WORLD RADIO TV HANDBOOK

WRTH

THE DIRECTORY OF GLOBAL BROADCASTING

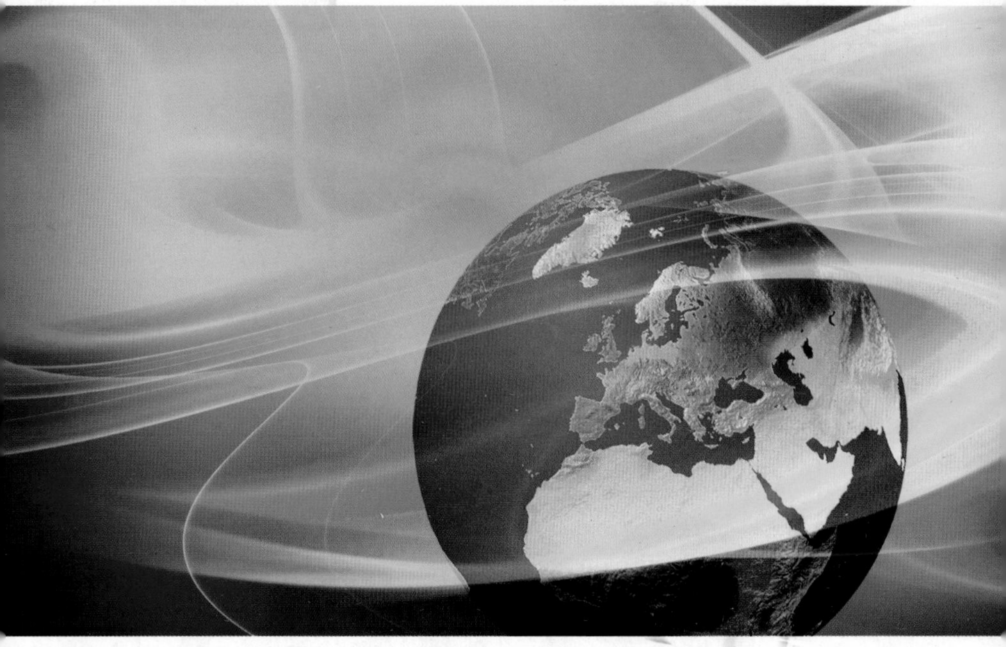

2011

BBC
WORLD
SERVICE

BBC
WORLD
NEWS

Never stop asking

Q Q Q Q Q
Q Q Q Q Q
Q Q Q Q Q
Q Q Q Q Q
Q Q Q & A

**Plus how and why from the
BBC's international news teams**

VOLUME 65 – 2011

Publisher
Nicholas Hardyman

International Editor
Sean Gilbert

Television Editor
Bernd Trutenau

Technical Editor
John Nelson

Contributing Editors
George Jacobs
Bengt Ericson
Dave Kenny
Mauno Ritola
Bernd Trutenau
Torgeir Woxen

Cover Design
Richard Boxall Design Associates

Advertising Sales Manager
Beth Leinbach

Published in the UK by:
WRTH Publications Limited
PO Box 290
Oxford OX2 7FT
United Kingdom
Tel & Fax: +44 (0) 1865 514405
Email: wrth@wrth.com
Web: www.wrth.com
ISBN 978-0-9555481-3-0

Distributed in the USA by:
Innovative Logistics
575 Prospect St.
Lakewood, NJ 08701
USA
Web: www.innlog.net
Email: lsucar@innlog.net
Tel US toll free: 866-289-2088
Fax US toll free: 877-372-8892
ISBN 978-0-9555481-3-0

Distributed in Germany by:
Gert Wohlfarth GmbH
Verlag Fachtechnik + Mercator-Verlag, Duisburg
Stresemannstrasse 20-22
47051 Duisburg
Tel: +49 (0) 203 3 05 27-50
Fax: +49 (0) 203 3 05 27-820
E-mail: info@wohlfarth.de
ISBN 978-3-87463-483-0

Printed and bound in the UK by CPI William Clowes, Beccles NR34 7TL

WORLD RADIO TV HANDBOOK

CONTENTS

Section Contents

Features & Reviews

National Radio

International Radio

Frequency Lists

Terrestrial Television

Reference

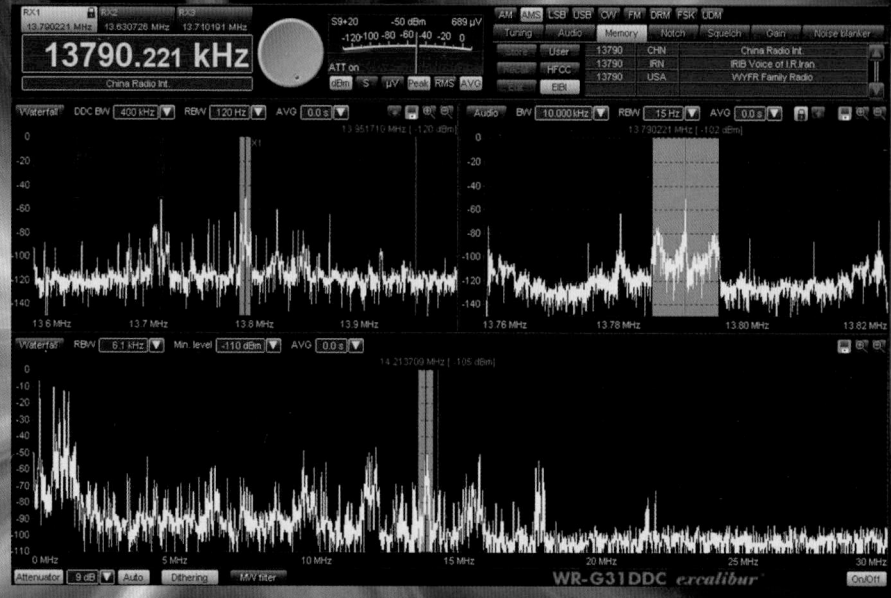

Editorial

NEW WAVES
Welcome to the 65th edition of *WRTH*. We bring you once again the most comprehensive and up-to-date information on global broadcasting that we and our network of diligent contributors can establish, and we are delighted at the evident good health of radio listening and DXing.

We were very pleased this year to learn that AOR had produced a new high-grade receiver and quickly requested a sample for review – we were not disappointed. We have also reviewed three new SDRs, and must make particular mention of the WinRadio Excalibur which is the finest SDR we have seen so far. Equally, we were very impressed with two of the ultralights we tested in our comparative review.

HF BROADCASTING
During the past year the debate about the future of international HF broadcasting has intensified. The financial strain placed on Western governments by the banking crisis has increased the attractiveness of reducing or abandoning this high-cost activity. This pressure on resources comes at a time when, as a spokesman for the BBC World Service pointed out, "...other countries are increasing, not reducing, their own investments in international broadcasting", a comment echoed by the Heritage Foundation in America.

Despite the reductions announced by large international broadcasters and the migration of services to FM feeds and the internet, analogue HF broadcasting is in surprisingly good health. As our readers will know, we strongly support international broadcasting. Analogue shortwave broadcasting remains a very effective method for transmitting signals around the world. However, reception quality is an unavoidable problem and in an age concerned with climate change and the effects of a warming planet, the ecological cost of running power-hungry shortwave transmitters is very high. Large transmitter sites consume enough electricity in a year to power a small town. The obvious answer to these two problems is to produce a digital solution which would provide clear audio at a comparatively low cost. This would safeguard the future of international broadcasting. This is what DRM should have done.

It is our view that the failure to grasp this opportunity, and to ensure the supply of inexpensive and readily available receivers capable of operating under the conditions experienced by analogue receivers, follows from a collective lack of will on the part on the states and organisations who should have been the keenest proponents of this technology. As a result of this failure we are reluctantly drawn to the sad conclusion that any hope that DRM will revitalise the prospects for international broadcasting can now be consigned to the waste-bin.

DIGITAL RADIO – A FOOTNOTE
We were delighted to hear that the proposed transition to DAB radio in the UK in 2015 is now an intention rather than a fixed schedule, suggesting that the inanites inherent in the decision to switch off analogue radio have been recognised.

WRTH FREQUENCY BARGRAPH
We have had many requests to publish something akin to the "Blue Pages" of the much-lamented *Passport to World Band Radio*. Doing this in *WRTH* would unfortunately mean increasing both its size and extent, but we are pleased to announce that we will be producing a CD of the full international schedules and domestic shortwave in bargraph form - see page 47.

WEBSITE UPDATES
We will be uploading pdf updates to the B10 season in February 2011 and the full schedules for the A11 season in May 2011. We will also continue to provide updates to the National Radio section on the *WRTHmonitor* page on the site.

PRIZE DRAW RESULTS
These winners were again drawn by members of the British DX Club.

RESULTS OF THE
WRTH 2010 PRIZE DRAW

First prize: the Etón G3

FIRST PRIZE
A Ferenc, UK

RUNNERS-UP PRIZES (copies of WRTH 2011)
N-E Johansen, Norway
G Serra , Italy
C Ghibaudo, France
S Obara, Japan
P Martin, USA

I hope you enjoy reading and using this edition of *WRTH*

Nicholas Hardyman
Publisher

WRTH Contributors 2011

The *WRTH* contributors come from all walks of life; yet share a common fascination with all aspects of global broadcasting. Each year we profile one of the people whose dedicated enthusiasm makes possible the enormous task of updating *WRTH* each year. This year it is the turn of Alan Davies, our contributor for many countries in South-East Asia, who gives us an insight into what makes a *WRTH* contributor.

Alan Davies

Writing this short piece has led me to reflect on how I ended up contributing a small part to this book. At the age of 12 or so, I received a National Panasonic radio as a Christmas present. It was nothing special, but it could receive shortwave programmes from across the world. I vividly remember listening to Australia, Japan, China and Grenada in our front room at home, and inflicting them on long-suffering family members. Actually *WRTH* shares some of the blame for this. I happened to buy a second-hand copy of the 1975 edition, and it was seeing the possibilities that the book promised in its shortwave listings that prompted me to say that the radio would be a welcome gift.

I later decided to study Asian languages and history at university, which began a long-lasting interest in this part of the world. After numerous visits, I made a permanent move to South East Asia ten years ago and have been working here since, most recently in education management.

These days I'm based in Indonesia, but I still try to make several visits each year to the nearby countries whose National Radio entries in *WRTH* constitute my "patch". Travelling as a radio enthusiast I find that listening to local broadcasting adds a sense of depth and purpose to any visit to a new place. I like to travel light, and these days a Sony SW7600GR, with its compact size, robustness and good performance on mediumwave, shortwave and FM alike, is my radio of choice.

Everywhere in South East Asia, a scan across the MW and FM bands will provide a satisfying blend of the familiar and the exotic. It's likely that at least one station will be playing western pop music and there is plenty of local fare on offer too: perhaps *lukthung* (Thailand's answer to country music), Malay, Chinese and Tamil hits in Malaysia and Singapore, chanted Buddhist liturgies in Cambodia, the Muslim call to prayer in Brunei, or the unmistakable sound of talk radio in the Philippines mixing words from Tagalog, English and Spanish into one effusive whole. Most distinctive of all are the unique tonalities of vocal music from the minority ethnic groups in Laos and Vietnam, also available to lucky DXers who can pick up the few domestic outlets still using SW.

Over the past decade, radio in South East Asia has experienced all kinds of changes, most noticeably the decline of AM radio on MW and SW alike, with many stations now closed or joining the exodus to FM. Notable too is the rise of community radio as an alternative to state and military-

owned radio stations, especially in Thailand, and the increasingly commercial and sophisticated sound of state-run national radio stations in countries such as Vietnam and Myanmar. Much else remains the same, however, and radio continues to be an important medium despite the rise of satellite and cable TV and the internet. The ties between radio stations – especially local ones – and their communities are strong enough to hold up against global competition.

So what of the future? At present FM is king, but shortwave still plays a role in countries where the domestic media is tightly controlled. Some countries in the region have experimented with digital radio, and one or two have officially launched DAB services. Nevertheless, analogue radio's dominance looks secure, and ten years from now FM will surely still be the place for stations to be if they want to be heard. Internet, cable and satellite are more of a threat to international shortwave broadcasting. I wonder how many voices will be left for the remaining shortwave listeners to hear.

Alan Davies

Official *WRTH* contributor for Brunei, Cambodia, Laos, Malaysia, Myanmar, Philippines, Singapore, Thailand and Vietnam

A large project such as WRTH could not be produced without the help of many people from all over the world. The following organisations and publications give invaluable help:

Asian Broadcasting Institute, BC-DX Top, British DX Club, Danish Shortwave Club International, DX-Listening Digest, DX Mix News, Electronic DX Press, Grupo Radioescucha Argentino, National Radio Club Inc (USA listings), Radio Heritage Foundation

our Country Contributors provide us with updated entries for the countries for which they are responsible:

Herman Boel, Christer Brunström, Luís Carvalho, Swopan Chakroborty, Svetomir Cuckovic, Alan Davies, Alok Dasgupta, Bengt Ericson, David Foster, Henrik Hargatai, Stig Hartvig Nielsen, Karel Honzik, Jose Jacob (vu2Jos), Richard Jary, Dave Kenny, Tetsuya Kondo, Vashek Korinek, Kai Ludwig, Dario Monferini, Svein Olav Pedersen, Andy Reid, David Ricquish, Mauno Ritola, Tony Rogers, Roberto Scaglione, Bernd Trutenau, Max van Arnhem, Thierry Vignaud, Tore B Vik, Torgeir Woxen

they and we are greatly aided by our other major contributors:

Teresa Beatriz Abreu, Carlos Benoit, Erich Bergmann, Dino Bloise, Héctor García Bojorge, Jordi Brunet, Mustafa Cancurt, Alfredo Cañote, Marcelo A. Cornachioni, Patricio R. de los Rios, Samir Elahcene, Jack FitzSimons, José Días Gómez, Victor Goonetilleke, Noel Green, Chris Greenway, Rudolf Walter Grimm, Alokesh Gupta, Wolf Harranth, Glenn Hauser, Aslam Javaid, Hans Johnson, Anatoly Klepov, Sergey Kolesov, Andrej Kuznecov, James MacDonell, Alexander Mak, Cláudio Rótolo de Moraes, Adán Mur, Michael Nevradakis, Horacio Nigro, Alexey Osipov, Samuel Ouma, Anker Petersen, Mieczyslaw Pietruski, Arnulf Piontek, Rimantas Pleikys, Patrick Robic, Rafael Rodríguez, Célio Romais, Daniel Rosenzweig, Ibrahim Rustamov, Victor Rutkovsky, Jari Savolainen, Zhang ShiFeng, Paulo Roberto e Souza, David Stanley, George Touliatos, Numan Vasquez, Tarek Zeidan

We thank them, and also all our readers who have written or emailed us with useful ideas and information. Please keep sending your thoughts and updates to:

wrth@wrth.com

or write to:
WRTH Publications Limited
PO Box 290
Oxford OX2 7FT
UK

WRTH Receiver Reviews 2011

In the intervening twelve months since the last edition of *WRTH* the world has suffered major economic recession and many international broadcasters are facing large-scale reductions in their operating budgets. One of the more striking indications of the degree of the changes afoot is that BBC Monitoring was facing the prospect of swingeing cuts. In July 2010 it emerged that the £25m annual government grant from the Cabinet Office, which provides the vast majority of the unit's funding, was set to be slashed in a spending review scheduled to be completed towards the end of the year. Its director, Chris Westcott, has said that the organisation is facing a "tipping point" and that the service could potentially be closed down.

Allied to the inexorably increasing cost of HF broadcasting as a result of rising energy charges and a continuing loss of the HF audience in favour of satellite transmission, local FM stations and internet-based radio and TV, we also wonder whether classical 'broadcasting' via the electromagnetic spectrum is also going to reach a tipping point at some time in the foreseeable future. But there are far too many variables in this particular equation and there is only one certainty, which is that no-one knows precisely what will happen. At a domestic level it is not yet clear anywhere in the world what – if anything – will replace analogue AM and FM radio and what the timescale will be. And it is not yet clear which of several incompatible standards will dominate digital television, if any.

One reason for the continuing uncertainty is the internet and the concomitant large-scale uptake of techniques such as streaming and podcasting. It is now clear that one way or another the internet will ultimately become the bearer of choice as far as broadcast 'reception' is concerned, at least in the developed world. The only issues which will ultimately prevent it from becoming dominant are political: issues around copyright and licensing seem to be less difficult to solve in practice than was expected a few years ago and there is clearly much more of a will to solve them. But where that leaves conventional broadcasting will continue to be an open question for the foreseeable future.

REVIEWS

It is pleasant to be able to record that this has been something of a vintage year for new receivers. At the top end of the market, the WinRadio G31DDC Excalibur has moved the state of the consumer SDR art onwards and the Medav LR2 has highlighted other aspects of what has become possible in this area. The AOR AR5001 proved itself a formidably good all-round performer, being in essence a 14-bit MF/HF SDR in the same box as a wideband scanner. In complete contrast, we have looked at some ultralight and very inexpensive portables and been astonished by how good two of them in particular turned out to be. The Tecsun PL-310 and PL-380 are extraordinarily capable and versatile and both entirely re-define the market for pocket-size receivers. Although they are not available via conventional distribution networks and have to be bought via internet auction sites (is this the way of the future?) this is a minor inconvenience to those well versed in on-line shopping. As we went to press there were indications that some of these receivers might be marketed via Amazon, which would make life even easier. Our reviews showed that not all ultralight portables are created equal and we hope to have clarified the issues of what to look for.

One of many praiseworthy points about the Excalibur is that it handles DRM very well indeed. In fact it makes the case for DRM being a viable broadcast mode better than any other receiver we have so far seen. It is therefore all the more regrettable that one class of receiver not represented this year is a stand-alone device with 'DRM' on the bandswitch. Last year we said that the Uniwave DiWave 100 was to be launched at the Amsterdam IBC show, which indeed it was after a fashion. A few retail outlets carried them in very limited numbers for a short time and they attracted generally lukewarm comments on the internet. We can only conclude that this radio will not now see the light of day, and that it is increasingly the case that manufacturers simply do not see DRM as a viable mass-market technology.

ABOUT THE AUTHOR
John Nelson is an author, editor and consultant specialising in audio, radio and communications technology. After graduation in 1974, John worked for the BBC and the Radio Society of Great Britain before forming his own company in 1986 and has written, edited and contributed to a wide variety of publications. He is now the managing director of Crew Green Consulting Ltd, which specialises in electronic and communications systems design and assessment for a wide international client base.

John's amateur radio callsign GW4FRX is often heard on the HF bands – where he currently has 322 DXCC countries confirmed – and on 144MHz. In his remaining spare time he enjoys aviation, music, literature and architecture. He lives in east Wales.

AOR AR5001D

US$3800   £2995   €3450

OVERVIEW

AOR's latest receiver is the AR5001D. The marketing literature states that it has been ". . . developed to meet the needs of security professionals and government agencies". This sounds suitably impressive although one would imagine that such users would be more likely to employ the very similar AR2300 (in essence an AR5001D with no manual controls) in conjunction with bespoke software. As the designation suggests, the receiver can be regarded simplistically as a development of the celebrated AOR5000, production of which ceased in April 2008. That receiver covered 10kHz-2036MHz whereas the 5001D caters for 40kHz-3.15GHz. There is in fact a marked physical and ergonomic resemblance between them – if you can already drive a 5000 you will have no difficulties with the 5001D – but the newer unit offers more functionality.

FEATURES

The AR5001D is a substantial device, measuring 305 x 220 x100mm and weighing about 5kg. The front panel is dominated by the lime green-backlit display which crams in a great deal of information. In frequency-display mode the upper three lines are small and closely spaced indications of control settings with the frequency, mode and IF bandwidth displayed in slightly more prominent characters. In bandscope mode various selected functions can be displayed. Below the S-meter are rotary stepped AF gain and squelch controls, the setting of the latter being shown on the display in terms of the squelched RF level threshold. Headphone and accessory sockets are adjacent to the rotary controls and the SD-card slot sits beneath. The accessory socket gives access to low-level audio, detector and +12V outputs and also offers a facility for a "GPS time pulse input".

No further reference to this could be found in the manual although the literature does mention an optional GPS unit which ". . .locks the reference oscillator of the AR5001D to (GPS) and provides 0.01ppm accuracy". Initially this sounds useful but in point of fact it only amounts to one part in 10^{-8} whereas a modern GPS-disciplined TCXO frequency standard should be several orders of magnitude better than this. Happily AOR has also provided an SMA socket on the rear panel for a 10MHz reference-frequency input. Something eminently suitable can be easily derived from a suitable GPS-based standard nowadays at low cost and should confer an accuracy of better than one part in 10^{-12} with no difficulty. The commendably large keypad sits below the display with the main tuning and "sub-dial" knobs to its right. The latter is used for various selections such as mode and IF bandwidth. All the keys have shifted functions, selected by a FUNC button to the left of the S-meter. The controls feel pleasantly precise and fall readily to hand.

Fundamentally the AR5001D can be considered as two architecturally distinct receivers in a common box. Between 40kHz and 25MHz it employs direct conversion with a 14-bit 65Ms/s sampling architecture. Between 25 and 200MHz the arrangement becomes that of a conventional double-conversion superheterodyne with IFs of 294.5 and 45.05MHz. An additional up-converting IF at 1.7045GHz is introduced between 200 and 420MHz with reversion to a double superhet above this range. The 45.05MHz IF appears on a rear-panel jack and its bandwidth is ±7.5MHz, so it could in principle be handled by a separate receiver or dedicated processor. Internally this IF is subjected to digital conversion such that all subsequent filtering and demodulation takes place in this domain. The 5001D's designers have

opted for switched rather than continuously variable bandwidths, which for a wide-range receiver is probably the optimum choice. Nine selectable bandwidths are available, namely 200 and 500 Hz and 3, 6, 15, 30, 100, 200 and 300kHz together with an IF shift function. The available modes are AM (conventional and synchronous), ISB, USB, LSB, CW, AIQ (AF-IQ), narrowband FM and wide FM which includes stereo. The AF-IQ provision is interesting in that it takes the form of a 12kHz IF output suitable for a PC sound card. This would be very useful for DRM reception. DTMF, CTCSS and DCS can be catered for. Both PAL and NTSC video demodulation modes are available and a composite video output is provided on the rear panel of the unit.

The memory and VFO facilities are very comprehensive, as might be expected. There are five separate VFOs and 2,000 memory channels organised by default as 50 channels in 40 banks; each of the latter can be arranged to contain anywhere between 5 and 95 channels as required. The claimed scanning rate is 100 per second and this is quite believable. Channels and banks can be individually named. Simultaneous reception of one frequency below 25MHz and one above it is possible, as is reception of two frequencies above 25MHz spaced within 10MHz of each other. Another useful feature is an on-board SD card slot, to which the AR5001D can directly record demodulated audio in .WAV format.

AOR states that the synthesiser is a low-noise DDS design and that the front end is "carefully designed by CAD to obtain optimum performance." In the absence of a schematic it was not possible to judge this claim although some brief comparative trials in the FM broadcast and civil aeronautical VHF bands suggested that the sensitivity was approximately on a par with that of the resident Icom IC-R8500. A few spurious responses, chiefly breakthrough from local FM broadcast and paging transmitters, were noted in the 120-130MHz region and were strong enough to break through squelch settings of -100dB or so. Given that the receiver is understood to contain extensive internal preselection and that the test site is not particularly demanding in respect of very strong local transmissions, this was a slightly disappointing result.

So far you might be forgiven for thinking that the AR5001D is a versatile and competent wideband scanning receiver, which indeed it is. What makes it far more than a mere "scanner" are two things. One is its ability to be controlled by a PC running Windows XP or higher, and the software provided gives access to all the receiver's functionality and also offers some very useful aids to managing locally used frequencies. But it also facilitates display of the results of the AR5001D's internal FFT analyser-cum-spectrum display, which is of course viewable via the normal front-panel display but looks considerably better on a PC screen. The analyser covers 400kHz-10MHz in 100kHz increments and worked very well indeed in our trials. For professional users the AR5001D receiver can be operated remotely using an optional LAN controller, and up to 1MHz of digital I/Q output can be recorded to a local computer for later analysis with the optional I/Q interface board installed.

PERFORMANCE

In performance terms the AR5001D acquitted itself very well. Almost every AOR receiver hitherto known to the author has exhibited what might be kindly called idiosyncratic ergonomics but the AR5001D is not as demanding as some of its predecessors; its interface was easily mastered and the unit thoroughly enjoyed. The claimed sensitivity figures were easily met and the receiver is always likely to be interference-limited rather than noise-limited in the HF region. The IF filters were excellent, as expected, with fine stopbands and shape factors. On HF the claimed spurious and image rejection is better than 70dB and this was met with ease. The claimed IPI_3 at 14.1MHz is +20dBm and this is certainly the case. Measured in the classical way the figure is about +24dBm although it is considerably worse than that in the VHF region. The top-mounted speaker did the unit no favours but good-quality headphones or the rear-mounted extension speaker socket produced very pleasant and acceptable audio. Listening for long periods was a pleasure.

CONCLUSION

All in all, the AR5001D is undeniably expensive but is also undeniably a tempting proposition for those wishing to have a thoroughbred HF receiver together with a fine wideband VHF/UHF scanner in a single unit. The quality of manufacture is exemplary and the performance and functionality are both excellent. We should like to thank the UK agent, Waters & Stanton, for the loan of the review unit.

Rating table AOR AR5001D

Mechanical design	★★★★★
Constructional quality	★★★★★
Firmware	★★★★★
Sensitivity	★★★★★
Dynamic range	★★★★
RF intermodulation	★★★★
Audio quality	★★★★
Versatility	★★★★★
VFM	★★★★

Overall rating   ★★★★★

Key:
★ = Poor   ★★ = Fair   ★★★ = Average
★★★★ = Good   ★★★★★ = Excellent
VFM = Value for money

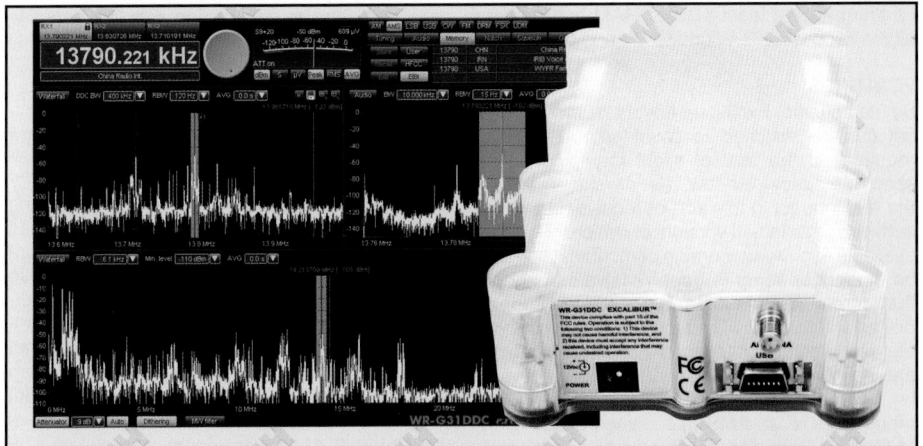

Winradio G31DDC Excalibur

US$850 £545 €630

OVERVIEW

In recent years a design approach which links the power of a modern personal computer (PC) to some of the functionality of a radio has produced some very interesting and capable receivers. The Icom PCR-1000 was an early example, as were WinRadio's first products.

Early receivers of this general form partially followed the architecture of the DSP radios discussed in our feature on page 22. A conventional RF amplifier and superheterodyne frequency changer was used to derive a baseband IF which was put into a digital format by an on-board analogue-to-digital (A-to-D) converter. In a standalone receiver this was further processed by on-board digital signal processing (DSP) hardware and software. In early PC-controlled receivers the processing was also carried out within the receiver, leaving the sound card to deal with the audio and with the PC handling control and display functionality. As hardware became more capable the A-to-D conversion function itself began to be carried out within the PC. This in turn opened the door to more extensive implementation of software-controlled digital signal-processing techniques. The keyboard and mouse of the PC continued to form the interface between the user and the electronics, and its display could be used in a variety of ways as required.

As always in electronics, hardware becomes more capable with the passage of time. In consequence the software-controlled approach has begun to be supplanted by another which is known as *software-defined*. In this the RF and mixer stages of the conventional superhet give way to what is referred to as direct down-conversion (DDC). Here the fixed IF generated by the RF and mixer stage is replaced by a technique in which the entire spectrum of interest is digitised

as a whole. Since it would require a vast amount of computing power to process all signals simultaneously, a smaller portion is selected by a process known as *decimation* and down-converted for the PC to perform the usual filtering and demodulation functions using digital processing techniques implemented in software. One very interesting consequence of this approach is that it

WinRadio Excalibur
BEST SDR

becomes possible to record and replay the decimated part of the spectrum and subject it to whatever analysis and processing is required. The massive advantages to the listener is that this need no longer take place in real time.

As SDRs have developed, the amount of the spectrum which can be handled in this way has gradually increased. Early radios such as the SDR-14 and FlexRadio 1000 could record and replay 192kHz segments. Initial versions of the Perseus offered 400, then 800 and 1600kHz which was enough to cater for the MF and several HF broadcast and amateur bands. The WinRadio WR-G31DDC Excalibur which is the subject of this review has a DDC bandwidth of no less than 2MHz, which as far as we are aware is the largest of any current consumer SDR. In passing, it goes without saying that the larger the DDC bandwidth, the more processing power is required in the connected PC. The minimum requirement for the Excalibur is a 2GHz dual-core

Pentium with at least 1GB of RAM and running either Windows XP, Windows 7 or Vista.

FEATURES

By default the Excalibur covers 10kHz-30MHz but can be switched to cover up to 50MHz if required, at the cost of slight cramping of the spectrum-space window. As always, however, the real interest in this sort of receiver lies in the software, the latest version of which is always available via www.winradio.com. Excalibur can be thought of as up to three receivers in one, in that three independent demodulators can be placed within the DDC spectrum. Each can independently demodulate signals of different modes and bandwidths and each can also be separately recorded and played back. The available modes are AM, AMS (i.e. synchronous AM), LSB, USB, CW , FM, FSK and a 'user definable' mode. Several bandwidth-selection buttons appear for each mode and can either be used as-is or serve as starting points to allow fine setting of the bandwidth as required by dragging the edges of the passband area with the mouse.

There is also DRM, which requires the appropriate plug-in. We should say at this stage that the DRM implementation in Excalibur is truly excellent and the best we have ever used. It produces almost immediate and consistent results from signals which most other receivers would either not process at all or suffer persistent dropouts. As long as there was enough signal to generate a stable S/N of 20dB or so, reliable audio would always appear. This is how DRM should be from the user point of view – a complete non-event. The user-definable mode amounts to a combination of mode, bandwidth, audio filter setting and any other parameters one might wish to use in association with it. We found it ideal for NDB and amateur-band MFSK16 reception in particular, the latter working superbly well in conjunction with the excellent waterfall display, but it would no doubt be useful for all sorts of other purposes. Synchronous AM generally seemed to work well but was sometimes slow to lock up.

Using Excalibur is very like using any other SDR. There are several ways of entering frequency, tuning and so on and a variety of keyboard shortcuts can be used, many of which are configurable. The memory functions are very comprehensive. Both the HFCC and EIBI databases are provided as text files and can be selected along with the user memories, the latter being tree-structured and only limited by the size of the available hard drive. Frequencies can also be stored in the twelve function keys.

There are many praiseworthy features of the Excalibur, all selectable via drop-down menus and tabs. One in particular is the fully configurable AGC. As supplied there are fast, medium and slow settings which are fixed but three user-definable settings are also provided. All work

very well, and in conjunction with the versatile squelch system they make reception of utility SSB as well as broadcasts very pleasant.

PERFORMANCE

In performance terms the Excalibur sets new standards in several areas. It is the most sensitive SDR we have yet measured, with MDS figures of between -119 and -122dBm depending on the tuned frequency. This is slightly less than one would expect with a high-grade conventional receiver and it is just possible that a low-noise wideband preamplifier would be beneficial in certain circumstances. Selectivity is generally excellent, as would be expected, and we encountered no problems at all during extensive testing with assorted long wires, dipoles and our resident Wellbrook loops. Indeed, the WinRadio seemed to like the latter best of all and we spent considerable tme in side-by-side tests against a Harris RF-590 and Racal RA1778 and RA1792 receivers. There were no circumstances in which one receiver could hear signals which the others could not but audio recovery of some weak and fading AM broadcast signals was occasionally better on the Excalibur than the others, chiefly because of the ability to tailor the IF bandpass and audio filters of the latter. That said, it was interesting that in some way which was rather difficult to define, both the Racal receivers sometimes sounded 'quieter' than the Excalibur, especially on the LF and MF bands. Utility SSB signals in particular often seemed to 'stand out' better against a quiet background. But the Excalibur's ease and fluidity of operation, massively versatile functionality and excellent RF performance – all from a small and unassuming box coupled to a commonplace PC – remind us that technology has inexorably moved on.

CONCLUSION

All in all, Excalibur is already the best SDR we have used – and knowing WinRadio we imagine that future software releases will only serve to make it even better.

Rating table for Winradio Excalibur

Constructional quality	★★★★★
Software	★★★★★
Sensitivity	★★★★★
Dynamic range	★★★★
RF intermodulation	★★★★★
Audio quality	★★★★★
Versatility	★★★★★
VFM	★★★★★

Overall rating ★★★★★

Key:
★ = Poor   ★★ = Fair   ★★★ = Average
★★★★ = Good   ★★★★★ = Excellent
VFM = Value for money

Ultralights

INTRODUCTION

In last years' edition of *WRTH* we examined a group of very low-cost portable receivers from assorted Chinese manufacturers, of the kind often purchased via internet auction sites and the like. At the quoted price of between £5 and £15 we were not expecting stellar performance and we certainly did not find it. Their claimed HF performance was marginal at best and all exhibited more or less serious shortcomings in various areas. We concluded that none of the receivers could really be recommended to the serious or semi-serious listener or *WRTH* reader.

For this edition we decided to go up-market and examine several slightly more expensive contemporary Chinese-made receivers. Three of them fall into what has become known in some quarters as the "ultralight" category but as we will see this does not necessarily imply chronic performance shortcomings. All are available via eBay and retail outlets in some parts of the world and we have seen several of them in European and Far Eastern airport duty-free shops. However, as far as we are aware there are no formally appointed European or North American distributors and the warranty position in particular is unclear, so there is no formal procedure for returning defective items and claiming a refund or replacement. This may be a point of concern for some readers.

Tecsun PL-310

Approx. US$55 £35 €40

TECSUN PL-310

The PL-310 is one of two Tecsun receivers in this group to embody the Silicon Labs Si4734 integrated circuit. This is a fully integrated CMOS-based LW/MW/SW/FM radio receiver incorporating some very clever digital signal processing (DSP) and capable of remarkably high performance. In fact the author of this review recently designed a receiver for a client using the RDS-capable version of this IC (the Si4735) and was pleasantly surprised by its price-per-formance ratio. Amongst the available functionality are five levels of switched AM bandwidth (1, 2, 3, 4 and 6kHz) and selectable AM tuning step size, several different tuning and frequency-entry methods, comprehensive memories and a display of signal strength in sensible units of dBµV. As if the latter was not enough there is also a display of instantaneous signal/noise ratio! The HF coverage is from 2.3 to 21.95MHz in 13 bands and the FM band depends on the market into which the radio is supplied. There is even a real-time clock and an alarm facility. The size and clarity of the display puts that of many receivers costing three times as much to shame.

Tecsun PL-310
BEST ULTRALIGHT

Looking rather like a smaller version of the Sangean ATS-909, the PL-310 measures a mere 140 x 86 x 28mm and weighs only 220g with its three AA cells fitted. In a nice touch, NiCd or NiMH cells can be used and recharged via any convenient USB socket using the supplied cable. There is a connection for an external FM or HF antenna although the supplied telescopic and internal ferrite antennas seem to work remarkably well overall.

Given its size and price, the performance of the PL-310 can only be described as astonishing. There are minor irritants; the strong-signal handling in certain circumstances is not outstanding and the "soft mute" feature of the Si4734 – which is enabled in this radio but not in the essentially similar PL-380 – occasionally leads to disconcerting changes in volume on HF when sharp fading is present. This is the one factor which limits the PL-310 for weak-signal working and SW DX enthusiasts may well prefer the PL-380. The variable-rate tuning is a little irritating until one acquires the knack of using it and some other aspects of the ergonomics are mildly idiosyncratic. There are some heterodynes and spurii in the MW band although commendably few elsewhere. But the switchable selectivity is very useful, the AM sensitivity is remarkable and the long-wave performance (which for some reason is not enabled by default) surprisingly good for a small

low-cost portable. The PL-310 handles external antennas with aplomb and coped particularly well with our Wellbrook wideband loop.

At the quoted price, the PL-310 is an extremely tempting proposition.

Tecsun PL-380

Approx. US$60　£40　€45

TECSUN PL-380

Similar in many ways to the PL-310 and identical in size, the PL-380 employs the same Si4734 DSP integrated circuit but omits the "soft mute" feature which for weak-signal AM reception is something of a problem. This function causes an abrupt change in volume with fading – it is rather like a fast-acting audio gain control – but also introduces what can best be described as "pumping" of the AGC proper, giving rise to disconcerting variations in apparent signal strength. On steady local signals this is not a problem but with weak and fading MF and HF transmissions it can make listening rather harder work than necessary. Anecdotally it is understood that Tecsun introduced the PL-380 chiefly as an alternative to the PL-310 in which this particular feature of the Si4734 was not enabled, thus making it rather more useful to those wishing to chase and log distant stations.

Apart from that, the only major physical difference between the PL-310 and the PL-380 are that the latter is tuned via a thumb-wheel instead of a tuning knob. It has 100 preset memories for FM, MW and LW and no less than 250 for the HF bands, a slightly more comprehensive alarm system and at least seven different tuning modes. Electrically the internal "loopstick" antenna appears to be slightly smaller than that in the PL-310 and this is possibly the reason for the slightly reduced AM sensitivity exhibited by the PL-380 in general use; our tests suggested that there was about 3-5dB difference between the two. Of course, a committed DXer would probably want to use some form of external antenna and the issue then disappears.

In general everyday use the variable-rate tuning is the only feature of the PL-380 which takes a little practice and it should also be mentioned that the confirming tones strike us as unneces-

sarily loud. Apart from that, the receiver worked extremely well. The FM performance was again very good indeed and the absence of the soft-mute feature made listening to weak AM signals much more pleasant, especially when it became necessary to tune away from the nominal carrier frequency slightly to assist in identification. We rather enjoyed ourselves one warm summer evening logging distant LF-band non-directional beacons in this way and there can be little wrong with a receiver capable of copying OZN (372kHz) in Prins Christian Sund, Greenland from the UK.

Neither the PL-310 nor the PL-380 are receivers for the hard-core HF-band listener, especially if SSB or CW facilities are required, but both do a very creditable job – which is emphatically not the case with most low-cost portables. On the MW and FM bands in particular both offer remarkably good performance. For the price it is difficult to imagine anything better.

Kchibo KK-D6110

Approx. US$45　£30　€35

KCHIBO KK-D6110

Another contender in the ultralight category, the Kchibo KK-D6110's packaging prominently displays "DSP Radio". One might initially think of it as a variation on the Tecsun theme of a modern portable employing the Silicon Labs Si4734 DSP integrated circuit. Unfortunately this does not seem to be the case if the measured and subjective performance is anything to go by. Evaluation of the receiver is not helped by the fact that all the control labelling and front-panel markings are in Chinese, as is the accompanying literature, and it may be that we have missed some salient features. However, there appears to be no switchable AM bandwidth and we could find no other specific functionality which would suggest the use of the Si4734.

The KK-D6110 measures 146 x 90 x 26mm and weighs about 320g with the battery installed. As far as could be determined the FM frequency range is 87-108MHz and the MF band either 522-1710kHz (in 9kHz steps) or 520-1710kHz in 10kHz steps. HF coverage is 2.30-21.85MHz. The receiver runs from three AA cells and a 5V

mains power supply-cum-charger was provided with our sample.

Compared with the two Tecsun receivers, the KK-D6110's overall performance was not as good. On both MF and HF the sensitivity was well down. Measurements with a signal generator were not helped by erratic results depending on lead positioning and so on but in very general terms the KK-D6110 was about 10-12dB down on the PL-310. Some mild tendencies to instability and hand-capacitance effects were noted in several HF bands together with a degree of digital "hash" on frequencies towards the upper and lower edges of each band. Touching the antenna usually had some effect on the problem but not consistently. Sensitivity on FM was between 6 and 8dB worse than in the PL-310 and during periods of low deviation there was what sounded very like old-style synthesiser whine in the background. Here again there were occasional tendencies to instability at the low-frequency end of the band and these became noticeably worse when the batteries were partially discharged.

Given the strength of the competition, it is difficult to recommend the Kchibo KK-D6110 for purchase. At the time of writing it also appeared to be less generally available than the other receivers in the tested group.

Degen DE-1103/Kaito KA 1103

Approx. US$80   £50   €60

DEGEN DE-1103

Generally comparable in price with the Tecsun PL-380, the Degen DE1103 (also marketed as the Kaito KA1103) is fractionally larger at 165 x 104 x 29mm and weighs slightly more at 330g. It is not a DSP-based receiver but a conventional up-converting superheterodyne with a first IF of 55.845MHz. The manual states that AM coverage is 0.1-29.99MHz but in our sample this was not the case; the AM tuning operated in 1kHz steps between 520 and 1710kHz. The FM band coverage is 76-108MHz and HF is handled in ten bands, namely 3.2-4, 4.6-5.4, 5.6-6.4, 6.6-7.4, 9.1-9.9, 11.5-12.4, 13.5-14.4, 15-15.9, 17.1-18 and 21-21.9MHz. There is provision for SSB reception with the aid of a "fine tuning" control although both the available filter bandwidths are rather too wide to be very useful in this mode. There is a total of 256 memories with autoscan and store. The DE1103 runs from four AA rechargeable cells and an AC adaptor is supplied although the cells are not, or at least were not supplied with our review sample. There is provision for an external antenna although for some reason this did not work on FM. Neither did the four-segment S-meter.

The single biggest drawback of the DE1103 is its ergonomics, which could charitably be described as unusual. The liquid-crystal display is large and clear and takes up half the panel but it mimics a travelling-bar analogue display that was passé by the 1960s. It is difficult to read, chronically imprecise and in our opinion represents poor engineering design. The display is coupled to a rotary tuning control which almost unbelievably doubles as the audio volume control and is switched between these duties by a button which is almost as far away from the control as it could be. The keypad – such as it is – is a row of small horizontally disposed buttons beneath the display and there are some function buttons beneath it. All have much higher breakout force than is commensurate with their size together with excessive tactile feedback, and the combination feels very clumsy. The button spacing is much too close for the average finger. The tuning knob is not detented and the 1kHz tuning steps are not ideal.

The performance of the DE1103 is reasonable enough but overall does not compare well with either of the Tecsun receivers. On the MF and HF bands there were some spurious responses, notably one at 900kHz below the wanted signal which was only about -28dB, and blocking was observed on several occasions when using the receiver with small external antennas. On side-by-side comparisons with the Tecsun receivers the Degen sounded rather muffled and lack-lustre despite its larger speaker and the FM sensitivity was clearly not as good. Reception on HF was pleasant on strong signals and sensitivity using the receiver's own antennas was very high but the switchable bandwidth of the Tecsun receivers gave them both a definite advantage on weak and fading signals.

All in all, the poor ergonomics of the DE1103 and its less than sparkling performance rule it out of contention in the tested group unless you have a particular requirement for SSB reception. The Tecsun PL-380 at the same price is superior in almost all respects.

CONCLUSION

The contrast between the low-cost receivers tested last year and those in the present group could hardly be greater, and the Tecsun models in particular offer extraordinarily good value for money. Either can be confidently recommended, especially the PL-310.

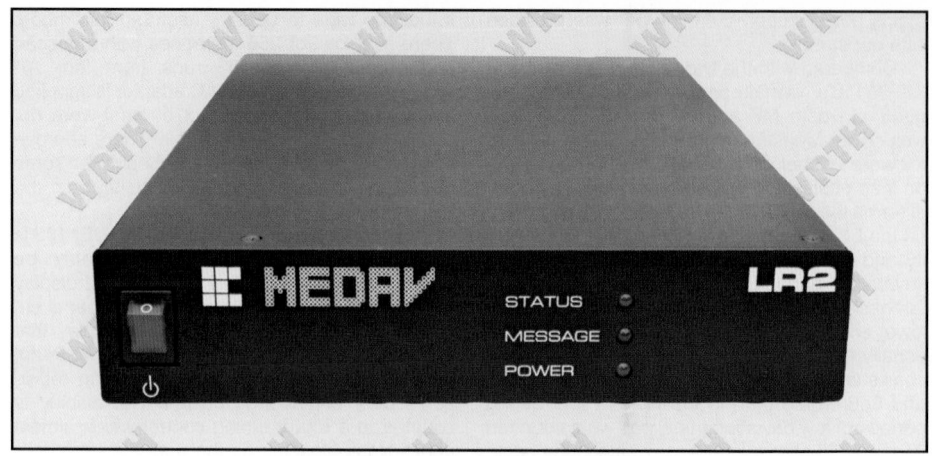

Medav LR2

US$4800   £3100   €3600

Medav may not be a familiar name to *WRTH* readers but the company is well known in its native Germany as a manufacturer of high-grade electronic systems for professional and military markets. The LR2 is a software-defined receiver which is available in three variants; D2 covers 100kHz-30MHz, D3 caters for 30MHz-3GHz and D4 handles the entire 100kHz-3GHz spectrum. We assessed the D2 version.

The LR2 is a conventional SDR with some interesting features and functionality. The rear drop has two N-type antenna input sockets (much superior to the usual BNC connectors in our opinion), a USB port and a standard RJ45 Ethernet socket. A standard IEC320 mains connector is provided and the internal PSU operates with any mains voltage between 85 and 264. The unit is beautifully made and finished. The requirements for the associated computer are quite stringent, namely a 3GHz CPU with at least 1GB of RAM and running either Windows XP or Vista.

As with any SDR, the most interesting aspects are internal. The tuning resolution is 1Hz and the modes available are LSB, USB, CW, AM and FM. The sampling rate is 66.66 Ms/sec, equating to a sampling frequency of just over 66MHz, with 16-bit resolution. Via the LAN output the selectable DDC bandwidths available are 8, 15, 30, 60, 120, 150, 250 and 500kHz. The latter two are not available at the USB port, giving a maximum bandwidth at that point of 150kHz. Given that several SDRs offer bandwidths in the megahertz region via USB, this is some way behind the state of the art. A 16-bit signed integer I/Q output at a data rate of about 2.5MB/sec is available via the LAN output, with about 750kB/sec at the USB port.

Some of the LR2's circuit design details are fascinating. The antenna inputs are selectable and are routed to one of 11 automatically switched filters with ranges carefully chosen to maximise the strong-signal performance. Up to 10MHz these are essentially half-octave, widening out slightly thereafter. Medav claims an IP_3 in excess of 35dBm (typically 40dBm) which is frankly beyond the capability of our test equipment to measure with any confidence in the result. A 30dB switchable attenuator with 10dB steps is provided but it is slightly difficult to imagine any real-world situation in which this might be necessary. The stated noise figure is 9dB, which no doubt reflects the use of a low-noise post-filter amplifier and should guarantee interference-limited results at even the highest frequencies. It also substantiates the claimed sensitivity figures, which are clearly conservative. The specification gives a figure of -108dBm for a 10dB S/N in a 6kHz bandwidth for 50% AM modulation at 1kHz. Our measurements suggested that the performance was about 4dB better than this, which makes the LR2 very nearly as sensitive as the Excalibur (see page 13) and hence as good as any SDR we have tested.

The software supplied with the LR2 is quite daunting at first glance and is clearly oriented towards measurement and analysis applications. However, a little persistence soon gave us fairly standard spectrum and waterfall displays. The LR2's strong-signal performance was superb and in practical terms it was never remotely threatened even by extremely strong 6/7MHz signals taken on a full-size dipole.

CONCLUSION

Overall, the LR2 is not for everyone: notwithstanding its excellent performance, the very high price and somewhat basic software facilities put it at some disadvantage. However Medav is clearly a company to watch.

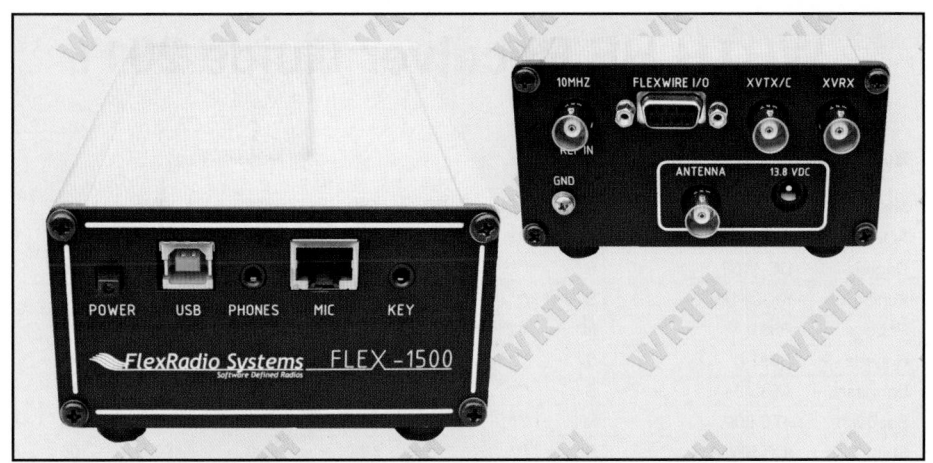

Flexradio FLEX-1500

US$650   £550   €650

In the 2007 edition of *WRTH* we looked at the then-new FlexRadio SDR-1000. This was a 1W transceiver used in conjunction with a PC and soundcard; the transmitter portion could easily be disabled, leaving the unit as an 11kHz-65MHz receiver. Licensed amateurs could add a 100W power amplifier, an auto ATU and a 144MHz transverter. The associated software was – and remains — entirely open-source and downloadable from the FlexRadio website (www.flexradio.com). The FLEX-1500 can be viewed as a development of the SDR-1000 and in fact is one of four products currently available from FlexRadio, the others being the 3000 and two versions of the 5000. One criticism of the SDR-1000 was that it required an associated 24-bit soundcard which by the standards of the time was exotic and expensive. However, the FLEX-1500 does not share this requirement, needing only a simple USB connection to the local PC. Neither does it need the elaborate set-up procedure associated with the SDR-1000. As a receiver the coverage is 490kHz-54MHz (the specification states that external filters must be used to eliminate images if the receiver is used at lower frequencies) and the supported modes are synchronous and conventional AM, USB, LSB, DSB and CW. Other modes could be used if required, and the 'Knowledge Base' on the manufacturer's web site gives a variety of useful information.

The receiver specifications with the preamplifier off claim an MDS at 14MHz of -116dBm in a 500Hz bandwidth, which is reasonable but not outstanding; a good conventional receiver could be expected to be somewhat better. However, the +20dB preamplifier naturally improves this and the claimed figure becomes -127dBm. The claimed image rejection is in excess of 100dB but

we occasionally found some strong stations to be audible at ±18kHz; for some reason it was not possible to duplicate this finding reliably in single-tone measurements.

At the time of writing the latest version of the PowerSDR software appeared to be version 2.0.8 and there were some known bugs with this. Cutting a long story short, by the end of our test period we were not quite convinced that this version was ready for non-technical users or those without at least a degree of familiarity with the PC. It seems that Windows 7, XP and Vista are supported and that Win 7 works best because of its use of a low-latency sound-card driver (WASAPI) instead of the ASIO4ALL driver in the other versions. Our testing took place using a fully patched version of Win XP SP3 and various minor difficulties became evident. Having said that, the basic receiver performance seemed to be very good. As hitherto, the most obvious visual element of the software is the splendid real-time spectrum display which offers histogram, waterfall and high-resolution real-time modes. As would be expected, the IF filters display classical DSP 'brick-wall' characteristics. The bandwidth can be continuously adjusted between 10Hz and 16kHz, making high-quality CW reception in particular an absolute pleasure since as with all digital filtering there is no ringing whatsoever.

CONCLUSION

In general terms we quite liked the FLEX-1500. But given that it occupies a similar price class to the WinRadio Excalibur (page 13), which offers a combination of exemplary performance and superbly versatile functionality, it does not appear to represent very good value for those interested only in listening to HF broadcasting.

WRTH HF Receiver Guide 2011

Budget, Hand-held & Travel Portables

Maker	Model	Size	SEL	DR	OV	US$	£	€
AOR	AR8200 MkIII	H	****	***	***	600	440	480
Degen	DE-1103	S	***	***	**	80	50	60
Kchibo	KK-D6110	S	***	***	***	45	30	35
Roberts	R861	M	****	***	****	280	185	210
Roberts	R9914	S	****	***	****	160	100	110
Sangean	ATS-404	S	***	***	***	75	55	60
Sangean	ATS-909	M	****	***	****	220	150	165
Sony	ICF-SW11	S	***	***	***	80	50	65
Sony	ICF-SW12	S	***	**	*	80	50	65
Sony	ICF-SW35	S	***	***	****	180	145	160
Sony	ICF-SW7600GR	S	****	****	****	200	130	150
Tecsun	PL-310	S	****	****	*****	55	35	40
Tecsun	PL-380	S	****	****	*****	60	40	45

PC Radios, SDRs, Serious Shortwave & Semi-pro Receivers

Maker	Model	Size	SEL	DR	OV	US$	£	€
AOR	AR5001D	M	*****	****	****	3800	3000	3450
AOR	AR8600	L	**	***	***	900	600	720
Elad	FDM77	C	****	****	****	640	400	450
Etón	Satellit 750	L	***	***	***	299	299	485
FlexRadio	FLEX-1500	C	****	****	***	650	550	650
FlexRadio	FLEX-5000A	C	*****	*****	*****	2700	2495	2720
Icom	IC-718	L	***	****	****	800	520	620
Icom	IC-7000	M	*****	****	*****	1300	1090	1375
Icom	IC-7600	L	*****	****	****	3800	3150	3800
Icom	IC-R9500	L	*****	*****	*****	13000	10000	12000
Medav	LR2	C	*****	*****	***	4800	3100	3600
Microtelecom	Perseus	C	*****	*****	*****	1200	750	820
Palstar	R30	M	*****	*****	****	680	600	660
RFSpace	SDR-IQ	C	****	****	****	500	470	560
Ten-Tec	RX320	C	***	***	***	370	200	240
Ten-Tec	RX340	L	*****	*****	*****	4250	3600	3925
WinRadio	G31DDC Excalibur	C	*****	****	*****	850	545	630
WinRadio	G303i	C	****	****	*****	500	460	450
WinRadio	G313i	C	*****	****	*****	1000	900	900
WinRadio	G313e	C	*****	*****	*****	1200	1090	1100
WinRadio	G305e	C	****	****	****	650	590	590
Yaesu	VR-5000	L	****	***	****	640	510	610

KEY: SEL = Selectivity, DR = Dynamic Range, OV = Overall Value. H = Hand-held, C = PC radio/SDR, S = Small, easily portable. M = Medium, suitcase size. L = Large, table top use. * = Avoid ** = Poor *** = Fair **** = Good ***** = Outstanding.
NOTE: Prices vary due to exchange rate fluctuations. Some models may be unavailable in certain markets.

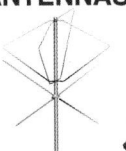

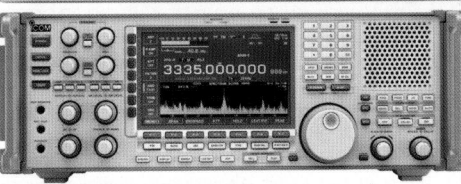

80s & 90s Classic Receivers

For the last few years we have published features on some classic professional and military receivers. This year we have looked at several other iconic and highly desirable professional and military classics of the later Cold War period and beyond.

The receivers we have examined for this feature represent the apotheosis of what might be called the "conventional" superheterodyne design in conjunction with extensive internal digital memory and processing circuitry or microprocessor control. All embody advanced synthesizers and all have extremely good signal-handling abilities. Two in particular feature digital signal processing (DSP). Progress in this field has been astonishingly rapid and what was state-of-the-art in 1990 looks quite limited today. But even early DSP systems offered functionality which was remarkable. At this date, what is usually referred to as the *front end* was fairly conventional insofar as there was an RF amplifier, mixer, local oscillator and IF amplifier – the latter often with some degree of AGC control applied. After that, however, the signal was fed to an analogue-to-digital converter (often abbreviated to A/D in the literature) and thence to whatever digital signal-processing implementation the manufacturer chose.

There are several major benefits to this type of signal-processing chain. One is that the IF filters exist solely as software and it is consequently not difficult to arrange for as many as are required, with properties which can be easily defined and specified. By the standards of high-grade and highly expensive conventional crystal or mechanical filters, the resulting shape factors of around 1.15 or better coupled with very low passband ripple (usually no worse than about 0.2dB) are extremely good, as is the excellent phase linearity. The importance of this latter point should not be overlooked. It implies that the group delay throughout the passband is constant and hence any signal – speech, music or even a digital pulse train – retains its original characteristics. Even more remarkably, signals with fast rise and fall times such as CW do not suffer from the 'ringing' introduced by conventional narrowband filters. To those whose ears have been wearied over the years by listening to Morse code in this way, the performance of modern digital filters is nothing short of a revelation.

Modern DSP systems add a variety of other features to the functionality of a receiver. Such delights as noise reduction, notch filtering, highly configurable AGC with individually selectable attack and release times and clever squelch systems are perfectly feasible. Early DSP receivers were not quite so capable but even by the mid-1990s it was fairly clear that the technology marked a major step forwards.

We mentioned the Rohde & Schwarz EK receiver family in last years' *WRTH*, and in some ways this and the Telefunken E1700 can be regarded as the high points of the conventional European superheterodyne with high-grade conventional filters and designed for the widest possible dynamic range. But other companies were still very active in the field; the post-Cold War wind-down in military and professional electronics had not quite taken effect and manufacturers still retained healthy and capable R & D departments and budgets. However, it was becoming evident that times were changing and that the money-no-object approach to receiver design would have to give way to something else. One can, for example, view the changes in the Racal product line between the 177x family of the 1970s and the 370x family of the 1980s as an exercise in maintaining and if possible improving performance and reliability but considerably reducing the cost of production. But it is striking that some long-term users of Racal receivers including BBC Monitoring, the UK Foreign Office and GCHQ changed their allegiance to Watkins-Johnson in the early 1990s, supposedly on the basis of procurement cost but also because of the steeply rising costs associated with maintenance. It is often forgotten that the important point about what might be called a 'professional' receiver is that not only must it display excellent performance but it must do so consistently under a variety of environmental and operating conditions. To take a simple example, most of us can switch off and disconnect our antennas when a thunderstorm is in the vicinity. A broadcast monitoring establishment or indeed a military or SIGINT site can hardly do the same. Its receivers also need to be capable of operation by users who might not be familiar with the finer points of receiver technology and who might also not be very mechanically or electrically sympathetic. So reliability and resistance to the slings and arrows of everyday 24/7 operation is vitally important. In this area, the BBC quickly found that its W-J receivers were an order of magnitude more reliable than their Racal predecessors.

The RA1217 family was Racal's solid-state follow-up to the seminal RA17 of the 1960s but unfortunately it exhibited several shortcomings. This was especially the case in the crucial area of strong-signal handling, where it quickly became clear that a fresh approach was required. The result was the mid-1970s RA1772 and its variants,

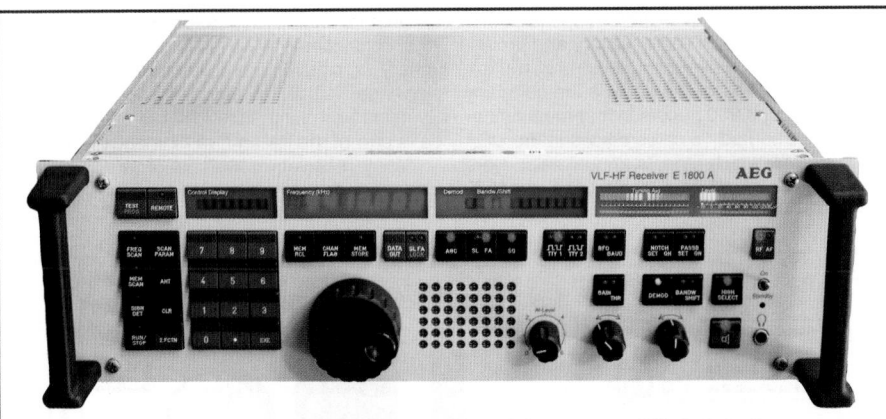

AEG-Telefunken E1700/E1800

Originating in the late 1970s, the Telefunken E1700 is generally stated to have been designed in response to a requirement from one of the German SIGINT agencies for a surveillance receiver to replace its predecessor, the E1500. It had a unique front end consisting of a Rafuse quad-FET followed by parallel-path discrete crystal filters and fascinating frequency-independent constant-impedance all-pass networks with strictly controlled phase shift, giving the mixer a constant 50ohm load impedance across the entire frequency range. The result was exceptionally good intermodulation performance, with the third-order intermodulation intercept being a very high +38dBm even without the optional (and highly expensive) preselector.

Many enthusiasts view the E1700 as the finest-performing HF receiver ever made although it is not at all intuitive to use and the learning curve is quite steep. Despite its rather 'spooky' origins the E1700 saw wide use in other applications, notably maritime coast stations, and examples are still in use by European broadcasters for quality-monitoring purposes. The E1800 and later E1800/3 variant, manufactured

between 1984 and 1992, employed microprocessor control and the E1800A (1993-98) embodies very capable digital signal processing. Amongst other things this features a notch filter which cleverly remains in the same part of the passband as the receiver is tuned and shows the result on a small display. In conjunction with passband tuning, the stepped variable-bandwidth filters cover a range of 100Hz to 10kHz and exhibit phenomenally low phase distortion. Curiously the AGC is rather rudimentary and tends to generate pops and clicks in the audio under certain conditions; the AGC performance is perhaps the only weak point in an otherwise formidably good receiver. There is no noise blanker but a squelch facility is provided. There are 100 memory channels and any combination of frequency, bandwidth and mode can be stored and recalled as required.

Unfortunately all these receivers are somewhere between uncommon and exceedingly rare and are regarded as highly desirable. Prices are consequently high; the E1800A in particular commands a five-figure sum on the extremely infrequent occasions one appears for sale.

notably the RA1778 with continuous tuning and 12 memories. All embodied both a high-grade wideband preamplifier and the then-new Rafuse switching mixer employing four high-current FETs and about +7dBm of local-oscillator injection. The later RA1792 family was largely a development of the 177x series but intended to be much less expensive, with a new fractional-*n* synthesiser and microprocessor control instead of the complex multiple-loop TTL-based synthesiser in the RA1772. With the possible exception of the RA1217, all these Racal receivers remain deservedly popular with enthusiasts today.

Some of our American readers have gently chastised us for so far making little mention of the Watkins-Johnson company in this feature series, arguing that any discussion of high-grade

receivers is incomplete without at least some reference to them. As long-standing enthusiasts for this illustrious brand, this is an omission we are very happy to rectify. The Watkins-Johnson company was founded in 1957 by Dean A Watkins and H Richard Johnson; Watkins was a professor of electrical engineering at Stanford University and Johnson was head of the microwave laboratory at Hughes Aircraft in southern California. Their declared aim was initially to manufacture electronic components and then to diversify into electronic systems and devices. In the 1960s several significant mergers and takeovers of companies heavily involved in radio receiving systems considerably increased the company's size and scope, and by 1970 its sales were worth about $65 million. By the mid-seventies Watkins-

Johnson is said to have been responsible for about 5% of the USA's entire military and professional market for receivers and electronic-warfare systems. By the end of the decade the company employed about 3,200 people in four manufacturing facilities and had 15 sales offices in the USA and Europe, by which time its annual sales were approaching $130 million. However, the company fell on hard times in the 1990s and in 1999 its various constituents were sold off piecemeal to a variety of buyers. The telecommunications division went initially to Marconi and then to BAE Systems and Signia-IDT. What remains is nowadays part of the Italian firm Finmeccanica.

Many receivers nowadays regarded as classics were produced by Watkins-Johnson during

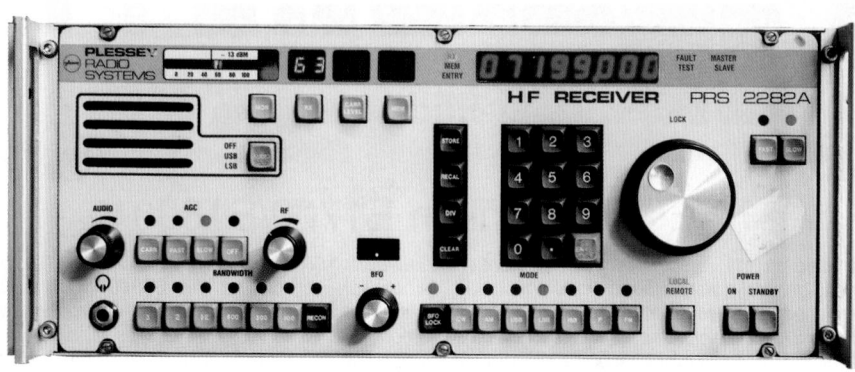

Plessey PRS2282A

It is often forgotten nowadays that the Plessey company was a major UK institution between its founding in 1917 and its takeover by GEC Siemens plc in September 1989. At its height in the 1970s the Plessey Group had about 85,000 employees worldwide and was represented in 136 countries, with some 130 establishments in the UK alone and 250 elsewhere in the world. The company was a major supplier to the British Post Office, the Ministry of Defence, civil and military aviation companies and many others. It was active in several fields including telecommunications, radio and radar, avionics, data processing and semiconductor design and manufacturing. Plessey's Communications Division at Ilford, UK designed and built a wide variety of HF, VHF and UHF radios for military and professional users, some of which are still in service today.

One of the last in a long line of illustrious HF receivers of which the PR155 was one of the best known, the PRS2282A was introduced in 1987 as a fairly conventional double-conversion superhet covering 10kHz-30MHz. As with its predecessors (the PRS2250/2280 family) the 2282A was fully modular and designed with ease of maintenance in mind. It was also beautifully engineered to the highest possible standards, and internally it is a work of art. Catering for AM, SSB, ISB, CW and FM, the PRS2282A offered IF bandwidths of 0.1, 0.3, 0.5, 1, 2.7 and 8kHz and had 100 memories with scan facilities. It is one of the last non-DSP receivers of its generation and its performance is excellent. There is a good deal of very careful and purposeful design in the

PRS2282A and it is a testament to the ability of the Ilford development team.

The PRS2282A was chiefly employed by the UK Ministry of Defence at several domestic and overseas HF sites. In particular it was used by the RAF at locations requiring SELCAL ground facilities, for which the receiver was specifically certified. It is also believed that the New China News Agency was a large-scale user. There is a prevailing view that a version of the PRS2282A was a component of the Plessey Pusher CDAA direction-finding system, also referred to by the US military as the AN/FRD13, of which about 25 were installed in various places around the world. Some receivers appearing on the surplus market have exhibited rather unusual IF and detector arrangements which are not covered in the extant literature and might indeed be associated with a DF application but the facts currently remain unclear and regrettably will probably always remain so.

It is rather unfortunate that relatively little service and maintenance information about the PRS2282A seems to be available. To make matters worse, the design made considerable use of Plessey's own SL-series integrated circuits for which some spares are virtually unobtainable nowadays and the extension cards necessary to work on the receiver's printed-circuit boards are exceedingly uncommon. Despite the difficulties associated with them Plessey HF receivers seem to have a cachet all their own and a PRS2282A generally finds a ready buyer on the infrequent occasions it appears on the market.

Racal RA3701

Although the RA1792 was an undoubted success, the company did not rest on its laurels and the RA370x family which replaced it in 1987 was a marked improvement in certain respects. There were four main models. The 3701 was the standard receiver and the 3702 was essentially two receivers in a single chassis. The 3703 was a remotely controllable variant of the 3701 and the 3704 was a dual 3703. Two others in the series (the 3705 and 3706) were for specialised direction-finding applications. All were entirely modular with individually shielded circuit elements linked together on a common backplane. Standard receivers comprised a front end, LO synthesiser, reference oscillator and BFO and IF/AF together with processor, power supply and front panel modules. However, a range of other modules was available and up to five could be fitted to a 3701. Coverage was 15kHz-30MHz and the standard filter set was 0.3, 1, 2.7, 6 and 12kHz although one of the optional (and enormously expensive) modules added a further seven.

The RA1792 had featured a very early 3850 microprocessor working in conjunction with the 3853 static memory interface. Its firmware was held in EPROMs which by today's standards had minuscule capacity. All these ICs were very fragile by today's standards and were replaced with rather more robust items in the 370x family. The microprocessor was changed to a Motorola 68000 16-bit device clocked at 4MHz. One of the minor problems presented to present-day 370x owners is that compatible 16-bit EPROMs are remarkably difficult to find and firmware changes

or maintenance can consequently be awkward. The RA1792 was notorious for problems with its liquid-crystal displays, many of which became unusable after a few years of service. Replacements were essentially impossible to find until a very enterprising Italian company (http://www.1792lcd.com) began production in 2009. The 370x uses dot-matrix displays which do not seem to give any problems apart from mildly unreliable backlighting.

As with the Plessey PRS2282A introduced in the same year, the RA370x did not feature digital signal processing. It is best considered as the acme of the Racal approach to conventional receiver design with its extremely quiet synthesiser, very robust front-end and excellent ergonomics, the latter always being a Racal strong point. With its control range of about 120dB, the AGC in the 370x is amongst the best of any HF receiver. The AGC threshold can be changed with the manual IF gain control, which incidentally can also act as a squelch and RF gain. The keypad incorporates four 'soft keys' which give access to a comprehensive menu system to set various infrequently changed parameters and also sets up the passband tuning and filter configurations. Like the 1792, the 370x has 100 memories and scan facilities. It also shares the latter's BITE but the facilities in the 370x are considerably more comprehensive. In general terms the performance of any of the 370x family is very much on a par with that of an RA1792. The filters have very good shape factors and the ergonomics are beyond reproach.

its lifetime although a high proportion of them cover frequencies rather higher than the HF broadcast bands. As any collector of these will ruefully confirm, the company was notorious for producing numerous variants of the same basic receiver; given the covert purposes to which many of them were put, there is often very little information available about them. Happily the sit-

uation is rather easier in respect of the HF receivers and devices such as the 8700, 8711, 8718, 8888 and the like are highly prized by collectors. We would be hard put to name a favourite amongst them but if forced to choose one it would probably be the 8711A.

Should you buy a classic receiver? As we said last year, the answer is a qualified yes. Much

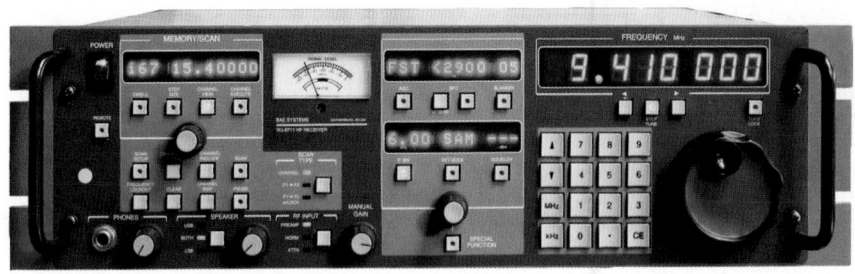

Watkins-Johnson 8711A

Many readers will know the Watkins-Johnson 8711 as the professional precursor of the consumer-market HF-1000 introduced in 1993 and discontinued in 1999 when the company's telecommunications group was sold. BBC Monitoring at Caversham was for many years a large-scale user of Racal receivers but in the 1990s it changed to Watkins-Johnson, initially using the 8718 and then acquiring the 8711. Although HF listening is nowadays a tiny part of Caversham's task, there are still some 8711As in service. The 8711A was a later variant with slightly different hardware (notably a different DSP IC) and some other changes to improve performance and address issues such as extremely harsh audio quality and a susceptibility to internally generated noise. Both receivers cover 5kHz-30MHz and handle AM, synchronous AM, ISB, USB, LSB, CW and FM. There are 100 memory channels and a versatile scanning facility. An optional sub-octave preselector is available and the BBC's experience was that this was useful if the receiver was used in conjunction with some of its larger antennas.

Visually the 8711A is impressive although it must be one of the lightest high-grade receivers ever made; the internal switching-regulator power supply was one of the first of its type to be used in a professional HF receiver and is both very lightweight in comparison with a conventional linear supply and more efficient and cool-running. Intended for mounting in a standard 19in rack, the front panel is dominated by the large and clear LED frequency display (reading to 1Hz) and the tuning knob and keypad beneath

it. The 'feel' of the tuning is pleasant and positive. Unlike some later DSP receivers with continuously variable filters, the 8711A has 66 discrete IF bandwidths ranging from 56 Hz to 16kHz. Initially this facility seems a little unusual, principally because not all of them are available to each mode; each has 'factory defaults' which may be easily changed if desired, although in practice they are quite well chosen. For use with teletype and other data modes the BFO frequency is adjustable in 10Hz increments. There is also a notch filter and adjustable noise blanker. The latter is simply superb and in our experience capable of dealing with practically any source of extraneous noise if it is correctly used.

The manufacturer's claimed performance is on a par with that of other receivers in its class, notably a +30dBm IPI_3 and very tightly defined IF filter shape factors. In use, an 8711A certainly sounds very clean and capable even when driven from a large antenna. The synchronous detector seems to work quite well although there have been reports that in earlier receivers it loses lock rather more easily than might be expected. Later firmware releases apparently addressed this issue with a degree of success, and in fact the 8711A saw several revisions and improvements to both hardware and firmware during its production lifetime.

Neither the 8711 and 8711A nor the HF-1000 derivative are particularly rare but they continue to command quite high prices. If you must have a Watkins-Johnson receiver in your collection but would prefer to pay rather less, the earlier 8718 is also something of a classic.

depends on whether you have the ability to maintain it yourself or know someone who can; many classic receivers of the period are complex and documentation is not always easily available. Some of them also lack modern features and functionality. But for receivers with external 455kHz IF outputs, the addition of a suitable

12kHz down-converter can bring DRM and the benefits of DSP together with noise reduction, notch filtering and so on.

But we repeat last years' word of warning; you may become so attached to the combination of wonderful ergonomics and superb performance that one may not be enough.

QSLing Then & Now

QSL cards, once a visible sign of the importance of verifying sucessful transmissions, are becoming a thing of the past. But Jerry Berg, renowned DXer and keen collector of QSLs, argues that QSLing can still be an important part of a DXer's life.

Thanks for writing. We always like to hear from our radio audience. Can we do anything for you? This verifies your reception.

THE RADIOVOX CO.
R. M. ROLAN,
Station Director

Broadcast stations began using form QSLs early. This card from WHK, Cleveland, Ohio, is from 1924

While many DXers of broadcast stations, both on shortwave and mediumwave, are happy just to log or record a station, others view QSLing as an essential step in their DXing.

Although broadcast station QSLing dates back to the earliest days of radio, the elements have remained constant over the years. The listener reports the date, time and frequency of reception, together with enough program details to prove that it was the reported station that he heard. The station is expected to check the report and send the listener a card or letter verifying the reception. Some of the details of the process have changed. The advent of home recording enabled the listener to submit a recording, all but eliminating the issue of what station was heard,

and DXers became more relaxed about the early requirement that, for a QSL to be valid it had to contain the date, time and frequency of reception.

There is less QSLing taking place now than in times past. The world has grown smaller, and long distance broadcasting, for many decades a technological marvel to both listener and station, now competes with numerous other content delivery vehicles. As the novelty of broadcast listening, particularly on shortwave, has lessened, so the number of QSL collectors has declined. But QSLs are still prized by those seeking tangible mementos of their DX experiences.

Technology has had both positive and negative impacts on QSLing. At the most fundamental level there are now many fewer stations, and thus

In the 1950s, many Scandinavian clubs arranged for special broadcasts, which the club then verified.

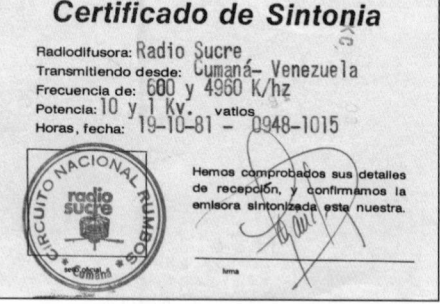

This prepared card from Radio Sucre is typical of the pre-pared cards that were used in the pre-computer days

Prepared cards can now easily be designed for a particular station. This one is from the IBB station on Saipan

fewer listening targets and fewer opportunities to obtain QSLs. Many stations also now have a variety of other methods of determining reception quality, and have to rely relatively little on listener reports. Receiving reception reports and QSLing them has long been more of a courtesy than an engineering activity, and this is truer now than ever. Add in the budgetary pressures that are faced by many stations today, and the active QSLer is likely to encounter a less enthusiastic station response than he once did. Some stations have stopped QSLing altogether.

However, for listeners and stations that still engage in QSLing, technology has greatly facilitated the process. Today, reception reports are often submitted by email rather than by postal mail and thus reach the station much more quickly. In addition, technology permits the attachment of a recording of reception with relatively little effort. Of course, electronic reports also lend themselves to being more easily ignored.

R. Nacional de Venezuela sent this QSL by email in 2004

While the sending of return postage is a problem with electronic reports, the need for it is eliminated if the station replies by email. Email replies are usually received more promptly than postal replies. However, the sameness in appearance of emails makes them less attractive for display purposes. Thus many listeners print out their email replies, sometimes on glossy or colored paper, and dress them up with station-related graphics taken from the internet. This makes them more presentable, but care must be taken not to alter the email to the point where it is amplified beyond its plain meaning, such as adding a "QSL" overprint to an e-mail that is just a thank you.

Today's computer technology also provides smaller stations that can't afford the cost of printing and mailing QSLs with a vehicle for designing

Radio Xoriyo

Date: Wed, 18 Jun 2003 17:53:50 -0700 (PDT)
From: International Ogaden Website <ogaden@yahoo.com>
Subject: Re: Reception of Radio Xoriyo in the U.S.
To: Jerry Berg <jsberg@rcn.com>

Dear Jerry Berg:

We thank you for your interest about Radio Xoriyo. The information you provided is correct.

Regards,
Ogaden Online staff

This Ethiopian clandestine station QSLed by email. The logo, found on the internet, was added to the email

a nice looking QSL and sending it as an email attachment at virtually no cost.

So while the anticipation that always accompanied a trip to the mailbox is now largely a thing of the past, there are offsetting pluses when one receives an email or a QSL attachment from a station that might not have replied by postal mail.

The process of sending prepared cards to stations that do not have their own QSLs is also greatly improved. With the most basic computer and printer the listener can prepare station-specific prepared cards that look much better than the generic prepared cards that have long been sent with postal reports.

The world of QSLing is changing, and while we might still long for the days of fancy envelopes from exotic places, the e-world does offer some compensating benefits.

Jerry Berg is a member of the Executive Council of the North American Shortwave Association, chair of the Committee to Preserve Radio Verifications, and co-producer of the www.ontheshortwaves.com website. He has been a DXer for over 50 years and has written three books about the history of shortwave broadcasting and DXing: On the Short Waves, 1943-1945, Listening on the Short Waves, 1945 to Today, *and* Broadcasting on the Short Waves, 1945 to Today.

Image 1 courtesy of Committee to Preserve Radio Verifications

AFN on Saddlebunch Key

The low-lying Florida Keys are home to a shortwave transmitting station that beams talk radio to American forces worldwide. Hans Johnson went to take a look.

The Granger Conical Monopole antenna, model AS-1974/FRC. The frequency range is 7-28mHz

The Florida Keys – turquoise seas, endless sunshine, and shortwave broadcast antennas. Antennas located on the southern isles of the Sunshine State? Well, on one key anyway. Saddlebunch Key, 15 miles (24 kilometers) east of Key West, is home to a U.S. Navy broadcast facility. The base carries out a number of duties. Some are secret, but the station's best known mission is unclassified. That mission is broadcasting U.S. Armed Forces Network (AFN) programming to American servicemen and women.

Official American forces broadcasts started during the Second World War. Shortwave, transmitted in AM mode within the recognized international shortwave broadcast bands, was used extensively. These transmissions ended in 1988 and were replaced by satellites. But as many other broadcasters have realized, the shortwave transmissions had ended too soon. Even in the satellite age, not all military personnel on board a ship or ashore could receive AFN programming.

A small ship equipped with a satellite dish might not have sufficient bandwidth to devote space to receive AFN programming, and personnel on a military exercise or temporarily deployed might be out of reach of existing stations.

As a result shortwave transmissions were restored in 1998 to fill this gap. The Navy already ran a global network of shortwave transmitters used for naval communications and the revived shortwave service used this same network for AFN broadcast transmissions.

At its peak, the network included stations in locations as diverse as Puerto Rico, Iceland, and Guam. There are just a few stations left in the network now and Saddlebunch (Key West) is one of them, the others being Pearl Harbor, Diego Garcia and Guam. A visit to the station starts with the Overseas Highway which connects the Florida Keys with the mainland. Locals use the highway mileage markers as a means of giving directions. Fifty meters past mile marker 15 is the entrance to the station. One can easily see the station's large antenna field to the north at this point. A locked, non-descript gate off the highway blocks access to the station.

The entrance road winds through the low mangroves and contains a second gate. Saddlebunch is a low-lying island with the base of the antennas not far above the high-tide line. The stout concrete station building was built on top of fill and is just a bit higher.

Andrews SPIRA cone antenna, model 3002-36HE. The frequency range is 2/4-30mHz

The 11m Whip antenna by Chu Associates Inc. The frequency range is 2-32mHz

A number of civilians working for the Navy run the station with their headquarters located in the northern Florida city of Jacksonville.

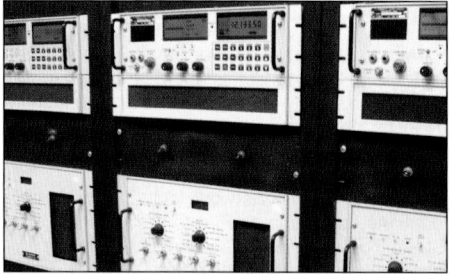

Signals are fed from the Navy base in Key West. The transmitter hall contains dozens of one kilowatt Harris AN/URT-119 and AN/URT-42 transmitters. Three of these transmitters running at full power carry the AFN voice channel service 24 hours a day on 5446.5, 7812.5, and 12133.5 kHz in the upper side band. Listeners will need a general coverage receiver capable of receiving single-side band to tune in.

A variety of antennas is used, including inverted cone (Granger AS-1976 series), conical monopole (Granger AS-1974 Series) with low and medium take-off angles, or SPIRA cone (Andrews 3002-36HE series). All these antennas are omni-directional.

Propagation from the island is excellent and

the station has received reports from as far away as Japan. The station is interested in distant reception and will verify reports sent to: Technical Director, PO Box 9045, NAS Key West, Key West, FL 33040-9405, USA. Return postage is not necessary. Listeners wanting to tune in can find a program schedule at the following web address: http://myafn.dodmedia.osd.mil/AFNRadio.aspx

The primary audience for the service remains, of course, the Navy personnel in the surrounding

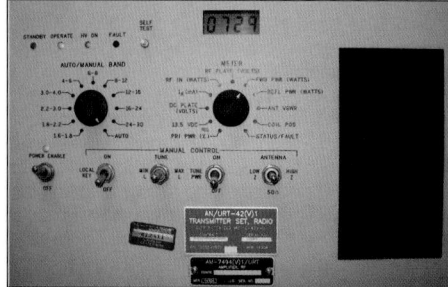

waters and as far away as the North Atlantic, for whom the Sports programs are the most popular. The need for the AFN shortwave stations remains and they have a bright future.

Hans Johnson is a freelance and technical writer based in Naples, Florida. He founded Cumbre DX in 1994.

Digital Update

Our regular round-up of what has been happening in the world of digital radio and TV over the past year

"Radio with pictures" was one of the issues promoted during the early days of DAB as an inherent feature and benefit of the technology. Much was made of this in broadcasting circles but realisations that it would not actually work very well in practice caused the notion to be quietly forgotten. On the other hand DRM has always been of some interest to broadcasters as a potential carrier for low-rate video and data as well as conventional radio. In *Diveemo* the video is encoded in H.264 format and the associated audio uses HE-AAC (high efficiency advanced audio coding) which is in essence the MPEG-4 system used in DRM. More than one audio stream can be carried alongside the video, allowing multi-language soundtracks. Diveemo is currently awaiting standardisation approval from the European Telecommunications Standards Institute (ETSI). Operating at a mere eight frames per second, Diveemo transmissions are clearly neither designed nor intended to compete with standard broadcast television. The service is instead being positioned for large-area distribution of education and news programmes where the video forms a supplement to the existing audio content.

The HF DRM standard has now become 'DRM30' and 'DRM+' refers to use of the mode on frequencies above 30MHz. There are, to be sure, some developments in DRM30 to report. In April 2010 the Indian government elected to fund a "digitalisation" programme worth about $200 million for All India Radio such that approximately 70% of the country would be covered. The associated tender involved bids for the supply of 34 new MW transmitters, the upgrade of 36 existing MW transmitters and purchase of five more for HF broadcasting. Malaysia and Australia have both acquired two DRM-capable HF transmitters, and in South Korea the Digital Radio Committee has "officially decided to examine DRM+ in their comprehensive comparison of digital radio technologies", whatever that may mean. The Russian Radio Frequency Centre announced the introduction of DRM in the MF and HF bands in March 2010, with an ". . .upgrade of the radio infrastructure to digital" apparently to start in 2012. In Brazil a series of DRM trials involving DRM30 and a 26MHz SFN was about to end as this was written. It was announced earlier in 2010 that Continental Electronics was to supply four 250 kW DRM-ready HF transmitters and associated equipment to the Saudi Arabia Ministry of Information (MOI) through First Gulf Company of

Riyadh. Apparently First Gulf is intending to construct ". . .an entirely new HF station where the transmitters, antennas, and other equipment will be installed at the existing Al Khumra site outside Jeddah." This site was originally constructed by Continental Electronics and its civil contractor between 1978 and 1980 and presently accommodates several high-power MW transmitters. The new HF transmitters will be delivered in the

> ## *". . .regrettable conclusion that mass-market manufacturers have no interest at all in DRM."*

latter part of 2010 and the station is planned to be fully operational by mid-2011.

All this is all very well. We know DRM30 works and we know that DRM+ works. We know that a variety of broadcasters are doggedly transmitting DRM and that with a suitable receiver it can give good results. We are well aware of the efficiency advantages; a DRM transmitter costs a good deal less to run than a conventional AM transmitter with equivalent coverage, which in an energy-conscious age is a powerful argument for the mode. But the questions are – who is listening and where are the receivers?

Consider some simple facts. Elsewhere in this issue we have reviewed some very low-cost portable radios of Chinese origin which use a modern and very capable DSP integrated circuit from a well-regarded American manufacturer. These receivers cost at most a few tens of dollars in the West – probably less than that in the Far East – and they work astonishingly well. They offer local AM and FM reception together with perfectly adequate HF. Radios such as this are affordable by even the lowest paid and they sell by the container-load in their native country and the less developed world.

Now consider the case of DRM. What mass-market receivers are available for this mode? None whatsoever. The RadioScape RS500 module which was used in the short-lived Morphy Richards DRM receiver is no longer in production. The NewStar WR608 DRM/DAB chipset was intended to be used in the Uniwave DiWave 100, which we have been mentioning in these pages for several years. Readers will no doubt raise wry

smiles when we report that production samples have not become available.

We are driven to the regrettable conclusion that mass-market manufacturers have no interest at all in DRM. And even if they had, it is inconceivable that a DRM receiver could be built down to the same order of price as a current Chinese low-cost portable within any realistically foreseeable timescale. Most *WRTH* readers will know that a high proportion of modern SDRs cater for DRM with no particular difficulty. The WinRadio WR-G31DDC Excalibur reviewed on page 13 is the best-performing DRM receiver we have encountered to date. It produces reliable audio from a variety of DRM transmissions and can hardly be faulted in this respect. But no SDR is a mass-market item. It is inevitably a discretionary purchase on the part of an enthusiast who can afford the required outlay. In fact it would be interesting to know how many DRM receivers are in use in the world during an average period of 24 hours but we suspect the answer would be depressing for all concerned.

There is also the fact that HF broadcasting overall continues to be in decline, increasingly supplanted by satellite TV and radio, local FM radio and the internet. And sadly there is little reason to expect the position to reverse. A recent estimate suggested that as far as major international broadcasters were concerned, something less than 40% of their audience relied on HF reception. However, overall HF transmission costs represented somewhere between 20 and 25% of the total annual budget. Given the depth of recession currently being experienced worldwide, it is extremely difficult to argue against the proposition that costly and power-hungry conventional HF transmitters seem ultimately destined to fade away. In this context, arguments over DRM uptake seem irrelevant.

To make matters worse, the future of domestic analogue radio broadcasting – certainly in Region 1 – remains very far from clear. An intriguing straw in the wind was that in mid-2010 the BBC's Chairman called for a review of its radio strategy, acknowledging in effect that DAB had not been a success. He said "The BBC's newer stations were designed in part to drive digital take-up. By 2010 we can see that take-up of DAB radio has been slower than expected ten years ago and the BBC's digital-only stations have not achieved the audiences or impact that was then expected, although the intention behind the Digital Economy Act was to provide new impetus". He added that "The BBC is already committed to playing a role in leading the UK radio industry to a fully digital future. A question remains about what that means in the longer term and ultimately what the potential is for internet-based radio platforms to evolve."

It seems clear now that any notion of an analogue switch-off before about 2020 is wide of the mark. It is also clear that when it does take place, DAB will not be the large-scale replacement. It seems that in most parts of the world – especially the developed world – "digital" listening will continue to take some other form than conventional off-air terrestrial broadcast reception; that is to say, podcasting, streaming or via satellite and the internet. Presumably at some stage there will be what might fancifully be called "Son of DAB" but it

> *"...we can see that take-up of DAB radio has been slower than expected..."*

is still not clear whether it will be DAB+, DVB-H, DRM+ or something else. This is an indictment of official decision-making.

The situation in other ITU regions is no clearer. In Region 2 there are now about 1,800 HD Radio outlets and about 90 different receivers were on the market late in 2010. But overall awareness of the mode remains remarkably low and it seems that for non-technical listeners HD Radio and satellite radio are often perceived as being the same thing. There appears to be a continuing groundswell of dissatisfaction with the audio quality of HD as well, and even high-grade FM tuners with clever post-detector filtering do not always perform as well in subjective terms as measurements suggest they might. Given these reservations together with the reluctance of iBiquity to release details of its proprietary codec, there is little doubt that HD Radio will not become a world standard, although it should be mentioned that there are a few stations in other Region 2 countries apart from the USA.

And finally, the situation in respect of television is little different from that we reported last year. The only development of note is that the European Commission has proposed requiring member countries to sell off analogue TV frequencies by 2013 and to allow wireless broadband elsewhere in the UHF spectrum by 2012. The proposal appears to have its origins in an EU commitment to some form of broadband being universally available by 2013 and at a minimum of 30Mb/s by 2020. It appears that several European countries will have to deregulate spectrum at 900 and 1800MHz currently allocated to GSM mobile telephones and sell off 2.6GHz and 3.5GHz by 2012. This process is supposed to run in parallel with the abandonment of analogue television in Band V (i.e. around 800MHz) by January 2013 – in what is referred to in some circles as the "digital dividend" – and thence the provision of broadband access to all EU citizens. To all of which the only possible rejoinder is that it will be very interesting to see what happens.

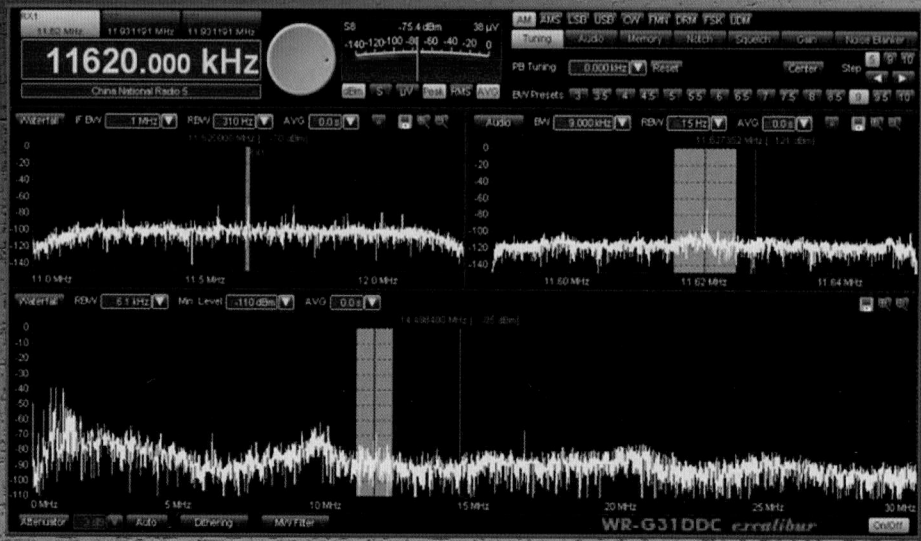

Radio St Helena

Radio St. Helena occupies a special place in the affections of DXers and listeners worldwide. Here Gary Walters, Tony Leo and Robert Kipp tell the story of the station, and of Radio St Helena Day

A view of the island with the Cable & Wireless station in the lower right foreground

The medium-wave station, Radio St Helena, was established in 1967 as a result of the initiative and hard work of George Lawrence (Information Officer for the Government of St Helena), George Barett (Diplomatic Wireless Relay Service on St. Helena (DWRS)) and Alan Johns (Education Officer).It was part of a ten-year plan for education, under the control of the Education Department, and was officially opened by Sir Dermond Murphy, the Governor of St Helena, on 25 December 1967. Prior to that, programmes for schools and for the adult population had been broadcast as test transmissions on a frequency of 1511 kHz for a period of about six months. The station had been registered with the International Registration Frequency Board in London as the "Saint Helena Broadcasting Station", and its iden-

tification signal was four evenly spaced trumpet calls followed by a few bars of "Life on the Ocean Waves" played by the band of the Royal Marines, which, in turn, was followed by the announcement: "This is the Saint Helena Government Broadcasting Station operating in the medium wave band on a frequency of 1511 kHz, 200 metres". The MW frequency was changed to 1548 kHz. in November 1978.

The Radio St. Helena (RSH) building was financed from Colonial Development & Welfare Funds provided by the UK Government and was designed and constructed by the Public Works Department in consultation with the engineering staff attached to the DWRS stationed on the island. A large proportion of the original equipment was of pre-World War II vintage and included

The QSL card for the "Final Transmission" in 1999

The QSL card for the Anniversaries in 2007

Tony Leo, Station Manager April 1973 to October 2001

the equipment single-handed. A total of 33 volunteers were busy at RSH in the 1980s, and today RSH has 21 active volunteers. The listening audience had a choice of light music interspersed with informative and educational items and with the inclusion of topical tapes from the BBC Transcription Services (RSH now uses pod casts with kind permission from the BBC). However, some 20 hours per week of locally-produced programmes were the biggest draw with the heterogeneous population of some 5,000 people. As they still do with current population of approximately 4,000 people on the island.

During a four-month period starting in August 1973, a series of shortwave test transmissions was carried out on frequencies of 11830 kHz and 6100 kHz to determine whether it was possible simultaneously to transmit local MW broadcasts to the 400 or so "Saints", or residents of St

two Marconi 500 Watt MW transmitters. The MW station now uses two transistorised 1 kW Gates One (Harris) transmitters. The MW antenna is a vertical radiator with capacity top-hat for tuning and was built by Larry Francis (station engineer at RSH in the 1990's).

In the early years, the Information Officer (also editor of the "St Helena News Review"), assisted by a clerk, was in charge of operations and programming and the Education Officer operated the station. There was a total of just three members of staff in the Information and Broadcasting Services office. Finally, in April 1973, Tony Leo was appointed as the first Broadcasting Officer (later renamed Station Manager). From the start, many programmes were produced and broadcast by voluntary helpers who sometimes operated all

Ralph Peters, Station Manager October 2001 to 2006

Helena, based on Ascension Island. This test failed, however, as transmissions could only be heard on professional receivers.

RSH's first live broadcast was of a local football match in 1974 and, in the early 1990s, RSH obtained a complete mobile studio in a 'bus' from which to cover local events such as the first visit of the liner Queen Mary II on 30 March 2010.

In 1990, John Ekwall, Jan Tunér, and Lennart Deimert in Sweden organised a broadcast by RSH, using mediumwave and shortwave in parallel, especially for participants of the NorDX radio listening contest. Cable & Wireless (C&W), located near Jamestown, allowed RSH to use a 1.5 kW Redifon G423B transmitter (with USB modulation) and a cage dipole antenna. On 6 October 1990, with Joy Lawrence at the microphone in Studio A, RSH began broadcasting on shortwave for the first time. These one-day-per-year short-

Laura Lawrence, Station Manager 2006 to Sept. 2008

The RSH building with the SW and MW antennas and satellite dish

wave broadcasts on 11092.5 kHz continued in 1992, and became world-famous as "Radio St. Helena Day" (RSD). After the transmission in 1999, C&W had to scrap the ancient Redifon transmitter and the transmissions ceased. The 1999 broadcast was thenceforth known as the "Final Transmission".

It was, however, not to be the final transmission for in late 2005 Robert Kipp in Germany, along with the considerable help of many radio clubs and individuals around the world (especially the Japan Short Wave Club), started a project to put RSH back on the air on shortwave. This project came to fruition on 4 November 2006 when RSH broadcast the RSD "Revival" programme using a 1 kW (USB mode) power amplifier and a 3-element Yagi directional antenna. In late 2007 RSH celebrated an important double anniversary with the 10th broadcast on SW and 40 years of broadcasting on MW.

RSH has had four station managers: Tony Leo, Ralph Peters, Laura Lawrence and Gary Walters (September 2008 to the present). Tony Leo, "Mr RSH", retired in 2001, after leading RSH for nearly 29 years and producing all the RSD programmes in the 1990s.

It is hoped that RSH can continue to develop and expand for its many listeners on the island and the thousands of supporters it has in all walks of life around the world.

RSH would like to thank the many Friends, Sponsors, Experts, and Radio Hams who have assisted or made donations to RSH and also all those radio listeners from around the world who have sent in reception reports and written so many encouraging and friendly letters.

Gary Walter writes: *Special praise and thanks go not only to my staff and producers: Jane John, Claire Bennett, Manfred Williams (handyman), and Bobby Ellick (who just retired as station engineer); but also to my 'Saint' wife Cherry and my daughters Jodie, Emily, and Helena who all help to make RSH and RSD a success.*

Gary Walters, the current Station Manager

Ears To Our World

The importance of radio shows most clearly in those countries without access to other media or even some of the basic tools of education. That is why Thomas Witherspoon set up his charity.

A teacher and his pupils admire a new wind-up radio in Uganda

For as long as I can remember, I've been passionate about radio. When I was a kid back in the early 1980s, our family room featured, alongside the cable TV, a vintage RCA 6K3, an elegant wooden floor radio that once belonged to my grandfather. He used it, I was told, for civil defense monitoring during World War II. My father would put the forty-year-old instrument through its paces every Sunday morning, tuning to WWV in order to set his watch by the atomic clock. Dad would sometimes allow me to tune the dial, too, and I would hear foreign voices and mysterious numbers drifting through the ether. No doubt about it, I was hooked.

My passion grew still further after my great-aunt presented me with a Zenith Transoceanic, an old radio she unearthed in her cluttered basement. I made a home for it in my bedroom; now I could listen whenever I liked to international broadcasters, and soon began to learn when they were on the air, to recognize their interval signals, and to remember which bands the stations frequented. I simply couldn't get enough of the music and voices I heard.

Radio captured my imagination as TV never could, and taught me early on that everyone has a story. Radio taught me, too, that each voice is different in the consideration of what's meaningful or newsworthy. I learned to understand – or at least appreciate – the diverse perspectives I heard in my radio journeys, and from these sprang my own opinions, hopes, beliefs. Radio became my teacher, a teacher who gave me, in my formative years, a global perspective.

Radio has shaped my life. That's probably why it has recently become a mission for me. Today, I am the founder and director of Ears To Our World (ETOW), a charitable organization with a simple objective: distributing self-powered world band radios to schools and communities in the less developed world, so that kids and those who teach them can learn about their world, too. I want others, especially children and young people, who lack reliable access to information, to have the world of radio within their reach.

ETOW works in rural, impoverished, and sometimes war-torn or disaster-ravaged parts of the world. Places that lack reliable access to electricity (let alone the internet) and where radio is often the only link to the world outside. The heart of our mission is to allow radio to be used as a tool for education, so we give radios to teachers, who, in turn, use the radios in the classroom and at home to provide real-life, up-to-date feedback about the world around them.

It is a budding organization. But through the

encouragement of our good friends at Universal Radio and the extraordinary magnanimity of Etón corporation, who donate our wind-up world band radios, we are honored to have, in just two years and on a budget of less than $3500, distributed radios to schools and communities in nine countries on three continents – in Africa, Eastern Europe, Central and South America, and the Caribbean – as well as to both Haiti and Chile, where the dissemination of information through radio has been life-saving.

We have managed to do this through partnerships with other reputable and established nonprofit agencies just like us. Ones that already help struggling schools throughout the world, and who believe, as we do, in *freedom of* and *access to* information. Our partner organizations have laid the groundwork in these regions, and have established reliable connections with communities in them. Their need is for resources – like radios.

By working cooperatively with other established organizations, we find we're able to distribute radios much more cost-effectively. Because of our strong partnerships, for example, money otherwise spent on travel can be put into shipping costs instead, thus getting more radios to more of the world with less donated funds.

So far, our scope is limited only by our financial resources. We are looking to place radios in other countries off the beaten path, but we're not simply focusing on expansion. ETOW is establishing strong, lasting bonds with our schools and teachers so as to better serve their needs long term. We would also like to develop on-air teacher training programs, and a new partnership with Oklahoma University has the aim of developing

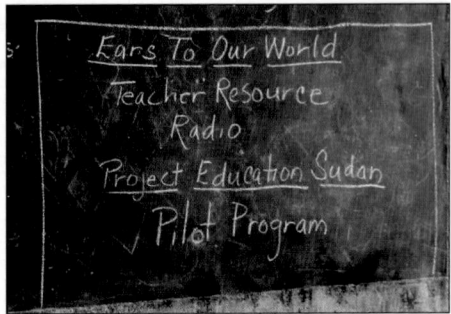

Helping teachers gain access to information which they can pass on to their students is a key aim of ETOW

and disseminating content on important subjects, such as literacy and health education.

We use radio instead of, say, computers for information access, because much of the world cannot access the world wide web and other dynamic media sources such as digital television or wireless networks. Political instability, meanwhile, can undermine the integrity of information provided by the written word. Radio, however, is simplicity itself: all one needs is a modest yet capable receiver and one has instant access to local and world media. Every teacher we've worked with so far already knows something about radio. But in these places it can take up to an entire week's wages to pay for a set of cells. ETOW's wind-up radios become vital: we effectively eliminate this cost, while giving them steady access to information. And the reports we have received back from the field have been overwhelmingly encouraging. Teachers in rural Cameroon are able to teach current events. Visually impaired children in rural Belize can listen to the outside world and hear information they had no knowledge of before. A remote community in southern Sudan was able to listen to reports of their country's first democratic election. Children in Haiti and families in Chile learned where to go to get food, medical care, and information about loved ones affected by the earthquakes.

As fellow radio enthusiasts and broadcasters, you already know that radio is a remarkable tool; allow me to convince you that it is a powerful teaching tool, as well. Just as radio taught me, and opened my young mind, I'm convinced that it can teach and open the minds of others. In some parts of our world, futures are still written on the airwaves. *Listen and learn.* It's a simple idea, but to some young people, it can mean the world.

Want to help us give the gift of radio? Visit ETOW online at earstoourworld.org or write us at PO box 3230, Cullowhee, NC 28723, USA. Your personal interest, or that of your local radio club or business, could put radios in a school or village in the most remote corner of the world.

A young boy in Belize is delighted with his radio

HF BROADCAST RECEPTION CONDITIONS EXPECTED DURING 2011

George Jacobs, MSEE, Fellow IEEE, Dean of WRTH Contibuting Editors, analyses likely listening conditions in the coming year

The Quiet Sun is Awakening

CYCLE 24 CONTINUES

During the course of an approximate eleven year sunspot cycle, the minimum phase, or quiet sun, is generally considered to exist during the time when the smoothed sunspot number (SSN) drops, and remains below 30. The smoothed sunspot number is a monthly index compiled by the Royal Observatory of Belgium for measuring solar cycle progress.

An unbroken string of smoothed sunspot numbers has been recorded since 1750. The present period of quiet sun began as declining Cycle 23 dropped below the SSN 30 level during April, 2005. NASA is predicting that the new Cycle 24 will rise above the 30 mark during early 2011, marking the end of the quiet sun period. A period of moderate solar activity is expected for the remainder of 2011, reaching a sunspot count on the order of 60 by year's end.

This six-year solar quiet period was the deepest and most persistent recorded in almost two hundred years. It mystified solar scientists, and it is another example of how little is yet known about sunspots and of the nature of the Sun itself.

To learn more about sunspots and the sunspot cycle I suggest that you visit this website solarscience.msfc.nasa.gov/SunspotCycle.shrml.

NASA'S SOLAR DYNAMICS OBSERVATORY

The Sun, our closest star, and upon which all life on Earth depends, is still a great mystery to scientists. On February 11 2010 NASA launched the Solar Dynamics Observatory (SDO), a bold attempt to solve many of the Sun's mysteries. The SDO is unlike any other satellite. It is taking a much closer, more penetrating look at the Sun and its atmosphere, specifically at sunspots, solar flares, solar eruptions and storms, the sun's magnetic field, the effects of solar radiation upon the Earth's atmosphere, including Its magnetic field and the ionosphere. The satellite is collecting huge amounts of solar data every day, along with solar images with definition much greater than HDTV. The Information is summarized daily along with images which can be viewed at this page on the NASA site: www.nasa.gov/sdo.

EXPECTED HF BROADCAST RECEPTION 2011

With the new Cycle 24 finally with us, a greater level of ultraviolet radiation from the increasing sunspot count is producing a stronger ionosphere in the earth's upper atmosphere. It is the ionosphere that reflects HF radio waves worldwide. A stronger ionosphere also extends the range of HF frequencies that will be reflected. This will be evident in 2011 with an increasing number of broadcast stations returning to the 17 and 21 MHz bands.

Table 1 is a summary of reception conditions expected in each HF broadcasting band during 2011. The 17, 15, 13 and 11 MHz bands are expected to be the most popular for worldwide reception during the daytime hours, with the 9, 7 and 6/5 MHz bands most popular during night time hours.

For readers who would like to learn more, an excellent introductory discussion on the ionosphere and HF propagation can be found at: en.wikipedia.org/wiki/Radio_propagation.

Good listening on the HF broadcast bands during the course of 2011.

TABLE 1

MHz	m.	%	Reception Characteristics
26	11	*	No significant HF broadcasting
21/19	13	2	Day: long distance, fall-spring seasons
17	16	5	Day: mid and long distance, all seasons
15	19	15	Day: mid and long distance, all seasons
			Eve: mid and long distance, not winter
13	22	10	Day: mid and long distance, all seasons
			Eve: mid and long distance, not winter
11	25	15	Day: short and mid distance, all seasons
			Eve: mid and long distance, not winter
9	31	18	Day: short and mid distance, all seasons
			Eve: mid and long distance, all seasons
			Night: mid and long distance, summer
7	41	16	Day: short and mid distance, all seasons
			Night: mid and long distance, all seasons
6/5	49	17	Day: short and mid distance, all seasons
			Night: mid and long distance, all seasons
4/3	75	2	Day: short distance, all seasons
			Night: short and mid distance, all seasons

m. = Metre Bands % = % of total band usage
* = used for local digital broadcasting in Europe
Short-distance: up to c. 1200 miles (2000 km)
Mid-distance: c. 1000-2400 miles (1600-4000 km)
Long-distance: over 2400 miles (4000 km)

ABOUT THE AUTHOR

George Jacobs is the dean of WRTH contributing editors, this being his 49th year of writing for the Handbook. He is a world-renowned innovative engineer and diplomat with a fierce belief in the free flow of information. George is a legend in the field of HF broadcasting, with 2011 marking his 70th year of practice. His biography can be found in the Marquis edition of "Who's Who in the World". He can be reached by e-mail at broadcaster@gjainc.com..

How to use *WRTH*

ORGANISATION OF THE BOOK

The book consists of three main areas: **Features**, consisting of equipment reviews, broadcasting predictions and informative radio-related articles; **Directory**, which is further divided into *National Radio*, *International Radio*, *Frequency Lists* (which includes Mediumwave lists by region, Shortwave Stations of the World and DRM broadcasts), and Terrestrial *Television*; and finally **Reference** where a full country index, abbreviations used in WRTH and transmitter site location tables, as well as other useful information related to the world of radio broadcasting can be found.

Each section is identified by a unique 'side-bar', which can be found both on the main contents page and on each individual page throughout the book. Each section starts with an alphabetical country listing.

In the Directory, countries are listed alphabetically within each section so that they may be easily located by flicking forward to the relevant location. Alternatively, the index in the Reference section may be used to find the exact page number for a specific country of interest.

Under each country in the National Radio section, state broadcasters are listed first followed by major networks and then other stations. Armed forces stations and local relays of international stations are at the end of the entry. For all stations, mediumwave is listed first, followed by shortwave and finally FM. Many stations now only broadcast on FM. Details are given of digital radio multiplexes where appropriate.

OPERATING TECHNIQUES

When operating their receivers, the majority of listeners tend to operate in one of two main modes, switching between them as and when they deem appropriate. One method is to 'target' a given station or country by monitoring known frequencies and the other is simply to 'cruise' a specific band and identify each station as they occur. We have designed WRTH in such a way that either of these methods can be accommodated.

TARGETING

When operating in the targeting mode there are two ways to find a particular country. The first option is to go to the main contents page and use the section 'side-bars' to direct you to the right area of the book. Once there, you then only have to flick forward a few pages to locate the country of interest. Alternatively you can use the country index at the back of the book, which will tell you the precise page number.

However, as you develop a 'feel' for the book and get used to the alphabetical layout, you will probably find that the side-bar method is simpler and quicker than using the country index.

BAND-SCANNING

Should you prefer to use band-scanning, there are listings of both medium wave and international shortwave broadcasts available in the Frequency Listings. These can also be useful for casual listening, but in either case can help to identify a station by frequency – whereupon further details can be obtained using the country entry to identify alternative frequencies for the station of interest.

RECEPTION REPORTS

WRTH recommends use of the simple SINPO code using the following scale:

S	I N P	O
5=Excellent	5=None	5=Excellent
4=Good	4=Slight	4=Good
3=Fair	3=Moderate	3=Fair
2=Poor	2=Severe	2=Poor
1=Barely Audible	1=Extreme	1=Worthless

S=Signal Strength, I=Interference, N=Noise, P=Propagation-disturbance, O=Overall Merit

It is courteous to enclose return postage when writing to small domestic broadcasters. This can be in the form of an International Reply Coupon (IRC) available from post offices. In all cases, when writing to radio stations you must write clearly. Remember, if the station cannot read your address, then you cannot expect to receive a reply!

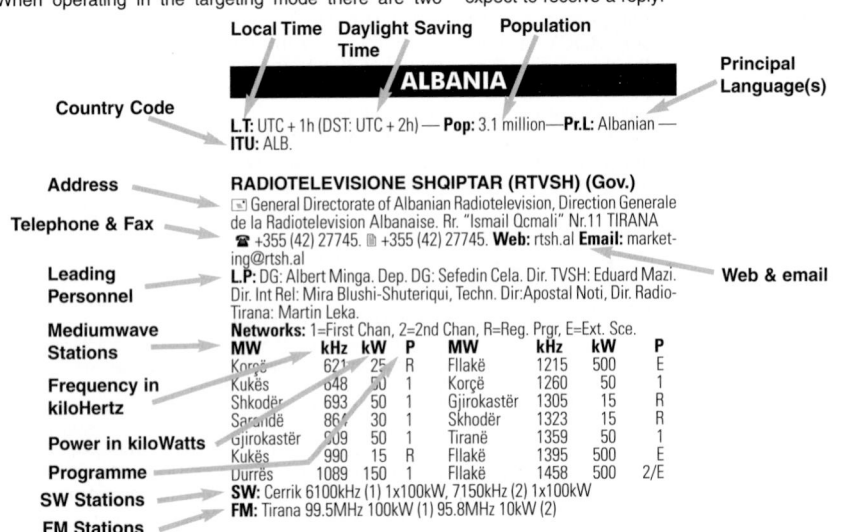

Local Time — Daylight Saving Time — Population — Principal Language(s)

ALBANIA

Country Code — **L.T:** UTC + 1h (DST: UTC + 2h) — **Pop:** 3.1 million — **Pr.L:** Albanian — **ITU:** ALB.

RADIOTELEVISIONE SHQIPTAR (RTVSH) (Gov.)

Address — ⊡ General Directorate of Albanian Radiotelevision, Direction Generale de la Radiotelevision Albanaise. Rr. "Ismail Qcmali" Nr.11 TIRANA

Telephone & Fax — ☎ +355 (42) 27745. 🖷 +355 (42) 27745. **Web:** rtsh.al **Email:** marketing@rtsh.al — Web & email

Leading Personnel — **L.P:** DG: Albert Minga. Dep. DG: Sefedin Cela. Dir. TVSH: Eduard Mazi. Dir. Int Rel: Mira Blushi-Shuteriqui, Techn. Dir:Apostal Noti, Dir. Radio-Tirana: Martin Leka.

Mediumwave Stations — **Networks:** 1=First Chan, 2=2nd Chan, R=Reg. Prgr, E=Ext. Sce.

MW	kHz	kW	P	MW	kHz	kW	P
Korçë	621	25	R	Fllakë	1215	500	E
Kukës	648	50	1	Korçë	1260	50	1
Shkodër	693	50	1	Gjirokastër	1305	15	R
Sarandë	864	30	1	Skhodër	1323	15	R
Gjirokastër	909	50	1	Tiranë	1359	50	1
Kukës	990	15	R	Fllakë	1395	500	E
Durrës	1089	150	1	Fllakë	1458	500	2/E

Frequency in kiloHertz / Power in kiloWatts / Programme

SW Stations — **SW:** Cerrik 6100kHz (1) 1x100kW, 7150kHz (2) 1x100kW

FM Stations — **FM:** Tirana 99.5MHz 100kW (1) 95.8MHz 10kW (2)

International Frequency Allocation Chart

Top face of each bar shows regional differences where appropriate

= Broadcast Band
= Radio Amateur Band
= Utility - Other Services
= Standard Time & Frequency Transmission

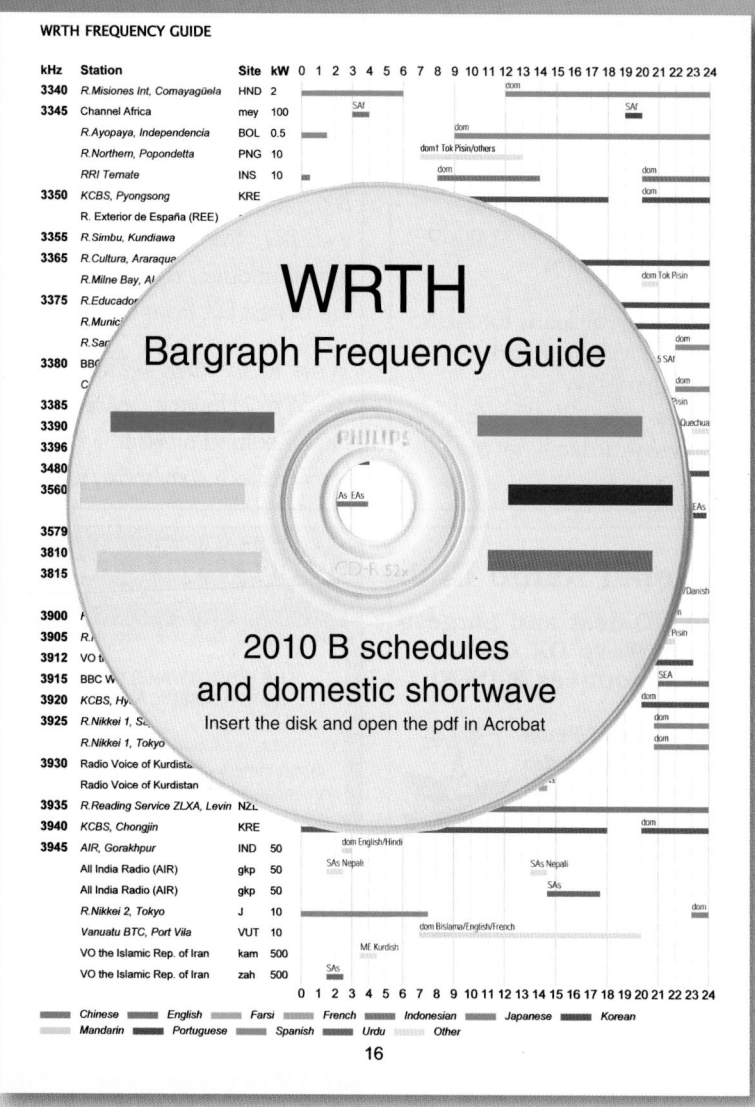

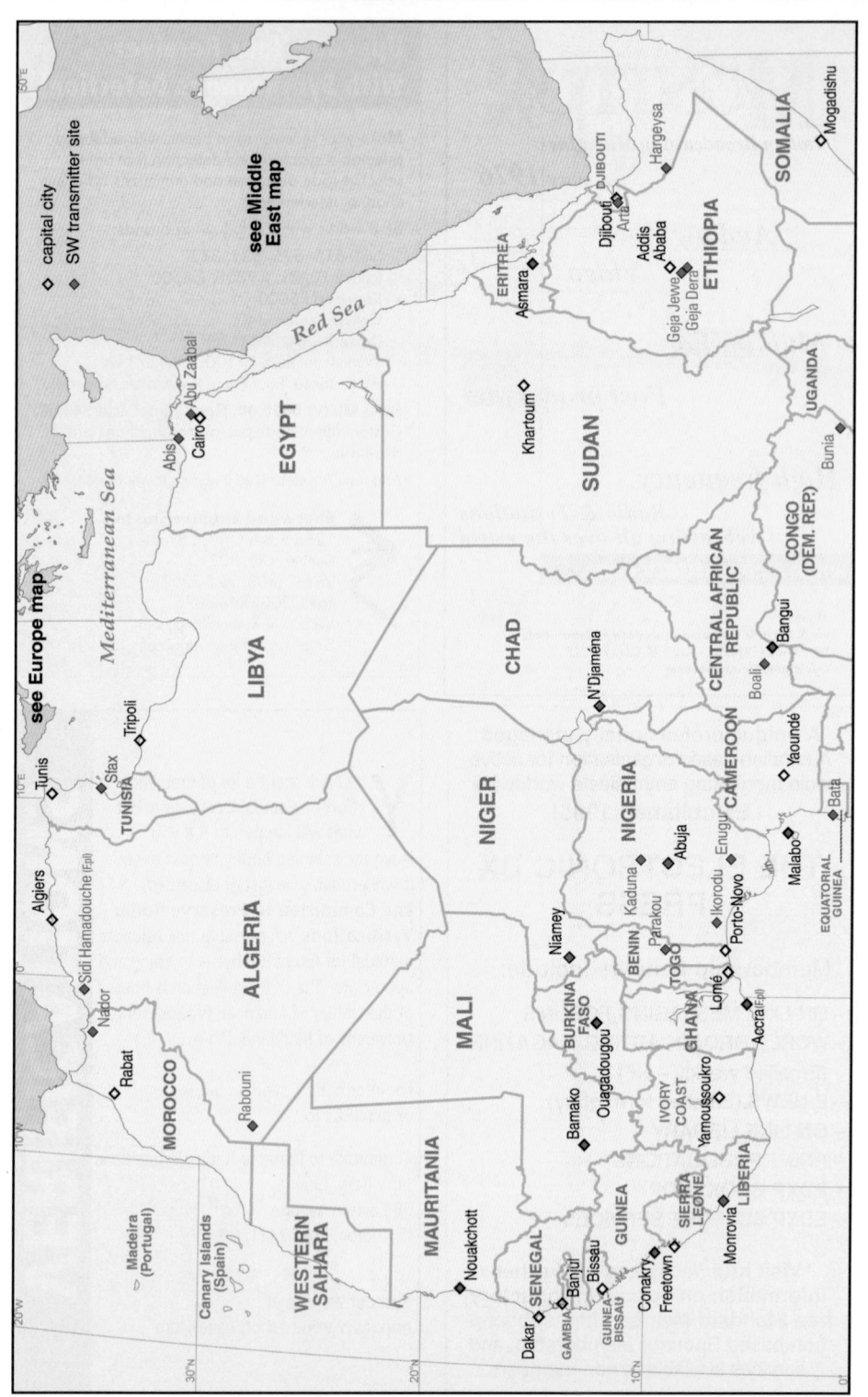

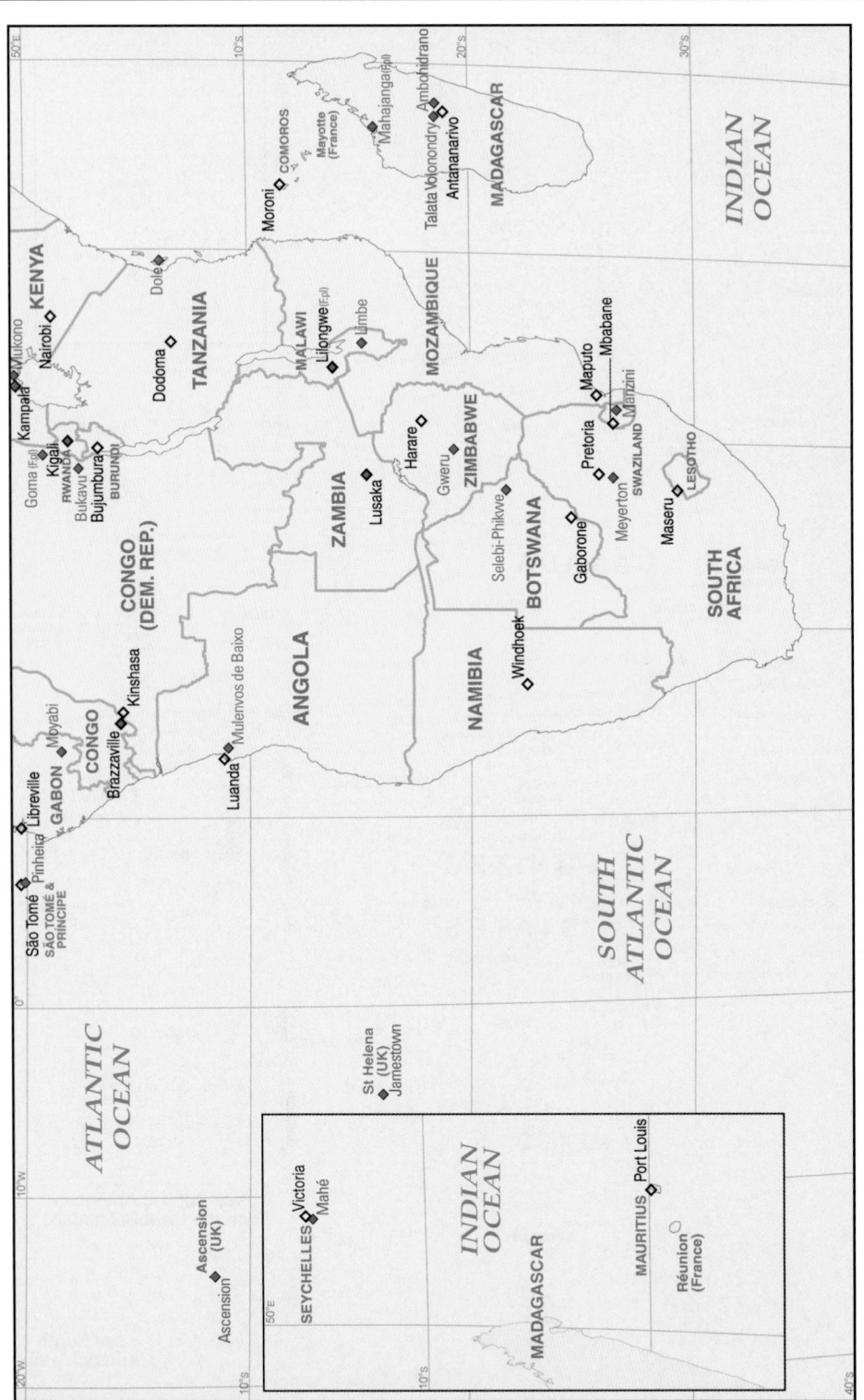

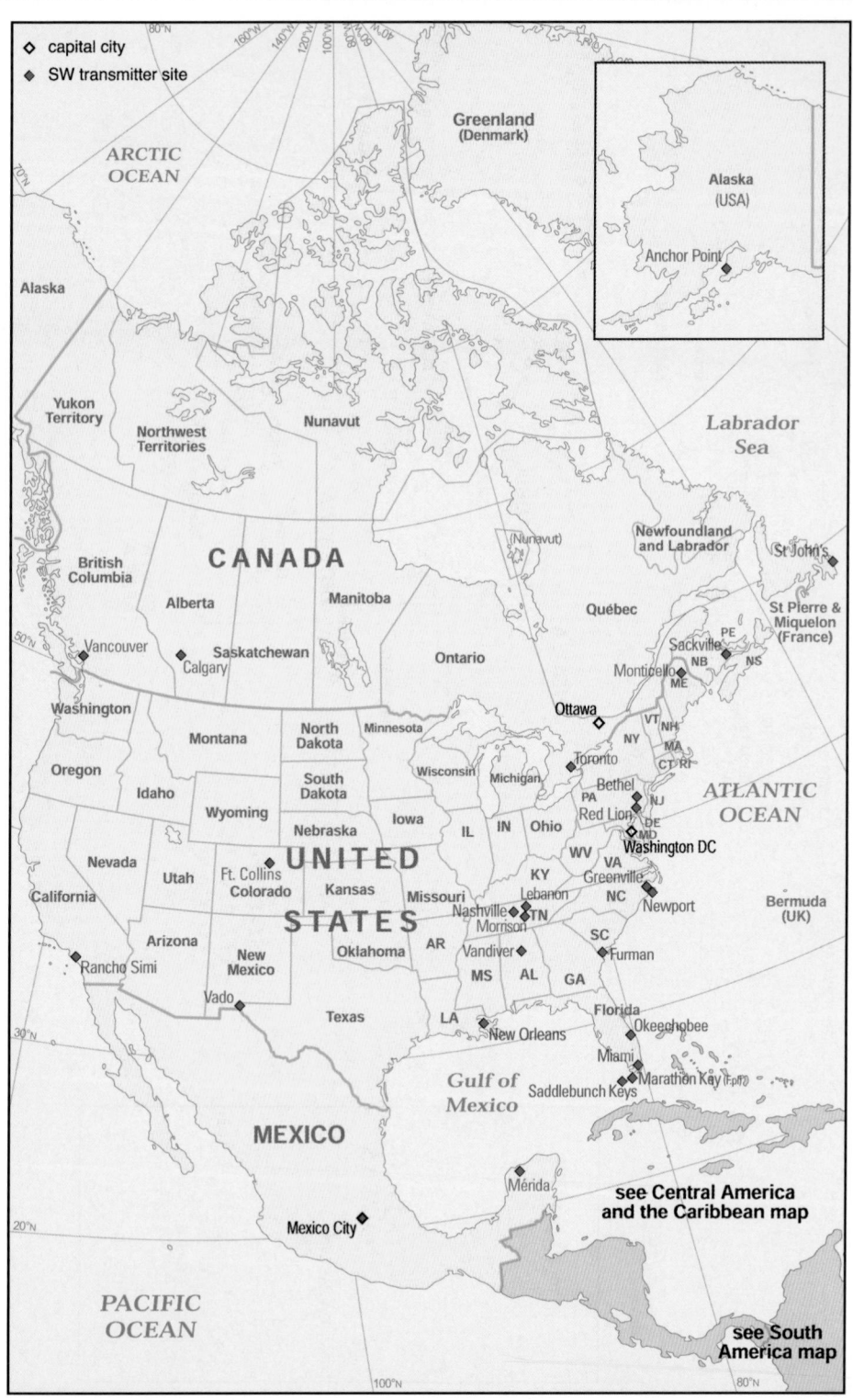

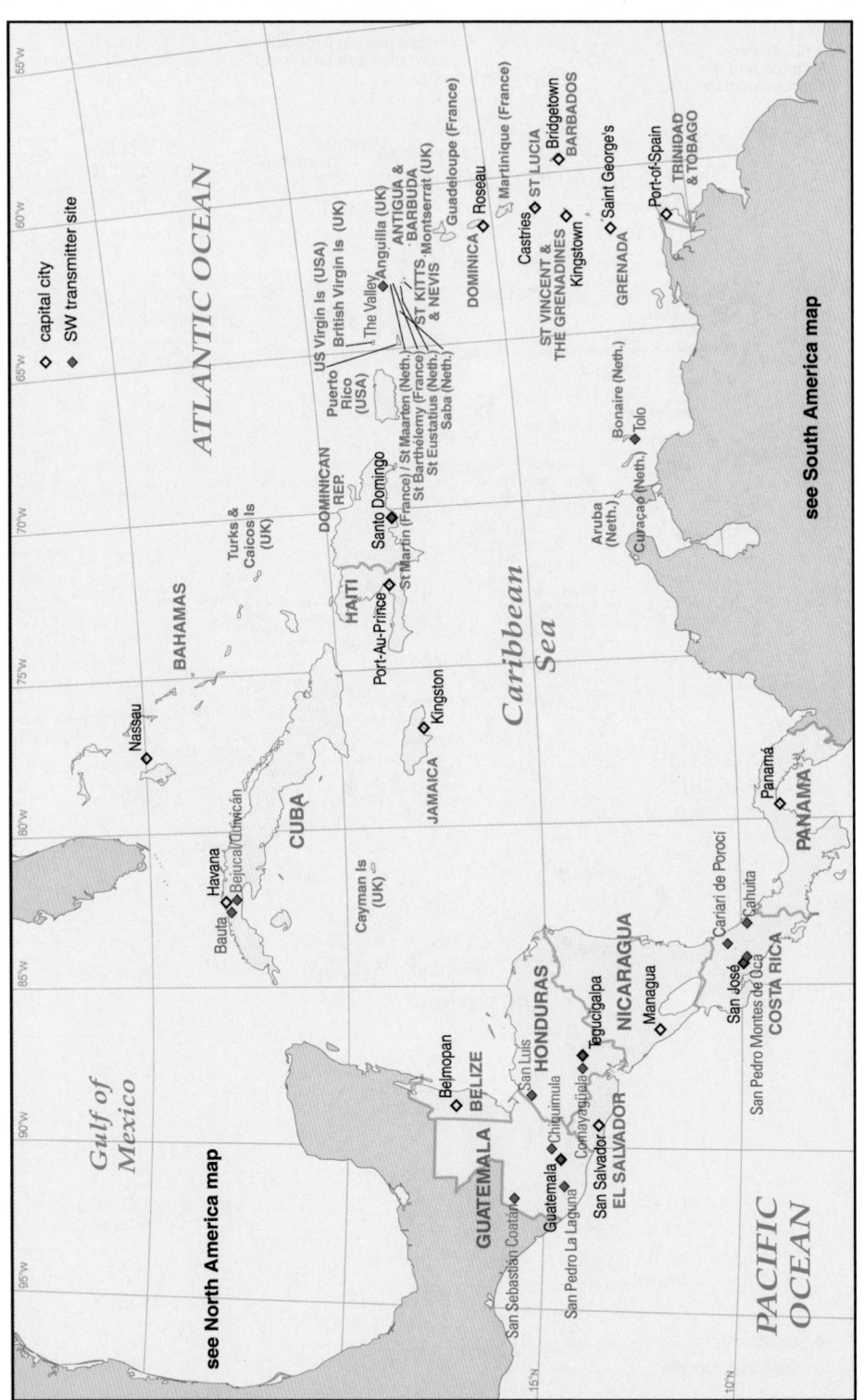

♦ capital city
♦ SW transmitter site

ATLANTIC OCEAN

US Virgin Is (USA)
British Virgin Is (UK)
The Valley △ Anguilla (UK)
ANTIGUA &
BARBUDA
Montserrat (UK)
Guadeloupe (France)
Martinique (France)
Roseau
DOMINICA
Castries ♦ ST LUCIA
Bridgetown
BARBADOS
ST VINCENT &
THE GRENADINES
Kingstown
Saint George's
GRENADA
Port-of-Spain
TRINIDAD
& TOBAGO

Puerto
Rico
(USA)
St Maarten (Neth.)
St Martin (France)
St Barthelemy (France)
St Eustatius (Neth.)
Saba (Neth.)
ST KITTS
& NEVIS

DOMINICAN
REP.
Santo Domingo

Turks &
Caicos Is
(UK)

HAITI
Port-Au-Prince

BAHAMAS

Nassau

Caribbean
Sea

Bonaire (Neth.)
Tolo
Aruba
(Neth.)
Curaçao (Neth.)

Kingston
JAMAICA

Cayman Is
(UK)

CUBA
Bauta
Havana
Bejucal/Quivicán

Panamá
PANAMA

see South America map

Gulf of
Mexico

see North America map

Belmopan
BELIZE
San Luis
HONDURAS
Tegucigalpa
NICARAGUA
Managua

San Sebastián Coatán
San Pedro La Laguna
GUATEMALA
Guatemala
Chiquimula
Comayagüela
San Salvador
EL SALVADOR

Cariari de Poroci
Cahuita
San José
San Pedro Montes de Oca
COSTA RICA

Panamá
PANAMA

PACIFIC
OCEAN

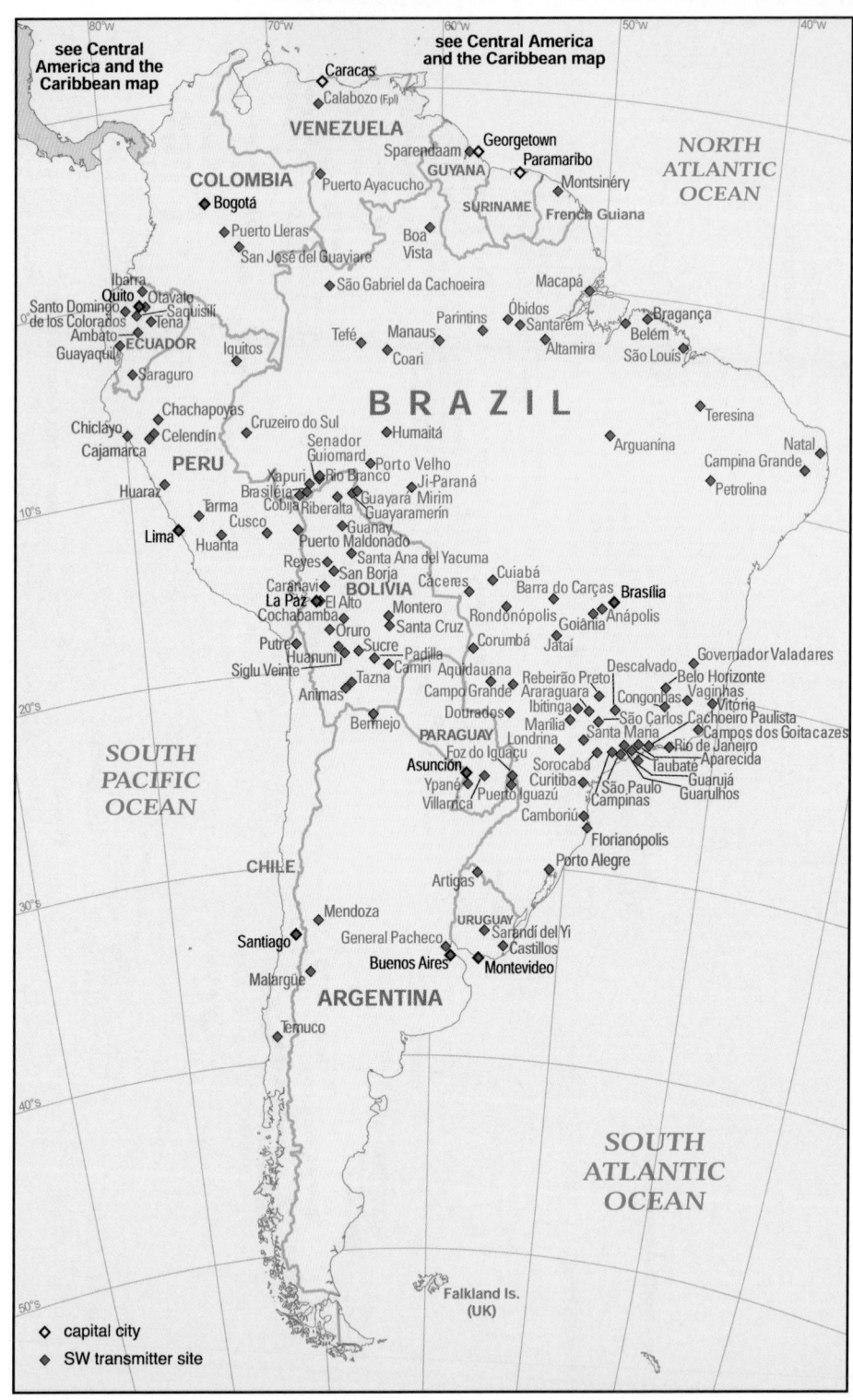

see Central America and the Caribbean map

see Central America and the Caribbean map

VENEZUELA
Caracas
Calabozo (F,pl)

Georgetown
Sparendaam
Paramaribo
GUYANA
Montsinéry
Puerto Ayacucho
SURINAME
COLOMBIA
French Guiana
Bogotá
Puerto Lleras
Boa
San José del Guaviare
Vista

Ibarra
Santo Domingo
de los Colorados
Quito
Otavalo
Saquisilí
Ambato
ECUADOR
Tena
Guayaquil
Iquitos
Saraguro

São Gabriel da Cachoeira
Macapá
Óbidos
Bragança
Parintins
Santarém
Belém
Manaus
São Louís
Tefé
Altamira
Coari

B R A Z I L

Chachapoyas
Cruzeiro do Sul
Humaitá
Teresina
Chiclayo
Celendín
Arguanína
Natal
Cajamarca
Senador
Campina Grande
PERU
Guiomard
Porto Velho
Petrolina
Xapuri
Rio Branco
Huaraz
Brasiléia
Ji-Paraná
Tarma
Cobija
Riberalta
Guayará Mirim
Cusco
Guayaramerín
Lima
Huanta
Guanay
Puerto Maldonado
Reyes
Santa Ana del Yacuma
Cuiabá
San Borja
Caranavi
Cáceres
Barra do Carças
Brasília
La Paz
BOLIVIA
El Alto
Montero
Anápolis
Cochabamba
Rondonópolis
Goiânia
Putre
Oruro
Santa Cruz
Corumbá
Jataí
Siglu Veinte
Huanuni
Sucre
Padilla
Governador Valadares
Camiri
Aquidauana
Descalvado
Belo Horizonte
Animas
Tazna
Campo Grande
Araraquara
Congonhas
Vaginhas
Bermejo
Dourados
Ibitinga
São Carlos
Cachoeiro Paulista
Vitória
PARAGUAY
Marília
Santa Maria
Campos dos Goitacazes
Londrina
Rio de Janeiro
Foz do Iguaçu
Sorocaba
Taubaté
Aparecida
Asunción
Curitiba
Guarujá
Ypané
Puerto Iguazú
São Paulo
Guarulhos
Villarrica
Campinas
Camboriú

Florianópolis
Porto Alegre

Rebeirão Preto

SOUTH PACIFIC OCEAN

CHILE
Artigas
URUGUAY
Mendoza
Sarandí del Yi
Santiago
General Pacheco
Castillos
Malargüe
Buenos Aires
Montevideo
ARGENTINA
Temuco

NORTH ATLANTIC OCEAN

SOUTH ATLANTIC OCEAN

Falkland Is. (UK)

◇ capital city
◆ SW transmitter site

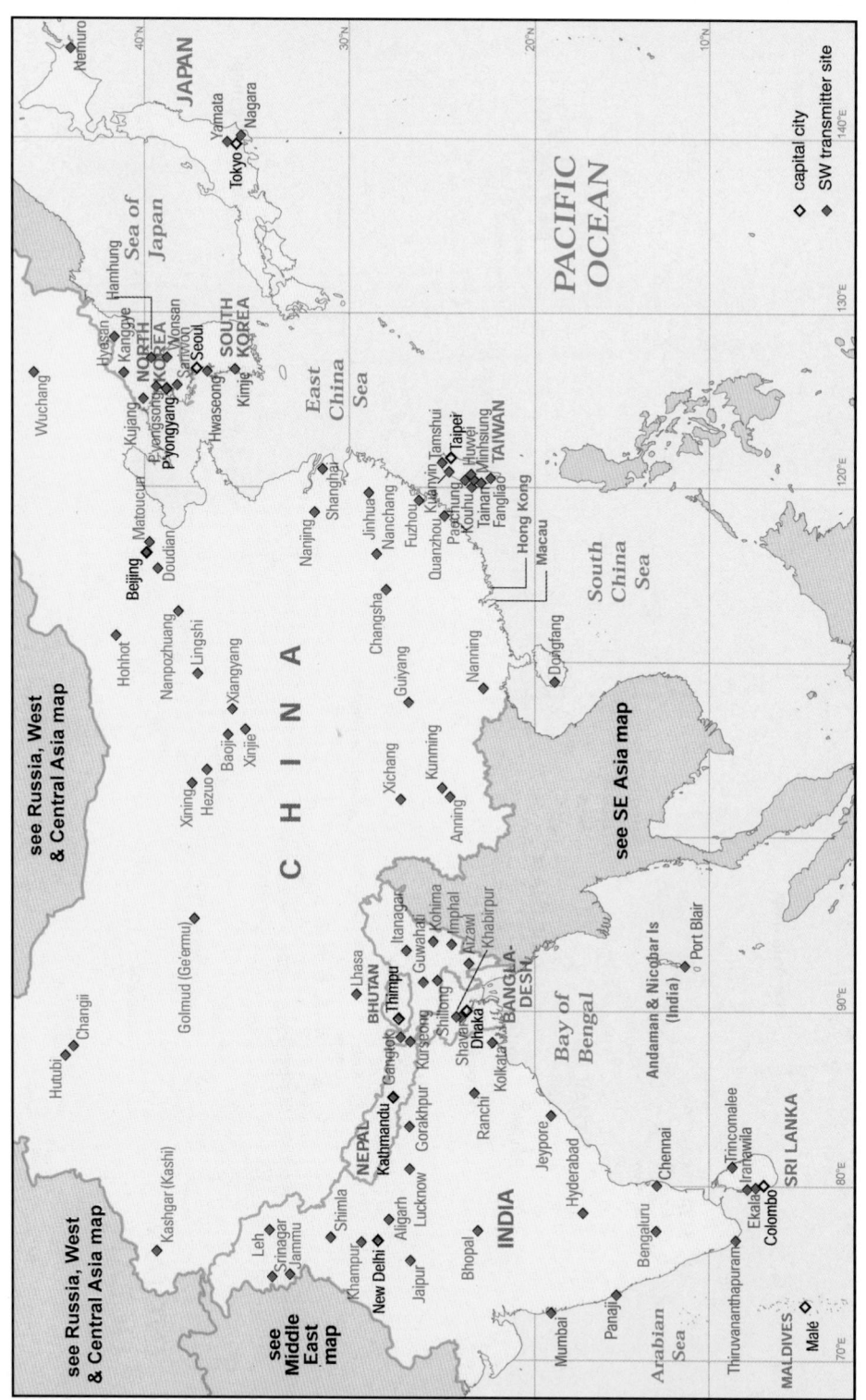

◇ capital city
◆ SW transmitter site

PACIFIC
OCEAN

JAPAN
Nemuro
Yamata
Nagara
Tokyo
Sea of
Japan
Hamhung
Hwasin
Kanggye
NORTH KOREA
Sariwon/Wonsan
Seoul
Hwaseong
SOUTH KOREA
Kimje
Kujang
Pyongsong
Pyongyang
East
China
Sea
Wuchang
Matoucun
Doudian
Beijing
Shanghai
Nanjing
Jinhua
Nanchang
Fuzhou
Quanzhou Kuanyin Tamshui
Paochung Taipei
Kouhu Huwei
Tainan Minhsiung
Fangliao TAIWAN
Hong Kong
Macau
South
China
Sea
Hohhot
Nanpozhuang
Lingshi
Xiangyang
Changsha
Guiyang
Nanning
Dongfang
Xining
Baoji
Hezuo
Xinjie
Xichang
Kunming
Anning
Changji
Hutubi
C H I N A
Golmud (Ge'ermu)
see SE Asia map
Lhasa
BHUTAN
Thimpu
Itanagar
Guwahati
Kohima
Imphal
Aizawl
Khabirpur
BANGLA-DESH
Shavar
Dhaka
Kolkata
Shillong
NEPAL
Kathmandu
Gangtok
Kurseong
Gorakhpur
Ranchi
Kashgar (Kashi)
Leh
Srinagar
Jammu
Shimla
Khampur
New Delhi
Aligarh
Lucknow
Jaipur
Bhopal
Jeypore
Hyderabad
INDIA
Bengaluru
Port Blair
Andaman & Nicobar Is
(India)
Bay of
Bengal
Trincomalee
Ekala Irahawila
Colombo
SRI LANKA
Chennai
Panaji
Mumbai
Arabian
Sea
Thiruvananthapuram
MALDIVES
Male

see Russia, West
& Central Asia map

see Russia, West
& Central Asia map

see
Middle
East
map

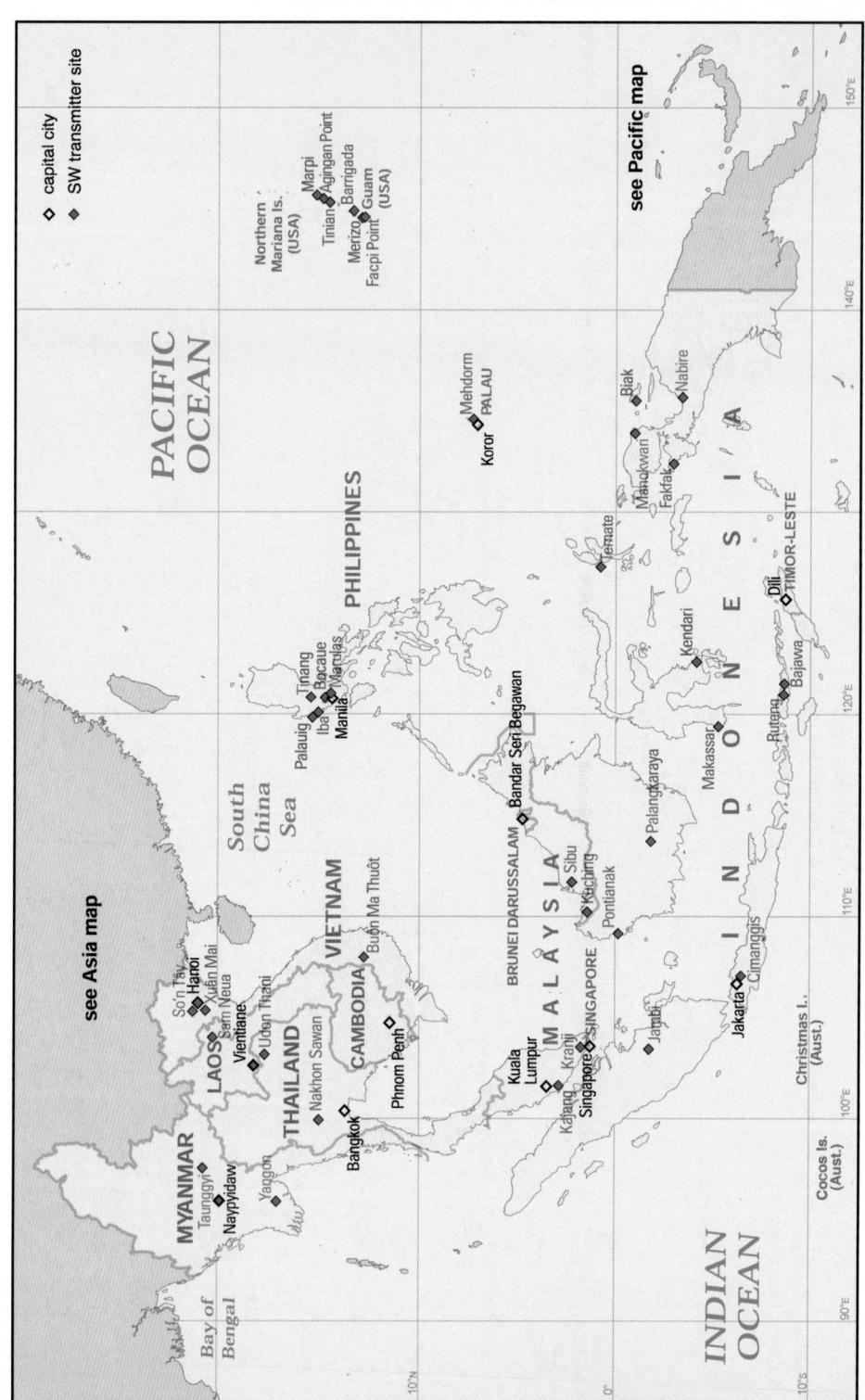

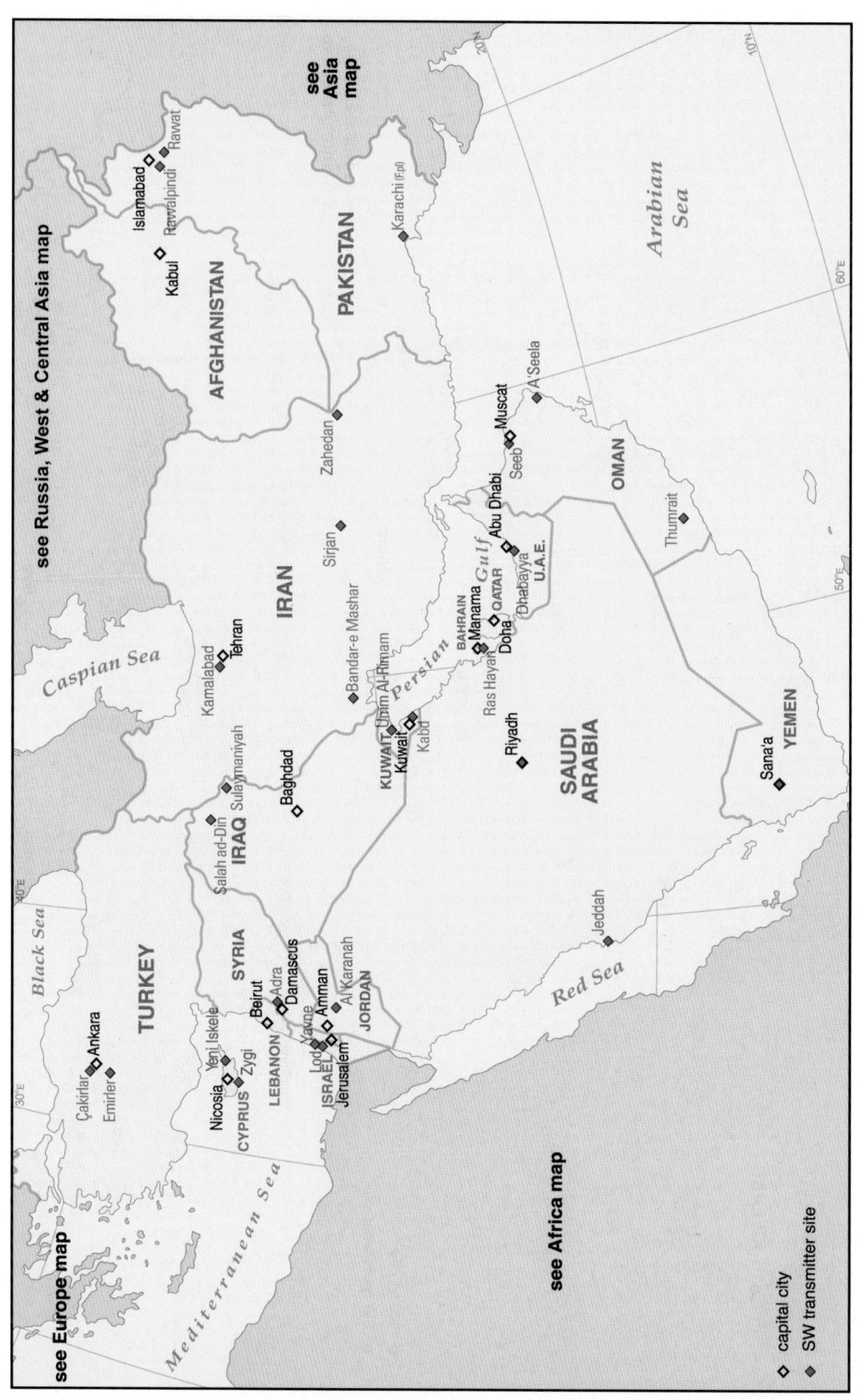

see Asia map

see Russia, West & Central Asia map

see Europe map

Rawat
Islamabad
Rawalpindi
Kabul
AFGHANISTAN

PAKISTAN

Karachi (Fpl)

Arabian Sea

Zahedan

Sirjan

IRAN

Caspian Sea

Kamalabad
Tehran

Bandar-e Mashar

Umm Al-Rimam
Persian Gulf

KUWAIT Sulaymaniyah
Kuwait
Kabd

Salah ad-Din
IRAQ
Baghdad

Black Sea

TURKEY

Çakırlar Ankara
Emirler
Yeni Iskele
Beirut
Adra
Damascus
Zygi
Lavrie
Nicosia Lod
CYPRUS LEBANON
ISRAEL
Jerusalem

Amman
Al Karanah
JORDAN
SYRIA

A'Seela
Muscat
Seeb

Abu Dhabi
Dhabayya
QATAR U.A.E.
Manama
BAHRAIN
Doha
Ras Hayan

OMAN

Thumrait

Riyadh

SAUDI
ARABIA

Sana'a YEMEN

Jeddah

Red Sea

Mediterranean Sea

see Africa map

◇ capital city
◆ SW transmitter site

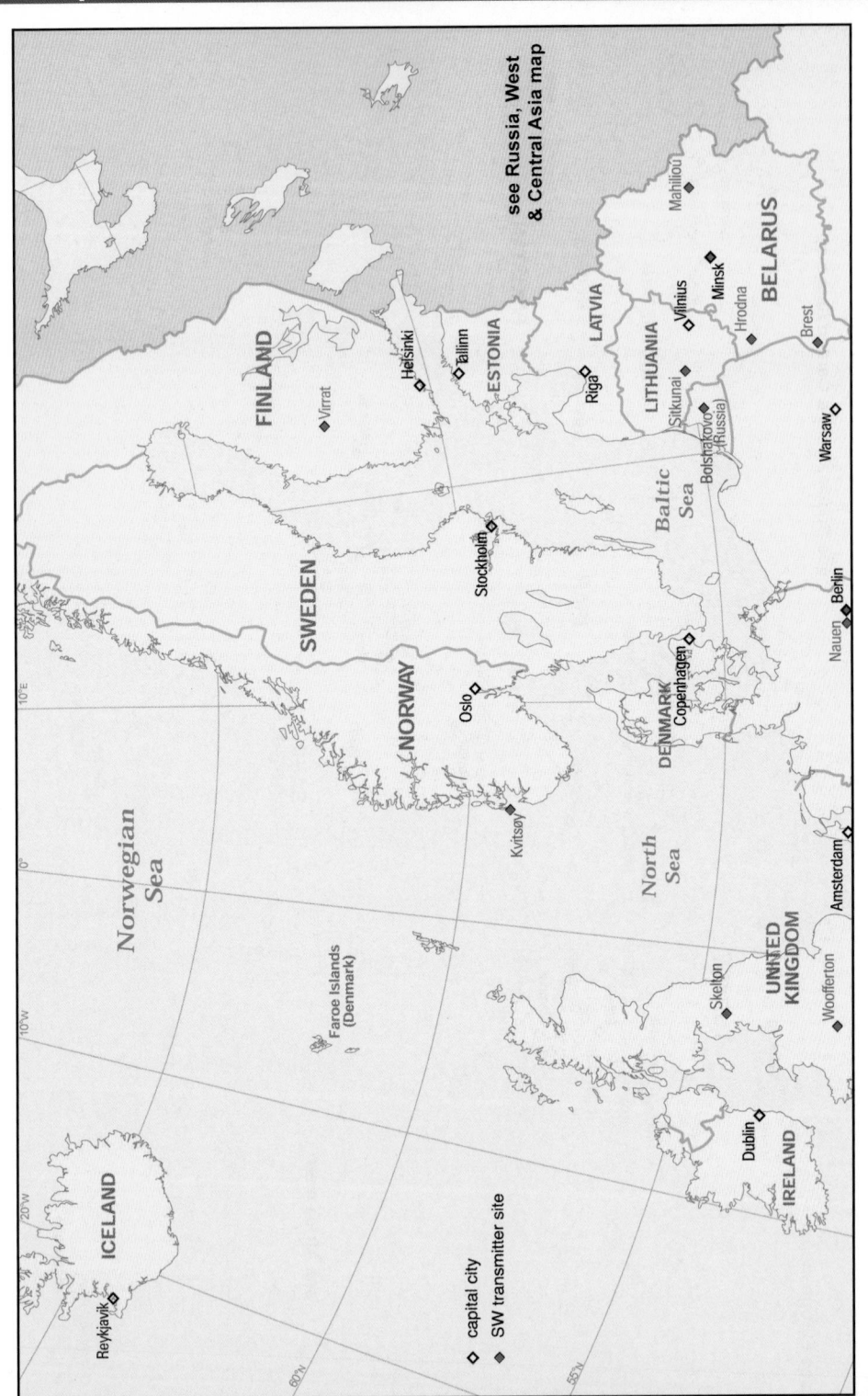

see Russia, West
& Central Asia map

BELARUS

Mahiliou

Minsk

LATVIA

Vilnius

Hrodna

Brest

LITHUANIA

FINLAND

ESTONIA

Tallinn

Helsinki

Riga

Sitkunai

Virrat

Bolshakovo
(Russia)

Warsaw

Baltic
Sea

SWEDEN

Stockholm

Berlin

Nauen

NORWAY

Oslo

DENMARK

Copenhagen

Norwegian
Sea

Kvitsøy

North
Sea

Amsterdam

Faroe Islands
(Denmark)

UNITED
KINGDOM

Skelton

Woofferton

ICELAND

Dublin

IRELAND

Reykjavík

◇ capital city

◆ SW transmitter site

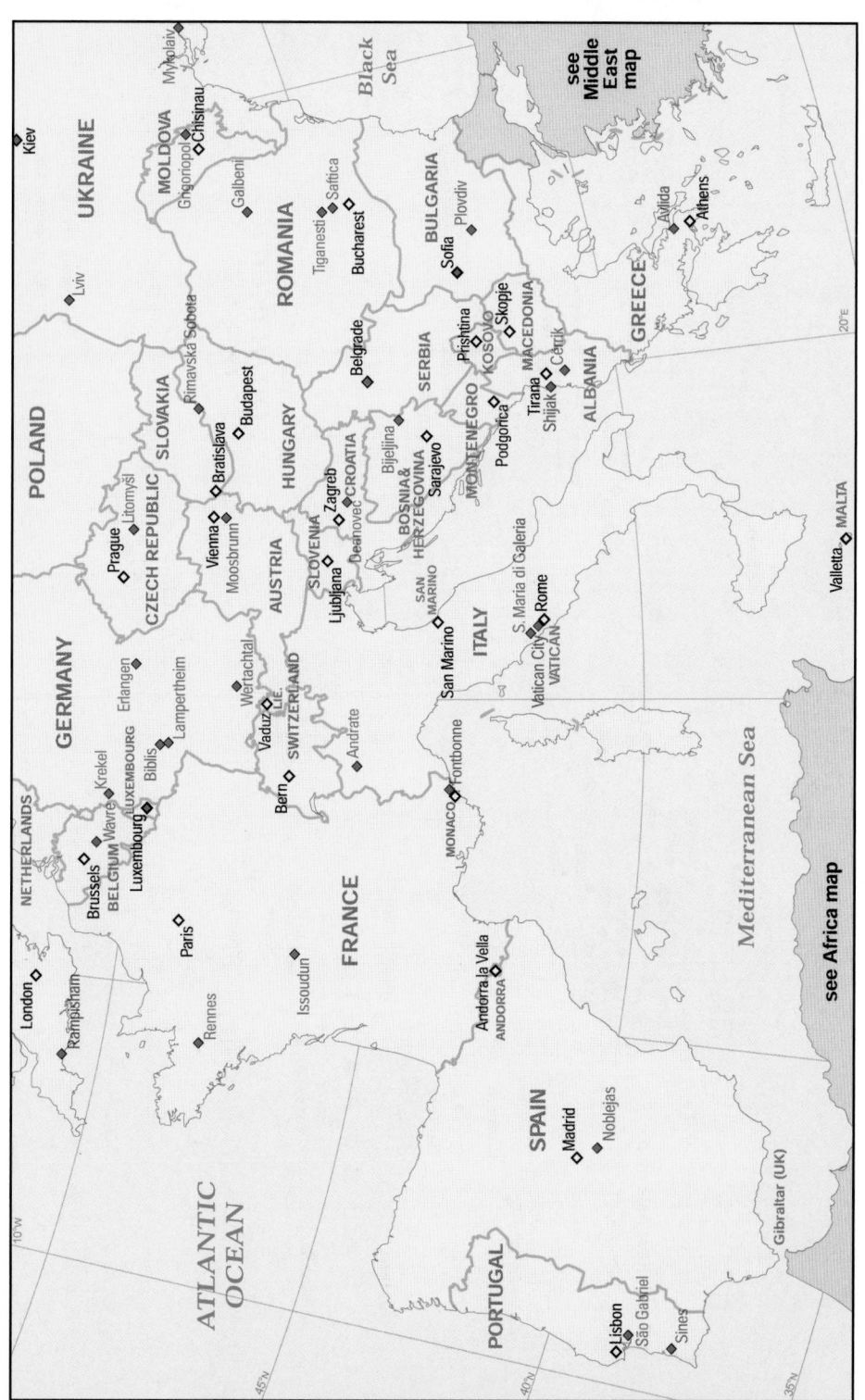

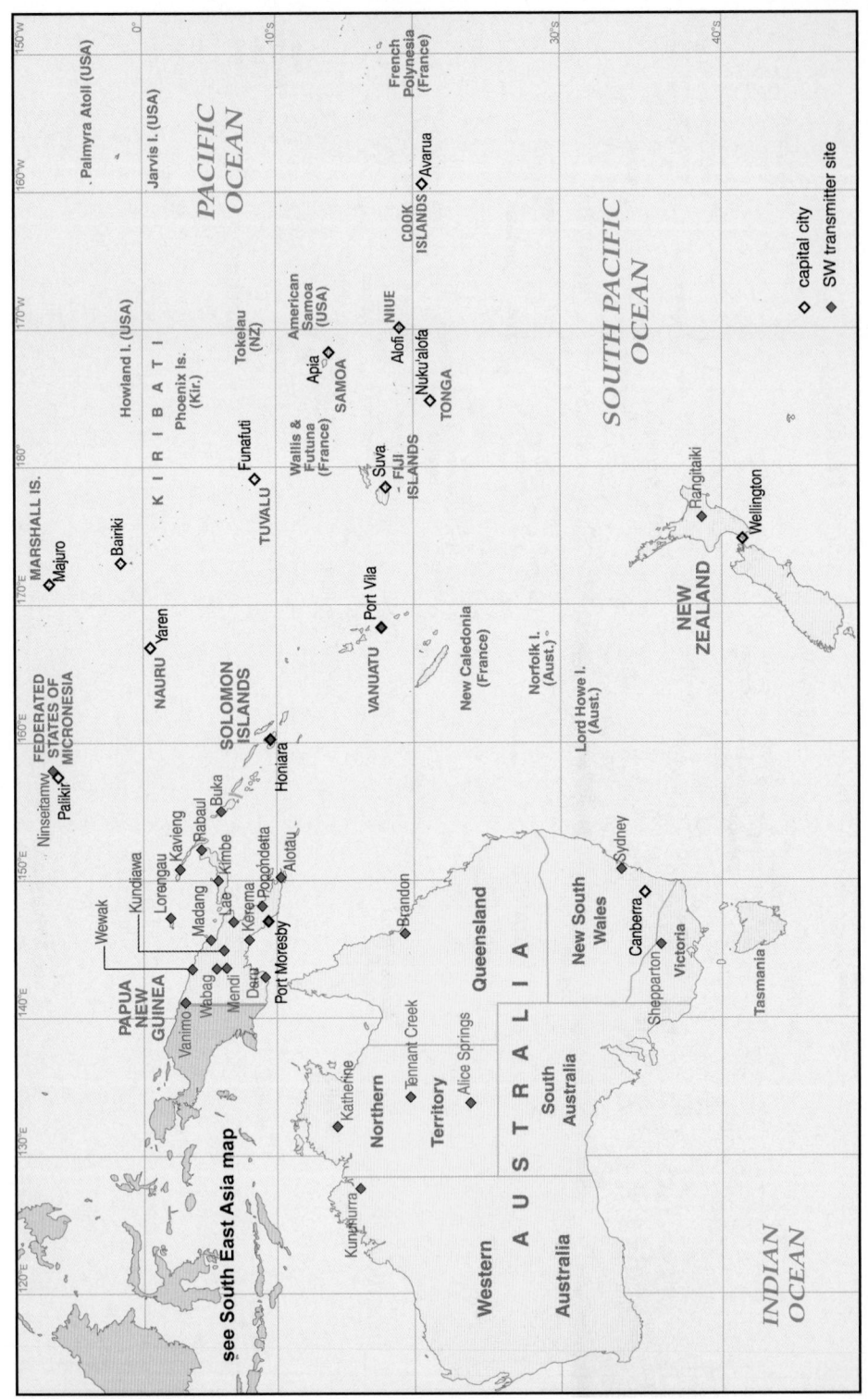

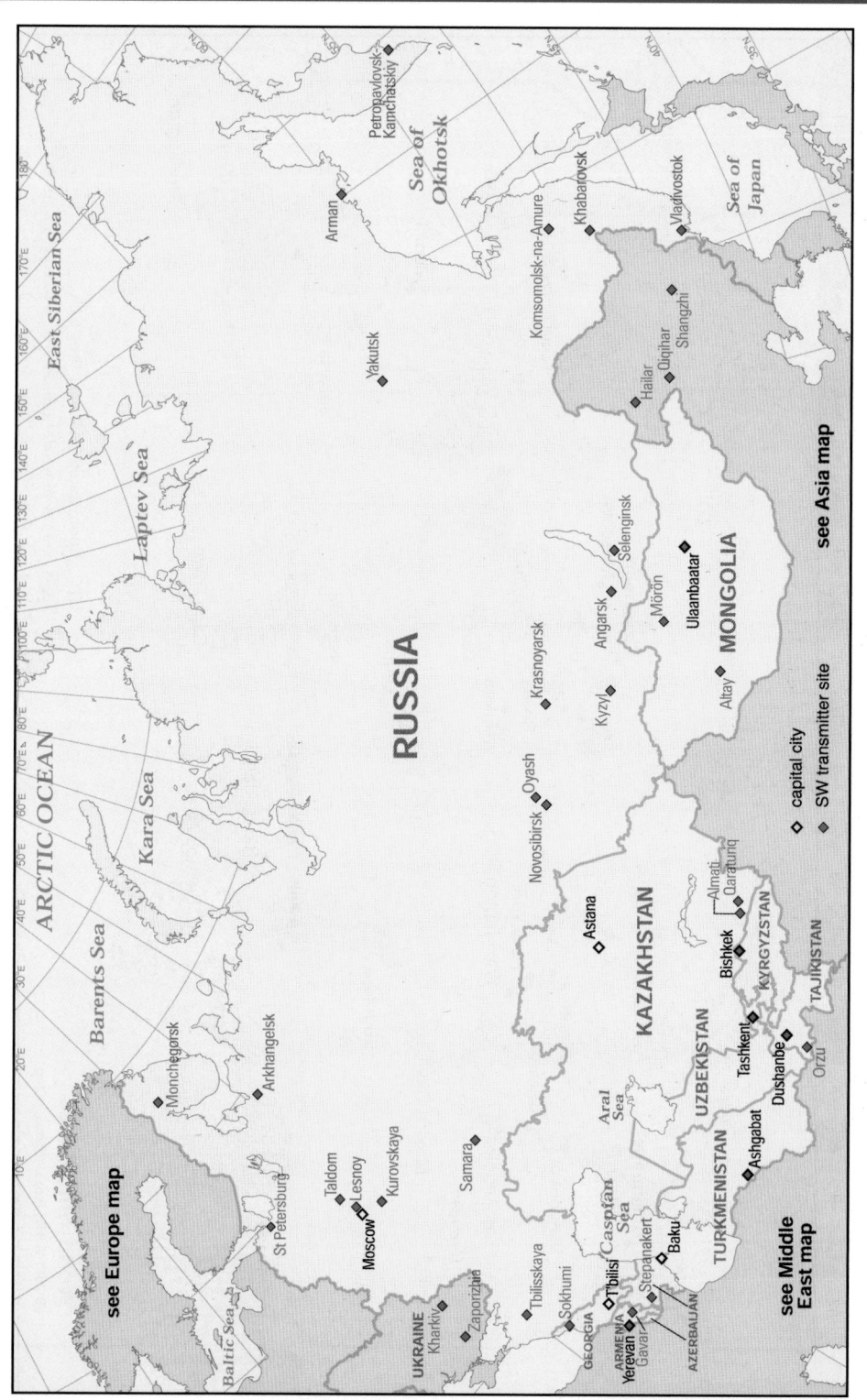

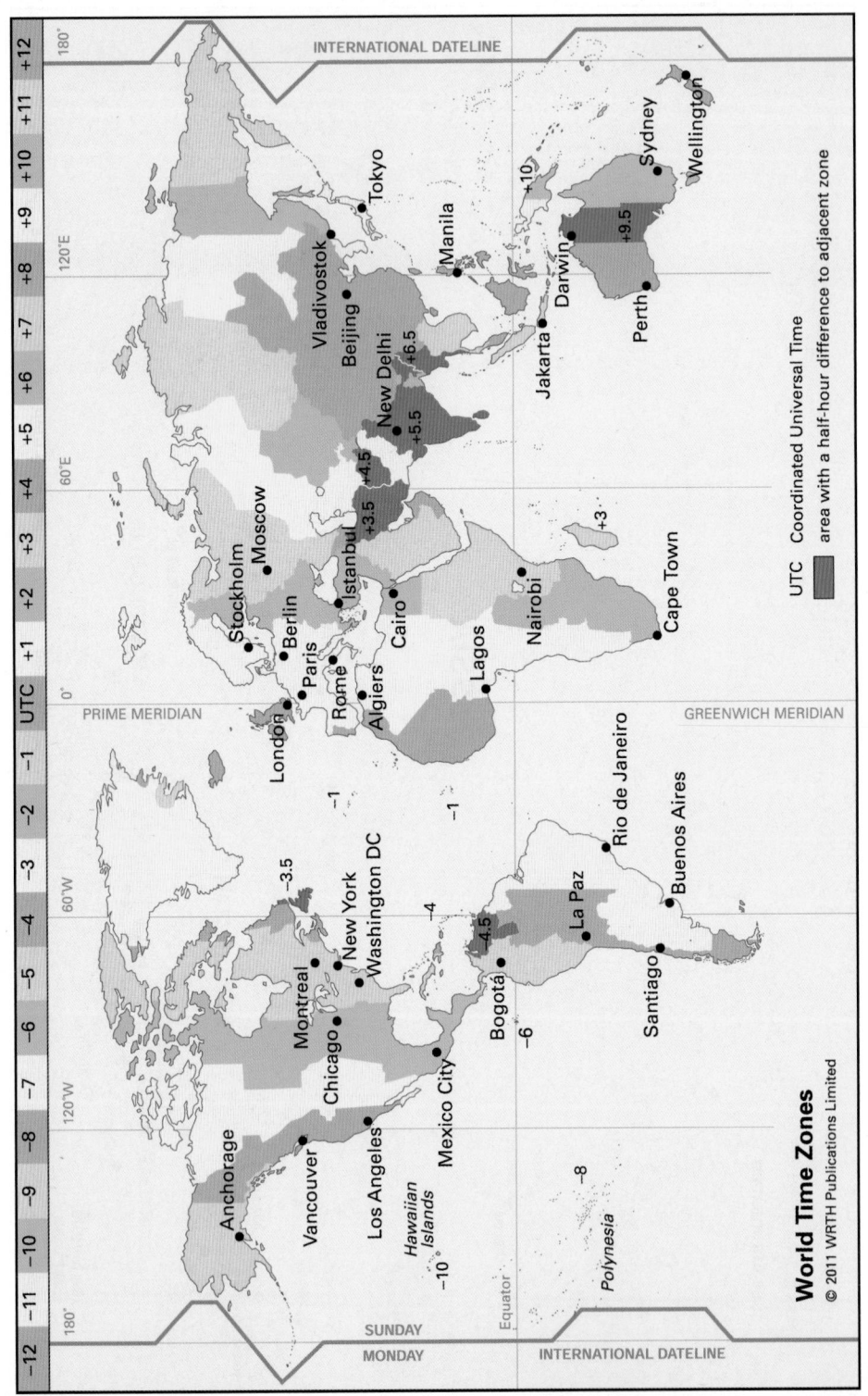

World Time Zones
© 2011 WRTH Publications Limited

UTC   Coordinated Universal Time

area with a half-hour difference to adjacent zone

INTERNATIONAL DATELINE

PRIME MERIDIAN   GREENWICH MERIDIAN

SUNDAY
MONDAY   INTERNATIONAL DATELINE

WORLD TIME TABLE

Differences marked + or - indicate the number of hours ahead of, or behind, UTC. Variations from Standard Time for part of the year (referred to as DST or Summer Time) are shown below; see the various country sections for the dates of operation. *) DST2011-2012 subj. to confirmation **N**=Normal (Standard) Time; **D**=Daylight Saving Time (DST) ¹) in parts of the territory ²) for other regions, see country section

Country	N	D
Afghanistan	+4½	+4½
Alaska	−9	−8
Aleutian Is	−10	−9
Albania	+1	+2
Algeria	+1	+1
American Samoa	−11	−11
Andorra	+1	+2
Angola	+1	+1
Anguilla	−4	−4
Antarctica		
(Argentinan)	−3	−3
(Chilenan)	−4	−3
(McMurdo)	+12	+13
Antigua	−4	−4
Argentina (B.Aires)[2]	−3	−3
Armenia	+4	+5
Aruba	−4	−4
Ascension I.	UTC	UTC
Australia		
We. Australia	+8	+8
No. Territory	+9½	+9½
So. Australia	+9½	+10½
Queensland	+10	+10
VIC, NSW, TAS	+10	+11
Austria	+1	+2
Azerbaijan	+4	+5
Azores	−1	UTC
Bahamas	−5	−4
Bahrain	+3	+3
Bangladesh*	+6	+6
Barbados	−4	−4
Belarus	+2	+3
Belgium	+1	+2
Belize	−6	−6
Benin	+1	+1
Bermuda	−4	−3
Bhutan	+6	+6
Bolivia	−4	−4
Bonaire	−4	−4
Bosnia & Herzegovina	+1	+2
Botswana	+2	+2
Brazil (Brasília)[2]	−3	−2
British Ind. Oc. Terr.	+6	+6
British Virgin Is	−4	−4
Brunei	+8	+8
Bulgaria	+2	+3
Burkina Faso	UTC	UTC
Burundi	+2	+2
Cambodia	+7	+7
Cameroon	+1	+1
Canada		
NL (SE Labr. & Is)	−3½	−2½
NL[1], NB, NS, PE, QC[1]	−4	−3
QC[1]	−4	−4
NT[1], NU[1], ON[1], QC[1]	−5	−4
NU[1]	−5	−5
MB, NU[1], ON[1]	−6	−5
SK[1]	−6	−6
AB, BC[1], NT, NU[1], SK[1]	−7	−6
BC[1]	−7	−7
BC[1], YT	−8	−7
Canary Is	UTC	+1
Cape Verde	−1	−1
Cayman Is	−5	−5
Ce. African Rep.	+1	+1
Chad	+1	+1
Chile	−4	−3
China (P.R.)	+8	+8
Christmas Is	+7	+7
Cocos (Keeling) Is	+6½	+6½
Colombia	−5	−5
Comoros	+3	+3
Congo (Kinshasa)[2]	+1	+1
Congo (Rep.)	+1	+1
Cook Is	−10	−10
Costa Rica	−6	−6
Côte d'Ivoire	UTC	UTC
Croatia	+1	+2
Cuba	−5	−4
Curaçao	−4	−4
Cyprus	+2	+3
Akrotiri & Dhekelia	+2	+3
Czech Rep.	+1	+2
Denmark	+1	+2
Djibouti	+3	+3
Dominica	−4	−4
Dom. Rep.	−4	−4
Easter I.	−6	−5
Ecuador	−5	−5
Egypt	+2	+3
El Salvador	−6	−6
Equatorial Guinea	+1	+1
Eritrea	+3	+3
Estonia	+2	+3
Ethiopia	+3	+3
Falkland Is	−4	−3
Faroe Is	UTC	+1
Fiji*	+12	+13
Finland	+2	+3
France	+1	+2
French Guiana	−3	−3
French Poly.(Tahiti)[2]	−10	−10
French So.& Ant. L.	+5	+5
Gabon	+1	+1
Galapagos Is	−6	−6
Gambia	UTC	UTC
Georgia (Tbilisi)[2]	+4	+4
Germany	+1	+2
Ghana	UTC	UTC
Gibraltar	+1	+2
Greece	+2	+3
Greenland (Nuuk)[2]	−3	−2
Grenada	−4	−4
Guadeloupe	−4	−4
Guam	+10	+10
Guatemala	−6	−6
Guinea	UTC	UTC
Guinea-Bissau	UTC	UTC
Guyana	−4	−4
Haiti	−5	−5
Hawaii	−10	−10
Honduras	−6	−6
Hong Kong	+8	+8
Hungary	+1	+2
Iceland	UTC	UTC
India	+5½	+5½
Indonesia (Jakarta)[2]	+7	+7
Iran	+3½	+4½
Iraq	+3	+3
Ireland	UTC	+1
Israel	+2	+3
West Bank & Gaza	+2	+3
Italy	+1	+2
Jamaica	−5	−5
Japan	+9	+9
Jordan	+2	+3
Kazakhstan (Astana)[2]	+6	+6
Kenya	+3	+3
Kiribati	+12	+12
Korea (North, DPR)	+9	+9
Korea (South, Rep.)	+9	+9
Kosovo	+1	+2
Kuwait	+3	+3
Kyrgyzstan	+6	+6
Laos	+7	+7
Latvia	+2	+3
Lebanon	+2	+3
Lesotho	+2	+2
Liberia	UTC	UTC
Libya	+2	+2
Liechtenstein	+1	+2
Lithuania	+2	+3
Lord Howe I.	+10½	+11
Luxembourg	+1	+2
Macau	+8	+8
Macedonia	+1	+2
Madagascar	+3	+3
Madeira	UTC	+1
Malawi	+2	+2
Malaysia	+8	+8
Maldives	+5	+5
Mali	UTC	UTC
Malta	+1	+2
Marshall Is	+12	+12
Martinique	−4	−4
Mauritania	UTC	UTC
Mauritius	+4	+4
Mayotte	+3	+3
Mexico (Mexico City)[2]	−6	−5
Micronesia		
Chuuk, Yap	+10	+10
Kosrae, Pohnpei	+11	+11
Moldova	+2	+3
Monaco	+1	+2
Mongolia	+8	+8
Montenegro	+1	+2
Montserrat	−4	−4
Morocco	UTC	+1
Mozambique	+2	+2
Myanmar	+6½	+6½
Namibia	+1	+2
Nauru	+12	+12
Nepal	+5¾	+5¾
Netherlands	+1	+2
New Caledonia	+11	+11
New Zealand	+12	+13
Nicaragua	−6	−6
Niger	+1	+1
Nigeria	+1	+1
Niue	−11	−11
Norfolk I.	+11½	+11½
No. Mariana Is	+10	+10
Norway	+1	+2
Oman	+4	+4
Pakistan	+5	+5
Palau	+9	+9
Panama	−5	−5
Papua N. Guinea	+10	+10
Paraguay	−4	−3
Peru	−5	−5
Philippines	+8	+8
Poland	+1	+2
Portugal	UTC	+1
Puerto Rico	−4	−4
Qatar	+3	+3
Réunion	+4	+4
Romania	+2	+3
Russia (Moscow)[2]	+3	+4
Rwanda	+2	+2
Saba	−4	−4
Samoa	−11	−10
San Marino	+1	+2
São Tomé & Prínc.	UTC	UTC
Saudi Arabia	+3	+3
Senegal	UTC	UTC
Serbia	+1	+2
Seychelles	+4	+4
Sierra Leone	UTC	UTC
Singapore	+8	+8
Slovakia	+1	+2
Slovenia	+1	+2
Solomon Is	+11	+11
Somalia	+3	+3
So. Africa	+2	+2
Spain	+1	+2
Sri Lanka	+5½	+5½
St. Barthélemy	−4	−4
St. Eustatius	−4	−4
St. Helena	UTC	UTC
St. Kitts & Nevis	−4	−4
St. Lucia	−4	−4
St. Martin	−4	−4
St. Pierre & Miq.	−3	−2
St. Vincent & Gren.	−4	−4
Sudan	+3	+3
Suriname	−3	−3
Swaziland	+2	+2
Sweden	+1	+2
Switzerland	+1	+2
Syria	+2	+3
Taiwan	+8	+8
Tajikistan	+5	+5
Tanzania	+3	+3
Thailand	+7	+7
Timor-Leste	+9	+9
Togo	UTC	UTC
Tokelau	-10	-10
Tonga	+13	+13
Trinidad	−4	−4
Tristan da Cunha	UTC	UTC
Tunisia	+1	+1
Turkey	+2	+3
Turkmenistan	+5	+5
Turks & Caicos Is	−5	−4
Tuvalu	+12	+12
Uganda	+3	+3
Ukraine	+2	+3
United Arab Em.	+4	+4
United Kingdom	UTC	+1
Uruguay	−3	−2
USA		
Eastern Time (CT, DE, FL, GA, IN[1], KY, MA, MD, ME, MI, NC, NH, NJ, NY, OH, PA, RI, SC, VA, VT, WV)	−5	−4
Central Time (AL, AR, IA, IL, IN[1], KS, LA, MN, MO, MS, ND, NE, OK, SD, TN, TX, WI)	−6	−5
Mountain Time[a] (N-E AZ, CO, ID, MT, NM, UT, WY)	−7	−6
[a]) exc. most of AZ	−7	−7
Pacific Time (CA, NV, OR, WA)	−8	−7
Uzbekistan	+5	+5
Vanuatu	+11	+11
Vatican City State	+1	+2
Venezuela	−4½	−4½
Vietnam	+7	+7
Virgin Is	−4	−4
Wallis & Futuna	+12	+12
Yemen	+3	+3
Zambia	+2	+2
Zimbabwe	+2	+2

NISTAN

n — **Pr.L:** Dari, Pashto, Turkmen,
AFG

ON AND CULTURE
side Spinzar Hotel, Kabul ☎+93 20
www.moic.gov.af
dom Raheen. Deputy Min.: Din

ANISTAN (RTA, Gov.)
Wazir Akbar Khan, Kabul ☎+93 20

i. Dir. Eng: Mr. Zarin Anzor.
Hz 400kW.
0W.
to 1430, Dari 1530.

kHz‡ 5kW, 105.1MHz‡ 30W – **R.**
z 6kW, 91.4MHz 250W. 0130-0330,
an, Pol-e-Khomri 103.0MHz, Baghlan:
ar-e-Sharif: 1584kHz 10kW, 105.1MHz
k: 0230-0430 (Fri 0430-0730), 1230-
R. Day Kundi: Nili 1200kHz‡ 0.5kW,
0.5kW – **R. Farah**, Farah: 1044kHz‡
0430 – **R. Faryab,** Maimana: 594kHz
– **R. Ghazni**: 1017kHz 10kW 92.4MHz
Ghor, Chaghcharan: 1584kHz 0.5kW,
ar Ga: 999kHz‡ 5kW, 96.1MHz 1kW.
rat: 1550kHz 0.1kW, 95.5MHz 250W.
wzjan, Sheberghan 106.6MHz 250W
90.6MHz 1kW. 0230-1430 – **R. Kapisa,**
V – **R. Khost:** 1602kHz 10kW, 91.2MHz
R. Kunar, Asadabad: 1575kHz 10kW,
0-1500 – **R. Kunduz**: 909kHz 10kW,
n, Mehtarlam: 88.2MHz‡ 0.5kW – **R.**
R. Nangarhar, Jalalabad: 1440kHz
730, 1030-1130 – **R. Nimroz,** Zaranj:
0330-0530, 1230-1500 (Fri 1930) – **R.**
0.1kW, 88.5MHz 300W – **R. Paktia,**
Hz 0.5kW – **R. Paktin Voice,** Zareh
, 89.8MHz 150W. 0230-1800. Web:
Bazarak: 88.0MHz 0.1kW – **R. Parwan,**
gan, Aybak: 1500kHz‡ 0.1kW, 90.4MHz‡
z – **R. Takhar**, Taloqan: 91.2MHz 30W
MHz 50W – **R. Wardak,** Meydan Shahr:
6kHz 10kW, 88.7MHz. ‡ inactive

STAN/R. MASHAAL/VOA ASHNA

Ashna R: MW: Kabul (Pol-e-Charkhi)
t/Jalalabad/Kabul/Kandahar/Mazar-e-
Pashto/E.
R: MW: Khost 621kHz 0100-1900. **FM:**
5MHz.
tails see International R. section (USA).

.Prgr: 24h in Pashto/Dari. Schedule v.
: "Da Sola Radyo day", Dari: "Inja Solh-e
g to Peace Radio".

stations:
045, Central PO, Kabul **W:** arman.fm **E:**
aad Mohseni. **FM:** 98.1MHz in Kabul
ad/Kandahar/Konduz/Lashkar Ga/ Mazar-
laman St. (near Ministry of Trade), Kabul
television.com **E:** marketing@arianatele-
Zubair. **FM:** Kabul: 93.5MHz – **R. City** ✉
High School, 2nd st. next to Uzbekistan
520094/77 7855900 **W:** citymedia.af **L.P:**
bul/Mazar-i-Sharif: 95.5MHz 3.5/1kW – **R.**
h-e-Awal, Kulula Pushta, Kabul ☎+93 77
E: info@killid.com **L.P:** Dir: Najiba Ayubi.
Herat 88.0MHz 2kW, Khost 88.2MHz.
– **R. Maiwand,** Kabul: 92.7MHz 1.5kW
g **L.P:** Dir: Mohammad Waqfi. **FM:** Kabul/
andahar/Konduz/Mazar e-Sharif/Paktia:
103.1MHz **R. Tapesh: W:** tapesh.fm **FM:** Kabul 97.1MHz.

Other stations (powers 50-500W if not stated otherwise):
Ayna R, Shebergan: 88.0MHz – **ERTV R,** (UNESCO), Kabul: 96.8MHz – **Gorbat FM,** Kabul: 91.8MHz – **R. Abasin,** Jalalabad: 93.7MHz – **R. Amo,** Faizabad: 91.5MHz – **R. Armangan,** Sheberghan: 89.9MHz – **R. Arzu,** Mazar-e-Sharif: 91.8MHz – **R. Azad Afgan,** Kandahar: 88.1MHz – **R. Baharak,** Badakshan prov: 95.3MHz – **R. Bamyan**: 88.0MHz – **R. Baran,** Herat: 98.7MHz 1kW – **R. Daikundi**, Nili: 88.5MHz – **R. Charchino,** Uruzgan prov: 88.1MHz – **R. Dehrawod,** Uruzgan: 87.5MHz – **R. Dunia,** Parwan: 88.8MHz – **R. Ejtima,** Logar: 88.5MHz – **R. Ertibat,** Malistan: 8.3MHz – **R. Faryad,** Herat: 87.8MHz – **R. Ghaznawiyan,** Ghazni: 89.3MHz – **R. Istiqlal,** Baraki-Barak: 89.5MHz – **R. Jaghuri,** Ghazni prov.: 87.8MHz – **R. Jaihoon**, Imam Shahib: 88.0MHz – **R. Javan,** Ghazni: 96.6MHz – **R. Jurm,** Badakhsan prov: 98.0MHz – **R. Kalagush,** Nuristan prov: 90.0MHz – **R. Kawoon,** Meterlam: 90.1MHz – **R. Khorasan,** Rokha: 91.3MHz – **R. Kishim,** Badakshan prov: 90.3MHz – **R. Milli e-Paygham,** Mohamad Agha: 94.0MHz –**R. Mowj,** Kabul: 105.5MHz 1kW – **R. Muram,** Jalalabad: 97.2MHz – **R. Nargis,** Jalalabad: 88.6MHz – **R. Nan:** Ghanikhel 99.9MHz, Khost 89.1MHz – **R. Naw-e-Bahar,** Balkh: 87.9MHz – **R. Nedaye Afghan,** Kabul: 99.6MHz – **R. Nedaye Solh,** Ghoreyan: 90.4MHz – **R. Nedaye Subh,** Ghoryan: 90.0MHz – **R. Paiwastoon,** Uruzgan: 89.2MHz – **R. Paktia,** 94.2MHz – **R. Paktika Voice,** Zareh Sharan: 92.9MHz – **R. Payman,** Kabul: 90.0MHz – **R. Pashtun Voice,** Sharana: 89.4MHz – **R. Payam,** Faizabad: 94.7MHz – **R. Qarabagh,** Kabul prov: 91.3MHz – **R. Quyash,** Maimana: 89.0MHz – **R. Rabia Balkhi,** Mazar-e-Sharif: 89.7MHz – **R. Roshani,** Kunduz: 89.5MHz – **R. Sabawoon**, Lashkar Ga: 88.0MHz 0.6kW – **R. Sahar,** Herat: 88.7MHz – **R. Samun,** Lashkar Ga: 88.6MHz – **R. Sharq,** Jalalabad: 87.6MHz – **R. Solh-e-Paygham,** Khost: 93.1MHz – **R. Spin Ghar**, Ghani Khel: 89.4MHz – **R. Tamana,** Faryab: 89.6MHz – **R. Takharistan,** Taluqan: 90.0MHz – **R. Tanin,** Shindan: 89.7MHz – **R. Tiraj Mir,** Pol-e-Khumri: 91.0MHz – **R. Watandar:** Herat/Kabul 87.5MHz 0.5/1kW – **R. VO Adalat,** Chagcharan: 90.3MHz – **R. VO Haqiqat,** Aibaq: 90.0MHz – **R. VO Jawan,** Herat University: 92.3MHz – **R. VO Kishim:** 90.3MHz – **R. VO Peace,** Naray: 94.0MHz – **R. VO Peace,** Jabul Saraj: 96.7MHz – **R. Wronga,** Kandahar: 92.5MHz 0.6kW –**R. Yawali Voice,** Sayedabad: 94.4MHz – **R. Zafar,** Paghman: 93.5MHz – **R. Zala Kunar,** Azadabad: 89.2MHz – **R. Zohra,** Kunduz: 89.8MHz – **Rana FM,** Kandahar: 88.5MHz. Web: www.ranafm.org – **Salam Watandar,** Kabul: 98.9MHz 1kW. Web: salamwatandar.com – **Shamshad FM,** Ghanikhel: 101.1MHz – Spogmai R, Kabul: 102.2MHz – **University R.** (UNESCO), Kabul: 106.7MHz 1.2kW, Herat 92.3MHz – **Voice of Women** (UNESCO)**:** Kabul 96..3MHz 10kW, Herat 88.7MHz – **Zala FM,** Naray: 89.2MHz 600W – **Zawon Voice R,** Khost: 99.7MHz - **Zenat R,** Pol-e Alam: 105.7MHz.
NB: many stations are Internews affiliates and relay their news prgr: "Salam Watandar". Web: internews.org

American Forces Network: Bagram 103.1/105.7/107.3MHz, Kabul: 105.7/107.3MHz, Kandahar 105.1/107.3MHz.
BBC World Sce: in English/Pashto/Dari/Uzbek/Farsi: Kunar 87.5MHz, Gardez 87.9MHz, Konduz 88.1MHz,, Ghazni/Saloquan/Taloqan 88.3MHz, Ghazni 88.3MHz, Faizabad 88.4MHz, Kabul/Bamian/Jalalabad/Pol-e-Khomri/Shebergan 89.0MHz, Herat/Lashkar-Ga/Mazar-e-Sharif 89.2MHz, Kandahar 90.0MHz, Khost 90.1MHz, Maimana 92.1MHz, Jabal-os-Saraj 92.2MHz, Kabul 101.6MHz.
BFBS R: Afghanistan: 102.1MHz at 2 sites & 102.4MHz at 6 sites; R. UK: Kabul 104.1/104.6MHz, 104.9MHz at 6 sites; Gurkha R: Kabul 103.2/103.5MHz, 106.3MHz at 5 sites.
Voice of Freedom, CJPOTF, HQ ISAF, Massoud Circle, Kabul. FM (1-5kW): Baghlan/Faizabad/Taloqan 87.5MHz, Kabul/Chekhcheran/Farah/Herat/Konduz/Mazar-e-Sharif/Maimana/Pol-e-Khomri/Qalaye Naw 88.5MHz, Kabul 91.6MHz (Turkish), Kabul 107.5MHz (German). Commando R, Kabul: 95.1MHz.
Deutsche Welle/Monte-Carlo Doualiya: Kabul: 90.5MHz 1kW.
R. France Int: Kabul 89.5MHz 0.2kW.
Many stns supporting the former Taliban gov. rep. to be in operation.

ALASKA (USA)

L.T: UTC -9h (13 Mar-6 Nov -8h) Aleutian Is. UTC -10h (13 Mar-6 Nov -9h) — **Pop:** 627,000 — **Pr.L:** English — **E.C:** 60Hz, 120/240V — **ITU:** ALS

FEDERAL COMMUNICATIONS COMMISSION (FCC)
see USA for details

ALASKA BROADCASTERS ASSOCIATION
✉ 700 W 41st Ave, Anchorage AK 99503 ☎+1 907 258 2424 🖷 +1 907 258 2414 **W:** www.alaskabroadcasters.org

THE WRTH 2011 QUESTIONNAIRE

TELL US WHAT YOU THINK *and* **WIN** A PRIZE!

This is your chance to tell us what you think of WRTH and a little bit about yourself so that we can continue to shape the book to your needs. Complete entries will be eligible to enter a prize draw. The first prize this year is a **Tecsun PL-380**. Five runners-up will each win a free copy of WRTH 2012

1. Name and address (use CAPITALS only please)

Name

Street

City State

Zip or Postal Code Country

2. Age: a. ❑ 15–30 b. ❑ 31–40 c. ❑ 41–50 d. ❑ 50+
3. Which category best describes your situation?
   a. ❑ broadcasting professional b. ❑ dedicated radio hobbyist/DXer c. ❑ radio listener
4. How many times have you bought WRTH? a. ❑ 0 b. ❑ 1–3 c. ❑ 4–6 d. ❑ 6+
5. How often do you buy WRTH?
   a. ❑ every year b. ❑ once every two years c. ❑ less often
6. How many other people read your copy of WRTH? a. ❑ 0 b. ❑ 1 c. ❑ 2 d. ❑ 3+
7. Are you a member of a DX club? a. ❑ Yes b. ❑ No
8. How much do you spend on radio equipment annually?
   a. ❑ less than $200 b. ❑ $200–$500 c. ❑ $500–$2000 d. ❑ $2000–$5000
9. Do you have plans to buy new equipment next year? a. ❑ Yes b. ❑ No
10. Are you connected to the Internet, or do you expect to be in the next six months?
    a. ❑ Yes b. ❑ No c. ❑ No PC
11. What is your main area of interest?
    a. ❑ International radio and SW broadcasts
    b. ❑ MW broadcasts (including both domestic and international)
    c. ❑ FM broadcasts
    d. ❑ TV DXing
12. Do you find all the information you need on radio listening in WRTH (aside from free internet resources)? a. ❑ Yes b. ❑ No
13. What information would you like to be added to WRTH, and are there any changes you think should be made to the content? ______

______

______

14. If you would like to be told by email when the new edition of WRTH is available please give your e-mail address here (please write very clearly):

______

Please tear this card out, fold it with the address showing, tape the *two* sides together (no staples please) and mail it. Please return this form BEFORE 1 JUNE 2011. The prize draw will be held by August and winners will be notified by mail or email. All entries become the property of WRTH and no correspondence will be entered into. WRTH reserves the right to list winners' names.

Visit www.wrth.com for
advertisers weblinks

Place
Stamp
Here

WORLD RADIO TV HANDBOOK
P.O. Box 290
OXFORD
OX2 7FT
U.K.

NATION

Section
Contents

Initial entries for each letter,
see Main Index for full details

AFGHA

L.T: UTC +4½h — **Pop:** 33 milli
Uzbek — **E.C:** 50Hz, 220V — **ITU**

MINISTRY OF INFORMATI
Mohammad Jan Khan Watt, be
210 1301 +93 20 229 0088 **W:**
L.P: Minister: Dr. Sayed Makh
Mohamad Mobariz Rashid.

RADIO TELEVISION AFGH
PO Box 544, Street 10, Lane 2
210 1085-7 **W:** rta.org.af
L.P: DG: Mr. Ghulam Rasol Hazra
MW: Kabul (Pol-e-Charkhi) 1107k
FM: Kabul 93.0 1kW, 105.2MHz
D.Prgr: 0100-1830. Main **N:** Pash
Ann: "Radyo Afghanistan, Kabul

PROVINCIAL STATIONS
R. Badakhshan, Faizabad: 584
Badghis, Qalay-e Naw: 1500kH
1330-1630, times vary – **R. Bagh**
106.6MHz 250W – **R. Balkh,** Ma
0.1kW. In Dari/Pashto/Tajik/Uzb
1530 – **R. Bamyan:** 96.0MHz
103.2MHz 30W, Qalat 99.0MHz
7kW, 88.5MHz 1kW (F.Pl.). 0300-
5kW, 104.3MHz 30W. 1230-1430
30W. 0230-0330, 1130-1530 – **R**
93.4MHz – **R. Helmand,** Lashk
0330-0730, 1130-1430 – **R. He**
0300-0500, 1130-1330 – **R. Jo**
– **R. Kandahar:** 1305kHz 10kW,
Mahmud-e-Raqi: 101.1MHz 600
1kW. 0230-0630, 1130-1530 –
100.5MHz 1kW.0330-0530, 093
94.4MHz 250W – **R. Laghma**
Logar, Pol-e-Alam: 92.7MHz
10kW, 93.5MHz 250W. 0230-0
1584kHz 2kW, 90.0MHz 1kW.
Nuristan, Nuristan: 1500kHz
Gardez: 909kHz 50W, 104.2M
Sharan, Paktika 1386kHz 5kV
paktinghag.com – **R. Panjshir,**
Charikar: 88.9MHz – **R. Saman**
150W – **R. Sar-e-Pol:** 89.9M
– **R. Uruzgan,** Tarin Kowt: 93.0
88.9MHz – **R. Zabul,** Qalat: 93

BBG–R. FREE AFGHAN
& DEEWA R. (US Gov.)
R. Free Afghanistan & VOA
1296kHz 400kW. **FM:** Hera
Sharif 100.5MHz. 24h in Dari/
R. Mashal & VOA Deewa
Asadabad/Gardez/Khost 100.
For SW broadcasts & more de

PEACE RADIO (Mil.)
W: cjtf82.com
SW: Orgun 6700kHz 1kW.
and operation irr. **Ann:** Pashto
Radyoe", E: "You are listenin

Independent commercial
Arman FM P.O. Box 1
info@arman.fm **L.P:** Dir:
(2kW)/Ghazni/Herat/Jalalab
e-Sharif – **Ariana R.** Da
+93 70 151515 **W:** arian
vision.com **L.P:** Dir: Ahmad
Karti 3, in front of Habibia
Embassy, Kabul +93 78
Nasir Totakhil, Pres. **FM:** Ka
Killid House 223, Estg
1088888 **W:** radiokillid.af
FM: Kabul 88.0MHz 4kV
Kabul Rock R: 108.0MH
– **Nawa R. W:** sabacent.o
Ghazni/Herat/Jalalabad/

MW	kHz	Call	kW	N	Location
2)	550	KTZN	5		Anchorage
3)	560	KVOK	1		Kodiak
4)	580	KRSA	5	d	Petersburg
5)	590	KHAR	5		Anchorage
6)	620	KGTL	5		Homer
7)	630	KJNO	5/1		Juneau
8)	630	KIAM	10/3.1		Nenana
9)	640	KYUK	10		Bethal
2)	650	KENI	50		Anchorage
11)	660	KFAR	10		Fairbanks
12)	670	KDLG	10		Dillingham
13)	680	KBRW	10		Barrow
14)	700	KBYR	10		Anchorage
15)	720	KOTZ	10		Kotzebue
5)	750	KFQD	50		Anchorage
17)	770	KCHU	9.7		Valdez
18)	780	KNOM	25/14		Nome
19)	790	KCAM	5		Glennallen
20)	800	KINY	10/7.6		Juneau
11)	820	KCBF	10		Fairbanks
22)	830	KSDP	1		Sand Point
24)	850	KICY	50	*	Nome
25)	870	KSKO	10		McGrath
26)	890	KBBI	10		Homer
27)	900	KZPA	5	r	Fort Yukon
28)	910	KIYU	5		Galena
29)	920	KSRM	5		Soldotna
30)	930	KTKN	5/1		Ketchikan
12)	930	KNSA	2.5	r	Unalakleet
32)	950	KSEW	1		Seward
33)	970	KFBX	10		Fairbanks
34)	1020	KOAN	10	d	Eagle River
34)	1080	KUDO	10		Anchorage
35)	1110	KAGV	10		Big Lake
29)	1140	KSLD	10		Soldotna
37)	1170	KJNP	50/21		North Pole
7)	1230	KIFW	1		Sitka
39)	1230	KVAK	1		Valdez
40)	1330	KXLJ	10/3		Juneau
41)	1430	KMBQ	1	+	Wasilla
42)	1450	KLAM	0.25		Cordova

d=directional *=directional 0800-1200 r=relay +=F.P.I.

FM	Call	MHz	kW	Location
	KAKL	88.5	11	Anchorage
	KATB	89.3	4.9	Anchorage
	KNBA	90.3	100	Anchorage
	KSKA	91.1	100	Anchorage
	KFAT	92.9	10	Anchorage
	KAFC	93.7	27	Anchorage
5)	KEAG	97.3	55	Anchorage
34)	KLEF	98.1	25	Anchorage
2)	KYMG	98.9	100	Anchorage
2)	KBFX	100.5	25	Anchorage
2)	KGOT	101.3	26	Anchorage
	KDBZ	102.1	23	Anchorage
5)	KMXS	103.1	100	Anchorage
5)	KBRJ	104.1	55	Anchorage
34)	KNLT	105.7	51	Anchorage
5)	KWHL	106.5	100	Anchorage
2)	KASH-FM	107.5	100	Anchorage
8)	KYKD	100.1	12	Bethel
	KCUK	88.1	6	Chevak
	K220CL	91.9	1	Chignik
11)	KTDZ	103.9	3	College
42)	KCDV	100.9	1.2	Cordova
	KRUP	99.1	6	Dillingham
	KDJF	93.5	20.5	Ester
27)	KUAC	89.9	38	Fairbanks
	KSUA	91.5	3	Fairbanks
11)	KXLR	95.9	25	Fairbanks
	KYSC	96.9	5.8	Fairbanks
11)	KWLF	98.1	28	Fairbanks
33)	KAKQ-FM	101.1	50	Fairbanks
33)	KIAK-FM	102.5	100	Fairbanks
33)	KKED	104.7	50	Fairbanks
	KEUL	88.9	1.4	Girdwood
17)	KXGA	90.5	3.2	Glennallen
	KHNS	102.3	3	Haines
	KHGO	89.9	3	Homer
6)	KWVV-FM	103.5	100	Homer
	KBBO-FM	92.1	10	Houston
34)	KZND-FM	94.7	15	Houston
	KXLW	96.3	10	Houston
	KXLL	100.7	6	Juneau

FM	Call	MHz	kW	Location
	KRNN	102.7	6	Juneau
	KTOO	104.3	1.4	Juneau
7)	KTKU	105.1	3.8	Juneau
20)	KSUP	106.3	10	Juneau
	KWJG	91.5	1	Kasilof
29)	KFSE	106.9	8	Kasilof
	KDLL	91.9	4.9	Kenai
29)	KWHQ-FM	100.1	25	Kenai
	KRBD	105.3	3.4	Ketchikan
	KMXT	100.1	3	Kodiak
3)	KRXX	101.1	3.1	Kodiak
8)	KAKN	100.9	3	Naknek
	KXBA	93.3	50	Nikiski
18)	KNOM-FM	96.1	1	Nome
24)	KICY-FM	100.3	1	Nome
37)	KJNP-FM	100.3	25	North Pole
	KFSK	100.9	2	Petersburg
32)	KKNI	105.9	3	Seward
38)	KSBZ	103.1	3.1	Sitka
	KCAW	104.7	3.6	Sitka
29)	KKIS-FM	96.5	10	Soldotna
6)	KPEN-FM	101.7	25	Soldotna
	KUHB-FM	91.9	15	St. Paul
	KWMD	90.1	1.2	Sterling
34)	KMVV	104.9	45	Sterling
	KTNA	88.5	7.2	Talkeetna
	K220AD	91.9	1.1	Valdez
39)	KVAK-FM	93.3	1.2	Valdez
41)	KMBQ-FM	99.7	51	Wasilla
	KAYO	100.9	50	Wasilla
	KSTK	101.7	3	Wrangell

NB: Txs below 1kW not listed

Addresses:
2) 800 E Dimond Blvd, Ste. #3-370, Anchorage AK 99515-2058 – 3) P.O.Box 708, Kodiak AK 99615-0708 – 4) P.O.Box 650, Petersburg AK 99833-0650 – 5) 301 Arctic Slope Ave #200, Anchorage AK 99518-3035 – 6) P.O.Box 109, Homer AK 99603-0109 – 7) 3161 Channel Drive #2, Juneau AK 99801-7815 – 8) P.O.Box 474, Nenana AK 99760-0474 – 9) P.O.Box 468, Bethel AK 99559-0468 – 11) 819 1st Ave #A, Fairbanks AK 99709-4449 – 12) P.O.Box 670, Dillingham AK 99576-0670 – 13) P.O.Box 109, Barrow AK 99723-0109 – 14) 1399 W 34th Ave #202, Anchorage AK 99503-3659 – 15) P.O.Box 78, Kotzebue AK 99752-0078 – 17) P.O.Box 467, Valdez AK 99686-0467 – 18) P.O.Box 988, Nome AK 99762-0988 – 19) P.O.Box 249, Glennallen AK 99588-0249 – 20) 1107 W 8th St #2, Juneau AK 99801-1896 – 22) P.O.Box 328, Sand Point AK 99661 – 24) P.O.Box 820, Nome AK 99762-0820. Russian 0800-0930 – 25) P.O.Box 70, McGrath AK 99627-0070 – 26) 3913 Kachemak Way, Homer AK 99603-7618 – 27) P.O.Box 50, Fort Yukon AK 99740-0050 – 28) P.O.Box 165, Galena AK 99741-0165. Mostly rel. 25) – 29) 40960 Kalifornsky Beach Rd, Kenai AK 99611-6445 – 30) 526 Stedman St, Ketchikan AK 99901-6629 – 32) P.O.Box 2414, Seward AK 99664-2414 – 33) 546 9th Ave, Fairbanks AK 99701-4902 – 34) 4700 Business Park Blvd #E-44A, Anchorage AK 99503-7176 – 35) 4723 King David St, Houston AK 99694 – 37) P.O.Box 56359, North Pole AK 99705-1359 – 39) P.O.Box 367, Valdez AK 99686-0367 – 40) 1105 W 9th St, Juneau AK 99801-1811 – 41) 2200 E Parks Hwy, Wasilla AK 99654-7355 – 42) P.O.Box 60, Cordova AK 99574-0060

EXTERNAL SERVICE: Radio Station KNLS
see International Broadcasting section

ALBANIA

L.T: UTC +1h (27 Mar-30 Oct +2h) — **Pop:** 3.6 million — **Pr.L:** Albanian — **E.C:** 50Hz, 220V — **ITU:** ALB

KËSHILLI KOMBËTAR I RADIOS DHE TELEVIZIONIT (KKRT) (NATIONAL COUNCIL OF RADIO & TELEVISION)
Rruga "Abdi Toptani", Ish Hotel Drini, Tirana. ☎ +355 4 233326 +355 4 226287 W: kkrt.gov.al E: kkrt@kkrt.gov.al LP: Chairman: Mesila Doda. Tech. Dir: Pirro Koci.

RADIOTELEVIZIONI SHQIPTAR (RTSH)
ALBANIAN RADIO & TELEVISION (Pub.)
Rruga Ismail Qemali 11, Tirana ☎ +355 4256 059 +355 422 7745 W: rtsh.al E: marketing@rtsh.al LP: DG: Petrit Beci. Tech. Dir: Agron Aranitasi, Dir Stns: Eng. Arben Mehilli.

FM	MHz	kW	Ch	FM	MHz	kW	Ch
Llogora	88.3	1	1	Mile	93.0	1	1
Korce	89.5	1	1	Ishem	94.1	1	1
Tarabosh	91.0	1	1	Petresh	95.4	1	1
Shkodra	92.0	2	R	Tirana(Dajt)	95.8	2	2

FM	MHz	kW	Ch	FM	MHz	kW	Ch
Mide	96.0	3	1	Kukesi	100.4	1	R
Cervenake	99.1	1	1	Qafe Prush	100.7	1	1
Tirana(Dajt)	99.5	10	1	Homesh	102.2	1	1
Zvernec	99.8	3	1	Gjirokastra	102.5	0.3	R
Erseke	100.2	1	1	Sopot	107.0	1	1

+11 more txs under 1kW.

D.Prgr: 1st Ch.: 24h, **2nd Ch.:** 24h, **Regional Ch.:** 0600-2000.

EXTERNAL SERVICES: R. Tirana + relays (CRI & TWR) on MW 1215/1395/1458kHz and SW: see International Broadcasting section.

Nationwide Private FM Stations:
+2 RADIO
✉ Rr. Aleksandër Moisiu Nr 76/1, ish kinostudjo Shqipëria e Re, Tirana ☎+355 4 368490 **W:** plus2radio.com.al **E:** info@plus2radio.com.al **LP:** Mgr: Leonard Gremi.

FM	MHz	kW	FM	MHz	kW
Fushe Dajt	89.8	63	Zvernec	96.3	0.3
Cervenake	90.3	4	Gllave	97.6	3
Mide	94.3	6	Tirana/Dajt	101.6	89

TOP ALBANIA RADIO
✉ Piramida QNK, Blvd. Dëshmorët e Kombit, Tirana. ☎+355 4 247492 🖷 +355 4 247493 **W:** topalbaniaradio.com **E:** contact@topalbaniaradio.com

FM	MHz	kW	FM	MHz	kW
Kerculle	93.0	1	Sarande	100.6	1
Shkoder	94.1	2	Mide	101.3	15
Korce	95.0	2	Elbasan	102.2	2
Gllave	96.0	1	Sopot	104.0	5
Dürres	99.0	1	Lezhe	104.3	1
Tirana/Dajt	100.0	15	Ardenice	104.5	2

Other Private FM Stations

FM Station	MHz	kW	FM Station	MHz	kW
1) R. Ngjallja	88.5	6	1) R. Koha	98.1	2
1) R. New Planet	89.0	3	1) R. Aldo 03	98.8	2
1) R. Kontakt	89.3	2	1) R. Super Star	99.0	-
1) R. Love	90.7	2	16) R.Armonia	99.2	0.4
2) R. Alfa	90.9	1	10) R. Saranda	100.0	0.5
3) R. Argjiropoli	91.0	1	1) R. Club FM	100.4	13
1) R. Skorpion	91.4	0.3	11) R. E Pare	100.4	0.1
4) R. +3	91.6	1	12) R. Prespa	100.8	0.3
1) R. Italia	92.4	0.4	1) R. Top Gold	100.8	56
5) R. Planet	93.0	2	1) R. Boom Boom	101.2	38
6) R. Klea	93.9	1	1) R. Alfa & Omega	102.6	6
7) R. Alpo	94.1	0.1	14) R. Val e Kalter	103.3	2
1) R. Eurostar	94.5	2	1) R. Club FM	104.3	1
6) R. Fantasy	94.7	0.3	1) R. Rock	104.6	25
8) R. Ruzvelt	94.8	0.1	1) R. Perla	105.0	1
9) R. Emanuel	95.7	4	1) R. Ime	105.4	0.5
1) R. Oxygen	96.1	2	4) R. Star	105.5	2
1) R. Real	96.4	2	1) R. Club Alsion	106.3	1
2) R. Eurostar	96.6	1	15) R. Perla	106.4	0.5
1) R. Rash	97.0	3	1) R. NRG	106.6	2
1) R. Muzika Jone	97.3	2	9) R. Stinet	106.9	1
1) R. +7	97.7	6	1) R. Magic Star	107.0	4
9) R. ABC	98.0	2	4) R. Fieri	107.2	1
			1) R. House of Arts	107.7	2

Locations: 1) Fushe Dajt, 2) Petresh, 3) Kerculle, 4) Fier, 5) Tirana, 6) Kavaje, 7) Kerculle, 8) Memaliaj, 9) Korce 10) Mile 11) Burrel, 12) Prespe, 13) Shkoder, 14) Zvernec, 15) Dürres, 16) Bularat.

Other FM stations:
R. France International: Fushe Dajt/Korca 102.0MHz 1kW
VOA Europe: Tirana (Dajt) 107.4MHz 0.8kW
BBC Europe: Tirana (Dajt) 103.9MHz 2.2kW
Deutsche Welle: Tirana 106.0MHz 6.3kW

ALGERIA

L.T: UTC +1h — **Pop:** 35 million — **Pr.L:** Arabic, French, Berber dialects — **E.C:** 50Hz, 230V — **ITU:** ALG

TÉLÉDIFFUSION D'ALGERIE (TDA)
✉ Direction Générale, B.P. 50, Bouzaréah, Route de Baïnam, 16340 Algér ☎+213 21 901717 🖷 +213 21 902424 **W:** tda.dz **E:** contact@tda.dz **LP:** DG: Abdelmalek Houyou. Dir. Tech. Sces: Mohamed Hacine Ladj.

RADIO ALGÉRIENNE (RA, Pub.)
✉ DCRR, 12 Rue Shakespeare, El Mouradia, Algér ☎+213 21 230805 🖷 +213 21 694620 **W:** radioalgerie.dz **E:** technique@algerian-radio.dz **LP:** DG: Tewfik Khelladi.

LW/MW	kHz	kW	Ch.	Hrs.
Béchar	153	2000	1	24h
Ouargla	198	2000	1	24h
Tipaza	252	*1500	3	24h
F'kirina	531	600	1	24h
Sidi Hamadouche	549	600	1	24h
20)Touggourt	558	10	1/L	24h
4) Béchar	576	*400	L	24h
31)Tindouf	666	10	1/L	24h
Aboudid (Ain el H.)	693	5	2	0400-2400
1) Reggane	693	10	1/L	24h
14)Laghouat	702	25	3/L	0500-0100
8) In Amenas	738	5	1/L	24h
8) Djanet	783	5	1/L	24h
24)El Oued	783	10	1/L	24h
Béchar	837	5	3	0500-0100
10)Ghardaïa	873	10	1/L	24h
Ouled Fayet	891	600	1	24h
Tamanrasset	909	10	1	24h
1) Timimoun	927	10	1/L	24h
Ouled Fayet	981	100	2	0400-2400
3) Batna	1017	10	1/L	24h
20)Hassi Messaoud	1026	10	1/L	24h
Illizi	1071	5	1	24h
1) Adrar	1089	10	1/L	24h
27)In Salah	1161	5	1/L	24h
10)El Golea	1287	5	1/L	24h
Ouled Fayet	1422	50	C	0400-0100

*) half-power 1900-0600. **F.PI:** New 50 kW at Aïn Beida for 648kHz, presumably for Chaîne 3.

FM (MHz)	1	2	3	I	kW(TRP)
Adrar			88.8		2
Aflou	90.7				10
Akfadou		91.8			10
Bains Romains				95.6	0.1
Bordj El Bahri	91.0		89.2	104.2	2
Chréa			88.4	101.5	10
Doukhane			91.0		2
Gara Djebilet	98.0				0.1
Kef El Akhal			87.6		10
Mahouna	97.6				2.5
M'cid	91.9				10
Mecheria	87.8				10
Meghriss	93.5				10
Nador		88.4	91.5		10
Tessala	102.7				2
Tiaret	89.4				2
Tizi Ouzou		88.0			0.1

Local station, location	MHz	kW(TRP)	Ch.
R. Laghouat, Aflou	87.6	10	3/L
R. Chlef, Ain N'sour	87.7	10	3/L
R. Tébessa, Doukhane	87.9	2	1/L
R. Batna, Metlili	88.1	10	2/L
R. Soummam, Akfadou	88.7	10	1/L
R. Annaba, M'cid	88.8	10	2/L
R. Béchar	89.3	5	1/L
R. El Hidhab, Megriss	90.4	10	1/L
R. Rélizane, Ain N'sour	90.8	10	1/L
R. Naama, Mecheria	90.9	10	1/L
R. Souhoub, Sbaa Mokrane	91.1	2	3/L
R. Biskra, Metlili	91.2	10	1/L
R. El Bahdja, Chréa	91.5	10	L
R. Adrar, Adrar	91.9	2	1/L
R. Ouargla	92.1	3	1/L
R Tiaret, Tiaret	92.5	2	B/L
R. Oran, Tessala	92.7	10	3/L
R. Illizi	93.5	0.1	1/L
R. Constantine, Kef El Akhal	93.9	10	1/L
R. El Bahdja, Bordj el Bahri	94.2	2	L
R. Hodna, M'sila	94.7	2	1/L
R. Mitidja, Chréa	94.7	10	1/L
R. Tlemcen, Nador	94.7	10	1/L
R. Jijel	94.8	0.1	1/L
R. Skikda	94.8	0.1	1/L
R. Souk Ahras, M'cid	95.1	10	3/L
R. Ghardaïa, El Golea/Guerara	98.0	3	1/L
R. Laghouat, Hassi R'Mel	98.0	2	1/L
R. Tamanrasset, 7 locations	98.0	1	1/L
R. Tindouf	98.0	0.1	1/L
R. Mascara, Chareb Rih	98.5	3	1/L
R. Sidi Bel Abbés, Tessala	99.2	10	1/L
R. Annaba, Bouzizi	99.3	2	3/L
R. El Bayadh	100.1	0.1	1/L
R. Annaba, Bouzizi	100.3	2	3/L
R. Mostaganem	107.2	0.5	1/L

+25 more stations under 1kW.

D.Prgr: 1=Chaîne 1 in Arabic: 24h. Satellite Eutelsat 702MHz. **2=Chaîne 2** in Tamazight: 0400-2400. Satellite Eutelsat 728MHz. **3=Chaîne 3** in French, 0500-0100. Satellite Eutelsat 738MHz. **I=R. Algérie Internationale** in Arabic, French, E. & Spanish: 0800-2300. **C=** incl. **R. Koran** 0400-0800, R. Mitidja 0800-1200, Chaîne 1 1200-1600, **R. Culture** 1600-2200 and Chaîne 3 2200-0100 on 1422kHz. **L=Local stations:** times of local prgrs vary by stn (may be longer in summer than winter), but at least between 0800-1600. Most local stns transmit Chaîne 1 at other times (666, 1161 & 1287kHz rep. with R. Algérie Int. relay). Many local stations carry the Arabic news from Chaîne 1 1100-1200 and at other times.

Addresses for local stations:
1) R. Adrar, B.P. 309, Adrar. Web: www.radio-adrar.net – **2) R. Annaba**, 7, Boulevard Radji Mokhtar, Quartier Annasr, Annaba. Web: www.annabafm.net – **3) R. Batna**, B.P. 453, Batna. Web: www.radio-batna.dz – **4) R. Béchar**, Cité Badr, B.P. 330, Béchar. Web: www.radiobechar.com – **5) R. Chlef**, Ain N'sour. Web: www.radio-chlef.com – **6) R. Constantine**, B.P. 28B El Koudia, Constantine. Web: www.radio-constantine.dz – **7) R. El Bahdja**. Web: msila28dz.site.voila.fr/nahdja.html – **8) R. El Bayadh**, B.P. 195, El Bayadh – **9) R. El Hidhab**, B.P. 54, Ain Tbinet, Sétif. Web: www.radio-setif.com – **10) R. Ghardaïa**, B.P. 17, Ghardaïa. Web: www.radioghardaia.dz – **11) R. Hodna**, B.P. 1400, Grand Poste, M'sila. Web: msila28dz.site.voila.fr/bostradiohodna.html – **12) R. Illizi**, Route de l'aéroport, B.P. 230, Illízi – **13) R. Jijel**, B.P. 48, Jijel. Web: www.radio-jijel.dz – **14) R. Laghouat**. Web: site.voila.fr/radiolagh – **15) R. Mascara**, Place Mostfa Ben Touhami, Mascara. Web: www.radiomascara.com – **16) R. Mitidja**, 12 Avenue William Shakespeare, Le Mouradia – **17) R. Mostaganem**, Place El Matemar, B.P. 1014, Mostaganem 027000. Web: www.radiomostaganem.com – **18) R. Naama**, Av. du 1 Novembre, BP 223, Naama. Web: www.radionaamafm.com – **19) R. Oran**. Web: www.radio-oran.com – **20) R. Ouargla**, Ruissat, B.P. 83, Ouargla – **21) R. Rélizane**, 15 Rue Ismail Mustapha, Maison de Culture, Rélizane. Web: www.radio-relizane.com – **22) R. Sidi Bel Abbés**, Gare de l'État, Sidi Bel Abbés 022000. Web: www.radiosba.net – **23) R. Skikda**, Porte des Aurès, B.P. 55, Skikda – **24) R. Souf**, Cité Reml El Oued, B.P. 172, El Oued Web: radio-souf.free.fr – **25) R. Souk Ahras**, Blvd. Messous Hamid, Souk Ahras. Web: www.radio-soukahras.dz – **26) R. Soummam**, Boulevard Youcef Bouchebah, Béjaia. Web: www.radiosoummam-bejaia.com – **27) R. Tamanrasset**, B.P. 1080, Tamanrasset. – **28) R. Tassili**, B.P. 230, Illizi – **29) R. Tébéssa**, Unité de Tébéssa, Parc des Loisirs, Tébéssa. Web: www.radio-tebessa.dz – **30) R. Tiaret**, Rue des Fréres Saim Tiaret, B.P. 671, Tiaret – **31) R. Tindouf**, B.P. 213, Agence Enasr, Tindouf – **32) R. Tlemcen**, B.P. 44K, Tlemcen. Web: www.radiotlemcen.org – **33) R. Ziban**, Av. Idriss Mohamed, Biskra Web: www.zibanfm.com

Ann: Chaîne 1: "Kanet al Oula min Idha'at al-Djazairiyah.", 2: "Radio Isnath", 3: "Chaine Trois, Radio Algérienne", C: "Idha'atul-Thaqafiyah", K:"Idha'atul-Koran al Karim". **IS:** Oriental Lute (Ud)

ANDORRA

L.T: UTC + 1h (27 Mar-30 Oct: + 2h) — **Pop:** 82,000 — **Pr.L:** Catalan, French, Spanish — **E.C:** 50Hz, 230V — **ITU:** AND

ANDORRA TELECOM
Av. Meritxell 112, Andorra la Vella, AD 500 ☎+376 875000 +376 821414 **E:** comunicacio@andorratelecom.ad **W:** www.andorratelecom.ad **L.P:** Admin Dir: Jaume Salvat Font. Dir. International relations & Television: Xavier Jimenez Beltran. **NB:** All stns located in Andorra la Vella except where noted.

FM	MHz	kW	Station
2)	87.8	0.3	R. María España (relay)
1)	88.1	0.3	Cadena Dial
2)	89.0	0.3	R7P – RAC 1
2)	89.5	0.3	Europa FM
2)	90.1	0.3	R. Tele Taxi
10)	90.9		iCAT FM
4)	91.4	1	R. Nacional d'Andorra
2)	92.1	1	Ona Andorra/Ona Catalana, Escaldes Engordany
1)	92.6	1	M80 R.
2)	93.3	0.3	R. Valira/Punto R.
5)	93.8	0.3	Flaix FM Vallnord
4)	94.2	1	R. Nacional d'Andorra
9)	94.6	0.3	Ona 7 R., Escaldes Engordany
2)	95.0	0.3	R. Valira/Punto R.
10)	95.6		Catalunya Informaciò
1)	96.0	1	R. Flaixbac
2)	96.5		Catalunya Música
4)	97.0	1	Andorra Música
2)	98.1	0.1	R. Valira/Onda Cero
2)	98.5	0.1	R. Valira/Onda Cero
6)	98.9	1	Kiss FM (relay)
2)	100.2		R. Valira/Punto R.
9)	100.2		Ona 7 R., Escaldes Engordany
6)	100.6	0.3	N.R.J. (relay)
8)	101.5	1	Andorra7R.
6)	101.8	1	France Inter (relay)
1)	102.3	0.5	R. SER Principat d'Andorra
6)	102.6	1	France Musiques (relay)
1)	103.3	0.3	Los 40 Principales
6)	104.0	1	France Culture (relay)
10)	104.6	1	Catalunya R. (relay)
6)	106.0	1.2	RNE R. 4 (relay)
6)	106.8	1.2	RNE R. 1 (relay)
7)	107.5	1.2	R. Principat/R. Estel
6)	107.9	1.2	RNE R. 3 (relay)

Addresses & other information:
1) C. Prat de la Creu, 32, AD-500 Andorra la Vella – **2)** Cadena Pirenaica de Ràdio i Televisió, SA, Av. Príncep Benlloch 24, AD-200 Encamp +376 732000. **E:** info@cadenapirenaica.com (R. Valira stations at C/ del Parnal 2, "Edifici Prat de Casa de Escaldes", AD-500 Andorra la Vella) – **3)** Parc de la Mola 10, Torre Caldea, planta 3, AD-700 Escaldes Engordany – **4)** Baixada del Molí 24, AD 500 Andorra la Vella +376 873777 +376 863242 **W:** www.rtva.ad **E:** rtva@rtva.ad **LP:** DG: Enric Castellet Pifarré – **5)** Av. Meritxell 75, AD-500 Andorra la Vella +376 862288 +376 862287 – **6)** Servei de Telecomunicacions d'Andorra, Av. Meritxell 112, AD-500 Andorra la Vella – **7)** Passeig del Parc 16-18, ES-25700 Seu d'Urgell, Spain. +973 354400. **E:** radio@radioprincipat.com Frequent relays of R. Estel, Barcelona – **8)** C. Bonaventura, Avda Riberaygua, 39, Edifici Alexandre, AD-500 Andorra la Vella **W:** www.andorra7radio.com **E:** laradio@andorra7radio.com – **9)** Carrer del Parnal 2, AD-700 Escaldes Engordany – **10)** Av. Diagonal 614-616, 08021 Barcelona, Spain.

ANGOLA

L.T: UTC +1h — **Pop:** 13 million — **Pr.L:** Portuguese + ethnic — **E.C:** 50Hz, 220V — **ITU:** AGL

MINISTÉRIO DA COMUNICAÇÃO SOCIAL (MCS)
Av. Comandante Valódia 1°&2° amdar , CP. 2608, Luanda ☎+244 22 443495 +244 22 2392649 **W:** www.comunicacaosocial.gv.ao **E:** mcs@netangola.com **L.P:** Dep. Min: Miguel de Carvalho.

RÁDIO NACIONAL DE ANGOLA (RNA, Pub.)
Av. Comandante Gika, CP. 1329, Luanda ☎+244 22 2323172/321258 +244 22 2324647/391234 **W:** www.rna.ao **E:** dgeral@rna.ao **L.P:** DG: Eduardo Magalhães. PD: Júlio Mendonça. TD: Cândido R. Pinto.

MW	kHz	kW	Prgr.	H of tr
Mulenvos	1088	25	A	24h
SW	**kHz**	**kW**	**Prgr.**	**H of tr.**
Mulenvos	4950	25	A	24h
Mulenvos	v7217	15	N/A	24h

FM (MHz): Luanda 4kW: 93.5 (A), 94.5 (5), 96.5MHz (FME), 99.9 (RL), 101.4MHz (N).
Ann: "Rádio Nacional de Angola". **F.PI:** new 100kW tx on MW.
Prgrs: A=Canal A in Portuguese (general coverage): 24h. **N:** on the h. **N=Rádio N'Gola Yetu** (ethnic): 0000-2400. **N:** rel. Canal A. **FME=Rádio FM Estéreo** (music): 2400-2400. **RL=Rádio Luanda** (capital channel): 24h. **5=Rádio 5** (sports): 0500-2300. **P=Emissora Provincial:** 0400-2300, rel. A at night.

PROVINCIAL STATIONS

MW	kHz	kW	Pr		MW	kHz	kW	Pr
6) Benguela	774	50	A	15)	Namibe	1314	10	P
3)Kuito	990	50	A	12)	Saurimo	1386	10	P
1)Mulenvos	1134	10	P	3)	Kuito	1404	v 10	P
17)Mbanza Congo	1152	10	P	1)	Dundo	1440	10	P
6)Huambo	1170	25	P	14)	Luena	1458	v 10	P
13)Malange	1197	v 10	P	8)	Menongue	1467	10	P
7) Lubango	1233	10	P	5)	Sumbe	1485	v 10	P
9) N'dalatando	1260	10	P	2)	Benguela	1503	10	P
4) Tenda	1278	25	A	4)	Tenda	1530	10	P
16)Uíge	1296	10	P					
FM	**MHz**	**kW**	**Pr**		**FM**	**MHz**	**kW**	**Pr**
9) Dondo	88.8	0.5	P	3)	Kuito	91.0	4	A
2) Benguela	90.4	1	S	16)	Uíge	91.0	4	P

FM	MHz	kW	Pr		FM	MHz	kW	Pr
7) Lubango	91.1	4	P		2) Lobito	93.5	0.25	P
4) Cabinda	91.3	4	P		13) Malange	93.7	4	P
1) Caxito	91.5	4	P		11) Xá Muteba	93.8		A
11)Capenda Cam.	91.5		A		16) Negage	94.4	0.25	P
10)Sumbe	91.7	4	P		15) Tômbwa	95.9	0.25	P
9) N'dalatando	91.8	0.15	M		14) Luena	96.0	2	P
6) Huambo	92.1	4	P		17) Mbanza Congo	96.1	2	A
5) Ondjiva	92.2	4	A		17) Mbanza Congo	97.7	0.5	P
15)Namibe	92.5	4	P		5) Ondjiva	98.0	5	P
12)Saurimo	92.5	0.25	P		9) Ambaca	98.0	0.5	P
16)Uíge	92.5	2	A		9) Golungo Alto	100.4	0.2	P
17)Soyo	92.7	0.1	P		2) Lobito	100.4	1	5
2) Benguela	92.9	4	P		2) Benguela	101.7	2	A
8) Menongue	93.1	0.25	P		2) Lobito	104.9	2	A
11)Dundo	93.3	4	P					

+30 more stations under 1kW.

Addresses & other information:
1) EP de Bengo, Caxito – **2)** EP de Benguela, C.P. 19, Benguela. Also rep. on 1200kHz – **3)** EP do Bié, C.P. 33, Kuito. Dir: Cordeiro Chimo – **4)** EP de Cabinda, Cabinda – **5)** EP de Cunene, Ondjiva – **6)** EP do Huambo, C.P. 125, Huambo – **7)** EP da Huíla, C.P. 111, Lubango – **8)** EP do Kuando-Kubango, C.P. 36, Menongue – **9)** EP do Kuanza Norte, C.P. 174, N'dalatando – **10)** EP do Kuanza Sul, C.P. 10, Sumbe – **11)** EP da Lunda Norte, Lucapa – **12)** EP da Lunda Sul, C.P. 116, Saurimo – **13)** EP de Malange, C.P. 83, Malange – **14)** EP do Moxico, C.P. 74, Luena – **15)** EP do Namibe, C.P. 174, Namibe – **16)** EP do Uíge, C.P. 140, Uíge – **17)** EP do Zaire, Mbanza Congo.
Affiliated community stations on FM in Buco Zau, Camabatela, Golungo Alto, Tombwa, Viana and Virei.

EXTERNAL SERVICE: see International Radio section.

RÁDIO ECCLESIA (Rlg.)
✉ Rua Comandante Bula 118, São Paulo, CP. 3579, Luanda ☎+244 22 2443041 🖷 +244 22 2443093 **W:** radioecclesia.org
LP: Exec. Dir: Father Maurício Kamutu. Adm. Dir: Sister Fátima Kavate. **FM:** 97.5MHz 5kW. **D.Prgr** in Portuguese: 0500-2030.

Private station:
Luanda Antena Comercial, Luanda. 95.5MHz 5kW. Web: lacluanda. com Email: lac@lacluanda.com

ANGUILLA (UK)

LT: UTC -4h — **Pop:** 16,000 — **Pr.L:** English — **E.C:** 50Hz, 230V — **ITU:** AIA

RADIO ANGUILLA (Gov. Comm.)
✉ P.O. Box 60, The Valley ☎+1 264 497 2218/0955 🖷 +1 264 497 5432 **E:** radioaxa@anguillanet.com **W:** radioaxa
LP: Dir.: Farrah Banks. PM: Keith Greaves. Eng.: Lester Richardson
FM: 95.5MHz, Crocus Hill. **D.Prgr:** Mon-Sat: 0925-0300, Sun: 1055-0200. N: Mon-Fri 1106 & 2306. Sat: 1106.

THE CARIBBEAN BEACON (Rlg.)
✉ P.O. Box 690, The Valley ☎+1 264 497 4340 🖷 +1 264 497 4311
LP: GM & CEN: Eddie Sutton
MW: 690kHz 50kW, 1610kHz 30kW
SW: 6090/11775kHz 100kW
FM: 100.1MHz 35kW
D.Prgr: 24h. Local prgrs 1000-1600 on 1610kHz
Owned by the University Network

PRIVATE FM STATIONS
GEM RADIO FM: 107.9MHz (relay Trinidad) – **KLASS FM**, Wilmot Estate Rock Farm, PO Box 339, The Valley ☎+1 264 497 3791 **W:** www.klass929. com **FM:** 92.9MHz – **KOOL FM**, North Side, The Valley ☎+1 264 497 0102 🖷 +1 264 497 0104 **W:** www.koolfm103.com **FM:** 103.3MHz. Format: Urban Caribbean – **NEW BEGINNING 99.3 RADIO GRACE FM**, Shoal Bay, PO Box 1122, The Valley ☎+1 264 497 0977 🖷 +1 264 497 7977 **W:** www.nbr993.com **FM:** 99.3MHz. Format: Rlg. – **PULSE FM**, The Valley AI-2640 ☎+1 264 498 1075. **FM:** 107.5MHz – **RAINBOW FM**, PO Box 603, The Valley ☎+1 264 498 9305. **W:** www.rainbowfm935. com **FM:** 93.3MHz – **UP BEAT RADIO**, Cedar Av., Rey Hills, PO Box 5045, The Valley ☎+1 264 497 3354 🖷 +1 264 497 5995 **W:** www.hbr1075.com **FM:** 97.7MHz – **VOICE OF CREATION RADIO J.E.S.U.S.**, Sachasses, The Valley ☎+1 264 497 0106 🖷 +1 264 497 0106 **FM:** 106.7MHz Format: Gospel – **VIBZ FM**, The Valley, 90.5MHz (relay Antigua) – **ZRON – TRADEWINDS RADIO**, 398 East Dania Beach Blvd. No. 210, Dania Beach, FL 33004, USA. **W:** www.tradewindsradio.com **FM:** 105.1MHz

ANTARCTICA

L.T: Antártida Argentina: UTC -3h; Antártida Chilena: UTC -4h (10 Oct 10-13 Mar 11, 16 Oct 11-11 Mar 12: -3h); Ross Dependency (NZL)/South Pole Station: UTC +12h (26 Sep 10-3 Apr 11, 25 Sep 11-1 Apr 12: +13h) — **Pop:** 4,120 (Su), 1,066 (Wi) — **ITU:** ATA

RADIO NACIONAL ARCANGEL SAN GABRIEL
✉ LRA36 Radio Nacional Arcangel San Gabriel, Base de Ejercito Esperanza, CP 9411-Antártida Argentina, Argentina. ☎+54 2974 44 5304. **E:** esperanzaantar@infovia.com.ar, lra36@infovia.com.ar
L.P: Dir: Nestor Arguello, Op.: Mario Gallardo
SW: (G.C: 63S24 056W59): LRA36 15476kHz 2kW. **Spanish.:** M-F 1900-2100 – **V.** by QSL-card & letter only. Re. with 1 IRC to Dir.
FM: 97.6MHz 24h.

SOBERANIA FM: ✉ Villa Las Estrellas, Antartica Chilena, Chile. **FM:** 90.5MHz 0.1kW.
ICE FM: ✉ McMurdo Station, Ross Dependency. **FM:** 104.5MHz 0.05kW
KOLD FM: ✉ South Pole Station, Ross Dependency. **FM:** 87.5MHz
88.7 FM: ✉ McMurdo Station, Ross Dependency. **FM:** 88.7MHz
Scott Base R.: ✉ Scott Base, Ross Dependency. **FM:** 97.0MHz

AMERICAN FORCES ANTARCTIC NETWORK (AFAN McMurdo)
✉ AFAN McMurdo, US Naval Support Force Antarctica, 651 Lyons Str, Port Hueneme, CA 93043-4345, USA.
FM: 93.9MHz 0.03kW, **D.Prgr:** 24h on both freqs. Rel AFRTS exc. some local prgrs on 104.5MHz

ANTIGUA & BARBUDA

L.T: UTC -4h — **Pop:** 88,000 — **Pr.L:** English — **E.C:** 50/60Hz, 110/220V — **ITU:** ATG

ANTIGUA & BARBUDA BROADCASTING SERVICE (Gov. Comm.)
✉ Cross Street, P.O. Box 590, St. John's ☎ +1 268 462 4427/0112 🖷 +1 268 462 2801
L.P: SM: Alex Nicholas. PM: Kenny Nibbs. CEN: Denis Leandro.
MW: 620kHz 10kW **FM:** 90.5/105.5MHz **D.Prgr:** ABS Radio 24h

CARIBBEAN RADIO LIGHTHOUSE (Rlg.)
✉ Jolly Hill, P.O. Box 1057, St.John's ☎ +1 268 462 1454 🖷 +1 268 462 7420/1452 **E:** info@radiolighthouse.org **W:** www.radiolighthouse. org **L.P:** Dir: Curt Waite. Asst. Dir. & Tech. Dir: Jerry Baker. Owned by Baptist International Missions Inc.
MW: 1160kHz 10kW **FM:** 92.5MHz 2kW.
D.Prgr: 0925-0145 on MW. 24h on FM.

GRENVILLE RADIO LTD. (Comm.)
✉ P.O. Box 1100, St. John's ☎ +1 268 462 1100 🖷 +1 268 462 1001/1100 **E:** mail@radiozdk.com **W:** www.radiozdk.com
L.P: PD: Ivor Bird
FM: ZDK Liberty Radio International 97.1MHz 1kW 24h – **Sun FM** 100.1MHz 2.5kW 24h – **Hit Radio Music Power** (see St. Lucia) 99.1MHz.

FAMILY RADIO NETWORK (Comm.)
✉ P.O. Box W1102, Belmont School of Business, St. John's ☎ +1 268 560 7578/9 🖷 +1 268 560 7577 **W:** www.vibzfm.com
L.P: MD: John Silcott. GM: Terence Dublin
FM: VIBZ FM: 92.9MHz – **Riddim FM**: 96.1MHz

CARIBBEAN RELAY CO. LTD.
✉ P.O. Box 1203, St. John's ☎ +1 268 462 0436/562 1128 🖷 +1 268 462 0487 **L.P:** SM: David Bones.
SW (G.C: 61.48W/17.06N): 4 x 250kW txs (inactive)
FM: 98.1MHz (rel. BBCWS & Caribbean Report)

Other FM stations:
ABUNDANT LIFE RADIO, Codrington Village, Barbuda ☎ +1 268 560 2676 🖷 +1 268 560 2676 **W:** www.abundantliferadio.com **E:** lifefm1031@ hotmail.com **L.P:** SM: Clifton Francois. Barbuda 103.1MHz 1kW, Antigua 103.9MHz 1kW. Format: Gospel – **CATHOLIC RADIO**, Michaels Mount, P.O. Box 836, St. John's ☎ +1 268 562 6868. **E:** catholicradio@candw. ag. 89.7MHz – **CRUSADOR RADIO**, Temple Street, P.O Box 2379, St. John's ☎ +1 268 562 4610 **W:** www.crusaderradio.com **E:** crusader-radio@candw.ag. L.P: SM: Conrad Pole. 107.3MHz – **HITZ 91.9FM**, High Street, P.O.Box 1318, St.John's ☎ +1 268 481 9190. **E:** info@hitz919fm.

com. 91.9MHz – **JAZZ FM** 97.5MHz – **Nice FM** 104.3MHz. **W:** www.nicafm1043.fm. Format: AC – **OBSERVER RADIO 911**, P.O. Box 1318, Ryans Place, High Street, St. John's. ☎ +1 268 460 8911 📠 +1 268 725 9125 **W:** www.antiguaobserver.com 91.1MHz. L.P.: PM Denise Francis. Ann: 'The Voice of the People - Radio 911FM – the Pulse of the Nation'. Format: 24h News/talk. – **RED HOT RADIO**. Carlisle Estate ☎ +1 268 562 9805 📠 +1 268 562 6693 **E:** redhotfm@hotmail.com 98.5MHz – **SECOND ADVENT RADIO**, P.O. Box 2962, St. John's ☎ +1 268 562 1015 **W:** www.secondadventradio.com **E:** radio101.5fm@hotmail.com 101.5MHz. L.P.: SM Maple Lake. Format: Rlg. – **POWER RADIO** (UNO Radio). 88.5MHz. Format: Latin – **VARIETY RADIO**, Cooks Hill, Harbour View, St. John's ☎+1 268 562 4835. **E:** varietyradio@hotmail.com L.P.: SM: Kelvin Carter. 102.3MHz 1kW

ARGENTINA

L.T: UTC -3h; exc. SC: -4h. — **Pop:** 41 million — **Pr.L:** Spanish — **E.C.:** 50Hz, 220V — **ITU:** ARG — **Int. dialling code:** +54

SECRETARIA DE COMUNICACIONES (SECOM)
✉ Tucumán 744, piso 4, (C1049AAP) Buenos Aires ☎11 5071 9412 **W:** www.secom.gov.ar
L.P: Lic. Arq.Carlos Lisandro Salas (Secretario de Comunicaciones).

COMISION NACIONAL DE COMUNICACIONES (C.N.C.)
✉ Perú 103, (C1067AAC) Buenos Aires 11 4347-9850 **W:** www.cnc.gov.ar **L.P:** Interventor: Ceferino Namuncurá

AUTORIDAD FEDERAL DE SERVICIOS DE COMUNICACIÓN AUDIOVISUAL (AFSCA)
✉ Suipacha 765, 9° piso, (C1008AAO) Buenos Aires ☎11 4320-4900 **W:** www.afsca.gov.ar **L.P:** Interventor: Lic. Juan Gabriel Mariotto. AFSCA controls certain technical aspects of broadcasting, and also controls the prgrs. trs over all kinds of broadcasting stns.

SECRETARIA DE PRENSA Y DIFUSION
Reports to the Presidencia de la Nación. Administers the media.

SISTEMA NACIONAL DE MEDIOS PÚBLICOS (SNMP)
✉ Balcarce 50, 1° piso, (C1064AAB) Buenos Aires ☎11 4344 3850 **W:** www.snmp.gov.ar **L.P:** Interventor: Tristan Bauer
All LRA stns belong to S.O.R. (incl. LRA36 in Antarctica). Common prgrs (originated from LRA1) in network called "Cadena Celeste y Blanca de Emisoras Argentinas".

° = on-air stn name not confirmed, * = inactive, v = varying freq.

MW Call		kHz	kW	Station, location and h of tr
CF34)		530		LV de las Madres, Buenos Aires
BA119)		540		R.Italia, Villa Martelli (n.f. 1610)
SF01)	LRA14	540	25/1	R. Nal., Santa Fé: 0900-0300
SA01)	LRA25	540	5	R. Nal.,Tartagal: 1000-0400
CB04)	LU17	540	10/5	R. Golfo Nuevo, Pto. Madryn: 24h
CB01)	LRA9	540	25/1	R. Nal., Esquel: 0900-0300
BA01)	LRA13	560	25/5	R. Nal., Bahía Blanca: 0900-0400
ER01)	LT15	560	10/5	R. del Litoral, Concordia: 0900-0500
SJ01)	LV1	560	25/5	R. Colón, San Juan: 24h
JU01)	LRA16	560	10/1	R. La Quiaca: 0855-0400
CF30)		570		R. Argentina, Buenos Aires
CB02)	LU20	580	10/5	R. Chubut, Trelew: 24h
CO01)	LW1	580	25/5	R. Univ. Nal. de Córdoba, Córdoba: 0800-0500
CF01)	LS4	590	25/5	R. Continental, Buenos Aires: 24h
RN01)	LRA30	590	25/1	R. Nal., San Carlos de Bariloche: 0855-0400
TU01)	LV12	590	4	R. Independencia, San Miguel de Tucumán: 24h
NE04)	LU5	600	20/5	R. Neuquén, Neuquén: W: 0900-0300, Sat/Sun: 1100-0300
SE03)	LRK201	610	1	R. Solidaridad, Añatuya: Mon-Sat: 1000-2200, Sun 1000-2400
BA91)		610	5/1	La Buena Radio, Villa Lynch: 24h
ME07)	LV4	620		R. San Rafael, San Rafael: 1000-0400
SC01)	LRA18	620	25/7	R. Nal., Río Turbio: 0845-0400
MS01)	LT17	620	25/5	R. Provincia de Misiones, Posadas: 0900-0500
CH03)	LRA26	620	25/5	R. Nal., Resistencia: 0830-0300
LR01)	LRA28	*620	25/5	R. Nal., La Rioja: 0900-0400
CF02)	LS5	630	25/5	R. Rivadavia, Buenos Aires: 24h
JU03)	LW8	630	25/5	R. San Salvador de Jujuy: 0900-0400
CB05)	LU4	630	10/5	R. Dif. Patagonia Argentina, Comodoro Rivadavia: 0900-0500
TF01)	LRA24	640	25/5	R. Nal., Río Grande: 24h
RN02)	LU18	640	10/5	R. El Valle, "640 AM", General Roca: 0900-0300

MW	Call	kHz	kW	Station, location and h of tr
SL01)	LV15	640	10/5	R. Villa Mercedes: 0900-0300
CF38)		650		R.Reporter, Buenos Aires
ER04)	LT41	660	1/5	R. LV del Sur Entrerriano, Gualeguaychú: 0800-0300
BA68)		660		R. Popular, Claypole
BA172)		660		R. Antartida, San Justo
CB03)	LRA11	670	25/5	R. Nal., Comodoro Rivadavia: 24h
NE01)	LRA52	670	1	R. Nal., Chos Malal: 0900-0300
MS02)	LT4	670	25/5	R. Dif. Misiones, Posadas: 0800-0200
BA02)	LRI209	670	25/5	R. Mar del Plata, Mar del Plata: 24h
BA132)		*670		R. Maranata,Lomas del Mirador
BA173)		680		R. Magna, San Martin
SF07)	LT3	680	25	R. Cerealista, Rosario: 24h
SC02)	LU12	680	25/5	R. Río Gallegos, Río Gallegos: 1000-0300
ME01)	LV6	680	25/5	R. Nihiul, Mendoza: 24h
CF42)		*690		AM Dakota, Buenos Aires
SA02)	LRA4	690	25/5	R. Nal. Salta: 0900-0400
RN03)	LU19	690	10/3	R. LV de Comahue, Cipolletti: 0900-0500
CO05)	LV3	700	25/5	R. Córdoba: 24h
NE02)	LRA17	710	25/1	R. Nal., Zapala: 0900-0300
MS03)	LRA19	710	25/5	R. Nal., Pto. Iguazú: 0900-0300
CF13)	LRL202	710	50	R. Diez, Buenos Aires: 24h
SC03)	LRA59	720	1	R. Nal., Gobernador Gregores: 1100-2300
ME02)	LV10	720	25/5	R. de Cuyo, Mendoza: 24h
LP01)	LRA3	730	25/5	R. Nal., Santa Rosa: 0900-0300
SC04)	LU23	730	10/1	R. Lago Argentino, El Calafate:1000-0300
CA01)	LRA27	730	20/5	R. Nal., Catamarca: 24h
BA153)		730		R. Guarani AM, San Justo
BA206)		730		R.General Gümes "La Radio Mundial", Buenos Aires
CB07)	LRA55	740	1	R. Nal., Alto Río Senguer: 0900-0300
SC05)	LRI200	740	10/1	R. Municipal: 24h
CH05)	LRH251	740	25/5	R. Chaco, Resistencia: 0800-0300
BA207)		750		R. AM 7-50, Lomas de Zamora
CO02)	LRA7	750	100/10	R. Nal., Córdoba: 0900-0500
BA09)	LU6	760	25/5	Emisora Atlántica, Mar del Plata: 24h
BA36)		*770		Amplitud 770, Lomas del Mirador
CF21)		770	5/1	R. Cooperativa, Buenos Aires: 24h
TF02)	LRA10	780	5/1	R. Nal., Ushuaía: 24h
CS01)	LRA12	780	5	R. Nal., Santo Tomé: 0900-0300
ME03)	LV8	780	25/5	R. Libertador, Mendoza: 24h
CB06)	LRF210	780	5	R. Tres, Trelew: 0900-0300
ME06)	LV19	790	5	R. Malargüe: 1100-0400
CF04)	LR6	790	25/5	R. Mitre "AM 80," Buenos Aires: 24h
JU02)	LRA22	790	25/5	R. Nal, San Salvador de Jujuy: 0945-0500
MS05)	LT46	*790	1/0,25	R. Provincial, Bernardo de Irigoyen: 0900-0300
RN04)	LU15	800	24/5	R. Viedma: 0900-0300
NE08)		800		R. AM 800 Wajzugun, San Martin de los Andes
ME04)	LRA23	800	1/0.25	R. Rio Atuel, General Alvear:1000-0400
CH01)	LT43	800	1/0.25	R. Mocoví, Charata: 0900-0300
BA182)		810		R. La Gauchita, Morón
CO16)		810	10/1	R. Mitre AM 810, Córdoba
FO01)	LRA8	820	25/5	R. Nal., Formosa: 0855-0400
BA04)	LU24	820	5/1	R. Tres Arroyos- "LV del Pueblo": 0900-0400
BA33)	LRI208	820	5/1	Estacion 820, Lomas de Zamora
JU04)	LRK221	820	1/0.25	R. Ciudad Perico, Perico: 1000-0300
BA117)		*830		R.Filadelfia, San Justo
CF24)		830	5	R. del Pueblo, Buenos Aires: 24h
SF02)	LT8	830	10/5	R. Rosario, Rosario: 24h
ME08)	LV18	*830	0.25	R. Municipal, San Rafael: 1030-0230
SC06)	LU14	830	25	R. Provincia de Santa Cruz,Río Gallegos: 0900-0500
CS05)	LT21	830	1/0.5	R. Municipal, Alvear
CS02)	LT12	840	25/5	R. General Madariaga, Paso de los Libres: 0900-0300
BA05)	LU2	840	25/5	R. Bahía Blanca, Bahía Blanca: 24h
SA03)	LV9	840	25/5	R. Salta, Salta: 1000-0500
CF18)		840	3	R. General Belgrano, Buenos Aires
BA126)		850	10	LV de América, San Miguel
BA139		860		R. Digital, Lanus: 24h
SC07)	LRA56	860	1	R. Nal., Perito Moreno: 0955-0300
LR02)		*860	5/1	R. Municipal, Chilecito: 1000-0400
CF05)	LRA1	870	100	R. Nal., Buenos Aires: 24h
SC06)	LU14	880	10	R. Provincia de Santa Cruz, Las Heras (//LU14 830): 1000-0300
BA183)		890		R.Libre, San Justo
LP02)	LU33	890	25/5	R. Pampeana, Santa Rosa: 24h
SE01)	LV11	890	25/5	Em. Santiago del Estero, Santiago del Estero: 0900-0500
CS03)	LT7	900	25/5	R. Provincia de Corrientes, Corrientes: 0900-0300
LP05)		900	1	R. Municipal, 25 de Mayo
SJ02)	LRA23	910	50/5	R. Nal.,San Juan: 0855-0400
CF06)	LR5	910	25/5	R. La Red, Buenos Aires: 24h
TU02)	LV7	930	25/5	R. Tucumán, San Miguel de Tucumán: 24h

MW	Call	kHz	kW	Station, location and h of tr
CO09)	LV28	930	5/1	R. Villa María, Villa María: 24h
BA138)		930		R. Nativa – "LV de Nuestra Gente", Ciudad Madero
SL03)	LRJ241	940	20/5	R. Dimensión, San Luís: 0900-0400
ER07)	LRH200	940	3/5	R. Chajarí, Chajarí: 0900-0300
CF07)	LR3	950	25/5	R.Belgrano, Buenos Aires: 24h
CH04)	LT16	950	25/5	RSP – R. Sáenz Peña (Cadena Eco), Roque Saénz Peña: 0900-0300
ME05)	LRA6	960	10/1	R. Nal., Mendoza: 1000-0500
BA06)	LU13	960	10/3	R. Necochea, Necochea: 0900-0400
CF35)		970		R. Genesis, Buenos Aires
CO03)	LV2	970	25/5	R. General Paz "AM 970", Córdoba: 24h
CS07)	LT25	970	1/0.25	R. Guaraní, Curuzú Cuatiá:0900-0300
LP03)	LU37	980	3/1	R. General Pico "Radio37",: 0930-0300
BA188)		980		R.Regional, San Miguel
ER06)	LT39	980	5	R. Victoria, Victoria: 0900-0400
RN11)	LRG387	980		R. Luján AM, Valcheta
CF12)	LR4	990	25/5	R. Splendid AM 990, Buenos Aires: 24h
SJ03)	LRJ201	990	1	R. Calingasta, Barreal: 1000-0500
F003)	LRH203	990	25/5	AM 990, Formosa: 24h
BA120)		1000		R.Sintonia, José C.Paz: 24h
RN05)	LU16	1000	1/0.25	R. Río Negro, Villa Regina: 0900-0300
CS06)	LT42	1010	1/0.25	R. Del Iberá, Mercedes: 0900-0300
CO04)	LV16	1010	20/10	R. Río Cuarto, Río Cuarto: 0830-0500
SA05)	LW2	1010	1/0.25	R. Emis. Tartagal: 1000-0500
CF15)		1010		R. Onda Latina, Buenos Aires: 24h
LR01)	LRA28	1010	1	R. Nacional, La Rioja // LRA28 – 620
SJ07)	LRJ214	1020	25/5	AM Mil 20 - La R. de la Gente, San Juan
SF03)	LT10	1020	10/5	R. Univ. Nal. del Litoral, Santa Fé: 0800-0500
CH02)	LRA58	1020	1	R. Nacional, Río Mayo: 1100-2300
CF08)	LS10	1030	25/5	R. del Plata, Buenos Aires: 24h
CF27)		1050	10/5	Concepto AM, Buenos Aires: 24h
CO08)	LV27	1050	10	R. San Francisco, San Francisco: 0900-0300
BA186)		1060		R. Las Naciones, Monte Grande
CF09)	LR1	1070	25/5	R. El Mundo, Buenos Aires: 24h
BA08)	LU3	1080	25/5	R. del Sur, Bahía Blanca (Cadena Eco),
BA127)		1080		R. Claridad, Monte Grande (r. 1230)
NE05)		1080	10/1	R. Departamento Minas, Andacollo: 1000(SS 1200)-2300
SA06)	LW4	1080	25/5	R. Orán/R.Maria: 1000-0600
BA141)		1090		R. Décadas, Hurlingham
BA79)		1090		R. Nuestras Raíces, Valentin Alsina
SF17)		1090	0.5	Libertad AM 1100, Rosario
BA66)		1100	1.3	R. Estilo, Glew: 24h
CF03)	LS1	1110	25/5	R. de la Ciudad"La Porteña", Buenos Aires:
BA184)		1120		R.Sudamericana, Victoria
BA198)		1120		Em. Santiago y Copla, Gregoria de Laferrere
CF33)		1120		AM Tango, Buenos Aires (r. 1540)
SJ04)	LV5	1120	25/5	R. Sarmiento, San Juan: 0900-0400
SE02)	LRA21	1130	25/5	R. Nal., Santiago del Estero: 0900-0400
CF22)		1130	10	R. Cadena Vida, Buenos Aires
BA128)		*1130		R. Carisma, El Talar
BA12)	LU22	1140	10/1	R. Tandil, Tandil: 0830-0300
BA185)		1140		R. Independencia, Remedios de Escalada
BA208)		1140		R. La Luna, El Palomar
RN06)	LRA2	1150	25/5	R. Nal., Viedma: 0900-0400
SJ05)	LRA51	1150	1	R. Nal., Jáchal: 1030-0300
SF04)	LT9	1150	10/5	R. Brigadier López, Santa Fé: 0700-0300
MS07)	LRH202	1150	10	R. Tupá Mbaé, Posadas: 0800-0300
BA154)		1150	5	R. Sagrada Familia (R.Maria), San Justo
BA56)		1160		R. Excelsior, Monte Grande
RN07)	LRA57	1160	1	R. Nal., El Bolsón: 0900-0300
MS04)	LRH253	1160	5/10	R. Cataratas, Pto. Iguazú: 1000-1600
BA10)	LU32	1160	10/2.5	R. Coronel Olavarría, Olavarría: 0900-0300
SL02)	LRA29	1170	25/3	R. Nal., San Luis: 1000-0500
BA77)		1170	5	R. Mi País, Hurlingham: 24h
BA155)	LRI357	1180		R. de la Sierra, Tandil
TU03)	LRA15	1190	50	R. Nal., San Miguel de Tucumán: 24h
CF10)	LR9	1190	25/5	R. América, Capital Federaf: 24h
ME05)	LRA6	1200	1	R. Nal. Mendoza (r. LRA6 960), Valle de Uspallata: 1000-0500
CS04)	LT6	1200	5/1.5	R. Goya, Goya: 0900-0300
BA29)	LRI229	1210	5/1	R. Las Flores, Las Flores: 24h
BA54)		1210		R. La Luz, Lomas de Mirador
BA200)		1210		R. Mailín, Gregorio de Laferrere (n.f. 1330)
BA83)	LRI224	1220	1	R. Onda Marina, Mar del Plata (Cad. Eco)
CF16)	LRL328	1220	5/1	LV del Aire (Cad.Eco), Buenos Aires: 24h
CH06)		1220		R. LRC – "La Radio de Chaco", Pres. Roque Sanez Peña
SF05)	LT2	1230	25/5	R. Gen. San Martín "R.Dos", Rosario: 24h
JU05)	LW5	1230	5/1	R. Libertador, General San Martín
BA89)		1230		R. Litoral, Isidro Casanova
BA127)		1230		R. Claridad, Monte Grande (nf. 1080)
CF44)		1230		R.Creativa, CA Buenos Aires
LP07)		1230		R. La Bendición, General Pico

MW	Call	kHz	kW	Station, location and h of tr
CF31)		1240		R. Cadena Uno, Buenos Aires: 24h
BA31)		1250	1	R. Estirpe Nacional., San Justo: 24h
BA131)		1260	2	R. Oasis, Victoria: 24h
CF36)		*1260		R. Olivia, CA Buenos Aires
CF39)		1260		R. Panamericana, CA Buenos Aires
ER02)	LT14	1260	10/5	R. General Urquiza, Paraná: 24h
F002)	LRA20	1270	5	R. Nal., Las Lomitas: 0900-0300
BA11)	LS11	1270	25/50	R. Provincia de Buenos Aires, La Plata: 24h
BA15)	LU11	1280	10/5	R. Trenque Lauquén, Tr. Lauquén: 0900-0300
CF17)		1280		R. Eco Porteña, CA Buenos Aires (n.f. 1530)
BA73)		1280		R. Mística, Libertad (n.f. 1320)
CF41)		1280		R. Punto, Buenos Aires
BA157)		*1280		El Sonido de la Gente, Gregorio de Laferrere
SF11)	LRI371	1290	1	R. Amanecer, Reconquista: 0800-0400
BA72)		1290		R. Provinciana, San Miguel
BA53)		1290		R. Interactiva, Ciudad Madero
ME10)	LRJ212	1290	5/1	R. Murialdo, Villa Nueva de Guaymallén: 0900-0300
SF06)	LRA5	1300	10/5	R. Nal., Rosario: 0853-0303
BA63)		1300		Plus Radio, Lanús
CF26)		1300		R. Identidad, Buenos Aires: 24h
BA199		1310		Gesell Radio, Villa Gesell
ER03)	LRA42v	1310	1	R. Nal., Gualeguaychú
NE06)		1310	1	R. Dr. Gregorio Alvarez (Cadena Eco) Piedra del Aguila: 24h
BA90)		1310		R. Imagen, Castelar
CF29)		1310		R. AM 13-10, Buenos Aires:24h
BA13)	LU10	1320	5/3	R. Azul, Azul: 0900-0400
BA43)		1320		R.Ciudad, Remedios de Escalada
BA73)		1320		R. Mística, Libertad (r. 1280)
BA187		1320		R. Máster, Luján
BA209		1320		R.Area 1, Caseros
BA210		1320		R. S'Combro, José de Paz
ME09)	LV24	1320	0.25	R. Río Tunuyán, Tunuyán: 1000-0600
BA200)		1330		R. Mailín, Gregorio de Laferrere (r. 1210)
BA202)		1340		AM Renacer, Moreno
CF11)	LS6	1350	25/5	R. Buenos Aires, Buenos Aires 24h
CO13)	LRJ747	1350	5/1	R. Sucesos, Córdoba: 24h
BA38)		1360		R. Nuestra Señora de Itatí – "R. Itati", Morón
RN08)	LRA54	1370	1	R. Nal., Ingeniero Jacobacci: 0900-0300
BA76)		1370	5/3	AM-1370, Isidro Casanova: 24h
BA143)		1380		R. Redentor, Claypole
BA144)		1380		R. Buenas Nuevas, Merlo
BA189)		1380		R. Los Toldos, Los Toldos
BA190)		1380	5/1	LV del Sudeste, Necochea
BA14)	LR11	1390	10	R. Univ., La Plata: 0800-0300
BA158)		*1390		R. Ribera Sur, Ingeniero Budge
NE07)	LRG202	1400	10/1	R. Cumbre, Neuquén
BA69)		1400		R. Gama, Valentin Alsina: 1000-0300
BA123)		1400		R. AM 1400, Luján
BA211)		1400		R. del Buen Ayre, Ituzaingó
BA42)		1410	5/1	R. Folclorismo, José Léon Suárez: 1100-0100
BA214)		1410		R. Fundacion, Rafael Calzada
LP06)	LRG203	1410	5/1	R. Capital "Antena 10", Santa Rosa
BA108)		1420		R. Génesis 2000, General Conesa
CF28)	LRI220	1420	1/0.25	AM La Marea, Buenos Aires
BA16)	LT24	1430	1/0.25	R. San Nicolás, San Nicolás: 24h
CO06)	LV26	1430	1/0.25	R. Río Tercero (Cad. 26), Río Tercero: 24h
BA17)	LRI235	1430	0.25	R. Balcarce, Balcarce: 0900-0300
BA65)		1430		R. José de S. Martín, "La Pionera",El Jagüel
BA152)		1430		R.Victoria, La Plata
BA201)		1430		R. Shekinah, Merlo
BA212)		1430		R. AM 1430 – "La Radio de los Cunumí Guazú"
BA146)		1440		R. Cristo Viene, Mar del Plata
NE03)	LRA53	1440	1	R. Nal., S. Martín de los Andes: 1000-0400
BA18)	LU36	1440	0.25	R. Coronel Suárez, Coronel Suárez: 1000-0300
CO07)	LV20	1440	1/0.25	R. Laboulaye, Laboulaye: 0900-0300
BA52)		1440		R. Impacto, Tapiales
SF12)	LRI221	1440	5/1	R. General Obligado, Reconquista
CF37)		1450	5/1	R. El Sol, Buenos Aires
SJ06)	LRI211	1450	5/1	R. Las Cuarenta, San Juan
SF08)	LT29	1460	1/0.25	R. Venado Tuerto, Venado Tuerto: 24h
BA19)	LU30	1460	0.25	R. Maipú (Cad. Eco), Maipú: 0900-2400
BA20)	LU34	1460	0.1	R. Pigüé, Pigüé: 1000-0300
BA41)		1460		R. Contacto, San Antonio de Padua: 24h
BA191)		1460		R. Jerusalem, Jerusalem
TU04)	LRK204	1460	1	R. 21 (Cadena Eco), Yerba Buena: 24h
BA21)	LT20	v1470	1/0.25	R. Junín: 0900-0300
ER05)	LT26	1470	1/0.25	R. Nuevo Mundo, Colón: 0900-0300
SF09)	LT28	1470	1/0.25	R. Rafaela, Rafaela: 0900-0300
BA22)	LU26	1470	0.25	R. Coronel Dorrego "La Dorrego", Coronel Dorrego: 1030-0300 (r. 1468)
RN09)		1470	1	R. Municipal, Luis Beltrán: 1200-0100
BA49)		1470		R. Mburucuya, José León Suarez
BA84)		1470		Cadena 1470, Lanús: 1000-0300

MW	Call	kHz	kW	Station, location and h of tr
BA50)	LU27	1480	1	R. Dolores, Dolores: 1100-0300
BA147)		1480		R. Sensaciones, Tapiales
BA174)		1480		R. Líder, Mariano Acosta
CO15)	LV22	1490	1	R. Huinca Renancó, Huinca Renancó: 1000-0300
BA24)	LU25	1490	0.1	R. Carhué, Carhué: 0900-0330
BA47)		1490		R. Cielo Nuevo, Isidro Casanova
BA67)		1490		R. Vida, Mar del Plata
BA07)	LRI214	1500	5/1	R. Bonaerense, Lavallol: 0900-0300
RN10)		1500	1/0.25	R. Municipal, Gral. Conesa: 1000-2400
BA23)	LT34	v1500	0.25	R. Nuclear, Zárate: 0900-0000
BA34)		1510		LV del Oeste,Libertad: 24h
CO11)	LV21	*1510	1	R. Champaqui (Cadena Eco), Villa Dolores:1000-0300
CF20)		1510		R. Urkupiña, Buenos Aires (n.f. 1550)
BA74)		1510		R. Alabanza, Guernica
SF14)	LRI253	1510	1/0,25	R. Belgrano, Suardi: 0900-0400
ER08)	LT38	1520	0.25	R. Gualeguay, Gualeguay: 0900-0300
BA40)		1520		R. Metropolitana "R.Metro", Ciudadela
BA92)		1520	5/1	R. Chascomús, Chascomús: 24h
BA145)		1520	3	R. Norteña, Los Polvorines
BA148)		1520		Cadena D, Monte Chingolo
BA161)		1520		R. AM Fortaleza, Ezeiza
CF17)		1530		R.Eco Porteña, CA Buenos Aires (r. 1280)
BA192)		1530		R.Esencia "LV del Litoral", San Miguel Oeste
CO12)	LRJ2001	1530	1	R. Centro Morteros, Morteros: 0900-0300
BA25)	LT35	1540	0.25	R. Mon, Pergamino: 0900-0400
BA26)	LU28	1540	0.25	Cadena Uno, Gen.Madariaga:1000-0300
BA100)		1540		R. AM Líder, Martinez
BA150)		1540		R. Cotidiana, Merlo
BA130)		*1540		R. Fuego, Longchamps
CF33)		1540		AM Tango, Buenos Aires (n.f. 1120)
CF43)		1540		R.Amanecer, CA Buenos Aires
CF20)		1550		R. Urkupiña, Buenos Aires (r. 1510)
SF10)	LT23	1550	5/0.25	R. Regional, San Jenaro Norte: 0900-0300
BA27)	LT32	1550	0.25	R. Chivilcoy, Chivilcoy: 1000-0400
ER09)	LT40	1550	0.25	R. LV de la Paz, La Paz: 0900-0100
BA101)		*1550		R. Trompeta de Diós, Isidro Casanova
BA175)		1550		R. Popular, José León Suárez
BA213)		1550		R. Esperanza, Gregorio de Laferrere
ER10)	LT11	1560	2.5/1.5	R. Gral. Francisco Ramírez, Villaguay
BA28)	LT33	1560	0.25	R. Nueve de Julio, 9 de Julio: 0900-0300
BA44)		1560		R. Castañares, Ituzaingó
BA94)		*1560		R. Ebenezer, Ezeiza
BA99)		1560		R. Restauración, Llavallol
BA194)		1560	0.5/0.3	AM 1560 "La R. de la Gente", Tandil
BA196)		1560		R. Antena Lobos, Lobos
BA55)		1570		R. Melody, Remedios de Escalada
BA71)		1570	2	R. AM Rocha, La Plata
BA162)	LRI223	1570	5/1	Lomas de Zamora (F.PI.)
BA163)		1570		R. La Morena de Itatí, Grand Bourg
ER11)	LT27	1580	1.25	R. LV del Montiel, Villaguay: 0900-0300
BA48)		1580		R. Tradición, San Martín
BA135)		1580	1	R. 26. de Julio, Longchamps
BA164)		1580		R. Tradición, Isidro Casanova
BA176)	LT36	1580	0.25	R. Chacabuco, Chacabuco
RN12)		1580		R. Provincial de Sierra Colorada, Sierra Colorada
BA39)		1590		R. Cristiana Adonal, Bánfield Oeste
BA124)		1590		R. Guaviyú, Gregorio de Laferrere (n.f. 1610)
BA178)		1590		R. Olivera, General Rodriguez
BA37)		1600	1.2	R. Armonia, José Ingenieros: 24h
BA57)		1600		R. Metropolitana, Luís Guillón
BA51)		1610	0.05	R. Luz del Mundo, Rafael Calzada
BA124)		1610		R. Guaviyú, Gregorio de Laferrere (r. 1590)
CO14)		1610	0.5	R. Buenas Nuevas, Laboulaye
SF18)		1610	0.2	R. Fósil, Rosario
BA45)		1620		R. Sión, Monte Grande
BA119)		1620		R.Italia, Villa Martelli (r. 540)
BA180)		1620		R. AM 16-20, Mar del Plata
BA134)		1630		AM Restauración, Hurlingham
BA179)		1630		R. AM Súpe Sport, Lomas de Zamora
BA177)		1640		Hosanna AM 1640, Isidro Casanova
CF23)		*1640		R. Nueva Bolivia , Buenos Aires
BA181)	LRI227	1650	1/0.5	Antares AM 1650 "La R. de la Familia", Pilar
BA195)		1650		R.Fenix. Temperley
BA156)		1650		R. Reivir, Isidro Casanova
BA203)		*1660		R.Esperanca, Virrey Del Pino
BA167)		v1670		R. Bethel, Buenos Aires (r. 1672-1675)
BA166)		1680		R.Hosanna Tropical, Ezeiza
BA107)		1690	1/0.25	R. Apocalipsis II, San Justo
BA204)		*1700		R. Cristiana Principe con Dios, Banfield Oeste
CF32)		1710		AM 1710 – R.Estudio ESBA, Buenos Aires

SW	Call	kHz	kW	Station, location and h of tr
CF05)	LRA31	6060	30	R. Nal., Buenos Aires: 2100-1500
ME06)	LV19	*6160	1	R. Malargüe, Malargüe
ME05)		*6180	7.5	R. Nal. Mendoza

ASOCIACION DE RADIODIFUSORAS PRIVADAS ARGENTINAS (ARPA)

✉ Tte. Gral. Juan D. Perón 1561, Piso 3, (C1037ACB) Buenos Aires ☎ 11 4371 5999 🖷 11 4382 4483 **W:** www.arpa.org.ar **E:** arpaorg.@arpa.org.ar **L.P:** Presidente: Carlos Maria Molina.
ARPA is an association of privately owned commercial stns.

ASOCIACION DE RADIODIFUSORES CATOLICOS ARGENTINOS (ARCA)

✉ Av. Juan D. Perón 3461, (S2003FYC) Rosario, Santa Fe ☎341 431 2872 **E:** info@radiodelrosario.com **L.P:** Presidente: Pbro. Osvaldo Bufarini

Addresses and other information:

BA00) BUENOS AIRES (PROV.):
BA01) Moreno 30, (B8000FWB) Bahía Blanca ☎ 291 453 2700 **W:** www.nacionalbahiablanca.com.ar **E:** bahiablanca@radionacional.gov.ar - **FM:** 99.3MHz – **BA02)** Hipolito Yrigoyen 2641, (B7600DPG) Mar del Plata. ☎223 492 2020 **W:** www.lu9mardelplata.com.ar **E:** lu9-adm@lacapitalnet.com.ar - **FM:** 103.3MHz FM 103 Universo – **BA04)** Av. Belgrano 457, (B7500EBE) Tres Arroyos ☎2983 42 3504 🖷2983 42 7000 **W:** www.lu24.com.ar **E:** lu24@lu24.com.ar - **FM:** 95.3MHz – **BA05)** Rodriguez 55 (B8000HSA) Bahía Blanca ☎291 459 0002 🖷291 455 5556 **W:** www.lu2.com.ar **E:** radio@lu2.com.ar - **FM:** 94.7MHz – **BA06)** Calle 64 No. 2946, Gran Galería Central, EP, (B7630CIR) Necochea ☎2262 42 0100 **W:** www.radionecochea.com.ar **E:** administracion@radionecochea.com.ar - **FM:** 88.1MHz – **BA07)** Doyhenard 316, (B1836EVH) Lavallol ☎11 4231 3225 **W:** www.am1500.com.ar **E:** radiobonarense@yahoo.com — **BA08)** Av. Lamadrid 116, (B8000FKD) Bahía Blanca ☎291 452 0382 **W:** www.lu3am1080.com.ar **E:** contacto@lu3am1080.com.ar - **FM:** 94.1MHz – **BA09)** Córdoba 1865, (B7600DVM) Mar del Plata ☎223 491 7047 🖷223 491 2355 **W:** www.lu6.com.ar **E:** radioa@lu6.com.ar - **FM:** 93.3MHz – **BA10)** Alsina 3377, (B7400COW) Olavarría ☎2284 41 0911 **W:** www.lu32.com.ar **E:** iadministracion@lu32.com.ar - **FM:** 98.7MHz – **BA11)** Calle 53 No. 810, (B1900BBQ) La Plata ☎221 424 9713 **W:** www.amprovincia.com.ar **E:** secretaria@radioprovinciaba.com.ar - **FM:** 97.1MHz – **BA12)** Gral. Rodriguez 762, PA, (B7000AOP) Tandil ☎2293 42 7493 **W:**www.lu22radiotandil.com.ar **E:** radiotandil@arnet.com.ar - **FM:** 97.1MHz – **BA13)** Av. Bartolomé Mitre 819/21, (B7300IKQ) Azul ☎2281 42 5628 **E:** lu10radioazul@latinmail.com - **FM:** 89.5MHz FM Celestial – **BA14)** Plaza Rocha 133, 2°piso, (B1900DVA) La Plata ☎221 422 0330 🖷221 422 4165 **W:** www.lr11.com.ar **E:**secretaria@lr11.com.ar - **FM:** 89.7MHz – **BA15)** Av. Pedro Garcia Salinas 1815, (B6400EIF) Trenque Lauquen ☎2392 42 5454 **W:** www.radiotrenquelauquen.com.ar **E:** radiolu11@speedy.com.ar or radiolu11@ciudad.com.ar - **FM:** 88.5MHz – **BA16)** Av. Moreno 124, (B2900GPO) San Nicolás ☎3461 42 5222 🖷3461 42 4479 **W:** www.lt24online.com.ar **E:** lt24@cablenet.com.ar - **FM:** 88, 88.3MHz – **BA17)** Av. San Martin 2200, (B7620) Balcarce ☎2266 43 0780 🖷2266 43 0779 **W:** //radiobalcarce.blogspot.com **E:** radiobalcarce@telefax.com.ar - **FM:** 89.7MHz – **BA18)** Garibaldi 71, (B7540DQA) Coronel Suárez ☎2926 43 2706 **W:** www.1440am.com.ar **E:** produccion@1440am.com.ar - **FM:** 100.5MHz "Frecuencia 36" – **BA19)** Lavalle Sud 312, (B7160BAH) Maipú ☎226 842 1774 **W:** www.lu30radiomaipu.com.ar **E:** lu30radiomaipu@hotmail.com – **BA20)** Rivdavia 382, (B8170ACH) Pigüé ☎2923 47 2535 🖷2923 47 2709 **W:** www.radiopigue.com.ar **E:** radiolu34@s8.coopnet.com.ar - **FM:** 96.3MHz – **BA21)** Hipólito Yrigoyen 86, (B600DDB) Junín ☎2362 44 3610 🖷2362 44 3474 **W:**www.lt20radiojunin.com.ar **E:** oyentes@lt20radiojunin.com.ar - **FM:** 89.1MHz – **BA22)** Uslenghi 592, 1° Piso, (B8150EGD) Coronel Dorrego ☎2921 45 3456 **W:** www.ladorrego.com.ar **E:** info@ladorrego.com.ar – **BA23)** Independencia 501, (B2800JIG) Zárate ☎3487 42 3116 🖷3487 43 9500 **W:** www.radionuclear.com.ar **E:** radionuclear@delta.com.ar - **FM:** 90.1MHz – **BA24)** Av. Colón 985, (B6430BHF) Carhué ☎2936 43 2660 🖷2936 43 2955 **E:** radiocarhue@yahoo.com.ar – **BA25)** Dr. Alem 340, (B2700LHH) Pergamino ☎2477 42 4022 **W:**www.lt35radiomon.com.ar **E:** lt35radiomon@speedy.com.ar - **FM:** 90.3MHz FM Mágica – **BA26)** Av. San Martín 366, (B7163EGQ) General Madariaga ☎ 2267 55 1540 **W:**www.radiotuyu.blogspot.com **E:** lu28@radiotuyu@telpin.com.ar - **FM:** 92.5MHz R.Tuy – **BA27)** Av. Mitre 924. (B6620BMW) Chivilcoy ☎2346 43 0690 - **FM:** 101.1MHz FM Sónica – **BA28)** Pte. Hipólito Irigoyen 969, (B6500DJQ) 9 de Julio ☎2317 52 1560 **W:** www.radio9dejulio.com.ar **E:** imagen@cadenanueva.com.ar - **FM:** 99.9MHz Maxima FM – **BA29)** Av. Avellaneda 773, (B7200AOH) Las Flores **W:** www.multimedialasflores.com.ar **E:** am1210@multimedialasflores.com.ar ☎2244 45 2320 🖷 2244 45 2838 - **FM:** 89.7MHz FM Condor – **BA31)** Juan Florio 3579, (B1754AJK) San Justo ☎ 11 4441 1400 **W:** www.estirpe1250.com.ar **E:** estirpe1250@yahoo.com.ar –

BA33) Antonio Sáenz 572, 2° piso, (B1832HUL) Lomas de Zamora ☎11 4243 7891 🖹11 4292 5559 **W:** www.radioestacion820.com **E:** 820am@speedy.com.ar – **BA34)** Isla Soledad 205 =ex 2560=, (B1716NXB) Libertad ☎220 494 1300 **W:** www.lavozdeloeste1510.com.ar **E:** oyentes@lavozdeloeste1510.com.ar - **FM:** 91.9MHz – **BA36)** Roque Sáenz Peña 3241, (B1752CBQ) Lomas del Mirador ☎11 4655 0400 **W:** www.am770.com.ar **E:** amplitud@am770.com.ar – **BA37)** Asuncion 3057, (B1703ATQ) José Ingenieros ☎11 4734 7100 **W:** www.am1600armonia.com.ar – **BA38)** Calle San Luís 989, (B1708JUE) Morón ☎11 4627 7439 – **BA39)** Arroyo Santa Catalina 4067, Barrio Juan Manuel de Rosas, (B1828) Bánfield Oeste ☎11 4693 2789 – **BA40)** Julio A. Roca 3414, (B1702BCL) Ciudadela ☎11 4488 3644. 🖹11 4657 4098 **W:** www.radiometro1520.com.ar **E:** amradiometro@yahoo.com.ar – **BA41)** Corre 275, (B1718BSE) San Antonio de Padua ☎220 482 4526 **W:** www.amcontacto.blogspot.com **E:** contacto1460@gmail.com – **BA42)** Lacroze 1871, (B1655LVS) José León Suárez ☎11 4720 2688 **W:** www.radiofolclorisimo.com **E:** info@radiofolclorisimo.com.ar – **BA43)** Fray Mamerto Esquiú 2855, (B1826GBO) Remedios de Escalada ☎11 4225 3823 **W:** www.radiociudad.com.ar **E:** radiociudad@movi.com.ar – **BA44)** 33 Orientales 1033, Villa Ariza, (B1714NOS) Ituzaingó ☎11 4623 4549 – **BA45)** Mariano Alegre 23, (B1842FSA) Monte Grande. ☎11 4281 4049 **W:** www.radiovidaam.com **E:** contacte@radiovidaam.com – **FM:** 104.9MHz – **BA47)** Juan Jofré 4243, (B1765MOY) Isidro Casanova. ☎11 4694 8131 **W:** www.canaancelestial1490.com.ar **E:** info@canaancelestial1490.com.ar – **BA48)** Pueyrredón 3846, (B1650CVP) San Martin. ☎11 4754 8784. 🖹11 4713 2517 **W:** amtradicion.com.ar **E:** amtradicion@tutopia.com – **BA49)** Santa Cruz 1312, (B1655IHD) José León Suárez ☎11 4720 0059.- **W:** www.radioam1470.com.ar **E:** radiomb@arnet.com – **BA50)** Faustino Brughetti 1329, (B7100) Dolores. ☎2245 44 3131 **W:** www.radiodolores.com **E:** radiodolores@hotmail.com.ar – **FM:** 94.9MHz Red 94 – **BA51)** Catamarca 2560, (B1847CXH) Rafael Calzada. ☎11 4219 1150 **W:** www.radioluzdelmundo.com.ar **E:** radioluzdelmundo@hotmail.com – **BA52)** Juncal 12, 1° piso, Of. "3", (B1770AOB) Tapiales. ☎11 4442 6333 **W:** www.am-1440.com **E:** impactoam@hotmail.com – **BA53)** Marquita Sánches de Thompson 1850, (B1768BDP) Ciudad Madero. ☎11 4622 1570. – **BA54)** Av.General Paz 13869, Villa Insuperable, (B1751BRG) Lomas del Mirador ☎11 4454 7799 **W:** fuentedeaguaviva.com.ar **E:** radiolaluz@gmail.com – **FM:** 105.7 MHz – **BA55)** Las Piedras 2447, (B1826DJO) Remedios de Escalada. ☎11 4249 6047 **W:** www.melody1570.com **E:** info@melody1570.com.ar – **BA56)** Gral. Martin Rodríguez 377, (B1842DIG) Monte Grande ☎11 4290 5245 **W:** www.am1160.com.ar **E:** radioexcelsior@ciudad.com.ar - **FM:** FM Malvinas 91.7MHz – **BA57)** Robertson 1249, 1° Piso "3", (B1838AIE) Luis Guillón ☎11 4296 3396 **W:** www.netmetro.com.ar **E:** am1600@netmetro.com.ar – **FM:** 96.9MHz FM Metro – **BA63)** Calle Eva Perón 1169, (B1824IBI) Lanús. ☎11 4247 3106 **W:** www.plusradioam1300.com.ar **E:** plusradioam1300@hotmail.com – **BA65)** Av. Enrique Santamarina 2642, (B1805AYF) El Jagüel, ☎11 4232 5635 – **BA66)** Florencio Sánchez 119, Barrio Los Alamos, (B1856FXE) Glew. ☎11 4233 1323 **W:** www.am1100estilo.blogspot.com **E:** amestilo@hotmail.com – **BA67)** Av. Jacinto Peralta Ramos 675, (B7608CFM) Mar del Plata. ☎223 482 0126.– **BA68)** Potrerillos 1246, (B1849DVX) Claypole. ☎11 4219 3850 **W:** www.am660popular.com **E:** info@am660popular.com.ar – **FM:** 89.1MHz - Popular – **BA69)** Choele Choel 1233, (B1822DPY) Valentín Alsina. ☎11 4218 4860 **W:** www.gama1400.com **E:** radiogama@hotmail.com – **BA71)** Calle 39 No. 256, (B1902APL) La Plata. ☎221 427 3360 **W:** www.radiorocha.com.ar **E:** director@radiorocha.com.ar – **BA72)** Domingo F.Sarmiento 2220, (B1663GFX) San Miguel. ☎11 4667 4460 – **BA73)** Congresales 570, 1° piso, (B1716) Libertad ☎220 495 0245 **E:** chmistica@hotmail.com – **BA74)** Santiago del Estero 73, (B1862SCA) Guernica. ☎2224 47 6963 – **BA76)** Cristianía 3049, (B1765HOG) Isidro Casanova. ☎11 4694 5434 🖹11 4694 7222 **W:** www.la1370.com.ar **E:** landproducciones@yahoo.com.ar - **FM:** 92.1MHz – **BA77)** Jauretche 1052, 1° piso "B", (B1686FCD) Hurlingham. ☎11 4662 9534 **W:** radiomipais1170.com.ar **E:** info@radiomipais1170.com.ar – **BA79)** Av. José María Moreno 1443, 1° Piso "A-B", (C1424ABB) CA Buenos Aires. ☎11 4296 1623 **W:** www.radio1090.com **E:** mensajes@radio1090.com.ar – **BA83)** España 468, (B7600CXJ), Mar del Plata. ☎223 475 1365 **W:** www.cadenaeco.com.ar **E:** radio@cadenaeco.com.ar **FM:** 89.1MHz**BA84)** Carlos Gardel 599. (B1824NTK) Lanús. ☎11 4225 7304 **W:** www.cadenaam1470.com **E:** info@cadenaam1470.com - **FM:**106.5MHz – **BA89)** José P. de Lafayette 549, (B1765GXC) Isidro Casanova ☎11 4485 7376 🖹11 4485 6516 **W:** www.litoral1230am.com.ar **E:** info@litoral1230am.com.ar – **BA90)** Madrid 3987, Barrio San Juan, (B1712NMO) Castelar ☎11 4692 4412 – **BA91)** Av. General Paz 3755, (B1672AMA) Villa Lynch. ☎11 4755 90 61 **W:** www.radioam610.com **E:** radio610@gmail.com – **BA92)** Libres del Sur 128, (B7130ACD) Chascomús. ☎2241 42 5367 **W:** www.rchradiochascomus.com.ar **E:** rch@radiochascomus.com.ar – **FM:** 90.9MHz – **BA94)** Catamarca 445, Villa Golf, (B1803DYI) Ezeiza. ☎11 4295 2246 – **BA99)** Av. Alte. Francisco Seguí 1059, (B1836BYK) Llavallol. ☎11 4293 9904 **W:** www.

restaurandote.com – **BA100)** Av. Santa Fe 2470, 1° piso, (B1640IFY) Martinez. ☎11 4578 2223 **W:** www.amlider.com.ar **E:** info@amlider.com.ar - **FM:** 99.3MHz – **BA101)** Juan Jofré 4243, (B1765MOY) Isidro Casanova. -☎11 4694 2538 – **BA107)** Av. Brig. Gral. Juan Manuel de Rosas 4357, (B1754FVB) San Justo. ☎11 4484 4517 **W:** www.cristolasolucionsj.com.ar **E:** contacto@cristolasolucionsj.com.ar - **FM:** 90.7MHz – **BA108)** Manuel Dorrego 292, (B7101) General Conesa ☎🖹2245 49 2140 **W:** www.radiotordillo.com – **BA117)** Balbastro 3681, 1° Piso, (B1754GPM) San Justo ☎11 4482 3230 **E:** fmfiladelfia@speedy.com.ar – **FM:** 97.1MHz – **BA119)** Gral. Martín Miguel de Güemez 5025, (B1603CUE) Villa Martelli ☎11 4709 1172 **W:** www.amitalia.com.ar **E:** radioitalia.am@hotmail.com – **BA120)** Domingo F.Sarmiento 4154, (B1665KON) José C.Paz. ☎2320 42 3306 **W:** www.sintonia1000.com.ar **E:** sintonia1000@yahoo.com.ar – **BA123)** 25 de Mayo 579, (B6700AKL) Luján ☎2323 44 2020 **E:** am1400lujan@yahoo.com.ar - **FM:** 91.1MHz FM Fantastica – **BA124)** Soberanía Nacional 2945, (B1757KHY) Gregorio de Lafferrere ☎11 4457 3674 **W:** www.guabiyu1610.com.ar **E:** oyentes@guabiyu1610.com.ar – **BA126)** Maestro Ferreyra 175, Barrio Trujui, (B1663CHC) San Miguel. ☎11 4455 1408 – **BA127)** Vicente López 235 2° Piso, PB A, (B1842AUE) Monte Grande. ☎11 4284 3186 **W:** www.radioclaridad.com.ar **E:** radio@radioclaridad.com.ar – **BA128)** Lavalle 1105, Dpto 1, (B1618CTU) El Talar. ☎11 4726 3556 **E:** amcarisma980@yahoo.com.ar – **BA130)** Florida 3168, (B1854HVH) Longchamps. ☎11 4233 3990 **W:** www.am1540.com.ar **E:** radiofuego@yahoo.com.ar – **BA131)** Av. Pte. Juan D. Perón 2514, (B1644CYP) Victoria. ☎11 4746 6856 **W:**www.radiooasis.com.ar **E:** radiooasis@mixmail.com - **FM:** 92.5MHz – **BA132)** Coronel Dorrego 346, (B1752DPH) Lomas del Mirador. ☎11 4699 0584.- **W:** www.radiomaranata.net **E:** radiomaranata@hotmail.com - **FM:** 89.3MHz – **BA134)** Av.Gral. Pedro Diaz 1460, (B1686IQH) Hurlingham. ☎11 4662 6387 **W:** www.radiorestauracion.com.ar **E:** radiorestauracionam@hotmail.com – **BA135)** San Martin 513 (B1854FEM) Longchamps. ☎11 4233 5560 **W:** www.radio26.com.ar **E:** radio26dejulio@gmail.com – **BA138)** Santander 5714, 2° Piso "D" (C1439ASZ) CA Buenos Aires . ☎11 4602 8553 **W:**www.amnativa.co.ar – **BA139)** Fray Mamerto Esquiú 1161, (B1824BFQ) Lanús ☎11 4225 2256 **W:** www.digital860.99k.org **E:** digital860@hotmail.com – **BA141)** Jauretche 1052, 1° "C", (B1686FCD) Hurlingham. ☎11 4452 8688 **W:** decadasam1090.com.ar **E:** info@decadasam1090.com.ar – **BA143)** Av. Monteverde 8158, (B1849HAX) Claypole. ☎11 4238 8427 **W:** //radioredenfor.mmmenargentina.com **E:** radioredentorradio@yahoo.com.ar – **BA144)** Santa Fe 2540, (B1722BGZ) Merlo. ☎220 485 6696 **W:**buenasnuevasradio.com.ar **E:** radiobuenasnuevasama1380@yahoo.com.ar – **BA145)** Ex. Combatientes de Malvinas 2053, (B1613ECO) Los Polvorines. ☎11 4663 5422 **W:** www.santiaguenañorteña.com.ar **E:** info@santiaguenanortena.com.ar – **BA146)** Jujuy 2928, (B7602BKD) Mar del Plata – **BA147)** Donovan 1433, (B1770AHK) Tapiales ☎11 4426 0185 **W:** www.am1480.com.ar **E:** sensacionesam@yahoo.com.ar – **BA148)** Victor Hugo 647, (B1825FBI) Monte Chingolo. ☎11 4220 6822 **W:** www.logdesignmedia.com.ar/socalo_am1520.swf **E:** am1520_suma@yahoo.com.ar - **FM:** 106.1MHz – **BA150)** San Martin 1337, (B1722LTK) Merlo. ☎220 482 1760 **E:** radiocotidiana@hotmail.com – **BA152)** Diagonal 74 No 1357, (B1900BZG) La Plata, ☎22 427 3227 – **BA153)** Dr. Ignacio Arieta 3950, (B1754AQT) San Justo ☎11 4482 2597 **W:** www.radioguarani.com.ar **E:** info@radio-guarani.com.ar – **BA154)** Salta 2641, (B1754IQS) San Justo. ☎11 4441 8196 - **FM:** 104.5MHz FM Sintonia – **BA155)** Gral. Belgrano 531, (B7000GEK) Tandil ☎2293 44 6383 **W:** www.am1180.com.ar **E:** am1180@speedy.com.ar - **FM:** 99.5MHz – **BA156)** Juan Sebastian Bach 3687, (B1756KKM) Isidro Casanova. ☎11 4694 6470 **W:** radioerevivir.com **E:** radio@radiorevivir.com – **BA157)** Av.Luro 4740, (B1757AAQ) Gregorio de Laferrere ☎11 4467 1134 – **BA158)** Azamor 2858, (B1827AIX) Ingeniero Budge. ☎11 4273 1229 **E:** la1390@argentina.com – **BA161)** Independencia 646, (B1804BCN) Ezeiza ☎11 4232 4355 – **BA162)** Lomas de Zamora – **BA163)** Juan F.Segui 895, (B1615MNA) Grand Bourg. ☎23 2041 4426 - **FM:** 105.3MHz – **BA164)** Elías Bedoya 2020/4, (B1765LXH) Isidro Casanova ☎11 4669 4925 **E:** stcom@uolsinectis.com.ar – **BA166)** Reconquista 27, (B1804CFA) Ezeiza ☎11 4232 0321 – **BA167)** Benito Pérez Galdós 688, Villa Florito, (B1821EON) Banfield. ☎11 4276 5194 – **BA172)** Juan Florio 3579, (B1754AJK) San Justo. ☎11 4441 8200 **W:** www.radiorepublica.com.ar – **BA173)** Belgrano 4033, (B1650CCS) San Martin. ☎11 4713 8808 **W:** www.am680.com.ar **E:** radio@am680.com.ar – **BA174)** Heredia 920, Augstín Ferrari, (B1724ETO) Marioano Acosta. ☎220 498 1498 – **BA175)** Av. Brigadier Juan Manuel de Rosas 2468, (B1655MSS) José León Suárez. ☎11 4729 1545 – **BA176)** Almirante Brown 135, (B6740DRB) Chacabuco. ☎2352 43 1136 **W:**lt36radiochacabuco.com.ar **E:** radiochacabuco@topmail.com.ar - **FM:** 91.7 MHz FM Universal – **BA177)** Zufriategui 871, (B1765CKQ) Isidro Casanova. ☎11 4467 2468 **W:** www.radiohosannaam1640.com **E:** hosannaam1640@hotmail.com - **FM:** 94.5MHz – **BA178)** Pedro Laurenz 237, Las Malvinas (B1748), General Rodríguez. ☎237 487 3200 **W:** http//radioolivera.es.tl **E:** radioolivera@hotmail.com– **BA179)** Bombero Ariño 1150, (B1834IAX) Temperly. ☎11 5290 0075 **W:**www.lasuper-

sport.com.ar **E:** lasuper@lasupersport.com.ar – **BA180)** Hipólito.Yrigoyen 2629, (B7600DPG) Mar del Plata. ☎22 3494 1428 **W:** www.1620.com.ar **E:** am1620@yahoo.com.ar – **BA181)** Cjal. Manuel Martitegui 598, Fátim, (B1629JGL) Pilar ☎2322 49 9899 – **BA182)** Salta 138, 2° Piso "C", (B1708JOD) Morón. ☎11 4489 2024 **W:** www.lagauchita810.com.ar **E:** amlagauchita@hotmail.com – **BA183)** Calle Juan Florio 3573, (B1754AJK) San Justo, Partido de La Matanza. ☎11 4651 1694 **W:** www.am890.com.ar **E:** radio@am890.com.ar – **BA184)** Santa Rosália 1465 (B1651CXE) San Andrés ☎11 4752 3245 **W:** www.sudamericana1120.com.ar **E:** info@sudamericana1120.com.ar – **BA185)** Fray Mamerto Esquiú 2855, (B1826GBO) Remedios de Escalada. ☎11 4225 3198 **W:** www.radioindependencia@hotmail.com – **BA186)** Calle Angel Rotta 168, (B1842AED) Monte Grande. ☎11 4296 0771 – **BA187)** Las Heras 1478, (B6700AUO) Luján. ☎2323 42 9595 **W:** www.radiomasterlujan.com.ar – **BA188)** Av. Pte. Juan Domingo Perón 1774, (B1663GHT) San Miguel ☎11 4664 0077 **E:** rafolk@yahoo.com.ar – **FM:** FM Integracion 99.3 MHz – **BA189)** Paso 1943, (B6015ASC) Los Toldos. ☎2358 44 3954 – **BA190)** Avenida 59 N° 2465, 1° Piso, (B730GYJ) Necochea. ☎2262 52 0003 **W:** am1380.com. ar **E:** info@am1380.com.ar – **BA191)** Fragate Heroína 2035, (B1842FQI) Monte Grande. ☎11 4284 2830 **W:** www.jerusalenradio.com **E:** info@jerusalen.com.ar – **BA192)** Paula Albarracin 3957, (B1663CPE) San Miguel de Oeste ☎2320 0649 – **BA194)** Av. Aristóbulo del Valle 1202 (B7000HLN) Tandil **W:** www.lavozdetandil.com.ar **E:** lavozdetandil.com.ar – **BA195)** Coronel Suárez 548/554, (B1834GHL) Temperley. ☎11 4244 1843 **W:** www.amfenix.com.ar **E:** amradiofenix@hotmail.com – **BA196)** Aristóbulo del Valle 23, (B7240IXA) Lobos. ☎2227 42 1211 – **BA198)** Luis Vernet 6654, (B1757MOB) Gregorio de Laferrere. ☎11 4467 4224 **W:** www.santiagoycopla.com.ar **Emai:** santiagoycopla@hotmail.com – **BA199)** Av. Buenos Aires s/n, av. Avenidas 6 y 7, Galeria Pinar, Local 11, (BXXXX) Villa Gesell. ☎2255 46 5686 **W:** www.am1310gesell.com. ar **E:** am1310gesell@hotmail.com – **FM:** 89.9MHz – FM Plus – **BA200)** Fournier 4075, (B1757IDW) Gregorio de Laferrere. ☎11 4457 7204 **W:** www.am1330radiomailin.com.ar **E:** radiomailin@am1330radiomailin.com.ar – **BA201)** Burela 560, (B1722PHL) Merlo. ☎220 489 0463 – **BA203)** Alekandro Volta e/Vesalio y Dubalia, (B1763) Virry Del Pino. ☎2202 49 3745.– **BA204)** Salta 4198, Banfield Oeste. ☎11 4693 4570 – **BA205)** Dr. Eugenio Asconape 371, (B1744FIG) Moreno ☎237 460 0878 – **BA206** Buenos Aires – **BA207)** Venezuela 370, 2° piso, (C1095AAH) CA Buenos Aires – **BA208)** Ramón I.Falcón 2193, (B1685BDY) El Paøomar. ☎11 4443 7424 **W:** www.radiolaluna.com.ar – **BA209)** Dr.Rebizzo- (Calle 626) – No 3917 (B1678BCC) Caseros ☎11 4578 5130 **W:** www.area1.am1320.com.ar **E:** radioarea1nail.com – **BA210)** Av Presidenteblllia – (ex. Ruta 8) – y Cnel. Suárez, (Bxxxx) José C.Pac ☎320 42 8351 – **BA211)** Ituzaingó – **BA212)** Marcelo T.de Alvear 650 (B1755JMN) Rafael Castillo ☎11 4697 4919 – **BA213)** Mñor. López May 3372, (B1757DHJ) Gregorio de Laferre ☎11 4467 3600 **W:** // radioesperanza.3a2.com **E:** radioesperanza@gmail.com – **BA214)** Calle Gen. Lavalle 2307, (B1847BQW) Rafael Calzada, Partido de Almirante Brown ☎11 4219 1903 **W:** www.am1410radiofundacion. org **E:** info@am1410radiofundacion.org

CA00) CATAMARCA
CA01) Chacabuco 762, (K4700BTP) S.F. del Valle de Catamarca ☎3833 42 4223 ☎3833 42 2251 **E:** catamarca@radionacional.gov.ar – **FM:** 103.3MHz

CB00) CHUBUT:
CB01) Av. Alvear 1180, (U9200AXY) Esquel ☎2945 45 1900 **W:** www. lra9.com.ar **E:** esquel@radionacional.gov.ar – **FM:** 88.7MHz – **CB02)** Av. Hipólito Yrigoyen 1735, (U9102BGM) Trelew. ☎2965 43 0580 ☎2965 42 5457 **W:** www.radiochubut.com.ar **E:** lu20@speedy.com.ar or info@lu20radiochubut.com.ar – **FM:** 95.7MHz Galaxia – **CB03)** 25 de Mayo 453, (U9000CXC) Comodoro Rivadavia, ☎297 447 2125 ☎297 446 2564 **E:** comodororivadavia@radionacional.gov.ar – **FM:** 94.5MHz – **CB04)** Estivariz 226, (U9120KET) Puerto Madryn. ☎2965 45 1600 **W:** www.lu17.com **E:** lu17@patagonia.net – **FM:** 100.3MHz **CB05)** Av. Rivadavia 198, (U9000AKP) Comodoro Rivadavia. ☎297 446 6561 **W:** www.lu4radio.com.ar **E:** direccion@lu4radio.com.ar – **FM:** 101.7MHz Paraiso FM – **CB06)** Calle 25 de Mayo 740, (U9100BRP) Trelew ☎2965 43 5221 **W:** www.radio3patagonia.com.ar **E:** radiotres@speedy.com.ar – **CB07)** Av. Comandante Fontana y Dr. Mariano Moreno, (U9033) Alto Rio Senguer. ☎2945 49 7050 **E:** altoriosenguer@radionacional.gov.ar - **FM:** 93.5MHz

CF00) CIUDAD AUTÓNOMA DE BUENOS AIRES (BUENOS AIRES):
CF01) Rivadavia 835, (C1002AAG) CA Buenos Aires ☎11 4999 1500 ☎11 4338 4250 **W:** www.continental.com.ar **E:** info@continental.com.ar – **FM:** 104.3 – 105.5MHz–**CF02)** Arenales 2467, (C1124AAM) CA Buenos Aires. ☎11 5219 4744 ☎11 5219 4760 **W:** www.rivadavia.com.ar **E:** info@rivadavia.com.ar – **CF03)** Sarmiento 1551, 8° piso (C1042ABC) CA Buenos Aires. ☎11 5371 4646 ☎11 4556 9056 **W:** www.radiodelaciudad.gov.ar **E:** ciudadam1110@gmail.com – **CF04)** Gral. Mansilla

2668, 1° piso (C1425BPD) CA Buenos Aires. ☎11 5777 1500 ☎11 5777 1504 **W:** www.radiomitre.com.ar **E:** info@radiomitre.com.ar – **CF05)** Maipú 555, (C1006ACE) CA Buenos Aires ☎11 4327 3021. ☎11 4325 9433 **W:** www.radionacional.com.ar **E:** buenosaires@radionacional.gov. ar - **International Sce:** see International Broadcasting section – **CF06)** Fitz Roy 1460, (C1414CHT) CA Buenos Aires. ☎11 5032 0400 **W:** www. radiolared.multimediosamerica.com.ar **E:** info@radiolared.com.ar – **CF07)** Tucumán 1, 19° Piso (C1041AAH) CA Buenos Aires ☎11 4318 7888 **W:** www.ambelgrano.com **E:** info@ambelgrano.com – **CF08)** Olleros 3551, (C1427EEA) CA Buenos Aires ☎11 4556 9000 **W:** www.amdelplata.com – **CF09)** Rivadavia 825, (C1002AAG) CA Buenos Aires ☎11 4121 8900 **W:** www.radioelmundo.com.ar **E:** info@radioelmundo.com.ar – **CF10)** José de Amenábar 23, (C1426AIA) CA Buenos Aires ☎11 4778 8500 **W:** www.estoesamerica.com – **CF11)** Av. Entre Ríos 1931, (C1133AAH) CA Buenos Aires ☎11 4307 2200 **W:** www.radiobuenosaires.com.ar **E:** am1350@radiobuenosaires.com.ar – **CF12)** San Martín 569, 2° piso "6", (C1004AAK) CA Buenos Aires. ☎11 4893 1701 **W:** www.amsplendid. com **E:** info@amsplendid.com.ar – **CF13)** Fitz Roy 1940, (C1414CID) CA Buenos Aires. ☎11 4535 4000 **W:** www.infobae.com **E:** radi10. com **E:** radio10@infobae.com – **CF15)** Av. Corrientes 1847, 19° Piso "A", (C1045AAA) CA Buenos Aires ☎11 4372 2841 **W:** www.am1010ondalatina.com.ar **E:** contactoam1010@yahoo.com.ar – **CF16)** Av. Rivadavia 10561, 3° piso, (C1408AAF) CA Buenos Aires ☎11 5631 1000 ☎11 5631 **W:** www.cadenaeco.com.ar **E:** radio@cadenaeco.com.ar – **CF17)** Av.Rivadavia 10561, 3° Piso, (C1408AAF) CA Buenos Aires. ☎11 5631 1000 **W:** www.cadenaecoa.com.ar **E:** am1530@cadenaeco.com.ar – **CF18)** Traful 3834, (C1437HML) CA Buenos Aires. ☎11 4912 0497 **W:** www.am840generalbelgrano.ar **E:** am840generalbelgrano@hotmail. com – **CF20)** Av. Saénz 459, (B1437DNE) CA Buenos Aires. ☎11 4912 0819 **W:** www.radiourkupina.com.ar **E:** www.urukpina.com.ar **W:** sergiocorrea.com.ar – **CF21)** Cerrito 242, PB "B", (C1010AAF) CA Buenos Aires, ☎11 5252 0741 **W:** www.am740.com.ar **E:** cooperativa@am740. com.ar – **CF22)** Av. San Juan 2461, (C1232AAG) CA Buenos Aires ☎11 4942 6913 **W:** www.cadenavidaonline.com – **CF23)** Av. Int. Francisco Rabanal 1465, PA, (C1437FPB) CA Buenos Aires. ☎11 4919 2994 **W:** www.radiobolivia.net – **CF24)** Montevideo 497, 3° Piso "B", (C1019ABI) CA Buenos Aires ☎11 4371 2597 **W:** www.amradiodelpueblo.com. ar **E:** gerencia@amradiodelpueblo.com.ar – **CF26)** Bonpland 1114, (C1414CMJ) CA Buenos Aires. ☎11 4856 8819 **W:** www.radioidentidad. com.ar **E:** radioidentidad@radioidentidad.com.ar – **CF27)** Maipu 267 7° piso, (C1084ABE) CA Buenos Aires. ☎11 4136 1050 **W:** www.conceptoam.com.ar **E:** radio@conceptoam.com.ar – **CF28)** Salguero 2745, 6° piso, Of.. 63, (C1425DEL) CA Buenos Aires ☎11 4803 4434 ☎11 4807 6006 **W:** www.amlamarea.com.ar **E:** amlamarea@amlamare.com.ar – **CF29)** Av.José Maria Moreno 1443, 1° piso, (C1424AAB) CA Buenos Aires. ☎11 4926 1661 **W:** www.radioam1310.com.ar **E:** radioam1310@ gmail.com – **CF30)** San Martin 569 2° piso "6" , (C1004AAK) CA Buenos Aires. ☎11 4893 1701 **W:** www.radioargentina.com.ar **E:** radio@ am570radioargentina.com.ar – **CF31)** Manzanares 4006, (C1430AEN) CA Buenos Aires. ☎11 4541 0303 **W:** www.cadenauno.com.ar **E:** mensajes@cadenauno.com.ar – **CF32)** Av. Triunvirato 4671, (C1431FBJ) CA Buenos Aires. ☎11 4521 3931 **W:** www.am1710.com **E:** mensajes@am1710.com.ar – **CF33)** Brasil 907, 1° Piso "B", (C1154AAO) CA Buenos Aires. ☎11 4307 1835 **W:** www.radiouno.com.ar **E:** amtango@ amtango.com.ar – **CF34)** Cerrito 242, PB, "A", (C1010AAF) CA Buenos Aires. 11 4382 9327 **W:** www.madres.org/am530/ **E:** radio@madres.org – **CF35)** Av. José María Moreno 1443, 1° Piso, (C1424ABB) CA Buenos Aires ☎11 4926 1622 **W:** www.radiogenesis970.com.ar **E:** mensajes@ radiogenesis970.com.ar – **CF36)** Fonrouge 76, (C1408HFB) CA Buenos Aires. ☎11 3979 5663 **W:** www.radioolivia.com **E:** radioolivia@hotmail. com – **CF37)** Alicia Moreau de Justo 2050, 1°P, Of. "132", (C1107AFP) CA Buenos Aires. ☎11 4893 7555 **W:** www.radioelsol.com.ar **E:** trd@trd-publicida.com – **FM:** 93.1MHz –**CF38)** San Martín 569, 2° Piso "6", (C1004AAK) CA Buenos Aires. ☎11 4893 1701 **W:** www.reporteram650. com.ar – **CF39)** Pje. Espinillo 1449, PA "1", (C1407ISA), CA Buenos Aires. ☎11 4683 3641. – **CF41)** Rivadavia 1615, 12° Piso, (C1033AAG) CA Buenos Aires. ☎11 4383 9773 **W:** www.amradiopunto.com.ar **E:** contacto@amradiopunto.com.ar – **CF42)** Yatay 977, (C1184AQD) CA Buenos Aires ☎11 4866 3883. – **CF43)** Cnel. Martiniano Chilavert 5875, (C1439CLM) CA Buenos Aires ☎11 4605 4857 **W:** www.siembraelpan. com.ar/amanecer.html **E:** radio_amanecer@hotmail.com – **CF44)** Av. Callao 441,1° piso "G", (C1022AAE) CA Buenos Aires ☎11 4372 5863 **W:** www.am1230creativa.com.ar **E:** am1230creativa@yahoo.com.ar

CH00) CHACO:
CH01) Av. General Güemes 1103, (H3730AML) Charata ☎3731 42 0150 ☎ 3731 42 0735 **W:** www.mocovi.com.ar **E:** am800@mocovi. com.ar - **FM:** 95.7MHz FM Lider – **CH02)** Acceso Ruta Nacional N° 40 s/n, Barrio Gendarmería, (U9030) Río Mayo ☎ 2903 42 0099 **E:** riomayo@radionacional.gov.ar – **FM:** 88.1MHz – **CH03)** Av. Sarmiento 1255, (H3502COE) Resistencia. ☎3722 43 2920. ☎3722 42 4937.- **W:** www.radionacional.chaco.com.ar **E:** resistencia @radionacional.gov.ar

Guaraní: Sat. 1800 - **FM:** 96.7MHz – **CH04)** Avellaneda Calle 19 No 151, (H3700ASC) Presidencia Roque Sáenz Peña. ☎3732 42 9651 **W:** www.lt16.net **E:** lt16am950@yahoo.com.ar - **FM:** 93.3MHz – **CH05)** Córdoba 710, (H3500APP) Resistencia. ☎3722 43 3900 ☐3722 43 3999 **W:** www.radiochaco.com.ar **E:** radiochaco740@yahoo.com.ar - **FM:** 101.5MHz FM Chaco – **CH06)** Urquiza Calle 6 No.949/51, (H3700HFS) Presidente Roque Sáenz Peña. ☎3732 42 7188 **W:** www.llrcradio.com.ar **FM:** 101.5MHz

CO00) CORDOBA:
CO01) Fray Miguel de Mojica 1600, Barrio Marquez de Sobremonte, (X5008CCN) Córdoba. ☎351 410 5000 **W:** www.580am.com.ar **E:** administracion@srtunc.com.ar - **FM:** 102.3MHz Power FM – **CO02)** Santa Rosa 241, (X5000ESE) Córdoba. ☎351 422 5664 ☐351 422 5665 **W:** www.adionacionalcba.com.ar **E:** cordoba@radionacional.gov.ar - **FM:** 91.3MHz – **CO03)** 27 de Abril 979, (X5000AES) Córdoba ☎351 526 5200 ☐351 526 5222 **W:** www.am970.com.ar **E:** info@am970.com.ar - **FM:** 99.7MHz R. Dos – **CO04)** Constitución 399, (X5800BBB) Río Cuarto. ☎358 463 8255 **W:** www.lv16.com **E:** lv16@lv16.com - **FM:** 93.9, 106.9MHz – **CO05)** Alvear 139, (X5000ILC) Córdoba. ☎351 526 0597 **W:** www.lv3.com.ar **E:** audencia@cadena3.com.ar - **FM:** 92.3, 100.5, 106.9MHz – **CO06)** Libertad 455 2° piso, (X5850KNI) Río Tercero. ☎3571 42 1019 **W:** www.lv26.com.ar **E:** lv26@itc.com.ar - **FM:** 94.5MHz FM Libra – **CO07)** Tucumán 159, (X6120EOC) Laboulaye. ☎3385 42 6259 ☐3385 42 5848 **W:** www.radiolv20.com.ar **E:** radiolv20@arnet.com.ar - **FM:** 89.9MHz – **CO08)** Córdoba 51,"Edificio Reggio II", (X2400PQA) San Francisco. ☎3564 42 2186 **W:** www.lavozdesanjusto.com.ar **E:** infolv27@cordoba.com.ar or infolv27@poraire.net - **FM:** 88.7MHz FM Galaxia – **CO09)** Santa Fe 1490, (X5900DTJ) Villa María. ☎353 452 2699 **W:** radiovillamaria.com **E:** lv28@radiovillamaria.com - **FM:** 98.5 MHz FM Record – **CO11)** Belgrano 33, Galeria Central,Local 8 y 10, (X5870ABA) Villa Dolores. ☎3544 42 3222 **W:** www.radiochampaqui.com.ar **E:** direccion@radiochampaqui.com.ar - **FM:** 100.1MHz – **CO12)** Blvd. 25 de Mayo 133, PB, (X2421ABB) Morteros ☎3562 42 2148 ☐3562 42 3176 **W:** www.radiomorteros.com **E:** radiomorteros@yahoo.com.ar - **FM:** 90.3MHz FM Selección – **CO13)** Av. Concepción Arenale 1174, (X5004AAY) Córdoba. ☎351 460 1010 **W:** www.radiosucesos.com **E:** audiencia@radiosucesos.com.ar - **FM:** 104.7MHz – **CO14)** Dr.Luis Tozzini 40 (X6120DDB) Laboulaye. ☎3385 42 6664 **W:** www.radiobuenasnuevas.com.ar **E:** radiobuenasnuevas@gmail.com – **CO15)** Santa Fé 804, (X6270CWR) Huinca Renancó. ☎2336 44 2007 **E:** LV22_1490@yahoo.com.ar – **CO16)** Av. Fernando Fader 3469, Barrio Cerro de Las Rosas, (X5009ABB) Córdoba. ☎3515 571 0810 **W:** www.cienradios.com.ar/argentina/mitre-cordoba - **FM:** 97.9MHz

CS00) CORRIENTES:
CS01) Chacra 46, La Tablada, (W3340) Santo Tomé. ☎3756 42 0090 **E:** santotome@radionacional.gov.ar - **FM:** 100.5MHz – **CS02)** Juan Sitja Nin 491, (W3230GEQ) Paso de los Libres. ☎3772 42 4332 **W:** www.radiolt12.com.ar **E:** contacto@radiolt12.com.ar - **FM:** 92.7MHz – **CS03)** La Rioja 743, (W3400BZG) Corrientes. ☎3783 42 3560 ☐3783 423149 **W:** www.radiolt7.com **E:** info@radiolt7.com - **FM:** 95.3MHz – **CS04)** Mariano I. Loza 231, (W3450BXE) Goya. ☎3777 43 3002 **W:** www.lt6noticias.com.ar **E:** lt6radiogoya@hotmail.com - **FM:** 98.3MHz FM Splendid – **CS05)** General Paz 903, (W3344AYQ) Alvear. ☎3772 47 0699 **E:** lt21radiomunicipal@hotmail.com – **CS06)** Av. Atlantico Aguirre Km 2, (W3470EHA) Mercedes. ☎3773 42 0087 **W:** www.radioliberia.com **E:** oyentes@lt42.net.ar - **FM:** 93.5MHz – **CS07)** San Martín 1380, (W3461AKA) Curuzú Cuatiá. ☎3774 42 2634 ☐3774 42 5873 **W:** www.lt25..com.ar **E:** info@lt25..com.ar - **FM:** FM Guarani 107.1MHz

ER00) ENTRE RIOS:
ER01) San Martín 371, (E3200FUG) Concordia. ☎345 421 5506 **W:** www.lt15concordia.com.ar **E:** lt15adm@arnet.com.ar - **FM:** 89.3MHz – **ER02)** Rivadavia 126, (E3100GNO) Paraná. ☎343 423 0101 **W:** www.radiolt14.com.ar **E:** lt14@radiolt14.com.ar - **FM:** 93.1MHz Radio 100 – **ER03)** Justo José de Urquiza al Oeste, Parada 12, (E2820) Gualeguaychú. ☎3446 42 6159 **E:** gualeguaychu@radionacional.gov.ar - **FM:** 98.7MHz – **ER04)** Carlos Pellegrini 106, (E2822EWD) Gualeguaychú. ☎3446 45 4660 ☐3446 42 7088 **W:** www.radiolt41.com **E:** info@radiolt41.com.ar – **FM:** 90.3 / 97.9MHz – **ER05)** Av. Pte. Juan D. Perón 117, (E3280CBS) **W:** www. nuevomundodigital.com.ar **E:** radionmundo@ colonred.com.ar - **FM:** 93.7MHz – **ER06)** Sarmiento 474, (E3153EZH) Victoria. ☎3436 42 1285 **W:** www.lt39am980. com **E:** lt39am980@arnet.com.ar - **FM:** FM Victoria 90.3MHz – **ER07)** Pablo Stampa 2430, (E3228FDD) Chajarí. ☎3456 42 0002 **W:** www.multimedioschajari.com.ar **E:** multimedioschajari.com - **FM** 107.7MHz – **ER08)** Chacabuco 38, 1° Piso, (E2840BFB) Gualeguay. ☎3444 42 4915 **W:** www.radiogueleguay.com.ar **E:** info@radiogueleguay.com.ar - **FM:** 104.3MHz FM Gama – **ER09)** Roque Sáenz Peña 1082, (E3190FZJ) La Paz. ☎3437 42 1568 **W:** www.lt40.com.ar **E:** info@lt40.com.ar - **FM:** 91.3MHz FM La Paz – **ER10)** Onésimo Leguizamón 469, (E2820FGO) Concepción del Uruguay. ☎3442 42 5661 **W:** www.lt11.net **E:** direccion@eleteonce.com.ar - **FM:** 92.9MHz FM Arenas – **ER11)** Av. Vélez Sársfield 1111, (E3240AUL) Villaguay. ☎3455 42 1717 **W:** www.lt27villaguay.com.ar **E:** lt27@clavis.

com.ar - **FM:** 88.7MHz R.Urbana.

FO00) FORMOSA:
FO01) Junín 655, (P3600IDM) Formosa. ☎3717 42 6197 **E:** formosa@radionacional.gov.ar - **FM:** 94.1MHz – **FO02)** Ruta Nacional 81 y Ruta Provincial 32, (F3630) Las Lomitas. ☎3715 43 2167 **E:** laslomitas@radionacional.gov.ar - **FM:** 93.5MHz – **FO03)** Av. 9 de Julio 165, (P3600BCB) Formosa. ☎3717 42 2590 **W:** www.am990formosa.com.ar **E:** info@am990formosa.com.ar - **FM:** 98.9MHz

JU00) JUJUY:
JU01) Av. España (Sur) 700, (Y4650ALN) La Quiaca. ☎3885 42 2356 **E:** laquiaca@radionacional.gov.ar - **FM:** 92.5MHz – **JU02)** Rio Bermejo y Olavarria, (Y4600) San Salvador de Jujuy. ☎388 422 2781 ☐388 422 6047 **E:** jujuy@radionacional.gov.ar - **FM:** 94.3MHz – **JU03)** Dr. Horacio Guzmán 496, (Y4600) San Salvador de Jujuy. ☎388 423 0035 **W:** radiovisionjujuy.com.ar **E:** rvj@arnet.com.ar - **FM:** 97.7MHz - Tropico – **JU04)** Av. Villafañe y Calilegua, (Y4608) Perico. ☎388 491 1465 **W:** radiovisionjujuy.com.ar **E:** rvj@arnet.com.ar – **JU05)** Jujuy 470, (Y4512DRJ) Libertador General San Martín. ☎38 8642 3399 **W:** radiovisionjujuy.com.ar **FM:** 104.5MHz

LP00) LA PAMPA:
LP01) Rivadavia 202, 4° piso, (L6300DWF) Santa Rosa. ☎2954 42 2456 ☐2954 42 5102 **E:** santarosa@radionacional.gov.ar - **FM:** 96.1MHz – **LP02)** Lisandro de la Torre 474, (L6300BQJ) Santa Rosa. ☎2954 43 3505 **W:** www.lu33pampeana.com.ar **E:** lu33noticias@hotmail.com - **FM:** 103.7MHz – **LP03)** Calle 40 No 1250, (L6360EVZ) General Pico. ☎2302 43 0055 **W:** www.radiolu37.com.ar **E:** radiolu37@radiolu37.com.ar - **FM:** 89.9MHz – **LP05)** General Pico 610, (L8201BIL) 25 de Mayo. ☎299 494 8086 **W:** 25demayo.gov.ar/comunicacion_y_prensa.htm **E:** radiomuni@hotmail.com - **FM:** 91.1MHz FM Rio – **LP06)** 25 de Mayo 383, (L6300DVG) Santa Rosa. ☎29 5442 1191 - **FM:** 103.1MHz R.Horizonte – **LP07)** Calle 39 No 1531, (L6360CNE) General Pico. ☎2302 43 4892 **W:** www.amorymiserocordia.org **E:** iglesiaamorymisericordia@gmail.com - **FM:** 106.3MHz

LR00) LA RIOJA:
LR01) Hipólito Yrigoyen 324, (F5300DIH) La Rioja ☎3822 42 7067 ☐3822 42 5396 **W:** www.lra28.com.ar **E:** larioja@radionacional.gov.ar - **FM:** 102.5MHz – **LR02)** Arturo Marasso 170, (F5360CPF) Chilechito. ☎3825 42 5025 – **FM:** 99.3 MHz

ME00) MENDOZA:
ME01) Manuel A.Sáez 2421, (M5539HSW) Las Heras. ☎261 430 1600 **W:** www.radionihuil.com.ar **E:** radionihuil@radionihuil.com - **FM:** 98.9MHz – **ME02)** Rioja 1093, (M5500ALU) Mendoza. ☎261 420 5100 **W:** www.lvdiez.com **E:** eleve10@infovia.com.ar or info@lvdiez.com - **FM:** 102.1, 100.9MHz Estacion del Sol – **ME03)** Rioja 1484, (M5500AMD) Mendoza. ☎261 423 8872 **W:** www.amlibertador780.blogspot.com **E:** radiorepublica@argentina.com.ar - **FM:** 92.7MHz – **ME04)** Bernado de Irigoyen 17, PA (M5620BDA) General Alvear. ☎2625 42 6566 **E:** lv23am800@yahoo.com.ar - **FM:** 88.9MHz – **ME05)** Emilio Civit 460, (M5502GVR) Mendoza. ☎261 438 0596 **E:** mendoza@radionacional.gov.ar - **FM:** 97.3MHz – **ME06)** Esquivel Aldao 350, (M5613AEH) Malargüe. ☎2627 47 1160 ☐2627 47 0658 **W:** www.lv19radiomalargue.com.ar **E:** radiomalargue03@yahoo.com.ar - **FM:** 88.1MHz – **ME07)** Av. Hipólito Yrigoyen 223, (M5602HBC) San Rafael. ☎2627 43 0055 ☐2627 43 0065 **W:** www.radiosanrafael.com **E:** lv14@radiosanrafael.com - **FM:** 97.3MHz – **ME08)** Cdte. Salas 287, (M5600DJE) San Rafael. ☎2627 42 2047 **E:** lv18@sanrafael.gov.ar - **FM:** 98.7MHz "FM Municipal" – **ME09)** Leandro N. Alem 1687, (M5560DAG) Tunuyán. ☎2622 42 5700 **E:** lv24am1320@yahoo.com.ar - **FM:** 104.5MHz – **ME10)** Av. Bandera de los Andes 4404, (M5521AXL) Villa Nueva de Guaymallén. ☎261 421 3992 **W:** www.radiomurialdo.com.ar **E:** mensajes@fmfamilia.com - **FM:** 90.5MHz FM Familia.

MS00) MISIONES:
MS01) Cristóbal Colón 1452 =128=, (N3300LXF) Posadas. ☎3752 43 3432 **W:** www.lt17.com.ar **E:** info@lt17.com.ar - **FM:** 107.3MHz FM Provincia – **MS02)** Félix de Azara 2440, (N3300LQZ) Posadas. ☎3752 43 0500 **W:** www.radiolt4.com - **FM:** 104.5MHz – **MS03)** Av. Victoria Aguirre Sur 809, (N3370AYI) Puerto Iguazú. ☎3757 42 0099 **E:** puertoiguazu@radionacional.gov.ar - **FM:** 99.1MHz – **MS04)** Av. Las Calandrias y Las Golondrinas, (N3370) Puerto Iguazú ☎3757 42 0600 **W:** www.radiocataratas.com **E:** info@radiocataratas.com - **FM:** 94.7MHz – **MS05)** Av. Independencia 156, (N3366xxx) Bernardo de Irigoyen. ☎3741 49 0002 - **FM:** 96.9MHz – **MS07)** Domingo F.Sarmiento 1847, 7° piso, (N3300HUM) Posadas. ☎3752 42 0203 ☐3752 42 2758 **W:**www.tupambaenoticias.com.ar **E:** tupambae@tupambaenoticias.com.ar or tupambae@arnet.com.ar - **Guaraní:** Sun 0900-1500. - **FM:** 105.9MHz

NE00) NEUQUÉN:
NE01) Gral. Paz 536, (Q8353CGL) Chos Malal. ☎2948 42 1198 **E:** chosmalal@radionacional.gov.ar - **FM:** 92.3MHz – **NE02)** Av. San Martín 324, (Q8340EYQ) Zapala. ☎2942 42 2960 **E:** zapala@radionacional.gov.ar - **FM:** 93.9MHz – **NE03)** Gral. Villegas 1320, (Q8370ELB) San Martín de los Andes. ☎☐2972 42 7766 **E:** sanmartindelosandes@radiona-

cional.gov.ar - **FM:** 92.5MHz – **NE04)** Juan B. Alberdi 189, (Q8300HLC) Neuquén. ☎299 443 2772 **W:** www.lu5am.com.ar **E:** info@lu5am.com.ar - **FM:** 94.7, 102.5, 103.3MHz – **NE05)** Ramón Elías Troitiño 649, (Q8354) Andacollo. ☎2948 49 4025. - **FM:** 99.5MHz – **NE06)** Calle Las Rosas 81, Barrio Jardín, (Q8315AYA) Piedra del Aguila. ☎2942 49 3216 - **FM:** 99.5MHz – **NE07)** Pte. Bernandino Rivadavia 609, (Q8300HDM) Neuquén. ☎299 443 0249 **W:** www.cumbream.com **E:** cumbre@cumbream.com.ar - **FM:** 89.9MHz – **NE08)** Sarmiento 340, (Q8370DOH) San Martin de los Andes. ☎2972 41 0669 **W:** www.am800wajzugun.com.ar **E:** info@am800wajzugun.com.ar

RN00) RIO NEGRO:
RN01) Av. 12 de Octubre 2421, (R8403AOH) San Carlos de Bariloche. ☎2944 42 2457 ▤2944 42 4035 **W:** www.lra3Obariloche.com.ar **E:** bariloche@radionacional.gov.ar - **FM:** 95.5MHz – **RN02)** Tucumán 1074, (R8332HQV) General Roca. ☎2941 430640 **W:** www.radioelvalle.com **E:** am640@radioelvalle.com - **FM:** 99.3MHz FM Color – **RN03)** Gral. Roca 365, 2° piso, (R8324BPG) Cipolletti. ☎299 477 6333 ▤299 477 6800 **W:** www.lu19lavozdelcomahue.com **E:** ricardodiluca@lu19lavozdelcomahue.com - **FM:** 102.9MHz FN 19 – **RN04)** Av. Alvaro Barros 1148, (R8500FFX) Viedma. ☎2920 42 7700 **W:** www.lu15radio.com.ar **E:** gerencia@lu15radio.com.ar - **Italian:** Sat 1500-1600. - **FM:** 94.3MHz – **RN05)** Remedios de Escalada 52, (R8336FED) Villa Regina. ☎2941 46 1102 **E:** info@lu16radio.com.ar - **FM:** Rio Negro 92.7MHz – **RN06)** Gral. Manuel Belgrano 710, (R8500FAP) Viedma. ☎2920 43 1697 **E:** viedma@radionacional.gov.ar - **FM:** 93.5MHz – **RN07)** Av. San Martín y Salta, (R8430) El Bolsón. ☎2944 49 2350 **E:** elbolson@radionacional.gov.ar - **FM:** 92.5MHz – **RN08)** Martín Coronado y José Hernández, (R8418) Ingeniero Jacobacci. ☎2940 43 2032 **E:** ingenierojacobacci@radionacional.gov.ar - **FM:** 93.5MHz – **RN09)** Casa de Tucumán 481, (R8361BKO) Luis Beltrán. ☎2946 48 0180 **W:** www.lamunicipalam1470.com.ar **E:** contacto@lamunicipalam1470.com.ar - **FM:** 93.9MHz – **RN10)** Pte. Julio A. Roca 570, (R8503BHL) General Conesa. ☎2931 49 8653 – **RN11)** Hipolito Yrigoyen y Remedios de Escalada, (R8536BBE) Valcheta. ☎2934 49 3283 - **FM:** 105.3MHz – **RN12)** Hipólito Yrigoyen 402, (R8534) Sierra Colorada. ☎2940 49 5176

SA00) SALTA:
SA01) Ruta Nal. 34,km. 1433, (A4560CJA) Tartagal. ☎3875 421600 **E:** tartagal@radionacional.gov.ar - **FM:** 92.3MHz – **SA02)** Dr.Carlos Pellegrini 715, 1° piso, (A4402FYO) Salta. ☎387 426 0109 **W:** www.radionacionalsalta.blogspot.com **E:** salta@radionacional.gov.ar - **FM:** 102.9MHz – **SA03)** Deán Funes 28 (⌂ Cas. 113), (A4400EDB) Salta. ☎387 431 3233 ▤387 431 1140 **W:** www.radiosalta.com **E:** contacto@radiosalta.com - **FM:** 96.9MHz – **SA05)** Gorriti 524, (A4560BRL) Tartagal. ☎3875 42 4141 **E:** lw2@fullnet.com.ar - **FM:** 96.9MHz – **SA06)** 9 de Julio 163, (A4530XBF) San Ramón de la Nueva Orán. ☎3878 42 1026 **W:** www.radiooran.com.ar **E:** oranradio@yahoo.com - **FM:** 90.9MHz

SC00) SANTA CRUZ:
SC01) Comodoro Py 342, Planta Baja, (Z9407BFH) Río Turbio. ☎2902 421131 **W:** www.lra18.com.ar **E:** rioturbio@radionacional.gov.ar - **FM:** 90.3MHz – **SC02)** Zapiola 25, (Z9400BCA) Río Gallegos ☎2966 42 0023 ▤2966 42 2608 **W:** www.lu12.com.ar **E:** lu_12am680@speedy.com.ar - **FM:** 92.9MHz – **SC03)** Av. San Martín 1114, (Z9311AVY) Governador Gregores ☎2962 49 1044 **E:** gobernadorgregores@radionacional.gov.ar - **FM:** 99.9MHz – **SC04)** Hermanos Vidal 261, (Z9405) El Calafate **W:** www.lu23..com.ar **E:** info@lu23.com.ar ☎2902 49 5686. - **FM:** 88.1MHz – **SC05)** Ramón Lista 36, (Z9050DLB) Puerto Deseado. ☎297 487 1211 **E:** lri200@deseado.com.ar - **FM:** 92.1MHz – **SC06)** Av. Pte. Julio A. Roca 823, 1°piso, (Z9400BAH) Río Gallegos **W:** www.santacruz.gov.ar/lu14 **E:** radiolu14@santacruz.gov.ar - **FM:** 92.1MHz – **SC07)** Saavedra 1318, (Z9040BQN) Perito Moreno. ☎2963 43 2233 **E:** peritomoreno@radionacional.gov.ar - **FM:** 93.9MHz

SE00) SANTIAGO DEL ESTERO:
SE01) 9 de Julio 390, (G4200DEH) Santiago del Estero. ☎385 421 3230 **W:** radiolv11.com.ar **E:** mensajeria@radiolv11.com.ar - **FM:** 88.1, 89.5MHz FM Total – **SE02)** Urquiza 392, 1° Piso, (G4300DHH) Santiago del Estero. ☎385 421 2565 **E:** santiagodelestero@radionacional.gov.ar - **FM:** 93.5MHz – **SE03)** Av. 25 de Mayo sur 69, (G3760AEA) Añatuya. ☎3844 42 1661 **W:** www.amradiosolidaridad.com.ar **E:** amsolidaridad@yahoo.com.ar

SF00) SANTA FE:
SF01) Mendoza 2430, 7° piso, (S3000CHB) Santa Fé. ☎342 453 3340. ▤342 452 8640 **W:** www.nacionalsantafe.ceride.gov.ar **E:** santafe@radionacional.gov.ar - **FM:** 94.9MHz – **SF02)** Córdoba 1843, (S2000AXC) Rosario. ☎341 410 0600. ▤34 1410 0637 **W:** www.lt8.com.ar **E:** lt8@lt8.com.ar - **FM:** 99.5MHz – **SF03)** 9 de Julio 3560, (S3002EXB) Santa Fé. ☎342 452 0187 **W:** www.lt10digital.com.ar **E:** correo@lt10digital.com - **FM:** 103.5MHz – **SF04)** 8 de Enero 2153, (S3000FHY) Santa Fé. ☎342 410 9999 **W:** www.lt9.ceride.gov.ar **E:** lt9@ceride.gov.ar - **FM:** 92.5MHz – **SF05)** Av. Pte. Juan Domingo Perón 8101, (S2010ACF) Rosario. ☎341 457 5415 **W:** www.rosario3.com **E:** radio2@rosario3.com - **FM:** 97.9MHz

– **SF06)** Córdoba 1331, 1° piso, (S2000AWS) Rosario. ☎341 440 2490 **E:** rosario@radionacional.gov.ar - **FM:** 104.5MHz – **SF07)** Balcarce 840, (S2000DNR) Rosario. ☎341 530 1190 **W:** www.lt3.com.ar **E:** lt3@lt3.com.ar - **FM:** 102.7MHz – **SF08)** Av. Casey 642, (S6400FJN) Venado Tuerto. ☎346 242 0777 **W:** www.radiovenadotuerto.com.ar **E:** lt29@radiovenadotuerto.com.ar - **FM:** 88.9MHz – **SF09)** Blvd. Lehman 245, (S2300GSC) Rafaela. ☎3492 50 1470 **W:** www.lt28rafaelaargentina.com or www.am1470.com.ar **E:** lt28@lt28rafaelaargentina.com - **FM:** 96.5MHz – **SF10)** Juan Chavarri 458, (S2147AUH) San Jenaro Norte. ☎3401 49 3069 **W:** www.radiolt23.com.ar **E:** lt23@co19set.com.ar - **FM:** 92.1MHz FM Concierto – **SF11)** Lucas Funes 1258, (S3560ETZ) Reconquista. ☎3482 42 8945 **W:** www.radioamanecer.com.ar **E:** radioamanecer@radioamanecer.com.ar - **FM:** 92.7MHz FM Amanecer – **SF12)** Patricio Diez 374, (S3560FUH) Reconquista. ☎3482 42 2005 **W:** www.radioobligado.com.ar **E:** info@am1440.com.ar - **FM:** 96.1MHz – **SF14)** Belgrano 470, (S2349AJJ) Suardi. ☎3562 47 7612 **W:** www.radiobelgranosuardi.com.ar **E:** belgrano@suardi.com.ar - **FM:** 104.9MHz – **SF17)** Corrientes 1172, 8° Piso "A", (S2000CTX) Rosario. ☎341 558 1090 **W:** www.amlibertad.com **E:** am1100@argentina.com – **SF18)** José Grevasio Artigas 253, (S2013ALA) Rosario ☎341 455 4827

SJ00) SAN JUAN:
SJ01) Mitre 50 Oeste, (J5402CXB) San Juan. ☎264 422 2344 ▤264-421 4950 **W:** www.lv1radiocolon.com.ar **E:** colonsa@speedy.com.ar - **FM:** 105.7, 106.3MHz – **SJ02)** Av. Ignacio de la Roza 293 Este, 2° Piso, (J5402DBC) San Juan. ☎264 421 4149. ▤264 421 4264 **E:** sanjuan@radionacional.gov.ar - **FM:** 91.3MHz – **SJ03)** Av. Belgrano y San Martin, (J5405AAA) Barreal. ☎2648 44 1260 - **FM:** FM Nuestra 103.5MHz – **SJ04)** Mendoza 452 Sur, (J5402GUJ) San Juan. ☎264 420 4028 **W:** www.lv5radiosarmiento.com.ar **E:** lv5@radiolv5.com - **FM:** 102.3, 103.7, 104.3, 104.7MHz– **SJ05)** General Paz 631, (J5460BBM) San José de Jáchal. ☎2647 42 0028. ▤2647 42 0561 **E:** jachal@radionacional.gov.ar - **FM:** 102.7MHz – **SJ06)** Mitre 11 este, (J5402CWA) San Juan, ☎264 427 2740. - **FM:** 105.1 – **SJ07)** Santa Fe 681 oeste, (J5402ACM) San Juan. ☎264 422 4646 ▤264 421 2443 **W:** www.am1020sj.com.ar **E:** info@am1020sj.com.ar - **FM:** 96.3

SL00) SAN LUIS:
SL01) Lavalle 291, Planta Alta, (D5732AEE) Villa Mercedes. ☎▤2657 42 4400 **W:** www. radiolv15.com.ar **E:** lv15@speedy.com.ar - **FM:** 95.5MHz FM Unica – **SL02)** Av. Lafinur 488, (D5700DCR) San Luís. ☎2652 43 1318 **E:** sanluis@radionacional.gov.ar - **FM:** 96.7MHz – **SL03)** Ruta Provincial No 3 s/n, (D5700) San Luís. ☎2652 45 8100 **W:** www.am940dimension.com **E:** info@am940dimension.com.ar - **FM:** Maxima FM 92.5//102.5MHz

TU00) TUCUMAN:
TU01) Lapride 530, (T4000IFL) San Miguel de Tucmán. ☎381 484 5100 **W:** www. lv12.com **E:** radiolv12@yahoo.com - **FM:** 99.1MHz – **TU02)** Mendoza 273, (T4000DAE) San Miguel de Tucumán. ☎381 497 5080 **W:** www.lv7.com.ar **E:** lv7@lv7.com.ar - **FM:** 102.7MHz – **TU03)** San Martín 241, (T4000CVE) San Miguel de Tucumán. ☎381 431 0131 ▤381 430 2409 **E:** tucuman@radionacional.gov.ar - **FM:** 93.3MHz – **TU04)** San Martín 610, Piso 6 "6", (T4000CVN) San Miguel de Tucumán. ☎381 400 0247 **W:** www.radio21tucuman.com.ar **E:** info@radio21tucuman.com.ar

TF00) TERRITORIO NACIONAL DE LA TIERRA DEL FUEGO, ANTARTIDA E ISLAS DEL ATLANTICO SUR:
TF01) Leonardo Rosales 490, (V9420CMJ) Río Grande. ☎2964 42 2176 ▤2964 43 0763 **W:** www.nacionalriogrande.com.ar **E:** direccion@nacionalriogrande.gov.ar - **FM:** 88.1MHz – **TF02)** Av. San Martín 331, (V9410BFD) Ushuaía. ☎2901 42 1670 **W:** www.rnacionalushuaia.com.ar **E:** ushuaia@radionacional.gov.ar - **FM:** 92.1MHz

FM in Buenos Aires: CF05) 93.7 R.Pop Nac – 96.7 FM Clásica Nacional – 98.7 Folclórica FM – **CF22)** 89.1 – **CF28)** 89.9 FM La Isla – **CF16)** 90.3 Eco Radio, 91.1 R.Abierta, 92.1 Mambo, 92.3 La Radio – **CF03)** 92.7 La Ciudad, 93.7, FM Federal – **CF09)** 94.3 Disney, 94.7 FM Palermo – **CF08)** 95.1 La Metro – **CF12)** 95.9 Rock & Pop, 96.3 R.Jai – **CF05)** 96.7 Clásica, 97.1 FM Europa, 97.3 Contacto FM, 97.9 R.Cultura – **CF13)** 98.3 Mega – **CF05)** 98.7 FM Folklorica, 99.1 Cadena 3 Argentina – **CF04)** 99.9 Cadena 100, 100.3 FM Cultural Musical – **CF07)** 100.7 Blue FM, 101.1 La Ciento Uno – **CF06)** 101.5 Pop Radio – **CF10)** 102.3 Aspen Classic – **CF02)** 103.1 R. Uno, 103.7 Amadadeus FM – **CF01)** 105.5 FM Hit – **CF11)** 106.3 R.Alfa, 106.7 X4, 107.3 Milenium, 107.9 Kabul Rock.
In the city area there are over 150 unlicensed LP FM stns, about 900 in the rest of the country.
FM in Córdoba: 88.1 FM Láser, 90.1FM Sur – **CO02)** 91.3 R.Nacional, 91.9 Hot FM – **CO05)** 92.3 R.Popular, 92.9 FM Logos, 93.7 R.Vital, 94.3 R.Universidad, 95.1 Radiocentro Bar, 96.1 FM Shopping Classics, 96.3 FM Norte, 96.5 R.Suquia, 96.9 CNI, 98.5 FM Latinoamericana, 99.1 FM Amistad, 99.3 FM Impacto – **CO03)** 99.7 Estacion Tierra – **CO05)** 100.5 FM Córdoba, 101.5 R.Maria – **CO01)** 102.3 Power 102, 102.7

FM Vision – **CO13)** 104.7, 105.5 FM Cielo, 107.5 Box Music Station, 107.9 FM Potencia

FM in Mar del Plata: 87.7 R.Urbana, 87.9 R. 87.9, 88.1 Graffiti FM, 88.3 LV del Puerto, 88.5 Onda Cero, 88.7 DeLaAzotea, 88.9 Mediterrano – **BA83)** 89.1 Red Impacto, 89.3 Láser, 89.7 d-Rock, 90.1 R. 90.1, 90.5 Kids, 90.7 Rural, 90.9 Concierto FM, 91.3 La Red, 91.7 K.L.A., 92.1 R. María, 92.3 Nova-92, 92.7 Líder – **BA09)** 93.3 Atlantica Latina, 93.7 Lisán, 94.1 R. 94-1 – **BA105)** 94.5 Latina, 94.9 Mega, 95.3 R. Uno, 95.9 Compacto, 96.5 Residencias, 96.9 Red 92, 97.1/ 97.5 R 97-1, 97.5 Popular, 97.6 R. 97.6, 97.7 Faro, 97.9 Estación 97, 98.1 R.Disney, 98.5 Brisas, 98.9 Rock'n Pop, 99.1 R. 99.1, 99.5 Más R., 99.9 Coast, 101.1 Cadena Musica, 100.3 Cadena Latina, 100.7 Del Sol, 101.1 Red Master, 101.5 Arena Sports, 101.7 La Ola, 101.9 Concierto, 102.1 Bristol, 102.1 R.10, 102.3 Municipal, 102.5 Nativa, 102.9 Ferimar, 102.9 Box, 102.9 La Nueva – **BA02)** 103.3 Universo, 103.7 Premium, 103.9 Canaan, 104.1 FM 104-1, 104.5 Via, 104.7 Urbana, 104.9 LV Amiga, 104.9 Cosmos, 105.1 Señal, 105.5 Inolviable, 105.9 Coast-Melody, 106.3 Five, 106.5 Argentian, 106.7 Sur, 106.9 Veronica, 107.1 Cielo, 107.5 Radioactiva, 107.7 R.107.7, 107.9 Trinidad.

FM in Rosario (Santa Fe): 89.5 R.Fisherton (CNN R.), 90.9 Uruguay – **SF05)** 92.3 FM Vida, 92.7 A-Z 927, 92.9 Radioactiva, 93.7 Cordial, 94.5 Latina, 95.5 Corazon, 96.6 Rio – **SF05)** 97.9 Vida, 98.5 Tango, 98.9 FM Si – **SF02)** 99.5 Estacion del Siglo, 100.5 Radiofónica, 100.9 Meridiano, 101.3 Hollywood – **SF02)** 102.3 FM No – **SF07)** 102.7 – **SF06)** 104.5 R.Nacional, 105.5 Tiempo Libre, 107.1 R.Universidad, 107.9 Cristal FM

FM in Santa Fe: 89.9 Federal, 90.7 Eclipse – **SF04)** 93.7 Láser, 93.1 Estacion Rock – **SF01)** 94.9 R.Nacional, 94.1 Santa Fe Capital, 101.7 Cielo, 104.3 Plenitud, 104.5 Sensación, 105.1 Hot 105, 105.5 News, 105.9 Ibiza FM – **SF03)** 107.3 FM X, 107.9 R.Antena (CNN Radio)

ARMENIA

LT: UTC +4h (27 Mar-30 Oct: +5h) — **Pop:** 3.2 million — **Pr.L:** Armenian — **E.C:** 50Hz, 220V — **ITU:** ARM

NATIONAL COMMISSION ON TV AND RADIO (NCTR)
✉ Saryan St. 22, 0002 Yerevan ☎ +374 10 539509 ▤ +374 10 539034 **E:** nctr@tvradio.am **W:** www.tvradio.am
LP: Chmn: Grigor Amalyan
NB. NCTR is the licensing body for broadcasting.

HAYASTANI HANRAYIN RADIO (Armenian Public Radio)
✉ A.Manoogian St. 5, 0025 Yerevan ☎ +374 10 551143 ▤ +374 10 554600 **E:** aa@arradio.am **W:** www.armradio.am
LP: DG: Armen Amiryan

MW	kHz	kW	Prgr
Yerevan	1395	150	AR1, F

FM: AR1: Yerevan 107.6MHz 1kW & nationwide network.
D.Prgr: AR1 (Armenian Pub. R. 1) 24h on FM; on MW the schedule varies and includes External Service (F) trs. **Local channel:** on Yerevan 103.8MHz.
External Service (Public R. of Armenia): see Int Radio section.

OTHER STATIONS
FM	MHz	kW	Location	Station
4)	90.2	1	Yerevan	Nor Radio
5)	90.7	1	Yerevan	Ararat FM
8)	100.6	1	Yerevan	R. Aurora
2)	102.0	2	Yerevan	AR R.Intercontinental
10)	102.4	1	Yerevan	FM-102.4
3)	103.5	1	Yerevan	R. Ardzaganq
1A)	104.1	1	Yerevan	Hay FM
7)	104.9	1	Yerevan	Russkoye R.
1B)	105.5	1	Yerevan	FM-105.5
9)	106.0	1	Yerevan	City FM
6)	107.0	1	Yerevan	R. Jazz

NB: Txs below 1kW not listed.
Addresses & other information:
1A,B) Pavstos Buzandi St. 1/3, 0010 Yerevan. **E:** haifm@haifm.am. 1A) via nationwide tx network; 1B) incl. rel. Europa+ (Russia). – **2)** Marash 13th St. 153, 0047 Yerevan. **E:** aa@arradio.am – **3)** Armeniak Armenakyan St. 250, 0047 Yerevan. **E:** ardzagank@ardzagank.com – **4)** A.Manoogian St. 5, 0025 Yerevan. – **5)** Acharyan St. 42, 0040 Yerevan. – **6)** Yeghvard Highway 1, 0054 Yerevan. – **7)** Khandjyan St. 13a, 0010 Yerevan. **E:** alfael@alfael.am – **8)** Nairi Zaryan St. 22, 0051 Yerevan. **E:** radio-aurora@mail.ru – **9)** Hyusisain Ave. 1, 0001 Yerevan. **E:** info@cityfm.am – **10)** Ovsepyan St. 95, 0047 Yerevan. Rel. RFI (France), Deutsche Welle (Germany).

Int. relays on MW (txs operated by Radio CJSC): Gavar 864/1350/1377kHz 1000kW. See Int Radio section.

ARUBA (Netherlands)

LT: UTC -4h — **Pop:** 103,065— **Pr.L:** Dutch (official), Papiamento, English, Spanish — **E.C:** 50+60Hz, 127/220V — **ITU:** ABW

DIRECTIE TELECOMMUNICATIE ZAKEN
✉ Rumbastraat 19, Oranjestad ☎ 297-582-6069 ▤ 297-582-5307
E: dirtelza@setarnet.aw **W:** www.dtz.aw

MW	kHz	kW	Station
2)	1270	2.5	Radio 1270 AM
3)	1320	2.5	Voz di Aruba

FM	MHz	kW	Station, location
12)	88.1	0.55	Mega 88FM, Oranjestad
11)	88.9	0.1	Bo Guia, Oranjestad
3)	89.9	0.25	Canal 90FM, San Nicolas
14)	90.7	0.45	Real FM
7)	91.5	0.125	Rumba 91.5 FM
10)	92.3	0.1	Latina 92.3, Oranjestad
1)	93.1	0.66	R. Victoria, Oranjestad (Rlg.)
12)	94.1	0.7	Hit 94 FM, Oranjestad
7)	95.1	0.325	Top 95.1 FM, Oranjestad
8)	96.5	0.2	Magic 96.5 FM, Oranjestad
6)	97.9	0.2	Easy 97.9 FM, Oranjestad
13)	98.9	0.2	Cool FM 98.9, Oranjestad
9)	99.9	0.5	R. Galactica 99.9 FM, Oranjestad
16)	100.9	0.85	Hit 100, Oranjestad
17)	101.7	0.7	Blizz FM
15)	105.3	0.5	Vision FM, Oranjestad
4)	106.7	0.44	R. Kelkboom, Oranjestad
5)	107.5	0.2	Mi FM, San Nicolas

Addresses and other information
1), Washington 23A (P.O.Box 5291) Oranjestad ☎/▤ +297 587 3444 **E:** radio.victoria@setarnet.aw **Mngr :** Nico Arts. Rlg: 24h in English, Spanish, Papiamento, Dutch, Creole, Tagalog and Cantonese www.radiovictoriaaruba.com– **2)** Balashi 62A, San Nicolas. MW: "Aruba Star" Mi FM: ☎+297 585 1075 ▤ +297 585 8622 Radio 1270 AM SM: J.A.C. Alders. Dir: F.A. Leauer.– **3)** Van Leeuwenhoekstraat 26, Oranjestad ☎+297 582 3355 ▤ +297 583 7340 **E:** info@canal90fm.aw **W** www.canal90fm.aw **MW:** in Dutch and Spanish – **4)** Bloemond 14, Paradera, Oranjestad ☎+297 582 1899 ▤ +297 584 3825 GM: Emile A.M. Kelkboom. 1000-0500 in Papiamento, Dutch, English, Spanish, Creole and Portuguese **W:** www.watapana-aruba.com **E:** radiokelkboom@setarnet.aw – **6)** Sabana Basora 31-D, Oranjestad ☎+297 593 3637 ▤ +297 585 2639 GM: W. Gesterkamp 24h in English, Papiamentu, Spanish and Dutch **E:** aruba@easyfm.com – **7)** Santa Cruz 110, Oranjestad ☎+297 585 9500 ▤+297 585 0951 **W:** www.top95fm.aw **W:** www.rumba91fm.com **8)** Caya G.F. Betico Croes 164 ☎+297 586 5353 ▤ +297 586 5354 24h **W:** www.magic965.com **E:** magic965fm@hotmail.com – **9)** Italiestraat 40, Oranjestad ☎+297 583 0999 ▤ +297 588 2536 Managing Director: Richard A. Arends Station Manager: Maikel Oduber **E:** gfmgalactica@gmail.com 24h in Papiamento, English and Dutch – **10)** L.G. Smith Boulevard 116 +297 583 3111 ▤ +297 583 3101 **11)** Tanki Flip 26B ☎+297 587 5000 ▤ +297 587 5889 **E:** radio.boguia@gmail.com – **12)** Caya Ernesto Petronia 68-A , Oranjestad Mega FM: ☎ +297 582 6888 ; Hit 94 FM: ☎ +297 583 9494 GM: Johnny Habibe Hit 94FM Stereo: 24h Top 40 radio in Dutch, Papiamento and Spanish **W:** www.hit94fm.com **E:** hit94@setarnet.aw ; Mega 88FM ☎+297 582 6888 ▤ +297 582 0494 **E:** mail@mega88fm.com **W:** www.mega88fm.com – **13)** Sabana Basora 31-D + 297 593 3637 ▤+ 297 583 3101 **W:** www.coolaruba.com **E:** coolaruba@gmail.com – **14)** John G. Emanstraat 124A ▤+ 297 583 1434 ▤+ 297 583 1515 **W:** www.real907.com **15)** Cumana 20, Oranjestad ☎+ 297 583 3939 ▤+ 297 582 5477 **W:** www.visionaruba.comb **16)** South Beach Mall, Palm Beach 51 lok. D-7 ☎+297 582 2384 **W:** http://hits100.fm **E:** contact@hits100.fm and inf@hits100.fm – **17)** Piedra Plat 44 C-D ☎+297 585 9031 **W:** www.blizz.aw **E:** info@blizz.aw

ASCENSION ISLAND (UK)

LT: UTC — **Pop:** 900 — **Pr.L:** English — **E.C:** 50Hz, 220V — **ITU:** ASC

VOLCANO RADIO (USAF)
✉ Ascension Radio Station, Ascension AAF, P.O. Box 4235, Patrick AFB, FL 32925-0235, USA.
FM: 98.7MHz 0.4kW 24h.
NB: MW service ZD8VR on 1602kHz currently inactive

BBC ATLANTIC RELAY STATION
✉ English Bay, Ascension Island, So. Atlantic.
Local Sce: FM: 93.2MHz 15W (24h relay of BBCWS in English plus occ. local prgrs).

See International section for details of SW relays.

BFBS: BFBS 2 on 105.3MHz and 107.3MHz
SAINT FM: 91.4MHz 25W, internet feed from St Helena

AUSTRALIA

L.T: See World Time Table. DST (where applicable): 3 Oct 10–3 Apr 11, 2 Oct 11-1 Apr 12 — **Pop:** 22 million — **Pr.L:** English — **E.C:** 50Hz, 230V — **ITU:** AUS

ABORIGINAL RESOURCE & DEVELOPMENT SERVICES
✉ Box 1671, Nhulunbuy NT 0881 ☎+61 (08) 8987 3910 🖷 +61 (08) 8982 3499 E: www.ards.com.au **E:** dale@ards.com.au
L.P.: Radio Services Manager: Dale Chesson
MW: 1530kHz Humpty Doo 2kW
SW: 5050kHz VKD963 Humpty Doo 400W (G.C. 131.05E 12.34 S) (currently off air)
F.PI: 1530kHz Nhulunbuy 400W, 1611kHz VKD883 Milingimbi, 1611kHz VKD884 Groote Eylandt, 1620kHz VKD885 Galiwin Ku (Elcho Island), 1629kHz VKD886 Gapuwiyak

AUSTRALIAN BROADCASTING CORP. (ABC)
HQ: Ultimo Centre, 700 Harris Str, Ultimo, NSW 2007.(✉ GPO Box 9994, Sydney NSW 2001) ☎+61 (02) 9333 1500 🖷 +61 (02) 9333 5305
MW: N = R. National, R = Regional, M = Metropolitan, P = Parliamentary & News Netw. **Call letters:** 2 = NSW (exc. Canberra = A.C.T.), 3 = Victoria, 4 = Queensland, 5 = So. Australia, 6 = We. Australia, 7=Tasmania, 8=Northern Territory.

MW Call	kHz	kW	Netw	Location
36) 6DL	531	10	R	Dalwallinu
37) 4QL	540	10	R	Longreach
38) 2CR	549	50	R	Orange (Cumnock)
39) 6WA	558	50	R	Wagin
40) 4JK	567	10(d)	R	Julia Creek
36) 6MN	567	0.1	R	Newman
36) 6PN	567	0.1	R	Pannawonica
36) 6PU	567	0.1	R	Paraburdoo
36) 6TP	567	0.1	R	Tom Price
2) 2RN	576	50	N	Sydney
6) 6PB	585	10	P	Perth
7) 7RN	585	10	N	Hobart
41) 3WV	594	50	R	Horsham
2) 2RN	603	10(d)	N	Nowra
37) 4CH	603	10(d)	R	Charleville
10) 6PH	603	2	R	Port Hedland
4) 4QR	612	50	M	Brisbane
6) 6RN	612	10	N	Dalwallinu
3) 3RN	621	50	N	Melbourne
2) 2PB	630	10	P	Sydney
40) 4QN	630	50	R	Townsville (Brandon)
13) 6AL	630	5	R	Albany
7) 7RN	630	0.4	N	Queenstown
14) 4MS	639	2	R	Mossman
15) 5CK	639	10	R	Port Pirie (Crystal Brook)
8) 8RN	639	2	N	Katherine
17) 2NU	648	10	R	Tamworth (Manilla)
18) 6GF	648	2	R	Kalgoorlie
38) 2BY	657	10(d)	R	Byrock
8) 8RN	657	2	N	Darwin
1) 2CN	666	5	M	Canberra ACT
20) 2CO	675	10	R	Albury (Corowa)
36) 6BE	675	5	R	Broome
21) 2KP	684	10	R	Kempsey (Smithtown)
22) 6BS	684	5	R	Busselton
8) 8RN	684	1	N	Tennant Creek
15) 5SY	693	2(d)	R	Streaky Bay
2) 2BL	702	50	M	Sydney
6) 6KP	702	10	R	Karratha
23) 4QW	711	10(d)	R	Roma/St.George
25) 2ML	720	0.4	R	Murwillumbah
2) 2RN	720	0.05	N	Armidale
26) 3MT	720	2(d)	R	Omeo
14) 4AT	720	4	R	Atherton
6) 6WF	720	50	M	Perth
5) 5RN	729	50	N	Adelaide
25) 2NR	738	50	R	Grafton
6) 6MJ	738	5(d)	R	Manjimup
23) 4QS	747	10	R	Toowoomba (Dalby)
7) 7PB	747	10(d)	P	Hobart (5.5kW night)
8) 8JB	747	0.2	R	Jabiru
21) 2TR	756	2(d)	R	Taree
3) 3RN	756	10(d)	N	Wangaratta
3) 3LO	774	50	M	Melbourne

MW Call	kHz	kW	Netw	Location
28) 8AL	783	2	R	Alice Springs
4) 4RN	792	25	N	Brisbane
14) 4QY	801	2	R	Cairns
29) 2BA	810	10	R	Bega
6) 6RN	810	20	N	Perth
17) 2GL	819	10	R	Glen Innes
36) 6KW	819	5	R	Kununurra
26) 3GI	828	10	R	Sale (Longford)
36) 6GN	828	10	R	Geraldton
30) 4RK	837	10	R	Rockhampton (Gracemore)
36) 6ED	837	1	R	Esperance
1) 2RN	846	10	N	Canberra
36) 6CA	846	2.5	R	Carnarvon
31) 4QB	855	10(d)	R	Pialba
31) 4QO	855	10	R	Eidsvold
6) 6DB	873	2	R	Derby
5) 5AN	891	50	M	Adelaide
4) 4PB	936	10	P	Brisbane
7) 7ZR	936	10(d)	M	Hobart
5) 5PB	972	2	P	Adelaide
3) 3RN	990	0.5	R	Albury-Wodonga
8) 8GO	990	0.5	R	Gove (Nhulunbuy)
32) 2NB	999	2(d)	R	Broken Hill
36) 6WH	1017	0.5	R	Wyndham
3) 3PB	1026	10	P	Melbourne
33) 2UH	1044	2(d)	R	Muswellbrook
14) 4WP	1044	0.5	R	Weipa
6) 6BR	1044	1	R	Bridgetown
14) 4TI	1062	2	R	Thursday Island
5) 5MV	1062	2	R	Renmark/Loxton
2) 2RN	1098	0.2	N	Goulburn
6) 6PNN	1152	10(d)	P	Busselton
6) 6RN	1152	10(d)	N	Manjimup
34) 5PA	1161	10(d)	R	Naracoorte
24) 7FG	1161	1(d)	R	Fingal
6) 6XM	1188	2	R	Exmouth
6) 6NM	1215	0.5	R	Northam
6) 6RN	1224	5	N	Busselton
11) 2NC	1233	10	M	Newcastle
6) 6RN	1269	5	N	Busselton
6) 6RN	1296	10	N	Wagin
5) 5RN	1305	2	N	Renmark/Loxton
38) 2LG	1395	0.2	R	Lithgow
2) 2RN	1431	2	N	Wollongong
2) 2PB	1458	2	P	Newcastle
5) 5MG	1476	1(d)	R	Mt. Gambier
2) 2RN	1485	0.1	N	Wilcannia
40) 4HU	1485	0.05	R	Hughenden
15) 5LN	1485	0.2	R	Port Lincoln
2) 2RN	1512	10	N	Newcastle
30) 4QD	1548	50	R	Emerald
12) 4GM	1566	0.2	R	Gympie
32) 2WA	1584	0.1	R	Wilcannia
15) 5WM	1584	0.05	R	Woomera
7) 7SH	1584	0.1	R	St. Helens
29) 2CP	1602	0.05	R	Cooma
16) 3WL	1602	0.25	R	Warrnambool
15) 5LC	1602	0.2	R	Leigh Creek South

FM stations (txs of greater than 1kW)
Networks: N=Radio National, R/M=Regional or Metropolitan Network, FM=Fine Music Network, JJJ=Triple J Network (alternative)

FM	Area	State	N	R/M	FM	JJJ
5)	Adelaide	SA			103.9	105.5
5)	Adel. Foothills	SA			97.5	95.9
3)	Alexandra	VIC	104.5	102.9		
17)	Armidale	NSW		101.9	103.5	101.1
3)	Bairnsdale	VIC	106.3			
42)	Ballarat	VIC		107.9	105.5	107.1
29)	Batemans Bay	NSW	105.1	103.5	101.9	
2)	Bega/Cooma	NSW	100.9		99.3	100.1
12)	Bendigo	VIC		91.1	92.7	90.3
29)	Bombala	NSW		94.1		
2)	Bourke	NSW	101.1			
4)	Brisbane	QLD			106.1	107.7
6)	Broken Hill	NSW	102.9		103.7	102.1
6)	Broome	WA	107.7			
6)	Bunbury	WA			93.3	94.1
24)	Burnie	TAS		102.5		
14)	Cairns	QLD	105.1	106.7	105.9	107.5
14)	Cairns North	QLD	93.9	95.5	94.7	97.1
1)	Canberra	ACT			102.3	101.5
6)	Cen.Agricult	WA			98.9	98.1
3)	Cen.Table'nds	NSW	104.3		102.7	101.9
38)	Cen. Western	NSW	107.9	107.1	105.5	102.3

	Area	State	N	R/M	FM	JJJ
4)	Darling Downs	QLD	105.7		107.3	104.1
8)	Darwin	NT		105.7	107.3	103.3
2)	Deniliquin	NSW	99.3			
7)	Devonport E	TAS		100.5		
2)	Dubbo City	NSW		95.9		
4)	Emerald	QLD	93.9		90.7	
6)	Esperance	WA	106.3		104.7	
6)	Geraldton	WA	99.7		94.9	98.9
8)	Glen Innes	NSW	105.1			
11)	Gold Coast	QLD	90.1	91.7	88.5	97.7
12)	Goulburn V.	VIC		97.7	96.1	94.5
25)	Grafton/Kemp.	NSW	99.5	92.3	97.9	91.5
31)	Gympie	QLD	96.9	95.3	93.7	
9)	Hay	NSW	88.9	88.1		
7)	Hobart	TAS			93.9	92.9
27)	Illawarra	NSW		97.3	95.7	98.9
2)	Jerilderie	NSW	94.1			
6)	Kalgoorlie	WA	97.1		95.5	98.7
28)	Katherine	NT		106.1		
5)	Keith	SA	96.9			
24)	King Island	TAS		88.5		
3)	Latrobe Valley	VIC		100.7	101.5	96.7
24)	Lileah	TAS	89.7	91.3		
25)	Lismore	NSW	96.9	94.5	95.3	96.1
4)	Longreach	QLD	99.1			
19)	Mackay	QLD	102.7	101.1	97.9	99.5
21)	Manning River	NSW	97.1	95.5	98.7	96.3
4)	Meandarra	QLD	104.3			
3)	Melbourne	VIC			105.9	107.5
35)	Mildura	VIC	105.9	104.3	102.7	101.1
4)	Mission Beach	QLD	90.9	89.3		
4)	Monto	QLD	101.9			
30)	Moranbah	QLD	106.5	104.9		
4)	Mossman	QLD	90.1			
4)	Mount Isa	QLD	107.3	106.5	101.7	104.1
5)	Mt Gambier	SA	103.3		104.1	102.5
3)	Murray Valley	VIC		102.1	103.7	105.3
20)	Murrumbidgee	NSW	98.9	100.5	97.3	96.5
2)	Muswellbrook	NSW		105.7		
12)	Nambour	QLD		90.3	88.7	89.5
6)	Narrogin	WA			92.5	
7)	NE Tas.	TAS	94.1	91.7	93.3	90.9
2)	Newcastle	NSW			106.1	102.1
3)	Nhill	VIC	95.7			
6)	Perth	WA			97.7	99.3
6)	Port Hedland	WA	95.7			
16)	Portland	VIC	98.5	96.9		
5)	Renmark	SA			105.1	101.9
4)	Rockhampton	QLD	103.1		106.3	104.7
6)	Roebourne	WA	107.5			
4)	Roma	QLD	107.3	105.7	97.7	
5)	Roxby Downs	SA	101.9	102.7	103.5	
4)	Salmon Gums	WA	100.7			
6)	S. Agricultural	WA	96.9		94.5	92.9
6)	South'n Cross	WA	107.9	106.3		
23)	South'n Downs	QLD	106.5	104.9	101.7	103.3
5)	Spencer Gulf N	SA	106.7		104.3	103.5
5)	Streaky Bay	SA	100.9			103.3
20)	SW Slopes	NSW	89.1	89.9	88.3	90.7
2)	Sydney	NSW			92.9	105.7
2)	Tamworth	NSW	93.9		103.1	94.7
4)	Townsville	QLD	104.7		101.5	105.5
5)	Tumby Bay	SA	101.9			
20)	Upper Murray	VIC		106.5	104.1	103.3
6)	Upper Namoi	NSW	100.7	99.1	96.7	99.9
3)	Warrnambool	VIC	101.7		92.1	89.7
8)	Western Vic.	VIC	92.5	94.1	93.3	94.9
31)	Wide Bay	QLD	100.9	100.1	98.5	99.3
4)	Winton	QLD	107.9			
5)	Wirrulla	SA	107.3			
5)	Wudinna	SA	107.7			105.3
6)	Wyndham	WA				98.9
3)	Yackandandah	VIC	99.9			

NB: Parliamentary News Network: 89.1 Emerald QLD (4.5), 89.3 Horsham VIC (20), 89.5 Bendigo VIC (10d), 90.5 Burnie TAS (1d), 90.9 Illawarra NSW (150d), 91.3 Warrnambool VIC (3.2d), 91.5 SW Slopes/E Riverina NSW (80), 91.5 Tumby Bay SA (2), 91.7 Tamworth NSW (10), 91.7 Western Victoria (80d), 92.1 Southern Agricultural WA (80), 92.5 NE Tasmania (192d), 93.5 Inverell NSW (10d), 93.9 Renmark SA (150d), 94.3 Townsville QLD (92d), 94.5 Gympie QLD (20d), 94.7 Manning River NSW (5d), 94.9 Port Hedland WA (2), 95.1 Latrobe Valley VIC (200d), 95.7 Gold Coast (26d), 95.9 Murray Valley VIC (20d), 96.3 Cairns North QLD (10d), 96.3 Wagin WA (5), 97.7 Portland VIC (2.6d), 98.1

Murrumbidgee NSW (100), 98.5 Lismore NSW (100d), 99.7 Central Agricultural WA (80), 100.3 Mildura VIC (150d), 100.3 Kalgoorlie WA (6), 100.5 Batemans Bay NSW (40d), 100.9 Deniliquin NSW (2d), 101.1 Cairns QLD (100), 101.3 Geraldton WA (10), 102.1 Devonport East TAS (1.2), 102.5 Darwin (32d), 102.7 Armidale NSW (4), 102.7 Spencer Gulf North SA (70), 103.1 Esperance WA (5), 103.9 Canberra ACT (80), 104.1 Alice Springs NT (1), 104.3 Mackay QLD (100d), 104.5 Broken Hill NSW (4), 104.7 Colac VIC (10d), 104.9 Muswellbrook NSW (16), 104.9 Mount Isa QLD (1), 105.3 Katherine NT (1), 105.5 Rockhampton QLD (80), 105.7 Mt Gambier SA (240d), 106.3 Central West Slopes NSW (220d), 106.9 Broome WA (2), 107.9 Bairnsdale VIC (2d).

Reports for Radio National, Parliament, ABC-FM and Triple J should go to the capital city ABC office in that state (Addresses 1-8)

ABC regional addresses:
1) GPO Box 365, Canberra ACT 2601 – **2)** PO Box 487 Sydney NSW 2001 – **3)** GPO Box 1686, Melbourne VIC 3001 – **4)** GPO Box 293, Brisbane QLD 4001 – **5)** GPO Box 1419H, Adelaide SA 5001 – **6)** GPO Box 190D Perth WA 6001 – **7)** GPO Box 9994, Hobart TAS 7001 – **8)** PO Box 9994, Darwin NT 0800 – **9)** 100 Fitzmaurice St, Wagga Wagga NSW 2650 – **10)** PO Box 387 Port Hedland WA 6721 – **11)** PO Box 217, Mermaid Beach QLD 4217 – **12)** PO Box 1922, Shepparton VIC 3630 – **13)** PO Box 489 Albany WA 6330 – **14)** PO Box 932 Cairns QLD 4870 – **15)** PO Box 289 Port Pirie SA 5540 – **16)** Kepler St., Warrnambool VIC 3280 – **17)** PO Box 558 Tamworth NSW 2340 – **18)** PO Box 125 Kalgoorlie WA 6430 – **19)** PO Box 127, Mackay QLD 4740 – **20)** PO Box 321 Albury NSW 2640 – **21)** PO Box 76 West Kempsey NSW 2440 – **22)** PO Box 242 Bunbury WA 6230 – **23)** PO Box 358 Toowoomba QLD 4350 – **24)** PO Box 201 Launceston TAS 7250 – **25)** PO Box 435 Grafton NSW 2460 – **26)** PO Box 330 Sale VIC 3850 – **27)** cnr Kembla & Market Sts, Wollongong NSW 2520, – **28)** PO Box 1144 Alice Springs NT 0871 – **29)** PO Box 336 Bega NSW 2550 – **30)** PO Box 911 Rockhampton QLD 4700 – **31)** PO Box 376 Maryborough QLD 4650 – **32)** PO Box 315 Broken Hill NSW 2880 – **33)** 47 Newcomen St, Newcastle NSW 2300 – **34)** PO Box 448 Mount Gambier SA 5290 – **35)** PO Box 5051, Mildura VIC 3502 – **36)** PO Box 387 Geraldton WA 6530 – **37)** PO Box 318 Longreach QLD 4730 – **38)** PO Box 863 Orange NSW 2600 – **39)** PO Box 242 Bunbury WA 6230 – **40)** PO Box 694 Townsville QLD 4810 – **41)** PO Box 506 Horsham VIC 3400 – **42)** PO Box 637 Bendigo VIC 3550.

EXTERNAL SERVICE: Radio Australia
See International Broadcasting section.

DAB: Now rolled out in capital cities: Channel 9A 202.928 MHz, 9B 204.64 MHz, 9C 206.352 MHz. All 50kW. Relays existing stations and some extra programming controlled by existing stations. All programming also available at http://www.digitalradioplus.com.au/. Sydney NSW, Melbourne VIC, Brisbane QLD all 9A, 9B and 9C. Adelaide SA and Perth WA 9B and 9C.

NORTHERN TERRITORY SHORTWAVE SERVICE
✉ Box 9994, Darwin NT 0801

SW: VL8A Alice Springs: 2310kHz (0830-2130), 4835kHz (2130-0830), 3230kHz (alt. freq) – **VL8T Tennant Creek:** 2325kHz (0830-2130), 4910kHz (2130-0830), 3315kHz (alt. freq) – **VL8K Katherine:** 2485kHz (0830-2130), 5025kHz (2130-0830), 3370kHz (alt. freq). Programming has been known to run over designated times.
D. Prgr. in English & Aboriginal languages:. Prgrs produced by Top End Aboriginal Bush Broadcasting Assoc. (TEABBA): VL8K: MF 2045-2330, 0730-0830. Rel. of ABC Alice Springs. All txs 50kW

COMMERCIAL RADIO AUSTRALIA
✉ Level 5, 88 Foveaux Street, Surry Hills NSW 2010 Australia ☎ +61 2 9281 6577 🖷 +61 2 9281 6599 **CEP:** Joan Warner.

Abbreviations: N-1: News on the hour. N-2: News on the half hour. N-3: News on the hour and half hour. The numeral preceding the call letters indicates the state: 2=New South Wales,. 3=Victoria, 4=Queensland, 5=South Australia, 6=Western Australia; 7=Tasmania, 8=Northern Territory.

News: Additional newscasts are often carried during breakfast and drive times. t=translator (relays main station).

MW	Call	kHz	kW	Location
1)	2PM	531	5(d)	Port Macquarie
2)	3GG	531	5(d)	Warragul
3)	4KZ	531	5(d)	Innisfail
4)	7SD	540	5(d)	Scottsdale
5)	4AM	558	5(d)	Atherton
6)	4GY	558	5(d)	Gympie
7)	7BU	558	2	Burnie
8)	2BH	567	0.5	Broken Hill
164)	6EL	621	2	Bunbury

MW	Call	kHz	kW	Location		MW	Call	kHz	kW	Location
133)	2HC	639	5(d)	Coffs Harbour		82)	3GV	1242	5(d)	Sale
33)	4CC(t)	666	2(d)	Biloela		85)	4AK	1242	2	Toowoomba
103)	4LM	666	2	Mount Isa		84)	5AU	1242	2(d)	Port Augusta
160)	6LN	666	1	Carnarvon		86)	2DU	1251	2	Dubbo
105)	3AW	693	5(d)	Melbourne		32)	3SR	1260	2	Shepparton
9)	4KQ	693	5(d)	Brisbane		88)	6KA	1260	1	Karratha
3)	4KZ(t)	693	0.5	Tully		89)	2SM	1269	5	Sydney
103)	4LM(t)	693	0.5	Cloncurry		90)	3EE	1278	5	Melbourne
31)	6FMS	747	1	Exmouth		91)	2TM	1287	2	Tamworth
131)	6SE	747	5(d)	Esperance		32)	3BT	1314	5(d)	Ballarat
166)	6TZ	756	2	Margaret River		40)	5DN	1323	2	Adelaide
10)	2EC	765	5(d)	Bega		98)	3SH	1332	2	Swan Hill
73)	4GC(t)	765	0.5	Hughenden		99)	4BU	1332	5(d)	Bundaberg
134)	5CC	765	2	Port Lincoln		102)	2LF	1350	5(d)	Young
88)	6SAT	765	0.1	Paraburdoo		37)	4WK(t)	1359	0.25	Toowoomba
88)	6SAT	765	0.1	Tom Price		104)	2GN	1368	2	Goulburn
147)	8HOT(t)	765	0.5	Katherine		105)	3MP	1377	5(d)	Melbourne
11)	4TO	774	5(d)	Townsville		106)	5AA	1395	5(d)	Adelaide
13)	6VA	783	2	Albany		108)	2PK	1404	2	Parkes/Forbes
14)	5RM	801	2	Berri		5)	4AM(t)	1422	1(d)	Port Douglas
73)	4GC	828	1	Charters Towers		111)	2MG	1449	5(d)	Mudgee
16)	7XS	837	0.5	Queenstown		32)	3ML	1467	2	Mildura
17)	4EL	846	5(d)	Cairns		115)	4ZR	1476	2	Roma
18)	4GR	864	2	Toowoomba		116)	2AY	1494	2	Albury
19)	6AM	864	2	Northam		117)	2BS	1503	5(d)	Bathurst
21)	2GB	873	5	Sydney		142)	6BAY	1512	5	Geraldton
22)	3YB	882	2(d)	Warrnambool		119)	2QN	1521	2	Deniliquin
24)	4BH	882	5(d)	Brisbane		120)	2VM	1530	2	Moree
23)	6PR	882	10	Perth		121)	2RE	1557	2	Taree
25)	2LM	900	5(d)	Lismore		122)	3NE	1566	5(d)	Wangaratta
107)	2LT	900	5(d)	Lithgow		10)	2EC(t)	1584	0.2	Narooma
58)	6BY	900	2	Bridgetown		33)	4CC(t)	1584	0.5	Rockhampton
27)	7AD	900	2	Devonport		153)	4VL(t)	1584	0.2	Cunnamulla
28)	8HA	900	2	Alice Springs						
29)	2XL	918	2	Cooma		**FM stations** (1kW and higher):				
153)	4VL	918	2/2.5	Charleville		FM	Call	MHz	kW	Location
164)	6NA	918	2	Narrogin		117)	2BS(t)	88.1	2(d)	Burraga
32)	3UZ	927	5	Melbourne		3)	4KZ(t)	88.5	1(d)	Mission Beach
33)	4CC	927	5(d)	Gladstone		165)	4RGC	88.5	1	Mossman
69)	4HI(t)	945	1(d)	Dysart		41)	3HFM	88.9	20(d)	Hamilton
36)	2UE	954	5	Sydney		56)	4KRY	89.1	15	Kingaroy
17)	4EL(t)	954	0.35	Gordonvale		117)	2BS(t)	89.3	1(d)	Blayney
38)	2RG	963	5(d)	Griffith		63)	7LAA	89.3	5(d)	Launceston
37)	4WK	963	5(d)	Warwick		46)	4TAB	89.7	5(d)	Beaudesert
93)	5SE	963	5(d)	Mt. Gambier		71)	5CCC	89.9	6(d)	Port Lincoln
164)	6TZ	963	2	Bunbury		48)	7EXX	90.1	5(d)	Launceston
86)	2DU(t)	972	0.3	Cobar		69)	4HIT(t)	90.3	1	Blackwater
39)	2MW	972	5(d)	Murwillumbah		130)	5SSA(t)	90.3	2(d)	Adelaide Foothills
112)	2NM	981	5(d)	Muswellbrook		154)	4SEA	90.9	25(d)	Gold Coast
41)	3HA	981	2	Hamilton		170)	8SAT	90.9	4(d)	Maitland SA
42)	6KG	981	2	Kalgoorlie		168)	4MCY	91.1	10(d)	Nambour
43)	4RO	990	5(d)	Rockhampton		150)	4RBL	91.1	1.5	Tara
45)	2ST	999	5(d)	Nowra		123)	2MAC	91.3	1	Campbelltown
46)	4TAB	1008	10(d)	Brisbane		120)	2NOW(t)	91.3	1	Lightning Ridge
49)	2KY	1017	5	Sydney		32)	3SRR(t)	91.3	1.2(d)	Mt. Buller
139)	4AA	1026	5(d)	Mackay		96)	4HIT	91.3	5	Moranbah
52)	6NW	1026	2	Port Hedland		148)	3PTV	91.5	56(d)	Melbourne
53)	5AU	1044	2	Port Pirie		29)	2SKI(t)	91.7	1	Bombala
54)	2CA	1053	5(d)	Canberra		45)	2ST	91.7	2(d)	St Georges Basin
55)	3EL	1071	5(d)	Maryborough		17)	4HOT(t)	91.7	1	Mossman
56)	4SB	1071	2	Kingaroy		52)	6HED	91.7	2	Port Hedland
151)	6WB	1071	2	Katanning		55)	3BDG	91.9	120(d)	Bendigo
57)	2MO	1080	2	Gunnedah		18)	4RGD	91.9	2	Warwick
167)	6IX	1080	2	Perth		15)	4SEE	91.9	10(d)	Nambour
59)	2EL	1089	5(d)	Orange		169)	5ADL	91.9	20(d)	Adelaide
60)	3WM	1089	5(d)	Horsham		150)	4BRZ	92.1	5(d)	Beaudesert
61)	4LG	1098	2	Longreach		16)	7AUS	92.1	20(d)	Queenstown/Zeehan
62)	6MD	1098	2	Merredin		29)	2XL(t)	92.5	1	Bombala
63)	7LA	1098	5(d)	Launceston		96)	4CCA(t)	92.5	1	Mossman
156)	3AK	1116	5(d)	Melbourne		155)	4GLD	92.5	25(d)	Gold Coast
65)	4BC	1116	6.3/17(d)	Brisbane		86)	2ZOO	92.7	10	Dubbo
135)	6MM	1116	2	Mandurah		15)	4SSS	92.7	10(d)	Nambour
113)	5MU	1125	5(d)	Murray Bridge		29)	2SKI(t)	92.9	1	Thredbo
66)	2AD	1134	2(d)	Armidale		91)	2TTT	92.9	20(d)	Tamworth
67)	3CS	1134	5(d)	Colac		120)	2VM(t)	92.9	1	Lightning Ridge
164)	6TZ(t)	1134	2	Collie		44)	6PPM	92.9	40(d)	Perth
68)	2HD	1143	2	Newcastle		143)	2GEE	93.1	10	Mudgee
69)	4HI	1143	5(d)	Emerald		70)	2WZD	93.1	80	Wagga Wagga
70)	2WG	1152	2	Wagga Wagga		154)	4RGB	93.1	3(d)	Bundaberg
30)	4FC	1161	2	Maryborough		14)	5RIV	93.1	10(d)	Renmark/Loxton
72)	2CH	1170	5	Sydney		163)	2DBO	93.5	10	Dubbo
75)	2NZ	1188	2	Inverell		1)	2PM(t)	93.5	3(d)	Port Macquarie
80)	2CC	1206	5(d)	Canberra		104)	2SNO	93.5	40	Goulburn
78)	2GF	1206	5(d)	Grafton		35)	3BBO	93.5	120(d)	Bendigo
69)	4HI(t)	1215	0.25	Moranbah		43)	4ROK	93.5	1(d)	Gladstone
						41)	3HFM(t)	93.7	2	Portland

FM	Call	MHz	kW	Location
87)	3SUN(t)	93.7	1(d)	Alexandra/Eildon
87)	3SUN(t)	93.7	1.2(d)	Mt. Buller
87)	3SUN(t)	93.7	1(d)	Yea
150)	4RBL	93.7	4	Tenterfield NSW
47)	6PER	93.7	40(d)	Perth
102)	2LFF	93.9	40	Young
101)	3BAY	93.9	55(d)	Geelong
99)	4RUM	93.9	3.2(d)	Bundaberg
83)	2BDR	94.1	1(d)	Falls Creek VIC
2)	3SEA(t)	94.3	7	Warragul
60)	3WWM(t)	94.5	2	Nhill/Lawloit
79)	6MIX	94.5	40(d)	Perth
29)	2SKI(t)	94.7	2	Jindabyne
69)	4HIT	94.7	5	Emerald
45)	2WSK	94.9	50(d)	Nowra
141)	4MIX	94.9	50(d)	Ipswich
88)	6KAN	94.9	5	Katanning
75)	2GEM	95.1	10	Inverell
43)	4RGK	95.1	1(d)	Gladstone
115)	4ROM	95.1	1	Roma
145)	2PTV	95.3	150(d)	Sydney
32)	3SRR	95.3	100(d)	Shepparton
162)	3YFM	95.3	20(d)	Warrnambool
13)	6AAY	95.3	50(d)	Albany
170)	8SAT	95.3	2(d)	Karoonda SA
108)	2ROK	95.5	10	Parkes/Forbes
101)	3CAT	95.5	55(d)	Geelong
170)	8SAT	95.5	3(d)	Kingscote (SA)
151)	6BUN	95.7	40(d)	Bunbury
73)	4CHT	95.9	1.5	Charters Towers
12)	2ONE	96.1	5	Katoomba
6)	4NNN	96.1	5(d)	Gympie
150)	4RBL	96.1	1	Weipa
93)	5SEF	96.1	20	Mount Gambier
125)	6NOW	96.1	40(d)	Perth
29)	2XL(t)	96.3	2(d)	Jindabyne
8)	2HIL	96.5	4(d)	Broken Hill
95)	2UUL	96.5	40(d)	Wollongong
142)	6GGG	96.5	30(d)	Geraldton
19)	6NAM	96.5	10	Northam
170)	8SAT	96.5	4(d)	Pinnaroo (SA)
40)	5ADD(t)	96.7	2(d)	Adelaide Foothills
161)	2SYD	96.9	150(d)	Sydney
118)	3SUN	96.9	100(d)	Shepparton
14)	5RIV	97.1	2.5(d)	Morgan
159)	4BFM	97.3	12	Brisbane
150)	4RBL	97.3	1	Inglewood
135)	6CST	97.3	5(d)	Mandurah
57)	2GGG	97.5	20(d)	Gunnedah
29)	2SKI	97.7	50(d)	Cooma
170)	8SAT	97.7	3(d)	Coonalpyn SA
136)	3RMR	97.9	12(d)	Mildura
5)	4AMM	97.9	5(d)	Atherton
42)	6KAR	97.9	6	Kalgoorlie
170)	8SAT	97.9	1.3(d)	Roxby Downs SA
112)	2VLY	98.1	20(d)	Muswellbrook
123)	2WIN	98.1	40(d)	Wollongong
142)	6BAY	98.1	30(d)	Geraldton
120)	2NOW	98.3	100(d)	Moree
11)	4TOO	98.3	2(d)	Bowen
3)	4ZKZ	98.3	20(d)	Innisfail
77)	5MMM(t)	98.3	2(d)	Adelaide Foothills
60)	3WWM(t)	98.5	1(d)	Ararat
98)	3SHI(t)	98.7	1	Kerang
158)	4RGM	98.7	100(d)	Mackay
113)	5EZY	98.7	20(d)	Murray Bridge
170)	8SAT	98.9	10(d)	Minlaton SA
169)	5ADL(t)	99.1	2(d)	Adelaide Foothills
117)	2BXS	99.1	10	Bathurst
170)	8SAT	99.3	4	Streaky Bay (SA)
150)	4RBL	99.4	2(d)	Mt. Tamborine
114)	3MDA	99.5	20(d)	Mildura
82)	3TFM	99.5	20(d)	Sale
150)	4RBL	99.5	1	Meandarra
157)	4RGC	99.5	10(d)	Cairns
170)	8SAT	99.5	1(d)	Kapunda SA
38)	2RGF	99.7	50	Griffith
113)	5EZY	99.7	1	Victor Harbour
132)	6CAR	99.7	10	Carnarvon
154)	7RGS	99.7	5(d)	Scottsdale
82)	3TFM(t)	99.9	5(d)	Bairnsdale
69)	4HI(t)	100.1	1(d)	Rolleston Mine
147)	8HOT	100.1	16	Darwin
66)	2NEB	100.3	10	Armidale
121)	2RE(t)	100.3	1.6	Forster

FM	Call	MHz	kW	Location
172)	3MEL	100.3	56(d)	Melbourne
51)	4MKY	100.3	100(d)	Mackay
113)	5EZY(t)	100.3	1	Mount Barker
170)	8SAT	100.3	5(d)	Padthaway East (SA)
116)	2AAY(t)	100.5	1(d)	Falls Creek VIC
88)	6BET	100.5	5	Bridgetown
164)	6NAN	100.5	5	Narrogin
150)	4BRZ	100.6	2(d)	Mt. Tamborine
1)	2PQQ	100.7	20(d)	Port Macquarie
18)	4RGD	100.7	10(d)	Toowoomba
11)	4RGR	100.7	100(d)	Townsville
25)	2ZZZ	100.7	32(d)	Lismore
144)	7TTT	100.9	36	Hobart
50)	3TTT	101.1	56(d)	Melbourne
140)	2CFM	101.1	16	Gosford
60)	3WWM	101.3	20(d)	Horsham
52)	6HED	101.3	2	Broome
43)	4RGK	101.5	10	Rockhampton
81)	2UUS	101.7	150(d)	Sydney
33)	4CCC	101.7	2	Charleville
20)	7HHO	101.7	36	Hobart
27)	7SEA(t)	101.7	20(d)	Burnie
120)	2NOW(t)	101.9	1	Collarenebri
126)	3FOX	101.9	56(d)	Melbourne
149)	4CEE	101.9	10(d)	Maryborough
139)	4MMK	101.9	100(d)	Mackay
122)	3NNN	102.1	25(d)	Wangaratta
1)	2ROX	102.3	20	Port Macquarie
94)	3RBA	102.3	20(d)	Ballarat
11)	4TOO	102.3	100(d)	Townsville
40)	5ADD	102.3	20(d)	Adelaide
131)	6SEA	102.3	5	Esperance
10)	2EEE	102.5	5	Bega
119)	2MOR	102.5	50	Deniliquin
150)	4BRZ	102.5	4	Tenterfield NSW
150)	4BRZ	102.5	1	Childers
170)	8SAT	102.5	3	Bourke
96)	4CCA	102.7	10(d)	Cairns
75)	2GEM	102.9	2	Warialda
109)	2KKO	102.9	20(d)	Newcastle
45)	2ST	102.9	2	Bowral
76)	4HTB	102.9	25(d)	Gold Coast
52)	6NW	102.9	2	Broome
94)	3BBA	103.1	20(d)	Ballarat
34)	4RAM	103.1	100(d)	Townsville
120)	2VM(t)	103.5	1	Collarenebri
96)	4HOT	103.5	10(d)	Cairns
149)	4MBB	103.5	10(d)	Maryborough
78)	2GF(t)	103.9	5(d)	Maclean
128)	2DAY	104.1	150(d)	Sydney
39)	2MW(t)	104.1	1(d)	Gold Coast QLD
10)	2EEE	104.3	20(d)	Batemans Bay/Moruya
25)	2LM(t)	104.3	1(d)	Kyogle
74)	3KKZ	104.3	56(d)	Melbourne
171)	2GOS	104.5	16	Gosford
61)	4LRE	104.5	1(d)	Longreach
127)	4MMM	104.5	12	Brisbane
78)	2CLR	104.7	20(d)	Grafton
138)	2ROC	104.7	20	Canberra
77)	5MMM	104.7	16(d)	Adelaide
116)	2AAY	104.9	100(d)	Albury
129)	2MMM	104.9	150(d)	Sydney
150)	4RBL	104.9	3	Bourke NSW
146)	8MIX	104.9	15	Darwin
120)	2NOW(t)	105.1	1	Walgett
59)	2OAG	105.1	5	Orange
1)	2ROX	105.1	10(d)	Kempsey
124)	3MMM	105.1	56(d)	Melbourne
62)	6MER	105.1	10	Merredin
139)	2NEW	105.3	20(d)	Newcastle
92)	4BBB	105.3	12	Brisbane
133)	2CSF	105.5	15	Coffs Harbour
10)	2EC(t)	105.5	1	Eden
120)	2VM(t)	105.5	1	Mungindi
83)	2BDR	105.7	100(d)	Albury
167)	6IX(t)	105.7	4(d)	Wanneroo
10)	2EC(t)	105.9	20(d)	Batemans Bay/Moruya
59)	2GZF	105.9	5	Orange
84)	5AUU	105.9	20	Spencer Gulf North
170)	8SAT	106.1	3	Ceduna/Smoky Bay (SA)
137)	1CBR	106.3	20	Canberra
133)	2CFS	106.3	15	Coffs Harbour
67)	3CCS	106.3	10(d)	Colac
11)	4RGT	106.3	100(d)	Townsville
64)	2WFM	106.5	150(d)	Sydney

FM	Call	MHz	kW	Location
88)	6RED	106.5	1	Karratha
1)	2PQQ	106.7	10(d)	Kempsey
45)	2ST(t)	106.7	1.6	Ulladulla
120)	2VM(t)	106.7	1	Walgett
150)	4RBL	106.7	1	Childers
100)	2XXX	106.9	20(d)	Newcastle
26)	4BNE	106.9	12	Brisbane
146)	8MIX	106.9	1	Katherine
120)	2NOW(t)	107.1	1	Mungindi
130)	5SSA	107.1	16(d)	Adelaide
121)	2MVB	107.3	10(d)	Taree
152)	7XXX	107.3	36	Hobart
170)	8SAT	107.3	2(d)	Kingston SE (SA)
97)	2GGO	107.7	16	Gosford
98)	3SHI	107.7	10	Swan Hill
27)	7DDD	107.7	7(d)	Devonport
107)	2ICE	107.9	10(d)	Lithgow
34)	4RAM(t)	107.9	2(d)	Bowen
43)	4ROK	107.9	10	Rockhampton

Addresses and other information (ARN:Australian Radio Network). **NB:** The term midnight-to-dawn refers to local time. Exact hours vary from stn to stn

1) PO Box 1161, Port Macquarie NSW 2444 (DMG). Supplementary stn. on 102.3MHz and 105.1MHz – **2)** PO Box 253, Warragul Vic. 3820. (N-2) – **3)** PO Box 19, Innisfail, Qld. 4860 **E:** zedamfm@4kz.com.au (N-1:) Translators: Tully 693kHz 0.5kW, Dunk Island 88.5MHz 0.5kW – **4)** PO Box 189, Scottsdale, TAS. 7254 (N-1).Part of TASmanian Broadcasting Network – **5)** PO Box 177, Mareeba, QLD 4880 (N-1) Translators: Port Douglas 1422kHz, Weipa 97.7MHz – **6)** PO Box 42, Gympie QLD 4370 (N-1) – **7)** PO Box 120, Burnie, TAS. 7320 (N-1) – **8)** 25 Garnet St, Broken Hill, NSW 2880 (N-3). Supplementary stn on 106.9MHz – **9)** PO Box 693, Newstead, QLD 4006 (N-1) – **10)** PO Box 471, Bega, NSW 2550. Translators: 1584=Narooma, 105.9MHz = Batemans Bay – **11)** PO Box 986, Townsville, QLD 4810 **E:** fourto@ultra.net.au **W:** www.ozemail. com.au/~aschter (N-1:) – **12)** PO Box 145, Penrith, NSW 2750 (N-1) – **13)** PO Box 293, Albany, WA 6330. (N-1) – **14)** PO Box 321, Berri SA 5343 **E:** fiverm@riverland.net.au **W:** www.riverland.net.au /~fiverm/ (N-1:) – **15)** PO Box 828, Nambour, QLD 4560 (N-1) – **16)** PO Box 315, Queenstown, TAS 7467 (N-3). Translators at Strahan 105.1MHz 25w & Rosebery 107.1MHz 0.3kW – **17)** PO Box 6110, Cairns, QLD 4870. (N-1: Sky Radio) – **18)** PO Box 111, Toowoomba, QLD 4350 (N-1) – **19)** PO Box 256 Northam, WA 6401 – **20)** GPO Box 542F, Hobart, TAS 7001 (N-3) – **21)** GPO Box 4290, Sydney 2001 (N-3) – **22)** PO Box 485, Warrnambool, Vic. 3280 – **23)** GPO Box 6072, Perth, W.A. 6000 (N-1) – **24)** GPO Box 906, Brisbane, QLD 4001 (N-1) – **25)** PO Box 44, Lismore, NSW 2480. (N-1) – **26)** Locked Bag 1069, Fortitude Value BC, QLD 4006 – **27)** PO Box 635, Launceston TAS 7310 – **28)** PO Box 2106, Alice Springs 0871 (N-1).Translator at Yularoon 100.5MHz with 100w. Supplementary st. 8SUN on 96.9MHz with 300w at Alice Springs – **29)** PO Box 651, Cooma, NSW 2630 (N-1) Relays 2UE 9:00-10:00 and AUSTEREO 16:00-18:00. Translators: Thredbo 92.1MHz 1kW, Jindabyne 96.3MHz 2kW and Perisher 98.7MHz 1kW – **30)** 625 Wyndham St, Shepparton, VIC 3630 – **31)** PO Box 665, Carnarvon WA 6701 – **32)** 3UZ Pty Ltd, PO Box 927, Carlton, VIC 3053 (N-1) ID's as "Sport 927" – **33)** PO Box 420, Gladstone, QLD 4680. (N-1). Translator at Rockhampton on 1584 with 500w and at Biloela on 666kHz with 2.5 Kw – **34)** PO Box 986, Townsville, QLD 4810 (FM **E:** hotfm@ultra.net.au) (N-1). 4RR: Racing format, prgrs 8.00-24.00, also relays 4TAB 1008. 4RAM: Translator at Mt Stuart 107.9MHz 1kW, ID's as "103.1 Hot FM" – **35)** PO Box 108, Golden Square, Vic. 3555 (N-1) – **36)** PO Box 950, North Sydney, NSW 2059 (N-3) – **37)** PO Box 195, Warwick, QLD 4370 (N-1) Rel 2TM 1287 7:00pm to 6:00am. Translator: Toowoomba 1359kHz 0.3kW – **38)** PO Box 493, Griffith, NSW 2680 (N-1 0) – **39)** PO Box 97, Coolangatta, QLD 4225 (N-1). Ids as "Radio 97" – **40)** 201 Tynte St, Nth Adelaide SA 5006. **W:** www.5dn.com.au (N-1) – **41)** PO Box 981, Hamilton, VIC 3300 (N-1) – **42)** PO Box 440, Kalgoorlie, WA 6430 (N-1) – **43)** PO Box 159, Rockhampton QLD 4700 (N-1) – **44)** PO Box 157, Subiaco, WA 6008 (N-1) – **45)** PO Box 540, Nowra 2540 (N-1). Translators: Uladulla 106.7MHz. Supplementary St. on 94.9MHz (N-1) – **46)** Radio 4TAB, PO Box 275, Albion, QLD 4010. Racing format – **47)** Level 1, 464 Hay St, Subiaco, WA 6008 – **48)** G.PO Box 572F, Hobart, TAS 7001 on 1008kHz & 1080kHz, 87.6MHz 1W narrowcast throughout Queenstown, Strahan, Zeehan, Roseberry, Tullah, Stanley& Smithton (N-1:Sky Radio). Racing format. Rel 2UE M-F – **49)** PO Box 1303, Parramatta, NSW 2150 (N-3) Provides relays to over 100 NSW stns carrying racing: – **50)** Private Bag 1011, Richmond Vic. 3121 (N-1) – **51)** PO Box 183, Mackay, QLD 4740 (N-1). Airlie Beach on 94.7MHz. Bowen on 107.9MHz – **52)** PO Box 2216, South Hedland, WA 6722 – **53)** PO Box 481, Pt. Pirie, SA 5540 (N-1) – **54)** G.PO Box 163, Canberra City, ACT 2601 **W:** www.2ca.village.com.au (N-3) – **55)** PO Box 178, Bendigo VIC 3550 – **56)** PO Box 305, Kingaroy, QLD 4610 (N-1) ID's

as "1071AM" and "Classic Gold" – **57)** PO Box 62, Gunnedah 2380 – **58)** 3 Gommes Lane, Yornup WA 6256 – **59)** PO Box 88, Orange, NSW 2800. (N-1:Sky Radio) – **60)** PO Box 606, Horsham, VIC 3400. (N-1) – **61)** PO Box 20, Longreach, QLD 4730 – **62)** PO Box 264, Merredin, WA 6415. (N-1) – **63)** PO Box 835G, Launceston, TAS 7250 (N-3) – **64)** PO Box 1107, Neutral Bay NSW 2089 (N-1). ID's as "Mix 106.5 FM" – **65)** G.PO Box 95, Brisbane, QLD 4001 (N-1) – **66)** PO Box 270, Armidale, NSW 2350. **E:** 2AD@mpx.com.au (N-1) – **67)** PO Box 63, Colac, Vic. 3250 (N-1) – **68)** PO Box 19, Mayfield, NSW 2304 (N-3) – **69)** PO Box 267, Emerald, QLD 4720. (N-1). Translators: 945kHz 1kW, 1215kHz 0.1kW, 88.1MHz 30W, 92.5MHz 10W, 98.2MHz 0.1kW, 102.1MHz 0.25kW. Rel 4AM 558kHz, 4ZR 1476kHz, 4CC 927kHz – **70)** PO Box 480, Wagga Wagga, NSW 2650. (N-1). Translator at Tumut on 107.9MHz with 10w. Supplementary stn on 93.1MHz. Both stns – **71)** PO Box 143, Maryborough, QLD 4650. (N-1) – **72)** PO Box 2516, Nth Sydney, NSW 2001 (N-1) – **73)** PO Box 381, Charters Towers, QLD 4820 Translator Hughenden 765kHz 0.5kW – **74)** Private Bag 1043, Richmond Vic. 3121 ID's as "Gold FM" – **75)** PO Box 770, Inverell, NSW 2360. (N-3) – **76)** PO Box 10290, Southport BC, QLD 4215 – **77)** PO Box 1047, Unley, SA 5061 (N-1) Translator in Adelaide city on 98.3MHz 0.5kW – **78)** PO Box 276, Grafton, NSW 2460. (N-1) – **79)** PO Box 945, Subiaco, WA 6008 (N-1: BBC) – **80)** PO Box 1499, Canberra City, ACT 2601 (N-1) – **81)** PO Box 234, Seven Hills, NSW 2147 (N-1 – **82)** PO Box 160, Sale, Vic. 3850 (N-1) – **83)** 490 David Street, Albury NSW 2640 – **84)** PO Box 496, Port Augusta, SA 5700 (N-1) – **85)** PO Box 783, Toowoomba, QLD 4350 (N-1) – **86)** PO Box 1221, Dubbo, NSW 2830 **E:** 2du@lisp.com.au. (N-1) FM station "ZOO FM" Dubbo 92.7MHz, Cobar 103.9MHz (N-1) – **88)** PO Box 153, Karratha, WA 6714. (N-1) – **89)** 8 Jones Bay Road, Pyrmont NSW 2009 **E:** contact@kick-am.com.au. **W:** www.kick-am.com.au (N-1) ID's as "Kick AM" – **90)** GPO Box 369F, Melbourne 3001. **W:** www.3aw.com.au/ – **91)** PO Box 497, Tamworth, NSW 2340 (N-1). Supplementary stn. on 92.9MHz – **92)** PO Box 105, Albion, QLD 4010 (N-1) ID's as "B105" – **93)** PO Box 500, Mt. Gambier, SA 5290 (N-1) – **94)** PO Box 360, Ballarat, VIC 3350. (N-1) – **95)** PO Box 1234, Wollongong, NSW 2500 **E:** mike@w151. aone.net.au (N-1) – **96)** 68 Abbott St Cairns QLD 4870 – **97)** PO Box 564, Gosford, NSW 2250 (N-1) – **98)** PO Box 504, Swan Hill, VIC 3585 (N-1) – **99)** PO Box 1059, Bundaberg, QLD 4670 (N-1) – **100)** PO Box 97, Charlestown, NSW 2290 (N-1) – **101)** PO Box 9550, Geelong, VIC 3220 (N-1) – **102)** PO Box 31, Young, NSW 2594 (N-1) – **103)** PO Box 780, Mount Isa, QLD 4825 (N-1). Relays to 4GC 828. Translator: Cloncurry 693kHz. Supplementary FM license at Mt. Isa. (N-1) – **04)** PO Box 115, Goulburn, NSW 2580 (N-1: Sky Radio) – **105)** PO Box 75, Frankston, Vic. 3199 **E:** magic@magic.com.au (N-1). 3EE ID's as "Magic" – **106)** GPO Box 5AA, Adelaide SA 5001 (N-1) – **107)** Mailbag 90, Lithgow, NSW 2790 **E:** 2lt@lisp.com.au. (N-1) (for QSL'ing purposes) c/o John Wright, 15 Olive Cres, Peakhurst NSW 2210 – **108)** PO Box 295, Parkes, NSW 2870. (N-1) – **109)** PO Box 606, Charlestown, NSW 2290. (N-1) – **111)** PO Box 17, Mudgee, NSW 2850 – **112)** PO Box 600, Muswellbrook, NSW 2333 (N-1) 2VLY 98.1 ID's as "Power FM" – **113)** PO Box 470, Murray Bridge, SA 5253 (N-1). Serves Murray Bridge, The Coorong and Meningie – **114)** PO Box 539, Mildura, VIC 3500 (N-1). 3MA 99.5 ID's as "Today's Music 99.5FM" – **115)** PO Box 22, Roma, QLD 4455. (N-1:Sky Radio) – **116)** PO Box 670, Albury, NSW 2640 **W:** www.albury.net.au/radio.albury.wodonga/2ay.html (N-1). Supplementary stn. on FM – **117)** PO Box 310, Bathurst, NSW 2795 **E:** stereo@2bs.ix.net.au or 2bs@csu.edu.au **W:** www.2bs.ix.net.au (N-1) FM service on 99.3MHz – **118)** PO Box 195, Shepparton, Vic. 3630 – **119)** PO Box 312, Deniliquin, NSW 2710. (N-1) 2MOR 102.5 ID's as "Classic Rock 102.5" – **120)** PO Box 389, Moree, NSW 2400. (N-1). Supplementary license on 98.3MHz. Translator on 88.7MHz with 250w r. (N-1) – **121)** PO Box 275, Taree, NSW 2430 (N-1). Translator: Gloucester 100.1MHz and Forster on 100.3MHz – **122)** PO Box 449, Wangaratta, VIC 3677 (N-1). 3NE Translators: Mt. Hotham 89.3MHz 0.02kW, Mt. Buffalo 105.3MHz 0.2kW, Mt. Beauty 90.3MHz 10w. 3NNN ID's as "Edge FM" – **123)** Locked Bag 6198 Sth Coast Mail Centre NSW 2521 (N-1) ID's as "98FM" – **124)** GPO Box 105, Melbourne, VIC (N-1) – **125)** 111 Wellington Str, East Perth, WA 6004. (N-1) – **126)** PO Box 1019, St. Kilda, Vic. 3182 (N-1) – **127)** GPO Box 1041, Brisbane, QLD 4001. (N-1) – **128)** PO Box 920, Crows Nest, NSW 2065 **W:** www.2dayfm. com.au (N-1) – **129)** GPO Box 442, Sydney, NSW 2001 (N-1). **W:** www. mrock.com.au – **130)** PO Box 1071, Unley, SA 5061.24h (N-1) Translator South Tce, Adelaide on 91.1MHz 1kW. ID's as "SAFM" – **131)** PO Box 527, Esperance, WA 6450. N-1. Rel 6PPM-FM 1000-2200 – **132)** PO Box 665, Carnarvon, WA 6701. 2200-1500 (N-1). Translator: Exmouth – **133)** PO Box 1950, Coffs Harbour, NSW 2450 (N-1). Rp – **134)** PO Box 483, Port Lincoln, SA 5606. (N-1) – **135)** 141 Mandurah Tce, Mandurah, WA 6210 (N-1) – **136)** PO Box 183, Canberra, ACT 2601. Belongs to 54). **F.PI:** translator for Tuggeranong area – **137)** PO Box 106, Dickson, ACT 2602. (N-1) ID's as "Mix 106.3" – **138)** GPO Box 163, Canberra, A.C.T. 2601 (N-1) – **139)** PO Box 185, Mackay QLD 4740 – **140)** PO Box 2101, Gosford, NSW 2250 (N-1) – **141)** PO Box 7, Ipswich, QLD 4305 (N-1) ID's

as "Mix 106.9 QFM" – **142)** PO Box 128 Geraldton, WA 6530. 24h (N-1) – **143)** 15 Puttabucca Rd, Mudgee NSW 2850 – **144)** G.PO Box 1800, Hobart, TAS. 7001 (N-1) – **145)** Locked Bag 5000, Broadway NSW 2007 – **146)** GPO Box 2510, Darwin NT 0801 – **147)** 4 Peary St., Darwin, NT 0800 (N-1) Translators: Katherine 765kHz 0.5kW – **148)** 678 Victoria St, Richmond VIC 3121 – **149)** 403 The Esplanade, Torquay QLD 4655 – **150)** PO Box 332, Beaudesert QLD 4285 – **151)** PO Box 148, Bunbury WA 6231 – **152)** GPO Box 1345, Hobart TAS 7001 – **153)** PO Box 84, Charleville, QLD 4470 (N-1) ID's as "Outback Radio". Translator: Cunnamulla 1584kHz 0.2kW – **154)** PO Box 5910 Gold Coast Mail Centre Bundall QLD 4217 r. (N-1) – **155)** Private Bag 925 Gold Coast Mail Centre QLD 4215. (N-1) – **156)** Paul Taylor, 41 Allards Crt., Clifton Springs VIC 3222 – **157)** Sea FM, 320 Sheridan St Cairns QLD 4870 – **158)** Sea FM, Suncorp/Metway Building Suite 3, Level 3, 123 Victoria St Mackay QLD 4740 – **159)** 444 Logan Rd, Stones Corner QLD 4120 – **160)** PO Box 665 Carnarvon WA 6701 – **161)** 33 Saunders Road, Pyrmont NSW 2009 – **162)** Regional Communications Pty Ltd, PO Box 7515, St Kilda Road VIC 3004 – **163)** 47 Wingewarra St Dubbo NSW 2830 – **164)** DMG Regional Radio, Locked Bag 5000, Broadway NSW 2007 – **165)** 68 Aboott St, Cairns QLD 4870 – **166)** PO Box 112, Bunbury WA 6230 – **167)** PO Box 33, Tuart Hill WA 6060 **168)** cnr Plaza Pde & Carnaby St, Maroochydore QLD 4558 – **169)** Locked Bag 919, Adelaide SA 5001 – **170)** PO Box 579, Lilydale VIC 3140 – **171)** PO Box 3535, Erina NSW 2250, – **172)** Level 2, 678 Victoria Street, Richmond, Vic, 3121.

COMMUNITY BROADCASTING ASSOCIATION OF AUSTRALIA

Suite One, Level Three, 44-54 Botany Rd. Alexandria, NSW 2015
☎+61 (2) 9310 2999 📠 +61 (2) 9319 4545
PRN: Public Radio Network, CBAA: Community Broadcasting Association of Australia. CBAA provides ComRadSat.

PUBLIC BROADCASTING STATIONS

	MW Call	Location	kHz	kW
1)	2WEB	Bourke	585	10(d)
2)	6WR	Kununurra	693	5
3)	3CR	Melbourne	855	2(d)
4)	7RPH	Hobart	864	2
210)	6FX	Perth	936	5
220)	6RPH	Perth	990	5
7)	1RPH	Canberra	1125	2(d)
9)	3RPH	Melbourne	1179	5
8)	4BI	Brisbane	1197	0.5(d)/1
10)	5RPH	Adelaide	1197	2
11)	2RPH	Sydney	1224	2
6)	4MW	Thursday Is.	1260	2
112)	4RPH	Brisbane	1296	5(d)
209)	3KND	Melbourne	1503	5(d)

	FM Call	MHz	kW	Location
151)	3MFM	88.1	2(d)	Leongatha
9)	3BPH	88.7	10	Bendigo
176)	3RUM	88.7	1	Walwa/Jingellic
13)	2RBR	88.9	1(d)	Coraki
14)	2YOU	88.9	1(d)	Tamworth
112)	7DBS	88.9	1(d)	Lileah
217)	4CCR	89.1	2(d)	Cairns
148)	5BBB	89.1	1	Barossa Valley
16)	4SDB	89.3	2	Warwick
17)	4CRB	89.3	25(d)	Gold Coast
149)	5EFM	89.3	1	Victor Harbour
150)	5GFM	89.3	10	Arhurton
151)	3MFM	89.5	1(d)	Foster
218)	2HIM	89.7	1(d)	Tamworth
19)	2TEN	89.7	4	Tenterfield
147)	5TCB(t)	89.7	1.5(d)	Naracoorte
158)	6TCR	89.7	2(d)	Wanneroo
152)	3TSC	89.9	56(d)	Melbourne
20)	4DDD	89.9	2	Dalby
153)	5GSFM	90.1	1	Victor Harbour
153)	3SYN	90.1	56(d)	Melbourne
21)	4CSB	90.7	5	Wondai
22)	5KIX	90.7	3(d)	Kangaroo Island
23)	1CMS	91.1	20	Canberra
24)	2CBD	91.1	5	Deepwater
1)	2WEB(t)	91.1	1(d)	Coonamble
25)	2MAX	91.3	10(d)	Narrabri
102)	4BRR	91.5	1	Gayndah
26)	4GCR	91.5	1	Gympie
27)	1WAY	91.9	20	Canberra
28)	4RGL	91.9	1	Gladstone
30)	2MFM	92.1	6	Sydney
31)	6RTR	92.1	10(d)	Perth
33)	2MCE	92.3	2	Bathurst
34)	3ZZZ	92.3	56(d)	Melbourne

	FM	Call	MHz	kW	Location
35)		1ART	92.7	20	Canberra
155)		5FBI	92.7	20(d)	Adelaide
36)		2NCR	92.9	6	Lismore
37)		4GOD	92.9	4	Toowoomba
147)		5TCB	92.9	2	Kingston SE
38)		2BBB	93.3	3.2	Dorrigo
156)		2MNO	93.3	2	Monaro
157)		2SNR	93.3	2	Gosford
9)		3RPH	93.5	1	Warragul
39)		2BAR	93.7	2	Bega
212)		2LND	93.7	50(d)	Sydney
40)		5DDD	93.7	6.3	Adelaide
226)		2CCM	94.1	2(d)	Gosford
41)		2LIV	94.1	4(d)	Wollongong/Nowra
42)		4JAZ	94.1	25(d)	Gold Coast
43)		2DCB	94.3	10	Dubbo
213)		2FBI	94.5	150(d)	Sydney
9)		3RPH	94.5	5(d)	Warrnambool
44)		5CCR	94.5	1	Ceduna/Smoky Bay
45)		8KNB	94.5	15	Darwin
33)		2MCE	94.7	1(d)	Orange
228)		3PLS	94.7	56(d)	Geelong
47)		4BCR	94.7	3(d)	Bundaberg
218)		2GCB	94.9	2	Gosford
160)		2MIA	95.1	5	Griffith
229)		2TRR	95.3	2(d)	Coolah
161)		6EBA	95.3	16(d)	Perth
162)		7HRT	95.7	1(d)	Nthn Midlands
12)		4BVR	95.9	1	Esk
229)		2TRR	96.1	1(d)	Dunedoo
32)		7THE	96.1	3	Hobart
48)		2CCC	96.3	2	Gosford
49)		3GGR	96.3	56(d)	Geelong
50)		2CHR	96.5	2	Cessnock/Maitland
51)		3EON	96.5	1(d)	Bendigo (city)
52)		4FRB	96.5	12	Brisbane
46)		4RFM	96.9	5	Moranbah
159)		2MAC	97.3	1	Lake Macquarie
54)		3HCR	97.3	1	Omeo
165)		7TAS	97.7	1	Tasman Peninsula
55)		8GGG	97.7	15	Darwin
56)		2LVR	97.9	4	Parkes/Forbes
57)		6DBY	97.9	2	Derby
58)		4EB	98.1	12	Brisbane
59)		1XXR	98.3	20	Canberra
60)		6MKA	98.3	1	Meekatharra
61)		2OOO	98.5	25(d)	Sydney
62)		3ONE	98.5	10(d)	Shepparton
167)		4YOU	98.5	1	Rockhampton
147)		5TCB	98.5	1	Padthaway
63)		6SON	98.5	16(d)	Perth
64)		2KRR	98.7	1	Kandos
65)		4CIM	98.7	10(d)	Cairns
66)		4AAA	98.9	9.5	Brisbane
29)		3SFM	99.1	1	Swan Hill
67)		3RPC	99.3	2(d)	Portland
168)		7EDG	99.3	1	Hobart South
68)		2RFM	99.7	10(d)	Newcastle
69)		2UUU	99.7	1	Ulladulla
70)		3MCR	99.7	1	Mansfield
215)		4ACR	99.7	1	Woorabinda
169)		4RED	99.7	2(d)	Redcliffe
71)		6GME	99.7	2	Broome
170)		2BAY	99.9	3	Byron Bay
72)		2PMQ	99.9	3	Port Macquarie
73)		3BBB	99.9	3	Ballarat
74)		4TCB	99.9	20	Townsville
171)		5MBS	99.9	2(d)	Adelaide Foothills
75)		2BCB	100.1	10	Bathurst
9)		3SPH	100.1	1	Shepparton
172)		4RIM	100.1	1	Boonah
214)		5GTR	100.1	1	Mt. Gambier
5)		6NR	100.1	6.5(d)	Perth
76)		2TLC	100.3	1	Maclean
216)		2YAS	100.3	2	Yass
77)		4BAY	100.3	4(d)	Wynnum/Redlands
11)		2RPH	100.5	4	Newcastle
11)		2RPH	100.5	1(d)	Sydney Eastern Suburbs
1)		2WEB	100.7	1(d)	Nyngan
78)		3CH	100.7	1	Kyneton
79)		4US	100.7	1	Rockhampton
211)		2PSR	100.9	1(d)	Port Stephens
80)		6CRA	100.9	10	Albany
166)		6NME	100.9	16(d)	Perth
81)		4CBL	101.1	4(d)	Logan

FM	Call	MHz	kW	Location
82)	3WPR	101.3	2(d)	Wangaratta
83)	2GLA	101.5	10(d)	Forster
230)	3BBS	101.5	1(d)	Bendigo
174)	4BSR	101.5	1	Beaudesert
84)	4OUR	101.5	3(d)	Caboolture
185)	5UV	101.5	20(d)	Adelaide
9)	2APH	101.7	2	Albury
231)	6SEN	101.7	5(d)	Perth
177)	2PAR	101.9	1	Ballina
86)	4ZZZ	102.1	12	Brisbane
87)	6WR	102.1	1	Wyndham
88)	2NIM	102.3	1	Nimbin
89)	2MBS	102.5	50(d)	Sydney
90)	3RRR	102.7	56(d)	Melbourne
178)	4DDB	102.7	4	Toowoomba
91)	2CVC	103.1	1	Grafton
92)	2WET	103.1	1	Kempsey
210)	3BBR	103.1	1	Warragul
93)	5EBI	103.1	20(d)	Adelaide
94)	2CBA	103.2	50(d)	Sydney
95)	2TLP	103.3	3(d)	Taree
97)	2CCB	103.5	5	Orange
98)	3MBR	103.5	4.8	Murrayville
99)	3MBS	103.5	56(d)	Melbourne
100)	2NUR	103.7	10(d)	Newcastle
53)	3WAY	103.7	5	Warrnambool
101)	4MBS	103.7	12	Brisbane
179)	7LTN	103.7	3.2	Launceston
103)	2WAY	103.9	3	Port Macquarie
104)	3BGR	103.9	3	Ballarat
209)	3GCB	103.9	10(d)	Latrobe Valley
105)	4TTT	103.9	20	Townsville
106)	6ESP	103.9	5	Esperance
107)	2CHY	104.1	1.6	Coffs Harbour
108)	8TOP	104.1	16	Darwin
69)	2UUU	104.5	2(d)	Nowra
147)	5TCB	104.5	1.6	Keith
180)	8KIN	104.5	1	Katherine
109)	2BOB	104.7	5	Taree
110)	3GCR	104.7	7.9(d)	Latrobe Valley
111)	3GRR	104.7	5(d)	Echuca
112)	7DBS	104.7	2	Devonport
181)	4SFM	104.9	3	Nambour
182)	5RCB	104.9	20	Mt. Gambier
113)	4WBR	105.1	10(d)	Maryborough
201)	5TRX	105.1	5	Port Pirie
114)	7WAY	105.3	3.2	Launceston
115)	4MET	105.7	10(d)	Gold Coast
24)	2CBD	105.9	3	Glen Innes
116)	2NVR	105.9	2(d)	Nambucca Heads
175)	4MUR	105.9	1	Mackay
147)	5TCB	106.1	1.6	Bordertown
112)	7DBS	106.1	10(d)	Wynyard
117)	2CUZ	106.5	10	Bourke
183)	4CLG	106.5	3	Nambour
96)	7HFC	106.5	36	Hobart
118)	3PBS	106.7	56(d)	Melbourne
119)	2VOX	106.9	2(d)	Wollongong
120)	3UGE	106.9	1	Alexandra/Eildon
4)	7RPH	106.9	3.2	Launceston
121)	4KIG	107.1	16	Townsville
122)	2REM	107.3	2	Albury
123)	2SER	107.3	14	Sydney
124)	4CAB	107.3	10(d)	Gold Coast
125)	2EAR	107.5	1.6(d)	Moruya
173)	2OCB	107.5	5	Orange
9)	3MPH	107.5	1	Mildura
126)	4CRM	107.5	1	Mackay
176)	3RUM	107.7	1	Tumbarumba NSW
127)	2AIR	107.9	5	Coffs Harbour
128)	2COW	107.9	1	Casino
129)	5RAM	107.9	20(d)	Adelaide
130)	6CCR	107.9	1(d)	Fremantle

HIGH POWER OPEN NARROWCAST STATIONS

These stns are licensed for a specific market such as horseracing or certain local audience that cannot be filled by commercial or other stns. Official callsigns are not issued for these stns but may be used. **NB:** many expanded band (1611-1701kHz) stns are licensed but may not be on air. Confirming their status is difficult.

MW	Call	kHz	kW	Location
188)	5RTI	531	0.5	Adelaide
141)	6——	657	2	Perth
187)	2RF	801	5(d)	Gosford
189)	4——	873	2	Innisfail

MW	Call	kHz	kW	Location
138)	4TAB	891	5(d)	Townsville
146)	3UZ	945	2	Bendigo
190)	2TAB	1008	0.3	Canberra
232)	7TAB	1008	5	Launceston
132)	6TAB	1017	1	Bunbury
141)	4——	1053	0.5	Brisbane
191)	7TAB	1080	5(d)	Hobart
132)	6TAB	1206	2	Perth
190)	2TAB	1215	0.35	Bowral
140)	8TAB	1242	2	Darwin
190)	2TAB	1314	5(d)	Wollongong
220)	1KIX	1323	0.4(d)	Canberra
190)	2TAB	1341	5(d)	Newcastle'
192)	3——	1341	5(d)	Geelong
193)	3——	1359	0.2	Mildura
132)	6TAB	1404	4	Busselton
194)	3——	1413	0.5(d)	Shepparton
195)	3XY	1422	5	Melbourne
200)	6GS	1422	0.4	Wagin
132)	6TAB	1431	2	Kalgoorlie
132)	6TAB	1449	2	Mandurah
186)	2——	1476	0.3(d)	Penrith
187)	2RF	1539	1	Sydney
139)	5TAB	1539	5(d)	Adelaide
139)	5TAB	1557	0.5(d)	Renmark/Loxton
187)	2RF	1575	5(d)	Wollongong
154)	2——	1593	0.2	Murwillumbah
156)	3RG	1593	5(d)	Melbourne
196)	2NTC	1611	0.4	Grafton
234)	2——	1611	0.001	Newcastle
235)	2——	1611	0.05	Wee Waa
233)	2——	1611	0.4	Western Sydney
196)	2NTC	1611	0.4	Tamworth
187)	2RG	1611	0.4	Griffith
236)	3——	1611	0.4	Mildura
164)	3UCB	1611	0.4	Hoppers Crossing
197)	4KIK	1611	0.05	Croydon
197)	4KZ	1611	0.4	Karumba
199)	6AY	1611	0.4	Albany
198)	6——	1611	0.4	Margaret River
200)	6GS	1611	0.4	Wagin
187)	6RF	1611	0.4	Esperance
187)	7RF	1611	0.4	Devonport
187)	7RF	1611	0.4	Launceston
187)	8RF	1611	0.4	Darwin
187)	1RF	1620	0.4	Canberra
190)	2KM	1620	0.4	Sydney
202)	3GB	1620	0.4	Melbourne
197)	4KZ	1620	0.4	Taylors Beach
187)	4RF	1620	0.4	Gladstone
187)	4RF	1620	0.4	Sunshine Coast
187)	4RF	1620	0.4	Toowoomba
221)	6——	1620	0.4	Perth
203)	2HRN	1629	0.1	Newcastle
196)	2NTC	1629	0.4	Bathurst
196)	2NTC	1629	0.4	Dubbo
187)	3RF	1629	0.4	Shepparton
187)	4RF	1629	0.4	Mackay
187)	5RF	1629	0.4	Adelaide
187)	5RF	1629	0.4	Mount Gambier
236)	6ABN	1629	0.4	Busselton
187)	6RF	1629	0.4	Albany
187)	6RF	1629	0.4	Mundaaring
187)	6RF	1629	0.4	Perth
222)	2ME	1638	0.4	Sydney
196)	2NTC	1638	0.4	Armidale
222)	3ME	1638	0.4	Melbourne
222)	8ME	1638	0.4	Darwin
222)	4ME	1647	0.4	Brisbane
198)	4——	1647	0.4	Mackay
222)	5ME	1647	0.4	Adelaide
225)	2——	1656	0.05	Broken Hill
238)	2MM	1656	0.4	Sydney
223)	4——	1656	0.4	Brisbane
222)	6ME	1656	0.4	Perth
206)	2MM	1665	0.4	Sydney
207)	2——	1683	0.4	Sydney
198)	4——	1692	0.4	Nanango
222)	2——	1701	0.4	Sydney
224)	3VMV	1701	0.4	Somerton
208)	4——	1701	0.1	Brisbane

SW	Call	kHz	kW	Location
85)	2——	2368.5	1	Sydney

FM	MHz	kW	Location
131)	88.7	2(d)	Atherton QLD

FM	MHz	kW	Location
132)	89.5	1.2	Esperance WA
187)	90.3	1(d)	Griffith NSW
142)	90.5	1(d)	Barossa Valley SA
133)	90.5	1(d)	Tamworth NSW
232)	90.5	5(d)	Perth WA
134)	90.9	1	Mossman QLD
133)	90.9	1	Mudgee NSW
135)	91.5	1	Toowoomba QLD
154)	91.9	8(d)	Latrobe Valley VIC
136)	92.3	10(d)	Maryborough QLD
132)	92.5	2	Port Hedland WA
133)	92.7	1	Inverell NSW
133)	92.7	3(d)	Port Macquarie NSW
164)	93.7	50(d)	Albany WA
133)	94.3	1	Goulburn NSW
137)	94.9	4	Broken Hill NSW
133)	95.5	10(d)	Wagga Wagga NSW
138)	95.5	3(d)	Bundaberg QLD
138)	95.5	4.5(d)	Emerald QLD
139)	95.5	5(d)	Renmark/Loxton SA
140)	95.9	1	Alice Springs NT
133)	95.9	20(d)	Gunnedah NSW
163)	96.5	1	Katherine NT
133)	96.9	1(d)	Cooma NSW
164)	97.5	2(d)	Bairnsdale VIC
138)	97.5	1	Blackwater QLD
165)	97.7	1(d)	Burnie TAS
136)	98.1	1	Inglewood QLD
184)	98.7	1	Alice Springs NT
141)	99.1	2(d)	Atherton QLD
133)	99.9	10	Parkes/Forbes NSW
138)	99.9	1	Rockhampton QLD
133)	100.5	4	Broken Hill NSW
132)	100.5	1	Wyndham WA
133)	100.9	10	Bathurst NSW
142)	101.1	10(d)	Nowra NSW
143)	101.3	2	Devonport TAS
133)	101.5	1	Grafton NSW
133)	101.5	1	Kempsey NSW
132)	101.7	1	Karratha WA
164)	102.1	1	Hamilton VIC
133)	102.7	2(d)	Jindabyne NSW
164)	102.9	2	Horsham VIC
133)	103.3	1	Muswellbrook NSW
138)	103.5	100(d)	Mackay QLD
140)	103.7	1	Katherine NT
133)	103.7	1	Moree NSW
133)	103.7	2(d)	Nowra NSW
133)	104.3	10	Armidale NSW
143)	104.3	10	Cairns QLD
164)	105.3	1	Portland NSW
144)	105.3	2(d)	Wollongong NSW
133)	105.7	5	Taree NSW
145)	106.1	1	Deniliquin NSW
133)	106.7	5	Orange NSW
146)	106.9	10	Swan Hill VIC
133)	107.1	1	Eden NSW
133)	107.5	3	Glen Innes NSW

NB FM stns below 1kW are not mentioned. There have been allocations for a number of years for HF outlets but only one has any ever made it to air.

Addresses and other information
1) Western Region Educational Broadc. Co. Ltd, PO Box 426, Bourke NSW 2840. Plus 5 FM translators. – **2)** Radio Station 6WR. PO Box 162 Kununurra WA 6743. (N-1:CBAA) Aboriginal programs from National Indigenous Radio Service – **3)** Community R. Federation Ltd, PO Box 277, Collingwood VIC 3066. Various foreign languages – **4)** Radio 7RPH Broadcasting Services for Handicapped Inc. 136 Davey st Hobart TAS. 7000. Information and reading service format. Relays BBCWS 11:00pm to 10:00am Mon-Sat, Sunday – **5)** Curtin Univ of Technology, GPO Box U1987, Perth WA 6001.. Rel. CBAA Network at times and BBCWS overnight – **6)** PO Box 385, Thursday Island QLD 4875 – **7)** Print-Handicapped Radio of ACT Inc, Barton Highway, Gungahlin, ACT 2912 – relays Switch FM, PO Box 173, Fortitude Valley QLD, 4006 – **9)** Assoc. for the Blind, 454 Glenferrie Rd, Kooyong 3144. Relays BBCWS overnight – **10)** Radio 5RPH, 231 Morphett St. Adelaide SA 5000. Relays BBCWS overnight – **11)** R. for the Print-Handicapped (NSW) Co-op Ltd, 2/252 Illawarra Rd, Marrickville NSW 2204 – **12)** PO Box 148, Toogoolawah QLD 4313 – **13)** PO Box 50 Houghwood Rd, Bora Ridge NSW 2471 – **14)** PO Box 998, Tamworth NSW 2340 – **15)** PO Box 891, Barham NSW 2732 – **16)** Rainbow FM, PO Box 473 WArwick QLD 4370 – **17)** PO Box 86, Burleigh Heads QLD 4220 – **18)** 120 McCrae St, Bendigo VIC 3550 – **19)** PO Box

93, Tenterfield NSW 2372 – **20)** PO Box 483, Dalby QLD 4405 – **21)** Crow FM, PO Box 171, Emerald QLD 4606 – **22)** PO Box 90, Kingscote SA 5223 – **23)** PO Box 3882, Weston ACT 2611 (Ethnic) – **24)** Gough St, Deepwater NSW 2371 – **25)** PO Box 94, Narrabri NSW 2390 – **26)** The Positive Alternative, PO Box 774, Gympie QLD 4570 (Christian) – **27)** Canberra Christian Radio, PO Box 927, Fyshwick ACT 2609 (Christian) – **28)** 257 Goondoon St WArwick QLD 4680 (Christian) – **29)** PO Box 998, Swan Hill VIC 3585 – **30)** Muslim Community Radio, PO Box 969, Bankstown NSW 1885 (Ethnic) – **31)** Arts Radio, PO Box 949, Nedlands WA 6009 – **32)** GPO Box 1324, Hobart TAS 7001 – **33)** Charles Sturt University, Locked Bag 30, Bathurst NSW 2795 – **34)** PO Box 1106, Collingwood VIC 3066 (Ethnic) – **35)** Artsound, PO Box 87, Curtin ACT 2605 – **36)** PO Box 5123, East Lismore NSW 2480 – **37)** The Light, PO Box 3367, Village Fair, Townsville QLD 4350 (Christian) – **38)** PO Box 304, Dorrigo NSW 2454 – **39)** Edge FM, PO Box 771, Bega NSW 2550 – **40)** 48 Nelson St, Stepney SA 5069 – **41)** Living Sound Broadcasters, PO Box 7, Coniston NSW 2500 (Christian) – **42)** Radio Hope Island, PO Box 16 SAnctuary Cove QLD 4212 – **43)** Radio Rhema, PO Box 1502, Dubbo NSW 2830 (Christian) – **44)** PO Box 271, Ceduna SA 5690 – **45)** Radio Larrakia, Shop 2, Alawa Shops, Alawa NT 0810 (Aboriginal) – **46)** PO Box 597, Moranbah QLD 4744 – **47)** PO Box 2678, Bundaberg QLD 4670 – **48)** PO Box 19, Gosford NSW 2250 – **49)** Rhema FM, PO Box 886, Belmont VIC 3216 (Christian) – **50)** PO Box 421, Cessnock NSW 2325 – **51)** Radio KLFM, PO Box 2997, Bendigo Delivery Centre VIC 3554 – **52)** Family Radio, PO Box 1700, Milton QLD 4064 – **53)** PO Box 752 WArrnambool VIC 3280 – **54)** PO Box 86, Omeo VIC 3898 – **55)** Darwin Christian Broadcasters, PO Box 43146, Casaurina NT 0810 (Christian) – **56)** Parkes Road, Forbes NSW 2871 – **57)** PO Box 655, Derby WA 6728 – **58)** 140 Main St, Kangaroo Point QLD 4169 – **59)** 2XX, GPO Box 812, Canberra ACT 2601 – **60)** PO Box 259, Meekatharra WA 6642 – **61)** Radio 2000, 2/25 Belmore Rd, Burwood NSW 2134 (Ethnic) – **62)** PO Box 6824, Shepparton VIC 3630 – **63)** Sonshine FM, PO Box 6340, Morley WA 6062 (Christian) – **64)** PO Box 99, Kandos NSW 2848 – **65)** PO Box 1856, Cairns QLD 4870 (Aboriginal) – **66)** Box 6229, Fairfield Gardens QLD 4103 (Aboriginal) – **67)** PO Box 450, Portland VIC 3305 – **68)** Rhema FM, PO Box 2000, Dangar NSW 2309 (Christian) – **69)** PO Box 884, Nowra NSW 2541 – **70)** PO Box 667, Mansfield VIC 3724 – **71)** PMB Turkey Creek, via Kununurra WA 6743 (Aboriginal) – **72)** Radio Rhema, PO Box 1537, Port Macquarie NSW 2444 (Christian) – **73)** Voice FM, PO Box 149, Ballarat VIC 3350 – **74)** Live FM, PO Box 332, Aitkenvale QLD 4814 (Christian) – **75)** Radio Rhema, PO Box 615, Bathurst NSW 2795 (Christian) – **76)** PO Box 210, Yamba NSW 2464 – **77)** PO Box 1003, Cleveland QLD 4163 – **78)** PO Box 26, Kyneton VIC 3444 – **79)** PO Box 663, Rockhampton QLD 4700 (Aboriginal) – **80)** 211-217 North Road, Albany WA 6330 – **81)** PO Box 2101, Logan City DC QLD 4114 – **82)** PO Box 605 Wangaratta VIC 3676 – **83)** PO Box 1015, Tuncurry NSW 2428 – **84)** PO Box 418, Caboolture QLD 4510 – **85)** Radio Symban, 867 New Canterbury Rd, Hurlstone Park NSW 2193 – **86)** PO Box 509, Fortitude Valley QLD 4006 – **87)** PO Box 815, Kununurra WA 6743 (Aboriginal) – **88)** PO Box 522, Nimbin NSW 2480 – **89)** 76 Chandos St, St Leonards NSW 2065 – **90)** PO Box 304, Fitzroy VIC 3065 – **91)** PO Box 115, Grafton NSW 2460 (Christian) – **92)** PO Box 200, West Kempsey NSW 2440 – **93)** 10 Byron Pl, Adelaide SA 5000 (Ethnic) – **94)** PO Box 54, Five Dock NSW 2046 – **95)** Ngarralinyi, The Listening Place, PO Box 657, Taree NSW 2430 (Aboriginal) – **96)** PO Box 1033, New Town TAS 7008 – **97)** Radio Rhema, PO Box 974, Orange NSW 2800 – **98)** PO Box 139, Murrayville VIC 3512 – **99)** 146 Cotham Road, Kew VIC 3101 – **100)** University Dr, Callaghan NSW 2308 – **101)** 384 Old Cleveland Rd, Coorparoo QLD 4151 – **102)** PO Box 915, Gayndah QLD 4625 – **103)** PO Box 603, Port Macquarie NSW 2444 – **104)** Good News Radio, PO Box 312, Ballarat VIC 3350 – **105)** PO Box 1033, Townsville QLD 4810 – **106)** PO Box 2154, Esperance WA 6450 – **107)** PO Box J233, Coffs Harbour NSW 2450 – **108)** PO Box 40146, Casaurina NT 0810 – **109)** PO Box 400, Taree NSW 2430 – **110)** PO Box 579, Morwell VIC 3840 – **111)** 1/15 Matong Rd, Echuca VIC 3564 – **112)** PO Box 333, Wynyard TAS 7325 – **113)** Rhema FM, PO Box 384, Hervey Bay QLD 4655 (Christian) – **114)** 93 Reatta Rd, Trevallyn TAS 7250 (Christian) – **115)** Radio Metro, PO Box 6530, GCMC QLD 9726 – **116)** PO Box 69, Bowraville NSW 2449 – **117)** PO Box 363, Bourke NSW 2840 (Aboriginal) – **118)** PO Box 2917, Fitzroy VIC 3065 – **119)** PO Box 1663, Wollongong NSW 2500 – **120)** PO Box 270, Alexandra VIC 3714 – **121)** PO Box 5483, Townsville QLD 4810 (Aboriginal) – **122)** Garland Ave, North Albury NSW 2640 – **123)** PO Box 123, Broadway NSW 2007 – **124)** Life FM, PO Box 948, Southport QLD 4125 (Christian) – **125)** PO Box 86, Moruya NSW 2537 – **126)** PO Box 1075, Mackay QLD 4740 – **127)** PO Box 2028, Coffs Harbour NSW 2450 – **128)** PO Box 1149, Casino NSW 2470 – **129)** Radio Alta Mira, PO Box 1079, North Adelaide SA 5006 (Christian) – **130)** Unit 4, 153 Rockingham Rd, Hamilton Hill WA 6163 – **131)** KIK FM, PO Box 1434, Edge Hill QLD 4870 – **132)** TAB WA, 14 Hasler Rd, Osborne Park WA 6017 – **133)** PO Box 1303, Parramatta NSW 2150 – **134)** 90.9 FM, Mossman & Port Douglas PO Box 383, Whakatane NEW ZEALAND – **135)** PO Box 111,

Toowoomba QLD 4350 – **136)** PO Box 1059, Bundaberg QLD 4670 – **137)** Cross FM, Broken Hill Church of Christ, 232 Lane St, Broken Hill NSW 2880 – **138)** Radio 4TAB, PO Box 275, Albion QLD 4010 – **139)** GPO Box 5AA, Adelaide SA 5001 – **140)** NT Racing Commission, PO Box 3170, Darwin NT 0800 – **141)** Rete Italia, PO Box 159, Clifton Hill VIC 3068 – **142)** Ambersky, PO Box 540, Nowra NSW 2541 – **143)** PO Box 5109, GCMC, Bundall QLD 9726 – **144)** 63 Minimbah Rd, Northbridge NSW 2063 – **145)** Rich Rivers Radio, PO Box 312, Deniliquin NSW 2710 – **146)** PO Box 927, Carlton VIC 3053 – **147)** PO Box 526, Bordertown SA 5268 – **148)** PO Box 654, Tanunda SA 5352 – **149)** PO Box 591 Victor Harbour SA 5211 – **150)** PO Box 390, Kadina SA 5554 – **151)** PO Box 144, Inverloch VIC 3996 – **151)** PO Box 899, Mont Albert VIC 3127 – **152)** PO Box 999 VICtor Harbour SA 5211 – **153)** PO Box 12013, A'Beckett St, Melbourne VIC 3000 – **154)** PO Box 519 Gold Coast Mail Centre Bundall QLD 4217 – **155)** Level 2, 230-232 Angas St, Adelaide SA 5000 – **156)** PO Box 28, Nimmitabel NSW 2631 – **157)** PO Box 2050, Gosford NSW 2250 – **158)** PO Box 281 WAnneroo WA 6946 – **159)** 30 Pillarapi Rd, Brightwaters NSW 2264 – **160)** PO Box 2122, Griffith NSW 2680 – **161)** PO Box 1005, Subiaco WA 6904 – **162)** c/- PO, Gordon St, Poatina TAS 7302 – **163)** PO Box 79, Earlwood NSW 2206 – **164)** Vision FM, Locked Bag 3, Springwood QLD 4127 – **165)** GPO Box 1345, Hobart TAS 7001 – **166)** PO Box 105, Bentley WA 6102 – **167)** PO Box 5035, North Rockhampton MC QLD 4701 – **168)** GPO Box 252-44, Hobart TAS 7001 – **169)** PO Box 139, Redcliffe QLD 4020 – **170)** PO Box 440, Byron Bay NSW 2481 – **171)** PO Box 7016, Hutt St, Adelaide SA 5000 – **172)** PO Box 243, Boonah QLD 4310 – **173)** PO Box 1031, Orange NSW 2800 – **174)** PO Box 235, Beaudesert QLD 4285 – **175)** PO Box 5337, Mackay MC QLD 4741 – **176)** 55 Main St WAlwa VIC 3709 – **177)** PO Box 612, Ballina NSW 2478 – **178)** PO Box 400, Toowoomba QLD 4350 – **179)** 43 Tamar St, Launceston TAS 7250 – **180)** CAAMA, PO Box 2608, Alice Springs NT 0871 – **181)** 5 Desiree Cl, Buderim QLD 4556 – **182)** Radio Rhema, PO Box 1465, Mt. Gambier SA 5290 – **183)** Radio Rhema, PO Box 200, Woombye QLD 4559 – **184)** 17 North St, Frewville SA 5063 – **185)** 228-230 North Tce, Adelaide SA 5000 – **186)** PO Box 259, Lane Cove NSW 1595 – **187)** c/- John Wright, 29 Milford Rd, Peakhurst NSW 2210 – **188)** GPO Box 1329, Adelaide SA 5001 (Italian) – **189)** PO Box 19, Innisfail QLD 4860 – **190)** Suite 1C, 9 Burwood Rd, Burwood NSW 2134 – **191)** GPO Box 572F, Hobart TAS 7001 – **192)** PO Box 9550, Geelong VIC 3220 – **193)** PO Box 1067, Mildura VIC 3502 – **194)** Radio Rhema, PO Box 980, Shepparton VIC 3630 – **195)** 1c Bell St, Preston VIC 3072 – **196)** 5 Macquarie St, Parramatta NSW 2150 – **197)** PO Box 177, Mareeba QLD 4880 – **198)** 46 Ord St, West Perth WA 6005 – **201)** Box 887, Port Pirie SA 5540 – **202)** PO Box 2160, Bayswater VIC 3153 – **203)** Hospital Radio Network, 70 Dawson St, Cooks Hill NSW 2300 – **204)** R. Salsa, PO Box 1141, Altona Meadows VIC 3028 – **205)** PO Box 630, Dalby QLD 4405 – **206)** Locked Bag 888, St. Peters NSW 2044 – **207)** 4/9 Mavis St, Revesby NSW 2212 – **208)** 5 Cheviot Pl, Sinnamon ark QLD 4073 – **209)** PO Box 124 Sale VIC 3853 – **210)** PO Box 995, Drouin VIC 3818 – **211)** PO Box 22 SAlamander Bay NSW 2317 – **212)** PO Box 966, Strawberry Hills NSW 2012 – **213)** PO Box 1962, Strawberry Hills NSW 2012 – **214)** PO Box 2161, Mt Gambier SA 5290 – **215)** Rankin St, Woorabinda QLD 4702 – **216)** PO Box 51, Yass NSW 2582 – **217)** PO Box 891, Manunda QLD 4870 – **218)** PO Box 1527, Tamworth NSW – **219)** GPO Box 568, Melbourne VIC 3001 – **220)** 11-17 Swanson St, Belconnen ACT 2617 – **221)** 12 Pickering Close Hoppers Crossing VIC 3029 – **222)** 5 Phoenix St, Castle Hill NSW 2154 – **223)** Suite 46-48, Pacific Centre, 223 Calam Rd, Sunnybank Hills QLD 4109 – **224)** PO Box 20, Cambellfield VIC 3061 – **225)** 622 Beryl St, Broken Hill NSW 2880 – **226)** PO Box 1042, Gosford NSW 2250 (country format) – **227)** PO Box 1120, Glenorchy TAS 7010 – **228)** 68-70 Little Ryrie Street, Geelong VIC 3220 – **229)** PO Box 1000, Dunedoo NSW 2844 – **230)** PO Box 1206, Bendigo Central VIC 3552 – **231)** PO Box 1388, Booragoon WA 6954 – 232) 15 Peneral Way, Bulleen VIC 3105 – 233) Traffic 1611, 5 Macquarie St, Parramatta NSW 2150 – 234) Francis Greenway High School, Lawson Ave, Beresfield NSW 2322 – 235) PO Box 2777, Ashmore QLD 4124 – 236) PO Box 644, Donnybrook WA 6239 – 237) PO Box 163, Dulwich Hill NSW 2203.

SPECIAL BROADCASTING SERVICE (SBS)
Locked Bag 028, Crows Nest, NSW 2065 ☎ +61 (02) 9430 2828 📠 +61 (02) 9430 3700

	MW	Call	kHz	kW	Service
1)	Wollongong	2EA	1035	2	National Prgr
1)	Sydney	2EA	1107	5	Sydney 1
2)	Melbourne	3EA	1224	5	Melbourne 1
1)	Newcastle	2EA	1413	5	National Prgr
1)	Canberra	1SBS	1440	2	National Prgr

	FM	Call	MHz	kW	Service
1)	Cairns	4SBS	90.5	1(d)	National Prgr
1)	Griffith	2SBS	92.7	1	National Prgr
2)	Melbourne	3SBS	93.1	100	Melbourne 2
1)	Brisbane	4SBS	93.3	96	National Prgr
1)	Leeton	2SBS	94.7	1	National Prgr

	FM	Call	MHz	kW	Service
1)	Adel. Hills	5SBS	95.1	2(d)	National Prgr
2)	Ballarat	3SBS	95.9	1	National Prgr
1)	Perth	6SBS	96.9	100	National Prgr
1)	Sydney	2SBS	97.7	150	Sydney 2
1)	Lismore	2SBS	98.9	1	National Prgr
1)	Wondai	4SBS	98.9	2(d)	National Prgr
1)	Renmark	5SBS	99.1	1	National Prgr
1)	Darwin	8SBS	100.9	32	National Prgr
1)	Sapphire	4SBS	103.5	1	National Prgr
1)	Canberra	2SBS	105.5	50	National Prgr
1)	Hobart	7SBS	105.7	56	National Prgr
1)	Adelaide	5SBS	106.3	32	National Prgr

Addresses and other information
1) Locked Bag 028, Crows Nest, NSW 2065
2) 2 Kavanagh St, South Melbourne, VIC 3205

AUSTRIA

LT: UTC +1h (27 Mar-30 Oct: +2h) — **Pop:** 7.8 million — **Pr.L:** German — **E.C:** 50Hz, 230V — **ITU:** AUT

ORF - ÖSTERREICHISCHER RUNDFUNK
ORF-Funkhaus, Argentinierstr. 30A, 1040 Wien ☎+43 1 50101 18699 📠 +43 1 50101 82500 **W:** www.orf.at
L.P: DG: Alexander Wrabetz. MD: Dr. Willy Mitsche. Tech. Dir: Peter Moosmann.

FM (MHz)	Ö-1	Ö-Reg	Ö-3	FM4	kW
Bad Gleichen		94.9a		6	
Bludenz	87.6	96.0h	98.8		4
Bregenz	93.3	98.2h	89.6	102.1	50
Bruck/Mur	87.6	93.2f	98.7	102.1	20
Graz	91.2	95.4f	89.2	101.7	67
Innsbruck	92.5	96.4g	88.5	101.4	45
Klagenfurt	92.8	97.8b	90.4	102.9	100
Kufstein	97.5	95.4y	103.9	99.9	5
Lienz	89.3	93.8b	99.3	101.0	2.6
		95.9g			2.6
Linz	97.5	95.2d	88.8	104.0	100
		90.1c			10
Mattersburg	89.0	96.2a	100.9	0.6/3/0.6	
Rechnitz	90.6	93.5a	87.9	97.4	6
		100.1f			3
Salzburg	90.9	94.8e	99.0	104.6	100
		101.2d			7
St. Pölten	97.0	91.5c	89.4	98.8	100
Schärding	92.5	99.5d	88.2		3/4/3
Schladming	94.3	96.3f	101.3	103.3	3
Semmering	90.3	95.8c	88.2	92.4	9
Spittal/Drau	91.6	100.4b	87.9	103.6	3
Weitra	92.7	95.7c	98.1	101.4	2
Wolfsberg	96.7	94.5b	99.5	102.3	1.5
Wien	92.0	89.9i	99.9	103.8	100
		97.9c			100
		94.7a			2.4

+ more than 500 low power txs

Österreich-1 (Ö1): 24h. **N:** on the h
Österreich 2 (Ö2): Regional services
a) Burgenland – Buchgraben 51, 7001 Eisenstadt. **W:** burgenland. orf.at **b)** Kärnten – Sponheimerstr. 13, 9010 Klagenfurt **W:** kaernten. orf.at **c)** Niederösterreich – Radioplatz 1, 3100 St. Pölten **W:** noe. orf.at **d)** Oberösterreich – Europaplatz 3, 4010 Linz **W:** ooe.orf.at **e)** Salzburg – Nonntaler-Haupstr. 49d, 5020 Salzburg **W:** salzburg.orf. at **f)** Steiermark – Marburgerstr. 20, 8042 Graz **W:** steiermark.orf.at **g)** Tirol – Rennweg 14, 6010 Innsbruck. **W:** tirol.orf.at **h)** Vorarlberg – Höchsterstrasse 38, 6851 Dornbirn **W:** vorarlberg.orf.at **i)** Wien – Argentinierstr. 30a, 1040 Wien **W:** wien.orf.at
Österreich 3 (Ö3): 24h. **N:** on the h.
FM4: Prgrs in English (0000-0500, 0900-1400) and German (0500-0900, 1400-2400) **W:** http://fm4.orf.at
Ann: "Österreich 1", "Ö2 (Wien, Niederösterreich, Tirol)", "Ö3"
IS: Österreich 1: composition by Werner Pirchner. Ö2: Composition by Bert Breit. Ö3: Electronic Music.
F.PI: Permission to use the following freq for commercial radio (kHz): 585, 630, 774, 891, 963, 1026, 1125, 1143, 1314, 1458, 1485, 1548 & 1602.

EXTERNAL SERVICES: Radio Ö1 International
see international radio section

PRIVATE STATIONS
KRONEHIT
Nationwide network with regional news windows
Daumegasse 1, A-1100 Wien **W:** www.kronehit.at

FM	MHz	kW	FM	MHz	kW
Weitra	90.2	3	Schärding	104.9	8
Linz	92.6	7	St. Pölten	105.3	100
Semmering	102.9	8	Schladming	105.6	2
B.Gleichenberg	103.2	8.5	Wien	105.8	100
Mattersburg	103.4	1	Innsbruck	106.5	32
Klagenfurt	103.7	1	Graz	107.5	1
Rechnitz	104.1	6			

+ 40 txs below 1kW

Other Private Stations by Area FM (MHz):
BURGENLAND: HiT FM, 106.3 Mattersburg, 1kW + 1rly
KÄRNTEN (Carinthia): Antenne Kärnten, 104.9 Klagenfurt, 100kW;
107.4 Spittal a.d.Drau, 3kW; 104.3 Wolfsberg, 2kW + 4rly – **Radio
Dva (ORF)-Agora** *(German/Slovenian progr)*, 105.5 Klagenfurt, 10kW;
106.8 Wolfsberg, 1kW + 7rly – **Radio Harmonie,** 95.2 Klagenfurt,
1kW + 4rly – **Radio Maria,** 99.3 Spittal a.d. Drau, 0.2kW
NIEDERÖSTERREICH (Lower Austria): HiT FM, 104.9 Weitra/
Nebelstein, 3kW; 103.3 Melk, 3kW; 100.8 St. Pölten, 2kW; 106.7
Hornstein, 1kW; 101.6 Horn 1kW + 9rly – **Campus Radio 94.4,** 94.4
St. Pölten, 0.2kW – **Radio Ypsilon,** 94.5 Hollabrunn, 0.1kW + 1rly
– **Radio Maria,** 104.7 Waidhofen a.d.Ybbs, 0.5kW – **Radio Arabella
99.4,** 99.4 Tulln-Judenau, 0.3kW + 1rly – **Radio Arabella 96.5,** 96.5
Ybbs a.d. Donau, 2kW + 2rly
OBERÖSTERREICH (Upper Austria): Life Radio, 100.5 Linz, 100kW;
102.6 Schärding, 3kW; 102.2 Bad Ischl, 0.4kW + 7rly – **Welle 1 Music
Radio,** 102.6 Steyr, 1,4kW + 3rly – **R. Arabella 96.7,** 96.7 Linz, 4kW
– **Radio FRO,** 105.0 Linz, 0.3kW – **LoungeFM,** 102.0 Linz, 2.2kW +
2rly – **Freies Radio Salzkammergut,** 100.2 Bad Ischl, 1kW; 107.3
Gmunden, 0.1kW + 4rly – **Antenne Wels,** 98.3 Wels, 0.1kW – **FR
107.1 Freies Radio Freistadt** 107.1 Freistadt, 0.6kW + 1rly
SALZBURG: Antenne Salzburg, 101.8 Salzburg/Gaisberg, 10kW;
105.9 Zell am See, 1kW; 102.5 St. Michael i. Lungau, 0.5kW + 15rly
– **Welle 1 Music Radio,** 106.2 Salzburg/Gaisberg, 2kW; 107.1 Zell
am See, 0.3kW; 107.5 St. Johann Pongau, 0.2kW – **Radiofabrik,** 107.5
Salzburg-Maria Plain, 0.5kW – **R. Arabella 102.5,** 102.5 Salzburg-
Nonntal, 0.3kW; **Energy Salzburg 94,0,** 94.0 Salzburg/Gaisberg,
0.3kW
STEIERMARK (Styria): Antenne Steiermark, 99.1 Graz/Schöckl,
80kW; 105.7 Bruck a.d. Mur, 20kW; 92.0 Schladming, 2kW; 106.1
Rechnitz, 3kW + 17rly – **Das Soundportal,** 100.4 Bad Gleichenberg
1.3kW; 97.9 Graz, 1kW + 5rly – **Radio Helsinki,** 92.6 Graz, 0.3kW
– **Radio Eins,** 89.6 Bruck a.d. Mur 8kW + 106.3 Schladming 2kW +
4rly – **Radio Grün-Weiss,** 106.6 Bruck a.d.Mur 1.2kW + 3rly – **Radio
West,** 107.3 Köflach, 0.1kW + 1rly – **Radio Freequenns,** 100.8
Liezen/Salberg, 1kW – **Radio Graz,** 94.2 Graz, 0.5kW
TIROL (Tyrol): Life Radio Tirol, 103.4 Inzing, 8kW; 101.8 Innsbruck,
1kW; 104.4 Lienz, 1kW; 105.4 Haiming, 1kW; 106.0 Landeck, 1kW;
106.8 Kufstein, 1kW + 8rly – **Antenne Tirol,** 100.8 Zirog (Italy), 1kW;
105.1 Innsbruck, 0.2kW; 106.4 Lienz 0.5kW; 104.6 Jenbach 0.2kW
+ 4rly – **Welle 1 Innsbruck,** 92.9 Innsbruck, 0.3kW – **Welle 1
Oberland,** 87.7 Ötz i. Tirol, 3kW; 103.9 Haiming (Telfs), 0.5kW; 104.3
Inzing, 0.4kW; 107.1 Landeck, 0.4kW + 5rly – **Welle 1 Ausserfern,**
104.0 Reutte, 0.3kW + 2rly – **Radio Freirad,** 105.9 Innsbruck, 0.3kW
– 97.0 Innsbruck, 1kW + 8rly – **Radio Osttirol,** 101.7 Matrei-Hopfgarten, 0.7kW; 107.2 Huben
0.6kW; 107.8 Lienz, 0.4kW; + 6rly – **Energy Innsbruck,** 99.9 Innsbruck
0.3kW – **Klassik Radio,** 95.5 Innsbruck, 0.1kW – **Radio Maria,** 107.9
Jenbach 0.2kW; 96.0 Mayrhofen 0.2kW
VORARLBERG: Antenne Vorarlberg, 106.5 Bregenz/Pfänder, 50kW;
101.1 Bludenz, 1.5kW; 105.1 Feldkirch 0.2kW + 2rly – **Radio Proton
104,6,** 104.6 Bludenz, 0.5kW; 95.9 Bregenz, 0.3kW; 104.3 Feldkirch
0.2kW – **Lokalradio Bregenz,** 103.2 Bregenz/Pfänder, 1kW (F.PI.)
WIEN (Vienna): Radio 88.6, 88.6 Wien/Kahlenberg, 10kW – **Antenne
Wien** 102.5, 102.5 Wien/Kahlenberg, 10kW – **R. Arabella 92.9,**
92.9 Wien-Donauturm, 3kW – **R. Energy 104.2,** 104.2Wien/RiFu-
Arsenal, 1kW – **R. Stephansdom,** 107.3Wien-Donauturm, 2kW – **R.
Orange 94,0,** 94.0Wien/Donauturm, 1kW; **98.3 superfly,** 98.3 Wien/
Donauturm, 0.4kW; **Lounge FM,** 103.2 Wien-Raiffeisenhaus, 0.3kW

AZERBAIJAN

L.T: UTC +4h (27 Mar-30 Oct: +5h) — **Pop:** 8.9 million — **Pr.L:** Azeri,
Russian, Armenian — **E.C:** 50Hz, 220V — **ITU:** AZE

MILLI TELEVISIYA VÄ RADIO SURASI (MTRS)
(National TV & Radio Council)
  Nizami küç. 105, AZ 1000 Baki ☎ +994 12 5983659 🖷 +994 12
4987668 **E:** office@ntrc.gov.az **W:** www.ntrc.gov.az
L.P: Chmn: Nusirävan Umud oglu Mähärrämov
NB. MTRS is the licensing body for broadcasting.

AZÄRBAYCAN TELEVIZIYA VÄ RADIO VERILISLÄRI
(Azerbaijan TV and Radio Broadcasting) (Gov)
  Mehdi Hüseyn küç. 1, AZ 1011 Baki ☎ +994 12 4923807 🖷 +994
12 4972020 **E:** info@aztv.az **W:** www.aztv.az
L.P: Chmn: Arif Nizam oglu Alisanov

MW	kHz	kW	MW	kHz	kW
Gäncä	549	70	Baki	891	30
Haciqabul	801	150	Sixli	1476	1
FM	**MHz**	**kW**	**FM**	**MHz**	**kW**
Gülüstan	88.0	5	Danaçi	103.0	2
Astara	90.0	1	Ordubad*	103.0	1
Poylu	90.0	5	Säki	104.0	1
Daskäsän	101.5	2	Babäk*	104.5	1.5
Lerik	101.5	2	Särur*	105.0	1
Yergüc	101.5	2	Baki	105.0	5

+ txs below 1kW. *) situated in the Naxçivan exclave
D.Prgr: AzR (Azärbaycan Radiosu) 24h.

ICTIMAI TELEVIZIYA VÄ RADIO YAYIMLARI SIRKETI
(Public TV and Radio Broadcasting Co.)
  Särifzadä küç. 241, AZ 1012 Baki ☎ +994 12 4313968 🖷 +994 12
4302958 **E:** info@itv.az **W:** www.itv.az
L.P: Dir: Ismayil Ömärov

FM	MHz	kW	FM	MHz	kW
Daskäsän	88.3	1	Danaçi	100.6	1
Lerik	88.3	1	Gülüstan	102.5	5
Yergüc	88.3	1	Poylu	103.0	5
Baki	90.0	5	Babäk*	103.0	1
Säki	91.6	1			

+ txs below 1kW. *) situated in the Naxçivan exclave
D.Prgr: Ictimai R. 24h.

OTHER STATIONS

FM	MHz	kW	Location	Station
1)	88.6	1	Danaçi	Bürc FM
7)	89.0	1	Gäncä	Xäzär FM
7)	89.0	1	Lerik	Xäzär FM
1)	91.0	1	Gülüstan	Bürc FM
3)	92.0	1	Lerik	R. ANS-ÇM
2)	100.0	1	Poylu	R. Antenn
2)	100.0	1	Lerik	R. Antenn
1)	100.7	1	Quba	Bürc FM
2)	101.0	1	Baki	R. Antenn
2)	101.2	1	Gülüstan	R. Antenn
3)	102.0	1	Baki	R. ANS-ÇM
3)	102.0	1	Säki	R. ANS-ÇM
1)	102.8	1	Lerik	Bürc FM
7)	103.0	2	Baki	Xäzär FM
4)	104.0	2	Baki	R. Space
3)	104.5	1	Xizi	R. ANS-ÇM
6)	105.0	1	Gäncä	Lider Jazz FM
5)	105.3	2	Daskäsän	R. Antenn
8)	105.5	2.5	Baki	Media FM
5)	106.3	2	Baki	R. Azad Azärbaycan
6)	107.0	1	Baki	Lider Jazz FM
6)	107.0	1	Imisli	Lider Jazz FM

NB: Txs below 1kW not listed.
Addresses & other information:
1) Atatürk pr. 28, AZ 1069 Baki. **E:** info@burc.fm – **2)** Särifzadä küç.
1, AZ 1072 Baki. **E:** info@antenn.az – **3)** Keçid 1128, 504-cü mähällä,
AZ 1073 Baki. **E:** info@ansradio.ws – **4)** C.Cabbarli küç. 33, AZ 1009
Baki. **E:** radio@spacetv.az – **5)** A.Abbaszadä küç. 8, AZ 1001 Baki. **E:**
azad_azerbaijan@mail.az – **6)** S. Mehdiyev küç. 83/23, AZ 1141 Baki. **E:**
mail@lidermedia.az – **7)** Atatürk pr. 28, AZ 1069 Baki. **E:** info@xazar.tv
– **8)** Teymur Äliyev küç. 25A, AZ 1130 Baki. E: mediafm@mediafm.az.

MOUNTAINOUS KARABAGH

ARTSAKH PUBLIC RADIO
  Tigran Mets St. 23a, Stepanakert ☎ +374 47 941733 **E:** artv@
ktsurf.net **L.P:** Dir: Slava Stepanyan
FM: Stepanakert 102.1MHz. **D.Prgr:** 0500-1700 in Armenian.

OTHER STATIONS

SW	kHz	kW	Location	Station
1)	9677v	5	Stepanakert	Ädalätin Säsi
FM	**MHz**	**kW**	**Location**	**Station**
6)	101.8	-	Stepanakert	Ekho Moskvy
4)	103.0	-	Stepanakert	R. Novoya volna
3)	104.3	-	Stepanakert	R. Pace
5)	105.0	-	Stepanakert	Mix FM
2)	105.5	-	Stepanakert	R. Hay
7)	107.5	-	Stepanakert	R. Vem

Addresses & other information:
1) Tigran Mets St. 23a, Stepanakert. In Azeri, Armenian, Russian:

Wed/Sat 0600-0630, Tue/Fri 1400-1430 (times may vary). – **2)** Azatamartikneri St. 18a, Stepanakert. Rel. Hay FM (Armenia) – **3)** Vazgen Sarkisyan St. 25, Stepanakert. **E:** pace@nk.am – **4)** A.Akopyan 30, Stepanakert. In Russian. – **5)** Azatamartikneri St. 18a, Stepanakert. **E:** radio@mix.am.m – **6)** Stepanakert. Rel. Ekho Moskvy (Russia) – **7)** Stepanakert. Rel. R. Vem (Armenia)

AZORES (Portugal)

LT: UTC -1h (27 Mar-30 Oct: UTC) — **Pop:** 245,000 — **Pr.L:** Portuguese — **E.C:** 50Hz, 220/380V — **ITU:** AZR

ANACOM-Autoridade Nacional de Comunicações, Delegação dos Açores
✉ Rua dos Valados, 9500-652 Relva (São Miguel) ☎+351 296 30 20 40 ▤ +351 291 30 20 41

RDP-RADIODIFUSÃO PORTUGUESA, S.A - Centro Regional da RDP-Açores
✉ Rua de Castelo Branco, 9500-761 Ponta Delgada ☎+351 296 201100 ▤ +351 296 201120 **E:** rdp.acores@rtp.pt **W:** www.rtp.pt
LP: Dir. Pedro Bicudo

RDP Antena 1 Açores

MW	kHz	kW	Island
Santa Bárbara	693	10	Terceira
Monte das Cruzes	828	1	Flores
Pico da Barrosa	837	10	São Miguel

FM (MHz)	Ant. 1	Ant. 2	Ant. 3	kW
Arrife	94.5	97.5		0.3
Cabeço Gordo	88.9	105.8		9.1
Cabeço Verde	98.1			1
Cascalho Negro	92.2			1
Espalamaca	93.8	101.4	102.7	0.03/0.5/1
Fajãzinha	100.4	103.7		1
Furnas	93.6			1
Lajes das Flores	102.6	97.0		0.2/0.5
Lajes do Pico	96.5	93.5		1
Macela	87.6	93.2		1
Monte das Cruzes	99.8	97.4		1
Morro Alto	93.5	91.9		1
Nordeste	104.6			0.1
Nordestinho	103.7	91.8		1
Pico Alto Sta. Maria	96.7			10
Pico Bartolomeu	92.7	89.9		0.5
Pico da Barrosa	97.9	101.7	87.7	33/33/30
Pico das Eguas	89.5			10
Pico do Geraldo	103.7	107.5		1
Pico do Jardim	97.0			0.9
Pico São Mateus	103.4			0.1
Ponta Delgada	94.1	100.8		0.3/1.3
Povoação	102.8	97.2		0.5
Santa Bárbara	90.5	98.9	103.0	35/35/30
Serra do Cume	99.7	89.2	103.9	0.9

D.Prgrs: all networks 24h
V: by QSL card via RDP in Lisboa
Prgr: RDP Antena 1 Açores carries its own prgrs M-F 0730-0200, Sat. 0700-0200, Sun. 0800-0200 LT; Antena 2 relays Lisboa 24h

DAB: RDP txs (9) at Ponta Delgada, Espalamaca, Stª Bárbara, Pico da Barrosa, Monte das Cruzes, Pico São Mateus, Serra do Cume, Cabeço Verde and Pico do Jardim, broadcasting Antena 1, 2, 3, RDP África and RDP Internacional on 225.648MHz, ch. 12, block B.

COMMERCIAL STATIONS:
RÁDIO RENASCENÇA – Emissora Católica Portuguesa (Rlg/Comm)
✉ (see Portugal) – **FM:** Pico da Barrosa 95.2MHz 50kW (RR), 100.0MHz 50kW (RFM)
Private stations (only active on FM, but owning MW licences):
RÁDIO CLUBE DE ANGRA – "A VOZ DA TERCEIRA" (Comm.)
✉ Av. Tenente Coronel José Agostinho, 4, 9700 Angra do Heroismo ☎+351 295 21 31 01 ▤ +351 295 21 31 02 **E:** administrativo@rcangra.com **W:** www.rcangra.com
FM: Santa Bárbara 101.1MHz 0.4kW, Serra do Cume (Terceira island) 94.7MHz 0.05kW
ESTAÇÃO EMISSORA DO CLUBE ASAS DO ATLÂNTICO (Comm.)
✉ Aeroporto de Santa Maria, 9580 Vila do Porto ☎+351 296 88 64 68 ▤+351 296 88 64 59 **E:** asasdoatlantico@clix.pt
FM: Pico Alto, 103.2MHz, 2kW. **D.Prgr:** 0700-2400 local time.

Other Stations
FM	Island	MHz	kW	Station, location
6)	Pico	100.2	0.5	R. Pico, Madalena do Pico

FM	Island	MHz	kW	Station, location
10)	São Miguel	102.4	0.5	Top FM, Pico da Barrosa
1)	Terceira	104.4	1	R. Horizonte Açores , Angra do Heroísmo
4)	Pico	104.7	0.5	R.Clube das Lajes do Pico, "A Voz da Montanha", Lajes do Pico
18)	São Miguel	105.0	0.5	R. Vila Franca, Pico da Barrosa
13)	São Miguel	105.5	2	R.Nova Cidade,Pico da Barrosa
15)	Flores	104.5	0.5	R.Flores/Canal FM
7)	São Miguel	106.0	0.5	R. Nordeste FM, Pico Bartolomeu
16)	Pico	106.1	0.5	R. Cais, São Roque do Pico
8)	São Miguel	106.3	0.8	R.Atlântida, Ponta Delgada
12)	Terceira	106.6	1	Top FM - Praia da Vitória, Santa Bárbara
17)	São Jorge	107.1	0.5	R. Lumena, Pico Rebineu

Addresses & other information (add +351 to tel/fax nos): **1)** Caminho do Meio, nº 51, S. Carlos, 9700 Angra do Heroísmo ☎295 216011/2/3/4 ▤295 216015 **E:** gerencia@horizonteacores.com **W:** www.horizonteacores.com – **2)** Rua Manuel Augusto Amaral 1-D 2-Direito 9500-222 Ponta Delgada ☎296 307 470 ▤296 307 479 **E:** radio@canal.fm **W:** www.canal.fm – **3)** Rua de São João, 38-B, 9900-129 Horta ☎292 29 33 90, ▤ 292 39 16 02 **E:** antenanove@iol.pt – **4)** R. S. Pedro, 9, 9930-129 Lajes do Pico ☎292 672299 ▤292 672950 **E:** radiomontanha@iol.pt **W:** www.radiomontanha.pt.to – **5)** (relays R. Horizonte Açores 104.4) Rua Nova da Misericórdia, 271, 9500-336 Ponta Delgada ☎296 653 911/2/3/4 ▤ 296 653 910 **E:** horizonte2@horizonteacores.com **W:** www.horizonteacores.com – **6)** Avenida Machado Serpa nº 54 9950-321 Madalena do Pico ☎292 622 727 ▤292 622 874 **E:** geral@radiopico.com, radiopico@sapo.pt **W:** www.radiopico.com – **7)** Rua da Assomada s/nº, 9630-070 São Pedro Nordestinho ☎296 098 480 **E:** nordeste.fm@gmail.com **W:** www.radionordestefm.net – **8)** Rua Bento José Morais 23 - 5º Sul, 9500-772 Ponta Delgada ☎296 201910 ▤296 629856 **E:** webmaster@radioatlantida.net **W:** www.radioatlantida.net – **9)** (relays TSF Lisboa) Rua Dr. Bruno Tavares Carreiro 34 - 2º - 9500 Ponta Delgada ☎296 202800 ▤296 202825 **E:** radioacores@acorianooriental.pt – **10)** Caminho do Meio, n.º 51, 9700-222 Angra do Heroísmo ☎295 216011/13 ▤295 216015 **E:** geral@mytop.fm ,topfm@live.com **W:** www.mytop.fm – **11)** Rua Nova da Misericórdia, 271-r/c, 9500-336 Ponta Delgada ☎ 296 65 39 11, ▤ 296 65 39 10 – **12)** (see nr. 10) – **13)** Rua Adolfo Coutinho de Medeiros Nº 24, Apartado 70 , 9600 Ribeira Grande ☎296 472738 / 296 472802 ▤296 472812 **E:** geral@radionovacidade.pt, radionovacidade@gmail.com **W:** www.radionovacidade.pt – **14)** R. do Corpo Santo, 37, 9880-368 Santa Cruz da Graciosa ☎295 732536 ▤295 712768 **E:** geral@radiograciosa.com **W:** www.radiograciosa.com – **15)** (see 2) – **16)** Largo do Museu da Indústria da Baleia ☎292 642930 ▤ 292 642934 **E:**geral@radiocais.com **W:** www.radiocais.com – **17)** Rua Cunha da Silveira, 25, Apartado 8, 9800-531 Velas ☎295 412 575/295 412 819/ ▤295 432458/459/460 ▤295 412810 **E:** radiolumena@radiolumena.com, radiolumena@iol.pt **W:** www.radiolumena.com – **18)** R. Nova Misericórdia, 271-R/C 9500-336 Vila Franca do Campo ☎296 654 112 / 296 653 053 ▤296 653 053 **W:** www.105fmazores.com

Military Stations:
RÁDIO LAJES – A VOZ DA FAP-FORÇA AÉREA PORTUGUESA (The Voice of the Portuguese Air Force)
✉ Comando da Zona Aérea dos Açores, 9760-290 Lajes, Terceira ☎+351 295 540891/2/3 ▤ +351 295 540791
E: fap.radiolajes@emfa.pt **W:** www.radiolajes.com/lajes
UNITED STATES AFRTS
✉ Lajes, Terceira, 9760 Praia da Vitoria ☎+351 295 57 34 97
W: www.lajes.af.mil
MW 1503kHz 100 watt (1kW nominal). **FM** 96.1MHz 150W
D.Prgr: Locally produced & relays of AFRTS

BAHAMAS

LT: UTC -5h (13 Mar–6 Nov: -4h) — **Pop:** 343,000 — **Pr.L:** English — **E.C:** 60Hz, 120/220V — **ITU:** BAH

ZNS – THE BROADCASTING CORPORATION OF THE BAHAMAS (Comm., Gov.)
✉ P.O. Box N-1347, Nassau ☎+1 242 502 6598 ▤ +1 242 322 3924 **W:** www.znsbahamas.com **LP:** Chairman: Michael Moss. GM: Anthony Foster. CEN: Donald Rolle.
MW: Nassau ZNS1 1540kHz 8kW, ZNS2 1240kHz 1kW Freeport ZNS3 810kHz 1kW **FM:** ZNS-FM: 104.5MHz 10kW. ZNS1: 107.1MHz 0.3kW So. Bahamas/107.7MHz 0.3kW No. Bahamas/107.9 0.3kW So. Bahamas
D.Prgr 24h. **N.** (ZNS1): 0800, 1300, 1830, 0000, 0300.
Ann. ZNS1: "This is Radio Bahamas" or "National Voice of Bahamas". ZNS2: "Inspiration 1240". ZNS3: This is Northern Service, Radio Bahamas". ZNS-FM: "Power 104.5"

TRIBUNE MEDIA GROUP (Comm.)

Radio House, P.O. Box N-3207, Nassau ☎+1 242 328 0950 🖷 +1 242 356 5343 **W:** www.100jamz.com **L.P:** PD Eric Ward.
Stations: Cool 96 FM: Yellow Pine Street, P.O. Box F-40773, Freeport ☎+1 242 352 7440 🖷 +1 242 352 8709 **W:** www.cool96fm.com **L.P:** GM: Andrea Gottlieb. **FM:** 96.1MHz – **100 Jamz. FM:** Nassau 100.3, Freeport 100.3, Abaco: 100.1 and Coopers Town: 100.5MHz **Joy FM,** P.O.Box N-1807, Nassau +1 242 356 5110. **L.P:** Steven Haughey. **FM:** Nassau 101.9MHz – **Y98.7. FM:** Nassau 98.7MHz

OTHER STATIONS:

RADIO ABACO, P.O. Box AB-20418, Abaco ☎+1 242 367 4935 **W:** www.radioabaco935.com **FM:** 93.5MHz – **BREEZE FM** 98.3MHz, Georgetown, Exuma. ☎+1 242 225 0775 **W:** www.thebreezefmexuma.com **L.P.** CEO Dwight Hart. **FM:** 98.3MHz – **GEMS RADIO,** P.O. Box SS-6094, Nassau ☎+1 242 326 4381 🖷 +1 242 326 4371 **W:** www.gemsbahamas.com **L.P.:** CEO Deborah Bartlett. **FM:** Nassau: 105.9MHz – **ISLAND FM,** P.O.Box N-1807, Nassau ☎+1 242 3322 8826 **W:** www.islandfmonline.com **FM:** Nassau 102.9MHz – **LOVE-97 FM,** P.O. Box N-3909, Nassau ☎+1 242 356 4960 🖷 +1 242 356 7256 **FM:** Nassau 97.5MHz – **MIX 102,** P.O.Box F-44008, Freeport ☎+1 242 373 2275 🖷 +1 242 373 2271 **L.P:** GM: Don Martin. **FM:** Freeport 102.1MHz 5kW – **MORE 94 FM – THE BAHAMAS SUPER STATION:** P.O. Box CR54245, Nassau ☎+1 242 361 2447 🖷 +1 242 361 2448 **W:** www.more94fm.com **FM:** Nassau 94.9MHz – **SPLASH FM,** P.O. Box EL-27495, Spanish Wells, Eleuthera ☎+1 242 333 4638 **W:** www.splash899fm.com **L.P:** Chris Forsythe **FM:** Eleuthera: 89.9MHz, Nassau: 92.5MHz. Abaco (95.5MHz) and South Eleuthera (98.5MHz)

BAHRAIN

LT: UTC +3h — **Pop:** 700,000 — **Pr.L:** Arabic — **E.C:** 50Hz, 230V(60/110 at Awala) — **ITU:** BHR

BAHRAIN RADIO & TV CORPORATION (BRTC, Gov.)

P.O.Box 1075, Manama ☎+973 17780780 🖷 +973 17780911 **W:** www.radiobahrain.fm **E:** info@bahrainradio.com **L.P:** CEO: Ahmed Najim. Dir. of Bc: Hamad Al-Manai. Act. Dir. Tech: Mr. Abdulla Ahmed Al-Balooshi.
General Prgr. in Arabic: 24h on **MW:** 801kHz 100kW, 1458kHz 10kW. **FM:** 90.9MHz 3kW.
Quran Prgr. in Arabic: Quran recitations & religious affairs 0300-2100 on **MW:** 612kHz 100kW.
2nd Prgr. in Arabic : 0500-1700 on **MW:** 1521kHz 10kW. **FM:** 93.3MHz 3kW.
Prgr. for the Indian community: 24h on **FM:** 104.2MHz.
English Sce (R. Bahrain): 0300-2100 on **MW:** 1584kHz 1kW, **FM:** 96.5MHz 2.5kW. 1600-2100 on 99.5MHz 0.5kW.
Shabab FM (Youth prgr.): Manama 98.4MHz. **W:** 984shabab.com
Ann: A: "Idha'atul-Bahrain". E: "Radio Bahrain".
IS: Local composition on guitar and violin.
Relays for abroad on shortwave: see International Radio section.

Other stations:
Sawt el-Ghad, Manama: 94.8MHz. Web: sawtelghad.com
Voice FM, Manama: 104.2MHz. Web: www.voicebahrainfm.com
Emarat FM, Manama: 92.3MHz. See main entry under UAE.
Panorama FM, Manama 103.0MHz. See main entry under UAE.
BBC World Sce, Manama: English 101.0MHz, Arabic 103.8MHz.
Deutsche Welle/Monte Carlo Doualiya, Manama: 90.9MHz 1kW.
R. Sawa, Manama 89.2MHz 1kW. 24h in Arabic.

BANGLADESH

LT: UTC +6h — **Pop:** 141 million — **Pr.L:** Bengali — **E.C:** 50Hz, 220/440V — **ITU:** BGD

BANGLADESH BETAR (Gov.)

National Broadcasting Authority, NBA House, 121 Kazi Nazrul Islam Ave, Dhaka-1000 ☎+880 2 ,8625538 8625904 🖷 +880 2 8612021 **E:** rrc@dhaka.net **W:** www.betar.org.bd
L.P: DG: A K M Shameem Chowhdhuri (DDG Prgrs): Abdur Rouf DDG (News) Nasimul Quadir Chowdhury: Sr. Engineer: Zia Hassan Dy.Station Engineer: Zakir Hussain Asst. Radio Engineer: Muzibur Rahman

MW	kHz	kW	Times
Khulna	558	100	0030-0400, 0600-1710
Dhaka-B	630	100	0000-0145, 0300-1710, 1800-2100
Dhaka-A	693	1000	0030-0610, 0830-1730
Rajshahi	846	100	0030-0400, 0600-1710
Chittagong	873	100	0030-0400, 0600-1710
Sylhet	963	20	0030-0400, 0800-1710
Thakurgaon	999	10	0950-1710

MW	kHz	kW	Times
Rangpur	1053	20	0030-0400, 0800-1710
Rajshahi	1080	10	0030-0400 0600-1710
Rangamati	1161	10	0530-1030
Dhaka-C	1170	10	0900-1100
Barishal	1287	10	0445-1115
Cox's Bazar	1314	10	0545-1045
Comilla	1413	10	1000-1710
Bandorban	1431	10	0530-1030

SW	kHz	kW	Times
Shavar	4750	100	0600-1500

NB: Relays Dhaka A prog

FM	MHz	kW	Rel	Times
Chittagong	105.4	2	b, d	0030-0400, 1300-1710
Comilla	101.2	2	b, d	0030-0400 1100-1710
Dhaka	97.6	5	d	0200-0600 0930-1330
Dhaka	103.2	5		0030-0400 1300-1710
Dhaka 100	100.0	3	b	0700-1000 1800-2100
Khulna	102.0	1	b, d	0030-0400 1300-1710
Rajshahi	104.0	5		0030-0400 1300-1710
Rajshahi	105.0	1	b, d	0030-0400 1300-1710
Rangpur	105.4	1	b, d	0030-0400, 1300-1710
Sylhet	105.0	1	b, d	0030-0400, 1300-1710
Thakurgaon	92.0	5		1000-1710
Traffic Channel	88.8	10		0200-1400

b) Rel. BBC World Service 0030-0100, 0130-0200, 1330-1400, 1630-1700. d) Rel. Deutsche Welle 0200-0230, 1400-1430
N. in English: 0200, 1100, 1530, 1805. **N. in Bengali:** 0100, 0300, 0400, 0500, 0600, 0900, 1000, 1200, 1430, 1600, 1700 **SAARC N. in Bengali:** 1235 **SAARC N. in English:** 1250 Every Mon
F.PI: , FM network covering entire country upgrading to 10kW. Set up of one 10kW FM tx in Dhaka. Set up of six FM 10kW txs for different regional stns. Replacement of one 250kW SW tx located in Kabirpur, Dhaka under implementation. China R Int to start broadcast Bengali & English via FM in Dhaka and Chittagong. 14 community radio stns to be set up. R. Japan will provide tech. support to Bangladesh Betar in exchange for broadcasting their prog via FM in Chittagong, Comilla , Dhaka, Khulna, Rajshahi, Rangpur and Sylhet.

EXTERNAL SERVICE: BANGLADESH BETAR
See International Radio section.

OTHER STATIONS:

RADIO TODAY
Radio Broadcasting FM (Bangladesh) Co. Ltd Awal Centre (13th & 19th floor), Kamal Atraturk Avenue, Banani, Dhaka 1213 ☎+880 2 8829293 8836491, 8836492 🖷 +880 2 8836494
E: info@radiotodaybd.fm **W:** www.radiotodaybd.fm
FM: 89.6MHz 10kW Dhaka 24h, 88.6MHz Chittagong
RADIO FOORTI
Radio Foortii Limited Landmark (8th flr), 12-14 Gulshan-2 North C/A, Dhaka 1212 ☎+880 2 8835747 8835748 **E:** info@radiofoorti.fm **W:** www.radiofoorti.fm **FM:** 88.0MHz 20kW Dhaka 24h, Chjittagong, Sylhet, Mymensingh, Rajshahi, Barishal, Khulna, Cox's Bazar
UNIWAVE BROADCASTING COMPANY LTD
L.P: CEO: Zulfiker Ahmed **E:** info@radioaamar.com **W:** www.radio-aamar.com **FM:** Radio Aamar 24h 88.4MHz Dhaka
AYENA BROADC. CORP.
Dhaka Trade Centre, 99 Kazi Nazrul Islam Avenue, Kawran Bazar, Dhaka **W:** www.abcradiobd.fm
L.P: Man. Dir.: Matiur Rahman Choudhury. **FM:** 89.2MHz 24h in Dhaka.

F.PI: , License to be issued for 8 new pvt. FM station: Peoples Radio Ltd, Dhaka FM Ltd, Media City Ltd, Asiatic Marketing Communication, Ayurvedio Pharmacy (Dhaka) Ltd, Next Wave Broadcasting Co. Ltd, Asian Radio Ltd and Gunchil Ltd.

BARBADOS

LT: UTC -4h — **Pop:** 283,000 — **Pr.L:** English — **E.C:** 50Hz 110V — **ITU:** BRB

CARIBBEAN BROADCASTING CORP. (Gov. Comm.)

The Pine, Wildey, St.Michael ☎+1 246 429 2041 🖷 +1 246 429 4795 **W:** www.cbc.bb **L.P:** Chairman: Leroy Parris. GM: Lars G. O. Söderström. Head of radio: Pearson Bowen.
MW: CBC Radio 900 AM 900kHz 5kW: 24h
FM: CBC Radio 900 AM 94.7MHz 5kW: 24h – **98.1 FM The One:** 98.1MHz 5kW: 24h – **Quality/Q FM 100.7** 100.7MHz 5kW

BARBADOS BROADCASTING SERVICE (Comm.)

Astoria, St George ☎+1 246 437 9550 🖷 +1 246 437 9203

L.P: MD: Anthony T. Brian. GM: Shery Anne Padmore.
FM: BBS-FM 90.7MHz: 24h – **Faith FM** (Rlg.) 102.1MHz: 24h

STARCOM NETWORK INC. (Comm.)

River Road, P.O. Box 1267, Bridgetown ☎+1 246 430 7300 📠 +1 246 426 5377 **W:** www.starcomnetwork.net & www.vob929.com
L.P: CEO: Victor Fernandez. Mgr Ops: Lennox Edwards. PM (Radio): Patrick Gollop
FM: Gospel 97.5 (Rlg.): 97.5MHz – **VOB Voice of Barbados:** 92.9MHz – **HOTT 95.3 FM:** 95.3MHz – **LOVE 104.1 FM:** 104.1MHz
All stns: 24h

Other Stations

BBC: FM 92.1. 24h relay of BBC World Service. — **EDUCATIONAL RADIO 91.1FM (Educ.):** Elsie Payne Complex, Constitution Rd., St. Michael **FM:** 91.1MHz. **D.Prgr:** Mon-Fri 0900-1213 & Tue + Thu 1400-1500 during school terms only and irr. at other times — **MIX 96.9 (Comm.):** 1st floor Sheraton Mall, Sargeants Village ☎ +1 246 228 4183 📠 +1 246 228 3550 **W:** www.mix969fm.com **L.P:** MD: Scott Weatherhead. **FM:** 96.9MHz. — **RADIO GED (Educ.):** Barbados Community College, General Education Dept., Eyrie Howells Cross Road, St. Michael ☎+1 246 426 3312 📠 +1 246 429 5935 **FM:** 106.1MHz (0.02kW) **D.Prgr:** 1500-1900MF during school terms only. — **SLAM FM (Comm.):** Hagget Hall, St. Michael ☎+1 246434 1011 📠 +1 246 437 7526. **W:** www.slam101fm.com. **FM:** 101.1MHz

BELARUS

L.T: UTC +2h (27 Mar-30 Oct: +3h) — **Pop:** 9.4 million — **Pr.L:** Belarusian, Russian — **E.C:** 50Hz, 220V — **ITU:** BLR

MINISTERSTVA KULTURY (Ministry of Culture)

pr. Peramozcaú 11, 220004 Minsk ☎ +375 17 2037574
E: ministerstvo@kultura.by **W:** www.kultura.by
L.P: Minister: Pavel Latruška
NB. The Ministry of Culture is issuing broadcasting licenses.

NACYJANALNAJA DZIARZAÚNAJA TELERADYJO-KAMPANIJA RESPUBLIKI BELARUS (Gov)
(State TV & Radio Co. of Belarus)

vul. Makajonka 9, 220807 Minsk ☎ +375 17 3896352 📠 +375 17 2678182 **E:** pr@tvr.by **W:** www.tvr.by
Belaruskaje Radyjo (BR): vul. Cyrvonaja 4, 220807 Minsk; exc. **Radyus FM:** vul. Cyhunacnaja 27/2, 220014 Minsk.
L.P: Chmn: Aliaksandr Zimóuski

LW/MW	kHz	kW	Prgr	MW	kHz	kW	Prgr*
Sasnovy	279	500	BR1	Salihorsk	1026	5	BR2
Hrodna	1008	7	BR2	Minsk (a)	1125	75	BR2
Brest	1026	7	BR2	Sasnovy	1170	800	F, BR1
Miadziel	1026	25	BR2	Brest	1278	10	BR1a
SW	**kHz**	**kW**	**Prgr***	**SW**	**kHz**	**kW**	**Prgr***
Brest	6010	5	BR1a	Mahilioú	6190	5	BR1d
Hrodna	6040	5	BR1c	Mahilioú	7235	5	BR1d
Brest	6070	5	BR1a	Hrodna	7265	2.5	BR2
Minsk (a)	6080	150	BR1	Hrodna	7280	5	BR1c
Minsk (a)	6115	75	BR1				

*) incl. reg. prgrs (see below), F=External Service (a) Kalodziscy
NB: BR1 relay on SW for listeners in Russia: see Int Radio section

FM (MHz)	*BR1[1]	BR2[1]	*BR1[2]	BR2[2]	RS	RFM	kW
Asipovicy	67.46d	71.69	91.0d		107.7	72.47	104.9	2x4/1/3x2
Asveja	-	-	103.5e	106.0	-	-	4
Babrujsk	71.45d	68.96	101.6d	106.0	73.01	104.1	2x4/2x2/4/2
Berazino	70.79	-	94.7	107.4	67.07	100.7	3x1/0.25/1
Baranavicy	-	-	-	102.5	71.18	-	0.5/0.25
Bycycha	-	-	101.0e	101.8	-	-	0.5
Brahin	67.37b	68.30	103.3b	105.8	69.11	100.8	2x4/2/2x2
Braslaú	69.08e	71.99	105.7e	107.7	73.49	102.3	2x4/2/2x4/2
Brest	70.91a	71.69	100.0a	88.5	72.47	103.7	2x4/2x1/2x4
Drahicyn	72.14a	-	101.8a	-	69.80	102.4	0.5/1/0.5/1
Heraniony	72.32c	68.39	105.8c	102.2	69.26	103.3	2x4/2/1/4/2
Homiel	67.76b	69.26	105.1b	91.5	66.20	100.1	2x2/2x1/21
Hrodna	66.98c	66.20	103.0c	95.0	68.90	100.5	3x4/1/4/2
Kapyl	-	70.97	102.3	103.6	73.22	103.9	01/4/05/01/05
Kascjukovicy	66.47	68.03	104.7	107.2	69.38	102.2	2x4/2/1/4/2
Krupki	-	-	103.8	-	106.3	-	4
Liubca	-	-	-	-	70.85	-	0.5
Luki	-	-	90.6c	94.7	-	104.3	1
Mahilioú	72.74d	71.96	105.9d	99.1	71.18	100.9	2x4/2/1/4/1
Mazyr	-	-	101.2	103.5	-	104.8	0.5
Miadziel	68.69	70.31	106.4	104.9	66.86	103.9	2x4/2x2/2x4
Minsk	71.33	70.43	106.2	102.9	72.89	103.7	2x4/2/1/2x4
Minsk	-	-	-	-	72.11	4	
Mscislaú	66.89d	-	106.7d	101.7	73.73	102.9	3x1/0.25/1

FM (MHz)	*BR1[1]	BR2[1]	*BR1[2]	BR2[2]	RS	RFM	kW
Pinsk	66.32a	67.10	104.5a	106.8	67.88	102.0	2x4/2/3x4
Pruzany	-	-	98.3a	103.1	-	-	0.1
Salihorsk	70.22	72.23	100.3	106.7	68.57	102.8	2x1/4/2/1/4
Slonim	66.56c	67.34	106.5c	97.3	-	104.0	3x4/1/4
Smarhon	67.97c	70.13	103.6c	-	66.38	101.4	2x2/3x1
Smiatanicy	67.22b	68.00	106.3b	104.6	70.28	103.8	4
Staryja Darohi	-	-	100.6	103.1	-	-	1
Strelciki	-	72.23	-	-	-	-	0.5
Stolin	-	73.79	104.1a	-	-	-	0.5
Svislac	-	66.08	105.9c	98.9	68.72	96.7	
2x4/0.5/4/0.1							
Trokeniki	-	70.76	104.5c	-	-	-	0.5/1
Ušacy	72.65e	66.74	106.7e	101.7	70.94	102.7	3x4/1/2x4
Valozyn	-	-	101.0	-	-	-	0.15
Vasilievicy	-	-	102.0b	106.5	73.19	-	1
Viciebsk	70.67e	69.92	100.5c	99.3	72.26	105.5	2x4/2x2/2x4
Vidzy	72.80e	74.00	-	-	-	-	0.03
Vorša	67.85e	69.14	107.0e	105.0	73.82	100.2	2/4/3x1/2
Zlobin	69.68b	71.03	105.5b	101.0	71.81	100.5	3x2/1/2x2

Simulcast via: [1]) OIRT FM netw. [2]) CCIR FM netw.;*) incl. reg.prgrs
D.Prgr: BR1 (Peršy nacyjanalny kanal): 24h in Belarusian, Russian on FM/LW; limited schedule* on MW/SW. Includes reg. prgrs (see below). Rel. R. Stalica (not via txs with regional prgrs): 0440-0500 (W), 1600-1700. NB: SW txs with reg. prgrs may rel. the separate regional channels at nighttime (irr.) (see under "BR regional stns"). – **BR2 (Kanal Kultura):** 0400-2400 in Belarusian on FM; limited schedule* on MW/SW. – **RS (Radyjo Stalica):** 0400-2400 in Belarusian. **RFM (Radyus FM):** 0300-0200 mainly in Russian. *) Generally 0300-0200 (BR1), 1600-2200 (BR2), but subject to changes.
External Service (Radio Station Belarus): see Int. Radio section. On FM (MHz): Brest 96.4 (0.5kW), Hrodna 96.9 (1kW), Heraniony 99.9 (1kW), Svislac 100.8 (1kW), Miadziel 102.0 (1kW), Braslaú 106.6 (1kW)

BR REGIONAL STATIONS (Gov)

D.Prgr: via BR1 txs (see tx table) Mon-Sat 0440-0500 & 1600-1700. Also fulltime via separate tx networks as shown below. All stns broadcast in Belarusian and Russian. TRK = teleradyjokampanija. **a) TRK "Brest"**, vul. Kujbyšava 64, 224030 Brest: "R. Brest". **E:** radio-brestgosti@tut.by. Via separate network: 24h on (MHz) 69.08 (Pinsk 4kW), 69.44 (Slonim), 69.68 (Brest), 94.6 (Pinsk 1kW), 101.1 (Baranavicy 0.5kW), 102.5 (Stolin 0.5kW), 104.2 (Drahicyn 2kW), 104.8 (Brest 2kW), 105.8 (Pruzany 0.1kW) – **b) TRK "Homiel"**, vul. Puškina 8, 246050 Homiel: "Homiel FM". **E:** radio@tvrgomel.by. Via separate network: 0300-0200 on (MHz) 66.44 (Smiatanicy 4kW), 66.98 (Homiel 2kW), 68.45 (Zlobin 2kW), 69.92 (Brahin 4kW), 101.3 (Homiel 1kW), 103.0 (Zlobin 2kW), 105.3 (Brahin 1kW), 107.8 (Vasilievicy 1kW). – **c) TRK "Hrodna"**, vul. Horkaha 85, 230015 Hrodna: "R. Hrodna". **E:** radio@tvr.grodno.by. Polish: Mon 1600-1640. Via separate network: 24h on (MHz) 67.76 (Hrodna 4kW), 68.12 (Slonim 4kW), 71.54 (Heraniony 4kW), 72.80 (Trokeniki 0.5kW), 101.2 (Hrodna 1kW), 103.8 (Luki 1kW), 102.5 (Slonim 1kW), 102.8 (Smarhon 1kW), 104.4 (Svislac 4kW), 107.8 (Heraniony 1kW). – **d) TRK "Mahilioú"**, vul. Peršamajskaja 83, 212030 Mahilioú: "R. Mahilioú". **E:** radiomogilev@tut.by. Via seperate network: 0440-2000 on (MHz) 66.02 (Babrujsk 4kW), 67.25 (Kascjukovicy 4kW), 70.10 (Mahilioú 4kW), 96.4 (Mahilioú 1kW), 99.4 (Kascjukovicy 1kW) 100.4 (Mscislaú 0.25kW), 102.3 (Asipovicy 1kW), 105.6 (Babrujsk 1kW). – **e) TRK "Viciebsk"**, vul. Kamunistycnaja 8, 210602 Viciebsk: "R. Viciebsk". **E:** info@radio.vitebsk.by. Via seperate network: 24h on (MHz) 68.30 (Ušacy 4kW), 71.48 (Viciebsk 4kW), 91.2 (Viciebsk 2kW), 102.0 (Asveja 1kW), 102.4 (Vorša 2kW), 104.0 (Bycycha 0.5kW), 104.6 (Braslaú 1kW) 107.8 (Ušacy 4kW).

OTHER REGIONAL STATIONS (Gov)

Minskaya volna: vul. Ckalova 5, 220039 Minsk. In Russian, Belarusian 0400-2200 on (MHz) 97.4 (Minsk 2kW), 102.4 (Miadziel 1kW), 102.6 (Krupki 0.5kW), 104.2 (Barysaú 0.5kW), 104.8 (Kapyl 1kW), 105.3 (Salihorsk 2kW), 105.5 (Berazino 1kW), 105.6 (Staryja Darohi 1kW), 107.4 (Maladzecna 2kW).

OTHER STATIONS

FM	MHz	kW	Location	Station
7)	67.70	1	Minsk	Avtoradio
1A)	91.0	1	Homiel	R. BA
5)	92.2	1	Viciebsk	Pilot FM
15)	92.4	2	Minsk	R. Minsk
5)	93.2	1	Mahilioú	Pilot FM
1A)	95.7	1	Hrodna	R. BA
1B)	96.2	1	Minsk	Melodii veka
16)	97.8	1	Viciebsk	Evropa plus Vitebsk
9)	98.4	1	Minsk	Novoye R.
11)	98.6	1	Mahilioú	Nashe R.
9)	98.7	1	Viciebsk	Novoye R.
3)	98.9	2	Minsk	Russkoye R.

FM	MHz	kW	Location	Station
6)	99.5	2	Minsk	R. Unistar
13)	100.4	1.5	Minsk	Hit FM
4)	100.8	2	Brest	Alfa-Radio
2)	101.2	1	Brest	R. ROKS
5)	101.2	2	Minsk	Pilot FM
1A)	101.5	1	Slonim	R. BA
14)	101.7	2	Minsk	R. ONT
8)	101.8	1	Viciebsk	R. Mir Belarus
5)	102.1	1	Hrodna	Pilot FM
2)	102.1	1	Minsk	R. ROKS
6)	102.3	1	Brest	R. Unistar
2)	102.6	1	Homiel	R. ROKS
5)	102.9	1	Brest	Pilot FM
2)	103.0	1	Viciebsk	R. ROKS
2)	103.4	1	Mahilioú	R. ROKS
8)	103.6	1	Babrujsk	R. Mir Belarus
1A)	104.5	1	Mahilioú	R. BA
12)	104.6	4	Viciebsk	Retro FM
1A)	104.6	4	Minsk	R. BA
7)	105.1	1	Minsk	Avtoradio
10)	106.1	1	Pinsk	Svaje R.
1A)	106.2	4	Brest	R. BA
15)	106.4	1	Viciebsk	R. Minsk
8)	106.6	1	Brest	R. Mir Belarus
9)	106.7	1	Homiel	Novoye R.
2)	106.9	1	Hrodna	R. ROKS
8)	107.1	4	Minsk	R. Mir Belarus
4)	107.6	1	Viciebsk	Alfa-Radio
8)	107.8	4	Mahilioú	R. Mir Belarus
4)	107.9	2	Minsk	Alfa-Radio

NB: Txs below 1kW not listed.

Addresses & other information:

1A,B) vul. Surhanova 26, 220010 Minsk. **E:** radioba@list.ru – **2)** vul. Starazoúskaja 8a, 220002 Minsk. **E:** radio@roks.com. In Russian. – **3)** vul. Starazoúskaja 8a, 220002 Minsk. E:info@rusradio.by. In Russian. – **4)** pr. Nezaleznasci 181, 220125 Minsk. **E:** alpha@alpha.by In Russian. – **5)** vul. K.Marksa 40, 220030 Minsk. **E:** pilot-fm@mail.ru – **6)** pr. Nezaleznasci 4, 220050 Minsk. **E:** radio@unistar.by. In Russian. – **7)** vul. Prytyckaha 62, 220050 Minsk. **E:** ar_minsk@mail.ru. In Russian. – **8)** vul. Kamunistycny 17, 220029 Minsk. **E:** info@radiomir.by. In Russian. – **9)** pr. Puškina 39, 220092 Minsk. **E:** reklama@novo-eradio.by In Russian. – **10)** vul. Karasiova 6, 225710 Pinsk. **E:** v_vizit@varjag.net. In Russian. – **11)** vul. Caljuskincaú 105, 212003 Mahilioú. **E:** nashe_mogilev@mail.ru – **12)** vul. Hoholia 11, 210601 Viciebsk. Rel. Retro FM (Russia). – **13)** vul. Kamunistycny 6a, 220029 Minsk. Rel. Hit FM (Russia). – **14)** vul. Kamunistycny 6, 220029 Minsk. **E:** fm@ont.by – **15)** zav. Kaliningradski 20a, 220012 Minsk. **E:** radio924fm@gmail.com. In Russian. – **16)** Maskaúski pr. 10, 210015 Viciebsk. In Russian.
Radio via DTT: see TV section

BELGIUM

L.T: UTC +1h (27 Mar-30 Oct: +2h) — **Pop:** 10.4 million — **Pr.L:** Flemish, French, German — **E.C:** 50Hz, 230V — **ITU:** BEL

FLANDERS Pop: 6.17 million — **Pr.L:** Flemish

VLAAMSE RADIO EN TELEVISIEOMROEP (VRT) (Pub) Flemish (Dutch) Language Network
Public Sce. grants by Flemish government.
✉ VRT, August Reyerslaan 52, B-1043 Brussels ☎+32 2 741 3111 🖷 +32 2 734 9351 **W:** www.vrt.be **E:** info@vrt.be
L.P: MD: Mieke Berendsen.
Regional Centres Radio 2:
Antwerpen: Jan Van Rijswijcklaan 157, 2018 Antwerpen ☎+32 3 2479111🖷+32 3 2378282 **E:** redactieantwerpen@radio2.be
Vlaams-Brabant: Dikke Lindelaan 2, 1020 Brussels ☎+32 2 7414147 🖷 +32 2 4780800 **E:** redactievlaamsbrabant@radio2.be
Oost-Vlaanderen: Martelaarslaan 232, 9000 Gent ☎+32 9 2247256 🖷 +32 9 2254903 **E:** redactieoostvlaanderen@radio2.be
West-Vlaanderen: Doorniksesteenweg 241B, 8500 Kortrijk ☎+32 56 247311 🖷 +32 56 221358 **E:** redactiewestvlaanderen@radio2.be
Limburg: Via Media 2, 3500 Hasselt ☎+32 11 249611 🖷 +32 11 242436 **E:** redactielimburg@radio2.be

MW		kHz		kW	Prgr	
Wolvertem		927		100	Radio Vlaanderen Info	

FM (MHz)	R1	R2	RK	SB	M	kW
Antwerpen	-	-	92.0	-	-	1
Brussegem	-	90.7	-	-	-	1
Brussels	-	-	-	88.3	-	1
Diest	-	92.4	-	-	-	1
Egem O.+W.-VI.	95.7	-	90.4	102.1	101.5	50/50/50/40

FM (MHz)	R1	R2	RK	SB	M	kW
Egem O.-VI	-	98.6	-	-	-	50
Egem W.-VI.	-	-	-	100.1	-	50
Genk	99.9	97.9	89.9	101.4	102.0	20/20/20/40/40
	-	-	-	-	93.0	3
Gent	-	-	-	94.5	-	0.5
Leuven	98.5	-	-	88.0	-	0.5/0.5
N'kerken Waas	-	89.8	-	-	-	1
Schoten	94.2	97.5	96.4	100.9	89.0	20/20/3/40/20
St-Pieters-Leeuw	91.7	93.7	89.5	100.6	97.0	50/50/50/50/2
Veltem	-	88.7	-	-	94.8	1/1

+3 txs under 1kW

R1=Radio Een (information and music), **R2**=Radio Twee (light & popular music), **RK**=Radio Klara (classical music), **SB**=Studio Brussel (youth stn), **M**=MNM (hit music stn)
Radio 2 Regional programmes: 0500-0700 (M-F), 1100-1200 (M-F), 1500-1700 (daily) on FM
Radio Vlaanderen Info: relay of Radio 1 and Radio 2 prgrs on 927kHz
Sporza: broadcasts on 927kHz and replaces normal Radio 1 prgrs during sports events **W:** www.sporza.be
D.Prgr: Night prgr on all frequencies.
DAB: 223.936MHz all services plus Sporza, MNM Hits, Klara continuo, Rvi, and Nieuws+.
Ann: R1:"Radio Een", R2: "Radio Twee", RK: "Radio Klara", SB: "Studio Brussel", M: "MNM"

EXTERNAL SERVICE: Radio Vlaanderen Internationaal
See International Broadcasting section

COMMERCIAL NETWORKS:
NB: For further information on all radio stns in Flanders visit **W:** www.radioinvlaanderen.info

FM	Mhz	kW	Location	Station
1)	87.6	3	Oostende	Nostalgie
1)	88.0	2	Kortrijk	Nostalgie
1)	88.1	1	Brugge	Nostalgie
2)	88.3	1	Oost-Vleteren	QMusic
2)	88.6	3	Gent	QMusic
4)	88.9	1	Diksmuide	Club FM
3)	89.1	1	Tongeren	Joe FM
2)	89.6	2	Brugge	VBRO
3)	90.6	5	Turnhout	Joe FM
6)	91.3	3	Brugge	Topradio
3)	92.2	3	Dendermonde	Joe FM
2)	92.2	2	Herentals	Qmusic
2)	92.7	2	Kortrijk	Exqi FM
3)	92.8	1	Gent	Joe FM
1)	92.8	1	Beringen	Nostalgie
3)	93.5	2	Sint-Niklaas	Joe FM
3)	93.5	2	Geel	Joe FM
3)	93.5	1	Bree	Joe FM
4)	93.6	5	Egem	Club FM
8)	94.0	2	Koolskamp	Mi Amigo
9)	95.1	1	Turnhout	Voodoo Radio
3)	95.5	1	StPietersLeeuw	Joe FM
3)	95.6	1	Brussegem	Joe FM
2)	95.8	1	Veltem	Qmusic
7)	96.3	2	Gent	Exqi FM
3)	96.7	2	Mechelen	Joe FM
1)	96.9	1	Sint-Truiden	Nostalgie
10)	98.0	1	Antwerpen	Minerva
1)	98.1	10	Brussel	Nostalgie
1)	98.2	3	Egem	Nostalgie
11)	98.8	3	Brussel	FM Brussel
12)	99.0	1	Mechelen	Randstad
2)	99.2	10	Antwerpen	Qmusic
13)	99.3	1	Bree	Hit FM
6)	99.4	4	Gent	Topradio
3)	99.7	1	Leuven	Joe FM
2)	100.0	1	Wuustwezel	Qmusic
9)	100.2	2	Antwerpen	Voodoo Radio
1)	101.0	20	Oost-Vleteren	Nostalgie
2)	102.5	1	Brussel	QMusic
2)	102.5	50	Genk	Qmusic
13)	102.6	2	Leuven	Hit FM
14)	102.6	1	Eeklo	Family Radio
15)	102.7	3	Aalst	City Music
16)	102.7	2	Brugge	Exclusief
4)	102.8	1	Brussel	Club FM
17)	102.8	1	Gent	Zen FM
1)	102.9	50	Schoten	Nostalgie
2)	103.0	20	Egem	Qmusic
1)	103.0	1	Bree	Nostalgie
2)	103.1	50	StPietersLeeuw	Qmusic
3)	103.3	1	Diest	Joe FM

FM	Mhz	kW	Location	Station
2)	103.3	1	Brugge	Qmusic
3)	103.4	10	Brussel	Joe FM
3)	103.4	5	Antwerpen	Joe FM
3)	103.4	20	Genk	Joe FM
1)	103.5	20	Gent	Nostalgie
4)	103.6	1	Oostende	Club FM
3)	103.7	1	Wuustwezel	Joe FM
3)	103.7	3	Lommel	Joe FM
1)	103.7	2	Sint-Niklaas	Nostalgie
1)	103.8	3	Leuven	Nostalgie
15)	103.8	2	Gent	City Music
9)	103.8	1	Geel	Voodoo Radio
14)	103.8	1	Hasselt	Family Radio
3)	104.1	50	Egem	Joe FM
4)	104.1	2	Hasselt	Club FM
7)	104.2	5	Leuven	Exqi FM
3)	104.2	1	Aalst	Joe FM
18)	104.2	2	Antwerpen	Crooze FM
1)	104.2	1	Overpelt	Nostalgie
5)	104.5	1	Oostende	VBRO
6)	104.5	1	Poperinge	Topradio
1)	104.5	1	Mechelen	Nostalgie
1)	104.5	1	Turnhout	Nostalgie
1)	104.6	2	Geel	Nostalgie
7)	104.6	2	Antwerpen	Exqi FM
19)	104.7	1	Hasselt	Viva FM
1)	104.8	3	Aalst	Nostalgie

+ 273 stns below 1kW

Addresses and other information

1) Katwilgweg 2, 2050 Antwerpen **W:** www.nostalgie.eu – **2)** Medialaan 1, 1800 Vilvoorde **W:** www.qmusic.be – **3)** Medialaan 1, 1800 Vilvoorde **W:** www.joe.be – **4)** Stationsstraat 68, 9900 Eeklo **W:** www.clubfm.be – **5)** Jan Miraelstraat 24, 8000 Brugge **W:** www.vbro.be – **6)** Nekkerputstraat 150, 9000 Gent **W:** www.topradio.be – **7)** Fabriekstraat 38, 2547 Lint **W:** www.exqifm.be – **8)** Dadizeelsestraat 9, 8890 Moorslede **W:** www.miamigomedia.be – **9)** Koningin Astridplein 38 bus 3, 2000 Antwerpen **W:** www.voodooradio.be – **10)** Wandeldijk 20, 2050 Antwerpen **W:** www.radio-minerva.be – **11)** Eugène Flageyplein 18 bus 18, 1050 Elsene **W:** www.fmbrussel.be – **12)** Hogeweg 211 2800 Mechelen **W:** www.randstad.fm – **13)** KAAI.16, Scheepvaartkaai 16 A bus 8, 3500 Hasselt **W:** www.hitfm.be – **14)** Leopoldlaan 98c, 9900 Eeklo **W:** www.familyradio.be – **15)** Geraardsbergsestraat 21, 9300 Aalst **W:** www.city-music.be – **16)** Emile Bethunelaan 5, 8200 Brugge – **17)** Einde Were 150, 9000 Gent **W:** www.zenfm.be – **18)** Diksmuidelaan 173, 2600 Berchem **W:** www.crooze.fm – **19)** Bremtstraat 22, 3511 Hasselt **W:** www.vivafm.be

WALLONIA Pop: 3.4 million — **Pr.L.:** French, German

RADIO-TÉLÉVISION BELGE DE LA COMMUNAUTE FRANCAISE (R.T.B.F.) (Pub.)
French Language Network
Public sce. Grants by French Parliament.
⌕ Cité de la Radio-Television, B-1044 Brussels ☎+32 2 737 2111 ▤ +32 2 737 4357 **W:** www.rtbf.be
L.P: Admin. Gen: Jean-Paul Philippot. Dir. Radio: Francis Goffin
Regional & Local Centres: Brussels: Reyerslaan 52, 1044 Brussel **Liège:** 203 rue de Verviers, 4821 Andrimont **Hainaut:** Anne-Charlotte de Lorraine, 7000 Mons **Namur:** Av. Golenvaux 8, 5000 Namur **Luxembourg:** Parc des Expositions, 6700 Arlon

MW	kHz	kW	Prgr	MW	kHz	kW	Prgr
Wavre	621	300	Ext.Svce	Liège	1233	0.2	5
Houdeng	1125	9	2	Marche	1305	7/9	5

FM (MHz)	1	2	3	4	5	kW
Anderlues	93.4	92.3			99.1	96.6 0.6/40/40/40
Bruxelles		99.3	91.2	93.2	88.8	3/40/1/?
Léglise	96.4	91.5	94.1	87.6		10/10/10/10
Liège	96.4	90.5	99.5	95.6	92.5	5/40/40/13/0.1
Malmédy	89.2	91.6				1/0.1
Marche	93.3	95.2				0.5/4
Profondeville	102.7		92.8	90.8		25/10/10
Tournai	106.0	101.8	102.6	104.6	90.6	25/30/30/30/?
Verviers	91.3	103.0			87.9	1/3/0.1
Wavre	96.1	97.3			101.1	10/35/50

+ many txs under 1kW

Network 1 (Première – information & musique)
Network 2 (Vivacité – light music)
Reg. Prgrs: W 0530-0800, 1200-1300, 1600-1800; Fri 1800-2100 – R. Hainaut (Mons) on 92.3/101.8MHz – R. Liège 90.5/103.0/94.6/89.1/89.4MHz – R. Namur on 97.3/92.8/89.3/91.5/89.4/90.2MHz – R. Bruxelles on 93.2MHz

Local Prgrs: R. Verviers (Radiolène) on 103.0MHz
Network 3 (Musique 3 – classical music)
Network 4 (Classic 21 – oldies & rock classics)
Network 5 (Pure FM – youth station)
DAB: 225.648MHz
Ann: "Vous écoutez La Première, Vivacité, Musique trois, Radio 21, Pure FM, Classic 21"

EXTERNAL SERVICE: RTBFi
See International Broadcasting section

BELGISCHES RUNDFUNK-UND FERNSEHZENTRUM DER DEUTSCHSPRACHIGEN GEMEINSCHAFT (BRF)
German Language Network (Pub)
Grants by RDG-Rat (German speaking community council).
⌕ Kehrweg 11, B-4700 Eupen ☎+32 87 59 1111 ▤+32 87 591199 **W:** www.brf.be **E:** info@brf.be
Regional: ⌕ Blvd. Reyers 52, B-1044 Brussels – Malmedyer Str. 25, B-4780 St. Vith **LP:** Pres. of Admin. Council: J-F Crucke; Dir. of BRF: H. Engels. Chief Editor: R Schroeder, PR:R. Ducomble.

FM	MHz	kW	FM	MHz	kW
Lüttich	88.5	50	Brussel*	95.2	2
Auel	92.2	0.16	Eupen	98.4	1
Recht	94.9	5	Recht	104.1	20

*: broadcasts joined prgrs of Deutschlandfunk and BRF
Ann: "Hier ist der Belgischer Rundfunk"

COMMERCIAL NETWORKS:
NB: For further information on all radio stns in Wallonia visit **W:** www.tuner.be/tuner.asp?content=radio

FM	Mhz	kW	Location	Station
14)	87.6	1	Ath	Sud Radio
14)	88.2	1	Charleroi	Sud Radio
11)	88.3	5	Bouillon	NRJ
11)	88.7	1	Arsimont	NRJ
1)	88.9	1	Fauquez	Bel RTL
10)	89.2	1	La Louvière	Radio Nostalgie
10)	90.0	1	Tournai	Sud Radio
10)	92.3	1	Verviers	Radio Nostalgie
11)	92.7	1	Malmédy	NRJ
7)	92.9	1	Bastogne	Fun Radio
10)	94.4	1	Bütgenbach	100.5 Das Hitradio
9)	94.7	5	Bouillon	Must FM
10)	95.0	1	Liège	Radio Nostalgie
15)	95.6	1	Houdeng	Twizz
12)	97.1	4	Bouillon	Radio Contact
14)	97.6	1	Braine-le-cte	Sud Radio
12)	97.6	2	Bütgenbach	Radio Contact (DE)
1)	99.0	2	Bouillon	Bel RTL
7)	99.0	1	Liège	Fun Radio
10)	100.0	5	Brussel	Radio Nostalgie
4)	100.1	1	Liège	Equinoxe FM
1)	100.2	1	Limal	Bel RTL
10)	100.4	4	Namur	Radio Nostalgie
2)	100.5	20	Eupen	100.5 Das Hitradio
10)	100.5	2	Couvin	Radio Nostalgie
10)	100.7	2	Dinant	Radio Nostalgie
8)	100.9	2	Liège	Maximum FM
1)	101.6	5	Marche	Bel RTL
12)	101.6	2	Verviers	Radio Contact
1)	101.7	2	Namur	Bel RTL
1)	101.7	1	Couvin	Bel RTL
1)	101.8	50	Meix Le Tige	Bel RTL
1)	101.9	1	Dinant	Bel RTL
14)	102.0	2	Mons	Sud Radio
12)	102.2	10	Brussel	Radio Contact
12)	102.2	5	Charleroi	Radio Contact
12)	102.2	1	Liège	Radio Contact
12)	102.3	5	Mons	Radio Contact
10)	102.4	1	Arlon	Radio Nostalgie
12)	102.5	1	Houffalize	Radio Contact
11)	103.2	10	Vlessart	NRJ
15)	103.2	2	Liège	Twizz
1)	103.4	5	Mons	Bel RTL
7)	103.5	1	Charleroi	Fun Radio
1)	103.6	1	Ath	Bel RTL
1)	103.6	50	Liège	Bel RTL
11)	103.7	1	Brussel	NRJ
6)	104.0	1	Brussel	Foo Rire
12)	104.0	2	Charleroi	Bel RTL
10)	104.1	1	Huy	Radio Nostalgie
14)	104.3	15	Brussel	Bel RTL
11)	104.3	5	Namur	NRJ
12)	104.5	1	Wavre	Radio Contact
12)	104.6	1	Marche	Radio Contact

FM	Mhz	kW	Location	Station
7)	104.7	1	Brussel	Fun Radio
12)	104.7	2	Malmédy	Radio Contact
12)	104.7	5	Namur	Radio Contact
2)	104.8	3	Sankt Vith	100.5 Das Hitradio
12)	104.8	2	Virton	Radio Contact
7)	105.5	1	Louvain-L-N	Fun Radio
12)	106.2	1	Florzé	Radio Ourthe Amblève
3)	106.3	2	Tubize	Radio Antipode
11)	106.7	1	Bastogne	NRJ
5)	106.8	1	Malmédy	Est FM
8)	106.9	1	Aywaille	Maximum FM
10)	100.2	50	Saint-Hubert	Radio Nostalgie
11)	104.5	5	Liège	NRJ
12)	107.0	1	Eupen	Radio Contact (DE)
7)	107.5	1	Arlon	Fun Radio
2)	107.6	1	Honsfeld	100.5 Das Hitradio
10)	107.6	5	Bouillon	Radio Nostalgie
11)	107.7	1	Tournai	NRJ
12)	107.8	1	Libramont	Radio Contact

+ many stns below 1kW

Addresses:
1) Avenue Georgin 2, 1030 Bruxelles 02-3376911 **W:** www.belrtl.
be – **2)** Kehrweg 11, B-4700 Eupen +32 87 591259 ✍+32 87 591249
W: www.hitradioworld.fm – **3)** Boîte postale 2, 1348 Louvain-La-
Neuve ☎010-451110 ✍ 010-451717 **W:** www.antipode.be – **4)** Rue
Montagne St Walburge, 261, 4000 Liège **W:** www.equinoxefm.be – **5)**
– – **6)** Avenue d'Hougoumont, 2, 1180 Bruxelles **W:** www.foorirefm.
be – **7)** Av. Telemaque 33, 1190 Bruxelles ☎02-3457575 **W:** www.
funradio.be – **8)** 22 Rue de la Chaudronnerie, 4030 Grivegnée **W:** www.
maximumfm.be – **9)** BP20, 1360 Perwez ☎081-655469 **W:** www.
mustfm.be – **10)** Quai au Foin 55, 1010 Bruxelles ☎02-2270450 ✍ 02-
2231455 **W:** www.nostalgie.eu – **11)** Chaussée de Louvain 467, 1030
Bruxelles ☎02-5137575 ✍ 02-5114859 **W:** www.nrj.be – **12)** Avenue
des Croix de Guerre 94, 1120 Bruxelles 02-2442711 ✍ 02-2442710 **W:**
www.radiocontact.be and www.derbestemix.be (DE) – **13)** Rue Armand
Binet 35B, 4140 Rouvreux (Sprimont) – **14)** 42, rue de la chaussée de
Mons, 7000 Mons ☎065-401010 ✍ 065-401011 **W:** www.sudradio.
net – **15)** Rue des Francs, 79, 1040 Bruxelles **W:** www.twizz.be

Military Stations:
AMERICAN FORCES NETWORK, SHAPE
✉ Box 7, 7010 SHAPE. (APO AE 09700) ☎+32 65 44 41 21
L.P: Officer-in-charge: Cpt. G. Martel. Broadc. Superv: SFC C. Kubicek.
Chief Eng: René Libre
Stations: Kleine Breugel 106.2MHz 0.1kW, Brussels 101.7MHz
0.9kW, SHAPE 104.2/106.5MHz 4kW, Florennes 107.7MHz 0.1kW,
Chievres 107.9.
D.Prgr: 24h on 101.7/104.2/107.7MHz. Own prgrs Mon-Fri 0500-
0800, 1400-1700; Sat 0800-1200. Other times rel. AFN Europe.
AFN-2: 24h easy listening stereo prgr on 106.5MHz
BRITISH FORCES BROADCASTING SERVICE
Stations: SHAPE BFBS 1, 107.7MHz Casteau 0.05kW
✉ Wentworth B., Listnstr., D-32049 Herford, Germany
D.Prgr: rel. BFBS Germany.

BELIZE

L.T: UTC -6h — **Pop:** 301,000 — **Pr.L:** English, Spanish — **E.C:** 60Hz,
110/220V — **ITU:** BLZ

FM	MHz	KW	Name and loc. (Hrs of tr usually 24h)
5)	88.9	40	Love FM, Ladyville
1)	90.5	0.25	Vibes R., Belize City
2)	92.3		Reef R., San Pedro
9)	93.7	0.5	My Refuge Christian R., Roaring Creek
3)	94.3		Our Lady of Mount Carmel R., Benque Viejo
4)	94.5		People's R. (Da Beat), Belize City
5)	95.1		Love FM, Belize City
6)	96.5		Krem FM, Belize City
7)	97.1		Integrity R., Belize City
	97.5		Estéreo Amor, Belize City
5)	99.5		More FM, Belize City
9)	100.5	0.2	My Refuge Christian R., Belize City (Rel. 93.7)
10)	101.3		R. Emanuel, San Pedro
1)	102.9		Vibes R., Belmopan
12)	104.5		Faith FM, Punta Gorda
13)	105.9		Wave 105.9, Belmopan
5)	107.1		More FM, Belmopan

+ several other local FM stns

Addresses and other information:
1) Belize City **E:** positivevibesbz@yahoo.com **2)** San Pedro **W:** www.
threeefradio.com **3)** Our Lady of Mount Carmel High School, Benque

Viejo del Carmen ☎+501 823-3177 **W:** www.carmelradio.org **E:**
general.manager@carmelradio.org **D. Prgr:** English and Spanish
1100-0500 **F.PI.** 1 kW FM transmitter at Pine Ridge **4)** 3321 Central
American Boulevard, Belize City **E:** modulation945@yahoo.com **5)**
7145 Slaughterhouse Road, PO Box 1865, Belize City ☎+501-203-
2098 ✍ +501-203-0529 **FM:** Belize City 95.1MHz + 9 FM repeaters **W:**
www.lovefm.com **E:** lovefmbelize@yahoo.com **6)** ✉ 3304 Partridge
Str, c/o P.O. Box 15, Belize City ☎+501 (2) 75929 ✍ +501 (2) 74079
D.Prgr: 1100-0500 **7)** Belize City **9)** ✉ P. O. Box 275, Belmopan ☎
+501-822-1080 ✍ +501 (8)22601 **E:** myrefuge@charter.net **W:** www.
myrefugebelize.org **D.Prgr:** 24h **10)** ✉ San Pedro, Ambergris Caye **12)**
Punta Gorda **13)** Belmopan **E:** waveradio105_9@yahoo.com

BRITISH FORCES BROADC. SERVICE BELIZE
✉ BFBS Belize, Airport Camp, BFPO 12 ☎ +501 225 2333 ✍ +501
225 2334 **W:** http://bfbs-radio.com **E:** bfbsbelize@bfbs.com
FM: BFBS Radio, 99.1MHz 0.5kW **BFBS Radio 2** 93.1MHz 0.015kW

BENIN

L.T: UTC +1h — **Pop:** 9 million — **Pr.L:** French + 18 ethnic — **E.C:**
50Hz, 220V — **ITU:** BEN

**HAUTE AUTORITÉ DE L'AUDIOVISUEL ET DE LA
COMMUNICATION (HAAC)**
✉ BP 3567, Ave. de la Marina, Face Hôtel du Port, 01 Cotonou ☎+229
21311743 ✍ +229 21311742 **W:** haacbenin.org

**OFFICE DE RADIODIFFUSION ET TÉLÉVISION DU
BENIN (ORTB, Gov.)**
✉ 01 B.P. 366, Cotonou ☎ +229 21360047 **W:** www.ortb.bj **E:** ortb@
intnet.bj **L.P:** DG: Hamado Ouangraoua. Chief Tech. Sces: Anastase
Adjoko.
Regional ✉ B.P. 128, Parakou ☎+229 23611096
MW: Parakou 936kHz 8kW.

SW	kHz	kW	Times
Parakou	5025v	10	MF0500-0900, 1100-1400, 1700-2300 SS 0700-2300

FM: Cotonou 94.7 10kW, Parakou 89.4/92.5MHz 2kW.
Radio Nationale: from Cotonou in French/ethnic.
N. in French: 0615MF, 0800SS, 1200SS, 1215MF, 1930, 2115.
R. Regionale Parakou: in French/ethnic.
(Cotonou & Parakou carry the same programme between 1900-2000).
Atlantic FM, Cotonou: 0700-2300 on 92.2MHz.
Ann: "Ici R. Bénin, Office de Radiodiffusion et Télévision du Bénin,
émettant de Cotonou". "Ici Parakou, Office de Radiodiffusion et
Télévision du Bénin, station regionale" **IS:** Bénin Tam-Tam

Other stations:
R. Adja Ouèrè FM, Cotonou: 92.6/100/100.6/107.6MHz – **R. Afrique
Espoir**, Porto-Novo: 99.1MHz – **Benin Culture**, Porto-Novo: 93.4MHz
– **CAPP FM**, Cotonou: 99.6MHz – **R. Carrefour**, Cotonou: 90.4MHz –
Cité Savalou Culture FM, Cotonou: 87.8MHz – **Deeman R**, Parakou:
90.2MHz – **FM Ahémé**, Possotome: 99.6MHz – **FM Alakétou**, Ketou:
95.8MHz – **FM Monts Kouffé**, Bassila: 103MHz – **FM Noon Sina**,
Bembereke: 90.8MHz – **FM Oré Ofé**, Tchetti: 102.1MHz – **Gerddes
FM**, Cotonou: 89.5 MHz – **Golfe FM**, Cotonou: 105.7MHz Web: eit.
to/golfefm.htm – **La Voix de la Lama**, Porto-Novo: 103.8MHz – **La
Voix de l'Islam**, Cotonou: 91.2MHz – **R. Immaculee Conception**
(Rlg.): Djougou 89.1MHz, Natitingou 93.1MHz, Parakou 93.3MHz,
Cotonou 98.7MHz, Bembérèkè 100.8MHz, Bohicon 100.9MHz, Allada
101.3MHz, Dassa-Zoumey 107.3MHz – **R. Liéma**, Cotonou: 104MHz –
R. Planète, Cotonou: 95.7MHz – **R. Maranatha**, Cotonou: 103.1MHz
W: eit.to/RadioMaranatha.htm – **R. Allodalome**, Cotonou: 97.4MHz
– **R. Star**, Cotonou: 94.3/96.3MHz – **R. Tokpa**, Cotonou: 104.3MHz
– **R. Tonassé**, Cové: 107.6MHz – **R. Wekê**, Cotonou: 107MHz – **R.
Rurale** stns in Tanguiéta 90, Ouessè 97.7, Dogbo 100, Ouaké 101 &
Banikoara 104.2MHz.
Africa No 1: Porto-Novo 102.6MHz (see main entry under Gabon).
BBC African Service: Cotonou 101.7MHz.
RFI Afrique: Cotonou 90MHz, Parakou 106.1MHz.
Trans World R, Parakou: 1566kHz 100kW 0300-0535, 1700-2115. (For
details see International Radio section)

BERMUDA (UK)

L.T: UTC -4h (13 Mar-6 Nov: - 3h) — **Pop:** 65,000 — **Pr.L:** English
— **E.C:** 60Hz, 115/230V — **ITU:** BER

BERMUDA BROADCASTING CO. LTD. (Comm.)
✉ 4 Fort Hill Road, Prospect, Devonshire DV 02, P.O. Box HM 452,
Hamilton HM BX ☎ +1 441 295 2828 ✍ +1 441 295 4282 **E:** zbmzfb@
bermudabroadcasting.com

L.P: CE: Ulric P Richardson. Ops. Mgr: E Delano Ingham. News Dir: Darlene Ming.
MW: ZFB 1230 AM Spirit AM: 1230kHz: 24h Christian Community Stn – **ZBM 1340 AM:** 1340kHz 1kW: 24h live & networked prgr.
FM: ZBM FM89 89.1MHz 15kW: 24h automated prgr. – **Power 95 FM:** 94.9MHz 1kW: 24h live & networked prgr. – **Government Emergency Broadcast Station:** 100.1MHz/1610kHz

DEFONTES BROADCASTING CO. LTD. (Comm.)
✉ P.O. Box HM 1450 (**studios:** 94 Reid Str.) Hamilton HM FX ☎ +1 441 292 0050 🖷 +1 441 295 1658 **E:** vsbnews@ibl.bm **W:** www.vsb-bermuda.com **L.P:** CEO: Kenneth De Fontes. SM: Mike Bishop. N. Dir: Chris Lodge. TD: Ed. Tucker & Fred Blanchette
MW: 1450 AM Gold 1450kHz 1kW 24h (oldies). – **Bible Broadcasting Network** 1280kHz 1kW 24h (rlg.) – **BBC World Service** 1160kHz 1kW 24h. Rel. BBCWS 24h except for special event prgrs
FM: Mix 106.1 FM 106.1MHz 2.5kW: 24h (CHR)

INTER-ISLAND COMMUNICATIONS (Comm.)
✉ 49 Union Square Mall, Hamilton ☎ +1 441 297 1075 🖷 +1 441 296 7680 **E:** feedback@hott1075.bm **W:** www.hott1075.com
L.P: CEO: Glenn Blakeney
FM: Magic 102.7: 102.7MHz – **Hott 107.5** 107.5MHz

LTT BROADCASTING LTD. (Comm.)
✉ P.O. Box 1564 HMGX, Hamilton HMEX ☎ +1 441 700 9810 🖷 +1 441 292 1593 **E:** jazz@bdaradio.com **W:** kjazzfm.xanga.com
L.P: Leo Trott **FM: KJAZ 98.1FM:** 98.1MHz

BHUTAN

L.T: UTC +6h — **Pop:** 2 million — **Pr.L:** Dzongha, Sharchopkha, Lhotsam(Nepali), English — **E.C:** 50Hz, 220V — **ITU:** BTN

BHUTAN BROADCASTING SERVICE (Corp.)
✉ P.O. Box 101, Thimpu ☎ +975 2 322866/322533/323071 🖷 +975 2 323073 **W:** www.bbs.com.bt **E:** request@bbs.com.bt
L.P: Exec. Dir: Sonam Tshong.Tech. MD: Pema Choden, Dir: Dorji Wangchuk. Prgr. Dir: Tashi Dhendup. News Dir: Thinley Tobgye
SW: 6035kHz 10/100kW **D.Prgr:** 0000-0700, 0800-1400 (Irr. Usually 0100/0300-1300/1330 and using 10kW at press date).
FM(MHz): 88.1 & 96.0 Thimphu, 92.0 Phuentsholing, 92.0 Paro, 93.0 Trongsa, 98.0 Chhukha, 90.0 Trashigang, also at Samtse, Haa and Tsirang
D.Prgr: 24h **English: Daily** 0500-0600, 0800-0900, 1500-1700. **N:** Daily: 0500, 0800, 1500 (15/20'); **English Music:** 1600-1700; **UN Radio Prgr:** Thurs 0520, **Bhutan This week:** Fri 0820; **Dzongkha:** Daily 0000-0300,0700-0800,1100-1500,1700-2400 (N:0100,0200,0700, 1100,1200,1300,1400; **Music:** 1700-2000, **Repeat b'cast:** 2000-2400) **Sharchhop:** 0300-0400, 0900-0700,0900-1000 (N:0300,0600,0900), **Lhotsam** (Nepali): Daily 0400-0500, 1000-1100 (N:0400,1000);
V. by QSL-card. 15 min prgr details req. Rp. (2 IRCs)

Other FM Stations
Kuzoo FM, ✉ P.O. Box 419, Thimpu **W:** www.kuzoo.net **DPrgr:** 24h in English on 104.0MHz and Dzongkha on 105.0MHz – **Radio Valley,** ✉ P.O. Box 224/225, Thimpu **Web:** www.radiovalley.bt **DPrgr:** 0200-1630 in English on 99.9MHz – **Sherubtse FM** 94.7MHz – **Centennial Radio** Thimpu 101.0 MHz

BOLIVIA

L.T: UTC -4h — **Pop:** 9.7 million — **Pr.L:** Spanish, Quechua, Aymara — **E.C:** 50Hz, La Paz 110/220V, Santa Cruz 220/380V — **ITU:** BOL — **Int. dialling code:** +591

SUPERINTENDENCIA DE TELECOMUNICACIONES (SITTEL)
✉ Calle 13 entre Sauces y Costanera No. 8280 y 8260, Calacoto, La Paz ☎ 2 2772266 🖷 2 2772299 **E:** informacion@sittel.gov.bo **W:** www.sittel.gov.bo **L.P:** Superintendente: Lic. Jóse Antonio Morales

CAMARA NACIONAL DE MEDIOS DE COMUNICACION
✉ Casilla 2431, La Paz
° = on-air stn name not confirmed, * = inactive, v = varying freq.

MW	kHz	kW	Station, location, h of tr
LP77)	540		Radiodifusora Victoria, La Paz
LP03)	*560	15	R. El Mundo, La Paz
LP01)	580	10	R. Panamericana, La Paz: 1000-0300 (Sun 1100-0100)
CH01)	600	10	R. ACLO, Sucre: 0900-0200
LP35)	600	1	Radioemisoras del Recobro, La Paz

MW	kHz	kW	Station, location, h of tr
LP02)	620	10	R. San Gabriel, La Paz: 0900 (Sun 1000)-0200
LP11)	650	15	R. Dif. Integración, El Alto: 0900-0130
SC10)	660	1	R. ABC, Santa Cruz: 0900-0100
LP27)	680	5	R. Andina, La Paz: 0900-0300
OR17)	710	10	R. Pío XII, Siglo Veinte: 0830-0230
LP06)	720	10	R. La Cruz del Sur, La Paz: 1300-2100(Sun. 1200-1600)
LP05)	720	2.5	R. Yungas, Chulumani: 0900-1700, 2000-0100
SC01)	*730	3	R. Mensaje, Montero
LP07)	760	50	R. Fides, La Paz: 1000 (Sun 1100)-0300(Sat 0500)
CO02)	760	5	R. Cosmos, Cochabamba: 1100-0300
CO56	760		R.Casachun Coca, Lauca Ñ
LP08)	*800	5	R. Libertad, La Paz: 1000-0200 (Sun. 2400)
LA09)	800	0.25	R. Santa Clara, Sorata: 0900-2400
CH12)	800	1	R. Churuquella, Sucre: 1000-0200
LP10)	820	10	R. Altiplano, La Paz: 24h
LP75)	840	3	R.Atipiri, El Alto
PO01)	850	1	R. 21 de Diciembre, Mina Catavi: 1000-0100
SC03)	850	5	R. María Auxiliadora, Montero: 0900-0100
LP12)	*860	10	R. Nueva América, La Paz: 1015-2400 (Sun 1100-0100)
BE01)	*860		R. Paitití, Guayaramerín: 1030-0300
CO31)	870	0.6	LV del Campesino, Sipe Sipe (r 972): 0700-2300
OR21)	875		R. Eucaliptos, Eucaliptos
LP42)	880		R. Inca, El Alto: 0900-0100
TA01)	900	0.25	R. LV Nacional, Tarija: 0100-2300
LP36)	900	5/0.1	R. Popular, La Paz: 0900-0100
CO33)	*902	1	R. Central Misionera, Cochabamba (n.f. 900):1100-0100
SC09)	905	1.5	R. Norte, Montero (n.f. 990): 0930-0200
CH11)	920	3	R. Encuentro, Santa Cruz: 1100-2200
LP65)	920	1	R. San Andres de Topohoco, Topohoco: 0900-0330
CH13)	940	1	R.Chuquisaca XXI, Sucre
LP13)	940		R. Metropolitana, La Paz: 0900-0500
CO49)	941	0.8	R. San Lorenzo, Colcapirhua (n.f. 940): 0900(Sun 1100)-0200
CO01)	950	3	R. Yurac Molino, Chimboata
PO02)	960	1	R. Kollasuyo, Potosí: 1000-0400 (Sun 1000-0200)
SC04)	960	10	R. Santa Cruz, Santa Cruz
LP43)	962	1	R. Huayna Potosí, Milluni
CO31)	972	0.6	LV del Campesino, Sipe Sipe (n.f.: 870): 0700-2300
LP14)	980	2.5	R. Mar Plus, La Paz: 1000-0200
CO04)	980	5	R. Esperanza, Aiquile: 0930-0200
BE02)	1000	3	R. Dif. Trópico, Trinidad: 1300-0030
SC33)	1000	1	R. Piraí, Santa Cruz
OR03)	1000	10	R. Bahá'í de Bolivia, Caracollo: 0800-0200
CO03)	*1000		R. LV de Sipe Sipe, Sipe Sipe 0900 (Sun 1000) - 0200
LP44)	1000	1	R. Mística, La Paz
LP15)	1020	10	R. Illimani - R.Patria Nueva, La Paz: 0900-0400
CH19)	1030		R. Em.Comunitaria Mojocoya (RPN), Mojocoya: 1000-0200
OR26)	1030	3	R. Orinaca (RPN), Orinaca
CO51)	1030		R.Totora (RPN) Totora
CO48)	1030	3	R. Independencia, Ayopaya
PO27)	1030		R. Em. Comunitaria Colquechacam (RPN), Colquechaya
BE20)	1030	3	R. Patria Nueva (RPN), Riberalta
CH18	1040		R. Em. Comunitaria 12 de Marzo (RPN), Tarabuco
OR28)	1040		R. Em. Comunitaria Qaqachaca (RPN), Qaqachaca
PO03)	1040	1	R. Villazón, Villazón
CO20)	1040	0.25	R. Sipe Sipe, Quillacollo: 1000-0300 (Sun 1100-0400)
TA11)	1040		R. Em.Comunitaria Libertad (RPN), Villamontes
SC37)	1040		R. Em.Comunitaria Camiri (RPN), Camiri
SC38)	1040		R. Em.Comunitaria San Juan (RPN), San Julián
SC30)	1040		R. San José (RPN), San José de Chiquitos
LP45)	1043	1	R. Bolivianísima, La Paz
CO52	1050		R. Em.Comunitaria (RPN), Independencia
SC06)	1050	5	R. El Mundo, Santa Cruz: 1000-0200
OR27)	1050	3	R. Sabaya (RPN) Sabaya
PO26)	1050		R. Em.Comunitaria Caizad (RPN), Caizad
OR01)	1060	1.5	R. Noticias, Oruro: 1000-2200, 0200-0600
SC07)	1060	0.5	R. LV de la Frontera, Pto. Suárez: 0900-0300 (Sat 0230, Sun 2300)
LP38)	1060	10	R. Eco Loyola, La Paz: 1100-0100
SC05)	1075	0.5	R. Agricultura, Portachuelo(n.f.: 1030): 1000-0400
CH02)	1080	1	R. Dif. Colosal, Sucre(n.f.: 1060): 0900-0300
LP76)	1080		LV de la Mayoria, Caranavi
CO06)	1090	3	R. Cultura, Cochabamba: 0900-0400 (Sun 1200-2300)
OR06)	1100	1	R. Universidad de Oruro (n.f.: 620): 1100 (Sun1200)-2300
LP04)	1100	4	R. Mundial, La Paz: 0900-0330
LP29)	1100		R. Chaka, Pucarani: 0900-1300, 2030-0130
PO18)	1115	1	R. Difusoras Independencia, Atocha: 0930-0300
BE03)	1120	1	R. Estación El Dorado, Trinidad: 1000-0100
CO53)	1120		R.Porvenir, Tiquipaya
LP48)	1120		R. Revelacion El Gran Yo Soy, El Alto: 0900-0100 (Sun -2400)
OR02)	1125	0.3	R. Em. Cooperativa Poopó, Poopó
SC08)	1125	0.5	R. Cruceña, Cotoca(n. 1530): 1130-2200

MW	kHz	kW	Station, location, h of tr
LP49)	1140	2	R. Pico Verde, Chulumani
CO45)	1140		R. San Isidro, Colami
SC31)	*1143	1	R. Colonia, Yapacani
LP17)	1145	1	R. Chuquiago Musical, La Paz: 1000-0200
OR22)	1150		R. 24 de Noviembre, Eucaliptos
TA02)	1150	0.2	R. Chaco, Yacuíba (n. 1100): 1000-2400, (Sun 1100-1600)
OR07)	1150	0.5	R. El Cóndor, Oruro: 1100-2030
LP50)	1150	0.3	R. Guaqui, Guaqui
SC11)	1160	5	R. Centenario, "La Nueva", Sta. Cruz
CO08)	1160	3/1	R. RTC, Cochabamba: 1030-2400
CH03)	1160	1	R. Nuevo Mundo, Sucre: 1000-0300
LP33)	1160	10	R. Continental, La Paz: 0930-2400
OR04)	1180	5	R. Central, Oruro: 0900-2400
LP18)	1180	1	R. Em. Ingavi, Viacha: 1000-0200 (Sun 1100-2400)
CO38)	1180	1	Radioem. 20 de Septiembre, Arbieto: 0900-0400 (Sun 1200-2300)
PO04)	1180		R. Kollasuyo, Potosí
LP73)	1180		R. LV de Dios, Al Alto
CO09)	*1195	1	R. Independencia, Quillacollo: (n. 1200)
SC12)	1200	5	R. Oriental, Santa Cruz: 0930-0200
CO10)	1200	0.25	R. 24 de Noviembre, Arani: 1030-0400
CH10)	1200		R. Mauro Nuñes, Villa Serrano
CO55)	1215		R. Tupuñan, El Paso
LP19)	1220	1	R. Splendid, La Paz: 0900-0100
OR09)	1220	1	R. Batallón Topáter, Oruro: 1000-0200 (Sat 1000-2300, Sun 1100-2300)
CO11)	r1220		R. El Cóndor, Arque
CO12)	1240	1	R. San Miguel, Arani: 1000-0300
TA03)	1240	2	R. Los Andes, Tarija: 1000-2200
LP51)	1240		R. Achocalla, Achocalla (r1255)
SC13)	1250	0.5	R. Sararenda, Camiri: 0900-0400
SC14)	1250	1	R. Amboró, Santa Cruz: 0845-0200 (Sun 1000-2400)
CO13)	*1250	0.5	R. Nacional, Cochabamba: 1200-2400
OR10)	1250	0.4	R. Oruro, Oruro: 1300-0400
PA01)	1250	0.1	R. Frontera, Cobija: 1000-1800
CH04)	°1250	2.5	R. La Plata, Sucre: 1000-0200
PO20)	1250		R. Uncia, Uncia
LP51)	v1255		R. Achocalla, Achocalla (nf. 1240)
CO54)	1260	1	R. LV de la Esperanca, Quillacollo
LP20)	1260	2	Radioemisora Unidas, La Paz: 0900-0100
OR20)	1260	10	R. Nacional de Huanuni, Huanuni
PO05)	1265	0.4	R. Uncía, Uncía: 1100-0300
LP21)	1270	1	R. Vanguardia, Colquiri: 1000-0400
SC15)	1275	0.5	R. Chané, Mineros: 1000-0400
LP68)	1280		R. Ondas del Titicaca, Huarina
OR12)	1290	1	Radiodifusoras Minería, Oruro: 1000-2400
CH05)	1300	2.5	R. Loyola, Sucre: 1000-2400. Sun: 1015-0200
PO06)	*1300	0.15	R. Juan XXIII, Uyuni
PO07)	1300	1	R. Fides, Potosí: 1000-0100 (Sun 1100-0500)
PO08)	1300	0.3	R. Chichas, Siete Suyos: 1100-0400
SC16)	1300	1	R. Coronel Eduardo Avaroa, Sta. Cruz: 0930-0230
LP23)	1300	15/6	R. Sol, "Poder de Diós", La Paz: 1000-0100
BE18)	1300	5	R. Bandera Beniana, Trinidad
CO14)	1310	10	R. San Rafael, Cochabamba
LP52)	1320		R. Panorama, Achocalla
OR08)	1330	3	R. América, Oruro
TA04)	1330	1	R. Frontera, Yacuíba: 0930-0330
SC17)	v1340	1	R. Grigotá, Santa Cruz: 1000(Sun 1100)-0100v
LP22)	1340	0.35	R. San Francisco, Apolo
LP39)	1340	0.5	R. Copacabana, Copacabana
LP40)	1340	0.5	R. Jach'a Suyu, Corocoro: 1000-1630, 2000-0100
CO05)	1350	2.5	R. Cochabamba, «CBA»: 1030-0200
SC18)	1350	2.5	R. Ichilo, Yapacaní: 0930-0135
CH06)	1350	1	América Radiodifusión, Sucre: 0800-0200
CO15)	v1355	0.25	R. Armonía, Cliza (n.f.: 1350)
LP16)	1360	5	Radiodifusoras Jiménez, El Alto: 0800-0100
PO24)	1370	0.5	R. LV de Minero, Siglo XX: 0900-1700,2200-2300
OR11)	1370	5/3	Radiodifusoras Coral, Oruro: 1000-0400 (Sun 1100-2400)
LP24)	1370	1	R. Agricultura, Achacachi: 1000-2400
CO16)	1370	0.15	R. Libertad, Cliza: 0930-0300
CO34)	1380	1.5	R. Bandera Tricolor, Cochabamba: 1100-0300
CO17)	*1380	0.25	R. 16 de Noviembre, Sacaba: 1000-2300
TA06)	1380	0.5	R. Luis de Fuentes, Tarija: 0930-0400
LP66)	1380		R. Misericordia, El Alto
PO11)	1390	0.25	R. LV Minera del Sud, Mina Telamayu: 1100-0300
CO50)	1390		R. Mancomunidad Andina, Cochabamba
CH20)	1400		R. Antena 2000, Sucre
LP25)	1400	5	R. Nacional de Bolivia, La Paz: 0900-0200
SC19)	*1400	1	R. Libertador, Santa Cruz
LP41)	1400		R. Comunidad, Patacamaya
OR14)	1410	0.25	R. Atlántida, Oruro: 1100-2400
SC20)	1410	0.25	R. Roboré, Roboré

MW	kHz	kW	Station, location, h of tr
TA05)	1420	1.5	R. Guadalquivir, Tarija: 0900-0100
CO18)	1420	1	R. Centro, Cochabamba: 1030-2300 (Sun 1100-0000)
CH15)	1420	1	R. Real Audiencia, Sucre: 0900-0200
CO19)	1430	0.25	R. Nuestra Señora de Burgos, Mizque: 1000-2300
PO17)	1430	0.15	R. Centinela, Tupiza
LP26)	1440	1	R. Batallón Colorados, La Paz: 1100-0100 (Sun 2400)
SC21)	1440	2/1	R. Yaguari, Vallegrande: 1000-0200
SC32)	1440	0.5	R. Oriente, Camiri
CO42)	*1440	0.25	R. Bolivia, Cochabamba: 1000-1830, 2200-0200
SC22)	1445	0.5	Super Broadcasting Alborada, "SBA", Santa Cruz (n.1475): 1000-0300
OR13)	1450	1	R. Em. Bolivia, Oruro: 0900-0130
SC24)	1450	1	R. Verde y Blanco, Santa Cruz: 1000-0700
OR23)	1450		R. Amanacer, Huari
PA03)	1450		R. Amazonia, Cobija: 1100-0200(Sun 2200)
CO39)	1455	0.5	R. Magnal, Capinota: 1600-2400, (Sun 1000-1800)
CO44)	r1460		R. Morochata, Morochata
CO36)	1461		R. LV del Pueblo de Dios, Cochabamba: 0900 (Sun 1100)-0100
CH07)	1470	0.25	R. Cordech, Alcalá: 1100-1600, 1900-0200
CO22)	r1475		R. Tiraque, Tiraque
PO12)	1480	0.1	Patrimonio Radiodifusión, Potosí: 0950-0100
CO32)	1480	1/0.8	R. Chiwalaki, Vacas: 0900-1400, 2100-0100
CO40)	1480		R. Domingo Savio, Villa Independencia (r:1495)
LP58)	1480		R. Amor de Diós, La Paz
CO23)	1485	1	R. LV del Valle, Punata (n.f.: 1580)
OR15)	1490	1	R. San José, San José, Oruro: 1000-1300, 0000-0400 (Sun 1200-1300)
BE05)	1490	0.25	R. Moxos, San Ignacio de Moxos
LP30)	1490	0.35	R. Pedro Domingo Murillo, Quime: 1000-1400
SC23)	1490	0.25	R. Mairana, Mairana: 1100-0300
CO21)	*1495	2.5	El Mundo Radiodifusión, Sacaba: 1000-0300
CO40)	1495		R. Domingo Savio, Villa Independencia (n: 1480)
SC25)	1500	1	R. Sagrado Corazón, Mineros
LP31)	*1500	5/1	R. Chuquisaca, El Alto
CO47)	1500		R. Comuicacion Cristiana, Quillacollo
LP71)	1500		R. Huaycheño, Puerto Acosta
TA08)	*1510	0.25	R. 27 de Diciembre, Villamontes 1000-0200 (Sun 1100-0100)
SC26)	1520	1	R. Petrolera, Sta. Cruz: 1200-0330, (Sun1400-2300)
LP32)	1520	0.25	R. LV del Cobre, Corocoro: 1000-0400
OR08)	1520		R. Melodía, Oruro: 0930-2400
LP59)	1520		R. La Luz del Tiempo, El Alto
BE07)	1530	0.5	R. Em. Ballivián, San Borja (r:1535): 1030-0100
PO10)	1530	0.25	R. Litoral, Llica
CO24)	1530	1	R. Don Bosco, Kami: 1900-0400
PO13)	1530	0.1	R. Horizonte, Huanuni
LP69)	1530		R. Huaycheño, La Paz
LP34)	1540	0.8	R. Sariri, Escoma: 1000-1300, 2200-0200
CO46)	1540		R. Wiña Kalpachaj, Tarata.
LP67)	1543		R. Bendita Trinidad Espíritu de Dios, El Alto
CO25)	1545	0.35	R. Mejillones, Tarata
TA09)	*1545		R. Emisoras Villamontes, Villamontes
LP28)	1550	10	R. Caranavi, Caranavi: 0930-1800, 2200-0200
SC27)	1550	1	R. Tamengo, Pto. Quijarro (n.f.: 1495)
OR19)	1560	1	R. Occidental, Oruro: 0930-2400
CO26)	1560	0.5	1° de Octubre, Capinota: 1030-0100
CO27)	1560	0.5	R. Urkupiña, Quillacollo: 1000-2400
LP53)	1560		R. Tawantinsuyo, Taraco
SC28)	1570	0.5	R. 1° de Mayo, 1° de Mayo: 1000-0200 (Sun 1800)
CO24)	1578		R. Don Bosco, Kami(n. 1530)
SC29)	1580	1	R. Andrés Ibáñez, Santa Cruz: 1000-0300
LP62)	1580		R. El Fuego del Espíritu Santo, El Alto: 1000-2400
LP61)	1590		R. Kollasuyo Marka, Tiawanaku
TA07)	1590	3	R. Bermejo, Bermejo: 0930-0100
OR16)	1590	0.5	R. Producciones Pusisuyu, Oruro
SC34)	1590		R. Globo, La Guardia
CO28)	1600	0.5	R. Continental, Punata: 1000-0200
LP63)	1600	1	R. La Voz del Espíritu Santo, El Alto: 0850-0200

SW	kHz	kW	Station, location
CO29)	3310	10	R. Mosoj Chaski, Cochabamba: 0900-1300, 2100-0100
CH08)	3390	1	R. Em. Camargo "LV del Valle Cinteño", Camargo
BE10)	4409	0.5	R. Eco, Reyes
BE17)	4422		R. Emisora Reyes, Reyes
BE04)	4451	1	R. Santa Ana, Santa Ana del Yacuma:
BE11)	v4472	1	R. Movima, Santa Ana del Yacuma
PA02)	4600	0.2	R. Perla del Acre, Cobija: 0930-2400
BE08)	v4699		R. San Miguel, Riberalta: 1030-0330
PO15)	v4717	1	R. Yatun Ayllu Yura, Yura (n. 4715)
PO20)	4722		R. Uncia, Uncia
PO25)	4728		R. Aripalca
PA04)	4732		R. Universitaria, Cobija
LP37)	*4762	1	R. Constelación, Guanay

SW	kHz	kW	Station, location
PO21)	4763		R. Chicha, Tocla
LP74)	4782		R. Tacana, Tumupasa
PO19)	v4796		R. Lípez, Uyuni: 1000-2330
PO22)	4835	0.5	R. Virgen de los Remedios, Tupiza (r)
LP70)	4845		R. Norteño, Caranavi
OR18)	*4864	1	R. Emisora 16 de Marzo, Mina Bolívar
SC36)	4865	5	R. Logos, Santa Cruz
LP06)	*4875	10	R. La Cruz del Sur, La Paz: 1300-2100 (Sun. 1220-1600
LP15)	*4945	10	R. Illimani, La Paz
BE02)	v4958	1.5	R. Tropico, Trinidad:
SC30)	5580	0.25	R. San José, San José de Chiquitos: 1100-1700, 2100-0200v
OR17)	v5952	5	R. Pío XII, Siglo Veinte: (n. 5955)
OR20)	5965	1	R.Nacional, Huanunui
CH05)	5996	2.5	R. Loyola, Sucre: 1000-2400, Sun. 1015-2200
LP15)	6025	10	R. Illimani - R. Patria Nueva, La Paz: 0930-0300
SC02)	v6054	3	R. Cultural Juan XXIII, San Ignacio de Velasco: (n. 6055): 1030-2300
CO56)	6075		R.Casachun Coca, Lauca Ñ
LP02)	6080	5	R. San Gabriel, La Paz: 0900 (Sun:1000)-0200
LP01)	v6105	10	R. Panamericana, La Paz
SC04)	6135	10	R. Santa Cruz, Santa Cruz: 1100-0200
LP07)	6155	10	R. Fides, La Paz: irr
SC36)	6165	1	R. Logos, Santa Cruz
LP07)	9625	15	R. Fides, La Paz: irr

NB: RPN = Red Patira Nueva

Addresses and other information

ERBOL (Educación Radiofónica de Bolivia), Calle Ballivián 1323, 4° piso (☞ Cas. 5946), La Paz ☎ 2 232 4606, 232 4768 📠 2 239 1985 – Pte.: Jorge Trias S.J. Secr. Ejecutivo: Jorge Aliaga Murillo
W: www.erbol.combo
UNESBO (Unión de Emisoras Sindicales de Bolivia), Yanacocha 689, La Paz ☎ 1 234 1881 Pte: Jorge Bustillo Burgos
NB: Whenever listed, Casilla addresses should preferably be used for mailing purposes.

BE00 (BENI)
BE01) Cas 172, Guayaramerín - **FM:** 100.1MHz – **BE02)** Cas 60, Trinidad – **BE03)** 18 de Noviembre 628 (☞ Cas 720), Trinidad – **BE04)** Calle Sucre 250, Santa Ana de Yacuma – **BE05)** San Ignacio de Moxos – **BE06)** Sucre 320, Guayaramerín – **BE07)** Oruro 52, San Borja ☎3 848 3020 – **BE08)** Calle Tomás Daney s/n, Barrio San José, Riberalta ☎591 3952 3363 📠591 3852 3268 - **FM:** 99.1MHz "Centenario" – **BE09)** Av Selim Majuli (☞ Correo Central), San Borja, Pcia BalliviáN – **BE10)** Reyes, Pcia Ballivián – **BE11)** Calle Baptista 24, Santa Ana de Yacuma – **BE12)** Calle Nicanor Gonzalo Salvatierra 249, Riberalta - **FM:** 91.1MHz **BE13)** Ballivián s/n, San Ignacio de Moxos **BE14)** Cas 395, Guayaramerín – **BE15)** Calle Beni s/n, Guayaramerín – **BE16)** Plaza Fr Martín Baltasar de Espinosa, Santa Ana del Yacuma – **BE17)** Reyes, Prov de Ballivián – **BE18)** Calle Santa Cruz esq Mamoré s/n, Trinidad – **BE19)** Avenida Primero de Mayo esquina Loreto, Guayaramerín **E:E:** ninafelima@hotmail.com – **BE20** Riberalta, Prov Vaca Diez.

CH00 (CHUQUISACA)
CH01) Loa 602 (or Cas 538), Sucre **W:** www.aclo.org.bo **E:E:** aclo@entelnet.bo ☎ 4 646 2213 📠 4 646 2618 Prgrs in **Quechua** except Spanish 1330-2030 0900-0200 – **CH02)** Cas 335, Sucre - **FM:** 90.7MHz – **CH03)** Cas 25, Sucre – **CH04)** Calle 276, Sucre ☎ 4 645 3231 - **FM:** 92.1MHz – **CH05)** Calle Ayacucho 161, Sucre ☎4 645 3677 📠4 644 2555 **W:** www.radioloyola.com **E:** loyola@radiofides.com - **FM:** 98.3MHz "Onda Joven" – **CH06)** Calle Nataniel Aguirre 560, Sucre ☎4 645 2500 📠4 645 350 **E:** america@radioamerica.combo - **FM:** 97.5MHz – **CH07)** Cas 156, Sucre – **CH08)** Cas 09, Camargo, Pcia Nor Cinti **W:** www.radiocamargo.cjb.net - **FM:** 100.0MHz – **CH09)** Alcaldía Municipal, Padilla – **CH10)** CEDEC, Cas 196, Sucre ☎4 645 5008 📠4 646 2628 - **FM:** 103.1MHz – **CH11)** Calle Loa No 41, Sucre ☎64 40904 **W:** www.encuentroradio.com - **FM:** 95.9MHz – **CH12)** Av Hernando Siles 614, Sucre – **CH13)** Calle Kantuta 3, Barrio Ferroviario, Zona San Matias, Sucre – **CH15)** Calle Avarioa 537, Sucre – **CH16)** Sucre – **CH17)** Sucre – **CH18)** Comunidad de Tarabco, Prov Yamparáez – **CH19)** Comunida de Mojocoya, Prov Zudáñez – **CH20)** Sucre.

CO00 (COCHABAMBA)
CO01) Chimboata – **CO02)** Av Heroinas O-467, Zona Sur (☞ Cas 1092), Cochabamba ☎4 425 0422 📠4 425 1173 **Quechua:** 0930-1030, 0000-0200 1100-0300 - FM: 95.1MHz "Fides" – **CO03)** Sipe Sipe – **CO04)** Loa Final s/n, Aiquile, or: Cas 5716, Cochabamba **E:** aiquile@pino.cbb.entelnet.bo **Quechua:** 8 hours daily - **FM:** 100.3MHz – **CO05)** Calle 25 de Mayo 230 entre Bolívar y Sucre(☞ Cas 5500), Cochabamba ☎4 425 1504 📠4 425 1561 **E:** ragarobol@yahoo.es - **FM:** 104.3MHz "Gaviota" – **CO06)** Santiváñez 172 Cas Junín (☞ Cas 719), Cochabamba **Aymara & Quechua:** 1130-1300 – **CO07)** Cochabamba – **CO08)** Lanza esq Ecuador N-0261 (☞ Cas 846), Cochabamba ☎4 425 7289 📠4 424

1414 – **CO09)** Cochabamba esq Heroes del Chaco, Quillacollo (or Cas 108), Cochabamba – **CO10)** Arani – **CO11)** Arque – **CO12)** Plaza Progreso 201 Arani – **CO13)** Calle 16 de Julio S-0435 (☞ Cas 4274), Cochabamba – **CO14)** Calle Calama E- 0315 (Cas 546), Cochabamba **Quechua/Aymara:** 0900-1100, 2300-0200 - **FM:** 92.3MHz – **CO15)** Calle 6 de Agosto 11, Cliza – **CO16)** Calle Santa Cruz 4, Cliza – **CO17)** Cas 2522, Cochabamba – **CO18)** Cas 839, Cochabamba - **FM:** 96.3MHz – **CO19)** Cas 893, Cochabamba – **CO20)** Plaza de Granos 44, Quillacollo - **FM:** 99.1MHz – **CO21)** Plaza 6 de Agosto 44, Sacaba (or Cas 3230, Cochabamba) ☎4 428 6150 **English:** 1900-2230 – **CO22)** Calle Junín s/n, Tiraque – **CO23)** Cas 1361, Cochabamba – **CO24)** Congregación Salesiana, Kami or: Cas 1151, Cochabamba ☎4 811 9295 Prgrs in Sp, **Aymara & Quechua** – **CO25)** Calle Esteban Arce 401, Tarata – **CO26)** Calle Bolívar 13, Capinota – **CO27)** Plaza Bolivar 25 acera oeste, Quillacollo – **CO28)** Ayacucho 138, Punata – **CO29)** Calle Abaroa S-0254 (Cas 4493), Cochabamba **E:** rmchaski@bo.net ☎4 422 0651 📠4 425 1041 – **CO31)** Calle Rafael Urquidi 238, Comunidad Huañakahua, Prov Sipe Sipe **FM:** 108.2MHz – **CO32)** Misuk'ani, Vacas, Prov de Arani **E:** chiwalak@pino.cbb.entelnet.bo ☎4 411 3111 or Calle Manko Kapaq 174 (or Cas 80), Cochabamba **E:** incca@pino.cbb.entelnet.bo ☎4 425 5390 Prgrs mainly in **Quechua** 0900-1400, 2100-0100 – **CO33)** Av Petrolera Km 0.5, Cochabamba – **CO34)** Av Oguendo N-0560 entre Pacieri y Pedro Blanco (Cas 3655), Cochabamba **E:** latripple999@yahoo.com ☎4 425 3299 📠4 411 9729 **Quechua:** 1000-1200 - **FM:** 99.9MHz "La Triple" – **CO36)** – **CO38)** Calle Sucre s/n, Arbieto, Provi de Estaban Arce – **CO39)** Augusto Larrain, Capinota – **CO40)** Villa Independencia, Prov Ayopaya – **CO42)** Calle Calama 0-0135, Cochabamba – **CO43)** Calle Junín 309, Tiraque, Prov Arani – **CO44)** Morochata – **CO45)** Colami – **CO46)** Calle Estéban Arce 40, Tarata – (Casilla 4310, Cochabamba) – **CO47)** Quillacollo – **CO48)** Provincia de Ayaypoya – **CO49)** Cajón 21, Cochabamba – **CO50)** Calle Tumusla esquina Ecuador No 310, Plaza Cobija, area Oeste (Casilla 1986) Cochabamba **W:** www.cepra-bo.bo **E:** cepra@supernet.combo – **CO51)** Totora, Prov Carrasco – **CO52)** Comunidad de Independencia, Prov de Ayopaya – **CO53)** Calle Pablo Jaimes 188, Tiquipaya, Prov Quillacollo - **FM:** 90.5MHz – **CO54)** Quillacollo – **CO55)** El Paso – **CO56)** Lauka Ñ, Chapare, Provincia de Tiraque **W:**http//radiokawsachuncoca.com **E:** radiokawsachuncoca@gmail.com – **FM:**99.7MHz

LP00 (LA PAZ)
LP01) Av 16 de Julio, Edificio 16 de Julio, Of 902, El Prado (☞ Cas 5263), La Paz **W:** www.panamericana-bolivia.com **E:** pana@panamericana.bo ☎2 233 4271 **N:** «El Panamericano» relayed by many stns – **LP02)** Gral Lanza 2001 (Cas 4792), Alto Sopocachi Cristo Rey, La Paz ☎2 241 4371 📠2 241 1174 Prgrs in **Aymara** exc Sp & Quechua 1400-1430, Sat 2100-2130 – **LP03)** 16 de Julio 1295, La Paz – **LP04)** Av Sánchez Lima 2554, P 2, La Paz – **LP05)** Calle Ballivian No 1277, Chulumani (Cas 4535, La Paz) ☎2220 3672 **Aymara:** 0930-1130, 2230-0030 0900-1700, 2000-0100 - FM: 104MHz – **LP06)** Nicaragua 1759(Cas 1408), La Paz ☎2 220541 📠2 243337 **W:** www.radiocruzdelsur.com **E:** contactos@radiocruzdelsur.com – **LP07)** Calle Jenaro Sanjinés 799, Zona Norte (Cas 9143), La Paz ☎2 240 6363 📠2 240 6320 **W:** www.radiofides.com **E:** sistemas@radiofides.com **N:** «La hora del país», relayed by many stns, at 1100, 1630, 2230, 0130 – **LP08)** Cas 5324 or: Av Sánchez Lima 2278, 3° piso (entre Fernando Guachalla y Rosendo Guttierez), La Paz ☎ 2 236 1591 📠 2 236 3069 **Aymara:** 0945-1015 1000-0200 (SS 2400) – **LP09)** Cas 2329, La Paz – **LP10)** Calle Abdon Saavedra Nº 2110 casi esquina Fernando Guachalla, or: Cas 8631, La Paz ☎2 426 742 **W:** altiplanoadvenir.org – **LP11)** Cas 312472, La Paz ☎2 810048 📠2 813424 **Aymara:** 0830-1200 0900-0130 – **LP12)** Calle Abdón Saavedra 1990 (or Cas 2431), La Paz ☎2 235 6622 – **LP13)** Juan de la Riva 1527 (☞ Cas 8704), La Paz ☎2 236 3745 📠2 237 6785 **Aymara:** 0900-1000 – **LP14)** Calle Jenaro Sanjinés 799 esqquina Calle Sucre, La Paz **E:** editor@radiofides.com – **LP15)** Av Camacho 1485, Edificio La Urbana P 6, La Paz ☎2 220 0282 **W:** www.patrianueva.bo **E:** illimani@comunica.gov.bo – **LP16)** Av Panamericana 93, Cd Satélite, El Alto (or Cas 6412, La Paz) – **LP17)** Calle Nueva York 140, 1er Pasaje Chijini (or Cas 8084), La Paz – **LP18)** Calle General Lanza 93, Viacha, Provincia Ingavi – **LP19)** Cas 1539, La Paz Prgr in **Aymara** 0900-0100 – **LP20)** Calle Tumusla 765, La Paz – **LP21)** Cas 154, Colquiri – **LP22)** Apolo, Pcia Franz Tamayo – **LP23)** Plaza del Estudiante 1905, La Paz – **LP24)** Achacachi, Pcia Omasuyos – **LP25)** Tumusla 612, 3° piso (or Cas 2532), La Paz. 0900-0200 Prgr. mainly in **Aymara** – **LP26)** Av Saavedra esq My Zubieta Estado Mayor (Miraflores), La Paz ☎2 237 2065 **Aymara:** 1100-1200 – **LP27)** Calle Chichas esq Litoral No 1292, Planta Baja(Cas 5303), La Paz – **LP28)** Liga de Oración en Misión Mundial, Cas 266, La Paz (or Correo Central, Caranavi) – **LP29)** Casilla 204, Colegio Don Bosco, Pucarani Prgrs mainly in **Aymara**, but also in Spanish 0900-1300, 2030-0130 – **LP30)** Quime, Inquisivi – **LP31)** Cas 3123, La Paz – **LP32)** Correo Central, Corocoro, Provincia Pacajes – **LP33)** Av República 872, La Paz – **LP34)** Colegio Don Bosco, Plaza, Escoma (or Cas 204, La Paz) **E:** escoma@caoba.entelnet.bo ☎📠2 213 5336 - **FM:** 104.7MHz – **LP35)**

Calle Murillo 1379, La Paz ☎2 235 0588 – **LP36)** Calle Panama 11-53, Eduficio Shopping Miraflores, 1 °piso, Oficina 1, La Paz – **LP37)** Calle Boston de Guanay 123, Guanay, Prov Larecaja (or Casilla 15012, La Paz) **E:** constelación@hotamil.com – **LP38)** Plaza Alonso de Mendoza 500 5° piso del Edificio Santa Anita, Ofic 501 (or Cas 4973), La Paz ☎2 239 0542 – – **LP39)** Calle Professor Pedro F Mejía 6, Esquina Plaza Central de la Basílica Nacional Nuestra Senora de Copacabana, Copacabana, Prov Manco Capac – **LP40))** Plaza 15 de Agosto, Corocoro, Prov Pacajes **E:** tricolor-jachasuyu@hotmail.com ☎2 283 0192 Prgr In **Aymara & Sp** 1000-1630, 2000-0100 – **LP41)** Patacamaya – **LP42)** Av Patriotica 3048 (ex Calle 16 de Octubre), 1° piso, Zona Bolívar Municipal, El Alto – **LP43)** Avenida del Ejercito No 30, Zona Central, La Paz – **LP44)** Calle Heroínas 1010 esq 10 de Julio, El Alto, La Paz – **LP45)** Calle Frederico Zuaso 1621, P-1 Z Central, La Paz – **LP46)** Comunidad Contorno Letania, Camino a Collana 30, Letania, Prov Ingavi – **LP47)** Av Manco Kapac 50, Tiawanaku, Prov Ingavi – **LP48)** Calle Tarija 233, El Alto, La Paz – **LP49)** Plaza Libertad s/n, Chulumani, Prov Ingavi – **LP50)** Puerto de Guaqui, Prov Ingavi – **LP51)** Achocalla – **LP52)** Av La Paz 90, Comunidad Pasajes, Localidad Achocalla – **LP53)** Plaza 16 de Julio s/n, Cantón Taraco, Taraco, Prov Ingavi – **LP54)** Calle Noel Kempf 140, El Alto, La Paz – **LP55)** Tiawanaku, Prov Ingavi – **LP56)** Calle Yanacocha 70, Achacachi – **LP57)** Plaza 4 de Octubre, Rosario, Corapata, Prov Los Andes – **LP58)** Edificio Hansa, piso 17, La Paz – **LP59)** Raúl Salmón 92 entre Calle 4 y 5, Zona 12 de Octubre, El Alto, La Paz ☎2 282 5269 – **LP60)** Plaza Principal, Cantón Villa Iquiaca, Vilaque, Prov Los Andes – **LP61)** Av Principal s/n, Tiawanaku, Prov Ingavi – **LP62)** El Alto – **LP63)** Av Héctor Ormachea 5671 entre Calle 10 y 11, La Paz – **LP64)** Calle Topater 830, La Paz – **LP65)** Plaza Abaroa, Camino a Viacha No 105, Topohoco – **LP66)** Calle Jotan Save 3132 entre Bluniel, Zona 16 de Julio, La Paz – **LP67)** El Alto – **LP68)** Huarina – **LP69)** Puerto Acosta – **LP70)** Caranavi – **LP71)** Puerto Acosta **LP73)** El Alto – **LP74)** Tumupasa, Prov de Iturralde – **LP75)** Zona Villa Amor de Dios, Calle Noaviri No 2105, El Alto – **LP76)** Caranavi, Prov Caranavi – **LP77)** Calle 11 de Calacote No 7837, La Paz.

OR00 (ORURO)
OR01) Calle Ayacucho 785 (Alto) (✉ Cas 670), Oruro ☎2 525 3534 ▤2 525 2515 **E:** cpi@coteor.net.bo **Aymara & Quechua:** 1130, 0230 – 1000-2200, 0200-0600 – **OR02)** Cooperativa Minera Poopó, F Fontanilla y Oblitas, Oruro ☎2 511 2113 **Quechua & Aymara:** 0900-1100 – **OR03)** Cas 1019, Oruro ☎2 511 2259 **W:** www.bahai. org.bo/oportunidades/servicio.htm **E:** radbahia@nogal.oru.entelnet. bo **Aymara & Quechua:** 11 hours daily – **OR04)** Montesimos 436 entre 6 de Octubre y Potosí, Oruro **Quechua & Aymara:** 0900-1230 0900-2400 - **FM:** 98.7MHz – **OR05)** Oruro – **OR06)** Calle Cochabamba esquina 6 de Octubre (✉ Cas 49), Oruro ☎2 525 0004 ▤2 524 2215 – **OR07)** Junín 508, P.2, Oruro– **OR08)** Calle Cochabamba 998 y Camacho (✉ Cas 41), Oruro ☎2 525 255 ▤2 524 0035 – **OR09)** Calle Junín y 6 de Agosto, Oruro ☎2 526 0200 **Aymara & Quechua:** 1000-1100 - **FM:** 98.3MHz – **OR10)** Ayacucho 663 y La Plata, Oruro – **OR11)** Av 6 de Octubre 1042 y Montecinos (✉ Cas 845), Oruro **E:** rdcoral@ coteor.net.bo ☎2 525 4143 ▤2 527 6645 **Quechua & Aymara:** 1000-1130 - **FM:** 97.1MHz – **OR12)** San Felipe 493 entre Tarapacá y Tejerina (Cas 247), Oruro **Aymara:** 2200-2400 - **FM:** 107.7MHz – **OR13)** Av Velasco Galvarro entre León y Rodriguez 1551, Oruro – Prgrs in **Sp., Aymara, Aymara & Poquina - FM:** 105.1MHz – **OR14)** Linares 1160 entre Cochabamba y Caro, Oruro – **OR15)** Caro 235 entre Pagador y Av Velasco Galvarro, Oruro – **OR17)** Campamento Siglo XX, Llallagua (or Cas 434, Oruro) **W:** www.radiopio12.org **E:** rpiodoce@ entelnet. bo – ☎2 582 0250 ▤ 2 582 0554 **Aymara & Quechua** 6 hrs daily - **FM:** 99.9MHz – **OR18)** Centro Minero Bolívar, Canadon Antequera, Provincia Poopo – **OR19)** Av Bakovic 1027 entre Caro y Montecinos, or Cas 326, Oruro 0930-2400 **Aymara & Quechua:** 0930-1030 - **FM:** 93.1MHz – **OR20)** Calle Sucre, Huanuni (✉ Cas 681), Oruro 2 552 0421 **E:** fstmb@hotmail.com – **OR21)** Eucaliptos, Provincia Tomas Baron – **OR22)** Eucaliptos, Provincia Tomas Baron – **OR23)** Huari – **OR24)** Calle Adolfo Mier 1231, Oruro – **OR25)** Huanuni – **OR26)** Orinoca, Prov Sud Carangas – **OR27)** Sabaya, Prov Sabaya – **OR28)** Comunidad de Oaqahaca, Prov Eduardo Abaroa

PA00 (PANDO)
PA01) Cas 179, Cobija – **PA02)** Av Prof Alguira Gutierrez (✉ Cas 209), Cobija – **PA03)** Calle 9 de Octubre 2, Barrio Conavi, Cobija – **PA04)** Radio y Televison Universitaria, Campus Universitario, Av Las Palmas, Cobija 3 842 2141 **E:** radiouap@hotmail.com

PO00 (POTOSI)
PO01) Plaza 6 de Agosto, Camamento Mina Catav - **FM:** 105.7MHz – **PO02)** Cobija 15, Zona Central, Potosí ☎62 22680 ▤62 26210 **E:** www.radiokollasuyo.net - **FM:** 105.1MHz – **PO03)** Cas 58, Villazón – **PO04)** Plaza Alonso de Ibañez entre Pasaje Boulevard, Potosí – **PO05)** Cas 15, Uncía – **PO06)** Cas 28, Uyuni – **PO07)** Cas 328, Potosí – **PO08)** Campamento Minero, Siete Suyos – **PO10)** Llica, Pcia Daniel Campos 1400-0300 – **PO11)** Campamento Minero, Telamayu – **PO12)** Victor

Flores 410, Potosi – **PO13)** Plaza Fermín López, Huanuni (or Cas 147, Oruro) - **FM:** 93.5MHz – **PO14)** Campamento Minero Tazna, Pcia Nor Chichas – **PO15)** Cas 326, Yura, Prov Antonio Quijarro **E:** radioyura@ hotmail.com – **PO16)** Dtto Minero de Animas – **PO17)** Cas 180, Tupiza – **PO18)** Mendez Arcos s/n, Atocha, Prov Sud Chichas **Quechua:** M-F 1100-1200 - **FM:** 102.1MHz – **PO19)** Calle Final Uruguay s/n (✉ Cas 16), Uyuni, Prov Antonio Quijarro ☎2 693 2145 **E:** max_nelson_t@ hotmail.com – **PO20)** Plaza 6 de Agosto y Calle Villazon, Uncía - **FM:** 105.3MHz – **PO21)** Tocla, Prov de Nor-Chicas – **PO22)** Casa Parroquial de Tupiza (Cas 198),Tupiza **E:** radiovirgenderemedios@hotmail.com ☎2 694 4662 - **FM:** 89.5MHz – **PO24)** Campamento Minero, Siglo XX – **PO25)** Comunidad de Aripalca, Municipio de Vitichi, Provincia de Nor Chicas ☎2 613 7068 **PO26)** Comunidad de Caiza, Prov Chayanta – **PO27)** Comunidad de Colquechaca, Prov de José María Linares.

SC00 (SANTA CRUZ)
SC01) Iglesia Metodista de Bolivia, Cas 434, Santa Cruz – **SC02)** Vicariato Apostólico, Calle Santa Cruz, San Ignacio de Velasco 1030-2300 - **FM:** 100.3MHz – **SC03)** Calle Potosí s/n (or Cas 38), Montero ☎4647 2469 **FM:** 105.7MHz "Concierto" – **SC04)** Calle Mario Flores esquina Güendá No 20 (Cas 672), Santa Cruz **E:** infacruz@entelnet. bo ☎3 353 1817 ▤3 353 2257 **Guaraní:** 1830-1900 - **FM:** 92.1MHz – **SC05)** Calle Warnes s/n, Portachuelo, Provincia Sarah – **SC06)** Parque Industrial Manzana No 7 (or Cas 1984), Santa Cruz **E:** diario@ mail.elmundo.combo or radio@mail.elmundo.combo **Guarani:** 1115, 2315 – **SC07)** Cas 18, Pto Suárez – **SC08)** Victoriano Gutiérrez 200, Cotoca – **SC09)** Warnes 195, Altos Cine Escorpio, Montero ▤3 992 0970 – **FM:** 99.9MHz – **SC10)** Warnes 334 (✉ Cas 629), Santa Cruz ☎3 336 3990 ▤3 336 3992 - **FM:** 92.7MHz – **SC11)** Cas 818, Santa Cruz ☎3 352 9265 ▤3 352 4747 **E:** mision.eplabol@scbbs-bo.com **Quechua & Guarani:** 0900-0945 - **FM:** 90.7MHz "R Super Color" – **SC12)** Independencia 372 (✉ Cas 186), Santa Cruz ☎ 3333 7194 ▤3 333 5778 - **FM:** 96.3MHz – **SC13)** Cas 7, Camiri – **SC14)** Cas 697, Santa Cruz – **SC15)** Mineros – **SC16)** Av Charcas 1051 lado octava División del Ejército, Santa Cruz ☎3 336 0447 ▤3 337 2242 - **FM:** 98.1MHz – **SC17)** Calle Colón 58, piso 5 Of 501-2, Cas 1399, Santa Cruz ☎ 3 332 2142 - **FM:** 90.3MHz – **SC18)** Calle Calama 118, Yapacani, Provincia Ichilo or Cas.463, Santa Cruz **Quechua:** - **FM:** 101.1MHz – **SC19)** Calle Cochabamba 255 (Cas 1333), Santa Cruz - **FM:** 107.5MHz – **SC20)** Santa Cruz – **SC21)** Florida esq Montes Claros 143, Vallegrande ☎3 942 2033 – **SC22)** Av Perimetral 284 esq Mutualista (or Cas 2024), Santa Cruz ☎3 347 0877 ▤3 332 9180 – **SC23)** Correo Central, Mairana, Prov de Florida – **SC24)** Av Paraguá 2465, Santa Cruz – **SC25)** Cas 507, Santa Cruz - **FM:** 89.5MHz. Prgrs also in **Quechua SC26)** Barrio Petrolero 21 de Diciembre fte Plaza Tarija, Santa Cruz – **SC27)** Puerto Quijarro, Prov Angel Sandoval – **SC28)** Calle 5 lado este, Villa 1° de Mayo – **SC29)** Calle España 572, 2°piso, Santa Cruz - **FM:** 97.9MHz – **SC30)** Cas 15, Santa Cruz de Chiquitos (Santa Cruz) – **SC31)** Carretera Internacional Santa Cruz – Cochabamba (or Correo Central), Santa Fé de Yapacani, Prov Ichilo (Santa Cruz) ☎ 3 933 6000 – **Quechua:** 0900-1100 – **FM:** 99.7MHz – **SC32)** Cas 30, Camiri – **SC33)** Cas 1766, Santa Cruz – **SC34)** El Toro s/n, La Guardia, Prov Andrés Ibañez – **SC35)** Calle Quijarro 74 esq Av Uruguay, Santa Cruz - **FM:** 105.5MHz – **SC36)** Santa Cruz – **E:** relacionespublicas@radiologosnetwork.com – **SC37)** Comunidad de Camiri, Prov Cordillera – **SC38)** Comunidad de San Julián, Prov Nuflo de Chávez.

TA00 (TARIJA)
TA01) Calle Virginio Lema 788 (Cas 404), Tarija ☎4664 3890 – **TA02)** Inst Politécnico Campesino, Cas 42, Yacuíba 1000-2400, (Sun 1100-1600) – **TA03)** Av Las Américas 963, Edif Radiofónico Los Andes (✉ Cas 344), Tarija - **FM:** 103.1MHz – **TA04)** Cas 24, Yacuíba – **TA05)** Daniel Campos 824, Tarija ☎4 663 4444 ▤4 663 5555 - **FM:** 91.5MHz – **TA06)** Bolívar 376, Edificio Borda (Cas 125), Tarija - **FM:** 93.1MHz – **TA07)** Av Barrientos esq Ameller, Bermejo ☎▤4 696 1584 - **FM:** 99.1MHz – **TA08)** Plaza del 15 de Abril, Villamonte – **TA09)** Av Méndez Arcos 157, Villamontes, Prov Gran Chaco – **TA10)** Av Bolívar 608, Bermejo – **TA11)** Comunidad de Villamontes, Prov Gran Chaco

FM in La Paz (MHz): 87.7 87.7 FM – 25) 88.5 Doble 8 Latina – 88.9 Gente – 89.3 Sistema Cristiano de Comunicaciones – 89.7 Salesiana – 16) 90.1 Caliente FM – 90.5 Panamericana Classica – 90.9 PCM – 91.3 Ciudad – 91.7 El Comercio – 92.1 Estudio 92 FM – 92.5 Estelar – 92.9 Galáctica – 93.3 Melodia – 93.7 Chacaltaya – 94.1 R La Voz de Bolivia – 94.5 Red Nuevo Tiempo – 94.9 Gigante – LP06) 95.3 – 95.7 Digital Sur – 1) 96.1 – 96.5 La Paz – 96.9 Diferente – 97.3 Stereo 97 – 6) 97.7 – 11) 98.1 Laser – 98.5 Andina – 98.9 Restauración – 25) 99.3 Melodía – 99.7 Cristo Viene – 41) 100.1 FM Cien – 100.5 Constelación – 100.9 R. Color – LP07) 101.3 – 101.7 Graffitti – 102.1 RRB – 102.5 Sintonia – 102.9 Cristal – 103.3 R. Deseo – 103.7 San Francisco de Asis – 104.1 Cadena CNT – 104.5 RCN – 104.9 Fantástica – 105.3 Nuevo Amanacer – 105.7 Majestad – 106.1 Pachamama – 17) 106.5 – 106.9 Paris-La Paz – 107.3 Nueva Cosmos – 107.7 Central FM

BONAIRE (Netherlands)

L.T: UTC -4h — **Pop:** 10,100 — **Pr.L:** Dutch (official), Papiamentu, English — **E.C:** 50Hz, 127/220V — **ITU:** ATN

BUREAU TELECOMMUNICATION AND POST
✉ Kaya Grandi 69, P.O. Box 791, Bonaire ☎ + 599 717 3140 🖷 +599 717 3554

MW Call	kHz	kW	Station, location
1) PJB	800	100	Trans World Radio, Kralendijk

FM	MHz	kW	Station, location
1)	89.5		Trans World Radio, Kralendijk
5)	91.1		Radio Digital 91.1 FM, Kralendijk
2)	93.1		Alpha FM 93.1, Kralendijk
2)	94.7	1	Voz di Bonaire, Kralendijk
3)	97.1	5	Ritmo FM, Kralendijk
4)	97.5		Dolfijn FM, Kralendijk
2)	101.1	0.4	Mega Hit FM, Kralendijk
6)	102.7		Bon FM, Kralendijk

Addresses and other information
1) Kaya Gob. N Debrot 64, Kralendijk, Bonaire (PO Box 388, Bonaire) ☎ +599 717 8800 🖷 +599 717 8808 **W:** www.twrbonaire.com **E:** 800am@twr.org, 895fm@twr.org **– 2)** Radiodifushon Boneriano NV, Kaya Gob. N. Debrot 2, Kralendijk, Bonaire ☎+599 717 5947 🖷 +599 717 8220 **W:** www.vozdibonaire.com www.megahitfm.com **E:** vdb@vozdibonaire.com **E:** info@megafm.com **– 3)** Kaya Korona 22, Bonaire ☎+599 717 5971 **– 4)** Sea Aquarium Beach, Bapor Kibra z/n, Willemstad ☎+599 9 465 9975 🖷 599 9 461 9975 **E:** info@dolfijnfm.com **W:** www.dolfijnfm.com **– 5)** Kaya Grandi z/n , Kralendijk, Bonaire ☎+599 717 9911 **– 6)** Kaya Irlanda 11, Kralendijk ☎+599 717 2102 🖷+599 717 2002 Manager: Carmo R. (Bubui) Cecilia 24 h in Papiamentu, news in Dutch and a music prgr in English **W:** www.bonfm.com **E:** bonfm@hotmail.com

TRANS WORLD RADIO (Rlg. Cult. Educ.)
RADIO NEDERLAND RELAY STATION):
see International Broadcasting section

BOSNIA & HERZEGOVINA

L.T: UTC +1h (27 Mar-30 Oct: +2h) — **Pop:** 4.6 million — **Pr.L:** Bosnian, Croatian, Serbian — **E.C:** 50Hz, 220V — **ITU:** BIH

REGULATORNA AGENCIJA ZA KOMUNIKACIJE (RAK)
✉ Mehmeda Spahe 1, 71000 Sarajevo ☎ +387 33 250600 🖷 +387 33 713080 **E:** info@rak.ba **W:** www.cra.ba
LP: DG: Kemal Huseinovic
NB. RAK is the licensing body for broadcasting.

RADIO-TELEVIZIJA BOSNE I HERCEGOVINE (BHRT) (Pub)
✉ Bulevar Meše Selimovica 12, 71000 Sarajevo ☎ +387 33 461101 🖷 +387 33 464061 **E:** lejla.babovic@bhrt.ba **W:** www.bhrt.ba
LP: DG: Nermin Durmo

MW	kHz	kW			
Sarajevo (a)	612	100 (a) Donje Moštre			

FM	MHz	kW	FM	MHz	kW
Capljina	87.8	-	Forteca	97.3	0.5
Ivornik	88.1	30	Blagaj	99.3	0.01
V. Gomila	88.8	1	Lisin	100.3	-
Lipik	93.7	3	Hum	101.7	0.4
Leotar	94.1	30	Kozara	103.1	30
Drvar	94.7	0.4	Trebevic	103.7	0.4
Marijin	96.2	-	Hadzica	107.3	0.1
Vlasic	97.0	5			

D.Prgr: BH Radio 1 in Bosnian, Croatian, Serbian: 24h.

FEDERACIJA BOSNE I HERCEGOVINE

RADIO-TELEVIZIJA FEDERACIJE BOSNE I HERCEGOVINE (RTV FBiH) (Pub)
✉ Bulevar Meše Selimovica 12, 71000 Sarajevo ☎ +387 33 461539 🖷 +387 33 461539 **E:** press@rtvfbih.ba **W:** www.rtvfbih.ba
LP: Chmn: Igor Soldo

FM	MHz	kW	FM	MHz	kW
Tuzla	88.5	24	Dreznica	95.7	-
Capljina	89.1	-	Hum	95.7	10
Vlasic	89.3	50	V. Gomila	95.7	30
Jablanica	90.0	24.8	Lipik	98.9	-
Forteca	91.7	30	Hadzica	99.5	5
Kladanj	94.5	-	Neum	102.5	-
Lisin	94.5	0.25	Blagaj	103.5	-

D.Prgr: Radio FBiH in Bosnian, Croatian: 24h.

OTHER STATIONS

MW	kHz	kW	Location	Station
15)	774	2	Tuzla	R. Tuzla
23)	792	1	Banovici	R. Banovici
24)	1503	1	Zavidovici	R. 1503 Zavidovici
25)	1584	1	Bos. Petrovac	R. Bosanski Petrovac

FM	MHz	kW	Location	Station
16)	88.2	2	Drvar	Radiopostaja Drvar
7)	90.3	5	Posušje	R. Livno
7)	90.8	5	Glamoc R.	R. Livno
13)	90.9	1	Sarajevo	R. Stari Grad
7)	91.5	5	Bos. Grahovo	R. Livno
5)	91.5	4	Sarajevo	R. Kalman
6)	91.8	2	Gradacac	R. Kameleon
20)	92.0	1.5	Sarajevo Centar	RTV Kometa
7)	93.1	5	Posušje	Hrvatski rp. Široki Brijeg
18)	94.7	1	Bihac	RTV Bihac
21)	96.2	4	Bihac	RTV USK
4)	96.2	1.5	Mostarsko Blato	R. Dobre vibracije
15)	96.5	3	Majevica Lipik	R. Tuzla
1)	96.9	3.5	Posušje	HRTV Mostar
13)	97.3	3.5	Zenica	R. Stari Grad
14)	97.6	2	Glamoc R.	R. Studio N
11)	98.1	30	Bjelašnica	R. Herceg-Bosne
8)	98.7	1	Sarajevo	R. M
3)	99.3	2.5	Zenica	R. BM
7)	100.9	1	Livno	R. Livno
1)	100.9	7.5	Mostar	HRTV Mostar
19)	101.5	1	Gorazde	RTV Gorazde
7)	101.8	1	Bugojno	R. Livno
4)	102.5	1.5	Neum	R. Dobre vibracije
10)	103.7	2	Osjecenica	R. Sana
11)	103.9	30	Velez	R. Herceg-Bosne
13)	104.3	16.5	Bjelašnica	R. Stari Grad
21)	105.1	31.5	Plješivica B.	RTV USK
9)	105.2	1	Fojnica	R. Q
12)	105.6	1	Sarajevo	R. Istocno Sarajevo
22)	106.6	10	Zenica Lisac	RTV Zenica
17)	106.7	3	Bosanska Dubica	R. Mir Medjugorje
17)	107.8	50	Bihac	R. Mir Medjugorje
17)	107.8	5	Petricevac	R. Mir Medjugorje

NB: Txs below 1kW not listed.

Addresses & other information:
1) Dubrovacka 4, 88000 Mostar. **E:** htvmostar@tel.net.ba **– 2)** Trg Gojka Šuška 5c, 88220 Široki Brijeg. **E:** radio-sirokibrijeg@tel.net.ba **– 3)** Talica brdo br.11 i 13, 72000 Zenica. **E:** urednik@bmradio.info **– 4)** Kralja P. Krešimira IV, 88000 Mostar. **E:** rdv@rdv.ba **– 5)** Varazdinska 18, 71000 Sarajevo. **E:** redakcija@kalmanradio.ba **– 6)** Dr. Milana Jovanovica 6, 75000 Tuzla. **E:** kameleon@kameleon.ba **– 7)** Kneza Mutimira 29, 80101 Livno. **E:** radio.livno@tel.net.ba **– 8)** Fra Andjela Zvizdovica 1, 71000 Sarajevo. **E:** hitradio@radiom.net **– 9)** Donje Rosulje 59, 71300 Visoko. **E:** radio@bih.net.ba **– 10)** Banjalucka 2, 75260 Sanski Most. **E:** radiosana@yahoo.com **–11)**Kralja Petra Krešimira IV bb, 88000 Mostar. **E:** radiohb@tel.net.ba **– 12)** Stefana Nemanje 8, 71123 Srpsko Sarajevo. **E:** radiois@sarajevo-rs.com **– 13)** Urijan Dedina 7, 71000 Sarajevo. **E:** rsg@rsg.ba **– 14)** Splitska bb., 80101 Livno. **E:** studio.n@tel.net.ba **– 15)** ul. Mirze Delibašica 4, 75000 Tuzla. **E:** radiotz@radiotuzla.com **– 16)** Ante Bruno Bušic 4, 80260 Drvar. **– 17)** Gospin Trg 1, 88266 Medugorje. **E:** radio-mir@medjugorje.hr **– 18)** Krupska bb, 77000 Bihac. **E:** rtvbihac@bih.net.ba **– 19)** Zaima Imamovica 2, 73100 Gorazde. **E:** martvgo@bih.net.ba **– 20)** Dedijerova 12, 71123 Istocno. **E:** kometa@paleol.net **– 21)** Dom Kulture, 77000 Bihac. **E:** rtvuskbi@bih.net.ba **– 22)** Bulevar Kralja Tvrtka I bb., 72000 Zenica. **E:** info@rtvze.com **– 23)** ul. 7 novembra 4, 75290 Banovici. **E:** radiobanovici@yahoo.com **– 24)** Maršala Tita 3, 72220 Zavidovici. **– 25)** Bosanska bb, 77250 Bosanski Petrovac.

REPUBLIKA SRPSKA

RADIO TELEVIZIJA REPUBLIKE SRPSKE (RTRS) (Pub)
✉ ul. Kralja Petra I Karadordevica 129, 78000 Banja Luka ☎ +387 51 339900 🖷 +387 51 301922 **E:** radio@rtrs.tv **W:** www.rtrs.tv
LP: DG: Dragan Davidovic

FM	MHz	kW	FM	MHz	kW
Kmur	87.3	1	Duge Njive	90.7	1
Trebevic	88.7	10	Banja Luka	90.9	1
Udrigovo	89.9	1	Kozara	92.7	30
Veliki Zep	90.3	30	Leotar	92.8	30
Duge Njive	90.7	1	Petrovo	93.5	-

D.Prgr: Radio RS in Serbian: 24h.

OTHER STATIONS

FM	MHz	kW	Location	Station
9)	87.7	5	Sarajevo	Nes R.
5)	88.4	6	Ugljevik	RTV Step

FM	MHz	kW	Location	Station
6)	89.9	2	Bosanska Dubica	RTV Vikom
12)	93.6	5	Banja Luka	Big R. 1
10)	94.3	2.5	Gracanica	Obiteljski R. Valentino
3)	94.7	10	Bosanska Dubica	R. Feniks
11)	95.3	23	Kozara	R. Bobar
1)	95.6	1	Banja Luka	R. Balkan
2)	96.3	2	Doboj	R. Doboj
9)	99.3	2	Velika Kladuša	Nes R.
4)	99.9	9	Banja Luka	RTV BN
11)	100.9	30	Majevica	R. Bobar
8)	102.7	3	Banja Luka	Hard Rock R.
11)	102.8	8	Trebevic	R. Bobar
11)	104.7	30	Vlašic	R. Bobar
6)	105.3	3	Banja Luka	RTV Vikom
11)	105.4	4.5	Tušnica	R. Bobar
11)	105.5	22	Leotar	R. Bobar
11)	105.9	39	Velez	R. Bobar
9)	106.4	22.5	Kozara	Nes R.
7)	106.7	1	Drvar	TMK Radio
11)	107.3	12.5	Trovrh	R. Bobar
11)	107.8	77	Plješivica B.	R. Bobar

NB: Txs below 1kW not listed.
Addresses & other information:
1) Kralja Petra I Karadordevica 113/11, 78000 Banja Luka. **E:** balkan@radio-balkan.com – **2)** Kneza Lazara 8, 74000 Doboj. **E:** rdoboj@doboj.net – **3)** Svetosavska 5/c, 79240 Kozarska Dubica. **E:** feniksfm@teol.net – **4)** Laze Kostica 146, 76320 Bijeljina. **E:** director@rtvbn.com – **5)** Naselje Gojsovac bb., 76320 Dvorovi. – **6)** Srpska 2/II, 78000 Banja Luka. **E:** vikom.tv – **7)** Bul. Vojvode Petra Bojevica bb., 78000 Banja Luka. **E:** tmk.oksigen@gmail.com – **8)** P.C. Krajina, Vidovdanska bb., 78000 Banja Luka. **E:** hardrockradio@blic.net – **9)** Brace Pišteljic 1, 78000 Banja Luka. **E:** nesradio@teleklik.net – **10)** Kolodvorska 108 a, 76204 Bijela. **E:** desk@valentinobh.com – **11)** Filipa Višnjica 211, 76320 Bijeljina. **E:** rbobar@teol.net – **12)** Vuka Karadzica 6, 78000 Banja Luka. **E:** redakcija@bigradiobl.com

BOTSWANA

L.T: UTC +2h — **Pop:** 2 million — **Pr.L:** English, Setswana — **E.C:** 50Hz, 220V — **ITU:** BOT

NATIONAL BROADCASTING BOARD (NBB)
206/207 Independence Ave, Private Bag 00495, Gaborone ☎+267 3957755 ▤ +267 3957976 **W:** www.bta.org.bw/nbb.htm **E:** info@bta.org.bw **L.P:** Chairman: Dr. Masego Mpotokwane.

RADIO BOTSWANA (Pub, Comm.)
Private Bag 0060, Gaborone ☎+267 3653000 ▤ +267 257138 **W:** www.mcst.gov.bw **E:** Rbeng@info.bw **L.P:** Dir. Broadc. Sces: Habuji Sosome. Chief Broadc. Officer: Mrs. Banyana Segwe. CE: Kingsley Reetsang. Mgr RB: Margaret Modise.

MW	kHz	kW	MW	kHz	kW
Maun	531	50	Mmathethe	945	25
Muchenje	558	50	Sebele‡	972	50
Selebi-Phikwe	621	100	Jwaneng	1071	25
Mopipi	648	50	Mahalapye	1215	50
Shakawe	693	25	Tshabong	1350	50
Gantsi	873	50	‡inactive		

FM(MHz)	RB1	RB2	kW	FM(MHz)	RB1	RB2	kW
Bobonong	95.9	102.7	0.5	Mabule	92.3	105.9	
Charleshill	93.5	103.5		Mabutsane	94.2	104.6	
Francistown	103.6	90.5	3	Mahalapye	96.6	107.0	2.5
Gaborone	89.9	103.0	5	Maun	94.2	104.6	0.5
Gantsi	94.0	100.8	0.1	Olifant's Drift	88.0	104.7	
Good Hope	94.6	101.6	0.5	Orapa	89.9	98.6	0.1
Hukuntsi	99.9	96.2	0.5	Palapye	94.5	101.5	0.5
Jwaneng	99.2	106.3	0.5	Sekakangwe Hill	91.9	101.9	2
Kanye	89.0	95.3	0.5	Selebi-Phikwe	94.2	104.6	0.1
Kang	89.3	98.9		Serowe	99.4	92.9	1
Kasane	94.4	104.8	0.5	Sojwe	87.7	90.7	
Lobatse	98.6	105.7	1	Tsabong	96.6	107.0	0.5

National Sce. (RB1) in Setswana/English on MW/FM: 24h.
N. in English: on the hour exc. **in Setswana:** 1100, 1600, 1900.
Commercial Sce. (RB2): 24h. **N:** rel. RB1.
Ann: E: "This is R. Botswana broadcasting from Gaborone", "RB", "RB1", "RB2". Setswana: "Se Ke Seromamowa Sa Botswana mo Gaborone".
IS: RB1: Bird chirps and first bars of the National Anthem.

Other stations:
Duma FM: Gaborone 93.0MHz, Francistown 93.6MHz – **Gabz FM,**

Private Bag BO 319, 2nd Floor, Beta House, Plot 17954, Old Lobatse Rd, Gaborone. Web: www.gabzfm.com **FM:** Maun/Selebi-Phikwe 91.0MHz, Mahalapye 93.4MHz, Lobatse 94.1MHz, Palapye 94.7MHz, Serowe 96.1MHz, Gaborone 96.2MHz, Francistown 96.8MHz – **Yarona FM:** Francistown 100.1MHz, Gaborone: 106.6MHz. **W:** www.yaronafm.co.bw
Voice of America relay station (MW):
Selebi-Phikwe, Moepeng Hill: 909kHz 600kW 0300-0700 ,1600-2200 & SW. For further details see International Radio section (USA)

BRAZIL

L.T: PE (Fernando de Noronha only): UTC -2h. AL, AP, BA, CE, DF*, ES*, GO*, MA, MG*, PA, PB, PE, PI, PR*, RJ*, RN, RS*, SC*, SP*, TO: UTC -3h (*=DST: UTC -2h). AC, AM, MS*, MT*, RO, RR: UTC -4h (*=DST: UTC -3h). *) DST: 17 Oct 10-20 Feb 11, 23 Oct 11-19 Feb 12 — **Pop:** 199 million — **Pr.L:** Portuguese — **E.C:** 60Hz, 220V — **ITU:** B— **Int. dialling code:** +55

AGÊNCIA NACIONAL DE COMUNICAÇÕES (ANATEL)
SAS Quadra 06 Bloco H, Ed. Ministro Sérgio Motta, 2° andar, 70313-900 Brasília, DF **W:** www.anatel.gov.br
LP: Dir. Gen. Dr. Rubens Bussacos. Dir. of Radio: Roberto Blois Montes de Souza. Dir. Dept. of Authorizations: Domingo Poty Chabalgoity.

ASSOCIAÇÃO BRASILEIRA DE EMISSORAS DE RADIO E TELEVISÃO (ABERT)
SCN Quadra 4 Bloco B-100, sala 501, Centro Empresarial Varig, 70714-900 Brasília, DF (C.P. 08780, 70312-970)
☎61 2104 4600 ▤61 2104 4626 **W:** www.abert.org.br
LP: Pres.: Paulo Machado de Carvalho Neto. Exec. Dir: Antonio Abelin

N.B: all stns carry «A Voz do Brasil» (official prgr.). Main tr. M-F 2200-2300. Stns may transmit at other times during the day.
MW: Call ZY—
The letters preceding the stn number indicate the state or territory.
* = inactive v = varying freq ff = future frequency

MW	Call	kHz	kW	Station, location, h. of tr.
BA01)	H481	540	1/0.25	R. Regional, Irecê: 24h
CE01)	H610	540	1/0.25	R. Jornal, Canindé
GO01)	H755	540	10/1	R. Riviera, Goiânia
MA01)	H894	540	1/0.25	R. Guajajara, Barra do Corda
MG151)	L331	540	1/0.5	R. Ipanema, Ipanema
PI31)	I914	540	1/0.25	R. Primeiro de Julho, Agua Branca
PR110)	J322	540	1/0.25	R. Nova Era, Borrazópolis: 0800-0000
RJ01)	J450	540	10/2.5	R. Fluminense, Niterói
RS01)	K226	540	1/0.5	R. Real, Canoas
RS02)	K322	540	10/1	R. São Sebast. Santo Ângelo: 24h
SC01)	J778	540	10/1	R. Mirador, Rio do Sul: 0800-0300
SE01)	J924	540	10/2.5	R. Jornal AM, Aracaju: 24h
SP01)	K697	540	1/0.25	R. Uirapuru, Biriguí
SP02)	K734	540	1/0.25	R. Nova Sumaré, Sumaré
CE59)	H644	550	1/0.25	R. Vale do Quinçoé, Acopiara
MG01)	L225	550	5/0.5	R. Cataguases, Cataguases
MG02)	L263	550	20/5	R. Soc. Norte de Minas (RBV), M. Claros
MT29)	I429	550	10/5	R. Banda, Sinop
PE01)	I796	550	5/1	R. Meridional, Garanhuns
PI01)	I902	550	1/0.25	R. Serra da Capivara, São Raimundo Nonato
PI22)	I907	550	10/0.5	R. Globo, Parnaíba: 24h
PR139)	J331	550	10/0.5	R. Banda B, Curitiba
RS03)	K287	550	2.5/0.25	R. Sta. Cruz do Sul, Sta. Cruz do Sul
SP03)	K578	550	5/0.5	R. Mantiqueira, Cruzeiro: 0800-0300
SP04)	K696	550	5/0.5	R. Boa Vontade, Sertãozinho
AM09)	H289	560	1/0.25	R. Educação Rural, Coari: 1000-0100
BA02)	H456	560	5/1	R. Jornal, Itabuna
CE41)	H604	560	1/0.25	R. Educ. Jaguaribana, Limoeiro do Nte: 0700-0100 Sun:0800-2300
GO25)	H769	560	5/0.25	R. Emissora Sul Goiana, Quirinópolis
MA02)	H887	560	25/5	R. Educadora do Maranhão, São Luís
MG05)	L277	560	5/0.25	R. Dif., Patrocínio
MT01)	I395	560	10/2.5	R. Em. Aruanã, Barra do Garças
MT24)	I419	560	10/1	R. Pioneira, Tangará da Serra
PB16)	I695	560	1/0.25	R. Maná, Mamanguape
PR01)	J214	560	1/0.5	R. Londrina Londrina
PR02)	J281	560	2.5/0.25	R. Cultura, Guarapuava
RJ02)	J496	560	5/0.25	R. Costa do Sol, Araruama
RS04)	K231	560	5/1	R. São Francisco Sat, Caxias do Sul
SP213)	K761	560	35/1	R. Paulista, Santa Isabel
AL01)	H244	570	5/1	R. Novo Nordeste, Arapiraca
CE02)	H613	570	5/0.25	R. Verde Vale, Juazeiro do Norte
CE03)	H614	570	1/0.25	R. Uirapuru, Itapipoca
GO02)	H750	570	2.5/0.5	R. Cultura, Catalão: 0700-0200

MW	Call	kHz	kW	Station, location
MA06)	H890	570	10/1	R. Imperatriz, Imperatriz
MG03)	L261	570	25/5	R. Gospa Mira, Belo Horizonte: 24h
MT30)	N407	570	1/0.25	R. Jornal, São José dos Quatro Marcos: 24h
PR146)	J349	570	1/0.5	R. Continental, Palotina: 24h
RS05)	K267	570	1/0.5	R. Diário da Manhã, Passo Fundo
SC02)	J735	570	5/0.5	R. Eldorado, Criciúma
SC99)	J794	570	1/0.5	R. Fronteira, Dionísio Cerqueira
SP05)	K595	570	1/0.25	R. Clube de Itapeva, Itapeva: 0900-2300, Sat: 0800-0100 Sun: 0800-2200
SP06)	K672	570	5/1	R. Dif., Taubaté
SP195)	K698	570	1/0.25	R. Jornal, Nhandeara
SP150)	K717	570	1/0.25	Bariri R. Clube, Bariri
BA03)	H477	580	1/0.25	R. Dif., Teixeira de Freitas
G044)	H799	580	1/0.5	R. Serra Azul, Caiapônia
MG04)	L328	580	7/0.5	R. América, Uberlândia
MS01)	I387	580	25/1	R. Inmaculada Conceicao, Campo Grande
PE02)	I776	580	20/10	R. Boas Novas, Recife: 24h
PI12)	I905	580	5/1	R. Itamaraty, Piripiri
PR105)	J327	580	2/0.25	R. Pitanga, Pitanga
PR03)	J330	580	2.5/0.25	R. Grande Lago, Santa Helena: 0800-0300 Sun:0900-0100
RJ03)	J465	580	50/5	R. Nova Relogio, Rio de Janeiro
RS06)	K299	580	2/0.5	R. São Gabriel, São Gabriel
RS07)	K318	580	10/5	R. Fátima, Vacaria: 24h
SP07)	K540	580	1/0.25	R. Você, Americana
SP08)	K724	580	1/0.25	R. Regional, Palmital
T001)	H785	580	10/2	R. Tocantins, Porto Nacional
BA04)	H445	590	10/5	R. Cruzeiro da Bahia, Salvador
CE04)	H627	590	10/1	R. Vale do Rio Poty, Crateús
ES01)	I213	590	1/0.5	R. Tribuna, Vitória
G003)	H798	590	10/1	R. Manchester, Anápolis
MG93)	L249	590	10/0.5	R. Cultura, João Monlevade
MT03)	I420	590	10/5	CBN Cuiabé, Cuiabá: 24h
PB25)	I692	590	5/0.25	R. Serrana, Araruna: 0730-0100
PR04)	J234	590	10/5	R. Difusora AM 590, Curitiba
PR38)	J240	590	2.7/0.75	R. Dif. Regional, Cruzeiro do Oeste
RR01)	O700	590	10	R. Dif. de Roraima, Boa Vista: 0800-0300
RS08)	K210	590	5/0.5	R. Alegrete, Alegrete: 0900-0300
SC03)	J809	590	2/1	R. Progresso, Descanso
SP09)	K534	590	10/1	R. Atlântica, Santos: 24h
SP10)	K612	590	1/0.25	R. Clube, Mirandópolis
SP11)	K696	590	5/1	R. 79, Ribeirão Preto
AM02)	H287	600	10	R. Municipal, São Gabriel da Cachoeira: 0900-0300 (Sun:1000-0100)
BA05)	H486	600	10/1	R. Vale do Rio Grande, Barreiras
BA64)	H…	600	1/0.25	R. Dif.de Rio Real, Rio Real
CE38)	H627	600	1/0.25	R. Cultura, Aracati
MA38)	H920	600	10/1	R. Mirante, São Luís
PE03)	I789	600	1/0.25	R. Cardeal Arcoverde, Arcoverde
RS09)	K278	600	100	R. Gaúcha, Porto Alegre
AL10)	H249	610	10/2	R. Imperial, Marechal Deodoro
AM18)	H321	610	50/14.5	Super R.Boa Vontade, Iranduba: 24h
GO10)	H786	610	25/0.5	R. Mega, Luziânia
MG06)	L268	610	100/25	R. Itaiaia, Belo Horizonte
MT04)	I425	610	10/5	R. Celeste, Sinop
PB01)	I678	610	10/1	R. Progresso, Sousa
PI02)	I899	610	10/1	R. Poty, Teresina
SC04)	J746	610	10/0.5	R. Super Condá , Chapecó
SP12)	K532	610	1/0.25	CBN - (R.Chamonix), Mogi Mirim
SP13)	K577	610	1/0.25	R. Bandeirantes, Catanduva
SP14)	K589	610	1/0.25	Super R. Piratininga, Guaratinguetá
SP15)	K726	610	1/0.25	R. Paranapanema, Piraju
SP113)	K502	610	1/0.25	R. Presidente Venceslau, Pres. Venceslau
CE05)	H590	620	10	R. Globo, Fortaleza
GO56)	H777	620	1/0.25	R.620 AM, Pires do Rio
MG145)	L357	620	1/0.25	R. Catuaí, Manhuaçu
PR05)	J332	620	2.5/0.25	R. Jandaia, Jandaia do Sul
RS10)	K270	620	10/1	R. Pelotense, Pelotas
RS11)	K315	620	1/0.25	R. Municipal, Tenente Portela: 0830-0300 Sat:0900-0000 Sun:0900-2200
SC05)	J779	620	5/0.25	Super-Radio Dif. AM, Rio do Sul: 0800-0100
SP16)	K521	620	50/10	R. Jovem Pan, São Paulo
AP01)	H422	630	25/10	R. Dif. de Macapá, Macapá: 24h
CE58)	H636	630	1/0.5	R. Cidade, Campos Sales (ff 1480): 0800-0300
GO04)	H777	630	1/0.25	R. Gospel 630 AM, Pires do Rio: 24h
MA16)	H924	630	10/0.5	R. Macaru, Viana
MG07)	L299	630	10/0.5	R. Jornal da Manha, Uberaba
MS32)	N603	630	10/1	R. Novo Tempo, Campo Grande
MT05)	I384	630	10/5	R. Dif. Bom Jesús, Cuiabá
PI03)	I904	630	1/0.25	R. Dif., Barras
PR06)	J284	630	10/0.5	R. Educativa, Curitiba
PR07)	J300	630	5/0.25	R. Educ. Marechal, Marechal Cândido Rondón
RJ04)	J466	630	25/10	R. Roquette Pinto, Rio de Janeiro
RS12)	K259	630	1/0.5	R. Cacique, Lagoa Vermelha: 24h
RS13)	K289	630	1/0.25	R. Santamariense, Santa Maria
SC06)	J800	630	1/0.25	R. Doze de Maio, São Lourenço d'Oeste
SE02)	J920	630	10/5	R. Aperipe, Aracaju
SP17)	K613	630	1/0.25	R. Difusora, Mirassol
SP18)	K635	630	5/0.25	R. Cidade, Presidente Prudente
BA12)	H458	*640	10/0.5	R. Dif. Sul da Bahia, Itabuna
ES02)	I204	640	10/0.5	R. Vitória, Vitória
GO05)	H757	640	50/5	R. Dif. Goiânia, Goiânia
MG08)	L308	640	3/0.25	R. Santa Cruz, Pará de Minas: 24h
MG125)	L320	640	10/0.25	R. Educadora, Porteirinha: 0800-0300
MT06)	I406	640	10/5	R. Progresso, Alta Floresta
MT18)	I424	640	10/1	R. Tangará, Tangará da Serra
PI28)	I924	640	1/0.25	R. Cruzeiro, Pedro II
PR08)	J262	640	5/1	R. Tupi, Londrina
RJ05)	J489	640	5/1	R. Agulhas Negras, Resende
RN01)	I590	640	10/5	R. Globo, Natal
RS14)	K277	640	50/10	R. Bandeirantes, Porto Alegre
SP19)	K547	640	5/1	R. Morada do Sol, Araraquara
BA06)	H462	650	5/0.5	R. Clube, Valença: 24h
GO06)	H790	650	1/0.25	R. Cultural do Araguaia, Jussara
MG09)	L200	650	10/0.5	R. Princesa, Lagoa Formosa
MG85)	L309	650	5/0.5	R. Veredas, Unaí: 24h
MG142)	L372	650	10/1	R. Itatiaia AM Vale do Aço, Timóteo
MT19)	I414	650	5	R. Educadora, Colider
PA20)	I540	650	10/1	R. Tropical, Santarém
PB02)	I672	650	5/0.5	R. Alto Piranhas, Cajazeiras
PI26)	I925	650	1/0.25	R. Tapuio, Miguel Alves
PR09)	J202	650	1/0.5	R. Banda B Norte Pionerio, Cambará
PR91)	J250	650	8/1	R. Colméia, Cascavel
RS15)	K238	650	5/0.5	Radiodif. Sul Riograndense, Erechim
SP20)	K508	650	1/0.25	R. Andradina, Andradina
SP22)	K518	650	5/1	R. Terra, Santos
SP21)	K524	650	5/0.8	R. Dif., Piracicaba
BA07)	H465	660	1/0.25	R. Jornal, Itapetinga
BA36)	H480	660	10/0.25	R. Bom Jesus, Bom Jesus da Lapa
BA66)	H510	660	1/0.25	R. Tribuna do Vale do São Francisco, Xique-Xique
BA65)	H518	660	1/0.25	R. Planalto, Euclides da Cunha
CE06)	H619	660	1/0.25	R. Rio das Garças, Itarema (Acaraú)
GO07)	H778	660	1/0.25	R. Primavera, Itapuranga
GO51)	H794	660	1/0.25	R. Alvorada, Quirinópolis
MG11)	L206	660	10/0.25	R. Clube, Curvelo
MT07)	I401	660	10/0.5	R. Juventud, Rondonópolis
PA21)	I552	660	1/0.25	R. Xinguara, Xinguara
PE04)	I787	660	5/1	R. Jornal, Limoeiro
PE05)	I795	660	1/0.25	R. Grande Serra, Araripina
PI34)	I925	660	1/0.25	R. Tacarijus, São Miguel do Tapuio
RJ06)	J472	660	5/1	R. Friburgo, Nova Friburgo: 24h
R001)	J673	660	10/5	R. Boas Novas AM, Porto Velho
RS16)	K286	660	1/0.25	R. Marajá, Rosário do Sul: 0900-0300 Sat:0830-0200 Sun:1100-0200
RS17)	K319	660	10/0.5	R. Esmeralda, Vacaria
SP23)	K639	660	10/0.5	R. Clube, Ribeirão Preto
SP112)	K656	°660	10/0.25	R. Mundial, São Paulo
AC11)	H…	670	0.25	R. Dif., Sena Madureira: 0900-0400
AM04)	H288	670	1	R. Nal. do Alto Solimões, Tabatinga: 24h
AM03)	H297	670	1/0.25	R. Vale do Rio Madeira, Humaitá: 0800-0400
AP02)	H420	670	10/1	R. CBN, Macapá: 24h
BA95)	H…	670	5/0.5	R. Cidade, Barreiras
CE07)	H606	670	1/0.25	R. Cultura, Várzea Alegre: 0800-0100
GO08)	H747	670	10/1	R. São Francisco, Anápolis
MG12)	L310	670	10/2.5	R. Educadora, Montes Claros
MG126)	L347	670	5/0.25	R. Montanhesa, Ponte Nova: 24h
MG123)	L361	670	5/0.25	R. Uberaba, Uberaba
MG135)	L310	670	1/0.25	R. Cidade, Bambuí: 0800-0200
MS23)	I408	670	1/0.25	R. Patriarca, Cassilândia
MS28)	N600	670	10	Super R. Fronteira, Ponta Porã
MT16)	I422	670	6/1	R. Transpantaneira, Poconé
MT44)	I436	670	1	R. Atitude, Lucas do Rio Verde
PA02)	I537	670	1/0.25	R. Rural, Altamira
PA22)	I539	670	5/0.25	R. Atalaia, Óbidos
PA27)	I546	670	1/0.25	R. Tropical, Paragominas
PI32)	I927	670	1/0.25	R. Livramento, José de Freitas
PR10)	J231	670	3/0.25	R. Canção Nova Esperança, Nova Esperança
PR11)	J248	670	10/2	R. Globo, Curitiba:24h
RS18)	K296	670	2.5/0.25	R. Cult. Jaguarão, Santa. Vitória do Palmar
RS19)	K370	670	10/0.5	R. Gazeta, Carazinho: 0900-0130
SE03)	J921	670	10/5	R. Cultura de Sergipe, Aracaju: 24h
SP24)	K574	670	1/0.5	R. Oceânica, Caraguatatuba
SP25)	K585	670	1/0.5	R. Centro Oeste, Garça
SP26)	K598	670	1/0.5	R. Convenção, Itu: 24h
BA08)	H471	680	5/0.5	R. Clube, Sto. Antônio de Jesús
GO09)	H765	680	10/0.5	R. Difusora, Jataí
GO49)	H787	680	5/1	R. Mantiqueira, Niquelândia
MA03)	H885	680	20	R. Dif. do Maranhão, São Luís

MW	Call	kHz	kW	Station, location
MG147)	L348	680	1/0.25	R. Ibiá, Ibiá
MG173)	L270	680	2/0.25	R. Difusora, Ouro Fino
MG13)	L326	680	1/0.25	R. União, João Pinheiro
MG71)	L296	680	5/0.25	R. Novo Tempo, Governador Valadares
MS02)	I389	680	5/1	R. Cultura, Campo Grande
PB26)	I683	680	2.5/0.25	R. Integração do Brejo, Bananeiras: 24h
PE06)	I793	680	10/1	R. Grande Rio, Petrolina: 0800-0300
PR155)	J362	680	5/0.25	R. Poema, Pitanga
RJ07)	J452	680	20/5	R. Copacabana, Rio de Janeiro
RS69)	K275	680	50	R. Farroupilha, Porto Alegre
SP27)	K576	680	1/0.25	R. Dif., Catanduva
SP28)	K628	680	2/0.25	R. Piratininga, Piraju
BA96)	H453	690	10/1	R.Cultura, Ilhéus
CE08)	H587	690	25/10	R. Shalom, Fortaleza: 0800-0300 Sat/Sun:24h
ES10)	I201	690	10/1	R. America, Vitoria
G048)	H780	690	50/1	R. Sociedade Ceres, Ceres
MG14)	L228	690	50/5	R. Mineira, Belo Horizonte
MS03)	I402	690	5/0.25	R. Cultura, Naviraí
MT31)	I451	690	5/1	R. Parecis, Diamantino
PA03)	I532	690	20/5	R. Clube do Pará, Belém: 24h
PR13)	J229	690	5/1	R. Dif., Londrina
PR14)	J252	690	1/0.25	R. Dif., Ponta Grossa
PR143)	J360	690	5/0.25	R. Voz do Sudoeste, Coronel Vívida: 24h
RS21)	K252	690	5/0.5	R. Progresso, Ijuí
SC07)	J772	690	5/1	R. Clube, Lages
SP29)	K561	690	1/0.25	R. Bebedouro, Bebedouro: 0800-2300
SP30)	K588	690	1/0.25	R. Clube, Guaratinguetá
SP31)	K625	690	1/0.25	R. Cidade, Pereira Barreto: 24h Sun.; 0900-0300
SP220)	K646	690	1/0.25	R. Brasil, Santa Bárbara d'Oeste
TO12)	N661	690	25/1.5	R. Canção Nova do Coração de Jesus, Palmas: 24h
BA10)	H500	700	25/1	R. Cultura, Feira de Santana
MT21)	I428	700	25/1	R. Sorriso, Sorriso
G047)	I489	700	25/0.5	R. Pouso Alto, Piracanjuba
PI04)	I890	700	10/5	R. Globo, Teresina
PR92)	J225	700	10/0.25	R. Capital do Papei, Telêmaco Borba
RJ56)	J507	700	5/0.4	R. Aliança, ItaÍva: 0800-2300, Sat/Sun:0800-2230
RS123)	K247	700	1/0.25	R. Sideral, Getúlio Vargas
RS22)	K356	700	1/0.25	R. Batovi, São Gabriel
SP32)	K686	700	50	R. Eldorado, São Paulo
AL02)	H240	710	5/1	R. Jornal, Maceió
BA46)	H490	710	10/0.25	R. Nova Jacarandá, Eunápolis:24h
CE09)	H628	710	1/0.25	R. Asa Branca, Boa Viagem
DF07)	H710	710	10/2.5	R. Aliançe, Brasília: 24h
MA12)	H891	710	1	R. Verdes Campos, Pinheiro: 0800-0100
MA27)	H910	710	10/0.5	R. Verdes Vales, Grajaú
MG79)	L219	710	1	R. Cancella, Ituiutaba
MG15)	L258	710	20/0.5	R. Manhuaçu, Manhuaçu
MG80)	L319	710	2/0.25	R. Planeta, Carmo do Paranaíba
MG16)	L333	710	2/0.25	R. Dif., Pouso Alegre
MT08)	I386	710	5/0.5	R. Cultura, Cuiabá
MT23)	I436	710	5/1	R. Nova Xavantina, Nova Xavantina
PA04)	I534	710	25/5	R. Rural, Santarém: 0800-0300
PB03)	I685	710	1/0.25	R. Educadora, Conceição
PI19)	I901	*710	1/0.25	R. Alvorada do Sertão, São João do Piauí
PI23)	I933	710	1/0.5	R. Clube, Barras
PR141)	J328	710	1/0.25	R. Alternativa, Cândido de Abreu
RJ09)	J451	710	10	R. Sucesso AM, Rio de Janeiro:24h
SC08)	J793	710	1/0.25	R. Fraiburgo, Fraiburgo: 0800-0300
SP33)	K559	710	10/0.25	R. 710, Bauru
AC01)	H202	720	10	R. Integração, Cruzeiro do Sul: 0900-0300
AM05)	H281	720	1	R.Difusora, Itacoatiara: 0900-0400
MG28)	L330	720	2.5/0.5	R. Divinópolis, Divinópolis
MS04)	I390	720	5/1	R. Clube, Dourados
MT20)	I411	720	5/1	R. Difusora, Barra do Garças
PE07)	I770	720	100	R. Clube AM 720, Recife: 24h
RS23)	K276	720	100	R. Guaíba, Porto Alegre
SP34)	K575	720	1/0.25	R. Difusora, Casa Branca: 0800-0000 Sun:0900-2300
SP35)	K701	720	1/0.25	R. Sentinela, Ourinhos
SP36)	K718	720	1/0.25	R. RC Vale, Cruzeiro
SP37)	K722	720	1/0.25	R. Menina, Olímpia
CE45)	H640	730	1/0.5	R. Sinal, Aracati
ES16)	I217	730	10/0.5	R. SIM, Vitória
G031)	H759	730	25/5	R. 730 do Brasil, Goiânia
MA04)	H896	730	1/0.25	R. Eldorado, Codó
MG17)	L287	730	5/1	R. Soc. Triângulo Mineiro, Uberaba
MG18)	L297	730	10/1	R. Manchester, Juiz de Fora
MS38)	I452	730	1/0.5	R. Princesa do Vale, Camapuã
MT09)	I410	730	10/2.5	R. Jornal, Cáceres
PE08)	I780	730	1/0.5	Em. Rural, A Voz do São Francisco, Petrolina
PR15)	J208	730	7/0.6	R. Marumby, Curitiba
PR16)	J323	730	1/0.25	R. Rural, Campo Mourão

MW	Call	kHz	kW	Station, location
PR147)	J353	730	5/0.25	R. Integração Oeste, Corbélia
RS24)	K268	730	5/1	R. Planalto, Passo Fundo
SC09)	J787	730	5/1	R. Tubá, Tubarão: 0800-0100
SP38)	K523	730	10/0.25	R. Cidade, Jundiaí
SP39)	K610	730	10/1	R. Dirceu, Marília
AC02)	H206	740	20/10	Super R. Alvorada, Rio Branco: 0900-0500 (Sun:0900-0300)
BA11)	H446	740	100	R. Soc. da Bahia, Salvador: 24h
MT25)	N403	740	1/0.5	R. Cidade, Alto Araguaia
PR17)	J259	740	1/0.25	R. Goioerê, Goioerê
PR135)	J354	740	1/0.25	R. Placar, Ortigueira
RS25)	K265	740	2.5/0.25	R. Palmeira, Palmeira das Missões
RS26)	K283	740	1/0.25	R. Nativa, Rio Grande
SC10)	J753	740	10/1	CBN – (R. Diário da Manha), Florianópolis
SP93)	K519	740	10/1	R. Assunção, Jales
SP40)	K553	740	1/0.25	R. Cultura, Bariri
SP41)	K650	740	25/0.5	R. Trianon, São Paulo
DF01)	H709	750	50/25	R. Jovem Pan, Brasília
MG19)	L213	750	100/5	R. América, Belo Horizonte
PA28)	I541	750	1/0.25	R. Ximango, Alenquer
PB04)	I682	750	1/0.25	R. Panati, Patos
PI05)	I897	750	1/0.25	R. Liberdade, Campo Maior
RS27)	K264	750	1/0.25	R. Osório, Osório
SC11)	J815	750	5/0.25	R. Aliança, Concórdia
SE04)	J927	750	10/0.25	R. Progresso, Lagarto: 0800-0300
SP42)	K516	750	5/1	R. Clube, Osvaldo Cruz
SP43)	K642	750	12/0.5	R. CMN, Ribeirão Preto
SP44)	K661	750	2.5/0.25	Super R Piratininga, São José dos Campos
SP283)	K696	750	5/0.5	R. Atual, Registro
TO03)	H792	750	1/0.25	R. Tocantins, Tocantinópolis
AL12)	H252	760	1/0.25	R. Pioneir (R.Delmiro), Delmiro Gouveia
AP03)	H424	760	5	Rede Amapaense de Radiodif., Macapá
BA44)	H461	760	5/0.5	R. Cidade, Vitória da Conquista (ff 1550)
CE10)	H588	760	10	R. Uirapuru, Fortaleza
G043)	H775	760	5/0.5	R. Rio Claro, Iporá
G011)	H783	760	5/0.5	R. Pousada do Rio Quente, Caldas Novas
MG83)	L257	760	2.5/0.25	R. Difusora, Machado
MG137)	L360	760	10/0.5	R. Terra, Monte Claros
MT32)	N408	760	10/5	R. Natureza, Chapada dos Guimarães
PR12)	J343	760	10/0.25	R. Cacique, Guarapuava
RJ11)	J478	760	25/1	R. Manchete AM, Niterói: 24h
RS28)	K222	760	2.5/0.25	R. Princesa do Jacuí, Candelária
RS29)	K351	760	2.5/0.25	R. Ametista, Planalto
SC12)	J742	760	25/2	R. Nereu Ramos, Blumenau: 0730-0100
SP149)	K541	760	10/0.25	R. Urubupungá, Andradina
SP45)	K560	760	10/0.5	R. Auri-Verde, Bauru
BA51)	H491	770	5/0.5	R. Rio Corrente, Santa Maria da Vitória
CE11)	H609	770	10/0.25	R. Vale do Salgado, Lavras da Mangabeira
ES03)	I211	770	10/1	R. Globo AM, Cachoeiro de Itapemirim
G012)	H745	770	5/1	R. A Voz do Coração Imaculado, Anápolis
MA28)	H922	770	10/1	R. Vitória, Coelho Neto
MG20)	L209	770	2.5/0.25	R. Cultura d'Oeste, Lavras: 0800-0200
MG21)	L302	770	10/0.5	R. Clube de Patos, Patos de Minas
MG108)	L315	770	5/0.5	R. Pontal do Triangulo, Iturama
MG22)	L337	770	1/0.25	R. Itabira, Itabira
MT28)	N404	770	1	R. Cidade de Matupá, Matupá
MT45)	I434	770	5/1	R. Xavantes, Jaciara
PA29)	I557	770	10/0.25	R. Clube, Marabá
PR131)	J344	770	5/0.25	R. Cidade, Cambé
SE05)	J922	770	10/5	R. Atalaia de Sergipe, Aracaju
SP46)	K506	770	5/0.5	R. Mix, Limeira
CE55)	H657	780	10/1	R. Dif., Nova Russas
G013)	H789	780	10/1	R. Soc. Vera Cruz, Goianésia
MA24)	H919	780	10/5	R. Alvorada, Zé Doca
MG23)	L246	780	1	R. Educadora Jovem Pan, Uberlândia
MG103)	L259	780	10/1	R. Manhumirim, Manhumirim
PE09)	I771	780	30/10	R. Jornal do Comércio, Recife: 24h
PR18)	J247	780	10/0.5	R. Porta Voz, Cianorte: 24h
PR19)	J305	780	5/0.25	R. Chopinzinho, Chopinzinho
RS30)	K229	780	5/2	R. Diário da Manhã, Carazinho
RS31)	K279	780	25	R. Jornal do Sul, Porto Alegre //970: 24h
SC13)	J788	780	1	R. Marconi, Urussanga
SC54)	J751	780	0.5	R. Brasil Novo, Jaraguá do Sul
SP161)	K619	780	1/0.25	R. Difusora, Monte Aprazível: 0800-0100
SP47)	K695	780	50/10	CBN, São Paulo
BA13)	H484	790	1/0.25	R. Barreiras, Barreiras
BA14)	H505	790	1/0.25	R. Regional, Serrinha
CE12)	H629	790	10/1	R. Jornal Centro Sul, Iguatu: 24h
G028)	H761	790	5/0.5	R. Xavantes, Ipameri: 0800-0200, Sat:-2100, Sun: 0900-0100
G014)	H771	790	1/0.25	R. Eldorado, Mineiros
MA29)	H904	790	1/0.25	R. Rio Turiaçu, Santa Helena
MA20)	H915	790	1/0.25	R. Cultura, Açailândia
MA30)	H899	790	1/0.25	R. Rio Flores,Tuntum
MG24)	L279	790	5/0.25	R. Soc. Ponte Nova, Ponte Nova
MG25)	L311	790	1/0.25	R. Treze de Junho, Mantena

MW	Call	kHz	kW	Station, location		MW	Call	kHz	kW	Station, location
MG26)	L314	790	5/1	R. Tropical, Lagoa da Prata		DF10)	H713	840	50/2.5	R. Senado, Brasília (CP)
MT22)	I456	790	1	R. Regional, Nortelândia		PI38)	I930	840	1/0.25	R. Ribeirão, Demerval Lobão
PB05)	I679	790	2.5/1	R. Cultura 790, Guarabira: 0700-0300		PI39)	I937	*840	1/0.25	R. Vitória, Batalha
PI36)	I931	790	1/0.25	R. Mafrense, Simplício Mendes		PR75)	J320	840	10/1.2	R. Inconfidência, Umuarama
PR130)	J316	790	2.5/0.25	R. Clube, Faxinal: 0800-2400		RS36)	K248	840	10	R. Capital, Porto Alegre
PR20)	J337	790	10/0.25	R. RCC, Curitiba: 24h		SC17)	J750	840	10/1	R. Rural, Concórdia
RS32)	K285	790	1/0.25	R. Rio Pardo, Rio Pardo		SP57)	K687	840	100/50	R. Bandeirantes, São Paulo
SC14)	J789	790	1/0.25	R. Videira, Videira		BA17)	H474	850	5/0.25	R. Caraíba, Senhor do Bonfim
SP48)	K538	790	1/0.25	R. Brasil, Adamantina		CE15)	H599	850	1	R. Iracema, Juazeiro do Nte
SP49)	K546	790	10/0.5	R. Cultura, Araraquara		GO17)	H776	850	5/1	R. Tropical, Porangatu
SP162)	K674	790	5/0.25	R. Cultura, Taubaté		MA31)	H923	850	10/0.5	R. Cidade, Vitória do Mearim
AC14)	H...	800	10	R. Líder Comunicações., Rio Branco		MG32)	L233	850	1/0.25	R. Difusora Formiguense, Formiga
AL17)	H256	800	10	R. Palmares, Maceió		MG33)	L254	850	10/0.5	R. Por um Mundo Melhor,
DF02)	H705	800	10/1	R. MEC, Brasília						Governador Valadares
PI08)	I921	800	10	R. Antares, Teresina		MG34)	L295	850	5/0.25	R. Tupaciguara, Tupaciguara: 0900-0300
RJ12)	J457	800	100	R. MEC, Rio de Janeiro		MS30)	I438	850	1/0.25	R. Difusora Nor'estado, São Gabriel
RS33)	K292	800	10	R. R.Universidade, Santa Maria		MT02)	I416	850	10/2	R. Cultura, Poxoréo
BA54)	H528	810	10/0.25	R. Guadalupe,		PA17)	I538	850	10/1	R. Itacaíunas, Marabá
				Riacho de Santana		PA06)	I555	850	1/0.25	R. Tocantins, Cametá
CE13)	H589	810	50/5	R. Verdes Mares, Fortaleza		PA18)	I557	850	5/1	R. Itaituba, Itaituba
GO15)	H767	810	5/0.5	R. Alvorada, Rialma		PB22)	I693	850	5/1	R. Rural, Guarabira
MG27)	L202	810	1	R. Aimorés, Aimorés		PI30)	I909	850	1/0.25	R. Grande Picos, Picos
MG92)	L252	810	1/0.25	R. Educadora, Ubá		PR50)	J254	850	1/0.25	R. Dif. Colméia, Campo Mourão: 24h
MG76)	L266	810	1/0.25	R. Clube, Nepomuceno		PR86)	J291	850	2/0.25	R. Alvorada do Sul, Rebouças: 24h
MG138)	L354	810	1/0.25	R. Cidade, Capinópolis		RJ31)	J470	850	10/0.5	R. Dif., Campos dos Goytacazes: 24h
MG156)	L366	810	5/0.5	R. Rainha de Paz, Patrocínio		RO03)	J675	*850	5/1	R. Ariquemes, Ariquemes
MT33)	N402	810	1/0.25	R. Integração, São José do Rio Claro		SC20)	J808	850	2.5/0.25	R. Cidade, Brusque
MT34)	N406	810	1/0.25	R. Floresta AM, Alta Floresta		SC102)	J807	850	5/1	R. Atalaia, Campo Erê
PR49)	J261	810	1/0.25	R. RCC Cornélio Procópio		SP59)	K563	850	1/0.5	R. Nova Clube, Biriguí
PR111)	J336	810	5/0.5	R. Esperança, Prudentópolis: 0800-0300		SP58)	K644	850	2.5/0.25	R. Jornal, Rio Claro
RS136)	K324	810	1.9/0.25	R. Cinderela, Campo Bom		CE16)	H592	860	25/10	R. Cidade, Maracanaú
SP50)	K604	810	5/0.25	R. Dif. Jundiaiense, Jundiaí		RJ14)	J459	860	100	R. CBN, Rio de Janeiro
SP89)	K655	810	1/0.5	R. Universal, Santos		RS37)	K288	860	10/1	R. Guarathan, Santa Maria
SP51)	K732	810	5/0.5	R. Cancao Nova São José do Rio Preto		AL03)	H245	870	5/1	R. Educ. Sampaio, Palmeira dos Indios
AC03)	H...	820	1/0.25	R. Educ. 6 de Agosto, Xapuri: 1000-0100		AM19)	H322	870	1/0.25	R. Cidade, Manacapuru: 1000-0200
AC12)	H207	820	0.25	R. Dif. de Tarauacá, Tarauacá		BA18)	H457	870	12/0.25	R. Nacional, Itabuna
AM06)	H294	820	1/0.25	R. Princesa, Manacapuru: 24h		BA84)	H499	870	5/1	R. Cidade, Juazeiro
BA15)	H534	820	20/1	R. Cultura, Utinga		CE17)	H591	870	1/0.25	R. Liberdade, Iguatu
CE14)	H624	820	1/0.25	R. União, Camocim		CE66)	H658	870	1/0.25	R. Tabajara, São Benedito
CE60)	H655	820	1/0.25	R. Sul Cearense, Brejo Santo		ES20)	I...	870	5/0.25	R. Globo. Linhares: 24h
ES04)	I212	820	10/2.5	R. Gazeta, Vitória		GO18)	H749	870	5/0.5	R. Lago Dourado, Uruaçu
GO16)	H752	820	50/5	R. Jornal, Goiânia:1000-2300		GO32)	H754	870	10/0.5	R. Universitária, Goiânia
MG29)	L255	820	5/0.25	R. Globo, Barbacena		MA05)	H903	870	10/0.5	R. Mirante, Codó
MG167)	L273	820	3/0.25	R. Bom Sucesso, Minas Movas		MG78)	L304	870	1/0.5	R. Juriti, Paracatu
MG30)	L291	820	1/0.25	R. Paraíso, São Sebastião do Paraíso:		MG38)	L318	870	1/0.5	R. Cultura, Diamantina
				0800-2330 Sat/Sun: 2100		MG66)	L324	870	5/0.25	R. Sacramento, Sacramento
MT10)	I400	820	10/1	R. Dif., Cáceres: 24h		MG96)	L247	870	1/0.25	R. Difusora, Ituiutaba: 0800-0300
PA05)	I543	820	5/1	R. Regional, Conceição do Araguaia:						Sun 0800-0100
				0900-2300		MG128)	L349	870	5/0.25	R. Atividade Muriaé
PE10)	I775	820	5/1	R. Universitária, Recife		MG127)	L350	870	5/0.25	R. Voz, Pioneira do Vale
PI06)	I912	820	5/0.25	R. Cacique Bruenque, Regeneração		MT35)	N409	870	1/0.5	R. RGB (Garça Branca), Guiratinga: 0830-2100
PR21)	J238	820	10/5	R. Cultura, Foz do Iguaçu		PA11)	I547	870	5/0.25	R. Marajó, Breves
PR150)	J357	820	1/0.25	R. Princesa, Roncador		PR25)	J243	870	5/0.25	R. Nova Ingá, Maringá
RJ13)	J477	820	5/0.25	R. Globo, Macaé		SC96)	J784	870	12/0.25	R. São Francisco, São Francisco do Sul
RS34)	K241	820	5/1	R. Alto Taquari, Estrela		SP60)	K620	870	1/0.25	R. Novo Horizonte, Novo Horizonte
SC15)	J738	820	10/5	CBN, Blumenau: 24h		SP61)	K705	870	5/1	R. Central, Campinas
SP52)	K542	820	1/0.25	R. Aparecida, Aparecida		TO04)	H762	870	1/0.25	R. Anhanguera, Araguaína
SP53)	K602	820	1/0.25	R. Jauense, Jaú		MG35)	L275	880	100	R. Inconfidência, Belo Horizonte: 24h
SP54)	K622	820	0.5/0.25	R. Clube, Ourinhos		PB18)	I680	880	5/0.25	R. Maringá, Pombal
SP55)	K624	820	5/0.25	R. Difusora, Penápolis		RS38)	K249	880	10/2.5	R. Itaí, Porto Alegre
BA67)	H506	830	5/0.25	R. Extremo Sul da Bahia, Itamaraju:		RS87)	K317	880	2.5/0.25	R. São Miguel, Uruguaiana
				0700-0100 Sun:.0000		RS20)	K363	880	5/0.25	R. Seberi, Seberi: 24h
CE65)	H659	830	1/0.25	R. Pioneira, Forquilha		CE46)	H642	890	1/0.25	R. Itatiaia, Santa Quitéria
GO50)	H805	830	1/0.25	R. Nova Cidade, Goiatuba		DF03)	H706	890	50/2.5	R. Clube AM 890, Brasília
MA26)	H905	830	10/1	R. Mirante do Maranhão, Imperatriz		MG36)	L250	890	10/1	R. Santa Cruz, Jequitinhonha
MA21)	H925	830	1/0.25	R. Boa Esperança, Esperantinópolis		MG154)	L370	890	5/0.25	R. Clube, Inhapim
MG31)	L244	830	50/5	R. Cultura, Belo Horizonte		MS33)	I453	890	10/0.5	R. Guaicurus, Fátima do Sul
MS07)	I396	830	5/0.5	R. Cidade Maracaju, Maracaju		PA13)	I536	890	5/1	R. Ponta Negra, Santarém
MT11)	I430	830	5/0.25	R. Xavantes, Jaciara		PE11)	I772	890	20/10	R. Tamandaré, Recife: 24h
MT26)	N401	830	1/0.25	R. Educadora, Juina: 0800-0100		PR117)	J287	890	5/0.25	R. Ubá, Ivaiporã
PA24)	I556	830	10/1	R. Guaraní de Marajó, Soure		PR26)	J338	890	5/0.25	R. Super Itapuã, Pato Branco
PI07)	I906	830	1/0.25	R. Primeira Capital, Oeiras		RJ59)	J499	890	10/0.5	R. Musical, Cantagalo
PI37)	I934	830	1/0.25	R. União, União		RS39)	K215	890	5/0.25	R. Difusora, Bento Gonçalves: 0800-
PR22)	J224	830	7.5/0.75	R. Iguassu, Araucária						0300 Sat/Sun:24h
PR24)	J266	830		CBN – (R.Tabajara), Londrina		RS40)	K295	890	2/0.5	R. Noroeste, Santa Rosa
PR23)	J311	830	1/0.25	R. Progresso, Clevelândia		SC52)	J745	890	2/0.25	R. Clube, Canoinhas: 0700-0300
RJ39)	J488	830	10/0.5	R. Tropical Solimos, Rio de Janeiro: 24h		SC18)	J755	890	1/0.25	R. Santa Catarina, Florianópolis
RN02)	J595	830	1/0.5	R. Rural do Caicó, Caicó		SP62)	K690	890	50/10	R. Gazeta, São Paulo
RS35)	K332	830	5/0.25	R. Independente, Cruz Alta		SP127)	K703	890	2.5/0.25	R. Cidade (R. Notícias), Matão
SC16)	J773	830	1/0.25	R. Cruz de Malta, Lauro Müller		SP178)	K562	890	1/0.25	R. Imaculada Conceicao, Bilac
SE06)	J926	830	20/1	R. Princesa da Serra, Itabaiana		BA19)	H488	900	1	R. Sisal, Conceição do Coité
SP56)	K681	830	1/0.25	R. Lider, Votuporanga		GO41)	H768	900	10/1	R. Rio Verde AM, Rio Verde: 0945-0300
SP227)	K746	830	5/1	R. Novo Tempo, Nova Odessa		MG86)	L207	900	10	R. Imbiara, Araxá
AM07)	H298	840	1/0.25	R. Rio Madeira, Manicoré		MG148)	L311	900	2.5/0.25	R. Onda Viva, Carangola
BA16)	H447	840	25/5	R. Excelsior da Bahia, Salvador		MG124)	L338	900	1/0.25	R. Vinícola, Andradas
CE51)	H648	840	1/0.5	R. Campo Maior, Quixeramobim		MT36)	I455	900	10/2.5	R. Difusora Arco-Íris, Araputanga

MW	Call	kHz	kW	Station, location
MT41)	I431	900	5/1	R. Integração, Primavera do Leste
PR27)	J272	900	5/0.25	R. Sant'Ana, Ponta Grossa
PR28)	J295	900	5/0.25	R. União, Toledo
RJ15)	J454	900	50/10	R. Tamoio, Rio de Janeiro: 24h
RN03)	J591	900	10	R. Nordeste Evangélica, Nata: 24h
RO04)	J672	900	5/1	R. Alvorada de Rondônia, Ji-Paraná
RS41)	K211	900	2.5/0.5	R. Aratiba, Aratiba: 24h
RS179)	K263	900	5/0.5	R. ABC 900, Nôvo Hamburgo
RS164)	K303	900	1/0.25	R. Municipal, São Pedro do Sul
SP63)	K511	900	5/0.25	R. Difusora, Presidente Prudente
SP64)	K664	900	10/0.8	R. Jovem Pan, São José do Rio Preto
SP65)	K742	900	1/0.25	R. Globo, Itapetininga
CE61)	H645	910	4/0.25	R. Caiçari, Sobral
GO20)	H763	910	5/0.25	R. Paranaíba, Itumbiara
GO23)	H804	910	10/0.5	R. Cidade, Jaraguá
MG37)	L292	910	1/0.25	R. Teófilo Otoni, Teófilo Otoni
MG132)	L346	910	1/0.25	R. Difusora, Nova Serrana
MG149)	N206	910	5/1	R. Globo, Juiz de Fora
PE12)	I785	910	5/1	R. Super Liberdade, Caruaru
PI41)	I935	910	10/1	R. CBN, Teresina
PR29)	J207	910	1/0.25	R. Nova AM, Apucarana
RS43)	K320	910	5/0.5	R. Venâncio Aires, Venâncio Aires
SC19)	J811	910	4/0.5	R. Difusora, Içara
SC90)	J824	910	1/0.25	R. Rainha das Quedas, Abelardo Luz
SP66)	K536	910	5/0.25	R. Onda Livre, Piracicaba
SP228)	K763	910	1/0.25	R. Princesa, Monte Azul Paulista
BA42)	H476	920	2.5/0.25	R. Educ. Santana de Caetité, Caetité: 24h
BA57)	H519	920	25/2	R. Novo Tempo, Salvador
ES05)	I207	920	10/5	R. Cultura, Linhares
GO33)	H788	920	5/1	R. Vale da Serra, São Luís de Monte Belos
MG39)	L271	920	5/0.5	R. Cultura, Visconde do Rio Branco
PB31)	I697	920	5/0.5	R. Maná, João Pessoa
PI11)	I893	*920	10/0.5	R. Educadora, Parnaíba
PI09)	I895	920	1/0.25	R. Difusora, Picos
RJ41)	J494	920	1/0.5	R. Stereo Sul, Volta Redonda
RN04)	J600	920	1/0.25	R. Currais Novos, Currais Novos
RS44)	K348	*920	20/2	R. Tramandaí, Tramandaí
SP67)	K584	920	10/0.25	R. Imperator, Franca
SP222)	K769	920	1/0.25	R. Bandeirantes, Penápolis
SP221)	K775	920	40/1	R. Nacional Gospel, Cotia
AM16)	H240	930	5/0.25	R. Boas Novas, Manaus: 24h
CE18)	H605	930	1/0.25	R. Cetama, Barbalha
CE52)	H646	930	7/0.25	R. Metropolitana, Fortaleza
MG41)	L220	930	2.5/0.25	R. Clube, Campo Belo
MG42)	L229	930	10/5	R. Araguari, Araguari
MG87)	L237	930	20/1	R. Globo, Governador Valadares: 24h
MS05)	I454	930	10/0.25	R. Capital, Campo Grande: 0900-0200
MT12)	I423	930	10/0.5	R. Clube, Rondonópolis
MT37)	N400	930	10/0.25	R. Jornal, Pontes e Lacerda
PA07)	I600	930	5/1	R. Liberal, Castanhal
PR30)	J227	930	1/0.25	R. Cultura, Rolândia
PR31)	J232	930	10/1	R. Cultura, Curitiba
PR69)	J235	930	10/1	R. Princesa, Francisco Beltrão
RS45)	K230	930	20/2.5	R. Caxias, Caxias do Sul
RS46)	K298	930	10/0.5	R. Santo Ângelo "R. Super", Santo Ângelo
SE07)	J923	930	20/5	R. 930AM – Liberdade AM, Aracaju
SP71)	K500	930	1/0.25	R. Dinâmica de Santa Fe, Santa. Fé do Sul
SP68)	K503	930	1/0.25	R. Clube, Itapira
SP69)	K652	930	10/1	R. Cultura, Santos
SP70)	K713	930	5/1	R. Canção Nova, Agudos
SP214)	K747	930	1/0.25	R. Jóia, Adamantina
AC04)	H204	940	10/1	R. Verdes Florestas, Cruzeiro do Sul: 0930-0200 (Sat/Sun 0930-0040)
PI25)	I911	940	10/0.25	R. Sete Cidades, Piracuruca
RJ16)	J453	940	100	Super R. Brasil (RBV), Rio de Janeiro
BA50)	H489	950	1/0.25	R. Bahia Noroeste, Paulo Afonso
CE19)	H593	950	5/1	R. Educadora do Nordeste, Sobral
GO21)	H764	950	5/0.25	R. Difusora, Itumbiara: 0800-0300, Sat:0800-2300, Sun:1000-0300
MA17)	H916	950	10/0.25	R. Capital, João Lisboa
MG43)	L212	950	25/10	R. Atalaia, Belo Horizonte
MG44)	L281	950	7/0.5	R. Indy, Bueno Brandão
MT17)	I439	950	5/1	R. Tucunaré, Juara
PB17)	I681	950	1/0.25	R. Jornal (Radio News), Sousa
PE13)	I782	950	25/5	R. Planalto, Carpina: 24h
PI20)	I915	950	10/0.25	R. São José dos Altos, Altos
PI42)	I923	950	1/0.25	R. Boa Esperança, Padre Marcos
PR114)	J239	950	5/0.25	R. Difusora, Irati
RS47)	K260	950	10/0.25	R. Independente, Lajeado
SC21)	J736	950	1/0.25	R. Vale, Tijucas
SP72)	K510	950	5/0.25	R. Clube de Vera Cruz, Marília
AL04)	I241	960	10	R. Difusora de Alagôas, Maceió
CE37)	H618	960	1/0.25	R. Cultura dos Inhamuns, Tauá
ES15)	I216	960	25/0.25	R. Diocesana, Cachoeiro de Itapemirim
GO45)	H802	960	50/0.25	R. Caraiba, Aparecida de Goiânia: 24h
PA26)	I551	960	5/1	R. Clube, Itaituba
PR109)	J217	960	2.5/0.25	R. Legendária, Lapa
PR32)	J257	960	1/0.25	R. Difusora, Maringá
RS48)	K291	960	10/1	R. Imembuí, Santa Maria
SC22)	J733	960	5/0.25	R. Guarujá, Orleães: 0800-0100 Sun: 2400
SC23)	J813	960	8/0.25	R. Super Difusora, Xanxerê
SP73)	K689	960	50/10	R. São Paulo, São Paulo
TO09)	H793	960	25/5	R. Jovem Palmas, Palmas
BA20)	H451	970	10/5	R. Sociedade, Feira de Santana
CE20)	H612	970	5/0.25	R. Monólitos, Quixadá
MG45)	L243	970	5/0.25	R. Caratinga, Caratinga
MG46)	L285	970	2.5/0.25	R. São João Del Rey, São João Del Rey
MG47)	L321	970	1/0.25	R. Central, Monte Alegre de Minas
MS18)	I399	970	5/0.5	R. Vale do Taquari, Coxim
PB06)	I684	970	1/0.25	R. Princesa Isabel, Princesa Isabel
PI43)	I910	970	10/0.5	R. Vale do Parnaíba, Luzilândia
PR33)	J260	970	7.5/1	R. Alvorada, Londrina
PR34)	J277	970	10/0.25	R. Difusora do Paraná, Marechal Cândido Rondón
RS49)	K201	970	50/10	R. Pampa, Porto Alegre //780
RS50)	K349	970	5/0.25	R. Alto Uruguai, Humaitá
SC24)	J730	970	5/0.25	R. Araguaia, Brusque: 24h
SP74)	K505	970	5/0.25	R. Transamerica, Itapetininga
SP75)	K529	970	5/0.25	R. Piratininga, São João da Boa Vista
SP76)	K684	970	5/0.25	R. Hertz, Franca
SP215)	K744	970	5/0.25	R. Alvorada, Estrela d'Oeste: 24h
DF04)	H707	980	50/300	R. Nacional, Brasília: 24h
AM23)	H299	990	1	R. Independência, Maués: 0900- 0200
BA21)	H483	990	1/0.25	R. Alvorada Gospel, Caravelas
PI44)	I922	990	1/0.25	R. Vale do Canindé, Oeiras
PR128)	J293	990	1/0.25	R. Najuá, Irati
PR121)	J321	990	5/0.25	R. Capital, Cianorte: 24h
RJ53)	J461	990	100/10	R. Record, Rio de Janeiro: 24h
RN05)	J596	990	10/1	R. Rural, Mossoró
RS51)	K314	990	2.5/0.5	R. Tupã, Tupanciretã
RS52)	K335	990	2.5/0.25	R. Sananduva, Sananduva: 24h
RS154)	K360	990	1/0.25	R. Clube, Pedro Osório
SC25)	J763	990	1/0.25	R. Itapiranga, Itapiranga
SP239)	K579	990	1/0.25	R. Cultura Regional, Dois Córregos
PB30)	I698	1000	2.5/0.5	R. Oeste da Paraíba, Cajazeiras
PE14)	I791	1000	5/1	R. Princesa Serrana, Timbaúba
SP77)	K522	1000	200	R. Record, São Paulo
BA22)	H448	1010	25/5	R. Bahia, Salvador
CE21)	H625	1010	12.5/2.5	R. CBN, Fortaleza
GO39)	H772	1010	10/1	R. Santelenense, Sta. Helena de Goiás
MG50)	L230	1010	10/1	R. Educadora, Coronel Fabriciano
MG48)	L264	1010	5/0.5	R. Solar, Juiz de Fora
MG49)	L325	1010	5/0.25	R. Estância, Jacutinga
MT13)	I421	1010	5/1	R. Dif., Mirassol d'Oeste: 0800-0200
PR35)	J263	1010	25/5	R. Celinauta, Pato Branco
RS53)	K232	1010	7.5/0.75	R. 1010, Caxias do Sul
RS54)	K344	1010	3/1	R. Missioneira, São Luíz Gonzaga
SC71)	J764	1010	10/0.3	R. Jaraguá, Jaraguá do Sul: 24h
SC75)	J758	1010	1/0.25	R. Bandeirantes, Imbituba
SP151)	K507	1010	5/0.5	R. Difusora, Lençóis Paulista
SP78)	K556	1010	1/0.5	R. Independente, Barretos
SP79)	K611	1010	1/0.25	R. Diario, Martinópolis
AL05)	H247	1020	25/1	R. Jovem Pan, Maceió
AP04)	H423	1020	1/0.25	R. Porto, Santana
CE22)	H600	1020	5/1	R. Educadora Cariri, Crato
CE79)	H664	1020	10/4	R. Macambira, Ipueiras: 0800-0100
ES06)	I205	1020	10/0.4	R. Difusora, Colatina
GO52)	H781	1020	10/1	R. Maranata, Firminópolis
MG55)	L224	1020	10/1	R. Congonhas, Congonhas: 0800-0100 Sun -2400
MG51)	L260	1020	10/0.25	R. Cultura, Uberlândia
MS06)	I381	1020	10/0.25	R. Independente, Aquidauana
PB19)	I686	1020	1/0.25	R. Cenecista, Picuí: 0900-2300
PR36)	J244	1020	1/0.25	R. Colombo do Paraná, Curitiba
PR37)	J307	1020	1/0.25	R. Independência, Medianeira
PR142)	J359	1020	1/0.25	R. Campo Aberto, Laranjeiras do Sul
RJ42)	J484	1020	5/0.25	R. Canção Nova, Salvador Campos dos Goytacazes
RO02)	J680	1020	5/1	R. Educadora, Rolim de Moura
RR04)	J702	1020	10/5	R. Folha, Boa Vista: 0900-0500
RS49)	K202	1020	25/5	R. Caiçara, Porto Alegre
SC27)	J805	1020	2.5/0.25	R. Continental, Coronel Freitas
SP80)	K513	1020	10/0.25	R. Canção Nova, Cachoeira Paulista
SP81)	K515	1020	5/0.5	R. Cultura, Assis: 0900-0300
SP82)	K531	1020	2.5/0.5	R. Educadora, Limeira
SP83)	K600	1020	5/0.25	R. Cultura, Jales
BA40)	H475	1030	10/1	R. Bahiana, Itaberaba
GO22)	H746	1030	10/1	R. Imprensa, Anápolis
MA07)	H892	1030	10/1	R. Jainara, Bacabal
PE15)	I777	1030	20/5	R. Olinda, Olinda
PR39)	J271	1030	5/0.25	R. Atalaia, Londrina
PR120)	J312	1030	2.5/0.25	R. Clube, Realeza

MW	Call	kHz	kW	Station, location
PR40)	J329	1030	1/0.25	R. Dif. do Xisto, São Mateus do Sul
RJ18)	J467	1030	100/5	R. Capital, Rio de Janeiro
RN31)	J612	1030	1/0.25	R. Em. Vale do Apodi, Apodi
RO10)	J683	1030	5/1	R. Rondônia, Ariquemes
RS129)	K224	1030	1/0.25	R. Cultura, Canguçu
RS55)	K253	1030	1/0.25	R. Repórter, Ijuí
SC28)	J771	1030	2.5/0.5	R. Princesa, Lages
SP84)	K525	1030	5/0.25	R. Difusora, Franca
SP85)	K554	1030	1/0.25	R. Emissora da Barra, Barra Bonita
SP86)	K606	1030	1/0.25	Lins Rádio Clube, Lins
T005)	H791	1030	1/0.25	R. Colinas, Colinas do Tocantins
SP87)	K537	1040	200/100	R. Capital, São Paulo
BA47)	H494	1050	25/0.75	CBN, Camaçari: 24h
CE54)	H647	1050	10/0.5	R. Primeira Capital, Aquiraz
ES07)	I203	1050	100/1	R. Capixaba, Vitória
GO40)	H760	1050	1/0.25	R. Jornal, Inhumas
MG52)	L236	1050	1/0.25	R. Rural, Tupaciguara
MS20)	I391	1050	10/0.5	R. Dif. Paranaibense, Paranaíba
PB07)	I676	1050	5/1	R. Caturité, Campina Grande
PR66)	J226	1050	1/0.25	R. Dif. Platinense, Sto. Antônio da Platine
PR99)	J286	1050	5/0.25	R. Club de Palmas, Palmas: 0900-0300 Sat:1030-0100
RJ19)	J497	1050	10/0.5	R. Angra, Angra dos Reis
SC26)	J830	1050	7/0.25	R. Verde Vale, Braço do Norte: 0720-0300
SP160)	K601	1050	10/0.5	R. Show Jardinópolis
BA23)	H460	1060	5/1	R. Clube de Conquista, Vitória da Conquista
BA68)	H520	1060	2.5/0.25	R. Clube, Itapicuru
MG53)	L278	1060	100/5	R. Grande Belo Horizonte "R. Grande Be Aga", Pedro Leopoldo
MG54)	L306	1060	1/0.25	R. Itajubá, Itajubá
MS39)	N604	1060	5/1	R. Imaculada Conceiçoa, Dourados
PR42)	J246	1060	10/0.5	R. Evangelizar, Curitiba
PR43)	J298	1060	10/0.5	R. Colorado, Colorado
PR44)	J306	1060	10/0.5	R. Educadora, Francisco Beltrão
RJ20)	J495	1060	30/1	R. Cancao Nova, Miguel Pereira
RN06)	J597	1060	3	R. Tapuyo, Mossoró (RPC)
RS56)	K220	1060	5/0.25	R. Camaqüense, Camaquã
RS57)	K302	1060	2/0.25	R. São Luís, São Luís Gonzaga
RS81)	K307	1060	2/0.5	R. Cristal, Soledade: 24h
SC103)	J830	1060	2/0.4	R. Regional, Florianópolis
SP88)	K533	1060	5/0.25	R. Educadora, Piracicaba
SP229)	K765	1060	5/0.25	R. Universitária, Garça
BA48)	H492	1070	5/0.5	R. Rural "R. Tropical", Ipiaú
MG56)	L316	1070	1/0.25	R. do Povo, Muzambinho
MG150)	L355	1070	5/0.25	Super R. Patos, Patos de Minas
MT14)	I427	1070	10/2.5	R. Bandeirantes, Cuiaba
PB08)	I673	1070	20/2.5	R. Dif. Cajazeiras, Cajazeiras: 0800-0300
PR45)	J203	1070	5/0.25	R. Difusora União, União da Vitória
PR46)	J319	1070	1/0.25	R. Guaraniaçu (Super RG), Guaraniaçu
RJ21)	J483	1070	10/0.25	R. Record, Campos dos Goitacazes
RS58)	K218	1070	2/0.25	R. Caçapava, Caçapava do Sul
RS59)	K343	1070	1/0.25	R. Metrópole, Crissiumal
RS60)	K357	1070	1/0.25	R. Bento, Bento Gonçalves
SC91)	J747	1070	1/0.25	R. Gralha Azul, Urubici
SP91)	K603	1070	1/0.25	R. Nova Piratininga, Jaú
SP145)	K615	1070	1/0.25	R. Metropolitana, Mogi das Cruzes
SP92)	K633	1070	10/1	R. Presidente Prudente, P. Prudente: 1000-2000
SP212)	K758	1070	1/0.25	R. Jornal, Barretos
BA24)	H470	1080	10/0.5	R. Subaé, Feira de Santana
BA25)	H485	1080	1/0.25	R. Fascinação, Itapetinga: 0800-0230
CE82)	H670	1080	2.5/0.25	R. Cultura, Quixadá
DF05)	H708	1080	25/5	R. Capital, Brasília
MG109)	L232	1080	2.5/0.5	R. Cultura, Dores do Indaiá
MG57)	L251	1080	5/0.7	R. Capital, Juiz de Fora
MT27)	I437	1080	2/0.5	R. Gaspar, Itiquira
PA32)	I540	1080	15/5	R. Novo Tempo, Belém
PE16)	I784	1080	10/0.5	R. Jornal do Comercio, Caruaru: 24h
PR47)	J201	1080	2.5/0.5	R. Clube Pontagrossense, Ponta Grossa: 24h
PR48)	J245	1080	1/0.25	R. Cultura do Norte, Paranavaí
RS61)	K254	1080	3/0.25	R. Marabá, Iraí
RS62)	K260	1080	10	R. da Universidad, Porto Alegre
SC29)	J759	1080	2/1	R. Clube, Indaial: 0800-0200
SP94)	K557	1080	5/1	R. Difusora, Batatais
SP95)	K607	1080	1/0.25	R. Alvorada, Lins
SP96)	K669	1080	10/1	R. Boa Nova, Sorocaba
SP97)	K704	1080	1/0.25	R. Monumental, Aparecida
SP190)	K710	1080	1/0.25	R. Alvorada, Cardoso
AL15)	H254	1090	5/0.5	R. Gazeta, Pão de Açucar
BA26)	H455	1090	1/0.25	R. Santa Cruz, Ilhéus
GO24)	H758	1090	25/1	R. 1090, Goiânia
MA08)	H893	1090	10	R. Rio Balsas, Balsas
PR51)	J283	1090	2.5/0.25	R. Vicente Palotti, Coronel Vivida
PR171)	J345	1090	1/0.25	R. Banda 1, Sarandi
RJ22)	J468	1090	25/5	R. Metropolitana, Rio de Janeiro: 24h
RN07)	J592	1090	10/5	R. Rural, Natal

MW	Call	kHz	kW	Station, location
RS63)	K216	1090	5/0.25	R. Cachoeira, Cachoeira do Sul
RS64)	K262	1090	1/0.25	R. Salete, Marcelino Ramos
RS65)	K341	1090	1/0.25	R. Giruá, Giruá
SC30)	J732	1090	1/0.25	R. Colón, Joinville
SC31)	J786	1090	5/0.5	R. Bandeirantes, Tubarão
SP98)	K609	1090	3/0.5	R. Clube, Marília
SP99)	K618	1090	1/0.25	R. Cultura, Monte Alto
SP233)	K768	1090	1/0.25	R. Canção Nova da Divina Providência, Paulína
CE67)	H638	1100	1/0.25	R. Dif. dos Inhamuns, Tauá: 0730-0030 Sun:0800-0200
CE73)	H668	1100	1/0.25	R. Difusora do Vale Acaraú, Acaraú
RN18)	J607	1100	1/0.25	R. A Voz do Seridó, Caicó
SP100)	K694	1100	150	R. Globo, São Paulo
CE32)	H620	1110	5/0.5	R. Litoral, Cascavel
GO46)	H782	1110	25/2	R. Redentor, Sto. Antônio do Descoberto
MG58)	L205	1110	1/0.25	R. Planalto, Araguari
MG59)	L267	1110	1	R. Aurilândia, Nova Lima
MS08)	I392	1110	1/0.25	R. Ponta Porã, Ponta Porã
PB09)	I678	1110	20/10	R. Tabajara, João Pessoa
PR52)	J241	1110	10/1	R. Paiquerê, Londrina
PR151)	J356	1110	1/0.25	R. Clube, Ubiratã
RJ23)	J471	1110	50/5	R. Cultura, Campos dos Goitacazes
RS66)	K257	1110	2.5/0.25	R. Cultura Jaguarão, Jaguarão
RS67)	K306	1110	1/0.25	R. Sobradinho, Sobradinho
RS68)	K325	1110	2.5/0.25	R. Cruzeiro do Sul, Itaqui: 0830-0200
RS152)	K364	1110	1/0.25	R. Solaris, Antônio Prado
SC74)	J743	1110	1/0.25	R. Caçanjure, Caçador: 0900-0300
SC32)	J752	1110	1/0.5	R. Cultura, Florianópolis: 0800-0300 Sun:1000-0100
SC33)	J812	1110	1/0.25	R. São Carlos, São Carlos
SP101)	K544	1110	1/0.25	R. Jovem Luz, Araçatuba: 1000-0000
SP102)	K592	1110	1/0.25	R. Ibitinga, Ibitinga
SP103)	K617	1110	1/0.25	R. Cultura, Mogi Mirim
BA72)	H658	1120	5/0.25	R. Belo Campo, Belo Campo
CE23)	H598	1120	5/1	R. Tupinambá, Sobral: 0800-0100 Sat:0800-2100 Sun:0900-2200
ES14)	I215	1120	10/1	R. Sim, São Mateus
MG10)	L272	1120	2.5/0.5	R. Itatiaia, Ouro Preto
MG60)	L301	1120	5/1	R. Sete Colinas, Uberaba
MG61)	L332	1120	1/0.25	R. Serra AM, Boa Esperança: 0900-0100
MS40)	N606	1120	25/1	R. Concordia, Campo Grande
PB10)	I687	1120	5/1	R. Independência, Catolé do Rocha
PE17)	I778	1120	5/1	R. Relógio Musical, Recife
PR53)	J253	1120	25/1	R. Mais, São José dos Pinhais: 24h
PR85)	J285	1120	5/0.5	R. Educadora, Laranjeiras do Sul
RS191)	K274	1120	50	R. Rural, Porto Alegre
RS156)	K367	1120	10/0.6	R. Querência, Santo Augusto: 0800-0300
SP104)	K631	1120	1/0.25	R. Nova Porto, Porto Feliz: 0900-0230 Sat: -2300 Sun:1000-
SP105)	K660	1120	10/1	R. Bandeirantes, São José dos Campos
SP106)	K671	1120	1/0.25	R. Clube Imperial, Taquaritinga: 0800-0000 Sun: 0800-2105
CE72)	H667	1130	10/0.25	R. Patu, Senador Pompeu
PA08)	I531	1130	10	R. Marajoara, Belém
PE18)	I783	1130	5/1	R. Cultura do Nordeste, Caruaru
RS55)	J220	1130	5/0.25	R. Castro, Castro: 0800-0300
PR54)	J333	1130	5/0.25	R. Ingamar, Marialva: 0900-2300 Sun:0900-2100
RJ17)	J460	1130	100/50	R. Nacional, Rio de Janeiro: 24h
RO06)	J677	1130	5/0.25	R. Ji-Parana, Ji-Paraná: 0800-0300
SR70)	K290	1130	5/1	R. Medianeira, Santa Maria
SC34)	J790	1130	5/1	R. Princesa d'Oeste, Xanxerê
SP107)	K676	1130	1/0.25	R. Tupã, Tupã
BA81)	H449	1140	10	R. Cultura da Bahia, Salvador: 24h
CE24)	H607	1140	1/0.25	R. Progresso, Russas
GO26)	H751	1140	5/0.25	R. Formosa, Formosa
MG63)	L248	1140	1/0.25	R. Doicesana, Campanha
MG64)	L253	1140	5/0.5	R. Muriaé, Muriaé: 0800-0300
MG129)	L362	1140	5/0.5	R. Clube, Bocaiúva
MS22)	I398	1140	10/0.5	R. Globo Regional, Fátima do Sul
MS36)	I418	1140	5/0.25	R. Cidade, Aparecida do Taboado
MT47)	I435	1140	5/0.25	R. Dif. Juara, Juara
PR144)	J352	1140	1/0.25	R. Difusora America, Chopinzinho: 0830-0300, Sat: 1015-0300
RS71)	K228	1140	2/0.25	R. Cruz Alta, Cruz Alta
RS72)	K316	1140	5/0.7	R. Charrua, Uruguaiana: 0900-2200
RS73)	K330	1140	5/0.25	R. Sobral, Butiá: 1000-2200
SC35)	J748	1140	10/0.5	R. Coroado, Curitibanos
SP108)	K550	1140	10/0.5	R. Difusora, Assis
SP109)	K555	1140	2/0.5	R. Barretos, Barretos
SP110)	K645	1140	1/0.25	R. Educação e Cultura, Rio Claro
SP111)	K709	1140	5/0.5	R. Costa Azul, Ubatuba
SP273)	K708	1140	1/0.25	R. Nova Regional, Registro: 24h
AL11)	H250	1150	20/1	R. Cultura, Arapiraca
CE47)	H643	1150	5/0.5	R. Moriá, Paracuru: 24h

MW	Call	kHz	kW	Station, location
MG65)	L283	1150	50/10	R. Globo, Belo Horizonte
PI10)	I891	1150	10/5	R. Pioneira, Teresina: 24h
RJ24)	J456	1150	10/0.5	R. Três Rios, Três Rios
RN25)	J617	1150	5/0.5	R. Cabugi do Seridó, Jardim do Seridó: 0800-0100
SP232)	K656	1150	100/50	R. Tupi, São Paulo: 24h
AM24)	H323	1160	4/0.5	R. Soc. TV Manauara, Boca do Acre
BA94)	H…	1160	5/0.25	R. São José Canção Nova, Itabuna
CE62)	H652	1160	1/0.25	R. Vale do Coreaú, Granja
CE80)	H660	1160	1/0.25	R. Montevidéu, Cedro: 0800-0200
DF09)	H714	1160	30/0.5	R. Globo Gama: 24h
ES08)	I202	1160	50/10	R. Espírito Santo, Vitória
GO19)	H803	1160	30/0.85	R. Bandeirantes, Santo Antônio do Descoberto
GO42)	H784	1160	1/0.25	R. Silvestre, Itaberaí: 0900-0100
MT15)	I385	1160	10/5	R. A. Voz d'Oeste, Cuiabá
PA30)	I558	1160	5/1	R. Guamá, São Miguel do Guamá
PB11)	I674	1160	1	R. Cariri, Campina Grande
PR56)	J258	1160	10/1	R. Manchete, Londrina: 0915-2315
RS74)	K242	1160	9/1	R. Miriam, Farroupilha
RS75)	K245	1160	5/1	R. Luz e Alegria, Frederico Westphalen
RS76)	K256	1160	2.5/0.5	R. Jaguari, Jaguari
RS77)	K273	1160	2.5/1	R. Universidade Católica, Pelotas
SC36)	J741	1160	9/0.7	R. Itaberá, Blumenau: 0900-0300
SC37)	J767	1160	1/0.25	R. Dif, Laguna: 0800-0000 (Sun:0700-)
SP114)	K517	1160	5/0.5	R. Cacique, Sorocaba
SP124)	K558	1160	2.5/1	R. Bandeirantes, Baurú
SP115)	K582	1160	1/0.25	R. Difusora, Fernandópolis
SP237)	K673	1160	1/0.25	R. Cacique, Taubaté
SP116)	K685	1160	1/0.25	R. Boa Nova, Mococa: 24h
AC15)	H205	1170	1/0.25	R. Dif. de Feijó, Feijó
AM08)	H284	1170	5/2.5	R. Guaranópolis, Maués: 1000-0200
BA27)	H473	1170	5/0.25	R. Jornal, Eunápolis
MG152)	L234	1170	1/0.25	R. Clube Fronteira
MG104)	L269	1170	5/0.25	R. Sociedade, Oliveira
MG75)	L327	1170	10/0.25	R. Vanguarda, Ipatinga
MG67)	L336	1170	5/0.25	R. Cidade, Araxá
PR57)	J273	1170	20/10	R. Atalaia, Curitiba
PR90)	J334	1170	2.5/0.45	R. Entre Rios, Sto. Antônio do Sudoeste
PR154)	J363	1170	8/1	R. Colméia, Mandaguaçu
RJ49)	J498	1170	5/0.25	R. Bom Jesus, Bom Jesus do Itabapoana: 0800-0300
RN08)	J598	1170	10/1	R. Difusora, Mossoró
RS78)	K207	1170	5/1	R. Itapuí, Santo Antônio da Patrulha
RS79)	K213	1170	5/1	R. Difusora A Voz de Bagé, Bagé
RS80)	K359	1170	5/1	R. Uirapuru, Passo Fundo
RS155)	K380	1170	2/1	R. Pitangueira, Itaqui
SP117)	K569	1170	10/5	R. Bandeirantes, Campinas
AL06)	H248	1180	1/0.25	R. Correio do Sertão, Santana do Ipanema
AM20)	H280	1180	10/2.5	R. Dif. do Amazonas, Manaus: 24h
MA09)	H889	*1180	10/5	R. Capital, São Luís
MG118)	L203	1180	10/0.25	R. Cultura, Alfenas
MS34)	I602	1180	10	R. Guanandi, Campo Grande
MT38)	N405	1180	5/0.25	R. Enauan, Guarantã do Norte
PB20)	I690	1180	1/0.25	R. Bonsucesso, Pombal: 0830-0100 Sun: 0930-2100
PE19)	I797	1180	1/0.25	R. Cult. de Vitória, Vitória de Sto. Antão
PR126)	J223	1180	2.5/0.5	R. Atalaia, Guarapuava
PR58)	J231	1180	10/0.5	R. Guaçu, Toledo: 0800-0200
PR81)	J314	1180	1/0.25	R. Educadora, São João do Ivaí: 0800-0100
RJ25)	J463	1180	50/10	R. Mundial (IMPD), Rio de Janeiro
RS128)	K340	1180	10/0.5	R. Gazeta, Santa Cruz do Sul
SC39)	J737	1180	5/0.25	R. Integração d'Oeste, São José do Cedro: 0830-1000
SC72)	J770	1180	1/0.5	R. Guri, Lages
SP260)	K567	1180	1/0.25	R. Brotense, Brotas
SP179)	K647	1180	2/0.5	R. Difusora, Santa Cruz do Rio Pardo
SP217)	K749	1180	5/0.25	R. Nova, Bebedouro
BA28)	H459	1190	10/1	R. Juazeiro, Juazeiro
CE68)	H663	1190	1/0.25	R. Guaraciaba, Guaraciaba do Norte: 24h
GO35)	H800	1190	10/0.25	R. Rio Vermelho, Silvânia
MG68)	L221	1190	10/1	R. Guarani, Belo Horizonte
MG40)	L276	1190	10/0.25	R. Mineira do Sul, Passa Quatro
PR59)	J309	1190	1/0.25	R. Pontal, Nova Londrina
PR136)	J355	1190	5/0.4	R. Cidade, Palmital
RN09)	J594	1190	10/1	CBN,Natal: 24h
RS82)	K234	1190	5/0.5	R. Cerro Azul, Cerro Largo
RS102)	K301	1190	2.5/0.25	R. São Lourenço, São Lourenço do Sul
RS83)	K354	1190	2.5/1	R. Rosário, Serafina Corrêa: 1000-0100
SC40)	J783	1190	1/0.25	R. Clube, São João Batista: 24h
SC87)	J817	1190	1/0.25	R. Planalto, Major Vieira: 24h
SC41)	J820	1190	2.5/0.25	R. Clube, São Domingos
SP199)	K512	1190	2.5/0.25	R. Clube Marconi, Paraguaçu Paulista
SP118)	K700	1190	5/0.25	R. Cidade, Votuporanga
SP119)	K729	1190	10/0.25	R. 31 de Março, Sta. Cruz das Palmeiras
SP120)	K741	1190	1/0.5	R. Regional, Taquarituba: 0800-0200
AL14)	H251	1200	20/1	R. Correio, Maceió
BA29)	H482	1200	5/0.5	R. Clube Rio do Ouro, Jacobina: 24h
CE26)	H585	1200	10	R. Clube AM 1200, Fortaleza: 24h
RS84)	K239	1200	5/1	R. Erechim, Erechim
RS85)	K342	1200	1/0.5	R. Fundação Cotrisel, São Sepé: 0800-0200 Sat/Sun:0800-0100
SP121)	K520	1200	100/20	R. Cultura, São Paulo
BA30)	H452	1210	10/1	R. Povo, Feira de Santana
BA58)	H498	1210	10/0.25	R. Canção Nova, Vitória da Conquista
CE50)	H637	1210	5/0.25	R. Príncipe Imperial, Crateús
CE48)	H641	1210	5/0.25	R. Boa Esperança, Barro: 0800-2100, Sun:1000-2200
DF08)	H711	1210	50/2.5	R. Brasília (RBV), Brasília
ES09)	I200	1210	25/1	R. Sim Cachoeiro, Cachoeiro de Itapemirim: 24h
MG69)	L238	1210	10/0.5	R. Clube, Varginha
MG70)	L305	1210	10/1	R. Itatiaia do Triângulo, Uberlândia
PE20)	I786	1210	10/1	R. Jornal, Garanhuns: 24h
PR60)	J219	1210	10/5	Super Rádio Deus é Amor, Curitiba
PR140)	J325	1210	1/0.5	R. Brotense, Porecatu
RN29)	J620	1210	5/0.5	R. Vale do Potengi, São Paulo do Potengi
RS86)	K240	1210	10/5	R. Record, Porto Alegre
RS88)	K353	1210	1/0.5	R. Blau Nunes, Santa Bárbara do Sul
SC42)	J785	1210	10/0.5	R. Super Santa, Tubarão
SP122)	K509	1210	10/1	R. Vida Nova, Jaboticabal
SP123)	K545	1210	5/0.25	R. Bandeirantes, Araçatuba
SP125)	K668	1210	5/0.25	R. Vanguarda, Sorocaba: 24h
RJ25)	J458	1220	150	R. Globo, Rio de Janeiro: 24h
BA59)	H532	1230	1/0.25	R. Povo, Ubatã
GO27)	H756	1230	10/2.5	R.Daqui, Goiânia
MA23)	H896	1230	1/0.25	R. Alecrim, Caxias
MG105)	L208	1230	5/0.5	R. Correio da Serra, Barbacena
MG102)	L216	1230	2.5/0.25	R. Passos , Passos
MG176)	ZYN203	1230	10/0.7	R.Estrela de Ibiúna, Campina Verde
PB12)	I670	1230	10/1	CBN, João Pessoa
PR170)	J350	1230	1/0.25	R. Nova Mensagem, Telêmaco Borba
RS146)	K297	1230	1/0.25	R. Santiago, Santiago
RS89)	K326	1230	2/0.25	R. Clube Nonoai, Nonoai
RS90)	K333	1230	2.3/0.35	R. Prata, Nova Prata
RS91)	K352	1230	1/0.5	R. Encruzilhadense, Encruzilhada do Sul
SC38)	J776	1230	5/0.65	R. Colméia, Porto União
SC88)	J816	1230	10/1	R. Guararema, São José: 24h
SP126)	K573	1230	1/0.5	R. Cacique, Capão Bonito
SP258)	K637	1230	10/0.25	R. Difusora, Rancharia
SP128)	K716	1230	1/0.5	R. Jequitibá, Campinas
SP266)	K766	1230	50/10	Super R. Boa Vontade, São Paulo
BA31)	H463	1240	10/0.5	R. Nova AM, Alagoinhas
CE49)	H654	1240	1/0.25	R. São Francisco, Canindé
MG97)	L294	1240	5/0.25	R. Cl. Três Pontas, Três Pontas
MG84)	L298	1240	1/0.5	R. Ubaense, Ubá
MG72)	L303	1240	10/0.35	R. Platina, Ituiutaba
MG116)	L317	1240	5/0.25	R. Pirapora AM, Pirapora: 0800-2200
MS09)	I388	1240	5/0.25	R. Dif. Pantanal, Campo Grande: 24h
PE21)	I774	1240	5	R. Capibaribe, Recife
PR112)	J215	1240	1/0.25	R. Arapongas, Arapongas: 24h
PR61)	J280	1240	2/0.25	R. Matelândia, Matelândia
PR94)	J301	1240	1/0.25	R. Dif. Ubiratanense, Ubiratã
RS92)	K200	1240	1/0.25	R. Aparados da Serra, Bom Jesus
RS93)	K251	1240	5/0.25	R. Ibirubá, Ibirubá
RS94)	K355	1240	1/0.65	R. São Jerônimo, São Jerônimo
SC43)	J774	1240	5/0.5	R. São José, Mafra: 24h
SC44)	J810	1240	1/0.25	R. Iracema, Cunhe Porã
SP129)	K565	1240	10/0.25	R. Municipalista, Botucatu: 0830-0200
SP130)	K621	1240	1/0.25	Orlândia R. Clube, Orlândia
SP131)	K653	1240	10/2.5	R. Clube, Santos
SP132)	K711	1240	1/0.25	R. Vale do Rio Tietê, José Bonifácio
CE27)	H594	1250	1	R. Educadora, Crateús
CE69)	H669	1250	1/0.25	R. Liberdade, Itarema
ES18)	I218	1250	10/1	R. Globo, Vitória
GO29)	H748	1250	1/0.25	R. Coração Fiel, Ceres
MG73)	L282	1250	5	R. Difusora, Poços de Caldas
MG153)	L367	1250	20/0.25	R. Metropolitana, Vespasiano
MS10)	I394	1250	1/0.25	R. Difusora, Três Lagoas
MS11)	I412	1250	5/0.5	R. Caiuás, Dourados
PB27)	I701	1250	1/0.25	R. Sociedade de Soledade, Soledade
PI47)	I932	1250	1/0.25	R. João de Paiva, Altos
PR62)	J211	1250	5/0.5	R. Difusora, Guarapuava
PR63)	J233	1250	5/0.4	R. Paranavaí, Paranavaí: 0800-0300, Sat1200-0300, Sun1030-0100
PR64)	J313	1250	1/0.25	R. Danúbio Azul, Sta. Isabel do Oeste: 24h
RJ50)	J500	1250	10/0.5	R. Litoral, Casimiro de Abreu
RS95)	K233	1250	10/0.5	R. Difusora Caxiense, Caxias do Sul
RS96)	K272	1250	1	R. Tupanci, Pelotas
RS142)	K361	1250	5/0.6	R. Aguas Claras, Catuípe
SC45)	J766	1250	5/0.25	R. Cultura Jovem Pan, Joinville

MW	Call	kHz	kW	Station, location
SE08)	J925	1250	10/1	R. Esperança, Estância
SP133)	K702	1250	5/0.5	R. Canção Nova, Caçapava
AL07)	H242	1260	50/5	R. Gazeta de Alagoas, Maceió: 24h
CE28)	H596	1260	1/0.25	R. Vale do Jaguaribe, Limoeiro do Nte: 0730-0100
RO09)	J670	1260	5	R. Educ., Guajará Mirim: 0900-0300
RS97)	K204	1260	1/0.25	R. Cultura, São Borja: 24h
RS98)	K327	1260	5/0.25	R. Fandango, Cachoeira do Sul
RS99)	K345	1260	1/0.25	R. Gaurama, Gaurama
SC46)	J740	1260	5/0.5	R. Blumenau, Blumenau
SP257)	K629	1260	1/0.25	Pirajau R. Clube, Pirajuí: 24h
SP134)	K688	1260	100/40	R. Morada do Sol, São Paulo
AM10)	H271	1270	5	R. Educação Rural, Tefé: 1000-0200
GO30)	H753	1270	100/10	R. Brasil Central, Goiânia
MG74)	L227	1270	5/1	R. Carijós, Conselheiro Lafaiete
MG107)	L300	1270	2.5/0.5	R. Estância, São Lourenço
MG155)	L240	1270	5/1	R. Libertas do Vale do Aço, Ipatinga
PA09)	I530	1270	10/2.5	R. R.RBN, Belém
PB28)	I696	1270	5/0.25	R. Cidade, Sumé
PR65)	J222	1270	5/0.5	R. Guairacá, Mandaguari: 0800-0100, Sat:0800-2300 Sun:0900-2300
PR67)	J236	1270	10/1	R. Continental, Curitiba
PR68)	J289	1270	5/0.5	R. Globo (R.Cidade) , Cascavel: 24h
RJ26)	J474	1270	5/0.5	R. Continental, Campos dos Goitacazes
RN10)	J593	1270	5/0.5	R. Clube AM 1270, Natal: 24h
RS131)	K206	1270	5/0.5	R. América, Montenegro
RS101)	K250	1270	5/0.5	R. Vera Cruz, Horizontina
SC47)	J765	1270	12/0.25	R. Catarinense, Joaçaba
SC48)	J768	1270	1/0.25	R. Garibaldi, Laguna: 0830-0200 Sun:1045-1600
SP136)	K678	1270	5/0.5	R. Brasil, Campinas: 24h
SP274)	K640	1270	10/0.5	R. Globo, Ribeirao Preto
PB21)	I688	1280	10/5	R. Sanhauá, Bayeux
RJ27)	J455	1280	100	R. Tupi, Rio de Janeiro
AM11)	H286	1290	10/2.5	R. Rio Mar, Manaus: 0900-0300
BA32)	H450	1290	10/1	R. Metropole, Salvador
MA10)	H888	1290	50/5	R. Timbira do Maranhão, São Luís: 0800-0230 Sun:1000
MG77)	L273	1290	25/5	R. Record Uberlandia, Uberlandia
MG164)	L345	1290	5/0.25	R. Cidade, Arcos
PR73)	J310	1290	10/0.5	R. Brasil Sul, Londrina
RN26)	J619	1290	5/0.25	R. Caicó, Caicó: 0600-0100 Sun:0700-2300
RS103)	K331	1290	5/2	R. Planetário, Espumoso
SC81)	J734	1290	5/1	R. Araranguá, Araranguá:0800-0200
SC49)	J804	1290	5/1	R. Camboriú, Balneário Camboriú: 0800-0300
SP240)	K662	1290	5/0.5	R. Difusora, São José do Rio Pardo
SP137)	K663	1290	5/1	R. Novo Tempo, São José do Rio Preto:24h
SP216)	K745	1290	1/0.5	R. Eldorado, São José dos Campos: 24h
CE29)	H586	1300	10	R. Iracema, Fortaleza
ES11)	I210	1300	5/0.25	R. Novo Tempo, Afonso Cláudio
MG143)	L339	1300	5/1	R. Eldorado, Sete Lagoas
PE22)	I799	1300	1/0.25	R. Guarany, Camaragibe
PR71)	J278	1300	1/0.25	CBN, Ponta Grossa
PR127)	J288	1300	5/0.5	R. Educadora, Dois Vizinhos
RS104)	K203	1300	50/13	Super R.Boa Vontade , Porto Alegre: 24h
RS105)	K337	1300	1/0.25	R. Regional, Santo Cristo: 0800-0100
RS106)	K347	1300	5/0.5	R. Maratan, Santana do Livramento: 24h
SC89)	J819	1300	1/0.25	R. Alvorada, Santa Cecília: 0900-0130 Sun:1000-0130
SP138)	K535	1300	50/1	R. Universo, São Bernardo do Campo
SP252)	K649	1300	30/0,25	R. Onda Viva, Santo Anastácio: 0855-2200
SP226)	K762	1300	1/0.25	R. Realidade, São Carlos
AP05)	H422	1310	1/0.25	R. Mazagão, Mazagão
BA33)	H454	1310	1/0.25	R. Nova Bahiana, Ilhéus: 0800-0300 Sat/Sun: 0900-0100
BA63)	H501	1310	5/0.25	R. Jaraguar, Jacobina: 0800-0200
CE30)	H602	1310	1	R. Progresso de Juazeiro, Juazeiro do Nte: 0800-2400, Sat:0900-0100 Sun:0830-2300
CE63)	H656	1310	1/0.25	R. Liberdade, Boa Viagem
MG144)	L359	1310	1/0.25	R. Montanhesa, Vazante
MS31)	I426	1310	5/1	R. Pindorama, Sidrolândia
MT65)		1310	1/0.25	R. Jauru, Jauru
PB23)	I691	1310	10/0.5	R. Cidade (Rede Esperança), Esperança
PR70)	J274	1310	10/0.5	R. Atalaia, Maringá
RJ28)	J504	1310	1/0.25	R. Difusora Coroados, São Fidélis
RO17)	J684	1310	10/5	R. Tropical, Porto Velho
RS107)	K305	1310	10/1	R. Sarandi, Sarandi: 0900-0000
RS124)	K329	1310	5/0.45	R. Integração, Restinga Seca
RS160)	K371	1310	5/0.5	R. Horizonte, Capão da Canoa
SC85)	J801	1310	10/0.5	R. Sintonia, Ituporanga
SP141)	K566	1310	10/0.25	R. Bragança, Bragança Paulista
SP139)	K596	1310	2/1	R. Difusora, Itápolis
AL08)	H243	1320	10/0.25	R. Milenio, Maceió
BA69)	H503	1320	5/0.5	R. Regional, Cícero Dantas
CE31)	H597	1320	1	R. Regional, Sobral
CE70)	H672	1320	1/0.5	R. Moriá, Aracati: 24h

MW	Call	kHz	kW	Station, location
MG62)	L204	1320	5/0.25	R. Minas, Divinópolis
MG136)	L322	1320	1/0.25	R. Mucuri, Teófilo Otoni
PE31)	I823	1320	1/0.25	R. Cultura, São José do Egito
PR72)	J255	1320	12/0.5	R. Tropical, Curitiba: 24h
PR145)	J351	1320	5/0.5	R. Foz, Foz do Iguaçu
RJ29)	J475	1320	25/5	R. Boas Novas, Petropôlis
RS108)	K223	1320	1/0.25	R. Clube, Canela
RS109)	K266	1320	3/0.25	R. Panambi, Panambi
RS110)	K271	1320	5/1	R. Cultura, Pelotas
SC68)	J762	1320	5/0.45	R. Litoral, Imaruí
SC104)		1320	5/0.45	R. Vitória, Videira
SP140)	K630	1320	1/0.25	R. Difusora, Pirassununga
SP241)	K675	1320	1/0.25	R. Clube, Tupã
BA34)	H468	1330	5/1	R. Continental, Serrinha
MS54)	N610	1330	1/0.25	R. Pantanal, Coxim
PA10)	I533	1330	25/5	R. Liberal, Belém: 24h
PR74)	J264	*1330	10/0.5	R. Jaguariaíva, Jaguariaíva
RN27)	J621	1330	10/0.5	R. Eldorado, Natal
RS111)	K236	1330	1/0.25	R. Upacaraí, Dom Pedrito
RS112)	K323	1330	1/0.5	R. Diplomata, São Marcos
SC50)	J739	1330	5/0.5	R. Clube, Blumenau: 0800(Sun:0900) -0100
SC51)	J749	1330	5/1	R. Chapecó, Chapecó
SP142)	K638	1330	30/0.25	R. Paulista, Regente Feijó
SP143)	K641	1330	5/1	R. Eldorado, Ribeirão Preto
SP187)	K736	1330	50/10	R. Terra, Osasco
CE71)	H661	1340	2.5/0.25	R. Pituaguary, Maracanaú: 0800-0300
MA11)	H886	1340	10/2	R. Cl. de São Luís, São Luís
MG81)	L241	1340	10/5	R. Cultura, Itabirito
MG139)	L352	1340	1/0.5	R. Globo, Passos
MS12)	I380	1340	3/0.25	R. Dif. 1340, Aquidauana
PB13)	I671	1340	5/1	R. Correio, João Pessoa
PR76)	J205	1340	2.5/0.25	R. Difusora, Rio Negro
PR77)	J249	1340	5/0.25	R. Cultura, Arapongas
PR41)	J368	1340	5/0.25	CBN, Cascavel
RJ40)	J490	1340	5/0.5	R. Tupi, Rio Bonito
RS113)	K227	1340	25/4	CBN (R. Educadora), Porto Alegre
RS173)	K377	1340	10/4	R. Journal da Manhã, Ijui
SP144)	K543	1340	5/1	R. Cultura, Araçatuba
SP203)	K571	1340	5/0.25	R. Em. Campos do Jordão, Campos do Jordão (ff 1560)
SP146)	K738	1340	1/0.25	R. Nova Canoa Grande, Igaraçu do Tietê
AC05)	H201	1350	50/5	R. Capital (RBV), Rio Branco
BA70)	H520	1350	50/10	Super R. Cristal, Salvador
CE56)	H662	1350	1/0.25	R. Liberal Jagoaribana, Morada Nova: 0800-2300 Sat/Sun:0830-2200
MG82)	L214	1350	10/5	R. Cultura, Poços de Caldas
PB14)	I675	1350	5/0.5	R. Clube AM 1350, Campina Grande
RS114)	K205	1350	2.5/0.25	R. Aurora, Guaporé: 1000-0100 Sat/Sun:0900
RS115)	K313	1350	5/1	R. Difusora, Três Passos
RS116)	K336	1350	1/0.25	R. Agudo, Agudo: 0900-0200
SC53)	J760	1350	1/0.25	R. Clube Bandeirantes, Itajaí
SP265)	K692	1350	10/0,25	R. Excelsior, Ibiúna
BA35)	H469	1360	10/1	R. Cultura, Paulo Afonso
CE57)	H650	1360	5/1	R. Iracema, Ipu: 0900-0200, Sa.:21hSun:13h
MS14)	I383	1360	2	R. Bandeirantes, Corumbá
PR165)	J265	1360	10/0.25	R. Cidade, Pato Branco
PR78)	J268	1360	5/1	R. Lider, Assaí
RJ30)	J464	1360	50/10	R. Bandeirantes, Rio de Janeiro
RN11)	J605	1360	1/0.5	R. Ouro Branco, Currais Novos
RS117)	K261	1360	5/0.5	R. Alvorada, Marau
RS151)	K281	1360	3/0.25	R. Navegantes, Porto Lucena
SC69)	J757	1360	25/0.4	R. Belos Vales, Ibirama
SP147)	K581	1360	5/1	R. Aguas Quentes, Fernandópolis: 24h
SP148)	K739	1360	5/0.25	R. Regional, Dracena: 24h
SP235)	K759	1360	1/0.25	R. Luzes da Ribalta, Santa Bárbara d'Oeste
BA82)	H555	*1370	0.25	R. Piquaraca, Monte Santo
CE81)	H628	1370	1/0.25	R. Vanguarda, Caridade
PE34)	I800	1370	1/0.25	R. Vale do Capibaribe, Sta. Cruz do Capibaribe
PI13)	I892	1370	10/1	R. Difusora, Teresina
PR80)	J267	1370	50/7	R. Canção Nova, Curitiba
RN28)	J618	1370	2.5/0.25	R. Difusora,São Miguel
RS173)	K374	1370	1/0.5	R. Jornal da Manhã, Ijuí
RS118)	K243	1370	25/0.5	R. Mãe de Deus, Flores da Cunha
RS119)	K334	1370	1/0.5	R. Gazeta, Alegrete: 0900-0400 Sat/Sun:0900
SC55)	J782	1370	10/0.5	R. Peperi AM, São Miguel do Oeste
SE09)	J929	1370	5/0.5	R. Capital do Agreste, Itabaiana: 0700-0300
SP223)	K766	1370	100/20	R. Iguatemi, São Paulo
AM12)	H283	1380	5/1	R. Alvorada, Parintins: 0900-0200
BA83)	H495	*1380	5/1	R. União, Gandu
ES21)	I...	1380	10/1	R. Itaí de Rio Claro, Rio Claro
MG172)	L218	1380	1/0.25	R. Cidade, Brasópolis
MG102)	L284	1380	5/0.25	R. Paranaíba, Rio Paranaíba: 0900-0100
MG130)	L323	1380	1/0.25	R. Gorutubana, Janaúba
MT49)	I...	1380	1/0.25	R. Educadora, Nova Brasilândia
PA11)	I547	1380	1/0.25	R. Marajó, Breves

MW	Call	kHz	kW	Station, location
PE23)	I773	1380	10/5	R. Continental, Recife
PR122)	J276	1380	2/0.25	R. Bom Jesus, Siqueira Campos: 0730-0200
PR152)	J367	1380	1/0.25	R. Integração, Toledo
RS120)	K293	1380	1/0.25	R. Cultura, Santana do Livramento
RS134)	K311	1380	5/0.25	R. Maristela, Torres: 24h
RS121)	K350	1380	6/0.25	R. Cultura, Tapera
RS165)	K372	1380	6/0.25	R. Chiru, Palmitinho: 0800-0200
SC56)	J821	1380	6/0.25	R. Cidade, Itaiópolis
SC93)	J827	1380	6/0.25	R. Barriga Verde, Capinzal
SC105)	J…	1380	1/0.25	R. Freguencia, Garopaba
SP152)	K616	1380	1/0.5	R. Difusora, Mogi Guaçu
SP247)	K623	1380	1/0.25	R. Cultura, Pederneiras
SP224)	K751	1380	5/0.25	R. Globo, Presidente Prudente
SP234)	K772	1380	1/0.25	R. República, Morro Agudo
ES13)	I209	1380	5/0.25	R. Educadora, Afonso Cláudio: 0700-0100
MG157)	I358	1390	2.5/0.25	R. Ouro Verde, São Sebastião do Paraíso
MG178)	N210	1390	10/0.5	R. Itaiaia Triângulo, Uberlândia
PA12)	I535	1390	10/1	R. Educadora, Bragança: 0830-0100
PE24)	I788	1390	5/1	R. Jornal O Povo, Pesqueira
PR82)	J242	1390	10/1	R. Cultura, Maringá
PR83)	J335	1390	5/0.25	R. Independência, Salto do Lontra
RJ32)	J473	1390	5/0.5	R. Sul Fluminense, Barra Mansa
RN32)	J599	1390	5/0.25	R. Farol, Touros
RO18)	J687	1390	5/1	R. Planalto, Ji-Piraná
RR02)	O701	1390	10/5	R. Roraima, Caracaraí
RS122)	K209	1390	10	R. Esperança, Porto Alegre: 24h
RS166)	K368	1390	8/0.25	R. Atlântica, Constantina: 0900-0200
SC57)	J769	1390	7.5/0.65	R. Globo, Lages
SP153)	K570	1390	5/0.25	R. Globo, Campinas: 24h
SP90)	K594	*1390	2.5/0.25	R. Anchieta, Itanhaém
SP154)	K636	1390	1/0.25	R. Cultura, Promissão
AC06)	H200	1400	10/1	R. Dif. Acreana, Rio Branco: 0900-0400
BA71)	H529	1400	5/0.25	R. Vale do Vasa-Barris, Jeremoabo
PB15)	I677	1400	5/1	R. Espinharas, Patos
PI27)	I926	1400	1/0.25	R. Cantagalo, Jaicós
PR84)	J256	1400	5/0.25	R. Globo, Londrina
PR87)	J299	1400	1/0.25	R. Fronteira d'Oeste, Terra Roxa
PR119)	J339	1400	10/1	R. Gospel, Balsa Nova
PR148)	J346	1400	1/0.45	R. Jornal São Miguel, São Miguel do Iguaçu: 0900-0200
RJ33)	J462	1400	50/5	R. Rio de Janeiro "R. Rio AM", Rio de Janeiro
RS192)	K376	1400	1/0,4	R. Educadora, São João da Urtiga: 0800-0130
SC58)	J775	1400	5/0.35	R. Entre Rios, Palmitos: 0800-0200 Sun.0900-2300
SP155)	K527	1400	1/0.25	R. Difusora, Lucélia
SP156)	K658	1400	5/0.25	R. Clube, São Carlos
SP157)	K682	1400	5/0.25	R. Metrópole AM, São José do Rio Preto: 0900-2300
TO08)	N660	1400	5/1	Radiodifusão Guaraí, Guaraí
BA37)	H467	1410	10/0.5	R. São Gonçalo, São Gonçalo dos Campos: 0800-2000 Sat/Sun:0800-0000
CE25)	N639	1410	10/1	R. Boa Nova, Pacajus
MS14)	I382	1410	5/1	Nova R. Clube, Corumbá: 0700-2300, Sat:-0100
RJ34)	J486	1410	10/0.5	R. Itaperuna, Itaperuna: 24h
RN21)	J614	1410	1/0.5	R. Santa Cruz, Santa Cruz: 0800-0200 Sun.0900-2300
RS137)	K246	1410	5/1	R. Garibaldi, Garibaldi
RS125)	K284	1410	1/0.25	R. Cassino, Rio Grande: 24h
RS126)	K294	1410	5	R. Santa Rosa, Santa Rosa
SC100)	J799	1410	2/0.5	R. Namba, Ponte Serrada
SP158)	K691	1410	50/25	R. América, São Paulo
SP264)	K683	1410	2.5/0.5	R. Excelsior, Rio Claro
BA60)	H504	1420	1/0.25	R. Cidade, Irecê
MG88)	L286	1420	1/0.25	R. Difusora, São João Nepomuceno
MG89)	L288	1420	5/1	R. Cultura, Sete Lagoas
MG90)	L313	1420	1/0.25	R. Montanhês Botelhos, Botelhos
MS16)	I397	1420	1/0.25	R. Difusora Cacique, Nova Andradina
PR88)	J269	1420	5/0.25	R. Cult, Umuarama: 0900 (Sun:1000)-2105
PR89)	J282	1420	1/0.25	R. Educadora, Jacarezinho
RN12)	J609	1420	1/0.25	R. Farol, Alexandria (RPC)
RS149)	K258	1420	10/0.25	R. 14 de Julho, Júlio de Castilhos
RS171)	K308	1420	0.5	R. Tapense, Tapes
SC60)	J744	1420	6/0.5	R. Cultura, Campos Novos: 0855-0100
SC59)	J754	1420	5/2.5	R. Guarujá, Florianópolis
SP159)	K597	1420	1/0.5	C.R.N., Itatiba: 24h
SP163)	K733	1420	1/0.25	R. Nova São Manuel, São Manuel
MG91)	L239	1430	1/0.25	R. Clube, Guaxupé
MG158)	L371	1430	1/0.25	R. Planalto, Perdizes
PE32)	I826	1430	5/0.25	R. Independência, Goiana
PR134)	J200	°1430	50/10	R. Clube B2, Curitiba: 24h
RN13)	J604	1430	10/0.5	R. Libertadora Mossoroense, Mossoró
RO11)	J671	1430	10	R. Caiari, Porto Velho: 0900-0300
RS167)	K366	1430	1/0.25	R. Guarita, Coronel Bicaco
SP164)	K666	1430	1/0.25	R. Serra Negra, Serra Negra
SP275)	K707	1430	25/0,25	R. Imaculada Conceição, São Roque
AM13)	H285	1440	10	R. Baré, Manaus: 24h
BA38)	H466	1440	50/1	R. Independência, Santo Amaro
CE33)	H603	1440	10/1	R. Araripe, Crato
MG159)	L365	1440	1/0.25	R. Som 2000, Santa Vitória: 24h
MS41)	I407	1440	1/0.25	R. Bela Vista, Bela Vista
RJ35)	J469	1440	20/5	R. Livre, Rio de Janeiro: 24h
RS168)	K221	1440	1/0.5	R. Ceres, Naõ Me Toque
RS130)	K328	1440	5/0.3	R. Excelsior, Gramado
RS153)	K362	1440	2.5/0.25	R. Caibaté, Caibaté: 0800 - 0200
SC61)	J792	1440	2.5/0.25	R. Difusora, Maravilha: 24h
SC82)	J797	1440	10/0,35	R. Educadora, Taió
SE11)	J930	1440	5/0.25	R. Educadora, Frei Paulo: 0700-0300
SP253)	K568	1440	5/0.25	R. Eldorado Centro Norte Paulista, Cajuru
SP165)	K634	1440	5/0.25	R. Comercial, Presidente Prudente: 24h
SP218)	K752	1440	1/0.25	R. Azul Celeste, Americana
BA73)	H531	1450	1/0.25	R. Ipirá, Ipirá
CE34)	H601	1450	1/0.25	R. Difusora Cristal, Quixeramobim
CE35)	H623	1450	1/0.25	R. Pinto Martins, Camocim
ES12)	I208	1450	1/0.5	R. Sim (R.Gaeta), Guarapari
MA32)	H900	1450	1/0.25	R. Boa Esperança, São João dos Patos
MA13)	H901	1450	1/0.25	R. Cultura, Pedreiras
MG94)	L312	1450	10/0.25	R. Diamante, Coromandel
MS13)	I417	1450	1/0.25	R. Difusora, Rio Brilhante: 0800-0100
PA33)	I559	1450	1/0.25	R. Juruá, São Felix do Xingu
PB29)	I699	1450	1/0.25	R. Itatiunga, Patos
PE25)	I794	1450	1/0.25	R. Cultura, Palmares
PI21)	I908	1450	5/0.25	R. Cultura do Gurguéia, Bom Jesus
PI35)	I917	1450	1/0.25	R. Confederação Valenciana "R. Valenciana", Valença do Piauí
PR93)	J279	1450	1/0.25	R. Cabiúna, Bandeirantes
PR95)	J317	1450	1/0.25	R. Rainha d'Oeste, Altônia
PR162)	J364	1450	1/0.25	R. Clube, Mallet
PR179)	J302	1450	1/0.25	R. Dif., Ubiratã
RJ36)	J480	1450	5/0.25	R. Comércio, Barra Mansa
RJ37)	J503	1450	1/0.25	R. Feliz, Santo Antônio de Pádua
RO12)	J674	1450	1/0.25	R. Vilhena AM, Vilhena
RR03)	O701	1450	1/0.25	R. Transamérca Hits, Alto Alegre
RS132)	K346	1450	5/0.6	R. Cassino, Rio Grande
RS177)	K338	1450	1/0.25	R. Cultura, Arvorezinha
SC62)	J802	1450	1/0.25	R. São Bento, São Bento do Sul
SC63)	J822	1450	10/0.25	R. Hulha Negra, Criciúma
SC97)	J828	1450	5/0.25	R. Belos Montes, Seara
SE12)	J932	1450	10/0.5	R. Abais, Estância
SP238)	K526	1450	2.5/0.25	R. Cultura, Ituverava
SP166)	K587	1450	1/0.25	R. Difusora, Guararapes
SP167)	K591	1450	50/5	R. Boa Nova, Guarulhos
SP168)	K657	1450	5/0.25	R. São Carlos, São Carlos
AL13)	H253	1460	10/0.25	R. Canaviero, União dos Palmares
AM17)	H300	1460	5	R. Clube, Parintins: 0830-0200
BA39)	H472	1460	1	R. Povo, Jequié
BA85)	H523	1460	1/0.25	R. Ferro Doido, Morro do Chapéu
BA74)	H536	1460	1/0.25	R. Alvorada, Cruz das Almas
CE53)	H595	1460	1/0.25	R. Ressurreição, Sobral: 0845-2200
CE36)	H616	1460	1/0.25	R. Uirapuru, Morada Nova
GO34)	H766	1460	1/0.25	R. Morrinhos, Morrinhos
MA33)	H917	1460	1/0.25	R. Vanguarda, Santa Luzia
MG95)	L201	1460	5/0.25	R. Cultura de Porto Novo, Além Paraíba
MG161)	L356	1460	5/0.25	R. Buritis, Buritis
MG131)	L363	1460	1/0.25	R. Soc. Entre Rios, Raul Soares
MS56)	I....	1460	1/0.25	R. Globo, Costa Rica: 0700-0100 (Sun-0400)
MT66)		1460	1/0,25	R. Vila Bela, Vila Bela da Santíssima Trinidade
PI14)	I903	1460	1/0.25	R. Cultura, Amarante
PR96)	J204	1460	2/0.25	R. Difusora, Paranaguá
PR97)	J228	1460	1/0.25	R. Central do Paraná, Ponta Grossa
PR98)	J251	1460	1/0.25	R. Cultura, Apucarana
PR100)	J297	1460	1/0.25	R. Guaíra, Guaíra
PR101)	J308	1460	5/0.25	R. Ampere, Ampere
PR102)	J318	1460	1/0.25	R. Guadalupe AM, Loanda: 24h
RN20)	J615	1460	10/0,25	R. Agreste, Santo Antônio
RS133)	K214	1460	1/0.25	R. Cultura, Bagé
RS135)	K312	1460	1/0.5	R. Colonial, Três de Maio
RS175)	K373	1460	1/0.25	R. Campinas, Campinas do Sul
RS176)	K378	1460	1/0.25	R. Mostardas, Mostardas
RS196)	K......	1460	0,25	R. Fronteira, Santiago
RS194)	K......	1460	1/0.25	R. Educadora, Bom Retiro do Sul
SC64)	J756	1460	1/0.25	R. Sentinela do Vale, Gaspar
SP242)	K548	1460	5/0.5	R. Clube Ararense, Araras
SP170)	K608	1460	1/0.25	R. Cultura, Lorena: 24h
SP171)	K707	1460	1/0.25	R. Universal, São Roque
TO02)	H774	1460	1/0.25	R. Independência do Tocantins, Paraíso do Tocantins
BA86)	H509	1470	0.25	R. Morro Verde, Mairi

MW	Call	kHz	kW	Station, location
CE64)	H665	1470	1/0.25	R. Guanancés de Itapajé, Itapajé: 0800-0100
ES17)	I214	1470	1/0.25	R. Globo, Barra de São Francisco
GO36)	H773	1470	10/0.25	R. Dif. Serra dos Cristais, Cristalina
GO37)	H779	1470	1/0.25	R. Cidade, Goiás
MA34)	H908	1470	1/0.25	R. Paranoá, Presidente Dutra
MA39)	H901	1470	1/0.25	R. Urbano Santos,Urbano Santos
MS29)	I413	1470	1/0.25	R. Alvorada, Dourados
PA25)	I548	1470	5/1.5	R. Moreno Braga, Vigia
PE35)	I822	1470	1/0.25	R. Educadora de Belém, Belém de São Francisco
PE37)	I827	1470	1/0.25	R. Papacaça, Bom Conselho
PI15)	I900	*1470	1/0.25	R. Difusora Vale do Uruçuí, Uruçuí
PI24)	I913	1470	1/0.25	R. Ingazeira, Paulistana
PI33)	I928	1470	1/0.25	R. Cidade, Castelo do Piauí
PR103)	J294	1470	1/0.25	R. Educadora, Ibaiti: 24h
PR104)	J304	1470	2/0.25	R. Jornal, Assis Chateaubriand
PR172)	J…	1470	1/0.25	R. Panorama, Itapejara d'Oeste
PR173)	J…	1470	3/0.25	R. Tradição, Rio Branco de Sul
RJ08)	J476	1470	1/0.25	R. Jornal Fluminense, Campos dos Goitacazes
RJ38)	J481	1470	1/0.25	R. Barra do Pirai, Barra do Piraí
RN23)	J616	1470	1/0.25	R. Rural de Parelhas, Parelhas: 0800-0100 Sat/Sun: 0900-
RO05)	J676	1470	1/0.25	R. Soc. Rondônia, Cacoal
RS138)	K208	1470	25/0.25	R. Assisense, São Francisco de Assis
RS169)	K219	1470	1/0.25	R. Cultura, Cacequi
SC65)	J781	1470	10/1	R. Mais Alegrie, São José
SC66)	J798	1470	1/0.25	R. Nova Líder, Herval d'Oeste
SP172)	K586	1470	1/0.25	R. Cultura, Guaíra
SP173)	K599	1470	1/0.25	R. Mensagem, Jacareí
SP174)	K632	1470	1/0.25	R. Primavera, Porto Ferreira
SP175)	K712	1470	1/0.25	R. Jornal, Indaiatuba: 24h
SP243)	K771	1470	1/0.25	R. Bastos AM, Bastos
BA61)	H508	1480	5/1	R. Alvorada, Guanambi
BA75)	H524	1480	1/0.25	R. Santana, Santana
CE74)	H671	1480	1/0.25	R. Princesa do Norte, Morrinhos
CE58)	H636	1480	1/0.5	R. Araripe (Cidade), Campos Sales
MA14)	H897	1480	1/0.25	R. Itapecuru, Colinas
MG98)	L235	1480	1/0.25	R. Nova Frutal AM, Frutal
MG99)	L265	1480	5/0.25	R. Difusora, Nanuque
MG100)	L307	1480	2.5/0.25	R. Emboabas, Tiradentes: 0900-2200 Sun:1530
MS17)	I393	1480	1/0.25	R. Caçula, Três Lagoas
PE26)	I790	1480	1/0.25	R. A Voz do Sertão, Serra Talhada: 0800-2300
PE36)	I825	1480	5/0.25	R. Cançao Nova, Gravatá
PI29)	I929	1480	1/0.25	R. Vale do Coroatá, Elesbão Veloso
PR79)	J302	1480	5/0.25	R. Cultura, Iporã
PR153)	J221	1480	1/0.25	R. Brotas, Piraí do Sul
PR106)	J230	1480	1/0.25	R. Astorga, Astorga
PR107)	J270	1480	1/0.25	R. Educadora, União da Vitória: 0900-0200
PR174)	J370	1480	1/0.25	R. Pérola, Pérola d'Oeste
RJ55)	J485	1480	10/0.5	R. Popular, Duque de Caxias
RN14)	J601	1480	1/0.25	R. Princesa do Vale, Açu
RO15)	J681	1480	1/0.25	R. Rondônia, Pimenta Bueno
RS100)	K244	1480	2.5/0.25	R. São Roque, Faxinal do Soturno
RS127)	K321	1480	5/0.25	R. Veranense, Veranópolis: 24h
RS178)	K255	1480	0.5/0.25	R. Guaramano, Guarani das Missões: 0900-0200 Sun:-0100
SC67)	J731	1480	1/0.25	R. Difusora, Joinville
SC94)	J826	1480	1/0.25	R. Caíbi, Caíbi
SE10)	J928	1480	1/0.25	R. Nova Cidade, Simão Dias: 24h
SP176)	K539	1480	1/0.25	R. Clube, Altinópolis
SP177)	K551	1480	1/0.25	R. Atibaia, Atibaia
SP255)	K767	1480	0.5/0.25	R. Nova America, Boituva
T006)	H795	1480	1/0.25	R. Cultura, Miracema do Tocantins
AL09)	H246	1490	5/1	Em. Rio São Francisco, Penedo
BA41)	H478	1490	1/0.25	R. Educadora, Ipaú: 0800-0300 Sat/Sun: 0900-2100
BA77)	H507	1490	1/0.25	R. Rio São Francisco, Bom Jesus da Lapa
BA87)	H512	1490	0.25	R. Planalto d'Oeste, Correntina
BA88)	H522	1490	2.5/0.5	R. Antena Um, Ribeira do Pombal
MG162)	L231	1490	0.25	R. Onda Viva, Araguari
MG163)	L234	1490	1/0.25	R. Paraisópolis, Paraisópolis
MG165)	L353	1490	1/0.25	R. Pirapetinga, Pirapetinga
MG179)	L209	1490	1/0.25	Momento de Comunicação,Santa Luzia
MS19)	I404	1490	1/0.25	R. Nova Paiaguás, Glória de Dourados
MT53)		1490	1/0.25	R. Vila, Vila Rica
PI40)	I918	1490	1/0.25	R. Lagoa, Buriti dos Lopes
PR108)	J210	1490	1/0.25	R. Cornélio, Cornélio Procópio
PR149)	J347	1490	1/0.25	R. Dif. São Jorge do Oeste: 0900-0100
RS140)	K309	1490	1/0.25	R. Taquara, Taquara
SC70)	J791	1490	2.5/0.25	R. Cultura, Xaxim
SP180)	K530	1490	1/0.25	R. Difusora, Olímpia
SP181)	K580	1490	1/0.25	R. Globo, Dracena: 24h
SP182)	K583	1490	1/0.25	R. Educadora Santa Rita, Fernandópolis
SP244)	K680	1490	1/0.25	R. Cult, Vargem Grande do Sul: 0900-0100
SP183)	K764	1490	25/0.5	R. Imaculada Conceição, Mauá
BA49)	H487	1500	5/0.5	R. Jacuípe, Riachão do Jacuípe: 0800-2000
CE39)	H615	1500	2.5/0.25	R. Macico, Baturité
MG101)	L215	1500	5/0.25	R. Montanhesa, Viçosa
MG140)	L340	1500	1/0.25	R. Aparecida do Sul, Ilicínea
PA19)	I542	1500	1/0.25	R. Floresta, Tucuruí
PE27)	I779	1500	1/0.25	R. Super Pajeu, Afogados da Ingazeira: 0800-0200
PI46)	I919	1500	1/0.25	R. Voz do Longa, Esperantina
PR163)	J366	1500	2.5/0.25	R. Aracauria, Margueirinha
RS139)	K225	1500	1/0.25	R. Liberdade, Canguçu: 0900-0200 (Sun:2300)
RS161)	K365	1500	3/0.25	R. Simpatia , Chapada: 0900-0100
SC106)	J……	1500	1/0.25	R. Guri AM, Balneário Camboriú
SP184)	K549	1500	2.5/0.25	R. Fraternidade, Araras
SP185)	K626	1500	1/0.25	R. Difusora, Pindamonhangaba
SP186)	K706	1500	1/0.25	R. Vale do Rio Grande, Miguelópolis
SP211)	K773	1500	1/0.25	R. Cumbica, Guarulhos
SP236)	K776	1500	1/0.25	R. Cidade, Apiaí
BA52)	H493	1510	5/0.5	R. Dif. do Descobrimento, Porto Seguro
CE40)	H608	1510	1/0.25	R. Nova Plan, São Benedito
CE84)	H630	1510	0.25	R. Trapiá, Pedra Branca
GO38)	H770	1510	1/0.25	R. Goiatuba, Goiatuba
PA01)	I544	1510	10/0.25	R. Oriente de Redenção, Redenção
PI16)	I894	1510	1/0.25	R. Difusora, Floriano
PI17)	I896	*1510	1/0.25	R. Progresso, Corrente
PI49)	I…	1510	1/0.25	R. Nordeste, Picos
PR113)	J216	1510	1/0.25	R. Educadora, Venceslau Bráz
PR115)	J326	1510	1/0.25	R. União, Céu Azul
RJ43)	J492	1510	1/0.25	R. Teresópolis, Teresópolis: 0900-0200 Sat:-0100 Sun:-2400
RN15)	J602	1510	1/0.25	R. Centenário, Caraúbas (RPC)
SC73)	J795	1510	1/0.25	R. Centro Oeste, Pinhalzinho: 0800-2300
SP188)	K654	1510	10/1	R. Cacique, Santos
SP189)	K665	1510	0.25	R. Clube Regional, São Manuel: 0700-2200
SP256)	K770	1510	0.5/0.25	R. Vale do Tietê, Salto
SP230)	K…	1510	1/0.25	R. Rural, Rinópolis
SP269)	K…	1510	1/0.25	R. Athenas Paulista, Jaboticabal: 0800-0300
BA78)	H530	1520	5/1	R. Povo, Poções
CE75)	H635	1520	1/0.25	R. Araripe, Ipu
CE83)	H653	1520	1/0.25	R. Cachoeira, Solonópole
GO53)	H806	1520	1/0.25	R. Nova RCB, Campos Belos: 0800-0300 Sat:1000-1900 Sun:1500-1700
MA15)	H899	1520	1/0.25	R. Ribamar, Pindaré-Mirim
MA35)	H428	1520	1/0.25	R. Mirante AM, Chapadinha
MG174)	L223	1520	0.25	R. Cultura, Cássia
MG106)	L245	1520	2.5/0.25	R. Clube, Itaúna
MS21)	I405	1520	1/0.25	R. Jornal, Amambaí: 0800-0300 Sat: 24h Sun: 1100-
MS42)	N605	1520	1/0.25	R. Campo Alegre, Rio Verde de Mato Grosso
PE28)	I801	1520	1/0.25	R. Surubim, Surubim
PR116)	J218	1520	2.5/0.25	R. Serra do Mar, Antonina: 0900-2100 Sat:0100 Sun:2300
PR118)	J292	1520	1/0.25	R. Nova Cultura, Palotina
PR132)	J340	1520	1/0.25	R. Internacional, Quedas do Iguaçu
PR156)	J358	1520	1/0.25	R. Guairacá, Terra Rica
RJ44)	J491	1520	10/0.5	R. Continental, Rio de Janeiro: 24h
RJ10)	J499	1520	1/0.25	R. Musical, Cantagalo
RN22)	J610	1520	1/0.25	R. Salinas, Macau
RS141)	K217	1520	1/0.25	R. Vale do Jacui, Cachoeira do Sul
RS177)	K338	1520	1/0.25	R. Cultura, Arvorezinha
SC76)	J806	1520	2.5/0.25	R. Cultura, Timbó: 0800-0200
SE13)	J931	1520	1/0.5	R. Ilha Aracaju, Tobias Barreto
SP191)	K614	1520	10/1	R. Iguatemi, Mogi das Cruzes
SP192)	K627	1520	1/0.25	R. Pinhal Clube, Espírito Sto. do Pinhal
SP225)	K760	1520	1/0.25	R. Manchester, Sorocaba
SP270)	K…	1520	1/0.25	R. Torre Forte, Buritama (Rede Amiga)
T007)	H797	1520	1/0.25	R. Cristal, Cristalândia
BA43)	H479	1530	10/0.5	R. Cultura, Guanambi
BA96)	H…	1530	0.25	Grupo Frajola de Comunicação, Capim Grosso
CE76)	H666	1530	1/0.25	R. Tres Fronteiras, Campos Sales: 0800-0000
CE96)	H…	1530	1/0.25	Rede Sol de Comunicação, Granja
MG175)	L262	1530	0.25	R. Progresso, Monte Santo de Minas
MG110)	L280	1530	1/0.25	R. Clube, Pouso Alegre
MT42)	I432	1530	1/0.25	R. Atual, Peixoto de Azevedo
PE29)	I781	1530	1/0.25	R. Bitury, Belo Jardim
PR157)	J348	1530	1/0.25	R. Vale do Iguaçu, Verê
RJ57)	J482	1530	1/0.25	R. Búzios, Cabo Frio
RJ45)	J502	1530	1/0.25	R. Princesinha do Norte, Miracema
RN16)	J603	1530	1/0.25	R. Curimataú, Nova Cruz
RO16)	J685	1530	5/1	R. Planalto, Vilhena: 0900-0330
RS143)	K235	1530	1/0.25	R. Sulina, Dom Pedrito

MW	Call	kHz	kW	Station, location
RS144)	K300	1530	5/0.25	R. Progresso, São Leopoldo
RS145)	K304	1530	1/0.25	R. Tapejara, Tapejara: 24h Sun:0800-2100
SC77)	J761	1530	1/0.25	R. Dif., Itajaí: 0800-0300 Sun:1000-0200
SC78)	J780	1530	2.5/0.25	R. Difusora, São Joaquim: 0900-0200
SC79)	J796	1530	1/0.25	R. Porto Feliz, Mondaí
SP193)	K677	1530	1/0.25	R. Difusora Digital, Tupi Paulista
SP194)	K699	1530	1/0.25	R. Noticias, Tatuí
SP231)	K755	1530	1/0.25	R. Universal, Teodoro Sampaio
BA79)	H511	1540	0.25	R. Jornal, Souto Soares: 0900-0100
BA89)	H...	1540	0.25	R. Sociedade, Itiruçu
CE42)	H611	1540	1/0.5	R. Sant'Ana, Tianguá
CE77)	H631	1540	1/0.25	R. Sertões, Mombaça
CE78)	H629	1540	1/0.25	R. Aratanha, Pacatuba
ES19)	I206	1540	0.25	R. Agricultura, Santa Teresa
MA36)	H921	1540	1/0.25	R. Santa Maura, Lago da Pedra
MG111)	L217	1540	1/0.25	R. Bomdespachense, Bom Despacho
MG112)	L226	1540	1/0.25	R. Cabana, Conselheiro Lafaiete
MG113)	L293	1540	1/0.25	R. Tropical, Três Corações: 24h
MG168)	L351	1540	10/0.25	R. Difusora, Salinas
MS35)	N601	1540	1/0.5	R. Nova Piravevê, Ivinhema
PA14)	I545	1540	1/0.25	R. Boa Vista, São Sebastião da Boa Vista
PB24)	I694	1540	1/0.5	R. Santa Maria, Monteiro
PE33)	I824	1540	1/0.25	R. Voluntários da Pátria, Ouricuri
PR168)	J306	1540	0.25	R. Litorânea, Guaratuba
RJ54)	J508	1540	1/0.25	R. Clube, Paraíba do Sul: 0900-0300
RN30)	J611	1540	1/0.25	R. Baixa Verde, João Câmara
RS157)	K282	1540	1/0.25	R. Quaraí, Quaraí
SC80)	J803	1540	1/0.25	R. Capinzal, Capinzal
SP245)	K514	1540	1/0.25	R. Cultura, Leme
SP196)	K564	1540	2/0.5	R. Em. Botucatu, Botucatu
SP197)	K723	1540	50/1	R. Nova Difusora, Osasco
SP246)	K737	1540	0.5/0.25	R. Central, Pompéia
BA44)	H461	1540	5/0.5	R. Cidade, Vitória da Conquista: (F.P.I. still on 760)
BA80)	H518	1550	5/0.25	R. Independ. do São Francisco, Juazeiro
MA37)	H926	1550	0.25	Sistema Janaína de Radiodifusão, Vargem Grande
MG114)	L211	1550	1/0.25	R. Cultura, Monte Carmelo
MG169)	L222	1550	1/0.25	R. Difusora, Carmo do Rio Claro
MG115)	L289	1550	1/0.25	R. Difusora Santarritense, Santa Rita do Sapucaí: 24h
MG195)	N211	1550		R. Cidade, Guanhães: 24h
PA34)	I550	1550	1/0.25	R. Cabano, Maracanã
PB32)	I700	1550	10/0.25	R. Jardim, Areia: 0800-0200
PR133)	J213	1550	1/0.25	R. Ipiranga, Palmeira: 24h
PR123)	J303	1550	1/0.25	R. Pioneira, Formosa do Oeste: 0800-2300 Sat/Sun: 0900-2200
PR124)	J315	1550	1/0.25	R. Cristal, Marmeleiro
PR169)	J217	1550	1/0.25	R. Itay, Tibagi
RJ46)	J479	1550	1/0.25	R. Imperial, Petrópolis
RN19)	J606	1550	1/0.25	R. Ivipanim, Areia Branca (RPC)
RO23)	J...	1550	0.25	R. Suprema, Cacoal
RS162)	K377	1550	1/0.25	R. Jornal, Capão do Leão
RS159)	K375	1550	1/0.25	R. Soledad, Soledade
SC92)	J814	1550	5/0.25	R. Imigrantes, Turvo: 0800-0100
SP198)	K501	1550	1/0.25	R. Clube, Itararé
SP259)	K528	1550	1/0.25	R. Tambaú, Tambaú: 24h
SP200)	K572	1550	1/0.25	R. Cacique, Capivari
SP201)	K590	1550	10/1	R. Guarujá AM, Guarujá
SP202)	K659	1550	1/0.25	R. São Joaquim, São Joaquim da Barra
SP219)	K740	1550	1/0.25	R. Nova Difusora, Auriflama
SP281)	K59	1550	1/0.25	R. Lidersom, Orlândia
AL16)	H257	1560	1/0.25	R. Princesa das Matas, Viçosa: 0830-0200 Sat/Sun:24h
BA90)	H526	1560	0.25	R. Povo Pombal, Ribeira do Pombal
CE43)	H622	1560	1/0.25	R. Difusora Vale de Curu, Pentecoste: 0800-0100 Sun:0900-0000
MA18)	H902	1560	1/0.25	R. Agua Branca, Vitorino Freire
MG117)	L256	1560	1/0.25	R. Jornal, Leopoldina
MT56)	I...	1560	1/0.25	R. Paranaíta. Paranaíta
PR125)	J275	1560	1/0.25	R. Capanema, Capanema
PR161)	J361	1560	0.25	R. Cultura Serpin, Ribeirão do Pinhal
PR184)		1560	10/0.25	R. Barigui, Almirante Tamandaré
RJ47)	J501	1560	5/0.25	R. Grande Rio, Itaguaí: 24h
RN17)	J608	1560	5/0.25	R. Cultura do Oeste, Pau dos Ferros
RS158)	K310	1560	2.5/0.25	R. Açoriana, Taquari
RS172)	K369	1560	5/0.25	R. Poata, São José do Ouro
SC95)	J825	1560	1/0.25	R. Cidade, São Miguel d'Oeste
SP261)	K593	1560	1/0.25	R. Show, Igarapava: 24h
SP248)	K679	1560	1/0.25	R. Valparaíso, Valparaíso
SP249)	K725	1560	0.25	R. Regional AM, Pedreira
SP250)	K778	1560	1/0.25	R. Vale do Rio Paraná, Presidente Epitácio
SP203)	K571	1560	5/0.25	R. Em. Campos do Jordão, Campos do Jordão: (F.P.I. still on 1340)
BA56)	H496	1570	1/0.25	R. Educadora, Jaguaquara
BA91)	H533	1570	1/0.25	R. Lider, Central

MW	Call	kHz	kW	Station, location
CE44)	H621	1570	1/0.25	R. Sertão Central,Senador Pompeu
MA22)	H907	1570	10/0.5	R. Cultura do Rio Jordão, Coroatá
MG146)	L242	1570	0.25	R. Universitária, Itajubá
MG141)	L344	1570	1/0.25	R. Cidade, Corinto
MG170)	L364	1570	10/0.25	R. Difusora, Piranga
MS24)	I409	1570	1/0.25	R. Nova Difusora, Caarapó
MS55)	I418	1570	1/0.25	R. Cidade, Aparecida do Taboado
PE30)	I798	1570	5/1	R. Asa Branca, Salgueiro
PR158)	J324	1570	0.25	R. Nova Brasileira, Bela Vista do Paraíso
PR137)	J341	1570	1/0.25	R. Club, Nova Aurora
PR159)	J365	1570	1/0.25	R. Arapoti Popular, Arapoti
PR166)	J209	1570	1/0.25	CBN, Paranaguá
RJ48)	J493	1570	1/0.25	R. Cultura, Valença: 24h
RO19)	J678	1570	5/0.25	R. Soc. Espigão, Espigão d'Oeste
RS147)	K358	1570	1/0.25	R. Metrópole, Cachoeirinha: 24h
SC83)	J777	1570	1/0.25	R. Rio Negrinho, Rio Negrinho
SC98)	J829	1570	1/0.25	R. Modelo, Modelo
SC107)	J...	1570	1/0.25	R. Tangrá, Tangará
SP204)	K552	1570	1/0.25	R. Avaré, Avaré
SP205)	K605	1570	1/0.25	R. Junqueirópolis, Junqueirópolis: 0900-2300 Sat/Sun:0900-2200
SP262)	K648	1570	1/0.25	R. Zequinha de Abreu, Santa Rita do Passa Quatro
SP206)	K651	1570	10/0.25	R. ABC, Santo André: 0800-0000
SP207)	K667	1570	1/0.25	R. Socorro, Socorro
SP208)	K670	1570	1/0.25	R. Clube, Tanabi
TO10)	N665	1570	1/0.25	R. Rio Bonito, Gurupi
BA53)	H497	1580	2.5/0.25	R. BM, Barra do Mendes
BA62)	H502	1580	1/0.25	R. Atalaia, Canavieiras
MG119)	L210	1580	1/0.25	R. Liberdade, Itapecirica
MG121)	L290	1580	1/0.25	R. Cult, Santos Dumont:24h Sun:0900-1500
MG122)	L329	1580	1/0.25	R. Educadora, Espinosa
MG133)	L335	1580	1/0.25	R. Nova Guaranésia, Guaranésia
MS25)	I415	1580	1/0.25	R. Laguna, Jardim: 0800-0200 Sat:0900
MS43)	N611	1580	1/0.25	R. Difusora, Ivinhema
PI18)	I898	1580	1/0.25	R. Santa Clara, Floriano
PR164)	J342	1580	2.5/0.25	R. São João do Sudoeste, São João
RJ51)	J487	1580	1/0.25	R. Popular Fluminense, Conceição de Macabu
RJ52)	J506	1580	5/0.25	R. Resende AM, Resende
RJ58)	J505	1580	0.25	R. Geração 2000, Teresópolis
RN24)	J613	1580	1/0.25	R. Verdade, Ceará Mirim
RS148)	K237	1580	1/0.25	R. Encantado, Encantado
RS150)	K339	1580	1/0.25	R. Dif. Fronteira, Arroio Grande
RS197)		*1580		R. Dif. das Missaõs, Palmeria das Missaõs
SC84)	J818	1580	1/0.25	R. Pomerode, Pomerode
SP251)	K504	1580	1/0.25	R. Difusora, Amparo
SP209)	K743	1580	0.25	R. Pedra Bonita, Itaporanga
BA55)	H...	1590	1/0.25	R. Vale do Jiquiriçá, Jiquiriça
CE80)	H660	1590	1/0.25	R. Montevidéu, Cedro
ES22)	I...	1590	1/0.25	R. Novo Tempo, Cachoeiro de Itapemirim
MG171)	L368	1590	1/0.25	R. Cidade Carinho, Ubá
MG134)	L369	1590	10/1	R. Guaicuí, Várzea da Palma
MG193)	N207	1590	0,25	R. Globo, Lambari
MS26)	I403	1590	1/0.25	R. Independência, Eldorado
PB34)	I703	1590	1/0.25	R. Correio do Vale, Itaporanga
PE40)	I...	1590	1/0.25	R. Dif. Rainha do Céu, Bezerros
PR129)	J290	1590	1/0.25	R. Nova Cultura, Andirá
PR160)	J296	1590	1/0.25	R. Havaí, Capitão Leônidas Marques
RS174)	K212	1590	0.25	R. Clube, Bagé
SC101)	J823	1590	10/0.5	R. Globo, Joinville
SP254)	K774	1590	10/0.5	R. Japi, Cabreúva
BA45)	H464	1600	10/1	R. Vox AM, Muritiba
SP263)	K779	°1600	100/20	R. Nove de Julho, São Paulo: 24h

SW	Call	kHz	kW	Station, location, h. of tr.
SP82)	G852	2380	0.25	R. Educadora, Limeira
AC02)	F204	2460	1	Super R. Alvorada, Rio Branco: 1000-2200
SP98)	G860	3235	0.5	R. Clube, Marília
AC03)	F...	3255	1	R. Educ. 6 de Agosto, Xapuri: 1000-0100
SP112)	G867	3235	2.5	R.Mundial, Guarulhos: irr
SP49)	G855	3365	1	R. Cultura, Araraquara
AM02)	F276	*3375	5	R. Municipal, São Gabriel da Cachoeira: 0900-1300, 2100-0100
RO09)	G792	3375	5	R. Educadora, Guajará Mirim: 0900-1300, 2100-0130
MS01)	F904	4755	10	R. Imaculada Conceição, Campo Grande
MG55)	G207	4775	1	R. Congonhas, Congonhas
RO11)	G790	*4785	10	R. Caiari, Porto Velho: 0900-1400, 1900-0300
SP136)	G857	4785	1	R. Brasil, Campinas
AM20)	F273	4805	10/5	R. Dif. do Amazonas, Manaus: 0930-1330, 1500-1800, 2000-0100
PR13)	G640	4815	10	R. Dif. Londrina
SP80)	G868	4825	10	R. Canção Nova, Cachoeira Paulista
PA12)	G364	4825	5	R. Educadora, Bragança: 0830-0300
SP102)	G869	4845	1	R. Meteorologia Paulista, Ibitinga:

SW	Call	kHz	kW	Station, location
				(relays R. Ternura FM)
AM14)	F278	4845	10	R. Cultura, Manaus: 1000-0200
AC04)	F203	4865	5	R. Verdes Florestas, Cruzeiro do Sul
PR33)	G641	4865	5	R. Alvorada, Londrina
RR01)	G810	4875	10	R. Roraima, Boa Vista: 0800-0300 (Sat/Sun:0900-0230)
GO12)	F692	4885	1	R. Maria, Anápolis
AC06)	F201	4885	5	R. Dif. Acreana, Rio Branco: 0900-0400
PA03)	G362	4885	10	R. Clube do Pará, Belém: 24h
AM13)	F274	*4895	5	R. Baré, Manaus
MS32)	R200	4895	5	R. Novo Tempo, Campo Grande
T004)	R193	4905	1	R. Anhanguera, Araguaína
AP01)	F360	4915	25	R. Dif. Macapá, Macapá: 24h
GO27)	F691	4915	10	R. Daqui, Goiânia
AM10)	F282	4925	5	R. Educação Rural, Tefé: 1000-1400, 2000-0200
ES07)	F641	4935	1	R. Capixaba, Vitória
SP187)	G865	*4975	1	R. Iguatemi, São Paulo
GO30)	F690	4985	10	R. Brasil Central, Goiânia
SP52)	G853	5035	10	R. Aparecida, Aparecida
AM09)	F272	5035	5	R. Educação Rural, Coari: irr 1000-0100
PA16)	G360	5045	10	R. Cultura do Pará, Belém: 24h
AM15)	F274	5055	5	R. Jornal A Crítica, Manaus: irr
MT10)	F901	5055	1	R. Difusora, Cáceres
SC86)		5940		R.Voz Missionario, Camboriú
SP62)	E965	5955	10	R. Gazeta, São Paulo
SP280)	E858	5965	7.5	R. Trans Mundial, Santa Maria
MG06)	E523	5970	10	R. Itatiaia, Belo Horizonte
SC59)	E891	5980	10	R. Guarujá, Florianópolis
DF10)	E773	5990	250	R. Senado, Brasília
RS23)	E852	6000	10	R. Guaíba, Porto Alegre
MG35)	E521	6010	5	R. Inconfidência, Belo Horizonte
RS09)	E850	6020	10	R. Gaúcha, Porto Alegre
PR60)	E726	6060	10	Super Rádio Deus é Amor, Curitiba
RJ18)	E765	6070	7.5	R. Capital, Rio de Janeiro
GO27)	E441	6080	5	R. Daqui, Goiânia
PR15)	E726	6080	10	R. Marumby, Curitiba
SP57)	E956	6090	10	R. Bandeirantes, São Paulo
SP80)	E971	6105	10	R. Canção Nova, Cachoeira Paulista
SP52)	E954	6135	25	R. Aparecida, Aparecida
SP77)	E950	6150	7.5	R. Record, São Paulo
AM11)	E245	6160	10	R. Rio Mar, Manaus: 0900-2100
RS104)	E854	6160	1	R. Boa Vontade (RBV), Porto Alegre
DF06)	E365	6185	250	R. Nal. da Amazônia, Brasília
SP77)	E951	9505	7.5	R. Record, São Paulo
PR15)	E726	9515	10	R. Marumby, Curitiba
SP280)	E858	9530	10	R. Trans Mundial, Santa Maria
RS104)	E855	9550	10	R. Boa Vontade (RBV), Porto Alegre
PR60)	E727	9565	20	Super Rádio Deus é Amor, Curitiba
SP52)	E954	9630	10	R. Aparecida, Aparecida:
SP57)	E957	9645	7.5	R. Bandeirantes, São Paulo
SC86)	E890	9665	10	R. Voz Missionária, Camboruú
SP80)	E971	9675	10	R. Canção Nova, Cachoeira Paulista
SP62)	E963	9685	7.5	R. Gazeta, São Paulo
AM11)	E245	9695	7.5	R. Rio Mar, Manaus: 0900-2100
SP263)	E...	9820	10	R. Nove de Julho, São Paulo: 24h
PR15)	E726	11725	10	R. Marumby, Curitiba
SP280)	E858	11735	50	R. Trans Mundial, Santa Maria
SC86)		11750	1	R. Voz Missionária, Camboriú
PR60)	E726	11765	20	Super Rádio Deus é Amor, Curitiba
DF06)	E365	11780	250	R. Nal. da Amazônia, Brasília
RS23)	E853	11785	7.5	R. Guaíba, Porto Alegre
PR60)	E776	11805	10	Super Rádio Deus é Amor, Rio de Janeiro
GO30)	E440	11815	7.5	R. Brasil Central, Goiânia
GO27)	E441	11830	10	R. Daqui, Goiânia
SP52)	E954	11855	1	R. Aparecida, Aparecida
RS104)	E856	11895	10	R. Boa Vontade (RBV), Porto Alegre
RS09)	E851	11915	10	R. Gaucha, Porto Alegre
SP57)	E958	*11925	10	R. Bandeirantes, São Paulo:
MG35)	E622	15190	5	R.Inconfidência, Belo Horizonte
SP121		26045		R. Cultura, São Paulo - DRM

RADIO NETWORKS

There are several radio networks. Below are listed just some of them. The affiliated outlets are often subject to alteration.

CENTRAL BRASILEIRA DE NOTICIAS – CBN: W: www.radioclick.globo.com/cbn
IGREJA PENTECOSTAL DEUS È AMOR: W: www.ipda.com.br
IGREJA UNIVERSAL DO REINO DE DEUS: W: www.igrejauniversal.org.br
JOVEM PAN Av. Paulista 807, 24º andar, 01311-915 São Paulo, SP **W:** www.jovempan.uol.com.br
RADIO BANDEIRANTES W: www.radiobandeirantes com.br
RADIO GLOBO: W: www.radioclick.globo.com
REDE BOA VONTADE - LBV Legião da Boa Vontade, Rua Doraci 90,

Bairro Bom Retiro, 01134-020 São Paulo, SP
W: www.redeboavontada.com
REDE BOAS NOVAS – RBN: W: www.rbn.org.br
REDE CANÇÃO NOVA DE RÁDIO Rua João Paulo II s/, Alto da Bela Vista, 12630-000 Cachoeira Paulista, SP **W:** www.cancaonova.com **E:** radio@cancaonova.com
REDE CATÓLICA DE RÁDIO – RCR: União de Radiodifusão Católica, Rua Vergueiro 3086, Conj. 91, Vila Mariana, 04102-001 São Paulo, SP **W:** www.rcrunda.com.br **E:** rcr@rcrunda.com.br
REDE DO ESTADO DE SÃO PAULO: W: www.redecbs.com.br
REDE GAÚCHA SAT Av. Erico Veríssimo 400, Edifício Maurício Sirotsky Sobrinho, 90169-900 Porto Alegre, RS **W:** www.rbs.clicrbs.com.br
REDE ESPERANÇA: W: www.redeesperança
REDE ITATIAIA: W: www.itatiaia.com.br/rede
REDE MILICIA SAT: W: www.milicia.org.br
REDE MINERIA DE RADIO; W: www.redemineriaderadio.com.br
REDE NOVO TEMPO: W: www. novotempo.org.br
REDE PAULUS SAT: Rua Doutor Pinto Ferraz 183, Vila Mariana, 04117-900 São Paulo, SP **W:** www.radioamericasp.com.br/paulussat.htm
REDE POTGUAR DE COMUNICAÇÃO (RPC): W: www.redepotiguar.com
REDE SUL DE RÁDIO: W: www.saofrancisco.am.br
REDE TRANSMARICA: W: www.transanet.uol.com.br
SISTEMA GLOBO DE RADIO: W: www.radioclick.globo.com/globobrasil
SISTEMA GUAÍBA SAT: Rua Caldas Jr. 219, 2º andar, 90019-900 Porto Alegre, RS **W:** www.guaiba.com.br
REDE SOMZOOM SAT: Av. Herois do Acre 590, Passaré, 60743-760 Fortaleza, CE **W:** www.somzoom.com.br **E:** somzoomsat@somzoom.com.br

Addresses and other information

AC00) ACRE
AC01) Rua de Alagoas, 270 - Colégio, 69980-000 Cruzeiro do Su ☎68 3322 4637 I **E:** radiointegracao@hotmail.com – **FM:** 99.9MHz – **AC02)** Av Ceará, 2150 Jardim Nazle, 69900-460 Rio Branco **E:** seve@jornalatribuna.com.br ☎68 3226 2660 – **AC03)** Rua Coronel Brandão, 1665 - Bairro Aeroporto, 69930-000 Xapuri ☎68 3542 2830 **E:** raimari.cardoso@hotmail.com – **AC04)** Travessa Mário Lobão 81, 69980-000 Cruzeiro do Sul ☎68 3322 3309 **E:** verdesflorestas@uol.com.br – **AC05)** Rua Epaminondas Jacome, 3121 - Base, 69908-420 Rio Branco ☎68 3224 2380 **W:** www.radiocapitalacre.com.br – **AC06)** Rua Benjamin Constant 1282, 69900-161 Rio Branco ☎68 3223 9696 **E:** comercial.difusora@ac.gov.br or producao.difusora@ac.gov.br – **AC07)** Av Castelo Branco 329, 69925-000 Senador Guiomard – **AC08)** Rua Genni Assis s/n, 69932-000 Brasiléia – **AC10)** Governo do Estado doAcre, 69960-000 Fejo – **AC11)** Av Brasil, 1800 – Jorge Alves Jr., 69940-000 Sena Madureira ☎68 3612 2626 **E:** rivaldosevero@hotmail.com - **FM:** 105.9MHz – **AC12)** Rua Nilo Freire de Albuquerge, Lotamente SEHAB Lotes 1 2 11 E 12 QD 16, 69970-000 Tarauacá **E:** railtonrodrigues@ac.gov.br – **AC14)** 69900-000 Rio Branco – **AC15)** Travasse Diamantino Macedo s/n, 69960-000 Feijo **E:** jocivaldogomes@bol.com.br

AL00) ALAGOAS
AL01) Av Coronel Wilson Santa Cruz 6, 57314-000 Arapiraca ☎82 3521 0570 **W:** www.novonordeste.com **E:** am@novonordeste.com – **AL02)** Via Expressa 4360, Serraria, 57080-000 Maceió – **AL03)** Rua José Maria Passos 25, 57600-030 Palmeiras dos Indios - **FM:** 92.5MHz – **AL04)** Rua Barão José Miguel, 400 - Farol, 57055-160 Maceió ☎82-315 1960 **W:** www.radiodifusora-al.com.br **E:** difusora@radiodifusora-al.com.br – **AL05)** Rua Miguel Palmeira 1513, 7º andar, Farol, 57055-330 Maceió – **AL06)** Praça Senador Eneas Araújo 61, 57500-000 Santana do Ipanema – **AL07)** Rua Saldanha da Gama s/n, Farol, 57051-580 Maceió ☎82 4009 7070 ▤82 4009 7719 **W:** http://gazetaWglobo.com/v2/radiogazeta/ 24h - **FM:** 94.1MHz – **AL08)** Rua Barão de Penedo 258, 57020-340 Maceió – **AL09)** C.P 6, 57201-970 Penedo– **AL10)** Quadra A lote 04, 57160-000 Marechal Deodoro – **AL11)** Rua Porcos s/n, 57300-000 Arapiraca – **AL12)** Praça Manoel Monteiro 72, 57480-000 Delmiro Gouveia - 82 3641 4047 82 3641 4061 **W:** www.radiodelmiro.com.br - **FM:** 89.9MHz – **AL13)** BR-104 Km 36, Bairro Roberto Correia de Arajuó, 57800-000 União dos Palmares – **AL14)** Rua Aldeir Lima Peixoto 123, 3º andar, Farol, 57051-110 Maceió ☎82 4009 0009 **W:** www.radiocorreioam1200.com **E:** ouvinte@radiocorreioam1200.com – **AL15)** Av Braulio Cavalcante 415, 57400-000 Pão de Açucar ☎82 3624 1238 – **AL16)** Rua Mota Lima, 19 – Centro, 57700-000 Viçosa ☎82 3283 1842 **W:** www.princesadasmatas.com – **AL17)** 57100-000 Maceió

AM00) AMAZONAS
AM02) Av Alvaro Maia s/n, 69750-000 São Gabriel da Cachoeira **E:** rmunicipalsgc@yahoo.com.br ☎97 3471 1768 – **AM03)** Rua Júlio de Oliveira, 1323 - Centro, 69800-000 Humaitá ☎97 3373 1600 **W:** www.portalvrm.blogspot.com – **AM04)** A/C Prefeitura Municipal de Tabatinga (⊠C.P 31), 69640-000 Tabatinga ☎97 3412 4078 **W:**www.ebc.com.br/ebc/canais/radios/radio-nacional-do-alto-solimoes **E:** lana.micol@radiobras.gov.br – **FM:** 96.1 MHz – **AM05)** Rua Solimões 809, 69100-000

Itacoatiara ☎92 📠92 3521 1237 **E:** radiodifusora_ita@hotmail.com - **FM:** 94.5MHz – **AM06)** Rua Joana D'Angelo s/n, 69400-000 Manacapuru ☎92 3361 2042 **E:** gadelha.silva@redeamazonica.com.br – **AM07)** Av Major Santana 2502, 69280-000 Manicoré – **AM08)** Estrada dos Morais 1455, 69190-000 Maués ☎92 3542 2264 **E:** radioguaranopolis@hotmail.com – **AM09)** Praça São Sebastião 228, 69460-000 Coari ☎97 3561 2474 **E:** radiocoari@hotmail.com – **AM10)** Rua Benjamin Constant 283 (✉ C.P 21), 69470-000 Tefé ☎97 3343 3017 📠97 3343 2663 **E:** rert@osite.com.br – **AM11)** Rua José Clemente, 500 - Centro, 69010-070 Manaus ☎92 3633 2295 📠92 3232 7763 **W:** www.riomaronline.com.br **E:** dianavieiramoreno@gmail.com – **AM12)** Rua Governador Leopoldo Neves 516, 69151-460 Parintins ☎92 3533 2002 📠92 3533 2004 **E:** sistemaalvorada@jurupari.com.br – **FM:** 100.1MHz – **AM13)** Av Tefe, 3025 - Japiim, 69078-000 Manaus ☎92 2101 5500 **W:** www.radiobare.com.br – **AM14)** Rua Barcelos, s/n Praça 14, 69020-200 Manaus ☎92 2101 4967 📠92 2101 4950 **E:** radiocultura@tvcultura.am.gov.br or radiocultura@hotmail.com – **AM15)** Av Andre Araujo, 1924 A Aleixo, 69060-001 Manaus ☎92 2123 1097 **E:** walteryallas@acritica.com.br – **FM:** 93.1MHz **AM16)** Av General Rodrigo Jordão Ramos, 1655 Anexo 3 - Japiim, 69077-000 Manaus ☎92 3671 3000 **W:** www.rbn.org.br **E:** franklincosta@gmail.com - **FM:** 100MHz – **AM17)** Av Amazonas 1958, 69151-000 Parintins ☎92 3533 1564 📠92 3533 2456 – **E:** rcparintins@jurupari.com.br or tadeudesouza@hotmail.com – **AM18)** Rodovia Manoel Urbano, km 2, 69405-000 Iranduba **W:** www.redeboavonde.com – **AM19)** Boulevard Pedro Rate, 176 - São José, 69400-000 Manacapuru.☎92 3361 2192 📠92 3361 2453 – **AM20)** Av Eduardo Ribeiro 639, Ed Palácio do Comércio, 20° andar, Centro, 69010-001 Manaus ☎92 3633 1001 **W:** www.difusoramanaus.com.br **E:** difusora@internext.com.br - **FM:** 96.9MHz – **AM21)** Estrada do Gavião, km 05 andar, 69500-000 Carauari – **AM23)** Praça Coronel João Vercosa, 47 - Centro, 69190-000 Maués ☎92 3542 1897 – **AM24)** Av Leopoldo Nelves 360, 69850-000 Bôca do Acre – **AM25)** Rua Anori/Anama 327, 69440-000 Anori – **AM26)** Praca Eduardo Ribeiro 2058, 69151-271 Parintins – **AM27)** 69670-000 Fonte Boa – **AM28)** 69140-000 Nhamundã – **AM29)** 69200-000 Borba – **AM30)** 69630-000 Benjamin Constant – **AM31)** 69750-000 Saó Gabriel da Cachoeira – **AM32)** 69680-000 Santo Antonio do Iça

AP00) AMAPÁ
AP01) Rua Cândido Mendes, 525 – Centro, 68900-100 Macapá ☎96 3131 2716 **W:** www.difusora.ap.gov.br **E:** rdm@rdm.ap.gov.br – **AP02)** Rua Eliezer Levy, 684 -Trem, 68901-090 Macapá ☎96 3222 3111 **E:** equatorialfm@uol.com.br – **FM:** 94.5MHz – **AP03)** Av Nações Unidas 256, 68906-100 Macapá – **AP04)** 68925-000 Santana – **AP05)** 68930-000 Mazagão – **AP06)** Oiapoque.

BA00) BAHIA
BA01) Praça Mario Dourado 78-A, 44900-000 Irecê ☎74 3641 3717 **W:** www.regionalam.com.br **Emai:** contato@regionalam.com.br – **BA02)** Av. Itajuipe No 1789, Barro Santo Antonio, , 45602-380 Itabuna. ☎73 3211 2385 **W:** www.radiojornaldeitabuna.com – **BA03)** Praça da Independência 244, 45995-000 Teixeira de Freitas – **BA04)** Rua Lima e Silva, N° 214 – Edif Dois Corações, Liberdade, 40325-281 Salvador ☎71 3326 3603 **W:** www.radiocruzeiro.com.br – **BA05)** Rua Luis Augusto Fernandes Borges 306, 47800-000 Barreiras ☎77 3611 3570 **W:** www.radiovale.com.br – **BA06)** Av Aurelino Ribeirao Noaves s/n, Graça, 45400-000 Valença ☎75 5575 3641 **W:** www.radioclubedevalenca.com.br **E:** clube@radioclubedevalenca.com.br – **BA07)** Praça Duque Caxias 3, 45700-000 Itapetinga– **BA08)** Rua Marieta Martins, 336 - São Benedito, 44573-250 Santo Antônio de Jesus ☎75 3631 5680 **W:** www.radioclube680.com.br **E:** contato@radioclube680.com.br – **BA10)** Rua Germiniano Costa 47, 44025-070 Feira de Santana – **BA11)** Rua Jardim Federaão 81, Federação, 40231-060 Salvador , ☎71 3486 3201 📠71 3486 3214 **W:** www.radiosociedadeam.com.br **E:** comercial@radiosociedadeam.com.br – **BA12)** C.P 19, 45601-970 Itabuna – **BA13)** Marechal Deodoro 639, 47800-000 Barreiras ☎77 3611 4545 **W:** www.rb.am.br – **BA14)** Praça Luiz Nogueira, 48700-000 Serrinha – **BA15)** Rua Antonio Neto No 27, 46810-000 Utinga ☎75 3337 1011 **W:** www.radiocultura820.com.br **E:** comercial@radiocultura820.com.br – **BA16)** Fundação Dom Avelar Brandão, Rua Martin Afonso de Souza, 270 - Garcia, 40100-050 Salvador ☎71 3328 5088 **W:** www.excelsiordabahia.com.br **E:** comercial@am840.com.br – **BA17)** Av Visconde do Rio Branco. 68 - Centro, 48970-000 Senhor do Bonfim ☎74 3541 4617 **W:** www.radiocaraiba.com.br – **BA18)** Travessa da Catedral s/n, 45600-000 Itabuna ☎73 3215 0909 **W:** www.radionacionalitabuna.com.br **E:** contato@radionacionalitabuna.com.br – **BA19)** Rua Wercelêncio da Mota 81, Centro, 48730-000 Conceição do Coité ☎75 3262 1010 **W:** www.radiosisal.com – **BA20)** Rua Frei Hermenegildo, 300 - Capuchinos, 44050-240 Feira de Santana 75 2101 9700 **W:** www.sociedadedefeiraam.com.br- **FM:** 96.9MHz – **BA21)** Gleba Fazenda Ouro Verde, 45900-000 Caravelas ☎73 3011 1299 **W:** www.radioalvoradagospel.com.br **E:** diretoria@radioalvoradagospel.com.br – **BA22)** Rua Gabriel Soares 23, 40060-040 Salvador – **BA23)** Praça Barão do Rio Branco 42, 45100-

000 Vitória da Conquista – **BA24)** Av Maria Quitéria, 223 - Mar de Tranquilidad, 44062-630 Feira de Santana 75 3623 1080 75 3623 2851 **W:** www.radiosubaeam.com.br **E:** radio@radiosubaeam.com.br - **FM:** 95.3MHz Nordeste FM – **BA25)** Rua José Bonifacio No 17 2° andar, Centro , 45700-000 Itapetinga ☎77 3261 2610 **W:** www.radiofascinacao.com.br **E:** comercial@radiofascinacao.com.br – **BA26)** Rua Marquês de Paranaguá 259, 45660-000 Ilhéus ☎73 3231 3612 **W:** www.santacruzam.com.br **E:** contato@santacruzam.com.br – **BA27)** C.P 29, 45825-000 Eunápolis – **BA28)** Rua Aprigo Duarte 4, 48900-000 Juazeiro **W:** www.radiojuazeiro.com.br **E:** geraldojoseba@hotmail.com – **BA29)** Rua Senador Pedro Lago, 54 - Centro, 44700-000 Jacobina ☎74 3621 9150 **W:** www.radiocluberiodoouro.com.br **E:** radio@radiocluberiodoouro.com.br – **BA30)** Rua Monte Castelo,45, Sobradinha, 44018-210 Feira de Santana ☎75 3221 8815 **W:** www.radiopovo.com.br **E:** cariocapovo@veloxmail.com.br – **BA31)** Rua Dom Pedro II 98, 48100-000 Alagoinhas **W:** www.novaam1240.com.br **E:** oliveirafm@oliveirafm.com – **BA32)** Rua Cde Pereire Carneiro, 226 -Pernambúes,41100-010 Salvador ☎71 3505 5000 📠71 3505 5040 **W:** www.radiometropole.com.br - **FM:** 101.3MHz – **BA33)** Av. Itabuna, 63 – Centro,, 45653-160 Ilhéus ☎73 3231 5462 **W:** www.novabahianaam.com.br **E:** radionovabahianadeilheus@bol.com.br or clinton.alves@hotmail.com – **BA34)** Praça Luiz Nogueira 385, 48700-000 Serrinha **W:** www.continentalam.com.br **E:** continental@continentalam.com.br – **BA35)** Rua São Francisco 159 - 163A, 48600-000 Paulo Afonso ☎73 3281 1266 **W:** www.redecultura.com.br **E:** redecultura@redecultura.com.br - **FM:** 92.7MHz – **BA36)** Praça da Bandeira s/n, Centro, 47600-000 Bom Jesus da Lapa ☎77 3481 5179 **W:** www.radiobomjesusam.com.br – **BA37)** Av Getúlio Vargas 394, 44330-000 São Gonçalo dos Campos – ☎77 3481 6161 **W:** www.radiobomjesusam.com.br – **BA38)** Rua Coronel Francisco Pinto, 44200-000 Santo Amaro ☎75 3241 4665 📠75 3241 1602 **W:** www.radioindependenciabahia.com.br – **BA39)** Rua 2 de Julho 20, 45200-000 Jequié ☎73 3527 4114 **W:** www.radiopovo.com.br **E:** radiopovojequie@gmail.com – **BA40)** Rua Sítio Escurinha s/n, BR-242, km 90, 46880-000 Itaberaba – **BA41)** Praça Virgilio Damasio, 140b 1° andar – Centro , 45570-000 Ipiaú ☎73 3531 3441 📠73 3531 3419 **W:** www.radioeducadoradeipiau.com.br **E:** contato@radioeducadoradeipiau.com.br – **BA42)** Av. Dom Manuel Raimundo de Mello, 607 - Bairro São José, 46400-000 Caetité ☎77 3454 1819 **W:** www.educadorasantana.com **E:** educadora@educadorasantana.com.br – **BA43)** Rua Otavio Mangabeira 1026, Bela Vista, 46430-000 Guanambi ☎77 3451 1348 **W:** www.radioculturagbi.com.br **E:** radiocutura@micks.com.br – **BA44)** Av Ascendino Melo 297, 267 – Sis 106/107, Shopping Itatiaia, Recreio, 45020-908 Vitória da Conquista ☎77 3472 0760 **W:** www.radiocidadeconquista.com.br **E:** radiocidade@clubenet.com.br – **BA45)** Tv Virgillo Gonzalves Pereira 196, Centro, 44340-000 Muritiba ☎73 3424 2048 – **BA46)** Av Porto Seguro 718, 1° andar, Centro, 45820-006 Eunápolis. ☎73 3281 5594 **W:** www.novaradiojacaranda.com **E:** jacaranda@euinanet.com.br – **BA47)** Rua da Bandeira 27, 42800-000 Camaçari **W:** www.cbnsalvador.com.br – **BA49)** Travessa do Contorno 26, 45570-000 Ipiaú – **BA49)** Rua Padre Argemiro Guimarães, 32 - Centro, 44640-000 Riachão do Jacuípe ☎75 3264 2189 **W:** www.radiojacuipeam.com.br **E:** radiojacuipe@yahoo.com.br – **BA50)** Av Getúlio Vargas 43, 48601-000 Paulo Afonso 75 3281 3009 **W:** www.radiobahianordeste.com.br **E:** rbm@radiobahianordeste.com.br – **BA51)** Rua Rio Corrente s/n, 47640-000 Santa Maria da Vitória – **BA52)** Rua Saldanha Marinho 30 Sala 23/24, Mesmo, 45810-000 Porto Seguro ☎73 3288 2136 – **BA53)** Rua Alvaro Campos 83, 44990-000 Barra do Mendes **W:** www.rbm.am.br **E:** rbm@holistica.com.br – **BA54)** Av Dom Avelar Brandão Vilella s/n, Sítio São Félix, 46470-970 Riacho de Santana **W:** www.micks.com.br/guadalupe **E:** guadalupe@micks.com.br – **BA55)** Rua Coronel Vicente s/n, 45470-000 Jiquiriça – **BA56)** Praça Guilherme Silva 85, 1° andar, 45345-000 Jaguaquara ☎73 3534 1422 📠73 3534 2220 **W:** www.radiopovo.com.br – **BA57)** Rua Gamboa de Cima, 18 - Campo Grande, 44000-008 Salvador **W:** www.novotemposalvador.com.br **E:** novotemposalvador@terre.com.br – **BA58)** Av Regis Pacheco 534, Centro, 45100-000 Vitória da Conquista **W:** www.cancaonova.com **E:** radioconquista@cancaonova.com – **BA59)** Rua Gonçalo Martins 19, 45550-000 Ubatã ☎73 3245 1233 **E:** povoam@gmail.com – **BA60)** Rua Antônio Otaviano Dourado 91, 44900-000 Irecê ☎74 3641 3111 – **BA61)** C.P 45, 46430-000 Guanambi ☎77 3451 1596 **W:** www.radioalvoradaam.com.br **E:** alvorada@alvoradaam.com.br – **BA62)** Rua General Pederneiras 62, 45860-000 Canavieiras – **BA63)** Rua Margem do Rio do Ouro 115, 44700-000 Jacobina **W:** www.radiojaraguar.com.br – **BA64)** Rua Farias Goes 164, 48330-000 Rio Real – **BA65)** Rua Otavio Mangabeira 13, 48500-000 Euclides da Cunha – **BA66)** Rua Rui Barbosa 119, 47400-000 Xique-Xique – **BA67)** Rua Jose de Anchita, 128 - 2° andar - Centro, 45836-000 Itamaraju. ☎73 3294 5455 **W:** www.extremosulam.com – **BA68)** Av José Candido dos Santos 20, Centro, Lagos Redonda, 49300-000 Itapicuru ☎79 3541 1067 **W:** www.rci1060am.com.br – **BA69)** Rua Frei Apolônio de Todi 10, 48410-000 Cícero Dantas **E:** regional@fallnet.com.br – **BA70)** Terreiro de Jesus 13, Centro Histórico, 40025-010 Salvador **W:** www.radio.boavontade.com/ba

– **BA71**) Rua Vicente Paula Costa 16, 48540-000 Jeremoabo – **BA72**) Rua 2 de Julho s/n, Centro, 45160-000 Belo Campo ☎☎773437 2233 – **BA73**) Praça São José 279 44600-000 Ipirá. – **BA74**) Rua Desidério Brandão, 15 – Centro, 44380-000 Cruz das Almas ☎☎75 3621 2716 **W:** www.radioalvoradaam1460.com.br **E:** alvoradaamcomercial@hotmail. com – **BA75**) Rua Teixeira de Freitas s/n, 47700-000 Santana – **BA76**) 47400-000 Xique-Xique – **BA77**) Rua Barão do Rio Branco s/n, 47600-000 Bom Jesus da Lapa – **BA78**) Rua Dulce Pazzi 6, Alto da Bela Vista, 45260-000 Poções ☎77 3431 1848 **W:** www.radiopovo.com.br **E:** povopocoes@hotmail.com – **BA79**) Rua Idalina Pinto 169, 46990-000 Souto Soares ☎75 3339 2328 **W:** www.radiojornal1540.com.br **E:** comercial@ radiojornal1540,com.br – **BA80**) Rua José Inácio 31, 48700-000 Juazeiro – **BA81**) Rua Carlos Gomes 980, Centro, 40285-280 Salvador ☎71 3329 6128 **W:** www.cultura1140am.com.br **E:** radio@ cultura1140am.com.br – **BA82**) Rua Desembargador Salvio Martins 321, 48800-000 Monte Santo – **BA83**) Parque Emilia Costa s/n, 45450-000 Gandu – **BA84**) Praça da Bandeira 47, 3° andar, Centro, 48900-000 Juazeiro ☎74 3611 5533 **W:** www.radiocidadeam870.com.br **E:** contato@radiocidadeam870. com.br – **BA85**) Rua Coronel Dias Coelho 249, 44850-000 Morro do Chapéu – **BA86**) Travessa Juracy Magalhães 4, 2° andar, 44630-000 Mairi **W:** www.radiomorroverde.com.br **E:** adm@radiomorroverde.com. br – **BA87**) Rua Dr Guerra 91, 47650-000 Correntina **W:** www.radioplanaltodooeste.com.br **E:** contato@radioplanaltodooeste.com.br – **BA88**) Rua Espírito Santo s/n, 48400-000 Ribeira do Pombal – **BA89**) Rua João Brandão 233, Centro, 45350-000 Itiruçu – **BA90**) Praça Getúlio Vargas 211, 48400-000 Ribeira do Pombal ☎75 3276 1164 **W:**www.radiopovopombal.com.br **E:** educadorapombalgerencia@hotmail.com – **BA91**) Rua do Comércio 31, 44940-000 Central – **BA94**) 45600-000 Itabuna – **BA96**) Rua Juana Angélica 125, Centro, 45653-640 Iléhus ☎73 3634 6940 **W:** www.radioculturadeilheus.com.br – **BA97**) 46100-000 Brumado **BA95**) 47800-000 Barreiras – **BA96**) Rua Joana Angélica 125, Centro, 45653-640 Ilhéhus ☎73 3634 6940 📠73 3634 2465 **W:** www.radioculturadeilheus.com.br

CE00) CEARÁ

CE01) Rua Romeu Martins, Centro S/N, Ed 29 de Julho, 62700-000 Canindé ☎85 3343 2233 📠85 3343 1602 **W:** www.radiojornal540.com **E:** admin@radiojornal540.com – **CE02**) Rua São Pedro 918, 63010-010 Juazeiro do Norte **W:** www.verdadocairi.com.br/radio.php **E:** raimundodantas10@yahoo.com.br – **CE03**) Av Monsenhor Tabosa, 2514 – Bairro das Madalenas, 62500-000 Itapipoca ☎88 3631 2173 📠88 3631 0469 **W:** www.radiouirapurudeitapipoca.com.br **E:** ribamar@bol.com.br – **CE04**) Rua Carlos Rolim/Praça da Matriz, 63700-000 Crateús – **CE05**) Av.Rui Barbosa, 1901 - Aldeota, 60115-221 Fortaleza **W:** www.lamb.com. br– **CE06**) Praça da Matriz s/n, 62590-000 Itarema – **CE07**) Rua Dep. Luiz Otacílio Correia, 221 - Centro, 63540-000 Várzea Alegre ☎88 3541 1055 📠88 3541 1072 **W:** www.radiocultura670.com **E:** radiocultura@radiocultura670.com – **CE08**) Shalom da Paz, Rua Maria Tomásia, 72 – Aldeota, 60150-170 Fortaleza ☎85 3261 4444 **W:** www.comshalom.com/radio/ **E:** benfeitordapaz@comshalom.org – **CE09**) Rua Agronomando Rangel 475, 63870-000 Boa Viagem **W:** www.radioasabranca.com.br **E:** asabranca@ radioasabranca.com.br ☎88 3427 1104 📠88 3427 1456 – **CE10**) Rua Marcondes Pereira, 426-Joaquim Távora, 60130-060 Fortaleza ☎88 3272 3733 📠88 3272 3749 – **CE11**) Rua Hilda Augusto 201, 63300-000 Lavras da Mangabeira ☎88 3536 1257 **W:** www.radiovaledosalgado. com.br **E:** contato@ radiovaledosalgado.com.br – **CE12**) Rua Floriano Peixoto 351, Centro, 63500-000 Iguatu ☎88 3581 1402 📠88 3581 0828 **W:** www.jornalam.com **E:** contato@jornalam.com.br – **CE13**) Praça de Imprensa s/n (C.P 851, 60001-970), 60135-900 Fortaleza 85 3266 9776 **W:** www.verdesmares.com.br – **CE14**) Praça Vicente Aguilar 16, Centro, 62400-000 Camocim ☎88 3621 1395 **W:** www.deputadosergioaguiar. com.br – **CE15**) Rua São Luís 68, 62040-450 Juazeiro do Norte ☎88 3512 3581 📠88 3511 5387 **W:** www.radioiracema.hpg.ig.com.br **E:** radioiracema@ig.com.br – **CE16**) Av Senador Virgilio Tavoar 2279, 60170-251 Fortaleza **W:** www.cidadeam860.com.br **Email** comercial860@yahoo. com.br – **CE17**) Rua Floriano Peixoto 358, 63500-000 Iguatu **W:** www.radioliberdadeam.com **E:** contato@radioliberdadeam.com – **CE18**) Rua Totonho Figueiras 244, 63180-000 Barbalha **W:** www.radiocetama.com. br – **CE19**) Praça Quirino Rodrigues 76/3, 62011- 280 Sobral. 88 3611 1550 **W:** www.educadora950.com **E:** radioeducadora@sobral.org – **CE20**) Rua Tabelião Enéas, 495 - 2° Andar – Centro, (✉ C.P 87, 63901-970) 63900-000 Quixadá ☎88 3414 5970 📠88 3412 0554 **W:** www.sistemamonolitos.com.br **E:**contato@sistemamonolitos.com.br– **FM:** 105.9 MHz – **CE21**) Av Aguanambi 282, 60055-402 Fortaleza ☎85 3264 3373 **W:** www.noolhar.com.br/amdopovo **E:** diretoria@radios.opovo.com.br – **CE22**) Rua Coronel Antônio Luiz 1068, Bairro do Pimenta, 63100-000 Crato – **CE23**) Rua Conselheiro José Júlio 126, 62010-820 Sobral **W:** www.radiotupinamba.com – **CE24**) Rua Raul Vieira 562, 62900-000 Russas – **CE25**) Rua Conego Eduardo Araripe, 1692 – Centro, 62870-000 Pacajus ☎85 3348 0725 **W:** www.comshalom.org/radio **E:** boanova1410@yahoo.com.br– **CE26**) Av Senador Virgilio Távora 2279, 60170-251 Fortaleza. ☎85 3264 2944 **W:** www.clube.am **E:** acavalcanti@

baydenet.com.br – **CE27**) Rua Coronel Zezé 1158, 63700-000 Crateús – **CE28**) Rua Luis Vicente Ferreira Lima 222, Bairro Populares, 62930-000 Limoeiro do Norte – ☎88 3423 4100 📠88 3423 2440 **W:** www.radiovaledojaguaribe.com.br**E:** comercial@radiovale.com – **CE29**) Av Santos Dumont 1687, Aldeota, 60150-160 Fortaleza – **CE30**) Rua São Francisco 374, 63010-210 Juazeiro do Norte ☎88 3511 2404 **W:** www.radioprogressoam.com.br – **CE31**) Rua Cel Joaquim Ribeiro , 405 s 2 - Centro, 62011-020 Sobral ☎88 3611 7888 – **CE32**) Av Dr Pedro de Queiroz Ferreira 2129, 62850-000 Cascavel – **CE33**) Rua São Francisco 139, 63100-000 Crato – **CE34**) Rua Monsenhor Salviano Pinto 71, 63800-000 Quixeramobim ☎88 3441 0014 **W:** contato@difusoracristal.com.br **E:** contato@difusoracristal.com.br – **CE35**) Praça Pinto Martins 260, 62400-000 Camocim – **CE36**) Av Manoel Castro 815, 62940-000 Morada Nova – **CE37**) Rua Moacir Pereira Gondim 333, 63650-000 Tauá – **CE38**) Rua Coronel Alexanzito 835, 62800-000 Aracati – **CE39**) Rua Hildo Furtado s/n, 62760-000 Baturité – **CE40**) Rua Italiano Júlio Filizola 551, 62370-000 São Benedito ☎88 3626 2142 **W:** www.novaplan.am.br – **CE41**) Rua Coronel Antônio Joaquim 2143, 62930-000 Limoeiro do Norte ☎88 3423 4225 **W:** www.radioeducadora.com.br **E:** educadora560@yahoo. com.br – **CE42**) Rua Maestro Quincas Bezerril s/n, 62320-000 Tianguá – **CE43**) Rua João Verçosa s/n, 62640-000 Pentecoste ☎85 3352 2554 **W:** www.radiovaledocuru.com – **CE44**) Rua Santos Dumont, 414 - Centro, 63600-000 Senador Pompeu ☎88 3449 0206 **W:** www.radiosertaocentralam.com.br **E:** radiocertaocentral@hotmail.com – **CE45**) Praça Adolfo Caminha 247, 62800-000 Aracati – **CE46**) Maria de Lourdes 545, 62280-000 Santa Quitéria – **CE47**) São João Evangeliste, 655 Campo de Aviação , 62680-000 Paracuru ☎58 3269 1469 **W:** www.deusespirito. org **E:** pr.francisco.oliveira@hotmail.com – **CE48**) Rua Justino Alves Feitosa, 364 - centro, 63380-000 Barro ☎88 3554 1166 **W:**www. rbe1210.com **E:** radioboaesperance1210am@hotmail.com – **CE49**) Rua Simão Barbosa,1290 - Centro, 62700-000 Canindé – **E:** rsf@sntralnet,com. br 📠85 33430 0574 **W:** www.radiosaofrancisco.com.br**E:**rsf@sntralnet,com. br – **CE50**) Rua Coronel Lucio 489, 63700-000 Crateús ☎88 811 0060 – **CE51**) Rua Monsenhor Salviano Pinto 507, 63800-000 Quixeramobim **E:** campo@disovernet.com.br – **CE52**) Rua Juaci Sampaio Pontes 695 – salas 25/56 2° andar, 61600-150 Caucaia ☎85 3342 1230 **W:** www. lamb.com.br **E:** radiometropolitana@fortalnet.com.br – **CE53**) Av da Ressurreiçao, 926 - Bairro Pe Ibiapina, 62000-000 Sobral ☎88 3611 3349 **W:** www.rsobral.com.br **E:** radioressurreicao@oi.com.br – **CE54**) Rua Tibúrcio Targino 155, 61700-000 Aquiraz – **CE55**) Rua Dr Almir Farias 446, 62200-000 Nova Russas. **W:** www.rd780.com.br – **CE56**) Rua Raimundo Nonato 81 – Centro, 62940-000 Morada Nova ☎88 3422 2561 **W:** www.portalliberal.com.br – **CE57**) Rua Dr Chagas Pinto, 3510 - Centro, 62250-000 Ipu ☎88 3683 2186 **W:** www.radioiracemadeipu. com.br **E:** radioiracemaipu@hotmail.com – **CE58**) Rua Francisco Gomes de Souza, 198 - Centro, 63150-000 Campos Sales. ☎88 3533 1188 **W:** www.cidadeam630.com.br **MSN:** r.cidade@hotmail.com – **CE59**) Rua Cazuzinha Marques, 87 - Centro, 63560-000 Acopiara ☎88 3565 0063 **W:** www.radiovaleacopiara.am.br **E:** contato@radiovaleacopiara.am.br – **CE60**) Rua Manoel Inacio de Lucena, 249 s 2 - Centro, 63260-000 Brejo Santo ☎88 3531 1093 **W:** www.radiosulcearense.com.br – **CE61**) Av. Mons Aloísio Pinto, 100 Dom Expedito, 62050-999 Sobral ☎88 3614 4043 **W:** www.radiocaicara.com **E:** radio@radiocaicara.com – **CE62**) Av. Senador Esmerino Arruda s/n , 62430-000 Granja ☎88 3624 1106 **W:** http://radiovaleam.com.br **E:**contato@radiovaleam.com.br – **CE63**) Rua Antônio Queiroz, 343 - Centro, 63870-000 Boa Viagem ☎88 3427 1064 **W:** www.amliberdade.com.br **E:** radioliberdadebv@yahoo.combr – **CE64**) Rua Major Barreto, 3000 - Centro, 62600-000 Itapajé ☎85 9104 5447 **W:** www.radioguanaces.com.br **E:** kekpubli@hotmail.com – **CE65**) Loteamento Pioneiro, Estrada Sobral Santa Quitéria km 03, 62115-000 Forquilha – **CE66**) Rua Capitão Carapeba 67, Centro, 62370-000 São Benedito **E:** radiotabajara1@hotmail.com ☎88 3626 2266 📠88 3626 2166 – **CE67**) Rua Monsenhor Joviniano Barreto 22, 2° andar, 63660-000 Tauá **W:** www.difusorataua.com **E:** contato@difusorataua.com – **CE68**) Rua Monsenhor Furtado, 149 - Centro, 62380-000 Guaraciaba do Norte ☎88 3652 2112 **W:** www.somzoom.com.br – **CE69**) Av Rios 92, 62590-000 Itarema – **CE70**) Av Coronel Alexanzito 369, 62800-000 Aracati **W:** www.deuseespirito.com **FM:** 105.3MHz– **CE71**) Av.7 N 260 Altos Conj. Jereissati - Maracanau, 61900-320 Maracanaú ☎85 3382 2222 **W:** www.radiopitaguaryam.com **E:** radiopitaguaryam@ibest.com.br – **CE72**) Rodovia BR-226 km 20, Distrito de Bonfim, 63600-000 Senador Pompeu – **CE73**) Avenida José Júlio Lousada 312, Centro, 62580-000 Acaraú ☎88 36611280 **W:** www.difusoraacarau.com – **CE74**) Rua Padre Antônio Tomaz s/n, 62550-000 Morrinhos – **CE75**) Rua Major Liberalino s/n, 62250-000 Ipu – **CE76**) Rua Joaquim Távora, 333 - Centro, 63150-000 Campos Sales ☎88 3533 1530 **W:** www.tresfronteirasam.com.br **E:** tresfronterasam@gmail.com – **CE77**) Rua Manoel Alencar 35, 63610-000 Mombaça – **CE78**) Rua Caio Prado Fontenele 61, Centro, 62230-000 Pacatuba – **CE79**) Rua Raul Catunda Fontenele 61, Centro, 62230-000 Ipueiras ☎88 3685 1368 **W:** www.radiomacambira.com.br – **CE80**) Rua Raimundo Guedes Martins 25, Centro,, 63400-000 Cedro ☎88 3564

1075 **W:** www.radiomontevideoam.com.br – **CE81)** Ro BR 020 s/n, Zona Rural, 62730-000 Caridade ☎85 3324 1292 **W:** www.radiovanguardaam. com.br **E:** studio@radiovanguardaam.com.br — **CE82)** Rua Francisco Brasileiro, Centro, 63900-000 Quixadá ☎88 3412 3047 **W:** www.radio-culturaquixada.com.br **E:** radioculturaquixada@gmail.com — **CE83)** Av. Rabelo, s/n – Alto Vistoso, 63620- 000 Solonópole ☎88 3518 1520 **W:** www.radiocachoeiraam.com.br **E:** radio.cachoeira.am@hotmail.com — **CE84)** Rua Augusto Vieira, 32-Centro, 63630-000 Pedro Branca ☎88 3515 2121 **W:** www.amtrapia1510.com.br **E:** amtrapia1510@amtra-pia1510.cm.br — **CE86)** 63140-000 Assaré — **CE88)** 63170-000 Araripe — **CE89)** 624000-000 Camocim — **CE96)** 62430-000 Granja — **CE98)** 61760-000 Eusebio

DF00) DISTRITO FEDERAL

DF01) SRTS, Qd 701, Ed Assis Chateaubriand, Bl 2, salas 701 a 716, 70340-906 Brasília ☎61 9316 9530 🖷61 9223 0532 — **DF02)** Setor de Rádio e TV Sul, Palácio do Rádio, Bloco 1 6° andar, 70340-901 Brasília **W:** www.radiomec.com.br — **DF03)** Sig Quadra.02 Lt 340 Bl.02, 1° andar,(✉) C.P. 8042, 70673-1080) 70610-901 Brasília ☎61 3214 1019 **W:** www. clube.am– **DF04)** SCRN 702/03, B1 «B», Edifício Radiobrás, 70710-750 Brasília ☎61 3327 4260 **W:** www.radiobras.gov.br – **DF05)** SRTV/Sul, Q-701, bloco E, Térreo, 70340-000 Brasília — **DF06)** C.P 070.747, 70359-970 Brasília **W:** www.radiobras.gov.br/ Pres: Marcelo Netto Dir Radio: Iolando Lourenço Dir News: Miriam Moura Dir Adm: Januário Procópio Dir Tec: Toshihiro Kanagae Dir SW: Renato Geraldo de Lima – **DF07)** SRTV-Sul, Q. 701, Conj «E», Bloco 2 e 4, sala 316, 70340-902 Brasília ☎61 2103 0710 **W:** www.redeboavontade.com.br – **FM:** 103.3MHz– **DF08)** SCS-Quadra 05, Bl B, Lotes 47 a 57, Ns 39/40, 70340-000 Brasília ☎61 3245 3747 **W:** www.redeboavontade.com.br **DF09)** Torre Televisi̇Dao s/n, 70070-300 Brasília ☎61 3321 6589 **W:** www.radiogloborasilia. com.br – **DF10)** Senado Federal, Praça dos Tres Poderes, Anexo II, Bloco B -Térreo, 70165-900 Brasília ☎61 3311 4691 🖷61 3311 4238 **W:** www. senado.gov.br/radio **E:** radio@senado.gov.br

ES00) ESPÍRITO SANTO

ES01) Rua Joaquim Plácido da Silva, 225, Ihla de Santa Maria, 29051-070 Vitória ☎27 3331 9000 🖷27 3223 7340 **W:** www.redetribuna. com.br/radio/am – **FM:** 99.1MHz – **ES02)** C.P 700, 29001-970 Vitória **W:** www.vitoria640am.com.br – **ES03)** Rua Dr Deolindo 65, Baiminas, 29305-440 Cacheiro de Itapemirim – **ES04)** Rua Chafic Murad 902, Ilha de Monte Belo, 29050-901 Vitória **W:** http://gazetaonline.globo.com - **FM:** 92.5MHz «Antena Um», 102.3MHz «Litoral FM» – **ES05)** C.P 125, 29900-971 Linhares - **FM:** 98.7MHz – **ES06)** C.P 178, 29790-971 Colatina ☎27 3721 1506 **W:** www.difusoracolatina.com.br **Email** radio@difu-soracolatina.com.br **ES07)** Av Santo Antônio, 366 al Caratoira, 29025-645 Vitória ☎27 3222 4376 🖷27 3222 7747 **W:** www.radiocapixaba. com.br **E:** radiocap@terra.com.br – **ES08)** Av NS da Penha 2141, (✉C. P 809, 29001-970) 29045-403 Vitória ☎27 3137 2475 **W:** www.radioe-spiritosanto.com.br **E:** falecom@radioespiritosanto.com.br – **ES09)** Rua Bernando Horta 324, Guandu, 29300-782 Cachoeiro de Itapemirim ☎28 3517 9696 **W:** www.radiosimcachoeiro.com.br – **ES10)** Rua Alberto de Oliveira Santos 42, 19° andar, salas 1916-1920, Centro, 29010-901 Vitória **W:** www.redeamericaes.com.br **E:** america@ebr.com.br ☎27 3222 2365 – **FM:** 101.5MHz «Cidade» – **ES11)** Rua José Cupertino 120, 29600-000 Afonso Cláudio – **ES12)** Rua da Matriz 85, 29200-000 Guarapari **W:** www.simamguarapari.com.br – **ES13)** Av Presidente Vargas 449, Centro, 29600-000 Afonso Cláudio ☎27 2735 1120 **W:** www.educadoraafonsoclaudio.com.br – **ES14)** C.P 132, 29930-000 São Mateus **W:** www.redesimatura.com.br/radio – **ES15)** Rua Costa Pereira 37, Centro, 29300-970 Cachoeiro de Itapemirim **W:** www.radiodiocesana com.br **E:** radio@radiodiocesana.com.br ☎27 3521 1960 – **Italian:** Sun 1300-1600 – **ES16)** Rua Graciano Neves 290, 29156-050 Cariacica **W:** www.redesimsat.com.br **ES17)** Av Perfeito Manoel Vila 660, 29800-000 Barra de São Francisco **W:** www.radiosaofranciscoam.com.br **E:** radiosaofranciscoam.com.br – **ES18)** Av Marechal Campos, 310 - Ed Milton Magnago, 2° pavimento, Consolação, 29045-460 Vitória ☎27 3322 1245 **W:** www.radionovaestacao.com.br **E:** falecom@radionovaestacao. com.br – **ES19)** 29650-000 Santa Teresa – **ES20)** Av João Felipe Calmon, 819 – Centro, 29900-010 Linhares ☎22 3371 0288 **W:** www.globolin-hares.com.br – **ES21)** Rodovia Mickel Chequer – Fazenda Vargem Alegre S/N. 293900-000 Iúna – **ES22)** 29300-000 Cachoeirp de Itapemirim – **ES23)** Baixo Guandu.

G000) GOIÁS

G001) Av. Goiás Q 10, 636 - Centro , 74010-010 Goiânia ☎62 3212 7872 **W:** http://radioriviera.webnode.com **E:** radioam.riviera@gmail. com – **G002)** Av. João XXIII, 381 - Centro, 75702-130 Catalão ☎64 3441 2700 **W:** www.radioculturaonline.com.br **E:** radiocultura@gmail. com – **G003)** Rua Rui Barbosa, 420 – Central, 75025-060 Anápolis **W:** www.radiomanchester.com.br **E:** fm@radiomanchester.com.br – **FM:** 93.3MHZ – **G004)** Av Egídio Francisco Rodrigues 54, Centro, 75200-000 Pires do Rio ☎64 3461 7346 **W:** www.gospelam.com.br **E:** am630gospel@ hotmail.com - **FM:** 102.3MHz «Rio FM»– **G005)** Av 24 de Outubro 1854, Campinas, 74505-016 Goiânia ☎62 3233 4000 **W:** www.difusora.am.br

E: difusora@netgo.com.br – **G006)** Av Marechal Rondón, Q 18 Lote 9, 76270-000 Jussara – **G007)** Rua 48, 1254 – Joaquim da Silva Moreira,, 76680-000 Itapuranga ☎62 3312 1548 **W:** www.radioprimavera.net **E:** radioprimavera@radioprimavera.net – **G008)** Av 1° de Maio, 30 - Centro, 75001-970 Anápolis ☎62 3328 1641 **W:** www.radiosaochico. com.br **E:** contato@radiosaochico.com.br - **FM:** 96.3MHz – **G009)** Rua José de Carvalho, 542 - Centro, 75800-447 Jataí ☎64 3631 1245 **W:** www.difusoraonline.com.br **E:** contato@difusoraonline.com.br – **G010)** Rua Evangelino Meireles 26, 72800-000 Luziânia – **G011)** Rua Coronel Gonzaga, 540 – Centro, 75690-000 Caldas Novas ☎64 3435 1100 **W:** www.radiopousada.com.br **E:** rpousada@itcn.com.br – **G012)** Rua Br Cotegipe s/n, S.Central, 75025-010 Anápolis ☎62 3098 3977 **W:** www.radiovozimaculada.franciscanosdaimaculada.org**E:** fras-tan@hotmail.com – **R.Maria:** Qna 05 Lote 20. 72110-050 Taguatinga Norte, DF ☎61 3562 8888 **W:** www.radiomaria.org.br **E:** radiomaria@ radiomaria.org.br – **G013)** Av Brasil 272, 76380-000 Goianésia – **G014)** Av. António C. Paniago, 1, S. Pecuári, 75830-000 Mineiros ☎62 3661 1316 **W:** www.eldoradomineiros.com.br **E:** radio@eldoradomineiros. com.br– **G015)** Av Bernardo Sayao 371, 76310-000 Rialma – **G016)** Rua Teixeira de Freitas Qd. 04 Lt 26, Setor Serrinha 74835-180 Goiânia ☎62 3945 3820 **W:** www.820am.com.br **E:** internet@820am.com.br – **G017)** Av Belém Brasilia Q5 10 lt 4, S Central, 76550-000 Porangatu ☎62 3362 4085 **E:** contato@radiotropicalcn.com.br – **G018)** Av Tocantins N° 65 1° andar, Centro, 76400-000 Uruaçu ☎62 3357 6626 **W:** www. radiolagodourado.com.br **E:** contato@radiolagodourado.com.br – **G019)** C.P 17, 72900-000 Santo Antônio do Descoberto (or Av.São João Batista 23, 77223-000 Cidade Eclética) – **G020)** Rua Minas Gerais, 135 - Central, 75503-190 Itumbiara ☎64 3431 8485 **W:**radioparanaiba.com.br - **FM:** 92.3MHz – **G021)** Rua Uberaba, 9 , 75500-000 Itumbiara ☎64 3431 7400 **W:** www.difusoraitumbiara.com.br **E:** contato@www.difusorai-tumbiara.com.br – **G022)** C.P 501, 75001-970 Anápolis– **G023)** Rua do Contorno 702, 76330-000 Jaraguá – **G024)** Av 4a Radial Q4, 847 lt 9 t R, Jardim das Esmeraldas, , 74830-130 Goiânia ☎62 3222 1090 **W:** www. radio1090.com.br – **G025)** Av Lazaro Xavier 18, 75860-000 Quirinópolis. - **G026)** Rua Herculano Lobo 80, 12573800-000 Formosa **E:** radioformo-saam@brturbo.com.br – **G027)** Rua Thomaz Edson Qd 07, Bairro Serrinha, 74835-900 Goiânia - **FM:** 92.7MHz «Executiva FM» and 97.1 «Araguaia» – **G028)** Av. Br. Rio Branco, 1 – S. Central, (✉C.P 34, 75781-970) 75780-000 Ipameri ☎64 3491 1657 **W:** www.xavantes.net **E:** radio_xavantes@ hotmail.com – **G029)** Rua 42, 50 – Rialma II,76310-000 Rialma ☎62 3397 2175 **W:** www.coracaofiel.com.br **E:**radio@coracaofiel.com.br – **G030)** Rua SC-1 No 299, Parque Santa Cruz, 74860-270 Goiânia ☎62 3201 7600 **W:** www.agecom.go.gov.br/RBCAM.php **E:** rbc@agecom. go.gov.br or fernandocozzao@yahoo.com.br - **FM:** 90.1MHz – **G031)** Av Goiás 174, Ed.São Judas Tadeu, 16° andar, Centro 74010-010 Goiânia ☎62 3216 0730 🖷62 3521 0412 **W:** www.radio730.com.br – **G032)** Alameda das Rosas 2200, Setor Oeste, 74126-010 Goiânia ☎62 3521 1707 **W:** www.radio.ufg.br **E:** tro@radio.ufg.br – **G033)** Av Amazonas, 367 – S.Central, 76100-000 São Luís de Montes Belos ☎64 3671 1621 **W:**www.valedaserraam.com.br **E:** radiovaleamfm@hotmail.com **FM:** 102.5MHz – **G034)** Rua Barão do Rio Branco, 989, - www.radiovale-mfm.com.br**www** Centro, 75650-000 Morrinhos **W:** www.intergracaofm. com.br - **FM:** 94.5MHz «Integração FM» – **G035)** Praca Rui Barbosa, 471 - Centro, 75180-000 Silvânia **W:** www.radioriovermelho.com.br **E:** atendimento@radioriovermelho.com.br – **G036)** Rua Kisleu Maciel 113, 73850-000 Cristalina ☎61 3612 2929 **W:** www.radioserradoscristais. com.br **E:**contato@radioserradoscristais.com.br – **G037)** C.P 60, 76601-970 Goiás.- **W:**www.cidadeam.com.br **E:** radiocidade@virtnet.com.br – **G038)** C.P 70, 75601-970 Goiatuba – **G039)** Praça Pres Médici, s/n, Central, 75920-000 Santa Helena de Goiás **W:** www.radiosantelenense. com.br **E:** radiosantelenense@globo.com ☎63 3641 1555 – **G040)** C.P 117, 75401-970 Inhumas **W:** www.rjiam.com.br **E:** jornalam@hotmail. com –**G041)** Av. Pauzanes Carvalho Q 25,s/n lt 7/9, S.Pauzanes (✉C. P 131, 75901-970) 75930-000 Rio Verde ☎64 3621 4433 **W:** www. rioverdeam.com.br – **FM:** 95.3MHz – **G042)** Rua Benedito Lemes, 45 – S Centro, 76630-000 Itaberaí ☎62 3375 2901 **W:** www.radiosilvestream. com.br **E:** joãosilvestre@radiosilvestream.com.br **FM:** 87.9MHz – **G043)** Rua Catalão 862, 76310-000 Iporã **W:** www.rioclaroamfelicidadefm.com. br **E:** contato@rioclaroamfelicidadefm.com.br – **G044)** Av Lindolfo Alves Dias, (C.P 50) 571 - Centro, 75850-000 Caiapônia ☎64 3363 1150 **W:** www.serraazulam.com.br **E:** atendimento@serraazulam.com.br – **G045)** Rua Lazer Q82, s/n lt 2 Res Village Garavelo, 74900-000 Aparecida de Goiânia ☎62 3283 1040 **W:** www.radiocaraibagoiania.com.br – **G046)** Quadra 33, Lotes 23/24, 72900-000 Santo Antônio do Descoberto (C.P 06-799, 71701-970 Brasília, DF) – **G047)** Rua 22, 150 – St.Aeroporto, 75640-000 Piracanjuba ☎64 3405 1919 **W:** www.radiopousaalto.com. br **E:** contato@radiopousoalto.com.br – **G048)** Rua 49 Q 53, 218 (✉C.P 185, 76301-970) 76300-000 Ceres ☎62 3307 3042 **W:** www.radiso-ciedadeceres.com.br – **G049)** Praça Silva Junior, 184, Centro76420-000 Niquelândia ☎62 3354 1430 **W:** www.radiomantiqueiraam.com.br **E:** radiomantiqueira@uol.com.br – **G050)** Rua Xingu 625, 75600-000

Goiatuba – **GO51**) Rua Francisco Corra Neves 100 – 2° andar, 75860-000 Quirinópolis – ☎️🖷64 3651 2106 **W:** www.radioalvoradaam.com.br **E:** radioalvorada@cultura.com.br **GO52**) Av Joaquim David Ferreira 1390, 76105-000 Firminópolis – **GO53**) Av Santana, Qd 55, lote 01, Sector Vila Baiana, 73840-000 Campas Belos ☎️62 3451 1209 **W:** www.rcbam. com.br – **GO54**) 78505-000 Caçu – **GO55**) 73900-000 Posse – **GO56**) 75200-000 Pires do Rio

MA00) MARANHÃO
MA01) Av Eliézer Moreira s/n, Incra, 65950-000 Barra do Corda – **MA02**) Praça Dom Pedro II s/n, 65030-000 São Luís **W:** www.educadora.elo. com.br – **MA03**) Av Camboa do Mato, 120, Camboa, 65020-260 São Luís – ☎️🖷98 3214 3000 **W:** www.sistemadifusora.com.br **E:** difusora@ sistemadifusora.com.br - **MA04**) Rua Henrique de Figueiredo 485, 65400-000 Codó – **MA05**) Av São Benedito 1075, Bairro São Benedito, 65400-000 Codó **MA06**) Rua Simplicio Moreira 1686, 1° andar, Centro, 65901-490 Imperatriz **E:** radioimp@aeronet.com.br **W:** www.radioimperatriz.com.br – **MA07**) Rua Manoel Alves de Abreu 373, 65700-000 Bacabal – **MA08**) Av Coronel Fonseca 200, 65800-000 Balsas **MA09**) Av Cel Colares Moreira, 1000 s11, Calhua, 65075-440 São Luís ☎️🖷98 3235 7676 – **MA10**) Palácio dos Leões, Av. Dom Pedro II, s/n, Centro, 65010-904 São Luís – ☎️98 3236 9419 🖷98 3226 8896 **W:** www.ma.gov.br/timbira **E:** radiotimbira.raimundofilho@bol.com.br – **MA11**) Av dos Africanos, 77 - Areinha, 65031-410 São Luís ☎️98 2109 7777 **W:** www.grupozildenifalco.com.br **E:** opecsaoluis@yahoo. com.br - **FM:** 102.5MHz – **MA12**) Rua 30 de Março, 627 - Centro, 65200-000 Pinheiro ☎️🖷98 3381 3215 **W:** www.sistemapericuma.com. br **E:** rvc@sistemapericuma.com.br – **MA13**) Av Rio Branco, 670,, 65725-000 Pedreiras **MA14**) Av Keened, 65690-000 Colinas – **MA15**) BR-316 km 115, 65370-000 Pindaré Mirim – **MA16**) Fazenda São João, BR-14 Km 37, 65215-000 Viana – **MA17**) Rua Guarani 3, Parque das Laranjeiras, 65922-000 João Lisboa – **MA18**) Rua A Bandeira 831, 65320-000 Vitorino Freire – **MA19**) 65922-000 João Lisboa – **MA20**) Rua Piauí 895, 65930-000 Açailândia – **MA21**) C Carneiro 177, 65750-000 Esperantinópolis – **MA22**) Travessa Tiradentes 338, Centro, 65415-000 Coroatá – **MA23**) Rua Aarão Reis s/n, 65604-060 Caxias – **MA24**) Av Coronel Stanley Fortes Batista 454, 65365-000 Zê Doca – **MA25**) 65278-000 Turiau – **MA26**) Rua Alagoas 497, 65900-490 Imperatriz – **MA27**) Av Amaral Raposo s/n, 65940-000 Grajaú – **MA28**) Rua Rui Barbosa s/n, 65620-000 Coelho Neto – **MA29**) Rua Dr Paulo Ramos 495, 65208-000 Santa Helena – **MA30**) Rua Frederico Coelho esquina com Av Frei Aniceto, 65763-000 Tuntum – **MA31**) Av do Puraqueu s/n, 65350-000 Vitória do Mearim – **MA32**) Parque da Bandeira 222, Edificio Ariana, Centro, 65665-000 São João dos Patos – **MA33**) Praça do Guarim s/n, 65390-000 Santa Luzia – **MA34**) Rua Terra esquina com Rua Jupiter s/n, 65760-000 Presidente Dutra – **MA35**) Praça Coronel Luis Vieira 25, 65500-000 Chapadinha – **MA36**) Rua Cel Pedro Bogéa 227, Centro, 65715-000 Lago da Pedra – **MA37**) Rua Hemeterio Leitão 103, 65430-00 Vargem Grande – **MA38**) Av Ana Jansen 200, 65000-000 São Luís ☎️98 3215 5124 **W:**http://imirante.globo.com/miranteam/_index.asp **E:** comercialam@mirante.com.br **FM:** 96.1MHz – **MA39**) 65530-000 Urbano Santos – **MA40**) 65600-000 Caxias – **MA41**) Praça Geutulio Vargas, 65800 000 Balsas

MG00) MINAS GERAIS
MG01) Rua Rabelo Horta, 39-Centro, (C.P 123, 36711-970) 36770-066 Cataguases ☎️32 3422 1724 **W:** www.sistemamultisom.com.br **E:** contato@sistemamultisom.com.br - **FM:** 89.5MHz **MG02**) Rua General Carneiro 10, Edificio Milinardo, s 200 à 305, 39400-095 Montes Claros – **MG03**) R.Júlia Lopes de Almeida 12, Santa Maria, 30525-100 Belo Horizonte ☎️31 3011 7900 **W:** www.gospamira.com.br **W:**www.vibeflog. com/jesusnanet/p/8339488 - **FM:** 105.7 – **MG04**) Praça Nossa Senhora Aparecida 134, Bairro Aparecida, 38400-726 Uberlândia E: america@ radioamerica.com.br **W:** www.radioamerica.com.br ☎️34 3214 4342 🖷34 3210 2298 - **FM:** 98.7MHz **MG05**) Av Padre Matias, 1089 - Bairro Marciano Brandão, 38740-000 Patrocínio ☎️34 3831 1546 🖷34 3831 1896 **W:** sistemadifusoraderadio.com.br - **FM:** 98.9MHz – **MG06**) Rua Itatiaia 117, Bairro Bonfim, 31210-170 Belo Horizonte ☎️31 3421 3588 🖷31 3422 8588 E: itatiaia@itatiaia.com.br **W:** www.itatiaia.com.br **FM:** 95.7MHz – **MG07**) Av. Dr. Fidélis Reis 820, Centro, , 38010-030 Uberaba. ☎️34 3331 7900 🖷34 3321 8200 **W:** www.jmonline.com.br **E:** jmonlin@ emonline.com.br – **MG08**) Av Presidente Vargas, 372 - Centro, 35661-000 Pará de Minas ☎️37 3232 1588 **W:** www.santacruzam.com.br **E:** contato@ padregabriel.com.br **MG09**) Rua Euripides Ribeiro, 739 - Centro, 38720-000 Lagoa Formosa ☎️34 3824 2628 **W:** www.princesaam.com.br **E:** contato@princesaam.com.br – **MG10**) Rua Xavier de Veiga 85, 35400-000 Ouro Preto ☎️31 3551 2166 🖷31 3551 2325 **W:** www.itatiaia.com. br/ouropreto/ **E:** ouropreto@itatiata.com.br **MG11**) C.P 30, 35791-970 Curvelo – **MG12**) Rua Prof Monteiro Fonseca 119, 39400-149 Montes Claros – **MG13**) Praça Coronel Hermógenes 292, 38770-000 João Pinheiro – **MG14**) Rua Entre Rios 33, Bairro Carlos Prates, 30710-080 Belo Horizonte ☎️31 3231 7151 **W:** www.radiomineira.com.br **E:** contato@ radiomineria.com.br – **MG15**) Praça 15 de Novembro 388, 36900-000

Manhuaçu – **MG16**) C.P 37, 37551-970 Pouso Alegre – **MG17**) Rua Guilherme Ferreira 650, salas 81/82, 38022-200 Uberaba E: sociedade@ ldc.com.br **MG18**) Rua Dr João Penido Filho 269, 36021-600 Juiz de Fora – **MG19**) Av Itaú, 515, Dom Cabral, 30730-910 Belo Horizonte **W:** www.americabh.com.br **E:** radioamerica-diretoria@ pucminas.br ☎️31 3469 2500 🖷31 3469 2551 – **MG20**) Praça Leonardo Venerando Pereira 200 - Centro, 37200-000 Lavras 🖷35 3822 5000 **W:** www.radiocultura770. com.br – **MG21**) Avenida Getúlio Vargas, 142 Centro, 38700 128 Patos de Minas ☎️34 3823 1100 **W:** www.clubeam.com **E:** clubeam@clubeam. com – **MG22**) Rua dos Cravos, 467 – Bairro São Pedro,(C.P 10, 35901-970) 35900-125 Itabira ☎️31 3831 4106 **W:** www.radioitabira.com.br **E:** euclideseder@yahoo.com.br - **FM:** 93.3MHz – **MG23**) Av Prof José Ignácio de Souza 2710, Umuarama, 38405-330 Uberlândia ☎️34 3212 0010 🖷34 3232 2044 **W:** www.ameducadora.com.br ☎️34 3232 2044 **MG24**) Ch. Vascobcelos, s/n - Chácara , 35430-017 Ponte Nova ☎️31 3817 1025 **W:** www.radiopontenova.com.br **E:** radiopontenova@radiopontenova.com. br – **MG25**) Rua Margarida Monteiro 125 (C.P 153), 35290-000 Mantena ☎️33 3241 1559 **W:** www.radio13dejunho.com.br **E:** radio13dejunho@ ralnet.com.br – **MG26**) Rua Luz 235, Bairro Américo Silva, 35590-000 Lagoa da Prata. ☎️37 3261 4500 **W:** www.tropical790.com **E:** thiagomartins@tropical790.com.br – **MG27**) Rua Esposalina Leal, 141 - Centro, 35200-000 Almorés ☎️33 3267 1021 **W:** www.radioam810.org.br - **FM:** 90.3MHz – **MG28**) Rua Maranhao 400, 35500-066 Divinópolis ☎️37 3222 7070 **W:** www.divinopolisvirtual.com.br - **FM:**100.5MHz – **MG29**) Praça Dom Pedro Teixeira 49, 36200-000 Barbacena **E:** rbq@ prower.com.br – **MG30**) Rua Dos Antunes, 1175, Ed São Sebastião (C.P 36), 37950-000 São Sebastião do Paraíso ☎️35 3531 2396 🖷35 3531 2332 **W:** www.paraisoam.com.br **E:** paraisoam@paraisoam.com.br – **MG31**) Rua Itatiaia 117, Bairro Bonfim, 31210-170 Belo Horizonte **W:** www.culturabh.com.br – **MG32**) Rua Barão de Piunhi 31, 35570-000 Formiga ☎️37 3322 2565 **W:** www.difusoraformiga.com.br – **MG33**) Av Brasil, 2770 2° andar, Centro, 35520-070 Governador Valadares. ☎️33 3271 7322 **W:** www.radiomundomelhor.com.br **E:** radiomundomelhor@ wkve.com.br - **FM:** 97.7MHz – **MG34**) Rua Coronel Joaquim Mendes 19 Ed Cine T Helena, 38430-000 Tupaciguara ☎️34 3281 4050 **W:** www. radiotupaciguara.com.br **E:** direcao@radiotupaciguara.com.br - **FM:** 91.9MHz **MG35**) Av Raja Gabáglia, 1666 – Gutierrez, 30441-194 Belo Horizonte ☎️31 3298 3434 🖷31 3298 3400 **W:** www.inconfidencia.com. br **E:** inconfidencia@inconfidencia.com.br or engenharia@ inconfidencia. com.br - **FM:** 100.9 MHz **MG36**) Rua Dr Olinto Martins 207, 39960-000 Jequitinhonha ☎️33 3741 1521 🖷33 3741 1332 – **MG37**) Av Getúlio Vargas, 420 Centro, 39800-015 Teófilo Otoni ☎️33 3522 3635 **W:** www. radioteofilotoni.com **E:** vanessa@radioteofilotoni.com.br – **MG38**) Praça Dom Joaquim 125, Centro, 39100-000 Diamantina 🖷38 3531 1408 – **MG39**) Praça 28 de Setembro 39, 36520-000 Visconde do Rio Branco – **MG40**) Rua Tenente Viotti, 131- Centro, 37460-000 Passa Quatro ☎️35 3371 3301 **W:** www.mineiradosul.com.br **E:** contato@mineiradosul.com. br – **MG41**) Av Afonso Pena, 795 2° andar - Centro, 37270-000 Campo Belo **W:** www.radioclubecampobelo.com.br **E:** radioam@ radioclubecampobelo.com.br ☎️🖷35 3832 2700 – **MG42**) Av Teodolino Pereira de Araújo 731, 38440-000 Araguari – **MG43**) Av Ferroviaria s/n, 30520-480 Belo Horizonte ☎️31 3361 0205 – **MG44**) Av Bom Jesus, Centro - 464, 37578-000 Bueno Brandão ☎️35 3463 1006 **W:** www. radioindy.com.br **E:** comercial@radioindy.com.br – **MG45**) Rua Radialista Hamilton Macedo, 204 - Limoero, 35300-000 Caratinga ☎️33 3329 0112 **W:** www.radiocaratinga.com.br **E:** contato@radiocaratinga.com.br **MG46**) Av Tirantendes, 209 - Centro, 36307-346 São João del Rei – ☎️32 3371 7777 **W:** www.radiosaojoaodelrei.am.br **E:** contato@ radiosaojoaodelrei.am.br **MG47**) Rua Rio Barbosa 259, 38420-000 Monte Alegre de Minas ☎️34 3281 2100 – **MG48**) Rua Espírito Santo, 95 – Poco Rico, 36020-000 Juiz de Fora ☎️32 3215 1620 🖷32 3215 4360 **W:** www.radiosolaram.com.br **E:** solarfm@radiosolaram.com.br - **FM:** 88.9MHz – **MG49**) Rua Afonso Pena, 340 - Centro, 37590-000 Jacutinga ☎️35 3443 2121 🖷35 3443 1625 **W:** www.radiojacutinga.com.br **E:** joao@radiojacutinga.com.br – **MG50**) Rua Manoel Joaquim Pires 43, 35170-082 Coronel Fabriciano ☎️31 3842 1400 **W:** www.educadoramg. com.br **E:** educadora@usinet.com.br - **FM:** 107.1MHz – **MG51**) Rua Rio Grande do Norte, 1096, Marta Helena, (C.P 557, 38409-970) 38402-039 Uberlândia ☎️34 3291 5500 **W:** www.radioculturafm.com.br - **FM:** 95.1MHz **MG52**) Rua Bueno Brandão 26, 38430-000 Tupaciguara – **MG53**) Rua Tamoios 200, 21° andar, 30140-000 Belo Horizonte – **MG54**) Rua Olegário Maciel 200, Bairro Vila Podis, 37500-000 Itajubá ☎️35 3623 2471 **W:** www.radioitajuba-am.com.br **E:** radioitajubaam1060@yahoo. com.br **MG55**) Praça da Basílica 100, (🖂 CP 05, 36415-000) Congonhas ☎️🖷31 3731 4217 **W:** www.radiocongonhas.com.br **E:** gerencia@ radiocongonhas.com.br – **MG56**) Av Dr Américo Luz 153, 37890-000 Muzambinho ☎️35 3571 2399 **W:** www.radiodopovo.com.br **E:** radiodopovoam@yahoo.com.br – **MG57**) Rua Halfeld 744 Sl 401, Centro,36010-003 Juiz de Fora ☎️32 3215 4477 **W:** www. radiocapitaljuizdefora.com.br **E:** contato@radiocapitaljuizdefora. com.br – **MG58**) C.P 85, 38441-970 Araguari ☎️34 3241 3030 **W:** www.

radioplanalto.com.br **E:** planaltoam@netsite.com.br – **MG59**) Rua Areião do Matadouro s/n, Parque da California, 34000-000 Nova Lima – **MG60**) Rua França, 506 – Boa Vista, (✉C.P 253), 38001-970 Uberaba ☎34 3313 3400 **W:** www.setecolinas.com.br **E:** contato@setecolinas.com.br - **FM:** 98.1MHz – **MG61**) Av Juscelino Kubstchek 740, 37170-000 Boa Esperança ☎35 831 1000 🖷35 851 1475 **W:** www.radioserraam1120.com **E:** contato@radioserraam1120.com – **MG62**) Av Antonio Olímpio de Morais 545, Centro, 35500-005 Divinópolis **W:** www.radiominasam.com.br **E:** sistemampa@radio94FM.com ☎37 3222 0001 🖷37 3222 0009 - **FM:** 95.3MHz – **MG63**) Rua João Bressane 1, 37400-000 Campanha ☎35 3261 1229 **W:** www.diocesedacampanha.org.br/radiodiocesana.html **E:** radiocesana@yahoo.com.br – **MG64**) Av Constantino Pinto, 90 - Centro, 36880-000 Muriaé ☎32 3729 2929 **W:** www.radiomuriae.com.br **E:** diretoria@radiomuriae.com.br - **FM:**) 96.3MHz «R 96» – **MG65**) Av Raja Gabáglia 3502, 4° andar, Estoril, 30350-540 Belo Horizonte ☎31 3298 9303 🖷31 3298 9305 **W:**http://radioclick.globo.com/globobrasil/ – **MG66**) Rua do Rádio 60, Bairro Pepétuo Socorro, 38190-000 Sacramento **E:** radiosac@sacranet.com.br ☎34 3351 1735 🖷34 3351 1432 – **MG67**) Rua Cassiano Lemos, 87 - Centro, 38180-000 Araxá ☎34 3661 2844 **W:** www.cidadeamfm.com.br **E:** cidade@radiocidadedearaxa.com.br - **FM:** 94.5 MHz – **MG68**) Av Assis Chateaubriand 499, Floresta, 30150-101 Belo Horizonte **W:** www.guarani.am ☎31 3237 6000 🖷31 3237 6699 – **MG69**) Praça Cleber de Holanda 111, Jardim Sion, 37048-370 Varginha **E:** sistemaclube@varginha.com.br - **FM:** 99.3MHz – **MG70**) Av Brasil 4460, Bairro Umuarama, 38405-312 Uberlândia ☎34 3212 0885 🖷34 3212 0190 **W:** www.itatiaia.com.br/triangulo/ **E:** triangulo@itatiaia.com.br – **MG71**) Rua Barao do Rio Branco 461, Conj 901 – Edifico Rio Branco, 35010-030 Governador Valadares. – **MG72**) Av. 13, 658 6° andar, Edifico Ituiutaba, Centro (✉C.P 110), 38301-970 Ituiutaba ☎34 3268 1813 **W:** www.sistemacancella.com.br – **MG73**) Rua Rio Grande do Sul, 631- Centro, 37701-001 Poços de Caldas **W:** www.difusorapocos.com.br **E:** comercial@difusorapocos.com.br ☎35 3722 1530 - **FM:**104.1MHz – **MG74**) Rua Capitão Henrique Albuquerque 55, 36400-000 Conselheiro Lafaiete **W:** www.radiocarijos.com.br ☎31 3763 1752 – **FM:** 89.9MHz – **MG75**) Rua Itajubá 80, 35160-035 Ipatinga **W:** www.vanguardaam.com.br **E:** producao@vanguardaam.com.br – **MG76**) Rua Ernane Vilela Lima, 114 - Centro, 37250-000 Nepomuceno ☎35 3861 1278 **W:** www.radioam810.com.br **E:** folha@agyonet.com.br – **MG77**) Rua Duque de Caxias 450, 16° andar, Edificio Chams, 38400-066 Uberlândia – **MG78**) Rua Alexandre Silva, 295 - Centro, 38600-000 Paracatu – ☎ 38 3671 3047 **W:** www.radiojuriti.com.br **E:** radiojuriti@ada.com.br – **MG79**) Av 13 No 658, 6° andar, Edificío Ituiutaba, Centro, 38300-140 Ituiutaba **W:** www.sistemacancella.com.br **E:** radiocancella@mgt.com.br ☎34 3268 1813 🖷34 3251 7071 - **FM:** 93.7MHz – **MG80**) Av Costa Junior, 467 Centro, 38840-000 Carmo do Paranaíba ☎34 3851 2066 **W:** www.sistemaplaneta.net **E:** rplaneta@sistemaplaneta.net – **MG81**) Rua José Benedito 441, Santa Efigenia, 35450-000 Itabirito – **MG82**) Av João Pinheiro 596, 1° andar (C.P 143), 37701-386 Poços de Caldas **W:** www.radioculturapocos.com.br **E:** cultura@radioculturapocos.com.br ☎35 3722 1687 🖷35 3722 2687 – **MG83**) Rua Cel José Paulino 257, Centro, 37750-000 Machado ☎35 3295 1361 **W:** www.difusoramachado.com.br **E:** gilson0408@gmail.com – **MG84**) Praça Guido Marliere 30, 36500-000 Ubá **W:** www.radioubaense.com.br – **MG85**) Rua Calixto Martins de Melo, 391, Centro, 38610-000 Unaí ☎38 3676 1490 **W:** www.radioveredas.com.br **E:** portalveredas@portalveredas.com.br - **FM:** 98.0MHz – **MG86**) Rua Calimeiro Guimarães 308, 38180-000 Araxá – **MG87**) Rua Antônio Dias Adorno 1290, 35045-040 Governador Valadares ☎33 3275 0930 **W:** www.radioglobogv.com.br **E:** comercial@radioglobogv.com.br - **FM:** 100.1MHz «Imparsom» – **MG88**) Rua Dr Péricles de Mendonça 91, 36680-000 São João Nepomuceno **W:** www.difusorasjn.com.br – **MG89**) Rua Niquel 457, 35701-107 Sete Lagoas ☎31 3773 3694 **E:** musirama@mrnet.com.br - **FM:** 92.1MHz «Musirama» – **MG90**) Av Major Antônio Alberto Fernandes 445, 37720-000 Botelhos – **MG91**) Av Conde Ribeirão do Vale 661, 37800-000 Guaxupé.- ☎35 3551 1245 **W:** amclube.com.br **E:** amclube@amclube.com.br – **MG92**) Rua XV de Novembro 62, 36500-000 Ubá ☎32 3531 1830 **W:** http://am.educadora.com.br - **FM:** 94.5MHz – **MG93**) Praça Minas Gerais 50, Satélite, 35930-259 João Monlevade ☎31 3851 6001 **W:** www.cultura590.com.br **E:** rcultura@robynet.com.br – **MG94**) Rua Gerson Coutinho da Silva 1001, 38550-000 Coromandel – **MG95**) Rua Juliano Marques Duarte 110, 36660-000 Além Paraíba ☎32 3462 7400 - **FM:** 95.5MHz «Juventude» – **MG96**) Av 15 895, 10° andar Sala 1002 e 1005, Edifico Executivo, 38300-000 Ituiutaba ☎34 3261 7118 **W:** www.difusoraituiutaba.com.br **E:** administracao@difusoraituiutaba.com.br – **MG97**) Av. Iparanga, 198 - Centro, 37190-000 Três Pontas ☎35 3265 2252 – **MG98**) Rua Coronel Domiciano Ferreira 314, 38200-000 Frutal – **MG99**) Av Belo Horizonte 108, 39860-000 Nanuque – **MG100**) Praça Dr. Antônio das Chagas Viegas, 130 2° andar - Centro, 36300-000 São João del Rei ☎32 3371 8394 🖷32 3371 8025 **W:** www.emboabasfm.com.br – **FM:** 96.9 MHz– **MG101**) Rua Floriano Peixoto, 31 - Centro, 36570-000 Viçosa ☎31 3891 1500 🖷31 3891 1242 **W:** www.montanhesa.am.br **E:**

montanhesavicosa@montanhesa.am.br - **FM:** 97.9MHz – **MG102**) C.P 28, 37901-970 Passos **W:** www.radiopassos.com.br – **MG103**) Rua Nunes Roas, Centro (✉C.P 61, 36971-970), 36970-000 Manhumirim ☎33 3341 1491 **W:** www.radiomanhumirim.com.br – **MG104**) Rua Dr Coelho de Moura 158, Centro, 35540-000 Oliveira **W:** www.radiosociedade.com.br **E:** geral@radiosociedade.com.br ☎37 3331 1170 🖷37 3331 1510 – **MG105**) Rua 13 de Maio 425, 36200-000 Barbacena ☎32 3332 5299 **W:** www.correiodaserra.com.br – **MG106**) Praça Dr Augusto Gonçalves 146, salas 411/412, Centro, 35680-054 Itaúna ☎37 3742 1910 **W:** www.clubeamfm.com.br **E:** comercial@clubeamfm.com.br - **FM:** 93.5 MHz – **MG107**) C.P 189, 37471-970 São Lourenço **W:** www.radioestancia.com.br **E:** estancia@radioestancia.com.br - **FM:** 94.3MHz – **MG108**) Av Juscelino Kubtschek, 1016 – Bairro Boa Vista, (C.P 99) 38280-000 Iturama ☎34 3411 3388 **W:** www.radiopontal.com **E:** contato@radiopontal.com – **MG109**) Av Magalhães Pinto 829, 35610-000 Dores do Indaiá ☎37 551 1622 – **MG110**) Rua Adalberto Ferraz 50, 2° andar, 37550-000 Pouso Alegre – **MG111**) Rua Dr José Gonçalves 17, 35600-000 Bom Despacho – **MG112**) Praça Getúlio Vargas 81, (C.P 123) Centro 36400-000 Conselheiro Lafaiete ☎31 3763 2466 – **MG113**) Rua Casemiro Avelar Filho 143, Centro, 37410-000 Três Corações ☎35 231 1000 **W:** www.radiotropical.net - **FM:** 95.5MHz – **MG114**) Praça Nossa Senhora do Carmo 224, 38500-000 Monte Carmelo – **MG115**) Rua Sancho Viderla 19, Centro, 37540-000 Santa Rita do Sapucaí ☎35 3473 4400 **W:** www.difusora1550.com.br **E:** comercial@difusora1550.com.br - **FM:** 95.3MHz – **MG116**) Av Brasil 508, Centro, 39270-000 Pirapora ☎38 3741 1400 🖷38 3741 2981 **W:** www.radiopirapora.com.br **E:** pirapora@pirapora.com.br – **MG117**) Praça João XXIII 15, salas 303/307/308, 36700-000 Leopoldina **W:** www.sistemamultisom.com.br – **MG118**) Rua Bias Fortes, 191 - Centro, 37130-000 Alfenas ☎35 3299 3887 🖷35 3299 3891 **W:** www.radioculturaalfenas.com.br **E:** radiocultura@unifenas.com.br – **MG119**) Rua JK 108, 35500-000 Itapecerica paranaiba@paranaibamaximus.com.br **FM:** 101.3MHz – **MG120**) Rua Anastácio José Gonçalves, 139 - Centro, 38810-000 Rio Paranaíba ☎34 3855 1433 **W:** www.paranaibamaximus.com.br **E:** rcultura@sdnet.com.br - **FM:** 101.5 MHz – **MG121**) Rua Sérgio Neves, 63/Sala 103 – Centro, 36240-000 Santos Dumont ☎32 3251 6534 **W:** www.radioculturasd.com.br **E:** contato.cultura@radiomineira.com – **MG122**) Av Minas Gerais 584, 39510-000 Espinosa – **MG123**) Praça Nossa Senhora da Abadia 490, Abadia, 38025-430 Uberaba ☎34 3322 6200 🖷34 3322 6430 **W:** www.670am.com.br – **MG124**) Av Hermenegildo Donatti 199, Jd. Nova Andrades, 37795-000 Andradas ☎35 3731 2291 **W:** www.radiovinicola.com.br **E:** vinicola@andradas-net.com.br - **FM:** 94.9MHz – **MG125**) Praça Coronel Odilon Coelho s/n, 39520-000 Porteirinha. ☎38 3831 1228 **W:** www.educadoraam640.com.br – **MG126**) Av. Dr. Otávo Soares 108, salas 707 a 712, Palmeiras, 35430-229 Ponte Nova ☎31 3881 8831 **W:** www.montanhesa.am.br **E:** montanhesapontenova@montanhesa.am.br – **MG127**) Travessa Dona Santinha, 20 - Centro, 39480-000 Januária ☎38 3621 1856 **W:** www.alternativafm **Email.:** alternativa@comnt.com.br - **FM:** 90.7 MHz – **MG128**) Rua Benedito Valadares 423, 36880-000 Muriaé – **MG129**) Rua Padre Pedro 53, Bonfim, 39390-000 Bocaiúva ☎38 3251 1995 **W:** www.radioclubebocaiuva.com.br – **MG130**) Rua Rui Barbosa 74, 39440-000 Janaúba – **MG131**) Av Getúlio Vargas 205, Centro Shopping Luziana, 35350-000 Raul Soares – **MG132**) Praça José Batista de Frieitas 78, 3° andar, 35519-000 Nova Serrana – **MG133**) Av. Deputada Humborta de Almeida 60, Centro, 37810-000 Guaranésia ☎35 3555 1350 🖷35 3555 2150 **W:** www.radioam1580guaranesia.com.br **E:** radioam1580@guaranesia.com.br – **MG134**) BR-496 Km 33, 39260-000 Várzea da Palma – **MG135**) Rua Getúlio Vargas, 4 - Centro, 38900-000 Bambuí ☎37 3431 3290 **W:** www.cidadeambambui.com **W:** **MG136**)Rua Jair Werneck, 330, Cidade Alta, 39800-000 Teófilo Otoni **W:** www.radiomucuri.com.br – **MG137**) Rua Major Honor Sarmento, 393 - São João, 39400-533 Montes Claros ☎38 3223 5666 🖷38 3221 5590 **W:** www.radioterra.am.br **E:** contato@radioterra.am.br – **MG138**) Rua 102 No 498, 38360-000 Capinópolis ☎34 3263 1481 – **MG139**) Praça Monsenhor Messias Bragança 80, sala 203, 37900-000 Passos – **MG140**) Rua Padre João Lourenço Leite 100, 37175-000 Ilicínea – **MG141**) Rua Astor Goulart de Moura 51, 39200-000 Corinto ☎38 3751 1858 **W:** www.radiocidadecorinto.com.br **E:**radiocidadecorinto@futuretec.com.br – **MG142**) Rodovia BR-381 Km 195, Bairro Cachoeira do Vale, 35180-001 Timóteo ☎31 3849 4000 **W:** www.itatiaia.com.br/valedoaco/ **E:** valedoaco@itatiaia.com.br - **FM:** 95.7MHz – **MG143**) Rua Coronel Américo Teixeira Guimarães 38, Centro, 35700-181 Sete Lagoas ☎31 3772 0244 **W:** www.eldorado1300.com.br **E:** comercial@eldorado1300.com.br – **MG144**) Rua 1° de Janeiro, N° 382 Serra Dourada, 38780-000 Vazante ☎34 3813 1113 **W:** www.montanheza.com.br **E:** radio@montanheza.com.br – **MG145**) Rua Leandro Gonçalves 88, 3° andar, 36900-000 Manhuaçu – **MG146**) Rua Coronel Rennó 7, 37500-000 Itajubá – **MG147**) Praça São Pedro 49, 38950-000 Ibiá – **MG148**) Praça Getúlio Vargas, 108 – Triângulo, 36800-000 Carangola ☎34 3241 5823 **W:** www.ondavivaaraguari.com.br **E:** comercial@radioondaviva.com.br - **FM:**

102.7MHz «Caparaó» – **MG149**) Rua Oscar Vidal, 416 - Centro, 36016-290 Juiz de Fora ☎32 2102 9500 **W:** www.radioglobojf.com.br **E:** atendimento@radioglobojf.com.br – **MG150**) Rua Getúlio Vargas, 254 - Centro, 38700-128 Patos de Minas ☎34 3823 1070 **W:** www.radiopatos.com.br **E:** radiopatos@radiopatos.com.br – **MG151**) Av 7 de Setembro 55-A, 36950-000 Ipanema – **MG152**) Rua Julio Cosi 5, 38230-000 Fronteira ☎34 3428 2099 – **MG153**) Av Francisco Epifâno Fagundes, 161, Fagundes, 33200-000 Vespasiano ☎31 3621 3811 – **MG154**) Rua Padre Vigilato, s/n Centro., 35330-000 Inhapim ☎33 3315 1299 **W:** www.radioclubedeinhapim.com.br – **MG155**) Rua João Valentim Pascoal 669, 5° andar, 35160-000 Ipatinga – **MG156**) Rua GervásioMarques da Silveira, 1498 – Barrio Marthina, 38740-000 Patrocínio ☎34 3831 7244 **W:** www.radiorainhadapaz.com.br **E:** contato@radiorainhadapaz.com.br – **MG157**) Praça Comendador José Honorio 100, 37950-000 São Sebastião do Paraiso – **MG158**) Rua Pref Terêncio Pereira Vale 10, 38170-000 Perdizes - **FM:** 96.1MHz – **MG159**) Av Joaquim Ribeiro de Gouveia, 1651 - Centro, 38320-000 Santa Vitória ☎34 3251 2000 **W:** www.radiosom2000.com.br **E:** radiosom2000@mgt.com – **MG161**) Rua das Acacias, 672 - Canaã, 38660-000 Buritis ☎38 9966 2412 **W:** www.radioburitis.com.br **E:** radioburitis@hotmail.com – **MG162**) Rua Silvino Brandão, 164 – Aeroporto, 38440-170 Araguari ☎34 3241 5823 **W:** www.ondavivaaraguari.com.br **E:** comercialmaisfm@yahoo.com.br – **MG163**) Travessa Cônego Benedito Profício, 95 - Centro, 37660-000 Paraisópolis. **W:** www.paraisopolisam.com.br – **MG164**) Av Progresso 177, Olaria, 35588-000 Arcos ☎37 3351 2100 **W:** www.radiocidadearcos.com.br **E:** radiocidadeam@gmail.com – **MG165**) Rua Antônio Ribeiro da Costa Junior 16, 36730-000 Pirapetinga – **MG167**) Praça Dr Badaró 112, 39650-000 Minas Novas ☎33 3764 1181 ☐33 3764 1185 **W:** www.radiobomsucesso.com.br **E:** lalado@radiobomsucesso.com.br – **MG168**) Rua Marcos Vinícius Ferreira 226, São Miguel, 39560-000 Salinas ☎38 341 1060 **W:** www.radiodifusoradesalinas.com.br **E:** fernandobrasilmg@gmail.com **FM:** 104.9MHz – **MG169**) Av Rondón Pacheco, 450 – Santo Antônio, 37150-000 Carmo do Rio Claro ☎35 3561 1967 **W:** www.radiodifusoraam.com.br **E:** contato@radiodifusoraam.com.br – **MG170**) Rua Vereador Mario Anselmo 33, 36480-000 Piranga – **MG171**) Rua Coronel Carlos Brandão 98, sala 07/08, 36500-000 Ubá –**MG172**) Av Dr Pedro Rosa s/n, 37530-000 Brasópolis ☎35 3641 1317 – **MG173**) Rua Silviano Brandão, 795, Centro, (C.P 100) 37570-000 Ouro Fino **W:** www.difusoraourofino.com.br **E:** radio@difusoraourofino.com.br ☎35 3441 1433 ☐35 3441 1800 – **MG174**) Praça Vital Brasil 56, 37980-000 Cássia – **MG175**) Praça Coronel Silverio de Melo 172, 37958-000 Monte Santo de Minas – **MG176**) Rua Artur de Vasconcelos 18, Centro, 38270-000 Campina Verde – **MF178**) Av.Brasil, 4460 Umuarama, 38405-312 Uberlândia ☎34 3212 0855 **W:** www.itatiaia.com.br/triangulo **E:** triangulo@itatiaia.com.br –**MG179**) 33100-000 Santa Luzia – **MF181**) 39270-000 Pirapoar – **MG182**) 39900-000 Almenara – **MG183**) 39740-000 Guanhãs – **MG186**) 38240-000 Itapagipe – **MG193**) Rua José M. Duarte, 301, Vila Rubens, 37480-000 Lambari **W:** www.globolambari.com.br ☎35 3521 - 201- Centro, 39740 000 Guanhães ☎33 3421 3503 **W:** www.cidadeam1550.com.br

MS00) MATO GROSSO DO SUL
MS01) Av Mato Grosso 530, Centro, 79002-233 Campo Grande **W:** www.miliciadaimaculada.org.br/v2/Rural580.asp **E:** 580am@miliciadaimaculada.org.br – **MS02**) Av Senador Felinto Müller 59, 79080-190 Campo Grande **W:** www.culturaam680.com.br – **MS03**) Rua Jamil Selem N° 27(C.P 104 79951-970) 79950-000 Naviraí **W:** www.cultura.com.br **E:** culturanav@terra.com.br – **MS04**) Rua Ciro Melo 2045, 79805-000 Dourados **W:** www.radioclubeam720.com.br – **MS05**) Rua Anchieta 871, Bairro Parati, 79081-180 Campo Grande ☎67 3346 2686 **W:** www.amcapital.com.br **E:** contato@amcapital.com.br – **MS06**) Rua 15 de Agosto 98, 79200-000 Aquidauana **W:** www.pantanalnews.com.br/radioindependente –**MS07**) Rua Melanio Garcia Barbosa 749, 79150-000 Maracaju **W:** www.rcmdigital.com.br **E:** rcmdigital@terra.com.br – **MS08**) Rua Joaquim Pereira Teixeira 135, 79900-000 Ponta Porã – **MS09**) Rua 15 de Novembro 2649, Jardim dos Estados, 79020-300 Campo Grande ☎67 3349 1240 **W:**www.difusorapantanal.com.br **E:** contato@difusorapantanal.com.br – **MS10**) Rua Plinio Alarcon, 901(C. P 37, 79601-970) 79630-213 Três Lagoas ☎67 3524 2129 **W:** www.radiodifusora1250.com.br **E:** radiodifusora1250@radiodifusora1250.com.br – **MS11**) Av Marcelino Pires 1404, 79801-002 Dourados – **MS12**) Rua Marechal Deodoro, 504, Bairro Guanandy, 79200-000 Aquidauana ☎67 3241 3957 **W:** www.difusora1340.com.br – **MS13**) Rua Antônio Lino Barbosa 961, 79130-970 Rio Brilhante **W:** www.difusorarb.com.br ☎67 3452 7451 – **MS14**) C.P 138, 79301-970 Corumbá ☎67 3232 3135 **W:** www.novaclubeam.com – **MS16**) Av Antonio Joaquim de Moura Andrade, 145 - Centro, 79750-000 **MS15**) Rua Delmare 1274, Centro 793330-040 Corumbá ☎67 3441 2136 **W:** www.radiocacique.com **E:** falecom@radiocacique.com – **MS17**) Av Aldair Rosa de Oliveira 1045, 79640-100 Três Lagoas – **MS18**) Rua Ferreira 69 B, Piracema, 79400-

000 Coxim **W:** www.radiovaledotaquari.com.br – **MS19**) Rua Angélica 455, Centro, 79730-000 Glória de Dourados ☎63 3466 2040 **W:** www.paiaguas.grupofeitosa.com.br – **MS20**) Rua Visconde de Taunay 895, 79500-000 Paranaíba **W:** www.radiodifusoraam.com.br **E:** difusoraparanaiba@hotmail.com – **MS21**) Rua General Câmara No 888 - Centro, 79990-000 Amambaí ☎67 3481 1373 ☐67 3481 1391 – **FM:** 102.5MHz – **MS22**) Av 9 de Julho 1557, 79700-000 Fátima do Sul – **MS23**) C.P 200, 79541-970 Cassilândia – **MS24**) Av Presidente Vargas, 669 - Centro, 79940-000 Caarapó ☎67 3453 1810 **W:** www.difusora.grupofeitosa.com.br– **MS25**) Rua/Av.Ra 7 de Setembro,740, Centro, 79240-000 Jardim ☎67 3251 1531 **W:** www.radiolaguna.com.br – **MS26**) Rua Rui Barbosa 753, 79970-000 Eldorado – **MS27**) C.P 129, 79301-970 Corumbá – **MS28**) C.P 199, 79901-970 Ponta Porã – **MS29**) C.P 68, 79804-970 Dourados – **MS30**) Rua São Paulo 1359, 79490-000 São Gabriel d'Oeste **W:** www.difusora850.com.br – **MS31**) Rua Thomás Cáceres 349, Bairro São Bento, 79170-000 Sidrolândia ☎67 3272 1541 **W:** www.pindorama.grupofeitosa.com.br – **MS32**) 135 Rua Amando do Oliveria, 135 - Bairro Amambai, 79005-380 Campo Grande (C.P.146 79002-970 Campo Grande) ☎67 3324 0630 **W:** www.radionovotempo.org.br **E:** ellen@radionovotempo.org.br – **MS33**) Rua Severino de Araujo Ferreira 1375, 79700-000 Fátima do Sul **W:** www.fifasul.com.br/guaicurus – **MS34**) Av Calogeras 1932, 79012-003 Campo Grande – **MS35**) Av Costa Rica 654, 79740-000 Ivinhema ☎67 3442 1450 **W:** www.radioatual1530.com.br – **MS36**) Rua João Pedro Pedrossian 4058, 79570-000 Aparecida do Taboado **W:** www.radioarcoiris.com.br – **English & Spanish:** 1100-1400 – **MS37**) 79300-000 Corumbá – **MS38**) Rua Candido Severino 462, 79420-000 Camapuã ☎67 286 1366 ☐67 286 1239 – **MS39**) Rua 01 No 1550, Altos do Indaiá, 79823-500 Dourados **E:** tepims@menthor.com.br - **Spanish & Guaraní:** Sat 1400-1600 – **MS40**) Rua José Antônio Pereira 1488, sala 23, 79010-190 Campo Grande – **MS41**) Rua Antônio Maria Coelho 289, Centro, 79260-000 Bela Vista ☎67 3439 1243 **W:** radiobelavista.com.br **E:** vozdoapa@vsp.com.br – **MS42**) Rua Porfirio Gonçalves 1240, 79480-000 Rio Verde de Mato Grosso ☎67 3292 1561 **W:** www.radiocampoalegre.com.br – **MS43**) Rua Atilio Reginato 355, 79740-000 Ivinhema – **MS44**) 79300-000 Corumbá – **MS46**) 79935-000 Sete Quedas – **MS47**) 79430-000 Bandeirantes.– **MS48**) 79780-000 Bataguassu.– **MS53**) 79100-000 Campo Grande – **MS54**) Av.Federal 69 – Piracema, 79400-000 Coxim – **MS55**) Av.Joao Pedro Pedrossian 4058, 79570-000 Aperecida do Taboado ☎67 3565 1075 **W:** www.cidade.grupofeitosa.com.br – **MS56**) Av José Ferreira da Costa, 771, 79550-000 Costa Rica ☎67 3247 2007 **W:** www.globocostarica.com.br **E:** radioglobo@globocostarica.com.br – **MS57**) 79560-000 Chapadao do Sul – **MS58**) 79220-000 Nivague – **MS59**) 79180-000 Ribas do Rio Pardo – **MS59**) 79220-000 Nioaque – **MS60**) 79180-000 Ribas do Rio.

MT00) MATO GROSSO
MT01) Rua Boróros 45, 78600-000 Barra do Garças – **MT02**) Av Brasil 27, 78600-000 Poxoréo – **MT03**) Professora Tereza Lobo 30, Concil, 78048-700 Cuiabá ☎65 3612 6530 **W:** www.cbncuiaba.com.br **E:** equipedeouro@gazetadigital.com.br – **MT04**) Rua das Primaveras 3971A, 78550-000 Sinop **W:** www.radioceleste.com.br **E:** administracao@radioceleste.com.br – **MT05**) Praça do Seminário 239, 78015-140 Cuiabá ☎65 3617 7917 **W:** www.arquidiocesecuiaba.org.br**E:** producaocuiaba@cancaonoca.com.br – **MT06**) Av Ludovico de Riva Netto, 3724 - Centro, 78580-000 Alta Floresta **W:** www.radioprogresso640.com.br **E:** contato@radioprogresso640.com.br – **MT07**) Rua Joþao Pessoa 453, Cebtro (C.P 401, 78700-970) 78700-000 Rondonópolis ☎66 3422 9894 – **MT08**) Rua Joaquim Murtinho, 1456, Centro,Centro Sul, 78025-000 Cuiabá **W:** www.radioculturadecuiaba.com.br – **MT09**) Rua São Pedro 806, 78200-000 Cáceres ☐65 223 1663 – **MT10**) Rua Tiradentes 979, Centro, 78200-000 Cáceres ☎65 3223 3820 **W:** www.radiosoracaceres.com.br - **FM:** 102.3MHz – **MT11**) C.P 227, 78821-970 Jaciara – **MT12**) Av Cuiabá 829, Edifício Mikerinos, 12° andar, 78700-090 Rondonópolis **W:** www.radioclubemt.com.br – **MT13**) Rua 28 de Outobro N° 3391, 78280-000 Mirassol d'Oeste ☎65 3241 1288 ☐65 3241 1770 **W:** www.difusoramirassol.com.br – **MT14**) Rua Benedito Monteiro 68, 78110-390 Várzea Grande – **MT15**) Rua Zulmira Canavarros 285, 78005-390 Cuiabá – **MT16**) Rua 2 No 32, 78175-000 Poconé – **MT17**) Rua Sorocaba 716 B, 78575-000 Juara **W:** www.radiotucunare.com.br – **MT18**) Av Brasil 780, 78300-000 Tangará da Serra ☎65 3326 2080 **W:** www.radiotangara.com.br **E:** radiotangara@radiotangara.com.br – **MT19**) Av T Neves 1682, 78500-000 Colider – **MT20**) Tv Pref Alexandrina Gomes 87, Centro, 78600-000 Barra do Garças ☎65 3401 6155 **W:** www.radiodifusorabarra.com.br**E:** radiodifusoramt@hotmail.com – **MT21**) Rua Criciuma 165, 78890-000 Sorriso **W:** www.radiosorriso.com.br – **MT22**) Praça Edgar de Araujo, Cx Postal 332, 78430-000 Nortelândia ☎65 3346 1729 **W:** www.radioregionalnortelandia.br **E:** edivaldodesa@reporter-news.com – **MT23**) Av Mato Grosso 133, 78690-000 Nova Xavantina ☎65 3438 1218 **W:** www.radionx.com.br **E:** webmaster@radionx.com – **MT24**) Rua 6 No 498, 78500-000 Tangará da Serra ☎65 726 1084 – **MT25**) Rua Benjamim Constant s/n, 78780-000 Alto Araguaia – **MT26**) Av Holmis Ioris 429, 78320-000 Juina ☎66 3566 1505 ☐66

3566 1228 **W:** www.rej.am.br **E:** rej@juina-fox.com.br – **MT27**) Av Mario Correa 350, 78790-000 Itiquira – **MT28**) Rua 01, 600, 78525-000 Matupá ☎66 3595 1144 **W:** www.anoticiadigital.com.br **E:** radiocidade@vsp.com.br – **MT29**) Rua das Aroeiras, 1557 – Setor Comercial, 78550-000 Sinop ☎66 3531 3550 **W:** www.bandsinop.com.br – **MT30**) Avenida Luiz Barbosa esq. c/ Sete de Setembro No 477, 78285-000 São José dos Quatro Marcos ☎65 3251 1062 **W:** www.radiojornal570.com.br **E:** radiojornal@vsp.com.br – **MT31**) Rua 6 s/n, 78400-000 Diamantino ☎65 736 1316 – **MT32**) Rua Perimetral s/n, Bairro Bom Clima, 78195-000 Chapada dos Guimarães – **MT33**) Av Roberto Valdecir Briante 99, Centro, 78435-000 São José do Rio Claro ☎66 3386 2216 **W:** www.integracao810.com.br – **MT34**) Rua U-2 s/n – Canteiro Central, 78580-000 Alta Floresta **W:** www.sistemarainha.com.br/radiofloresta/ – **MT35**) Rua Jovino Lopes 1292, 2° andar, Bairro Santa Maria Bertila, 78760-000 Guiratinga ☎66 9615 9449 **W:** www.radiogarcabrancaam.com.br **E:** radiorgb@hotmail.com or radiorgb@uol.com – **MT36**) Rua Joaquim Nabuco 450, 78260-000 Araputanga **W:** www.radioarcoiris.com.br **E:** radioarcoiris@terra.com.br – **MT37**) Rua São Paulo 1440, 78250-000 Pontes e Lacerda – **MT38**) Rua do Burtis s/n, 78520-000 Guarantã do Norte ☎66 3552 1114 **W:** www.anoticiadigital.com.br **E:** enauen@vsp.com.br – **MT39**) Rua Dom Antônio Malan 674, 78015-600 Cuiabá – **MT40**) Rua Carajás 69, 76800-000 Barra do Garças – **MT41**) 78850-000 Primavera do Leste – **MT42**) Rua Filinto Muller, 1804 Morada do Sol, 78043-500 Cuiaba **E:** radioatual_brasil1530am@hotmail.com.br – **MT44**) Av Universitária 540W, 78455-000 Lucas do Rio Verde ☎65 3549 1550 **W:** www.radioatual1530.com.br **E:** atendimento@atitudeam.com.br – **MT45**) Rua Potiguaras, 809 – Centro Edifico Santa Fé - 2° andar, (✉ C.P. 227) , 78820-000 Jaciara ☎66 3461 1966 **W:** www.radioxavantes.com.br **E:** radioxavantes@vsp.com.br or radioxavantes@hotmail.com – **MT46**) 78700-000 Rondónopolis – **MT47**) Rua Araçuaí 1105, Centro, 78575-000 Juara ☎66 3556 1478 **W:** www.difusorajuara.com.br **E:** difusoraaovivo@hotmail.com – **MT49**) 78860-000 Nova Brassilândia – **MT50**) 78635-000 Água Boa. – **MT51**) 78785-000 Alto Taquari – **MT52**) 78325-000 Aripuanã – **MT53**) Rua Padre Rolin No 282, Barrio Inconfidencia, 78645-000 Vila Rica ☎3554-1723 **W:** www.radiovilaam.com.br **E:** am@radiovilaam.com.br – **MT54**) 78505-000 Terra Nova do Norte – **MT55**) 78390-000 Barra do Bugres – **MT56**) 78590-000 Paranaita – **MT62**) 78840-000 Campo Verde – **MT63**) 78625-000 Novo São Joaquim – **MT64**) Campo Novo do Parecis – **MT65**) 78255-000 Jauru – **MT66**) Vila Bela da Santíssima Trinidade.

PA00) PARÁ
PA01) Av Araguaia 247, 68551-100 Redenção **E:** roriente@realonline.com.br – **PA02**) C.P 119, 68371-970 Altamira – **PA03**) Av Almirante Barroso 2190 3° andar, Marco, 66095-000 Belém **W:** www.radioclubedopara.com.br **E:** alderio@rbadecomunicacao.com.br – **PA04**) Av São Sebastião, 622-A Bloco A - Centro, 68005-090 Santarém ☎93 3523 1066 ☐93 3523 2685 **W:** www.radiorruraldesantarem.com.br **E:** rural@radiouralsantarem.com.br or edilrural@gmail.com.br – **PA05**) Av. Juscalino Kubitschek Oliveira s/n, Centro , 68540-000 Conceição do Araguaia ☎94 3421 1576 **W:** www.radioregionaldoaraguaia.com.br – **PA06**) Praça dos Notáveis 1006, 68400-000 Cametá – **PA07**) Rodovia BR-316 Km 58, 68741-740 Castanhal – **PA08**) Travessa Campos Sales 370, 66015-080 Belém ☎91 4005 4400 **W:** www.supermarajoara.com.br **E:** comercial@supermarajoara.com.br - **FM:** 100.9MHz – **PA09**) Travessa Vileta 2193, 66093-380 Belém **W:** www.rbn.org.br **E:** rbncontato@rbn.org.br – **PA10**) Av Bráz de Aguiar, No 351, Nazaré, 66035-000 Belém **W:** www.radioliberal.com.br **E:** dirgel@radioliberal.com.br ☎91 3213 1500 **FM:** 97.5MHz – **PA11**) C.P 038, 68801-970 Breves – **PA12**) Rua 13 de Maio, s/n - Centro, 68600-000 Bragança ☎91 3425 1774 ☐91 3425 1702 **W:** www.fundacaoeducadora.com.br **E:** contato@fundacaoeducadora.com.br – **FM:** 106.7MHz – **PA13**) Av Mendonça Furtado, 1481 Santa Clara 68005-100 Santarém **W:** www.rtvpontanegra.com.br **E:** am890@rtvpontanegra.com.br ☎93 3523 3348 – **PA14**) Av Coronel Monfredo 42, 68820-000 São Sebastião da Boa Vista – **PA16**) Av Almirante Barroso 735, 66093-020 Belém ☎91 4005 7700 **W:** www.portalcultura.com.br **E:** radiocultura@gmail.com.br - **FM:** 93.7MHz – **PA17**) Margem da Rodovia PA -130 Km 8/9, Nova Marabá, 68500-000 Marabá ☎94 3322 2220 **W:** www.tveldoradosbt.com.br/radio.html **E:** belemfm@hotmail.com – **PA18**) Rodovia Transamazonica Km 01, 68180-010 Itaituba – **PA19**) Rua Lauro Sodré 722, 68456-000 Tucuruí – **PA20**) Av Rui Barbosa 825, 68005-080 Santarém – **PA21**) Av Xingu s/n, 68555-010 Xinguara – **PA22**) Travessa E Simões 230, 68250-000 Óbidos – **PA23**) Travessa Dr.Lauro Sodré 299, 68250 000 Obidos **W:** www.kaleb.hpg.ig.com.br ☎☐93 3547 1699 – **PA24**) Travessa 18 No 1863, entre 4 e 5 ruas, 68870-000 Soure – **PA25**) Av Visconde de Sousa Franco 116, Centro, 68780-000 Vigia – **PA26**) Av.Fernando Guilhon, 358 Bela Vista, 68180-000 Itaituba ☎93 9121 6510 **W:** www.960rci.com.br **E:** radioclube@zzum.com.br – **PA27**) Av Tropical s/n, 68625-000 Paragominas – **PA28**) Rua 2 de Outobro s/n, 68200-000 Alenquer – **PA29**) Rodovia Transamazônica Km 04, 68502-290 Marabá – **PA30**) Rodovia BR-010 Km 09, Bairro Industrial, 68660-000 São Miguel do Guamá – **PA32**) Travessa

Mauriti 1006, Bairro da Pedreira, 66080-650 Belém – **PA33**) Av Antonio Marques Ribeiro, 242, 68380-000 São Felix do Xingu ☎91 4351 1243 – **PA34**) Av Augusto Montenegro s/n, 68710-000 Maracanã – **PA35**) Rua Barao Rio Branco 1562, 68440-000 Abaetetuba – **PA36**) 68230-000 Almeirim – **PA37**) 68430-000 Igarapé Miri – **PA40**) 68220-000 Monte Alegre – **PA41**) 68300-000 Gurupa – **PA42**) 68620-000 Visen – **PA47**) 68530-000 Rio Maria – **PA48**) Oriximineâ – **PA49**) Tailândie – **PA50**) Portel – **PA51**) Prainha – **PA52**) Barcarena – **PA53**) Beiao – **PA54**) Almeirien – **PA55**) Tomé Açu,

PB00) PARAÍBA
PB01) Rua Pres. João Pessoa 25, 58800-010 Sousa ☎83 3521 2116 **W:** www.portalprogresso.com – **PB02**) C.P 26, 58900-970 Cajazeiras **W:** www.diariodosertao.com.br **E:** diariodosertao@gmail.com ☐83 3531 1334 – **PB03**) Rua Padre Manoel Otaviano 340, 58970-000 Conceição **W:** www.radioeducadordeconceicao.com – **PB04**) Rua Presidente Epitácio Pessoa 242, 58700-020 Patos ☎83 3421 3932 **W:** www.radiopanati.com.br – **PB05**) Rod PB 075 s/n km 1,25, Zona Rural (✉C.P 40, 58200-970) 58200-000 Guarabira **W:** www.cultura790.com.br – **PB06**) Praça Pres Epitácio Pessoa 167, 58755-000 Princesa Isabel – **PB07**) Rua Joao Pessoa 313 1° andar (✉ C.P 134 58100-970), Campina Grande ☎83 3349 2101 **W:** www.radiocaturite.com.br **E:** radiocaturite@radiocaturite.com.br – **PB08**) Rua Coronel Juvêncio Carneiro, 160 - Centro, 58900-000 Cajazeiras ☎88 3531 4530 **W:** www.radiocajazeiras.com.br **E:** radiocajazeiras@hotmail.com – **PB09**) Av D.Pedro II s/n, Torre, (C.P 1089, 58001-970) 58013-420 João Pessoa ☎83 3218 7900 **W:** www.radiotabajara.pb.gov.br – **PB10**) Rua Manoel Pedro s/n, 58884-000 Catolé do Rocha ☎83 3441 2013 **W:** www.portalprogresso.com **E:** portalprogreso@yahoo.com **PB11**) Av.Pres Getúlio Vargas. 566 Centro, , 58101-200 Campina Grande ☎83 3342 0046 **W:** www.radiocaririam.com.br – **PB12**) Av Pedro II 523, Centro, 58013-420 João Pessoa ☎83 3216 5044 **W:** www.correiosat.com.br/?radio=3 – **PB13**) Rua das Trincheiras 198, Centro, 58011-000 João Pessoa – **PB14**) C.P 160, 58100-970 Campina Grande **W:** www.clube.am **E:** martha@clubeampb.com.br – **PB15**) Rua Rui Barbosa, 53 – Centro (✉C.P 57, 58790-970) 58700-060 Patos **W:** www.radioespinharas.com.br **E:** antoniosilva.1@hotmail.com.br ☎83 3421 3791 ☐83 3221 3795 – **PB16**) Rua Orlando Soares de Oliveira 299 Bairro Miramar, 58032-083 Joao Pessoa – ☎83 3226 2876 **W:** www.comunidademana.com.br **E:** mana@comunidademana.com.br – **PB17**) Rua Dr Carlos Pires 17, Centro, 58804-200 Sousa ☎83 3522 1525 **W:** www.950news.com.br **E:** contato@950news.com.br – **PB18**) Rua Monsenhor Valeriano s/n, 58840-000 Pombal – **PB19**) Rua Antônio Firmino 344, 58187-000 Picuí ☎83 3371 2400 **W:** www.radiocenecistapicui.com.br – **PB20**) Rua Cândido de Assis 421, Centro, 58840-000 Pombal ☎83 3431 3558 **W:** www.bonsucessoam.com.br **E:** contactos@bonsucessam.com.br – **PB21**) Rua Conselheiro Henrique, 17 – Centro, 58010-690 João Pessoa ☎83 3241 5657 **W:** www.portalsanhaua.com.br **E:** radioshanua@ig.com.br – **PB22**) Rua Epitácio Pessoa, 8 - Centro, 58200-000 Guarabira ☎83 3321 8437 **W:** www.radiorruralam850.com.br **E:** radio.rural@oi.com.br – **PB23**) Rua Monsenhor Palmeira 471, Centro, 58135-000 Esperança ☎83 3322 6327 **W:** www.redeesperanca.com.br **E:** contato@redeesperanca.com – **PB24**) Rua Getúlio Vargas 129, 58500-000 Monteiro ☎83 3351 2612 – **PB25**) Rua Coronel Pedro Targino s/n, 58233-000 Araruna ☎83 3370 1102 ☐83 3370 1104 **W:** www.radioserrana.com **E:** radioserrana@gmail.com or radioserranaam@hotmail.com 0730-0100 – **PB26**) Rod. PB 105, 105 km 33, Zona Rural Solânea, 58220-000 Bananeiras ☎83 3367 1599 – **PB27**) Rua Prefeito Inacio Claudino 121, 58155-000 Soledade – **PB28**) Rua João Sabiá 56, 58540-000 Sumé ☎83 3533 2456 **W:** www.radiocidadesume.com – **PB29**) Praça Frei Martinho s/n, 1° andar, Centro, 58700-100 Patos ☎83 3421 3132 - **FM:** 102.9MHz – **PB30**) Rua Cel Guimarães, 6 - Centro, 58900-000 Cajazeiras ☎83 3531 3715 **W:** www.oeste1000.com.br **E:** oesteam@hotmail.com – **PB31**) Rua Orlando Soares de Oliveira 299 Bairro Miramar, 58032-083 Joao Pessoa ☎83 3226 2876 **W:** www.comunidademana.com. br **E:** mana@comunidademana.com.br – **PB32**) Rua Epitácio Pessoa, 184 - Centro, 58397-000 Areia ☎63 3362 2778 **W:** www.radiojardim.com.br **E:** radiojardim@hotmail.com – **PB34**) Av Ananias Conserva 18, Centro, 58780-000 Itaporanga ☎83 3451 3879 **W:** www.correiosat.com.br/?radio=5 – **PB35**) 58100-000 Campina Grande.

PE00) PERNAMBUCO
PE01) Av Santo Antônio 324, 55290-000 Garanhuns – **PE02**) Rua Floriano Peixoto 780, São, José, 50020-060 Recife ☎81 3799 9512 ☐81 3797 9511 **W:** www.redebrasildecomunicacao.com.br **E:** boasnovas@radioboasnovas.net – **PE03**) A Joaquim Nabuco 322, São Crístovão,56503-150 Arcoverde ☎87 3821 0664 **W:** www.radiocardealam.com **E:** rcardeal@arconet.com.br **MSN:** cardealam@hotmail.com – **PE04**) Praça da Bandeira s/n, 55700-000 Limoeiro – **PE05**) Rodovia Araraipina – Picos Km 3, 56280-000 Araripina ☎87 3873 1366 **W:** www.radiograndeserra.com.br **FM:** 94MHz – **PE06**) Av Sete de Setembro s/n, Bairro Km 02, 56300-000 Petrolina ☎87 3861 4744 **W:** www.granderioam.com **E:** granderioam@uol.com.br – **PE07**) Rua do Veiga 590, Santo Amaro, 50040-915 Recife ☎81 3421 4244 **W:** www.radioclubeam.com.br **E:** car-

los.miguel@radioclubeam.com.br – **PE08**) Praça Maria Auxiliadora, 205 - Centro, 56302-335 Petrolina ☎87 3862 1522 **W:** www.emissorarural.com.br - **F.PI**: FM – **PE09**) Rua Capitão Lima 250, Santo Amaro, 50040-080 Recife **W:** www.radiojornal.com.br – **PE10**) Av.Norte, 68 – Santo Amaro, 50040-200 Recife ☎81 2126 8063 **W:** www.tvu.ufpe.br **E:** radioam@ufpe.br – **PE11**) Av Pres Kennedy 3092, Peixinhos, 53260-640 Olinda ☎81 3444 8282 **W:** www.radiotamandare.com.br – **PE12**) Rua da Conceição 16/22, 2° andar, Centro, 55000-000 Caruaru ☎81 3722 5201 **W:** www.liberdade.com.br **E:** programacaoam@liberdade.com.br - **FM:** 94.7MHz – **PE13**) Av Padre Rocha s/n, 55810-000 Carpina **W:** www.radioplanaltoam950.com.br – **PE14**) Av Maria Emília Cavalcanti 570, 55870-000 Timbaúba – **PE15**) Estrada do Passarinho 1415, Caixa D´água, 53170-110 Olinda ☎81 3444 7855 ▤81 3444 7858 **W:** www.radiolindaam.com.br **E:** comercial@radioolindaam.com.br – **PE16**) C.P 88, 55001-970 Caruaru – **PE17**) Rua Floriano Peixoto 780, 1° andar, 50020-060 Recife **W:** http://jc3.uol.com.br/radiojornal/canal.php?canal=49 **E:** gus@jc.com.br – **PE18**) Av Rui Barbosa, 65 - Divinópolis, 55010-540 Caruaru **W:** www.radioculturadonordeste.com.br **E:** jornalismo@radio-culturadonordeste.com.br **PE19**) Rua dos Ferreiros s/n, Granja Fazenda Nova, 55600-000 Vitória de Santo Antão – **PE20**) Av. Rui Barbosa 1236, Heliópolis 55293-300 Garanhuns ☎87 3762 7244 **W:** http://jc3.uol.com.br/radiojornal – **PE21**) Rua Coronel Urbano Ribeiro de Sena 956, Cajueiro, 52221-000 Recife ☎81 3444 2566 **W:** www.radiocapibaribe.com.br **E:** radiocapibaribd@radiocapibaribe.com.br – **PE22**) Av. Timbi, 672 - Timbi, 54765-240 Camarajibe ☎81 3485 1322 **W:** www.radioguarany.com.br – **PE23**) Rua Capitão do Lima 50, Santo Amaro, 50040-080 Recife – ☎81 3413 6374 ▤81 3413 6384 – **PE24**) Av F Pessoa de Queiróz s/n, 55200-000 Pesqueira **W:** www.radiojornal.com.br – **PE25**) Rua Manoel Paulino dos Santos, s/n 55540-000 Palmares ☎81 3662 1020 **W:** www.rcpalmares.com.br **PE26**) Rua Inocencio Gomes de Andrade, 619 - Centro, 56903 906 Serra Talhada ☎83 3831 1700 **W:** www.radioavozdosertao.com **PE27**) Rua Newton César, 5 - Centro, 56800-000 Afogados da Ingazeira ☎87 3838 1213 **W:** www.superpajeu.com.br **E:** contato@superpajeu.com.br 0800-0200 – **PE28**) Rua Benjamin Constant 16, 55750-000 Surubim – **PE29**) Rua José Lopes da Silva s/n, São Pedro, 55150-000 Belo Jardim – **PE30**) Rua Antônio Figueira Soares s/n, 56000-000 Salgueiro **W:** www.asabranca.com.br **E:** asabranca@asabranca.com.br – Rua Jão Pessoa, 25 - Centro, 56700-000 São José do Egito ☎87 3844 1024 **W:** www.cultura.am.br **Email;** cultura@cultura.am.br – **PE32**) Praça Duque de Caxias 818, 55900-000 Goiana – **PE33**) Av Fernando Bezerra 1123, Centro, 56200-000 Ouricuri ☎87 3874 1559 – **PE34**) Rua Manoel Balbino 184, 55190-000 Santa Cruz do Capibaribe – **PE35**) Av Coronel Trapia, s/n – Centro,, 56440-000 Belém de São Francisco ☎87 3876 1380 **W:** www.radioeducadoradebelem.com.br **E:** educadoradebelem@yahoo.com– **PE36**) Rua São Francisco de Assis s/n, Centro , 55641-190 Gravatá (C.P 64, 55641-970) ☎81 3533 4764 **W:** www.cancaonova.com or blog.cancaonova.com/gravata **E:** radiogravata@cancaonova.com – **PE37**) Rua Conselheiro João Alfredo, 22 Centro, 55330-000 Bom Conselho – **PE38**) 56460-000 Petrolândia – **PE39**) 56180-000 Cabrobó – **PE40**) 55660-000 Bezerros.

PI00) PIAUÍ
PI01) Trav Franisco Antônio da Silva, N° 115, Centro, 64770-000 São Raimundo Nonato ☎89 3582 1497 ▤89 3582 1649 **W:** www.radioserradacapivara.com.br **E:** capivara550@yahoo.com.br – **PI02**) Rua Alvaro Mendes 972, 64000-060 Teresina - **FM:** 94.1MHz – **PI03**) Rua Taumaturgo de Azevedo 995, 64100-000 Barras – **PI04**) Av Valter Alencar 2120, Monte Castelo, 64017-500 Teresina ☎86 2107 6600 - **FM:** 99.1MHz – **PI05**) Av Heróis do Jenipapo 37, 64280-000 Campo Maior – **PI06**) Rua Pres Kennedy 233, 64490-000 Regeneração – **PI07**) Praça de Comércio 400, 1° andar, 64500-000 Oeiras – **PI08**) Av.Prof Valter Alencar 2021, 64017-500 Teresina **W:** www.antares.com.br – **PI09**) Rua Joaquim Baldoíno 40, 64600-000 Picos – **PI10**) Rua 24 de Janeiro, 150, Sul, 64001-230 Teresina **W:** www.radiopioneira.am.br **E:** rosemiro@radiopioneira.am.br ☎86 2107 8121 ▤86 2107 8122 – **PI11**) Av Presidente Getúlio Vargas 266, 64200-000 Parnaíba – **PI12**) Rua 18 de Setembro 678, 64260-000 Piripiri – **PI13**) Av Miguel Rosa 3775 Sul, 64001-490 Teresina – **PI14**) Av Prefeito J de Carvalho, 64400-000 Amarante – **PI15**) Av Rio Branco 314, 64860-000 Uruçuí – **PI16**) Rua Clementino Ribeiro 56, 2° andar, 64800-000 Floriano – **PI17**) Praça Emilio Cavalcante 29, 64980-000 Corrente – **PI18**) Rua Antônio Neto 1065, 64800-000 Floriano – **PI19**) Rua Sabino Paulo 696, 64760-000 São João do Piauí – **PI20**) Av João de Paiva 94, 64290-000 Altos – **PI21**) Av Dr Raimundo Santos, 537 – Centro , 64900-000 Bom Jesus – **PI22**) Praca Sto. Antônio, 1019 ap 101, Centro, 64200-361 Parnaíba ☎86 3322 3550 **W:** www.radiogloboparnaiba.com.br – **PI23**) Rua General Taumaturgo de Azevedo 800, 64100-000 Barras – **PI24**) Av Marechal Deodoro, 203, 64750-000 Paulistana – **PI25**) Rua Fernando Bacelar 480, 64240-000 Piracuruca – **PI26**) Av José de Deus Lacerda 534, 64130-000 Miguel Alves – **PI27**) Av Governador Chagas Rodrigues s/n, 64575-000 Jaicós – **PI28**) Rua Corinto Andrade, s/n - Centro, 64255-000 Pedro II ☎86 3271 1186 **W:** http://radiocruzeiroam.com.br **E:** radiocruzeirop2@hotmail.com – **PI29**) Praça da Independência

69, 64325-000 Elesbão Veloso – **PI30**) Rua Joaquim Baldoíno 48, 64600-000 Picos - **FM:** 94.5MHz – **PI31**) Av João Ferreira 199, 64460-000 Agua Branca – **PI32**) Rua Hugo Napoleão 940, 64110-000 José de Freitas – **PI33**) Rua Pedro II 695, 64340-000 Castelo do Piauí – **PI34**) Rua Pedro II s/n, 64330-000 São Miguel do Tapuio – **PI35**) Rua Euripedes Martins, 500 - Centro, 64300-000 Valença do Piauí – **PI36**) Rua Matias Gomes 510, 64700-000 Simplício Mendes – **PI37**) Rua Coronel Narciso 728, 64120-000 União – **PI38**) Rua Padre Joaquim Nonato s/n, 64390-000 Demerval Lobão – **PI39**) Rua Coronel Messeas Melo s/n, 64190-000 Batalha – **PI40**) Estrada Barra do Longa, Periferia de Cidade, 64230-000 Buriti dos Lopes – **PI41**) Av Antonino Freire 1356, 64001-040 Teresina – **PI42**) Praça Waldemar Leal 42, 64680-000 Padre Marcos – **PI43**) Rua Sete de Setembro 471, 64160-000 Luzilândia – **PI44**) Praça da Bandeira 93, 64500-000 Oeiras – **PI45**) Rua Prof Alceu Brandão 2397, Barrio Monte Castelo, 64016-150 Teresina – **PI46**) Rua Coronel José Fortes 549, 64180-000 Esperantina – **PI47**) Rodovia BR-343 s/n, 64290-000 Altos – **PI48**) Av Professor Alceu Brandão 2750, 64016-150 Teresina **W:** radioclick.globo.com - **FM:** 91.1MHz – **PI49**) Rua São VicenteS/N, Bomba, 64600-000 Picos ☎89 3415 900 ▤89 3415 5152 **W:** www.grnordeste.com.br/radionordeste

PR00) PARANÁ
PR01) Rua Quintino Bocaiuva, 41 - Centro, 86020-100 Londrina ☎43 3344 2038 **W:** www.radiolondrina.com.br **E:** radiolondrina@onda.com.br – **PR02**) Rua XV de Novembro 3466, Centro, 85010-000 Guarapuava **E:** cultura@gol.psi.br **W:** www.centralcultura.com.br ☎42 3623 6423 ▤42 3723 7269 - **FM:** 93.7MHz – **PR03**) Av Brasil, 1720(⊠C.P 10, 85892-970), 85892-000 Santa Helena **W:** www.radiograndelago.com.br **E:** grandelago@rgl.com.br ☎45 3268 1212 ▤ 45 3268 1135 – **PR04**) Av Mal Humberto de Alencar Castelo Branco 590, Cristo Rei 82 530-020 Curitiba ☎41 3263-3311 **W:** www.difusoraam590.com.br – **PR05**) Praça de Café N° 1100, 86900-000 Jandaia do Sul ☎43 3432 9797 **W:** www.radiojandaia.com.br **E:** contato@radiojandaia.com.br **FM:** 103.3MHz – **PR06**) Rua Julio Perneta, 695, Mercês, 80810-110 Curitiba ☎41 3331 7400 ☎41 3331 7449 **W:** www.pr.gov.br/rtve/ – **PR07**) Rua 7 de Setembro, 520, Centro, (C.P 1026), 85960-000 Marechal Cândido Rondón **W:** www.radioeducadora@rondonet.com.br **E:** educadora@rondonet.com.br ☎45 3284 1212 – **PR08**) Rod João Carlos Strass s/n, Heimtal (⊠C.P. 337, 86001-970) 86084-610 Londrina ☎43 3339 6244 **W:** www.radiotupilondrina.com.br – **PR09**) Rua Antônio Costa N° 529, Bairro Bela Vista Alegre das Mercâs, 86390-000 Cambará ☎41 3240 7500 **W:** www.radiobandab.com.br/norte_pioneiro.php **E:** nortepioneiro@radiobandab.com.br ▤43 3532 4050 – **PR10**) Rua Lord Lovat 497, Centro, 87600-000 Nova Esperança ☎44 3252 4533 **W:** www.cancaonova.com **E:** radionovaesperanca@cancaonova.com – **PR11**) Rua Oyapock, 649, Cristo Reiç, 80050-450 Curitiba ☎41 3318 5860 **W:** www.radiogloboucuritiba.com.br – **PR12**) Rua Saldanha Marinho 1581, Apto B - Ventro, 85010-290 Guarapuava ☎42 3035 7010 **W:** www.radiocaciqueam.com.br **E:** comercialcacique@brturbo.com.br – **PR13**) Rua Sergipe, 843 - sala 05, (C.P 916) 86010-380 Londrina **W:** www.radiodifusoradelondrina.com.br **E:** radiodifusora690@aol.com ☎43 3324 7369 – **PR14**) Rua 15 de Novembro 433, Centro, 84010-905 Ponta Grossa. ☎42 3025 7211 **W:** www.difusora690.com.br **Email.** suporte@difusora690.com.br – **PR15**) Av Paraná 1885, Bairro Boa Vista, 82510-000 Curitiba ☎41 3251 2410 **W:** http://radioevangelismo.com.br **E:** radio@radioevangelismo.com.br – **PR16**) Av. Capitão Índio Bandeira, 1400 5° Andar – Centro Empresarial Antares, 87300-005 Campo Mourão ☎44 3523 5248 **W:** www.portalradiorural.com.br/2009/ – **PR17**) Av 19 de Agosto 522, 1° andar, 87360-000 Goio-Erê ☎44 3522 7777 ▤44 3522 1162 **W:** www.radiogoioere.com.br **E:** rgam@goioere.com.br – **PR18**) Av. Maranhão 62 2° piso, Centro, 87200-000 Cianorte ▤44 3629 1514 **W:** www.radioportavozam.com.br **E:** radioportavozam@radioportavozam.com.br – **PR19**) Rua Frei Everaldo 445, 85560-000 Chopinzinho **W:** www.radiochopinzinho.com.br **E:** radiochopinzinho@chnet.com.br – **PR20**) Rua Bruno Filgueira 1210, 80440-220 Curitiba ☎42 3623 7565 **W:** www.centralcultura.com.br - **FM:** 93.7 – **PR21**) Rua Marechal Floriano Peixoto 1123, 85851-020 Foz do Iguaçu **W:** www.equipelegal.com.br – **PR22**) Rodovia do Xisto BR 475 Km 20 No 2018, 83702-560 Araucária ☎41 3642 1010 **W:** www.radioiguassu.com.br – **PR23**) Rua Coronel Manoel Ferreira Bello 64, 85530-000 Clevelândia **W:** www.rdprogresso.com.br **E:** rdprogresso@pinet.com.br – **PR24**) Rua Anita Garibaldi, 43 - Centro, 86020-500 Londrina ☎43 3326 1400 ▤43 3321 3501 **W:** www.cbnlondrina.com.br **E:** cbnlondrina@cbnlondrina.com.br **FM:** 93.5MHz – **PR25**) Av. Euclides Cunha 455, Zona 04, 87014-250 Maringá ☎44 3225 1765 **W:** www.pingafogonoticias.com.br/radio-novainga.html – **PR26**) Rua Iguaçu 808, Centro, 85501-270 Pato Branco ☎46 3220 0890 **W:** www.radioitapua.com.br – **PR27**) Praça Marechal Floriano Peixoto 581 - 3° andar, (C.P 090 84001-970) 84010-910 Ponta Grossa ☎42 3025 1900 ▤42 3027 2112 **W:** www.radiosantana.com.br **E:** adm@radiosantana.com.br – **PR28**) Av Largo São Vicente de Paulo 1085, 85900-215 Toledo **W:** www.radiouniaodetoledo.com.br **E:** contato@radiouniaodetoledo.com.br ☎45 3055 2841 ▤45 3055 2488 – **PR29**) Rua Sao Paulo, 910 - Centro, 86808-070 Apucarana **W:** www.novaam.

com.br **E:** novaam@uol.com.br☎43 3423 1100 – **PR30**) C.P 178, 86600-970 Rolândia E: radiocultura@onda.com.br – **PR31**) Rua João Negrão 558, 80010-200 Curitiba ☎41 3339 2900 **W:** www.radiocultura930.com. br **E:** radiocultura@radiocultura930.com.br – **PR32**) Rua Joubert de Carvalho 623, 87013-200 Maringá – **PR33**) Rua Dom Bosco, 145 – Jardim Dom Bosco, 86060-340 Londrina **W:** www.radioalvorada.am.br **E:** alvorada@radioalvorada.am.br ☎43 3347 0606 ☐43 3347 0303 – **PR34**) Rua Santa Catarina 970, 85960-000 Marechal Cândido Rondón 45 3284 8080 **W:** www.radiodifusora.net **E:** comercial@radiodifusora.net - **FM:** 91.5MHz – **PR35**) Rua Tocantins 1991, 85505-140 Pato Branco **W:**radiocelinauta.com.br **German:** Sun 2200-2300 – **PR36**) Praça Generoso Marques 90, , Galeria Andrade, Ed Claudia 1° andar, Centro, 80020-230 Curitiba **W:** www.radiocolombo.com.br – **PR37**) Av Pedro Soccol 542, São Cristovão, 85884-000 Medianeira **W:** www.independenciaam.com. br **E:** contato@independenciaam.com.br – **PR38**) Rua Paraná 650, Centro, 874-000 Cruzeiro do Oeste **W:** www.difusoraregional590.com.br **E:** alodifusora@bol.com.br – **PR39**) Rua Visconde de Mauá 123, Jardim Shangrilá, 86070-540 Londrina ☎43 3328 1030 – **PR40**) Rua Ulisses Faria 1077, 83900-000 São Mateus do Sul – **PR41**) Rua Maranhão, 2955 – Alto Alegre, 85805-220 Cascavel ☎45 3321 7000 ☐45 3226 5565 **W:** www.cbncascavel.com.br **E:** cantini@cbncascavel.com.br - **FM:** 102.7MHz – **PR42**) Praça Senador Corrêa 128, 80230-130 Curitiba ☎41 3221 6060 **W:** www.padreginaldomanzotti.org.br – **PR43**) Rua Bahia 667, 86690-000 Colorado – **PR44**) Rua Porto Alegre 21, 1° andar, 85601-480 Francisco Beltrão **W:** www.radioeducadorafb.com.br **E:** comercialeducadora@wmail.com.br ☎ 46 3524 2255 – **PR45**) Rua Dario Antônio Bordin, 313 - Centro, 84600-000 União da Vitória ☎42 3522 3596 **W:** www.radiouniaoam.com.br **E:** gerente@radiouniaoam.com.br – **PR46**) Av Ivan Ferreira do Amaral, 331-Centro,85400-000 Guaraniaçu ☎45 3232 2722 **W:** www.superrg.com.br **E:** superrq@superrg.com.br – **PR47**) Rua 15 de Novembro 344, Centro, 84010-020 Ponta Grossa **W:** www.prj2.com.br – **PR48**) Rua Edson Martins, 1935, Esquina c/Avenida Parigot de Souza - Centro, 87703-420 Paranavaí ☎44 3423 6565 **W:** www.culturaparanavai.com.br **E:** am@culturaparanavai.com.br – **PR49**) C.P 101, 86300-000 Cornélio Procópio **E:** educa1080@uol.com.br – **PR50**) Av Brasil 1407, 87302-230 Campo Mourão ☎44 3525 1413 **W:** www.radiocolmeiaam.com.br – **PR51**) Rua das Américas 255, 85550-000 Coronel Vivida ☎46 3232 1398 **W:** www.radiovicentepallotti.com.br **E:** atendimento@radiovicentepallotti.com.br – **PR52**) Av Higienópolis 2100, 86015-905 Londrina **W:** www.paiquere.com.br **E:** paiquere@paiquere.com.br ☎43 3323 5500 ☐43 3339 1175 - **FM:** 98.9MHz – **PR53**) Quince de Novembro 2175 - 8° andar, 83005-000 São José dos Pinhais ☎41 3568 1718 **W:** www.radiomais.am.br **E:** radio@radiomais.am.br - **FM:** 97.9MHz – **PR54**) Av Cristóvão Colombo 1055, 86990-000 Marialva ☎44 3232 1115 **W:** www.ingamar.com.br **E:** empresamartiniradio@hotmail.com – **PR55**) Praça Manoel Ribas 112, Centro, 84165-510 Castro ☎42 3232 2224 **W:** www.radiocastro.com.br **E:** radiocastro@convoy.com.br – **PR56**) Rua Fidelcino Dourado, 225, Centro, 86200-000 Londrina **W:** www.radiomanchetelondrina.com.br – **PR57**) Rua João Negrão 595, Centro, 80010-200 Curitiba – **PR58**) Rua Raimundo Leonardi 1301, 85900-110 Toledo ☎45 3378 3161 **W:** www.rioguacu.com.br **E:** radioguacu@uol.com.br – **PR59**) Av Londrina 500, 87970-000 Nova Londrina – **PR60**) Rua João Negrão 595, Centro, 80010-200 Curitiba ☎41 3324 3849 **W:** www.superradiodeuseamor.com.br – **PR61**) Av Paraná 596, 85887-000 Matelândia **E:** radiomatelandia@matelnet.com. br – **PR62**) Rua Afonso Alves de Camargo 1175. Alta da XV, 85010-320 Guarapuava ☎42 3035 8000 **W:** www.difusoraguarapuava.com.br **E:** difusora@mattosleao.com.br – **PR63**) Av Paeaná, 271 an 1, 87704-100 Paranavaí ☎44 3422 3322 **W:** www.radioparanavai.com.br **E:** contato@radioparanavai.com.br – **PR64**) Av. Dos Pineiros s/n, 85650-000 Santa Isabel do Oeste ☎46 3542 1239 **W:** www.radiodanubioazul.com.br **E:** contato@radiodanubioazul.com.br – **PR65**) Rua José Ferreira 262, 86975-000 Mandaguari ☎44 3233 1180 **W:** www.radioguairaca.com.br **E:** guairaca@bwnet.com.br – **PR66**) Rua Marchal Deodora 172, Centro, (☐ C.P 91, 86430-970) 86430-000 Santo Antônio da Platina ☎43 3534 4321 **W:** www.difusoraplatinense.com.br **E:** radiofmvaledosol@bol.com.br - **FM:**. 100.5MHz – **PR67**) Rua Pedro Eloy de Souza 51, 82820-130 Curitiba ☎41 3323 3467 **W:** www.radiocontinental1270.com.br **E:** comercial@radiocontinental1270.com.br – **PR68**) Rua Rio Grande do Sul 1110, 85806-010 Cascavel ☎45 3224 2717 **W:** www.radioglobocascavel.com.br **E:** radiocidade@certto.com.br – **PR69**) C.P 71, 85601-600 Francisco Beltrão **W:** www.seleski.com.br **E:** seleski@win.com.br or 105@wmail.com.br - **FM:** 105.1MHz «FM Super Jovem» – **PR70**) Av Pedro Taques 1864, Jardim Alvorada (☐C.P 1300, 87001-970) 87033-000 Maringá ☎44 3267 3000 **E** radioatalaia@turbo.com.br – **PR71**) Rua XV de Novembro, 591, Centro, 84010-020 Ponta Grossa ☎42 3028 1300 **W:** www.cbnpg.com.br **E:** cbnpg@cbnpg.com.br– **PR72**)Rua Desembargador Westphalen 295, 80010-110 Curitiba ☎41 3366 5657 **W:** http://tropicalam1320.com.br **E:** tropicalam1320@tropicalam1320.com.br – **PR73**) Rua Ébio de Carvallo, 699 – Rui Barbosa, 86031-720 Londrina ☎43 3378 2100 **W:** www.radiobrasilsul.com **E:** brasilsul@radiobrasilsul.com

– **PR74**) TV Silvério Carneiro 3 (☒C.P 26, 84201-970), 8420-970 Jaguariaíva ☎43 3535 1144 **W:** www.radiojaguariaiva.com.br – **PR75**) C.P 13, 87502-970 Umuarama **E:** inconfidencia@fenixnet.com.br – **PR76**) Rua Exp Adir Jorge 511, Centro, 83880-000 Rio Negro ☎47 3642 3969 **W:** www.difusorarionegro.com.br – **PR77**) Rua Flamingos 357, 86701-390 Araponas **W:** www.transnorte.com.br – **PR78**) Av Paul Harris, 50 – Conjunto Paraiso, 86220-000 Assaí ☎43 3262 1367 **W:** www.lideram.com.br **E:** contato@lideram.com.br – **PR79**) Rua Pedro Alvares Cabral 1609, 87560-000 Iporã – **PR80**) Rua Fioravante Dalla Stela 66, Barrio Cristo Rei, 80050-150 Curitiba **W:** www.cancaonova.com/curitiba/radiocancao-nova-nossa-am-1370 – **PR81**) Rua Paraiba 168, 86930-000 São João do Ivaí ☎43 3477 1117 **W:** www.educadora1180.com.br**E:** radioeducadora@ligbr.com.br – **PR82**) Av.Mauá 1988, Vila Operária, 87050-020 Maringa (C.P 76, 87001-970 Maringá) ☎44 3221 3221 ☐44 3222 4969 **W:** www.culturamaringa.com.br **E:** radio@culturamaringa.com.br - **FM:** 102.5MHz – **PR83**) Av Bertino Warmling, 1110 sala 02 - Centro, 85670-000 Salto do Lontra ☎46 3538 1320 **W:** www.rinet.com.br **E:** ri_ouvinte@slnet.com.br – **PR84**) Rua Pref Hugo Cabral, 192 - Centro, 86020-110 Londrina **W:** www.radioglobolondrina.com.br **E:** radioglobolondrina@radioglobolondrina.com.br ☎43 3373 5500 – **PR85**) Av Dep Ivan Ferreira do Amaral Filho, 86 - Centro, 85301-070 Laranjeiras do Sul ☎42 3635 1396 **W:** www.radioeducadora1120.com.br **E:** comercial@radioeducadora1120.com.br – **PR86**) Rua Simão Domingues, 26 - Centro, 84550-000 Rebouças ☎42 3457 1150 **W:** www.alvoradanoar.com.br **E:** comercial@alvoradanoar.co.br – **PR87**) Rua Azauri Guedez Pereira n°1351, 85990-000 Terra Roxa ☎44 3645 1135 **W:** www.radiofronteiradoeste.com **E:** radiofronteiraam@hotmail.com – **PR88**) Rua Nicanor dos Santos Silva 4465,, 87501-120 Umuarama ☎44 3624 4664 **W:** www.culturaumuarama.com.br – **PR89**) Rua Antônio Lemos, 807 - Centro, 86400-000 Jacarezinho **W:** www.educadora1420.com.br **E:** educadora@uol.com.br ☎43 3525 0773 ☐43 3527 2029 – **PR90**) Av Brasil 702, 85710-000 Santo Antônio do Sudoeste **W:** http//radioentrerios.com **E:** contato@radioentrerios.com ☎46 3563 1541 – **PR91**) Rua Mato Grosso 2229, (C.P 66) 85812-020 Cascavel ☎45 3220 1717 **W:** www.radiocolmeia.com.br **E:** radiocolmeia@brturbo.com.br – **PR92**) Av Horácio Klabin 383, 2° andar, 84261-000 Telêmaco Borba - **FM:** 92.9MHz – **PR93**) Rua Vicente Inácio Filho 241, 86360-000 Bandeirantes ☎43 3542 3233 **W:** www.radiocabiuna.com.br – **PR94**) Rua Pedro Oliveira s/n, 85440-000 Ubiratã **E:** difusora@ubinet.com.br ☎44 3543 1717 – **PR95**) Rua Mal Deodoro da Fonseca 711,(☐C.P 171, 87550-970) 87550-000 Altônia ☎44 3659 3444 **W:** www.radiorainha.com.br **E:** radiorainha.com.br – **PR96**) Rua Prof Cleto, 281 - Centro, 83221-320 Paranaguá ☎41 3423 4322 **W:** www.difusoraam1460.com.br **E:** administracao@difusoraam1460.com.br – **PR97**) Rua XV de Novembro 522, 84010-908 Ponta Grossa ☎42 3225 2144 ☐3222 7115 **W:** www.centraldoparana.com.br **E:** central@centraldoparana.com.br 24h – **PR98**) Rua Sao Paulo, 910 – Vila Feliz (☐C.P 777 86800-970), 86808-070 Apucarana ☎43 3033 2216 **W:** www.radioculturaapucarana.com.br **E:** contato@radioculturaapucarana.com.br or amcultura@net21.com.br – **PR99**) Rua Jesuino Alves da Rocha Loures 1964 (C.P 66) 85555-000 Palmas ☎46 3263 1818 **W:** www.redebomjesus.com.br **E:** comercial@radioclubeamfm.com.br 0900-0300 Sat.:1030-0100 - **FM:** 96.5MHz «Horizonte» – **PR100**) Rua Acácio Nunes, 1065 (C.P 217 85980-970), 85980-000 Guaíra ☎44 3642 2068 – **PR101**) Rua Londrina 410, 85640-000 Ampére **W:** www.ampernet.com.br **E:** radioampere@ampernet.com.br – **PR102**) Av. Belo Horizonte 497,87900-00 Loanda ☎44 3425 5252 **W:** www.guadalupeam.com.br **E:** guadelupeam@guadalupeam.com.br – **PR103**) Rua Nilo Sampaio 531, Centro, 84900-000 Ibaiti ☎43 3546 1291 **W:** www.radioeducadora1470.com – **PR104**) Praça Nossa Senhora do Carmo 99 (C.P 54), 85935-000 Assis Chateaubriand ☎44 3528 4477 **E:** radiojornal@visaonet.com.br **W:** www.radiojornalam.com.br – **PR105**) Rua Ebano Pereira 157, 85200-000 Pitanga **W:** www.radiopitanga.com.br – **PR106**) C.P 45, 86730-970 Astorga – **PR107**) Rua Ipiranga 91, 84600-000 União da Vitória ☎42 3522 1098 **W:** www.educadora-uv.com.br **E:** contato@educadora-uv.com.br – **PR108**) Rua João Carlos Farias 85, (☐C.P 230), 86300-000 Cornélio Procópio ☎43 3524 2333 **W:** www.radiocornelio.com.br **E:** contato@radiocornelio.com.br – **PR109**) Rua 7 de Setembro 42, 83750-000 Lapa **W:** www.legendaria.am.br **E:** am960@matrix.com.br ☎41 3622 1918 ☐41 3622 1428 – **PR110**) Avenida Paraná,540, 86925-000 Borrazópolis ☎43 3425 1233 **W:** www.radionovaera.com.br **E:** radionovaera@brturbo.com.br – **PR111**) Av. São João No. 1952 (☒C.P 121, 84400-970) 84400-000 Prudentópolis ☎42 3446 1547 **W:** www.radioesperancaam.com.br **E:** radioesperancaam@hotmail.com – **PR112**) Rua Rouxinol 752, 86701-150 Arapongas ☎43 3055 4535 ☐43 3055 2133 **W:** www.radioarapongas.com.br**E:** radio@arapongas.com.br – **PR113**) Rua João Ramos Piedad 120, 84950-000 Venceslau Brás ☎43 3528 1105 **W:** www.educadoraonline.com - **FM:**95.7MHz – **PR114**) Rua Dr Correis, 289, Centro, 84500-000 Irati **W:** www.radiodifusoradeirati.com **E:** difusoraam@hotmail.com – **PR115**) Rua Florianópolis 1636, 85840-000 Céu Azul **E:** uniao@netceu.com.br – **PR116**) Travessa Vale Porto 240, 83370-000 Antonina ☎41 3432 1362

W: www.serradomaram1520.com.br E: radioserradomar@yahoo.com.br – PR117) Av Souza Naves 1265, 86870-000 Ivaiporã – PR118) Rua 5 de Julho 1065, 85950-000 Palotina – PR119) Rua D Pedro II 1889, 834601-610 Campo Largo E: radiorbn@uol.com.br – PR120) Rua Mauá 2518, (C.P 66), 85770-000 Realeza ☎46 3543 1196 W: radioclube.ampernet.com.br E: radioclube@ampernet.com.br – PR121) Rua Florianópolis 1813, Zona 02, 87200-000 Cianorte ☎44 3629 1317 W: www.radiocapital990.com. br E: adm@radiocapital990.com.br – PR122) Praça Alfredo João Lazzarotto, s/nº (Caixa Postal 61), 89490-000 Siqueira Campos ☎43 3571 1125 W: www.radiobomjesus.com.br E: bomjesus@hotmail com – PR123) Av. Redife, 43485830-970 Formosa do Oeste ☎41 3243 0950 W: www.pioneiraam.com.br – PR124) Av Dambros e Piva, 946(C.P 10, 85615-970) 85615-000 Marmeleiro ☎46 3525 1183 ▤46 3525 1142 W: www.cristal.seleski.com.br E: radiocristal@wim.com.br – PR125) Av Brasil 502, 85760-000 Capanema – PR126) Rua Senador Pinheiro Machado 1536, Centro, 85010-100 Guarapuava ☎42 3035 8000 W: http://difusoraguarapuava.com.br/atalaia E: atalaia@mattosleao.com.br – PR127) Rua do Comércio 654, 85660-000 Dois Vizinhos W: www.educadoradv.com.br E: radio@educadoradv.com.br ☎46 3536 3131 ▤46 3536 3003 – FM: 100.7MHz «Vizinhança» – PR128) Rua Benjamin Constant, 440 - Centro, 84500-000 Irati ☎42 3423 1331 W: www.radio-najua.com.br E: radionajua@radionajua.com.br FM: 106.9MHz – PR129) Rua Sao Paulo 180, 86380-000 Andirá ☎43 3538 4150 W: www.cultu-raandira.com.br E: comercial@culturaandira.com.br – PR130) Rua São Paulo 489, 86840-000 Faxinal ☎43 3461 1129 W: www.radioclubde-faxinal.com.br E: radioclub@folnet.com.br – PR131) Rua Noruega 98, 86182-000 Cambé – PR132) Praça São Pedro 999, 85460-000 Quedas do Iguaçu ☎46 3532 1416 W: www.radiointernacional.com.br E: ademir@radiointernacional.com.br – PR133) Praça Marechal Floriano 108, 84130-000 Palmeira ☎42 3252 3669 W: www.radiopipiranga.com.br E: radioipiranga@br10.com.br – PR134) Rua Amauri Lange Silvéro, No 300 - Pilarzinho, 82120-000 Curitiba ☎41 3271 4700 W: www.clubeb2.com.br E: comercial@clubeam.com 24h – FM: 101.5MHz – PR135) Av Brasil 740, 84350-000 Ortigueira – PR136) Av Maximiliano Vicentin 240, 85270-000 Palmital ☎42 3657 1442 W: www.radiocidadepalmital.com.br E: radiopalmital@hotmail.com – PR137) Rua Melissa 520, 85410-000 Nova Aurora E: raclubna@sercopa.com.br – PR138) Avenida Brasil 531, Sala 74, Centro, 85851-000 Foz do Iguaçu ☎46 3552 2410 W: www.radiofiladelfia.com.br E: radioculturamorenafiladelfia@gmail.com – PR139) Rua Antônio Costa, 529-Vista Alegere das Mercês, 80820-020 Curitiba W: www.radiobandab.com.br E: bandabad@radiobandab.com.br ☎41 3240 7000 – PR140) Rua Urbano Lunardelli 875, 86160-000 Porecatu E: radiobrotense@com.br – PR141) Av Paraná, 220 - Centro, 84470-000 Cândido de Abreu W: www.alternativa710.com.br E: radio@alternativa710.com.br ☎43 3476 1244 – PR142) Av Santos Dumont, 2505-Centro, 85301-040 Laranjeiras do Sul W: www.radiocampoaberto.com.br E: radiocampoaberto.com.br – PR143) Av Genoroso Marques 599, 2º andar, Centro, 85550-000 Coronel Vivida ☎46 3232 1191 W: www.radiovozdosudoeste.com.br/2008 E: radiovoz@win.com.br – PR144) Rua Sete de Setembro 3910, 85560-000 Chopinzinho ☎46 3242 1435 W: www.redebomjesus.com.br E: difusoraaamerica@chnet.com.br - FM: 98.7MHz – PR145) Av Paraná, 201 – Jardim Itajubá, 85857-970 Foz do Iguaçu W: www.radiofoz.com.br E: radiofoz@compubras.com.br ☎45 3523 2211 – PR146) Av Presidente Kennedy, 170 - Norte, 85950-000 Palotina ☎44 3649 7700 ▤44 3649 7713 W: www.radiocontinental-am.com E: comercialmidia@gmail.com 24h - FM: 93.5MHz – PR147) Rua Gârdenia 08, 85420-000 Corbélia ☎45 3242 1799 W: www.radiointegracao.net E: integracao@realplus.com.br – PR148) Rua Farroupilha 80, 2º andar, 85877-000 São Miguel do Iguaçu ☎43 3565 1033 W: www.radiojornalsaomiguel.com.br E: rjcomercial@brturbo.com.br – PR149) Av Iguaçu, 288 - Centro, 85575-000 São Jorge d'Oeste ☎☎46 3534 1184 W: www.difusorasaojorge.com.br E: radiodifusora1490@hotmail.com – PR150) Av Santo Antônio 826, 87320-000 Roncador W: www.princesa820.com.br – PR151) Av Yolanda Loureiro de Carvalho 1021, 87350-000 Ubiratã – PR152) Rua Dom Pedro II, 1581 – Conjunto Residencia, 85901-270 Toledo. ☎45 3252 7095 W: www.radiointegracaoam.com.br E: contato@radiointegracaoam.com.br – PR153) Rua Perfeito Rolim de Moura 104, 84240-000 Piraí do Sul – PR154) Rua Castro Alves 39, (C.P 20) 87160-000 Mandaguaçu W: www.colmeia1170am.com E: contribuinto@colmeia1170am.com.br ☎44 3245 1776 – PR155) Rua Rosalvo Petrechem 551, 85200-000 Pitanga ☎42 3646 3366. – PR156) Av Euclides da Cunha s/n, 87890-000 Terra Rica E: guairaca@vsp.com.br – PR157) Av Iguaçu 858, Ed Palanea, 85585-000 Verê E: rvaledoiguacu@gualinet.com.br – Italian: Sat mornings – PR158) Av Independência s/n, Prox Escola de Aplica, 86130-000 Bela Vista do Paraíso – PR159) Rua Luiz Pinheiro 1446, 84990-000 Arapoti – PR160) Av Iguaçu 366, 85790-000 Capitão Leônidas Marques E: hawai@certto.com.br – PR161) Rua Antonio Rosa 1170, 86490-000 Ribeirão do Pinhal E: am@radioserpin.com.br – PR162) Rua Vicente Machado 385, 84570-000 Mallet ☎42 3542 2004 W: www.radioclubemallet.com E: contato@

radioclubemallet.com – PR163) Rua Marechal Deodoro 22, 85540-000 Mangueirinha E: radioaraucaria@qualinet.com.br – PR164) Rua São Miguel 577, 85570-000 São João ☎46 35331474 W: www.radiosaojo-ao.com.brE: radiosj@radiosaojoao.com.br – PR165) Rua Guarani 829, sala 01, 85501-050 Pato Branco ☎46 3225 4000 W: www.radiopatobranco.com.br E: ouvinte@radiopatobranco.com.br – PR166) Av.Artur de Abreu 29, Ed Palácio do Café, 11° andar, conj 6, Centro, 83203-480 Paranaguá – PR167) 85850-000 Foz do Iguau – PR168) Rua Guilherme Pequeno 413, Centro, 83230-000 Guaratuba ☎41 3472 3275 ▤41 3472 3019 W: www.radiolitoranea.com.br E: amlitorane@onda.com.br – PR169) Rua Ana Beje, 84300-000 Tibagi – ☎42 3275 3247 W: www.radioitay.com.br E: radioitay@radioitay.com.br – PR170) Av Vice-Prefeito Reginaldo Guedes Nocera 84260-000 Telêmaco Borba – PR171) Av Londrina, 523, Centro, 87111-220 Sarandi ☎44 3035 7476 W: www.banda1am.com.br – PR172) 85580-000 Itapejara d´Oeste – PR173) 83540-000 Rio Branco do Sul – PR174) Rua Parigot de Souza, 47, Centro, 86740 Pérola d´Oeste ☎46 3556 1048 W: www.radioperola.com.br E: admin@radioperola.com.br – PR175) 87530-000 Icaraíma – PR176) 86180-000 Cambé – PR178) 84430-000 Imbituva – PR179) Rua Pedro de Oliveira c/Rua Santos Dumont, 85440-000 Ubiratã ☎44 3543 1717 W: www.radiodifusoraubirata.com.br – PR182) 86400-000 Jacarezinho – PR183) 86400-000 Jacarejinho – PR184) Rua José Carlos Colodel, Vila Santa Terezina 306, 83501-140 Almirante Tamandaré W: www.radioba-rigui.com.br

RJ00) RIO DE JANEIRO

RJ01) Rua Benedito Hipolito 1, Centro, 20211-130 Niterói ☎21 2509 6525 ▤21 2245 2028 W: www.ofluminense.com.br – RJ02) Rua Costa Rica 151, Parque Hotel, 28970-000 Araruama ☎22 2665 4119 W: www.radiocostadosol560.com E: radiocostadosol@araruama.com.br – RJ03) Rua Paramopama 131, Ribeira, Ilha do Governador, 21930-110 Rio de Janeiro W: www.radiorelogio.com.br – RJ04) Av.Erasmo Brage 118, 11° andar, Centro 20020-000 Rio de Janeiro – RJ05) Rodovia Presidente Dutra Km 303, Fazenda Barra, 27365-000 Resende – ☎24 3544 1222 ▤24 3355 0733 W: www.radioagulhasnegras.com.br E: agulhasnegrasam@resenet.com.br – FM: 93.9MHz – RJ06) Praça Demerval Barbosa Moreira 28, Centro , 28610-160 Nova Friburgo ☎22 2523 3034 W: www.radiofriburgoam.com.br E: contato@novafriburgoam.com.br – RJ07) Av Dom Helder Camara 4242, 20771-000 Rio de Janeiro – RJ08) Av Alair Ferreira 201A, 28015-020 Campos dos Goitacazes – ☎22 2735 3815 – RJ09) Rua México 111 slj, Centro, 20031-145 Rio de Janeiro ☎21 2220 3656 W: www.redesucesso.com – RJ10) Av Djalma Beda Coube 719, 28500-000 Cantagalo – RJ11) Rua de Assembléia, 10/3401- Centro, 20011- 901 Rio de Janeiro ☎☎ 21 2531 0070 W: www.radiomanchete.com.br E: comercial@radiomanchete.com.br – RJ12) Praça da República 141-A, 3° andar, sala 306, Centro 20211-350 Rio de Janeiro ☎21 2508 8295 W: www.radiomec.com.br – RJ13) Av Rui Barbosa 749, 3° andar, 27910-260 Macaé ☎22 3311 3142 W: www.radio820.com.br E: marcelo820@gmail.com – RJ14) Rua do Russel 434, Glória, 22210-010 Rio de Janeiro – RJ15) Av Portugal 96, Urca, 22291-050 Rio de Janeiro ☎21 4002 3599 W: www.radiotamoio.com.br – RJ16) Av. Marchal Floriano, 114 - Centro , 20540-004, Rio de Janeiro W: www.redeboavontade.com.br – RJ17) Rua de Relação, 18 – Lapa , 20231 110 Rio de Janeiro ☎21 2117 6209 W: www.ebc.com.br/canais/radios/radio-nacional-am-rio-de-janeiro – RJ18) Rua da Conceião 178 an 3, Centro, 20080-032 Rio de Janeiro ☎21 2253 5499 W: www.radiocapitalrio.com.br – RJ19) Travessa Santa Luiza 91, 23900-900 Angra dos Reis ☎24 3365 1352 E: jangra@uol.com.br – RJ20) Rua Buenos Aires N° 68, 19° andar, Centro, 20070-020 Rio de Janeiro ☎21 3171 1067 W: http://blog.cancaonova.com – RJ21) Av Alberto Torres 164, 28035-582 Campos dos Goitacazes ☎22 2733 1082 – RJ22) Estrada Adhemar Bebiano (ex-Estr. Velha da Pavuna) 3517Inhaúma, 20765-170 Rio de Janeiro ☎21 2176 8276 W: www.metropolitana1090.com.br E: contato@metropolitana1090.com.br – RJ23) Av Alair Ferreira 201, Turf-Club, 28022-000 Campos dos Goitacazes ☎22 2734 8544 – RJ24) Rua Presidente Vargas, 541 - Centro, 25802-200 Três Rios ☎24 2252 1797 W: www.radiotresrios.com.br E: rtr@radiotresrios.com.br - FM: 89.7MHz – RJ25) Rua do Russel 426/434, Glória, 22210-010 Rio de Janeiro W: www.mundial.am.br – RJ26) Rua dos Andrades, 109 - 3° andar, Centro, 28010-300 Campos dos Goitacazes ☎22 2722 5699 W: www.radiocontinentalam.com E: continental@viacabo.com.br – RJ27) Rua do Livramento 189, 8° andar, 20221-191 Rio de Janeiro ☎21 2126 2421 W: www.tupi.com – RJ28) Rua Alberto Torres 410, 3° andar, 28400-000 São Fidelis ☎22 2758 1275 – RJ29) Av Boulevard Vinte e Oite de Setembro, 258 - Vila Isabel, 20551-031 Rio de Janeiro ☎21 2576 3049 W: www.boasnovas.com.br E: comercial@radioboasnovas.com.br – RJ30) Rua Alvaro Ramo, 350, Botafogo, 22280-110 Rio de Janeiro ☎21 2586 9400 – RJ31) Rua Carlos de Lacerda 52, 2° andar, 28013-030 Campos de Goytacazes ☎22 2733 0102 W: www.radiocamposdifusora.com.br E: angeladifusora@yahoo.com.br – RJ32) Av.Joaquim Leite 465, 1° andar, 27345-391 Barra Mansa ☎24 3323 3300 W: www.sulfluminense.com.br FM: 99.3MHz

– **RJ33**) Estrada do Dendê, 659, Tauá, Ilha do Governador, 21920-000 Rio de Janeiro ☎21 3396 5252 **W:** www.radioriodejaneiro.am.br **E:** marketing@radioriodejaneiro.am.br **Esperanto:** Wed 2330, Sun 1100 – **RJ34**) Av Cardoso Moreira 422, Sobrado, Centro, 28300-000 Itaperuna ☎24 3824 1410 **W:** www.itaperunaam.com.br **E:** thiago@radioita-perunaam.com.br or radioitaperunaam@gmail.com – **RJ35**) Rua do Mercado 34 Grupo 802, Praça XV, 20010-120 Rio de Janeiro ☎21 2203 2288 **W:** www.radiolivre1440.com.br **E:** contato@radiolivre1440.com. br – **RJ36**) Av Joaquim Leite, 279 - Centro, 27330-042 Barra Mansa **W:** www.radiodocomercio.com.br **E:** atendimento@radiodocomercio.com. br ☎24 3323 3848 – **RJ37**) Rua Dr Temistocles de Almeida 97, 28470-000 Santo Antônio de Pádua ☎22 3853 3173 **E:** radiofeliz@ig.com.br – **RJ38**) Rua Ana Nery, 120 9° andar Centro, 27123-150 Barra do Piraí ☎24 2443 1470 **W:** www.rbpfm.com.br **E:** rbpam@terra.com.br **FM:** 89.9MHz –**RJ39**) Rua Senador Dantas 117, cob 02, Nova Iguaçú ☎21 2767 3333 **W:** www.tropical830am.com.br **E:** tropical@tropical830am. com.br – **RJ40**) Rodovia BR-101 s/n km 270, Zona Rural, 28800-000 Rio Bonito ☎21 2734 8688 – **RJ41**) Av.Fransisco Torres 12, 27286-440 Volta Redonda **E:** diretoria105fm@aleluianet.com.br – **RJ42**) Rua Cardoso 357, 28013-460 Campos dos Goytacazes - ☎22 2733 9072.– **RJ43**) Rua Nilza Chiapeta Fadigas, 275 Sobrado - Várzea, 25963-150 Teresópolis ☎21 2742 1040 ☎21 2742 1920 **W:** www.radioteresopolis.com.br **E:** radioteresopolisam@gmail.com – **RJ44**) Rua Carolina Machado, 173 - Madureira (☒ CP 17030, Madureira, 21310-971), 21351-021 Rio de Janeiro ☎21 3390 1422 **W:** www.continental1520.com.br http://continental1520.com.br **E:** contatos@continental1520.com.br – **RJ45**) Rua Paulino Padilha 80, 28460-000 Miracema – ☎22 3852 0899 – **RJ46**) Rua Marechal Deodoro 46, 9° andar, 25620-150 Petrópolis ☎24 2237 6000 - **FM:** 88.5MHz – **RJ47**) Rua Vereador D.Teixeira Fontes 566, 23815-270 Itaguaí ☎21 688 2267 ☎21 688 1684 **W:** www.radiogranderio1560. com.br 24h – **RJ48**) Rua Carneiro de Mendonça 29-A, Centro, 27600-000 Valença ☎24 2453 4418 **W:** www.radioc.culturadovale.com.br – **RJ49**) Rua Tenente José Teixeira 147, 28360-000 Bom Jesus do Itabapoana ☎22 3831 1570 ☒22 3831 1295 **W:** www.bomjesusam.com.br **E:** radiobjam@acessototal.com.br – **RJ50**) Av Amaral Peixoto 366, 28860-000 Casimiro de Abreu **E:** jornalismo@radiolitoralam.com.br – **RJ51**) Rua Frei Valerio 58, 28740-00 Conceição de Macabu ☎22 2779 2100 **W:** www.popularfluminens.com.br **E:** radiopo-pular@facilite.com.br – **RJ52**) Rua Nilo Peçanha 329, 3° andar, 27542-210 Resende ☎24 3355 2266 – **RJ53**) Rua Gal.Gustavo C.Farias 84, 20910-220 Rio de Janeiro **W:** www.radiorecordrj.com – **RJ54**) Rua Barao Piabanha, 107 - Centro, 25850-000 Paraíba do Sul ☎24 2263 2343 **W:** www.radioclube1540.com.br – **RJ55**) Rua Gal Dionísio, 327 - Guaratiba , 23025-330 Duque de Caxias ☎21 3336 2580 **W:** www.radiopopularam. com.br **E:** rdcradio@yahoo.com.br – **RJ56**) Rua Figueiras de Barros 100, 28210-000 Italva ☎22 2783 1777 **W:** www.aliancaam.com.br **E:** superaliancaam@bol.com.br – **RJ57**) Praça Porto Rocha 56, Apt 102, Centro, 28905-250 Cabo Frio ☎22 2645 4000 **W:** www.cabofriobuzios. com.br **E:** radiocabofrio@mar.com.br – **RJ58**) Rua Coronel Santiago 250, 25950-000 Teresópolis – **RJ59**) Av.Djalma Beda Combe 719, 28500-000 Cantagalo ☎22 2555 4455 – **RJ60**) 28400-000 São Fidelis

RN00) RIO GRANDE DO NORTE

RN01) Av Duque de Caxias, 106 - Ribeira, 59010-200 Natal **W:** www. radioglobonatal.com.br **E:** rcmidia@terra.com.br ☎84 4006 6186 ☒84 4006 6178 – **RN02**) Praça Dom José de Madeiros Delgado s/n, 59300-000 Caicó **W:** www.radiorural.com **E:** comercial@radiorural.com ☎84 3421 2270 ☒84 3421 1229 - **FM:** 95.0MHz – **RN03**) Rua Luiz XV, Bairro Nordeste, 59042-070 Natal ☎84 3653 3780 **W:** www.nordeste-evangelica.com.br **E:** sthenio@nordesteevangelica.com.br 24h – **RN04**) Rua João Pessoa 22, 1° andar, 59380-000 Currais Novos **W:** www. radiocurraisnovosam.com.br – **RN05**) Praça Vigário Antonio Joaquim 39, Centro, 59600-160 Mossoró ☎84 3314 7256 **E:** rrural@serv1000.com.br – **RN06**) Rua Prof Antonio Campos, s/n, Pres Costa e Silva, 59619-218 Mossoró ☎84 3312 4618 **W:** www.redepotiguar.com **E:** boanoite@ rederpc.com.br – **RN07**) Rua Açú 335, Tirol, 59020-110 Natal ☎84 3201 1690 **W:** http://blog.cancaonova.com/natal **E:** radionatal@cancaonova. com – **RN08**) Rua Dr Cunha Mota, s/n - Pereiros, 59600-160 Mossoró ☎84 3317 6167 **W:** www.difusoramossoro.com **E:** gravacao@difu-soramossore.com – **RN09**) Rua Romualdo Galvão, 973 - Lagoa Seca, 59056-100 Natal **E:** cbn@redetropical.com **W:** www.redetropical. com.br ☎84 3211 6400 **FM:** 103.9MHz –**RN10**) Av Deodoro 245, 59012-600 Natal **W:** www.clube.am **E:** nilsonpinheiro.rn@diariosas-sociados.com.br 24h – **RN11**) Praça Des Tomáz Salustino, 42, Centro 1596 ☎84 3431 1266 – **RN12**) Rua Francisca Delfina 30, Centro, 59860-620 Alexandria ☎84 3381 2320 **W:** www.redepotiguar.com **E:** rpcfarol@risanet.com.br – **RN13**) Praça Bento Praxedes, 104 - Centro, 59600-620 Mossoró – **RN14**) Rua Otávio Amorim 643, 59650-000 Açu ☎84 3331 1222 **W:** www.radioprincesadovale.com.br – **RN15**) Rua Nero Nazareno Fernandes, 250 - Centro, 59780-300 Caraúbas ☎84 3337 2297 **W:** http://radiocentenarioam.blogspot.com **E:** centenario@ brisanet.com.br **RN16**) Rua Frei Alberto Cabral 08 - Centro, 59215-000

Nova Cruz ☎ ☒ 84 3281 2123 – **RN17**) Rua Getúlio Vargas, 1296-Centro, 59900-000 Pau dos Ferros ☎84 3351 2388 – **RN18**) Rua Augusto Monteiro, 415 -Centro, 59300-000 Caicó ☎84 3421 1988 – **RN19**) Rua Avinida Rio Branco, 173 - Centro, 59655-000 Areia Branca ☎84 3312 4618 **W:** www.redepotiguar.com **E:** comecial@rederpc.com.br – **RN20**) Rua Ana de Pontes 419, 59255-000 Santo Antônio ☎84 3282 2347 ☒84 3282 2346 – **RN21**) Rua Odorico Ferreira de Souza 70, Bairro DNR, 59200-000 Santa Cruz **W:** www.radiosantacruzam.com.br **E:** contato@ radiosantacruzam.com.br ☎84 3291 2300 ☒84 3291 2201 – **RN22**) Rua Experidião Coimbra, 22 - Centro, 59500-000 Macau ☎84 3521 1765 **W:** www.redetropical.com.br – **RN23**) Rua Cícero Tomáz de Azevedo 1052, Cruz do Monte, 59360-000 Parelhas ☎84 3471 2401 **W:** www. ruralam.com.br **E:** ruraldeparelhas@hotmail.com – **RN24**) Rua Bela Vista 1420, Centro, 59570-000 Ceará Mirim ☎84 3274 2794 ☒84 3274 2119 – **RN25**) Rua Sebastião Guilherme Caldas, s/n – Bairro Baixa da Beleza Jardim do Seridó, 59343-000 Jardim do Seridó ☎84 3472 2587 **W:** www.cabugidoserido.com **E:** cabugidoserido@hotmail.com or cabu-gidoserido@yahoo.com.br – **RN26**) A. Cel Martiniano, 1077 - Centro, 59300-000 Caicó ☎84 3421 4181 ☒84 3417 1112 **W:** www.radiocaico. com.br **E:** suerdamedeiros@uol.com.br – **RN27**) Rua São Jorge 1290, Vale Dourado, 59104-200 Natal ☎84 3664 1330 – **RN28**) Rua Padre Cosme 45, Centro, 59930-000 São Miguel ☎ 84 3353 2166 ☒ 84 3353 2112 – **RN29**) Rua Coronel Freire 242, Centro, 59460-000 São Paulo do Potengi ☎84 3251 2263 ☒84 3251 2381 – **RN30**) Av 21 de Abril 460, BR-460, 59550-000 João Câmara ☎ ☒84 3262 2189 – **RN31**) Rua Joel do Amaral Gurgel s/n, Bairro Cohab, 59700-000 Apodi ☎84 3333 2528 – **RN32**) Rua do Chafariz 1390, Bairro Novo Horizonte, 59584-000 Touros ☎84 3263 2121 ☒84 3263 2526

RO00) RONDÔNIA

RO01) Rua Joaquim Nabuco 1573, 79815-350 Porto Velho ☎69 3224 1081.– **RO02**) Av Rondônio s/n, 78987-000 Rolim de Moura – **RO03**) Av Jamari 4218, 78932-000 Ariquemes ☎69 3536 3385 **W:** www. radioariquemes.com.br **E:** amauri@ariquemes.com.br – **RO04**) Rua Dom Augusto, 1491 - Centro, 78961-380 Ji-Paraná **W:** www.sgcrondonia. com.br **E:** rd-alvorada@ulbrajp.com.br – **RO05**) Av 02 de Junho, 2224, 78975-000 Cacoal ☎69 3441 5316 **W:** www.radio-rondonia.com – **RO06**) Rua Feijo, 2930 Cafézinho, 78963-085 Ji-Paraná ☎69 3424 0406 **W:** www.radiojiparana.com.br **E:** radiojiparana@uol. com.br – **RO07**) Rua Dom Augusto, 1297 Centro, 78964-140 Ji-Parana – **RO08**) Rua Ricardo Catanhede esquina com a Rua Goiás s/n, 78941-000 Jaru – **RO09**) Praça Mário Correa 90, Cristo Rei, 76850-000 Guajará Mirim ☎69 3541 6333 **W:** www.radioeducadoraam.com.br **E:** adm@ radioeducadoraam.com.br - **FM:** 93.7MHz **RO10**) Rua Dourados 4, Setor Industriales, 78930-000 Ariquemes ☎69 3535 3000 **W:** www. radiorondonia.com – **RO11**) Rua das Crianças, 4646 - Areal da Floresta, 78912-210 Porto Velho ☎69 3210 3621 **W:** www.radiocaiari.com **E:** comercialcaiari@gmail.com 0800-0300 SW: 0900-1400, 1900-0300 – **RO12**) Rua Princesa Isabel 128, 78995-000 Vilhena ☎69 3321 3309 **W:** www.radiovilhena.com.br **E:** radiovilhena@brturbo.com.br – **RO14**) Loteamento Monte Alegre, Quadras 35, 36, 38, 39, 41 e 42, 78987-000 Rolim de Moura – **RO15**) Rua Carlos Doneje 1308, Ctg, 78984-000 Pimenta Bueno ☎69 3222 5308 **W:** www.radiorondonia.com **E:** comerci-alpb@radiorondonia.com – **RO16**) Rua 1005, Nro. 1522 – Setor Pionero, (☒ C.P. 105), 76980-000 Vilhena ☎69 3322 2589 **W:** www.plansol.com. br **E:** plansol@hotmail.com – **RO17**) Rua Miguel Chakian 1300, Bairro Embratel, 78906-300 Porto Velho – **RO18**) Rua 6 de Maio No 1811, Bairro Casa Preta(☒ C.P. 163), 76980-000 Ji-Paraná ☎69 3421 1390 **W:** www. plansol.com.br **English:** 0000-0200 – **RO19**) Rua Sergipe 1766, Morada do Sol, 78983-000 Espigão D'Oeste ☎69 3481 3348 **W:** www.radiosoci-edadeespigao.com.br **E:** radio@radiosociedadeespigao.com.br – **RO20**) 78949-000 Ouro Preto do Oeste – **RO21**) 78941-000 Jaru – **RO22**) 78984-000 Pimenta Bueno – **RO23**) Rua Anel Viario, 1782, Parque Brizon, 78975-000 Cacoal ☎69 3443 2928 **W:** www.radiosuprema.com.br **E:** estudio@radiosuprema.com.br – **RO24**) 78900-000 Porto Velho.

RR00) RORAIMA

RR01) Av Capitão Ene Garcez, 888 - São Francisco, 69301-160 Boa Vista ☎95 3623 2259 **E:** direcao@radiororaima.com.br **W:** www.radior-oraima.com.br – **RR02**) Rua Sebastião Diniz 363, 69360-000 Caracaraí – **RR03**) 69350-000 Alto Alegre – **RR04**) Rua Lobo D'Almada, 43 - Sao Francisco, 69035-050 Boa Vista ☎95 3623 8801 ☒95 3623 8803 **W:** www.folhabv.com.br **E:**radiofolha@folhabv.com.br - **FM:** 91.9MHz – **RR05**) 69380-000 Bonfim.

RS00) RIO GRANDE DO SUL

RS01) Av Victor Barreto 3056, Conj 207, 92010-901 Canoas ☎51 3466 1200 ☒51 3476 5077 **W:** www.radiorealam.com.br **E:** radioreal@terra. com.br– **RS02**) Av Antunes Ribas 1535, 98801-630 Santo Angelo ☎55 3313 3666 **W:** www.radiosepe.com.br **E:** contato@radiosepe.com.br – **RS03**) Rua Marechal Deodoro 1157, 96810-110 Santa Cruz ☎51 3711 3908 **W:** www.radiosantacruz.com.br **E:** gerencia@radiosantacruz.com. br – **RS04**) Rua General Sampaio, 161 - Bairro Rio Branco, 95097-000 Caxias do Sul **W:** www.radiosaofrancisco.am.br **E:** redsul@saofrancisco.

am.br ☎54 2101 5222 📠54 2101 5236 - **FM:** 98.5MHz – **RS05**) Av. Scarpelini Guezzi 353, 99072 000 Passo Fundo **E:** info@radioplanalto. com ☎54 3313 2587 📠54 3314 3280 – **RS06**) Rua Mascarenhas de Morães 586, 97300-000 São Gabriel **W:** www.redetche.com.br/saoga-briel/radio.asp **E:** amir_comercial@hotmail.com amir_comercial@hot-mail.com ☎55 3232 6336 – **RS07**) Avenida Moreira Paz 726 (✉ C.P. 67), 95200-000 Vacaria ☎54 3231 7500 **W:** www.redesul.am.br/index. php?emissora=29 **E:** gerente@fatima.am.br – **FM:** 101.5MHz «R Mais Nova FM» – **RS08**) Praça Oswaldo Aranha 39, Centro, 97540-000 Alegrete ☎55 3422 1600 **W:** www.radioalegrete.com.br – **RS09**) Av Ipiranga, 1075/2° andar-Azenha, 90160-093 Porto Alegre **E:** gaucha@ rdgaucha.com.br **W:** www.clicrbs.com.br ☎51 3218 6600 📠51 3218 6680 - Satellite signal downlinked via 165 stns in southern Brazil forming Rede Gaúcha Sat – **RS10**) Rua Andrade Neves, 2316-Centro (C.P 284, 96001-970) 96020-080 Pelotas ☎53 3222 4334 📠53 3222 7407 **W:** www.radiopelotense.com.br **E:** ppelotense@terra.com.br – **RS11**) Rua Suécia 255, 98500-000 Tenente Portela ☎55 3551 1395 📠55 3551 1211 **W:** www.radiomunicipal.com.br **E:** radiomunicipal@uol.com.br – **RS12**) Rua 14 de Julho, 588 – Centro, (✉ C.P 54), 95300-000 Lagoa Vermelha ☎54 3358 1788 **W:** www.redesul.am.br/index.php?emissora=28 **E:** cacique@cacique.am.br – **RS13**) Rua Venâncio Aires 1851-1° andar, 97010-003 Santa Maria **E:** radiosantamariense@terra.com.br – **RS14**) Rua Delfino Riet, 183 - Santo Antonio, 90660-120 Porto Alegre ☎51 2101 0010 **E:** direcaogeral@bandrs.com.br **W:** www.bandrs.com.br – **RS15**) Av Mauricio Cardoso 88, 1° andar, Centro, 99700-000 Erechim ☎55 3321 2243 **W:** www.radiodifusaosul.com.br **E:** radiodifusao@cli-calpha.com.br – **FM:** 94.9MHz – **RS16**) Rua Voluntários da Patria 1432, 97590-000 Rosário do Sul ☎55 3231 2533 📠55 3231 4141 **W:** www. radiomaraja.com.br **E:** radiomaraja@rfc.com.br – **RS17**) Rua Ramiro Barcelos 800, sala 201, Centro, 95200-000 Vacaria **W:** www.radioesme-ralda.com.br **E:** comercial@radioesmeralda.com.br **FM:** 93.1MHz–**RS18**) Rua Neita Ramos 217, 96230-000 Santa Vitória do Palmar ☎53 3263 1660 **W:** www.redemeridional.com/sta.html **E:** culturasantavitoria@ redemeridional.com – **RS19**) Rua Domingos Secchi 35, Bairro Boa Vista, 99500-000 Carazinho **W:** www.gazeta670.com.br **E:** comercial@gaze-ta670.com.br ☎54 3330 3143 📠54 3330 2800 – **RS20**) Travessa 4 de Junho 84, 98380-000 Seberi ☎55 3746 1040 📠55 3746 1004 **W:** www. seberiam.com.br **E:** diretor@seberiam.com.br – **RS21**) Rua XV de Novembro 275, 9° andar, 98700-000 Ijuí **E:** contato@radioprogresso.com. br **W:** www.radioprogresso.com.br ☎55 3332 8888 📠55 3332 9999 – **RS22**) Rua Mascarenhas de Morães 298, 97300-000 São Gabriel ☎55 3232 2244 📠55 3232 5920 **E:** rbatovi.comerciais@terra.com.br – **RS23**) Rua Caldas Jr., 219- 2° andar-Centro, 90019260 Porto Alegre **E:** diretor@ radioguaiba.com.br **W:** www.radioguaiba.com.br ☎51 3215 6333 📠51 3215 6317 - **FM:** 101.3MHz –**RS24**)Rua Coronel Chicuta 436, 5° andar,– Ed Nossa Senhora Aparecida - Centro, 99010-051 Passo Fundo ☎54 3045 2914 **W:** www.rplanalto.com **E:** comercial@rplanalto.com – **FM:** 105.9MHz –**RS25**) Av Júlio de Castilhos, 435 - Bairro Vista Alegre, 98300-000 Palmeira das Missões **W:** www.radiopalmeira.com.br **E:** direcao@radiopalmeira.com.br ☎55 3742 1082 📠55 3742 2626 - **FM:** 101.7MHz –**RS26**) Av Silva Paes 363-A , 96200-340 Rio Grande ☎53 3232 2303 **W:** radionativa.com.br **E:** radionativa@vetorial.net – **RS27**) Av Marechal Floriano 920, Sala 301, 95520-000 Osório **W:** www.radio-sorio.com **E:** radioosorio@ com.br ☎51 3663 3435 – **RS28**) Rua Botucaraí 911, 96930-000 Candelária **E:** radioprincesa@terra.com.br ☎51 3743 1031 – **RS29**) Av Gal Daltro Filho 1000, 98470-000 Planalto ☎55 3794 1249 📠55 3794 1025 **E:** patyzanella@bol.com.br – **RS30**) Rua Santiago Matiotti 670, 99500-000 Carazinho – **RS31**) Rua Orfanatrófio 711, 90840-440 Porto Alegre **W:** www.redepampa.com.br **E:** prin-cesaam@pampa.com.br ☎51 3218 2525http://www.redepampa.com.br – **RS32**) Rua São João 485, 96640-000 Rio Pardo ☎57 3731 1390 **W:** redecartario.com.br/rccam790 **E:** contatosam790@redecartario.com.br ☎51 3731 2199 📠51 3731 1390 – **RS33**) Prédio da Reitoria, 10° andar, Campus Universitário de Camobi, 97105-900 Santa Maria ☎51 3220 8550 📠 51 3220 8390 **W:** www.radio.ufsm.br **E:** radio800am@ ahoo.com.br – **RS34**) Rua Fernando Abott 427, 2° andar, 95880-000 Estrela – ☎51 3720 5076 📠51 3712 1259 **W:** www.radioaltotaquari. com.br **E:** administracao@radioaltotaquari.com.br – **RS35**) Av Presidente Vargas 892, 98005-160 Cruz Alta ☎55 3322 6499 📠55 3322 6100 **E:** radioindependente@diarioserrano.com.br – **RS36**) Rua Silveiro 1321, 90850-000 Porto Alegre ☎51 3227 6809 – **RS37**) Caladão Salvador Isaia 1330 - 3° andar 97010-902 Santa Maria **W:** www.guarathan.com.br **E:** guarathan@terra.com.br ☎55 3025 5755 – **RS38**) Rua Correa Lima 1831, 90850-250 Porto Alegre ☎51 3024 2401 **W:** www.radioitai.com. br – **RS39**) Rua Osvaldo Aranha, 808 Sala 102B – Bairro Juventud, 95700-000 Bento Gonçalves ☎54 3452 7777 **W:** www.radio890am.com. br **E:** contato@radio890am.com.br - **FM:** 94.5MHz – **RS40**) Praça da Bandeira 36, 2° andar, 98900-000 Santa Rosa **E:** noroeste@viabrazil.com. br ☎55 3512 5757 – **RS41**) Rua XV de Novembro 336, 99770-000 Aratiba ☎54 3376 1170 📠54 3376 1270 **W:** www.radioaratiba.com.br **Emai:** atendimento@radioaratiba.com.br – **RS43**) Rua sete de Setembro,

1441 - Centro, 95800-000 Venâncio Aires ☎51 3741 2000 📠51 3741 2130 **W:** www.radiovenancioaires.com.br **E:** rva@gruporva.com.br - **FM:** 105.1MHz – **RS44**) Av Fernandes Bastos, 314, 2° andar - Centro, 99590-000 Tramandaí 51 3684 5132 **W:** www.radiotramandai.com.br –**RS45**) Rua Garibaldi 789-21° andar, 95080-150 Caxias do Sul 54 3289 3000 **W:** www.radiocaxias.am.br **E:** guilherme@radiocaxias.am.br – **RS46**) Av Getúlio Vargas, 1797 - Centro, 98801-590 Santo Ângelo **W:** www.radio-santoangelo.com.br **E:** contato@radiosantoangelo.com.br ☎55 3313 2440 📠55 3313 2467 – **RS47**) Av Alberto Müller 242, 95900-000 Lajeado **W:** www.independiente.com.br **E:** comercial@independente.com.br – **RS48**) Av. Walter Jobim, 222 s 106 – Patronato,, 97020-426 Santa Maria **E:** www.imembui.com.br **W:** www.imembui.com.br radioimembui@via-rs.nrt – **RS49**) Rua Orfanatrófio 711, 90840-440, Canoas 51 3218 2610 – **RS50**) Av Getúlio Vargas 412, 98670-000 Humaitá **E:** radioaltouruguai@hns.com.br ☎55 3525 1212 📠55 3525 1222 – **RS51**) Rua Otacílio Tupanciretã de Azevedo 2, 98170-000 Tupanciretã ☎55 3272 1763 📠55 32721763 **W:** www.tupa.am.br **E:** contato@tupa.am.br - **FM:** 92.5MHz – **RS52**) Av Fiorentino Bachi 791, 99840-000 Sananduva ☎54 3343 1662 📠54 3343 1438 **W:** radiosananduva.com.br **E:** radiosan@terra.com.br - **FM:** 97.7MHz – **RS53**) Rua Garibaldi 789, 23° andar - Centro, 95084-900 Caxias do Sul 54 3289 3000 **W:** www.radio1010.am.br **E:** 1010@ radio1010.am.br – **RS54**) Rua Julio de Castilhos 2236, 97800-000 São Luís Gonzaga **W:** www.radiomissioneira.com.br **E:** missioneira@viacom. com ☎55 3352 4141 – **RS55**) Av David José Martins 1206, 98700-000 Ijuí **W:** www.radioreporter.com.br **E:** radioreporter@san.psi.br ☎📠55 3332 8000 - **FM:** 101.5MHz «Iguatemi» – **RS56**) Rua General Zeca Netto 1396, 96180-000 Camaquã ☎51 3671 0962 **W:** www.redemeridional. com **E:** radiocamaquense@redemeridional.com – **RS57**) Rua São João, 1894-Centro, 98700-000 São Luís Gonzaga **W:** www.radiosaoluiz.com.br **E:** ouvinte@radiosaluiz.com.br ☎55 3352 4444 – **RS58**) Rua XV de Novembro 236, 96570-000 Caçapava do Sul ☎55 3281 1495 **W:** www.redemeridio-nal.com/cacapava.html **E:** radiocacapava@redemeridional.com – **RS59**) Rua Tucunduva 758, 98640-000 Crissiumal **E:** metrópolo@virnet.com.br ☎55 3524 1212 📠55 3524 1223 – **RS60**) Rua Marechal Deodoro 101,Galeria Central, 7° andar, 98700-000 Bento Gonçalves **W:** www. radiobento.com.br **E:** gera1070@radiobento.com.br ☎54 3455 3999 - **FM:** 92.5MHz «Serrana» – **RS61**) Rua João Carlos Machado 645, 98460-000 Iraí **E:** maraba@speedrs.com.br ☎55 3745 1474 – **RS62**) Rua Sarmento Leite 426, Campus Central da UFRGS 90046-900 Porto Alegre **W:** www.ufrgs.br/radio **E:** radio@ufgrs.br ☎51 3316 3435 📠51 3316 3192 24h - **Spanish:** Fri 2400 – **RS63**) Rua Ramiro Barcelos 2092, 96508-070 Cachoeira do Sul **W:** www.radiocachoeira.com.br **E:** radiocachoei-ra@radiocachoeira.com.br ☎51 3722 4022 - **German & Italian:** 3h weekly – **RS64**) Praça Pe Basso 95, 99800-000 Marcelino Ramos – **RS65**) Av Bento Gonçalves 733, 98870-000 Giruá **E:** radio@srgirua.com.br – **RS66**) Rua Osvaldino Barbosa Silveira, prolongamento da Júlio de Castilhos 2470, 96300-000 Jaguarão ☎53 3261 2933 **W:** www.redeme-ridional.com/jaguarao.html **E:** culturajaguarao@redemeridional.com – **RS67**) Rua Padre Oswaldo Stracke 56, 96900-000 Sobradinho ☎51 3742 1090 📠51 3742 1833 **W:** www.radiosobradinho.com.br **E:** radiosobra-dinho@terra.com.br - **FM:** 97.3MHz "R.Jacuí" – **RS68**) Rua Borges do Canto 1056, 97650-000 Itaqui ☎55 3433 8409 **W:** www.radiocruzeiro-dosul.com.br **E:** cruzeiro@bnet.com.br – **RS69**) Rua Corrêa Lima, 1960 – Morro Santa Tereza, 90850-250 Porto Alegre **W:** www.clicrbs.com.br/ especial/rs/farroupilha/home,0,2204,Home.html **E:** farroupilha@rdfar-roupilha.com.br ☎51 3218 5781 📠51 3218 5789 – **RS70**) Av Rio Branco 809, 97010-122 Santa Maria ☎55 3223 7231 📠55 3222 9500 **W:** www.radiomedianeiraam.com.br **E:** ouvinte@radiomedianeira.com.br – **RS71**) Rua Pinheiro Machado 628, 98005-000 Cruz Alta **W:** www.radiocruzalta. com.br **E:** gravadora-rca@comnet.com.br ☎55 3322 7222 📠55 3322 7292 - **FM:** 105.1MHz – **RS72**) Rua Domingos de Almeida 2194, 97500-004 Uruguaiana ☎55 3412 1731 📠55 3412 3046 **W:** www.radiochar-ru.com.br **E:** amfm@radiocharru.com.br **FM:** 97.7MHz – **RS73**) Rua André Kopaeff 02, 96750-000 Butiá **W:** www.radiojornalsobral.com.br **E:** radiojornalsobral.com.br ☎51 3652 1140 📠51 3652 1752 – **RS74**) Rua Rui Barbosa 96, 95180-000 Farroupilha **W:** www.radiomi-riam.com.br **E:** radiomiriam@radiomiriam.com.br ☎54 3042 1160 📠54 3260 5151 – **RS75**) Rua Tenente Lira 950, 98400-000 Frederico Westphalen **E:** www.luzealegria.com.br **E:** radioluzealegriaam@hot-mail.com ☎55 3744 3500 📠55 3744 3508 - **Italian & Polish:** Sun 0930 & 1600 - **FM:** 95.9MHz – **RS76**) Rua General Osório, 1160 - Centro, 97760-000 Jaguari **W:** www.radiojaguari.com.br **E:** radiojaguari@brtur-bo.com.br ☎55 3255 1474 – **RS77**) Rua Félix da Cunha 328, 3° andar, 96010-000 Pelotas **W:** www.radiouniversidadeam.com.br **E:** opecalfa@ via-rs.net ☎53 3222 1160 – **RS78**) Av Coronel Victor Villa Verde 491, 95500-000 Santo Antônio da Patrulha 51-662 1255 **W:** www.jwm.br/ita-pui/index **E:** radio.itapui@terra.com.br – **RS79**) Av 7 de Setembro 1115, 96400-001 Bagé– **W:** www.difusorabage.com.br **E:** difusora@difusora-bage.com.br ☎53 3242 5211 – **RS80**) Rua 7 de Setembro, 366 – Centro,(C.P 326) 99010-121 Passo Fundo **W:** www.radiouirapuru.com.br **E:** uira-

puru@rduiarapuru.com.br ☎54 2104 1600 – **RS81**) Av Maurício Cardoso 697, 99300-000 Soledade ☎54 3381 1144 ▤54 3381 1781 **W:** www.redesul.am.br/index.php?emissora=24**E:** gerente@cristal.am.br - **FM:** 99.1MHz –**RS82**) Rua da Anunciação, 480 - Morro do Convento, 97900-000 Cerro Largo - ☎55 3359 2022 ▤55 3359 1291 **W:** www.radiocerroazul.com.br **E:** fale@radiocerroazul.com.br – **German:** Tuesday 1230-1330 - FM: 105.9MHz «Shamballa» – **RS83**) Rua Mons. Scalabrini, s/n - Centro, 99250-000 Serafina Corrêa ☎54 3444 1212 **W:** www.rsradios.com.br **E:** rosario@net11.com.br – **RS84**) Rua Torres Gonçalves 33, 99700-000 Erechim ☎54 3522 1389 **W:** www.radioerechim.com.br **E:** toligado@radioerechim.com.br – **RS85**) BR 392 - Km 232, 97340-000 São Sepé **W:** www.radiocotrisel.com.br **E:** radfunco@plugnet.psi.br ☎55 3233 1113 ▤55 3233 1163 – **RS86**) Travessa Francisco de Leonardo Truda 40, 90010-050 Porto Alegre ☎51 3221 8711 ▤51 3221 2752 **E:** radiorec@zaz.com.br – **RS87**) Rua Domingos de Almeida 2194, 97500-003 Uruguaiana ☎55 3412 1731 ▤55 3412 3046 **W:** www.portaluruguaiana.com.br/smiguel **E:** radiosanmiguel@uol.com.br – **RS88**) Rua Coronel Vitor Dumoncel 1756, 98240-000 Santa Bárbara do Sul **E:** sales5@ligbr.com.br ☎55 3372 1453 ▤55 3372 1136 – **RS89**) Rua Rui Barbosa 373, Centro, 99600-000 Nonoai **W:** www.cluberadio.com.br ☎54 3362 1384 – **RS90**) Av Adolfo Schneider No 85 - 2° andar, Ed Elias, 95320-000 Nova Prata **W:** www.radioprata.com.br **E:** radioprata@netprata.com.br ☎54 3242 1684 ▤54 3242 1212 – **RS91**) Praça Silvestre Corréa 77, Centro, 96610-000 Encruzilhada do Sul ☎53 3537 1168 **W:** www.radioencruzilhadense.com **E:** faleconosco@radioencruzilhadense.com – **RS92**) Rua Júlio de Castilhos 605 2° andar, 95290-000 Bom Jesus E: radioaparados@m2net.com.br ☎54 3237 1247 ▤54 3237 1755 – **RS93**) Rua General Osório, 1134 - Centro, 98200-000 Ibirubá **W:** www.sistemaepu.com.br **E:** iatendimento@sistemaepu.com.br ☎54 3324 1758 ▤54 3324 1083 - **FM:** 96.6MHz –**RS94**) Rua Ponciano Ramos 74, 96700-000 São Jerônimo **E:** radio.saojeronimo@terra.com.br ☎51 3651 4228 – **RS95**) Av Júlio de Castilhos 1511, 8° andar, salas 81/84, Centro, 95010-003 Caxias do Sul **W:** www.radiodifusoracaxiense.com.br **E:** radio@www.radiodifusoracaxiense.com.br ☎54 3223 6788 – **RS96**) Rua 15 de Novembro 717, 96015-000 Pelotas **W:** www.radiotupanci.com.br **E:** tupanci@terra.com.br ☎53 3225 0930 ▤53 .3222 6167 – **RS97**) Rua Riachuelo, 928 - Centro, 97670-000 São Borja ☎55 3431 2244 ▤55 3431 1993 **W:** www.radiofronteirafm.com.br **E:** radio@gpsnet.com.br – **FM:**97.1MHz "Radio Fronteira" –**RS98**) Rua Moron, 1520 - Centro, 96508-030 Cachoeira do Sul ☎51 3722 3033 ▤51 3722 3633 **W:** www.radiofandango.com.br **E:** radiofandango@radiofandango.com.br - **FM:** 102.5MHz –**RS99**) Rua José Sponchiado 418, 99830-000 Gaurama **E:** radiogaurama@awo.com.br ☎54 3391 1104 - **Italian & Polish:** 1300-1400 – **RS100**) Rua Benjamim Santo Zago 601, 97220-000 Faxinal do Soturno ☎55 3263 1021 ▤55 3263 1335 **W:** www.radiosaoroque.com.br **E:** falecom@radiosaoroque.com.br – **RS101**) Rua Balduino Schneider, 254-Horizontina, 98920-000 Horizontina **W:** www.radioveracruz.com.br **E:** radioveracruz@radioveracruz.com.br ☎55 3537 1212 ▤55 3537 1414 –**RS102**) Rua Dr Pio Ferreira, 453, Centro, 96170-000 São Lourenço do Sul ☎53 3251 1303 **W:** www.radiosaolourenco.com.br **E:** radio@cybersul.com.br - **German:** Sun 1100 – **RS103**) Av Angelo Macalós 246, 99400-000 Espumoso **W:** www.radioplanetario.com **E:** falecom@radioplanetario.com ☎54-3383 1082 - **German:** Sun 1600 - **FM:** 95.3MHz –**RS104**) Av São Paulo 722, 3° andar, Bairro São Geral, 90230-160 Porto Alegre **W:** www.redeboavontade.com.br ☎51 3325 7000 –**RS105**) Rua 25 de Julho 39, Centro, 98960-000 Santo Cristo ☎55 3541 1188 **W:** www.radioregional1300.com.br – **RS106**) Rua dos Andrades, 663 - Centro, 97573-000 Santana do Livramento **W:** radiomaratan.webnode.com **E:** estudiomaratan@hotmail.com ☎55 3241 1300 – **RS107**) Av Duque de Caxias 1320, 99560-000 Sarandi ☎54 3361 1455 **W:** www.rsradios.com **E:** sarandiradio@viaradiointernet.com.br – **RS108**) Av Júlio de Castilhos 232, 95680-000 Canela ☎54 3282 8822 ▤54 3282 0900 **W:** www.radioclubedecanela.com.br **E:** contato@radioclubecanela.com.br - **FM:** 88.5MHz –**RS109**) Rua General Osório 1276, 98280-000 Panambi **E:** sulbrasileira@profnet.com.br ☎55 3375 3580 - **Italian:** 1300-1330 - **German:** Sun 1230-1255 –**RS110**) Rua Sete de Setembro 393, 96015-300 Pelotas **E:** radiopel.sul@terra.com.br ☎53 3229 3174 ▤53 3227 2382 –**RS111**) Av Rio Branco 401, 96450-000 Dom Pedrito ☎53 3243 3400 ▤53 3243 1257 **W:** www.radioupacarai.com.br **E:** upacarai@radioupacarai.com.br – **RS112**) Rua Pe Feijo 833 s42, 95190-000 São Marcos **W:** www.radiodiplomata.am.br **Emai:** diplomata@nsol.com.br ☎54 3291 2422 ▤54 3291 1563 – **RS113**) Av Ipiranga 1075, 2° andar, 90160-093 Porto Alegre ☎51 3218 6751 **W:** www.rbs.com.br **E:** cbn@rbsradios.com.br – **RS114**) Av Scalabrini 777, 99200-000 Guaporé **W:** www.radioaurora@tl.com.br ☎54 3443 4488 **Italian:** Mon 1300-1500 – **RS115**) Av Santos Dumont 240, 98600-000 Três Passos ☎55 3522 1011 **W:** www.difusoraceleiro.com.br – **RS116**) Av Concordia, 1480 - Centro, 96540-000 Agudo **E:** ☎55 3265 2045 W:www.radioagudo.com.br **E:** radioagudo@terra.com.br –**RS117**) Rua Lauro R.Bortolon 402, 99150-000 Marau **W:** www.alvorada.am.br **E:** gerencia@alvorada.am.br ☎54 3342 3300 - **FM:** 94.7MHz «Kosmos» –

RS118) Rua John Kennedy, 2220 sala 18 CxP 199, 95270-000 Flores da Cunha ☎54 3292 2311 ▤54 3028 3888 **E:** am1370@comunidadeoasis.org.br – **RS119**) Rua Gaspar Martins 55-3° andar, 97542-000 Alegrete ☎55 3422 1590 **W:** www.radiogazetadealegrete.com.br **E:** radiogazetadealegrete@brturbo.com.br – **RS120**) Rua Conde de Porto Alegre 521, 97573-581 Sant´Ana do Livramento **W:** www.culturalivramento.com.br **E:** culturalivramento@brturbo.com.br ☎55 3242 3066 – **RS121)** Rua XV Novembro, 1369-2° andar – Centro, (C.P 57) 99490-000 Tapera 543385 1166 543385 1855 **E:** cultura@sistemaepu.com.br – **RS122**) Rua Chaves Barcellos 36, conj 1205, Centro, 90030-120 Porto Alegre ☎51 3226 1390 **W:** www.radioesperanca.com.br **E:** adm@radioesperanca.com.br – **RS123**) Rua Pedro Toniollo, 529 - Centro, 99900-000 Getúlio Vargas **W:** www.radiosideral.com.br **E:** fale@radiosideral.com.br ☎54 3341 1555 ▤54 3341 1554 – **RS124**) Rua Augusto Rossi 316, 97200-000 Restinga Sêca **W:** www.radiojornalintegracao.com.br **E:** radio@integracao-rs.com.br ☎55 3261 1270 ▤55 3261 1030 – **RS125**) Rua Benjamin Constant, 327, 96200-090 Rio Grande ☎53 3231 3084 **W:** www.radio-cassino.com.br **E:** cassinoam@vetorial.net – **RS126**) Rua São Francisco 246, Cruzeiro, 98900-000 Santa Rosa ☎55 3512 5265 **W:** www.radi-osantarosa.com.br **E:** estudiob107@viabrazil.com.br ☎55 3512 5265 – **RS127**) Rua 24 de Maio 671, 95330-000 Veranópolis ☎53 3441 3200 **W:** http://www.redesul.am.br/index.php?emissora=22 **E:** gerencia@radioveranense.am.br – **RS128**) Rua Ramiro Barcelos 1206 - Centro, 96901-900 Santa Cruz do Sul **E:** gazeta1180@viaradio.com.br ☎51 3711 2211.- **FM:** 101.7MHz –**RS129**) Rua Teófilo Conrado de Matos, 96600-000 Canguçu **W:** www.radiocultura1030.com **E:** cultura@supersul.com.br ☎53 3252 1144 –**RS130**) Av das Hortencias 78, 95670-000 Gramado ☎54 3286 5516 ▤54 3286 1902 **W:** www.radioexcelsior.com.br **E:** excelsioram@serragaucha.com.br – **RS131**) Rua São João, 1637 - Centro, 95780-000 Montenegro ☎51 3632 1867 **W:** www.radioamerica-am.com.br **E:** radio@radioamerica.com.br – **RS132**) Rua Benjamin Constant 377, 96200-400 Rio Grande ☎53 3231 3048 **W:** www.radio-cassino.com.br **E:** cassinoam@vetorial.net – **RS133**) Av Sete de Setembro 672, Centro, 96400-003 Bagé ☎53 3242 1471 ▤53 3242 1211 **W:** www.radioculturabage.com.br **E:**radioculturabage@hotmail.com – **RS134**) Rua Borges de Medeiros 401, 95560-000 Torres ☎51 3664 4188 **W:** www.radiomaristela.com.br **E:** zyk311maristela@terra.com.br – **RS135**) Rua Dr Bruno Dockhorn 18, 98910-000 Três de Maio **W:** www.radiocolonial.com.br **E:** radiocolonial@mksnet.com.br ☎55 3535 1022 ▤55 3535 2488 –**RS136**) Rua Santos Inacio de Loiola 253, sl 203, 93700-000 Campo Bom ☎51 3585 1470 **W:** www.radiocinderela.com.br **E:** radiocinderela@gmail.com.br – **RS137**) Rua Julio de Castilhos 325, 95720-000 Garibaldi **W:** www.garibaldi.com.br **E:** gerencia@garibaldi.am.br ☎54 3462 1557 ▤54 3462 1911 - **Italian:** Sat 1800-1900, Sun 1030-1200 - **FM:** 88.1MHz –**RS138**) Rua Gabriel Machado 1590, 3° andar, 97610-000 São Francisco de Assis **E:** radiodifusoram@terra.com.br ☎55 3252 1455 ▤55 252 1166 – **RS139**) Rua General Osorio 943, 96600-000 Canguçu ☎53 3252 1515 **W:** radioliberdadeam.com.br **E:** falecoanosco@radioliberdadeam.com.br – **RS140**) Rua Rio Branco, 1006 - Centro, 95600-000 Taquara **W:** www.jornalpanorama.com.br **E:** radiotaquara@faccat.br ☎51 3542 2288 ▤51 3542 2511 – **RS141**) Rua São Vicente 345, 96501-180 Cachoeira do Sul ☎51 3723 7534 **W:** www.radiovaledojacui.com.br **E:** rvj@radiovaledopjacui.com.br – **RS142**) Av Rio Branco, 616 - Centro, 98770-000 Catuípe **W:** www.radioaguasclaras.com.br **E:** ricardocatuipe@hotmail.com ☎55 3336 1328 ▤55 3336 1042 – **RS143**) Rua José Bonafácio 1128, 96450-000 Dom Pedrito ☎53 3243 3110 ▤53 3243 1434 **W:** www.radiosulina.com.br **E:** sulina@radiosulina.com.br – **RS144**) Rua Dom João Becker, 604 - Centro, 93010-010 São Leopoldo ☎51 3568 8680 ▤51 3554 2894 **W:** www.redetche.com.br/progresso **E:** rdsaoleo@zaz.com.br – **RS145**) Rua Cel Amâncio Cardoso 596, 99950-000 Tapejara ☎54 3344 1185 ▤54 3344 1185 **W:** www.radiotapejara.com.br **E:** contato@radiotapejara.com.br - **Italian & German:** Sat 1230-1400, Sun 1300-1400 – **RS146**) Trav Jaime Pinto 136, 97700-000 Santiago **W:** www.radiosantiago.com.br **E:** zyk297@radiosantiago.com.br ☎55 3251 3972 ▤55 3251 1487 – **RS147**) Av Flores da Cunha, 4283 - Centro, 949150-001 Cacheoirinha ☎51 3421 1922 **W:** www.radiometropoleam.com.br **E:** radiometropoleam@terra.com.br – **RS148**) Rua 7 de Setembro 792, 95960-000 Encantado ☎51 3751 1903 **W:** www.rdencantado.com.br **E:** ouvinte@encantoam.com.br - **FM:** 97.7 MHz –**RS149)**Av Assis Brasil 263, 98130-000 Júlio de Castilhos **W:** www.radio14dejulho.com.br **E:** radio14@jcvirtual.com.br ☎55 3271 1414 – **RS150**) Rua José Bonifácio 41, 96330-000 Arroio Grande **W:** www.difusora1580.com.br **E:** radiodifusoraam@terra.com.br ☎53 3262 1008 – **RS151**) Rua Paraguai 42, 98980-000 Porto Lucena **E:** radionavegantes@san.psi.br ☎53 3260 1200 ▤55 3565 1221 – **RS152**) Av Valdomiro Bocchese 872, apt 01, 95250-000 Antônio Prado **W:** www.radiosolaris.com.br **E:** radiosolaris@nol.com.br ☎54 3293 1110 ▤54 3293 1733 – **RS153**) Av Padre Reus 1344, 2° andar, 97930-000 Caibaté **W:** www.radiocaibate.com.br **E:** radiocaibate@radiocaibate.com.br ☎55 3355 1335 ▤55 3355 1349 – **RS154**) Rua Rui Barbosa 46, 1° andar, 96360-000 Pedro Osório.53-255 – **RS155**) Av Borges de Medeiros, 1462

- Chacara, 97650-000 Itaqui **W:** www.radiopitangueira.com.br **E:** contato@radiopitangueira.com.br ☎56 3433 2301 🖷55 3433 2157 – **RS156)** Rua. Pe. Roque Gonzáles, 08, Centro (✉C.P. 241 98590-970), 98590-000 Santo Augusto ☎55 3781 1255 **W:** www.radioquerenciaonline.com **E:** radio@querenciaonline.com – **RS157)** Rua Baltazar Brum 343, 97560-000 Quaraí **E:** quarai@terra.com.br ☎55 3423 3001 – **RS158)** Rua Osvaldo Aranha 179 (C.P. 80), 95860-000 Taquari **E:** radioacoriana@taquari.com.br – **RS159)** 99300-000 Soledade – **RS160)** Rua Dom Luiz Guanella N° 2313, 95555-000 Capão da Canoa ☎51 3625 2300 **W:** www.radiohorizonte.com.br **E:** radiohorizonte@radiohorizonte.com.br – **RS161)** Rua da República 220, 99530-000 Chapada ☎54 3333 1338 **W:** www.radiosimpatia.com.br **E:** simpatia@radiosimpatia.com.br – **RS162)** Av Narciso Silva 1791, 96160-000 Capão do Leão – **RS164)** Rua Floriano Peixoto 222, 97400-000 São Pedro do Sul ☎55 3276 1311 🖷55 3276 4335 **W:** www.saopedrodosul.org/radio-municipal **E:** radiomunicipal@hotmail.com 1– **RS165)** Rua Duque de Caxias 255, Centro, 98430-000 Palmitinho ☎55 3791 1175 **W:** www.radiochiru.com.br **E:** radiochiru@radiochiru.com.br – **RS166)** Rua João Maffesoni, 10 - centro, 99680-000 Constantina ☎54 3363 1330 🖷54 3363 1122 **W:** www.radioatlantica.net.br **E:** radio.atlantica@hotmail.com – **RS167)** Rua Francisco Gobbi 545, 98580-000 Coronel Bicaco. ☎55 3557 1195 🖷55 3557 1220 **W:** www.radioguarita.com.br **E:** radioguarita@hotmail.com – **RS168)** Av Alto Jacuí 435, Terreo, 99470-000 Não Me Toque **E:** radioceres@dgnet.com.br ☎54 3332 1488 🖷54 3332 1498 – **RS169)** Rua Brasil 806, 97450-000 Cacequi ☎55 3254 1366 🖷55 3254 1157 **E:** socradiocccacequienselt-da@brturbo.com.br **E:** daltro@brturbo.com.br – **RS170)** Rua São João da Urtiga ☎54 3532 1247 – **RS171)** Rua Luiz Vieira 525, 96760-000 Tapes ☎51 3672 1031 🖷51 3672 1031 **E:** rt@conectsul.com.br – **RS172)** Av Antônio Finco 700 (C.P. 19), 99870-000 São José do Ouro ☎54 3352 1008 🖷54 3352 1108 **E:** radiopoata@ouronetonline.com.br – **RS173)** Rua Albino Brendler, 122 – Centro, 98700-000 Ijuí ☎55 3331 0300 🖷55 3331 0303 **W:** www.jornaldamanhaijui.com **E:** edmundo@jornaldamanhaijui.com – **RS174)** Rua Consórcio, s/n, Conj 09, 96400-970 Bagé ☎53 3242 4668 **W:** www.radioclubebage.com.br **E:** akucera@globo.com – **RS175)** Rua Pedro Alvares Cabral 164, 99660-000 Campinas do Sul **E:** radiocampinas@toirs.com.br ☎51 3366 1266 🖷54 3613 3366 – **RS176)** Rua XV de Novembro 609, 96270-000 Mostardas ☎51 3673 2062 🖷51 3673 2062 **E:** radiomostardas.brtdta.com.br – **RS177)** Av Barão do Triunfo, 584 - Centro, 95995-000 Arvorezinha ☎51 3772 2443 🖷51 3772 2129 **W:** www.radiocultura.inf.br **E:** cultura@radiocultura.inf.br – **FM:** 92.3MHz – **RS178)** Rua Comandaí, 615 - Centro, 97950-000 Guarani das Missões **W:** www.radioguaramano.com.br **E:** faleconosco@comunidadeoasis.org.br or radioguaramano@brturbo.com.br ☎55 3353 1721 🖷55 3353 1722 – **RS179)** Rua Jornal NH, 99-Bairro Ideal, 93334-350 Novo Hamburgo ☎51 3299 0999 **W:** www.radioabc900.com.br **E:** gerenciabc900@gruposinos.com.br – **RS180)** Rua 15 de Novembro 236, 96570-000 Caapava do Sul ☎53 3281 1495 **E:** radiocapava@farrapo.com.br – **RS191)** Rua Rádio e TV Gaúcha, 189 - Morro Santa Tereza, 90859-900 Porto Alegre ☎51 3218 5260 🖷51 3218 5285 **W:** www.clickrbs.com.br **E:** rrural@rbsradios.com.br – **RS192)** Rua Sananduva, 178-Centro, 99855-000 São João da Urtiga ☎54 3532 1247 **W:** www.radioeducadoraurtiga.com.br **E:** rdeducadora@brturbo.com.br – **RS193)** 95880-000 Estrele – **RS194)** 95870-000 Bom Retiro do Sul – **RS195)** 93180-000 Portao – **RS196)** 97700-000 Santiago. – **RS197)** Rua Rui Barbosa, 349 – Vila Velha, 98300-000 Palmeita das Missões. ☎55 3742 1000

SC00) SANTA CATARINA

SC01) Almeda Aristiliano Ramos 36 1°/2° andar (✉C.P. 61), 89160-000 Rio do Sul **W:** www.radiomiradorr.com **E:** am540@radiomirador.com.br ☎47 3531 2100 🖷47 3531 2102 - **FM:** 93.3MHz – **SC02)** Av Centenario 6050, Próspera, (C.P. D2, 88801-970) 88815-000 Criciúma ☎48 3439 5111 **W:** www.radioeldorado.net – **SC03)** Av Martin Piaseski 25, Centro, 89910-000 Descanso. ☎49 3633 0300 **W:** www.progresso.am.br – **SC04)** Rua Benjamin Constant 286-D, 3 e 4 andares, Centro, 89801-970 Chapecó ☎49 3332 5177 **W:** www.superconda.com.br **E:** jornalismoconda@zipway.com.br – **SC05)** Rua Carlos Gomes 12, Centro, 89160-000 Rio do Sul **W:** www.amanda.fm.br **E:** difusora@superdifusora.am.br ☎47 3521 1155 - **FM:** 94.9 «Amanda FM» – **SC06)** Rua João Beux Sobredinho, 350 Centro, 89990-000 São Lourenço d'Oeste **W:** www.radiodoze.com.br **E:** radiodoze@brturbo.com.br ☎49 3344 1544 🖷49 3344 1748 – **SC07)** Rua Carlos Jofre do Amaral 67, 88501-010 Lages **E:** jota@iscc.com.br – **SC08)** Av Sete de Setembro 109, Centro, 89580-000 Fraiburgo ☎49 3256 1010 **W:** www.radiofraiburgo.am.br **E:** estudio@radiofraiburgo.am.br – **SC09)** Rua Senador Gustavo Richard, 90 - Centro, 88701-220 Tubarão ☎48 3626 5177 **W:** www.radiotuba.com.br **E:** radiotuba@radiotuba.com.br – **SC10)** Rua General Vieira da Rosa, 1570 – Morro da Cruz, 88020-420 Florianópolis **W:** www.rbs.com.br **E:** cbndiario@rbsradios.com.br ☎48 3216 2500 🖷48 3216 2675 – **SC11)** Rua Leonel Mosele 275, 89700-000 Concórdia ☎49 3442 1366 **W:** www.radioalianca.com.br - **Italian & German:** Sun 1130, 1600 – **SC12)** Rua Buenos Aires 145, Edifico Senador Evelásio Vieira 145, Ponta Agude, 89051-050 Blumenau ☎47 3222 9000 **W:** www.radionereuramos.

br **E:** neruam@terra.com.br – **SC13)** Rua da Criança, 171, Centro, 88840-000 Urussanga **W:** www.radiomarconi.net **E:** radiomarconi@radiomarconi.net ☎48 3465 1055 – **SC14)** Rua Venereanos dos Passos 385, 89560-000 Videira **W:** www.radiovideira.com.br – **SC15)** Rua Angelo Dias 207, 6° andar, 89010-020 Blumenau **W:** www.cbnblumenau.com **E:** cbn@rfc.com.br ☎47 3041 8020 – **SC16)** Rua Itagiba, 215 - Centro, 88880-000 Lauro Müller **W:** www.radiocruzdemalta.com.br **E:** radiocruzdemalta@netlm.com.br ☎48 3464 3762 - **Italian:** Sat 1600 – **SC17)** Rua João Suzin Marini, 64 - Centro (C.P 71, 89700-970) 89700-000 Concórdia ☎48 3437 4602 **W:** www.radiorural.com.br/2005/ - **FM:** 96.3MHz – **SC18)** Av do Adão 1784, Morro da Cruz, 88025-150 Florianópolis **W:** www.bandeirantes890.com.br **E:** lucio.jornalismo@radiosantacatarina.com.br - **FM:** 101.7MHz «Transamérica» – **SC19)** Rua Rodovia SC 444 - km3, 88820-000 Içara ☎48 3461 0700 **W:** www.difusora910.com.br **E:** difusoranoticia@hotmail.com – **SC20)** Rua Conselheiro Rui Barbosa, 50 1° andar - Centro, 88350-000 Brusque **W:** www.radiocidadeam.com.br **E:** diretoria@radiocidadeam.com.br ☎47 3351 4611 – **SC21)** Rua Jardim Portobello, 50 - Centro, 88200-000 Tijucas ☎48 3263 0303 **W:** www.radiovaletj.com.br **E:** contato@radiovaletj.com.br – **SC22)** Rua Aristiliano Ramos, 134 Ed.Regina Sala 02 - Centro, 88870-000 Orleáns ☎48 3466 0533 **W:** www.guarujaam.com.br **E:** contato@guarujaam.com.br – **SC23)** Av.Brasil 260 Centro Comercial Tiradentes-3° andar, 89820-000 Xanxerê 49 3433 0171 **W:** www.superdifusora.com.br **E:** difusora@superdifusora.com.br – **SC24)** Rua Mathilde Hoffman 66, sala 21 e 22 (C.P 96, 88350-970), 88353-120 Brusque ☎47 3351 1744 **W:** www.araguaia970am.com.br **E:** contato@araguaia970am.com.br - **FM:** 107.7MHz – **SC25)** Rua São Bonifacio 280 89896-000 Itapiranga ☎49 3622 1877 **W:** www.peperi.com.br – **SC26)** Rua Severiano Francisco Sombrio 732, Centro (✉C.P 67), 88750-000 Braço do Norte ☎48 3658 2178 **W:** www.verdevale.com.br **E:** radio@verdevale.com.br 0720-0300 – **SC27)** Rua Pernambuco 329, 89840-000 Coronel Freitas – **SC28)** Rua Otacilio Vieira da Costa 40, Centro, 88501-050 Lages ☎49 3222 3011 **W:** www.rfc.com.br/princesa **E:** radioprincesa@rfc.com.br - **FM:** 95.7MHz «Amizade» – **SC29)** Rua Manoel Simão, 177-Salas 24 e 25 - Bairro das Nações, 89130-000 Indaial ☎47 3333 0499 **W:** www.radioclubeindaial.com.br **E:** falecom@radioclubeindaial.com.br – **SC30)** Rua Rolf Colin 80 (C.P 25, 89201-970), 89204-070 Joinville – **SC31)** Rua Vidal Ramos 519, 88701-160 Tubarão ☎48 3626 5488 - **FM:** 98.9MHz «Band FM» – **SC32)** Rua Padre Schoereder 1, Agronômica, (C.P 1477, 88010-900), 88025-090 Florianópolis ☎48 3201 1110 🖷48 3228 4950 **W:** www.divinooleiro.com.br – **SC33)** Av Santa Catarina 828, Edifico Dona Olivia - 2° andar, Centro, 89885-000 São Carlos **W:** www.radiosaocarlos.com.br **E:** am1110@sancasnet.com.br ☎49 3325 4355 🖷49 3325 4483 – **SC34)** Travessa João Winkler 15, 89820-000 Xanxerê ☎49 3433 1110 🖷49 3433 0682 **W:** www.redeprincesa.com.br **E:** studio@redeprincesa.com.br - **FM:** 101.3MHz – **SC35)** Rua Cel. Vidal Ramos, 861 - Centro, 89520-000 Curitibanos **E:** coroado@coroado.am.br **W:** www.coroado.am.br ☎49 3241 0923 🖷49 3241 0499 - **FM:** 98.9MHz – **SC36)** Rua 15 de Novembro 600, sala 401, Edifico Visconde de Mauá, 89010-000 Blumenau ☎🖷47 3322 9773 **W:** www.radioitabera.com.br **E:** contato@radioitabera.com.br – **SC37)** Rua Conselheiro Jeronimo Coelho 48, Centro, 88790-000 Laguna ☎48 3644 0025 **W:** www.difusoralaguna.com.br **E:** radio_difusora@yahoo.com.br – **SC38)** Rua Siqueira Campos 33, 89400-000 Porto União ☎42 3522 2245 **W:** www.colmeia.com.br **E:** colmeia@colmeia.am.br – **SC39)** Rua Padre Aurélio 240, 89930-000 São José do Cedro ☎49 3604 0151 **W:** www.radiointegracaoam1180.com.br **E:** comercial@radiointegracaoam1180.com.br – **SC40)** Rua Otavianpo Dadan, 355-Centro, 88240-000 São João Batista ☎48 3265 0222 **W:** www.radioclubeam.com **Email** contato@radioclubeam.com – **SC41)** Rua São Cristóvão 393 (C.P 59, 89835-970), 89835-000 São Domingos ☎49 3443 0139 **W:** www.clubesd.com.br **E:** contato@clubesd.com.br – **SC42)** Av Patricio Lima, 3073 - Bairro São Bernardo, 88708-201 Tubarão ☎48 3628 0658 🖷48 3628 1356 **W:** www.radiosc.com.br **E:** radiosc@radiosc.com.br – **SC43)** Rua Tenente Ary Rauen, 1361 - Alto, 89300-000 Mafra ☎47 3642 3955 **W:** www.saojoseam.com.br **Email.** radionovaera@netuno.com.br - **FM:** 104.5MHz «Nova Era» – **SC44)** Porã – **SC45)** Rua 9 Março 737, 8° andar, Ed.Turim 8° andar, 89201-400 Joinville ☎47 3026 4111 **W:** www.jovempanjoinville.com.br - **FM:** 91.1 MHz – **SC46)** Rua XV de Novembro, 550 - 6° andar, Sala 601 - Centro, 89012-060 Bluemau ☎47 3340 1260 **W:** www.radioblumenau.com.br **E:** radioblu@bnu.nutecnet.com.br – **SC47)** Av XV de Novembro, 608-Centro, 89600-000 Joaçaba **W:** www.radiiocatarinense.com.br **E:** radiocatarinense@radiocatarinense.com.br ☎49 3551 2424 🖷49 3551 2426 – **Italian:** Sun 1215-1500 - **FM:** 92.3MHz – **SC48)** Rua Osvaldo Cabral 68 - 1° andar, Centro, 88790-000 Laguna, ☎48 3646 0337. **W:** www.garibaldilaguna.com.br **E:** raduiogaribaldi@brturbo.com.br – **SC49)** Av Alvin Bauer 585, Centro, 88330-000 Balneário Camboriú ☎47 3367 1044 🖷47 3367 4949 **W:** www.radiocamboriu.com.br **E:** radiocamboriu@radiocamboriu.com.br – **SC50)** Rua Buenos Aires 145, Edifico Senador Evelásio Vieira, Ponta Aguda, 89051-050 Blumenau ☎47 3222 9070 **W:** www.radioclubeblumenau.com.br 0800 (Sun.:0900) - 0100 – **SC51)** Rua Marechal Floriano

Peixoto, 161-0 - Centro, 89800-000 Chapecó **W:** www.radiochapeco.com.
br **E:** soesca@redamp.com.br ☎49 3322 0688 🖷49 3322 0429 - **FM:**
107.1MHz – **SC52**) Rua Vereador Guilherme Prust, 311 - Campo d'Água
Verde, 89460-000 Canoinhas ☎47 3622 7000 **W:** www.radioclubedeca-
noinhas.com.br – **SC53**) Av Gov Adolfo Konder 1500, Bairro São Vicente,
88308-000 Itajaí ☎47 3248 1350 **W:** www.radioclubebandeirantes.com.
br **E:** jornalismo@radioclubebandeirantes.com.br– **SC54**) Rua Olivio D
Brugnago, 181 - Vila Nova, (C.P 405, 89251-970) 89259-260 Jaraguá do
Sul 🖷 47 3371 0444 **W:** www.radiobrasilnovo.com.br **E:** rbn@radio-
brasilnovo.com.br – **SC55**) Rua Duque de Caxias 1302, 2° andar, 89900-
000 São Miguel d'Oeste **W:** www.peperi.com.br **E:** rede@peperi.com.br
☎49 3622 1877 – **FM:** 104.9MHz – **SC56**) Rua José Gonçalves, 333 -
Lucena, 89340-000 Itaiópolis ☎ 🖷47 3652 2279 **W:** www.cidade1380.
am.br– **SC57**) Rua Carlos Jofre do Amaral 67, 88501-010 Lages ☎49
3221 3000 🖷49 3221 3030 – **SC58**) Rua Visconde do Rio Branco, 1028 -
Centro, 89887-000 Palmitos ☎49 3647 0292 **W:** www.radioentrerios.
com.br **E:** entrerios@promitos.com.br – **SC59**) Rua Nunes Machado 94,
10° andar, 88010-460 Florianópolis ☎48 3222 5555 **W:** www.radio-
guaruja.com.br **SC60**) Rua Marechal Deodoro, 298 Ed Pe Quintílio
Costini - Centro, 89620-000 Campos Novos ☎🖷49 3541 0391 **W:** www.
rsradios.com.br **Email:** culturaam@rsradios.com.br – **SC61**) Rua 7 de
Setembro 341, Centro, 85505-030 Maravilha ☎46 3664 0029 **W:** www.
difusoramaravilha.com.br **E:** atendimento@difusoramaravilha.com.br–
SC62) Rua Ervino Rank 37, Serra Alta, 89291-695 São Bento do Sul ☎47
3633 0572 🖷47 3634 2497 **W:** www.radiosaobento.com **E:** comercial@
radiosaobento.com – **SC63**) Av Centenario, 6050 - Bairro Prospera,
88815-000 Criciúma ☎48 3478 5659 **W:** www.hulhanet.com.br – **SC64**)
Rua São Pedro, 245, Centro, 89110-000 Gaspar **W:** www.sentine-
ladovale.com.br **E:** radiosentinela@terra.com.br ☎47 3332 0783 🖷47
3332 1200 – **SC65**) Rua Jornalista Bento Silvério 906, Kobrasol 88102-
090 São José ☎48 3244 1240 **W:** www.radiomaisalegria.com **E:**
amgazeta@hotmail.com – **SC66**) Rua Santos Dumont, 204 - Centro,
89610-000 Herval d'Oeste ☎49 3544 1817. - **Italian:** Sat 1530-1730 **W:**
www.radiolider.am.br– **SC67**) Rua Coronel Procópio Gomes 1155,
89202-300 Joinville ☎ 🖷48 3801 6533 **W:** www.difusora.net **E:** difusora@difusora.net –
SC68) Rua Capitão Jerônimo Luiz de Bittencourt 103, sala 01, Centro,
88770-000 Imaruí ☎48 3643 0000 **W:** www.litoralam.com.br **E:** lito-
ralam@litoralam.com.br – **SC69**) Rua Tiradentes, 283, sala 21, Centro,
89140-000 Ibirama – ☎47 3357 2236 **W:** www.belosvales.com.br **E:**
belosvales@ibnet.com.br – **SC70**) Av Plínio Arlindo de Nes 476, 89825-
000 Xaxim ☎49 3353 2425 **W:** www.radioculturaxaxim.com.br **E:** cultur-
axaxim@brturbo.com.br – **SC71**) Rua Max Wilhelm, 373 - Baependi,
89256-000 Jaraguá do Sul ☎ www.jaraguaam.com.br **E:** jaraguaam@
jaraguaam.com.br ☎47 3371 1010 🖷47 3275 0304 – **SC72**) Av Luis de
Camões 1370 , 88523-000 Lages ☎43 3222 8222 **W:** www.radioguri.
com.br **E:** rco@rco.com.br or faleconosco@radioguri.com.br – **SC73**) Av
Belém, 500 - Centro, 89870-000 Pinhalzinho ☎49 3366 1111 **W:** www.
rco.com.br – **SC74**) Rua Altamiro Guimarães 480, Centro, 89500-000
Caçador ☎49 3536 2211 **W:** www.am1110.com.br – **SC75**) Praça
Henrique Lage 797, 88780-000 Imbituba **W:** www.bandeirantes1010.
com.br – **SC76**) Rua Equador 245, 89120-000 Timbó ☎47 3382 3888 **W:**
www.radioculturaam.com **E:** radiocultura@tpa.com.br – **SC77**) Rua
Manoel Vieira Garcao 3, 88301-010 Itajaí ☎47 3348 2992 **W:** www.
difusoraitajai.com.br – **SC78**) Rua Boanerges P de Medeiros 205, 2° e 3°
andares, 88600-000 São Joaquim ☎49 3233 0021 **W:** www.difuso-
ra1530.com.br **E:** difusora@iscc.com.br – **SC79**) Av Porto Feliz 151,Centro,
89893-000 Mondaí ☎49 3674 0122 **W:** www.portofeliz.com.br – **German
& Spanish:** Sun 1300-1400 – **SC80**) Rua Carmelo Zocoli 205, 89665-000
Capinzal ☎49 3555 1333 **W:** www.radiocapinzal.am.br **E:** radiocapin-
zal@radiocapinzal.com.br – **SC81**) Av Getúlio Vargas, 429 - Centro, 88900-
000 Araranguá ☎🖷48 3524 0137 **W:** www.radioararangua.com.br **E:**
contato@radioararangua.com.br – **FM:** 92.5MHz – **SC82**)Rodovia SC -
422 km 3 (C.P 3, 89190-970), 89190-000 Taió **W:** www.educadora.am.br
E: comercial@educadora.am.br– **SC83**) Rua Carlos Weber 228, 89295-
000 Rio Negrinho – **SC84**) Av 21 de Janeiro, 966 - Centro, 89107-000
Pomerode ☎47 3395 1580 **W:** www.radiopomerode.com.br **Email**
radiopomerod@radiopomerode.com.br – **SC85**) Rua João Steffens 260 (
C.P 100), 88400-000 Ituporanga ☎47 3533 8310 **W:** www.sintonia.am.br
E: radio@sintonia.am.br – **SC86**) Rua Joaquim Nuns 244, Centro, (🖃
C.P 2004) 888340-000 Camboriú ☎47 3261 3232 **W:** www.gmuh.com.
br/radio/sintonia.htm **E:** contato@gideos.com.br – **SC87**) Rua João
Florentino de Souza 700, 89480-000 Major Vieira ☎47 3635 1177 **W:**
www.radioplanaltodemajorvieira.com.br – **SC88**) Rua Renato Ramos da
Silva 239, Barreiros, (C.P 1477, 88103-970), 88110-015 São José ☎47
3041 4103 **W:** www.radioguararema.com.br **E:** grasiele@radioguarare-
ma.com.br 24h - **FM:** 103.5/107.7MHz – **SC89**) Rua Sargento Juvenil
Pereira de Souza. 476 - Centro, 89540-000 Santa Cecília ☎49 3244
2188 **W:** www.radioalvoradasc.com – **SC90**) Av Getúlio Vargas 860,
89830-000 Abelardo Luz – **SC91**) Rua Ricardo Kruger 140, sala 02,
88650-000 Urubici – **SC92**) Rua Rui Barbosa, 1321 - Cemtro, 88930-000
Turvo **W:** www.radioimigrantes.com.br **E:** imigrantes@radioimigrantes.

com.br 🖷48 3525 0321 – **SC93**) Rua Professor João Sobotka 222,
Bairro São Cristovão, 89665-000 Capinzal ☎ 49 3555 1799 **W:** www.
radiobarrigaverde.am.br - **Italian:** Sun 1500-1600 – **SC94**) Av Progresso
569, 89888-000 Caibi **E:** radiocaibi@cbi.cpnet.com.br – **SC95**) Rua Duque
de Caxias 1302, 2° andar, 89900-000 São Miguel d'Oeste **W:** www.
peperi.com.br ☎49 3622 1877 – **SC96**) Rua Rafael Pardinho, 249 -
Centro, 89240-000 São Francisco do Sul **W:** www.radiosaofranciscosc.
com.br **E:** radio.saofrancisco@ilhanet.com.br ☎47 3444 2733 🖷47 3444
0450 – **SC97**) Rua do Comercio 215, 89770-000 Seara **W:** www.belos-
montes.com.br – **SC98**) Rua Nereu Ramos 2222, Centro , 89872-000
Modelo ☎49 3365 3294 **E:** radiomodelo@mhnet.com.br – **SC99**) Rua 7
de Setembro, 496, Centro, 89950-000 Dionísio Cerqueira ☎49 3644
1042 **W:** www.radiofronteira.com.br - **FM:** 94.3 – **SC100**) Rua Marechal
Floriano, 505 – Ponte Serrada, 89683-000 Ponte Serrada ☎49 3435
0171 **W:** www.radionamba.com.br **E:** atendimento@radionamba.com.br
– **SC101**) Av Dr Albano Schultz, 925 - 2° andar Centro, 89201-220
Joinville ☎47 3481 3030 **W:** www.radioglobojoinville.com.br **E:** com-
ercial@radioglobojoinville.com.br – **FM:** 103.1MHz – **SC102**) Rua
Maranhão 700, sala 02, 89980-000 Campo Erê **W:** www.peperi.com.br
– **SC103**) Av. Presidente Kennedy, 222 - 4° andar – Campinas, 88101-000
São José ☎48 3879 2011 **W:** www.radiocentralam.com.br – **FM:** 106.5
MHz – **SC104**) Rua XV de Novembro 495, 89560-000 Videira ☎49 3650
2500 **W:** www.vitoriaam.com.br **E:** adm@vitoriaam.com.br – **SC105**) Rua
João Lino da Silva Neto 621, 88495-000 Garopaba ☎48 3254 3055 **W:**
www.radiofrequencia.net **E:** frequencia@radiofrequencia.net – **SC106**)
88330-000 Balneário Cambouri – **SC107**) 89642-000 Tangará – **SC108**)
88950-000 Jacinte Machado.

SE00) SERGIPE

SE01) Rua Claudio Batista 334, Santo Antônio, 49060-100 Aracaju ☎79
3234 3232 **W:** www.radiojornal540.com **E:** jornal@radiojornal540.
com.br – **SE02**) Rua Propria 124, 49010-020 Aracaju **W:** www.aperipe.
se.gov.br/ - **FM:** 104.9MHz – **SE03**) Rua Simão Dias, 643 - Centro,
49010-430 Aracaju ☎79 3226 8710 **W:** www.radiocultura670.com.br
E: cultura@cultura670.com.br – **SE04**) 49400-000 Lagarto ☎79 3631
8500 **W:** www.radioprogressoam.com **E:** contato@radioprogressoam.
com – **SE05**) C.P 409, 49001-970 Aracaju **W:** amatalaia.com.br – **SE06**)
Rua 13 de Maio 119, 49500-000 Itabaiana ☎79 3431 1762 – **SE07**) Rua
Pacatuba 254, Ed Paulo Figueiredo, sala 1116, Centro 49010-900 Aracaju
☎79 3213 1174 **W:** www.930am.com.br **E:** jornalismo@930am.
com.br **SE08**) Praça Coronel Gonçalo Prado s/n, 49200-000 Estância
☎79 3522 1411 🖷79 3522 2327 – **SE09**) Av Dr Luíz Magalhães 346,
49500-000 Itabaiana ☎79 3431 7928 **W:** www.capitaldoagreste.com.
br **E:** contato@capitaldoagreste.com.br – **SE10**) Rodovia Lourival Batista
2153, 49480-000 Simão Dias ☎79 3611 1488 **W:** www.novacidadeam.
com.br **E:** novacidadeam@hotmail.com – **SE11**) Av. Napoleão Emifio
Costa, 1052 – Centro, , 49514-000 Frei Paulo ☎79 3447 1745 **W:** www.
radioeducadorafreipaulo.com.br **E:** radioeducadorafreipaulo@yahooa.
– **SE12**) Rua Barão do Rio Branco 262, 49000-000 Estância
– **SE13**) Travessa Santa Luzia 69, 49300-000 Tobias Barreto ☎79 3541
1548 **W:** www.redeilha.com **E:** am1520@redeilha.com)

SP00) SÃO PAULO

SP01) Rua Padre Geraldo Goseling 798, 16200-000 Birigui – **SP02**) Rua
Antônio do Vale Mello 807, Centro, 13170-011 Sumaré – **SP03**) Av
Nesralla Rubez 353, 12700-000 Cruzeiro ☎12 3144 0606 🖷12 3144 3688
W: www.radiomantiqueira.com.br **E:** atendimento@mantiqueira.com.br
- **FM:** 100.7MHz – **SP04**) Rua José Bonini 1415, 14160-000 Sertãozinho
W: h//radio.boavontade.com – **SP05**) Prefeito João Benedito
Barbosa,161, Vila Nova, 18400-000 Itapeva ☎15 3522 2000 **W:** www.
radioclubeitapeva.com.br **E:** radioclube@dioclubeitapeva.com.br- **FM:**
93.5MHz «Cristal» – **SP06**) Rua Dr Sousa Alves 960, 12020-030 Taubaté
☎12 3632 8122 **W:** www.difusora570.com.br – **SP07**) Rua Rui Barbosa,
580-Centro, 13465-280 Americana **E:** radiovoce@radiovoce.com.br **W:**
www.radiovoce.com.br ☎19 3475 8801 🖷19 461 7081 - **FM:** 88.9MHz
"Notícia" – **SP08**) Av Rotary 85, 19970-000 Palmital ☎18 3351 2601 **W:**
www.radioregionalpalmital.com.br – **SP09**) Rua Pedro Lessa 1640, sala
809 – Embaré, 11025-002 Santos **W:** www.radioatlantica.com.br **E:**
radioatlantica@radioatlantica.com.br ☎13 3273 6900 24h – **SP10**) Rua
das Nações Unidas 127, 16800-000 Mirandópolis **E:** clubeam@express-
net.com.br ☎18 3701 4084 🖷18 3701 4143 – **SP11**) Av Jerônimo
Gonçalves 640, 14010-040 Ribeirão Preto **W:** www.radio79.com.br –
SP12) Av Luíz Gonzaga de Amoedo Campos 28, 13800-000 Mogi Mirim
– **SP13**) Rua Pará 155, Centro, 15800-000 Catanduva **W:** www.rbcatan-
duva.com.br - **FM:** 94.9MHz – **SP14**) Rua Conselheiro Rodrigues Alves,
104 - 3° andar - Centro, 12560-020 Guaratinguetá ☎12 3122 3155 **W:**
superradiopiratininga.com.br **E:** ouvintes@superradiopiratininga.com.br
– **SP15**) C.P 150, 18800-970 Piraju ☎14 3351 1066 **W:** www.winf.com.
br/paranapanema/ **E:** paranapanema@winf.com.br – **SP16**) Av Paulista
807, 24° andar, 01311-915 São Paulo **E:** info@jovempan.com.br **W:**
www.jovempan.com.br – **SP17**) Rua Capitão Neve 1840, 15130-000
Mirassol ☎17 3242 3076 – **SP18**) Av Marcondes Filho 1130, 19013-160
Presidente Prudente – **SP19**) Rua Nove de Julho 1300, 14804-295

Araraquara ☎16 3303 3622 🖷16 3303 0114 E: radiomorada@uol.com. br W: www.radiomorada.com.br - FM: 98.1MHz – SP20) Rua Homero Rodrigues Silva 1090, 16900-000 Andradina - FM: 97.9MHz «Cidade Andradina» – SP21) Praça José Bonifácio 815, 13400-340 Piracaba ☎19 2105 6600 W: www.rdifusora.com.br E: contato@difusorapiraci-caba.com.br - FM: 102.3MHz – SP22) Rua Tolentino Figueiras, 119 7° andar - cj 71/72, Gonzaga, 11060-471 Santos ☎13 3289 5259 W: www. criacaoconsultoria.com.br - FM: 105.5MHz – SP23) Av Nove de Julho 606, 14025-000 Ribeirão Preto W: www.clube.com.br - FM: 100.5MHz – SP24) Rua Teotonio Tibiriçá Pimenta, 380 - Centro, 11660-230 Caraguatatuba ☎12 3882 5000 W: www.radiooceanicaam.com.br E: radio.oceanica@uol.com.br – SP25) Rua Pref Salviano Pereira de Andrade 20, 17400-000 Garça ☎14 3471 2241 🖷14 3471 0396 W: www.rco670am.com.br E: rcostudioam1@pop.com.br – SP26) Rua Quintino Bocaiúva, 37 - Centro, 13300-135 Itu.- ☎11 4023 2363 W: www.radioconvencao.com.br E: radioconvencao@hotmail.com – SP27) Rua 13 de Maio 720, 15800-000 Catanduva – SP28) Av Dr Alvaro Schmidt Gallo 317, 18800-000 Piraju ☎14 3351 1680 W: www.winf. com.br/piratininga E: pirat680@winf.com.br – SP29) Rua Francisco Inácio 257, 14700-000 Bebedouro ☎17 3342 2484 W: www.radi-obebedouro.com.br E: gerencia.rb@mdbrasil.com.br – SP30) Praça Conselheiro Rodrigues Alves 170, Centro, 12500-020 Guaratinguetá E: rclube@provale.com.br - FM: 97.1MHz – SP31) Rua Humberto Liedtke 1936, 15370-000 Pereira Barreto ☎18 3704 6677 W:www.radioci-dadeam690.com E: contato@radiocidadeam690.com MSN: radioci-dadeam690@hotmail.com – SP32) Av.Eng Caetano Alvares, 55 – Limão, 02598-900 Sao Paulo ☎11 2108 6700 W: www.radioeldoradoam.com. br E: am@radioeldorado.com.br or spaulo@radioeldoradi.com.br - FM: 92.9MHz – SP33) Rua 1 de Agosto 927, 17010-011 Bauru W: www. radio710bauru.com – SP34) Rua dos Pelegrini 11, Bairro do Desterro, 13700-000 Casa Branca ☎19 3671 2101 W: www.radiodifusoracasa-branca.com.br – SP35) Rua Antonio Carlos Mori 288 (C.P 355 19900-970), 19900 080 Ourinhos ☎ Rua Dr Carlos Varela 104 (🖂C.P 25,12700-970),12701-301 Centro, Cruzeiro ☎12 3143 6894 W: www. rcvale.com.br – SP37) Rua Dr Antônio 227, Centro 15400-000 Olímpia ☎17 3281 3045 W: www.meninaam.com.br E: ammenina@uol.com.br – SP38) Rua Siqueira de Morães 578, 10° andar, Ed Marijú, 13201-803 Jundiaí E: cidade@radiojundiai.com.br W: www.radiojundiai.com.br ☎11 4586 0969 – SP39) Rua Coronel Galdino de Almeida, 55 - Centro, bloco 3 sala 1, (🖂 C.P 324, 17500-970) 17500-100 Marília ☎14 3402 5128 🖷14 3402 5127 W: www.dirceu.am.br Emal: radio@dirceu.am.br – SP40) C.P 88, 17250-970 Bariri – SP41) Av Paulista 900, 01310-100 São Paulo – SP42) Rua Itapura 06, Jardim América, 17700-000 Osvaldo Cruz W: www.radioosvaldocruz.com.br E: calfm@cruz.com.br – FM: 97.3MHz «California FM» – SP43) Rua Ramos de Azevedo, 622 – Jardim Paulista, 14090-180 Ribeirão Preto ☎16 3624 2848 W: www.radiocmn. com.br – SP44) Rua Euclides Miragaia 394 – 18° andar, Centro, 12245-901 São José dos Campos ☎12 3909 3941 🖷12 3941 1999 W: www. superradiopiratininga.com.br E: ouvintes@superradiopiratininga.com.br - FM: 99.7MHz – SP45) Rua Virgilio Malta, 6-78, 17015-220 Bauru ☎14 3104 0761 W: www.auriverde.am.br E: auriverde@auriverde.am.br – SP46) Rua Santa Cruz 655, 13480-041 Limeira ☎19 3404 4000 W: www.mixam.com.br - FM: 100.7MHz – SP47) Rua das Palmeiras 315, Vila Buarque, 01226-901 São Paulo ☎11 3824 3200 🖷11 3825 8844 W: www.radioclick.globo.com.br/cbn/ E: contato@globonoar.com.br – SP48) Alameda Dr Armando de Sales Oliveira, 575 – Centro, 17800-000 Adamantina 18 3521 1242 W: www.radiobrasilam.com.br E: contato@radiobrasilam.com.br – SP49) Av Bento de Abreu, 789 - Fonte, (C.P 59 14801-970) 14802-396 Araraquara ☎16 3303 7799 W: www.radiocul-tura.net E: comercial@radiocultura.net- FM: 97.3MHz – SP50) Rua Barão de Jundiaí, 1041 – 9° andar - Centro, 13201-906 Jundiaí ☎11 4586 2020 W: www.radiodifusorajundiai.com.br E: radio@radiodifusora-jundiai.com.br – SP51) Rua Benjamin Constanst, 3327 - Centro, 15015-600 São José do Rio Preto - Blog: http://blog.cancaonova.com/riopreto E: radioriopreto@cancaonova.com.br – SP52) Av Getúlio Vargas, 185 - Centro, (C.P 02, 12570-970 Aparecida) 12570-000 Aparecida W: www. radioaparecida.com.br ☎12 3104 4400 🖷12 3104 4451 DX-prgrm – Encontro DX, Saturdays at 22.00 UTC - FM: 90.9MHz – SP53) Rua Tenente Lopes, 191-Centro, (C.P.3) 17201-460 Jaú – ☎14 3622 2800 W: www.radiojauense.com.br E: radiojauense@netsite.com.br - FM: 101.1MHz – SP54) Rua José Galvão 359, (C.P 94) 19900-260 Ourinhos ☎14 3322 2997 🖷14 3322 6255 W:radioclube820.com.br E: clube@ radioclube820.com.br – SP55) Av.Antonieta Vilela Ferreira 900 - Violage, 16300-000 Penápolis ☎18 3652 0027 🖷18 3522 1577 W: www.difuso-radepenapolis.com.br E: difusora@difusoradepenapolis.com.br – SP56) Rua Sao Paulo 1091, 15500-000 Votuporanga ☎17 3422 3301 W: www. lider830.com.br E: club92fm@votuporanga.com.br - FM: 92.1MHz – SP57) Rua Radiantes 13, Morumbi, 05699-900 São Paulo W: www. bandeirantes.com.br E: rbnoar@band.com.br - FM: 90.9MHz – SP58) Av Visconde do Rio Claro 2128, 13500-580 Rio Claro – SP59) C.P 154, 16200-970 Birigui – SP60) Rua Prudente de Morães 418, 14960-000

Novo Horizonte – SP61) Rua Romualdo Andreazzi 516, Jd Leonor, 13041-030 Campinas W:radiocentral.com.br E: radiocentral@radiocentral.com. br ☎19 3272 1400 - FM: 103.7MHz «Nova» – SP62) Av.Paulista, 900 4° andar, Bairro Bela Vista, 01310-940 Sao Paulo ☎11 3170 5757 🖷11 3170 5630 W: www.radiogazeta.com.br E: fuba@radiogazeta.com.br – SP63) Rua Rui Barbosa 273, 19015-000 Presidente Prudente ☎18 222 2500 W: www.difusoraprudente.com.br – SP64) Rua Siqueira Campos 3223, 15010-210 São José do Rio Preto - FM: 102.1MHz «R Onda Nova FM» – SP65) Rua Dr Virgilio de Rezende 400, Centro, 18200-180 Itapetininga – SP66) Rua S. José 412, 13400-120 Piracicaba ☎19 3432 3000 W: www.ondalivre.com.br E: comercial@ondalivre.com.br - FM: 105.3 – SP67) Rua Monsenhor Rosa, 1561 - Cetro, 14400-670 Franca ☎16 3713 3977 🖷16 3713 3905 W: www.radioimperador.com.br E: contato@radioimperador.com.br – SP68) Av Brasil, 31 – Prados, (🖂 C.P 52) , 13973-255 Itapira W: www.radioclubeitapira.com.br E: radioclube@ dglnet.com.br ☎19 3843 5198 🖷19 3813 3948 - FM: 91.1MHz «Clube FM» – SP69) Av Ana Costa 532 - 5° andar, Gonzaga, 11060-002 Santos W: www.radiocultura.com.br E: cultura@radiocultura.com.br ☎13 3289 5757 🖷13 3289 4758 - FM: 106.7MHz – SP70) Av Aviador Marques Penedo 11-13, Bairro Jardim Europa, 17045-460 Bauru ☎14 3223 9433 W: www.cancaonova.com E: radio@cancaonova.com – SP71) Rua Doce, 303 - Centro, 15775-000 Santa Fé do Sul ☎17 3631 4859 W: www.radiosantafe.com.br E: comercial@radiosantafe.com.br - FM: 104.7MHz – SP72) Av Sampaio Vidal, 185 - Centro, 17501-040 Marília ☎14 3453 2145 W: www.radio950.com.br E: Webmaster@radio950. com.br– SP73) Av João Dias 1800, Santo Amaro, São Paulo ☎11 5641 4499 – SP74) Rua Quintino Bocaiuva 330/340 (C.P 56, 18200-970) 18200-014 Itapetininga W: www.difusoratransamerica.com.br E: ouvintes@ difusoratransamerica.com.br – SP75) Rua Floriano Peixoto 64, Santo André, 13870-060 São João da Boa Vista W: www.piratininga970am. com.br E: radio970@dglnet.com.br – SP76) Al Vicente Leporace, 4583 – Parque dos Pinhas, (C.P 34, 14400-970) 14405-610 Franca – ☎16 3724 6651 W: www.radiohertz.com.br - FM: 96.5MHz «R 10» – SP77) Rua da Váras, 240 –Barra Funda, 01140-080 São Paulo W: www.radi-orecord.com.br E: radio@rederecord.com.br ☎11 3661 6727 🖷11 2184 4971 – SP78) Praça Joel Waldo Dal Moro 1, 14781-574 Barretos ☎17 3322 9411 W: www.odiariodebarretos.com.br – SP79) Rua Kametaro Morishita, 95 – 3°andar - cidade Universitaria, 19050-700 Presidente Prudente – SP80) Rua João Paulo II s/n, Alto da Bela Vista (C.P 57), 12630-000 Cachoeira Paulista W: www.cancaonova.com E: radio@can-caonova.com ☎12 3186 2600 "Além Fronteiras" (Beyond Boundaries in SS, PP and EE) Sat.: 2200-2300 - FM: 96.3MHz – SP81) Rua Benjamin Constant 33, 10° andar, Centro, 19806-130 Assis ☎18 3322 8811 🖷18 3322 1319 W: www.culturadeassis.com.br E: cultura@culturadeassis. com.br - FM: 100.1MHz – SP82) Rua Profa Aparecida M.Faveri 988, Jd. Fumagalli, 13485-316 Limeira (C.P 105, 13480-970 Limeira) W: www. educadoraam.com.br E: radio@educadoraam.com.br ☎19 3441 3760 – SP83) Rua 24, 2442 (C.P 16, 15700-970), 15700-000 Jales ☎17 3622 5508 W: www.radioculturadejales.com.br – SP84) Rua Ouvidor Freire, 1986 – Centro, (CP 243), 14400-630 Franca ☎16 3713 8800 🖷16 3722 1214 E: administracao@comerciodafranca.com.br or W: www. comerciodafranca.com.br – SP85) Av.Pedro Ometto 2420, 17340-000 Barra Bonita - FM: 97.7MHz – SP86) Rua Floriano Peixoto, 1840 - Labate, 16400-101 Lins - FM: 103.1MHz – SP87) Praça Rodrigues de Abreu, 228 - Paraiso, 04040-080 São Paulo W: www.radiocapital-1040.com.br E: administra@radiocapital.am.br ☎11 3053 1040 – SP88) Rua Boa Morte, 1122 - Centro, 13400-140 Piracicaba ☎19 3422 1060 W: www. educadora1060.com.br E: ouvinte@educadora1060.com.br - FM: 103.1MHz – SP89) Rua Rangel Pestana 147, 11031-551 Santos – SP90) C.P 31, 11740-970 Itanhaém – SP91) Rua Marechal Bitencourt 346, 17201-430 Jaú W: radiopiratininga.am.br – SP92) Av Washington Luiz, 1250-Centro, (C.P 704, 19015-970) 19015-150 Presidente Prudente ☎81 2104 6000 W: www.prudente.am.br E: contato@fm1010fm.com.be - FM: 101.1MHz – SP93) Rua 20 3011 15700-000 Jales ☎17 3622 5505 W: www.radioassuncao.com.br E: comercial@regionalfm.com.br - FM: 103.5MHz «Regional FM» – SP94) Rua Santos Dumont, 239 - Centro, 14300-000 Batatais W: www.difusoraam.com.br E: diretoria@difusor-aam.com.br ☎16 3761 3600 🖷16 3761 3623 – SP95) Rua Olavo Bilac 693, Centro, 16400-000 Lins ☎14 3522 4644 W: www.radioalvora-dadelins.com.br E: alvorada@superig.com.br – SP96) C.P 565, 18001-970 Sorocaba W: www.radioboanova.com.br – SP97) Av Marginal Beira Rio, 13 - Apt 02, Ponte Alta, 12570-000 Aparecida ☎12 3105 1492 W: www.rmpansat.com E: rmpansat@rmpansat.com – SP98) Rua Carlos Artêncio 117 (C.P 326), 17519—255 Marília W: www.radioitaipu.com.br E: itaioufm@terra.com.br – SP99) Rua Jeremias de Paulo Eduardo 916, 15910-000 Monte Alto – SP100) Rua das Palmeiras 315, 01221-010 São Paulo W:radioclick.globo.com/globobrasil E: contato@globonoar.com.br – SP101) Rua Marechal Deodoro da Fonseca 675, Sobre Loja, 16011-000 Araçatuba ☎18 3624 9797 W: www.jovemluz.com.br – SP102) Rua Capitão Joao Marques 89, Jardim Centenarioi – 14940-000 Ibatinga (C.P 91, 14940-970) 16 3341 9900 W: www.radioibitinga.com.br E: radio.

ibatinga@ibinet.com.br - **FM:** 99.3MHz «Ternura FM» – **SP103)** Av Luíz Gonzaga de Amoêdo Campos 28, 13800-000 Mogi Mirim - **FM:** 93.9MHz – **SP104)** Rua Bandeirantes 104 18540 000Porto Feliz ☎15 3562 1215 **W:** www.radionovaporto.com.br **E:** ouvinte@radioportofelicense.com – **SP105)** Av Dr Mário Galvão 463, Jardim Bela Vista, 12209-004 São José dos Campos ☎12 3925 7000 **W:** www.radiobandeirantes1120. com.br **E:** contato@bandvale.com.br **E:** ouvinte@bandvale.com.br - **FM:** 97.5MHz «Nativa FM» – **SP106)** Rua Duque de Caxias 260 cj 22 an 2, 15900-000 Taquaritinga ☎16 3252 2999 **W:** www.radioimperial.com.br – **SP107)** Rua Cherentes 250 – 13° andar (C.P 258, 17600-970) 17600-090 Tupã ☎14 3496 3255 🖳14 3496 6835 **W:** www.radiotupa.com.br **E:** tupaam@radiotupa.com.br - **FM:** 97.7MHz – **SP108)** Rua Gonçalves Dias 208, 19800-110 Assis ☎18 3322 3833 🖳18 3322 8477 **W:** www.difusora-assis.com.br **E:** difusora@difusoraassis.com.br – **SP109)** Praça Joel Waldo Dal Moro 1, Centro, 14781-574 Barretos – **SP110)** C.P 139, 13500-970 Rio Claro **SP111)** Rua Bernardino 546, (C.P 153, 11680-970) 11680-000 Ubatuba ☎12 3832 2993 **W:** www.radiocostaazul.com.br – **SP112)** Av Paulista 2198, Térreo, 01310-300, Sao Paulo **W:** www.radiomundial. com.br **E:** radio@radiomundial.com.br - **FM:** 95.7MHz – **SP113)** Rua Almirante Barroso 456, 19400-000 Presidente Venceslau ☎18 3271 1213 **W:** www.venceslauam.com.br **E:** contato@venceslauam.com.br - **FM:** 95.1MHz «R Jovem Som» – **SP114)** Rua Saldanha de Gama, 184 - Centro, 18035-040 Sorocaba ☎15 3234 3444 **W:** www.radiocacique. com.br **E:** contato@radiocacique.com.br - **FM:** 96.5MHz – **SP115)** Av. Manoel Marques Rosa, 1075 – Ed. Atlântis - Térreo, 15600-000 Fernandópolis ☎17 3442 2666 **W:** www.radiodifusorafernandopolis. com.br - **FM:** 99.0MHz – **SP116)** Rua Barão de Monte Santo, 1211, 3° andar , Centro, 13730-000 Mococa ☎19 3656 6534 – **SP117)** Rua Eng Antonio Francisco de Paula Souz 2799, Jd.São Gabriel, 13044-370 Campinas ☎19 3779 7404 – **SP118)** Rua Pernambuco 4006, (C.P 380, 15500-970), 15500-000 Votuporanga ☎17 3421 2113 **W:** www.radiocidade1190.com.br **E:** contato@radiocidade1190.com.br – **SP119)** Av XV de Novembro 715, (C.P 75) 13650-000 Santa Cruz das Palmeiras – **SP120)** Rodovia Taquarituba Avare s/n km 384, 18740-000 Taquarituba ☎14 3762 1487 🖳14 3762 1009 **W:** www.radioregional1190.com.br **E:** regionalam@yahoo.com.br – **SP121)** Rua Vladimir Herzog, 75 - Agua Branca, 05036-900 São Paulo **W:** www.tvcultura.br **E:** dpt@tvcultura.com.br ☎11 2182 3080 🖳11 3611 1914 – **SP122)** Rua Rui Barbosa 546, 4° andar, 14870-000 Jaboticabal – **SP123)** Rua Tupinambás 115, Bairro São João, 16025-180 Araçatuba – **SP124)** Av Dr Nunu de Assis 550 (C.P 209, 17001-970), 17010-120 Bauru ☎14 3232 3572 – **SP125)** Av. Roberto Simonsen 280, Jd, Santa Rosalia,18090-000 Sorocaba ☎15 3224 5300 🖳15 3231 4938 - **FM:** 94.9MHz – **SP126)** Rua Floriano Peixoto 375, 18300-000 Capão Bonito – **SP127)** Av Mato Grosso 496, Jardim de Bosque, 15997-122 Matão.**W:** radionoticias.am.br **E:** radionoticias@radionoticias.am.br – **SP128)** Rua Dr Miguel Penteado 585, Jardim Chapadão, 13073-180 Santos – **SP129)** Rua Dr. Cardoso Almeida, 1000 ap 131 – Centro, , 18600-005 Botucatu ☎14 3815 3025 **W:** www.radiomunicipalista.com **W:** Rua 8 No 472, 14620-000 Orlândia ☎16 3826 3000 🖳16 3826 3006 **W:** www.orc.com.br **E:** orc@orc.com.br – **SP131)** Rua José Caballero 60, 11055-300 Santos – **SP132)** Av Nove de Julho 265, 15200-000 José Bonifácio **W:** www.radiovaledotiete.com.br **E:** contato@radiovaledotiete.com.br – **SP133)** Rua Vilaa 195, Sala 23, Centro, 12210-000 São Jose dos Campos ☎12 3923 7000 **W:** www.cancanova.com.br **E:** radiosjc@cancaonova.com – **SP134)** Av Prof Alceu Maynard Araújo, 153, 7° andar - Santo Amaro, 04726-160 São Paulo **W:** www.radiomorada.com.br **E:** radiomorada@uol.com.br – **SP136)** Av. Benjamin Constant 1214, 5° andar, Centro, 13010-141 Campinas ☎19 3231 5322 **W:** www.brasilcampinas.com.br **E:** radio@brasilcampinas.com.br – **SP137)** Rua Jamil Khauan19A, Vila Imperial, 15015-620 São José do Rio Preto ☎17 3233 3322 **W:** www.novotempoam.com.br **E:** comercial@novotempoam.com.br – **SP138)** Rua Carina 05, 09732-060 São Bernardo do Campo **W:** www.ipda.com.br – **SP139)** Rua Barao do Rio Branco 559 (C.P 66), 14900-000 Itápolis – **SP140)** Av Cap Antonio Joaquim Mendes, 790 - Jardim Carlos Gomes, 13633-030 Pirassununga ☎19 3561 2200 **W:** www.minhacidadetem.com.br/difusora/– **SP141)** Rua Coronel Osório 84, 12900-000 Bragança Paulista ☎11 4034 0442 **W:** www.radiobraganca.com.br **E:** contato@radiobraganca.com.br – **SP142)** Rua Brigadeiro Tobias 911, 19570-000 Regente Feijó – **SP143)** Rua Costabile Romano 2201, Ribeirania, 14096-380 Ribeirão Preto – **SP144)** Rua Osvaldo Cruz 67, (C.P 246 16010-971) 16010-040 Araçatuba. ☎18 3623 8726 🖳18 3622 6024 **W:** www.culturaam1340.com.br - **FM:** 95.5MHz – **SP145)** Rua Barão de Jacequai 468, 08710-905 Mogi das Cruzes – **SP146)** C.P 20, 17350-970 Igaraçu do Tietê ☎14 3644 1122 **W:** www.novacanoa.com.br **Email:** contato@novacanoa.com.br – **SP147)** Rua Sao Paulo 1708 (✉ C.P 173, 15600-970) 15600-000 Fernandópolis ☎17 3442 6639 **W:**www.aguasquentes1360.com.br - **FM:** 90.5MHz – **SP148)** Av Orlando Fruchi No 97, Distrito Industrial, 17900-000 Dracena ☎18 3821 2593 **W:** www.radioregionaljp.com.br **E:** contato@radioregionaljp.com – **SP149)** Rua Cuiabá 2790, 16900-000 Andradina – **SP150)** Av Sérgio Forein 230, 17250-000, Jardim Santa Rosa, Bariri ☎14 3662

6444 **W:** www.baririradioclube.com.br **E:** contato@baririradioclube.com. br – **SP151)** Rua Pedro Natalia Lorenzetti 172, 18680-030 Lençóis Paulista **W:** www.difusora.lpnet.com.br – **SP152)** Rua Guanabara 144, 13840-000 Mogi Guaçu – **SP153)** Rua Benjamin Constant 1214, 3° andar, 13010-141 Campinas **W:** www.globocampinas.com.br **E:** cbncampinas@globo.com ☎19 3731 5100 - **FM:** 99.1MHz – **SP154)** Rua Dr Erico de Abreu Sodré 542, 16370-000 Promissão – **SP155)** Av Brasil 1119, 17780-000 Lucélia – **SP156)** Rua Salomao Shevs 670, (C.P 96, 13560-970) 13560-270 São Carlos ☎16 3375 3046 Email.: info@clube.com.br - **FM:** 104.7MHz – **SP157)** Rua Jorge Tibiriçá, 2253 - Boavista, 15025-060 São José do Rio Preto ☎17 3212 7012 **W:** www.radiometropole1400.com.br**E:** metropoloam@terra.com.br – **SP158)** Rua Doutor Pinto Ferraz 183, Vila Mariana, 04117-040 São Paulo **W:** www.radioamericasp.com.br ☎11 5579 2916 – Satellite signal downlinked to 44 stations forming Paulus Sat Network – **SP159)** Ladeira Prof Irineu Lopes de Lima, 418 - Centro, 13250-241 Itatiba ☎11 9214 7078 **W:** www.crnitatiba.com.br – **SP160)** Rua Cerqueira Cesar 481, 14010-130 Ribeirão Preto – **SP161)** Rua Mato Grosso 37, Vila Aparecida, 15150-000 Monte Aprazível ☎17 3275 1772 **W:** www.difusoraaparecida.com.br **E:** difusoraaparecida@bol.com.br – **SP162)** Praça Barao do Rio Branco 30, 12010-090 Taubaté – **SP163)** Rua Coronel Joaquim Floriano 287, 18650-000 São Manuel – **SP164)** Praça Lourenço Franco de Oliveira 81, 13930-000 Serra Negra – **SP165)** Av Manoel Goulart, 291 1° andar - Centro, 19010-270 Presidente Prudente ☎18 3221 2900 **W:** www.comercialam.com.br **E:** radio@comercialam.com.br – **SP166)** Praça Nossa Senhora da Conceição 434, 16700-000 Guararapes – **SP167)** Fundação Espírita André Luiz, Av André Luís, 723 - Picanço, 07082-050 Guarulhos. ☎11 2458 321 🖳11 6457 8085 **W:** www.radioboanova.com.br – **SP168)** C.P 115, 13560-970 São Carlos – **SP170)** Av Peixto de Castro, 539 - Vila Celeste 12630-610 Lorena ☎12 3153 1691 **W:** www.cultural1460.com.br **E:** tecnica@cultural1460com.br 24h – **SP171)** Rua Onorio Mendes de Moraes 23, 18130-000 São Roque – **SP172)** Av 15 No 225, 14790-000 Guaíra – **SP173)** Av Rui Barbosa, 229,- Centro, 12308-520 Jacareí ☎12 3954 3000 🖳12 3954 3009 **W:** www.radiomensagem.am.br **E:** mensagem@radiomensagem.am.br – **SP174)** Rua Sao Sebastiao, 33 - Centro, 13660-000 Porto Ferreira ☎19 3581 1552 **E:** primavera@linkway.com.br – **SP175)** Rua 13 de Maio 2680, Jardim Avaí, 13333-080 Indaiatuba (C. P.297 13330-970) ☎19 3875 9141 🖳19 3875 6270 **W:** www.radiojornalindaiatuba.com.br **E:** contato@radiojornalindaiatuba.com.br – **SP176)** Rua Renato Jardim 511, 14350-000 Altinópolis – **SP177)** Rua Adolfo André 478, 2° andar, 12940-280 Atibaia – **SP178)** Praça Oswaldo Martins 218, 16210-000 Bilac – **SP179)** Rua Catarina Estuco Umezo, 171 - Centro, 18900-000 Santa Cruz do Rio Pardo ☎18 3372 1996 **W:** www.difusorasantacruz.com.br **E:** contato@difusorasantacruz.com.br – **SP180)** Av Governador Dr Ademar Pereira de Barros 134, 15400-000 Olímpia **W:** www.difusoraolimpia.com.br – **SP181)** Rua Monte Castelo. 941 - Centro, 17900-000 Dracena ☎18 3821 6492 **W:**www.radioglobodracena.com.br – **SP182)** Av Libero Almeida Silvares 3168, 15600-000 Fernandópolis ☎17 3442 1945 **W:** www.educadora.hd1.com.br **E:** educadorasr@acif.com.br – **SP183)** Rua Padre Moro Grande, 870 - Barrio dos Finco, 09830-670 São Bernardo do Campo ☎11 4354 0059 **W:** www.sagradocoracaojesus.com.br/radio_imaculada.php ou www.miliciadaimaculada.org. br **E:** sam@miliciadaimaculada.org.br - **FM:** 107.1MHz – **SP184)** Av Guerino Turatti, 200, D.Industrial III, 13600-970 Araras ☎19 3541 8322 **W:** www.fraternidade.br **E:** fraternidade@fraternidade.com.br - **FM:** 97.9MHz – **SP185)** Rua Rubião Júnior, 192 - Centro, 12400-450 Pindamonhangaba ☎12 243 1566 **W:** www.rededifusora.com.br/am1500/index.htm **E:** difusorapinda@rededifusora.com.br – **SP186)** Av Leopoldo Carlos de Oliveira 1038, 14530-000 Miguelópolis – **SP187)** C.P 66, 06001-970 Osasco **W:** www.radioterra.am.br – **SP188)** Rua Silva Jardim, 480 - Macuco, 11015-020 Santos ☎13 3221 1846 **W:** www.radiocacique1510.com **E:** ouvintes@radiocacique1510.com – **SP189)** Rua Epitácio Pessoa, 440 – Centro, (C.P 13), 18650-000 São Manuel ☎14 3841 2522 **W:** www.cluberegional.com.br – **SP190)** Av Romeu Viana Romaneli 1510, 15570-000 Cardoso ☎17 3453 1330 **W:** www.radioalvoradacardoso.com – **SP191)** Rua Princesa Izabel de Braganca, 235 - cj 1401, Centro, 08710-460 Mogi das Cruzes ☎11 4796 1478 **W:** www.radioiguatemi.com.br – **SP192)** C.P 66, 13990-970 Espírito Santo do Pinhal - **FM:** 102.7MHz – **SP193)** C.P 125, 17930-970 Tupi Paulista – **SP194)** Rua Capitão Lisboa 1080, 18270-000 Tatuí ☎15 3251 3840 **W:** www.radionoticias.com.br **E:** ouvinte@radionoticias.com.br - **FM:** 93.9MHz «Ternura» – **SP195)** Rua Benedito Carlos dos Reis 700, 15190-000 Nhandeara – **SP196)** Rua Marechal Deodoro 320, 18600-320 Botucatu ☎14 3882 1535 **W:** www.prf-8.com.br **E:** contato@prf-8.com.br - **FM:** 103.1MHz – **SP197)** Av Diogo Antonio Feijo N° 1185, 06114-029 Osasco ☎11 3681 1134 🖳11 3683 0034 **W:** www.novadifusora.com.br **E:** faleconosco@novadifusora.com.br – **SP198)** Rua Dom José Carlos Aguirre 567, 18460-000 Itararé ☎15 3532 4050 🖳15 3532 4499 **W:** www.radioclube.cjb.net **E:** radioclubeam@terracom.br – **SP199)** Rua Pedro de Toledo, 205 - Centro, 19700-000 Paraguaçu Paulista ☎18 3361 1268 🖳18 3361 1197 **W:** www.radiomarconi.com.br **E:** comercial@radi-

omarconi.com.br – **SP200**) Rua Regente Feijó 121, 13360-000 Capivari – **SP201**) Rua Montenegro 196, 11410-040 Guarujá ☎ 13 3269 1010 **W:** www.radioguarujaam.com.br **E:** radioguarujaam@ radioguarujaam.com. br - **FM:** 104.5MHz – **SP202**) C.P 135, 14600-970 São Joaquim da Barra – **SP203**) Av Dr Januario Miraglia 2818, Vila Jaguaribe, 12460-000 Campos do Jordão **W:**radiosaojoaquim. com.br - **FM:** 94.9MHz – **SP204**) Rua A.G Guerra 175 18700-000 Avaré – **SP205**) Rua Belo Horizonte N°930 17890-000 Junqueirópolis ☎18 3841 1465 **W:** www.radiojunqueiropolis.com **E:** comercial@radiojun-queiropolis.com – **SP206**) Av Pereira Barreto, 1200 Vila Gilda, 09190-210 Santo André ☎11 4435 9000 ▤11 4435 9001 **W:** www.radioabc.com.br – **E:** radioabc@radioabc.com.br – **SP207**) Rua Dr Vicente D´Anna 473, 13960-000 Socorro ☎19 3895 1444 **W:** www.radiosocorro.com.br **E:** radiosocorro@terra.com.br – **SP208**) Rua Capitão da Daniel da Cunha Moras 344, 15170-000 Tanabi ☎17 3272 2967 **W:** www.radioclubetan-abi.com.br **E:** radioclubetanabi@hotmail.com – **SP209**) Rua Dr Felipe Vita 1616, 18480-000 Itaporanga – **SP210**) Rua José Bonifácio, 765 - Centro, 13690-000 Descalvado **E:** rscapin@gmail.com – **SP211**) Rua Joaquim Moreira, 12 Parque São Miguel, 07260-220 Guarulhos ☎11 6499 2708 **W:** www.radiocumbica.com.br **E:** webmaster@radiocumbica. com.br – **SP212**) Av 17 No 560, 14780-000 Barretos – **SP213**) Av Paulista 2202, 8° andar, Conj 81/82, 01310-300 São Paulo – **SP214**) Av Capitão José Antônio de Oliveira, 544 - Centro, 17800-000 Adamantina ☎18 3521 3547 **W:** www.radiojoia.com.br **E:** contato@radiojoia.com - **FM:** 93.7MHz «Antena 1» – **SP215**) Rua Brasil 648, 15650-000 Estrela d'Oeste ☎17 3833 1389 **W:** www.alvorada970.com.br – **SP216**) Rua Rubião Junior, 84 – Cj.89 – Shopping Centro, 12210-180 São José dos Campos ☎12 3018 4889 **W:** www.radioeldoradosjc.com.br **E:** contato@ radioeldoradosjc.com.br – **SP217**) Rua Brandão Veras 1274, 14700-000 Bebedouro – **SP218**) Rua Antônio Lobo, 237 1° andar - Centro, 13465-000 Americana ☎19 3462 3992 **W:** www.azulceleste.com.br **E:** azul-celeste@azulceleste.com.br – **SP219**) Estrada Serrinha Km 200, 15350-000 Auriflama **W:** A Monte Castelo, 225 - Centro, 13450-285 Santa Bárbara d'Oeste ☎19 3463 5255 **W:** radiobrasilsbo.com.br - **E:** radio-brasil@radibrasilsbo.com.br – **SP221**) Rua Professor Maximo Ribeiro Nunes 75, 05535-000 Rondopolis(24h Gospel prgrs) ☎11 3721 8010 **W:** www.pazevida.com.br **E:** radio@nacionalgospel.com.br – **SP222**) Rua Dr Mário Sabino 131, 16300-000 Penápolis – **SP223**) Avenida Paulista 2200-5°andar, Bela Vista, 01310-300 São Paulo – **W:** www.radioiguatemi.com.br **E:** radioiguatemi@radioiguatemi.com.br – **SP224**) Rua Kametaro Morishita 95, 2° andar, 19050-700 Presidente Prudente - **FM:** 106.7MHz – **SP225**) Rua Paula Ney 79, 18110-000 Votorantim – **SP226**) Rua Bento Carlos 61, 13560-660 São Carlos - **FM:** 96.9MHz «Jovem Pan FM» – **SP227**) Rua Duque de Caxais 33, Centro, 13460-000 Nova Odessa ☎19 3466 5127 **W:** www.novotempocampi-nas.com.br **E:** comercial@novotempocampinas.com.br – **SP228**) Rua Américo Vespúcio 20, 14730-000 Monte Azul Paulista ☎17 3361 2215 ▤17 3361 2717 **W:** www.radioprincesaam.com.br **E:** radioprincesa@ monteazul.com.br – **SP229**) Av Dr Labiano da Costa Machado 1735 (C.P 235), 17400-000 Garça **W:** www.unimidianet.com – **SP230**) Rua Professor Sud Menucci 464, 17740-000 Rinópolis – **SP231**) Al Trifon Infante Algarim 1914, 19280-000 Teodoro Sampaio – **SP232**) Rua Vergueiro 2045, Liberdade, 04101-000 São Paulo ☎11 5081 579 **W:** www.radiotupiam.com.br **E:** comercial@radiotupiam.com.br– **SP233**) Av 9 de Julho, 304 Nova Paulina, 13140-000 Paulínia **W:** www.cancaonova. com **E:** admpaulina@cancaonova.com ☎19 3844 8500 – **SP234**) Rua Carlos Gomes, 534 - Centro, 14640-000 Morro Agudo ☎16 3851 2414 **W:** www.radiorepublica.com.br **E:** republica@com4.com.br – **SP235**) Rua General Câmara 733, 2° andar, 13450-029 Santa Bárbara d'Oeste radioluzes@netsbo.com.br ▤19 463 3490 – **SP236**) Rua Joaquim Eliziário de Campos 126, 18320-000 Apiaí ☎15 552 1968 ▤15 552 1060 – **SP237**) Rua 5 No 170, Bairro Cristo Redentor, 12100-000 Taubaté – **SP238**) Rua Soares de Oliveira 2070, 14500-000 Ituverava ☎16 3839 7739 **W:** www.nossasenhoradocarmo.com.br/radiocultura – **SP239**) Av Frederico Ozanan 554, 17300-000 Dois Córregos ☎14 3652 2166 – **SP240**) Av Olinda Ralston 411B, Vila Formosa, 13720-000 São José do Rio Pardo **E:** cidlivreadm@vd.com.br - **FM:** 88.7MHz «R 88» – **SP241**) Rua Bororos 344, 17600-020 Tupã **W:** www.radioclubeshow.com.br – **SP242**) Av Washington Luíz, 214 – Centro, 13600-720 Araras ☎19 3541 1265 ▤19 3541 0477 **E:** radioclube@radioclube.com.br **W:** www.radio-clube.com.br– **SP243**) Av 18 de Junho 367, 17690-000 Bastos – **SP244**) Rua Santana 440, Centro, 13880-000 Vargem Grande do Sul ☎19 3641 1152 **W:** http//radioculturaam.com.br **E:** r.cultura@itelefonica.com.br – **SP245**) Rua Newton Prado,344 - Centro, 13610-120 Lemé ☎19 3554 6933 **W:** www.radioculturadeleme.com.br **E:** ouvinte@radiocultur-adeleme.com.br – **SP246**) Rua Francisco Geraldino 71, 17580-000 Pompéia ☎14 9631 0974 **W:** www.sitenc.com.br **E:** leitor@sitenc.com. br – **SP247**) Rua 7 de Setembro S-73, 17280-000 Pederneiras - **FM:** 88.3MHz – **SP248**) Rua Tenente Adolfo Padilha 157, 16800-000 Valparaíso – **SP249**) Rua 15 de Novembro 52, 13920-000 Pedreira – **SP250**) Av Presidente Vargas 2-44, 19470-000 Presidente Epitácio ☎18

3281 8333 **W:** www.radiovaledorioparana.com.br– **SP251**) Av Dr Carlos Burgos 1680, 13901-300 Amparo – **SP252**) Rua Padre João Goetz 370, Jardim Esplanada, 19061-460 Predisete rudente ☎18 3918 5300 **W:** www.ondaviva.com.br **E:** radioondaviva@stetnet.com.br – **SP253**) Rua 7 de Setembro 911, 14240-000 Cajuru **W:** www.miviu.com/eldorado/ index.html – **SP254**) Av São Paulo 1220, 13310-000 Cabreúva – **SP255**) Alameda dos Lírios 111, Bairro Fazenda Castelo (✉ C.P. 789), 18550-000 Boituva ☎15 3268 7202 **W:** www.aesp.org.br/radioamerica1480 **E:** radionovamerica@fasternet.com.br – **SP256**) Rua José Revel 477, Centro, 13320-020 Salto – **SP257**) Rua 9 de Julho 666, 16600-000 Pirajui ☎14 3572 1352 ▤14 3572 1941 **W:** www.pirajuiradioclube.net.br – **SP258**) Rua dos Operários 1441, Vila Guaçu, 19600-000 Rancharia – **SP259**) Rua Coronel João de Carvalho, 39 1° andar - Centro, 13710-000 Tambaú ☎19 3673 1729 **W:** www.radiotambauam.com.br– **SP260**) Av. Professor Jesuíno, 352 - Centro, 17380-000 Brotas ☎14 3653 1306 **W:** www.radiobrotense.com.br – **SP261**) Rua Benjamim Constant 543, 14540-000 Igarapava ☎16 3172 2918 **W:** www.radioshowam.com.br **E:** radioshowam@yahoo.com.br – **SP262**) Rua Inácio Ribeiro 592, 13670-000 Santa Rita do Passa Quatro – **SP263**) Rua Manoel de Arzão 85, Freguesia do Ó, 02730-030 São Paulo ☎11 3935 0108 **W:** www.radi-o9dejulho.com.br **E:** radio9dejulho@terra.com.br **H.of tr,**; 24h – **SP264**) Rua 6 No 1460, 4° andar-Edf São Lucas, 13500-151 Rio Claro ☎19 3532 5507 **W:** www.radioexcelsiorrioclaro.com.br **E:** excelsior@radioexcel-siorrioclaro.com.br – **SP265**) Av São Sebastião 162, 3° piso, sala 1, 18150-000 Ibiúna **W:** www.radioexcelsiorad.com.br – **SP266**) Rua Sergio Tomaz, 740 - Bairro Bom Retiro, 01131-010 Sao Paulo **W:** www.rede-boavontade.com.br – **SP267**) 14100-000 Ribeirão Preto – **SP268**) 18730-000 Itaí – **SP269**) Av Carlos Berchieri, 390 - Centro, 14870-010 Jaboticabal ☎16 3203 5355 **W:** www.radioathenas.com.br **E:** adm@ radioathenas.com.br – **SP270**) Rua Dos Pereiras, 1197, Bairro Palmeiras, 15290-000 Buritama ☎18 3691 3279 **W:** www.radioamiga.com.br **E:** tor-reforteam@uol.com.br – **SP271**) 14740-000 Virdouro – **SP273**) Av Clara Gianotti de Suza 1124, 1 andar, 11900-000 Registro ☎13 3821 1606 **W:** www.radionovaregionalam.com.br **E:** niltonfrancorochaotmail.com – **SP274**) Av Maurilio Biagi 2103 – Ribeirania, 14096-170 Ribeirao Preto. – **SP275**) Rua Honório Mendes de Moraes 23, Esplanada Mendes Moares, 18130-760 São Roque ☎ 11 4712 496 **W:** www.milicia.com.br – **SP276**) Rua barao do Rio Branco, 18550-000 Boituva – **SP277**) Capão Bonito – **SP280**) Rua Épiró 110, (C.P 18113, 04626-970) 04635-030 São Paulo **W:** www.transmundial.com.br **E:** rtm@transmundial.com.br – **SP281**) Rua Minas Gerais 1225, Liberson, 14620-000 Orlândia **W:** www. lidersom.com.br – **SP282**) 14160-000 Sertaozinho – **SP283**) Praca Osvaldo Cruz, 124 – Conjuto 116, 11900-000 Registro.

TO00) TOCANTINS

TO01) Av Joaquim Aires 2393, 77500-000 Porto Nacional – **TO02**) Praça Joe Tôrres, 3 – St Central, 77600-000 Paraíso do Tocantins ▤63 3602 1135 – **TO03**) Av Nossa Senhora de Fátima 894, 77900-000 Tocantinópolis – **TO04**) BR-157 Km 1103, Zona Rural, 77804-970 Araguaína -**FM:** 99.7MHz «Araguaia» – **TO05**) Rua Raul do Espírito Santo 1334, 77760-000 Colinas do Tocantins **E:** elgb@zipmail.com.br – **TO06**)Rua Justianio Borpa, 344 – Setor Santa Filomena, 77650-000 Miracema do Tocantins ☎63 3366 1264 **W:** www.rcmmiracema.com **E:** contato@rcmmiracema.com.br – **TO07**) Almeda João Pires Querido 07, 77490-000 Cristalândia – **TO08**) Av Bernardo Sayão 2201, 77700-000 Guaraí **W:** www.radioguaraito.com.br **E:** gersonnk@hotmail.com – **TO09**) 77054-970 Palmas – **TO10**) 77402-970 Gurupi – **TO11**) 77300-000 Dianópolis – **TO12**) C.P 09, 77054-970 Palmas ☎63 3218 8585 **W:** www.arquidiocesedepalmas.org.br/canais/690am/ **E:** radiopalmas@ cancaonova.com.br – **TO13**) 77950-000 Araguatins – **TO14**) 77700-000 Guarai.– **TO19**) 77770-000 Goiatins – **TO20**) 77700-000 Guarai

FM stations in principal cities. All MHz.
Belo Horizonte: 88.7 Scala FM – 90.7 Cidade – 91.7 Horizontes de Minas – 94.9 Alvorada – MG06) 95.7 – MG68) 96.5 – MG53) 97.3 Altaneira FM – 98.3MHz 98 (Del Rey) – 99.9MHz Terra – MG35) 100.9–102.1 BH FM – MG06) 103.9–105.1 Antena Um – MG65) 106.1–107.5 FM
Brasília: 89.9 Brasília Super FM – 91.7 Brasília Comunicação – 93.7 Atlântida FM – DF01) 95.3 – DF04) 96.1–96.9 Dest Cámara Deputados 97.7 Manchete FM – 99.3 Antena 9 – 100.1 Transamérica – DF02) 100.9 Cultura FM – 101.7 R Jornal de Brasília – 105.5 FM 105 (Planalto) – 106.3 Sigma Radiodifusão – DF03) 107.1 Atividade.
Curitiba: PR15) 88.5 – PR139) 90.1–91.3 Transamerica Hit – 92.3 Scala FM – 93.9 Capital – 95.1 Transamérica Light – 96.3 Studio 96 – PR06) 97.1 – 97.9 Melodia – 98.7 FM 98 – PR42) 99.5 Paraná FM – 100.3 Transamérica – PR134) 101.5 – 102.3 Caioba – 103.9 Jovem Pan – 105.5 Ouro Verde – 106.5 R Novo Tempo.
Fortaleza: 88.9 FM Jangadeiro – 89.9 Capital – 92.9 Tropical – 93.9 FM 93 – 94.7 Jovem Pan – CE21) 95.5 FM do Povo – 99.1 Cidade FM – CE08) 99.9 Dragão FM – 100.9 Pajeu – 101.7 FM Casablanca – 103.9 FM O Tempo – 105.7 Atlântico Sul FM – 106.7 Hoje – 107.9 Universitária.

Porto Alegre: 89.3 Antena Um – 90.3 Transamérica FM – 92.1 Cidade – 92.9 Alegria FM – 93.7 Metropolitana – 94.3 Atlántida – 94.9 Ipanema – 95.9 Liberdade FM – 96.7 Eldorado – 97.5 Jovem Pan FM – 98.3 Continental – 99.3 Band FM – 99.9 Novo Tempo FM – 100.5 Capital – RS23) 101.3 – RS09) 102.3–104.1 FM 104 (Rede Pampa) – 106.3 Aliança – 107.7 Cultura.
Recife: 88.7 Antena Um – 90.3 J.C. FM – 91.9 Rede Aleluia – 92.7 Transamérica – 94.3 Manchete – 95.9 Cidade – 97.5 Recife FM Stereo – 99.1 Caetes – PE10) 99.9–100.7 Evangélica do Brasil – 103.9 Maranata FM – 106.9 CBN – 107.9 JMB Empreendimentos.
Rio de Janeiro: 88.5 Tribuna – RJ11) 89.3–90.3 M.P.B. FM – RJ25) 98.1 FM 98 – RJ12) 98.9 – RJ16) 99.7 – RJ35) 100.5 FM O Dia – 101.3 Transamérica FM – 102.1 Jovem Pan – 102.9 Cidade – 91.1 Diário – RJ25) 92.5–93.3 El Shaddai FM – RJ04) 94.1–94.9 Jovem Rio FM – 95.7 Alvorada FM RJ27) 96.5–97.3 Melodia FM – 103.7 Antena Um – 104.5 Tropical – RJ16) 105.1 105 FM – 106.3 Universidade – 106.7 Catedral – 107.1, 107.9 Universidade.
Salvador: BA57) 90.1 Globo FM – 91.3 Itaparica – 92.3 Salvador FM – 94.3 Piata – 95.9 FM 96 (Aratu) – 97.5 Itapuã – 99.1 Bandeirantes – 100.1 Transamérica – 101.3 BA32) Metrópole – BA04) 102.3–103.9 104 FM (R FM a Tarde) – 104.7 Manchete – BA55) 107.5.
São Paulo: SP62) 88.1–89.1 FM 89 – SP77) Nova FM 89.7 – SP47) 90.5 CBN – 91.3 Manchete FM – 92.1 Lider FM – 92.5 R.Cançao Nova FM – SP32) 92.9 – 93.7 R. USP – 94.1 Deus e Amor – 94.7 Antena Um FM – 95.3 Nativa FM – 95.7 Scala FM – SP57) 96.1 Band FM – 96.9 Cidade FM – 97.7 FM 97 – SP232) 98.1–98.5 Metropolitana FM – 99.3 99 FM – 100.1 Transamérica – SP16) 100.9–101.7 Alpha FM – 102.1 Kiss FM – 102.5 Imprensa – SP121) 103.3–104.1 Apolo FM – 104.7 Transcontinental – 105.1 105 FM – 105.7 Musical FM – SP41) 106.3 Mix FM – 106.9 Nova Omega FM – 107.3 Brasil 2000 – 107.5 Antena 1 – 107.9 Tropical FM

BRITISH INDIAN OCEAN TERRITORY

L.T: UTC + 6h — **Pr.L:** English — **Pop:** variable (US & British military personnel). The original population of approx. 3000 was removed to Mauritius — **E.C:** 60Hz, 110/220V — **ITU:** BIO **Diego Garcia ITU:** DGA

ARMED FORCES RADIO AND TELEVISION SERVICE (U.S. Mil.)
✉ Naval Media Center Detachment-Diego Garcia, PSC 466 Box 14, FPO, AP 96595-0014. ☎+246 370 3680/3685 🖷+246 370 3681
Email: Dgar@mediacen.navy.mil
MW: Island Talk, 1485kHz 200W, news, sports & talk
FM: Power 99, 99.1MHz 200W, weekdays 0600-1400, rock & roll, live DJ. **Island Variety,** 101.9MHz 200W, mixture of rock, alternative, urban & country. – **D.Prgr:** 24h – **V.** by letter.

BRUNEI DARUSSALAM

L.T: UTC +8h — **Pop:** 428,000 — **Pr.L:** Malay, English, Chinese, Gurkha — **E.C:** 50Hz, 240V — **ITU:** BRU

RADIO TELEVISION BRUNEI - RTB (Gov.)
✉ Prime Minister's Office, Jalan Elizabeth II, Bandar Seri Begawan BS8610 ☎ +673 2 243111 🖷 +673 2 241882 **E:** (International and Public Relations): rtbipro@brunet.bn. **W:** www.rtb.gov.bn.
L.P: Acting Dir: Haji Idris bin Haji Md. Ali. Acting Deputy Dir: Pg. Hj. Mahari Pg Hj Abd Rajak. Acting Head of Radio Prgrs: Dayang Hjh Zalinar binti Hj Abdullah. Superintendent of Engineering: Madam Lim Soh Kwang.

MW	kHz	kW	Netw.	H of tr
1) Tutong	594	200	RN	24h

NB: 594kHz r. inactive.

FM (MHz)	RN	RPi	RPe	RH	NI	kW
1) Andulau	93.8	96.9	91.0	97.7	94.9	5
2) Bukit Subok	92.3	95.9	91.4	94.1	93.3	5/0.5

DAB: on 225.648MHz
1) Kuala Belait & Tutong areas. 2) Bandar Seri Begawan (BSB) area.
RN = Rangkaian Nasional FM in Malay: 24h **RPi** = R. Pilihan. English: 0300-0800 (Sat 0300-0700), 1200-0100. Chinese: 0100-0300, 0800-1100. Gurkha: daily 1100-1200, Sat 0700-0800. **RPe** = Rangkaian Pelangi (Pelangi FM, prgrs for young people): 24h Additional FM freqs: 88.5MHz in BSB area, 96.3MHz in Kuala Belait area. **RH** = Rangkaian Harmoni (music sce.): 24h **NI** = Rangkaian Nur Islam (rlg. talk channel): 24h. **Ann:** (RN in Malay) "Nasional FM, Radio Brunei".

KRISTAL MEDIA SDN. BHD. (subsidiary of DST Group, DataStream Technology Sdn Bhd) (Comm.)
✉ Unit 1-345, 1st Fl., Gadong Properties Centre, Gadong BE 4119 11

☎ +673 2 456828 🖷 +673 2 420682 **W:** www.dst-group.com
E: kristalfm@dst-group.com

FM (MHz)	KFM	RQ
1) Andulau	98.7	99.7
2) Bukit Subok	90.7	89.1

KFM=Kristal FM. RQ=Recital of Al-Quran. 1), 2) as above.
D.Prgr: Kristal FM: 24h in English/Malay. RQ: 24h in Arabic.
N: (Kristal FM) rel. RTB and BBC WS.

BRITISH FORCES BROADCASTING SERVICE
✉ BFBS Brunei, BFPO 11 ☎ +673 3 223424 🖷 +673 3 224113 **E:** bfbsbrunei@bfbs.com **W:** www.ssvc.com/bfbs/radio/brunei/
L.P: Station Manager: Mr. Steve Britton **Prgr.:** 24h in English on 101.7MHz 0.25kW, in Nepali (Gurkha) on 89.5MHz 0.25kW. Location: Brunei Garrison HQ, Tuker Lines, Seria, Belait District.

BULGARIA

L.T: UTC +2h (27 Mar-30 Oct: +3h) — **Pop:** 7.2 million — **Pr.L:** Bulgarian, Turkish — **E.C:** 50Hz, 220V — **ITU:** BUL

SAVET ZA ELEKTRONNI MEDII
(Council for Electronic Media)
✉ bul. Shipchenski prohod 69, 1574 Sofiya ☎ +359 2 9708810 🖷+359 2 9733769 **E:** office@cem.bg **W:** www.cem.bg
L.P: Chmn: Georgi Lozanov
NB. The Council for Electronic Media is the regulatory authority for broadcasting.

BALGARSKO NATSIONALNO RADIO (BNR) (Pub)
✉ bul. Dragan Tsankov 4, 1040 Sofiya ☎ +359 2 93361 **E:** bnr@bnr.bg **W:** www.bnr.bg **L.P:** DG: Valeri Todorov

LW/MW	kHz	kW	Prgr	MW	kHz	kW	Prgr
Sofiya (a)	261	75	1+P	Shumen	963	75	R
Vidin	*576	500	2	Kardzhali[1]	963	50	1+T
Pleven	*594	250	1	Sofiya (c)	963	40	1
Plovdiv	*648	30	R	Kresna	*963	40	1
Kresna	*702	10	2	M.Tarnovo	963	5	1+T
Petrich[4]	747	300	F	Varna	*1143	40	1
Shumen	747	10	1+T	St.Zagora	*1161	500	1
Varna	774	75	R	Targovishte	1161	10	1+T
Shumen	*828	500	2	Dulovo	1161	10	1+T
Sofiya (b)	*828	50	2	Vidin[3]	1224	300	1/F
Blagoevgrad	864	75	R	Kardzhali[2]	1296	150	2
Samuil	864	10	1+T	Pleven	*1296	30	2
St. Zagora	873	60	R				

(a) Vakarel (b) Stolnik (c) Dragoman *) inactive at editorial deadline R= reg. sce (see below) T= incl. Turkish sce P= incl. Parliamentary ch. F= External Service [1]=0300-0100, [2]=0400-2200, [3]=0600-2200, [4]=1400-2300

FM (MHz)	BNR1	BNR2	kW	FM	BNR1	BNR2	kW
Belogradchik	102.3	88.2	10/1	Pleven	102.7	100.2	1
Botev vrah	100.9	92.2	10/1	Plovdiv	88.1	91.7	1
Burgas	102.5	95.3	10/1	Popovo	-	95.7	1
Burgas (a)	90.2	96.1	0.5/1	Razgrad	103.5	-	1
Dobrich	104.3	102.3	3/1	Ruse	103.0	95.7	10/1
Dupnitsa	104.1	87.8	1	Silistra	103.3	107.2	1
Gabrovo	103.2	95.4	1	Sliven	87.8	98.7	1
G.Delchev	100.3	98.5	10/1	Smolyan	101.6	96.0	5/1
Kardzhali	105.0	99.2	10/1	Sofiya	103.0	92.9	10/1
Karnobat	103.2	95.0	1	St.Zagora	-	98.3	1
Kavarna	88.1	90.1	1	Svilengrad	99.7	94.9	3
Kyustendil	102.1	99.3	10/1	Shumen	102.0	100.4	10/1
M.Tarnovo	90.2	106.1	1	Varna	100.9	104.8	5/1
Montana	101.4	-	10	V.Tarnovo	89.1	-	1
Nikopol	96.4	98.2	0.1/1	Vratsa	103.4	97.8	1
Oryakhovo	99.8	-	1	Yablanitsa	89.2	50.3	1

NB: Sites with only txs below 1kW not listed. (a) downtown
D.Prgr: BNR1 (Horizont): 24h. On SW (DRM): Fri 0400-0700 on 9400; Sat 0600-0900, Mon-Thu 0900-1200 on 11900kHz. – **BNR2 (Hristo Botev):** 24h. – **Parliamentary Channel:** parliamentary broadcasts 0800-1700 (Tue-Thu), 0900-1200 (Fri) exc. August. On 261kHz + Sofiya 94.5MHz (1kW). – **Service for Turkish minority (Bulgaristan Ulusal Radyosu):** 0600-0630 (SS 0700), 1300-1330, 1800-1900, freqs see MW table. – **Local station: Radiostantsiya Sofiya** on 94.5MHz (1kW): 24h (incl. Parliamentary ch. relays).
External Service (R. Bulgaria): see International Radio section.

BNR REGIONAL STATIONS (Pub)
a) R. Blagoevgrad: ul. Ivan Mihaylov 56, 2700 Blagoevgrad. **E:** reklama@radioblg.com On 864kHz + (MHz) 90.9 (Yakoruda 0.1kW), 102.3 (Gotse Delchev 1kW), 103.2 (Blagoevgrad 0.25kW), 105.2 (Kresna 1kW), Kyustendil 106.6 (1kW): 0400-2200. – **b) R. Plovdiv:**

ul. Dondukov korsakov 2, 4000 Plovdiv. **E:** director@radioplovdiv. bg. On (MHz) 94.0 (Plovdiv 1kW), 100.1 (Velingrad 0.25kW), 100.6 (Dospat 0.25kW), 103.1 (Smolyan 1kW): 0400-2200. – **c) R. Stara Zagora:** ul. Knyaz Boris 75, 6000 Stara Zagora. **E:** rsz@radio-sz.net. On 873kHz + (MHz) 88.3 (Stara Zagora 1kW), 97.2 (Sliven 1kW), 100.8 (Svilengrad 1kW): 0400-2200. – **d) R. Shumen:** ul. Dobro Voynikov 7, 9700 Shumen. **E:** news@radioshumen.net. On 963kHz + (MHz) 87.6 (Shumen 10kW), 90.3 (Silistra 0.25kW), 104.5 (Isperih): 0400-2200. – **e) R. Varna:** bul. Primorski 12, 9000 Varna. **E:** bnr@radiovarna.com. On 774kHz (0200-2400) + (MHz) 88.5 (Burgas 1kW), 88.7 (Dobrich 1kW), 98.2 (Kavarna 0.5kW), 88.9 (Provadiya 0.1kW), 103.4 (Varna 1kW): 24h. – **f) R. Vidin:** ul. Gradinska 1, 3700 Vidin. **E:** office@radio-vidin.com. On (MHz) 94.4 (Vratsa 0.3kW), 97.1 (Vidin 0.25kW), 97.5 (Belogradchik 1kW), 103.9 (Montana 1.5kW): 0400-2200.

OTHER STATIONS

FM	MHz	kW	Location	Station
2)	87.6	1	Nesebar	R. N-Joy
1)	87.9	1	Sandanski	Darik R.
1)	89.7	2	Samokov	Darik R.
4B)	90.6	1	Burgas	NRJ
1)	91.0	1	Belogradchik	Darik R.
8)	91.1	1	Burgas	R. Bravo
1)	91.5	1	Shumen	Darik R.
3A)	91.9	1	Petrich	R. Vega+
1)	93.2	1	Kardzali	Darik R.
6)	94.1	1	Burgas	R. FM+
5A)	94.8	1	Burgas	R. Veselina
1)	96.7	1	Yablanitsa	Darik R.
5B)	96.7	1	Burgas	R. Vitosha
7)	96.8	1	Smolyan	R. Fokus
1)	99.3	1	Kavarna	Darik R.
4A)	99.9	1	Burgas	R. Veronika
1)	100.6	1	Kyustendil	Darik R.
2)	100.6	1	Primorsko	R. N-Joy
1)	100.7	1	Silistra	Darik R.
1)	101.2	1	Sliven	Darik R.
2)	101.8	1	Burgas	R. N-Joy
3B)	103.4	1	Sandanski	R. Ultra
1)	104.0	1	Svilengrad	Darik R.
1)	104.5	1	Burgas	Darik R.
1)	105.0	2	Sofiya	Darik R.
1)	105.4	1.5	Plovdiv	Darik R.
1)	106.2	1	Goce Delchev	Darik R.
1)	106.6	1	V.Tarnovo	Darik R.
1)	106.8	1	Varna	Darik R.
1)	107.0	1	Smolyan	Darik R.
1)	107.7	1	Dobrich	Darik R.
7)	107.7	1	Vidin	R. Fokus

NB: Txs below 1kW not listed.

Addresses & other information:
1) bul. Knyaz A.Dondukov 82, 1504 Sofiya. E: reklama@darik.net – **2)** ul. Panayot Volov 3, 1504 Sofiya. E: jazzfm@netbg.com – **3A,B)** ul. Todor Aleksandrov 3, 2700 Blagoevgrad. **E:** 3A) vega_plus@abv.bg, 3B) radio_ultra@abv.bg – **4A,B)** bul. Tsar Boris III 23, 1612 Burgas. E: 4A) reklama@radioveronika.bg, 4B) reklama@nrj.bg – **5A,B)** ul. Srebarna 21, 1407 Sofiya. E: office@sbsbroadcasting.bg – **6)** bul. Erusalim 51, Zhilishen Kompleks Mladost 1, 1784 Sofiya. E: fmplus@fmplus.net – **7)** ul. Filip Stanislavov 6, 1505 Sofiya. E: focus@focus-news.net – **8)** bul. Stefan Stambolov 2, 8000 Burgas. E: burgas@radiobravo.com.
Radio via DTT: see TV section

BURKINA FASO

L.T: UTC — **Pop:** 15 million — **Pr.L:** French + 16 ethnic — **E.C:** 50Hz, 220V — **ITU:** BFA

CONSEIL SUPÉRIEUR DE LA COMMUNICATION (CSC)
🖃 BP 6437, Ouagadougou ☎+226 50301124 🖷 +226 50301133 **W:** www.csi.bf **E:** infos@csc.bf **L.P:** Dir: Luc Adolphe Diao.

RADIODIFFUSION TÉLÉVISION DU BURKINA (Gov.)
🖃 BP 7029, Ouagadougou 01 ☎+226 50324302 🖷 +226 50310441 **W:** radio.bf **E:** radio@rtb.bf **L.P:** MD: Marcel Toe. Head of Tr. Centre: Marcel Teho. Prgr.Dir: Pascal Goba.

SW	kHz	kW	Times
Ouagadougou	5030	100	0530-0800, 1700-2400 ‡
Ouagadougou	7230	100	0800-1700 ‡

FM: 88.5/92.0/99.9MHz 0.02kW
D.Prgr in French/Ethnic: 0530-2400. **N. in French:** 0630MF, 1000SS + Thurs, 1245 (regional), 1300, 1900, 2200. **N. in English:** W1920 (approx.) **Ann:** "RTV Burkina", "RTB". **IS:** Balafon.

CANAL ARC-EN-CIEL, 03 BP 7045, Ouagadougou. **L.P:** Alphousseini

Bassolet. **FM:** Ouagadougou 96.6MHz, Bobo-Dioulasso 89.8MHz.

REGIONAL STATIONS
Radio Bobo, BP 392, Bobo-Dioulasso. **FM:** 92.0MHz 0.02kW. **D.Prgr:** MF 0600-0800, 1200-1400, 1600-2400, SS 0800-2400 – **R. Gaoua,** Gaoua. **FM:** 90.1MHz – **R. Rurale:** FM txs in Diapaga, Djibasso, Gassan, Kongoussi, Orodara & Poura.

Other stations:
Al Houda FM, Ouagadougou: 98.5MHz – **R. de l'Alliance Chrétienne,** Bobo-Dioulasso: 95.9MHz – **R. Balafon,** Bobo-Dioulasso: 102.7MHz – **Bankuy FM,** Dédougou: 107.7MHz – **R. FM Boulgou,** Garango: 101.1MHz – **R. Cascade,** Banfora: 98MHz – **R. Daande Sahel,** Dori: 104.6MHz – **R. Djongo,** Pô: 106.4MHz – **Echo des Cotonniers,** Solenzo: 95.1MHz – **R. Djibasso:** 94.6MHz – **R. Energie,** 01 BP 6437, Ouagadougou - **FM:** Kaya 92.2MHz, Yako 94.9MHz, Fada N'Gourma 98.8MHz – **R. de l'Espoir,** Réo: 102.8MHz – **R. Évangile Développement,** 04 BP 8050, Ouagadougou. **FM:** Ouagadougou 93.4MHz 0.1kW, Koudougou 95.4MHz, Leo 97.8MHz, Bobo-Dioulasso 103.6MHz, Ouahigouya 104.6MHz **W:** www.autre.net/red – **R. Evangile du Sud-Ouest,** Gaoua: 99.7MHz – **R. Fréquence Espoir:** Dédougou 96.8MHz 1kW, Tougan 101.4MHz 0.1kW – **R. Frontière,** Tenkodogo: 97.6MHz – **R. Gambidi,** Ouagadougou: 97.7MHz – **R. Gassan:** 105.5MHz – **R. du Grand Nord,** Dori: 97.5MHz – **Horizon FM,** 01 BP 2710, Ouagadougou - **FM:** Tenkodogo 97.6MHz, Banfora 98MHz, Koudougou 98.7MHz, Ouayigouya 100.4MHz, Dédougou 102.7MHz, Ouagadougou 104.4MHz, Dori 104.6MHz – **R. Kadoadb,** Ziniare: 107.7MHz – **R.Kongoussi:** 93.2MHz – **R. Kouritta,** Koupela: 93.7MHz – **R. Lumière,** 01 BP 108, Ouagadougou: 98.1MHz – **R. Manegda,** Kaya: 99.4MHz – **R. Maria,** 01 BP 90, Ouagadougou. **FM:** Ouagadougou 91.6MHz 1kW, Kaya 99.4MHz, Koupela 96.9MHz 1kW. **W:** www.radiomaria.org – **Media Star,** Bobo-Dioulasso: 96.7MHz – **R. Munuy FM,** Banfora: 94.7MHz – **R. Naboswende,** Pouytenga: 103.7MHz – **R. Natigmeb Zanga,** Yako: 98.2MHz – **R. Nostalgie,** Ouagadougou, 94.4MHz – **R. Notre Dame:** Kaya 102.9MHz, Kouhogou 105.8MHz, Ouahigouya 102.6MHz – **R. Ouaga FM,** Avenue Loudun-Immeuble Obouf, Ouagadougou: 105.2MHz – **R. Palabre,** B.P. 196, Kougougou: 92.2MHz – **R. Poura:** 98.2MHz – **R. Pulsar,** 01 BP 5976, Ouagadougou: 94.8MHz 0.4kW – **R. Salankoloto,** 01 BP 1716, Ouagadougou: 97.3MHz – **R. Sanmentenga,** Kaya: 96.1MHz – **R. Savane,** Ouagadougou: 103.4MHz – **La Voix du Sud-Ouest,** Diébougou: 101.5MHz – **R. Taanba,** Fada N'Gourma: 98.8MHz 1kW – **R. Tapao,** Diapaga: 95.8MHz – **R. Unitas,** Diébougou: 94.7MHz – **R. Vive le Paysan,** B.P. 74, Saponé: 107MHz – **R. la Voix du Passoré,** Yako: 105.3MHz – **R. la Voix du Paysan,** Ouahigouya: 97MHz – **R. La Voix du Verger,** Orodara: 91.2MHz – **R. Zoodo,** Ouahigouya: 100.4MHz.

Africa No. 1: Ouagadougou 90.3MHz (see main entry under Gabon).
BBC African Sce: Ouagadougou 99.2MHz.
RFI Afrique: Banfora 91.5MHz, Koudougou 93MHz, Ouagadougou 94MHz, Ouahigouya 94.3MHz, Bobo-Dioulasso 99.4MHz.
Voice of America: Ouagadougou 102.4MHz

BURUNDI

L.T: UTC +2h — **Pop:** 9 million — **Pr.L:** Kirundi, Swahili, French, English — **E.C:** 50Hz, 220V — **ITU:** BDI

CONSEIL NATIONAL DE LA COMMUNICATION(CNC)
+257 22223742 🖷 +257 22226547 **W:** www.burundi.gov.bi **L.P:** Chairman: Vestine Nahimana.

RADIO-TÉLÉVISION NATIONALE DU BURUNDI (RTNB, Gov.)
🖃 B.P. 1900, Bujumbura ☎+257 22223742 🖷 +257 22226547 **W:** rtnb.bi **E:** rtnb@cbinf.com **L.P:** D.G.: Innocent Muhozi

FM(MHz)	RTNB1	RTNB2	FM(MHz)	RTNB1	RTNB2
Birime	94.2	98.9	Kaberenge	94.7	98.0
Bujumbura	102.9	92.9	Manga	95.6	98.9
Inanzerwe	88.4	91.4	Mutumba	88.8	91.9

D.Prgr: W 0300-0700 & 0900-2100, Sun 0300-2100. (RTNB1 in Kirundi, RTNB2 in French/Swahili/English). **N.** in French: 0530, 1200, 1500, 1900. **N.** in Swahili: 0630, 1245, 1800. **N.** in Kirundi: 0500, 0700, 1130, 1800, 2000. **N.** in English: 0445, 1230, 1600, 1845.
Ann: "Ici Bujumbura, Radio-Télévision Nationale du Burundi". **IS:** Drums.

Other stations:
Bonesha FM: Manga 87.7MHz, Bujumbura: 96.8MHz, Jenda 102.4MHz – **R. CCIB FM,** Bujumbura: 99.4MHz, nationwide 102.4MHz – **R. Culture,** Bujumbura: 88.2/99.9MHz – **R. Isanganiro:** Bujumbura 89.7MHz, Bururi 93.3/95.1MHz, Kirundo 90.6MHz, Ruyigi 90.7MHz, Manga 101Mhz. **W:** www.web-africa.org/isanganiro **E:** isanganiro@

yahoo.fr – **R. Ivyizigiro**, B.P. 6445, Bujumbura: 90.9/104.8MHz – **R. Public Africaine**, Bujumbura: 91.5MHz – **R. Renaissance**, Bujumbura 101.4MHz – **R. Scolaire Nderagakura**, Bujumbura 87.9MHz – **Rema FM**, Bujumbura: 88.0/107.5MHz. Web: remafm.com.
RFI Afrique: Manga 103.7MHz in F/E/Swahili.
BBC African Sce: Bujumbura 90.2MHz, Mount Manga 105.6MHz.

CAMBODIA

L.T: UTC +7h — **Pop:** 14.8 million — **Pr.L:** Khmer (Cambodian) — **E.C:** 50Hz, 230V — **ITU:** CBG

NATIONAL RADIO OF KAMPUCHEA (RNK)
✉ No 6 Street 19 (Corner Street 102), Sangkat Wat Phnom, Khan Daun Penh, Phnom Penh 12202 ☎ +855 23 725522 🖷 +855 23 427319
W: www.rnk.gov.kh
L.P: Dir. Gen: HE Tan Yan. Dir. Radio Programming Dept: Mr Bou Vannarith. Dir. Radio Tech Dept: Oum Phin. Dep. Dir. Gen, AM 918: Som Sarun. Dep. Dir. Gen. FM 96: Touch Sareth

MW	kHz	kW
Phnom Penh a)	918	200

a) Location: Steung Meanchey (G.C: 11N32 104E53)
National sce AM 918: 2230v-1600 on 918kHz.
FM-96 (comm.): ✉ Steung Meanchey, Phnom Penh 12352. **FM:** 96.0MHz 20kW
Wat Phnom FM: Phnom Penh 105.75MHz. Prgrs for young people & relays of 918kHz.
Cambodia-China Friendship Radio (joint service with China Radio International): 2300-1700 in Chaozhou, English, Khmer and Mandarin. **FM:** Phnom Penh 96.5MHz.
Provincial sces: Battambang **FM:** 92.7MHz. Pursat **FM:** 98.5MHz.
Ann: (Khmer): "Thini Sathani Vithayu Cheat Kampuchea"

Other stations

FM	Location	MHz	kW	Station
1)	Phnom Penh	88.0		Sweet FM (Chinese/Khmer)
42)	Phnom Penh	88.25		R. Meanchey FM
2)	Siem Reap	88.25	2	R. Mahanakor Khemara
3)	Kompong Thom	88.5		Steung Sen R.
4)	Phnom Penh	89.5		VO New Life R. (Samlang Chivit Thmey)
5)	Ratanakiri	89.5		Provincial R.
6)	Phnom Penh	90.0	10	FM90 (Reach Sey)
7)	Battambang	90.25		R. Khlaing Meoung
8)	Pailin	90.5		Provincial R. (Sweet FM)
8)	Phnom Penh	90.5		Ta Phrom Radio
9)	Battambang	91.0		R. FM Khemara
10)	Kampot	91.0	2	R. Bayon FM
10)	Kompong Thom	91.0	2	R. Bayon FM
11)	Phnom Penh	91.0		R. New Phnom Penh (Phnom Penh Thmey)
10)	Takhao	91.0		R. Bayon FM
43)	Phnom Penh	91.25		Sleuk Mas FM
10)	Oddar Meanchey	91.5	2	Provincial R.
10)	Pursat	91.5	2	R. Bayon FM
12)	Phnom Penh	92.0		R. France Int.
10)	Sihanoukville	92.0	3	R. Bayon FM
12)	Siem Reap	92.0		R. France Int.
10)	Stung Treng	92.0	2	R. Bayon FM
5)	Kompong Cham	92.5		Provincial R. (Sweet FM)
10)	Siem Reap	93.0	10	R. Bayon FM 93
5)	Kampot	93.25	1	Provincial R. (Sweet FM)
13)	Phnom Penh	93.5		FM 93.5 (Metropolitan FM)
14)	Svay Rieng	93.75		R. WMC
10)	Banteay Meanchey	94.0	4	R. Bayon FM
12)	Battambang	94.5		R. France Int.
12)	Kompong Cham	94.5		R. France Int.
15)	Phnom Penh	94.5		ABC Traffic Radio
12)	Sihanoukville	94.5		R. France Int.
10)	Phnom Penh	95.0	24	R. Bayon FM
16)	Siem Reap	95.5		R. Angkor Ratha (R. Sarika Angkor)
10)	Svay Rieng	95.5	2	R. Bayon FM
5)	Sisopohon	96.5	10	Banteay Meanchey Provincial R.
17)	Phnom Penh	97.0	10	R Apsara
1)	Phnom Penh	97.5	1	Love FM (English/Khmer)
1)	Siem Reap	97.5		Love FM (English/Khmer)
18)	Phnom Penh	98.0	10	FM98 (Armed Forces R.)
18)	Siem Reap	98.0		FM98 (Armed Forces R.)
19)	Phnom Penh	98.25		Farmers Radio (Kasekor FM)
18)	Sihanoukville	98.5		FM98.5 (Armed Forces R.)
20)	Phnom Penh	99.0	10	FM99
21)	Sisophon	99.0	2	Meanchey FM
5)	Preah Vihear	99.0		Provincial R. (Sweet FM)
11)	Kompong Chhnang	99.25		R. New Phnom Penh (Phnom Penh Thmey)
22)	Siem Reap	99.25		BBC World Service (E.): 24h
23)	Sisophon	99.5	1	My FM

FM	Location	MHz	kW	Station
24)	Phnom Penh	99.5	10	KRUSA FM (FEBC/Family FM)
22)	Phnom Penh	100.0		BBC World Service (E.): 24h
5)	Kompong Cham	100.5	1	FM 100.5
1)	Pursat	100.5		Provincial R. (Sweet FM)
5)	Siem Reap	100.5	1	Provincial R. (Sweet FM)
5)	Sihanoukville	100.5		Provincial R. (Sweet FM)
5)	Stung Treng	100.5		Provincial R. (Sweet FM)
25)	Phnom Penh	101.3		Motherland R.
18)	Battambang	101.5		FM 101.5 (Armed Forces R.)
26)	Phnom Penh	101.5	1	Radio Australia (E.): 24h
26)	Siem Reap	101.5	0.5	Radio Australia (E.): 24h
14)	Phnom Penh	102.0	10	R. WMC
27)	Phnom Penh	102.0		R. Tonle FM
28)	Siem Reap	102.5		Sathani Vithayu Krom Siem Reap (Siem Reap City R. St., FM 102.5)
1)	Phnom Penh	103.0	10	Municipality R. (Khmer)
5)	Battambang	103.25		Provincial R. (Sweet FM)
5)	Bokor Hill	103.5	1	Sweet FM (rel. Kampot)
29)	Sisophon	103.5		Sweet FM
5)	Svay Rieng	103.75		Provincial R. (Sweet FM)
30)	Phnom Penh	104.0		R. Sovann Phum
14)	Kompong Thom	104.3		R. WMC
31)	Phnom Penh	104.5		R. Hang Meas FM
32)	Siem Reap	105.0	5	Sombok Ka Mum (R. Beehive)
33)	Phnom Penh	105.5		KCS Radio
34)	Siem Reap	105.5		R. Mongkul Sovann
35)	Phnom Penh	106.0		South East Asia FM
36)	Siem Reap	106.25		R. For Buddhism
37)	Phnom Penh	106.5		R. Sarika FM (V of Democracy)
41)	Sisophon	106.5	1	U FM
38)	Phnom Penh	107.0	1	Khmer FM (Smile R.)
5)	Kompong Penh	107.25		Provincial R.
39)	Phnom Penh	107.5		ABC Cambodia Radio
40)	Phnom Penh	108.0		R. Solida (Soft FM)

Addresses and other information
1) No 02, Confédération de la Russie, Sangkat Monorom, Khann 7 Makara, Phnom Penh **W:** www.tv3.com.kh/radio103.htm. Prgr in Cham: 1300-1400 on 103.0MHz – **3)**.Slaket Village, Prey Tahou Commune, Steung Sen District, Kompong Thom – **4)** PO Box 1426 Phnom Penh. Operated by Final Frontiers Foundation – **5) Provincial R. Stations (Gov.):** TV3 Relay Station, Battambang; Phum Andong Chen, Khum Ochar, Battambang; Phum Sovansako, Kompong Kandal, Kampot; Khum 3, Veal Vong, Kompong Cham; Phum Chamka Kafe, Toul Lavea, Pailin. Phum Sra Em; Khum Kantout, Chamksan, Preah Vihear; Phum Mondol 2, Khum Svay Dangkhum, Siem Reap; Phum 1, Sangkat 3, Khan Mittapheap, Sihanoukville; Phum Svay Hill, Kou Than Village, O Ampil Commune, Sispohon, Banteay Meanchey Province; Phum Mepleang, Svay Rieng. NB: Most provincial stns carry local Sweet FM prgrs – **6)** Chamkadong, Phnom Penh 12401 – **8)** No 27B Street 472, Phnom Penh 12312. Owned by Funcinpec Party – **10)** HQ: 3 Street 466, Phnom Penh 12310; Phnom Penh stn: 22 Street 106, Toul Krasang, Takhmao, Kandal; Siem Reap: Kasekam Village, Sro Nge, Siem Reap; reg. stns relay 95.0MHz **W:** www.bayontv.com.kh – **11)** 99.25MHz relays Phnom Penh 91.0MHz. English: Sat/Sun 0900-1100 – **12)** Centre Culturel Français, No 218, Keo Chea (Street 184), Phnom Penh. **D.Prgr:** 24h in French exc.1200-1300 in Khmer – **13)** Cheung Ek Village, Phnom Penh 12415 – **14)** Women's Media Center of Cambodia, 30 Street 488, Sangkat Phsar Demthkov, Khan Chamcar Morn, Phnom Penh 12307. 94.5 & 104.3MHz mostly rel. Phom Penh 102.2MHz, with limited local prgrs at times. **W:** www.wmc-cambodia.org/radio102_en.html – **15)** Chamkadong, Phnom Penh 12401 (Affiliated with Station 37) – **16)** Affiliated with stn 25) – **17)** No 69, Street No 57 (Corner Street No 370), Phnom Penh **W:** www.apsaratv.com.kh/radio97program – **18)** Street 169, Borei Keila, Phnom Penh 12253. – Module 3 Slorkram Village, Siem Reap. Owned by the Royal Cambodian Armed Forces, operated by MICA Media Co Ltd, part of Kantana Group (Bangkok, Thailand). Sts carry prgrs of Virgin Hitz R. – **20)** No 69, Street No 57 (Corner Street No 370), Phnom Penh –**23)** Sisophon, Banteay Meanchey Province – **24)** No 8D Street 355, Phnom Penh 12105 – **26)** Preah Monivong (corner Kampuchea Krom), Phnom Penh. – **27)** No 246, Street 63, Sangkat Beong Kengkang, Chamkarmorn, Phnom Penh 12258 – **29)**.Sisophon, Banteay Meanchey Province – **30)** 29 Street 210, Phnom Penh 12158 – **31)** 33 Street 115, Phnom Penh – **32)** No 44G, Street 360, Sangkat Boeung Keng Kang III, Khan Chamkarmorn Phnom Penh 12304 **W:** www.sbk.com.kh – **34)**.National Highway 6, Ta Tean Village, Sala Kamkreuk, Siem Reap –**35)** Sleng Roleung Village Khan Sen Sok, Phnom Penh – **36)** Wat Po, Siem Reap – **37)** No 14A Street 392, Boeung Keng Kang 1, Khan Chamkarmorn, Phnom Penh 12302 (operated by Cambodian Center for Independent Media, CCIM) – **38)** No 18, Rd. 562, Phnom Penh 12151 – **39)** 50A Russian Boulevard, Sangkat Tektla, Khan Rusey Keo, Phnom Penh **W:** www.abccambodia.com – **40)** TS3 Thmey Village, Chamkardong, Phnom Penh 12401 – **42)** Thmey Village, Chamkardong, Phnom Penh 12410 – **43)** Chamkardong, Phnom Penh 12401.

CAMEROON

L.T: UTC +1h — **Pop:** 19 million — **Pr.L:** French, English, ethnic — **E.C:** 50Hz, 220V — **ITU:** CME

NATIONAL COMMUNICATIONS COUNCIL (CNC)
✉ Yaoundé. **L.P:** Prof. Laurent Charles Boyomo.

CAMEROON RADIO TELEVISION (CRTV, Gov.)
✉ B.P. 1634, Yaoundé **☎** +237 22214077 **📠** +237 22204340 **W:** www.crtv.cm **E:** infos@crtv.cm **L.P:** DG: Amadou Vamoulke. Deputy DG: Francis Wete. Dir of Inf. Radio: Michel Ndjock Abanda.

CRTV R. Nationale: Yaoundé 88.8MHz 10kW, Douala 89.2MHz 10kW, Buéa 98.6MHz, Bertoua 89.8MHz 10kW, Bafoussam 91.1MHz 10kW.
Regional stations:
CRTV Yaoundé FM: 94MHz, **CRTV Centre,** Yaoundé: 101.9MHz – **CRTV Littoral,** B.P. 986, Douala: 91.3MHz, **Suelaba FM:** 104.9MHz 10kW – **CRTV Sud-Ouest,** Buea 94.5MHz, **CRTV Mont Cameroun,** Buea 98.0MHz – **CRTV Nord,** B.P. 103, Garoua: 101.2MHz 10kW – **CRTV Est,** B.P. 230, Bertoua: 92.9MHz 10kW – **CRTV Ouest,** B.P. 970, Bafoussam: 93.5MHz 10kW, Pouala FM: 104.5MHz – **CRTV Nord-Ouest,** B.P. 4049, Bamenda: 93.5MHz 10kW – **CRTV Sud,** Ebolowa 97.6MHz 10kW, **Kaze FM** 91.1MHz – **CRTV Adamaoua,** Ngaoundéré: 102.5MHz 10kW – **CRTV Extrême-Nord,** Maroua 94.8MHz, Kousseri 95.5MHz.
SW: CRTV Buea rep. on 6005kHz between 0700-1700.

Other stations:
Dynamic FM, Douala: 103.9MHz – **Magic FM,** Yaoundé: 100.1MHz – **R. Bon Berger,** Kaélé 99.0MHz – **R. Bonne Nouvelle:** Yaoundé 97.7MHz, Ngaoundéré 98.4MHz, Douala 102.5MHz, Ebolowa 102.7MHz – **R. Environnement,** Yaoundé: 107.7MHz – **R. Equinoxe,** Douala: 93MHz Web: www.lanouvelleexpression.net – **R. Le Lauréat,** Douala: 90.5MHz – **R. Lumière,** Yaoundé: 91.9MHz – **R. Nostalgie,** Douala: 96MHz – **R. Reine,** Yaoundé 103.7MHz 1kW, Buéa 97.7MHz 1kW – **R. Salaaman,** Garoua: 89.0MHz – **R. Sawtu Linjiila,** Ngaoundéré: 92.0MHz 1kW (SW relays see Target Broadcasts Section) – **R. Siantou,** Yaoundé: 90.5MHz – **R. Venus,** Yaoundé: 95.4MHz – **R. Veritas,** Douala: 96.8MHz – **R. Vie Nouvelle,** Douala: 100.5MHz – **Real Time Music,** Douala: 103.5MHz, Yaoundé 106.0MHz – **Sky One R,** Yaoundé: 104.5MHz, Douala 100.1MHz. Web: www.skyonecameroun.com – **Sweet FM,** Douala: 88.7MHz – **TBC FM,** Yaoundé: 93.0MHz.
BBC African Sce: Garoua 94.4MHz, Bamenda 95.7MHz, Yaoundé 98.4MHz, Douala 101.3MHz.
RFI Afrique: Yaoundé 105.5, Douala 97.8, Bafoussam 101.1MHz.
Africa No 1: Douala 102MHz, Yaoundé 106.7MHz (see Gabon)

CANADA

L.T: See World Time Table (DST where applicable 13 Mar-6 Nov) — **Pop:** 33 million — **Pr.L:** English, French — **E.C:** 60Hz, 120V. — **ITU:** CAN

CANADIAN RADIO-TELEVISION AND TELECOMMUNICATIONS COMMISSION - CRTC
✉ Ottawa, ON K1A 0N2 **☎** +1 819 997 0313 **📠** +1 819 994 0218 **W:** www.crtc.gc.ca
L.P: Chair: Konrad W. von Finckenstein. Vice Chair: Broadcasting: Michel Arpin.
The CRTC is an independent public organization that regulates and supervises Canadian broadcasting and telecommunications systems.

Provinces & Territories: AB=Alberta, BC=British Columbia, MB=Manitoba, NB=New Brunswick, NL=Newfoundland & Labrador, NS=Nova Scotia, NT=North West Territories, NU=Nunavut, ON=Ontario, PE=Prince Edward Island, QC=Québec, SK=Saskatchewan, YT=Yukon

CBC/RADIO-CANADA (Pub)
✉ Box 3220 Stn C, Ottawa ON K1Y 1E4 **☎** +1 613 288 6033 **W:** www.cbc.radio-canada.ca
L.P: Chair, Board of Dir: Tim W. Casgrain. Pres. and CEO: Hubert T. Lacroix. VP Brand, Comm. and Corp. Affairs: William B. Chambers. VP and CFO: Suzanne Morris. VP Real Estate, Legal Sces. and General Counsel: Maryse Bertrand. VP, People and Culture: Katya Laviolette. Sen. VP Corp. Strategy and Business Partnerships: Michel Tremblay. VP and Chief Regulatory Officer: Steven Guiton. Sen. Dir. Corp. Comm: Martine Ménard.
English Networks: ✉ Box 500 Stn A, Toronto ON M5W 1E6 **W:** www.cbc.ca

L.P: Interim Exec. VP, Eng Sces. and General Manager CBC TV: Kirstine Stewart. GM and Editor in Chief, CBC News, English Sces: Jennifer McGuire. Exec. Dir. Marketing Comms, English Sces: Bridget Hoffer.
French Networks: ✉ Box 6000, Montréal PQ H3C 3A8 +1 514 597 6000 **W:** www.radio-canada.ca
L.P: Exec. VP, French Sces: Sylvain Lafrance. Exec, Dir, News and Current Affairs, French Sces: Alain Saulnier. Exec. Dir, Comm. and Branding: Guylaine Bergeron.

English Radio
CBC Radio One: c=moving to FM *=also on SW +=F.Pl.

MW	Location	Prov.	kHz	kW	Call
1)	Grand Falls	NL	540	10	CBT
2)	Watrous	SK	540	50	CBK
3)	Whitehorse	YT	570	5/1	CFWH
15)	St. Anthony	NL	600	10	CBNA
4)	St. John's	NL	640	10	CBN
5)	Vancouver	BC	690	50	*CBU
7)	Edmonton	AB	740	50	CBX
24)	Bonavista Bay	NL	750	10	CBGY
10)	Prince Rupert	BC	860	10/2.5	CFPR
11)	Inuvik	NT	860	1	CHAK
15)	Winnipeg	MB	990	50/46	CBW
16)	Corner Brook	NL	990	10	CBY
17)	Calgary	AB	1010	50	CBR
18)	Sydney	NS	1140	10	CBI
21)	Iqaluit	NU	1230	1	CFFB
11)	Yellowknife	NT	1340	2.5	CFYK
24)	Gander	NL	1400	4	CBG
25)	Windsor	ON	1550	10	CBE

SW	Location	Prov.	kHz	kW	Call	Relays
5)	Vancouver	BC	6160	0.5	CKZU	CBU
22)	St. John's	NL	6160	1	CKZN	CFGB-FM

FM	Location	Prov.	MHz	kW	Call
17)	Calgary	AB	99.1	7	CBR-1-FM
17)	Edmonton	AB	93.9	3.9	CBX-2-FM
17)	Lethbridge	AB	100.1	100	CBRL-FM
30}	Kamloops	BC	94.1	4.8	CBYK-FM
30)	Kelowna	BC	88.9	5.2	CBTK-FM
33)	Prince George	BC	91.5	100	CBYG-FM
5)	Vancouver	BC	88.1	19.5	CBU-2-FM
31)	Victoria	BC	90.5	6.3	CBCV-FM
16)	Brandon	MB	97.9	90	CBWV-FM
16)	Winnipeg	MB	89.3	2.8	CBW-1-FM
18)	Allardville	NB	97.9	50	CBAA-FM
18)	Fredericton	NB	99.5	3.2	CBZF-FM
18)	Moncton	NB	106.1	69.5	CBAM-FM
32)	St. John	NB	91.3	80	CBD-FM
22)	Goose Bay	NL	89.5	4.5	*CFGB-FM
9)	Halifax	NS	90.5	91	CBH-FM
9)	Middleton	NS	106.5	93.4	CBHM-FM
9)	Mulgrave	NS	106.7	93.4	CBHB-FM
6)	Huntsville	ON	94.3	70	CBLU-FM
6)	Kingston	ON	107.5	100	CBCK-FM
6)	Kitchener/Waterloo	ON	89.1	13.5	CBLA-FM-2
6)	London	ON	93.5	100	CBCL-FM
6)	Ottawa	ON	91.5	84	CBO-FM
6)	Peterborough	ON	98.7	19.2	CBCP-FM
20)	Sudbury	ON	99.9	50	CBCS-FM
6)	Thunder Bay	ON	88.3	23.7	CBQT-FM
6)	Toronto	ON	99.1	98	CBLA-FM
23)	Charlottetown	PE	96.1	100	CBCT-FM
29)	Chicoutimi	QC	102.7	30	CBJE-FM
13)	Montréal	QC	88.5	25	CBME-FM
13)	Québec	QC	104.7	65.8	CBVE-FM
35)	Sept îles	QC	96.9	15	CBSE-FM
13)	Sherbrooke	QC	91.7	25	CBMB-FM
2)	Regina	SK	102.5	2.7	CBKR-FM
2)	Saskatoon	SK	94.1	4.1	CBK-1-FM

+approx 375 relay txs **NB:** Stns identify as 'CBC Radio One'

CBC Radio Two:

FM	Location	Prov.	MHz	kW	Call
17)	Calgary	AB	102.1	100	CBR-FM
7)	Edmonton	AB	90.9	100	CBX-FM
17)	Lethbridge	AB	91.7	100	CBBC-FM
5)	Kamloops	BC	105.3	4.8	CBU-FM-4
5)	Kelowna	BC	89.7	5	CBU-FM-3
5)	Prince George	BC	90.3	0.2	CBU-FM-5
5)	Vancouver	BC	105.7	50	CBU-FM
31)	Victoria	BC	92.1	87	CBU-FM-1
16)	Brandon	MB	92.7	90	CBWS-FM
16)	Winnipeg	MB	98.3	160	CBW-FM
32)	Fredericton/St. John	NB	101.5	81	CBZ-FM
18)	Moncton	NB	95.5	77	CBA-FM
15)	Corner Brook	NL	91.1	3	CBN-FM-2
1)	Grand Falls	NL	90.7	100	CBN-FM-1

FM	Location	Prov.	MHz	kW	Call
4)	St. John's	NL	106.9	100	CBN-FM
9)	Halifax	NS	102.7	92	CBH-FM
9)	Middleton	NS	93.3	16.6	CBH-FM-1
9)	Mulgrave	NS	103.1	100	CBH-FM-2
19)	Sydney	NS	105.1	100	CBI-FM
11)	Yellowknife	NT	95.3	0.1	CFYK-FM
9)	Huntsville	ON	106.9	70	CBL-FM-1
6)	Kingston	ON	92.9	1.6	CBBK-FM
6)	Kitchener/Waterloo	ON	90.7	10.6	CBL-FM-2
6)	London	ON	100.5	22.5	CBBL-FM
12)	Ottawa	ON	103.3	84	CBOQ-FM
6)	Peterborough	ON	103.9	26	CBBP-FM
20)	Sudbury	ON	90.1	50	CBBS-FM
8)	Thunder Bay	ON	101.7	23.5	CBQ-FM
6)	Toronto	ON	94.1	38	CBL-FM
25)	Windsor	ON	89.9	100	CBE-FM
23)	Charlottetown	PE	104.7	100	CBCH-FM
13)	Montréal	QC	93.5	24.6	CBM-FM
13)	Québec	QC	96.1	0.8	CBM-FM-2
13)	Sherbrooke	QC	89.7	25	CBM-FM-1
2)	Regina	SK	96.9	100	CBK-FM
2)	Saskatoon	SK	105.5	98	CBKS-FM
3)	Whitehorse	YT	104.5	0.5	CBU-FM-8

+approx 20 relay txs **NB:** Stns identify as 'CBC Radio Two'
NB: Full list of CBC English freqs at **W:** www.cbc.ca/frequency

French Radio – Radio Canada
Première Chaîne:

MW	Location	Prov.	kHz	kW	Call
25)	Windsor	ON	540	2.5/5	CBEF
28)	New Carlisle	QC	540	10	CBGA-1
7)	Edmonton	AB	680	10	CHFA
2)	Gravelbourg	SK	690	5	CBKF-1
6)	Toronto	ON	860	50	CJBC
2)	Saskatoon	SK	860	50	CBKF-2
16)	Winnipeg	MB	1050	10	CKSB

FM	Location	Prov.	MHz	kW	Call
7)	Edmonton	AB	101.1	3.9	CHFA-10-FM
5)	Vancouver	BC	97.7	100	CBUF-FM
16)	Winnipeg	MB	90.5	2.8	CKSB-10-FM
18)	Allardville	NB	105.7	10	CBAF-FM-2
18)	Fredericton/St. John	NB	102.3	84	CBAF-FM-1
18)	Moncton	NB	88.5	50	CBAF-FM
9)	Halifax	NS	92.3	91	CBAF-FM-5
12)	Ottawa	ON	90.7	84	CBOF-FM
20)	Sudbury	ON	98.1	50	CBON-FM
25)	Windsor	ON	105.5	2.4	CBEF-2-FM
29)	Chicoutimi	QC	93.7	50	CBJ-FM
28)	Matane	QC	102.1	42.9	CBGA-FM
13)	Montréal	QC	95.1	100	CBF-FM
27)	Québec	QC	106.3	52.5	CBV-FM
26)	Rimouski	QC	89.1	38.8	CJBR-FM
36)	Rouyn-Noranda	QC	90.7	25	CHLM-FM
33)	Sept îles	QC	98.1	96.8	CBSI-FM
38)	Sherbrooke	QC	101.1	35	CBF-FM-10
37)	Trois-Rivières	QC	96.5	100	CBF-FM-8
2)	Regina	SK	97.7	13.7	CBKF-FM

+approx 150 relay txs
Espace musique:

FM	Location	Prov.	MHz	kW	Call
7)	Edmonton	AB	90.1	100	CBCX-FM-1
5)	Vancouver	BC	90.9	2.8	CBUX-FM
16)	Winnipeg	MB	89.9	61	CKSB-FM
18)	Allardville	NB	101.9	25	CBAL-FM-1
18)	Fredericton/St. John	NB	88.1	78.5	CBAL-FM-4
18)	Moncton	NB	98.3	77	CBAL-FM
9)	Halifax	NS	91.5	77.5	CBAX-FM
12)	Ottawa	ON	102.5	84	CBOX-FM
20)	Sudbury	ON	90.9	50	CBBX-FM
6)	Toronto	ON	90.3	10	CJBC-FM
29)	Chicoutimi	QC	100.9	50	CBJX-FM
28)	Matane	QC	107.5	31.7	CBRX-FM-1
13)	Montréal	QC	100.7	100	CBFX-FM
27)	Québec	QC	95.3	64.6	CBVX-FM
26)	Rimouski	QC	101.5	100	CBRX-FM
36)	Rouyn-Noranda	QC	89.9	26.7	CBFX-FM-4
35)	Sept îles	QC	96.1	84.8	CBRX-FM-2
38)	Sherbrooke	QC	90.7	25	CBFX-FM-2
37)	Trois-Rivières	QC	104.3	100	CBFX-FM-1
2)	Regina	SK	88.9	96.4	CKSB-FM-1
2)	Saskatoon	SK	88.7	100	CKSB-FM-2

+approx 15 relay txs
NB: Full list of CBC French freqs at **W:** www.radio-canada.ca/radio

CBC North-Québec SW

SW	Location	Prov.	kHz	kW	H of tr & language
39)	Sackville	NB	9625	100	1200-0610 E/F/Inuktitut/Cree

Addresses:
1) 2 Harris Ave, Grand Falls-Windsor NL A2A 2Y2 **W:** www.cbc.ca/thecentralmorningshow – **2)** 2440 Broad St, Regina SK S4P 4A1 **W:** www.cbc.ca/sask **W: (F):** www.radio-canada.ca/regions/saskatchewan – **3)** 3103 3rd Ave, Whitehorse YT Y1A 1E5 **W:** www.cbc.ca/north – **4)** P.O.Box 12010 Stn A, St. John's NL A1B 3T8 **W:** www.cbc.ca/nl – **5)** P.O.Box 4600, Vancouver BC V6B 4A2 **W:** www.cbc.ca/bc **W: (F):** www.radio-canada.ca/regions/colombie-britannique – **6)** P.O.Box 500 Stn A, Toronto ON M5W 1E6 **W:** www.cbc.ca/toronto **W: (F):** www.radio-canada.ca/regions/ontario – **7)** P.O.Box 555, Edmonton AB T5J 2P4 **W:** www.cbc.ca/edmonton **W: (F):** www.radio-canada.ca/regions/alberta – **8)** 213 East Miles St, Thunder Bay ON P7C 1J5 **W:** www.cbc.ca/thunderbay – **9)** P.O. 3000, Halifax NS B3J 3E9 **W:** www.cbc.ca/ns – **10)** 222 3rd Ave W, Prince Rupert BC V8J 1L1 **W:** www.cbc.ca/daybreaknorth – **11)** P.O.Box 160, Yellowknife NT X1A 2N2 **W:** www.cbc.ca/north –**12)** P.O.Box 3220 Stn C, Ottawa ON K1Y 1E4 **W:** www.cbc.ca/ottawa **W: (F):** www.radio-canada.ca/regions/ottawa – **13)** P.O.Box 6000, Montréal QC H3C 3A8 **W:** www.cbc.ca/montreal **W: (F):** www.radio-canada.ca/regions/montreal – **14)** P.O.Box 2200, Fredericton NB E3B 5G4 **W:** www.cbc.ca/informationmorningfredericton – **15)** 541 Portage Ave, Winnipeg MB R3B 2G1 **W:** www.cbc.ca/manitoba **W: (F):** www.radio-canada.ca/regions/manitoba – **16)** P.O.Box 610, Corner Brook NL A2H 6G1 **W:** www.cbc.ca/thewestcoastmorningshow – **17)** P.O.Box 2640, Calgary AB T2P 2M7 **W:** www.cbc.ca/calgary – **18)** 250 University Ave, Moncton NB E1C 8N8 **W:** www.cbc.ca/informationmorningmoncton – **19)** 285 Alexandra St, Sydney NS B1S 2E8 **W:** www.radio-canada.ca/regions/acadie – **20)** 15 MacKenzie St, Sudbury ON P3C 4Y1 **W:** www.cbc.ca/sudbury – **21)** P.O.Box 490, Iqaluit NU X0A 0H0 **W:** www.cbc.ca/north – **22)** P.O.Box 1029 Stn C, Happy Valley – Goose Bay NL A0P 1C0 **W:** www.cbc.ca/labradormorning – **23)** P.O.Box 2230, Charlottetown PE C1A 8B9 **W:** www.cbc.ca/pei – **24)** P.O.Box 369, Gander NL A1V 1W7 **W:** www.cbc.ca/thecentralmorningshow – **25)** 825 Riverside Dr W, Windsor ON N9A 5K9 **W:** www.cbc.ca/windsor – **26)** 273, rue Saint-Jean-Baptiste Ouest, Rimouski QC G5L 4J8 **W: (F):** www.radio-canada.ca/regions/bas-st-laurent – **27)** 888, rue Saint-Jean, Québec QC G1R 5H6 **W: (F):** www.radio-canada.ca/regions/quebec – **28)** 155, rue Saint-Sacrement, Matane QC G4W 1Y9 **W: (F):** www.radio-canada.ca/regions/gaspesie-lesiles – **29)** 500, rue des Saguenéens, Chicoutimi QC G7H 6N4 **W: (F):** www.radio-canada.ca/regions/saguenay-lac – **30)** 243 Lawrence Ave, Kelowna BC V1Y 6L2 **W:** www.cbc.ca/daybreaksouth – **31)** 1025 Pandora Ave, Victoria BC V8V 3P6 **W:** www.cbc.ca/ontheisland – **32)** P.O.Box 2358, St. John NB E2L 3V6 **W:** www.cbc.ca/informationmorningsaintjohn – **33)** Unit 1, 890 Victoria St, Prince George BC V2L 5P1 **W:** www.cbc.ca/daybreaknorth – **35)** 350, rue Smith Bureau 30, Sept îles QC G4R 3X2 **W: (F):** www.radio-canada.ca/regions/cote-nord – **36)** 70, rue Principale, Rouyn-Noranda QC J9X 4P2 **W: (F):** www.radio-canada.ca/regions/abitibi – **37)** 225 des Forges suite 101, Trois-Rivières QC G9A 2G7 **W: (F):** www.radio-canada.ca/regions/mauricie – **38)** 1335, rue King Ouest, Sherbrooke QC J1J 2B8 **W: (F):** www.radio-canada.ca/regions/estrie – **39)** P.O.Box 6000 17th flr, 1726 Montréal QC H3C 3A8 **W:** www.cbc.ca/north

EXTERNAL SERVICE: Radio Canada International (RCI)
See International Broadcasting section.

PRIVATE STATIONS English unless: F=French m=multilingual/ethnic c=moving to FM r=relay *=also on SW +=F.Pl.
NB: Txs below 100W not listed

MW	kHz	Call	kW	N	Location, Prov.
705)	530	CIAO	1/0.25	m	Toronto, ON
202)	560	CHTK	1/0.25		Prince Rupert, BC
701)	560	CFOS	7.5/1		Owen Sound, ON
204)	570	CKWL			Williams Lake, BC
500)	570	CFCB	1		Corner Brook, NL
702)	570	CKGL	10		Kitchener, ON
912)	570	CKSW	10		Swift Current, SK
100)	580	CKUA	10		Edmonton, AB
703)	580	CFRA	50/10		Ottawa, ON
706)	580	CKWW	0.5		Windsor, ON
207)	590	CFTK	1		Terrace. BC
302)	590	CFAR	10/1		Flin Flon, MB
402)	590	CJCW	1/0.25		Sussex, NB
501)	590	VOCM	10		St. John's, NL
707)	590	CJCL	50		Toronto, ON
708)	600	CKAT	10/5		North Bay, ON
900)	600	CJWW	25/8		Saskatoon, SK
101)	610	CKYL	10		Peace River, AB

MW	kHz	Call	kW	N	Location, Prov.
209)	610	CHNL	25/5		Kamloops, BC
303)	610	CHTM	1		Thompson, MB
709)	610	CKTB	10/5		St. Catharines, ON
950)	610	CKRW	1		Whitehorse, YT
501b)	620	CKCM	10		Grand Falls-Windsor, NL
901)	620	CKRM	10		Regina, SK
102)	630	CHED	50		Edmonton, AB
711)	630	CFCO	10/6		Chatham-Kent, ON
712)	640	CFMJ	50		Toronto, ON
238)	650	CISL	10/9		Richmond-Vancouver, BC
501d)	650	CKGA	5		Gander, NL
908)	650	CKOM	10		Saskatoon, SK
108)	660	CFFR	50		Calgary, AB
305)	680	CJOB	50		Winnipeg, MB
707)	680	CFTR	50		Toronto, ON
501e)	710	CKVO	10		Clarenville, NL
716)	710	CJRN	5/2.5		Niagara Falls, ON
213)	730	CHMJ	50		Vancouver, BC
306)	730	CKDM	10/5		Dauphin, MB
804)	730	CKAC	50	F	Montréal, QC
501a)	740	CHCM	10		Marystown, NL
717)	740	CFZM	50		Toronto, ON
911)	750	CKJH	25		Melfort, SK
224a)	760	CFLD	1	r	Burns Lake, BC
103)	770	CHQR	50		Calgary, AB
104)	790	CFCW	50		Camrose, AB
500a)	790	CFNW	1	r	Port au Choix, NL
206)	800	CKOR	10/0.5		Penticton, BC
502)	800	VOWR	10/2.5		St. John's, NL
704)	800	CJBQ	10		Belleville, ON
706)	800	CKLW	50		Windsor, ON
810)	800	CJAD	50/10		Montréal, QC
811)	800	CHRC	50	F	Québec, QC
902)	800	CHAB	10		Moose Jaw, SK
307)	810	CKJS	10	m	Winnipeg, MB
403)	810	CJVA	10	F	Bathurst, NB
724)	820	CHAM	50/10		Hamilton, ON
128)	830	CKKY	10/3.5		Wainwright, AB
204a)	840	CKBX	1/0.5		100 Mile House, BC
206d)	870	CKIR	1/0.25	r	Invermere, BC
224)	870	CFBV	1/0.5		Smithers, BC
500b)	870	CFSX	0.5		Stephenville, NL
102)	880	CHQT	50		Edmonton, AB
312)	880	CKLQ	10		Brandon, MB
230)	890	CJDC	10		Dawson Creek, BC
216)	900	CKMO	10		Victoria, BC
603)	900	CKDH	1	c	Amherst, NS
725)	900	CHML	50		Hamilton, ON
903)	900	CKBI	10		Prince Albert, SK
106)	910	CKDQ	50		Drumheller, AB
308)	920	CFRY	25/15		Portage la Prairie, MB
728)	920	CKNX	10/1		Wingham, ON
107)	930	CJCA	50		Edmonton, AB
405)	930	CFBC	50		St. John, NB
501)	930	CJYQ	25		St. John's, NL
904)	940	CJGX	50/10		Yorkton, SK
309)	950	CFAM	10		Altona, MB
406)	950	CKNB	10/1		Campbellton, NB
108)	960	CFAC	50		Calgary, AB
213)	980	CKNW	50		Vancouver, BC
731)	980	CFPL	10/5		London, ON
905)	980	CJME	10/5		Regina, SK
819)	990	CKGM	50		Montréal, QC
733)	1010	CFRB	50	*	Toronto, ON
130)	1020	CKVH	1/0.4	c	High Prairie, AB
215)	1040	CKST	50		Vancouver, BC
820)	1040	CJMS	10/5	F	Saint-Constant, QC
734)	1050	CHUM	50		Toronto, ON
906)	1050	CJNB	10		North Battleford, SK
111)	1060	CKMX	50	*	Calgary, AB
220)	1070	CFAX	10		Victoria, BC
736)	1070	CHOK	10		Sarnia, ON
221)	1130	CKWX	50		Vancouver, BC
122)	1140	CHRB	50/46		High River, AB
222)	1150	CKFR	10		Kelowna, BC
724)	1150	CKOC	50		Hamilton, ON
907)	1190	CFSL	10/5		Weyburn, SK
225)	1200	CJRJ	25	m	Vancouver, BC
703)	1200	CFGO	50		Ottawa, ON
505)	1210	VOAR	10		St. John's, NL
910a)	1210	CFYM	1/0.25		Kindersley, SK
309a)	1220	CJRB	10		Boissevain, MB
825)	1220	CKSM	10/2.5	F	Shawinigan, QC
209c)	1230	CJNL	1	r	Merritt, BC
500c)	1230	CFGN	0.25		Ch.-Port aux Basques, NL
223)	1240	CKMK	1		Mackenzie, BC
206b)	1240	CJOR	1		Osoyoos, BC
233)	1240	CFNI	1		Port Hardy , BC
302a)	1240	CJAR	1		The Pas, MB
501c)	1240	CKIM	1	r	Baie Verte, NL
747)	1240	CJCS	1		Stratford, ON
826)	1240	CFLM	1	F	La Tuque, QC
309b)	1250	CHSM	10		Steinbach, MB
721)	1250	CJYE	10/5		Oakville, ON
119)	1260	CFRN	50		Edmonton, AB
410)	1260	CKHJ	10		Fredericton, NB
605)	1270	CJCB	10		Sydney, NS
830)	1270	CFGT	10/5	F c	Alma, QC
833)	1280	CFMB	50	m	Montréal, QC
909)	1280	CJSL	10		Estevan, SK
311)	1290	CFRW	10		Winnipeg, MB
751)	1290	CJBK	10		London, ON
127)	1310	CHLW	10	c	St. Paul, AB
753)	1310	CIWW	50		Ottawa, ON
229)	1320	CHMB	50	m	Vancouver, BC
721)	1320	CJMR	20	m	Oakville, ON
910)	1330	CJYM	10		Rosetown, SK
106a)	1340	CIBQ	1	c	Brooks, AB
209a)	1340	CINL	1	r	Ashcroft, BC
210)	1340	CFKC	0.25		Creston, BC
211)	1340	CIVH	1		Vanderhoof, BC
507)	1340	CKHV	1		Happy Valley, NL
612)	1350	CKAD	1		Middleton, NS
812)	1350		1/0.18	F+	Gatineau, QC
129)	1370	CFOK	10	c	Westlock, AB
758)	1380	CKPC	25		Brantford, ON
106b)	1400	CKSQ	1		Stettler, AB
209b)	1400	CHNL-1	1	r	Clearwater, BC
206c)	1400	CKGR	1		Golden, BC
206a)	1400	CIOR	1	r	Princeton, BC
215)	1410	CFTE	50		Vancouver, BC
751)	1410	CKSL	10		London, ON
836)	1410	CJWI	10	F+	Montréal, QC
612a)	1420	CKDY	1		Digby, NS
764)	1430	CHKT	50	m	Toronto, ON
121)	1440	CKJR	10		Wetaskiwin, AB
214)	1450	CHOR	1		Summerland, BC
612b)	1450	CFAB	1		Windsor, NS
834)	1450	CHOU	2/1	m	Montréal, QC
768)	1460	CJOY	10		Guelph, ON
234)	1470	CJVB	50	m	Vancouver, BC
912a)	1490	CJSN	1		Shaunavon, SK
772)	1510	CKOT	10/-		Tillsonburg, ON
773)	1540	CHIN	50/30	m	Toronto, ON
310)	1570	CKMW	10		Winkler, MB
835)	1570	CJLV	10	F	Laval-Montréal, QC
756)	1580	CKDO	10		Oshawa, ON
774)	1610	CHHA	10/1	m	Toronto, ON
836)	1610	CJWI	1	F	Montréal, QC
729)	1650	CINA	1/0.68	m	Mississauga, ON
837)	1650	CJRS	1	m	Montréal, QC
823)	1670	CJEU	1	F	Gatineau, QC
720)	1690	CHTO	3/1	m	Toronto, ON
831)	1690	CJLO	1		Montréal, QC

SW	kHz	Call	kW	Location	Relays
111)	6030	CFVP	0.1	Calgary, AB	CKMX
733)	6070	CFRX	1	Toronto, ON	CFRB

NB: Affiliates of stns that broadcast a common prgr during part of the day have a letter as part of the reference no.

Alberta
100) 4th flr -10526 Jasper Ave NW, Edmonton AB T5J 1Z7 – **101)** P.O.Box 300, Peace River AB T8S 1T5 – **102)** 5204 84th St NW, Edmonton, AB T6E 5N8 – **103)** Shaw Court, 105 630 - 3rd Ave SW, Calgary AB T2P 4L4 – **104)** 5708-48 Ave, Camrose AB T4V 0K1 – **106)** P.O.Box 1480, Drumheller AB T0J 0Y0 – **106a)** 8-403 2nd Ave W, Brooks AB T1R 0S3. Own prgrs Mon-Fri 1300-2000, Sat 1400-1900 – **106b)** 4812A 50 St, Stettler AB T0C 2L2. Own prgrs Mon-Fri 1300-0100, Sat 1400-1900 – **107)** 5316 Calgary Trail NW, Edmonton AB T6H 4J8 – **108)** 2723 37th Ave NE #240, Calgary AB T1Y 5R8 – **111)** 300-1110 Centre St NE, Calgary AB T2E 2R2 – **119)** 100-18520 Stony Plain Rd NW, Edmonton AB T5S 2E2 – **121)** 5214A-50th Ave, Wetaskiwin AB T9A 0S8 – **122)** 11-5th Ave SE, High River AB T1V 1G2 – **127)** 201-4341 50th Ave, St. Paul AB T0A 3A3 – **128)** 1037 2nd Ave 2nd flr, Wainwright AB T9W 1K7 – **129)** 17-10030 106th St, Westlock AB T7P 2K4 – **130)** 4833 52nd St, High Prairie AB T0G 1E0
British Columbia
202) 215 Cow Bay Rd #212, Prince Rupert BC V8J 1A2 – **204)** 83 First

Ave S, Williams Lake BC V2G 1H4. Own prgrs 1400-1800 – **204a)** P.O.Box 1834, 100 Mile House BC V0K 2E0. Own prgrs Mon-Fri 1500-2100, Sat 1500-1700, 2000-2100 – **206)** 33 Carmi Ave, Penticton BC V2A 3G4 – **206a)** P.O.Box 1400, Princeton BC V0X 1W0 – **206b)** 203 – 8309 Main St, Osoyoos BC V0H 1V0 – **206c)** P.O.Box 1403, Golden BC V0A 1H0 – **206d)** Invermere BC – **207)** 4625 Lazelle Ave, Terrace BC V8G 1S4 – **209)** 611 Lansdowne St, Kamloops BC V2C 1Y6 – **209a)** Ashcroft BC – **209b)** Clearwater BC – **209c)** P.O.Box 1630 Stn Main, Merritt BC V1K 1B8. Own prgrs 1200-2100, 2000-2100 – **210)** P.O.Box 310, Creston BC V0B 1G0. Own prgrs 1200-0600. Rel: CJAT-FM – **211)** P.O.Box 1370, Vanderhoof BC V0J 3A0 – **213)** 2000-700 W Georgia St, Vancouver BC V7Y 1K9 – **214)** P.O.Box 1170, Summerland BC V0H 1Z0 – **215)** 300-380 2nd Ave W, Vancouver BC V5Y 1C8 – **216)** 3100 Foul Bay Rd, Victoria BC V8P 5J2 – **220)** 1420 Broad St, Victoria BC V8W 2B1 – **221)** 2440 Ash St, Vancouver BC V5Z 4J6 – **222)** 300-435 Bernard Ave, Kelowna BC V1Y 6N8 – **223)** 2nd flr – 1810 3rd Ave, Prince George BC V2M 1G4. Rel: CKDV-FM – **224)** P.O.Box 335, Smithers BC V0J 2N0 – **224a)** Burns Lake BC – **227)** 110-3060 Norland Ave, Burnaby BC V5B 3A6 (Lic to Vancouver) – **229)** 100-1200 73rd Ave W, Vancouver BC V6P 6G5. Mostly Chinese – **230)** 901 102nd Ave, Dawson Creek BC V1G 2B6 – **233)** 7035A Market St, Port Hardy BC V0N 2P0 – **234)** 2090 Aberdeen Centre 4151 Hazelbridge Way, Richmond BC V6X 4J7 (Lic to Vancouver) – **238)** #20-11151 Horseshoe Way, Richmond BC V7A 4S5

Manitoba
302) P.O.Box 430 Stn Main, Flin Flon MB R8A 1N3 – **302a)** P.O.Box 2980 Stn Main, The Pas MB R9A 1R7 – **303)** 103 Cree Rd, Thompson MB R8N 0B9 – **305)** 930 Portage Ave, Winnipeg MB R3G 0P8 – **306)** 27 3rd Ave NE, Dauphin MB R7N 0Y5 – **307)** 520 Corydon Ave, Winnipeg MB R3L 0P1 – **308)** 350 River Rd, Portage la Prairie MB R1N 0N6 – **309)** P.O.Box 950, Altona MB R0G 0B0 – **309a)** Boissevain MB – **309b)** 105-32 Brandt St, Steinbach MB R5G 2J7 – **310)** P.O.Box 399 Stn Main, Winkler MB R6W 4A6 – **311)** 1445 Pembina Hwy, Winnipeg MB R3T 5C2 – **312)** P.O.Box 880, Brandon MB R7A 6N6

New Brunswick
402) P.O.Box 5900 Stn Main, Sussex NB E4E 5M2 – **403)** 195 rue Main 2nd flr, Bathurst NB E2A 1A7 – **405)** P.O.Box 930 Stn Main, St. John NB E2L 1B1 – **406)** 74 Water St, Campbellton NB E3N 1B1 – **410)** 206 Rookwood Ave, Fredericton NB E3B 2M2

Newfoundland & Labrador
500) P.O.Box 570 Stn Main, Corner Brook NL A2H 6H5 – **500a)** Port au Choix NL – **500b)** 60 West St, Stephenville NL A2N 1C6. Own prgrs 0930 Sat 1030, Sun 1630)-2100 – **500c)** P.O.Box 1230, Ch.-Port aux Basques NL A0M 1C0. Own prgrs Mon-Fri 1230-2130 – **501)** P.O.Box 8590 Stn A, St. John's NL A1B 3P5 – **501a)** P.O.Box 560, Marystown NL A0E 2M0. Own prgrs 0730-1630 – **501b)** P.O.Box 620 Stn Main, Grand Falls-Windsor NL A2A 2K2. Own prgrs 0730-2230 – **501c)** Baie Verte NL – **501d)** P.O.Box 650 Stn Main, Gander NL A1V 1X2. Own prgrs 0700-1900 – **501e)** Gen. Delivery, Clarenville NL A1Y 1A2. Own prgrs 0730-1630 – **502)** P.O.Box 7430 Stn C, St. John's NL A1E 3Y5 – **505)** 1041 Topsail Rd, Mt. Pearl NL A1N 5E9 (Lic to St. John's) – **507)** P.O.Box 160, Nain NL A0P 1L0 (Lic to Happy Valley)

Nova Scotia
603) P.O.Box 670 Stn Main, Amherst NS B4H 4B8 – **605)** P.O.Box 1270 Stn A, Sydney NS B1P 1C8 – **612)** P.O.Box 550, Middleton NS B0S 1P0 – **612a)** P.O.Box 1420, Digby NS B0V 1A0 – **612b)** 169-A Water St, Windsor NS B0N 2T0

Ontario
701) P.O.Box 280 Stn Main, Owen Sound ON N4K 5P5 – **702)** 305 King St W #1101, Kitchener ON N2G 4E4 – **703)** 87 George St, Ottawa ON K1N 9H7 – **704)** P.O.Box 488 Stn Main, Belleville ON K8N 5B2 – **705)** 5312 Dundas St W, Toronto ON M9B 1B3 (Lic to Brampton). Mostly langs – **706)** 1640 Ouellette Ave, Windsor ON N8X 1L1 – **707)** 777 Jarvis St, Toronto ON M4Y 3B7 – **708)** P.O.Box 3000, North Bay ON P1B 8K8 – **709)** P.O.Box 977 Stn Main, St. Catharines ON L2R 6Z4 – **711)** P.O.Box 100 Stn Main, Chatham-Kent ON N7M 5K1 – **712)** Corus Quay 25 Dockside Dr, Toronto ON M5A 0B5 (Lic to Richmond Hill) – **716)** 4668 St. Clair Ave, Niagara Falls ON L2E 6X7 – **717)** 550 Queen St E Suite 205, Toronto ON M4K 1K2 – **720)** 437 Danforth Ave Suite 204, Toronto ON M4K 1P1. Mostly Greek – **721)** 284 Church St, Oakville ON L6J 7N2 – **724)** 883 Upper Wentworth St Suite 401, Hamilton ON L9A 4Y6 – **725)** 875 Main St W #900, Hamilton ON L8S 4R1 – **728)** 215 Carling Terrace, Wingham ON N0G 2W0 – **729)** 1515 Britannia Rd Suite 315, Mississauga ON L4W 4K1. Mostly langs – **731)** P.O.Box 2580 Stn B, London ON N6A 4H3 – **733)** 2 St. Clair Ave W 2nd flr, Toronto ON M4V 1L6. Rpt: **E:** cfrx@ymail.com – **734)** 250 Richmond St W, Toronto ON M5V 1W4 – **736)** 1415 London Rd, Sarnia ON N7S 1P6 – **747)** 376 Romeo St S, Stratford ON N5A 4T9 – **751)** 743 Wellington Rd S, London ON N6C 4R5 – **753)** 2001 Thurston Dr, Ottawa ON K1G 6C9 – **756)** 207-1200 Airport Blvd, Oshawa ON L1J 8P5 – **758)** 571 West St, Brantford ON N3T 5P8 – **764)** 8-135 East Beaver Creek Rd, Richmond Hill ON L4B 1E2 (Lic to Toronto). Mostly Chinese – **768)** 75 Speedvale Ave E, Guelph

ON N1E 6M3 – **772)** P.O.Box 10 Stn Main, Tillsonburg ON N4G 4H3. Daytime only: Jan. 1300-2215, July 1000-0100 – **773)** 622 College St, Toronto ON M6G 1B6. Mostly Italian – **774)** 22 Wenderly Dr, Toronto ON M6B 2N9. Mostly Spanish

Québec
804) 800 rue de la Gauchetière Ouest Bureau 1100, Montréal QC H5A 1M1 – **810)** 300-1411 Du Fort St, Montréal QC H3H 2R1 – **811)** 250W boulevard Wilfrid-Hamel Colisée Pepsi, Entrée nord-ouest, 4e étage, Québec QC G1L 5A7 – **812)** 4020 rue St-Ambroise, Montréal QC H4C 2C7 (Lic to Gatineau). Rel: CIRA-FM – **819)** 300 - 1310 Greene Ave, Westmount QC H3Z 2B5 (Lic to Montréal) – **820)** 143 rue Saint-Pierre, Saint-Constant QC J5A 2G9 – **823)** 855 boul. de Gappe pièce 310, Gatineau QC J8T 8H9 — **825)** 6183 boul. Royal Bureau 130, Shawinigan QC G9N 8O3 – **826)** C.P. 850 Stn Bureau-CHEF, La Tuque QC G9X 3P6 – **830)** 460 SacréCoeur Ouest Bureau 200, Alma QC G8B 1L9 – **831)** 7141 Sherbrooke St Ouest Room CC430, Montréal QC H4B 1R6 – **833)** 35 rue York, Westmount QC H3Z 2Z5 (Lic to Montréal). Mostly Italian and other langs – **834)** 11876 rue demeulles, Montréal QC H4J 2G6 – **835)** 2040 Autoroute Laval, Laval QC H7S 2M9 – **836)** 3733 rue Jarry E 2e etage, Montréal QC H1Z 2G1 – **837)** 4835 Côte St. Catherine Rd #2, Montréal QC H3W 1M4. Mostly langs.

Saskatchewan
900) 345 3rd Ave S, Saskatoon SK S7K 1M5 – **901)** 1900 Rose St, Regina SK S4P 0A9 – **902)** P.O.Box 800 Stn Main, Moose Jaw SK S6H 4P5 – **903)** P.O.Box 900 Stn Main, Prince Albert SK S6V 7R4 – **904)** Broadc Place 120 Smith St E, Yorkton SK S3N 3V3 – **905)** 210-2401 Saskatchewan Dr, Regina SK S4P 4H8 – **906)** P.O.Box 1460 Stn Main, North Battleford SK S9A 2Z5 – **907)** P.O.Box 340 Stn Main, Weyburn SK S4H 2K2 – **908)** 715 Saskatchewan Cres W, Saskatoon SK S7M 5V7 – **909)** P.O.Box 1280 Stn Main, Estevan SK S4A 2H8 – **910)** P.O.Box 490, Rosetown SK S0L 2V0 – **910a)** P.O.Box 1330, Kindersley SK S0L 1S1 – **911)** P.O.Box 750, Melfort SK S0E 1A0 – **912)** 134 Central Ave N, Swift Current SK S9H 0L1 – **912a)** P.O.Box 1176, Shaunavon SK S0N 2M0

Yukon Territory
950) 203-4103 4th Ave, Whitehorse YT Y1A 1H6

FM	Prov.	MHz	kW		Call
Airdrie	AB	106.1	6		CFIT-FM
Athabasca	AB	94.1	9		CKBA-FM
Bonnyville	AB	99.7	50		CFNA-FM
Bonnyville	AB	101.3	27		CJEG-FM
Calgary	AB	88.1	27	m	CKAV-FM-3
Calgary	AB	88.9	100		CJSI-FM
Calgary	AB	90.3	100		CKMP-FM
Calgary	AB	90.9	4		CJSW-FM
Calgary	AB	92.1	100		CJAY-FM
Calgary	AB	92.9	100		CFEX-FM
Calgary	AB	93.7	100		CKUA-FM-1
Calgary	AB	94.7	65		CHKF-FM
Calgary	AB	95.9	100		CHFM-FM
Calgary	AB	96.9	100		CJAQ-FM
Calgary	AB	97.7	100		CIGY-FM
Calgary	AB	98.5	100		CIBK-FM
Calgary	AB	101.5	100		CKCE-FM
Calgary	AB	103.1	100		CFXL-FM
Calgary	AB	105.1	100		CKRY-FM
Calgary	AB	107.3	100		CFGQ-FM
Camrose	AB	98.1	50		CFCW-FM
Cold Lake	AB	95.3	100		CJXK-FM
Drayton Valley	AB	92.9	100		CIBW-FM
Drumheller	AB	91.3	100		CKUA-FM-13
Edmonton	AB	89.3	100	m	CKAV-FM-4
Edmonton	AB	91.7	96		CHBN-FM
Edmonton	AB	92.5	100		CKNG-FM
Edmonton	AB	94.9	100		CKUA-FM
Edmonton	AB	95.7	100		CKEA-FM
Edmonton	AB	96.3	100		CKRA-FM
Edmonton	AB	97.3	100		CIRK-FM
Edmonton/Spruce Grove	AB	98.5	9.3	m	CFWE-FM-4
Edmonton	AB	99.3	100		CHMC-FM
Edmonton	AB	100.3	97		CFBR-FM
Edmonton	AB	101.7	100	m	CKER-FM
Edmonton	AB	102.3	100		CKNO-FM
Edmonton	AB	102.9	100		CHDI-FM
Edmonton	AB	103.9	98		CISN-FM
Edmonton	AB	104.9	100		CFMG-FM
Edmonton	AB	105.9	100		CJRY-FM
Edmonton	AB	107.1	40		CJNW-FM
Edson	AB	94.3	20		CFXE-FM
Fort McMurray	AB	93.3	43.5		CJOK-FM
Fort McMurray	AB	94.5	23.5	m	CFWE-FM-5
Fort McMurray	AB	97.9	43.5		CKYX-FM
Fort McMurray	AB	100.5	50		CHFT-FM
Fort McMurray	AB	103.7	50		CFVR-FM

FM	Prov.	MHz	kW		Call
Grande Prairie	AB	93.1	100		CJXX-FM
Grande Prairie	AB	96.3	70		CJGY-FM
Grande Prairie	AB	97.7	100		CFGP-FM
Grande Prairie	AB	98.9	100		CIKT-FM
Grande Prairie	AB	100.9	100		CKUA-FM-4
Grande Prairie	AB	104.7	100		CFRI-FM
High Level	AB	102.1	34		CKHL-FM
High Level	AB	106.1	34		CFKX-FM
High Prairie	AB	93.5	25		CKVH-FM
High River/Okotoks	AB	99.7	16		CFXO-FM
High River/Okotoks	AB	100.9	100		CKUV-FM
Joussard	AB	91.7	4.2	m	CFWE-FM-1
Lacombe	AB	94.1	55		CJUV-FM
Lethbridge	AB	94.1	100		CJOC-FM
Lethbridge	AB	95.5	100		CHLB-FM
Lethbridge	AB	98.1	20		CKVN-FM
Lethbridge	AB	99.3	100		CKUA-FM-2
Lethbridge/Taber	AB	106.7	100		CJRX-FM
Lethbridge	AB	107.7	100		CFRV-FM
Lloydminster	AB	95.9	100		CKSA-FM
Lloydminster	AB	106.1	100		CKLM-FM
Medicine Hat	AB	94.5	100		CHAT-FM
Medicine Hat	AB	96.1	100		CFMY-FM
Medicine Hat	AB	97.3	100		CKUA-FM-3
Medicine Hat	AB	102.1	40		CJCY-FM
Medicine Hat	AB	105.3	100		CKMH-FM
Moose Hills	AB	96.7	100	m	CFWE-FM-3
Olds	AB	96.5	35		CKLJ-FM
Olds	AB	104.5	35		CKJX-FM
Peace River	AB	96.9	100		CKUA-FM-5
Peigan/Blood River	AB	89.3	10.2	m	CFWE-FM-2
Pincher Creek	AB	92.7	6		CJPV-FM
Red Deer	AB	95.5	100		CKGY-FM
Red Deer	AB	98.9	100		CIZZ-FM
Red Deer	AB	100.7	100		CKRI-FM
Red Deer	AB	101.3	50		CKIK-FM
Red Deer	AB	105.5	100		CHUB-FM
Red Deer	AB	106.7	100		CFDV-FM
Red Deer	AB	107.7	100		CKUA-FM-6
Slave Lake	AB	92.7	5		CHSL-FM
Suffield	AB	104.1	4.3		CKBF-FM
Taber	AB	93.3	100		CJBZ-FM
Wainwright	AB	93.7	100		CKWY-FM
Westlock	AB	97.9	48	+	CKWB-FM
Wetaskiwin	AB	93.5	5.1		CIHS-FM
Whitecourt	AB	96.7	9		CFXW-FM
Whitecourt	AB	105.3	42.3		CIXM-FM
Campbell River	BC	99.7	6		CIQC-FM
Chilliwack	BC	98.3	5		CKSR-FM
Courtenay	BC	97.3	11.6		CKLR-FM
Courtenay	BC	98.9	5		CFCP-FM
Duncan	BC	89.7	3.5		CJSU-FM
Fort St. John	BC	98.5	50		CHRX-FM
Fort St. John	BC	100.1	20		CKFU-FM
Fort St. John	BC	101.5	40		CKNL-FM
Gibsons	BC	107.5	4.6		CISC-FM
Houston	BC	105.5	3.5		CJFW-FM-7
Kamloops	BC	97.5	4.3		CKRV-FM
Kamloops	BC	98.3	4.3		CIFM-FM
Kamloops	BC	100.1	3.5		CKBZ-FM
Kamloops	BC	103.1	5		CJKC-FM
Kelowna	BC	96.3	31		CKKO-FM
Kelowna	BC	99.9	35		CHSU-FM
Kelowna	BC	101.5	33.3		CILK-FM
Kelowna	BC	103.1	35		CKQQ-FM
Kelowna	BC	103.9	10		CJUI-FM
Kelowna	BC	104.7	36		CKLZ-FM
Nanaimo	BC	101.7	3		CHLY-FM
Nanaimo	BC	102.3	3		CKWV-FM
Nanaimo	BC	106.9	3		CHWF-FM
Penticton	BC	100.7	14.1		CIGV-FM
Port Alberni	BC	93.3	6		CJAV-FM
Powell River	BC	95.7	5.8		CFPW-FM
Prince George	BC	94.3	11.5		CIRX-FM
Prince George	BC	97.3	12		CJCI-FM
Prince George	BC	99.3	9.3		CKDV-FM
Prince George	BC	101.3	9.1		CKKN-FM
Squamish	BC	107.1	30		CISQ-FM
Terrace	BC	103.1	3.2		CJFW-FM
Trail	BC	95.7	13.5		CJAT-FM
Vancouver	BC	93.1	8	m	CKYE-FM
Vancouver	BC	93.7	75		CJJR-FM
Vancouver	BC	94.5	90		CFBT-FM
Vancouver	BC	95.3	71.3		CKZZ-FM
Vancouver	BC	96.1	100	m	CHKG-FM
Vancouver	BC	96.9	75		CKLG-FM
Vancouver	BC	99.3	75		CFOX-FM
Vancouver	BC	100.5	11		CKPK-FM
Vancouver	BC	101.1	75		CFMI-FM
Vancouver	BC	102.7	5.5		CFRO-FM
Vancouver	BC	103.5	100		CHQM-FM
Vancouver	BC	104.3	10		CHHR-FM
Vancouver	BC	104.9	31		CFUN-FM-2
Vancouver	BC	106.3	9	m	CKAV-FM-2
Vernon	BC	105.7	100		CICF-FM
Vernon	BC	107.5	100		CKIZ-FM
Victoria	BC	91.3	3.5		CJZN-FM
Victoria	BC	98.5	100		CIOC-FM
Victoria	BC	100.3	100		CKKQ-FM
Victoria	BC	103.1	20		CHTT-FM
Victoria	BC	107.3	20		CHBE-FM
Brandon	MB	91.5	100		CIWM-FM
Brandon	MB	94.7	100		CKLF-FM
Brandon	MB	96.1	88.7		CKX-FM
Brandon	MB	101.1	100		CKXA-FM
Neepawa	MB	97.1	3.2	+	CJBP-FM
Portage La Prairie	MB	93.1	27		CFRY-1-FM
Portage La Prairie	MB	96.5	24		CJPG-FM
Selkirk	MB	104.1	100		CFQX-FM
Selkirk	MB	105.5	100		CICY-FM
St. Boniface	MB	91.1	61	F	CKXL-FM
Steinbach	MB	96.7	100		CILT-FM
Swan Lake	MB	90.5	3.7		CISF-FM
Winkler/Morden	MB	93.5	100		CJEL-FM
Winnipeg	MB	92.1	140		CITI-FM
Winnipeg	MB	94.3	100		CHIQ-FM
Winnipeg	MB	95.1	100		CHVN-FM
Winnipeg	MB	97.5	310		CJKR-FM
Winnipeg	MB	99.1	100		CJGV-FM
Winnipeg	MB	99.9	100		CFWM-FM
Winnipeg	MB	100.7	80		CHNK-FM
Winnipeg	MB	102.3	100		CKY-FM
Winnipeg	MB	103.1	100		CKMM-FM
Winnipeg	MB	104.7	3	m	CIUR-FM
Bathurst	NB	92.9	100	F	CKLE-FM
Bathurst	NB	104.9	33.5		CKBC-FM
Campbellton	NB	103.9	15	F	CIMS-FM
Edmundston	NB	92.7	40.8	F	CJEM-FM
Fredericton	NB	92.3	93		CFRK-FM
Fredericton	NB	105.3	78		CFXY-FM
Fredericton	NB	106.9	78		CIBX-FM
Grand Falls	NB	93.5	5.3		CIKX-FM
Grand-Sault	NB	105.1	3	F	CFAI-FM-1
Inkerman/Pokemouche	NB	97.1	44.4	F	CKRO-FM
Kedgwick	NB	90.1	3	F	CFJU-FM
Miramichi City	NB	93.7	11	F	CKMA-FM
Miramichi City	NB	99.3	17.8		CFAN-FM
Moncton	NB	90.7	30	F	CFBO-FM
Moncton	NB	91.9	70		CKNI-FM
Moncton	NB	94.5	19		CKCW-FM
Moncton	NB	96.9	100		CJXL-FM
Moncton	NB	99.9	9.5	F	CHOY-FM
Moncton	NB	103.1	46.8		CJMO-FM
Moncton	NB	103.9	70		CFQM-FM
Saint John	NB	88.9	79		CHNI-FM
Saint John	NB	94.1	100		CHSJ-FM
Saint John	NB	97.3	100		CHWV-FM
Saint John	NB	98.9	12		CJYC-FM
Saint John	NB	100.5	100		CIOK-FM
Shediac	NB	89.5	38	F	CJSE-FM
St. Stephen	NB	98.1	40		CHTD-FM
Woodstock	NB	104.1	10		CJCJ-FM
Argentia	NL	100.3	3.7		CFOZ-FM
Bonavista	NL	92.1	6.7		CJOZ-FM
Carbonear	NL	103.9	30		CHVO-FM
Clarenville	NL	100.7	4.1		VOCM-FM-1
Corner Brook	NL	92.3	50		CKOZ-FM
Corner Brook	NL	103.9	40		CKXX-FM
Gander	NL	98.7	6		CKXD-FM
Grand Falls	NL	102.3	24		CKXG-FM
Marystown	NL	96.3	31.3		CIOZ-FM
Rattling Brook	NL	95.9	50		CKMY-FM
St. John's	NL	94.7	100		CHOZ-FM
St. John's	NL	97.5	100		VOCM-FM
St. John's	NL	99.1	100		CKIX-FM
St. John's	NL	101.1	20		CKSJ-FM
Stephenville	NL	98.5	4.3		CIOS-FM
Amherst	NS	101.7	40	+	CKDH-FM
Amherst	NS	107.9	6.5	+	CFTA-FM
Antigonish	NS	98.9	75.4		CJFX-FM

FM	Prov.	MHz	kW		Call	FM	Prov.	MHz	kW		Call
Barrington	NS	96.3	5.5		CJLS-FM-2	Kingston	ON	98.3	95.5		CFLY-FM
Bridgewater	NS	98.1	32		CKBW-FM	Kingston	ON	98.9	15		CKLC-FM
Bridgewater	NS	100.7	10		CJHK-FM	Kingston	ON	101.9	3		CFRC-FM
Cheticamp	NS	106.1	3	F	CKJM-FM	Kingston	ON	104.3	8		CKWS-FM
Glace Bay	NS	89.7	6		CKOA-FM	Kingston	ON	105.7	50		CIKR-FM
Halifax	NS	89.9	100		CHNS-FM	Kirkland Lake	ON	101.5	23		CJKL-FM
Halifax/Dartmouth	NS	92.9	100		CFLT-FM	Kitchener/Waterloo	ON	91.5	10		CKBT-FM
Halifax/Dartmouth	NS	93.9	5		CJLU-FM	Kitchener	ON	96.7	100		CHYM-FM
Halifax	NS	95.7	65		CJNI-FM	Kitchener/Waterloo	ON	98.5	27		CKWR-FM
Halifax	NS	96.5	100		CKUL-FM	Kitchener	ON	99.5	4.3		CKKW-FM
Halifax	NS	100.1	100		CIOO-FM	Kitchener	ON	105.3	100		CFCA-FM
Halifax	NS	101.3	100		CJCH-FM	Kitchener/Waterloo	ON	106.7	5		CIKZ-FM
Halifax	NS	101.9	91		CHFX-FM	Leamington	ON	92.7	4		CJSP-FM
Halifax	NS	103.5	100		CKHZ-FM	Leamington	ON	96.7	27		CHYR-FM
Halifax/Dartmouth	NS	104.3	100		CFRQ-FM	Lindsay	ON	91.9	11.4		CKLY-FM
Halifax	NS	105.1	45		CKHY-FM	Little Current	ON	100.7	27.5		CFRM-FM
Inverness	NS	102.5	10		CJFX-FM-1	London	ON	92.7	50		CJBX-FM
Kentville	NS	89.3	30		CIJK-FM	London	ON	94.9	6		CHRW-FM
Kentville	NS	94.9	100		CKWM-FM	London	ON	95.9	300		CFPL-FM
Kentville	NS	97.7	18		CKEN-FM	London	ON	97.5	50		CIQM-FM
Liverpool	NS	94.5	8.7		CKBW-1-FM	London	ON	102.3	12.1		CHST-FM
New Glasgow	NS	94.1	80		CKEC-FM	London	ON	106.9	3		CIXX-FM
New Tusket	NS	93.5	3		CJLS-FM-1	Marathon	ON	93.1	50		CFNO-FM
Petit-de-Grat	NS	104.1	5.8	F	CITU-FM	Midland	ON	104.1	20		CICZ-FM
Port Hawkesbury	NS	101.5	38.1		CIGO-FM	Napanee	ON	88.7	11.1		CKYM-FM
Shelburne	NS	93.1	8.6		CKBW-2-FM	New Liskeard	ON	104.5	10		CJTT-FM
Sydney	NS	94.9	61		CKPE-FM	Newmarket	ON	88.5	30		CKDX-FM
Sydney	NS	98.3	100		CHER-FM	Niagara Falls	ON	105.1	15		CFLZ-FM
Sydney	NS	101.9	58		CHRK-FM	North Bay	ON	100.5	100		CHUR-FM
Sydney	NS	103.5	26.5		CKCH-FM	North Bay	ON	101.9	100		CKFX-FM
Truro	NS	99.5	16.8		CKTY-FM	North Bay	ON	106.3	10		CFXN-FM
Truro	NS	100.9	50		CKTO-FM	Orangeville	ON	103.5	30.7		CIDC-FM
Weymouth	NS	103.3	3		CKDY-1-FM	Orillia	ON	105.9	20		CICX-FM
Yarmouth	NS	95.5	18		CJLS-FM	Oshawa	ON	94.9	50		CKGE-FM
Yarmouth	NS	104.1	39.3	F	CIFA-FM	Ottawa	ON	88.5	12		CILV-FM
Ajax	ON	95.9	50		CJKX-FM	Ottawa	ON	89.1	18.1		CHUO-FM
Bancroft	ON	97.7	50		CHMS-FM	Ottawa	ON	89.9	27		CIHT-FM
Barrie	ON	93.1	100		CHAY-FM	Ottawa	ON	93.1	12		CKCU-FM
Barrie	ON	95.7	100		CFJB-FM	Ottawa	ON	93.9	95		CKKL-FM
Barrie	ON	100.3	40		CJLF-FM	Ottawa	ON	95.7	9.1	m	CKAV-FM-9
Barrie	ON	101.1	7.5		CIQB-FM	Ottawa	ON	97.9	6.8	m	CJLL-FM
Barrie	ON	107.5	50		CKMB-FM	Ottawa	ON	99.1	66		CHRI-FM
Belleville	ON	91.3	3.4		CJLX-FM	Ottawa	ON	99.7	100		CJOT-FM
Belleville	ON	95.5	64		CJOJ-FM	Ottawa	ON	100.3	100		CJMJ-FM
Belleville	ON	97.1	50		CIGL-FM	Ottawa	ON	101.9	4.5		CIDG-FM
Belleville	ON	100.1	32		CHCQ-FM	Ottawa	ON	105.3	84		CISS-FM
Belleville	ON	102.3	15		CKJJ-FM	Ottawa	ON	106.1	100		CHEZ-FM
Bluewater	ON	91.7	6		CIBU-FM-1	Ottawa	ON	106.9	84		CKQB-FM
Bracebridge	ON	99.5	12		CFBG-FM	Owen Sound	ON	92.3	9.4		CJOS-FM
Brantford	ON	92.1	80		CKPC-FM	Owen Sound	ON	93.7	22		CKYC-FM
Brockville	ON	103.7	100		CJPT-FM	Owen Sound	ON	106.5	28		CIXK-FM
Brockville	ON	104.9	7.7		CFJR-FM	Paris	ON	88.3	10.6		CJIQ-FM
Cambridge	ON	107.5	6		CJDV-FM	Parry Sound	ON	103.3	46.6		CKLP-FM
Chatham	ON	89.3	18.7		CKGW-FM	Pembroke	ON	96.7	100		CHVR-FM
Chatham	ON	94.3	50		CKSY-FM	Pembroke	ON	99.9	7.5		CKQB-FM-1
Chatham	ON	95.1	42		CKUE-FM	Penetanguishene	ON	88.1	4.5	F	CFRH-FM
Cobourg	ON	93.3	15.5		CKSG-FM	Peterborough	ON	99.7	11		CKPT-FM
Cobourg	ON	103.1	86.7		CFMX-FM	Peterborough	ON	100.5	15		CKRU-FM
Cobourg	ON	107.9	20		CHUC-FM	Peterborough	ON	101.5	15.2		CKWF-FM
Cornwall	ON	92.1	45.6	F	CHOD-FM	Peterborough	ON	105.1	7.5		CKQM-FM
Cornwall	ON	101.9	3.2		CJSS-FM	Port Elgin	ON	97.9	9		CFPS-FM
Cornwall	ON	104.5	28.2		CFLG-FM	Renfrew	ON	98.7	20		CJHR-FM
Dryden	ON	92.7	39		CKDR-FM	Sarnia	ON	99.9	50		CFGX-FM
Elliot Lake	ON	94.1	90		CKNR-FM	Sarnia	ON	103.3	6		CKCI-FM
Fort Erie	ON	101.1	50		CKEY-FM	Sarnia	ON	106.3	50		CHKS-FM
Fort Frances	ON	93.1	21		CFOB-FM	Sault Ste. Marie	ON	100.5	13.9		CHAS-FM
Goderich	ON	104.9	12.6		CHWC-FM	Sault Ste. Marie	ON	104.3	100		CJQM-FM
Guelph	ON	106.1	50		CIMJ-FM	Simcoe	ON	98.9	50		CHCD-FM
Haldimand	ON	92.9	10		CKJN-FM	Smiths Falls	ON	92.3	17		CJET-FM
Haliburton	ON	93.5	6		CFZN-FM	Smiths Falls	ON	101.1	100		CKBY-FM
Haliburton	ON	100.9	3.4		CKHA-FM	St. Catharines	ON	97.7	50		CHTZ-FM
Hamilton/Burlington	ON	94.7	100		CIWV-FM	St. Catharines	ON	105.7	50		CHRE-FM
Hamilton	ON	95.3	100		CING-FM	St. Thomas	ON	103.1	50		CFHK-FM
Hamilton	ON	102.9	40.3		CKLH-FM	Stratford	ON	107.7	6		CHGK-FM
Hamilton/Burlington	ON	107.9	26.1		CJXY-FM	Sudbury	ON	91.7	50		CICS-FM
Hearst	ON	91.1	5.5	F	CINN-FM	Sudbury	ON	92.7	100		CJRQ-FM
Huntsville	ON	105.5	43.4		CFBK-FM	Sudbury	ON	93.5	100		CIGM-FM
Kapuskasing	ON	89.7	3	F	CKGN-FM	Sudbury	ON	95.5	8.1		CJTK-FM
Kapuskasing	ON	93.7	3.4	F	CHYX-FM	Sudbury	ON	103.9	100		CHNO-FM
Kapuskasing	ON	100.9	12		CKAP-FM	Sudbury	ON	105.3	100		CJMX-FM
Kenora	ON	89.5	50		CJRL-FM	Sunderland	ON	89.9	5		CJKX-FM-1
Kenora	ON	100.5	40		CIKN-FM	Thunder Bay	ON	91.5	100		CKPR-FM
Kincardine	ON	95.5	5.7		CIYN-FM	Thunder Bay	ON	94.3	93		CJSD-FM
Kingston	ON	93.5	7.5		CKXC-FM	Thunder Bay	ON	105.3	100		CKTG-FM
Kingston	ON	96.3	28		CFMK-FM	Tillsonburg	ON	101.3	26		CKOT-FM

FM	Prov.	MHz	kW		Call
Tillsonburg	ON	107.3	7.8		CJDL-FM
Timmins	ON	92.1	40		CJQQ-FM
Timmins	ON	93.1	3.6		CHMT-FM
Timmins	ON	99.3	40		CKGB-FM
Timmins	ON	104.1	3.5	F	CHYK-FM
Toronto	ON	88.9	4.2	m	CIRV-FM
Toronto	ON	89.5	15		CIUT-FM
Toronto	ON	91.1	40		CJRT-FM
Toronto	ON	92.5	13		CKIS-FM
Toronto	ON	93.5	3.7		CFXJ-FM
Toronto	ON	96.3	60		CFMZ-FM
Toronto	ON	97.3	28.9		CHBM-FM
Toronto	ON	98.1	44		CHFI-FM
Toronto	ON	99.9	40		CKFM-FM
Toronto	ON	100.7	8.5	m	CHIN-FM
Toronto	ON	102.1	35.4		CFNY-FM
Toronto	ON	104.5	40		CHUM-FM
Toronto	ON	107.1	40		CILQ-FM
Trenton	ON	107.1	15		CJTN-FM
Welland	ON	91.7	50		CIXL-FM
Windsor	ON	88.7	100		CIMX-FM
Windsor	ON	93.9	100		CIDR-FM
Windsor	ON	95.9	11.8		CJWF-FM
Windsor	ON	100.7	9		CKUE-FM-1
Wingham	ON	94.5	75		CIBU-FM
Wingham	ON	101.7	100		CKNX-FM
Woodstock	ON	103.9	52		CKDK-FM
Woodstock	ON	104.7	20		CIHR-FM
Charlottetown	PE	93.1	75		CHLQ-FM
Charlottetown	PE	95.1	100		CFCY-FM
Charlottetown	PE	100.3	88		CHTN-FM
Charlottetown	PE	105.5	88		CKQK-FM
Elmira	PE	99.9	3.4		CHTN-FM-1
Elmira	PE	103.7	3.4		CKQK-FM-1
St. Edward	PE	89.9	5		CHTN-FM-2
St. Edward	PE	91.1	5		CKQK-FM-2
Summerside	PE	102.1	50		CJRW-FM
Alma	QC	95.7	100	F	CKYK-FM
Alma	QC	104.5	50	F +	CFGT-FM
Amos/Val d'Or	QC	103.5	100	F	CHOA-FM-1
Amos/Val d'Or	QC	104.3	100	F	CHGO-FM
Amos	QC	105.3	32.2	F	CHOW-FM
Amqui	QC	99.9	23.8	F	CFVM-FM
Asbestos	QC	99.3	11.1	F	CJAN-FM
Baie-Comeau	QC	97.1	4.2	F	CHLC-FM
Bécancour-Nicolet	QC	90.5	60	F	CKBN-FM
Cabano	QC	98.3	3	F	CIEL-FM-3
Cap-aux-Meules	QC	92.7	6.3	F	CFIM-FM
Carleton	QC	94.9	37.6	F	CIEU-FM
Chandler	QC	96.3	22.9	F	CFMV-FM
Chibougamau	QC	93.5	56.2	F	CKXO-FM
Chicoutimi	QC	94.5	100	F	CJAB-FM
Chicoutimi	QC	96.9	100	F	CFIX-FM
Chicoutimi	QC	98.3	100	F	CKRS-FM
Chicoutimi	QC	106.7	46.2	F	CION-FM-2
Chisasibi	QC	101.1	3		CHFG-FM
Dégelis	QC	95.5	12.5	F	CFVD-FM
Dolbeau	QC	100.3	50	F	CHVD-FM
Drummondville	QC	92.1	3	F	CJDM-FM
Drummondville	QC	105.3	5.3	F	CHRD-FM
Forestville	QC	100.5	6	F	CFRP-FM
Fort Coulonge	QC	101.7	11.9	F	CHIP-FM
Gaspé	QC	94.5	6	F	CJRG-FM
Granby	QC	104.9	4.3	F	CFXM-FM
Hull	QC	94.9	84	F	CIMF-FM
Hull/Gatineau	QC	97.1	11.2	F	CHLX-FM
Hull/Gatineau	QC	104.1	19	F	CKTF-FM
Hull/Gatineau	QC	104.7	100	F	CJRC-FM
Joliette	QC	103.5	4.5	F	CJLM-FM
La Baie	QC	105.5	6	F	CKGS-FM
La Pocatière	QC	97.5	25.2	F	CHOX-FM
La Sarre	QC	102.1	4.1	F	CJGO-FM
Lac-Etchemin	QC	100.5	9.6	F	CFIN-FM
Lac-Mégantic	QC	106.7	4.3	F	CJIT-FM
Lachute	QC	104.9	3	F	CJLA-FM
Les Escoumins	QC	94.9	5	F	CHME-FM
Longueuil	QC	98.5	40.8	F	CHMP-FM
Matane	QC	95.3	30	F	CHOE-FM
Matane	QC	105.3	30	F	CHRM-FM
Mistassini	QC	95.3	50	F	CINI-FM
Mont-Joli	QC	93.3	27.3	F	CFYX-FM
Mont-Laurier	QC	104.7	16.9	F	CFLO-FM
Montmagny	QC	90.3	41.6	F +	CIQI-FM
Montréal	QC	89.3	10	F	CISM-FM
Montréal	QC	90.3	5		CKUT-FM

FM	Prov.	MHz	kW		Call
Montréal	QC	91.3	36.2	F	CIRA-FM
Montréal	QC	91.9	4.7	F	CKLX-FM
Montréal	QC	92.5	41.4		CFQR-FM
Montréal	QC	94.3	41.4	F	CKMF-FM
Montréal	QC	95.9	41.2		CJFM-FM
Montréal	QC	96.9	307	F	CKOI-FM
Montréal	QC	97.7	41.2		CHOM-FM
Montréal	QC	99.5	8.7	F	CJPX-FM
Montréal/Laval	QC	105.7	41	F	CFGL-FM
Montréal	QC	107.3	42.9	F	CITE-FM
Natashquan	QC	104.1	6.6	F	CKNA-FM
New Carlisle	QC	107.1	5.5	F	CHNC-FM
Pikogan	QC	100.1	3.7	m	CKAG-FM
Port-Cartier	QC	99.1	45	F	CIPC-FM
Québec	QC	90.9	5.7	F	CION-FM
Québec	QC	91.9	31	F	CJEC-FM
Québec	QC	93.3	33	F	CJMF-FM
Québec	QC	98.1	40	F	CHOI-FM
Québec	QC	98.9	41	F	CHIK-FM
Québec/Lévis	QC	102.1	78	F	CFEL-FM
Québec/Lévis	QC	102.9	32.8	F	CFOM-FM
Québec	QC	105.7	37	F	CITF-FM
Rimouski	QC	96.5	6.4	F	CKMN-FM
Rimouski	QC	98.7	100	F	CIKI-FM
Rimouski	QC	102.9	33.6	F	CJOI-FM
Rivière-du-Loup	QC	103.7	60	F	CIEL-FM
Rivière-du-Loup	QC	107.1	100	F	CIBM-FM
Roberval	QC	99.5	50	F	CHRL-FM
Rouyn-Noranda	QC	88.7	3.4	F	CHIC-FM
Rouyn-Noranda	QC	95.7	44	F	CHGO-FM-1
Rouyn-Noranda	QC	96.5	61.1	F	CHOA-FM
Rouyn	QC	99.1	3	F	CJMM-FM
Saguenay	QC	92.5	14.2	F	CKAJ-FM
Sainte-Foy	QC	94.3	6	F	CHYZ-FM
Sept-îles	QC	94.1	11.3	F	CKCN-FM
Sherbrooke	QC	93.7	25.5	F	CFGE-FM
Sherbrooke	QC	102.7	92	F	CITE-FM-1
Sherbrooke	QC	104.5	50	F	CKOY-FM
Sherbrooke	QC	107.7	25	F	CHLT-FM
Sorel	QC	101.7	3	F	CJSO-FM
St-Gabriel-de-Brandon	QC	99.1	9.8	F	CFNJ-FM
St-Georges-de-Beauce	QC	99.7	100	F	CHJM-FM
St-Georges-de-Beauce	QC	103.5	15	F	CKRB-FM
St-Hyacinthe	QC	106.5	3	F	CFEI-FM
St-Jérôme	QC	103.9	39.3	F	CIME-FM
Ste-Marie-de-Beauce	QC	101.5	100	F	CHEQ-FM
Thetford Mines	QC	97.3	100	F	CFJO-FM
Thetford Mines	QC	105.5	6	F	CKLD-FM
Trois-Rivières	QC	89.1	3	F	CFOU-FM
Trois-Rivières	QC	89.9	6	F	CIRA-FM-2
Trois-Rivières	QC	94.7	100	F	CHEY-FM
Trois-Rivières	QC	100.1	64.1	F	CJEB-FM
Trois-Rivières	QC	102.3	5.8	F	CIGB-FM
Trois-Rivières	QC	106.9	100	F	CHLN-FM
Valleyfield	QC	103.1	3	F	CKOD-FM
Val-d'Or	QC	102.7	63.1	F	CJMV-FM
Ville-Marie	QC	93.1	34	F	CKVM-FM
Waskaganish	QC	92.5	7.1		CJRH-FM
Waswanipi	QC	93.9	6.2		CFNE-FM
Wemindji	QC	99.9	4.8		CHPH-FM
Dafoe	SK	100.3	100		CJVR-FM-1
Estevan	SK	102.3	100		CHSN-FM
Meadow Lake	SK	102.3	45		CJNS-FM
Meadow Lake	SK	104.5	45		CJCQ-FM-1
Melfort	SK	105.1	100		CJVR-FM
Moose Jaw	SK	100.7	100		CILG-FM
Moose Jaw	SK	103.9	100		CJAW-FM
Nipawin	SK	94.7	14.8		CJNE-FM
North Battleford	SK	93.3	100		CJHD-FM
North Battleford	SK	95.5	28		CJLR-FM-6
North Battleford	SK	97.9	100		CJCQ-FM
Okanese First Nation	SK	95.3	50		CHXL-FM
Prince Albert	Sk	88.1	49	m	CJLR-FM-3
Prince Albert	SK	90.1	3	F	CKSF-FM
Prince Albert	SK	99.1	100		CFMM-FM
Prince Albert	SK	101.5	100		CHQX-FM
Regina	SK	90.3	43	m	CJLR-FM-4
Regina	SK	92.1	100		CHMX-FM
Regina	SK	92.7	100		CHBD-FM
Regina	SK	94.5	100		CKCK-FM
Regina	SK	98.9	100		CIZL-FM
Regina	SK	104.9	100		CFWF-FM
Saskatoon	SK	92.9	100		CKBL-FM
Saskatoon	SK	95.1	100		CFMC-FM
Saskatoon	SK	96.3	100		CFWD-FM

FM	Prov.	MHz	kW	Call
Saskatoon	SK	98.3	100	CJMK-FM
Saskatoon	SK	102.1	100	CJDJ-FM
Swift Current	SK	94.1	100	CIMG-FM
Swift Current	SK	97.1	100	CKFI-FM
Wapella	SK	102.9	14	CFGW-FM-2
Waskesiu Lake	SK	106.3	11	CJVR-FM-2
Weyburn	SK	103.5	100	CKRC-FM
Yorkton	SK	94.1	100	CFGW-FM
Whitehorse	YT	98.1	4.3	CHON-FM

NB1: Txs below 3kW not listed
NB2: Most stns identify using a slogan rather than calls. Industry Canada stn list database **W:** www.strategis.ic.gc.ca/eic/site/smt-gst.nsf/eng/h_sf01842.html. Stn history & info **W:** www.broadcasting-history.ca (university site)

BRITISH FORCES BROADC. SCE. Suffield AB
✉ BFBS Canada, BATUS, BFPO 14, UK ☎ +1 403 544 4104 **W:** www.bfbs-radio.com/pages/extranet/bfbs-canada-i-1239.php **BFBS 1: FM:** 98.1MHz, 104.1MHz

WEATHERADIO CANADA
The Meteorological Service of Canada operates a network of Weatheradio stns throughout Canada. These stns broadcast current conditions, forecasts and alerts in English and French. Weatheradio stns broadcast on 7 freqs in the VHF 162MHz range as well as low power MW & FM. Stn lists and info **W:** www.ec.gc.ca/meteo-weather/default.asp?lang=En&n=792F2D20-1

CANARY ISLANDS (Spain)

LT: UTC (27 Mar-30 Oct: +1h) — **Pop: 2** million — **Pr.L:** Spanish — **E.C:** 50Hz, 220V — **ITU:** CNR

MW	kHz	kW	Net	Location	Island
1)	576	25	RNE-1	Las Palmas	GC
2)	621	300	RNE-1	Santa Cruz	TF
2)	720	25	RNE-5	Santa Cruz	TF
1)	747	25	RNE-5	Las Palmas	GC
3)	837	10	COPE	Las Palmas	GC
4)	882	20	COPE	La Laguna	TF
5)	1008	25	PR	Las Palmas	GC
6)	1179	25	SER	Tenerife	TF
7)	1269	25	ECCA	Las Palmas	GC

Abbreviations: GC=Gran Canaria, GCF=Fuerteventura, GCL=Lanzarote, TF=Tenerife, TFP=Isla de La Palma, TFG=Isla de la Gomera, TFH=Hierro. (For network abbreviations refer to Spain).

Addresses and other information
1) R. Nacional de España, Plazoleta Milton 1, 35005 Las Palmas de Gran Canaria ☎ +34 928 364 088 ▤ +34 928 362 754 **2)** R. Nacional de España, San Martín 1, 38001 Sta. Cruz de Tenerife ☎ +34 (922) 288400 ▤ +34 922 283363 **R.1:** 24h on 621kHz **N:** On the h. **R.2:** (classical music) 24h. **R.3:** 24h. **R.5:** 24h - **3)** R. Popular de Las Palmas, Av. Escaleritas 60, Las Palmas 35011 ☎ +34 928 286970 **E:** laspalmas@cadenacope.net Dir: Antonio Miguel Díaz **D.Prgr:** 24h. **4)** R. Popular de Tenerife, Darías y Patrón, 1-2°-38003 Santa Cruz de Tenerife ☎ +34 922 236900/05/09 ▤ +34 922 2369121 **E:** tenerife@cadenacope.net Dir: José Carlos Marrero Gonzá. **D.Prgr:** 24h - **5)** Radio Las Palmas, C/ Profesor Lozano 5, 2:a planta, Urb. Industrial El Sabadal, 35008 Las Palmas de Gran Canaria **W:** www.radiolaspalmas.com ☎ +34 928 462052 ▤ +34 928 462057 Dir:María Enma Hernández Martín. **D.Prgr:** 24h - **6)** R. Club Tenerife, Av. de Anaga 35, Santa Cruz de Tenerife 38001 ☎ +34 922 270400 ▤ +34 922 281043 **E:** radioclubtenerife@unionradio.es Dir: Mª José Pérez. **7)** Av. Escaleritas 64, 35011 Las Palmas de Gran Canaria **W:** www.radioecca.org

Major FM-Networks
Gran Canaria (MHz): RNE5 88.6 – COPE 90.1 – ECCA 90.4 – COPE 91.0 – Punto R 91.2 – Canaras R. 91.4 – C100 91.8 – RNE1 92.8 – Cadena 40 94.4 – RNE2 95.1 – RNE3 98.5 – Rock & Gol 99.0 – SER 99.8 – Canarias R. 99.9 – Canarias R. 100.8 – Dial 101.4 – SER 102.4 - Radio Canarias 103.0 – RNE5 104.8 – R. Las Palmas 105.1 - M80 105.4 – OCR 106.8
Fuerteventura (MHz): RNE2 87.7 – OCR 90.7 - COPE 91.2 – Canarias R. 92.0 – ECCA 93.0 – RNE1 94.6 – Canarias R. 96.9 – Onda Cero 97.7 – RNE3 100.6 – Canarias R. 104.4 – RNE5 104.8
Lanzarote (MHz): SER 89.7 – COPE 90.7 – RNE1 92.5 – ECCA 93.0 – RNE2 94.9 – Punto 96.4 - RNE5 100.2 – Canarias R. 101.2 – Punto R 102.0 - RNE3 102.8 – Canarias R. 103.5 – Cadena 40 104.0
Tenerife (MHz): Dial 87.8 – Canarias R. 88.1 – RNE5 88.8 - Dial 88.9 – ECCA 89.6 – RNE3 90.0 – Dial 91.6 – Punto R 91.8 – RNE1 92.3

– RNE2 92.6 – Cadena 40 93.2 – OCR 94.0 – RNE1 94.8 – RNE3 95.4 – RNE2 96.2 – COPE 97.1 – RNE5 98.1 – Cadena 40 99.1 – Punto R 99.5 – M80 100.1 – SER 101.1 – COPE 101.4 – RNE2 102.1 – RNE5 104.0 – Canarias R. 104.2 – Canarias R. 104.7 – RNE1 105.6 – RNE3 105.7
Isla de la Palma (MHz): RNE5 89.6 – Onda Cero 91.7 – COPE 95.1 – Cadena 40 97.4 – Canarias R. 97.5 – RNE5 98.4 – ECCA 99.5 – Canarias R. 100.5 – SER 101.6 – RNE1 102.7 – Dial 104.1 - RNE2 104.5 – RNE3 106.1 – OCR 106.4
La Gomera (MHz): RNE5 91.7 – RNE1 94.3 – Canarias R. 96.7 – SER 98.8 – RNE2 101.8 – RNE3 105.2
El Hierro (MHz): RNE5 89.8 – SER 92.0 – RNE1 92.5 – RNE2 93.9 – RNE3 96.4 – RNE2 97.0 – RNE5 98.2 – RNE1 101.2 – Canarias R. 102.3 – Canarias R. 103.2 – RNE3 104.9
There are hundreds of low power FM stns throughout the islands.

Tourist Radio FM Stations
These stns broadcast in German, English and other languages to tourists visiting the Canary Islands. Most operate 24h
Atlantis FM (GCF) 99.3 (GCL) 101.7MHz **W:** www.atlantisfm.de – Coast FM (TF) 89.2, 93.8MHz **W:** www.coastmusicradio.co.uk – **Happy Radio** (TF) 98.7MHz **W:** www.happyradioteneriffa.com – **Holiday FM** (GC, TF) 98.2, 99.0, 100.0, 105.5MHz **W:** www.holidayfm.com – **Mix 101 FM Radio** (GC) 101.0MHz **W:** www.mix101.net – **Oasis FM** (TF) 101.0, 100.2MHz **W:** www.oasisfm.com – QFM (TF) 94.3, 94.6MHz **W:** www.qmusica.com – **R. Europa/R. Syd** (TF) 101.5MHz **W:** www.radioeuropaweb.net/radioeuropa.htm – **R. Megawelle** (TF) 88.3MHz, 103.7, 104.7MHz (GC) 88.3, 101.7MHz **W:** www.megawelle.com – **Sands FM** (GC) 104.8MHz **W:** www.sandsfm.com – **UK Away FM** (GCL) 99.4, 99.9MHz **W:** www.ukawayfm.com

CAPE VERDE

LT: UTC -1h — **Pop:** 400,000 — **Pr.L:** Portuguese, Crioulo — **E.C:** 50Hz, 220V — **ITU:** CPV.

AGENCIA NACIONAL DAS COMUNICAÇÕES (ANAC)
✉ C.P. 892, Edifício MIT, Ponta Belém, Praia ☎ +238 2604400 ▤ +238 2613069 **W:** www.anac.cv **E:** info.anac@anac.cv

RÁDIOTELEVISÃO DE CABO VERDE (RCV, Gov.)
✉ Rua 13 de Janeiro 1-A, Achada de Santo António, Praia ✉ C.P. 29, Av. Marginal, Mindelo, São Vicente ✉ C.P. 40, Espargos, Ilha do Sal ☎ ▤ +238 2411444 **W:** www.rtc.cv **E:** rtc@cvtelecom.cv
L.P: Dir: Marcos Oliveira. PD: Giordano Custodio. Dir. Inf: Mario Almeida. Dir. Tec: Francisco Lopes Monteiro.
FM: Monte Verde 87.6MHz 1kW, Morro Curral 89.7MHz 0.25kW, Monte Tchota 91.6MHz 1kW, Mindelo 95.6MHz 0.5kW, Praia 98.1MHz 0.1kW + 12 relays below 0.1kW. **D.Prgr:** 24h.

Other stations:
R. Nova, C.P. 426, Mindelo, São Vicente. **E:** radionova@cvtelecom.cv FM (MHz): Pinhão 91.8, Monte Vermelho 94.1, Cachaço 95.1, Sal Rei 97.0, Monte Tropetona 99.1, Pedra Rachada 99.9, M. Tchota 101.6 0.25kW, Mindelo 102.3, M- Verde 104.3 0.5kW, Morro do Curral 106.4 – **R. Comercial:** Santiago 92.9 1kW, Ponta Rachada 96.1MHz, Praia/M. Verde 99.9MHz – **R. Crioula:** M. Tchota 88.5MHz, M. Barro/M. Verde 89.6MHz, Praia 94.9MHz, Morro Curral 98.9MHz – **R. Educativa:** Monte Verde 101.5MHz, M. Tchota 102.3MHz, Praia & 4 sites 103.1MHz – **Mosteiros FM:** M. Chota 96.1MHz, São Filipe/Mosteiros 97.3MHz
RDP Africa: Monte Verde 93.9MHz 3kW, Monte Tchota/Pedra Rachada 105.2MHz 3/1kW, Pedra Rachada 105.2MHz 1kW + 4 trs under 1kW.
RFI Afrique, Praia/Santo Antão 99.3MHz 1kW, Fogo/Mindelo/Sal 100.7MHz 0.25kW in French and Portuguese.

CAYMAN ISLANDS (UK)

LT: UTC -5h — **Pop:** 56,000 — **Pr.L:** English — **E.C:** 60Hz, 110V — **ITU:** CYM

RADIO CAYMAN (Gov. Comm.)
✉ 71C Elgin Av, Box 1110, George Town, Grand Cayman KY1-1102 ☎ +1 345 949 7799 ▤ +1 345 949 6536 **W:** www.radiocayman.gov.ky **L.P:** Dir: Norma McField. Sales Mgr: Paulette Conolly-Bailey
FM: R. Cayman One: Grand Cayman 89.9MHz 5kW/ Cayman Brac: 89.9MHz 0.25kW: music and news – 24h Relays BBCWS 0500-1100 – **Breeze FM:** Grand Cayman 105.3MHz 5kW/ Cayman Brac: 105.3MHz 0.25kW: music and news – **Cayman Weather R.:** 107.9MHz

HURLEYS ENTERTAINMENT CORPORATION (Comm.)
✉ 256 Crewe Road, Box 30110 SMB, George Town, Grand Cayman

KY1-1201 ☎ +1 345 945 1166 🖷 +1 345 945 1006 **E:** info@z99.ky **W:** www.z99.ky and www.rooster101.ky
L.P: Pres. & GM: Randy Merren. PD (Z99): Scott Hamilton. PD (Roster 101.9): Keith Michaels.
FM: Z99: George Town 99.9MHz 5kW: CHR – **Rooster 101.9:** 101.9MHz 5kW/Cayman Brac 101.9MHz 1 kW: Country

PARAMOUNT MEDIA SERVICES (Comm.)
🖳 Rankin's Plaza, 21 Eclipse Drive, P.O.Box 10734, George Town, Grand Cayman KY1-1007 ☎ +1 345 949 8423 🖷 +1 345 946 9867 **E:** info@vibefm.ky **W:** www.spinfm.ky and www.vibefm.ky
FM: Spin FM: George Town 94.9MHz 1kW: dance music – **Vibe 98.9:** George Town 98.9MHz 2 kW/Cayman Brac 98.9MHz 0.3kW: urban Caribbean

ICCI-FM (Educ.)
🖳 International College of the Cayman Islands, Newlands, P.O. Box 136 SAV, Savannah Post Office, Grand Cayman KY1-1501. College ☎ +1 345 947 1100 🖷 +1 809 947 1230 **L.P:** College Pres. & GM: Elsa M. Cummings **FM:** 101.1MHz 0.5kW
D.Prgr: 24h. Locally produced prgrs in English & Spanish for Grand Cayman residents, or continuous music, acc. to availability of student volunteers. **Ann:** "This is ICCI-FM, Newlands", "FM 101.1 ICCI".

CHRISTIAN COMMUNICATIONS ASSOCIATION (Rlg.)
🖳 Box 31481 SMB, George Town, Grand Cayman KY1-1206 ☎ +1 345 945 2797 🖷 +1 345 945 2707 **E:** heaven97@candw.ky **W:** www.heaven97.com.
L.P: SM: Steve Faucette.
FM: Gospel 88.7: George Town 88.7MHz 2kW: Gospel – **Heaven 97:** George Town 97.7MHz 2kW continuous Christian music

DMS BROADCASTING LTD. (Comm.)
🖳 Box 31910 SMB, Grand Cayman KY1-1208 ☎ + 1 345 943 1367 🖷 +1 345 943 1368 **E:** info@dmsbroadcasting.ky **W:** www.dmsbroadcasting.ky **L.P:** MD: Don Seymour. GM: Steve Jones.
FM: CayRock George Town: 96.5MHz (1kW) – **HOT** George Town: 104.1MHz – **KISS FM** George Town 106.1MHz – **X 107.1** George Town 107.1MHz – **New FM** Cayman Brac 96.5MHz (0.3kW)

WESTPOINT RADIO (Rlg.)
🖳 Goulds Estate, Box 349, West Bay, Grand Cayman ☎ + 1 345.926 2100 🖷 +1 345 949 2166 **W:** www.westpointradio.ky **L.P:** MD: Locksley Gould **FM:** 94.3MHz 3kW

PRAISE 87.9 RADIO (Rlg.)
🖳 37 Barnes Rd, Box 152, George Town, Grand Cayman KY1-1501.
FM: 87.9MHz **Format:** Adventist religion

CAYMAN BROADCASTING LTD. (Comm.)
🖳 125 Eastern Av.,P.O.Box 1336, George Town, Grand Cayman KY1-1108. **Love FM:** 103.1MHz 2kW.

L.T: UTC +1h — **Pop:** 4.5 million — **Pr.L:** French, Sango — **E.C:** 50Hz, 220V — **ITU:** CAF.

MINISTÈRE DE LA COMMUNICATION
🖳 B.P. 940, Bangui ☎+236 612766/615247 🖷 +236 615985
L.P: Minister: Abdou Karim Meckassolia

RADIO CENTRAFRIQUE (Gov.)
🖳 B.P. 940, Bangui ☎+236 75 503632 **W:** radiocentrafique.org **E:** radio.centrafrique@yahoo.fr **L.P:** DG: Aimé-Christian Ndotah. PD: Mrs. Pauline Gbianza.

MW: Bangui 1440kHz 50kW 0700-1700.
SW: Bangui 720kHz 50kW 0500-1700 (irreg.), 5035kHz 1kW 0500-1905
FM: 106.9MHz 24h.
D.Prgr in French/Sango: 24h. **N. in French:** 0600, 0700, 1300, 1800.
Ann: F: "Ici Bangui, Radio Centreafricaine". **IS:** Repeated piano chord. Opens and closes with National Anthem.

RADIO ICDI (Integrated Community Development International, Rlg.)
🖳 B.P. 362, Bangui ☎+236 508622 **W:** www.icdinternational.org/radio.html **E:** radioicdi@gmail.com **L.P:** Josue Mbami, Coordinator.
SW: Boali 6030 & 3390kHz 1kW. **D.Prgr.** in French, Sango, Bayaka and Fulfulde: 0400-1800 on 6030kHz and 1800-0400 on 3390kHz. (3390 kHz not yet in operation at the editorial deadline).

RADIO NDEKE LUKA
(joint initiative between the UN Development Programme, CAF government and Hirondelle Foundation)
🖳 c/o PNUD, Av. de l'Indépendance, B.P. 872, Bangui ☎+236 610652 **W:** radiondekeluka.org **L.P:** Dir: Patrick Stéphane Akibata.
FM: Bangui 100.8MHz 1kW. **D.Prgr:** 24h in Sango/French.
F.PI: nationwide coverage.

Other stations:
R. Be Oko, Bambari: 103.5MHz – **R. ESCA La Voix de la Grâce,** Bangui: 98.5MHz. Email: radiovoixdelagrace@yahoo.fr – **R. Évangile Néhémie,** Bangui: 104.4MHz – **R. Maigaro,** Bouar: 98MHz – **R. MKA,** Berbérati: 105.9MHz – **R. Ndoye,** Bossangoa: 98MHz 1kW – **R. Notre Dame,** Bangui: 103.3MHz 1kW – **R. Songo,** M'Baiki: 97.2MHz 1kW – **R. Voix de la Sangha,** Nola: 98.0MHz.
Africa No. 1, Bangui: 94.5MHz (see main entry under Gabon).
BBC African Sce, Bangui: 90.2MHz.
RFI Afrique, Bangui: 99.8MHz

L.T: UTC +1h — **Pop:** 10 million — **Pr.L:** French, Arabic, 8 ethnic — **E.C:** 50Hz, 220V — **ITU:** TCD

HAUT CONSEIL DE LA COMMUNICATION(HCC)
🖳 N'Djamena. **L.P:** Moussa Mahamat Dago, president.

OFFICE NATIONAL DE RADIO ET TÉLÉVISION DU TCHAD (Gov.)
🖳 B.P. 892, N'Djamena ☎+235 514253 **L.P:** Dir: Nguérébaye Adoum Saleh. Dir. of Tech. Sces: Raphael Mbaissane.
Station: N'Djamena-Gredia.
MW: 840kHz 20kW **SW:** 6165kHz 250kW, alt. fqs 4905/7120kHz.
FM: 94.5MHz 0.1kW
D.Prgr: in French/Arabic/others: 0430-2230, Sat -2300.
Ann: "Ici N'Djamena, Office National de Radio et Télévision du Tchad". **IS:** Balafon.

REGIONAL STATIONS
R. Moundou, B.P. 122, Moundou. **FM:** 94.05/98.3MHz 200/450W.
R. Sarh, B.P. 270, Sarh. ☎+235 681361/681422 - **FM:** 94MHz.
R. Abéché, B.P. 105, Abéché. **L.P:** Dir: Sanoussi Saïd. **FM:** 101MHz.
R.Faya-Largeau: FM 99.1MHz.

Other stations:
Al-Bayan FM, N'Djamena: 93.7MHz – **Al-Nasr,** N'Djamena: 102.1MHz – **Al-Quran Al-Karim,** N'Djamena: 88.3MHz – **Dja FM,** N'Djamena: 96.9MHz 0.5kW – **R. Duji Lokar,** Mondou: 101.8MHz 0.5kW – **R. Évangile Développement,** Pala: 88.5MHz – **R. Lotiko:** Koumra 100.1MHz, Sarh 97.6MHz (also r. BBC). **W:** www.lotiko.org/fr/save/lotiko.htm **E:** lotiko@intnet.td – **R. Terre Nouvelle,** Bangor 99.4MHz 1.1kW – **La Voix du Paysan,** Doba: 96.2MHz 1kW.
Africa No. 1: N'Djamena 103MHz (see main entry under Gabon)
BBC African Sce: N'Djamena 90.6MHz
RFI Afrique: N'Djamena 100.2MHz

L.T: UTC -4h (10 Oct 10-13 Mar 11, 16 Oct 11-11 Mar 12: -3h) — **Pop:** 17 million— **Pr.L:** Spanish — **E.C:** 50Hz, 220V — **ITU:** CHL — **Int. dialling code:** +56

SUBSECRETARIA DE TELECOMUNICACIONES
Offices: Amunátegui 139, Santiago 🖳 Clasificador 120, Correo 21, Santiago ☎02 672 6503 🖷02 699 5138 **W:** www.subtel.cl **L.P:** Subsecr of Telecommunocations: Roberto Pliscoff Vásquez.
ASOCIACION DE RADIODIFUSORES DE CHILE
🖳 Cas. 10476, Santiago de Chile ☎2 6398755 🖷2 6394205 **W:** www.galeon.com/redarca
L.P: Pres: César Molfino Mendoza. Dir: Alfredo Matte L.
STATIONS: MW: Call letters CA, CB, CC and CD indicate: A=No. Zone, B=Central Zone, C=So. Zone and D=Antarctic Zone. The figures indicate the freq. in kHz minus one cipher, f. inst. CB82 = Central Zone 820kHz. SW: Call letters CE are used for all zones.
° = on-air stn name not confirmed * = inactive v = varying freq.

MW	Call	kHz	kW	Station, location, h of tr
MS01)	CB54	540	1	R. Ignacio Serrano, Melipilla: 1100-0400
LL01)	CD54	540	1	R. R.Ainil, Valdivia: 1000-0400
BB01)	CC55	550	2	R. Concepcion, Penco
LA02)	CD55	550	1	R. LV. de la Tierra, Angol
MS02)	CB57	*570	50	R. Agricultura, Santiago
AN01)	CA59	590	1	R. Horizonte, Antofagasta

MW	Call	kHz	kW	Station, location, h of tr
BB02)	CC59	590	1	R. Portales, Concepción: 24h
MC01)	CC59	590	10	R. Patagonica, Punta Arenas
MS03)	CB60	600	10	R. Monumental, Santiago: 24h
GS01)	CD61	*610	5	R. Puerto Aysen, Puerto Aysén
CO01)	CA62	620	1	R. Norte Verde, Ovalle: 1100-0400
BB03)	CC62	620	10	R. Bío-Bío, Concepción: 24h
VA01)	CD63	630	10	R. Stela Maris, Valparaíso: 1100-0500
LA03)	CD64	640	10	R. Cooperativa AM, Temuco: 1030-0230
MA13)	CC64	640	0,25	R. Portales, Curico
MS26)	CB66	640	50	R. UC, Santiago
BB04)	CC68	680	1	R. Cooperativa, Concepción
MS05)	CB69	690	10	R. Santiago, Santiago: 1000-0600
LL02)	CD69	690	10	R. Estrella del Mar, Ancud: 1100-0330
AT01)	CA70	700	1	R. Nibsan, Copiapó: 1030-0300
LL03)	CD70	700	1	R. Valdivia, Valdivia: 1030-0500
MC02)	CD70	700	5	R. Magallanes, Punta Arenas
TA09)	CA72	720	1	R. Portales, Iquique
VA02)	CD73	730	10	R. Cooperativa AM, Valparaíso: 1000-0430
BB19)	CD73	730	1	R. Angelina, Los Angeles
GS02)	CD73B	730	1	R. Aysén, Pto. Aysén: 1100-0400
MS06)	CB76	760	50	R. Cooperativa, Santiago: 24h
LA10)	CD76	770	1	R. Agricultura, Temuco: 1000-0400
LL04)	CD77	770	1	R. Cooperativa, Castro
LL05)	CD78	780	10	R. Sago AM, Osorno: 1000-0400
VA03)	CB80	800	5/1	R. Maria, Viña del Mar
CO02)	CA82A	820	10/1	R. Portales, La Serena: 1000-0400
MS07)	CB82	820	10/5	Radioem. Carabineros de Chile, Santiago: 24h
BB05)	CC82	820	1	R. Maria Inmaculada, Concepción: 24h
LL06)	CD82	820	1	R. Concordia, La Unión: 1100-2330
AN02)	CA82	820	0.25	R. Pampa, Pedro de Valdivia
VA04)	CB84	840	10	R. Portales, Valparaíso: 1000-0500
GS03)	CD84	840	10	R. Santa María, Coyhaique: 1030-0230
BB06)	CC86	860	10	R. Inés de Suárez, Concepción: 24h
MS08)	CB88	880	10	R. Colo Colo, Santiago: 24h
TA07)	CA89	*890	10	R. León XIII, Pozo Almonte: 1100-0200
BB07)	CC89	890	20	R. Interamericana, Concepción: 24h
MC03)	CD89	890	20	R. Nal., Punta Arenas
AT02)	CA90	900	5	R. Universidad, Copiapó
VA05)	CB90	900	1	Cablenoticias, Valparaíso: 1100-0500
BB08)	CC90	900	1	R. Mayor, Chillán: 1100-0400
LL07)	CD90	900	1	R. LV de la Costa, Osorno: 1030-0400
LA04)	CD92	920	1	R. 920, Temuco
MS09)	CD93	930	10	R. Nuevo Mundo, Santiago: 1000-0530
LL08)	CD93	930	10	R. Reloncaví, Puerto Montt: 1100-0400
VA06)	CB94	940	1	R. Valentín Letelier, Valparaíso
VA21)	CC94	940	1	R. Armonia, Viña del Mar
AT03)	CA94	*940	1	R. 9-40, Copiapó
MS10)	CB96	960	10	R. Carrera, Santiago: 1100-0400
MC04)	CD96	960	1	R. Polar, Punta Arenas
AN04)	CA97	970	1	R. Calama, Calama: 1000-0400
MA01)	CC97	970	1	R. Lautaro, Talca: 1000-0500
GS04)	CD97A	970	1	R. Patagonia Chilena, Coyhaique: 1000-0400
LL09)	CD97	970	1	R. Austral, Valdivia: 1030-0400
VA07)	CB98	980	5	R. Agricultura, Valparaíso
MA12)	CC99	990	1	R. El Roble, Parral (n. 1590)
TA08)	CA98	*980	1	R. Univ. de Tarapaca, Arica
MS11)	CB100	1000	10	R. BRB, Santiago: 1030-0500
LA21)	CD101	1010	10	R. Nielol, Temuco
MA02)	CC102	1020	10	R. Amiga, Talca: 1000-0500
MS12)	CB103	1030	1	R. Progreso, Talagante
LL10)	CD103	1030	1	R. Chiloé, Castro:1100-0330
MC05)	CD103A	1030	1	R. Payne AM, Puerto Natales
LA06)	CD104	1040	1	R. Raíces, Curacautín: 1100-0100
LL11)	CD105	1050	1	R. Armonía, Osorno: 1030-0630
MS13)	CB106	1060	50	R. Maria, Santiago
CO03)	CA108	1080	1	R. Río Elqui, Vicuña
LA07)	CD108	1080	1	R. Los Confines, Angol: 1100-0300
AN05)	CA110	*1100	1	R. La Portada, Antofagasta
VA08)	CB110	1100	10	R. Integridad, Viña del Mar: 1100-0500
LA08)	CD111	1110	10	R. La Frontera, Temuco: 1000-0400
MS14)	CB114	1140	75	R. Nal., Santiago: 1000-0600
MA04)	CC116	1160	1	R. Ancoa, Linares: 1000-0400
CO10)	CB116	1160	1	R. El Espectador de America, La Serena
LA01)	CD116A	1160	1	R. Baha'i, Temuco: 1030-0230
MC06)	CD117	1170	3	R. Natales, Puerto Natales: 1200-0400
MS15)	CB118	1180	50	R. Portales, Santiago: 1030-0430
BB20)	CD120	1200	10	R. Agricultura, Los Angeles: 1000-0400
AN09)	CA121	1210		R. Universidad de Antofagasta, Antofagasta
LL12)	CD121	1210	5	R. Armonia, Puerto Montt
MA05)	CC121	1210	1	R. Universidad de Talca, Talca:1100-0400
VA20)	CB121	1210		R. Valparaíso, Valparaíso
CO04)	CA122	1220	1	R. La Caribeña, La Serena
LA09)	CD122	*1220	10	R. Santa Maria de Guadalupe, Temuco
MS16)	CB124	1240	10	R. Universidad de Santiago, Santiago: 1100-0400
AN06)	CA124	1240	0.25	R. Principal Chuquicamata, Calama: 0950-0500
CO05)	CA125	1250	1	R. Santa Maria de Guadalupe, La Serena
LL13)	CD125	*1250	10	R. Armonía, Valdivia: 1100-0400
TA02)	CA126	*1260	10	R. Nal., Arica
MA06)	CC126	1260	2	R. Condell, Curicó: 1100-0500
MC07)	CD126	*1260	10	R. Santa Maria de Guadalupe, Punta Arenas
VA09)	CB127	v1270	10	R. Festival, Viña del Mar: 1000-0700
BB10)	CC128	1280	1	R. Arturo Prat Chacón AM, San Carlos: 1050-0405
LL14)	CD128	1280	10	R. del Sur «En Voz Alta», Osorno
AN07)	CA129	1290	0.25	R. Coya, María Elena
BB24)	CC129	1290	5	R. Doce Noventa, Los Angeles
MS25)	CB130	1300	5	R. Tierra, Santiago: 1700-0100
AT04)	CA132	1320	0.25	R. Estrella del Norte, Vallenar
BB21)	CD132	1320	1	R. Lincoyan, Mulchén: 1100-0300
MS17)	CB133	1330	3	La Mexicana, Santiago: 24h
LL16)	CD133	1330	3/1.5	R. Vicente Pérez Rosales, Puerto Montt: 1055-0400
VA10)	CB134	1340	10	R. Colo Colo, Valparaíso: 0930-0300
BB11)	CC134	v1340	1	R. La Discusión, Chillán: 24h
LL17)	CD134	1340	1	R. Panguipulli, Panguipulli: 1200-0100
CO06)	CA135	1350	1	R. Riquelme, Coquimbo: 1030-0430
BB12)	CC136	1360	5	R. Universidad del Bío-Bío, Concepción: 24h
LA11)	CD137	1370	1	R. Portales ,Temuco: 1100-0400 (Sat 0800)
MS18)	CB138	1380	50	R. Corporación, Santiago
BB22)	CD140	1400	5	R. La Amistad, Los Angeles: 1030-0400
TA03)	CA140	*1400	0.25	R. Altisima, Iquique: 1045-0600
LL18)	CD140A	1400	5	R. Viento del Sur, Puerto Montt
VA11)	CB141	1410	3	R. Quinta Región, Valparaíso
LA12)	CD141	1410	1	R. Loncoche, Loncoche: 1100-0330
MS19)	CB142	v1420	1	R. Panamericana, Santiago: 1100-0400
MA07)	CC142	1420	1	R. Maule, Cauquenes: 1050-0430
TA04)	CA144	*1440	1	R. Santa Maria de Guadalupe, Arica: 1000-0430
CO07)	CA144A	1440	1	R. Agricultura, La Serena: 1000-0400
BB13)	CC144	1440	1	R. El Sembrador, Chillán
VA12)	CB145	1450	1	R. Universidad Técnica "Federico Santa María", Valparaíso: 1100-0300
MA08)	CC145	*1450	5	R. Libertad, Curicó: 1000-0400
LL19)	CD145	*1450	1	R. Sta Maria de Guadalupe, Puerto Varas: 24h
AN08)	CA146	1460	10	R. Antofagasta, Antofagasta: 1130-0300
MS20)	CB146	1460	1	R. Yungay, Santiago: 1100-0400
BB14)	CC146	1460	1	R. Armonía, Concepción
VA13)	CB147	1470	1	R. Romantica, San Antonio
CO08)	CA148	1480	1/0.25	R. Amanecer, Ovalle: 1100-0400
BB15)	CC148	1480	1	R. La Amistad AM, Tomé: 24h
LL20)	CD148	1480	1	R. General Baquedano, Valdivia
AT06)	CA149	1490	1	R. Chañaral, Chañaral
MS21)	CB149	1490	1	R. El Canelo de Nos AM, San Bernardo: 1100-0400
LA13)	CD149A	1490	5	R. Malleco, Victoria
TA05)	CA150	*1500	1	R. Santa Maria de Guadalupe, Iquique: 1100-0500
MA09)	CC150	1500	1	R. Centenario, San Javier: 1100-0300
MC08)	CD150	*1500	1	R. Tierra del Fuego, Puerto Porvenir: 1100-0400
VA14)	CB150	1500	1	R. Trasandina, Los Andes: 1100-0400
CO09)	CA151	1510	1/0.5	R. Luís Alvarez Sierra, Illapel: 1100-0400
GB02)	CC151	1510	1	R. Rancagua, Rancagua
LA20)	CD151	1510	0.05	R. La Trompeta de Dios, Loncoche: 1100-0400
MA10)	CC152	1520	1	R. Nueva Soberanía, Linares: 1030-0430
LA14)	CD152	1520	0.1	R. Aníbal Pinto, Lautaro
VA15)	CB152	v1520	1	R. Integración, San Antonio: 0900-0500
AT05)	CA153	1530	1	R. Portales, Copiapó
VA16)	CB153	1530	1	R. Nexo, Quillota: 1100-0600
BB16)	CC153	1530	1	R. Portales (R. Corporación), Lota
LL21)	CD153	1530	1	R. Nuvo Mundo, Puerto Montt
MS22)	CB154	1540	1	R. Sudamérica, Santiago:1000-0030
BB17)	CC154	1540	1	R. Portales (R. Centra), Chillán
LL22)	CD154	1540	1	R. San José de Alcudia, Río Bueno: 0955-0300
VA17)	CB155	1550	1	R. Provincial AM, Putaendo
GB03)	CC155	v1550	1	R. Manuel Rodríguez, San Fernando:(r 1555:
LA15)	CD155	1550	0.25	R. Regional, Traiguén
TA06)	CA156	1560	5/3	R. Parinacota, Putre: 24h
MS23)	CB156	1560	1	R. Manantial, Talagante: 1100-0400
LA16)	CD156	1560	1	R. Parque Nacional, Villarrica: 1100-0400
GB04)	CC157	v1570	1	R. Niebla, Rancagua:
MA11)	CC157A	1570	7	R. Familia de Talca, "La Voz de la Región", Talca: 24h
LA17)	CD157	1570	0.25	R. Acuarela, Nueva Imperial: 1155-1810, 2200-0200
GB05)	CC158	v1580	1	R. Colchagua, Santa Cruz: 1000-0400

MW	Call	kHz	kW	Station, location, h of tr
BB23)	CD158	1580	0.25	R. Millaray, Cañete
LA18)	CD158A	1580	0.5	R. Continental, Collipulli: 1100-0430
VA18)	CB159	v1590	1	R. Aconcagua, San Felipe: 1100-0430
GB06)	CC159	v1590	0.25	R. Rengo, Rengo
MS24)	CB160	1600	0.25	R. Nuevo Tiempo, Santiago
VA19)	CB160A	1600	0.25	R. Radiocable, Viña del Mar
BB18)	CC160	v1600	0.25	R. Llacolén, Concepción

SW	Call	kHz	kW	Name, location, h of tr
TA06)	CE601	6010	1	R. Parinacota, Putre: 24h
LA19)	CE609	6090	10	R. Esperanza, Temuco: 24h

Addresses and other information

A complete list of Chilean radio stns is available as part of the Chilean yellow pages, with links to the Websites of individual stns and a free Internet Fax to each stn at **W:** www.chilnet.cl/rubros/radioeo1

AN00 (ANTOFAGASTA):
AN01) Washington 2562, Depto. 204 (⬚ Cas. 1060), Antofagasta – **AN02)** Of. SOQUIMICH, Pedro de Valdivia. – **AN04)** Rafael Vargas 1875, Calama ☎55 364353 - **FM:** 104.7MHz «Aurora FM». – **AN05)** Cas. 410, Antofagasta. – **AN06)** Cas. 127, Calama ☎55 342712 **E:** soc-integral@hotmail.com - **FM:** 99.7MHz "Sencación El Abra FM" – **AN07)** Of. SOQUIMICH, María Elena. – **AN08)** Gloria Postrera, Colectivo Peru, Ofic. 1, Antofagasta ☎55 229561 **W:** www.radio-antofagasta.cl 1130-0300 – **AN09)** Angamos No 0601, Antofagasta - **FM:** 99.9MHz

AT00 (ATACAMA):
AT01) Vallejos 650, Departamento 11, Copiapó ☎52 214133 – **AT02)** Rodríguez No 660 – 2do Piso , Copiapó ☎52 212650. **W:** www. radiouda.cl **E:** info@radiouda.cl - **FM:** 96.5MHz – **AT03)** Maipú 370, Copiapó. – **AT04)** Cas. 13, Vallenar ☎ 51 913847 ▤51 613739 **E:** radioestrelladelnortevallenar@yahoo.es – **AT05)** Colipí 371, Copiapó ☎ 52 212031. – **AT06)** Montandon No 916, Chañaral

BB00 (BIO BIO):
BB01) Membrillad N° 230, Penco. – **BB02)** Castellon N° 477 piso 5, Concepción – **BB03)** O'Higgins 680, Concepción ☎41 620620 **W:** www.radiobiobio.cl **E:** biobio@laradio.cl - **FM:** 98.1MHz – **BB04)** Paicavi 119, 2° piso, (Paza Peru) (or Cas. 2337), Concepción ☎41 223207 ▤41 234697 **W:** www.cooperativa.cl **E:** cgomez@coopertativa. cl – **BB05)** Barras Arana 544 – 3er piso, Concepción ▤41 2626168 **W:** www.radiomi.cl **E:** radiomi@radiomi.cl – **BB06)** Castellón 477, 3° piso (or Cas. 862), Concepción ☎41 2938972 **W:** www.radioinesdesuarez.cl **E:** contacto@radioinesdesuarez.cl – **BB07)** Calle Barros Arana 871, 5° piso, Of. 51, Concepción ☎41 2214450 **W:** www.radiointeramericana. net **E:** contacto@radiointeramericana.net – **BB08)** 5 de Abril 655 (⬚ Cas. 267), Chillán **W:** www.radiomayor900am.cl **E:** radiocontigo@gmail. com ☎42 237820 - **FM:** 89.7MHz "R.Nuble" – **BB09)** Aníbal Pinto 215, of 801, Concepción ☎32 259129 – **W:** www.radioagricultura.cl/ - **FM:** 90.1MHz «Galaxia», 106.5MHz «Aurora» – **BB10)** Cas. 265, San Carlos **W:** www.radiosancarlos.cl – **BB11)** 18 de Septiembre 721 (or Cas. 479), Chillán **W:** www.diarioladiscusion.cl **E:** radiotv@ladiscusion. cl ☎42 211667 ▤42 213578 - **FM:** 94.7MHz – **BB12)** Avenida Collao 1202, sector Puchacay (⬚Cas. 5-C), Concepción ☎41 273104 **W:** www.radioubb.cl/ **E:** nino@ubiobio.cl – **BB13)** Arauco 447 (⬚Cas. 336), Chillán ☎42 224603 **W:** www.radioelsembrador.cl **E:** administracion@radioelsembrador.cl - **FM:** 104.7MHz «Aurora FM» – **BB14)** Av.Los Carrera N° 464, Concepcion ☎ 41 2854594 **W:** www.armonia. cl – **BB15)** Sotomayor 952, Tomé ☎41 2653629 ▤41 2650657 **W:** www.radiolaamistad.com **E:** contacto@radiolaamistad.com – **BB16)** Pedro Aguirre Cerda N° 377 (⬚Cas. 66), Lota – **BB17)** Bulnes 220 (⬚ Cas. 35), Chillán ☎42 225220 **E:** radiocontigo@gmail.com – **BB18)** Cas. 2311, Concepción – **BB19)** Colo-Colo 451 Of. 120, Nivel 2, Los Angeles ☎43 349920 **E:** contacto@radiocamila.cl – **BB20)** Colón 143, Los Angeles ☎43 312538 ▤43 322663 **W:** www.radioagricultura.cl **E:** director@radioagricultura.cl - **FM:** 97.5, 100.5MHz «San Cristóbal – **BB21)** Gana 360, Mulchén ☎43 562739 – **BB22)** Cas. 541, Los Angeles ▤43 313964 **E:** prensalaamistad@gmail.com – **BB23)** Arturo Prat 399, Cañete ☎41 61108 **W:** www.radiom.cl - **FM:** 98.5MHz «Cañete» – **BB24)** Almagro N° 595 Oficina 4, Piso 2, Los Angeles

C000 (COQUIMBO):
C001) Cas. 355, Ovalle ☎53 620359 ▤53 621509 – **C002)** Los Carrera 525, 3° piso, Departamento C, La Serena **W:** www. radiogabrielamistral.8m.com **E:** radiomistral@mixmail.com ☎51 221659 - **FM:** 98.5MHz «Intima» – **C003)** San Marín 14, Vicuña ☎51 412867 **W:** radiorioelquiam.blogspot.com **E:** radiorioelqui108@yahoo. cl – **C004)** O'Higgins No 519, Piso 2, Oficina 09, La Serena – **C005)** Matta 591, La Serena – **C006)** Aldunate 1619, Coquimbo ☎51 321051 **W:** www.radioriquelme.cl **E:** radioriquelme@portalquimbo.com0 – **C007)** Cas. 536, La Serena. ☎51 240291 – **C008)** Libertad 786 (⬚ Cas. 34), Ovalle. ☎53 620651 – **C009)** Independencia 175, Illapel ▤53 522831 **E:** lradios@gmail.com - **FM:** 100.9MHz – **C010)** O´Higgins No 519, Piso 2, Oficina 09, La Serena **E:** radiosamericachile@yahoo.cl

GB00 (LIBERTADOR GENERAL BERNARDO O´HIGGINS RIQUELME):
GB02) Pasaje Hoffman 61, Rancagua – ☎72 234 999 **W:** www. radiorancagua.cl **E:** radio@radiorancagua.cl – **GB03)** Chacabuco Esq. España, San Fernando ☎72 714267 – **GB04)** Calvo 447, Rancagua - **FM:** 101.3MHz «FM San Fernando».– **GB05)** Rafael Casanova 146 (⬚Cas.170), Santa Cruz ▤72 822193 **W:** www.radiocolchagua.com **E:** contacto@radiocolchaque.com - **FM:** 105.5MHz «Ensueño» – **GB06)** Urriola 485, Rengo ▤34 510198

GS00 (GENERAL CARLOS IBANEZ DEL CAMPO):
GS01) Puerto Aysén – **GS02)** Carrera 545, 2° piso, Puerto Aysén **E:** radioaysen@yahoo.com ☎67 332626 – **GS03)** Francisco Bilbao 691, Coyhaique ☎67 232398 ▤67 231306 **W:** www.radiosantamaria.cl **E:** contacto@radiosantamaria.cl - **FM:** 102.3MHz – **GS04)** Simón Bolívar 26, Coyhaique ☎67 232240 ▤67 233287 - **FM:** 99.3MHz «Acro Iris».

LA00 (LA ARAUCANIA):
LA01) Cas. 56-D, Temuco **E:** kalimat@telsur.cl ☎ 45 375142 ▤45 323657 - **Mapuche:** 1030-1400, 1600-2100, 0000-0230 – **LA02)** Av. Bernardo O'Higgins 294, piso 2 (⬚ Cas. 268), Angol **E:** vozdelatierra@123click.cl ☎45 714706 – **LA03)** Portales 775, Temuco ☎43 311015 **W:** www. cooperativa.cl **E:** cgomez@cooperativa.cl - **FM:** 93.5MHz «Temuco Rock & Pop FM» – **LA04)** Gral. Cruz 551 (or Cas. 1499), Temuco ☎45 212707 – **LA06)** Cas. 136, Curacautín, Malleco – **LA07)** Lautaro 124 (⬚ Cas. 211), Angol. – **LA05)** ☎ 45 413647 - **FM:** 94.9MHz – **LA08)** Claro Solar 536, Temuco. – ☎45 213166 **E:** araucnayfrontera@entelchile. net - **FM:** 95.9MHz «La Araucana» – **LA09)** Antonio Varas 920, Temuco – **LA10)** Lynch 646, Temuco ☎45 213854 – **W:** www.radioagricultura.cl/ - **FM:** 105.7MHz «San Cristóbal – **LA11)** Rudecindo Ortega 691(⬚Cas. 12), Temuco ☎45 748702 – **LA12)** Ignacio Serrano 264 (⬚ Cas. 61), Loncoche ☎45 411567 **E:** radiocd141@gmail.com - **FM:** 105.9MHz «Vibración» – **LA13)** Cas. 267, Victoria – **LA14)** O'Higgins 828, 2° piso, Of. 5 (⬚Cas. 15), Lautaro – **LA15)** General Lagos 662 (⬚ Cas. 186), Traiguén. - **FM:** 101.3MHz «Granero» – **LA16)** Vicente Reyes 753 (⬚ Cas. 110), Villarrica ☎45 411567 – **LA17)** Cas. 18, Nueva Imperial 1155-1810 – **LA18)** Alcázar 1158, 2° piso, Collipulli ☎45 811623 – **LA19)** Luis Durand 3057 (or Cas. 830), Temuco ☎45 367070 ▤45 213790. - **English:** 0800-0830. **German:** Sun 1230-1300 - **FM:** 106.9MHz – **LA20)** Sector Elecoyan, Loncoche ☎45 471052 – **LA21)** Sector Pircunche, Localidad de Cajon, Temuco

LL00 (LOS LAGOS):
LL01) Ismael Valdez 500, piso 2, Validivia ▤63 230101 W:.: www. ainil.cl – **LL02)** Ramírez 207, or Cas. 260, Ancud ☎65 622905 ▤65 622722 – **W:** www.cooperativa.cl – **LL03)** Caupolicán 597, of. 31, Valdivia – **LL04)** Thompson 25 (or Cas. 174), Castro – **LL05)** Juan Mackenna 904, entrepiso (or Cas. 35-0), Osorno **W:** www.sago.cl **E:** sago@telsur.cl ☎64 233881 - **FM:** 94.5MHz – **LL06)** Arturo Prat 466 (or Cas. 312), La Unión ☎64 322275 ▤64 322322 **E:** radioconcordia@ surnet.cl – **LL07)** Cochrane 746 (⬚Cas. 5-0), Osorno ▤64 268911 **W:** www.radiovozdelacosta.cl **E:** contacto@radiovozdelacosta. cl – **LL08)** Illapel 60 (⬚ Cas. 67), Puerto Montt **W:** www.radiore-loncavi.cl **E:** radio-rr@telsur.cl ☎65 252946 ▤65 256523 – **LL09)** Arauco 363 3° piso, Valdivia ☎63 213601 **E:** radioaustral@surnet.cl – **LL10)** Bernardo O' Higgins 486 (⬚Cas. 106), Castro ▤65 632260 **W:** www.radiochiloe.cl **E:** gerencia@ www.radiochiloe.cl - **FM:** 90.1MHz «Martin Ruiz de Gamboa» – **LL11)** Ramirez N° 816 2° piso Oficina 6, Osorno ☎64 643650 **W:** www.armonia.cl – **LL12)** La Serena 971, piso 4, Puerto Montt ☎65 254997 **W:** www.armonia.cl – **LL13)** Arauco N° 340, Valdivia ☎66 333280 **W:** www.armonia.cl – **LL14)** Patricio Lynch 1814-B, Osorno ☎64 330400 ▤64 33041 - **FM:** 101.5MHz «La Palabra» – **LL16)** Concepcion 110 (⬚Cas. 166), Puerto Montt ▤65 258439 **E:** adona@telsur.cl – **LL17)** Bernard O'Higgins 793, Panguipulli. ▤63 310796 **E:** radio2@surnet.cl – **LL18)** Benavente 385, tercere piso, Puerto Montt ▤65 258048 **W:** www.radiovientodelsur.cl **E:** contacto@ radiovientodelsur.cl - **FM:** 92.3MHz «Aurora» – **LL19)** San Francisco 248, Of. 2, Puerto Varas – **LL20)** Av. Ramón Picarte 4215 (⬚Cas. 35), Valdivia – **LL21)** Sector Alto Bonito, Puerto Montt – **LL22)** Pedro Lagos 295, Río Bueno ▤64 341531 **W:** www.radiosanjosedealcudia.cl **E:** radio@radiosanjosedealcudia.cl

MA00 (MAULE):
MA01) 6 Oriente No 928 2 y 3 Sur (⬚Cas. 214), Talca ☎71 231344 **W:** www.radiolautaro.cl **E:** radiolautaro@tie.cl – **MA02)** Diagonal Isidoro del Solar 285, Talca **W:** www.magica.cl **E:** chb@entelchile.net ☎73 210917 ▤73 217143 - **FM:** 100.7MHz «Futura FM», 107.1MHz «Logika FM» – **MA04)** Independencia 631 (or Cas. 500), Linares **W:** www. radioancoa.cl **E:** cecili.rojas@tv5linares.com - **FM:** 90.7MHz – **MA05)** Casa 2 Norte 685, Talca ☎71 233019 **W:** http://radioemisoras.utalca. cl **E:** radioemisoras@utalca.cl - **FM:** 94.1MHz – **MA06)** Cas. 492, Curicó. ☎75 310023 **W:** http://radiocondell.cl - **FM:** 89.9MHz «Futura – **MA07)** Claudina Urrutia 707, Interior (⬚Cas. 196), Cauquenes ☎73 514303 - **FM:** 101.9MHz «Dinastia» – **MA08)** Cármen 714, Curicó. ☎75 543523 **W:** www.radiocondell.cl **E:** amfm@radiocondell.cl - **FM:**

103.1MHz (rel. AM), 92.7MHz «Opus» – **MA09)** Cas. 18-D, San Javier ☎73 322529 📠73 321226 - **FM:** 105.5MHz «Musical FM» – **MA10)** Diputado Dario Dueñas 340 (or Cas. 67), Linares ☎73 210277 **E:** soberania@hotmail.com – **MA11)** 1 Poniente 1239 (or Cas. 516), Talca **E:** radiofamilia@gmail.com ☎71 227255 – **MA12)** Cas. 37, Parral – **MA13)** Villouta N° 558, Curico

MM00 (MAGALLANES Y LA ANTARCTICA CHILENA):
MC01) Errazuriz 675, 2° piso, Punta Arenas 📠61 225958 **W:** www.radiopatagonica.com/portal – **FM:** 96.9MHz – **MC02)** José Nogueira 1370, Punta Arenas ☎61 243551 **W:** radiomagallanes.cl **E:** prensa@radiomagallanes.cl – **MC03)** Rocha No 931, 2° piso, Punta Arenas. ☎61 617 115 **W:** www.radio-nacional.cl **E:** radionacional@tie.cl – **MC04)** Bories 871, Punta Arenas **W:** www.radiopolar.com **E:** informaciones@radiopolar.com ☎61 241417 📠61 249001 - **FM:** 96.5-98.5-105.7MHz «Finísima» – **MC05)** Puerto Eberhard 229, Puerto Natales **W:** www.radiopayne.cl – **MC06)** Eberhard 212, Puerto Natales ☎61 410157 📠61 410157 – **MC07)** Faguano 548 A, Punta Arenas – **MC08)** Bulnes 449, Puerto Porvenir ☎61 580100 📠61 580094 **W:** www.radiotierradelfuego.cl **E:** director@radiotierradelfuego.cl -

MS00 (METROPOLITANA DE SANTIAGO):
MS01) Avenida Ortuzar N° 935 (Cas.110), Melipilla ☎832 3193 📠832 34440 **E:** radioserrano@123mail.cl - **FM:** 104.5MHz «Caricia FM» – **MS02)** Av. Manuel Rodriguez 15, Santiago ☎392 3000 📠392 23072 – **W:** www.radioagricultura.cl **E:** director@radioagricultura.cl – **MS03)** Av. Condell 910, Santiago ☎2 2224500 📠2 2223093 – **W:** www.monumental.cl **E:** contacto@monumental.cl – **MS05)** Triana 868, Providencia (or Cas. 10195), Santiago. - ☎2 2360096 📠2 363495 **W:** radiosantiago.cl **E:** gerenciageneral@radiosantiago.cl - 1000-0600 – **MS06)** Antonio Bellet 353, Providencia, Santiago **W:** www.cooperativa.cl **E:** info@cooperativa.cl ☎23 648000 📠23 648010 – **MS07)** Av. Presidente Bulnes 80, Of. 127, Santiago **W:** www.carabinerosdechile.cl **E:** radio@carabineros.cl ☎02 698 8141 – **MS08)** Alameda 43623 (or Cas. 56042), Santiago **W:** www.radiocolocolo.cl ☎📠2 6642353 – **MS09)** Estados Unidos 246, Santiago **W:** www.radionuevomundo.cl **E:** gerencia@radionuevomundo.cl ☎2 460 8211 – **MS10)** Eleodoro Flores 2475, Nunoa, Santiago ☎02 2692255 📠02 2692257 **W:** www.nuevaradiocarrera.cl **E:** info@radiocarrera.cl – **MS11)** Av. Bulnes 120, Oficina 89 (or Cas. de Correo 14351), Santiago – **MS12)** Enrique Alcalde 1081, Talagante ☎28 151666 **W:** nuevaprogresoam@gmail.com - **FM:** 103.9MHz «Contacto» – **MS13)** Alcalde Dávalos 124 - Providencia, Santiago ☎22 7322344 – **MS14)** Cas. 244-V, Santiago 1000-0600 – **MS15)** Fanor Velasco 11, Santiago ☎26 723288 📠26 980664 – **W:** www.radioportales.net **E:** contacto@radioportales.net - 1000-0430 – **MS16)** Cas. 442, Correo 2, Santiago **W:** www.radiousach.cl **E:** radio@usach.cl – **FM:** 94.5MHz – **MS17)** Los Leones 668, Providencia, Santiago ☎25 836602 **W:** www.lamexicana.cl **E:** contacto@lamexicana.cl – **MS18)** Portugal 810, Santiago ☎26 650673 📠26 651032 **W:** www.radiocorporacion.cl – **MS19)** Gran Avenida Jose Miguel Carrera 5848, 4° piso, Santiago ☎5 242868 **W:** www.radiopanamericanadechile.cl **E:** director@radiopanamericanadechile.cl – **MS20)** Irarrazaval 2821, Of. 427, Edif. Century, Torre B Ñuñoa, Santiago ☎22 746596 **E:** cb146yungay@hotmail.com – **MS21)** Av. Portales 3020 (📠 Cas. 380), San Bernardo ☎28 414135 📠28 571160 **W:** www.radiocanelo.cl **E:** canelo@rdc.cl – **MS22)** Cas. 1346, Santiago **W:** www.radiosudamerica.cl **E:** director@radiosudamerica.cl ☎2 527 3999 – **MS23)** Av. Lib.Bdo. O'Higginsa 854, (Cas. 223), Talagante ☎81 51374 - **FM:** 102.9MHz «Embrujo FM» – **MS24)** María Luisa Santander 0292, Providencia, Santiago ☎22 844921 📠22 087252 **W:** www.nuevotiempo.cl **E:** contactos@nuevotiempo.cl – **MS25)** Purisima 251, Barrio Bellavista, Recoleta, Santiago ☎27 323748 **W:** www.radiotierra.com – **MS26)** Alameda 340, Santiago ☎2354 2020 📠2354 2054 **E:** radio@uc.cl

TA00 (TARAPACA):
TA02) Baquedano 575, (📠 Cas. 49-D), Arica. - **FM:** 90.9MHz «Pukara FM» – **TA03)** Luis Uribe N° 355, Piso 2, Iquique ☎57 510750 – **TA04)** General Lagos 678 (📠 Cas. 225), Arica – **TA05)** Esmeralda 594 (📠 Cas. 290), Iquique ☎57 422693 – **TA06)** Calle José Miguel Carrera 350 esquina Av. Circulación O'Higgins, Putre (📠 Cas 82, Arica) 📠58 252803 **E:** prensaputre@hotmail.com - **FM:** 94.5MHz – **TA07)** Ap. 6, Pozo Almonte, Iquique ☎58 222380 📠58 222278 **E:** radiodir@uta.cl - **FM:** 95.9MHz – **TA09)** Iquique

VA00 (VALPARAISO):
VA01) Pedro Montt 1766 (Cas. 3304), Valparaíso ☎32 274 5537 📠32 259 6064 **W:** www.radiostellamaris.cl **E:** direccion@radiostellamaris.cl – **VA02)** Morris No 106, Depto. 155, Piso 15, Valparaíso – **W:** www.cooperativa.cl **E:** cgomez@cooperativa.cl – **VA03)** 5 Norte 168, Viña del Mar ☎32 971201 – **VA04)** Cas. 89-V, Valparaíso ☎32 258699 – **W:** www.radioportales.cl/portal **E:** radioportalsger@adsl.tie.cl - **FM:** 98.9MHz «Carolina» – **VA05)** Valparaíso 633, piso 3, Valparaíso ☎32 695485 – **VA06)** Av. Errazuriz 2120, Valparaíso **W:** www.radiovalentinletelier.cl - **FM:** 97.3MHz – **VA07)** Victoria

B° 2321(📠Cas. 90), Valparaíso ☎39 23000 📠39 23072 **W:** www.radioagricultura.cl **E:** director@radioagricultura.cl - **FM:** 97.3MHz – **VA08)** Plaza Vergara 172, Oficina 22, Viña del Mar **E:** vina@rrb.org ☎32 885524 – **VA09)** Paseo Cousiño 8, Viña del Mar **W:** www.radiofestival.cl **E:** servicios@festival.cl - 1000-0700 – **VA10)** Plaza de la Justicia 45, Of. 702, Valparaíso ☎32 2312297 📠32 256509 **W:** www.radiocolocolo.cl – **VA11)** Chacabuco 2370, Valparaíso ☎32 681756 **W:** www.radiofe.cl **E:** leopoldmoreno@hotmail.com – **VA12)** Av. España 1680, Valparaíso **W:** radio.utfsm.cl ☎32 654137 - **FM:** 99.7MHz – **VA13)** Las Colinas 284(📠 Cas. 68, Correo 2), San Antonio ☎35 211321. - **FM:** 90.9MHz «Cristalina» – **VA14)** Papudo 155 (or Cas. 307), Los Andes ☎34 421425 **E:** gerencia@radiosaconcagua.cl – **VA15)** Cas. 33, Llo-Lleo, San Antonio ☎35 215257 📠35 283543 **E:** radiointegracion@hotmail.com – **VA16)** Cas. 529, Quillota ☎33 470003 **W:** www.radiolibra.cl/NEXO **E:** lpardo@radiolibra.cl - **FM:** 104.7MHz «Libra Stereo FM» – **VA17)** Cas. 75, Putaendo ☎34 501428 📠34 504040 – **VA18)** Santo Domingo 99, Oficina 4 (📠Cas. 100), San Felipe ☎34 510198 **W:** www.radioaconcagua.cl **E:** ecornejo@radiosaconcagua.cl - **FM:** 91.7MHz «Colunquén FM» – **VA19)** Cas. 972, Viña del Mar – **VA20)** Eusebio Lillo 520, local 12, edifico Torre Valparaiso, Valparaiso ☎32 296 3793 **W:** www.radiovalparaiso.cl **E:** prensa@radiovalparaiso.cl – **VA21)** Calle Antofagasta N° 131, Paradero 2, Nueva Aurora, Viña del Mar ☎32 296 2478 – **W:** www.armonia.cl

FM in Santiago. **Power** 1-10kW **Slogans:** Name + «FM»
MHz: 11) 88.1 Aurora – 88.5 Concierto – 61) 88.9 R. Futuro – MS13) 89.3 R.Maria – 11) 89.7 Duna – 90.5 Pudahuel – 91.3 El Conquistador – 91.7 Amistad – 3) 92.1 – 92.5 Radioactiva – 92.9 Romance – 93.3 La Cooperativa – 93.7 Universo – 16) 94.1 Rock & Pop – 68) 94.5 – 95.3 40 principales – 76) 95.9 Tiempo – 96.5 Beethoven – 97.1 Caracol – 97.7 Zero – 98.5 FM 2 – 29) 99.3 Carolina – 99.7 Bío Bío – 76) 100.1 Infinita – 11) 100.9 – 101.3 Corazón – 101.7 Pet Hit – 102.1 Oasis – 102.5 Univ. de Chile – 103.3 Horizonte – 103.9 Maria – 104.1 Romantica – 104.9 Nina – 105.7 Para ti – 106.3 Armonía – 46) 106.9 Sintonía – 107.5 Fantasía

CHINA (People's Rep. of)

L.T: UTC +8h — **Pop:** 1,336 million —**Pr.L:** Mandarin, Amoy, Cantonese, Chaozhou, Hakka, Kazakh, Korean, Mongolian, Tibetan, Uighur, Zhuang, a.o. — **E.C:** 50Hz, 220V — **ITU:** CHN

MINISTRY OF INDUSTRY AND INFORMATION TECHNOLOGY
📧 13 Xi Chang'an Jie, Beijing 100804 **L.P:** Minister: Li Yizhong.

THE STATE ADMINISTRATION OF RADIO, FILM AND TELEVISION (SARFT) (Gov.)
📧 2 Fuxingmenwai Dajie, Beijing 100866 or P.O.Box 4501, Beijing ☎ +86 10 6809 2707 📠 +86 10 6851 2174
W: www.sarft.gov.cn **L.P:** Dir: Wang Taihua.

Official P.R.C Abbreviations: The 31 regions of the People's Republic of China, with their abbreviations and names in Pinyin (Chinese Phonetic Alphabet) version followed by the old spelling in brackets:
AH: Anhui (Anhwei) – BJ: Beijing M. (Peking) – CQ: Chongqing M. (Chungking) – FJ: Fujian (Fukien) – GD: Guangdong (Kwangtung) – GS: Gansu (Kansu) – GX: Guangxi Zhuang A.R. (Kwangsi) – GZ: Guizhou (Kweichow) – HAN: Hainan (Hainan) – HB: Hubei (Hupeh) – HEB: Hebei (Hopeh) – HEN: Henan (Honan) – HL: Heilongjiang (Heilungkiang) – HN: Hunan (Hunan) – JL: Jilin (Kirin) – JS: Jiangsu (Kiangsu) – JX: Jiangxi (Kiangsi) – LN: Liaoning (Liaoning) – NM: Nei Menggu A.R. (Inner Mongolia) – NX: Ningxia Hui A.R. (Ningsia) – QH: Qinghai (Tsinghai) – SC: Sichuan (Szechwan) – SD: Shandong (Shantung) – SH: Shanghai M. (Shanghai) – SN: Shaanxi (Shensi) – SX: Shanxi (Shansi) – TJ: Tianjin M. (Tientsin) – XJ: Xinjiang Uighur A.R. (Sinkiang) – XZ: Xizang A.R.(Tibet) – YN: Yunnan (Yunnan) – ZJ: Zhejiang (Chekiang).

Regional Services: Add "Renmin Guangbo Diantai" (People's Broadcasting Station) to the stn name shown in the table below to obtain the full name in Standard Chinese.
Abbreviations: 1 = 1st prgr, 2 = 2nd prgr, 3 = 3rd prgr; EBS = Economic Broadcasting Station, LBS = Literary Broadcasting Station.
Languages: Standard Chinese (Putonghua), based on the Beijing dialect, is used in broadcasts throughout China. Various dialects and minority languages are included in the relevant regional services and in broadcasts to Taiwan.
Abbreviations: Ch = Standard Chinese, Kg = Kirghiz, Ko = Korean, Kz = Kazakh, Mo = Mongolian, Tb = Tibetan, Ug = Uighur.

MW	kHz	kW	Station	Tx Location
ZJ1)	531	10	Zhejiang	Jinhua
1)	540	50	CNR 1	
NM18)	540		Genhe	
QH4)	540		Haixi	Da Qaidam
1)	549	1200	CNR 5	Putian, FJ
EN2)	549	25	Zhengzhou	
NM12)	549	10	Alxa	Bayanhot
NM5)	549	10	Chifeng	
EB1)	558		Hebei	Shijiazhuang
FJ1)	558	50	Fujian	Jianyang
NM17)	558	1	Zalantun	
NM3)	558	10	Baotou	
XJ1)	558	120	Xinjiang	Hutubi
YN16)	558		Nujiang	Lushui
1)	567	10	CNR 1	Lianyungang, JS
EN17)	567	10	Zhoukou	
TJ1)	567	20	Tianjin	
EN4)	576	10	Luoyang	
FJ5)	576		Quanzhou	
YN1)	576	10	Yunnan	Dali/Wenshan
ZJ1)	v576		Zhejiang	Linhai
14)	585	200	Southeast BC	Fuzhou, FJ
EB11)	585	10	Langfang	
EN14)	585	10	Nanyang	
EN1A)	585		Henan	Anyang
GS3)	585	3	Jinchang	
HB8)	585	10	Jingzhou	
HL3)	585		Qiqihar	
JL2)	585	10	Changchun	
JL10)	585		Yanbian	
JS1)	585	50	Jiangsu	Nanjing
JX5)	585	10	Xinyu	
LN16)	585	1	Chaoyang	Beipiao
SX6)	585	10	Jincheng	
SD1)	594	50	Shandong	Jinan
XZ1)	594	300	Xizang	Lhasa
13)	603		VO Pujiang	SH
AH1)	603	10	Anhui	Hefei
AH4)	603		Huaibei	
BJ1)	603	25	Beijing	
EB1)	603		Hebei	Shijiazhuang
EB12)	603		Hengshui	
EB6)	603		Zhangjiakou	
EN13)	603	10	Sanmenxia	
EN1A)	603	200	Henan	Zhengzhou
GD1)	603	10	Guangdong	Guangzhou
GZ1)	603	10	Guizhou	Guiyang
HB3)	603	10	Wuhan	
HL5)	603		Shuangyashan	
JL10)	603		Yanbian	Dunhua
JL3A)	603		Jilin-shi EBS	
JS1)	603		Jiangsu	Yangzhou
JS12)	603		VO Jiangnan	
JS9)	603	1	Nantong	
JX1)	603		Jiangxi	Shangrao
JX1)	603		Jiangxi	Jiujiang
JX9)	603	10	Ji'an	
LN10)	603		Yingkou	
NM10)	603	10	Ordos BS	
NM19)	603		Morin Dawa	
NM8)	603	50	Hulun Buir	Hailar
SD3)	603	10	Qingdao	
SD10)	603	10	Jining	
SD5)	603	10	Zaozhuang	
SH2)	603		Dongfang	
SN1)	603	25	Shaanxi	Xi'an
SN7)	603		Yan'an	
SX1)	603	30	Shanxi	Taiyuan
SX4)	603	1	Yangquan	
XJ10)	603		Shihezi	
XJ9)	603	1	Ili	Yining
ZJ1)	603		Zhejiang	Hangzhou
ZJ4)	603	10	Ningbo	
FJ1)	612	100	Fujian	Ningde
LN1)	612	10	Liaoning	Dandong
SC1)	612	10	Sichuan	Neijiang/Yibin
HB9)	621	10	Yichang	
HL1)	621	200	Heilongjiang	Harbin
QH4)	621	20	Haixi	Da Qaidam
SC9)	621	3	Guangyuan	
SD1)	621	10	Shandong	Liaocheng
YN11)	621	1	Zhaotong	
1)	630	200	CNR 2	Nanchang, JX
1)	630	100	CNR 2	Yingyang, HEN
1)	639	200	CNR 1	BJ

MW	kHz	kW	Station	Tx Location
AH3)	648	1	Huainan	
GD1)	648	150	Guangdong	Guangzhou
JL2)	648		Changchun	
LN14)	648		Liaoyang	
LN16)	648	3	Chaoyang	
SH1)	648	10	Shanghai	
XJ7)	648		Kashi	
EN1)	657	300	Henan	Zhengzhou
JL7)	657	1	Baishan	
ZJ6)	657		Jiaxing	
11)	666	600	VO Strait	FJ
AH2)	666	10	Hefei	
GZ5)	666	1	Anshun	
HL10)	666	10	Jiamusi	
JL4)	666	10	Siping	
LN8)	666	2	Jinzhou	
QH1)	666	200	Qinghai	Xining
SD10)	666	1	Jining	
TJ1)	666	50	Tianjin	
YN10)	666	1	Dongchuan	
ZJ5)	666	7.5	Wenzhou	
JX5)	675	1	Xinyu	
NM1)	675	200	Nei Menggu	Hohhot
XJ1)	675		Xinjiang	Altay
YN12)	675	1	Gejiu	
YN15)	675	10	Diqing	Shangri-la
ZJ9)	675		Jinhua	
1)	684	1200	CNR 6	Putian, FJ
AH1)	684		Anhui	Xuancheng
AH13)	684		Suzhou	
EB8)	684	10	Tangshan	
GS1)	684		Gansu	Jinchang
HB2)	684	10	Chutian	Jingmen
HL9)	684	50	Mudanjiang	
LN5)	684	10	Fushun	
XJ1)	684	3	Xinjiang	Hotan
ZJ11)	684	10	Zhoushan	
HL3)	693	10	Qiqihar	
SN1)	693	300	Shaanxi	Xianyang
2)	702		CRI DS	Zhuhai, GD
JL3)	702	10	Jilin-shi	
JS1)	702	200	Jiangsu	Nanjing
LN16)	702	3	Chaoyang	Lingyuan
NM15)	702	1	Manzhouli	
NM5)	702	10	VO Tongliao	
NM6)	702	10	Ulanqab	Jining
SC11)	702	1	Neijiang	
XJ1)	702	10	Xinjiang	Urumqi
YN5)	702	10	Honghe	Gejiu
AH12)	711	3	Fuyang	
AH16)	711	1	Lu'an	
EN2)	711	10	Zhengzhou	
QH1)	711	10	Qinghai	Golmud
SC5)	711	1	Panzhihua	
SC8)	711	1	Mianyang	
ZJ10)	711	3	Quzhou	
ZJ12)	711		Lishui	
1)	720	200	CNR 2	BJ
SC7)	720	1	Deyang	
AH1)	720		Anhui	
EN16)	729	10	Shangqiu	
JX1)	729	200	Jiangxi	Nanchang
HN1)	738	200	Hunan	Changsha
JL1)	738	150	Jilin	Changchun
XJ1)	738	120	Xinjiang	Hutubi
ZJ8)	738	5	Shaoxing	
1)	747		CNR 12	BJ
AH2)	747	1	Hefei	
EB1)	747		Hebei	Shijiazhuang
EB5)	747	10	Baoding	
EB12)	747		Hengshui	
EN1)	747	25	Henan	Huangchuan
EN5)	747	10	Pingdingshan	
FJ6)	747		Longyan	
HB1)	747	10	Hubei	Qichun
JS11)	747	3	Changzhou	
JS6)	747		Yancheng	
JX8)	747		Ganzhou	
LN10)	747		Yingkou	
LN16)	747	3	Chaoyang	Jianping
LN5)	747		Fushun	
NM4)	747	1	Wuhai	
NX1)	747	10	Ningxia	Yinchuan
SC1)	747		Sichuan	Chengdu
SC14)	747		Nanchong	

MW	kHz	kW	Station	Tx Location	MW	kHz	kW	Station	Tx Location
SD11)	747		Rizhao		2)	846	10	CRI DS 3	BJ
SD13)	747		Linyi		AH1)	846	10	Anhui	Suzhou
SD5)	747		Zaozhuang		AH14)	846	1	Chaohu	
SN1)	747	50	Shaanxi	Xianyang	AH4)	846	1	Huaibei	
SN6)	747	1	Weinan		EB10)	846	10	Cangzhou	
TJ1)	747		Tianjin Binhai		EB11)	846	10	Langfang	
YN6)	747	100	Xishuangbanna	Jinghong	EB3)	846		Handan	
ZJ11)	747		Zhoushan		EN1)	846	100	Henan	Zhengzhou
ZJ4)	747		Ningbo		EN5)	846	3	Pingdingshan	
SX8)	750	1	Xinzhou		EN7)	846		Hebi	
1)	756	150	CNR 1	Harbin, HL	GX1)	846	10	Guangxi	Qinzhou
1)	765	600	CNR 5	Fuzhou, FJ	HB1)	846	15	Hubei	Qichun
AH7)	765	1	Bengbu		HB2)	846	10	Chutian	Xianning/Yichang
EN24)	765	10	Gongyi		JL1)	846	10	Jilin	Changchun
GD8)	765	10	Shaoguan		JS11)	846	10	Changzhou	
GZ1)	765	10	Guizhou	Zunyi	JS13)	846	5	Suzhou	
GZ3)	765	1	Liupanshui		LN13)	846	10	Fuxin Mo BS	
NM1)	765	10	Nei Menggu	Baotou	LN8)	846		Jinzhou	
BJ1)	774	10	Beijing		SD15)	846	10	Binzhou	
HB1)	774	100	Hubei	Wuhan	SD2)	846		Jinan	
LN8)	774		Jinzhou		SD3)	846	10	Qingdao	
SX2)	774		Taiyuan		SD7)	846	5	Weifang	
XJ6)	774	10	Hotan		SD9)	846	5	Weihai	
11)	783		VO Strait	Zhangpu, FJ	SX1)	846	20	Shanxi	Changzhi
EB1)	783		Hebei	Chengde	XJ1)	846	3	Xinjiang	Hotan
EB1)	783	100	Hebei	Baoding	XZ1)	846	10	Xizang	Lhasa
GD10)	783		Meizhou		1)	855	50	CNR 2	Anning, YN
EN19)	792	1	Xinmi		XJ1)	855	10	Xinjiang	Urumqi
GS5)	792	1	Jiayuguan		AH1)	864	50	Anhui	Hefei
GX1)	792	200	Guangxi	Nanning	EB20)	864		Renqiu	
LN2)	792	10	Shenyang		EN20)	864		Qinyang	
NM10)	792		Ordos BS	Otog	ZJ1)	864		Zhejiang	Ninghai
SC3)	792		Chengdu		ZJ15)	864	1	Jiangshan	
SH2)	792	50	Dongfang		15)	873	200	China Huayi BC	Fuzhou, FJ
XJ2)	792		Urumqi		EB13)	v873		Xinji	
AH1)	801	10	Anhui	Hefei	EN3)	v873	10	Kaifeng	
AH12)	801	10	Fuyang		GS1)	873	50	Gansu	Lanzhou
AH15)	801	1	Chizhou		HB3)	873	50	Wuhan	
EB10)	801	25	Cangzhou		HL1)	873	200	Heilongjiang	Harbin
EB8)	801	10	Tangshan		SD13)	873	10	Linyi	
EN8)	801		Xinxiang		XJ8)	873		Changji	
FJ3)	801		Xiamen		ZJ7)	873		Huzhou	
GD2)	801	50	Zhujiang EBS	Maoming	EB2)	882	10	Shijiazhuang	
GS1)	801		Gansu	Lanzhou	EN23)	882		Ruzhou	
HB1)	801	10	Hubei	Jingmen/Macheng	EN9)	882		Anyang	
JS2)	801		Nanjing		FJ1)	882	100	Fujian	Fuzhou
JS3)	801		Xuzhou		GZ6)	882	1	Qiannan	Duyun
JS5)	801	10	Huai'an		LN2)	882	10	Shenyang	
JS7)	801		Yangzhou		LN3)	882	50	Dalian	
LN16)	801	1	Chaoyang	Lingyuan	NM2)	882	10	Hohhot	
NX2)	801	10	Yinchuan		QH3)	882	10	Yushu	
SD10)	801	1	Jining		XJ4)	882		Karamay	
SD13)	801		Linyi		XJ9)	882	1	Ili	Yining
SD14)	801		Liaocheng		LN7)	891	10	Dandong	
SD4)	801	1	Zibo		NM13)	891	10	Hinggan	Ulanhot
SD8)	801	10	Yantai		NX1)	891	200	Ningxia	Yinchuan
SN2)	801		Xi'an		XJ10)	891		Shihezi	
XJ7)	801		Kashi		1)	900	10	CNR 2	Golmud, QH
ZJ5)	801	10	Wenzhou		2)	900		CRI DS 4	BJ
EN18)	810		Zhumadian		AH1)	900		Anhui	Bengbu
JL5)	810	10	Liaoyuan		EB1)	900		Hebei	Shijiazhuang
LN1)	810	5	Liaoning	Panjin	EB6)	900		Zhangjiakou	
LN15)	810		Tieling		EB7)	900	1	Chengde	
LN16)	810	10	Chaoyang		EB8)	900		Tangshan	
SN2)	810	50	Xi'an		EB9)	900		Qinhuangdao	
ZJ1)	810	200	Zhejiang	Hangzhou	EN1)	900	100	Henan	Zhengzhou
SX1)	819	200	Shanxi	Taiyuan	EN12A)	900		Luohe EBS	
XJ11)	819		Korla		FJ1)	900	1	Fujian	Yongding
XJ12)	819	1	Kuytun		GD6)	900		Zhuhai	
XJ13)	819		Bayingolin	Korla	HB1)	900	1	Hubei	Yingcheng
BJ1)	828	50	Beijing		HB2)	900	10	Chutian	Xiangfan
EN1)	828	10	Henan		HL1)	900	50	Heilongjiang	Bei'an/Jiamusi
EN17)	v828	10	Zhoukou		HN1)	900		Hunan	Changsha
EN6)	828		Jiaozuo		JL16)	900	1	Yanji	
GD1)	828	50	Guangdong	Heyuan	JL2)	900		Changchun	
HB23)	v828	1	Xiantao		JL4)	900	1	Siping	
HB8)	828	10	Jingzhou		JS10)	900		Zhenjiang	
1)	837		CNR 1		JS12)	900		Wuxi	
1)	837		CNR 5	FJ	JS2)	900		Nanjing	
EN15)	837		Xinyang		JS6)	900		Yancheng Huanghai	
FJ1)	837	3	Fujian	Fuding	LN19)	900	1	Haicheng	
HL2)	837	20	Harbin		LN6)	900	1	Benxi	
LN14)	837	1	Liaoyang		NM5)	900	10	Chifeng	
XJ1)	837	10	Xinjiang	Urumqi	SD3)	900	10	Qingdao	
12)	846	3	Jiangsu	Nanjing, JS	SN1)	900	30	Shaanxi	Xi'an

MW	kHz	kW	Station	Tx Location
SN4)	900	1	Baoji	
SX1)	900		Shanxi	Taiyuan
SX3)	900	10	Datong	
YN9)	900	100	Dehong	Luxi
ZJ11)	900		Zhoushan	
1)	909	100	CNR 6	Quanzhou, FJ
HL8)	909	7.5	Yichun	
JL6)	909	10	Tonghua	
QH1)	909	10	Qinghai	
SC1)	909	10	Sichuan	Jiange
TJ1)	909	50	Tianjin	
XJ1)	909		Xinjiang	
SD1)	918	200	Shandong	Jinan
1)	927		CNR 6	FJ
BJ1)	927	50	Beijing	
EB4)	927	12.5	Xingtai	
EN11)	927		Xuchang	
EN14)	927	10	Nanyang	
EN16)	927	10	Shangqiu	
GD1)	927	10	Guangdong	Guangzhou
GZ1)	927	200	Guizhou	Kaili
HB1)	927	10	Hubei	Xianning
HB12)	927	1	Xiaogan	
HB2)	927		Chutian	
HB2)	927	10	Chutian	Suizhou
HL1)	927	10	Heilongjiang	Shuangyashan
HL2A)	927	1	Harbin EBS	Hulan
JL19)	927	1	Hunchun	
JL3)	927	10	Jilin-shi	
JS11)	927	3	Changzhou	
JS16)	927	1	Changshu	
JS8)	927		Taizhou	
JX1)	927	10	Jiangxi	Jiujiang
LN1)	927	12.5	Liaoning	Shenyang
NM7)	927	10	Xilingol	Xilinhot
SH1)	927		Shanghai	
XJ3)	927		Urumqi EBS	
YN17)	927	1	Lufeng	
ZJ7)	927		Huzhou	
ZJ1)	930		Zhejiang	
AH1)	936	200	Anhui	Hefei
NM10)	936	10	Ordos BS	
1)	945	400	CNR 1	Jiaohe, JL
HB2)	945	10	Chutian	Qichun
HB2)	945	10	Chutian	Jingzhou
HL1)	945	50	Heilongjiang	Harbin/Jiamusi
SD10)	945	1	Jining	
EB12)	954	1	Hengshui	
GS2)	954	10	Lanzhou	
HA1)	954	30	Hainan	Haikou
LN4)	954	10	Anshan	
NM8)	954	50	Hulun Buir	Hailar
SC1)	954	10	Sichuan	Chengdu
SC6)	954		Luzhou	
ZJ2)	954	25	Hangzhou	
EB3)	963	10	Handan	
HB5)	963	10	Huangshi	
LN1)	963	50	Liaoning	Dalian
XJ1)	963		Xinjiang	Gulja
ZJ1)	963	10	Zhejiang	
EN1)	972	150	Henan	Zhengzhou
HL2A)	972	10	Harbin EBS	
XJ1)	972		Xinjiang	Altay
1)	981	200	CNR 1	Changchun, JL
1)	981	200	CNR 1	Nanchang, JX
SD7)	981	5	Weifang	
EB9)	990		Qinhuangdao	
SH1)	990	100	Shanghai	
YN1)	990	10	Yunnan	Gejiu
AH19)	999		Bozhou	
GD1)	999	10	Guangdong	Guangzhou
GZ2)	999	10	Guiyang	
LN1)	999	200	Liaoning	Shenyang
SC1)	999		Sichuan	Chengdu
SD1)	999	1	Shandong	Jining
XJ1)	999	10	Xinjiang	Hami
XZ1)	999		Xizang	Lhasa
1)	1008	200	CNR 1	Anning, YN
2)	1008		CNR DS 3	BJ
AH1)	1008	10	Anhui	Hefei/Wuhu
EB11)	1008		Langfang	
EB3)	1008	10	Handan	
EN13)	1008		Sanmenxia	
EN2)	1008		Zhengzhou	
FJ1)	1008	1	Fujian	Zhangping
GD22)	1008		Chenghai	
HB2)	1008	50	Chutian	Jingmen
HB22)	1008	10	Suizhou	
HN7)	v1008	1	Yueyang	
HN9)	v1008		Yiyang	
JS12)	1008		Wuxi	
JS2)	1008	10	Nanjing	
NX1)	1008	1	Ningxia	Guyuan
SD12)	1008	10	Dezhou	
SD3)	1008		Qingdao	
SN1)	1008	10	Shaanxi	Hanzhong/Yan'an
SN1)	1008		Shannxi	Xi'an
TJ1)	1008	50	Tianjin	
1)	1017		CNR 1	Dongtou, ZJ
1)	1017	200	CNR 8	Changchun, JL
EB5)	1017	10	Baoding	
GD1)	1017		Guangdong	Shaoguan/Shantou
1)	1026		CNR 6	FJ
BJ1)	1026	50	Beijing	
GZ1)	1026	200	Guizhou	Guiyang
JS14)	1026		Yizheng	
JS6)	1026		Yancheng	
LN10)	1026	2	Yingkou	
XJ6)	1026	10	Hotan	
1)	1035	50	CNR 1	
XJ1)	1044	1	Xinjiang	Yiwu
XJ1)	1044	10	Xinjiang	Urumqi
YN8)	1044	1	Dali	
ZJ1)	1050		Zhejiang	
1)	1053		CNR 10	BJ
AH2)	1053	1	Hefei	
EB10)	1053	10	Cangzhou	
EB15)	1053	1	Shahe	
EB17)	1053	1	Zhuozhou	
EN18)	1053	10	Zhumadian EBS	
EN3)	1053	10	Kaifeng	
EN4)	1053	10	Luoyang	
HB1)	1053	50	Hubei	Qianjiang
HN7)	1053		Yueyang	
JL10)	1053	20	Yanbian	Yanji
JS1)	1053	10	Jiangsu	Nanjing
LN1)	1053	50	Liaoning	Shenyang
SD2)	1053	10	Jinan	
YN4)	1053	50	Wenshan	
GD2)	1062	150	Zhujiang EBS	Guangzhou
HL11)	1062		Qitaihe	
AH13)	1071	1	Suzhou	
FJ1)	v1071		Fujian	
GX1)	1071	10	Guangxi	Ningming
LN4)	1071	2	Anshan	
SD16)	1071	10	Heze	
SN4)	1071	10	Baoji	
TJ1)	1071	50	Tianjin	
XJ3)	1071		Urumqi	
ZJ1)	1071	10	Zhejiang	Hangzhou
GD7)	1080	5	Shantou	
HL7)	1080	1	Daqing	
HL12)	1080	1	Suihua	
JS13)	1080	10	Suzhou	
YN1)	1080		Yunnan	
ZJ1)	v1080		Zhejiang	
1)	1089	600	CNR 6	Fuzhou, FJ
HN3)	1089	1	Zhuzhou	
LN1)	1089	200	Liaoning	Shenyang
AH1)	1098		Anhui	Wuhu
AH18)	1098		Dangtu Xian	
AH7)	v1098	1	Bengbu	
EN1A)	1098		Henan	Zhengzhou
EN5)	1098		Pingdingshan	
GD11)	1098	1	Huizhou	
GD19)	1098	5	Maoming	
GD4)	1098		Guangzhou	
HB1)	1098	10	Hubei	Jingzhou
HB8)	1098	10	Xiangfan	
HN5)	1098	10	Hengyang	
JS17)	1098	10	Zhangjiagang	
JS3)	1098		Xuzhou	
LN8)	1098		Jinzhou	
NM1)	1098	10	Nei Menggu	
SD12)	1098	10	Dezhou	
TJ1)	1098		Tianjin	
XJ5)	1098	1	Hami	
ZJ11)	1098	10	Zhoushan	
1)	1098		CNR 11	BJ
AH6)	1107	1	Tongling	

MW	kHz	kW	Station	Tx Location	MW	kHz	kW	Station	Tx Location
EN7)	1107	10	Hebi		GD1)	1206		Guangdong	Shenzhen/Zhaoqing
FJ3)	1107	10	Xiamen		HN1)	1206	10	Hubei	Xiangfan
HA1)	1107	10	Hainan	Wuzhishan	JL10)	1206	150	Yanbian	Yanji
JL1)	1107	1	Jilin	Siping/Fuyu	JS1)	1206	1	Jiangsu	Nanjing
JX4)	1107	1	Pingxiang		NX1)	1206	1	Ningxia	Zhongning
XJ1)	1107	120	Xinjiang	Hutubi	SD9)	v1206	1	Weihai	
ZJ6)	1107	10	Jiaxing		1)	1215	20	CNR 2	Shenyang, LN
1)	1116	120	CNR 2	Harbin, HL	1)	1215	50	CNR 7	Zhuhai, GD
1)	1116	600	CNR 5	Shaowu, FJ	HB1)	1215		Hubei	Yichang
AH12)	1116	10	Fuyang		HL14)	1215		Heihe	
HA1)	1116		Hainan		XJ1)	1215	10	Xinjiang	Urumqi
SC1)	1116	200	Sichuan	Chengdu	1)	1224		CNR 6	FJ
SD10)	1116		Jining		GX1)	1224	100	Guangxi	Nanning
EB1)	1125		Hebei	Shijiazhuang	JS10)	1224	10	Zhenjiang	
HB3)	v1125	50	Wuhan	Changjiang	HN1)	1233	10	Hunan	Yueyang
1)	1134	1200	CNR 1	Golmud, QH	JS9)	1233	10	Nantong	
GD18)	1134	10	Zhanjiang		XJ1)	1233		Xinjiang	Bortala
GS9)	1134	10	Yumen		XJ1)	1233	120	Xinjiang	Hutubi
SN3)	1134	10	Tongchuan		AH16)	v1242		Lu'an	
XJ9)	1134	1	Ili	Yining	HB20)	1242	1	Macheng	
ZJ1)	v1134		Zhejiang	Ningbo	HB24)	v1242	1	Qianjiang	
1)	1143	10	CNR 8		JX9)	1242		Ji'an	
BJEB18)	1143	1	Dingzhou		LN9)	1242	1	Huludao	
EB8)	1143		Tangshan		ZJ10)	1250		Quzhou	
EN1)	1143	100	Henan	Zhengzhou	2)	1251		CRI DS 1	BJ
EN5)	1143	3	Pingdingshan		EB1)	1251		Hebei	Qinhuangdao
GD1)	1143		Guangdong	Zhanjiang	EB2)	1251	25	Shijiazhuang	
GS4)	1143	10	Tianshui		EN10)	1251	10	Puyang	
GZ5)	1143		Anshun		EN12)	1251	10	Luohe	
HA1)	1143		Hainan	Haikou	EN22)	1251	1	Yima	
HB1)	1143	10	Hubei	Shiyan	EN6)	1251		Jiaozuo	
HL10)	1143		Jiamusi		EN9)	v1251		Anyang	
JL17)	1143	1	Tumen		HB1)	1251	5	Hubei	Jingmen
JL3)	1143	10	Jilin-shi		HN2)	1251		Changsha	
JL8)	1143		Songyuan		JL3A)	1251		Jilin-shi EBS	
JS11)	1143	3	Changzhou		JS12)	1251	10	Wuxi	
JS2)	1143		Nanjing		JS2)	1251		Nanjing	
JS9)	1143		Nantong		JS4)	1251		Lianyungang	
LN10)	1143		Yingkou		JS5)	1251	10	Huai'an	
LN14)	1143		Liaoyang		LN4)	1251		Anshan	
LN5)	1143		Fushun		QH1)	1251	100	Qinghai	Xining
NM5)	1143		Chifeng		SD1)	1251		Shandong	Jinan
NM16)	1143	1	Yakeshi		SD18)	1251		Longkou	
QH1)	1143		Qinghai	Xining	SD3)	1251	10	Qingdao	
SC11)	1143		Neijiang		SD7)	1251		Weifang	
SC16)	1143	1	Dazhou		SN8)	1251	10	Hanzhong	
SC9)	1143		Guangyuan		YN14)	1251		Yuxi	
SD13)	1143	10	Linyi		ZJ1)	1251		Zhejiang	Jinhua
SD14)	1143	10	Liaocheng		ZJ4)	1251		Ningbo	
SD4)	1143	1	Zibo		ZJ7)	1251		Huzhou	
SN9)	1143	1	Yulin		HN8)	1260	1	Changde	
XJ1)	1143		Xinjiang	Kashi	LN1)	1260		Liaoning	Tieling
ZJ1)	v1143		Zhejiang	Yuhuan	XZ1)	1260	1	Shannan	Nedong
HN1)	1152	150	Hunan	Changde	11)	1269	200	VO Strait	FJ
LN3)	1152	10	Dalian		JL18)	1269	1	Dunhua	
NM11)	1152	10	Bayannur	Linhe	JS3)	1269		Xuzhou	
NM13)	1152	10	Hinggan	Ulanhot	SX1)	1269	10	Shanxi	Xinzhou
1)	1161		CNR 1		EB1)	1278	100	Hebei	Shijiazhuang
GD2)	1161		Zhujiang EBS	Taishan	HL13)	1278	7.5	Daxing'anling	Jagdaqi
GX1)	1161	7.5	Guangxi	Beihai	JX2)	1278	10	Nanchang	
HB10)	1161	10	Jingmen		1)	1287		CNR 1	
JS12)	1161		Wuxi		EB20)	v1287	1	Renqiu	
SD7)	1161	10	Weifang		EN11)	1287	1	Xuchang	
AH16)	1170		Lu'an		GD5)	1287	30	Shenzhen	
AH17)	1170	3	Xuancheng		JS12)	1287		Wuxi	
AH2)	1170	10	Hefei		LN12)	1287	20	Fuxin	
GD)	1170		Guangzhou		NX1)	1287	10	Ningxia	Guyuan
GS1)	1170		Gansu	Lanzhou	SD7)	1287	5	Weifang	
SD15)	1170	10	Binzhou		YN7)	1287	10	Chuxiong	Chuxiong
SD5)	1170	1	Zaozhuang		ZJ1)	1287		Zhejiang	Dongtou
HB2)	1179	100	Chutian	Wuhan	EB16)	1296	1	Qinghe	
HL5)	1179	10	Shuangyashan		LN20)	1296		Xingcheng	
JS7)	1179		Yangzhou		LN6)	1296	10	Benxi	
XJ4)	1179		Karamay		SC10)	1296		Suining	
EB19)	1188	1	Botou		SH2)	1296	25	Dongfang	
EB4)	1188	10	Xingtai		SN5)	1296	10	Xianyang	
JL10)	1188		Yanbian		1)	1305		CNR 2	
FJ5)	v1197		Quanzhou		SD2)	1305		Jinan	
HL3)	1197	10	Qiqihar		CQ1)	1314	50	Chongqing	
SD16)	1197	10	Heze		HB14)	1314	3	Xianning	
SH1)	1197	10	Shanghai		HB6)	1314		Xiangfan	
12)	1206	1	VO Jinling	Nanjing, JS	JS1)	1314	10	Jiangsu	Suzhou
EB10)	1206		Cangzhou		SD8)	1314	10	Yantai	
EB3)	1206	10	Handan		ZJ1)	1314		Zhejiang	
EN21)	1206	1	Huixian		HN2)	1323	10	Changsha	

MW	kHz	kW	Station	Tx Location
JL9)	1323	10	Baicheng	
LN17)	1323	1	Wafangdian	
SD16A)	1323	10	Mudan	Heze
SN1)	1323	10	Shaanxi	Xi'an
ZJ4)	1323	20	Ningbo	
EN1)	1332	100	Henan	Zhengzhou
FJ1)	1332	10	Fujian	Yunxiao
FJ2)	1332	10	Fuzhou	
GS6)	1332	10	Gannan	Hezuo
JL2)	1332	10	Changchun	
1)	1341	100	CNR 1	GD
HB19)	1341	1	Yingcheng	
HB21)	1341	1	Chibi	
HL1)	1341	100	Heilongjiang	Heihe
JS8)	1341	10	Taizhou	
LN2)	1341	10	Shenyang	
SD12)	1341	1	Dezhou	
SD19)	1341	1	Qufu	
JX1)	v1350	50	Jiangxi	Ji'an
LN19)	1350		Haicheng	
NM9)	1350	50	Tongliao	
YN2)	1350	50	Kunming	
1)	1359		CNR 1	
FJ1)	v1368	1	Fujian	Changding
HB18)	1368	1	Guangshui	
HL6)	1368	10	Jixi	
1)	1377	600	CNR 1	Yingyang, HEN
15)	1377	200	China Huayi BC	Fuzhou, FJ
AH11)	1377	1	Chuzhou	
FJ1)	1377		Fujian	Nanping
NX5)	1377	1	Qingtongxia	
QH1)	1377		Qinghai	Xining
SD3)	1377	10	Qingdao	
XZ1)	1377		Xizang	
FJ1)	1386		Fujian	Quanzhou
GX3)	1386	5	Liuzhou	
HB11)	1386	1	Ezhou	
HB17)	1386	1	Shishou	
JS15)	1386	1	Jiangyin	
TJ1)	1386	50	Tianjin	
AH1)	1395	50	Anhui	Hefei
FJ1)	1395		Fujian	Hui'an
NM1)	1395		Nei Menggu	
NM7)	1395	10	Xilingol	Xilinhot
FJ1)	1404	50	Fujian EBS	Fuzhou
HB1)	1404	10	Hubei	Suizhou
LN7)	1404	10	Dandong	
ZJ1)	1404		Zhejiang	Wenling
HL4)	v1413	1	Hegang	
JS1)	1413	1	Jiangsu	Xuzhou/Yancheng
LN15)	1413		Tieling	
NX4)	1413	1	Wuzhong	
XJ1)	1413	5	Xinjiang	Kunes
1)	1422	600	CNR 1/8	Kashi, XJ
13)	1422	20	VO Pujiang	SH
SC4)	1422	10	Zigong	
SH2)	1422	20	Dongfang	
SX2)	1422	10	Taiyuan	
AH10)	v1431	2	Huangshan	
AH4)	v1431	10	Huaibei	
EB2)	1431	10	Shijiazhuang	
HB16)	1431	1	Danjiangkou	
HN10)	1431	1	Jinshi	
JL8)	1431		Songyuan	
NM14)	1431	1	Fengzhen	
GX1)	1440	50	Guangxi	Bose
LN18)	1440		Zhuanghe	
NM12)	1440	10	Alxa	Bayanhot
NM5)	1440	50	Chifeng	
FJ1)	1449		Fujian	Dongshan
JX1)	1449	20	Jiangxi	
SD11)	1449	10	Rizhao	
SD6)	1449	10	Dongying	
JS4)	1458	10	Lianyungang	
LN4)	1458		Anshan	
NM1)	1458	200	Nei Menggu	Hohhot
EB5)	1467	10	Baoding	
JX3)	1467	7.5	Jingdezhen	
SD1)	1467	1	Shandong	Dezhou
1)	1476	200	CNR 2	Shuangyashan, HL
HB15)	v1476	1	Laohekou	
HL1)	1476		Heilongjiang	
HL9)	1476		Mudanjiang	
JL14)	1476	1	Qian Gorlos	
JX5)	1476		Xinyu	
QH2)	1476	10	Xining	
SC12)	1476		Leshan	
SD4)	1476		Zibo	
ZJ1)	1476		Zhejiang	Leqing
GS1)	1485		Gansu	
GX1)	1485	1	Guangxi	Lingshan
GX4)	1485	1	Guilin	
GX5)	1485	1	Wuzhou	
HB7)	1485	10	Shiyan	
HL6)	1485	1	Jixi	
JL11)	1485	1	Gongzhuling	
JX6)	1485	1	Jiujiang	
LN11)	1485	1	Panjin	
SC3)	1485	1	Chengdu	
SD1)	1485	1	Shandong	Weihai
SX10)	1485	1	Shuozhou	
XJ12)	1485	1	Kuytun	
XJ5)	1485	1	Hami	
YN13)	1485	1	Chuxiong	
YN5)	1485	1	Honghe	Jinping
AH5)	v1494	1	Wuhu	
FJ1)	1494		Fujian	Lianjiang
NM1)	1494		Nei Menggu	
XJ1)	1494	1	Xinjiang	Yiwu
XJ11)	1494		Korla	
AH12)	v1503	1	Fuyang	
HN4)	1503	10	Xiangtan	
LN14)	1503		Liaoyang	
ZJ1)	1503	1	Zhejiang	Xinchang
GS7)	1512	10	Linxia	
NM5)	1512	1	Chifeng	Lindong
SD2)	1512		Jinan	
EB1)	1521		Hebei	Baoding
EB11)	1521		Langfang	
EN1)	1521		Henan	Zhengzhou
EN5)	1521	3	Pingdingshan	
EN8)	1521	10	Xinxiang	
FJ3)	1521		Xiamen	
GD20)	1521	1	Zhaoqing	
GZ5)	1521		Anshun	
HB25)	1521	10	Xiangyang	
HL1)	1521	1	Heilongjiang	Jingbohu
JL15)	1521	1	Taonan	
JS11)	1521	3	Changzhou	
JS12)	1521		Wuxi	
JS13)	1521		Suzhou	
JS17)	1521		Zhangjiagang	
JS5)	1521	10	Huai'an	
JS7)	1521		Yangzhou	
NM6)	1521		Ulanqab	Jining
SC1)	1521	10	Sichuan	Chengdu
SD2)	1521		Jinan	
SN1)	1521	1	Shaanxi	Shangluo
SX5)	1521	1	Changzhi	Qinxian
YN3)	1521	1	Qujing	
YN5)	1521	1	Honghe	
ZJ1)	1521		Zhejiang	
ZJ7)	1521		Huzhou	
JL1)	1530		Jilin	Yanji
SX7)	1530	1	Jinzhong	
ZJ1)	1530	50	Zhejiang	Hangzhou
1)	1539	10	CNR 1	
HN6)	1548	10	Shaoyang	
SD1)	1548	200	Shandong	Weifang
EB10)	1557	25	Cangzhou	
EB14)	1557	1	Nangong	
LN12)	1557		Fuxin	
EB6)	1566	10	Zhangjiakou	
JL10)	1566		Yanbian	Songjiang
GS8)	1566		Pingliang	
SX9)	1566	1	Yuncheng	
JL12)	1575		Lishu	
LN3)	1575	2	Dalian	
AH8)	v1584	1	Ma'anshan	
AH9)	v1584	1	Anqing	
EB7)	1584	1	Chengde	
GZ4)	1584	1	Zunyi	
JL13)	1584	1	Meihekou	
SX1)	1584		Shanxi	Taiyuan
SX5)	1584	10	Changzhi	
ZJ14)	1584		Rui'an	
1)	1593	600	CNR 1	Changzhou, JS
HL1)	1593	10	Heilongjiang	
HL12)	1593	1	Suihua	
XJ1)	1593		Xinjiang	Korla

MW	kHz	kW	Station	Tx Location
JS1)	1602	1	Jiangsu	Hongze

SW	kHz	kW	Station	Tx Loc.	Times
13)	3280	15	VO Pujiang	Shanghai	1130-1600
NM8)	3900	2	Hulun Buir	Hailar	as 603kHz
XJ1)	*3950	100	Xinjiang	Urumqi	Nov-Apr only
1)	3985	100	CNR 2	Golmud	1300-1605
GS6)	3990	15	Gannan	Hezuo	as 1332kHz
XJ1)	*3990	50	Xinjiang	Urumqi	Nov-Apr only
QH1)	4220	15	Qinghai	Xining	as 1251kHz
XJ1)	*4330	100	Xinjiang	Urumqi	Nov-Apr only
1)	4460	100	CNR 1	Beijing	1955-2330, 1100-1735
XJ1)	*4500	50	Xinjiang	Urumqi	Nov-Apr only
1)	4750		CNR 1		1955-1735
QH1)	4750	50	Qinghai	Xining	as 666kHz
XZ1)	4820	100	Xizang	Lhasa	as 999kHz
11)	*4900	50	VO Strait	Fuzhou	Winter only
XZ1)	4905	50	Xizang	Lhasa	as 594kHz
XZ1)	*4920	50	Xizang	Lhasa	
11)	4940	50	VO Strait	Fuzhou	2230-2400, 1200-1600
13)	4950	15	VO Pujiang	Shanghai	1130-1600
XJ1)	*4980	50	Xinjiang	Urumqi	Nov-Apr only
HN1)	4990	10	Hunan	Changsha	as 738kHz
FJ1)	v5005	10	Fujian	Fuzhou	2250-2320, 0925-1030
FJ1)	5040	10	Fujian	Fuzhou	2250-2325, 0925-1035
11)	*5050	50	VO Strait	Fuzhou	Winter only
15)	5050	15	China Huayi BC		1000-1300
XJ1)	*5060	50	Xinjiang	Urumqi	Nov-Apr only
13)	*5075	15	VO Pujiang	Shanghai	Winter only
XZ1)	*5240	50	Xizang	Lhasa	
12)	5860	50	VO Jinling	Nanjing	1445-1705
1)	5925	50	CNR 5	Beijing	0955-0005
XZ1)	*5935	100	Xizang	Lhasa	
1)	5945	100	CNR 1	Beijing	1955-0100, 1300-1735
1)	5955	50	CNR 8	Beijing	1200-1300, 2155-2400
XJ1)	5960	50	Xinjiang	Urumqi	2300-0300, 1200-1800
GS6)	5970	15	Gannan	Hezuo	as 1332kHz
1)	5975	100	CNR 8	Beijing	2155-2300
QH1)	5990	15	Qinghai	Xining	as 1251kHz
1)	6010	100	CNR 11	Xi'an	2155-2400, 1030-1605
XJ1)	6015	100	Xinjiang	Urumqi	2330-0300, 1200-1800
1)	6030	100	CNR 1	Beijing	1955-1735
1)	6040	150	CNR 2	Beijing	2055-2300
NM1)	6040	50	Nei Menggu	Hohhot	2150-1605
XZ1)	6050	100	Xizang	Lhasa	as 999kHz
SC1)	6060	10	Sichuan	Xichang	2155-1515
1)	6065	150	CNR 2	Beijing	2055-2330, 1200-1605
1)	6080	100	CNR 1	Golmud	1955-2300, 1200-1735
NM8)	6080	2	Hulun Buir	Hailar	as 954kHz
1)	6090	100	CNR 2	Golmud	2055-0100, 1000-1605
XZ1)	*6110	100	Xizang	Lhasa	
11)	6115	50	VO Strait	Fuzhou	2230-1600
XJ1)	6120	50	Xinjiang	Urumqi	2300-0300, 1200-1800
1)	6125	100	CNR 1	Beijing	1955-2400, 0900-1735
1)	6125	100	CNR 1	Shijiazhuang	1955-2300, 1100-1735
XZ1)	6130	100	Xizang	Lhasa	as 594kHz
1)	6140	50	CNR 8	Beijing	1600-1705
QH1)	6145	15	Qinghai	Xining	2200-0230, 0900-1605
1)	6155	150	CNR 2	Beijing	2055-0100, 1000-1605
1)	6165	50	CNR 6	Beijing	2055-0105, 0900-1805
1)	6175	100	CNR 1	Beijing	1955-2400, 0900-1735
15)	6185	15	China Huayi BC	Fuzhou	2230-1000
1)	6190	100	CNR 2	Golmud	2055-2300
XJ1)	6190	100	Xinjiang	Urumqi	2330-0330, 1210-1800
XZ1)	6200	100	Xizang	Lhasa	
XJ1)	7205	50	Xinjiang	Urumqi	2300-0200, 1400-1800
YN1)	7210	20	Yunnan	Kunming	1055-1500
1)	7215	100	CNR 1	Shijiazhuang	1955-2400
1)	7220	100	CNR2	Golmud	2300-1300
1)	7220	50	CNR 8	Beijing	1400-1500
SC1)	7225	10	Sichuan	Xichang	2155-1515
1)	7230	100	CNR 1	Xi'an	1955-1735
XJ1)	7230	50	Xinjiang	Urumqi	as 1233kHz
XZ1)	7240	100	Xizang	Lhasa	2000-0200, 0900-1800
1)	7245	150	CNR 2	Beijing	2055-2300, 1300-1605
1)	7255	100	CNR 2	Xi'an	2055-0100
XZ1)	*7255	100	Xizang	Lhasa	
XJ1)	7260	100	Xinjiang	Urumqi	as 738kHz
1)	7265	100	CNR 2	Xi'an	1230-1605
NM1)	7270	50	Nei Menggu	Hohhot	2150-1605
1)	7275	100	CNR 1	Beijing	1955-2300, 1100-1735
XJ1)	7275	50	Xinjiang	Urumqi	as 558kHz
11)	7280	50	VO Strait	Fuzhou	2230-1600
1)	7290	100	CNR 1	Beijing	1955-2400, 1100-1735
XJ1)	7295	50	Xinjiang	Urumqi	Nov.-Apr.only
1)	7305	100	CNR 1	Shijiazhuang	1955-2200, 1000-1735
XJ1)	7310	50	Xinjiang	Urumqi	2300-0300, 1400-1800
1)	7315	150	CNR 2	Xi'an	2055-0100, 1100-1605
1)	7335	100	CNR 2	Baoji-Sif.	2055-0030, 1000-1605
XJ1)	7340	100	Xinjiang	Urumqi	as 1107kHz
1)	7345	100	CNR 1	Beijing	1955-2400, 1100-1735
1)	7350	100	CNR 11	Baoji-Sif.	0900-1605
1)	7360	100	CNR 11	Baoji-Sif.	2155-2400
1)	7365	100	CNR1	Shijiazhuang	1200-1735
1)	7370	100	CNR 2	Beijing	1300-1605
1)	7375	150	CNR 2	Beijing	2055-2300, 1200-1605
XZ1)	7385	100	Xizang	Lhasa	2100-0200, 0930-1805
1)	7410	100	CNR 8	Beijing	1000-1100
NM1)	7420	50	Nei Menggu	Hohhot	2150-1605
1)	7425	150	CNR 2	Xianyang	2055-2400, 1300-1605
1)	7445	100	CNR 8	Beijing	1500-1705, 2300-2400
XZ1)	*7450	100	Xizang	Lhasa	
1)	7620	50	CNR 5	Beijing	0955-0005
1)	7935	50	CNR 8	Beijing	1600-1705
1)	9170	50	CNR 6	Beijing	2055-0105, 1100-1805
1)	9410	50	CNR 5	Beijing	0955-2200
1)	9420	100	CNR 1/8	Lingshi	1100-1705
1)	9440	50	CNR 8	Beijing	0400-0500
1)	9455	100	CNR 1/8	Lingshi	1955-0200
XJ1)	9470	15	Xinjiang	Urumqi	0300-1200
1)	9480	100	CNR 11	Baoji-Sif.	2155-0100, 0800-1605
XZ1)	*9490	100	Xizang	Lhasa	
1)	9500	100	CNR 1	Shijiazhuang	1955-1735
11)	9505	50	VO Strait	Fuzhou	0000-1200
XJ1)	9510	50	Xinjiang	Urumqi	0530-1030
1)	9515	150	CNR 2	Beijing	2055-2400, 1200-1605
NM1)	9520	50	Nei Menggu	Hohhot	2150-1605
1)	9530	100	CNR 11	Baoji-Sif.	0000-1030
XJ1)	9560	50	Xinjiang	Urumqi	0300-1200
1)	9570	100	CNR 2	Golmud	0100-1000
XZ1)	9580	100	Xizang	Lhasa	0200-0930
XJ1)	9600	50	Xinjiang	Urumqi	0300-1400
1)	9610	50	CNR 8	Beijing	0300-0500, 1200-1300
1)	9620	150	CNR 2	Beijing	2300-1300
1)	9630	100	CNR 1	Golmud	2300-1200
1)	9630	100	CNR 1/8	Lingshi	1200-1600
1)	9645	100	CNR 1/8	Beijing	2330-1100, 1300-1705
1)	9655	100	CNR 1	Lingshi	1955-2400
1)	9665	50	CNR 5	Beijing	2200-0005
1)	9675	100	CNR 1	Beijing	2300-1000
1)	9685	50	CNR 5	Beijing	0055-0615
1)	9690	50	CNR 8	Beijing	1100-1200
13)	9705	15	VO Pujiang	Shanghai	1130-1600
XJ1)	9705	50	Xinjiang	Urumqi	0330-0530, 1030-1230
1)	9710	100	CNR 1	Shijiazhuang	1955-2330, 1100-1735
1)	9720	150	CNR 2	Xi'an	0000-1000
NM1)	9750	50	Nei Menggu	Hohhot	2150-1605
1)	9755	100	CNR 2	Xi'an	2055-0200, 1000-1605
1)	9775	150	CNR 2	Beijing	2055-0200, 0900-1605
QH1)	9780	15	Qinghai	Xining	0230-0900
1)	9785	50	CNR 8	Beijing	1000-1100
1)	9810	100	CNR 1	Nanning	1955-2300, 1300-1735
1)	9810	100	CNR 2	Xi'an	0100-1230
1)	9820	150	CNR 2	Xi'an	2055-0100, 1100-1605
1)	9830	100	CNR 1	Beijing	1955-0100, 0730-1735
XJ1)	9835	50	Xinjiang	Urumqi	0300-1200
1)	9845	100	CNR 1	Beijing	1955-0100, 1030-1735
QH1)	9850	15	Qinghai	Xining	0230-0830
1)	9860	100	CNR 1	Beijing	1200-1735
1)	9890	100	CNR 1/8	Lingshi	1955-2400, 1400-1705
1)	9900	100	CNR 1	Beijing	1955-2300
1)	11610	150	CNR 2	Beijing	2300-1300
1)	11620	50	CNR 5	Beijing	0055-0615
1)	11630	100	CNR 1/8	Lingshi	1955-2400, 0200-1600
1)	11630	100	CNR 8	Beijing	0000-0200
1)	11660	150	CNR 2	Xi'an	0100-1100
1)	11670	150	CNR 2	Beijing	2330-1200
1)	11685	100	CNR 11	Baoji-Sif.	0000-0900
1)	11710	100	CNR 1	Beijing	1955-0030, 1000-1735
1)	11720	100	CNR 1/8	Shijiazhuang	2330-1200
1)	11740	50	CNR 2	Xi'an	2055-0100, 1100-1605
1)	11750	100	CNR 1	Shijiazhuang	2200-1000
1)	11760	100	CNR 1	Shijiazhuang	0000-1200
XJ1)	11770	50	Xinjiang	Urumqi	as 738kHz
1)	11780	100	CNR 8	Beijing	0500-1000
1)	11800	150	CNR 2	Beijing	2300-1200
1)	11810	100	CNR 8	Lingshi	0000-0200
1)	11810	100	CNR 8	Beijing	0200-0300
1)	11815	50	CNR 8	Beijing	0300-0400, 0700-0800

SW	kHz	kW	Station	Tx Loc.	Times
1)	11835	150	CNR 2	Xi'an	0000-1300
1)	11845	150	CNR 2	Xi'an	0000-1100
XZ1)*	11860	100	Xizang	Lhasa	
XJ1)	11885	50	Xinjiang	Urumqi	as 558kHz
1)	11905	50	CNR 6	Beijing	0355-0900
1)	11915	100	CNR 2	Xi'an	0030-1000
1)	11925	100	CNR 1	Lingshi	1955-2330 0900-1735
1)	11935	50	CNR 5	Beijing	0055-0615
XZ1)	11950	100	Xizang	Lhasa	0200-0900
1)	11960	100	CNR 1	Beijing	0000-0900
XJ1)	11975	15	Xinjiang	Urumqi	0330-0530, 1030-1230
1)	12045	100	CNR 1	Beijing	2300-1200
1)	12055	100	CNR 1/8	Lingshi	0200-1200
1)	12080	100	CNR 2	Xi'an	0200-1000
1)	13610	100	CNR 1	Nanning	2300-1300
XJ1)	13670	50	Xinjiang	Urumqi	0200-1400
1)	13700	100	CNR 1/8	Lingshi	2300-1400
1)	15270	150	CNR 2	Beijing	0200-0900
1)	15370	100	CNR 1	Shijiazhuang	0100-1100
1)	15380	100	CNR 1	Beijing	2300-1100
1)	15390	100	CNR 1/8	Lingshi	0100-1100
1)	15415	100	CNR 8	Beijing	0500-1000
1)	15480	100	CNR 1	Beijing	0100-1300
1)	15500	100	CNR 2	Xi'an	0100-1100
1)	15540	50	CNR 2	Xi'an	0100-1100
1)	15550	100	CNR 1	Beijing	0000-1100
1)	15570	100	CNR 11	Baoji-Sif.	0100-0800
1)	15710	50	CNR 6	Beijing	0355-1100
1)	17550	100	CNR 1	Beijing	0100-1030
1)	17565	100	CNR 1	Beijing	0100-0730
1)	17580	100	CNR 1	Lingshi	2330-0900
1)	17595	100	CNR 1	Shijiazhuang	0100-1100
1)	17605	100	CNR 1	Beijing	0030-1000
1)	17625	150	CNR 1	Beijing	0000-1200
1)	17725	100	CNR 1	Shijiazhuang	0000-1200
1)	17890	100	CNR 1	Beijing	0000-1000

NB: Baoji-Sif. = Baoji-Sifangshan *) inactive.

FM(MHz)	CNR 1	CNR 2	CNR 3	Prov.E	Prov.M	City.E	City.M
Beijing	106.1	96.6	90.0	-	-	107.3	97.4
Changchun	99.1	104.7	91.6	-	92.7	90.0	99.6
Changsha	95.0	105.9	107.7	90.1	97.5	-	106.1
Chengdu	103.7	-	107.6	94.0	95.5	-	105.6
Chongqing	92.0	102.9	88.9	-	-	101.5	88.1
Dalian	107.8	-	107.8	87.6	-	99.1	89.1
Daqing	103.9	96.7	-				
Fuzhou	93.5	-	92.6	96.1	91.3	87.6	89.3
					99.6a		
Guangzhou	89.3	106.6	87.4	95.3	99.3	-	102.7
				97.4b			
Guiyang	93.6	-	107.3	98.9	91.6	104.0	102.7
Haikou	105.8	87.8	105.8	103.8	-	-	91.6
Hangzhou	90.2	97.9	103.2	95.0	96.8	91.8	105.4
Harbin	89.9	88.1	100.9	-	95.8	92.5	98.4
Hefei	93.5	104.7	94.3	97.1	89.5	-	87.6
Hengyang	95.0	105.9	-	91.0	96.9	98.9	-
Hohhot	97.1	104.5	99.1	101.4	93.6	-	-
Jinan	95.7	-	89.8	96.0	99.1	90.9	88.7
Jinzhou	104.9	101.2	106.0				
Kunming	96.0	100.0	94.0	88.7	97.0	-	-
Lanzhou	94.8	90.3	88.3	93.4	-	-	99.5
Lhasa	89.2	104.0	96.1	98.0	-	-	-
Nanchang	89.1	96.9	87.2	-	-	-	95.1
Nanjing	95.8	107.5	98.9	93.7	89.7	-	105.8
Nanning	106.2	104.0	103.6	97.0	95.0	-	107.4
Ningbo	95.7	92.8	107.7	103.3	-	102.9	93.9
Qingdao	93.1	89.7	98.0	-	106.0	102.9	106.6
Quanzhou	96.9	98.3	102.5	95.5	-	92.3	-
Shanghai	99.0	91.4	107.7	-	-	97.7	96.8
Shenyang	94.8	93.5	99.8	89.5	102.9	90.4	87.6
Shenzhen	95.8	-	101.2	105.7	-	-	97.1
Shijiazhuang	95.6	97.2	105.1	97.2	90.7	100.9	106.7
Suzhou	100.0	-	107.7				
Taiyuan	97.0	99.0	89.3	93.8	101.5	104.4	-
Tangshan	93.2	107.4	-	90.8	94.8	95.5	102.0
Tianjin	102.9	98.0	92.5	-	-	101.4	99.0
Urumqi	88.7	90.6	-	92.9	94.9	97.4c	100.7
Weifang	96.7	-	-	97.9	101.1	93.3	88.7
Wenzhou	90.2	-	94.9	-	-	88.8	100.3
Wuhan	95.6	97.8	90.7	99.8	103.8	100.6	101.8
					105.8d		
Xiamen	102.6	93.5	105.2	98.6	-	107.0	90.9
Xi'an	96.4	103.0	95.5	89.6	98.8	-	93.1
Xining	91.6	105.6	100.6	-	97.2	-	104.3
Yinchuan	96.4	107.8	99.7	92.8	-	95.0	100.6

FM(MHz)	CNR 1	CNR 2	CNR 3	Prov.E	Prov.M	City.E	City.M
Zhangjiakou	88.9	-	-	94.8		98.6	100.0
Zhengzhou	102.6	98.3	-	98.6	-	-	99.1
Zhengzhou	101.2	96.7	100.2	103.2	88.1	92.6	91.8
Zhuhai	99.1	-	101.2	103.8	93.9	-	87.5

Prov.E=Provincial economic stn **Prov.M**=Provincial music stn **City. E**=City economic stn **City.M**=City music stn a) V.O. the Strait, b) Zhujiang EBS, c) Traffic music ch, d) Chutian BS

Addresses and other information:
1) CHINA NATIONAL RADIO (CNR)
✉ 2 Fuxingmenwai Dajie, Xicheng Qu, Beijing 100866 ☎ +86 10 8609 2636 **W:** www.cnr.cn
LP: Gen. Dir: Wang Qiu. CE: Qian Yuelin.
V.O. China (1st Prgr "News Radio"): 24h (exc. Tues 0600-0850) on MW/SW/FM. **V.O. the Economy (2nd Prgr "China Business Radio"):** 2055-1605 (exc. Wed 0600-0900) on MW/SW/FM. **V.O. the Music (3rd Prgr "Music Radio"):** 2200-1600 (exc. Tues 0600-0855) on FM. **V.O. the City (4th Prgr "Fresh FM"):** 2155-1600 (exc. Tues 0500-0855) on 101.8MHz **V.O. Zhonghua (5th Prgr "Zhonghua News Radio"):** 0055-0615, 0955-0005 on MW/SW/102.3MHz(Fuzhou). **V.O. Shenzhou (6th Prgr "Shenzhou Easy Radio")** in Chinese, Amoy and Hakka: 2055-0105, 0355-1805 on MW/SW/106.2MHz(Fuzhou). **V.O. Huaxia (7th Prgr "Huaxia Radio")** for the Zhujiang Delta: Chinese Ch. on 87.8/92.3MHz 2055-1805 (exc. Tues 0600-0855), Bilingual Ch. on 1215kHz/104.9MHz 2055-1805 (exc. Tues 0600-0855) in Chinese and Cantonese. **V.O. the Literary (9th Prgr "Story Radio"):** 2155-1805 on 106.6MHz **V.O. Old Age (10th Prgr.):** 1955-1735 (exc. Tues 0605-0855) on 1053kHz. **V.O. the Entertainment (12th Prgr):** 2155-1805 (exc. Tues 0505-0855) on 747kHz.

V.O. Minorities (8th Prgr "Ethnic Minority Radio"): 2155-1705 on MW/SW/104.5MHz(Hohhot)/105.7MHz(Lhasa) – +) relayed by regional stns

Kazakh			
0200-0300		15670, 15390, 13700, 12055, 11810, 11630, 1422, 1143	
0500-0600+	XJ	15415, 15390, 13700, 12055, 11780, 11630, 1422, 1143	
1100-1200		15415, 15390, 13700, 12055, 11780, 11630, 1422, 1143	
1300-1400		13700, 11630, 9645, 9630, 9420, 7445, 1422, 1143	
1400-1500+	XJ	11630, 9890, 9645, 9630, 9420, 7220, 1143	
Korean			
2155-2300		5975, 5955, 1143	
0400-0500		9610, 9440, 1143	
1000-1100+	JL/HL	9785, 7410, 1143, 1017	
Mongolian			
2300-2400		7445, 5955, 1143	
0300-0400+	NM	11815, 9610, 1143	
1200-1300	NM	9610, 5955, 1143	
Uighur			
0000-0200		15670, 15390, 13700, 11810, 11630, 9455, 1422, 1143	
0600-0700+	XJ	15415, 15390, 13700, 12055, 11780, 11630, 1422, 1143	
0700-0800		15415, 15390, 13700, 12055, 11780, 11630, 1422, 1143	
0800-1000		15415, 15390, 13700, 12055, 11780, 11630, 1422, 1143	
1500-1600+	XJ	11630, 9890, 9645, 9630, 9420, 7445, 1143	
1600-1705		9890, 9420, 7935, 7445, 6140, 1143	

11th Prgr. Tibetan Service 2155-1605 on MW/SW/105.7MHz (Lhasa)

Tibetan	
2155-0000	9480, 7360, 6010, 1098
0000-0100	11685, 9530, 9480, 1098
0100-0800	15570, 11685, 9530, 1098
0800-1300	11685, 9530, 9480, 1098
1300-1605	9480, 7350, 6010, 1098

2) CHINA RADIO INTERNATIONAL (CRI)
(Zhongguo Guoji Guangbo Diantai)
✉ Jia 16, Shijingshan Lu, Shijingshan Qu, Beijing 100040 ☎ +86 10 6889 1001 **W:** www.chinabroadcast.cn **LP:** Gen. Dir: Wang Gengnian
Domestic Sce:
Beijing 1 "**Easy FM**" (1251kHz/91.5MHz): 24h in English – **Beijing 2** "**Hit FM**" (88.7MHz): 24h in English – **Beijing 3** (846/1008kHz): 24h in English – **Beijing 4** "**News Radio**" (900kHz/90.5MHz): 2200-1600 in Chinese, 1600-2200 in English – **Tianjin 1** "**Hit FM**" (105.4MHz): 24h in English – **Tianjin 2** "**News Radio**" (90.6MHz): 2200-1700. –**Shanghai 1** "**Easy FM**" (87.9MHz): 0000-0300. – **Shanghai 2** "**Hit FM**" (87.9MHz): 0000-2400, 0300-1730. – **Hefei 1** "**Easy FM**" (92.4MHz): 2200-1700. – **Hefei 2** "**News Radio**" (90.1MHz): 2200-1700. – **Wuhu** "**News Radio**" (89.8MHz): 2200-1700. – **Xiamen** "**Easy FM**" (95.8MHz): 2200-1700. – **Guangzhou 1** "**Hit FM**"

(88.5MHz): 24h – **Guangzhou 2** "News Radio" (702kHz/107.1MHz): 24h – **Chongqing** "Hit FM" (91.2MHz): 24h – **Lhasa** "Easy FM" (105.7MHz): 1930-1600. – **Lanzhou** "Easy FM" (98.5MHz): 2200-1600.

DAB: Radio, Film & TV Bureau of Guangdong on 209.936MHz (test transmission).

EXTERNAL SERVICES: China Radio International, Voice of Beibu Bay Radio, Yunnan Broadcasting Station
See International Broadcasting section

BROADCASTS TO TAIWAN

11) Voice of the Strait (Haixia zhi Sheng), Xindian, Fuzhou or P.O.Box 187, Fuzhou, Fujian 350012. Operated by the People's Liberation Army of China. **W:** www.vos.com.cn News Sce. on 666/1269/4940/9505kHz 2230-1600 (exc. Wed 0400-0955) in Ch. English Prgr. "Focus on China": Sun1500-1525. - Automobile Life Sce. on 5050/7280kHz/90.6MHz 24h (exc. Wed 0400-0955) in Ch. - Dialect Sce. on 783/4900/6115kHz 2230-1600 (exc. Wed 0400-0953) in Amoy. - Bus Sce. on 99.6MHz 2230-1600 (exc. Wed 0400-0953). – **12)** Voice of Jinling (Jinling zhi Sheng), P.O.Box 268, Nanjing, Jiangsu 210002. On 1206/5860kHz 1445-1705. – **13)** Voice of Pujiang (Pujiang zhi Sheng), 1376 Hongqiao Lu, Shanghai or P.O.Box 518, Shanghai 200051. On 603/1422/3280/4950/5075/97 05kHz 1130-1600. – **14)** Southeast Broadcasting Company, 2 Gutian Lu, Fuzhou, Fujian 350001 **W:** www.sebc.com.cn On 585kHz/97.6/ 106.2MHz 0955-1600 in Ch and Amoy. – **15)** China Huayi Broadcasting Corporation, P.O.Box 251, Fuzhou, Fujian 350001 **W:** www.chbcnet.com On 873/1377/4830/6185kHz/107.1MHz for Taiwan, Hong Kong, Macao and Southeast Asia. 24h (exc. Wed 2125-1000).

ANHUI PROVINCE

AH1) 355 Tongcheng Nanlu, Hefei, Anhui 230065 **W:** www.ahradio. com.cn News General Sce. on 936/846kHz/95.5MHz 2000-1700(Tues 1500). - Economic Sce. on 864/1098kHz/97.1MHz 24h (exc. Tues 1500-2000). - Travel Sce. on 900kHz/96.1/106.5MHz 2130-1700(Tues 1500). - Traffic Sce. on 90.8MHz 24h (exc. Tues 1500-2150). - Life Sce. on 603/684kHz/105.5MHz 2100-1800 (exc. Mon 1500-1800). - Farm Sce. on 720/1008kHz/FM on 2100-1700 (exc. Tues 1500-1950). - Music Sce. on 801kHz/89.5MHz 24h. - Novel and Storytelling Sce. on 1395kHz/107.4MHz 2030-1800 (exc. Tues 1500-1800). - Chinese Opera Sce. on 99.5MHz 24h (exc. Tues 1500-2000). – **AH2)** 114 Rongshida Dadao, Hefei, Anhui 230001. News General Sce. on 666kHz/91.5MHz 2120-1600. - Traffic St. on 1053kHz/102.6MHz 24h (exc. Tues 0600-0850). - Literary St. on 747kHz/87.6MHz 2200-1700. - Story Sce. on 1170kHz/98.8MHz 2100-1700. - Charm Music Sce. on 88.6MHz – **AH3)** 11 Dongshan Zhonglu, Huainan, Anhui 232001. News General Sce. on 648kHz/103.7MHz 2125-1500. - Traffic and Literary Sce. on 97.9MHz 2140-1500. – **AH4)** 336 Huaihai Donglu, Xiangshan Qu, Huaibei, Anhui 235000. News General Sce. on 1431kHz/94.9MHz 2115-1500. - Traffic and Music Sce. on 100.4MHz. - Economic Life Sce. on 603kHz/89.3MHz 2110-1600. – **AH4A)** Huaibei Experimental BS, Renmin Lu, Xiangshan Qu, Huaibei, Anhui 235000. – **AH5)** 197 Beijing Donglu, Wuhu, Anhui 241000. News Sce. on 100.4MHz 2130-1600. - Life Sce. on 1494kHz. - Traffic and Economic Sce. on 96.3MHz 2130-1630. - Music and Story Sce. on 98.2MHz 2200-1600. – **AH6)** Yi'an Beilu, Tongling, Anhui 244000. On 1107kHz/100.0MHz 2120-0005, 0155-0510, 0855-1305. - Traffic and Life Sce. on 88.7MHz 2255-1100. – **AH7)** Xuehua Shan, Shengli Donglu, Bengbu, Anhui 233000. News Sce. on 765kHz/107.9MHz 2200-1400. - Economic Sce. on 1098kHz/104.2MHz 2150-1430. – **AH8)** 46 Yushan Zhonglu, Ma'anshan, Anhui 243011. News General Sce. on 1584kHz/ 105.1MHz 2130-1500 (exc. Tues 0600-0850). - Traffic and Music Ch. on 92.8MHz 2130-1500 (exc. Tues 0600-0850). – **AH9)** 23 Guanyue Miao, Anqing, Anhui 246004. News General Sce. on 1584kHz/90.3MHz 2200-0530, 0930-1400. – **AH10)** 9 Tiandu Dadao, Tunxi Qu, Huangshan, Anhui 245000. On 1431kHz/87.5/93.3MHz 2200-0600, 0920-1500. - Traffic and Travel Sce. on 100.4MHz 2200-1500. – **AH11)** 225 Langxie Lu, Chuzhou, Anhui 239000. News General Sce. on 95.0MHz 2125-1440 (exc. Tues 0500-0930). - Traffic and Music Sce. on 105.4MHz 2125-1530. - Literary and Story Sce. on 1377kHz/97.0MHz 2100-1500. – **AH12)** Nan 2 Huan Lu, Fuyang, Anhui 236034. News Sce. on 1116kHz/91.6MHz 2130-1610. - Economic Sce. on 711/801/1503kHz/94.1MHz 2145-1550 (exc. Tues 0700-0830). - Traffic Sce. on 90.0/103.5MHz 2130-1530. – **AH13)** Baihuiyuan, Huaihai Lu, Suzhou, Anhui 234000. On 1071kHz/105.0MHz 1015-?. - Traffic and Music St. on 96.1MHz – **AH14)** 436 Dongfeng Xilu, Juchao Qu, Chaohu, Anhui 238000. News General Sce. on 846kHz/ 88.1MHz 2125-1500. - City Traffic Sce. on 93.8MHz 2200-1500. – **AH15)** Qiupu Donglu, Guichi Qu, Chizhou, Anhui 247100. 2120-2325, 0315-0515, 0950-1310. – **AH16)** Meishan Nanlu, Lu'an, Anhui 237001. News General Sce. on 711/1242kHz/102.1MHz 2155-1505. - Traffic and Music Sce. on 1170kHz/96.4MHz 2155-1505. – **AH17)** 10 Zhuangyuan Lu,

Xuancheng, Anhui 242000. News General Sce. on 1170kHz/100.6MHz 2155-1500. - Traffic and Literary Sce. on 106.1MHz – **AH18)** Dangtu Xian, Anhui 243100. 0250-0450. – **AH19)** Renmin Zhonglu, Bozhou, Anhui 236800. - News General Sce. on 999kHz/88.2MHz - Traffic and Music Sce. on 107.2MHz

BEIJING MUNICIPALITY

BJ1) 14 Jianguomenwai Dajie, Chaoyang Qu, Beijing 100022 **W:** www. bjradio.com.cn News Sce. on 828kHz/100.6MHz 24h (exc. Tues 1630-2100 Thurs 0700-0800). – Public Service Sce. on 1026kHz/107.3MHz 2100-1700. - Sports Sce. on 102.5MHz 24h. – Communication Sce. on 103.9MHz 24h. - Story Sce. on 603kHz/89.1MHz 2100-1730(Mon 1630). - Foreign Language Sce. "Radio 774" on 774kHz 24h (exc. Mon 1600-2200, Thurs 0700-0800). - Literary Sce. on 87.6MHz 2130-1730. - Music Sce. on 97.4MHz 24h (exc. Mon 1600-2100). - "i Home Radio" on 927kHz 2130-1600 (exc. Thurs 0700-0800).

CHONGQING MUNICIPALITY

CQ1) 159 Zhongshan 3 Lu, Yuzhong Qu, Chongqing 400015 **W:** fm968. cbg.cnNews Sce. on 1314kHz/96.8MHz 24h. - Economic Sce. on 101.5/107.7MHz 24h. - Traffic Sce. on 95.5/88.9/92.7MHz 24h. - City Sce. on 93.8MHz 2130-1800. - Music Sce. on 88.1MHz 24h. – **CQ2)** V.O. Jialing, 6 Nanjing Lu, Beibei Qu, Chongqing 400700. On 97.4/102.1MHz 2255-1600.

HEBEI PROVINCE

EB1) 63 Yuhua Donglu, Shijiazhuang, Hebei 050012. **W:** www.hebradio. com News Sce. on 1278/783kHz/FM 2030-1800 (exc. Tues 0600-0900). - Economic Sce. on 1125/1251kHz/FM 2130-1800 (exc. Tues 0530-0900). - Life Sce. on 747/783kHz/FM 24h. - Traffic Sce. on 99.2MHz 2130-1700(exc. Tues 0600-0900). - Literary Sce. on 900kHz/FM 2130-1700. - Music Sce. "i Radio": on FM 2130-1700. - Farm Sce. on 558kHz/98.1MHz 2130-1800. - Travel Culture Sce. on 603/1521kHz/ 88.1MHz 2200-1600. - Children Music Sce. "M Radio" on 89.5MHz – **EB2)** 302 Tiyu Nan Dajie, Shijiazhuang, Hebei 050021. News Sce. on 882kHz/88.2MHz 2125-1700. - Economic Sce. on 100.9MHz 2125-1600 (exc. Tues 0600-0825). - Music Sce. "NuStar Radio" on 106.7MHz 2155-1600. - Storytelling and Entertainment Sce. on 1431kHz 2125-1600. - Farm Sce. on 1251kHz/96.1MHz 2100-1600. - Traffic Sce. on 94.6MHz 2130-1700. – **EB3)** 246 Renmin Lu, Handan, Hebei 056002. News General Sce. on 963kHz/96.4MHz 2100-1600. - Economic Life Sce. on 1206kHz/102.8MHz 2100-1600. - Traffic Sce. on 1008kHz/106.8MHz 2100-1600. – Chinese Opera and Storytelling Sce. on 846kHz/104.8MHz 2100-1600. – **EB4)** 15 Yejin Lu, Xingtai, Hebei 054000. News Sce. on 1188kHz/90.3MHz 2125-1600 (exc. Tues 0530-0930). – Economic Life Sce. on 927kHz/102.0MHz 2120-1500 (exc. Tues 0530-0930). - Traffic and Music Sce. on 91.8/101.2MHz 2225-1500 (exc. Tues 0530-0930). – **EB5)** 1620 Yangguang Bei Dajie, Baoding, Hebei 071051. News Sce. on 1467kHz/90.9/93.7MHz 24h (exc. Tues 0600-0855). - Economic Sce. on 1017kHz/99.7MHz 2145-1600 (exc. Tues 0600-0930). - Traffic Sce. on 747kHz/104.8MHz 1850-1600 (exc. Tues 0600-0900). - City Service Sce. on 101.6MHz 2200-1400. - City and Country Alliance Sce. on 101.3/103.2/105.6MHz - Traffic and Music Ch. on 105.8MHz 2200-1630. – **EB6)** 17 Jianguo Lu, Qiaodong Qu, Zhangjiakou, Hebei 075000. News General Sce. on 1566kHz/101.0,107.4MHz 2155-1505. - Literary Sce. on 900kHz/94.5MHz 2155-1500. - V.O. the Earth: on 603kHz/103.6MHz 2155-1500 (exc. Tues 0500-0900). – **EB7)** 120 Guangdian Lu, Shuangqiao Qu, Chengde, Hebei 067000. News General Ch. on 1584kHz/93.8MHz 2155-0540, 0950-1400. - Traffic Sce. on 900kHz/97.6MHz 2255-1500. Rel. CRI English prgr. 2300-0500, 1300-1500. - Storytelling Ch. on 105.8MHz – **EB8)** 1 Guangda Jie, Wenhua Lu, Tangshan, Hebei 063000. News General Sce. on 684kHz/91.7MHz 2030-1605 (exc. Tues 0705-0855). - Economic Sce. on 801kHz/95.5MHz 2130-1530 (exc. Tues 0700-0830). - Traffic and Literary Sce. on 1143kHz/96.8MHz 2135-1505. - Music Sce. on 94.0MHz 2200-1600. – "V.O. Cao Jidian" Novel Sce. on 900kHz 2200-1600 (exc. Tues 0800-0900). - Cultural and Entertainment Sce. on 105.9MHz – **EB9)** 9 Yingbin Lu, Haigang Qu, Qinhuangdao, Hebei 066000. On 990kHz/89.1MHz 2055-1600. - V.O. Qinhuangdao: on 900kHz/103.8MHz 2055-1600. - Communication Sce. on 99.5/100.4MHz 2155-1600. - Sports and Music Sce. on 97.3MHz 2200-?. - Story Sce. on 89.9MHz – **EB10)** 12 Jiefang Xilu, Cangzhou, Hebei 061001. News General Sce. on 1557kHz/97.0MHz 2057-1500. - Agricultural Economic Sce. on 1053kHz - Traffic and Music Sce. on 1206kHz/93.8MHz 2200-1500. - Literary Sce. on 846kHz/103.6MHz 2200-1500. - Storytelling Sce. on 801kHz 2200-1500. – **EB11)** 8 Yongfeng Dao, Langfang, Hebei 065000. News General Ch. on 1008/846kHz/95.1MHz 2055-1700. - Storytelling Ch. on 585kHz/100.3MHz 24h. - Chinese Opera Ch. on 1521kHz/104.9MHz 2055-1700. – **EB12)** 49 Hongqi Dajie, Hengshui, Hebei 053000. On 954kHz 2225-0535, 0825-1630. - Traffic and Information St. on 603kHz. - Literary St. on 747kHz/87.7MHz – **EB13)** 167, Bei Duan, Xinghua Lu, Xinji, Hebei 052360. 2225-2355, 0255-0500, 1025-1250. – **EB14)** Xitou, Shengli Dajie, Nangong, Hebei 055750. 2225-0045, 1005-1400. – **EB15)** 36 Yingxin Dajie, Shahe, Hebei 054100. 2200-1600. – **EB16)** Sanyang

Dongjie, Qinghe Xian, Hebei 054800. 2210-0330, 0910-1230. – **EB17)** Beiguan, Zhuozhou, Hebei 072750. – **EB18)** Zhongshan Xilu, Dingzhou, Hebei 073000. 2200-1500. – **EB19)** 393 Xiguan Xijie, Botou, Hebei 062150. 2225-2355, 0345-0450, 1025-1235. – **EB20)** 12-1 Xihuan Lu, Renqiu, Hebei 062550. General Ch. on 1287kHz/92.8MHz 2225-1600, Storytelling Ch. on 864kHz 2255-1400.

HENAN PROVINCE

EN1) 2 Jing 5 Lu, 18 Zhenghua Lu, Zhengzhou, Henan 450003. **W:** www.radiohenan.com News Sce. on 657kHz/FM. 24h. - Economic Sce. on 972/846kHz/103.2MHz 24h. - Traffic Sce. on 900kHz/104.1MHz 24h. - Farm Sce. on 846/107.4MHz 24h. - Travel Sce. on 1332kHz/99.9MHz 24h. - Music Sce. "Meili (charm) 881": on 88.1MHz/FM 24h. - Visual Sce. "My Radio": on 1521kHz/90.0MHz 24h. - Chinese Opera Sce. on 1143kHz/97.6MHz 24h. – **EN1A)** 2 Wei 1 Lu, Zhengzhou, Henan 450003. **W:** www.hnir.com Information Sce. on 603kHz/96.7MHz 24h. - Infromation Sce. FM Prgr. on 585kHz/96.2/103.4/105.6MHz 24h. - "Binfen 1098": on 1098kHz. 24h. – **EN2)** 67 Huaihe Lu, Zhengzhou, Henan 450052. News General Sce. on 549kHz/88.9/98.6MHz 24h (exc. Tues 0600-1000, Thurs 1600-2200). - Economic Sce. on 711kHz/93.1MHz 24h. - City Sce. "Automobile FM": on 91.2MHz 24h. - Music Sce. "Simul Radio": on 94.4MHz 24h. - Literary Sce. on 1008kHz/91.8MHz 24h. – **EN3)** 78 Songcheng Lu, Kaifeng, Henan 475004. General Sce. on 873kHz/98.6MHz 24h. - News Sce. on 101.4MHz 2200-1530. - Economic Sce. on 100.2MHz 2155-1530. - Traffic Sce. on 105.1MHz. - New Farm Sce. on 1053kHz. – **EN4)** 67, Jiudu Lu, Luoyang, Henan 471009. News Sce. "V.O. Heluo": on 576kHz/88.1MHz 2150-1600. - Economic Sce. on 1053kHz/106.5MHz 2155-1600. - Traffic Sce. on 92.7MHz 2155-1600. – **EN5)** Zhong Duan, Jianshe Lu, Pingdingshan, Henan 467000. News Ch. on 747kHz/98.9MHz 2055-1500. – Economic Ch. on 1143kHz 2155-1600. –Literary Ch. on 846kHz 2200-1600, on 99.6MHz 24h. - Traffic Sce. on 1521kHz/96.4MHz 24h. – **EN6)** 217 Jiefang Zhonglu, Jiaozuo, Henan 454002. News General Sce. on 828kHz/103.0MHz 2200-1700. - Traffic and Travel Sce. on 99.5MHz - Life and Literary Sce. on 1251kHz/89.4MHz 2200-1700 (exc. Tues 0600-0955). – **EN7)** Zhong Duan, Huashan Lu, Hebi, Henan 458030. On 1107kHz/100.3MHz 2155-0535, 0955-1430. - Economic Ch. on 846kHz 2155-0535, 0955-1330. – **EN8)** 73 Renmin Lu, Xinxiang, Henan 453003. News General Sce. on 801kHz/92.9MHz 2125-1600 (exc. Tues 0530-0955). - Traffic Sce. on 1521kHz/99.1MHz 2155-1500 (exc. Tues 0500-0955). – **EN9)** Zhong Duan, Wenfeng Dadao, Anyang, Henan 455000. News General Sce. on 882kHz/94.2MHz 2155-1530. - Traffic Sce. on 1251kHz/104.3MHz 2200-1400. - Life Sce. on 89.0MHz 2155-1600. – **EN10)** 379 Zhongyuan Lu, Puyang, Henan 457000. News General Sce. on 1251kHz/100.1MHz 2130-1535 (exc. Tues 0600-0900). - Economic Life Sce. on 91.0MHz 2130-1530. - Traffic and Music Sce. on 93.7MHz 2100-1600. – **EN11)** 72 Balong Lu, Xiao Nanhai, Xuchang, Henan 461000. News Sce. on 1287kHz/93.8MHz 2120-1500. - Economic and Literary Sce. on 927kHz 2200-0530, 1000-1345. - Xuchang Literary and Information BS: on 92.6MHz – **EN12)** 152 Daxue Lu, Luohe, Henan 462000. News Sce. on 1251kHz/89.0MHz 2050-1620. - Traffic and Music Sce. on 106.7MHz 2155-1600. - City Sce. on 98.1MHz 2155-1500. – **EN12A)** 1 Wenhua Lu, Luohe, Henan 462000. – **EN13)** Zhong Duan, Jianshe Lu, Sanmenxia, Henan 472000. News General Ch. on 603kHz/90.8/98.9MHz 2155-1605 (exc. Tues 0530-0955). - Literary and Traffic Ch. on 1008kHz/104.0MHz 2255-1500 (exc. Tues 0530-1000). - Story Ch. on 100.0MHz 24h. – **EN14)** Zhong Duan, Funiu Lu, Nanyang, Henan 473056. News Ch. on 104.2MHz 2130-1605. - General Ch. on 585kHz/93.6MHz 2130-1605. - Literary and Life Ch. on 927kHz/106.0MHz 2130-1605. - Traffic and Music Ch. on 97.7MHz 2155-1605. – **EN15)** 19 Dongfanghong Dadao, Xinyang, Henan 464000. News Sce. on 837kHz/88.8MHz 2155-1600. - Traffic and Music Sce. on 94.8MHz - Literary Sce. on 106.8MHz – **EN16)** 35 Xinjian Nanlu, Shangqiu, Henan 476000. News General Ch. on 729kHz/89.0MHz 2100-1500. - City Ch. on 927kHz/100.7MHz 2155-1505. - Traffic Ch. on 94.5MHz 2200-1600. – **EN17)** 10, Dong Duan, Jianshe Lu, Zhoukou, Henan 466000. News Sce. on 828kHz 2050-1515. - Economic Sce. on 567kHz 2100-1500. - Traffic Sce. on 89.3MHz - Music Sce. on 96.0MHz – **EN18)** 209 Wenhua Lu, Zhumadian, Henan 463000. On 810kHz/97.2MHz 2125-?. - Traffic St. on 102.4MHz - Zhumadian EBS: on 1053kHz. – **EN19)** Qingping Lu, Xinmi, Henan 452370. – **EN20)** Lianmeng Xiaoqu, Chengguan Zhen, Qinyang, Henan 454550. – **EN21)** 25 Xi Dajie, Huixian, Henan 453600. – **EN22)** 10 Qianqiu Lu, Yima, Henan 472300. – **EN23)** 30 Guangyu Lu, Ruzhou 467500. – **EN24)** Dufu Lu. Gongyi, Henan 451200. On 765kHz/98.2MHz 2155-1600. - Sunshine Ch. on 107.5MHz 2155-1530.

FUJIAN PROVINCE

FJ1) 2 Gutian Lu, Fuzhou, Fujian 350001. **W:** www.fjgb.com News General Sce. on 558/612/882kHz/SW/FM. 24h (exc. Tues 0630-0855) in Ch and Amoy. - City Life Sce. on 98.7/101.5MHz 24h. - Traffic Sce. on 100.7MHz 24h (exc. Tues 0600-0850). - Music Sce. on 91.3MHz 24h (exc. Tues 0600-0900) - Fujian EBS "Interactive FM": on 1404kHz/FM. 24h (exc. Tues 0600-0850) in Ch and Amoy. – **FJ2)** Fuzhou Radio and TV, 1 Yuanyang Lu, Fuzhou, Fujian 350004. News Ch. on 1332kHz/94.4MHz 24h (exc. Wed 0605-0925) in Ch and Fuzhou dialect. - Music Ch. on 89.3MHz 24h (exc. Thurs 0600-0900). - Business and Traffic Ch. on 87.6MHz 24h. – **FJ3)** 123 Hubin Beilu, Xiamen, Fujian 361012. News Sce. on 1107kHz/99.6MHz 2130-1700 in Ch and Amoy. - Economic and Traffic Sce. on 107.0MHz 2200-1700 (exc. Tues 0600-0900). - V.O. Minnan: on 801kHz/101.2MHz 2200-? in Amoy. - Music Sce. on 90.9MHz 24h. – **FJ4)** Putian PBS, 416 Puyang Qu, Chengxiang Qu, Putian, Fujian 351100. General St. on 93.7MHz 2155-1800 in Ch and Puxian dialect. - Music St. on 103.0MHz – **FJ5)** 1 Guangdian Lu, Quanzhou, Fujian 362000. News Sce. on 576kHz/88.9MHz 2200-1600. - Qise (Colorful) FM: on 92.3MHz 24h. - V.O. the Traffic: on 90.4MHz 24h (exc. Tues 0500-0900). - V.O. Citong: on 105.9MHz 24h in Quanzhou dialect. – **FJ6)** 62 Heping Lu, Longyan, Fujian 364000. 2158-1620 (exc. Tues 0600-1000). – **FJ7)** Zhangzhou PBS, Shengli Donglu, Zhangzhou, Fujian 363000. News General Sce. on 89.6/96.2MHz 2153-1700. - Literary Sce. "Traffic FM": on 96.6/92.7MHz - Music Sce. on 99.1MHz – **FJ8)** Sanming PBS, 32 Zhuang, Liedong Shuangyuan Xincun, Sanming, Fujian 365000. News General Ch. on 87.6/103.4MHz - City Life Ch. on 97.5MHz

GUANGDONG PROVINCE

GD1) 686 Renmin Beilu, Guangzhou, Guangdong 510012 **W:** www.radio-gd.com Satellite Sce. (News Ch.) on 648/828/1017/1143/1206kHz/91.4MHz 24h. - V.O. the City: on 103.6/90.0MHz 24h. - Yangcheng Traffic St. on 105.2MHz 24h. - Southern Life Sce. on 999kHz/93.6MHz 24h (exc. Mon 0400-1000). - Stock Sce. "Caijing 927": on 927kHz/95.3MHz 24h. - V.O. the Music: on 99.3/93.9/96.8MHz 24h. - V.O. Nanyue "Liuxing 1057": on 105.7MHz 24h. - Literary and Sports Sce. on 603kHz/107.7MHz 24h. – **GD2)** Zhujiang EBS, 686 Renmin Beilu, Guangzhou, Guangdong 510012. On 1062/801/1161kHz/97.4MHz 24h in Cantonese. – **GD4)** 231 Huanshi Zhonglu, Guangzhou, Guangdong 510010. News Information Sce. "Fengyun 962": on 96.2MHz 24h (exc. Sun 1700-2200) in Cantonese. English Prgr: Fri 1300-1400. - Golden Hit Sce. "Jinqu 1027": on 102.7MHz 24h (exc. Sun 1600-2100) in Cantonese. - Traffic Sce. "Jiaotong 1061": on 1098kHz/106.1MHz 2200-1600 (exc. Mon 1600-2200). - Youth Sce. on 1170kHz/88.0MHz 24h (exc. Mon 1600-2100). – **GD5)** 1 Pengcheng 1 Lu, Futian Qu, Shenzhen, Guangdong 518026. News Ch. on 89.8MHz 24h (exc. Tues 0530-0930) in Ch and Cantonese. - Life Ch. on 94.2MHz 24h in Ch and Cantonese. - Music Ch. "Feiyang 971" on 97.1MHz 24h in Ch and Cantonese. - Traffic Ch. on 1287kHz/106.2MHz 2230-1800. – **GD6)** 1129 Dong, Jiuzhou Dadao, Xiangzhou Qu, Zhuhai, Guangdong 519015. V.O. the City: on 95.1MHz 2225-1700 in Ch and Cantonese. - Traffic Music Sce. "Feiyue 875" on 900kHz/87.5MHz 2225-1700 in Ch and Cantonese. – **GD7)** Chaoshan Lu, Shantou, Guangdong 515021. News Ch. on 1080kHz/99.3MHz 2200-1600 in Ch and Chaozhou dialect. - V.O. the Life and Economy: on 102.5MHz 2300-1600. - Music Ch. on 107.1MHz 24h (exc. Wed 0600-0900). – **GD8)** 57 Huimin Beilu, Shaoguan, Guangdong 512026. Chinese Ch. on 105.7MHz 2225-1600. - Cantonese Ch. "V.O. Beijiang" on 765kHz/95.2MHz 2225-1700. – **GD9)** Heyuan PBS, 1 Xingyuan Donglu, Yuancheng Qu, Heyuan, Guangdong 517000. On 92.2/97.8MHz in Ch and Cantonese. – **GD10)** 42 Dong Jiaochang Bei, Meizhou, Guangdong 514011. Life Sce. on 94.8/97.8MHz 2155-1600 in Ch and Hakka. - Happy Sce. on 100.3MHz – **GD11)** Ehu Lu, Huicheng Qu, Huizhou, Guangdong 516001. News General Ch. on 100.0/88.3MHz 2230-1630 (exc. Tues 0030-0830) in Ch and Cantonese. - Environment and Traffic Sce. on 1098kHz 2230-1830.. - Traffic Sce. on 98.8MHz – **GD12)** Shanwei PBS, Zhong Duan, Shanwei Dadao, Shanwei, Guangdong 516600. News Ch. on 90.0/103.5MHz 0940-? in Ch and Hakka. – **GD13)** Dongguan PBS, 35 Xizheng Lu, Cheng Qu, Dongguan, Guangdong 523000. General Ch. on 100.8MHz 2225-1600 in Cantonese. - Music Ch. on 106.9MHz 0100-1600 in Cantonese. – **GD14)** Zhongshan BS, 4 Xingzhong Dao, Dong Qu, Zhongshan, Guangdong 528403. On 96.7MHz 2200-1800 in Cantonese. - V.O. the Environment and Travel: on 88.8MHz 2200-1800. - Music St. on 89.3MHz – **GD15)** Jiangmen PBS, 19 Jianshe Lu, Jiangmen, Guangdong 529000. On 100.2MHz 2200-1600 in Cantonese. - Travel and Music St. on 93.3MHz 2200-1600. – **GD16)** Foshan PBS, Jihua 6 Lu, Chancheng Qu, Foshan, Guangdong 528000. Zhen'ai Ch. (Love FM) on 94.6MHz 24h in Cantonese. - Qianse Ch. (Color FM) on 98.5MHz 24h in Cantonese. - Travel Ch. on 88.3MHz - Feiyue 92.4 "Simul Radio": on 92.4MHz – **GD17)** Yangjiang PBS, 114 Mojiang Lu, Jiangcheng Qu, Yangjiang, Guangdong 529500. V.O. the City: on 95.6MHz in Ch and Cantonese. – **GD18)** 93 Yuejin Lu, Chikan Qu, Zhanjiang, Guangdong 524038. 1st St. on 1134kHz/95.1MHz 2220-1700 in Ch, Cantonese and Leizhou dialect. - 2nd St. on 89.3MHz 2220-1700. - V.O. the Traffic and Music: on 98.1/102.4MHz - Zhanjiang EBS: on 104.6MHz 2200-0600, 0800-1600. – **GD19)** 13 Gaoliang Zhonglu, Hedong Qu, Maoming, Guangdong 525000. News St. on 106.1MHz 2230-1500 in Ch and Cantonese. - Music St. on 1098kHz/96.7MHz 2230-1700. – **GD20)** Xinghu Dadao, Zhaoqing, Guangdong 526060. Information Sce. on 1521kHz/92.9MHz 2200-1600 in Ch and Cantonese. - Music Sce. on 90.9MHz 2200-1600 in Ch and Cantonese. – **GD21)** Qingyuan PBS, 18 Xincheng, Yinquan Lu,

Qingyuan, Guangdong 511515. On 88.7/96.7MHz 2225-1600 in Ch and Cantonese. – **GD22)** Wenci Donglu, Chenghai Qu, Shantou, Guangdong 515800. 2250-1600 in Ch and Chaoshan dialect. – **GD23)** Jieyang PBS. Radio and TV Center, Jinxianmen Dadao, Jieyang , Guangdong 522000. 1st prgr. on 103.9MHz - 2nd prgr. on 106.5MHz

GANSU PROVINCE

GS1) 561 Zhangsutan, Chengguan Qu, Lanzhou, Gansu 730010. **W:** www.gstv.com.cn News Sce. on 684/873kHz/FM 2150-1605 (exc. Tues 0600-0850). - City FM: on 102.2/106.6MHz 2200-1800 (exc. Tues 0600-0855). - Economic Sce. "V.O. Yellow River": on 801kHz/93.4MHz 2255-1700. - Traffic Sce. on 103.5/104.8MHz 2150-1800 (exc. Tues 0600-0855). - Youth Sce. "Sunshine FM": on 104.8MHz 0000-1600. - Farm Sce. "Voice of Country" on 1170kHz/92.2MHz 2225-1700 (exc. Tues 0600-0850). – **GS2)** 92 Qingyang Lu, Lanzhou, Gansu 730030. News Sce. on 954kHz/97.3MHz 2125-1700. - Traffic and Music Sce. on 99.5MHz 2200-1900 (exc. Mon 0600-1000). - Life and Literary Sce. on 100.8MHz 2230-1800. – **GS3)** 6 Yan'an Xilu, Jinchang, Gansu 737100. News General Sce. on 585kHz/101.4MHz 2150-1600. - Traffic and Literary Sce. on 103.8MHz on – **GS4)** 11-5 Huancheng Zhonglu, Qincheng Qu, Tianshui, Gansu 741000. News General Sce. on 1143kHz/98.2MHz 2220-1600. - Music and Literary Sce. on 93.7MHz – **GS5)** 10 Fuqiang Xilu, Jiayuguan, Gansu 735100. – **GS6)** 49 Xi 2 Lu, Hezuo, Gansu 747000. On 1332/3990/5970kHz/97.2MHz 2220-0100, 0420-0620, 0950-1400 in Ch and Tb. – **GS7)** 45 Tuanjie Lu, Linxia, Gansu 731100. 2255-0130(Sun 0230). – **GS8)** 45 Hongqi Jie, Kongtong Qu, Pingliang, Gansu 744000. 2200-1500. – **GS9)** Gongyuan Lu, Zhongping Qu, Yumen, Gansu 735200.

GUANGXI ZHUANG AUTONOMOUS REGION

GX1) 75 Minzu Dadao, Nanning, Guangxi 530022. **W:** www.gxradio. com Satellite Sce. on 792/1071/1440/1485kHz/FM. 24h (exc. Tues 0500-0930). - Economic Sce. on 846/1161/1224kHz/FM 24h (exc. Tues 0500-0830) in Ch and Guangxi dialect. - Education and Life Sce. "City 930": on 88.5/90.1/93.0MHz 2100-1900 (exc. Tues 0500-0930) in Ch and Zhuang. - Traffic St. on 100.3/89.5/106.3MHz 2200-1700 (exc. Tues 0500-0930). - Literary Sce. on 95.0/105.0MHz 24h (exc. Tues 0500-0930). – **GX2)** Nanning PBS, 25 Gecun Lu, Nanning, Guangxi 530012. News General Ch. on 101.4MHz 2055-1700 in Ch and Guangxi dialect. - Country and Life Sce. on 104.9MHz 2300-1700 - Traffic and Music Sce. on 107.4MHz 2240-1600. – **GX3)** 1 Guizhou Dadao, Liuzhou, Guangxi 545006. News Sce. on 1386kHz/102.9MHz 24h in Ch and Liuzhou dialect. - Traffic and Life Sce. on 99.1MHz - Music Sce. on 94.5MHz – **GX4)** 1 Anxin Beilu, Xiangshan Qu, Guilin, Guangxi 541002. News General Sce. on 1485kHz/97.7MHz 2225-0525(Sun 0350), W0910-1400, Sun1025-1410. - Travel and Music Sce. on 88.3MHz – **GX5)** 69 Xinxing 3 Lu, Wuzhou, Guangxi 543002. News Sce. on 1485kHz/100.8MHz 2200-1630 in Ch and Guangxi dialect. - V.O. the Music and Traffic: on 107.5MHz 2200-1300 (exc. Mon). – **GX6)** Beihai PBS, 36 Guizhou Nanlu, Beihai, Guangxi 536000. On 93.5MHz 2200-1700 in Ch and Guangxi dialect. – **GX7)** Qinzhou PBS, 18 Liqiao Jie, Qizhou, Guangxi 535000. News General Sce. on 98.6MHz 2220-1645. - Music Sce. on 88.9MHz 2300-1645. – **GX8)** Yulin PBS, 1 Guangdian Lu, Yulin, Guangxi 537000. News Sce. on 97.8MHz 2200-1700 (exc. Tues 0600-0900). - Traffic and Music St. on 99.2MHz 2200-1700.

GUIZHOU PROVINCE

GZ1) 302 Qingyun Lu, Guiyang, Guizhou 550002. **W:** www.gzbs.cn Satellite Sce. (General Ch.) on 765/927/1026/7275kHz/FM 2150-1705 (exc. Tues 0600-0900).–City Sce. "Simul Radio" on 97.2MHz 24h. - Traffic Sce. on 95.2MHz 24h. - Health Sce. on 106.2MHz 24h (exc. Tues 0700-1000). - Music Sce. on 91.7MHz 2300-1600 (exc. Tues 0600-0900). – **GZ2)** 15 Zunyi Lu, Guiyang, Guizhou 550002. News St. on 999kHz/88.9MHz 2150-1800. - Female Ch. on 104.0MHz 2150-1800. - Traffic and Travel Sce. on 101.1MHz 2250-1800. – **GZ3)** 31 Minghu Lu, Zhongshan Qu, Liupanshui, Guizhou 553001. News General Ch. on 765kHz/99.8MHz 2225-1800. - Safe Traffic Ch. on 96.8MHz 2225-1800. - Star Music FM: on 102.1MHz 2225-1800. – **GZ4)** 11 Wenmiao Xiang, Fenghuang Lu, Honghuagang Qu, Zunyi, Guizhou 563000. – **GZ5)** 14 Waihuan Xinan Lu, Anshun, Guizhou 561000. – **GZ6)** Qiannan PBS, 267 Huandong Zhonglu, Duyun, Guizhou 558000. News General Ch. on 882kHz/98.0MHz 2200-1530. - Traffic and Travel Ch. on 93.3MHz/92.2MHz 2225-1600.

HAINAN PROVINCE

HA1) Hainan Radio and TV St, 61 Nansha Lu, Haikou, Hainan 570206 **W:** www.hnwtv.com News Sce. on 954/1107/1116kHz/FM 24h in Ch and Hainan dialect. - Economic Sce. on 99.0/103.8/106.8MHz 2225-1705. - Traffic Sce on 1143kHz/89.3/100.0MHz 2255-1805. – **HA2)** Haikou Radio and TV St, 15 Zhongsha Lu, Haikou, Hainan 570206. News General Sce. on 101.8MHz 2155-1800 in Ch and Hainan dialect. - City and Country Sce. on 95.4MHz - Music Sce. "Simul Radio" on 91.6MHz 2155-1630. – **HA3)** Sanya PBS, Jiefang 4 Lu, Sanya, Hainan 572000. General Ch. on 104.6MHz - Traffic Ch. on 100.3MHz

HUBEI PROVINCE

HB1) 1237 Jiefang Dadao, Hankou, Wuhan, Hubei 430022. **W:** www.

hbtv.com.cn News General Sce. on 774/1404kHz/FM 2000-1730 (exc. Tues 0700-0850). - Economic Ch. on 1053/1251kHz/FM 1940-1800. - Life Ch. on 801/846/900/927/1098/1143/1215kHz/96.6MHz 2000-1800. - Traffic Sce. "Simul Radio" on 107.8/88.0/90.4MHz 2055-1700. - Women and Children St. "Sunshine FM": on 747/1206kHz/102.6MHz 2000-1700. - Music Ch. on 103.8MHz 2000-1700. – **HB2)** Chutian BS, 1237 Jiefang Dadao, Hankou, Wuhan, Hubei 430022 **W:** www.ctbs.cn News St. on 1179/927/945/1008kHz 1955-1700 (exc. Tues 0630-0855). - Satellite St. on 684/846/900/927/945kHz/FM 1955-1700 (exc. Tues 0630-0855). - Traffic and Sports St. on 92.7MHz 2125-1630 (exc. Tues 0630-0855). - Music Sce. on 105.8MHz 24h (exc. Tues 0630-0855). – **HB3)** 620 Jianshe Dadao, Hankou, Wuhan, Hubei 430015. On 873kHz/ 88.4MHz 2030-1700 (exc. Wed 0600-0925). - Changjiang Economic Sce. on 1125kHz/100.6MHz 2100-1700 (exc. Thurs 0600-0900). - Traffic Sce. on 603kHz/89.6MHz 2100-1700 (exc. Tues 0500-0900). - Music Sce. on 101.8MHz 2100-1700 (exc. Tues 0600-1000). - Children and Storytelling Sce. on 93.6MHz 2100-1700. – **HB5)** 188 Wuhan Lu, Huangshi, Hubei 435000. News St. on 963kHz/101.2/101.8MHz 2200-1600. - Traffic and Music St. on 103.3MHz 2200-1600. - V.O. Cihu (Ci Lake): on 105.0MHz 2145-1600. – **HB6)** 78 Zhongshan Houjie, Fancheng Qu, Xiangfan, Hubei 441021. General Sce. on 1314kHz/90.9MHz 2120-1635 (exc. Tues 0700-0830). - News Sce. on 104.0MHz 2120-1635 (exc. Tues 0700-0830). - Automobile Sce. on 1098kHz/105.3MHz 2155-1635 (exc. Tues 0700-0830). - Traffic Sce. on 89.0MHz 2155-1635 (exc. Tues 0700-0830). – **HB7)** 4 Renmin Beilu, Shiyan, Hubei 442000. News St. on 1485kHz/106.2/107.3MHz 2100-1600. - V.O. Checheng (Mobile City): on 99.1MHz 2200-1600. - Music and Traffic St. on 101.9MHz 2200-1700 (exc. Tues 0600-1000). – **HB8)** 266 Jiangjin Xilu, Shashi Qu, Jingzhou, Hubei 434000. News St. on 828kHz/98.4MHz 2030-1630. - Traffic and Music St. on 96.3MHz 2200-1700 (exc. Tues 0700-0900). - 901 Automobile St. on 90.1MHz 2200-1600. - Health Life St. on 585kHz/97.2MHz 2100-1600. – **HB9)** 2 Guoyuan 1 Lu, Yichang, Hubei 443000. News General Sce. on 621kHz/95.6MHz 2100-1615 (exc. Tues 0600-0700). - Economic Life Sce. on 100.6MHz 2130-1600 (exc. Tues 0600-0700). - Traffic and Music Sce. on 105.9MHz 2225-1645 (exc. Tues 0600-0700). – **HB10)** 100 Xiangshan Dadao, Dongbao Qu, Jingmen, Hubei 448000. News and Economic St. on 1161kHz/89.7MHz 2120-1600. - Health and Music St. on 93.2MHz 2120-0500, 0930-1500. - Traffic and Literary St. on 90.3MHz 2120-1600. – **HB11)** 157 Binhu Lu, Ezhou, Hubei 436000. 2100-1600. – **HB12)** 116 Changzheng Lu, Xiaogan, Hubei 432100. News General Ch. on 927kHz/91.2MHz 2155-1530 (exc. Tues 0500-1000). - Traffic and Music Ch. on 87.7MHz 2255-1505. – **HB13)** 169 Dongmen Lu, Huangzhou Qu, Huanggang, Hubei 438000. On 91.4MHz 2220-1540. - Traffic Sce. on 107.6MHz 2220-1530. – **HB14)** 38 Wenquan Lu, Xianning, Hubei 437100. ?-0800, 0930-?. – **HB15)** 32 Xuefu Lu, Laohekou, Hubei 441800. 2220-0500, 0800-1400. – **HB16)** 4 Renmin Lu, Danjiangkou, Hubei 441900. News General Ch. on 1431kHz 2220-1600. – **HB17)** 2 Shannan Xiaoqu, Shishou, Hubei 434400. 2200-0005, 0955-1235. – **HB18)** 56 Guang'an Lu, Yingshan Zhen, Guangshui, Hubei 432700. 2155-0115, 0955-1305. – **HB19)** 146 Puyang Dadao, Yingcheng, Hubei 432400. 2200-0600, 0900-1305. – **HB20)** 199 Nanhuan Lu, Macheng, Hubei 436100. Educational and Music St. on 1242kHz/92.5/105.0MHz 2155-1430. – **HB21)** 50 Chunchuan Daqiao Lu, Chibi, Hubei 437300. 0950-1340. – **HB22)** 359 Lieshan Dadao, Suizhou, Hubei 441300. News General Ch. on 1008kHz 2130-2330, 0330-0500, 1030-1305. - Traffic and Music St. on 96.2MHz – **HB23)** 117 Mianyang Dadao, Xiantao, Hubei 433000. 2130-1600. – **HB24)** 16 Jianghan Lu, Yuanlin Zhen, Qianjiang, Hubei 433100. 2205-0445, 0930-1600. – **HB25)** 201 Hangkong Lu, Xiangyang Qu, Xiangfan, Hubei 441104. On 1521kHz/96.5MHz 2155-1600.

HEILONGJIANG PROVINCE

HL1) 333 Hanshui Lu, Nangang Qu, Harbin, Heilongjiang 150090 **W:** www.hljradio.com News St. on 621/900/927/1341/94.6MHz 2000-1600. - Life St. on 1476kHz/FM 24h (exc. Tues 1600-2100). - Traffic St. on FM 24h (exc. Tues 1600-2100). - City Women St. on 102.1MHz 2100-1600. - "FM 97" on 97.0MHz 2200-1600. EG Prgr: 1400-1500. - Music St. on FM 24h exc. Tues 1600-2100). - Country St. on 945kHz/94.3MHz 2100-1500. - Heilongjiang Korean BS: on 873kHz/96.1MHz W2100-2300, Sun0100-0400, W0400-0500, D0900-1100 and on 95.8MHz 2100-2400, 1000-1300 in Ko. - University St. "Radio Young" on 99.3MHz 2200-?. - Sanya BS "V.O. Tianya" on 104.6MHz 2200-?. – **HL2)** 1 Huashan Lu, Xiangfang Qu, Harbin, Heilongjiang 150036. News Ch. on 837kHz/105.6MHz 24h. - Literary Sce. on 98.4MHz 24h. – **HL2A)** 2 Wenzheng Jie, Dongli Qu, Harbin, Heilongjiang 150040. People Life Ch. on 972kHz. 24h (exc. Tues 0500-0900). - Traffic Sce. on 92.5/95.3MHz 24h. - Music Sce. on 927kHz/88.8/90.9/103.0MHz 24h. – **HL3)** 99 Yong'an Dajie, Longsha Qu, Qiqihar, Heilongjiang 161005. News Sce. on 1197kHz/97.8MHz 2000-1600(Tues 1405). - Life and Literary Sce. on 693kHz/89.4MHz 2020-1505. - Traffic Sce. on 94.1MHz 2050-1605. - Country Sce. on 585kHz/103.4MHz – **HL4)** Jiuma Lu, Xiangyang Qu, Hegang, Heilongjiang 154100. News Sce. on 1413kHz/97.2MHz 2055-

1400. - Traffic and Literary Sce. on 106.1MHz 2145-1400. - Life Sce. on 93.3MHz – **HL5)** 240 Xinxing Dajie, Jianshan Qu, Shuangyashan, Heilongjiang 155100. On 1179kHz 2120-0120, 0320-0530, 0905-1230. - Traffic and Literary Sce. on 98.1MHz - Storytelling Sce. on 603kHz/ 88.6MHz – **HL6)** 11 Diantai Lu, Jiguan Qu, Jixi, Heilongjiang 158100. News General Sce. on 1368kHz 2130-0600, 0850-1350. - Traffic Sce. on 1485kHz/95.9MHz - Literary and Life Sce. on 98.6MHz - Storytelling Sce. on 103.9MHz 2055-1500. – **HL7)** Jia 1, Dongfeng Lu, Sa'ertu Qu, Daqing, Heilongjiang 163311. News Sce. on 1080kHz/96.7MHz 24h. - Traffic Sce. on 95.0MHz 24h. - Music Sce. on 106.0MHz 1955-1600.- Story Sce. on 103.9MHz 24h. - V.O. Baihu: on 91.9MHz 2000-1600. – **HL8)** 16 Linshan Lu, Yichun Qu, Yichun, Heilongjiang 153000. News General Sce. on 909kHz/92.4/102.1MHz 2130-0810 (exc. Mon 0725-1000) - Traffic and Life Sce. on 98.5MHz 2200-1400. – **HL9)** 138 Taiping Lu, Mudanjiang, Heilongjiang 157000. News Sce. on 684kHz/87.9MHz 2105-1400 (exc. Tues 0800-0855). - City Life Sce. on 1476kHz/91.6MHz 2200-1530. - Traffic and Literary Sce. on 98.2MHz 2300-1300. – **HL10)** 35 Shunhe Lu, Jiamusi, Heilongjiang 154002. News General Sce. on 666kHz/101.7MHz 2055-0530, 0855-1400. - Economic Sce. on 1143kHz/ 95.0MHz 2055-1600. - Traffic and Literary Sce. on 90.0/93.8MHz 2225-1600. – **HL11)** 2 Shanhu Dajie, Taoshan Qu, Qitaihe, Heilongjiang 154600. News General Ch. on 1062kHz/98.8MHz 0820-1100. - Traffic Ch. on 89.1MHz – **HL12)** Xizhi Lu, Suihua, Heilongjiang 152054. - Traffic Sce. on 90.7MHz 2155-1400. - Music Sce. on 1080kHz – **HL13)** 2 Xing'an Dajie, Jagdaqi Zhen, Heilongjiang 165000. 2120-0155, 0255-0535, 0825-1430. – **HL14)** 93 Hailan Jie, Aihui Qu, Heihe, Heilongjiang 164300. On 1215kHz/103.8MHz 2100-1400.

HUNAN PROVINCE

HN1) 167 Yuhua Lu, Changsha, Hunan 410007 **W:** www.hnradio. com Satellite Sce. (News Ch) on 738/1152/1233/4990kHz/FM 2100-1700 (exc. Tues 0500-0900). - V.O. the Country "Green 938": on 900kHz/93.8/100.7MHz 2100-1705. - Economic Ch. "E-FM": on FM 2130-1700 (exc. Tues 0500-0900). - Literary Ch. "Kuaile (Happy) 975": on 97.5/87.5/90.6/90.8/96.9MHz 2130-1700 (exc. Tues 0500-0900). - Traffic Ch. on FM 24h. - V.O. the Music "Super 893": on 89.3/102.1MHz - Travel Sce. on 106.9MHz – **HN2)** 237 Laodong Xilu, Changsha, Hunan 410015. News Sce. on 1323kHz/105.0MHz 24h (exc. Tues 0600-0900). - Music Ch. on 1251kHz/106.1MHz 24h (exc. Tues 0600-0900). - V.O. Jinying (Golden Vox): on 95.5MHz 24h. - V.O. Rongcheng (Legend 886): on 88.6MHz 24h. – Sound of City "City FM": on 101.7MHz 24h. – **HN3)** Caotangba Xiang, Jianshe Zhonglu, Zhuzhou, Hunan 412007. News Ch. on 1089kHz/101.2MHz 2150-1600. - Traffic Ch. on 98.4MHz – **HN4)** Donghu Lu, Xiangtan, Hunan 411104. General Ch. on 1503kHz/104.2MHz 2200-1700. - V.O. the Music: on 98.6MHz – **HN5)** 114 Xianfeng Lu, Hengyang, Hunan 421001. On 1098kHz/98.9MHz 2130-1700. – **HN6)** 373 Zhangshulong, Baoqing Xilu, Daxiang Qu, Shaoyang, Hunan 422000. On 1548kHz 2225-0530, 0955-1600. - Traffic Ch. on 95.4MHz – **HN7)** Nanhu Dadao, Yueyang, Hunan 414000. News and Traffic Sce. on 1053kHz/100.1/104.1MHz 2200-0530, 0925-1340. - Music Sce. on 1008kHz/106.1MHz 2155-1700. – **HN8)** 69 Wuling Dadao, Changde, Hunan 415000. On 1260kHz 2200-2355, 0400-0500, 0925-1100. - News Ch. on 105.6MHz 2225-1705. - Traffic Ch. on 97.1MHz – **HN9)** Chaoyang Lu, Yiyang, Hunan 413000. Economic Sce. on 1008kHz/99.7MHz 2220-1600 (exc. Wed 0800-1000). – **HN10)** 51 Renmin Lu, Jinshi, Hunan 415400. – **HN11)** Chenzhou PBS, 7 Li Dadao, Chenzhou, Hunan 423000. Politics and General Ch. on 99.2/89.9MHz 2200-1700. - Music and Traffic Ch. on 102.8MHz 2225-1700. – **HN12)** Huaihua PBS, Tianxing Lu, Huaihua, Hunan 418000. News Ch. on 97.2MHz - Traffic and Literary Ch. on 103.8MHz

JILIN PROVINCE

JL1) 242 Xi'an Dalu, Changchun, Jilin 130051 **W:** jlradio.chinajilin. com.cn News General Sce. on 738/1107/1530kHz/FM 24h (exc. Tues 0500-0900). - Economic Sce. on 846kHz/95.3MHz 24h. - Health and Entertainment Sce. on 101.9MHz 24h (exc Tues 1500-1800). - Traffic Sce. on 103.8MHz 24h. - Infromation Sce. on 100.1MHz 24h (exc. Tues 1500-1800). - Music Sce. on 92.7MHz 24h (exc. Tues 1500-1800). - Country Sce. on FM 24h. - Story Sce. on 103.3MHz 2130-1600. – **JL2)** 3 Baicao Lu, Changchun, Jilin 130061. On 585kHz/88.9MHz 24h. - Country Sce. on 90.0MHz - Young and Old Life Sce. on 648kHz. - Country Sce. on 900kHz/88.0/105.8MHz 24h. - V.O. the Traffic on 96.8MHz 24h. - Health Life Sce. on 107.9MHz - City Moving "Top Radio": on 106.4MHz - Changchun EBS: on 1332kHz 2125-1600. - Changchun LBS: on 99.6MHz 24h. – **JL3)** 90 Jilin Dajie, Jilin-shi, Jilin 132011. News General St. on 927kHz/100.8MHz 24h. - Public Life St. on 702kHz 2045-1600. - Traffic St. on 105.3MHz 24h. - Novel and Storytelling St. on 1143kHz. 2100-1400 - V.O. Songhuajiang (Songhua River): on 88.3MHz 24h. – **JL3A)** Jilin-shi EBS, 181 Jiefang Dalu Xi, Jilin-shi, Jilin 132011. "Dushi (city) 110" St. on 1494kHz/90.3MHz 24h. - "Qin'ai (dear) 603" St. on 603kHz 2020-1600. - "Luyou (Travel) 893" St. on 89.3MHz 2030-1700. - "Jiankang (Health) 1251" St. on 1251kHz – **JL4)** 39 Nan Xinhua Dajie, Siping, Jilin 136000. News General Ch. on 666kHz/93.9MHz 2135-

1300. - Traffic and Literary Ch. on 99.5MHz 24h. - Public Storytelling Ch. on 900kHz/90.5MHz 2155-1300. – **JL5)** 20 Hebin Lu, Longshan Qu, Liaoyuan, Jilin 136200. On 810kHz/100.0MHz. - Traffic and Literary St. on 96.2MHz 2125-1500. – **JL6)** 199 Cuiquan Lu, Longquan Jie, Tonghua, Jilin 134001. News General Sce. on 909kHz 2150-1505 (exc. Tues 0705-0855). – Urban Sce. on 90.9MHz 2200-1400 - Traffic and Literary St. on 93.8MHz 2140-1700. - Storytelling St. on 97.9MHz 2150-1400. – **JL7)** 36 Hunjiang Dajie, Badaojiang Qu, Baishan, Jilin 134302. News General St. on 657kHz/107.7MHz 2100-1530. - Traffic Sce. on 98.4MHz 2100-1500. – **JL8)** 71 Linjiang Lu, Ningjiang Qu, Songyuan, Jilin 131200. News General Ch. on 1431kHz 2100-1500. - Economic St. on 89.9MHz - Traffic St. on 1143kHz/99.9MHz 2100-1500. - Storytelling and Entertainment St. on 102.5MHz 2055-1600. – **JL9)** 18 Chunyang Lu, Baicheng, Jilin 137000. News General Sce. on 1323kHz/103.0MHz 2100-1400 (exc. Tues 0630-0940). - Traffic and Literary Sce. on 96.5MHz - City Life Sce. on 105.8MHz - Storytelling Sce. on 98.5MHz – **JL10)** 166 Juzi Jie, Yanji, Jilin 133000. Ch Satellite Sce. on 1053/603/1566kHz/FM 2130-1630. - Ch News Sce. on 88.2/98.3MHz - Ko News General Sce. on 1206kHz 2040-1600 (exc. Tues 0540-0900). - Ko Satellite Sce. on 585/1188kHz/ FM 2130-1510. - Traffic and Literary Sce. on 105.9MHz 2130-1600 in Ch. – **JL11)** 45 Dong Huancheng Lu, Gongzhuling, Jilin 136100. V.O. the Public: on 1485kHz 2050-1420. - V.O. the Traffic: on 101.3MHz – **JL12)** 18 Nan Dalu, Lishu Xian, Jilin 136500. – **JL13)** 70 Henan Jie, Meihekou, Jilin 135000. On 1584kHz/95.7MHz 2155-1130. – **JL14)** Yucai Jie, Qian Gorlos, Jilin 131100. 2125-2330, 0325-0500, 0955-1230 in Ch and Mo. – **JL15)** 29 Gushu Nanjie, Taonan, Jilin 137100. – **JL16)** 7 Yongle Jie, Yanji, Jilin 133000. Ch Prgr. on 900kHz. 24h. - Ko Prgr. "Arirang Radio": on 88.0MHz 2100-1700. - Yanji V.O. the Traffic BS: on 93.5MHz – **JL17)** 12 Xiangshang Jie, Tumen, Jilin 133100. 2155-2400, 0330-0500, 0855-1230 in Ch and Ko. – **JL18)** 1-8 Xinhua Xilu, Dunhua, Jilin 133700. 2130-1500 in Ch and Ko. – **JL19)** Jinghe Jie, Hunchun, Jilin 133300. Storytelling St. on 927kHz 2030-1530 in Ch and Ko - North East Asia V.O. Hunchun on 101.0MHz.

JIANGSU PROVINCE

JS1) Jiangsu Prov. Radio and TV Headquarters, 8 Xi Citang Xiang, Zhongshan Donglu, Nanjing, Jiangsu 210002 **W:** www.jsbc.com News General Ch. on 702/1314/1413/1602kHz 2000-1700 (exc. Tues/Thurs 0600-0850). - News Sce. on FM 2100-1600. - Home Sce. on 107.1MHz 24h (exc. Tues 0600-0900, Wed 1700-2100). - Health Sce. on 846/603kHz. 24h (exc. Tues 0600-0900). - Financial Sce. on 1206kHz/95.2MHz 2100-1600 (exc. Tues 0600-0900). - Traffic Sce. on 101.1MHz 24h (exc. Tues 1800-2000). - Story Sce. on 585kHz 2000-1700 (exc. Tues 0600-0900). - Music St. "City FM": on 89.7/107.8MHz 24h (exc. Tues 0600-0900). - Classic Music St. on 97.5MHz 24h. - Chinese Opera Sce. on 1053kHz 2100-1600. – **JS2)** Nanjing Radio and TV Bldg., 358 Baixia Lu, Nanjing, Jiangsu 210001. News St. on 1008kHz 1950-1800 (exc. Tues 0600-0800). - News FM: on 96.6MHz 2200-1400. - Economic St. on 900kHz. 24h. - Economic St. "City FM": on 101.7MHz 2200-1600. - City Control Sce. on 1143kHz 2100-1600. - Traffic St. on 102.4MHz 2130-1800. - Automobile Music Sce. on 105.8MHz 24h. - Sports St. on 1251kHz/104.3MHz 24h. - Sports St. Entertainment Sce. "Gandong (move) 801": on 801kHz 2100-1600 (exc. Tues 0600-0800). – **JS3)** 223 Zhongshan Nanlu, Xuzhou, Jiangsu 221003. News Sce. on 1269kHz/93.0MHz 2000-1730. - Life Sce. on 801kHz/91.6MHz 2025-1730. - Traffic Sce. on 103.3MHz 2030-1700. - Literary Sce.on 1098kHz/89.6MHz 2015-1600. - Music Sce. on 99.6MHz 24h. – **JS4)** 221 Jiefang Xilu, Xinpu Qu, Lianyungang, Jiangsu 222003. News Sce. on 1458kHz/97.2/98.3MHz 2100-1600 (exc. Tues 0600-0855). - Economic Sce. on 1251kHz/90.2/90.7MHz 2115-1600. - Traffic Sce. on 900kHz/96.0/101.4MHz 2100-1600. - Story Sce. on 98.1MHz – **JS5)** 6 Dazhi Lu, Huai'an, Jiangsu 223001. News General Sce. on 801kHz/106.7MHz 2000-1600 (exc. Tues 0600-0900). - Economic Life Sce. on 1251kHz/105.0MHz 2000-1600 (exc. Tues 0600-0840). - Traffic and Literary Sce. on 1521kHz/94.9MHz 2100-1600. - Public (Chengshi Guanli) Sce. on 96.8MHz 2055-1600. – **JS6)** Yancheng Radio and TV St, 4 Shengyuan Lu, Yancheng, Jiangsu 224001. News and Financial Ch. on 1026kHz/99.6MHz 2100-1600. - City and Traffic Ch. on 747kHz/105.3MHz 2100-1600. - Huanghai Mingzhu (Pearl) Sce. on 900kHz/93.9MHz – **JS7)** 8 Meiling Donglu, Yangzhou, Jiangsu 225002. News Sce. on 1179kHz/98.5/105.5MHz 2120-1800 (exc. Tues 0600-0930). - Traffic Sce. on 1521kHz/103.5MHz 2120-1800. - Life Sce. on 801kHz/94.9MHz 2130-1600 (exc. Tues 0645-0800). – **JS8)** 20 Qingnian Lu, Taizhou, Jiangsu 225300. News Sce. on 1341kHz/103.7MHz 2120-1525 (exc. Tues 0530-0855). - City Sce. on 927kHz/97.3MHz 2145-1600. - Traffic Sce. on 92.1MHz 2200-1600. – **JS9)** 100 Renmin Zhonglu, Nantong, Jiangsu 226001. News Information Ch. on 1233kHz/97.0MHz 2130-1600. - Economic Life Ch. on 603kHz/103.0MHz 2130-1600. - Music and Traffic Ch. on 1143kHz/92.9MHz 2125-1600. – **JS10)** 94 Zhongshan Xilu, Zhenjiang, Jiangsu 212004. News Ch. on 99.4MHz 2055-1600. - Economic Ch. on 104.5MHz 2215-1600. - Health Life Ch. on 1224kHz/94.0MHz 2055-1600. - Traffic and City Sce. "City Radio": on 96.3MHz 2200-1600. - Music FM "Aiting 905": on 90.5MHz

2200-1700. - Story Sce. on 900kHz/102.7MHz 2100-1600. – **JS11)** 88 Guangshi Lu, Changzhou, Jiangsu 213016. General Ch. "AM846":on 846kHz 2059-1600 (exc. Tues 0600-0850). - General Ch. "FM103.4": on 103.4MHz Economic Ch. "AM1143": on 1143kHz 2200-1600. - Economic Ch. "FM105.2": on 105.2MHz - Traffic Ch. on 90.0MHz 2100-1800. – Traffic and Literary Ch. on 747kHz 2110-1600 (exc. Tues 0600-0900). - Music Ch. on 1521kHz/93,5/100.1MHz 2150-1600. - Classic Music Ch. on 927kHz 2200-1600. – **JS12)** Wuxi Radio and TV St, 4 Hubin Lu, Wuxi, Jiangsu 214061. News Ch. News Sce. on 1161kHz/89.4MHz 2020-1700 (exc. Tues 0500-0900). - News Ch. Information Sce. on 93.7MHz 2020-1700. - Economic Sce. on 1251kHz/104.0MHz 2030-1800. - Story and Chinese Opera Sce. on 1008kHz 24h. Traffic St. "Automobile Ch." on 106.9MHz 24h. - Music Ch. on 900kHz/91.4MHz 2130-1800. - City Life Ch. on 1521kHz/98.7MHz 2130-1800 (exc. Tues 0500-0700). - V.O. Jiangnan: on 603kHz/92.6MHz 2100-1800 (exc. Tues 0500-0800). – **JS13)** Suzhou Radio and TV Headquarters, 4 Gongyuan Lu, Suzhou, Jiangsu 215006. News General Ch. on 1080kHz 2030-1600 (exc. Tues 0600-0730) in Ch and Suzhou dialect. - V.O. the City "My Radio": on 91.1MHz 2030-1630. - Traffic Sce. on 1521kHz/104.8MHz 2130-1600. - Life Sce. on 96.5MHz 2130-1600(Tues 1525). - Music Sce. on 94.8MHz 24h (exc. Tues 0600-1000). - Chinese Opera Sce. on 846kHz - Fortune Sce. on 102.8MHz 2200-1600. – **JS14)** 43 Gongnong Lu, Yizheng, Jiangsu 211400. On 1026kHz/94.3MHz 2155-0535, 0725-1350. – **JS15)** 79 Zhongshan Nanlu, Jiangyin, Jiangsu 214400. Happy Life Ch. on 1386kHz/106.0MHz 2200-1530. - T Automobile Ch. on 90.7MHz 2200-1500. – **JS16)** 29 Haiyu Beilu, Changshu, Jiangsu 215500. News General Ch. on 1116kHz 2130-1400 (exc. Sat 0630-0830). - Economic Service Ch. on 927kHz 2150-1400 (exc. Sat 0630-0830). - Traffic and Music Ch. on 747kHz/100.8MHz 2130-1400 (exc. Sat 0600-0800). – **JS17)** Chenjiachang Nong, Yangshe Zhen, Zhangjiagang, Jiangsu 215600. News Sce. on 1098kHz 2140-1455 (exc. Wed 0600-0830). - Traffic Sce. on 102.0MHz 2155-1500 (exc. Wed 0530-0955). - Music Sce. on 1521kHz 2155-1500.

JIANGXI PROVINCE

JX1) 207 Hongdu Zhong Dadao, Nanchang, Jiangxi 330046 **W:** www. jxgdw.com/jxgd/jxgbdt/ News Sce. on 729/1350/1449kHz/FM 2000-1700 (exc. Tues 0600-0855). - City Sce. on 927kHz/FM 2200-1700 (exc. Tues 0600-0900). - People Life Sce. on 603/927kHz/101.9/94.7/ 101.2MHz 2200-1600. - Scientific Education and Farm Sce. on 603kHz/ 98.5/88.3MHz 2200-1600 (exc. Tues 0600-0900). - Traffic Sce. "My FM": on 96.9/105.4MHz 24h. - Literary and Music Ch. on 103.6/94.9/9 7.9/100.2/101.6/103.8/107.6MHz 24h. – **JX2)** 241 Ruzi Lu, Nanchang, Jiangxi 330009. News General Ch. on 1278kHz/91.7MHz 2030-1700 (exc. Tues 0500-0900). - V.O. the Traffic and Music: on 95.1MHz 24h (exc. Tues 0500-0900). - V.O. Fortune: on 89.7MHz 24h. - Music and Story Sce. on 90.6MHz 24h. - V.O. the Beauty: on 87.9MHz 2200-1600. – **JX3)** 1073 Cidu Dadao, Jingdezhen, Jiangxi 333000. News General Sce. on 1467kHz/96.5/107.3MHz 2200-1600. – **JX4)** Jiangwan Li, Binhe Xilu, Pingxiang, Jiangxi 337005. News General Ch. on 1107kHz/96.8/ 106.8MHz 2155-1500. - Traffic and Literary Ch. on 88.8MHz 24 h. – **JX5)** 49 Xianlai Zhong Dadao, Xinyu, Jiangxi 338000. General Ch. on 675kHz 2130-1800. - City Sce. on 585kHz/94.0MHz - Health Sce. on 1476kHz 2130-1800. - Traffic Sce. "Love Radio": on 96.2MHz 2200-1800. – **JX6)** 84 Changhong Dadao, Jiujiang, Jiangxi 332000. News General Sce. on 90.0/91.6MHz 2155-1600. - Traffic Sce. on 88.4/88.9MHz 2255-1500 (exc. Tues 0530-0855). - City Life Sce. on 1485kHz 2155-1600. – **JX7)** Yingtan PBS, 3 Jianshe Lu, Yingtan, Jiangxi 335200. V.O. Xinjiang (Xin River): on 103.2MHz 2200-1605. - V.O. the Traffic and Music: on 95.6MHz 2200-1800. – **JX8)** 58 Hongqi Dadao, Ganzhou, Jiangxi 341000. News General Ch. on 747kHz/93.7/101.8MHz 2130-1700 (exc. Tues 0600-0830) in Ch and Hakka. - V.O. the City: on 94.5MHz 2200-1700. - Traffic Sce. on 99.2MHz – **JX9)** 19 Beimen Jie, Ji'an, Jiangxi 343000. - V.O. Jinggang: on 603/1242kHz/95.6/102.1MHz 2150-1600. - Traffic and Entertainment Sce. on 100.6MHz – **JX10)** Shangrao PBS, 51 Qingfeng Lu, Shangrao, Jiangxi 334000. News General Ch. on 93.4MHz 2200-1630. - Traffic and Music Ch. on 96.6MHz 2200-1630.

LIAONING PROVINCE

LN1) Liaoning Radio and TV St, 10 Guangrong Jie, Heping Qu, Shenyang, Liaoning 110003 **W:** www.lnradio.cn General Sce. on 612/963/1089/1260kHz/102.9MHz 24h (exc. Tues 0605-0855). - News Sce. on 88.8/99.5/105.5MHz - Economic Sce. on 999kHz/FMMHz 24h (exc. Tues 0540-0855). - Country Sce. on 927kHz/96.9MHz - Traffic Sce. on 97.5MHz 24h (exc. Tues 0540-0850). - Literary Sce. on 810kHz/95.9/99.5MHz 24h (exc. Tues 0540-0900). - Story Sce. on 1053kHz/101.8MHz 24h. - Information Sce. (Dalian Blanch): on 90.6MHz 24h. – **LN2)** Shenyang Radio and TV St, 89 Sanhao Jie, Heping Qu, Shenyang, Liaoning 110004. News Sce. on 792kHz/104.5/107.0MHz 24h. - Ecomonic Sce. on 882kHz/90.4MHz 24h (exc. Tues 0500-0855). - City Sce. on 103.4MHz 24h. - Traffic Sce. on 98.6MHz 24h (exc. Thurs 0500-0855). - Sports and Leisure Sce. on 1341kHz/105.9MHz 24h. - Literary Sce. on 92.1MHz 24h. - Music Space Sce. on 87.6MHz

2130-1600. – **LN3)** 162 Minquan Jie, Shahekou Qu, Dalian, Liaoning 116022. News Sce. on 882kHz/103,3MHz 1955-1605 (exc. Tues 0600-0800). - Financial Sce. on 93.1MHz 24h (exc. Tues 0630-0800). - City Sce. on 1152kHz/99.1MHz 2025-1605 (exc. Tues 0630-0800). - Traffic Sce. on 100.8MHz 24h (exc. Tues 0600-0800). - Sports and Leisure Sce. on 105.7MHz 2025-1605 (exc. Tues 0600-0800). - New City and Country Sce. on 1575kHz/95.6MHz 24h. - Children Sce. "Easy Radio": on 106.7MHz 24h (exc. Tues 0600-0800). – **LN4)** 3, 219 Lu, Tiedong Qu, Anshan, Liaoning 114002. News St. on 954kHz/101.0MHz 24h. - Economic St. on 1071kHz/89.7MHz 24h. - Traffic St. on 1458kHz/ 105.1MHz 24h. - Storytelling St. on 1251kHz/87.9MHz – **LN5)** 2 Hunhe Beilu, Shuncheng Qu, Fushun, Liaoning 113006. News Sce. on 684kHz/93.0/93.8MHz 2000-1500. - Traffic Sce. on 747kHz/106.1MHz 24h. - Music Sce. on 100.6MHz 2030-1600. - Storytelling St. on 1143kHz/88.2MHz – **LN6)** 15 Tiyu Lu, Mingshan Qu, Benxi, Liaoning 117000. On 1296kHz 2125-1500. - Traffic and Economic St. on 900kHz/ 107.4MHz 2155-1600. - Life and Entertainment St. on 96.4MHz ?-1500. – **LN7)** 1 Shanshang Jie, Zhenxing Qu, Dandong, Liaoning 118000. News Ch. on 1404kHz/103.6MHz 24h. - Traffic Ch. on 891kHz/101.7MHz 2000-1600. - Life and Entertainment Ch. on 104.3MHz 2000-1600. - V.O. the Yellow Sea: on 88.0MHz 2100-1500. – **LN8)** 3, 4 Duan, Beijing Lu, Jinzhou, Liaoning 121000. News St. on 666kHz 2125-1500 (exc. Tues 0530-0855). - Economic St. on 774kHz 2125-1500 (exc. Tues 0530-0855). - Public Life St. on 1098kHz/97.7MHz 2125-1500 (exc. Tues 0530-0855). - Traffic and Literary Sce. on 846kHz/100.3MHz 2125-1500 (exc. Tues 0530-0855). – **LN9)** Huludao Radio and TV St, 23 Haixing Lu, Longwan Dajie, Huludao, Liaoning 125000. News General Sce. on 1242kHz/93.1/95.2MHz 2130-1535. - Economic Sce. on 106.3MHz - Traffic and Literary St. on 87.8MHz 2150-1330 (exc. Tues 0540-0955). – **LN10)** 10, Dong, Bohai Dajie, Zhanqian Qu, Yingkou, Liaoning 115000. News General Sce. on 1026kHz/88.4/106.2MHz 2055-1500. - Economic Life Sce. on 747kHz/91.1/92.8MHz 2100-1500. - Traffic and Literary Sce. on 1143kHz/95.1MHz 2130-1600. - Storytelling and Entertainment Sce. on 603kHz/94.1MHz 2125-1500. – **LN11)** 7 Shifu Dajie, Xinglongtai Qu, Panjin, Liaoning 124010. News General Sce. on 1485kHz/88.2MHz 2115-0155, 0330-0540, 0925-1400. - Traffic and Literary Sce. on 90.1MHz 2100-1500. - Life and Entertainment Sce. on 104.2MHz 2100-1600. - Storytelling and Chinese Opera Sce. on 101.8MHz 2100-1550. – **LN12)** 61 Zhonghua Lu, Haizhou Qu, Fuxin, Liaoning 123000. On 1287kHz/100.9MHz 2115-0625, 0755-1245. - Economic and Storytelling Sce. on 1557kHz/89.3MHz 2055-1600. - Literary Sce. on 105.3MHz - Traffic Sce. on 88.7MHz – **LN13)** Fuxin Mongolian BS, 52-2 Yan'an Lu, Haizhou Qu, Fuxin, Liaoning 123000. 2155-0015, 0325-0530, 1040-1300 in Mo. – **LN14)** Liaoyang Radio and TV St, 59 Qingnian Dajie, Taizihe Qu, Liaoyang, Liaoning 111000. News General Sce. on 837kHz 2030-1530 (exc. Tues 0600-0800). - Life Sce. on 1143kHz/102.0MHz 2025-1530 (exc. Tues 0540-0755). - Traffic and Literary Sce. on 1503kHz/107.8MHz 2025-1530. - Storytelling and Chinese Opera Sce. on 648kHz/106.0MHz 2030-1530. – **LN15)** Tieling Radio and TV St, 45 Gongren Jie, Yinzhou Qu, Tieling, Liaoning 112000. News Sce. on 1413kHz/90.8MHz 2150-1300. - Traffic Sce. on 95.2MHz 2200-1300. - Country Sce. on 810kHz/101.2MHz 2130-1430. - Literary Sce. on 95.9MHz – **LN16)** 88, 1 Duan, Xinhua Lu, Shuangta Qu, Chaoyang, Liaoning 122000. News General Sce. on 585kHz/96.1/101.1MHz 2125-1600. - New Farm Sce. on 810/702/747kHz/99.5MHz 1955-1600 (exc. Tues 0600-0900). - Traffic and Entertainment Sce. on 93.8/103.1MHz 1955-1600 (exc. Tues 0600-0900). – Economic Life Sce. on 648/801kHz/106.5MHz 2000-1600 (exc. Tues 0600-0900). – **LN17)** 67 Jinluan Lu, Wafangdian, Liaoning 116300. 2125-1345. – **LN18)** 385, 1 Duan, Huanghai Dajie, Zhuanghe, Liaoning 116400. 2100-0100, 0855-1200. – **LN19)** Haicheng Radio and TV St, 14 Huancheng Xilu, Haicheng, Liaoning 114200. News General Sce. on 900kHz/90.4MHz 2135-1500. Traffic and Entertainment Sce. on 1350kHz/106.9MHz 2135-1500. – **LN20)** 18, 2 Duan, Xinghai Beilu, Xingcheng, Liaoning 121600. – **LN21)** 6 Qingnian Lu, Nanshan Jie, Beipiao, Liaoning 122100. V.O. Beipiao on 91.2MHz 2125-1500 (exc. Tues 0500-0930)

NEI MENGGU AUTONOMOUS REGION

NM1) 55 Xinhua Dajie, Hohhot, Nei Menggu 010058 **W:** www.nmrb. cn Chinese News General Sce. on 675/765/1494/7420/9520kHz 2150-1605 (exc. Tues 0600-0950). - Ch News Sce. on 89.0MHz 2150-1600. - Mongolian News General Sce. on 1458/1098/1395/6040/7270/97 50kHz/FM 2150-1605 (exc. Tues 0600-0950). - Economic Life Sce. on 101.4MHz 2150-1700. - V.O. the Traffic: on 89.6/89.8/90.6/95.7/101 .9/105.6MHz 2155-1700. - V.O. the Music: on 93.6MHz 2150-1600. - Storytelling and Art Sce. on 102.8MHz 2150-1700. - V.O. the Green Field: on 91.9MHz 2150-1605 (exc. Tues 0600-0950). – **NM2)** 159 Gongyuan Xilu, Hohhot, Nei Menggu 010035. On 882kHz 2200-0130, 0150-0530, 0830-1400 in Ch and Mo. - Traffic Sce. on 107.3MHz 2250-1600. - City and Life Sce. on 90.1MHz 24h. – **NM3)** 12 Gangtie Dajie, Hondlon Qu, Baotou, Nei Menggu 014030. News General Ch. on 558kHz/94.9MHz 2045-1600 (exc. Tues 0600-0955). - Life and Entertainment Ch. on

105.9MHz 1955-1605 (exc. Tues 0600-0915). - Traffic and Literary Ch. on 89.2MHz 24h (exc. Tues 0600-0920). - Urban and Rural Music Ch. on 100.1MHz 24h (exc. Tues 0730-0930). – **NM4)** 17 Ordos Dongjie, Haibowan Qu, Wuhai, Nei Menggu 016000. W2225-0025, Sun0025-0515, W0325-0520, D1025-1305(SS 1405). – **NM5)** 12, Xi Duan, Gangtie Xijie, Hongshan Qu, Chifeng, Nei Menggu 024001. Ch General Sce. on 1143/549/1512kHz/96.0MHz 1958-1730. - Mo General Sce. on 1440kHz/89.4MHz 2000-1730. - Traffic Sce. on 102.4MHz 24h. - Farmers and Herdsmen Literary Sce. on 900kHz/101.8MHz 2100-1600. - "Lark" FM Stereo Sce. on 89.4MHz 2220-1400. – **NM6)** 86 Qiaoxi Shahe Lu, Jining Qu, Ulanqab, Nei Menggu 012000. Ch News General Sce. on 702kHz/93.3/98.7MHz 2150-1500. - Mo Prgr. on 1521kHz 2125-1505. - Traffic Sce. on 90.7/99.0MHz in Ch. – **NM7)** 89 Xilin Dajie, Xilinhot, Nei Menggu 026000. Ch General Sce. on 1395kHz/99.4MHz 2225-1455. - Mo General Sce. on 927kHz 2220-1505. - General Literary Sce. on 106.9MHz 2255-1505. - Traffic and Literary Sce. on 97.5MHz 2225-1455. – **NM8)** 43 Manzhouli Lu, Hailar Qu, Hulun Buir, Nei Menggu 021008. Chinese News General Ch. on 603/3900kHz/99.9MHz 2130-0700(Tues 0210), 0900-1440. - Mongolian News General Ch. on 954/6080kHz 2150-2400, 0355-0600, 0920-1500. - Life and Literary Ch. on 104.6MHz. – **NM9)** 29 Heping Lu, Horqin Qu, Tongliao, Nei Menggu 028001. V.O. Tongliao: on 702kHz/97.2MHz 2150-0520, 0930-1330. - Ch 2nd Prgr. on 1233kHz/90.4MHz. - Mo Prgr. "V.O. Horqin": on 1350kHz/93.7/100.3MHz 2145-0500, 0925-1500. – **NM10)** Ordos BS, Manduhai Xiang, Dongsheng Qu, Ordos, Nei Menggu 017000. Ch Prgr. on 936/792kHz/98.9MHz 2220-0020, 0320-0600, 1000-1320. - Mo Prgr. on 603kHz. - Traffic Sce. on 100.8MHz – **NM11)** 26 Xinhua Xijie, Linhe Qu, Bayannur, Nei Menggu 015000. News General Sce. (V.O. Hetao): on 1152kHz/107.0MHz. - Traffic and Literary Sce. (V.O. the Yellow River): on 97.7MHz 2200-1600. - Life and Entertainment Sce. (V.O. the City): on 95.8MHz. – **NM12)** 1 Elute Donglu, Bayanhot Zhen, Alxa Zuoqi, Nei Menggu 750306. Ch Prgr. on 549/6025kHz 2230-1600. - Mo Prgr. on 1440kHz 2230-1600. – **NM13)** 73 Hinggan Bei Dalu, Ulanhot, Nei Menggu 137400. V.O. Hinggan: on 891kHz/89.1MHz 2125-1430 in Ch. - V.O. Alateng Hinggan: on 1152kHz/90.5MHz 2130-1430 in Mo. - V.O. the Traffic: on 99.0MHz 2125-1500 (exc. Tues 0600-0800). - V.O. the City: on 106.8MHz 2200-1500. – **NM14)** Xuegang Shan, Xinchengwan Xiang, Fengzhen, Nei Menggu 012100. 2225-0020, 0355-0505, 0955-1215. – **NM15)** 1 Dianshi Jie, Manzhouli, Nei Menggu 021400. 2225-1500. – **NM16)** 1 Xing'an Dongjie, Yakeshi, Nei Menggu 022150. – **NM17)** 3 Shengli Lu, Shiqiao Jie, Zalantun, Nei Menggu 162650. ?-0635, 0925-?. – **NM18)** Zhongyang Dajie, Genhe, Nei Menggu 022350. 2130-0700, 0900-1430. – **NM19)** 129 Nawenxi Dajie, Nirji Zhen, Morin Dawa, Nei Menggu 162850.

NINGXIA HUI AUTONOMOUS REGION

NX1) 66 Xingqing Zhonglu, Jinfeng Qu, Yinchuan, Ningxia 750001. **W:** www.nxtv.cn/radio/ News General Ch. on 891/1206/1287kHz/FM 2115-1900 (exc. Tues 0640-0930). - Economic Ch. on 747kHz/92.8MHz 2200-1600 (exc. Tues 0600-1000). - City Ch. on 103.7MHz 2225-1635 (exc. Tues 0630-0955). - Traffic Ch. on 98.4MHz 24h(exc. Tues 0600-0955). – **NX2)** 11 Zhongshan Beijie, Xingqing Qu, Yinchuan, Ningxia 750004. On 801kHz W2225-0145, Sun0000-0500, W0355-0530, D1025-1310(Sun 1330). - Traffic Sce. on 100.6MHz 2225-1800. – **NX2A)** Yinchuan City Economic St, 11 Zhongshan Beijie, Xingqing Qu, Yinchuan, Ningxia 750001. On 95.0MHz 0000-1705. – **NX3)** Shizuishan PBS, 363 Youyi Xijie, Dawukou Qu, Shizuishan, Ningxia 753000. – **NX4)** 54 Yumin Dongjie, Litong Qu, Wuzhong, Ningxia 751100. On 1413kHz 2230-1600. – **NX5)** Wenhua Jie, Xiaoba Zhen, Qingtongxia, Ningxia 751600.

QINGHAI PROVINCE

QH1) 81 Xiguan Dajie, Xining, Qinghai 810008 **W:** www.qhradio.com News General Ch. (Satellite Sce.) on 666/711/909/4750/6145/9780kHz/ 91.6MHz 2200-1605 (exc. Tues 0600-0855). - Tb Sce. on 1251/4220/5990/9850kHz/98.3MHz 2250-1600. - Economic Ch. on 1143kHz/107.5MHz 2255-1600 (exc. Tues 0600-0855). - Traffic and Music Sce. on 1377kHz/97.2MHz 2255-1600. – **QH2)** 43 Nanguan Jie, Xining, Qinghai 810000. News General Sce. on 1476kHz 2200-1630. - Health Sce. on 95.6MHz - Music Sce. on 104.3MHz 2230-1905. – **QH3)** 139 Hongwei Lu, Jiegu Zhen, Yushu Xian, Qinghai 815000. On 882/6075kHz 2255-0100, 1025-1230 in Ch and Tb. Rel. CNR 1: 1135-1230. – **QH4)** Haixi PBS, 7 Changjiang Lu, Delingha, Qinghai 817000. Ch Prgr. on 621kHz. Mo/Tb Prgr. on 540kHz.

SHANDONG PROVINCE

SD1) 81 Jing 11 Lu, Lixia Qu, Jinan, Shandong 250062. **W:** www. sdgb.cn News Ch. on 918/1467/1485/1548kHz/FM 1940-1700 (exc. Tues 0530-0900). English Prgr: 1650-1700. - Economic Ch. "Fortune Media" on 594kHz. 24h. - Economic Ch. "FM96" on 96.0MHz 24h. - Life Ch. on 105.0/88.6/104.7/104.9/107.8MHz 24h. - V.O. the Traffic and Music "Love FM": on 101.1/106.0/106.9MHz 24h. - Entertainment FM: on 97.5MHz 24h (exc. Tues 0500-0900). - Country Ch. "V.O. Green": on 1251/621/999kHz/FM 24h. - Music Ch. "City FM": on 99.1/106.6/107.8MHz 24h. - Sports and Leisure Sce. "Star FM" on

102.1MHz – **SD2)** 32 Jing 11 Lu, Lixia Qu, Jinan, Shandong 250014. News Sce. on 1053kHz/89.3/106.6MHz 24h (exc. Tues 0410-0850). - Economic Sce. on 846kHz/90.9MHz 2055-1700. - Traffic Sce. on 103.1MHz 24h (exc. Tues 0400-0850). - Music Sce. on 88.7/105.8MHz 24h. - Literary Sce. on 1305kHz/93.6MHz 24h (exc. Tues 0400-0900). - Story Sce. on 1512kHz/101.6/104.4MHz 24h. - Folk Art Sce. on 99.7MHz – **SD3)** 200 Ningxia Lu, Qingdao, Shandong 266071. News Life Prgr. on 1377kHz/104.1MHz - News Sce. on 107.6MHz 2030-1630. - Automobile Life Sce. on 1251kHz/102.9MHz 2035-1800. - Economic Sce. "Happy 603": on 603kHz/100.7MHz 2040-1800. - Traffic Sce. on 900kHz/89.7MHz 24h. - Literary Sce. "Your FM": on 846/1008kHz/96.4MHz 2100-1800. - Music and Sports Sce. "Simul Radio" on 91.5MHz 24h. – **SD4)** Zibo Radio and TV Headquarters, 52 Huaguang Lu, Zhangdian Qu, Zibo, Shandong 255047. News Sce. on 1143kHz/89.0MHz 2155-1700. - Economy Sce. on 801kHz/106.7MHz 2155-1700. - Traffic and Literary Sce. on 1476kHz/100.0MHz 2145-1700 (exc. Tues 0500-0900). - V.O. the City (Vanguard FM): on 92.6MHz 2145-1700. – **SD5)** 88 Guangming Xilu, Zaozhuang, Shandong 277102. News General Sce. on 1170kHz/97.6MHz 2155-1600. - Economic Life Sce. on 603kHz/103.7MHz 2200-1600. - Traffic and Literary Sce. on 747kHz/105.2MHz 2200-1600. – **SD6)** 1229 Dongcheng Nan 1 Lu, Dongying, Shandong 257091. News Ch. on 1449kHz/102.2MHz 2155-1430. - Economic Ch. on 105.3MHz 2150-1435. - Traffic and Music Ch. on 94.9MHz 2150-1435. – **SD7)** 248 Dongfeng Dongjie, Kuiwen Qu, Weifang, Shandong 261041. News Sce. on 1161kHz/100.2MHz 2055-1600. - i Home Sce. on 1287kHz/93.3MHz 2100-1700. - Traffic and Music Sce. on 846kHz/95.9MHz 24h. - Health and Entertainment Ch. on 98.3MHz 2055-1700. - Music Sce. "Simul Radio": on 88.7MHz 24h. - Story and Chinese Opera Sce. on 981kHz/107.1MHz 2055-1700. – **SD8)** 32 Wenhua Xiang, Zhifu Qu, Yantai, Shandong 264000. News Ch. on 1314kHz/101.0/98.6MHz 2055-1600. Rel. CRI Korean news: 1300-1315. - Economic Life Ch. on 801kHz/105.9/102.7MHz 2055-1600. - Traffic Sce. on 103.0/89.1/95.3MHz 2055-1600. Rel. CRI English prgr: 1500-1600. - Music Ch. "I Radio": on 91.2MHz 24h. - Storytelling Sce. on 88.4/96.6MHz 2055-1800. Rel. CRI "News Radio": 2300-2400, 0400-0500, 1000-1100, 1500-1600. – **SD9)** 66 Wenhua Zhonglu, Weihai, Shandong 264200. News General Ch. on 1206kHz/99.6/105.1MHz 2100-1600 (exc. Tues 0600-0825). Ko Prgr: 0530-0600, 1430-1500. - Traffic and Literary Ch. on 846kHz/96.1/101.7MHz 2125-1500 (exc. Tues 0600-0855). - Story Ch. on 95.0MHz 2100-1600. – **SD10)** 11 Hongxing Zhonglu, Jining, Shandong 272037. News Sce. on 666kHz/104.2MHz 2200-1700. - Economic Life Sce. on 1116kHz/99.3MHz 2200-1700. - Traffic and Literary Sce. on 801kHz/101.8MHz 2200-1700. - Entertainment Sce. on 945kHz/103.1MHz 2200-1700. – **SD11)** Beishou, Yantai Lu, Rizhao, Shandong 256618. News General Ch. on 1449kHz/95.0MHz 2130-1600 (exc. Tues ?-0945). - Traffic and Life Ch. on 747kHz/88.1MHz 2130-1530. - Literary and Sports Ch. on 104.0MHz 2130-1530. – **SD12)** 64 Dongfanghong Lu, Dezhou, Shandong 253012. News General Sce. on 1098kHz/92.9MHz 2150-1600. - Traffic and Music Sce. on 1341kHz/94.1MHz 2150-1600. - Literary and Life Sce. on 1008kHz/104.1MHz 2150-1600. – **SD13)** Linyi Radio and TV St, 21 Jinqueshan Lu, Lanshan Qu, Linyi, Shandong 276004. News Sce. on 873kHz/97.6MHz 2125-1600 (exc. Tues 0530-1020). - Literary Sce. on 1143kHz/93.2MHz 2125-1600. - City Sce. on 747kHz/101.0MHz 2155-1600. - Traffic and Music Sce. on 801kHz/89.9MHz 2155-1600 (exc. Tues 0600-0950). - Story Sce. on 104.5MHz 2155-1600. – **SD14)** 41 Liuyuan Beilu, Liaocheng, Shandong 252000. News Sce. on 1143kHz/96.8MHz 2125-1500. - Economic Sce. on 801kHz 2130-1600. - Traffic Sce. on 98.9MHz 2130-1600. - Music Sce. on 92.4MHz – **SD15)** 358 Huanghe 5 Lu, Binzhou, Shandong 256618. News General Sce. on 846kHz/107.6MHz 2130-1535. - Economic Life Sce. on 1170kHz/99.4MHz 2200-1530. - Traffic Sce. on 93.1MHz – **SD16)** 28 Zhonghua Donglu, Heze, Shandong 274033. News Ch. on 1197kHz/92.7MHz 2055-1600. - Traffic Ch. on 94.8MHz 2055-1600. - Chinese Opera Ch. on 1071kHz/96.8MHz 2055-1500. – **SD16A)** Mudan PBS, 2093 Changjiang Lu, Heze, Shandong 274000. V.O. Heze on 1323kHz/97.2MHz 2155-1700. - Heze V.O. the City: on 104.0MHz 2125-1600. – **SD17)** Qingzhou PBS, 21 Fangongting Xilu, Qingzhou, Shandong 262500. On 95.4MHz 2125-1600. – **SD18)** Huangcheng Xihuan Lu, Longkou, Shandong 265701. On 101.6MHz - Yantai Longkou Economic and Literary BS: on 1251kHz 2228-0200, 0500-0700. – **SD19)** 4 Gulou Beijie, Qufu, Shandong 273100. On 1341kHz/98.7MHz 2155-0510(SS0450), 0955-1430(SS1410).

SHANGHAI MUNICIPALITY

SH1) Shanghai Radio and TV St, 1376 Hongqiao Lu, Shanghai 200051 **W:** www.smg.cn News Ch. on 990kHz/93.4MHz 24h (exc. Thurs 1705-2100). - Traffic Ch. on 648kHz/105.7MHz 24h (exc. Fri 1700-2100). - Chinese Opera Ch. on 1197kHz/97.2MHz 2150-1600 (exc. Wed 0530-0830) in Ch and Shanghai dialect. - New Entertainment FM "Happy Radio" on 927kHz/107.2MHz 2200-1600 (exc. Wed 0530-0830). - Sports Ch. on 94.0MHz 2155-1600 – **SH2)** Shanghai Dongfang BS (Eastern Radio), 1376 Hongqiao Lu, Shanghai 200051. **W:** www.smg.cn News

St. on 1296kHz/90.9MHz 24h (exc. Thurs 1600-2100). - "City 792": on 792kHz/89.9MHz 24h (exc. Thurs 1600-2100). - First Financial and Ecomonic Ch. on 1422/603kHz/97.7MHz 2158-1600. - Popular Music Ch. "Donggan 101": on 101.7MHz 2200-1800 (exc. Fri 0600-0800). - Popular Music Ch. "Love Radio": on 103.7MHz 24h (exc. Thurs 1600-2200). - Classical Music Ch. on 94.7MHz 2200-1700.

SHAANXI PROVINCE

SN1) 336 Chang'an Nanlu, Xi'an, Shaanxi 710061 **W:** www.sxradio.com.cn News Sce. on 693/1008/1521/6176kHz/FM 24h (exc. Tues 0600-0900). - Economic (Fortune) Sce. on 89.6MHz 1930-1830. - City Sce. on 1008kHz/101.8MHz 24h. - Traffic Sce. on 1323kHz/91.6MHz 24h. - Farm Sce. on 900kHz 1930-1700. - Youth Sce. "Young Radio": on 105.5MHz 24h. - Chinese Opera Sce. on 747kHz/99.4MHz 24h. - Music Sce. on 98.8/94.8/97.5MHz 24h. - Story Sce. on 603kHz 24h (exc. Wed 1700-1945). - Qin Melody Sce. on 101.1MHz 24h. – **SN2)** 100, Zhenxing Lu, Xi'an, Shaanxi 710068. News Sce. on 810kHz/102.1MHz 2100-1600. - Information Sce. on 106.1MHz 24h (exc. Tues 1730-2100). - Traffic and Travel Sce. on 104.3MHz 2150-1800. - Music Sce. on 801kHz/93.1MHz 24h. – **SN3)** Miaopu Lu, Hongqi Jie, Tongchuan, Shaanxi 727000. 2210-0015, 0330-0515, 0915-1405. – **SN4)** 47 Hongqi Lu, Baoji, Shaanxi 721000. On 1071kHz 2145-2400, 0325-0610, 0930-1500. - Music and Storytelling Sce. on 105.3MHz 2255-1700. - Economic and Traffic Sce. on 900kHz/102.8MHz 2155-0600, 0955-1400. – **SN5)** Nan Duan, Fu'an Lu, Xianyang, Shaanxi 712000. News General Sce. on 1296kHz/100.7/107.6MHz 2150-1740. - City Music Sce. on 99.9MHz 2200-1740. – **SN6)** Xi Duan, Dongfeng Jie, Weinan, Shaanxi 714000. News Sce. on 747kHz/101.3/102.6MHz - Traffic Sce. on 90.9MHz – **SN7)** Dongguan Jie, Yan'an, Shaanxi 716000. News Sce. on 603kHz/100.1/104.6MHz 2210-1500 (exc. Wed 0630-0910). - Traffic Sce. on 98.7MHz – **SN8)** 14 Dong Jianshe Xiang, Hanzhong, Shaanxi 723000. News Sce. on 1251kHz/95.6MHz 2130-1620. - Music Sce. on 97.1/99.5MHz 24h (exc. Wed 0700-0930). - Traffic and Travel Sce. on 93.0/94.3/101.8MHz – **SN9)** 7 Zhonglou Xiang, Yulin, Shaanxi 719000. News Sce. on 1143kHz/99.4MHz - Traffic and Literary Sce. on 95.9MHz – **SN10)** Ankang PBS, 113 Bashan Zhonglu, Ankang, Shaanxi 725000. News Sce. on 89.7MHz - Traffic Travel and Music Sce. on 95.9MHz 2155-1600.

SHANXI PROVINCE

SX1) Shanxi Radio and TV Headquarters, 318 Yingze Dajie, Taiyuan, Shanxi 030001 **W:** www.sxrtv.com News General Sce. on 819/846/900/1269kHz/FM 2159-1700 (exc. Tues 0600-0900). - Changcheng Economic Sce. on 95.8MHz 2200-1700 (exc. Tues 0600-0900). - V.O. the Health: on 1584kHz/105.9MHz 2200-1700 (exc. Mon 0600-0900). - Traffic Sce. on 88.0MHz 24h. - Farm Sce. on 603kHz 2200-1700. - Music Sce. on 94.0MHz 2200-1700. – **SX2)** 2 Yifen Jie, Taiyuan, Shanxi 030024. News Ch. on 91.2MHz 24h. – V.O. Old Age: on 1422kHz 2155-1600. - Economic Ch. on 774kHz/104.4MHz 24h. - Traffic Ch. on 107.0MHz 24h. – **SX3)** 178 Yingbin Xilu, Datong, Shanxi 037006. General St. on 900kHz/103.8MHz 2200-1605. - Health St. on 91.1MHz 2200-1605. - Traffic St. on 99.6MHz 2200-1605. – **SX4)** Ningbo Lu, Yangquan, Shanxi 045000. 2150-2400, 0300-0535, 0955-1355. – **SX5)** 87 Yingxiong Zhonglu, Changzhi, Shanxi 046000. News General Sce. on 1584kHz/98.8MHz 2120-0600, 0915-1530. - Traffic and Literary Sce. on 94.9/101.1MHz 2225-1600 (exc. Tues 0500-0900). – **SX6)** Fengtai Xijie, Jincheng, Shanxi 048000. News General Sce. on 585kHz/89.8MHz 2155-1600. - Traffic and Health Sce. on 93.5MHz 2155-1600. – **SX7)** 3 Xiaoyuan Lu, Yuci Qu, Jinzhong, Shanxi 030600. On 1530kHz/92.5MHz 2200-1600. – **SX8)** Cangcheng Xijie, Xinzhou, Shanxi 034000. – **SX9)** 233 Hongqi Dongjie, Yuncheng, Shanxi 044000. News General Sce. on 1566kHz/93.2MHz 2200-1600. – Traffic and Literary Sce. on 101.9MHz 2200-1600. – **SX10)** 1 Minfu Xijie, Shuozhou, Shanxi 036002. – **SX11)** Linfen PBS, 10 Guangxuan Jie, Linfen, Shanxi 041000. News General Ch. on 94.1MHz - Traffic and Literary Sce. on 88.9MHz

SICHUAN PROVINCE

SC1) 119-1 Hongxing Zhonglu, Chengdu, Sichuan 610017. News Ch. on 612/909/1116kHz/98.1/90.0/93.7/95.7/103.9/106.6MHz 24h. - Economic Ch. "Times Broadcast" on 88.4/94.0MHz 2300-1700. - Economic Ch. "Public Broadcast" on 89.4MHz 24h. - Traffic Sce. on 101.7MHz - V.O. the Health: on 999kHz 2200-1700 (exc. Tues 0600-0800). - City Life Sce. on 97.0MHz 2155-1705. - Minority Sce. on 954/6060/7225kHz 2155-1705 in Ch, Tb and Yi. - Educational Prgr. on 1521kHz/98.1MHz 0900-1030, 1330-1500. - Minjiang Music St. on 95.5MHz 2230-1700 (exc. Tues 0600-1000). - Sound of City "City FM": on 102.6MHz 2200-1600. - Sichuan Women & Children BS: on 747kHz 2200-1640 (exc. Tues 0600-1000). – **SC3)** 99 Shuanglin Lu, Chengdu, Sichuan 610021. News Sce. on 792kHz/90.8MHz 2130-1700. - Traffic Sce. on 1485kHz/91.4MHz 2200-1700 (exc. Tues 0500-0800). - Music Sce. "Love Radio": on 105.6MHz 2200-1700. - Cultural and Leisure "V.O. Feiyang": on 94.6MHz – **SC4)** 1 Wenhua Lu, Huidong Xinqu, Zigong, Sichuan 643000. On 1422kHz/100.9/103.0MHz 2220-1505 (exc. Tues 0600-1000). - Yandu Music St. on 97.7MHz 2150-1530. – **SC5)** 338 Linjiang Lu, Dong Qu, Panzhihua, Sichuan 617000. On 711kHz/88.5MHz

2150-1605. – **SC6)** Datong Lu, Chengbei Xinqu, Luzhou, Sichuan 646000. News General Sce. on 954kHz/89.8/97.0MHz 2155-1600. - Traffic and Music Sce. on 96.0/100.6MHz 2155-1600. – **SC7)** 63, 1 Duan, Taishan Nanlu, Deyang, Sichuan 618000. News Sce. on 720kHz/95.9MHz 2200-1600. Music and Traffic Sce. on 107.8MHz 2300-1600. – **SC8)** 232, Nan Duan, 1 Huan Lu, Fucheng Qu, Mianyang, Sichuan 621000. News Sce. on 711kHz/96.7/102.0MHz 2200-1600. - Scientific and Life Sce. on 91.2/92.6MHz 2200-1600 (exc. Tues 0700-0900). – **SC9)** 585, Xi Duan, Hezhou Donglu, Guangyuan, Sichuan 628017. 2135-1700. – **SC10)** 358 Suizhou Zhonglu, Suining, Sichuan 629000. 2200-1530. – **SC11)** 33, 1 Xiang, Xianglong Lu, Neijiang, Sichuan 641000. – Economic Sce. on 1143kHz/101.4MHz– **SC12)** 300, Nan Duan, Chunhua Lu, Shizhong Qu, Leshan, Sichuan 614000. News General Sce. on 1476kHz/102.8MHz 2225-1735. – **SC13)** Jiazhou EBS, 40 Dingdong Jie, Leshan, Sichuan 614000. On 95.7MHz 2200-1405 (exc. Wed 0500-1000) – **SC14)** 6 Sichou Lu, Nanchong, Sichuan 637000. On 747kHz 2155-0015, 0355-0520, 1155-1415. - V.O. Nanchong: on 91.5MHz – **SC15)** Yibin PBS, 106 Renmin Lu, Yibin, Sichuan 644000. News General Ch. on 92.8/97.0/101.4MHz 2210-1500 (exc. Tues 0630-1000). - Traffic and Music Ch. on 94.2/105.9MHz 2200-1700. - Farm and Literary Ch. on 104.2MHz – **SC16)** 92 Zhangjiawan, Tongchuan Qu, Dazhou, Sichuan 635000. News General Ch. on 1143kHz 2200-1600.

TIANJIN MUNICIPALITY

TJ1) 143 Weijin Lu, Heping Qu, Tianjin 300070 **W:** www.radiotj.com News FM Sce. on 97.2MHz 2055-1800(Tues 1600). - News MW Sce. on 909kHz 2055-1800(Tues 1600). - Economic Sce. on 1071kHz/101.4MHz 2055-1800(Tues 1600). - Economic Sce. "V.O. Hangu": on 101.8MHz - Traffic Sce. "Chinese Comic Dialogue": on 567kHz 2155-1800(Tues 1600) - Traffic Sce. on 106.8MHz 24h (exc. Tues 1600-2100). - Life Sce. on 1386kHz/91.1MHz 2055-1800(Tues 1600). - Literary Sce. on 1098kHz/104.6MHz 2155-1800(Tues 1600). - Music FM Sce. on 99.0MHz 24h (exc. Tues 1600-2055). - Music MW Sce. on 1008kHz 2055-1800(Tues 1600). - Binhai Sce. on 747kHz/92.0MHz 2055-1800(Tues 1600). - Entertainment Sce. on 87.8MHz 2155-1800(Tues 1600). - Novel Sce. on 666kHz 2200-1800(Tues 1600).

XINJIANG UIGHUR AUTONOMOUS REGION

XJ1) 84 Tuanjie Lu, Urumqi, Xinjiang 830044 **W:** www.xjbs.com.cn Ch Prgr. on 702/738/999/1494/5960/7260/7310/9600/9835/11770kHz 2300-1800 (exc. Tues 0800-1100). - Ug Prgr. on 558/855/1044/1413/612 0/7205/7275/9560/11885/13670kHz 2300-1800 (exc. Tues 0800-1100). - Kz Prgr. on 963/1107/6015/7340/9470kHz 2330-1800 (exc. Tues/Thurs 0800-1100). - Mo Prgr. on 1233/6190/7230/9510kHz 2330-0330, 0530-1030(Tues/Thurs 0800), 1230-1800. - Kirghiz Prgr. on 1233/6190/7230 /9705/11975kHz 0330-0530, 1030(Tues/Thurs 1100)-1230. - Ch News Sce. on 96.1MHz 2300-1800. - City Sce. on 837/1215kHz/92.9MHz 2300-1800. - Traffic Sce. on 94.9/101.8MHz 2330-1800. - Literary Sce. on 101.7MHz 2300-1800 (exc. Tues 0800-1100) in Ug. - Story Sce. on 102.8MHz 2300-1800. – **XJ2)** 28 Xinmin Lu, Urumqi, Xinjiang 830002. News Ch. on 792kHz 2345-1720. - Traffic and Music Ch. on 100.7MHz 2345-1730. - Travel and Entertainment Sce. on 106.5MHz 2345-1725. - Uighur General Ch. on 1071kHz 2345-1630. – **XJ3)** Urumqi EBS, 11 Nanchang Lu, Urumqi, Xinjiang 830002. Health and Life Ch. on 927kHz 2345-1800. - Traffic and Music Ch. on 97.4MHz 2345-1800. – **XJ4)** 42 Tianshan Xilu, Karamay, Xinjiang 834000. Ch News General Ch. on 1179kHz 2355-1800. - Ug Ch. on 882kHz 2355-1800. - City FM: on 92.6MHz 0000-2000. – **XJ5)** 2 Hongxing Xilu, Hami, Xinjiang 839000. Ch News Information St. on 1485kHz 2200-1800. - Ug Prgr. on 1098kHz/107.9MHz 2300-1600. - City FM: on 103.5MHz 2300-1800. – **XJ6)** 13 Urumqi Nanlu, Hotan, Xinjiang 848000. Ch Prgr. on 1026kHz. - Ug Prgr. on 774kHz/92.2MHz 2300-1800. – **XJ7)** Keziduwei Lu, Kashi, Xinjiang 844000. Ch Prgr. on 648kHz 2355-0215, 0455-0710, ?-1335. - Ug Prgr. on 801kHz 2355-?. – **XJ8)** 15 Nan Gongyuan Xilu, Changji, Xinjiang 831100. News General Ch. "V.O. Wuchang": on 873kHz 2300-1730. - Legal FM: on 105.3MHz – **XJ9)** Ili PBS, 1 Hongqi Lu, Yining, Xinjiang 835000. Ch Prgr. on 1134kHz/105.9MHz 2350-0200, 0550-0700, 1150-1600. - Ug Prgr. on 603kHz/96.3MHz 2350-0200, 0550-0700, 1150-1600. - Kz Prgr. on 882kHz 2350-0200, 0550-0700, 1220-1500. – **XJ10)** 184 Bei 2 Lu, Shihezi, Xinjiang 832000. News Ch. on 891kHz/103.5MHz 0030-0730, 1130-1600. - Literary Ch. on 603kHz/89.3MHz – **XJ11)** Renmin Donglu, Korla, Xinjiang 841000. – **XJ12)** Korla Donglu, Kuytun, Xinjiang 833200. Ch Prgr. on 1485kHz W2355-0230, Sun0025-0335, Sun0528-0720, W0558-0740, D1123-1425. - Kz Prgr. on 819kHz. – **XJ13)** Bayingolin PBS, 1 Jianguo Lu, Korla, Xinjiang 841000

XIZANG AUTONOMOUS REGION

XZ1) 41 Beijing Zhonglu, Lhasa, Xizang 850000 **W:** www.tibetradio.cn Chinese News General Ch. on 999/1377kHz/SW/93.3MHz 2000-1800 (exc. Tues 0600-1000). - Tibetan News General Ch. on 594/846kHz/SW/101.6MHz 2050-1805 (exc. Tues 0600-1000). English Prgr. "Holy Tibet": 0700-0730, 1630-1700. - Kham (Tibetan dialect) Ch. on 594kHz 2200-1605 (exc. Tues 0600-1000). - City Life Ch. on 98.0MHz 2300-1700 (exc. Tues 0600-1000). – **XZ2)** Lhasa PBS, Lhasa, Xizang 850000.

On 91.4MHz 2350-1410 in Tb and Ch. – **XZ3)** 25 Nedong Lu, Zetang Zhen, Nedong, Xizang 856000. 2335-0135, 0405-0535, 1005-1340 in Ch and Tb.

YUNNAN PROVINCE

YN1) 182 Renmin Xilu, Kunming, Yunnan 650031. News Sce. on 576/990/1080kHz/94.4/105.8MHz 2200-1700. - V.O. Shangri-la: on 99.0MHz - Life FM: on 88.7MHz 2250-1700. Rel. CRI English prgr: 1300-1500. - Minority Language Sce. on 7210kHz 1055-1500 in Lahu, Jingpo, Lisu, Dehong Dai and Xishuangbanna Dai. - V.O. the Traffic: on 91.8MHz 2240-1600. - Automobile Sce. on 95.4MHz - V.O. the Music "Binfen 97": on 97.0MHz 2300-1600 (exc. Tues 0600-0900). - Fashion FM: on 100.0MHz - Youth and Entertainment FM: on 101.7MHz 2250-1700. – **YN2)** 198 Danxia Lu, Kunming, Yunnan 650118. Sunlight Ch. on 1350kHz/100.8/95.4MHz 24h (exc. Tues 0400-0800). - City FM: on 102.8MHz 24h. – **YN3)** Qilin Nanlu, Qujing, Yunnan 655000. – **YN4)** Wolong Xiaoqu, Panzhihua Zhen, Wenshan Xian, Yunnan 663000. Minority Ch. on 1053kHz 2225-0030, 0355-0530, 0955-1400 in Ch, Zhuang, Miao and Yao. - News General Ch. on 103.0/105.2MHz 2220-1500. – Qihua FM: on 97.3MHz 2220-1600. – **YN5)** Honghe PBS, 31 Jianshe Donglu, Gejiu, Yunnan 661000. News General Ch. on 1521/1485kHz 2200-1800. - Minority Language Ch. on 702kHz/101.4MHz 2000-1800 in Ch, Hani and Yi. - City FM: on 92.9MHz - Music Sce. on 97.5MHz 2200-1900. – **YN6)** Xishuangbanna PBS, 2 Nonglin Xilu, Jinghong, Yunnan 666100. 2210-0100, 0250-0600, 1030-1540 in Ch, Xishuangbanna Dai and Hani. – **YN7)** Chuxiong Autonomous Prefecture PBS, 144 Lucheng Donglu, Chuxiong, Yunnan 675000. News General St. on 1287kHz/93.9MHz 2225-0050, 0310-0625, 0955-1355. - Economic and Music Sce. on 96.7MHz 2300-1600. – **YN8)** Wanhua Lu, Xiaguan Zhen, Dali, Yunnan 671000. News General Ch. on 1044kHz. - Cang'er FM: on 99.9/105.5MHz 2200-1600. – **YN9)** Dehong PBS, 30 Yingjian Lu, Mangshi Zhen, Luxi, Yunnan 678400. Minorities Ch. on 900kHz 2230-0110, 0330-0700, 1030-1530 in Ch, Dehong Dai, Jingpo and Zaiwa. - V.O. the Peacock: on 104.3MHz 2245-1600. – **YN10)** Donghuan Lu, Dongchuan Qu, Kunming, Yunnan 654100. – **YN11)** 6 Longquan Lu, Zhaotong, Yunnan 657000. News General Sce. on 621kHz 2225-1600. – **YN12)** Baohua Lu, Gejiu, Yunnan 661400. – **YN13)** 38 Xueqiao Jie, Chuxiong, Yunnan 675000. W2225-2400, Sun2325-0200, D0325-0600, D0955-1405. - Dianzhong FM: on 106.1MHz – **YN14)** 29 Guihua Lu, Yuxi, Yunnan 653100. Green FM on 1251kHz/102.4MHz 2225-1600. – **YN15)** Diqing PBS, 37 Changzheng Lu, Jiantang Zhen, Shangri-la Xian, Yunnan 674400. – **YN16)** Nujiang PBS, 96 Xiangyang Xilu, Liuku Zhen, Lushui Xian, Yunnan 673100. – **YN17)** Lufeng, Yunnan. 2225-1230.

ZHEJIANG PROVINCE

ZJ1) 111 Moganshan Lu, Hangzhou, Zhejiang 310005 **W:** www.cztv.com.cn News St. "V.O. Zhejiang": on MW/FM. 24h (exc. Tues 0600-0858). - Economic St. "Fortune Sce." on 95.0MHz 24h. - V.O. the City "Private Car 107": on 1530kHz/FM 24h. - V.O. the Traffic: on 93.0/93.6MHz 24h. - Music FM "Moving 968": on 1071kHz/96.8/88.6/89.8MHz 24h (exc. Tues 0600-0800). - Music FM "People Life 996": on 930/1050/1314kHz/99.6MHz 24h. - V.O. the Travel "Anchorwomen BS": on 603/1251/1521kHz/FM 24h. – **ZJ2)** 86 Moganshan Lu, Hangzhou, Zhejiang 310005. **W:** www.radiohz.com News General Ch. (Jinqiu Ch.) on 954kHz 2000-1600. News Sce. on 89.0MHz 24h. – **ZJ2A)** Hangzhou Traffic and Economic Sce, 5 Qingchun Donglu, Hangzhou, Zhejiang 310016. On 91.8MHz 24h. – **ZJ3)** Automobile Sce. "V.O. Xihu", 86 Moganshan Lu, Hangzhou, Zhejiang 310005. On 105.4MHz 24h in Ch and Hangzhou dialect. – **ZJ4)** 109 Heyi Lu, Ningbo, Zhejiang 315000. News Sce. on 1323kHz/92.0MHz 2055-1610 (exc. Tues 0600-0800). English N: D1600-1610. - Old and Young Sce. on 1251kHz/90.4MHz 2200-1400 (exc. Tues 0600-0900). - Economic Sce. on 747kHz/102.9MHz 2100-1830 (exc. Tues 0600-0730). - Traffic Sce. on 603kHz/93.9MHz 24h (exc. Tues 0500-0900). - Music Sce. on 98.6MHz 2300-1600 (exc. Tues 0500-0900). – **ZJ5)** Xincheng Dadao, Lucheng Qu, Wenzhou, Zhejiang 325027. News Sce. on 666kHz/94.9/103.9MHz 24h (exc. Tues 0600-0900) in Ch and Wenzhou dialect. - Economic Life Sce. on 801kHz/88.8MHz 24h (exc. Tues 0600-0900). - Traffic Sce. on 103.9MHz 24h. - V.O. the Music: on 100.3MHz 24h (exc. Tues 0600-0900). - Air Shopping Prgr. on 97.2MHz – **ZJ6)** Jiaxing Radio and TV Headquarters, 6 Dongsheng Lu, Jiaxing, Zhejiang 314001. News Sce. "V.O. Nanhu": on 1107kHz/104.1MHz 2125-1505. - Traffic and Economic Sce. on 657kHz/92.2MHz 2130-1700 (exc. Tues 0530-0730). - City and Country Life Sce. on 88.2MHz 2130-1430. – **ZJ7)** 628 Xinhua Lu, Huzhou, Zhejiang 313000. News General Ch. on 873kHz/105.0MHz 2155-1600 (exc. Tues 0600-0730). - Traffic and Economic Ch. on 927/1521kHz/103.5MHz 2155-1600. - City Literary Ch. on 1251kHz/98.5MHz 2110-1600. – **ZJ8)** 508 Yan'an Donglu, Shaoxing, Zhejiang 312000. News General Ch. on 738kHz/96.0MHz 2100-1600. - Traffic Sce. on 94.1MHz 2130-1600 (exc. Tues 0600-0830). - Chinese Opera Ch. on 102.5MHz 2130-1300. - Music Ch. "i Music": on 103.5MHz 2100-1500 (exc. Tues 0600-0900). – **ZJ9)** 238 Renmin Xilu, Jinhua, Zhejiang 321000. News Sce. on 675kHz/104.4MHz 2100-1600 (exc. Tues 0600-0900).

- Fortune and Music Sce. on 101.4MHz 2200-1700. - Traffic Sce. on 94.2MHz 24h. – **ZJ10)** Quzhou Radio and TV Headquarters, 35 Nanjie, Quzhou, Zhejiang 324000. News Sce. on 711kHz/105.3MHz 2155-1600 (exc. Tues 0500-0725). - Traffic and Music Ch. on 1250kHz/97.5MHz 2200-1700. – **ZJ11)** Zhoushan Radio and TV Headquarters, 137 Changguo Lu, Dinghai Qu, Zhoushan, Zhejiang 316000. News General Ch. on 684kHz/99.8MHz 2130-1500 (exc. Tues 0530-0855). - Traffic and Economic Sce. on 1098kHz/97.0MHz 2155-1500 (exc. Tues 0500-0900). - Literary and Educational Ch. on 900kHz/91.0/102.6MHz 2155-1500 (exc. Tues 0530-0855). – **ZJ12)** Lishui Radio and TV Headquarters, 2 Huayuan Lu, Liandu Qu, Lishui, Zhejiang 323000. News General Ch. on 711kHz/94.0/96.4MHz 2155-1600. - Traffic and Music Ch. on 106.9MHz 2155-1600. – **ZJ13)** Xiaoshan PBS, Nanduan, Yucai Lu, Xiaoshan Qu, Hangzhou, Zhejiang 311200. On 107.9MHz 2155-1400. – **ZJ14)** Xishan, Chengguan, Rui'an, Zhejiang 325200. On 1584kHz/91.0MHz ?-1305. – **ZJ15)** 121 Zhongshan Lu, Jiangshan, Zhejiang 324102. – **ZJ16)** Taizhou PBS, 355 Donghuan Dadao, Jiaojiang Qu, Taizhou, Zhejiang 318000. News Sce. on 98.7/88.0MHz 24h. - Traffic Sce. on 102.7MHz 24h. - Music St. "Easy Radio" on 100.1/104.9MHz 2200-1600.

CHRISTMAS ISLAND (Australia)

L.T: UTC +7h — **Pop:** 1,402 — **Pr.L:** English, Malay, Cantonese, Hokkien, Mandarin — **E.C:** 50Hz, 240V — **ITU:** CHR

MW	kHz	Call	kW	MW	kHz	Call	kW
1)	1422	6ABCRN	0.5	2)	1620	-	0.4
FM	**Mhz**	**Call**	**kW**	**FM**	**Mhz**	**Call**	**kW**
1)	97.3	6ABCRN	0.02	5)	102.1	6RCI	0.02
3)	98.9	6FMS	0.02	5)	105.3	6RCI	0.02
4)	100.5	6JJJ	0.02	3)	106.9	6FMS	0.04

Addresses & other information
1) 24hr satellite relay ABC Radio National – **2)** Promo Radio Pty Ltd, 12 Pickering Close, Hoppers Crossing VIC 3029 (currently inactive) – **3)** 24h satellite relay RED FM, Perth WA – **4)** 24hr satellite relay ABC Triple J – **5)** Radio Christmas Island, Broadcast House, Murray Road, Drumsite (PO Box 474) Christmas Island WA 6798. ☎ +61 8 9164 8316/8422 ▤ +61 8 9164 8315 (local community stn).

COCOS (KEELING) ISLANDS (Australia)

L.T: UTC +6½h — **Pop:** 596 — **Pr.L:** English, Cocos Malay — **E.C:** 50Hz, 220V — **ITU:** ICO

MW	kHz	Call	kW	Station	Location
1)	1611	-	0.4	Promo Radio	West Island
2)	1629	-	0.4		West Island
FM	**Mhz**	**Call**	**kW**	**Station**	**Location**
3)	96.0	6CKI	0.1	VO the Cocos	West Island
4)	100.5	6FMS	0.1	RED FM	West Island
5)	102.1	6ABCRR	0.1	ABC Kimberley	West Island
3)	102.7	6CKI	0.02	VO the Cocos	Home Island

Addresses & other information
1) Promo Radio Pty Ltd, 12 Pickering Close, Hoppers Crossing VIC 3029 (currently inactive) – **2)** E. Ardstrom, 2/33 Musgrave St, Mosman NSW 2088 (currently inactive) – **3)** Voice of the Cocos (Keeling) Islands, PO Box 1093, Cocos (Keeling) Islands WA 6799. ☎+61 8 9162 6700. Local community stn 24h with local news 0700 UTC M-F. – **4)** 24h satellite relay from Perth WA – **5)** 24h satellite relay from Broome WA.

COLOMBIA

L.T: UTC -5h — **Pop:** 45 million — **Pr.L:** Spanish — **E.C:** 60Hz, 110V — **ITU:** CLM

MINISTERIO DE COMUNICACIONES Dirección General de Telecomunicaciónes y Servicios Postales ▤ Edificio Murillo Toro, Cras 7 y 8, Calles 12A y 13, Santafé de Bogotá, DC ☎+57 1 286 6911 **W:** www.mincomunicaciones.gov.co Call HJ-, ° also on shortwave, * = inactive, rel. = relay v = varying freq, SF de Bogotá = Santa Fe de Bogotá. The letters preceding the stn number indicate the departamento. Addresses are listed by departamento in alphabetical order.
Hr of tr. usually 24h – see address section for variations

MW	Call	kHz	kW	Station, location
DC01)	KA	540	20	R. Auténtica Básica, SF de Bogotá
DC02)	HF	550	50	R. Nal., Marinilla (rel: 570)
DC02)	ZQ	*550	50	R. Nal., Neiva (rel: 570)
DC02)	GS	*560	10	R. Nal., Tunja (rel: 570)
GU01)	PF	560	25/10	LV de la Pampa, Maicao
DC02)	ND	570	100	R. Nal de Colombia, SF de Bogotá

MW	Call	kHz	kW	Station, location
DC02)	HP	580	50/10	R. Nal., Cali (rel: 570)
AN01)	CR	590	50	W Radio, Medellín
AT01)	HJ	600	50	R. Libertad, Barranquilla
NA13)	Z95	600	1	LV de los Awas, Ricaurte el Diviso
DC02)	D90	610	50	R. Nal., Uríbia
DC03)	KL	610	30	La Cariñosa, SF de Bogotá
B001)	VP	620	15	Colmundo, Cartagena
VA01)	EL	620	50/20	Colmundo, Cali
CL01)	FD	630	10	R. Manizales, Manizales
GN01)	WC	630	10	LV del Guainía, Puerto Inírida
MA01)	BJ	640	10	RCN, Santa Marta
DC03)	KH	650	100	RCN Antena 2, SF de Bogotá
NS01)	QS	660	25	Colmundo, Cúcuta
VA02)	JM	660	20	R. Auténtica, Cali
AN02)	PL	670	50	RCN Antena 2, Medellín
SS28)	R33	670	10	R. U.I.S - Universidad Industrial de Santander, Bucaramanga
DC02)	ZO	680		R. Nal., Barranquilla (Sabanagrande)
AN56)	Z73	690	1	LV Indígena de Uberaba, Apartadó
DC04)	CZ	690	50/12	R. Recuerdos, SF de Bogotá
VA03)	CX	700	120	W Radio, Cali
AN03)	NX	710	10	R. Super, Medellín
BY14)	YD	710	5	R. La Paz, Paipa
AT01)	AN	720	30	Emisoras Unidas, Barranquilla
DC02)	ZX	*720	50	R. Dif. Nal., Rionegro (rel: 570)
QU01)	VO	720	25	Transmisora Quindío, Armenia
C003)	TJ	730	15	R. Uno, Montería
DC05)	CU	°730	100	Cad. MelodíaRadio Lider, SF de Bogotá
CE01)	NS	740	50	R. Guatapurí, Valledupar
NA01)	HB	740	10	Ecos de Pasto, Pasto
AN01)	DK	750	50	Caracol Colombia, Medellín
CS01)	LH	750	5	LV de Yopal, Yopal
AT02)	AJ	760	30/10	RCN, Barranquilla
DC03)	JX	770	100	RCN, SF de Bogotá
GU02)	ZW	780	10/5	R. Almirante, Riohacha
SS30)	C21	780	10	Antena del Río, Barrancabermeja
VA04)	ZG	780	10	LV del Valle, Cali
AN01)	DC	790	50	R. Caracol, Medellín: (Múnera Eastman R.)
DC02)	BU	*790	50	R. Nal., Zambrano (rel: 570)
DC02)	ZR	*790	50	R. Nal., Villavicencio (rel: 570)
T003)	NC	790	1	Ecos del Combeima, Ibagué
SS01)	BW	800	100	RCN, Bucaramanga
DC04)	CY	810	200	Caracol Colombia, SF de Bogotá
B002)	AD	820	10	R. Vigía, Cartagena
VA03)	ED	820	50	Caracol Colombia, Cali
AN01)	DM	830	25	R. Reloj, Medellín
HU01)	KK	840	10	H J Doble K, Neiva
MA02)	BI	840	30	Ondas del Caribe, Santa Marta
DC04)	KC	850	50	W Radio, SF de Bogotá
CE02)	NJ	860	12	LV del Cañaguate, Valledupar
VA05)	FP	860	10	Voces de Occidente, Buga
AN09)	ZH	870	5	Vida AM, Medellín
AT03)	SB	870	25	R. Mar Caribe Int., Barranquilla
BY17)	GD	870	5	Em. Reina de Colombia, Chiquinquirá
T001)	LA	870	10	LV del Tolima, Ibagué
CL04)	FH	880	10	R. Regional Independiente, Anserma
SS02)	GE	880	20	Caracol Colombia, Bucaramanga
AT13)	HKO93	890	0.25	R. Ecos de Soledad, Soledad
DC06)	CE	890	10	R. Continental, SF de Bogotá
MA03)	PM	890	20	R. Galeón, Santa Marta
NS02)	DD	900	15/5	R. Super, Cúcuta
VA04)	EY	900	10	LV de Cali, Cali
AN04)	DO	910	10	LV del Rio Grande, Medellín
BY12)	TT	910	1	Ondas del Porvenir, Samacá
DC24)	S52	910	15	Colombia Estereo, Florencia
IS01)	MY	910	30	RCN, San Andrés (rel. 770 Bogotá)
B003)	AA	920	30	Em. Fuentes, Cartagena
NA02)	JN	920	10	Ondas del Mayo, Pasto
T002)	SJ	920	10	Colmundo, Ibagué
DC07)	CS	930	30	LV de Bogotá, SF de Bogotá
NS03)	TL	940	25	RCN, Cúcuta
VA04)	GB	940	10	R. Calima, Cali
BY19)	UJ	950	8	Armonias Boyacenses, Tunja
RI01)	FN	950	15	Caracol Colombia, Pereira
B008)	HN	960	50	Caracol Colombia, Magangué
IS05)	R31	960	15	Candela, San Andrés (Rel: Candela 101.9 Bogotá)
SS23)	HX	960	1	Candela AM, Bucaramanga
CA01)	VK	970	30	Armonias del Caquetá, Florencia
DC08)	CI	970	10	R. Super, SF de Bogotá
GU02)	ME	970	10	RCN Guajira, Maicao
QU09)	HKX59	970	0.25	R. Quimbaya, Calarca
NS04)	JV	980	10	La 980 Sensacional, Cúcuta
VA06)	ES	980	100	RCN, Cali

MW	Call	kHz	kW	Station, location
AN02)	DB	990	100	RCN, Medellín
BY07)	HI	990	5	LV de Garagoa, Garagoa
B004)	AQ	1000	15	RCN, Cartagena
DC02)	ZP	*1000	50	R. Nal., Yopal (rel: 570)
CC01)		1000	0.8	R. Panamericana, Cajibío
DC02)	JG	1000	10	R. Nal., Manizales (rel: 570)
AT04)	OP	1010	10	Oxígeno R., Barranquilla
C001)	ZD	1010	15	R. Panzenú, Montería
DC04)	CN	1010	10	R. Reloj/W R., SF de Bogotá
HU02)	JR	1010	15	Caracol Colombia, Neiva
NA03)	BN	1010	10/5	LV del Galeras, Pasto
SS03)	IX	1010	10	R. Yarima, Barrancabermeja
AN04)	DQ	1020	10	Emisora Claridad, Medellín
ME01)	KS	1020	15	LV del Llano, Villavicencio
RI02)	FQ	1020	10	RCN, Pereira
SS04)	DZ	1020	15	R. Primavera, Bucaramanga
T003)	FT	1020	10	R. Super, Ibagué
BY01)	DJ	1030	10	RCN LV de los Libertadores, Duitama
CE03)	RF	1030	15	Ondas del Cesar, Aguachica
CO02)	GX	1030	1	R. Progreso de Córdoba, Lorica
VA06)	ER	1030	30	RCN Antena 2, Cali
VP01)		1030	5	Ondas del Vaupés, Mitú
AT05)	AI	1040	15	R. Tropical, Barranquilla
CC02)	SY	1040	10	R. 1040/La Caucana 10-40, Popayán
DC10)	CJ	1040	15	Colmundo, SF de Bogotá
NA04)	UB	1040	15	Colmundo, Pasto
NS05)	BF	1040	15	LV del Norte, Cúcuta
QU02)	FM	1040	15	LV de Armenia, Armenia
AN04)	DR	1050	10	R. Unica, Medellín
AR01)	LZ	1050	10	LV del Cinaruco/Caracol, Arauca
CE04)	BB	1050	15	Caracol Colombia, Valledupar
CS03)	S62	1050	15	Cusiana R., Yopal
ME02)	IO	1050	5	LV de la Conquista, Granada
SS05)	GU	1050	10	R. Bucarica, Bucaramanga
T004)	FZ	1050	10	La Cariñosa del Centro, Antena 2, Espinal
VA07)	NG	1050	10	R. Palmira, Palmira
AN05)	MG	1060	1	R. Litoral, Turbo
BY02)	MV	1060	10	R. Furatena, Chiquinquirá
CL02)	FJ	1060	15	RCN Caldas, Manizales
GU04)	LY	1060	5	R. Delfín, Riohacha
HU03)	OV	1060	15	R. Surcolombiana, Neiva
SU11)	YX	1060	1	Caracoli, Sincelejo
AT06)	AH	1070	20	Em. Atlántico, Barranquilla
CC03)	VR	1070	15	R. Super, Popayán
DC11)	CG	1070	30	R. Santa Fé, SF de Bogotá
AN01)	AX	1080	10	La 1080, Medellín
CL03)	JS	1080	15	R. Pontoná, La Dorada
CO04)	AW	1080	10	LV de Montería, Montería
ME03)	KT	1080	10	R. Autentica/R. Macarena, Villavicencio
SS06)	MH	1080	10	Melodía AM, Floridablanca
VA04)	JF	1080	10	R. Eco, Cali
B005)	OM	1090	5	R. Bucanero, Cartagena
BY03)	IH	1090	10	Caracol Colombia, Sogamoso
CA02)	IG	1090	10	R. Autentica, Florencia
CL01)	IA	1090	10	Oxígeno R., Manizales
NS06)	BC	1090	15	Caracol Colombia, Cúcuta
T005)	JB	1090	10	LV de los Pijaos, Guamo
AN06)	GQ	1100	5	Transmisora Surandes, Andes
AT04)	AT	1100	15	Caracol Colombia, Barranquilla
C005)	MK	1100	5	Emisora Ideal, Planeta Rica
DC27)	CN	1100	10	BBN R., SF de Bogotá
HU04)	YZ	1100	10	R. Super, Neiva
SS07)	GI	1100	5/1	LV de Colombia, Socorro
VI01)	EF	*1100	2	LV del Vichada, Puerto Carreño
AN07)	DI	1110	10	R. Bolivariana, Medellín
AR02)	GP	1110	5	LV del Río Arauca, Arauca
IS02)	PA	1110	1	LV de las Islas, San Andrés
ME04)	JP	1110	10	RCN, Villavicencio
SU02)	ZE	1110	15	R. Piragua, Sincelejo
VA03)	EW	1110	10	Oxígeno 1110, Cali
BY04)	KQ	1120	15	J.B. R., Tunja
DC24)	Q92	1120	5	Colombia Mía, Yopal, CS
NS01)	TI	1120	10	Vox Dei, Cúcuta
RI03)	JC	1120	5	R. Matecaña, Pereira
SS02)	GH	1120	15	Oxígeno R., Bucaramanga
AT07)	AC	1130	10	Em. Riomar, Barranquilla
B006)	NN	1130	1	Ondas del Río, Magangué
DC09)	VA	1130	15	Vida AM Básica, SF de Bogotá
NA05)	QQ	1130	10	Oxígeno R., Pasto
AN02)	DL	1140	10	R. Paisa "La Cariñosa de Medellín", Medellín
B007)	KO	1140	10	R. Esperanza, Cartagena
CC12)		1140		R. Piendamo, Piendamo
CU01)	CL	1140	10	R. Panamericana, Girardot
ME05)	RW	1140	10	Caracol Villavicencio, Villavicencio

MW	Call	kHz	kW	Station, location
SS08)	RN	1140	10	RCN, Barbosa
BY05)	GJ	1150	1	JB Radio, Duitama
CH01)	TE	1150	1	LV del Chocó, Quibdó
HU05)	FP	1150	5	RCN, Neiva
NS07)	BT	1150	10	R. Catatumbo, Ocaña
QU03)	FI	1150	15	Caracol Colombia, Armenia
AT01)	BL	1160	10	R. Aeropuerto, Barranquilla
CA03)	AU	1160	15	Ondas del Orteguaza, Florencia
CO06)	AZ	1160	5	Frecuencia Bolivariana "tu emisora", Montería
DC13)	OC	1160	15	Ecos de Colombia, SF de Bogotá
NA06)	ZV	1160	5	RCN R. Las Lajas, Ipiales
NS08)	EC	1160	10	R. San José de Cúcuta, Cúcuta
RI04)		1160		Ondas del Puerto, La Virginia
SS09)	S31	1160	10	Colombia Mía, Barrancabermeja
VA04)	EV	1160	10	R. Unica, Cali
AN04)	KW	1170	10	R. Nutibara, Medellín
AR04)	E74	1170	10	Meridiano 70, Arauca
B008)	NW	1170	10	Caracol Colombia, Cartagena
BY04)	GA	1170	10	R. Recuerdos, Tunja
CE06)	PB	1170	10	Ondas de Macondo, Valledupar
ME01)	BX	1170	10	Ondas del Meta, Villavicencio
VA08)	JE	1170	1	RCN, Tuluá
AN08)		1180		Em. Coorpurabá, Apartadó
CL05)	FX	1180	15	Caracol Colombia, Manizales
GV01)	WA	°1180	5	LV del Guaviare, San José del Guaviare
SS10)	GK	1180	20	R. Santander 2, Bucaramanga
T006)	JT	1180	10/5	RCN, Ibagué
AT05)	CT	1190	10	LV de la Costa, Barranquilla
CO07)	KI	1190	1	R. Barají, Sahagún
DC07)	CV	1190	10	R. Cordillera, SF de Bogotá
NA07)	KG	1190	10	R. Mira, Tumaco
VA09)	EO	1190	10	Ondas del Valle, Cartago
GU05)		1195		Ondas del Ranchería, Barrancas
AN49)	IJ	1200	15	R. 1200 "LV de la Raza", Medellín
B017)	BV	1200	10	R. Príncipe, Cartagena
BY06)	LR	1200	10	La Cariñosa, Antena2, Sogamoso
CU02)	CD	1200	10	Em. Nueva Epoca, Fusagasugá
GU06)	BZ	1200	10	Ondas del Riohacha, Riohacha
VA10)	NF	1200	10	R. Super, Cali
HU02)	FR	1210	10	R. Recuerdos, Neiva
NS03)	BE	1210	10	La Cariñosa, Antena 2, Cúcuta
RI02)	BQ	1210	10	La Cariñosa, Pereira
AT05)	FF	1220	15	R. Reloj, Barranquilla
CO08)	AV	1220	10	RCN, Montería
DC22)	KR	1220	5	R. María, "LV Católica de su Hogar", SF de Bogotá
NA08)	NM	1220	10	R. Viva Cultural Bolívar, Ipiales
SS11)	MT	1220	10	RCN La Radio, San Gil
AN10)	IL	1230	10	Minuto de Dios, Medellín
BY04)	BR	1230	10	Oxígeno R., Tunja
CU03)	TP	1230	1	R. Colina, Girardot
GU03)	MJ	1230	1	RCN Antena 2, Maicao
SS12)	GV	1230	15	Colmundo, Bucaramanga
VA06)	KL	1230	10	R. Calidad "La Cariñosa", Cali
AR03)	GO	1240	1	R. Caribabare, Saravena
QU04)	FG	1240	10	RCN, Calarcá
SS13)	GN	1240	5	R. Barrancabermeja, Barrancabermeja
VA11)	JA	1240	5	R. Buenaventura, Buenaventura
AT07)	OK	1250	10	Em. ABC, Barranquilla
DC14)	CA	1250	10	Capital Radio, SF de Bogotá
NA15)	FV	1250	10	R. Viva/R. María, Pasto
NS06)	HS	1250	15	R. Reloj, Cúcuta
SU03)	EM	1250	1	LV de Corozal, Corozal
AM01)	OU	1260	2	Ondas del Amazonas, Leticia
AN11)	DA	1260	5	R. Auténtica, Medellín
BY05)	NO	1260	5	Bésame AM, Duitama
CE08)	OH	1260	5	RCN Cesar, Valledupar
IS03)	HU	1260	1	Caracol Colombia, San Andrés (rel 810 Bogotá)
ME06)	LX	1260	5	Minuto de Dios Eco Llanero, Villavicencio
NS10)	TM	1260	5	R. Sonar, Ocaña
T007)	DV	1260	5	Caracol Colombia, Ibagué
VA29)	ET	1260	5	R. María, Cali
BO04)	AR	1270	2	La Cariñosa, Antena 2,Cartagena
CE05)	KJ	1270	1.5	LV de Curumaní, Curumaní
CU04)	XQ	1270	1	LV Amiga, Ubaté
DC24)	Q99	1270	5	Colombia Mía, San José del Guaviare
PU01)	SV	1270	1	LV de Orito, Orito
RI05)	IM	1270	5	Colmundo, Pereira
SS02)	TX	1270	5	Bésame AM, Bucaramanga
T012)	BM	1270	5	R. Internacional, Honda
AN12)	MB	1280	5	R. Suroeste, Concordia
AT01)	SO	1280	5	R. Playa Mendoza, Barranquilla
DC07)	KN	1280	5	R. Única, SF de Bogotá
GU07)	HO	1280	5	Impacto Popular, San Juan del Cesar
HU06)	CM	1280	5	R. Sur, Pitalito
NA05)	LR	1280	5	Caracol Colombia, Pasto
NS11)	RP	1280	5	Ecos de Tibú, Tibú
SS14)	NQ	1280	1	LV del Río Suárez, Barbosa
VA12)	TK	1280	5	R. Super, Caicedonia
AN13)	TH	1290	5	LV de las Estrellas, Medellín
DC24)	SZ	1290	5	Colombia Mía, Saravena, AR
CU05)	KY	1290	5	RCN, Girardot
MA04)	EB	1290	5	LV del Turismo, Santa Marta
ME07)	NE	1290	5	LV del Ariari, Granada
SU04)	OI	1290	5	R. Chacurí, Sampués
VA13)	MC	1290	5	R. Viva 12-90, Cali
B010)	OG	1300	5	LV de las Antillas, Cartagena
BY08)	RB	1300	5	CRB Cadena Radial Boyacense, Tunja
CC04)	IN	1300	5	R. Eucha, Belalcázar
PU02)	UA	1300	5	R. Sindamanoy, Mocoa
RI01)	LD	1300	5	Oxígeno R., Pereira
SS02)	NB	1300	5	Onda 5, Bucaramanga
T008)	EA	1300	5	R. Lumbí, Mariquita
AN14)	LM	1310	5	R. Santa Bárbara
AN15)	IR	1310	5	RCN Urabá, Apartadó
AT08)	AK	1310	7	LV de la Patria Celestial, Barranquilla
CO09)	DG	1310	5	Caracol Colombia, Monteria
DC09)	JZ	1310	5	R. Manantial, SF de Bogotá
HU07)	WD	1310	5	Micrófono Cívico, Palermo
NS12)	TQ	1310	5	R. Tasajero, Cúcuta
AN16)	TA	1320	1	R. María, Medellín
BY09)	HT	1320	5	R. Guateque, Guateque
CU06)	NO	1320	5	La Cariñosa, Girardot
IS04)	QI	1320	10	R. Leda Int., San Andrés
MA05)	LV	1320	5	R. Onda Fantastica, Fundación
SS15)	MS	1320	5	R. Fiesta, Barrancabermeja
VA14)	NK	1320	1	R. Luna, Palmira
AN17)	RD	1330	1	R. Fénix de Oriente 1330 AM, El Peñol
B002)	AP	1330	5	R. Auténtica, Cartagena
CE09)	MP	1330	1	LV de Aguachica, Aguachica
CC05)	LS	1330	5	Caracol Colombia, Popayán
CL17)	HKR33	1330	0.25	Alcaldía de Salamina, Salamina
RI02)	FE	1330	5	Antena 2, Pereira
SS16)	NR	1330	5	La Caliente 13-30, San Gil
AN18)	NP	1340	1	R. Comunal, Nariño
AT03)	FA	1340	5	R. Olímpica AM, Barranquilla
DC03)	FB	1340	5	Fiesta 13-40, SF de Bogotá
HU05)	KD	1340	5	La Cariñosa/Antena 2, Neiva
NA10)	HA	1340	5	RCN Nariño, Pasto
NS04)	PY	1340	5	R. Lemas, Cúcuta
NS13)	VL	1340	0.5	Brisas del Catatumbo, Tibú
SS05)	NY	1340	1	R. Unica, Bucaramanga
SU05)	HY	1340	5	RCN Sucre, Sincelejo
VA15)	IS	1340	5	R. El Sol, Buenaventura
AN19)	OS	1350	5	R. Ondas de la Montaña, Medellín
AN20)	LO	1350	5	RCN Antena 2/La Cariñosa, Caucasia
BY10)	HW	1350	5/1	Em. Ecos del Río, Puerto Boyacá
CE10)	MN	1350	1	R. Perijá, Codazzi
CE12)		1350	1	R. Cultural 2001, Pailitas
MA01)	OC	1350	5	La Cariñosa, Ant. 2, Santa Marta
T009)	HL	1350	5	R. Reloj, Ibagué
VA16)	EN	1350	10	R. Armonía, Cali
VA30)	HKZ98	1350	0.25	Alcaldía de Caicedonia, Caicedonia
AN21)	PK	1360	10/5	LV de Abejorral, Abejorral
AN22)		1360	0.5	R. Segovia, Segovia
B008)	TU	1360	5	Oxígeno R., Cartagena
RI06)	RA	1360	1	Eco 13-60 "La Superestación", Pereira
SS17)	NX	1360	1	R. Láser, Zapatoca
T018)	MI	1360	5	R. Auténtica, Melgar
AN23)	NU	1370	2.5	RCN, Rionegro
AT09)	BO	1370	5	Minuto de Dios, Barranquilla
CC06)	EQ	1370	10	RCN Cauca, Popayán: 24h
DC01)	KX	1370	5	R. Mundial, SF de Bogotá
NS15)	BD	1370	1	La Nueva R. Guaimaral, Cúcuta
SU14)	NI	1370	1	R. Sabana, Sincelejo
VA17)	JQ	1370	1	RCN Antena 2, Zarzal
AN57)	JD	1380	3	R. Nuestra Señora del Encuentro con Dios, Medellín
BY11)	EE	1380	5	RCN, Tunja
CE13)	MM	1380	5	R. Recuerdos, Valledupar
CL06)	LG	1380	3	LV de La Dorada, La Dorada
HU08)	ID	1380	5	R. Potencia Latina, La Plata
VA18)	EJ	1380	1	Armonías del Palmar, Palmira
AN25)		1390	0.1	R. Ciudad de Antioquia, Santa Fé de Antioquia
CL07)	FO	1390	5	Red de los Andes, La Voz de Siempre, Manizales
CU07)	YW	1390	5	R. Auténtica, Pacho

NATIONAL RADIO

MW	Call	kHz	kW	Station, location
SS18)	ZY	1390	1	La Primera, Bucaramanga
TO10)	FY	1390	5	R. Avenida, Espinal
AN26)	LL	1400	1	RCN Antena 2, Santa Bárbara
AT02)	AS	1400	5	R. Uno/RCN Antena 2, Barranquilla
CC07)	WY	1400	1	LV de los Samanes: Quilichao
CC13)		1400	0.45	R. Cañaveral, Morales
CH02)	IT	1400	1	Ecos del Atrato, Quibdó
CO10)		1400	0.25	Brisas del Sinú, Tierralta
CO11)	DF	1400	5	LV de Niquel, Montelíbano
DC16)	KM	1400	5	Em. Mariana de Bogotá "R.Multi-Cultural" SF de Bogotá
NA11)	JJ	1400	5	R. Ipiales, Ipiales
NA12)		1400	1.5	LV de Samaniego, Samaniego
NS16)	BK	1400	1	Voz Grancolombia, Cúcuta
QU04)	HM	1400	5	La Cariñosa de Armenia, Calarcá
SS19)	D31	1400	2.5	LV de Cimitarra, Cimitarra
SS20)	TY	1400	1	Caracol Colombia, Vélez
SU12)	HKZ25	1400	0.25	Alcaldía de Ovejas, Ovejas
SU13)	HKZ22	1400	0.25	Alcaldía de Majagual, Majagual
AN27)	DU	1410	5	Em. Cultural Universidad de Antioquia, R. Universidad, Medellín
BY18)	HKP79	1410	1	R. Universidad, Tunja
BY23)	HKP86	1410	0.25	Alcaldía de Chiquinquira, Chiquinquira
GU08)	P79	1410	2	R. Evangélica, Uribia
TO11)	FS	1410	5	RCN, Honda
VA19)	EI	1410	5	R. Guadalajara, Buga
AN28)	D23	1420	1	Ecos de Frontino, Frontino
BO02)	AP	1420	5	R. Autentica, Cartagena
CL05)	HK	1420	5	R. Recuerdos, Manizales
MA06)	BH	1420	5	Caracol Colombia/R. Magdalena, Santa Marta
SS21)	SN	1420	1	R. Lenguerque, Zapatoca
TO06)	LE	1420	1	La Cariñosa, Antena 2, Ibagué
AN29)	CK	1430	1	R. Sensación, Yarumal
AN30)	MF	1430	2	R. Venus, Puerto Berrío
AN47)	G42	1430	0.5	R. Alejandría, Alejandría
AT10)	PW	1430	5	Colmundo, Barranquilla
CC08)	EG	1430	1	LV de Belalcázar, Popayán
CL08)	IU	1430	5	Armonías del Ingrumá, Riosucio
DC17)	KU	1430	5	1430 AM, SF de Bogotá (Rel: Sonríele a Jesús R.)
NS17)	BP	1430	2	R. Cariongo, Pamplona
PU03)	HKK38	1430	0.5	R. Manantial, Sibundoy
QU08)	X61	1430	0.25	R. Dif. Cultural del Quindío, Armenia
RI08)	HKX73	1430	0.25	R. Ciudad de Pereira, Pereira
SU07)	QX	1430	5	R. Majagual, Corozal
AN46)	NZ	1440	5	Colmundo, Medellín
BY06)	GM	1440	5	RCN, Sogamoso
CA04)	IB	1440	1	RCN Caquetá, Florencia
CU19)	HKT58	1440	0.25	Alcaldía de Ubala, Ubala
TO12)	BM	1440	5	R. Internacional, Honda
VA20)	EK	1440	5	R. Reloj, Tuluá
CU08)		1445	0.5	Em. R. Únion, La Palma
AN31)	E20	1450	1	R. María, Urrao
AN32)		1450	0.2	R. LV del Nordeste, Remedios
BO11)	MX	1450	1	R. Mancomoján, Carmen de Bolívar:
BY12)	TT	1450		Ondas del Porvenir, Samacá
CC09)		1450	0.5	LV del Cauca, El Bordo
CL02)	NL	1450	5	La Cariñosa, Ant. 2, Manizales
SS22)	HH	1450	5	R. Católica Metropolitana, Bucaramanga
TO13)	BY	1450	5	Oxígeno R., Flandes
AN33)	TF	1460	5	R. María, Turbo
AN34)	MN	1460	5	LV de Amalfi "La Primera", Amalfi
AN45)	E26	1460	1	R. Capiro, La Ceja
AT02)	VH	1460	5	R. Uno/RCN Antena 2, Barranquilla
CL18)	HKR44	1460	0.25	Alcaldía de Victoria, Victoria
DC18)	JW	1460	5	Em. Nuevo Continente, SF de Bogotá
HU09)	FL	1460	2.5	Agustiniana Minuto de Dios, San Agustín
NA10)	ZU	1460	5	RCN Antena 2, Pasto
NS18)	IW	1460	1	R. Monumental, Cúcuta
QU06)	JH	1460	5	R. Ciudad Milagro, Armenia
SS29)	HKY73	*1460	0.25	Alcaldía de San Andrés, San Andrés
SU08)	AL	1460	1	R. Sincelejo, Sincelejo
AN04)	IM	1470	5	R. Popular, Medellín
AT14)	HKO96	1470	5	Alcaldía de Baranoa, Baranoa
BO12)	PX	1470	5	Colmundo, Cartagena
BY13)	HJB63	1470	1	R. Uno, Iza
CU09)	HQ	1470	5	R. Futurama, Pacho
PU04)	JIF	1470	1	R. Tres Fronteras, Puerto Asís
TO14)	TB	1470	5	Ondas de Ibagué, Ibagué
TO21)	JS20	1470	0.25	Ecos de Palo Cabildo, Palo Cabildo
VA26)	NT	1470	1	R. Huellas 1470, Cali
AN35)	TC	1480	1	R. Sonsón, Sonsón (n.f.1490)
MA07)	OD	1480	2.5	R. Rodadero, Santa Marta
NS14)		1480	0.25	LV del Samán, Bochalema
RI03)	FC	1480	5	R. Unica, Pereira
SS10)	TZ	1480	5	RCN Antena 2, Bucaramanga
TO15)	VB	*1480	1	R. Guayabal, Armero, Guayabal
AT11)	AY	1490	5	R. Vida Nueva "Te acerca a Dios", Barranquilla
BO14)	J76	1490	0.2	Alcaldía de El Peñon, El Peñon
DC19)	BS	1490	5	Em. Punto Cinco, SF de Bogotá
HU10)	AG	v1490	1	R. Garzón, Garzón
NA18)	HKW24	1490	0.2	Alcaldía de Guaitarilla, Guaitarilla
SU09)	JO	1490	1	LV de San Marcos, San Marcos
VA21)	ZB	1490	5	Robles 14-90, La Nueva, Tuluá
BY15)	SH	1500	5	R. Reloj, Moniquirá
CL09)	UW	1500	5	R. María, Manizales
CU10)	TW	1500	5	R. Sumapaz, Fusagasugá
CU16)	HKT71	150	1	Macheta
VA22)	LJ	1500	5	Sonora, La Voz de la Red, Cali
AN37)	D24	1510	5	LV de La Unión, La Unión
BY16)	A22	1510	2	LV de San Luis, San Luis de Gaceno
DC24)	HKY41	1510	1	Colombia Mía, Barrancabermeja, SS
QU07)	ZA	1510	1	R. Cristal, Armenia:
SS23)	HX	1510	1	Candela AM, Bucaramanga
TO16)		1510	0.5	LV de los Cedros, Libanó
VA28)	HKZ98	1510	0.25	Alcaldía de Caicedonia, Caicedonia
VA31)	HKZ94	1510	0.25	Alcaldía de Buenaventura, Buenaventura
VA32)	HKZ93	1510	0.25	Alcaldía de Versalles, Versalles
AN38)		1520	0.3	Brisas del Palmar, Caucasia
AN39)	MA	1520	1	LV de Suroeste, Jericó
AT03)	LQ	1520	5	R. Minuto, Barranquilla
CC11)	HKS24	1520	0.5	R. Cristalares Timbío, Timbío
CL10)		1520		Sonoradio 1520 AM, Viterbo
CO13)	HKT20	1520	0.25	Alcaldía de Montería, Montería
CU14)	V37	1520	1	R.Pueblo Viejo, Zipacon
DC09)	LI	1520	5	J-C R. Pasión Extrema, SF de Bogotá
DC24)	T21	1520	0.25	Colombia Mía, Tierralta, CO
NA16)	HKW37	1520	1	R. Universidad, Pasto
NA19)	HKW43	1520	0.1	Alcaldía de Tangua, Tangua
NS19)	J98	1520	1	Em. Una Voz de la Frontera, Puerto Santander
RI07)	RL	1520	1	Antena de los Andes, Santa Rosa de Cabal
SU10)	MZ	1520	1	Ecos de la Sierra Flor, Sincelejo
TO17)	AM	1520	1	R. Altamizal, Dolores
AN58)	DN	1530	5	Yeshu'a LV de Jesucristo, Medellín
AN50)	HKN57	1530	0.25	Alcaldía de San Juan de Uraba, San Juan de Uraba
AN53)	HKN85	1530	0.25	Alcaldía de Anza, Anza
AN55)	HKN79	1530	0.25	Alcaldía de Uramita, Uramita
CE11)	HKS56	1530		Fascinación AM, Becerril
CE15)	HKS58	1530	0.1	Alcaldía de El Copey, El Copey
CE14)		1530		R. Integración, Morales
DC24)	HKN65	1530	0.25	Colombia Mía, Caucasia, AN
GU09)	OZ	1530	5	LV de la Prov. de Padilla, San Juan del Cesar
ME10)	V82	°1530	0.25	Alcaraván Radio, Puerto Lleras
VA23)	JB	1530	1	Caracol Sevilla, Sevilla
VA25)	HKR73	1530	1	Ecos del Pacífico, Guapí
AN40)		1540	0.25	LV Dorada, Segovia
AN41)	B89	1540	1	Em. Brisas del Río Chico, Belmira
BO15)	HKP50	1540	0.25	Alcaldía de Arjona, Arjona
CL11)	ZF	1540	1	R. Cóndor "Em. Universitaria", Manizales
CS02)	HKR80	1540	0.15	Alcaldía de Sacama, Sacama
DC24)	HKZ52	1540	1	Colombia Mía, Chaparral, TO
NA09)	RQ	1540	2	R. Austral, Túquerres
SS24)		1540	0.25	R. El Sur, San Vicente de Chucurí
SS25)	HD	1540	1	LV del Petróleo, Barrancabermeja
AN36)		1550	0.5	Ondas del Nechí, Campamento
AT02)	CB	1550	5	R. El Sol "La Cariñosa", Barranquilla
CL16)	UN	1550	5	LV del Río Arma, Aguadas
DC21)	ZI	1550	5	MCI Radio 15-50, SF de Bogotá
DC24)	HKV38	1550	1	Colombia Mía, Pitalito, HU
DC24)	HKX29	1550	5	Colombia Mía, Tibú, NS
NA20)	HKW53	1550	0.1	Alcaldía de El Tablón, El Tablón
NA21)	HKW55	1550	0.1	Alcaldía de Guachucal, Guachucal
NA22)	HKW50	1550	0.25	Alcaldía de Mallama, Mallama
QU03)	QD	1550	5	R. Bésame, Armenia
VA33)	NC	1550	1	Em. Revivir en Cristo, Cali
AN42)		1555	0.5	R. Parroquial, El Santuario
AN52)	XZ	1560	1	Santa María de la Paz R., Medellín
AN54)	HKO35	1560	0.25	Alcaldía de Cañasgordas, Cañasgordas
CE07)	HKS65	1560	0.5	R. Tamalameque, Tamalameque
CE14)	PZ	1560	1	R. Codazzi, Codazzi
CU11)	CP	1560	5	RCN Antena 2, Arbelaez
ME11)	HKV90	1560	0.25	Alcaldía de Villavicencio, Villavicencio
SS26)	HE	1560	5	Voces Rovirenses, Málaga
VA08)	LP	1560	5	La Cariñosa, Antena 2, Tuluá

MW	Call	kHz	kW	Station, location
AN43)	C22	1570	1	R. Ciudad Dabeiba, Dabeiba
BO16)	HKP58	1570	0.25	Alcaldía de Sta Rosa Sur, Sta Rosa Sur
BY24)	HKQ83	1570	0.25	Alcaldía de Maripi, Maripi
BY25)	HKQ82	1570	0.25	Alcaldía de Sta María, Sta María
CA05)	HKR66	1570	0.2	R.Universidad de la Amazonia, Florencia
CA06)	HJR66	1570	0.5	Timbiqui Estéreo, Timbiqui
CL12)	ZT	1570	1	R. Auténtica, Manizales
CU18)	HKU42	1570	0.15	Alcaldía de Cajica, Cajica
DC22)	TG	1570	1	R. María, Machetá
DC24)	E96	1570	1	Colombia Mía, Palmira, VA
DC24)	HKX52	1570	2	Arc. Armada de Colombia, Pto Leguizamo
NS09)		*1570		LV de Fomeque, Fomeque
RI09)	HKX80	1570	0.1	R. Marsella, Marsella
RI11)	HKX78	1570	0.25	Alcaldía de Balboa, Balboa
AT12)	QZ	1580	5	R. María, Barranquilla
CC16)	HKS46	1580	0.15	R. Alcaldía de Padilla, Padilla
CH01)	TE	1580		LV de Chocó, Quibdó
CO12)	HKT34	1580	0.25	Alcaldía de San Antero, San Antero
CU17)	HKU42	1580	0.25	Alcaldía de Cajica, Cajica
DC25)	QT	1580	5	Sonríele a Jesús R., SF de Bogotá
HU11)		1580		Alcaldía de Yaguará, Yaguará
MA08)	LC	1580	1	LV del Banco, El Banco
NA23)	HKW741	1580	0.1	Alcaldía de Pupiales, Pupiales
NS20)	KB	1580	1	R. Zulima, Villa del Rosario
SU01)	RM	1580	5	Caracol Colombia, Sincelejo
TO19)	DE	1580	1	R. Miraflores, Rovira
VA24)	SQ	1580	5	R. Robledo/RCN Antena 2, Cartago:
AN44)	IP	1590	5	BBN 15-90 R., Envigado
CE16)	HKS72	1590		Alcaldía de La Gloria, La Gloria
CL13)	QM	1590	1	Ecos de la Miel, Samaná
CU14)		*1590		Ondas del Rioseco, Rioseco
SS27)	WB	1590	5	Em Nuestra Sra del Socorro, Socorro: 0500-2300
VA25)		1590		R. Espacial, Andalucía
AN51)	HKO63	1600	0.25	Alcaldía de Jardín, Jardín
BY20)		*1600		R. Fortaleza, Sogamoso
BY22)		*1600		R. Bello Horizonte, Pesca
CC15)		*1600	0.25	R. Impacto Cristiano, Popayán
CL14)		*1600	0.25	LV de Aranzazu
CL15)	HKR52	1600	0.25	LV de Colina, Risaralda
CO14)	HKT39	1600	0.25	Alcaldía de Valencia, Valencia
CU13)	HV	1600	0.1	Armonías Zipaquireñas, Zipaquirá
CU15)		1600	1	LV del Rosario, Junín
DC24)	HKO72	1600	5	Colombia Mía, Carepa, AN
RI10)	HKX84	*1600		Em. Mundial, Dosquebradas
RI12)	HKX83	1600	0.25	Alcaldía de La Celia, Celia
TO20)	HKZ79	1600	0.15	Alcaldía de Cajamarca, Cajamarca
TO22)	HKZ77	1600	0.15	Alcaldía de Venadillo, Venadillo
VA27)	F33	1600	5	R. Restauración, Cali
AN48)		*1610		Armonías de Occidente, Medellín
RI13)		*1610		R. Estelar, Santuario
BY26)		v1613	1	R. Ideal, Úmbita (nom 1600)

SW	Call	kHz	kW	Name and h of tr
DC26)	DH	5910	5	Marfil Estéreo, Pto Lleras (Meta): 24h (rel. 88.8MHz FM)
DC26)	DH	6010	5	LV de tu Conciencia, Pto Lleras (Meta): 24h
GV01)	OY	6035	5	LV del Guaviare, San José del Guaviare
DC05)	QE	*6140	5	Cad. Melodía R Lider, SF de Bogotá

Stns with (*) are reported to be inactive, but may occasionally be reactivated for variable periods of time.

Major Networks:
CARACOL (Primera Cadena Radial Colombiana)
✉ Calle 67 N° 7-37, Santafé de Bogotá, DC ☎+57 1 348 7600 📠+57 1 337 7126 **W:** www.caracol.com.co
E: caracolcolombia @caracol.com.co
RCN (Radio Cadena Nacional)
✉ Cra. 13A N° 37-32, Santafé de Bogotá, DC ☎+57 1 314 7070 📠+57 1 314 7070 **W:** www.rcn.com.co
Regularly all "La Cariñosa" stations relay sport trs from Antena 2.
SUPER RADIO
✉ Calle 39A N° 18-12 (or: Ap. 23316), Santafé de Bogotá, DC ☎+57 1 338 2166 📠+57 1 287 8678 **W:** www.cadenasuper.com
TODELAR (Circuito Todelar de Colombia)
✉ Ap. 27344 (Av. Cra 20, N° 83-64), Santafé de Bogotá, DC ☎+57 1 621 6621 📠+57 1 616 0056 **W:** www.todelar.com **E:** todelar@ telesat.com.co
COLMUNDO
✉ Diagonal 58 N° 26A-29, Santafé de Bogotá, DC ☎+57 1 217 8911 📠+57 1 348 2746 **W:** http://colmundoradio.com.co **E:** correo@ colmundoradio.com
CADENA RADIAL AUTENTICA DE COLOMBIA (RIg.)
✉ Ap. 18350, (Calle 32 N° 16-12), Santafé de Bogotá, DC. Carrera

38D # 1-52, Barrio Santa Isabel, Cali. ☎+57 1 285 3360 📠+57 1 285 2505 **W:** www.cmbflorestacali.org
RTVC (Radio Televisión de Colombia) (Publ)
✉ Avenida El Dorado – CAN, Bogotá, D.C. **W:** www.rtvc.gov.co

State abbreviations: (Departamentos) AM = Amazonas, AN = Antioquia, AR = Arauca, AT = Atlántico, BO = Bolívar, BY = Boyacá, CA = Caquetá, CC = Cauca, CE = Cesar, CH = Chocó, CL = Caldas, CO = Córdoba, CS = Casanare, CU = Cundinamarca, DC = Distrito Capital, GN = Guainía, GU = Guajira, GV = Guaviare, HU = Huila, IS = Islas San Andrés y Providencia, MA = Magdalena, ME = Meta, NA = Nariño, NS = Norte de Santander, PU = Putumayo, QU = Quindío, RI = Risaralda, SS = Santander del Sur, SU = Sucre, TO = Todelar, VA = Valle del Cauca, VI = Vichada, VP = Vaupés.
N.B: These abbreviations are not officially recognized by the Colombian Post Office. Letters should therefore carry full name.

Addresses and other information:
AM00) AMAZONAS
AM01) Cra. 6A N° 10-104 (or: Ap. 236), Leticia 1100-0500
AN00) ANTIOQUIA
AN01) Cra. 79A N° 39-45, Medellín **W:** www.radiomunera.com **E:** munera@munera.eastman.com – **AN02)** Edificio Coltejer, Calle 52 #47-42, Medellín. – **AN03)** Calle 50 Colomb N° 67-141, Medellín. – **AN04)** Av.13 N° 84-42 (or: Ap. 1431), Medellín – **AN05)** Cra. 19 N° 20-66, Turbo **W:** www.1060radiolitoral.com – **AN06)** Ap. 1431, Andes - 1000-0200 – **AN07)** Circular 1a N° 70-01, Bloque 6, P7 U.P.B. Laureles, Medellín **W:** www.ubicar.com/radiobolivariana - **FM:** 92.4MHz – **AN08)** Apartadó. – **AN09)** Cra. 77B N° 48-144, Medellín ☎+57 421 0102 **W:** www.vidaam.com **E:** info@vidaam. com – **AN10)** Calle 56 N° 41-57, Medellín. **W:** www.infonetway. com/minuto/ or www.1230amradio.com **E:** minuto1230@amradio. com – **AN11)** Calle 41 N° 80B-46, P2, Medellín. – **AN12)** Cra. 3 Calles 2 y 3, Concordia. – **AN13)** Ap. 4300, Medellín. – **AN14)** Cra. 51 N° 51-38 (or Ap. 3854), Medellín - 1000-0500 – **AN15)** Calle 94 N° 99-51, Apartadó. – **AN16)** Calle 50 N° 67-141 (or: Ap. 65103), Medellín. – **AN17)** Centro Cooperativo, Parque Principal, El Peñol. – **AN18)** Cra. 11 N° 10-34, Nariño - 1100-0100 – **AN19)** Calle 44 N° 94-15, P3, Medellín **E:** ondasm@cis.net.co – **AN20)** Cra. 2 N° 21-54, Caucasia. – **AN21)** Cra. 51 N° 50-09, Abejorral - 0900-0500 – **AN22)** Segovia - 1100-0300 – **AN23)** Cra. 51 N° 49-09, Rionegro. – **AN24)** Calle 48B N° 79-38, Medellín. – **AN25)** Casa de la Cultura, Santa Fé de Antioquia - 1100-2300 – **AN26)** Cra. Bolívar, Calle López, Santa Bárbara. – **AN27)** Ap. 1226 (or: Cra. 44 N° 48-72), Medellín - 1100-0500 **W:** www. udea.edu.co/emisora/ - **FM:** 101.9MHz – **AN28)** Cra. 32 N° 30-05, Frontino. – **AN29)** Cra. 20 N° 20-21, Yarumal. – **AN30)** Calle 6 N° 1-23, Puerto Berrio. – **AN31)** Urrao - 0900-0300 – **AN32)** Remedios - 1100-2400 – **AN33)** Ap. 1289, Medellín - 1000-0400 – **AN34)** Cra. 19 Restrepo N° 19-61, Amalfi - 1000-0300 **W:** http://lavozdeamalfi. com - **FM:** 103.9MHz – **AN35)** Calle 8 N° 6-60, Sonsón. – **AN36)** Casa Cural, Campamento - 1100-2300 – **AN37)** Calle 10 N° 9-37, La Unión - 0900-0500 **W:** Ap. 4897, Medellín) - 1000-0200 **E:** emivozunion@epm.net. co – **AN38)** Batallón de Infantería N° 29 "Rifles", Barrio El Palmar, Caucasia - 1130-0400 – **AN39)** Calle 7, Cras. 3 y 4, Jericó - 1000-0500 – **AN40)** Batallón Bombona, Segovia. – **AN41)** Cra. 20 N° 20-14, Belmira. – **AN42)** Parroquia de Nuestra Señora de Chiquinquirá, El Santuario - 1230-1500, 2200-2400 – **AN43)** Edif.Restrepo, P3, Plaza Principal, Dabeiba - 0900-0300 – **AN44)** Ap. 81095 (or: Cra. 44A N° 31 Sur-16,Barrio San Marcos, Medellín), Envigado. – **AN45)** Calle 20 N° 27-20, La Ceja - 1100-0300 (Sun -0100) **E:** radiocapiro@yahoo. com ☎+57 4 5531528 📠+57 4 5530785. – **AN46)** Cra. 80 N° 46-74, Medellín - 1100-0600 – **AN47)** Junta de Acción, Comunal Central, Alejandría – **AN48)** Calle 100 N° 14-06, Turbo. **AN49)** Cra.73 N° 47-35, Medellín. – **AN50)** Palacio Municipal de San Juan de Uraba, San Juan de Uraba. – **AN51)** Palacio Municipal de Jardín, Jardín.– **AN52)** Calle 10 N° 42-22, Medellín **W:** www.santamariadelapaz.org/index.html **E:** webmaster@santamariadelapaz.org – **AN53)** Palacio Municipal de Anza, Anza. – **AN54)** Palacio Municipal de Cañasgordas. – **AN55)** Palacio Municipal de Uramita, Uramita. – **AN56)** Calle 105F No. 51-16, Barrio 20 de Enero, Apartado. – **AN57)** Calle 43 No. 67a-16, Barrio San Joaquín, Medellín **E:** nsemedellin@testimonio.net. – **AN58)** Cra 81A No. 48-B – 71, Barrio Calasanz, Medellín.
AR00) ARAUCA
AR01) Calle 19 N° 19-62 P2, Arauca. – **AR02)** Cra. 20 N° 19-09, P5, Arauca (or: Ap. 16555, SF de Bogotá).– **AR03)** Calle 20, Cra. 27 (or: Ap. 6558), Saravena –**AR04)** Cra 20 N° 17-57, P3, Arauca **W:** meridiano70. com **E:** info@meridiano70.com
AT00) ATLANTICO
AT01) Cra. 53 N° 15-166 (or: Ap. 3143), Barranquilla. **FM:** 96.9. – **AT02)** Barranquilla - 1000-0300 – **AT03)** Calle 73 N° 41B-106, Barranquilla - 0900-0400 ☎+575 368 2832 **W:** www.radiominuto.org **E:** info@ radiominuto.org – **AT04)** Ap. 1688, Barranquilla. – **AT05)** Cra. 53 N°

82-132, Barranquilla 1030-0200 **W:** http://emisoralavozdelacosta.net – **AT06)** Organización Radial Olímpica, Calle 72 No 48-37, Barranquilla - 1000-0500 – **AT07)** Cra. 48 N° 72-25, Ofc. 306, (or: Ap. 2010), Barranquilla - 0930-0500 – **AT08)** Cra. 45 No. 76-125, Barranquilla - 0900-0500 ☎+57 5 3692208 **E:** vozdelapatriacelestial1310@gmail.com – **AT09)** Calle 53 N° 50-11, P2, Barranquilla **W:** www.minutodedios.org or www.1370am.org/ **E:** aj1370@latinmail.com – **AT10)** Cra. 44 N° 70-61, Barranquilla. – **AT11)** Cra. 26, No 75B-07, Barranquilla - 1100-0500 **W:** www.radiovidanueva.net **E:** contacto@radiovidanueva.com – **AT12)** Calle 60 N° 47-70, Centro Cultural Santa Catalina, Barranquilla - 0930-0130 ☎ +575 344 2032. – **AT13)** Palacio Municipal de Soledad, Soledad. – **AT14)** Palacio Municipal de Baranoa, Baranoa.

B000) BOLÍVAR
BO01) Av. Venezuela, Edif. Banco Internacional, La Matuna 8B-05, Cartagena. – **BO02)** Calle Real 20-217, Cartagena - 1000-0500 – **BO03)** Calle Mayor N° 6-34 (or: Ap. 1771), Cartagena - 1000-0420. – **BO04)** Ap. 246, Cartagena. – **BO05)** Banco Popular, Ofc 1103 Mutana, Cartagena. – **BO06)** Ap. 180, Maganguvé. – **BO07)** Calle Sta Fe, N° 13-113, Torices, Cartagena ☎+ 57 5 666 2072 **W:** www.radioesperanza1140. net **E:** info@radioesperanza1140.net – **BO08)** Matuna, Calle 32 No. 8-21, Of. 1106, Edificio Banco Popular, Cartagena. – **BO09)** Av. 3 N° 21-62, La Manga, Cartagena. – **BO10)** Cra. 21 N° 29B-10, Cartagena. – **BO11)** Calle 56 N° 26-01, Carmen de Bolívar - 1030-0400 – **BO12)** Av. Venezuela, Edif. Suramericana, Of. 801, Cartagena. – **BO14)** Palacio Municipal de El Peñon, El Peñon. – **BO15)** Palacio Municipal de Arjona, Arjona. – **BO16)** Palacio Municipal de Sta Rosa Sur, Sta Rosa Sur. **BO17)** Manzana H, Lote 20, La Consolata, Cartagena.

BY00) BOYACÁ
BY01) Calle 16 N° 15-21, P8, Edif.Camara de Comercio, Duitama. – **BY02)** Cra. 10 N° 16-36, Chiquinquirá - 0900-0600 – **BY03)** Ap. 282, Sogamoso. – **FM:** 88.5MHz, 107.3MHz – **BY04)** Edif. Camol, Piso 11, Cra. 10 No. 21-15, Tunja - 1100-0500. – **BY05)** Cra. 15 N° 14-47, Duitama. – **BY06)** Ap. 019, Maganguvé - 1000-0500 – **FM:** 106.1MHz – **BY07)** Cra. 9 No. 8-65, Garagoa - 1000-0330 – **BY08)** Calle 20 N° 10-64, Tunja. – **BY09)** Cra. 7 N° 9-57, Guateque (or: Ap. 17387, SF de Bogotá) - 1000-0300 – **BY10)** Cra. 3 N° 13-74, P2, Puerto Boyacá - 0900-0400 – **BY11)** Cra. 10 N° 17-50, P5, Tunja. – **BY12)** Calle 5,N° 5-25, P2, Parque Santander, Samacá - 0900-0300 – **BY13)** Iza (or: Cra. 7 N° 17-51, Of. 610, SF de Bogotá). – **BY14)** Cra. 6 N° 6-93, Paipa - 1000-0400 – **BY15)** Calle 7 N° 3-61, Moniquirá 1030-0300 – **BY16)** Calle 6 N° 5-42, San Luis de Gaceno - 0900-0300 – **BY17)** Calle 18 N° 12-81, P2, Chiquinquirá. – **FM:** 92.6MHz – **BY18)** Universidad Pedagógica y Técnico de Colombia, Tunja. – **BY19)** Calle 20 N° 10-64, Ofc.307, Tunja ☎+57 8 744 1562 **E:** armoniasboyacenses@hotmail. com – **BY20)** Cra. 10 N° 1495, Sogamoso. – **BY22)** Pesca. – **BY23)** Palacio Municipal de Chiquinquirá, Chiquinquirá. – **BY24)** Palacio Municipal de Maripi, Maripi. – **BY25)** Palacio Municipal de Sta María, Sta María. – **BY26)** Calle 16A N° 3-58, Umbita.

CA00) CAQUETÁ
CA01) Cra.14 N° 12-129, Casa Episcobal, P2 (Ap. 285), Florencia - 1000-0300 – **CA02)** Ap. 465, Florencia. – **CA03)** Calle 17 N° 10-40, P2, (Ap. 209), Florencia - 1030-0300 – **CA04)** Ap. 150, Florencia. – **CA05)** Ap. 192, Florencia **W:** www.uniamazonia.edu.co – **FM:** 98.1MHz – **CA06)** Timbiqui.

CC00) CAUCA
CC01) Barrio El Porvenir, Cajibío (or: Ap. 945, Popayán) - 1300-2300 – **CC02)** Cra 8 N° 3-17 (or: Ap. 1321), Popayán - 1000-0400 **W:** www. caucana1040.am – **CC03)** Cra. 8 N° 5-41, Popayán. – **CC04)** Casa Cural, Parque Principal, Belalcázar (or: Ap. 987, SF de Bogotá) - 1000-2400 – **CC05)** Calle 5A N° 11-25, Popayán. – **CC06)** Ap. 535, Popayán. – **CC07)** Cra. 13 N° 9-20, Santander de Quilichao - 1100-2400 – **CC08)** Calle 2a N° 1-06 (or: Ap. 759), Popayán - 0930-0400 – **CC09)** Batallón José Hilario López, Bordo. – **CC11)** Calle 15 Cra. 17 Esq. Casa de la Cultura, Timbío (or: Calle 12B N° 13B-22, Popayán) - 1200-2400 – **CC12)** Cra. 4 N° 9-42, Piendamo. – **CC13)** Barrio Sagrada Familia, Cra. 3 esq., Morales - 1130-1700, 1900-2200 – **CC14)** Casa de la Cultura, Morales. – **CC15)** Ap. 789, Popayán - 1000-0500 – **CC16)** Palacio Municipal de Padilla.

CE00) CESAR
CE01) Calle 17 N° 15-67, Valledupar - 0900-0300 – **CE02)** Cra. 5 N° 13-52, Valledupar. – **CE03)** Calle 7 N° 16-39, Aguachica. – **CE04)** Ap. 22, Valledupar - 0900-0500 – **CE05)** Calle 6 N° 19-66, Curumaní - 1000-0100 – **FM:** 95.7MHz – **CE06)** Calle 16B N° 13-74, Valledupar. – **CE07)** Casa de la Cultura, Tamalameque. – **CE08)** Ap. 250, Valledupar. – **CE09)** Cra. 10a N° 4-38, P2, Aguachica. – **CE10)** Cra. 16 N° 11-102, Codazzi. – **CE11)** Becerril. – **CE12)** Pailitas. – **CE13)** Cra. 9 N° 5-02, Valledupar. – **CE14)** Calle 12 N° 15-08, Codazzi. – **CE15)** Palacio Municipal de El Copey, El Copey. – **CE16)** Palacio Municipal de La Gloria.

CH00) CHOCÓ
CH01) Calle 28 N° 1-04, P2 (or: Ap. 482), Quibdó. – **CH02)** Cra. 4 N° 25-18, P2, (or: Ap. 196), Quibdó - 1000-0400 – **CH03)** Choco.

CL00) CALDAS
CL01) Ap. 67, Manizales. – **CL02)** Ap. 244, Manizales - 1000-0500 – **CL03)** Cra. 2 N° 13-31, P3, La Dorada - 1000-0500 – **CL04)** Cra. 4 N° 8-58, P3, Anserma - 0900-0300 – **CL05)** Ap. 2000, Manizales. – **CL06)** Calle 11 N° 3-58 (or: Ap. 34), La Dorada - 0930-0500 – **CL07)** Calle 22 N° 21-40, Plaza Bolívar, Manizales **W:** www.laredsonoraradio.com – **CL08)** Cra. 5 N° 11-102, Av. Los Fundadores, Riosucio - 1000-0500 – **CL09)** Cra. 23 N° 71-03 (or: Ap. 990), Manizales 1015-0500 – **CL10)** Viterbo. – **CL11)** Antigua Estación del Ferrocaril, Manizales - 1200-0400 **W:** www.autonoma.edu.co/emisora.html **E:** rcondor@autonoma. edu.co – **CL12)** Cra. 23 N° 71-03, Av.Sant, Manizales. – **CL13)** C. A. M, Samaná - 1000-0100 – **CL14)** La Parroquia de Nuestra Señora del Rosario, Aranzazu - 1100-1300, 1700-1900, 2100-2300 – **CL15)** Av. Joaquín 1-09, Salida a San José, Risaralda. – **CL16)** Cra. 3 N° 7-31, Aguadas - 1000-0300 – **CL17)** Palacio Municipal de Salamina, Salamina. – **CL18)** Palacio Municipal de Victoria, Victoria.

CO00) CÓRDOBA
CO01) Cra. 3A N° 30-12, P2, Montería - 1000-0400 – **CO02)** Av. Olaya Herrera, Edif. Jatin, Lorica - 1000-0400 – **CO03)** Calle 23 N° 1-53, Montería - 1000-0100 – **CO04)** Cra 2 N° 28-53, P2, (or: Ap. 497), Montería - 1000-0400 – **CO05)** Cra. 8 N° 17-56, Planeta Rica. – **CO06)** Ap. 148, Montería. – **CO07)** Calle de Comercio, Sahagún - 1000-0500 – **CO08)** Calle 27 N° 8-25, Montería. – **CO09)** Ap. 364, Montería. – **CO10)** Brigada N° 11, Tierralta. – **CO11)** Cra. 5 N° 14-35, Montelibano. – **CO12)** Palacio Municipal de San Antero, San Antero. – **CO13)** Palacio Municipal de Montería, Montería. – **CO14)** Palacio Municipal de Valencia, Valencia.

CS00) CASANARE
CS01) Calle 9 N° 22-63, Edif. Cine Casanare, P2, Yopal - 1000-0500 **W:** www.lavozdeyopal.com.co **FM:** 97.7MHz – **CS02)** Palacio Municipal de Sacama, Sacama. – **CS03)** Yopal.

CU00) CUNDINAMARCA
CU01) Calle 14 N° 11-23, P2, Ofc.202, Girardot. – **CU02)** Av. Las Palmas N° 5-08, P5, Fusagasugá - 0900-0400 – **CU03)** Terminal de Transportes, Girardot. – **CU04)** Cra. 6 N° 6-38, Ubaté. – **CU05)** Ap. 416, Girardot. – **CU06)** Calle 16 N° 10-38, P3, Girardot - 1030-0300 – **FM:** 93.6MHz – **CU07)** Calle 7 N° 14-83, Pacho. – **CU08)** La Palma. – **CU09)** Calle 3 N° 16-39, Pacho - 0930-0400 – **CU10)** Calle 8 N° 5-59, Fusagasugá - 0900-0300 – **CU11)** Cra. 3 N° 2-36, Arbeláez (or: Av. 37 N° 75-84, SF de Bogotá) – **CU13)** Calle 3 N° 7-56, Zipaquirá. – **CU14)** Zipacon. – **CU15)** Alcaldia Municipal, Junín. – **CU16)** Macheta. – **CU17)** Palacio Municipal de Cajica, Cajica. – **CU18)** Palacio Municipal de Cajica, Cajica. – **CU19)** Palacio Municipal de Ubala.

DC00) DISTRITO CAPITAL
DC01) Calle 32 N° 16-12, SF de Bogotá ☎+57 1 285 3360 🖷 +57 1 285 2505 – **DC02)** Cra. 45 No. 26-33, SF de Bogotá **W:** www.radionacional-decolombia.gov.co – **DC03)** Cra. 13A N° 37-32, SF de Bogotá. – **DC04)** Calle 67 No. 7-37, SF de Bogotá – **DC05)** Calle 45 N° 13-70 (or: Ap. 19823), SF de Bogotá ☎+57 1 323 1500 🖷+57 1 288 4020 – E: radiolider@cadenamelodia.com **DC06)** Calle 48 N° 18-77, SF de Bogotá. – **DC07)** Av. 13 N° 84-42, SF de Bogotá – **DC08)** Calle 39A N° 18-12, SF de Bogotá ☎+57 1 338 2166 – **DC09)** Sistema Vida Colombia, Avenida Calle 13 N° 79-70, SF de Bogotá, DC - 1100-0300 – **DC10)** Diagonal 58 N° 26A-29, SF de Bogotá - 1100-0300 – **DC11)** Calle 5 N° 17-48, SF de Bogotá ☎+57 1 345 6781 🖷+ 57 1 345 7080 **W:** www. radiosantafe.com – **DC12)** Cra. 16 N° 43-09, SF de Bogotá ☎+57 1 288 3766. 🖷+57 1 288 7720 **E:** radiok@multiphone.net.co – **DC13)** Cra. 13 N° 46-72, SF de Bogotá – **DC14)** Cra. 30 N° 91-84 (or: Ap. 250649), SF de Bogotá ☎+57 1 610 2079 🖷+57 1 218 0312 – **DC15)** Ap. 9291, SF de Bogotá. – **DC16)** Calle 6 N° 7-22, (or: Ap. 3201), SF de Bogotá (alt.address: Calle 385 N° 75-31, Cd. Kennedy, SF de Bogotá) - 1100-0130 (Frequent relays of Sonríele a Jesús R.) **W:** www.emisoramari-ana.com – **DC17)** Av. Cd. Kennedy 72825, SF de Bogotá - 1100-0300 **W:** www.1430amradio.com – **DC18)** Cra. 27 N° 49-48, SF de Bogotá. – **DC19)** Av. 15 N° 123-61, Of. 408, SF de Bogotá - 1100-2300 – **DC20)** Diagonal 88 bis N° 26-40 (or: Ap. 90883), SF de Bogotá. – **DC21)** Calle 22C N° 31-01, SF de Bogotá ☎+57 1 368 7431 **E:** mciradio@latino.net. co – **DC22)** Carrera 21A, No. 151-23, SF de Bogotá ☎+57 1 216 9839 🖷 +57 1 614 3730 **W:** www.radiomariacol.org **E:** info.col@radiomaria. org – **DC24)** Escuela de Cadetes José María Cordoba, Calle 80 N° 38-00, SF de Bogotá 🖷+57 1 240 7374 – **DC25)** Fundación Sonríele a Jesús, Calle 72a No. 86-64, SF de Bogotá ☎+57 1 4362014 **W:** www. sonrieleajesusradio.com **E:** sonriele1580@sonrieleajesus.com – **DC26)** Librería Colombia para Cristo, Calle 46 N° 13-56, Blg C, Ap.to 215, (or Apartado Aéreo 67751) SF de Bogotá. (Reports c/o Rafael Rodríguez R., Apartado Aéreo 67751, Bogotá, DC. 2 IRC's required for QSL reply) **E:** libreria@fuerzadepaz.com or (for reports) rafaelcoldx@yahoo.com ☎+57 1 338 4716. – **DC27)** Av. Boyacá 48 A 11, Edificio Castillo Dorado, Of. 301, SF de Bogotá.

GN00) GUAINÍA
GN01) Casa Cultura, Calle 6 con Cra. 3, Puerto Inírida. - **FM:** 88.9MHz

Super Estación.
GU00) GUAJIRA
GU01) Cra. 9 N° 12-31, Maicao - 0900-0300 – **GU02)** Cra. 8 N° 3-27, Riohacha - 1000-0300 **E:** mercorio@col3.telecom.com.co – **GU03)** Ap. 125 & 256, Maicao. – **GU04)** Calle 15, Salida a Maicao, Riohacha - 0930-0400 – **GU05)** Barrancas. – **GU06)** Cra 8A N° 3-27 (or: Ap. 3), Riohacha. – **GU07)** Cra. 6 N° 6-60, San Juan del César. – **GU08)** Cra. 18 N° 13-54, Uribia. – **GU09)** Calle 1 N° 5-63, San Juan del Cesar - 1000-0400
GV00) GUAVIARE
GV01) Cra 22 con Calle 9, San José del Guaviare **E:** mercorio@col3.telecom.com.co
HU00) HUILA
HU01) Calle 7 N° 10-36, Neiva - 1000-0300 – **HU02)** Ap. 150, Neiva. – **HU03)** Ap. 496 (or: Cra. 7, Calles 21 y 22), Neiva - 1000-0530 – **HU04)** Cra. 13 N° 3A-24, Neiva. – **HU05)** Cra. 4 N° 2-21, Of. 501-502, Neiva. – **HU06)** Calle 6 N° 5-36, P4, Pitalito - 0900-0400 – **HU07)** Cra 8 N° 8-60, P2, Palermo - 0900-0300 – **HU08)** Calle 4a N° 5-59, La Plata - 1000-0100 – **HU09)** Cra. 14 N° 2-47, San Agustín. **E:** – **HU10)** Cra. 7 N° 7-05, Garzón - 1000-0300 – **HU11)** Palacio Municipal de Yaguará, Yaguará.
IS00) ISLAS SAN ANDRÉS Y PROVIDENCIA
IS01) Ap. 354, San Andrés Isla. – **IS02)** Avenida Los Libertadores No 3a – 73, Oficina 204, San Andrés Isla 1030-0300 ☎+ 57 8 51 29138 – **IS03)** Edif. Bermuda, P2, Av. de las Américas, San Andrés Isla. – **IS04)** Av. Providencia N° 1A-48, (or: Ap. 665), San Andrés Isla - 1100-0500 – **IS5)** San Andrés Isla.
MA00) MAGDALENA
MA01) Av. Libertadores 27-101, Santa Marta - 1000-0300 – **MA02)** Cra. 5 N° 18-32 (or: Ap. 757), Santa Marta 0945-0500 – **MA03)** Calle 17 N° 5-83 (or: Ap. 103), Santa Marta. – **MA04)** Calle 18 N° 5-58, Santa Marta - 1000-0200 – **MA05)** Cra. 9 N° 14-13, Fundación. – **MA06)** Ap. 1240, Santa Marta. – **MA07)** Calle 11 C N° 18a-34, Santa Marta - 1000-0300 – **MA08)** Ap. 45, El Banco - 1000-0400
ME00) META
ME01) Calle 41B N° 30-11, Barrio La Grama, Villavicencio. – **ME02)** Cra. 13 N° 15-52, Granada. – **ME03)** Calle 38 N° 32-41, P7, Edif. Prollano, Ofc 702, Villavicencio **W:** www.cmb.org.co/cra/index.html **E:** cmbvillavo@andinet.com – **ME04)** Cra. 30 N° 36-14, P4, Villavicencio. – **ME05)** Cra. 31 N° 37-71, Of.1001, (Ap. 2472), Villavicencio - 0900-0500 – **ME06)** Cra. 40 N° 34-34, Baltazar Alto, Villavicencio. – **ME07)** Calle 10 N° 28-05 (Ap. 001), Granada. – **ME10)** (See DC26). **W:** www.fuerzadepaz.com – **FM:** 88.8MHz Marfil Stereo. – **ME11)** Palacio Municipal de Villavicencio, Villavicencio.
NA00) NARIÑO
NA01) Cra. 29 N° 17-30 (or: Ap. 375), Pasto - 1000-0200 – **NA02)** Cra.20A N° 16-73, P2, Pasto - 1000-0100 **E:** Ondasdelmayo@latinmail.com – **NA03)** Ap. 454, Pasto. – **NA04)** Calle 20 N° 24-73, Of 603, P6, Pasto - 0900-0500. – **NA05)** Cra. 27 N° 19-30, Pasto. **E:** caracol@telesat.com.co **NA06)** Ap. 1005, Ipiales - 1100-0200 – **NA07)** Parque Colón (or: Ap. 165), Tumaco - 1100-0400 – **NA08)** Cra. 8 N° 4-48, Ipiales - 1100-0200 – **NA09)** Calle 20 N° 15-13, Túquerres - 0900-0400 – **NA10)** Ap. 516, Pasto. – **NA11)** Cra. 6A N° 9-14, P2, Ipiales. – **NA12)** Cra. 5 N° 3-15, Samaniego. – **NA13)** Fundación Tomás Cipriano de Mosquera, Ricaurte el Diviso 1300-2300 – **NA14)** Nevado Cumbal. – **NA15)** Calle 15 N° 14-24 , Pasto - 1100-0500 **E:** sannicolas@organizacionsolarte.com – **NA16)** Universidad de Nariño, Cra. 25 N° 19-12, Pasto. – **NA17)** Cra. 1 N° 21-36, Pasto. – **NA18)** Palacio Municipal de Guaitarilla, Guaitarilla. – **NA19)** Palacio Municipal de Tangua, Tangua. – **NA20)** Palacio Municipal de El Tablón, El Tablón. – **NA21)** Palacio Municipal de Guachucal, Guachucal. – **NA22)** Palacio Municipal de Mallama, Mallama. – **NA23)** Palacio Municipal de Pupiales, Pupiales.
NS00) NORTE DE SANTANDER
NS01) Calle 5 N° 3-26 (or: Ap. 1650), Cúcuta. – **NS02)** Centro Comercial Bolívar, Local E4 y E5, Cúcuta - 1000-0400 – **NS03)** Ap. 400, Cúcuta. – **NS04)** Calle 5A N° 0-45, Cúcuta. – **NS05)** Av. O. N° 10-54, P2 (or: Ap. 624), Cúcuta. – **NS06)** Ap. 519, Cúcuta. – **NS07)** Cra. 13 N° 9-10, P7, Ocaña. – **NS08)** Calle 7N N° 4-117 (or: Ap. 2284), Cúcuta. – **NS09)** Cra. 4 Calle 5 junto Almacén Fotorubio, Fomeque - 1000-1700, 2200-0100 – **NS10)** Calle 11 N° 15-24, Ocaña - 1030-0300 – **NS11)** Calle 7 N° 4-50, Tibú - 1000-2400 – **NS12)** Calle 16 N° 2-17 (or: Ap. 473), Cúcuta - 1000-0500 – **NS13)** Base Militar "San Jorge", Tibú - 1030-1700, 2100-0200 – **NS14)** Av. 2 N° 4-11, Bochalema - 1000-0500 – **NS15)** Calle 12 N° 4-19, Ofc. 214, (or: Ap. 2582), Cúcuta - 1000-0500 – **NS16)** Av. 0A N° 12-75, Ofc. 101 (or: Ap. 1303), Cúcuta. – **NS17)** Cra. 6 N° 4-59, P3 (or: Ap. 1074), Pamplona - 0900-0300 – **NS18)** Av. N° 11-17, Ofc. 303, Cúcuta - 1000-0400 – **NS19)** Cra. 2 N° 1-10, Puerto Santander - 1000-2200 – **NS20)** Av. 5 N° 9-58, P2, Edif. Mut.Aux (or: Ap. 151), Villa del Rosario.
PU00) PUTUMAYO
PU01) Calle Principal, Orito - 1100-2300 – **PU02)** Calle 10 N° 6-01 (or: Ap. 011), Mocoa - 0900-0300 – **PU03)** 19A Barrio Oriental, Sibundoy - 1300-2300 - **FM:** 107.3MHz – **PU04)** Calle 11 N° 17-18 (or: Ap. 9),

Puerto Asís - 1100-0200
QU00) QUINDÍO
QU01) Cra 16 N° 19-23, P10, Armenia. – **QU02)** Calle 9 N° 13-50 (or Ap. 2361), Armenia. – **QU03)** Ap. 2481, Armenia. – **QU04)** Ap. 556, Calarcá. – **QU06)** Cra. 14 N° 21-26, P2, (or: km 2 via al Aeropuerto), Armenia - 1000-0500 - **FM:** 104.7MHz Robles FM Stereo. – **QU07)** Calle 21 N° 16-31, Ofc 702, (or: Ap. 617), Armenia. – **QU08)** Universidad del Quindío, Av.Bolívar Cra.15 Calle 12 Norte, Armenia **W:** www.uniquindio.edu.co **E:** uq@uniquindio.edu.co - **FM:** 102.1MHz – **QU09)** Cra.24 N° 39-52, Calarcá. – **QU10)** Palacio Municipal de Armenia, Armenia.
RI00) RISARALDA
RI01) Ap. 354, Pereira. – **RI02)** Ap. 045, Pereira. – **RI03)** Ap. 221, Pereira - 1000-0500 – **RI04)** La Virginia. – **RI05)** Crra. 7a N° 18-80, Of. 705, Edificio Centro Financiero, Pereira - 0930-0515 – **RI06)** Cra. 7 N° 15-10, P3 (or: Ap. 1262), Pereira. – **RI07)** Cra. 15 N° 11-80, Santa Rosa de Cabal (or: Calle 19 N° 8-74, Pereira) - 1000-0300 – **RI08)** Palacio Municipal, Pereira. – **RI09)** Calle 17 N° 9-10, Marsella. – **RI10)** Centro Administrativo Municipal, Dosquebradas. – **RI11)** Palacio Municipal de Balboa, Balboa. – **RI12)** Palacio Municipal de La Ceila La Ceila. – **RI13)** Santuario.
SS00) SANTANDER DEL SUR
SS01) Ap. 915, Bucaramanga. – **SS02)** Ap. 223, Bucaramanga. – **SS03)** Calle 50 N° 17-71, P3, Barrancabermeja. – **SS04)** Cra. 27 N° 45-80, Bucaramanga - 0900-0400 – **SS05)** Ap. 007, Bucaramanga - 1000-0200. – **SS06)** Calle 36 N° 14-58, Ofc.707, Floridablanca. – **SS07)** Calle 16 N° 15-01, Esquina, Socorro - 0930-0300 **E:** vozdecolombia1100am@gmail.com – **SS08)** Transv. 6 N° 9-56, Barbosa. – **SS09)** Batallón de Artillería de Defensa Aerea N° 2 "Nueva Granada" (or: Ap. 036), Barrancabermeja. – **SS10)** Ap. 1100, Bucaramanga. – **SS11)** Calle 11 N° 9-80, p. 3, San Gil. – **SS12)** Calle 48 N° 35A-25, Bucaramanga. – **SS13)** Edif.Súper Estrellas, Ofc.409 (or: Ap. 23), Barrancabermeja - 1000-0300 – **SS14)** Calle 7 N° 17-44, Barbosa 0730-2330 – **SS15)** Ap. 578, Barrancabermeja - 0900-0400 – **SS16)** Calle 12 N° 10-30, Centro, San Gil - 0900-0300 **W:** www.lacaliente1330.com – **SS17)** Calle 16 N° 4-47, Zapatoca. – **SS18)** Calle 35 N° 20-39 (or: Ap. 3104), Bucaramanga. – **SS19)** Cra. 4 N° 4-118, P2, Cimitarra - 0900-0300 – **SS20)** Cra. 3 N° 3-42, Vélez. – **SS21)** Calle 20 N° 6-36, Zapatoca. – **SS22)** Av. 36, No. 19-76, Piso 9, Bucaramanga. **E:** emisora@radiocatolicametropolitana.com – **SS23)** Calle 41 N° 19-87, Bucaramanga - 1000-0300 – **SS24)** Batallón Luciano D'Ahuyar, San Vicente de Chucurí. – **SS25)** Calle 12 N° 17-10, Ofc.302 (or: Ap. 250), Barrancabermeja - 1000-0400 – **SS26)** Calle 11 N° 6A-11, Edif. San Gabriel, P2, Málaga - 1000-0200 – **SS27)** Diócesis del Socorro y San Gil, Cra. 13 N° 34, Esquina Socorro. – **SS28)** Cra.27, Calle 9, Televis, Bucaramanga. – **SS29)** Palacio Municipal de San Andrés, San Andrés. **SS30)** Diócesis de Barrancabermeja, Calle Octava, entre Carreras 15 y 16, Barrancabermeja.
SU00) SUCRE
SU01) Ap. 167, Sincelejo - 1000-0430 – **SU02)** Cra. 18 N° 20-48 (or: Ap. 448), Sincelejo – **SU03)** Cra. 24 N° 29-50 (or: Ap. 100), Corozal - 1100-0500 – **SU04)** Cra. 20 N° 16-40 (or: Ap. 191), Sincelejo - 1100-0300 – **SU05)** Calle 20 N° 24-93, Av. las Penitas, Sincelejo – **SU07)** Cra. 20 N° 25-82 (or: Ap. 542), Corozal. – **SU08)** Cra. 20 N° 21-46 (or: Ap. 303), Sincelejo - 1000-0400 – **SU09)** Cra. 28 Calle 18, San Marcos. – **SU10)** Calle 25A N° 18, Sincelejo - 1030-0430 – **SU11)** Cra. 20 N° 25-92, P2, Sincelejo - 1030-0200. – **SU12)** Palacio Municipal de Ovejas, Ovejas. – **SU13)** Palacio Munivipal de Majagual, Majagual. – **SU14)** Calle 24 No 18-31, Sincelejo.
T000) TOLIMA
T001) Calle 12 N° 1-17, P5, Ibagué - 0900-0400 - **FM:** 96.3MHz – **T002)** Calle 14 N° 2A-14, P2, Ibagué. – **T003)** Parque Murillo Toro N° 3-29, P4, Ibagué - 1000-0400 – **T004)** Cra 7 con Calle 10, Espinal. – **T005)** Calle 11 N° 10-36, Guamo - 1100-0400 – **T006)** Ap. 2419, Ibagué. – **T007)** Ap. 1094, Ibagué. **FM:** 93.9MHz – **T008)** Calle 5 N° 6-25, Mariquita. – **T009)** Calle 9 N° 1-124, P3, Ibagué. – **T010)** Calle 11 N° 4-26 (or: Ap. 64), Espinal - 1000-0400 – **T011)** Ap. 536, Honda. – **T012)** Ap. 509, Honda - 1000-0300 – **T013)** Cra. 2 N° 11-27, Flandes. – **T014)** Cra 3 N° 12-76, Ofc.801 (or: Ap. 589), Ibagué - 1000-0400 – **T015)** Armero, Guayabal. – **T016)** Cra. 3 N° 5-61, Líbano - 1000-0300 – **T017)** Cra. 7a N° 5-36, Dolores - 0945-0300 – **T018)** Calle 7a No. 20-70, Melgar - 0900-0500 – **T019)** Cra. 2 N° 3-74, Rovira. – **T020)** Palacio Municipal de Cajamarca, Cajamarca. – **T021)** Palacio Municipal de Palo Cabildo, Palo Cabildo. – **T022)** Palacio Municipal de Venadillo, Venadillo.
VA00) VALLE DEL CAUCA
VA01) Cra. 26 N° 5C-25, San Fernando, Cali. – **VA02)** Cra. 38D Diagonal 37A-52B/Santa Isabel, Cali. **VA03)** Ap. 1941, Cali **E:** Gerencia: caracolra@emcali.net.co – Dir.Técnico: caracol@emcali.net.co – **VA04)** Ap. 4666, Cali - 1100-0400 – **VA05)** Cra. 14 N° 2-25, P2 (or: Ap. 96), Buga - 1100-0500 **W:** www.pablus.net/vocesdeoccidente – **VA06)** Av. 5B Norte N° 21-02, Cali ✉ +57 92 667 5536 – **VA07)** Cra. 33 N° 28-51 (Ap. 280), Palmira - 1000-0500 **E:** radiopalmira@

uni **Web**.net.co – **VA08)** Ap. 126, Tuluá - 1000-0300 – **VA09)** Cra. 4A
N° 10-75 (or: Ap. 145), Cartago. – **VA10)** Calle 21 Nte N° 3N-49, P5,
Cali **E:** superam@telesat.com.co –**VA11)** Calle 12-39, Ofc 301, Edif.
R.Buenaventura (Ap 383), Buenaventura 1030-0500 ☎ +57 92 242
4387 ▤+57 092 242 2969 **E:** fdradbue@col2.telecom.com.co – **VA12)**
Cra. 16 N° 6-22, P2, Caicedonia. – **VA13)** Cra 19 N° 2N-29, Ofc 21B,
Cali. – **VA14)** Cra. 30 N° 29-09, Palmira - 1000-0400 – **VA15)** Cra 6 N°
54-08, Av. Simon Bolívar, Buenaventura. – **VA16)** Carrera 66B, No. 6-68,
Barrio El Limonar, Cali – **VA17)** Cra. 11 N° 11-43, P2, Zarzal. – **VA18)**
Cra. 29 N° 32-88/90 (or: Ap. 201), Palmira - 1130-0300 – **VA19)** Cra. 14
N° 5-77, Buga. – **VA20)** Cra. 26 N° 28-72, Tuluá - 1000-0500 – **VA21)**
Calle 27 N° 33-35, Tuluá - 1100-0500 – **VA22)** Av. Roosevelt N° 34-37,
Cali **E:** redsonora@hotmail.com – **VA23)** Cra. 51 N° 49-21, Sevilla.
– **VA24)** Calle 10 N° 6-87, P3, Cartago - 1100-0500 – **VA25)** Calle 14
N° 4A-63, Andalucía 1300-2300 – **VA26)** Calle 13, No. 19-59, Barrio
Guayaquil, Cali - 1100-0500 **W:** www.radiohuellas1470.com – **VA27)**
Cra. 13 N° 10-58, Cali. – **VA28)** Palacio Municipal de Caicedonia,
Caicedonia. –**VA29)** Av.Roosevelt N° 28, Cali. (Or: Transversal 34 N°
149-23, Cedro Golf, SF de Bogotá) ☎+57 52 514 2641 ▤+57 52 558
1113 – **VA30)** Palacio Municipal de Caicedonia, Caicedonia. – **VA31)**
Palacio Municipal de Buenaventura, Buenaventura. – **VA32)** Palacio
Municipal de Versalles, Versalles. – **VA33)** Cra. 13 N° 10-62, Cali
- 1100-0400 ☎+57 2 880 4660

VI00) VICHADA
VI01) Av. Orinoco, Puerto Carreño.
VP00) VAUPÉS
VP01) Mitú.

FM in Santafé de Bogotá (MHz): 88. 9 R. Uno (RCN) – 89. 9 40
Principales (Caracol) – 90.4 La UD (University) – 90. 9 La Mega
(RCN) – 91. 9 Javeriana Estereo (University))– 92. 4 Policía Nacional
(Colombian Police) – 92. 9 La Z (Todelar) – 93.4 Colombia Estéreo
(Colombian Army) – 93. 9 RCN La Radio (rel. 770) – 94. 9 La FM (RCN)
– 95. 9 R. Nacional de Colombia (RTVC) – 96. 9 Melodía FM Estéreo –
97.4 Bésame (Caracol) –97.9 Radioactiva (Caracol) – 98. 5 Universidad
Nacional (University) – 99. 1 Radionica (RTVC) – 99. 9 W Radio
(Caracol) – 100.4 Oxígeno (Caracol) – 100. 9 Caracol Colombia (rel.
810kHz) – 101. 9 Candela (W Radio) – 102. 9 Tropicana (Caracol) – 103.
9 La X (Todelar) – 104.4 Amor Estéreo – 104. 9 Vibra Bogotá (W Radio)
– 105.4 Rumba Stereo (RCN) – 105. 9 Olímpica – 106. 9 Universidad
Jorge Tadeo Lozano (University) – 107. 9 Minuto de Dios (Rlg).

COMOROS

L.T: UTC +3h — **Pop:** 700,000 — **Pr.L:** French, Comorian, Arabic
— **E.C:** 50Hz, 220V — **ITU:** COM

**OFFICE DE RADIO TÉLÉVISION DES COMORES
(ORTC, Gov.)**
▤ BP 452, Moroni, Grand Comoro ☎+269 7732531 ▤ +269 7730303
W: radiocomores.km **LP:** Tech. Dir: Abdulkader Radjab.
FM: Moroni, R.Studio 1 101.2MHz. Nkazi, R.Nkazi 107.0MHz. 6 x 1kW,
3 x 0.5kW txs. **D.Prgr:** 0300-1900. **Ann:** F: "Ici Radio Comoro".
Other stations:
R. Dziyalandze, Anjouan: 90.0MHz (rel. RFI 1700-1030).
R. Ocean Indien, Ngazidja: 100.5MHz. Web: radioceanindien.km
R. France Int: 103.0MHz

CONGO (Dem. Rep.)

L.T: Kinshasa & western part: UTC +1h, eastern part: UTC +2h — **Pop:**
67 million — **Pr.L:** French, Lingala, Swahili, Tshiluba, Kikongo — **E.C:**
50Hz, 220V — **ITU:** COD

HAUTE AUTORITÉ DES MÉDIAS (HAM)
▤ Boulevard Sendwe n°5058, Commune de Kalamu, Kinshasa **W:**
www.ham-rdc.org **E:** info@ham-rdc.org **L.P:** Pres.: Modeste Mutinga.

**RADIO-TÉLÉVISION NATIONALE CONGOLAISE
(RTNC, Gov.)**
▤ B.P. 3171, Kinshasa-Gombe **L.P:** DG: E. K. Mukambilwa, Deputy
DG: M. Makuala.
FM: Kinshasa: **National channel:** 100.0MHz, **Kinshasa channel:**
91.8MHz, **Channel for national languages:** 97.0MHz.
D.Prgr: 24h in French/Swahili/Lingala/Tshiluba/Kikongo. Also relayed by
other stns. **Ann:** "RTNC, Radio-Télévision Nationale Congolaise, émettant
de Kinshasa".
Provincial stations:
FM: 3) 88.9/92MHz 1.5kW – **4)** 93.3MHz 1.5kW/90MHz 0.05kW – **6)**
89.1MHz 50kW – **7)** 90.0MHz – **9)** 94.8MHz – **10)** 90.1MHz – **11)**
93.7MHz.
Addresses: 2) B.P. 7296, Lumbumbashi – **3)** RTNC Kivu, B.P. 475, Bukavu

– **4)** B.P. 1061, Mbandaka – **5)** B.P. 1232, Mbuji-Mayi – **6)** B.P. 708,
Kananga, Western Kasai – **7)** B.P. 704, Matadi – **8)** B.P. 1745, Kisangani
– **9)** Butembo, Nord-Kivu – **10)** Goma, Nord-Kivu. **E:** rtncnordkivu@yahoo.
fr – **11)** Ulvira, Sud-Kivu.
F.PI: 10kW SW txs in Goma and Bukavu in the 60 metre band and
100kW FM txs in all main cities.

RADIO TÉLÉ CANDIP
▤ B.P. 373, Bunia ☎+243 987 46297 **L.P:** Henri d'Odz Korr, Dir.
SW: Bunia v5066kHz 1kW. **FM:** 98MHz 1kW.
D.Prgr in French/Ethnic: 0400-0700, 1300-1620.

RADIO KAHUZI (Rlg.)
▤ Ave. Masikits 2, Muhumba, Bukavu or B.P. 42, Cyangugu, Rwanda.
W: radiokahuzi.com **E:** radiokahuzi@gmail.com **L.P:** Dir: Richard
McDonald, St. Mgr: Barbara Smith.
SW: Bukavu 6210kHz 0.8kW. **FM:** 91.1/102.1MHz 0.2kW. **D.Prgr:**
0530-2010 in French, English, Kikongo, Kinyarwanda, Lingala, Mashi,
Swahili and Tshiluba. Also rel. VOA. **F.PI:** FM transmitters in Fizi,
Shabunda and Wamazi.

RADIO OKAPI
(joint initiative between the UN Mission in the DRC [MONUC] and
Hirondelle Foundation)
▤ QG Monuc, 12 Av. des Aviateurs, Kinshasa-Gombe ☎+243-81-
890-6747 **W:** www.radiookapi.net **E:** info@hirondelle.org
L.P: Dir: Yves Laplume.
FM (MHz): (powers 1-5kW): Isiro 90.1, Beni 92.0, Butembo 92.9,
Gbadolite/Kananga/Lisala 93.0, Mbuji-Mayi 93.8, Kisangani 94.8,
Bukavu 95.3, Gemena 95.4, Lubumbashi 95.8, Kanyabayonga/
Mahagi 96.0, Aru 98.0, Bundundu 99.0, Matadi 102.0, Baraka/Kindu/
Mbandaka 103.0, Kikwit/Kinshasa/Mbuji Mayi 103.5, Kamina 104.3,
Manono 104.5, Bunia/Walikale 104.9, Kalemie 105.0, Goma/Uvira
105.2, Shabunda 105.4, Tshomo Ini 106.5.
D.Prgr: 0430-2200 in in French/Lingala/Swahili/Tshiluba.
NB: For relays via txs outside Congo DR: see Clandestines & other
target broadcasts section.

Other stations:
Business R. Africa, Kinshasa: 98.6MHz. **W:** www.brt-africa.com
E: redaction@brt-africa.com – **Canal Congo pour Christ,** Bukavu:
97.3MHz – **Canal Futur,** Kinshasa-Gombe: 107.4MHz – **CEBS,**
Kinshasa: 93.7MHz – **RATELKI,** Kinshasa: 90.2MHz – **R. Artemis,**
Bunia: 90.2MHz – **R. Boboto,** Isiro: 100.6MHz 100W – **R. Butembo:**
100MHz – **R. Canal CVV,** Kinshasa: 102.3MHz – **R. Canal Révélation,**
Bunia: 100.7MHz 0.3kW – **R. Congo FM,** Kinshasa: 96.4MHz – **R. ECC,**
Kinshasa: 104.0MHz – **R. Elikya,** Kinshasa: 97.5MHz – **R. Liberté
Kinshasa (RALIK),** 96.8MHz – **R. Maendeleo,** 15 Ave. Kolwezi,
Commune d'Ibanda, B.P. 3133, Bukavu. FM: Bukavu 88.7MHz 1kW,
Chomuhini 103.3MHz 1kW. Web: www.radiomaendeleo.org – **R.
Malebo Broadcast Channel** (MBC), Kinshasa: 98.3MHz – **R. Maria
Malkia wa Amani** (Rlg.): Bukavu 94 & 97MHz. **W:** www.radiomaria-
bukavu.best.cd – **R. Méthodiste Lokole,** Kinshasa-Gombe: 100.8MHz
– **R. Moto** (Rlg.): Kivu 103MHz 1.2kW, Butembo 106MHz (also rel.
RFI) – **R. Neno la Uzima,** Bukavu: 100.2MHz – **R. Parole Eternelle,**
Kinshasa: 103.8MHz – **R. Raga FM,** Kinshasa-Gombe 90.5MHz 1.1kW.
(Also rel. BBC & VOA). **W:** www.raga.net/ragafm.htm **E:** info@raga.
cd – **R. Rehema,** Chamuhini 89.5MHz 0.25kW, Bukavu 99.7MHz 1kW
– **R. Réveil FM,** Kinshasa-Gombe: 105.4MHz **W:** reveilfm.itgo.com
– **R. Sango Malamu:** Boma 102.5MHz, Kinshasa 104.5MHz – **R.
Tangazeni Kristo** (Rlg.): Bunia 88.6MHz, Aru/Kwandruma 90.0MHz **E:**
buero@diguna.de – **R. Télé Armée de l'Eternel,** Kinshasa: 94.5MHz
– **R. Télé Amani,** Kisangani; 100.1MHz 25W, 103.1MHz 0.5kW – **R.
Téle Boma** (RTB), Boma: 98.0MHz – **R. Télé Graben,** Beni/Butembo
98MHz – **R. Télé Groupe l'Avenir (RTGA),** Kinshasa 88.1MHz –**R.
Télé Kin Malebo** (RTKM): Kinshasa 95.9MHz, Kananga 97.5MHz (also
rel. RFI) – **R. Télé Kintuadi** (RTK): Boma 91.1MHz, Kinshasa 97.1MHz,
Mbanza Ngungu 103.4MHz, Matadi 107.5MHz – **R. Télé Message de
Vie,** Kinshasa: 88.7MHz – **R. Télé Mosaïque,** Likasi: 88.5MHz (also
rel. RFI) – **R. Télé Puissance,** Kinshasa 101.0MHz – **RTV Bukavu
Liberté,** Ibanda 107.3MHz – **RTV Mulangane,** Bukavu: 100.1MHz
0.25kW – **R. Sentinelle,** Kinshasa: 97.1MHz – **R. Tomisa,** Kikwit:
97.5MHz 0.5kW (also rel. RFI) – **R. Veritas** (Rlg.), Kabinda: 105.0MHz
– **R. Univ. of Kinshasa,** fq. not known – **R. Vuvu Sence,** Mbanza
Ngungu: 101.0MHz – **RCLS,** Kirumba: 91MHz – **REB,** Butembo:
90.7MHz – **RTIV,** Kisangani: 89.4MHz 0.5kW – **Sauti ya Mkaaji,**
Makongo: 87.85MHz – **Top Congo,** Kinshasa: 88.4MHz.
Africa No. 1 Kinshasa 102MHz (see main entry under Gabon).
BBC African Sce: Kinshasa 92.7MHz, Kisangani/Lubumbashi 92MHz.
RFI Afrique: Bunia 90.2MHz, Bukavu/Kisangani/Lubumbashi/ Matadi
92MHz, Kinshasa 105MHz.

RTBFi (Belgium), Kinshasa: 99.2MHz

CONGO (Rep.)

L.T: UTC +1h — **Pop:** 4 million — **Pr.L:** French, Lingala, Kikongo — **E.C:** 50Hz, 230V — **ITU:** COG

CONSEIL SUPÉRIEUR DE LA LIBERTÉ DE LA COMMUNICATION (CSLC)
▣ Brazzaville. **L.P:** President: Jacques Banangadzala.

TELEDIFFUSION DU CONGO - RADIO CONGO (Gov.)
▣ Direction Générale, B.P. 2912, Brazzaville.
L.P: DG: Jean Médard Bokatola. Dir. Radio: Sylver Sandi Ibambo.
SW: Brazzaville 6115kHz 50kW 1700-2130 (irreg.)
FM: 90.1/94/96.4MHz
National Network: 0420-2300 in French & ethnic. **N. in English:** 1900 (approx.) **Ann:** "Radio Congo, Chaîne Nationale".
IS: Zansi solo. Opens and closes with National Anthem.

Other stations:
Digital R. N° 1, Brazzaville: 92.2MHz – **R. Brazzaville:** 98.0MHz – **R. Mucodec,** Brazzaville: 88.4MHz. Web: mucodec.com – **R. Rurale Congolaise,** Brazzaville: 99.3MHz – **R. Liberté,** Brazzaville 106.0MHz.
Africa No 1, Brazzaville 89.6MHz (see main entry under Gabon).
BBC African Service: 103.8MHz.
RFI Afrique: Brazzaville/Pointe-Noire 93.2MHz

COOK ISLANDS

L.T: UTC-10h — **Pop-**11,870 — **Pr.L:** English, Cook Island Maori — **E.C:** 50Hz, 220V — **ITU:**CKH

MW	kHz	kW	Station	Location
1)	630	2.5	R. Cook Islands AM	Rarotonga
FM	**MHz**	**kW**	**Station**	**Location**
2)	88.1		88 FM	Rarotonga
1)	88.1		Araura FM	Aitutaki
1)	89.0		R. Cook Islands	Mitiaro
1)	89.0		R. Cook Islands	Pukapuka
1)	89.9		R. Cook Islands	Rarotonga
1)	90.6		R. Cook Islands	Mangaia
1)	90.6		R. Cook Islands	Rakahanga
1)	90.6		R. Cook Islands	Palmerston
4)	91.9		Matariki FM	Rarotonga
1)	92.2		R. Cook Islands	Atiu
1)	92.2		R. Cook Islands	Penrhyn
3)	93.0		R. Australia	Rarotonga
1)	93.8		R. Cook Islands	Mauke
1)	93.8		R. Cook Islands	Nassau
1)	95.4		R. Cook Islands	Aitutaki
1)	95.4		R. Cook Islands	Manihiki
4)	96.7		Matariki FM	Rarotonga
9)	97.9		Marantha FM	Rarotonga
5)	98.7		Rarotonga's Gospel & Comm. R. TK3ANA	Rarotonga
4)	99.9		Matariki FM	Rarotonga
1)	100.0		R. Cook Islands	Aitutaki
1)	101.1		Ocean & Earth HITZ FM	Rarotonga
6)	103.3	1	R. Ikurangi KCFM	Rarotonga
7)	105.0		R. Atiu	Atiu

Addresses & other information
1) Elijah Communications, PO Box 126, Avarua, Rarotonga ☎ +682 29460 🖷 +682 21907 **W:** www.radio.co.ck **E:** jeanne@oyster.net.ck. **R.Cook Islands AM Hours:** M-F 1600-0900 [Fri 1000] Sat 1600-1000 Su 1700-0900 **N:** Local news hourly M-F 1700-0200 **RNZI** 1600, 1700, 1800 M-F **Prgr:** Talkback, news, information, religious services and music in English & Cook Isl Maori. **Ocean & Earth HITZ FM** 24h **Prgr:** contemporary hit music. **Outer Island Network:** Txs outside Rarotonga are owned by the Cook Islands Government and relay Radio Cook Islands AM and in some cases also originate prgrs as local community stns – **2)** The Digital Factory, Avarua, Rarotonga ☎: +682 22836 **L.P:** Nicholas Henry **ID:** '88FM Raro's Hottest Hits' 24h – **3)** 24h satellite rel. – **4)** Matariki FM Ltd, PO Box 511, Avarua, Rarotonga ☎: +682 25997 **W:** www.matarikifm.co.ck **E:** onair@matarikifm.co.ck **L.P:** William Franheim – **5)** Cook Islands Seventh Day Adventist Mission, Avarua, Rarotonga ☎: +682 22851 **L.P:** Priest Uma Kalu **E:** uma@adventist.org.ck **Prgr:** Religious. Rptd inactive July 2010 – **6)** Kia Orana Country Radio, PO Box 521, Avarua, Rarotonga ☎ +682 23203 – **7)** Enuamanu School, Mapumai, Atiu. Mgr: Bazza Ross ☎ +682 33264 **E:** rossb@oyster.net.ck **-8)** Aitutaki. **ID:** '88FM Aitutaki's Hottest Hits' 24h. Same family ownership as #2 with joint marketing and sales with the Rarotonga stn **-9)** Rarotonga. **Prgr:** Religious.

COSTA RICA

L.T: UTC -6h — **Pop:** 4.2 million — **Pr.L:** Spanish — **E.C:** 60Hz, 120V — **ITU:** CTR

CONTROL NACIONAL DE RADIO (CNR)
▣ Ministerio de Gobernación y Policia, Ap.10006, 1000 San José ☎ +506 2221 0992, 2221 9910

CAMARA NACIONAL DE RADIO (CANARA)
▣ Ap.1583, 1002 San José ☎ +506 2256 2338 🖷 +506 2255 4483
E: info@canara.org

Hrs of tr. usually 24h

MW	Call	kHz	kW	Station, location and h. of tr
1)	RI	530	10	R. Sinfonola, Cartago: (r: FM 90.3)
2)	SCL	550	5	R. Santa Clara, Cd. Quesada: 1100-0130
3)	ELR	570	5	R. Libertad, San José: 1200-0400
4)	RN	590	5	R. Nacional, San José
6)	RMV	610	15	R. María, San José
5)	ALY	640	20	R. Rica, San José: 1130-0400
7)	TNT	670	10	R. Managua, San José
12)	JC	700	10	FCNRADIO.COM, San José
68)		730	1	R. Pacífico, Puntarenas: 1400-0200
70)	HB	*730	20	Sin Fronteras, Desamparados
9)	LX	760	5	R. Columbia, San José
10)	RA	780	10	R. América, San José: 1000-0500
8)	SD	800	5	R. La Gigante, San José
11)	GC	820	2.5	R. Centro AM, San José: 1130-0600
62)	RDR	850	2	R. Cartago, Cartago: 1100-0400
13)	UCR	°870	10	R. 870 UCR, San Pedro Montes de Oca
7)	BAS	890	10	R. Heredia, Heredia
14)	UM	910	5	BBN, San José
31)	RCR	930	5	R. Costa Rica, Guadalupe
71)	SD	960	5	R. Actual 960, San José: 1100-0600
72)	RC	980	10	R. Alajuela, Alajuela: 1100-0200
17)	MIL	1000	1	100.7/Mil FM, San José
29)	TIC	*1020	5	LV de la Liberación, San José: 1100-0500
18)	AC	1045	5	R. Fides, San José
60)	HG	1040	5	R. Nosara, Hojancha: 1100-1400 & 2100-2300
9)	LX	1060	1	R. Columbia, San Isidro del General (r: 760)
19)	FC	1080	1	Faro del Caribe, San José
69)	SBC	1100	5	R. Guápiles, Guápiles
20)	SCR	1100	5	R. Chorotega, Santa Cruz: 1315-0000
54)	ACE	1120	1	R. Miel, Alajuela
15)	DKN	1140	1	R. Nueva, Guápiles
9)	CA	1160	1	R. Columbia, Puntarenas (r: 760)
22)	PJ	1180	5	R. Victoria, Heredia: 1100-0400
23)	TQ	1200	5	R. Cucú, San José: 1000-0600
73)	Q	1220	1	R. Fe y Poder, Limón
63)	WC	1240	1	R. Corobici, Cañas
24)	DIO	1260	5	R. Emaús, San Vito de Coto Brus: 1100-0300
25)	GV	*1280	2	Visión 1280, San José
26)	GL	1300	1	La Fuente Musical, Cartago
9)	LX	1320	1	R. Columbia, San José (r: 760)
27)	HR	1340	5	R. Sideral, San Ramón: 1000-0400
31)	DS	*1360	5	R. Radio 1360, San José
28)	MS	1380	1	R. Guanacaste, Liberia: 1000-0500
61)	GJ	1400	5	R. Sinaí, San Isidro del General: 1000-0400
32)	RPN	1420	1	R. Pampa, Liberia: 1100-0100
30)	RDVC	1430	3	R. San Carlos, Cd. Quesada: 1100-0300
9)	LX	1460	1	R. Columbia, Ciudad Quesada (r: 760)
47)	AW	1480	2	R. El Sol, Puntarenas: 1200-0400
55)	ASF	1500	1	R. Radio1500, Sarapiqui: 1100-0300
9)	LX	1520	1	R. Columbia, Cartago (r: 760)
57)	CUB	*1540	1	Enlace Radio, Pavas
32)	OAR	1560	5	R. Nicoya, Nicoya: 1000-0300
44)	RCVT	1580	0.25	LV de Talamanca, Talamanca
34)	RCLS	1580	0.25	R. Cultural Los Santos
35)	RCC	1580	0.25	R. Cultural de Corredores,
36)	RCLC	1580	0.25	R. Sistema Cultural de La Cruz
37)	RSCM	1580	0.25	R. Cultural Maleku
38)	RCL	1580	0.25	R. Sistema Cultural de Los Chiles
45)	RCP	1580	0.5	R. Cultural Pejibaye, Pérez Zeledón
46)	RCS	1580	0.5	R. Cultural Santiago
47)	RCT	1580	0.5	R. CulturalTilarán
64)	LG	*1580	1	R. Casino, Siguirres, Limón
64)	LGJ	1590	1.5	R. 16, Grecia: 1100-0400
39)	RSCN	1600	0.25	R. Sistema Cultural Nicoyano
40)	RCT	1600	0.25	R. Cultural de Turrialba
41)	RCBA	1600	0.25	R. Cultural de Buenos Aires
42)	RCP	v1600	0.25	R. Cultural de Pital
43)	RCU	1600	0.25	R. Cultural de Upala
48)	RCCH	1600	2.5	R. Cultural Chirripó
49)	RCSG	1600	0.5	R. Cultural San Gabriel

MW	Call	kHz	kW	Station, location and h. of tr
50)	RCPV	1600	2.5	R. Cultural Puerto Viejo
65)	CC	1600	2.5	R. Radio Cima, Pto Golfito
66)	MQ	1600	1.5	R. Pococí, Guápiles: 1100-0400
51)	RPQ	1600	0.5	R. Más, Pto Quepos

Call: TI–, ° = also on SW, * = inactive, (r) = repeater, v = varying fq.

SW	Call	kHz	kW	Station, location and h of tr
19)	FC	*5054	5	Faro del Caribe, San José: irr
13)	UCR	*6105	10	R. Universidad de Costa Rica, San Pedro Montes de Oca
19)	FC	*6175	2.5	Faro del Caribe, San José
19)	FC	*9645	5	Faro del Caribe, San José

All current SW stns are reported to be inactive, but may occasionally be reactivated for variable periods of time.

Addresses and other information:
1) Interamericana Sur, Km 19, Taras, Cartago **E:** rumbo@racsa.co.cr **☎**+506 2537 1002 – **2)** Ap. 221, 4400 Cd. Quesada **W:** www.radiosantaclara.org **E:** radio@radiosantaclara.org **☎**+506 2460 6666 – **3)** Cadena Radial Costarricense, 100m oeste de Taca, La Uruca, 1000 San José or Ap. 301-2400 Desamparados **☎**+ 506-2232 3672 **☖**+506 2232 9750 – **4)** Ap. 7-1980 (or: La Uruca 1 km Oeste Parque Diversionales), 1000 San José **☎**+506 2231 3331. **☖**+506 2231 6604 **W:** www.sinart.go.cr **E:** rnacional@sinart.go.cr **5)** Ap. 1695, 1000 San José **☎**+506 2258 5806 **☖**+506 2258 5803 **E:** radiorica@racsa.co.cr – **6)** Escuela Pilar Jimenez 25 Sur, Goicoechea, 1000 San José **E:** info.crc@radiomaria.org **☎**+506 2234 1676 – **☖**+506 2225 5795. – **7)** Ap. 800-1000 (or: Costado Oeste del Puente Juan Pablo II), 1000 San José. **E:** carias@monumental.co.cr **W:** www.monumental.co.cr **☎**+506 2296 6093. **☖**+506 2296 0413 – **8)** Calles 15-13, Av. 11, Barrio Aranjuez (or: Ap. 1735) 1000 San José **☎**+506 2257 3131 - **☖**+506 2221 9679 **W:** www.radiogigante800am.com– **9)** Ap. 708, 1000 San José **W:** www.columbia.co.cr – **10)** Edificio de la Prensa Libre, Calle 4, Avenida 4 (or Ap. 177-1009) San José **E:** radioamerica@780am.com **☖**+506 2255 3712 **W:** www.780america.com. – **11)** Ap. 6133, 1000 San José **☎**+506 2240 7591 **☖**+506 2236 3672 **E:** info@radiocentrocr.com – **12)** Family Christian Network, Ap. 60-2020, Zapote **W:** www.fcnradio.com - **E:** info@fcnradio.com **☖**+506 2293 7993– **13)** Cd. Universitaria Rodrigo Facio, San Pedro Montes de Oca, 2060-1000 San José **W:** http://radiosucr.com **E: info@**radiosucr.com **☎** +506 2511 3721. **☖** +506 2511 4832. – **14)** De la Municipalidad de Tibas 100 mtrs al Norte y 75 metrs al Oeste, casa blanca a mano derecha (Ap. 2006), 1100 San José. **W:** www.bbnradio.org **☎**+506 240 2900 – **15)** Ap. 266, 7210 Guápiles **☎** +506-2711 1140 **☖** +506 2710 4011 **E:** radionueva@gmail.com– **17)** Ap. 10. 001, 1000 San José **E:** radiomil@racsa.co.cr **W:** www.cienpuntosiete.com **☎**+506 2225 1000 **☖**+506 2234 6198 – **18)** Ap. 5079, 1000 San José **☎** +506 2258 1415 **☖** +506 2233 2387 **W:** www.radiofides.co.cr **E:** rafides@racsa.co.cr – **19)** Ap. 2710, 1000 San José. **☎**+506 2227 1725 **W:** www.farodelcaribe.org **E:** tifc@farodelcaribe.org – **20)** 700 mts este de Almacén Jiménez y Chaverrí, (or Ap. 92), 5175 Santa Cruz **☎** +506 2680 0447 **☖**+506-2680 2435 **E:** ugiocr@hotmail.com – **21)** Ap. 287, 7300 Puerto Limón **☎** +506 2758 0029 **☖** +506 2758 3029 - **FM:** 98.3MHz **W:** www.radiocasinodelimon.com **E:** radiocasino@racsa.co.cr – **22)** Ap. 298, 3000 Heredia **☎**+506 2260 2323 **☖**+506 2237 5736 **E:** gpiedra@gmail.com – **23)** Ap. 1128, 1000 San José. **E:** gerencia@radiocucu.com **☎**+506-2221 8620 **☖**+506 2221 8636 **W:** www.radiocucu.com – **24)** Ap.262, 8257 San Vito de Coto Brus **E:** radioemaus@racsa.co.cr **☎** +506 2773 3101 **☖**+506 2773 4035 – **25)** De la Nissan de Paseo Colón 100 al sur, (or: Ap. 1851, 4050 Alajuela) 1000 San José **W:** www.estereovision.com **E:** estereovision@racsa.co.cr **☖**+506 2256 6361 – **26)** 1 km este de la Basilica de los Angeles, Carr. a Paraíso, 7050 Cartago **E:** lafuentemusical@ice.co.cr **☎**+506 2553 2389 **☖**+506-2591 1090 – **27)** Ap. 73, 4250 San Ramón **E:** rsideral@racsa.co.cr **☎**+506 2245 5046 **☖**+506 2445 5130 – **28)** Residencial Las Brisas, Casa #11A, Buscando la quebrada (Ap. 27), 5600 Liberia, (or: Ap. 6462, 1000 San José) **☎** +506 8387 3133 **☖**+506 2235 3704. – **29)** Ap. 8130, 1002 San José – **30)** 500 Sur 25 Este del Parque de Ciudad Quesada (Ap. 25), 4400 Cd. Quesada **W:** www.radiosancarlos.com **E:** radiosancarlos@ice.co.cr **☎**+506 2460 0339 **☖**+506 2460 0358 – **31)** Barrio Córdoba, Autos Bohío 100 sur y 100 este, 894-2200 Coronado. **W:** www.radiocr.net **E:** radiocostarica@gmail.com **☎**+506 905 2 930 930 **☖**+506 2294 1479. – **32)** Ap. 50, 5200 Nicoya **E:** micoya1560@racsa.co.cr **☎**+506 2685 5757 **☖**+506 2685 5543. – **33)** Ap. 248, 5000 Liberia **E:** pamparamirez@costaricense.cr **☎**+506 2666 4933 **☖**+506-2666 5989 **W:** www.radiopampa.com – **34-50)** Stns are affiliated to Instituto Costarricense de Enseñanza Radiofónica, Ap.132, 2050 San Pedro Montes de Oca (Ministerio de Educación Pública) **W:** www.icer.co.cr **☎**+506 2225 9252. –**34)** Edificio Municipal, Barrio de las Tres Marías, San Marcos de Tarrazú –**35)** Frente al Parque Central, Ciudad Neilly, Corredores. – **36)** Costado sur del Comando Norte, La Cruz, Guanacaste. –**37)** Palenque Tonjibe, frente a la plaza de fútbol, Tonjibe, San Rafael de Guatuso, Prov. de Alajuela. – **38)** Costado Oeste de Edificio Municipal, Los Chiles, Prov. de Alajuela. – **39)** De la esquina noreste de la Iglesia Nueva, 200m Norte, B:o Santa Lucia, Nicoya, Prov. de Guanacaste. – **40)** Palacio Municipal, 132-2050 Turrialba, Prov. de Cartago. – **41)** 300 metros al Norte del Cuerpo de Bomberos, Buenos Aires de Puntarenas. – **42)** Edificio de la Asociación de Desarrollo, Pital, San Carlos, Prov. de Alajuela. – **43)** Frente a la Sucursal Banco Nacional de Costa Rica, Upala, Prov. de Alajuela. – **44)** Frente de la Plaza de Futbol de Amubri, Talamanca. – **45)** Pérez Zeledón, San José - **46)** Puriscal, San José - **47)** Tilarán, Guanacaste - **48)** Chirripó, Turrialba - **49)** San Gabriel de Aserrí, San José - **50)** Puerto Viejo, Cachuita, Limón – **51)** 6350 Quepos – **52)** Ap. 23, 1200 Pavas **W:** www.enlace.org **E:** radio@enlace.org – **54)** 300 metros Norte y 50 Oeste del Antiguo Hospital de Alajuela, Alajuela (or Ap. 233-4060, Moll International, Alajuela) **E:** nelson@enlace.org **☎**+506-2442 9764 **☖**+506 2443 1922 – **55)** Ap. 827-8000 San Isidro Pérez Zeledón **☎**+ 506 2460 7900 – **60)** Casa Cultural de Hojancha, Hojancha, Guanacaste **☎** +506 659 9028 **☖** +506 659 9038 – **61)** Ap. 262, 8000 San Isidro del General **E:** radiosinai@ice.co.cr **☎** +506 2771 4367 **☖** +506 2771 4367 – **62)** Altos de Apolo, frente al Palacio Municipal, Cartago **W:** www.radiocartago.net **E:** radiocartago@ice.co.cr **☎** +506 2591 0542 **☖**+506 2552 4497 – **63)** Frente a la Central de Hielo Frío, Cañas, Guanacaste **☎**+506-2669 2023 - **64)** Centro Comercial San Francisco, Loc. 5 y 6, (or Ap. 16), 4100 Grecia **☎** +506 2494 0016 **☖**+506 2494 2031 **W:** www.radio16.com **E:** gerencia@radio16.com – **65)** Barrio El Invú, La Rotonda, Pto Golfito) **☎** +506-2775 0068 **☖** +506-2775 3303 – **66)** Costado Oeste del Estadio de Guápiles (or Ap. 160), 7210 Guápiles **☎**+506 2710 1600 **☖**+506 2710 9884 **E:** jorgepococi@yahoo.com - **67)** Ap. 421, 2020 Zapote- **☖**+506 253 4887.- **68)** Puntarenas - **69)** Guápiles, Pococí, Limón - **70)** Desamparados, San José – **71)** Calle 13-15, Av. 11, (or: Apartado 1735-1000), San José. – **72)** 150 metros al sur del antiguo Seguro, Alajuela. – **73)** Iglesia Maranatha, 7300 Puerto Limón.

FM in San José and vicinities (MHz): 88.7 Lira – 89.1 R. 89.1 La Super Estación – 89.5 Sendas de Vida – 89.9 R. 8 99 – 90.3 Sinfonola – 90.7 R. Ritmo 90.7 – 91.1 911 La Radio – 91.5 R. 915 – 91.9 Puntarenas – 92.3 Onda Radial – 92.7 Columbia Stereo – 93.1 Fides – 93.5 Monumental – 93.9 Sonido Latino – 94.3 Reloj – 94.7 "94.7" – 95.1 Z-FM – 95.5 95 Cinco Jazz – 95.9 R. 95.9 – 96.3 Centro – 96.7 Universidad – 97.1 Faro del Caribe – 97.5 Musical – 97.9 "979" – 98.3 R. Estéreo Visión – 98.7 Columbia – 99.1 La Mejor FM – 99.5 R. Dos – 99.9 R. Azul – 100.3 La Paz del Dial – 100.7 R.100.7 – 101.1 R. Disney – 101.5 R. Nacional FM – 101.9 "U" – 102.3 Súper – 102.7 Exa FM – 103.1 "103" – 103.5 Best FM – 103.9 Sinai – 104.3 Oxígeno – 104.7 Hit – 105.1 Omega – 105.5 Ten Fifty-Five/Omega – 105.9 Beatz 106 – 106.3 R. Peninsular – 106.7 Premium – 107.1 Estéreo Actual – 107.5 R. 107.5 Real Rock

CROATIA

L.T: UTC +2h (27 Mar-30 Oct: +3h) — **Pop:** 4.5 million — **Pr.L:** Croatian — **E.C:** 50Hz, 220V — **ITU:** HRV

HRVATSKA AGENCIJA ZA POŠTU I ELEKTRONICKE KOMUNIKACIJE (HAKOM)
☐ Jurišiceva 13, 10002 Zagreb **☎** +385 1 4896000 **☖** +385 1 4920227 **W:** www.hakom.hr **LP:** Dir: Drazen Lucic
NB. HAKOM is the regulatory authority for broadcasting.

HRVATSKI RADIO (HR) (Pub)
☐ Prisavlje 3, 10000 Zagreb **☎** +385 1 6342634 **☖** +385 1 6343712 **E:** hrt@hrt.hr **W:** www.hrt.hr **LP:** Dir (Radio) Zoran Mihajlovic

MW	kHz	kW	Prgr			
Zadar	1134	600	HR GH (1700-2345)			

FM (MHz)	HR1	HR2	HR3	kW
Belje	93.3	98.1	-	50
Biokovo	89.7	98.9	-	80
Borinci	88.3	96.1	-	3
Brac	99.8	-	88.8	3
Buje	91.3	103.7	93.2	1
Celavac	95.1	98.1	-	3
Drenovci	92.1	104.4	-	3
Gruda	101.7	106.1	-	2
Ivanščica	102.4	106.4	96.1	15/15/30
Kalnik	90.8	105.8	107.8	15
Labinština	91.3	96.1	-	30
Licka Plješivica	87.7	90.5	100.3	50
Limski kanal	90.2	102.6	-	1
Mirkovica	91.3	93.3	-	30
Pag	98.5	103.4	-	3
Papuk	94.9	106.8	97.7	10
Psunj	97.3	99.7	-	80
Pula	91.4	102.1	94.2	5

FM (MHz)	HR1	HR2	HR3	kW
Slavonski Brod	91.3	105.1	107.9	15
Sljeme	92.1	98.5	-	120
Srdj	88.9	98.5	-	30
Stipanov Gric	102.3	97.5	89.7	15
Subicevac	94.0	90.0	102.3	1
Ucka	99.3	105.3	100.5	80
Ugljan	91.6	87.6	-	5
Uljenje	95.1	103.0	105.6	3

NB: Txs below 1kW not listed.
D.Prgr: HR1 (Prvi program): 24h. – **HR2 (Drugi program):** 24h.
– **HR3 (Treci program):** 24h. – **HR GH (Glas Hrvatske):** 24h. Own
prgrs, rel. of HR1 & news in English, Hungarian, Italian, Spanish.
Details see International section: **Voice of Croatia (Glas Hrvatske)**

HR REGIONAL STATIONS (Pub)

D.Prgr: all stns 24h (incl. rel. of HR1). **HR R. Dubrovnik:** Branitelja
Dubrovnika 21, 20000 Dubrovnik. **E:** radiodubrovnik@hrt.hr. On (MHz)
88.2 (Rota), 89.5 (Ilija), 97.2 (Blato), 101.1 (Vela Luka), 103.7 (Slano),
103.8 (Korcula), 105.0 (Srd), 106.2 (Lastovo), 106.5 (Lopud & Ston).
– **HR R. Knin:** Kralja P. Krešimira IV 30, 23300 Knin. **E:** radio.knin@
hrt.hr. On (MHz) 88.1 (Subicevac), 90.2 (Knin), 94.4 (Promina). – **HR R.
Osijek:** Šamacka 13, 31000 Osijek. **E:** radioosijek@hrt.hr. On (MHz)
102.0 (Psunj), 102.4 (Drenovci & Osijek), 102.8 (Beli Manastir), 105.3
(Borinci), 105.6 (Zlatarevac), 105.8 (Ilok). Incl. Hungarian ("Eszéki Rádió")
1730-1800. – **HR R. Pula:** Riva 10, 52100 Pula. **E:** radiopula@hrt.hr. On
(MHz) 93.8 (Novigrad), 93.9 (Limski kanal), 96.3 (Koromacno), 96.4 (Buje),
100.0 (Pula & Vrsar), 101.3 (Ucka), 103.8 (Raša). Incl. Italian ("R. Pola"):
MF 1000-1005, W 1300-1305, 1530-1630. – **HR R. Rijeka:** Korzo 24,
51000 Rijeka. **E:** redakcija@radio-rijeka.com. On (MHz) 94.5 (Brgud),
95.1 (Pulac), 97.9 (Cres), 98.1 (Kupjacki Vrh), 100.3 (Licka Plješivica),
101.7 (Prezid), 102.7 (Mirkovica), 104.0 (Fuzine), 104.7 (Ucka), 107.4
(Mali Lošinj II), 107.5 (Mrkopalj). Incl. Italian ("R. Fiume"): 0900-0903,
1100-1103, 1400-1403, 1500-1530. – **HR R. Sljeme:** Prisavlje 3, 10000
Zagreb. **E:** radio_sljeme@hrt.hr. On 88.1MHz (Sljeme). – **HR R. Split:**
Mazuranicevo šetalište 24a, 21000 Split. **E:** radio.split@hrt.hr. On
(MHz) 88.4 (Komiza), 100.2 (Hvar), 101.0 (Labinštica), 102.0 (Biokovo),
104.5 (Brac), 105.3 (Orlovaca), 105.8 (Vrlika). – **HR R. Zadar:** Poljana
Šime Budinica 3, 23000 Zadar. **E:** radio_zadar@hrt.hr. On (MHz) 101.8
(Ugljan), 103.0 (Celevac), 105.9 (Pag).

OTHER STATIONS

FM MHz	kW	Location	Station
13) 87.8	1.85	Brac	R. Dalmacija
7) 88.0	3	Beli Monastir	R. Baranja
26) 88.6	3	Slavonski Brod	R. Slavonija
45) 89.0	1	Valpovo	Hrvatski R. Valpovština
49) 89.3	5.5	Zadar	Novi R.
25) 89.4	2	Sisak	R. Sisak
11) 89.6	1	Porec	R. Centar Porec
50) 89.7	4.7	Sljeme	Antena Zagreb
34) 90.2	1	Vinkovci	Radio postaja Vincovci
28) 90.2	1	Slavonska Pozega	R. Vallis aurea
22) 90.3	1	Dugo Selo	R. Martin
44) 90.4	1	Vrlika	Hrvatski R. Sinj
48) 90.5	1	Komiza	Nautic R. Vis
35) 91.0	1	Dakovo	Slavonski R. Osijek
31) 91.6	5	Vinkovci	R. VFM
19) 91.7	5	Koprivnica	R. Koprivnica (RKC)
44) 92.2	1	Krizice	Hrvatski R. Sinj
37) 92.4	2	Alaginci	Zupanijski R. Pozega
2) 92.6	2	Zagreb	Otvoreni R.
32) 92.9	1.9	Virovitica	R. Virovitica
11) 93.6	7	Rusnjak	R. Centar Porec
18) 93.8	1	Jastrebarsko	R. Jaska
42) 94.4	6	Cetingrad	Hrvatski R. Karlovac
47) 94.7	1	Hvar	Megamix R. Hvar
24) 94.9	1	Rovinj	R. Rovinj
30) 94.9	5	Velika Gorica	R. Velika Gorica (RVG)
1) 95.3	3	Osijek	Narodni R.
45) 95.4	5	Drenovci	Hrvatski R. Vukovar
23) 95.4	5	Duga Resa	R. Mreznica
19) 95.5	5	Kalnik	R. Koprivnica (RKC)
3) 95.5	1	Ugljan	Hrvatski Katolicki R.
27) 95.6	2	Donja Stubica	R. Stubica
21) 96.4	1	Zagreb	R. Marija
52) 96.5	1	Rijeka	Primorski R.
4) 96.5	1	Varazdin Breg	R. 042
17) 96.9	5	Ucka	R. Istra
20) 97.6	4	Bogomolje	R. Makarska Rivijera (RMR)
47) 97.9	1	Split	Hrvatski Katolicki R.
51) 98.0	5	Zagreb	Plavi R.
17) 98.0	5	Pula	R. Istra
43) 98.1	1	Nova Gradiska	Hrvatski R. Nova Gradiska

FM MHz	kW	Location	Station
36) 98.4	1	Rijeka	Svid R.
3) 98.6	1	Osijek	Hrvatski Katolicki R.
19) 98.8	5	Zabno	R. Koprivnica (RKC)
15) 99.1	2	Bijele Vode	R. Glina
40) 99.1	1	Osijek	Gradski R. Osijek
33) 99.5	1	Zapresic	R. Zapresic
41) 99.5	1	Sveta Marija	Hrvatski R. Cakovec
39) 100.1	5	Moslavacka Gora	Bjelovarsko-Bilogorski R.
14) 100.2	1	Djakovo	R. Djakovo
35) 100.6	1	Osijek	Slavonski R. Osijek
38) 100.7	3	Zirje	Zupanijski R. Sibenik
53) 101.0	120	Sljeme	R. 101
1) 101.2	3	Metkovic	Narodni R.
10) 101.3	1	Slavonski Brod	R. Brod
19) 101.5	5	Sedlarica	R. Koprivnica (RKC)
22) 101.8	1	Zagreb	R. Martin
37) 102.4	2	Pakrac	Zupanijski R. Pozega
9) 102.7	1	Brac	R. Brac
3) 103.5	120	Sljeme	Hrvatski Katolicki R.
3) 103.9	3	Psunj	Hrvatski Katolicki R.
6) 104.0	1	Sveti Martin	R. 105
3) 104.1	1	Licka Plješivica	Hrvatski Katolicki R.
45) 104.1	1	Zupanja	Hrvatski R. Vukovar
2) 104.4	5	Papuk	Otvoreni R.
12) 104.5	1.85	Zagreb	R. Cibona
38) 104.9	5	Sibenik	Zupanijski R. Sibenik
8) 105.1	1	Okucani	R. Bljesak
5) 105.6	1.5	Cakovec	R. 1
16) 105.7	1.5	Stipanov Gric	R. Gospic
35) 106.2	50	Beli Manastir	Slavonski R. Osijek
2) 106.5	3	Vidova Gora	Otvoreni R.
3) 106.7	3	Ucka	Hrvatski Katolicki R.
21) 106.8	1	Zagreb	R. Marija
27) 106.9	2	Ostri Hum	R. Stubica
13) 106.9	4	Labinstica	R. Dalmacija
29) 107.1	4	Varazdin Breg	R. Varazdin
2) 107.3	80	Celevac	Otvoreni R.
46) 107.2	5	Vinkovci	Hrvatski R. Vukovar
3) 107.9	80	Biokovo	Hrvatski Katolicki R.

NB: Txs below 1kW not listed.
Addresses & other information:
1) Avenija Dubrovnik 15, 10000 Zagreb. **E:** marketing@narodni.hr – **2)**
Cebini 28, Buzin, 10000 Zagreb. **E:** otvoreni@otvoreni.hr – **3)** Vocarska
cesta 106, 10000 Zagreb. **E:** hkr@hkr.hr – **4)** Trstenjakova 3, 42000
Varazdin. **E:** marketing@radio1.hr – **5)** Nova ulica 7, 40000 Cakovec.
E: marketing@radio1.hr – **6)** B. Radica 23, 40314 Selnica. **E:**
radio105.hr – **7)** Trg slobode 32/III, 31300 Beli Manastir. **E:** radio@radio-
baranja.hr – **8)** Blazenog kardinala A. Stepinca 24, 35430 Okucani. **E:**
radio-bljesak@sb.hinet.hr – **9)** Mladena Vodanovica 3, 21400 Supetar.
E: radio-brac@st.htnet.hr – **10)** Dr. Mile Budaka 1, 35000 Slavonski
Brod. **E:** radio-brod@sb.hinet.hr – **11)** Vitomira Širole - Paje 18, 52440
Porec. **E:** radio-centar-studio-porec@pu.tel.hr – **12)** Palmoticeva 7/I,
10000 Zagreb. **E:** marketing@radio-cibona.hr – **13)** Kralja Zvonimira
14/2, 21000 Split. **E:** marketing@radiodalmacija.hr – **14)** Bana Jelacica
6/5, 31400 Djakovo. **E:** radio-djakovo@os.hinet.hr – **15)** Antuna i
Stjepana Radica 8, 44400 Glina. **E:** radio-glina@sk.htnet.hr – **16)**
Trg Stjepana Radica 4, 53000 Gospic. **E:** radio-gospic1@gs.htnet.hr
– **17)** Jurja Dobrile 6, 52000 Pazin. **E:** radioistra@radioistra.hr – **18)**
Strossmayerov trg 5, 10450 Jastrebarsko. **E:** radio-jaska@zg.hinet.
hr – **19)** Zagrebacka b.b., 48000 Koprivnica. **E:** radio-koprivnica@
kc.htnet.hr – **20)** Don Mihovila Pavlinovica 1, 21300 Makarska. **E:** radio-
makarska-rivijera@st.hinet.hr – **21)** Jordanovac 110, 10000 Zagreb.
E: info@radiomarija.hr – **22)** Josipa Zorica 17, 10370 Dugo Selo. **E:**
radio.martin@zg.hinet.hr – **23)** Jozefinska cesta 8, 47250 Duga Resa.
E: radio-mreznica@ka.htnet.hr – **24)** Carducci 13, 52210 Rovinj. **E:**
arting-radio-rovinj@pu.tel.hr – **25)** Antuna i Stjepana Radica 2, 44000
Sisak. **E:** radio-sisak@sk.htnet.hr – **26)** Mile Budaka 1, 35 000 Slavonski
Brod. **E:** radioslavonija@radioslavonija.hr – **27)** Toplicka 5, 49240 Donja
Stubica. **E:** radio-stubica@ht.hinet.hr – **28)** Cehovska 8/I, 34000 Pozega.
E: radio-vallis-aurea@po.tel.hr – **29)** Zagrebacka 3, 10410 Velika Gorica. **E:**
radio@globalnet.hr – **30)** Zagrebacka 3, 10410 Velika Gorica. **E:**
radio@globalnet.hr – **31)** Trg Franje Tudjmana 2, 32100 Vinkovci. **E:**
vfm@vfm.hr – **32)** F. Rusana 1/IX, 33000 Virovitica. **E:** rtv@icv.hr – **33)**
Trg zrtava fašizma 6, 10290 Zapresic. **E:** info-centar-zapresic@zg.tel.hr
– **34)** Jurja Dalmatinca 29, 32100 Vinkovci. **E:** radio-vinkovci@vk.htnet.
hr – **35)** Hrvatske Republike 20, 31000 Osijek. **E:** slavonski-radio@glas-
slavonije.tel.hr – **36)** Trpimirova 3, 51000 Rijeka. **E:** svid-radio@hi.hinet.
hr – **37)** Antuna Kaniziica 3/I, 34000 Pozega. **E:** radio.pozega@inet.hr
– **38)** Trpimirova 3, 22000 Sibenik. **E:** info@radiosibenik.hr
– **39)** Trg E. Kvaternika bb, 43000 Bjelovar. **E:** bbr@bbr.hr – **40)** Trg Ante
Starcevica 7/1, 31000 Osijek. **E:** marketing@eter.hr – **41)** Trg republike
5, 40000 Cakovec. **E:** info@radio-cakovec.hr – **42)** Ambroza Vraniczanya

2, 47000 Karlovac. **E:** radio.karlovac@ka.tel.hr – **43)** Relkoviceva 4, 35400 Nova Gradiška. **E:** hr-nova-gradiska@sb.hinet.hr – **44)** Glavicka 29, 21230 Sinj. **E:** radio-sinj@st.tel.hr – **45)** Kralja Petra Krešimira IV br.1, 31550 Valpovo. **E:** hrvatski-radio-valpovstina@os.htnet.hr – **46)** Trg Drazena Petrovica 1, 32010 Vukovar. **E:** hrvatski-radio-vukovar@vk.hinet.hr – **47)** Šime Ljubica 30, 21000 Split. **E:** megamix@st.htnet.hr – **48)** V. Nazora 19, 21480 Vis. **E:** nautic-radio@st.tel.hr – **49)** Opatice Vekenege 2, 23000 Zadar. **E:** noviradio@noviradio.hr – **50)** Avenija Veceslava Holjevca 29, 10000 Zagreb. **E:** antena@antenazagreb.hr – **51)** Slavonska avenija 2, 10000 Zagreb. **E:** marketing@plaviradio.hr – **52)** Barciceva 4a, 51000 Rijeka. **E:** marketing@primorski-radio.hr – **53)** Gajeva 10, 10000 Zagreb. **E:** marketing101@radio101.hr.

DAB (Trial): Sljeme ch12C (227.360MHz) 0.8kW. **Operator:** OIV

CUBA

LT: UTC -5h (13 Mar-30 Oct: -4h); Guantánamo Bay: UTC -5h (13 Mar-6 Nov: -4h) — **Pop:** 11.4 million — **Pr. L:** Spanish — **E.C:** 60Hz, 110/120V — **ITU:** CUB

MINISTERIO DE COMUNICACIONES (MC)
Dirección General de Telecomunicaciones
✉ Plaza de la Revolución, Ciudad de la Habana

INSTITUTO CUBANO DE RADIO Y TELEVISION (ICRT)
✉ Edif.Radiocentro, Av. 23 N° 258, Vedado, Habana 4 ☎ +53 7 8324648. Radio Cubana has links to most national and local stns **W:** www.radiocubana.cu
Hrs of tr. usually 24h – see address section for variations. Call CM-

MW	Call	kHz	kW	Primary network, location
N5)	BQ	530	10	R. Enciclopedia, HA
N1)	BA	540	1	R. Rebelde, Sancti Spíritus, SS
N1)	BA	550	10	R. Rebelde, Guantánamo, GU
N1)	BA	550	30	R. Rebelde, Pinar del Río, PR
N1)	BA	550	1	R. Rebelde, Manzanillo, GR
N1)	BA	560	30	R. Rebelde, Moa, HO
N2)	BD	570	30	R. Reloj, Santa Clara, VC
N1)	BA	580	5	R. Rebelde, Baracoa, GU
N1)	BA	580	10	R. Rebelde, Mantua, PR
N3)	BF	590	30	R. Musical, La Habana, CH
N1)	BA	600	150	R. Rebelde, Urbano Noris, HO
N1)	BA	610	1	R. Rebelde, Bahía Honda, PR
N2)	BD	610	1	R. Reloj, Trinidad, SS
N1)	BA	620	1	R. Rebelde, Moa, HO
N1)	BA	620	25	R. Rebelde, Colón, MA
N4)	BC	630	5	R. Progreso, Pinar del Río, PR
N4)	BC	640	50	R. Progreso, Guanabacoa, HA
N4)	BC	640	10	R. Progreso, Las Tunas, LT
N4)	BC	640	5	R. Rebelde, Las Mercedes, GR
N1)	BA	650	1	R. Rebelde, Stgo de Cuba, SC
N1)	BA	650	1	R. Rebelde, Media Luna, GR
N1)	BC	660	30	R. Progreso, Santa Clara, VC
N1)	BA	670	50	R. Rebelde, Arroyo Arenas, CH
N2)	BD	670		R. Reloj
N1)	BA	680	10	R. Rebelde, Ciego de Ávila, CA
N4)	BC	680	1	R. Progreso, Cienfuegos, CI
N4)	BCDB	680	1	R. Progreso, Stgo de Cuba, SC
N4)	BC	690	20	R. Progreso, Jovellanos, MA
N5)	BQ	700	1	R. Enciclopedia, Guantánamo, GU
N4)	BC	700		R. Progreso, Baracoa
N1)	BA	700	1	R. Rebelde, Sancti Spíritus, SS
N1)	BA	710	150	R. Rebelde, La Julia, HA
N1)	BA	710	10	R. Rebelde, Holguín, HO
N1)	BA	710	50	R. Rebelde, Santa Clara, VC
N1)	BA	710	30	R. Rebelde, Camagüey, CM
N1)	BA	720	1	R. Rebelde, Cienfuegos, CI
N4)	BC	730	10	R. Progreso, Nueva Gerona, IJ
N4)	BC	740	20	R. Progreso, Camagüey, CM
HO01)	KO	740	10	R. Angulo, Sagua de Tanamo, HO
N4)	BC	750	10	R. Progreso, Palmira, CI
N4)	BC	750	1	R. Progreso, Trinidad, SS
N2)	BD	760	10	R. Reloj, Las Mercedes, GR
N2)	BD	760	1	R. Reloj, La Habana, HA (emergency tx)
N1)	BA	770		R. Rebelde
N1)	BA	780		R. Rebelde
N2)	BD	790	30	R. Reloj, Pinar del Río, PR
N4)	BC	800	1	R. Progreso, Manzanillo, GR
N4)	BC	810	10	R. Progreso, Guantánamo, GU
N4)	BC	820	10	R. Progreso, Ciego de Avila, CA
CH01)	BE	820	10	R. Ciudad de la Habana, Santa Catalina, CH
N2)	BD	820	10	R. Reloj, Contramaestre
N2)	BD	830	5	R. Reloj, Holguín, HO
N4)	BC	840	1	R. Progreso, Las Tunas, LT

MW	Call	kHz	kW	Primary network, location
N5)	BQ	840	1	R. Enciclopedia, La Fé, IJ
VC01)	HW	840	10	Doblevé, Santa Clara, VC
SC01)	KC	840	1	R. Revolución, Stgo de Cuba, SC
N4)	BC	850	1	R. Progreso, Trinidad, SS
N2)	BD	850	1	R. Reloj, Nueva Gerona, IJ
N2)	BD	860	1	R. Reloj, Baracoa, GU
N2)	BD	860	5	R. Reloj, Colón, MA
N1)	BA	860	10	R. Rebelde, Arroyo Arenas, CH
N2)	BD	870	1	R. Reloj, Sancti Spíritus, SS
N4)	BC	880	12.5	R. Progreso, Pinar del Río, PR
N4)	BC	890	1	R. Progreso, Santa Clara, VC
N4)	BC	890	25	R. Progreso, Chambas, CA
N4)	BC	900	25	R. Progreso, Cacocum, HO
N2)	BD	910	5	R. Reloj, Bolondrón, MA
CH02)	BL	910	10	R. Metropolitana, La Lisa, CH
CM01)	HA	910	25	R. Cadena Agramonte, Camagüey, CM
N2)	BD	920	1	R. Reloj, Moa, HO
N4)	BC	920	1	R. Progreso, Bayamo, GR
N2)	BD	930	1	R. Reloj, La Jaiba, MA
CA01)	IP	930	25	R. Surco, Ciego de Ávila, CA
N2)	BD	930	1	R. Reloj, Stgo de Cuba, SC
N2)	BD	940	10	R. Reloj, Central España, MA
N2)	BD	940	10	R. Reloj, Holguín, HO
N4)	BC	940	1	R. Progreso, Sancti Spíritus, SS
N2)	BD	950	10	R. Reloj, La Habana, HA
N2)	BD	950	1	R. Reloj, Mayarí Arriba, SC
N5)	BQ	960	1	R. Enciclopedia, Matanzas, MA
N2)	BD	960	10	R. Reloj, Guantánamo, GU
N2)	BD	960	1	R. Reloj, Cienfuegos, CI
N3)	BF	960	0.25	R. Musical, Ciego de Avila, CA
N2)	BD	980	1	R. Reloj, Bayamo, GR
CO03)	B	980	5	R. COCO, Sapo, CH
HO04)	KV	*980	1	La Voz del Níquel, Moa, HO
PR01)	AM	990	1	R. Guamá, San Luís, PR
N3)	BF	1000	1	R. Musical, Sancti Spíritus, SS
N3)	BF	1000	1	R. Musical, Camagüey, CM
GR02)	NM	1000	1	R. Granma, Manzanillo, GR
HA01)	CH	1000		R. Cadena Habana
N3)	BF	1010	10	R. Musical, Holguín, HO
N2)	BD	1020	5	R. Reloj, Jorobo, LT
GU01)	M	1020	10	Cadena CMKS, Guantánamo, GU (Cfr. 1070)
HA01)	CH	1020		R. Cadena Habana
PR01)	AM	1020	10	R. Guamá, Bahía Honda, PR
PR01)	AM	1020		R. Guamá, Los Palacios, PR
PR01)	AM	1030	1	R. Guamá, La Palma, PR
N3)	BF	1030		R. Musical
LT01)	LL	1040	1	R. Victoria, Puerto Padre, LT
LT01)	LL	1050	10	R. Victoria, Las Tunas, LT
LT01)	L	1060	1	R. Victoria, Amancio Rodríguez, LT
GU01)	M	1060	5	Cadena CMKS, Baracoa, GU
MA01)	DL	1060		R. 26, Matanzas
PR01)	AM	1070	1	R. Guamá, Pinar del Río, PR
GU01)	M	1070	10	Cadena CMKS, Guantánamo, GU (Cfr. 1020)
HA01)	CH	1080	10	R. Cadena Habana, Güines, MB
PR01)	AM	1090	1	R. Guamá, Santa Lucia, PR
CH01)	CH	1090	1	R. Cadena Habana, La Salud, CH
HO01)	KO	1090	1	R. Angulo, Moa, HO
HO01)	KO	1100	1	R. Angulo, Banes, HO
HA01)	CH	1100	1	R. Cadena Habana, La Habana, CH
HA01)	CH	1110		R. Cadena Habana
HO01)	KO	1110	10	R. Angulo, Holguín, HO
N1)	BA	1120		R. Rebelde
HA01)	CH	1120	5	R. Cadena Habana, Artemisa, AR
HO01)	KO	1120	1	R. Angulo, Mayarí, HO
N5)	BQ	1130	1	R. Enciclopedia, Santa Clara, VC
HO05)	KA	1130	5	R. Angulo/Ecos del Sagua, Sagua de Tánamo, HO
N5)	BQ	1140	5	R. Enciclopedia, Loma de la Cruz, HA
GR01)	NL	1140	1	R. Bayamo, Media Luna, GR
HA01)	CH	1140	5	R. Cadena Habana, La Habana, CH
MA02)	DP	1140		R. Ciudad Bandera, Cárdenas, MA
N3)	BF	1140		R. Musical, Villa Clara, VC
CM02)	HC	1140	1	R. Camagüey, Camagüey, CM
GR01)	NL	1150	10	R. Bayamo, Entronque Bueycito, GR
GR01)	NL	1160	1	R. Bayamo, Pilón, GR
GU01)	M	1170	10	Cadena CMKS, Maisí, GU
N1)	BA	1170		R. Rebelde
N1)	BA	1180	50	R. Rebelde, Villa María, HA
N1)	BA	1180	1	R. Rebelde, Mayarí Arriba, SC
N2)	BD	1180		R. Reloj, Nueva Gerona, IJ
GU03)	MN	1180		La Voz del Toa, Baracoa
SC04)	JD	1190		R. Coral, Guamá, SC
SS01)	GL	1190	1	R. Sancti Spíritus, Trinidad, SS
MA01)	DL	1190	1	R. 26, La Caridad, MA
SC01)	KC	1200	10	R. Revolución, Palma Soriano, SC
SS01)	GL	1200	1	R. Sancti Spíritus, Sancti Spíritus, SS

MW	Call	kHz	kW	Primary network, location
SS01)	GL	1210	10	R. Sancti Spíritus, Sancti Spíritus, SS
SC01)	J	1210	1	R. Revolución, Chivirico, SC
SC01)	J	1210	1	R. Revolución, Mayarí Arriba, SC
IJ01)	BY	1220	5	R. Caribe, IJ
MA01)	DL	1220	10	R. 26, Central España, MA
MA01)	DL	1230	3	R. 26, Unión de Reyes, MA
MA01)	DL	1240	10	R. 26, Bolondrón, MA
GU02)	KS	*1250	0.25	R. Playitas, Imías, GU
N5)	BF	1260	5	R. Enciclopedia, Arroyo Arenas, CH
N2)	BD	1270	10	R. Reloj, Camagüey, CM
N5)	BQ	1270	1	R. Enciclopedia, Varadero, MA
SC03)	KW	1280	1	R. Mambí, Stgo de Cuba, SC
N4)	BC	1290	5	R. Progreso, La Pastora, HA
N5)	BQ	1290	5	R. Enciclopedia, La Habana, HA
VC01)	E	1290	1	CMHW, Rancho Veloz, VC
N5)	BQ	1300	1	R. Enciclopedia, Las Tunas, LT
VC01)	E	1300	1	CMHW, Sagua La Grande, VC
AR01)	CW	1320	0.5	R. Artemisa, Artemisa, AR
N5)	BQ	1320	1	R. Enciclopedia, Stgo de Cuba, SC
N5)	BQ	1320	1	R. Enciclopedia, Sancti Spíritus, SS
CI01)	FL	1340	10	R. Ciudad del Mar, Palmira, CI
CI01)	FL	1350	1	R. Ciudad del Mar, Aguada de Pasajeros, CI
LT03)	LM	1350	10	R. Libertad, Puerto Padre, LT (rel. CMLL R. Victoria 0300-1200)
CM01)	HA	1360	1	R. Cad. Agramonte, Rodolfo Ramírez Esquível, CM
CM01)	HA	1370	1	R. Cad. Agramonte/R. Nuevitas, Nuevitas, CM
CM01)	HA	1380	1	R. Cad. Agramonte, Central Brasil, CM
HA04)	BT	*1390	1	R. Jaruco, Jaruco, HA
N3)	BF	1400	1	R. Musical, Matanzas, MA
N5)	BQ	1410	1	R. Enciclopedia, Pinar del Río, PR
CM01)	HA	1410	1	R. Cadena Agramonte, Sta Cruz, CM
CA01	JY	1430	10	R. Surco/R. Amanecer, Primero de Enero, CA
CA01)	JP	1440	10	R. Surco, Ciego de Avila, CA
LT02)	LN	1450	1	R. Maboas, Amancio Rodríguez, LT
MB01)	CL	1450	1	R. Güines, Güines, MB
CM01)	HA	1460	1	R. Cadena Agramonte, Sola, CM
MA02)	GE	1470	1	R. Ciudad Bandera, Cárdenas, MA
HO03)	KN	1490	1	R. Mayarí, Mayarí, HO
N5)	BQ	1500	1	R. Enciclopedia, Holguín, HO
N5)	BQ	1510	1	R. Enciclopedia, Moa, HO
VC02)	ES	1540	1	R. Sagua, Sagua La Grande, VC
N1)	BA	1550	1	R. Rebelde, Nuevitas, CA
N5)	BQ	1560	1	R. Enciclopedia, Ciego de Avila, CA
N5)	BQ	1570	1	R. Enciclopedia, Las Tunas, LT
CM01)	HA	1580	1	R. Cadena Agramonte,Santa Cruz del Sur, CM
N4)	BQ	1590	1	R. Progreso, Manzanillo, GR
N2)	BD	1610		R. Reloj
N1)	BA	1620		R. Rebelde
SW	**Call**	**kHz**	**kW**	**Primary network, location**
N1)	BA	5025	50	R. Rebelde, Bauta, AR

v = varying freq. * = inactive

Provinces: AR=Artemisa CA=Ciego de Avila CH=Ciudad Habana CI=Cienfuegos CM=Camagüey GR=Granma GU=Guantánamo HA=Habana HO=Holguín IJ=Isla de laJuventud LT=Las Tunas MA=Matanzas MB=Mayabeque PR=Pinar del Río SC=Santiago de Cuba SS=Sancti Spíritus VC=Villa Clara
N.B.: Some stns relay different networks at different times of day. This applies especially to the three major networks; Progreso, Rebelde and Reloj. Radio Rebelde carries sports events which are relayed by many stns. National stations operate 24h.

FM in La Habana (MHz): 90.3 R. Progreso – 91.7 CMCK R. COCO – 93.3 R. Taíno – 94.1 R. Enciclopedia – 94.9 R.Ciudad de la Habana – 96.7 R. Rebelde – 98.3 Metropolitana – 99.1 R. Musical Nacional – 99.9 R. Cadena Habana – 100.9 Habana FM –101.5 R. Reloj – 104.7 R. Rebelde – 106.3 R. Progreso – 106.9 Habana R – 107.9 R. Rebelde.

National networks: N1) R. Rebelde, Ap. 6277, La Habana 10600 (or: Edif. del ICRT, Av. 23 N° 258, Vedado, La Habana 10400) **W:** www.radiorebelde.com.cu – **N2)** R. Reloj, Ap. 6277, Ciudad de La Habana (or Ed. Radiocentro, Calle 23 No. 258, (8avo piso), entre Ly M, Vedado, La Habana 10400 **W:** www.radioreloj.cu – **N3)** R. Musical Nacional, Edificio N, Calle N, entre 23 y 21, Vedado La Habana 10400 **W:** www.cmbfradio. cu – **N4)** R. Progreso, Ap. 4042, La Habana 10300 (or Infanta 105, Esq. A 25, Centro Habnana) **W:** www.radioprogreso.cu – **N5)** R. Enciclopedia, Edificio N, Calle N. N° 266 (bajos), entre 21 y 23, Vedado, La Habana 10400 **W:** www.radioenciclopedia.cu – **N6)** R. Taíno, Ap. 6277, La Habana 10400 (or Av. 23 N° 258, Vedado, La Habana 10400) – FM only

Provincial and municipal stations
AR00) Artemisa (AR)
AR01) Calle 50 No. 2310, entre 23 y 25, Artemisa 33800 **W:** www.

artemisaradioweb.cu
CA00) Ciego de Ávila (CA)
CA01) Ap. 183 (or Chicho Valdés 66), Ciego de Ávila 65100 **W:** www. radiosurco.cu
CH00) Ciudad de La Habana (CH)
CH01) Ap. 6599, La Habana 10600 (or Calle N No. 266 (5to piso), entre 21 y 23, Vedadado, Plaza de la Revolución, La Habana 10400) **W:** www.habanaenlinea.cu **CH02)** Ed. Focsa, Calle N No. 301 (1er piso), esq. A 17, Vedado, Plaza de la Revolución, La Habana 10400 **W:** www.radiometropolitana.cu **CH03)** Ed. Focsa, Calle N No. 301, esq. A 17, Vedado, Plaza de la Revolución, La Habana 10400 **W:** www. radiococo.cu
CI00) Cienfuegos (CI)
CI01) Ap. 290 (or Calle 37 No. 3602, entre 36 y 38, Cienfuegos 55100 **W:** www.rcm.cu
CM00) Camagüey (CM)
CM01) Calle Cisneros # 310 entre Ignacio Agramonte y General Gómez, Camagüey 70100 **W:** www.cadenagramonte.cu **CM02)** Same address as CM01.
GR00) Granma (GR)
GR01) Ap. 74 (or Calle General Calixto García 156, entre Figueredo y Luz Vásquez 74) Bayamo 85100 **W:** www.radiobayamo.co.cu **GR02)** Ap. 220 (or Calle Martí 341, entre Quintin Banderas y León), Manzanillo 87510 **W:** www.radiogranma.co.cu
GU00) Guantánamo (GU)
GU01) Ap. 96 (or Donato Mármol 409, entre José Martí y Pedro A. Pérez), Guantánamo 95100 **W:** www.radioguantanamo.cu **GU02)** Calle B No. 2050, Imías 97500. **GU03)** LV del Toa (CMDX), Martí 122, Baracoa 97310 **H of tr:** 1000-0200 **W:** www.radiobaracoa.icrt.cu
HA00) La Habana (HA)
HA01) Calle 15, esq. a J, No. 210, Vedado, Plaza de la Revolución, La Habana 10400 **W:** www.cadenahabana.cu
HO00) Holguín (HO)
HO01) Ap. 14 (or Calle Máximo Gómez 298 (3er piso) entre Frexes y Martí), Holguín 80100 **W:** www.radioangulo.cu **HO03)** Calle Martí 46, Mayarí 83000. **HO04)** Calle 9na s/n, Reparto Rolo Monterrey, Moa 83330. **HO05)** Sagua de Tánamo 83200.
IJ00) Isla de la Juventud (IJ)
IJ01) Calle 26, entre 41 y 43, Nueva Gerona 25100 **W:** www.radio-caribe.icrt.cu
LT00) Las Tunas (LT)
LT01) Ap. 211 (or Calle Colón 157, entre Julián Santana y Francisco Vega), Las Tunas 75100 **W:** www.tiempo21.cu **LT02)** Avenida Sergio Reynó 19, Amancio Rodriguez 77700 **W:** www.radiomaboas.cu **LT03)** Ap. 45 (or Avenida de La Libertad 95), Puerto Padre 77200 **W:** www. radiolibertad.cu
MA00) Matanzas (MA)
MA01) Ap. 51 (or Milanés final, esq. a Guachinango), Matanzas 40100 **W:** www.radio26.icrt.cu **MA02)** Calzada, esq. a Calvo, Cárdenas 42100.
MB00) Matabeque (MB)
MB01) Calle 76 No. 7707, entre 77 y 81, Güines 33900. 1000-0400 **W:** www.radioguines.icrt.cu
PR00) Pinar del Río (PR)
PR01) Calle Colón 14, entre Adela Azcuy y Juan Gualberto Gómez, Pinar del Río 20100 **W:** www.rguama.icrt.cu
SC00) Santiago de Cuba (SC)
SC01) Ap. 232 (or Aguilera 554, entre San Augustín y Barnada), Santiago de Cuba 90100 **W:** www.cmkc.cu **SC03)** Calle 8 No. 56, entre A e Independencia, Reparto Sueño, Santiago de Cuba 90900 **W:** www.radiomambi.icrt.cu. **SC04)** R. Coral, Calle C No. 64, Chivirico, Guamá 92800
SS00) Sancti Spíritus (SS)
SS01) Circunvalación s/n, Los Olivos 1, Sancti Spíritus 60100 **W:** www.radiosanctispiritus.cu
VC00) Villa Clara (VC)
VC01) Ap. 376 (or Parque Leoncio Vidal 4, entre Martha Abreu y Pao Chao), Santa Clara 50100 **W:** www.cmhw.cu **VC02)** Libertadores 100, esq. a Carmen Ribalta, Sagua la Grande, Villa Clara 52310 **W:** www. radiosagua.cu

Guantánamo Bay (leased to USA)

AFRTS (US Navy)
✉ Naval Media Center Broadcasting Detatcment, Guantánamo Bay, Cuba, PSC 1005, Box 22, FPO AE 09593, USA **E:** gitmo@mediacen. navy.mil
MW: Guantánamo Bay: 1340kHz 0.25kW
FM: 102.1MHz 0.5kW (stereo), 103.1MHz 0.5kW – **D.Prgr:** 24h on 1340kHz/102.1MHz. Rel AFRTS satellite sce on 103.1MHz

CURAÇAO (Netherlands)

L.T: UTC -4h — **Pop:** 133,600 — **Pr.L:** Dutch (official), Papiamentu, English, Spanish — **E.C:** 50Hz, 127/220V — **ITU:** ATN

Bureau Telecommunication and Post
✉ Beatrixlaan 9, P.O. Box 2047, Curaçao ☎ +599 9 463 1700 🖷 +599 9 736 5265 **E:** gen.affairs@burtel.an **W:** www.btnp.org

MW Call	kHz	kW	Station, location
1) PJZ-86	860	10	R. Curom, Willemstad

FM	MHz	kW	Station, location
1)	88.3		Rockorsou, Willemstad
18)	88.9		Radio Vishon, Willemstad
12)	90.1		Radio Krioyo, Willemstad
5)	91.5		Gold 91.5 , Willemstad
11)	92.1		Direct Life 92.1 FM, Willemstad
9)	92.7		R. Edukativo, Deltha 92, Willemstad
17)	93.3		Radio Top FM, Willemstad
3)	93.9	20	R. Korsou FM, Willemstad
8)	95.1	0.5	ClazzFM, Willemstad
1)	95.7	4	Mi 95Z FM, Willemstad
13)	96.5		New Song, Willemstad
10)	97.3		Dolfijn FM, Willemstad
8)	97.9	0.5	Easy 97.9 FM, Willemstad
4)	98.5	2.5	Radio Semiya, Willemstad
15)	99.7		Radio MAS, Santa. Maria
	100.3		Hit 100.3, Willemstad
3)	101.1	5	Laser 101, Willemstad
2)	101.9	5	Radio Hoyer 1, Willemstad
5)	103.1		Paradise FM, Willemstad
8)	103.9	0.5	Radio One FM, Willemstad
14)	104.5		Radio Active FM, Willemstad
2)	105.1	5	Radio Hoyer 2, Willemstad
16)	106.3		Fiesta FM, Willemstad
11)	107.1		Radio Direkt, Willemstad
6)	107.9	1	Rumbera Network, Willemstad

Addresses and other information
1) Roodeweg 64, Willemstad, Curaçao ☎ +599 9 462 2020 🖷 +599 9 462 5796. **E:** mi95@curom.com, z86@curom.com, 88rockorsou@curom. com **W:** www.curom.com, www.rockorsou.fm – **2)** Plasa Horacio Hoyer 21, Willemstad, Curaçao ☎ +599 9 461 1678 🖷 +599 9 461 6528 **E:** management@radiohoyer.com **W:** www.radiohoyer.com MD: Ms. Helen Hoyer. W,Sun: 1000-0400. R. Hoyer 1 in Papiamento, R. Hoyer 2 in Dutch -3) Bataljonweg 7, Willemstad, Curaçao ☎ +599 9 737 3012 🖷 +599 9 737 2888. Dir Hans Oosterhof. PD: Alan H. Evertsz. 24h. 24h. **N.** in Dutch: 1000, 2300. **N.** in Papiamento: 2200. English: Tues 0000. Portuguese: Wed 2330. Sranan Tongo: Fri 0000. Relaying Radio Nederland Wereldomroep at 09.30 **W:** www.korsou.com **E:** studio@ korsou.com Separate prgrs ("Laser 101") on 101.1MHz. **E:** studio@ laser101.com – **4)** Parmantierweg 2, Willemstad, Curaçao ☎ +599 9 462 4000 🖷 +599 9 462 4004. Dir: Ferris Thode.Rlg programs in English and Papiamento **W:** www.radiosemiyafm.org – **5)** De Rouvilleweg 7, ingang Klipstraat , Willemstad ☎ +599 9 462 8103 🖷 +599 9 461 9103 **W:** www. paradisefm.an www.gold915.com **E:** @paradisefm.an Paradise FM: Dutch with every hour and half hour Dutch news. Gold 915: English non stop The Golden Hits and English news; Owner: Cees Baas – **6)** Caracasbaaiweg, Willemstad ☎ +599 9 461 5027 🖷 +599 9 461 5028 **W:** www.rumberanetwork.org **E:** contacto@rumberanetwork. com.ve – **7)** Compleho Deportivo Casa Grandi Z/N Willemstad ☎ +599 9 747 3333 🖷 +599 9 747 7265. Manager: Elmer Cijntje. Prgrs in Papiamento, Dutch, Spanish and English **E:** hit100.3fm@cura.net **W:** www.hit24.com – **8)** Arikokweg 19A, Willemstad ☎ +599 9 462 3162 🖷 +599 9 462 8712. GM: Quintus Flashvoet ClazzFM **E:** info@clazzfm.com **W:** www.clazzfm.com 24h light music & jazz in E, Papiamento & Dutch, Radio One FM **E:** info@radioone.net **W:** www.radioone.nl 24h dance & Top 40 music in E, & Dutch. Easy FM: **W:** www.easyfm.com **E:** radio@ easyfm.com – **9)** Suffisantweg 16, Willemstad ☎ +599 9 888 0120 +599 9 888 0155– **10)** Sea Aquarium Beach, Bapor Kibra z/n, Willemstad ☎ +599 9 465 9975 🖷 +599 9 461 9975 **E:** info@dolfijnfm.com **W:** www. dolfijnfm.com – **11)** F.D Rooseveltweg 214, Tesoro Shopping Center, Willemstad ☎ +599 9 888 4107 🖷 +599 9 888 8407 **E:** studio@ direct107.com **W:** www.direct107.com; Direct Life 92.1 FM ☎ +599 9 888 8092 – **12)** Gosieweg 133,Willemstad ☎ +599 9 736 0901 🖷 +599 9 736 4914 **E:** kantoor@radiokrioyofm.com **W:** www.radiokrioyofm.com – **13)** Winkelcentrum Muizenberg z/n (Edificio New Song) ☎ +599 9 888 0965 🖷 +599 9 888 0561 **E:** newsong@cura.net **W:** www.biblevoice.org – **14)** Kaya Simon Pieters, Willemstad, Curaçao ☎+599 9 660 3302 🖷 +599 9 869 4109 **W:** www.active.fm **E:** active1045fm@hotmail.com –**15)** Fosfaatweg 8, Sta. Maria ☎+599 9 888 8997 🖷 +599 9 888 6997 –**16)** Fatimaweg 2, Suffisant ☎+599 9 868 3379 🖷 +599 9 869 6613 –**17)** F.D. Rooseveltweg 32-w ☎+599 9 888 7933 **E:** info@radiotopfmcuracao. com **W:** www.radiotopfmcuracao.com

---

CYPRUS

L.T: UTC +2h (27 Mar-30 Oct: +3h) — **Pop:** 800,000 — **Pr.L:** Greek, Turkish, Armenian — **E.C:** 50Hz, 240V — **ITU:** CYP

CYPRUS RADIO-TELEVISION AUTHORITY
✉ 32 Nikis Ave, P.O.Box 23377, 1682 Nicosia ☎+357 22 512468 🖷 F+357 22 512473 **W:** www.crta.org.cy **E:** crtauthority@cytanet.com.cy

CYPRUS BROADCASTING CORPORATION (semi-gov)
✉ CyBC Street, Athalassa, P.O. Bxo 24824, CY-1397 Nicosia ☎+357 22 862000 🖷 +357 22 314050 **W:** www.cybc.com.cy **E:** rik@cybc.com. cy**L.P:** DG: Themis Themistocleous. Deputy DG: Michael Stylianou.

MW	kHz	kW	Ch.	MW	kHz	kW	Ch.
Paphos	558	10	1	Paphos	918	10	3
Nicosia	603	100	3	Nicosia	963	100	3
Limassol	693	10	1	Limassol	1044	10	3

FM (MHz)	Ch. 1	Ch. 2	Ch. 3	Ch.4	kW
Larnaca	90.2	92.4	96.0		7
Mt. Olympos	97.2	91.1	94.8	88.2	30
Paphos	93.3	96.5	99.8		7
Paralimni	91.4	94.2	97.9		4

Ch. 1 (Proto) in Greek: 24h – **Ch. 2 (Deutero)** Multilingual: 24h. Programs in **English** 1030-1040, 1500-0300. **Turkish** 0300-1400. **Armenian:** 1400-1500 – **Ch. 3 (Trito)** in Greek: 24h – **Ch. 4 (R. Love)** in Greek: 24h.
Ann: Greek: "Radiofonikon Idryma Kyprou". Turkish: "Burasi Kibris Radyo Yayin Korporasyonu". **IS:** "Avkoritssa" (guitar).

EXTERNAL SERVICE: see International Radio section.

Other stations:
ANT1 FM: Larnaca 102.7MHz, Paphos 103.7MHz. **W:** www.ant1fm. com.cy – **Energy 107.6:** Mount Olympos 107.6MHz. **W:** energy1076. com – **Kanali 6:** Limassol 98.6MHz, Nicosia 106.0MHz, Mount Phanos 107.0MHz. **W:** kanali6.com – **Kiss FM,** Nicosia: 89.0MHz. **W:** kissfm. com.cy – **Klik FM,** Limassol: 105.6MHz. **W:** klikfm.com.cy – **Logos R:** Limassol 100.7MHz, Mount Olympos 101.1MHz, Paphos/Larnaca 101.6MHz, Mount Kykkos 102.4MHz. **W:** logosradio.com.cy – **R. Astra:** Mount Olympos 92.8MHz, Larnaca 105.3MHz **W:** astra.com.cy – **R. Athina:** Limassol 88.7MHz, Nicosia 100.7MHz. **W:** radioathina.com – **R. Proto:** Larnaca 89.3MHz, Mount Olympos 99.3MHz. **W:** www. radioproto.com.cy – **R. Sfera:** Paphos 96.8MHz, Limassol 106.4MHz. **W:** www.radiosfera.com.cy – **Super FM:** Larnaca 95.7MHz, Mount Olympos 104.7MHz. **W:** superfmradio.com
BBC World Sce: 1323kHz English 0200-2300 + SW. See Int. R. section.
R. Monte Carlo & Trans World R. relays on 1233kHz 0200-2115; for further details see International Radio section.
R. Sawa, 990kHz 24h. see International Radio section (USA).

NORTHERN CYPRUS

SUPREME BROADCASTING BOARD (YYK)
✉ Memduh Asaf St. 9, Kösklüçiftlik, Lefkosa, Northern Cyprus ☎+90 392 228 1368 🖷+90 392 228 1272
W: kktcyyk.org **E:** info@kktcyyk.org

BAYRAK RADYO TELEVIZYON KURUMU (BRTK, Gov.)
✉ BRT Sitesi, Dr. Fasil Küçük Bulvari, Lefkosa, Northern Cyprus, via Mersin 10, Turkey ☎+90 392 225 5555 🖷 +90 392 225 4991
W: www.brtk.cc **E:** brt@brtk.net
L.P: DG: Ahmet Okan. Head Tr. Dept: Mustafa Tosun.
MW: Iskele (Trikomo) 1098kHz 100kW.

FM	R.1	B.FM	B.Int.	Klasik	B.T.	kW
Kantara	90.6	98.1	87.8	93.4		10/10/10/1
Selvilitepe	102.0	92.1	105.0	88.4/102.5	94.6	5/5/5/1/5
Lefkosa		94.2				0.05

Radyo 1 in Turkish: 24h on 1098kHz & FM.
Bayrak FM in Turkish: 24h.
Bayrak International in Greek/English/Arabic/German: 24h.
Radyo Klasik: 24h.
Bayrak Türk Müzigi: 24h.

OTHER STATIONS FM (MHz):

Station	W	E	kW
1) Odtu FM	103.1		1
2) Cool FM	92.5	97.5	1
3) As FM	97.7		1
4) Sim FM	98.6	89.5	1/0.3
5) Süper FM	98.9		1

Station	W	E	kW
6) Metro FM	104.0		1
7) Kral FM	106.9		1
8) Kibris FM	103.4	100.2	10/2.5
9) First FM	90.0	96.6	1/0.2
10) Akdeniz FM	88.6		1
11) Günes FM	91.3		1
12) R. Vatan Türkü	104.5	94.4	5/5
13) R. Nihavent	100.4	89.8	1/3
14) Dance FM	95.5	95.8	2/0.3
15) Yakin Dogu FM	88.0		1
16) Kuzey FM	106.7		2
17) Radyo T	96.65		2
18) Güven FM	90.4	90.8	5
19) Plus FM	106.2		1
20) Laü FM	97.4		2
21) Mayis FM	96.0	99.5	2/0.05
22) Gaü FM	105.8		1
23) Ukü FM	107.2		1
24) Daü FM		106.5	2.5
25) Enerji FM		93.0	0.5
26) InterFirst FM	100.9		0.25
27) Güven Nostalji	102.8	91.3	
28) Avrasya FM	107.9	106.7	1
29) Ada FM	96.2	93.8	1

Tx sites: W (west) = Selvilitepe, E (east)= Kantara-Sinan Dagi. All stns 24h.
Location: All Lefkosa exc.: 1) Kalkanli, 2), 19) & 24) Magosa, 20) Lefke & 22) Gime.

AKROTIRI & DHEKELIA (UK)

Pop: 15,700 — **Pr.L:** English, Greek — **E.C:** 50Hz, 240V — **ITU:** CYP

BFBS RADIO, CYPRUS (Mil.)

✉ BFBS Akrotiri, BFPO 57, UK ☎ +357 2527 8518 ▤ +357 2527 8580
W: www.ssvc.com/bfbs/radio/cyprus **E:** cyprus@bfbs.com
L.P: GM: Tess Turner; Eng Mgr: J. Dunlop

FM (MHz)	BFBS1	BFBS2	kW
Akrotiri	92.1	89.9	25
Dhekelia	99.6	95.3	25
Nicosia	91.7	89.7	1.5

D.Prgr: 24h. **Ann:** "This is BFBS Radio"

Other stations:
BBC World Sce: Zakaki 639 & 720kHz: Arabic 0300-2200.

CZECH REPUBLIC

L.T: UTC +1h (27 Mar-30 Oct: +2h) — **Pop:** 10.5 million — **Pr.L:** Czech — **E.C:** 50Hz, 230V — **ITU:** CZE

CESKÉ RADIOKOMUNIKACE, a.s.

✉ U nákladového nádrazí 4, 130 00 Praha 3 ☎ +420 267 005 111
Operates the TV and radio transmission facilities.

CESKY ROZHLAS (CZECH RADIO)

✉ Vinohradská 12, 120 99 Praha 2 ☎ +420 221 551 111 ▤ +420 221 551 300 **E:** info@rozhlas.cz **W:** www.rozhlas.cz
L.P: DG: Peter Duhan. PD: Lukas Hurnik. TD: Martin Zadrazil

LW & MW:	kHz	kW	Prgr.
Uherské Hradiste	270	650	CRo 1
Praha (Liblice)	639	1500	CRo 2 + CRo 6
Ostrava-Svinov	639	30	CRo 2 + CRo 6
Brno (Dobrochov)	954	200	CRo 2 + CRo 6
Ceské Budejovice	954	30	CRo 2 + CRo 6
Karlovy Vary	954	20	CRo 2 + CRo 6
Moravské Budejovice	1332	50	CRo 2 + CRo 6

FM (MHz)	CRo 1	CRo 2	CRo 3	CRo 5	kW
9) As	107.9			96.7	0.1/0.2
1b) Benesov				99.0	1(5)
6) Brno	95.1		102.0	106.5	72/91/72
6) Brno (city)		92.6	90.4	93.1	6/6/2
2) C. Budejovice	91.1	103.7	96.1	106.4	80/1/40/80
Cheb			106.2		1
4) Chomutov	98.9	94.2	96.3	103.1	10
3) Domazlice	98.0			105.3	10
13) Frydlant			97.4		0.2
6) Hodonín	106.2	107.8	100.4	93.6	9/3/9
5) Hradec Králové				95.3	1
11) Hradec Králové			104.7		10
8) Hulín			101.6		1
9) Jáchymov			103.4		1
8) Jeseník	91.3	88.7	98.2	106.8	20/0.2/20/20
Jicin		106.9			1
10) Jihlava	90.7	107.1	95.4	87.9	20/10/20/10

FM (MHz)	CRo 1	CRo 2	CRo 3	CRo 5	kW
Kaplice			105.9		0.2
9) Karlovy Vary	102.6		105.7	91.0	0.1/0.2/1
Kasperské Hory			107.2		0.5
1b) Kladno				100.5	0.2
3) Klatovy	99.8	90.3	88.6	102.4	10
1b) Kutná Hora		102.2		100.5	1/3
13) Liberec	95.9	89.9	103.9	102.3	20/20/20/1
13) Liberec			91.3		0.5
8) Lipník n.Becvou			88.7		0.1
9) Marián. Lázne	97.6		100.8		1
1b) Mladá Boleslav			100.3		0.5
Nové Hrady		102.2			1
8) Olomouc		107.2	92.8		1
7) Opava		101.7		102.6	1/0.5
7) Ostrava	101.4	101.9	104.8	107.3	43/0.5/43/3
11) Pardubice	89.7	100.1	102.7	101.0	90/90/90/1
Písek	97.0	98.9	105.2		1
Plzen (North)	89.1	101.7	95.6		80
3) Plzen (East)	99.2		93.3	106.7	10
3) Plzen (city)				91.0	1
Praha		100.7			50
1a) Praha (city)	94.6	91.2	105.0	92.6	5/3/5/7
1b) Príbram	102.2	107.0		100.0	0.4/1/1
3) Prosec n.N.				102.3	1
1b) Rakovník				100.4	1
5) Rychnov n.K.				96.5	1
2) Slavonice		103.3		88.2	1
3) Sokolov	94.3			98.2	0.4
Susice	90.6				1
11) Svitavy				102.4	1
3) Tachov				106.3	0.4
10) Trebíc				90.1	0.2
7) Trinec	92.1			105.3	1
5) Trutnov	88.5		101.9	90.5	10/10/20
10) Uher. Hradiste				99.1	0.2
6) Uhersky Brod	93.0			107.3	1
12) Uhersky Brod				107.3	1
4) Ustí nad Labem	90.9		104.5	88.8	80
Ustí n.L. (city)		98.6			1
11) Usti n. Orlicí				98.6	1
7) Val. Mezirící	92.5	89.9	96.8	99.0	7/1/7/7
4) Varnsdorf			88.4	98.5	0.2
Votice	93.1	103.2			95
13 Vratislavice				91.3	0.5
7) Vrbno pod Prad.		103.6		95.5	1
7) Vsetín	92.1	102.9	98.3	89.5	0.1
12 Vsetín				89.5	0.1
2) Zelezná Ruda				95.8	0.2
6) Zlín	99.5	107.7	94.8	97.5	6
12) Zlín				97.5	6
2) Znojmo	101.2	89.6	99.2	97.3	1/3/3/1

CRo 1 (Radiozurnál): 24h (LW: Mon-Sat 0400-2300, Sun 0500-2300). **N:** on the h – **CRo 2 (Praha):** 24h (MW: Mon-Fri 0300-1700, Sat+Sun 0400-1700) – **CRo 3 (Vltava):** 24h – **CRo 4 (Radio Wave):** 24h (on internet only: www.rozhlas.cz/audio/vysilani) – **CRo 5 REGIONAL STATIONS** – own prgrs as listed, otherwise relays of CRo 2 – **CRo 6:** 1700-2300. **N:** on the h.
Addresses:
1a) CRo Regina Praha, Hybesova 10, 186 72 Praha 8: 24h **W:** www.rozhlas.cz/regina – **1b)** CRo Region - Strední Cechy, Hybesova 10, 186 72 Praha 8: 24h **W:** www.rozhlas.cz/strednicechy – **2)** CRo Ceské Budejovice, U Trí Ivu 1, 370 29 Ceské Budejovice: 24h **W:** www.rozhlas.cz/cb – **3)** CRo Plzen, Nám Míru 10, 320 70 Plzen: 24h **W:** www.rozhlas.cz/plzen – **4)** CRo Sever (=North), Na schodech 10, 400 91 Ustí nad Labem: 24h **W:** www.rozhlas.cz/sever – **5)** CRo Hradec Králové, Havlíckova 292, 501 01 Hradec Králové: 24h **W:** www.rozhlas.cz/hradec – **6)** CRo Brno, Beethovenova 4, 657 42 Brno: 24h **W:** www.rozhlas.cz/brno – **7)** CRo Ostrava, Dr Smerala 2, 729 91 Ostrava: 24h (Polish: Mon-Fri 1800-1900, Sun 1730-1800) **W:** www.rozhlas.cz/ostrava – **8)** CRo Olomouc, Horní námestí 21, 771 06 Olomouc: 24h **W:** www.rozhlas.cz/ol – **9)** CRo Karlovy Vary, Zitkova 3, 360 00 Karlovy Vary: Mon-Fri 1400-1630, otherwise CRo Plzen **W:** www.rozhlas.cz/plzen – **10)** CRo Region - Vysocina, Masarykovo nám 42, 586 01 Jihlava: 0400-1700 1700-0400: relay of CRo Region (1b) **W:** www.rozhlas.cz/vysocina – **11)** CRo Pardubice, Sv Anezky Ceské 29, 530 02 Pardubice: 0400-1700 1700-0400: relay of CRo Region (1b) **W:** www.rozhlas.cz/pardubice – **12)** CRo Zlín, Osvoboditelu 187, 760 01 Zlín **W:** www.rozhlas.cz/brno – **13)** CRo Liberec, Modrá 1048, 460 06 Liberec **W:** www.rozhlas.cz/liberec/portal/

EXTERNAL SERVICE: Radio Prague
See International Broadcasting section.

MAJOR PRIVATE STATIONS/NETWORKS:

RADIO IMPULS (Comm.)
✉ Ortenovo nám. 15a, 170 00 Praha 7 ☎ +420 255 700 700 🖷 +420 255 700 727 **E:** impuls@radioimpuls.cz **W:** www.radioimpuls.cz
FM: see list below **D.Prgr:** 24h

RADIO FREKVENCE 1 (Comm.)
✉ Wenzigova 4, 120 00 Praha 2 ☎ +420 257 001 111 🖷 +420 257 314 183 **E:** frekvence1@frekvence1.cz **W:** www. frekvence1.cz
FM: see list below **D.Prgr:** 24h

EVROPA 2 (Comm.)
✉ Wenzigova 4, 120 00 Praha 2 ☎ +420 257 001 111 🖷 +420 257 001 807 **E:** info@evropa2.cz **W:** www.evropa2.cz
FM: see list below **D.Prgr:** 24h

RADIO KISS FM (Comm.)
✉ Rícanská 3, 101 00 Praha 10-Vinohrady ☎ +420 267 009 800 🖷 +420 267 009 811 **E:** radio@kiss.cz **W:** www.kiss.cz
FM: see list below **D.Prgr:** 24h
Regional branches: R. KISS 98 FM, ✉ Rícanská 3, 101 00 Praha 10-Vinohrady ☎ +420 267 009 800 🖷 +420 267 009 811 **W:** www. kiss98.cz – **R. KISS Hády,** ✉ Stefánikova 38, 612 00 Brno 12 ☎ +420 541 221 143 🖷 +420 541 211 117 **W:** www.kisshady.cz – **R. KISS Jizní Cechy** ✉ U Vystaviste 15A, 370 05 Ceské Budejovice ☎ +420 385 510 888 🖷 +420 385 510 990. **W:** www.kissjiznicechy. cz – **R. KISS Morava** ✉ Starobelská 13, 700 30 Ostrava-Zábreh ☎ +420 596 708 401 🖷 +420 596 708 400 **W:** www.kissmorava.cz – **R. KISS ProTon** ✉ Husova 58, 301 24 Plzen 1 ☎ +420 377 235 808 🖷 +420 377 235 810 **W:** www.kissproton.cz – **R. KISS Publikum** ✉ Bartosova 45, 760 01 Zlín ☎ +420 577 009 036. 🖷 +420 577 009 033 **W:** www.kisspublikum.cz – **R. KISS Delta** ✉ Jana Palacha 1025, 293 01 Mladá Boleslav 1 ☎ +420 326 720 000. 🖷 +420 326 721 342 **W:** www.kissdelta.cz

RADIO PROGLAS (Relig.)
✉ Barvicova 85, 602 00 Brno ☎ +420 543 217 241-3. 🖷 +420 543 217 245 **E:** radio@proglas.cz **W:** www.proglas.cz
FM: see list below **D.Prgr:** 24h

COUNTRY RADIO (Comm.)
✉ Ricanská 3, 101 00 Praha 10-Vinohrady. ☎ +420 251 024 111. 🖷 +420 251 024 224 **E:** info@countryradio.cz **W:** www.countryradio.cz
MW: Praha 1062kHz 20kW (0500-1800), 1kW (1800-0500) **FM:** Kutná Hora 87.7, Praha 89.5MHz, Ceské Budejovice 94.7, Beroun 98.3MHz, Tábor 101.8, Plzen 102.1, Ceská Lípa 102.6MHz, 106.8 Usti nad Labem **D.Prgr:** 24h

FM Stations:

MHz	kW	Station	Location
87.6	70	R. Impuls	Brno
87.8	1	R. Blaník	Praha
87.8	1	R. Cerná hora	Králíky
88.1	1	R. Evropa 2	Liberec
88.1	10	Hitrádio Orion	Jeseník
88.2	5	R. Evropa 2	Praha
88.3	10	R. Kiss Hády 88 FM	Brno
88.4	1	R. Blanik - JC	Ceské Budejovice
88.7	1	R. Proglas	Tábor
88.9	10	R. Jih	Breclav
89.0	1	R. Práchen	Písek
89.0	45	R. Impuls	Ostrava
89.3	5	R. Novy Preston	Benešov/Kozmice (F.Pl.)
89.5	1	R. Cas	Trinec
89.5	5	Country Radio	Praha
89.6	1	R. Frekvence 1	Plzen
89.6	7	Rock Max	Zlín
89.8	1	BBCWS/R.C.	Ceské Budejovice
90.0	1	Hitradio Dragon	Cheb
90.0	1	R. Rubi	Sumperk
90.0	10	R. Kiss ProTon	Plzen
90.2	1	R. Kiss Delta	Kutná Hora
90.3	5	R. Expres	Praha
90.3	3	R. Kiss Publikum	Zlín
90.6	4	Hitrádio FM Most	Chomutov
90.6	1	R. Proglas	Bystrice pod Hostynem
91.0	70	R. Frekvence 1	Ostrava
91.0	1	R. Evropa 2	Mariánské Lázne
91.1	1	R. Kiss Delta	Pardubice
91.4	66	R. Impuls	Plzen
91.6	1	R. Blanik - SevC	Decín
91.6	5	Fajn rádio Life	Opatovice
91.7	4	R. Zlín	Zlín
91.9	1	R. 1	Praha
92.1	10	R. Impuls	Trutnov
92.3	1	R. Relax	Kladno
92.3	5	R. Haná	Pohorany
92.5	1	R. Egrensis	Mariánské Lázne

MHz	kW	Station	Location
92.8	5	R. Cas	Ostrava
92.8	1	Hitradio Magic	Náchod
92.9	1	R. Kiss Delta	Mladá Boleslav
93.2	1	R. Egrensis	Cheb
93.3	20	R. Proglas	Jeseník
93.4	1	R. Frekvence 1	Jihlava
93.5	50	R. Frekvence 1	Ustí nad Labem
93.6	1	Hitrádio Faktor	Písek
93.7	5	R. City	Praha
93.7	45	R. Hellax	Ostrava
93.8	1	R. Evropa 2	Karlovy Vary
93.9	80	R. Blaník V.Cechy	Pardubice
94.0	10	R. Impuls	Klatovy
94.1	10	R. Frekvence 1	Valasské Mezirící
94.1	50	R. Frekvence 1	Ceské Budejovice
94.3	10	Hitrádio Vysocina	Jihlava
94.7	1	R. Hey Ostrava	Ostrava
94.9	1	R. Hey Ostrava	Opava
95.0	95	R. Blaník	Votice
95.2	5	Fajn North Music	Ustí nad Labem
95.2	1	R. Sumava	Klatovy
95.3	5	R. Beat	Praha
95.7	2	R. Hey	Praha
95.8	1	Hitrádio Vysocina	Trebíc
96.2	1	R. Spin	Praha
96.2	1	R. Zlín	Uherský Brod
96.4	4	Hitrádio Orion	Ostrava
96.5	1	R. Kiss Morava	Sumperk
96.6	5	R. Impuls	Praha
96.7	5	BBCWS/R.C.	Jihlava
96.8	1	R. Hey Brno	Brno
96.9	1	R. Hey Profil	Pardubice
97.1	5	R. Rubi	Pohorany
97.1	1	R. Hey! Sever	Liberec
97.2	5	Fajn rádio	Praha
97.4	50	R. Frekvence 1	Pardubice
97.7	50	R. Kiss Jizní Cechy	Votice
97.7	1	Evropa 2	Ostrava
97.9	20	R. Proglas	Liberec
98.1	1	Fajn rádio Agara	Chomutov
98.1	1	R. Kiss 98 FM	Praha
98.3	1	R. Cas	Trinec
98.4	20	R. Frekvence 1	Trutnov
98.4	5	R. Impuls	Kasperské Hory
98.6	1	BBCWS/R.C.	Plzen
98.7	1	Hitrádio Orion	Trinec
98.7	5	R. Classic FM	Praha
99.0	1	Hitradio Brno	Brno
99.1	1	BBCWS/R.C.	Pardubice
99.2	1	BBCWS/R.C.	Liberec
99.3	1	Kiss Jizní Cechy	Cesky Krumlov
99.3	10	R. Evropa 2	Jeseník
99.3	1	R.France Int./Fr.Mus.	Praha
99.5	1	R. Evropa 2	Pardubice
99.7	1	Hitradio Dragon	Karlovy Vary
99.7	1	R. Gold	Ceské Budejovice
99.7	5	R. Bonton	Praha
99.8	5	Hitradio Apollo	Valasské Mezirící
99.9	1	Hitrádio Crystal	Ceská Lípa
100.3	20	R. Impuls	Jihlava
100.5	7	R. Impuls	Valasské Mezirící
100.6	2	R. Blaník Sever	Teplice
100.8	1	R. Beat	Slavonice
100.8	20	R. Impuls	Jeseník
101.1	1	R. Kiss Morava	Frydek-Místek
101.1	3	BBCWS/R.C.	Praha
101.3	10	R. Evropa 2	Plzen
101.4	20	R. Contact (RCL)	Liberec
101.8	1	Country R.	Tábor
102.0	50	R. Impuls	Ustí nad Labem
102.5	5	R. Frekvence 1	Praha
102.8	5	Hitradio Dragon	Mariánské Lázne
102.8	1	Hitrádio FM Labe	Ustí nad Labem
102.9	50	R. Impuls	Ceské Budejovice
103.0	10	R. Krokodyl	Brno
103.4	1	R. Blaník V.Cechy	Hradec Králové
103.4	5	R. Petrov	Brno
103.6	1	R. Hey Profil	Chotebor
103.7	1	Oldies R. Olympic	Praha
103.8	10	R. Frekvence 1	Klatovy
103.9	7	Hitrádio Orion	Valasské Mezirící
104.1	50	R. Frekvence 1	Plzen
104.2	1	R. Blanik - Jiz. Morava	Znojmo
104.3	20	R. Frekvence 1	Jeseník
104.3	32	Hitrádio Faktor	Ceské Budejovice

MHz	kW	Station	Location
104.5	50	R. Frekvence 1	Brno
104.7	10	R. Blanik - Západ	Plzen
105.0	10	R. Frekvence 1	Zlín
105.3	3	R. Cerná hora	Trutnov
105.4	1	R. Rubi	Vrbno pod Pradedem
105.5	95	R. Evropa 2	Votice
105.5	1.5	R. Evropa 2	Brno
105.7	1	R. Jizera	Mladá Boleslav
105.8	8	Hitrádio FM Plus	Klatovy
105.9	1	R. Cas	Frenstát p. Radh.
106.0	50	R. Impuls	Pardubice
106.1	3	Hitrádio FM Plus	Plzen
106.3	1	BBCWS/R.C.	Ostrava
106.4	1	R. Evropa 2	Vrchlabí
106.5	10	R. Blaník - Sever	Chomutov
106.6	1	Fajn rádio	Kutná Hora
106.7	1	R. Evropa 2	Znojmo
107.2	1	R. Evropa 2	Ustí nad Labem
107.4	1	Hitrádio FM Plus	Jáchymov
107.5	3	R. Proglas	Brno
107.5	2	R. Proglas	Nové Hrady

+ more than 70 txs below 1kW
NB: BBCWS/R.C. = BBCWS (in English) + Radio Cesko (in Czech)

DENMARK

L.T: UTC +1h (27 Mar–30 Oct: +2h)— **Pop:** 5.5 million— **Pr.L:** Danish — **EC:** 50Hz, 230/380V — **ITU:** DNK

BROADCAST SERVICE DENMARK A/S
Banestrøget 21, DK-2630 Taastrup ☎+45 70118011 ☏ +45 43711143
Broadcast Service Denmark is responsible for the operation of the txs carrying the prgrs. of DR and TV 2.

DR RADIO (Gov.)
DR Byen, Emil Holms Kanal 20, DK-0999 Copenhagen C ☎+45 35203040 **W:** www.dr.dk
L.P: DG: Kenneth Plummer. Media Dir.: Mikael Kamber.
LW: Kalundborg 243kHz 0.2kW (DRM tests. Inactive). **MW:** Kalundborg 1062kHz 250kW.

FM	P1	P2	P3	P4	kW
Bornholm	96.2	103.5	90.0	99.3	30
Copenhagen	90.8	102.3	93.9	96.5	60
Funen	89.0	100.5	92.6	96.8	60
Holstebro	90.2	100.3	92.9	98.5	60
Nakskov	89.4	98.8	94.1	92.2	8/15/15/30
Næstved	94.8	101.9	99.6	97.5	100
Skamlebæk	88.4	101.1	94.3	92.0	3/5/3/3
So. Jutland	95.1	102.1	97.2	99.9	60
Thisted	91.4	101.3	99.2	95.6	2/3/2/2
Tolne, N.Jutland	91.0	100.7	96.6	94.4	8/10/8/8
Ølgod	88.7	102.5	92.3	97.7	10
Varde				97.7	10
Vejle	95.5	100.9	90.7	94.0	10
Aalborg	93.3	102.7	89.7	98.1	60
Århus	88.1	103.0	91.7	95.9	60

+ 25 FM txs below 1kW. A full list is available at http://itst.dk
DAB: DAB1: ch.12C (227.360MHz). DAB2: ch.11C (220.352MHz) on Sealand & Funen and ch.13B (232.496MHz) in Jutland.
P1 on FM + DAB1. **N:** on the h (except Su 0900, 1000). N in Danish from KNR, Greenland: MF 1825-1830 – **P2** on FM + DAB1: Classical and jazz music and cultural prgrs. – **P3** on FM + DAB1: Popular music, news and sport. N: on the h. + MF: 0530, 0630, 0730 – **P4** on FM. News, entertainment and regional prgrs. N: national news on the h and regional news on the half h.
DR Boogieradio on DAB1: Continuous current chart hits – **DR Dansktop** on DAB2: Continuous C&W + schlager – **DR Hit** on DAB1: Continuous A/C chart hits – **DR Jazz** on DAB1: Continuous jazz – **DR Klassisk** on DAB2: Continuous classical music – **DR Nyheder** on DAB2: All news – **DR Oline** on DAB1: For children 3-6 years old – **DR P5** on DAB1: Oldies etc.; at times relays from P4 – **DR P5000** on DAB2: CHR – **DR Rock** on DAB2: Continuous rock music – **DR Unga Bunga** on DAB1: Continuous indie/alternative.
All prgrs are available on the internet. P1, P2, P3, P4 København and DR Klassisk are also available via satellite.
F.pl.: The FM frequencies of P2 will be taken away from DR late 2011, and instead allocated to a new private, non-commercial, gov. funded News/Talk station. DR is due to leave DAB2 as soon as the space has been awarded to new commercial stations.
Regional stations:
MF: 0507-0600, 0607-0700, 0707-0800, 0807-0900, 1130-1132, 1403-

1500, 1510-1550 & 1610-1700. Sat 0603-0700, 0707-0800, 0807-0900 & 1130-1132. Sun: 0603-0700, 0703-0800, 0807-0900 & 1130-1132. P4 Trekanten, P4 Esbjerg and P4 Nordvestsjælland are on the air at a reduced schedule. At other times national P4 prgrs are carried.
DR Nordjylland, Frederik Bajers Vej 9, DK-9220 Aalborg: on 89.1/94.4/96.7/98.1MHz – **DR Midt- & Vest**, Vestergade 1, DK-7500 Holstebro: on 95.6/ 97.7/98.5/102.2MHz – **DR Østjylland,** Olof Palmes Alle 10-12, DK-8200 Aarhus N: on 88.9/95.9/96.4/102.0MHz – **DR Trekanten,** Den Hvide Facet 4., DK-7100 Vejle: on 94.0MHz – **DR Syd,** H.P. Hansensgade 11, DK-6220 Aabenraa: on 94.0/96.6/99.0/99.9/103.7MHz – **DR Esbjerg,** Torvegade 8, 6700 Esbjerg on 99.0/103.7MHz – **DR Fyn,** Lille Tornbjergvej 10, DK-5220 Odense S: on 96.4/96.8MHz – **DR Sjælland,** Vadestedet 1, DK-4700 Næstved: on 92.0/102.3/107.4MHz – **DR Nordvestsjælland,** DR Sjælland, Ahlgade 3 F, 4300 Holbæk on 92.0MHz – **DR København,** Emil Holms Kanal 20, DK-0999 Copenhagen: on 96.5MHz – **DR Bornholm,** Aakirkebyvej 52, DK-3700 Rønne: on 93.7/99.3MHz
P6 – Kalundborg mellembølgesender on 1062kHz: 0445-0507, 0730-0807, 1045-1135, 1645-1716, 2145-2205. Special progrs.: Wrp. 0445-0500, 0745-0800, 1045-1100 & 1645-1700, gymnastics 0730-0745 & navigational warnings (repeated twice) 1700-1716. Also news from P4 at 0500-0507, 0800-0807 & 1100-1120MF/1100-1115SS.
Ann: FM: "Du lytter til P et/to/tre/fire" (1st, 2nd, 3rd & 4th prgr.). MW: "Du lytter til DRs mellembølgesender på 1062 kHz"

NEW RADIO (Comm.)
Rådhuspladsen 45, DK-1550 Copenhagen V ☎ +45 33378900 ☏ +45 33378967 **E:** info@radio100fm.dk **W:** www.radio100fm.dk & www.radiosoft.dk
LP: MD: Jim Receveur.
Radio 100FM: Hot AC. Hove (Copenhagen): On 27 low power FM txs in most major cities – **Radio Soft:** A/C Soft. On 7 low power FM txs in the larger Copenhagen area.

SBS RADIO (Comm.)
Magstræde 10, DK-1204 Copenhagen K ☎ +45 33376666 ☏ +45 33930807 **E:** info@thevoice.dk **W:** www.sbsradio.com/da, www.novafm.dk, www.thevoice.dk and www.popfm.dk.
LP: MD: Frederik Meyer. CEN: Jan Andersen
Nova FM: AC. Ølgod 87.8MHz 10kW, So. Jutland 89.3MHz 3kW, Copenhagen 91.4MHz 11kW, Bornholm 92.2MHz 1kW, Funen 93.4MHz 1kW, Vejle 99.3MHz 1kW, Tolne N. Jutland 102.4MHz 1kW, Holstebro 103.4MHz 60kW, Næstved 103.9MHz 100kW, Aalborg 106.0MHz 5kW + 20 stns below 1kW. Also nationwide on DAB2 – **The Voice:** CHR. On 26 low power FM txs in major cities. – **Pop FM:** Classic hits. Copenhagen 100.0MHz 60kW, Randers 99.9MHz 0.5kW + nationwide on DAB2.

NRJ (Comm.)
Bispevej 4, 1. DK-2400 Copenhagen NV ☎ +45 38168200 ☏ +45 28168202 **E:** info@nrj.dk **W:** www.nrj.dk
Radio Energy - NRJ: CHR. On 5 low power txs in Copenhagen, Holbæk, and Odense.

Private stations:
Approx. 200 organizations are operating low-powered FM txs. (0.16kW-0.5kW at 40m. height). Currently aprrox. 500 txs are on the air. Major stns in the main cities are as follows (only main frequency/frequencies mentioned):
Aabenraa: Radio Mojn, Box 44, 6200 Aabenraa: 102.6/104.5MHz
Aalborg: ANR, Box 7089, 9200 Aalborg SV: 87.6/102.0/103.8MHz – Radio Aura, Box 7089, 9200 Aalborg SV: 105.4/106.9MHz – The Voice: 100.2MHz – Various grassroots/community stns: 92.2/106.5/107.4MHz
Aarhus: go!FM, Vesterport 3, 4., 8000 Århus C: 92.2/94.6/106.5MHz – The Voice: 90.9/93.1/93.7MHz – Radio 100FM: 87.6MHz – Various grassroots/community stns: 98.7MHz
Copenhagen: The Voice: 91.8/96.1/104.4/104.9/105.4MHz – NRJ: 88.6/107.1MHz – Radio 100FM: 97.2/103.6/104.1/105.6MHz – Radio Soft: 95.0MHz – Various grassroots/ community stns: 87.6/90.2/90.4/92.9/94.5/95.2/95.5/97.7/98.9/100.9/102.9/103.4/105.9/106.3/107.4MHz
Esbjerg: Skala FM, Banegårdspladsen, 6700 Esbjerg: 101.7/106.8MHz – Radio 100FM: 106.3MHz
Frederikshavn: ANR, Tordenskjoldsgade 4, 9900 Frederikshavn: 107.5MHz – Radio Aura: 89.0MHz.
Herning: Radio M, Østergade 21, 7400 Herning: 105.8MHz – Radio Aura, Østergade 21, 7400 Herning: 89.5MHz
Hjørring/Hirtshals: Skaga FM, P. Rimmersgade 40, 9850 Hirtshals: 105.6MHz – ANR: 104.7MHz – Radio Aura: 89.0MHz
Holstebro: Holstebro Favorit FM, Lægårdvej 86, 7500 Holstebro: 105.1/106.2MHz
Horsens: Radio Horsens, Nørregade 42, 8700 Horsens: 91.1MHz – Horsens Classic: 105.3MHz – The Voice: 105.0MHz

Kolding: Skala FM, Dalbygade 40, 6000 Kolding: 94.4/105.2MHz – VLR: 103.2MHz – Globus Guld: 100.3MHz – Radio 100FM: 91.3/102.7MHz

Køge: Radio Køge, Box 222, 4600 Køge: 98.2/106.2/106.8MHz –The Voice: 93.6MHz

Nykøbing F: Radio Sydhavsøerne, Tværgade 18, 4800 Nykøbing Falster: 87.8MHz

Nykøbing M: Radio Limfjord, Gasværksvej 10, 7900 Nykøbing Mors: 104.7/106.9/107.8MHz – Radio Limfjord Plus: 94.7/107.7MHz

Næstved/Ringsted/Slagelse: Radio SLR, Dania 38, 4700 Næstved: 91.6/100.7/101.0/106.5MHz

Odense: Radio 3, Box 312, 5100 Odense C: 91.1MHz – The Voice: 104.2/105.1/107.6MHz – NRJ: 103.5MHz – Radio 100FM: 101.2MHz – Various grassroots/community stations: 107.1MHz

Randers: Radio ABC Box 174, 8900 Randers: 95.3/105.7MHz – Radio Alfa, Box 174, 8900 Randers: 91.3/102.4MHz – Radio ABC Solo FM, Box 174, 8900 Randers: 96.4MHz

Roskilde: The Voice: 106.6MHz – Radio 100FM: 103.6/104.3MHz – Radio Soft: 107.7MHz

Rødding: Radio Globus, Herrefogedvej 2, 6630 Rødding: 104.4MHz – Globus Guld, Herrefogedvej 2, 6630 Rødding: 93.0MHz

Silkeborg: Radio 1, Fredensgade 1, 8600 Silkeborg: 107.7MHz – Silkeborg Guld: 94.5/101.2MHz – Radio 100FM: 96.9MHz

Skive: Radio Skive, Nordbanevej 1A, 7800 Skive: 104.0MHz – Radio Alfa, Nordbanevej 1A, 7800 Skive: 101.8MHz

Svendborg: Radio Diablo, Voldgade 9,1., 5700 Svendborg: 107.7MHz – Radio Alfa Sydfyn: 106.5MHz

Vejle: VLR, Bugattivej 8, 7100 Vejle: 101.7MHz – The Voice: 105.9MHz

Viborg: Radio Viborg, Box 501, 8800 Viborg: 105.0MHz – Viborg Favorit FM: 93.8MHz

DJIBOUTI

L:T: UTC +3h — **Pop**: 500,000 — **Pr.L**: Arabic, French (official), Somali, Afar — **E.C**: 50Hz, 220V — **ITU**: DJI

MINISTÈRE DE LA COMMUNICATION ET DE LA CULTURE CHARGÉ DES POSTES ET DE TÉLÉCOMMUNICATIONS (MCCPT)

1 Rue de Moscou, B.P. 32, Djibouti ☎+253 355 672 +253 353 957 **W**: www.mccpt.dj **E**: mccpt@intnet.dj **LP**: Minister: Rifki Abdoulkader Bamakhrama.

RADIODIFFUSION TÉLÉVISION DE DJIBOUTI (Gov.)

1 Rue St. Laurent du Var, B.P. 97, Djibouti ☎+253 350484 +253 356502 **W**: www.rtd.dj **E**: rtd@intnet.dj
LP: DG: M. Abdi Atteyeh Abdi. Dir. Tec: Mohamed Moussed Yaya. PD: Nabil Dorani. Dir. Inf: Ms. Hasna Maki.
MW: Djibouti (Dorale) 1116kHz 50kW, 1539kHz 50kW.
SW: Djibouti (Dorale) 4780kHz 50kW. Site to be relocated.

FM (MHz)	1	2	kW	FM (MHz)	1	2	kW
Arta	93.5	89.5	5	Ballembaley	95.3	91.3	1/0.1
Djibouti	91.3	95.3	1	Dikhil	96.6	98.8	0.5/0.25
Ali Sabieh	90.3	94.2	0.5/0.25				

Channel 1 in Afar/Arabic/Somali: 0300-2100 on 1539 & 4780kHz + FM. **Channel 2** in Afar/French: 0300-2100 on 1116kHz & FM. French: 0700-1100 & 1400-1800.
Ann: "Radio Djibouti".

Other stations:
BBC African Sce: Djibouti 99.2MHz 1kW.
RFI Afrique: Arta 92.0MHz 2kW, Djibouti 104.0MHz 0.5kW.
Deutsche Welle/Monte-Carlo Doualiya: Arta 97.2MHz 5kW.
R. Sawa: Djibouti (Dorale) 1431kHz 600kW 1630-0400 & Arta 100.8MHz 5kW 24h.
Voice of America: Djibouti 102.0MHz 1kW

DOMINICA

L:T: UTC -4h — **Pop**: 71,000 — **Pr.L**: English, Patois — **E.C**: 50Hz, 240V — **ITU**: DMA

DOMINICA BROADCASTING CORP. (Gov. Comm.)

Victoria Str, PO Box 148, Roseau ☎ +1 767 448 3282/3 +1 767 448 2918 **E**: dbsradio@cwdom.dm **W**: www.dbcradio.net
LP: Chairman: Aurelius Jolly. Acting GM: Shermaine Green-Brown. CEN: Kurt Matthew
FM: Eggleston, Roseau 88.1MHz 1kW, Grand Fond 88.5MHz 0.03kW, Marigot 103.5MHz 0.3kW, Petite Soufriere 103.1MHz 0.1kW, Grand Bay 103.5MHz 0.1kW, Portsmouth 104.7MHz
D.Prgr: Own prgrs: 0900-0300. BBC relay: 0300-0900 **Patois**: 1800-1930MF **Ann**: "DBS Radio".

Other stations:
HIT RADIO MUSIC POWER (Comm.) P.O. Box 931, Roseau. FM: 93.5/96.5/100.1MHz (see St. Lucia) – **KAIRI FM (Comm.)** P.O. Box 931, Roseau ☎ +1 767 448 7330/7331 +1 767 448 7332 **W**: www. kairifm.com **L:P**: Mgr: Steve Vidal. FM: **Kairi FM**: 88.7/93.1/107.9MHz **Hot FM**: 91.1MHz – **Q95 (Comm.)** 15 Hanover Str., PO Box 861, Roseau ☎ +1 767 448 5822 +1 767 448 5828 **W**: www. wiceqfm.com L:P: CEO: Sheridan G. Gregoire FM: 89.7/95.1/105.7MHz – **RADIO CARIBBEAN INTERNATIONAL (Comm.) FM**: 98.1MHz 24h (see St. Lucia) – **VOICE OF LIFE RADIO – ZGBC RADIO (Rlg.)** P.O. Box 205, Madrelle, Loubiere, Roseau ☎ +1 767 448 7017 +1 767 440 0551 **W**: www.voiceofliferadio.dm L:P:CEO: Mc Kenzie Mitchelle. SM: Clementina Munro. CEN: Francis Guiste **FM**: 24h: Portsmouth 90.7MHz, Roseau 102.1MHz, Marigot 106.1MHz – **RADIO EN BA MANGO (Community)** Grand Bay. ☎ +1 767 446 3207 **W**: www.grandbayagain.com/radioanbamango.htm **FM**: 93.5MHz (3 kW). **D.Prgr**: 2200-0300

DOMINICAN REPUBLIC

L:T: UTC -4h — **Pop**: 9.5 million — **Pr.L**: Spanish — **E.C**: 60Hz, 110V — **ITU**: DOM

INDOTEL - INSTITUTO DOMINICANO DE LAS TELECOMUNICACIONES

Abrahan Lincoln N° 962, Edif. Osiris 1, Planta, 10148 Santo Domingo ☎ +1 809 732 5555 +1 809 732 3904 **W**: www.indotel. org.do **L:P**: Dir.Gen: Dr. David Pérez Taveras

Hrs of tr. usually 24h unless stated. Call HI–

	MW Call	kHz	kW	Station, location and hr of tr.
1)	CM	540	5	R. ABC, Sto Domingo: 0900-0400
60)	AA*	560	3	R. Ritmos, Santiago
52)	MS	570	10/5	R. Crystal, Sto Domingo
71)	FS	580	5	R. Montecristi, Montecristi
4)	DV	590	10/5	R. Santa María, La Vega: 0900-0300
7)	SD*	600		R. Santo Domingo, El Seybo (r: 620)
119)		600		Celestial 600, Santo Domingo
60)	JR	610	5	R. Amanecer, Santiago (r: 1580)
7)	SD*	610	1	R. Santo Domingo, Pedernales (r: 620)
7)	SD	620	10	R. Santo Domingo, Sto Domingo: 0900-0400
7)	SD*	630	1	R. Santo Domingo, San Juan (r: 620)
7)	SD*	640	10	R. Santo Domingo, Santiago
9)	AT	650	15/5	R. Universal, Sto Domingo
62)	AM	660	3	R. Visión Cristiana, Santiago: (r: 1330)
59)	BS	670	5	R. Dial, San Pedro de Macorís
7)	SD*	670	1	R. Santo Domingo, Barahona (r: 620)
11)	JX	*680	3	R. Zamba, San Ignacio de Sabaneta: 0900-0400
22)	AW	690	10	R. Guarachita "La Poderosa", Sto Domingo: 0900-0400
13)	DC	*700	0.6	R. Mao, Mao, Valverde
104)	WP	710		Onda del Caribe, San Cristóbal
14)	AQ	720	1.8	R. Norte, Santiago: 0900-0500
87)	EF	720	5	R. Cayacoa, Hainey: 0900-0400
15)	Z	730	10	R. HIZ, Broadc. Nac., Sto Domingo: 1100-0500
16)	DB	750	5	R. Jesús es el Señor, Santiago
17)	CO	*760	5	R. Cordillera, Sto Domingo
18)	MD*	770	5	R. Activa, Santiago: 3 daily news only
19)	BO	780	0.5	R. Constanza, Constanza: 1100-0200
20)	L	790	5	R. Centro, Sto Domingo (occ r. Romántica FM 107.7)
70)	VM	800	1	R. Bonao, Bonao: 1000-0400
24)	AV	810	5	R. Salvación, Baní: 1100-0300
21)	AZ	820	2.5	R.Santiago, Santiago
22)	JB	830	10	R. HIJB, Sto Domingo: 1100-0300
23)	AB*	840	1	R. Isabel de Torres, Puerto Plata
72)	GA	850	5	R. Guarocuya, Barahona: 1000-0400
5)	UA*	850	5	R. Clarín, Santiago (r: 860):1100-0300
5)	UA	860	10	R. Clarín, Sto Domingo
25)	VG	870	4	R. La Vega, La Vega: 1000-0400
26)	OR	890	3	La Consentida, Mao, Valverde: 1000-0400
27)	PJ	*890	4/5	R. Continental, Sto Domingo: 1000-0500
28)	EN	900	5/1	R. Puerto Plata, Puerto Plata: 0900-0400
60)	FK	900		R. Amanecer, Neiba (r: 1580)
29)	LB	910	5	R. 91 "La Grande", Bonao: 0930-0300
9)	BA	920	10	R. 9-20 AM-Stereo "Power", Sto Domingo
31)	CK*	930	10	Ondas del Yaque, Santiago: 0945-0400
8)	AS	940	3	R. Metro, Montecristi (F.PI)
32)	IG	950	10	R. Popular, Sto Domingo
33)	FF	960	5	LV del Atlántico, Puerto Plata: 1000-0500
50C)	CV	970	5/1	R. Barahona, Barahona
25)	VP	970	6	R. Olímpica, La Vega
36)	SA	990	1	R. Cibao HI-SA 9-90, Santiago
37)	HG*	1000	5/1	R. Beller, Dajabón: 1000-0300
38)	JA	1010	10	R. Comercial, Sto Domingo: 1100-0600
38)	JA	1010		R. Comercial, Salcedo (r: HIJA Sto Domingo 1010)

MW	Call	kHz	kW	Station, location and hr of tr.
38)	JA	1010		R. Comercial, San Juan de la Maguana (r: HIJA Sto Domingo 1010)
30)	TS	1020	10	R. Enriquillo, Neyba: 0900-0400
39)	DL	1030	5	R. Novedades, Santiago
40)	ON	1040	10	CDN Radio, Sto Domingo
14)	CB	1050	1.5	R. Hispaniola, Santiago
60)	AJ	1060		R. Amanecer, San Pedro de Macorís (r: 1580)
42)	XF	1060	1	R. Azua, Azua: 0900-0200
44)	BI	1070	5/1	HIBI R. 1070, San Francisco de Macorís: 0900-0400
45)	MC	1080	1	R. RPQ Sport, Sto Domingo
46)	JM	1090	3	R. Amistad, Santiago
50A	RB	1100	1	R. Jimaní, Jimaní
47)	HD	1100	1	R. Oriente, San Pedro de Macorís: 0900-0400
48)	MP	1100	1	R. Ocoa, San José de Ocoa: 1200-0200
49)	PS	1100	1	R. Comercial, Nagua: 0900-0400
51)	TC	1110	2.5	R. Jarabacoa, Jarabacoa: 1000-0400
95)	OS*	1110	1/0.5	R. Marién, Dajabón
52)	CN	1120	10	R. Metro Hit, Sto Domingo
52)	CN	1120	10	R. Metro Hit, Samaná (r: 1120): 1000-0400
109		1120		R. Antillas, Barahona
40)	RL	1130	10/1	CDN Radio, Santiago (r: 1040)
54)	RA	1140	5	R. Anacaona, San Juan de la Maguana: 1100-0400
55)	AS	1150	5	Onda Musical, Sto Domingo: 1100-0500
56)	BE	1160	1	Radiolandia, Santiago (Occ r: 1180kHz): 0900-0400
110)	AS	1170		Cadena Espacial, Azua
57)	BE	1180	10	R. Mil, Sto Domingo: 1000-0500
58)	AG	1190	10	Azul 11-90 Bachatisima, Santiago
50B)	MR	1200	1	R. Caracol, Azua
98)	AH	1200		R. VEN - Voz Evangelica Nacional, Sto Domingo
61)	CJ	1210	5	R. Merengue, San Francisco de Macorís
100)		1220		R. HIN, Sto Domingo (rel La Z 101)
63)	PM*	1230	1	R. Moca, Moca: 1000-0300
64)	AU	1240	1	R. Vida, Puerto Plata: 0900-0300
53)		1240	10	R. María, Santo Domingo
66)	BC	1250	5	LV del Progreso, San Francisco de Macorís: 1000-0400
67)	RJ	1250	5	R. Juventud, La Romana: 0930-0430
38)	T	*1260	1	R. Recuerdos, Sto Domingo
52)	DA	1270	1.2	R. Metro-Hit 12-70, Santiago
69)	TA	1270	1	R. Ambiente, Baní: 1000-0400
110	JH*	1280		Cadena Espacial, Sto Domingo
6)	BD*	1290	0.5	R. Jánico, Santiago
74)	KQ*	1300	1	Radio Radio/La Doz de HIZ, Sto Domingo
75)	MH	1310	1	R. Real, La Vega: 1100-0400
76)	BZ	1320	1/0.5	R. Centro, San Juan de la Maguana
62)	VC	1330	3	R. Visión Cristiana, Sto Domingo
77)	PM	1350	1	R. Rutas Musical, La Romana: 1100-0400
102)	JD	1350	4	Ondas del Yuna, Bonao
108)	XZ	1360		R. Tropical, Sto Domingo
79)	RP	1370	5	R. Seybo, El Seybo
80)	SC	1380	1	R. Nacional, Santiago: 1000-0300
81)	AR	1390	1	R. San Cristóbal, 1100-0300
82)	AC	1400	1	Ondas del Valle, La Vega: 1100-0200
65)	AE	1410	1	R. Tricolor, Sto Domingo
85)	JU	1410	1/0.5	R. Grí-Grí, Río San Juan: 1000-0400
50D)	CH	1410	3/0.5	R. 14-10, Barahona
86)	FD	1420	15	R. Oro, Cotuí
34)	JC	1430	6	R. Emanuel, Santiago
89)	AD	1440	5	R. San Juan, San Juan de la Maguana: 1000-0300
90)	AK	1440	5	R. Impactante, Sto Domingo
83)		*1450		R. Alfa y Omega, Sto Domingo
91)	AC	1450	10	R. Util, Salcedo: 0900-0400
92)	AN	1460	0.5	R. Renacimiento, Hato Mayor del Rey
93)	DE	1470	1	LV de la Alabanza, San Francisco de Macorís: 1000-0400
50C)	CH	1470		R. Vibra –La Deportiva, Barahona
50D)	CV	1470		R. Barahona, Provincia Independencia (r: R. Barahona 970)
68)	AH°	1480	5	R. Villa, Sto Domingo: 1000-0400 (Sun 1100-2300)
96)		*1490	3	R. Universal, Santiago (rel. 650)
97)	PA	1500	0.5	R. Higüey, Higüey: 0900-0400
111)	RD	1500		R. Juan Pablo Duarte, Elías Piña
98)	BL°	1510	10/3	R. Pueblo, Sto Domingo: 1000-0400 (Sun 1100-2300)
99)	WJ	1520	1	R. Samaná "R. 15-20", Samaná: 0930-0400
112)	JN*	1530	1	R. 1530, Santiago (irr. Channel 25 UHF audio)
38)	FP	1540	1	R. Criolla Comercial, Sto Domingo
41)	BUv	1540	1	LV de La Romana, La Romana (r): 0930-0400
50E)	PZ	1560	1/0.5	R. Pedernales, Pedernales
117)		*1560	1	R. Universidad UASD, Santo Domingo (CP)
101)	GL*	1560	1	R. Unica, Santiago
60)	AJ°	1580	10	R. Amanecer, Sto Domingo: 1000-0400
50F)	PK	1580	1	R. Neiba, Neiba: 0900-0400
101)	SF	1590	1	R. Libertad, Santiago
65)	FG	1600	5	R. Revelación en América, Sto Domingo: 1200-0200
103)	SR	1620		R. Taina/Planeta, San Pedro de Macorís
10)		1640	1/0.5	R. Juventus Don Bosco, Sto Domingo

MW	Call	kHz	kW	Station, location and hr of tr.
115)		*1650	5/3	RADECO, Santiago (CP)
116)		*1660	5/1	Fundación Lama, Sto Domingo (CP)
114)	SV	1680	1	R. Senda 1680 AM, San Pedro de Macorís
113)		1700	5/1	R. Eternidad, Sto Domingo: 1100-2400

° = also on SW, * = inactive, (r) = repeater, v = varying fq.

SW	Call	kHz	kW	Station, location and h of tr
118)		4730	0.1	R. Descubrimiento, Santo Domingo: irr
68)	VR*	4960	5	R. Villa/Cima 100/Super Q FM, Santo Domingo: irr
3)	MI	5010	1	R. Cristal Int., Sto Domingo: irr (rel: R.Pueblo 1510)
60)	IJ	*6025	1	R. Amanecer Internac., Sto Domingo: 0900-0300

Stns with a (*) are reported inactive, but may be reactivated for variable periods of time.

Addresses and other information

1) Av Rómulo Betancourt N° 2078, (or: Ap 517), Sto Domingo ☎+1 809 684 2888 – **2)** Calle Félix María Ruiz N° 6, La Trinitaria (or: Ap 581), Santiago ☎+1 809 724 9050 **FM:** 95.5MHz **W:** www.digital95fm.com – **3)** Ap 894 (or: Calle Pepillo Salcedo 18, Altos) Sto Domingo **E:** cristal-internacional@hotmail.com – **4)** Avenida Pedro A Rivera km 1.5 (or: Ap 55), La Vega ☎+1 809 573 2722 ☎+1 809 573 6200 **W:** www.rsan-tamaria.com - **FM:** 97.9MHz – **5)** Av Prolongación México, esquina Clarín,Sto Domingo – **6)** Santiago ☎+1 809 724 9050 - **FM:** 90.5MHz – **7)** Ap 869 (or: Dr.Tejada Florentino N° 8), Sto Domingo **W:** www.certvdominicana.com **E:** certvdominicana@gmail.com ☎+1 809 689 2121 ▤+1 809 688 6208 – **8)** Duarte N° 1 (or: Ap 52), Montecristi ☎+1 809 579 2421 **W:** www.microondasnacionales.com/radiomontecristi.htm **E:** montecristi@microondasnacionales.com **FM:** 97.1MHz – **9)** Av 27 de Febrero, Edificio Kira, Sto Domingo ☎+1 809 381 0531 ▤+1 809 381 0542 **W:** www.radiouniversalfm.com – **FM:** 98.1MHz – **10)** Calle Manuela Diez 67 del Barrio María Auxiliadora, (or Apartado Postal 4848), Sto Domingo **W:** www.radiojuventusdonbosco.com **E:** radioju-ventusdonbosco@yahoo.com ☎+1 809 538 4647 – **11)** Calle Restauración N° 60 (or: Ap 2), San Ignacio de Sabaneta **E:** t.sabaneta@verizon.net.do ☎+1 809 580 2455 ▤ +1 809 580 2808 - **FM:** 92.3MHz – **12)** Calle Palo Hincado 302, Sto Domingo ☎+1 809 541 2886 ▤+1 809 541 2885 – **13)** Calle Duarte N° 49 (or: Ap 20), Valverde Mao (or: Ap 789, Santiago) few hours on weekends ☎+1 809 572 3322 – **14)** Urb Las Hortensias (or: Ap 454), Santiago ☎+1 809 543 4321 ☎+1 809 971 4308 - **FM:** 103.5MHz – **15)** Calle El Conde Esq Sánchez, Edif Copelic (or: Ap 68), Sto Domingo **W:** www.hiz730.com ☎+1 809 682 2839 – **16)** Calle Sánchez Esq Pedro F Bonó, Santiago **W:** www.misionerosdejesus.org/radiojesus.html – **17)** Calle Emilio Á Morel esq Luis Pérez, Ensanche La Fé, Sto Domingo ☎+1 809 566 5875 – **18)** Calle El Sol 51, 3a Planta, Edif Lamarche Alvarez (or: Ap 1636), Santiago ☎+1 809 583 2610 - **FM:** 97.1MHz – **19)** Calle V.M de Robiou N° 18, Constanza ☎+1 809 539 2468 – **20)** Abraham Lincoln N° 58 (or: Ap 335) , Sto Domingo ☎+1 809 566 9696 – **21)** Av Estrella Sadhalá N° 3, Plaza Alejo, 3er piso (or: Ap 282), Santiago ☎+1 809 582 5725 **FM:** 99.1MHz **W:** www.radiosantiago820.com – **22)** Edif Teleantillas, Carr Duarte km 7.5, Sto Domingo ☎+1 809 567 5400 ▤+1 809 540 4912 - **FM:** 95.7MHz – **23)** Ap 146, Puerto Plata 0930-0330 ☎+1 809 586 2974 – **24)** Calle Mella esquina Calle 27 de Febrero, Baní **W:** www.radiosalvacion.com ☎+1 809 380 3800 FM 95.7 Baní **FM:** – **25)** Av Pedro A Rivera, Edif Microondas Nacionales (or: Ap 203), La Vega: microondasnacionales.com/radiolavega.htm **E:** rlavega@microondasnacionales.com ☎+1 809 573 2872 ▤+1 809 573 2317 - **FM:** 104.9MHz – **26)** Calle 27 de Febrero Esq Agustin Cabral (or Ap 80), Valverde Mao ☎+1 809 572 3234 - **FM:** 106.7MHz – **27)** Calle Dr Delgado N° 206 (or: Ap 156), Sto Domingo ☎+1 809 565 4621 – **28)** Av 26 de Agosto N° 38,, Puerto Plata ☎+1 809 586 5051 - **FM:** 99.7MHz – **29)** Calle Mella 50, Bonao ☎+1 809 525 2235 – **30)** Calle A Reyes N° 3 (or: Ap 99,) Neyba ☎+1 809 527 0475 - **FM:** 93.7 – **31)** Calle Restauración Esq 30 de Marzo (or ap 225), Santiago ☎+1 809 582 5855 – **FM:** 92.1MHz – **32)** Av Charles Summer N° 33, Los Prados (or: Ap 928), Sto Domingo ☎+1 809 566 6125 - **FM:** 97.3MHz – **33)** Av John F Kennedy N° (altos) Puerto Plata ☎+1 809 586 8940 ▤+1 809 320 8128 - **FM:** 97.3MHz – **34)** Calle Cuba No. 46, 3ra planta, Los Pepines, (or: Apartado Postal 897) Santiago ☎+1 809 587 5892 **W:** www.radioemanuel.org **E:** radioemanuelcabina@hotmail.com **FM:** 89.1MHz – **36)** Av Imbert, Gurabito (or Ap 141), Santiago ☎+1 809 575 2174 - **FM:** 95.1MHz – **37)** Av Pablo Reyes N° 1, Dajabón ☎+1 809 579 8294 ▤+1 809 533 4518 - **FM:** 91.7MHz – **38)** E A Morel 27 (or: Ap 1322), Sto Domingo ☎+1 809 565 2008 **W:** www.radiocomercial1010.com - **FM:** 106.5MHz – **39)** Av Estrella Sadhalá, Plaza Alejo (3era planta) , Santiago ☎+1 809 583 1030 - **FM:** 92.7MHz **W:** www.megamedios.net – **40)**Calle Dr Delfilló N° 4, Los Prados, Sto Domingo **W:** www.elcaribecdn.com ☎+809 683 8100 ▤+1 809 544 4003 – **41)** Av Gregorio Luperón N° 10-A, (or: Ap 213), La Romana ☎+1 809 556 4616 ▤+1 809 550 5167 – **42)** Calle Emilio Prud'homme 17A, Azua ☎+1 809 521 3317 ▤+1 809 521 4949 - **FM:** 97.1MHz – **44)** Av 27 de Febrero N° 51 (or:Ap 201), San Francisco de Macorís ☎+1 809 588 2450 ▤+1 809 588 2818 **W:** www.hibiradio.com - **FM:** 102.3MHz – **45)** Edif Jaar, Calle

El Conde esq Espaillat, Sto Domingo ☎+1 809 221 7925 – **46)** Av Texas Esq Calle 12, Jardines Metropolitanos (or: Ap 561), Santiago ☎+1 809 581 5067 ⊒+1 809 581 6614 **W:** www.radioamistadfm.com – **FM:** 101.9MHz – **47)** Calle Mariano Soler Merino N° 19 (altos) (or: Ap 64), San Pedro de Macorís ☎+1 809 529 7111 – **48)** Calle Canada, San José de Ocoa **W:** www.radioocoa.com – **49)** Calle Narciso Minaya N° 36, Nagua ☎+1 809 584 2382 – **50A-F)** Empresas Radiofónicas SA, Ap 20339, Sto Domingo **W:** www.suprafm.com/informativo.htm **E:** f.suprafm@codetel.net.do 50A) 27 de Febrero 1, Jimaní; 50B Félix del Rosario 1, Azua;50C-D) Edificio Rodulfo Lama, Calle María Montés #24 (Ap 20339), Barahona; 50E) Duarte 1, Pedernales; 50F) Cambronal 8, Neiba – **51)** Calle Domingo Sabio N° 1 (or: Ap 10), Jarabacoa ☎+1 809 549 7439 - **FM:** 98.7MHz – **52)** Urbanización Las Hortensias, Santiago ☎+1 809 971 4302 ⊒+1 809 971 4308 **W:** www.microondasnacionales.com **E:** rhit@microondasnacionales.com **FM:** 98.3MHz – **53)** Ave 27 de Febrero # 238, Edificio Rodríguez Sandoval, 5to piso, Santo Domingo **W:** http://radiomaridominicana.org – **54)** Calle Club de Leones N° 175 (or: Ap 37), San Juan de la Maguana ☎+1 809 557 2623 – **55)** Calle Palo Incado N° 161, Sto Domingo ☎+1 809 682 2078 **W:** www.ondamusical1150.com – **56)** Calle Restauración N° 64 (or: Ap 187), Santiago ☎+1 809 582 5414 ⊒+1 809 241 5845 - **FM:** 93.1MHz – **57)** Av Máximo Gómez N° 65 (or: Ap 1372), Sto Domingo ☎+1 809 565 5366 ⊒+1 809 566 9486 - **FM:** 103.3MHz – **58)** Calle Restauración Esq 30 de Marzo (or: Ap 79), Santiago ☎+1 809 582 5855 - **FM:** 94.3MHz – **59)** Av. Independencia No. 169, San Pedro de Macorís - **FM:** 90.7MHz Sultana + 98.7 Estéreo 98 **W:** www.radiodial670am.com.do – **60)** Juan Sánchez Ramírez #40, Gazcue (or: Ap 4680), Sto Domingo (Owned and operated by the Seventh Day Adventist Church) **E:** cabina@tradioamanecer **W:** www.radioamanecer.org ☎+1 809 688 8067 – F.PI: 10kW SW – **61)** Calle 27 de Febrero N° (or: Ap 57), San Francisco de Macorís ☎+1 809 588 2734 ⊒+1 809 588 6522 **W:** www.circuitomerengue.com - **FM:** 94.7MHz – **62)** Calle Sánchez #74, Casi Esquina Calle del Sol Centro, Santiago ☎+1 809 582 6009 Calle César Dargán #26, El Vergel (Frente a la Plaza Criolla), Sto Domingo ☎+1 809 472 4150 (or: P O Box 2908, Paterson, NJ 07509-2908, USA) **E:** radiovision@sprintmail.com – **63)** Ave 27 de Febrero #265, Suite 202, Piantini, Santo Domingo– **64)** Av Circunvalación Norte, Puerto Plata – **65)** Av 25 de Febrero 144, Ensanche Las Américas, P3 Hostal Hostal Puerto Rico, Sto Domingo ☎+1 809 592 ⊒+1 809 592 5311 **W:** www.radiorevelacionenamerica.org.do – **66)** Ap 264 (or: Calle San Francisco 50), San Francisco de Macorís – **67)** Calle Santa Rosa N° 18 (or:Ap 151), La Romana ☎+1 809 556 2891 **FM:** 107.5MHz – **68)** Av.27 de Febrero N° 265, Ofc.201, Ensanche, Sto Domingo ☎+1 809 543 1957 ⊒+1 809 685 2476 **W:** www.cima100fm.com cima100.htm – **69)** Sánchez esq Mella, Baní **FM:** 96.7MHz – **70)** Calle Libertad N° 15, Bonao ☎+1 809 525 5858 **FM:** 88.7MHz Latina 88 – **71)** C/ Proecto No 11, Las Colinas, Montecristi ☎ +1-809-579 2421 – **72)** Padre Billini esq Jaime Mota, Barahona– **73)** Conde esq 19 Marzo, Edif El Palacio, Sto Domingo ☎ +1 809 682 6526 **W:** www.la2dehiz.com – **75)** Juan Rodríguez 76-A, La Vega ☎+1 809 573 2833 ⊒+1 809 242 1258 **W:** www.radioreal.net – **76)** C/ Duarte Esq Mella, Sto Domingo, Santiago ☎ +1-809-525 5535 – **85)** Calle Sánchez N° 45 (or: Ap 003), Río San Juan ☎+1 809 589 2595 - **FM:** 105.9MHz – **86)** Calle Sánchez N° 48 , Cotuí ☎+1 809 585 2456 - **FM:** 97.3MHz – **87)** Diócesis de la Alta Gracia, Calle General Santana 65, Higüey **W:** http://lvozdelaaltagracia.com – **88)** Av 27 de Febrero No 265, Suite 202, Piantini, Santo Domingo – **89)** Calle Santomé N° 27 (or Ap 88), San Juan de la Maguana **FM:** 90.3MHz – **90)** Ave Sarasota esquina Winston Churchill, Plaza Universitaria, Local 9B, Sto Domingo **E:** carlos@radioimpacto.org ⊒ +1 809 508 2754 – **91)** Calle Mella N° 90 (altos) (or: Ap 2), Salcedo ☎+1 809 577 2576 **W:** www.radioutilfm.com - **FM:** 106.5MHz – **92)** Calle Felipe de Castro Esq Santana N° 4, Hato Mayor del Rey ☎+1 809 553 2222 – **93)** Carr.salida a Nagua al lado del Hospital del Seguro Social, San Francisco de Macorís – **95)** Pres Henriquez 53, Dajabón - **FM:** 105.1MHz – **96)** Plaza Alejo, Av.Estrella Sadhala, Santiago **FM:** Comando 88 – **97)** Calle Altagracia N° 70, Higüey – **98)** Ap 2217 (or: Calle Paseo de los Aviadores No. 3, Ensanche Miraflores), Santo Domingo **W:** www.radioven.com **E:** radioven@codetel.net.do – **99)** Av Malecón, Samaná ☎+1 809 538 2530 – **100)** Sto Domingo – **101)** Ap 1091, Santiago ☎+1 809 971 9084 **W:** www.radiopoder.com/index.html – **102)** Calle Duarte Esq Mella, Edif Fantino (2da planta), Bonao ☎+1 829 525 2221 ⊒+1 829 525 5444 – **103)** Circuito Telesonido, Mella N° 177, San Pedro de Macorís – **104)** San Cristóbal – **107)** Calle Proyecto, Neiba ☎+1 809

549 3362 ⊒+1 809 549 3362 – **108)** C/Paseo de Los Periodistas N° 52, Sto Domingo ☎+1 809 689 9772 – **109)** Barahona – **110)** Av Pasteur N° 204, Sto Domingo ☎+1 809 687 9161 – **111)** C/La Lira N° 18, Ens. Vergel, Elias Piña ☎+1 809 535 8290 – **112)** Calle General López Esq 16 de Agosto, Santiago **FM:** 91.3MHz – **113)** Luís Amiama Tió # 105, Arroyo Hondo, Santo Domingo ☎+1 809 566 1707 **W:** www.radioeternidad.org **114)** Calle René del Risco Bermúdez No.17, Villa Progreso, San Pedro de Macorís ☎+1 809 246 1680 **W:** www.radiosenda.net **E:** radiosenda@hotmail.com – **115)** Av San Cristóbal esq L Pérez García, Santiago – **116)** C/Fantino Falco No 47, 1er piso, Plaza Naco, Santo Domingo – **117)** Universidad Autonoma de Santo Domingo, Alma Mater, Santo Domingo **W:** www.uasd.edu.do – **118)** Santo Domingo (c/o WRMI, 175 Fontainbleau Blvd, Suite 1N4, Miami, FL 33172, USA) – **119)** Avenida Las Américas Esquina España, Santo Domingo

FM in Sto Domingo (MHz):

88.1 Primera FM – 88.5 Estudio Rock – 88.9 Escape – 89.3 Neon – 89.7 Renuevo FM – 90.1 Fuego 90 – 90.5 Estrella 90 – 90.9 Alianza Francesa - 91.3 La 91 FM – 91.7 La Roka FM – 92.1 Hits 92 – 92.5 CDN La Radio - 92.9 Pura Vida 92.9 – 93.3 Independencia FM – 93.7 Latidos FM – 94.1 Fidelity – 94.5 KQ-94.5 FM – 94.9 Kiss 95 – 95.3 Radeco – 95.7 La Nota Diferente – 96.1 Quisqueya FM (CERTV) – 96.5 Ritmo 96 – 96.9 Espacio 96.9 FM – 97.3 Disney – 97.7 R. Higo/Emociones FM – 98.1 Universal – 98.5 Rumba FM – 98.9 Dominicana FM (CERTV) – 99.3 Sonido Suave – 99.7 Listín – 100.1 Antena 100 – 100.5 Cima – 100.9 Super Q – 101.3 Z-101 – 101.7 Supra FM – 102.1 La X 102 – 102.5 Tentación – 102.9 Raíces FM - 103.3 Milenium FM – 103.7 Power FM – 104.1 R. Cordillera – 104.5 La Super Potente – 104.5 Mortal FM - 105.3 ABC – 105.7 Fiesta FM – 106.1 Disco – 106.5 Zol 106 – 106.9 LV de las FF AA – 107.3 Cadena Espacial – 107.7 Romántica R. Millón

EASTER ISLAND (Chile)

L.T: UTC -6h (10 Oct 10-13 Mar 11, 16 Oct 11-11 Mar 12: -5h) — **Pop:** 3,500 — **Pr.L:** Spanish, Rapanui — **E.C:** 50Hz, 220V — **ITU:** PAQ

FM	MHz	kW	Station	FM	MHz	kW	Station
1)	88.9	1	R. Manukena	2)	104.3	1	R. Activa

Addresses and other information
1) La Misma Municipalidad de Isla de Pascua, Calle Atamu Tekena, Hangaroa. Correo Isla de Pascua, Chile ☎+5632 255 1245 **LP:** Dir Responsable: Nelson Ramon Zapata Orellana. Volunteer operated community radio 24h. – **2)** 24h satellite relay from Santiago. **W:** www.radioactiva.cl

ECUADOR

L.T: UTC -5h — **Pop:** 13.9 million — **Pr.L:** Spanish, Quichua — **E.C:** 60Hz, 110/127 V — **ITU:** EQA

SUPERINTENDENCIA DE TELECOMUNICACIONES DEL ECUADOR

⊠ 9 de Octubre 1645 y Berlin, Quito ☎+593 22 2221500 **W:** www.supertel.gov.ec **E:** info@supertel.gov.ec

Hrs of tr. 24h unless stated below. Call HC-

MW	Call	kHz	kW	Station, location, hr. of tr.
PI01)	DC1	530	1	R. Iris/530 AM "LV de la Comunidad", Quito: 1000-0500
GU01)	FA2	540	25	R. Tropicana "Canal 540", Guayaquil: 1100-0600
PI02)	GM1	550	50	R. Reloj "5-50", Quito: 1100-0400
GU01)	RN2	560	25	C. R. E. Satelital, Guayaquil
PI03)	CE1	570	10	R. El Sol, Quito
GU02)	PC2	580	10	R. Uno, Guayaquil
PI04)	SP1	590	10	R. Carrousel, Quito: 1100-0200
GU03)	XY2	600	50	R. Nal. del Ecuador, Guayaquil: 1100-0400
PI05)	MJ1	610	10	R. Caravana AM, Quito
LO01)	XY3	620	50	R. Nal. del Ecuador, Loja: 1100-0400
LR01)	HA2	630	10	Ondas Quevedeñas, Quevedo
GU04)		640		R. Morena AM, Guayaquil
PI06)	XY1	640	50	R. Nal. del Ecuador, Quito: 1100-0400
MA01)	FD4	650	5	R. Visión Manta, Manta: 0900-0500
GU05)	LG2	660	30	R. Carrousel, Guayaquil: 1200-0500
PI07)	FF1	670	12/5	R. Jesús del Gran Poder, Quito: 0945-0500
GU06)	VP2	680	25/12	R. Atalaya, Guayaquil: W 0900-0500, Sun 1000-0300
MA02)	FA4	690	5	Sucre Portoviejo, Portoviejo
PI08)	JB1	°690	50d	LV de los Andes, Quito: 1030-0500
GU07)	RS2	700	50	Sucre Guayaquil, Guayaquil
CR01)	ER5	710	8	Escuelas Radiofónicas Populares, Riobamba: 0900-0500

MW	Call	kHz	kW	Station, location, hr. of tr.
EO01)	UE3	720	10	R. Única, Machala
LO02)	MO3	720	5	R. Matovelle "HCM-3", Loja: 1000-0200
MA03)	GB4	v720	10	LV de Portoviejo, Portoviejo: 1000-0400
PI09)	IC1	720	5	R. Municipal, Quito
GU08)	MG2	730	10	R. Guayaquil, Guayaquil
MA04)	SE4	v740	10	R. Libertad, Chone: 1100-0600
PI15)	GC1	740	10	R. Melodía "Canal 7-40", Quito: 1100-0400
GU09)	RC2	750	30	Caravana AM, Guayaquil
PI10)	QR1	°760	25	R. Quito "LV de la Capital", Quito
GU10)	MF2	770	25/12	R. El Telégrafo, Guayaquil: 1000-0500
PI20)	CM1	780	10/2	R. Colón AM, Quito
PI12)		790		R. Paraíso, Maldonado
IM01)		790		Su Radio 790 AM, Otavalo
GU05)	ML2	800	25	K 800, Guayaquil
PI13)	FB1	800	5	R. Sensación 800, Guayaquil: 1000-0300
GU11)	VT2	810	5	R. Atalaya, El Milagro: 2300-0300
TU01)		810		Sucre Ambato, Ambato
CA01)	VI5	820	5	LV de Ingapirca, Cañar: 0900-0330
MA06)	RF4	820	1	Canal Manabita, Portoviejo
PI54)	UP1	820	25	R. Unión, Quito: 1100-0100
CR02)	RP5	830	4.5	R. Promoción, Riobamba: 0900-1400, 2200-0200
GU12)	RM2	830	25	R. Huancavilca, Guayaquil
MA07)	EM4	840	1	R. Costa Azul, Portoviejo: 1100-0500
PI16)	PN1	840	50	R. Vigía "LV del Tránsito Nacional", Quito: 1100-0400
GU13)	VS2	v850	20/12	R. San Francisco, Guayaquil: M-F 0945-0500, Sat -0300, Sun - 0100
PI17)	PC1	v860	10	R. Positiva, Quito: 1015-0400
GU14)	NY2	870	20	R. Cristal "RCQ", Guayaquil: 1000-0600
TU02)	GS6	870	1	R. Píllaro, Píllaro: 1100-0400
PI18)	RP1	880	50/40	R. Católica Nacional, Quito: 1000-0200
CR03)	TL5	890	1	Ondas del Chimborazo, Riobamba: 1100-0100
EO02)	RS3	v890	25/20	R. Superior, Machala: 0900-0500
AZ01)	RR5	900	1	R. Reloj, Cuenca: 1100-0200
MA08)	OF4	v900	5	R. Chone, Chone: 1100-0400
PI19)	VA1	v900	10	Sucre Quito, Quito: 1100-0400
CR04)	GE5	910	5	R. Mundial, Riobamba: 1000-0400
GU15)	BO2	910	2	Nueva R. Colón "Espectáculo",Guayaquil
EO03)	RU3	920	10	CRO - Compañía Radiofónica Orense, Machala: 0930-0430
PI40)	AB1	920	1	R. Democrácia "La Cariñosa", Quito: 1000-0400
GU12)	VI2	v930	5	Canal Tropical, Guayaquil
TU03)	BA6	930	5	R. Ambato, Ambato
AZ21)		940		R. Austral del Ecuador, Cuenca
PI21)	BZ1	940	5	R. Dif. de la Casa de la Cultura Ecuatoriana, Quito: -0200
CR05)	UE5	950	3	R. Colta "LV de la Asociación", Colta: 0900-0200
GU17)	DE2	950	10	GRD R. Internacional, Guayaquil
IM02)		°950		Chaskis del Norte, Ibarra
AZ02)	SA5	960	1	Sono Onda Internacional, Cuenca: W0925-0430, Sun1200-0400
PI22)	NC1	v960	1	R. Cosmopolita, Quito: 1100-0500
TU04)	JX6	960	1	LV del Santuario, Baños: 1000-0300
SD01)	OT1	965	10	R. Católica Nacional, Sto Domingo de los Colorados (r: 880)
GU18)	AW2	970	20	R. Católica Nal. del Ecuador, Guayaquil: 1000-0500
IM03)	MB1	970	1	R. Imperio, Ibarra: 1030-0300
CR06)	JI5	980	1	R. El Prado, Riobamba
LO11)	CL3	980	5	R. Cariamanga, Cariamanga: 1000-0400
PI24)	GH1	990	25	R. Tarquí, Quito: 1015-0400
GU19)	EW2	990	15	Frecuencia Mil, Guayaquil
LO03)	NT3	*1000	1	Dinamita Mil AM, Catamayo: 1030-2330
AZ04)	RV5	1010	2.5	R. Visión, Cuenca
GU20)	RZ2	1010	3	R. Amiga, Guayaquil: 1100-0500
TU05)	NR6	1010	15	TSB R. Líder, Ambato: 0945-0300
BO01)	CR6	1020	5/3	R. Surcos, Guaranda: 1030-0100
EO04)	GO3	*1020	3	Canal Estelar, Santa Rosa
PI26)	HR1	1020	5	RTU (Radio y Televisión Unidas), Quito
GU21)	RF2	1030	5	R. Punto 1030/Ecuantena, Guayaquil: 1100-0500
AZ05)	EV5	v1040	10/5	R. Splendit, Cuenca
PI27)	CW1	1040	3	LV del Valle, Machachi: 1130-0100
TU06)	GB6	1040	3	R. Colosal, Ambato: 0930-0500
GU49)	RO2	1050	5	R. Águila, Guayaquil: 1030-0400
IM04)	IM1	1050	5/3	LV de Imbabura, Ibarra: 1000-0100
CP01)	MG6	1060	5	R. Ecos del Pueblo, Saquisilí: 1045-0330
EO19)		1060		R. Fiesta, Machala
LR02)		1060		R. Richi, El Empalme
AZ06)	CJ5	1070	5	R. LV de Tomebamba, Cuenca: 1000-0500
PI28)	VP1	1070	1	R. Libertad, Quito
SD02)	RS1	1070	5	R. Lubakán, Santo Domingo de los Colorados: 0950-0200 (Sun -2300)

MW	Call	kHz	kW	Station, location, hr. of tr.
CP02)	BH6	1080	10	R. Latacunga, Latacunga: 0900-0230
GU22)	KD2	1080	10	Sistema 2, Guayaquil
MA11)	AB4	v1080	1	R. Contacto, Manta: 0900-0300
PI30)	VI1	v1090	5	R. Irfeyal "Fe y Alegría", Quito
CP03)	GR6	1100	5/2	R. Novedades, Latacunga: 1000-0500
NA02)	LE7	°1100	1.5	R. Oriental, Tena: 0900-0400
AZ07)	JC5	1110	5	R. Ondas Azuayas, Cuenca: 1100-0200
PI31)	JR1	1110	10	R. Clásica, Quito
TU07)	RP6	v1110	5	R. Pelileo, Pelileo: 1100-0400
CC01)	EB1	1120	2	Canal 1120, San Gabriel
GU24)	FV2	1120	5	Estación Intercontinental, Guayaquil: 1100-0500
PA02)	AS7	1120	3	R. Variedades del Puyo, El Puyo
SD05)	LE1	*1120	10	R. Dif. Marañon, Sto Domingo de los Colorados
EO20)		1130		Romántica AM, Machala
IM05)	RD1	1130	5/3	R. Punto, Ibarra: 1000-0400
LR03)		1130		R. Sibimbe AM, Ventanas
TU08)	PV6	°1130	5	R. Centro, Ambato
AZ08)	AZ5	1140	1	R. Alfa Musical, Cuenca: 1100-0600
GU25)	FB2	1140	1.5	R. Cóndor, Guayaquil: 1130-0500
MA12)	MF4	1140	4	R. Rumbos, Portoviejo
PI33)	IR1	v1140	5	Raíz 11-40, Quito: 1130-0400
CR07)	GB5	v1150	10	LV de Riobamba "Antena 1", Riobamba
LO06)	AV3	1150	10	R. Luz y Vida, Loja: 1000-0330, Sat -0400, Sun -0700
CA02)		1160		LV del Pueblo, Azogues
CP04)	UR6	1160	1	R. Runatacuyaj "LV de la Asociación", Latacunga: W 1000-0200
EO05)	VR3	1160	2	R. Vía, Machala
MA13)	WD4	1160	1	R. Cenit, Portoviejo: 1200-0500
PI34)	CP1	1160	5	Super Auténtica, La Radio 11-60, Quito
CR08)	JV5	1170	5	R. Central, Riobamba: 0900-0500
EO06)		*1170	5	R. Trébol AM, Zaruma: 1130-0200
GU26)	RV2	1170	5	R. Filadelfia, Guayaquil
AZ09)	DP5	1180	4	R. Cuenca "LV de los 4 Ríos", Cuenca: 1200-0900
CC02)	RV1	°1180	1.2	R. Familiar, Julio Andrade: 1100-1800, 2200-0200
MA26)		1180		LV del Volante, Portoviejo
PI35)	LR1	1180	12.5	Nueva Em. Central, Quito: 1100-0400
CP05)	RF6	1190	1	R. El Sol, Pujilí: 1100-0200
GU22)	DE2	1190	2	Estudio Universidad Católica, Guayaquil: 1100-0500
AZ10)	RM5	1200	5	R. El Mercurio, Cuenca: 0900-0500
EO07)		1200		R. U Cadena Sur, Sta Rosa
LR04)	RE2	1200	5	LV del Trópico, Quevedo: 1000-0400
PI36)	CS1	v1200	5	R. Super K, La Líder, Sangolquí: 1000-0100
GU27)	BJ2	1210	20	R. El Mundo, Guayaquil: 1200-0300, Sat -0100, Sun -0400
LO07)	VC3	°1210	10	R. Centinela del Sur "CDS", Loja: 1100-0300
TU09)	JM6	1210	3	R. Sira, Ambato: 1000-0700
BO03)	EB6	1220	3/5	Ecos de Bolívar, Guaranda: 0930-0130
PI32)	AP1	1220	10	R. Marañon, Quito: 1300-0200
AZ11)	MV5	1230	3	R. Popular, Cuenca: 1045-0500
CP06)	RL6	°1230	1	LV de Saquisilí y Libertador, Saquisilí: 1045-0300
ES02)	FG4	°1230	5	Sucre Esmeraldas, Esmeraldas
GU48)	FV2	1230	15	R. Galáctica, Guayaquil: 1000-0400
IM06)	RI1	°1230	3	CRI-Centro Radiofónico de Imbabura, Ibarra: 1100-0300
CR09)	LA5	°1240		R. Musical, Riobamba (r. Alianza Cristiana)
EO08)	RF3	1240	5	R. Fenix, Zaruma: 1000-0100
PI37)	PA1	1240	1	R. Metropolitana, Quito: 1200-0300
CC03)	EM1	1250	10	Ondas Carchenses, Tulcán: 1000-0400
GU28)	HB2	1250	10	R. Tricolor, Guayaquil
SD03)	MY1	1250	3	LV del Triunfo, Sto Domingo de los Colorados: 1000-0500
AZ12)	PB5	1260	2	R. Contacto XG, Cuenca: 1100-0300
EO09)	RB3	1260	1	R. Benemérita, Sta Rosa: 1030-0100
PI39)	MO1	1260	10	LV del Santuario del Quinche, Quito: 1100-0300
TU10)	RO6	1260	3	R. Calidad, Ambato: 0930-0600
GU22)	UM2	1270	15	R. Universal, Guayaquil
MA15)	LD4	1270	3	R. Junín, Junín: 1100-0500
CR10)	NW5	1280	1	R. Canal Tropical, Riobamba: 1100-0100
MA16)	IN4	1280	1	LV del Sur de Manabí, Jipijapa: 1000-0500
AZ13)	JA5	1290	3	LV del Río Tarquí, Cuenca: 0900-0300
CP08)	VM6	1290	0.5	R. Once de Noviembre, Latacunga: 1200-0400
GU29)	OF2	1290	1	Canal Milagreño, El Milagro
IM07)	NS1	1290	1	R. Popular, Atuntaqui: 1100-0300
BO04)		1300		R. La Paz, Guaranda
GU30)	DC2	1300	5	R. Cenit, Guayaquil: 1200-0400
SD04)	RV1	1300	5	R. Festival, Sto Domingo de los Colorados: 0930-0300

MW	Call	kHz	kW	Station, location, hr. of tr.
SU02)	RS7	1300	2/1	R. Sucumbios, Nueva Loja: 1100-2400
CA03)	CI5	1310	3	T. V. O. "El Poder Mágico de la Fé", Biblián
CR20)	AI5	1310	0.5	Eco de los Andes, Cumanda: 1000-0200
EO11)	CP3	v1310	1	LV de El Oro, Pasaje
PI58)	GB1	°1310	20	R. Nal. Espejo, Quito
LR05)	FR2	1320	1	R. Guayaquil, Babahoyo: 1030-0300
MA24)	VO4	°1320	1	R. Stéreo Carrizal, Calceta: 1100-0300
TU11)	JD6	1320	10	R. Continental, Ambato: 0930-0400
AZ14)	LW5	1330	2	R. Visión Cristiana, Cuenca
CC04)	OV1	*1330	3	GRC AM-Grupo Radial Carisma, El Angel
EO12)	RV3	1330	5	Nacional El Oro, Machala: 1000-0600
GU31)		1330		Lomas Stereo 2000, Guayaquil
PI42)		v1330	3	R. Visión Cristiana, Quito
ES03)		1340		LV de su amigo "Esté Musical", Esmeraldas
L008)		1340	1	Ondas de Esperanza, Loja: 1100-0300
TU12)	RT6	1340	5	R. Paz y Bien, Ambato: 0930-0130
AZ15)	SF5	1350	2/1	LV de San Fernando, San Fernando: 1000-0300
GU47)	VP2	1350	2	Teleradio 13-50 AM Digital , Guayaquil
CR12)	RJ5	1360	1	R. América, Riobamba: 1100-0300
EO13)	HG3	1360	5	R. Jerusalem AM, Machala
PI44)	MT	v1360	3	Oyambaro AM, Tumbaco: 1000-0300
CA04)		v1370		R. El Rocio, Biblián
GU32)	VO2	1370	5	LV del Milagro, El Milagro
IM08)	JS1	1370	5	Ecos Andinos, Pimampiro
L009)	ER3	1370	5	R. Progreso, Loja: 1000-0315
PA03)	RP7	*1370	2	R. Pastaza, El Puyo: 1100-0100 (Sun 1200-2300)
EO14)	OA3	1380	1	La Mejor, Balsas
PI45)	CV1	1380	5	R. Cristal "RCQ", Quito: 0830-0300
TU13)		v1380	5	R. Mera, Ambato
CR13)	EA5	1390	5	R. Tropicana "Canal 13-90", Cuenca: 1200-0300
CR13)	DN5	*1390	3	R. Atenas, Riobamba: 0900-0500
ES04)	HE4	1390	1	LV de Esmeraldas, Esmeraldas
IM09)	IE1	1390	1.5	R. Uno, Urcuquí
CP09)		1400		Impacto 1400 AM, Latacunga
GU02)	)FL2	v1400	10	R. Z Uno, Guayaquil
AZ17)	GC5	1410	1	R. Centro Gualaceo, Gualaceo
CR14)		1410	1	Ondas Cisnerinas, Riobamba: 2000-2300
ES05)	FR4	1410	1	LV de Quinindé, Quinindé
GU33)	CQ2	1410	1	R. Net AM, El Milagro
PI59)	EC1	1410	1	R. El Tiempo "Em.del Amor", Quito
CP10)	MA6	1420	1	R. Alternativa, Salcedo: 1130-0300
EO15)		1420		Corazón AM, Machala
IM10)	RN1	°1420	3	R. Bahá'í, Otavalo: 0900-1500, 1930-0300
NA06)	VN7	1420		LV del Napo, Tena
BO05)	JC6	1430	5	R. Guaranda, Guaranda: 1100-0500
GU34)	MB2	1430	10	R. Federal, Virgen de Fátima
L010)	CV3	1430	5	Ondas del Zamora, Canal Juvenil, Loja: 1130-0330
PI46)	GF1	1430	3.5	R. Futura 14-30, Quito: 1300-0200
CA05)	OV5	1440	2.8	Ondas del Volante, Azogues: 1000-0400
CP11)	AQ6	1440	3/5	R. Fenix, Latacunga
EO10)		1440		Mi Radio AM, Machala
ES06)	DY4	1440	2.5	R. Iris, Esmeraldas; 1000-0400
IM11)	DF1	v1440	5	R. Panorama, Ibarra: 1030-0300
CR15)	SC5	1440	10	R. Calidad, Riobamba: 0800-0400
GU35)	DR	v1450	1	R. Minutera, Guayaquil
SE01)	SE2	1450	1	R. Santa Elena, Santa Elena: 2200-0200
PI47)	SC1	1450	1	AS La Radio, Tabacundo
CP12)	IC6	1460	5	R. Nuevos Horizontes, Latacunga: 1000-0200
MS04)	AA7	1460	5	LV de Gualaquiza, Gualaquiza: 1000-0300
GU37)	LD2	1470	1.5	R. Ecos de Naranjito, Naranjito
PI48)	JC1	1470	5	Ecos de Cayambe, Cayambe (occ. rel: Colón FM Guayaquil 92.9MHz)
CR16)	WP5	1480	3	R. Atlántida, Alausí: 1000-0400
CP13)	CY6	1480	5	R. Popular de La Maná, La Maná
EO16)	BS3	v1480	3	Oro Radio AM, Machala
IM12)	MC1	1480	1	R. Municipal, Cotacachi
MA20)	JV4	1480	3	R. LV de Jipijapa, Jipijapa: 1100-0400
CA06)	SM5	1490	5	R. Santa María, Azogues: 0930-0330
ES07)	AE4	1490	2.5	R. Unión, Esmeraldas: 1000-0300
GU38)	VY2	1490	1	La R. Dinámica, Guayaquil
PI60)		1490		Poderosa 14-90, Quito
TU14)	AI6	1490	3	R. Moderna, Píllaro: 1300-0200
IM13)	RO1	1500	1	R. Otavalo, Otavalo: 1200-0300
LR09)	HG2	1500	5	LV del Río Vinces, Vinces: 1100-0500
MA21)	AD4	*1500	5	R. Satélite, El Carmen: 1000-0500
BO07)	RY6	1510	1	R. Runacunapac Yachana "R. El Saber del Hombre", Simiátug
CA07)	RC5	1510	2	R. Punto C 1510 AM, Cañar: 1030-0400
GU39)	HD2	°1510	0.5	Inst. Oceanográfico de la Armada, Guayaquil: time signals 24h
LO13)	UC3	*1510	10	R. Unión Calvense, Cariamanga

MW	Call	kHz	kW	Station, location, hr. of tr.
PI56)		1510	5	R. Monumental, Quito: 1300-0400
SU03)	JV7	1510	3	R. Ecos del Oriente, Lago Agrio: 1030-0100
TU19)		1510		R. Net, Ambato
CR18)	RI5	1520	2.5	LV de Guamote, Guamote
GU40)	RN2	1520	1	LV de Naranjal, El Naranjal
IM14)	TI1	1520	1	R. Ibarra, Ibarra: 1000-0400
CA08)	CC5	1530	5	Ondas Cañaris AM, R. Universitaria Católica, Azogues
CR19)	VP5	v1530	3	R. LV de Pallatanga, Pallatanga: 1100-0300
SE02)	MP2	1530	5	LV de la Península, La Libertad: 1000-0300
TU15)	MZ6	1530	1	R. Deportes 15-30, Pelileo: 1130-0230
CP14)	MH	v1540	0.5	Cotopaxi Digital, Latacunga: 1000-0400
EO18)		1540		R. Flecha AM, Machala
LR10)	FM2	1540	3	R. Cristal de Ventanas, Babahoyo
MS05)	VB7	°1540	0.25	LV del Upano, Macas: 1030-0300
PI49)	DP1	1540	1	R. Caracol, Quito: 1000-0400
AZ18)	AD5	1550	5	LV de Chaguarurco, Santa Isabel: 1200-0300, Sun 1100-0300
GU42)	AD2	1550	2	LV del Triunfo, El Triunfo: 1100-0400
TU16)	EI6	1550	2	R. Montalvo, Ambato: 1130-0300
EO17)	TR3	1560	2	LV del Guabo, El Guabo: 1000-1300, 2300-0400
GU43)	CS2	1560	2	R. Sideral, Daule: 1300-0500
IM15)	ZD1	1560	1.5	Ecos Culturales de Urcuquí, Urcuquí
MA23)		1570	1	R. LV Espíritu Santo de Dios, Manta: 1100-0100
PI51)	PG1	1570	10	R. Nucanchic, Maldonado
TU20)		1570	0.5	Ondas Quereñas, Quero: 1100-0300
AZ19)	TP5	v1580	3	Ecos del Portete, Girón: 1200-0330
ES09)	VA4	1580	5	Estación de la Alegría, Esmeraldas
LO14)	AB3	*1580	0.25	Ondas de Paltas, Catacocha
PI52)	LF1	1580	1	Ecos de Orellana, Machach: 1030-0230
SE03)	AS2	1590	0.25	R. Record, La Libertad
PI53)	RZ1	1590	1	R. Mensaje, Cayambe: 1000-1400, 2130-0130
TU17)	QT6	1590	1	R. Panamericana, Quero: 1000-0200 (Sun -2400)
BO09)		v1600		Ondas de Caluma "R.del Pueblo", Caluma
PI57)		1600	5	R. Ilusión 1600 AM, Puembo: 0900-0500

° = also SW, * = inactive, (r) = repeater, v = varying fq.

SW	Call	kHz	kW	Name and h of tr
PI08)	JB1	3220	8	LV de los Andes/TWR, Quito
NA06)	VN7	3280	2.5	LV del Napo, Tena: 0900-1115, 1300-1400, 2200-0300 Prgrs: R. María
IM06)		*3380	1	Centro Radiofónico de Imbabura, Ibarra
NA02)	LE7	*4781	3	R. Oriental, Tena: irr
LO15)	AX3	4815	1	R. Buen Pastor, Saraguro 1000-1600, 2100-0355 (occ. rel. of R. Internacional/LV de los Andes)
CP06)	RL6	*4900	1	LV de Saquisilí, Saquisilí: irr
IM16)		4910		R. Chaskis, Otavalo (rep. on 4909.3kHz)
PI10)	QR1	*4919	12	R. Quito "LV de la Capital", Quito: irr
MS05)	VB7	°6000		LV del Upano, Lago Agrio, Sucumbíos: irr
PI08)		6050	10	HCJB, Quito 1030-0503

Stns with a (*) are reported to be inactive, but may occasionally be reactivated for variable periods of time.

Province-abbreviations: AZ=Azuay BO=Bolívar CA=Cañar CC=Carchi CP=Cotopaxi CR=Chimborazo EO=El Oro ES=Esmeraldas GU=Guayas IM=Imbabura LO=Loja LR=Los Ríos MA=Manabí MS=Morona Santiago, NA=Napo PA=Pastaza PI=Pichincha SD=Santo Domingo de los Tsáchilas SE=Santa Elena SU=Sucumbíos TU=Tungurahua ZC=Zamora Chinchipe **N.B.:** These abbreviations are not recognized by the Ecuadorian Post Office. Letters should carry the full name.
Addresses and other information:
AZ00) AZUAY
AZ01) Bolívar 368, Cuenca – **AZ02)** Av.Remigio Crespo y Calle La Libertad, Cuenca – **AZ04)** Cas 198, Cuenca – **AZ05)** Cas 01-01-1352, Cuenca - **FM:** 90.5MHz 92.5MHz – **AZ06)** Cas 01-01-0493, Cuenca **W:** www.tomebamba.satnet.net **E:** tomebamba@cue.satnet.net - **FM:** 94.9MHz 102.1MHz – **AZ07)** Cas 01-01-4980 (or: Av Héroes de Verdeloma 9-15), Cuenca ☎+593 7 823911 📠+593 7 839067 **E:** oazuayas@cue.satnet.net - **FM:** 93.7MHz Sunny – **AZ08)** Simon Bolívar 226, Cuenca – **AZ09)** Bomboiza 1-83, entre Loja-Pastaza, Cuenca – **AZ10)** Av.de las Américas, Edif.Mercurio, Cuenca – **AZ11)** La Gloria de Nanuncay, Av.Loja 2408, Cuenca – **AZ12)** J Dávila y C Merchán, Cuenca – **AZ13)** Manuel Vega 653 y Presidente Córdova, Cuenca – **AZ14)** Edif.Alfa, P4, Gran Colombia 739 y A Borrero, Cuenca – **AZ15)** Av José María Quito y Santiago de San Fernando, San Fernando ☎+593 7 279187 – **AZ16)** Cas 830 (or: Pumapungo 5-50), Cuenca – **AZ17)** Gran Colombia y 9 de Octubre 3102, Frente al Parque Central, Gualaceo – **AZ18)** Cas 01-01-46 (or: Calle Bolívar 7-64), Aperado (or: Calle 24 de Mayo y Abdon Calderón, Cuenca) **E:** chaguarurco60@hotmail.com – **AZ19)** Antonio Flor 6-57, Girón – **AZ21)** J Roldos 480, Edif El Consorcio, Cuenca.

B000) BOLÍVAR
BO01) Johnson City 204 y Sucre, Parraquia San Vicente, Guaranda - **FM:** 97.3MHz – **BO03)** 10 de Agosto 612, Guaranda - **FM:** 93.9MHz – **BO04)** G Moreno y 7 de Mayo, Guaranda – **BO05)** Federico Paez, Frente al Parque Cen, (or: Cas 86), Guaranda – **BO07)** Simiátug – **BO09)** Av La Naranja 169, Atras-Coliseo, Caluma.

CA00) CAÑAR
CA01) Av Ingapirca, Cdla El Vergel, Cañar (or: Cas 01-01-0447, Cuenca) **Quichua:** 0900-1300 - **FM:** 94.3MHz – **CA02)** General Vintimilla 1-10 y Oriente, Azogues – **CA03)** Mariscal Sucre 722 y B Ochoa, Biblián (or: Cas 729, Azogues) – **CA04)** Calle Mariscal Sucre 202 y Tarquí, Biblián – **CA05)** Bolivar y Azuay, Azogues ☎+593 27 2240274 ☎+593 27 2240898 – **CA06)** Cas 03-01-730, Azogues ☎+593 27 2240616 🖷+593 27 2243247 **W:** www.radiosantamaria.com **E:** stamaria@satnet.net – **CA07)** Bolívar y Borrero (Junto Parque Central), Cañar -0415 – **CA08)** Calle Rivera 613, Azogues.

CC00) CARCHI
CC01) Atahualpa 166 y Aristizava, San Gabriel – **CC02)** Calle 13 de Abril, Convento parroquial, Julio Andrade – **CC03)** Olmedo 52-025 y Ayacucho (or: Cas 30), Tulcán – **CC04)** Olmedo s/n y Bolívar, El Angel ☎+593 26 2977075

CP00) COTOPAXI
CP01) Imbabura 2333 y 9 de Octubre, Saquisilí – **CP02)** Cas 05-01-392 (or: Calle Quito 14-56, Pasaje La Catedral), Latacunga ☎+593 3 810287 🖷+593 3 802329 **E:** latacunga@andinanet.net - **FM:** 97.1MHz +102.1MHz – **CP03)** 2 de Mayo 438, entre Tarquí y General Maldonado, Latacunga – **CP04)** Bel.Quevedo Caserio Illuchi, Latacunga – **CP05)** B Quevedo 555, Pujilí – **CP06)** Av.24 de Mayo 669, Saquisilí – **CP08)** Calle Felix Valencia 432, Plaza El Salto, Latacunga **W:** www.radio11denoviembre.com **E:** programacion@radio11.net or radio11@radio11denoviembre.com – **CP09)** General Maldonado 379 y 2 de Mayo, Latacunga – **CP10)** Calle Bolívar 1509 y Sucre, Salcedo **W:** www.radio-alternativa.net **E:** informacion@radio-alternativa.net – **CP11)** Juan Abel Echeverria 6-56 y Quito, Latacunga ☎/🖷 +593 23 2812258 **E:** ehquintana@hotmail.com – **CP12)** Faustino Sarmiento 5046 y Vela, Latacunga – **CP13)** Enrique Gallo 164 y Av 19 de Mayo, La Maná – **CP14)** António Clavijo, P3, Latacunga.

CR00) CHIMBORAZO
CR01) Cas 06-01-693 (or: Juan de Velasco N° 20-60 y Guayaquil), Riobamba ☎+593 23 2961608 🖷+593 23 2961625 **W:** www.ferpe.org **Quichua:** 0900-1100, 2300-0300 - **FM:** 91.7MHz – **CR02)** Cas 06-01-0242, Riobamba – **CR03)** Pichincha 1363 y Cardondelet, Riobamba – **CR04)** Cas 06-01-572 (or: Av.Daniel León Borja 30-44), Riobamba ☎+593 23 2960101 🖷+593 23 2940464 **E:** radiomundial@soccer.com or camelosg@yahoo.com – **CR05)** Cas.87A, Majipamba, Colta Prgrs un **Quichua** only – **CR06)** Francia 1857 y Villaroel, Riobamba – **CR07)** Cardondelet 2952 y J.Montalvo, Riobamba – **CR08)** 10 de Agosto 1742 y Benalcazar, Riobamba – **CR09)** M E Flor 4009, Riobamba – **CR10)** Ayacucho 3234 (or: Cas 06-01-0471), Riobamba – **CR12)** Calle Pichincha 24-26 y Veloz (or: Cas 82), Riobamba - **FM:** 100.1 – **CR13)** Av.C.Norte y Av Circunvalación, Riobamba – **CR14)** Cas 334 (or: La Paz y México Esq.), Riobamba ☎+593 23 2961331 🖷+593 23 2961330 – **CR15)** Cas 06-01-0376, Riobamba – **CR16)** Cas 06-03-0805, Alausí – **CR18)** Comunidad Sta Cruz, Guamote – **CR19)** Panamericana y Eloy Alfaro, Pallatanga – **CR20)** 1 Constituyente y G Rendon, Cumanda.

E000) EL ORO
EO01) Bolívar Madero 1313, via Pto Bolívar, Machala ☎+593 47 2929845 – **EO02)** Cas 221, Machala – **EO03)** Bolívar 601, Edif.Encasa, Machala – **EO04)** Libertad y Vega Davila, Santa Rosa – **EO05)** Cas 07-01-0086, Machala (or: 9 de Octubre y Páez), Machala – **EO06)** Av Honorato Márquez, Zaruma ☎+593 47 2972565 🖷+593 47 2972165 – **EO07)** 9 de Octubre y 1 Diagonal, Santa Rosa – **EO08)** San Francisco 114 y Sucre, Zaruma – **EO09)** El Oro y Cuenca, Sta Rosa ☎/🖷 +593 7 943139 – **EO10)** 9 de Mayo y Rocafuerte, esq, piso 2, Machala – **EO11)** San Martín 720, Entre Municipalidad y Och, Pasaje – **EO12)** 9 de Octubre y Sta Rosa, Machala – **EO13)** Calle Pasaje s/n y Costa Oeste, Machala – **EO14)** Av 10 de Agosto 1303 y 23 de Febrero, Balsas – **EO15)** Av Buena Vista 742 y 4ta Norte, Machala – **EO16)** Machala (see also GU07) – **EO17)** Av del Ejército, El Guabo – **EO18)** Av del Periodista y Calle Jon, Machala – **EO19)** Av 9 de Octubre y 23 de Abril, Machala ☎+593 47 2962300 – **EO20)** Av 12va Norte y Buena Vista, Machala.

ES00) ESMERALDAS
ES02) Malecón 805 y Cañizares, Esmeraldas (se also GU07) – **ES03)** Manuela Cañizares y Olmedo, Esmeraldas - **FM:** 96.3MHz – **ES04)** Edif.Mutualisat Vargas Torres, Esmeraldas – **ES05)** Simon Plata Torres y Maclovio Velazco, Quinindé – **ES06)** Bolívar s/n, Esmeraldas – **ES07)** Gustavo Becerra y Piedrahita, Esmeraldas – **ES09)** Bolívar 513 y Piedrahita, Esmeraldas.

GU00) GUAYAS
GU01) Cas 4144 (or: Boyaca 642 y Padre Solano), Guayaquil **W:** www.

cre.com.ec/cre htm **E:** cresat@gye.satnet.net ☎+593 44 2564290 🖷+593 44 2560386 - **FM:** 105.7 – **GU02)** Cas 2119, Guayaquil (or: Amazonas 743 y Veintemilla, P8, Quito) **W:** www.radiocadenauno.com/ – **GU03)** Quisquis 316 y Garaicoa, Edif Huancavelica, Guayaquil – **GU04)** Quisquis 316 y Garaicoa, Guayaquil – **GU05)** Cas 9974 (or: Av de Las Américas junto Canal 10), Guayaquil **W:** http://superk800.com/ – **GU06)** Rumichaca 934 y Velez, Guayaquil – **GU07)** Cas 11714 (or: Av.Francisco de Orellana y Juan Tanca Marengo), Guayaquil ☎+593 44 2680588 🖷+593 44 2680592 **W:** www.radiosucre.com.ec **E:** rsucre@radiosucre.com or: info@radiosucretv.com - **FM:** 95.3MHz – **GU08)** Cas 2440 (or: Escobedo 1504 y Aguirre, P9), Guayaquil – **GU09)** Cas 716, (or: Av Juan Tanga Marengo km 3), Guayaquil ☎+593 44 2561220 🖷+593 44 564570 **E:** caravana@gye.satnet.net - **FM:** 88.1MHz – **GU10)** Cas 09-01-4203 (or: Colón 548 y Boyacá, P7), Guayaquil – **GU11)** Juan Montalvo 1042, El Milagro – **GU12)** Cas 856 (or: Edif Gran Pasaje Of 906/908), Guayaquil – **GU13)** Cas 09-01-5762, Guayaquil **E:** sanfrancisco850@hotmail.com ☎+593 44 2530058 – **GU14)** Cas 5062 (or: Laque 1407 y Antepara), Guayaquil **E:** rcristal@ecua.net.ec – **GU15)** Malecón 206 entre Juan Montalvo y Loja, Guayaquil - **FM:** 92.9MHz – **GU17)** García Moreno y Hurtavho, en los Altos, Ofc.Delgado Travel P3, Guayaquil – **GU18)** 10 de Agosto 504 y Chimborazo, P3), Guayaquil ☎+593 44 2322495 ☎+593 44 2329695 **E:** servidor1000@hotmail.com – **GU19)** Urdesa, Av.Circunvalación Sur 111-B, frente al parque, Guayaquil ☎+593 44 2885449 – **GU20)** José de Antepara 4415 y Nicolas González, Guayaquil – **GU21)** Los Ríos 609, Cond Orellana, P4, Ofc 2, Guayaquil – **GU22)** Miguel H Alcivar y Luis Orrantia s/n, Guayaquil – **GU24)** Aguirre 931 y L.de Garaycoa, Guayaquil – **GU25)** Febres Cordero 315 y Chile, Guayaquil – **GU26)** Veléz 905, Edif.Forum, P16 (or: Cas 8729), Guayaquil ☎+593 44 2530288 ☎+593 44 2530059 **E:** ife@interactive.net.ec – **GU27)** Jiguas 500 y V.Emilio Estrada, Guayaquil – **GU28)** Lorenzo de Garaycoa 2615, Guayaquil ☎+593 44 2412533 🖷+593 44 2412533 – **GU29)** Laurel y Guayacanes, El Milagro – **GU30)** Luis Urdaneta 202 y Cordoba, Guayaquil – **GU31)** Simon Bolívar, entre Gonzales y Telégrafo, Guayaquil – **GU32)** Av 17 de Septiembre, El Milagro – **GU33)** Calle García Moreno y Bolívar 1013, El Milagro – **GU34)** Km 26.5 vía Duran-Tambo, Virgen de Fátima – **GU35)** Quito 1520 entre Sucre y Colón, Guayaquil – **GU37)** Av 5 de Octubre 150, Naranjito ☎+593 44 2720212 – **GU38)** Av 25 de Julio cdla 7 Lagos C, Guayaquil – **GU39)** Cas 5940, Guayaquil ☎+593 44 2481300 🖷+593 44 2485166 **W:** www.inocar.mil.ec **E:** inocar@inocar.mil.ec – **GU40)** Pastaza y 15 de Octubre, El Naranjal **E:** jpinoargote@hotmail.com – **GU42)** Jaime Roldos 700 y Av.8 de Abril, El Triunfo – **GU43)** Cdla.Belén Piedrahita y 1era, Daule – **GU47)** 9 de Octubre y Baquerizo Moreno, Edif.Plaza, P1, Guayaquil ☎+593 44 2565615 – **GU48)** Edif El Forum, P5, Ofic 508, Guayaquil 🖷 +593 44 2533885 – **GU49)** Eloy Alfaro Duran en la Av Samuel Cisneros, via al Secap, Guayaquil.

IM00) IMBABURA
IM01) Morales 408 y Sucre, Otavalo – **IM02)** Celiano Aguinaga y Panamericana Sur, Atuntaqui, Ibarra – **IM03)** Cas 413 (or: Olmedo 1178 y Av.Peréz Guerrero), Ibarra – **IM04)** Cas 10-01-0179 (or: Bolívar y García Moreno, Ibarra - **FM:** 98.5MHz – **IM06)** Río Chinchipe 397 y Río Daule, Ibarra – **IM07)** Cas 3, Atuntaqui – **IM08)** Bolívar 10020 y Espejo, Pimapiro – **IM09)** Matovelle s/n, Urcuquí – **IM10)** Cas 10-02-1464, Otavalo **Quichua:** 0900-1200, 1930-2300, **Sp:** 1200-1500 – **IM11)** Juan José Flores 11-26 y Jaime Rivadeneira, Ibarra ☎+593 26 2956008 🖷+593 26 2950828 **W:** www.imbanet.net/panorama/panorama.html - **FM:** 93.7MHz – **IM12)** Av.Reales Tamarindos y Calle Tenis Club, Portoviejo – **IM13)** Rocafuerte 1-10 y Guayaquil, Otavalo – **IM14)** Calle Oviedo y Bolívar, Edif Way, P2, Ibarra – **IM15)** Antonio Ante s/n, Urcuquí – **IM16)** Jirón Roldos Aguilera y Panamericana Norte, Otavalo **E:** radiochaskis@hotmail.com

L000) LOJA
LO01) Av.J.A Eguuigurren y Bolívar, Loja – **LO02)** Cas 474 (or: Bernardo Valdiviezo 1054, entre Miguel Riofrio y Azuay), Loja - **FM:** 100.3MHz – **LO03)** 24 de Mayo y Eloy Alfaro, Catamayo ☎/🖷 +593 47 2677067 - **FM:** 93.7MHz – **LO06)** Cas 11-01-222, Loja **FM:** 88.3MHz **E:** Luzvida@easynet.net.ec – **LO07)** Cas 196 (or: Olmedo 11-56 y Mercadillo), Loja ☎+593 47 2561166 🖷+593 47 2562 270 - **FM:** 88.9MHz – **LO08)** Olmedo 1146 entre Azuay y Mercadillo, Loja – **LO09)** Av Gran Colombia 2663 y Ibarra, Loja – **LO10)** B Valdiviezo 08-59, Loja – **LO11)** Sector Colinas de San Juan, Cariamanga ☎+593 47 2687320 🖷+593 47 2687322 **E:** rcmga@loja.telconet.net – **LO13)** Sucre y García Moreno, Cariamanga – **LO14)** Coop Ahorro y Crédito 3 de Diciembre, Isidro Ayora 235, Catacocha – **LO15)** Asociación Cristiana de Indigenas Saraguros, Saraguro **Quichua:** 1200-1400, 2100-2300, **Sp:** 1400-1600, 2300-0100 ID in Quichua: "R Alli Michic" - **FM:** 93.1MHz

LR00) LOS RÍOS
LR01) 12 Calle N° 207 y 7 de Octubre, Quevedo – **LR02)** Av Manabí y Juan León Mera, El Empalme, Quevedo – **LR03)** Av Velasco Ibarra 1012, Ventanas – **LR04)** Av.7 de Octubre 727, Quevedo – **LR05)** Cdla El

Mamey, Babahoyo – **LR09)** Olmedo 109, Vinces – **LR10)** 28 de Mayo 1412 y 6 de Octubre, Babahoyo.

MA00) MANABÍ
MA01) Av.10ma y Calle 17, P2, Manta – **MA02)** 10 de Agosto 609 y Olmedo, Portoviejo (see also GU07) – **MA03)** Ricaurte y P.Moreira, Portoviejo – **MA04)** Av Lascano, Chone – **MA06)** Pedro Gual, Edif. Servicentro, Portoviejo – **MA07)** Colón 180, Portoviejo – **MA08)** 18 de Octubre 404, Chone – **MA11)** 9 y Malecón, Edif "Jacob Vera", P1 Ofc 7, Manta ☎+593 45 2622714 ▤+593 45 2628718 – **MA12)** C Central, Portoviejo – **MA13)** Bolívar y Espejo, Portoviejo – **MA15)** 10 de Agosto 180 y Eloy Alfaro, Junín – **MA16)** Cas 13-04-0705 (or: 9 de Octubre en Mejía y Rocafuerte), Jipijapa ☎+593 45 2600679 ▤+593 45 2601477 – **MA20)** Noboa y Colón, Jipijapa – **MA21)** 4 de Diciembre y Alfaro, El Carmen – **MA23)** 306 Entre Las Avenidas 204 y 205, Manta ☎+593 45 2923666 ▤+593 45 2921635 **E:** diosvenami@aol.com or esdvami@hotmail.com – **MA24)** Flavio Alfaro 718 Ciudadela San Bartolo, Calceta ☎+593 45 2685169 ▤+593 45 2685126 – **MA25)** Cas 13-02-0629 (or: Montufar N° 1014 y Aguilera), Bahía de Caráquez ☎+593 45 2690370 ▤+593 45 2690305 - **FM:** 95.3 – **MA26)** Morales 104 y Colón, Ed Sind Choferes Man, Portoviejo.

MS00) MORONA SANTIAGO
MS04) Luia Casiragui s/n y Amazonas, Gualaquiza ☎+593 27 2780227 - **FM:** 91.7MHz 96.5MHz – **MS05)** Misión Salesiana de Oriente, Calle 10 de Agosto s/n, Macas ☎+593 27 2700186 ▤+593 27 2701838 **E:** radioupano@cue.eolnet.net or radioupano@easynet.net.ec **Shuar:** 1200-1230, 2230-2300 on SW - **FM:** 90.5MHz For Radio María del Ecuador address, see NA06 – **MS06)** Federación de Centros Shuar, Domingo Comín 17-38, Sucúa (or: Cas 17-01-4122, Quito) Educational prgr in Shuar only ID in Shuar: "Shuar Achuara Tuntuiri".

NA00) NAPO
NA02) Cas 260 (or: Av.Jumandy 536, Barrio 2 Rios), Tena – **NA06)** Misión Josefina, Juan Montalvo s/n y P Central, Tena **E:** coljav20@yahoo.es Radio Maria del Ecuador: Calles Baquerizo Moreno 281 y Leonidas Plaza, Quito ☎+593 22 2564714 ▤+593 22 2237630 **W:** www.radiomaria ecuador.org **E:** radiomaria@andina.net or info.ecu@radiomaria.org

PA00) PASTAZA
PA02) Vía Macas km 1 5, El Puyo – **PA03)** Cas 728 (or: Ceslaos Marín 391), El Puyo

PI00) PICHINCHA
PI01) Ulloa 611 y Acuña, La Fincha, P1, Quito ☎+593 22 2551423 – **PI02)** Panamericana Sur km 14.5 (teléfono 2 691 573), Quito ☎+593 22 2691573 – **PI03)** Av.Maldonado 688 y Calvas, Quito – **PI04)** Conde Ruíz de Castilla 997 y Muregeón, Quito ☎+593 22 2442650 – **PI05)** Pasaje A 689 y Vasco de Contrera, Quito ☎+593 22 2442951 ☎+593 22 2443147 **W:** www.geocities.com/~crespo – **PI06)** Cas 60 (or: Mariano Echeverria y Brasil), Quito – **PI07)** Cuenca 477 y Sucre (El convento de San Francisco), Quito ☎+593 22 2953077 – **PI08)** Cas 17-17-691(or: Villalengua 884 y Av.10 de Agosto), Quito ☎+593 22 2266808. Rel. HCJB in Spanish and local languages on 6050kHz. International service: See Int Broadcasting section – **PI09)** García Moreno 751 entre Sucre y Bolívar, P3, Quito **W:** www.quito.gov.ec/homequito municipio.com – **PI10)** Cas 17-21-1971 (or: La Coruña 2104 y Whimper, Edif.Aragones) Quito ☎+593 22 2508301 **E:** radioquito@elcomercio.com ☎+593 22 2508301 – **PI12)** Av Principal s/n, Maldonado – **PI13)** Amazonas 1638 y La Pinta, Quito ☎+593 22 2559383 – **PI15)** Panamericana Sur km 14.5 (teléfono 2 678 989), Quito ☎+593 22 2678989 – **PI16)** Ramírez Dávalos 612 y 10 de Agosto, Quito – **PI17)** Av.Amazonas y Colón, Edif. España, P4, Ofc.42, Quito ☎+593 22 2905471 – **PI18)** Cas 17-03-540 (or: Av.América 1830 y Mercadillo),Quito ☎+593 22 2245770 **W:** www.radiocatolica-ecuador.org **E:** buenanoticia@radiocatolica.org.ec – **PI19)** Palacio 303 y Av La Gasca, Quito (see also GU07) – ☎+593 22 2484591 **PI20)** Avellanias E5-107 y Av.Eloy Alfaro, Quito ☎+592 22 2484574 **W:** www.colonfm.com **E:** escucha@colonfm.com – **PI21)** Cas 17-01-67, Quito – **PI22)** Morales 1224 y García Moreno, Quito ☎+593 22 2283096 – **PI24)** García Moreno 1315 y Olmedo, Quito– **PI26)** Edif Sevilla, P9, J L Mera 565 y Carrión, Quito – **PI27)** García Moreno 446, Machachi ☎+593 22 2315025 ▤+593 22 2315 73 – **PI28)** Tarqui 785 y Estrada,Edif.de Cosi, P2, Quito ☎+593 22 2903306 – **PI30)** Cas 17-03-31 (or: Carrión 1288 y Av 10 de Agosto), Quito **E:** lrfeval@ecuanex.net.ec or Radioirf@ecuanex.net.ec ☎+593 22 2903756 – **PI31)** Av.América 4829 y Naciones Unidas, Quito ☎+593 22 2563560 ▤+593 22 2543625 **W:** www.explored.com.ec/radio/ index.htm – **PI32)** Cas 17-111-2263 (or: Bolivar 359 entre García Moreno y Venezuela), Quito ☎+593 2 2950 060 ▤+593 2 2951 018 – **PI33)** Cas 17-01-638 (or Av Amazonas N35-89 y Corea, P4), Quito ☎+593 22 2255999 ▤+593 22 2462562 **W:** www.radioeres.com - **E:** izurieta@ecuafast.com – **PI34)** Marquesa de Solanda 722, Quito ☎+593 22 2583942 – **PI35)** Central Roca 331 y Av 6 de Diciembre, Quito ☎+593 22 2524158 – **PI36)** Cas 17-23-47 (or: Av General Enriquez N° 29-35 y Río Chinchipe), Sangolquí ☎+593 22 2331064 ▤+593 22 2330736 – **PI37)** 12 de Octubre 227,

Quito – **PI39)** Cas 17-01-3386 (or: García Moreno N 11-184 y Carchi), Quito **E:** hcmunomat@hotmail.com – **PI40)** Edif Doral Mariscal, Of 86, Páez y Mercadillo, Quito **W:** www.radiodemocracia.com – **PI42)** Reina Victoria 447 y Roca, Quito – **PI44)** Carvajal e Interoceania, Barrio Sta Rosa, Tumbaco - **FM:** 104.1MHz – **PI45)** Av de la Prensa N°60-22 y Av.de la Prensa, Quito ☎+593 22 2595 19 ▤+593 22 2532262 **E:** rcq_1380@yahoo.com – **PI46)** Av.Amazonas 3911 y Corea, Unicormio 2, P10, Ofc.1008, Quito – **PI47)** Calle Bolívar y Alfredo Boada (sobre el Banco del Pichincha), Tabacundo ☎+593 22 2365556 – **PI48)** Cas.17-25-5 (or: Terán 409 y Av 10 de Agosto, Cayambe ☎+593 22 2360047 – **PI49)** Venezuela 701 y Espejo, Quito ☎+593 22 2956679 – **PI51)** Pedro Vicente, Maldonado (or: Concejo Provincial de Pichincha, Manuel Larrea y Antonio Ante, Cas 298, Quito) – **PI52)** Luis Cordero 557 y J.Mejia, Machachi – **PI53)** Av.Natalia Jarrín 2-77 y Vivar, Cayambe ☎+593 22 2360516 **E:** acayambe@uio.satnet.net – **PI54)** Iñaquito 133-E2 y Unión Nacional de Periodistas, Quito ☎+593 22 2254782 – **PI56)** Manuel Cajias E 14-09 y Toribio Hidalgo, Quito ☎+593 22 2234234 – **PI57)** Pasaje Santa Rosa y Av. Interoceánica, Puembo ☎+593 22391 287 **E:** radioilusion1600@hotmail.com – **PI58)** Panamericana Sur km 14.5 (teléfono 2 245 300), Quito ☎+593 22 2245800 – **PI59)** Gonzalo Díaz de Pineda 290 y Pedro del Alfaro, Quito ☎+593 22 2660580 – **PI60)** Av Colón OE3-331 y Versalles, Edificio Villarre, Quito ☎+593 22 2238231.

SD00) SANTO DOMINGO DE LOS TSÁCHILAS
SD01) Calle Ibarra y Babahoyo esq, Sto Domingo de los Colorados – **SD02)** Guayaquil 124 y Tsáchilas, Sto Domingo de los Colorados – **SD03)** Cas 17-24-0043 (or: Ibarra 905 y Av 29 de Mayo, Edif Dueñas), Sto Domingo de los Colorados – **SD04)** Quito e Ibarra, Santo Domingo de los Colorados – **SD05)** Av Quevedo 405, Santo Domingo de los Colorados

SE00) SANTA ELENA
SE01) Guayaquil s/n y 9 de Octubre, Santa Elena – **SE02)** 4a Av 619 y Robles, La Libertad ☎+593 4 2785129 ▤+593 44 2786296 **E:** lvp@porta.net – **SE03)** 12 de Octubre 1032, La Libertad

SU00) SUCUMBIOS
SU02) Cas 21-01-14 (or: Venezuela y Progreso), Nueva Loja ☎+593 26 2830423 ▤+593 26 2830425 **E:** radiosuc@andinanet.net - **FM:** 105.3 – **SU03)** Cas 40 (or: Mariscal Sucre y 12 de Febrero), Lagos Agrio ☎+593 26 2830201 - **FM:** 99.3MHz

TU00) TUNGURAHUA
TU01) Cevallos 345, Ambato (see also GU07) – **TU02)** Bolívar 537 y Fund.del Canton, Píllaro (or: Cas 18-01-244, Ambato)– **TU03)** Cas 18-01-181 (or: Sucre 09-42 y Quito), Ambato ☎+593 23 2822450 ▤+593 23 2822450 **W:** www.radioambato.com **E:** radioambato com - **FM:** 96.7MHz R. Amor – **TU04)** 12 de Noviembre y Ambato, Edif. El Pelegrino, Baños **E:** radiosantuario@yahoo.es ☎+593 23 2740962 – **TU05)** Cas 18-01-0674 (or: Av.Cevallos 15-57 y Mera, P10, Ofc 1001), Ambato ☎+593 23 2823128 ▤+593 23 2823097 **E:** radiolider@uio. telconet.net – **TU06)** Bolívar y Martinez, Ambato – **TU07)** Cas 005, (or: Av 22 de Julio y Jorge Chacón 4-47), Pelileo ☎/▤ +593 23 2871155 **E:** radiopelileo@hotmail.com – **TU08)** Cas 18-01-0574 (or: Castillo entre 12 de Noviembre y Olmedo, Edif.R.Centro), Ambato - **FM:** 93.7MHz – **TU09)** Cevallos 1624 y Maldonado, Ambato – **TU10)** Cevallos 754 y Martinez (or: Cas 18-01-0198), Ambato – **TU11)** Cotacachi 176 e Iliniza, Ambato – **TU12)** Cas 18-01-115 (or: Fray Fausto Suárez, Francisco Flor 321), Ambato - **FM:** 92.9MHz 104.5MHz 106.9MHz – **TU13)** Cas 618 (or: Calle Ayllón 1753 y Darquea), Ambato – **TU14)** Barrio El Censo via San Miguelito, Píllaro – **TU15)** Av Padre Chancon s/n y Juan Velasco, Pelileo **W:** www.radiodeportesambato. com – **TU16)** Av.El Rey, Ciudadela Oriente, Ambato – **TU17)** Montalvo 106, Quero – **TU19)** Calle Montalvo y Av Cevallos, Ambato ☎+593 23 2421789 – **TU20)** Sector Kiambe, Quero.

FM in Quito (MHz): 88.1 Latina FM – 88.5 Metro – PI08) 89.3 HCJB – 89.7 Majestad – 90.1 Tropicalida – 90.5 Disney – 90.9 Platinum – 91.3 Sabormix – PI17) 91.7 Visión – 92.1 Contacto Nuevo Tiempo – 92.5 Genial Exa FM – 92.9 Música y Sonido - PI33) 93.3 Eres 93.3 – 93.7 Galaxia – PI18) 94.1 Católica Nacional FM – 94.5 Rumba – 94.9 La Gitana – 95.3 Universal – 95.7 G.R.D.Int – 96.1 Joya – 96.5 BBN - 96.9 Paraíso – PI31) 97.3 La Otra FM – Un producto de Hoy – 97.7 Centro – 98.1 Proyección – 98.5 Alfa – 98.9 Colón – 99.3 La Luna – 99.7 Añoranza La Rumbera – 100.1 María – PI23) 100.5 Stereo Zaracay – 100.9 Nacional del Ecuador – R. Pública - 101.3 Onda Azul – 101.7 Sucesos – 102.1 R.La Red – PI07) 102.5 Francisco Estéreo – 102.9 Armonía – 103.3 Onda Cero FM - 103.7 Sonorama – 104.1 Cobertura – 104.5 América – 104.9 Ecuashyri – 105.3 Kiss – 105.7 CRE – 106.1 Hot 106 R. Fuego – 106.5 Canela – 106.9 R. Urbana – 107.3 JC – 107.7 Más Candela
FM in Guayaquil (MHz): 88.1 María – 88.5 Galaxia Stereo – 88.9 Di Blu – 89.3 R. City – 89.7 Punto Rojo FM – 90.1 Romance FM – 90.5 Canela – 90.9 Kiss – 91.3 Tropicalida Stereo – 91.7 Antena Tres – 92.1

Estrella – 92.5 Forever Music FM – 92.9 Colón FM – 93.3 Majestad – 93.7 Disney – 94.1 Onda Positiva – 94.5 Platinum FM – 94.9 Sol 95 – 95.3 Cupido – 95.7 Metro Stereo – 96.1 Onda Cero FM – 96.5 Pasión – 96.9 Más Candela – 97.3 Nuevo Tiempo – 97.7 Centro FM – 98.1 Morena – 98.5 J C R. – 98.9 Impacto FM – 99.3 Sabormix FM – 99.7 Elite – 100.1 R. La Prensa – 100.5 RSN FM Stereo – 100.9 Mundial – 101.3 La Estación Musical – 101.7 Telequil R. Stereo – 102.1 WQ Dos – 102.5 HCJB – 102.9 Armonía Musical – 103.3 Joya Stereo – 103.7 Sonorama FM – 104.1 Alfa Stereo – 104.5 Corazón – 104.9 Once Q FM – 105.3 Nacional del Ecuador, R. Pública – 105.7 Fabustereo – 106.1 BBN – 106.5 Fuego – 106.9 Francisco Stereo – 107.3 Rumba – 107.7 Visión FM

EGYPT

L.T: UTC +2h (29 Apr-29 Sep: +3h). DST may end earlier (at start of Ramadan) — **Pop:** 70 million — **Pr.L:** Arabic — **E.C:** 50Hz, 220V — **ITU:** EGY

EGYPTIAN RADIO & TV UNION (Gov)
P.O. Box 1186, Cairo 11511 (Street: Radio & TV Building, Cornish El Nil, Cairo) ☎+20 2 25757715, 25789145 ▤ +20 2 25789461
Email: freqmeg@yahoo.com **Web:** www.ertu.org (Arabic)
L.P: Pres: Mr Ahmed Anis, Chmn Eng. Sector: Eng. Hamdy Mounir, Chmn Broadc. Sector: Intesar Shalaby

MW	kHz	kW	P	Times
Cairo	558	100	2k	1200-2400
Sohag	603	50	4	0200-2200
Batra	621	1000	6a	24h
Asswan	702	10	2e	0400-2000
			4	0200-0400, 2000-2200
El Kharga	702	10	2h	0400-1000, 1130-2000 (Fri 0400-2000)
			4	2000-2200
			10	1000-1130 (not Fri)
Tanta	711	100	1a	24h
Qena	756	10	2e	0400-2000
			4	0200-0400, 2000-2200
Abis	774	500	5	24h
Batra	819	1000	1a	24h
Santah	864	500	4	24h
Matruh	882	10	1a	1100-0700 (Fri 24h)
Bawti	918	10	1a	24h
Hurghada	918	10	2k	1300-2000
			4	0200-0500, 2000-2200
			10	0500-1300
Cairo	936	50	11	1500-2000
Salum	936	10	1a	24h
Abu Simbel	981	1	1a	24h
Assiut	981	10	2d	0400-2000
			4	0200-0400, 2000-2200
Baris	981	1	1a	0300-2400
El Arish	1008	100	6c	0600-1500
			7	1500-2200
El Farafra	1008	1	1a	24h
El Fayoum	1008	10	2d	0400-2000
Cairo	1071	100	1b	0300-1500
El Minya	1080	10	1a	0300-2400
Luxor	1080	10	1a	0300-2400
Batra	1107	600	6b	1700-2300
			6c	0600-1700
Tanta	1161	100	2b	0400-2200
Quena	1179	10	1a	0300-2400
Ras Gharib	1188	10	1a	0300-2400
Asswan	1278	10	1a	0300-2400
Assiut	1305	10	1a	0300-2400
Abu Simbel	1314	1	2e	0400-2000
			4	0200-0400, 2000-2200
Hurghada	1314	10	1a	0300-2400
Nag Hamadi	1314	1	1a	0300-2400
Cairo	1341	100	3c	1700-0100
			8	0500-1700
Bawiti	1341	10	2k	1300-2000
			4	0200-0500, 2000-2200
			10	0500-1300
Idfu	1341	10	1a	0300-2400
Siwa	1341	10	1a	24h
Quseir	1350	10	1a	0300-2400
El Farafra	1368	1	2k	1300-2000
			4	0200-0500, 2000-2200
			10	0500-1300
El Kharga	1368	10	1a	0300-2400
Barnis	1386	1	1a	0300-2400
Luxor	1386	10	2e	0400-2000
			4	0200-0400, 2000-2200

MW	kHz	kW	P	Times
Ras Gharib	1422	10	2k	1300-2000
			4	0200-0500, 2000-2200
			10	0500-1300
Salum	1422	10	2i	0400-2000
			4	0200-0400, 2000-2200
El Minya	1476	10	2d	0400-2000
			4	0200-0400, 2000-2200
El Tur	1485	1	2k	1300-2000
			10	0200-0500, 2000-2200
			1	10	0500-1300
El Arish	1503	25	2f	0400-2200
Quseir	1575	10	2k	1300-2000
			4	0200-0500, 2000-2200
			10	0500-1300
Baris	1584	1	2h	0400-1000, 1130-2000 (Fri 0400-2000)
			4	2000-2200
			10	1000-1130 (not Fri)
Idfu	1584	10	2e	0400-2000
			4	0200-0400, 2000-2200
Matruh	1593	10	2i	0400-2000
			4	0200-0400, 2000-2200
Nag Hamadi	1602	10	2e	0400-2000
			1	4	0200-0400, 2000-2200
Siwa	1602	10	2i	0400-2000
			4	0200-0400, 2000-2200

MW Prgrs: 1a=General Prgr, 1b=Adults Prgr, 2=Local Prgrs (2b=Mid Delta, 2d= North Upper Egypt, 2e= South Upper Egypt, 2f= North Sinai, 2h=El Wady El Gadid, 2i=Matruh, 2k=Educational), 3c=Cultural Prgr, 4=Holy Koran Prgr, 5=Middle East Prgr, 6a=Voice of the Arabs, 6b=Wadi El Nile, 6c=Palestine Prgr, 7=Hebrew Prgr, 8=Songs Prgr, 10=Youth and Sports Prgr, 11=Om Kalthoum Prgr.

FM (MHz)
Site	D	E	G	K	M	N	R	S	Y
Abu				90.6					
Abh							95.7		
Alx		94.3	104.7	90.1	88.0		101.1	97.6	
Al F			98.2	88.6			91.7		
Asy	99.1	102.6	99.1	95.8	89.1			92.6	99.1
Asw	98.6	92.1	98.6	95.3	89.0				98.6
Baris				88.8					
Bawiti				87.6					
Ben							91.4		
Cairo†		95.4	107.4	98.2	98.8	88.7	102.2	105.8	108.0
Dahab			98.5	92.0					
Dum				93.8			87.6		
El A	87.8	94.1		87.8	90.9		97.4		87.8
El D			91.1	88.0			94.3		
El K				88.4					
El M	91.0	94.2	91.0	101.0	87.9			104.6	91.0
El Tur	89.4		95.7	89.4	92.5		99.0		89.4
El Z				88.4					
Hal				96.7			107.5		
Hga	101.7	94.9	101.7	91.7	88.6			98.2	101.7
Idfu				101.7					
Ism		93.5			90.4		96.7		
Isna				90.3					
Kat			90.0	87.6					
Kom				92.8					
Lux	93.1	96.3	93.1	103.1	90.0				93.1
Mah				99.6	93.1		89.2		
Man				96.3					
Mat				99.1	95.8		102.6	92.6	
Nag			90.9	87.8				94.1	
Nuw	99.1		92.6	99.1	89.5		95.8		99.1
Pt S		98.0			101.5		91.5		
Qena	100.1		100.1	90.5	93.6			96.8	100.1
Qus				97.2					
Rafah				103.9					
Saf		96.1	92.9				89.8		
SeS‡	97.6		91.1	97.6	88.0		94.3		97.6
Sha		103.5	93.5						
Sid				101.2					
Siwa				90.6					
Soh	99.3		96.0	89.7	104.8		102.8	92.8	99.3
Suez		91.2		94.4	88.1				

†=Also Cultural Prgr on 91.5MHz 11.9kW & Middle East Prgr on 89.5MHz at 100kW. ‡=K Prgr also on 101.1MHz at 0.3kW.

FM Prgrs: C=Cultural Prgr, D=Educational Prgr, E=European Prgr, G=General Prgr, K=Koran Prgr, M=Musical Prgr, N=Radio Misr, R=Regional Prgr, S=Songs Prgr, T=Middle East Prgr, Y=Youth & Sport

Stations & powers: Abu=Abu Simbel 0.3kW, Abh=Abu Homus 4kW, Alx=Alexandria 58.6kW, Al F=Al Farfra 8kW/R, G 0.3kW, Asy=Assyout

11.2kW, Asw=Aswan 11.9kW, Baris 0.3kW, Bawiti 0.3kW, Ben=Beni Suef 0.3kW, Cairo 100kW, Dahab 0.3kW, Dum=Dumyat 0.3kW, El A=El Arish 54.5kW, El D=El Dakhla 4kW/K 0.3kW, El K=El Kharga 0.3kW, El M=El Minyah 18kW/S 4kW, El Tur 11.9kW, El Z=El Zayat 0.3kW, Hal=Halayeb 4kW/K 0.3kW, Hga=Hurghada 28.3kW/M, K, E 7.96kW, Idfu 0.3kW, Ism=Ismailia 61.5kW, Isna 0.3kW, Kat=Katherina 0.3kW, Kom=Kom Ombo 0.3kW, Lux=Luxor 11.7kW, Mah=Mahalla 155kW, Man=Managem 0.3kW, Mat=Matruh 9.77kW, Nag=Nag Hamady 10kW, Nuw=Nuweiba 9.53kW, Pt S=Port Said 10kW, Qena 28.6kW, Qus=Quseir 4kW, Rafah 0.3kW, Saf=Safaga 10kW, SeS=Sharm El Sheikh 7.41kW/K 0.3kW, Sha=Shalatin 0.3kW, Soh=Sohag 38kW/M 4kW/M 0.3kW, Sid=Sidi Barani 0.3kW, Siwa 0.3kW, Suez 8.71kW

Ann: General Prgr: "Idha'atu jumhuriya misr al'arabbiya min al-qahira". Voice of the Arabs: "Saut al-'arab, min al-qahira". Holy Koran prgr: "Idha'atu-l-Quran min al-qahira"

Other FM Stations
Nogoom FM, Cairo 100.6MHz 100kW. Arabic music, 24h
Nile FM, Cairo 104.2MHz 100kW. Mainly English pop & rock, 24h
Web: www.nilefmonline.com

F.PI.: Reports that four new FM stations to be launched: Mega FM (92.7MHz); Hits FM (88.2MHz), Radio Drama and Comedy Radio.

EXTERNAL SERVICES: Radio Cairo
see International Broadcasting Section.

Other Stations
AFRTS Low-power broadcasts of NPR and AFN to US contingent of UN MFO in Sinai rep. on wide range of freqs from 92.7 to 106.1. Also 107.0 at Gebel Musa.

EL SALVADOR

LT: UTC -6h — **Pop: 7.1** million—**Pr.L:** Spanish — **E.C:** 60Hz, 115V — **ITU:** SLV

SUPERINTENDENCIA GENERAL DE ENERGÍA Y TELECOMUNICACIONES (SIGET)
Sexta Décima Calle Poniente y 3°Av.Sur N° 2001, Colonia Flor Blanca, San Salvador ☎ +503 2257-4438 **W:** www.siget.gob.sv

ASOCIACION SALVADORENA DE RADIODIFUSION (ASDER) Calle La Ceiba # 261, Col. Escalon, San Salvador. **W:** www.asder.com.sv

	MWCall	kHz	kW	Station, location
1)	HV	540	5	La Estación de la Palabra, San Salvador
18)	FG	550	2	R. Cristo Te Llama, Sonsonate (r: 900)
3)	KT	570	10	R. Exus "YXR Radio", San Salvador
4)	NK	*600	3	Vox FM, San Salvador (r: 94.5)
64)	LN	630	10	R. Promesa, San Salvador: 1130-0400
7)		700	12	R. Mi Gente, San Salvador
7)		700	12	R. Mi Gente, San Miguel (r. 700)
9)	RA	720	1	Qué Buena, San Salvador (r:88.9)
10)	KL	760	5	YSKL La Poderosa, San Miguel (r:770)
10)	KL	760	1	YSKL La Poderosa, Sonsonate (r:770)
10)	KL	760	1	YSKL La Poderosa, Zacateluca (r: 770)
10)	KL	770	10	YSKL La Poderosa,San Salvador: 1030-0530
10)	KL	780	1	YSKL La Poderosa, Usulután (r:770)
10)	KL	780	1	YSKL La Poderosa, Sta Ana (r:770)
11)	AX	800	12	R. María El Salvador, San Salvador: 1230-0600
12)	FA	810	2	R. Lorenzana, San Vicente
44)	DA	810	1.5	R. Imperial, Sonsonate: 1100-0300
13)	PX	*830	5	R. Pax, San Miguel
14)	FB	840	10	R. Santa Biblia, San Salvador: 1030-0400
15)	RC	*860	1	R. Tecana, Sta Ana
16)	AR	870	10	R. Renacer, San Salvador
8)	CD	880	1	R. Ritmo, Stgo de María
17)	LA	890	3	R. Renacimiento, Sta Ana: 1000-0500
18)	QJ	900	2	R. Cristo Te Llama, San Salvador
59)		930		R. Rey de Gloria, San Salvador
62)	HG	*950	1	R. Cristo Te Llama, San Miguel
21)	TW	*960	0.5	R. Centro, Sonsonate
47)	MS	*970	5	R. UTEC–R. Universidad Tecnológica, San Salvador: 1200-0400
24)	HH	*1000	1	Estación H, Sta Ana
25)	CA	v1020	5	R. Int. /La Máxima, San Salvador: 1100-0400
27)	RM	1030	1	R. Frontera, Ahuachapán: 1200-0400
26)	AN	*1070	1	LV de los Ausoles, Ahuachapán
61)	ME	1080	6	R. CRET, San Salvador
61)		1090	1	R. CRET, Sta Ana
28)	MG	1090	3	R. 1090, Atiquizaya

	MWCall	kHz	kW	Station, location
29)	RF	1100	3	R. Universidad Don Bosco, San Salvador
30)	CL	*1110	2.5	R. Horizonte, San Miguel (r.1160)
58)	LR	1120	3	Una Voz que Clama en el Desierto, San Salvador: 1045-0500
20)	LG	1130	1	R. Chaparrastique, San Miguel
31)	AJ	1130	1	R. Moderna, Sta Ana: 1200-0400
11)	CF	1150	1	R. María Zona Oriental, San Miguel (r: 800)
48)	RG	*1160	1	R. Corporación, Sta Ana
55)	CR	*1170		R. Cristo Viene, San Miguel
68)	CB	1170	0.5	R. Pentecostés, Sonsonate
33)	VG	1180	5	R. VEA–Voz Evangélica de América, San Salvador: 1200-0300
34)	KJ	*1200	1	R. Sirama, San Miguel
22)	CG	1210	1	R. América/R. La Paz, Zacatecoluca
49)	MT	1240	0.5	R. Metapán, Metapán
50)	QN	1240	1	R. Norteña, San Miguel
35)	AA	*1260	12	R. Abba, San Salvador
34)	QZ	*1270	1	R. W "LV de la Verdad en Oriente", San Miguel
61)	QV	*1280	1	R CRET, Sta Ana
57)		1280		R. Emaús, San Vicente
37)	MA	1290	1	R. Chalatenango, Chalatenango: 1000-0300
38)	LV	1300	6	W-LV de la Verdad, San Salvador
56)	KG	1300		R. Llanera "La Campechana", San Miguel
51)	RV	1310	5	R. Veritas, Stgo de María
52)	AH	*1320	1	R. Emanuel, La Unión
39)	HQ	*1330	5	R. Cristo Te Llama, San Salvador
40)	XW	*1340	1	R. Novedades, Usulután
46)	FM	*1360	5	Super Radio, San Salvador
53)	KO	1370	1	R. Lluvias de Bendición, San Miguel: 1100-0300
63)		1390		R. Getsemani, La Unión
		1390		R. Fraternidad de Jesucristo, Chalchuapa
41)	JI	1400	1	LV del Litoral, Usulután: 1100-0400
54)	KR	*1450	1	R. Restauración, San Miguel: 1000-0400
67)	CS	1500	1	R. Pentecostal, Usulután
60)		1550	5	R. Sanidad Divina, San Salvador: 1000-0600
65)		*1580		R. Poder y Gloria, Santa Ana

Call YS–, * = inactive, (r) = repeater, v = varying fq

Addresses and other information:
1) Ap.2854 (or: Calle al Matazano N° 1, Final Col.Sta Lucía, Ilopango), San Salvador **W:** www.elim.org.sv – **2)** Carretera a Santa Ana, Colonia Monte Carlos, Calle Principal (or Apartado Postal 10), Sonsonate – **3)** Jardines de la Cima polígono "N" Calle las Begonias, Pasaje los Lirios #14, San Salvador ☎+503 2248 1683 📠+503 2248 1683 – **4)** Edif.TV2, Alameda Dr.Manuel E.Araujo, San Salvador – **6)** 65 Av S y Av.Olipica, 192 Edif.Corporación YSKL, San Salvador ☎+503 2223 9267 **W:** www.radiofmmonumental.com – **7)** 14 Calle Poniente, entre 43 y 45, Avenida Sur No. 2309, Col. Flor Blanca, San Salvador ☎+503 2245 4148 **W:** www.RadioCadenaMiGente.net – **8)** 2a Av.Norte 24, Stgo de María, Usulután – **9)** Ap.720, San Salvador – **10)** 65 Av S y Av.Olimpica, 192 Edif.Corporación YSKL, San Salvador **W:** www.radioyskl.com – **11)** Urb. General Escalon, Pasaje Beethoven 8/E, San Salvador ☎+503 2262 0800 📠+503 2262 0692 **W:** www.radiomaria.org.sv –**12)** Carretera a Tecoluca, Col. Najarro, San Vicente – **13)** 10 Calle Oriente 102 Bis, San Miguel – **14)** Iglesia San Pablo, Final 5a Calle Poniente, Colonia Escalón, San Salvador ☎+503 2263 0666.– **15)** Altos del Cine Tecana, Sta Ana – **16)** 27 Calle Poniente 544, San Salvador ☎+503 2248 1683 **W:** www.renace.org – **17)** 4a Av.Sur, Entre 7a y 9a Calle Poniente, Edif. Plaza de Vidrio, Sta Ana – **18)** Colonia San Miguel, CI Principal Pasaje Castillo, San Ramón Mejicanos, San Salvador **W:** www.cristotellama. org.sv – **19)** Calle al Trapiche, Chalchuapa, Santa Ana ☎+503 2408 2957 – **20)** 4a Av.Sur 303 bis, San Miguel ☎+503 2661 3640 📠+503 2661 7644 – **21)** 5a Calle Oriente 44 (or Apartado Postal 115), Sonsonate – **22)** 2a Calle Poniente 22, Zacatecoluca – **24)** 9a Calle Poniente 25, Sta Ana – **25)** Av.España y 23 Calle Oriente, Ex Cine Fausto, San Salvador – **26)** Av.Morazán km 101, Ahuachapán – **27)** Av.2 de Abril y 8a Calle Poniente, Ahuachapán – **28)** 5° Calle Oriente 3-204, Atiquizaya – **29)** Edificio 2, CITT, Universidad Don Bosco, Soyapango, San Salvador ☎+503 2251 5000 ext 1840 **W:** http://radiob.udb.edu.sv – **31)** 8a Calle Poniente 11A, Sta Ana – **33)** Carr Panamericana km 18.5, Cantón La Palma, San Martín, San Salvador **W:** www.dowlos.org – **34)** Carr.Litoral km 134, Cantón Jalacatal, San Miguel – **35)** Col.San Benito, Pasaje Las Palmas 182, San Salvador – **37)** Calle a San Francisco Lempa, Col. Veracruz, Chalatenango ☎+503 2335 2030 **W:** www.radiochalatenango.com.sv – **38)** 17 Calle Oriente 143, Barrio San Miguelito, San Salvador **W:** www.radiolavoz. org – **39)** Misión Evangelístyica Cristo Te Llama, Ap.855, San Salvador **E:** cristotellama@navegante.sv – **40)** 1a Oeste 18, Usulután – **41)** 12 Av.Sur y final 5a Calle Oriente, Col.Sta Rosa, Usulután **W:** www.lavozdellitoral. com **FM:** 90.1MHz – **44)** Ap.56, Sonsonate 📠+503 2450 0189 – **46)** Boulevard de los Héroes, Edificio Los Heroes, Local 8B, San Salvador – **47)** Universidad Tecnológica, 17 Av.Norte 130, San Salvador – **48)** Sta Ana – **49)** Calle Principal, Costado Norte Centro Judical, Col Lomas de

Montecristo, Metapán, Sta Ana – **50)** Col.Hirleman 14 C P Block 6 N°
9, San Miguel – **51)** Bo El Centro, C.Bolivar y 4 Av.S, Stgo de María,
Usulután – **52)** La Unión – **53)** Carr.Panamericana, Crio El Alto, 300 mts
al Norte, El Jalacatal, San Miguel ☎+503 2669 8303 **W:** www.lluvias-
debendicionradio.com.sv – **54)** Ap.210, San Miguel – **55)** San Miguel
– **56)** Col.Hirleman, 14 Calle Poniente, Bloque 6, N° 9, Col.Hirleman,
San Miguel ☎+503 2669 5151 📠+503 2669 5150 – **57)** 2 Av N N° 10,
San Vicente – **58)** San Salvador **W:** www.iglesiacuerpodecristo.org.sv
– **59)** Carretera Antigua a Plan del Pino, 1 Cuadra, Antes de la Ciudadela
Don Bosco, Soyapango, San Salvador ☎+503 2534 0967 – **60)** Calle
25 de Abril Poniente, Barrio San José # 22B, San Marcos, San Salvador
☎+503 2220 4849 **W:** www.radiosanidaddivina.net – **61)** Barrio La
Cruz, 10 Av Norte N° 203-Bis, San Miguel **W:** www.radiocret.net – **62)**
Misión Evangelística Cristo Te Llama, Barrio El Calvario, 4a Av.Sur 303
bis, San Miguel ☎+503 2661 3283 – **63)** Col. La Paz, La Unión (part of
Radio CRET Network) – **64)** 75 Av Norte, Prolongación Juan Pablo II,
Col Jardines de Escalón, final Pasaje KL, San Salvador **W:** www.radio-
promesa.org – **65)**Santa Ana – **67)** Kilómetro 112½, Carretera El Litoral,
Frente a Desvío Esmora, Usulután – **68)** Colonia Monte Carmelo, Calle a
Los Naranjos, Frente Antena de la YSU, Sonsonate.

FM in San Salvador (MHz): 72.5 Metroaudio – 72.9 – R. Selectos –
87.75 Canal 6 – 88.5 Paz – 88.9 Qué Buena – 89.3 Cool – 89.7 Bautista
– 90.1 Láser (español) – 90.5 Progreso – 90.9 UPA – 91.3 Exa – 91.7
YSUCA – 92.1 La Klave – 92.5 La Nueva – 92.9 Láser (inglés)– 93.3
Globo – 93.7 El Mundo – 94.1 Super Estrella – 94.5 Vox – 94.9 Astral
– 95.3 R. Eco – 95.7 Verdad – 96.1 Scan – 96.5 R. Roca – 96.9 R. El
Salvador – 97.3 Corazón – 97.7 Luz – 98.1 Gospel FM – 98.5 Cuscatlán
– 98.9 La Mejor FM – 99.3 Mesías – 99.7 Guapa – 100.1 ABC – 100.5
Retro – 100.9 La Chévere – 101.3 Monumental – 101.7 Mil 80 –102.1
102 Uno –102.5 Femenina – 102.9 102 Nueve – 103.3 Clásica – 103.7
Cadena Central – (6) 104.1 YSKL La Poderosa – 104.5 Sonora – 104.9
Fiesta – 105.3 UFG Radio – 105.7 YXY – 106.1 El Camino – 106.5
Ranchera – 106.9 Maya Visión – 107.3 YSU – 107.7 Fuego.
FM in San Miguel (MHz): 90.1 Stereo Caliente - 90.5 Siglo 21– 90.9
Popular – 91.7 YSUCA – 92.5 Monseñor Romero – 94.1 Cadena
Central – 96.5 Agape R. – 97.3 Carnaval – 98.1 La Pantera – 99.7 Mi
Consentida - 102.9 102 Nueve – 104.1 YSKL La Poderosa – 106.1 La
Grande.
FM in Santa Ana (MHz): 90.5 Supra Stereo - 91.7 YSUCA - 92.1 Fe
y Alegría – 92.5 R. Doremix – 93.3 Shabach – 95.3 Amor – 97.3 Uno
– 97.9 La Campirona – 99.7 R. RX FM - 102.9 Doble H – 104.1 YSKL La
Poderosa – 105.3 Soda Stereo – 106.1 Bautista

EQUATORIAL GUINEA

L.T: UTC +1h — **Pop:** 600,000 — **Pr.L:** Spanish, French, ethnic
— **E.C:** 50Hz, 220V — **ITU:** GNE

MINISTERIO DE INFORMACIÓN, TURISMO Y CULTURA
✉ Barrio Nzalang (antiguo África 2000), Malabo ☎+240 078221 📠
+240 072444 **W:** www.ceiba-guinea-ecuatorial.org/guineees/mininfo.
htm **E:** ibcc-net@ceiba-equatorial-guinea.org
L.P: Minister: Purificacion Opo Barila. Dir R & TV: Hermenesildo
Moliko Djele.

RADIODIFUSION DE GUINEA ECUATORIAL (Gov.)
✉ Ap. 749, Bata ☎+240 082592 📠+240 082093 ✉ Av. 3 de Agosto
90, Ap. 195, Malabo ☎+240 072260 📠+40 72097/3122

SW:	kHz	kW	Times
Bata	5005	50	0500-2300 (occasionally)
Malabo(Semu)	6250	20	0500-2200 (times vary)

FM: Bata 98/99.9MHz 1kW, Malabo 102MHz.
D.Prgr: in Spanish/French/ethnic. N: 0600, 1415, 2100.
Ann: "Esta es Radio Bata" or " ... Malabo".

Other stations:
Africa No. 1: Malabo 102MHz (see main entry under Gabon).
RFI Afrique: Malabo 88/97.5MHz in French/Spanish.
R. Asonga, Malabo & Bata: freq. not known.
Rural radio: La Voz de Kie Ntem at Ebibeyín, Ecos de Wele Nzás at
Mongomo and La Voz de Centro Sur at Evinayong

ERITREA

L.T: UTC +3h — **Pop:** 6 million — **Pr.L:** Afar, Amharic, Arabic,
Tigrinya, Tigre, others — **E.C:** 50Hz, 230V — **ITU:** ERI.

MINISTRY OF INFORMATION
✉ P.O. Box 872, Asmara ☎+291 1 120478/201820 📠+291 1 126747
W: shabait.com **E:** nesredin@tse.com.er

VOICE OF THE BROAD MASSES OF ERITREA (Gov.)
✉ P.O. Box 242, Asmara ☎+291 1 117111/118711 📠+291 1 124847
LP: DG: Mahmud Chirum. TD: Mehreteab Tesfagiorgis. PD: Abdu Heji.
Dir. Radio Eng.: Berhane Gerezgiher.
Station: Asmara (Selai Dairo).
MW: 837kHz 100kW (Prgr. 2), 945kHz 100kW (Prgr. 1).
SW: 7210 kHz 100kW (Prgr. 1), 7175 kHz 100kW (Prgr.2). Prgr. 2
heard also via three low power transmitters between 1500-1900 on
5060/6170/7120/9710kHz. **NB:** Frequencies highly variable to escape
Ethiopian jamming.
Prgr. 1 in Tigrinya/Tigre/Kunama: 0400-1000, 1300-2000 – **Prgr. 2** in
Arabic/Afar/Amharic/Oromo/Saho/Bilen: 0400-1000, 1300-2000.
F.PI: nationwide FM network.
Zara FM: 100MHz + others.
Ann: Amharic:"Yeh be Asmera ketema yemigegne yesifiw Yeritrea
hezeb demts yeamarigna agelgilot new". Arabic: "Huna Asmara,
Idha'at Sawt al-Jamahir al-Iritriyyah". Tigrigna: "Ezi kab Asmara
Zemehalalef Medeber Radio Demtsi Hafash Eritrea Eyu".
R. Sawa, west Eritrea: **FM** (fq. not known). Op. by Sawa National
Youth Training Centre.

ESTONIA

L.T: UTC +2h (27 Mar-30 Oct: +3h) — **Pop:** 1.3 million — **Pr.L:**
Estonian, Russian — **E.C:** 50Hz, 230V — **ITU:** EST

KULTUURIMINISTEERIUM (Ministry of Culture)
✉ Suur-Karja 23, 15076 Tallinn ☎ +372 6282208 📠 +372 6282200
E: peeter.sookruus@kul.ee **W:** www.kul.ee
LP: Head Media & Copyright Dept.: Peeter Sookruus
NB. The Ministry of Culture issues broadcasting licenses.

EESTI RAHVUSRINGHÄÄLING (ERR) (Pub)
✉ Gonsiori 21, 15020 Tallinn ☎ +372 6284100 📠 +372 6114457
✉ Studios (exc. ERR2): Kreutzwaldi 14, 10124 Tallinn
E: err@err.ee **W:** www.err.ee **LP:** Chmn: Andres Jõesaar

FM (MHz)	ER1	ER2	ER3	ER4	kW
Alu	-	99.5	89.1		0.26
Essu	106.0	95.4	104.5	94.1	0.1
Haapsalu	105.3	102.9	106.3	93.6	0.7
Haljala	106.0	95.8	104.5	94.2	0.1
Koeru	105.1	102.6	107.6	93.4	30/30/30/7.8
Kohtla-Nõmme	105.4	102.9	90.4	95.3	11.2
Kuressaare	105.6	103.1	107.0	-	1
Kõrgessaare	91.2	99.1	94.9		1
Möksi	-	-	-	99.9	3
Narva	104.7	102.3	89.4	100.9	0.5
Orissaare	105.9	103.4	107.8	-	20/10/10
Pärnu	104.8	102.3	107.3	94.8	10
Tallinn	104.1	101.6	106.6	94.5	30
Tartu	106.7	96.3	106.7	94.4	0.5
Valga	-	-	-	92.5	1.8
Valgjärve	106.1	103.6	105.7	-	40/40/12.5
Viiratsi	105.8	103.3	107.0	95.5	1

D.Prgr: ERR1 (Vikerraadio): 24h. – **ERR2 (Raadio 2):** 24h. – **ERR3
(Klassikaraadio):** 24h. – **ERR4 (Raadio 4/Radio 4)** in Russian: 24h.
– **ERR Raadio Tallinn** 103.5MHz (1kW): 24h. Own prgrs 0700-1700;
1700-0700 rel. BBC World Sce, RFI, Deutsche Welle.

OTHER STATIONS

MW	kHz	kW	Location	Station
6B)	1035	200	Tartu (a)	R. Eli

FM	MHz	kW	Location	Station
18)	87.7	1	Paldiski	Paldiski R.
3A)	88.1	2.6	Paide	Star FM
6A)	88.2	3	Kohtla-Nõmme	Tartu Pereraadio
2D)	88.2	3	Muhu	Raadio 3
1C)	88.3	2	Tallinn	Spin FM
4)	88.6	3	Pärnu	Raadio 7
5)	88.8	1.3	Tallinn	R. Mania
6A)	89.0	3	Rõõmu	Tartu Pereraadio
8)	89.0	2.4	Vanamõisa	Kuressaare Pereraadio
8)	89.4	1	Kõnnu	Kuressaare Pereraadio
10)	89.6	1	Tallinn	Tallinna Pereraadio
1B)	89.8	1.1	Vinni	Raadio Uuno
1A)	89.9	1.1	Pärnu	Raadio Kuku
12)	90.1	1	Kärdla	Raadio Kadi
1E)	90.2	1	Tallinn	Dinamit FM
12)	90.5	1	Kuressaare	Raadio Kadi
2C)	90.6	1.3	Tallinn	Russkoe R.
1B)	91.0	3	Pärnu	Raadio Uuno
7A)	91.2	6.5	Valgjärve	Raadio Elmar
7A)	91.5	1	Kuressaare	Raadio Elmar

FM	MHz	kW	Location	Station
7A)	91.7	7.5	Koeru	Raadio Elmar
7A)	92.2	3	Linnamäe	Raadio Elmar
3A)	92.2	3	Pada	Star FM
1B)	92.3	1	Parksepa	Raadio Uuno
14)	92.5	1	Vinni	Raadio Viru
11)	92.7	1.5	Seljametsa	Raadio Pärnu
3A)	92.9	3	Linnamäe	Star FM
2E)	93.2	1.5	Tallinn	Energy FM
3A)	93.3	3	Kuressaare	Star FM
2B)	93.8	3	Holsta	Sky Plus
2B)	95.2	1	Rõõmu	Sky Plus
2B)	95.4	3	Tallinn	Sky Plus
4)	96.1	3	Tamsalu	Raadio 7
2B)	96.3	1.9	Vätta	Sky Plus
3A)	96.6	1.5	Maardu	Star FM
17)	96.6	1	Sangaste	Ruut FM
2B)	96.8	1.6	Audru	Sky Plus
2B)	96.9	1.5	Palade	Sky Plus
1B)	97.2	3	Tallinn	Raadio Uuno
1B)	97.2	1	Rõõmu	Raadio Uuno
1B)	97.4	3	Sikassaare	Raadio Uuno
1B)	97.4	2.5	Koeru	Raadio Uuno
2D)	97.8	2	Tallinn	Raadio 3
2D)	98.1	1	Audru	Raadio 3
2A)	98.4	3	Tallinn	Sky Radio
2D)	98.6	2.5	Rõõmu	Raadio 3
7A)	99.0	3	Pärnu	Raadio Elmar
3A)	99.4	3	Rõõmu	Star FM
1B)	99.8	2.5	Linnamäe	Raadio Uuno
1F)	100.0	1	Narva	Narodnoe R.
7B)	100.2	2	Rõõmu	Raadio Tartu Kuku
1A)	100.3	1.9	Seljametsa	Star FM
1A)	100.5	2.2	Paide	Raadio Kuku
13)	100.7	3	Põlva	Raadio Marta
1A)	100.7	3	Tallinn	Raadio Kuku
1A)	100.8	1.5	Viljandi	Raadio Kuku
1A)	100.9	2.5	Linnamäe	Raadio Kuku
9)	101.0	3	Paide	Kuma Raadio
15)	101.7	1	Võru	Ring FM
3A)	101.9	2	Jõgeva	Star FM
3B)	102.1	1.5	Maardu	Power Hit R
3A)	103.2	3	Parksepa	Star FM
1B)	104.5	1	Liiva	Raadio Uuno
15)	104.7	1	Rõõmu	Ring FM
16)	104.9	1	Tallinn	Euro FM

NB: Txs below 1kW not listed. (a) Kavastu

Addresses & other information:
1A-F) Veerenni 58a, 11314 Tallinn. **E:** 1A) kuku@kuku.ee; 1B) uuno@uuno.ee; 1D), 1F) in Russian. – **2A-E)** Pärnu mnt. 139c, 11317 Tallinn. **E:** skymedia@sky.ee. In Russian, exc. 2B,2E. **E:** 2E) info@raadio3.ee – **3A,B)** Peterburi 81, 11415 Tallinn. **E:** 3A) starfm@starfm.ee, 3B) info@power.ee – **4)** Välja 18, 10506 Tallinn. **E:** raadio7@raadio7.ee – **5)** Tartu mnt. 80d, 10112 Tallinn. **E:** raadio@mania.ee – **6A)** Annemõisa 8, 50708 Tartu. **E:** tartu@pereraadio.ee; **6B)** Vabaduse 20, 20306 Narva. Religious prgrs in Russian (incl. TWR relays): 24h. **E:** am1035@bk.ru – **7A,B)** Õpetaja 9a, 51003 Tartu. **E:** 7A) elmar@elmar.ee; 7B) raadio@tartukuku.ee – **8)** Tallinna mnt. 45, 93811 Kuressaare. **E:** kuressaare@pereraadio.ee – **9)** Pärnu mnt. 57, 72712 Paide. **E:** kuma@kuma.ee – **10)** Endla 29, 10129 Tallinn. **E:** tallinn@pereraadio.ee – **11)** Esplanaadi 10, 80010 Pärnu. **E:** raadio@pfm.ee – **12)** Pikk tn. 62, 93811 Kuressaare. **E:** raadio@kadi.ee – **13)** Kesk 42, 63308 Põlva. **E:** martafm@martafm.ee – **14)** Rägavere 35, 44312 Rakvere. **E:** info@radioviru.ee – **15)** Teguri 37, 50107 Tartu. **E:** info@ringfm.ee – **16)** Kristiina 15, 15026 Tallinn. **E:** info@eurofm.ee – **17)** Pikk tn. 3a, 68206 Valga. **E:** ruutfm@ruutfm.ee – **18)** Sadama 21-12, 76806 Paldiski. **E:** prl04@hot.ee

ETHIOPIA

LT: UTC +3h — **Pop:** 83 million — **Pr.L:** Amharic, Oromo, Sidamo, Somali, Tigrinya — **E.C:** 50Hz, 220V — **ITU:** ETH

ETHIOPIAN BROADCASTING AUTHORITY (EBA)
✉ P.O. Box 43142, Hailalem Bldg. Kazanchis, Addis Ababa ☎+251 11 5538755 🖷 +251 11 5536767 **W:** www.eba.gov.et **E:** e.b.a1@ethionet.et

RADIO ETHIOPIA (Gov.)
✉ P.O. Box 1020, Addis Ababa ☎+251 11 5516977
W: www.erta.gov.et **E:** etv2@ethionet.et **L.P:** GM: Ato Fikadu Yimeru. SM: Kasa Miloko. CE: Kebede Gobena. Head of English Prgrs: Melesse Edea Beyi.

MW	kHz	kW	MW	kHz	kW
Bahir Dar	594	100	Metu	684	100

MW	kHz	kW	MW	kHz	kW
Arba Minch	828	100	Robe (Bale)	972	100
Harar	855	100	Addis Ababa	989	1
Addis Ababa	873	100	Mekele	1044	200
Dese	891	100			

SW: Addis Ababa (Geja): 5990kHz‡, 7110kHz‡, 9705kHz 100kW.
FM: Addis Ababa 93.2MHz 2.5kW, 97.1MHz.
National Prgr. in Amharic/Others: 0300-2100. **In English:** 1200-1300. **FM Addis** on 97.1MHz. **Reg. prgrs** and **BBC relays** at times. External Sce. relay: on 989kHz. **Ann:** Amharic: "Yeh Ye-Ethiopia Radio Naw". E: "This is the English service of R. Ethiopia". **IS:** Electronic keyboard.

EXTERNAL SERVICE: see International Radio section.

RADIO FANA (Priv.)
✉ P.O.Box 30702, near Black Lion Hospital, in front of Sweden Embassy, Addis Ababa. **W:** www.radiofana.com **L.P:** GM: Woldu Yemessel. Tech. Dir: Mulugeta Mehari.
MW: Addis Ababa 1080kHz 3kW.
SW: Addis Ababa 6110 & 7210kHz 100kW. (Irr., alt. fq 6890kHz 10kW).
FM: Addis Ababa 98.2MHz (separate programmin).
D.Prgr. in Afar/Amharic/Oromo/Somali/Tigrinya: 0300-2100.

Regional government stations:
RADIO OROMIYA (Oromiya Radio & TV, ORTV)
✉ P.O. Box 2919, Adama. **W:** oromiagov.org **L.P:** Abarra Hailu, Mgr.
MW: Robe (Bale) 837kHz 100kW, Adama (Nazret) 1035kHz 10kW, Nekemte 1053kHz 100kW (planned).
SW: Addis Ababa: 6030kHz 100kW.
D.Prgr. in Oromo: 0330-0600, 0900-1100, 1530v-1900.
Ann: Oromo: "Kun Radio Oromiya".

VOICE OF TIGRAY REVOLUTION (Gov.)
✉ P.O.Box 450, Mekele, Tigray ☎+251 34 4410544/5 **W:** www.dimtsiwoyane.com **E:** vort@ethionet.et **L.P:** Dir: Abera Tesfay
MW: Addis Ababa 1359kHz 100kW.
SW: Addis Ababa 5950kHz 100kW.
F.PI: FM transmitter in Mekele.
D.Prgr in Tigrinya/Afar: MF 0300-1900, SS 0300-1730. **Ann:** Tigrinya: "Dimtsi Woyane Tigray". **IS:** Melody played on washint (Ethiopian flute).

AMHARA STATE REGIONAL RADIO (Gov.)
✉ Amhara Mass Media Agency, Bahir Dar. **L.P:** Dir: Mr. Dereje Moges **W:** www.amma.gov.et **E:** ammawebmaster@yahoo.com
MW: Bahir Dar 801kHz 100kW. **SW:** Geja 6090kHz 100kW. **FM:** Bahir Dar 96.9MHz. **D. Prgr:** 0300-0600, 0900-1100, 1400-1900.

Addis Ababa Region R.: Addis Ababa 96.3MHz 4kW – **Debub R,** Awassa: 100.6MHz – **City FM,** Addis Ababa: 94.7MHz – **FM Dire:** Dire Dawa: 106.0MHz. – **FM Harar:** Harar: 101.4MHz – **R. Jigjiga,** Jigjiga 95.2MHz 2.5kW. Web: radiojigjiga.com

Other stations:
Afro 105.3 FM, Addis Ababa: 105.3MHz 2.5kW. Web: afro105fm.com
Ravos FM, Awassa: 100.9MHz.
R. Sidama, c/o Furra Institute of Development Studies, P.O. Box 69, Yirgalem: 954kHz 2.5kW; MF 0500-1400, SS 0400-1700.
Sheger FM, Addis Ababa: 102.1MHz. Web: shegerfm.com
Zami R, Addis Ababa: 90.7MHz 2kW. Web: www.zami.com.et

FALKLAND ISLANDS (UK)

L.T: UTC -4h (5 Sep 10-17 Apr 11, 4 Sep 11-1 Apr 12: -3h) — **Pop:** 3,100 (excl. military personnel) — **Pr.L:** English — **E.C:** 50Hz, 240V — **ITU:** FLK

FALKLAND ISLANDS RADIO SERVICE (Pub)
✉ John Str, Stanley FIQQ 1ZZ. ☎+500 27277. 🖷 +500 27279.
Web: www.firs.co.fk **Email:** cgoss@firs.co.fk **L.P:** Stn Man.: Corina Asbridge, Prgr Contr.: Elizabeth Elliot Senior Reporter: Stacy Bragger
MW: 530kHz 15kW. **FM**(MHz): Pt Stanley 88.3 15W, Sussex Mountains 88.0 5W, March Ridge 90.0 1kW, Sapper Hill 96.5 0.25kW, Mt Maria 102.0 2kW, Mt Alice 105.0 0.25kW, Mt Kent 105.0 0.25kW

BRITISH FORCES BROADCASTING SERVICE
✉ Rockhopper Road, RAF Mount Pleasant. BFPO 655. ☎+500 32179. 🖷 +500 32193. **Email:** adriana@bfbs.com **L.P:** SM: Steve. Eng. Mgr: Adrian J. Almond.
MW: BFBS 2: Bush Rincon 550kHz 4kW **FM**(MHz): **BFBS 1:** 88.0, 90.0, 93.8, 96.5, 98.5, 102.0, 105.0, **Gurkha** on 96.0. **D.Prgr:** 24h N: Every hour from Independent Radio News by satellite from London. **Ann:** "This is BFBS in the Falklands". **V.** by QSL-card. Rp

RADIO NOVA

☞ KTV Ltd, 68 Dean St., Stanley. ☎+500 22349 ▤ +500 21049
Email: kmzb@horizon.co.fk **Web:** www.ktv.co.fk

BBC World Service rel.: 106.5MHz (also on FIRS 530kHz)
Deutsche Welle rel.: 101.1MHz
Saint FM rel.: 95.5MHz. (rebroadcast from St Helena).

FAROE ISLANDS (Denmark)

L.T: UTC (27 Mar-30 Oct: +1h) — **Pop:** 48,000 — **Pr.L:** Faroese
— **E.C:** 50Hz, 220/380V — **ITU:** FRO

KRINGVARP FØROYA ÚTVARPIÐ (Gov.)

☞ Norðari Ringvegur, P.O.Box 1299, FR-100 Tórshavn ☎ +298 347500
▤ +298 347501 **E:** kringvarp@kringvarp.fo **W:** www.kringvarp.fo
L.P: SM: Annika Mittún Jacobsen. PD: Jógvan Arge. TD: Hans Andor
Johannesen.
MW: Akraberg 531kHz 50/100kW
FM: Tórshavn 89.9MHz 31kW, Klaksvík 94.3MHz 41kW, Hesturin
Suðurðy 97.5MHz 27kW, 100.0MHz Støðlafjall 3kW + 15 lp stns
D.Prgr: 24h All prgrs are in Faroese, except wrp. in English at approx.
0855 LT during four summer months. **Ann:** 'Útvarpið'

RÁS 2 (Comm.)

☞ Vágsbotnur, P.O.Box 76, FR-100 Tórshavn ☎ +298 359999 ▤ +298
359990 **E:** ras2@ras2.fo **W:** www.ras2.fo
L.P: SM: Jonhard Hammer.
FM (MHz): Tórshavn 102.0MHz, Suðuroy 102.6MHz, Streymoy
106.0MHz, Skarvanes 106.3MHz, Klaksvik: 107.0MHz + 9 lp stns.
D.Prgr: MS: 0700-2400, Sun 0915-2230

LINDIN KRISTILIGT KRINGVARP (Rlg.)

☞ Bøkjaragøta 9, P.O.Box 2063, FR-165 Argir (Tórshavn) ☎ +298
321377 ▤ +298 321379 **E:** lindin@lindin.fo **W:** www.lindin.fo
L.P: Chairman: Preben Hansen
FM: Streymoy 98.0MHz, Tórshavn 101.0MHz, Klaksvik: 103.0MHz,
Hestin Há Vági 105.5MHz + 7 lp stns. **D.Prgr:** 24h

FIJI

L.T: UTC + 12h (24 Oct 10-16 Mar 11 +13h) DST 2011-2012 subject to
confirmation — **Pop:** 944,720 — **Pr.L:** English, Fijian, Hindi — **E.C:**
50Hz, 240V — **ITU:** FJI

DEPARTMENT OF COMMUNICATIONS

☞ 1st Floor, Credit Corporation Building, Suva ☎ +679 330 0766 ▤
+679 331 5167 **LP:** Dep. Secretary: Josua Turaganivalu

FIJI BROADCASTING COMMUNICATIONS LTD (Pub)

☞ PO Box 334, Suva ☎ +679 331 4333 ▤ +679 330 1643 **W:** www.
radiofiji.com.fj **E:** via website form **LP:** CEO: Francis Herman.

MW	kHz	kW	Netw.	MW	kHz	kW	Netw.
Suva	558	10	RF1	Labassa	810	1	RF2
Lautoka	639	10	RF1	Sigatoka	927	2.5	RF1
Labassa	684	10	RF1	Rakiraki	1152	2.5	RF1
Namara	774	2.5	RF2	Rakiraki	1467	2.5	RF2

FM	RF1	RF2	Bula FM	R.Mirchi	RFGOLD	2dayFM
Ba		105.0	91.4	90.6	94.6	
Tavua/Rakiraki	92.2			93.0	94.6	
Suva		105.2	102.0	98.0	100.4	104.0
Sigatoka/Taveuni		105.0	103.0	98.2	100.6	
Lautoka/Nadi		105.4	102.4	98.4	100.0	107.4
Labassa/Savusavu		105.4	102.4	98.4	100.0	

Networks: RF1: R.Fiji One (Fijian), **RF2:** R.Fiji Two (Hindi), **Bula
FM** (Fijian), **R.Mirchi** (Hindi), **RFGOLD:** R.Fiji GOLD (English),
2dayFM (English). All 24h

Private and Commercial Stations

FM	MHz	Station, location		FM	MHz	Station, location
1)	88.2	BBCWS, Suva	11)	94.0	Mai Mix FM, Lautoka/Nadi
8)	88.2	BBCWS, Lautoka/Nadi	11)	94.0	Mai Mix FM, Suva
11)	88.6	Mai Mix FM, Ba/Tavua	7)	94.8	R. Naya Jiwan, Suva
2)	88.8	R. Pasifik Triple 8, Suva	12)	95.0	Bula Namaste FM,
3)	89.2	femTALK 89.2, Suva			Ba/Tavua/Rakiraki
4)	91.8	RFI, Suva	8)	95.4	FM96, Lautoka/Nadi
5)	91.8	Hope FM, Lautoka/Nadi	8)	95.4	FM96, Rakiraki
5)	92.2	Hope FM, Taveuni	8)	95.4	FM96, Labassa
6)	92.6	R. Australia, Suva	8)	95.4	FM96, Levuka
6)	92.6	R. Australia, Lautoka/Nadi	8)	96.0	FM96, Suva
7)	93.6	R. Light, Sigatoka	8)	96.6	FM96, Sigatoka

FM	MHz	Station, location		FM	MHz	Station, location
8)	96.6	FM96, Ba	8)	102.8	Viti FM, Suva
8)	96.6	FM96, Savusavu	7)	103.4	Nai Talai FM, Suva
9)	97.0	FM97, Suva	8)	103.8	Viti FM, Ba
8)	97.4	Navtarang, Lautoka/Nadi	8)	103.8	Viti FM, Sigatoka
8)	97.4	Navtarang, Rakiraki	8)	103.8	Viti FM, Savusavu
8)	97.4	Navtarang, Labassa	8)	104.2	R.Sargam, Lautoka/Nadi
5)	97.6	Hope FM, Suva	8)	104.2	R.Sargam, Rakiraki
8)	98.8	Navtarang, Suva	8)	104.2	R.Sargam, Labassa
8)	99.2	FM96, Ba	8)	104.6	R.Sargam, Suva
12)	99.2	Bula Namaste FM,	8)	105.6	R.Sargam, Sigatoka
		Sigatoka/Nadi Lautoka	8)	105.8	R.Sargam, Ba
12)	99.4	Bula Namaste FM, Suva	8)	105.8	R.Sargam, Savusavu
8)	99.6	Viti FM, Lautoka/Nadi	8)	105.8	R.Sargam, Levuka
8)	99.6	Viti FM, Labassa	7)	106.0	R.Light, Pacific Harbor
8)	99.6	Viti FM, Rakiraki	8)	106.4	Legend FM, Lautoka/Nadi
8)	99.6	Viti FM, Levuka	8)	106.8	Legend FM, Suva
8)	101.6	Navtarang, Ba	8)	106.8	Legend FM, Labassa
8)	101.6	Navtarang, Sigatoka	8)	107.2	Legend FM, Sigatoka
8)	101.6	Navtarang, Savusavu	8)	107.2	Legend FM, Savusavu

Addresses and other information

1) 24hr satellite relay. – **2)** USP Student's Association, University of
the South Pacific, PO Box 1168, Suva ☎+679 313900 ▤+679 312
591 – **3)** PO Box 2439, Government Buildings, Suva. Mgr: Sharon
Bhagwan Rolls. ☎+ 679 331 6290 **E:** femlinkpac@connect.com.fj
Mobile UNESCO funded community stn for women, operates across
Viti Levu island – **4)** 24hr satellite relay – **5)** currently reported inac-
tive – **6)** 24hr satellite relay. Currently silent after closure by military.
– **7)** Evangelical Bible Missions Trust Board, Studio:15 Tower Street,
Suva. Networks: R.Light (English), R.Naya Jiwan (Hindi), Nai Talai FM
(Fijian) – **8)** Communications Fiji Ltd, 231 Waimanu Road (Private Mail
Bag), Suva ☎+679 331 4766 ▤+679 330 3748 **W:** www.fijivillage.
com **E:** info@fm96.com.fj **LP:** Man Dir: William Parkinson. Networks:
FM96 (English), Legend FM (English), Viti FM (Fijian), Navtarang
(Hindi), R.Sargam (Hindi). All 24h – **9)** Trinity Broadcasting Network,
Khalsa Road, Kinoya, Nasinu [Suva] – **11)** 9 Nasoki Street, Lautoka.
☎+679 666 8900 **ID:** "Tune in to Mai Mix" – **12)** Level 2, Pacific
House, Cnr Butt & MacArthur Streets, Suva ☎+679 330 4025 ▤+679
331 0685 **E:** info@bulamastefm.com **W:** www.bulamastefm.com
– **Other: Galoa Radio,** Galoa Island, Nabouwalu ☎+679 820 1499,
R. Pachim, 218 Kings Road, Ba ☎+679 667 8233

FINLAND

L.T: UTC +2h (27 Mar-30 Oct: +3h) — **Pop:** 5.3 million — **Pr.L:** Finnish,
Swedish — **E.C:** 50Hz, 230V — **ITU:** FIN

VIESTINTÄVIRASTO
(FICORA, Finnish Communications Regulatory Authority)

☞ PL 313, FI-00181 Helsinki ☎+358 9 69661 ▤ +358 9 6966410
W: www.ficora.fi **E:** info@ficora.fi **LP:** DG: Rauni Hagman. Dir. of
Radio Adm.: Kari Koho.

DIGITA OY (programme distributor)

☞ Jämsänkatu. 2, FI-00520 Helsinki ☎+358 20411711 ▤ +358
204117234 **W:** www.digita.fi **E:** info@digita.fi
L.P: DG: Sirpa Ojala. Vice Pres, Netw. & Site Sces: Ilari Anttila.

YLEISRADIO (YLE, Pub.)

☞ FI-00024 Yleisradio ☎+358 9 14801 ▤ +358 9 14803216 **W:** yle.fi
E: fbc@yle.fi **LP:** DG: Lauri Kivinen

FM (MHz)	1	2	3	4	5	6	7	kW
Aavasaksa	87.9	89.8	94.7					3
Ahvenanmaa			100.3		104.9	93.1		10/3
Anjalankoski	88.5	92.8	96.9	91.4	99.5b			30/3
Enontekiö	88.5	91.4	98.7	104.6			101.2	5
Espoo	87.9	91.9	94.0	103.7	98.9	101.1		60
Eurajoki	87.7	92.0	94.8	103.5	99.4	103.0		30/3
Fiskars	90.9	93.1	97.0	105.0	102.5	99.7		3/1
Haapavesi	89.0	96.1	98.4	101.9				30/3
Hanko					101.9			5
Hämeenlinna			99.2					1
Iisalmi	87.7	92.8	96.5	107.9				2
Inari	88.4	92.8	98.8	105.3			101.9	50/3
Joensuu			106.9					1
Joutseno	88.0	90.9	98.5	100.7				30/3
Jyväskylä	89.9	92.5	99.3	87.6	103.5			30/3
Karigasniemi	89.5	93.4	96.8	103.7			100.8	2
Kerimäki	90.5	95.8	99.1					30
			97.7	103.2				6/3

FM (MHz)	1	2	3	4	5	6	7	kW
Kiihtelysvaara	88.4	94.9	97.2	100.4				5
Koli	90.2	93.4	99.6	106.4	102.4b			30/5
Kruunupyy	91.4	94.0	97.6	88.8	99.7	102.7		60/3
Kuopio	91.6	93.9	98.1	88.1	100.2b			50
Kuttanen	94.1	97.2	99.6	105.6			102.2	3
Lahti	93.2	95.5	97.9	90.3	100.6b			50/0.2
Lapua	88.2	90.1	93.1	97.5	95.2	101.5		60/2
Lohja			96.1	105.0				3
Mikkeli	88.9	92.1	94.6	101.8				30/3
Nilsiä				90.8				
Nuorgam	88.6	93.9	97.7				101.2	3
Oulu	90.4	93.2	97.3		100.3b			50/3
Parikkala			95.1					
Pello	90.2	97.0	99.7	103.4				3
Pernaja	89.5	92.3	95.0	96.4	102.2	98.3		3/1
Pieksämäki	89.4	95.3	97.4	104.9				2/0.5
Pihtipudas	88.6	91.1	97.0	94.7	100.8b			50/2
Posio	87.6	91.5	98.6	104.0				30/3
Pyhätunturi	91.0	97.6	99.9	102.4				50
Pyhävuori	88.9	91.0	94.2/97.2	104.2	98.6	102.6		30/1
Rovaniemi	88.2	94.0	96.7	106.8		103.0		30/3
Ruka	90.7	92.8	95.1	104.3				3/2
Sodankylä	87.8	90.1	94.3	106.5		101.3		3
Taivalkoski	89.2	91.9	99.2/103.6	106.5				60/3
Tammela	89.2	91.3	96.0	105.4				5
Tampere	90.7	93.7	99.9	88.3	102.1b			60/6
Tenola(NOR)	89.0	94.1	95.8			100.5		0.02
Tervola	88.6	92.6	95.6	101.6				30/3
Turku	89.8	92.6	94.3	96.7	98.2	101.4		60/6
Utsjoki	90.7	93.1	99.4	107.1			102.6	2/5
Vaasa	87.8	89.6	94.8	105.2	97.3	101.0		1
Vuokatti	92.3	94.3	98.9	101.2				60/3
Ylläs	92.2	95.3	98.1	100.7		103.8		50
Ähtäri	91.9	94.6	96.2	102.9				3

+30 transmitters under 1kW not mentioned.
b = "FSR Mix", mixture of FM5 & FM6.
NB: Current actual powers for YLE Puhe are in many cases lower than the ones listed.
D.Prgr: FM1 "YLE Radio 1" (classical music, culture, actualities): 24h. **N:** 0400, 0500, 0600, 0900, 1100, 1400, 1600, 1700, 2000, 2200. **N. in English:** 1525. **N. in Russian:** 2055. **N. in Latin:** Fri 0755, Sat 1055 – **FM2 "YleX"** (rock & pop culture for youth): 24h (r. FM1 W00-04).**N:** on the h – **FM3 "Radio Suomi"** (news, sports, popular music and regional prgrs): 24h. **N:** on the h – **FM4 "YLE Puhe"** (news & talk prgr.): 24h. – **FM5 "Radio Extrem"** (Swedish language prgr for young people). 24h (simultaneous night prgr. with R Vega) – **FM6 "Radio Vega"** (Swedish language prgr for elderly people and regional prgrs). 24h – **"FSR Mix"** carries R. eXtrem MF 0400-0650, 1415-1700, 2000-2200 Sat 1500-0000 Sun 1500-2200. At other times R. Vega – **FM7 "Sámiradio"** (Sámi language network). 24h. Carries YLE, SR & NRK Sámiradio: MF: 0515-0830, 1100-1130, 1300-1630 Sat 1700-1800 Sun 1700-1830, at other times FM3 – **"YLE Mondo"** (digital network carried also via Espoo 97.5MHz 5kW): 24h.

Regional & local prgrs:
In Finnish on FM3: MF 0430-1535, Sat 0503-0955 excl. nationwide news on the h. – **Ylen aikainen**, 00240 Helsinki: 94.0MHz. – **R. Itä-Uusimaa**, Porvoo: 90.3/95.0MHz. (also r. 94.0MHz) – **Ylen läntinen**, Lohja: 97.0/105.0MHz. (also r. 94.0MHz) – **Tampereen R**, Tampere: 99.9MHz. – **Lahden R**, Lahti: 97.9MHz. – **R. Häme**, Hämeenlinna: 96.0/97.3/99.2/107.1MHz. – **Turun R**, Turku: 94.3/100.3/107.1MHz. – **Satakunnan R**, Pori: 94.8/97.2MHz. – **R. Keski-Suomi**, Jyväskylä: 87.6/97.0/99.3MHz. – **Kymenlaakson R**, Kouvola: 96.9MHz. – **Etelä-Karjalan R**, Lappeenranta: 89.1/97.2/98.5/103.2MHz. – **Pohjois-Karjalan R**, Joensuu: 97.2/97.7/99.6MHz. – **Etelä-Savon R**, Mikkeli: 94.6/97.4/99.1MHz. – **R. Savo**, Kuopio: 96.5/98.1MHz. – **Pohjanmaan R**, Vaasa: 93.1/94.2/94.8/96.6MHz. – **R. Keski-Pohjanmaa**, Kokkola: 97.6MHz. – **Oulu-R**, Oulu: 95.1/97.3/98.4/99.2/102.5MHz. – **Kainuun R**, Kajaani: 98.9/103.6MHz. – **R. Perämeri**, Kemi: 94.7/95.6MHz. – **Lapin R**, Rovaniemi: 96.7MHz + 15 more freqs.
In Swedish on FM6: MF 0430-1000, 1330 & 1430. **R. Vega Mellannyland**, Helsingfors: 101.1MHz. – **R. Vega Östnyland**, Borgå: 91.4/98.3MHz. – **R. Vega Västnyland**, Ekenäs: 99.7/101.9MHz. – **R. Vega Åboland**, Åbo: 93.1/101.4/103.0MHz. – **R. Vega Österbotten**, Vasa: 101.0/101.5/102.6/102.7MHz.

Digital: YLE's digital radio broadcasts are carried on Digital Video Broadcasting (DVB) network within the digital TV multiplexes. They include Ylen 1, Ylen klassinen, YleX, R. Extrem, YLE Puhe, R. Suomi, R. Vega, YLE World, YLE Mondo, YLE Multifoorumi, YLE FSR+ and commercial channels Uusi Kiss, Iskelmä and Harju ja Pöntinen.

Other stations; main networks:

FM (MHz)	1)	2)	3)	4)	5)	6)	7)	8)	9)
Alajärvi		104.3							
Anjalankoski	105.7		102.7	90.0	89.3		104.9	96.2	
Espoo	106.2					92.5			
Eurajoki	106.0		101.7				101.7	95.7	
Forssa		98.5		107.5	103.3		90.1		
Haapavesi	104.1	100.1					106.1		
Hanko		107.5	96.2		104.5				
Harjavalta						93.9			
Heinola		87.6							
Helsinki		96.2	104.6	94.9	98.1	96.8	90.0	89.0	92.9
Huittinen		93.0							
Hyvinkää					95.7				
Hämeenlinna	100.2	101.7	106.5	92.3			97.3	105.9	88.1
Iisalmi		89.5	103.1	104.7	89.1				
Ikaalinen		99.0							
Imatra		105.3				102.5			
Inari						104.1			
Inkoo							105.5		
Joensuu		92.8	87.9	103.7	102.9	96.4			
Joutseno	103.8		94.2					96.0	
Juuka		103.3							
Jyväskylä	105.8	107.1	101.6	97.7	104.9	97.3	101.0	94.1	96.2
Jämsä		100.3	94.4						
Järvenpää							101.8		
Kajaani			102.8	107.0	96.3	93.7			
Kalajoki					104.6				
Kemi			105.2	98.8					
Kemijärvi	104.7								
Kerimäki	107.7							91.3	
Kitee		102.2							
Kokkola				99.1	106.3				
Koli	104.3							107.4	
Kotka		87.7			101.5				
Kouvola		100.1				93.8	96.2		
Kristiinankaup.		105.1							
Kruunupyy	107.2					104.9		104.3	
Kuopio	106.7	96.7	93.0	100.9	107.3	101.6	89.1	106.1	94.8
Kurikka		92.3							
Köyliö					107.9				
Lahti	104.4	103.0	105.0	89.7	102.8	96.6	94.2	106.4	107.4
Lappeenranta		93.5			94.8	96.5		96.0	
Lapua	106.5	96.9	105.4					89.4	
Lempäälä						102.8			
Lohja		96.5				88.8		107.2	
Loimaa		98.5							
Loviisa					104.6	105.2			
Luumäki								96.0	
Mikkeli	106.3	89.7	100.5	93.0	104.8		100.9	87.8	
Mäntsälä					103.4				
Mäntyharju				93.0					
Nilsiä					97.5				
Orivesi		103.8	101.2						
Oulu	104.8	89.4	101.4	95.8	96.4	99.1	106.2	106.9	99.6
Outokumpu		101.7							
Pieksämäki		102.2	101.3						
Pihtipudas	105.1	107.0				104.5		102.3	
Pohja		95.1							
Pori		100.4	104.5	96.5	90.4	98.7		95.7	
Porvoo		99.8				93.5			90.8
Pyhätunturi	105.8					106.2			
Pyhävuori	107.6								
Raahe		92.5	107.0	89.9		87.7			
Rauma		105.1	103.6						
Riihimäki			99.6				94.7		
Rovaniemi	105.5	89.3	102.0	107.0	103.1	101.1		93.4	
Ruka	100.8					96.3			
Ruovesi		103.8							
Salo			99.1				107.7		
Savonlinna		96.7	104.2			105.2		91.3	100.0
Seinäjoki		96.9		100.4	103.3	91.2		89.4	
Sievi		107.7							
Siilinjärvi		102.0							
Sonkajärvi		107.1							
Taivalkoski	106.5					94.6			
Tammisaari				91.4	89.4		107.0		104.3
Tampere	104.7	100.9	89.6	104.2	91.6	90.0	105.6	97.2	92.2
Tervola	107.5					100.1			
Tornio				92.0					
Turku	103.9	100.1	98.7	97.6	106.8	104.6	102.4	107.3	106.8
Uusikaupunki		96.2							
Vaasa			104.4	91.6	102.0	93.9			
Valkeakoski		95.0							
Varkaus		92.7				91.0	105.5		
Vihti		105.6							

FM (MHz)	1)	2)	3)	4)	5)	6)	7)	8)	9)
Vilppula		95.4							
Vuokatti	105.7				88.8				
Ylivieska			88.3						
Ylläs	107.9					106.0			
Ähtäri		97.8	102.9						

1) R. Nova, Ilmalank. 2C, PL 123, 00241 Helsinki. **W:** www.radionova.fi Powers 1-60kW – **2) Iskelmä,** Kehräsaari B5. 33200 Tampere. **W:** www.iskelma.fi Powers 0.1-3kW – **3) The Voice,** Tallbergink. 1C 7. krs, 00180 Helsinki **W:** www.voice.fi Powers 0.1-30kW – **4) R. Rock,** PL 350, Tehtaankatu 27-29 A, 00151 Helsinki **W:** www.radiorock.fi Powers 0.1-1kW – **5) R. Suomipop,** Lintulahdenk. 10, 00500 Helsinki **W:** www.radiosuomipop.fi Powers 0.1-1kW – **6) R. NRJ (Energy),** Kiviaidankatu 2 i, 00210 Helsinki. **W:** www.nrj.fi Powers 0.1-1kW – **7) R. Aalto,** PL 350, Tehtaankatu 27-29 A, 00151 Helsinki. **W:** www.radioaalto.fi Powers 0.1-10kW – **8) R. Dei (Rlg.),** Ilmalankuja 2 i, 00240 Helsinki. **W:** www.radiodei.fi Powers 0.2-5kW – **9) Rondo FM,** Ilmalankuja 2L, 00240 Helsinki. **W:** rondofm.fi Powers 0.2-1kW. About 50 more stations are in operation.

ÅLAND (autonomous province)

SVERIGES RADIO cf. Sweden

FM (MHz)	P1	P2	P3	P4	kW
Mariehamn	95.0		102.3	10	

Steel FM, Mariehamn: 95.9MHz 0.2kW. **W:** www.steelfm.net – **Rix FM** (cf. Sweden), Mariehamn: 101.8MHz 3kW – **R. Harmonica,** Mariehamn: 102.8MHz 1kW – **Soft FM,** Mariehamn: 107.2MHz 0.2kW. **W:** www.softfm.net – **Ålands R,** Mariehamn: 91.3MHz 10kW. **W:** www.radiotv.aland.fi

FRANCE

L.T: UTC +1h (27 Mar-30 Oct: +2h) — **Pop:** 63 million — **Pr.L:** French — **E.C:** 50Hz, 220V — **ITU:** F

CONSEIL SUPÉRIEUR DE L'AUDIOVISUEL (CSA)
✉ 39/43 quai André Citroën, 75739 Paris cedex 15 ☎ +33 1 40583800 📠 +33 1 45790006 **W:** www.csa.fr
LP: Pres: Michel Boyon
The CSA regulates TV and radio, and issues broadcast licenses.

TDF
✉ 106 avenue Marx Dormoy, 92541 Montrouge cedex ☎ +33 1 55951000 📠 +33 1 55952000 **W:** www.tdf.fr
LP: DG: Olivier Huart
TDF operates the majority of radio txs used by Radio France, Radio France International, RFO, TV txs and around 2000 private FM txs

TOWERCAST
✉ 46/50 avenue Théophile Gautier, 75016 Paris ☎ +33 1 40714071 **W:** www.towercast.fr
LP: Pres. : Jacques Roques
Towercast operates 1000 radio FM txs used by Radio France or private FM stns and a part of DTT txs

RÉSEAU FRANCE OUTRE-MER (RFO) (Pub)
✉ 35/37 rue Danton, 92240 Malakoff ☎ +33 1 55227100 **W:** www.rfo.fr **LP:** DG: Yves Garnier.
RFO is a part of TDF and produces public service prgrs (radio & TV) in the French overseas territories.

RADIO FRANCE (Pub)
✉ 116 Av. du Président Kennedy, 75220 Paris cedex 16 ☎ +33 1 56402222 **W:** www.radiofrance.fr **LP:** Pres. & DG: Jean-Luc Hess

HOME SERVICES:

LW & MW	N	kHz	kW	MW	N	kHz	kW
Allouis	A	162	*2000	Lille		1377	300
Paris	F	585	5	Ajaccio	B+L	1404	20
Lyon	I	603	300	Brest	I+L	1404	20
Rennes	I+L	711	300	Dijon	I	1404	5
Limoges	I	792	300	Grenoble	I	1404	20
Nancy	I	837	200	Pau	I	1404	20
Paris	B+L	†1864	300	Bastia	B+L	1494	20
Toulouse	I+L	945	300	Bayonne	I	1494	4
Bordeaux	I	1206	300	Besançon	I	1494	5
Marseille	I	1242	150	Clermont-Fd	I	1494	20
Strasbourg	B+L	1278	300	Nice	I	1557	300

N=Networks: A=France Inter, B=France Bleu, F=FIP, I= France Info, L=rel. local stns at certain times. †=AM stereo C-QUAM (*)1000kW 1700-0500 (Wi. time) 1900-0400 (Su. time)

FM: Station	C	D	E	F	kW
Abbeville	93.1	97.4	89.8		2.5
Ajaccio	92.4	97.6	88.0		10
Ajaccio (La Punta)	88.6	103.9		105.6	4
Albi				105.5	1
Alençon	93.0	88.0	91.0		13
Ales	87.6	96.1	98.6	105.1	1
Amiens (St Just)	95.4	102.5	99.4		20
Amiens (Dury)	92.6	89.0	89.3	105.5	2
Angers	93.2	91.4	97.4		10
Angers (La Ballue)				105.5	1
Angoulême	92.4	87.6	95.1	105.5	2
Arcachon	88.3	97.0	91.0	105.5	1.2
Argenton sur Creuse	101.9	89.8	97.2		1
Arles				105.0	1
Arnay le Duc	94.6	90.3	100.3		3
Aurillac	94.5	98.0	91.9		7
Autun	88.1	97.3	94.1		10
Auxerre	99.5	89.5	92.8		5
Auxerre (Venoy)				105.5	1
Avignon	97.4	90.7	93.2		4
Avignon (Sorgues)				105.2	2
Bar le Duc	90.9	88.4	92.7	104.5	10
Bastia	95.9	89.2	93.9	105.5	10
Bayonne	89.0	96.1	92.7	105.5	16
Beaucaire				105.2	1
Beauvais				105.5	1
Bergerac	92.3	94.0	97.1		26
Besançon (Montfaucon)	98.7	89.3	95.0	10	
Besançon (Lomont)	90.0	97.7	92.9		18
Beziers				105.1	1
Bonifacio				103.2	1
Bordeaux	89.7	97.7	93.5	105.5	6
Boulogne sur Mer	103.3	99.9	89.4	106.5	1
Bourges	94.9	88.5	91.8		74
Bourges (town)				105.5	1
Brest	95.4	97.8	89.4		200
Brest (town)				105.5	3
Briançon	91.5	97.8	89.5	105.4	1
Brignoles	106.7	104.0	105.5		1.5
Caen	99.6	91.5	95.6		100
Caen (town)				105.5	1
Calais	104.7			105.6	1
Cannes				105.9	1
Carcassone	88.3	96.5	90.9		80
Castres				105.5	2
Chambéry	93.5	90.5	98.6		8
Chambéry (town)				105.1	1
Champagnole	88.5	91.7	98.3		1
Charleville-Mézières	95.8	90.1	93.5	105.9	10
Chartres	94.6	98.1	89.7		32
Chartres (town)				105.7	4
Chateaubriant				105.5	1
Châteauroux				105.5	1
Chaumont	96.5	90.4	93.3		15
Chaumont (town)				105.5	1
Cherbourg-Octeville	94.1	89.2	92.3	105.6	1
Cholet				105.9	1
Clermont-Ferrand	90.4	98.4	95.5		35
Compiègne				105.3	1
Corse (East)	968	92.3	99.8		17
Corte	98.2	91.0	94.8		1.3
Creil	87.6	93.3	91.9	105.6	1
Dijon	95.9	93.7	99.2		25
Dunkerque				106.5	1
Epinal	98.6	92.4	89.4	106.5	10
Evreux	88.5	98.9	97.3	105.5	1
Fontainebleau				105.5	2
Gap	98.3	88.5	95.3	105.5	5
Gex	94.4	96.7	89.6	101.1	25
Grenoble (Chamrousse)	99.4	88.2	91.8	1	
Grenoble (T. s. Venin)	89.9	92.8	95.5	105.1	12
Guéret	100.7	98.8	90.8	105.5	12
Hirson	94.4	99.7	97.2		5
Hyères	91.6	97.5	94.5	107.1	1.5
L'Ile Rousse				105.2	1
Laon				105.3	1
Laval	95.1	88.3	92.1	105.5	5
La Rochelle				105.5	1
Le Havre	88.9	93.3	98.5	105.5	1
Le Mans	92.6	89.0	97.0	105.5	128
Le Puy	99.3	89.3	92.8		10
Lesparre	92.4	90.3	95.1		1.6
Lille (Bouvigny)	103.7	98.0	88.7	105.2	125
Limoges	93.0	89.5	97.5		150
Limoges (town)				105.5	2
Longwy	98.1	88.3	91.0	104.3	5

FM: Station	C	D	E	F	kW
Lourdes				105.3	3.5
Lyon (Mont Pilat)	99.8	88.8	92.4	103.4	150
Lyon (Town)	101.1	94.1	98.0	105.4	1
Mantes la Jolie	95.0	92.4	97.1		5
Marseille	91.3	99.0	94.2		400
Marseille				105.3	13
Marseille (town)	91.7	98.6	94.7		1
Maubeuge				106.2	2
Melun				105.7	1
Mende	90.1	96.9	93.7		10
Menton	97.0	89.6	91.7	105.5	5
Metz	99.8	94.5	89.7	106.8	145
Millau	94.9	99.2	88.9		6
Mont de Marsan				105.5	6
Montargis	102.9	98.8	94.1	105.5	1
Montauban				105.7	1
Montereau				105.7	1
Montlieu la Garde	88.3	104.8	98.8		3
Montluçon				105.5	1
Montpellier	89.4	97.8	92.9		18
Montpellier (Town)			96.4	105.1	1
Morosaglia	97.1	88.8	93.4		
Mulhouse	95.7	88.6	91.6	105.5	100
Nancy	96.9	88.7	91.7	105.9	5
Nantes	90.6	94.2	98.9	105.5	125
Neufchateau	96.3	100.3	91.5		1
Neufchatel-en-Bray	92.7	96.0	90.2		5
Nevers				105.5	1
Nice	100.2	101.9	92.2	105.7	100
Nimes	88.7			105.1	5
Niort	99.4	96.4	91.1		190
Niort (town)				105.5	1
Orléans	99.2	95.8	90.7		4
Orléans (town)				105.5	1
Paris	87.8	93.5	91.7	105.5	10
Parthenay	93.8	87.9	98.5	105.5	12
Pau				105.5	1
Perpignan	92.1	99.8	97.2	105.1	10
Poitiers	97.7	92.3	95.5	105.5	1
Porto Vecchio (Col de Mela)	96.8	90.8	98.9		1.5
Porto Vecchio (Punto di a Varra)	92.6	87.9	94.6		1
Privas	89.8	96.5	94.7	105.2	1
Redon				95.8	1
Reims	96.8	98.8	89.2	105.5	135
Rennes	93.5	98.3	89.9		100
Rennes (town)				105.5	2.5
Roanne				105.5	1
Rouen	96.5	94.0	92.0		100
Rouen (town)				105.7	2.7
Ruffec				105.2	1
Saint Brieuc				105.5	1
Saint Etienne	88.0	91.7	97.1	105.6	2
Saint-Nazaire	95.2	92.2	102.6	105.5	1.5
Saint-Quentin				105.6	1
Saint-Raphaël	96.3	88.7	99.6		40
Saint-Raphaël (town)				106.0	1
Sainte Foy la Grande				105.5	1
Sarrebourg	93.1	99.4	90.3		10
Sens	96.3	98.5	93.8		10
Sens (town)				94.3	1
Soissons				105.7	1
Strasbourg	97.3	87.7	95.0	104.4	48
Toulon	92.0	97.1	94.9	105.8	5
Toulouse (town)	88.1	96.3	91.1	105.5	2
Toulouse (Pic du Midi)	87.9	95.7	91.5		72
Tours	99.9	97.8	92.2		8
Tours (town)				105.5	1
Troyes	97.7	97.9	91.4		50
Troyes (town)				105.5	1
Ussel	96.0	88.2	99.7		10
Valence				105.4	3
Vannes	88.6	96.0	91.8	105.5	20
Verdun	92.1	99.3	97.4	106.3	6
Villebon sur Yvette	95.4	98.0	97.1		1
Villers-Cotterets	91.1	89.6	92.9		13
Vittel	98.2	89.0	94.0		8
Voiron	91.5	89.2	107.2	105.4	1

+ 1466 stns under 1kW

C=France Inter (stereo), D=France-Culture (stereo), E=France-Musique (stereo), F=France Info (mono). RDS on all txs.

France Inter Network A on **LW**, C on **FM**; Allouis 162kHz: **D.Prgrs**:24h exc. Tues 0005-0358. FM txs: 24h. **N**: Hourly, plus 0430, 0530, 0630 – **France Culture** (Network D) (stereo) **D.Prgrs**:24h. **N**:

0600, 0700, 0800, 1130, 1700, 2100 – **France Musique** (Network E) (stereo): **D.Prgrs**:24h **N**: 0600, 0700, 0800, 1200, 1700 – **France Info** (Network F) News and informations **D.Prgrs**:24h

Le Mouv'

📧 78 allée Jean Jaurès, 31009 Toulouse Cedex 6 ☎ +33 5 34417000
📠 +33 5 34417006

Station	MHz	kW	Station	MHz	kW
Ajaccio	92.0	4	Lyon	87.8	4
Amiens	91.0	1	Marseille	96.8	2.5
Angers	96.0	1	Marseille (town)	96.4	1
Besançon	93.5	1	Mende	107.2	0.2
Bordeaux	87.7	1	Montpellier	102.7	3
Brest	94.0	3	Nantes	96.1	3
Caen	87.8	1	Nice	101.0	1
Cannes	101.0	0.1	Paris	92.1	8
Carcassonne	90.0	1	Reims	101.1	0.5
Clermont-Fd	97.5	2	Rennes	107.3	2
Dijon	88.9	1	Rouen	95.8	1
Lille	91.0	2	Toulouse	95.2	5
Lorient	103.3	0.5	Tours	94.1	2
Limoges	107.6	2	Valence	100.7	0.5

D.Prgrs: 24h RDS on all txs. (stereo)

Local Stations "FIP"

FIP Bordeaux, 95 rue Judaïque, 33000 Bordeaux ☎ +33 5 56241313 - Bordeaux 96.7MHz 2.5kW, Arcachon 96.5 0.5 kW
FIP Nantes, 2 bis quai François Mitterrand, 44200 Nantes ☎ +33 2 40731414 - Nantes 95.7MHz 2.5kW, St Nazaire 97.2MHz 1.5kW
FIP Paris, 116 avenue du Président Kennedy, 75220 Paris Cedex 16 ☎ +33 1 42201234 - 585kHz 5kW, 105.1MHz 10kW
FIP Strasbourg, 4 rue Joseph Massol, 67000 Strasbourg ☎ +33 3 88352400 - 92.3MHz 4kW
Sts without local news: Marseille 90.9MHz 4kW, Montpellier 99.7MHz 1kW, Rennes 101.2MHz 1kW, Toulouse 103.5MHz 2kW
RDS on all txs **D.Prgrs:** 24h. Prgrs consist of music and news.

France Bleu

📧 116 av. du Président Kennedy, 75220 Paris Cedex 16
☎ +33 1 56401111 **D.Prgrs:** 24h uninterrupted music 2200-0400
Stations: **MW:** Network B + **FM**
France Bleu Local Stations (F.B = France Bleu) - At certain times, local stns relay national France Bleu prgrs.
F.B 107.1, 116 av du Président Kennedy, 75220 Paris Cedex 16 ☎ +33 1 56402222: **FM:** Paris 107.1MHz 10kW, Chartres 97.3MHz 4kW, **MW:** Paris 864kHz 300kW (C-QUAM stereo)
F.B Alsace, 4 rue Joseph Massol, 67000 Strasbourg ☎ +33 3 88762000 **FM:** Strasbourg 101.4MHz 48kW, Mulhouse 102.6MHz 100kW, **MW:** 1278kHz 300kW
F.B Armorique, 14 av Jean Janvier, 35031 Rennes Cedex ☎ +33 2 99674321 **FM:** Vannes 101.3MHz 20kW, Rennes 103.1MHz 100kW
F.B Auxerre, 12 place Saint Amâtre, B.P 101, 89002 Auxerre Cedex ☎ +33 3 86723456 **FM:** Sens 100.5MHz 10kW, Auxerre 101.3MHz 5kW, Nevers 104.0MHz 1kW
F.B Azur, 2 place Grimaldi, 06012 Nice Cedex 1 ☎ +33 4 97033636 **FM:** Nice 103.8MHz 100kW, Menton 94.8MHz 5kW, Saint Raphaël 100.7MHz 10kW
F.B Basse Normandie, 75 rue Basse, 14053 Caen Cedex ☎ +33 2 31471414 **FM:** Le Havre 102.2MHz 2.5kW, Caen 102.6MHz 10kW
F.B Béarn, 2 rue O'Quin, BP 211, 64002 Pau Cedex ☎ +33 5 59983030 **FM:** Oloron Sainte Marie 93.2MHz 1.5kW, Pau 102.5MHz 10kW
F.B Belfort Montbéliard, 10 rue des Capucins, 90008 Belfort Cedex ☎ +33 3 84579090 **FM:** Belfort 106.8MHz 2kW
F.B Berry, 10/12 rue de la République, 36000 Châteauroux ☎ +33 2 54606060 **FM:** Argenton 93.5MHz 5kW, Bourges 103.2MHz 19kW
F.B Besançon, 2 Place Granvelle, BP 591, 25027 Besançon Cedex ☎ +33 3 81212525 **FM:** Besançon 101.4MHz 18kW + 102.8MHz 10kW
F.B Bourgogne, 29 rue Guillaume Tell, BP 11888, 21018 Dijon Cedex ☎ +33 3 80592121 **FM:** Troyes 87.8 60kW, Arnay le Duc 103.4MHz 3kW, Dijon 103.7MHz 25kW
F.B Breizh Izel, 12 esplanade François Mitterrand, 29000 Quimper ☎ +33 2 98552929 **FM:** Brest 93.0MHz 200kW
F.B Champagne, 28 bd du Maréchal Joffre, BP 1094, 51054 Reims Cedex ☎ +33 3 26845151 **FM:** Charleville-Mézières 100.9MHz 10kW, Reims 95.1MHz 2kW, Châlons en Champagne 94.8MHz 1kW, Troyes 100.8MHz 1kW
F.B Cotentin, Hôtel Atlantique, rue Piedagnel, 50100 Cherbourg-Octeville ☎ +33 2 33885050 **FM:** Cherbourg-Octeville 100.7MHz 4kW
F.B Creuse, 7 rue de la République, 23000 Guéret ☎ +33 5 55612323 **FM:** Guéret 94.3MHz 12kW
F.B Drôme Ardèche, 7 rue Poncet, BP 519, 26005 Valence Cedex ☎ +33 4 75813333 **FM:** Valence 87.9MHz 9kW, Privas 98.4MHz 1.5kW, Vals les Bains 103.8MHz 1kW

F.B Frequenza Mora, 4 rue Favalelli, BP 130, 20289 Bastia Cedex ☎ +33 4 95329532 **FM:** Corse (east) 88.2MHz 17kW, Ajaccio 100.5MHz 10kW, + 97.0MHz 4kW + 1404kHz 20kW, Corte 100.0MHz 1.33kW, Bastia 101.7MHz 10kW + 1494kHz 20kW, Porto Vecchio 101.8MHz 1.5kW + 105.4MHz 1kW, Morosaglia 104.6MHz 1kW

F.B Gard Lozère, 10 bd des Arènes, 30020 Nîmes Cedex 1 ☎ +33 4 66363030 **FM:** Nîmes 90.2MHz 5kW, Alès 91.6MHz 2kW, Mende 104.9MHz 10kW

F.B Gascogne, 13 place Jean Jaurès, BP 289, 40005 Mont de Marsan Cedex ☎ +33 5 58854040 **FM:** Mont de Marsan 98.8MHz 20kW, Bayonne 100.5MHz 38kW, Mimizan 103.4MHz 20kW

F.B Gironde, 95 rue Judaïque, BP 585, 33006 Bordeaux Cedex ☎ +33 5 57812020 **FM:** Bordeaux 100.1MHz 6kW, Lesparre 101.6MHz 1.6kW

F.B Haute Normandie, 45 bd des Belges, 76000 Rouen ☎ +33 2 35073107 **FM:** Le Havre 95.1MHz 1kW, Rouen 100.1MHz 100kW, Neufchâtel en Bray 101.6MHz 5kW, Evreux 89.5MHz 1kW

F.B Hérault, 474 allée Henri II de Montmorency, 34034 Montpellier ☎ +33 4 67066565 **FM:** Montpellier 101.1MHz 18kW + 100.6MHz 1kW

F.B Isère, 27 av Félix Viallet, BP 154, 38003 Grenoble Cedex ☎ +33 4 76031838 **FM:** Chambéry 99.1MHz 5kW, Lyon 101.8MHz 25kW, Grenoble 102.8MHz 1kW + 98.2MHz 1.2kW

F.B La Rochelle, 5 av Michel Crépeau, 17025 La Rochelle Cedex 01 ☎ +33 5 46351717 **FM:** Royan 101.8MHz 1kW, Saintes 103.9MHz 60kW, Angoulême 101.5MHz 2kW, La Rochelle 98,2MHz 1kW

F.B Limousin, 23 bd Gambetta, BP 3603, 87036 Limoges Cedex 1 ☎ +33 5 55113811 **FM:** Chateauponsac 92.5MHz 1kW, Ussel 101.4MHz 10kW, Limoges 103.5MHz 150kW

F.B Loire Océan, 2 bis quai François Mitterrand, 44200 Nantes ☎ +33 2 40444546 **FM:** Saint Nazaire 88.1MHz 1.5kW, Nantes 101.8MHz 200kW, Angers 88.5MHz 1k W

F.B Lorraine Nord, 5, rue d'Austrasie, B.P 50071, 57003 Metz cedex 03 ☎ +33 3 87682222 **FM:** Metz 98.5MHz 1kW

F.B. Maine, 17 rue Pierre Mendès France, 72000 Le Mans ☎ +33 2 43297272 **FM:** La Flèche 91.7MHz 1 kW, Le Mans 96MHz 1 kW, Sablé sur Sarthe 105.7MHz 1kW

F.B Mayenne, 41 av Robert Buron, 53000 Laval ☎ +33 2 43495050 **FM:** Laval 96.6MHz 5kW

F.B Nord, 14 rue Léon Trulin, 59002 Lille Cedex 01 ☎ +33 3 20135962 **FM:** Lille (town) 87.8MHz 1kW, Lille (Bouvigny) 94.7MHz 125kW, Boulogne sur Mer 95.5MHz 1kW, Etaples 97.8MHz 2kW, Calais 106.2MHz 1kW

F.B Orléans, 8 rue d'Illiers, 45057 Orléans Cedex 1 ☎ +33 2 38714545 **FM:** Blois 93.9MHz 1kW, Orléans 100.9MHz 4kW, Montargis 106.8MHz 1kW

F.B Pays Basque, 46 allées Marines, 64116 Bayonne Cedex ☎ +33 5 59466464 **FM:** Bayonne 101.3MHz 15kW

F.B Pays d'Auvergne, 80 bd François Mitterand, BP 277, 63008 Clermont-Ferrand Cedex 01 ☎+33 4 73346363 **FM:** Clermont-Ferrand 102.5MHz 37kW, Aurillac 100.2MHz 1kW, Montluçon 96.7MHz 1kW

F.B Pays de Savoie, 256 rue de la République, 73000 Chambéry ☎ +33 4 79707374 **FM:** Annecy 95.2MHz 1kW, Chambéry 103.9MHz 8kW, Gex 106.1MHz 20kW

F.B Périgord, 1 cours Saint Georges, BP 3033, 24003 Périgueux Cedex ☎ +33 5 53062000 Limoges **FM:** 91.7MHz 100kW, Bergerac 99.0MHz 26kW

F.B Picardie, Rue du Maréchal de Lattre de Tassigny, 80000 Amiens ☎ +33 3 22711515 **FM:** Amiens 100.2MHz 2kW, Abbeville 100.6MHz 5kW, Hirson 101.3MHz 5kW, Sailly Saillisel 102.8MHz 15kW

F.B Poitou, 27, bd de Solférino, 86000 Poitiers ☎ +33 5 49605000 **FM:** Parthenay 106.4MHz 12kW, Niort 101.0MHz 1kW

F.B Provence, 560 av Mozart, 13617 Aix en Provence.Cedex 01 ☎ +33 4 42991313 **FM:** Brignoles 102.1MHz 1.5kW, Hyères 102.5MHz 1.5kW, Toulon 102.9MHz 5kW, Marseille 103.6MHz 200kW

F.B Roussillon, 24 av du Général Leclerc, 66000 Perpignan ☎ +33 4 68519000 **FM:** Perpignan 101.6MHz 1kW

F.B Sud Lorraine, 21/23 bd du Recteur Senn, 54042 Nancy Cedex ☎ +33 3 83195488 **FM:** Epinal 100.0MHz 1kW, Nancy 100.5MHz 5kW, Vittel 102.6MHz 1kW, Neufchateau 103.0MHz 1kW

F.B. Toulouse, 78 allée Jean Jaurès, BP 50901, 31009 Toulouse ☎ +33 5 34417010 **FM:** Toulouse 90.5MHz 2kW

F.B Touraine, place Gaston Pailhou, BP 3231, 37032 Tours Cedex 1 ☎ +33 2 47363737 **FM:** Tours 105.0MHz 8kW +98.7MHz 1kW

F.B Vaucluse, 25 rue de la République, BP 320, 84021 Avignon Cedex ☎ +33 4 90411312 **FM:** Avignon 100.4MHz 2kW

+ 324 txs less than 1kW not mentioned. Stereo and RDS on all txs.

Special Programmes (MW)
Lyon 603kHz (Rlg) Sun 1700-1800. ✉ Foyer Notre Dame des Ondes, 24 rue Paul Sisley, 69003 Lyon – **Strasbourg** 1278kHz (Rlg) Sun 0700-0900. First Sun 1200-1300 prgr in cooperation with SWF4. Prgr in Alsatian language 0600-1130 + 1300-1600 ✉ See under F.B.

Alsace – **Rennes** 711kHz + Brest 1404kHz Prgr in Breton language. Sat 1106 -1200 ✉ See under F.B. Armorique – **Toulouse** 945kHz Prgr in Occitan language. Sat 1045-1145 ✉ Passejada Occitana, France 3, 24 chemin de la Cépière, 31081 Toulouse Cedex ☎ +33 5 62 23 97 97 **W:** http://sud.france3.fr

RADIO FRANCE INTERNATIONALE (Pub)
✉ 116 av. du Président Kennedy, 75016 Paris ☎ +33 1 56401212 **W:** www.rfi.fr **LP:** Pres. & DG: Alain de Pouzilhac
RFI1 (French service): Paris **FM** 89MHz 10kW (stereo).
EXTERNAL SERVICE see International Broadcasting section.

PRIVATE MW STATIONS

MW	Location	kHz	kW		MW	Location	kHz	kW
1)	Toulouse	819	1		2)	Nice	1350	10
2)	Nîmes	1602	1					

Addresses
1) Sud Radio, see entry under FM – **2)** Radio Orient, see entry under FM.

PRIVATE FM STATIONS

FM	Station	MHz	kW		FM	Station	MHz	kW
25)	Auxerre	87.6	1		17)	Lyon	88.4	4
23)	Bayonne	87.6	3		6)	Mont de Marsan	88.4	1
17)	Bernay	87.6	1		19)	Nantes	88.4	2
17)	Besançon	87.6	1		21)	Sarreguemines	88.4	1
23)	Castres	87.6	1		5)	Thouars	88.4	1
20)	Le Havre	87.6	1		5)	Tonnerre	88.4	1
22)	Niort	87.6	1		10)	Compiègne	88.5	1
17)	Orléans	87.6	2		6)	Nogent le Rotrou	88.5	1
17)	Romilly sur Seine	87.6	1		20)	Quimper	88.5	1
22)	Vannes	87.6	1		17)	Annecy	88.6	1
8)	Yssingeaux	87.6	1		17)	Châlons en Champ.	88.6	1
23)	Bourges	87.7	1		6)	Châteaubriant	88.6	1
13)	Clermont Ferrand	87.7	1		6)	Chaumont	88.6	1
5)	Corte	87.7	1		19)	Confolens	88.6	1
5)	Figeac	87.7	1		16)	Paris	88.6	4
19)	Nice	87.7	2		23)	Porto Vecchio	88.6	1
21)	Saint Omer	87.7	1		22)	Vichy	88.6	1
9)	Tours	87.7	2		10)	Alençon	88.7	1
19)	La Flèche	87.8	1		25)	Avallon	88.7	1
7)	Le Blanc	87.8	1		23)	Bastia	88.7	1
18)	Mayenne	87.8	1		6)	Caen	88.7	2
23)	Mazamet	87.8	1		18)	Chartres	88.7	1
7)	Montluçon	87.8	1		18)	Châteauroux	88.7	1
7)	Verdun	87.8	1		19)	Étampes	88.7	1
6)	Dijon	87.9	1		25)	Ghisonaccia	88.7	4
21)	Menton	87.9	1		6)	Gray	88.7	1
6)	Montreuil	87.9	1		18)	Saint Flour	88.7	1
17)	Reims	87.9	2		13)	Saintes	88.7	1
5)	Saint Raphaël	87.9	1		22)	Toulouse	88.7	5
2)	Toulon	87.9	2		6)	Bonnières sur Seine	88.8	2
9)	Yvetot	87.9	1		12)	Clermont Ferrand	88.8	1
17)	Aubusson	88.0	1		8)	Nantes	88.8	3
10)	Calais	88.0	1		6)	Reims	88.8	2
23)	Châteauroux	88.0	1		17)	Saint Dizier	88.8	1
23)	Colmar	88.0	1		8)	Bagnères de Bigorre	88.9	1
17)	St Gilles Croix de Vie	88.0	1		17)	Bordeaux	88.9	1
2)	Vesoul	88.0	1		13)	Montluçon	88.9	1
3)	Villefranche sur Saône	88.0	1		23)	Rennes	88.9	1
3)	Vitry le François	88.0	1		6)	Aurillac	89.0	1
17)	Angers	88.1	1		6)	Avignon	89.0	1
3)	Avignon	88.1	1		6)	Avranches	89.0	1
6)	Brive la Gaillarde	88.1	1		17)	Brest	89.0	3
18)	Châtellerault	88.1	1		17)	Clamecy	89.0	1
23)	Dole	88.1	1		13)	Moulins	89.0	1
6)	Nice	88.1	5		23)	Bernay	89.1	1
17)	Rouen	88.1	1		1)	Bourges	89.1	1
6)	Soissons	88.1	1		5)	Gien	89.1	1
3)	Fontenay le Comte	88.2	1		18)	Perpignan	89.1	3
4)	Laval	88.2	1		8)	Saint Nazaire	89.1	1
17)	Le Havre	88.2	1		22)	Saint Quentin	89.1	1
10)	Nancy	88.2	1		22)	Valenciennes	89.1	1
6)	Saint Quentin	88.2	1		19)	Brive la Gaillarde	89.2	1
6)	Strasbourg	88.2	4		23)	Châteaubriant	89.2	1
22)	Tours	88.2	2		9)	Châtellerault	89.2	1
13)	Bonifacio	88.3	1		9)	Decazeville	89.2	1
17)	Brioude	88.3	1		22)	Lille	89.2	2
17)	Dijon	88.3	1		18)	Marseille	89.2	10
5)	L'Ile Rousse	88.3	1		13)	Montbard	89.2	1
8)	Lorient	88.3	1		17)	Nevers	89.2	1
8)	Moulins	88.3	1		23)	Ussel	89.2	1
17)	Roanne	88.3	1		17)	Vichy	89.2	1
18)	Saint Flour	88.3	1		17)	Castres	89.3	1
5)	Laon	88.4	1		17)	Cholet	89.3	1
9)	Luxeuil les Bains	88.4	1		13)	Longwy	89.3	1

FM	Station	MHz	kW
4)	Niort	89.3	1
9)	Nogaro	89.3	1
23)	Rouen	89.3	2
25)	Arras	89.4	1
17)	Aurillac	89.4	1
18)	Bayeux	89.4	1
18)	Bayonne	89.4	2
6)	Chambéry	89.4	1
23)	Marmande	89.4	1
6)	Roanne	89.4	1
6)	Saint Dizier	89.4	1
15)	Toulon	89.4	2
5)	Saintes	89.4	1
23)	Chaumont	89.5	1
6)	Strasbourg	89.5	4
6)	Ajaccio	89.6	8
14)	Angers	89.6	2
6)	Auch	89.6	1
6)	Clermont Ferrand	89.6	2
10)	La Rochelle	89.6	1
23)	Le Havre	89.6	1
4)	Marseille	89.6	4
21)	Mende	89.6	1
7)	Vierzon	89.6	1
5)	Aubusson	89.7	1
23)	Bastia	89.7	1
13)	Nevers	89.7	1
10)	Nîmes	89.7	1
23)	Perpignan	89.7	3
18)	Saint Nazaire	89.7	1
2)	Tours	89.7	2
6)	Troyes	89.7	1
6)	Agen	89.8	1
6)	Brioude	89.8	1
23)	Corte	89.8	1
23)	Gray	89.8	1
18)	Quimper	89.8	1
21)	Roanne	89.8	1
6)	Sablé sur Sarthe	89.8	1
18)	Toulon	89.8	4
18)	Alès	89.9	1
7)	Cognac	89.9	1
18)	Douai	89.9	1
19)	Montpellier	89.9	3
28)	Paris	89.9	6.3
3)	Périgueux	89.9	1
6)	Saint Dizier	89.9	1
20)	Saint Girons	89.9	1
18)	Saint Raphaël	89.9	1
20)	Bagnères de Bigorre	90.0	1
17)	Bayeux	90.0	1
22)	Brest	90.0	3
13)	Cosne Cours s Loire	90.0	1
23)	Marseille	90.0	10
21)	Quimperlé	90.0	1
9)	Royan	90.0	1
3)	Vichy	90.0	1
6)	Béthune	90.1	1
22)	Evreux	90.1	1
18)	Nantes	90.1	3
19)	Perpignan	90.1	3
9)	Poligny	90.1	1
7)	Toul	90.1	1
7)	Angoulême	90.2	1
23)	Bar le Duc	90.2	1
3)	Bergerac	90.2	1
9)	La Ferté s. Jouarre	90.2	1
13)	Melun	90.2	1
18)	Mimizan	90.2	1
14)	Pau	90.2	1
20)	Porto Vecchio	90.2	1
9)	Thionville	90.2	1
17)	Vannes	90.2	2
5)	Bastia	90.3	4
13)	Compiègne	90.3	1
18)	Decazeville	90.3	1
17)	Montargis	90.3	1
4)	Montmorillon	90.3	1
24)	Pamiers	90.3	1
3)	Saumur	90.3	1
19)	Valence	90.3	2
20)	Abbeville	90.4	1
18)	Auch	90.4	1
10)	Caen	90.4	2
5)	Calvi	90.4	1
13)	Châteaudun	90.4	1
22)	Dinan	90.4	1
21)	Longwy	90.4	1
13)	Paris	90.4	10
23)	Sablé sur Sarthe	90.4	1
3)	Alès	90.5	1
6)	Bourges	90.5	1
26)	Brest	90.5	1
13)	Chartres	90.5	1
23)	Le Mans	90.5	2
9)	Mont de Marsan	90.5	1
6)	Narbonne	90.5	1
13)	Rodez	90.5	1
23)	Tours	90.5	2
6)	Lourdes	90.6	3
7)	Maubeuge	90.6	1
20)	Melun	90.6	1
8)	Millau	90.6	1
7)	Dijon	90.7	1
18)	Figeac	90.7	1
4)	Laon	90.7	1
21)	Périgueux	90.7	1
6)	Soustons	90.7	1
13)	Troyes	90.7	1
1)	Bastia	90.8	1
9)	Château Thierry	90.8	1
9)	La Flèche	90.8	1
23)	Vannes	90.8	1
23)	Vesoul	90.8	1
17)	Annonay	90.9	1
19)	Brest	90.9	3
6)	Brive la Gaillarde	90.9	1
18)	Montreuil	90.9	1
14)	Poitiers	90.9	1
17)	Segré	90.9	1
7)	Villefrnch s. Saône	90.9	1
5)	Ajaccio	91.0	8
19)	Besançon	91.0	1
6)	Bourges	91.0	1
10)	Chambéry	91.0	1
13)	Colmar	91.0	1
13)	Fleurance	91.0	1
24)	Le Puy en Velay	91.0	1
15)	Limoges	91.0	2
5)	Sarrebourg	91.0	1
13)	Sens	91.0	1
10)	Vichy	91.0	1
25)	Boulogne sur Mer	91.1	1
23)	Dunkerque	91.1	1
19)	Nancy	91.1	1
19)	Orange	91.1	1
23)	Pau	91.1	1
3)	Villeneuve sur Lot	91.1	1
7)	Aubusson	91.2	1
7)	Grenoble	91.2	1
7)	Guéret	91.2	1
22)	Laval	91.2	1
22)	Mulhouse	91.2	1
17)	Orléans	91.2	2
5)	Saint Tropez	91.2	1
19)	Agen	91.3	1
19)	Cahors	91.3	1
5)	Cambrai	91.3	1
5)	Dinan	91.3	1
3)	Paris	91.3	10
27)	Reims	91.3	1
6)	Valence	91.3	1
19)	Amiens	91.3	1
13)	Bastia	91.4	4
18)	Béziers	91.4	1
17)	Brive la Gaillarde	91.4	1
18)	Jonzac	91.4	1
21)	Morlaix	91.4	1
6)	Beaune	91.5	1
13)	Blois	91.5	1
6)	Boulogne sur Mer	91.5	1
22)	Le Puy en Velay	91.5	1
17)	Clermont Ferrand	91.6	1
13)	Corte	91.6	1
5)	Dunkerque	91.6	1
17)	Epernay	91.6	1
17)	La Châtre	91.6	1
5)	Lens	91.6	1
5)	Perpignan	91.6	3
13)	Royan	91.6	1
3)	Tours	91.6	2
10)	Agen	91.7	1
18)	Bourgoin Jallieu	91.7	1
23)	Cholet	91.7	1
17)	Mrtgne au Perche	91.7	1
17)	Villefrnch s. Saône	91.7	1
7)	Amiens	91.8	1
7)	Bordeaux	91.8	5
8)	Brioude	91.8	1
3)	Castres	91.8	1
18)	La Rochelle	91.8	1
21)	Montélimar	91.8	1
7)	Montpellier	91.8	3
7)	Saint Dizier	91.8	1
7)	Saint Malo	91.8	1
7)	Saint Quentin	91.8	1
2)	Vichy	91.8	1
7)	Bressuire	91.9	1
9)	Chalon sur Saône	91.9	1
19)	Chaumont	91.9	1
4)	Civray	91.9	1
7)	Le Puy en Velay	91.9	1
7)	Lessay	91.9	1
7)	Porto Vecchio	91.9	1
8)	Salon de Provence	91.9	1
8)	Aix en Provence	92.0	1
7)	Albi	92.0	1
7)	Arcachon	92.0	1
23)	Auxerre	92.0	1
6)	Lille	92.0	2
23)	Montargis	92.0	1
7)	Pamiers	92.0	1
8)	Saint Affrique	92.0	1
7)	Saint Brieuc	92.0	1
10)	Saintes	92.0	1
19)	Soissons	92.0	1
3)	Brive la Gaillarde	92.1	1
21)	Cambrai	92.1	1
6)	Menton	92.1	1
7)	Nontron	92.1	1
19)	Troyes	92.1	1
13)	Amiens	92.2	1
22)	Béthune	92.2	1
5)	Colmar	92.2	1
7)	Dunkerque	92.2	1
4)	Lannemezan	92.2	1
7)	Laon	92.2	1
7)	Limoges	92.2	2
22)	Metz	92.2	1
22)	Mont de Marsan	92.2	1
5)	Montélimar	92.2	1
28)	Carcassonne	92.3	1
7)	Mimizan	92.3	1
10)	Rennes	92.3	1
13)	Vitry le François	92.3	1
23)	Albi	92.4	1
20)	Brest	92.4	1
14)	Montpellier	92.4	3
22)	Romilly sur Seine	92.4	1
5)	Saint Quentin	92.4	1
6)	Vannes	92.4	1
23)	Avignon	92.5	1
7)	Brive la Gaillarde	92.5	1
4)	Cahors	92.5	1
7)	Fontenay le Comte	92.5	1
17)	Issoudun	92.5	1
7)	Le Havre	92.5	1
5)	Lille	92.5	2
7)	Lourdes	92.5	3
19)	Rodez	92.5	1
1)	Aix en Provence	92.6	1
3)	Calvi	92.6	1
3)	Charolles	92.6	1
10)	Clermont Ferrand	92.6	1
23)	Nîmes	92.6	1
17)	Quimper	92.6	1
7)	Saint Raphaël	92.6	1
7)	Bastia	92.7	4
23)	Boulogne sur Mer	92.7	1
1)	Dreux	92.7	1
26)	Lorient	92.7	1
13)	Montélimar	92.7	1
7)	Montluçon	92.7	1
22)	Rennes	92.7	1
9)	Béthune	92.8	1
10)	Blois	92.8	1
22)	Castres	92.8	1
8)	Châteauroux	92.8	1
25)	Marseille	92.8	4
22)	Nice	92.8	5
23)	Vannes	92.8	1
3)	Cambrai	92.9	1
20)	Château Gontier	92.9	1
13)	Lyon	92.9	10
22)	Menton	92.9	1
7)	Montauban	92.9	3
10)	Orléans	92.9	2
7)	Rochefort	92.9	1
21)	Vichy	92.9	1
13)	Ajaccio	93.0	8
22)	Bonifacio	93.0	1
5)	Courtenay	93.0	1
18)	Hirson	93.0	1
21)	Lille	93.0	2
9)	Lourdes	93.0	1
22)	Saint Raphaël	93.0	1
7)	Verdun	93.0	1
13)	Annecy	93.1	1
11)	Arcachon	93.1	1
7)	Bayeux	93.1	1
13)	Bourg en Bresse	93.1	1
21)	Châlons en Champ.	93.1	1
6)	Châteaudun	93.1	1
7)	Coutances	93.1	1
13)	Dole	93.1	1
8)	Ernée	93.1	1
23)	Fontenay le Comte	93.1	1
23)	Royan	93.1	1
13)	Saint Étienne	93.1	2
7)	Toulon	93.1	4
19)	Avallon	93.2	1
13)	Bergerac	93.2	1
9)	Commercy	93.2	1
5)	Evreux	93.2	1
5)	Guéret	93.2	1
23)	Nevers	93.2	1
21)	Provins	93.2	1
5)	Romilly sur Seine	93.2	1
22)	Saint Tropez	93.2	1
5)	Ussel	93.2	1
13)	Arras	93.3	1
5)	Dreux	93.3	1
13)	Grenoble	93.3	1
7)	Lyon	93.3	1
6)	Marmande	93.3	1
4)	Montauban	93.3	3
8)	Montluçon	93.3	1
18)	Orléans	93.3	2
3)	Pamiers	93.3	1
3)	Poitiers	93.3	1
17)	Argentan	93.3	1
25)	Bourges	93.4	1
1)	Epernay	93.4	1
6)	Le Chambon s. Lignon	93.4	1
13)	Lille	93.4	1
8)	Marseille	93.4	4
7)	Moulins	93.4	1
10)	Narbonne	93.4	1
7)	Niort	93.4	1
23)	Quimper	93.4	1
9)	Vic Fezensac	93.4	1
22)	Ajaccio	93.5	8
13)	Béthune	93.5	1
18)	Dax	93.5	1
7)	Amiens	93.6	1
10)	Angers	93.6	1
8)	Brest	93.6	1
4)	Calvi	93.6	1
18)	Evreux	93.6	1
23)	La Roche sur Yon	93.6	1
19)	Laon	93.6	1
17)	Mazamet	93.6	1
6)	Montbard	93.6	1
6)	Pau	93.6	1
21)	Saint Gaudens	93.6	1
22)	Grenoble	93.7	1
10)	Le Havre	93.7	1
8)	Lyon	93.7	1
17)	Nancy	93.7	1
13)	Orléans	93.7	2
25)	Reims	93.7	1

FM	Station	MHz	kW
20)	Saint Nazaire	93.7	1
6)	Saintes	93.7	1
13)	Toulon	93.7	4
17)	Alençon	93.8	1
6)	Chaumont	93.8	1
2)	Marseille	93.8	4
18)	Montauban	93.8	3
7)	Orange	93.8	1
21)	Avallon	93.9	1
17)	Bourg en Bresse	93.9	1
19)	Bourges	93.9	1
13)	Carcassonne	93.9	1
13)	Château Gontier	93.9	1
20)	Condom	93.9	1
9)	Epernay	93.9	1
9)	Guéret	93.9	1
20)	Jonzac	93.9	1
3)	Lille	93.9	1
25)	Nogent le Rotrou	93.9	1
13)	Saint Brieuc	93.9	1
20)	Vannes	93.9	1
7)	Verdun	93.9	1
7)	Avignon	94.0	1
6)	Pau	94.0	1
6)	Rochefort	94.0	1
22)	Saint Flour	94.0	1
1)	Troyes	94.0	1
21)	Ussel	94.0	1
6)	Arcachon	94.1	1
25)	Chartres	94.1	1
19)	Decazeville	94.1	1
18)	Grenoble	94.1	1
6)	Mayenne	94.1	1
24)	Mont de Marsan	94.1	1
13)	Montmorillon	94.1	1
13)	Narbonne	94.1	1
13)	Reims	94.1	1
4)	Saint Gaudens	94.1	1
13)	Soissons	94.1	1
13)	Châlons en Champ.	94.2	1
3)	Chaumont	94.2	1
5)	Saint Omer	94.2	1
23)	Tarbes	94.2	1
6)	Bordeaux	94.3	5
4)	La Côte St. André	94.3	1
18)	Le Mans	94.3	1
23)	Lille	94.3	1
23)	Lorient	94.3	1
15)	Paris	94.3	4
23)	Saint Dizier	94.3	2
1)	Saint Étienne	94.3	1
10)	Saint Raphaël	94.3	1
21)	Sens	94.3	1
8)	Clermont Ferrand	94.4	1
8)	Le Puy en Velay	94.4	1
24)	Loches	94.4	1
19)	Orléans	94.4	1
4)	Parthenay	94.4	1
4)	Pau	94.4	1
6)	St. Gilles Croix de Vie	94.4	1
6)	Saintes	94.4	1
19)	Toulouse	94.4	1
5)	Gournay en Bray	94.5	1
7)	La Rochelle	94.5	1
7)	Laval	94.5	1
9)	Mazamet	94.5	1
5)	Montmorillon	94.5	1
9)	Nevers	94.5	1
9)	Rennes	94.5	2
17)	Arcachon	94.6	1
13)	Chambéry	94.6	1
7)	Colmar	94.6	1
18)	Lannemezan	94.6	1
22)	Perpignan	94.6	3
6)	Pouzauges	94.6	1
23)	Romilly sur Seine	94.6	1
7)	Saint Lô	94.6	1
7)	Béziers	94.7	1
18)	La Ferté Macé	94.7	1
13)	Le Havre	94.7	1
2)	Limoges	94.7	2
7)	Mende	94.7	1
6)	Nantes	94.7	3
4)	Poitiers	94.7	1
7)	Quimper	94.7	1
17)	Saint Étienne	94.7	2
3)	Vesoul	94.7	1
6)	Angers	94.8	2
20)	Chalon sur Saône	94.8	1
9)	Chaumont	94.8	1
22)	Forbach	94.8	1
5)	Longwy	94.8	1
5)	Mulhouse	94.8	1
22)	Nancy	94.8	1
5)	Nîmes	94.8	1
20)	Riscle	94.8	1
5)	Avignon	94.9	1
14)	Bordeaux	94.9	1
3)	Caen	94.9	1
13)	Hirson	94.9	1
9)	La Roche sur Yon	94.9	1
13)	Le Puy en Velay	94.9	1
19)	Lyon	94.9	1
5)	Montpellier	94.9	3
8)	Rennes	94.9	1
24)	Alès	95.0	1
21)	Autun	95.0	1
7)	Chambéry	95.0	1
5)	Cholet	95.0	1
19)	Clermont Ferrand	95.0	1
8)	Dinan	95.0	1
9)	Grenoble	95.0	1
10)	Lorient	95.0	1
13)	Mimizan	95.0	1
9)	Montauban	95.0	3
7)	Nice	95.0	5
13)	Niort	95.0	1
13)	Porto Vecchio	95.0	1
3)	Aubusson	95.1	1
17)	Douai	95.1	1
9)	Mâcon	95.1	1
24)	Marseille	95.1	4
3)	Pithiviers	95.1	1
10)	Saint Étienne	95.1	2
9)	Saint Raphaël	95.1	1
3)	Ussel	95.1	1
3)	Béziers	95.2	1
21)	Dunkerque	95.2	1
21)	Fontenay le Comte	95.2	1
3)	Jussey	95.2	1
9)	Périgueux	95.2	1
24)	Tarascon	95.2	1
21)	Argentan	95.3	1
3)	Bordeaux	95.3	5
20)	Chartres	95.3	1
20)	Château Thierry	95.3	1
7)	Dax	95.3	1
3)	Évreux	95.3	1
10)	Le Puy en Velay	95.3	1
21)	Lisieux	95.3	1
2)	Lyon	95.3	10
3)	Mirande	95.3	1
13)	Nancy	95.3	1
3)	Tarbes	95.3	1
3)	Toulon	95.3	4
18)	Cahors	95.4	1
18)	Chambéry	95.4	1
19)	Le Mans	95.4	2
17)	Orléans	95.4	2
17)	Pouzauges	95.4	1
17)	Ruffec	95.4	1
8)	Angers	95.5	1
24)	Bergerac	95.5	1
21)	Besançon	95.5	1
13)	Calvi	95.5	1
5)	Corte	95.5	1
17)	La Rochelle	95.5	1
9)	Mâcon	95.5	1
19)	Marseille	95.5	10
9)	Millau	95.5	1
19)	Nogent le Rotrou	95.5	1
13)	Niort	95.6	1
11)	Paris	95.6	1
9)	Saint Tropez	95.6	1
19)	Vannes	95.6	1
23)	Le Puy en Velay	95.7	1
19)	Lorient	95.7	1
22)	Lyon	95.7	4
3)	Nancy	95.7	1
6)	Perpignan	95.7	3
17)	St Amand Montrond	95.7	1
22)	Chambéry	95.8	1
6)	Guéret	95.8	1
6)	Montpellier	95.8	3
3)	Nice	95.8	5
21)	Saint Brieuc	95.8	1
6)	Toulon	95.8	4
23)	Beauvais	95.9	1
18)	Béthune	95.9	1
10)	Bourges	95.9	1
17)	Bourgoin Jallieu	95.9	1
13)	Brioude	95.9	1
18)	Cavaillon	95.9	1
23)	Commercy	95.9	1
10)	Limoges	95.9	2
22)	Mazamet	95.9	1
3)	Saint Étienne	95.9	2
3)	Annecy	96.0	1
9)	Brignoles	96.0	1
23)	Châteaudun	96.0	1
13)	Châtillon sur Seine	96.0	1
3)	Verdun	96.0	1
23)	Grenoble	96.0	1
3)	Lille	96.0	2
6)	Lisieux	96.0	1
13)	Marseille	96.0	4
23)	Paris	96.0	10
23)	Valence	96.0	2
23)	Auxerre	96.1	1
22)	Béziers	96.1	1
23)	Chartres	96.1	1
6)	Le Puy en Velay	96.1	1
23)	Lyon	96.1	4
13)	Montauban	96.1	3
19)	Moulins	96.1	1
23)	Nancy	96.1	1
13)	Saint Dizier	96.1	1
6)	Tours	96.1	2
17)	Vire	96.1	1
25)	Aix en Provence	96.2	1
19)	Alençon	96.2	1
18)	Brive la Gaillarde	96.2	1
22)	Châteauroux	96.2	1
23)	Clermont Ferrand	96.2	1
23)	Compiègne	96.2	1
23)	Douai	96.2	1
23)	Dunkerque	96.2	1
23)	Montbard	96.2	1
6)	Saint Brieuc	96.2	1
13)	Sedan	96.2	1
12)	Amiens	96.3	1
23)	Annonay	96.3	1
6)	Bourg en Bresse	96.3	1
7)	Caen	96.3	2
7)	L'Aigle	96.3	1
9)	Montluçon	96.3	1
7)	Morlaix	96.3	1
7)	Nogent le Rotrou	96.3	1
17)	Rennes	96.3	1
7)	Rodez	96.3	1
7)	Saint Étienne	96.3	2
22)	Bastia	96.4	4
22)	Calvi	96.4	1.3
20)	Cosne Cours sur Loire	96.4	1
7)	Épernay	96.4	1
13)	Granville	96.4	1
2)	Lille	96.4	2
6)	Lorient	96.4	1
2)	Paris	96.4	4
13)	Mont de Marsan	96.4	1
22)	Sarrebourg	96.4	1
20)	Bourges	96.5	1
6)	Brest	96.5	3
10)	Lyon	96.5	4
20)	Marmande	96.5	1
23)	Saint Flour	96.5	1
23)	Saint Nazaire	96.5	1
23)	Saint Omer	96.5	1
5)	Auxerre	96.6	1
6)	Châteauroux	96.6	2
7)	Clermont Ferrand	96.6	2
17)	Nice	96.6	1
3)	Nîmes	96.6	1
9)	Toulon	96.6	4
6)	Yssingeaux	96.6	1
23)	Abbeville	96.7	1
23)	Limoges	96.7	2
13)	Montargis	96.7	1
17)	Saint Lô	96.7	1
22)	Thionville	96.7	1
17)	Angoulême	96.8	1
23)	Brive la Gaillarde	96.8	1
6)	Caen	96.8	2
6)	Cahors	96.8	1
17)	Cannes	96.8	1
20)	Châtellerault	96.8	1
14)	Dreux	96.8	1
7)	Lille	96.8	1
18)	Mont de Marsan	96.8	1
13)	Nantes	96.8	3
13)	Redon	96.8	1
13)	Roanne	96.8	1
13)	Rochefort	96.8	1
13)	Valence	96.8	1
7)	Arras	96.9	1
13)	Guéret	96.9	1
7)	Montbard	96.9	1
17)	Montpellier	96.9	3
18)	Rennes	96.9	1
7)	Toulouse	96.9	1
25)	Calvi	97.0	1
21)	Chambéry	97.0	1
9)	Condom	97.0	1
3)	Mazamet	97.0	1
3)	Albi	97.1	1
9)	Bar le Duc	97.1	1
21)	La Ferté s. Jouarre	97.1	1
6)	Montluçon	97.1	1
7)	Pouzauges	97.1	1
9)	Bagnères de Bigorre	97.2	1
13)	Bourg en Bresse	97.2	1
18)	Chaumont	97.2	1
7)	Pithiviers	97.2	1
3)	Propriano	97.2	1
18)	Saint Omer	97.2	1
22)	Vichy	97.2	1
20)	Alençon	97.3	1
7)	Auch	97.3	1
13)	Bordeaux	97.3	5
22)	Le Havre	97.3	1
25)	Lyon	97.3	1
8)	Poitiers	97.3	1
8)	Rodez	97.3	1
3)	Vire	97.3	1
21)	Agen	97.4	1
22)	Bayeux	97.4	1
13)	Brest	97.4	1
21)	Grenoble	97.4	1
23)	Mont de Marsan	97.4	1
7)	Morhange	97.4	1
21)	Nice	97.4	5
19)	Paris	97.4	4
18)	Saint Malo	97.4	1
23)	Argentan	97.5	1
18)	Béziers	97.5	1
23)	Carmaux	97.5	1
13)	Corte	97.5	1
13)	Dijon	97.5	1
5)	Mayenne	97.5	1
3)	Nogent le Rotrou	97.5	1
7)	Rouen	97.5	2
18)	Avallon	97.6	1
23)	Caen	97.6	2
5)	Chambéry	97.6	1
22)	Fontenay le Comte	97.6	1
3)	Le Mans	97.6	2
7)	L'Ile Rousse	97.6	1
18)	Menton	97.6	1
13)	Metz	97.6	1
13)	Montauban	97.6	3
7)	Pamiers	97.6	1
3)	Perpignan	97.6	1
13)	Rennes	97.6	1
7)	Bayonne	97.7	1
9)	Castres	97.7	1
9)	Compiègne	97.7	1
4)	Dreux	97.7	1

FM	Station	MHz	kW
4)	Figeac	97.7	1
28)	Laval	97.7	1
5)	Maubeuge	97.7	1
9)	Montargis	97.7	1
22)	Nantes	97.7	3
8)	Ussel	97.7	1
24)	Brive la Gaillarde	97.8	1
6)	Chalon sur Saône	97.8	1
1)	Grenoble	97.8	1
17)	Porto Vecchio	97.8	1
6)	Reims	97.8	2
6)	Saint Étienne	97.8	2
6)	Vichy	97.8	1
4)	Bastia	97.9	1
13)	Parthenay	97.9	1
4)	Toulouse	97.9	5
9)	Angers	98.0	2
6)	Montélimar	98.0	1
23)	Montluçon	98.0	1
23)	Moulins	98.0	1
9)	Vannes	98.0	1
23)	Ajaccio	98.1	8
18)	Besançon	98.1	1
5)	Dax	98.1	1
28)	Nice	98.1	5
5)	Périgueux	98.1	1
21)	Ruffec	98.1	1
9)	Saint Flour	98.1	2
20)	Saintes	98.1	1
13)	Samatan	98.1	1
13)	Sens	98.1	1
3)	Auxerre	98.2	1
9)	Avignon	98.2	1
22)	Bernay	98.2	1
9)	Bordeaux	98.2	1
22)	Bourges	98.2	1
9)	Compiègne	98.2	1
19)	Limoges	98.2	2
6)	Lourdes	98.2	3
9)	Mâcon	98.2	1
17)	Narbonne	98.2	1
23)	Nevers	98.2	1
12)	Paris	98.2	4
9)	Quimperlé	98.2	1
3)	Sablé sur Sarthe	98.2	1
1)	Toulon	98.2	1
3)	Aix en Provence	98.3	1
18)	Bar le Duc	98.3	1
8)	Gien	98.3	1
17)	Montpellier	98.3	3
22)	Rouen	98.3	2
21)	Saint Affrique	98.3	1
23)	Saint Quentin	98.3	1
23)	Amiens	98.4	1
23)	Chaumont	98.4	1
3)	La Flèche	98.4	1
18)	Mazamet	98.4	1
5)	Mirande	98.4	1
8)	Royan	98.4	1
8)	Agen	98.5	1
3)	Alençon	98.5	1
18)	Bastia	98.5	1
8)	Beauvais	98.5	1
17)	Béziers	98.5	1
21)	Hirson	98.5	1
3)	Laval	98.5	1
5)	Albi	98.6	1
22)	Bergerac	98.6	1
9)	Cognac	98.6	1
20)	Dax	98.6	1
18)	La Roche sur Yon	98.6	1
18)	Vannes	98.6	1
9)	Auch	98.7	1
21)	Brioude	98.7	1
10)	Chartres	98.7	1
13)	La Rochelle	98.7	1
9)	Le Puy en Velay	98.7	1
9)	Niederbronn l. Bains	98.7	1
5)	Argentan	98.8	1
20)	Cannes	98.8	1
23)	Castres	98.8	1
7)	Grenoble	98.8	1
20)	Lorient	98.8	1
20)	Nice	98.8	1
19)	Toulon	98.8	4
7)	Valence	98.8	1
3)	Arcachon	98.9	1
6)	Auxerre	98.9	1
7)	Brest	98.9	3
6)	Le Creusot	98.9	1
3)	Lyon	98.9	1
9)	Mende	98.9	1
3)	Montauban	98.9	3
6)	Sens	98.9	1
3)	Vierzon	98.9	1
9)	Amiens	99.0	1
22)	Bayonne	99.0	2
10)	Boulogne sur Mer	99.0	1
5)	La Ferté Macé	99.0	1
22)	Poitiers	99.0	1
7)	Royan	99.0	1
17)	Saint Raphaël	99.0	1
6)	Ussel	99.0	1
3)	Aurillac	99.1	1
9)	Cervione	99.1	2
7)	Châteauroux	99.1	1
5)	Châtillon sur Seine	99.1	1
9)	Limoges	99.1	2
18)	Toulouse	99.1	5
3)	Abbeville	99.2	1
25)	Alençon	99.2	1
3)	Aubusson	99.2	1
7)	Bourges	99.2	1
3)	Brive la Gaillarde	99.2	1
22)	Calais	99.2	1
4)	Châtellerault	99.2	1
7)	Condom	99.2	1
7)	Mont de Marsan	99.2	1
20)	Narbonne	99.2	1
9)	Nice	99.2	5
25)	Saint Lô	99.2	1
23)	Vichy	99.2	1
13)	Argentan	99.3	1
22)	Cambrai	99.3	1
13)	L'Aigle	99.3	1
14)	Laval	99.3	1
18)	Montpellier	99.3	3
9)	Saint Nazaire	99.3	1
10)	Avignon	99.4	1
9)	Bastia	99.4	4
7)	Calvi	99.4	1
9)	Charolles	99.4	1
25)	Clermont Ferrand	99.4	1
5)	Fontainebleau	99.4	2
3)	Mâcon	99.4	1
3)	Mazamet	99.4	1
13)	Mulhouse	99.4	1
9)	Saint Malo	99.4	1
3)	Châteaudun	99.5	1
13)	Eauze	99.5	1
13)	Toulouse	99.5	1
3)	Abbeville	99.6	1
25)	Aurillac	99.6	1
18)	Bordeaux	99.6	5
6)	Bourges	99.6	1
3)	Carcassonne	99.6	1
17)	Carmaux	99.6	1
22)	Cholet	99.6	1
18)	Dijon	99.6	1
5)	Île de Ré	99.6	1
17)	Limoges	99.6	1
5)	Porto Vecchio	99.6	1
9)	Quimperlé	99.6	1
7)	Salon de Provence	99.6	1
9)	Vichy	99.6	1
6)	Bagnères de Bigorre	99.7	1
7)	Brest	99.7	3
13)	La Flèche	99.7	1
7)	Marseille	99.7	4
22)	Montauban	99.7	3
6)	Nevers	99.7	1
9)	Orléans	99.7	1
17)	Troyes	99.7	1
3)	Ajaccio	99.8	8
21)	Chartres	99.8	1
9)	Guéret	99.8	1
17)	Lavaur	99.8	1
27)	Le Havre	99.8	1
9)	Menton	99.8	1
18)	Montargis	99.8	1
6)	Mulhouse	99.8	1
21)	Noyon	99.8	1
21)	Parthenay	99.8	1
19)	Argentan	99.9	1
17)	Chaumont	99.9	1
9)	Mimizan	99.9	1
9)	Quimper	99.9	1
7)	Belfort	100.0	1
23)	Béziers	100.0	1
9)	Fontenay le Comte	100.0	1
2)	Laval	100.0	1
3)	L'Ile Rousse	100.0	1
13)	Limoges	100.0	2
9)	Montélimar	100.0	1
9)	Pithiviers	100.0	1
9)	Poitiers	100.0	1
22)	Porto Vecchio	100.0	1
3)	Rodez	100.0	1
13)	Romilly sur Seine	100.0	1
23)	Toulouse	100.0	1
18)	Angers	100.1	1
6)	Bagnères de Bigorre	100.1	1
3)	Bayonne	100.1	3
9)	Carcassonne	100.1	1
13)	Châteauroux	100.1	1
3)	Marseille	100.1	4
9)	Melun	100.1	1
22)	Reims	100.1	2
20)	Saint Brieuc	100.1	1
9)	Sens	100.1	1
23)	Alès	100.2	1
14)	Brest	100.2	1
7)	Chinon	100.2	1
5)	Coutances	100.2	1
3)	Gien	100.2	1
6)	Guéret	100.2	1
9)	La Rochelle	100.2	1
23)	Montpellier	100.2	3
3)	Orthez	100.2	1
9)	Troyes	100.2	1
9)	Valence	100.2	2
14)	Agen	100.3	1
9)	Angoulême	100.3	1
7)	Le Mans	100.3	2
9)	Lyon	100.3	4
13)	Mende	100.3	1
3)	Mont de Marsan	100.3	1
9)	Narbonne	100.3	1
9)	Paris	100.3	10
19)	St Gilles Croix de Vie	100.3	1
3)	Soissons	100.3	1
24)	Arcachon	100.4	1
6)	Besançon	100.4	1
21)	Bourges	100.4	1
9)	Chartres	100.4	1
9)	Chaumont	100.4	1
4)	Corte	100.4	1
6)	Lens	100.4	1
9)	Limoges	100.4	2
9)	Niort	100.4	1
9)	Orléans	100.4	1
22)	Royan	100.4	1
21)	Toulon	100.4	4
9)	Toulouse	100.4	5
17)	Tours	100.4	2
9)	Alençon	100.5	1
6)	Annecy	100.5	1
9)	Argentan	100.5	1
9)	Bourgoin Jallieu	100.5	1
21)	Brive la Gaillarde	100.5	1
9)	Château Gontier	100.5	1
9)	Compiègne	100.5	1
12)	Marseille	100.5	1
9)	Nogent le Rotrou	100.5	1
18)	Rodez	100.5	1
9)	Rouen	100.5	2
9)	Ruffec	100.5	1
22)	Saint Étienne	100.5	2
9)	Saint Nazaire	100.5	1
20)	Vichy	100.5	1
9)	Albi	100.6	1
9)	Avallon	100.6	1
7)	Blois	100.6	1
3)	Bourg en Bresse	100.6	1
3)	Brioude	100.6	1
9)	Carcassonne	100.6	1
9)	Dijon	100.6	1
9)	Douarnenez	100.6	1
9)	Ghisonaccia	100.6	4
9)	Parthenay	100.6	1
9)	Reims	100.6	2
17)	Saint Brieuc	100.6	1
20)	Villeneuve sur Lot	100.6	1
13)	Béziers	100.7	1
6)	Decazeville	100.7	1
13)	Laon	100.7	1
13)	Laval	100.7	1
13)	Le Mans	100.7	1
3)	Le Puy en Velay	100.7	1
4)	Paris	100.7	10
13)	Bastia	100.8	4
9)	Calvi	100.8	1
20)	Castres	100.8	1
23)	Chambéry	100.8	1
9)	Clermont Ferrand	100.8	1
9)	Grenoble	100.8	1
5)	Lisieux	100.8	1
9)	Nîmes	100.8	1
12)	Perpignan	100.8	1
7)	Saint Gaudens	100.8	1
5)	Thouars	100.8	1
20)	Vire	100.8	1
6)	Alençon	100.9	1
9)	Bayonne	100.9	5
9)	Besançon	100.9	1
9)	Marseille	100.9	10
13)	Nancy	100.9	1
22)	Rodez	100.9	1
9)	Amiens	101.0	1
24)	Aurillac	101.0	1
9)	Avignon	101.0	1
13)	Bergerac	101.0	1
13)	Château Thierry	101.0	1
9)	L'Aigle	101.0	1
6)	Lourdes	101.0	3
9)	Agen	101.1	1
9)	Aubusson	101.1	1
9)	Bar le Duc	101.1	1
10)	Beauvais	101.1	1
6)	Châteauroux	101.1	1
6)	La Roche sur Yon	101.1	1
10)	Laval	101.1	1
11)	Le Havre	101.1	1
13)	Metz	101.1	1
10)	Paris	101.1	10
23)	Poitiers	101.1	1
9)	Saint Malo	101.1	1
9)	Ajaccio	101.2	8
24)	Albi	101.2	1
9)	Arras	101.2	1
13)	Blois	101.2	1
21)	Chaumont	101.2	1
9)	Clermont Ferrand	101.2	1
7)	Épinal	101.2	1
9)	Bourgoin Jallieu	101.3	1
23)	Chambéry	101.3	1
21)	Châtellerault	101.3	1
22)	Dreux	101.3	1
9)	Dunkerque	101.3	1
4)	Forbach	101.3	1
3)	Jonzac	101.3	1
22)	La Rochelle	101.3	1
21)	Le Blanc	101.3	1
3)	Lille	101.3	1
20)	Orange	101.3	1
19)	Saint Étienne	101.3	2
9)	Sarlat la Canéda	101.3	1
10)	Amiens	101.4	1
9)	Caen	101.4	2
17)	Longwy	101.4	1
21)	Marseille	101.4	10
9)	Nice	101.4	5
21)	St Gilles Croix de Vie	101.4	1
24)	Toulouse	101.4	1
9)	Vic Fezensac	101.4	1
18)	Alès	101.5	1
5)	Cahors	101.5	1
14)	Évreux	101.5	1
21)	Guéret	101.5	1
5)	Montbard	101.5	1
18)	Nevers	101.5	1

FM	Station	MHz	kW
14)	Paris	101.5	10
13)	Poitiers	101.5	1
21)	Redon	101.5	1
21)	Rodez	101.5	1
17)	Valence	101.5	1
18)	Cambrai	101.6	1
18)	Issoudun	101.6	1
10)	Le Mans	101.6	2
5)	Mâcon	101.6	1
20)	Montargis	101.6	1
23)	Périgueux	101.6	1
10)	Quimper	101.6	1
21)	Saint Malo	101.6	1
13)	Valence	101.6	1
21)	Albi	101.7	1
5)	Bayeux	101.7	1
5)	Bayonne	101.7	5
18)	Compiègne	101.7	1
6)	Cosne Crs s. Loire	101.7	1
17)	Le Puy en Velay	101.7	1
21)	Limoges	101.7	2
21)	Mazamet	101.7	1
18)	Montluçon	101.7	1
22)	Montpellier	101.7	1
23)	Morlaix	101.7	1
9)	Provins	101.7	1
9)	Reims	101.7	2
9)	Romilly sur Seine	101.7	1
22)	Tarascon	101.7	1
6)	Aubusson	101.8	1
21)	Auxerre	101.8	1
21)	Bergerac	101.8	1
23)	Brest	101.8	3
7)	Fontainebleau	101.8	1
7)	Laon	101.8	1
6)	Le Havre	101.8	1
4)	Aix en Provence	101.9	1
3)	Cahors	101.9	1
9)	Châtillon s. Seine	101.9	1
9)	Cognac	101.9	1
9)	Coutances	101.9	1
7)	Evreux	101.9	1
9)	Martigues	101.9	1
9)	Mayenne	101.9	1
18)	Moulins	101.9	1
7)	Paris	101.9	10
9)	Tonnerre	101.9	1
9)	Abbeville	102.0	1
6)	Bar le Duc	102.0	1
17)	Beaune	102.0	1
6)	Falaise	102.0	1
23)	La Rochelle	102.0	1
4)	Metz	102.0	1
22)	Quimper	102.0	1
9)	Rennes	102.0	3
21)	Saint Quentin	102.0	1
24)	Toulouse	102.0	60
9)	Annecy	102.1	1
20)	Avallon	102.1	1
20)	Calvi	102.1	1
9)	Charolles	102.1	1
18)	Limoges	102.1	2
29)	Melun	102.1	1
9)	Mulhouse	102.1	1
6)	Nîmes	102.1	1
9)	St Amand Mntrnd	102.1	1
18)	Strasbourg	102.1	4
9)	Arcachon	102.2	1
9)	Blois	102.2	1
21)	Dole	102.2	1
17)	La Ferté Macé	102.2	1
7)	Montargis	102.2	1
7)	Thouars	102.2	1
21)	Troyes	102.2	1
9)	Avranches	102.3	1
3)	Cahors	102.3	1
17)	Chambéry	102.3	1
7)	Forbach	102.3	1
21)	Le Puy en Velay	102.3	1
6)	Marseille	102.3	10
7)	Montbard	102.3	1
18)	Nancy	102.3	1
21)	Nevers	102.3	1
29)	Paris	102.3	4
21)	Quimperlé	102.3	1
9)	Saint Brieuc	102.3	1
9)	Saint Omer	102.3	1
10)	Tours	102.3	2
7)	Auxerre	102.4	1
9)	Bordeaux	102.4	5
3)	Brest	102.4	3
6)	Castres	102.4	1
3)	Chalon sur Saône	102.4	1
20)	Chaumont	102.4	1
3)	Decazeville	102.4	1
10)	Grenoble	102.4	1
6)	Haguenau	102.4	1
5)	Montmorillon	102.4	1
9)	Nantes	102.4	3
7)	Perpignan	102.4	3
2)	Rennes	102.4	1
9)	Romorantin Lnthny	102.4	1
6)	Toulouse	102.4	5
3)	Vienne	102.4	1
22)	Angers	102.5	1
18)	Calais	102.5	1
24)	Carmaux	102.5	1
5)	Chartres	102.5	1
5)	Commercy	102.5	1
18)	Gourdon	102.5	1
8)	Les Sbls d'Olonne	102.5	1
5)	Melun	102.5	1
8)	Niort	102.5	1
3)	Dijon	102.5	1
18)	Angoulême	102.6	1
13)	Bergerac	102.6	1
21)	Carcassonne	102.6	1
21)	Montauban	102.6	3
27)	Orléans	102.6	1
3)	Quimper	102.6	1
9)	Saint Gaudens	102.6	1
20)	Troyes	102.6	1
9)	Abbeville	102.7	1
22)	Avallon	102.7	1
3)	Limoges	102.7	2
5)	Morlaix	102.7	1
3)	Nérac	102.7	1
8)	Paris	102.7	10
20)	Rochefort	102.7	1
9)	Saint Dizier	102.7	1
5)	Saint Flour	102.7	1
13)	Alès	102.8	1
5)	Annecy	102.8	1
13)	Avignon	102.8	1
23)	Bordeaux	102.8	5
9)	Bourg en Bresse	102.8	1
22)	Brive la Gaillarde	102.8	1
17)	Dax	102.8	1
20)	Lorient	102.8	1
20)	Parthenay	102.8	1
9)	Saint Étienne	102.8	2
2)	Saint Raphaël	102.8	3
21)	Tours	102.8	2
9)	Vitré	102.8	1
13)	Charolles	102.9	1
14)	Clermont Ferrand	102.9	1
21)	Confolens	102.9	1
5)	Guéret	102.9	1
1)	Le Mans	102.9	2
21)	Lourdes	102.9	3
9)	Lunéville	102.9	1
23)	Nantes	102.9	3
9)	Saint Lô	102.9	1
24)	Villeneuve sur Lot	102.9	1
17)	Carcassonne	103.0	1
9)	Chambéry	103.0	1
9)	Châteaudun	103.0	1
13)	Colmar	103.0	1
9)	Condom	103.0	1
18)	Le Puy en Velay	103.0	1
9)	Lyon	103.0	10
3)	Metz	103.0	1
21)	Moulins	103.0	1
17)	Neufchâtel en Bray	103.0	1
12)	Poitiers	103.0	1
9)	Tonnerre	103.0	1
23)	Angoulême	103.1	1
21)	Arcachon	103.1	1
20)	Bergerac	103.1	1
17)	Charensat	103.1	1
20)	Mont de Marsan	103.1	1
20)	Paris	103.1	10
5)	Roanne	103.1	1
24)	Saint Affrique	103.1	1
3)	Saint Dizier	103.1	1
13)	Saint Flour	103.1	1
10)	Toulouse	103.1	5
22)	Amiens	103.2	1
21)	Belfort	103.2	1
21)	Cervione	103.2	1
17)	Dole	103.2	1
6)	Douarnenez	103.2	1
9)	Grenoble	103.2	1
9)	Mirande	103.2	1
20)	Montmorillon	103.2	1
20)	Niort	103.2	1
20)	Nogent le Rotrou	103.2	1
24)	Perpignan	103.2	10
20)	Albi	103.3	1
18)	Aurillac	103.3	1
7)	Avesnes sur Helpe	103.3	1
6)	Carpentras	103.3	1
6)	Chartres	103.3	1
20)	Compiègne	103.3	1
20)	La Rochelle	103.3	1
20)	Lille	103.3	2
9)	Provins	103.3	1
7)	Nérac	103.3	1
20)	Orthez	103.3	1
3)	Rouen	103.3	2
20)	Sarlat la Canéda	103.3	1
5)	Strasbourg	103.3	4
10)	Toulon	103.3	1
7)	Vichy	103.3	1
2)	Bastia	103.4	1
25)	Cahors	103.4	1
13)	Carcassonne	103.4	1
7)	Nantes	103.4	3
12)	Orléans	103.4	1
6)	Rodez	103.4	1
25)	Tours	103.4	2
6)	Beauvais	103.5	1
7)	Dinan	103.5	1
6)	Épinal	103.5	1
6)	Fontainebleau	103.5	2
19)	Le Havre	103.5	1
6)	Le Mans	103.5	2
5)	Morlaix	103.5	1
6)	Paris	103.5	10
9)	Saint Affrique	103.5	1
23)	Beaune	103.6	1
21)	Blois	103.6	1
3)	Longwy	103.6	1
5)	Montluçon	103.6	1
20)	Saint Gaudens	103.6	1
7)	Saint Nazaire	103.6	1
23)	Alençon	103.7	1
17)	Grenoble	103.7	1
23)	Laval	103.7	1
20)	Mirande	103.7	1
21)	Niort	103.7	1
21)	Bastia	103.8	4
21)	Bergerac	103.8	1
21)	Chinon	103.8	1
24)	Figeac	103.8	1
21)	Lourdes	103.8	1
20)	Lourdes	103.8	3
4)	Nantes	103.8	3
5)	Saint Brieuc	103.8	2
18)	Troyes	103.8	1
9)	Ussel	103.8	1
24)	Bayonne	103.9	5
18)	Beauvais	103.9	1
13)	Montpellier	103.9	2
21)	Paris	103.9	10
9)	Rennes	103.9	3
19)	Saint Dizier	103.9	1
5)	Saint Flour	103.9	1
5)	Saint Lô	103.9	1
3)	Saint Quentin	103.9	1
21)	Toulouse	103.9	5
21)	Vierzon	103.9	1
20)	Arcachon	104.0	1
9)	Avignon	104.0	1
21)	Besançon	104.0	1
3)	Cervione	104.0	2
3)	Mauriac	104.0	1
18)	Millau	104.0	1
21)	Romorantin Lanth.	104.0	1
18)	St Gilles Crx de Vie	104.0	1
21)	Tours	104.0	2
24)	Villefr. de Rouergue	104.0	1
21)	Abbeville	104.1	1
21)	Alençon	104.1	1
9)	Bressuire	104.1	1
18)	Chartres	104.1	1
20)	Compiègne	104.1	1
17)	Confolens	104.1	1
20)	Laval	104.1	1
20)	Mâcon	104.1	1
24)	Mazamet	104.1	1
21)	Melun	104.1	1
8)	Menton	104.1	1
20)	Montauban	104.1	3
20)	Montélimar	104.1	1
21)	Montluçon	104.1	1
20)	Nancy	104.1	1
6)	Rouen	104.1	1
20)	Annecy	104.2	1
20)	Bordeaux	104.2	5
21)	Dijon	104.2	1
20)	Grenoble	104.2	1
20)	Lyon	104.2	4
20)	Mende	104.2	1
4)	Mirande	104.2	1
21)	Nogent le Rotrou	104.2	1
21)	Troyes	104.2	1
21)	Ajaccio	104.3	8
21)	Amiens	104.3	1
21)	Angers	104.3	1
20)	Arles	104.3	1
20)	Bastia	104.3	4
20)	Bayonne	104.3	5
20)	Béziers	104.3	1
21)	Brest	104.3	3
21)	Clermont Ferrand	104.3	2
13)	Épinal	104.3	1
21)	La Ferté Macé	104.3	1
21)	La Rochelle	104.3	1
21)	Le Havre	104.3	1
21)	Le Mans	104.3	2
20)	Limoges	104.3	2
21)	Lorient	104.3	1
20)	Marseille	104.3	10
17)	Montbard	104.3	1
20)	Montpellier	104.3	3
21)	Nantes	104.3	3
20)	Nîmes	104.3	1
21)	Orléans	104.3	2
21)	Paris	104.3	10
21)	Pau	104.3	1
21)	Péronne	104.3	1
21)	Perpignan	104.3	3
21)	Poitiers	104.3	1
21)	Quimper	104.3	1
21)	Rennes	104.3	1
20)	Saint Affrique	104.3	1
21)	St Amand Mntrnd	104.3	1
21)	Saint Nazaire	104.3	1
21)	Soissons	104.3	1
21)	Toulon	104.3	4
20)	Toulouse	104.3	5
20)	Valence	104.3	1
21)	Vannes	104.3	1
21)	Aubusson	104.4	1
21)	Auxerre	104.4	1
21)	Bourg en Bresse	104.4	1
21)	Jonzac	104.4	1
21)	Le Puy en Velay	104.4	1
19)	Montargis	104.4	1
2)	Nice	104.4	5
21)	Reims	104.4	1
24)	Rodez	104.4	1
18)	Romorantin Lnthny	104.4	1
21)	Ruffec	104.4	1
20)	Saint Étienne	104.4	2
20)	Agen	104.5	1
5)	Alençon	104.5	1

FM	Station	MHz	kW
20)	Avignon	104.5	1
7)	Baccarat	104.5	1
20)	Chambéry	104.5	1
11)	Chartres	104.5	1
21)	Compiègne	104.5	1
5)	Forbach	104.5	1
19)	Gien	104.5	1
17)	La Roche sur Yon	104.5	1
5)	Laval	104.5	1
9)	Le Creusot	104.5	1
5)	Melun	104.5	1
20)	Provins	104.5	1
17)	Redon	104.5	1
21)	Rouen	104.5	2
5)	Tours	104.5	1
5)	Alès	104.6	1
22)	Alta Rocca	104.6	1
3)	Avallon	104.6	1
5)	Bayeux	104.6	1
5)	Bordeaux	104.6	5
5)	Grenoble	104.6	1
5)	L'Aigle	104.6	1
5)	Lyon	104.6	4
5)	Nevers	104.6	1
5)	Nogent le Rotrou	104.6	1
20)	Saint Flour	104.6	1
21)	Saint Raphaël	104.6	1
5)	Amiens	104.7	1
5)	Angers	104.7	1
5)	Beauvais	104.7	1
5)	Brest	104.7	3
24)	Carcassonne	104.7	80
5)	Cholet	104.7	1
5)	Clermont Ferrand	104.7	2
5)	Dijon	104.7	1
22)	Ghisonaccia	104.7	2
5)	La Rochelle	104.7	1
5)	Le Mans	104.7	2
5)	Limoges	104.7	2
5)	Lorient	104.7	1
24)	Montpellier	104.7	1
5)	Nantes	104.7	3
5)	Orléans	104.7	2
5)	Paris	104.7	10
5)	Poitiers	104.7	1
5)	Quimper	104.7	1
5)	Rennes	104.7	2
5)	Saint Nazaire	104.7	1
5)	Soissons	104.7	1
5)	Toulon	104.7	4
5)	Troyes	104.7	1
5)	Vannes	104.7	1
5)	Arcachon	104.8	1
3)	Argentan	104.8	1
20)	Aubusson	104.8	1
20)	Auxerre	104.8	1
5)	Bernay	104.8	1
5)	Cambrai	104.8	1
5)	Châlons en Chmpgn	104.8	1
4)	Gourdon	104.8	1
5)	Marseille	104.8	10
21)	Metz	104.8	1
18)	Neufchâteau	104.8	1
5)	St Amand Mntrnd	104.8	1
5)	Saint Étienne	104.8	2
18)	Saint Lô	104.8	1
21)	Valence	104.8	1
5)	Abbeville	104.9	1
5)	Agen	104.9	1
5)	Besançon	104.9	1
7)	Chartres	104.9	1
5)	Compiègne	104.9	1
7)	La Roche sur Yon	104.9	1
22)	Laval	104.9	1
5)	Mont de Marsan	104.9	1
7)	Mont. Fault Yonne	104.9	1
5)	Moulins	104.9	1
7)	Parthenay	104.9	1
24)	Périgueux	104.9	1
5)	Rouen	104.9	1
21)	Royan	104.9	1
20)	Angoulême	105.0	1
20)	Auch	105.0	1
21)	Bar le Duc	105.0	1
21)	Caen	105.0	2
24)	Cahors	105.0	1
21)	L'Aigle	105.0	1
13)	Luxeuil les Bains	105.0	1
21)	Lyon	105.0	4
7)	Morlaix	105.0	1
23)	Reims	105.0	1
22)	Alençon	105.1	1
3)	Angers	105.1	5
21)	Bayonne	105.1	1
9)	Bonifacio	105.1	1
21)	Bordeaux	105.1	5
21)	Charolles	105.1	1
21)	Clermont Ferrand	105.1	1
9)	Dinan	105.1	1
25)	Le Puy en Velay	105.1	1
20)	Limoges	105.1	1
21)	Nancy	105.1	1
20)	Niort	105.1	1
17)	Toulon	105.1	4
7)	Ajaccio	105.2	8
20)	Brive la Gaillarde	105.2	1
9)	Issoudun	105.2	1
3)	Lons le Saunier	105.2	1
5)	Montauban	105.2	3
21)	Saint Étienne	105.2	2
21)	Saint Lô	105.2	1
21)	Vitré	105.2	1
3)	Chartres	105.3	1
19)	Cholet	105.3	1
5)	Metz	105.3	1
13)	Rouen	105.3	2
22)	Sens	105.3	1
13)	Strasbourg	105.3	4
13)	Dole	105.4	1
5)	Nancy	105.5	1
9)	Béziers	105.7	1
23)	Bonifacio	105.7	1
7)	Lannemezan	105.7	1
21)	Le Creusot	105.7	1
17)	Loches	105.7	1
5)	Marseille	105.7	1
9)	Redon	105.7	1
5)	Saint Flour	105.7	2
21)	Sancerre	105.7	1
21)	Strasbourg	105.7	4
9)	Vesoul	105.7	1
17)	Argenton s. Creuse	105.8	1
5)	Carcassonne	105.8	1
10)	Dijon	105.8	1
3)	Grenoble	105.8	1
23)	Nîmes	105.8	1
23)	Segré	105.8	1
5)	Bourges	105.9	1
18)	Brest	105.9	3
5)	Caen	105.9	2
22)	Clermont Ferrand	105.9	1
21)	Corte	105.9	1
20)	Ghisonaccia	105.9	2
9)	Le Mans	105.9	1
18)	Mende	105.9	1
22)	Paris	105.9	10
7)	Pau	105.9	1
20)	Périgueux	105.9	1
13)	Perpignan	105.9	1
13)	Saint Nazaire	105.9	1
21)	Saintes	105.9	1
7)	Toulouse	105.9	5
23)	Troyes	105.9	1
5)	Valence	105.9	1
5)	Ajaccio	106.0	8
22)	Besançon	106.0	1
22)	Blois	106.0	1
24)	Bordeaux	106.0	5
7)	Cahors	106.0	1
17)	L'Aigle	106.0	1
8)	Limoges	106.0	2
9)	Lorient	106.0	1
3)	Marseille	106.0	1
20)	Mauriac	106.0	1
21)	Mayenne	106.0	1
21)	Montargis	106.0	1
21)	Niort	106.0	1
20)	Rennes	106.0	1
9)	Roanne	106.0	1
13)	St Gilles Crx de Vie	106.0	1
23)	Agen	106.1	1
3)	Albi	106.1	1
27)	Amiens	106.1	1
13)	Angers	106.1	2
8)	Aubusson	106.1	1
19)	Bastia	106.1	4
5)	Brive la Gaillarde	106.1	1
19)	Calvi	106.1	1
22)	Chartres	106.1	1
22)	Melun	106.1	1
9)	Montpellier	106.1	3
10)	Rouen	106.1	2
5)	Saint Dizier	106.1	1
5)	Sarrebourg	106.1	1
5)	Tarbes	106.1	1
7)	Argentan	106.2	1
19)	Avignon	106.2	1
5)	Bergerac	106.2	1
18)	Étampes	106.2	1
20)	Laval	106.2	1
13)	Morlaix	106.2	1
3)	Nantes	106.2	2
5)	Neufchâteau	106.2	1
18)	Tonnerre	106.2	1
22)	Toulon	106.2	1
13)	Vendôme	106.2	1
21)	Arras	106.3	1
5)	Bourges	106.3	1
5)	Castres	106.3	1
6)	Laon	106.3	1
20)	Moulins	106.3	1
21)	Pau	106.3	1
13)	Quimper	106.3	1
22)	Saint Brieuc	106.3	1
5)	Sarlat la Canéda	106.3	1
5)	Toulouse	106.3	5
13)	Tours	106.3	1
8)	Vannes	106.3	1
5)	Avallon	106.4	1
15)	Bordeaux	106.4	5
5)	Caen	106.4	2
8)	Chambéry	106.4	1
18)	Clermont Ferrand	106.4	2
5)	Marseille	106.4	10
5)	Montargis	106.4	1
10)	Troyes	106.4	1
10)	Valence	106.4	1
7)	Agen	106.5	1
5)	Aurillac	106.5	1
5)	Blois	106.5	1
5)	Châteauroux	106.5	1
5)	Évreux	106.5	1
9)	Fougères	106.5	1
3)	Lons le Saunier	106.5	1
22)	Lourdes	106.5	1
13)	Nogent le Rotrou	106.5	1
22)	Périgueux	106.5	1
23)	Saint Étienne	106.5	2
17)	Yvetot	106.5	1
22)	Albi	106.6	1
10)	Brest	106.6	3
20)	Cahors	106.6	1
5)	Châtellerault	106.6	1
5)	Châtillon s. Seine	106.6	1
15)	Dreux	106.6	1
9)	Gournay en Bray	106.6	1
18)	La Flèche	106.6	1
6)	La Rochelle	106.6	1
5)	Le Puy en Velay	106.6	1
20)	Montluçon	106.6	1
23)	Quimperlé	106.6	1
13)	Saint Malo	106.6	1
8)	Toulon	106.6	1
9)	Vire	106.6	1
18)	Alençon	106.7	1
5)	Angoulême	106.7	1
5)	Bourges	106.7	1
6)	Calvi	106.7	1
10)	Carcassonne	106.7	1
5)	Chalon sur Saône	106.7	1
5)	Condom	106.7	1
25)	Lisieux	106.7	1
15)	Lyon	106.7	1
5)	Mende	106.7	1
10)	Nantes	106.7	4
1)	Paris	106.7	4
3)	Roanne	106.7	1
5)	Royan	106.7	1
5)	Saint Gaudens	106.7	1
20)	Ussel	106.7	1
6)	Vitry le François	106.7	1
8)	Alès	106.8	1
8)	Avallon	106.8	1
22)	Bordeaux	106.8	5
20)	Brioude	106.8	1
4)	Chartres	106.8	1
13)	Château Renault	106.8	1
20)	Grasse	106.8	1
17)	La Côte St André	106.8	1
12)	Marseille	106.8	4
5)	Niort	106.8	1
5)	Pau	106.8	1
13)	Perpignan	106.8	1
3)	Rennes	106.8	3
5)	Rethel	106.8	1
18)	Saint Affrique	106.8	1
23)	Béthune	106.9	1
19)	Bourg en Bresse	106.9	1
19)	Châteauroux	106.9	1
22)	Compiègne	106.9	1
9)	Grenoble	106.9	1
8)	La Roche sur Yon	106.9	1
8)	Le Havre	106.9	1
8)	Le Mans	106.9	2
7)	Lorient	106.9	1
9)	Mantes la Jolie	106.9	2
3)	Mazamet	106.9	1
23)	Melun	106.9	1
13)	Mers les Bains	106.9	1
21)	Montpellier	106.9	3
6)	Périgueux	106.9	1
18)	Poligny	106.9	1
20)	Propriano	106.9	1
5)	Provins	106.9	1
2)	Romilly sur Seine	106.9	1.2
29)	Saint Lô	106.9	1
2)	Strasbourg	106.9	1
5)	Bar le Duc	107.0	1
8)	Bressuire	107.0	1
8)	Cahors	107.0	1
7)	Château Thierry	107.0	1
2)	Clermont Ferrand	107.0	1
23)	La Ferté Macé	107.0	1
6)	L'Aigle	107.0	1
19)	Le Puy en Velay	107.0	1
21)	Mont de Marsan	107.0	1
8)	Montauban	107.0	1
17)	Montluçon	107.0	1
23)	Nice	107.0	5
21)	Nîmes	107.0	1
18)	Porto Vecchio	107.0	1
7)	Rouen	107.0	1
18)	Saint Quentin	107.0	1
1)	Valence	107.0	1
19)	Abbeville	107.1	1
10)	Arcachon	107.1	1
5)	Bourges	107.1	1
2)	Caen	107.1	2
21)	Carmaux	107.1	1
23)	Dijon	107.1	1
8)	Laval	107.1	1
21)	Mulhouse	107.1	1
9)	Nancy	107.1	1
10)	Poitiers	107.1	1
5)	Quimper	107.1	1
18)	Saint Étienne	107.1	1
5)	Saint Flour	107.1	1
5)	St Méen le Grand	107.1	1
21)	Alès	107.2	1
3)	Angers	107.2	1
21)	Avignon	107.2	1
21)	Bastia	107.2	1
20)	Blois	107.2	1
21)	Figeac	107.2	1
14)	Limoges	107.2	1
19)	Mâcon	107.2	1
2)	Nantes	107.2	3
21)	Pau	107.2	1
10)	Pau	107.2	1
8)	Rochefort	107.2	1
7)	Soissons	107.2	1
2)	Toulouse	107.2	1

FM Station	MHz	kW		FM Station	MHz	kW
20) Tours	107.2	2		9) Metz	107.3	1
24) Ussel	107.2	1		5) Millau	107.3	1
7) Arnay le Duc	107.3	1		10) Montpellier	107.3	3
2) Arras	107.3	1		20) Orléans	107.3	1
10) Auxerre	107.3	1		9) Parthenay	107.3	1
2) Bordeaux	107.3	1		21) Perpignan	107.3	3
2) Brest	107.3	3.2		23) Saint Brieuc	107.3	1
10) Brive la Gaillarde	107.3	1		23) Saint Raphaël	107.3	1
1) Carcassonne	107.3	1		23) Dreux	107.4	1
10) Chantilly	107.3	4		5) Granville	107.4	1
2) Châteauroux	107.3	1		7) Lisieux	107.4	1
21) Colmar	107.3	1		8) Nevers	107.4	1
8) Dax	107.3	1		18) Provins	107.4	1
23) Évreux	107.3	1		4) Château Gontier	107.5	1
2) Lens	107.3	1		20) Châteaudun	107.5	1
18) Lorient	107.3	1		21) Cherbourg Octeville	107.5	1
18) Lyon	107.3	1		20) Saint Dizier	107.5	1
6) Mazamet	107.3	1		20) La Ferté Macé	107.9	1
2) Menton	107.3	1		20) L'Aigle	107.9	1

NB: Stns under 1kW not mentioned.

As of September 2010, 4765 licenses (txs) were allocated to private commercial and non-commercial FM stns. Approx. 3250 stns are affiliated to one of the following private commercial national networks.
Addresses:
1) Beur FM ☎ 89 rue Oberkampf, B.P. 249, 75524 Paris Cedex 11 ☎ +33 1 49294337 ▤+33 1 48060662 **W:** www.beurfm.net. + 4 tx less than 1kW – **2)** BFM ▣ 12 rue d'Oradour sur Glane, 75740 Paris Cedex 15 ☎ +33 1 71191181 ▤ +33 1 71191180 **W:** www.radiobfm.com + 3 txs less than 1kW – **3)** Chérie FM ▣ 22 rue Boileau, 75016 Paris ☎ +33 1 40714200 ▤ +33 1 40714124 **W:** www.cheriefm.fr+ 54 txs less than 1kW – **4)** COFRAC 11 rue Rosenwald, 75015 Paris ☎+33 1 56232298 **W:** www.cofrac-media.com + 18 txs less than 1kW – **5)** Europe 1 ▣ 26 bis rue François 1er, 75008 Paris ☎ +33 1 44319000 ▤ +33 1 47231900 **W:** www.europe1.fr **LW:** 183kHz 2000kW see Germany. + 125 txs less than 1kW – **6)** Virgin Radio ▣ 26 bis rue François 1er, 75008 Paris ☎ +33 1 47231000. ▤ +33 1 47231071 **W:** www.virginradio.fr + 95 txs less than 1kW– **7)** Fun Radio ▣ 20 rue Bayard, 75008 Paris ☎ +33 1 40704848. ▤ +33 1 40704800 **W:** www.funradio.fr + 92 txs less than 1kW – **8)** MFM ▣ 104 avenue du Président Kennedy, 75016 Paris ☎ +33 1 55745570 ▤ +33 1 55745588 **W:** www.mfm.fr + 33 txs less than 1kW – **9)** NRJ ▣ 22 rue Boileau, 75016 Paris ☎ +33 1 40714000 ▤ +33 1 40714040 **W:** www.nrj.fr + 136 txs less than 1kW – **10)** Radio Classique ▣ 12 bis place Henri Bergson, 75382 Paris. Cedex 08 ☎ +33 1 40085000 ▤ +33 1 40085080 **W:** www.radioclassique.fr + 24 txs less than 1kW – **11)** Radio Courtoisie ▣ 61 bd Murat 75016 Paris ☎ +33 1 46510085 ▤ +33 1 46512182 **W:** www.radiocourtoisie.net + 3 txs less than 1kW – **12)** Radio FG ▣ 51 rue de Rivoli, 75001 Paris ☎ +33 1 40138800 ▤ +33 1 40138001 **W:** www.radiofg.com - 6 txs less than 1kW – **13)** Nostalgie ▣ 22 rue Boileau, 75016 Paris ☎ +33 1 40714000. ▤ +33 1 40714040 **W:** www.nostalgie.fr + 104 txs less than 1kW – **14)** Radio Nova ▣ 127 avenue Ledru Rollin, 75011 Paris ☎ +33 1 53333335 **W:** www.novaplanet.com + 4 txs less than 1kW – **15)** Radio Orient ▣ 98 bd Victor Hugo, 92110 Clichy ☎ +33 1 41061600 ▤ +33 1 41061619 **W:** www.radioorient.com + 9 txs less than 1kW – **16)** Radio Soleil ▣ 57 rue Avron, 75020 Paris ☎ +33 1 43488974 ▤ +33 1 43485558 **W:** www.radio-soleil.com + 2 txs less than 1kW – **17)** RCF ▣ 7 place Saint Irénée, 69321 Lyon Cedex 05 ☎ +33 4 72382022. ▤ +33 4 72382057 **W:** www.rcf.fr + 113 txs less than 1kW – **18)** RFM ▣ 28 rue François 1er, 75008 Paris ☎ +33 1 42322000 ▤ +33 1 47232466 **W:** www.rfm.fr + 81 txs less than 1kW – **19)** Rire et Chansons ▣ 22 rue Boileau, 75016 Paris ☎ +33 1 40714000 ▤ +33 1 40714040 **W:** www.rireetchansons.fr + 41 txs less than 1kW – **20)** RMC ▣ 12 rue d'Oradour sur Glane, 75740 Paris Cedex 10 ☎ +33 1 71191191 ▤ +33 01 71191190 **W:** www.rmc.fr **LW:** Roumoules 216kHz 1400kW See Monaco. + 112 txs less than 1kW – **21)** RTL ▣ 22 rue Bayard, 75008 Paris. ☎ +33 1 40704450 ▤ +33 1 40704450 **W:** www.rtl.fr **LW:** 234kHz 2000kW see Luxembourg. + 93 txs less than 1kW – **22)** RTL 2 ▣ 22 rue Bayard, 75008 Paris ☎ +33 1 40704000 ▤ +33 1 40704800 **W:** www.rtl2.fr + 61 txs less than 1kW – **23)** Skyrock ▣ 37 bis rue Greneta, 75002 Paris ☎ +33 1 44888200 **W:** www.skyrock.fm + 89 txs less than 1kW – **24)** Sud Radio ▣ Im. Les Allées du Lac, bât B, rue du Lac, 31681 Labège Cedex ☎ +33 5 61632020 ▤ +33 5 61632037 **W:** www.sudradio.fr + 28 txs less than 1KW – **25)** Jazz Radio ▣ 40 quai Rambaud, 69002 Lyon ☎ +33 4 72101535 **W:** www.jazzradio.fr + 13 txs less than 1 kW – **26)** La radio de la mer ▣ 12 rue Laugier, 75017 Paris ☎ +33 1 46220006 **W:** www.laradiodelamer.com + 6 txs less than 1kW – **27)** France Maghreb 2 ▣ 84 rue des Couronnes, 75020 Paris ☎ +33 1 40339081 **W:** www.francemaghreb2.fr + 2 txs less than 1kW – **28)** TSF Jazz ▣ 127 avenue Ledru Rollin, 75011 Paris ☎ +33 1 53332280 **W:** www.tsfjazz.com + 4 txs less than 1kW – **29)** Ouï FM ▣

2 rue de la Roquette, 75011 Paris ☎ +33 1 55281414 **W:** www.ouifm. fr + 5 txs less than 1 kW.

DAB: F.P.I.: New licences to be granted on VHF band III using T-DMB technology, at the end of 2009, were postponed to an unknown date. Tests in Paris ch9A and 11C, Lyon 8D, Nantes 9A (irr.).

FRENCH GUIANA

L.T: UTC -3h — **Pop:** 210,000 — **Pr.L:** French — **E.C:** 50Hz, 127/220V — **ITU:** GUF — **Int. dialling code:** +594

R.F.O. GUYANE (RADIODIFFUSION FRANÇAISE D'OUTRE-MER)
▣ B.P. 7013, 97305 Cayenne ☎ 594 301500 ▤ 594 302649
L.P: Dir: Anastasie Bourquin. Dir. Tec: Serge Sulpice-Timothe. PD: Jean-Pierre Karam
MW: Matoury 1070kHz 10kW, St. Laurent du Maroní 1060kHz 0.05kW
SW: Matoury (G.C: 04N54 052W20): 5055kHz 10kW
NB: Both MW and SW reported inactive
FM (MHz): Ouanary 90.0 – Sinnamary, Régina, Saint Laurent du Maroni 91.0 – Cayenne 92.0 – Kaw, Mana, Kourou, Saint-Georges de l'Oyapock 94.0 – Papaichton, Saül, Maripasoula, Grand Santi 95.0
D.Prgr: 24h **Ann:** "Ici Cayenne, RFO Guyane"
IS: "Nos richesses" on guitar. **V.** by QSL-folder. Rec. acc.

Local Radio: R. Mosaique, Cayenne 88.1MHz – R.2000, Cayenne 96.9MHz – Nostalgie Guyane, Cayenne 99.6MHz – Vinyl Radio, Cayenne 102.9MHz – RTM, Cayenne 103.3MHz

RADIO FRANCE INTERNATIONALE RELAY STATION
▣ TDF Montsinery, B.P. 97307, Cayenne Cedex
FM: Cayenne 98.7MHz, 102.0MHz. Sinnamary 104.0MHz

FRENCH POLYNESIA

L.T: Tahiti: UTC-10h, **Marquesas Is:** -9½h, **Gambier Is:** -9h, — **Pop:** 287,032 — **Pr.L:** French, Tahitian — **E.C:** 60Hz, 220V — **ITU:** OCE

RADIO POLYNESIE (Pub/Comm)
▣ RFO Polynesie Francaise, Centre Pamatai, FAAA BP 60125-98702, Papeete, Tahiti, French Polynesia ☎ +689 689861616 ▤ +689 689861611 **W:** www.polynesie.rfo.fr

MW	kHz	kW
Mahina,Tahiti	738	20

FM	MHz	FM	MHz
Atuona, Hiva Oa	88.2	Uturoa, Raiatea	94.0
Mahatea,Moorea	89.0	Otepa, Hao	94.4
Mont Muake, Nuku Hiva	89.0	Pouheva, Tuamotu	94.4
Papeete, Tahiti	89.0	Rapuarava, Reao	94.4
Tapeata, Hiva Oa	89.5	Rikitea, Gambier Islands	94.4
Moerai, Rurutu	89.6	Rotoava, Fakarava	94.4
Papetoai, Moorea	89.6	Tarione, Fakahina	94.4
Punaauia, Tahiti	89.6	Turipaoa, Manihi	94.4
Rairua, Raivavae	89.6	Fakamaru, Tureia	94.8
Rautini, Arutua	90.5	Fakatopatere, Takapoto	94.8
Taiohae, Nuku Hiva	90.5	Teana, Fangatau	94.8
Tiarei, Tahiti	90.5	Tuherahera, Tikehau	94.8
Vaipaee, Ua Huka	91.0	Tukuhora, Anaa	94.8
Hakahau, Ua Pou	91.5	Tumukuru, Tatakoto	94.8
Mont Marau, Tahiti	91.8	Mahaena, Tahiti	95.2
Arutua, Raitahiti	93.6	Marautangaroa, Pukarua	95.2
Pahua, Mataiva	93.6	Niutahi, Apataki	95.2
Raitahiti, Kaukura	93.6	Papara, Tahiti	95.2
Teavaroa, Takaroa	93.6	Apataki, Tuamotu	95.5
Tepukamaruia, Napuka	93.6	Vaitape, Bora Bora	96.6
Aeroport, Rangiroa	94.0	Mahina, Tahiti	99.0
Hitianau, Faaite	94.0	Taravao, Tahiti	99.0
Manihi, Tuamotu	94.0	Ahurei, Rapa	99.4
Pouheva, Makemo	94.0	Amaru, Rimatara	99.4
Pukapuka, Tuamotu	94.0	Mataura, Tubuai	99.4
Tavana, Nukutavake	94.0		

Other Stations

FM Location		MHz	kW	Station
1)	Taravao, Tahiti	87.6	1	R.Maria No Te Hau
2)	Taputapuatea, Raitea	88.0	0.5	R.Paofai
3)	Mont Marau, Tahiti	88.2	3	R.Maohi
4)	Maiao, Moorea	88.6	4	NRJ
4)	Faaa, Tahiti	88.6	3	NRJ
5)	Tiarei, Tahiti	88.7	0.3	R.Maohi
6)	Uturoa, Raiatea	88.8	0.2	R.Bora Bora
8)	Mahina, Tahiti	89.4	0.3	R.Te Reo o Tefana

FM	Location	MHz	kW	Station
5)	Papeete, Tahiti	89.9	0.2	R.Paofai
8)	Uturoa, Raiatea	90.0	0.5	R.Te Reo o Tefana
4)	Taiarapu, Tahiti	90.1	0.5	NRJ
5)	Tahaa	90.8	0.5	R.Paofai
9)	Taiarapu, Tahiti	90.9	1	Radio 1
10)	Mont Marau, Tahiti	91.4	3	R.Te Vevo o Te Tiaturiraa
11)	Atuona, Hiva Oa	92.0	6	R.Te Oko Nui
14)	Moerai, Rurutu	92.0	0.05	R.Rurutu
3)	Maatea, Moorea	92.3	3.6	R.Maohi
11)	Taiohae, Nuku Hiva	92.5	0.3	R.Te Oko Nui
1)	Uturoa, Raiatea	92.6	0.3	R.Maria No Te Hau
8)	Mont Marau, Tahiti	92.8	0.5	R.Te Reo o Tefana
12)	Taiohae, Nuku Hiva	93.5	3	R.Te Vevo Te Tiaturiraa
10)	Taiarapu, Tahiti	93.5	0.5	R.Te Vevo o Te Tiaturiraa
1)	Papeete, Tahiti	93.8	0.2	R.Maria No Te Hau
5)	Taiarapu, Tahiti	93.9	0.5	R.Paofai
11)	Mont Muake, Nuku Hiva	94.5	6	R.Te Oko Nui
13)	Uturoa, Raiatea	94.5	0.5	R.La Voix de l'Esperance
3)	Taravao, Tahiti	94.8	1	R.Maohi
14)	Manureva, Rurutu	95.0	0.3	R.Rurutu
9)	Uturoa, Raiatea	95.0	0.5	R.Tiare
6)	Vaitape, Bora Bora	95.4	0.2	R.Bora Bora
13)	Mont Marau, Tahiti	95.6	3	R.La Voix de l'Esperance
19)	Moorea	95.8		Taui FM
15)	Ahurei, Rapa	96.0	0.3	R.Kotokoto
16)	Marutea Sud	96.0	1	R.Marutea Sud
6)	Vaitape, Bora Bora	96.2	0.2	R.La Voix de l'Esperance
1)	Mont Marau, Tahiti	96.4	3	R.Maria No Te Hau
3)	Papeete, Tahiti	96.8	0.6	R.Maohi
17)	Mataura, Tubuai	97.0	0.2	R.Te Reo No Tubuai
10)	Uturoa, Raiatea	97.2	0.5	R.Te Vevo o Te Tiaturiraa
8)	Afareaitu, Moorea	97.4	3	R.Te Reo o Tefana
8)	Maatea, Moorea	97.4	4.25	R.Te Reo o Tefana
8)	Niau, Tuamotu	97.4	0.1	R.Te Reo o Tefana
19)	Maatea, Moorea	97.8	12	Taui FM
1)	Aeroport, Rangiroa	98.0	1	R.Maria No Te Hau
18)	Papeete, Tahiti	98.0	0.3	R.Poroi
7)	Uturoa, Raiatea	98.0	0.5	Star FM
9)	Taravao, Tahiti	98.3	1	R.Tiare
9)	Uturoa, Raiatea	98.4	0.4	Taui FM
20)	Otepa, Hao	98.5	0.1	R.Hao
21)	Pouheva, Makemo	98.5	0.2	R.Tanginui
22)	Punaauia, Tahiti	98.5	0.3	R.Nono
9)	Papeete, Tahiti	98.8	0.2	R. 1
13)	Maatea, Moorea	99.5	3	R.La Voix de l'Esperance
3)	Nunue, Bora Bora	99.7	0.5	R.Maohi
1)	Faaone	99.8	0.5	R.Maria No Te Hau
9)	Afareaitu, Moorea	100.0	3	Radio 1
1)	Mangareva	100.0	0.7	R.Maria No Te Hau
23)	Taugaraufara, Manihi	100.0	0.1	R.Poe Rava
5)	Taiarapu, Tahiti	100.3	0.5	R.Paofai
24)	Papeete, Tahiti	100.5	0.8	R.Fara
7)	Taravao, Tahiti	100.8	0.5	Star FM
9)	Uturoa, Raiatea	100.9	0.5	Radio 1
25)	Aeroport, Rangiroa	101.0	1	R.Te Reo Tuamotu
26)	Muake, Nuku Hiva	101.3	0.3	R.Marquises
27)	Taiarapu, Tahiti	101.3	0.5	R.Tiarapu
1)	Afareaitu, Moorea	101.5	2	R.Maria No Te Hau
3)	Uturoa, Raiatea	101.7	0.5	R.Maohi
10)	Aeroport, Rangiroa	102.0	1	R.Te Vevo Te Tiaturiraa
32)	Papara, Tahiti	102.2	0.3	R.Te Vevo No Papara
30)	Papeete, Tahiti	102.2	1	R.Tahiti Nui FM
9)	Nunue, Bora Bora	102.4	0.3	Radio 1
4)	Mont Marau, Tahiti	103.0	3	NRJ
26)	Taiohae, Nuku Hiva	103.3	0.05	R.Marquises
9)	Papeete, Tahiti	103.4	0.2	R.Tiare
9)	Mont Marau, Tahiti	103.8	3	Radio 1
9)	Mont Marau, Tahiti	104.2	3	R.Tiare
31)	Nuku Hiva	104.5	0.05	R.Te Tau Vae'ia
5)	Afareaitu, Moorea	104.7	2	R.Paofai
13)	Taravao, Tahiti	105.1	0.8	R.La Voix de l'Esperance
1)	Uturoa, Raiatea	105.4	1	R.Maria No Te Hau
9)	Maatea, Moorea	105.5	3	R.Tiare
19)	Taravao, Tahiti	105.8	0.4	Taui FM
9)	Papeete, Tahiti	106.0	0.5	R.Tiare
2)	Maiao, Moorea	106.4	4.25	Pacifique FM
27)	Vairau, Tahiti	106.6	0.5	R.Taiarapu
8)	Taravao, Tahiti	107.0	0.4	R.Te Reo o Tefana
19)	Faaa, Tahiti	107.3	1	Taui FM

Addresses and other information
1) BP 94-98713 Papeete. Dir: Mme Irene Paofai ☎ +689 689420011 🖷 +689 689420635 **E:** contactmnth@radiomarianotehau.pf **W:** www.radiomarianotehau.com – **2)** BP 14150-98701, Arue. Dir: Thierry Demary ☎ +689 689583747 🖷 +689 689429164 **E:** pacificfm@cara-

mail.com – **3)** Maison des Jeunes, BP 5038 Pirae ☎ +689 689433101 🖷 +689 689451650 (currently believed to be operating limited service only at some locations and reportedly in receivership) – **4)** BP 50-98713 Papeete ☎ +689 689421042 🖷 +689 689464346 **W:** www.nrj.pf **E:** nrj@mail.pf **LP:** GM: Nadine Richardson – **5)** BP 113-98713 Papeete, Dir: Maea Tematua Tech: Maurice Tupea Sec: Iteata Tevaarauhara ☎+689 689460624/689460606 🖷 +689 689419357 **E:** radiopaofai@epm.pf **W:** www.radio.radiopaofai.org – **6)** Vaitape, Bora Bora, Dir: Jean Claude. ☎ +689 689605873/689605888 – **7)** Papeete [believed silent and most frequencies reallocated to other stations in mid-2010] – **8)** BP 6295 Faaa, Papeete ☎ +689 689819797 🖷 +689 689825493 **E:** tereo@mail.pf. **LP:** GM: Terimateatea Mana – **9)** BP 3601-98713 Papeete. Dir: Mme Sonia Aline ☎ +689 689434100/689436100 🖷 +689 689422421/689423406 (Radio 1/R.Tiare) Radio 1/aline@aline.pf **W:** Radio 1: www.radio1.pf R.Tiare: www.tiare.pf and www.facebook.com/pages/tiarefm – **10)** BP 1817-98713 Papeete, 51, rue Dumont D'Urville, Orovini, Papeete. Dir: Christian Bradai ☎ +689 689412341 🖷 +689 689412322 **E:** contacts@mail.pf – **11)** Mission Catholique, Taiohae, Nuku Hiva – **12)** BP 1817 Papeete ☎ +689 689412341 🖷 +689 689412322. **E:** contacts@mail.pf – **13)** BP 95-97813 Papeete, Dir: Hubert Terorotua ☎ +689 689508259 🖷 +689 689451427 – **14)** Moerai, Rurutu ☎ +689 689940468 – **15)** ☎+689 689957272 – **16)** ☎ +689 689946151 – **17)** ☎ +689 689950821 – **18)** Catholic Mission, Papeete – **19)** BP 60076 Faa'a-Centre, Papeete ☎+689 689854747 🖷 +689 689412155 **W:** www.taui-fm.com **E:** taui@mail.pf – **20)** Association Jeunesse et Developpement de Hao [reported inactive] – **21)** Makemo, iles Tuamotu-Gambier – **22)** Association Pacifique Sound, Punaauia, Tahiti [inactive] – **23)** Manihi, iles Tuamotu-Gambier – **24)** Servitude Graffe, Taumoa, Dir: Mme Roti Make ☎ +689 689419125 – **25)** Cultural Association Iva Manu-Manu Arii, Avatoru, Rangiroa, Tuamotu-Gambier [reported silent] – **26)** BP 338,Taiohae, Nuku Hiva 98742 ☎ +689 689920790 🖷 +689 689920729 **W:** www.facebook.com/pages/radio-marquises **E:** radiomarquises@mail.pf – **27)** "Le Rythme de la Presque"ile", Vairao-Taiarapu, Tahiti ☎ +689 689575208 **LP:** Denis Tarovo – **28)** SARL Tahiti CD-Tahiti Pub, Papeete, Tahiti [inactive] – **31)** College de Taiohae, Taiohae, Nuku Hiva 98742 **E:** direction@clgtaio.ensec.edu.pf – **32)** Mairie de Papara, pk 34,800, Papara 98712, Tahiti ☎ +689 689279574

FRENCH SOUTHERN & ANTARCTIC LANDS

L.T: UTC+5h — **Pop:** 150 (wi), 310 (su) — **Pr.L:** French — **E.C:** 50Hz, 220V — **ITU:** none (WRTH: FSA); Isles Kerguelen: ITU: KER

FM	MHz	Station	FM	MHz	Station
1)	98.0	Radio Ker	3)	100.0	RTL
2)	100.0	France Inter			

Addresses & other information:
1) Port-aux-Francais, District de Kerguelen, Terres Australes & Antarctiques Francaises [via Reunion, Indian Ocean]. 24h local community station. – **2)** 24h satellite relay from Paris, Mon-Fri. – **3)** 24h satellite relay from Paris, weekends

GABON

L.T: UTC +1h — **Pop:** 1.5 million — **Pr.L:** French, Fang, Bopounou, Obamba, Djebi — **E.C:** 50Hz, 220V — **ITU:** GAB

CONCEIL NATIONAL DE LA COMMUNICATION(CNC)
🖃 B.P. 6437, Libreville ☎+241 762796
L.P: Pierre-Marie Dong, President.

RADIODIFFUSION-TÉLÉVISION GABONAISE(RTG,Gov.)
🖃 B.P. 10150, Libreville ☎+241 732459 🖷 +241 739775
LP: DG RTG-1: Willy Kombény. DG RTG-2: Jules Legnongo. Asst. DGs: Radio: Gilles Terence Nzoghe. Tech: Claude Nganga. Provincial Stns: Robert Aloli.

MW	kHz	kW	N	Times
Oyem	549	20	2	0430-0630, 1030-1430, 1600-2230

SW relays via Moyabi on 4777/7270kHz not heard recently., irr.
FM(MHz): Libreville 87.7/96.54 (1), 92.5 (2), Franceville 87.86 (2), Makokou 100.5 (2), Oyem 87.94 (2), Pt. Gentil 88.03 (2), Tchibanga 91.04 (2).
(1) **RTG Chaîne 1** in French (2) **RTG Chaîne 2** (provincial netw.) in French & ethnic languages. **FM Prgr:** 0500-2305 on FM only.
Ann: 1: "Ici Libreville, vouz écoutez Radio Gabon, chaîne 1".
IS: Indigenous instruments. Opens and closes with National Anthem.

AFRICA No.1 (Comm.)
🖃 B.P. 1, Libreville ☎+241 760001 🖷+241 742133 🖃 **in France:** 33

Rue du Faubourg Saint Antoine, F-75011 Paris +33 1 55075801 🖷 +33 1 55079748 **W:** www.africa1.com
LP: DG: Louis Barthélém Mapangou.
FM: Libreville 94.5MHz + rel. in other countries.
Africa Plus Gabon 99.5MHz. **SW:** see International section

Other stations:
R. Émergence, B.P. 06, Libreville: 91.6MHz 30W **W:** f-i-a.org/emergence – **R. Génération Nouvelle**, B.P. 727, Libreville: 97.4MHz – **R. Mandarine**, B.P. 511, Libreville: 106.6MHz – **R. Nostalgie**, B.P. 13050, Libreville: 93.0MHz – **R. Notre-Dame de Sainte-Marie**, B.P. 20348, Libreville: 99MHz – **R. Soleil FM**, B.P. 5420, Libreville: 107.7MHz – **Top FM**, B.P. 6554, Libreville: 105.5MHz (also rel. VOA) – **R. Unité**, B.P. 2676, Libreville: 100.5MHz.
RFI Afrique in Franceville, Libreville & Port-Gentil on 104MHz

GALAPAGOS ISLANDS (Ecuador)

L.T: UTC -6h — **Pop:** 19,000 — **Pr.L:** Spanish — **E.C:** 60Hz 110/220V — **ITU:** EQA (**WRTH:** GAL)

LA VOZ DE GALAPAGOS (Rlg)
Prefectura Apostólica de Galápagos, Puerto Baquerizo Moreno
☎ +593 5 459435
MW: La Voz de Galápagos 530kHz 5kW (inactive)
FM: Galápagos Stereo 97.1MHz **V.** by QSL card
FM in Pto Baquerizo Moreno (MHz): 91.1 Nacional del Ecuador FM – 94.7 R. Mar – 97.1 LV de Galápagos FM – 100.7 R. María – 101.9 Encantada FM – 104.3 Telegalápagos FM
FM in Pto Ayora (MHz): 88.7 R. Santa Cruz – 89.9 Caravana AM – 93.5 Pacífica FM – 94.7 R. Mar – 95.9 Antena 9 FM – 98.3 Stereo Zaracay – 99.5 Deportes solo deportes – 101.9 Encantada FM
FM in Pto Villamil (MHz): 106.7 Isabela FM

GAMBIA

L.T: UTC — **Pop:** 1.8 million — **Pr.L:** English, Mandinka, Fula, Wolof, Jola, Serahuleh, Manjago, Aku — **E.C:** 50Hz, 230V — **ITU:** GMB

MINISTRY OF COMMUNICATIONS, INFORMATION & INFORMATION TECHNOLOGY (MOCIIT)
🖃 New GRTS Building, MDI Road, Kanifing +220 4378000 🖷 +220 4378029 **W:** www.doscit.gm **E:** doscit@gamtel.gm
LP: S.S. Jallou, Perm. Sec.

GAMBIA RADIO AND TELEVISION SERVICE (GRTS)
🖃 Mile 7 Studios, P.O. Box 387, Banjul ☎+220 4495101/4497419 🖷 +220 4495102 **W:** www.grts.gm
LP: DG: Mr. Modou Sanyang. Deputy DG: Mr. Alhaji Modou Joof.
MW: Bonto 648kHz 50kW, Basse 747kHz 10kW.
FM: Bonto 91.4MHz, Serrekunda 96.0, Banjul 98.6MHz.
D.Prgr: Mon-Thurs 0530-1400, 1655-2400. Fri-Sun 0550-2400.
Ann: "You are tuned to Gambia Radio & Television Service from Banjul".**IS:** Cora (harp).

Other stations:
City Limits R., Serrekunda: 93,6MHz 0.25kW – **Hill Top R.**, Serrekunda: 104.7MHz – **Paradise FM**, Farafenni/Serrekunda: 105.5MHz, both 1kW. Web: paradisefm.gm – **R. KWT** (Kids With Talent), Banjul: 107.6MHz – **Unique FM**, Banjul/Basse: 101.7MHz. Web: uniquefm.gm – **West Coast R.**, Serrekunda: 92.1MHz. Web: westcoast.gm . Also rel. BBC.
RFI Afrique: Banjul 89.0MHz

GEORGIA

L.T: UTC +4h; Akhazia & South Ossetia: UTC +3h (27 Mar-30 Oct: +4h) — **Pop:** 4.6 million — **Pr.L:** Georgian, Russian, Abkhaz, Ossetic — **E.C:** 50Hz, 220V — **ITU:** GEO

GEORGIAN NATIONAL COMMUNICATIONS COMMISSION (GNCC)
🖃 Ave. Ketevan Tsamebuli/Bochorma St. 50/18, Tbilisi 0144 ☎ +995 32 921667 🖷 +995 32 921625 **E:** post@gncc.ge **W:** www.gncc.ge
LP: Chmn: Irakli Chikovani.
NB. GNCC is the regulatory authority for broadcasting.

SAKARTVELOS SAZOGADOEBRIVI MAUTS'Q'EBELI (Georgian Public Broadcasting)
🖃 M.Kostava St. 68, Tbilisi 0171 ☎ +995 32 409477 🖷 +995 32 409477 **E:** info@gpb.ge **W:** www.gpb.ge
LP: Chmn: Levan Gakheladze.

FM (MHz)	P1	P2	kW	FM	P1	P2	kW
Akhaltsikhe	102.4	-	-	Lentekhi	102.4	-	-
Ambrolauri	102.4	-	-	Sachkere	102.4	-	-
Batumi	102.4	-	5	Telavi	100.6	-	-
Chiatura	102.4	-	-	Tbilisi	102.4	100.9	10
Gori	100.6	-	0.25	Zugdidi	101.3	-	5
Kutaisi	100.3	-	0.5	P=Prgr			

D.Prgr: Prgr 1 (Sazogadoebrivi R.): 24h.– **Prgr 2 (Kartuli R.):** 0400-2200.

OTHER STATIONS

FM	MHz	kW	Location	Station
14)	93.8	1	Tbilisi	5 Lines
15)	94.3	1	Tbilisi	R. GIPA
22)	94.7	1	Tbilisi	R. Natsnobi
23)	95.1	1	Tbilisi	Avtoradio
3)	95.5	1	Tbilisi	R. Komersant
6)	95.9	1	Tbilisi	R. Muza
17)	96.3	5	Tbilisi	R. Jako
5C)	96.7	1	Tbilisi	R. Ar Daidardo
18)	97.1	1	Tbilisi	R. Akhali talga
10)	98.1	1	Tbilisi	R. Utsnobi
21)	98.8	1	Tbilisi	Golos Abkhazii
21)	98.8	1	Gori	Golos Abkhazii
21)	98.8	1	Kutaisi	Golos Abkhazii
25)	99.2	1	Tbilisi	Folk FM
5D)	99.7	1	Tbilisi	Guru FM
9)	100.0	1	Gori	R. Imedi
9)	100.1	1	Batumi	R. Imedi
7)	100.3	1	Tbilisi	Beat FM
9)	100.9	5	Qutaisi	R. Imedi
2)	101.4	5	Tbilisi	R. Monte-Karlo
12)	101.5	1	Gori	R. Mtsvane talga
19)	101.9	1	Tbilisi	R. Kalaki
5C)	102.7	2.5	Qutaisi	R. Ar Daidardo
5B)	103.4	1	Tbilisi	R. Fortuna+
5B)	103.4	5	Gori	R. Fortuna+
5B)	103.4	5	Batumi	R. Fortuna+
5B)	103.4	1	Qutaisi	R. Fortuna+
12)	103.6	1	Zugdidi	R. Mtsvane talga
24)	103.9	1	Tbilisi	R. Palitra
9)	104.2	1	Zugdidi	R. Imedi
4)	104.3	1	Tbilisi	R. Sindikat
16)	104.8	1	Gori	R. Trialeti
20)	105.0	1	Qutaisi	R. Nostalgia
8)	105.0	1	Tbilisi	R. 105
1)	105.4	5	Batumi	R. Iveria
1)	105.4	1	Qutaisi	R. Iveria
1)	105.4	1	Tbilisi	R. Iveria
9)	105.9	1	Tbilisi	R. Imedi
11)	106.4	10	Tbilisi	Pirveli R.
11)	106.4	1	Gori	Pirveli R.
5A)	106.9	1	Tbilisi	R. Fortuna
5A)	106.9	1	Gori	R. Fortuna
5A)	106.9	1	Batumi	R. Fortuna
21)	107.2	5	Zugdidi	Golos Abkhazii
12)	107.4	10	Tbilisi	R. Mtsvane talga
12)	107.5	2	Qutaisi	R. Mtsvane talga
13)	107.4	1	Tbilisi	R. Saqartvelos khma

NB: Txs below 1kW not listed.
Addresses & other information:
1) Erekle II square 1, Tbilisi 0105. **E:** iveria105.4@yahoo.com – **2)** M.Kostava St. 14, Tbilisi 0169. In Russian. – **3)** Nadiradze St. 8, Tbilisi 0102. **E:** info@commersant.ge – **4)** Ateni St. 18a, Tbilisi 0179. **E:** syndicate@radiosyndicate.ge. In Georgian & English. – **5A,B,C,D)** Marshal Gelovani St. 2, Tbilisi 0179. **E:** 5A,B) thc@radio.fm; 5C) tamara@fortuna.ge – **6)** Zandukeli St. 12, Tbilisi 0108. **E:** info@radiomuza.ge. In Georgian & English. – **7)** Tsinamdzgvrishvili 95, Tbilisi 0102. **E:** info@beatfm.ge – **8)** Agladze St. 31, Tbilisi 0119. **E:** n1001@geo.net.ge – **9)** Lubliana St. 5, Tbilisi 0159. **E:** info@radio-imedi.ge – **10)** M.Kostava St. 68, Tbilisi 0171. **E:** radio@ucnobifm.ge – **11)** Aleksidze St. 1, Tbilisi 0193. **E:** 106.4@radioone.ge – **12)** Vazha-Pshavela Ave. 45, Tbilisi 0177. **E:** gwave@greenwave.ge – **13)** Tashkenti St. 51, Tbilisi 0160. – **14)** Amaghleba St. 11, Tbilisi 0105. **E:** 5lines@5linesradio.ge – **15)** Marie Brosset St. 2, Tbilisi 0108. – **16)** Chavchavadze St. 45, Gori 1400. **E:** contact@trialeti.ge – **17)** Tbilisi. – **18)** Tbilisi. – **19)** M.Kostava St. 68, Tbilisi 0171. **E:** info@radio-kalaki.ge – **20)** Tamar Mepe Ave. 56, Qutaisi 4600. **E:** nostalgiarad@posta.ge – **21)** Vazha-Pshavela Ave. 76b, Tbilisi 0186. **E:** info@amc.ge – **22)** M.Kostava St. 68, Tbilisi 0171 – **23)** Tbilisi. – **24)** M.Kostava St. 77, Tbilisi 0161. **E:** info@radiopalitra.ge – **25)** Tbilisi.

ABKHAZIA

APSUA XÖYNTKARRATÄ TELERADIOKOMPANIA (Abkhaz State Radio & TV Co.)
🖃 Lasuria St. 16, Sokhumi ☎ +7 840 2224867 🖷 +7 840 2221144

E: apsuaradio1@mail.ru **L.P:** Dir: Zurab Argun

MW	kHz	kW			
Sokhumi	1350	30			
SW	kHz	kW	SW	kHz	kW
Sokhumi	9495v	5	Sokhumi	9535v	5

FM: Sokhumi 68.45/103.3MHz.
D.Prgr: Apsua R. 0300 (SS 0500)-2100v in Abkhaz, Russian on FM. On MW/SW: limited schedules with breaks, changing frequently. Outside of own prgrs, various other stns may be relayed: R. Rossii from Russia (incl. reg. reg. prgrs of GTRK "Kuban", Krasnodar & GTRK "Sochi", Sochi), commercial stns (e.g. Avtoradio).

OTHER STATIONS
FM	MHz	kW	Location	Station
A)	100.7	-	Ochamchire	Golos Rossii relay
A)	104.4	-	Sokhumi	Golos Rossii relay
A)	106.1	-	Tkvarcheli	Golos Rossii relay
2)	107.1	-	Sokhumi	Avtoradio
1)	107.9	0.3	Sokhumi	R. Soma

Addresses & other information:
1) Zvanba St. 9, Sokhumi. **E:** info@radiosoma.com – **2)** Sokhumi. Rel. Avtoradio (Russia).– **A)** Rel. Golos Rossii (Russia).

SOUTH OSSETIA

STATE RADIO & TV CO. "IR"
✉ Geroev St. 48, Tskhinvali ☎ +7 9974 451218 **E:** rsoradio@yandex.ru **W:** www.radioir.org **L.P:** Dir: Robert Kulumbegov
FM: Tskhinvali 102.3MHz.
D.Prgr: R. Ir in Russian, Ossetic: 24h.

OTHER STATIONS
FM	MHz	kW	Location	Station
A)	104.5	-	Kvaysa	Vesti FM relay
1)	105.9	-	Tskhinvali	Volna FM
B)	106.3	-	Tskhinvali	R. Mayak relay
2)	107.3	-	Tskhinvali	R. Yuzhnyy gorod

Addresses & other information:
1) Tskhinvali. – **2)** Tskhinvali. E: info@yugfm.ru – **A)** Rel. Vesti FM (Russia) – **B)** Rel. R. Mayak (Russia)

GERMANY

L.T: UTC +1h (27 Mar-30 Oct 2011: UTC +2h) — **Pop:** 82 million— **Pr.L:** German — **E.C:** 50Hz, 230V — **ITU:** D

BUNDESNETZAGENTUR
Authority responsible for frequency allocation matters.
✉ Postfach 8001, 53105 Bonn (office location: Tulpenfeld 4) ☎ +49 (228) 14 0 ▤ + 49 228 14 8872 **W:** www.bundesnetzagentur.de

NB: Due to the complexity of the broadcasting system in Germany, AM-transmitters (public service, commercial and military) are listed in a combined frequency table below. FM stations and other info can be found under the respective public radio station (section I), federal state (section II) or military station (section III).

LW/MW
Stn	kHz	kW	Site	Prgr
A)	153	500/250°	Donebach (Mudau)	DLF (1)
A)	177	500	Zehlendorf (Oranienbg.)	DK (1)
U)	183	2000°	Felsberg (Saarlouis)	Europe 1 (2)
A)	207	500/250°	Aholming (Deggendorf)	DLF
A)	549	100	Nordkirchen	DLF
A)	549	100	Thurnau-Tannfeld	DLF
I)	576	100	Mühlacker	SWR cont.ra
M)	603	20	Zehlendorf (Oranienbg.)	OldieStar Radio
R)	630	100/16	Scheppau (Braunschw.)	Voice of Russia (3)
I)	666	150	Rohrdorf	SWR cont.ra (4)
M)	693	250	Zehlendorf (Oranienbg.)	Voice of Russia (3)
E)	702	5	Flensburg	NDR Info Spezial
I)	711	5	Heilbronn	SWR cont.ra
I)	711	5	Ulm	SWR cont.ra
J)	720	85	Langenberg	WDR 2/VERA
	729	–	Putbus (Rügen)	(closed down)
B)	729	0.2	Hof	on3radio
B)	729	1	Würzburg	on3radio
J)	774	5	Bonn	WDR 2/VERA
A)	756	200/200°	Scheppau (Braunschw.)	DLF
A)	756	100	Ravensburg	DLF
D)	783	100	Wiederau (Leipzig)	MDR Info
E)	792	5	Lingen	NDR Info Spezial
B)	801	100	Ismaning (München)	on3radio
B)	801	10	Dillberg (Nürnberg)	on3radio
E)	828	20/5	Hemmingen (Hannover)	NDR Info Spezial

Stn	kHz	kW	Site	Prgr
I)	828	10	Freiburg	SWR cont.ra (4)
A)	855	10#	Berlin-Britz	DRadio Wissen
D)	873	150°	Weißkirchen (Oberursel)	AFN Power Netw.
D)	882	20°	Wachenbrunn (Themar)	MDR Info (5)
	936	–	Bremen Oberneuland	(closed down)
E)	972	100	Hamburg-Billwerder	NDR Info Spezial
A)	990	100	Berlin-Britz	DK (1)
I)	1017	100	Wolfsheim (Mainz)	SWR cont.ra
D)	1044	20	Wilsdruff (Dresden)	MDR Info (5)
	1107	10	Kaiserslautern	AFN Power Netw.
	1107	10	Vilseck	AFN Bavaria
	1143	10	Stuttgart-Hirschlanden	AFN Power Netw.
	1143	1	Mönchengladbach	AFN Benelux
	1143	1	Spangdahlem/Bitburg	AFN Power Netw.
	1143	1	Heidelberg	AFN Power Netw.
	1143	0.3	Bamberg	AFN Bavaria
	1143	0.01	Schweinfurt	AFN Bavaria
H)	1179	10	Heusweiler	SR Antenne Saar
D)	1188	3	Reichenbach (Görlitz)	MDR Info
A)	1269	300°	Arpsdorf (Neumünster)	DLF
X)	1323	1000°/150°	Wachenbrunn (Themar)	Voice of Russia (3)
A)	1422	400	Heusweiler	DLF
V)	1431	250/150	Wilsdruff (Dresden)	Voice of Russia (3)
	1485	–	Baden-Baden	(closed down)
M)	1485	0.5#	Berlin Schäferberg	OldieStar Radio
	1485	0.3	Ansbach	AFN Power Netw.
	1485	0.3	Hohenfels	AFN Bavaria
P)	1539	700	Mainflingen	ERF Radio (6)
J)	1593	20#	Langenberg	Kiraka

SW	kHz	kW	Site	Prgr
O)	5980	1	Krekel (Euskirchen)	Hamb. Lokalr. (8)
S)	6005	1	Krekel (Euskirchen)	Radio 700 (7)
O)	6045	100	Wertachtal (Buchloe)	Hamb. Lokalr. (8)
	6085	–	Ismaning (München)	(closed down)
A)	6190	17	Berlin-Britz	DLF
L)	15896	0.1#	Erlangen	bit eXpress
	26045	–	Hannover university	(closed down)

Powers day/night (usually 0500-1800/1800-0500). ° = directional. # = DRM (digital modulation).
Stn: Public stations **A-J** see section I. – Commercial and other stations see section II, chapters **M-X**, Voice of Russia schedules and contact details see under Russia in International Broadcasting Section – **AFN** see section III. Txs on air continuously unless otherwise stated.
Notes: 1) Also Dokumentation&Debatten. 177kHz 0100-0400 DRM tests – **2)** 0300-2400 – **3)** 0500-2300 – **4)** Mon-Fri 0400-2200, Sat/Sun 0600-2200 – **5)** Also Erfurt/Dresden parliament coverage – **6)** 0400-0900 and 1830-2200 (subject to change), NVIS antenna – **7)** 0600-1800 (subject to change) – **8)** 5980kHz daily 0800-0900 (subject to change), 6045kHz first Sun each month 1000-1100.

I. PUBLIC STATIONS

A) DEUTSCHLANDRADIO
Operates on behalf of all German states for nationwide coverage.
Cologne seat: ✉ Raderberggürtel 40, 50968 Köln ☎ +49 221 345 0 ▤ +49 221 345 4803
Berlin seat: ✉ Hans-Rosenthal-Platz, 10825 Berlin ☎ +49 30 8503 0 ▤ +49 30 8503 6168 **W:** www.dradio.de

FM (MHz)	DLF	DK	kW	FM (MHz)	DLF	DK	kW
Baden-Württemberg				Tübingen	93.9	99.4	0.5/1
Baden-Baden	-	107.9	0.1	Ulm	103.5	91.5	0.5/1
Biberach	100.5	-	0.5	Witthoh	100.6	-	40
Blauen	105.1	-	10	Wörth	-	96.6	0.2
Esslingen	96.7	-	0.1	**Bayern**			
Freiburg	-	90.6	0.2	Amberg	-	107.9	0.1
Geislingen	-	87.7	0.2	Ansbach	92.7	102.7	0.2
Göppingen	99.8	-	0.1	Aschaffenbg.	-	94.8	0.1
Heidelberg	106.5	-	0.4	Augsburg	101.5	100.0	0.5/15
Heidenheim	-	100.8	0.1	Berchtesgd.	91.6	103.4	0.1
Heilbronn	91.3	97.3	0.1	Brotjacklrieg.	100.1	-	100
Hornisgrinde	106.3	-	80	Burgbernhm.	106.3	94.30.2/0.3	
Kirchheim	91.3	-	0.1	Burglengenf.	-	107.3	0.1
Konstanz	-	94.5	0.2	Cham	-	101.4	0.1
Lörrach	-	95.0	0.1	Freilassing	100.3	-	15
Ludwigsburg	94.1	-	0.5	Füssen	87.6	103.6	0.1
Pforzheim	89.2	95.20.1/0.5		Hof Waldst.	-	89.3	20
Rottweil	106.0	-	0.1	Hohenpeißbg.	94.7	-	0.1
Schwäb. Hall	95.8	-	0.1	Ingolstadt	107.0	88.6	0.5
Schw. Gmünd	-	95.9	0.2	Kaufbeuren	-	107.3	0.1
Stuttgart	96.0	87.9	0.5/1	Kempten	89.3	98.8	0.1

FM (MHz)	DLF	DK	kW
Lansberg	90.3	107.9	0.1
Landshut	95.9	100.5	0.2
Marktoberdf.	101.9	-	0.2
Mittenwald	91.9	105.2	0.1
München	101.7	96.8	0.3
Nürnberg	90.1	105.6	0.1
Oberstdorf	92.0	96.5	0.1
Ochsenkopf	100.3	-	100
Passau	-	97.7	0.5
Pfronten	96.5	-	0.02
Regensburg	95.5	101.3	0.2
Rhön	103.3	-	100
Rosenheim	97.2*	96.2	0.1
	97.7*		0.1
Starnberg	87.9	94.7	0.1
Straubing	-	88.7	0.4
Traunstein	-	88.3	0.1
Weiden	-	103.7	0.1
Weilheim	94.7	-	0.05
Würzburg	100.3	101.3	0.1

Berlin & Brandenburg

FM (MHz)	DLF	DK	kW
Berlin A'platz	97.7	-	100
Berlin-Britz	-	89.6	100
Calau	-	90.8	10
Casekow	105.2	-	6
Cottbus	88.6	-	3
Eisenhütt.st.	100.2	-	1
Frankfurt (Bo.)	-	92.7	5
Herzberg/Els.	94.5	-	0.3
Rhinow	-	103.7	0.2

Bremen

FM (MHz)	DLF	DK	kW
Bremen	107.1	100.3	100/1
Bremerhaven	103.4	106.2	0.5/5

Hamburg

FM (MHz)	DLF	DK	kW
Hamburg	88.7	89.1	3/0.1

Hessen

FM (MHz)	DLF	DK	kW
Alsfeld	104.0	-	0.1
Bad Camberg	99.8	-	0.2
Bad Hersfeld	102.9	-	0.3
Darmstadt	102.0	91.1	0.4/0.1
Eschwege	100.6	-	0.5
Frankfurt/M.	97.6	91.2	0.3
Friedberg	89.9	-	0.3
Fritzlar	-	96.0	0.1
Fulda	-	90.7	0.3
Gelnhausen	93.9	-	0.2
Gießen	103.1	107.5	0.6/0.3
Hanau	92.4	107.7	0.3
Hofgeismar	106.9	-	0.3
Kassel	92.7	-	0.1
Korbach	92.8	-	0.1
Limburg	103.3	105.1	0.3
Mainz-Kastel	-	107.2	0.4
Marburg	103.5	93.3	0.5/0.1
Michelstadt	100.5	107.2	0.2
Oberursel	103.5	101.8	0.1
Rimberg	91.3	-	50
Wetzlar	103.7	97.3	0.5/0.3
Wiesbaden	103.7	-	0.5

Mecklenburg-Vorpommern

FM (MHz)	DLF	DK	kW
Anklam	107.4	-	1
Barth	100.3	-	0.1
Dargun	89.8	-	0.5
Greifswald	104.3	106.8	0.2
Güstrow	106.0	-	0.8
Helpterberg	96.5	97.1	10/30
Heringsdorf	98.4	107.1	0.5
Marlow	-	96.7	30
Neukloster	90.6	-	0.3
Neustrelitz	97.9	-	1
Ribn.-Damg.	102.1	-	0.2
Röbel	102.4	90.0	3
Rostock	106.5	-	1
Sassnitz	104.0	101.4	8
Schwerin	106.3	95.3	2/100
Stralsund	89.3	-	0.3
Waren/Mü.	91.3	-	0.1

Niedersachsen

FM (MHz)	DLF	DK	kW
Aurich	101.8	106.9	100/1
Cloppenbg.	-	95.5	0.1
Cuxhaven	101.6	107.7	2/20
Damme	95.4	97.5	0.3
Emden	-	93.4	1
Göttingen	101.0	-	0.1
Hannover	94.0	-	0.1
Hann. Münd.	98.5	-	0.5
Höhbeck	102.2	-	95
Jever	-	89.0	0.5
Leer	-	91.5	0.5
Lingen	102.0	-	25
		91.6*	0.4
		102.9*	0.3
Lübbecke	-	97.7	0.2
Meppen	-	100.7	0.1
Norden	-	105.3	0.3
Nordhorn	-	97.1	0.2
Oldenburg	-	102.8	1
Osnabrück	101.8	-	0.5
Seesen	88.0	-	0.1
Soltau	89.3	-	0.1
Stadthagen	106.1	-	1
Tecklenburg	-	101.1	0.5
Torfh./Harz	103.5	-	100
Uelzen	107.5	-	0.5
Visselhövede	-	88.8	1
Warendorf	107.2	-	1

Nordrhein-Westfalen

FM (MHz)	DLF	DK	kW
Aachen	102.7	-	0.5
B. Oeynhsn.	93.9	-	0.1
Beckum	91.5	-	0.2
Bielefeld	95.5	106.2	0.1
Bonn	89.1	98.9	5/0.1
Eifel-Bärbelk.	-	106.1	20
Gronau	-	94.6	0.2
Kleve	-	90.1	1
Köln	91.3	-	0.1
Langenberg	-	96.5	35
Lemgo	92.2	88.9	0.3
Lennestadt	-	98.9	0.1
Lübbecke	-	97.7	0.2
Münster	104.5	97.5	0.3/0.1
Nordhelle	102.7	-	20
Olpe	-	89.0	0.2
Olsberg	-	106.1	10
Paderborn	94.5	-	0.2
Schwerte	104.4	-	0.2
Siegen	94.2	100.2	0.1
Stadthagen	106.1	-	1
Steinfurt	-	91.0	0.2
Tecklenburg	-	101.1	0.5
Warendorf	107.2	-	1
Wesel	-	102.8	50

Rheinland-Pfalz

FM (MHz)	DLF	DK	kW
B. Kreuznach	106.5	-	0.1
Bingen	-	106.3	0.2
Bitburg	-	95.3	0.1
Boppard	90.5	88.9	0.1
Idar-Oberst.	89.5	94.7	0.2
Kaiserslaut.	105.1	98.1	0.2
Koblenz	99.8	105.3	0.5
Limburg	103.3	105.1	0.3
Linz	-	98.3	0.1
Lorch	88.1	-	1
Ludwigshafen	-	97.3	0.1
Mayen	100.8	-	0.2
Pirmasens	106.1	94.4	0.4
Prüm	95.4	-	0.1
Saarburg	104.6	105.3	20/0.1
Traben-Trarb.	88.7	106.2	0.3
Trier	-	94.3	0.2
Wörth	-	96.6	0.2

Saarland

FM (MHz)	DLF	DK	kW
Lebach	-	107.9	0.1
Neunkirchen	-	105.0	5
Oberperl	-	106.2	5
Saarbrücken	90.1	107.5	1/0.4
Saarlouis	-	96.3	0.1
Völklingen	-	88.6	0.1

Sachsen

FM (MHz)	DLF	DK	kW
Bad Düben	-	99.4	0.2
Bärenstein	-	104.3	1
Belgern	-	101.1	1
Chemnitz	-	106.3	0.5
Collmberg	-	96.1	0.3
Döbeln	-	101.3	1
Dresden	97.3	93.2	100/1
Eilenburg	-	92.0	0.2
Freiberg	-	100.7	1
Geyer (Erzg.)	97.0	-	100
Grimma	-	91.6	0.1
Hoyerswerda	-	89.7	0.5
Leipzig-Holzh.	-	100.4	2
Löbau	99.5	103.0	5/2
Pulsnitz	-	106.7	0.5
Schöneck	94.5	-	3
Weißwasser	-	97.7	2
Wiederau	96.6	-	100
Zwickau	-	104.6	0.2

Sachsen-Anhalt

FM (MHz)	DLF	DK	kW
Brocken/Harz	-	97.4	100
Dessau	107.1	-	0.3
Dequede	-	96.9	7
Eisleben	103.8	-	0.5
Schönebeck	102.0	-	20
Wittenberg	89.3	107.7	1/0.5
Zeitz	-	91.8	0.5

Schleswig-Holstein

FM (MHz)	DLF	DK	kW
Bungsberg	101.9	103.1	95/0.2
Flensburg	103.3	92.1	20/1
Garding	102.3	101.7	0.2/0.5
Güby	-	105.0	0.2
Heide	104.4	92.2	0.4/0.1
Helgoland	107.4	103.0	0.1
Husum	-	101.0	0.1
Itzehoe	102.2	97.5	0.4/0.1
Kaltenkirchen	-	105.5	0.1
Kiel	-	104.7	0.2
Neumünster	-	107.8	0.5
Niebüll	-	104.2	0.6
Rendsburg	-	95.2	0.3
Schleswig	-	105.0	0.2
Sylt	90.3	103.9	0.2

Thüringen

FM (MHz)	DLF	DK	kW
Bleßberg	-	94.2	100
Eisenach	106.5	-	0.5
Erfurt	103.1	-	2
Gera	94.3	93.6	0.2/0.3
Gotha	94.0	-	0.1
Ilmenau	99.9	-	0.1
Inselsberg	-	97.2	100
Jena	104.5	98.2	0.3/0.2
Mühlhausen	107.0	-	1
Saalfeld	98.7	-	0.1
Suhl	98.8	-	0.1
Weimar	89.7	-	0.5

*Directional, different beams.

DAB: See section II.
Satellite: Astra 1H, 11.954GHz h.
Deutschlandfunk: From Köln studios, emphasis on current affairs.
Deutschlandradio Kultur: From Berlin studios, emphasis on culture.
DRadio Wissen: From Köln studios, for young audiences, via digital distribution platforms only.
Dokumentation&Debatten: Parliament coverage, audio of TV talk-shows and other special prgr.; on 153/177/990kHz and via webstream. Sea weather forecasts and nautical warnings: 0005, 0540, 1005, 2005 on 177/1269/6190kHz.

ARBEITSGEMEINSCHAFT DER ÖFFENTLICH-RECHTLICHEN RUNDFUNKANSTALTEN DEUTSCHLANDS (ARD)

Umbrella organization of the public broadcasting institutions
✉ Arnulfstraße 42, 80335 München ☎ +49 89 5900 3344
W: www.ard.de
Overnight programming: ARD-Nachtexpress (light music, from 0300 as ARD-Radiowecker); ARD-Popnacht (pop); ARD-Nachtkonzert (classical music); produced on a rota system.
Satellite radio: Astra 1H, 12.266 GHz; carrying the vast majority of ARD radio stns.
DAB: See section II.

B) BAYERISCHER RUNDFUNK (BR)

Public broadcasting institution of Bavaria.
✉ Bayerischer Rundfunk, 80300 München (headquarter location: Rundfunkplatz 1) ☎ +49 89 5900 01 🖷 +49 89 5900 2375 **W:** www.br-online.de

FM (MHz)	B1	B2	B3	BR K	B5	kW
Augsburg	–	–	–	–	105.3	0.5
Bad Reichenhall	91.8	89.9	96.7	98.3	105.0	0.3
Bamberg	94.8N	98.6	99.8	102.9	97.4	25/5
Berchtesgaden	90.4	99.6	96.9	94.2	106.4	0.3/0.1
Brotjacklriegel	92.1R	96.5	94.4	100.9	106.9	100/50
Büttelberg	91.4N	98.2	99.3	95.5	104.0	25/10
Coburg	93.5N	88.3	99.2	97.7	92.8	5/0.3
Dillberg	88.9N	92.3	97.9	87.6	102.0	25
	104.5R					5
Eichstätt	101.6	90.5	97.6	89.0	106.1	25/10
Garmisch-Partenk.	89.2	93.5	97.7	95.9	104.9	0.1
Grünten (Allgäu)	90.7U	88.7	95.8	101.0	106.9	50/100
Herzogstand	88.1	97.0	91.0	–	106.7	0.1
Hochberg-Traunst.	98.0	91.5	95.9	97.0	107.1	5/0.5
Hohenpeißenberg	92.8	94.2	99.2	100.4	–	25
Hoher Bogen	96.8R	91.6	94.7	88.3	104.4	50/5
Hühnerberg	91.9U	96.1	99.5	93.1	107.6	25/11
Kreuzberg (Rhön)	98.3W	93.1	96.3	107.9	105.3	100/50
Landshut	92.0R	97.8	95.3	93.2	106.6	0.1
Lindau	88.1U	92.0	94.0	87.6	100.4	0.5/0.1
München-Ismaning	91.3	88.4	97.3	103.2	90.0	25
Ochsenkopf	90.7N	96.0	94.0	102.3	107.1	100/50
	91.2R					20

FM (MHz)	B1	B2	B3	BR K	B5	kW
Passau	87.7R	93.2	90.4	95.6	105.9	0.5/0.3
Pfaffenberg	95.6W	88.4	93.4	98.0	106.4	25/1
Regensburg	95.0R	93.0	99.6	97.0	105.0	25/5
Untersb. Geiereck*)	87.8	92.9	96.1	100.7	–	0.1
Wallberg	94.0	87.7	99.7	97.9	101.8	0.1
Wendelstein	93.7	89.5	98.5	102.3	105.7	100
Würzburg	90.9W	90.0	97.6	89.0	105.7	5/0.2

*) Site in Austria.

Bayern 1: Oldies format, at night rel. ARD-Nachtexpress, Mon-Fri 1105-1200 and 1805-1855 regional prgr. from Nürnberg (N), Regensburg (R), Würzburg (W) and Ulm (U) – **Bayern 2**: Various prgr., at night rel. ARD-Nachtkonzert – **Bayern 3**: AC, 24 hours – **BR Klassik**: Classical music, at night rel. ARD-Nachtkonzert – **B5 aktuell**: News, mono signal, rel. 2300-0500 MDR Info – **on3radio**: On 729/801kHz, satellite and DAB, 24 hours, alternative youth format.
Bayern plus, Bayern2plus, BR Verkehr, Bayern5plus: On DAB, Bayern plus (German light and folk music format) and Bayern5plus (coverage of parliament, sports and other events) also on satellite.

C) HESSISCHER RUNDFUNK (HR)

Public broadcasting institution of Hessen.
✉ 60222 Frankfurt am Main (headquarter location: Bertramstraße 8)
☎ +49 69 155 1 📠 +49 69 155 2900 **W:** www.hr-online.de

FM (MHz)	hr1	hr2	hr3	hr4	kW
Alsfeld-Homberg	–	–	105.6	–	0.1
Bad Hersfeld	88.9	–	102.9	–	0.3
Biedenkopf	91.0	99.6	87.6	104.3M	100
Bingen	–	–	91.1	–	0.3
Feldberg (Taunus)	94.4	96.7	89.3	102.5R	100
Frankfurt (HR headq.)	–	87.9	–	–	0.1
Fulda	–	106.6	88.5	103.9N	0.3
Habichtswald	–	–	101.2	103.2N	20
Hardberg (Odenw.)	90.6	95.3	92.7	101.6R	50
Heidelstein (Rhön)	104.8	–	106.2	107.3N	50
Hoher Meißner	99.0	95.5	89.5	101.7N	100
Kassel	94.3	93.7	–	–	0.5
Limburg	–	100.8	–	97.1M	0.3/0.2
Marburg	–	–	–	102.8M	1
Rimberg	–	95.0	–	91.9N	50/20
Rotenburg	–	–	105.7	–	0.3
Schlüchtern	–	–	88.9	–	0.3
Weilburg	–	–	–	97.9M	0.1
Wetzlar	–	–	–	90.5M	0.3
Wiesbaden	98.3	93.1	–	–	0.1
Würzburg (Odenw.)	88.1	97.4	89.7	103.8R	5

FM (MHz)	You FM	hr-info	kW
Alsfeld-Homberg	–	104.0	0.1
Bad Hersfeld	–	106.9	0.3
Badd Orb	–	89.8	0.3
Bensheim	90.2	91.2	0.2/0.1
Biedenkopf	–	102.3	10
Bingen	92.3	–	0.3
Darmstadt	104.3	107.0	0.8/5
Eltville	96.2	–	0.5
Eschwege	106.6	–	0.1
Frankfurt/Main	90.4	103.9	0.5
Friedberg	94.0	92.1	0.3
Fritzlar	–	106.6	0.1
Fulda	93.6	89.7	0.3/0.2
Gelnhausen	99.4	–	0.3
Gießen	97.9	99.2	0.5/0.3
Kassel-Wilhelmsh.	100.1	107.5	0.5/1
Korbach	–	102.6	1
Limburg	90.7	99.2	0.2/0.3
Marburg	93.9	98.5	1/0.3
Michelstadt	91.0	–	0.2
Reinhardshain	–	92.9	0.2
Rimberg	97.7	–	50
Rotenburg	–	96.8	0.3
Schlüchtern	88.2	91.5	0.3
Seeheim	–	88.2	0.1
Sontra	–	90.8	0.1
Wetzlar	105.5	–	0.2
Wiesbaden	99.7	97.2	0.2/0.1
Witzenhausen	91.1	–	0.3

hr1: Oldies format, at night rel. ARD-Popnacht – **hr2**: Culture and classical music, 2305-0500 rel. ARD-Nachtkonzert – **hr3**: AC format, at night rel. ARD-Popnacht – **hr4**: Produced at Kassel (Wilhelmshöher Allee 347, 34131 Kassel), light music format, 2305-0500 rel. ARD-Nachtexpress, regional prgr. Nordhessen (N; Kassel/Fulda), Mittelhessen (M; Gießen) and Rhein-Main (R; Frankfurt/Darmstadt) Mon-Fri only 0830-0835, 1105-1200 and 1505-1600 – **You FM**: CHR format, 24 hours – **hr-info**: News, 2305-0500 rel. ARD-Nachtkonzert..

D) MITTELDEUTSCHER RUNDFUNK (MDR)

Public broadcasting institution of Sachsen, Sachsen-Anhalt and Thüringen.
✉ Kantstraße 71-73, 04360 Leipzig (administration) **W:** www.mdr.de
✉ Gerberstraße 2, 06110 Halle/Saale ☎ +49 345 300 0 📠 +49 345 300 5544. (radio, except MDR 1 stns, see below)

FM (MHz)	MDR 1	JumpFigaro	Info	Sputnik	kW
Txs in Sachsen:					
Altenburg	–	–	101.5	–	1
Annaberg-Buchholz	–	–	91.2	–	0.2
Aue	–	–	95.1	–	1
Auerbach	–	–	101.7	–	0.4
Bautzen	–	98.8	87.9	–	0.2/0.1
Chemnitz-Reichenh.	–	–	94.7	–	0.5
Collmberg	101.8L	103.7	98.9	105.9	2x5/0.5/30
Döbeln-Mockritz	–	–	99.6	–	0.5
Dresden-Wachwitz	92.2	90.1	95.4	106.1	3x100/0.5
Eilenburg	–	–	92.4	–	0.2
Freiberg	99.1C	–	93.7	–	1/0.2
Freital	–	–	95.9	–	0.2
Geyer (Erzgebirge)	92.8C	89.8	87.7	–	100
Grimma-Hohnstädt	–	–	100.6	–	0.2
Görlitz	–	–	106.9	–	1
Hoyerswerda	93.0B	89.0	94.7	94.2	1/0.5/1/1
	100.4				30
Klingenthal	93.7C	–	98.4	–	0.2
Leipzig city	–	–	–	95.6	0.5
Löbau	98.2B	91.8	96.2	–	5
Markneukirchen	104.8C	–	106.4	–	0.5
Meißen-Korbitz	–	–	94.9	–	1
Neustadt	–	–	89.6	–	0.2
Plauen	–	–	102.0	–	1
Raschau	–	–	91.6	–	0.2
Seifhennersdorf	94.5B	96.9	103.4	–	0.25/0.3
Schöneck	88.7C	101.2	98.7	–	3/30/3
Stollberg	–	–	89.3	–	0.1
Torgau	88.9L	–	93.0	–	0.5/0.2
Weißwasser	–	–	90.5	–	1
Wiederau (Leipzig)	93.9L	90.4	88.4	–	100
	106.5H				*30
Zittau	87.7B	107.1	95.4	106.4	0.2/0.5
Zschopau	–	–	99.5	–	0.2
Zwickau	–	–	91.4	–	1
Txs in Sachsen-Anhalt:					
Aschersleben	–	–	102.8	–	1
Brocken	94.6	91.5	107.8	–	60/100/10
Burg	–	–	89.6	–	1
Dequede	94.9St	98.9	89.4	–	10
Dessau-Mildensee	–	–	90.0	–	0.3
Fleetmark	–	–	90.1	105.0	2/1
Gernrode	–	–	91.0	–	0.1
Haidberg	–	–	–	100.7	5
Haldensleben	–	–	99.1	–	1
Halle Petersberg	100.8H	–	95.3	104.4	5/2/10
Halle city	–	89.6	107.3	–	0.1
Hergisdorf	92.9H	–	–	–	1
Jerichow	–	–	–	90.5	1
Jessen	–	–	87.6	–	1
Klötze	–	–	–	100.7	5
Köthen	–	–	106.4	–	0.3
Magdeburg	96.1	107.4	–	–	10/30
Naumburg	92.3H	–	–	93.1	1/0.5
Sangerhausen	101.1H	–	99.9	–	0.1/1
Schneidlingen	–	–	106.7	–	0.5
Schönebeck	–	–	91.1	105.2	2/1.5
Stendal-Borstel	–	–	87.8	104.8	1
Weißenfels	–	–	88.8	–	1
Wernigerode	–	–	98.6	–	1
Wittenberg	88.1D	101.6	104.0	–	30/2x55
Zeitz-Hainichen	–	–	–	89.4	0.5
Txs in Thüringen:					
Apolda	–	–	91.2	–	1
Arnstadt	–	–	106.1	–	0.5
Bad Salzungen	–	–	94.0	–	0.1
Bleßberg	91.7S	96.9	–	–	100/20
Eisenach	–	–	100.0	–	0.2
Erfurt	94.4	–	97.8	–	2/1
Gera	–	–	91.1	–	1
Gotha	–	–	88.8	–	0.1
Greiz	–	–	93.3	–	0.1
Heiligenstadt	93.6He	–	90.5	–	1
Ilmenau	–	–	93.0	–	1
Inselsberg	92.5	90.2	87.9	–	100/100/60
Jena-Oßmaritz	88.2G	101.9	96.4	89.5	1/0.2
Keula	98.5He	–	–	–	20
Lobenstein	95.5G	–	101.8	–	2/0.5

FM (MHz)	MDR 1	JumpFigaro		Info	Sputnik	kW
Magdala	92.9	–	–	99.2	–	0.01/0.05
Meiningen	–	–	–	94.7	–	0.2
Mühlhausen	–	–	–	105.8	–	0.1
Nordhausen	88.3He	–	–	93.7	–	0.1
Pößneck	–	–	–	101.6	–	0.2
Remda	103.6	105.6	100.7	–	–	60
Ronneburg	97.8G	100.9	103.9	–	–	10/30/30
Saalfeld	–	–	–	104.6	–	0.1
Schleiz	–	–	–	105.1	–	0.2
Schmalkalden	–	–	–	100.0	–	0.1
Schmölln	–	–	–	107.9	–	0.2
Sondershausen	100.1He	–	–	95.1	–	0.05/0.1
Sonneberg	–	–	–	105.8	–	0.1
Suhl Erleshügel	93.7S	91.1	89.8	97.5	–	1/0.1/0.2/5
Weimar Ettersberg	93.3	–	–	–	–	5
Weimar Belvedere	–	–	–	102.6	–	2

*) Directional, to north and west only.

MDR 1 Radio Sachsen: Königsbrücker Str. 88, 01099 Dresden, regional prgr. from studios Bautzen (freq. marked (B), Chemnitz (C) and Leipzig (L); **MDR 1 Radio Sachsen-Anhalt:** Stadtparkstr. 8, 39114 Magdeburg, regional prgr. Dessau (D), Halle (H) and Stendal (St); **MDR 1 Radio Thüringen:** Gothaer Str. 36, 99094 Erfurt; regional prgr. Gera (G), Heiligenstadt (He) and Suhl (S). 2200-0400 on all MDR 1 stn's common prgr. – **Jump:** Modern Rock, 24 hours – **MDR Figaro:** Culture, rel. 2305-0500 ARD-Nachtkonzert – **MDR Info:** News, 24 hours, also on MW, at night rel. by BR and SWR – **Sputnik:** CHR, 24 hours – **MDR Klassik:** On satellite and DAB only, mostly rel. Figaro – **Serbske Rozhlas:** MDR, Studio Bautzen, Am Postplatz 2, 02607 Bautzen. Prgr. in Upper Sorbian on 14.0MHz Mon-Fri 0405-0700, Sat 0505-0800, Sun 1000-1130. "Radio Satkula" for young listeners Mon 1900-2100.

E) NORDDEUTSCHER RUNDFUNK (NDR)
Public broadcasting institution of Hamburg, Mecklenburg-Vorpommern, Niedersachsen and Schleswig-Holstein.
✉ Rothenbaumchaussee 132, 20149 Hamburg ☎ +49 40 4156 0 📠 +49 40 447 602 **W:** www.ndr.de **N.B** Adresses for regional NDR 1 services see below.

FM (MHz)	NDR 1	NDR 2	NDR-K	Info	N-Joy	kW
Txs in Hamburg:						
Moorfleet	90.3	87.6	99.2	92.3	94.2	80/5/1
	89.5No					10
Txs in Mecklenburg-Vorpommern:						
Anklam	94.6Gr	–	–	–	103.0	6.3/1.25
Bad Doberan	94.3R	–	–	–	103.7	0.2/5
Barth	87.6Gr	–	–	–	95.0	0.4/0.3
Dömitz	88.3	–	–	–	–	1
Garz/Rügen	102.5Gr	99.8	91.5	88.6	95.5	50/10
Greifswald	101.0Gr	–	–	–	–	0.16
Grevesmühlen	100.7W	–	–	–	103.4	0.5/5
Güstrow-Strentz	92.5R	–	–	–	104.4	1.25/0.63
Helpterberg	90.5N	99.1	96.0	101.8	103.2	100/1.25
	94.2Gr					6.3
Heringsdorf	97.6Gr	94.0	102.7	100.5	92.3	1
Marlow	91.0R	93.5	88.2	102.8	–	100/30/100
Malchin	–	–	–	103.5	94.4	1
Neubrandenburg	–	–	–	–	89.5	1
Pasewalk	93.7Gr	–	–	–	94.8	2.5/1.25
Ribnitz-Damgarten	–	–	–	–	99.4	0.3
Röbel	88.5N	107.0	94.7	100.4	97.4	10/60/4
Rostock	95.8R	–	–	–	88.9	0.16/2
Schwerin	92.8	98.5	89.2	105.3	99.5	30/100/2
Stralsund	92.1Gr	–	–	–	–	1
Ueckermünde	90.1Gr	–	–	–	104.1	4/1.5
Wismar	96.2W	–	–	–	–	0.2
Wolgast-Moeckow	89.0Gr	–	–	–	93.2	0.4/0.3
Txs in and for Niedersachsen:						
Alfeld	87.8B	93.6	96.5	91.1	92.9	0.05
Aurich-Popens	95.8Ol	98.1	90.0	96.4	92.7	25/10/1
Bad Pyrmont	88.6	92.6	95.7	98.5	–	0.05
Bad Rothenfelde	–	–	–	97.9	91.2	0.2/0.1
Braunlage	–	–	–	–	96.1	0.02
Braunschweig	–	–	–	–	100.3	15
Bremen-Walle	–	–	–	95.0	–	1
Bremerhaven	–	–	–	98.9	92.8	0.5/0.05
Cloppenburg	–	–	–	103.7	93.5	1
Cuxhaven	105.4Ol	97.9	94.6	93.1	91.6	20/10/1/10
	98.4					1
Damme	–	–	–	106.5	105.0	0.5/1
Dannenberg	91.2L	96.4	93.3	90.7	94.0	25/10/3/1
Goslar	88.2B	93.7	95.1	96.0	96.5	0.1
Göttingen	88.5B	94.1	96.8	99.9	95.9	5/0.5/5/0.5
Hann. Münden	88.2B	96.1	90.8	92.9	94.8	0.05
Hannover-Hemm.	90.9	96.2	98.7	88.6	92.6	155/150/525
Hildesheim	–	–	–	–	95.7	0.5

FM (MHz)	NDR 1	NDR 2	NDR-K	Info	N-Joy	kW
Holzminden	92.7B	96.0	98.4	88.6	99.7	0.5/0.1
Jever	–	–	–	–	97.3	0.3
Königslutter-Elm	–	–	–	88.7	–	0.2
Lingen	92.80	97.8	90.2	88.9	96.6	15/0.2/0.5
Meppen	–	–	–	–	93.3	0.05
Osnabrück	92.4Ol	89.2	98.8	87.6	96.4	8/2x0.2
Rinteln	–	–	–	95.3	105.2	0.1/0.04
Rosengarten	103.2L	–	–	–	91.4	20/0.3
Seesen	–	–	–	90.4	96.6	0.2/0.05
Stadthagen	100.8	102.6	104.4	98.2	91.3	25/1
Steinkimmen	91.1Ol	99.8	94.4	98.6	92.9	100/3/1
Torfhaus	98.0B	92.1	89.9	99.5	–	100/50
Visselhövede	91.8L	95.9	87.8	98.4	90.1	5/2/5/1
Wedel	–	–	–	–	95.6	0.2
Wolfsburg	–	–	–	88.2	–	0.1
Txs in Schleswig-Holstein:						
Bungsberg	97.8Lb	91.9	89.9	96.6	99.0	50/1/0.5
Flensburg	89.6F	93.2	96.1	87.7	91.0	25/10/0.5
Garding-Katingsiel	–	–	–	–	88.8	0.5
Heide-Welmbüttel	90.5H	96.3	99.4	87.9	94.9	15/0.5
Helgoland island	88.9H	93.4	97.0	92.5	91.5	0.01
Husum	–	–	–	–	93.7	0.05
Kiel-Kronshagen	91.3	98.3	95.7	99.7	94.5	15/1/0.4/15
Lauenburg	94.7Lb	–	–	96.8	99.8	0.3
Lübeck	93.1Lb	90.7	88.0	95.9	94.0	0.5/0.1/0.5
Mölln	104.5Lb	–	–	–	90.9	20/0.5
Neumünster	106.4No	–	–	90.8	98.7	20/1/0.5
Niebüll-Süderlügum	–	–	–	–	91.5	0.2
Sylt	90.9F	98.7	94.3	92.7	95.6	5
Wedel	–	–	–	–	95.6	0.2

NDR 90,3: from Hamburg studios, on 90.3/98.4MHz; **NDR 1 Radio MV:** Schloßgartenallee 61, 19061 Schwerin; via txs in Mecklenburg-Vorpommern, regional prgr. Greifswald (Gr), Neubrandenburg (N), Rostock (R) and Wismar (W, from Schwerin studios); **NDR 1 Niedersachsen:** Rudolf-von-Bennigsen-Ufer 22, 30169 Hannover; via txs in Niedersachsen, regional prgr. Braunschweig (B), Göttingen (G), Lüneburg area (L, from Hannover studios), Oldenburg (Ol) and Osnabrück (O); **NDR 1 Welle Nord:** Postfach 34 80, 24033 Kiel (studio location: Eggerstr. 16); via txs in Schleswig-Holstein and 89.5MHz; regional prgr. Flensburg (F), Heide (H), Lübeck (Lb) and Norderstedt (No). 2110-0430 common prgr. on all NDR 1 stns – **NDR 2:** AC format, 24 hours – **NDR Kultur:** Classical music, rel. 2305-0500 ARD-Nachtkonzert – **NDR Info:** Mon-Fri 0500-1850 and Sat 0500-1700 news format, remaining time diverse prgr. – **NDR Info Spezial:** On 702/792/828/972kHz, rel. Mon-Fri 1500-2000 Funkhaus Europa (see J), Sun 0500-0700 NDR 90,3 for Hamburger Hafenkonzert prgr., broadcast since 1929. Only on 702/972kHz: Sea weather forecasts at 2305 (also via NDR Info FM txs in Mecklenburg-Vorpommern), 0730 and 2105. – **N-Joy:** CHR format, 24 hours.

F) RADIO BREMEN (RB)
Public broadcasting institution of Bremen.
✉ Diepenau 10, 28195 Bremen ☎ +49 421 246 0 📠 +49 421 246 1010 **W:** www.radiobremen.de

FM (MHz)	Eins	NWRadio	Vier	Europa	kW
Bremen-Walle	93.8	88.3	101.2	96.7	100/50
Bremerhaven	89.3	95.4	100.8	92.1	25

Bremen Eins: Oldies format, rel. 2305-0400 (Sun til 0500) SWR1 – **Nordwestradio:** Culture, in cooperation with NDR. Rel. 2305-0500 ARD-Nachtkonzert – **Bremen Vier:** AC format, at night rel. ARD-Popnacht – **Funkhaus Europa:** See J).

G) RUNDFUNK BERLIN-BRANDENBURG (RBB)
Public broadcasting institution of Berlin and Brandenburg, operating from two main seats:
Potsdam: ✉ Marlene-Dietrich-Allee 20, 14482 Potsdam-Babelsberg ☎ +49 331 731 0 📠 +49 331 731 3571
Berlin: ✉ 14046 Berlin (studio/office location: Masurenallee 8-14) ☎ +49 30 3031 0 📠 +49 30)3015 062 **W:** www.rbb-online.de

Txs in Berlin:

MHz	kW	Site	Program
88.8	80	Scholzplatz	Radio Berlin
92.4	80	Scholzplatz	Kulturradio
93.1	25	Scholzplatz	Inforadio
95.8	100	Alexanderplatz	radioeins
96.3	80	Scholzplatz	Funkhaus Europa
99.7	100	Alexanderplatz	Antenne Brandenburg
102.6	20	Alexanderplatz	Fritz

Txs in Brandenburg:

FM (MHz)	Ant.B.	Eins	Fritz	Kultur	Info	kW
Belzig-Lütte	106.2	99.3	91.9	100.2	–	100/10

FM (MHz)	Ant.B.	Eins	Fritz	Kultur	Info	kW
Booßen	87.6F	89.1F	101.5	96.8	102.0	5/30/1.5
Calau	98.6C	95.1C	103.2	104.4	93.4*)	100/30
Casekow	91.1Pr	106.1	100.1	104.4	–	60/10
Cottbus	–	–	–	–	99.9	1
Guben	100.9C	–	–	–	–	6
Lübben	–	–	–	–	92.4	0.4
Perleberg	–	–	–	–	92.3	1
Prenzlau	99.4Pr	–	–	–	98.6	0.5
Pritzwalk	106.6Pe	99.9	103.1	91.7	–	100/10
Wittstock	–	–	–	–	97.7	1.3
Zehlendorf	90.8F	–	–	–	–	1.3

From Berlin studios: Radio Berlin, Berlin city prgr., 2305-0400 rel. ARD-Popnacht – **Inforadio**, news, 24 hours – **Kulturradio**, culture, 2305-0500 rel. ARD-Nachtkonzert.
From Potsdam studios: Antenne Brandenburg, Brandenburg state prgr., regional prgr. from studios Perleberg (Pe), Prenzlau (Pr), Frankfurt/Oder (F) and Cottbus (C), 2100-2305 common prgr. with Radio Berlin as "Pop nach zehn vom RBB", 2305-0400 rel. ARD-Nachtexpress – **radio-oeins**, progressive-style rock/pop and information, 24 hours, regional prgr. from Frankfurt/Oder and Cottbus – **Fritz**, youth, 24 hours.
*) Bramborske Serbske Radio: RBB, Studio Cottbus, Berliner Straße 155, 03046 Cottbus. Prgr. in Lower Sorbian Mon-Fri 1100-1200 and repeat at 1800-1900, Sundays and holidays 1130-1300, otherwise rel. Inforadio.
Funkhaus Europa: See J). Txs in Brandenburg no longer in use.

H) SAARLÄNDISCHER RUNDFUNK (SR)
Public broadcasting institution of the state of Saarland.
✉ Funkhaus Halberg, 66100 Saarbrücken ☎ +49 681 602 0 📠 +49 681 602 3874 **W:** www.sr-online.de

FM (MHz)	SR 1	SR 2	SR 3	UnserDing	kW
Bliestal-Webenheim	92.3	98.0	89.1	–	5
Göttelborner Höhe	88.0	91.3	95.5	–	100
Homburg	–	–	–	98.6	0.2
Merzig-Hilbringen	89.3	92.1	98.0	–	0.1
Neunkirchen	–	–	–	–	5
Oberperl	91.9	88.6	96.1	–	5
Saarbr. Schocksberg	–	–	–	103.7	100
Sankt Wendel	–	–	–	90.3	0.1

SR 1 Europawelle Saar: AC format, at night rel. ARD-Popnacht – **SR 2 KulturRadio:** Culture, at night rel. ARD-Nachtkonzert – **SR 3 Saarlandwelle:** Light music, at night rel. ARD-Nachtexpress, news in French at 0805 – **Unser Ding:** Prgr. for teenagers, at times rel. Das Ding (SWR) – **SR Antenne Saar:** On 1179kHz and DAB, rel. of SR 2, SWR cont.ra and Radio France Internationale.

I) SÜDWESTRUNDFUNK (SWR)
Public broadcasting institution of Baden-Württemberg and Rheinland-Pfalz.
✉ 76522 Baden-Baden (Location: Hans-Bredow-Straße)
☎ +49 7221 929 0 📠 +49 7221 929 2010
Broadcasting house Mainz: ✉ Postfach 3740, 55122 Mainz (Location: Am Fort Gonsenheim 39) ☎ +49 6131 929 0
Broadcasting house Stuttgart: ✉ Postfach 106040, 70049 Stuttgart (Location: Neckarstraße 230) ☎ +49 711 929 0

FM (MHz)	SWR1	SWR2	SWR3	SWR4	DasDing	kW
Txs in and for Baden-Württemberg:						
Aalen Braunenberg	95.1	91.1	98.1	96.9U	–	50/5
Albstadt-Mahlesfeld	–	–	–	99.5Tü	–	0.1
Bad Bellingen	–	–	–	96.6F	–	0.1
Bad Mergentheim	87.8	93.2	99.7	105.5H	100.5	10/0.04
Baden-Baden	90.9	98.9	99.6	88.5Ka	91.7	0.8/0.4
Baiersbronn	–	–	–	87.9O	–	0.1
Basel St. Crischona*)	87.9	92.0	98.3	89.5L	–	5
Blauen-Hochblauen	89.2	92.6	97.0	–	–	8.4
Buchen	91.9	97.1	94.1	107.5M	100.6	0.1/25
Elzach Hörnleberg	–	–	–	101.8F	–	0.1
Feldberg	89.8	97.9	93.8	104.0F	–	5
Freiburg-Lehen	107.0	91.1	99.2	100.7F	–	0.1/1
Freudenstadt	90.3	97.2	94.9	91.6H	–	0.01
Geislingen	93.0	88.5	95.5	107.9	–	0.5/0.1
Grünten*	98.7	–	103.0	–	–	30
Hausach Brandkopf	95.4	–	99.7	97.6O	–	0.5/0.1
Heidelberg Königstuhl	97.8	88.8	99.9	104.1M	–	100
Heilbronn	–	–	–	99.5H	–	2
Hornisgrinde	93.5	96.2	98.4	94.0O	–	80/5
Karlsruhe-Ettlingen	–	–	–	97.0Ka	–	20
Klettgau	95.1	92.8	98.5	87.7Lö	–	2.6
Lichtenstein	99.1	–	–	89.0Tü	–	0.1
Mannheim	–	–	–	–	91.5	4
Mötzingen	–	–	97.2	87.6Tü	90.5	1
Mühlacker	–	–	–	95.7B	–	2

FM (MHz)	SWR1	SWR2	SWR3	SWR4	DasDing	kW
Pforzheim	92.9	88.1	99.3	87.6Ka	–	5/0.2/0.5
Raichberg	88.3	91.8	94.3	107.3Tü	–	40/25
Ravensburg	99.0	–	87.9	–	107.2	0.1
Reutlingen	–	–	–	–	97.7	2
Schiltach-Simonsberg	90.8	–	94.5	99.2O	–	0.1
Schwäbisch Gmünd	–	–	–	100.9U	–	0.1
Sigmaringen	–	–	–	101.2Fr	–	0.1
Strasbourg*	–	–	–	88.9O	–	1
Stuttgart-Degerloch	94.7	105.7	92.2	90.1	90.8	100/2
Stuttgart (town)	99.6	93.1	–	–	–	0.5/0.2
Tübingen	–	–	–	–	†91.5	0.3
					97.3	2
Ulm Kuhberg	92.6	89.2	97.4	94.5U	98.9	10/1
Vaihingen	–	98.6	–	–	–	0.1
Villingen-Schwenningen	–	–	–	91.1F	–	1
Waldenburg	98.8	93.8	96.5	106.6H	–	100/50
Waldburg	–	94.9	–	99.5H	–	60
				91.2Fr	–	25
Weinheim	97.1	–	99.5	100.7M	–	0.04/0.1
Wertheim	96.9	91.8	94.6	101.2H	–	0.1
Witthoh	92.4	90.4	97.1	89.0Fr	–	40/5
Zell Hohe Möhr	87.6	–	96.8	100.2F	–	0.1
Zwiefalten	93.7	–	92.8	87.6Fr	–	0.1

†) SWR cont.ra. *Basel site in Switzerland, Strasbourg site in France, Grünten site in Bayern.

FM (MHz)	SWR1	SWR2	SWR3	SWR4	DasDing	kW
Txs in and for Rheinland-Pfalz:						
Bleialf-Buchet	88.3	99.7	–	98.5	94.6T	–	0.1
Daun	91.1	–	–	98.5	93.6T	–	8
Diez-Geisenberg	88.4	93.4	98.2	87.9K	–	0.01/0.1
Donnersberg	99.1	92.0	101.1	105.6K	–	60
Haardtkopf	97.7	93.0	90.0	107.1T	–	50/25
Hohe Wurzel	–	–	–	107.9M	–	6.2
Idar-Oberstein	88.5	95.1	98.1	106.4T	–	0.01/1
Kaisersl. Bomberg	90.8	93.9	97.5	99.6Kl	–	25/0.5
Koblenz-Waldesch	96.1	94.0	91.6	107.4K	99.4	10/40/0.2
Kreuzweiler	–	–	–	97.3T	–	0.3
Linz	92.4	–	94.8	97.4K	–	50
Mainz-Kastel*	87.7	103.2	93.7	91.4M	105.2	1
Mainz-Wolfsheim	–	–	–	94.9M	–	5
Marienberger Höhe	89.8	95.4	92.8	106.3K	91.3	25/0.1
Niersteim-Oppenheim*	–	–	–	92.9M	–	0.1
Pirmasens Kettrichhof	100.8	–	107.2	104.2Kl	–	5
Rüdesheim*	–	99.4	93.3	88.6M	–	0.1/0.5
Saarburg	99.2	93.8	90.6	101.2T	–	5
Trier	94.9	89.4	98.2	98.8T	91.7	0.1/0.3
Tübingen Herrenberg	–	–	97.2	87.6Tü	90.5	1
Weinbiet	89.9	102.2	–	95.9L	–	25
Zweibrücken	–	–	–	90.5Kl	–	0.2

*) Site in Hessen. +20 stns below 0.1kW

From Stuttgart studios, via txs in Baden-Württemberg: **SWR1 Baden-Württemberg**, oldies format, at night common SWR1 prgr. from Baden-Baden; **SWR4 Baden-Württemberg**, light music format, Mon-Sat 0500-0700, 0900-1000, 1130-1300, 1530-1700, Sun 1200-1300, 1600-1700 local prgr. from Freiburg (F), Friedrichshafen (F), Heilbronn (H), Karlsruhe (Ka), Lörrach (Lö), Mannheim (M), Offenburg (O), Tübingen (T) and Ulm (U), at night rel. ARD-Nachtexpress.
From Mainz studios, via txs in Rheinland-Pfalz: **SWR1 Rheinland-Pfalz**, oldies format, at night common SWR1 prgr. from Baden-Baden; **SWR4 Rheinland-Pfalz**, light music format, Mon-Sat 1100-1200 local prgr. from Kaiserslautern (Kl), Koblenz (K), Ludwigshafen (L) and Trier (T), at night rel. ARD-Nachtexpress.
From Baden-Baden studios: SWR2, culture, 1740-1800 prgr. from Mainz/Stuttgart, at night rel. ARD-Nachtkonzert; **SWR3**, AC format, 24 hours; **Das Ding**, youth, 24 hours; **SWR cont.ra**, News/talk, on 576/666/711/828/1017/1485kHz, rel. Mon-Fri 2200-0500, Sat/Sun 2200-0656 MDR Info.

J) WESTDEUTSCHER RUNDFUNK (WDR)
Public broadcasting institution of Nordrhein-Westfalen.
✉ 50600 Köln (Location: Appellhofplatz 1) ☎ +49 221 220 1 📠 +49 221 220 4800 **W:** www.wdr.de

FM (MHz)	ELive	WDR 2	WDR 3	WDR4	WDR5	kW
Aachen-Stolberg	106.4	100.8A	95.3	91.9	101.9	20
Arnsberg	96.0	99.4S	97.5	91.7	88.5	0.1
Bad Oeynhausen	107.7	99.1B	92.7	90.1	87.7	0.1
Bergheim	–	88.4K	–	–	–	0.5
Bonn Venusberg	102.4	100.4K	93.1	90.7	88.0	50
Dortmund	–	87.8D	–	–	–	0.1
Ederkopf	107.2	101.8S	–	100.7	95.8	15/20
Eifel-Bärbelkreuz	105.5	101.0	96.3	104.4	89.6	20/10/20/10
Gummersbach	–	91.8W	–	–	–	10
Hallenberg	105.7	–	–	96.1	88.3	0.1
Höxter Hasselberg	107.3	96.4B	95.2	87.8	93.9	0.5
Ibbenbüren	102.5	96.0M	93.0	91.0	99.5	0.5

FM (MHz)	ELive	WDR 2	WDR 3	WDR4	WDR5	kW
Klever Berg	103.7	93.3Dü	97.3	101.7	99.7	2
Köln	87.6	98.6K	–	–	–	0.3/0.5
Langenberg	106.7	99.2Dü	95.1	101.3	88.8	100
					*103.3	100
Lübbecke	93.6	96.0B	91.7	99.6	88.6	0.1
Münster-Baumberge	107.9	94.1M	89.7	100.0	92.0	25
Nordhelle	104.7	93.5S	98.1	103.8	90.3	35
Olsberg	107.0	102.1S	–	104.1	98.6	10
Remscheid	–	95.7W	–	–	–	
Schmallenberg	100.1	93.8S	97.8	101.1	90.0	0.1
Siegen	107.5	97.1S	98.4	101.2	97.6	0.5/1/0.5/1
Teutoburger Wald	105.5	93.2B	97.0	100.5	90.6	100
Warburg	98.2	91.8B	94.3	104.5	88.4	0.5
Wittgenstein	–	92.3S	88.7	–	–	15
Wuppertal	–	99.8W	–	–	–	1

*Funkhaus Europa.

1 Live: CHR format, 24 hours – **WDR 2:** Also on 720/774kHz, pop and information, 24 hours, incl. local news from Aachen (A), Bielefeld (B), Köln (K), Dortmund (D), Düsseldorf (Dü), Münster (M), Siegen (S), Wuppertal (W) – **WDR 3:** Culture, at night rel. ARD-Nachtkonzert – **WDR 4:** Light music, at night rel. ARD-Nachtexpress – **WDR 5:** Information, 24 hours with repeats overnight – **Funkhaus Europa:** Multicultural, also via RBB and RB txs (see F/G) and with some prgr. from their Bremen/Berlin studios – **VERA:** Continuous traffic jam information, 24 hours on satellite and DAB, at times also on 720/774kHz – **1 Live Diggi:** Via satellite and DAB, continuous CHR music – **Kiraka:** Via satellite, DAB and 1593kHz (DRM mode), repeats of childrens prgr. from other WDR networks.

II. COMMERCIAL AND OTHER STATIONS

Note: In Germany supervision and frequency allocation for commercial broadcasting services is the responsibility of the federal states. Each state (listed below) has its own media institution, with the exceptions of Berlin and Brandenburg as well as Hamburg and Schleswig-Holstein, respectively, who have common institutions. As a result of this situation most commercial stations broadcast on the territory of one state only.

K) BADEN-WÜRTTEMBERG

Media institution: Landesanstalt für Kommunikation (LfK) ✉ Postfach 102927, 70025 Stuttgart (office location: Reinsburgstraße 27 ☎ +49 711 669910 🖷 +49 711 6699111 **W:** www.lfk.de

Commercial stations:

FM	MHz	kW	Site	Station
2)	87.8	1	Mannheim	big FM
6)	88.6	2	Langenburg	Radio Ton
3)	89.1	0.5	Heilbronn	Hit-Radio Antenne 1
3)	89.3	0.1	Bad Urach	Hit-Radio Antenne 1
3)	89.5	10	Stuttgart Frauenkopf	big FM
3)	89.5	0.1	Wertheim	Hit-Radio Antenne 1
2)	89.7	1	Tübingen	big FM
11)	90.4	2	Karlsruhe	Klassik Radio
9)	90.5	2	Achern	Hitradio Ohr
2)	90.9	0.1	Heidelberg city	big FM
15)	91.4	3	Lützenhardt	R. TV Radio
7)	91.4	0.5	Pforzheim	die neue welle
18)	92.4	1	Hockenheimring	Rennradio
2)	92.7	1	Horb	big FM
9)	93.0	0.1	Haslach	Schwarzwald R.
16)	93.1	1	Rottweil-Zimmern	Radio Neckarburg
15)	94.7	0.5	Freiburg-Lehen	baden.fm
17)	95.4	0.1	Stuttgart SWR bldg.	Metropol FM
6)	95.6	1	Balingen	Radio Ton
6)	96.0	0.1	Künzelsau	Radio Ton
10)	96.4	1	Überlingen	Radio Seefunk
6)	96.8	0.3	Eppingen	Radio Ton
4)	96.9	0.1	Schussental	Radio 7
12)	97.2	1	Stuttgart-Münster	Motor FM
2)	97.2	0.5	Sinsheim-Dühren	big FM
13)	97.5	0.5	Esslingen	Die Neue 107.7
8)	97.6	0.3	Rudersberg	Energy Stuttgart
2)	99.0	0.5	Rottweil	big FM
6)	99.0	0.1	Bad Urach	Radio Ton
15)	99.2	0.2	Herrenberg	R. TV Radio
9)	99.2	0.1	Oberkirch	Hitradio Ohr
10)	99.3	5	Friedrichshafen	Radio Seefunk
2)	99.7	1	Ulm	big FM
3)	100.1	50	Schwäbisch Hall	Hit-Radio Antenne 1
6)	100.1	0.1	Hechingen	Radio Ton
2)	100.3	5	Geislingen	big FM
1)	100.4	80	Hornisgrinde	Radio Regenbogen
8)	100.7	20	Güglingen	Energy Stuttgart
6)	100.9	1	Tübingen	Radio Ton
7)	100.9	0.8	Baden-Baden	die neue welle
1)	101.1	8.4	Blauen-Müllheim	Radio Regenbogen
4)	101.2	0.1	Villingen-Schwenningen	Radio 7

FM	MHz	kW	Site	Station
3)	101.3	75	Stuttgart Frauenkopf	Hit-Radio Antenne 1
10)	101.6	0.5	Brandenkopf	Hitradio Ohr
7)	101.8	25	Karlsruhe	die neue welle
4)	101.8	10	Ulm-Ermingen	Radio 7
10)	101.8	10	Konstanz	Radio Seefunk
8)	101.8	1	Backnang	Energy Stuttgart
10)	101.9	0.1	Schopfheim	Radio Seefunk
16)	102.0	3	Villingen-Schwenningen	Radio Neckarburg
5)	102.1	25	Mudau	sunshine live
10)	102.4	0.2	Laufenburg [Switzerl.]	Radio Seefunk
4)	102.5	40	Witthoh-Tuttlingen	Radio 7
6)	102.6	0.5	Schwäbisch Hall	Radio Ton
8)	102.6	0.3	Ravensburg	Radio Seefunk
8)	102.6	0.1	Bad Wildbad	Energy Stuttgart
8)	102.7	0.1	Nagold	Energy Stuttgart
1)	102.8	50	Heidelberg.Königstuhl	Radio Regenbogen
2)	102.8	0.5	Freiburg	big FM
11)	103.0	1	Göppingen	Klassik Radio
8)	103.0	0.3	Calw	Energy Stuttgart
3)	103.1	5	Rheinfelden	Radio Seefunk
3)	103.1	0.1	Reutlingen	Hit-Radio Antenne 1
6)	103.2	25	Heilbronn	Radio Ton
3)	103.4	50	Raichberg	Hit-Radio Antenne 1
6)	103.5	20	Bad Mergentheim	Radio Ton
4)	103.7	50	Aalen	Radio 7
16)	103.7	0.1	Schramberg	Radio Neckarburg
2)	103.8	2	Baden-Baden	big FM
10)	103.9	10	Iberger Kugel	Radio Seefunk
11)	103.9	2	Stuttgart-Münster	Klassik Radio
10)	104.2	5	Sigmaringen	Radio Seefunk
8)	104.2	0.1	Heidenheim	Radio Ton
8)	104.3	2	Sindelfingen	Energy Stuttgart
10)	104.3	0.1	Lörrach	Radio Seefunk
8)	104.5	2	Waiblingen	Energy Stuttgart
8)	104.5	1	Winnenden	Energy Stuttgart
16)	104.6	1	Oberndorf	Radio Neckarburg
14)	104.6	0.3	Biberach	Donau 3 FM
1)	104.6	1	Buchen	Radio Regenbogen
7)	104.7	0.2	Heilbronn	big FM
6)	104.7	0.1	Wertheim	Radio Ton
13)	104.7	1	Geislingen	Die Neue 107.7
10)	104.8	1	Reutlingen	Radio Ton
9)	104.9	1	Offenburg-Ohlsbach	Hitradio Ohr
10)	104.9	1	Stuttgart-Münster	sunshine live
4)	105.0	50	Grünenbach	Radio 7
2)	105.1	0.2	Aalen	big FM
10)	105.2	20	Pforzheim	Radio Seefunk
10)	105.3	0.5	Singen	Radio Seefunk
3)	105.4	1	Geislingen	Hit-Radio Antenne 1
3)	105.4	0.3	Balingen	Hit-Radio Antenne 1
10)	105.4	0.1	Waldshut-Tiengen	Radio Seefunk
9)	105.5	0.5	Bühl	Hitradio Ohr
14)	105.9	5	Ulm-Ermingen	Donau 3 FM
15)	106.0	8.4	Blauen-Müllheim	baden.fm
3)	106.0	0.1	Bad Mergentheim	Hit-Radio Antenne 1
5)	106.1	1	Heidelberg-Königstuhl	sunshine live
13)	106.1	1	Göppingen	Die Neue 107.7
14)	106.2	0.5	Riedlingen	Donau 3 FM
13)	106.5	1	Kirchheim	Die Neue 107.7
15)	106.6	0.1	Titisee-Neustadt	baden.fm
13)	106.8	1	Nürtingen	Die Neue 107.7
3)	106.9	0.1	Leonberg	Hit-Radio Antenne 1
3)	107.0	5	Wannenberg-Klettgau	Radio Seefunk
3)	107.0	1	Pforzheim	Hit-Radio Antenne 1
6)	107.1	20	Aalen	Radio Ton
5)	107.1	0.1	Wiesloch	sunshine live
7)	107.3	0.1	Bruchsal	big FM
9)	107.4	5	Lahr	Hitradio Ohr
13)	107.4	0.1	Gosbach	Die Neue 107.7
13)	107.7	4	Stuttgart Frauenkopf	Die Neue 107.7
15)	107.7	0.5	Freiburg-Littenweiler	baden.fm
5)	107.7	0.1	Weinheim	sunshine live
1)	107.9	1	Sickingen	Radio Ton
5)	107.9	0.1	Mosbach	sunshine live
7)	107.9	0.1	Bretten	die neue welle

Addresses and other information:
1) P.O.-Box 10 26 55, 68026 Mannheim (studio location: Dudenstr. 12-26); **W:** www.regenbogenweb.de AC – **2)** Kronenstr. 24, 70173 Stuttgart; **W:** www.bigfm.de CHR, further txs see T), U) – **3)** Plieningerstr. 150, 70567 Stuttgart; **W:** www.antenne1.de AC – **4)** Gaisenbergstr. 29, 89073 Ulm; **W:** www.sunshine-live.de Techno, also via Astra 1H, 12.148GHz. Templin tx see M) – **6)** Allee 2, 74072 Heilbronn; **W:** www.radio-ton.de AC – **7)** Albert-Nestler-Str. 26, 76131 Karlsruhe; **W:** www.meine-neue-welle.de AC – **8)** Anton-Schmidt-Str. 36, 71332 Waiblingen; **W:** www.energy-stuttgart.de

CHR – **9)** Postfach 20 80, 77610 Offenburg (studio location: Hauptstr. 83a); **W:** www.hitradio-ohr.de www.schwarzwaldradio.com AC – **10)** Konzilstr. 1, 78462 Konstanz; **W:** www.radio-seefunk.de AC – **11)** see O) – **12)** see M), stn. 4 – **13)** Königstr. 2, 70173 Stuttgart; **W:** www.dieneue1077.de Rock – **14)** Basteistr. 37, 89073 Ulm; **W:** www.donau3fm.de AC – **15)** Munzingerstr. 1, 79111 Freiburg; **W:** www.baden.fm AC – **16)** August-Schuhmacher-Str. 10, 78664 Eschbronn-Mariazell; **W:** www.radio-neckarburg.de – **17)** see M), stn. 15 – **18)** during Hockenheimring races only

Non-commercial stations:

FM	MHz	kW	Site	Station
10)	88.4	0.3	Freiburg university	echo-fm
6)	88.6	1	Stuttgart-Münster	Hochschulr. Stuttg.
9)	89.2	0.1	Horb	Freies R. Freudens.
1)	89.6	0.1	Mannheim	bermuda.funk
4)	91.2	0.1	Bruchsal	LernRadio
8)	96.6	1	Tübingen	Wüste Welle
7)	97.5	0.1	Schwäbisch Hall	Radio StHörfunk
5)	99.2	0.3	Stuttgart-Münster	Freies R. f. Stuttg.
9)	100.0	0.5	Freudenstadt	Freies R. Freudens.
10)	102.3	1	Freiburg Vogtsberg	Radio Dreyeckland
7)	102.6	1	Ulm-Ermingen	Radio FreeFM
9)	104.1	0.1	Baiersbronn	Freies R. Freudens.
12)	104.5	0.5	Hohe Möhr	Radio Kanal Ratte
3)	104.8	1	Karlsruhe	Querfunk
2)	104.8	0.1	Crailsheim	Radio StHörfunk
1)	105.4	0.1	Heidelberg Königstuhl	bermuda.funk

Addresses and other information:
1) Brückenstr. 2-4, 68167 Mannheim; **W:** www.bermudafunk.org. Also rel. Radio Aktiv (Universität Mannheim, Postfach 144, 68131 Mannheim); **W:** www.radioaktiv-online.de; Mon-Wed 0600-1000 and 1700-1900, Thu-Fri 2300-1000 and 1700-1900, Sun 1900-2100 – **2)** Haalstr. 9, 74523 Schwäbisch Hall; **W:** www.sthoerfunk.de – **3)** Steinstr. 23, 76133 Karlsruhe; **W:** www.querfunk.de rel. Mon-Fri 0600-1100 and Mon-Thu 1600-2100 stn. 4), arrangement subject to change – **4)** Hochschule für Musik, Postfach 6040, 76040 Karlsruhe (studio location: Wolfartsweierer Str. 7a); **W:** www.lernradio.de – **5)** Freies Radio für Stuttgart, Rieckestr. 24, 70190 Stuttgart; **W:** www.freies-radio.de – **6)** Hochschulradio Stuttgart (Nobelstr. 10, 70569 Stuttgart; **W:** www.horads.de – **7)** Söflinger Str. 206, 89077 Ulm; **W:** www.freefm.de – **8)** Hechinger Str. 203, 72072 Tübingen; **W:** www.wueste-welle.de Rel. Sun 0900-1300 Tübingen university prgr. and Tue-Thu 0700-0800 Helle Welle (religious) – **9)** Freies Radio Freudenstadt, Forststr. 23, 72250 Freudenstadt; **W:** www.radio-fds.de – **10)** Adlerstr. 12, 79098 Freiburg; **W:** www.rdl.de – **11)** Georges-Köhler-Allee Geb. 076, 79110 Freiburg; **W:** www.echo-fm.uni-freiburg.de – **12)** Bahnhofstr. 3, 79650 Schopfheim; **W:** www.kanalrattefm.de
DAB: 16 txs on 225.7MHz (ch. 12B) DLF (mono), DK (128k), DRadio Wissen (mono), SWR1 Baden-Württemberg (128k), SWR2 (128k), SWR3 (128k), DasDing (128k), SWR cont.ra (mono).

L) BAYERN

Media institution: Bayerische Landeszentrale für Neue Medien (BLM) ✉ Heinrich-Lübke-Straße 27, 81737 München ☎ +49 89 638 080 🖷 +49 89 63808140; **W:** www.blm.de

FM networks:

Location	Ant.B.	Rock.	Klass	egoFM	Galaxy	kW
Amberg	–	–	–	–	105.5	0.1
Ansbach	–	–	–	–	105.8	0.1
Aschaffenburg	103.0	–	–	–	91.6	25/0.1
Augsburg	104.2	87.9	92.2	94.8	–	0.1/0.3
Bad Reichenhall	103.7	–	–	–	–	0.1
Bamberg	101.1	–	–	104.7	25/0.5
Bayreuth	–	–	–	–	92.7	0.1
Bayrischzell	106.7	–	–	–	–	0.1
Berchtesgaden	107.9	–	–	–	–	0.3
Breithart	101.5	–	–	–	–	25
Brotjacklriegel	103.5	–	–	–	–	100
Coburg	103.8	–	–	–	90.4	5/0.2
Dillberg	100.6	–	–	–	–	25
Eichstätt	100.2	–	–	–	–	25
Enterbach	101.1	–	–	–	–	0.5
Grünten	104.4	–	–	–	–	50
Heidelstein	101.9	–	–	–	–	100
Herzogstand	102.0	–	–	–	–	0.1
Hochries	107.7	–	–	–	–	50
Hof	–	–	–	–	94.0	0.2
Högl-Freilassing	105.3	–	–	–	–	1
Hohenpeißenb.	103.8	–	–	–	–	25
Hoher Bogen	101.9	–	–	–	–	50
Ingolstadt	–	–	–	–	107.9	0.1
Kempten	–	–	–	–	88.1	0.3
Konradsreuth	–	–	–	–	98.1	0.1
Landshut	99.3	–	–	–	99.8	0.1

Location	Ant.B.	Rock.	Klass	egoFM	Galaxy	kW
Lindau	99.0	–	–	–	–	0.5
Münchberg	–	–	–	–	98.1	0.1
München	101.3	–	107.2	104.0	–	0.3/1/0.1
Naila	–	–	–	–	96.5	0.1
Nördlingen	103.3	–	–	–	–	25
Nürnberg	–	–	105.1	103.6	–	0.5/0.3
Oberaudorf	94.6	–	–	–	–	0.3
Ochsenkopf	103.2	–	–	–	–	100
Passau	102.1	–	–	–	91.7	1/0.2
Pfaffenhofen	92.6	–	–	–	–	0.5
Regensburg	103.0	–	91.1	107.5	–	25/0.3/0.3
Reit im Winkel	101.6	–	–	–	–	0.1
Rosenheim	–	–	–	–	106.6	0.1
Selb	–	–	–	–	93.4	0.1
Sonthofen	93.6	–	–	–	–	0.1
Traunstein	103.7	–	–	–	–	5
Ulm	104.8	–	–	–	–	0.1
Weiden	–	–	–	–	89.8	0.1
Weiler Simm.	106.0	–	–	–	–	0.1
Wunsiedel	–	–	–	–	97.3	0.1
Würzburg	104.4	–	92.1	95.8	–	5/0.3
Zugspitze	102.7	–	–	–	–	2

Addresses and other information:
Antenne Bayern (AC), **Rockantenne** (rock): Münchener Straße 101c, 85737 Ismaning, **W:** www.antenne.de www.rockantenne.de – **Klassik Radio:** see O) – **egoFM:** Leopoldstraße 254, 80807 München; **W:** www.egofm.de Alternative – **Radio Galaxy:** Lilienthalstraße 3c, 93049 Regensburg, **W:** www.radiogalaxy.de CHR. Mon-Fri 1400-1800 local prgr., produced by stns 17), 26), 29), 30/31), 32), 33), 35) (R. Euroherz), 36), 41), 42), 43) and 48) listed below. Antenne Bayern, Rockantenne, egoFM also via Astra 1H, 12.148GHz.

Local stations:

FM	MHz	kW	Site	Station
39)	87.9	0.3	Straubing Bogenberg	R. AWN
44)	87.9	0.1	Erding	Hitwelle Erding
35)	88.0	5	Großer Waldstein	extra-rad. / Euroherz
19)	88.1	0.1	Krumbach-Kirchberg	R. Prima 1
18)	88.2	0.2	Kaufbeuren	R. Ostallgäu
51)	88.2	0.1	Bad Reichenhall	R. Untersberg
32)	88.5	0.5	Bamberg Rothof	R. Bamberg
36)	88.5	0.1	Tirschenreuth	R. Ramasuri
27)	88.6	0.1	Karlstadt	R. Charivari
5)	89.0	0.3	München Olympiaturm	2DAY/Neues Europa
42)	89.0	0.1	Dingolfing	R. Trausnitz
51)	89.0	0.1	Högl-Freilassing	R. Untersberg
26)	89.1	0.1	Wassertrüdingen	R. 8
31)	89.2	0.5	Coburg Eckardtsberg	R. EINS
17)	89.3	0.1	Oberstdorf-Steinach	RSA R.
40)	89.3	0.2	Regen Geiskopf	Unser R. Deggendorf
26)	89.4	0.5	Ansbach Ludwigshöhe	R. 8
24)	89.7	0.1	Dillingen	RT.1 Nordschwaben
38)	89.7	0.3	Regensburg Ziegetsberg	gong fm
41)	89.7	0.3	Bad Griesbach	Unser R. Passau
26)	89.8	0.1	Dinkelsbühl	R. 8
45)	89.8	0.1	Landsberg-Stoffen	R. 106.4
31)	90.0	0.1	Kronach-Neuses	R. EINS
19)	90.2	0.32	Bad Grönenbach	R. Prima 1
26)	90.2	0.1	Gunzenhausen	R. 8
47)	90.2	0.1	Miesbach-Bergham	R. Alpenwelle
21)	90.3	0.1	Günzburg	Hitradio X
26)	90.4	0.2	Neuestadt / Aisch	R. 8
27)	90.4	0.1	Gemünden / Lohr	R. Charivari
49)	90.4	0.1	Mühldorf	Inn-Salzach-Welle
30)	90.5	0.1	Bad Kissingen	R. PrimaTon
29)	90.8	0.2	Alzenau	R. Primavera
15)	91.0	0.2	Fürth	Vil R.
47)	91.7	0.1	Holzkirchen Jasberg	R. Alpenwelle
42)	91.8	0.2	Pfeffenhausen-Stollnried	R. Trausnitz
47)	92.0	0.1	Wolfratshausen	R. Alpenwelle
6)	92.4	0.3	München Olympiaturm	(shared freq.)
16)	92.7	0.1	Weiler Simmerberg	Welle Bodensee
37)	92.7	0.4	Hoher Bogen	Charivari Regensbg.
49)	92.7	0.3	Reichertsheim	Inn-Salzach-Welle
13)	92.9	0.3	Nürnberg	Hi R. N1
17)	93.0	0.1	Immenstadt	RSA R.
49)	93.1	0.1	Burgkirchen-Gendorf	Inn-Salzach-Welle
2)	93.3	0.1	München Olympiaturm	Energy 93.3
33)	93.3	0.1	Pegnitz	R. Mainwelle
23)	93.4	0.3	Augsburg	R. Fantasy
14)	93.6	0.3	Erlangen	Energy Nürnberg
36)	93.6	0.1	Waidhaus Fischerberg	R. Ramasuri
19)	93.9	0.3	Mindelm-Altensteig	R. Prima 1
41)	93.9	0.3	Vilshofen-Otterkirchen	Unser R. Passau
30)	94.0	0.1	Bad Brückenau	R. PrimaTon
37)	94.0	1	Seubersdorf Göschberg	Charivari Regensbg.

FM	MHz	kW	Site	Station
7)	94.5	0.1	München Blutenburgstr.	M 94,5
11)	94.5	0.3	Nürnberg	R. F / Jazztime
43)	94.6	0.1	Schrobenhausen	R. IN / R. ND1
47)	95.0	0.2	Bad Tölz	R. Alpenwelle
35)	95.1	0.1	Marktredwitz	extra~r. / Euroherz
36)	95.3	1	Hirschberg Rothbühl	R. Ramasuri
31)	95.4	0.3	Lichtenfels	R. EINS
43)	95.4	0.1	Ingolstadt	R. IN
3)	95.5	0.3	München Olympiaturm	Charivari 95.5
24)	95.6	1	Harburg Hühnerberg	RT.1 Nordschwaben
30)	95.7	0.1	Haßfurt/Main	R. PrimaTon
39)	95.7	0.1	Mallersdorf-Hofkirchen	R. AWN
14)	95.8	0.3	Nürnberg	R. Z
4)	96.3	0.3	München Olympiaturm	R. Gong 96,3
38)	96.3	0.32	Burglengenfeld	gong fm
32)	96.6	0.1	Forchheim Pinzberg	R. Bamberg
45)	96.6	0.1	Starnberg	R. 106.4
17)	96.7	0.1	Kempten town	RSA R.
22)	96.7	0.1	Augsburg	Kit R. RT.1
48)	96.7	0.3	Flintsbach Dandlberg	Charivari Rosenheim
12)	97.1	0.3	Nürnberg	Gong 97.1
24)	97.1	0.1	Donauwörth	RT.1 Nordschwaben
41)	97.2	0.1	Grafenau Liebersberg	Unser R. Passau
26)	97.3	0.3	Feuchtwangen	R. 8
38)	97.3	0.1	Schwandorf Weinberg	gong fm
46)	97.5	0.1	Weilheim	R. Oberland
17)	97.6	1	Kempten Blender	RSA R.
18)	98.0	0.1	Füssen	R. Ostallgäu
51)	98.1	0.1	Berchtesgaden	R. Untersberg
37)	98.2	0.3	Regensburg Ziegetsberg	Charivari Regensb.
41)	98.3	0.2	Passau-Haidenhof	Unser R. Passau
10)	98.6	0.3	Nürnberg	Charivari 98.6
40)	98.7	0.1	Deggendorf-Hochobernd.	Unser R. Deggendorf
37)	98.8	0.5	Burglengenfeld	Charivari Regensb.
34)	98.9	0.1	Stadtsteinach	R. Plassenburg
25)	99.0	0.2	Lauf Moritzberg	star fm
27)	99.0	0.1	Marktheidenfeld	R. Charivari
43)	99.1	0.1	Eichstätt-Seuversholz	R. IN
50)	99.4	0.3	Haslach-Einham	R. Chiemgau
36)	99.9	0.2	Weiden Fischerberg	R. Ramasuri
47)	99.9	0.3	Herzogstad	R. Alpenwelle
29)	100.4	1	Aschaffenburg	R. Primavera
30)	100.5	0.5	Schweinfurt	R. PrimaTon
1)	100.8	0.1	München Blutenburgstr.	R. Arabella
26)	100.8	0.1	Burgbernheim	R. 8
43)	101.2	0.2	Neuburg/Donau	R. IN / R. ND1
46)	101.2	0.1	Oberammergau	R. Oberland
46)	101.4	0.3	Sindelsdorf	R. Oberland
30)	101.5	0.3	Bad Neustadt-Unsleben	R. PrimaTon
41)	101.5	0.1	Freyung Geyersberg	Unser R. Passau
50)	101.5	0.3	Trostberg	R. Chiemgau
34)	101.6	5	Kulmbach Rehberg	R. Plassenburg
27)	102.4	0.3	Würzburg	R. Charivari
37)	102.6	0.32	Waldmünchen Perlhütte	Charivari Regensb.
16)	103.6	0.5	Lindau Hoyerberg	Welle Bodensee
36)	103.9	0.1	Amberg Eisberg	R. Ramasuri
37)	103.9	0.5	Kelheim Leitenberg	Charivari Regensb.
42)	104.1	1	Landshut	R. Trausnitz
48)	104.2	0.3	Oberaudorf-Hölzelsau	Charivari Rosenheim
33)	104.3	10	Oschenberg	R. Mainwelle
47)	104.3	0.5	Enterbach-Ringberg	R. Alpenwelle
46)	104.6	0.1	Herzogstad	R. Oberland
43)	104.8	0.2	Pfaffenhofen Wolfsberg	R. IN
36)	105.1	0.5	Wiesau-Fuchsmühle	R. Ramasuri
1)	105.2	25	München-Isen	R. Arabella
18)	105.2	0.1	Obergünzburg	R. Ostallgäu
43)	105.4	0.1	Beilngries	R. IN
37)	105.5	0.3	Lam-Koppenhof	Charivari Regensb.
42)	105.5	0.32	Landau	R. Trausnitz
20)	105.9	5	Ulm-Ermingen	R. Donau 1
37)	105.9	0.32	Nabburg Galgenberg	Charivari Regensb.
32)	106.1	0.1	Burglesau Reisberg	R. Bamberg
8)	106.2	0.1	Erlangen	afk max
46)	106.2	0.3	Garmisch-Partenkirchen	R. Oberland
47)	106.2	0.1	Schliersbergalm	R. Alpenwelle
18)	106.3	0.5	Eisenberg Schloßberg	R. Ostallgäu
36)	106.4	0.1	Königstein Gr. Ossinger	R. Ramasuri
45)	106.4	2	Fürstenfeldbruck	R. 106.4
49)	106.4	0.3	Lohkirchen	Inn-Salzach-Welle
8)	106.5	0.1	Nürnberg	afk max
28)	106.9	5	Würzburg	R. Gong 106,9
9)	106.9	0.3	Nürnberg	Energy Nürnberg
42)	107.4	0.1	Pfarrkirchen-Postm.	R. Trausnitz
25)	107.8	0.2	Schwabach Heidenberg	star fm
40)	107.8	0.2	Brotjacklriegel	Unser R. Deggendorf

+ 28 txs less than 0.1kW

Adresses and other information:
Dienstleistungsgesellschaft für Bayerische Lokal-Radioprogramme (BLR) ✉ Rosenheimer Straße 145c, 81671 München **W:** www.blr.de and www.radiodienst.de Program supplier for many of the above listed stns. Nationwide content delivery under the brand RadioDienst.
1) Paul-Heyse-Str. 2-4, 80336 München, **W:** www.radioarabella.de – **2)** Pestalozzistr. 15-19, 80469 München, **W:** www.energy.de/muenchen – **3)** Postfach 20 16 09, 80016 München (studio location as stn. 1), **W:** www.charivari.de – **4)** Franz-Joseph-Str. 14, 80801 München, **W:** www.radiogong.de – **5)** Schneemanstr. 25, 81369 München, **W:** www.radio2-day.de Tr. rel. instead Sat 2300-Mon 0500 R. Neues Europa: Konviktstr. 1, 85049 Ingolstadt – **6) Radio Horeb,** Postfach 1165, 87501 Immenstadt; **W:** www.radiohoreb.de Religious. On 92.4MHz Mon-Fri 2300-1500, Sat/Sun 2300-0500, Sun 0900-1200 and 1300-2000. 24 hours via Astra 1C, 10.832GHz. **Christliches Radio München,** Postfach 310201, 80102 München; **W:** www.christlichesradiomuenchen.de Religious. Mon-Fri 1500-1600, Sun 0800-0900 and 1200-1300. **Lora München,** Gravelottestr. 6, 81667 München, **W:** www.lora924.de Non-commercial. Mon-Fri 1600-2300. **Feierwerk München,** Hansastr. 39, 81373 München; **W:** www.feierwerk.de Non-commercial. Sat 0500-2300, Sun 0600-0800 and 2000-2200. **N.B** FM sce. of Net.FM has been closed down. – **7)** Schwere-Reiter-Str. 35, 80797 München, 80538 München, **W:** m945.afk.de Journalist training stn. – **8)** Fürther Str. 212, 90429 Nürnberg, **W:** www.afkmax.de Journalist training stn. – **9)** Ostendstr. 100, 90482 Nürnberg, **W:** www.energy.de/nuernberg – **10),11),12),13)** Funkhaus Nürnberg, Senefelder Str. 7, 90409 Nürnberg, **W:** www.funkhaus.de 92.0MHz also rel. Camillo 92.9 (Mon, Tue, Sun 2000-2200), R. AREF (Sun 0900-1100), Pray 92.9 (Sun 1100-1200), R. Meilensteine (Sun 0800-0900), 94.5MHz also rel. Jazztime München (Mon 2100-2200, Thu 2000-2100). – **14)** Kopernikusplatz 12, 90459 Nürnberg, **W:** www.radio-z.net. 1300-0100 only, other times rel. stn 25) (**N.B** R. Aladin has been closed down) – **15)** Platnersgasse 1, 90403 Nürnberg, **W:** www.vilradio.de – **16)** **W:** www.welle-bodensee.de – **17)** Rottachstr. 17, 87439 Kempten, **W:** www.allgaeuseite.de/rsa_radio – **18)** Rottachstr. 17, 87439 Kempten, **W:** www.allgaeuseite.de/rsa_radio – **18)** **W:** www.roal.de – **19)** Hirschgasse 1, 87700 Memmingen, **W:** www.prima1.de – **20)** Leipzigstr. 26, 88400 Biberach, **W:** www.radiodonau1.de – **21)** Augsburger Str. 1 ½, 89312 Günzburg, **W:** www.hitradiox.de – **22)** Curt-Frenzel-Str. 4, 86167 Augsburg, **W:** www.radio-rt1.de – **23)** Ludwigstr. 1, 86150 Augsburg, **W:** www2.fantasy.de Rel. Mon 2100-2400 Kanal C (university stn.): Eichleitnerstr. 30, 86159 Augsburg, **W:** www.kanal-c.de – **24)** Artur-Proeller-Str. 1, 86609 Donauwörth, **W:** www.rt1-nordschwaben.de – **25)** O'Brien Str. 2, 91126 Schwabach; **W:** www.rocksender.de/rocksender_nuernberg/ – **26)** Postfach 8, 91510 Ansbach (studio location: Schalkhäuser Landstr. 5), **W:** www.radio8.de – **27), 28)** Semmelstr. 15, 97070 Würzburg, **W:** http://charivari.fm and www.gong.fm Also rel. Radio Opera – **29)** Am Funkhaus 1, 63743 Aschaffenburg, **W:** www.radio-primavera.de – **30),31)** Seifartshofstr. 21, 96450 Coburg, **W:** www.radioeins.com – **32)** Gutenbergstr. 5, 96050 Bamberg, **W:** www.radio-bamberg.de – **33)** Postfach 10 11 61, 95411 Bayreuth (studio location: Richard-Wagner-Str. 33), **W:** www.mainwelle.de – **34)** E.C.-Baumann-Str. 5, 95326 Kulmbach, **W:** www.radio-plassenburg.de – **35)** 0900-1000, 1200-1300 and 1800-2000 **extra~radio**, Postfach 1745, 95016 Hof (studio location: Kreuzsteinstr. 2-6), **W:** www.extra-radio.de; otherwise: **R. Euroherz,** Pfarr 1, 95028 Hof, **W:** www.euroherz.de – **36)** Unterer Markt 35, 92637 Weiden, **W:** www.ramasuri.de – **37),38)** Lilienthalstr. 3c, 93049 Regensburg, **W:** www.radiocharivari.de and www.gongfm.de – **39),40)** Bahnhofstr. 28, 94469 Deggendorf, **W:** www.unserradio.de – **41)** Medienstr. 5, 94036 Passau, **W:** as stn. 40) – **42)** Postfach 381, 84028 Landshut, **W:** www.radio-trausnitz.de – **43)** Donaustr. 11, 85049 Ingolstadt, **W:** www.radio-in.de, rel. 0500-0900 on 94.6/101.2MHz R. ND1 – **44)** Postfach 1155, 84420 Isen, **W:** www.hitwelle.de – **45)** Schöngeisingerstr. 11, 82256 Fürstenfeldbruck, **W:** www.radio1064.de – **46)** Postfach 1752, 82467 Garmisch-Partenkirchen (studio location: Marienplatz 17), **W:** www.radio-oberland.de – **47) W:** www.radio-alpenwelle.de – **48)** Hafnerstr. 5, 83022 Rosenheim, **W:** www.radio-charivari.de – **49)** Mozartstr. 3a, 84508 Burgkirchen/Alz, **W:** www.inn-salzach-welle.de – **50)** Rupertistr. 40-42, 83278 Traunstein, **W:** www.radio-chiemgau.de – **51)** www.untersberg.de **N.B** stns 49), 50), 51) also rel. prgr. of independent producers.
DAB: 39 txs on 229.1MHz (ch. 12D) BR Klassik (192k), on3radio (128k), Bayern plus (128k), Bayern2plus (128k), B5 plus (mono), BR Traffic News (mono), Rockantenne (192k), Radio Galaxy (160k).
München and Nürnberg txs on 222.1MHz(ch. 11D) Bayern 1 (160k), Bayern 2 (160k), Bayern 3 (160k), B5 aktuell (mono), BR Traffic News (mono). München txs on 220.4MHz (ch. 11C) and Ingolstadt txs on 216.9MHz (ch. 11A) DLF (mono), DK (128k), DRadio Wissen (mono), Fantasy (192k), Digital Classix (160k), R. DeLuxe (160k), R. Gong Mobil (160k).
Nürnberg/Erlangen txs on 213.4MHz (ch. 10C) DLF (mono), DK (128k), DRradio Wissen (mono), Energy Nürnberg (160k), Fantasy (192k), Vil R. (160k), Pirate R. (160k).
Augsburg tx on 206.4MHz (ch. 9C) DLF (mono), DK (128k), DRadio

Wissen (mono), R. Kö (160k), Fantasy Bayern (192k), Fantasy Aktuell (192k), R. Augsburg (192k), Smart R. (160k).
Regensburg txs on 223.9MHz (ch. 12A) Charivari (128k), Antenne Bayern (AAC 128k), Galaxy (AAC 128k), Gong FM (AAC 128k).
DVB-H: On Erlangen 706MHz (ch. 50; 0.1kW) bit eXpress (run by Erlangen university; Am Wolfsmantel 33, 91058 Erlangen; W: www.bitexpress.de; also on 15896kHz).

M) BERLIN & BRANDENBURG
Media institution: Medienanstalt Berlin-Brandenburg (MABB) ☒ Kleine Präsidentenstraße 1, 10178 Berlin ☎ +49 30 264 9670 ▤ +49 30 264 96730 **W:** www.mabb.de
Berlin sites:

FM	MHz	kW	Site	Station
14)	87.9	1	Alexanderplatz	Star FM
16)	88.4	0.5	Hallesches Ufer	88vier
19)	90.2	4	Scholzplatz	R. Teddy
16)	90.7	0.1	Schäferberg	88vier
2)	91.4	100	Alexanderplatz	Berliner Rundfunk
10)	93.6	3	Alexanderplatz	JAM FM
3)	94.3	20	Alexanderplatz	rs2
12)	94.8	16	Alexanderplatz	BBCWS
13)	97.2	0.1	Hallesches Ufer	Russkij / Blu
9)	98.2	8	Scholzplatz	(see note)
11)	98.8	1	Alexanderplatz	KISS FM
26)	99.1	0.1	Schönhauser Allee	(special stns)
4)	100.6	13	Alexanderplatz	100,6 Motor FM
8)	101.3	5	Alexanderplatz	Klassik R.
15)	101.9	4	Schäferberg	Metropol FM
5)	103.4	10	Alexanderplatz	Energy Berlin
18)	104.1	0.2	Hallesches Ufer	NPR FM Berlin
6)	104.6	10	Alexanderplatz	104.6 RTL
7)	105.5	5	Alexanderplatz	Spreeradio
17)	106.0	1	Alexanderplatz	RFI
12)	106.8	0.5	Alexanderplatz	Jazz Radio
1)	107.5	13	Schäferberg	BB Radio

Brandenburg sites:

FM	MHz	kW	Site	Station
8)	87.6	0.4	Brandenburg/Havel	Klassik Radio
5)	87.6	0.2	Prenzlau	Energy Berlin
6)	88.0	1	Crinitz	104.6 RTL
20)	88.3	0.5	Neuruppin	Power Radio
6)	89.5	0.5	Elsterwerda-Hohenl.	104.6 RTL
23)	90.3	0.5	Spremberg	94.5 Radio Cottbus
9)	90.4	0.2	Guben-Reichenbach	(see note)
1)	90.9	0.8	Rhinow	BB Radio
8)	91.0	0.5	Booßen (Frankf./O.)	Klassik Radio
3)	91.3	1	Lauchhammer West	rs2
5)	91.6	1.3	Casekow	Energy Berlin
	91.6	0.5	Cottbus-Klein Oßnig	(to be allocated)
5)	91.7	0.1	Herzberg/Elster	Energy Berlin
20)	91.8	1.3	Zehlendorf	Power Radio
23)	92.1	1	Guben-Reichenbach	94.5 Radio Cottbus
22)	93.9	3	Fürstenwalde	Sender KW
20)	94.4	1.3	Perleberg	Power Radio
23)	94.5	0.3	Cottbus-Madlow	94.5 Radio Cottbus
3)	94.7	3	Booßen (Frankf./O.)	rs2
25)	94.9	0.5	Templin	sunshine live
20)	95.2	0.4	Belzig-Lütte	Power Radio
20)	95.3	0.1	Fürstenwalde	Power Radio
1)	95.4	1.3	Zehlendorf	BB Radio
9)	95.5	0.2	Eisenhüttenstadt	(see note)
3)	95.6	1.3	Cottbus-Klein Oßnig	rs2
5)	96.6	0.5	Wittstock	Energy Berlin
3)	96.7	1	Crinitz	rs2
21)	96.7	0.5	Pausin	OldieStar Radio
20)	97.0	0.3	Erkner	Power Radio
	99.3	0.8	Booßen (Frankf./O.)	(to be allocated)
3)	100.1	3	Lübben	rs2
2)	100.9	5	Casekow	Berliner Rundfunk
1)	102.1	20	Casekow	BB-Radio
2)	102.2	3	Cottbus-Klein Oßnig	Berliner Rundfunk
23)	102.7	0.5	Forst	94.5 Radio Cottbus
1)	103.7	0.6	Eisenhüttenstadt	BB Radio
24)	103.8	1.5	Großräschen	Elsterwelle
3)	103.9	6	Forst	rs2
2)	104.2	20	Booßen (Frankf./O.)	Berliner Rundfunk
1)	104.3	100	Pritzwalk-Buchholz	BB Radio
21)	104.9	1.3	Zehlendorf	OldieStar Radio
1)	105.0	3	Brandenburg-Krahne	BB Radio
22)	105.1	0.8	Königs Wusterh.	Sender KW
9)	105.9	1.6	Booßen (Frankf./O.)	(see note)
3)	106.3	4	Spremberg	rs2
1)	107.3	100	Calau	BB Radio
3)	107.3	12	Casekow	rs2
1)	107.8	30	Booßen (Frankf./O.)	BB Radio

FM	MHz	kW	Site	Station
1)	107.9	5	Zehlendorf	BB Radio

Addresses and other information:
1) Großbeerenstr. 185, 14482 Potsdam; **W:** www.bbradio.de AC, with short local insertions (different ones on both Zehlendorf freq.) – **2)** Grunewaldstr. 3, 12165 Berlin; **W:** www.berliner-rundfunk.de Oldies – **3)** as stn. 2); **W:** www.rs2.de AC – **4)** Brunnenstr. 24, 10119 Berlin; **W:** www.motorfm.de Alternative. Stuttgart tx see K) – **5)** Hardenbergstr. 4-5, 10623 Berlin; **W:** www.energy.de/berlin. CHR – **6), 7)** Kurfürstendamm 207-208, 10719 Berlin; **W:** www.104.6rtl.com (CHR), www.spreeradio. de (oldies) – **8)** see O) – **9)** Freq. reallocated to stn. 21), at time of editing subject of legal action by former user R. Paradiso (addr. as stn. 10, www.paradiso.de) – **10)** Am Kleinen Wannsee 5, 14109 Berlin; W: www.jamfm.de Black – **11)** as stn. 2); **W:** www.kissfm.de CHR – **12)** At time of editing subject of allocation procedures due to insolvency of Jazz R. BBC World Service seeks to use 101.9MHz as replacement for 90.2MHz – **13) R. Russkij Berlin**, Kochstr. 54, 10969 Berlin; W: www.radio-russkij-berlin.de In Russian, on 97.2MHz 0600-1800. **Blu FM**, Sophienstr. 8, 10178 Berlin; **W:** www.blu.fm. Dance music, on 97.2MHz 1800-0600. **N.B** Rel. of WRN Deutsch ceased – **14)** Dircksenstr. 48, 10178 Berlin; **W:** www.starfm.de Rock – **15)** Markgrafenstr. 11, 10969 Berlin, **W:** www.metropolfm.de; prgr. in Turkish. Further txs see K) and T) – **16)** c/o ALEX, Voltastr. 5, 13355 Berlin; **W:** www.alex-berlin.de Run by MABB, own citizen radio b'casts and prgr. from various small ventures – **17)** See International Broadcasting section under France. Freq. subject of reallocation in 2012 – **18)** See National Public Radio under USA; **W:** www.nprberlin.de – **19)** August-Bebel-Str. 26-53, 14482 Potsdam; **W:** www.radioteddy.de; childrens prgr., also via Astra 1H (12.246GHz). Kassel tx see P) – **20)** Potsdamer Str. 131, 10783 Berlin; **W:** www. powerradio918.de Oldies – **21)** Pfalzburger Str. 43-44, 10717 Berlin; **W:** www.oldiestar.de – **22)** Am Funkerberg, Senderhaus 1, 15711 Königs Wusterhausen; **W:** www.sender-kw.de Oldies – **23)** Schloßkirchplatz 3, 03046 Cottbus; **W:** www.radiocottbus.de AC – **24)** see V), stn. 8) **25)** See K), stn. 5) – **26)** Frequently in use for special event stn's.
MW: On 603kHz OldieStar, on 693kHz Voice of Russia, on 1485kHz OldieStar in DRM mode.
DAB: Britz (1kW); Scholzplatz (1kW), Marzahn (0.5kW) txs on 199.4MHz (ch. 8C) DLF (mono), DK (128k), DRadio Wissen (160k), WDR 2 (192k). Alexanderplatz and Michendorf txs (1kW each) on 229.1MHz (ch. 12D) DLF (64k), DK (128k), DRadio Wissen (mono), OldieStar (128k).

DVB-T: Alexanderplatz (10kW) and Schäferberg (20kW) txs on 775.25MHz (ch. 59) 104.6RTL, 104.6 RTL Best of Modern Rock & Pop, Spreeradio, R. Paloma, ERF, R. Horeb, place2be. Pilot project, multiplex usage changes frequently.
Booßen (50kW) and Calau (100kW) txs on 759.25MHz (ch. 57) experimental sce. with Antenne Brandenburg, R. Eins, Fritz, Kulturr., Infor.

N) BREMEN
Media institution: Bremische Landesmedienanstalt (Brema) ☒ Grünenweg 26, 28215 Bremen ☎ +49 421 334940 ▤ +49 421 323533 **W:** www.bremische-landesmedienanstalt.de

FM	MHz	kW	Site	Station
1)	89.8	1	Bremen-Walle	Energy Bremen
3)	90.7	0.2	Bremerhaven	Radio Wester TV
3)	92.5	0.2	Bremen Neuenstr.	Radio Weser TV
	97.2	–	Bremen-Walle	(allocated to Motor FM)
1)	104.3	8	Bremerhaven	Energy Bremen
2)	104.8	0.1	Bremen-Walle	Hit-Radio Antenne
2)	107.9	0.3	Bremerhaven	Hit-Radio Antenne

Addresses and other information:
1) Erste Schlachtpforte, 28195 Bremen; **W:** www.energy.de/bremen CHR – **2)** see R) – **3)** Richtweg 14, 28195 Bremen; **W:** www.radioweser.tv Citizen radio.

O) HAMBURG & SCHLESWIG-HOLSTEIN
Media institution: Medienanstalt Hamburg / Schleswig-Holstein (MA HSH) ☒ Rathausallee 72-76, 22846 Norderstedt ☎ +49 40 3690050 ▤ +49 40 36900555 **W:** www.ma-hsh.de

FM	R.SH	delta	Nora	Klass.	kW
Ahrensburg	–	96.5	–	–	2
Bredstedt	–	–	98.1	–	0.1
Bungsberg (Eutin)	100.2	104.1	106.2	97.2	2x50/0.2
Flensburg-Freienwill	101.4	105.6	–	–	20
Flensburg-Harrislee	–	–	88.5	106.5	0.5
Garding	–	–	94.1	91.7	0.5
Hamburg-Bergedorf	–	107.7	–	–	0.1
Hamburg Hertz-T.	100.0	93.4	–	98.1	2x2/0.1
Heide-Welmbüttel	103.8	100.4	–	–	15
Heide (town)	–	–	96.9	–	0.1
Helgoland (island)	100.0	103.5	101.6	89.8	0.1
Husum	–	92.0	–	–	0.1

FM	R.SH	delta	Nora	Klass.	kW
Itzehoe	–	–	104.9	92.7	1/0.5
Kaltenkirchen	102.9	107.4	101.1	–	20
Kiel	102.4	105.9	97.0	97.4	2x15/0.3
Lauenburg	102.5	105.6	97.4	–	1/1/0.3
Lübeck	–	–	91.5	–	0.3
Mölln-Berkenthin	101.5	107.9	91.5	93.6	2x20/0.3
Neumünster	–	–	88.9	–	0.5
Niebüll	–	–	107.2	94.7	0.2
Rendsburg	–	–	93.6	92.9	0.5
Schleswig (town)	–	–	92.4	100.8	1/0.5
Schleswig-Borgwedel	–	–	–	93.9	0.5
Westerland (Sylt)	102.8	104.8	89.1	89.8	5/5/1/0.5

Addresses and other information:
R.SH (AC), **delta radio** (CHR), **Radio Nora** (oldies)**:** Wittland 3, 24109 Kiel; **W:** www.rsh.de www.deltaradio.de www.radionora.de – **Klassik Radio:** Postfach 57 03 60, 22772 Hamburg (studio location: Planckstr. 15); **W:** www.klassikradio.de Classical music. Further txs see K), L), M), P), R), X). Also via Astra 1F, 11.273GHz.

Hamburg only:

FM	MHz	kW	Site	Station
1)	88.1	0.1	Bergedorf	Oldie 95
1)	88.5	2	Otterndorf*)	R. Hamburg
2)	91.7	0.1	H.-Hertz-Turm	Alsterradio
2)	93.6	2	Otterndorf*)	Alsterradio
1)	95.0	0.1	H.-Hertz-Turm	Oldie 95
3)	97.1	0.1	H.-Hertz-Turm	Energy 97.1
3)	100.9	0.1	Bergedorf	Energy 97.1
1)	103.6	80	Moorfleet	R. Hamburg
1)	104.0	0.2	H.-Hertz-Turm	R. Hamburg
4)	106.0	0.1	H.-Hertz-Turm	(temporary use)
2)	106.8	40	Rahlstedt	106!8 rock'n pop

*) Tx in Niedersachsen, serving Neuwerk and Scharhörn islands (belonging to Hamburg).

Addresses and other information:
1) Postfach 10 01 23, 20001 Hamburg (studio location: Spitalerstraße 10); **W:** www.radiohamburg.de www.oldie95.de – **2)** Rödingsmarkt 29, 20459 Hamburg; **W:** www.106acht.de **F.PI.:** Separate prgr. on 91.7MHz – **3)** Winterhuder Marktplatz 6-7, 22299 Hamburg; **W:** www.energy.de/hamburg – **4)** Temporary stn's during cultural events.

Non-commercial stations:

FM	MHz	kW	Site	Station
1)	93.0	0.1	Hamburg Hertz-Turm	Freies Sender Kombinat
1)	96.0	0.1	Hamburg Hertz-Turm	TIDE 96.0 / HLR
4)	97.6	0.5	Garding	OK Westküste
5)	98.8	0.5	Lübeck-Stockelsdorf	OK Lübeck
4)	98.8	0.1	Husum	OK Westküste
3)	101.2	0.1	Kiel	Kiel FM
5)	105.2	0.1	Heide	OK Westküste

Addresses and other information:
1) Schulterblatt 23c, 20357 Hamburg; **W:** www.fsk-hh.org – **2) TIDE 96.0**, Uferstraße 2, 22081 Hamburg; **W:** www.tidenet.de Run by Hamburg Media School. Mon 0500-2300 and thorough Tue 0500 til Sun 0500. **Hamburger Lokalradio**, Kulturzentrum LOLA, Lohbrügger Landstraße 8, 21031 Hamburg; **W:** www.hhlr.de On 96.0MHz Sun 0500 til Mon 0500 and night Mon/Tue 2300-0500. On 5980kHz (low power tx) 0800-0900. Via WRN Deutsch Sun 1000-1100. On 6045kHz first Sun each month 1000-1100 – **3)** Hamburger Chaussee 36, 24113 Kiel; **W:** www.kielfm.de – **4)** Landvogt-Johannsen-Str. 11, 25746 Heide; **W:** www.okwestkueste.de – **5)** Kanalstr. 42-48, 23554 Lübeck; **W:** www.ok-luebeck.de
DAB: Hamburg tx (0.5kW) on 227.4MHz (ch. 12C), Kiel tx (0.4kW) on 229.1MHz (ch. 12D) rel. Niedersachsen ensemble.

P) HESSEN

Media institution: Hessische Landesanstalt für Privaten Rundfunk (LPR) Wilhelmshöher Allee 262, 34131 Kassel ☎ +49 561 935860 +49 561 9358630; **W:** www.lpr-hessen.de

FM	FFH	plan.	Kla.	Bob	Ener	harm	kW
Alsfeld	88.1	–	–	101.5	–	94.1	4/0.1
Bad Camberg	–	–	–	–	–	105.4	0.2
Bad Hersfeld	95.9	–	93.8	99.8	–	88.4	0.1/0.3
Bad Nauheim	–	104.6	–	106.6	–	100.4	0.5/1
Bensheim	–	–	–	103.3	–	107.5	0.2
Bingen	106.9	–	103.4	–	–	101.8	0.2/0.3
Butzbach	–	96.0	–	–	–	–	0.1
Darmstadt	–	–	92.4	100.8	–	–	0.2/0.5
Dieburg	–	90.1	–	99.5	–	104.7	1/0.2
Dillenburg	100.0	–	–	–	–	–	30
Driedorf	106.8	–	–	–	–	–	30
Eisenberg	–	100.3	–	–	–	–	50
Eltville	90.3	–	–	–	–	–	0.2
Eschwege	–	104.6	–	103.0	–	88.3	0.5/0.3
Feldberg	105.9	–	–	–	–	–	100
Frankfurt	–	100.2	107.5	101.4	95.1	105.4	1/0.1

FM	FFH	plan.	Kla.	Bob	Ener	harm	kW
Fritzlar	–	–	88.4	–	–	97.1	0.1
Fulda	–	99.9	102.8	105.7	–	95.7	0.2/0.3
Gießen	–	93.7	88.0	92.6	105.2	102.0	0.5/0.1
Glashütten	–	–	–	–	–	93.2	0.5
Habichtsw.	103.7	–	–	–	–	–	20
Hanau	–	–	–	97.3	106.8	–	0.5
Heidelstein*	100.9	–	–	–	–	–	50
Hofgeismar	–	–	88.8	–	–	–	0.1
Hoherodskopf	–	–	94.7	–	–	–	0.1
Homberg	–	–	99.3	–	–	–	0.1
H. Meißner	105.1	–	–	–	–	–	100
Idstein	–	–	–	–	–	93.2	0.5
Kassel	–	104.6	104.1	99.4	–	96.6	0.5/0.2
Krehberg	105.0	–	–	–	–	–	20
Korbach	107.7	94.0	–	96.5	–	107.4	20/0.2
Limburg	–	97.6	102.0	90.2	–	92.1	0.5/0.2
Marburg	–	101.0	104.9	103.9	–	96.2	0.3/0.1
Michelstadt	96.1	–	–	98.5	–	104.6	0.1/1
Offenbach	–	–	–	–	–	99.3	0.3
Rimberg	–	–	90.5	–	–	–	0.1
Rotenburg	–	–	93.5	–	–	–	0.1
Schlüchtern	–	–	101.3	–	–	–	0.2
Schotten	–	–	94.7	–	–	–	0.1
Vogelsberg	–	–	94.7	–	–	–	0.1
Wetzlar	–	100.5	88.2	105.0	–	101.3	0.3/0.5
Wiesbaden	102.0	90.1	–	101.4	95.1	88.2	0.1/0.5

Addresses and other information:
Hit-Radio FFH (AC), **planet radio** (black/CHR), **harmony.fm** (oldies)**:** FFH-Platz 1, 61111 Bad Vilbel; **W:** www.ffh.de www.planet-radio.de, www.harmonyfm.de; also via Astra 1L, 12.633GHz (harmony.fm using two freq. at Frankfurt due to interference situation) – **Klassik Radio** see O) – **Radio Bob**, Friedrich-Ebert-Str. 2, 34117 Kassel; **W:** www.radiobob.de also via Astra 1KR (10.936GHz, proprietary ADR system) – **Energy Rhein-Main** (former Main FM, relaunched in 2010), Rüsselsheimer Str. 22, 60326 Frankfurt am Main; **W:** www.energy.de/rhein-main
Evangeliums-Rundfunk (German branch of Trans World Radio): Postfach 1444, 35573 Wetzlar (studio location: Berliner Ring 62) ☎ +49 6441 9570, +49 6441 957120; **W:** www.erf.de
ERF Radio: 0400-0900 and 1830-2200 on 1539kHz, 24 hours via Astra 1H (12.148GHz) and at studio site on 90.0MHz (20W). 0400-0445 prgr. in foreign languages.

Other stations:

FM	MHz	kW	Site	Station
5)	90.1	0.1	Marburg-Lahnberge	R. Unerhört
3)	90.9	0.3	Rüsselsheim	R. Rüsselsheim
8)	91.7	0.2	Kassel Tannenwäldchen	R. Teddy
1)	91.8	0.1	Frankfurt-Ginnheim	R. X
2)	92.5	0.1	Wiesbaden	R. RheinWelle 92,5
7)	96.5	0.3	Witzenhausen	RundFunk Meißner
9)	99.2	0.3	Fulda	Domradio
7)	99.4	0.1	Sontra	RundFunk Meißner
7)	99.7	0.5	Eschwege	RundFunk Meißner
7)	102.6	0.3	Hessisch Lichtenau	RundFunk Meißner
4)	103.4	0.3	Darmstadt	R. Darmstadt
6)	105.8	0.5	Kassel Tannenwäldchen	Freies R. Kassel

Adresses and other information:
1) Schützenstr. 12, 60311 Frankfurt; **W:** www.radiox.de – **2)** Postfach 49 20, 65039 Wiesbaden; **W:** www.rheinwelle.de – **3)** Ludwigstr. 13-15, 65428 Rüsselsheim; **W:** www.radiok2r.de – **4)** Steubenplatz 12, 64293 Darmstadt; **W:** www.radiodarmstadt.de – **5)** Rudolf-Bultmann-Str. 2b, 35039 Marburg; **W:** www.radio-rum.de – **6), 7)** Niedonher Str. 1, 37269 Eschwege; **W:** www.eschwege.de/rfm – **8)** see M) – **9)** see S)
DAB: Feldberg (0.5kW) and Frankfurt-Ginnheim (1kW) txs on 227.4MHz (ch. 12C) DLF (mono), DK (128k), DRadio Wissen (mono).

Q) MECKLENBURG-VORPOMMERN

Media institution: Landesrundfunkzentrale Mecklenburg-Vorpommern, Bleicheufer 1, 19053 Schwerin ☎ +49 385 5588 10 +49 385 5588 130; **W:** www.lrz-mv.de

Commercial stations:

Location	Ant.MV	Osts	103.3	Klass	kW
Ahrenshoop	–	–	103.3	–	0.3
Garz (Rügen)	105.1	107.6	–	–	50
Grevesmühlen	105.8	94.7	–	–	0.2/0.1
Güstrow	107.7	98.0	–	–	1/0.4
Helpterberg	103.8	105.8	–	–	100
Heringsdorf	105.4	103.3	–	–	10/2
Marlow	100.8	104.8	–	–	100
Röbel	93.8	92.2	–	–	50/0.1
Rostock	97.3	105.6	–	–	2
Schwerin	101.3	107.3	–	90.1	2x100/0.2
Waren	98.3	93.0	–	–	0.2/0.1

Location	Ant.MV	Osts	103.3	Klass	kW
Wismar	88.7	93.7	–	–	0.2/0.1
Wolgast	–	100.0	–	–	0.5

Adresses and other information:
Antenne MV, 19086 Plate; **W:** www.antennemv.de AC – **Ostseewelle**, Warnowufer 59a, 18057 Rostock; **W:** www.ostseewelle.de CHR – **103.3 Ihr Lokalradio** (predecessor of Radio FDZ, closed down in 2010), Cubanzestr. 19b, 18211 Kühlungsborn (office; studio in Wieck); **W:** www.radio1033.de Oldies – **Klassik Radio** see O).

Non-commercial stations:

FM	MHz	kW	Site	Station
1)	88.0	0.8	Neubrandenburg	NB-Radiotreff
3)	90.2	0.1	Rostock-Stadtweide	LOHRO
2)	98.1	0.2	Greifswald	Radio 98eins
1)	98.7	0.1	Malchin	NB-Radiotreff

Adresses and other information:
1) Treptower Str. 9, 17033 Neubrandenburg; **W:** www.nb-radiotreff. de; run by Landesrundfunkzentrale, also prgr. from Malchin studio – **2)** Domstr. 12, 17489 Greifswald; **W:** www.98eins.de; run by university, Mon-Fri 1800-2200 only, otherwise rel. stn. 1) – **3)** Margaretenstr. 43, 18057 Rostock; **W:** www.lohro.de
DAB: Schwerin tx (1kW) on 225.7MHz (ch. 12B), rel. Niedersachsen ensemble.

R) NIEDERSACHSEN

Media institution: Niedersächsische Landesmedienanstalt für privaten Rundfunk (NLM), ✉ Seelhorststraße 18, 30175 Hannover ☎ +49 511 28477 0 🖷 +49 511 28477 36 **W:** www.nlm.de

FM	ffn	Ant.	R. 21	Klass	kW
Aurich	103.1	104.9	100.6	–	2x25/1
Bad Rehburg	–	–	89.4	–	0.5
Barsinghausen	101.9	103.8	–	–	25
Braunschw.-Broitzem	103.1	106.9	104.1	–	15/13/1
Celle	–	–	93.5	–	0.2
Cuxhaven-Otterndorf	102.6	104.6	–	–	20
Dannenberg-Zernien	102.7	106.1	–	–	25
Delmenhorst	–	–	107.6	–	0.1
Goslar	–	–	87.7	–	0.5
Göttingen	102.8	106.0	93.4	–	2x5/1
Hannoversch Münden	100.7	106.7	–	–	0.5
Hannover	–	–	104.9	107.4	0.5/0.2
Helmstedt	–	–	94.1	–	0.5
Hildesheim	–	–	105.8	–	1
Holzminden	102.2	105.7	–	–	0.5
Leer-Nüttermoor	–	–	104.5	–	0.3
Lingen-Damaschke	101.5	104.3	106.9	–	2x15/0.5
Oldenburg	–	–	104.1	–	0.24
Osnabrück	103.4	105.9	95.3	–	2x10/0.1
Rosengarten	100.6	105.1	–	–	20
Seesen	–	100.9	–	–	0.1
Steinkimmen	102.3	105.7	–	–	100
Torfhaus (Harz)	102.4	106.3	–	–	100
Visselhövede	101.7	104.2	–	–	10
Wilhelmshaven	–	–	99.1	–	0.3
Wolfsburg	–	–	95.1	–	0.5

N.B R. Hamburg / 106!8 rock'n pop txs at Cuxhaven see O).

Addresses and other information:
radio ffn, Stiftstraße 8, 30159 Hannover; **W:** www.ffn.de AC – **Hit-Radio Antenne**, Goseriede 9, 30159 Hannover; **W:** www.antenne.com. AC – **Radio 21**, An der Feuerwache 3-5, 30823 Garbsen; **W:** www.radio21.de Rock, cooperates with Rockland Radio, see T) – **Klassik Radio** see O).

Non-commercial stations:

FM	MHz	kW	Site	Station
4)	87.7	0.2	Emden	Radio Ostfriesland
3)	87.8	1	Wilhelmshaven	Radio Jade
1)	88.0	1	Uelzen	Radio ZuSa
1)	89.7	0.5	Dannenberg-Zernien	Radio ZuSa
4)	94.0	1	Aurich-Haxtum	Radio Ostfriesland
8)	94.8	1	Bad Pyrmont	Radio Aktiv
2)	95.2	0.2	Nordhorn	Ems-Vechte-Welle
1)	95.5	1	Lüneburg	Radio ZuSa
5)	95.6	1	Lingen-Schepsdorf	Ems-Vechte-Welle
5)	99.3	1	Molbergen-Cloppenburg	Ems-Vechte-Welle
8)	100.0	0.3	Hameln	Radio Aktiv
12)	100.0	0.1	Hannover Bettfedernf.	Radio Flora
4)	103.9	0.2	Leer	Radio Ostfriesland
10)	104.6	0.5	Braunschweig-Broitzem	Radio Okerwelle
6)	104.8	1	Osnabrück	OS Radio 104,8
11)	105.3	1	Hildesheim	Radio Tonkuhle
6)	106.5	1	Oldenburg-Wahnbek	Oldenburg Eins
7)	106.5	0.3	Hannover Telemaxx tower	Leinehertz
9)	107.1	1	Göttingen	StadtRadio Gött.

Adresses and other information:
1) Ilmenaaufer 47, 29525 Uelzen and Scharnhorststr. 1, 21335 Lüneburg; **W:**

www.zusa.de – **2)** Bahnhofstr. 11, 26122 Oldenburg; **W:** www.uni-oldenburg.de/ok_ol/ – **3)** Kieler Str. 31, 26382 Wilhelmshaven; **W:** www.radio-jade.de – **4)** VHS Emden, An der Berufsschule 3, 26721 Emden; **W:** www.radio-ostfriesland.net – **5)** Halle IV, Kaiserstr. 10a, 49809 Lingen; **W:** www.emsvechtewelle.de – **6)** Lohstr. 45a, 49074 Osnabrück; **W:** www.os-radio.de – **7)** Hildesheimer Str. 29, 30169 Hannover; **W:** www. leinehertz.de – **8)** Hefehof 23, 31785 Hameln; **W:** www.radio-aktiv.de – **9)** Groner Str. 2, 37073 Göttingen; **W:** www.stadtradio-goettingen.de – **10)** Rebenring 18, 38106 Braunschweig; **W:** www.okerwelle.de – **11)** Andreas-Passage 1, 31134 Hildesheim; **W:** www.tonkuhle.de – **12)** Zur Bettfedernfabrik 3, 30451 Hannover; **W:** www.radioflora.de On 100.0MHz during special events, otherwise via webstream only.
Permanent special stns: Radio SWS (**W:** www.radio-sws. de), Norderney 104.0MHz; **Radio S.A.S.** (**W:** www.radio-sas.de), Stadthagen 94.5MHz; **Lamberti-Kirchenfunk** (**W:** www.soerenkoenig. com/Radlam) Aurich 106.0MHz; **Kirchenfunk Esterwegen**, 106.6MHz; **Kirchenfunk Lorup**, 107.6MHz; **Kirchenfunk Herzlake**, 106.1MHz; **Pfarrfunk Breitenberg**, 98.4MHz.; **Kirchenfunk Meppen**, 95.0MHz.
MW: On 630kHz Voice of Russia.
DAB: Hannover, Braunschweig, Visselhövede, Steinkimmen and Aurich txs on 223.9MHz (ch. 12A) DLF (192k), DK (192k), DRadio Wissen (mono), NDR 1 Niedersachsen (128k), NDR Info (mono), NDR 2 Plus (128k), NDR Musik Plus (128k), NDR Traffic Channel (mono).
Hildesheim tx (0.5kW) on 227.4MHz (ch. 12C) rel. Bayern ensemble as engineering test.

S) NORDRHEIN-WESTFALEN

Media institution: Landesanstalt für Medien Nordrhein-Westfalen (LfM) ✉ Postfach 10 34 43, 40025 Düsseldorf (office location: Zollhof 2) ☎ +49 211 77 007 0 🖷 +49 211 727 170 **W:** www.lfm-nrw.de

Radio NRW, Essener Str. 55, 46047 Oberhausen; **W:** www.radionrw.de The following stns are affiliates with some hours of own prgr. per day, other times rel. Radio NRW with local ID's inserted automatically.

FM	MHz	kW	Site	Station
5)	87.7	0.2	Krefeld-Oppum	Welle Niederrhein
26)	88.1	4	Eggegebirge	R. Hochstift
19)	88.2	0.5	Lüdinghausen	R. Kiepenkerl
37)	88.2	0.5	Siegen	R. Siegen
35)	88.3	0.1	Meinerzhagen	R. MK
16)	88.4	1	Bocholt	Westmünsterlandw.
36)	89.1	0.2	Schmallenberg	R. Sauerland
7)	89.4	1	Düsseldorf Rheinturm	NE-WS 89.4
6)	90.1	0.3	Mönchengladbach	R. 90,1
31)	90.8	0.1	Herne	Herne 90acht
30)	91.2	0.2	Dortmund	R. 91.2
39)	91.2	0.2	Siegburg	R.Bonn/Rhein-Sieg
42)	91.4	0.1	Bergheim	R. Erft
35)	91.5	0.1	Altena	R. MK
33)	91.5	0.1	Hattingen-Schierken	R. en
3)	91.7	0.1	Moers-Meerbeck	R. K.W.
23)	91.7	0.1	Vlotho	R. Herford
4)	92.2	0.1	Duisburg	R. Duisburg
35)	92.5	0.3	Iserlohn	R. MK
20)	92.6	1	Sendenhorst	R. WAF
43)	92.7	0.5	Düren-Hürtgenwald	R. Rur
13)	92.9	0.6	Mülheim-Saarn	R. Mülheim
29)	92.9	0.1	Selm	antenne unna
16)	93.0	0.5	Ahaus	Westmünsterlandw.
26)	93.7	0.1	Paderborn	R. Hochstift
39)	94.2	0.1	Much-Wersch	R.Bonn/Rhein-Sieg
9)	94.3	0.2	Solingen	R. RSG
15)	94.6	0.1	Recklinghausen	Hit Radio Vest
20)	94.7	0.2	Warendorf	R. WAF
36)	94.8	0.1	Marsberg	R. Sauerland
23)	94.9	0.5	Herford	R. Herford
24)	95.1	0.1	Rahden	R. Westfalica
18)	95.4	0.2	Münster	Antenne Münster
15)	95.6	0.1	Berghaltern	Hit Radio Vest
24)	95.7	0.5	Minden Jakobsberg	R. Westfalica
20)	95.7	0.3	Beckum	R. WAF
14)	96.1	0.1	Gelsenkirchen	REL
36)	96.2	0.4	Olsberg-Antfeld	R. Sauerland
20)	96.3	0.2	Oelde	R. WAF
38)	96.9	0.5	Leverkusen-Oplalen	R. Berg
1)	97.2	0.1	Simmerath	(see note)
35)	97.2	0.1	Werdohl	R. MK
37)	97.3	0.1	Bad Laasphe	R. Siegen
29)	97.4	0.5	Lünen	Antenne Unna
11)	97.6	4	Langenberg	R. Neandertal
16)	97.6	1	Borken	Westmünsterlandw.
22)	97.6	0.4	Friedrichsdorf	R. Bielefeld
39)	97.8	0.5	Bonn Venusberg	R.Bonn/Rhein-Sieg
2)	98.0	1	Kleve	Antenne Niederrhein

FM	MHz	kW	Site	Station
22)	98.3	0.1	Bielefeld	R. Bielefeld
32)	98.5	0.5	Bochum	R. 98.5
14)	98.7	0.5	Bottrop	REL
37)	98.9	0.1	Neunkirchen	R. Siegen
35)	99.5	0.1	Plettenberg	R. MK
44)	99.7	0.1	Euskirchen	R. Euskirchen
38)	99.7	0.5	Gremberg	R. Berg
39)	99.9	0.5	Bonn-Königswinter	R.Bonn/Rhein-Sieg
1)	100.1	0.4	Aachen Karlshöhe	(see note)
35)	100.2	0.5	Lüdenscheid	R. MK
5)	100.6	1	Viersen	Welle Niederrhein
27)	100.9	1	Soest-Möhnesee	Hellweg R.
25)	101.0	0.5	Schieder-Schwalenbg.	R. Lippe
7)	102.1	0.3	Grevenbroich	NE-WS 89.4
12)	102.2	0.3	Essen-Werden	R. Essen
29)	102.3	1	Schwerte Sommerberg	Antenne Unna
5)	102.5	0.3	Viersen Süchtelner Höhe	Welle Niederrhein
16)	103.6	0.1	Gronau	Westmünsterlandw.
27)	103.6	0.1	Lippstadt	Hellweg R.
17)	104.0	1	Tecklenburg	R. RST
8)	104.2	1	Düsseldorf	Antenne Düsseldorf
33)	104.2	1	Witten-Stockum	R. en
26)	104.8	0.5	Neuhaus-Hasselberg	R. Hochstift
26)	104.8	0.1	Büren	R. Hochstift
36)	104.9	0.1	Meschede	R. Sauerland
1)	105.0	0.1	Monschau	(see note)
12)	105.0	1	Essen-Holsterhausen	R. Essen
28)	105.0	0.2	Hamm	R. Lippewelle
38)	105.2	4	Lindlar	R. Berg
17)	105.2	1	Schöppingen	R. RST
15)	105.2	0.1	Dorsten	Hit Radio Vest
37)	105.4	4	Aue-Kirchhundem	R. Siegen
38)	105.7	1	Waldbröl	R. Berg
3)	105.7	0.5	Geldern	Antenne Niederrhein
33)	105.7	0.1	Gevelsberg	R. en
42)	105.8	1	Köln-Ehrenfeld	R. Erft
16)	106.2	0.1	Oberhausen	R. Oberhausen
19)	106.3	0.2	Dülmen	R. Kiepenkerl
36)	106.5	0.5	Hallenberg	R. Sauerland
36)	106.5	0.3	Arnsberg	R. Sauerland
25)	106.6	1	Lemgo	R. Lippe
24)	106.6	0.1	Lübbecke	R. Westfalica
21)	106.6	4	Borgholzhausen	R. Gütersloh
45)	106.9	0.4	Schleiden (Eifel)	R. Euskirchen
41)	107.1	0.5	Köln Neumarkt	R. Köln
33)	107.2	0.1	Herdecke	R. en
27)	107.3	0.2	Wickede	Hellweg R.
19)	107.4	1	Coesfeld	R. Kiepenkerl
25)	107.4	1	Linderhofe-Dörenbeg	R. Lippe
10)	107.4	0.5	Wuppertal	R. Wuppertal
44)	107.4	0.1	Bad Münstereifel	R. Euskirchen
21)	107.5	1	Oelde	R. Gütersloh
43)	107.5	0.1	Linnich	R. Rur
36)	107.6	0.5	Sundern	R. Sauerland
3)	107.6	0.2	Wesel-Büderich	R. K.W.
40)	107.6	0.1	Leverkusen-Wiesdorf	R. Leverkusen
27)	107.7	0.2	Belecke-Sennhöfe	Hellweg R.
34)	107.7	0.2	Hagen	R. Hagen
1)	107.8	0.4	Aachen Stolberg	(see note)
39)	107.9	0.5	Herchen-Rosbach	R.Bonn/Rhein-Sieg
9)	107.9	0.1	Remscheid	R. RSG

Adresses and other information:
1) Radio Aachen closed down in 2010, Antenne AC to be incorporated into planned new stn. for Aachen area – 2) Stechbahn 2-8, 47533 Kleve, **W:** www.antenneniederrhein.de – 3) Rheinstr. 24-26, 47495 Rheinberg, **W:** www.radiokw.de – 4) Ruhrorter Str. 187, 47119 Duisburg, **W:** http://medien.freepage.de/guidojansen – 5) Uerdinger Str. 543, 47800 Krefeld, **W:** www.welleniederrhein.de – 6) Lüpertzander Str. 159, 41061 Mönchengladbach, **W:** www.radio901.de – 7) Moselstr. 16, 41464 Neuss, **W:** www.news894.de – 8) Kaistr. 7, 40221 Düsseldorf, **W:** www.antenneduesseldorf.de – 9) Postfach, 42621 Solingen (studio location: Alleestr. 1) **W:** www.radiorsg.de – 10) Friedrich-Engels-Allee 426, 42283 Wuppertal, **W:** www.radiowuppertal.de – 11) Elberfelder Str. 81, 40804 Mettmann, **W:** www.radioneandertal.de – 12) Sachsenstr. 36, 45128 Essen **W:** www.radio-essen.de – 13) Essener Str. 99, 46047 Oberhausen **W:** www.106.2.radiooberhausen.de and www.92.9.radiomuelheim.de – 14) Hochstr. 68, 45894 Gelsenkirchen, **W:** www.radio-emscher-lippe.de – 15) Schaumburgstr. 14, 45657 Recklinghausen **W:** www.hitradiovest.de – 16) Heinrich-Hertz-Str. 6, 46325 Borken **W:** www.radiomw.de –17) Postnstr. 3, 48431 Rheine, **W:** www.radiorst.de – 18) Nevinghoff 14/16, 48147 Münster, **W:** www.antennemuenster.de – 19) Tiberstr. 21, 48249 Dülmen, **W:** www.radio-kiepenkerl.de – 20) Am Schweinemarkt 3, 48231 Warendorf, **W:** www.radiowaf.de – 21) Feldstr. 14, 33330 Gütersloh **W:** www.radioguetersloh.de – 22) Niederstr. 21-27, 33602 Bielefeld **W:** www.radiobielefeld.de – 23) Berliner Str. 30, 32052 Herford **W:** www.radioherford.de – 24) Johanniskirchhof 2, 32423 Minden **W:** www.radiowestfalica.de – 25) Lagesche Str. 17, 32756 Detmold **W:** www.radiolippe.de – 26) Frankfurter Weg 22, 33106 Paderborn **W:** www.radio-hochstift.de – 27) Jakobistr. 46, 59494 Soest **W:** www.hellwegradio.de – 28) Königstr. 39, 59065 Hamm **W:** www.lippewelle.de – 30) Karl-Zahn-Str. 11, 44141 Dortmund **W:** www.radio912.de – 31) Bahnhofstr. 45, 44623 Herne **W:** www.radio-herne.de – 32) Westring 26, 44787 Bochum, **W:** www.ruhrwelle-bochum.de – 33) Mühlenstr. 55, 58285 Gevelsberg **W:** www.radio-en.de – 34) Rathausstr. 23, 58095 Hagen, **W:** www.radio-hagen.de – 35) Vinckestr. 9-13, 58636 Iserlohn, **W:** www.radio-mk.de – 36) Steinstr. 32, 59872 Meschede, **W:** www.radio-sauerland.de – 37) Postfach 10 02 42, 57002 Siegen (studio location: Obergraben 33), **W:** www.radio-siegen.de – 38) Friedrich-Ebert-Str., 51429 Bergisch Gladbach, **W:** www.radioberg.de – 39) Kennedybrücke 4, 53225 Bonn, **W:** www.radio-bonn.de – 40) Bismarckstr. 71, 51373 Leverkusen, **W:** www.radioleverkusen.de – 41) Stolberger Str. 374, 50933 Köln, **W:** www.radiokoeln.de – 42) Hürth Park, 50354 Hürth, **W:** www.radioerft.de – 44) August-Klotz-Str. 21, 52349 Düren, **W:** www.radiorur.de – 45) Rheinstr. 55, 53881 Euskirchen, **W:** www.radioeuskirchen.de

Other stations:

FM	MHz	kW	Site	Station
12)	87.9	0.05	Bielefeld	Hertz 87,9
13)	89.4	0.03	Paderborn	L'Unico
9)	90.0	0.2	Bochum	CT das radio
1)	92.0	0.05	Pulheim	Domradio
8)	93.0	0.05	Dortmund university	Eldoradio
3)	94.3	0.05	Bielefeld-Bethel	Antenne Bethel
14)	94.7	0.05	Meschede	Radio FH
7)	96.8	0.5	Bonn	(shared freq.)
6)	97.1	0.04	Düsseldorf-Bilk	Hochschulr. Düsseld.
2)	97.2	0.1	Simmerath	Antenne AC
4)	99.1	0.1	Aachen	Hochschulr. Aachen
5)	100.0	0.1	Köln Sternengasse	Kölncampus
1)	101.7	0.03	Köln Sternengasse	Domradio
11)	103.9	0.5	Steinfurt college	Radio Q
10)	104.5	0.2	Essen university	Campus FM
10)	105.6	0.05	Essen university	Campus FM
2)	107.8	0.4	Aachen-Stolberg	Antenne AC

Adresses and other information:
1) Domkloster 3, 50667 Köln; **W:** www.domradio.de; further txs on P) and T), also via Astra 1H (12.460GHz). Run by Catholic church – 2) Merzbrück 214, 52146 Würselen; **W:** www.antenne-ac.de Commercial – 3) Quellenhofweg 25, 33617 Bielefeld-Bethel; **W:** www.antenne-bethel.de Run by diacony – 4) Wüllnerstr. 5, 52056 Aachen; **W:** www.hochschulradio-aachen.de – 5) Albertus-Magnus-Platz, 50923 Köln; **W:** koelncampus.com – 6) Universitätsstr. 1, 40225 Düsseldorf; **W:** www.hochschulradio.uni-duesseldorf.de – 7) shared by six groups – 8) Vogelpothsweg 74, 44227 Dortmund; **W:** www.eldoradio.de – 9) 44780 Bochum (studio location: Ruhr university, room 04/452); **W:** www.radioct.de – 10) Universitätsstr. 2, 45141 Essen; **W:** www.campusfm.info – 11) Bismarckallee 3, 48151 Münster (lp. tx on 90.9MHz here); **W:** www.radioq.de – 12) Universitätsstr. 25, 33615 Bielefeld; **W:** www.radio-hertz.de – 13) Warburger Str. 100, 33098 Paderborn; **W:** www.l-unico.de – 14) Jahnstr. 23, 59872 Meschede; **W:** www.radiofh.de – 15) Radio Triquency, Liebigstr. 87, 32657 Lemgo; **W:** www.triquency.de Via lp. txs on 95.9/96.1/99.4MHz **N.B** Stn's 4)-15) university/college.

DAB: 25 txs on 229.1MHz (ch. 12D) DLF (mono), DK (128k), DRadio Wissen (mono), Eins Live (128k), WDR 2 (mono), Funkhaus Europa (128k), Eins Live diggi (128k), Kiraka (128k), Vera (mono), WDR Event (mono), Domradio (mono).

DMB: Köln tx (10kW) on 221.1MHz (ch. 11D) Eins Live.

T) RHEINLAND-PFALZ
Media institution: Landesanstalt für Medien und Kommunikation (LMK) ✉ Postfach 21 73 63, 67072 Ludwigshafen (office location: Turmstraße 8) ☎ +49 621 5252 0 🖷 +49 621 5252 152 **W:** www.lmk-online.de

FM	RPR 1	bigFM	Rockl.	Metrop	kW
Bad Bergzabern	103.3	–	–	–	0.3
Bad Dürkheim	98.1	96.4	–	–	0.1
Bad Kreuznach	89.7	104.8	–	–	0.1/0.2
Bad Marienberg	102.9	–	–	–	25
Bernkastel-Kues	–	100.5	–	–	0.1
Betzdorf	–	107.7	–	–	0.5
Bitburg	–	–	107.9	–	0.1
Bornberg-Eßweiler	103.1	107.6	–	–	25
Daun (Eifel)	102.1	106.6	–	–	20
Diezer Hain	101.2	100.4	–	–	0.1
Grünstadt/Mertesh.	103.3	–	–	–	0.1
Haardtkopf	100.1	–	–	–	50
Heckenbach	103.5	104.9	–	–	30

FM	RPR 1	bigFM	Rockl.	Metrop	kW
Hohe Wurzel	–	–	107.9	–	6
Idar-Oberstein	100.3	101.9	–	–	1
Kalmit	103.6	106.7	–	–	25
Kirchheimbolanden	–	–	97.1	–	0.2
Kleinkarlbach	91.1	–	–	–	0.1
Koblenz Kühkopf	101.5	104.0	–	–	40
Koblenz-Bendorf	–	–	88.3	107.8	0.3
Linz	–	–	96.9	–	0.2
Ludwigshafen	–	–	–	88.4	0.1
Mainz Ober-Olm	100.6	104.5	–	–	20
Mainz (city)	98.1	106.6	–	96.0	0.2/0.4
Mannheim	–	–	93.2	–	1
Pirmas. Kettrichhof	104.7	–	–	–	5
Pirmasens (town)	–	96.7	–	–	0.4
Rivenich	–	95.8	–	–	0.2
Saarburg	102.6	96.5	–	–	20
Trier Petrisberg	102.9	106.4	105.8	–	0.1/0.5
Zweibrücken	103.3	106.6	–	–	2/0.1

Addresses and other information:
RPR 1, Turmstr. 8, 67059 Ludwigshafen; **W**: www.rpr1.de – **bigFM**: see K), stn.2); rel. of adopted version in responsibility of RPR – **Rockland Radio**, Wallstr. 1-5, 55122 Mainz; **W**: www.rockland.de cooperates with Radio 21, see R) – **Metropol FM** see M), stn. 15. **N.B** Antenne West closed down in 2010.

Local stations:

FM	MHz	kW	Site	Station
3)	87.6	0.2	Idar-Oberstein	Radio Idar-Oberstein
7)	87.8	0.1	Welschbillig	Antenne Trier
9)	87.9	0.1	Bretzenheim (church)	Studio Nahe
2)	88.3	0.1	Bad Kreuznach	Antenne Bad Kreuznach
9)	88.3	0.1	Ludwigshafen	Metropol FM
7)	88.4	0.5	Trier Petrisberg	Antenne Trier
6)	88.4	0.3	Pirmasens	Radio Pirmasens
8)	94.1	0.3	Mommenheim	97eins
5)	94.2	1	Neustadt/Weinstr.	Antenne Pfalz
7)	94.7	0.1	Trierweiler	Antenne Trier
5)	94.8	0.1	Landau	Antenne Landau
4)	96.9	0.5	Kaiserslautern	Antenne Kaiserslautern
8)	97.1	0.1	Bodenheim	97eins
1)	98.0	1	Koblenz Moselw. Str.	Antenne 98.0
1)	98.0	1	Neuwied	Antenne 98.0
1)	98.0	1	Koblenz-Bendorf	Antenne 98.0

Addresses and other information:
1) Friedrich-Ebert-Ring 54, 56068 Koblenz; **W**: www.akoblenz.de – **2)** Kreuzstr. 31-33, 55543 Bad Kreuznach; **W**: www.antenne-kh.de – **3)** Auf der Idar 2a, 55743 Idar-Oberstein; **W**: www.radio-io.de – **4)** Am Altenhof 11-13, 67655 Kaiserslautern; **W**: www.antenne-kl.de – **5)** Europastr. 3, 67433 Neustadt/Wstr.; **W**: www.antenne-pfalz.de – **6)** Schloßstr. 44, 66953 Pirmasens; **W**: www.radio-pirmasens.de – **7)** run by stn. 4) – **8)** Am Kümmerling 21-25, 55294 Bodenheim; **W**: www.97eins.de – **9)** Obere Grabenstr. 29, 55450 Langenlonsheim; **W**: www.studio-nahe.de Run by Catholic church, mostly rel. Domradio, see S).
DAB: 9 txs on 223.9MHz (ch. 12A) DLF (mono), DK (128k), DRadio Wissen (mono), SWR1 Rheinland-Pfalz (160k), SWR2 (160k), SWR3 (160k), DasDing (128k), SWR cont.ra (mono).

U) SAARLAND
Media institution: Landesmedienanstalt Saar (LMS) ✉ Postfach 11 01 64, 66070 Saarbrücken (office location: Nell-Breuning-Allee 6) ☎ +49 681 389880; 🖷 +49 681 3898820; **W**: www.lmsaar.de

FM networks:

Location	Salü	C.Ro.	bigFM	Saar.	(F.PI.)	kW
Merzig	103.0	–	92.6	–	105.1	0.1/0.5
Mettlach	104.2	–	–	–	106.1	0.1
Neunkirchen	–	–	–	94.6	–	0.6
Oberperl	100.3	–	–	–	–	5
Saarbr. Schoksbg.	101.7	–	–	–	–	100
Saarbr. Halberg	–	–	94.2	–	–	1
Saarbr. Winterberg	–	92.9	–	–	–	1
Saarbr. Schwarzenbg.	–	–	–	99.6	–	0.1
St. Ingbert	–	100.6	–	–	–	1
Sulzbach	–	–	96.8	–	–	0.1
Webenheim	100.0	–	–	–	–	5

Addresses and other information:
Radio Salü, Classic Rock Radio: Postfach 10 08 44, 66008 Saarbrücken (studio location: Richard-Wagner-Str. 58-60); **W**: www.salue.de, www.classic-rock-radio.de Run by Groupe Lagardère, also operates Felsberg tx on 183kHz for Europe 1 prgr. from Paris (stn. details see under France) – **bigFM Saarland**: Gutenbergstr. 11-23, 66103 Saarbrücken; **W**: www.bigfm-saarland.de; mostly rel. Stuttgart prgr. (see K), stn. 2) – **Radio Saarbrücken**: Nell-Breuning-Allee 6, 66115 Saarbrücken; **W**: www.radio-sb.de **F.PI.:** Affiliated stn's on 94.6/105.1/106.1MHz. **N.B** Antenne West closed down in 2010.

DAB: 5 txs on 197.6MHz (ch. 8B) DLF (192k), DK (192k), DRadio Wissen (mono), SR 1 (160k), SR 2 (192k), SR 3 (192k), Unser Ding (160k), Antenne Saar (mono), R. Salü (128k).

V) SACHSEN
Media institution: Sächsische Landesanstalt für privaten Rundfunk und neue Medien (SLM) ✉ Postfach 10 16 62, 04016 Leipzig ☎ +49 341 22 59 0 🖷 +49 341 22 59 199; **W**: www.slm-online.de; office location: Ferdinand-Lassalle-Straße 21.

FM	PSR	R.SA	RTL	Radio	Energ.	kW
Annaberg-Buchholz	–	104.8	–	–	–	0.5
Auerbach	–	107.9	–	–	–	0.1
Bärenstein	–	–	107.2E	–	–	0.2
Beilrode	–	99.6	–	–	–	1
Borna	–	–	–	99.5L	–	0.1
Chemnitz-Reichenh.	–	91.0	–	102.1C	97.5	3
Collmberg	98.0	–	104.7	–	–	5/10
Döbeln	–	107.9	–	–	98.3	1/0.2
Dresden-Gompitz	–	–	–	91.1D	–	1
Dresden-Wachwitz	102.4	89.2	105.2	103.5D	100.2	100/2
Ebersbach	–	106.1	–	–	–	0.5
Elsterberg	–	99.7	–	–	–	0.2
Flöha	–	98.4	–	99.0C	–	0.1
Freiberg	–	90.6	–	104.2D	96.4	0.2/0.5
Freital	–	88.3	–	107.0D	–	0.2
Geyer (Erzgebirge)	100.0	–	105.4	–	–	100
Görlitz	–	105.1	–	–	–	1
Grimma	–	107.4	–	90.9L	93.3	2/0.3
Hoyerswerda-Zeißig	–	96.9	–	–	87.6	0.2/0.3
Leipzig-Holzhausen	–	–	–	91.3L	99.8	4
Leipzig-Reudnitz	–	98.2	–	–	–	1
Leisnig	–	100.5	–	–	–	0.2
Limbach-Oberfrohna	–	–	–	107.3C	–	0.1
Löbau Schafberg	101.0	–	105.6	107.6G	–	30
Löbau town	–	87.6	–	–	–	0.1
Markneukirchen	–	89.6	–	–	–	0.2
Meerane	–	–	–	89.2Z	–	0.1
Meißen-Korbitz	–	–	–	107.5D	–	0.2
Mittelherwigsdorf	–	100.0	–	94.3G	–	0.5/0.3
Mügeln	–	91.2	–	–	–	0.5
Neukirchen	–	–	–	95.8C	–	1
Niederschöna	–	94.4	–	–	–	0.5
Nossen	–	91.4	–	–	–	0.2
Oelsnitz (Vogtland)	–	91.5	–	–	–	0.1
Olbernhau	–	101.0	–	–	–	0.5
Oschatz	–	89.1	–	–	–	0.3
Pirna	–	–	–	96.4D	–	1
Plauen	–	93.5	–	–	–	1
Reichenbach/Vogtl.	–	92.4	–	–	–	0.2
Riesa	–	106.4	–	–	91.7	2/1
Rothenburg	–	100.0	–	–	–	0.5
Schöneck	92.0	–	106.0	–	–	10/30
Sohland	–	107.0	–	–	–	1
Stollberg	–	93.4	–	–	–	1
Torgau	–	91.1	–	–	–	0.2
Werdau	–	–	–	90.9Z	–	0.3
Wiederau (Leipzig)	102.9	–	106.9	–	–	100
Wilkau-Haßlau	–	92.3	–	103.4Z	–	0.5
Wilthen	–	106.5	–	–	104.9	1/0.5
Wurzen	–	95.0	–	–	–	0.4
Zschopau	–	–	–	91.7C	–	0.3
Zwickau-Ebersbrunn	–	–	–	96.2Z	98.2	0.5/0.3

Addresses and other information:
Radio PSR (AC), **R.SA** (oldie-based AC), **Energy Sachsen** (CHR): Thomasgasse 2, 04102 Leipzig; **W**: www.radiopsr.de www.rsa-sachsen.de www.nrj.de – **Hitradio RTL** (AC), **Radio Chemnitz / Dresden / Erzgebirge / Lausitz / Leipzig / Zwickau** (AC, on freq. marked C, D, E, G, L, Z, with some content from local studios): Ammonstr. 35, 01067 Dresden; **W**: www.bcs-sachsen.de

Other stations:

FM	MHz	kW	Site	Station
6)	88.2	0.4	Auerbach	Vogtland R.
9)	89.2	1	Weißwasser	R. WSW
1)	89.2	0.1	Leipzig-Reudnitz	R. Blau
1)	94.4	0.3	Leipzig-Stahmeln	R. Blau
9)	94.9	0.2	Wilthen	R. WSW
6)	95.4	2	Plauen	Vogtland R.
4)	97.6	4	Leipzig-Holzhausen	mephisto 97.6
2)	98.4	0.1	Dresden-Gompitz	coloRadio
1)	99.2	0.5	Leipzig-Connewitz	R. Blau
2)	99.3	0.1	Freital (Dresden)	coloRadio
5)	99.3	0.1	Mittweida	R. Mittweida
6)	100.5	1	Reichenbach/Vogtland	Vogtland R.
3)	102.7	1	Chemnitz-Reichenhain	Radio T
8)	102.8	0.5	Hoyerswerda-Zeißig	Elsterwelle
6)	103.8	0.5	Markneukirchen	Vogtland R.

FM	MHz	kW	Site	Station
7)	107.7	2	Fichtelberg	R. Erzgebirge

Addresses and other information:
Apollo Radio: Carolastr. 4-6, 09111 Chemnitz; **W**: www.apolloradio. de; jazz and classical music, Mon-Fri 2200-1700, Sat-Sun 2300-1100 via txs of stn's 1), 2), 3).

1) Paul-Gruner-Str. 62, 04107 Leipzig; **W**: www.radioblau.de – **2)** Jordanstr. 5, 01099 Dresden; **W**: www.coloradio.net – **3)** Karl-Liebknecht-Str. 19, 09111 Chemnitz; **W**: www.radiot.de; rel. 1700-1800 Chemnitz university prgr. – **4)** Ritterstr. 9-13, 04109 Leipzig; **W**: mephisto976.uni-leipzig.de, run by Leipzig university; Mon-Fri 0900-1100 and 1700-1900, other times rel. R.SA – **5)** Leisniger Str. 9, 09648 Mittweida; **W**: www.radio-mittweida. de; run by Mittweida college – **6)** Haselbrunner Str. 114, 08225 Plauen; **W**: www.vogtlandradio.de – **7)** Vierenstr. 11, 09484 Oberwiesenthal; **W**: www.radioerzgebirge-online.de – **8)** Walther-Rathenau-Str. 27, 02977 Hoyerswerda; **W**: www.elsterwelle.de, 103.8MHz tx see M) – **9)** Werner-Seelenbinder-Str. 54a, 02943 Weißwasser; **W**: www.radiowsw.de
MW: On 1431kHz Voice of Russia.
DAB: Bautzen tx on 178.4MHz (ch. 5C); Dresden, Freiberg, Chemnitz, Schöneck, Leipzig, Collmberg txs on 223.9MHz (ch. 12A) DLF (mono), DK (128k),DRadio Wissen (mono), MDR Klassik (192k), MDR Info (AAC 96k), MDR Sputnik (AAC 96k), 90elf (AAC 64k).

W) SACHSEN-ANHALT
Media institution: Medienanstalt Sachsen-Anhalt (MSA) ⌨ Reichardtstraße 9, 06114 Halle/Saale ☎ +49 345 52550 🖷 +49 345 5255 121 **W**: www.msa-online.de

FM	R Bro	RTL	SAW	Rock	kW
Bernburg	–	–	–	95.0	1
Blankenburg	99.9	–	95.7	–	0.3/0.1
Brocken	–	89.0	101.4	–	60/100
Dequede	101.0	–	95.6	–	60/1
Dessau-Mildensee	90.6	–	92.6	94.1	0.8/2/0.3
Eisleben	93.7	–	–	–	1
Fleetmark-Lüge	–	–	103.9	–	5
Halle Petersberg	93.5	–	103.3	–	5
Halle city	–	–	–	98.3	0.5
Hergisdorf-Wolferode	93.7	–	–	–	1
Köthen	–	–	–	97.1	1
Magdeburg-Buckau	–	–	–	98.7	0.2
Naumburg	98.8	–	95.1	99.6	10/0.5/1
Sangerhausen	107.1	–	99.4	–	0.1
Schneidlingen	–	–	–	107.2	2.5
Schönebeck	105.7	–	100.1	–	15/20
Stendal Tucholsky-Str.	–	–	100.5	–	1
Weißenfels	–	–	–	88.0	1
Wernigerode	105.4	–	90.8	–	0.5/1
Wiederau (Leipzig)	–	–	104.9	–	*90
Wittenberg-Gallun	102.3	–	98.4	–	4/5
Zeitz-Hainichen	99.1	–	–	–	0.5
Ziesar	–	–	102.8	–	2

*) Tx in Sachsen, sharply directional towards Sachsen-Anhalt.
Addresses and other information:
Radio Brocken (oldie-based AC), **89.0 RTL** (CHR): Große Ulrichstr. 60D, 06108 Halle; **W**: www.brocken.de, www.89.0rtl.de – **Radio SAW** (AC), **Rockland Sachsen-Anhalt:**: Hansapark 1, 39116 Magdeburg; **W**: www.radiosaw.de, www.rockland-digital.de
Non-commercial stations:

FM	MHz	kW	Site	Station
2)	92.5	1	Aschersleben	radio hbw
1)	95.9	0.6	Halle Petersberg	R. Corax

Adresses and other information:
1) Unterberg 11, 06108 Halle; **W**: www.radiocorax.de – **2)** Herrenbreite 9, 06449 Aschersleben; **W**: www.radio-hbw.de
DAB: 13 txs on 227.4MHz (ch. 12C) DLF (mono), DK (128k), DRadio Wissen (mono), MDR Klassik (160k), 89.0 RTL (128k), SAW (128k), Rockland Sachsen-Anhalt (128k).

X) THÜRINGEN
Media institution: Thüringer Landesmedienanstalt (TLM) ⌨ P.O.-Box 90 03 61 (office location: Steigerstraße 10), 99096 Erfurt ☎ +49 361 211770 🖷 +49 361 2117755 **W**: www.tlm.de

FM	Ant.T	LW	Top 40	Klass	kW
Altenburg	–	–	98.4	107.5	0.5
Apolda	–	–	–	99.5	0.2
Arnstadt	–	–	–	96.5	0.1
Bleßberg	102.7	106.7	–	–	60
Dingelstädt	103.9	–	–	–	5
Eisenach	–	–	93.5	90.9	0.2
Erfurt-Windischh.	100.2	99.7	–	–	3/0.5
Erfurt-Hochheim	–	–	88.6	–	0.5
Gera	98.3	105.8	95.3	104.5	0.2/1
Gotha	–	–	90.8	99.3	0.1/0.2

FM	Ant.T	LW	Top 40	Klass	kW
Heiligenstadt	–	88.7	–	–	0.1
Ilmenau	–	–	94.8	–	0.1
Inselsberg	102.2	104.2	–	–	100
Jena-Oßmaritz	90.9	106.1	–	–	1
Jena Kernberge	–	–	94.8	–	0.2
Keula	–	104.5	–	–	10
Kulpenberg	104.7	96.8	–	–	3
Lobenstein	93.2	98.5	–	107.4	1/2/0.2
Meiningen	–	–	99.5	90.6	0.2/0.1
Mühlhausen	–	–	93.8	102.9	0.2/0.5
Nordhausen	106.8	105.8	103.0	–	0.1
Pößneck	–	–	98.9	–	0.2
Remda Kalmberg	107.6	95.7	–	–	60/10
Ronneburg	102.5	94.9	–	–	30/3
Saalfeld	–	–	97.6	–	0.1
Schleiz	–	–	–	92.4	0.2
Sömmerda	–	91.0	–	–	0.1
Sondershausen	–	90.7	–	–	0.2
Sonneberg	–	–	88.8	–	0.1
Suhl	101.3	88.6	92.1	–	2x1/0.1
Weimar Ettersberg	107.2	89.2	–	–	0.25
Weimar Belvedere	–	–	97.9	88.7	0.1

Addresses and other information:
Antenne Thüringen (AC), **Top 40** (rock): Belvederer Allee 25, 99425 Weimar; **W**: www.antennethueringen.de, www.radiotop40.de; Top 40 also via Astra 1H, 12.633GHz – **LandesWelle Thüringen** (AC): Mehringstr. 5, 99086 Erfurt; **W**: www.landeswelle.de – **Klassik Radio** see O)
Non-commercial and other stations:

FM	MHz	kW	Site	Station
1)	96.2	0.6	Erfurt-Hochheim	Funkwerk, F.R.E.I.
4)	96.5	2	Eisenach	Wartburg-R.
3)	98.1	0.1	Ilmenau	hsf Studentenradio
6)	100.4	0.1	Nordhausen	Offener Kanal Nordh.
7)	101.4	0.1	Saalfeld	SRB
5)	103.4	0.3	Jena-Oßmaritz	Radio OKJ
2)	106.6	2	Weimar Belvedere	Funkwerk, Lotte, b11

Addresses and other information:
1) Funkwerk, Juri-Gagarin-Ring 96, 99084 Erfurt; **W**: www.funkwerk. de Mon-Fri 1200-2000, Fri 2300- Sat 2300. **F.R.E.I.**, Gotthardstr. 21, 99084 Erfurt; **W**: www.radio-frei.de Mon-Thu 0600-1200 and 2000-2400, Fri 0600-1200 and 2000-2300, Sat 2300- Sun 2400 – **2) Radio Lotte**, Herderplatz 14, 99423 Weimar; **W**: www.radiolotte.de Mon 0600-1200 and 2300-2400, Tue-Thu 0600-1200 and 2000-2400, Fri 0600-1200 and 2000-2300, Sat 2300- Sun 2400. **studio b11**, Bauhaus-Universität, Bauhausstr. 11, 99421 Weimar; **W**: www.radiostudio.org. Mon 1900-2300 only. Also rel. Funkwerk from Erfurt – **3)** Postfach 100 565, 98684 Ilmenau; **W**: www.hsf.tu-ilmenau.de – **4)** Georgenstr. 43, 99817 Eisenach; **W**: www.wartburgradio.com – **5)** Helmboldstr. 1, 07749 Jena; **W**: www.radio-okj.de – **6)** August-Bebel-Platz 6, 99734 Nordhausen; **W**: www.ok-nordhausen.de – **7)** Tiefer Weg 7, 07318 Saalfeld; **W**: srb.fm. **N.B** Some stns rel. BBC World Service at times.
MW: On 1323kHz Voice of Russia.
DAB: 7 txs on 225.6MHz (ch. 12B) DLF (mono), DK (128k), Dradio Wissen (mono), MDR Klassik (192k), MDR Info (AAC 96k), MDR Sputnik (AAC 96k).

III. ARMED FORCES STATIONS

FM	MHz	kW	Site	Station
Baden-Württemberg				
2)	102.3	100	Stuttgart Frauenkopf	AFN Heidelberg
2)	104.6	0.4	Heidelberg-Wieblingen	AFN Heidelberg
2)	107.3	0.1	Mannheim-Käfertal	AFN Heidelberg
Bayern				
3)	87.7	0.1	Schweinfurt	AFN Bavaria
1)	89.4	0.2	Hohenfels	AFN Power Network
1)	90.0	0.2	Amberg	AFN Power Network
3)	90.3	0.1	Garmisch-Partenk.	AFN Bavaria
3)	98.5	0.1	Grafenwöhr	AFN Bavaria
3)	98.9	0.1	Bamberg	AFN Bavaria
1)	101.4	0.2	Grafenwöhr	AFN Power Network
3)	104.9	0.4	Illesheim	AFN Bavaria
3)	107.3	1	Ansbach-Katterbach	AFN Bavaria
1)	107.6	0.2	Vilseck	AFN Power Network
Hessen				
4)	98.7	50	Feldberg (Taunus)	AFN Wiesbaden
Niedersachsen				
7)	93.0	40	Braunschweig	BFBS Germany
7)	95.2	0.1	Fallingbostel	BFBS Radio 2
7)	95.4	0.2	Celle	BFBS Germany
7)	97.6	30	Visselhövede	BFBS Germany
7)	99.3	0.1	Hameln	BFBS Germany
7)	104.7	0.2	Bergen-Belsen	BFBS Radio 2

FM	MHz	kW	Site	Station
7)	106.8	0.1	Hameln	BFBS Radio 2

Nordrhein-Westfalen

7)	91.3	0.1	Rheindahlen	BFBS Germany
7)	91.7	0.3	Gütersloh	BFBS Radio 2
7)	92.5	0.8	Dülmen	BFBS Germany
7)	101.6	0.3	Bielefeld	BFBS Radio 2
7)	101.9	7	Wulfen	BFBS Germany
7)	102.2	0.3	Münster	BFBS Radio 2
7)	103.0	70	Bielefeld Hünenburg	BFBS Germany
7)	104.0	2.4	Niederkrüchten	BFBS Germany
7)	104.3	0.3	Rheindahlen	BFBS Radio 2
7)	105.0	0.3	Paderborn	BFBS Radio 2
7)	105.1	0.5	Rheinberg	BFBS Germany
7)	106.0	3	Dortmund	BFBS Germany
1)	107.6	0.3	Bonn (US consulate)	AFN Power Network

Rheinland-Pfalz

5)	100.2	1	Kaiserslautern	AFN Kaiserslautern
8)	101.9	0.1	Ramstein	CFN/RFC
5)	103.0	0.5	Pirmasens	AFN Kaiserslautern
6)	105.1	1	Spangdahlem	AFN Eifel
5)	106.1	0.1	Baumholder	AFN Kaiserslautern

Schleswig-Holstein

7)	88.4	0.1	Kiel-Holtenau	BFBS Germany

N.B Geilenkirchen airbase served by AFN, BFBS and CFN txs at Brunssum, see under Netherlands.

Addresses and other information:
1) AFN Europe, Coleman Barracks, 68307 Mannheim-Sandhofen; ▤ +49 (621) 46085 335; **W:** www.afneurope.net. Produces AFN Power Network (talk format, includes rel. of NPR and commercial US stns) and The Eagle (AC format) for distribution by affiliates 2)...5), AFN Benelux (see under Belgium, also rel. by Mönchengladbach tx on 1143kHz) and AFN South stn's (see under Italy) as well as non-affiliated stn. 6) – **2) AFN Heidelberg**, studio location as stn. 1); **W:** heidelberg.afneurope. net Own prgr. Mon-Fri 0400-0800 and 1400-1700. **N.B** To be moved and accordingly renamed AFN Stuttgart until 2013, Mannheim and Heidelberg txs to be closed down. AFN Europe expected to move to Wiesbaden – **3) AFN Bavaria**, Rose Barracks, 92249 Vilseck; **W:** bavaria.afneurope.net Own prgr. Mon-Fri 0500-0800 and 1400-1700 – **4) AFN Wiesbaden**, Würgelstr. 1217, Flugplatz Erbenheim, 65205 Wiesbaden; **W:** wiesbaden. afneurope.net. Own prgr. Mon-Fri 0500-0900 and 1300-1700 – **5) AFN Kaiserslautern**, Vogelweh, Bldg. 2058, 67661 Kaiserslautern; **W:** kaiserslautern.afneurope.net Own prgr. Mon-Fri 0500-1700, Sat 0700-1100. 103.0MHz rel. Sat 1100-Mon 0500 Power Network instead – **6) AFN Eifel**, Spangdahlem Air Base, 54529 Spangdahlem; **W:** www.spangdahlem.af.mil/afn.asp Own prgr. Mon-Fri 0500-0900 and 1300-1600, Sat 0800-1100 – **7) BFBS Germany**, Bergen-Hohne Garrison, Lager Hohne, 29303 Lohheide; **W:** bfbs-radio.com **N.B** New use of 96.5/97.6MHz see A), S). Further tx closures to be expected, 93.0MHz could close at yearend 2010 – **8)** Rel. CFN/RFC Brunssum, see under Netherlands.

Radio Andernach (German forces broadcasting sce.): Bundeswehr, Zentrum Operative Information, Kürrenberger Steig 34, 56727 Mayen; **W:** www.radio-andernach.de At present on air in Bosnia (Rajlovac 97.7MHz), Serbia (Suva Reka 89.9MHz, Prizren 106.9MHz) and Afghanistan (Kabul 107.5MHz), carrying satellite feeds from Mayen and local shows.

GHANA

L:T: UTC — **Pop:** 23 million — **Pr.L:** English, Akan, Dagbani, Ga, Ewe, Hausa, Nzema, others — **E.C:** 50Hz, 230V — **ITU:** GHA

NATIONAL COMMUNICATIONS AUTHORITY (NCA)
▤ P.O. Box CT1568, 1st Rangoon Close, Cantonments, Accra ☎+233 21 776621 ▤ +233 21 763449 **W:** www.nca.org.gh **E:** nca@nca.org.gh
L:P: Acting DG: Major J. R. K. Tandoh.

GHANA BROADCASTING CORPORATION (GBC, Pub.)
▤ P.O. Box 1633, Broadcasting House, Ring Road Central, Kanda, Accra ☎+233 21 786567 ▤ +233 21 768975
W: www.gbcghana.com **E:** info@gbcghana.com **L:P:** DG: William Ampem-Darko. Dir. Radio: Theo Agbam. Dir. Eng: Mrs. Sarah Boye.
SW: F.PI: new 50kW transmitter.
FM: D.Prgr in **English & ethnic:** 0525-2400. **French:** MF 1330-1500. **Network N. in E** (rel. by all GBC stations)**:** 0600, 0700, 0900, 1100SS, 1300, 1400, 1800, 2000, 2200, 2345.
Ann: "This is Radio Ghana in Accra". **IS:** Sign on with drum beat.
GBC Regional & partnership stations:

FM	MHz	Name	Web/Addr./Area
Bolgatanga	89.8	URA R.	Upper East
Han	90.1	Upper West R.	Upper West
Tamale	91.2	R.Savanna	North
Ho	91.5	Volta Star R.	Volta

FM	MHz	Name	Web/Addr./Area
Kumasi	92.1	Garden City R.	Ashanti
Cape Coast	92.5	R.Central	Central
Sunyani	93.5	R. Bar	Brong Ahafo
Accra	93.7	R. Ada	P.O. Box 9482, K.I.A
Wa	93.9	Upper West R.	Upper WEst
Sekondi-Takoradi	94.7	Twin City R.	West
Dormaa-Ahenkro	94.9	R. Dormaa	Brong Ahafo
Accra	96.5	Obonu FM	Greater Accra
Apam	96.5	Apam R.	Central
Accra	97.5	Unique FM	Greater Accra
Swedru	98.6	Swedru R.	Central
Kumasi	99.5	Luv FM	P.O. Box 17207, Accra
Accra	99.7	Joy FM	www.myjoyonline.com
Koforidua	106.1	Sunrise FM	East

Other FM stations in Accra:
Asempa FM, P.O. Box 17013, Accra-North: 94.7MHz – **Atlantis R,** P.O. Box 14629, Accra: 87.9MHz 5kW – **Channel R,** P.O. Box AN 8135, Accra-North: 92.7MHz – **Choice FM,** Accra: 102.3MHz. **W:** www.choice-fmghana.com – **Citi FM,** P.O. Box 30211, K.I.A, Accra: 97.3MHz – **Happy FM,** P.O. Box 1538, Dansoman, Accra: 98.9MHz – **Hot FM,** P.O. Box KD594, Kanda, Accra: 93.9MHz – **Peace FM,** Accra: 104.3MHz 5kW. **W:** www.peacefmonline.com – **R. Gold FM,** P.O. Box 17298, Accra: 90.5MHz – **R. Hit,** P.O. Box 17013, Accra-North: 103.7MHz – **R. Universe,** P.O. Box 25, Legon: 105.7MHz – **Sunny FM,** Box CT 3850, Cantonments, Accra: 88.7MHz – **Top R,** P.O. Box CT 4748, Cantonments, Accra: 103.1MHz – **Vibe FM,** Priv. Mailbag CT 183, Accra 91.9MHz.
+ 75 more stations elsewhere.
BBC African Service: Accra 101.3MHz.
RFI Afrique: Accra 89.5MHz, Kumasi 92.9MHz in French/English.
VOA Africa: Accra 98.1MHz

GIBRALTAR (UK)

L:T: UTC +1h (27 Mar-30 Oct: +2h) — **Pop:** 28,000 — **Pr.L:** English, Spanish — **E.C:** 50Hz, 240V — **ITU:** GIB

GIBRALTAR BROADCASTING CORP.
▤ Broadcasting House, 18 South Barrack Rd, Gibraltar ☎ +350 20079760 ▤ +350 20076432 **W:** www.gbc.gi **E:**info@gbc.gi
L:P: GM: Vacant, Head of Radio: Gerard Teuma, Head of Engineering: John Tewkesbury.
MW: 1458kHz 2kW
FM: 91.3MHz 0.2kW, 92.6MHz 1.0kW, 100.5MHz 0.2kW
D. Prgr: 24h. Spanish 1300-1500. 1700/1900-0600 Latest hits, classic songs. **Ann:** "Radio Gibraltar"

BRITISH FORCES BROADC. SCE. GIBRALTAR
▤BFBS Gibraltar, BFPO 52 ☎ +350 20055389 ▤ +350 20055528
Managing editor: Mario Chrisostomou
W: http://bfbs-radio.com **E:** gib@bfbs.com
FM: BFBS Radio 1: North Mole 93.5MHz; O'Hara's Battery 97.8MHz 1kW
BFBS Radio 2: North Mole 89.4MHz; O'Hara's Battery 99.5MHz 0.25kW
D. Prgr: 24h **Ann:** "BFBS Community Radio on the Rock, BFBS 1/2 FM"

GREECE

L:T: UTC +2h (27 Mar-30 Oct: +3h) — **Pop:** 11 million — **Pr.L:** Greek — **E.C:** 50Hz, 220V — **ITU:** GRC

ETHNIKO SIMVOULIO RADIOTILEORASIS (ESR, National Council for Radio & Television)
▤ Panepistimiou & Amerikis 5, 10564 Athens ☎+30 210 3354500 ▤ 210 3319881 **W:** www.esr.gr **E:** ncrtv@otenet.gr
L:P: President: Ioannis Laskaridis.

ELLINIKI RADIOFONIA (ERA, Greek Public Radio)
▤ Leof. Mesogeion 432, 15342 Agia Paraskevi, Athens ☎210 6066000. ▤ 210 7292826 **W:** tvradio.ert.gr/radio **E:** ntheleriti@ert.gr
L:P: DG: Antonis Andrikakis. Dir. Network Op: Mihalis Tzouvelekis. Dir. Int. Rel: Evi Demiri. Head of Eng. Dept.: Kostantinidis Tsiakalos.

MW	kHz	kW	Prgr.		MW	kHz	kW	Prgr.
	Athens	666	100	F	7)	Thessaloniki1	1044	150	M1
	Athens 1	729	100	1	8)	Orestias	1080	10	R
2)	Ioannina	765	10	R	7)	Thessaloniki2	1179	50	M2,4
5)	Thessaloniki	792	100	4	9)	Florina	1278	10	R
3)	Zakynthos	927	50	1,2,4	10)	Tripolis	1314	10	R
4)	Larisa	945	5	R	11)	Komotini	1404	100	R
	Heraklion	954	10*	R	12)	Patras	1485	1	R
	Athens 4	981	200	4	13)	Volos	1485	1	R
6)	Kerkyra	1008	100	R	14)	Rhodes	1494	100	R
					15)				

MW	kHz	kW	Prgr.	MW	kHz	kW	Prgr.
16) Chania	1512	100	R	19) Kavala	1601	1	R
18) Serres	1584	1	R	20) Kozani	1602	1	R

Prgrs: 1=ERA NET, 4=ERA Spor, 5=Voice of Greece, M1=Macedonia, M2=Macedonia 2, R=Regional prgrs & rel. ERA 1,2 & 4. F= Filia Radio (mainly foreign language programmes). * - Inactive.

FM(MHz)	NET	ERA2	ERA3	ERASp	Reg.	kW
17) Agios Ioannis					96.4	1
5) Ahentrias	94.4	96.4			103.6	3/10
3) Ainos	96.9	98.9	104.2	106.8	93.2	10
2) Akarnanika	88.9	91.3	102.5	97.3	100.3	10/5
Amfissa		107.0				3
9) Assea	88.3	103.5	90.3	101.5	95.3	10/3
Borsa	106.6	90.5				3
1) Bournias		104.8		106.8	89.7	10/2
Didima	101.2	99.4		103.2		2
4) Dovroutsi					98.3	2
12) Erateini		96.5		94.5	89.9	10/2
10) Frangopidima					102.4	10
Geraneia	97.9	99.9		105.0		3
Hamezi	89.9	89.0				1
Hlomo		101.5		107.4		2
Hortiatis	88.0	90.0	92.0	93.9		5
1) Ikaria					89.1	1
Imittos	105.8	103.7	90.9	101.8		100
12) Kalavrita					93.9	10
19) Kastania		103.6	88.2	105.6	100.2	10
17) Kefalohori					101.5	3
Lefkes		98.9		102.7		10
2) Ligiades	97.8	99.8	103.3	106.1	88.2	10/3
Lihada	104.2	88.8				3
Makrovouni		99.4				10
15) Malaxa					100.6	1/3
2) Manoliassas					102.1	10
1) Merovigli				102.1	100.1	1
1) Olympos	92.3	94.3	106.4		104.4	3/10
18) Paggaio	89.2	91.2	97.5	107.3	96.3	2
12) Panahaiko		102.3	104.3	87.9	92.5	3/10
6) Pantokratoras	91.8	93.8	89.8	101.1	99.3	10/3
Parnitha	91.6	102.9	95.6	100.9		100
16) Petalidi	92.2	94.2	89.3	100.4	105.4	3
13) Pilio	92.8	94.8	96.8	107.1	101.2	10
1) Pithion	98.9	93.8	88.1	89.4	101.0	2
1) Platanos		87.7			91.7	1
1) Plaka		90.6			98.1	4/2
14) Prof. Ilias(R)	88.4	90.4	103.4	101.4	92.7	10/3
14) Prof. Ilias(T)					93.3	10
Reihea		91.0		93.0		3
14) Rhodes town					93.1	1
1) Rogdia	104.8	99.2	91.3	93.9	97.5	3
15) Skloka	92.9	94.9	106.0	90.1	104.0	3/10
3) Skopos					95.2	2
1) Smerna					103.7	10
13) Soros					100.7	2
5) Stavros					105.3	1
14) Sympetro					98.4	1
1) Thanos					96.5	1
18) Thasos Isl.		95.1		104.7	106.7	10
1) Tholo Potami					95.2	2
1) Vathi					89.7	1
8) Vitsi	88.6	90.6	103.1	106.1	96.6	3/10
Xanthi (Pilima)		101.4		101.1		10

+14 stations under 1 kW.

Other ERT Stations:
Athens: Kosmos Radio. Parnitha 93.6, Imittos 107.0 100kW W: www.kosmos936.gr.
Athens: Filia Radio 666kHz & 106.7 MHz (Parnitha), rel. VO Greece Mon-Thu 0500-0600, Fri 0500-0600 and 1700-2200, SS 2000-2300, rel. Kosmos MF 2200-0500, Sat 2300-0700, Sun 1800-0500).
D.Prgr: All 24h. **ERA (1) (NET):** News, talk, current affairs. **N:** every hr. 0400-2300 except 0700 & 1600, common prgr. with ERA 2 0300-0400. **ERA 2:** Mainly music. **N:** every h, common prgr. with NET 0300-0400. **ERA 3:** Classical music, arts & drama. **ERA (4) Spor: N:** every half hour, Sport N: every hour, common prgr. with ERA 2, 0100-0400. **Regional programmes:** typically MF 0500-1200, 1400-1700 + sometimes SS. At other times rel. NET, ERA 2, ERA 3.

Regional station addresses:
1) Northern Aegean: E. Bostani 69, GR-81100 Mitilini **2) Ioannina:** N. Papadoupulou 2, GR-45444 Ioannina **3) Zakynthos:** Ampelokipoi, GR-201 00 Zakynthos **4) Larissa:** Iroon Politehniou 1, 1h Stratia, GR-412 22 Larissa **5) Heraklion:** Maxis Kritis 161, GR-71303 Iraklio **6)**

Kerkira: Ethniki Lefkimis, GR-49100 Kerkira **7) Orestiada:** Euripidou 15, GR-68200 Orestiada **8) Florina:** Megarovou 20, GR-53100 Florina **9) Tripoli:** Erithrou Staurou 1, 221 00 Tripoli **10) Pirgos:** Olympion 70, GR-27100 Pirgos **11) Komotini:** P.O. Box 5, Kosmiou Terma, GR-69100 Komotini **12) Patra:** Riga Feraiou 104, GR-26221 Patra **13) Volos:** Pl. Agiou Konstantinou, GR-32222 Volos **14) Southern Aegean:** 30 km. Leof. Kallitheas, GR-85100 Rhodes **15) Chania:** Ellis 40, GR-73200 Chania **16) Kalamata:** Anataliko Kentro 10-11, GR-24100 Kalamata **17) Serres:** P.O. Box 91, Stratopedou Kolokotroni, GR-62100 Serres **18) Kavala:** Sof. Venizelou & Iokastis, Ag. Paraskevi,, GR-65100 Kavala **19) Kozani:** I. Tranta 19, GR-50100 Kozani. **W:** tvradio.ert.gr/radio/localprofil.asp?id=22
IDs: The opening notes of the Greek folk song "Tsopanakos Imouna" (Once I Was A Shepherd Boy) played on flute and sheep bells.

EXTERNAL SERVICES: The Voice of Greece (ERA 5th Prgr.): see International Radio section.

RADIOFONIKOS STATHMOS MAKEDONIAS (Gov.)
✉ Aggelaki 14, 546 36 Thessaloniki ☎2310 299451 📠 2310 299451 **W:** www.ert3.gr **E:** pr@ert3.gr **LP:** Dir.: Klearhos Tsaousidis, Head of Int. Rel.: Lefty Kongalides, Tech. Dir.: Papagiannis Vouras.
Makedonia 1: MW: 1044kHz, **FM:** Hortiatis 102.0MHz 10kW, 24h., Agios Ioannis (Serres) 89.6MHz 2kW, Metaxas (Kozani) 89.1MHz 2kW, Poligiros 102.0MHz, Thasos 100.8MHz 10kW Vitsi (Kastoria) 100.6MHz 10kW.
Makedonia 2: MW: 1179kHz, **FM:** Hortiatis 95.8MHz 10kW, Poligiros 95.8MHz, 24h.
Relays on shortwave: see International Radio section.
ANN: "Elliniki Radiophonia, Radiofonikos Stathmos Makedonias"

RADIOFONIKOS STATHMOS AMALIADAS (Priv.)
✉ Ag. Trifonos 5, 27200 Amaliada. **W:** www.rsafm.gr **D.Prgr:** 24h. **MW:** Amaliada 1584kHz 1kW. **FM:** 92.7MHz 2kW.

1431 AM (Educ.)
✉ Aristotle University of Thessaloniki. 1os Orofos Ptergas THMMHY, Politehniki Sholi, 54124 Thessaloniki. **W:** www.1431am.gr **D.Prgr:** 24h. **MW:** Amaliada 1431kHz 350W.

PRIVATE FM STATIONS in Athens and Thessaloniki
Athens

FM	MHz	Station	kW	FM	MHz	Station	kW
1)	87.5	Kriti FM	2	31)	97.5	Love Radio	10
2)	87.7	En Lefko	10	32)	97.8	Real FM	10
3)	88.0	Oasis 88	10	33)	98.0	Liquid FM	10
4)	88.3	VFM 88,3	17	34)	98.3	Athena 9,84	10
5)	88.6	Fresh 88,6	10	35)	98.6	Derti 98,6	10
6)	88.9	Freedom	14	36)	98.9	Radio 9	10
7)	89.2	Music 89,2	10	37)	99.2	Melodia FM	10
8)	89.5	Ekklesia Ell.	19	38)	99.5	Vima FM	10
9)	89.8	Dromos 89,8	10	39)	99.8	99,8 FM	10
10)	90.1	902 Aristera	10	40)	100.3	Skai 100,3	12
11)	90.4	Kanali 1	10	41)	101.3	Diesi 101,3	10
12)	90.6	Radio Asty	5	42)	101.6	FM1	10
13)	91.2	Peiraiki Ekkl.	10	43)	102.2	Sfera 102,2	10
14)	91.4	Kritiki Radiof.	5	44)	102.5	Nitro Radio	10
15)	92.0	Galaxy 92	10	45)	102.8	Top FM	2
16)	92.3	Lampsi 92,3	10	46)	103.2	R. Blackman	5
17)	92.6	Best 92,6	10	47)	103.3	Sentra FM	10
18)	92.9	Kiss FM	17	48)	104.0	Parea FM	10
19)	93.2	Happy 93,2	10	49)	104.2	Minore FM	5
20)	94.0	Epikinonia FM	5	50)	104.4	Athens Int'l R.	10
21)	94.3	Xenios FM	5	51)	104.6	Hot 104,6	10
22)	94.6	NovaSpor FM	10	52)	104.8	Styl FM	5
23)	94.9	Rythmos 949	10	53)	105.2	Atlantis FM	10
24)	95.2	Athens Deejay	10	54)	105.5	Sto Kokkino	10
25)	95.8	Omega FM	5	55)	106.2	Mad Radio	10
26)	96.0	Flash 96	10	56)	107.0	Future Radio	10
27)	96.3	Red 96,3	10	57)	107.4	New Radio	10
28)	96.6	Difono FM	10	58)	107.6	Aktina FM	10
29)	96.9	Rock FM	10	59)	107.8	Star FM	5
30)	97.2	Antenna 97,2	10	60)	108.0	Ihorama FM	10

Thessaloniki

FM	MHz	Station	kW	FM	MHz	Station	kW
61)	87.6	Laikos FM	5	69)	91.4	Ola FM	15
62)	88.5	88miso	10*	70)	91.7	RSO 91,7	5
63)	89.0	89 Rainbow	15	71)	92.4	Radio Ekfrasi	5
64)	89.4	Thes. Deejay	10	72)	92.8	Aris FM	15
65)	89.7	Imagine 89,7	3	73)	93.1	Heart FM	5
66)	90.4	904 Aristera	30	74)	93.4	Mythos FM	3
67)	90.8	Zoo Radio	5	75)	93.7	Radio Gnomi	2
68)	91.1	Radio 91.1	5	76)	94.2	Radio Lydia	18

FMMHz	Station	kW	FM	MHz	Station	kW
77) 94.5	R. Thessaloniki	5	96)	101.3	POPS 101,3	5
78) 94.8	Eroticos FM	5	97)	101.7	Kalamaria FM	25
79) 95.1	Cosmoradio	20	98)	102.3	Radio Akrites	10
80) 95.5	Metropolis FM	5	99)	102.6	Plus Radio	10
81) 96.1	S'Agapo 96,1	10	100)	103.0	Sport 103	5
82) 96.5	Palmos 96,5	5	101)	103.6	Paradosiakos	5
83) 96.8	Velvet 96,8	5	102)	104.0	Rythmos 104	20
84) 97.1	Star FM	30	103)	104.4	Radiokymata	5
85) 97.5	Antenna 97,5	25	104)	104.7	Rock Radio	10
86) 98.0	Oasis 98	20	105)	104.9	R. D.E.Th.	15
87) 98.4	Panorama 9,84	20	106)	105.5	1055 Rock	5
88) 98.7	Athlitika Nea	5	107)	105.8	Mou. Galaxias	2
89) 99.0	Radio 1	20	108)	106.1	City Int'l	2
90) 99.4	Flash 99,4	15	109)	106.5	Foni tis Toumpas	5
91) 99.7	Radio Ekrixi	5	110)	106.8	Iera Mt. Langada	2
92)100.0	FM 100	20	111)	107.1	Real FM	3
93)100.3	Republic 100,3	15	112)	107.4	Libero 107,4	5
94)100.6	FM 100,5	20	113)	107.7	Peiratikos FM	5
95)101.0	FM 101	20	* - Inactive			

Patras FM: Greca FM 87.6 – Relax by MTV 88.2 – Apostolic Church 88.5 – Melody FM 88.8 – Politia FM 89.1 – R. Ena 89.4 – Oasis FM 89.7 – Omega R. 90.0 – Mythos FM 90.6 – Star FM 90.9 – Yes R. 91.2 – Radio 91,5 91.5 – Heaps Radio 91.7 – Kiss FM 92.2 – Top FM 93.0 – Max FM 93.4 – R. Gamma 94.0 – Alpha Patras 94.4 – Studio Patras 94.9 – Oxygen 95.3 – Ionion FM 95.7 – Spor FM 96.3 – Sfera 96.6 – Wave R. 97.4 – Light FM 97.6 – R. Messatida 98.0 – R. Blackman 98.5 – Flash Patras 98.7 – R. Aigio 99.2 – Fasma FM 99.7 – You FM 100.1 – Melodia FM 100.4 – Smart FM 100.7 – Studio 20 FM 101.1 – Hroma FM 102.1 – M FM 102.7 – 902 Aristera sta FM 103.0 – Lampsi FM 103.3 – Dytikos FM 103.7 – Palmos FM 104.1 – Klik FM 104.8 – Peiraiki Ekkl. 105.0 – Antenna Patras 105.3 –Beautiful R. 105.7 – Galaxy FM 106.1 – R. Patra 106.5 – Enjoy FM 107.0 – Hristianismos FM 107.2 – Mojo R. 107.7. Powers 1–5kW.

+ approx 1100 additional private stns nationwide.

NB: no official information available about powers of Athens stations and Thessaloniki powers are mostly based on estimates.

Addresses & other information:
1) Peloponissou 42, 18121 Koridallos Athens. **W:** www.875.gr – 2) Perikleos 49, 15451 Neo Psihiko Athens. **W:** www.877.gr – 3) Leof. Siggrou 174, 17671 Kallithea Athens. **W:** www.oasis88.gr – 4) Perikleous 49, 15154 Neo Psihiko Athens. **W:** www.vfm883.gr – 5) Leof. Siggrou 174, 17671 Kallithea Athens. **W:** www.freshmusic.gr – 6) Eth. Makariou & Delta Falireos 2, 18547 Neo Faliro Athens. **W:** www.freedomfm. gr – 7) **Papanikoli 24**, 152 32 Halandri Athens. **W:** www.music892. gr – 8) Iasiou 1, 11526 Athens. **W:** www.ecclesia.gr – 9) Viltanioti 36, 14564 Kato Kifisia Athens. **W:** www.dromosfm.gr – 10) Leof. Irakliou 145, 14231 Nea Ionia Athens. **W:** www.902.gr – 11) Evripidou 79, 18532 Piraeus. **W:** www.kanaliena.gr – 12) Praxitelous 58, 17674 Kallithea Athens. **W:** www.radioasty.gr – 13) Deligiorgi 47, 18535 Piraeus. **W:** www.pe912fm.com – 14) Athens. **W:** www.radiocreta.gr – 15) Pirronos 12, 16346 Ilioupoli Athens. **W:** www.galaxy92.gr – 16) Androutsou 24 & Kifisou 89, 18233 Rentis Athens. **W:** www.lampsifm.com – 17) Perikleos 49, 15451 Neo Psihiko Athens. **W:** www.bestradio.gr – 18) Vas. Sofias 85, 15124 Maroussi Athens. **W:** www.kiss.gr – 19) Dimitros 31, 17778 Tavros Athens. **W:** www.932happy.gr – 20) S. Karagiorgi 2 & M. Antypa, 14121 Iraklio Athens. **W:** www.94fm.gr – 21) Plateia Iroon 1, 13341 Ano Liosia Athens. **W:** www.xenios943.gr – 22) Davaki 58, 17672 Kallithea Athens. **W:** www.sport-fm.gr – 23) Theotokopoulou 4, 15124 Maroussi Athens. **W:** www.rythmosfm.gr – 24) Leof. Kifisias 215, 15124 Maroussi Athens. **W:** www.athensdeejay.gr – 25) Athens. **W:** www.floga-radio.gr 26) Leof. Kifisias 64, 15125 Maroussi Athens. **W:** www.flash.gr – 27) Eth. Makariou/Delta Falireos 2, 18547 Neo Faliro Athens. **W:** www.redfm.gr – 28) Athens. **W:** www.difono.gr – 29) Paradeisou 14, 15125 Maroussi Athens. **W:** www.rockfm.gr – 30) Leof. Kifisias 10-12, 15125 Maroussi Athens. **W:** www.972.gr – 31) Dimitros 31, 17778 Tavros Athens. **W:** www.loveradio.gr – 32) Leof. Kifisias 215, 15124 Maroussi Athens. **W:** www.realfm.gr – 33) Athens. **W:** www. liquidfm.gr – 34) Leof. Peiraios 100, 11854 Athens. **W:** www.athina984. gr – 35) N. Plastira 172, 13561 Ag. Anargiroi Athens. **W:** www.derti. gr – 36) 40o km. Attikis Odou, SEA Mesogeion, Ktirio 6, 19002 Paiania Athens. **W:** www.radio9.gr – 37) Eth. Makariou/Delta Falireos 2, 18547 Neo Faliro Athens. **W:** www.melodia.gr – 38) Mihalakopoulou 80, 11528 Athens. **W:** www.vimafm995.gr – 39) Athens. – 40) Eth. Makariou/Delta Falireos 2, 18547 Neo Faliro Athens. **W:** www.skai. gr/1003/ – 41) Leof. Mesogeion 411, 15343 Agia Paraskevi Athens. **W:** www.diesi.gr – 42) Athens. – 43) M. Antypa 41-45, 14121 Neo Iraklio Athens. **W:** www.sfera.gr – 44) M. Antypa 41-45, 14121 Neo Iraklio Athens. **W:** www.nitroradio.gr – 45) Athens. – 46) Papanastasiou 25, 18755 Keratsini Athens. **W:** www.blackman.gr – 47) Benaki 1, 15235

Metamorfosi Halandriou Athens. **W:** www.sentrafm.gr – 48) Thiseos 218, 17675 Kallithea Athens. **W:** www.pareafm.gr – 49) Athens **W:** www.minorefm.gr – 50) Leof. Peiraios 100, 11854 Athens. **W:** www. athina984.gr/1044fm **D.Prgr.:** Programming in 13 languages: English, Russian, German, Spanish, Italian, Arabic, French, Chinese, Bulgarian, Polish, Romanian, Filipino – 51) Mitropoleos 43, 15124 Maroussi Athens. **W:** www.hotfm1046.gr – 52) Athens. **W:** www.stylfm.gr – 53) Ag. Konstantinou 11, 18544 Piraeus. **W:** www.atlantisfm.gr – 54) Sarri 19, 10554 Athens. **W:** www.stokokkino.gr – 55) Eth. Antistaseos 253 & E. Kostopoulou, 15351 Pallini Athens. **W:** www.madradio.gr – 56) Athens. – 57) Athens. – 58) Paraskevopoulou 4, 12133 Peristeri Athens. **W:** www.aktinafm.gr – 59) Athens. – 60) Athens. **W:** www. hxorama.gr – 61) Kalapothaki 4, 54624 Thessaloniki. **W:** www.laikos. gr – 62) Parodos I. Koletti 20B, 54627 Thessaloniki. **W:** www.88miso. gr – 63) Leof. Karamanli 62, 54642 Thessaloniki. **W:** www.89rainbow.gr – 64) Aristotelous 7, 54624 Thessaloniki. **W:** athensdeejay.gr – 65) A. Papagou 41, 56334 Kordelio Thessaloniki. **W:** www.imagine897.gr – 66) Egnatias 69, 54631 Thessaloniki. 67) Aristotelous 3, 54624 Thessaloniki. **W:** www.zooradio.gr – 68) Ostovou 39, 54453 Thessaloniki. **W:** www.911.gr – 69) 1o km. Filiro-Langada, 57010 Filiro Thessaloniki. **W:** www.olafm.gr – 70) 1o km. Filiro-Langada, 57010 Filiro Thessaloniki. **W:** www.rso.gr – 71) I. Korovagou 3, 4os Orofos, 54627 Thessaloniki. **W:** www.fm-ekfrasi.gr – 72) Aristotelous 3, 54624 Thessaloniki. **W:** www. arisfm.gr – 73) Aristotelous 5, 54624 Thessaloniki. **W:** www.heartfm. gr 74) K. Karamanli 84, 54644 Faliro Thessaloniki. **W:** www.mymythos. eu – 75) Ag. Sofias 43, 54623 Thessaloniki. **W:** www.gnominet.gr – 76) Eleftherias 15, 56123 Ambelokipi Thessaloniki. **W:** www.radiolydia.gr – 77) 17o km. Moudianon, Kombos Risiou, 57001 Thermi Thessaloniki. **W:** www.rthess.gr – 78) Tsimiski 19, 54624 Thessaloniki. **W:** www. eroticos.gr – 79) Tsimiski 51, 6os Orofos, 54623 Thessaloniki. **W:** www. cosmoradio.gr – 80) K. Palama 6A, 54352 Thessaloniki. **W:** www.met-ropolisradio.gr – 81) Mitropoleos 34, 8os Orofos, 54623 Thessaloniki. **W:** www.sagapo961.gr – 82) K. Kristalli 30, 54630 Thessaloniki. **W:** www.palmos965.gr – 83) Aristotelous 7, 54624 Thessaloniki. **W:** www. velvet968.gr – 84) Aristotelous 3, 54624 Thessaloniki. **W:** www.starfm. gr – 85) 26hs Oktovriou 30, 54627 Thessaloniki. **W:** www.ant1fm. gr – 86) Leof. Karamanli 60, 54623 Thessaloniki. **W:** www.oasisfm. gr – 87) Mitropoleos 34, 54623 Thessaloniki. **W:** www.panorama984. gr – 88) Mitropoleos 61, 54623 Thessaloniki. **W:** www.athlitikanea. gr – 89) K. Karamanli 175, 54249 Thessaloniki. **W:** www.99fm.gr – 90) P.O. Box 680, 57001 Neo Risio Thessaloniki. **W:** www.europe-one.gr – 91) Melenikou 31A, 56224 Evosmos Thessaloniki. **W:** www. ekrixifm.gr – 92) Aggelaki 16, 54621 Thessaloniki. **W:** www.fm100. gr – 93) Tsimiski 60, 54622 Thessaloniki. **W:** www.republicradio. gr – 94) Aggelaki 16, 54621 Thessaloniki. **W:** www.fm100.gr – 95) Aggelaki 16, 54621 Thessaloniki. **W:** www.fm100.gr – 96) Vas. Irakliou 4, 54623 Thessaloniki. **W:** www.popsnet.gr – 97) Andrianoupoleos 4 & Epanomis 26, 55133 Kalamaria Thessaloniki. **W:** www.kalamariafm. gr – 98) Vas. Othonos 12, 54629 Stavroupoli Thessaloniki. **W:** www. akritestoupontou.gr – 99) Aristotelous 7, 54624 Thessaloniki. **W:** www. plusradio.gr – 100) Monastiriou 85, 54627 Thessaloniki. **W:** www. sport103.gr – 101) Palamidou 4, Ano Poli, 54633 Thessaloniki. **W:** www. studio3.gr – 102) 26hs Oktovriou 30, 54627 Thessaloniki. **W:** www. rythmosfm.gr – 103) A. Papandreou 27, 56334 Thessaloniki. **W:** www. radiokymata.gr – 104) Kouskoura 5, 54625 Thessaloniki. **W:** www.tif. gr – 105) Egnatia 154, 54636 Thessaloniki. **W:** www.tif.gr – 106) Aggelaki 31, 54621 Thessaloniki. **W:** www.1055rock.gr – 107) Kromnis 10, 54453 Toumpa Thessaloniki. **W:** www.g-radio.gr – 108) Mitropoleos 34, 54623 Thessaloniki. **W:** www.cityinternational.gr – 109) Aggelaki 31, 54621 Thessaloniki – 110) Iera Mitropoli Langada, 57200 Langadas Thessaloniki. **W:** www.imlagada.gr/Radio.htm – 111) Navmahias Ellis 4, Thessaloniki. **W:** www.realfm.gr – 112) Aristotelous 10, 54624 Thessaloniki. **W:** www.libero.gr – 113) Aristotelous 4, 54624 Thessaloniki. **W:** www.peiratikos.fm

AMERICAN FORCES RADIO & TV SERVICE (Mil.)
W: myafn.dodmedia.osd.mil **FM:** "107.3 The Odyssey": Souda Bay 107.3MHz 0.5kW.

GREENLAND (Denmark)

LT: UTC -3h (DST*: -2h) Qaanaaq & Thule Air Base: UTC -4h (DST*: -3h; not Thule AB) Ittoqqortoormiit: UTC -1h (DST*: UTC) Danmarkshavn: UTC. *) 27 Mar-30 Oct — **Pop:** 56,000 — **Pr.L:** Greenlandic, Danish — **E.C:** 50Hz, 220V — **ITU:** GRL

KALAALIT NUNAATA RADIOA – KNR (Gov. Comm.)
✉ Kissarneqqortuunnguaq 15, PO Box 1007, DK-3900 Nuuk ☎ +299 361500 🖷 +299 361502 **W:** www.knr.gl **E:** info@knr.gl
LP: Chairman: Peter Jensen. MD (radio & TV): Ivalo Egede. Head of prod: Frederik Lund. Head of radio: Thora H. Nielsen.

MW	kHz	kW	MW	kHz	kW
Nuuk	570	5	Upernavik	810	5
Qeqertarsuaq	650	5	Uummannaq	900	5
Simiutaq	720	10			

FM	MHz	kW	FM	MHz	kW
Nuuk*	90.5	0.05	Aasiaat	95.5	0.1
Sisimiut	95.0	0.1	Ilulissat	96.0	0.05
Kangerlussuaq	96.0	0.01	Tasillaq	96.0	0.05
Uummannaq	95.0	0.05	Sanderson Hope	96.0	0.1

+ 60 additional stns 0.05kW or less. *) = stereo

SW: 3815kHz 0.2kW (USB) Tasiilaq: 1500-1615, 2100-2215. **N:** Greenlandic: 1515, 2130. Danish: 1530, 2200. **D.Prgr:** 24h
Ann: "Kallaallit-Nunaata Radioa", "Grønlands Radio" **IS:** "Sunnia Kalippoq" (The Whaleboat "Sonja" drags whale) played on celeste.
F.pl.: All MW transmitters will be closed January 2011. News and weather forecasts are to be carried via coastal stn Aasiaat Radio.
DR P1, Denmark. Sat. relay 24h: Nuuk 98.0MHz 0.1kW
RÚV Rás 2, Iceland. Sat. relay 24h: Narsaq 88.0MHz

Private stations (local radio):
Akisuasoq Radio, Box 29, 3912 Maniitsoq: 90.5MHz (0.3kW), 93.0MHz (0.1kW), 99.0MHz (0.1kW). **W:** www.akisuasoq.gl/radio.html – **Inuunerup Nipaa**, Ilivinnguaq 1, Box 67, 3900 Nuuk: 88.5MHz. **W:** www.ino.gl. Format: Rlg. – **Kangaatsiap Tusaataa**, 3955 Kangaatsiaq: 103.0MHz – **Kap York Radio**, B157, 3971 Qaanaaq: 93.5MHz – **Lokal Radio Ilulissat**, Box 1004, 3952 Ilulissat: 99.0MHz – **Nanortalik Lokalradio**, 3922 Nanortalik – **Nuuk FM**, Box 1462, 3900 Nuuk: 93.0MHz (0.1kW) – **Qasigiannguit Lokalradio**, B20, 3951 Qasigiannguit: 103.0MHz – **Radio 5OZ20**, Den Danske Radio, SPE, Box 139, Thule Air Base, 3970 Pituffik: 97.1MHz (0.1kW) – **Radio Grønnedal**, Grønlands Kommando, 3930 Kangilinnguit: 91.5MHz (0.015kW) – **Radio Narsaq TV**, Josifip aqq. 543, Box 74, 3962 Upernavik: 93.0MHz (0.025kW). **W:** www.radionarsaq.gl – **Radio Upernavik**, Box 244, 3962 Upernavik: 93.0MHz – **Seekon Radio**, Box 361, 3920 Qaqortoq: 93.0MHz – **Sisimiut Tusaataat**, Box 312, 3911 Sisimiut: Sarfannguit 91.0MHz (0.02kW), Sisimiut 93.0MHz (0.05kW), Kangerlussuaq 93.0MHz (0.02kW), Sarfannguit 98.0MHz (0.02kW). **W:** www.sistus.gl – **Tusaataat Paamiut**, Box 533, 3940 Paamiut: 93.0MHz (0.02kW) – **Tusaat TV Aasiaat**, Box 20, 3950 Aasiaat: 93.0MHz (0.1kW) – **Uummannap Tusaataa**, Box 195, 3961 Uummannaq: 98.2MHz (0.075kW).

GRENADA

L.T: UTC -4h — **Pop:** 108,000 — **Pr.L:** English — **E.C:** 50Hz, 230/400V — **ITU:** GRD

GRENADA BROADCASTING NETWORK – G.B.N. Radio (Gov, Comm.)
Observatory Road, PO. Box 535, St. George's ☎ +1 473 440 3033 ▤ +1 473 444 4180 **W:** www.klassicgrenada.com
Email: gbn@spiceisle.com **LP:** GM: Ruel Edwards. Op.mgr.: Clarence Cosmos Baker. CEN: Kennedy Bowen
MW: Klassic AM: 540kHz 10kW: 0900-0200. Rel. BBC 0200-0900.
FM: HOTT FM: 98.5/105.5MHz 1000-0300

HARBOUR LIGHT OF THE WINDWARDS (Rlg.)
Carriacou ☎ +1 473 443 7628 ▤ +1 473 443 7628 **W:** www.harbourlightradio.org & www.lastchanceministries.com/harbourlight.htm
E: harbourlight@spiceisle.com **LP:** SM: Randy Cornelius
MW: 1400kHz 5kW **FM:** 92.3MHz 0.25kW, 94.5MHz 0.25kW
D.Prgr: MW: 0953-0245. FM: 24h. **N:** rel. BBC & VOA
Ann: "This is the Harbour Light of the Windwards broadcasting from beautiful and friendly Carriacou"

PRIVATE STATIONS:
BOSS FM, Sauteurs, St Patricks ☎ +1 473 442 1177 **W:** www.bossfmgrenada.com **FM:** 104.1/104.3MHz – **CITY SOUND**, River Road, St George's ☎ +1 473 440 9616 ▤ +1 473 440 7838 **W:** www.citysoundfm.com **L.P.:** Mgr Alphonses Strachan. **FM:** 97.5MHz – **CRFM COMMUNITY RADIO:** Morne Jaloux, St George's ☎ +1 473 440 4848 ▤ +1 473 440 4991 **L.P:** Mgr Rawl Ghatts. **FM:** 89.5MHz – **FUNCITY FM,** Central Depradive St., Gouyare, St John's ☎ +1 473 417 0433. **W:** www.funcity909.webs.com **FM:** 90.9MHz – **GFN – GRENADA FAMILY NETWORK**, PO Box 2747, St George's ☎ +1 473 435 4297. **W:** www.globalfamilynetwork.net **L.P.:** Pres. David Gates. **FM:** 91.3MHz. Format: Rlg. (Adventist) – **GN FM - GOOD NEWS FM**, Marrast Hill, St George's ☎ +1 473 435 1301 ▤ +1 473 435 1278 **FM:** 96.3MHz. Format: Rlg., regional – **GNG GOOD NEWS GRENADA RADIO**, Box 224, St George's ☎ +1 473 435 0143 L.P.: Cyril Hopkin. **FM:** 99.5MHz. Format: Rlg – **KYAK 106 FM**, Church

Street, Hillsborough, Carriacou ☎ +1 473 443 6262 ▤ +1 473 443 6262 **W:** www.kyak106.com **FM:** 106.3MHz – **REAL FM GRENADA**, St George's. **FM:** 91.9MHz – **SAC FM**, St Andrews Connection, Grenville ☎ +1 473 442 4745 ▤ +1 473 438 0338 L.P.: Mgr Bernard La Mothe, **FM:** 104.7MHz – **SGU 107.5**, Office of University Communications, 2nd floor, Chancellery, St George's University, St George's ☎ +1 473 444 4175 ext. 2191 ▤ +1 473 444 3153 **W:** www.sgu.edu . **FM:** 107.5MHz. Format: Non-commercial community radio – **SISTER ISLE RADIO**, Fort Hill, Hillsborough, Carriacou ☎ +1 473 443 8141/8142. **W:** www.sisterisleradio.com **FM:** 92.9MHz – **SPICE CAPITAL RADIO**, PO Box 90, St George's ☎ & ▤ +1 473 435 3563 **W:** www.spicecapitalradio.net **L.P.:** Mgr Paul Roberts. **FM:** 90.1MHz – **VOG VOICE OF GRENADA**, Lagoon Road, St George's ☎ +1 473 440 8171 ▤ +1 473 440 8505 **W:** www.vogfm.com **FM:** 88.9/95.7/103.3MHz – **WEE FM**, Grenada Wireless Comm Network, Lower Depradine St, PO Box 555, Gouyave, St John's ☎ +1 473 440 4933 ▤ +1 473 440 8724 **W:** www.weefmgrenada.com L.P.: GM: Alvin Dabreo. **FM:** 93.3/93.9MHz

GUADELOUPE (France)

L.T: UTC -4h — **Pop:** 394,000 — **Pr.L:** French, Créole Patois — **E.C:** 50Hz, 230V — **ITU:** GLP

RADIO GUADELOUPE
Morne Bernard-Destrellan, B.P. 180, F-97122 Baie-Mahault. ☎+590 590939696. ▤ +590 590939682. **L.P:** Dir. R.Surjus. Editor-in-Chief: Philippe Goudé. PD: L.Francil. Head Communications Dept: Sonia Gémieux
MW: Point-à-Pitre 640kHz 40kW
FM: Point-à-Pitre 88.9MHz, Marie Galante 89.1MHz, Deshaies 96.8MHz, Basse-Terre 97.0MHz, Pointe-Noire 97.4MHz.
D.Prgr: 24h. **N:** 1100, 1700, 2230, plus relays of France-Inter.
Ann: "Ici Point-à-Pitre, RFO Guadeloupe" or "RFO".
IS: "Biguin" (guitar) **V.** by QSL-card. Rp.

RCI - RADIO CARAÏBES INTERNATIONAL GUADELOUPE (Comm.)
RCI Guadeloupe, B.P. 1309, F-97187 Point-à-Pitre Cédex. ☎ +590 590839696 ▤ +590 590839697
FM: 95.1/98.6/106.6MHz. **D.Prgr:** 24h. **N:** on the h. (rel. Europe 1).
RCI 2 (Comm.)
FM: 96.3/100.6/102.6MHz (rel. RCI2 St. Martin) **D.Prgr:** 24h.
Europe 2: Basse Terre 96.6MHz, Point-à-Pitre 103.4MHz.
R. France Internationale: via 98.2MHz and 103.0MHz.

RADIO MASSABIELLE (Rlg)
B.P. 607, 97168 Point-à-Pitre ☎+590 590 832521 ▤ +590 590 834861. **L.P:** Pres: José Colat-Jolivière
W: www.radiomassabielle.fr **E:** contact@radiomassabielle.fr
FM: 97.8MHz 0.6kW, 101.8MHz 1kW

RADIO SAPHIR FM
rue Bel Air Bourg, 97170 Petit-Bourg ☎+590 690 352274
E: saphirfm@live.fr **W:** www.radiosaphirfm.com
FM: 89.4

Other stations: 14 other FM stns are r. to be operating.

GUAM (USA)

L.T: UTC +10h — **Pop:** 178,430 — **Pr.L:** English, Chamorro, Filipino — **E.C:** 60Hz, 110/220V — **ITU:** GUM

FEDERAL COMMUNICATIONS COMMISSION (FCC)
see USA for details

	MW	kHz	kW		MW	kHz	kW
1)	KGUM	567	10	13)	-	1170	0.25
2)	KUAM	630	10	13)	-	1350	0.25
3)	KTWG	801	10	14)	KVOG	1530	0.25
12)	-	1017	4.0				

	FM	MHz	kW		FM	MHz	kW
4)	KHMG	88.1	8	6)	KOLG	90.9	5.7
5)	KPRG	89.3	6.6	7)	KSDA	91.9	3.8
3)	K225AN	92.9	0.01	8)	KISH	102.9	25
2)	KUAM	93.9	5.2	11)	KIJI	104.3	12.5
8)	KSTO	95.5	25	1)	KGUM	105.1	12
1)	KZGZ	97.5	40	15)	KGCA	106.9	0.07
9)	KOKU	100.3	50	13)	K300AV	107.9	0.023
10)	KTKB	101.9	46				

Addresses and other information
1) 111 Chalan Santa Papa, Suite 800; Hagatna, GU 96910-5193 ☎+1

671 477-5700, 808 524-6495, 📠+1 671 477-3982 **W**: www.radiopacific. com www.k57.com Format: KGUM-AM Talk, news – **2)** 600 Harmon Loop Road, Suite 102; Dededo, GU 96929-6536 ☎+1 671 637-KUAM (637-5826) 📠+1 671 637-9865 **W**: www.kuam.com Format: KUAM-AM Island contemporary music – **3)** 1868 Halsey Drive; Asan, GU 96910-1505 ☎+1 671 477-5894 📠+1 671 477-6411 **W**: www.ktwg.com Format: Protestant Christian talk and instruction, gospel music **NB:** Korean Mon & Fri 0800-0830, Tagalog Wed 0800-0830, Chamorro Thu 0800-0815 & Sun 0700-0730, Japanese Thu 0815-0830 – **4)** c/o Harvest Christian Academy, 170-C Machaute St., Barrigada, GU 96913-1193 – **5)** c/o University of Guam, 303 University Drive; UOG Station; Mangilao, GU 96923-1871 **NB:** BBCWS Daily 0700-0800, Sun 1900-2100, Mon 1900-2000, Tue 1400-2000, Wed & Thu 1400-1800 & 1900-2000, Fri 1400-1800, Sat 1700-2000 – **6)** Chalan Santo Papa; P.O. 23006, Guam Mail Facility, Barrigada, GU 96921-3006 – **7)** 290 Chalan Palasyo, Hagatna Heights, Piti, GU 96910-6405 – **8)** Nimitz Hill, 1868 Halsey Drive, Piti, GU 96910-1505 – **9)** 107 Julale Center, 424 West O'Brien Drive, Hagatna, GU 96910-5078 – **10)** 177-B Ilipog Drive, Suite 203; Tamuning, GU 96913-4107 – **11)** 543A N Marine Dr, Tamuning, GU 96913-4217 **LP:** SM: Rich de Vera ☎+1 671 478-0104 📠+1 671 647-7480 **W**: www. kijifm104.com – **12)** (CP) Powell Meredith Communications Co, 813 Ventura Park, Irving TX 75061 ☎+1 915 695-9898 – **13)** Management Advisory Services Inc, 125 Tun Jesus, Crisotomo Street #308, Tamuning GU 96913 T: +671 648-4262. Two new AM/FM stations under construction – **14)** Guam Power II Inc, 1100 Alakea #1800, Honolulu HI 96813-2839 ☎+1 808 521-4711 – **15)** 1ST Fl, J. Perez Bldg, 138 Seaton Blvd, Hagatna GU 96913 ☎+1 671 648-4262

ADVENTIST WORLD RADIO - ASIA (Rlg.) and TRANS WORLD RADIO - ASIA (Rlg.): See International Radio section

GUATEMALA

L.T: UTC -6h – **Pop:** 13 million – **Pr.L:** Spanish – **E.C:** 60Hz, 120V – **ITU:** GTM

SUPERTEL
✉ 14 Calle N° 3-51, Z-10, Edif. Murano, Nivel 16, Guatemala ☎ +502 2366 5880 **W**: www.sit.gob.gt

MW	Call	kHz	kW	Station, location, h. of tr.
AV01)		540		R. Cobán, Cobán
SO03)		540	0.02	R. Amistad, San Pedro de Laguna
GU01)	RV	560	10	R. 560, Guatemala: 1200-0500
SM01)		560	1	R. Quetzal, Malacatán
ES01)	PA	570	1	R. Palmeras, Escuintla
GU02)	Y	580	5	R. Progreso, Guatemala: 1100-0600
QU01)	RQ	590	5	R. Quiché, Sta Cruz del Quiché: 1100-0400
ES02)	RC	*600	1	R. Campesina, Escuintla
GU03)	GA	610	5	R. Alianza, Guatemala: 1000-0300
TO01)	PQ	620	5	R. 6-20, San Cristóbal: 1200-0400
PE01)	EL	630		R. Cultural Porvenir, Sta Elena: (r. 730)
QE01)	Q	660	3	La Voz de Quetzaltenango, Quetzaltenango: 1100-0400
AV02)	VP	680	10	R. Norte, Cobán: 1000-0500
JU01)	VB	690	1	R. Tamazulapa, Jutiapa
ES03)	AJ	700	1	R. Inspiración, Escuintla
GU06)	HR	700	15	R. Mundial, Guatemala
QE02)	XL	710	1	R. Tecún Umán, Quetzaltenango (r. 730)
IZ01)	RO	*720	1	R. Corona, Morales
GU07)	N	°730	10	R. Cultural, Guatemala
GU08)	HB	*760	5	Nueva R. Super, Guatemala: 1000-0500
QE03)	BX	770	1	R. Nueva Fratemidad, Quetzaltenango: 1000-0600
ZA01)	CK	780	1	Sultana La Cristiana, Zacapa
GU09)	O	*790	3	R. Festival, Guatemala: 1100-0400
SR01)	YZ	*800	1	R. Rosa, Chiquimulilla
PE02)		810		R. Moapán, Sta Elena
SA01)		810		R. Circuito San Juan, San Juan
SM06)	END	810		R. Constelación, San Marcos: 1200-2400
GU10)	TO	820	10	R. Kyrios, Guatemala: 1000-0600
SU01)	AV	830	5	R. Satélite, Mazatenango: 1100-0400
AV06)		840	2.5	R. Luz, San Pedro Carchá
JU04)		840		R. Idea 840, Jutiapa
GU11)	X	*850	1	R. Ciro, Guatemala
SU02)	L	870	0.5	R. Victoria, Mazatenango
GU12)	J	*880	10	R. Nuevo Mundo, Guatemala: 1030-0500
ES04)	HU	890	1	R. Escuintla, Escuintla
IZ02)	MA	900	1	R. Amatique, Puerto Barrios
GU30)	KL	910	10	R. Fe y Esperanza, Guatemala: 1130-0600
ES05)	RS	920	0.2	R. Cultural, Escuintla (r. 730)
GU13)	TL	940	5	Eventos Católicos R., Guatemala: 1200-0500
GU13)	TL	940	1	Eventos Católicos R., Sacatepeque (r: 940)
SU03)	AF	950	1	R. Indiana, Mazatenango
GU14)	AX	970	5	R. Continental, Guatemala: 1200-0430

MW	Call	kHz	kW	Station, location, h. of tr.
SM04)	MQ	980	1	R. Retama, San Marcos: 1200-0500
CH01)	AL	*990	1	R. Perla de Oriente, Chiquimula
CM02)		1000		R. Cultural y Educativa, Patzún
GU32)		1000		R. Revelación y Verdad, Guatemala: 1055-0500
IZ06)		1010	1	R. Caribe, Izabal
QU03)	XI	1010	1	R. Ixil, Nebaj: 1100-0200
SM05)	CM	1020	5	R. Frontera, Pajapita: 1100-0400
GU15)	UX	1030	10	R. Panamericana, Guatemala: 1100-0500
JA01)	JP	1040	1	R. Oriental, Jalapa
HU01)	SL	1050	5/1	LV de los Cuchumatanes, Huehuetenango: 1100-0600
GU16)		*1060	10	R. Favorita, Guatemala: 1100-0600
QE04)	D	1070	3/2	LV de Occidente, Quetzaltenango: 1200-0400
ZA02)	LU	*1080	1	R. Novedad, Zacapa
QE05)	SR	*1100	1	R. Superior, Coatepeque
AV04)	MK	1110	1	R. Verapaz, Cobán
GU17)	C	1120	0.5	R. Poderosa "La Voz de la Liberación", Guatemala: 1100-0600
RE01)	VR	1130	1	Em. Unidas LV de la Costa Sur, Retalhuleu
GU17)	T	1150	10	R. Sonora, Guatemala: 1100-0600
IZ03)	RI	1160	1	R. Izabal, Morales (r: 730): 1300-0300
QE06)	RL	1170	5	R. Cadena Landívar, Quetzaltenango: 0900-0300
GU33)		1180		R. 10, Guatemala
JU02)	RJ	1200	12	R. Unción, Jutiapa
GU19)	MX	1210	10/5	R. Miel, Guatemala
IZ04)	AT	1230	1	R. Atlántida, Puerto Barrios: 1130-0500
SU04)		1230		R. América, Cuyotenango
GU20)	K	1240	5	R. Luz, Guatemala
CH02)	PY	1250	1	R. Payakí, Esquipulas: 1100-0300
TO04)		1250	1	LV Cristiana, Totonicapán
GU21)	CQ	1270	2.5	R. Exclusiva, Guatemala
BV01)	VY	*1280	2.5	R. Zamaneb, Salamá: 1100-0200
ZA03)		1290		R. Miramundo "LV del Ejercito", Zacapa
QE07)	AN	1310	1	R. LV de los Altos, Quetzaltenango: 1100-0700
JU03)	ME	1320	1	R. Quezada, Jutiapa
GU22)	MU	1330	5.5	Unión R. "La Voz de la Esperanza", Guatemala
AV05)	MC	1350	1	R. Monja Blanca, Cobán
GU15)	LK	1360	10	R. Tic Tac "LV del Evangelio Pentecostés", Guatemala
QE09)	AC	1370	1	LV de Colomba, Colomba
TO03)	EB	1380	1	R. Momostenango Educativa, Momostenango: 1100-0300
IZ05)	RB	*1400	1	R. Porteña, Puerto Barrios
QE10)	GH	1410	5	R. Xelajú, Quetzaltenango
GU24)	RP*	1420	1	R. Capital, Guatemala: 1130-0600
HU02)	AG	1430	1.2	LV de Huehuetenango: 1100-0400
SU05)	MS	1440	0.5	R. Nacional, Mazatenango: 0000-0400
GU06)	LG	1450	1	R. Hosanna, Guatemala: 1000-0600
PE04)	RN	1460	2.5	R. Petén, Flores
GU25)	HB	1480	5	R. Horizontes, Guatemala: 1030-0200
RE02)	RE	1490	1	R. Modelo, Retalhuleu
GU32)	DX	1510	5	R. Centroamericana del Amor, Guatemala: 1055-0500
PE05)		1520		R. Taysal, Sta Elena de la Cruz
QE11)	RS	*1520	1	R. Superior, Coatepeque
GU29)		*1540	1	R. Cultura y Deportes, Guatemala
QE12)		1560		R. Inspiración, Coatepeque
GU27)	VE	1570	10	VEA - Voz Evangélica de América, Guatemala: 1030-0600
CM01)	XC	1590	1	R. Triunfadora, Chimaltenango

Call TG—° = also on SW, * = inactive, (r) = repeater, v = varying fq.

SW	Call	kHz	kW	Station, location & h. of tr
GU07)	NC	*3300	10	R. Cultural, Guatemala
CH04)	AV	*4052.5	0.5	R. Verdad, Chiquimula: 1130-0500
SO03)		*4699	0.5	R. Amistad, San Pedro La Laguna
HU04)	CT	*4780	1	R. Cultural Coatán, San Sebastián Coatán: 1100-1500, 2200-0230
HU05)	MI	*4800	1	R. Buenas Nuevas, San Sebastián, Huehuetenango: 1000-0430 (r: 102.1)
GU07)	NA	*5955	0.5	R. Cultural, Guatemala: irr

Stns with a (*) are reported to be inactive, but may occasionally be reactivated for variable periods of time.

State abbreviations: (Departamentos) AV = Alta Verapaz, BV = Baja Verapaz, CH = Chiquimula, CM = Chimaltenango, ES = Escuintla, GU = Guatemala, HU = Huehuetenango, IZ = Izabal, JA = Jalapa, JU = Jutiapa, PE = Petén, QE = Quetzaltenango, QU = Quiché, RE = Retalhuleu, SA = Sacatepéquez, SR = Santa Rosa, SM = San Marcos, SO = Solola, SU = Suchitepequez, TO = Totonicapán, ZA = Zacapa. **N.B:** These abbreviations are not recognized by the Post Office. Letters should therefore carry the full name.
Addresses and other information:
AV00) ALTA VERAPAZ
AV01) 5 Calle 1-06, Z-3, 16001Cobán – **AV02)** 2 Calle 5-57, Z-3,

16001Cobán – **AV04)** 2 Calle 5-57, Z-3, 16001Cobán – **AV05)** Edif Municipalidad, 5a Calle 1-06, 16001Cobán – **AV06)** 11 Av Zona 1, Colonia Cuatro Caminos, San Pedro Carchá (or Apartado Postal 14, 16001 Cobán) - 1100-0400.

BV00) BAJA VERAPAZ
BV01) Inst de Educación Básica, Barrio Abajo San Jerónimo, 15001 Salamá. Prgrs. in Spanish, Achi and Q'eqchí

CH00) CHIQUIMULA
CH01) 7 Calle Av 4-00, Z-1, 20001Chiquimula (or: 6 Av 0-60, Z-4, Torre Prof II, Of 904, 01004 Guatemala) – **CH02)** 5 Av 6-37, Z-1, 20007 Esquipulas – **CH04)** Estación Educativa Evangélica, Ap 5 (or: 4 Av 2-24, Z-1), 20901 Chiquimula **W:** www.radioverdad.com **E:** radioverdad5@yahoo.com ☎ +502 79425689 📠+502 79420362 **FM:** 102.7MHz

CM00) CHIMALTENANGO
CM01) 2 Calle 3-33, Z-3, 04001Chimaltenango – **CM02)** 6ta Calle 3-88, Zona 5, Patzún 050, Chimaltenango.

ES00) ESCUINTLA
ES01) 15 Calle 2-48, Z-3, 05001Escuintla – **ES02)** Col 15 de Junio, Z-3, Tiquisate, 05001Escuintla – **FM:** 92.3MHz – **ES03)** 4 Av 12-27, Z-1, 05001Escuintla **E:** radioinspiracion@gmail.com – **ES04)** 4 Av 11-38, Z-1, 05001Escuintla – **ES05)** Central American Benevolent Association, 05001Escuintla – **FM:** 96.3MHz

GU00) GUATEMALA
GU01) 8 Calle 1-11, Z-1, 01001Guatemala – **GU02)** 9 Av 0-32, Z-2, 01002 Guatemala ☎ +502 22542440 📠+502 22542541 – **GU03)** 34 Av "A" 7-60 Tikal 2, Z-7, 01007 Guatemala **W:** www.radioalianza.com – **GU04)** 18 Calle 6-72, Z-1, 01001Guatemala – **GU06)** 8 C 10-54, Zona 11, Col. Roosevelt, 01011Guatemala **W:** www.lamisionera1450.com – **GU07)** Ap 601 (or: 4 Av 30-09, Z-3), 01901 Guatemala – **English:** 0300-0430 on 730 kHz ☎+502 24721745 📠+502 24400260 **W:** www.radiocultural.com **E:** tgn@radiocultural.com – **GU08)** 30 Av 3-86, Z-11, Utatlán II, 01011 Guatemala – **GU09)** 11 Calle 2-43, Z-1, 01001Guatemala – **GU10)** 3a Av 0-54, Z-13, Guatemala – **GU11)** Calzada San Juan 7-90, Edif.Acuario, Z-7, 01007 Guatemala – **GU12)** 6a Av 0-60, Zona 4, Torre Profesional 1, Niv. 9, Of. 911, 01004 Guatemala – **GU13)** 10a Avenida "A" 2-57, Z-1, 01001 Guatemala ☎+502 23820222 **W:** www.eventoscatolicos.net – **GU14)** 15 Calle 3-45, Z-1, 01001Guatemala – **GU15)** 1 Av 35-48, Z-7, Col Toledo, 01007 Guatemala ☎+502 55958504 📠+502 55912293 – **GU16)** 10 Calle 5-20, Z-1, 01001 Guatemala – **GU17)** 2 Calle 18-07, Zona 15, Vista Hermosa 1, 01015 Guatemala **W:** www.sonora.com.gt – **GU19)** 4 Av 1-14, Z-1, 01001 Guatemala – **GU20)** Ap 281, 01901 Guatemala – **GU21)** 7 Av. 15-13, Zona 1, Edificio Ejecutivo, Niv. 8, 01001Guatemala **E:** radioexclusiva@motivacioncristiana.org – **GU22)** Ap 51-C, 01015 Guatemala - 1100-2330. Owned and operated by Adventist World Radio **W:** www.unionradiogt.org – **GU24)** 4 Av 0-60, Z-4, 01004 Guatemala – **GU25)** 17 Av.21, Cnt.Com Las Pergolas, Z-11, 01011Guatemala – **GU27)** Ap 1213 (or: 30 Av "A" 7-33, Z-7, Col Tikal, 01007 Guatemala) 01901 Guatemala **W:** www.radiovea.org **E:** radiovea@radiovea.org – **GU29)** Guatemala – **GU30)** 10a Avenida 0-61, Z-19, Colonia La Florida, 01019 Guatemala – **GU32)** 17 Av. 5-47, Zona 11, Col. Miraflores, 01011 Guatemala **W:** www.corporacionverdad.org – **GU33)** 6a Avenida 11-77, Zona 10, Pent House, Guatemala **W:** www.radio10.com.gt.

HU00) HUEHUETENANGO
HU01) 2 Calle 4-42, Z-1, 13001 Huehuetenango – **HU02)** Ap 13, 13901 Huehuetenango **W:** www.lavozdehuehue.comlu.com – **HU04)** 13025 San Sebastián Coatán Programming in Spanish & Chuj Coatán – **FM:** 92.5MHz – **HU05)** 13020 San Sebastián H, Huehuetenango **W:** http://tgmiradiobuenasnuevas.com

IZ00) IZABAL
IZ01) Calle Principal, Morales – **IZ02)** Ruta Atlántico km 291, 18001 Puerto Barrios – **IZ03)** Barrio El Carrizal, Morales – **IZ04)** Ap 425, 18901Puerto Barrios – **IZ05)** 8 Av 15 y 16 Calle, 18001 Puerto Barrios – **IZ06)** Izabal

JA00) JALAPA
JA01) Av Chipilapa "A" 1-03, Z-2, 21001 Jalapa.

JU00) JUTIAPA
JU01) Calle 15 Septiembre, Sta Cruz, 22001 Jutiapa – **JU02)** Carr Interamericana km 117, 22001 Jutiapa – **JU03)** Quezada – **JU04)** 22001 Jutiapa

PE00) PETÉN
PE01) Sta Elena de la Cruz **FM:** 96.9MHz – **PE02)** Sta Elena de la Cruz – **PE04)** Isleta Sta Bárbara, 17001 Flores (or: 1 Av 1-22, Z-1, Guatemala) 1100-0500 **E:** radiopeten@hotmail.com ☎/📠+502 22515516 **FM:** 105.3MHz – **PE05)** Ministerio de la Defensa Nacional, Sta Elena de la Cruz

QE00) QUETZALTENANGO
QE01) Ap 113 (or 13 Av 8-19, Z-1), 09901 Quetzaltenango – **QE02)** 6 Av 6-41, Z-1, 09001 Quetzaltenango – **QE03)** 5 C 13-56, Zona 3, Xelajú (Ap 90), 09901 Quetzaltenango – **FM:** 99.1MHz– **QE04)** 7 Av 0-26, Z-2, 09002

Quetzaltenango ☎+502 77610582 📠+502 77612062 – **QE05)** 3 Calle 3-38, Z-1, Coatepeque – **QE06)** 14 Av "A" 0-78, Z-1, 09002 Quetzaltenango – **QE07)** Ap 107, 09901 Quetzaltenango – **QE09)** Calle Principal, Z-2, Colomba ☎+502 77723050 📠+502 77723075 –**FM:** 99.1MHz – **QE10)** 4 Calle 15A-62, Z-1, 09002 Quetzaltenango - 1200-0600 – **QE11)** 3 Calle 3-38, Z-1, Coatepeque, Retalhuleu – **QE12)** Quetzaltenango.

QU00) QUICHÉ
QU01) 7 Calle 3-67, Z-5, 14001 Sta Cruz del Quiché **W:** www.radioscatolicasdequiche.com – **FM:** 91.5MHz – **QU03)** 5 Av 1-32, Canton Batzbaca, 14013 Nebaj

RE00) RETALHULEU
RE01) Ap 84, 11901Retalhuleu – **RE02)** 7 Av 6-72, 11001 Retalhuleu (or: Ap 183-A, Guatemala) - 0900-0300

SA00) SACATEPÉQUEZ
SA02) San Juan Sacatepéquez **W:** www.radiocircuitosanjuan.com

SR00) SANTA ROSA
SR01) Edif Municipal, Chiquimulilla.

SM00) SAN MARCOS
SM01) 5 Calle 3-58, Z-1, Malacatán – **SM04)** 5 Calle 8-21, Z-1, San Pedro – **SM05)** Pajapita, 12001 San Marcos – **SM06)** 12001 San Marcos

SO00) SOLOLA
SO03) Iglesia Bautista Getsemani, San Pedro La Laguna (or: International Mission Board, SBC, Ap 25, Bulevares, MX 53140, México) - **FM:** 97.6MHz

SU00) SUCHITEPEQUEZ
SU01) 10001 Mazatenango - 1100-0400 – **SU02)** La Libertad 9-91, Z-1, 10001 Mazatenango – **SU03)** 6 Av 10-54, Z-1, 10001 Mazatenango – **SU04)** 13 Av 23-60, Z-12, 10012 Coyotenango – **SU05)** Calle 30 de Junio 1a y 2a, Z-5, 10001 Mazatenango

T000) TOTONICAPAN
T001) Barrio La Cienaga, 08002 San Cristóbal Totonicapán – **T003)** Momostenango, 08001 Totonicapán – **T004)** Totonicapán

ZA00) ZACAPA
ZA01) 4 Calle 12-54, Z-1, 19001 Zacapa – **ZA02)** 4 Calle 10-34, Z-1, 19001 Zacapa – **ZA03)** Zona Militar N° 7, 19001 Zacapa

FM in Guatemala City (MHz): 88.1 Fabuestereo - 88.5 Galaxia La Picosa – 88.9 Fabulosa 88.9 – 89.3 Estrella – 89.7 Em.Unidas – 90.1 Yo Sí Sideral – 90.5 Punto – 90.9 Exitos – 91.3 Furia Musical – 91.7 Fiesta – 92.1 Universidad – 92.5 40 Principales – 92.9 Disney – 93.3 FM Joya – 93.7 Mía – 94.1 94 FM – 94.5 La Sabrosita – 94.9 Nueve Cuatro Nueve – 95.3 Viva – 95.7 Ranchero – 96.1 Nuevo Mundo – 96.5 Atmósfera - 96.9 Sonora – 97.3 Alfa – 97.7 Kiss FM - 98.1 Doble S – 98.9 Globo – 99.3 La Grande - 99.7 Conga – 100.1 Infinita – 100.5 Cultural – 100.9 La Hit FM – 101.3 R. Extrema – 101.7 Exa – 102.1 Stereo 102 – 102.5 FM Fama – 102.9 Caliente – 103.3 R. María – 103.7 R. Fiesta – 104.1 Stereo Visión – 104.9 Tropicalada – 105.3 Shock FM – 105.7 Union – 106.1 Red Deportiva – 106.5 Clásica – 106.9 Internacional – GU04) 107.3 TGW LV de Guatemala – 107.7 Mega

FM in Quetzaltenango (MHz): 88.1 Dinámica – 88.5 La Consentida – 89.5 Emisoras Unidas – 89.9 Prisima FM – 90.3 Tropicalidad – 90.7 María – 91.1 La Nueva Mega – 91.7 La Rubia – 92.3 R. Cadena Sonora – 92.7 Cadena Caliente – 93.1 Nahual Estereo – 93.7 Fiesta – 94.3 Diamante – 94.7 Punto – 95.1 Ke Buena – 95.5 Evolución – 95.9 FM Globo – 96.3 FM Intima – 97.1 Exa FM – 97.5 Gaviota FM – 98.3 La Grande – 98.7 Yo Sí Sideral – 99.1 RTVA Arqueocesana – 99.5 Génesis – 99.9 Galaxia – La Picosa – 100.3 Stereo Cien – 100.7 R. Culturas – 101.1 R. Estéreo Tulán – 101.5 Estéreo Alegre – 102.3 Precencias R. – 102.9 Cristal – 103.3 La Voz de Dios – 104.3 Emisoras Unidas – 104.7 Razón – 105.3 La Voz del Evangelio – 105.9 FM Luna – 106.3 La Visión F – 106.7 Alfa – 107.1 R. Exitos – 107.5 TGQ La Voz de Quetzaltenango – 107.9 R. Estéreo Vida

GUINEA

LT: UTC — **Pop:** 10 million — **Pr.L:** French, Fulah, Maninké, Soussou — **E.C:** 50Hz, 220V — **ITU:** GUI

CONSEIL NATIONAL DE LA COMMUNICATION (CNC)
📧 Conakry **W:** www.guinee.gov.gn **L.P:** Chmn: Mounir Camar.

RADIO TÉLÉVISION GUINÉE (RTG, Gov.)
📧 B. P. 391, Conakry ☎+224 30 41 55 19. **W:** www.rtg- conakry.com **L.P:** DG: Alpha Kabinet Keita. Dir. Tech: Aladji Touré.
SW: Conakry (Sonfonia): 7125kHz 50kW (irreg.).
FM: Conakry 88.55/91.7MHz.
D.Prgr. in French/Others: W 0555-2400, Sun 0800-2400. **N:** French: 0645, 0915Sun, 1200Sun, 1245W, 1300Sun, 1615W, 1945W, 2000Sun, 2200, 2350. **English:** 1845 (irr.).
Ann: F: "R. Conakry", "R. Guineé". **IS:** Guitar.

RADIO RURALE (RTG rural stations)
R. Rurale de la Moyenne Guinée ("R. Fouta Internationale [RFI]"),
B.P. 169, Labé. **MW**: Labé (Dayabé) 1386kHz 50kW (inactive). **FM**:
87.6MHz. **D.Prgr**: 0600-2330v in French/Pular/Maninke/others. Rel.
RTG 0600-1800 and news 1945 – **R. Rurale de la Haute Guinée**:
Mandiana 88.2MHz, Kankan 92.1MHz, Dabadou 93MHz, Siguiri
97MHz, Douabou 99MHz – **R. Rurale de Kindia**: 88.3MHz, Kakoulima
98.7MHz, Koliadi 99.9MHz – **R. Rurale de N´Zerekore**: 89MHz.

RADIO FAMILIA (Rlg, operated by Actualité Féminine en Guinée)
✉ Timbi **☎**+224 62412893 **W:** familiafm.com **E:** directiontech.
familiafm@yahoo.fr **LP:** DG: Mrs. Colette Baudais. Dir. Tech: Laurent
Koulemou.
SW: Timbi-Madina 4900kHz 1kW. **FM:** Conakry 105.3MHz.
D.Prgr. in French/Susu/Kpèlè/Pular/Maninka: FM: 0600-2400, SW
1800-2400v. Different prgr. on SW and FM.

Private stations:
R. Djiguii, Conakry: 105.7/107.7MHz. **W:** djiguii.com – **R. FM
Liberté:** Conakry 101.3MHz – **R. Nostalgie Guinée:** Conakry:
98.2MHz – **Soleil FM:** Conakry 101.7MHz.
R. France Int: Conakry/Labé 89.9MHz.

GUINEA-BISSAU

L.T: UTC — **Pop:** 1.5 million — **Pr.L:** Portuguese, Crioulo, others
— **E.C:** 50Hz, 220V — **ITU:** GNB

INSTITUTO DAS COMUNICAÇÕES DA GUINÉ-BISSAU(ICGB)
✉ Av. Domingos Ramos 53, C.P. 1372, Bissau **☎**+245 3204873/74 🖷
+245 3204876 **E:** icgb@mail.bissau.net

RADIODIFUSÃO NACIONAL (RDN, Gov.)
✉ C.P. 191, Bissau **☎**+245 3212426 **L.P:** DG: Hipolito José Mendes.
FM: 88/91.5/93.7/98MHz. **D.Prgr:** 0600-1330, 1530-2400.
Ann: "Escutam a Radiodifusão Nacional da República da Guiné-Bissau".

Other stations:
R. Bombolom, Bissau: 106.2MHz. Also rel. BBC & DW – **R. Mavegro,**
Bissau: 100.0MHz. Also rel. BBC – **R. Nossa** (Rlg.), Bissau: 98.9MHz
– **R. Pindjiguiti,** Bissau: 95.0MHz. Also rel. VOA – **R. Sol Mansi**
(Rlg.) Mansoa: 90.0MHz 4kW, Bissau/Bafatá 101.8MHz 1kW. 0630-
2300. Also rel. Vatican R. and UN prgrs.
RFI Afrique: Bissau 94.7MHz in French/Portuguese.
RDP África: Nhacra 88.4MHz 25kW, Gabú 100MHz 1kW +1tx under 1kW.
+20 community radio stations

GUYANA

L.T: UTC -4h — **Pop:** 772,000— **Pr.L:** Creole, English, Hindi, Urdu,
Amerindian dialects — **E.C:** 50Hz, 240V — **ITU:** GUY

PUBLIC UTILITIES COMMISSION
✉ Parliament Buildings, Brickdam, Georgetown **☎** +592 227 3293
🖷+592 227 3534

**NATIONAL COMMUNICATIONS NETWORK INC.
(ex. GUYANA BROADCASTING CORP)**
✉ Broadcasting House, P.O. Box 10760, Georgetown **☎**+592 227
5166 🖷+592 226 2253 **W:** www.ncnguyana.com
L.P: SEO: Mohammed Sattaur GM: Mazrul Bacchus Prod. Mgr: Martin
Goolsarran
MW: Georgetown 560/760kHz 10kW, Linden 700kHz 1kW
NB: 560kHz & 760kHz txs r. *inactive
SW: Sparendaam (G.C: 06N49 058W10): 3290/5950kHz 10kW
FM: Georgetown 100.1/102.5MHz, Linden 106.5MHz
R. Roraima: 0800-0200 on (760kHz*) + 100.1MHz. **N:** 0900, 1000,
1100, 1330, 1500, 1900, 2100, 2230 (Sun), 2300 (W), 0100.
Voice of Guyana: 24h on (560*)/700kHz + 102.5/106.5MHz, 0900-
2200 on 5950kHz, 2200-0900 on 3290kHz **N:** as R. Roraima.
V. by letter.

HAITI

L.T: UTC -5h — **Pop:** 8.9 million — **Pr.L:** Créole, French — **E.C:**
50+60Hz, 110V — **ITU:** HTI

**CONSEIL NATIONAL DES TELECOMUNICATIONS
(CONATEL)**
✉ B.P.2002 (or: Cité de l'Exposition 16), Port-au-Prince **☎** +509

25163325 🖷+509 22239229 **W:** www.conatel.gouv.ht **E:** info@cona-
tel.gouv.ht

MW	kHz	kW	Station, location
4)	630	1	Rdif. Jérémienne, Jérémie
6)	660	5	R. Lumiere, P-au-P
6)	720	1	R. Lumière, Petite Riv.
6)	740	1	R. Lumière, Pignon
6)	760	2	R. Lumière, Cayes
6)	780	0.5	R. Lumière, Jérémie
8)	780	10	Eben-Ezer, Mirebalais
9)	810	0.05	R. Atlantique, Gonaives
11)	840	10	R. 4VEH, Cap Haitien
13)	860	3	R. Men Kontre, Cayes
14)	870	1	R. Express, Jacmel
15)	880	0.3	R. Independance, Gonaives
16)	890	0.5	R. Trans Artibonite, Gonaives
17)	890	1	Voix du Nord'est, Forte Liberte
20)	930	5	R. Cap Haitien, Cap Haitien
21)	930	0.1	R. Echo 2000, Val. de Jacmel
22)	940	0.25	R. St Marc, St marc
23)	940	0.2	Rdif. Jacmelienne, Jacmel
66)	1030		R. Ginen, P-au-P
27)	1080	20	R. Nationale, P-au-P
30)	1170		R. Tropicale Internationale, Jérémie
31)	1190	0.3	R. Grand Anse, Jérémie
32)	1200		Voix de la Paix, Port de Paix
33)	1220	1	Voix du Plateau Central, Hinche
34)	1230	1	Voix de L'ave Maria, Cap Haitien
36)	1280		R. Transcaribbean International, Jean Rabel
37)	1330	10	R. Haiti Inter, P-au-P
38)	1350	0.25	R. Dame Marie, Dame Marie
40)	1370	0.5	R. Citadelle, Cap Haitien
41)	1370	1	Rdif. Cayenne, Cayes
43)	1410	3	Voix de Nord-ouest, Port de Paix
44)	1420	0.5	R. Messie Continental, Dessalines
46)	1460	0.2	Voix du Nord, Cap Haitien

Abbreviations: P-au-P = Port-au-Prince, * = inactive, v = varying fq.
Addresses and other information:
4) 82, Rue Eugene Magron, Jérémie – **6)** Côte Plage 16, Carrefour,
P-au-P (or BP 1050) **H of tr:** 1000-0200 **W:** www.radiomlumiere.org
– **8)** 27, Rue Clair Heureuse, Mirebalais – **9)** Rlle Laporte, Gonaive
– **11)** Box 1, Cap-Haitien (or: Radio 4VEH, P.O.Box 24638, West Palm
Beach, FL 33416, USA) **☎** +509 454 1334 **W:** www.radio4veh.org **E:**
contact@radio4veh.org – **13)** 137, Rue Simon, Cayes – **14)** 31, Rue
Stenio Vincent, Jacmel – **15)** Rue Egalite, Gonaives – **16)** Rue du Quai,
Gonaives – **17)** Rue Bourbons, Forte Liberte – **20)** 30 Rue 10A, Cap
Haitien – **21)** 15, rue Alcius Charmant, Jacmel – **22)** 20 Rue A Thoby,
St Marc – **23)** 32, Rue D'Orleans, Jacmel – **26)** Vaudreuil, Cap Haitien
– **27)** Rue du Magasin d'État (or BP 1143), P-au-P – **30)** Jérémie – **31)**
54, Rue Eugene Magron, Jérémie – **32)** Eveché de P.de.P, Rue L'Hôpital,
Port de Paix – **33)** 657, Rue Toussaint L'Ouverture, Hinche – **34)** Rue
19H, Cap Haitien **E:** radiovoixavemaria@hotmail.com – **36)** 28, Cité
James, Coicou, Jean Rabel – **37)** Delmas 66A, P-au-P – **38)** 252, Rue
Frere Portier, Dame Marie – **40)** Rue 10-11-E, Cap Haitien – **41)** 77, Rue
Duvivier-Hall, Cayes – **43)** 84, Rue Christophe, Port de Paix – **44)** 15,
Rue Jacques 1er, Marchands Dessalines, Dessalines – **46)** Rue 20-A-B,
Cap Haitien – **66)** 9 Bis, Delmas 31, P-au-P
FM in Port au Prince (MHz): 88.1 R. Visa FM – 88.5 R. Kyskeya
– 88.9 R. Indigène – 89.3 RFI – 89.7 LV de l'Espérance – 90.1 R.
One – 90.5 R. Signal FM – 90.9 R. Timoun – 91.3 Tropic FM – 91.7
R. Etoile – 92.1 R. Lumière – 92.5 R. Commerciale – 92.9 R. Ginen
– 93.3 Antilles Internationales – 93.7 R. Vasco – 94.1 R. Nouvelle
Génération – 94.5 Caraïbes FM Stéréo – 94.9 R. MBC – 95.3 LV de
l'Evangile – 95.7 R. Horizon 2000 – 96.1 R. Communauté Haïtienne
2000 – 96.5 R. Sky FM – 96.9 R. Antilles Internationales – 97.3 R.
Mega Star – 97.7 R. Lumière – 98.1 R. Maxima FM – 98.5 R. Ibo – 98.9
R. Maximum Power – 99.3 R. Vision 2000 – 99.7 Sweet FM – 100.1
R. Métropole – 100.5 R. Eclair – 100.9 R. Magic Stéréo – 101.3 R.
Univers FM – 101.7 Energie FM – 102.1 R. Nationale – 102.5 Zenith
FM – 102.9 R. Super Star – 103.3 R. Mélodie FM – 103.7 R. Lakensyèl
– 104.1 R. SODEC Service – 104.5 R. Galaxie – 104.9 R. RFM – 105.1
R. Nationale – 105.5 R. Soleil – 106.1 R. Haïti Inter – 106.5 R. Planet
Kreyol – 106.9 R. Kadans – 107.5 R. Solidarité – 107.7 R. Vertières

RADIO FRANCE INTERNATIONALE
✉ B.P. 1126, Port-au-Prince **☎** +509 22224724 🖷 +509 22229140
E: ablanc@acn2.net **FM:** Port-au-Prince 89.3MHz, Cap Haïtien
100.5MHz

HAWAII (USA)

L.T: UTC -10h — **Pop:** 1.28 million — **Pr.L:** English, Japanese, Filipino
— **E.C:** 60Hz, 120V — **ITU:** HWA

FEDERAL COMMUNICATIONS COMMISSION (FCC)
see USA for details

THE HAWAII ASSOCIATION OF BROADCASTERS, INC.
P.O. Box 22112, Honolulu HI 96823-2112 **W:** www.hawaii-broadcasters com **E:** stephanieueyeda@hawaii.rr.com **L.P:** Pres: Chris Leonard, Exec. Dir.: Stephanie Uyeda

MW	kHz	kW	Call	Location
1)	550	5	KMVI	Wailuku, Maui
37)	570	1	KQNG	Lihue, Kauai
3)	590	7.5	KSSK	Honolulu, Oahu
4)	620	5	KHNU	Hilo, Hawaii
4)	620	10	KHNU	Kalaoa, Hawaii
4)	620	10	KHNU	Naalehu, Hawaii
5)	650	10	KRTR	Honolulu, Oahu
6)	670	5	KPUA	Hilo, Hawaii
7)	690	5	KHNR	Honolulu, Oahu
37)	720	5	KUAI	Eleele, Kauai
35)	740	5		Kihei, Maui (CP)
7)	760	10	KGU	Honolulu, Oahu
1)	790	5	KKON	Kealakekua, Hawaii
3)	830	10	KHVH	Honolulu, Oahu
1)	850	5	KHLO	Hilo, Hawaii
1)	880	2	KHCM	Honolulu, Oahu
1)	900	5	KNUI	Kahului, Maui
5)	940	10	KKNE	Honolulu, Oahu
5)	990	5	KIKI	Honolulu, Oahu
8)	1010	50/10		Honokaa, Hawaii (CP)
1)	1040	10	KLHT	Honolulu, Oahu
10)	1060	5	KIPA	Hilo, Hawaii*
11)	1080	5	KWAI	Honolulu, Oahu
2)	1110	5	KAOI	Kihei, Maui
12)	1130	1	KPHI	Honolulu, Oahu
36)	1170	0.33/0.14	KORL	Honolulu, Oahu (CP)
36)	1180	1	KORL	Honolulu, Oahu*
13)	1210	1	KZOO	Honolulu, Oahu
14)	1270	5	KNDI	Honolulu, Oahu
15)	1320	5	KEWA	Ewa Beach, Oahu*
6)	1340	10		Honaunau, Hawaii (CP)
16)	1370	6.2	KUPA	Pearl City, Oahu
17)	1420	5	KKEA	Honolulu, Oahu
18)	1460	5	KHRA	Honolulu, Oahu
37)	1500	10	KUMU	Honolulu, Oahu
19)	1540	5	KREA	Honolulu, Oahu
20)	1570	15	KUAU	Haiku, Maui

FM	MHz	kW	Call	Location
21)	88.1	35	KHPR	Honolulu, Oahu
21)	88.1	1	KHPR-FM1	Makaha, Hawaii
21)	88.7	5	KIPH	Kailua, Hawaii (CP)
9)	88.9	21	KHJC	Lihue, Kauai
21)	89.3	38.5	KIPO	Honolulu, Oahu
21)	89.3	1	KIPO-FM1	Makaha, Hawaii (CP)
21)	89.7	36.9	KIPM	Hana, Maui (CP)
22)	90.1	3	KTUH	Honolulu, Oahu (CP)
38)	90.1	4		Lihue, Kauai (CP)
22)	90.3	3	KTUH	Honolulu, Oahu
23)	90.3	5	KCIF	Hilo, Hawaii
25)	90.5	2.2		Hauula, Oahu (CP)
24)	90.5	9	KPHL	Ocean View, Hawaii*
21)	90.7	56	KKUA	Wailuku, Maui
26)	90.9	0.9	KKCR	Hanalei, Kauai
21)	91.1	30	KANO	Hilo, Hawaii
27)	91.7	1	KAHU	Pahala, Hawaii*
26)	91.9	1.75	KAQA	Kilauea, Kauai
10)	92.1	4.5	KHWI	Holualea, Hawaii
3)	92.3	100	KSSK-FM	Waipahu, Oahu
1)	92.5	1.7	KLHI-FM	Kahului, Maui
10)	92.7	7.5	KHBC	Hilo, Hawaii
37)	93.1	100	KQMQ-FM	Honolulu, Oahu
3)	93.1	10	KMWB	Captain Cook, Hawaii
1)	93.5	72	KPOA	Lahaina, Maui
37)	93.5	51	KQNG-FM	Lihue, Kauai
3)	93.9	100	KIKI-FM	Honolulu, Oahu
1)	93.9	7.3	KLUA	Kailua-Kona, Hawaii
37)	94.7	100	KUMU-FM	Honolulu, Oahu
6)	94.7	51	KWXX-FM	Hilo, Hawaii
2)	95.1	100	KAOI-FM	Wailuku, Maui
2)	95.5	100	KAIM-FM	Honolulu, Oahu
1)	95.9	39	KPVS	Hilo, Hawaii
1)	95.9	1	KPVS-FM1	Ocean View, Hawaii (CP)
37)	95.9	51	KSRF	Poipu, Kauai
5)	96.3	75	KRTR-FM	Kailua, Oahu
29)	96.9	100	KFMN	Lihue, Kauai
28)	97.1	38	KNWB	Hilo, Hawaii
1)	97.3	1.5	KRKH	Wailea-Makena, Maui
37)	97.5	80	KHCM-FM	Honolulu, Oahu
1)	97.9	51	KKBG	Hilo, Hawaii
1)	97.9	1	KKBG-FM1	Ocean View, Hawaii (CP)
12)	98.1	51	KJMQ	Lihue, Kauai
1)	98.3	9.4	KJMD	Pukalani, Maui
3)	98.5	51	KDNN	Honolulu, Oahu
12)	98.9	51	KITH	Lihue, Kauai
1)	99.1	7.3	KAGB	Waimea-Kamuela, Hawaii
7)	99.5	100	KHUI	Honolulu, Oahu
1)	99.9	72	KJKS	Kahului, Maui
12)	99.9	51	KTOH	Kalaheo, Kauai
1)	100.3	35	KAPA	Hilo, Hawaii
1)	100.3	7.3	KAPA-FM1	Puueo, Hawaii
5)	100.3	100	KCCN-FM	Honolulu, Oahu
12)	101.1	100	KORL-FM	Waianae, Oahu
6)	101.1	6.5	KAOY	Kealakekua, Hawaii
3)	101.9	100	KUCD	Pearl City, Oahu
37)	102.1	50	KTBH-FM	Kurtistown, Hawaii
2)	102.3	1.9	KMKK-FM	Kaunakakai, Molokai
37)	102.7	61	KDDB	Waipahu, Oahu
30)	102.9	1.5	KLZY	Paia, Maui*
2)	103.3	51	KSHK	Kekaha, Kauai
31)	103.5	2.2	KHAI	Wahiawa, Oahu
2)	103.7	100	KNUQ	Paauilo,Maui
5)	104.3	75	KPHW	Kaneohe, Oahu
1)	104.7	72	KONI	Lanai City, Maui
5)	105.1	100	KINE	Honolulu, Oahu
4)	105.3	28	KBGX	Keaau, Hawaii
4)	105.3	1	KBGX-FM1	Naalehu, Hawaii (CP)
32)	105.5	59	KPMW	Haliimaile, Maui
37)	105.9	100	KPOI	Honolulu, Oahu
1)	106.1	7.3	KLEO	Kahaluu-Kona, Hawaii
28)	106.5	72	KRYL	Haiku, Maui
33)	106.7	25	KNAN	Nanakali, Oahu*
34)	106.9	5.5	KWYI	Kawaihae, Hawaii
4)	107.7	28	KKOA	Volcano, Hawaii
4)	107.7	1	KKOA-FM5	Naalehu, Hawaii
7)	107.9	100	KKOL	Aiea, Oahu

Txs on air or CP less than 1kW not mentioned. *) currently off-air

Addresses and other information:
Addresses: Add state abbreviation HI between location and zip code as appropriate. **Prgr:** All stns 24h unless otherwise stated
1) Pacific Radio Group, Maui: KMVI **W:** www.espn550.com KNUI **W:** www.foxnews900.com KLHI-FM **W:** www.x925.fm KPOA **W:** www.kpoa.com KJMD **W:** www.dajam983.com KJKS **W:** www.kiss99fm.com 311 Ano Street, Kahului, 96732-1304; **Hawaii:** KKON **W:** www.espnhawaii.com KHLO **W:** www.espnhawaii.com KLUA KPVS KKBG **W:** www.kbigfm.com KAGB **W:** www.kaparadio.com KAPA **W:** www.kaparadio.com KLEO **W:** www.kaparadio.com 913 Kanoelehua Ave, Hilo 96720-5116 – **2) Visionary Related Entertainment LLC, Molokai:** KMKK-FM **W:** www.vremaui.com/KMKK 130 Kamehameha V Highway, Kaunakakai 96748; **Maui:** KAOI **W:** www.kaoi1110.com KAOI-FM **W:** www.kaoifm.com KNUQ **W:** www.q103maui.com 1900 Main Street, Wailuku 96793-1900 – **3) Capstar TX Ltd Partnership, Oahu:** KSSK **W:** www.ksskradio.com KHVH **W:** www.khvhradio.com KHBZ **W:** www.khbz.com KSSK-FM KIKI-FM [www.hot939.com KDNN **W:** www.island985.com KUCD **W:** www.star1019.com 650 Iwilei Rd #400, Honolulu 96817-5319 – **4) Mahalo Broadcasting LLC, Hawaii:** KHNU KBGX **W:** www.lava1053.com KKOA 74-5605 Luhia St #B-7, Kailua-Kona 96740-1678 – **5) Cox Radio Inc, Oahu:** KRTR **W:** www.650amhawaii.com KKNE **W:** www.am640hawaii.com KRTR-FM **W:** www.krater96.com KCCN-FM **W:** www.kccnfm100.com KPHW **W:** www.power1043.com KINE **W:** www.hawaiian105.com 900 Fort St Mall #700, Honolulu 96813-3797 – **6) New West Broadcasting Corporation, Hawaii:** KPUA **W:** www.kpua.net K*** [1340] KWXX-FM **W:** www.kwxx.com KNWB **W:** www.b97hawaii.com KMWB **W:** www.b97hawaii.com KAOY **W:** www.kwxx.com 1145 Kilauea Ave, Hilo 96720-4203 – **7) Salem Media of Hawaii Inc, Oahu:** KHNR **W:** www.khnrtownhall.com KGU **W:** www.kguradio.com KHCM [**Prgr:** 24h. China R. International relay in English, Chinese, Korean & Japanese] KAIM-FM **W:** www.thefishhawaii.com KHCM-FM **W:** www.975countrykhcm.com KHUI **W:** www.khuiradio.com KKOL **W:** www.oldies1079honolulu.com 1160 N King St #200, Honolulu 96817-330 – **8) Family Worship Center Church Inc, Hawaii:** K*** [1010]: PO Box 262550, Baton Rouge LA 70810 – **9) Calvary Chapel of Honolulu Inc, Oahu:** KLHT **W:** www.klight.org 98-106 Komo Mai Drive, Aiea 96701-1901 ; **Kauai:** KHJC: 2970 Kele St #117, Lihue 96766-1803 – **10) Parrott Broadcasting Ltd Partnership, Hawaii:** KIPA KHWI KHBC 688 Kinoole St #112, Hilo 96720-3877 – **11) Radio Hawaii Inc, Oahu:** KWAI **W:** www.kwai1080am.com 100 N Beretania St #401, Honolulu 968174724 – **12) Hochman-McCann Hawaii Inc, Oahu:** KPHI KORL-FM 900 Fort St Mall #450, Honolulu 96813-3713; **Maui:** KRKH KONI 300 Ohukai Road #C-318, Kihei 96753-7050; **Kauai:** KJMQ KITH KTOH

4334 Rice St #204-B, Lihue 96766-1801 **W:** www.hhawaiimedia.net – **13) Polynesian Broadcasting Inc, Oahu:** KZOO **W:** www.kzoohawaii.com **Prgr:** Japanese. ☒ 2752 Woodlawn Drive #5-204, Honolulu 96822-1855 – **14)Broadcast House of the Pacific Inc, Oahu:** KNDI **W:** www.kndi.com **Prgr:** multicultural, religious. ☒ 1734 S King St, Honolulu 96826-2042 – **15) KM Communications Inc, Oahu:** KEWA: 3654 W Jarvis Ave, Skokie IL 60076 – **16) Broadcasting Corporation of America, Oahu:** KUPA: 4766 Holladay Blvd, Holladay UT 84117 – **17) Blow Up LLC, Oahu:** KKEA **W:** www.sportsradio1420.com ☒ 1088 Bishop St #LL2, Honolulu 96813-3113 – **18) RK Media Group, Oahu:** KHRA **Prgr:** Korean. ☒ 1311 Kapiolani Blvd #204, Honolulu 96814-4513 – **19)** JMK Communications Inc, Oahu: KREA **Prgr:** Korean. ☒ 1839 S King St #203, Honolulu 96814-2137 – **20) First Assembly King's Cathedral & Chapel, Maui:** KUAU **W:** www.kingscathedral.com – **21) Hawaii Public Radio Inc, Oahu:** KHPR **W:** www.hawaiipublicradio.org] KIPO [includes BBC relay] ☒ 738 Kaheka St #101, Honolulu 96814-3726; **Simulcast: Maui:** KKUA **F.PL:** KIPM (KIPO relay) **Hawaii:** KANO **F.PL:** KIPH (KIPO relay) – **22) The University of Hawaii, Oahu:** KTUH **W:** www.ktuh.org ☒ 202 Hemenway Hall, University of Hawaii, 2445 Campus Road, Honolulu 96822-2216 – **23) Hilo Christian Broadcasting Corporation, Hawaii:** KCIF **W:** www.kcifhawaii.org ☒ 180 Kinoole St, Hilo 96720-2827 – **24) Vineyard Christian Fellowship of Honolulu Inc, Hawaii:** KPHL ☒ 250 Kawaihea St #15E, Honolulu 96825 – **25) Halau Lokahi Public Charter School, Oahu:** K**** [90.5]: 401 Waikakamilo Road #1A, Honolulu 96817 – **26) Kekahu Foundation Inc, Kauai:** KKCR **W:** www.kkcr.org **F.PL:** 20kW] KAQA **W:** www.kkcr.com **F.PL:** 4.5kW] ☒ 4520-D Hanalei Plantation Road, PO Box 825, Hanalei 96714-0825 – **27) Haola Inc, Hawaii:** KAHU: PO Box 5024, Hilo 96720-1054 – **28) Captain Cook Broadcasting Inc, Hawaii** KMWB [see #6] **Maui:** KRYL. 8215 Birch St, New Orleans LA 70118 – **29) FM97 Associates – Kauai:** KFMN **W:** www.kfmn97.com ☒ 1860 Leleiona Road, Lihue 96766-9000 – **30) Chaparral Broadcasting Inc, Maui:** KLZY: c/o Jerrold T Lundquist, 14 Cockenoe Drive, Westport CT 06880-6908 – **31) Educational Media Foundation, Oahu:** KHAI: 5700 W Oaks Blvd, Rocklin CA 96765 [**F.PL:** 53kW] – **32) Rey-Cel Broadcasting Inc, Maui:** KPMW **W:** www.myspace.com/wild105 ☒ 230 Hana Hwy, Kahului 96732 – **33) Big D Consulting Ltd, Oahu:** KNAN: 3800 Howard Hughes Parkway 17th Floor, Las Vegas NV 89109 – **34) Colin H Naito, Hawaii:** KWYI ☒ 64-1040 Mamalahoa Hwy #4, Kamuela 96743-6540 **-35) IHR Educational Broadcasters, Maui:** K*** PO Box 180, Tahoma CA 96142 – **36) Centro Cristiano Vida Abundante Inc, Oahu:** KORL 121 W Alvin St, Santa Maria CA 93421 – **37) Ohana Broadcast Company LLC, Kauai:** KQNG KQNG-FM **W:** www.kongradio.com KUAI KSRF **W:** www.surf959fm.com KSHK **W:** www.shaka103.com 4271 Halenani St, Lihue 96766-1312 **Oahu:** KQMQ-FM **W:** www.931thezone.net KUMU KUMU-FM **W:** www.kumu.com KDDB KPOI **W:** www.kpoifm.com 765 Amana St #206, Honolulu 96814-3248 **Hawaii:** KTBH – **38) Calvary Chapel Kauai, Kauai:** K*** [90.1] PO Box 1062, Kapa'a 96746

HONDURAS

L.T: UTC -6h — **Pop: 7.**6 million — **Pr.L:** Spanish — **E.C:** 60Hz, 110V — **ITU:** HND

COMISION NACIONAL DE TELECOMUNICACIONES (CONATEL)
☒ Ap. 15012 (or Edificio CONATEL, Colonia Modelo, Sexta Avenida Suroeste Contigua a Hondutel), Tegucigalpa ☎ +504 234 8600 ☒ +504 236 8611 **W:** www.conatel.hn **E:** conatel@conatel.gob.hn

ASOCIACION NACIONAL DE RADIODIFUSORES DE HONDURAS (ANARH)
☒ Ap. 4039, Tegucigalpa
Hrs of tr usually 24h – see address section for variations. Call HR–

MW	Call	kHz	kW	Station, location
155)	XT	550	1	R. X, Tegucigalpa
2)	XD	550	0.5	R. Manantial, Sta Rosa de Copán
76)	VF	560	1	R. Valladolid, Comayagua
3)	RZ	560	5	VRZ R. Juticalpa, Juticalpa
4)	KL	560	1	R. Reloj, San Pedro Sula
157)	OX	570	1	R. El Triunfo, Choluteca
5)	LP	570		R. América, Tela (r: 610)
3)	ZQ	580	3	R. Noticias STC, Tegucigalpa
112)	EO	580		Super Estrella de Occidente, Sta Rosa de Copán
5)	LP3	590		R. América, San Pedro Sula (r. 610)
227)	RE	590		R. Renacer, Catacamas
209)	EK	600		R. Orion, La Ceiba
5)	LD	610	10	R. América, Tegucigalpa
5)	LP	610	10	R. América, Sta Rosa de Copán (r:610)
5)	LP5	620	1	R. América, Comayagua (r:610)
5)	LP	620	1	R. América, Juticalpa (r:610)

MW	Call	kHz	kW	Station, location
28)	LP17	620	1	R. Continental, San Pedro Sula
218)	LO	620		R. Litoral, Tocoa
5)	LP	630	1	R. América, Choluteca (r:610)
5)	LP7	630	1	R. América, La Ceiba (r:610)
8)	UP	640	1	R. Centro, Tegucigalpa
185)	JT	640		R. Jerusalen, Sta Bárbara
241)	VS	650	25	Nuestra Señora de La Esperanza, San Pedro Sula
5)	LP	650	15	R. América, Danlí (r:610)
182)	TA	650		R. Turquesa, Siguatepeque
198)	VS	650	1	R. Católica Olancho, Olanchito
8)	NN18	660	3	LV de Honduras, La Ceiba (r:670)
239)	KV	*660		R. Betania, Choluteca
8)	N	670	10	LV de Honduras, Tegucigalpa
8)	NN20	670	1	LV de Honduras, Sta Rosa de Copán (r:670)
8)	NN8	680	10	LV de Honduras, San Pedro Sula (r:670)
8)	NN11	680	10	LV de Honduras, Tocoa (r:670)
8)	NN2	680	10	LV de Honduras, Siguatepeque (r:670)
8)	NN7	680	1	LV de Honduras, Danlí (r:670)
8)	NN10	680	1	LV de Honduras, Juticalpa (r:670)
8)	NN9	680	1	LV de Honduras, Tela (r: 670)
8)	NN3	690	1	LV de Honduras, Choluteca (r:670)
203)	KL	700	5	R. Reloj, Tegucigalpa
8)		700		LV de Honduras, Olanchito (r: 670)
9)	RH	710	3	LV de Occidente, Sta Rosa de Copán
14)	UP3	710	1	R. Rock 'n Pop, San Pedro Sula
10)	LK	710	2	R. Comayagua/LV Católica, Comayagua
11)	KN	710	2.5	LV de Olancho, Catacamas
8)	NN13	710	1	LV de Honduras, Yoro (r:670)
79)	NN3	720	1	R. Caribe, La Ceiba
238)	ZN	720		R. San Lorenzo, San Lorenzo
8)	NN4	730	1	R. Exitos, Tegucigalpa
162)	XG	730	0.25	R. Cadena Dial, Sta Bárbara
12)	QQ	740	1	R. Intibuca, La Esperanza
13)	IH	740	1	7-40 La Super Grande, Juticalpa
14)	TG2	740	1	R. Satélite, San Pedro Sula.
90)	VC	740		LV Evangélica, Olanchito (r: 1390)
18)	XW	760	2.5	R. Comayagüela/Stereo Azul, Comayagüela
231)	IJ	760		R. Jicatuyo, San José de Colinas
14)	NN21	770	10	R. Norte, San Pedro Sula
19)	MV	770	0.5	R. Aguán, Olanchito
20)	PI	770		R. Sui Generis, Comayagua
135)	RD	770	1	R. Majestad "LV del Guayape", Juticalpa
5)	QN	780		R. Sonora, La Ceiba
163)	SE	780	1	Alabanza Estéreo, Choluteca
20)	FI	790	1	R. Feliz, Sta Bárbara
8)	TG	790	3	R. Satélite, Tegucigalpa
21)	DL	v800	1	R. Corporación, Comayagua
5)	QN	800	1	R. Sonora, Danlí
217)	MD	800		R. Yoro, Jocon
170)	GW	800		R. Patria, Catacamas
17)	MA	800	3	R. Moderna, San Pedro Sula
90)	VC	810	6	LV Evangélica, La Ceiba (r:1390)
25)	LP24	810	3	R. Valle, Choluteca
5)	LP16	820	5	R. Moderna, Tegucigalpa
84)	KW	820	7/3	R. Sultana, Sta Rosa de Copán
24)	RU	830	1	R. Uno, San Pedro Sula
26)	JB	830	1	Cadena Radial Impacto, Comayagua
27)	VQ	830	1	R. Excelsior, Juticalpa
216)	TB	830		R. Colón, Tocoa
18)	CR	840	1	Dif. Cristiana de Radio "DCR", Choluteca
8)	UP	850	10	R. Televisión, Tegucigalpa
165)	IF	850	0.5	R. Inspiración, La Entrada
28)	BS	860	10	R. San Pedro, San Pedro Sula
110)	LS	860	0.5	R. Dinorama, La Paz
215)	NZ	860		La Respuesta es la Cruz, Olanchito
225)	BV	860		R. Piedra Blanca-LV de Nuestra Gente, Catacamas
1)	H9	870	6	R. Nacional de Honduras, La Ceiba (r:880)
1)	H10	870	5	R. Nacional de Honduras, Puerto Lempira (r:880)
1)	H4	870	3	R. Nacional de Honduras, Nacaome (r:880)
1)	H	880	10	R. Nacional de Honduras, Tegucigalpa
1)	H5	880	5	R. Nacional de Honduras, Sta Rosa de Copán
1)	H3	890	10	R. Nal de Honduras, San Pedro Sula (r:880)
1)	H7	890	3	R. Nacional de Honduras, Juticalpa (r:880)
1)	H9	890	10	R. Nacional de Honduras, Siguatepeque (r:880)
1)	H2	890	1	R. Nacional de Honduras, Comayagua (r:880)
1)	H6	890	5	R. Nacional de Honduras, El Paraíso (r:880)
1)	H8	890	5	R. Nacional de Honduras, Olanchito (r:880)
1)	H	890	1	R. Nacional de Honduras, Danlí (r: 880)
8)	UP6	900	1	R. Satélite, La Ceiba
8)	UP	900	1	R. Centro, Choluteca
29)	VS	910	10	R. Católica "LV de Suyapa", Tegucigalpa
151)	NM	910	2.5	R. Comunidad, Ocotepeque
21)	RM	920	1	R. Sistema, Comayagua

MW	Call	kHz	kW	Station, location
31)	ZV	920	1	Una Voz que clama en el desierto, San Pedro Sula
32)	SK	920	5	R. Catacamas, Catacamas
1)	H11	920	1	R. Nacional de Honduras, Danlí (r:880)
214)	VS	920		R. Católica, Tocoa
208)	CQ	930		Cadena R. Samaritano, La Ceiba
237)	LD	930		R. Estéreo Leed, Nacaome
18)	CR	940	1	R. Dif. Cristiana de R. "DCR", Tegucigalpa
15)	BO	940	1	R. Cadena Occidental, La Entrada
34)	QL	950	1	Centro Radial Hondureño, Siguatepeque
35)	ZE	*950	1.5	R. Cortés AM, Puerto Cortés
213)	XI	950		R. El Camino, Olanchito
224)	QJ	950		R. Agalta, San Esteban
246)		950	6	R. Choloma, Choloma
36)	YF	960	1	R. Fergusón, Choluteca
203)	XB	960		R. Bautista Buenas Nuevas, Puerto Lempira
38)	LY	970	2	R. Milenium, Tegucigalpa
230)	KI	970		La Picosa, N. Ocotepeque
41)	AO	980	1	R. Tocoa, Tocoa
39)	ZC	980	2	R. Monumental, San Pedro Sula
90)	VC	980		LV Evangélica, Siguatepeque (r:1390)
223)	UI	*980		Super 10, Catacamas
140)	PR	990	3.5	R. Paz, Choluteca
207)	OJ	990		R. Vida Cristiana, La Ceiba
44)	XZ	1000	1	R. Alfa, Tegucigalpa
256)		1000	6	R. Río de Piedras, Lempira
5)	QN	1010		R. Sonora, San Pedro Sula
89)	CD	1010	1	R. Constelación. Juticalpa
228)	AE	1010		R. Apaguiz, Danlí
255)		1010		R. Visión Cristiana, Tocoa
236)	PN	1020		R. Visión Cristiana Roca de Salvación, Marcovia
152)	RJ	1030	1	R. Ticante, Ocotepeque
8)	UP3	1030	1	R. Rock 'n Pop, Tegucigalpa
14)	NNY	1040	3	Exitos, San Pedro Sula
33)	MJ	1040	1	R. Renovación, Comayagua
90)	VC	1040		LV Evangélica, Juticalpa (r:1390)
90)	VC	1040		LV Evangélica, Danlí (r: 1390)
206)	OK	1050		R. Roatán, Roatán
7)	KT	1060	2	La Catracha, Tegucigalpa
50)	FA	1060	0.5	R. Peña Blanca, Sta Barbara
52)	GR	1070	3	Cadena Guaymuras, El Paraíso
53)	LE	1070	1	R. Unica AM, San Pedro Sula
54)	QN	1070	2.5	R.Sonora, Siguatepeque
180)	BN	1070		R. Unidad Evangélica, Catacamas
56)	ID	1080	1	R. Miramar, Tela
235)	IE	1080		R. Evangélica Senda de Vida, Nacaome
8)	LB	1090	1	R. La Mejor, Sta Rosa de Copán
90)	CQ	1090	1	Cad. Radial Samaritano, Tegucigalpa
58)	ND	1100	1	R. Esperanza, La Esperanza
59)	VA	1100	1	R. Tiempo/R. Fama, San Pedro Sula
60)	FQ	1100	1	R. Máxima, Olanchito
222)	AJ	1100		R. Antena 5, Catacamas
5)	QN	1110		R. Sonora, Choluteca
38)	TL	1120	2	R. Fiesta, Tegucigalpa
196)	VR	1120		R. Marchala, Ocotepeque
61)	PL	1130	5	R. Progreso, El Progreso
63)	BT	1130	1	R. San Francisco, San Francisco de la Paz
99)	HP	1130	1	R. Pinares, Siguatepeque
249)		1130		R. Pirata, Sonaguera
65)	UL	1140	1	R. Pico Bonito1140 AM, La Ceiba
90)	VC	1140		LV Evangélica, Choluteca (r: 1390)
5)	LP12	1150		R. Universal, Tegucigalpa
66)	AV	1150	5	Ondas del Ulúa, Sta Bárbara
68)	GF	v1160	0.5	R. El Paraíso, El Paraíso
34)	VZ	1160	1	R. Juan Pablo II, Siguatepeque
204)	FJ	1160		R. País "LV del Valle de Sula", Progreso
212)	HZ	1160		R. Liberación, Tocoa
221)	BJ	1160		R. Nueva Palestina, Nueva Palestina
45)	AF	1170	2	R. Campeonísima, Choluteca
149)	VS	1180	1/0.8	R. Congolon, Gracias: 1100-0500
188)	AZ	1180	1	R. La Tigre, Tegucigalpa
134)	GK	1190	1	R. Brassabola, Minas de Oro
240)	ZQ	1190		R. Notícias STC, El Progreso
259)		1190		R. Ecológica de Olancho, Catacamas
72)	SI	1200	1	R. Impacto, Tela
73)	RO	1210	1	R. Capital, Comayagua
114)	MY	1210	1	LV Evangélica, La Entrada (r: 1390)
74)	OP	1220	1	R. Costeña Ebenezer, San Pedro Sula
75)	YS	1220	1	R. Suari, Marcala
148)	SD	1220	3	R. Destellos de Luz, Sabá
260)		1220		R. Sintonía, Juticalpa
56)	QW	1230	10	R. Tela, Tela
171)	CQ	1230	0.25	R. Samaritano, San Marcos de Colón
133)	ZC	1240	1	R. Vanguardial, Tegucigalpa
172)	VN	1240	1	R. Venus, Sta Bárbara
51)	YF	1250	1	R. Cristiana 1250, Comayagua
115)	DG	1250	1	R. Oriental, Danlí
200)	YL	1250	1	R. Sonaguera, Sonaguera
253)		1250		Super R., San Pedro Sula
77)	FP	1260	1	R. Amistad, San Marcos de Colón
5)	QN	1270	1	R. Sonora, Tegucigalpa
117)	OF	1270	1	Ecos del Celaque, Gracias
78)	AM	1280	1	R. Unción AM, Olanchito
136)	RF	1280	1	R. Cadena de Notícias, San Pedro Sula
107)	BN	1280	1	R. San Miguel, Marcala
251)		1280		R. Armonía, Juticalpa
81)	NN26	1290	1	R. Choluteca, Choluteca
118)	GS	1290	1	R. HRGS/Bay Island Christian Network, Utila
82)	LR	1300	5	R. Sta Rosa, Sta Rosa de Copán
83)	IV	1300	1	R. C.C.I., Tegucigalpa
90)	VC	1310	2.5	LV Evangélica, San Pedro Sula (r:1390)
103)	RL	1310	1	R. Libertad, Marcala, La Paz
258)	CM	1310	5	R. Universidad de Agricultura, Catacamas
119)	MG	1320	1	R. Bahía "La Super Grande", La Ceiba
173)	FL	1330	1	R. Florida, La Entrada
262)		1330		R. Emisora Evangélica, Tegucigalpa
153)	TQ	1340	10	R. 1430/R. El Mundo, San Pedro Sula
120)	CQ	1340	1	Cadena Radial Samaritano, Comayagua
245)		1340		Telecolor R, Catacamas
143)	JV	1350	1	R. Henecan, San Lorenzo
193)	EL	1350		R. Estelar, La Ceiba
28)	BS	1360	1	R. San Pedro, Tegucigalpa (r:860)
197)	BH	1360	5	R. Sta Bárbara, Sta Bárbara
175)	SQ	1370	1	R. El Shaddai R., Siguatepeque
88)	ST	1370	1	R. Fraternidad, San Pedro Sula
220)	ZG	*1370		R. Guayapeña, Catacamas
127)	AH	1380	0.5	R. Redención, Jutiapa
176)	EJ	1380	1	R. Monjaras, Choluteca
90)	VC	°1390	10/5	LV Evangélica, Tegucigalpa
90)	VC	1390	1	LV Evangélica, Sta Rosa de Copán (r:1390)
80)	YT	1400	1	R. Estrella de Oro, San Pedro Sula
177)	AU	1400	1	R. Alegre, Sava Colón
187)	BO	1400		R. Punto, Comayagua
219)	UV	1400		R. Universitaria, Catacamas
94)	SL	1420	1	R. Stereo Actualidad, Trinidad
244)	GB	1420		R. Sabanagrande, Sabanagrande
257)	FO	1430	1	R. Shekina, Puerto Cortés
97)	VM	1430	1	R. Maranatha, La Paz
211)	QV	1430		La Nueva Potencia, Olanchito
250)		1430		Ministerios Cristianos Fuente de Vida, Juticalpa
98)	RD	1440	5	R. Belén, La Ceiba
121)	RY	1440	0.5	R. Ekklesia Int., San Marcos de Colón
144)	BR	1450	1	R. Cultural, La Entrada
244)	GB	1450		R. Sabanagrande, Tegucigalpa
202)	GC	1460	2.5	R. Conga, San Pedro Sula
122)	CX	1460	0.5	LV de Patuca, Catacamas
113)	KS	1460	0.5	R. Ministerio Bautista, Yoro
242)	FR	1460		R. Firmamento, La Paz
234)	XH	1470		R. Globo Grupera, Nacaome
145)	WP	1480	1	R. Soberanía, San Marcos, Ocotepeque
102)	EZ	°1480	1	LV de Misiones "R. MI", Comayagüela
35)	GO	1490	1.2	R. Porteña, Puerto Cortés
104)	OM	1490	1	R. Omega "Sonido Internacional", La Esperanza
210)	OE	1490		R. Pijol, Morazán
105)	TX	1500	1	R. Victoria, Choluteca
109)	VP	1500		R. Sion, La Ceiba
261)		1500		R. MI-EL, Sabá
124)	PG	1510	1	R. Gualcho, Tegucigalpa
106)	EM	1510	1	R. Emanuel, Ocotepeque
108)	CR	1520	1	Dif. Cristiana de R. "DCR", San Pedro Sula
131)	HJ	1520	1	R. Santiago, Yoro
183)	DF	1520		R. Rios de Agua Viva, Siguatepeque
192)	MQ	1520		R. Manantial de Vida Eterna, Juticalpa
247)		1530		Super Q, Choluteca
194)	VK	1540		R. Nuevo Mundo "Cadena Radial Reloj", Tegucigalpa
132)	JX	1550	1	R. Cristiana Nueva Vida, San Pedro Sula
101)	JO	1550	1	R. Campeona, Comayagua
248)		1550		R. Wuampu, Dulce Nombre de Culmi
233)	FD	1560	1	R. Mi Preferida, Choluteca
57)	RF	v1570	2.5	R. Cad Nal de Noticias "RCN", Tegucigalpa
229)	TF	1570		Difusora Cristiana Torre Fuerte, Gracias
252)		1580		R. La Voz Lenca, La Esperanza
181)	BX	1590		R. Perla, El Progreso
232)	ZL	1590		R. Zol, Choluteca
111)	PC	°1600	1	R. Luz y Vida, San Luís
243)	PQ	1600		R. Poderosa, Tegucigalpa

° = also on SW, * = inactive, (r) = repeater, v = varying fq.

SW	Call	kHz	kW	Name and h of tr
111)	PC	v3250	1	R. Luz y Vida, San Luís: 1100-1600, 2200-0400
102)	EZ	3340		R. Misiones Int. "R.MI", Comayagüela
90)	VC	*4819	5	LV Evangélica, Tegucigalpa
102)	EZ	*5010	1	R. Misiones Int., Comayagüela: 1200-0500

Stns with a (*) are reported to be inactive, but may occasionally be reactivated for variable periods of time.

Addresses and other information:
1) Ap 403 (or Avenida Miguel de Cervantes, Ed Lucía, frente a la Iglesia Menonita, Colonia: Av La Paz), Tegucigalpa **W:** www.rnh.hn – **2)** 1a Av 439, Barrio San Martín, Sta Rosa de Copán - 1115-0300 ☎ +504 662 0318 – **3)** Ap 3, Juticalpa - 1100-0400 - **FM:** 97.9MHz – **4)** Ap 24, San Pedro Sula – **5)** Edif Audio Video, Ap 259, Tegucigalpa - 1030-0400 **W:** www.radioamerica.hn ☎+504 290 4950 圖+504 232 2923 – **6)** Blvd Morazán, Edificio Classic, 2ndo piso, Frente a Banco Ficohsa, Tegucigalpa **W:** www.radiocadenavoces.com – **8)** Emisoras Unidas, Col Florencia, Blv Suyapa (or: Ap 642), Tegucigalpa - 1045-0500 **W:** www.radiohrn.hn **E:** noticias@radiohrn.hn – **9)** Ap 206, Sta Rosa de Copán - **FM:** 92.1MHz – **10)** Ap 347, Comayagua - **FM:** 90.3MHz – **12)** Barrio El Way, Calle Principal, La Esperanza, Intibucá - 1100-0100 ☎ +504 783 0171 – **13)** Ap 9, Juticalpa 1200-0400 – **14)** Emisoras Unidas, Ap 163, San Pedro Sula - 1100-0400 – **15)** Atras de Gasolinera Shell, La Entrada, Copán ☎ +504 661 2425 – **17)** Av New Orleans 20C, San Pedro Sula – **18)** Ap 3448 (or: Iglesia Amor Viviente, Col Godoy frente a F.H.I.S.), Tegucigalpa - 1200-0400 – **19)** Coyoles Central, Olanchito, Yoro – **20)** Ap 26, Sta Bárbara - **FM:** 90.9MHz – **21)** Barrio San Francisco, Fte Parque, Comayagua - 1100-0400 - **FM:** 99.9MHz – **22)** Danlí, El Paraíso – **24)** Edif.Maranata, Calle 8 y 9, San Pedro Sula ☎ +504 350 4614 – **25)** Ap 29, Choluteca - 1000-0400 – **26)** Ap 33, Comayagua – **27)** Ap 28, Juticalpa - 1100-0400 ☎ +504 885 1277 – **28)** Ap 364 (or Av.New Orleans), San Pedro Sula – **29)** Ap 480 (or: Edif Radio Católica, Av Paz Barahona), Tegucigalpa **E:** rcatolica@unete.com ☎+504 237 2848 圖+504 237 2017 – **30)** Comayagua – **31)** Ap 2918 (or: 5 Calle, 10 y 11 Av S.O 91), San Pedro Sula **E:** radiofabulosa@sigmanet.hn - **FM:** 102.1MHz Radio Fabulosa ☎ +504 553 1228 – **32)** Ap 50, Catacamas - 1200-0400 - **FM:** 104.5MHz ☎ +504 899 4891 – **33)** Ap 10, 12101 Comayagua 1100-2400 ☎ +504 772 6581 圖+504 772 1926 - **FM:** 89.1MHz R.Vida – **34)** Barrio Abajo, 2da Ave, 2 y 3 Calle, Siguatepeque - 1100-0300 - **FM:** 95.7MHz ☎ +504 773 1632 – **35)** 3 Av entre 7 y 8 Calles N° 772, Puerto Cortés - 1200-0300 ☎ +504 665 2810 **E:** radio-cortes@lemaco.hn - **FM:** 88.9MHz +105.7MHz – **36)** Calle Vicente Williams, Edif Fergusón, Choluteca **W:** http://hometown.aol.com/djrubbik/stereoF.html - **FM:** 103.3MHz – **38)** Ap 2821, Tegucigalpa ☎ +504 237 7927 – **39)** Ap 996 (or: 9 Calle, S.O 44, Entre 8 y 9 Av), San Pedro Sula - 1200-0600 - **FM:** 98.5MHz Estéreo Mass – **41)** Tocoa, Colón- **44)** Ap 614 (or: Barrio Abajo, 6 Calle), Tegucigalpa - 1000-0600 – **45)** Ap 78, Choluteca ☎ +504 990 3796 - **FM:** 97.3MHz – **50)** Peña Blanca, 10 km al norte de Las Vegas, Sta Bárbara – **51)** Barrio Costado Norte Cine Valladolid, Barrio Sta Clara, El Paraíso - 1100-0400 – **53)** 9 Av 4 Calle, Edif Las Fuentes, San Pedro Sula - **FM:** 88.3MHz – **54)** Barrio Fatima, Edif Audiovideo, Siguatepeque 1055 – **56)** Av Panamá, Edif Canales N° 861, Tela, Atlántida - 1200-0400 - **FM:** 104.7MHz ☎ +504 448 2957 – **57)** Ap 2250, Tegucigalpa – **58)** Ap 25, La Esperanza, Intibucá 1100-0300 ☎+504 783 0025 圖+504 783 0644 – **59)** Ap 906, San Pedro Sula -**FM:** 97.9MHz – **60)** Calle El Calvario Frente Al Parque, Edif Plaza, Olanchito, Yoro - **FM:** 88.7MHz – **61)** Ap 20, El Progreso, Yoro - **FM:** 103.3MHz Stereo Alegría – **63)** San Francisco de la Paz, Olancho - 1100-0200 – **65)** Barrio La Isla, La Ceiba – **66)** Ap 004, Sta Bárbara ☎+504 643 2406 圖+504 643 2940 - 0900-0400 - **FM:** 97.5MHz – **68)** Barrio San Isidro, El Paraíso - **FM:** 93.3MHz – **72)** Calle José Trinidad Cabañas, Edif Hotel Presidente, Tela - 1200-0400 - **FM:** 89.9 – **73)** Col El Prado 1C-107A, Comayagüela ☎ +504 239 2228 – **74)** Iglesia de Cristo, Ministerio Ebenezer, 14 Calle A, Costado Sur de Wendy's Circunvalación (or: Ap 34-76), San Pedro Sula - 1100-0600 – **W:** www.ebenezer.hn - **FM:** 91.9MHz +93.7MHz – **75)** Calle Principal, Marcala, La Paz – **76)** Barrio Abajo, Comayagua – **77)** San Marcos de Colón, Choluteca – **78)** Olanchito, Yoro – **79)** Emisoras Unidas, Solares Nuevos, Av República, La Ceiba – **80)** Ap 303, San Pedro Sula - 1100-0200 **E:** efmhonduras@globalnet.hn - **FM:** 97.3MHz – **81)** Barrio Campo Luna, Choluteca - 1050-0400 – **82)** Ap 203, Sta Rosa de Copán - **FM:** 94.5MHz – **83)** Ap 955, Tegucigalpa – **84)** Ap 204, Sta Rosa de Copán - 1100-0400 - **FM:** 90.3MHz Rosa de Copán – **88)** Colonía Fesitran, San Pedro Sula – **89)** Juticalpa **1200-0400** – **90)** Ap 3252, Tegucigalpa – (Owned and operated by Conservative Baptist Home Mission Society, Box 828, Wheaton, IL 60187, USA) ☎ +504 234 6640 **W:** www.hrvc.org - **E:** gerencia @ hrvc.org – **94)** Barrio El Centro, 22115 Trinidad, Sta Bárbara ☎+504 664 1706 圖+504 664 1663 - **FM:** 105.3MHz – **95)** 12 Calle 2a Ave 206, Barrio La Curva, Puerto Cortés – **97)** Santiago de la Paz, La Paz (or: Col.21 de Octubre, Sector 3, Bloque 2, Casa 5, Tegucigalpa) – **98)** Av San Isidro, Entre Calles 9 y 10, La Ceiba – **99)** Casa 269, Barrio Abajo, Siguatepeque

1155- - **FM:** 91.5MHz – **101)** Barrio Abajo 229, Comayagua – **102)** Ap 20583, Comayagüela (or: IMF World Misiones, PO Box 6321, San Bernardino, CA 92412, USA) - 1100-0300 – **103)** Barrio San Miguel, Calle Principal, Marcala, La Paz ☎ +504 764 5377 - 1200-0500 – **104)** Av España, La Esperanza, Intibucá ☎ +504 898 2063 - 1100-0400 – **105)** Barrio La Cruz, Calle Chorotega, Choluteca - **FM:** 96.2 – **106)** Barrio San Andrés, Ocotepeque – **107)** Barrio Concepción, Marcala, La Paz (or: Palacio Arzobispal, Av.Cervantes, Barrio El Centro, Tegucigalpa) - 1000-0400 – **108)** Ap 2017, San Pedro Sula - 1200-0400 – **109)** La Cruzada del Evangélico de Honduras, La Ceiba – **110)** Parque Central, La Paz – **111)** Barrio Luz y Vida, San Luis, Sta Bárbara (or: Ap 303, San Pedro Sula) - 1200-0300 **English:** Sat 0300-0400, Sun 0230-0400 **E:** efmhonduras@globalnet.hn – **112)** Sta Rosa de Copán – **113)** Ap.23301, Olanchito, Yoro – **114)** Barrio El Progreso, 2da y 3ra Calle, Av.La Entrada, La Entrada, Copán – ☎ +504 661 2049 – **117)** 2a Av 9C N° 9, Barrio Rosario No 9, Gracias, Lempira ☎ +504 686 1087 – **118)** Col de Jerico, Utila – **W:** www.ibnet.org – **119)** Barrio La Bara, Calle Pavimentada, Casa 1185, La Ceiba ☎ +504 443 2481 – **120)** Calle del Comercio 12A, Comayagua – **121)** San Marcos de Colón, Choluteca – **122)** Barrio La Mora, Catacamas, Olancho - 1000-0400 - **FM:** 99.1MHz – **124)** Col 21 de Octubre, Sector 3, Bl 1, Casa 4, Tegucigalpa – **127)** 1 Av Calle Principal, Jutiapa, Atlántida - 1200-0600 - **FM:** 95.7 ☎ +504 898 4918 – **128)** Radio Ensenanzas Evangelicas, Puerto Lempira – **131)** Yoro. Dir: Jamil N Hawit Castro – **132)** Ap 2424, San Pedro Sula - 1000-0200 – **133)** Ap 914, Tegucigalpa – **134)** Barrio La Manzana, Minas de Oro, Comayagua – **135)** Ap 15, 16101 Juticalpa - **FM:** 106.3MHz Prgrs in Sp and E – **136)** Col Río Piedras, 5 Calle 26 Av., San Pedro Sula – **140)** Ap 40, Choluteca - 1000-0400 - **FM:** 95 5MHz – **143)** San Lorenzo – **144)** La Entrada, Copán – **145)** Barrio San Sebastián 2 Calle, San Marcos, Ocotepeque – **147)** Ap 888, (or: Centro Comercial San José), La Ceiba ☎+504 441 5973 **W:** www.applegatefellowship.org/missions/honduras.asp – **E:** radiolitoral@psinet.hn **English:** Weekends 0400-0500 – **148)** Barrio La Pava, Sabá, Colón ☎ +504 424 82 49 – **149)** Frente al Parque "Lempira", Gracias, Lempira (or: Ap.1579, Tegucigalpa) - 1100-0500 - **FM:** 95.1MHz R Galaxia 21 FM Stereo ☎ +504 656 1068 – **151)** Ocotepeque, Ocotepeque - 1100-0300 **W:** www.cauaguanca.com/radio-comunidad **E:** radiocomunidad910@yahoo.es ☎+504 653 3994 – **152)** Media Cuadra Al Norte del ParWque, B:o El Centro, Ocotepeque - 1200-0400 - **FM:** 92.1MHz – **153)** Ap 210 (or: 5 Calle, 10 y 11 Av S.O., Barrio Beuque), San Pedro Sula - **FM:** 90.7MHz – **155)** Col Miraflores, Tegucigalpa ☎ +504 661 2327 – **157)** Calle Vicente Williams 345, Choluteca – **162)** 2 Av Calle 38, Trinidad, Sta Bárbara ☎ +504 664 1681 – **163)** Barrio La Esperanza 4A N° 142, Choluteca - **FM:** 98.5MHz – **165)** Barrio El Banco, La Entrada - 1100-0300 ☎ +504 661 2327 - **FM:** 103.5MHz – **170)** Calle del Estadio, Barrio El Campo, Catacamas – **171)** San Marcos de Colón, Choluteca – **172)** Av Independencia, Sta Bárbara - **FM:** 89.7MHz – **173)** Barrio Miraflores, La Entrada – **175)** Barrio El Centro, Siguatepeque – **176)** Barrio Guadalupe, Choluteca - **FM:** 100.9MHz – **177)** Barrio El Coyol, Sava Colón – **180)** Barrio La Cruz, Contiguo a la Iglesia el Encuentro, Catacamas, Olancho ☎ +504 899 4329 – **181)** 4 y 5 Ave, 3 Calle 442, Barrio Las Delicias, El Progreso, Yoro ☎ +504 898 4803 – **182)** Barrio El Campo, Frente a la Ferreteria San António, Sigiatepeque – **183)** Barrio El Centro, Valle del Boulevard, Contiguo a la Iglesia Adventista, Siguatepeque – **185)** Barrio El Centro, Contiguo al Banco Atlántida, Las Vegas, Sta Bárbara ☎ +504 659 3156 – **187)** Balneareo Pasada del Sol, Barrio Arriba, Comayagua ☎ +504 772 0565 – **188)** Tegucigalpa – **192)** Barrio de Jesús, Casa 7, Calle Principal, Juticalpa – **193)** Col.Irias, Primera Calle, 5 Casas a Mano Izquierda, La Ceiba ☎ +504 441 0238 – **194)** Radio Industrias de Honduras, Tegucigalpa – **196)** Barrio San José, Ocotepeque ☎ +504 443 0435 – **197)** Apartado 004, Sta Bárbara 1200-0400 – **198)** Olanchito, Yoro – **200)** Sonaguera, Colón – **202)** Ap 534, San Pedro Sula – **203)** Puerto Lempira – **204)** Progreso, Yoro ☎504 647 1717 – **205)** Puerto Lempira, Gracias a Dios - **206)** Roatán, Islas de la Bahía – **207)** La Ceiba, Atlántida – **208)** La Ceiba, Atlántida – **209)** La Ceiba, Atlántida – **210)** Morazán, Yoro – **211)** Olanchito, Yoro – **212)** Tocoa, Colón – **213)** Olanchito, Yoro – **214)** Tocoa, Colón – **215)** Olanchito, Yoro – **216)** Tocoa Colón – **217)** Jocón, Yoro – **218)** Tocoa, Colón – **219)** Catacamas, Olancho – **220)** Catacamas, Olancho – **221)** Nueva Palestina, Olancho – **222)** Catacamas, Olancho – **223)** Catacamas, Olancho – **224)** San Esteban, Olancho – **225)** Catacamas, Olancho – **227)** Catacamas, Olancho – **228)** Danlí, El Paraíso – **229)** Lempira, Gracias a Dios – **230)** Nueva Ocotopeque, Ocotepeque – **231)** San José de Colinas, Santa Bárbara – **232)** Choluteca – **233)** Choluteca – **234)** Nacaome, Valle – **235)** Nacaome, Choluteca – **236)** Marcovia, Choluteca – **237)** Nacaome, Valle – **238)** San Lorenzo, Valle – **239)** Choluteca, Choluteca – **240)** El Progreso, Yoro – **241)** San Pedro Sula – **242)** La Paz, La Paz – **243)** Tegucigalpa – **244)** Sabanagrande, Francisco Morazán – **245)** Catacamas, Olancho – **246)** Choloma, Cortés – **247)** Choluteca – **248)** Dulce Nombre de Culmi, Olancho – **249)** Intibucá – **250)** Juticalpa, Olancho – **251)** Juticalpa, Olancho ☎ +504 785 2609 **E:** manantial1520@yahoo.com –

252) Copinh, La Esperanza, Intibucá – **253)** San Pedro Sula – **254)** Sonaguera, Colón – **255)** Tocoa, Colón ☎+504 444 3972 – **256)** Lempira – **257)** Puerto Cortés, Cortés – **258)** Universidad Nacional de Agricultura, Catacamas, Olancho – **259)** Barrio El Centro, Esquina opuesta a Edificio Villatoro, Catacamas, Olancho – **260)** Juticalpa, Olancho – **261)** Iglesia de Cristo Elim, Sabá, Colón – **262)** Asociación de Iglesias Evangélicas Centroamericanas, Tegucigalpa

FM in Tegucigalpa (MHz): 88.1 Ke Buena – 88.7 Globo Grupera – 89.3 Power – 89.9 R. Red de Radiodifusión Bíblica – 90.5 R. Corazón 90.5 – 91.1 R. Kairos FM – 91.7 R. Buenísima – 92.3 Rock & Pop – 92.9 La Voz de Honduras – 93.3 Estéreo Fantasía – 93.5 R. Notícias STC – 94.1 FM 94 – 94.7 América – 95.3 Digital – 95.9 R. Panamericana – 96.5 R. Estéreo Fiel – 97.1 EstéreoTic Tac – 97.7 Azul – 98.3 Estéreo Concierto – 98.9 Estéreo Fe – 99.5 Suprema – 100.1 Super 100 – 100.7 R. Exa FM – 101.3 R. Nacional de Honduras – 101.9 Vox – 102.5 Suave – 103.1 FM 103.1 – 103.7 Luz – 104.3 Momentos FM – 104.9 Estéreo Amor – 105.5 Musiquera – 106.1 Romántica – 106.7 Stereo Rumba – 107.3 W107 Energía Estéreo – 107.9 Top Music con La Onda del Nuevo Mundo.

AFRTS (Air Force)
✉ JTF-B, APO AA 34042, USA **E:** PAO@jtfb-emh1.army.mil
FM (MHz): 106.5MHz Soto Cano Air Base, 0.25kW **D.Prgr:** 24h

HONG KONG (China, SAR)

L.T: UTC +8h — **Pop:** 7.0 million — **Pr.L:** Cantonese, English — **E.C** 50Hz, 200/220V — **ITU:** HKG

RADIO TELEVISION HONG KONG (Gov.)
✉ Broadcasting House, 30 Broadcast Drive, Kowloon, Hong Kong ☎ +852 2339 6300 🖷 +852 2336 9314 **E:** ccu@rthk.org.hk **W:** www.rthk.org.hk **L.P:** Dir. of Broadc: Gordon Leung Chung-tai, Asst. Dir. of Broadc. (Radio): Tai Keen-man

MW (kHz)	Network	Location	kW
567	Radio 3	Golden Hill	20
621	P. Ch	Golden Hill	20
675	Radio 6	Peng Chau	10
783	Radio 5	Golden Hill	20
1584	Radio 3	Chung Hom Kok	0.1

P. Ch = Putonghua Channel

FM(MHz)	Network	kW	Tx Location	Target Div.
92.3	Radio 5	0.038	Tin Shui Wai	
92.6	Radio 1	3	Mt. Gough	Kowloon
92.9	Radio 1	0.1	Golden Hill	Tsuen Wan
93.2	Radio 1	0.5	Cloudy Hill	Fan Ling
93.4	Radio 1	0.7	Castle Peak	Tuen Mun
93.5	Radio 1	0.15	Beacon Hill	Sha Tin
93.6	Radio 1	0.5	Lamma Island	HK Island south
94.4	Radio 1	1	Kowloon Peak	HK Island north, Sai Kung
94.8	Radio 2	3	Mt. Gough	Kowloon
95.3	Radio 2	0.5	Cloudy Hill	Fan Ling
95.6	Radio 2	0.1	Golden Hill	Tsuen Wan
96.0	Radio 2	0.5	Lamma Island	HK Island south
96.3	Radio 2	0.15	Beacon Hill	Sha Tin
96.4	Radio 2	0.7	Castle Peak	Tuen Mun
96.9	Radio 2	1	Kowloon Peak	HK Island north, Sai Kung
97.6	Radio 4	3	Mt. Gough	Kowloon
97.8	Radio 4	0.5	Cloudy Hill	Fan Ling
97.9	Radio 3	0.02	Mt. Nicholson	Jardine's Lookout
98.1	Radio 4	0.15	Beacon Hill	Sha Tin
98.2	Radio 4	0.5	Lamma Island	HK Island south
98.4	Radio 4	0.1	Golden Hill	Tsuen Wan
98.7	Radio 4	0.7	Castle Peak	Tuen Mun
98.9	Radio 4	1	Kowloon Peak	HK Island north, Sai Kung
99.4	Radio 5	0.015	Tseung Kwan O	Jank Bay
100.9	P. Ch	0.005	Jardine's Lookout	Happy Valley
100.9	P. Ch	0.003	Castle Peak	Tuen Mun
103.3	P. Ch	0.015	Tseung Kwan O	Jank Bay
103.3	P. Ch	0.038	Tin Shui Wai	
106.8	Radio 5	0.01	Castle Peak	Tuen Mun
106.8	Radio 3	0.06	Chung Hom Kok	HK Island south
107.8	Radio 3	0.015	Tseung Kwan O	Jank Bay
107.8	Radio 3	0.038	Tin Shui Wai	

RTHK Radio 1 in Cantonese/Chinese: 24h – **RTHK Radio 2** in Cantonese: 24h – **RTHK Radio 3** in English: 24h – **RTHK Radio 4** in English/Cantonese: 24h – **RTHK Radio 5** in Cantonese/Chinese: 24h – **RTHK Radio 6** Relay BBCWS English: 24h – **RTHK Putonghua Channel** in Chinese: 24h
Ann: Cantonese: "Hoenggong dintoi dai (number) toi".

HONG KONG COMMERCIAL BROADC. CO. LTD
✉ 3 Broadcast Drive, Kowloon, Hong Kong ☎ +852 2336 5111

🖷+852 2338 0021 **E:** cs@881903.com **W:** www.881903.com

MW (kHz)	kW	Location	Prgr.
864	10	Peng Chau	Quote AM

FM(MHz)	Network	kW	Tx Location	Target Div.
88.1	CR1	3	Mt.Gough	Kowloon
88.3	CR1	0.5	Cloudy Hill	Fan Ling
88.6	CR1	0.7	Castle Peak	Tuen Mun
88.9	CR1	0.1	Golden Hill	Tsuen Wan
89.1	CR1	0.5	Lamma Island	HK Island south
89.2	CR1	0.15	Beacon Hill	Sha Tin
89.5	CR1	1	Kowloon Peak	HK Island north, Sai Kung
90.3	CR2	3	Mt.Gough	Kowloon
90.7	CR2	0.5	Cloudy Hill	Fan Ling
90.9	CR2	0.1	Golden Hill	Tsuen Wan
91.1	CR2	0.15	Beacon Hill	Sha Tin
91.2	CR2	0.7	Castle Peak	Tuen Mun
91.6	CR2	0.5	Lamma Island	HK Island south
92.1	CR2	1	Kowloon Peak	HK Island north, Sai Kung

HKCR CR1 (Supercharged 881) in Cantonese. 24h **N:** half-hourly.
HKCR CR2 (Ultimate 903) in Cantonese. 24h **N:** hourly. **Ann:** "Cikzak gaulingsaam".
HKCR AM864 in English. 24h **N:**On the hour from 2200-1500.

METRO BROADCAST CORPORATION LTD.
✉ Basement 2, Site 6, Whampoa Gardens Hunghom, Kowloon, Hong Kong ☎ +852 2123 9888 🖷+852 2123 9877 **E:** tech@metroradio.com.hk **W:** www.metroradio.com.hk

MW (kHz)	Network	kW	Location
1044	Metro Plus	10	Peng Chau

FM(MHz)	Network	kW	Tx Location	Target Div.
99.7	Metro Showbiz	3	Mt.Gough	Kowloon
100.0	Metro Showbiz	0.5	Cloudy Hill	Fan Ling
100.4	Metro Showbiz	0.7	Castle Peak	Tuen Mun
100.5	Metro Showbiz	0.15	Beacon Hill	Sha Tin
101.0	Metro Showbiz	0.01	Stanley	
101.6	Metro Showbiz	0.1	Golden Hill	Tsuen Wan
101.8	Metro Showbiz	1	Kowloon Peak	HK Island north
102.1	Metro Showbiz	0.5	Lamma Island	HK Island south
102.4	Metro Finance	0.15	Beacon Hill	Sha Tin
102.5	Metro Finance	0.7	Castle Peak	Tuen Mun
102.6	Metro Finance	0.01	Stanley	
104.0	Metro Finance	3	Mt.Gough	Kowloon
104.5	Metro Finance	0.5	Lamma Island	HK Island south
104.7	Metro Finance	0.5	Cloudy Hill	Fan Ling
105.5	Metro Finance	0.1	Golden Hill	Tsuen Wan
106.3	Metro Finance	1	Kowloon Peak	HK Island north

Metro Plus in English (Partly Cantonese, Mandarin, Filipino and Indonesian). 24h. Music, news and information. **Metro Showbiz** in Cantonese. 24h. **Ann:** "Sanseng dintoi zisoen toi". **Metro Finance** in Cantonese. 24h

F.PI: Wave Media 24h on 220.352MHz (DAB tr) planned for 2010.

HUNGARY

L.T: UTC +1h (27 Mar-30 Oct: UTC +2h) — **Pop:** 10.02 million — **Pr.L:** Hungarian — **E.C:** 50Hz, 220V — **ITU:** HNG

MAGYAR RÁDIÓ
✉ Bródy Sándor u. 5-7, H-1800 Budapest ☎ +36 1 3287000 or 3287878 🖷 36 1 3287406 **W:** www.radio.hu **L.P:** Pres.: not elected
MR1 Kossuth Rádió ☎ +36 1 3287945 **MR2 Petőfi Rádió** ☎ +36 1 3288555 **MR3 Bartók Rádió** ☎ +36 1 3288772

MW	kHz	kW	Prg	MW	kHz	kW	Prg
Solt	540	2000	1	Marcali	1188	300	4
Lakihegy	873	20	4	Szolnok	1188	100	4
Pécs	873	20	4	Szombathely	1251	25	6c
Miskolc	1116	15	6b	Nyíregyháza	1251	25	6a
Mosonmagyaróvár	1116	5	6c	Györ	1350	5	6c

FM (MHz):	MR1	MR2	MR3	MR6+7	kW
Budapest	107.8	94.8	105.3		10
Csávoly		89.4			3
Debrecen	99.7	89.0	106.6	91.4a	1
Györ		93.1	106.8		3/1
Kabhegy	102.3	93.9	105.0		10/3/10/10
Kékestető		102.7	90.7		3
Kiskörös		95.1	105.9		3/1
Komádi	89.9	96.7	105.1		3
Miskolc	103.8	102.3b	107.5	102.3b	3/3/3/1
Nagykanizsa	106.7	94.3	104.7		10
Pécs	104.6	103.7	107.6	101.7e	10/10/10/2
Sopron	91.8	99.5	107.8		10/10/1

FM (MHz):	MR1	MR2	MR3	MR6+7	kW
Szeged	101.9	104.6	105.7	93.1d	0.5/2/1/1
Szentes	91.6	98.8	107.3		3/1/10/
Szolnok				101.2a	2
Tokaj	88.3	92.7	105.5		10/5/10/
Úzd		90.3	106.9		3
Vasvár	103.6	98.2	106.9		3

+ 15 MR1 txs below 1kW

Prgrs: 1= **MR1 Kossuth** (news-talk) 2=**MR2 Petöfi** (pop) 3=**MR3 Bartók**.(classical) 4=**MR4 Nemzetiségi** (Ethnic), **MR5 Parlament** (internet/satellite only), 6a-e=**MR6** (regional) 7=**MR7** (folk+operetta)
D.Prgr: 24h exc **MR1+MR6** MW: M-Sa 0330-2125 Sun: 0400-2130 (**NB**: tr. is cut on MR1/MR6 MW, continues on FM); **MR4** MW: 0600-1900
ANN: **MR1**: "MR egy Kossuth Rádió, a szavak ereje" **MR2**: "MR kettő Petöfi Rádió, nagyon zene", **MR3**: MR három Bartók Rádió, a klasszikus zene rádiója" **MR6**: "Itt az MR hat, a régió rádiója." **MR7**: "Dalok és dallamok". **S/on**: Rákóczi March (only MR1 MW), S/off 2125 MR1 MW and 2300 MR2: Nat. Anthem, 2258 on MR3: "Szózat"; 1100 MR1, MR2, MR4, MR6: midday church bell toll
MR4 Nemzetiség Ethnic Prgrs: +36 (1) 328-8672 +36 (1) 328-8682. **W**: http://mr4.radio.hu/ Croatian 0700-0900, German: 0900-1100, Roma/Gipsy: 1103-1200; Mo: Slovenian, Tu:Rusyn/Ruthenian , W: Greek, Th: Bulgarian, F: Ukranian, Sa: Armenian 1200-1230; M-F: Minority music, Sa: Polish 1230-1300; Serbian 1300-1500; Romanian: 1500-1700, Slovak: 1700-1900
MR6 Regional Network: **W**: www.mr6.hu **a) MR6 Debrecen** 4024 Debrecen, Piac u. 28/c +36 52 525-325 36 52 525-333 **b) MR6 Miskolc** 3527 Miskolc, Bajcsy-Zsilinszky u. 15 +36 46 502-719 36 46 502-728 **c) MR6 Györ** 9027 Györ, Nagy Imre u. 28 +36 96 514-222 36 96 514-224 **d) MR6 Szeged** 6720 Szeged, Stefánia 7 +36 62 554-814 36 62 554-804 **e) MR6 Pécs** 7621 Pécs, Szent Mór u. 1 +36 72 518-313 36 72 518-320**Local Prgr**: 0500-1630. rel MR7 (see MR7), other times on AM rel MR1, Miskolc on FM rel MR2
MR7: mr7.hu 24h online, AM/FM 0400-0500, 1700-2130, FM only 2130-0400 on the MR6 network

EXTERNAL SERVICE: Radio Budapest
See International Broadcasting Section

National Media and Communications Authority (NMHH)
1015 Budapest, Ostrom utca 23-25 +36 1 3757777 +36 1 3565520 **W**: www.nhh.hu
National Radio and Television Commission (ORTT)
1088 Budapest, Reviczky u. 5 +36 1 4298600, 2672590 +36 1 2672612 **W**: www.ortt.hu
(to be replaced by the Media Council in 2010/2011)
Antenna Hungária Broadc. & Radiocommunications Ltd (AH Zrt.)
119 Budapest, Petzvál József u. 31-33 +36 1 2036060 +36 1 4642525 **W**: www.ahrt.hu
Hungarian Federation of Free Radios (SZARÁMASZER)
1077 Budapest Rózsa utca 34 fsz 9. / +36 1 3111855 **W**: www.szabadradio.hu
National Association of Local Radios (HEROE)
8000 Székesfehérvár, Donát u. 92 +36 22 505310 +36 22 505 312 **W**: www.heroe.hu
DX data: www.radiosite.hu, tv.tvnet.hu/radio, www.frekvencia.hu, /www.helyiradiok.hu

National stations and networks
CLASS FM (National, Comm.)
1089 Budapest, Üllöi út 102. +36 1 5555500
NEO FM (National, Comm.)
1025 Budapest, Csévi u. 7/B. +36 1 5551000

FM (MHz):	CLA	NEO	kW	FM (MHz):	CLA	NEO	kW
Budapest	103.3	100.8	10	Nagykanizsa		90.2	10
Csávoly		96.7	3	Pécs	105.5	95.9	10
Debrecen	101.1	87.6	1	Sopron		102	96.8	3/10
Györ	101.4	87.6	3	Szeged	94.9	90.3	0.5
Kabhegy	100.5	107.2	10	Szentes		100.4	3/0.25
Kékes	104.7	95.5	3	Tokaj	103.5	97.5	10
Kiskörös		88.4	1	Uzd		101.5	3
Komádi	101.6	103.0	3	Vasvár		91.6	3
Miskolc	98.3	97.1	3				

MAGYAR KATOLIKUS RÁDIÓ (MKR) (National Relig.)
1062 Budapest, Délibáb u. 15-17 +36 1 255-3366 +36 1 255-3399 **W**: www.katolikusradio.hu

MW	kHz	kW	MW	kHz	kW
Lakihegy	810	12	Szolnok	1341	150
Balatonszabadi	1341	150			

Other Stations

MW	kHz	kW	Location	Station
22)	1485	0.2	Mohács	Régió Rádió (0455-1900)

FM	MHz	kW	Location	Station
1)	88.1	1.9	Budapest	Rise FM (dance)
3)	89.5	10	Budapest	Juventus R. (AC)
4)	88.8	0.1	Budapest	Rádió C (ethnic)
5)	90.3	1	Budapest	Tilos R. (community)
31)	90.6	1.6	Veszprém	Klubrádió (mono, talk)
34)	90.9	2	Budapest	Jazzy (smooth jazz)
6)	91.5	1	Györ	Kék Duna R.
28)	91.7	1.2	Kiskörös	Kunság R./Mária R.
7)	91.8	1	Eger	Szent István R. (rlg)
38)	92.1	2.2	Budapest	Klasszik R. (classical)
3)	92.6	1.3	Siófok	Juventus R. (AC)
29)	93.4	1	Dabas	Dabas R.
31)	93.5	1	Debrecen	Klubrádió (mono, talk)
8)	93.6	1	Gödöllö	Rádióaktív
2)	94.2	1	Budapest	Mária Rádió (rlg)
7)	95.1	1	Miskolc	Szent István R. (rlg)
31)	95.3	1	Budapest	Klubrádió (talk)
21)	95.6	1	Debrecen	Gazdasági R.(financ. talk)
21)	95.7	1	Balassagyarmat	Gazdasági R.
9)	95.8	1.3	Budapest	Info R. (mono, news)
10)	96.3	1.8	Miskolc	Juventus (AC)
11)	96.4	2.5	Budapest	Roxy R. (dance)
33)	96.5	2	Kecskemét	Gong R.
31)	97.7	1	Kecskemét	Klubrádió (mono, talk)
18)	98.0	1	Budapest	Civil R. (community)
14)	98.6	0.8	Budapest	Radiocafé (alt. rock)
37)	98.9	0.9	Tihany	Rise FM (dance)
40)	98.9	1	Szigetvár	N-Joy R.
15)	99.5	3	Budapest	Rádió Q
23)	100.2	1	Szeged	R. Plusz
21)	100.7	1	Eger	Gazdasági R. (financ. talk)
17)	100.3	1	Budapest	Lánchíd R. (newstalk)
32)	100.6	3	Telkibánya	Híd R.
39)	101.8	1	Székesfehérvár	Rádió 1 (CHR)
24)	101.9	1	Tamási	Tamási R.
16)	102.1	1	Budapest	Sztár FM (oldies)
23!)	103.0	0.3	Miskolc	Sztár FM
25)	103.9	1	Nyíregyháza	Retro R.
39)	103.9	5	Budapest	Rádió 1 (CHR)
20)	104.0	1	Békéscsaba	Csaba R.
21)	105.9	1	Budapest	Gazdasági R (financ. talk)
26)	106.5	1	Dunaföldvár	Alisca R.
27)	106.5	1	Miskolc	Gold FM
26)	107.5	1	Paks	Alisca R.

+ 200 additional FM txs below 1kW
Addresses and other information
1) 1134 Budapest, Róbert Károly körút 54-58. +36 30 9002086 **– 2)** 1054 Budapest, Szabadság tér 2 +36 1 3730701 +36 1 3730702 **– 3)** 1134 Budapest, Róbert Károly körút 82/84 +36 1 237 5300 +36 1 320-1299 **– 4)** 1086 Budapest, Teleki tér 7 +36 1 4590095 +36 1 4590094 **– 5)** 1092 Budapest, Kinizsi u 28 +36 1 4768491 +36 1 4768492 **NB:** Some English prms **– 6)** 9021 Györ, Szent István út 10/a. +36 96 524444 +36 96 524445 **– 7)** 3301 Eger, Pf 86 +36 36 510-610 +36 36 510-614 **W:** www.mkr.hu **– 8)** 2100 Gödöllö, Dózsa Gy. út 39. +36 28 510745 +36 28 510746 **– 9)** 1088 Budapest, Múzeum u 9 +36 1 4832950 +36 1 4832625 **– 10)** 3530 Miskolc, Kisavas Alsósor 22 **– 11)** Budapest, Fény u +36 1 466 03 44 +36 1 466 06 73 **–13)** 1116 Budapest, Sztregova utca 3 +36 1 489-0997 **– 14)** 1064 Budapest, Vörösmarty u 67 +36 1 301 2170 +36 1 301 2190 **– 15)** Budapest 1051 József A u 12 +36 1 5053900 **–17)** 1089 Budapest, Üllöi út 102 +36 1 8148730 +36 1 8148754 **– 19)** 4400 Nyíregyháza, Dózsa Gy u 3 +36 42 508530 +36 42 507158 **– 20)** 5600 Békéscsaba, Teleki utca 5 +36 66 441-111 +36 66 441-112 **– 21)** 1133 Budapest, Váci út 78/B +36 1 8873500 +36 1 8873501 **– 22)** 7700 Mohács, Bakács u. 5. +36 69 511555 **– 23)** 6723 Szeged, Római körút 23 +36 62 635635 +36 62 630200 **– 24)** 7100 Szekszárd Szent László u 19 NB: 0900-0500 Rel R Alisca **– 25)** 4400 Nyíregyháza, Széchenyi u 9/A +36 42 444022 +36 42 401035 **– 26)** 7100 Szekszárd, Szent László u 19 +36 74 410191, +36 74 410191 **– 27)** 3530 Miskolc, Széchenyi út 72 +36 46 999940 **– 28)** 6200 Kiskörös, Petöfi tér 10-11 +36 78 414030 +36 78 414020 **– 29)** 2370 Dabas, Szent István tér 1/b +36 29 562562 +36 29 562560 **– 31)** 1037 Budapest, Kunigunda u 64 +36 1 2406953 **– 32)** 3713 Arnót, Rákóczi u 22 **– 33)** 6000 Kecskemét, Petöfi Sándor u 1/b +36 76 414020 **– 34)** 1024 Budapest Ady Endre út 24 +36 1 7876992 **– 37)** 1134 Budapest, Róbert Károly körút 54-58 +36

30 5807899 – **38)** ✉ Budapest Ady Endre u. 24. I/1 ☎ +36 1 7866464 – **39)** ✉ 1016 Budapest, Mészáros utca 58/b ☎ +36 1 473-2600 – **40)** ✉ 7400 Kaposvár, Arany János u. 97. ☎ +36 82 814700

Kisközösségi rádió (micropower community radio)
By 2010, 68 non-profit low-power stns (0.1–10W) had been granted licences in cities and country villages. Info **W:** www.szabadradio.hu Low power txs in Budapest **FM(MHz):** 87.6, 93.5, 93.7, 97.0, 101.4, 107.3

DAB+: Budapest. MuxA: MR1-3, MKR, Inforádio, Klubrádió, Juventus, Gazdasági + 3 Swiss satellite stns on 222.064MHz (11D) 2x250W.

ICELAND

L.T: UTC — **Pop:** 307,000 — **Pr.L:** Icelandic — **E.C:** 50Hz, 230V — **ITU:** ISL

ÚTVARPSRÉTTARNEFND
✉ Kringlan 4-12, 103 Reykjavík ☎ +354 5512114 🖷 +354 5335578 **E:** utvarpsrettarnefnd@utvarpsrettarnefnd.is **W:** www.utvarpsrettarnefnd.is **L.P:** Chmn: Benedikt Bogason
NB. Útvarpsréttarnefnd issues broadcasting licenses.

RÍKISÚTVARPIÐ (RÚV) (Pub)
✉ Efstaleiti 1, 150 Reykjavík ☎ +354 5153000 🖷 +354 5153010 **E:** isradio@ruv.is **W:** www.ruv.is **L.P:** Dir (Radio): Páll Magnússon

LW	kHz	kW	Prgr	LW	kHz	kW	Prgr
Gufuskálar	189	300	Rás 1/2	Eiðar	207	100	Rás 1/2

FM (MHz)	Rás 1	Rás 2*	Rondó	kW
Almannaskarð	90.3	104.8a	-	1
Auðsholt	91.3	95.3d	-	2.5
Borgarland	88.0	96.3	-	3/3.5
Gagnheiði	98.8	87.7b	-	5
Girðisholt	92.9	-	-	3.5
Háfell	93.8	98.7d	-	14/34
Hegranes	90.6	98.8b	-	3.1/5
Hnjúkar	89.1	95.5b	-	6/6.2
Skálafell	92.4	99.9	-	24
Vaðlaheiði	91.6	96.5b	-	9.3
Vatnsendi	93.5	90.1	87.7	3.4/2/2
Vestmannaeyar	97.1	88.1d	-	17/24
Viðarfjall	88.1	96.1b	-	3.3

NB: Txs below 1kW not listed. *) Carries reg. prgrs (see below)
D.Prgr: Rás 1: 24h. — **Rás 2:** 24h. — **Rondó:** 24h.
On LW: MF: 0000-0700 Rás 1, 0700-0900 Rás 2, 0900-1400 Rás 1, 1400-1600 Rás 2, 1600-1615 Rás 1, 1615-1930 Rás 2, 1930-1945 Rás 1, 1945-2200 Rás 2, 2200-2400 Rás 1. SS: 0000-1615 Rás 1, 1615-1930 Rás 2, 1930-1945 Rás 1, 1945-2200 Rás 2, 2200-2400 Rás 1.
Reg. Prgrs (via Rás 2 txs): **a)** Svæðisútvarp Austurlands, Miðvangi 2-4, 700 Egilsstaðir: Tue-Fri 1725-1755. – **b)** Svæðisútvarp Norðurlands, Kaupvangsstræti 1, 602 Akureyri: 1725-1755. – **c)** Svæðisútvarp Vestfjarðar, Aðalstræti 22, 400 Ísafjörður: Tue-Fri 1725-1755. – **d)** Svæðisútvarp Suðurlands, Austurvegi 4, 800 Selfoss: Tue-Fri 1725-1755.

OTHER STATIONS

FM	MHz	kW	Location	Station
5)	88.5	1	Reykjavík	XA Radíó
1E)	90.4	2	Vestmannaeyjar	Xið
1C)	90.9	2	Mosfellsbær	GullBylgjan
1A)	92.7	2	Vaðlaheiði	Bylgjan
A)	94.3	2	Mosfellsbær	BBCWS relay
1A)	94.5	2	Háfell	Bylgjan
1B)	94.7	1	Egilsstaðir	FM957
1B)	95.1	1	Hegranes	FM957
1B)	95.7	2	Reykjavík	FM957
8)	96.3	1	Selfoss	Suðurland FM 96,3
1D)	96.7	2	Mosfellsbær	LéttBylgjan
6)	97.2	2	Reykjavík	Útvarp Flensborg
1E)	97.7	2	Mosfellsbær	Xið
1A)	97.9	1	Hegranes	Bylgjan
4)	98.7	1	Vaðlaheiði	Voice 987
1A)	98.9	2	Vatnsendi	Bylgjan
1A)	98.9	1	Hnjúkar	Bylgjan
3)	100.5	1	Bláfjöll	Kaninn
1A)	100.9	2	Vestmannaeyjar	Bylgjan
1B)	101.7	2	Vestmannaeyjar	FM957
1F)	102.2	2	Vatnsendi	Barnaútvarpið
1B)	102.5	1	Skáneyjarbunga	FM957
2)	102.9	2	Reykjavík	Lindin
1B)	103.2	1	Selfoss	FM957
1A)	103.3	1	Skáneyjarbunga	Bylgjan

NB: Txs below 1kW not listed.

Addresses & other information:
1A-1F) Skaftahlíð 24, 105 Reykjavík. **E:** 1A) bylgjan@bylgjan.is, 1B) fm957@fm957.is, 1C) gull@bylgjan.is, 1D) lettbylgjan@lettbylgjan.is – 2) Krókhálsi 4a, 110 Reykjavík. **E:** lindin@lindin.is – **3)** Grænásbraut 619, 235 Reykjanesbæ. E: kaninn@kaninn.is – **4)** Ráðhústorg 7, 2 hæð, 600 Akureyri. E: voice@voice.is – **5)** c/o Mörður Ingólfsson, Brávallagötu 18, 101 Reykjavík. Mainly in English. – **6)** Pósthólf 240, 222 Hafnarfjörður. – **7)** Skipholti 50D, 105 Reykjavík. **E:** frettir@frettavakt.is – **8)** Hrísmýri 6, 800 Selfoss. E: 963@963.is – **A)** Lynghálsi 5, 110 Reykjavík. Rel. BBCWS (UK).

DAB (Trial): Reykjavík ch11C (220.352MHz). **Operator:** RÚV

INDIA

L.T: UTC +5½h — **Pop:** 1,166 million — **Pr.L:** Assamese, Bangla, Bodo, Dogri, English, Gujarati, Hindi, Kannada, Kashmiri, Maithili, Marathi, Malayalam, Nepali, Oriya, Punjabi, Santhali, Sindhi, Tamil, Telugu & Urdu — **E.C:** 50Hz 220/400V — **ITU:** IND

MINISTRY OF INFORMATION & BROADCASTING
Main Secretariat: ✉ A-Wing, Shastri Bhawan, New Delhi-110001 **W:**www.mib.nic.in
L.P: Minister for Info. & Broadcasting: Ms.Ambika Soni

PRASAR BHARATI (BROADCASTING CORPORATION OF INDIA) (Public Corporation)
✉ 2nd Floor, PTI Building, Parliament Street, New Delhi-110001 ☎ +91 11 23382094/5/7/8/9 🖷 +91 11 23386507 **L.P:** Chairman: Ms. Mrinal Pande ☎91 11 23753687 🖷91 11 23737589, CEO: Baljit Singh Lalli ☎91 11 23737603, 23352558 🖷91 11 23352549 **E:** ceobci@yahoo.com

AKASHVANI – ALL INDIA RADIO
Administration/Engineering: ✉ Directorate General, All India Radio, Akashvani Bhawan, Parliament Street, New Delhi-110001 ☎ +91 11 23421006, 23715413 🖷+91 11 23711956 **E:** airlive@air.org.in **W:** www.allindiaradio.gov.in
L.P: DG: Ms. Noreen Naqvi ☎+91 11 23710300/23421061 🖷 91 11 23421956 **E:** dgair@air.org.in Engineer in Chief: M.C.Aggarwal ☎+91 11 23421058, 23421459 🖷 +91 11 23421459 **E:** einc@air.org.in
Spectrum Management & Synergy: Room No.204, All India Radio, Akashvani Bhavan, Parliament Street, New Delhi-110001 **L.P:** Dir. M.S.Ansari, Dy.Dir (Eng) B.K. Oberoi ☎+91 11 23421062, 23421145 **E:** spectrum-manager@air.org.in
Programming: ✉ New Broadcasting House, 27 Mahadev Road, New Delhi-110 001 ☎ +91 (11) 23421218 ✉ Akashvani Bhawan, Parliament Street, New Delhi-110001 ☎+91 11 23715411
News Services Division: ✉ New Broadcasting House, 27 Mahadev Road, New Delhi-110 001. Newsroom ☎ +91 11 23421100, 23421101 🖷+91 11 23711956 **E:** nbhnews@air.org.in **W:** www.newsonair.nic.in **L.P:** Dir.General (News): G.Mohanty **E:** dgn@air.org.in News on phone: English ☎ +91 11 2332-4343/1259, Hindi +91 11 2332-4242/1258
Commercial Service (Vividh Bharati): All India Radio, ✉ Gorai Road, Borivli West, Mumbai-400 091, Maharashtra ☎ +91 22 28692698 **E:** vbsmumbai@gmail.com
National Channel: ✉ All India Radio, Todapur, New Delhi 110012 ☎ +91 11 25843207 **E:** senchair@yahoo.com
Research & Development: ✉ Office of the Chief Eng R & D, All India Radio, 14-B, Indra Prashta Estate, Ring Road, New Delhi-110002 ☎ +91 11 23378211, 23378212 **E:** researchdelhi@yahoo.co.in
Monitoring: ✉ International Monitoring Stn., All India Radio, Dr. K.S. Krishnan Rd, Todapur, New Delhi-110097 ☎ +91 11 25842939 ✉ Central Monitoring Stn, All India Radio, Ayanagar, New Delhi-110047.
Audience Research: ✉ Audience Research Unit, AIR, Akashvani Bhavan, Parliament Street, New Delhi 110001 ☎ +91 11 23421022 **L.P:** Dir: M.N.Jha

Regional Headquarters: (Office of the Chief Engineer)
North Zone: AIR, Jamnagar House, Shahjahan Road, New Delhi-110011 ☎ +91 11 23389000
East Zone: AIR, 4th Floor, Akashvani Bhawan, Eden Garden, Kolkata-700001 ☎ +91 33 22480158
North-East Zone: AIR, Dr P Kakati's Building, G.S. Road, Guwahati-781006, Assam ☎ +91 361 2230326
West Zone: AIR, 101 M.K.Road, Mumbai-400020, Maharashtra ☎ +91 22 22014287
South Zone: AIR, Swami Sivanada Salai, Chepauk, Chennai-600005 ☎ +91 44 25383253

NB: Thiruvananthapuram is referred to throughout as Trivandrum.

MW: c) Vividh Bharati, e) ext.sce., n) national channel, r) relay stn
*) off air due to installation of new transmitter

KHz	Station	kW	reg	KHz	Station	kW	reg
531	Jodhpur A	300	N	1170	Hyderabad (St'by)	1	S
540	Aizawl	20	NE	1179	Rewa	20	W
549	Ranchi A	100	E	1188	Mumbai C	50	W,c
558	Mumbai B	100	W	1197	Shillong (St'by)	1	NE
567	Dibrugarh	300	NE	1197	Tirunelveli	20	S
576	Alappuzha	200	S, r	1206	Bhawanipatna	200	E
585	Nagpur A	300	W	1215	New Delhi	20	N, n
594	Chinsurah	1000	E, er	1215	Pudducherri	20	S
603	Ajmer	200	N, r	1224	Srinagar C	10	N
612	Bengaluru A	200	S	1233	Tura	20	NE
621	Patna A	100	E	1242	Varanasi	100	N
630	Thrissur	100	S	1251	Sangli	20	W
639	Kohima	100	NE	1260	Ambikapur	20	W
648	Indore A	200	W	1269	Agartala	20	NE
657	Kolkata A	200	E	1269	Madurai	20	S
666	New Delhi B	100	N	1278	Lucknow C	10	N, c
675	Bhadravathi	20	S	1287	Panaji A	100	W
675	Chhatarpur	20	W	1296	Darbhanga	10	E
675	Itanagar	100	NE	1305	Parbhani	20	W
684	Kozhikode A	100	S	1314	Bhuj	20	W
684	Port Blair	100	S	1314	Cuttack B	1	E, c
684	Kargil A	200	N	1323	Kolkata C	20	E, c
702	Jalandhar A	200	N, e	1332	Tezu	10	NE
711	Siliguri	200	E	1341	Kohima	1	NE
720	Chennai A	200	S	1350	Jalandhar C	1	N, c
729	Guwahati A	20	NE	1350	Kupwara	20	N, r
738	Hyderabad A	200	S	1368	New Delhi C	20	N, c
747	Lucknow A	300	N	1377	Hyderabad B	20	S
756	Jagdalpur	100	W	1386	Gwalior	20	W
765	Dharwad A	200	S	1395	Bikaner	20	N
774	Shimla	100	N	1404	Gangtok	20	NE
783	Chennai C	20	S, c	1413	Kota	20	N
792	Pune A	100	W	1440	Kurseong	1	E
801	Jabalpur	200	W	1449	Kanpur	1	N, c
810	Rajkot A	300	W	1458	Barmer	20	N
819	New Delhi A	200	N	1458	Bhagalpur	20	E
828	Panaji B	20	W, c	1467	Jeypore	100	E
828	Silchar	20	NE	1476	Jaipur A	1	W
837	Vijayawada A	100	S	1485	Adilabad	1	S
846	Ahmedabad A	200	W	1485	Ahwa	1	W
864	Shillong	100	NE	1485	Chamoli	1	N
873	Jalandhar B	300	N	1485	Drass	1	N, cr
882	Imphal	300	NE	1485	Joranda	1	E
891	Rampur	20	N	1485	Khaltsi	1	N, cr
900	Kadapa	100	S	1485	Nongstoin	1	NE
909	Gorakhpur	100	N	1485	Nyoma	1	N, cr
918	Suratgarh	300	N	1485	Pithoragarh	1	N, r
927	Visakhapatnam	100	S	1485	Soro	1	E
936	Tiruchirapalli A	100	S	1503	Vijayawada B	1	S, c
945	Sambalpur	100	E	1512	Kokrajhar	20	NE
954	Najibabad	200	N	1521	Aurangabad	1	W
963	Jalgaon	20	W	1521	Tawang	10	NE
972	Cuttack A	300	E	1530	Agra	20	N
981	Raipur	100	W	1566	Nagpur	1000	W, nr
990	Jammu A	300	N	1584	Diphu	1	NE
999	Almora	1	N	1584	Himmat Nagar	1	W
999	Coimbatore	20	S	1584	Jamshedpur	1	E
1008	Kolkata B	100	E	1584	Kalpa	1	N, r
1017	Chennai B	20	S	1584	Kargil B	1	N
1017	New Delhi	10	N	1584	Kavaratti	1	S
1026	Allahabad A	20	N	1584	Keonjhar	1	E
1035	Guwahati B	10	NE	1584	Mathura	1	N
1044	Mumbai A	100	W	1584	Mon	1	NE
1053	Leh	20	N	1584	Padam	1	N,c
1053	Tuticorin	200	S, e	1593	Bhopal A	10	W
1062	Pasighat	10	NE	1602	Diskit	1	N cr
1071	Rajkot *	1000	W, e	1602	Pauri	1	NE
1089	Udipi	20	S, r	1602	Saiha	1	NE
1089	Naushera	20	N, r	1602	Solapur	1	W
1107	Gulbarga	20	S	1602	Tiesuru	1	N,cr
1116	Srinagar A	300	N	1602	Tuensang	1	NE
1125	Tezpur	20	NE	1602	Udagamangalam	1	S
1134	Chinsurah*	1000	E, ner	1602	Uttarkashi	1	N, r
1143	Ratnagiri	20	W	1602	Varanasi B	1	N, c
1143	Rohtak	20	N	1602	William Nagar	1	NE
1161	Trivandrum	20	S	1602	Ziro	1	NE

Regional Domestic SW stations

kHz	kW	Station	H. of tr.
3945 #	50	Gorakhpur	0230-0300
4760	10	Leh	s0128/w0213-0400/0413/Sun 0430, 1200-1700
4760	8.5	Port Blair	2355-0300 1030-1700/1730
4775	50	Imphal	s0000/w0030-0215 1030-1700/1730
4800	50	Hyderabad	0020-0215 1130-1742
4810	50	Bhopal	0025-0215 1130-1742
4820	50	Kolkata	0025-0215 1220-1744
4830	50	Jammu*	0025-0445/Sun 0450 1030-1741
4835v	10	Gangtok	0100-0401/ 1030-1600 (currently on 4837.2)
4840	50	Mumbai	2355-0400 1230-1730
4850	50	Kohima *	0000-0415 1000-1600/1630/1700
4860 #	50	Delhi (Kingsway)	0025-0440, 1220-1330
4870	100	Delhi (Kingsway)	0230-0330 1430-1530 (R. Sadaye Kashmir)
4880	50	Lucknow	0025-0430 (Sun 0415), 1215-1741
4895	50	Kurseong	0055-0400 1130-1700(Sat, Sun 1741)
4910	50	Jaipur	0025-0430(Sun 0530) 1130-1741
4920	50	Chennai	0015-0245 1200-1739
4940	50	Guwahati	0000-0415(Sun 0450) 1135v-1700 (Sat 1741)
4950	50	Srinagar	s0000/w0120-0215 1120-1739 (2220v-2315v during Ramadan)
4960	50	Ranchi*	0025-0445 1100 (Sun 1130) -1741
4965	50	Shimla	0025-0200 1235 v-1730 (Sat, Sun 1741)
4970	50	Shillong	0025-0400 1056-1630
4990	50	Itanagar*	0020-0400 1000-1630
5010	50	Trivandrum	0020-0215 1130-1740
5015	50	Delhi (Kingsway)	1220-1841
5040	50	Jeypore	0025-0435 (Sun 0445), 1130 (Sun 1115)-1741
5050	10	Aizawl	0025-0400 1130-1630
5965	50	Jammu*	0630-0930
5985	50	Ranchi*	0630-1000
6000	10	Leh	0700 (Sun 0630)-0930
6020	50	Shimla	0215-0410, 0700 (Sun 0415-1000)-0930, 1130-1230
6030	50	Delhi (Kingsway)	0200-0310
6030	100	Delhi (Kingsway)	1215-1430
6040	50	Jeypore*	0700-0936
6065	50	Kohima*	0430-0510 0700-0900
6085	50	Delhi (Kingsway)	1220-1310, 1330-1340, 1345-1420, 1430-1440, 1445-1615/1630/1700/1730v, 1730-1740
6085	10	Gangtok	For special broadcasts in day time
6090		Delhi	0900-1200 (Vividh Bharati, Tests)
6100	100	Delhi (Kingsway)	0730-0830 (Radio Sadaye Kashmir)
6100	100	Delhi (Khampur)	0900-1200 (Vividh Bharati, DRM)
6110	50	Srinagar	0225-0509 (Sun 1115), 0600-1115
6150	50	Itanagar*	0700-0900
6190	50	Delhi (Kingsway)	0730-1030
7210	50	Kolkata	0230-0401 (SS 0501), 0700-1000 (Sun 1030)
7230	50	Kurseong	0620-1030
7235	50	Delhi (Kingsway)	0215-0320 0330-0340 (Sun 0355)
7240	50	Mumbai	0530-1035
7250 #	50	Gorakhpur	1130-1140
7270 #	100	Chennai	0130-0430
7280	50	Guwahati	0600-0930 0945-1130 (Sun 0530-1145)
7290	50	Trivandrum	0230-0430 (Sat , Sun 1030), 0630-1000
7295	10	Aizawl	0700-1000
7315	50	Shillong	0656-0931
7325	50	Jaipur	0630-0931
7335	50	Imphal	0225-0400 (Sun 0430), 0630 (Sun 0600)-1000
7340 #	100	Mumbai	1130-1140
7370	50	Delhi (Kingsway)	0030-0040
7370 #	100	Delhi (Kingsway)	1550-1615/1630/1700/1730v 1730-1740
7380	50	Chennai	0300-0430 (Sun 0500), 0610-0930 (Sun 1130)
7390	8.5	Port Blair	0315-0400 (Sat 0415, Sun 0500), 0700-0931 (Sun 1000)
7420	50	Hyderabad	0225-0400 (Sat 0505, Sun 0500), 0545/0610-0930 (Sun 0530-1030)
7420 #	50	Guwahati	0230-0300,1515-1600,1730-1740
7430	50	Bhopal	0225-0447 (Sun 0531), 0630/0700-0932, (Sun 0700-1031)
7440	50	Lucknow	0700 (Sun 0430)-1000, 1005-1006
9425	500	Bengaluru	1320-0043 (National Channel)
9470	250	Aligarh	1320-0043 (National Channel)
9575 #	50	Delhi (Kingsway)	1330-1420, 1430-1440, 1445-1615/1630/1700/1730v, 1730-1740
9595 #	50	Delhi (Kingsway)	0810-0830, 1130-1140
9810 #	50	Delhi (Kingsway)	For special broadcasts
9835 #	50	Delhi (Kingsway)	1330-1420, 1430-1440, 1445-1615/1630/1700/1730v, 1730-1740

D.Prgr: Varies from stn to stn. Most stns have 3 transmissions daily ie Morning/Noon/Evening. Some smaller stns have only 1 or 2 transmissions. Extended coverage during sports or special events.
National Ch.: 1325-0043 on 1215, 1566, 9425, 9470 kHz
F.PI: DRM at 40 MW stations. **New:** 1kW at Dharmanagar, Dungarpur (1485 kHz). **Current: stns:** 10kW Kavaratti, 1000kW Chinsurah, Rajkot.

kHz	kW	Station	H. of tr.
9870	500	Bengaluru	0025-0435, 0900-1200, 1245-1740
			(Vividh Bharati)
11620	#250	Delhi (Khampur)	1130-1140
11710	# 50	Delhi (Kingsway)	1115-1140
11830	50	Delhi (Kingsway)	0125-0340 (Sun 0355)
15050#	250	Delhi (Khampur)	For special broadcasts
15135	50	Delhi (Kingsway)	0125-0205 0215-0310 (Sun 0355)
15185	# 50	Delhi (Kingsway)	0730-0930 1115-1140
15260	# 50	Delhi (Kingsway)	0700-0930
17860	#100	Delhi (Kingsway)	1220-1245

s = summer, w = winter, v = timing/frequency varies. * = irregular / off air, #= Frequency also used by External Services at other times

N in English originating in New Delhi and relayed by most stns: 0035-0040, 0245-0300, 0335-0340, 0435-0440, 0630-0635, 0730-0735, 0830-0900, 0935-0940, 1030-1035, 1135-1140, 1230-1235, 1430-1435, 1435-1440(Sports), 1530-1545, 1730-1735. Extended broadcasts for special events, important Parliament sessions, sports and on January 26 (Republic Day) and August 15 (Independence Day) **V.** by QSL-card. Reception Reports for SW stations only to: Director (Spectrum Management & Synergy), All India Radio, Room No.204, Akashvani Bhavan, New Delhi-110001 or online form: www.allindiaradio.gov.in/recepfdk.html 91-11-23421062, 23421145 **E**: spectrum-manager@air.org.in In charge of processing reception reports: Director: M.S.Ansari. Local stns also verify directly in many cases by letter or email. No return postage is necessary. **F.P.I.**: DRM at Aligarh, Bengaluru, New Delhi

Addresses of SW stations: (Reception reports may be addressed to the Station Engineer)
1) Aizawl: Radio Tila, Tuikhuahtlang, Box 13, Aizawl-796001 ☎+91 389 2322415 **E**: aizawl@air.org.in – **2) Aligarh**: Anoopshahar Road, Aligarh-202001, Uttar Pradesh ☎+91 571 2700972 **E**: aligarh@air.org.in – **3) Bengaluru**: Super Power Transmitters, Yelahanka New Town, Bengaluru-560064, Karnataka ☎+91 80 27601149 **E**: sptairynk@rediffmail.com– **4) Bhopal**: Shyamla Hills, Bhopal-462002, Madhya Pradesh ☎+91 755 2660088 **E**: se_airbpl@dataone.in – **5) Chennai**: S.M.Nagar PO, Avadi, Chennai-600062, Tamilnadu. Tel. 91 44 2638 3204. email : airavadi@dataone.in – **6) Gangtok**: Old MLA Hostel, Gangtok-737101, Sikkim ☎+91 3592 202636 **E**: seairgtk@yahoo.co.in – **7) Gorakhpur**:Town Hall, Gorakhpur-273001, Uttar Pradesh ☎+91 551 2337401 **E**: seairgkp@rediffmail.com – **8) Guwahati**: Chandmari, Guwahati-781003 Assam ☎+91 361 2660235 **E**: guwahati@air.org.in – **9) Hyderabad**: Rocklands, Saifabad, Hyderabad-500004, Andhra Pradesh ☎+91 40 23234904. **E**: airhyderabad@rediffmail.com – **10) Imphal**: Palace Compound Imphal-795001, Manipur ☎+91 385 2220534. **W**: http://cicmanipur.nic.in/html/air_imp.htm **E**: airimfal@sancharnet.in – **11) Itanagar**: 'C' Sector, Itanagar-791111, Arunachal Pradesh ☎+91 360 2212881 **E**: itanagar@air.org.in – **12) Jaipur**: 5 Park House, Mirza Ismail Road, Jaipur-302001, Rajasthan ☎.91 141 2366263. **E**: jaipur@air.org.in – **13) Jammu**: Radio Kashmir, Begum Haveli, Old Palace Road, Jammu-180001, Jammu & Kashmir ☎+91 191 2544411. **E**: jammu@air.org.in – **14) Jeypore**:764005, Orissa ☎+91 6854 232524 **E**: airjeypore@rediffmail.com – **15) Kohima**:797001, Nagaland ☎+91 370 2245556 **E**: kohima@air.org.in – **16) Kolkata**:Eden Gardens, Kolkata-700001, West Bengal ☎+91 33 22481705 **E**: sgeairkolkata@rediffmail.com – **17) Kurseong**: Mehta Club Bldg, Kurseong-734203, Darjeeling Dist., West Bengal ☎+91 354 2344350 **E**: kurseong@air.org.in – **18) Leh**: Leh-194101, Ladakh Dist., Jammu & Kashmir ☎+91 1982 252080 **E**: seairleh@rediffmail.com – **19) Lucknow**:18 Vidhan Sabha Marg, Lucknow-226001, Uttar Pradesh ☎+91 522 2237476 **E**: lucknow@air.org.in – **20) Mumbai**: Marve Road, Malwani, Malad West, Mumbai 400095, Maharashtra. Tel. 91 22 28821867. **W**: www.hptmald.org.in **E**: malad@air.org.in – **21A) New Delhi**: High Power Transmitters, Khampur, New Delhi -110036 ☎+91 11 27203560 **E**: hptkhampur@yahoo.co.in –**21B)** High Power Transmitters, Kingsway, New Delhi-110009 ☎+91 11 27436661 **E**: hptkingsway@yahoo.com – **22) Panaji**: Goa University PO, Panaji-403206 ☎+91 832 2222311 **E**: airtrgoa@sancharnet.in – **23) Port Blair**: Haddo Post, Dilanipur, Port Blair-744102, Andaman & Nicobar Islands ☎+91 3192 230682 **E**: airportblair@rediffmail.com – **24) Ranchi**: 6 Ratu Rd, Ranchi-834001, Jharkhand ☎+91 651 2283310 **E**: ranchi@air.org.in – **25) Shillong**: North Eastern Service, Pomdngiem, Opposite GPO, Shillong-793001, Meghalaya **W**: www.airshillong.org ☎+ 91 364 2224443 **W**: www.airshillong.org **E**: shillong@air.org.in – **26) Shimla**: Choura Maidan, Shimla-171004, Himachal Pradesh ☎+91 177 2811355 **E**: airshimla@yahoo.com – **27) Srinagar**: Radio Kashmir, Sherwani Rd, Srinagar-190001, Jammu & Kashmir ☎+91 194 2452100 **E**: srinagar@air.org.in – **28) Thiruvanathapuram :** Bhakti Vilas, Vazuthacaud, Thiruvanathapuram-695014, Kerala ☎+91 471 2325009 **W**: www.airtvm.com **E**: se_airtvm@rediffmail.com

FM: b) FM Rainbow c) Vividh Bharati g) FM Gold r) relay stn

MHz	location	kW	reg	MHz	location	kW	reg
93.9	Vadodara	10	W	102.2	Kathua	10	N
96.7	Ahmedabad	10	W, c	102.2	Murshidabad	6	E
100.1	Ahmednagar	6	W	102.2	Vijayawada	1	S, b,c
100.1	Bengaluru	3	S	102.3	Chennai	20	S, g
100.1	Gorakhpur	1	N, c	102.3	Daman	3	W
100.1	Kothagudem	6	S	102.3	Guna	6	W
100.1	New Delhi	5	N	102.3	Hissar	6	N
100.2	Haflong	6	NE	102.3	Karwar	3	S
100.2	Kolkata	20	E, g	102.3	Kochi A	6	S
100.2	Patiala	6	W	102.3	Kurseong	5	E, b
100.2	Shivpuri	6	W	102.3	Purnea	6	E
100.3	Allahabad	10	N, c	102.4	Akola	6	W
100.3	Asansol	6	E, r	102.4	Kurnool	6	S
100.3	Jaipur B	6	W	102.4	Rajkot	10	W, c
100.3	Jammu A	3	N	102.5	Dharmapuri	10	S
100.3	Karaikal	6	S	102.5	Kullu	6	N, r
100.3	Mangalore	10	S	102.5	Patna	6	E, c
100.4	Bareilly	6	N	102.6	Chitradurga	6	S
100.4	Mandla	1	W, c	102.6	New Delhi	20	N, b
100.5	Dhule	6	W	102.6	Rourkela	6	E
100.5	Hospet	10	S	102.6	Sagar	5	W
100.5	Kodaikanal	10	S, b	102.6	Srinagar	6	N, c
100.6	Berhampur	6	E	102.7	Jalandhar	10	N,b
100.6	Mysore	10	S	102.7	Kolhapur	6	W
100.6	Nagpur	6	W, c	102.7	Manjeri	6	S, b
100.6	Varanasi	1	N,c	102.7	Nagaon	6	NE
100.7	Aizawl	6	NE	102.7	Obra	6	N
100.7	Churu	6	W	102.7	Yeotmal	6	W
100.7	Lucknow	6	N, b	102.8	Hyderabad	6	S, c
100.7	Mumbai	20	W, g	102.8	Puducherry	5	S
100.7	Poonch	6	N	102.9	Saraipalli	1	W, c
100.7	Raigarh	6	W, c	102.9	Baripada	6	E
100.7	Rajgarh	3	W	102.9	Beed	6	W
100.8	Guwahati	10	NE, c	102.9	Bengaluru	10	S, c
100.8	Jamshedpur	6	E, c	102.9	Chittorgarh	6	W
100.9	Mukokchung	6	NE	102.9	Jabalpur	10	W, c
100.9	Port Blair	10	S, c	103.0	Chandrapur	6	W
100.9	Shimla	1	N, c	103.0	Coimbatore	10	S, b,c
101.0	Bhaderwah	6	N	103.0	Daltonganj	6	E
101.0	Nagercoil	6	S	103.0	Dharwad	10	S
101.0	Pune	6	W, c	103.0	Jhansi	6	N
101.1	Bathinda	6	N	103.0	Kohima	6	NE
101.1	Jowai	6	NE	103.1	Alwar	6	N
101.1	Surat	6	W	103.1	Betul	6	W
101.2	Khandwa	6	W	103.1	Chandigarh	6	N
101.2	Aligarh	6	N, br	103.1	Itanagar	10	NE
101.3	Balaghat	6	W	103.1	Macherla	3	S
101.3	Banswara	6	W	103.1	Madikeri	6	S
101.3	Bengaluru	10	S, b	103.1	Satara	6	W
101.3	Cuttack	6	E, b	103.1	Shanthi Nikethan	3	W
101.3	Osmanabad	6	W	103.2	Bilaspur	6	E
101.4	Chennai	20	S, b	103.2	Jhalawar	6	N
101.4	Devikulam	6	S	103.2	Kailashahar	6	NE
101.4	Kurukshetra	6	N	103.2	Nizamabad	6	S
101.4	Nasik	6	W	103.2	Tirupati-I	10	S
101.5	Siliguri	10	E, c	103.3	Bellary	1	S
101.5	Kannur	6	S	103.3	Dhubri	6	NE, r
101.5	Markapur	6	S	103.3	Madurai	1	S
101.6	Sawai Madhapur	6	W	103.3	Ranchi	6	E, c
101.6	Agartala	10	NE, c	103.4	Dharamsala	10	N
101.6	Indore	6	W, c	103.4	Jorhat	6	NE
101.6	Raipur	1	W	103.4	Puri	3	E
101.7	Anantapur	6	S	103.4	Sasaram	6	E
101.7	Aurangabad	1	W, c	103.5	Bhopal	6	W, c
101.7	Chaibasa	6	E	103.5	Churachandpur	6	NE
101.7	Udaipur	1	N, c	103.5	Imphal	10	NE, c
101.8	Bijapur	6	S	103.5	Mount Abu	6	W
101.8	Hamirpur	6	N	103.5	Rohtak	1	N, c
101.8	Jaisalmer	10	W	103.6	Warangal	10	S, c
101.9	Bolangir	6	E	103.6	Kozhikode	20	S, c
101.9	Faizabad	6	N	103.6	Oros	6	S
101.9	Hyderabad	5	S, b	103.6	Shillong	10	NE, b
101.9	Lungleh	6	NE	103.7	Belonia	6	NE
101.9	Rajouri	10	N, r	103.7	Gulbarga	1	S,c
101.9	Trivandrum	10	S, c	103.7	Kanpur	1	N
102.0	Shahdol	6	W	103.7	Nagaur	6	N
102.0	Visakhapatnam	10	S, b,c	104.5	Jammu B	10	N, c
102.1	Hazaribagh	6	E	105.4	Panaji	6	W, b
102.1	Jodhpur	6	N, c	106.4	New Delhi	20	N, g
102.1	Mussoorie	10	N, br	107.0	Kolkata	20	E, b
102.1	Raichur	6	S	107.1	Mumbai	20	W, b
102.1	Tiruchirapalli	10	S, b,c	107.2	Kasauli	10	N, bcr
102.2	Chindwara	6	W	107.5	Kochi B	10	S, bc
102.2	Godhra	6	W	107.5	Tirupati -II	3	S
102.2	Hassan	6	S	+ 51 Stns of 100W			

F.PI: DRM+ in about 170 FM stations. Relay stations of 100 watts in about 200 locations. **1 kW:** Anini, Bhadravati, Changlang, Chempai, Cuddapah, Daporijo, Dawki, Dibrugarh, Gairsen, Goalpara, Jeypore, Karimganj, Khonsa. Kalasib, Kota, Longtherai, Lumding, New Tehri, Nutan Bazar, Parbhani, Phek, Rairangpur, Rampur, Ratnagiri, Sangli, Srikakulam, Suryapet, Tamenglong, Tezpur, Thrissur, Tuipang, Tuticorin, Udaipur(Tripura), Ukhrul, Wokha, Zunheboto **5 kW:** Ambikapur, Bageshwar, Bhuj, Chhatarpur, Gwalior, Silchar. Karimnagar, Ujjain.**10 kW:** Adliababd, Alwar, Amravati, Balurghat, Banda, Banswara, Bardhaman, Bellary, Bhawanipatna, Bikaner, Chandigarh(Kasauli), Chittorgarh, Coochbehar, Darjeeling, Dehraduni(Mussoorie), Dhanbad, Gangtok, Gorakhpur, Gulbarga, Haldwani, Hyderabad (2 Nos), Jalandhar, Jamshedpur, Jorhat, Junagadh, Kanpur, Kochi, Kurukshetra, Lakhimpur Kheri, Lucknow, Madurai, Maunath Bhanjan, Mehaboobnagar, Patna, Pondicherry, Raipur, Ranchi, Rohtak, Shimla, Srinagar, Tirunelveli, Udaipur,Varanasi, Vijayawada. **20 kW:** Amristsar, Chautan Hill, Fazilka **Replacement of 6kW to 10kW:** Nagpur, Surat, Pune. **Replacement of 1kW MW to 10kW FM:** Aurangabad, Solapur.

Addresses of MW & FM stations (See also SW stn addresses):
Adilabad-504002, Andhra Pradesh – Palace Compound, North Gate – **Agarthala**-799001, Tripura – Vivbhav Nagar, **Agra**-282001, Uttar Pradesh – Ashram Rd, Navarangpura, **Ahmedabad**-380009, Gujarat – **Ahmednagar**-414001, Maharashtra – **Ahwa**-394710, Dangs Dist., Gujarat – 21/10 Vaishali Nagar, **Ajmer**-305001, Rajasthan – **Akola**-444001, Maharashtra – Pathirapally, **Alappuzha**-688521, Kerala – Z-9 Dayanand Marg, **Allahabad**-211001, Uttar Pradesh – Almora–263601, Kumaon Dist., Uttarakhand – Scheme No 6, Mangal Vihar, **Alwar**-301001, Rajasthan – Kumar Palace, **Ambikapur**-497001, Surguja Dist., Chhatisgarh – Near, Collectorate, **Anantapur**-515001, Andhra Pradesh – **Asansol**-713301, Burdwan Dist., West Bengal – Jalna Rd, **Aurangabad**-431005, Maharashtra – **Aurangabad**-842101, Bihar–**Balaghat**-481001, Madhya Pradesh – Raj Bhavan Rd, **Bengaluru**-560001, Karnataka - **Banswara**-327001, Rajasthan – No 15, Lal Phatak, Badaun Road, **Bareilly**-243004, Uttar Pradesh – **Baripada**-757001, Mayurbhanj Dist., Orissa – Laxmi Nagar, **Barmer**-344001, Rajasthan – Khandeshwari Road, **Beed**-431122, Maharashtra – **Bellary**-583101, Karnataka – **Belonia**-799155, Tripura – **Berhampur**-760001, Ganjam Dist., Orissa – **Betul**-460001, Madhya Pradesh – J.P.S.Colony, Paper Tower, **Bhadravati**-577302, Karnataka – Port Campus, **Bhagalpur**-812001, Bihar – **Bathinda**-151005, Punjab – **Bhaderwah**-182202, Doda Dist., Jammu & Kashmir – **Bhawanipatna**-766001, Nektiguda, Kalahandi Dist., Orissa – **Bhuj**-370001, Kutch Dist., Gujarat – **Bijapur**-586101, Karnataka – **Bikaner**-334001, Rajasthan – Nutan Colony, **Bilaspur**-495001, Chhattisgarh – **Bolangir**-767001, Orissa – Tungri Maidan, **Chaibasa**-833201, Singhbhum Dist., Jharkhand – **Chamoli**-246424, Gopeshwar, Uttarakhand – **Chandrapur**-442401, Maharashtra – Sector-19B, **Chandigarh**-160019 – 7, Kamarajar Salai, Mylapore, **Chennai**-600004, Tamilnadu – **Chhatarpur**-471001, Madhya Pradesh – **Chindwara**-480001, Madhya Pradesh – **Chinsurah**-712102, West Bengal – **Chitradurga**-577501, Karnataka – Sector 4, Gandhi Nagar, **Chittorgarh**-312001, Rajasthan – **Churu**-331001, Rajasthan – **Churachandpur**- 795128, Manipur - Trichy Rd, Ramanathapuram, **Coimbatore**-641045, Tamilnadu – Madhupur House, Bakshi Bazar, Cantonment Rd, **Cuttack**-753001, Orissa – **Daltonganj**-822101, Jharkhand – Opp. Varkunt, Mota Fliya, **Daman**-396210, Daman & Diu – **Darjeeling**-734101, West Bengal – **Darbhanga**-846004, Bihar – **Deogarh**-768108, Orissa – **Devikulam**-685613, Idukki Dist., Kerala – **Dharmapuri**-636701, Tamilnadu – **Dharmasala**-176215, Kangra Dist., Himachal Pradesh – Saptapur, **Dharwar**-580008, Karnataka – **Dhubri**-783301, Assam – **Dhule**-424001, Maharashtra – Malakhubasa, **Dibrugarh**-786001, Assam – **Diphu**-782460, Kabri Anglong Dist., Assam – **Diskit**-194401, Leh Dist., Jammu & Kashmir– **Drass**-194102, Kargil, Jammu & Kashmir – Begumganj Garahiya, **Faizabad**-224001, Uttar Pradesh – **Godhra**-389001, Gujarat – Aiwan-e-Shahi, Municipal Garden, **Gulbarga**-585103, Karnataka – **Guna**-473001, Madhya Pradesh – Gandhi Rd, **Gwalior**-474002, Madhya Pradesh – **Haflong**-788819, Assam – **Hamirpur**-177001, Himachal Pradesh – Salagame Road, **Hassan**-573201, Karnataka – Jail Road, **Hazaribagh**-825301, Jharkhand – **Himmat Nagar** – 383001, Gujarat – **Hissar**-125001, Haryana – **Hospet**-583201, Karnataka – Malwa House, Residency Area, **Indore**-452001, Madhya Pradesh – 373 Napier Town, **Jabalpur**-482001, Madhya Pradesh – Collectorate Rd, **Jagdalpur**-494 001, Bastar Dist., Chhattisgarh – Vyas Colony, **Jaisalmer**-345001, Rajasthan – Jalandhar-144001, Punjab –Jilhapet, **Jalgaon**-425001, Maharashtra – Adityapur, Gamharia Rd, **Jamshedpur**-831013, Jharkhand – Jungle Road, **Jhalawar**-326001, Rajasthan – Kanpur Road, **Jhansi**-284128, Uttar Pradesh – Paoata 'C' Road, **Jodhpur**-342006, Rajasthan – **Joranda**-759014, Dhenkanal Dist., Orissa – **Jorhat**-785001, Assam – **Jowai**-793150, Jaintia Hills, Meghalaya – Cooperative Colony, **Kadapa**-516001, Andhra Pradesh – **Kailashahar**-799277, Tripura –

Kalpa-172108, Kinnaur Dist., Himachal Pradesh – **Kannur**-670001, Kerala – **Kanpur**-208001, Uttar Pradesh – **Kargil**-194103, Jammu & Kashmir – Radio Avenue, Nehru Ngr., **Karaikal**-609606, Puducherri – **Karwar**-581301, Karnataka – **Kasauli**-173204, Solan Dist., Himachal Pradesh – **Kathua**-184104, Jammu & Kashmir – **Kavaratti**-682555, Lakshadeep – **Keonjhar**-758001, Orissa –**Khaltsi**–194106, Leh, Jammu & Kashmir – **Khandwa**-450001, Nimar Dist., Madhya Pradesh – BMC PO, **Kochi**-682021, Ernakulam Dist., Kerala – Anandagiri, **Kodaikanal**-624101, Tamilnadu – **Kokrajhar**-783370, Assam – Sardar Cly, Taravai Park, **Kolhapur**-416003, Maharashtra – Jawahar Rd, **Kota**-324001, Rajasthan – Ramavaram, **Kothagudam**-507118, Khammam Dist., Andhra Pradesh – Beach Rd, **Kozhikode**-673001, Kerala – **Kulu**-175101, Himachal Pradesh – **Kupwara** – 193222, Jammu & Kashmir – Bellary Road, **Kurnool**-518003, Andhra Pradesh – **Kurushetra**-132118, Haryana – **Lungleh**-796701, Mizoram – **Macherla**-522426, Guntur Dist., Andhra Pradesh – **Madikeri**-571201, Kodagu Dist., Karnataka – Lady Doak College Rd, Chokkikulam, **Madurai**-625002, Tamilnadu – Kadri Hills, **Mangalore**-575004, Dakshin Kanara Dist., Karnataka – **Manjeri**-676121, Kerala – **Mandla**-481661, Madhya Pradesh - **Markapur**-523316, Prakasam Dist., Andhra Pradesh – Vrindavan Rd, Gayatri Tapobhumi, **Mathura**-281003, Uttar Pradesh – **Mokokchung**-798601, Nagaland – **Mount Abu** -307501, Sirohi Dist., Rajasthan – **Mumbai:** Broadcasting House, Backbay Reclamation, Mumbai-400020, Maharashtra – **Murshidabad**-742101, West Bengal – **Mussoorie**-248179, Dehradun Dist., Uttarakhand – Yadavagiri, **Mysore**-570020, Karnataka – **Nagaon**-782002, Assam – Basni Rd, **Nagaur**-341001, Rajasthan – Konam, **Nagercoil**-629004, Kanya Kumari Dist., Tamilnadu – Civil Lines, Palam Rd, **Nagpur**-440001, Maharashtra (National Channel: Seminary Hills, **Nagpur** 440006, Maharashtra) – Kotwali Rd, **Najibabad**-246763, Bijnor Dist., Uttar Pradesh – Vasrania, **Nanded**-431601, Maharashtra – **Naushera**-193125, Jammu & Kashmir – **Nizamabad**-503012, Andhra Pradesh – **Nongstoin**-793119, West Khasi Hills, Meghalaya – **Nyoma**-194101, Leh Dist, Jammu & Kashmir – **Obra**-231219, Uttar Pradesh – Tambri Vibhag, **Oros** 416812, Sindhudurg Dist, Maharashtra – **Osmanabad**-413501, Maharashtra –**Padam**, Jammu & Kashmir - Altinho, **Panaji**-403001, Goa – Jamakar Colony, Nawa Mondha, **Parbhani**-431401, Maharashtra – **Pasighat**-791102, East Siang Dist., Arunachal Pradesh – Phase-I, Urban Estate, Rajpura Rd, **Patiala**-147002, Punjab – Frazer Road, Chhaju Bagh, **Patna**-800001, Bihar – **Pauri**-246001, Uttarakhand – **Pithorgarh**-262501, Uttarakhand – **Poonch**-185101, Jammu & Kashmir – 24 Coubert Avenue, Gorimedu, **Puducherri**-605001– University Rd, Shivaji Nagar, **Pune**-411005, Maharashtra – **Puri**-751001, Orissa – **Purnea**-854302, Bihar – **Raichur**-584101, Karnataka – Chote Atarmude, **Raigarh**-496001, Chhattisgarh – Kamla Nehru Marg, Civil Lines, **Raipur**-492001, Chhattisgarh – **Rajgarh** -465661, Madhya Pradesh – Opposite Race Course, Sitaram Pandit Marg, **Rajkot**-360001, Gujarat – **Rajouri**-185131, Jammu & Kashmir – **Rampur**-244901, Uttar Pradesh – Thiba Palace Rd, **Ratnagiri**-415612, Maharashtra – 6 Civil Lines, **Rewa**-486001, Madhya Pradesh – Subhash Rd, **Rohtak**-124001, Haryana – **Rourkela**-769001, Orissa – **Sagar**-470001, Madhya Pradesh – **Saiha**-796901, Chhimtuipui Dist., Mizoram – 3, Kuchery Rd, **Sambalpur**-768001, Orissa – Market Yard, Kolhapur Rd, **Sangli**-416416, Maharashtra – **Saraipalli**-493558, Raipur, Chhatisgarh - **Sasaram**-821115, Rohtas Dist., Bihar – **Satara**-415001, Maharashtra – Pali Road, **Shahdol**-484001, Madhya Pradesh – **Shanthi Nikethan**, West Bengal – Physical College, **Shivpuri**-473551, Madhya Pradesh – **Silchar**-788001, Cachar Dist., Assam – 2 Mile Sevoke Rd, **Siliguri**-734401, Darjeeling Dist., West Bengal – **Solapur**-413006, Maharashtra – **Soro**-756045 , Balasore Dist, Orissa – **Surat**-395001, Gujarat – **Suratgarh**-335804, Sriganganagar Dist., Rajasthan – **Swai Madhopur**-322001, Rajasthan – **Tawang**-790104, Arunachal Pradesh – **Tezpur**-784001, Sonitpur Dist., Assam – **Tezu**-792001, Lohit Dist., Arunachal Pradesh – Ramavarmapuram, **Thrissur**-680631, Kerala – 28-3 Promenade Rd, **Tiruchirapalli**-620001, Tamilnadu – Sarojini Park, Palayamkottai, **Tirunelveli**-627006, Tamilnadu – **Tirupati**-517501, Andhra Pradesh – **Tuensang**-798612, Nagaland – Lower Chandmari, **Tura**-794001, Meghalaya – Millerpuram, Playamkottai Road, **Tuticorin**-628008, Tamilnadu – Chetak Circle, **Udagamandalam**-643001, Nilgris, Tamilnadu, **Udaipur**-313001, Rajasthan – Brahmavar, **Udipi**-576213, Dakshina Kanara Dist., Karnataka – **Uttar Kashi**-249193, Uttarakhand – Makarpura Rd, **Vadadora**-390009, Gujarat – Mahmoorganj, **Varanasi**-221010, Uttar Pradesh – Bandar Rd, Punnammathota, **Vijayawada**-520010, Andhra Pradesh – Siripuram, **Visakhapatnam**-530003, Andhra Pradesh – **Waranagal**-506002, Andhra Pradesh – **William Nagar**, Meghalaya – **Yeotmal**-445001, Maharashtra – **Yercaud**-636601, Salem, Tamilnadu – **Ziro**-791120, Lower Subansiri Dist., Arunachal Pradesh.

EXTERNAL SERVICES: All India Radio
see International Broadcasting section.

Private FM Stations:

Location	MHz	Station	Location	MHz	Station
Agartala	91.9	R. Ooo La La	Jaipur	93.5	Red FM
Agra	91.9	Radio Mantra	Jaipur	94.3	My FM
Agra	92.7	Big FM	Jaipur	98.3	Radio Mirchi
Ahmedabad	91.1	Radio City	Jalandhar	91.9	Radio Mantra
Ahmedabad	93.5	Red FM	Jalandhar	92.7	Big FM
Ahmedabad	94.3	My FM	Jalandhar	94.3	My FM
Ahmedabad	95.0	Radio One	Jalandhar	98.3	Radio Mirchi
Ahmedabad	98.3	Radio Mirchi	Jalgaon	91.1	Radio City
Ahmednagar	91.1	Radio City	Jalgaon	106.4	Dhamaal 24
Ahmednagar	106.4	Dhamaal 24	Jammu	92.7	Big FM
Ajmer	92.7	Big FM	Jamshedpur	92.7	Big FM
Ajmer	94.3	My FM	Jamshedpur	93.5	Red FM
Akola	91.1	Radio City	Jamshedpur	104.8	Radio Dhoom
Aligarh	92.7	Big FM	Jhansi	92.7	Big FM
Allahabad	92.7	Big FM	Jodhpur	92.7	Big FM
Allahabad	93.5	Red FM	Jodhpur	94.3	My FM
Amritsar	92.7	Big FM	Jodhpur	104.8	Oye FM
Amritsar	94.3	My FM	Kannur	91.9	Radio Mango
Amritsar	104.8	Oye FM	Kannur	93.5	Red FM
Asansol	92.7	Big FM	Kannur	94.3	Club FM
Asansol	93.5	Red FM	Kannur	95.0	Best FM 95
Aurangabad	93.5	Red FM	Kanpur	92.7	Big FM
Aurangabad	98.3	Radio Mirchi	Kanpur	93.5	Red FM
Bareilly	91.9	Radio Mantra	Kanpur	98.3	Radio Mirchi
Bareilly	92.7	Big FM	Karnal	91.9	Radio Mantra
Bengaluru	91.1	Radio City	Karnal	106.4	Dhamaal 24
Bengaluru	91.9	Radio Indigo	Kochi	91.9	Radio Mango
Bengaluru	92.7	Big FM	Kochi	93.5	Red FM
Bengaluru	93.5	Red FM	Kochi	94.3	Club FM
Bengaluru	94.3	Radio One	Kolhapur	94.3	Tomato FM
Bengaluru	98.3	Radio Mirchi	Kolhapur	98.3	Radio Mirchi
Bengaluru	104.0	Fever FM	Kolkata	91.9	Friends FM
Bhopal	92.7	Big FM	Kolkata	92.7	Big FM
Bhopal	93.5	Red FM	Kolkata	93.5	Red FM
Bhopal	94.3	My FM	Kolkata	94.3	Radio One
Bhopal	98.3	Radio Mirchi	Kolkata	98.3	Radio Mirchi
Bikaner	92.7	Big FM	Kolkata	104.0	Fever FM
Bilaspur	94.3	My FM	Kolkata	104.8	Oye FM
Chandigarh	92.7	Big FM	Kolkata	106.2	Amar FM
Chandigarh	94.3	My FM	Kolkata	107.8	Power FM
Chennai	91.1	Radio City	Kota	92.7	Big FM
Chennai	91.9	Aahaa FM	Kota	94.3	My FM
Chennai	92.7	Big FM	Kota	95.0	Tadka 95 FM
Chennai	93.5	Suryan FM	Kozhikode	91.9	Radio Mango
Chennai	94.3	Radio One	Kozhikode	93.5	Red FM
Chennai	98.3	Radio Mirchi	Lucknow	91.1	Radio City
Chennai	104.8	Chennai Live	Lucknow	92.7	Big FM
Chennai	106.4	Hello FM	Lucknow	93.5	Red FM
Coimbatore	91.1	Radio City	Lucknow	98.3	Radio Mirchi
Coimbatore	93.5	Suryan FM	Madurai	93.5	Suryan FM
Coimbatore	98.3	Radio Mirchi	Madurai	98.3	Radio Mirchi
Coimbatore	106.4	Hello FM	Madurai	106.4	Hello FM
Cuttack	92.7	Big FM	Mangalore	92.7	Big FM
Cuttack	93.5	Red FM	Mangalore	93.5	Red FM
Cuttack	104.0	R. Choklate	Mangalore	98.3	Radio Mirchi
Dhule	106.4	Dhamaal 24	Mumbai	91.1	Radio City
Gangtok	91.9	Nine FM	Mumbai	92.7	Big FM
Gangtok	93.5	Red FM	Mumbai	93.5	Red FM
Gangtok	95.0	Radio Misty	Mumbai	94.3	Radio One
Gorakhpur	91.9	Radio Mantra	Mumbai	98.3	Radio Mirchi
Gulbarga	93.5	Red FM	Mumbai	104.0	Fever FM
Guwahati	91.9	R. Ooo La La	Mumbai	104.8	Oye FM
Guwahati	92.7	Big FM	Muzzafarpur	106.4	Dhamaal 24
Guwahati	93.5	Red FM	Mysore	92.7	Big FM
Guwahati	94.3	Gup Shup	Mysore	93.5	Red FM
Gwalior	91.9	Suno Lemon	Nagpur	91.1	Radio City
Gwalior	94.3	My FM	Nagpur	93.5	Red FM
Gwalior	95.0	Radio Chaska	Nagpur	94.3	My FM
Hissar	98.3	Dhamaal 24	Nagpur	98.3	Radio Mirchi
Hissar	91.9	Radio Mantra	Nanded	91.1	Radio City
Hissar	92.7	Big FM	Nasik	93.5	Red FM
Hyderabad	91.1	Radio City	Nasik	98.3	Radio Mirchi
Hyderabad	92.7	Big FM	New Delhi	91.1	Radio City
Hyderabad	93.5	Red FM	New Delhi	92.7	Big FM
Hyderabad	98.3	Radio Mirchi	New Delhi	93.5	Red FM
Indore	92.7	Big FM	New Delhi	94.3	Radio One
Indore	93.5	Red FM	New Delhi	95.0	Hit FM
Indore	94.3	My FM	New Delhi	98.3	Radio Mirchi
Indore	98.3	Radio Mirchi	New Delhi	104.0	Fever FM
Itanagar	91.9	R. Ooo La La	New Delhi	104.8	Oye FM
Jabalpur	93.5	Red FM	Panaji	91.9	Radio Indigo
Jabalpur	94.3	My FM	Panaji	92.7	Big FM
Jabalpur	98.3	Radio Mirchi	Panaji	98.3	Radio Mirchi
Jabalpur	106.4	Dhamaal 24	Patiala	92.7	Big FM
Jaipur	92.7	Big FM	Patiala	104.8	Oye FM

Location	MHz	Station	Location	MHz	Station
Patiala	106.4	Dhamaal 24	Surat	91.1	Radio City
Patna	98.3	Radio Mirchi	Surat	92.7	Big FM
Pondicherry	92.7	Big FM	Surat	94.3	My FM
Pondicherry	93.5	Suryan FM	Surat	98.3	Radio Mirchi
Pondicherry	106.4	Hello FM	Trivandrum	92.7	Big FM
Pune	91.1	Radio City	Trivandrum	93.5	Red FM
Pune	93.5	Red FM	Trivandrum	94.3	Club FM
Pune	94.3	Radio One	Trivandrum	98.3	Radio Mirchi
Pune	98.3	Radio Mirchi	Thrissur	91.9	Radio Mango
Raipur	94.3	My FM	Thrissur	93.5	Red FM
Raipur	95.0	Tadka 95 FM	Thrissur	95.0	Best FM 95
Raipur	98.3	Radio Mirchi	Thrissur	104.8	Club FM
Raipur	104.8	Rangila FM	Tiruchirapalli	106.4	Hello FM
Rajahmundry	93.5	Red FM	Tirunelveli	93.5	Suryan FM
Rajkot	92.7	Big FM	Tirunelveli	106.4	Hello FM
Rajkot	93.5	Red FM	Tirupati	92.7	Big FM
Rajkot	98.3	Radio Mirchi	Tirupati	93.5	Red FM
Ranchi	91.9	Radio Mantra	Tuticorin	93.5	Suryan FM
Ranchi	92.7	Big FM	Tuticorin	106.4	Hello FM
Ranchi	104.0	Radio Tarang	Udaipur	92.7	Big FM
Ranchi	104.8	Radio Dhoom	Udaipur	94.3	My FM
Ranchi	106.4	Dhamaal 24	Udaipur	95.0	Tadka 95 FM
Rourkela	92.7	Big FM	Vadodara	91.1	Radio City
Rourkela	104.0	R. Choklate	Vadodara	92.7	Big FM
Sangli	91.1	Radio City	Vadodara	93.5	Red FM
Shillong	91.9	R. Ooo La La	Vadodara	98.3	Radio Mirchi
Shillong	93.5	Red FM	Varanasi	91.9	Radio Mantra
Shimla	91.9	Big FM	Varanasi	93.5	Red FM
Shimla	104.8	Oye FM	Varanasi	98.3	Radio Mirchi
Shimla	106.4	Dhamaal 24	Vijayawada	93.5	Red FM
Siliguri	91.9	Nine FM	Vijayawada	98.3	Radio Mirchi
Siliguri	92.7	High FM	Visakhapatnam	91.1	Radio City
Siliguri	93.5	Red FM	Visakhapatnam	92.7	Big FM
Siliguri	94.3	Radio Misty	Visakhapatnam	93.5	Red FM
Solapur	91.1	Radio City	Visakhapatnam	98.3	Radio Mirchi
Solapur	92.7	Big FM	Warangal	93.5	Red FM
Srinagar	92.7	Big FM	**F.PI.**	More stns in more areas	

Web addresses: Aahaa FM: www.aahaafm.com **Amar FM:** http://aamar106fm.com **Big FM:** www.big927fm.com **Chennai Live:** www.chennailive.fm **Club FM:** http://www.clubfm.in **Dhamaal 24:** http://dhamaal24.com **Fever FM:** www.fever.fm **Friends FM:** http://www.abp.in/30002.html **Hello FM 106.4 FM:** www.hello.fm **My FM:** www.myfmindia.com **Oye FM:** http://oyefm.in **Power FM:** http://power-er1078fm.com **Radio Chaska:** http://www.radiochaska.com **Radio Choklate:** http://www.radiochoklate.com **Radio City:** www.radiocity.in **Radio Indigo:** radioindigo.fm, **Radio Mango:** www.radiomango.co.in **Radio Mirchi:** www.radiomirchi.com **Radio Misty:** www.radiomisty.co.in **Radio One:** www.radioone.in **Rangila FM:** www.rangilafm.com **Red FM:** www.sunnetwork.org/redfm/Index.htm **Suno Lemon:** www.sunolemon.com **Suryan FM:** www.sunnetwork.org/suryanfm

Gyan Vani (Educational FM Channel)

✉ Electronic Media Production Centre, Sanchar Kendra, Indira Gandhi National Open University (IGNOU), Maidan Garhi, New Delhi-110068 ☎ 91-11-29533079 📠 91-11-29534299 **E:** gyandarshan@ignou.ac.in **W:** www.ignou.ac.in/gyandarshan/gvani.html

Location	MHz	kW	Location	MHz	kW
Agra	105.6	10	Lucknow	105.6	10
Ahmedabad	105.4	10	Madurai	105.6	10
Allahabad	107.4	10	Mumbai	105.6	10
Aurangabad	104.8	10	Mysore	105.2	10
Bengaluru	106.4	10	Nagpur	107.8	10
Bhopal	105.6	10	New Delhi	105.6	10
Chandigarh	105.4	10	Panaji	107.8	10
Chennai	105.6	10	Patna	105.6	10
Coimbatore	91.9	10	Pune	105.6	10
Cuttack	105.6	10	Raipur	105.6	10
Hyderabad	105.6	10	Rajkot	107.8	10
Indore	105.6	10	Shillong	103.6	10
Guwahati	107.8	10	Srinagar	107.8	10
Jabalpur	105.6	10	Trivandrum	105.6	10
Jaipur	105.6	10	Tiruchirapalli	105.6	10
Jalandhar	105.6	10	Tirunelveli	105.6	10
Kanpur	105.6	10	Varanasi	105.6	10
Kochi	105.6	10	Visakhapatnam	106.4	10
Kolkata	105.4	10			

NB: Txs located at and maintained by AIR.

Community FM Radio Stations: Run by Educational Institutions and others with 50W on FM 90.4MHz, 90.8MHz, 91.2MHz, 96.9MHz, 107.2MHz, 107.4MHz and 107.8MHz. **F.PI.** More community radio stns by different institutions.

INDONESIA

LT: We. Indonesia (Java, Sumatra, We. & Ce. Kalimantan): UTC +7h; Ce. Indonesia (So. & Ea. Kalimantan, Sulawesi, Bali, Nusa Tenggara): UTC +8h; Ea. Indonesia (Maluku, Papua): UTC +9h — **Pr.L:** Bahasa Indonesia (Indonesian) — **Pop:** 243 million — **E.C:** 50 Hz, 230V — **ITU:** INS

DIRECTORATE GENERAL OF POSTS & TELECOMMUNICATIONS (Direktorat Jenderal Pos dan Telekomunikasi)

✉ Gedung Sapta Pesona, Medan Merdeka Barat 17, Jakarta 10110 ☎ +62 21 3835955 📠 +62 21 3860754 **W:** www.postel.go.id **E:** admin@postel.go.id

INDONESIAN BROADCASTING COMMISSION (Komisi Penyiaran Indonesia, KPI)

✉ Gedung Sekretariat Negara Lt VI, Jl. Gajah Mada 8, Jakarta 10120 ☎ +62 21 6340713 📠 +62 21 6340667 **W:** www.kpi.go.id **LP:** Head: Mr Dadang Rahmat Hidayat. Deputy Head: Ms Nina Mutmainnah.

RADIO REPUBLIK INDONESIA (RRI) (Gov.)

National Station: RRI, Jakarta ✉ Jl. Medan Merdeka Barat 4-5, Jakarta 10110, or Tromolpos 1157 (or Kotak Pos 356), Jakarta 10001 ☎ +62 21 3842083 📠 +62 21 3457132 **W:** www.rri.co.id **E:** info@rri.co.id **Programa 1 (Prosatu):** Information and entertainment on 92.1MHz **Programa 2 (Produa):** Music and information on 105.0MHz **Programa 3 (Protiga):** News and information on 999, 92.8MHz 24h, also relayed in full on FM by most regional stns. N: on the h. Sports N. (Berita Olahraga): 0400, 0800. **Programa 4 (Proempat):** Culture and sport on 1332, 9680kHz, 88.8MHz 24h Relays Protiga 1700-2200.

MW	kHz	kW	Station	MW	kHz	kW	Station
JB01	540	2/10	Bandung	LA01	1035	1/5	Bandar Lampung
JT01	585	50	Surabaya	SH01	1035		Palu
SL01	630	50	Makassar	PA04	1044	10	Biak
PB01	702	2/10	Manokwari	ST02	1044	10	Tahuna
MA01	720	10	Ambon	SU02	1044	10	Sibolga
PA06	729		Nabire	PA01	1053	10	Jayapura
BE01	747	10	Bengkulu	BA02	1080	2/10	Singaraja
JH03	756	2/10	Purwokerto	JA01	1098	2/10	Jambi
MA02	765	1	Tual	JT05	1098	2/10	Sumenep
PB02	774		Fak-Fak	NT01	1107	1/5	Kupang
NT02	783	10	Ende	YG01	1107	1/10	Yogyakarta
JH01	801	10	Semarang	KS01	1134	1/25	Banjarmasin
SU01	801	1	Medan	AC02	1152		Lhokseumawe
PA02	810	7.5	Merauke	JH01	1170		Semarang
NB01	855	2/10	Mataram	SB01	1179	2/10	Padang
SU01	855	10	Medan	ST01	1188	1	Manado
JB03	864	2/10	Cirebon	KH01	1197	10	Palangkaraya
SG01	882		Kendari	KT01	1215	0.5/10	Samarinda
JT02	891	10	Malang	KB01	1233	02/1/5	Pontianak
MU01	891	10	Ternate	JB02	1242	10	Bogor
PB03	909	5/10	Sorong	AC01	1251	10	Banda Aceh
RI01	927		Pekanbaru	SS01	1287	20/25	Palembang
SG01	954	10	Kendari	JK01	1332	10	Jakarta
JT03	963	2/10	Jember	KR01	1341	1/5	Tanjung Pinang
JH02	972	50	Surakarta	KT02	1350	10	Tarakan
JK01	999	1/150	Jakarta	SH02	1377	10	Tolitoli
G001	1008		Gorontalo	PA05	1395	1	Wamena
JT04	1008	10	Madiun	BB01	1413	5	Sungai Liat
PA03	1026	5	Serui	SB02	1512		Bukittinggi

SW	kHz	kW	Station, h. of tr.
KH01	3325	10	Palangkaraya: 2200-0100, 0900-1610
MU01	3345	10	Ternate: 2000-0030, 0750-1500
KB01	3976		Pontianak: 2200-0030, 1000-1700
PB01	3987		Manokwari*
SG01	3995	5	Kendari: 2100-0100, 0750-1600
PA03	4605	1	Serui: 2000-2315, 0800-1500 irr.
SL01	4750	20	Makassar: 2030-0000, 0745-1500
PB02	4790		Fak-Fak: 2000-2230, 0700-1500
PA05	4870		Wamena: 2000-2315, 0800-1400
PA04	4920		Biak*
JA01	4925	10	Jambi: 2200-0205, 0900-1600 irr.
PA06	6125		Nabire*
PA06	7290		Nabire: 2200-2315, 0500-0815v
JK01	9680	250	Jakarta (Cimanggis): 2200-1500
PB03	9743	10	Sorong*

NB: During the Muslim month of Ramadan several stns begin morning transmissions as early as 1800.

Addresses (Jl = Jalan)

AC01) Jl Sultan Iskandar Muda 13, P.O Box 112, Banda Aceh 23423, Nanggroe Aceh Darussalam **E:** sekretariat@rribandaaceh.net -**FM:** 88.6/92.6/97.7MHz + 90.5MHz (Tapaktuan), 91.9MHz (Langsa), 92.0MHz (Sinabang), 92.3MHz (Kutacane), 93.0MHz (Subulussalam), 95.1MHz (Lamno), 97.3MHz (Jantho), 97.5MHz (Calang), 99.7MHz (Beuneuruen) — **AC02)** Jl Peutua Ibrahim 75, Teumpok Teungoh, Lhokseumawe 24352, Nanggroe Aceh Darussalam - **FM:** 89.3/94.4/95.2/101.9MHz — **AC03)** RRI Sabang, Jl Yos Sudarso 65, Cot Bak U, Kecamatan Sukajaya, Sabang, Nanggroe Aceh Darussalam - **FM:** 94.0MHz — **AC04)** RRI Takengon, Takengon, Aceh Tengah, Nanggroe Aceh Darussalam - **FM:** 93.0MHz — **AC05)** RRI Meulaboh, Meulaboh, Nanggroe Aceh Darussalam - **FM:** 88.7/97.0MHz

BA01) Jl Hayam Wuruk 70, Keladis, Denpasar 80233 (Kotak Pos 31, Denpasar 80001), Bali - **FM:** 88.6/93.4/95.3/102.5MHz + Paradise FM 100.9MHz: English — **BA02)** Jl Gajah Mada 144, Tromolpos 153, Singaraja 81113, Bali - **FM:** 97.9/102.0/103.7MHz + 99.5MHz (Bukit Kutul)

BB01) Jl Jend Ahmad Yani, Sungai Liat 33211, Bangka, Bangka Belitung - **FM:** 96.4/97.2/103.9MHz + 90.4MHz (Toboali), 95.4MHz (Mentok), 95.5MHz (Tanjung Pandan), 99.8MHz (Pangkalpinang)

BE01) Jl Let Jend S Parman 25, Kotak Pos 13, Bengkulu 38227, Bengkulu - **FM:** 90.9/92.5/105.1MHz + 95.4MHz (Muko-Muko), 97.0MHz (Bintuhan), 98.0MHz (Curup), 101.3MHz (Ipuh)

G001) Jl Jenderal Sudirman 30, Gorontalo 96128, Gorontalo **E:** layanan@rrigorontalo.com - **FM:** 92.4/96.7/101.8MHz + 92.5MHz (Baroko), 97.0MHz (Marisa)

JA01) Jl Jendral A Yani 5, Telanaipura, Jambi 36122, Jambi **E:** rrijambi@rri.co.id - **FM:** 88.5/90.9/94.4MHz + 95.8MHz (Bangko), 99.0MHz (Kualatungkal), 99.0MHz (Sarolangun), 99.8/101.0MHz (Sungai Penuh), 99.8MHz (Tungkal Ilir), 101.0MHz (Muara Bungo)

JB01) Jl Diponegoro 61, Bandung 40122 (Kotak Pos 1055, Bandung 40001), Jawa Barat - **FM:** 96.0/97.6/102.0MHz + 95.0MHz (Gunung Malang), 97.0MHz (Purwakarta/Subang), 97.8MHz (Tasikmalaya), 98.0MHz (Bayah), 98.2MHz (Puncak Surangga), 98.9MHz (Saketi), 102.5MHz (Cikuray), 103.3MHz (Garut) — **JB02)** Jl Pangrango 30, P.O Box 232, Bogor 16161, Jawa Barat - **FM:** 93.7/106.8/107.1MHz — **JB03)** Jl Brigjen Dharsono/By Pass, Cirebon 45132, Jawa Barat **E:** rricirebon@rricirebon.info - **FM:** 93.5/94.8/97.5MHz

JH01) Jl Ahmad Yani 144-146, Kotak Pos 1307, Semarang 50241, Jawa Tengah - **FM:** 89.0/90.6/95.3/99.4MHz + 94.2MHz (Colo), 96.7MHz (Batang), 97.7MHz (Gunung Gantungan), 99.4MHz (Gunung Depok), 99.5MHz (Gunung Periksa) — **JH02)** Jl Abdul Rahman Saleh 51, Kotak Pos 40, Surakarta 57133, Jawa Tengah - **FM:** 97.0/101.8/105.5MHz + 96.3/102.0MHz (Tawangmangu) — **JH03)** Jl Jendral Sudirman 427, Kotak Pos 5, Purwokerto 53116, Jawa Tengah - **FM:** 93.1/99.0/107.3MHz

JK01) Jl Medan Merdeka Barat 4-5, Jakarta 10110 (Tromolpos 1157, Jakarta 10001).

JT01) Jl Pemuda 82-90, Kotak Pos 239, Surabaya 60271, Jawa Timur - **FM:** 95.2/99.2/106.3MHz + 91.1MHz (Cemoro Lawang), 97.9MHz (Pacitan), 99.2MHz (Alas Malang), 99.2MHz (Pare), 99.2MHz (Sidoarjo), 102.3MHz (Pulau Bawean) — **JT02)** Jl Candi Panggung 58, Kotak Pos 78, Mojolangu, Malang 65142, Jawa Timur - **FM:** 91.9/94.6/99.4/105.3MHz — **JT03)** Jl D.I Panjaitan 61, Jember 68110 (Kotak Pos 166, Jember 68101), Jawa Timur - **FM:** 87.9/89.5/95.4MHz — **JT04)** Jl Mayjen Panjaitan 10-12, Madiun 63133, Jawa Timur **E:** rri@rrimadiun.net - **FM:** 97.7/99.7/104.0MHz +96.3MHz (Pare-Kediri), 96.3MHz (Kemiri) — **JT05)** Jl Urip Sumoharjo 26, Sumenep 69411, Madura, Jawa Timur - **FM:** 93.0/98.5/101.3MHz

KB01) Jl Jendral Sudirman 7, Kotak Pos 6, Pontianak 78111, Kalimantan Barat - **FM:** 90.3/101.8/104.2MHz +95.0MHz (Nangamerakai), 96.8MHz (Ketapang), 97.0MHz (Sanggau), 97.7MHz (Sambas), 97.7MHz (Singkawang), 98.0MHz (Kendawangan), 98.2MHz (Semitau), 99.3MHz (Sanggau Ledo), 100.2MHz (Balaikarangan) — **KB02)** RRI Sintang, Jl Oevang Oeraya, Baning, Sintang, Kalimantan Barat - **FM:** 90.7/96.6/102.5MHz — **KB03)** RRI Entikong, Jl Lintas Negara Indonesia-Malaysia, Entikong — Sanggau, Kalimantan Barat - **FM:** 100.2MHz

KH01) Jl M Husni Thamrin 1, Palangkaraya 73112, Kalimantan Tengah - **FM:** 89.2/92.4/95.9MHz + 93.6MHz (Kuala Kapuas), 93.6MHz (Sampit), 96.0MHz (Muara Teweh), 97.1MHz (Pulang Pisau), 97.3MHz (Buntok), 99.2MHz (Pangkalan Bun)

KR01) Jl Ahmad Yani, Kotak Pos 8, Tanjung Pinang 29133, Bintan, Kepulauan Riau - **FM:** 92.1/98.3/101.2MHz +96.6MHz (Karimun), 99.6MHz (Tarempa) — **KR02)** RRI Ranai, Jl Sepempang, Ranai, Pulau Natuna Besar 29183, Kepulauan Riau - **FM:** 90.0/104.0/105.9MHz — **KR03)** RRI Batam, Gedung Sumtra Expo Lantai III Batam Centre, Batam, Kepulauan Riau **E:** rri.batam@yahoo.co.id - **FM:** 105.1MHz

KS01) Jl Jenderal A. Yani Km 3.5 No 234, Kotak Pos 117, Banjarmasin 70234, Kalimantan Selatan - **FM:** 87.7/92.5/95.2/97.6MHz + 89.4MHz (Batu Licin), 90.2MHz (Kotabaru), 90.7MHz (Amuntai), 105.7MHz (Kandangan)

KT01) Jl Moh Yamin 8, P.O Box 45, Samarinda 75110, Kalimantan Timur **E:** layananusaha@gmail.com - **FM:** 88.5/97.6/98.4MHz + 95.5MHz (Pulau Sebatik), 96.0MHz (Penajam), 96.7MHz (Berau), 96.8MHz (Tanah Grogot), 97.0MHz (Balikpapan), 97.1MHz (Nunukan), 97.4MHz (Melak), 97.4MHz (Bontang/Sangata), 99.0MHz (Tenggarong) – **KT02)** Jl Sungai Mahakam 10, Kampung Empat, Tarakan Timur 77125, Kalimantan Timur - **FM:** 88.8/97.9/101.9MHz – **KT03)** RRI Malinau, Jl Pusat Perkantoran, Tg. Belimbing – Malinau, Kalimantan Timur - **FM:** Freq. unknown

LA01) Jl Gatot Subroto 26, Kotak Pos 24, Pahoman, Bandar Lampung 35213, Lampung **E:** rri_bdl@yahoo.com - **FM:** 87.7/90.9/92.5MHz + 93.0MHz (Kalianda), 95.8MHz (Kotabumi), 97.0MHz (Kota Agung), 99.0MHz (Simpang Pematang), 99.4MHz (Liwa), 99.7MHz (Padang Cermin), 100.2MHz (Tulungbawang).

MA01) Jl Jendral Akhmad Yani 1, Ambon 97124, Maluku - **FM:** 90.3/98.4/102.0MHz + 92.0MHz (Amahai/Masohi), 94.3MHz (Saumlaki) – **MA02)** Jl Sukarno-Hatta, Kec Wat Deh, Tual 97661, Pulau Kai, Maluku - **FM:** 93.2/97.6/103.6MHz

MU01) Jl Sultan Khairun, Kedaton, Ternate 97720, Maluku Utara - **FM:** 90.4/95.3/104.1MHz + 92.8MHz (Pulau Morotai), 93.7MHz (Soasiu)

NB01) Komplek Perumahan RRI Mataram, Jl Majapahit, P.O Box 2, Mataram, Lombok, Nusa Tenggara Barat **E:** pro1mataram@gmail.com - **FM:** 89.2/94.3/104.2MHz + 89.1MHz (Bima), 89.1MHz (Dompu), 89.1MHz (Lombok Timur), 96.3MHz (Lombok Tengah), 98.1MHz (Tanjung), 104.0MHz (Sumbawa Besar)

NT01) Jl Tompello 8, Kupang 85225, Timor, Nusa Tenggara Timur - **FM:** 90.9/94.4MHz + 88.8MHz (Soe), 90.7MHz (Kefamenanu), 91.5MHz (Atambua), 92.5MHz (Baa) – **NT02)** Ende, Flores, Nusa Tenggara Timur - **FM:** 92.2/101.0/105.0MHz

PA01) Jl Tasangkapura 23, Kotak Pos 1077, Jayapura 99200, Papua - **FM:** 90.1/96.0/97.6/105.9MHz +93.3MHz (Sentani), 94.5MHz (Timika), 96.5MHz (Sarmi), 96.7MHz (Sorendiweri), 100.0MHz (Genyem) – **PA02)** Jl Jendral A Yani, Mopa Baru, Merauke 99611 (Kotak Pos 11, Merauke 99601), Papua - **FM:** 90.0/95.2/98.1/105.0/107.0MHz – **PA03)** Jl Pattimura, Serui 98213, Papua - **FM:** 92.5/94.8/97.4/103.0MHz – **PA04)** Jl Majapahit, Kotak Pos 505, Biak 98117, Papua - **FM:** 90.5/9 2.5/95.3/95.8/96.9MHz + 96.3/97.6MHz (Numfor) – **PA05)** Jl Jendral A Yani 64, Wamena 99511 (Kotak Pos 10, Wamena 99501), Papua - **FM:** 93.5/94.7/96.3/97.8MHz – **PA06)** Jl Merdeka 74, Nabire 98811 (Kotak Pos 110, Nabire 98801), Papua - **FM:** 93.55/94.5/96.3/98.1MHz – **PA07)** RRI Boven Digul, Tanah Merah, Boven Digul, Papua - **FM:** 88.3/96.6MHz

PB01) Jl Merdeka 68, Manokwari 98311, Papua Barat - **FM:** 92.4/93.5/94.3/97.75MHz + 95.9MHz (Bintuni) – **PB02)** Jl Kapt P Tendean, Kotak Pos 154, Fak-Fak 98612, Papua Barat - **FM:** 89.9/93.3/97.2/99.0MHz + 96.3MHz (Kaimana), 97.2/98.1MHz (Kokas) – **PB03)** Jl Sam Ratulangi 4, Kotak Pos 146, Sorong 98414, Papua Barat - **FM:** 95.1/96.7/100.2MHz + 95.1/96.3MHz (Teminabuan)

RI01) Jl Jend Sudirman 440, Kotak Pos 51, Pekanbaru 28115, Riau **E:** admin@rripekanbaru.com - **FM:** 88.4/91.2/99.1MHz +90.6MHz (Selat Panjang), 90.6MHz (Dumai), 92.6 MHz (Pasir Pegarayan), 95.7MHz (Sungai Pakning), 98.5MHz (Baserah), 99.3MHz (Tembilahan), 99.9MHz (Siak)

SB01) Jl Jendral Sudirman 12, Kotak Pos 77, Padang 25124, Sumatera Barat **E:** rripadang@rri.co.id - **FM:** 88.4/90.8/97.5MHz + 88.4MHz (Pandai Sikek Padang Pariaman), 89.5MHz (Bukit Gompong Solok), 92.0MHz (Bungkit Palakat), 96.8MHz (Pasaman Barat), 97.1MHz (Pariaman), 97.9MHz (Bukit Langkisau Painan), 97.9MHz (Dharma Seraya), 97.9MHz (Kota Pariaman), 98.5MHz (Mentawai) – **SB02)** Jl.Prof Muhammad Yamin 199, Kotak Pos 3, Aurkuning, Bukittinggi 26131, Sumatera Barat - **FM:** 90.5/94.8/97.2MHz

SG01) Jl Laute Mandonga 44, Kotak Pos 7, Kendari 93111, Sulawesi Tenggara - **FM:** 90.8/96.7/107.0MHz + 93.5MHz (Boepinang), 97.0MHz (Raha), 99.4MHz (Bau-Bau), 99.5MHz (Lasolo)

SH01) Jl R.A Kartini 39, Palu 94112, Sulawesi Tengah - **FM:** 90.5/94.8/105.0MHz + 95.4MHz (Ampana), 95.5MHz (Tanjung Santigi), 96.0MHz (Banggai), 96.2MHz (Poso), 97.1MHz (Toboli), 99.2MHz (Luwuk) – **SH02)** Jl Jenderal Sudirman, Tolitoli 94514, Sulawesi Tengah - **FM:** 90.2/93.0/94.5/102.0MHz

SL01) Jl Riburane 3, Kotak Pos 103, Makassar 90111, Sulawesi Selatan.- **FM:** 92.9/94.4/96.8/106.3MHz + 90.6MHz (Bontu Tabang), 94.0MHz (Baraka), 96.0MHz (Mamuju), 99.0Mhz (Parepare), 99.0MHz (Bantaeng)

SS01) Jl Radio 2 Km 4, Palembang 30128, Sumatera Selatan **E:** rripalembang@yahoo.co.id - **FM:** 88.4/91.6/92.4/97.1MHz + 90.3MHz (Sekayu), 90.5MHz (Baturaja), 90.5MHz (Pagar Alam), 95.1MHz (Lubuklinggau), 97.7MHz (Prabumulih), 99.9MHz (Muara Enim)

ST01) Jl Radio 1, Kotak Pos 1110, Tikala Ares, Manado 95124, Sulawesi Utara - **FM:** 92.0/92.5/94.5/97.7/104.4MHz + 92.0MHz (Lirung), 95.5MHz (Paguyaman) – **ST02)** Jl Tona, Tahuna, Sangihe, Sulawesi Utara - **FM:** 92.0/98.7/99.5MHz

SU01) Jl Jend Gatot Subroto Km 5.6, Medan 20123, Sumatera Utara

- **FM:** 88.8/92.4/94.3MHz + 90.0Mhz (Natal), 90.3MHz (Teluk Dalam), 90.6MHz (Rantau Prapat), 91.9MHz (Kotanopan), 92.0MHz (Prapat), 92.0MHz (Sidikalang), 94.5MHz (Simar Jarunjung), 96.1MHz (Pematang Siantar), 96.3MHz (Tarutung), 99.1MHz (Sibuhan), 99.3MHz (Pulau Raja) – **SU02)** Jl Ade Irma Suryani Nasution 11, Sibolga 22513, Sumatera Utara - **FM:** 94.8/97.2/103.0MHz + 93.0/99.9MHz (Padangsidempuan) – **SU03)** RRI Gunungsitoli, Desa Iraonogeba, Gunungsitoli, Nias, Sumatera Utara - **FM:** 90.3/96.2/101.3MHz

YG01) Jl Ahmad Jazuli 4, Tromolpos 18, Kotabaru, Yogyakarta 55224, Daerah Istimewa Yogyakarta - **FM:** 91.1/102.5/102.9MHz

FEDERATION OF INDONESIAN NATIONAL COMMERCIAL BROADCASTERS (Persatuan Radio Siaran Swasta Nasional Indonesia)

✉ Pengurus Pusat, Persatuan Radio Siaran Swasta Nasional Indonesia, Jl. Raya Pondok Gede 96, Jakarta 13810 ☎ +62 21 8414311 🖷 +62 21 8414314 **W:** www.radioprssni.com **E:** radioprssni@ radioprssni.com or ppjkt @indosat.net.id **LP:** Chmn: Shidki Wahab. **PRSSNI** has 758 members. Commercial station permitted power: 1 kW (MW) and 10 kW (FM).

LOCAL PUBLIC BROADCASTING STATIONS (Lembaga Penyiaran Publik Lokal)

Local government stations are in the process of making the transition to local government owned but autonomous public broadcasters. As a result, the names of former local government radio stations (Radio Siaran Pemerintah Daerah) are being changed. Where still used, these station headings apply: **RKIP:** Radio Khusus Informasi Pertanian – **RKPD:** Radio Khusus Pemerintah Daerah – **RPD:** Radio Pemerintah Daerah – **RPD Kotamadya:** Radio Pemerintah Daerah Kotamadya (only intended for particular cities) – **RPK:** Radio Pemerintah Kabupaten – **RSPD:** Radio Siaran Pemerintah Daerah – **RSPK:** Radio Siaran Pemerintah Kabupaten.

INDONESIAN COMMUNITY RADIO NETWORK (Jaringan Radio Komunitas Indonesia)

✉ Sekretariat, Jaringan Radio Komunitas Indonesia, Jl Dwi Sri 10, Bandung, Jawa Barat ☎ +62 22 5224205 **W:** http://jrki.wordpress.com **E:** suara.jrki@gmail.com or jrk_kongres04@yahoo.com **LP:** Chmn: Bowo Usodo.

The majority of community stations operate from 107.7 to 108.0 MHz.

MW	kHz	kW	Station
BN01	531		R. Palanta, Tangerang
LA02	540		R. Dei Marganusa, Trimurjo
JB04	549		Inyong R., Depok
BN02	558		R. Swara Angkasa Megah, Pandeglang
JH04	558		R. Diantara Vita Kharisma (D.V.K.), Kebumen
JK02	576		R. Vineyard Indonesia (Ravii), Jakarta*
BN03	603		R. Suka, Tangerang
KH02	603		R. Riwut Melawen, Buntok
JH05	v607	0.25	R. Suara Banyumas Asli (Subali), Purwokerto
BN04	612	0.25	R. Gema Bahari Selatan, Rangkasbitung
JB05	612		R. Geswara Pamanukan (G.S.P.), Pamanukan
KB04	612		R. Kijang Berantai (Kiber) Perkasa, Sambas
NB02	612		R. Yayasan Attohiriyah Alfadiliyah (R. Yatofa), Bodak-Praya
SS02	612		R. Swara Betung Indah, Betung-Musi Banyuasin
JB06	630		R. Citra Suara Sukapura, Tasikmalaya
JK03	630		R. Samhan, Jakarta
LA03	630		R. Gema Swarna Dwipa (Slendro), Terbanggi Besar
JA02	648		R. Batanghari Permai (B.H.P.), Muarabulian
JH06	648		R. Santo Bernadus D.S., Pekalongan
JH07	648		R. Roro Djonggrang B.S., Prambanan
JH08	648		R. Aji Satria, Ajibarang
JK04	648		R. Rahmat Emmanuel Ministries, Jakarta
NB03	648		R. Suara Hamzanwadi, Selong
JB07	666	0.22	R. Linggarjati Utama (Rasilima), Kuningan
JH09	666	0.25	R. Ramakusala, Surakarta
JH10	666		R. Tunggul Suara Dirgantara, Purbalingga
JK05	666		R. Dwi Karya 69 (R. Sekuntum Bunga Yonina) (SBY), Jakarta
SU04	666		R. Gelora Remaja Sibolga (R. Gresia), Sibolga
JB08	684		R. Swara Pakusarakan Pratika, Sawangan*
JB09	684		R. Angkasa Media, Kadipaten
KS02	684		R. Purnama Nada, Kandangan
JB10	702		R. Bravo, Bandung
JK06	702		R. Tona, Jakarta
PA08	702		R. Suara Kasih Agung, Jayapura
YG02	702		R. Suara Konco Tani, Sidokarto
JH11	720	0.25	R. Lusiana Namberwan, Semarang
JH12	720		R. Gagah Sehat Berbobot (Gasebo), Majenang
JK07	720		R. Gracia, Jakarta*

MW	kHz	kW	Station
LA04)	720		R. Bhara Kharisma Suryajaya, Talangpadang*
NT03)	720		RSPD Sumba Timur, Waingapu
JB11)	729		R. Viriaana Nusa Karya (VNK), Tomo - Sumedang
NB04)	729		R.Dewi Anjani, Selong*
BN05)	738		R. Bharata Bhakti Nusa, Tangerang
JH13)	738		R. POP (Konservatori), Surakarta
KB05)	738		R. Swara Pinohperkasa, Sintang
NT04)	738		RSPD Timor Tengah Selatan, Soe
SL02)	738		R. Rina Bestari, Rantepao
SS03)	738		R. Aditya Nada Jaya, Indralaya
JA03)	740		RSPD Batanghari, Muarabulian
NT05)	743	0.3	RPD Sumba Barat, Waikabubak
JT06)	v747		R. Swara Nabawiy, Pasuruan
JT07)	747		R. Canka Bhalaria, Ngawi
JB12)	756		R. Rodja, Cileungsi - Bogor
BN06)	774		R. Klasik Galih Lestari (Gless R.), Tangerang
JH14)	774	0.2	R. Leonardus Buana Suara, Salatiga
JT08)	774		R. Pesona 2000, Sumenep
KS03)	774		R. Ruhui Rahayu, Rantau
SB03)	774	0.35	RSPD Kotamadya Payakumbuh
SL03)	774		R. Suara Adyafiri, Watansoppeng
YG03)	774		R. Swara Kenanga, Yogyakarta
KS04)	v783		R. Dakwah Masjid Raya Sabilal Muhtadin Banjarmasin
JB13)	792		R. Cempaka Angkasa, Ciamis
JB14)	792		R. Swara Citra Cianjur Mandiri, Cianjur
JH15)	792		R. Bayu Sakti, Kroya
JK08)	792		R. Assyafi'yah, Jakarta
LA05)	792		R. Suara Dwi Amanda, Gadingrejo
NB05)	792		R. Mitra Idola Kita, Pancor
SS04)	792		R. Suara Ria Jaya Sentosa (S.R.J.S.), Baturaja
AC06)	810		R. Amanda Rasisonia, Takengon
JB15)	810		R. Indraswara Cakrawala Nada, Majalengka
JB16)	810		RSPD Bandung (R. Swara Bale Endah)
JH16)	810		R. Suara Maung Sakti, Banjarnegara
JK09)	810		R. Universitas Mercu Buana, Jakarta
SL04)	810		R. Megapesona, Enrekang
SU05)	810		R. Suara Tanjung Berjaya, Tanjungbalai
JH17)	819		R. Pancabayu Madugondo (Suara R.P.M.), Sukoharjo
JB17)	828	0.25	R. Leidya Swara Utama (R. Kharisma), Bandung
JB18)	828		R. Adhika Pariwara, Pelabuhanratu
JH18)	828		R. S.B.S., Purbalingga
JK10)	828		R. Berita Klasik (RBK), Jakarta
KB06)	828		R. Mahkota Ngabang Gemaswara, Ngabang
SL05)	828		R. Swara Christy Ria, Makassar
YG04)	828		R. Suara Parangtritis, Parangtritis
JA04)	837		R. Kelapa Indah (R. KIN), Tanjung Jabung Barat
JH19)	837		R. Bima Sakti, Boyolali
JK11)	837		R. Garis Visi (R. AM Moslem), Jakarta*
JB19)	846		R. Suara Galunggung Giri Sakti, Tasikmalaya
JH20)	846		R. Cipta Bentala Swara (CBS), Magelang
JH21)	846		R. Immanuel, Surakarta
JH22)	846		R. Suara Karangbolong, Gombong
JH23)	846		R. Swara Anggada Senatama, Purbalingga
JH24)	846		R. Suara Tegal Agung Raya (Star), Tegal
JT09)	846	0.5	RKPD Ponorogo (R. Suara Ponorogo)
JT10)	846		R. Miniwatt Pesona Indah, Surabaya
KS05)	846		RSPD Kotabaru
JK12)	864		R. Hana Citra Swara Jakarta (Suara Jakarta), Jakarta
JT11)	864		R. Menara III, Surabaya *
SL06)	864		Suara AsAdiyah, Sengkang
SR01)	864		R. Manakara, Mamuju
YG05)	864		R. Satunama, Gunung Kidul
JT12)	879	0.5	RSPK Sidoarjo (R. Suara Delta)
JB20)	882		R. Suara Anggada Senatama (S.A.S.), Banjarsari
JB21)	882		R. Pelangi Nusantara, Bekasi
JH25)	882		R. Swara Kranggan Persada, Temanggung
JT13)	882		R. Gema Panca Arga, Pacitan
SL07)	882	0.5	R. Bambapuang, Pangkajene
AC07)	900		R. Siaran Cempaka Nadacitra, Desa Tonjong
AC08)	900		R. Siaran Nada Karya Semesta, Lhok Sukon
JA05)	900	0.25	R. Gema Nugraha, Sungai Penuh
JH26)	900		R. Bintoro Karya, Demak
JH27)	900		R. Suara Sendang Mas, Banyumas
JK13)	900		R. Sindajaya, Jakarta
JT14)	900		R. Suara Al Iman, Surabaya
KB07)	900		R. Aries Sanggau Perkasa, Sanggau
LA06)	900		R. Swara Alfina Shakti, Kalianda*
SB04)	900		R. Elkartika Angkasa Niaga, Padang*
SU06)	v900		R. Bethany (Voice of Victory), Medan
JH28)	909		R. Blora Sakti (R.B.S.), Cepu - Blora
JB22)	918		R. Gema Nury (El Nury), Bogor
JB23)	918	1.5	R. Citra Wahana Indonesia (R. Debora), Bandung*
JH29)	918		R. Suara Selomanik (R.S.S.), Banjarnegara
JT15)	918	0.5	RKPD Gresik
NT06)	918		R. Balistik, Kupang*
SU07)	918		R. Gelora Pertiwi, Medan
LA07)	927		R. Primanada, Gisting*
JH30)	936		R. Widya Bhakti, Magelang
JH31)	936		R. Kelana Sumbangsihku (Kasihku), Bumiayu
JK14)	936	0.25	R. Puspa Dwi Swara Cipta (P2SC), Jakarta
JT16)	936		R. Suara Fiskarama, Bondowoso
KB08)	936		R. Swara Dermagaria Persada Cakrawala, Sekadau
SL08)	936		R. Suara Viktori, Makassar
JH32)	945		R. Swara Buana Asri, Wonosobo
SB05)	945		R. Galundi Pradana, Gando Sulit Air
SU08)	945		R. Tuah Swara Murni, Lubukpakam
JB24)	954	0.25	R. Sena Bahana Cakrawala, Sukabumi
JB25)	954		R. Suara Terunajaya, Pemeungpeuk
JH33)	954		R. Gita Nusantara Perkasa (Studio 99), Purbalingga
JT17)	954	0.25	R. El Bayu, Gresik
SS05)	954	0.15	R. Garuda Kenten Jaya, Palembang
BN07)	972		R. Pusako Minangkabau, Tangerang
JB26)	972		R. Antares, Garut
JT18)	972		R. Suara Harmoni, Situbondo*
SS06)	972		R. Nada Santika, Pagar Alam
JH34)	985	0.25	RSPD Magelang*
AC09)	990		R. Cakra Donya Multi Swara, Lhokseumawe*
JH35)	990		R. Gita Lestari, Brebes
JH36)	990		R. Pesona Bahari, Weleri
KS06)	990		R. Bahana Al-Mursyidul Amin, Martapura
NT07)	990		R. Gema Suara Gloria (G.B.S.), Kupang*
SU09)	990		R. Nias Mitra Dharma, Gunungsitoli
KT04)	1005		RPD Kutai, Tenggarong
AC10)	1008		R. Siaran Dwieka Swara, Beureunuen
KB09)	1008		R. Suara Pemangkat, Pemangkat
SU010)	1008		R. Citra Tebingtinggi Idola Nada, Tebingtinggi
JH37)	1017		R. Suara Gajah Mungkur, Ngadirejo
SU11)	1017		R. Kardopa, Medan
NT08)	1024	0.25	RPD Belu, Atambua
AC11)	1026		R. Gema Cakrawala Utama, Kuala Simpang
BA03)	1026		R. Diva, Denpasar
JB27)	1026		R. Ummat, Bandung*
JK15)	1026		Suara Multazam, Jakarta Utara
SS07)	1026		R. Suara Enim Jaya Perkasa, Muara Enim
JB28)	1029	0.5	RPK Ciamis
NT09)	1034		RSPK Ngada, Bajawa
JH38)	1035		RSPD Temanggung
JH29)	1044		R. Duta Angkasa, Pangandaran
JB30)	1044		R. Lima Swara Mandiri (Purnayudha, "Country Stn"), Bekasi
JH39)	1044	0.25	R. Raka, Tegal
KB10)	1044		R. Ramagentara, Sungai Pinyuh
RI02)	1044		R. Soreram Indah, Pekanbaru
RI03)	1044		R. Bagan Batu Citra Nuansa, Bagan Batu
AC12)	1050		RPD Aceh Timur, Langsa
JB31)	1062		R. Swakarya Niaga (SKN), Cianjur
JH40)	1062		R. P.T.D.I. Unisa 205, Semarang
JK16)	1062	0.25	R. Cendrawasih Pusat, Jakarta
JT19)	1062	1	R. Sangkakala, Surabaya
PA09)	v1062		R. Swara Lembah Baliem, Wamena
SS08)	1062	0.25	R. Gema Mutiara, Palembang
SU12)	1062		R. Tembang Perbauangan Indah, Perbauangan
YG06)	1062		R. Erbe (RB), Yogyakarta*
JT20)	1071		RKPD Pacitan (Suara Pacitan)
AC13)	1080		R. Katalina, Sigli
JB32)	1080		R. K.C.B.S., Losali
JK17)	1080		R. Swara Mega Asri (R. Safari), Jakarta
KH03)	1080		R. Bahana Nusantara, Ampah
KH04)	1080		R. Citra Barito, Muarateweh
LA08)	1080		R. Idola Nada, Tulang Bawang
SG02)	1080		R. Gema Gersamata, Kolaka
SU13)	1080		R. Sukma Aroma, Rantau Prapat
JB33)	v1088		R. Mahendra, Subang
JK18)	1098		R. Media Mahasiswa Tarumanegara, Jakarta
BN08)	1107		R. Swara Mitra, Tangerang
SU14)	1107	0.25	R. Bintang Niaga, Kisaran
JB34)	1116		R. Lisma, Bandung
JH41)	1116	0.25	R. Bhakti Dirgantara Suara Batang, Batang
JH42)	1116		R. Indah Sragen Asri, Sragen
SL09)	1116		R. Mitra Niaga Suara Utari, Bantaeng
SS09)	1116		R. Dian Bahagia Sentosa, Prabumulih Barat
SU15)	1116	0.25	RPD Kotamadya Binjai
JT21)	v1117	0.25	R. Carolina Arjuno, Surabaya
SL10)	1125	0.25	RPD Luwu, Palopo
SU16)	1127	0.2	RPD Kotamadya Medan*
JT22)	1134		R. Duta Nusantara Suara Ponorogo, Ponorogo

MW	kHz	kW	Station
AC14)	1140		R. Siaran Niaga dan Budaya Milanda, Meulaboh
JH43)	v1143		R. Swara Delanggu (Swadesi), Delanggu
AC15)	1145		R. Daerah Perdjaya Bebas Sabang, Sabang*
AC16)	1152		R. Citraganda Kencanaswara, Bireuen
BA04)	1152		RPK Jembrana, Negara
BE02)	1152		R. Gita Buana Suara, Ipuh - Bengkulu
JB35)	1152		R. Rama Sutra, Sukamandi-Subang
JB36)	1152		R. Pasundan Citra Angkasa (PAS), Cianjur
JH44)	1152		R. Sangkakala Pertiwi, Semarang
JT23)	1152		R. Yasmara, Surabaya
SS10)	1152	0.25	R. Enes Duabelas Ulu, Palembang
YG07)	1152		R. Suara Istana, Yogyakarta
BE03)	1160		R. Ratu Anda Swara, Argamakmur
AC17)	1170		R. Kazuma Bawana Swara, Lhokseumawe
AC18)	1170		R. Swara Fatali Nusajaya, Blangpidie
JB37)	v1170		R. Dios (R. Paksi), Bandung
JT24)	1170		R. Rajawali, Surabaya*
PA10)	1170		R. Suara Nusa Bahagia, Jayapura
SR02)	1170		R. Lariang Indah, Mamuju
JH45)	v1180	0.5	RSPD Wonogiri
NT10)	1185		Suara Kelimutu (RSPD), Ende
AC19)	1188		R. Rapeja, Lamno
BE04)	1188	0.25	R. Namora Swara Pratama, Curup
JB38)	1188		R. Swara Selabintana Permai (SSP), Sukabumi
JB39)	1188		R. Duta Swara Parahyangan (DSP), Bekasi
JH46)	1188		R. Suara Ayukarya Banjaran Adiwerna (RSA-Abadi), Tegal
JT25)	1188		R. Swara Perak Jaya P.T.D.I., Surabaya
YG08)	1188		R. Anak Pemkot Yogyakarta
SU17)	1197	0.5	RSPD Labuhan Batu, Rantau Prapat
SU18)	1205	0.2	RPD Simalungun, Pematang Siantar
AC20)	1206		R. Geunta Suara, Geudong
AC21)	1206		R. Mariba Raya (Maya), Kuala Simpang
SB06)	1206		R. Suara Dikara Bawana (Dirgan Bravo), Padang
SU19)	1206		R. Suara Langkat Tanjung Persada, Tanjungpura
JH47)	v1213	0.25	R. Suara Perwira (RSPD), Purbalingga
JB40)	1224		R. Sonata 47, Bandung
JH48)	1224		R. Angkasa Bahana Citra (A.B.C.), Surakarta
JH49)	1224		RSPD Banyumas, Purwokerto
JK19)	1224		R. Metro, Jakarta
JT26)	1224		RKPD Sumenep (R. Dinamika Suara Pariwisata)
SU20)	1224		R. Barisan Nauli, Sidikalang
SU21)	1224		R. Alnoria Dirgantara, Tebingtinggi
SU22)	1224		R. Cipta Anindya Guna, Binjai
AC22)	v1233	0.3	RPD Kotamadya Sabang*
JH50)	1224		R. Duta Suara Garuda Sakti, Blora
SL11)	1242		R. Suara Bulusaraung, Pangkep
SU23)	1242		R. Langkat Jaya, Binjai
YG09)	1251		R. Edukasi, Yogyakarta
JH51)	1260		R. P.T.D.I. Suara Kaliwungu Dirgantara, Kaliwungu
JT27)	1260		R. Gabriel, Madiun
SB07)	1260		R. Gitamitra Suara Perdana, Lubuk Basung
SL12)	1260		R. Molina Indah Pesona, Sinjai
SU24)	1260		R. Swara Jupti Indah, Sibolga
JT28)	1278		R. Antariksa Radang IV, Surabaya
PA11)	1278		R. Pikonane, Yahukimo
SG03)	1278		R. Ringan Mutiara, Kendari
SU25)	1278		R. Suaratama Citra Mitra, Bandar Pulau
SU26)	1278		R. Cempaka Selaras Silindung, Tarutung
KH05)	1287		R. Gema Kahayan, Pangkalan Bun
SU27)	1290	0.2	RPD Kotamadya Pematang Siantar
KH06)	1296		R. Merak Jaya, Muarateweh
NB06)	1296	0.5	R. Duta Gita Bhyomantara Sinta Rama, Cakranegara
SL13)	1296		R. Suara Kelandka, Palopo
SU28)	1296		R. Begita, Kabanjahe
JT29)	1304	0.5	RKPD Nganjuk
JA06)	1305		R. Kerinci Giri Swara (K.G.S.), Sungaipenuh
JB41)	1314		R. Mutiara, Bandung
JH52)	1314		Suara Sion Perdana, Karanganyar
JH53)	1314		R. Gema Sritanjung Mediatama (G.S.M.), Jatibarang
SG04)	1314		R. Buana Sutra, Kendari
JT30)	1332		RKPD Ngawi (Suara Ngawi)
SG05)	1332		R. Suara Bhakti Nusantara, Bau-Bau*
SH03)	1341		R. Bittara Indah, Tolitoli
JT31)	1350		RKPD Kotamadya Surabaya (R. Gelora Surabaya)
JT32)	1350	0.5	RKPD Situbondo
SS11)	1350		R. Baturaja Mutiara Wahana (B.M.W.), Baturaja
SU29)	1350	0.75	RPD Tapanuli Utara, Balige
SU30)	1351		R. Delijaya, Tebingtinggi
JB42)	v1368		R. Suara Citra Aditama, Bekasi
SU31)	1368		R. Swaratama Jatayu, Limapuluh-Asahan
SU32)	1368		R. Kharisma Swararia, Balige
JB43)	1385	0.25	RPK Majalengka
SH04)	1386		R. Swara Maya Prastha, Poso
SU33)	1395		R. Deli Indah Swara Diah, Tebingtinggi

MW	kHz	kW	Station
SS12)	1404		R. Puspa Irama, Belitang Oku
SH05)	v1415		R. Swara Magaga, Tolitoli
JT33)	1422		R. Perkasa Muda Agung (P.M.A.), Kraksaan
KR03)	1422		R. Karastina, Tanjung Pinang*
SU34)	1431		R. Buana Serdang, Dolok Masihul
BN09)	v1440		R. Edukasi, Tangerang
JH54)	1440		R. Dian Sindoro Suara Semesta, Temanggung
JT34)	1440		R. Nada Kemala Jaya, Sumenep
ST03)	1440		Suara Totabuan Ria, Kotamobagu
JT35)	1449		RKIP Wonocolo, Surabaya
JB44)	1458		R. Fajri, Bandung
SS13)	1458	0.5	R. Lematang Indah, Bandar Agung
JH55)	1476	0.25	R. Siaran Niaga Hiukencana (R.H.K.), Semarang
PA12)	1476		R. Wagadei, Paniai
JK20)	v1490		R. Karya Bersama, Jakarta
SH06)	1494	0.25	R. Toddo Puli (Topsi), Palu*
SU35)	1494		R. Al Rona Bahana, Padangsidempuan
JT36)	1503		R. Pendidikan Jawa Timur, Surabaya
KT05)	1512	0.25	R. Swara Mitra Dirgantara (Rasmira), Balikpapan
AC23)	1521		R. Dirgantara, Langsa
JH56)	1521	0.1	RSPD Klaten
JH57)	1521	0.3	RSPD Wonosobo
SS14)	1521		R. Suara Musijaya Pratama, Sekayu
JK21)	1530		R. Islam Sabili (R.I.S.), Jakarta
RI04)	1539	0.25	R. Esti Elita, Pekanbaru
JB45)	1565		R. Swara Primadona Mahardika, Cikampek
KS07)	1584		Swara Al Karomah Pratama, Martapura
KT06)	1584		R. Pangkalan Remaja Derap Bhakti (PRDB), Samarinda

NB: * = r. inactive or moved to FM.

EXTERNAL SERVICES: The Voice of Indonesia
see International Broadcasting section.

Addresses (Jl = Jalan)

AC00) NANGGROE ACEH DARUSSALAM (State of Aceh)
AC06) Takengon – **AC07)** Jl Teuku Umar Km 10, Desa Tonjong, Lho'nga Leupeung 23353 – **AC08)** Jl Teuku Cik Ditiro 1, Lhok Sukon – **AC09)** Jl Veteran 18, Kp Jawabaru, Kec Bandasakti, Lhokseumawe 24351 – **AC10)** Jl Letkol Abdullah Basah 3, Bandar Mutiara, Beureunuen 24173 – **AC11)** Jl Mayjen Sutomo 31, Kuala Simpang 24475 – **AC12)** Langsa – **AC13)** Jl Mawar 25, Sigli 24112 – **AC14)** Jl Ujung Kalak 16, Meulaboh 23613 – **AC15)** Jl.Diponegoro, Sabang, Weh – **AC16)** Jl Gayo 137, Bireuen 24211 – **AC17)** Jl Rel Kereta Api 14, Lhokseumawe 24310 – **AC18)** Blangpidie – **AC19)** Lamno – **AC20)** Jl Kreung Pase 12, Geudong, Lhokseumawe 24374 – **AC21)** Jl Rantau Dusun Jawa, Rantau, Kuala Simpang 24474 – **AC22)** Jl Diponegoro 53, Sabang, Weh – **AC23)** Jl Rantau 19, Langsa.

BA00) BALI
BA03) Jl Imam Bonjol, Gang Gunung Sabha 6, Banjar Abian Timbul, Denpasar – **BA04)** Jl Jend Sudirman 25, Negara.

BE00) BENGKULU
BE02) Muko-Muko - Ipuh, Bengkulu – **BE03)** Argamakmur, Bengkulu Utara – **BE04)** Jl D.I Panjaitan 99, Curup, Bengkulu 39118.

BN00) BANTEN
BN01) Jl Gatot Subroto Km 8, Jatake, Tangerang – **BN02)** Jl Raya Serang Km 2, Kesambi, Pandeglang 42213 – **BN03)** Tangerang – **BN04)** Jl Raya Km 4, Kaduagung Barat – Rangkasbitung 42311 – **BN05)** Jl Radeh Fatah, Perum Lembang Baru I/3, Ciledug, Tangerang 15151 – **BN06)** Tangerang – **BN07)** Cipondoh, Tangerang – **BN08)** Tangerang – **BN09)** Pusat Teknologi Informasi dan Komunikasi (PUSTEKKOM), Departemen Pendidikan Nasional (DEPDIKNAS), Ciputat, Tangerang.

JA00) JAMBI
JA02) Jl Sultan Thoha 1, Komplek Airpanas, Muarabulian 36613 – **JA03)** Jl.Gajah Mada, Muarabulian 36610 – **JA04)** Jl Panglima H Saman 297B, Kuala Tungkal, Tanjung Jabung Barat 36513 – **JA05)** Jl Yos Sudarso 55, Sungai Penuh, Kerinci – **JA06)** Jl Sisingamangaraja 23, Sungaipenuh, Kerinci 37113.

JB00) JAWA BARAT (West Java)
JB04) Depok – **JB05)** Jl Ion Martasismata 24, Pamanukan, Subang 41254 – **JB06)** Jl Raya Karangnunggal 157, Karangnunggal, Tasikmalaya 46186 – **JB07)** Jl Radio 9, Cirendang, Kuningan – **JB08)** Jl Raya Bojongsari 17, Sawangan, Bogor 16516 – **JB09)** Jl Brawijaya, P.O Box 12, Kadipaten, Majalengka 45452 – **JB10)** Jl. Raya Batujajar 288, Bandung Barat 40561 – **JB11)** Tomo – Sumedang – **JB12)** Cileungsi - Bogor – **JB13)** Jl Batulawang 1, Banjar, Ciamis 46133 – **JB14)** Cianjur – **JB15)** Jl Pramuka 10, Majalengka 45418 – **JB16)** Jl Adikusumah, Babe Endah, Dayeuh Kolot, Bandung – **JB17)** Jl Siliwangi 5, Bandung 40132 – **JB18)** Jl Siliwangi 103, Pelabuhanratu, Sukabumi 43164 – **JB19)** Jl Raya Timur 12, Singaparna, Tasikmalaya – **JB20)** Jl Raya Barat 98, Banjarsari, Ciamis 46383 – **JB21)** Gedung Sasana Kriya TMII, Bekasi 13560 – **JB22)** Jl Raya Kedung Halang 2, Waru Jambu,

Bogor 16710 – **JB23)** Jl Soekarno-Hatta 613B, Bandung 40257 – **JB24)** Jl Perintis Kemerdekaan 86, Cibadak, Sukabumi – **JB25)** Jl Satria 22, Pemeungpeuk, Garut 44175 – **JB26)** Jl Merdeka 92 Lt 2, Garut 44151 – **JB27)** Jl Gegerkalong Girang 67, Bandung 40154 – **JB28)** Jl Ir H Juanda 128, Ciamis 46211 – **JB29)** Jl Pramuka 653, Pangandaran, Ciamis – **JB30)** Jl Cendana 70, Bekasi 17100 – **JB31)** Jl Raya Bandung Km 15, Ciranjang, Cianjur 43282 – **JB32)** Jl Ky Dulngali 6, Losali, Cirebon – **JB3o)** Subang – **JB34)** Bandung – **JB35)** Jl A Yani 56, Ciasem, Sukamandi-Subang – **JB36)** Cianjur – **JB37)** ITC Kosambi Blok G-16 Lt I, Jl Baranangsiang, Bandung 40112 – **JB38)** Jl Selabintana 146, P.O Box 59, Sukabumi – **B39)** Bekasi – **JB40)** Bandung – **JB41)** Jl Cikamiri 7, Cisadea, Bandung – **JB42)** Tambun Selatan, Bekasi – **JB43)** Jl Raya Timur, Majalengka – **JB44)** Bandung – **JB45)** Jl Siswa 56, Cikampek, Karawang 41373.

JH00) JAWA TENGAH (Central Java)
JH04) Jl Kutoarjo 60, Kebumen 54312 – **JH05)** Jl Margantara Tanjung, P.O. Box 45, Purwokerto 53143 – **JH06)** Jl Barito 4, Pekalongan 51116 – **JH07)** Jl Pamukti Baru 9, Prambanan, Klaten 57454 – **JH08)** Jl Pancurendang 26, Ajibarang, Banyumas 53163 – **JH09)** Jl Purworejo VI/10, Surakarta – **JH10)** Jl Mayjen Sungkono 89, Purbalingga – **JH11)** Jl Raung 7, Candi Baru, Semarang – **JH12)** Jl Pang Diponegoro 18, Majenang, Cilacap 53257 – **JH13)** Jl K.H Agus Salim 22, Surakarta 57147 – **JH14)** Jl Kemuning 30, P.O Box 48, Salatiga 50724 – **JH15)** Jl Kendeng (Pesayangan) 55, Kroya, Cilacap – **JH16)** Jl Letjend S Parman 28, Banjarnegara – **JH17)** Jl Madugondo 15, Grogol, Sukoharjo 57552 – **JH18)** Jl Overste Isdiman 22, Purbalingga 53313 – **JH19)** Boyolali – **JH20)** Jl Pahlawan 99, Magelang – **JH21)** Jl D.I. Panjaitan 3, Surakarta – **JH22)** Jl Yos Sudarso 171, Gombong, Kebumen 54411 – **JH23)** Purbalingga – **JH24)** Jl Raya Kramat Km 7, Tegal 52181 – **JH25)** Jl Kanjengen C-308, Kranggan, Temanggung 56271 – **JH26)** Jl Kyai Jebat 1, Demak – **JH27)** Jl Kompleks Kawedanan Lama 296, Banyumas 53192 – **JH28)** Jl Pemuda 55, Cepu, Blora 58312 – **JH29)** Jl D.I Panjaitan 3, Banjarnegara 53415 – **JH30)** Jl Pahlawan 134A, Magelang 56116 – **JH31)** Jl Pasar Hewan 75, Bumiayu 52273 – **JH32)** Wonosobo – **JH33)** Jl Kemuning 3, Purbalingga 53316 – **JH34)** Jl Pemuda Pucungrejo, Muntilan, Magelang – **JH35)** Jl Pesantren 19, Ketanggungan, Brebes 52263 – **JH36)** Jl Bahari 325, Weleri, Kendal 51355 – **JH37)** Jl Raya Nagadirejo, Ngadirejo, Wonogiri – **JH38)** Jl Jenderal Ahmad Yani 32, Temanggung – **JH39)** Jl Tentara Pelajar 52, Tegal 52122 – **JH40)** Yayasan Badan Wakaf Sultan Agung (YBWSA), Universitas Islam Sultan Agung, Jl Raya Kaligawe Km 4, Semarang 50012 – **JH41)** Jl Ahmad Yani 186, Batang 51215 – **JH42)** Jl Raya Sukowati 530, Sragen 57215 – **JH43)** Jl Raya Delanggu Utara 53, Delanggu, Klaten 57471 – **JH44)** Jl Pandanwangi Selatan A-88. Kel. Kedungmundu, Tembalang, Semarang – **JH45)** Komplek Perluasan Kota, Jl Plongkowati, Wonogiri – **JH46)** Jl Raya Banjaran 34B, Adiwerna, Tegal 52194 – **JH47)** Jl Dipokusumo, Purbalingga – **JH48)** Jl Kapt Mulyadi 117, Surakarta 57113 – **JH49)** Purwokerto – **JH50)** Blora – **JH51)** Jl Raya Kramat 1, Kaliwungu, Kendal 51372 – **JH52)** Jl Dr Muwardi 47, Badranasri, Karanganyar 5771 – **JH53)** Jl Syah Alibahayar Salamah 2, Jatibarang, Brebes 52261 – **JH54)** Jl Kartini 34, Temanggung 56216 – **JH55)** Jl H Kimar III/5, Semarang 50249 – **JH56)** Jl Pemuda Tengah 56, Kotak Pos 113, Klaten – **JH57)** Komplek Kabupaten Wonosobo, Jl Merdeka 1, Wonosobo.

JK00) JAKARTA
JK02) Jakarta – **JK03)** Jl Swadaya Raya 26/143, Raden Inten, Jakarta – **JK04)** Apartemen Robinson Lt 6, Jembatan Dua Raya 2, Jakarta – **JK05)** Jl Matraman 39, Jakarta – **JK06)** Jl Bintaro Rosali IV/10, Bumi Pintaro Permai, Jakarta – **JK07)** Lantai 4, Pusat Perdagangan Senen, Jakarta – **JK08)** Jl Masjid Al Barkah 17, Tebet, Jakarta Selatan – **JK09)** Universitas Mercu Buana, Meruya Selatan, Jakarta Barat – **JK10)** Sunter Agung, Jakarta – **JK11)** Jl Kampung Melayu Kecil III/40, Tebet, Jakarta – **JK12)** Jakarta Selatan – **JK13)** Kampung Beting, Jakarta Utara – **JK14)** Jl Dakota V/1, Kemayoran, Jakarta 10630 – **JK15)** Jakarta Utara – **JK16)** Jl Batu Ceper V/52, Jakarta Pusat 10120 – **JK17)** Jakarta – **JK18)** Jl Letjen S Parman 1, Jakarta Barat – **JK19)** Jakarta – **JK20)** Kemang, Jakarta – **JK21)** Graha Sabili, Jl Cipinang Cempedak II/11A, Polonia, Jakarta.

JT00) JAWA TIMUR (East Java)
JT06) Desa Sungi Wetan, Pasuruan – **JT07)** Ngawi – **JT08)** Jl Yos Sudarso 173, Sumenep – **JT09)** Jl Alun-Alun Utara 3, Ponorogo 63413 – **JT10)** Jl Dharmhusada Indah Blok A75, Surabaya 60285 – **JT11)** Jl Simolawang I/96, Surabaya – **JT12)** Wisma Sarinadi, Kawasan GOR, Sidoarjo – **JT13)** Jl Gatot Subroto 107, Pacitan – **JT14)** Komplek STAI Ali Bin Abi Thalib, Jl Sitopo Kidul 51, Surabaya – **JT15)** Jl K.H Wakhid Hasyim 9, Gresik – **JT16)** Jl Veteran 6B, Bondowoso 68211 – **JT17)** Jl Aipda Karel Sasuit Tubun 15, Gresik 61114 – **JT18)** Jl W.R Supratman 29, Situbondo – **JT19)** Kompleks Manyar Indah Plaza, Jl Ngagel Jaya Selatan, Surabaya – **JT20)** Jl Jaksa Agung Suprapto 8, Pacitan 63512 – **JT21)** Jl Ngagel Jaya Utara IV/21, Surabaya 60283 – **JT22)** Jl Sidoluhur 2A, Ponorogo 63410 – **JT23)** Jl Amir Hamzah 18, Surabaya

60241 – **JT24)** Jl Panglima Sudirman 72, Surabaya 60272 – **JT25)** Jl Teluk Aru 68, Surabaya 60165 – **JT26)** Jl Dr Cipto, Sumenep 69410 – **JT27)** Jl Pesanggrahan V Taman, Madiun 63131 – **JT28)** Jl Kusuma Bangsa 4, Surabaya 60241 – **JT29)** Jl Dr Sutomo 60, Nganjuk – **JT30)** Jl Teuku Umar 12, Ngawi – **JT31)** Humas Gelora 10 Nopember, Jl Tambaksari, Surabaya 60136 – **JT32)** Jl P.B Sudirman 14, Situbondo 68416 – **JT33)** Jl P Sudirman 62, Kraksaan, Probolinggo 67282 – **JT34)** Sumenep – **JT35)** Jl Ahmad Yani 112, Wonokromo, Surabaya – **JT36)** Jl Gentengkali 33, Surabaya.

KB00) KALIMANTAN BARAT (West Kalimantan)
KB04) Jl Raya Sambas Bukitluwing 1, Sambas 79162 – **KB05)** Jl Kelam Akcaya I/18, Sintang 78611 – **KB06)** Jl Raya Ngabang 72, Ngabang, Pontianak – **KB07)** Jl Kom Yos Sudarso 9, Sanggau 78582 – **KB08)** Jl Kawak 26, Sekadau 78582 – **KB09)** Jl Pembangunan RT 003/XIV, Desa Harapan, Pemangkat 79153 – **KB10)** Jl Pendidikan II, Sungai Pinyuh 78353.

KH00) KALIMANTAN TENGAH (Central Kalimantan)
KH02) Jl Merdeka Raya 21, Buntok 73711 – **KH03)** Jl Pongsongteleng 47, Ampah 73652 – **KH04)** Muarateweh – **KH05)** Jl Ahmad Yani 45, Pangkalan Bun 74113 – **KH06)** Jl Merak 34, Muarateweh 73810.

KS00) KALIMANTAN SELATAN (South Kalimantan)
KS02) Jl Pahlawan 33, Kandangan 71211 – **KS03)** Jl Brigjen H Hasan Basry 56/58, Rantau 71111 – **KS04)** Banjarmasin – **KS05)** Kotabaru – **KS06)** Jl Barintik 35, P.O Box 48, Martapura 70613 – **KS07)** Jl Jend A Yani, Pesayangan Utara, Martapura 70619.

KT00) KALIMANTAN TIMUR (East Kalimantan)
KT04) Jl Mulawarman 66, Tenggarong – **KT05)** Jl A Yani 50, Balikpapan 76123 – **KT06)** Jl Anggur 33, Samarinda 75123.

LA00) LAMPUNG
LA02) Jl Veteran 475, Purwodadi, Trimurjo 34114 – **LA03)** Jl Raya Simpang Agung 58, Terbanggi Besar, Lampung Tengah – **LA04)** Jl Batu Tegi 40, Talangpadang 35377 – **LA05)** Jl Raden Intan 188, Wonodadi, Gadingrejo 35372 – **LA06)** Jl Indra Bangsawan 188, Kalianda 35513 – **LA07)** Jl Tanggamus 69, Gisting – **LA08)** Jl Raya Lintas Timur 148, Unit II, Tulang Bawang.

NB00) NUSA TENGGARA BARAT (West Nusa Tenggara)
NB02) Bodak, Praya, Lombok – **NB03)** Jl Pahlawan 70, Pancor, Selong, Lombok – **NB04)** Selong, Lombok – **NB05)** Jl Jend Sudirman 10, Pancor, Selong 83611, Lombok – **NB06)** Jl Miru 72, Cakranegara, Mataram 83511, Lombok.

NT00) NUSA TENGGARA TIMUR (East Nusa Tenggara)
NT03) Waingapu, Sumba – **NT04)** Soe, Timor – **NT05)** Waikabubak, Sumba – **NT06)** Jl Nusa Indah 21, Oepura, Kupang 85117, Timor – **NT07)** Jl Untung Suropati 2B, Kupang 85119, Timor - **FM:** 103.5MHz – **NT08)** Jl Basuki Rahmat 2, Atambua 85711, Timor – **NT09)** Jl Sukarno-Hatta, Bajawa, Flores – **NT10)** Jl Panglima Sudirman, Ende, Flores.

PA00) PAPUA (formerly Irian Jaya)
PA08) Jayapura – **PA09)** Jl Bhayangkara, Wamena – **PA10)** Jl Skyline, Jayapura – **PA11)** Anyelma, Kurima, Yahukimo – **PA12)** Enarotali, Paniai.

RI00) RIAU
RI02) Jl Putri Nilam 51, Sukajadi, Pekanbaru 28128 – **RI03)** Jl Jend Sudirman 674, Bagan Batu, Bengkalis – **RI04)** Jl Teratai 17, Pekanbaru.

SB00) SUMATERA BARAT (West Sumatra)
SB03) Jl Jend Sudirman 18, Payakumbuh 26211 – **SB04)** Jl Sisingamangaraja 1, Padang 25122 – **SB05)** Jl Limo Singke Baringin, Gando Sulit Air, Solok – **SB06)** Jl W.R Mongonsidi 4B, Lantai 2, Padang – **SB07)** Lubuk Basung, Agam

SG00) SULAWESI TENGGARA (South-East Celebes)
SG02) Kolaka – **SG03)** Kendari – **SG04)** Kendari – **SG05)** Bau-Bau, Buton.

SH00) SULAWESI TENGAH (Central Celebes)
SH03) Jl Magamu 33, Tolitoli 94514 – **SH04)** Jl Pulau Kalimantan 45, Poso 94610 – **SH05)** Jl Saputan Raya 117, Tolitoli – **SH06)** Jl Setia Budi 20A, Palu 94111.

SL00) SULAWESI SELATAN (South Celebes)
SL02) Jl Ratulangi 17, Rantepao 91831 – **SL03)** Jl Poros Cabenge 1, Watansoppeng – **SL04)** Jl Abubakar Lambogo 11, Enrekang 91711 – **SL05)** Jl Manggis 16, Makassar 90112 – **SL06)** Jl Mesjid Raya 100, Sengkang, Wayo 90914 – **SL07)** Jl Andi Naboang 1, Pangkajene – **SL08)** Kompleks Ruko Somba Opu Blok B/19, Tanjung Bunga - Makassar – **SL09)** Jl Gelatik 2, Kel Pallanting, Bantaeng 92411 – **SL10)** Jl Mangga 1, Palopo 91921 – **SL11)** Jl Sultan Hasanuddin 94, Pangkep – **SL12)** Sinjai – **SL13)** Jl Mannennungeng Kav 33, Palopo 91922.

SR00) SULAWESI BARAT (West Celebes)
SR01) Mamuju – **SR02)** Jl Pasar Sentral 48, Mamuju.

SS00) SUMATERA SELATAN (South Sumatra)
SS02) Jl Raya Betung 281, Betung, Musi Banyuasin – **SS03)** Jl Simpang Tiga Tanjung Seteko, Indralaya, OKI 30662 – **SS04)** Jl Cut Nyak Din 3,

Baturaja, OKU 32111 – **SS05)** Jl Dr M Isa 38, 8 Ilir, Palembang 30114 – **SS06)** Pagar Alam - Lahat – **SS07)** Jl Pramuka I/15, Muara Enim – **SS08)** Jl D.I Panjaitan 3/41, Plaju, Palembang 30265 – **SS09)** Jl Jend Sudirman 182/IV, Prabumulih Barat 31123 – **SS10)** Jl K.H.A Azhari 136, 12 Ulu, Palembang 30262 – **SS11)** Jl Mayor Iskandar 427, Baturaja – **SS12)** Jl Sakura 103, RT 04, Bedilan, Belitang 32182 – **SS13)** Jl Raya Bandar Agung 4, Bandar Agung, Lahat 31414 – **SS14)** Jl Kol Wahid Udin 565, Lingkungan 7, Sekayu, Musi Banyuasin.
ST00) SULAWESI UTARA (North Celebes)
ST03) Jl Teuku Umar 155, Kotamobagu.
SU00) SUMATERA UTARA (North Sumatra)
SU04) Jl K.H Zainal Arifin 13, Sibolga 22521 – **SU05)** Jl M.T Haryono 64, Tanjungbalai 21311 – **SU06)** Medan – **SU07)** Jl Cirebon 3, Belawan, Medan 20412 – **SU08)** Jl Galang 9, Lubukpakam 20510 – **SU09)** Jl Diponegoro 69, Gunungsitoli, Nias – **SU10)** Jl Imam Bonjol 16, Tebingtinggi 20610 – **SU11)** Jl Iskandar Muda 117A, Medan 20119 – **SU12)** Jl Deli Gg Kereta Api 6, Perbaungan, Deli Serdang 20586 – **SU13)** Rantau Prapat, Labuhanbatu – **SU14)** Jl Cokroaminoto 171, Kisaran 21216 – **SU15)** Jl Ismail 5A, Binjai – **SU16)** Medan – **SU17)** Jl W.R Supratman 37, Rantau Prapat – **SU18)** Jl Merdeka 1, PO Box 25, Pematang Siantar – **SU19)** Jl Pemuda, Gg Singadua 29A, Tanjungpura, Langkat 20853 – **SU20)** Jl Dr F.L Tobing 59, Sidikalang 22212 – **SU21)** Jl Raya Medan 52, Tebingtinggi 20610 – **SU22)** Jl Hasanudin 35, Binjai 20713 – **SU23)** Jl Palembang 36, Binjai 20721 – **SU24)** Jl Letjend Suprapto 101, Sibolga 22351 – **SU25)** Jl Beringin 1, Tanjung Gading, Asahan 21257 – **SU26)** Jl Kol Liberti Malau, Pasar Baru, Tarutung – **SU27)** Jl Merdeka, Pematang Siantar – **SU28)** Jl Jend Sudirman 35, Kabanjahe 22113 – **SU29)** Komplek Monumen D.I Panjaitan 1, P.O Box 4, Balige – **SU30)** Tebingtinggi – **SU31)** Jl Besar Sumberpadi, Limapuluh 21255 – **SU32)** Jl Sisingamangaraja 188, Balige 22316 – **SU33)** Jl Jenderal Sudirman, Tebingtinggi – **SU34)** Jl Besar 159, Dolok Masihul, Deli Serdang – **SU35)** Jl Kamboja 1, Padangsidempuan 22730.
YG00) DAERAH ISTIMEWA YOGYAKARTA (Yogyakarta Special Region)
YG02) Sidokarto, Sleman – **YG03)** Jl Panti Wreda 5, Giwangan, Umbulharjo, Yogyakarta 55163 – **YG04)** Parangtritis – Yogyakarta – **YG06)** Yogyakarta – **YG07)** Jl Puro Pakualaman, Yogyakarta 55112 – **YG08)** Taman Pintar, Jl Panembahan Senopati, Gondomanan – Yogyakarta – **YG09)** Balai Pengembangan Media Radio, Pusat Teknologi Informasi dan Komunikasi Pendidikan, Departemen Pendidikan Nasional, Jl Sorowajan Baru 367, Banguntapan, Yogyakarta 55198.

FM: A large number of FM stns operate throughout the country. See RRI address list for RRI FM freqs.
Jakarta FM(MHz): 87.6 Antarnusa Jaya (Hard Rock FM) – 88.0 Mustang Utama – 88.4 Arief Rahman Hakim (ARH/Global R.) – 89.2 Metro Jaya Kartika (R. 68H/Green R.) – 89.6 Mustika Abadi (I Radio) – 90.0 R. Elshinta – 90.4 Muara Abadi Nusa (Cosmopolitan) – 90.8 Suara Gema Pembangunan Utama (Oz R. Jakarta) – 91.6 Indika Millenia – 92.0 Sonora – 92.4 Primaswara Adi Spirit Semesta (PAS FM/R. Bisnis Jakarta) – 93.2 Merpati Dharmawangsa (MD R.) – 93.6 Gema Wargakarya Satnawa (Gaya) FM, Bekasi – 93.9 Swara Mersidiona (Mersi), Tangerang – 94.3 Gardia Asia Bumi (Woman FM) – 94.7 Agustina Yunior (U FM) – 95.1 Kirana Indah Suara (KIS) – 95.5 R. Siaran Alaikassalam Sejahtera (RASfm) – 95.9 Smart Media Utama – 96.3 R. Pelita Kasih (RPK) – 96.7 Swara Rhadana Dunia (Rhadio A) – 97.1 Suara Monalisa (R. Dangdut TPI) – 97.5 Safari Bina Budaya (Motion R.) – 97.9 Bahana Sanada Dunia (Female), Tangerang – 98.3 Cakrawala Gita Swara – 98.7 Attahiriyah (Gen FM) – 99.1 Delta Insani – 99.5 Kayumanis – 99.9 Draba (Ninetyniners FM Jakarta) – 100.3 Elgangga, Bekasi – 100.6 Jati Yaski Mandiri (Heartline), Tangerang – 101.0 Suara Irama Indah (Jak FM) – 101.4 Suara Kejayaan (Trax FM) – 101.8 Terik Matahari Bahana Pembangunan (Bahana FM) –102.2 Prambors – 102.6 Camajaya Surya Nada – 103.4 Taman Mini (DFM) – 103.8 Brava – 104.2 Media Suara Trisakti (MS-Tri) – 104.6 Trijaya Sakti – 105.4 Niaga Chakti Bhudi Bhakti (CBB FM) – 105.8 Ramako Jaya Raya (Lite FM) – 106.2 Bergaya Nyanyian Irama Sejati, Tangerang (Bens R.) – 106.6 Sabda Sosok Sohor (M) – 107.0 Nada Komunikasi Utama (Dakta), Bekasi – 107.3 Suara Tunggal Angkasa Raya (Star), Tangerang – 107.5 R. Mitra Carita Gemabelas (Music City/MCFM), Bogor – 107.7 Islamic R. – 107.8 Suara Metro (Jakarta Police).

Denpasar (Bali) FM (MHz): 87.8 R. Baturiti Menara Swara (Hard Rock R.) – 88.2 R. Rock FM – 89.4 Gema Sunari Indah – 89.8 Organik Lestari Sejahtera (Pak Oles FM), Tabanan – 90.2 R. Swara Sanathana Dharma – 90.6 R. Gema Megantara Pratama (R. Megantara Bali), Tabanan – 91.0 Gita Bakti Persada (R. Phoenix) – 91.45 R. Pemerintah Kota Denpasar – 91.8 Flamboyant Bali Indah (FBI) – 92.2 Gema Megantara Pesona (Heartline) – 92.6 R. Suara Yudha – 93.0 R. Gita Semara Persada (GSP),

Semarapura – 93.8 R. Suara Calvary, Klungkung – 94.5 Citra Dharma Bali Satya (CDBS) – 94.9 Click R. Bali, Bangli – 96.1 R. Genta Suara Bali – 96.5 R. Swara Kinijani (Global FM Bali), Tabanan – 96.9 R. Elang Kosa Gagana (Elkoga) – 97.3 R. Sonata Indah (Soni FM) – 97.7 R. Gema Merdeka – 98.1 Gia FM, Gianyar – 98.5 R. Plus, Sanur – 98.9 R. Bali Perkasa (Balisa FM), Gianyar – 99.3 R. Duta Dewata (Duta Female) – 99.7 R. Semarapura FM, Klungkung – 100.5 R. Dunia Bokashi Raya, Klungkung – 101.2 R. Bali Swara Mitragama (Oz R.), Kuta – 101.6 R. Ujung Bali, Amlapura – 102.0 R. Suara Denpasar Chakti (Cassanova) – 102.4 R. Balina Citra (BFM), Bangli – 102.8 R. Menara – 103.2 R. Mega Nada, Tabanan – 103.6 R. Pinguin – 104.4 Aneka Rama (AR) – 104.8 R. Jegeg Bali, Gianyar – 105.2 Surya Permai (Super R.) – 105.6 R. Bali Mandala Perkasa, Gianyar – 106.0 Swara Kreasi Utama (Kuta R.), Kuta – 106.4 R. Gelora Gianyar, Gianyar – 107.2 R. Dian Mandiri – 107.5 R. Suara Udayana.

IRAN

L.T: UTC +3½h (22 Mar–22 Sept: +4½h) — **Pop:** 66 million — **Pr.L:** Farsi (Persian) — **E.C:** 50Hz, 230V — **ITU:** IRN

ISLAMIC REPUBLIC OF IRAN BROADCASTING (Gov.)
⌨ P .O. Box 19395-333, Tehran. (Int. Tech. Affairs: P.O. Box 15875-4344, Tehran) ☎ +98 21 2204 1093 ▤ +98 21 2222 1508
W: www.irib.com/radio **E:** radio@irib.ir
LP: President: Ezatollah Zarghami. DG: Gholamali Ramezani.
MW:

Prov & Location	kHz	kW	N		Prov & Location	kHz	kW	N
3 Azarshahr	531	500	I		24 Zahedan	1017	100	I
24 Iranshahr	531	600	I		3 Tabriz	1026	100	R
29 Mashhad	540	200	I		26 Yazd	1035	50	R
13 Sirjan	549	400	I		12 Dehloran	1044	50	I
21 Gheslagh	558	1000	F		18 Khorramabad	1053	100	I
15 Mahshahr	576	750	R/E		24 Saravan	1053	25	R
25 Tehran	585	600	Q		13 Kerman	1062	50	I
7 Shiraz (Dehnow)	594	400	R		22 Alborz (Qom)	1071	100	M
24 Zahedan	603	50	I		15 Mahshahr	1080	750	E
Unknown loc.	603		R		23 Biarjmand	1089	50	I
14 Qasr-e-Shirin	612	600	E		24 Zabol	1098	100	R
11 Bandar Abbas	621	50	R		29 Sabzevar	1107	50	R
30 Birjand	621	50	R		26 Ardekan	*1116	200	I
3 Bonab	639	400	I		21 Qazvin	1125	50	R
5 Shahr-e-Kord	648	50	I		30 Nehbandan	1125	10	R
24 Zahedan	657	100	R		28 Bojnurd	1134	50	I
29 Shushtar	666	50	I		16 Yasuj	1143	50	I
10 Hamadan	675	50	R		3 Tabriz	*1152	100	R
29 Gaem (Mashhad)	684	100	R		14 Qasr-e-Shirin	1161	600	E
11 Bandar Abbas	693	100	R		15 Abadan	1170	750	I
4 Bushehr	702	100	I		23 Damghan	1170	50	I
12 Kiashahr	702	500	E		9 Gonbad	1179	50	R
15 Ahwaz	711	200	R/E		24 Chabahar	1179	50	I
14 Mahidasht	720	750	R/E		25 Tehran	1188	300	P
29 Tayebad	720	400	R/E		14 Moghan	1197	50	R
4 Dayyer	738	50	R		7 Dasht	1197	10	R
9 Gonbad	747	150	I		30 Unknown loc.	1206	50	R
13 Kerman	747	100	R		20 Chalus	1215	50	R
8 Rasht	756	50	R		Unknown loc.	1224		
24 Chabahar	765	1000	E		13 Kerman	1224	400	I
5 Shahr-e-Kord	765	100	R		7 Abadeh	1233	50	R
19 Arak	774	100	R		27 Zanjan	1242	50	I
24 Iranshahr	783	150	R		13 Kiashahr	1251	100	R
27 Sohravard (Zanjan)	792	50	R		6 Khur	1260	10	I
29 Kashmar	801	50	I		1 Khalkhal	1269	50	I
18 Khorramabad	810	100	I		14 Kermanshah	1278	100	R
10 Sari	819	30	R		7 Lar	1287	100	R
26 Tabas	828	50	R		21 Qazvin	1296	50	R
6 Habibabad	837	300	R		24 Zabol	1296	10	R
3 Mianeh	846	10	I		4 Bushehr	1305	50	R
14 Qasr-e-Shirin	864	50	I		1 Ardabil	1314	50	I
10 Sari	873	100	I		3 Jolfa	1323	50	R
28 Bojnurd	873	50	R		25 Tehran	1332	300	T
10 Mahabad	882	60	R		8 Bam/Sirjan	1341	10	I
16 Yasuj/Dehdasht	891	50	R		7 Darab	1359	50	I
25 Tehran	900	600	I		9 Gorgan	1368	100	R
7 Lar	909	50	S		24 Chabahar	1377	50	I
13 Jiroft	918	50	I		14 Paveh	1377	10	R
18 Dorud	927	50	I		10 Hajiabad	1395	50	I
2 Urumiyeh	936	100	R		7 Dasht	1404	10	I
17 Dehgolan	945	50	I		11 Estahban	1413	10	R
30 Birjand	963	50	I		14 Kermanshah	*1422	100	I
12 Ilam	972	10	R		6 Habibabad	1431	200	I
10 Hamadan	981	100	I		7 Lamerd	1431	50	o
7 Shiraz (Dehnow)	990	400	I		9 Bandar-e-Torkaman	1449	400	I
17 Baneh	999	50	R		30 Birjand	1458	50	R
23 Semnan	1008	100	R		22 Alborz	1467	100	R

Prov & Location	kHz	kW	N
17 Marivan	1476v	20	R
7 Khoy	1485	10	I
7 Jahrom	1485	10	R
15 Abadan	1485	10	R
23 Damghan	1485	1	I
2 Maku	1494	20	I
4 Bushehr	1503	200	R
1 Ardabil	1512	50	R
8 Kiashahr	*1521	100	F
26 Yazd	1530	50	R
9 Gorgan	1539	50	R
16 Gachsaran	1548	10	R
20 Larijan	1548	10	R

Prov & Location	kHz	kW	N
17 Sanandaj	1548	10	I
30 Ferdows	1548	10	I
24 Zabol	1557	50	I
11 Bandar Abbas	1566	100	I
29 Qayen	1575	10	I
2 Maku	1584	10	I
15 Dezful	1602	1	R
7 Estahban	1602	1	I
7 Kazerun	1602	1	I
26 Bahabad	1602	1	I
23 Semnan	1602	10	R

*inactive

FM:

Pr. Location	I	R	Q	P	M	J	F	V/Tj
1 Ardabil	88.8	101.6	94.8	100.9	98.3	102.4	89.3	105.2
1 Khalkhal		94.4	99.2	90.1	103.9	95.8		94.0
1 Parsabad		92.4	92.8	106.5	98.0		99.3	
2 Khoy			95.0	93.8	98.2		101.8	
2 Maku	100.0	102.6			92.2	104.9		
2 Urumiyeh	106.5	91.1	106.1	95.8		92.6	102.6	99.1
2 Maragheh			94.7	95.3	101.5		98.0	
3 Tabriz	98.7	96.3	90.7	102.9	96.1	93.9	94.2	99.4
4 Bushehr	94.2	92.2	99.0	104.4	96.2		94.2	106.6
5 Ardal		106.9	99.8	93.2	90.2	96.5	103.2	
5 Ben				94.4	97.7	91.2		
5 Borujen		92.0	106.7	96.2	103.1	99.6	93.1	90.0
5 Farsan		99.5	94.5	98.5	100.5	106.5		92.5
5 Lordegan	90.8		100.0	95.8	102.6	91.2		95.8
5 Saman		98.8		89.3	95.5	92.3	102.3	105.9
5 Shahr-e-Kord	97.1	90.0	96.0	104.3	93.6	98.0	87.6	
5 Shalamzar		88.2		97.8	101.3	94.5	91.3	104.9
6 Fereydunshahr	101.8	108.0			96.6		93.3	103.3
6 Golpayegan	106.0			93.8			97.0	107.2
6 Isfahan	88.1	96.8	107.2	88.5	96.8		98.6	
6 Kashan	98.0	101.1	102.8		105.5		96.8	
6 Khonsar			99.3	107.3	102.8		106.0	
6 Nain	98.1			93.3	92.8		91.1	
6 Semirom	99.0	105.6		90.8	102.0		93.8	
7 Marvdasht			96.2		93.0		99.5	107.3
7 Shiraz	88.3	94.7	95.6	93.7	107.3	96.0	91.6	96.5
8 Rasht	98.5	92.5	101.3	104.9	104.1	91.3	106.0	107.7
8 Rudbar	91.8		97.6	108.0	91.8			
9 Gorgan		97.5	96.8	91.0	101.0	103.6	97.5	94.2
10 Hamadan	96.0	101.4	91.4	97.8	103.7	90.0	106.1	88.4
10 Mahabad	91.1		97.0	92.2				93.4
10 Malayer			96.1	96.3	93.1			96.0
10 Nahavand			94.2	94.7	101.0		101.7	
11 Bandar Abbas	93.4	97.0	90.1		95.2	94.2	104.6	105.6
11 Kish Island	103.9		103.9					107.1
12 Dehloran	100.7	102.9			94.7		101.4	
12 Ilam	92.1	106.1	105.0	95.3	100.0	93.5	98.6	102.1
12 Malekshahy	98.6	104.3	95.7	96.8	92.5			
13 Jiroft			103.2	96.4	99.7		94.5	
13 Kerman	92.8	90.2	90.2	93.3	96.5	93.9	95.1	99.8
13 Sirjan			90.7	100.7	104.3		97.2	
13 Zarand			103.5	93.3	93.5		101.1	
14 Kermanshah			106.9	93.3	90.2	93.0	99.8	103.3
15 Abadan	89.7	88.0	107.6	87.6	104.5		106.4	93.8
15 Ahwaz	106.0	93.3	103.0	93.7	97.9	90.0	99.6	107.4
15 Dezful			104.2	103.0	107.3		93.7	
15 Masjed Soleiman			105.4	97.5	104.5		99.3	
16 Yasuj	97.7	100.0	106.5	96.1	89.8	99.4	92.9	102.9
16 Baneh	107.0		93.2	99.7		96.4	103.2	
17 Bijar			100.2	90.6	103.7	96.9	107.3	93.7
17 Dehgolan			95.2			100.0	102.0	98.5
17 Divandareh	90.0		93.5	103.5		96.7	100.0	
17 Kamyaran	87.5		93.0	99.5		96.2	103.0	
17 Marivan	93.5		95.7	102.5		99.0		92.5
17 Qorveh			102.8		96.0	92.8	99.3	
17 Sanandaj	89.4	97.1	95.7	92.5		102.5	99.0	106.1
18 Saqez	106.1		92.5	99.2		95.7	102.5	
18 Sarvabad			92.9	106.5		100.2	99.4	89.8
18 Borujerd	89.3	92.9	94.9	89.0	98.2		93.6	
18 Khorramabad			93.2	91.0	101.0	100.1	94.2	
18 Kuhdasht			106.2	94.6	102.6		92.4	
19 Arak	90.1	88.5	94.7	105.8	93.2		106.8	96.4
19 Ashtian	101.0		97.5	87.9	91.0	94.2	97.3	104.6
19 Delijan	95.7	96.0	102.5	92.5	99.0	89.4	100.6	106.1
19 Komejan	91.0		94.2	97.5	104.6	101.0		87.9
19 Khomein	87.7		97.3	94.0	90.8	103.8	100.8	104.4
19 Khondab	106.6		91.4	97.8	102.0	104.0	101.4	103.0
19 Mahallat	94.9		88.6	101.7	91.7	98.2	107.6	105.3

Pr. Location	I	R	Q	P	M	J	F	V/Tj
19 Saveh	90.7		101.2	94.4	97.2	107.0	87.8	103.4
19 Shazand	93.6		90.5	96.8	100.1	103.6	107.2	89.4
19 Tafresh	91.0	89.1	101.0	100.0	97.5	92.2	103.2	94.2
20 Behshahr			96.8	87.7	96.2		93.0	
20 Chalus	97.5		98.2		88.6		91.7	
20 Sari	101.1	96.5	94.2	98.6	96.2	105.7	102.1	92.1
21 Abgarm				90.9	87.8	94.1	100.9	97.4
21 Moallem Kalayeh		92.0	102.0	95.2	88.9	98.5		105.6
21 Ghazvin	94.3	100.1	107.2	103.7	93.6	96.8	91.1	97.6
22 Qom				96.6	93.4			
22 Biarjmand			95.9	98.5	98.5		88.9	
23 Semnan	92.7	94.5	94.0	89.6	102.7		101.5	99.2
23 Shahrud	90.8		97.3	100.8		104.4	90.8	94.0
24 Saravan	95.5	92.2	95.4		92.2			
24 Zabol	92.0		96.0	87.9	97.5	91.0		94.2
24 Zahedan	93.7	90.5	93.7	90.5	100.1	100.1	103.7	107.3
25 Tehran			91.3	104.7	99.6	88.1	106.7	102.5
26 Ardakan	100.0		99.4	92.2	102.9	101.5	92.9	92.7
26 Bafgh			90.8	95.7	99.0		92.5	
26 Tabas	101.0	100.0		102.7	102.4		98.9	
26 Yazd	101.5	97.1	96.9	94.4	96.3	94.7	93.1	103.1
27 Mahnshan	92.3		102.0	94.1				99.8
27 Zanjan	99.7	93.2	105.4	96.4	102.3	90.1	93.2	106.8
28 Bojnurd	88.1	101.0	103.0		93.5	90.1	95.0	
29 Dorud	94.5		104.8	102.9	101.2		98.6	
29 Ferdows	89.0		94.3	96.0	96.2			93.6
29 Torbat	94.4	100.9	92.6				98.0	
29 Mashhad	94.8	88.1	95.2	101.6	105.8	98.7	102.2	91.6
29 Shabzevar	92.9		96.0	96.4		91.1		
30 Birjand	102.3	96.0	91.2	105.6	94.5	101.3	98.2	

Provinces: 1) Ardabil 2) West Azerbayjan **E:** 162-waz@irib.ir 3) East Azerbayjan **E:** tabriz@irib.ir 4) Bushehr **E:** prbushehr@irib.ir 5) Chaharmahal & Bakhtiari 6) Isfahan **E:** isfahan@irib.ir 7) Fars **E:** fars@irib.ir 8) Gilan **E:** gilan@irib.ir 9) Golestan 10) Hamadan 11) Hormozgan **E:** modirkol-klf@irib.ir, kish@irib.ir 12) Ilam 13) Kerman 14) Kermanshah 15) Khozestan **E:** abadan@irib.ir 16) Kohgiluyeh & Boyerahmad 17) Kurdistan 18) Lorestan 19) Markazi **E:** 162-mzn@irib.ir 20) Mazandaran 21) Qazvin 22) Qom **E:** khoshi@irib.ir 23) Semnan **E:** semnan@irib.ir 24) Systan & Baluchestan **E:** zahedan@irib.ir 25) Tehran 26) Yazd 27) Zanjan 28) North Khorasan **E:** kh-shomali@irib.ir 29) Razavi Khorasan **E:** infoplanning-ksnr@irib.ir 30) South Khorasan **E:** birjand@irib.ir

Networks:
I=Radio Iran: 24h, but hrs. of operation vary by station. Frequencies for R. Iran and provincial prgrs at the same site can often be swapped. **N:** on the half hour – **R=Regional (Provincial) network.** Studios in 32 centres producing prgrs in Farsi and local langs, including some locally produced Ext. Sce. prgrs. Regional prgrs are usually between 0230-1730, in some cases 24h, and they may r. R. Iran network 1630-2030 or overnight – **Q=R. Quran**(rlg.): 24h on MW 585kHz and FM – **P=R. Payam** ("Message", actualities): 24h on MW 1188kHz + FM. **M=R. Ma'aref** ("Presentation", rlg.): 24h on MW 1071kHz & FM **English: Call of Islam R.** on satellite and Internet: mms://62.220.122.10/maarefeng – **J=R. Javan** (Youth): 24h on MW 1206kHz & FM – **F=R. Farhang** (cultural): 24h on MW 558/1521kHz & FM – **V=R. Varzesh** (sports) and **Tj=R. Tejarat** (R. Trade) on FM. Partly sharing their frequencies – **T=Tehran City Prgr.** 24h on MW 1332kHz & FM 95.0MHz – **R. Salamat** (Health R): 0230-1430 in Tehran on 103.9MHz – **R Ava/Nava,** Tehran: 24h on 107.2MHz. – **R. Goftegoo,** Tehran: 0230-2030 on 103.9MHz.
Foreign Language prgrs in Tehran: 100.7MHz: various 24h. On 106.7MHz: **English:** 2130-2230, 2030-2130. **Russian:** 1930-2030.
Ann: S: "Inja Tehran ast, Radio Iran, Sedaye Jomhuriye Islamiye Iran". Farhang: "Inja Tehran ast, Sedaye Jomhuriye Islamiye Iran, shabakeye Farhang". **M:** "Inja Qom ast, shabakeye Ma'aref, Sedaye Jomhuriye Islamiye Iran". **Q:** "Radio Qur'an". **R:** "Inja (capital) ast, Sedaye Jomhuriye Islamiye Iran, shabakeye/markaze (province)."

EXTERNAL SERVICE: Voice of the Islamic Republic of Iran; see International Radio section

IRAQ

LT: UTC +3h — **Pop:** 28 million — **Pr.L:** Arabic, Kurdish, Assyrian, Turkoman — **E.C:** 50Hz, 230V — **ITU:** IRQ

IRAQI NATIONAL COMMUNICATION AND MEDIA COMMISSION (INCMC)
CPA Media Regulation Office, Convention Center, 3rd floor, Baghdad ☎ +964 1 7180009 **W:** www.ncmc-iraq.org **E:** enquiries@ncmc-iraq.org **L.P:** CEO: Siyamend Othman

IRAQI MEDIA NET - REPUBLIC OF IRAQ RADIO (Gov)
✉ near Al-Mansoor Melia Hotel, Salihiya, Baghdad ☎+964 790 1
325260 🖷 +964 15 377070 **W:** english.iraqimedianet.net **E:** rir.info@
iraqimedianet.net **LP:** DG: Hassan Al-Musawi. Dir. Eng: Emad Aziz.
Dep. DG & Dir. Admin: Amer Selman.

MW	kHz	kW	Prgr.
Mosul	603	20	R. Nineva/Main
Kirkuk	657	1	Main
Baghdad	675	1	Main
Nasiriya	846	20	Main‡
Ramadi	864	10	Main‡
Basra	909	25	Provincial
Hilla	1071	20	R. Babil
Tikrit	1215	10	Main
‡inactive			

FM	MHz	kW	Prgr.
Nasiriya	88.8		Main
Hilla	92.2/94.2		Main
Sinjar	90.5		Main
Diwaniya	92.2	10	Main
Samawa	92.2/92.7		Main
Karbala	94.0	1	Main
Baquba	94.8	10	R. Diyala
Basra	96.0	1	Quran
Tikrit	96.0/97.9/98.2		Main
Najaf	96.5	5	Quran
Basra	98.1	1	Shahrazad
Baghdad	98.3	1	Quran
Nasiriya	99.0		Quran
Faluja	99.9	1	Main
Basra	100.0	5	Main
Kut	100.5	1	Main
Mosul	88.7/103.4		Main
Najaf	101.0	5	Provincial
Baghdad	103.3		Shahrazad
Amara	104.1		Quran
Baghdad	105.0		Al Jel
Amara	106.0		Main

Main Prgr (Republic of Iraq R.): 24h in Arabic on on MW and FM
except for Provincial programmes on some transmitters during the
day. **Quran prgr:** 24h.
Shahrazad R. (for women and children). **R. Al Jel** (for youth). **R.
Nineva:** daytime on Mosul trs. **R. Babil:** on 1071kHz and FM. **Ann:**
Main prgr: "Idha'at Jumhuriyah al-Iraq min Baghdad".

Other stations:

	MW	kHz	kW	Location	Station	H of tr
1)		594	20	Baghdad	R. Al-Nas	0400-1500
2)		610	0.5	Nasiriya	R. Kull al-Iraq	
3)		756		No. Iraq	Information R.	
10)		756	3	Basra	R. Dar as-Salam	0400-2100
8)		810	5/3	Baghdad	R. Om Al-Qura	0400-1830
	)	819	10	Basra	R. Al-Amal	
3)		846	unk. loc.		Information R.	
10)		882	5	Mosul	R. Dar as-Salam	0400-2100
41)		918		Baghdad	R. Shafaq	F.P.I.
18)		936	20	Basra	R. as-Safir	
5)		999	20	Baghdad	R. Bilad	0400-1600
6)		1008	20	Najaf	Sowt al-Fadhila	
50)		1017	10	Karbala	R. Karbala	F.P.I.
29)		1044		Basra	Vo the South	
7)		1053	3	Baghdad	R. As-Salam	0700-1700
10)		1116	20	Baghdad	R. Dar as-Salam	0400-2100
11)		1179	30	Baghdad	R. Voice of Iraq	0400-1800
10)		1197	1	Kirkuk	R. Dar as-Salam	0400-2100
12)		1206		Sulaimaniya	VO People of Kurdistan	03-06,17-20
13)		1341		Kirkuk	Voice of Komal	
4)		1395	0.5	Basra	R. Shanasheel	0400-2300
16)		1588	0.1	Baghdad	R. Shrara	0600-1000

	FM	MHz	kW	Location	Station	H of tr
48)		87.5		Kirkuk	R. Vision	
9)		87.8		Kirkuk	Vo Kurdistan	0300-2000
19)		88.0	2	Basra	BBC English	24h
20)		88.0		Sulaimaniya	R. Sawa	24h
21)		88.0	1	Baghdad/Mosul	DW/Monte-Carlo Doualiya	24h
43)		88.2		Baghdad	R. Dijla	0500-0100
43)		88.4			R. Dijla	0500-0100
22)		88.4	1	Sulaimaniya	R. Free Iraq/VOA	24h
23)		88.6	1	Baghdad	Panorama FM	24h
38)		88.6	1	Halabja	R. Dênge Nwe	0400-1600
10)		88.8		Kirkuk	R. Dar as-Salam	0500-1800
10)		89.0	2	Baghdad +2 stns	BBC Arabic	24h
39)		89.0		Kirkuk	R. Ashur	
47)		89.0		Amediye/Duhok	R. Duhok	
46)		89.1		Penjwin	R. Nawa (Kurdish)	
40)		89.2		Arbil	Sumer FM	24h

	FM	MHz	kW	Location	Station	H of tr
45)		89.3		Amara	Al-Mirbad R.	24h
46)		89.3		Halabja	R. Nawa (Kurdish)	
10)		89.4		Mosul	R. Dar as-Salam	0500-2100
18)		89.4		Basra	R. As-Safir	
21)		89.4		Basra	Monte-Carlo Doualiya	24h
46)		89.5		Kirkuk	R. Nawa (Kurdish)	
47)		89.5		Arbil	R. Duhok	
31)		89.6		Karbala/Al-Kut	Al-Huda Islamic R.	24h
46)		89.7		Kalar	R. Nawa, (Arabic)	
46)		89.9		Baghdad	R. Nawa (Kurdish)	
19)		90.0		Basra	BBC Arabic	24h
24)		90.0		Kirkuk	Turkoman FM	0510-2200
39)		90.0		Mosul	R, Ashur	
27)		90.3		Baghdad	R. Al-Noor	(inactive)
20)		90.4	1	Hillah	R. Sawa	24h
26)		90.4		Baghdad	R. Al-Yauwm	-1500
4)		90.6	0.2	Basra	R. Shanasheel	0400-2300
17)		90.6	0.1	Kirkuk	R. Lawani Kurdistan	
46)		90.6		Suleimaniya	R. Nawa, (Arabic)	
10)		91.0	1	Baghdad	R. Dar as-Salam	0500-2100
39)		91.1		Nineva	R. Ashur	
9)		91.4		Salah al Din	Vo Kurdistan	0300-2000
28)		91.5	5	Baghdad	R. Al-Rashid	0300-2300
9)		91.5		Erbil	VO Kurdistan	0300-2000
47)		91.5		Zakho	R. Duhok	
29)		91.6		Basra	Vo the South	
7)		92.0	0.3	Baghdad	R. As-Salam	0700-1700
46)		92.0		3 locations	R. Nawa (Kurdish)	
17)		92.0	0.1	Ranye	R. Lawani Kurdistan	
40)		92.1		Kirkuk	Sumer FM	24h
30)		92.3	1	Baghdad	AFN "Voice Channel"	24h
46)		92.4		Saidsadq	R. Nawa (Kurdish)	
19)		92.8	2	Kirkuk	BBC Arabic	24h
31)		92.8	1	Karbala	Al-Huda Islamic R.	24h
32)		92.8	0.6	Basra	Al-Nakhil R.	0300-2100
40)		92.8		Sulaimaniyah	Sumer FM	24h
43)		93.0			R. Dijla	0500-0100
46)		93.0		Kalar	R. Nawa (Kurdish)	
45)		93.3		Basra	Al-Mirbad R.	24h
30)		93.3	1	Qayara/Tikrit	AFN "Freedom R."	24h
9)		93.3		Dohuk	Vo Kurdistan	0300-2000
31)		89.6		Karbala/Al-Kut	Al-Huda Islamic R.	24h
43)		94.0			R. Dijla	0500-0100
16)		94.0		Dohuk	R. Shrara	0600-1000
8)		94.5	3	Baghdad	R. Om Al-Qura	0400-1830
46)		94.6		Kirkuk	R. Nawa, (Arabic)	
4)		94.6		Basra	R. Nahrain	
21)		95.0		Tikrit	Monte-Carlo Doualiya	24h
14)		95.5	10	Baghdad	R. Al-Mustaqbal	0500-1700
12)		95.5		Kirkuk	VO People Kurdistan	0500-2100
20)		95.7	10	So. Iraq	R. Sawa	24h
33)		96.0	5	Baghdad	R. Al-Mahaba	
19)		96.0	2	Erbil/Mosul	BBC Arabic	24h
20)		96.0	1	Tikrit	R. Sawa	24h
34)		96.1		Najaf	Al-Ghadeer R.	
35)		96.3		No. Iraq	Guven R.	
46)		96.4		Qaladze	R. Nawa (Kurdish)	
1)		96.6	5	Baghdad	R. Al-Nas	0400-1500
22)		96.9	1	Baghdad	R. Free Iraq	24h
19)		96.9	2	Baghdad	BBC English	24h
46)		97.1		Sara	R. Nawa (Kurdish)	
37)		97.3		Baghdad	Sowt al-Jam'ah	
46)		97.7			R. Nawa (Kurdish)	
12)		97.9		Baghdad	VO People Kurdistan	0500-2100
24)		98.0	0.1	Erbil	Turkoman FM	0510-2200
46)		98.0		Duhok	R. Nawa (Kurdish)	
46)		98.5		Zakho	R. Nawa (Kurdish)	
43)		98.8			R. Dijla	0500-0100
36)		98.8	5	Baghdad	Ur FM	24h
20)		98.8	1	Kirkuk	R. Sawa	24h
25)		99.1		Karbala	Karbala FM	0300-1500
15)		99.3	3	Bahrez	Ind. RTV Netw.	0500-2100
39)		99.4		Baghdad	R. Ashur	0600-1700
40)		99.8	5	Baghdad	Sumer FM	24h
19)		100.0	2	Nasiriya	BBC Arabic	24h
30)		100.1	0.2	Tikrit	AFN "Voice Channel"	24h
20)		100.4	10	Baghdad	R. Sawa	24h
4)		100.4		Basra	R. Nahrain	
42)		101.0		Baghdad	R. Al-Ahd	
45)		101.4		Nasiriya	Al-Mirbad R.	24h
17)		101.5		Baghdad	VO Iraqi National Congress	
41)		102.0	3	Baghdad	R. Shafaq	0500-1700

FM	MHz	kW	Location	Station	H of tr
22)	102.4	10	Baghdad	R. Free Iraq/VOA	24h
21)	103.0		Erbil	Monte-Carlo Doualiya	24h
20)	103.2	1	Erbil	R. Sawa	24h
49)	103.3		Kirkuk	R. Justice	
12)	104.0		Kirkuk	R. Kirkuk	
30)	104.1		Baghdad	AFN "Voice Channel"	
22)	104.5	1	Erbil	R. Free Iraq/VOA	24h
22)	104.6	5	Mosul	R. Free Iraq/VOA	24h
22)	105.0	10	Basra	R. Free Iraq/VOA	24h
30)	105.1	1	Mosul	AFN "Freedom R."	24h
20)	105.8	1	Sulaimaniya	R. Sawa	24h
44)	106.0	1	Baghdad	As-Salam 106 FM	24h
20)	106.6	5	Mosul	R. Sawa	24h
20)	107.0	10	Basra	R. Sawa	24h
30)	107.1		Taqaddum	AFN	
30)	107.3	0.25	Balad/Kirkuk	AFN "Freedom R."	24h
30)	107.7	1	Baghdad	AFN "Freedom R."	24h
50)	107.7		Zakho	R. Hizal	
30)	107.9	0.25	Sinjar	AFN "Freedom R."	24h

Addresses and other information:

1) R. Al-Nas ("People"). Web: www.radioannas.com Email: radioal-nas@yahoo.com – **2)** R. Kull al-Iraq ("All of Iraq") – **3)** Operated by the US forces – **4)** R. Shanasheel ("Balcony"), Basra – **5)** R. Bilad ("Lands"). Operated by the Islamic Virtue Party. Web: albilad.org Email: albilad@albilad.org – **6)** Sowt al- Fadhila ("Voice of Virtue"), Najaf – **7)** R. As-Salam ("Peace")Email: safa565@hotmail.com Alt. freq. 1030/1035kHz – **8)** Web: heyetnet.org – **9)** Operated by the Kurdistan Democratic Party Web: kurdistanradio.net Email: info@kurdistanradio.net Prgrs in Sorani Kurdish/Arabic – **10)** R. Dar As-Salam ("Haven of Peace"), The Voice of the Iraqi Islamic Party. Web: www.darusalam.org Email: darusalam@darusalam.net – **11)** Operated by Imam Al-Shirazi International Association. Web: www.voiraq.com Email: voiceiraq@yahoo.com Prgrs in Arabic/English/Turkmen – **12)** Operated by the Patriotic Union of Kurdistan Web: www.puk.org Email: puk@puk.org . In Sorani Kurdish/Arabic – **13)** Operated by the Kurdistan Islamic Group. Web: www.komall.org Email: islamicjamaa@maktoob.com prgrs in Arabic/Kurdish/Turkish – **14)** R. Al-Mustaqbal ("The Future"). Operated by the Iraqi National Accord Movement. Web: www.wifaq.com – **15)** Web: www.irtniraq.com Email: kahoofy2005@yahoo.com – **16)** Operated by the Assyrian Patriotic Party. Web: www.atranaya.org Email: app@atranaya.org. In Assyrian/Arabic – **17)** Kurdistan Youth R. Web: www.mosy-krg.org – **18)** Basra. Web: aliraqnews.com – **19)** BBC Arabic Service Email: arabicservice@bbc.co.uk – **20)** R. Sawa ("Together"). Also on MW via Kuwait 1548kHz 600kW 24h. For more details see International Radio section (USA) – **21)** R. France Internationale & Monte-Carlo Doualiya. prgrs in Arabic/French. For details see International radio section (France). **22)** Iraqi Sce. of R. Free Europe/R. Liberty. Also on MW via Kuwait 1593kHz 150kW 1400-0700. In Arabic incl. VOA in Kurdish/English. For more details see International Radio section (USA) – **23)** See MBC entry under UAE – **24)** Web: www.kerkuk.net Prgrs in Turkoman/Arabic – **25)** Shammasyia St. 29, Quarter 318, Line 55, House 31, Adhadmyia, Baghdad – **26)** R. Al-Yauwm ("Today Radio") – **27)** R. Al-Noor ("Light"). Email: alnoor903fm@yahoo.com – **28)** Web: www.aljanabigroup.ae Email: alrasheedfm@yahoo.com – **29)** Web: www.almannarah.com Email: almannarah@almannarah.com – **30)** American Forces Network Iraq. Web: www.afniraq.army.mil Email: freedomradio@iraq.centcom.mil – **31)** Web: radio.alhodaonline.com Email: alhodaonline@gmail.com – **32)** Operated by the Islamic Supreme Council of Iraq. Web: www.almejlis.org Email: info@almejlis.org – **33)** R. Al-Mahaba ("Friendship"), Voice of Iraqi Women. Supported by the United Nations Development Fund for Women (UNIFEM). Web: www.okiinc.org/vow_radio.html – **34)** Al-Ghadeer R, Najaf – **35)** Operated by the Turkish army – **36)** Web: www.ur.fm Email: info@ur.fm F.PI: txs in Mosul, Sulaimaniya, Arbil, Kirkuk and Basra – **37)** Sawt al-Jam'ah ("Voice of the University") – **38)** R. Dênge Nwe ("New Voice"). Web: www.wadinet.de/projekte/newiraq/radio/index_en.htm Email: radiodangenwe@yahoo.com – **39)** Operated by the Assyrian Democratic Movement (ADM/ZOWAA) Web: www.zowaa.org Email: info@zowaa.org .In Assyrian/Arabic. – **40)** Web: www.sumerfm.com Email: sumerfm@sumerfm.com . – **41)** R. Shafaq ("Twilight"). Web: www.shafaaq.com . In Kurdish/Arabic – **42)** R. Al-Ahd (Oath), Baghdad – **43)** R. Dijla (Tigris). Web: www.radiodijla.com . Also r. Deutsche Welle – **44)** As-Salam (Peace) 106 FM. Web: peace106fm.com E-mail: peace106fm@yahoo.com – **45)** Web: almirbad.com E-mail: info@almirbad.com – **46)** Web: www.radionawa.com Email: info@radionawa.com – **47)** Web: duhokradio.org Email: info@duhokradio.net – **48)** Operated by the Iraqi Turkmen Brotherhood Party – **49)** Operated by the Iraqi Turkmen Justice Party – **50)** Zakho

IRELAND

LT: UTC (27 Mar-29 Oct: +1h) — **Pop:** 4.5 million — **Pr.L:** Irish Gaelic, English — **E.C:** 50Hz, 230V — **ITU:** IRL

RAIDIÓ TEILIFÍS EIREANN (Statutory Corporation)

Donnybrook, Dublin 4 ☎ +353 1 208 3111 ▯ +353 1 208 3080
E: info@rte.ie **W:** www.rte.ie
LP: DG: Cathal Goan; Ch.Financial Offr.: Conor Hayes. MD TV: Glen Killane. MD Radio: Clare Duignan; MD News: Ed Mulhall. Dir Corp. Dev.: Brian Dalton
Raidió Na Gaeltachta: Casla, Conamara, Co Galway ☎ +353 91 506677 ▯ +353 91 506666 **E:** rnag@rte.ie **W:** www.rte.ie/rnag
Lyric FM: Cornmarket Square, Limerick ☎ +353 61 207300 ▯ +353 61 207390 **E:** lyric@rte.ie **W:** www.rte.ie/lyricfm **Pub.:** RTE Guide
Networks: 1=R1, 2=2FM, 3=Raidió Na Gaeltachta, 4=Lyric FM

LW	kHz	kW	N			
Summerhill	252	300	1			
FM (MHz)	**1**	**2**	**3**	**4**		**kW**
Achill	89.9	92.1	94.3	99.5		2
Aranmore	89.6	91.8	94.0	99.2		3
Ballybofey	89.7	91.9	94.1	99.3		1
Bantry	88.7	90.9	93.1	98.3		1
Cahirciveen	89.5	91.7	93.9	99.1		2
Cairn HI (Longford)	89.8					16
Casla	88.4	90.6	92.8	98.0		2
Castlebar	89.3	91.5	93.7	98.9		3
Castletownbere	88.3	90.5	92.7	97.9		1
Clermont Carn	87.8	97.0	102.7	95.2		40
Clifden	89.5	91.7	93.9	99.1		3
Clonmel	88.3	90.5	92.7	97.9		1
Cnoc an Oir	89.2	91.4	93.6	98.7		1
Cork (Spur Hill)	89.2	91.4	93.6	98.8		5
Crosshaven	88.2	90.4	92.6	97.8		3
Dungarvan	88.5	90.7	92.9	98.1		3
Fanad	89.8	92.0	94.2	99.4		4
Greystones	89.5	91.7	93.9	99.1		1
Holywell Hill	89.2	91.4	93.6	98.8		2
Kippure	89.1	91.3	93.5	98.7		40
Knockmoyle	88.4	90.6	92.8	98.0		1
Limerick City	89.4	91.6	93.8	99.0		2.5
Maghera	88.8	91.0	93.2	98.4		160
Malin	89.9	91.1	93.3	98.5		1
Monaghan	88.9	91.1	93.3	98.5		3
Moville	88.3	90.5	92.7	97.9		10
Mt. Leinster	89.6	91.8	94.0	99.2		100
Mullaghanish	90.0	92.2	94.4	99.6		160
Suir Valley	89.0	91.2	93.4	98.6		3
Three Rock	88.5	90.7	92.9	96.7		10
Truskmore	88.2	90.4	92.6	97.8		125

+ 9 relays below 1kW

1) RTE Radio 1: 24h in English & Irish on LW, FM, satellite and internet. **N. in English:** on the h **N. in Irish Gaelic:** 2150 – **2)** 2FM: 24h in English and Irish on FM, satellite and internet. – **3)** Raidió Na Gaeltachta: 24h in Irish Gaelic on FM, satellite and internet – **4)** Lyric FM: 24h on FM, satellite and internet.

EXTERNAL SERVICE (RTE Radio Worldwide): see Int. section

DIGITAL RADIO (DAB): DAB trs are on Band 3 operated by RTE. Two multiplexes. **RTE national multiplex** Block 12C 227.360 MHz trs. in Dublin, NE Ireland, Cork, Limerick carrying 10 RTE services (R1, 2FM, Lyric FM, R Na Gaeltachta, 2XM, Chill, Choice, Junior, Gold, R1 Extra, Pulse).
Independent multiplex: Trial in SE Ireland Block 9B commenced April 2010 with simulcast of several commercial stns; Block 12A 223.936 MHz trs in Dublin & NE Ireland ceased operation Nov 2008. Further information **W:** www.digitalradio.ie

BROADCASTING AUTHORITY OF IRELAND (BAI)

2-5 Warrington Place, Dublin 2 ☎ +353 1 644 1200 ▯ +353 1 644 1299 **E:** info@bai.ie **W:** www.bci.ie **LP:** Chief Exec: Michael O'Keeffe. Dep Ch Exec & Dir Broadcasting: Celene Craig.
Responsible for regulation of commercial broadcasting in the Irish Republic. Full list of licensed stns can be found on BCI website

TODAY FM (Comm.)

Marconi House, Digges Lane, Dublin 2 ☎ +353 1 804 9000 **W:** www.todayfm.com

FM	MHz	kW	FM	MHz	kW
Crosshaven	100.0	6	Clonmel	100.1	2
Truskmore	100.0	250	Knockmoyle	100.2	2

FM	MHz	kW	FM	MHz	kW
Moville	100.1	2	Woodcock Hill	101.2	5
Dungarvan	100.3	6	Greystones	101.3	2
Maghera	100.6	320	Clifden	101.3	6
Monaghan	100.7	5	Kilkeaveragh	101.3	6
Suir Valley	100.8	6	Mt. Leinster	101.4	400
Kippure	100.9	100	Fanad	101.6	8
Holywell Hill	101.0	12	Achill	101.7	6
Spur Hill, Cork	101.0	10	Mullaghanish	101.8	320
Knockanore	101.0	2	Three Rock	101.8	2
Castlebar	101.1	6	Clermont Carn	105.5	80

D.Prgr:24h **N:** on the h, also on the half h at peak times.

NEWSTALK (Comm.)

✉ Marconi House, Digges Lane, Dublin 2 ☎ + 353 1 644 5100 📠 + 353 1 644 5101 **W:** www.newstalk.ie
LP: CE Frank Cronin.

FM	MHz	kW	FM	MHz	kW
Monaghan	103.3	2.5	Mullaghanish	107.4	80
Capard	105.8	4	Truskmore	107.4	80
Three Rock	106.0	10	Mohercrom	107.4	10
Holywell Hill	106.9	12	Waterford	107.2	2
Longford	106.9	5	Maghera	107.6	32
Limerick City	107.0	2	Saggart	107.6	2
Nagles	107.0	0.5	Dungarvan	107.6	5
Ballyguile	107.0	2	Cork City	107.8	10
Kilitimagh	107.2	6	Kilduff	107.8	2
Mt Leinster	107.2	2	Gorey	107.8	1
Knockmoyle	107.2	4			

Local Stations:

	FM MHz	kW	Station, tx location
31)	87.8	1	Connemara Community R
35)	94.6	1	4FM, Saggart
32)	94.7	5	Spin South West, Clifden
18)	94.8	4	Northern Sound, Slieve Glah
35)	94.8	3	4FM, Churchfield (Mallow)
1)	94.9	9	East Coast FM, Avoca
35)	94.9	4	4FM, Three Rock
2)	95.0	10	Limerick's Live 95 FM, Woodcock Hill
16)	95.1	10	WLR FM, Faha, Dungarvan
10)	95.2	2	Highland R, Aran Mor
35)	95.4	9	4FM, Nowen Hill
7)	95.5	2	Clare FM, Kilrush
17)	95.6	1	Cork's 96 FM, Kilworth,NE Cork
3)	95.6	4	South East R, Mt.Leinster
4)	95.8	10	LM FM, Mt. Oriel
17)	95.8	10	Cork's 96 FM, Nowen Hill
7)	95.9	2	Clare FM, Woodcock Hill
8)	96.0	1	KCLR, Corbally Wood
5)	96.1	10	MWR FM,Kiltimagh
17)	96.1	1	Cork's 96 FM, Mount Hillary
20)	96.2	6	R. Kerry, Cahirciveen
1)	96.2	10	East Coast FM, Bray
18)	96.3	10	Northern Sound, Monaghan
7)	96.4	10	Clare FM, Maghera
17)	96.4	2	Cork's 96 FM, Holly Hill
8)	96.6	10	KCLR , Johns Well
9)	96.8	10	Galway Bay FM, Knockroe
8)	96.9	4	KCLR, Rossmore
33)	96.9	9	i102-104, Scalp Mountain
20)	97.0	40	R. Kerry, Mullaghanish
11)	97.1	10	Tipp FM, Scrouthea
5)	97.1	3	MWR FM, Achill
12)	97.3	5	KFM, Rossmore
35)	97.4	4	4FM, Bweeng Mountain
9)	97.4	2	Galway Bay FM, Redmount Hl
16)	97.5	10	WLR FM, East Waterford
20)	97.6	2	R. Kerry, Knockanore
12)	97.6	4	KFM, Slieve Thuile
13)	98.1	9	98 FM, Three Rock
1)	99.9	9	East Coast FM, Saggart Hill
36)	100.3	11	R. Nova, Three Rock
24)	102.0	13	Beat 102-103 FM, Mount Leinster
33)	102.1	9	i102-104, South Galway
10)	102.1	1.3	Highland R., Feirn Hill
24)	102.2	6	Beat 102-103 FM, West Waterford
26)	102.2	5	Q 102, Three Rock
32)	102.3	1	Spin South West, Ennistymon
24)	102.4	10	Beat 102-103 FM, Clonmel
21)	102.5	4	Ocean FM, Truskmore
32	102.5	5	Spin South West, Knockmoyle
17)	102.6	2	C103, Cork City
32	102.7	9	Spin South West, Maghera
24)	102.8	10	Beat 102-103 FM, East Waterford

	FM	MHz	kW	Station, tx location
1)		102.9	16	East Coast FM, Ballyguille
1)		102.9	2	East Coast FM, Baltinglass
17)		102.9	1	C103, NE Cork
32)		102.9	9	Spin South West, Cahirciveen
32)		103.0	2.5	Spin South West, Woodcock Hill
33)		103.1	9	i102-104, Longford
33)		103.1	4	i102-104, Achill
33)		103.1	1	i102-104, Senafaistin
22)		103.2	0.5	Dublin City FM, Three Rock
10)		103.3	10	Highland R, Scalp Mountain
17)		103.3	10	C103, Nowen Hill
33)		103.3	4	i102-104, Clifden
6)		103.5	2.6	Midlands Radio 3, Sliabh Bloom
32)		103.5	1	Spin South West, Knockanore
17)		103.7	5	C103, Mt. Hillary
33)		103.7	9	i102-104, Castlebar
25)		103.8	5	Spin 103.8, Three Rock
11)		103.9	3.2	Tipp FM, Kilduff
33)		104.0	2	i102-104, Aranmore
14)		104.1	5	Shannonside 104FM, Sliabh Bawn
35)		104.2	9	4FM, Limerick
15)		104.4	10	FM 104, Three Rock
33)		104.4	9	i102-104, Sligo (Truskmore)
28)		104.5	10	Red FM, WCork (Nowen Hill)
10)		104.5	2	Highland R, Back Mountain
35)		104.6	9	4FM, Maghera,Co Clare
34)		104.7	1	i105-107, Saggart
19)		104.8	10	Tipperary Mid-West R, Dangandargan
34)		104.8	2.5	i105-107, Cavan
35)		104.9	9	4FM, Galway City
21)		105.0	5	Ocean FM, Mt.Charles
34)		105.0	5	i105-107, Mt Oriel (Louth)
30)		105.0	4	Inishowen Community R, Malin
28)		105.0	1	Red FM, Newmarket
29)		105.2	4	Phantom FM, Three Rock
28)		105.7	5	Red FM, North Cork (Nagles)
28)		106.1	2	Red FM, Cork City
34)		106.2	5	i105-107, Capard
23)		106.4	2	Raidió Na Life, Three Rock
34)		106.7	10	i105-107, Monaghan
27)		106.8	4	Dublin's Sunshine 106.8, Three Rock

+ approx 110 additional txs of less than 1kW

Addresses and other information:

1) Radio Centre, Killarney Rd, Bray, Co Wicklow **E:** mail@eastcoast. fm **W:** www.eastcoast.fm – **2)** Radio House, Dock Rd, Limerick **E:** mail@live95fm.ie **W:** www.live95fm.ie – **3)** Custom House Quay, Wexford Town **E:** info@southeastradio.ie **W:** www.southeastradio. ie – **4)** Broadcasting House, Rathmullen Rd, Drogheda, Co Louth **E:** info@lmfm.ie **W:** www.lmfm.ie – **5)** Clare Str, Ballyhaunis, Co Mayo **E:** info@midwestradio.ie **W:** www.midwestradio.ie – **6)** Tindle House, Axis Business Park, Tullamore, Co Offaly **E:** info@midlandsradio.fm **W:** www.midlandsradio.fm – **7)** Abbeyfield Centre, Francis Str, Ennis, Co Clare **E:** info@clarefm.ie **W:** www.clarefm.ie – **8)** The Broadcast Centre,Carlow Rd, Kilkenny **E:** info@kclr96fm.com **W:** www.kclr96fm. com – **9)** Unit 13, Sandy Rd, Galway **E:** info@galwaybayfm.ie **W:** www. galwaybayfm.ie – **10)** Pine Hill, Letterkenny, Co Donegal **E:** enquries@ highlandradio.com **W:** www.highlandradio.com – **11)** Broadcast Centre, 4A Gurtnafleur Business Park, Clonmel, Co Tipperary **E:** sales@tippfm. com **W:** www.tippfm.com – **12)** KFM Broadcast Centre, M7 Business Park,Newhall, Naas, Co Kildare **E:** info@kfmradio.com **W:** www. kfmradio.com – **13)** South Block, The Malt House, Grand Canal Quay, Dublin 2 **E:** website@98fm.ie **W:** www.98fm.ie – **14)** Unit 1E Master Tech Business Park, Athlone Rd, Longford **E:** info@shannonside.ie **W:** www.shannonside.ie – **15)** Macken House, Mayor Str Upper, Dublin 1 **E:** sales@fm104.ie **W:** www.fm104.ie – **16)** Broadcast Centre, Ardkeen, Dunmore Rd, Waterford **E:** reception@wlrfm.com **W:** www.wlrfm.com – **17)** Broadcasting House, Patrick's Place, Cork **E:** info@96fm.ie, info@ C103.ie **W:** www.96fm.ie www.c103.ie – **18)** Unit 3 Milltown Business Park, Monaghan & Thomas Ashe St.,Cavan **E:** info@norththernsound.ie **W:** www.northernsound.ie – **19)** St Michael's St, Tipperary **E:** tippmid-westradio@eircom.net **W:** www.tippmidwestradio.com – **20)** Maine St, Tralee, Co Kerry **E:** info@radiokerry.ie **W:** www.radiokerry.ie – **21)** Ocean FM Broadcasting Centre, North West Business Park, Collooney, Co Sligo **E:** sales@oceanfm.ie **W:** www.oceanfm.ie – **22)** Docklands Innovation Park, Unit 6, 128-130 Eastwall Rd, Dublin 3 **E:** info@dublincityfm.com **W:** www.dublincityfm.com – **23)** 7 Merrion Square, Dublin 2 **E:** eolas@raidi-onalife.ie **W:** www.raidionalife.ie (Irish language stn) – **24)** Broadcast Centre, Ardkeen, Dunmore Rd, Waterford **E:** info@beat102103.com **W:** www.beat102103.com – **25)** Level 3, South Block, Malt House, Grand Canal Quay, Dublin 2 **E:** info@spin1038.com **W:** www.spin1038. com – **26)** Macken House, 39-40 Upper Mayor Str, Dublin 1 **E:** info@

q102..ie **W:** www.q102.ie – **27)** Radio Centre, Killarney Rd, Bray, Co Wicklow **E:** mail@countrymix.ie **W:** www.countrymix.ie – **28)** 1 University Technology Centre, Bishoptown, Cork **E:** info@redfm.ie **W:** www.redfm.ie – **29)** 73 North Wall Quay, Dublin 1 **E:** info@phantom.ie **W:** www.phantom.ie – **30)** Pound Str., Carndonagh, Inishowen, Co Donegal **E:** studio@icrfm.ie **W:** www.icrfm.ie – **31)** Connemara West Centre, Letterfrack, Co Galway **E:** info@connemarafm.com **W:** www.connemarafm.com – **32)** 2nd Floor Landmark Building, Raheen, Limerick **E:** info@spinsouthwest.com **W:** www.spinsouthwest.com – **33)** 19 Wellpark Leisure Block, Galway **E:** info@iradio.ie **W:** www.i102104.ie. – **34)** iRadio, NEM Unit 3, Monksland Business Park, Athlone. **E:** info@iradio.ie **W:** www.i105107.ie – **35)** Latin Hall, Golden Lane, Dublin 8. **E:** info@4fm.ie **W:** www.4fm.ie – **36)** 1ˢᵗ Floor Castleforbes House, Castleforbes Rd, Dublin 1. **E:** info@nova.ie **W:** www.nova.ie.

Community/special interest stns: 21 stns in operation at October 2010. **Hospital/Institutions:** 5 stns. **Temporary/Special Event services:** see BCI **W:** www.bai.ie. **Wireless Public Address System (WPAS):** religious and other sces broadc. to housebound via CB radio 26.7–27.99MHz

ISRAEL

LT: UTC +2h (1 Apr - 2 Oct: +3h) — **Pop:** 7.6 million — **Pr.L:** Hebrew, Arabic — **E.C:** 50Hz, 230V — **ITU:** ISR

ISRAEL BROADCASTING AUTHORITY (IBA)
⌨ 161 Jaffa Road, P.O. Box 28080, Jerusalem 91280 ☎+972 2 501 5555 📠 +972 2 501 5504 **W:** iba.org.il **E:** webmaster@iba.org.il
LP: Chairman: Amir Gilat. DG: Moti Sklar. Transmission Mgr: David Gombosh.

KOL ISRAEL – THE VOICE OF ISRAEL (Pub.)
⌨ Heleni Hamalka 21, P.O. Box 1082, Jerusalem 91010 ☎1-599-509-510 **W:** iba.org.il/kolisrael **E:** reception@iba.org.il
LP: Dir. & PD: Yonatan Ben-Menachem. Dir. of Eng: Efraim Porat. Liaison & Coordination: Raphael Kochanowski.

MW	kHz	kW	Prgr	MW	kHz	kW	Prgr.
Yavne	531	100	A	Yavne	1080	10	D
Yavne	657	200	B	Acre	1206	50	D
She'ar-Yeshuv	882	10	B	She'ar-Yeshuv	1458	10	A
Eilat	927	10	B	Eilat	1458	10	A
Acre	927	50	B				

FM (MHz)	A	B	C	D	M	X	R
Menara	104.8	100.5	97.7		95.7		106.8
Safed	105.1	92.0	97.8	88.8	98.5	87.6	101.8
Haifa	104.8	89.5	105.5	88.8	100.2	88.0	100.5
		94.5		92.4			
				99.3			
Kohav Hayarden		95.0	88.9		97.2		100.7
Netanya		95.2		93.7			100.5
Tel Aviv	104.8	95.0	89.7		97.2	88.0	101.2
Central area	105.3	95.2	97.5	88.8		87.6	
				93.7			
Jerusalem	104.8	95.0		99.3		88.0	100.3
Eitanim	105.1	95.5	97.8		91.3	87.6	101.3
Kalya		95.5	97.8	90.3	105.0		
Beersheba	100.7	103.3	106.9	93.3	90.2	88.0	107.3
		94.4					
Sha'ar HaNegev	104.8	106.2		92.4			
Merkaz Sapir		104.1	89.5				
Mitzpe Ramon		95.5	104.3		97.2		
Arava		95.5	98.9		92.4		
Eilat		90.7	100.5		97.0		

Prgrs (in Hebrew if not mentioned otherwise):
A: "Reshet Alef": Talk & cultural programming 24h excl. times listed in Reshet Moreshet below. **N.** in Hebrew: rel. Prgr. B. – **B: "Reshet Bet":** 24h. News, current affairs & sports. **N:** on the h. – **C: "Reshet Gimmel":** 24h Israeli popular music. **N:** rel. Prgr. B. – **D: "Reshet Dalet":** (Arabic). 24h. – **X: "88 FM":** 24h. **N:** rel. Prgr. B. Light music, traffic reports – **R. REQA (Reshet Qlitat Aliya):** immigrants network, 24h in Russian, Amharic, French, Yiddish, Ladino, Romanian, Spanish, Moghrabi, Bukharian, Georgian, Hungarian and English 0430-0445, 1030-1045 & 1830-1845 – **M: VO Music. "Kol Ha Musica":** 24h, classical music and drama. – **"Reshet Moreshet"** (Heritage Network). Religious programming on Reshet Alef. Sun-Thurs 1400-2200, Fri 0600-1500, Sat 1900-2200 – **Local education prgrs "Tachana Chinuchit"** are transmitted on 106.0MHz in Holon, Tiberias, Beer Sheva, Haifa, Bet El and more LP trs in colleges around Israel. Complete station listing in Hebrew at Web: eduradio.iba.org.il – **External service:** see International Radio section.

GALEI TZAHAL (Israel Defence Forces R, Mil.)
⌨ Military Post Office Box 01005. ☎+972 3 5126666

W: www.glz.co.il **E:** glz@galatz.co.il
L.P: Commander: Yitzchak Tunik.

MW	kHz	kW	MW	kHz	kW
Yavne	945	50*	Rosh-Pina	1305	50
Beersheba	1224	20	Shivta	1368	20

*F.PI: 100kW.

FM (MHz)	Main	kW	GalGalatz	kW
Kiryat Shmona	-	-	104.1	1
Nebi Yesha	93.9	5	-	-
Bet Shean	99.8	5	140.0	0.3
Haifa	102.3	10	107.0	10
Jerusalem	96.6	10	93.9	0.5
Tel Aviv	104.0	3.5	91.8	10
Beer Sheva	102.3	10	99.8	10
N. Dead Sea	96.9	0.25	93.6	0.25
Mitzpe Ramon	104.0	1	106.4	1
Eilat	104.0	0.25	106.4	0.25

Main Prgr on MW&FM: 24h. (news, talk show, music). **N:** on the h.
GalGalatz on FM(traffic reports and music): 24h.
Ann: Main Prgr: "Galei Tzahal, Shidure Tsva Hagana Le'Yisrael".
Relays on shortwave for Europe: see International Radio section.

Regional commercial FM radio (All in Hebrew except as noted):
Radio A´shams (The Sun) in Arabic at Nazareth-Ein Hahoresh area. 98.1 & 101.1MHz. Web: ashams.com – **R. Darom** (Southern R.): Beersheba 97.0MHz, Kiryat Gat, Ashkelon. Aravah & Dead Sea settlements: 95.8MHz. Web: dromi.co.il/radio/index.asp **R. Darom** (Southern R.) :101.5MHz. Web: dromi.co.il/1015/index.asp – **R.Emtza Ha Derech**, (Middle of the Road), Tel Aviv: 90.0, 94.7MHz. Web: 90fm.co.il – **R.Haifa**, Haifa: 99.5, 107.5MHz. Web: 1075.fm – **R. Kol Barama**, Haifa 92.1MHz, Beersheva 104.3MHz, Jerusalem 105.7MHz. Web: kol-barama.co.il – **R. Kol Chai**: Bene Brak 92.8, Jerusalem 93.0MHz. Web: www.93fm.co.il – **R. Kol Rega**: Galilee 96.0MHz, Tiberias 91.5MHz. Web: 96fm.co.il – **R. Lev Hamedena**: Shfela 91.0MHz, Beersheba 93.3MHz. Web: 91fm.co.il – **Kol Ha Yam Ha Adom** (VO the Red Sea): 101.1, 102.0MHz 1kW. Web: fm102.tapuz.co.il – **R. Jerusalem:** Jerusalem 101.0MHz, Bet Shemesh 89.5MHz. Web: 101fm.tapuz.co.il – **R. L'Lo Hafsaka** (Nonstop): Upper Galilee 101.5MHz, Ramat Gan 103.0MHz, Lower Galilee 104.5MHz. Web: 103.fm – **Pervoye R,** Rishon Le'Zion. In Russian: Ashdod 89.1MHz. Web: www.891fm.co.il – **Radius 100 FM**, Tel Aviv: 100.0MHz. Web: 100fm.co.il – **R. Tel Aviv**: 102.0MHz. Web: 102fm.co.il – **ECO99fm**, Hertzliyah: 99.0MHz. Web: eco99.fm – **Galey Israel**: south 102.5MHz, Dan area 106.5 MHz, Benjamin area 89.3MHz. Web: galeyisrael.co.il

WEST BANK & GAZA STRIP
(Palestinian territories)
LT: UTC +2h (26 Mar-2 Sep: +3h), subject to change — **Pop:** 4 million — **Pr.L:** Arabic — **E.C:** 50Hz, 230V — **ITU:** XWB (West Bank), XGZ (Gaza Strip)

PALESTINIAN BROADCASTING CORPORATION (Gov)
⌨ P.O. Box 984, Al-Bireh, Ramallah, West Bank ☎+970 2 2988888 📠 +970 2 2959891 **W:** www.pbc.gov.ps **E:** pbcinfo@pbc.gov.ps
L.P: Chmn: Radwan Abu Ayyash.
FM: Ramallah 90.7MHz, Gaza 99.4MHz, Jenin 102.2MHz.
D.Prgr. in Arabic: 0400-2300 **Ann:** "Sawt Filastin".

Private FM stations:
Al-Balad FM, Jenin: 104.8/105.8MHz. Web: albaladfm.com – **Al-Manar R,** Gaza: 92.0MHz. Web: manarfm.com – **Al-Qamar R,** Jericho: 89.4MHz. Web: maannet.org – **Ajyal R:** Ramallah 103.4MHz, Hebron 107.1MHz. Web: radioajyal.com – **Amwaj R.,** Ramallah 91.5MHz, south 99.4MHz, north 104.8MHz. Web: amwaj.ps – **Angham R.,** Ramallah: 92.3MHz. Web: radioangham.com – **Cool FM**: 100.3MHz, all English. Web: coolfm.ps – **Gaza FM**: 100.9MHz. Web: gazafm.net – **Hebron R:** 90.4MHz. Web: hebronradio.com – **Holy Qur'an R,** Jerusalem 88.4MHz, Nablus 96.0MHz. Web: quran-radio.com – **Iman R,** Gaza: 96.2MHz. Web: imanradio.com – **Kul Al-Nas R:** Tulkarem: 107.3MHz – **Najah FM,** Nablus: 88.4MHz. Web: najah.edu/fm – **Quds R,** Gaza: 102.7MHz. Web: www.qudsradio.ps – **R. Al-Shabab,** Gaza: 105.0MHz. Web: 105fm.ps – **R. Al-Shamal,** Qalqiliya: 96.6MHz. Web: alshamal.net – **R. All for Peace,** Ramallah: 107.2MHz (Hebrew), 89.8MHz (Arabic). Web: allforpeace.org Also rel. VOR R. Japan and Polish R. – **R. Bethlehem 2000,** Bethlehem: 89.6/106.4MHz 5kW. Web: www.radiobethlehem2000.net Also rel. BBC&DW – **R. Isis,** Bethlehem: 87.5MHz. Web: radioisis.net – **R. Manbar Al-Hurriya,** Hebron: 92.7MHz – **R. Marah,** Hebron: 100.4MHz. Web: marah-fm.ps – **R. Mawwal,** Bethlehem: 101.7MHz. Web: mawwal.ps – **R. Nablus:** 93.5MHz. Web: radionablus.com – **R. Nagham,** Qalqiliya: 99.6MHz. Web: radionagham.com – **R. Nisaa,** Ramallah: 96.0MHz. Web: radionisaa.net – **R. Tariq al-Mahabeh,** Nablus: 97.7/108.0MHz. Web: tmfm.net – **Sawt al-Aqsa,** Gaza: 106.7MHz. Web: alaqsavoice.ps – **Sawt**

al-Sha'ab, Gaza: 106.0MHz. Web: sha3bvoice.com – **Sawt el-Ghad,** Salfit: 105.0MHz. Web: sawtelghad.ps – **Sirajj R,** Hebron: 105.7MHz. Web: sirajjfm.ps – **VO Love and Peace,** Ramallah: freq. not known Web: www.volpfm.com – **Monte-Carlo Doualiya:** Ramallah 94.6MHz, Nablus 97.3MHz, Hebron 99.7MHz – **R. Sawa:** Bethlehem: 94.2MHz.

ITALY

L.T: UTC +1h (27 Mar-30 Oct: +2h) — **Pop:** 58 million — **Pr.L:** Italian — **E.C:** 50Hz, 220V — **ITU:** I

RAI-RADIOTELEVISIONE ITALIANA (Pub.)

☑ Viale Mazzini 14,IT- 00195 Roma ☎ +39 06 38781 🖷 +39 06 3622621 ☎ (Listeners) Centro Corrispondenza, C.P. 320, IT-00100 Roma ☎ +39 06 3317 2591 🖷 +39 06 3317 1895 **E:** service@rai. it **W:** www.rai.it Tech. Dept: Rai Teche: Via Cernaia 33, IT-10121 Torino, Dir.Barbara Scaramucci **W:** www.teche.rai.it **E:** teche@rai.it Rai Way:Centro Ascolto e Qualità Controllo Servizio RAI Monza, Via Parco Mirabellino 1, IT-20052 Monza. Pres.:: Francesco Di Domenico **W:** www.raiway.rai.it **E:** raiway@rai.it **V:** QSL-card. No Rp. Sedi Regionali: **W:** www.sediregionali.rai.it **E:** sedi.regionali@rai.it
L.P: Pres.: Paolo Garimberti GM: Mauro Masi , Dir.Reg.Radio: , Alberto Maccari Dir.Rai International: Daniele Renzoni .

Regional Centres: 1 **Abruzzo:** Viale de Amicis 27, IT-65123 Pescara – 2 **Alto Adige:** Piazza Mazzini 23, IT-39100 Bolzano/Bozen – 3 **Basilicata:** Via dell'Edilizia 2, IT-85100 Potenza – 4 **Calabria:** Viale G. Marconi 1,IT- 87100 Cosenza – 5 **Campania:** Via Marconi 11,IT-80125 Napoli – 6 **Emilia-Romagna:** Viale della Fiera 13, IT-40127 Bologna – 7 **Friuli-Venezia-Giulia:** Via Fabio Severo 7, IT-34133 Trieste – 8 **Lazio:** Largo Willy de Luca 4, IT-00188 Roma – 9 **Liguria:** Corso Europa 125, IT-16132 Genova – 10 **Lombardia:** Corso Sempione 27, IT-20145 Milano – 11 **Marche:** Piazza della Repubblica 1,IT- 60121 Ancona – 12 **Molise:** Viale Principe di Piemonte 59, IT-86100 Campobasso – 13 **Piemonte:** Via G.Verdi 16, IT- 10121 Torino – 14 **Puglia:** Via Dalmazia 104, IT-70121 Bari – 15 **Sardegna:** Via Barone Rossi 27, IT-09125 Cagliari – 16 **Sicilia:** Viale Strasburgo 19, IT-90146 Palermo – 17 **Toscana:** Largo Alcide de Gasperi 1, IT-50136 Firenze – 18 **Trentino:** Via Fratelli Perini 141, IT-38100 Trento – 19 **Umbria:** Via L. Masi 2, IT- 06121 Perugia – 20 **Valle d'Aosta:** Loc.Grande Charriere 70, IT- 11020 Saint Cristophe – 21 **Veneto:** Palazzo Labia, Campo S. Geremia,Sestiere Cannaregio 275, IT- 30121 Venezia.

MW Station		kHz	kW	Prgr
6)	Bologna (Budrio)	567	60	R1
16)	Caltanissetta (St.Anna)	567	20	R1 (b)
2)	Bolzano (Monticolo)	657	25	R1 (e)
5)	Napoli (Marcianise)	657	120	R1
7)	Pisa (Coltano)	657	55	R1
3)	Potenza	693	20	R1
10)	Milano (Siziano)	693	100	Test DRM
7)	Trieste (Monte Radio)	819	20	R1 (a)
8)	Roma (Santa Palomba)	846	50	Test DRM inactive
14)	Taranto	873	1	R1
10)	Milano (Siziano)	900	600	R1
16)	Trapani	936	10	R1
21)	Venezia (Campalto)	936	20	R1 (+a)
7)	Trieste(Monte Radio)	981	10	S
4)	Vibo Valentia (CapoVatic)	999	2	R1
19)	Perugia(Torgiano)	999	2	R1
6)	Rimini (Viserba)	999	20	R1
13)	Torino (Volpiano)	999	2	R1
1)	Pescara (SanSilvestro)	1035	10	R1
14)	Lecce/Salento Specchia)	1035	2	R1
11)	Ancona (Montagnolo)	1062	10	R1
15)	Cagliari (Sestu)	1062	25	R1 (d)
16)	Catania (Barriera del Bosco)	1062	2	R1 (c)
18)	Trento (Villazzano)	1062	2	R1
8)	Roma (Monte Ciocci)	1107	10	R1
20)	Aosta (Gerdaz)	1116	2	R1 (f)
14)	Bari (Ceglie d, Campo)	1116	2	R1
13)	Cuneo (Tetti Pesio)	1116	20	R1
16)	Palermo (Mte Pellegrino)	1116	10	R1 (c)
3)	Matera	1314	2	R1
14)	Foggia	1431	2	R1
13)	Biella (S.Paolo)	1449	2	R1
2)	Bolzano (Bressanone)	1449	2	R1 (e)
2)	Bolzano (Brunico)	1449	2	R1 (e)
10)	Sondrio	1449	2	R1
12)	Campobasso	1575	2	R1
9)	Genova (Portofino)	1575	50	R1
7)	Gorizia (Piuma)	1575	2	R1 (a)
16)	Nuoro (S.Onofrio)	1575	1	R1 (d)
19)	Terni (S.Lorenzo)	1584	2	R1

FM (MHz)		R1	R2	R3	R4	GRP	kW
6)	Bertinoro	90.8	93.4	99.6		89.7	30
6)	Bologna	89.5	91.7	93.9		93.6	60
2)	Bolzano	91.5	93.7	97.1	99.6	95.1	14
6)	Ca' del Vento	92.1	96.5	98.5		90.6	40
8)	Canepina-PNibbio	93.7	99.4				12
4)	Capo Spartivento	95.6	97.6	99.7		104.2	10
21)	Col Visentin	91.1	93.1	95.5			30
4)	Crotone	94.9	97.9	99.9		97.4	10
17)	Firenze	87.8	91.1	98.4		88.0	10
4)	Friscano	88.4	90.5	94.1			10
4)	Gambarie	95.3	97.3	99.3			40
			103.9				40
9)	Genova	89.5	91.9	95.1		104.5	30
5)	Golfo di Policastro	88.5	90.5	92.5			10
5)	Golfo di Salerno	95.1	97.1	99.1			20
7)	Gorizia	89.5	92.3	94.6	98.3	106.8	10
14)	Martina Franca	89.1	91.1	93.1		90.3	100
10)	Milano	90.6	93.7	99.4	102.2	88.3	60
17)	Monte Argentario	90.1	92.1	94.3		99.6	70
			89.0				16
9)	Monte Beigua	91.5	94.6	98.9		100.5	40
14)	Monte Caccia	94.6	96.7	99.2		98.2	100
16)	Monte Cammarata	94.1	95.9	99.9		98.3	100
6)	Monte Canate		95.9				24
8)	Monte Cavo	87.6	91.2	98.4		99.3	80
11)	Monte Conero	88.3	90.3	92.3		105.2	100
5)	Monte Faito	94.1	96.1	98.1		91.0	100
16)	Monte Lauro	94.7	96.7	98.7		89.0	100
18)	Monte Limbara	88.9	95.3	99.3			60
17)	Monte Luco	88.1	92.5	96.2		103.2	30
14)	Monte Nerone	94.7	96.6	98.7		88.1	100
19)	Monte Miranda	95.7	97.7	99.7		102.1	60
			88.3				30
10)	Monte Penice	94.2	97.4	99.9		88.2	120
			103.0				120
3)	Monte Pierfaone	88.1	90.1	92.1		91.2	45
19)	Monte Sambuco	88.6	90.7	93.5			100
			100.7				100
4)	Monte Scuro	88.5	90.5	92.5		98.4	30
15)	Monte Serpeddi	90.7	92.7	96.3		106.5	70
17)	Monte Serra	88.5	90.5	92.9		88.2	70
16)	Monte Soro	89.9	91.9	93.9		104.2	30
19)	Monte Subasio	89.3	91.4	93.5		104.6	30
21)	Monte Venda	88.1	89.0	89.9			160
5)	Monte Vergine (AV)	87.9	90.3	92.3		93.0	20
5)	Napoli Camaldoli	89.3	91.3	93.3	103.9	101.0	12
3)	Nova Siri			89.5			10
16)	PalermoMtePellegri	94.9	96.9	98.9		90.3	40
1)	Pescara S. Silvestro	89.2	94.3	96.4		100.2	70
3)	Pomarico	88.7	92.7	95.7			10
15)	Punta Badde Urbara	91.3	93.3	97.3			70
8)	Roma M. Mario	89.7	91.7	93.7		100.3	100
4)	Roseto Capo Spulico	94.4	96.5	98.5			10
14)	Salento Turrisi	90.7	95.5	97.5		91.0	60
17)	San Cerbone	95.3	97.3	99.3			12
21)	SanZenodiMontagna	93.2	96.5	98.5		89.5	10
18)	Selva Piana	88.4	90.3	92.4			20
13)	Torino Eremo	92.1	95.6	98.2	101.8	88.2	100
16)	Trapani Erice	88.4	90.5	92.5		90.8	60
7)	TriesteMteBelvedere	91.5	93.6	95.8	103.9	106.7	30
7)	Udine	94.9	97.2	99.8			60
8)	Velletri	88.7	90.7	92.7			15

+ over 6000 stns below 1kW not mentioned.

D.Prgr:All stns transmit from 0500 to 2300, except for Milano 900kHz, Roma 846kHz (DRM Tests) inactive ,Napoli 657kHz all 24h
R1=Radiouno, **R2FM**=Radiodue, **R3FM**=Radiotre, **S**=Special Prgrs.
Regional Prgrs: 0620-0628 Mon/Sat RAI 1; 1110-1127 Mon/ Sun . **(a)** Regional service : Friuli 0620-0657 Mon-Sat RAI 1, 1003-1157 Mon-Sat RAI1, 1130-1157 Sun RAI 1,1300-1415 Mon-Fri RAI 1, 1330-1400 Sat RAI 1, 1730-1756 Mon/Fri RAI 1, 1715-1756 Sat RAI 1, 0740-0910, 1108-1157, 1730-1756 Sun sport RAI 1 ; **(+a)**: "L'ora della Venezia Giulia" 1445-1545 Mon/Sat RAI 1, 1330-1400 Sun RAI 1,.Summertime 1h earlier **(b)** Regional service : Sicilia 1230-1245, Mon/Sat RAI 1 Arabic service only over FM stns.Summertime 1h earlier. **(c)** Regional service: Sicilia .0630- 0657.1110-1127,1315-1400, 1730-1756 Mon/ Sat;RAI 1; 1140-1157,1730-1756 Sun sport RAI1 Summertime: 1h earlier. **(d)** Regional service: Sardegna 0630-0657,1315-1400, 1730-1756 Mon/Sat RAI 1; 1730-1756 Sun sport RAI 1 Summertime: 1h earlier. **(f)** Regional service: Valle D'Aosta 1315-1400 Mo/Sat, 1730-1756 Sun sport RAI 1.(Bilingual) Summertime 1h earlier. **(e)** Regional service Alto Adige 0630-0657,1315-1400,1730-1756 Mon/Sat; 1730-1756 Sun sport RAI 1. Summertime 1h earlier.

SPECIAL PRGRS
ISO Radio: 24h sce. for motorway users on 103.3MHz **FM** (220 txs of 5kW or less);103.2MHz Milano,Como,Lecco area; 103.5MHz, Roma area **W**: www.isoradio.it
GR Parlamento: 24h sce. Italian Parliament channel. 2000-0455. . **FM** (150 txs of 5kW or less) **W**: www.radio.rai.it/grparlamento
Sender Bozen (Bolzano): Prgrs in **German** on FM (46 txs of 1kW or less) **DPrgr**: 0500 (Sun 0600)-2300. **N.** 0615 (W), 0800 (Sun), 1000 (W), 1100, 1200, 1300, 1700 (W), 1930. **W**: www.senderbozen.rai.it
Regional Prgr. in Slovene: Trieste 981kHz 10kW + 103.9MHz 20kW (and 22 additional FM-stns). **D. Prgr**: 0500 (Sun 0600)-2300. **N**: W 0500, 0700, 0900, 1200, 1300, 1600, 1800; Sun 0700, 1200, 1300, 1800.
N. in German: 0900 (W). **Night** : Relay V channel Filodiffusione or Notturno Italiano 2300-0500
ANN: Home Sce: "RAI Radiouno", "RAI Radiodue", "RAI Radiotre" as appropriate. Night Prgr: "RAI-Radiotelevisione Italiana stazioni a onda media di Milano kHz 900, di Roma kHz 1107, e di Napoli kHz 657 Notturno Italiano".
Notturno Italiano Notturno Italiano: musical and cultural prgr with I/E/F/ news. irr.schedule

R.A.S.
✉ Europaallee 164/A,IT- 39100 Bozen ☎ +39 0471 546666 🖷 +39 0471 200378 **W**: www.ras.bz.it **E**: info@ras.bz.it
LP: Pres . Rudi Gamper MD: Georg Plattner . Dir. Tec: Dr .Johann Silbernagl RAS is a public body of the autonomous Region of Southern Tyrol whose purpose is to relay TV and Radio from Germany, Austria and Switzerland to the German-speaking population.

FM (MHz)	RAS 1	RAS 2	RAS 3	kW
Kronplatz	100.7	103.0	104.7	2
Meransen	101.3	103.9	107.3	1
Obervinschgau	100.5	103.0	106.1	0.6
Penegal	103.3	100.3	104.7	2
Perdonig	101.8	104.0	106.0	1
Plose	99.8	102.0	105.6	1
Vinschgau	101.1	102.9	105.0	2

+ 880 low power stns
RAS 1: rel. OE-3 (Austria) - **RAS-2**: rel. OE-R (Austria) - **RAS-3**: rel. OE-1 (Austria)
DAB: RAI & RAS on Blocks 12A-12DA, 223.936MHz - 229.072MHz
Consorzio DAB Italia on block 9D. 208.064MHz **W**: www.dab.it

PRIVATE STATIONS
Only stns with MW/SW broadcasts and FM networks are listed. A number of other stns are heard irr. There are approx. 600 FM stns

MW	kHz	kW	Station, location and h of tr.
4)	1368	0.4	Challenger R., Villa Estense (testing)
1)	1404	1	Radio Luna, Chiozza di Scandiano 0600-1700
2)	1485	0.5	Broadcast Italia 24h
3)	1512	0.5	Onda Media Broadcast 0500-2130 irr
4	1566	0.5	Challenger Radio, Villa Estense 24h test DRM, prgrs Universal Welle
5)	1584	10	Radio Studio X, Momigno: 24h
6)	1584	0.5	Radio Verona,Verona 24h
SW	**kHz**	**kW**	**Station, h of tr.**
7)	26000	0.1	Radio Maria 24h DRM test AM mode
7)	26010	0.1	Radio Maria 24h DRM test

Addresses and other information:
1) Via Brolo Sotto 52, 42019 Chiozza di Scandiano)RE) ☎ +39 0522 856598 🖷 +39 0522/857084 **W**: www.radioluna.com **E**: info@ radioluna.com **FM**: 105.9MHz 5kW **SM**: Battista Francia – **2) W**: www.broadcastitalia.it **E**: maurizioamici@tin.it **SM**: Maurizio Amici

– **3)** 40018 San Pietro Iin Casale)BO) **E**: ondamediabroadcast@ gmail.com **SM**: Roberto Furlan ☎ +39 3927558283 **4)** Via Legnaro 6, IT-35040 Villa Estense (PD) 🖷 +39 0429 662280 **W**: www.challenger. it **E**: challenger@challenger.it **V.** by letter Rp. Rpt :Vita Universale Italia, C.P. 16068, IT-20158 Milano (MI) **SM**: Maurizio Anselmo **5)** Via Mammianese 687, IT-51030 Momigno (PT) **W**: www.radiostudiox.it **E**: info@radiostudiox.it SM: Luca Betti.i- **FM**: 96.55MHz 5kW **V.** by letter Rp – **6)** Via della Scienza 25, IT-37139 Verona (VR) ☎ +39 045 8063260 🖷 +39 045 8063261 **W**: www.radioverona.it **E**: radioverona@ radioverona.it **SM**:.M.Puliero **FM**: 103/103.90MHz 10kW **V.** by letter. Rp – **7)** DRM tests Radio Maria, c/o Claudio Re, Via Mazzini 15,IT-21020 Casciago (VA) **W**: www.mediasuk.org/archive/radio_maria. html **E**: qsl@radiomaria.org **V.** by letter. Rp **QSL Mgr.**: Giampiero Bernardini, Via Tertulliano 35, IT-20137 Milano (MI) Rp

FM NETWORKS IN MAJOR CITIES (MHz):

	Network	To	Mi	Ve	Bo	Ge
1)	Circuito Margherita	91.8	89.5	-	-	90.1
2)	Kiss Kiss	92.4	97.6	-	101.8	104.9
3)	InBlu Radio	89.0	101.7	94.6	92.3	88.8
4)	Latte Miele	88.5	92.2	106.2	105.0	-
5)	m2o	90.3	91.0	87.8	89.0	88.6
6)	Popolare Network	97.6	107.6	99.1	96.2	-
7)	Radio Capital	93.0	90.0	98.5	99.4	93.9
8)	Radio 105	99.6	99.1	98.9	103.5	99.5
9)	Radio 101	101.0	100.9	107.3	96.0	105.2
10)	Radio Classica	98.7	94.0	-	-	-101.1
11)	Radio Cuore	95.0	92.4	-	-	105.8
12)	Radio Deejay	106.9	99.7	94.8	99.7	96.9
13)	Radio RDS	96.4	94.4	99.8	104.2	95.7
14)	R. Italia Anni 60	103.7	106.3	99.5	102.1	91.3
15)	R. Italia S.M.I	106.6	98.4	101.5	100.6	106.3
16)	Radio Maria	107.7	89.0	106.5	90.5	106.6
17)	Radio Mater	105.7	95.3	100.1	-	102.0
18)	R. Padania Libera	106.0	103.5	93.8	106.2	96.0
19)	Radio Radicale	102.8	87.9	105.4	92.0	95.4
20)	Radio RMC1	105.5	105.3	100.4	101.3	104.2
22)	Radio 24	105.0	104.8	106.8	107.0	97.2
24)	RTL 102.5	102.5	102.5	102.5	101.6	102.4
25)	Virgin Radio	90.9	104.5	93.1	106.5	105.5
	Network	**Fi**	**Rm**	**Na**	**Ba**	**Pa**
1)	Circuito Margherita	96.7	-	100.7	95.2	95.2
2)	Kiss Kiss	92.8	97.9	89.0	100.8	103.0
3)	InBlu Radio	95.4	96.3	101.45	100.0	88.0
4)	Latte Miele	91.4	93.1	101.2	93.5	94.6
5)	m2o	105.8	97.0	98.3	87.6	107.8
6)	Popolare Network	93.6	103.3	-	97.3	-
7)	Radio Capital	97.6	95.5	104.6	99.5	103.3
8)	Radio 105	105.0	96.1	99.7	87.9	105.1
9)	Radio 101	94.9	92.0	96.6	91.2	97.2
10)	Radio Classica	99.4	89.5	-	-105.0	99.5
11)	Radio Cuore	100.3	-	105.3	-	89.1
12)	Radio Deejay	100.6	101.0	92.3	93.2	107.5
13)	Radio RDS	88.3	103.0	107.5	89.1	106.6
14)	R. Italia Anni 60	96.2	89.1	104.1	89.6	95.8
15)	R. Italia S.M.I	107.6	104.2	96.8	103.5	104.8
16)	Radio Maria	88.8	95.1	98.8	102.0	89.4
17)	Radio Mater	93.9	93.5	-	95.4	-
18)	Radio Norba	-	-	92.7	105.5	-
19)	R. Padania Libera	-	-	-	-	-104.0
20)	Radio Radicale	97.0	88.6	101.6	89.3	92.0
21)	Radio RMC1	106.6	106.3	91.6	92.0	90.0
22)	Radio Subasio	94.5	94.5	106.5	-	-
23)	Radio 24	103.8	107.9	103.5	88.2	104.5
24)	RTL 102.5	100.9	102.1	102.6	102.8	102.3
25)	Virgin Radio	107.2	98.7	93.1	106.5	102.1

To=Torino **Mi**=Milano **Ve**=Venezia **Bo**=Bologna **Ge**=Genova **Fi**=Firenze
Rm=Roma **Na**=Napoli **Ba**=Bari **Pa**=Palermo
Reference to Italian frequency on **W**: www.fmdx.altervista.org

Addresses and other information
1) Via Marchese di Villabianca 82, IT-90143 Palermo (PA) ☎ +39 091 302712 🖷 +39 091 8724835 **W**: www.radiomargherita.com **E**: info@ radiomargherita.com **SM**: Giuseppe Orobello **V.** by letter. Rp. –**2)** Via Sgambati 61, IT- 80131 Napoli (NA) ☎ +39 081 5461212 🖷 +39 081 5467789 **W**: www.kisskiss.it **E**: info@kisskiss.it info@kisskissnetwork. it **SM**: Antonio Niespolo **V.** by letter. Rp. – **3)** Via Aurelia 796, IT- 00165 Roma (RM) ☎ +39 06 6650851 🖷+39 06 66508516 **W**: www.radioin-blu.it **E**: info@radioinblu.it **TM**: Marco Giubileo. **V.** by letter. Rp – **4)** Via Andrea Costa 10, IT- 40013 Castelmaggiore (BO) ☎ +39 051 70928 🖷 +39 051 6325710 **W**: www.lattemiele.com **E**: info@lattemiele.com **SM**: Franco Magnani – **5)** Piazza della Repubblica 23/c, IT- 00185 Roma (RM) ☎ +39 06 492311 🖷 +39 06 4453758 **W**: www.m2o.it **E**: contatti@m2o.it **PM**: Fabrizio Tamburini **V.** by letter. Rp – **6)** Via U.Olleano

5, IT- 20155 Milano (MI) ☎ +39 02 392411 ▤ +39 02 39273125 **W:** www.radiopopolare.it **E:** Radiopop@radiopopolare.it **SM:** Stefano Di Blasio **V.** by QSL-card. Rp – **7)** Via C. Colombo 90, IT- 00147 Roma (RM) ☎ +39 06 494321 ▤ +39 06 44702290 **W:** www.capital.it **E:** infoline@capital.it **SM:** Vittorio Zucconi. **V.** by QSL-card. Rp – **8)** Largo G. Donegani 1, IT- 20121 Milano (MI) ☎ +39 02 6596116 ▤ +39 02 6592272 **W:** www.105.net **E:** diretta@105.net **SM:** Alberto Hazan **V.** QSL-Card. Rp – **9)** Via Giovanni Ventura 3, IT- 20134 , Milano (MI) ☎ +39 02 210831 ▤ +39 02 21083210 **W:** www.r101.it **E:** infor101@r101.it **SM:** Francesco Perilli **V.** by QSL-card. Rp.. – **10)** Via M.Burigozzo 5, IT-20122 Milano (MI) ☎ +39 02 58219600 ▤ +39 02 58219407 **W:** www.radioclassica.fm **E:** radioclassica@class.it **SM:** Carla Signorile **V.** by letter. Rp. – **11)** Via Giovanni da Verrazzano 16, Localita Le Melorie, IT- 56038 Ponsacco (PI) ☎ +39 0587 2861 ▤ +39 0587 733861 **W:** www.mediahit.it/ **E:** info@mediahit.it **SM:** Italo Bessi **V.** by letter. Rp. –.**12)** Via Massena 2,IT-20154 Milano(MI) ☎ +39 02 342522 ▤ +39 02 342888 **W:** www.deejay.it **E:** segnalazioni@deejay.it **SM:** Linus **V.** by QSL-card. Rp..– **13)** Via G.Mazzini 119, IT- 00195 Roma (RM) ☎ +39 06 377051 ▤ +39 06 3725336 **W:** www.rds.it **E:** ufficiotecnico@rds.it **SM:** Stefano Montefusco **V.** by letter. Rp. – **14)** Via Lussimpiccolo 3, 00177 Roma (RM) ☎ +39 010 6196417 ▤ +39 02 70036123 **W:** www.radioitaliaanni60.it **E:** info@radioitaliaanni60.it **SM:** Francesco Nisi **V.** by letter. Rp. – **15)** Viale Europa 49, IT- 20093 Cologno Monzese (MI) ☎ +39 02 25441 ▤ +39 02 25444220 **W:** www.radioitalia.it **E:** info@radioitalia.it **SM:** Mario Volanti **V.** by letter. Rp. – **16)** Via F.Turati 7, IT- 22036 Erba (CO) ☎ +39 031 610600 ▤ +39 031 611288 **W:** www.radiomaria.it **E:** info.ita@radiomaria.org **SM:** Don Livio Fanzaga **V.** by QSL-card. Rp – **17)** Via G.Marconi 85, IT- 22036 Arcellasco d'Erba (CO) ☎ +39 031 645214 ▤ +39 031 6490527 **W:** www.radiomater.com **E:** info@radiomater.com **SM:** Don Mario Galbiati **V.** by letter. Rp. – **18)** Via Foggia 29, IT- 70014 Conversano (BA) ☎ +39 80 4951229 ▤ +39 80 4953079 **W:** www.radionorba.com **E:** radionorba@radionorba.it **SM:** Annamaria Fantasia **V.** by letter. Rp. – **19)** Via C.Bellerio 41, IT- 20161 Milano (MI) ☎ +39 02 66203529 ▤ +39 02 66220964 **W:** www.radiopadania.net **E:** direzione@radiopadania.net **SM:** Cesare Bosetti **V.** by letter. Rp. – **20)**Centro di Produzione, Via Principe Amedeo 2, IT- 00185 Roma (RM) ☎ +39 06 488781 ▤ +39 06 4880196 **W:** www.radioradicale.it **E:** staff@radioradicale.it **SM:** Massimo Bordin **V.** by letter. Rp. – **21)** Via Principe Amedeo 2, IT- 20121 Milano (MI) ☎ +39 02 29001636 ▤ +39 02 6551451 **W:** www.radiomontecarlo.net **E:** rmc@radiomontecarlo.net **SM:** Paolo Del Forno **V.** by QSL-card. Rp–**22)** Localita Colle de Bensi, 06081 Assisi (PG) ☎ +39 075 8060 ▤ +39 075 8065419 **W:** www.radiosubasio.it **E:** subasio@radiosubasio.it **SM:** Mario Settimi **V.** by letter. Rp. – **23)** Via Monte Rosa 91, IT- 20149 Milano (MI) ☎ +39 02 30221 ▤ +39 02 30224462 **W:** www.radio24.it **E:** info@radio24.it **SM:** Elia Zamboni **V.** by letter. Rp – **24)** Viale Piemonte 61/63, IT- 20093 Cologno Monzese (MI) ☎ +39 02 251515 ▤ +39 02 25096201 **W:** www.rtl.it **E:** qualita@rtl.it **SM:** Luigi Tornari **V.** QSL-card. Rp.– **25)** Largo Donegani 1, IT-20121 Milano (MI) ☎ +39 02 6596116 ▤ +39 02 62537460 **W:** www.virginradioitaly.it **E:** guastivirgin@virginradio.it **SM:** Francesco Migliozzi **V.** QSL-Card. Rp.

NEXUS - INTERNATIONAL BROADCASTING ASSOCIATION
See International Broadcasting section.

AMERICAN FORCES NETWORK EUROPE (U.S. Mil.)
W: www.afneurope.net **E:** harringtonj@afns.vicenza.army.mil
1st Prgr. The Eagle on 106.0MHz **Key Stations: Vicenza** (10kW) AFN,C/o Caserma Ederle,Via della Pace 100, IT-36100 Vicenza (VI) **W:** http://vicenza.afneurope.net **V.** by letter. NoRp.
Livorno (10kW) AFN Livorno,UNIT 31301,Box 64,APO AE,09613,USA. Local prgr 0500-0800,1000-1200,1400-1700,2000-2200 Mon-Fr **W:** http://livorno.afneurope.net/ **Other stns: Aviano/San Vito** (all 5kW).
Napoli "LAVA 106" (10kW), PSC 817,Box 31,FPO AE 09622,USA **W:** http://naples.afneurope.net **E:** Ask.NSA@nsa.naples.navy.mil **2nd Prgr. Power Network** on 107.0MHz **Key Station: Aviano** (10kW) AFN,C/o Base USA, Dept.8.AFBS/ XOOR, IT-33081 Aviano (PN) ☎+39 0434 664634 **W.** by letter. No Rp. Other stns: **Vicenza/Livorno/Sigonella** (all 5kW) **E:** jeffrey.wells@afn.sicily.army.mil **Napoli** (10kW), **San Vito** (0.5kW)

IVORY COAST

L.T: UTC — **Pop**: 21 million — **Pr.L:** French, Diola, 12 ethnic — **E.C:** 50Hz, 220V — **ITU:** CTI

CONSEIL NATIONAL DE LA COMMUNICATION AUDIOVISUELLE (CNCA)
▣ Place de la République, B.P. V 56, Abidjan ☎+225 20 311580 **W:** lecnca.net **E:** infos@lecnca.net **L.P:** Gen. Secr: Franck Anderson Kouassi. Chmn: Jerome Diegou Bailly

RADIODIFFUSION-TÉLÉVISION IVOIRIENNE(RTI, Gov.)
▣ B.P. 191, Abidjan ☎+225 20 214800 ▤ +225 20 215038 **W:** www.rti.ci **L.P:** Interim DG: Pierre Brou Armessan. Deputy DG for Tech: Issa Yéresso Sangaré. Dir. Broadc: Robert Kossi N'Da.

FM (MHz)	1	2	kW	FM (MHz)	1	2	kW
Abobo-Abidjan	90	92	5	Man	96.9	100.2	-
Bouaflé	99	102.6	-	Naingbo	93	103	-
Dabakala	91	101	-	Niangue	93	95.9	-
Dimbokro	99	102.9	-	Séguéla	89	95	-
Divo	88	90.8	10	Tengréla	96.3	99.6	-
Grabo	88	91	0.5	Tiémé	88	91	5
Kouakoussikro	89.3	92.4	-	Touba	94.7	101.5	-
Koun Fao	94.2	101	-				

R. Côte d'Ivoire (1): 0500-2400. **Fréquence Deux (2):** 24h.
Ann: "R. Côte d'Ivoire" or "Fréquence Deux". **IS:** s/on with clock chimes.

Other stations:
City FM, Abidjan: 106.1MHz – **Cocody FM**, Abidjan: 98.5MHz 1kW. **W:** radiococodyfm.com – **Fréquence Vie**, Abidjan: 89.4MHz 1kW. **W:** frequencevie.en-fete.org – **La Voix de l'Évangile**, Abidjan: 102.5MHz – **N'Gowa FM**, Abidjan: 89.7MHz – **ONUCI FM:** Daloa 91.4MHz, Bangolo/Guiglo/Zouénoula 93.7MHz, Yamoussoukro 94.4MHz, Abengourou 94.7MHz, Bouaké/Korhogo/Man/Odiénné/Séguéla 95.3MHz, Abidjan 96.0MHz, Danané 97.6MHz, Bondoukou 100.1MHz, Bouna 102.8MHz, Duékoue/Ferkéssédougou 104.4MHz Sanpédro 106.3MHz. **W:** www.onuci.org/onucifm – **R. Al Bayane**, Abidjan: 95.7MHz. **W:** radio-albayane.info – **R. Arc-en-ciel**, Abidjan: 102.0MHz. – **R. Espoir**: Abidjan: 102.8MHz. **W:** radioespoir.ci – **R. Jam:** Yamoussokro 88.1MHz 1kW, Abidjan 99.3MHz 3kW. **W:** radiojam.ci – **R. Nostalgie**, Abidjan: 101.1MHz. **W:** nostalgie.ci – **Zenith FM**, Abidjan: 92.8MHz
Africa No 1: Abidjan 91.1MHz (see main entry under Gabon).
BBC African Sce: Abidjan 94.3MHz.
RFI Afrique: Abidjan/Bouaké/Korogho on 97.6MHz.
Voice of America, Abidjan: 99.0MHz

L.T: UTC -5h — **Pop:** 2.7 million — **Pr.L:** English — **E.C:** 50Hz, 110/220V — **ITU:** JMC

RJR COMMUNICATIONS GROUP - RADIO JAMAICA LTD. (Comm.)
▣ 32 Lyndhurst Road, P.O. Box 23, Kingston 5 ☎ +1 876 926 1100 ▤ +1 876 929 7467 **E:** webmaster@radiojamaica.com **W:** www.radiojamaica.com
L.P: Chmn. & MD: J.A. Lester Spaulding. CEN: Carroll Lawrence. Exec. Prod. RJR 94: Henry Stennett. Exec. Prod. FAME FM: Francois St. Juste. Media Srvces Mgr. (radio): Donald Topping. Publ. Rel. (radio): Norma Brown-Bell
FM: RJR94 94.1/94.3/94.5/94.7/94.9MHz **Fame95**: 95.1/95.3/95.5/9 5.7/95.9MHz **Hitz92**: 92.1/92.3/92.5/92.2/92.9MHz
MW: 550/580/700/720/770kHz (all inactive)

Other stations:
BBC 104 FM. FM: 104.1/104.3/104.5/104.7/104.9MHz D.Prgr: 24h relay of the BBC World Service – **BESS FM**, 4 East Bloomsbury Rd, Kingston 10 ☎ +1 876 754 1898 ▤ +1 876 920 4749 **W:** www.bessfm.com **FM:** 100.1/100.3/100.5/100.7/100.9 – **FREE FM**, General Penitentiary, South Camp Rd, Kingston. **FM:** Kingston 88.9MHz – **HOT 102FM**, 37 St. James St., Montego Bay ☎ +1 876 952 3056 **W:** www.hot102.fm. **L.P:** Andrea Wilson-Messam Prgs/operations mgr: Tomlin Ellis. **FM:** 102.1/102.3/102.5/102.7/102.9MHz. Format: Talk & hit radio – **IRIE FM**, P.O. Box 282, Coconut Grove, Ocho Rios ☎ +1 876 974 5051/968 5023 ▤ +1 876 974 5943 **E:** info@iriefm.net **W:** www.iriefm.net **FM:**101.1/ 101.3/107.5/107.7/107.9MHz. **L.P:** MD Chad Young. Format: Reggae – **JET FM**, Hills of St Mary. **FM:** 88.9. Format: Community radio operated by Jeffrey Town Farmers Association Ltd. – **KLAS FM**, 17 Haining Rd, Kingston 5 ☎ +1 876 929 1344 ▤ +1 876 906 7604 **W:** www.klassportsradio.com **FM:** 89.1/89.3/89.5/89.9MHz. Format: Sport – **KOOL 97 FM**, 1 Braemar Ave, Kingston 10 ☎ +1 876 978 4037 ▤ +1 876 978 3346 **W:** www.kool97fm.com **L.P:** GM: Linvol Stephens. **FM:** 97.1/97.3/97.5/97.7/97.9MHz – **LOVE 101**, 81 Hagley Pk Rd, Kingston 11 ☎ +1 876 968 9596 ▤ +1 876 968 7545 **W:** www.love101.org **L.P:** GM: Winston Ridgard. **FM:** 101.1/101.3/101.5/101.7 /101.9MHz. Format: Rlg. – **LINKZ FM**, 8 Beckford St., Savanna-la-mar ☎ +1 876 955 4650 ▤ +1 876 955 4321 **W:** www.linkzfm.com **L.P:** CEO Roger Allin. **FM:** 96.5/96.9MHz – **MEGA JAMZ**, 20 Ballater Av., Kingston 10 ☎ +1 876 631 5269 ▤ +1 876 929 9566 **W:** www.megajamz98fm.com. **L.P:** MD Katherine Chong. **FM:** 98.1/98.3/98.5/9 8.7/98.9MHz. Format: Oldies. – **MELLO FM**, 63 Barnett St, Montego Bay ☎ +1 876 971 4163 ▤ +1 876 940 6397 **L.P:** CEO Al Robinson.

Ops Mgr Edwin George. **FM:** 88.1MHz – **MUSIC 99,** 6 Bradley Av, Kingston 10 ☎ +1 876 968 4880 🖹 +1 876 968 9165 L.P: MD Newton James. **FM:** 99.1/99.3/99.5/99.7/99.9MHz – **NATIONWIDE RADIO,** Mannings Hill Rd, Kingston 8 ☎ +1 876 755 6397 🖹 +1 876 924 5375 **Web:** www.nationwideradiojm.com L.P: CEO: Chris Hughes. Op.mgr: Lennie Gordon. **FM:** 90.3/90.5/90.7MHz – **NEWSTALK 93FM,** Universal Media Company, 18 Ring Rd., Mona, Kingston 7 ☎ +1 876 970 2345 🖹 +1 876 970 2472 **W:** newstalk.com.jm. L.P: Chairman/MD: Gordon Shirley. **FM:** 93.1/93.3/93.5/93.7/93.9MHz – **POWER 106 FM,** 6 Bradley Av., Kingston 10 ☎ +1 876 968 4880 🖹 +1 876 968 9165 **E:** power106@cwjamaica.com **W:** www.go-jamaica.com/power L.P: MD Newton James. **FM:** 106.1/106.3/106.5/106.7/ 106.9MHz – **RADIO FRANCE INTERNATIONALE, FM:** 96.5MHz – **ROOTS FM,** 1 Mahoe Drive, Kingston 11 ☎ +1 876 923 6488 🖹 +1 876 923 6000 **W:** http://www.mustardseed.com/community/roots_fm.html. L.P: Chm: Trevor Gordon Somers. **FM:** Kingston 96.1MHz – **STYLZ FM,** 4 Boundbrooke Ave, Port Antiono p.o., Portland. ☎ +1 876 453 1444 🖹 **FM:** 96.1MHz – **SUN CITY RADIO,** Mother In Crisis, Shop#10 Port Henderson Rd, Portmore ☎ +1 876 938 8132. **W:** suncityradio.fm. **FM:** 104.9MHz – **TBC FM (The Breath of Change),** 51 Molynes Rd, Kingston 10 ☎ +1 876 754 5120 🖹 +1 876 968 9159 L.P: GM Gary Callam. **FM:** Kingston 88.5MHz. Format: Rlg (Baptist). – **VYBZ FM,** 98 Great George Street, Savanna-la-Mar, Westmoreland ☎ +1 876 918 2521 🖹 +1 876 918 2394 **W:** www.vybzfm.net. L.P: CEO Viannie Bedward-Morgan. **FM:** Westmoreland **FM:** 96.3MHz – **WIC RADIO,** Northern Caribbean University, Manchester Rd.,Mandeville. **FM:** 91.1MHz. Format: Rlg. (Adventist) – **ZIP 103,** 1B Derrymore Road, Kingston ☎ +1 876 929-2748/6233 🖹 +1 876 960 0523 **W:** www.zipfm.net. L.P: MD Chadl Young. FM: 103.1/103.3/103.5/ 103.7/103.9MHz. Format: Techo/dance/alternative.

JAPAN

LT: UTC +9h — **Pop:** 127.3 million — **Pr.L:** Japanese — **EC:** 50 & 60Hz, 100V — **ITU:** J

INFORMATION AND COMMUNICATIONS POLICY BUREAU, MINISTRY OF INTERNAL AFFAIRS AND COMMUNICATIONS (SOUMU SHO)
🖥 1-2, Kasumigaseki 2-chome, Chiyoda-ku, Tokyo 100-8926 ☎ +81 3 5253 5111 **W:** www.soumu.go.jp **L.P:** Minister: K.Haraguchi

NIPPON HOSO KYOKAI (NHK)
(The Japan Broadcasting Corporation)
🖥 2-1, Jinnan 2-chome, Shibuya-ku, Tokyo 150-8001 ☎ +81 3 3465 1111 **W:** www.nhk.or.jp
L.P: Chmn. (Board of Governors): S.Komaru. Pres: S.Fukuchi. Exec. Vice-Pres: Y.Imai. MD & Exec. Dir. Gen: K.Nagai. MD's: S.Kaneda, H. Hyuga, A. Mizoguchi, K. Yahata, N.Onishi, T.Imai, T.Kuroki, H.Tsukada, K.Yoshikuni. **Pub:** NHK Nenkan (Japanese) and NHK Update (English).

MWLoc. & Prgr	Call	kHz	kW		MWLoc. & Prgr	Call	kHz	kW
F2) Morioka 1	QG	531	10		D1) Hiroshima 2	FB	702	10
E2) Nago 1		531	1		C1) Nagoya 2	CK	729	50
F3) Yamagata 1	JG	540	5		G1) Sapporo 2	IB	747	500
E3) Miyazaki 1	MG	540	5		E8) Kumamoto 1	GK	756	10
E4) Kitakyushu 1	SK	540	1		F4) Akita 2	UB	774	500
A2) Matsumoto 1		540	1		C5) Takayama 1		792	1
C2) Nanao 1		540	1		G4) Enbetsu 1		792	1
E2) Ishigaki 1		540	1		A3) Takada 1		792	1
E2) Okinawa 1	AP	549	10.0		E5) Naze 1		792	1
G1) Sapporo 1	IK	567	100.0		A2) Nagano 1	NK	819	5
E5) Kagoshima 1	HG	576	10.0		B1) Osaka 2	BB	828	300
C3) Hamamatsu 1	DG	576	1		A3) Niigata 1	QK	837	10
G2) Kushiro 1	PG	585	10		G4) Nayoro 1		837	1
A1) Tokyo 1	AK	594	300		F5) Koriyama 1		846	5
D2) Okayama 1	KK	603	5		H1) Uwajima 1		846	1
G3) Obihiro 1	OG	603	5		E8) Hitoyoshi 1		846	1
E1) Fukuoka 1	LK	612	100		E8) Kumamoto 2	GB	873	500
G4) Asahikawa 1	CG	621	3		C3) Shizuoka 1	PK	882	10
B2) Kyoto 1	OK	621	1		F1) Sendai 1	HK	891	20
A2) Iida 1		621	1		C1) Nagoya 2	CB	909	10
E3) Nobeoka 1		621	1		A4) Kofu 1	KG	927	5
E6) Oita 1	IP	639	5		C6) Fukui 1	FG	927	5
C3) Shizuoka 2	PB	639	10		G4) Wakkanai 1		927	1
C4) Toyama 1	IG	648	5		D2) Tsuyama 1		927	1
B1) Osaka 1	BK	666	100.0		H2) Tokushima 1	XK	945	1
G5) Hakodate 1	VK	675	5		I2) Muroran 1	IQ	945	3
D3) Yamaguchi 1	UG	675	1		B3) Hikone 1	QP	945	1*
E7) Nagasaki 1	AG	684	1		E7) Fukue 1		945	1
A1) Tokyo 2	AB	693	500		F6) Aomori 1	TG	963	5
G6) Kitami 2	KD	702	10		H1) Matsuyama 1	ZK	963	1

MWLoc. & Prgr	Call	kHz	kW		MWLoc. & Prgr	Call	kHz	kW
E9) Saga 1	SP	963	1		F6) Hachinohe 2		1377	1
D4) Yonago 1		963	1		F2) Morioka 2	QC	1386	10
D3) Hagi 1		963	1		C2) Kanazawa 2	JB	1386	10
A2) Kisofukushima 1		981	1		E5) Kagoshima 2	HC	1386	10
E7) Sasebo 1		981	1		D2) Okayama 2	KB	1386	5
H3) Kochi 1	RK	990	10		G5) Hakodate 2	VB	1467	1
F6) Hachinohe 1		999	1		A2) Nagano 2	NB	1467	1
D1) Fukuyama 1		999	1		E6) Oita 2	ID	1467	1
H3) Nakamura 1		999	1		E3) Miyazaki 2	MC	1467	1
E1) Fukuoka 2	LB	1017	50		G4) Wakkanai 2		1467	1
C4) Toyama 2	IC	1035	1		A2) Iida 2		1476	1
H4) Takamatsu 2	HD	1035	1		F4) Akita 1	UK	1503	10
F3) Tsuruoka 2		1035	1		E8) Aso 1		1503	1
D1) Hiroshima 1	FK	1071	20		H1) Matsuyama 2	ZB	1512	5
F1) Sendai 1	HB	1089	10		F5) Koriyama 2		1512	1
E2) Okinawa 2	AD	1125	10		A2) Matsumoto 2		1512	1
D3) Obihiro 2	OC	1125	1		F3) Yamagata 2	JC	1521	1
G7) Muroran 2	IZ	1125	1		F6) Aomori 2	TC	1521	1
D4) Tottori 2	LC	1125	1		C3) Hamamatsu 2	DC	1521	1
G4) Nayoro 2		1125	1		C6) Fukui 2	FC	1521	1
D3) Hagi 2		1125	1		D4) Yonago 2		1521	1
C5) Takayama 2		1125	1		H3) Nakamura 2		1521	1
G2) Kushiro 2	PC	1152	10		E2) Ishigaki 2		1521	1
H3) Kochi 2	RB	1152	10		A3) Niigata 2	QB	1593	10
G6) Kitami 1	KP	1188	10		D5) Matsue 2	TB	1593	10
C2) Kanazawa 1	JK	1224	10		G4) Asahikawa 2	CC	1602	1
D5) Matsue 1	TK	1296	10		A4) Kofu 2	KC	1602	1
F5) Fukushima 1	FP	1323	1		E4) Kitakyushu 2	SB	1602	1
F2) Yamada 1		1323	1		F5) Fukushima 2	FD	1602	1
E8) Minamata 1		1341	1		D1) Fukuyama 2		1602	1
F5) Iwaki 1		1341	1		G4) Enbetsu 2		1602	1
H4) Takamatsu 1	HP	1368	5		H1) Uwajima 2		1602	1
D4) Tottori 1	LG	1368	1		E8) Hitoyoshi 2		1602	1
F3) Tsuruoka 1		1368	1		E3) Nobeoka 2		1602	1
D3) Yamaguchi 2	UC	1377	5		E5) Naze 2		1602	1
E7) Nagasaki 2	AC	1377	1					

+ approx 240 stns below 1kW

1: NHK Radio One, **2:** NHK Radio Two. **Call:** JO(call). *stn announces its callsign as "JOBK".

FM Location	Call	MHz	kW		FM Location	Call	MHz	kW
A5) Utsunomiya	BP	80.3	1		E1) Fukuoka	LK	84.8	3
A6) Chiba	MP	80.7	5		E2) Miyakojima		85.0	1
C4) Toyama	IG	81.5	1		A10) Saitama	LP	85.1	5
A7) Maebashi	TP	81.6	1		G1) Sapporo	IK	85.2	5
C7) Tsu	NP	81.8	3		F5) Fukushima	FP	85.3	1
A8) Yokohama	GP	81.9	5		E8) Kumamoto	GK	85.4	1
F3) Yamagata	JG	82.1	1		A4) Kofu	KG	85.6	1
C2) Kanazawa	JK	82.2	1		E5) Kagoshima	HG	85.6	1
A3) Niigata	QK	82.3	1		D5) Hamada		85.8	1
A1) Tokyo	AK	82.5	10		F6) Aomori	TG	86.0	3
C1) Nagoya	CK	82.5	10		H4) Takamatsu	HP	86.0	1
F1) Sendai	HK	82.5	5		F4) Akita	UK	86.7	3
B2) Kyoto	OK	82.8	1		H1) Matsuyama	ZK	87.7	1
F2) Morioka	QG	83.1	1		B1) Osaka	BK	88.1	10
A9) Mito	EP	83.2	1		E2) Okinawa	AP	88.1	1
C6) Fukui	FG	83.4	1		G4) Nayoro		88.2	1
H2) Tokushima	XK	83.4	1		D1) Hiroshima	FK	88.3	1
A3) Yamato		83.5	1		D2) Okayama	KK	88.7	1
C5) Gifu	OP	83.6	1		C3) Shizuoka	PK	88.8	1
B3) Otsu	QP	84.0	1		E6) Oita	IP	88.9	1
B4) Himeji		84.2	1		G4) Chikoma		89.1	1
E5) Tanegashima		84.4	1		G2) Nakashibetsu		89.9	1

+ approx 481 stns below 1kW **Call:** JO(call)-FM

Addresses of regional HQs:
A) Kanto-Koshinetsu area = Tokyo A1): same as NHK general HQ address. **B)** Kinki area = Osaka B1): 1-20, Otemae 4-chome, Chuo-ku, Osaka 540-8501. **C)** Tokai-Hokuriku area = Nagoya C1): 13-3, Higashisakura 1-chome, Higashi-ku, Nagoya 461-8725. **D)** Chugoku area = Hiroshima D1): 11-10, Otemachi 2-chome, Naka-ku, Hiroshima 730-8672. **E)** Kyushu area = Fukuoka E1): 1-10, Ropponmatsu 1-chome, Chuo-ku, Fukuoka 810-8577. **F)** Tohoku area = Sendai F1): 11-1, Nishiki-machi 1-chome, Aoba-ku, Sendai 980-8435. **G)** Hokkaido area = Sapporo G1): 1-1, Odori Nishi, Chuo-ku, Sapporo 060-8703 **H)** Shikoku area = Matsuyama H1): 5, Horinouchi, Matsuyama 790-8501.

NHK R. One (General prgr): 24h **N:** every h(exc Sun 0000). Also at 2030(exc Sat), 2140(exc Sat), 2230(exc Fri&Sat), 2330(exc Sat), 0030(exc Sat&Sun), 0130(exc Sat&Sun), 0230(exc Sat&Sun), 0430(exc Sat&Sun), 0535(exc Sat&Sun), 0630(exc Sat&Sun), 0730(exc Sat&Sun), 1130(exc Sat&Sun). **Regional and local prgrs** (the amount of local prgrs varies between stns) 2055wrp, 2125N/wrp/inf, 2155 wrp/inf,

2215(Fri&Sat 2210)N/wrp, 2240(Fri&Sat 2255)N/wrp/inf, 2355N/wrp/inf, 0055(exc Sun)N/wrp/inf, 0155N/wrp/inf, 0250wrp/inf, 0315(Sat & Sun 0310)N/wrp, 0355(exc Sun)wrp/inf, 0455N/wrp/inf, 0555N/wrp/inf, 0655N/wrp/inf, 0755N/wrp/inf, 0855N/wrp/inf, 0950N/wrp/inf, 1015(Sat & Sun)N/wrp, 1045(exc. Sat &Sun)N/wrp/inf, 1055(Sat&Sun)N/wrp/inf, 1155(exc. Sat & Sun)N/wrp/inf, 1255N/wrp/inf, 1355(Sat)N/wrp, 1410(exc. Sat)N/wrp. **IS:** Original music played by Celesta. **Ann:** "JO(call), NHK (location) Daiichi Hoso desu". Local ID's with call letters, network & location given by studio stns just before: 2000, 0300, 1000.

NHK R. Two (Educational prgr): 2100-1540(variable). No regional and local prgrs. **Foreign language N** (rel. NHK World - R. Japan): **Chinese:** 0900-0915(Sat&Sun 0910). **Korean:** (0915-0930(Sat&Sun 0910-0920). **English:** 0500- 0530(Sat&Sun 0510). **Portuguese:** 0930-0945(Sat&Sun 0920-0930). **Spanish:** 0400-0415(Sat&Sun 0410). Weather map: 0010, 0700, 1300 (all 20 mins.) **IS:** Original music played by Celesta. Nat. Anthem at s/on on national holidays & s/off. **Ann:** "JO(call), NHK (location) Daini Hoso desu". Local IDs (as 1st Netw) just before 2100, 0030, 0720, 1320 and sign off. Portuguese prgr (produced by NHK World): 1540-1600(variable), 2040-2100 on 639, 909, 1125 and 1521 kHz (see C1, C3) & C5)).

NHK FM Netw: 24h. 1600-2000 relays R. One, **N:** 2200, 0300, 0950(local), 1000. **Ann:** "JO(call)-FM, NHK (location) FM Hoso desu". Local IDs just before 2000, 0300, 1000.

V: NHK officially has no organised QSL sce. However, many local stns verify by QSL card or letter for DX reports.

EXTERNAL SERVICES:
RADIO JAPAN, NHK WORLD NETWORK
See International Broadcasting section

THE NATIONAL ASSOCIATION OF COMMERCIAL BROADCASTERS IN JAPAN (NIPPON MINKAN HOSO RENMEI)

3-23, Kioi-cho, Chiyoda-ku, Tokyo 102-8577 ☎ +81 3 5213 7711 +81 3 5213 7703 **W:** www.nab.or.jp
L.P: Pres: M. Hirose, Vice-Presidents: H. Shinkura, H. Inoue, K. Manabe, H. Hayakawa, K. Toyoda, S. Sugaya, K.Takada, Y. Nagamori. Exec. Dir: T. Fukuda. **Pub:** Nippon Minkan Hoso Nenkan (Japanese), Gekkan Minpo (Japanese), Minkan Hoso (Japanese) and NAB Handbook (English) etc.

MW Call	kHz	kW	ID	Station, location & h of tr
1) CR	558	20	CRK	R. Kansai, Kobe
2) WN	639	5	STV	STV Radio, Hakodate
3) DF	684	5	IBC	Iwate Hoso, Morioka
3) LO	684	1	IBC	Iwate Hoso, Ofunato
4) IL	720	1	KBC	Kyushu Asahi Hoso, Kitakyushu
5) LR	738	5	KNB	Kita Nihon Hoso, Toyama
5)	738	1	KNB	Kita Nihon Hoso, Takaoka
6) RR	738	10	RBC	Ryukyu Hoso, Naha
7) JF	765	5	YBS	Yamanashi Hoso, Kofu
8) PF	765	5	KRY	Yamaguchi Hoso, Shunan
9) XR	864	10	ROK	R. Okinawa, Naha: 2000-1900(Sat -1800, Sun -1500)
10) SO	864	1	SBC	Shin'etsu Hoso, Matsumoto
11) HE	864	3	HBC	Hokkaido Hoso, Asahikawa
11) QF	864	3	HBC	Hokkaido Hoso, Muroran
11)	864	1	HBC	Hokkaido Hoso, Enbetsu
12) PR	864	1	FBC	Fukui Hoso, Fukui
13) XN	864	1	CRT	Tochigi Hoso, Nasu
2) WS	882	3	STV	STV Radio, Kushiro
2)	882	1	STV	STV Radio, Esashi
11) HO	900	5	HBC	Hokkaido Hoso, Hakodate
14) HF	900	5	BSS	San'in Hoso, Yonago: (off air Sat1800-1855, Sun1500-1855)
15) ZR	900	5	RKC	Kochi Hoso, Kochi
2) VX	909	5	STV	STV Radio, Abashiri
16) EF	918	5	YBC	Yamagata Hoso, Yamagata
16)	918	1	YBC	Yamagata Hoso, Tsuruoka
16)	918	1	YBC	Yamagata Hoso, Yonezawa
16)	918	1	YBC	Yamagata Hoso, Shinjo
8) PM	918	1	KRY	Yamaguchi Hoso, Shimonoseki
8) PN	918	1	KRY	Yamaguchi Hoso, Iwakuni
17) TR	936	5	ABS	Akita Hoso, Akita
18) NF	936	5	MRT	Miyazaki Hoso, Miyazaki
18)	936	1	MRT	Miyazaki Hoso, Nobeoka
18)	936	1	MRT	Miyazaki Hoso, Nichinan
18)	936	1	MRT	Miyazaki Hoso, Kobayashi
18)	936	1	MRT	Miyazaki Hoso, Takachiho
19) KR	954	100	TBS	TBS Radio, Tokyo: (S)
20) NR	1008	50	ABC	Asahi Hoso, Osaka
21) AR	1053	50	CBC	Chubu Nippon Hoso, Nagoya: (S)
2) WM	1071	5	STV	STV Radio, Obihiro
10) SR	1098	5	SBC	Shin'etsu Hoso, Nagano
10) SW	1098	1	SBC	Shin'etsu Hoso, Iida

MW Call	kHz	kW	ID	Station, location & h of tr
22) MF	1098	1	NBC	Nagasaki Hoso, Sasebo
23) GF	1098	5	OBS	Oita Hoso, Oita
24) WO	1098	5	RFC	R. Fukushima, Koriyama
25) CF	1107	20	MBC	Minami Nihon Hoso, Kagoshima
25)	1107	1	MBC	Minami Nihon Hoso, Akune
25)	1107	1	MBC	Minami Nihon Hoso, Oguchi
25)	1107	1	MBC	Minami Nihon Hoso, Sendai
26) MR	1107	5	MRO	Hokuriku Hoso, Kanazawa
26)	1107	1	MRO	Hokuriku Hoso, Nanao
27) AF	1116	5	RNB	Nankai Hoso, Matsuyama
27) AL	1116	1	RNB	Nankai Hoso, Niihama
27) AM	1116	1	RNB	Nankai Hoso, Uwajima
28) DR	1116	5	BSN	Niigata Hoso, Niigata
29) QR	1134	100	NCB	Bunka Hoso, Tokyo: (S)
30) BR	1143	20	KBS	KBS Kyoto, Kyoto
31) OR	1179	50	MBS	Mainichi Hoso, Osaka
15)	1197	1	RKC	Kochi Hoso, Nakamura
32) FO	1197	5	RKB	RKB Mainichi Hoso, Kitakyushu
33) BF	1197	10	RKK	Kumamoto Hoso, Kumamoto
33)	1197	1	RKK	Kumamoto Hoso, Hitoyoshi
33)	1197	1	RKK	Kumamoto Hoso, Aso
33)	1197	1	RKK	Kumamoto Hoso, Goshoura
34) YF	1197	5	IBS	Ibaraki Hoso, Mito: 2045 (Sat 2100)- 2000 (Sun 1530)
2) WL	1197	3	STV	STV R., Asahikawa
2)	1197	1	STV	STV R., Wakkanai
2)	1197	1	STV	STV R., Nayoro
2)	1197	1	STV	STV R., Enbetsu
30) BO	1215	2	KBS	KBS Kyoto, Maizuru
30a) BW	1215	1	KBS	KBS Shiga, Hikone
22) UR	1233	5	NBC	Nagasaki Hoso, Nagasaki
35) GR	1233	5	RAB	Aomori Hoso, Aomori
36) IR	1242	100	NBS	Nippon Hoso, Tokyo: (S)
37) IR	1260	5	TBC	Tohoku Hoso, Sendai
11) HW	1269	5	HBC	Hokkaido Hoso, Obihiro
11) FM	1269	1	HBC	Hokkaido Hoso, Esashi
38) JR	1269	5	JRT	Shikoku Hoso, Tokushima
38)	1269	1	JRT	Shikoku Hoso, Ikeda
32) FR	1278	50	RKB	RKB Mainichi Hoso, Fukuoka
11) HR	1287	50	HBC	Hokkaido Hoso, Sapporo
39) UF	1314	50	OBC	R. Osaka, Osaka: (S)
40) SF	1332	50		Tokai R. Hoso, Nagoya: (S)
41) ER	1350	20	RCC	Chugoku Hoso, Hiroshima: (S)
11) TS	1368	1	HBC	Hokkaido Hoso, Wakkanai
1) CE	1395	1	CRK	R. Kansai, Toyooka
24) WE	1395	1	RFC	R. Fukushima, Wakamatsu
11) QL	1404	5	HBC	Hokkaido Hoso, Kushiro
42) VR	1404	10	SBS	Shizuoka Hoso, Shizuoka
42) VO	1404	1	SBS	Shizuoka Hoso, Hamamatsu
4) IF	1413	50	KBC	Kyushu Asahi Hoso, Fukuoka
43) RF	1422	50	RF	RF R. Nippon, Yokohama
14) HL	1431	1	BSS	San'in Hoso, Tottori: (as 900kHz)
14)	1431	1	BSS	San'in Hoso, Izumo: (as 900kHz)
22)	1431	1	NBC	Nagasaki Hoso, Fukue
24) WW	1431	1	RFC	R. Fukushima, Iwaki
44) VF	1431	5	WBS	Wakayama Hoso, Wakayama: (S)
45) ZF	1431	5	GBS	Gifu Hoso, Gifu: 2030-1600.
2) WF	1440	50	STV	STV Radio, Sapporo
2)	1440	3	STV	STV Radio, Muroran
2)	1440	1	STV	STV Radio, Tomakomai
11) QM	1449	5	HBC	Hokkaido Hoso, Abashiri
46) KF	1449	1	RNC	Nishi Nippon Hoso, Takamatsu
46)	1449	1	RNC	Nishi Nippon Hoso, Marugame
22a) UO	1458	1	NBC	Nagasaki Hoso, Saga
24) WR	1458	1	RFC	R. Fukushima, Fukushima
34) YL	1458	1	IBS	Ibaraki Hoso, Tsuchiura
34)	1458	1	IBS	Ibaraki Hoso, Sekiyo
41)	1458	1	RCC	Chugoku Hoso, Shobara
8) PL	1485	1	KRY	Yamaguchi Hoso, Hagi
35) GO	1485	1	RAB	Aomori Hoso, Hachinohe
11) TL	1494	1	HBC	Hokkaido Hoso, Nayoro
47) YR	1494	10	RSK	Sanyo Hoso, Okayama: (S)
47)	1494	1	RSK	Sanyo Hoso, Takahashi: (S)
47)	1494	1	RSK	Sanyo Hoso, Tsuyama
47)	1494	1	RSK	Sanyo Hoso, Niimi
47)	1494	1	RSK	Sanyo Hoso, Bizen
47)	1494	1	RSK	Sanyo Hoso, Ochiai
28) DO	1530	5	BSN	Niigata Hoso, Joetsu
13) XF	1530	5	CRT	Tochigi Hoso, Utsunomiya
41) EO	1530	1	RCC	Chugoku Hoso, Fukuyama
41)	1530	1	RCC	Chugoku Hoso, Mihara

Relay stns below 1kW (approx 125 stns) not included.

Call: JO(call). (S): AM Stereo (C-QUAM System). **Schedule:** 24h, unless otherwise indicated above. Most 24h stns are off the air for 1 to 5 hours until 1900 or 2000 on Sun unless mentioned. All other days

a network prgr is aired 1600 or 1800 to 2000 on most stns. Network prgrs may also be broadcast at other times of day. **ID:** Company initials are usually used as stn identification.

Addresses and other information:
1) R Kansai Co., Ltd., 5-7, Higashi Kawasaki-cho 1-chome, Chuo-ku, Kobe 650-8580 **W:** http://jocr.jp – **2)** The STVradio Broadcasting Co., Ltd, 1-1, Nishi 8-chome, Kita 1-jo, Chuo-ku, Sapporo 060-8705 **W:** www.stv.jp – **3)** Iwate Broadc Co., Ltd., 6-1, Shike-cho, Morioka 020-8566 **W:** www.ibc.co.jp – **4)** Kyushu Asahi Broadc Co., Ltd, 1-1, Nagahama 1-chome, Chuo-ku, Fukuoka 810-8571 **W:** www.kbc.co.jp – **5)** Kita-nihon Broadc Co., Ltd.,10-18 Ushijima-machi, Toyama 930-8585 **W:** www.knb.ne.jp – **6)** Ryukyu Broadc Corp., 3-1, Kumoji 2-chome, Naha 900-8711 **W:** www.rbc-ryukyu.co.jp – **7)** Yamanashi Broadc System, 6-10, Kitagouchi 2-chome, Kofu 400-8525 **W:** www.ybs.ne.jp – **8)** Yamaguchi Broadc Co., Ltd., Koen-ku, Shunan 745-8686 **W:** http://kry.co.jp – **9)** R Okinawa Corp., 4-8, Nishi 1-chome, Naha 900-8604 **W:** www.rokinawa.co.jp – **10)** Shin-etsu Broadc Co., Ltd., 1200, Toigoshomachi, Nagano 380-8521 **W:** http://sbc21.co.jp – **11)** Hokkaido Broadc Co., Ltd., 2, Nishi 5-chome, kita 1-jo, Chuo-ku, Sapporo 060-8501 **W:** www.hbc.jp – **12)** Fukui Broadc., Ltd., 37-1-1 Owada-cho, Fukui 910-8588 **W:** www.fbc.jp –**13)** Tochigi Broadc Co., Ltd., 12-11 Honcho, Utsunomiya 320-8601 **W:** www.crt-radio.co.jp – **14)** Broadc System of San-in, 1-1-71, Nishi-Fukubara, Yonago 683-8670 **W:** http://bss.jp – **15)** Kochi Broadc Co., Ltd., 2-15, Hon-machi 3-chome, Kochi 780-8550 **W:** www.rkc-kochi.co.jp – **16)** Yamagata Broadc Co., Ltd., 2-5, Hatago-machi, Yamagata 990-8555 **W:** www.ybc.co.jp – **17)** Akita Broadc System, 9-42, Sanno 7-chome, Akita 010-8611 **W:** www.akita-abs.co.jp – **18)** Miyazaki Broadc Co., Ltd., 4-6-7, Tachibanadori-nishi, Miyazaki 880-8639 **W:** www.mrt.jp – **19)** TBS Radio & Communications, Inc., 3-6, Akasaka 5-chome, Minato-ku, Tokyo 107-8006 **W:** www.tbs.co.jp/radio – **20)** Asahi Broadc Corp., 1-30, Fukushima 1-chome, Fukushima-ku, Osaka 553-8503 **W:** http://asahi.co.jp/ – **21)** Chubu-Nippon Broadc Co., Ltd., 1-2-8, Shinsakae, Naka-ku, Nagoya 460-8405 **W:** http://hicbc.com – **22)** Nagasaki Broadc Co., Ltd., 1-35, Uwa-machi, Nagasaki 850-8650 **W:** www.nbc-nagasaki.co.jp – **22a)** Nagasaki Broadc Co., Ltd Saga station, 1249, Honjo-machi, Saga 840-0027 **W:** www.nbc-saga.jp – **23)** Oita Broadc System, 1-1, Imazuru 3-chome, Oita 870-8620 **W:** www.e-obs.com – **24)** R Fukushima Broadc Co., Ltd., 8, Shimoarako, Fukushima 960-8655 **W:** www.rfc.jp – **25)** Minaminihon Broadc Co., Ltd., 5-25, Korai-cho, Kagoshima 890-8570 **W:** www.mbc.co.jp – **26)** Hokuriku Broadc Co., Ltd., 3-2-1, Honda-machi, Kanazawa 920-8560 **W:** www.mro.co.jp – **27)** Nankai Broadc Co., Ltd., 1-1-1, Honmachi , Matsuyama 790-8510 **W:** www.rnb.co.jp – **28)** Broadc System of Niigata, Inc., 3-18, Kawagishi-cho, Chuo-ku, Niigata 951-8655 **W:** www.ohbsn.com – **29)** Nippon Cultural Broadc., Inc.,1-31, Hamamatsu-cho, Minato-ku, Tokyo 105-8002 **W:** www.joqr.co.jp – **30)** Kyoto Broadc System Co., Ltd., Kamichojamachi, Karasumadori, Kamigyo-ku, Kyoto 602-8588 **W:** www.kbs-kyoto.co.jp – **30a)** KBS Shiga Station, 13-1, Daito-cho, Hikone 522-0074 – **31)** Mainichi Broadc System, Inc., 17-1, Chayamachi, Kita-ku, Osaka 530-8304 **W:** www.mbs.jp – **32)** RKB Mainichi Broadc Corp., 2-3-8, Momochihama, Sawara-ku, Fukuoka 814-8585 **W:** www.rkb.ne.jp – **33)** Kumamoto Broadc Co., Ltd., 30, Yamasaki-machi, Kumamoto 860-8611 **W:** www.rkk.co.jp – **34)** Ibaraki Broadc System, 2084-2, Senba-cho, Mito 310-8505 **W:** www.ibs-radio.com – **35)** Aomori Broadc Corp., 8-1, Matsumori 1-chome, Aomori 030-8655 **W:** www.rab.co.jp – **36)** Nippon Broadc System, Inc., 9-3, Yurakucho 1-chome, Chiyoda-ku, Tokyo 100-8439 **W:** www.jolf.co.jp – **37)** Tohoku Broadc Co., Ltd., 26-1, Kasumi-cho, Yagiyama, Taihaku-ku, Sendai 982-8668 **W:** www.tbc-sendai.co.jp – **38)** Shikoku Broadc Co., Ltd., 5-2, Nakatokushima-cho 2-chome, Tokushima 770-8573 **W:** www.jrt.co.jp – **39)** Osaka Broadc Corp., 2-4, Benten 1-chome, Minato-ku, Osaka 552-8501 **W:** www.obc1314.co.jp – **40)** Tokai Radio Broadc Co., Ltd., 14-27, Higashisakura 1-chome, Higashi-ku, Nagoya 461-8503 **W:** www.tokairadio.co.jp – **41)** RCC Broadc Co., Ltd., 21-3, Moto-machi, Naka-ku, Hiroshima 730-8504 **W:** www.rcc.co.jp – **42)** Shizuoka Broadc System, 3-1-1, Toro, Suruga-ku, Shizuoka 422-8680 **W:** www.digisbs.com – **43)** RF Radio Nippon Co., Ltd., 5-85, Choja-machi, Naka-ku, Yokohama 231-8611 **W:** www.jorf.co.jp – **44)** Wakayama Broadc System, 3-3, Minato-honmachi, Wakayama 640-8577 **W:** www.wbs.co.jp – **45)** Gifu Broadc System, 52, Hashimotocho 2-chome, Gifu 500-8588 **W:** www.zf-web.com – **46)** Nishi-nippon Broadc Co., Ltd., 8-15, Marunouchi, Takamatsu 760-8575 **W:** www.rnc.co.jp – **47)** Sanyo Broadc Co., Ltd., 1-3, Marunouchi 2-chome, Okayama 700-8580 **W:** www.rsk.co.jp.
V: Most stns verify by QSL-card. Rec acc. Rp.

NIKKEI RADIO BROADCASTING CORPORATION (RADIO NIKKEI)

📠 9-15, Akasaka 1-chome, Minato-ku, Tokyo 107-8373 ☎ +81 3 3583 8151 📠 +81 3 3583 7441 **W:** www.radionikkei.jp

SW	kHz	kW	Prgr	SW	kHz	kW	Prgr
JOZ	3925	50	1	JOZ6	6115	50	2
JOZ4	*3925	10	1	JOZ3	9595	50	1
JOZ5	3945	10	2	JOZ7	9760	50	2
JOZ2	6055	50	1				

*) Nemuro; others Nagara (Chiba)
1st Prgr: Sun-Wed2225-1330 (Thu2225-Fri1415, Fri2155-Sat1200, Sat2155-Sun1200) on 3925/ 6055/9595kHz; as above except 2300-0750 on 3925kHz (Nemuro).
2nd Prgr: Sun-Thu 2300-0605, Fri & Sat 2300- 0900 (9760kHz: 0800)
IS: Slow tempo chime with Japanese instrument "Koto" at sign on and sign off - **V.** by QSL card. Rp.

COMMERCIAL FM STATIONS:

FM	Call	MHz	kW	Station, location & h of tr
1)	QU	76.1	1	FM Iwate, Morioka
2)	LU	76.1	1	FM Fukui, Fukui
3)	DW	76.1	10	Inter FM, Tokyo
4)	FW	76.1	1	Love FM, Fukuoka
5)	SV	76.4	1	R. Berry, Utsunomiya
6)	AW	76.5	10	FM CO-CO-LO, Osaka
7)	VW	76.8	1	FM Okayama, Okayama
8)	UV	77.0	1	E-Radio, Otsu
9)	JU	77.1	5	Date FM, Sendai
10)	SU	77.4	1	FM Kumamoto, Kumamoto
11)	VU	77.4	0.5	V-air, Matsue
12)	XU	77.5	1	FM Niigata, Niigata
13)		77.6	1	Kiss-FM, Himeji
14)	QV	77.8	10	ZIP FM, Nagoya
15)	NV	77.9	0.5	FM Saga, Saga
16)	GV	78.0	5	bayfm, Chiba
17)	GU	78.2	1	Hiroshima FM, Hiroshima
18)	YU	78.6	1	FM Kagawa, Takamatsuh
19)	RV	78.7	3	CROSS FM, Kitakyushu (Fukuoka)
20)	NU	78.9	3	Radio Cube, Tsu
21)	WV	79.0	1	FM Port, Niigata: (off air SS 1600-2100)
22)	KU	79.2	1	K-MIX, Hamamatsu (Shizuoka)
23)	UU	79.2	1	FM Yamaguchi, Yamaguchi
24)	HU	79.5	1	FM Nagasaki, Nagasaki
25)	DV	79.5	5	NACK 5, Saitama
26)	EU	79.7	1	FM Ehime, Matsuyama: 2057-1803 (Fri -2003, Sat -1703, Sun -1533)
27)	ZU	79.7	1	FM Nagano, Matsumoto (Nagano)
28)	OV	79.8	1	µ FM, Kagoshima
29)	WU	80.0	1	FM Aomori, Aomori
30)	AU	80.0	10	Tokyo FM, Tokyo
31)	XV	80.0	1	Radio 80, Ogaki (Gifu)
32)	FV	80.2	10	FM 802, Osaka
33)	FU	80.4	5	AIR-G', Sapporo
34)	EV	80.4	1	Rhythm Station, Yamagata
35)	HV	80.5	1	FM Ishikawa, Kanazawa
36)	CU	80.7	10	FM Aichi, Nagoya
37)	MV	80.7	1	FM Tokushima, Tokushima
38)	DU	80.7	3	FM Fukuoka, Fukuoka
39)	AV	81.3	10	J-WAVE, Tokyo
40)	LV	81.6	0.5	FM Kochi, Kochi
41)	TV	81.8	1	Fukushima FM, Koriyama(Fukushima)
42)	PV	82.5	5	FM North Wave, Sapporo
43)	OU	82.7	1	FM Toyama, Toyama
44)	PU	82.8	3	FM Akita, Akita
45)	CV	83.0	1	FM Fuji, Kofu 2000-1715 (Fri -1800, Sat -1700, Sat1930-Sun1530)
46)	MU	83.2	1	Joy FM, Miyazaki
47)	TU	84.7	5	FM Yokohama, Yokohama
48)	BU	85.1	10	FM Osaka, Osaka
49)	RU	86.3	1	FM Gunma, Maebashi
12)		86.5	1	FM Niigata, Yamato
11)		86.6	1	V-air, Hamada
50)	IU	87.3	1	FM Okinawa, Naha
51)	JV	88.0	1	FM Oita, Oita
52)	KV	89.4	3	Alpha-Station, Kyoto
13)		89.9	1	Kiss-FM, Kobe

NB: Relay stns below 1kW and community stns are not included.
Call: JO(call)-FM. Schedule: 24h unless otherwise indicated above. Most 24h stns are off the air for 2 to 5 hours until 1900, 2000 or 2100 on Sun.

Addresses and other information:
1) FM Iwate Broadc Co., 2-10, Uchimaru, Morioka 020-8512 **W:** www.fmii.co.jp – **2)** Fukui FM Broadc Co., Ltd., 1-1, Miyuki 1-chome, Fukui 910-8553 **W:** www.fmfukui.jp – **3)** FM Inter-wave Inc., 3-3, Higashishinagawa 1-chome, Shinagawa-ku, Tokyo 140-0002 - Prgr in English & foreign languages **W:** www.interfm.co.jp – **4)** Kyushu International FM Inc., 5-35, Tenjin 2-chome, Chuo-ku, Fukuoka 810-8565 Prgr in English, Chinese and Korean etc **W:** http://lovefm.co.jp – **5)** FM Tochigi Brordc co.,ltd., 2-1, Chuo 1-chome, Utsunomiya 320-8550 **W:** www.berry.co.jp – **6)** Kansai Intermedia Corp., 14-16, Nanko-Kita 1-chome, Suminoe-ku,

Osaka 559-8522 Foreign language prgr in English, Chinese, Korean, etc **W:** www.cocolo.co.jp – **7)** Okayama FM Broadc Co., Ltd., 1-8-45, Nakasange, Okayama 700-0821 **W:** www.fm-okayama.co.jp – **8)** FM Shiga Co., Ltd., 19-10, Nishinosho, Otsu 520-0818 **W:** www.e-radio. co.jp – **9)** Sendai FM Broadc., Inc., 10-28, Honcho 2-chome, Aoba-ku, Sendai 980-8420 **W:** www.datefm.co.jp – **10)** FM Kumamoto Broadc Co., Ltd., 5-50, Chibajomachi, Kumamoto 860-0001 **W:** http://fmk. fm – **11)** FM San-in Co., Ltd., 383, Tono-machi, Matsue 690-8508 **W:** www.fm-sanin.co.jp – **12)** FM Radio Niigata Co., Ltd., 3-5, Saiwainishi 4-chome, Chuo-ku, Niigata 950-8581 **W:** www.fmniigata.com – **13)** Kiss-FM KOBE Inc., 5-4 Hatoba-cho, Chuo-ku, Kobe 650-8589 **W:** www. kiss-fm.co.jp – **14)** ZIP-FM Inc., 20-17, Marunouchi 3-chome, Naka-ku, Nagoya 460-8578 **W:** http://zip-fm.co.jp – **15)** FM Saga Co., Ltd., 286-5, Fukuro, Honjo-machi, Saga 840-0023 **W:** www.fmsaga.co.jp – **16)** bayfm78 Co., Ltd., 2-6, Nakase, Mihama-ku, Chiba 261-7127 **W:** www. bayfm.co.jp – **17)** Hiroshima FM Broadc Co., Ltd., 8-2, Minamimachi 1-chome, Minami-ku, Hiroshima 734-8511 **W:** www.hfmweb.jp – **18)** FM Kagawa Broadc Co., Ltd., 4-23, Saiho-cho 1-chome, Takamatsu 760-8584 **W:** www.fmkagawa.co.jp – **19)** Cross FM Co., Ltd, 1-1, Kyomachi 3-chome, Kokurakita-ku, Kitakyushu 802-8570 **W:** www. crossfm.co.jp – **20)** Mie FM Broadc co., Ltd., 1043-1, Kannonji-cho, Tsu 514-8505 **W:** www.fmmie.jp – **21)** Niigata Kenmin FM Broadc Co., Ltd., 1-1, Bandai 2-chome, Chuo-ku, Niigata 950-8579 **W:** www.fmport. com – **22)** Shizuoka FM Broadc Co., Ltd, 133-24, Tokiwa-cho, Naka-ku, Hamamatsu 430-8575 **W:** www.k-mix.co.jp – **23)** FM Yamaguchi Co., Ltd., 3-31, Midori-cho, Yamaguchi 753-8521 **W:** www.fmy.co.jp – **24)** FM Nagasaki Co., Ltd., 5-5, Sakae-machi, Nagasaki 850-8550 **W:** www. fmnagasaki.co.jp – **25)** FM Nack 5 Co Ltd., 682-2, Nishiki-cho, Omiya-ku, Saitama 330-8579 **W:** www.nack5.co.jp – **26)** FM Ehime Broadc Co., 1-10-7, Takewara-machi, Matsuyama 790-8565 **W:** www.joeufm.co.jp – **27)** Nagano FM Broadc Co., Ltd, 13-5, Honjo 1-chome, Matsumoto 390-8520 **W:** www.fmnagano.co.jp – **28)** FM Kagoshima Co., Ltd., 1-38, Higashisengoku-cho, Kagoshima 892-8579 **W:** www.myufm.jp – **29)** Aomori FM Broadc Co., Ltd., 7-19, Tsutsumi-machi 1-chome, Aomori 030-0812 **W:** www.afb.co.jp – **30)** Tokyo FM Broadc Co., Ltd., 1-7 Kojimachi, Chiyoda-ku, Tokyo 102-8080 **W:** www.tfm.co.jp – **31)** Gifu FM Broadc Co., Ltd., 35-10, Kono 4-chome, Ogaki 503-8580 **W:** www. radio-80.com – **32)** FM 802 Co., Ltd., Kita 2-6, Tenjinbashi 2-chome, Kita-ku, Osaka 530-8580 **W:** www.funky802.com – **33)** FM Hokkaido Broadc Co., Ltd., 1, Nishi 2-chome, kita 1-jo, Chuo-ku, Sapporo 060-8532 **W:** www.air-g.co.jp – **34)** FM Yamagata Co., Ltd., 14-69, Matsuyama 3-chome, Yamagata 990-9543 **W:** www.rfm.co.jp – **35)** FM Ishikawa Broadc Co., Ltd., 1-45, Hikoso-machi 2-chome, Kanazawa 920-8605 **W:** http://hellofive.jp – **36)** FM Aichi Broadc Co., Ltd., 15-18, Chiyoda 2-chome, Naka-ku, Nagoya 460-8388 **W:** http://fma.co.jp – **37)** FM Tokushima Broadc Co., 1-6, Saiwai-cho, Tokushima 770-8567 **W:** www. fm807.jp – **38)** Fukuoka FM Broadc Co., Ltd., 9-19, Kiyokawa 1-chome, Chuo-ku, Fukuoka, 810-8575 **W:** www.fmfukuoka.co.jp – **39)** J-WAVE Inc., Roppongi Hills Mori Tower 33F, 6-10-1, Roppongi,, Minato-ku, Tokyo 106-6188 **W:** www.j-wave.co.jp – **40)** FM Kochi Broadc Co., Ltd., 1-5, Takashocho 2-chome, Kochi 780-8532 **W:** www.fmkochi.com – **41)** FM Fukushima Inc., 4-4 Shinmei-cho, Koriyama, 960-8013 **W:** www. fmf.co.jp – **42)** FM North Wave Co., Ltd., 1, Nishi 4-chome, Kita 7-jo, Kita-ku, Sapporo 060-8557 **W:** http://825.fm/northwave – **43)** Toyama FM Broadc Co., Ltd., 2-11, Okuda-machi, Toyama 930-8567 **W:** www. fmtoyama.co.jp – **44)** FM Akita Broadc Co., Ltd., 7-10, Yabase-Honcho 3-chome, Akita 010-0973 **W:** www.fm-akita.co.jp – **45)** FM Fuji Co Ltd., Aria 105, Kawadamachi, Kofu 400-8550 **W:** www.fmfuji.co.jp – **46)** Miyazaki FM Broadc Co., Ltd., 2-78, Gion, Miyazaki 880-8583 **W:** www.joyfm.co.jp – **47)** Yokohama FM Broadc Co., Ltd., 2-2-1, Minato-Mirai, Nishi-ku, Yokohama 220-8110 **W:** www.fmyokohama.co.jp – **48)** FM Osaka Co., Ltd., 3-1, Minatomachi 1-chome, Naniwa-ku, Osaka 556-8510 **W:** http://fmosaka.net – **49)** FM Gunma Broadc Co., Ltd., 4-8, Wakamiyacho 1-chome, Maebashi 371-8533 **W:** www.fmgunma. com – **50)** FM Okinawa Broadc Corp., 40, Kowan, Urasoe, Okinawa 901-2525 **W:** www.fmokinawa.co.jp – **51)** FM Oita Broadc., Co., Ltd., 17-19, Higashikasuga-machi, Oita 870-8558 **W:** www.fmoita.co.jp – **52)** FM Kyoto, Inc., CoCon Karasuma 8F, 620, Suiginya-cho, Karasuma-dori Shijo-sagaru, Shimogyo-ku, Kyoto 600-8566 **W:** www.fm-kyoto.jp
V. Most stns verify by QSL card. Rec acc. Rp. **NB:** Aichi International Broadc Co., Ltd. (Radio-I) 79.5MHz discontinued since Oct. 1, 2010.

THE OPEN UNIVERSITY OF JAPAN (HOSO DAIGAKU)
✉ Hoso Daigaku, 2-11, Wakaba, Mihama-ku, Chiba 261-8586 ☎ +81 43 276 5111 **W:** www.ouj.ac.jp
FM: JOUD-FM 77.1MHz 10kW, Tokyo. 78.8MHz 1kW, Maebashi **D. Prgr:** 2100-1500 **V.** by QSL card. Rp.

Digital Radio: Digital system testing in Tokyo by Digital Radio Promotion Association (DRP) since October 2003 on 190.214286MHz.

Service is not DAB, but the ISDB-T system (Integrated Services Digital Broadcasting for Terrestrial): Tokyo (JOAZ-FM) on 2.4kW. Txs to be phased out July 2011.

AMERICAN FORCES NETWORK (AFN) (U.S. Mil.)
The network serves the members of the US forces. The stns in Japan broadcast by authority of Commander, US Forces, Japan, in cooperation with the Information and Communications Policy Bureau in Japan. Stns are linked by land line and microwave.
⊡ **AFN Tokyo,** Det 10, Unit 5091 Bldg 3266, Yokota Air Base, Fussa, Tokyo 197-0001 or Det 10, Unit 5091 Bldg 3266, APO/AP 96328-5091 ☎ +81 42 552 2511 ext 52374 🖷 +81 42 552 2511 ext 52386 **E:** AFN. Eagle810@yokota.af.mil **W:** www.afntokyo.com
Other stns: AFN Okinawa: Okinawa **E:** AFNRadio@kadena.af.mil **W:** www.kadenaforcesupport.com/afn – **AFN Misawa:** Misawa, Aomori **E:** afn@misawa.af.mil **W:** http://myafn.dodmedia.osd.mil/affiliateinfo. aspx?a=221 **AFN Iwakuni:** Iwakuni, Yamaguchi **E:** det13bg@iwakuni. usmc.mil **W:** http://myafn.dodmedia.osd.mil/affiliateinfo.aspx?a=220 – **AFN Sasebo:** Sasebo, Nagasaki **E:** roger.dutcher@sasebo.navy.mil **W:** www.cnic.navy.mil/sasebo/AFN/index.html

MW	kHz	kW	MW	kHz	kW
Okinawa	648	10	Misawa	1575	0.6
Tokyo	810	50	Sasebo	1575	0.25
Iwakuni	1575	1			

FM: Okinawa 89.1MHz 20kW.
D. Prgr: 24h **N:** on the h. **Ann:** "This is the American Forces Network" **V.** by QSL card or letter.

L.T: UTC +2h (31 Mar-28 Oct: +3h) — **Pop:** 6.4 million — **Pr.L:** Arabic — **E.C:** 50Hz, 230V — **ITU:** JOR

AUDIOVISUAL COMMISSION (AVC)
✉ P.O.Box 142515, Amman 11814 ☎+ 962 6 5549720 🖷 + 962 6 5535093 **W:** www.avc.gov.jo **LP:** Dir: Hussein Bani Bani.

JORDAN RADIO & TELEVISION CORP. (JRTV, Gov.)
✉ Al-Shara Al-Musharrafah St, P.O.Box 1041, JO-11118 Amman ☎+962 6 4773111 🖷+962 6 4778 578 **W:** jrtv.gov.jo **E:** rj@jrtv.gov.jo
L.P: CEO: Saleh Al-Kallab. Dir. Radio: Mazen Majali, Dir. Eng: Sufian Nabulsi. Dir. Int. Rel: Mrs. Jihan Al-Haeyk.

MW	kHz	kW	Prgr	Times
Amman	612	200	Arabic	0330-0015
Ruweished	693	50	Arabic	0330-0015
Ajlun	‡801	*2000	Arabic	0330-0015
Amman	855	10	Arabic	0430-2000
Amman	1035	20	Arabic	0430-2000
Aqaba	1485	10	Arabic	0330-0015
Al Karanah	‡1494	1000	Arabic	0330-0015

*200kW 0530-1845. ‡ inactive

FM	Main	Amman FM	English	Quran	kW
Ajlun	106.7	95.8	90.9		10/5
Amman	90.0	99.0	96.3	93.1	5/10
Aqaba	101.5	105.6	99.7	92.0	5/1
Irbid	103.8	95.4		98.7	1
Kerak	103.6				1
Salt		105.0			1
Tafeleh		90.8			1
Zarqa	100.6	88.0			1

Main Arabic sce: 24h. **N:** on the h. (not 0400, 1100 Fri, 1300, 1700). Jordan Armed Forces R: 1400-1600. Also relayed on SW. **Amman FM,** in Arabic: 0430-2000 (Irbid with some local prgr). **English prgr:** 0300-2400. **French:** 1400-1600 on English prgr fqs. **Quran Prgr:** 0500-2200 on FM & MW 855kHz.
Hawa FM, Amman: 105.9MHz. Web: www.ammancity.gov.jo
Ann: Arabic: "Huna Amman, Idha'atu-l-Mamlaka al-Urdoniya al-Hashemiya". Armed forces R: "Idha'at Al-Quwaat Al-Musala al-Urdoniya, al-Gayish al-Arabi". E: "This is R. Jordan broadcasting from Amman".

EXTERNAL SERVICE: R. Jordan see International Radio section.

Other stations:
Amen FM: Amman/Aqaba 89.5MHz, Irbid 89.7MHz. **W:** amenfm. jo – **Ayyam FM:** Amman 91.5MHz, Irbid 91.9MHz, Petra 92.1MHz. **W:** ayaamfm.jo – **Beat FM,** Amman: 102.5MHz. English. **W:** www. mybeat.fm – **Energy FM,** Amman: 97.7MHz. English. **W:** energyradio. jo – **Hayat FM:** Irbid 94.7MHz, Amman 104.7MHz. **W:** www.hayat. fm – **Mazaj FM:** Amman 95.3MHz, Irbid 101.7MHz. **W:** mazajfm. com – **Melody FM:** Amman 91.1MHz, Zarqa 105.5MHz. – **Mood FM,** Amman: 92.0MHz. English. **W:** mood.fm – **Play FM,** Amman: 99.6MHz. English. **W:** play.jo – **R. Al-Balad,** Amman: 92.4MHz **W:** ammannet.

net – **R. Fann:** Aqaba 91.1MHz, Irbid/Ruweished 91.3MHz, Ajlun/Karak 94.3MHz, Salt/Tafileh 94.7MHz, Amman 102.1/104.2MHz, Petra/Azraq 105.4MHz. **W:** radiofann.com – **R. Farah Al-Nas,** Amman: 98.5MHz. **W:** farahalnas.jo – **Rotana R:** Irbid 90.5MHz, Amman 99.9MHz. **W:** rotana.net – **Sawt Al-Janoub,** Ma'an: 90.5MHz. – **Sawt al-Madina:** Amman: 88.7MHz. **W:** sawtalmadenah.net – **Sawt el-Ghad:** Amman 101.5MHz. **W:** sawtelghad.com – **Spin Jordan:** Irbid 88.3MHz, Ma'an 88.5MHz, Amman 94.1MHz, Aqaba 103.5MHz. English. **W:** spin.jo – **Sunny,** Amman: 105.1MHz. English. **W:** sunny.jo – **Watan FM:** Irbid 92.7MHz, Amman 100.3MHz. **W:** watanonline.net – **Yarmouk FM,** Irbid: 105.7MHz. **W:** www.yu.edu.jo
BBC Arabic Sce: Amman 103.1MHz 5kW, Ajlun 89.1MHz 10kW.
Monte Carlo Doualiya/DW: Amman 97.4MHz, Ajlun 106.2MHz.
R. Sawa: Amman 98.1MHz 10kW, Ajlun 107.4MHz.
F.PI: Virgin R. Jordan, Amman: 93.7MHz. Web: virginradiojordan.com

KAZAKHSTAN

L.T: UTC +6h (exc. West Kazakhstan: +5h) — **Pop:** 15.4 million — **Pr.L:** Kazakh, Russian — **E.C:** 50Hz, 220V — **ITU:** KAZ

MÄDENÏET MÏNÏSTRLIGI (Ministry of Culture)
✉ House of Ministries, 010000 Astana ☎ +7 7172 740251
W: www.mki.gov.kz **L.P:** Minister: Muxtar Abraruli Qul-Muxammed
NB. The ministry's Committee of Information and Archives is responsible for issuing broadcasting licenses.

"QAZAQSTAN" RESPWBLĪKALIQ TELERADÏO-KORPORACÏYASI (Gov)
(Republican Broadcasting Corp. "Qazaqstan")
✉ Almati broadcasting house: Jeltoqsan kös. 175a, 050013 Almati ☎ +7 727 2721360 ✉ +7 727 2613777 ✉ Astana broadcasting house: Moskovskaya kös. 55a, 010032 Astana ☎ +7 7172 394969 ✉ +7 7172 393159 **W:** www.kaztrk.kz; www.kazradio.kz
L.P: Dir (Radio): Galimjan S. Meldesov

MW	kHz	kW	Prgr			
Aqtaw	1341	25	QR			
FM (MHz)	**QR**	**S**	**FM**		**QR**	**S**
Almati	101.0	106.5	Qizilorda		102.0	101.0
Aqtaw	100.1	102.1	Öskemen		104.0	105.6
Aqtöbe	102.2	105.7	Pavlodar		101.0	-
Astana	106.8	100.4	Petropavl		106.8	-
Atiraw	101.2	102.8	Semey		100.1	104.4
Köksetaw	101.0	-	Simkent		100.0	102.7
Oral	101.2	103.2	Taraz		100.8	102.6
Qaragandi	104.3	102.3	Türkistan		101.0	-
Qostanay	105.4	-	Ülken Sagan		101.0	-

+ translators.
D.Prgrs: Qazaq Radïosi (QR) in Kazakh, Russian: 24h. For ethnic minorities ("Dostiq"): Mon-Sat 1120-1140 (Mon German, Tue Uighur, Wed Korean, Thur Turkish, Fri Tatar, Sat Azeri) – **Salqar (S)** in Kazakh: 0000-1800. – **Local station: Astana Radïosi** on Astana 101.4MHz in Kazakh, Russian: 24h. **E:** 317577@mail.ru.

"QAZAQSTAN" RTRK REGIONAL STATIONS (Gov)
The regional branches (oblistiq fïlïali) of "Qazaqstan" RTRK are broadcasting on own FM frequencies at various times in Kazakh and languages of ethnic minorities.
a) Aqmola oblistiq fïlïali: Awezov kös. 230, 020000 Köksetaw. E: akmol_tv@mail.ru – **b) Aqtöbe oblistiq fïlïali:** Axtanov kös. 54, 030000 Aqtöbe. E: atrk@aktobe.kz – **c) Atiraw oblistiq fïlïali:** Moldagalïev kös. 29, 060005 Atiraw. E: bayan-baha@mail.ru – **d) Batis Qazaqstan oblistiq fïlïali:** Amanjolov kös. 104, 090000 Oral. E: zapad_tv@mail.ru – **e) Jambil oblistiq fïlïali:** Swleymenov kös. 6, 080000 Taraz. E: otrk@taraz.kz – **f) Mañgistaw oblistiq fïlïali:** 24 iqsam awdani, 130000 Aqtaw. E: gulek.aktau@mail.ru – **g) Pavlodar oblistiq fïlïali:** Derïbas kös. 21, 140000 Pavlodar. E: oblradio@nursat.kz . On Pavlodar 100.5MHz + translators. – **h) Qaragandi oblistiq fïlïali:** Jaw-ïnternacïonalïstov kös. 14, 100000 Qaragandi. E: obltrk@nursat.kz – **i) Qostanay oblistiq fïlïali:** Puskïn kös. 54, 110000 Qostanay. E: office@otrk.kst.kz – **j) Qizilorda oblistiq fïlïali:** Jeltoqsan kös. 11, 120014 Qizilorda. E: o14@kazakstan.kz – **k) Semey oblistiq fïlïali:** Sugaev kös. 157, 071403 Semey. On Semey 106.9MHz (R.7). E: semey-tv@kazakstan.kz – **l) Soltüstik Qazaqstan oblistiq fïlïali:** Brwsïlovskïy kös. 1, 150000 Petropavl. E: otrk@inbox.ru – **m) Sigiz Qazaqstan oblistiq fïlïali:** Staxanov kös. 70, 070020 Öskemen. E: vktrk@ukg.kz – **n) Öñtüstik Qazaqstan oblistiq fïlïali:** Qazibek bi kös. 20, 160000 Simkent. E: uktv_aha@mail.ru.

OTHER STATIONS
MW	kHz	kW	Location	Station
B)	*1098	-	Almati	BBC relay

MW	kHz	kW	Location	Station
B)	*1188	-	Sariagas	BBC relay
B)	*1197	10	Astana	BBC relay
A)	1341	30	Almati	RFE/RL relay
B)	*1440	-	Qizilorda	BBC relay

*) lease by BBC in 2011 subject to confirmation. Txs may carry Qazaq R. as filler when no BBC prgr is relayed.

FM	MHz	kW	Location	Station
B)	66.80	4	Simkent	BBC relay
A)	70.19	1	Simkent	˙RFE/RL relay
A)	71.21	1	Semey	RFE/RL relay
4)	101.2	1	Simkent	R. Yumaks
5)	101.4	1	Aqtaw	Tengri FM
5)	101.4	1	Semey	Tengri FM
2)	102.8	1	Almati	R. Xabar
6)	103.2	1	Astana	RDV
3)	103.5	2	Almati	R. 31
5)	103.5	1	Öskemen	Tengri FM
1)	104.0	1	Qaragandi	Europa+ Kazakhstan
5)	104.5	1	Astana	Tengri FM
5)	104.7	1	Simkent	Tengri FM
5)	105.0	1	Astana	Europa+ Kazakhstan
5)	105.8	1	Oral	Tengri FM
5)	107.0	1	Almati	Europa+ Kazakhstan
5)	107.6	1	Köksetaw	Tengri FM
5)	107.7	1	Jambil	Tengri FM

NB: Txs below 1kW not listed.
Addresses & other information:
1) Respwblïk alana 13, 050013 Almati. **E:** program@europaplus.kz – **2)** Jeltoqsan kös. 185, 050013 Almati. **E:** radio@khabar.kz – **3)** Minbaev kös. 53, 050057 Almati. **E:** radio@31.kz – **4)** Sairam kös. 198, 160008 Simkent. **E:** reklama@umax.kz – **5)** Begalïn kös. 148, 050051 Almati. – **6)** Astana. – **A)** Rel. RFE/RL (USA) – **B)** Rel. BBC (UK).
NB: There are more than 40 independent radio stations in Kazakhstan. They generally use txs with a power of less than 1kW

KENYA

L.T: UTC +3h — **Pop:** 38 million — **Pr.L:** English, Swahili, Kikuyu, Luhya, Luo, Kalenjin, Somali, others — **E.C:** 50Hz, 240V — **ITU:** KEN

COMMUNICATIONS COMMISSION OF KENYA (CCK)
+ P.O. Box 14448, Nairobi 00800 ☎+254 20 4242000 ✉ +254 20 4451866 **W:** www.cck.go.ke **E:** info@cck.go.ke

KENYA BROADCASTING CORPORATION (KBC, Pub.)
✉ P.O. Box 30456, Nairobi 00100 ☎+254 20 318823 ✉ +254 20 2229658 **W:** www.kbc.co.ke **E:** md@kbc.co.ke **L.P:** Chmn: Charles Musyoki Muoki.

MW	kHz	kW	Netw.	MW	kHz	kW	Netw.
Voi	540	100	S	Malindi	*927	100	E
Kapsimotwa+	558	25	W	Nyamninia+	954	100	E
Garissa	567	50	S	Voi	981	100	E
Ngong++	612	100	S	Malindi	*1044	100	E
Garissa	639	50	E/N	Maralal	1107	100	E
Marsabit	675	50	S	Kitale	1134	50	E
Marania+++	702	100	S	Wajir	1152	50	S
Ngong++	747	100	E/C	Marsabit	1233	50	E
Nyamninia+	846	100	S	Wajir	1305	50	E/N
Kitale	882	50	S	Maralal	1386	100	E
Marania+++	900	100	E				

+) near Kisumu, ++) near Nairobi, +++) near Meru. *=inactive.

FM	S	E	W	Metro	Coro	Pwani
Limuru**	92.9	95.6	-	101.9	99.5	-
Malindi	90.1	93.3	-	96.5	-	-
Nyambene	90.4	103.5	-	-	100.3	-
Nyeri	87.8	100.7	-	97.0	102.3	-
Mombasa	100.8	104.4	-	89.1	-	103.1
Timboroa	88.6	91.5	-	-	-	-
Nakuru	104.1	-	-	94.5	-	-
Eldoret	-	-	-	97.9	-	-
Kapsimotwa	-	-	-	100.9	-	-
Kisumu	104.5	-	100.2	87.7	-	-
Kisii	103.3	-	-	-	-	-
Nyadundo	-	-	-	-	99.7	-

**Limuru txers serve the Greater Nairobi area.
Networks (from Nairobi studios unless stated):
S=Swahili Sce ("KBC Idhaa ya Taifa): 0200-2110. Also rel. China R. Int. – **E=English Sce:** 0200-2105 (on FM non-stop music 2105-0200). Includes schools prgrs & relays of China R. Int. at 1630 and other times. Some MW freqs carry N & C sces at times – **N=(North) Eastern Sce.** in local langs: MF variable between 0900-1905 – **C=Central Sce.** in local langs: Mon-Sat variable times between 0200-2010 – **W=Western Sce.** from Kisumu studios in local langs

– **Metro FM** ("House of Reggae") in English/Swahili: 24h – **Coro FM** in Kikuyu – **Pwani FM** from Mombasa studios.
F.PI: Metro FM, Coro FM and Pwani FM to be privatized and run by separate commercial entity.
Ann: E: "This is KBC, Nairobi". **IS:** Flute & drum melody. Nat. Anthem at s/on and s/off.

ROYAL MEDIA SERVICES LTD. (RMS)
✉ P.O. Box 7468, Nairobi 00300 ☎+254 20 2721415/6 🖷 +254 20 2724211 Web: royalmediaservices.co.ke Email: info@royalmedia.co.ke **L.P:** Owner: Samuel K. Macharia. MD: Wachira Waruru.
FM: R. Citizen in Swahili/Eng: Nairobi 106.7MHz, Chuka 93.2MHz, Eldoret 90.4MHz, Garissa 95.7MHz, Kapenguria 94.5MHz, Kisii 95.1MHz, Kisumu 97.6MHz, Kitui 89.9MHz, Malindi 97.4MHz, Meru 94.3MHz, Mombasa 97.3MHz, Nakuru 100.5MHz, Namanga 106.7MHz, Nyeri 104.3MHz, Voi 91.8MHz – **Hot 96 FM** in Eng/Sheng/Swahili: Nairobi 96.0MHz, Eldoret 87.6MHz, Kisumu 103.1MHz, Mombasa 90.4MHz, Nakuru 102.5MHz, Nyeri 88.6MHz.
RMS also operates the following stns for specific lang. communities (freqs in Nairobi unless stated): **Bahari FM** in Swahili & coastal langs: Mombasa 94.2MHz – **Chamgei FM** in Kalenjin: 90.4MHz, Nakuru 95.0MHz, Eldoret 97.5MHz – **Egesa FM** in Kisii: 103.2MHz, Kisii 94.6MHz – **Inooro FM** in Kikuyu: 98.9MHz, Chuka 102.0MHz, Eldoret 107.0MHz, Muranga 96.9MHz, Meru 95.1MHz, Mombasa 99.2MHz, Nakuru 89.8MHz, Nyadundo 88.9MHz, Nyeri 97.8MHz – **Mulembe FM** in Luhya: 97.9MHz, Rift Valley 94.0MHz, Western Province 89.6MHz – **Musyi FM** in Kamba: 102.2MHz, Kitui 103.6MHz – **Muuga FM** in Meru: Meru 88.9MHz – **Ramogi FM** in Luo: 107.1MHz, Kisumu 107.6MHz, Nakuru 95.4MHz, Siaya 98.6MHz – **Wimwaro FM** in Embu: Embu 93.0MHz.

RADIO AFRICA LTD.
✉ P.O. Box 74497, Nairobi ☎+254 20 4244000 🖷 +254 20 4447410 **W:** kissfm.co.ke **E:** info@kissfm.co.ke **L.P:** MD: Patrick Quarcoo.
FM: Kiss 100 in Eng/Swahili: Nairobi 100.3MHz, Eldoret 89.1MHz, Kisumu 92.5MHz, Meru 93.5 MHz, Mombasa 88.7MHz, Nakuru 98.1MHz, Nyeri 100.1MHz, Webuye 104.7MHz – **Classic 105 FM** in Eng/Swahili: Nairobi 105.2MHz, Mombasa 107.5MHz, Nakuru 94.9MHz, Meru 105.5MHz – **East FM** (Asian): Nairobi 106.3MHz, Mombasa 89.5MHz – **Jambo FM** (sports): Nairobi 97.5MHz & relays elsewhere – **X FM** (rock music): Nairobi 105.5MHz – **Smooth FM** (R&B music): Nairobi 103.5MHz.

NATION MEDIA GROUP LTD.
✉ P.O. Box 49010, Nairobi 00100 ☎+254 20 3288000 **W:** www.nationmedia.com **L.P:** Chmn: Dr Martin Aliker. CEO: Linus Gitahi. MD Broadc. Div: Ian Fernandes.
FM: Easy FM in Eng/Swahili: Nairobi 96.3MHz, Eldoret 102.7MHz, Kisumu 102.1MHz, Meru 93.9MHz, Mombasa 101.5MHz, Nakuru 97.7MHz, Nyeri 104.9MHz – **QFM** in Swahili: Nairobi 94.4MHz, Eldoret 96.7MHz, Meru 107.1MHz, Mombasa 87.9MHz, Nakuru 103.3MHz, Nyeri 90.9MHz.

OTHER FM STNS IN NAIROBI (including relays elsewhere; freqs are in Nairobi unless stated): **Biblia Husema Broadcasting** (Christian): 90.7MHz, Eldoret 96.3MHz, Lokichokio 102.5MHz, Nakuru 102.9MHz, Machakos 96.7MHz, Timboroa 101.5MHz – **Capital FM** (in Eng): 98.4MHz, Garissa 102.7MHz, Kitui 106.5MHz, Malindi 104.5MHz, Meru 103.9MHz, Mombasa 98.4MHz, Nakuru 98.5MHz, Nyeri 98.5MHz, Timboroa 93.0MHz, Voi 104.9MHz – **ECN FM** (Kenya Institute of Mass Communication): 104.7MHz – **East Africa R.** (in Eng/Swahili - relay of Tanzanian stn): 94.7MHz – **R. 316** (Christian, formerly Family FM): 103.9MHz, Kisumu 96.5MHz, Mombasa 97.9MHz, Nakuru 102.1MHz – **Frontier FM** (in Somali, also known as Garissa FM): 88.7MHz, Garissa 107.5MHz, Wajir 88.9MHz. Also rel. VOA – **Ghetto R.:** 89.5MHz – **Homeboyz R.:** 91.5MHz – **Hope FM** (Pentecostal Church): 93.3MHz, Mombasa 101.9MHz, Timboroa 93.9MHz – **Iqra FM** (Islamic): 95.1MHz – **Kameme FM** (mainly in Kikuyu): 101.1MHz, Eldoret 101.9MHz, Nakuru 99.3MHz, Nyeri 92.3MHz, Meru 88.3MHz. Also rel. BBC – **Kass FM** (in Kalenjin): Nairobi 89.1MHz, Eldoret 90.0MHz, Kisii 99.3MHz, Kisumu 91.0MHz, Mombasa 102.7MHz, Nakuru 92.5MHz – **Milele FM** (in Swahili): 93.6MHz, Eldoret/Kapenguria 88.3MHz, Kibwezi 104.3MHz, Kisumu 99.7MHz, Malindi 101.3MHz, Meru 101.5MHz, Mombasa 96.7MHz, Nakuru/Nyahururu 90.7MHz, Nyeri 91.7MHz, Taita-Taveta 88.7MHz, Webuye 92.7MHz – **R. Maisha** (owned by Standard newspaper): 102.7MHz, Kisumu 105.3MHz, Mombasa 105.1MHz, Meru 105.1MHz, Nakuru 104.5MHz, Nyeri 105.7MHz – **Sound Asia:** 88.0MHz, Mombasa 89.9MHz – **Star FM** (in Somali/Swahili/Eng): 105.9MHz, Daadab/Garissa 97.1MHz, Mandera 97.5MHz, Wajir 97.3MHz. Also rel. BBC. Web: starfm.co.ke – **R. Umoja:** 101.5MHz, Kisumu 97.3MHz, Mombasa 94.7MHz, Nakuru 87.7MHz – **R. Waumini** (Catholic Church):

88.3MHz. Also rel. Vatican R.
NB: 99.9MHz is assigned for use in Nairobi by very low-powered community stns: There are many private FM stns outside Nairobi. A full list of allocations is at: **W:** cck.go.ke/licensing/broadcasting/register_radio.html
RELAYS OF INT. STNS: BBC WS (Eng/Swahili): Nairobi 93.9MHz, Mombasa 93.9MHz, Kisumu 88.1MHz – **VOA** (Eng/Swahili): Nairobi 107.5MHz – **RFI Afrique** (F/E/Swahili): Nairobi 89.9MHz, Mombasa 105.5MHz – **China R. Int.** (Eng/Swahili/Chinese): Nairobi 91.9MHz

KIRIBATI

L.T: UTC +12h — **Pop:** 112,850 — **Pr.L:** I-Kiribati, English, Gilbertese — **E.C:** 50Hz, 240V — **ITU:** KIR

RADIO KIRIBATI (Broadcasting and Publications Authority)
✉ PO Box 78, Bairiki, Tarawa **Mgr:** Tibwere Bobo ☎ +686 21187 🖷 +686 21096. **E:** bpa@tskl.net.ki. **W** streaming: http://radio-tarawa.tskl.net.ki
MW: Bairiki 846kHz 10kW (sometimes silent because of lack of staff, parts and funds) **FM:** 99.0MHz 0.1kW
D.Prgr: I-Kiribati (90%) English (10%): 1800-2000, 0000-0130, 0530-0930 **N. in English:** 0600 (rel.BBC, Radio Australia, Radio NZ International) followed by local news bulletin.
Ann: "This is Radio Kiribati, the national broadcasting service of Kiribati in the Central Pacific" "Aio bwanaan Kiribati te botaki ni kanako bwanaa I bukin Kiribati I nukan te Betebeke."
F.PI: Extension of service to eastern islands when funding available.

KIRITIMATI RADIO (Ministry for Line & Phoenix Island Development)
✉ Ronton, Kiritimati Island, Kiribati, Central Pacific ☎ +686 81211 🖷 +686 81278 **E:** ttemoku@hotmail.com **L.P:** Hon. Tarvita Temoku
FM: Ronton (London) 93.5MHz 0.5kW **Prgr:** Satellite feed from Radio Kiribati 99.0 FM and some local prgrs in Gilbertese language.

Other Stations

FM	Locatin	MHz	Station
3)	Ambo, Betio	89.0	Newair FM89
1)	Bairiki	90.0	Radio Australia
2)	Bairiki	95.0	BBC World Service
2)	Tarawa	100.0	BBC World Service
3)	Bairiki	101.0	Newair FM101

1) 24h satellite relay from Melbourne – **2)** 24h satellite relay from London – **3)** PO Box 204, Bairiki, Tarawa. **L.P:** Ieremia Tabai. **D.Prgr:** Local commercial prgrs in English & I-Kiribati ☎ +686 21671 **E:** newairfm89kiribati@gmail.com

Community FM: Fanning Island, Washington Island, Bairiki, Bairiki: St.Paul's Roman Catholic Church. (Awaiting frequency details)

KOREA (North, DPR)

L.T: UTC +9h — **Pop:** 26 million —**Pr.L:** Korean — **E.C:** 60Hz, 100/200/220V — **ITU:** KRE

THE RADIO AND TELEVISION BROADCASTING COMMITEE OF THE DEMOCRATIC PEOPLE'S REPUBLIC OF KOREA
✉ Jonsong-dong, Moranbong District, Pyongyang ☎ +850 2 816035 **L.P:** Chairman: Cha Sung Su.

**KOREAN CENTRAL BROADCASTING STATION
(Joson Jung-ang Pangsong)**
✉ Jonsung-dong, Moranbong District, Pyongyang ☎ +850 2 812301

MW	kHz	kW	Prgr	MW	kHz	kW	Prgr
Chongjin	702	50	C/R	Wonsan	882	250	C/R
Wiwon*	720	500	C/R	Hwangju+	927	50	C/R
Hyesan	765	50	C/R	Hamhung	999	250	C/R
Kaesong	810	50	C/R	Haeju	1080	1500	C/R
Pyongyang	819	500	C	Pyongyang	1368	2	E
Sinuiju	873	250	C/R				

SW	kHz	Prgr	SW	kHz	Prgr
Sariwon	2350	C/R	Kanggye	3960	C/R
Pyongyang	2850	C	Wonsan	3970	C/R
Hamhung	3220	C/R	Kanggye	6100	C
Pyongsong	3350	C/R	Pyongyang	9665	C
Hyesan	3920	C/R	Kanggye	11680	C
Chongjin	3940	C/R			

*= Kanggye, += Sariwon, C = Central Broadcast from Pyongyang, R = Regional Sce, E = rel. Ext. Sce.
NB: all freqs variable **FM:** Kaesong 102.3MHz
D.Prgr in Korean: 2000-1800 on all freqs exc. 6100 (2000-0630 &

1330-1800). **N:** 2100, 2200, 0100, 0300, 0600, 0800, 1100, 1200, 1300. Regional Prgrs: W0500-0600. Rel. Pyongyang Broadc. St: 1500-1800 on 702/720/864kHz. 1500-2000 on 102.3MHz. 1800-2000 on 3220/3940/3970kHz

Ann: "Joson Jung-ang Pangsong-imnida". Reg. Prgrs: "(location) Pangsong-imnida". **IS:** Song of General Kim Il Sung. Opening & closing music: Nat. Anthem. **V:** not verified.

EXTERNAL SERVICES: Voice of Korea, Pyongyang Broadcasting Station, Pyongyang Branch of the Anti-Imperialist National Democratic Front – See International Broadcasting section

PYONGYANG FM BROADCASTING STATION
(Pyongyang FM Pangsong)

FM	MHz	kW	FM	MHz	kW
Pyongsong	90.1	2	Komdok	102.1	1
Kaesong	92.5	2	Sariwon	103.0	2
Kanggye	93.3	5	Haeju	103.7	10
Hyesan	93.8	2	Pyongyang	105.2	20
Wonsan	95.1	5	Chongjin	105.5	10
Heaju	97.8	10	Hamhung	106.1	20
Sinuiju	101.3	5	Nampo	107.2	2

D.Prgr: 0700-2000, 2100-2400 (National holidays: 2100-2030) (music, drama and novel).

Ann: "Pyongyang FM Pangsong-imnida". **IS:** Song of General Kim Jong Il. Opening music: Pyongyang Is My Heart

FRONTLINE SOLDIERS RADIO
(Jonyon Chobyongdurul Wihan Pangsong)

MW & SW: 1613vkHz(irr.), 3025kHz (irr.) **D.Prgr in Korean:** 2030-2230 and 0730-0930. Almost relay Korean Central Broadcasting Station. Times and freqs are variable.

Ann: "Jonyon Chobyongdurul Wihan Pangsong-imnida"

KOREAN PEOPLE'S ARMY FM BROADCASTING STATION
(Josong Inmingun FM Pangsong)

FM: 95.5MHz

Ann: "Josong Inmingun FM Pangsong-imnida".

KOREA (South, Rep.)

L.T: UTC +9h — **Pop:** 47 million — **Pr.L:** Korean — **E.C:** 60Hz, 110/220V — **ITU:** KOR

KOREAN BROADCASTING SYSTEM (KBS)
(Hanguk Bangsong Gongsa) (Public Corporation)

✉18, Yeouido-dong, Yeongdeungpo-gu, Seoul 150-790 ☎ +82 2 781 1000 🖷 +82 2 761 2499 **W:** www.kbs.co.kr

L.P: Pres & CEO: Kim In-Kyu. Auditor General: Lee Gil-Yung. Exec. Vice Pres: Cho Dae-Hyun, Kim Young-Hae, Exec. Man. Dirs: Ji Yun-Ok (Audience Relations), Lee Jung-Bong (N & Sports), Gil Hwan-Young (Content), Kim Seon-Kwon (New Media & Tech), Lee Dong-Shik (Policy Planning). Dir. Int. Rel. Div:Min Eun-Gyung.

MW	Location	Call	kHz	kW	MW	Location	Call	kHz	kW
9)	Jangsu	-	540	1	13)	Gumi	-	909	10
13)	Jeomchon	-	540	1	N1)	Yeoncheon*	-	918	50
8)	Hongseong	-	540	10	8)	Buyeo	-	927	10
10)	Jangheung	-	540	10	18)	Hadong	-	927	1
13)	Daegu+2	QH	558	250	5)	Hongcheon	-	927	1
3)	Jeonju+	KF	567	100	16)	Changwon 3	-	936	10
12)	Suncheon 3	-	576	1	8)	Boeun	-	945	10
14)	Yeongju	-	594	10	19)	Jeju+	KS	963	10
N2)	Namyang*	SA	603	500	14)	Andong+	CR	963	10
4)	Taebaek	-	621	10	K1)	Danjin*	CA	972	1500
19)	Seogwipo	-	621	10	4)	Gangneung 3-		1008	50
6)	Yeongdong	-	621	1	3)	Hwacheon	-	1026	1
3)	Inje	-	630	5	18)	Geochang	-	1026	1
5)	Yeosu	-	630	10	15)	Pohang+	CP	1035	10
10)	Boseong	-	648	1	4)	Samcheok	-	1044	10
3)	Chuncheon+	KM	657	50	7)	Jecheon	-	1044	1
3)	Jeonju 3	-	675	10	6)	Cheongju+	KQ	1062	50
N1)	Sorae*	KA	711	500	18)	Chungju+	CH	1089	10
13)	Daegu+	KG	738	100	18)	Jinju+	CJ	1098	20
10)	Gwangju+	KH	747	100	N3)	Hwaseong*	KC	1134	500
N1)	Yeoju*	-	756	100	5)	Wonju+	CW	1152	10
5)	Yeongwol	-	783	10	K2E)	Gimje*	SR	1170	500
3)	Yanggu	-	846	5	5)	Jeongseon	-	1206	1
4)	Gangneung+	KR	864	100	14)	Cheongsong	-	1206	10
8)	Daejeon+	KI	882	20	10)	Gwangju 3	-	1224	20
2)	Busan+	KB	891	250	5)	Pyeongchang-		1233	1

MW	Location	Call	kHz	kW	MW	Location	Call	kHz	kW
9)	Namwon	-	1260	10	14)	Bonghwa	-	1458	1
N1)	Yangju	-	1269	10	11)	Mokpo+	KN	1467	50
9)	Gurye	-	1269	1	12)	Goheung	-	1485	1
16)	Hapcheon	-	1278	1	8)	Gongju	-	1485	1
15)	Uljin	-	1305	10	13)	Gimcheon	-	1503	1
10)	Yeonggwang	-	1323	1	19)	Gosan	-	1539	1
15)	Ulleung	-	1323	1	7)	Danyang	-	1584	1
9)	Muju	-	1368	1	18)	Sancheong	-	1584	1
N1)	Cheorwon*	-	1395	10	8)	Geumsan	-	1584	1
17)	Ulsan+	QB	1449	10	5)	Sabuk	-	1602	1
18)	Hamyang	-	1458	1					

MW: N1 = KBS R. One, N2 = KBS R. Two, N3 = KBS R. Three, K1 = Global Korean Network 1, K2 = Global Korean Network 2, E = also used for Ext. sce., KBS WORLD R, N = Netw. or local stn. area, *) Key stn, + = Regional key St, 2 = rel N2 exc. for local prgrs, 3 = rel N3 (other local st take N1), Call: HL(call).

NB: Global Korean Network stns and FM-stns do not use call letters (even if assigned). Other stns without call letters use the calls from their regional key stns.

	FM(MHz)	I	II	III	kW
1)	Namsan		93.1	89.1a	-/10/10
1)	Gwanaksan	97.3*		106.1b	10/-/10
1)	Gwanaksan			104.3c	-/-/2
1)	Yongmunsan	90.3*			1
2)	Yeongdo	103.7	92.7	97.1b	3/5/3
3)	Hwaaksan	99.5*	91.1	98.7b	5/5/3
4)	Gwaebangsan	98.9*	89.1	102.1b	1/5/5
5)	Baegunsan	97.1	89.5		1/3
5)	Taegisan	95.5*			1
4)	Hambaeksan	93.7*	97.3		1/3
6)	Sikchangsan		102.1		-/3
8)	Sikchangsan			100.9b	-/-/3
6)	Heukseongsan	89.9*			1
6)	Uamsan	89.3	94.1		1/1
6)	Gayeopsan			90.9b	-/-/3
7)	Gayeopsan	92.1*	100.3		1/3
8)	Gyeryongsan	94.7*	98.5		1/5
9)	Moaksan	96.9*	100.7	92.9b	5/5/?
9)	Nogodan	88.3*	104.5		1/3
10)	Mudeungsan	90.5*	92.3	95.5b	5/5/3
11)	Yangulsan		98.3		-/1
12)	Daedunsan	105.9			1
12)	Suncheon			102.7b	-/-/3
12)	Mangunsan	95.7*	94.5		1/3
13)	Palgongsan	101.3*	89.7	102.3b	5/5/3
14)	Ilwolsan	90.5*			1
14)	Hakkasan		88.1		-/3
15)	Johangsan	95.9*	93.5		1/3
16)	Bulmosan	91.7*	93.9	106.1b	1/1/3
17)	Muryongsan	90.7*	101.9		1/3
18)	Gamaksan		92.1		-/3
18)	Mangjinsan	90.3	89.3		1/1
18)	Gyeonwolak	99.1*	96.3	91.9b	5/3/3
19)	Sammaebong		99.9	89.9b	-/3/1

+ low power relay stn

Reg = region in MW section. I-Standard FM(R. One); II-KBS FM One; III a = FM Two, b = R. Two, c=R. Three. *) are also SCA (R. Three).

KBS R. One (KBS Je-il Radio, HLKA): 24h. Non-commercial nationwide news sce. Key freqs 711/756kHz, 90.3/97.3MHz. Also rel. by Standard FM stns and most reg. stns. Reg. stns may broadcast local prgrs at designated times. **N:** hourly 2000-1600 except 1100(W). Local N: 2205(Sun), 2210(W), 0000(Sun), 0005(w), 0310(Sun), 0315(w), 0605, 0805(Mon-Fri), 0900(Sun), 0905(W).

KBS R. Two (KBS Je-i Radio, Happy FM, HLSA): 2000-1800 (558kHz to 1500). Commercial. Key freq's 603kHz/106.1MHz. Reg. stns may broadcast local prgrs at designated times. **N:** hourly 2000-1200. Local N: 2300, 0400, 0700, 1200. Global Korean Network prgr 1700-1800.

KBS R. Three (KBS Je-sam Radio, Sarang-ui Sori Bangsong, HLKC): 2100-1800. Non-comm. sce. **N:** 0000(W), 0100(W), 0300(W), 0800(Mon-Fri).

KBS FM One (KBS Je-il FM Bangsong, Classic FM, HLKA-FM): 24h. Mainly Korean traditional and western classical music.

KBS FM Two (KBS Je-i FM Bangsong, Cool FM, HLKC-FM): 24h. Mainly Korean and western popular and light classical music.

NB: Regional FM One stns relay FM Two 2100-2200.

KBS Global Korean Network (Hanminjok Bangsong)
See International Broadcasting section

Ann:N1: "AMChilbaek-sib-il(711)kHz, FMGusib-chil-jeom-sam(97.3)MHz, Je-il Radiomnida. HLKA". N2: "KBS Je-i Radiomnida". N3: "AM Yukbeak-samsib-gu(639)kHz, KBS Je-sam Radio, Sarang-ui Sori Bangsong-imnida.

HLKC". Global Korean Network 1:"Jungpa Gubaek-chilsib-i(972)kHz, Hanminjok Neteuwokeu Chaeneol, KBS Hanminjok Je-il Bangsong-imnida". Global Korean Network 2:"Jungpa Cheonbaek-chilsip(1170)kHz, Daehan Mingook Seoureseo Bonae Deurineun Hanminjok Neteuwokeu Chaeneol, KBS Hanminjok Je-i Bangsong-imnida".

Addresses of regional key stations:
2) 63, Namcheon-dong, Suyeong-gu, Busan 608-790 – **3)** 86-1, Nagwon-dong, Chuncheon-si, Gangwon-do 200-100 – **4)** 62-5, Yonggang-dong, Gangneung-si, Gangwon-do 210-070 – **5)** 79-1, Won-dong, Wonju-si, Gangwon-do 220-060 – **6)** 417, Gaesin-dong, Cheongju-si, Chungcheongbuk-do 361-790 – **7)** 417, Munhwa-dong, Chungju-si, Chungcheongbuk-do 380-790 – **8)** 300, Mannyeon-dong, Seo-gu, Daejeon 302-790 – **9)** 523-3, Geumam 2-dong, Deokchin-gu, Jeonju-si, Jeollabuk-do 560-790 – **10)** 1206-1, Chipyeong-dong, Seo-gu, Gwangju 502-270 – **11)** 1188-3, Yongdang 1-dong, Mokpo-si, Jeollanam-do 530-360 – **12)** 91-3, Seokhyeon-dong, Suncheon-si, Jeollanam-do 540-100 – **13)** 245, Beomeo 4-dong, Suseong-gu, Daegu 706-790 – **14)** 666, Taehwa-dong, Andong-si, Gyeongsangbuk-do 760-790 – **15)** 655, Sangdo-dong, Nam-gu, Pohang-si, Gyeongsangbuk-do 790-790 – **16)** 97-1, Sinwol-dong, Changwon-si, Gyeongsangnam-do 641-790 – **17)** 416-7, Dal-dong, Nam-gu, Ulsan 680-790 – **18)** 13-22, Sinan-dong, Jinju-si, Gyeongsangnam-do 660-790 – **19)** 302-3, Yeon-dong, Jeju-si, Jeju 690-170.
Local identifications: Within local prgrs. **N1**: just before the h. at 2000, 2200(Sun), 2300, 0000(W), 0200, 0300, 0500, 0700(Mon-Fri), 0800, 0900(Sun), 1000(W), 1100(Sun), 1300, 1400, 1500(Sun), 1600. **N2**: just before the h. 2000-1700. **N3**: just before the h. 2100-1700. **FM One**: just before the h. at 2000-2200, 0000, 0200, 0300, 0500, 0700-0900, 1100, 1300, 1500, 1600, 1800. **FM Two**: just before the h.
F.PI:FM Two nationwide network in 2010.

EXTERNAL SERVICES: KBS WORLD RADIO
See International Broadcasting section

KOREA EDUCATIONAL BROADCASTING SYSTEM (EBS)
(Gyoyuk Bangsong) (Pub.)
✉ 463, Dogok 2-dong, Gangnam-gu, Seoul 135-854 ☎ +82 2 526 2000 🖷 +82 2 526 2419 **W:** www.ebs.co.kr
Call letters HLQL used for all the stns.

FM	Tx location	MHz	kW
Chungju	Gayeopsan	104.1	5
Changwon	Bulmosan	104.3	5
Seoul	Gwanaksan	104.5	5
Jinju	Gamaksan	104.7	3
Gangneung	Gwaebangsan	104.9	3
Wonju	Baegunsan	104.9	3
Seogwipo	Sammaebang	104.9	3
Daegu	Palgongsan	105.1	5
Gwangju	Mudeungsan	105.3	5
Daejeon	Gyeryongsan	105.7	5
Ulsan	Muryongsan	105.9	3
Yeosu	Mangunsan	106.3	1
Chuncheon	Hwaaksan	106.5	5
Pohang	Johangsan	106.7	3
Jeonju	Moaksan	106.9	5
Taebaek	Hambaeksan	107.1	3
Jeju	Gyeonwolak	107.3	3
Namwom	Nogodan	107.5	3
Andong	Hakkasan	107.7	3
Busan	Yeongdo	107.7	3
Cheongju	Sikjangsan	107.9	3

+ low power relay stns
D.Prgr: 2000-1700 **Ann:** "EBS, Gyoyuk Bangsong-imnida"

GUGAK FM BROADCASTING SYSTEM
(Gugak Bangsong) (Pub.)
✉ DMS Bldg., C-3 Area, DMC Block, Sang-am-dong, Mapo-gu, Seoul 120-803 ☎+82 2 300 9990 🖷 +82 2 300 9959
W: www.gugakfm.co.kr
Stations: Seoul HLQA-FM 99.1MHz 5kW: 24h, Namwon 95.9MHz 1kW: 24h, Namdo 94.7MHz 0.5kW, Gwangju/Pohang 107.9MHz 3kW: 24h **Ann:** "Gugak Bangsong-imnida."
F.PI: Relay stns in Busan, Chuncheon

MUNHWA BROADCASTING CORP. (MBC)
(Munhwa Bangsong) Nationwide comm. netw.
✉ 31, Yeouido-dong. Yeongdeungpo-gu, Seoul 150-728 ☎ +82 2 784 2000 **W:** www.imbc.com

MW	Call	kHz	kW	Station	MW	Call	kHz	kW	Station
1)	CQ	765	10	Daejeon MBC	5)	CN	819	20	Gwangju MBC
2)	AJ	774	10	Jeju MBC	6)	AU	846	10	Ulsan MBC
3)	AN	774	10	Chuncheon MBC	7)	CX	855	10	Jeonju MBC
4)	CT	810	20	Daegu MBC	8)	KV	900	50	Seoul MBC
9)	AP	990	10	Changwon MBC	15)	SB	1242	10	Wonju MBC
10)	AW	1017	10	Andong MBC	16)	AF	1287	10	Gangneung MBC
11)	AT	1080	10	Yeosu MBC	17)	AX	1287	10	Cheongju MBC
12)	AV	1107	10	Pohang MBC	18)	AO	1332	10	Chungju MBC
13)	KU	1161	20	Busan MBC	19)	AQ	1350	10	Samcheok MBC
14)	AK	1215	10	Jinju MBC	20)	AM	1386	10	Mokpo MBC

D.Prgr: All 24h

		Music FM		Standard FM	
FM	Location	MHz	kW	MHz	kW
8)	Seoul	91.9	10	95.9	10
13)	Busan	88.9	5	95.9	3
4)	Daegu	95.3	5	96.5	5
5)	Gwangju	91.5	5	93.9	5
	Gwangju	95.1	3	-	-
1)	Daejeon	97.5	5	92.5	3
7)	Jeonju	99.1	5	94.3	2
	Jeonju (Namwon)	-		101.7	3
9)	Changwon	100.5	1	98.9	3
3)	Chuncheon	94.5	3	92.3	3
17)	Cheongju	99.7	1	107.1	1
2)	Jeju	90.1	3	97.9	1
	Jeju(Seogwipo)	102.9	3	97.1	1
	Ulsan	98.7	3	97.5	1
16)	Gangneung	94.3	5	96.3	3
14)	Jinju	97.7	1	91.1	3
	Jinju	96.1	3	93.5	1
20)	Mokpo	102.3	1	89.1	2
11)	Yeosu	98.3	2	100.3	1
10)	Andong	91.3	3	100.1	3
15)	Wonju	98.9	3	92.7	1
	Wonju	-		102.5	1
18)	Chungju	88.7	3	96.1	1
19)	Samcheok	98.1	3	101.5	1
	Samcheok	99.9	1	93.1	3
12)	Pohang	97.9	3	100.7	3
	Pohang(Uljin)	94.9	1	102.7	1

+low power rel. stns
NB: Standard FM stns simulcast with the MW stn in the same city. A separate sce. is provided to the Music FM stns. All regional stns broadcast a combination of a feed from Seoul and their own local prgrs. Standard FM stns follow the same schedule as their corresponding MW outlet. Music FM of Seoul MBC sched: 24h.
Ann: "(freq. and location) Munhwa Bangsong-imnida. (call letters)" or "Munhwa Bangsong-imnida" or "MBC". Seoul: "Jungpa Gubaek (900)kHz, Pyojun FM Gushib-o-jeom-gu 95.9MHz Munhwa Bangsong-imnida"

Addresses and other information
Add "(location) Munhwa Broadc. Corp." to addr.
1) 4-5, Doryong-dong, Yuseong-gu, Daejeon 305-740 **W:** www.tjmbc.co.kr – **2)** 321-22, Yeon-dong, Jeju-si, Jeju Teukbyel Jachido 690-170 **W:** www.jejumbc.co.kr – **3)** 238-3, Samcheon-dong, Chuncheon-si, Gangwon-do 200-200 **W:** www.chmbc.co.kr – **4)** 1, Beomeo-dong, Suseong-gu, Daegu 706-728 **W:** www.tgmbc.co.kr – **5)** 300, Wolsan-dong, Nam-gu, Gwangju 503-728 **W:** www.kjmbc.co.kr – **6)** 409-1, Hakseong-dong, Jung-gu, Ulsan 681-728 **W:** www.ulsanmbc.co.kr – **7)** 151-9, Junghwasan-dong 2-ga, Wansan-gu, Jeonju-si, Jeollabuk-do 560-728 **W:** www.jmbc.co.kr – **8)** National addr. – **9)** 525-1, Yangdeok-dong, Masan Hoewon-gu, Changwon-si, Gyeongsangnam-do 630-713 **W:** www.changwonmbc.co.kr – **10)** 709-1, Taehwa-dong, Andong-si, Gyeongsangbuk-do 760-290 **W:** www.andongmbc.co.kr – **11)**50-15, Yeomun-ro, Yeosu-si, Jeollanam-do 550-728 **W:** www.ysmbc.co.kr – **12)** 907-4, Daejam-dong, Pohang-si, Gyeongsangbuk-do 790-728 **W:** www.phmbc.co.kr – **13)** 316-2, Millak-dong, Suyeong-gu. Busan 613-728 **W:** www.busanmbc.co.kr – **14)** 700-1, Gajwa-dong, Jinju-si, Gyeongsangnam-do 660-728 **W:** www.jinjumbc.co.kr – **15)**1023-70, Hakseong 1-dong, Wonju-si, Gangwon-do 220-031 **W:** www.wjmbc.co.kr – **16)** 126, Gajak-ro, Ponam 2-dong, Gangneung-si, Gangwon-do 210-112 **W:** www.gnmbc.co.kr – **17)** 352-8, Gagyeong-dong, Heungdeok-gu, Cheongju-si, Chungcheongbuk-do 361-855 **W:** www.cjmbc.co.kr – **18)** 680, Hoam-dong, Chungju-si, Chungcheongbuk-do 380-130 **W:** www.cjmbc.co.kr – **19)** 111, Galcheon-dong, Samcheok-si, Gangwon-do 245-090 **W:** www.scmbc.co.kr – **20)** 1096-1, Yongdang-dong, Mokpo-si, Jeollanam-do 530-728 **W:** www.mokpombc.co.kr

CHRISTIAN BROADCASTING SYSTEM (CBS)
(Gidokkyo Bangsong)

MW Call	kHz	kW	Station and h.of tr.
1) KY	837	50	CBS Seoul: 24h
2) CL	999	10	CBS Gwangju: 2000-1600
4) KT	1251	10	CBS Daegu: 2000-1600
5) CM	1314	10	CBS Jeonbuk: 2000-1600
6) KP	1404	10	CBS Busan: 2000-1600

CBS FM:
1) CBS-FM Seoul HLKY-FM 93.9MHz 7kW 24h (Music FM)
1) CBS Seoul HLKY-SFM 98.1MHz 10kW 24h
2) CBS Gwangju HLCL-SFM 103.1MHz 5kW 2000-1600
3) CBS Jeonnam HLCL-FM 102.1MHz 2kW 2000-1600
4) CBS Daeju HLKT-SFM 103.1MHz 5kW 2000-1600
5) CBS Jeonbuk HLCM-SFM 103.7MHz 5kW 2000-1600
6) CBS Busan HLKP-SFM 102.9MHz 5kW 2000-1600
7) CBS Cheongju HLAC-FM 91.5MHz 3kW 2000-1600
8) CBS Chuncheon HLDC-FM 93.7MHz 3kW 2000-1600
9) CBS Daejeon HLDX-FM 91.7MHz 5kW 2000-1600
10) CBS Pohang HLCB-FM 91.5MHz 3kW 2000-1600
11) CBS Gyeongnam HLCC-FM 106.9MHz 5kW 2000-1600
12) CBS Jeju HLKO-FM 93.3MHz 3kW 2000-1600
12) CBS Jeju (Seogwipo relay st) 90.9MHz 1kW 2000-1600
13) CBS Yeongdong HLCO-FM 91.5MHz 3kW 2000-1600
14) CBS Ulsan HLKP-FM 100.3MHz 1kW 2000-1600
+low power relay stns

Addresses and other information:
1) 917-1, Mok 1-dong, Yangcheon-gu, Seoul 158-701 ☎ +82 2 2650 7000 W: www.cbs.co.kr **Ann:** "Jeongjikhan Sesang-eul Gakkuneun AM Palbaek-samsip-chil(837)kHz, Pyojun FM Gusip-pal-jeom-il(98.1)MHz, CBS-mnida. HLKY." – **2)** 721-2, Geumho-dong, Seo-gu, Gwangju 506-154 ☎ +82 62 376 8500 – **3)** 117-5, Maegok-dong, Suncheon-si, Jeollanam-do 540-947 ☎ +82 61 902 1000 – **4)** 3-7, Chimsan 2-dong, Buk-ku, Daegu 702-703 ☎ +82 53 426 8001 – **5)** 114-8, Daga-dong, Wansan-gu, Jeollabuk-do 560-053 ☎ +82 63 281 0430 – **6)** 1155-2, Beomchon 4-dong, Busanjin-gu, Busan 614-024 ☎ +82 51 636 0050 – **7)** 1010, Sugok-dong, Heungdeok-gu, Cheongju-si, Chungcheongbuk-do 361-150 ☎ +82 43 292 4100 – **8)** 174-3, Ungyo-dong, Chuncheon-si, Gangwon-do 200-080 ☎ +82 33 255 2001 – **9)** 1-13, Munhwa-dong, Jung-gu, Daejeon 301-130 ☎ +82 42 259 8888 – **10)** 640-7, Daedo-dong, Nam-gu, Pohang-si, Gyeongsangbuk-do 790-824 ☎ +82 54 277 5500 – **11)** 323-3, Sanho-dong, Masan Happo-gu, Changwon-si, Gyeongsangnam-do 630-811 ☎ +82 55 224 5600 – **12)** 271, Yeon-dong, Jeju-si, Jeju Teukbyeol Jachido 690-813 ☎ +82 64 744 0933 – **13)** 935-1, Gyo 1-dong, Gangneung-si, Gangwon-do 210-923 ☎ +82 33 642 9131 - **14)** 186-11, Sinjeong 3-dong, Nam-gu, Ulsan 680-822 ☎ +82 52 256 3333
Ann: stns 2)-8): "Jeongjikhan Sesang-eul Kakkuneun (freq.), CBS (location) Bangsong-imnida. (call)" or "Maeumgwa Maeumi Mannaneun Bangsong, (freq.), CBS (location) Bangsong-imnida. (call)"
F.PI: Relay stns in Chungju, Wonju, Jinju, Gongju, Seosan. Music FM in Daejeon, Gwangju, Jeju, Ulsan, Jeonbuk(Jeonju), Gyeongnam(Changwon), Busan, Daegu.

SEOUL BROADCASTING SYSTEM (SBS)
⊡ 920 Mok-dong, Yangcheon-gu, Seoul 158-725 ☎ +82 2 2061 0006 🖷 +82 2 2113 3169 **W:** www.sbs.co.kr
MW: HLSQ Goyang (near Seoul) 792kHz 50kW. **D.Prgr:** 24h.
Standard FM (Love FM): 103.5MHz HLSQ-SFM 10kW: 24h.
Music FM (Power FM): 107.7MHz HLSQ-FM 10kW: 24h.
Ann: "AM Chilbaek-gusib-i 792kHz, FM Baek-sam-jeom-o 103.5MHz, SBS Love FM-imnida. HLSQ", "FM Baek-chil-jeom-chil 107.7MHz, Yeoreobune SBS Power FM-imnida. HLSQ"

FAR EAST BROADCASTING CO., KOREA (Rlg.)
MW	kHz	kW	Station, location
1)	1188	100	HLKX, Seoul
2)	1566	250/100	HLAZ, Jeju
FM	MHz	kW	Station, location
1)	106.9	5	HLKX-SFM, Seoul
3)	93.3	5	HLAD-FM, Daejeon
4)	98.1	5	HLDD-FM, Changwon
5)	90.1	3	HLDY-FM, Yeongdong
6)	100.5	1	HLKW-FM, Mokpo
7)	90.3	3	HLDZ-FM, Pohang
8)	107.3	3	HLQR-FM, Ulsan
9)	93.3	1	HLQO-FM, Busan
10)	91.9	1	HLCU-FM, Daegu

+ low power relay stns
Addresses and other information
1) Far East Broadc. Co.(Geukdong Bangsong), 89, Sangsu-dong, Mapo-gu, Seoul 121-707 ☎ +82 2 320 0114 🖷 +82 2 320 0229 **W:** www.febc.net
D.Prgr: 1900-1700. Korean: 1900-1100, 1600-1700(Standard FM: 1900-1700) **English:** 1100-1200(1188kHz) **Chinese:** 1500-1600(1188kHz). **VOA Relay in Korean:** 1200-1500(1188kHz). **Ann:** Korean "Jungpa Cheonbaek-palsip-pal(1188)kHz, Pyojun FM Paeng-nyuk-jeom-gu(106.9)MHz, Areumdaun Chanyanggwa Gibbeun Sosigeul Jeonhaneun Geukdong Bangsong-imnida.". English: "This is HLKX Radio broadcasting with 100,000 watts of power on 1188kHz" **FI:** by contributions & free

---

will offerings. – **2)** Jeju Geukdong Bangsong, 2761, Hagwi-ri, Aewol-up, Bukjeju-gun, Jeju Teukbyel Jachido 695-750 ☎ +82 64 799 8100 **D.Prgr:** 24h. **Korean:** 1900-1100. **Chinese:** 1100-1230, 1345-1730, 1800-1900. **Japanese:** 1230-1345. **Russian:** 1730-1800. – **3)** Daejeon Geukdong Bangsong, 233-15, Chijok-dong, Yuseong-gu, Daejeon 305-711 ☎ +82 42 828 9330. **D.Prgr:** 24h. – **4)** Changwon Geukdong Bangsong, 117, Jungang-dong, Changwon-si, Gyeongsang-nam-do 641-030 ☎ +82 55 269 9810 **D.Prgr:** 24h. – **5)** Yeongdong Geukdong Bangsong, 500-1, Jangsa-dong, Sokcho-si, Sokcho-si, Gangwon-do 217-130 ☎ +82 33 638 9000 **D.Prgr:** 1900-1700. – **6)** 878-9, Sang-dong, Mokpo-si, Jeollanam-do 530-822 ☎ +82(61)284 9000 **D.Prgr:** 1900-1700. – **7)** 122-4, Deoksan-dong, Buk-gu, Pohang-si, Gyeongsangnam-do 791-020 ☎ +82 54 256 3000 **D.Prgr:** 24h. – **8)** 589-3, Dal-dong, Nam-gu, Ulsan-si 680-080 ☎ +82 52 256 2000 **D.Prgr:** 24h. – **9)** 4th Floor, Centum Venture Town, 1475, U-dong, Haeundae-gu, Busan612-020 ☎ +82 51 759 6000 **D.Prgr:** 24h. – **10)** 1326-3, Manchon 1-dong, Suseong-gu, Daegu, ☎ +82 53 745 1187 **D.Prgr:** 24h.**F.PI:** Regional stns in Gwangju, Yeosu. Standard FM in Jeju. Relay stn in Taebaek.

PYEONGHWA BROADCASTING CORP. (PBC)
(Pyeonghwa Bangsong) Endowment by the Catholic Church.
Stations:
1) Seoul HLQP-FM 105.3MHz 5kW: 1957-1702 – **2)** Gwangju HLDL-FM 99.9MHz 5kW, 99.5MHz 1kW(rel. stn in Yeosu): 1957-1702 – **3)** Deagu HLDK-FM 93.1MHz 3kW, 96.9MHz(rel. st. in Pohang), 100.7MHz (rel. stn in Andong): 1957-1702 – **4)** Busan HLDW-FM 101.1MHz 3kW, 105.5MHz(rel. st. in Changwon): 1957-1702 – **5)** Daejeon HLQO-FM 106.3MHz 3kW: 1957-1702.

Addresses
1) 2-3, Jeo-dong 1-ga, Jung-gu, Seoul 100-031 ☎ +82 2 2270 2114 🖷 +82 2 2270 2210 W: www.pbc.co.kr **Ann:** "Saengmyeng Sarang, FM Baeg-o-jeom-sam(105.3)MHz, Gibbeun Sosik, Balgeun Sesang, PBC Pyeonghwa Bangsong-imnida. HLQP." – **2)** 3-5, Geumnam-ro 3-ga, Dong-gu, Gwangju 501-023 – **3)** 71, Gyesan-dong 2-ga, Jung-gu, Daegu 700-082 – **4)** 81-1, Daecheong-dong 4-ga, Jung-gu, Busan 600-094 – **5)** 189, Daeheung-dong, Jung-gu, Daejeon 301-802 ☎ +82 42 250 3200.

BUDDHIST BROADCASTING SYSTEM (BBS)
(Bulgyo Bangsong) Owned and operated by the Buddhistns.
Stations:
1) Seoul HLSG-FM 101.9MHz 5kW: 2000-1700 – **2)** Gwangju HLDB-FM 89.7MHz 3kW: 2000-1700 – **3)** Busan HLDA-FM 89.9MHz 3kW: 2000-1700 – **4)** Daegu HLDI-FM 94.5MHz 3kW, 105.5MHz(rel. stn in Pohang), 97.7MHz(rel. stn in Andong): 2000-1700 – **5)** Cheongju HLDJ-FM 96.7MHz 3kW: 2000-1700 – **6)** Chuncheon HLQM-FM 100.1MHz 3kW: 2000-1700 – **7)** Ulsan HLQU-FM 105.3MHz 1kW: 2000-1700.
Addresses:
1) Dabo Building; 140, Mapo-dong, Mapo-gu, Seoul 121-050 ☎ +82 2 705 5114 🖷 +82 2 705 5229 **W:** www.bbsfm.co.kr – **2)** Daesaeng Bldg, 78-2, Im-dong, Buk-gu, Gwangju. 500-010 ☎ +82 62 520 1114 – **3)** Bosaeng Bldg, 833-13, Beomil 2-dong, Dong-gu, Busan 601-060 ☎ +82 51 520 5114 – **4)** Jingak Hoegwan, 156-1, Daebong-dong, Jung-gu, Daegu 700-430 ☎ +82 53 427 5114 – **5)** 1646, Yongam-dong, Sangdang-gu, Cheongju-si, Chungcheongbuk-do 360-181 ☎ +82 43 294 5114 – **6)** 4-1, Yoseon-dong, Chuncheon-si, Gangwon-do 200-030 ☎ +82 33 250 2114 – **7)** 1359-11, Dal-dong, Nam-gu, Ulsan ☎ +82 52 279 8114.
Ann: 1) "FM Baeg-il-jeom-gu (101.9)MHz, BBS Bulgyo Bangsong-imnida. HLSG." **F.PI:** Relay stn in Gangneung

SEOUL TRAFFIC BROADCASTING SYSTEM (TBS)
(Gyotong Bangsong)
Municipal Station. This stn is operated by the Seoul Municipal Traffic Broadcast Headquarters to provide traffic information and education to the citizens of Seoul and surroundings.
⊡ 3-8, Yejang-dong, Jung-gu, Seoul 100-250 ☎ +82 2 311 5114 🖷 +82 2 311 5219 **W:** www.tbs.seoul.kr
Station: HLST-FM(Live FM) 95.1MHz 5kW: 24h in Korean. HLSW-FM(Soul FM) 101.3MHz 1kW: 2000-1700 in English.
Ann: "FM Gusib-o-jeom-il(95.1)MHz, TBS Gyotong Bangsong-imnida","You're listening to 101.3 tbs-eFM"

TRAFFIC BROADCASTING NETWORK (TBN)
(Hanguk Gyotong Bangsong)
⊡ 171, Sindang-dong, Jung-gu, Seoul 100-789 ☎ +82 2 2230 6114 🖷 +82 2 2230 6269 **W:** www.tbn.or.kr
Stations:
1) Busan 94.9MHz HLDN-FM 3kW, 100.1MHz 1kW(rel. st. in Jinju): 24h – **2)** Gwangju 97.3MHz HLDM-FM 3kW, 103.5MHz 1kW(rel. st. in Gwangyang): 24h – **3)** Daejeon 102.9MHz HLDT-FM 3kW: 24h – **4)**

Daegu 103.9MHz HLDU-FM 3kW: 24h – **5)** Incheon 100.5MHz HLSU-FM 1kW: 24h – **6)** Gangwon(Wonju) 105.9MHz HLSV-FM 3kW: 24h, Gangwon(Chuncheon) 103.7MHz 3kW: 24h , Gangwon(Gangneung) 105.5MHz 1kW: 24h – **7)** Jeonju 102.5MHz HLCM-FM 1kW: 24h + low power relay stns

Addresses and other information
1) 580-8, Daeyeon 3-dong, Nam-gu, Busan 608-023 ☎ +82 51 6105 114 **Ann:** "FM Gusib-sa-jeom-gu(94.9)MHz, Busan Gyotong Bangsong-imnida. HLDN-FM" – **2)** 665-2, Ssangam-dong, Gwangsan-gu, Gwangju 506-303 ☎ +82 62 9701 114 **Ann:** "FM Gusib-chil-jeom-sam(97.3)MHz, Gwangju Gyotong Bangsong-imnida. HLDM" – **3)** 152-7, Nae-dong, Seo-gu, Daejeon 302-181 ☎ +82 42 6001 114 **Ann:** "FM Baeg-i-jeom-gu(102.9)MHz, Dallineun Radio Daejeon Gyotong Bangsong-imnida." – **4)** 1679-2, Daemyeong-dong, Nam-gu, Daegu 705-031 ☎ +82 53 6060 114 **Ann:** "FM Baek-sam-jeom-gu(103.9)MHz, Daegu Gyotong Bangsong-imnida. HLDU-FM" – **5)** 401-74, Hagik-dong, Nam-gu, Incheon 402-865 ☎ +82 32 4531 114 **Ann:** "FM Baek-jeom-o(100.5)MHz, TBN Incheon Gyotong Bangsong-imnida. HLSU" – **6)** 1400, Bangok-dong, Wonju-si, Gangwon-do ☎ +82 33 7490 114 **Ann:** "Haengbogui Giljabi, Ggumi Inneun Bangsong, FM Baeg-o-jeom-gu(105.9)MHz, Gangwon Gyotong Bangsong-imnida." – **7)** 410-1, Jinbuk-dong, Deokjin-gu, Jeonju 561-162. ☎ +82 63 2593 114 **Ann:** "FM Baeg-i-jeom-chil(102.7)MHz, TBN Jeonju Gyotong Bangsong-imnida. HLCM"
F.PI: Regional stns in Jeju, Ulsan, Changwon.

KOREA NEW NETWORK CORP. (KNN)
🖃 603-8, Yeonsan-4-dong, Yeonje-gu, Busan 611-084 ☎ +82 1 850 9000 **W:** www.knn.co.kr **Station:** HLDG-FM 99.9MHz 3kW: 24h.
Ann: "Gushib-gu-jeom-gu (99.9), KNN Radiomnida.HLDG"

TAEGU BROADCASTING CORPORATION (TBC)
(Daegu Bangsong)
🖃 201-9, Tusan-dong, Susong-gu, Daegu 760-080 ☎ +82 53 760 1900 **W:** www.tbc.co.kr
Station: HLDE-FM(Dream FM) 99.3MHz 5kW: 24h. Relay stn: Pohang 99.7MHz. **Ann:** "HLDE-FM TBC Dream FM-imnida."

KWANGJU BROADCASTING CO., LTD. (KBC) (Gwangju Bangsong)
🖃 111-14, So-dong, Nam-gu, Gwangju 503-010 ☎ +82 62 650 3114 **W:** www.ikbc.co.kr **Station:** HLDH-FM(MY FM) 101.1MHz 5kW: 24h. Relay stn: Yeosu 96.7MHz. **Ann:** "HLDH, FM 101.1MHz, 96.7MHz, Yeollin Sesang, Joheun Chingu, KBC MY FM."

TAEJON BROADCASTING CO., LTD. (TJB)
(Daejeon Bangsong)
🖃 122-1, Hyo-dong, Tong-gu, Daejeon 300-722 ☎ +82 42 281 1101 **W:** www.tjb.co.kr
Station: HLDF-FM(Power FM) 95.7MHz 5kW: 24h Relay stn: Seosan 96.5MHz **Ann:** "Gusib-o-jeom-chil(95.7), Gusim-nyuk-jeom-o(96.5)MHz, TJB Power FM-imnida. HLDF"

JEONJU TELEVISION CORPORATION (JTV)
(Jeonju Bangsong)
🖃 656-3, Seonosong-dong, Deokjin-gu, Jeonju-si, Jeollabuk-do 561-090 ☎ +82 63 250 5200 **W:** www.jtv.co.kr
Station: HLDQ-FM(Magic FM) 90.1MHz 5kW: 24h.
Ann: "FM Gusib-jeom-il(90.1)MHz, JTV Magic FM-imnida. HLDQ"

CHEONGJU BROADCASTING CORPORATION (CJB)
(Cheongju Bangsong)
🖃 12-16, Sajik 2-dong, Hongdeok-gu, Cheongju-si, Chungcheongbuk-do 361-102 ☎ +82 43 265 7000 **W:** www.cjb.co.kr
Station: HLDI-FM(Joy FM) 101.5MHz 5kW: 24h.
Ann: "FM Baeg-il-jeom-o(101.5)MHz, CJB Joy FM-imnida. HLDI"

ULSAN BROADCASTING CORPORATION (UBC)
(Jeonju Bangsong)
🖃 1521-1, Samsan-dong, Nam-gu, Ulsan 680-732 ☎ +82 52 228 6000 **W:** www.ubc.co.kr
Station: HLDP-FM(Green FM) 92.3MHz 5kW: 24h.
Ann: "Gusib-i-jeom-sam(92.3)MHz, UBC Green FM Bangsong-imnida. HLDP"

JEJU FREE INTERNATIONAL CITY BROADCASTING SYSTEM (JIBS) (Jeju Gukje Jayu Dosi Bangsong)
🖃 2750, Ora 3-dong, Jeju-si, Jeju Teukbyeol Jachido 690-163 ☎ +82 64 740 7800 **W:** www.jibstv.com
Station: HLQC-FM(Power FM) 101.5MHz 3kW: 24h. Relay stn: Seogwipo 98.5MHz
Ann: "JIBS New Power FM Bangsong-imnida."

GANGWON TELEVISION BROADCASTING CO., LTD (GTB)
(Gangwon Minbang)
🖃 635, Janghak-ri, Dong-myeon, Chuncheon-si, Gangwon-do 200-853 ☎ +82 33 248 5000 **W:** www.igtb.co.kr
Station: HLCG-FM(Fresh FM) 105.1MHz 3kW: 24h. Relay stn: Gangneung 106.1MHz 3kW, Wongju 103.1MHz 1kW.
Ann: "Chuncheon Baeg-o-jeom-il(105.1)MHz, Gangneung Baeng-ryuk-jeom-il(106.1)MHz, GTB Fresh FM, HLCG"

KYONGGI BROADCASTING CO. (KFM) (Gyeonggi Bangsong)
🖃 961-17, Yeongtong-dong, Yeongtong-gu, Suwon-si, Gyeonggi-do 443-810 ☎ +82 31 210 0999 **W:** www.kfm.co.kr
Station: HLDS-FM 99.9MHz 5kW: 24h.
Ann: "FM Gusib-gu-jeom-gu(99.9)MHz, Gyeonggi Bangsong-imnida. HLDS"

Kyung-In Broadcasting SUNNY FM
🖃 1, Aam 5-gil, Nam-gu, Incheon 402-773 ☎ +82 32 830 1000 **W:** www.sunnyfm.co.kr
Station: HLDO-FM 90.7MHz 1kW: 24h **Ann:** "Gusib-jeom-chil(90.7)MHz, Gyeong-In Bangsong, Sunny FM-imnida"

YTN RADIO(YTN FM)
🖃 YTN Tower, 6-1, Namdaemun-ro 5-ga, Jung-gu, Seoul 100-800 ☎ +82 2 398 8000 **W:** www.ytnfm.co.kr
Station: HLQV-FM 94.5MHz 3kW: 24h **Ann:** "FM Gusib-sa-jeom-o(94.5)MHz, YTN FM-imnida. HLQV"

WON-BUDDHISM BROADCASTING SYSTEM (WBS)
(Woneum Bangsong)
🖃 **1)** 1-3, Heukseok 1-dong, Dongjak-gu, Seoul 156-856 ☎ +82 2 2102 7700 **W:** www.wbsfm.com - **2)** 38-6, Sinchang-dong 1-ga, Jung-gu, Busan 600-061 ☎ +82 51 247 3844 - **3)** 344-2, Sinyongdong, Iksan-si, Jeollabuk-do 570-754 ☎ +82 63 837 0979- **4)** 1286, Ssangchon-dong, Seo-gu, Gwanju
Stations:
1) Seoul HLQK-FM 89.7MHz 1kW: 24h – **2)** Busan HLQJ-FM 104.9MHz 3kW: 24h – **3)** Jeonbuk(Iksan) HLDV-FM 97.9MHz 3kW: 24h.
Ann:1) FM Palsib-gu-jeom-chil(89.7)MHz, WBS Woneum Bangsong-imnida. HLQK" – **2)** "FM Baeg-sa-jeom-gu(104.9)MHz, WBS Busan Woneum Bangsong-imnida. HLQJ" – **3)** "FM Gusib-chil-jeom-gu(97.9)MHz, WBS Jeonbuk Woneum Bangsong-imnida. HLDV" – **4)** "FM Baek-chil-jeom-gu(107.9)MHz, WBS Gwangju Woneum Bangsong-imnida. HLQN"
F.PI: Regional stns in Daegu

KOREA INTERNATIONAL BROADCASTING FOUNDATION (Arirang Radio)
🖃 Arirang Tower, 1467-80, Seocho-dong, Seocho-gu, Seoul 137-878 ☎ +82 2 3475 5000 **W:** www.arirang.co.kr
Station: Jeju HLQE-FM 88.7MHz: 24h in English. Relay stn: Seogwipo 88.1MHz. **Ann:** "You're listening to Arirang Radio"

GFN Foundation
🖃 177-39, Sa-dong, Nam-gu, Gwangju 503-030 ☎ +82 62 460 0987 **W:** www.gfn.or.kr
Station: HLSY-FM 98.7MHz 1kW: 2000-1700 in English. **Ann:** "Listen more Feel more! GFN 98.7 FM"

Busan e-FM
🖃 15, Jeongbo Town 5-ro, Yeonje-gu, Busan 611-711 ☎ +82 51 861 8601 **W:** www.befm.co.kr
Station: HLSX-FM 90.5MHz 1kW: 2000-1700 in English. **Ann:** "Now you're listening to Busan e-FM 90.5"

KOREAN FORCES NETWORK (Friends FM) (Gukkun Bangsong)
🖃 San 2, Yongsan-dong 2-ga, Yongsan-gu, Seoul 140-022 **W:** www.dema.mil.kr/web/fm.do
Stations: FM (operated by KBS): Hwaaksan HLSE-FM 96.7MHz 5kW, Namsan 96.7MHz 2kW, Yongmunsan 101.1MHz 3kW, Gwaebangsan 92.5MHz 3kW + 5 lp stns
D.Prgr: 24h. Own prgrs 2100-1400, other times relay KBS R. One (HLKA). prgrs for soldiers located near the demilitarized zone. Also 0805-0900(Sun) via KBS R. One network.
Ann: "Hamggehaeyo Seonjin Ganggun, Silcheonhaeyo Noksaek Seongjang, Friends FM Gukkun Bangsong Radio"".

AMERICAN FORCES NETWORK KOREA (AFN)
🖃 As below ☎ +82 2 7914 6495/6 **W:** afnkorea.com

MW & FM Stations	kHz	kW	MHz	kW
1) Seoul/Yongsan	1530	5	102.7	5
2) Munsan/Western Corridor	576	5	88.5	0.05
3) Daegu/Camp Walker	1080	5	88.5	1

Korea (South) — continued

MW & FM Stations	kHz	kW	MHz	kW
4) Busan/Camp Hialeah	1260+	5	88.1	0.25
Chuncheon/Camp Page	1044	1	88.5	0.1
Uijeongbu/Camp Red Cloud	1161	0.25	88.5	0.1
5) Dongducheon/Camp Casey	1197+	1	88.3	0.25
Chuncheon/Camp Page	1260 F.PI	1	88.5	0.1
6) Songtan/Osan Air Base	1359*	1	88.5	0.05
7) Pyeongtaek/Camp Humphroys	1440+	1	88.3	0.05
8) Gunsan/Gunsan Air Base	1440	1	88.5	0.25
Wonju/Camp Long	1440	0.25	88.3	0.05
Waegwan/Camp Carrol	1440	0.25		
2) Munsan/Western Corridor	1440 F.PI	5	88.5	0.05
Pohang/Camp Libby	1512	0.25		
Jinhae/Naval St.	1512	0.25	88.5	0.05
Kotar Range	1512			

Low power: 1512kHz (Sangdong, Jeju Teukbyel Jachido); 88.5MHz Gwangju Air Base.

+= local prgrs 2005-0000 Mon-Fri; otherwise rel.1).

*= local prgrs 2005-0000 & 0605-0900 Mon-Fri; otherwise rel.1).

D.Prgr: 24h (MW/FM sep. prgrs). N. on the h. Formal sign on at 1505.

Ann: AM: "American Forces Network Korea", FM (Seoul): "This is Eagle FM"

Addresses
1) Headquarters, American Forces Network Korea, Unit #15324, APO AP 96205-0097, USA (+82(2) 7914 6495. Commanding Officer: LTC Chad C. Starr — **2)** Unit #15325, APO AP 96251-0098, USA – **3)** Unit #15029, APO AP 96218-0186, USA – **4)** Unit #15184. APO AP 96259-0274, USA – **5)** Unit #15116, APO AP 96224-0380, USA – **6)** Unit #2034. APO AP 96278-5000, USA – **7)** Unit #15473. APO AP 96271-0543, USA – **8)** Unit #2011, APO AP 96264-5000, USA

KOSOVO

L.T: UTC +1h (27 Mar-30 Oct: +2h) — **Pop:** 1.8 million — **Pr.L:** Albanian, Serbian, Bosnian, Turkish, Romany — **E.C:** 50Hz, 220V — **ITU:** pending (**WRTH:** RKS)

KOMISIONI I PAVARUR PËR MEDIA (KPM)
(Independent Media Commission)
🖃 Rr. "Gazmend Zajmi" nr. 1, 10000 Prishtinë ☎ +381 38 245031 🗎 +381 38 245034 **E:** info@imc-ko.org **W:** www.imc-ko.org
L.P: Agim Sopi
NB. KPM is the licensing body for broadcasting.

RADIO TELEVIZIONE KOSOVËS (RTK) (Pub)
🖃 Rr. "Xhe Prishtina" nr. 12, 10000 Prishtinë ☎+381 38 230102 🗎 +381 38 235336 **E:** post@rtklive.com **W:** www.rtklive.com
L.P: Chmn: Rrahman Paçarizi
🖃 R.Kosova/R.Blue Sky: Rr. "Nëna Tereze" pa numër, 10000 Prishtinë ☎ +381 38 249077 (R.Kosova); +381 38 226553 (R.Blue Sky) **E:** radiokosova@rtklive.com; radiobluesky@rtklive.com.

MW	kHz	kW	Prgr		
Prishtinë	549	10	1		

FM	Prgr 1	Prgr 2	kW	FM	Prgr 1	Prgr 2	kW
Cërnusha	87.6	91.5	0.5	Prishtinë	91.9	93.3	0.5
Maja e Gjelbërt	88.5	90.5	0.5	Zatriqi	88.9	92.4	0.5
Goleshi	95.7	97.7	5				

D.Prgr: Prgr 1 (R. Kosova): 24h in Albanian. – **Prgr 2 (R. Blue Sky):** 24h in Albanian; exc. 1300-1500 Serbian, 1500-1520 Turkish.

OTHER STATIONS

FM	MHz	kW	Location	Station
A)	88.6	1	Prishtinë	Deutsche Welle relay
D)	89.6	1	Mitrovicë	RFI relay
2)	92.7	1	Maja e Gjelbërt	R. Dukagjini
2)	94.5	1	Zatriqi	R. Dukagjini
1)	94.8	1	Maja e Gjelbërt	R. 21
B)	96.2	1	Prishtinë	RFE/RL & VOA relay
C)	98.6	5	Goleshi	BBC relay
2)	99.7	1	Goleshi	R. Dukagjini
D)	101.0	1	Prishtinë	RFI relay
1)	102.8	1	Goleshi	R. 21
1)	103.9	1	Zatriqi	R. 21

NB: Txs below 1kW not listed.
Addresses & other information:
1) Pallati i mediave, aneks II, 10000 Prishtinë. **E:** radio21@rtv21.tv – **2)** Rr. "Ismail Qemajli" nr. 7, 30000 Pejë. **E:** radio@radio-dukagjini.com – **A)** Rel. Deutsche Welle (Germany) – **B)** Rel. RFE/RL & VOA (USA) – **C)** Rel. BBC (UK) – **D)** Rel. RFI (France)

KUWAIT

L.T: UTC +3h — **Pop:** 2.7 million — **Pr.L:** Arabic — **E.C:** 50Hz, 240V — **ITU:** KWT

MINISTRY OF INFORMATION
🖃 P.O. Box 193, 13002 Safat ☎+965 22415301 🗎 +965 22434511
RADIO OF THE STATE OF KUWAIT
🖃 P.O. Box 967, 13010 Safat ☎+965 22423773 🗎 +965 22456660
W: www.moinfo.gov.kw **E:** info@media.gov.kw
L.P: Mr. Hani Al-Naqi, Dir. Freq. Mgmt.

MW(kHz)	kW	Prgr.	Times
540	600	Main Arabic	24h
630	100	Holy Quran	24h
963	20	Main Arabic	1200-1600, 2100-0500
		Multilingual	0500-1200, 1600-2100
1134	100	Main Arabic &Sports	24h
1269	100	Classical Arab Music	24h
1341	100	Holy Quran	2100-0700
		2nd Arabic	0700-2100

F.PI: 25kW transmitter on 1530 kHz.

FM(MHz)	kW	Prgr.	Times
87.9		Classical Arab Music	24h
89.5		Main Arabic	24h
92.5	5	Easy FM	24h
93.3		National Assembly	24h (also Classical mx)
96.3		Main Arabic	1200-1600, 2100-0500
		Multilingual	0500-1200, 1600-2100
97.5	2	Holy Quran	2100-0700
		2nd Arabic	0700-2100
98.9	20	Holy Quran	24h
99.7		Super Station	24h
100.5	-	TV sound (Prgr. 1)	24h
103.7	-	Modern Arab Music	24h

Main Arabic prgr: 24h. **N:** 0300, 0500, 1000, 1700, 2100 – **2nd Arabic prgr:** 0700-1700 – **Classical Arab Music prgr:** 24h – **Modern Arab Music prgr:** 24h – **Multilingual prgr:** English 0500-0800, 0900-1000, Persian 0800-1000, Filipino 1000-1200, Urdu 1600-1800 – **Quran prgr:** 24h – **"Easy FM" in English:** 24h – **"FM Super Station"** in English: 24h. **N:** on the hour.
Ann: "Idha'at al-Dawlat Al Kuwait".
External service on shortwave: see International Radio section.

Other stations:
Marina FM, 88.8MHz. Web: www.marinafm.com
Mix FM, 98.4MHz. Web: www.mix984.com
AFN: Al-Jabber/Camp Doha 101.5/107.9MHz 50W/5kW.
BBC World Sce: Kuwait City: Arabic 90.1MHz, English 100.1MHz.
Panorama FM: 91.4MHz. See main entry under UAE.
DW/Monte Carlo Doualiya: 107.3MHz 1kW. **RFI:** 106.3MHz.
R. Sawa: 1548kHz 300/600kW, 95.7MHz 5kW, both 24h.
VOA: 96.9MHz 1kW. **VOA/RFERL** MW: 1593kHz 150kW 1400-0700

KYRGYZSTAN

L.T: UTC +6h — **Pop:** 5.4 million — **Pr.L:** Kyrgyz, Russian — **E.C:** 50Hz, 220V — **ITU:** KGZ

MADANIYAT ZHANA MAALYMAT MINISTRILIGI
(Ministry of Culture and Information)
🖃 Pushkin St. 78, 720040 Bishkek ☎ +996 312 620482 **E:** min-cultkr@mail.ru **L.P:** Minister: Sadyk Sher-Niyaz
NB. The Ministry of Culture and Information is responsible for issuing broadcasting licenses.

KYRGYZ RESPUBLIKANYN ULUTTUK TELERADIO BERÜÜ KORPORATSIYASY (Gov)
(Kyrgyz National Broadcasting Corp.)
🖃 Jash Gvardiya blvd. 59, 720010 Bishkek ☎ +996 312 655677 🗎 +996 312 651064 **E:** ntrk@ntrk.kg **W:** www.ntrk.kg
L.P: GD: Kubat Otorbaev

MW	kHz	kW	Prgr	MW	kHz	kW	Prgr
Bishkek (a)	612	150	KGR2	Haidarkan	1404	7	KGR1
Bishkek (a)	1287	150	KGR1	Jojomel	1404	20	KGR1
Naryn	1404	7	KGR1	Orgochor	1404	-	KGR1
Cholpon-Ata	1404	1	KGR1				

SW	kHz	kW	Prgr	SW	kHz	kW	Prgr
Bishkek (a)	4010	100	KGR1	Bishkek (a)	4795	15	KGR1

FM (MHz)	KGR1	KGR2	kW
Bishkek	104.1	106.9	1

+ nationwide network (CCIR-FM). (a) Krasnaya Rechka

D.Prgr: KGR1 (Kyrgyz radiosu): 2300-1800 in Kyrgyz, Russian. – **KGR2 ("XXI kylym" radiosu):** 0000-1800 (Sun 1200) in Kyrgyz, Russian. – **Parliamentary channel:** Bishkek 90.2MHz.

OTHER STATIONS

MW	kHz	kW	Location	Station
B)	1467	150	Bishkek (a)	TWR relay
A)	4050	100	Bishkek (a)	R. Rossii relay
FM	**MHz**	**kW**	**Location**	**Station**
A)	66.26	17	Karakol	R. Rossii relay
A)	67.94	17	Bishkek	R. Rossii relay
A)	68.66	17	Kara-Kul	R. Rossii relay
A)	69.92	17	Osh	R. Rossii relay
A)	69.95	17	Kazarman	R. Rossii relay
7)	70.24	4	Karakol	Avtoradio (R. LW)
A)	70.07	17	Arstanbap	R. Rossii relay
A)	70.40	17	Suluktu	R. Rossii relay
A)	70.82	17	Narin	R. Rossii relay
A)	72.20	17	Jalal-Abad	R. Rossii relay
A)	72.44	17	Suluktu	R. Rossii relay
6)	87.5	1	Bishkek	R. Mir
2)	88.0	1	Bishkek	R. NS
4)	89.0	1	Bishkek	Love R.
9)	101.4	1	Cholpon-Ata	Hit FM
10)	101.7	1	Bishkek	Evropa Plus
5)	102.9	1	Bishkek	R. Manas FM
1)	103.2	1	Cholpon-Ata	Russkoye R.
1)	104.5	1	Bishkek	Russkoye R.
3)	105.0	1	Bishkek	R. Piramida
8)	106.3	1	Kara-Balta	R. Tatina

NB: Txs below 1kW not listed. (a) Krasnaya Rechka.

Addresses & other information:
1) Almaty St. 4b, 720082 Bishkek. **E:** rusradio@europa.kg – **2)** Bishkek. **3)** Jantoshev St. 70, 720005 Bishkek. **E:** pyramid@mail.elcat.kg – **4)** Ibraimov St. 24, 720031 Bishkek. **E:** office@loveradio.kg – **5)** Mir pr. 56, 720044 Bishkek. **E:** manasfm@manas.kg – **6)** Bishkek. – **7)** Gebze St. 120, 722360 Karakol. **E:** lw@pari.issyk-kul.kg – **8)** Gvardeyskaya St. 18, 722030 Kara-Balta. **E:** tatina@infotel.kg – **9)** Bishkek. – **10)** Bishkek. – **A)** Rel. R. Rossii (Russia): 2300-1900 – **B)** Rel. TWR (USA)

LAOS

L.T: UTC +7h — **Pop:** 7.0 million — **Pr.L:** Lao (Lao Soung, Lao Theung dialects), Hmong, Khmu — **E.C:** 50Hz, 230V — **ITU:** LAO

LAO NATIONAL RADIO – LNR (Gov.)

✉ B. P. 310, Vientiane; Phainam Rd, Ban Sisakhet, Chantabouly District, Vientiane ☎ +856 21 212468 🖷 +856 21 212430 **W:** www.lnr.org.la **L.P:** DG: Mr Sipha Nonglath. Dep. DG: Keungkham Vilayasith. Tech. Dir: D. Sisombath

City and Provincial sces: These are operated by the local governments ✉ Sisavangvong Rd, Ban Pakhame, Luang Prabang – Km 2 Route 13 South, Oudomsavane Village, Pakse, Champassak Province - Manthatulat Road, Vientiane – Houamouangtai Village, Savannakhet, Khantabouly – Nongbouakham Village, Tha Khek, Khammouane.

MW	kHz	kW	S	H of tr
Vientiane+	567	200	N	2200-0800, 0900-1600
Khantabouly, S	585	20	P	2230-1300
Vientiane	640	10	C	2330-1000, r. inactive
Luang Prabang	705	10	P	2200-0800, 1025-1500
Tha Khek, Kh	765	10	P	unconfirmed, r. inactive
SW	**kHz**	**kW**	**S**	**H of tr**
Sam Neua, HP	v4413	1	P	2300-0130, 0925-1230
Vientiane	6130	50	N	2200-0800, 0900-1600

S=Sce., N=National, P=Provincial, C=City
HP=Houa Phan prov. Kh=Khammouane prov. O=Oudomxay prov. S=Savannakhet prov. XK=Xiang Khouang prov. Reg. stns genrally rel. national news at 0000, 0500, 1200 +) Transmitter loc.: Kilometre 49 (GC: 18N20 102E27); alt. freq: 580kHz
National Sce in Lao: 2300-0600, 0900-1600; **Hmong:** 2200-2230, 0600-0700; **Khmu:** 2230-2300, 0700-0800; **N:** 2300, 0000, 0500, 0800, 1200, 1400. **English/French LL:** Mon-Thurs 1415-1430.
Ann: LNR: "Thini Sathani Vitthayou Krachaisiang Hengsat". Sam Neua: "Thini Vitthayou Krachaisiang Houa Phan, krachaisiang chak Muang Sam Neua".
IS: Music on Khéne (mouth organ) & Solo (bamboo instrument).

LNR FM sces: Vientiane FM1 103.7MHz 20kW: 2300-1700. Vientiane FM2 97.25MHz 2kW: 2230-1700 rel. Int. Sce, and at other times carries entertainment prgrs in Lao. **Vientiane City FM sce:** Vientiane 105.5MHz: 2330-1700. Vientiane 98.8MHz: operated by Butterfly Media Co. Ltd. **Provincial FM sces:** Attapeu: 102.7MHz 100W, r.

inactive. Bolikhamsay prov: 101.5MHz 5kW Houai Xay, Bokeo prov: 102.75MHz 1kW Khantabouly, S: 100.75MHz 1kW Luang Namtha prov.: 98.MHz 1kW Luang Prabang: 103.5MHz 300W. Muang Hay, O: 100MHz 100W. Muang Khong, C: 97.2MHz. Pakse, Champassak prov: 103.7MHz 1kW Phonsavan, XK: 97.5MHz 5kW Phongsali prov: 102MHz 100W. Sam Neua, HP: 102.75MHz 100W (ann. freq): As 4640kHz. Saravane: 101.2MHz 300W. Saiyabouly: 96.5MHz 5kW Saysomboun Special Reg.: 100MHz 5kW Sekong: 102.7MHz 100W. Tha Khek, Kh: 95.5MHz 100W.

EXTERNAL SERVICES: see International Broadcasting section.

OTHER STATIONS:
China Radio International: Vientiane 93.0MHz 10kW **D.Prgr:** 0300-1530 rel. CRI from Beijing in Chinese, English & Lao.
Radio Australia: Vientiane 96.0 MHz **D.Prgr:** 24h in English.
Radio France Internationale: Vientiane 100.5 MHz **D.Prgr:** 24h rel. RFI from Paris in French.

LATVIA

L.T: UTC +2h (27 Mar-30 Oct: +3h) — **Pop:** 2.2 million — **Pr.L:** Latvian, Russian — **E.C:** 50Hz, 220V — **ITU:** LVA

NACIONALA RADIO UN TELEVIZIJAS PADOME (NRTP) (National Radio and TV Council)

✉ Smilšu iela 1/3, 1939 Riga ☎ +371 67221848 🖷 +371 67220448 **E:** nrtp@nrtp.lv **W:** www.nrtp.lv **L.P:** Chmn: Abrams Kleckins
NB. NRTP is the licensing body for broadcasting.

LATVIJAS RADIO (Pub)

✉ Doma laukums 8, 1505 Riga ☎ +371 67206722 🖷 +371 67206709 **E:** radio@latvijasradio.lv **W:** www.latvijasradio.lv
L.P: DG: Dzintris Kolats

FM (MHz)	LR1	LR2	LR3*	LR4	kW
Aluksne	106.8	104.3	-	-	3.5
Auce	99.6	-	-	-	1
Cesvaine	102.5	105.0	103.5	107.9	20/20/5/5
Dagda	102.6	-	-	-	1.7
Daugavpils	106.1	100.7	90.6	104.0	6.5/10/3.2
Dundaga	91.1	106.7	-	-	4
Kuldiga	95.9	101.3	92.0	-	10/16.6/3.3
Liepaja	107.1	101.0	104.6	97.9	10/10/1/5
Rezekne	104.2	101.0	101.8	107.5	20/20/20/5
Riga	90.7	91.5	103.7	107.7	15/3.5/5/6.3
Valmiera	104.0	101.5	87.6	-	18
Ventspils	99.2	103.0	89.8	95.3	0.3/0.3/0.3/1
Viesite	107.6	104.7	102.2	-	5
Vitrupe	105.5	-	-	-	1.6
Zilupe	-	-	-	100.1	1.7

*) time-shared with LR5 (exc. Riga 103.7)

D.Prgr: LR1 (Latvijas R. Viens): 24h. – **LR2 (Latvijas R. Divi):** 24h. **LR3 (Klasika):** 24h. – **LR4 (Doma laukums/Domskaya ploshchad):** 24h, mainly in Russian. For various other ethnic minorities: MF 1215-1245. – **LR5 (R. NABA/Saemas kanals):** 24h. On Riga 93.1MHz (5kW) & parttime via LR3 network. ✉R. NABA: Aspazijas bulv. 5, 1050 Riga. **E:** naba@radionaba.lv.

OTHER STATIONS

FM	MHz	kW	Location	Station
25)	1485	1.25	Riga	R. Merkurs
FM	**MHz**	**kW**	**Location**	**Station**
2)	87.9	2	Madona	Star FM
2)	88.1	1	Selpils	Star FM
3)	88.4	2.2	Liepaja	Kurzemes R.
4)	88.4	1	Gulbene	European Hit R.
14C)	88.6	1.1	Ulbroka	Jumor FM
1C)	89.2	2.8	Riga	R. SWH Rock
1A)	89.3	1	Dundaga	R. SWH
23)	90.3	1	Matisi	Latviešu R.
5)	90.8	1	Ventspils	Kristigais R.
23)	90.9	2.1	Lautere	Latviešu R.
2)	91.0	2.2	Liepaja	Star FM
12)	91.6	1	Viški	R. 1
2)	91.9	2	Rezekne	Star FM
2)	92.0	1	Sipolkalni	Star FM
23)	92.1	1.1	Ošani	Latviešu R.
10)	92.2	1.25	Rugaji	R. Vidzeme
19)	92.3	1	Liepaja	City R.
3)	92.4	1.3	Tukums	Kurzemes R.
10)	92.8	1	Limbazi	R. Vidzeme
18)	93.5	2	Liepaja	R. 101
14D)	93.9	9.3	Riga	R. Baltcom

FM	MHz	kW	Location	Station
23)	94.3	1.6	Talsi	Latviešu R.
1A)	94.7	1.4	Saldus	R. SWH
22)	95.2	5	Daugavpils	Latgolys Radeja
23)	95.2	1	Liepaja	Latviešu R.
22)	95.8	1.6	Jekabpils	Latgolys Radeja
4)	96.1	1.1	Liepaja	European Hit R.
11)	96.1	2.1	Kraslava	Novoye R. (Europa+)
9)	96.2	2	Riga	Krievu Hitu R. (Russkoye R.)
23)	96.8	1	Riga	Latviešu R.
10)	97.0	1	Valmiera	R. Vidzeme
17)	97.5	1.6	Liepaja	R. Skonto Liepaja
2)	97.7	1.1	Pure	Star FM
20)	98.1	1	Valmiera	R. Valmiera
5)	98.5	3.3	Kuldiga	Kristigais R.
13)	99.0	2	Jurmala	R. Jurmala
24)	99.4	1.2	Daugavpils	R. Daugavai
14B)	99.5	4.5	Riga	Fit FM
10)	99.8	2	Cesvaine	R. Vidzeme
5)	99.9	1.1	Daugavpils	Kristigais R.
16)	100.0	2.5	Riga	R. PIK
1A)	100.3	5	Cesvaine	R. SWH
15)	100.5	1	Ventspils	Ventspils R.
A)	100.5	1.2	Riga	BBC relay
5)	100.6	1.2	Liepaja	Kristigais R.
4)	100.8	1	Talsi	European Hit R.
7)	101.0	2	Iecava	Capital FM
1A)	101.2	1.4	Jekabpils	R. SWH
5)	101.3	1.2	Kraslava	Kristigais R.
6)	101.6	2.5	Daugavpils	Alise Plus
5)	101.8	5.1	Riga	Kristigais R.
2)	102.0	2	Saldus	Star FM
4)	102.0	1	Madona	European Hit R.
1A)	102.2	4	Talsi	R. SWH
14A)	102.7	6.9	Riga	R. Mix FM
5)	102.8	1	Jekabpils	Kristigais R.
22)	103.0	1.8	Rezekne	Latgolys Radeja
23)	103.2	1	Kuldiga	Latviešu R.
2)	103.2	1	Svente	Star FM
2)	103.8	8	Kuldiga	Star FM
4)	104.3	4.6	Riga	European Hit R.
18)	104.7	1	Cesis	R. 101
2)	105.0	1.6	Pope	Star FM
4)	105.1	1.3	Rezekne	European Hit R.
1A)	105.1	4.5	Liepaja	R. SWH
1A)	105.2	13.2	Riga	R. SWH
1A)	105.2	2.2	Daugavpils	R. SWH
1A)	105.4	1	Ventspils	R. SWH
1B)	105.7	4.5	Riga	R. SWH Pluss
5)	105.9	4	Cesvaine	Kristigais R.
2)	106.2	6.3	Riga	Star FM
3)	106.4	11.5	Kuldiga	Kurzemes R.
1A)	106.5	4.6	Valmiera	R. SWH
1A)	106.5	4	Rezekne	R. SWH
2)	106.6	1	Bauska	Star FM
17)	107.2	4.6	Riga	R. Skonto
8)	107.2	2.5	Daugavpils	R. Maksimums (Europa+)
4)	107.4	1	Kuldiga	European Hit R.
2)	107.4	2	Valmiera	Star FM
21)	107.6	1.6	Liepaja	R. Liepaja (Russkoye R.)
3)	107.9	1	Ventspils	Kurzemes R.

NB: Txs below 1kW not listed.

Addresses & other information:

1A,B,C) Skanstes iela 13, 1013 Riga. 1B) in Russian. **E:** 1A) radio@radioswh.lv, 1B) swhplus@radioswh.lv – **2)** Dzelzavas iela 120G, 1021 Riga. **E:** info@starfm.lv – **3)** Pilsetas laukums 4, 3300 Kuldiga. **E:** studija@kurzemesradio.lv – **4)** Elijas iela 17, 1050 Riga. **E:** radio@europeanhitradio.com – **5)** Lacpleša iela 37, 1011 Riga. In Latvian & Russian. **E:** lkr@lkr.lv – **6)** Raina iela 28, 5403 Daugavpils. In Russian & Latvian. **E:** radio@aliseplus.lv – **7)** L.Nometnu iela 62, 1002 Riga. **E:** info@capitalfm.lv – **8)** S.Mihoelsa iela 9, 5403 Daugavpils. In Russian, with Europa+ (Russia) as filler. **E:** radiomax@mbox.latg.lv – **9)** Elijas iela 17, 1050 Riga. In Russian, with Russkoye R. (Russia) as filler. – **10)** Rigas iela 19, 4201 Valmiera. – **11)** Latgales iela 20, 4600 Rezekne.In Russian, with Europa+ (Russia) as filler. **E:** radio.latgalei@rezekne.lv – **12)** Brivibas iela 116, 5200 Jekabpils. **E:** info@radio1.lv – **13)** Brivibas bulv. 30, 1050 Riga. – **14A,B,C)** K.Valdemara iela 8, 1010 Riga. 14A,C,D in Russian; 14C with Yumor FM (Russia) as filler. **E:** 14A) radio@mixfm.lv, 14B) fitfm@fitfm.lv, 14D) radio@radiobaltkom.lv – **15)** Saules iela 10, 3601 Ventspils. – **16)** Brivibas bulv. 30, 1050 Riga. In Russian. **E:** info@pik.lv – **17)** Kr.Valdemara 100, 1013 Riga. On Liepaja 97.5MHz: mostly local prgrs. **E:** reklama@radioskonto.lv – **18)** Elizabetes iela 55, 1050 Riga. **E:** radio@radio101.lv – **19)** Zivju iela 3, 3401 Liepaja. **E:** radio@cityradio.lv – **20)** Rigas iela 13, 4201 Valmiera. **E:** info@radiovalmiera.

lv – **21)** Klaipedas iela 19/21, 3401 Liepaja. In Russian, with Russkoye R. (Russia) as filler. **E:** radioliepaja@netlogs.lv – **22)** Atbrivošanas aleja 81/5, 4601 Rezekne. **E:** radeja@lr.lv – **23)** Elijas iela 19, 1050 Riga, **E:** info@latviesuradio.lv – **24)** Zakusalas krastmala 3, 1509 Riga. – **25)** P.O.Box 371, 1010 Riga. E: rni@apollo.lv – **A)** Rel. BBC (UK)

LEBANON

LT: UTC +2h (27 Mar–30 Oct: +3h) — **Pop:** 4 million — **Pr.L:** Arabic, French, English, Armenian — **E.C:** 50Hz, 110/220V — **ITU:** LBN

MINISTRY OF INFORMATION
Hamra, Beirut ☎+961 1 754400 **W:** www.ministryinfo.gov.lb

RADIO LEBANON (Gov.)
Rue Lyon, Sanayeh, P.O. Box 4848, Beirut ☎+961 1 743531 **W:** www.96-2.com **E:** mykee@cyberia.net.lb
L.P: Dir: Fuad Hamdan. Tech. Dir: Nazih Chahine. Chief, Prgr. Dept: Waheed Jalal. Chief, Public Rel: Faouzi Fehmy.
1st Prgr. in Arabic: 0330-2330 on 98.1/98.5MHz. **2nd Prgr. in French/ English/Armenian:** 24h on 96.2MHz. **Rel. R. France Int:** 13h daily. **Ann:** A: "Iza'at Loubnan min Beirut". F: "Ici Radio Liban émettant de Beyrouth" **IS:** Opening notes from the Lebanese National anthem played on guitar.

OTHER STATIONS:

FM	MHz	Name	FM	MHz	Name
14)	87.5	VO Charity	11)	96.9	Sawt el-Ghad
32)	87.4	R. Sawa	12)	97.4	Voice of Faith
20)	87.8	R. Nostalgie	26)	97.7	R. Strike
4)	88.5	R. Orient	5)	99.0	NRJ
13)	89.1	Holy Quran R.	12)	99.3	Voice of Faith, Beit Meri
7)	89.4	Risala R.	15)	99.7	Fame FM
16)	90.0	Voice of Liberty	24)	100.0	R. Scope, Zahle
13)	90.1	Holy Quran R.	23)	100.4	R. Lebanon Star
12)	90.3	Voice of Faith	25)	101.1	R. Scope
17)	90.6	R. Light FM	31)	101.4	R. Sevan
31)	90.8	R. Sevan	22)	102.0	R. Delta
8)	91.3	The Call of Islam	9)	102.5	R. Free Lebanon
3)	92.0	R. Al-Nour	21)	103.0	Pax R.
28)	92.7	Sawt el-Mada	29)	103.3	Monte-Carlo Doualiya
1)	93.3	Voice of Lebanon	2)	103.9	Voice of the People
6)	94.0	Radio One	19)	104.6	Mix FM, Achrafieh
13)	94.1	Holy Quran R.	6)	105.3	Radio One
6)	94.5	Radio One	14)	106.0	Voice of Charity
10)	94.8	Voice of Van	27)	106.7	Sound of Music
5)	95.1	NRJ, Farya Mzaar	18)	107.4	Master Broadc. Station
12)	95.5	Voice of Faith	14)	107.7	Voice of Charity
30)	95.9	Sawt el-Noujoum	30)	107.8	Sawt el-Noujoum

NB: the stns have been allocated 400kHz frequency range, of which the centre freq. is listed above. In many cases the trs from various sites are placed on both upper and lower limits of this range.

Addresses & other information:

1) P.O. Box 165271, Ashrafieh, Bachir el Gemayel Ave, Beirut. Web: www.vdl.com.lb – **2)** Jabal el Arab St, Wata el Mousaitbeh, P.O.Box 14/5425, Beirut. 0400-2300 – **3)** Al-Nour Bldg, Abdel Nour St, Haret Hreïk, P.O.Box 25-197, Ghbeiry, Beirut. Web: www.alnour.com.lb . Also relayed via Tartus, Syria on 1071 kHz – **4)** Annajah Centre, Mar Elias St, Karakol Druz, P.O. Box 11-6362, Beirut. Web: www.radioorient.com.lb – **5)** Studiovision Bldg,Naccache, Metn, Beirut. Web: nrjlebanon. com 24h in English. – **6)** Zakhem Bldg, Beit Meri El Metn. Web: www.radioonev5.com 0400-2300 in English – **7)** Fraiha Bld. 3rd Floor, Barbour Beirut. Web: risalaradio.com – **8)** Shaykh Ahmad Iskandarani Centre, Bourj Abi Haidar, Beirut – **9)** Kebbe Bldg, Adonis, Zouk Mosbeh, P.O.Box 110, Zouk Mekhael, Jounieh. Web: www.rll.com.lb 0340-2300 – **10)** 2nd floor, Shaghzoyan Centre, Borj Hammoud, P.O. Box 80-860, Beirut. Web: www.voiceofvan.net 24h in Armenian & Arabic – **11)** 2nd floor, Disco Samir Bldg, Zalka Rd, El Metn. Web: www.sawtelghad.com – **12)** Near Riyad el Solh Palace, Bir Hassan, P.O. Box 83/25, Ghbeiri, Beirut – **13)** Dar al Fatwa, P.O. Box 14-5380, Al Mazraa-Beirut. Web: www.darelfatwa.gov.lb – **14)** Couvent St. Jean, Fouad Chehab St, P.O. Box 850, Jounieh. Web: www.radiocharity.org 24h in Arabic/French/others. Also rel. by Vatican R. 0530-0555 11715kHz – **15)** 3rd floor, La Perla Centre, Sabra Highway, Jounieh. Web: www. famefm.com – **17)** 8th floor, Mansour Bldg, Independence St, Sassine Square, Achrafieh, Beirut. Web: www.radiolightfm.com 0500-2200 in English/French – **18)** St. Paul bldg, facing La Cite, Jounieh – **19)** Alfred Naccache Ave, P.O. Box 166-815, Achrafieh, Beirut. Web: www.mixfm.com.lb 24h in English – **20)** Mont Liban Bldg, Ave. Fouad Chehab, Fassouh, P.O.Box 16-6000, Achrafieh, Beirut. Web: www.nostalgie.com.lb 24h in French – **21)** P.O. Box 116-5104, Beirut. 24h in English. Web: paxradio.page.tl – **22)** Kahalé Bldg, Old St, P.O.Box 1306, Beit Meri el Metn. Web: www.4com.

net.lb/delta – **23)** 1st floor, Nehme Bldg, Sin el Fil main rd, Hayek – **24)** facing Tal-Shiha Hospital, Tal-Shiha, Zahlé. 0400-2200 in Arabic – **25)** 4th Floor, Hawa Chicken Building, Damascus Highway, Hazniyé – **26)** 2nd floor, Abi Jaber Bldg, Al Saideh St, Sin El-Fil, Beirut. Web: www.radiostrike.com 0600-2200 in Arabic – **27)** Centre Nasrallah, Rue Al-Anwar, Jdeideh, P.O. Box 90-1119, Beirut. Web: www.sawtelmousika.com 0430-2400 in Arabic – **28)** Mirna el Chalouhi Centre 2nd floor, Sin el-Fil, Beirut. Web: sawtlemada.com – **29)** txs in Beirut/Tripoli/Tyros/Sidon. For details see IntRad under France – **30)** Kreshet Bldg. 7th floor, Suyoufi St, Algazlep, Achrafieh, Beirut. Web: sawtelnoujoum.com – **31)** Khatchadurian Street, Khederlarian Building, Ground Floor, Beirut. Web: radiosevan.com – **32)** see in International section under USA

LESOTHO

L.T: UTC +2h — **Pop:** 2.2 million — **Pr.L:** Sesotho, English — **E.C:** 50Hz, 220V — **ITU:** LSO

LESOTHO COMMUNICATIONS AUTHORITY (LCA)
✉ P.O. Box 15896, 6th Floor, Moposo House, Kingsway Road, Maseru ☎ +266 22224300 🖷 +266 22310984 **W:** lca.org.ls **E:** lca@lca.org.ls

LESOTHO NATIONAL BROADCASTING SERVICE (LNBS, Pub.) - Radio Lesotho
✉ P. O. Box 552, Lerotholi St, Opposite Royal Palace, Maseru 100 ☎ +266 22321460 🖷 +266 22313980 **W:** www.radioles.co.ls **L.P:** Dir. of Broadc: Mr. Lebohang Dada Mokasa. CE: Mr. Motlatsi Monyane. Principal Tech. Officer: 'Masekoala Ratia.
MW: Maseru (Lancer's Gap): 639kHz 100kW, 891kHz 50kW.

FM	MHz	kW	FM	MHz	kW
Mokhele	90.5	1	Likhoele	97.2	1
Maseru (Berea)	93.3	1	Thaba-Ntso	98.9	0.25
Chafo	96.0	1	Sheep Stud Hill	102.4	1
Maseru (Ponoare)	96.2	1	Popa	103.6	0.25
Matshoana	96.8	1	Souru	105.4	

D.Prgr in Sesotho/English: 24h. **F.PI:** four new FM trs.
Ultimate FM in English: Lancer's Gap 891kHz & 99.8MHz, 24h.
Ann: E: "This is Radio Lesotho" or "This is the Lesotho National Broadcasting Service, Maseru". Sesotho: "Se-ea-le-moea sa Lesotho, Maseru". **IS:** native horn instruments.

OTHER STATIONS:
Catholic R. FM: Qoatsaneng, Maseru 103.3MHz – **Dope FM,** Maseru: 103.6 MHz – **Fill the Gap (Jesu ke Karabo):** Mafeteng 87.6MHz, Leribe 102.8MHz, Lancer's Gap 105.3MHz – **Harvest FM:** Lancer's Gap, Maseru 98.9MHz – **Joy FM,** Private Bag A68, Maseru 100: 106.9MHz 1kW **W:** www.joyfm.co.ls Also rel. VOA. F.PI: trs in Mafeteng and Maputsoe – **Lesotho Evangelical Church R.** Lancer's Gap, Maseru: 102.4MHz – **MoAfrika FM:** Leribe 89.7MHz, Mafeteng 90.7MHz, Lancer's Gap 99.3MHz. **W:** www.moafrika.co.ls – **People's Choice FM,** Development House, Block D, Floor 9, Kingsway Str, Maseru: 95.6MHz. **W:** www.pcfm.co.ls – **Thaha-Khube FM,** Ha Ts'osane, Maseru: 97.6MHz.
Family R, Maseru. **MW:** 1197kHz 100kW. **D.Prgr:** English 0300-0500, 1600-1900, 2000-2300, Portuguese 1900-2000. For further details see International Radio section (USA).
BBC African Sce: Lancer's Gap, Maseru 90.2MHz.
RFI Afrique: Lancer's Gap, Maseru 96.5MHz in French/English

LIBERIA

L.T: UTC — **Pop:** 3.5 million — **Pr.L:** English, 18 ethnic — **E.C:** 60Hz, 120V — **ITU:** LBR

LIBERIA TELECOMMUNICATIONS AUTHORITY (LTA)
✉ National Investment Commission Annex, 12th Street, Sinkor, Tubman Boulevard, Monrovia ☎ +231 770 54054 🖷 +266 770 00825 **W:** www.lta.org.lr

LIBERIA BROADCASTING SYSTEM (LBS, Pub.)
✉ P.O. Box 594, Paynesville **W:** liberiabroadcastingsystem.com **E:** info@liberiabroadcastingsystem.com **L.P:** Dir: Charles Snetter.
FM: R. Liberia: 99.9MHz 1kW.
D.Prgr: FM 0455-1005 & 1200-2400, Sat 0450-2400, Sun 0650-2400.
F.PI: building modern FM & shortwave services.

RADIO ELWA (Rlg.)
✉ P.O. Box 192, Monrovia **Web:** www.elwaministries.org **E:** elwaradio@yahoo.com **L.P:** GM: Moses T. Nyantee.
SW: Monrovia 4760kHz 1kW, 6070kHz 2kW (both inactive).
FM: Monrovia 94.5MHz 0.25kW.
D.Prgr in English/local lang's: 0600-1000, 1730-2300 (SS -2230).

RADIO VERITAS (Rlg.)
✉ P.O. Box 3569, Catholic Media Centre, Monrovia ☎ +231-77777987 🖷 231-6516906 **E:** radioveritas@hotmail.com **L.P:** Stn. Mgr: Ledgerhood Rennie.
SW: Monrovia 10kW: 0445-0900 on 6090kHz, 1745-2300v on 5470kHz. (inactive, but expected to return).
FM: 97.8MHz 5kW 0445-2300. **Prgr:** in English/local langs.
Ann: "This is R. Veritas, the Voice of the Truth, broadcasting from Monrovia. Liberia"

STAR RADIO (Priv.)
✉ 12 Broad St, Snapper Hill, Monrovia ☎ +231 77104411 **W:** www.starradio.org.lr **L.P:** James Morlu, Mgr.
SW: 3960kHz 2.5kW **FM:** Monrovia: 104.0MHz 1kW. **D.Prgr** in English/Local languages: 0500-0905, 1700-2105.

UNMIL RADIO (United Nations Mission in Liberia)
✉ UNMIL Force HQ, Star Building, Monrovia **W:** www.unmil.org **E:** webmaster@unmil.org **L.P:** Dir: Joseph Roberts-Mensah.
FM: Gbarnga 90.5MHz, Harper/Monrovia/Zwedru 91.5MHz, Sanniquellie 95.1MHz. Greenville/Voinjama 97.1MHz (Harper/Sanniquelle 1kW, others 5kW). **D.Prgr:** 24h in English/local langs

Other stations:
City FM, Monrovia: 90.2MHz – **Crystal FM,** Monrovia: 95.5MHz – **DC 101.1 FM,** Monrovia: 101.1MHz. Also rel. BBC African Sce. – **King's FM,** Monrovia: 88.5MHz. Also rel. VOA. – **Liberian Christian Broadcasting Network,** Monrovia: 102.3MHz – **Love FM,** Monrovia: 105.5MHz – **Power FM,** Monrovia: 93.3MHz – **Magic FM,** Monrovia: 99.2MHz – **Sky FM,** Monrovia: 107.0MHz. Email: skyliberia@yahoo.com – **Truth FM,** Monrovia 96.1MHz. Web: truthfm.com.lr – **United Metodist Church R,** Monrovia: 98.7MHz 0.3kW.
BBC African Sce, Monrovia: 103.0MHz
RFI Afrique: Monrovia 106.0MHz in French/English.
About 35 community radio stations are in operation

LIBYA

L.T: UTC +2h — **Pop:** 6.5 million — **Pr.L:** Arabic — **E.C:** 50Hz, 127/230V — **ITU:** LBY

LIBYAN JAMAHIRIYA BROADCASTING CORPORATION (LJBC, Gov.)
✉ El Fath Rd, P.O. Box 80237, Tripoli ☎ +218 21 3402153 🖷 +218 21 340 3458 **Web:** ljbc.net **Email:** info@ljbc.net **L.P:** DG: Ibrahim El Zawam. Dir. Tech, Dept: Ammar El Mahjoub.

MW	kHz	kW	N	MW	kHz	kW	N
Tobruk	‡ 648	300	V	Tripoli	1053	100	M
Benghazi	675	100	M	Kufra	‡ 1080	10	M
Jeffren	‡ 711	50	M	El Beida	1125	500	M
Sirt	‡ 792	20	M	Tripoli	1251	400M/V	
Sabha	‡ 828	300	M	Tripoli	‡ 1404	20	M
Ghiagboub	‡ 909	20	M	Al-Assah	1449	500	M
Sirt	972	50	M	Brach	‡ 1485	1	Q
FM	MHz		N	FM	MHz		N
Al Beida	87.9		M	Tripoli	97.5		M
Benghazi	89.3		L	Al Khums	97.9		M
Misurata	90.0		M	Tobruk	98.0		L
Misurata	91.9		L	Benghazi	98.8		Q
Beni Walid	92.3		L	Misurata	99.9		L
Misurata	93.4		H	Al-Zawia	101.3		L
Tripoli	94.0		L	Tripoli	103.4		L
Ghadames	95.3		M	Tripoli	104.0		M
Tripoli	95.5		M	Misurata	105.0		Q
Zultan	95.6		M	Al-Ujilat	105.1		Q
Baniwalid	95.8		M	Gharian	105.5		Q
Sirt	96.0		L	Tripoli	107.0		Q
Tripoli	96.6		M				

(Based mostly on monitoring reports. No info available from LJBC).
Networks: M=Main Sce, V=VO Africa, Q=Quran prgr, L=Local R.
Main Sce: 24h. Also on 1251kHz 1000-1700. **Voice of Africa:** 1700-0400 in Arabic incl. short newscasts in English/French/others at variable times. **Holy Quran Prgr:** 24h on FM. **Tripoli R. of the Arabs** on 104.0MHz: 0700-1100 & 1600-1800. **Ann:** M: "Idha'at al-Jamahiriya al-Ozma". **IS:** Prgrs open and close with National Anthem.

EXTERNAL SERVICE: V. of Africa: see International Radio section.

AL-LIBIYA FM
✉ One Nine Media Company, P.O. Box 91367, Tripoli ☎ +218 21 3339974 🖷 +218 21 4442390 **Web:** www.allibiya.fm **Email:** info@allibiya.fm **L.P:** DG: Abdussalam Meshri.
FM: Tripoli 93.4MHz. D.Prgr: 0600-2400

LIECHTENSTEIN

L.T: UTC +1h (27 Mar-30 Oct: +2h) — **Pop:** 34,000 — **Pr.L:** German, Alemannic — **E.C:** 50Hz, 230V — **ITU:** LIE

LIECHTENSTEINISCHER RUNDFUNK (Pub)
✉ Dorfstr. 24, 9495 Triesen, Fürstentum Liechtenstein ☎ +423 3991313 ▤ +423 3991366 **E:** admin@radio.li **W:** www.radio.li
L.P: CEO: Alois Ospelt
FM: Buchs 89.2 (0.5kW), Trübbach 95.0 (0.02kW), Steg 96.6 (0.25kW), Vaduz 96.9 (0.1kW), Nendeln I 100.2 (0.05kW), Vilters-Targön 103.4 (0.1kW), Nendeln II 103.7 (0.1kW), Rüthi 106.1 (1kW).
D.Prgr: Radio L 24h

LITHUANIA

L.T: UTC +2h (27 Mar-30 Oct: +3h) — **Pop:** 3.5 million — **Pr.L:** Lithuanian, Polish, Russian — **E.C:** 50Hz, 220V — **ITU:** LTU

LIETUVOS RADIJO IR TELEVIZIJOS KOMISIJA (LRTK)
✉ Vytenio g. 6/23, 03113 Vilnius ☎ +370 5 2330660 ▤ +370 5 2647125 **E:** lrtk@rtk.lt **W:** www.rtk.lt **P:** Chmn: Jonas Liniauskas
NB. LRTK is the regulatory authority for broadcasting.

LIETUVOS NACIONALINIS RADIJAS (Pub)
✉ S.Konarskio g. 49, 03123 Vilnius ☎ +370 5 2363209 ▤ +370 5 2363208 **E:** lrt@lrt.lt **W:** www.lrt.lt **L.P:** Dir (Radio): Jurgita Litviniene

FM (MHz)	LNR1	LNR2	LNR3	kW
Alytus	104.6	102.8	-	0.5
Anykščiai	101.9	104.4	106.5	17.4/4/0.5
Birzai	100.8	87.5	-	10/5
Bubiai	100.9	103.4	90.5	1.3/10/0.5
Dieveniškes	107.2	106.4	-	0.25
Druskininkai	102.3	103.7	-	5
Ignalina	92.3	99.6	-	1/2
Joniškes	-	94.4	-	0.25
Juragiai	102.1	96.2	98.0	20/10/1
Kalvarija	104.8	-	-	2
Klaipeda	102.8	105.3	91.9	17/17/0.5
Mazeikiai	93.3	101.8	-	5/0.5
Nida	106.8	103.3	-	0.5
Pazagieniai	107.5	105.3	93.7	1.5/1.5/1
Plunge	88.0	105.0	-	1/2
Raseiniai	-	-	87.7	0.2
Rokiškis	95.3	-	92.6	0.5
Skuodas	99.3	103.5	-	0.7/4
Taurage	98.8	107.4	104.2	4/1/0.5
Telšiai	93.0	107.0	-	0.2
Ukmerge	102.8	101.7	-	0.2/0.1
Utena	107.4	107.8	-	0.5
Varena	100.7	105.3	-	0.5
Vilnius	102.6	105.1	-	5
Visaginas	102.9	100.4	98.3	11/11/0.5

D.Prgr: LNR1 (Lietuvos radijas): 0300-2100. Russian: 1430-1500. – **LNR2 (Klasika):** Mon-Fri 0900-2000, Sat/Sun 0600-2100. For ethnic minorities: 1300-1330 Belarusian (Tue/Sat[1]), Russian (Wed/Sat[2]/Sun), Ukrainian (Thu), Yiddish (Mon); 1330-1400 Polish. – **LNR3 (Opus 3):** 24h via Internet, 0500-1600 on FM. [1] last Sat, [2] exc. last Sat

OTHER STATIONS

MW	kHz	kW	Location	Station
22A)	612	100	Vilnius	Relays (RBW)
22B)	1386	500*	Sitkunai	Relays (RBWI)

*) run with 150kW at times (upon customer's request)

FM	MHz	kW	Location	Station
22B)	68.24	6.3	Visaginas	Relays (RBWI)
2B)	88.2	3.2	Bubiai	ZIP FM
14)	88.7	1.6	Palanga	Hot FM
3B)	90.1	2	Klaipeda	Pukas 2
2C)	90.6	2.3	Klaipeda	Russkoje R. Baltija
2A)	91.2	4	Tryškiai	Radiocentras
2B)	91.6	2	Skuodas	ZIP FM
7)	91.8	2	Šiauliai	Marijos radijas
16)	92.2	3.5	Bubiai	Relax FM
3B)	92.4	3.2	Kaunas	Pukas 2
2B)	92.5	4.5	Klaipeda	ZIP FM
2B)	92.7	5	Mazeikiai	ZIP FM
1A)	92.8	5	Kedainai	M-1
4)	93.4	2.5	Marijampole	Ziniu radijas
3A)	94.0	2.5	Alytus	Pukas
3A)	94.2	5	Raseiniai	Pukas
3A)	94.6	2	Ukmerge	Pukas
3A)	94.8	3.2	Marijampole	Pukas

FM	MHz	kW	Location	Station
15A)	94.9	2	Klaipeda	Laluna
7)	95.0	2	Viešintos	Marijos radijas
2A)	95.2	1	Krakes	Radiocentras
19)	95.4	4	Telšiai	Zemaitijos radijas
A)	95.5	3.2	Vilnius	BBC relay
1C)	95.6	1	Rokiškes	Lietus
3A)	95.7	2	Šiauliai	Pukas
9)	95.9	2	Vilnius	Power Hit R.
4)	96.3	2	Birzai	Ziniu radijas
4)	96.4	2	Mazeikiai	Ziniu radijas
11)	96.4	1.4	Vilnius	A2
10)	96.6	1	Panevezys	Pulsas
9)	96.7	2.5	Klaipeda	Power Hit R.
4)	97.3	1	Vilnius	Ziniu radijas
3B)	97.4	1	Šiauliai	Pukas 2
1B)	97.6	4	Juragiai	M-1 Plius
2A)	97.7	2	Plunge	Radiocentras
21)	97.8	2	Šiauliai	Antroji radijo stotis
1B)	98.3	3	Klaipeda	M-1 Plius
1B)	98.7	3	Utena	M-1 Plius
12)	99.0	3.1	Alytus	FM 99
13)	99.7	1	Vilnius	European Hit R.
6)	99.8	3.2	Klaipeda	Kelyje
2A)	99.9	3	Raseiniai	Radiocentras
2B)	100.1	12.6	Vilnius	ZIP FM
1B)	100.2	1	Pazagieniai	M-1 Plius
8)	100.4	4	Mazeikiai	Mazeikiu aidas
2C)	100.4	4	Kaunas	Russkoje R. Baltija
1B)	100.5	2	Bubiai	M-1 Plius
2B)	100.8	1	Marijampole	ZIP FM
15B)	100.8	1.3	Klaipeda	Raduga
3B)	100.9	1.1	Vilnius	Pukas 2
2A)	101.0	2	Utena	Radiocentras
2A)	101.1	2	Alytus	Radiocentras
2A)	101.3	1	Jonava	Radiocentras
2A)	101.4	6	Pazagieniai	Radiocentras
2A)	101.5	3.5	Vilnius	Radiocentras
2A)	101.5	2.5	Klaipeda	Radiocentras
3A)	101.6	1.6	Taurage	Pukas
2A)	101.6	2	Druskininkai	Radiocentras
2A)	101.7	3.2	Bubiai	Radiocentras
2A)	101.8	2	Marijampole	Radiocentras
3A)	102.0	1	Visaginas	Pukas
4)	102.2	1.6	Klaipeda	Ziniu radijas
17)	102.3	3.4	Bubiai	Saules radijas
3A)	102.6	1	Skuodas	Pukas
2A)	102.7	1.5	Taurage	Radiocentras
2A)	102.7	1	Ignalina	Radiocentras
18)	102.9	4	Juragiai	Tau
1C)	103.0	3	Pazagieniai	Lietus
1C)	103.0	1.1	Tryškiai	Lietus
1C)	103.1	2	Vilnius	Lietus
1C)	103.1	1.7	Taurage	Lietus
1C)	103.3	2	Alytus	Lietus
1C)	103.3	2	Birzai	Lietus
1C)	103.4	1.2	Utena	Lietus
1C)	103.5	3.4	Juragiai	Lietus
1C)	103.7	1.2	Klaipeda	Lietus
4)	103.7	1.1	Visaginas	Ziniu radijas
20)	103.8	4	Vilnius	Znad Willi
1C)	103.9	2.2	Bubiai	Lietus
5)	104.1	3.2	Klaipeda	Laisvoji banga
2B)	104.1	20	Juragiai	ZIP FM
1B)	104.3	3.2	Marijampole	M-1 Plius
5)	104.3	1.8	Šiauliai	Laisvoji banga
4)	104.4	2.2	Utena	Ziniu radijas
5)	104.5	4	Juragiai	Laisvoji banga
5)	104.7	1	Vilnius	Laisvoji banga
5)	104.8	1.1	Pazagieniai	Laisvoji banga
4)	104.8	2.5	Taurage	Ziniu radijas
4)	104.9	1	Kaunas	Ziniu radijas
2B)	105.2	1.3	Raseiniai	ZIP FM
14)	105.4	4	Kaunas	Hot FM
2B)	105.4	10	Visaginas	ZIP FM
2A)	105.5	2	Birzai	Radiocentras
2B)	105.7	2	Taurage	ZIP FM
2C)	105.8	4.2	Šiauliai	Russkoje R. Baltija
1A)	105.9	3.2	Ignalina	M-1
6)	105.9	1	Kaunas	Kelyje
1A)	106.0	2	Pazagieniai	M-1
1A)	106.0	2	Tryškiai	M-1
1A)	106.0	2	Alytus	M-1
1A)	106.2	1.6	Taurage	M-1
1B)	106.2	2.5	Vilnius	M-1 Plius
1A)	106.3	2.5	Marijampole	M-1

FM	MHz	kW	Location	Station
1A)	106.3	2.5	Bubiai	M-1
1A)	106.3	2	Utena	M-1
1A)	106.4	2	Raseiniai	M-1
1A)	106.5	2.5	Klaipeda	M-1
1A)	106.6	2.5	Juragiai	M-1
2B)	106.7	2	Laukuva	ZIP FM
1A)	106.8	1	Vilnius	M-1
3A)	107.1	1	Varena	Pukas
2A)	107.1	2.5	Juragiai	Radiocentras
3A)	107.3	1	Vilnius	Pukas
10)	107.3	3,2	Birzai	Pulsas
3A)	107.6	4	Kaunas	Pukas
6)	107.7	1	Vilnius	Kelyje
3A)	107.8	3	Klaipeda	Pukas
4)	107.9	1.2	Panevezys	Ziniu radijas

NB: Txs below 1kW not listed.

Addresses & other information:
1A,B,C) Laisves pr. 60, 05120 Vilnius. **E:** 1A) m-1@m-1.fm, 1B) pliusas@pliusas.fm, 1C) lietus@lietus.fm – **2A,B,C)** Laisves pr. 60, 05120 Vilnius. 2C) in Russian. **E:** 2A) programa@rc.lt, 2B) info@zipfm.lt; 2C) rusradio@rc.lt – **3A,B)** Šaldytuvu 25, 45123 Kaunas. **E:** radio@pukas.lt – **4)** Laisves pr. 60, 05120 Vilnius. **E:** biuras@ziniur.lt – **5)** Gedimino pr. 50/2, 01110 Vilnius. **E:** info@laisvojibanga.lt – **6)** Savanoriu pr. 151, 50174 Kaunas. **E:** kelyje@takas.lt – **7)** M.Daukšos g. 21, 44282 Kaunas. **E:** direktorius@marijosradijas.lt – **8)** Sodu g. 13-93, 89116 Mazeikiai. **E:** info@mazeikiuaidas.lt – **9)** Kalvariju g. 143, 08221 Vilnius. **E:** info@powerhitradio.lt – **10)** Respublikos g. 28, 35174 Panevezys. **E:** pulsas@elektra.lt – **11)** Sausio 13-osios g. 10, 04347 Vilnius. **E:** a2@a2.lt – **12)** Rotušes a. 2a, 62141 Alytus. **E:** fm99@fm99.lt – **13)** Jasinskio g. 16g, 01108 Vilnius. **E:** info@ehr.lt – **14)** Virbališkes takas 3, 00127 Palanga. **E:** info@hotfm.lt – **15A,B)** Taikos pr. 81, 94114 Klaipeda. 15B) in Russian. **E:** 15A) laluna@laluna.lt; 15B) info@raduga.lt – **16)** Laisves pr. 60, 05120 Vilnius. **E:** info@relaxfm.lt – **17)** Aušros al. 64, 76240 Šiauliai. **E:** info@saulesradijas.lt – **18)** Draugystes g. 19, 51230 Kaunas. **E:** info@tau.lt – **19)** Zemaites g. 26-3, 87101 Telšiai. **E:** zemaitijos@radijas.lt – **20)** Laisves pr. 60, 05120 Vilnius. **E:** radio@znadwilii.lt. In Polish. – **21)** Varpo g. 22, 76297 Šiauliai. **E:** admin@2ra.lt – **22A,B)** Švitrigailos g. 11a-211, 03228 Vilnius. **E:** radio@balticwaves.cjb.net. Rel. on 612kHz: 0400-0600 RFE-RL, 1300-1500 VO Russia*, 1600-2200 RFE-RL, 2200-2230 R.Polonia (all in Belarusian exc.* in Russian). Rel. on 1386kHz: see Int R. section. Rel. on 68.24MHz: European R. for Belarus in Belarusian: 24h. – **A)** Rel. BBC (UK).

DAB (Trial): Vilnius ch13A (230.784MHz) 0.5kW. **Operator:** LRTC

LORD HOWE ISLAND (Australia)

L.T: UTC +10½ (3 Oct 10-3 Apr 11, 2 Oct 11-1 Apr 12: +11h) — **Pop:** 350 — **Pr.L:** English — **E.C:** 50Hz, 240V. — **ITU:** AUS **(WRTH:** LHW)

MW	kHz	Station	Call	kW
1)	1620	Promo Radio	-	0.4
FM	MHz	Station	Call	kW
2)	100.1	Lord Howe Island R.	-	0.02
3)	104.1	ABC Classic FM	2ABCFM	0.02
4)	105.3	ABC Triple J	2JJJ	0.02
3)	106.1	ABC Classic FM	2ABCFM	0.02

Addresses & other information
1) Promo Radio Pty Ltd, 12 Pickering Close, Hoppers Crossing VIC 3029 (currently inactive) – **2)** The Shack, New Jetty Complex, Lagoon Road [PO Box 52], Lord Howe Island NSW 2898. ☎+61 2 6563 2123 ▤ +61 2 6563 2127 Volunteer community radio **Prgr:** Thu night only, irregular at other times – **3)** 24h satellite relay ABC Classic FM – **4)** 24h satellite relay ABC Triple J.

LUXEMBOURG

L.T: UTC +1h (27 Mar-30 Oct: +2h) — **Pop:** 486,000 —**Pr.L:** Luxembourgish, French, German — **E.C:** 50Hz, 110/220V — **ITU:** LUX

CLT-UFA (Comm.)
▤ 45 blvd. Pierre Frieden, L-1543 Luxembourg ☎ +352 4214 22175 ▤ +352 4214 22756 **W:** www.clt-ufa.com
L.P: Pres:Jacques Santer.
Luxembourg Sce: RTL Radio Lëtzebuerg: ☎ +352 4214 23 ▤ +352 4214 22737 **W:** www.rtl.lu
German Sce: RTL Radio – die Grössten Oldies: ☎ +352 4214 23500 ▤ +352 4214 22738 **W:** www.rtlradio.de **LP:** PD: Holger Richter
French Sce: RTL: ☎ 22 rue Bayard, F-75008 Paris ☎ +33 1 4070 4070 ▤ +33 1 4070 4272 **RTL2:** ☎+33 (1) 4070 4000 ▤ +33 1 4070 4350 **W:** www.rtl2.fr **L.P:** CEO: Jacques Rigaud. VP & PD: Philippe Labro. Prgr. Mgr RTL2: Frédéric Jouve

LW/MW: See International Broadcasting section
FM: Marnach 88.9/92.5/97.0/107.0MHz 100kW. Dudelange 93.3MHz 100kW. Also FM relays in France & Germany.
RTL Radio Lëtzebuerg in English/German/Luxembourgish: 24h on 92.5MHz (2000-0500 rel. German Sce.)
RTL: 24h FM network in France
RTL2: 24h on FM network in France
RTL Radio – die Grössten Oldies: 24h on 93.3 and 97.0MHz & satellite and via txs in Germany.

OTHER STATIONS
Den Neie Radio, P.O. Box 1522, 1015 Luxembourg **W:** www.dnr.lu 102.9/ 104.2/107.7MHz – **Eldorado**, B.P. 1344, 1013 Luxembourg **W:** eldoradio.lu 24h on 105.0/107.2MHz – **Honnert.7** (non-comm.), b.p. 1833, 1018 Luxembourg **W:** www.100komma7.lu 100.7MHz – **R. Ara,** 2 rue de la Boucherie, 1247 Luxembourg **W:** www.ara.lu 103.3/105.2MHz – **R. Latina**, 2 rue Astrid, 1143 Luxembourg: 101.2/103.1MHz **W:** www.radiolatina.lu – **R. LRB**, B.P. 8, 3201 Bettembourg **W:** www.lrb.lu 24h on 103.9MHz – **Radio WAKY Power FM 107**, P. O. Box 70, 5801 Hesperange, Luxembourg: 24h on 107.0MHz in English & Luxembourgish **W:** www.waky.lu/home.html – **Sunshine Radio,** 29 ave. de la Financerie, 1510 Luxembourg: 24h on 102.2MHz in English & Luxembourgish

MACAU (China, SAR)

L.T: UTC +8h — **Pop:** 538,000 — **Pr.L:** Portuguese, Cantonese — **E.C:** 50Hz, 220V — **ITU:** MAC

TELEDIFUSÃO DE MACAU, SARL (Priv. Comm.)
▤ Avenida Dr. Rodrigo Rodrigues, No. 223-225, Edif. "Nam Kwong" 7 Andar, Macau ☎ +853 28335888 ▤ +853 28343199 **E:** rmacau@tdm.com.mo **W:** www.tdm.com.mo
D.Prgr in **Portuguese** 24h 98.0MHz 2.5kW, in **Cantonese** 24h: 100.7MHz 2.5kW

RÁDIO VILAVERDE LDA (Priv. Comm.)
▤ Hipódromo da Taipa, Macau ☎ +853 28820338 ▤ +853 28820337 **E:** am738@am738.com **W:** www.am738.com
D.Prgr in **Cantonese** 24h: 738kHz 10kW

MACEDONIA

L.T: UTC +1h (27 Mar-30 Oct: +2h) — **Pop:** 2.1 million — **Pr.L:** Macedonian, Albanian, Turkish — **E.C:** 50Hz, 220V — **ITU:** MKD

SOVET ZA RADIODIFUZIJA NA REPUBLIKA MAKEDONIJA (Broadcasting Council of the Republic of Macedonia)
▤ bul. Marks i Engels 4, 1000 Skopje ☎ +389 2 3103400 ▤ +389 2 3103401 **E:** sovet@srd.org.mk **W:** www.srd.org.mk
L.P: Pres: Zoran Stefanovski
NB. SRD is the regulatory authority for broadcasting.

MAKEDONSKO RADIO (MR) (Pub)
▤ bul. Goce Delcev bb, 1000 Skopje ☎ +389 2 3112200 ▤ +389 2 3112200 **E:** mkrtvir@mt.net.mk **W:** www.mr.com.mk
L.P: Dir (Radio): Grigori Popovski

MW	kHz	kW	Prgr	
Sveti Nikole	810	1200	MR1, (External Service)	
FM (MHz)	**MR1**	**MR2**	**kW**	
Belasica	91.5	97.8	10	
Boskija	95.3	98.1	10	
Bukovic	89.2	95.9	1	
Cocon	88.8	93.8	1	
Crn Vrv	97.3	94.1	100	
Gevgelija	99.2	102.4	10	
Golak	94.5	97.0	10	
Mali Vlaj	93.3	97.7	10	
Pelister	92.3	96.1	20	
Popova Šapka	88.8	96.3	5	
Stogovo	95.3	101.0	3	
Tepavci	94.9	103.4	1	
Turtel	93.3	90.5	50	
Vodno	98.9	92.4	10	

NB: Txs below 1kW not listed.
D.Prgr: MR1 (R. Skopje): 24h. – **MR2 (R. 2):** 24h in Macedonian, Albanian, Turkish, Bosnian, Romany, Vlakh, Serbian. – **Local Station "Kanal 103":** 24h on Vodno (Skopje) 103.0MHz (0.5kW).
External Service (R. Makedonija): see Int Radio section.

OTHER STATIONS

MW	kHz	kW	Location	Station
4)	639	1	Štip	R. Štip
FM	**MHz**	**kW**	**Location**	**Station**
2)	89.8	3	Pelister	R. Bitola
A)	91.3	1	Skopje	RFI relay
1)	92.9	1	Pelister	Antena 5
1)	95.5	1	Vodno	Antena 5
4)	96.6	1	Štip	R. Štip
3)	98.8	1	Ohrid	R. Ohrid
1)	104.8	1	Turtel	Antena 5
1)	106.3	1	Boskija	Antena 5

NB: Txs below 1kW not listed.
Addresses & other information:
1) ul. Tetovska 35, 1000 Skopje. **E:** mail@antenna5.com.mk — **2)** ul. Tomaki Dimitrovski 7, 7000 Bitola. **E:** radiobt@freemail.com.mk — **3)** ul. Sv. Klement Ohridski 2, 6000 Ohrid. **E:** radiooh@sonet.com.mk — **4)** ul. Vanco Prke bb, 2000 Štip. E: radiostip@mt.net.mk — **A)** Rel. RFI (France)

MADAGASCAR

L.T: UTC +3h — **Pop:** 21 million — **Pr.L:** Malagasy, French — **E.C:** 50Hz, 220V — **ITU:** MDG

RADIO MADAGASIKARA - RADIO NATIONALE MALAGASY (RNM, Pub.)
✉ BP 4422, Anosy, 101 Antananarivo ☎+261 20 2221745 🖷 +261 20 2232715 **W:** rnm.mg **L.P:** Dir: Johary Ravoajanarahy.

MW	kHz	kW	H of tr
Fenoarivo	630	75	0300-1900
SW	**kHz**	**kW**	**H of tr**
Ambohidrano	3289	10	0300-0500, 1500-1900
Ambohidrano	*5010	100	0300-0500, 1500-1900
Ambohidrano	6135	30	0500-1500
Ambohidrano	7105	20	0500-1500

*at times AM/USB.
FM: Antananarivo 99.2MHz (0.5kW) & relay txs.
D.Prgr: 0300-1900 (SS 2200) in Malagasy & French.
Ann: Malagasy: "R. Madagasikara"; F: "R. Nationale Malagasy".

OTHER STATIONS

SW	kHz	kW	Location	Station
1)	3215	50	Talata-Volonondry	R. Feon'ny Filazantsara
FM	**MHz**	**kW**	**Location**	**Station**
2)	88.6		Antananarivo	R. Fahazavana
	89.2		Antananarivo	BBCWS relay
3)	92.0		Antananarivo	Alliance FM (also r. RFI)
4)	93.4		Antananarivo	R. Don Bosco
5A)	94.4		Antananarivo	R. Tana
	96.0		Antananarivo	RFI
6)	96.6		Antananarivo	R. Des Jeunes
7)	97.6		Antananarivo	R. Antsiva
8)	98.2		Toamasina	R. Voanio
5B)	102.0		Antananarivo	R. 102
9)	105.2		Antananarivo	Ma FM
10)	106.0		Antananarivo	R. Lazan larivo
11)	107.4		Antananarivo	R. FMFOI

NB: Unlicensed stns are operating in many parts of the country.
Addresses & other information:
1) BP 95, 110 Antsirabe. **E:** flm@wanadoo.mg. In Malagasy: 1630-1700. Tr. is relayed by RNW — **2)** BP 623, Lot II J 11, Faravohitra, Rue Joël Rakotomalala, 101 Antananarivo. **W:** radiofahazavana. agilityhoster.com — **3)** Enceinte Maison Laborde, Andohalo, 101 Antananarivo. **W:** alliancefr.mg/institution/part_dg.htm#alliance92 — **4)** BP 60, Maison Don Bosco, Ivato Airport, 105 Antananarivo. **W:** radiodonbosco.mg — **5A-B)** Enceinte Sitram, Ankorondrano, 101 Antananarivo. **W:** www.rta.mg . A) in Malagasy, B) in French –**6)** BP 4370, Immeuble Vitasoa, Analakely, 101 Antananarivo. **W:** rdeejay. net — **7)** BP 12170, Zone Zital Ankorondrano, Enceinte RTA, 101 Antananarivo. **W:** www.antsiva.mg –**8)** BP 489, 11 Rue Grandidier, 501 Toamasina. **W:** voanio.com — **9)** BP 1414, Ankorondrano, 101 Antananarivo. **W:** matv.mg – **10)** BP 6319, V.A 49 Andafiavaratra, 101 Antananarivo. **W:** rli106fm.com — **11)** Rue Docteur Ralarosy V W01, Ambohipotsy, 101 Antananarivo. **W:** fmfoi.ifrance.com

MADEIRA (Portugal)

L.T: UTC (27 Mar-30 Oct: +1h) — **Pop:** 300,000 — **Pr.L:** Portuguese — **E.C:** 50Hz, 220/380V — **ITU:** MDR

ANACOM-Autoridade Nacional de Comunicações, Delegação da Madeira
✉Rua do Vale das Neves, 19, São Gonçalo, 9050-332 Funchal ☎ +351 291 79 02 00 🖷+351 291 79 02 01

RADIODIFUSÃO PORTUGUESA, S.A.
Centro Regional da RDP-Madeira
✉ Rua Tenente Coronel Sarmento, 15, 9000-020 Funchal ☎ +351 291 20 20 00 🖷 +351 291 23 07 53 **W:** www.rtp.pt **E:** rdpmadeira@rtp.pt
L.P: Dir: Tito de Freitas

MW (Antena 1 Madeira)

Location	kHz	kW	Location	kHz	kW
Pico do Areeiro	*603	10	Ponta do Pargo	1125	1
Monte	1332	1			

*) inactive because the tower collapsed; return is uncertain.

FM (MHz)	Ant. 1	Ant.2	Ant. 3	kW
Achada da Cruz	104.3		105.0	0.8
Cabo Girão	96.7	99.4	94.8	1/3/1
Calheta	105.4		107.5	0.1
Caniço	101.6	99.0	89.3	0.5
Encumeada	93.1		90.8	0.06
Gaula	98.5	106.3	91.3	1/0.7/1
Maçapez	92.0		95.7	0.1
Monte	104.6	102.4	89.8	1/1/0.7
Paúl da Serra	101.9		93.3	1
Pico do Areeiro	95.5		94.1	13/15
Pico do Facho	93.1		90.8	0.03
Ponta do Pargo	90.2		94.6	1
Porto Santo	100.5	103.3	96.5	10
Ribeira Brava	105.6		103.1	1
Santa Clara	104.6	102.4	89.8	

D.Prgr: all networks 24h. Antena 1 of RDP Madeira provides regional prgrs M-F 0700-2000, Sat 0700-1800, Sun 0900-1800 LT; Antena 2 relays Lisboa 24h; Antena 3 Madeira carries own prgrs. M-F 0700-2400, Sat 0000-0300 & 0600-2400, Sun 0800-2400 LT **DAB:** RDP has 7 T-DAB txs at Cabo Girão, Gaula, Monte, Pico do Facho, Pico do Areeiro and Monte, all in Madeira island and in Porto Santo, all broadc. Ant. 1, 2, 3, RDP África and RDPi on 225.648MHz, ch. 12, block B. Technical info. may be obtained from gabinete.tecnologias@rtp.pt
V. by QSL-card via RDP Lisboa.

RÁDIO RENASCENÇA – Em. Católica Portuguesa (Rlg/Comm)
✉ (see Portugal) – **FM:** Pico do Silva 88.0MHz 44kW (RR), 93.6MHz 44kW (RFM)

PEF – Posto Emissor de Radiodifusão do Funchal (Priv., comm.)
✉Rua Ponte de São Lázaro 3, 9000-027 Funchal ☎ +351 291 23 03 93 🖷 +351 291 22 17 97 **E:** pef@netmadeira.com **W:** www.pef.pt
MW: Funchal 1530kHz 10kW, Santana 1017kHz 1kW
FM: Funchal 92.0MHz 2kW **D.Prgr:** 24h. Relays R.Renascença, Lisboa, at certain times **Ann:** "PEF – a sua rádio regional"

Local FM stations:

FM	Island	Station & location	MHz	kW
4)	Madeira	R. Jornal da Madeira, Funchal	88.8	1
13)	Madeira	R. São Vicente, São Vicente	89.2	0.5
6)	Madeira	R. Zarco, Machico	89.6	1
9)	Pto Santo	R. Praia, Porto Santo	91.6	0.5
12)	Madeira	R. Santana/Santana FM, Santana	92.5	0.5
11)	Madeira	R. Palmeira, Santa Cruz	96.1	0.5
10)	Madeira	Girão FM, Ribeira Brava	98.4	0.5
1)	Madeira	R. Calheta, Calheta	98.8	0.5
5)	Madeira	R. Notícias/TSF, Funchal	100.0	2
2)	Madeira	R. Popular da Madeira, Câmara de Lobos	101.0	2
8)	Madeira	R. Porto Moniz, Porto Moniz	102.9	0.5
7)	Madeira	R. Sol, Ponta do Sol	103.7	0.5
3)	Madeira	R.Clube da Madeira, Funchal	106.8	0.4

+ five 50W repeaters used by three stns
Addresses and other information (add +351 to tel/fax nos):
1) Edifício Ondaparque, Av.ª D. Manuel I, 9370-133 Calheta ☎ 291 82 01 32/6, 🖷 291 82 01 38 **E:** radiocalheta@gmail.com **W:** www. radiocalheta.pt – **2)** Rua dos Netos, 23, 900-084 Funchal ☎291 76 41 24 🖷291-20 23 86 – **3)** Rua dos Estados Unidos da América, 146-150, 9000-090 Funchal ☎291 761068 🖷291 703 471 E: 106.8@radioclube. pt, ssfranco@radioclube.pt **W:** www.radioclube.pt – **4)** Rua Dr. Fernão de Ornelas, 35-r/c, 9054-528 Funchal ☎291 210 400 🖷291 210 409 **E:** radio@jornaldamadeira.pt **W:** www.jornaldamadeira.pt – **5)** Rua Fernão de Ornelas, 56-3°, 9050-021 Funchal ☎ 291 20 23 94/5/6 🖷 291 20 23 87; relays TSF Lisboa **E:** rmoliveira@dnoticias.pt **W:** www.dnoticias.pt/ tsfmadeira – **6)** (see nr.2) –**7)** (see nr. 2) –**8)** 9240 São Vicente ☎291 84 21 35 🖷291-84 26 66 –**9)** Rua Goulart de Medeiros, 1, 9400 Porto Santo ☎ 291 98 01 30 🖷 291 982484 **W:** www.radiopraia.pt – **10)** (see nr. 2) – **11)** (see nr. 2) – **12)** Rua da Igreja, 8-3°, 9325-031 Estreito de Câmara de Lobos ☎ 291 94 75 73 🖷 291 94 87 94 **E:** radio.santana@clix.pt – **13)** Bombeiros Voluntários de São Vicente e Porto Moniz, Vila de São Vicente, 9240 São Vicente ☎291 84 26 94 & 291 84 26 61 🖷 291 84 23 93 **E:** dpe_drci@netmadeira.com **W:** www.miradouro.pt

MALAWI

LT: UTC +2h — **Pop:** 14 million — **Pr.L:** English, Chichewa, Tumbuka, Lomwe, Sena, Yao, Nkhonde, Tonga — **E.C:** 50Hz, 230V — **ITU:** MWI

MALAWI COMMUNICATIONS REGULATORY AUTHORITY (MACRA)

✉ Salmon Amour Rd, Private Bag 261, Blantyre ☎+265 1 623611 🖷 +265 1 623890 **LP:** DG: Allexon Chiwaya. Ag. Dir of Bc: Kelton Masangano. **W:** www.macra.org.mw

MALAWI BROADCASTING CORPORATION (MBC, Pub.)

✉ P.O. Box 30133, Chichiri, Blantyre 3 ☎+ 265 1 871461 **E:** dgmbc@malawi.net **LP:** DG: Patrick D. Khoza. Deputy DG: Bright Malopa. Dir. of N. & Current Affairs: Maxwell Kasinja. Dir. of Eng: Joseph Chikagwa.

MW	kHz	kW	MW	kHz	kW
Mangochi	540	10	Bangula	810	10
Karonga	558	10	Nkhota Kota	1107	1
Lilongwe	594	30	Chitipa	1404	10
Ekwendeni	675	50	Matiya	1422	10
Blantyre	756	10			

FM	R1	R2	FM	R1	R2
Blantyre	95.4	92.2	Lilongwe	94.7	91.5
Chikangawa	105.9	103.0	Livingstonia	90.1	104.5
Chitipa	90.7	100.5	Mzuzu	97.8	91.3
Dedza	90.1	104.5	Nkhotakota	95.6	
Dwangwa	103.6	93.6	Ntchisi	100.5	92.4
Karonga	95.5	98.7	Zomba	94.1	96.8
Kasungu	94.5	96.2			

MBC Radio 1 in English/Chichewa/others on MW/FM: 0253-2200. **N. in English:** 0300, 0500W, 0600Sun, 0700W, 0800Sun, 0900W, 1000Sun, 1030MF, 1100SS, 1200MF, 1400MF, 1600, 1800, 2000W, 2100, 2200.
MBC Radio 2 in English/Chichewa on FM: 24h.
Ann: E: R1: "MBC Radio 1", R2: "Radio 2 FM". Chichewa: "Kuno ndi ku ya MBC". **IS:** 0253 Cock crow and rapid drum beat.

TRANS WORLD RADIO MALAWI (Rlg.)

✉ P.O. Box 52, Lilongwe ☎ 🖷 +265 1 751763 **E:** twr@malawi.net **LP:** Dir: Patrick Semphere. Dir. Prgr: Victor Kaonga.
FM: (alll 2kW): Blantyre 89.1MHz, Ntchis 90.7MHz, Mvera 91.1MHz, Dedza 96.4MHz, Yawo 106.2MHz, Chikangawa/Zomba 106.4MHz, Lilongwe 106.5MHz, Thyolo 107.1MHz.
D.Prgr: 24h in English/Chichewa.

Other stations:
R. Alinafe, Lilongwe: 97.1MHz. **E:** radioalinafe@sdnp.org.mw – **Calvary Family Church R,** Blantyre 3: 105.8MHz 0.25kW. **E:** calvaryministries@hotmail.com – **Capital R:** Blantyre/Mzuzu 102.5MHz, Dedza 105.2MHz, Lilongwe 102.8MHzm Zomba 96.1MHz. **W:** www.capitalradiomalawi.com – **Channel For All Nations,** Lilongwe: 101.5MHz – **FM 101 Power:** Ntcheu 101.5MHz txs 0.5/1kW): Ntcheu 88.1MHz, Livingstonia 93.2MHz, Chintheche 98.6MHz, Mzuzu 99MHz, Nkhoma 100.3MHz, Blantyre/Lilongwe/Nkhota-kota 101MHz, Dedza 103.9MHz, Ntchisi 104MHz, Dwangwa 107.2MHz. **W:** www.fm101.malawi.net – **Joy R,** Blantyre: 89.6MHz – **MIJ R,** Blantyre 3: 90.3MHz **W:** www.mijmw.net/radio.htm – **R. Maria Malawi:** Mangochi 88.5Mhz 1kW, Dowa 94.0MHz 1kW, Blantyre 99.2MHz 1kW, Zomba 99.4MHz 2kW, Dedza 99.7MHz 2kW. **W:** www.radiomaria.mw – **Star FM,** Blantyre: 89MHz – **Zodiak BS:** Chitipa 89.5, Dedza 89.0, Dowa 92.9, Karonga 93.7, Lilongwe 95.1, Livingstonia 93.0, Mpingwe 97.0, Mzuzu 95.1, Namwera 103.3, Zomba 89.3MHz. **W:** zbsmw.com
BBC African Sce: Mzuzu 87.9, Lilongwe 98.0, Blantyre 98.1MHz

MALAYSIA

LT: UTC +8h — **Pop:** 26.2 million — **Pr.L:** Bahasa Malaysia (Malay), English, Chinese. In West Malaysia also Tamil and various Orang Asli languages, East Malaysia also 12 local languages or dialects — **E.C:** 50Hz, 240V — **ITU:** MLA

MALAYSIAN COMMUNICATIONS AND MULTIMEDIA COMMISSION (MCMC) (Suruhan Komunikasi dan Multimedia Malaysia, SKMM)

Regulatory body for the communications & multimedia industries.
✉ 63000 Cyberjaya, Selangor ☎ +60 3 8688 8000 🖷 +60 3 8688 1000 **W:** www.skmm.gov.my **LP:** Chairman: Yg Bhg Tan Sri Khalid Bin Ramli

JABATAN PENYIARAN MALAYSIA (Dept of Broadcasting of Malaysia) (Gov.)

Parent body of RTM. ✉ Angkasapuri, 50614 Kuala Lumpur ☎ +60 322825333 🖷 +60 2282 5103

RADIO TELEVISION MALAYSIA - RTM (Gov.)

✉ Dept. of Broadcasting, Angkasapuri, Bukit Putra, 50614 Kuala Lumpur ☎ +60 322825333 🖷 +60 322824735 **W:** www.rtm.gov.my **E:** teknikalradio@rtm.net.my
LP: DG: Yg Bhg Datuk Ibrahim Yahaya. Prgr Dir (Radio): Mohammad Bin Mat Hussin. Dir. Technical Sces: Hj Ab. Wahid Bin Ab. Hamid.

MW	kHz	kW	Sce.	H of tr
Gerik	657	20	Perak FM	2200-1600

SW: Kajang:

kHz	kW	Sce.	H of tr	kHz	kW	Sce.	H of tr
5965v	100	1	24h	7295	100	4	24h
6050v	100	†	0000-1600	9750	100	VOI	1400-1600
6175	100	VOI	1400-1600				

† = Rel 7) 0000-1400, rel. VOI 1400-1600

FM (MHz)	Site	1	2	3	4	5	6	kW
Alor Setar	a	94.9	100.5	98.7	101.3	96.7		5
Balik Pulau	b	99.5	93.9	90.1	92.1	98.9		0.1
Baling	c	88.7	89.7	91.7	92.5	93.3		1
Besut	d	94.3	98.8	97.0	97.8	95.3		0.1
Cameron	e	89.1	93.1	101.1	103.5	104.3		0.1
Dungun	f	95.9	96.9	98.9	99.7	100.7		1
Gerik	g	97.8	95.4	98.4	100.8	100.0		0.1
Ipoh	h	88.3	90.9	90.1	92.1	98.9		1
Jeli	i	88.4	89.2	90.8	91.6	92.4		0.1
Jerantut	j	88.1	93.5	89.9	90.7	91.9		0.1
Johor Bahru	k	106.7	105.7	102.9	104.9	101.1		5
Kota Bharu	l	101.1	101.9	104.7	105.7	106.7		1
Kuala Lumpur	m	87.7	88.5	90.3	89.3	92.3		1
KL2	n	98.3	95.3	100.1	106.7	96.3		1
KT	o	92.5	91.7	89.7	90.5	87.9		1
Kuantan	p	107.9	107.1	105.3	106.1	103.3		1
Machang	q	95.5	96.5	98.5	99.3	100.9		2
Melaka	r	93.6	96.6	97.4	100.4	103.3		0.5
Mersing	s	90.1	90.9	92.9	89.1	88.3		1
Seremban	t	87.9	91.7	88.7	89.7	90.5		0.1
Seremban	u	107.9	-	-	-	-		0.1
Sik	v	99.5	102.7	105.9	106.7	107.5		0.1
Taiping	w	103.3	107.2	105.3	106.1	107.9		0.1

1-6: see RTM national sces below. **Asyik FM/VOI/Salam FM:** Cameron Highlands 105.1MHz,, Gunung Ulu Kali 102.5MHz, Kuala Lumpur 91.1MHz, Ulu Tembeling 89.3MHz. **FM:** All powers are TRP. KL2=KL/Selangor/Pahang (West). KT=Kuala Terengganu.
Sites: a) Gunung Jerai, b) Bukit Genting (Penang) c) Bukit Palong, d) Bukit Bintang, e) Gunung Berinchang, f) Bukit Bauk, g) RTM Gerik h) Bukit Keledang, i) Bukit Tangki Air, j) Bukit Istana, k) Gunung Pulai, l) Telipot, m) Menara KL (Bukit Nanas), n) Gunung Ulu Kali, o) Bukit Besar, p) Bukit Pelindung, q) Bukit Bakar, r) Gunung Ledang (Mt Ophir), s) Bukit Tinggi, t) Bukit Telapa Burok, u) RTM Seremban, v) Bukit Dedap, w) Bukit Larut (Maxwell Hill).

RTM national services

(1) Klasik Nasional FM: 24h news, information & Malay oldies presented in Malay. **(2) Muzik FM:** 24h Music presented in Malay. **(4) Traxx FM:** 24h News, music and travel sce in English. **(5) Ai FM:** 24h General Sce in Chinese (Mandarin exc. news at 0200 in Hakka, 0500 Cantonese, 0700 Hakka & 1300 Chaozhou). **(6) Minnal FM:** 24h General sce in Tamil. **Asyik FM:** 0000-1400 for Orang Asli in Jakun, Malay, Semai, Temiar & Temuan. **VOI, Voice of Islam (Suara Islam):** Rlg. prgrs in Malay/Indonesian 1400-1600. **Salam FM:** Rlg. prgrs in Malay from Jabatan Kemajuan Islam Malaysia 1600-2400 **V.** occasionally by letter or Email. **Ann:** names of networks and regional sces are sometimes preceded by the words "Radio Malaysia".

RTM regional services in West Malaysia:

Most sces. operate 24h in Malay (exceptions include Langkawi, which carries local & tourist information in English and Malay, and FMJB 107.5, which operates 0000-1200). Some sces relay RTM Klasik Nasional overnight. Refer to above lists for tx sites and powers for frequencies marked a-w. **FM:** All powers are TRP.
Johor: Johor FM (JFM), Karung Berkunci 716, 80990 Johor Bahru, Johor **W:** johor.dapat.fm **E:** rmjb@rtmjb.net.my State sce On 92.1MHz s, 101.9MHz k, 105.3MHz r Johor Bahru city sce:**FMJB 107.5** on 107.5MHz RTM Johor Bahru 1kW – **Kedah:** Kedah FM, Kompleks Penerangan dan Penyiaran Sultan Abdul Halim, KM 3, Jalan Kuala Kedah, 05400 Alor Setar, Kedah **W:** www.kedahfm.gov.my **E:** rtmas@rtm.net.my On 88.5MHz Selama-Bandar Baharu (site Bukit Sungai Kecil Hilir) 0.25kW, 90.5MHz b, 97.5MHz a, 105.1MHz v, 105.7MHz Gunung Raya 1kW, 107.0MHz Kuah – **Kelantan:** Kelantan FM, Peti Surat 143, 15720 Kota Bharu, Kelantan **E:** rtmkb@rtm.net.my On 88.1MHz FELDA Paloh 1kW, 97.3MHz q 102.9MHz l, 89.1MHz Gua Musang 0.1kW, 92.0MHz j, 88.9MHz Taman Wangi 0.1kW, 107.1MHz d – **Kuala Lumpur:** KL.fm

On 97.2MHz m **W:** www.kl.fm – **Langkawi (Kedah):** Langkawi FM, Tingkat 2, Bangunan Tabung Haji, Jalan Padang Mat Sirat, 07000 Kuah, Langkawi **W:** www.langkawifm.gov.my On 87.5MHz Kuah 0.1kW, 104.8MHz Gunung Raya 1kW **English:** 0100-0400, 0700-1000 Malay/ English 1300-1600 – **Melaka:** Melaka FM (MFM), Jalan Taming Sari, 75614 Melaka **W:** www.melakafm.webs.com **E:** rtmmlk@rtm.net.my On 102.3MHz r – **Negeri Sembilan:** Negeri FM, Jalan Raja Ali, 71000 Seremban, Negeri Sembilan **E:** rtmnegeri Fm On 92.5MHz t 96.3MHz Gemencheh (Batang Melaka) 0.1kW, 101.3MHz Port Dickson 0.1kW, 95.7MHz Gunung Tampin 0.1kW, 107.7MHz r – **Pahang:** Pahang FM, Peti Surat 152, 25710 Kuantan, Pahang **W:** www.pahangfm.gov.my **E:** rtmktn@rtm.net.my On 104.1MHz p, 107.5MHz n 100.3MHz e, 92.7MHz j, 92.0MHz Maran (Bukit Senggora) 0.25kW, 91.9MHz Rompin 0.25kW – **Perak:** Perak FM, Jalan Dairy, 31400 Ipoh, Perak **W:** www.perakfm.my **E:** rtmipoh@rtm.net.my On 657kHz, 89.6MHz Bukit Asa, 94.2MHz Lenggong (Bukit Ladang Teh) 0.025kW, 94.7MHz e, 95.6MHz g, 96.1MHz h, 97.3MHz Changkat Rembian 0.5kW, 104.1MHz w z – **Perlis:** Perlis FM, Tingkat 6, Bangunan WSP, Jalan Bukit Lagi, 01000 Kangar, Perlis **E:** rtm-kgr@rtm.net.my On 102.9MHz Pauh 2kW – **Pulau Pinang (Penang):** Mutiara FM, Jalan Burmah, 10350 Pulau Pinang **W:** www.mutiarafm.gov.my **E:** rtmpp@rtm.net.my On 90.9MHz b, 93.9MHz a, 95.7MHz Bukit Penara 1kW – **Selangor:** Selangor FM Bangunan Sultan Salehudin Abdul Aziz Shah, 40000 Shah Alam, Selangor **W:** rms.mmu.edu.my On 100.9MHz n – **Terengganu:** Terengganu FM, Peti Surat 63, 20914 Kuala Terengganu, Terengganu **W:** tfm.gov.my **E:** rtmkt @rtm.net.my On 88.7MHz o, 96.2MHz c, 97.7MHz f.

RADIO TELEVISION MALAYSIA SABAH (Gov.)
✉ 2.4km, Tuaran Road, Beg Berkunci 2022, 88614 Kota Kinabalu ☎ +60 88213444 ◻ +60 88223493 **W:** www.rtmsabah.gov.my
Addresses of local stns: ✉ Tingkat 6, Wisma Persekutuan, W.D.T. 52, 90500 Sandakan - Peti Surat 606, 91008 Tawau.
L.P: Dir. Broadcasting: Encik Zubad Ibrahim. Dir. Tech. (Radio): Abdul Jalani bin Mahmud. Dep. Dir. (Radio Prgr): Tuan Haji Hashim Jaffrey.

RADIO TELEVISION MALAYSIA LABUAN
✉ 5004 Tanjung Taras, Peti Surat 299, 87008 WP Labuan ☎ +60 87415677 ◻ +60 87416658 **W:** www.labuanfm.gov.my **E:** kejuruter-aan@labuanfm.net.my

MW	kHz	kW	Netw.	H of tr
Tenom	565	10	SF	1000-1500
Kudat	801	10	SF	2130-0800
Kudat	v1197	10	SV	2030-0800
Tuaran	d1475	700	SM	1100-1330

FM (MHz) Tx		SF	SV	1	2	4	5	kW
FELDA S	a	104.1	106.7	99.9	102.9	104.9	105.7	0.1
Gadong	b	89.3	92.6	88.0	88.9	90.7	91.6	0.1
Kota Belud	c	101.5	104.1	99.9	100.7	102.5	103.3	1
K. Kinabalu	d	89.9	92.7	88.1	88.9	90.7	91.9	1
Kudat	e	95.9	98.9	94.1	94.9	96.7	98.1	1
Labuan	f		93.3	87.6	88.5	90.3	92.3	0.1
Lahad Datu	g	89.7	92.6	87.9	88.7	90.5	91.7	1
Langkon		91.7	91.1	101.6	90.1	89.0	87.7	0.1
Layang-L.	h	104.5	107.1	99.5	100.3	105.3	106.3	1
Luasong			87.7	88.5	89.3	90.1	0.1
Sandakan	j	92.9	96.1	91.1	92.1	94.3	95.1	1
Sipitang	k	97.9	102.9	95.5	96.5	99.1	99.9	1
Tawau	l	95.7	99.3	93.9	94.7	97.1	98.1	1
Tenom	m	90.3	93.1	88.5	89.3	91.7	92.3	1

Sites: , a) FELDA Sahabat, b) Bukit Gadong, c) Bukit Pompoda d) Kota Kinabalu (Bukit Lawa Mandau) e) Bukit Kelapa, f) Bukit Timbalai, g) Gunung Silam, h) Layang-Layang (Mount Kinabalu) j) Bukit Trig, k) Bukit Tampulagus, l) Gunung Andrassy, m) Bukit Sigapon. Tx powers are TRP.
National networks: 1, 2, 4, 5: see RTM national sces above.
State networks: SF= Sabah FM in Malay 24h. **Reg. N:** 2200, 2330, 0400, 0530, 0830, 1400. **SV=** Sabah V FM in English 0300-0700 & 1500-1800, Mandarin 0030-0300 inc. news in Hakka at 0110, Bajau 1800-2000 & 0830-1100, Dusun 2300-0030, 1300-1500, Kadazan 2100-2300 & 0600-0830, Murut 2000-2100 & 1100-1300. **N. (English):** 0500, 1500. **SM=**Suara Malaysia (Overseas Sce) in Tagalog to the Philippines 1100-1330 from studios in Kota Kinabalu, also on 94.7MHz (studio link to MW tx).
Local Sce's: Labuan FM on 89.4MHz 0.1kW (Bukit Timbalai) & 103.7MHz 0.1kW (RTM Labuan): 2145-1200 incl. **English** 0100-0300 - Tawau FM on 100.1MHz 1kW (Gunung Andrassy): 2145-1100 – Sandakan FM on 90.1MHz 1kW (Bukit Trig): 2150-1000 – Keningau FM on 89.4MHz: 2300-0900 in Malay, Dusun and Murut.

RADIO TELEVISION MALAYSIA SARAWAK (Gov.)
✉ Broadcasting House, Jalan P. Ramlee, 93614 Kuching ☎ +60 82248422 ◻ +60 82241914 **W:** www.rtmsarawak.gov.my **E:** rtmsar@ rtm.gov.my

Addresses of local stns: Bangunan Penyiaran, 98700 Limbang – Bangunan Penyiaran, Jalan Brighton, 98000 Miri – Bangunan Penyiaran, 96009 Sibu – Bangunan Penyiaran, 95000 Sri Aman.
L.P: Dir. Broadcasting:Tuan Haji Monshi Abdullah.

Stations: Kuching (Stapok: G.C: 110.20E/01.33N)

SW		kHz	kW	Netw.		SW	kHz	kW	Netw.	
Kuching	5030	10		SF		Kuching	7270	100/10	W/L	

FM (MHz) Tx		SF	Red	1	2	4	5	kW
Belaga		105.4	107.8	103.8	104.6	106.2	107.0	0.1
Betong	a	94.4	97.8	92.8	93.6	95.2	96.0	0.1
Bintulu	b	93.7	100.5	87.9	90.3	98.5	99.3	1
Bintulu	c	94.7	96.7					1
Dalat			96.9					0.1
Kapit	e	92.7	89.9	90.7	91.9	88.1	88.9	0.1
Kuching	f	88.9	91.9	92.9	88.1	89.9	90.7	10
Lambir Hills	g	88.1	90.7	91.9	92.7	88.9	89.9	1
Lawas	h	97.5	100.5	94.7	96.7	98.5	99.3	0.1
Limbang	i	101.5	104.1	97.1	98.1	102.3	103.3	1
Limbang	m	100.0	107.7	95.3	99.2	106.0	106.8	0.1
Marudi	n	-	-	102.9	-	-	-	0.1
Miri	o	100.3	106.3	107.1	99.3	104.5	105.3	0.1
Mukah		89.9	92.3	88.3	89.1	90.7	91.5	0.5
Sarikei	p	91.5	89.2	87.9	90.3	92.3	93.6	0.1
Serian	j	94.8	97.2	98.0	94.0	95.6	96.4	0.5
Sibu	k	101.5	104.1	95.5	98.5	102.5	103.3	0.1
Song		95.7	99.0					
Sri Aman	l	100.3	106.3	107.3	98.9	92.3	105.3	1
Stapong	d	95.1	101.1	93.3	94.1	95.9	97.1	1

SF=Sarawak FM, W=Wai FM, Red=Red FM L=Limbang FM S=Sibu
FM: All powers are TRP. Additional local sce. freqs: Gunung Serapi (Kuching) 101.3MHz 10kW (Wai FM), Bukit Ampangan 106.9MHz 0.5kW (Wai FM), Bukit Lima (Sibu FM) 87.6MHz 0.1kW, Bukit Kayu Malam 94.6MHz 1kW (Sibu FM), Bukit Song 99.8MHz (Sibu FM), Bukit Kapit 94.3MHz 0.1kW (Sibu FM), Belaga 103.0MHz 0.1kW (Sibu FM), Bukit Singgalang 102.1MHz 1kW (Sibu FM), Mukah 98.7MHz 0.5kW (Sibu FM), Bukit Nyabau (Bintulu): 97.5MHz 1 kW (Bintulu FM), Lambir Hills (Miri FM) 95.7MHz 1kW, Miri (Miri FM) 98.0MHz 0.1kW, Bukit Temunduk (RaSa FM) 89.5MHz 1kW, Bukit Mas (Limbang FM) 104.9MHz 1kW.
Sites: a) Off. Spaoh b) Bukit Setiam c) Bukit Nyabau d) Bukit Singgalang e) Bukit Kapit f) Gunung Serapi g) Bukit Lambir h) Bukit Tiong i) Bukit Mas j) Bukit Ampangan k) Bukit Lima l) Bukit Temunduk m) Bukit Sagan Rudang, n) Bukit Kayu Malam, o) RTM Miri, p) Bukit Dabei, q) Bukit Song.
National networks: 1, 2, 4, 5: see RTM national sces above.
State networks: Sarawak FM in Malay 24h. **N.** (Kuching): 2200, 0400, 1000, 1400. **Red FM** 2200-1600, in Chinese: 2200-0200, 0700-1300; **English:** 0200-0700, 1300-1600. FM txs relay Sarawak FM overnight. Educational prgrs during school terms: MF 0100-0300. **N.** (Kuching): English 0400, 0700, 1300; Chinese 0600, 0801, 1000, 1245; Hakka 1030; Hokkien 1045. **Wai FM:** 2200-0100, 0400-0700 (FM only), 1000-1600 (exc. 101.3 & 106.9MHz) in Iban, 0100-0400 in Bidayuh, 0700-1000 in Kayan/Kenyah. Additional Wai FM sce in Bidayuh for Kuching area: 1300-1600 on 101.3 & 106.9MHz. FM txs relay Sarawak FM overnight. Iban sce relays Limbang FM Mon/Thurs 1300-1400.
Local sce's: Bintulu FM: 0100-1100 in Malay and Iban. **Limbang FM:** in Malay 0100-0400, 1000-1300; Lun Bawang (Murut) 0400-0700, also relayed via Kuching 7270kHz; Bisaya 0700-1000; Iban Mon/Thurs 1300-1400 also relayed via Kuching 7270 kHz. **Miri FM:** in Malay 0100-0400, 1000-1300, Chinese 0700-1000 (Tues/Thurs 0900), Iban 0400-0700, Tues/Thurs 0900-1000. **Sibu FM:** in Malay 0100-0400, 1000-1300, Chinese 0700-1000, Iban 0400-0700. **Sri Aman (RaSa FM):** in Malay, Iban. Local sces also relay Wai FM in Iban 2200-0100 and from close of local prgrs until 1600, and Sarawak FM overnight.
SW: 5030kHz: 2200-1600 (rel. Sarawak FM). 7270kHz: 2200-1600 (rel Wai FM Kuching, exc.0400-0700 and Mon/Thurs 1300-1400 rel Limbang FM).
IS: A musical phrase (played on a native instrument, the Sape), alternating between A and F.

EXTERNAL SERVICE: RTM OVERSEAS SERVICE
see International Broadcasting section.

AMP RADIO NETWORKS SDN. BHD. (ASTRO) (Comm.)
✉ All Asia Broadcast Centre, Technology Park Malaysia, Bukit Jalil, 57000 Kuala Lumpur **W:** www.ampradio.net
L.P: Exec. Dir.: Datuk Borhanuddin Osman.

FM (MHz)	MY	ERA	Lite	Mix	Hitz	Sin	XFM
Alor Setar a	99.7	104.5	104.4	91.0	92.8	97.1	106.5
Ipoh b	100.6	103.7	101.5	94.3	92.7	96.9	98.5
Johor Bahru c	95.4	104.5	94.6	99.1	97.6	87.5	98.4
Johor Bahru d	-	-	-	-	-	-	103.3

FM (MHz)	MY	ERA	Lite	Mix	Hitz	Sin	XFM
Kota Bharu e	102.3	103.3	104.3	94.6	92.8	93.8	99.8
KK f	104.0	102.4	103.2	101.6	100.8	104.9	98.6
KT g	101.2	102.8	105.9	94.8	98.3	97.5	104.0
KL/Selangor h	101.8	103.3	105.7	94.5	92.9	96.7	103.0
Kuantan i	101.1	98.0	104.7	94.1	93.2	97.2	100.0
Kuching j	96.9	96.1	100.1	97.7	95.3	102.1	103.7
Melaka k	106.4	90.3	92.2	91.1	93.0	96.0	107.3
Miri l	103.2	101.3	-	102.4	105.8	87.7	-
Seremban m	100.6	103.6	104.6	94.2	95.0	96.9	97.9
Taiping n	100.2	95.2	89.3	91.3	93.6	96.4	104.9

Prgrs: MY FM: Music channel in Mandarin & Cantonese. **ERA:** Contemporary Malaysian music channel in Malay. **Lite FM:** Easy listening music in English. **Mix FM:** Music and variety in English. **Hitz.fm:** Top 40 presented in English. **Sinar FM:** Malay oldies. **Xfresh FM (XFM):** Programming for teenagers in Malay. **THR.fm:** see Radio Lebuhraya Sdn Bhd below. **Sites:** a) Gunung Jerai b) Bukit Keledang c) Gunung Pulai d) Metropolis Tower, JB e) Bukit Panau f) Kota Kinabalu (Bukit Kokol) g) Kuala Terengganu (Bukit Jerung) h) Ulu Kali i) Bukit Pelindong j) Muara Tabuan k) Gunung Ledang l) Tanjong Lobang m) Telapa Barok n) Bukit Larut. **TRP:** 2kW, exc. Sinar FM at sites e, f, j, k and l: 0.25kW.

BFM MEDIA (Comm.)
L.P: Exec. Dir: Mr. Malek Ali.
✉ 5.01 Wisma BU8, 11 Lebuh Bandar Utama, 47800 Petaling Jaya ☎ +60 3 7629 7112 **W:** www.bfm.my
BFM: Kuala Lumpur/Klang Valley (Gunung Ulu Kali) 89.9MHz, 24h business prgrs and music in English and Malay.

CAPITAL FM SDN BHD (Comm.)
✉ Concorde Hotel, Jalan Sultan Ismail, Kuala Lumpur **W:** capitalfm.com.my
Capital FM: Kuala Lumpur/Klang Valley (Gunung Ulu Kali) 88.9MHz, 24h in English and Malay.

DIGITAL MEDIA BROADCASTING SDN. BHD. (Bernama News Agency) (Gov.)
✉ 15th Fl, Wisma Bernama, 28 Jalan 1/65A, off Jalan Tun Razak, 53300 Kuala Lumpur **W:** www.radio24.com.my
Bernama Radio24: Kuala Lumpur/Klang Valley (Bukit Nanas), 93.9MHz 1kW TRP, 24h in English and Malay.

HUSA NETWORK SDN. BHD. (Comm.)
✉ 4213C Tingkat 2, Lot 51 & 52, Seksyen 27, Jalan Kebun Sultan, 15350 Kota Bharu, Kelantan ☎ +60 97436661 🖷 +60 97436664 **W:** www.manis.fm
Manis FM: Kota Bharu (Bangunan Billion) 90.6MHz, Kuantan (Bukit Pelindong) 95.1MHz Kuala Terengganu (Bukit Jerung) 102.0MHz. Prgrs in Malay. **TRP:** all sites 2kW.

INSTITUT KEFAHAMAN ISLAM MALAYSIA (Institute of Islamic Understanding) (Gov., Rlg.)
✉ No 2, Langgak Tunku, Off Jalan Duta, 50480 Kuala Lumpur ☎ +60 3 62046200 🖷 +60 3 62014189 **W:** ikim.gov.my/ikim.fm/
L.P: Dir. of Radio: Mr. Asa'ari Bin Mat Noh.

FM	Tx	MHz	FM	Tx	MHz
Alor Setar	a	89.0	Kuantan		89.5
Ipoh	b	102.7	Kuching	h	F.P.I.
Johor Bahru	c	106.2	Klang Valley (KL)	i	91.5
Kota Bharu	d	89.9	Melaka		89.5
Kota Kinabalu	e	93.9	Negeri Sembilan	k	102.7
Kuala Terengganu	f	100.2			

Radio Ikim (IKIM.FM): 24h in Malay with limited Arabic and English. **Sites:** a) Gunung Jerai, b) Bukit Keledang, c) Gunung Pulai, d) Bukit Panau, e) Bukit Kokol, f) Bukit Besar, g) Bukit Pelindong 2, h) F.P.I., i) Bukit Cincin, j) Gunung Ledang, k) Gunung Telapa Burok. **TRP:** all sites 2kW.

KRISTAL HARTA SDN. BHD. (CATS RADIO) (Comm.)
✉ 5th Floor, Wisma Ting Pek Khiing, No.1 Jalan Padungan, 93100 Kuching, Sarawak ☎ +60 82 410933 🖷 +60 82 254993 **W:** www.catsfm.my **L.P:** Chmn: Tan Sri Datuk Amar Haji Bujang Mohd Nor. CEO: Tuan Haji Affandi Tahir.

FM	Tx location	MHz	FM	Tx location	MHz
Bintulu	Bukit Setiam	88.3	Mukah	Mukah	97.9
Kuching	Gunung Serapi	99.3	Sarikei	Bt. K. Malam	96.7
Limbang	Bukit Mas	88.7	Sibu	Bukit Lima	88.4
Miri	Lambir Hills	93.3	Sibu	Bt. Singgalang	99.9

Prgr.: 24h in Malay, English and Iban. **TRP:** all sites 1kW.

MEDIA PRIMA BHD. (Comm.)
✉ Tingkat 2, South Wing, Sri Pentas, Persiaran Bandar Utama, 47800

Petaling Jaya, Selangor Darul Ehsan ☎ +60 3 77105022 🖷 +60 3 77107098 **W:** www.hotfm.com.my or www.flyfm.com.my or www.onefm.com.my
L.P: Head of Radio Networks: Mr Ahmad Izham Omar

FM(MHz)	Tx	Hot	Fly	One	FM(MHz)	Tx	Hot	Fly	One
Alor Setar	a	88.2	91.7	87.8	Melaka	j	104.3	94.0	88.1
Ipoh	b	104.5	87.9	87.6	Kota Bharu	h	105.1	-	-
Johor Bahru	c	90.1	102.5	-	KL/Selangor	i	97.6	95.8	88.1
Kuantan	d	92.4	87.6	100.4	KT		105.0	107.5	-
Kota Kinabalu	e	87.7	-	95.7	Penang	k	-	89.9	-
Kuching	f	94.3	-	-	Seremban	l	99.5	98.6	-

Hot FM: 24h in Malay. Hot FM freqs are licensed to Synchrosound Studios Sdn Bhd. **Fly FM:** 24h in English/Malay. FlyFM freqs are licensed to Malaysian Airports (Sepang) Sdn. Bhd. **One FM:** 24h in Mandarin and Cantonese. **Sites:** a) Gunung Jerai b) Bukit Keledang c) Gunung Pulai d) Bukit Pelindong e) Bukit Kokol f) Muara Tabuan g) Gunung Ledang h) Peringat i) Bukit Cincin exc. Hot FM: Ulu Kali j) Kuala Terengganu k) Bukit Penara l) Bukit Telapa Burok. **TRP:** Fly FM 0.25kW exc. Bukit Cincin: 2kW

RADIO LEBUHRAYA SDN. BHD. (THR.FM) (Comm.)
✉ 20th Flr, Plaza Berjaya, 12 Jalan Imbi, 55100 Kuala Lumpur ☎ +60 (3) 2433088 **W:** www.thr.fm Operated by AMP Radio Networks.

Coverage Area	Tx	MHz	Coverage Area	Tx	MHz
Kota Bharu	a	88.1+	Negeri Sembilan	g	101.5
Kuantan	b	88.8+	No. Perak (Taiping)	h	102.1
Ce. Perak (Ipoh)	c	97.9	Kedah (Alor Setar)	i	102.4
KL/Selangor	d	99.3	Johor Bahru	j	103.7
So. Penang	e	99.3	Kuala Terengganu	k	106.8+
Melaka	f	99.7			

D.Prgr: THR Raaga: Music and traffic information presented in Tamil: 24h. **THR Gegar:** separate prgrs in Malay for East Coast.on frequencies marked with + **Sites:** a) Bukit Panau b) Bukit Pelindong c) Upper Keledang 0.5kW d) Gunung Ulu Kali e) Bukit Penara 0.5kW f) Gunung Ledang g) Bukit Telapa Burok h) Bukit Larut 0.5kW i) Gunung Jerai j) Gunung Pulai k) Bukit Jerung. Affiliated with AMP Radio Networks (see above). **TRP:** 1kW, exc. where indicated under sites.

RIMAKMUR SDN. BHD. (Comm.)
✉ Bangunan AMDB, Jalan Lurut, Kuala Lumpur ☎ +60 3 40440784 **W:** suriafm.com.my
L.P: Chief Operating Officer: Engku Emran Engku Zainal Abidin

FM	Tx	MHz	FM	Tx	MHz
Alor Setar	a	106.9	Kota Kinabalu	f	105.9
Ipoh	b	96.0	Kuala Terengganu	g	102.4
Johor Bahru	c	101.4	Kuantan	h	96.1
Klang Valley (KL)	d	105.3	Seremban	i	107.0
Kota Bharu	e	106.1	Taiping	j	91.7

Suria FM: 24h in Malay. **Sites:** a) Gunung Jerai b) Bukit Keledang c) Gunung Pulai d) Bukit Cincin 2kW TRP e) Bukit Panau f) Bukit Kokol g) Bukit Besar h) Bukit Pelindong 2 i) Gunung Telapa Burok j) Bukit Larut.

STAR RFM SDN. BHD. (Comm.)
✉ 19th Floor, Bangunan AMDB, 1, Jalan Lumut, 50400 Kuala Lumpur ☎ +60 340481988 🖷 +60 340439988 **W:** www.988.com.my or www.red1049.com **E:** rfm988@silicon.net.my

FM(MHz)	Tx	Red	988	FM(MHz)	Tx	Red	988
Alor Setar	a	98.1	96.1	Melaka	f	98.9	98.3
Ipoh	b	106.4	99.8	Penang	g	107.6	94.5
Johor Bahru	c	92.8	99.9	Seremban	h	106.0	93.3
KL/Selangor	d	104.9	98.8	Taiping		98.2	101.0
Kuantan	e	91.6	90.4				

Red FM: 24h in English & Malay. **988** (jiu ba ba): 24h in Mandarin & Chinese dialects. **Sites:** a) Gunung Jerai b) Gunung Keledang c) Gunung Pulai d) Gunung Ulu Kali e) Bukit Pelindong f) Gunung Ledang g) Bukit Penara h) Bkt Telapa Burok i) Bukit Larut

SUARA JOHOR (Comm.)
✉ Bukit Pelangi, Jalan Pasir Pelangi, 80050 Johor Bahru, Johor
L.P: CEO: Haji Bakhtiar Haji Arshad.
☎ +60 7 3316104 🖷 +60 7 3351104
BEST 104: Melaka & Segamat (Gunung Ledang) 94.8MHz, Johor Bahru (Gunung Pulai) 104.1MHz 10kW TRP, Kuala Lumpur/Selangor (Gunung Ulu Kali) 104.1MHz, Mersing (Bukit Tinggi) 102.5MHz.
D.Prgr: 24h (Malay & English music).

University stations:
Putra FM
✉ Tingkat 2 Jabatan Komunikasi, Fakulti Bahasa Moden dan

Komunikasi, Universiti Putra Malaysia, 43400 UPM Serdang, Selangor **W:** www.putrafm.upm.edu.my
Station: 90.7MHz 1kW: Mon-Fri 0200-1600 in Malay & English.
Radio UiTM (UFM)
📧 Level 13, Menara Ilmu Universiti Teknologi MARA, 40450 Bandaraya Shah Alam, Selangor **W:** www.uitm.edu.my/ufm
Station: 93.6MHz 1kW TRP.

MALDIVES

LT: UTC +5h — **Pop:** 400,000 — **Pr.L:** Dhivehi (Maldivian) — **E.C:** 50Hz, 230V — **ITU:** MLD

MALDIVES NATIONAL BROADCASTING CORPORATION (MNBC, Gov.)
📧 MNBC One Bldg, Buruzu, Magu, Male ☎ +960 3000 200 📠 +960 3325 083 **W:** mnbc..com.mv **E:** info@mnbc.com.mv **LP:** MD: Ibrahim Khaleel. Asst. Eng: Mohammed Hashim.
MW: 1449kHz 10kW.
FM: Male 91.0MHz 1kW, 103.8MHz 20W, Addu 90.0MHz 500W, Foahmula 89.0MHz 0.5kW.
"Raajje R." 1449kHz in Dhivehi: 24h. English: 1200-1400.
"Rajjee FM" (music channel) on 91.0MHz.
R. Eke (music, sports & entertainment) on 103.8MHz: 1745-0020.
Ann: MW: "Mee Dhivehi Raajjeyge Adu".

Other stations:
Capital R, Male: 95.6MHz 1 kW. Web: capital956.fm
Dhi FM, Male: 95.2MHz. Web: dhifm.com
Faraway FM, Male: 96.9MHz.
H FM, Male: 92.6MHz. Web: hfm.com.mv

MALI

LT: UTC — **Pop:** 13 million — **Pr.L:** French, Bambara, Peuls, Sonrhai, Sarakolé, Bobo, others — **E.C:** 50Hz, 220V — **ITU:** MLI

CONSEIL SUPÉRIEUR DE LA COMMUNICATION (CSC)
📧 B.P. 116, Bamako ☎+223 20232101

OFFICE DE RADIODIFFUSION TÉLÉVISION DU MALI (ORTM, Gov.)
📧 B.P. 171, Rue del Marne 287, Bamako ☎+223 20212019 📠 +223 20214205 **W:** www.ortm.ml **Email:** ortm@ortm.ml
LP: DG: Sidiki Konate. DG Adj: Nouhoum Traore. Dir. Radio: Seydou Baba Traore. Dir. Rural Radio: Mme Gnouma Keita. Dir. Research & New Tech.: Gaoussou Singare.

SW: Bamako (Kati) 50/100kW

kHz	Times	kHz	Times
5995	1800-2400	9635	0555-1800

FM:
National R. Bamako: 92.0MHz 1 kW + 47 txs of 0.5/0.25kW
Regional R. (Channel 2)

Location	MHz	kW	Location	MHz	kW
Mopti	94.4	10	Ségou	96.8	1
Bamako	95.2	1	Sikasso	98.3	1
Kayes	95.4	10			

National R. (Radio Mali) in French/Arabic/English/Bambara/others: SW & FM. **D.Prgr:** 0555-2400. **N. in English:** Sat 1905-1920.
Regional R. (Channel 2) on FM only: **D.Prgr:** 0800-1945.
Ann: "Vous écoutez l'office de Radiodiffusion-Télévision Malienne émettant de Bamako". E: "This is Bamako, Mali Radio Telecommunications". **IS:** Guitar.
R. Rurale on FM in Kayes 89.1MHz, Kolondieba 93.7Mhz, Koutiala and Macina.

Other stations in Bamako:
R. Patriote FM 88.1MHz – **R. Canal 2000:** 90.7MHz. **W:** membres. lycos.fr/canal2000 – **R. Mirador** 91.1MHz – **La Voix de la Verité** 91.5MHz – **Fréquence 3** 93.8MHz – **R. Tabalé** 94.3MHz – **R. Guintan** 94.7MHz – **R. Benkan** 97.1MHz – **R.Liberté** 97.7MHz **W:** www. comfm.com/live/radio/radioliberte E: liberte@mtelecom-mali.net – **R. Bamankan** 100.3MHz – **R. Klédu** FM 101.2MHz – **R Jakafo** 100.7MHz – **R. Kayira FM** 104.4MHz – **R. Voix de l'Islam** 107.4MHz.
RFI Afrique: Bamako 98.5MHz. Gao 92.1MHz, Kayes 102.2MHz, Mopti 97.7MHz, Segou 93.6MHz, Sikasso 95.0MHz.
Africa No. 1: Bamako 102MHz (see main entry under Gabon).
BBC African Service: Bamako 88.9MHz.
China R. Int relay station: see International Radio section

MALTA

LT: UTC +1h (27 Mar-30 Oct: +2h) — **Pop:** 405,000 — **Pr.L:** English, Maltese — **E.C:** 50Hz, 240V — **ITU:** MLT

MALTA BROADCASTING AUTHORITY (Regulatory Authority)
📧 7 Mile-end Rd, Hamrun HMR1719 ☎ +356 21221281, 21247908 📠 +356 21240855 **E:** info@ba-malta.org **W:** www.ba-malta.org
L.P: Chairman: Joseph Scicluna, Chief Exec: Dr. Pierre Cassar

PUBLIC BROADCASTING SERVICES LTD
📧 75, St. Luke's Road, Gwardamangia MSD 09 ☎ +356 21225051 📠 +356 21244601 **E:** info@pbs.com.mt **W:** www.pbs.com.mt
LP: Head of Radio: Natalino Fenech. CE: A. Psaila, Exec Engineer: Costantino Abela (cabela@pbs.mt)
RADIO MALTA
MW: Bizbizja 999kHz 5kW
FM: Bizbizja 93.7MHz 8kW, 107.5MHz 0.025kW
D.Prgr: 24h. **N:** 0600, 0700, 0900, 1100, 1400, 1700, 2130 (BBC WS night relay). BBC News Mo-Fr 0600-0640, 0900, 1100, 1500 - Sat 0600-0640, 0900, 1500 - Sun 0600-0640, 1500
MAGIC 91.7: FM: 91.7MHz 8kW, 24h.
MALTIN BISS: FM: 106.6MHz 8kW, 24h (live coverage of debates in the Maltese Parliament and indigenous music)

DIGI B NETWORK LTD.
📧 136, Alwetta Street, Mosta MST4508 ☎ +356 27420570 **E:** info@ digibnetwork.com **W:** http://www.dab.com.mt **LP:** Managing director: Sergio D'Amico **DAB+:** 6A, 6C, 12A, LP. Bouquet includes local, gov-ernative and international stations

COMMERCIAL STATIONS:
89.7 BAY, Eden Place, St. George's Bay STJ3310 ☎ +356 23710800 📠 +356 23710845 **E:** 897@bay.com.mt **W:** www.bay.com.mt. - **FM:** 89.7MHz 8kW. **LP:** Station Manager: Simon Lumsden - Sales Manager: Marthese Azzopardi — **CAMPUS FM,** University Broadcasting Services, Old Humanities Building, University of Malta, Tal-Qroqq Msida MSD 06. ☎ +356 21333313 📠 +356 21314485. **E:** campusfm@um.edu. mt **W:** http://campusfm.um.edu.mt - **FM:** 103.7MHz 8kW. **LP:** Station Manager: Fr. Joe Borg. Also relay of BBC WS, DW, UN Radio, Radio Netherland. — **CALYPSO 101.8,** Triq il-Gifen, Bugibba. ☎ +356 21578022 📠 +356 21578026 **E:** info@calypsoradio.com **W:** www. calypsoradio.com - **FM:** 101.8MHz 8kW. **LP:** Director: Frank Camilleri — **R. 101,** 2 Triq Herbert Ganado, Pieta' PTA1450. ☎ +356 25965407 📠 +356 21240261 **E:** news@media.link.com.mt **W:** www.radio101. com.mt - **FM:** 101.0MHz 8kW, 95.5MHz 300W. (Operated by Maltese Nationalist Party's) — **R. MARIJA,** Kunvent Patrijiet Dumnikani, Misrah San Duminku, Rabat RBT 06. ☎ +356 21453105, 21453106. 📠 +356 21453103. **E:** info.mal@radiomaria.org **W:** http://www.radjumarija. org - **FM:** 102.3MHz 8kW, 107.8MHz 200W. **LP:** Director: Fr. Charles Fenech — **RTK, MEDIA CENTRE,** Archdiocese of Malta and Diocese of Ghawdex, Triq Nazzjonali, Blata-Badja HMR02 ☎ +356 2569 9100, +356 2569 9158 📠 +356 2569 9151, +356 2569 9160 **E:** info@rtk.org. mt **W:** www.rtk.org.mt - **FM:** 103.0MHz 8kW, Ghawdex 97.8MHz 400W, Malta 97.6MHz 250W. **LP:** Chairman: Franco Azzopardi (fazzopardi@ mediacentre.org.mt). General Manager: Michael Francalanza (mfranca-lanza@mediacentre.org.mt). Mktg mngr: Sylvana Magro. Format: news, educational, entertainment — **SMASH R.,** 4 Thistle Lane, Paola PLA 19. ☎ +356 21667777 📠 +356 21697830 **E:** smash@vol.net.mt **W:** www. smashmalta.com - **FM:** 104.6MHz 8kW. **LP:** Head: Jesmond Saliba. — **SUPER 1,** A28B, Industrial Estate, Marsa, LQA 06. ☎ +35625682568 📠 +35621248249 **E:** radio@super1.com **W:** www.one.com.mt - **FM:** 92.7MHz 8kW, 98.2MHz 8kW, 88.0MHz 25W. **LP:** Managing Director: Dr. Michael Vella-Haber, Senior Manager Broadcasting: Ms. Ruth Vella, Senior Manager Radio: Mr. Ray Azzopardi. (Operated by Maltese Labour Party) — **VIBE FM,** Triq Tas-Sliema, Kappara, San Gwann ☎ +356 21385887 📠 +356 21383826 **E:** info@vibefm.com **W:** www.vibefm. com.mt. - **FM:** 88.7MHz 8kW. **LP:** Head: Justin Chircop — **XFM 100.2,** Grima Communications Ltd, 24, A.Cuschieri Str. Fleur-de-Lys, Birkirkara BKR 4916. ☎ +356 21376385 📠 +356 21378167 **E:** news@xfmmalta. com **W:** www.xfmmalta.com - **FM:** 100.2MHz 8kW. **LP:** Station man-ager: Donny Hughes. Marketing Manager: Antonella Vassallo

COMMUNITY STATIONS:
Temporary licences for up to 2 years and powers of 0.25-1W: 96.1 Vilhena FM 96.1MHz, Banda Fgura FM 93.1MHz, Bastjanizi FM 95.0MHz, Big FM 107.1MHz, BKR Radio 94.5FM 94.5MHz, Christian Light 105.4MHz, Deejays Radio 956FM 95.6MHz, District Convention of Jehovah's Witnesses 108.0MHz, Elenjani FM 95.8MHz, Energy FM 964

96.4MHz, Kiss FM Radio 91.3MHz, Kottoner 98 FM 98.0MHz, La Salle Radio 99.4MHz, Lehen il-Belt Gorgjana 105.6MHz, Lehen il-Belt Victoria 104.0MHz, Lehen il-Karmelitani 101.4MHz, MMG FM 97.5MHz, Power FM 90.4MHz, Pure-Gold Christian Radio 97.8MHz, Radio 12th May 96.5MHz, Radju 15 t'Awwissu 98.3MHz, Radju Bambina 98.3MHz, Radju Bartilmew 103.3MHz, Radju Belt Rebbieha 97.0MHz, Radju Elenjani 95.8MHz, Radju Festa 99.2MHZ, Radio Galaxy 105.0MHz, Radju Gilju Rebbieh 105.5MHz, Radju Hal Tarxien 99.0MHz, Radju Hompesch 90.0MHz, Radju Katidral 90.9MHz, Radju Kazin Banda San Filep 106.3MHz, Radju Lauretana 96.5MHz, Radju Lehen il-Guzeppini 89.1MHz, Radju Lehen il-Qala 106.3MHz, Radju Leonardo 105.2MHz, Radju Luminaria 106.9MHz, Radju Margerita 96.1MHz, Radju Marija Assunta 98.9MHz, Radju Maria Bambina 90.2MHz, Radju MMG FM 97.5MHz, Radju Pawlin 97.2MHz, Radju Prekursur 99.3MHz, Radju Sacro Cuor Sliema 105.2MHz, Radju San Gwann 96.9MHz, Radju Sant'Andrija 88.4MHz, Radju Santa Katarina 90.6MHz, Radju Santa Venera 91.0MHz, Radju Sokkors 95.1MHz, Radju Vilhena 106.0MHz, Radju Vizitazzjoni 92.4MHz, Radju Xeb-er-ras 90.8MHz, Tal-Gilju FM 95.4MHz, Trinitarji Fm 89.3MHz, VSB FM 103.40 103.4MHz.

MARSHALL IS (USA associated)

L.T: UTC +12h — **Pop:** 64,522 — **Pr.L:** English, Kajin Majol— **E.C:** 60Hz, 110/220V — **ITU:** MHL

RADIO MARSHALLS (Gov/Comm)
✉ PO Box 19, Majuro 96960 ☎+692 625 8413. Studio ☎ + 692 625 8411.**E:** v7ab@ntamar.net
L.P: GM: Antari Elbon, PD: Nixon Elisha, CE: Jambre Ralpho
MW: V7AB 1098kHz 25kW **FM:** 97.9MHz **D.Prgr:** 1900 (Sun 2000)-1130 **News:** Local bulletins and BBC hourly.

Other Stations:

MW	kHz	kW	Station
1) Kwajalein	1224	1	Armed Forces Network
FM	**MHz**	**kW**	**Station**
2) Majuro	95.5		V7MI
6) Majuro	96.5	0.03	WSO-FM
5) Majuro	98.5		BBC World Service
8) Majuro	98.5		R.Australia
1) Kwajalein	99.9	1	Armed Forces Network
7) Majuro	99.9		V7BNJ
1) Kwajalein	101.1	1	Armed Forces Network
1) Kwajalein	102.1	1	Armed Forces Network
4) Majuro	102.5		V7DJ
3) Majuro	104.1		V7AA
9) Majuro			VOA

Addresses and other information
1) Box 23 APO San Francisco CA 96555 **MW:** National Public Radio via satellite from Washington DC **FM:** Country Music (99.9), Active Rock Music (101.1) Hot AC Music (102.1) via satellite from AFRS. **D.Prgr:** 24h Local studio facilities are available, local breakfast show on 101.1 FM. – **2)** Pacific Media Services, Majuro 96960 ☎ +692 625 2911 **E:** v7emon@ntamar.net **L.P:** Mgr: Fred Pedro, CE: Benitito Kom **ID:** "V7Emon" [translated as 'V7Good'] – **3)** Majuro Independent Baptist Church, PO Drawer H, Majuro 96960-1008 ☎ +692 625 3141 🖷 +692 625 3141 **E:** v7aafm@ntamar.net – **4)** Ace Broadcasting, Majuro 96960 **L.P:** Mgr: Harry Doulatram ☎ +692 247 8735 -**5)** 24h satellite relay from London. This service is only funded until early 2011 when it will be replaced by R.Australia. -**6)** National Weather Radio, Majuro 96960. Live and recorded local weather and emergency information for the Majuro atoll area. 24h -**7)** Bukot Non Jesus Church [Assembly of God Part Two], Majuro 96960 **T:** +692 625 7914 **E:** eagle1@ntamar. net **L.P:** Pastor Paul & Laura Hensene **ID:** 'V7Eagle' -**8)** **F.PL:** 24h satellite relay from Melbourne from early 2011 -**9)** **F.PL:** 24h satellite relay from Washington DC planned from 2011.

MARTINIQUE (France)

L.T: UTC -4h — **Pop:** 418,000 — **Pr.L:** French, Créole Patois — **E.C:** 50Hz, 220V — **ITU:** MRT

RADIO MARTINIQUE
✉ RFO Martinique, B.P. 662, 97263 Fort de France. ☎+596 596595200 🖷 +596 596595280 **L.P:** Dir: Fred Jouhoud. CE: Jean Claude Arrivé.
MW: Fort-de-France 1310kHz 5kW
FM: Fort-de-France 92.0/94.5kHz, Morne-Rouge 94.3MHz, Marin 93.2MHz, Trinité 94.0MHz, Macouba 92.0MHz, St-Pierre 94.0MHz.
D.Prgr: 0800-0400. **N:** 1000, 1030, 2000, 2300 + rel. France-Inter.
IS: Piano **V.** by QSL-card.

RADIO CARAÏBES INTERNATIONAL MARTINIQUE (Comm)
✉ 2 Boulevard de la Marne, F-97200 Fort de France. ☎+596 596639870. 🖷 +596 596632659. **Web:** www.fwinet.com/rci
L.P: Dir: Yann Duval. Editor-in-Chief: Jean Philippe Ludon (**Email:** 100444.2371@compuserve.com). CE: Daniel Toussaint
FM: 91.2/88.5/98.7/104.6MHz.
D.Prgr: 24h. **N:** on the h. (rel. Europe 1) **V.** by letter. Rp.

R. France Internationale: rel. via R. Intertropical 99.9MHz, R. 105 Canal Antilles 105.0MHz & R. AS 106.2MHz.
Other stations: approx. 40 FM stns are operating.

MAURITANIA

L.T: UTC — **Pop:** 3.3 million — **Pr.L:** Arabic, French, Poular, Soninké, Wolof — **E.C:** 50Hz, 220V — **ITU:** MTN

HAUTE AUTORITÉ DE LA PRESSE ET DE L'AUDIOVISUEL (HAPA)
✉ Nouakchott **W:** www.mauritania.mr **L.P:** Dir: M. Imam Cheikh Ould Ely.

RADIO MAURITANIE (RM, Gov.)
✉ Av. Gamal Abdel Nasser 387, BP 5522 , Nouakchott ☎+222 5253 266 🖷 +222 525 4069 **E:** rm@mauritania.mr **L.P:** DG: Yeslem Ben Abdem
MW: Nouakchott 783kHz 50kW.
SW: Nouakchott: 4845/7245kHz 100kW (irreg.)
FM: Nouakchott 93.3MHz 2kW, Nouadhibou 94.7MHz + 98.0MHz (local st.), Néma 98.5MHz, Aïoun 94.7MHz, Kiffa 96.7MHz, Akjoujt 98.7MHz, Aleg 96.1MHz 1kW + 90.8MHz 1kW (local stn), Atar 98.5MHz, Kaedi 98.3MHz, Rosso 96.7MHz + 98.0MHz (local st.), Sélibabi 97.7MHz, Tidjikja 98.5MHz, Zouérate 96.7MHz, Barkéol 100MHz. Where no power level is shown, stns are 0.1kW.
Youth R: Nouakchott 98.0MHz. **F.PL:** Quran prgr.
D.Prgr. in Arabic/French/others: 24h on MW/FM, 0630-0830, 1700-0800 on 4845kHz, 0800-1700 on 7245kHz. **N:** Arabic: 0700, 1100(not Fri), 1200, 1300, 1500(not Fri), 1600(Fri), 2200, 2400. French: 1330(Fri), 1430, 1800v. **Ann:** A: "Huna Nouakchott, Idha'at al-Gumhuriyati al-Islamiyya al-Mauritaniya". F: "Ici Nouackchott, R. Mauritanie". **IS:** Mauritanian guitar.

BBC Arabic Sce: Nouakchott 106.9MHz, Nouadhibou 102.4MHz.
DW/Monte-Carlo Doualiya, Nouakchott: 90.2MHz 1kW.
RFI Afrique: Nouakchott 88.0MHz
Monte-Carlo Doualiya: Nouakchott 90.2MHz

MAURITIUS

L.T: UTC +4h — **Pop:** 1.3 million — **Pr.L:** English, French, 6 Indian langs, Chinese — **E.C:** 50Hz, 240V — **ITU:** MAU (Rodrigues: ROD)

INDEPENDENT BROADCASTING AUTHORITY
✉ 5 De Courson Str, Curepipe Rd, Forest Side **W:** iba.gov.mu **E:** iba@intnet.mu

MAURITIUS BROADCASTING CORPORATION (MBC, Pub.)
✉ 1 Louis Pasteur Str, Forest Side ☎+230 6021200 🖷 +230 6757332 **W:** mbc.intnet.mu **E:** mbc@mbc.intnet.mu **L.P:** DG: Dhanjay Callikan. Ag. CE: Cyril Nankoo.
MW: RM1, R.Maurice, Malberhes: 684kHz 10kW **D.Prgr:** 24h in French. Relay of KOOL FM during daytime and VOA English during local nighttime. **N** on the h. **English:** 0500-0515.
RM2, R.Mauritius, Malherhes: 819kHz 10kW **D.Prgr:** 24h in Indian languages.
R. Rodrigues, Citronelle: 1206kHz 1kW **Prgr:** Relay of RM1. Local prgr. 1400-1415. **FM**(MHz): 97.3 0.5kW 24h.

MBC FM	Location	MHz	kW
Kool FM	Signal Mt.	91.7	0.5
Kool FM	Plaine Wilhems	97.3	1.0
Kool FM	Jurançon	89.3	0.5
Taal FM	Signal Mt.	98.2	0.5
Taal FM	Plaine Wilhems	94.0	1
Taal FM	Jurançon	95.6	0.5
One World FM	Signal Mt.	94.9	0.5
One World FM	Plaine Wilhems	90.8	1
One World FM	Jurançon	92.4	0.5

Other stations:
R. Plus, Labourdonnais Str, Port Louis. **FM**(MHz): Centre 87.7, North 88.6, South 98.9.

R. One, Brown Sequard Str, Port Louis. **FM**(MHz): Centre 100.8, North 101.7, South 102.4.
Top FM Skywave: FM(MHz): Centre 104.4, North 105.7, South 106.0.
BBC World Sce: Bigara 1575kHz 2kW. 24h.
RFI Afrique: Port-Louis 100.8MHz, Rodrigues 93.2MHz

MAYOTTE (France)

L.T: UTC +3h — **Pop:** 200,000 — **Pr.L:** French, Mahorian — **E.C:** 50Hz, 220V — **ITU:** MYT

RADIO MAYOTTE (Gov)
RFO Mayotte, B.P. 103, Rue de jardins, 97600 Mamoudzou ☎+262 269601017 📠 +262 269601852 **W:** mayotte.rfo.fr
L.P: DG: Georges Chow-Toun.
MW: Pamandzi 1458kHz 5kW
FM: Dzaoudzi 91.0MHz 0.1kW,M'lima Combani 92.0MHz 0.5kW, Kanikeli-Choungui 101.3MHz 0.5kW, Mtsanboro-Madjabalini 103.3MHz 0.5kW.
D.Prgr in French/Mahorian: Local prgr. Mon-Sat 0045-1900, Sun 0145-1830. Relays RFI overnight.
Ann: "Vous êtes à l'écoute de RFO-Mayotte". "Radio Mayotte".
IS: Melody on guitar.

Europe 2: Boueni 90.2MHz, Mamoudzou 99.1MHz, Pamandzi 97.7MHz
France-Inter: Dzaoudzi 101.0MHz 24h

MEXICO

L.T: UTC -6h (DST*: -5h). Baja California Sur, Chihuahua, Nayarit, Sinaloa, Sonora: UTC -7h (DST*: -6h; not Sonora); Baja California Norte: UTC -8h (DST*: -7h) *) 3 Apr-30 Oct — **Pop:** 110 million — **Pr.L:** Spanish — **E.C:** 60Hz, 127V — **ITU:** MEX — **Intl. dialling code:** 52

COMISION FEDERAL DE TELECOMUNICACIONES (COFETEL)
Unidad de Sistemas de Radio y Televisión
Bosque de Radiatas # 44, Col. Bosques de las Lomas, 05120 Del. Cuajimalpa, México, D.F ☎ +52 55 5015 4000

DIRECCION DE RADIO
Departamento de Asignación de Frecuencias
Eugenia 197, Col.Narvarte, 03020 Delg. Benito Juárez México, D.F ☎ +52 55 5015 4785
Call XE–,° = also on SW, * = inactive, (r) = repeater, v = varying fq, d = daytime operation. The letters preceding the stn number indicate the state. Addresses are listed by state in alphabetical order. Hrs of tr usually 24h – see address section for variations.

MW Call		kHz	kW	Station, location
BC19)	SURF	540	0.1	540 AM, Tijuana: (rel. KGIL 1260)
CH15)	TX	540	5	La TX/La Ranchera de Paquime, Nuevo Casas Grandes
CS22)	MIT	540	5/1	LV de BalúnCanán, Comitán
ME09)	WF	540	20/2.5	La Poderosa del Oriente/La Ke Buena AM 540, Ixtapaluca
NL01)	WA	540	1	W R., Monterrey (rel: XEW 900kHz)
SL10)	WA	540	150	W R., San Luis Potosí (rel: XEW 900kHz)
SN13)	HS	540	5/2.5	La Norteñita, Los Mochis
CH01)	PL	550	5/0.15	La Super Estación, Cd. Cuauhtémoc
GR01)	ACD	550	1	Los 40 Principales, Acapulco
JL23)	ZK	550	2.5/1	Poder 55, Tepatitlán
NA14)	TNC	550	2.5/0.15	R. Aztlán, Tepic
OX01)	HLL	550	1.5/0.25	Los 40 Principales, Salina Cruz (rel: XHHLL 97.1)
VE01)	KL	550	5/0.25	W R., Jalapa
YU01)	QW	550	2/0.35	Q-W La Poderosa, Mérida
C007)	GIK	560	1.4/0.25	La Acerera, Monclova
CS06)	IN	560	2/0.25	LV del Valle, Cintalapa
DF01)	OC	560	0.75/0.5	La Mejor, México
DG05)	SRD	560	10/0.1	La Tremenda, Santiago Papasquiaro
JL33)	MZA	560	10/1	Fórmula Melódica del Pacífico, Manzanillo
QR13)	QAA	560	5/1	La Poderosa, Chetumal
S037)	YO	560	1/0.5	R. Lobo, Huatabampo
ZC12)	XZ	560	5/1	Ke Buena, Zacatecas
C001)	TJ	570	1	La Mexicana, Torreón
MI01)	LQ	570	2/1.7	R. 5-70, Morelia
NA10)	TD	570	0.25	R. Red/La Z, Tecuala
NL02)	BJB	570	5/0.5	La Sabrosita, Monterrey (rel: XHRK 95.7)
OX02)	OA	570	5/2.5	O-A R. Mexicana, Oaxaca
PU13)	VJP	570	0.5	R. Xicotepec, Xicotepec de Juárez
SO26)	UK	570	0.5/0.25	La U-K/XEUK, Caborca
TB01)	VX	570	10/1	Mass R./La Grande de Tabasco, Villahermosa
YU09)	ME	570	2.5	El Poder del Oriente", Valladolid (XHME 91.9)
CH08)	FI	580	5/0.7	R. Mexicana, Chihuahua

MW Call		kHz	kW	Station, location
CO02)	MU	580	5/0.5	La Rancherita del Aire, Piedras Negras
CS01)	UE	580	1/0.25	La Invasora, Tuxtla Gutiérrez
JL02)	AV	580	10/1	Canal 58, Guadalajara
QE08)	UAQ	580	0.25	R. Universidad, Querétaro(rel: XHUAQ 89.5)
QR01)	YI	580	5/0.5	Mix FM, Cancún: (rel: XHYI 93.1)
SO06)	HO	580	1/0.25	La Fuerza de la Palabra, Cd.Obregón
TM01)	HP	580	1	La Más Prendida, Cd.Victoria
VE02)	DZ	580	1/0.5	R. Ondas, Córdoba
CS02)	ZZZ	590	5/1	Triple Z, Tapachula
DF02)	PH	590	25/10	Sabrosita 590 , México
DG01)	E	590	1	R. Fórmula, Primera Cad., Durango (rel: XHE 105.3)
GJ10)	GTO	590	10/0.25	Tu Recuerdo, León (rel XHGTO 95.9)
JL31)	CJU	590	10/1	La Explosiva 590, Puerto Vallarta
SO01)	BH	590	1	La Mejor, Hermosillo (rel: XHBH 98.5)
TM02)	FD	590	5/0.5	La Mejor, Reynosa
VE07)	OM	590	1	R. Fórmula, Coatzacoalcos
CO01)	DN	600	1	R. Notícias, Torreón
CS24)	OCH	600	10/1	K'in Radio, Ococingo
GR03)	BB	600	5/1	La Comadre, Puros Éxitos, Acapulco (rel. ZHBB 101.5)
JL36)	LAZ	600	2.5	La Mejor, Cd.Guzmán
MI02)	TA	600	1	600 Solo Hits, Zitácuaro
NL03)	MN	600	1/0.5	La Regiomontaña, Monterrey
SL16)	CV	600	5/1	La Gran Compañía, Cd.Valles
SN20)	HW	600	5/1	La Fiera Digital, Rosario
YU02)	Z	600	20/1	R. Fórmula, Segunda Cad., Mérida (rel: XHZ 105.1)
CO03)	BX	610	5/0.5	La Primera, Sabinas
CO26)	SAC	610	1	R. Lobo, Saltillo
MI03)	UF	610	5/1	Variadísima, Uruapan (re XHUF 100.5)
OX03)	KZ	610	1/0.5	La Poderosa, Tehuantepec
SN02)	GS	610	1/0.5	La GS/La Ley, Guasave
VE01)	JA	610	1/0.5	Conexión 610/R. Fórmula, Jalapa
ZC01)	EL	610	5/0.1	El Super Canal 610, Fresnillo
YU04)	UM	610	10	Candela FM, Valladolid (rel: XHUM 92.7)
BC09)	SS	620	5	LaTremenda, Ensenada (rel. XESDD 1030)
CH36)	BU	620	5/1	La Norteñita, Chihuahua
DF03)	NK	620	10	R. 6-20, México
DG02)	CK	620	1/0.5	R. 6.20, Durango
NA01)	OO	620	5/1	W R., Tepic
SL14)	WZ	620	2.5/0.5	R. Novedades, San Luis Potosí
TB13)	HGR	620	2.5/1	R. Fórmula 620, Villahermosa
TM30)	GH	620	1/0.25	La Lupe, Reynosa (rel: XHCAO 89.1)
GR27)	JR	630	5	Coral 630, Zihuatanejo
JL34)	JB	630	10/0.5	Jalisco R., Guadalajara
NL03)	FB	630	10	F-B La Estación que da las notícias, Monterrey
QR08)	CCQ	630	0.5	Frecuencia Turquesa, Cancún (rel: XHCCQ 91.5)
SN03)	OPE	630	5/0.25	Exa, Mazatlán (rel:XHOPE 89.7)
SO02)	FX	630	1/0.25	Doble X, Guaymas
TM27)	ERO	630	1/0.5	R. Tamaulipas, Altamira
VE03)	FU	630	10/0.75	La Nueva Voz, Cosamaloapan (rel: XHFU 103.3)
CH27)	JUA	640	5	R. Recuerdo, Canal 640, Cd.Juárez
CH29)	HHI	640	10/1	R. Uno/La Número Uno, Hidalgo del Parral
CS03)	WM	640	5/1	Suprema 64, San Cristóbal de las Casas
HG04)	NQ	640	10	N-Q La Superestación, Tulancingo: (rel. XHNQ 90.2)
OX26)	HDL	640	5/1	Aro-AM, Huajuapán de León
TM19)	TAM	640	1/0.25	Ke Buena, Cd.Victoria (rel: XHTAM 96.1)
ZC09)	YQ	640	5/1	R. Uno, Fresnillo
CO13)	RCG	650	0.5d	R. Vida, Cd. Acuña
GR21)	CHH	650	5	Capital Máxima, Chilpancingo (rel: XHCHH 97.1)
JL03)	EJ	650	10/2.5	La Z, Puerto Vallarta
MI04)	ZM	650	5/1	La Zamorana, Zamora
OX12)	PX	650	2/0.2	LV de Angel/R. Fórmula, Puerto Ángel
SL04)	IY	650	1	Espectacular, Río Verde
SN04)	TNT	650	5/1	W R., R. 65, Los Mochis
SO32)	VSS	650	1/0.25	R. 13, Hermosillo
TB15)	VILL	650	1/0.5	La Comadre, Villahermosa
YU13)	VG	650	1	R. Fórmula, Primera Cad, Mérida (rel: XHVG 94.5)
AG01)	EY	660	50/10	6-60 La Consentida, Aguascalientes
BS06)	SJC	660	2.5/0.25	Cabo 6-60, San José del Cabo
CH04)	ACB	660	5	Radio 6-60/La Tremenda Número Uno, Cd.Delicias
DF04)	DTL	660	50	R. Ciudadana, México
DG01)	WX	660	10/1	R. Mexicana, Durango
NL02)	FZ	660	10/1	Noti-Radio 6-60, Monterrey
OX04)	YG	660	1/0.5	R. 660/R. Fiesta Mexicana, Matías Romero
QR09)	CPR	660	1	R. Chan Santa Cruz-LV de los Mayas, Felipe Carillo Puerto
TM20)	AR	660	5	La Mexicana, Tampico
CO04)	TOR	670	5/0.25	X-E-Tor R. Ranchito, Torreón
CS10)	OB	670	5/0.5	La Máquina Musical, Pichucalco
JL01)	IS	670	1	La Rancheríta Consentida, Cd.Guzmán
NA08)	LH	670	5/0.1	La Zeta 670, Acaponeta
QE01)	OG	670	1/0.1	ABC R. 670, Querétaro
VE38)	SIC	670	1	La Romántica, Córdoba

MW Call	kHz	kW	Station, location
CH08) FO	680	1/0.25	Éxtasis Digital, Chihuahua
CS04) KQ	680	5/3	La Mexicana, Tapachula
GJ25) LG	680	10/3	LG, La Grande, León
GR15) CHG	680	5/2.5	Ke Buena, Chilpancingo: (rel: XHCHG 107.1)
OX27) OAX	680	5	Aro AM "la radio que une a Oaxaca", Oaxaca
PU01) FJ	680	1/0.1	La Consentida/R. Fórmula, Teziutlán
SN05) ORO	680	1/0.5	La Mera Jefa, Guasave
SO35) SON	680	1	La Mexicana, Hermosillo
YU03) PY	680	2.5/1	Foro 6-80, Mérida
BC01) WW	690	78/50	W R. América, LV del Pueblo, Tijuana
CL03) CS	690	5/1	La Mejor, Manzanillo
DF08) N	690	100/5	La 69, México
MI10) XL	690	2.5	La Ley, Pátzcuaro
NL04) RG	690	10/1	La Deportiva 6-90/La R-G, Monterrey
SN19) ST	690	2/0.25	La Invasora, Mazatlán
VE07) AFA	690	2.5	Ke Buena, Coatzacoalcos
ZC03) MA	690	50/2	M-A/La Madre de Todas, Fresnillo
CA09) XPUJ	700	5	LV del Corazón de la Selva, X'pujil
CH25) GD	700	5/0.25	La Poderosa, Hidalgo del Parral
JL12) DKR	700	1	R. Red, Guadalajara
MI02) LX	700	5	La Ke Buena, Zitácuaro
SO44) ETCH	700	5d	LV de los Tres Ríos, Etchojoa
TB12) RV	700	2.5/0.5	R. Villa, Villahermosa
VE02) VC	700	2.5/0.1	Ke Buena, Córdoba
CH05) DP	710	7/0.1	La Ranchera de Cuauhtémoc, Cd.Cuauhtémoc
CL01) RL	710	1	La R-L de Colima, Colima
CO04) LZ	710	1	La Reina, Torreón (rel: XHLZ 103.5)
CS05) ON	710	4.5/1	R. Mexicana, Tuxtla Gutiérrez
DF04) MP	710	10	Interferencia 7 Diez, Mexico
GR03) MAR	710	1	Amor, Acapulco (rel. XHMAR 98.5)
NA02) RK	710	1	R. Korita, Tepic
OX05) RPO	710	5/0.5	La Ley 710, Oaxaca
SL19) SMR	710	1/0.25	R. Fórmula, San Luis Potosí
SN01) BL	710	5/0.25	La Ke Buena, Culiacán (rel: XHBL 91.9)
SO03) PS	710	1/0.25	La Super Grupera, Guaymas
TM10)OLA	710	1	Huasteca, Tampico
YU14) YK	710	5/0.25	La Z, Mérida
CO19) DE	720	8/0.25	La Kaliente, Saltillo
JL05) QZ	720	1/0.25	Ritmo 720/La Máquina Musical, San Juan de los Lagos
QR05) CPQ	720	2d	La Estrella Maya Que Habla, Felipe Carrillo Puerto
SN19) VU	720	1/0.5	Magia, Mazatlán (rel: XEHVU 97.1)
VE49) AVR	720	10/0.25	R. Fórmula, Primera Cadena Nacional, Veracruz (rel: XERFR 970kHz)
BC21) EBC	730	1/0.25	Ke Buena, Ensenada
BS07) LBC	730	10/1	R. La Giganta 730 AM, Loreto
CH07) HB	730	50/1	R. Viva Villa, Hidalgo del Parral
CO05) PQ	730	10/1	La 73/La Sabrosita, Cd.Muzquiz
CS14) VF	730	10/5	R. Villaflores, Villaflores
DF06) X	730	100	Estadío W, México
JL30) GDL	730	5/1	La Explosiva, Guadalajara
SO39) SOS	730	10	R. Uno, Agua Prieta
YU10) PET	730	10d	LV de los Mayas, Peto
AG05) LTZ	740	1	Globo 740/R. Fórmula , Aguascalientes
CO28) QN	740	10/1	R. Fórmula Primera Cadena, Torreón (rel: XERFR 970kHz)
GJ04) OF	740	5/1	Romántica, Celaya
JL25) VAY	740	1	Amor, Puerto Vallarta (rel: XHVAY 92.7)
OX19) POR	740	5/1	La Explosiva/R. Fórmula, Putla de Guerrero
QR11) CAQ	740	20/10	R. Fórmula QR Cancún, Cancún: (rel: XHCAQ 92.3)
SN18) CW	740	10/1	R. Variedades, Los Mochis
TB14) KV	740	10/1	Exa FM, Villahermosa (rel: XHKV 88.5)
VE29) GF	740	2/0.25	R. Fiesta, Gutiérrez Zamora
CH21) OH	750	1/0.75	La Pantera, Camargo
CS11) MG	750	1/0.25	La Ke Buena, Arriaga
GR01) KOK	750	5/0.25	La Poderosa, Acapulco
MI29) URM	750	10/1	Fiesta Mexicana, Uruapán
NA09) JMN	750	10d	LV de los Cuatro Pueblos, Jesús María
OX06) CORO	750	1/0.1	Ke Buena, Loma Bonita
SL17) RASA	750	1/0.1	Candela 750/Candela Pasión Grupera, San Luis Potosí
SN23) CSI	750	1/0.25	Vida 750, Culiacán
VE31) TI	750	10/0.25	R. Fiesta, La Más Picuda, Tempoal
CH03) ES	760	1/0.5	Antena Musical 7-60, Chihuahua
CS07) RA	760	5/0.5	R. Uno, San Cristóbal las Casas
DF07) ABC	760	70/10	ABC Radio, México
DG03) DGO	760	5/05	La Mejor, Durango (rel: XEDGO 103.7)
JL15) ZZ	760	5/1	R. Gallito , Guadalajara
SO07) EB	760	5/1	R. Fiesta, Cd.Obregón
SO17) NY	760	5/0.1	R. Geny, Nogales
YU01) YW	760	2.5/0.5	Mexicanísima, Mérida
GR20) SUR	770	5/1	Tu Ritmo Musical, Chilapa

MW Call	kHz	kW	Station, location
MI05) ML	770	5/1.5	La Ranchera, Apatzingán
NL07) ACH	770	25/1	R. Fórmula, Primera Cadena, Monterrey
OX28) MRO	770	1d	Aro-AM, Matias Romero
OX29) HUA	770	1d	Aro-AM, Sta Cruz Huatulco
SL13) ANT	770	10	LV de las Huastecas, Tancanhuitz de los Santos
SN04) REV	770	1/0.1	Los 40 Principales, Los Mochis
VE49) QRV	770	5/0.5	Ultra 770, La Radio, Veracruz
ZC01) IH	770	10/1	La Unica, Fresnillo
CO31) WGR	780	10/0.25	Exa FM, Monclova (rel: XHWGR 101.1)
CS02) TS	780	5/1	La Máquina Musical, Tapachula
GJ04) ZN	780	5/1	EXA FM, Celaya (rel: XHZN 104.5)
GR04) XY	780	2.5/1	LV del Balsas, Cd.Altamirano
JL07) LD	780	5/1	R. Costa, Autlán
OX15) GLO	780	10d	LV de la Sierra Juárez, Guelato de Juárez:
TM32)MTS	780	2.3/0.25	R. Fórmula, Tampico
TM25)SFT	780	5/1	La Triple T/La Caliente, San Fernando
AG03) BI	790	10/5	R. B-I, La Estación que da las Noticias, Aguascalientes
BC06) SU	790	1/0.25	R. 790/La Dinámica, Mexicali (rel: XHSU 105.9)
BS01) NT	790	5/0.75	La Paz/R. Fórmula, La Paz
CH02) RPC	790	5/0.4	R. Ranchito, Chihuahua
CO01) GZ	790		W R., Torreón
DF08) RC	790	50/1	Formato 21, México
JL28) GAJ	790	0.25d	R. Fórmula, Primera Cadena, Guadalajara
TB03) VA	790	25/5	R. Tabasco, La Em. del Hogar, Villahermosa
TM03)FE	790	1/0.5	Mi Radio 790 - La Fiesta, Nuevo Laredo
VE11) COV	790	1/0.5	R. Lobo, Poza Rica
YU06) UP	790	1	Candela, Tizimín (rel: XHUP 96.3)
BC03) SPN	800	0.5/0.25	ESPN 800, Tijuana
CH12) ROK	800	150	R. Cañon, Cd.Juárez
CO29) ZR	800	2/0.25	La Traviesa de Coahuila, Zaragoza
CS13) UI	800	5/1	R. Comitán, Comitán
GJ19) GX	800	5/1	Fiesta Mexicana , San Luis de la Paz
GR09) ZV	800	5d	LV de la Montaña, Tlapa de Comonfort
JL27) AN	800	1/0.1	R. Alegría, Ocotlán
NL10) DD	800	10/2.5	La Tremenda, Montemorelos
VE09) QT	800	1	La Poderosa, Veracruz (rel: XHQT 106.9)
CA12) IC	810	0.1	R. I-C, Campeche
CH07) SB	810	1d	R. Mexicana/La S B, Santa Bárbara
CL05) MAX	810	3/0.25	Radiomax, Tecomán
CO06) IM	810	1/0.5	Fiesta Mexicana, Saltillo
CS04) OE	810	2.5/1.5	Romántica, Tapachula
GJ12) EMM	810	1/0.5	R. La Salmantina, Salamanca
GR01) AGR	810	7/0.6	R. Fórmula, Primera Cadena, Acapulco (rel. XHAGR 105.5)
NA03) UX	810	10/0.25	La Legendaria, Tepic
QR04) RB	810	2.5/0.25	Sol Estéreo, Cozumel (rel: XHRB 89.9)
SO04) RSV	810	5/0.25	Tribuna R., Cd.Obregón
TM04)RI	810	1/0.1	R. Rey, Reynosa
TM05)FW	810	50/1	R. Estrella, Tampico
TX01) HT	810	5/1	R. Huamantla, Huamantla
YU03) MQ	810	2/0.25	Yóol lik/R. Mayub, Mérida
ZC05) ZC	810	5/1	R. Felicidad, Río Grande
BC07) ABCA	820	3.5/0.5	R. Frontera, Mexicali
CA01) ESC	820	0.75d	R. Escárcega, Escárcega
DG01) DRD	820	10/0.5	W R. , Durango
GR17) GRC	820	1d	Soy Guerrero, Coyuca de Catalán
JL15) BA	820	10/1	La Consentida, Guadalajara
OX23) YN	820	10/1	Romántica 8-20, Oaxaca
SL19) BM	820	10/1	La Mera Mera, San Luis Potosí (rel: XHBM 105.7)
SN14) UDO	820	10/1	R. Universidad de Occidente, Los Mochis
VE02) KG	820	2.5/0.1	Golden Hits/ABC R., Córdoba
CO15) IK	830	5	La Norteñita 8-30AM/R. Fórmula, Piedras Negras
DF19) ITE	830	10/5	R. Capital, México
MI06) PUR	830	8d	LV de los P'urhepechas, Cheran
NL13) LN	830	10/0.25	La Caliente 830 AM, Linares (rel: XHLRS 95.3)
OX20) TLX	830	6	La Poderosa/R. Tlaxiaco, Tlaxiaco
SN12) VQ	830	5/1	La Superestación, La Grande de Sinaloa, Culiacán
SO03) DR	830	5/1	Digital 99, Guaymas (rel: XHDR 99.5)
TB09) ZQ	830	5/1	R. Futurama, Villahermosa
VE26) DQ	830	1	R. Alegría/ LV Amiga de los Tuxtlas, San Andrés Tuxtla
ZC12) LK	830	10/0.5	R. Mexicana, Zacatecas
CS01) IO	840	10/2.5	La Más Picuda, Tuxtla Gutiérrez
GJ04) FG	840	5/1	La Pachanga, Celaya
JL16) XXX	840	5/1	Fiesta Mexicana/Fiesta Digital, Tamazula
NA02) TEY	840	1/0.25	R. Sensación, Tepic (rel: XHTEY 93.7)
PU) FJ	840	1/0.1	La Consentida, Teziutlán
TM18)MY	840	1d	La Jefa, Cd.Mante
VE18) PV	840	2.5/0.1	La Fiera Grupera, Papantla
BC04) ZF	850	1	La Rancherita Contenta, Mexicali
CH03) M	850	5/1	Renacimiento 850 , Chihuahua
JL08) MIA	850	3/1	850 Noticias, Información Que Sirve,

MW Call	kHz	kW	Station, location
			Guadalajara
MI14) ZI	850	1d	Maxistar, Zacapu
QE02) JAQ	850	1	R. Felicidad, Jalpan
SO05) US	850	1/0.2	R. Universidad de Sonora, Hermosillo (rel: XHUSH 107.5)
TB07) RTM	850	5/0.5	La Zeta, Macuspana
VE04) TQ	850	10/1	La Q Orizabeña/R. Fórmula, Orizaba
AG05) PLA	860	2.5	La Mexicana, Aguascalientes
BC05) MO	860	10/7.5	La Poderosa 860, Tijuana
CH33) ZOL	860	5/1	R. Noticias 860, Cd.Juárez
CL02) AL	860	5/0.1	R. Mundo/R. Fórmula, Manzanillo
CS08) DB	860	5/0.25	Canal 86, Tonalá
DF10) UN	°860	45/10	R. UNAM, México
DG03) DU	860	1/0.5	D-U la que le gusta a Usted, Durango
MI12) IW	860		1 R. 860, Uruapan
NL04) NL	860	5/2	R. Recuerdo, Monterrey
QR06) CTL	860	5/1	R. Chetumal, Presencia Mexicana en el Caribe, Chetumal
QR07) CCN	860	5	R. Caribe, Cancún (rel: XHCBJ 106.7)
SN07) NW	860	1/0.25	Máxima 103.3, Culiacán (rel: XHNW 103.3)
SO40) HX	860	5/0.25	La Mia, Ciudad Obregón
TB04) ZX	860	1/0.15	LV de Usumacinta, Tenosique
TM05) TW	860	1/0.25	R. Fiesta, Tampico
YU14) RRF	860	5/0.5	860 AM, Mérida
CH18) TAR	870	10d	LV de la Sierra Tarahumara, Guachochi
GJ05) AMO	870	1/0.5	AMO 870, Irapuato
GR14) GRO	870	1	Soy Guerrero, Chilpancingo (rel: XHGRC 97.7)
MI01) LY	870	1/0.1	R. Fórmula, Morelia
OX07) ACC	870	5/0.25	R. Fórmula/LV del Puerto, Puerto Escondido
PU06) NG	870	0.5	R. Huauchinango, Huauchinango
SN19) FIL	870	1/0.25	R. Noticias, Mazatlán
CH14) V	880	5/0.25	R. Fórmula, Primera Cadena Nacional, Chihuahua: (rel: XERFR 970kHz)
CO16) TC	880	10/1	880 AM, Torreón
GR10) IG	880	2.5/1	Los 40 Principales, Iguala
JL09) AAA	880	20/1	R. 880/La Triple A, Guadalajara
SL04) EM	880	5/1	La M Mexicana, Río Verde
SN18) PNK	880	10/2	Canal 88/Superestación, Los Mochis
TB10) QQQ	880	10/0.5	Ke Buena, Villahermosa
VE19) YV	880	1/0.5	La Invasora, Córdoba
CS18) FRT	890	10/1	R. Frontera, Comitán
GJ11) AK	890	5/0.5	R. Consentida, Acámbaro
NA02) PNA	890	1/0.25	R. Joya/R. Fórmula, Tepic
SN01) NZ	890	10/0.5	La Sinaloense, Culiacán
VE24) BY	890	5/1	R. Fórmula, Tuxpan
ZC06) PC	890	5/1	Sonido Estrella, Zacatecas
CH31) DT	900	5/1	La Reina, Cd.Cuahtémoc
CS04) TAK	900	1/0.75	Radiorama Siglo XXI, Tapachula
DF06) W	900	250	W R., México
JL21) ED	900	1	La Líder 900 AM, Arneca
VE05) WB	900	50/10	W R., Veracruz (rel: XEW 900kHz)
NL05) OK	900	10/2.5	OK Notícias/R. Tráfico, Monterrey
BC06) AO	910	0.25	R. Mexicana, Mexicali
GJ06) ACN	910	5/0.1	R. Fórmula León/R. Uno, León: (rel: XEDF 1500kHz)
NA01) NAY	910	10/1	La Poderosa, Puerto Vallarta
PU02) OL	910	10/2.5	R. Impacto, Teziutlán
TB06) ACM	910	5/1	R. Exitos, Cárdenas
CA11) TEB	*920	1.5/0.5	R. Mar, Campeche
CH36) QD	920	1/0.25	R. Noticias 920, Chihuahua
CO04) RCA	920	5/0.2	Planeta, Torreón (rel: XHRCA 102.7)
CO08) MJ	920	1/0.25	La Fronteriza del Aire, Piedras Negras
CS05) VV	920	10/0.5	La Poderosa/R. Fórmula, Tuxtla Gutiérrez
GJ07) RE	920	5/1	La Comadre, Puros Éxitos, Celaya: (rel: XHRE 88.1)
JL15) LT	920	10	R. María, Tlaquepaque
MI27) LCM	920	5/2.5	R. La Mexicana, Cd.Lázaro Cárdenas
OX24) PNX	920	1/0.15	R. Costa/Ke Buena, Santiago Pinotepa Nal
PU14) ZAR	920	1	La Z, Puebla
SN08) CQ	920	5/0.5	C-Q/La Ranchera de Culiacán, Culiacán
SO01) HQ	920	5/1	R. Capital, Hermosillo
TM32) LE	920	10	La Preferida, Tampico
CL01) TTT	930	1	Magia 930, Colima
CO17) SHT	930	1/0.25	La Poderosa, Saltillo
CS12) MK	930	5/2.5	M-K R. Mexicana, Huixtla
HG02) CY	930	2/1	R. Diversión, Huejutla
MI16) ZU	930	1	La Explosiva, Zacapu
OX17) TLA	930	5d	LV de la Mixteca, Tlaxiaco
VE06) U	930	1/0.5	La U de Veracruz, Veracruz
YU03) UL	930	2.5/0.2	La Picosita, Mérida
ZC03) QS	930	10/3	Romance en Radio/R. Fórmula, Fresnillo
BC07) MMM	940	1/0.1	940 AM Oldies/R. Fórmula, Mexicali
BS08) RLA	940	1	R. Santa Rosalía, Santa Rosalía
CO09) YJ	940	15/0.5	Mix 9-40, Sabinas
DF06) Q	940	50	La Q 9-40/Bésame 9-40, México

MW Call	kHz	kW	Station, location
JL20) HE	940	1d	La Melódica, Atotonilco
PU) OL	940	10/2.5	R. Impacto, Teziutlán
TB18) REC	940	1/0.25	W R., Villahermosa
TM07) RKS	940	1d	La Poderosa, Reynosa
AG05) CAA	950	1	La 950, Aguascalientes
BC08) KAM	950	20/5	R. Fórmula Californias, Tijuana
CA02) MAB	950	3/0.9	La Poderosa, Cad. del Carmen (rel: XHMAB 101.7)
CH02) FA	950	1/0.25	La Poderosa, Chihuahua (rel: XHFA 89.3)
CS01) TUG	950	1/0.25	Radiorama Siglo XXI, Tuxtla Gutiérrez
GJ24) CEL	950	10/1	R. Lobo Bajío, Celaya
GR02) ACA	950	5/1	R. Fórmula, Segunda Cadena, Acapulco
JL22) MEX	950	5/0.5	La Mexicana, Cd.Guzmán (rel: XHMEX 104.9)
NA07) ZE	950	2.5/1	La Poderosa, Santiago Ixcuintla
NL06) RN	950	5/1	R. Naranjera, Monterrey
OX16) OJN	950	10d	LV de la Chinantla, San Lúcas Ojitlán
SN18) ORF	950	5/1	R. Exitos, Los Mochis
SO12) PB	950	10/0.1	La Grande/R. Amor, Hermosillo
TM20) TO	950	5/2	Romántica, Tampico
CH11) FAMA	960	10/1	R. Fama, Cd.Camargo
CO10) KS	960	0.5/0.1	XEKS 960/LV del Tiempo, Saltillo
CS04) TAP	960	5/1	Imperio/La Poderosa, Tapachula
GR05) UQ	960	1/0.5	R. Variedades, Zihuatanejo
GR12) XC	960	1	ABC R. 960, Taxco
JL10) HK	960	10/2.5	LV de Guadalajara, Guadalajara
MI07) MM	960	1	960 Notícias, Morelia
QR02) ROO	960	5/0.5	La Guadalupana, Chetumal
SL03) CZ	960	1	ABC R., San Luis Potosí
SO07) IQ	960	1/0.5	R. Norteña, Cd.Obregón
TM08) K	960	5/1	La Radio 9-60/La Estación Grande, Laredo
VE07) GB	960	1/0.5	R. Fiesta, Coatzacoalcos
VE08) OZ	960	1/0.25	Amor, Jalapa (rel: XHOZ 91.7)
CH22) SW	970	1/0.5	R. Madera/La Mera Mera, Cd. Madera
CH30) J	970	1/0.5	La J Mexicana, Cd.Juárez
CO31) MF	970	1/0.5	La Mejor, Monclova
DF11) RFR	970	50/4	R. Fórmula, Primera Cadena, México
GJ08) UG	970	1	R. Universidad de Guanajuato, Guanajuato
MI08) CJ	970	1/0.25	R. Apatzingán, Apatzingán (rel: XHCJ 94.3)
SN10) VOX	970	10/1	Fiesta Mexicana, Mazatlán
SO08) EZ	970	5/0.25	La Mejor, Caborca
TB05) VT	970	10/5	970 AM Stereo, Villahermosa
TM01) BJ	970	1	R. 9-70, Cd. Victoria
TM09) O	970	1	R. Gallito, Matamoros
YU03) MH	970	5/0.5	Candela FM, Mérida (rel: XHMH 95.3)
ZC12) ZAZ	970	5/0.5	De Mil Amores 9-70, Zacatecas
CH24) JK	980	1	La Poderosa, Cd.Delicias
CO12) NR	980	5/0.5	R. 980, Nueva Rosita
MI09) LC	980	5/0.2	Dual Stereo, La Piedad (rel: XELC FM 92.7)
NA04) XT	980	1	R. Capital/Capital Máxima, Tepic
PU09) FS	980	1	R. Matamoros, Código 9-80, Izúcar de Matamoros
SO09) FQ	980	2.5/0.5	LV de la Ciudad del Cobre, Cananea
SO10) KE	980	1/0.5	KE-98, Solo para tí, Navojoa
TM10) TU	980	5	R. Tampico, Tampico
VE03) QO	980	5	R. Romance, Cosamaloapan
BC04) CL	990	1.4/3	Rockola 990, Mexicali
BS02) HZ	990	5/0.25	HZ La Pura Sabrosura, La Paz
CH13) ER	990	5/0.25	R. Lobo, Cd.Cuauhtémoc
CS09) TG	990	10/1	La Grande del Sureste, Tuxtla Gutiérrez
GR28) PI	990	20/5	W R., Chilpancingo
JL01) BC	990	1	La Buena Onda, Cd.Guzmán (rel: XHBC 95.1)
MI28) ATM	990	1	A Toda Máquina, Morelia
NL04) T	990	50	La T Grande, Monterrey
OX08) IU	990	1	Cristal, Oaxaca (rel: XHIU 105.7)
VE15) ID	990	10/2.5	R. Álamo, Álamo
ZC10) FP	990	1/0.3	R. Alegría, Xalpa
CH33) FV	1000	1	La Rancherita, Cd.Juárez
CH29) HPC	1000	1/0.5	R. Mil/R. Fórmula, Hidalgo del Parral
CS02) TAC	1000	10/1	Exa FM, Tapachula (rel: XHTAC 91.5)
DF02) OY	°1000	50/20	R. Mil, México
GJ23) RZ	1000	1/0.5	W R., León
SN11) MIL	1000	1/0.25	Planeta Mil, Los Mochis
SN24) MMS	1000	1	Ke Buena, Mazatlán
TM24) NLT	1000	1/0.1	R. Fórmula/Laredo R., Nuevo Laredo
VE30) CSV	1000	1	Máxima FM, Coatzacoales (rel: XHCSV 93.1)
YU01) MYL	1000	5/0.25	Los 40 Principales, Mérida (rel: XHMYL 92.1)
BC22) DX	1010	2/0.5	CBC R., Ensenada
CH14) LO	1010	5/0.5	La X, Chihuahua
CO01) VK	1010	1	La Poderosa 10-10 AM, Torreón
CO13) KD	1010	0.5/0.25	La Mejor, Cd.Acuña
HG11) HGO	1010	1d	R. Hidalgo, Huejutla
JL15) HL	1010	50/5	Estadio VI, Guadalajara
MI32) TUMI	1010	5d	LV Mazahua Otomi/LV de la Sierra Oriente, Tuxpán
PU19) PA	1010	10/1	Punto 10 R., Cholula

MW Call	kHz	kW	Station, location
SN12) WS	1010	5/1	Romántica, Culiacán
SO11) XN	1010	0.5/0.2	R. Ures, Ures
VE09) FM	1010	5/0.5	La Máquina Tropical, Veracruz
CL09) VE	1020	1	W R., Colima
NA02) PIC	1020	1	R. Hits, Tepic
OX14) OU	1020	5/1	Sensación Estéreo, Huajuapan de León
QE06) KH	1020	1	R. Centro, Querétaro
QR03) WO	1020	1/0.1	97.7, Chetumal (rel: XHWO 97.7)
VE14) PR	1020	5/0.5	Los 40 Principales, Poza Rica: (rel: XEPR 102.7)
BC09) SSD	1030	10	La Tremenda, Ensenada
CA05) BCC	1030	1/0.25	Los 40 Principales, Cad. del Carmen (rel: XHBCC 100.5)
CH32) YC	1030	5/0.5	R. Fórmula, Cd.Juárez
CS15) VFS	1030	10/0.25	LV de la Frontera Sur, Las Margaritas
DF08) QR	1030	50/5	R. Centro, México
GR01) VP	1030	1	W R., Acapulco
JL11) LJ	1030	20/2	La Ke Buena, Lagos de Moreno
OX11) TEKA	1030	1/0.5	R. T-K, Juchitán
QR12) NKA	1030	5d	LV del Gran Pueblo, Felipe Carillo Puerto
SL07) IE	1030	5/1	Stereo 1030 AM, Matehuala
SN13) MPM	1030	10/1	R. Fama, Los Mochis
TM10) PAV	1030	1/0.5	La Picosita, Tampico
CH08) HES	1040	5/0.25	Radiorama Siglo XXI, Chihuahua
CS27) PLE	1040	1/0.25	R. Palanque, Palenque
GJ26) SAG	1040	1/0.25	R. Lobo, Irapuato
JL09) BBB	1040	10/1	R. Mujer, Guadalajara
ME05) CH	1040	5/0.75	R. Capital, Toluca
SO42) GYS	1040	5/0.25	La Primera, Guaymas
VE01) GR	1040	2.5/1	Imagen, Jalapa
AG06) DC	1050	1	1050 Notícias, Aguascalientes
BC06) D	1050	10	Radiorama Siglo 21/W R., Mexicali
BS05) BCS	1050	10/1	R. Cultura Surcalifornia, La Paz
GR29) ZUM	1050	15	ABC R., Chilpancingo
MI29) IP	1050	1/0.5	La Poderosa, Uruapán (rel. XHIP 89.7)
NA12) RIO	1050	5	La Poderosa, Ixtlán del Río
NL08) G	1050	100	La Ranchera 1050 , Monterrey
QR10) QOO	1050	35/2.6	R. Imagen, Cancún (rel: XHQOO 90.7)
TB03) TAB	1050	10/5	LV de Tabasco, Villahermosa
VE25) JF	1050	5d	R. Max/R. Sensación, Tierra Blanca
DF12) EP	°1060	100/20	R. Educación, México
CA02) IT	1070	1/0.25	Exa FM, Cad. del Carmen (rel: XHIT 99.7)
CS01) RPR	1070	2.5	Extasis Digital, Tuxtla Gutiérrez
GR03) AGS	1070	1/0.2	Digital 101.3,/Solo Exitos Acapulco: (rel: XHAGS 101.3)
JL12) SP	1070	10/1	10-70 R. Notícias, Guadalajara
PU03) GY	1070	1/0.25	R. Lobo, Tehuacán
SL02) EI	1070	5/0.25	Ke Buena, San Luis Potosí
SO18) OBS	1070	1/0.25	R. Fórmula, Cd.Obregón
VE10) MI	1070	1/0.1	La Poderosa, Minatitlán
BS09) PAB	1080	0.5/0.25	R. Celebridad, La Paz
CL04) UU	1080	5/0.5	La Mejor, Colima (rel. XEUU 92.5)
GJ05) CN	1080	5/0.5	Los 40 Principales, Irapuato
JL32) JLV	1080	5d	Sistema Jalisciense, Puerto Vallarta
ME07) TUL	1080	5/0.25	R. Mexiquense Valle de México, Tultitlán
OX09) AX	1080	5/0.5	Magía, Oaxaca
SO31) DY	1080	5/0.5	R. Gallo, San LuisRíoColorado
VE11) XK	1080	10/0.25	R. Fórmula, Poza Rica: (rel XERFR 970kHz)
BC11) PRS	1090	50	XX 1090 AM, Rosarito
JL13) LB	1090	5/1	La Buenísima, La Barca
NL04) AU	1090	5/0.5	Milenio TV, Monterrey
PU16) HR	1090	10/2.5	La HR, al Servicio de Puebla/R. Fórmula, Puebla
QE07) XE	1090	2.5/1	R. Grupo Fórmula Querétaro, Querétaro (rel: XERFR 970kHz)
TM08) WL	1090	1d	La Romántica, Nuevo Laredo
VE50) MCA	1090	10	R. 1090, La Grande de las Huastecas, Pánuco
VE13) IL	1090	1	La Comadre, Veracruz
YU01) FC	1090	10/0.25	XEFC 1090, Mérida
BS10) BAC	1100	1	R. Asunción/R. Sur California, Bahía Asunción
GJ09) BV	1100	5	R. Alegría, Moroleón
GR24) GRM	1100	1d	Soy Geurrero, Ometepec
QR15) CAN	1100	4d	R. Mundo Maya Turquesa, Cancún
SL19) PO	1100	1/0.25	Imagen, San Luis Potosí
SO43) NAS	1100	1/0.5	Única 1100 AM, Navojoa
ZC11) TGO	1100	5/0.5	R. Cañón, Tlaltenango
CH33) WR	1110	5/1	R. Guadalupana, Cd.Juárez
CO33) PU	1110	0.25d	Patronato Cultural Monclova, Monclova
DF08) RED	1110	50	R. Red, México
GJ25) LEO	1110	5/1	La Rancherita, León
JL35) PVJ	1110	1/0.2	Ke Buena, Puerto Vallarta (rel: XHPVJ 94.3)
OX22) TEO	1110	0.4d	La Señal de Oaxaca, Teotitlán de Flores Magon
OX30) TUX	1110	0.5d	La Señal de Oaxaca, Tuxtepec
SO32) VS	1110	10/0.25	Maxima 96, Hermosillo (rel: XHVU 96.3)
TM02) OQ	1110	1d	Notigape 11-10/R. Fórmula, Reynosa
VE23) HTY	1100	10/1	La Tremenda, Tlapacayan
BC07) MX	1120	0.4/0.1	Sonido 1120, Mexicali

MW Call	kHz	kW	Station, location
JL14) UNO	1120	0.5	R. Uno La Popular , Guadalajara
OX09) ZB	1120	2/0.25	R. Oro/La Tremenda, Oaxaca
PU16) POP	1120	1/0.1	Fórmula 11-20 AM, Puebla
QE04) GV	1120	1/0.5	11-20 Notícias, Querétaro
SL05) TR	1120		1R. Panorámica, Cd.Valles
TB17) TQE	1120	5/0.5	La Morena 1230 AM, La Más Choca de Todas, Tenosique (rel: XETVH 1230kHz)
YU08) RUY	1120		1 R. Universidad, Mérida
AG03) YZ	1130	10/2.5	La Poderosa, Aguascalientes (rel: XHYZ 107.7)
ME10) TOL	1130	1/0.5	11-30 Notícias, Toluca
MI11) FN	1130	1/0.1	R. Moderna, Uruapan: (rel: XHFN 91.1)
NA05) LUP	1130	1d	R. Lupita, Las Varas
SN27) MOS	1130	10/1	La Invasora, Los Mochis
SO36) HN	1130	1d	Ke Buena/Mariachi Estéreo, Nogales
VE08) ZL	1130	10/1	Capital 11-30, Jalapa
CS28) TEC	1140	10/1	R. Tecpatán, Tecpatán
GJ23) XF	1140	5/1	R. Felicidad, León
HG12) PEC	1140	1d	Hidalgo R., San Bartolo Tutotepec
MI22) LIA	1140	5/0.5	La Tremenda, Morelia
NL02) MR	1140	5/1	M-R Deportes, Monterrey
PU04) TE	1140	1	1140 Punto Digital, Tehuacán
BC06) RM	1150	1	R. Fórmula, Mexicali
CH16) JS	1150	1/0.5	R. Exitos/JS Digital, Hidalgo del Parral
CO14) BF	1150	2.5/1	R. Extremo, San Pedro
DF08) JP	1150	50/10	El Fonógrafo, México
JL06) AD	1150	50/1	R. Metrópoli, Guadalajara
OX10) XP	1150	10/1	La Super Buena, Tuxtepec
QE10) QUE*	1150	1/0.5	R. Querétaro, (now on FM 100.3MHz)
SN16) UAS	1150	10/0.15	R. Universidad/ R. UA Sinaloa, Culiacán (rel: XHUAS 96.1)
SO18) SO	1150	5/0.3	La Poderosa, Cd.Obregón
VE24) TVR	1150	1.5/0.5	La Nueva Azul, Tuxpán (rel: XETVR 106.9)
ZC08) XM	1150	5/1	R. Jerez, Jerez de García Salinas
BC20) QIN	1160	10	LV del Valle, San Quintín
GJ11) VW	1160	2.5/0.5	R. Sensación, Acámbaro
MI33) IW	1160	2.5	Canal Stereo Juvenil, Aruapan .
SL12) GI	1160	1/0.1	R. Reyna, La Gigante del Cuadrante, Tamazunchale
VE16) BE	1160	1/0.1	R. Perote, Perote
AG03) UVA	1170	10/2.5	La Rancherita, Aguascalientes
CO11) MDA	1170	1/0.5	La Ley 11-70, Monclova
JL19) JTF	1170	1/0.1	Prisma La Poderosa/Prisma Musical, Zacoalco de Torres
ME01) RLK	1170	1/0.1	Super Stereo Miled, Atlacomulco
PU05) CD	1170	10/2.5	R.Oro, Puebla
SO33) IB	1170	1	La Primera, Caborca
SO35) FEM	1170	1/0.5	R. Manantial, Hermosillo
TM07) RT	1170	5d	Voz 1170/R. Formula, Reynosa
VE46) ZS	1170	2.5/1	R. Hit/La Explosiva, Coatzacoalcos (rel: XHZS 92.3)
BS05) UBS	1180	10d	R. Universidad Autonoma de Baja California Sur, La Paz
CH35) DCH	1180	5/1.5	Romántica 11-80, Cd.Delicias
DF14) FR	1180	5/0.5	R. Felicidad, Los Éxitos de Siempre, México
GJ05) YA	1180	1/0.8	La Picosa, Irapuato
OX11) AH	1180	0.5	Ke Buena, Juchitán
VE34) GN	1180	10/1	La Gigante, Piedras Negras
BC07) MBC	1190	0.25/0.1	CBC R., Mexicali
CH30) PZ	1190	5/0.1	R. Norteña, Cd.Juárez
JL15) WK	1190	50/10	W R./W Guadalajara, Guadalajara
MI13) SOL	1190	5/1	R. Sol, la pura ley, Cd.Hidalgo
MO01) JPA	1190	5	La Grande, Cuernavaca
NL08) CT	1190	10/0.1	Contacto 11-90, Monterrey
SL08) XQ	1190	25/1	R. Universidad, San Luís Potosí
TM33) TOT	1190	1	ABC R., Tampico
VE32) PP	1190	5/0.25	La Comadre, Puros Exitos (rel: XHPP 100.3)
AG05) AGA	1200		La Bonita, Aguascalientes
BS11) PAS	1200	1d	R. Punta Abreojos, Punta Abreojos
ME02) QY	1200	2.5	Uno Más Uno R., Toluca
QE11) QJAL	1200	5	R. Querétaro, Jalpán
SN06) WT	1200	1/0.25	W R., Culiacán
SO29) YF	1200	1/0.25	R. Fórmula Hermosillo, Hermosillo (rel: XERFR 970kHz)
VE11) PW	1200	1/0.3	W R., Poza Rica
CS26) COPA	1210	5d	LV de los Vientos, Copainalá
GJ27) ITC	1210	1	R. Tecnológico, Celaya
PU16) PUE	1210	5/1	Méxicana/R. Fórmula, Puebla
VE27) BD	1210	10/0.25	R. Centro, Jalapa
VE33) VZ	1210	5/1	R. La Veraz, Acayucan
CO32) SAL	1220	2.5d	R. Universidad, Saltillo
DF04) B	1220	100	La B Grande, México
JL28) DKN	1230	1/0.25	R. Fórmula, Segunda Cadena, Guadalajara
MI30) LP	1230	1	R. Pía, La Piedad
NL07) IZ	1230	1	R. Fórmula Cadena 3, Monterrey
PU20) TCP	1230	1	W R., Tehuacan
SN12) EX	1230	10/2	R. Fórmula, Culiacán

MW Call	kHz	kW	Station, location
TB11) TVH	1230	20/1	La Morena 1230, La Más Choca de Todas, Villahermosa
AG03) RO	1240	10/2.5	R. Recuerdo, Aguascalientes
CH12) WG	1240	1	Cambio 1240, Cd.Juárez
CH17) BN	1240	1	Radiola, Cd.Delicias
CO15) VM	1240	1	Amor 107, Piedras Negras (rel: XHPNS 107.1)
CS01) LM	1240	2.5	Romántica 12-40, Tuxtla Gutiérrez
HG08) RD	1240	3	R. Lobo, Pachuca (rel: XHRD 104.5)
MI15) RPA	1240	5/0.5	R. Ranchito, Morelia
NA06) SI	1240	1	R. Positiva, Santiago Ixcuintla
OX25) CE	1240	2.5/1	Ke Buena, Oaxaca
SO36) CG	1240	1	Romántica, Nogales
SO14) BQ	1240	1	FM 105, Guaymas (rel. XHBQ 105.3)
TM10) S	1240	1/0.25	W R., Tampico
VE04) OV	1240	2.5/0.5	La Picosa/R. Fórmula, Orizaba
CH19) AT	1250	5/0.25	R. Imagen/Nueva Imagen, Hidalgo del Parral
CO03) SC	1250	1/0.5	La Pantera/R. 1250, Sabinas
CO30) SJ	1250	5/0.5	R. Saltillo, Saltillo
JL26) DK	1250	10/1	DK 12-50, Guadalajara
ME07) TEJ	1250	1/0.25	Sistema XEGEM "R. Mexiquense", Tejupilco (rel: XEGEM 1600kHz)
PU17) ZT	1250	5/0.5	R. Tribuna, Puebla
QE07) JX	1250	5/1	Cadena R. Uno, Grupo Fórmula, Querétaro (rel: XEDF 1500kHz)
SO32) DL	1250	1/0.5	R. 13/DL/Fuerza de la Palabra, Hermosillo (rel: XEDA 1290kHz)
VE49) TF	1250	10	R. Fórmula, Segunda Cadena, Veracruz
CH20) OG	1260	5/0.25	R. Ranchito, Ojinaga
DF14) L	1260	20/10	La 12-60 AM, México
GJ13) ZH	1260	1/0.25	La Estación que se Escucha, Salamanca
JL24) JY	1260	5/1	La Mejor, El Grullo
MI04) QL	1260	1	Catedral de la Música, Zamora
NL13) R	1260	1	Hits 12-60, Linares
OX18) JAM	1260	10d	LV de la Costa Chica, Santiago Jamiltepec
SL16) XR	1260	5/1	R. Mensajera, Cd.Valles
SN26) SA	1260	5/0.5	La Mexicana, Culiacán
SO15) MW	1260	1/0.25	R. San Luis/Sonido 2, San Luis Río Colorado
VE22) MTV	1260	1	R. Lobo de Mina, Minatitlán (rel: XHMTV 100.9)
VE48) TBV	1260	1	La Poderosa, Tierra Blanca
BC12) AZ	1270	0.5	Canal 1270, Zeta 13, Tijuana
CO04) WM	1270	0.5/0.15	El Fonógrafo del Recuerdo, Torreón
DG04) HD	1270	1.5/0.5	R. Universidad, Durango
GJ01) RPL	1270	10/0.15	La Poderosa RPL, León (rel: XHRPL 93.9)
HG05) QH	1270	3	Milenium R., Ixmiquilpán
SO16) GL	1270	1/0.5	Digital 12-70, Navojoa
TB18) VHT	1270	1	Bésame 12-70, Villahermosa
TM10) RRT	1270	2/0.5	Sport R., Cd.Madero
VE51) RRR	1270	1/0.25	Romántica, Papantla
CA03) CAM	1280	2.5/1	Kiss FM, Campeche (rel: XHCAM 101.9)
CH36) BW	1280	1/0.1	Palabra Viva, Chihuahua
CS12) KY	1280	1/0.1	Romántica 12-80, Huixtla
GJ14) SQ	1280	2.5/1.15	R. San Miguel, San Miguel de Allende
JL28) BON	1280	0.5/0.25	R. Fórmula, Tercera Cadena, Guadalajara
NL04) AW	1280	10/1	A-W Notícias, Monterrey
PU18) EG	1280	1/0.5	ABC Radio, Puebla (rel: XEABC 760kHz)
TM12) TUT	1280	1	R. Tamaulipas, Tula
VE38) AG	1280	2/1	La Poderosa, Córdoba
CA04) TH	1290	0.25d	R. Palizada, Palizada
DF09) DA	1290	10/1	R. Trece, La Fuerza de la Palabra, México
GJ03) FAC	1290	5/0.25	La Mera Mera, Salvatierra
MI17) IX	1290	1/0.5	Enlace Digital 12-90/La Pantera, Sahuayo
SN10) NX	1290	10/1	R. Mujer, Mazatlán
SO18) AP	1290	1/0.25	Romántica 1290, Cd.Obregón
CH30) P	1300	50	R. 13/R. Centro, Cd.Juárez
GJ15) XV	1300	10/0.75	La Z, León (rel: XHXV 88.9)
HG13) AWL	1300	1/0.25	R. Jacala/Hidalgo R.,Jacala
MI07) KW	1300	1	La Guadalupana, Morelia
SN15) JL	1300	1/0.25	La 130, La Ley, Guamuchil
SO13) XW	1300	1/0.1	W R., Nogales
VE23) HU	1300	1	La Que Manda, Martínez de la Torre
BC13) C	1310	1	R. Enciso, Tijuana
BS12) BTS	1310	1d	R. Bahía de Tortugas, Bahía de Tortugas
BS13) LPZ	1310	1	R. La Paz, La Paz
CH23) RU	1310	1/0.25	R. Universidad, Chihuahua (rel: XHRU 105.3)
GR23) GRT	1310	1d	Soy Guerrero, Taxco
JL06) TIA	1310	10/1	R. Vital, Guadalajara
NL02) VB	1310	5/0.25	Digital 102.9, Monterrey (rel: XHMB 102.9)
PU14) HIT	1310	5/1	R. Felicidad, Puebla
QE06) NY	1310	5	Stereo Joya, la Música de tu Vida, Querétaro
SO19) FH	1310	1/0.1	R. Plan de Agua Prieta, Agua Prieta
TM14) AM	1310	5/0.25	La M Grande, Matamoros
VE06) HV	1310	2.5/1	H-V 1310, Veracruz
AG07) NM	1320	1/0.5	R. 1320, Estación sin fronteras, Aguascalientes
BS03) SR	1320	0.5/0.25	R. Cachanía, Santa Rosalia
CH06) JZ	1320	2.5/0.25	La Campera/R. Fórmula, Cd.Jimenez:

MW Call	kHz	kW	Station, location
CO25) CPN	1320	10/0.1	1320 Notícias La Mexicana, Piedras Negras
DF20) NET *	1320	20	México
MI18) NI	1320	10/1	Romántica 13-20, Uruapán
OX10) UH	1320	10/2	X R., Tuxtepec
SN03) RJ	1320	10/1	RJ 1320, La Ranchera de Mazatlán, Mazatlán
TB19) PAR	1320	2.5/1	La Buena, Villahermosa
CL08) MAC	1330	10	La Poderosa, Manzanillo
CO18) WQ	1330	4/0.25	R. Triunfadora, Monclova
CO27) AJ	1330	5/0.9	1330 AM R., Saltillo
GJ16) BO	1330	5/1	R. Variedades, Irapuato
PU07) EV	1330	0.5d	R. Festival, Izúcar de Matamoros
TM10) RP	1330	1/0.1	La Tremenda, Cd.Madero
BC24) AA	1340	1	13-40 AM, Mexicali
CH20) RCH	1340	1/0.5	R. Exitos, Ojinaga
CO13) DH	1340	1	R. Amistad, Cd.Acuña
GR03) CI	1340	1	Romántica 13-40, Acapulco
HG03) QB	1340	1	La Divertida/R. Fórmula, Tulancingo
JL26) DKT	1340	5/1	R. Ranchito, Guadalajara
MI05) APM	1340	1	Candela, Apatzingán (rel. XHAPM 95.1)
MI19) CR	1340	1	La Zeta, Morelia (rel: XHCR 96.3)
M001) ASM	1340	5	Romántica 13-40, Cuernavaca
NL02) NV	1340	1	91X La Experiencia, Monterrey (rel: XHXL 91.7)
PU08) LU	1340	10/5	Ke Buena Puebla, Cd.Serdán
SL02) SL	1340	2	Señal 13-40, Golden Music, San Luis Potosí
SN17) QE	1340	1d	La Mera Mera, Escuinapa
SO40) OS	1340	1	R. Mujer, Cd.Obregón
TM01) RPV	1340	1	La Cotorra, Cd.Victoria (rel: XHVIR 101.7)
TM14) MT	1340	1	Mi Radio 13-40 - Nostalgia, Matamoros
TM15) BK	1340	1	Mega ¡Sí pega!, Nuevo Laredo (rel: XHBK 95.7)
CO04) TB	1350	5/0.5	R. Laguna, Torreón
CS16) CAH	1350	5/1	La Popular 13-50/La Voz de Soconusco, Cacahoatán
DF04) QK	1350	5/1	Tropicalísima 13-50, México
PU15) CTZ	1350	10d	LV de la Sierra Norte, Cuetzalán
SO15) LBL	1350	8	R. Centro, San Luis Río Colorado
TM16) ZD	1350	1/0.25	Mi Radio 1350, La Preferida, Camargo
CH02) DI	1360	1/0.4	La Nueva, Chihuahua (rel: XHDI 88.5)
CS09) UD	1360	5/0.5	La Ley de Tuxtla/La Máquina Musical, Tuxtla Gutiérrez
GJ07) Y	1360	1/0.25	R. Fiesta Retro, Celaya
GR08) KF	1360	1	Canal 13-60, Iguala
VE42) ZON	1360	10d	LV de la Sierra de Zongolica, Zongolica
BC06) HG	1370	0.5	Romántica, Mexicali
CA06) A	1370	1	Ke Buena, Campeche
DG07) RPU	1370	1/0.25	La Z, Durango (rel: XERPU 102.9)
GJ17) JE	1370	1/1.5	R. Reyna, Dolores Hidalgo
JL08) PJ	1370	10/1	Frecuencia Deportiva, Guadalajara
MI20) SV	1370	1/0.5	R. Nicolaíta, Morelia
NL07) MON	1370	10	R. Fórmula, Segunda Cadena, Monterrey
SO36) HF	1370	5	R. Fórmula, Nogales
TM31) GNK	1370	5/0.5	Fiesta Mexicana, Nuevo Laredo
CO01) RS	1380	1/0.5	Romántica 1380, Torreón
CO20) VD	1380	1/0.1	R. Sensación, Allende
DF13) CO	1380	50/5	Radio 13-80, México: (rel: XEQ-FM 92.9)
TM01) GW	1380	5/1	Mazz W, Cd.Victoria
VE27) TP	1390	10/1	Sensación FM, Jalapa (rel: XHTP 95.5)
BC14) KT	1390	5/0.1	La Súper Estación, Tecate
CL06) TY	1390	10/2.5	Los 40 Principales, Tecomán (rel: XHTY 91.3)
GJ06) RW	1390	10/0.25	R. Fórmula, León (rel: XERFR 970kHz)
HG06) ZG	1390	0.5d	R. Mezquital y Huasteca Hidalguense, Ixmiquilpán
M003) CTA	1390	1	R. Cauatla, Cuautla (rel: XHVAC 102.9)
SO21) QC	1390	1/0.15	LV de Pto Peñasco/La Reyna del Mar, Pto Peñasco
TM18) XO	1390	5/1	La Super Buena, Cd.Mante
TM02) OR	1390	1	La Papaya Tropicalísima, Reynosa
VE24) TU	1390	5/1	R. Ola, Tuxpán (rel: XHTL 91.5)
AG02) AC	1400	1	Ke Buena de Aguascalientes, Aguascalientes
BC15) PF	1400	1	La Efectiva/La Rancherita, Ensenada
GR01) KJ	1400	1	Mariachi Stereo, Acapulco
ME03) XI	1400	2.5/1	La I de Ixtapan, Ixtapan de la Sal
MI21) OJ	1400	5/1	R. Horizonte/R. Fórmula, Cd.Lázaro Cárdenas
MI22) I	1400	5	R. Trece, Morelia
NL09) SH	1400	5	R. Sabinas, Cd.Sabinas
OX21) UBJ	1400	1	R. Universidad Benito Juárez, Oaxaca
QE03) VI	1400	1	EXA FM 99.1, San Juan del Río (rel: XHVI 99.1)
SL06) WU	1400	0.25	La Poderosa, Matehuala
SO23) AB	1400	0.25	R. Santa Ana, Santa Ana
CA13) CUA	1410	1	R. Universidad, Campeche
CO01) YD	1410	1/0.1	La Grande de Madero, Torreón
DF02) RS	1410	25/1	Sinfonola, La Mas Perrona, México
GR25) ZHO	1410	2/1	Aquamarina R., Zihuatanejo
JL17) KB	1410	25/10	Canal 14-10, Guadalajara
SL11) IR	1410	5/0.5	XEIR, La Señal Perfecta, Cd.Valles

MW Call	kHz	kW	Station, location
SN13) CF	1410	10/0.5	La Mexicana, Los Mochis
TM08)AS	1410	1/0.25	Ke Buena, Nuevo Laredo (rel: XHAS 101.5)
BC16) XX	1420	2	R. Mexicana/R.Fórmula 1420, Tijuana
CH33) F	1420	5/0.5	Línea Deportiva, Cd.Juárez
GJ05) WE	1420	10/1	La Estación Familiar, Irapuato
HG08) PK	1420	1	1 R. Felicidad 14-20, Pachuca
JL18) KMX*	1420	1d	La Super X, Sayula
NL03) H	1420	5/1	La H, Antología Vallenata, Monterrey
PU03) WJ	1420	1	WJ Fórmula, Tehuacán
TM21)EW	1420	1	W1420/LV del Bajo Bravo, Matamoros
VE10) AFQ	1420	1d	Romántica, Minatitlan
CA06) RAC	1430	0.25	La Número Uno en Campeche/R. Fórmula, Campeche
CL09) COC	1430	1	Inolvidable, Colima
SO24) OX	1430	5/0.5	Exa FM 106.5, Cd.Obregón (rel: XHOX 106.5)
TM22)WD	1430	5/0.15	La Grande de Ciudad Miguel Alemán, Cd. Miguel Alemán
TX02) TT	1430	5/1	R. Tlaxcala, Tlaxcala
VE06) LL	1430	5/1	Latido 14-30 AM, Veracruz
BS04) VSD	1430	1/0.15	La Señal del Progreso, Cd. Constitución
DF16) EST	1440	25/5	Cambio 14-40, México
JL02) ABCJ	1440	10/1	ABC Radio/Corazón , Guadalajara (rel: XEABC 760kHz)
CH26) ARE	1450	1/0.25	R. Pegüís/R. Lobo, Ojinaga
CO01) BP	1450	1	Bonita, Torreón
GR11) RY	1450	2/1	La Poderosa V del Sur, Arcelia
MI17) GC	1450	1	R. Impacto, Sahuayo
NL04) JM	1450	5/1	La Caliente, Monterrey (rel: XET 94.1)
OX32) PNO	1450	0.4d	La Señal de Oaxaca, Santiago Pinotepa Nal
QE05) NA	1450	1	R. Capital, Querétaro
SN18) CU	1450	10/1	La Rancherita, Los Mochis
SO25) DJ	1450	0.5	R. Clave, Magdalena
SO45) CB	1450	10/1	R. Ranchito, San Luis Río Colorado
TM18)CM	1450	1	Bonita, Cd.Mante
TM34)RDO	1450	5/1	La Radio 14-50, Reynosa
VE20) JD	1450	1	R. Mundo 14-50, Poza Rica
VE22) KM	1450	1	R. Mina/R. Fórmula, Minatitlán
OX05) KC	1460	5/0.5	Estéreo Estés, Oaxaca (rel: XHKC 100.5)
VE08) JH	1460	1	ABC R., Jalapa
BC05) RCN	1470	10/5	R. Hispana 14-70 San Diego y Tijuana, Tijuana
CA08) BAL	1470	2.5/0.5	R. Voz Maya de México, Bécal
DF11) AI	1470	50/5	R. Fórmula, Tercera Cadena, México
DG03) CAV	1470	5/1	Play 14-70, Tocando Tu Memoria, Durango
GJ26) IRG	1470	1	La Campirana, Irapuato
HG07) IND	1470	1/0.5	LV Sierra Hidalguense, Tlanchinol: (occ. Rel: XHBCD 98.1 Hidalgo R.)
SN21) ACE	1470	1/0.1	R. Fórmula Mazatlán, Mazatlán (rel: XHACE 91.3)
TM04)HI	1470	10/0.25	Mi Radio 1470, Puro Cañonazo, Ciudad Miguel Alemán
CO21) XU	1480	1/0.1	La Poderosa, Monclova
CH28) HM	1480	1/0.5	H-M Radio, Cd.Delicias
HG10) CARH	1480	2.5d	LV del Pueblo Hña-hñu, Cárdonal
JL10) ZJ	1480	2/1	Ciudad 1480, Guadalajara
NL04) TKR	1480	10/1	TKR Rancherita y Regional, Monterrey
SO10) NS	1480	5/0.25	Z14, Solo Exitos, Navojoa
TM28)VIC	1480	5/0.15	R. Tamaulipas, Cd.Victoria
CH10) CJC	1490	1	R. Net, Cd.Juárez
MI04) GT	1490	5/1	W R., Zamora
MI23) KN	1490	1	R. Variedades, Huetamo
NA11)SK	1490	1/0.25	La Super K/La Costeñita, Cd.Ruiz
SL06) FF	1490	1/0.25	R. Norteña, Matehuala
SO27) AQ	1490	1	La Caliente, Agua Prieta
TM29)MS	1490	1	R. Mexicana, Matamoros
VE28) YT	1490	1	R. Teocelo, Teocelo
CO22) JQ	1500	0.4d	La Explosiva, Parras
DF11) DF	1500	50	R. Fórmula, Segunda Cadena, México
GJ20) FL	1500	1/0.5	R. Santa Fe, Guanajuato (rel: XHFL 90.7)
HG09) HUI	1510	2	R. Huichapán, Huichapán: (occ. Rel: XHBCD 98.1 Hidalgo R.)
NL11) QI	1510	50d	La Nueva Radio, Monterrey
CH30) JCC	1520	5	La 1520, Cd.Juárez
CO23) VUC	1520	1d	La Norteñita, Allende
ME07)ATL	1520	1	Sist. XEGEM "R. Mexiquense", Atlacomulco (rel: XEGEM 1600kHz)
MO02)ART	1520	2	Señal 152, Jojutla
SO46) EH	1520	1	R. Exitos, San Luis Río Colorado
TM18)YP	1520	1d	Mazz 15-20, Cd. Mante
VE35) VO	1530	1	La Furia, San Rafael
DF13) UR	1530	50/1	Fiesta 15-30, México
GJ21) SD	1530	10/0.1	Los 40 Principales, Silao (rel: XESD-FM 99.3)
MI24) GQ	1530	1	La Reyna de los Reyes, Los Reyes
GJ28) NC	1540	1/0.25	La Auténtica 15-40, Celaya
NL04) STN	1540	1/0.5	R. Red, Monterrey (rel: XERED 1110kHz)
PU12) RTP	1540	2.5	La Poderosa/Impacto, San Martín Texmelucán
SO29) HOS	1540	5	La Poderosa, Hermosillo
BC03) BG	1550	1	CBC R., Tijuana
MI26) REL	1550	1	R. Michoacán, Morelia (rel: XHREL 106.9)
TM31)NU	1550	5/0.25	La Rancherita, Nuevo Laredo
VE36) RUV	1550	10	R. Universidad Veracruzana, Jalapa
CA07) SE	1560	5d	LV de Campeche, Champotón
CH33) JPV	1560	1d	R. Viva, Cd. Juárez
CS29) CHZ	1560	20/0.15	R. Lagarto, La Voz Viva de Chiapas, Chiapa de Corzo
DF20) INFO*	1560	50/10	México
GJ12) MAS	1560	1/0.25	Ke Buena, Salamanca
MI25) LAC	1560	5/1	R. Azul/LV del Balsas, Cd.Lázaro Cárdenas
CO24) RF	1570	100	La Poderosa, Cd.Acuña
GJ28) AF	1580	1/0.5	La Temeraria 15-80, Celaya
GR13) LI	1580	1/0.5	Super 94.7, Chilpancingo (rel: XELI 94.7)
ME06)VAB	1580	20	Stereo Miled, Valle del Bravo
SO35) DM	1580	10	DM Notícias, Hermosillo
BC10) HC	1590	1	R. Bahía, Ensenada
CH24) BZ	1590	1/0.25	Extasis Digital, Cd.Delicias
DF14) VOZ	1590	20/10	R. Mexicana, México
VE39) PT	1590	1/0.1	La Nueva Misantla R., Misantla
GR26) TPA *	1600	1d	Soy Guerrero, Tlapa de Comonfort
ME07)GEM	1600	5	Sist. XEGEM "R. Mexiquense", Metepec
ME08)UACH	1610	0.25d	R. Chapingo, Chapingo
BC25) UT	1630	10/1	R. Universidad UABC, Mexicali
BC02) PE	1700	1	San Diego 1700, Tecate

SW Call	kHz	kW	Station, location & h of tr
DF18) RTA	4800	1/0.4	XERTA R. Transcontinental de América, México
DF02) OI	6010	1	R. Mil Onda Corta, México
SL08) XQ	6045	1	R. Universidad, San Luis Potosí: - 1300-0500
YU03) QM	6105	0.25	RASA, Mérida: irr (rel. different Mérida stns)
DF12) PPM	6185	10	R. Educación, México: 0000-0600 (0600-1200 rel: XEEP 1060kHz)
DF10) YU	*9600	1	R. UNAM-Universidad Nacional Autónoma de México, México; 24h

Stns with an (*) are reported to be inactive, but may occasionally be reactivated for variable periods of time.

State abbreviations: AG = Aguascalientes; BC = Baja California; BS = Baja California Sur; CA = Campeche; CH = Chihuahua; CL = Colima; CO = Coahuila; CS = Chiapas; DF = Distrito Federal; DG = Durango; GJ = Guanajuato; GR = Guerrero; HG = Hidalgo; ME = Estado de México; MI = Michoacán; MO = Morelos; NA = Nayarit; NL = Nuevo León; OX = Oaxaca; PU = Puebla; QE = Querétaro; QR = Quintana Roo; SL = San Luis Potosí; SO = Sonora; SN = Sinaloa; SO = Sonora; TA = Tabasco; TM = Tamaulipas; TX = Tlaxcala; VE = Veracruz; YU = Yucatán; ZC = Zacatecas. **N.B:** These abbreviations are not officially recognized by the Mexican Post Office. Letters should therefore carry the abbreviations in brackets or full state name.

Addresses and other information:
AG00) AGUASCALIENTES (Ags.)
AG01) Av Universidad N° 1001, Desp 614, Edif.Torre Plaza Bosques, 20127 Aguascalientes – **AG02)** Bahía No 201, Fracc.La Fuente, 20239 Aguascalientes – **AG03)** Morelos 222, Col Centro, 20000 Aguascalientes **W:** www.radiogrupo.com.mx – **AG05)** Madero 333, 1er piso, Col.Centro, 20000 Aguascalientes - 1200-0600 – **AG06)** San Miguel 117-A, Col.Salud, 20240 Aguascalientes - 1200-0400 ☎ (449) 9182370 🖷 (449) 9182371 – **AG07)** Av.28 de Agosto s/n, 2020259 Aguascalientes - 1200-0600 ☎+52 449 994 6470 **W:** www.aguas-calientes.gob.mx/ryta/default.aspx **E:** ryta@aguascalientes.gob.mx
BC00) BAJA CALIFORNIA (B.C.)
BC01) Ap.100, 22000 Tijuana (or: c/o Noble Broadcasting of San Diego, 4891 Pacific Highway, San Diego, CA 92110, USA) **W:** xtrasports.com/ – **BC02)** 3655 Nobel Drive, Suite 470, San Diego, CA 92122-1005, USA **W:** http://sandiego1700am.com – **BC03)** Av.de los Olivos 3401, Fracc. Cubillas, 22410 Tijuana – **BC04)** Pasaje Vallarta 1128 Altos, Centro Cívico, 21010 Mexicali (or: Box 1014, Calexico, CA 92231, USA) - 1300-0100 – **BC05)** Gral.Manuel Márquez de León 950, Zona Río, 22320 Tijuana (or: 713 Broadway, Suite "F", Chula Vista, CA 91910, USA) - 1300-0800 – **BC06)** Av.Calafia 519, Centro Cívico, 21000 Mexicali - 1400-0300 – **BC07)** Francisco L.Montejano 2200, Fracc.Fovisste, 21030 Mexicali (or: P.O.Box 872125, Calexico, CA 92232, USA) - 1400-0800 ☎ 6556 0600 🖷 6556 0662 **W:** www.mvs.com.mx – **BC08)** Carr Escenica Tijuana-Ensenada km 22.5, 22440 Tijuana – **BC09)** Calle 3a N° 1323-15, Plaza Elva, 22800 Ensenada - 1400-0800 – **BC10)** Ap.777, 22800 Ensenada - 1400-0700 – **BC11)** Blvd.Agua Caliente 10535-506, Fracc.Chapultepec 22420 Tijuana (or: 3655 Nobel Drive, Suite 470, San Diego, CA 92122-1005, USA) – **BC12)** Baja California 1310, Zona Norte, 22100 Tijuana (or: Box 430233, San Ysidro, CA 92073, USA) – **BC13)** Ap.23, 22000 Tijuana – **BC14)** Ap.19, 21400 Tecate – **BC15)** Ap.123, 22800 Ensenada – **BC16)** Carlos Robirosa 3110, Fracc.Aviación, 22420 Tijuana – **BC19)** Blvd.Lázaro Cárdenas 10183, Desp.201, 22450 Tijuana – **BC20)** Calle Octava n° 139, Fracc Cd San Quintín, 22930 San Quintín

- 1200-0200 (Sun – 2200) ☎ 6165 2023 Prgrs in Sp., Mixteco, Triqui and Zapateco – **BC21)** Calle 16, N° 159, Centro, 22800 Ensenada – **BC22)** Ap.526, 22800 Ensenada - 1400-0800 **W:** www.bajanet.com.mx/cbc – **BC23)** Lázaro Cárdenas, Esq.Colegio Central, Centro Comercio, Villa Fontana, Loc 33 y 34, 21180 Mexicali – **BC24)** Boulevard Benito Juárez No 1990, Local 12, Plaza Fimbres, Col Jardines del Valle, 21270 Mexicali – **BC25)** Edif.Rectoria, Av.Alvaro Obregón y Calle Julián Carillo s/n, Col. Nueva, 21100 Mexicali (or: UABC Radio, 233 Paulin Avenue, P O Box MSC 5163, Calexico, CA 92231-2646, USA) - 1400-0800 **W:** www.uabc. mx/RadioU/radio.htm - **FM** 104.1MHz

BS00) BAJA CALIFORNIA SUR (B.C.S.)
BS01) Ap.105, 23010 La Paz - 1300-0700 – **BS02)** Hidalgo 314-B, Centro, 23000 La Paz – **BS03)** Av.Las Flores 1, 23920 Santa Rosalía - 1200-0600 – **BS04)** Ap.279, 23600 Cd.Constitución - 1300-0700 – **BS05)** Ap.19-B, 23010 La Paz - 1300-0700 – **BS06)** Blvd.Mauricio Castro, Dorada's Plaza 4, 23400 San José del Cabo - 1200-0700 – **BS07)** 23880 Loreto – **BS08)** Av de Las Flores 1, 23920 Santa Rosalía – **BS09)** 23010 La Paz – **BS10)** 23960 Bahía Asunción – **BS11)** 23970 Punta Abreojos – **BS12)** 23950 Bahía de Tortugas – **BS13)** 23010 La Paz.

CA00) CAMPECHE (Camp.)
CA01) Calle 44 y 21 s/n, 24350 Escárcega1200-2400 – **CA02)** Calle 22 N° 131, 24100 Cd.del Carmen - 1200-0400 – **CA03)** Av.Luis Álvarez Barret 11, 24000 Campeche 1155-0500 – **CA04)** Ap.22, 24200 Palizada - 1200-2400 – **CA05)** Calle 32 N° 23-2 P.B., Centro, 24100 Cd.del Carmen - 1200-0500 – **CA06)** Tamaulipas 15, Col.Santa Ana, 24050 Campeche - 1200-0600 – **CA07)** Tamaulipas 15, Col Santa Ana, 24050 Campeche - 1200-2400 – **CA08)** Ap.1, 24930 Bécal - 1200-0600 – **CA09)** Domicilio Conocido, 24640 X'pujil - 1100-1600, 2000-0000 - Prgrs in Sp., Maya and Chol – **CA10)** 24000 Campeche – **CA11)** Prol. Calle 53, Esq.Av.16 de Septiembre s/n, 24000 Campeche - 1200-0600 – **CA12)** Instituto Campechano, 24000 Campeche – **CA13)** Universidad Autónoma de Campeche, Orquidea y Narcisos s/n, Col Jardines, 24000 Campeche - 1200-0200.

CH00) CHIHUAHUA (Chih.)
CH01) Calle Agustín Melgar 473, 31500 Cd.Cuauhtémoc – **CH02)** Julián Carrillo No 701, 31000 Chihuahua – **CH03)** Boulevard Ortíz Mena No 3406, Col Lomas del Santuario 2ª Etapa, 31240 Chihuahua - 1300-0500 – **CH04)** Calle 4a Poniente 606, 33000 Cd.Delicias - 1200-0600 – **CH05)** Calle 2A N° 437 (or: Ap.271), 31500 Cd.Cuauhtémoc - 2300-0700 – **CH06)** Allende 613, 33980 Cd.Jiménez - 1300-0200 – **CH07)** Boulevard Ortíz Mena No 54, Centro, 33800 Hidalgo del Parral - 1300-0100 **E:** larancherita@red-sat.com.mx – **CH08)** Julián Carrillo 705-A, 31000 Chihuahua - 1200-0200 – **CH09)** Coronado 71, 33580 Santa Bárbara – **CH10)** José Borunda 1178 Oriente, 32030 Cd.Juárez **W:** www.radionet1490.com – **CH11)** Gonzáles Ortega 1130, Centro, 33700 Cd.Camargo ☎+52 648 462 0527 ▤ +52 648 462 3333 –**CH12)** Av.Insurgentes 2127, Col.Ex-Hipódromo, 32330 Cd Juárez **E:** readiocanon800@latinmail.com – **CH13)** Ap.1771, 31500 Cd.Cuauhtémoc - 1300-0600 – **CH14)** Cuauhtémoc 2000, Col.Centro, 31020 Chihuahua - 1200-0600 – **CH15)** Jesús Urueta 504, 31700 Nuevo Casas Grandes - 1200-0400 ☎ +52 636 694 0083 – **CH16)** Ap.125, 33800 Hidalgo del Parral - 1245-0500 – **CH17)** Ap.222, 33000 Cd.Delicias - 1300-0400 – **CH18)** Francisco M Blancarte y Felipe Angeles, Colonia El Salto, 33180 Guachochi - 1200-0100 ☎/▤ 1543 0168 –Prgrs in Sp., Tarahumara, Tepehuáno and Guarijío – **CH19)** Ap.122, 33800 Hidalgo del Parral - 1200-0500 – **CH20)** Calle de la Paz 822, 32080 Ojinaga - 1200-0400 – **CH21)** Av.Mariano Negrete 8, Fracc Los Pinos, 33700 Cd.Camargo ☎ +52 648 462 1316 – **CH22)** Calle 3a N° 1204, 31940 Cd.Madero - 1300-0400 – **CH23)** Universidad de Chihuahua, 31000 Chihuahua – **CH24)** Ap.250, 33000 Cd.Delicias - 1300-0500 – **CH25)** Ap.190, 33800 Hidalgo del Parral - 1200-0600 – **CH26)** Juárez y 2a 201, 32881 Ojinaga (or: Box 276, Presido, TX 79845, USA) - 1200-0500 – **CH27)** Avenida Tecnológico 1770, Colonia Fuentes del Valle, Galería E, Local D-07, 32000 Cd.Juárez - 1200-0700 – **CH28)** Av.del Parque Sur 6, 33000 Cd.Delicias - 1300-0300 – **CH29)** Blvd.Ortíz Mena 54, P3, 33800 Hidalgo del Parral – **CH30)** Av Vicente Guerrero 2329, Col.Partido Romero, 32280 Cd.Juárez - 1200-0500 – **CH31)** Agustín Melgar 602, Niños Heroes, 31500 Cd.Cuauhtémoc - 1300-0100 – **CH32)** José Borunda 1178, Col.Partido Romero, 32030 Cd.Juárez – **CH33)** Chapultepec 316, Col Cauahtémoc, Edificio NAFTA Center, 32000 Cd.Juárez - 1200-0200 **W:** http://radioguadelupana.org (or: Box 17718, El Paso, TX 79917-7718, USA) ☎ +52 656 614 2869 – **CH35)** Calle 2a Norte N° 309, Interior 107, Col.Centro, 33000 Cd.Delicias - 1200-0700 – **CH36)** Ignacio Allende No 2211, Colonia Zarco, 31020 Chihuahua - 1200-0700.

CL00 COLIMA (Col.)
CL01) Calzada la Armonía 270, 28020 Colima - 1100-0600 ☎ (3) 313 1940 ▤ (3) 313 1500 **W:** www.radiolevy.com/xetttinfo.htm **E:** grlevy@col1.telmex.net.mx – **CL02)** Boulevard Costera Miguel de la Madrid No 801-3, Crucero Las Brisas, 28200 Manzanillo - 1100-0700 – **CL03)** Carretera Manzanillo-Minatitlán, km 0.2, 28200 Manzanillo – **CL04)** Ignacio Sandoval 13, 28000 Colima - 1200-0600 – **CL05)**

Allende 408-102, 28100 Tecomán - 1100-0400 ☎ (3) 324 1950 ▤ (3) 324 1616 **W:** radiolevy.com/xemaxinfo.htm **E:** grlevy@prodigy net.mx – **CL06)** Av.Antonio Leaño del Castillo 663, 28160Tecomán - 1100-0300 – **CL07)** Ap.2-1690, Suc.A, 28950 Colima – **CL08)** Lote 4, Manzana B, Parque Industrial Fondeport, 28200 Manzanillo - 1100-0600 – **CL09)** Av.Félipe Sevilla del Río 585, Col.Jardines Cista Hermosa, 28017 Colima - 1200-0400.

CO00) COAHUILA (Coah.)
CO01) Blvd.González de la Vega 195, 27000 Torreón – **CO02)** Ap.3, 26000 Piedras Negras (or: Box 196, Eagle Pass, TX 78853-0196, USA) **W:** www.larancherita.com.mx **E:** xemu@larancherita.com.mx – **CO03)** Ap.60, 26700 Sabinas - 1100-0600 – **CO04)** Priv.Eulogio Ortiz y Pamanes, Col.Ampl.Los Angeles, 27140 Torreón - 1200-0600 – **CO05)** Ap.71, 26340 Cd.Múzquiz - 1200-0400 – **CO06)** Piedras Negras 1812, 25280 Saltillo - 1200-0700 – **CO07)** De la Fuente 223 Pte, 25700 Monclova **E:** – **CO08)** Rassini 617, Col.Bravo, 26000 Piedras Negras - 1200-0500 – **CO09)** Zaragoza Pte 1270, Del Valle, 26788 Sabinas ☎ (861) 614 - 1200 – **CO10)** Gral.Manuel Pérez Trevino 839, Pte Interior, Centro, 25000 Saltillo - 1100-0600 ☎ +52 018 414 8149 **W:** xeksradio.com.mx/set-b01.html **E:** xeks@infosel.com – **CO11)** Venustiano Carranza 612-2 Ote, 25700 Monclova - 1300-0600 – **CO12)** Pte Carranza 1000, Col.Comercial, 26850 Nueva Rosita - 1155-0600 – **CO13)** Madero 274 Pte. (Ap.10), 26200 Cd.Acuña - 1200-0600 ☎ 877 772 51 21 – **CO14)** Pedro G Garza s/n, Col Magisterial, 35000 San Pedro – **CO15)** Av Carranza 1104, Col.Roma, 26000 Piedras Negras (or: Box 1261, Eagle Pass, TX 78852, USA) - 1100-0600 – **CO16)** Acuña 276 Sur, P2, 27000 Torreón – **CO17)** América Latina y Alaska s/n, Col.Virreyes, 25230 Saltillo – **CO18)** De la Fuente 304 Ote, 25700 Monclova – **CO19)** Av.Universidad No 1035, Col Universidad, 25260 Saltillo – **CO20)** Juárez - 1400 Sur, 26530 Allende - 1200-0600 – **CO21)** Pte Carranza Carr 4 Ciénegas, 25700 Monclova - 1200-0600 – **CO22)** Fco.I.Madero 501 Pte, 27980 Parras - 1100-0100 – **CO23)** Boulevard Leonides Guadarrama No 890 Norte, Centro, 26170 Nava - 1200-0200 – **CO24)** Madero # 600, 26200 Cd.Acuña **W:** www.lapoderosa.imer.com.mx **E:** lapoderosa1570@imer.com.mx – **CO25)** Lerdo 1612, Col.Nísperos, 26020 Piedras Negras – **CO26)** Chihuahua 151, P1, Col.Reública, 25280 Saltillo - 1200-0600 – **CO27)** Av Universidad 1035, Col.Universidad, 25260 Saltillo ☎ (844) 438 8108 **E:** 1330radio@mail.com – **CO28)** Av.Morelos 1320-204, Edif.Monterrey, 27000 Torreón – **CO29)** 505 Sur Alto, Ap 26850, 26450 Zaragoza - 1200-0400 – **CO30)** América Latina y Alaska, Col. Virreyes Residencial, (Ap.27), 25230 Saltillo – **CO31)** Cipres 321, Col. Guadalupe, 25750 Monclova – **CO32)** Universidad Autónoma Agraria, "Antonio Narro", Buenavista, 25315 Saltillo

CS00 CHIAPAS (Chis.)
CS01) Ap.59, 29000 Tuxtla Gutiérrez - 1100-0600 – **CS02)** 2ª Calle Poniente No 4, Centro, 30700 Tapachula - 1100-0500 **W:** www.radionucleo.com – **CS03)** Ap.74, 29250 San Cristóbal de Las Casas - 1200-0600 – **CS04)** Ap.76 (or: 1a Av.Sur N° 2), 30700 Tapachula - 1100-0400 – **CS05)** Av.Central Pte 554-4, 29000 Tuxtla Gutiérrez - 1200-0600 – **CS06)** Ap.60, 30400 Cintalapa - 1100-0500 – **CS07)** Avenida Benito Juárez No 48, Interior Altos, 29200 San Cristóbal las Casas – **CS08)** CrR Tonalá-Arriaga 1500, 30500 Tonalá - 1100-0500 – **CS09)** Blvd. Belisario Domínguez 4820, 29000 Tuxtla Gutiérrez - 1100-0700 – **CS10)** Aurora No 31 Centro, 29520 Pichucalco - 1100-0500 – **CS11)** Ap.28, 30450 Arriaga - 1200-0400 – **CS12)** Av.Central Norte 8, 30640 Huixtla - 1100-0300 – **CS13)** 2a Norte N° 2, 30000 Comitán - 1200-0400 – **CS14)** 1a Av.Norte Pte N° 53, 30470 Villaflores - 1200-0400 – **CS15)** 14 Sur-Poniente s/n, Barrio San Sebastián, 30180 Las Margaritas - 1200-0030 (SS – 2400) Prgrs in Sp., Tojobal, Mame, Tzeltal and Tzotzil – **CS16)** Km 1.5 Carr Cacahoatán-Unión Juárez, Ejido Rosario Ixtal, 30890 Cacahoatán - 1100-0700 – **CS17)** Carr Boca del Limókm 2.5, 29500 Reforma – **CS18)** Primera Calle Norte Pte 7, 30000 Comitán - 1100-0500 – **CS22)** Av.Chichimá 405, 30000 Comitán - 1100-0700 – **CS24)** Gobierno del Estado de Chiapas, 29950 Ococingo – **CS26)** Primera Oriente s/n, Barrio Siete Hescos, 29650 Copainalá 1230-2230 Prg in Sp., Zoque and Tzotzil – **CS27)** 29960 Palenque 1000-0400 (belongs to Gobierno del Estado de Chiapas) – **CS28)** 2ª Sur y 1ª s/n, 29610 Tecpatan (belongs to Gobierno del Estado de Chiapas) –**CS29)** Km 14 Libramiento Norte, 29160 Chiapa de Corzo.

DF00) DISTRITO FEDERAL (D.F.)
DF01) Av.Chapultepec 473, P7, Col.Juárez, 06600 México - 1100-0700 ☎ +52 55 3547 4600 – **W:** www.lamejor560.com **E:** contacto@lamejor560.com – **DF02)** NRM Comunicaciónes, Prolongación Paseo de la Reforma 115, Col.Paseo de las Lomas, 01330 México ☎ +52 (55) 5258 - 1200 ▤ +52(55) 5663 0739 **W:** www.nrm.com.mx **E:** info@nrm.com. mx – **Radio Mil sw:** Ap.21-1000, 04021 México **E:** radiomil@nrm. com.mx – **DF03)** Radiodifusoras Asociadas, Durango 341, Planta Baja, Col.Roma, 06700 México ☎ + 52 (55) 5553 9620, ▤ +52 (55) 5286 2774 – **W:** www.rasa.com.mx **E:** info@rasa.com.mx – **DF04)** Instituto Mexicano de la Radio, Real de Mayorazgo 83, Barrio Xoco, 03330 México ☎ +52 (55) 5628 1700 ▤ +52 (55) 5628 1693 **W:** www.

imer.gob.mx – **DF06)** Televisa Radio, Calzada de Tlalpan 3000, Col Espartaco, 04870 México ☎+52 (55) 5327 2000 🖹+52 (55) 5679 7996 **W:** www.televisa.com.mx/radio www.wradio.com.mx www.esmas. com/radio **DF07)** México Radio, Basilio Vadillo 29, Col Tabacalera, 06030 México ☎ +52 (55) 5518 3293 🖹+ 52 (55) 5535 0295 **W:** www.oem.com.mx/abcradio www.760.com.mx – **DF08)** Grupo Radio Centro, Av.Constituyentes 1154, Col.Lomas Altas, 11590 México ☎ +52 (55) 5728 4800-10 **W:** www.radiocentro.com.mx – **DF09)** Radio S.A., Rodolfo Emerson 412, Col.Chapultepec Morales, 11570 México **W:** www.radio13.com.mx – **E:** radio13@radio13.com.mx ☎ +52 (55) 5203 5577 🖹 +52 (55) 5545 2078 – **DF10)** Universidad Nacional Autónoma de México, Adolfo Prieto 133, Col.del Valle, Del. Benito Juárez, 03100 México ☎ +52 (55) 5523 2633 **W:** www.radiounam. unam.mx **E:** contacto@radiounam.unam.mx – **DF11)** Organización R Fórmula, Av.Universidad 1273, Col.Del Valle, 03100 México ☎+52 (55) 5279 2207 **W:** www.radioformula.com.mx – **DF12)** Radio Educación, Ángel Urraza 622, Col.del Valle, 03100 México (Radio Educación Onda Corta: Apartado Postal 21-465, 04021 México) ☎ +52 (55) 5559 6169 🖹 +52(559) 5575 6566 **W:** www.radioeducacion.edu.mx **E:** informes@ radioeducacion.edu.mx – **DF13)** Radiorama, Paseo de la Reforma 56, P1, Col.Juárez, 06000 México ☎ +52 (55) 5566 0299/5566 0471 ☎ +52 (55) 5566 1454 **W:** www.radiorama.com.mx – **DF14)** Grupo ACIR, Pirineos 770, Lomas de Chapultepec, - 11000 México **W:** www. grupoacir.com.mx/acir/default.htm – ☎ +52 (55) 5201 1700 🖹 +52 (55) 5540 4106 – **DF15)** Grupo ACIR, Blvd.de los Virreyes 1030, Lomas de Chapultepec, - 11000 México ☎ +52 (55) 5520 1956 – **DF16)** Grupo 7 División Radio, Montecito 59, Col.Nápoles, 03810 México ☎ +52 (55) 5669 1421 🖹 +52 (55) 5569 0047 **W:** www.gruposiete.com.mx **E:** cambio@spin.com.mx – **DF17)** MVS Radio, Mariano Escobedo 532, Anzures, 11300 México ☎ +52 (55) 5263 2100 🖹 +52 (55) 5203 4574 **W:** www.mvsradio.com.mx – **DF18)** Radio Transcontinental de América S.A de C.V., Gabriel Guerra 13, Col Zona Escolar Oriente, 07230 México 75 ☎ +52 (55) 5306 4668 **W:** http://misionradio.com www.xertaradio. com **E:** info@xertaradio.com – **DF19)** Grupo Radiodifusoras Capital, Montes Urales 425, Col Lomas de Chapultepec, - 11000 México ☎ +52 (55) 5202 3370 **W:** www.gruporadiocapital.com.mx – **DF20)** Info-Red, La Presa 212, San Jerónimo Lidice, 10200 México ☎+52 (55) 5329 - 1100 ☎ +52 (55) 5681 5000 **DF21)** Universidad Ibero-Americana, Av.Prol.Paseo de la Reforma 880, Col.Lomas de Santa Fe, México 01210 ☎ +52 (55) 5267 4239 **DF22)** Instituto Politécnico Nacional, Av.Santa Ana 1000, San Fco.Culhuacán, México 04260 ☎ +52 (55) 5624 2012 **W:** www.radioipn.xs3.com

DG00) DURANGO (Dgo.)
DG01) Manuel Rangel 100, P3, 34270 Durango - 1200-0600 – **DG02)** Av.20 de Noviembre 1918 Ote, Col.Guillermina, 34270 Durango - 1200-0600 – **DG03)** Negrete 405-B Oriente, 34270 Durango - 1100-0500 – **DG04)** Universidad Juárez del Estado de Durango, 34270 Durango – **DG05)** Fco.I.Madero y Heroico Colegio Militar s/n, 34600 Santiago Papasquiaro - 1200-0400 – **DG07)** Capitán de Ibarra 1203, Farcc.del Lago, 34080 Durango – **DG08)** Blvd.González de la Vega 195, Sur Zona Industrial, 35000 Gómez Palacio.

GJ00) GUANAJUATO (Gto.)
GJ01) Cañada 310, Esq.Roca, Col Jardines de Moral, 37160 León W: www.radioramabajio.com ☎+052 477 773 3606 🖹 +52 477 773 2470 – **GJ02)** Ap.301, 37160 León – **GJ03)** Morelos 704, Centro, 38900 Salvatierra 1230-0130 ☎ +52 466 663 0365 – **GJ04)** Blvd.López Mateos Ote 1117, 38070 Celaya - 1200-0600 **E:** telradio@mail.mindvox. ciateg.mx ☎ +52 461 613 4400 🖹 +52 461 612 1164 – **GJ05)** Morelos 110, 36500 Irapuato - 1200-0600 **W:** www.intercom.net.mx/radio **E:** radiogpo@intercom.net.mx ☎ +52 462 626 - 1200 – **GJ06)** Blvd. Mariano Escobedo Pte 4206, Col.Flores Magón, 37350 León ☎+52 477 7777 1943 – **GJ07)** Corporación ACIR Celaya, Guanajuato 106, Col Alameda, 38090 Celaya - 1300-0600 ☎ +52 461 612 4710 – – **GJ08)** Palacio Federal, Casa de Moneda, Sopeña 1m, P2, 36000 Guanajuato - 1300-0500 **W:** http://radiogto.ugto.mx ☎ +52 473 732 1684 – **GJ09)** Elodia Ledezma 658, 38890 Moroleón - 1200-0600 – **GJ10)** Av Roma 910, Col Andrade, 37370 León ☎ +52 477 714 0002 **FM:** 95.9 – **GJ11)** Allende 17, 38600 Acámbaro - 1200-0400 ☎ +52 417 172 1960 – **GJ12)** Ap.300, 36700 Salamanca - 1300-0600 ☎ +52 464 648 9200 – **GJ13)** Ap.24, 36700 Salamanca - 1300-0500 ☎ +52 464 648 0227 – **GJ14)** Calle Solano 4, 37700 San Miguel de Allende - 1200-0400 ☎ +52 415 152 0227 – **GJ15)** Ap.13, 37000 León - 1200-2400 ☎/🖹 +52 477 770 0468 – **GJ16)** Ap.72, 36500 Irapuato 1030-0600 ☎+52 462 626 3733 - – **GJ17)** Ap.43, 37800 Dolores Hidalgo - 1200-0600 ☎ +52 418 182 0413 – **GJ19)** Ap.67, 37900 San Luis de la Paz - 1300-0300 ☎ +52 468 688 2849 – **GJ20)** Municipio Libre 8, 36080 Guanajuato - 1300-0300 ☎ +52 473 732 9909 – **GJ21)** Ap.60, 36100 Silao ☎+52 472 722 0302 – **GJ22)** Ap.528, 38000 Celaya – **GJ23)** Ap.311, 37530 León - 1200-0600 **E:** acir_leon@infosel.net.mx ☎ +52 477 711 7388 🖹 +52 477 711 7374 – **GJ24)** Corporación Celaya Radio, Privada Venustiano Carranza 119, P1, 38000 Celaya – ☎+52 461 613 0977 🖹 +52 461 613

9410 – **GJ25)** Ap.642, 37160 León - 1300-0100 ☎ +52 477 712 2000 – **GJ26)** Av.Guerrero y Francisco Sarabia, Centro Plaza Magna, Local 3-B, 36500 Irapuato - 1300-0100 **W:** www.radioramabajio.com ☎ +52 462 624 4665 – **GJ27)** Av Tecnológico y García Cubas s/n, 38110 Celaya - 1200-0600 **E:** xeite@ite.mx ☎ +52 461 611 8040 – **GJ28)** Grupo Radiocomunicación Trébol, Privada Ronavicienovación 135, Floresta del Sur, 38090 Celaya - 1200-0800 ☎+52 461 613 1580.

GR00) GUERRERO (Gro.)
GR01) Calle de la Paz 190, P2, Edif.Nick, 39300 Acapulco – **GR02)** Ap.60, 39390 Acapulco – **GR03)** Av.La Suiza 19, Fracc.Las Playas, 39390 Acapulco - 1200-0600 – **GR04)** Fray Bautista Moya 410, Centro, 40660 Cd.Altamirano - 1200-0430 – **GR05)** Paseo de Zihuatanejo Pte No 143, Col.Limón, 40880 Zihuatanejo - 1200-0300 – **GR06)** Avenida Insurgentes No 125, 39170 Tixtla de Guerrera – **GR08)** Juan N.Álvarez 3 Altos, 40000 Iguala 1155-0500 – **GR09)** Av Heroico Colegio Militar No 234, Col Aviación, 41304 Tlapa de Comonfort - 1200-0100 (SS – 2000) ☎ 7476 0156 Prgrs in Sp., Náhuatl, Mixteco and Tlapaneco – **GR10)** Av Bandera Nacional 51-A, P1, (or: Ap.52) 40000 Iguala - 1200-0600 – **GR11)** Avenida Lázaro Cárdenas No 54, Col Héroes Surianos, 40500 Arcelia - 1200-0300 – **GR12)** Cerro de la Bermeja s/n, Taxco de Juan Ruíz de Akarcón, 40200 Taxco - 1200-0300 – **GR13)** Ap.40, 39000 Chilpancingo – **GR14)** Palacio de la Cultura "Ignacio Manuel Altamirano", tercer piso, Plaza Cívica 1er Congreso de Anahuac, Col Centro, 39000 Chilpancingo - 1200-0700 – **GR15)** Av.del Sur 14, Col. Margarita Vigurí, 39000 Chilpancingo – **GR17)** Av Revolución 6, 40700 Coyuca de Catalán – **GR19)** Morelos 6-3, 39000 Chilpancingo - 1200-0400 – **GR20)** Calle 5 Sur 305, 41100 Chilapa - 1200-0200 – **GR21)** Av.Guerrero 10-B, P1, Desp.2, Centro, 39000 Chilpancingo - 1200-0400 **W:** www.pmp.com.mx/radio.html – **GR23)** Hacienda del Cernillo, Casa Gallos s/n, 40200 Taxco - 1200-0400 – **GR24)** Benito Juárez 19-A, Barrio del Carmen, 41700 Omotepec (belongs to Gobierno del Estado de Guerrero) – **GR25)** Avenida Benito Juárez No 21-A, Col Centro, 40880 Zihuatanejo – **GR26)** 41300 Tlapa de Comonfort (belongs to Gobierno del Estado de Guerrero) **GR27)** Paseo de la Boquita No 53, Col Centro, 40880 Zihuatanejo ☎ +52 (755) 554 9520 – **GR28)** Eje Central No 3, entre Avenida Rufo Figueroa y Circuito Poniente, Col Burócratia, 39090 Chilpancingo – **GR29)** Zapata 28, 2do piso, Col. Centro, 39000 Chilpancingo - 1200-0400

HG00 HIDALGO (Hgo.)
HG01) Ap.123, 42000 Pachuca – **HG02)** Ap.35, 43000 Huejutla - 1100-0200 – **HG03)** Hidalgo Ote.209, 43600 Tulancingo - 1200-0600 – **HG04)** Plaza Constitución y Manuel F Soto (or: Ap.96), 43600 Tulancingo – **HG05)** Carr a Cardonal km 2.689, Barrio San Nicolás, 42300 Ixmiquilpán - 1200-0500 – **HG06)** Félipe Ángeles s/n, 42300 Ixmiquilpán - 1300-0100 – **HG07)** 43150 Tlanchinol - 1200-0200 – **HG08)** Plaza Juárez 103 (or: Ap.123), 42000 Pachuca - 1200-0600 **E:** acirpachuca@netpac.net.mx – **HG09)** Chávez Macotela 8,, 42400 Huichapan – **HG10)** Domicilio Conocido, Col Buenos Aires, 42370 Cárdonal - 1300-2300 Prgrs in Sp., Otomí and Náhuatl – **HG11)** Radio y Televisión de Hidalgo, 43000 Huejutla - 1100-0300 – **HG12)** Radio y Televisión de Hidalgo, 43440 San Bartolo Tutotepec – **HG13)** Radio y Televisión de Hidalgo, 42200 Jacala - 1200-0300.

JL00) JALISCO (Jal.)
JL01) Hidalgo 158, Centro, 49000 Cd.Guzmán - 1200-0600 – **JL02)** c/o El Periodico el Occidental, Calzada Independencia Sur 324, Col Centro, 44100 Guadalajara – **JL03)** Paseo de las Gaviotas 198, Fracc. Las Gaviotas, 48328 Puerto Vallarta – **JL05)** Carr Tampico-Barra de Navidad km 695, 47000 San Juan de los Lagos - 1300-0300 – **JL06)** Av.México 3150, Activa de Centro SA de CV, 44670 Guadalajara – **JL07)** Avenida Hidalgo 111-C, 48900 Autlán – **JL08)** Av.Lázaro Cárdenas 2820, Jardines del Bosque, 44520 Guadalajara - 1200-0600 – **JL09)** Av.Mariano Otero 3405, Fracc.Verde Valle, 45060 Guadalajara – **JL10)** Vidrio 2056, 44100 Guadalajara **W:** http://ciudad1480.com – **JL11)** Constituyentes 262, 47400 Lagos de Moreno – **JL12)** Pablo Casal 567, Prados Providencia, 44670 Guadalajara – **JL13)** Km 6.5 Carr La Barca-Guadalajara, 47910 La Barca - 1200-0400 – **JL14)** Hidalgo 2055 Esq Tomas de Gómez, Col Arcos Sur, 44500 Guadalajara **E:** xkguad@mail. udg – **JL15)** Televisa Radio, Rubén Dario 158, Circunvalación vallarta, 44680 Guadalajara Radio María México-address: Av Cruz del Sur 3195, P3, Lomas de Victória, 44580 Tlaquepaque, Jalisco - 1230-0600 – **JL16)** Portal Hidalgo 13, Int.10, Centro, 49650 Tamazula - 1200-0600 – **JL17)** Av Francia 1783, Col.Moderna, Sector Juárez, 44190 Guadalajara - 1200-0600 – **JL18)** Ap.36, 49300 Sayula - 1200-2400 – **JL19)** Fco. I.Madero 77, 45750 Zacoalco de Torres - 1300-0100 – **JL20)** Centro Comercial del Valle de Atotonilco, Local 17, Centro, 44750 Atotonilco - 1300-0100 ☎ (391) 917 1358 – **JL21)** Ap.16, 46600 Ameca - 1200-0600 – **JL22)** Primero de Mayo 126-8, 49000 Cd.Guzmán - 1200-0600 – **JL23)** Lerdo de Tejada 184, 47600 Tepatitlán - 1200-0300 – **JL24)** 18 de Marzo No 55 Int 9 7 10, Centro, 48740 El Grullo - 1200-0300 – **JL25)** Paseo de las Gaviotas, 48328 Puerto Vallarta - 1200-0600 – **JL26)** Studios: Av.Lázaro Cárdenas 3126, Col.Chapalita, 45040 Guadalajara

Commercial sve: Av.de los Niños Heroes 1555, P6, Ofc 602, Plaza Tolsa, Col Moderna, 44100 Guadalajara – **JL27)** Monterrey 190, Fracc. Camino Real, 47820 Ocotlán - 1200-0400 – **JL28)** Av.México 3370, Plaza Bonita, Local Subanda P, 45120 Guadalajara - 1300-0700 – **JL29)** Montezuma 68, 49000 Cd.Guzmán – **JL30)** Avenida Enrique Díaz de León No 285-2, Col Jesús, 44200 Guadalajara - 1300-0100 – **JL31)** Blvd.Francisco Medina Asencio km 7.5, Plaza Marina Local 101, Col. Marina Vallarta, 48300 Puerto Vallarta – **JL32)** Oceano Pacífico 201, Palmar de Aramara, 48300 Puerto Vallarta – **JL33)** Av.México 3150, 44670 Guadalajara - 1200-0500 **W:** www.unidifusion.com.mx – **JL34)** Av Constituyentes 21, Nucleo Agua Azul, 44190 Guadalajara - 1300-0700 – **JL35)** Honduras 309 Int 161, Hotel Paloma del Mar, Col.5 de Diciembre, 48350 Puerto Vallarta – **JL36)** Moctezuma 68, Centro, 49000 Cd.Guzmán - 1200-0400 ☎ (341) 412 5710

ME00) ESTADO DE MÉXICO (Edo.Méx.)
ME01) Carretera Panamericana km 24, 50450 Atlacomulco – **ME02)** Paseo Tollocan 613, Oriente, Col. Valle Verde, 50130 Toluca – **ME03)** José María Morelos 948, 51900 Ixtapan de la Sal - 1200-0600 ☎+52 721 143 1111– **ME04)** Independencia 19, 50600 El Oro – **ME05)** Ernesto Monroy, Lote 7 Manzana 3, parque Industrial Exportec II, 50200 Toluca - 1100-0500 ☎ +52 (722) 276 0715 ▤ +52 (722) 276 0700 – **ME06)** Independencia 506, 51200 Valle del Bravo – **ME07)** Av.Estado de México Km 1, Col La Virgen, 52140 Metepec - 1200-0600 – **ME08)** CarrCarr México-Texcoco km 38.5, 56235 Chapingo 1800-0200 ☎ +52 (595) 952 1610 – **ME09)** (RVM) Radiorama del Valle de México, Av Cuauhtémoc # 3, Col Santa Bárbara, 56530 Ixtapaluca - 1200-0600 **W:** www.radiorama.com.mx **E:** grupo@radiorama.com.mx **E10)** Paseo Tollocan Poniente 300, Col. Univerisdad, 50130 Toluca ☎+52 722 212 7128.

MI00) MICHOACÁN (Mich.)
MI01) Aqua 78, Col.Prados del Campestre, 58297 Morelia - 1200-0300 ☎+52 443 315 1810 – **MI02)** Av Revolucion Sur 66 (or: Ap.50), 61500 Zitácuaro - 1200-0600 (SS -0400) – **MI03)** Ap.61, 60100 Uruapan – **MI04)** Av.5 de Mayo 501 Sur, Jardines de Catedral, 59670 Zamora - 1200-0600 – **MI05)** Av.Constitución de 1814 Norte 10 Altos, 60600 Apatzingán - 1200-0400 – **MI06)** Domicilio Conocido, Predio INI, 60270 Cheran - 1300-0020 Prgrs in Sp and Purépecha – **MI07)** Laguna de Parras 630, Col.Ventura Puente, 58020 Morelia ☎+52 443 314 3518 – **MI08)** Av.Constitución de 1814 Norte 2 Altos, 60600 Apatzingán - 1200-0400 – **MI09)** Ap.10, 59300 La Piedad - 1230-0500 – **MI10)** Ap.244, 61600 Pátzcuaro - 1200-0600 – **MI11)** Ap.132, 60000 Uruapan E: moderna@mail.compuscp.com – **MI12)** Mazatlán 30, 60050 Uruapan - 1300-0400 **MI13)** Altos Mercado Emiliano Zapata, 61100 Cd.Hidalgo - 1200-0400 ☎+52 715 154 0212 – **MI14)** Avenida Morelos No 529, Plaza Ruíz, Centro, (Ap.65), 58600 Zacapu - 1200-0200 – **MI15)** Av.Madero Pte 644, 58000 Morelia - 1200-0300 ☎+52 443 317 2158 – **MI16)** Ap.50, 58600 Zacapu – **MI17)** Ap.60, 59000 Sahuayo - 1300-0200 – **MI18)** Venezuela 116, Col.Ángeles, 60160 Uruapan – **MI19)** Aquiles Serdán 548 , 58020 Morelia ☎+52 443 312 3214 – **MI20)** Universidad Michoacana de San Nicolás de Hidalgo, Cd Universitaria,, 58000 Morelia - 1200-0600 – **MI21)** Av.Río Balsas 7, 60950 Cd.Lázaro Cárdenas - 1200-0400 **MI22)** 20 de Noviembre 358, 58000 Morelia - 1200-0400 ☎+52 443 312 0903 – **MI23)** Madero Norte 15, 61940 Huetamo – **MI24)** Mariano Jiménez Norte 8-1, Centro, 60300 Los Reyes - 1200-0200 – **MI25)** Ap.430, 60950 Cd.Lázaro Cárdenas - 1100-0600 – **MI26)** Camino de los Gatos 200, 58000 Morelia W: www.smrtv.michoacan.gob.mx – **MI27)** Carr Lázaro Cárdenas-La Mira, 5 de Mayo, 60990 Lázaro Cárdenas – **MI28)** Dulcamara s/n, 58254 Morelia - 1200-0900 – **MI29)** Macarena 32, Inhuambo, 60130 Uruapan – **MI30)** Ap.73, 59300 La Piedad - 1300-0300 – **MI32)** Carretera Federal N° 15 Morelia-Zitácuaro km 125.6, 61420 Tuxpan - 1200-2330 Prgrs in Sp., Mazahua, Otomí and Matlatzinca – **MI33)** Mazatlán # 30, Col. La Magdalena, 60080 Uruapan

M000) MORELOS (MoR)
M001) Av.Morelos 309, Col.Centro, 62000 Cuernavaca - 1200-0600 ☎ +52 777 312 8872 ▤ +52 777 312 4388 **W:** www.radioramamorelos.com.mx – **M002)** Plaza Yuliana, P2, 62900 Jojutla - 1200-0100 ☎ +52 734 342 1778 ▤ +52 734 342 1777 – **M003)** 62746 Cuautla (alt.address: Hidalgo 105, 62220 Ocotepec, Cuernavaca) ☎+52 777 313 6123.

NA00) NAYARIT (Nay.)
NA01) Puebla 64, Sur Centro, 63060 Tepic – **NA02)** Insurgentes 1046 Pte, Col.El Rodeo, 63060 Tepic - 1200-0800 – **NA03)** Av.Juarez N° 160 Ote, Col.Centro, 63000 Tepic **E:** XEUXAM@yahoo.com.mx – **NA04)** Calle María Curie No 67, entre Madero y Sócrates, Colonia Burócrata, 63180 Tepic - 1300-0700 – **NA05)** López Mateos 61, 63715 Las Varas - 1300-0100 – **NA06)** P Sánchez 87 (or: Ap.22), 63310 Santiago Ixcuintla - 1300-0500 – **NA07)** Amado Nervo 106-B Ote, (or: Ap.4), 63310 Santiago Ixcuintla - 1200-0300 – **NA08)** Morelos No 54 Poniente, Planta Alta, Depto 1, 63440 Acaponeta - 1300-0100 – **NA09)** Domicilio Conocido, 63530 Jesús María - 1200-2000 Prgrs in Sp., Cora, Huichol, Tepehuáno and Náhuatl – **NA10)** Ap.7, 63440 Tecuala - 1200-2400

– **NA11)** Puebla 3, 63600 Cd.Ruíz - 1200-0300 – **NA12)** Justo Barajas No 50 Norte, Centro, 63940 Ixtlán del Río - 1300-0100.

NL00) NUEVO LEÓN (N.L.)
NL01) Carr a Verde km 6.5, Col.Juárez, 78000 San Luis Potosí, S.L.P – **NL02)** Avenida Madero Oriente No 1110, 64000 Monterrey - 1100-0600 – **NL03)** Juan Ignacio Ramón 506 Oriente, P20, Edif Latino, 64000 Monterrey – **NL04)** Paricutín Sur 316, Col.Roma, (or: Ap.203) 64700 Monterrey - 1200-0700 – **NL05)** Calle Monterrey 698, Esq. Cerralvo, Col.Libertad, 64130 Cd Guadalupe – **NL06)** Privada Rhin 647, 64000 Monterrey – **NL07)** Av Paseo de los Leones No 2935, Local C, Col Cumbres, 64000 Monterrey - 1200-0000 – **NL08)** Ap.118, 64000 Monterrey – **NL09)** Calle Reforma s/n, Col Lozano, 65290 Cd.Sabinas Hidalgo - 1200-0600 – **NL10)** Capitán Alonso de León s/n o Antigua Carretera Nacional km 904, Barrio Zaragoza, (Ap.45), 67500 Montemorelos – **NL11)** Av San Francisco y Loma Grande, Col.Loma Grande, 64000 Monterrey ☎ (81) 8347 6573 – **NL12)** Ap.62, 67701 Linares – **NL13)** Ap.81, 67701 Linares - 1200-0600

OX00) OAXACA (Oax.)
OX01) M.Ávila Camacho 514, P3, 70600 Salina Cruz - 1200-0500 – **OX02)** Ap.175, 68000 Oaxaca – **OX03)** Ap.21, 70760 Tehuantepec - 1200-0600 – **OX04)** Áquiles Serdán y Mina, 70301 Matías Romero - 1200-0100 – **OX05)** Netzahualcóyotl 216, Col Reforma, 68050 Oaxaca - 1200-0600 – **OX06)** Carr Cd.Alemán a Sayula km 27, 68400 Loma Bonita - 1200-0400 – **OX07)** Carr Puerto Escondido-Pochutla km 143, 71980 Puerto Escondido - 1300-0600 – **OX08)** Jazmines 907, 68000 Oaxaca - 1200-0600 ☎ 9513 4344 – **OX09)** Gómez Farias 113, 68000 Oaxaca - 1200-0600 – **OX10)** Abasolo 37, Centro, 68300 Tuxtepec - 1200-0100 – **OX11)** Carr Cristóbal Colón km 819.5, 70030 Juchitán 1130-0630 ☎ 9711 1233 (Ap.60), 70030 Juchitán - 1200-0200 – **OX12)** Ap.35, 70900 Puerto Angel - 1200-0200 – **OX14)** Venustiano Carranza 74-A, Col Altavista de Juárez, (Apartado 48), 69005 Huajuapan de León - 1300-0100 ☎ +52 953 532 2004 **W:** www.xeouradio.com – **OX15)** Lázaro Cardenas s/n, 68770 Guelato de Juárez - 1200-0130 –Prgrs in Sp., Zapoteco, Mixe and Chinanteco – **OX16)** Domicilio Conocido, 68470 San Lúcas Ojitlán - 1400-2200 –Prgrs in Sp., Mazateco, Cuicateco and Chinanteco – **OX17)** Carr Yucudaa km 54.5, 69899 Tlaxiaco - 1200-2400 Prgrs in Sp., Mixteco and Triqui –**OX18)** Plaza de la Constitución y Negrete s/n, 71700 Santiago Jamiltepec Prgrs in Sp., Mixteco, Amuzgo and Chatino - 1200-2400 – **OX19)** Morelos 6-2, 71000 Putla de Guerrero - 1200-0130 ☎/▤ +52 (953) 533 0000 – **OX20)** José Inés Dávila 2, Centro, 69800 Tlaxiaco - 1200-0500 – **OX21)** Universidad AutónomaBenito Juárez, 68000 Oaxaca – **OX23)** Macedonio Alcala 915, 68000 Oaxaca - 1200-0600 – **OX24)** Av.Alfonso Perez Gasca 504, 71600 Pinotepa Nacional - 1200-0600 – **OX25)** Valerio Trujano 708, 68000 Oaxaca - 1130-0600 – **OX26)** 69000 Huajuapán de León - 1200-0600 (belongs to Gobierno del Estado de Oaxaca) – **OX27)** Corporación Oaxaqueña de Radio y Televisión, Madero s/n, Centro Cultural, 68000 Oaxaca **OX28)** 70300 Matías Romero (belongs to Gobierno del Estado de Oaxaca) – **OX29)** 70989 Sta Cruz Huatulco (belongs to Gobierno del Estado de Oaxaca) – **OX30)** 68300 Tuxtepec (belongs to Gobierno del Estado de Oaxaca) – **OX31)** 68500 Huautla de Jiménez – **OX32)** 71600 Santiago Pinotepa Nacional – **OX33)** Carr Ixtepec-Juchitán km 2, 70110 Ixtepec

PU00) PUEBLA (Pue.)
PU01) Allende 507, Col. Centro, 73800 Teziutlán ☎ +52 231 312 0268 **W:** www.xefj.com.mx 1200-0400 – **PU02)** Av. Juárez 1002, Col. Centro, 73800 Teziutlán - 1200-0400 ☎ +52 231 312 0389 – **PU03)** Uno Sur 112, Mezanine (or: Ap.84), 75700 Tehuacán - 1200-0600 ☎ +52 238 2 3240 ▤ +52 238 2 0777 – **PU04)** Primera Norte 101-103, Desp.6, 75700 Tehuacán - 1300-0600 ☎ +52 238 24452 ▤ +52 238 21093 – **PU05)** Teziutlán Sur 17, Col La Paz, 72160 Puebla ☎ +52 2230 3499 ▤ + 52 2249 4199 **W:** www.radiooro.com **E:** Organización@radiooro.com – **PU06)** Ap.54, 73160 Huauchinango – **PU07)** Prof. Emilio Carranza 122, Barrio Santiago Mihuacán, 74420 Izúcar de Matamoros - 1200-0600 ☎ + 52 243 61026 – **PU08)** Prol.Manuel M.Flores s/n, 75520 Cd.Serdán - 1300-0100 **W:** www.kebuenapuebla.com **PU09)** Zaragoza 31-A, Centro, 74400 Izúcar de Matamoros ☎+52 243 6 0594 ▤ +52 243 6 1325 – **PU12)** Edificio Impacto, Tlaxcala #3, Col. La Santísima, (Ap.4), 74000 San Martín Texmelucan - 1200-0600 ☎ +52 248 4 0520 ▤ + 52 2484 4040 – **PU13)** Plaza de la Constitución 102, altos 1, 73080 Xicotepec de Juárez - 1100-0500 – **PU14)** Av.15 Pte.1306, Col.Santiago, Santiago, 72000 Puebla ☎ +52 22 2243 0100 ▤ 2237 0738 – **PU15)** Miguel Alvarado 45, 73560 Cuetzalán - 1200-0100 ☎+52 233 1 0382 Prgrs in Sp., Náhuatl and Totonaco – **PU16)** Av.15 de Mayo 2939, Frac. Las Hadas, 72070 Puebla ☎ +52 2249 6840 ▤ +52 2249 6830 – **PU17)** 3 Sur N° 107, P3, 72000 Puebla ☎ +52 2232 8000 **W:** www.radiotribuna.com **E:** general@radiotribuna.com – **PU18)** San Martín Texmelucan 57, Col.La Paz, 72160 Puebla ☎ +52 2242 0911 ▤ +52 2242 3322 – **PU19)** Calle San Judas Tadeo No 4901, Col Santa Cruz, Buenavista Sur, 72170 Puebla - 1200-0600 ☎+52 22 2467 1012 – **PU20)** 1 Sur 108, Desp.307, Col.Centro, 75700 Tehuacan☎+52 238 2 7281.

QE00) QUERÉTARO (Qro.)
QE01) Avenida Constituyentes Oriente No 122, Colonia Los Arquitos, 76048 Querétaro ☎/🖷 +52 442 212 2466 – **QE02)** Carr San Juan del Rio-Xilitla km 181, Col.San José, 76340 Jalpan - 1200-0600 **QE03)** Av.Juárez 38 Pte, 76800 San Juan del Río - 1200-0400 ☎ +52 427 272 2664 🖷 +52 427 272 0588 **W:** www.xeviradio.com – **QE04)** Paseo del Prado 102, Desp 401, Fracc.del Prado, 76030 Querétaro ☎ +52 442 216 5556 🖷 +52 442 216 5555 – **QE05)** Ing José Antonio Septién 28, Col Alameda, 76000 Querétaro - 1200-0600 ☎ +52 442 215 1615 🖷 +52 442 192 0045 – **QE06)** Av.Carrizal 28-F2, Fracc.Ampliación Carrizal, 76030 Querétaro - 1200-0600 **E:** rrdir@rR com.mx - **W:** www.rR com.mx - ☎+52 442 215 0333 🖷 +52 442 215 5842 – **QE07)** Av.Tecnológico Sur 2, Local 106, Col.Niños Héroes, 76010 Querétaro ☎ +52 442 215 2236 – **QE08)** Centro Universitario, 76010 Querétaro - 1200-0600 🖷/☎ 442 216 3481 **E:** radiouaq@sunserver.uaq.mx **W:** www. uaq.mx/servicios/cultural/radio.html – **QE09)** Av.Tecnológico 100, Desp.306-307, Edif.Tec 100, 76000 Querétaro – **QE10)** Luis Pasteur 6 Norte, Col.Centro, 76000 Querétaro ☎ +52 442 212 0910 🖷 +52 442 212 2786 - **E:** radiogro@grol.telmex.net.mx –**W:** http://members.xoom. com/ RadioQro – **QE11)** Camino de Piedras Anchas No. 100, Cabecera Municipal de Jalpán de Serra, 76000 Jalpán ☎+52 442 238 5111.

QR00) QUINTANA ROO (Q.Roo.)
QR01) Calle 63, Super Manazana 61, Mzna 7, Lote 1, 77500 Cancún - 1100-0600 ☎ +52 (988) 9884 1068 – **QR02)** Avenida Quinana Roo No 221, Col Venustiano Carranza, 77012 Chetumal - 1200-0500 – **QR03)** Prol.Av.Heroes 680, 77000 Chetumal - 1200-0400 – **QR04)** Ap.299, 77600 Cozumel ☎ +52 (987) 9872 0948 – **QR05)** Ap.13, 77200 Félipe Carillo Puerto - 1300-0100 – **QR06)** Av.Miguel Hidalgo 201, 77000 Chetumal - 1100-0700 – **QR07)** Av.Uxmal s/n, 77500 Cancún - 1100-0500 – **QR08)** Región 92, Manzana 62, Lotes 21-23, Z-7, 77500 Cancún - 1200-0300 ☎ +52 (988) 9888 7163 – **QR09)** Av.Lázaro Cárdenas 46, 77200 Félipe Carillo Puerto – **QR10)** Av.Náder 25, SM2 Mzna 13, Desp.401, 77500 Cancún ☎ +52(988) 9887 4550 **W:** www.radiopirata. com – **QR11)** Nader 27, Desp.1043, 77500 Cancún ☎ +52 (988) 9887 6660 – **QR12)** Carr Carillo Puerto-Cancún km 1, Col Emiliano Zapata, 77229 Félipe Carillo Puerto - 1200-1700 Prgrs in Sp and Maya – **QR13)** André Quintana Roo No 221, Esq Primo de Verdad, 77000 Chetumal – **QR15)** 77500 Cancún.

SL00) SAN LUIS POTOSÍ (S.L.P.)
SL01) Carr a Verde km 6.5, Col.Juárez, 78000 San Luis Potosí – **SL02)** Capitan Caldera 315, Col.Tequisquiapan, 78250 San Luis Potosí - 1200-0600 – **SL03)** Los Bravo 445 Altos, 78000 San Luis Potosí – **SL04)** Hidalgo 7-A, 79600 Río Verde - 1300-0100 – **SL05)** Avenida México Laredo No 29, Sur, 79050 Cd.Valles – Ap.80, 78700 Matehuala - 1200-0400 – **SL07)** Betancourt No 401, 78700 Matehuala 1155-0405 – **SL08)** General Mariano Arista 245, Centro Hist´rico, 78000 San Luis Potosí ☎ (444) 4826-13-48 **W:** uaslp.mx/rtu – **SL10)** Saturnino Cedillo No 200, Col Benito Juárez, 78437 San Luis Potosí – **SL11)** Carretera México-Laredo Sur s/n, Lomas Poniente, 79099 Cd.Valles – **SL12)** Privada Pemex 3 Barrio San Rafael, 79690 Tamazunchale - 1200-0800 – **SL13)** Josefa Oríz de Domínguez s/n, 79800 Tancanhuitz de Santos - 1200-0700. Prgrs in Sp., Náhuatl, Pame and Huasteco – **SL14)** Fausto Nieto 220, 78000 San Luis Potosí - 1200-0400 – **SL15)** Venustiano Carranza 460, 78000 San Luis Potosí - 1200-0100 – **SL16)** Londres y atenas s/n, Fracc.Lomas, 79090 Cd.Valles - 1200-0500 – **SL17)** Carranza 1408-interior, 78250 San Luis Potosí – **SL19)** Av Dr Salvador Nava Martínez No 278, Col El Paseo, 78320 San Luis Potosí.

SN00) SINALOA (Sin.)
SN01) Av.Lazaro Cardenas 750 sur altos, Local I-1, Plaza Palacio, 80129 Culiacán – **SN02)** Ap.61, 81000 Guasave - 1230-0800 – **SN03)** Av.Juan Carrasco y Pequeira, Local 5-D, Plaza Las Américas, 82010 Mazatlán – **SN04)** Sinaloa 442 Pte, 81000 Los Mochis ☎ 6812 7879 **W:** www.promored.com.mx –**SN05)** Ap.68, 81000 Guasave - 1200-0600 – **SN06)** Av.Álvaro Obregón 24 Sur, L-50, Centro, 80000 Culiacán –**SN07)** Paseo Niños Héroes No 802 Poniente, Centro, 80000 Culiacán - 1300-0600 – **SN08)** Ap.233, 80000 Culiacán – **SN10)** Av Miguel Alemán No 312, Centro (Ap.148), 82006 Mazatlán – **SN11)** Antonio Rosales 223 Norte, 81200 Los Mochis - 1200-0700 –**SN12)** Av.Álvaro Obregón 650-1 Norte, 80000 Culiacán - 1300-0200 –**SN13)** Aquiles Serdán 860 Pte, 81200 Los Mochis - 1300-0700 – **SN14)** Universidad de Occidente, Blvd.Macario Gaxiola y Carr Internacional, 81200 Los Mochis - 1300-0500 – **SN15)** Blvd Antonio Rosales 509 Ote, Col.Morelos, 81460 Guamuchil - 1200-0800 – **SN16)** Universidad de Sinaloa, Rosales 284 Pte, 80000 Culiacán - 1300-0400 – **SN17)** Hidalgo 408, 82400 Escuinapa - 1200-0200 – **SN18)** Hidalgo 755 Pte, 81200 Los Mochis - 1200-0400 **W:** oirmochis.com.mx/ – **SN19)** Av.Miguel Aleman 619 Ote, 82000 Mazatlán - 1300-0700 – **SN20)** Ap.35, 82800 Rosario - 1200-0700 – **SN21)** Av.Del Mar 548, 82000 Mazatlán – **SN23)** Insurgentes 334 Sur, Centro Sinaloa, 80129 Culiacán – **SN24)** Av.del Mar 80, 82010 Mazatlán - 1300-0700 – **SN26)** Insurgentes Sur No 346, 2º Piso, Centro (Ap.113), 80129 Culiacán – **SN27)** Sinaloa 442

Pte, Col.Centro, 81200 Los Mochis.

SO00) SONORA (Son.)
SO01) Yáñez 5, entre Zacatecas y San Luis Potosí, 83000 Hermosillo - 1200-0700 ☎ (6) 215 1522 – **SO02)** Ap.630, 85480 Guaymas – **SO03)** Calle 19 N° 81, entre 15 y 16, Centro, 85400 Guaymas (tx-site Empalme) - 1300-0700 – **SO04)** Durango 901 Sur Altos, 85160 Cd.Obregón - 1300-0200 ☎ 6416 3878 – **SO05)** Ap.1817, 83000 Hermosillo - 1200-0620 – **SO06)** Blvd.Rodolfo Elias Calles 252 Ote, 85000 Cd.Obregón – **SO07)** Sinaloa Sur N° 408, 85000 Cd.Obregón – **SO08)** Obregón Este 184, Col. Centro, 83600 Caborca - 1400-0600 – **SO09)** Av.Juárez y 11ª Este 226, 84620 Cananea - 1200-0700 – **SO10)** Ap.226, 85800 Navojoa - 1200-0700 –**SO11)** Ap.6, 84900 Ures - 1300-0500 – **SO12)** Heriberto Aja 96, 83000 Hermosillo - 1300-0800 – **SO13)** Vasquez 127, P.PA, Local 1, Col. Fundo Legal, 84000 Nogales ☎ +52 (631) 312 0960 – **SO14)** Ap.371, 85400 Guaymas – **SO15)** Calle 5 No 470, Col Comercial (Ap.44), 83400 San Luis Río Colorado - 1300-0100 ☎ 6534 1901 – **SO16)** Av.Morelos y Ramón Corona s/n, Col.Constitución, 85830 Navojoa - 1200-0700 – **SO17)** Ap.256, 84000 Nogales (or: Box 1472, Nogales, AZ 85628, USA) – **SO18)** Guerrero y California, Plaza Tutuli, Local E17, 85000 Cd.Obregón - 1200-0800 – **SO19)** Ap.28, 84200 Agua Prieta - 1400-0300 – **SO21)** Ap.66, 83550 Pto Peñasco - 1200-0100 – **SO22)** Allende 914 Oriente, Esq.con Sinaloa, 85000 Cd Obregón ☎ +52 (644) 413 0607 – **SO23)** Ap.44, 84600 Santa Ana - 1400-0500 – **SO24)** Ap.158, 85000 Cd.Obregón – **SO25)** Ap.63, 84160 Magdalena - 1300-0300 – **SO26)** Morelos y Calle Obregón, 83600 Caborca – **SO27)** Ap.28, 84200 Agua Prieta 1500-0600 – **SO29)** Blvd Navarrete 38, Local 2, Col.Valle Hermoso, 83209 Hermosillo ☎ +52 (662) 215 4900 🖷 +52 (662) 215 4940 –**SO31)** Ap.148, 21960 Cd Morelos, BC - 1200-0700 - (transmitter site: San Luís Colorado, Sonora) – **SO32)** Heriberto Aja 96 y Nayarit, 83000 Hermosillo – **SO33)** Av.13 de Julio 5, 83600 Caborca – **SO35)** Ap.285, 83000 Hermosillo - 1300-0700 – **SO36)** Padre Nacho No 712, Col Fondo Legal, 84030 Nogales - 1200-0200 – **SO37)** Juarez 33 Ote, 85900 Huatabampo - 1300-0700 – **SO39)** Calle Internacional y Av.5 Int 8-C, 84200 Agua Prieta ☎ 6338 2017 – **SO40)** Veracruz 230 Sur Altos, 85000 Cd.Obregón – **SO42)** Edif.Leo, Abelardo Rodríguez 180, Desp.45, Col.Centro, 85400 Guaymas – **SO43)** Blvd Álvaro Obregón No 216-1, Col Reforma, 85830 Navojoa ☎ +52 (6) 422 2999 – **SO44)** Carr a Novojoa km 27, 85280 Etchojoa - 1300-0100. Prgrs in Sp., Mayo, Yaqui and Guarijīo – **SO45)** Calle 5 No 470, Col Comercial, 83400 San Luís Río Colorado – **SO46)** Prolongación Madero y Carretera Mexicali, Col Cuauhtémoc, 83400 San Luís Río Colorado - 1200-0600.

TB00) TABASCO (Tab.)
TB01) Paseo de la Ceiba 102, P1, Col.3 de Mayo, 86190 Villahermosa - 1000-0600 – **TB02)** J.Alvarez 301, 86000 Villahermosa – **TB03)** Paseo Usumacinta y Ayuntamiento, 86100 Villahermosa - 1000-0600 – **TB04)** Calle 28 N° 117, 86900 Tenosique - 1200-0200 – **TB05)** Rosendo Taracena s/n, Local 97, 86030 Villahermosa – **TB06)** Leandro Adriano y Rogelio Ruiz Rojas, 86500 Cárdenas - 1200-0600 – **TB07)** Santa María No 214, 9, Esquina Vicente Guerrero, Local 3, Centro, 86700 Macuspana - 1200-0600 ☎ 9302 2644 – **TB08)** José Julián Dueñas 201, Parque Hidalgo, 86800 Teapa – **TB09)** Av.Méndez 1407, P1, Col. Nueva Villahermosa, 86070 Villahermosa - 1200-0300 – **TB10)** José Martí 101-107, 86000 Villahermosa - 1100-0500 – **TB11)** Prolongación 27 de Febrero 1001, Col.F.Galaxias, 86035 Villahermosa - 0620 ☎ (993) 316 3317 – **TB12)** Sánchez Mármol 408, 86000 Villahermosa - 1100-0500 – **TB13)** José Pagés Ilergo No 116, Col Fracc Nueva Villahermosa, 86070 Villahermosa - 1200-0600 – **TB14)** Av Ruíz Cortines No 1228, Fracc Oropeza, 86030 Villahermosa – **TB15)** Calle de la Ceiba 102-3, Primero de Mayo, 86190 Villahermosa – **TB16)** Constitución 1011, Centro, 86000 Villahermosa – **TB17)** 86900 Tenosique (belongs to Gobierno del Estado de Tabasco) – **TB18)** Antonio Rullán Ferrer No 201, Col Mayito, 86090 Villahermosa – **TB19)** Av Rullán Ferrer No 201, Esquina Arista, Zona Centro, 86000 Villahermosa - 1200-0100.

TM00) TAMAULIPAS (Tamps.)
TM01) Gaspar de la Garza 170 Sur, 87000 Cd.Victoria - 1200-0600 – **TM02)** Ap.134, 88500 Reynosa - 1300-0100 – **TM03)** Ap.4, 88000 Nuevo Laredo 1230-0600 – **TM04)** Séptima No 233, Altos Centro, 88300 Ciudad Miguel Alemán - 1155-0500 – **TM05)** Ap.797, 89160 Tampico - 1200-0400 – **TM06)** Ap.79, 89901 Cd.Mante – **TM07)** Tiburcio Garza Zamora No 1245, Col Beatty, 88630 Reynosa - 1200-0400 – **TM08)** Paseo Colón No 3822, Plaza Cristal, Local 20, Col Jardin, 88260 Nuevo Laredo – **TM09)** Calle 14 y Abasolo Esq No 76, 87300 Matamoros - 1200-0600 – **TM10)** Valentín Gómez Farías, 89150 Tampico - 1200-0600 – **TM11)** Gonzales y Mendoza 747, Col. Centro, 88000 Nuevo Laredo (or: 1510 Calle del Norte, Suite 2, Laredo, TX 78041, USA) **E:** xe2xpk@nld.bravo.net – **TM12)** Diego Acuña, 87900 Cd.Tula - 1200-0600 – **TM13)** Aquiles Seldan 119 Sur, 89000 Tampico – **TM14)** Ap.540, 87300 Matamoros - 1200-0700 – **TM15)** Ap.232, 88000 Nuevo Laredo - 1155-0600 (Sat -0800, Sun -0200) – **TM16)** Carretera Ribereña KM 62, 88440 Cd.Camargo - 1200-0400 – **TM18)** Av.Juárez 703 Ote, 89800 Cd.Mante - 1200-0100 – **TM19)**

Carretera Victoria-Mante Km 2 s/n, Col Las Brisas, 87180 Cd.Victoria – **TM20)** Benito Juárez 506-A, Col.Tolteca, 89160 Tampico **E:** mexicana660tampico@terra.com – **TM21)** Av.Cuauhtémoc y Calle 12, Col. San Francisco, 87350 Matamoros - 1155-0600 ☎ 8812 0202 – **TM22)** Ap.13, 83000 Cd.Miguel Alemán 1155-0400 – **TM23)** Ap.1, 83000 Cd.Miguel Alemán – **TM24)** Morelos 2513, Juarez, 88209 Nuevo Laredo - 1200-0600 – **TM25)** Zaragoza 85, 87600 San Fernando - 1200-0600 – **TM27)** Altamira Calle Principal de Estereos, Carr Tampico-Gonzalez, 89600 Altaimra - 1200-0400 – **TM28)** Calle 8 y Cuauthémoc 125, Col.Pedro Sosa, 87120 Cd Victoria - 1200-0800 – **TM29)** Sexta y Fuerza Aéreo, Edif María Rebeca , 87300 Matamoros – **TM30)** Lázaro Cárdenas 210, Local 19,20 y 21, Col.Centro, 88500 Reynosa – **TM31)** González y Mendoza No 3848, Centro, 88000 Nuevo Laredo - 1200-0200 – **TM32)** Boulevard Adolfo López Mateos No 3205, Local 9 y 10, Santo Niño, 89160 Tampico - 1200-0600 – **TM33)** Altaimra No 311, Poniente, Zona Centro, 89000 Tampico – **TM34)** Carretera Monterrey-Reynosa No 210, Conjunto Inlosa, Planta Alta, Local 16, Col Portal de S Miguel, 88500 Reynosa - 1200-0600.

TX00) TLAXCALA (Tlax.)
TX01) Av.Juárez Norte 203, 90500 Huamantla - 1200-0600 – **TX02)** Calle Uno 420, 90070 Tlaxcala - 1200-0600.

VE00) VERACRUZ (VeR)
VE01) Plaza Crystal, Loc.26, 91150 Jalapa - 1300-0400 – **VE02)** Av.Tres 425, 94500 Córdoba - 1200-0600 **W:** www.rogsa.com.mx – **VE03)** Ap.18, 95400 Cosamaloapan - 1200-0600 – **VE04)** Av.Oriente 6 No 261-210, 94300 Orizaba - 1300-0900 – **VE05)** Playa Aventura s/n, Col Playa Linda, 91890 Mérida – **VE06)** Melchor Ocampo 119, P7, Edif.Pazos, 91700 Veracruz – **VE07)** Av.Hidalgo 1117-A Altos, 96400 Coatzacoalcos – **VE08)** Carr Jalapa-Veracruz 200, 91190 Jalapa - 1200-0600 – **VE09)** Bwnhamin Franklin 4, 91700 Veracruz - 1200-0600 – **VE10)** Juárez 100, 96700 Minatitlán - 1100-0600 – **VE11)** Blvd.Adolfo Ruiz y Heriberto Kehoe, Obrera, 93260 Poza Rica - 1200-0600 – **VE12)** Ignacio de la Llave 38, 92000 Pánuco – **VE13)** Av.Salvador Díaz Mirón 2625, Esq. Heroico, Col.Militar, 91700 Veracruz – **VE14)** Ap.4, 93300 Poza Rica – **VE15)** Esq.Comunicación, Gabino Gonzales, 92730 Alamo - 1200-0400 – **VE16)** Humboldt Sur 36, 91270 Perote - 1200-0300 – **VE17)** Zaragoza 300, Local 14, Galería Margón, 96400 Coatzacoalcos – **VE18)** Av.González Ortega 200, Centro, 93400 Papantla - 1200-0600 – **VE19)** Av.1 N° 211, Centro, 94500 Córdoba - 1200-0200 – **VE20)** Ap.4, 93300 Poza Rica – **VE21)** Bravo 1103 N° 201, 91700 Veracruz – **VE22)** Eulalio Vela 15, 96700 Minatitlán - 1200-0100 – **VE23)** Ferrer No 300, 93650 Tlapacayan - 1200-0600 – **VE24)** Garizurieta No 25, 92800 Tuxpán - 1200-0400 – **VE25)** Libertad y Morelos 301, 95100 Tierra Blanca - 1200-0200 – **VE26)** Francisco González Bocanegra No 10-B, Centro, 95700 San Andrés Tuxtla - 1200-0600 – **VE27)** Plaza Crystal, Loc.20, 91150 Jalapa - 1200-0600 – **VE28)** Ap.15, 91615 Teocelo - 1200-0200 – **VE29)** Av.Manuel Avila Camacho 11, Col.Centro, 93550 Gutiérrez Zamora - 1200-0300 – **VE30)** Lázaro Cárdenas No - 1200, Esq Melchor Ocampa, Local 1 a 3 P.A., 96410 Coatzacoales – **VE31)** Ap.1, 92060 Tempoal - 1200-0400 – **VE32)** Sur 31 N° 336, 94300 Orizaba - 1200-0600 – **VE33)** Ap.26, 96000 Acayucan - 1200-0600 – **VE34)** Av.Libertad 201, 95220 Piedras Negras - 1200-0600 – **VE35)** Carr Nacional 38, 93620 San Rafael - 1300-0100 – **VE36)** Universidad Veracruzana, Ap.629, 91000 Jalapa - 1100-0700 – **VE37)** Zamora 364-Altos, Centro, 91700 Veracruz – **VE38)** Calle 8 N° 119, Entre Av 1 y 3, 94500 Córdoda – **VE39)** 5 de Mayo No 212, 93820 Misantla - 1200-0400 – **VE41)** Fernando Siliceo 801, 91970 Veracruz – **VE42)** Azueta No 8, Col Centro, 95000 Zongolica - 1300-2400 Prgrs in Sp And Maya – **VE43)** Banderas 4, 92800 Tuxpam – **VE46)** Av.Vicente Guerrero Sur 202, 96400 Coatzacoalcos - 1200-0600 – **VE48)** Libertad No 315, Altos, entre Juárez y Madero, Centro, 95100 Tierra Blanca **VE49)** Avenida Jiménez Sur No 4286, Col Pascual Ortíz Rubio, 91750 Veracruz - 1200-0600 – **VE50)** Ignacio de la Llave No 36, Centro 92000 Pánuco – **VE51)** Avenida González Ortega No 200, Centro, 93400 Papantla - 1200-0300.

YU00) YUCATÁN (Yuc.)
YU01) Calle 62 N° 465, Entre 53 y 55, 97000 Mérida - 1130-0100 – **YU02)** Ap.152, 97001 Mérida – **YU03)** Edificio Publicentro, Calle 62, No 508 Altos, (Ap.217), 97001 Mérida - 1200-0600 (XEQM-6105 rel XHMH 95.3 "Candela FM" and also XEMH 970kHz) – **YU04)** Km 1 Carr Valladolid-Carillo Puerto, 97780 Valladolid - 1100-0500 ☎ (985) 9856 2101 – **YU05)** Calle 56 N° 447, 97000 Mérida – **YU06)** Ap.5, 97700 Tizimín - 1200-0600 – **YU07)** Ap.78, 97320 Progreso – **YU08)** Universidad Autonoma de Yucatán, Ap.63-B, 97000 Mérida - 1200-0600 – **YU09)** Calle 42 N° 194-A, Entre 35 y 37, 97000 Valladolid - 1200-0400 – **YU10)** Domicilio Conocido, 97930 Peto - 1100-0100 (Sun - 1300-2200) Prgrs in Sp And Maya – **YU11)** Calle 60 N° 451, Entre 49 y 51, 97000 Mérida – **YU12)** 97780 Valladolid – **YU13)** Calle 33-B, No 513 x 6, Col García Ginerés, 97000 Mérida - 1200-0400 – **YU14)** Calle 7, No 91 x 20 y 22, Col San Antonio Cinta, 97139 Mérida - 1200-0400.

ZC00) ZACATECAS (Zac.)
ZC01) Carr Panamericana km 724.6, 99030 Fresnillo - 1200-0600

E: xeelxeih@logicnet.com.mx – **ZC03)** Av.Hidalgo 316, P1, Centro, 99000 Fresnillo **E:** gpb15@gauss.logicnet.com.mx – **ZC05)** DR Gilberto Delgadillo 18-3, 98400 Río Grande - 1200-0500 –**ZC06)** Radio S.A.Julián Aguirre 110, Col.Lomas de la Soledad, 98040 Zacatecas – **ZC08)** Ramón López Velarde No 43, Colonia Centro, 99300 Jerez de García Salinas - 1200-0400 – **ZC09)** Ap.324, 99000 Fresnillo - 1200-0700 – **ZC10)** Ocampo No 622, Colonia Sagrado Corazón, 99601 Xalpa – **ZC11)** Josefa Ortiz de Dominguez 51, P3, 99700 Tlaltenango - 1200-0600 – **ZC12)** Juan de Tolosa 402, Col.Sierra de Alica, 98000 Zacatecas - 1100-0800

FM in México City (MHz): DF08) 88.1 Red FM – DF14) 88.9 Noticias – DF02) 89.7 Oye 89.7 Siempre Hits – DF19) 90.5 Imagen – DF21) 90.9 Ibero – DF08) 91.3 Alfa – DF08) 92.1 Universal – DF06) 92.9 La Ke Buena – DF08) 93.7 Stereo Joya – DF04) 94.5 Opus 94 – DF14) 95.3 La Nueva Amor – DF22) 95.7 El Politécnico en Radio – DF10) 96.1 UNAM – DF06) 96.9 W Radio – DF08) 97.7 Stereo 97-7 – DF19) Reporte 98.5 – DF14) 99.3 Digital 99 – DF02) 100.1 Stereo Cien – DF02) 100.9 Beat 100.9 – DF06) 101.7 Los 40 Principales – DF17) 102.5 MVS 102.5 – DF11) 103.3 R. Fórmula FM 103 Cadena – DF11) 104.1 R. Uno – DF17) 104.9 Exa-FM – DF04) 105.7 Reactor 105 – DF14) 106.5 Mix 106 – DF08) 107.3 La Z – DF04) 107.9 Horizonte 108

MICRONESIA (USA associated)

L.T: Chuuk, Yap: UTC +10h; Kosrae, Pohnpei: UTC +11h — **Pop:** 108,000 — **Pr.L:** Yapese, Trukese, Ponapean, Kosraean, English — **E.C:** 60Hz, 110/220V — **ITU:** FSM

FEDERATED STATES OF MICRONESIA BROADCASTING SERVICE (Gov.)
✉ Public Information Office, P.O. Box 34, Palikir Station, Pohnpei State FSM 96941 ☎ +691 320 2548 🖷 +691 320 4356

CHUUK STATE
MW Call	kHz	kW	MW Call	kHz	kW
1) V6A	1350	1	2) V6AK	1593	5
FM Call	**MHz**	**kW**	**FM Call**	**MHz**	**kW**
3) V6BC	88.1	0.1	4) V6CWS	89.5	
5)	89.1	0.2	4) V6CWS	98.5	
2) BWXX	89.5				

Addresses and other information
1) Baptist Church, Weno, Chuuk State FSM 96942 **L.P:** Pastor Jody Colson ☎ +691 330 3453 [reported inactive] **-2)** FSMBS Radio Chuuk, PO Box 189, Weno, Chuuk State FSM 96942 ☎ +691 330 2593 🖷 +691 330 2777 **W:** www.fm/chuuk/radio **ID:** 'Ach nenien appio V6AK ion Chuuk' **D.Prgr:** 2000-1400, 24h during adverse weather. BWXX-FM is repeater [reported inactive]. **-3)** Baptist Church, Weno, Chuuk State FSM 96942 **L.P:** Rev.Tom Phillips **Prgr:** conservative religious music and supplied paid prgrs **-4)** National Weather Radio [WSO FM], PO Box A, Weno, Chuuk State, FSM 96942. Live and recorded local weather and emergency information for Chuuk Lagoon area 24h **-5)** New Shine Radio, New Shine Church, Weno, Chuuk State FSM 96942 [r. inactive].

KOSRAE STATE
MW	Call	kHz	kW
1)	V6AJ	1503	1

Addresses and other information
1) FSMBS Radio Kosrae, PO Box 147, Tofol, Kosrae State FSM 96944 ☎ +691 370 3040 🖷 +691 370 3880 **W:** www.fm/kosrae/radio **E:** kosraebroadcast@yahoo.com **ID:** "Painge station V6AJ, fwin an Kosrae" **D.Prgr:** 2000-1400, 24h during adverse weather.

POHNPEI STATE
MW Station	kHz	kW	MW Station	kHz	kW
1) V6AF	999	10	2) V6AH	1449	10
FM Station	**MHz**	**kW**	**FM Station**	**MHz**	**kW**
3) 88.1FM	88.1		7) Magic FM	100.3	
4) V6MAThe Cross	88.5	0.3	8) V6AV	101.0	
5) V6CR	88.9		1) V6AF	104.1	
6) Paradise FM	89.5				

Addresses and other information
1) Independent Baptist Church, PO Box H, Kolonia, Pohnpei State FSM 96941. **L.P:** Dave Arthurs **E:** arthurs@mail.fm **D.Prgr:** 24h - **2)** FSMBS Radio Pohnpei, PO Box 1086, Kolonia, Pohnpei State FSM 96941 ☎ +691 320 2296 **F:** +691 320 5212 **W:** www.fm/pohnpei/radio **E:** v6ah@mail.fm **ID:** "Met Station V6AH nan Pohnpei" **D.Prgr:** 2000-1400, 24h during adverse weather **-3)** Bernard's Enterprises, Kolonia, Pohnpei State FSM 96941 ☎ +691 320 2441 🖷 +691 320 2444 **- 4)** Pacific Missionary Aviation, Radio Station, PO Box 517, Kolonia, Pohnpei State FSM 96941 ☎ +691 320 1122/2496 **W:** www.pmapacific.org **E:** radio@pmapacific.org **SW:** V6MP 4755 kHz 1kW located

at Ninseitamw, Kolonia simulcast of 88.5FM [reported inactive] - **5)** College of Micronesia, Media Studies Program, PO Box 159, Kolonia, Pohnpei State FSM 96941 ☎ +691 320 2480 🖷 +691 320 2479 **E:** national@comfsm.fm - **6)** Paradise Media, Kolonia, Pohnpei State FSM 96941 - **7)** Kolonia, Pohnpei State FSM 96941, joint ownership with KWAW Saipan CNM -**8)** Satellite rel. of BBC WS from London 24h.

YAP STATE

MW Station	kHz	kW
1) V6AI	1494	5

FM Station	MHz	kW	FM Station	MHz	kW
1) KUTE FM	88.1		3) V6AA	89.7	
2) V6JY	88.9	0.25	4) YEC-FM	101.1	0.25

Addresses and other information
1) FSMBS Radio Yap, PO Box 117, Colonia, Yap State FSM 96943 ☎ +691 350 2174 🖷 +691 350 4426 **W:** www.fm/yap/radio **E:** petergar@ mail.fm **ID:** "Pary e radio station V6AI nu Waab" **D.Prgr:** 2000-1400, 24h during adverse weather. KUTE-FM is repeater. - **2)** Joy Family Radio, Colonia, Yap State FSM 96943 ☎ +691 350 8483 **D.Prgr:** 24h religious - **3)** Voice of Hope, Colonia, Yap State FSM 96943 **Prgr:** religious - **4)** Yap Evangelical Church, Colonia, Yap State FSM 96943 ☎ +691 350 6101 **Prgr:** religious.

MOLDOVA

L.T: UTC +2h (27 Mar-30 Oct: +3h) — **Pop:** 4.3 million — **Pr.L:** Moldovan (Romanian), Russian, Gagauz — **E.C:** 50Hz, 220V — **ITU:** MDA

CONSILIUL COORDONATOR AL AUDIOVIZUALULUI (CCA) (Coordinating Audio-Visual Council)
🖃 str. Vlaicu Parcalab 46, 2012 Chisinau ☎ +373 22 277551 🖷 +373 22 277471 **E:** office@cca.md **W:** www.cca.md **L.P:** Pres: Gheorghe Gorincioi **NB.** CCA is the licensing authority for broadcasting.

TELERADIO-MOLDOVA (Pub)
🖃 str. Miorita 1, 2028 Chisinau ☎ +373 22 721388 🖷 +373 22 723537 **E:** rel@trm.md **W:** www.trm.md **L.P:** Dir (Radio): Alexandru Dorogan

MW	kHz	kW	MW	kHz	kW
Chisinau (a)	873	75	Edinet	1494	20
Cahul	1494	30	(a) Codru		

FM	MHz	kW	FM	MHz	kW
Sanatauqa	100.4	0.03	Mascauti	103.9	0.03
Caplani	100.5	0.03	Mîndrestii Noi	104.9	4
Cosauti	100.5	0.03	Crocmaz	105.1	0.03
Straseni	100.5	4	Naslavcia	105.2	0.03
Vasilcau	100.5	0.03	Soroca	105.8	0.25
Cahul	100.7	4	Causeni	106.8	4
Edinet	101.3	2	Ciobalaccia	106.8	0.03
Ungheni	102.0	4	Leova	106.8	0.25
Trifesti	103.3	2	Lipcani	107.9	0.03
Cimislia	103.5	5			

D.Prgr: R. Moldova 24h in Moldovan, exc. N. in Russian: 0700-0712, 0900-0912 (exc. Sat), 1300-1315, 1600-1615.

OTHER STATIONS
FM	MHz	kW	Location	Station
A)	67.46	17	Edinet	BBC relay
2)	67.58	17	Straseni	Antena C
8)	68.99	1	Edinet	R. Sanatate
A)	69.14	20	Cahul	BBC relay
A/C)	69.53	17	Ungheni	BBC/RFE-RL relay
B)	70.31	8	Edinet	RFE-RL relay
10)	71.57	2.5	Chisinau	Vocea Basarabiei
3)	87.6	1	Chisinau	R. Stil
2)	88.0	1	Chisinau	Antena C
14)	89.1	2	Chisinau	Retro FM
17)	89.6	1	Chisinau	City FM
15)	90.4	1.8	Chisinau	Pro FM Chisinau
2)	90.5	2	Balti	Antena C
9)	90.9	2.8	Causeni	Serebryanyy dozhd
10)	91.9	2	Causeni	Vocea Basarabiei
2)	92.3	1.25	Ungheni	Antena C
2)	92.6	2	Cahul	Antena C
16)	96.7	2	Chisinau	R. Alla
A)	97.2	3.2	Chisinau	BBC relay
6)	99.7	1.4	Chisinau	R. Noroc
6)	99.9	3.2	Causeni	R. Noroc
18)	99.9	3.2	Glodeni	R. Prim
8)	100.1	3.2	Chisinau	R. Sanatate
-10)	100.3	1	Glodeni	Vocea Basarabiei
5)	100.7	1.3	Mîndrestii Noi	Micul Samaritean
12)	100.9	2	Chisinau	Kiss FM
6)	100.9	1.6	Iagara	R. Noroc
1)	101.1	1.25	Proteagailovca	Hit FM
19)	101.4	1	Cahul	Univers FM

FM	MHz	kW	Location	Station
A)	101.5	15	Causeni	BBC relay
1)	101.7	1	Chisinau	Hit FM
10)	101.9	5	Taraclia	Vocea Basarabiei
5)	102.0	1	Causeni	Micul Samaritean
10)	102.3	10	Straseni	Vocea Basarabiei
5)	102.5	1	Glodeni	Micul Samaritean
13)	102.7	1	Chisinau	R. 21
A)	102.9	15	Mîndrestii Noi	BBC relay
4)	103.0	3.2	Causeni	Russkoye R.
3)	103.5	1.6	Balti	R. Stil
4)	103.7	1.6	Chisinau	Russkoye R.
5)	103.8	10	Edinet	Micul Samaritean
5)	104.2	1	Chisinau	Micul Samaritean
6)	104.3	5	Floresti	R. Noroc
1)	104.5	20	Ungheni	Hit FM
1)	105.2	20	Cahul	Hit FM
5)	105.4	10	Trifesti	Micul Samaritean
11)	105.6	1.6	Balti	Megapolis FM
10)	105.7	5	Nisporeni	Vocea Basarabiei
7)	105.9	5	Chisinau	Fresh FM
5)	107.0	20	Ungheni	Micul Samaritean
B)	107.3	10	Straseni	RFI relay
1)	107.6	20	Mîndrestii Noi	Hit FM
5)	107.7	20	Cahul	Micul Samaritean
9)	107.9	1	Chisinau	Serebryanyy dozhd
8)	107.9	10	Edinet	R. Sanatate

NB: Txs below 1kW not listed.
Addresses & other information:
1) str. Bucuresti 68, 2012 Chisinau. Rel. Hit FM (Russia). – **2)** str. Veronica Micle 10, 2012 Chisinau. – **3)** str. Sciusev 93, 2012 Chisinau. Rel. Vzrosloye Radio Shanson (Ukraine). – **4)** sos. Hîncesti 59/1, 2028 Chisinau. – **5)** str. Bucuresti 68, 2012 Chisinau. – **6)** bd. Negruzzi 6, 2001 Chisinau. – **7)** str. Bucuresti 68, 2012 Chisinau.– **8)** str. Mihai Vitezul 1, 2004 Chisinau. – **9)** sos. Hîncesti 59/1, 2028 Chisinau. Rel. Serebryanyy dozhd (Russia). – **10)** str. A.Puskin 20a, 2012 Chisinau. – **11)** str. Alba Iulia 75, 2028 Chisinau. – **12)** str. Ismail 33, 2011 Chisinau. Rel. Kiss FM (Romania). – **13)** str. Alecu Russo 1, 2068 Chisinau. Rel. R. 21 (Romania). – **14)** str. Frumusica 1, 2002 Chisinau. Rel. Retro FM (Russia). – **15)** str. Maior Petru 7, 2001 Chisinau. Rel. Pro FM (Romania). – **16)** str. Bucuresti 68, 2012 Chisinau. Rel. R. Alla (Russia). – **17)** str. Chisinaului 72, 3100 Balti. – **18)** str. Suveranitati 5, 2901 Glodeni. – **19)** str. I.Spirin 106, 3909 Cahul. – **A)** Rel. BBC (UK) – **B)** Rel. RFI (France) – **C)** Rel. RFE-RL (USA).

GAGAUZIA

GAGAUZIYA RADIO TELEVISIONU (GRT) (Pub)
🖃 str. Lenin 164, 3805 Comrat ☎ +373 298 23086 🖷 +373 298 26934 **E:** gagauztv@gagauztv.com **W:** www.gagauztv.com
L.P: Pres: Semion Lazarev

FM	MHz	kW	FM	MHz	kW
Comrat	102.1	5	Baurci	104.6	1.25
Vulcanesti	103.6	0.2			

D.Prgr: GRT FM in Gagauz, Russian: 0500-2200, incl. relay TRT (Turkey).

OTHER STATIONS
FM	MHz	kW	Location	Station
6)	99.5	2.5	Comrat	R. Noroc
20)	100.3	1	Comrat	PRO 100 R.
5)	103.2	1	Ciadîr-Lunga	Micul Samaritean
1)	106.6	5	Comrat	Hit FM
2)	107.5	1	Ciadîr-Lunga	Antena C

NB: Txs below 1kW not listed. **Addresses & other information:** see main table. **20)** str. Novaia 23, 3801 Comrat.

TRANSNISTRIA

RADIO PMR (Gov)
🖃 ul. Pravdy 31, 3300 Tiraspol ☎ +373 533 60701 🖷 +373 533 77758 **E:** radiopmr@inbox.ru **W:** www.radiopmr.org **L.P:** Dir: Anatoliy A. Kirsa

MW*	kHz	kW			
Maiac	549	150	*) irr.		

FM	MHz	kW	FM	MHz	kW
Slobozia	74.00	0.1	Camenca	104.0	-
Dnestrovsc	100.3	-	Maiac	105.0	1.5
Slobozia	100.7	1.5	Slobozia	105.8	0.2
Pervomaisc	103.4	-	Voroncovo	106.0	1
Tiraspol	104.0	0.2	Maiac	106.5	0.2

D.Prgr: R. Pridnestrovya in Russian, Ukrainian, Moldovan: 24h on FM. Own prgrs and rel. Golos Rossii (Russia). From time to time also on MW, with limited schedule.
External Service (R. PMR): see International Radio section.

OTHER STATIONS
FM	MHz	kW	Location	Station
4)	88.3	-	Tiraspol	Tiraspol FM

FM	MHz	kW	Location	Station
3)	88.8	3	Tiraspol[1]	R. Shanson
6)	89.3	-	Varnita	R. Novaya volna
2)	89.6	-	Slobozia	Hit FM
7)	90.1	-	Tiraspol	Retro FM
14)	90.5	-	Slobozia	Relax FM
4)	91.2	-	Dubasari	Tiraspol FM
9)	91.5	-	Slobozia	Yumor FM
15)	92.5	-	Tiraspol	R. Tochka
13)	98.7	1	Ribnita	R. Zhelannoye
10)	100.3	0.03	Dubasari	Dubosarsskoye R.
A)	101.1	0.1	Slobozia	R. Rossii relay
A)	104.6	-	Ribnita	R. Rossii relay
11)	104.6	0.1	Slobozia	Ekho Moskvy
5)	105.4	2	Tiraspol	Dorozhnoye R.
12)	106.3	-	Slobozia[1]	R. Rekord
9)	106.7	-	Ribnita	Yumor FM
16)	107.1	-	Slobozia	Avtoradio
1)	107.7	1	Tiraspol[1]	R. Inter FM

[1]) Synchro-network with txs in several towns.

Addresses & other information:
1) ul. K.Libnikhta 1/2, 3300 Tiraspol. **E:** reklama@inter-fm.idknet.com. – **2)** Rel. Hit FM (Russia). – **3)** ul. K.Libnikhta 1/2, 3300 Tiraspol. Rel. R. Shanson (Russia). – **4)** Tiraspol. – **5)** Tiraspol. Rel. Dorozhnoye R. (Russia). – **6)** ul. Internatsionalistov 13, Bender. E: nv893@mail.ru Rel. Love R. (Ruissia). – **7)** Rel. Retro FM (Russia). – **8)** Rel. R. 7 (Russia). – **9)** Re. Yumor FM (Russia). – **10)** ul. Dzerzhinskogo 4, 4501 Dubasari. – **11)** Rel. Ekho Moskvy (Russia). – **12)** Rel. R. Rekord (Russia). – **13)** ul. Kirova 130a, 5500 Ribnita. **E:** radio987@lan-rybnitsa.com – **14)** Rel. Relax FM (Russia). – **15)** Tiraspol. – **16)** Rel. Avtoradio (Russia). – **A)** Rel. R. Rossii (Russia).

Intermational rel. on MW: (operated by Pridnestrovskiy radio-teletsentr) Grigoriopol (Maiac) 621kHz 150kW, 999/1413/1548kHz 1000kW. See Int Radio section.

MONACO

L.T: UTC +1h (27 Mar-30 Oct : +2h) — **Pop:** 33,000 — **Pr.L:** French — **E.C:** 50Hz, 220V — **ITU:** MCO

MONTE CARLO RADIODIFFUSION (Comm.)
☎ 10 Quai Antoine 1er, MC-98000 Monaco ☎ +377 97974799 📠 +377 97974707 **W:** www.mcr.mc **E:** mcradiodiffusion@mcr.mc **L.P:** Patrick Jean.

MW	kHz	kW	Prgr.
Col de la Madone (France)	702	200	R.Chine Int. (0800-2300)*
Roumoules (France)	1467	1000	TWR relay(1915-2245) **
Col de la Madone (France)	1467	40	R. Marie France (0500-1900)***

*in French & Italian, **mainly Arab & Eng. Prgs, ***Radio Marie France , BP 42, F-06341 La Trinite , France **W:** www.radiomaria.fr
FM: Monaco 94.5MHz 1kW Jardin exotique **V.** by Letter Rp.

RMC INFOS (Comm.)
📠 HQ: 12 Rue d'Oradour sur Glane, 75740 Paris Cedex 15, France ☎ +33 1 71191191 📠 +331 71191190 **W:** www.rmc.fr **E:** technique@rmc.fr **L.P:** Pres: Alain Weill, GD: Franck Lanoux
LW: Roumoules (France) 216kHz 1400kW (reduced to 900 kW.), RMC INFOS 0330-2305 (Id as "RMC" only)
FM: Mont Agel 98.5MHz 50kW; Monaco Jardin Exotique 98.8MHz 1kW **V.** by QSL-card. Rp.

RADIO MONACO (Comm.)
📠 HQ: 17 Rue du Gabian, Gildo Pastor Centre, MC-98000 Monaco ☎ +377 97700700 📠 +377 97700701 **W:** www.radio-monaco.com
FM: Monaco Jardin Exotique 98.2MHz 0.5kW; Mont Agel 95.4MHz 40kW, Grasse (France) 103.2MHz 0.5kW **E:** info@radio-monaco.com
SW: see International Radio section for details
NB: News relays via Monaco Radio utility stn for seamen.weather forecast in French & English 0530, 1103. 1630 on 8728 kHz SSB News bulletin from Radio Monaco relay over Monaco Radio inFrench 1100-1103 (M-F) on 4363, 8728, 13246. 17260 kHz SSB

RADIO MONTE CARLO ITALIE(Comm.)
📠 8 Quai Antoine 1er, MC-98000 Monaco ☎ +377 97976666 📠 +377 97708661 **W:** www.radiomontecarlo.net **E:** rmc@radiomonter-carlo.net
FM: RMC1 RMC1 Jardin Exotique 91,3MHz 0.1kW;Monaco 106.8MHz 1kW; Mont Agel 107.4MHz 50kW, RMC2 : RMC2 Musée Ocean. 92.7MHz 0.05kW, Mont Agel 101.6MHz 10kW- **V.**by QSL-card. Rp.

RIVIERA RADIO (Comm)
📠 10 Quai Antoine 1er, MC-98000 Monaco ☎ +377 97979494

📠 +377 97979495 **W:** www.rivieraradio.mc **E:** info@rivieraradio.mc
L.P.: MD: Paul Kavanagh. Tech. Manager: Peter Miller
FM: Monaco 106.3MHz 10kW; Mont Agel 106.5MHz 50kW
D.Prgr. in English: 24h. Rel. BBCWS **N** every hour

Other FM stations (all sces. 24h):

FM	MHz	kW	Station, location
-)	88.2	0.1	R. 105, Monaco Jardin Exotique
-)	90.3	50	MFM, Mont Agel
-)	93.5	50	R. Nostalgie, Mont Agel
-)	93.8	1	R. Nostalgie, Monaco Jardin Exotique
-)	96.0	1	R.Chine Int., Monaco Jardin Exotique
1)	96.4	0.1	FG Radio, Monaco Jardin Exotique
-)	99.0	1	R. Capri, Monaco Jardin Exotique
-)	101.1	1	Virgin Radio Italia, Monaco Jardin Exotique
-)	102.1	1	RTL 2, Monaco Jardin Exotique
2)	102.4	1	One Sud, Monaco Musée Ocean.
-)	102.7	50	R .Classique, Mont Agel
-)	103.0	10	RFM, Mont Agel
-)	103.3	1	RDS Italia,Monaco Jardin Exotique
-)	104.5	1	Cherie FM, Monaco Jardin Exotique

Addresses:
– **1)** 51 rue de Rivoli,F-75001, Paris **W:** wwww.radiofg.com – **2)** ☎ +33 619212776 **W:** www.onesud.com

MONGOLIA

L.T: UTC +8h — **Pop:** 3.1 million — **Pr.L:** Mongolian — **E.C:** 50Hz, 220V — **ITU:** MNG

HUUL EÜY DOTOOD HERGIYN YAM
(Ministry of Justice and Home Affairs)
📧 Hudaldaanï gudamj 6/1, Ulaanbaatar 210646 ☎ +976 11 267014 📠 +976 11 325225 **E:** admin@mojha.gov.mn **W:** www.mojha.gov.mn
L.P: Minister: Tsend Nyamdorj
NB: The Ministry of Justice & Home Affairs issues broadc. licenses.

MONGOLÏN ÜNDESNIY OLON NIYTIYN RADIO TELEVIZ (MÜONRT)
(Mongolian National Public Radio & Television)
📧 Huvsgalïn zam 3, Ulaanbaatar 31 ☎ +976 11 328334 📠 +976 11 328334 **E:** mr@mongol.net **W:** www.mnb.mn
L.P: Pres/CEO: Naranbaatar Myanganbuu

LW	kHz	kW	Prgr	LW/MW		kHz	kW	Prgr
Ulaanbaatar (a)	164	500	1	Altay		227	75	1
Choybalsan	209	75	1	Mörön		882	75	1
Dalanzadgad	209	75	1	Ulaanbaatar (a)		990	500	F
Ölgiy	209	30	1	Choybalsan		1350	500	1*
(a) Hönhör; F=External Service; *) local evening hours								

SW	kHz	kW	Prgr	SW		kHz	kW	Prgr
Altay	4830	10	2	Ulaanbaatar (a)		7260	°50	2
Mörön	4895	10	2	(a) Hönhör; °) 0655-1500: 250kW				

FM (MHz)	MR1	MR3	kW
Ulaanbaatar	106.0	100.9	1

+ low power txs in provincial capitals
D.Prgr: MR1 (Mongolïn R.): 2200-1500. May incl. reg. trs. – **MR2 (Höh tenger):** 2300-0500, 0655-1500. – **MR3 (R3):** 2300-1600.
External Service (Voice of Mongolia): See Int Radio section.

OTHER STATIONS

MW	MHz	kW	Location	Station
C)	† 1350	500	Choybalsan	R.Free Asia relay

FM	MHz	kW	Location	Station
11)	93.0	0.05	Erdenet	Avtoradio
24)	96.3	-	Ulaanbaatar	Avto FM
27)	98.1	-	Ulaanbaatar	Formula
25)	98.5	-	Ulaanbaatar	Best R.
26)	98.9	-	Ulaanbaatar	Hifi R.
13)	99.3	1	Ulaanbaatar	Ineemselgel R.
28)	99.7	-	Ulaanbaatar	Ih Mongol
8)	100.1	-	Ulaanbaatar	R. Elgen Nutag
12)	100.4	0.05	Erdenet	Nomin FM
9)	100.5	1	Ulaanbaatar	R. Miniy Mongol
30)	101.5	-	Saynshand	Miniy nutag
10)	101.7	-	Ulaanbaatar	R. Höh Mongol
14)	102.1	-	Ulaanbaatar	Eh Oron R.
2)	102.5	-	Ulaanbaatar	Ulaanbaatar R.
15)	102.5	-	Baruun-Urt	Talïn Tsurai
A)	103.1	-	Ulaanbaatar	BBC Relay
29)	103.5	-	Baruun-Urt	Kiss You
16)	103.6	-	Ulaanbaatar	TV FM
17)	103.6	-	Dalanzadgad	Govïn Dolgion
18)	103.7	-	Darhan	Darhan FM

FM	MHz	kW	Location	Station
3)	104.0	-	Ulaanbaatar	Life FM
19)	104.0	-	Sühbaatar	Ögrön Selenge
7)	104.5	-	Ulaanbaatar	Ger Büülin R.
31)	104.5	-	Saynshand	Songodog FM
4)	105.0	-	Ulaanbaatar	Tany Degred
1)	105.5	-	Ulaanbaatar	Hotïn Hõgjim
20)	106.0	-	Darhan	Orhon R.
21)	106.2	-	Önderhaan	Dölgöön Herlen
22)	106.4	-	Choybalsan	Züün Büsïn Olon Hiytiyn
23)	106.5	-	Mörön	Möröngïn Dolgion
B)	106.6	1	Ulaanbaatar	VOA relay
6)	107.0	-	Ulaanbaatar	Shine Zuuny R.
5)	107.5	-	Ulaanbaatar	Shine Dolgion R.
32)	107.5	-	Baruun-Urt	Ertöntsiyn ayalguu
33)	107.5	-	Saynshand	Saynshand FM

† = tx also carries prgrs of Mongolian Public R.

Other information:
11) Rel. Avtoradio (Russia) – **A)** Rel. BBC (UK) – **B)** Rel. VOA (USA) – **C)** Rel. R. Free Asia (USA)

MONTENEGRO

L.T: UTC+1h (27 Mar-30 Oct: +2h) — **Pop:** 672,180 — **Pr.L:** Serbian — **E.C:** 50Hz, 220V — **ITU:** MNE

RADIO TELEVIZIJA CRNE GORE
Cetinjski put bb, 81000 Podgorica ☎ +382 20 245595 **W:** www.rtcg.me **E:** marketing@rtcg.org **LP:** DG: Z. Jokovic

MW	kHz	kW	Station, h. of tr.
Podgorica	882	5	R. Podgorica 1, 24h

FM (MHz)	I	R. 98I	kW	FM (MHz)	I	R. 98I	kW
Bjelasica	92.1	99.3	54	Podgorica	95.5	-	1
Durmitor	91.3	-	1	Sudj. Glava	88.0	98.9	10
Lovcen	94.9	98.0	54	Velji Grad	89.6	99.7	10
Mozura	-	93.4	1				

+ 11 txs.less than 1kW

Local/private stations
R. Antena M, Podgorica 87.6MHz + 5 relays – **R. Cetinje** 94.5MHz + 1 relay – **R. Elmag,** Podgorica 96.0MHz + 7 relays. – **R. 98,** Podgorica 98.0MHz + 2 relays. – **R. Bar** 92.9MHz + 1 relay – **R. Berane** 88.2MHz + 1 relay – **R Bijelo,** Polje 105.8MHz + 1 relay – **R. Budva** 98.7MHz + 1 relay – **R. Corona,** Bar 88.9MHz +1 relay – **R. D,** Podgorica 88.6MHz + 2 relays. – **R. Danilovgrad** 92.9MHz – **R. Fokus,** Bijelo Polje 93.9MHz – **R. Free Montenegro,** Podgorica 103.0.MHz – **R. Glas** Plava, Plav 102.9.MHz – **R. Gorica,** Podgorica 93.3MHz – **R. Herceg** Novi 90.0MHz +1 relay – **R. Jupok,** Rozaje 98.7MHz + 1 relay – **R. Kotor** 95.3MHz + 1 relay – **R. Max,** Danilovgrad 107.5MHz + 1 relay – **R. Mir,** Tuzi 106.1MHz + 1 relay – **R. Mojkovac** 92.8MHz – **R. Montena,** Podgorica 105.7MHz + 5 relays – **R. Niksic** 88.0MHz + 2 relays – **R. Ozon,** Kolasin 97.6MHz – **R. Panorama,** Pljevlja 89.2MHz – **R. Pljevlja** 94.8MHz + 1 relay – **R. Rozaje** 104.4MHz – **R. Svetigora,** Cetinje 101.0MHz – **R. Tivat** 88.5.MHz – **R. Ulcinj** 91.3MHz + 1 relay – **R. Zeta,** Podgorica 93.8MHz

MONTSERRAT (UK)

L.T: UTC -4h — **Pop:** 6,000 — **Pr.L:** English — **E.C:** 60Hz, 220V — **ITU:** MSR

RADIO MONTSERRAT (Gov. Comm.)
P.O. Box 51, Sweeneys ☎ +1 664 491 2885/6349/7242 🖷 +1 664 491 9250 **W:** www.zjb.gov.ms **E:** zjb@gov.ms **LP:** SM: Herman Sargeant. Technician: Ivor Grenaway
FM: 88.3 (Isles Bay Hill: 0.1kW), 95.5MHz (Silver Hills 5kW)
D.Prgr: 24h BBC relay at night 0400-0930
Ann: "ZJB Radio Montserrat, the Voice of Montserrat"

Other stations:
ETERNAL LIFE RADIO Cavalla Hill ☎ +1 664 496 6982.FM: 106.1MHz – **GEM RADIO 93.9** P.O Box W939, St.John's, Antigua ☎ +1 268 720 6017 - **FM:** 93.9/94.5MHz (Relay Trinidad) – **VIBZ FM - Family Radio Network** P.O Box 350, Baker Hill ☎ +1 664 491 7331 **W:** www.vibzfm.com - **FM:** 89.9/90.9MHz (Relay Antigua)

MOROCCO

L.T: UTC DST in 2011 (May-Aug: +1h) subject to confirmation) — **Pop:** 35 million — **Pr.L:** Arabic, French, Spanish, English, Berber languages, Hassania — **E.C:** 50Hz, 127/220V — **ITU:** MRC

HAUTE AUTORITÉ DE LA COMMUNICATION AUDIOVISUELLE (HACA)
Espace les Palmiers, Lot 26,Angle Avenues Anakhil et Mehdi Ben Barka, B.P. 20590, Rabat Ryad ☎ +212 53 7579600 🖷 +212 53 7714274 **W:** www.haca.ma **E:** info@haca.ma **LP:** DG: Ahmed Akhchichine.

SOCIÉTÉ NATIONALE DE RADIODIFFUSION ET DE TÉLÉVISION (SNRT) - RADIO MAROCAINE (Pub.)
1, Rue El Brihi, B.P. 1042, MA-10000 Rabat ☎ +212 53 7700 319 🖷 +212 53 772 2047 **W:** www.snrt.ma **Reg.** B.P. 459, Laayoune. **LP:** DG: Mohamed Ayad. Dir. Tech: Allal Kacimi.

LW/MW	kHz	kW	N	LW/MW	kHz	kW	N
Azilal	207	400	A	Laâyoune	711	300	R
Tahadart	540	300	A/R	Agadir	936	100	C/R
Oujda	*595	50	A/R	Sebaa-Aioun...	1044	300	C
Sebaa-Aioun	612	300	A/R	*alt. on 595 kHz.			

FM (MHz)	A	B	C	Q&R	kW
Agadir	91.0	94.2	97.5	87.9	
Beni Mellal	89.8	92.9	96.1		10
Casablanca	96.0	90.0	95.3	98.6	
Dakhla	93.5	91.8			
El Houceima	105.7	92.1	95.3		8
El Jadida	90.4				
Errachidia	91.3	97.8	94.5		
Essaouira	97.9	91.4			
Fès	88.8	95.1	101.9	98.4	
Figuig	91.9	95.1	98.4		
Ifrane	90.5	93.6	96.8		
Khenifra	91.6	87.9	104.6	94.2	10
Lâayoune	93.9	97.9	91.1	94.2	10
Marrakech	94.9	98.8		91.7	30
Meknès	88.8	95.1	101.9	98.4/92.5	10
Nador	87.6	93.9	97.2		
Ouarzazate	90.3	93.4	96.6		
Oujda	89.9	99.4		96.1	10
Oum Dreiga	97.9				
Rabat	91.0	87.9	104.6	94.2	40
Safi	90.9		94.1		
Settat	92.1	89.0			
Tanger	88.7	91.8	95.0	98.3/104.0	
Tantan	90.3				
Taza	91.7	96.9			10
Tétouan	90.6	100.2		93.7	12

A: National Network in Arabic: 0500-0100. **N:** on the h.
Netw. B, Chaine Inter : 0600-0100. **English:** W 1000-1200, Sun 1400-1500v. **Spanish:** 0900-1000. Other times in French.
Netw. C in Berber/Arabic dialects: 1200-2400 (incl. relays of Netw. A 0600-1200).
Netw. Q: Quran R. "Mohammed VI": 24h, MW 1800-0600.
Regional Prgrs (FM on Q network): **Agadir,** Avenue Hassan II, Agadir: Mo 0900-1200 in Arabic, 2000-2400 in Tachelhit dialect. Tues.-Sat. 0900-1300 2 h Arabic + 2 h Tachelhit dialect on 936kHz. **Casablanca,** Ain Chock, Casablanca: 1500-1600 & 1620-1800. **Laâyoune/Dakhla:** 0800-0200. Spanish Sat 2230-2400 on 711kHz & 91.8/91.1MHz. **Marrakech,** 40, Avenue Yugoslavie, Marrakech: 1500-1600 & 1620-1800 on 1593kHz. **Meknès-Fès:** 1400-1900 on 612kHz. **Oujda,** Avenue Omar Errifi, Oujda: 1200-1300 & 1330-1500 on 594kHz. **Tangier:** 33, Avenue Amir Moulay Abdallah, Tanger: 1500-1600 & 1620-1800 on 540kHz. **Tetouan,** 30, Avenue Mohammed V, Tetouan: 0800-1200 in Arabic and Rifain dialect on 1053kHz. **Ann:** Arabic: "Huna Ribat, Idha'atu-I-Mamlaka al Maghribiyya" ot "Idha'at al-Wataniya". French: "Ici Rabat, Radiodiffusion Télévision Marocaine". Berber: "Dahab Rbad Lidaa Attalfaza Li Mamlaka L'Maghrib".

EXTERNAL SERVICE: R. Marocaine, see International Radio section.

RADIO MEDITERRANÉE INTERNATIONALE - MEDI 1 RADIO (Comm, Semi-Gov.)
B.P. 2055, 3 rue Emsallah, 9000 Tanger ☎ +212 89 936363 🖷 +212 89 935755 **W:** www.medi1.com **E:** medi1@medi1.com
LW: Nador 171kHz 1600kW & SW: see International Radio section.
FM (MHz):Agadir 104.6 20kW, Al Hoceima/El Jadida 96.7, Beni Mellal 102.9, Casablanca 99.6 9kW, Dakhla 96.4, Enjil 97.0, Essaouira 94.6 12kW, Fès 101.4 2kW, Laâyoune 101.0 10kW, Marrakech 105.3 1kW, Meknès 105.5 10kW, Merchiche 87.6, Nador 105.3 10kW, Ouarzazate/Zaio 99.9, Oujda 102.9 12kW, Rabat 97.5 20kW, Safi 97.0, Slokia 95.3, Taliouine 92.2, Tanger 101.0 1kW, Tantan 93.4 12kW, Taroudante 95.4, Tetouan 103.7MHz 12kW, Zagora 97.0.

D.Prgr. in Arabic/French: 24h. **N. in Arabic:** 0600, 0700, 0800, 1200, 2000, 2300. **N. in French:** 0630, 0730, 0830, 1230, 1700, 1930, 2200.**Ann:** "Médi 1".

Other stations:

FM	1)	2)	3)	4)	5)	6)	7)	8)	9)	10)
Agadir	100.4	93.1	96.5		95.6		102.4	103.7		
Al Hoceima	97.7	93.3		102.1						
Béni Mellal	94.0					98.1		94.7		105.1
Casablanca	104.3	93.1	88.7		100.3	92.5	99.2	100.8	91.2	88.2
Dakhla	89.7				99.7		88.0			
El Jadida	95.1	93.1			94.5	97.3	98.0	96.2	91.5	
Errachadia	102.5				104.1		105.6			
Essaouira	92.8	93.3	99.9		96.1	102.0				
Fès	103.9				94.1	98.8			89.4	103.2
Figuig		93.1	105.5							
Gharb	99.3									
Guelmin					98.5					
Goulmima		93.1								
Ifrane	103.6									
Khenifra	102.4									
Laâyoune	104.6		107.1		91.6		98.6			
Larache					92.8					
Marrakech	100.6	93.8	97.7		94.4	90.5	101.2	98.5		
Meknès	99.9	93.7	102.5			97.2			90.7	92.9
Nador	104.3		101.0	90.7						101.4
Oujda	102.0			92.9	98.5		106.5			
Ouarzazate	91.2				92.0		103.4			
Rabat	95.7	93.5			99.8	106.9	107.4	103.7	96.5	97.0
Safi	103.6									
Settat	103.8				98.9	106.4		97.9	94.7	96.4
Skhour	102.2		102.6		95.8					
Taza	95.8				98.6					
Tafraoute			99.2							
Tamanar			98.4							
Tantan		93.1	99.9				101.3			
Tanger	102.3	93.3		105.4	96.4	103.3				91.1
Targuist				95.8						
Taroudante	101.3						88.1			
Tétouan	105.9	93.9		104.5	97.8					
Tiznit			104.2		91.5					
Zagora								105.9		

1) Aswat FM: Ghandi Mall, Imm 9, Bd. Ghandi, Casablanca. **W**: aswat.ma – **2) Radio 2M** (Semi-Gov.): Km 7300 route de Rabat Ain Seeba, Casablanca. **W**: www.2m.tv/radio2M – **3) MFM Atlas/Oriental/Sahara/Saïss/Souss & Casa FM**: Groupe New Publicity, 58 Av. des FAR, Tour des Habous, 18ème étage, Casablanca. **W**: radiocasafm.ma – **4) Cap Radio**: Zone industrielle, Route de Tétouan, Allée principale lot n°123, Tanger. **W**: capradio.ma – **5) Hit Radio**: 3 rue Assouhaili, Agdal, Rabat. **W**: hitradio.ma – **6) R. Atlantic**: Eco-Médias, 70 Bd. Massira Khadra, Casablanca. **W**: atlanticradio.ma – **7) Luxe Radio**: 48 Blvd. Rahal Elmeskini, Casablanca. **W**: luxeradio.ma – **8) R. Chada FM**: Société R. Kolinass, 42 Bd. Idriss 1er quartier des Hôpitaux, Casablanca. **W**: chadafm.net – **9) R. Mars**: 30 Ave. des Far 13ème étage, Casablanca 20000. **W**: radiomars.ma – **10) Medina FM:** Rue Oued ziz imm 51 appt 4 agdal, Rabat.
R. Sawa: Rabat & Agadir 101.0MHz 20kW, Casablanca 101.5MHz 10kW, Marrakech 101.7MHz 12kW, Meknès 91.9MHz, Fès 97.9MHz 2kW, Tetouan 92.1/101.8MHz 20kW.

CEUTA (Spain)
L.T: see Spain — **Pop**: 80,000 — **Pr.L**: Spanish

R. Nacional de España, Real 90, E-51001 Ceuta. **FM**: RNE-1 97.2MHz, R. Clásica 100.8MHz, RNE5TN 101.9MHz, RNE-3 106.8MHz, all 1kW.
SER Radiolé - R. Ceuta, Poblado Marinero, Local 32, E-51001 Ceuta. **MW**: 1584kHz 5kW 24h rel. of Radiolé netw. **FM**: 96.2MHz R. Ceuta
COPE, Sargento Mena 8,1°izq, E-11701Ceuta. **FM**: 89.8MHz COPE Ceuta/Cadena 100.
Onda Cero R, Grupos Alfa 4,3°nz, E-11701Ceuta. **FM**: 101.4MHz Onda Cero R. 3kW.
RTV Ceuta: 99.0MHz. Web: rtvce.es

MELILLA (Spain)
L.T: see Spain — **Pop**: 70,000 — **Pr.L**: Spanish

R. Nacional de España, Altos de la Vía 3, E-52004 Melilla. **MW**: RNE1 972kHz 5kW. **FM**: (0.3kW): 97.7MHz (R1), 100.1MHz (RNE5TN) ,105.3MHz (R3), 107.6MHz (R. Clásica).
SER Radiolé - R. Melilla, Muelle Ribera s/n, E-52005 Melilla. Email: radiomelilla@unionradio.es .**MW:** 1485kHz 1kW 24h. **FM:** (MHz): 96.3 Cadena 40 Melilla, 101.1 Dial Melilla.
COPE, C/ Pablo Vallescá 6 "Edificio Ánfora"2° - 1, E-52001 Melilla. **FM:** 91.3MHz Cadena 100, 98.4MHz COPE Melilla.
Onda Cero R, General Mola 26 E-29804 Melilla. **FM:** (MHz): 89.6 Onda Cero R, 98.4 Europa FM.
esRadio, Melilla: 92.2MHz.

MOZAMBIQUE

L.T: UTC +2h — **Pop**: 22 million — **Pr.L**: Portuguese, 20 ethnic languages — **E.C**: 50Hz, 220V — **ITU**: MOZ

INSTITUTO NACIONAL DAS COMUNICAÇÕES (INCM)
Av. Eduardo Mondlane, 123/127, PO Box 848, Maputo ☎+258 21 490131 ▤ +258 21 494435. **Web**: www.incm.gov.mz **E-mail**: info@incm.gov.mz

RÁDIO MOÇAMBIQUE (Pub.)
Rua da Rádio n.° 2, C.P. 2000, Maputo ☎+258 21 431687 ▤ +258 21 321816 **W**: www.rm.co.mz **E**: caprimoe@zebra.uem.mz
L.P: Chmn/CEO: Ricardo Malate. TD: Mr Hermenegildo Basílio Mula. Int. Rel. Dir: Ms. Maria Cremilda Massingue. Fin. Dir: Arlindo Piedade de Sousa.

MW	kHz	kW	N	MW	kHz	kW	N
1) Maputo	738	50	N	4) Chimoio	1026	50	N
3) Nampula	765	50	EP	7) Quelimane	1179	50	EP
10) Xai-Xai	810	50	EP	5) Inhambane	1206	50	EP
2) Beira	873	50	EP	8) Pemba	1224	50	EP
9) Tete	963	50	EP	6) Lichinga	1260	50	EP
1) Maputo	1008	50	EP				

FM	MHz	kW	N	FM	MHz	kW	N
10) Xai-Xai	87.8	-	N	1) Maputo	97.9	5	C
5) Massinga	89.9	0.12	N	9) Tete	100.7	-	EP
3) Tete	90.7	-	N	5) Inhambane	101.6	-	N
10) Xai-Xai	90.9	-	EP	6) Lichinga	101.7	-	N
2) Beira	91.6	-	N	1) Maputo	102.3	-	M
7) Quelimane	92.1	10	N	4) Chimoio	102.5	-	EP
1) Maputo	92.3	10	N	5) Inhambane	105.1	-	EP
1) Maputo	93.1	-	D	2) Beira	105.2	-	C
3) Nampula	95.1	-	N	3) Nampula	105.5	0.25	EP
8) Pemba	95.3	-	N	1) Maputo	105.9	-	E

Antena Nacional (N) in Portuguese: 24h.
Cidade FM (C) in Portuguese: 24h. Also rel. BBC.
RM Desporto (D) in Portuguese: 0300-2200.
Maputo Corridor R. (E) in English: 1000-2200. Also rel. BBC.
Emissão Provincial (EP) in Portuguese/ethnic: Mostly provincial prgrs on MW/FM 0250-2200. Also rel. network N news at 1600 and 2130 and "RM Jornal" at 1030-1100 & 1730-1800.
1) EP de Maputo – **2)** EP de Sofala, C.P. 1942, Beira – **3)** EP de Nampula, C.P. 93, Nampula – **4)** EP de Manica, C.P. 390, Chimoio – **5)** EP de Inhambane, C.P. 196, Inhambane – **6)** EP do Niassa, C.P. 171, Lichinga – **7)** EP de Zambézia, C.P. 333, Quelimane – **8)** EP de Cabo Delgado, C.P. 45, Pemba – **9)** EP de Tete, C.P. 384, Tete – **10)** EP de Gaza, C.P. 130, Xai-Xai.
Ann: "Rádio Moçambique, Antena Nacional", EP: "Rádio Moçambique, (province)". **IS:** Mbira (indigenous xylophone). Opens and closes with National Anthem.

Other stations:
R. A Voz do Islão, Maputo: 96.3MHz – **R. Capital,** Maputo: 90.7MHz (also rel. TWR) – **R. Maria Moçambique**: Maputo 103.1MHz, Villankulo/Xai Xai 102MHz, Chokwe 101.4MHz, Govure 102.5MHz, Quissico 106.4MHz, Maxixe 104.2MHz, Nova Mambone 104MHz Web: www.radiomaria.org mz – **R. Miramar**, Maputo :101.4MHz – **R. N'tyana,** Maputo: 93.5MHz – **R. SFM,** Maputo: 94.6MHz – **R. 99,** Maputo: 99.3MHz – **RTV Klint**, Maputo: 88.3MHz – **R. Terra Verde,** Maputo: 98.6MHz – **R. TGV 9FM,** Maputo: 99.3MHz Web: www.99fm.co.mz – **R. Viva,** Maputo: 99.6MHz 1kW. Web: radioviva.fm – **Top R,** Maputo: 104.2MHz
BBC African Service: Tete 87.8MHz, Nampula 88.3MHz, Beira 88.5MHz, Quelimane 95.3MHz, Maputo 95.5MHz 1kW, Xai-Xai 100.9MHz – **RDP África:** Beira 94.8MHz, Maputo 89.2MHz, Nampula 91.9MHz, Quelimane 89.0MHz (all 50kW) – **RFI Afrique:** Maputo 105.0MHz 1kW in French/English/ Portuguese.
In addition about 100 community radio stations are in operation.

MYANMAR

L.T: UTC + 6½h — **Pop**: 53.4 million — **Pr.L**: Burmese (Bamar), English Major minority languages: Kachin, Kayah, Kayin (Po & Sakaw), Chin, Mon, Rakhine, Shan — **E.C**: 50Hz, 230V — **ITU**: BRM

MINISTRY OF INFORMATION
Yaza Thingaha Rd, Zeya Theiddhi Ward, Nay Pyi Taw ☎ +95 67 412323.

MYANMA RADIO AND TELEVISION DEPT, MRTV (Gov.)
MYANMA RADIO
Tatkon Township, Nay Pyi Taw ☎ +95 67 79483. **Yangon centre.:** Pyay Rd, Kamayut-11041, Yangon ☎ +95 1 527122 ▤ +95 1 525428

E: mrtv@mptmail.net.mm **L.P:** DG: U Khin Maung Htay. Dir. (Broadc.): U Hla Swe. CE: U Tin Wan. Dir (TV): U Phone Myint.

MW (kHz)kW		Loc	Pr	H. of tr.
576	200	Y	P	2300-0010v
		N		0030-0730, 0930-1630
594	200	N	N	0030-0730, 0930-1630
711	400	N	P	2300-0130, 0730-1000, 1130-1530
729	?200	Y	Y	2330-0530, 0730-1330
			E	1330-1500

SW (kHz)	kW	Loc	Pr	H. of tr.
5915a	50	N	Mi	2330-0730, 0930-1330
			E	1330-1500
v5985	50	Y	P	2300-0130
			N	0930-1630
7200	50	Y	N	0040v-0300, 1010v-1230
v9731	50	Y	N	0215v-0730
			P	0730-1000

a=alt. freq.: 5920kHz

FM: Yangon area 98.0MHz (N prgr), 100.0MHz (P prgr). Other FM freqs may be in operation, details not available.
Loc=Location: N=Nay Pyi Taw. Y=Yangon. SW tx locations shown are based on monitoring and other research and are not yet confirmed.
Pr=Prgr: N=National prgr. in Burmese, English. Mi=minorities prgrs in Kayah, Kayan, Gekho, Geba, Chin, Kokang, Mindat Chin and Wa. Y=Yangon prgr. in Burmese. E=Educational prgr. in Burmese and English. P="Padauk Myay" music prgr in Burmese. **English** (in N prgr) 0230-0330, 0700-0730, 1530-1630. **N:** Generally 30 mins past the UTC hr on N and P prgrs; in English on N prgr at 0230, 0700, 1530.
Ann: E: "This is Myanma R" **IS:** Myanma Orchestral Music.

Other Stations:

Cherry FM (Comm.) Operated by Zaykabar Co. ☐ Taunggyi. Yangon office: 1 No 3, Main Rd, Mingalardon Garden City, Yangon. **FM:** Taunggyi 89.8MHz, Shan State (east) 88.0/88.6/91.3MHz – **City FM (FM-89) (Gov.)** ☐ Yangon City Development Committee, City Hall, Mahabandoola St corner Sule Pagoda Rd, Yangon. **FM:** Yangon 89.0MHz – **FM Bagan (Comm.)** Operated by Htoo Co. **FM:** 89.8MHz (may not yet be operaerting) – **Mandalay FM (Gov.)** ☐ Mandalay. Yangon office: Rm 1402-3, Olympic Twr, Bo Aung Kyaw St, Yangon. Joint venture of Forever Group and Mandalay City Development Committee (MCDC). **FM:** Mandalay (Sagaing Hill)/Taungoo/Yangon 87.9MHz, unk. loc. 88.3MHz – **Paddamya FM (Comm.) (Ruby FM.).** ☐ Sagaing. **FM:** Monywa 88.6MHz, Myitkyina/Sagaing 88.9MHz – **Pyinsawaddy FM (Comm.)** Operated by Forever Group. ☐ Sittwe, Rakhine State. **FM:** Sittwe 88.9MHz – **Shwe FM (Comm.)** Operated by Shwe Thanlwin Co. ☐ Bago. **FM:** Bilin/Nyaunglaybin 89.5MHz, Bago/Kyaikto/Mon State 89.8MHz, Yangon 90.0MHz – **Defence Forces Broadcasting Unit (Mil.)** ☐ Taunggyi, Shan State **SW:** 5770kHz 10kW **D.Prgr.** in Burmese and minority langs: 0030-0430, 0830-0930, 1130-1530.

NAMIBIA

LT: UTC +1h (5 Sep 10-3 Apr 11, 4 Sep 11-1 Apr 12: +2h) — **Pop:** 2.1 million — **Pr.L:** English, Afrikaans, German, local languages — **E.C:** 50Hz, 220V — **ITU:** NMB

NAMIBIAN COMMUNICATIONS COMMISSION (NCC)
☐ Communication House, 56 Robert Mugabe Ave, Private Bag 13309, Windhoek ☎+264 61 222666 ▤ + 264 61 222790 **W:** www.ncc.org.na **L.P:** Dep. Dir: Jan Kruger. Chief Eng. Tech: Barthos Hara-Gaeb.

NAMIBIAN BROADCASTING CORPORATION (Pub)
☐ P.O. Box 321, Cullinan Str, Northern Industrial Area, Windhoek 9000 ☎+264 612 913111 ▤ + 264 612 913325 **E:** pr@nbc.com.na **W:** www.nbc.com.na **L.P:** Chmn: Mr. Ponhele Ya France. Ag. DG: Yvonne Boois. Tech. Mgr: Ruben Prinz.

FM (MHz)	Afr.	Nat.	Ger.	Ova.	Her.	D/N.	Kav.	Lozi.	Tsw.	San.
Aminuis	88.9	92.0			95.2					98.5
Andara		92.5		95.7		99.0	106.1	102.5		
Arendsnes	88.7	90.1	91.8	96.4	99.7	106.8	93.2	95.0	98.3	103.2
Aroab	87.9	94.2			104.6					
Aussenkjer	92.5	95.7		98.7		102.5				
Bethanien	88.6	91.7	98.2		101.7					
Brukkaros	90.2	96.5			106.9					
Ekuli		91.5		88.4	101.5		94.7	98.0		
Epukiro	91.6	98.1			101.6					105.2
Erongo	90.6	93.7	96.9	100.2	103.7	107.3				
Gam		92.6	95.8	102.6	99.1					
Gibeon					100.7					
Gobabis	87.6	90.7	93.9	102.9	100.7	104.3	106.5	92.9	95.6	

FM (MHz)	Afr.	Nat.	Ger.	Ova.	Her.	D/N.	Kav.	Lozi.	Tsw.	San.
Gross-Herzog	88.6	91.7	94.9	98.2		101.7	105.3			
Kamanjab	89.7					106.4				
Katima Mulilo	89.5	95.8	90.9	99.1		94.1	106.2	100.9	92.6	87.8
Keetmanshoop	87.6	90.7	93.9	97.2		89.3	104.3			
KL. Waterberg	89.6	92.7	95.9	99.2		102.7	106.3			
Koes	88.8	95.1					105.5			
Kongola		88.3		91.4				94.6	97.9	
Luderitz	89.7	92.8	96.0	99.3		100.2	103.7			
Maltahohe	88.5		94.8				105.2			
Mariental	87.7	90.8	94.0	101.8		105.4	104.4			
Nakop	90.6	93.7		100.2			103.7			
Nkurenkuru		90.7		87.6		97.2		105.1	93.9	
Noordoewer	87.7	90.8		97.3			100.8			
Okongo		89.0		92.1			95.3	98.6		
Omega		90.6		93.7				96.9	100.2	
Omuthiya		89.2		98.8		102.3	105.9			
Opuwo		91.1		97.6		101.1				
Oranjemund	90.0	93.1		99.6			106.7			
Oshakati	89.2	87.8	96.4	97.4		98.8	105.9	93.2	90.9	99.7
Otjimbingwe						102.3	105.9			
Otjinene	90.2	93.3				103.3				106.9
Paresis	88.7	91.8	95.0	98.3		101.8	105.4			
Renosterkop	87.9	91.0				101.0	104.6			
Rietfontein	89.1	92.2				95.4				98.7
Rosh Pinah	90.3	93.4		96.6			99.9			
Rossing	89.7	92.8	96.0	99.3		98.4	106.4			
Rundu		89.6						95.9		
Shamvura		91.3		97.8			101.3	94.5	104.9	
Signalberg	87.7	90.8	94.0	97.3		100.8	104.4			
Stampriet	89.7	92.8	96.0				106.4			
Terrace Bay		104.3								
Tsumeb	88.6	91.7	94.9	98.2			105.3			
Tsumkwe		90.4		100.0		93.5				103.5
Ur		89.8	92.9	96.1	99.4	102.9	106.5			
Windhoek	89.5	92.6	95.8				107.1	93.5	90.4	

National Sce. in English: 24hrs — **Afrikaans Sce:** MF 0900-1600 & 1700-2000, SS 0600-2000. Relayed on other services overnight. – **German Sce:** MF 0900-1600 & 1700-2000, SS 0500-2000 –**Oshiwambo Sce in Ovambo/Kwanyama:** 0900 (SS 0500)-2200 – **Otjiherero Sce in Herero/Setswana:** MF 0900-1600 & 1700-2000, SS 0500-2000 – **Damara/Nama Sce:** MF 0900-1600 & 1700-2000, SS 0500-2000 – **Rukavango Sce in Kwangali:** 0900 (Sat/Sun 0500)-2000 – **R. Opuwo – San R.**
Ann: National Sce: "National Radio". On all NBC trs overnight: "Here is the National Sce. of the NBC Nationwide". G: "Hier ist das Deutsche Hörfunkprogramm der NBC". Damara/Nam: "Nes ge Damara/Nama Gowab loabas NBC's disa". A: "Dit is die Afrikaanse diens van die NBC". Otjiher: "Indji oradio ja Namibia morupa rueraka Otjiherero" .
IS: at s/on: National Anthem with choir/orchestra.

Other stations:

R. 99, P.O. Box 11849, Windhoek **FM:** Windhoek 99MHz, Walvis Bay & Swakopmund 96.5, Tsumeb 101.7, Oshakati & Ondangwa 104.5, Otjiwarango 99.9MHz. Also rel. VOA – **Kanaal 7,** Windhoek: 102.3MHz + 18 FM fq's. **W:** www.k7.com.na **E:** channel7@k7.com.na – **R. Energy,** P.O. Box 676, 17 Bismarck St, Windhoek West 9000 **W:** www.energy100fm.com **E:** energy@iway.na **FM:** Walvis Bay 88.8MHz 0.5kW, Windhoek 100MHz 0.5kW, Oshakati 100.9MHz 2kW – **R. Cosmos,** Windhoek: 94.1MHz. Mainly in Afrikaans – **R. Kudu,** P.O. Box 5369, Windhoek **W:** www.radiokudu.com.na **E:** radiokudu@radiokudu.com.na **Prgrs** in English. **FM:** Otjiwarongo 90.9MHz 0.1kW, Tsumeb 92.6MHz 0.1kW, Rundu 92.7MHz 0.1kW, Swakopmund 94.3MHz 0.5kW, Karibib 94.6MHz 0.1kW, Lüderitz 94.7MHz 0.1kW, Walvis Bay 95.1MHz 0.1kW, Grootfontein/Oshakati 95.5MHz 1kW, Gobabis/Keetmanshoop 95.6MHz 0.1kW, Mariental 97.3MHz 0.1kW, Rosh Pinah 103.4MHz 0.1kW, Okahandja/Rehoboth/Windhoek 103.5MHz 1kW, Katima Mulilo 107.4MHz 0.1kW – **Omulunga R,** P.O. Box 40789, Windhoek. Prgrs in Oshiwambo. **W:** www.omulunga.com.na **E:** omulunga@omulunga.com.na **FM:** Otjiwarongo/Tsumeb 89.2MHz 0.1kW, Grootfontein 92MHz 0.1kW, Rundu 92.7MHz 0.1kW, Mariental 95MHz 0.1kW, Oranjemund 96.3MHz 0.1kW, Windhoek 100.9MHz 1kW, Oshakati 102.3MHz 1kW, Walvis Bay/Swakopmund/Henties Bay/Arandis 104.7MHz, WalvisBay/Swakopmund/Henties Bay/Arandis 105.5MHz 0.1kW, Keetmanshoop 106MHz 0.1kW, Lüderitz 106.4MHz 0.1kW, Gobabis 107.5MHz 0.1kW – **R. Wave,** P.O. Box 9953, 30 Simpson St, Windhoek West **W:** www.radiowave.com.na **E:** radiowave@radiowave.com.na **FM:** Central Namibia 87.8MHz 1.5kW, Grootfontein 88.9MHz 0.1kW, Luderitz 90.6MHz 0.1kW, Henties Bay/Swakopmund/Rossing 91.1MHz 250W, Walvis Bay 91.9MHz 0.1kW, Keetmanshoop 92.4MHz 0.1kW, Tsumeb 95.8MHz 100W, Windhoek 96.7MHz 1.5kW, Otjiwarongo 100.9MHz 0.1lW, Katima Mulilo 104.5MHz 350W, Rundu 105.4MHz 350W, Erongo/Ondangwa/Oshakati 106.8MHz 350W.
R. France Int, Windhoek: 107.1MHz

NAURU

L.T: UTC +12h — **Pop:** 14,019 — **Pr.L:** English, Nauruan — **E.C:** 50Hz, 110/240V — **ITU:** NRU

NAURU BROADCASTING SERVICE (Gov)
⌨ P O Box 429, Rep. of Nauru, Ce. Pacific ☎ +674 555 6066 🖷 +674 4443195 **E:** radionaurufm@hotmail.com
L.P: SM: Miss Rin Tsitisi. TD: Malcom Aroi
FM: 88.8MHz **D.Prgr:** 1900-1130. Upgraded and local community prgrs reintroduced. Rel. R. Australia at other times **V.** by letter

RADIO PASIFIK NAURU Triple 9 FM (Edu)
⌨ University of the South Pacific Campus, Private Bag, Nauru PO, Nauru 00674 ☎ +674 4443774
L.P: Alamanda Lauti **E:** lauti_a@usp.ac.fj
FM: 99.9MHz 0.03kW **D.Prgr:** Mon-Sat 3 hrs daily includes lectures and tutorials from USP Suva, Fiji as well as local community prgrs. Tx is solar powered.

NEPAL

L.T: UTC +5¾h — **Pop:** 25.3 million — **Pr.L:** Nepali, English — **E.C:** 50Hz, 220V — **ITU:** NPL

RADIO NEPAL (Semi-Gov, Comm.)
⌨ Radio Broadcasting Service, G.P.O. Box 634, Singha Durbar, Kathmandu ☎ +977 1 4231804 🖷 +977 (1) 4221952 **W:** www.radionepal.org **E:** radio#rne.wlink.com.np (Eng. div: +977 1 4241923 **E:** radio#engg.wlink.com.np)
L.P: Exec. Dir: Er. R.S. Karki. Dep. Exec. Dirs: Mr Rajendra Prasad Sharma & Mr Sushil Koirala. Chief Eng.: Er. Ramesh Jung Karkee.

MW	kHz	kW	MW	kHz	kW
Surkhet	576	100	Kathmandu	792	100
Dhankuta	648	100	Dipayal	810	10
Pokhara	684	100	Bardibas	1143	10

SW: Khumaltar 5005kHz 100kW/5kW (irreg.)
D.Prgr on MW/SW: 2315-1715 (SW except 0515-0715 m-f). **N.** in **Nepali** at 0115, 0315, 0515, 0715, 0915, 1315, 1515, 1710; in **English:** 0215, 0815, 1415 & Brief **N** at other times on the hour; in **Hindi:** 1615; in **Sanskrit** 0010; in **Newari** 0330; in **Maithili** 1215
Variation at Regional Centres: 0400-0415 & 1145-1200
Ann: Nepali: "Yo Radio Nepal Ho"; English: "This is Radio Nepal"
IS: Instruments used are conch shell, violin, piano and jal tarang. **V.** by QSL-card.

FM STATIONS:

FM	MHz Station	FM	MHz Station
Baglung	91.6 Saypatri FM	Dhanusa	100.8 Radio Mithila ‡
Baglung	96.4 Baglung FM	Dhanusa	101.8 Janakpur FM ‡
Baglung	98.6 Dhawalgiri FM †	Dharan	95.6 Star FM
Bajhang	100.6 Saipal Radio FM †	Dharan	98.8 Bijaypur FM
Banepa	89.8 Radio ABC	Dolakha	100.6 Gaurishankar FM
Banke	94.6 Bageshwari FM †	Dolakha	106.4 R. Kalinchowk
Banke	95.6 Bheri Awaz FM	Gaur	90.8 Rautahat FM
Bhairahawa	96.1 R.Kantipur	Gorkha	92.8 Gorkha FM
Bhairahawa	99.2 Rupandehi FM	Gorkha	106.4 Deurali FM
Bharatpur	91.6 Synergy FM †	Gulariya	100.6 Phoolbari FM
Bharatpur	95.2 Kalika FM †	Hetauda	92.9 Manakamana FM
Bharatpur	96.1 R.Kantipur	Hetauda	95.4 Bindabasini FM
Bharatpur	103.0 R.Nepal	Hetauda	96.6 Hetauda FM
Biratnagar	88.2 Kankai Sa Geet	Hetauda	98.0 R.Nepal
Biratnagar	91.2 Birat FM	Humla	100.0 R.Nepal
Biratnagar	94.3 Koshi FM	Ilam	90.6 R.Fikkal FM
Biratnagar	105.6 Saptakoshi FM	Ilam	93.0 Ilam FM †
Biratnagar	106.6 Sky FM	Ilam	100.0 R.Nepal
Birgunj	91.4 Gadhina FM †	Itahari	90.0 Saptakoshi FM
Birgunj	96.1 R.Kantipur	Jaleshwar	106.6 R.Appan Mithila
Birgunj	99.0 Birganj FM ‡	Janakpur	91.0 R.Today
Birgunj	100.0 R.Nepal	Janakpur	106.0 Janaki FM
Butwal	94.4 Butwal FM †	Jhapa	92.6 Kanchanjhanga FM
Butwal	96.8 Radio Lumbini †	Jhapa	93.6 Pathibara FM ‡
Butwal	98.2 Tinau FM ‡	Jhapa	101.6 Saptarangi FM †
Damauli	94.2 Damauli FM †	Jomsom	100.0 R.Nepal
Dang	91.4 R.Madhya Paschim	Jumla	105.2 Radio Karnali †
Dang	101.6 Tulsipur FM	Kailai	93.8 Dinesh FM ‡
Dang	102.8 Swargadwari FM †	Kailai	100.4 Ghodaghodi FM
Daunne	100.0 R.Nepal	Kalyanpur	106.8 Nuwakot FM
Dhangadhi	101.8 R.Kantipur	Kanchanpur	92.0 Kanchanpur FM
Dhangadi	91.4 Khaptad FM	Kanchanpur	96.2 R.Mahakali
Dhangadi	93.8 Dinesh FM	Kanchanpur	94.0 Skuklapanda FM ‡
Dhankuta	92.2 Community R	Kaski	90.6 R.Gandki
Dhankuta	96.1 R.Kantipur	Kaski	92.2 Himchuli FM †
Dhankuta	106.2 R. Dhankuta	Kathmandu	87.6 R.Upatyaka

FM	MHz Station	FM	MHz Station
Kathmandu	88.8 Nepaliko Radio	Kathmandu	107.0 TU FM
Kathmandu	89.4 R.Mirmire	Kathmandu	107.4 R. Abhiyan
Kathmandu	90.0 Ujyaalo 90 Netw	Kavre	106.7 R.Nanabuddha
Kathmandu	90.6 Times FM	Kawasoti	101.0 R. Madhya Bindu
Kathmandu	91.2 Hits FM	Lampjung	95.0 R.Marsyangdi †
Kathmandu	91.8 Nepal FM	Makwanpur	96.6 Hetauda FM †
Kathmandu	92.4 Capital FM	Nawalparasi	90.2 R.Parasi
Kathmandu	93.0 Gorkha FM	Nawalparasi	101.6 Vijaya FM
Kathmandu	93.5 Mero FM	Nepalganj	101.8 R.Kantipur
Kathmandu	94.0 HBC FM	Nepalganj	105.5 Nepal Press Inst.
Kathmandu	94.6 Metro FM	Nilkantha	89.4 R.Loktantra FM
Kathmandu	95.2 Star FM	Nilkantha	106.0 Dhading FM
Kathmandu	96.1 R.Kantipur	Palpa	90.8 Muktinath FM
Kathmandu	96.8 V of Youth	Palpa	93.2 Shreenagar FM
Kathmandu	97.2 Headlines &	Palpa	99.4 R.Paschimanchal
	Music FM	Palpa	103.6 R.Rampur
Kathmandu	97.9 Image FM	Palpa	106.9 R.Madanpokhra †
Kathmandu	98.3 Keeps FM	Palung	107.2 Palung FM
Kathmandu	98.8 R. City	Parasi	101.6 Vijaya FM †
Kathmandu	99.4 Maitri FM	Parsa	103.8 Narayani FM †
Kathmandu	100.0 R.Nepal	Pokhara	91.0 Machhapuchhre
Kathmandu	100.6 ABC News R.	Pokhara	92.2 Himachuli FM
Kathmandu	101.2 Classic FM	Pokhara	93.4 R. Annapurna †
Kathmandu	101.8 Radio Filmy	Pokhara	95.8 Pokhra FM
Kathmandu	102.4 R. Sagarmatha †	Pokhara	97.9 Image Channel
Kathmandu	103.0 R.Nepal #	Pokhara	99.2 R.Barahi FM
Kathmandu	103.6 Image News	Pokhara	101.8 R.Kantipur
Kathmandu	104.2 ECR FM	Pokhara	104.6 R.Sarangkot
Kathmandu	104.8 Adhyatma Jyoti	Siraha	105.4 Fulbari Commu. R
Kathmandu	105.1 Good News FM	Solokhumbu	102.2 Solu FM
Kathmandu	105.4 Bhaktapur FM	Sunsari	90.0 Saptakoshi FM ‡
Kathmandu	106.0 CJMC FM	Surkhet	104.4 Bheri FM †
Kathmandu	106.3 R.Audio	Surkhet	107.6 Bulbule FM
Kathmandu	106.6 Newa FM	Tamghash	106.2 R.Resunga

†-carries BBC Nepali Service; ‡-carries BBC Nepali & Hindi Service
#-carries BBC WS 24hrs (except Nepali 1500-1530 hrs).

BBCWS daily MW relay via R. Nepal: Surkhet 576kHz 1630-1700 Hindi (rep. of 1430-1500 hrs), 1700-1730 Hindi, 1730-1800.
Guru-Baba FM: Bansgadi, 106.4MHz 0.1kW. Prgrs in Tharu
BFBS: Gurkha Radio, Kathmandu **FM:** English 92.1MHz, Gorkha 99.6MHz

NETHERLANDS

L.T: UTC +1h (27 Mar-30 Oct: +2h) — **Pop:** 16.7 million — **Pr.L:** Dutch — **E.C:** 50Hz, 230V — **ITU:** HOL

NEDERLANDSE OMROEP STICHTING (NOS)
⌨ Sumatralaan 45, 1217 GP Hilversum; Postbus 26600, 1202 JT Hilversum ☎ +31 35 6779222 🖷 +31 35 6772649
W: www.nos.nl **E:** publieksreacties@nos.nl

NEDERLANDSE PROGRAMMA STICHTING (NPS)
⌨ P.O. Box 29000, 1202 MA Hilversum ☎ +31 (35) 6774959 🖷 +31 (35) 6774959 **E:** publiek@nps.nl **W:** www.nps.nl (for all national public broadcasters).
Dutch national public prgrs are provided by the **NOS**, **NPS** and the following broadcasting organisations: **AVRO, BNN, EO, KRO, NCRV, TROS, VARA** and **VPRO**.

MW	kHz	kW	Prgr		
Zeewolde	747	400	Radio 5 (200kW nightime)		
Emmaberg	1251	10	Radio 5		
FM	**Radio 1**	**Radio 2**	**3FM**	**Radio 4**	**kW**
Amsterdam	98.6	92.3	96.5	94.5	0.09/0.09/0.09/0.09
Arnhem	98.6	92.9	96.5	92.1	0.07/0.07/0.07/0.07
Eys	-	97.2	-	-	13
Goes	104.4	94.4	99.8	95.0	50/0.1/15/15
Emmaberg	105.3	93.4	103.9	98.7	10/13/10/10
Hulst	-	107.1	-	-	0.1
Loon op Zand	-	-	-	98.2	55
IJsselstein	98.9	92.6	96.8	94.3	100/100/100/83
Markelo	98.4	104.6	96.2	91.4	100/100/100/100
Philippine	-	97.8	-	-	95
Roermond	104.8	88.2	90.9	94.5	100/100/100/100
Rotterdam	98.6	92.9	97.1	94.7	0.1/0.1/0.04/13
Hoogersmilde	91.8	88.0	88.6	94.8	100/100/38/100
Wieringerwerf	95.0	92.9	97.1	101.6	22/22/26/45

+ 20 low-power relays
Ann: "Dit is de VARA", "Dit is de VPRO" etc. **V:** by QSL-card.
Radio 1: news, sport; Radio 2 and 3 FM: music; Radio 4: classical music; Radio 5: light music, information, service and games

Regional stations:

FM	Mhz	kW	Location	Station
2)	87.6	8	Mierlo	Omroep Brabant
4)	87.9	15	Goes	Omroep Zeeland
9)	88.9	10	Amsterdam	Radio Noord-Holland
6)	89.1	5	Megen	Radio Gelderland
13)	89.3	10	Rotterdam	Radio 89 3 West
10)	89.4	10	De Lutte	Radio Oost
12)	89.8	25	Lelystad	Radio Flevoland
6)	90.4	10	Ruurlo	Radio Gelderland
5)	90.8	4	Hoogersmilde	Radio Drenthe
2)	91.0	15	Roosendaal	Omroep Brabant
2)	91.9	2	Loon op Zand	Omroep Brabant
3)	92.2	25	Irnsum	Omrop Fryslân
7)	93.1	4	IJsselstein	Radio M
11)	93.4	10	Rotterdam	Radio Rijnmond
9)	93.9	11	Wieringerwerf	Radio Noord-Holland
1)	95.3	10	Emmaberg	L1 Radio
10)	95.6	5	Markelo	Radio Oost
2)	95.8	5	Megen	Omroep Brabant
8)	97.5	40	Groningen	Radio Noord
7)	97.9	3	Rhenen	Radio M
10)	99.4	25	Zwolle	Radio Oost
1)	100.3	100	Roermond	L1 Radio
6)	103.5	20	Ughelen	Radio Gelderland

+5 low-power relays

1) Postbus 31, 6200 AA Maastricht ☎ +31 43 3467777 📠 +31 43 3467715 **E:** redactie@L1.nl **W:** www.l1.nl – **2)** Postbus 108, 5600 AC Eindhoven ☎ +31 40 2949494 📠 +31 40 2949320 **W:** www.omroepbrabant.nl – **3)** Postbus 7600, 8903 JP Leeuwarden ☎ +31 58 299 7799 📠 +31 58 2997778 **E:** direksje@omropfryslan.nl **W:** www.omroepfryslan.nl – **4)** Postbus 1090, 4388 ZH Oost-Souburg ☎ +31 118 499900 📠 +31 118 499929 **W:** www.omroepzeeland.nl – **5)** Postbus 999, 9400 AZ Assen ☎ +31 592 338080 📠 +31 592 331048 **E:** redactie@rtvdrenthe.nl – **6)** Postbus 747, 6800 AS Arnhem ☎ +31 26)3713713 📠 +31 26 3713710 **E:** rtv@omroepgelderland.nl **W:** www.omroepgelderland.nl – **7)** Postbus 1012, 3500 BA Utrecht ☎ +31 30 8500600 📠 +31 30 8500601 **E:** info@radiom.nl **W:** www.rtvutrecht.nl – **8)** Postbus 30101, 9700 RP Groningen ☎ +31 50 3199999 📠 +31 50 3185147 **E:** radio@rtvnoord.nl – **9)** Postbus 9823, 1006 AM Amsterdam ☎ +31 20 8505050 📠 +31 20 8505850. **E:** info@rtvnh.nl **W:** www.rtvnh.nl – **10)** Hazenweg 25, 7556 BM Hengelo (Ov) ☎ +31 74 2456456 📠 +31 74 2437148 **E:** info@rtvoost.nl **W:** www.rtvoost.nl – **11)** Postbus 1515, 3000 BM Rotterdam ☎ +31 10 4400600 📠 +31 10 4400698 **E:** info@rijnmond.nl **W:** www.rijnmond.nl – **12)** Postbus 567, 8200 AN Lelystad ☎ +31 320 285085 📠 +31 320 285099 **E:** rtv@omroepflevoland.nl **W:** www.omroepflevoland.nl – **13)** Postbus 24025, 2490 AA Den Haag ☎ +31 70 3078888 📠 +31 70 3078844 **E:** west@rtvwest.nl

Public local stations in major cities: FM(MHz):

Amersfoort 107.9 Bingo FM – **Amsterdam** 96.1 FUN X, 99.4 Wereld FM, 105.2 Radio Zuid Oost (RAZO), 106.8 Stads FM, 107.9 Caribbean FM – **Den Haag** 92.0 Den Haag FM, 98.4 FUN X – **Rotterdam** 91.8 FUN X, 93.9 Megastad FM – **Utrecht** 96.1 FUN X, 105.7 Bingo FM, 107.7 Bingo FM

NB: FUN X is an initiative of SALTO Omroep Amsterdam, Slor Rotterdam, Stadsomroep Den Haag and Omroep RTV Utrecht.

± 350 local stns

EXTERNAL SERVICE: Radio Nederland Wereldomroep (RNW)

See International Broadcasting Section

OTHER STATIONS

MW	kHz	kW	Location	Station
50)	675	120	Lopik	R. Maria Nederland (Rlg.)
56)	828	20/5	Heienoord	Radio 10 Gold (Comm.)
14)	891	20/5	Hulsberg	Radio 538 (Comm.)
53)	1008	200	Zeewolde	Groot Nieuws R. (Rlg.)
41)	1116	0.5	Bloemendaal	R. Bloemendaal (Rlg.)
8)	1332	2	Nieuwegein	Hot Radio Plus (Comm.)
55)	1395	20	Trintelhaven	Big L Int., The Mighty 1395 (Comm.)
52)	1584	0.15	Utrecht	R. Paradijs (Comm.)
54)	1557		Amsterdam	Magic Jazz (Comm.)
51)	1602	1	Pietersbierum	R. Waddenzee/R. Seagull (Comm.)

FM	MHz	kW	Location	Station
20)	87.6	1	Enschede	City FM
23)	88.2	1	Ugchelen	100%NL
23)	88.4	43	Roosendaal	Slam FM
3)	88.6	26	Mierlo	BNR Nieuwsradio
23)	88.7	1	Apeldoorn	Hot Radio Plus
15)	88.9	1	Den Bosch	Radio 8FM
23)	89.0	3	Lochem	100%NL
11)	89.1	1	Groningen	Radio NL

FM	MHz	kW	Location	Station
20)	89.2	3	Zwolle	City FM
9)	89.2	1	Venlo	Slam FM
15)	89.3	1	Eindhoven	Radio 8FM
23)	89.5	10	Alkmaar	100%NL
23)	89.5	5	Utrecht	100%NL
11)	89.6	2	Huissen	Radio NL
3)	89.6	5	Hoogersmilde	BNR Nieuwsradio
30)	89.6	1	Nieuwbergen	Maasland Radio
26)	89.9	1	Emmen	Waterstad FM
23)	90.0	3	Breskens	100%NL
23)	90.0	37	Loon op Zand	100%NL
9)	90.1	7	Den Helder	Slam FM
23)	90.2	50	Roosendaal	100%NL
27)	90.3	4	Eindhoven	XFM
27)	90.5	3	Helmond	XFM
9)	91.0	10	Irnsum	Slam FM
9)	91.0	1	Markelo	Slam FM
9)	91.1	40	Hilversum	Slam FM
48)	91.1	1	Gemert	Omroep Centraal
18)	91.3	1	Hoogezand	Simone FM
3)	91.3	70	Rotterdam	BNR Nieuwsradio
3)	91.3	1	Tilburg	BNR Nieuwsradio
3)	91.5	3	Breda	BNR Nieuwsradio
3)	91.5	3	Driewegen	BNR Nieuwsradio
3)	91.5	10	Eys	BNR Nieuwsradio
19)	91.6	5	Amsterdam	Radio Veronica
23)	92.1	10	Emmaberg	100%NL
39)	92.3	1	Rijssen	Radio 350
15)	92.4	3	Westdorpe	Radio Hollandio
47)	92.9	1	Wellerooi	Maasland Radio
44)	93.0	1	Meppel	Radio Meppel
3)	93.1	1	Emmen	Slam FM
26)	93.2	20	Irnsum	Waterstad FM
11)	93.3	1	Enschede	Radio NL
11)	93.5	9	Markelo	Radio NL
19)	93.6	1	Amsterdam	Wild FM Hitradio
22)	93.6	5	Eindhoven	Royaal FM
9)	93.6	6	Zwolle	Slam FM
9)	93.7	5	Enschede	Slam FM
9)	93.7	5	Groningen	Slam FM
40)	93.7	2	Leiden	Sleutelstad FM
9)	93.8	17	Megen	Slam FM
15)	93.9	4	Roosendaal	Radio Hollandio
19)	94.0	1	Emmen	Radio Veronica
35)	94.1	1	Den Bosch	Radio Mexico
11)	94.1	2	Tjerkgaast	Radio NL
24)	94.9	12	Mierlo	100%NL
23)	95.0	2	Amersfoort	100%NL
23)	95.0	2	Nijmegen	100%NL
9)	95.2	25	Alphen aan de Rijn	Slam FM
15)	95.2	1	Weert	Radio 8FM
3)	95.3	20	Zwolle	BNR Nieuwsradio
43)	95.3	1	Bedum	Regio FM
3)	95.4	1	Emmen	BNR Nieuwsradio
3)	95.5	30	Tjerkgaast	BNR Nieuwsradio
6)	95.6	1	Rijswijk	Fresh FM
6)	95.7	10	Amsterdam	Fresh FM
11)	95.7	2	Meppel	Radio NL
6)	95.9	3	Alphen aan de Rijn	Fresh FM
46)			Zoetermeer	ZFM
29)	96.3	1	Alkmaar	Wild FM Hitradio
19)	96.3	30	Loon op Zand	Radio Veronica
19)	96.6	1	Goes	Radio Veronica
11)	96.6	1	Leeuwarden	Radio NL
29)	97.3	2	Haarlem	Wild FM Hitradio
23)	97.6	1	Hengelo	100%NL
11)	97.6	2	Maastricht	Radio NL
36)	97.6	37	Rotterdam	Radio Decibel
19)	97.7	10	Arnhem	Radio Veronica
19)	97.7	6	Mierlo	Radio Veronica
11)	97.7	15	Landgraaf	Radio NL
19)	97.8	4	IJsselstein	Radio Veronica
36)	98.0	12	Amsterdam	Radio Decibel
11)	98.1	2	Eys	Radio NL
36)	98.3	5	Alkmaar	Radio Decibel
20)	98.5	5	Groningen	City FM
11)	98.5	1	Weert	Radio NL
20)	98.7	30	Hoogersmilde	City FM
23)	99.1	10	Enschede	100%NL
23)	99.1	4	Groningen	100%NL
23)	99.1	5	Tjerkgaast	100%NL
32)	99.1	1	Geleen	Streekradio START
9)	99.2	12	Vlissingen	Slam FM
9)	99.4	30	Mierlo	Slam FM
36)	99.4	2	Den Haag	Radio Decibel

FM	MHz	kW	Location	Station
3)	99.6	5	Dordrecht	BNR Nieuwsradio
9)	99.6	25	Hoorn	Slam FM
9)	99.6	17	Hoogersmilde	Slam FM
3)	99.9	1	Dedemsvaart	BNR Nieuwsradio
3)	99.9	3	Ugchelen	BNR Nieuwsradio
3)	99.9	27	Wormer	BNR Nieuwsradio
3)	100.1	100	IJsselstein	BNR Nieuwsradio
3)	100.1	4	Nijmegen	BNR Nieuwsradio
3)	100.2	12	Lochem	BNR Nieuwsradio
13)	100.4	15	Westdorpe	QMusic
13)	100.4	7	Roosendaal	QMusic
13)	100.4	3	Rotterdam	QMusic
13)	100.4	90	Hoogersmilde	Qmusic
13)	100.4	10	Doetinchem	QMusic
13)	100.5	5	Wieringerwerf	QMusic
13)	100.7	30	Breskens	QMusic
13)	100.7	10	Enschede	QMusic
13)	100.7	100	IJsselstein	QMusic
13)	100.7	10	Lichtenvoorde	Qmusic
24)	101.0	100	Hoogersmilde	Sky Radio 101 FM
24)	101.1	10	Nijmegen	Sky Radio 101 FM
24)	101.2	13	Hengelo	Sky Radio 101 FM
24)	101.2	200	Hilversum	Sky Radio 101 FM
24)	101.3	5	Roosendaal	Sky Radio 101 FM
24)	101.4	10	Deventer	Sky Radio 101 FM
24)	101.5	1	Arnhem	Sky Radio 101 FM
24)	101.5	7	Den Bosch	Sky Radio 101 FM
24)	101.5	8	Rotterdam	Sky Radio 101 FM
24)	101.6	5	Mierlo	Sky Radio 101 FM
24)	101.6	3	Roermond	Sky Radio 101 FM
24)	101.7	10	Breda	Sky Radio 101 FM
18)	101.7	1	Emmen	Simone FM
24)	101.9	2	Tilburg	Sky Radio 101 FM
24)	101.9	50	Goes	Sky Radio 101 FM
8)	101.9	1	Zieuwent	Hot Radio Plus
14)	102.1	100	Hilversum	Radio 538
14)	102.2	10	Groningen	Radio 538
14)	102.3	20	Alkmaar	Radio 538
14)	102.3	100	De Mortel	Radio 538
14)	102.3	15	Lochem	Radio 538
14)	102.3	2	Roermond	Radio 538
14)	102.4	1	Arnhem	Radio 538
14)	102.4	20	Westdorpe	Radio 538
14)	102.5	8	Tilburg	Radio 538
14)	102.5	50	Tjerkgaast	Radio 538
14)	102.5	1	Utrecht	Radio 538
14)	102.6	5	Enschede	Radio 538
14)	102.6	2	Nijmegen	Radio 538
14)	102.7	10	Emmen	Radio 538
14)	102.7	100	Rotterdam	Radio 538
19)	103.0	40	Lelystad	Radio Veronica
19)	103.1	20	De Lutte	Radio Veronica
19)	103.1	5	Megen	Radio Veronica
19)	103.2	40	Rotterdam	Radio Veronica
19)	103.2	13	Hoogersmilde	Radio Veronica
19)	103.3	6	Terneuzen	Radio Veronica
19)	103.4	5	Groningen	Radio Veronica
19)	103.5	2	Roosendaal	Radio Veronica
15)	103.6	2	Tilburg	Radio 8 FM
11)	104.2	5	Alkmaar	Radio NL
23)	104.4	50	Hilversum	100%NL
26)	104.4	2	Groningen	Waterstad FM
23)	104.6	87	Rotterdam	100%NL
45)	104.8	1	Zuidwolde	De Streekradio
34)	106.7	1	Lochem	Achterhoek FM

+ 419 stns below 1kW

Addresses:
3) Postbus 651, 1000 AR Amsterdam ☎ +31 20 5158515 🖷 +31 20 5898755 **E:** operations@bn.nl **W:** www.bn.nl – **4)** Postbus Vervaartlaan 6, 2288GM Rijswijk ☎ +31 20 5849999 and 070-3072520 🖷 +31 20 5849980 **W:** www.cityfm.nl **E:** info@cityfm.nl – **6)** Darwinstraat 20, 2722 PX Zoetermeer ☎ +31 79 3434491 🖷 +31 79 3434492 **W:** www.fresh.nl – **8)** Postbus 700, 7550 AS Hengelo ☎ +31 74 2509090. 🖷 +31 74 2509099 **W:** http://plus.hotradio.nl **E:** plus@hotradio.nl – **9)** Rhoneweg 54, 1043 AH Amsterdam **E:** radio@id-t.com **W:** www.id-t.com – **11)** Postbus 248, 8600AE Sneek ☎ +31 515 432360 🖷 +31 35 432986 **W:** www.radionl.fm **E:** info@radionl.fm – **12)** Postbus 36, 4450 AA Heinkenszand ☎ 0900 5105100 🖷 +31 113 567670 **E:** info@maximaal.nl **W:** www.maximaal.nl – **13)** Postbus 102, 1200 AC Hilversum ☎ +31 35 655 2 655 🖷 +31 35 655 2 656 **E:** info@q-music.nl **W:** www.q-music.nl – **14)** Postbus 2538, 1200 CM Hilversum ☎ +31 35 5385538 🖷 +31 35 6283538 **W:** www.radio538.nl – **15)** Postbus 8, 5201 AA Den Bosch ☎ +31 73 6312003 🖷 +31 73 6313311 **W:** www.radio8fm.nl **E:** info@radio8fm.nl **W:** www.radiohollandio.

nl – **17)** Postbus 1111, 6201 BC Maastricht ☎ +31 045 3215566 🖷 +31 045 3216677 – **18)** Hoogveen 2, 9501 XK Stadskanaal ☎ +31 599 312183 🖷 +31 599 312187 **E:** info@simone.nl **W:** www.radiosimone. nl – **19)** Postbus 1007, 1400 BA Hilversum ☎ +31 35 5277555 🖷 +31 35 5277557 **W:** www.radioveronica.nl – **20)** Hoofdstraat 80, 7941AL Meppel ☎ +31 522 242624 **W:** www.cityfm.nl **E:** info@cityfm.nl – +31 522 252485. – **21)** Postbus 77, 9640 AB Veendam ☎ +31 598 633055 🖷 +31 598 633308 **E:** info@touchradio.nl **W:** www.touchradio. nl – **22)** Dukaathof 3, 5551 VG Valkenswaard ☎ +31 40 2017283 🖷 +31 40 2040066 **E:** redactie@royaal.fm **W:** www.royaal.fm – **23)** Postbus 813, 1200AV Hilversum **W:** www.100p.nl – **26)** Postbus 248, 8600 AE Sneek ☎ +31 515 432360 🖷 +31 515 432 986 **E:** info@ waterstadfm.nl **W:** www.waterstadfm.nl – **27)** Postbus 170, 5660 AD Geldrop ☎ +31 40 2233303 🖷 +31 40 2960610 **W:** www.xfm.nl – **29)** Gyroscoopweg 144, 1042 AZ Amsterdam ☎ +31 20 4470808 🖷 +31 20 4118344 **W:** www.wildfm.nl **E:** spam@wildfm.nl – **30)** Raadhuisstraat 7, 5854AX Nieuwbergen ☎ +31 485 341234 🖷 +31 485 343118 **W:** www.maaslandradio.nl **E:** studio@maaslandradio.nl – **32)** **W:** www. streekomroepstart.net – **34)** Postbus 115, 7250AC Vorden ☎ +31 575 556560 🖷 +31 575 556564 **W:** www.achterhoekfm.nl – **35)** ☎ +31 73 6444887 **W:** www.radiomexicodenbosch.nl – **36)** Wilgenweg 16A, 1031HV Amsterdam ☎ +31 909 5008000 **W:** www.radiodecibel.nl **E:** studio@decibel.nl – **39)** Postbus 234, 7460AE Rijssen ☎ +31 548 681 010 🖷 +31 548 544 117 **W:** www.radio350.nl **E:** info@radio350. nl – **40)** Postbus 937, 2300AX Leiden ☎ +31 71 523 5907 🖷 +31 71 523 5908 **W:** www.sleutelstad937.nl **E:** info@sleutelstad.nl – **43) W:** www.regiofm.info **E:** info@regiofmradio.nl ☎ +31 598 4232 00 🖷 +31 8 42298 700 – **44)** Emmastraat 10, 7941HR Meppel ☎ +31 522 259319 🖷 +31 522 240694 **W:** www.omroepmeppel.nl **E:** secretari-aat@omroepmeppel.nl – **45)** Postbus 8, 7920AA Zuidwolde ☎ +31 528 373444 🖷 +31 528 372272 **W:** www.streekradio.com **E:** info@ streekradio.com – **46)** Postbus 841, 2700AV Zoetermeer ☎ +31 79 3317287 🖷 +31 79 3434193 **W:** www.zoetermeerfm.nl **E:** studio@ zoetermeerfm.nl – **47)** Raadhuisstraat 7, 5854AX Nieuw Bergen ☎ +31 485 341234 🖷 +31 485 343118 **W:** www.maaslandradio.nl **E:** redac-tie@maaslandradio.nl – **48)** St. Annastraat 60, 5421KC Gemert ☎ +31 492 366833 🖷 +31 492 366822 **W:** www.omroepcentraal.nl **E:** info@ omroepcentraal.nl – **49)** Vijverweg 14, 2061GX Bloemendaal ☎ +31 23 5250471 **E:** bureau@radiobloemendaal.nl **W:** www.radiobloemendaal. nl **D.Prgr:** Sun & Christian holidays 0800-2000 & Tues 1100-1130v. reception reports to: PA0WDG@amsat.org – **50)** Postbus 5045, 5201GA 's-Hertogenbosch **W:** www.radiomaria.nl **E:** info@radiomaria.nl; recep-tion reports to qsl@radiomaria.org – **51)** Postbus 24, 8860 AA Harlingen ☎ +31 06 28580161 🖷 +31 58 2662204 **E:** info@radiowaddenzee.nl **W:** www.radiowaddenzee.nl **D.Prgr:** 06.00-18.00 for tourists. After 1800: Radio Seagull **W:** www.radioseagull.com **E:** office@radioseagull. com – **52)** very irregular txs **E:** radiocaroline@wxs.nl – **53)** Einsteinlaan 41b, 3902 HN Veenendaal, Managers: Evert ten Ham and Arjan de Heer ☎+31 318 584 384 🖷 +31 318 584 380 **W:** www.grootnieuwsradio. nl **E:** info@grootnieuwsradio.nl **V:** by letter – **54) W:** www.magicfm. nl/home/6 – **55)** 1-3 Colmore Crescent, Moseley, Birmingham B19 9SJ, United Kingdom Reception Reports : Big L QSL, PO Box 1536, 1000BM Amsterdam ☎ +44 121 449 5051 **W:** www.bigl.co.uk **E:** dutch@bigl. co.uk; prgrs in English 2300-2000; Dutch 0600-0800: Transportradio **W:** www.transportradio.nl; sa/su 2300-0100 Dutch rlg. – **56)** Postbus 1056, 1200BB Hilversum ☎ +31 35 75 05 910 🖷 +31 35 628 35 38 **E:** info@ radio10gold.nl **W:** www.radio10gold.nl

Military stations

FM	MHz	kW	Location	Station
2)	87.7	0.1	Maastricht	BFBS 1
1)	89.2	1	Brunssum	AFN Benelux
2)	90.2	0.1	Brunssum	BFBS 1
1)	107.9	0.1	Volkel	AFN Benelux
3)	96.9	10	Brunssum	CFN/RFC
3)	99.7	0.5	Brunssum	CFN/RFC

Addresses and other information:
1) E: spannb@afn.shape.army.mil **W:** www.afneurope.net **D.Prgr:** 24h relay AFN SHAPE (Belgium) – **2) D.Prgr:** Relays BFBS 1 prgrs Germany **W:** www.ssvc.com/bfbs/index.htm – **3)** Slot 6041, 2nd 5053 Stn Forces, Belleville, Ontario K8N 5W6, Canada D.Prgr:24h ☎ +31 45 5263791 🖷 +31 45 5263792 **E:** mail@cfnradio.com **W:** cfnradio.com

NEW CALEDONIA (France)

L.T: UTC +11h – **Pop:** 227,436 – **Pr.L:** French, Kanak and other Melanesian-Polynesian dialects — **E.C:** 50Hz, 220V — **ITU:** NCL

RADIO NOUVELLE CALEDONIE (Pub)
📠 1 rue Marechal Leclerc, Mt Coffyn, B.P. G3 - 98848 Noumea Cedex

☎ +687 687274327 🖷 +687 687281252 **W:** www.nouvellecaledonie.rfo.fr (live streaming) **LP:** Reg. Dir: Benoit Saudeau. **D.Prgr:** 24h in French (local and RFO common prgr satellite feed) and Kanak (local). **Ann:** "Ici RNC, Radio Nouvelle Caledonie".
MW: Noumea 666kHz 20kW, Touho 729kHz 5kW

FM	Location	kW	FM	Location	kW
88.0	Bouloupari	2.2	90.0	Kone	3.5
88.0	Kaala-Gomen	1.2	90.0	Noumea-Mont Koghis	5
88.0	Mont Dore	3	90.0	Poum	0.01
88.0	Touho	0.6	90.0	Yate	0.09
88.5	Canala	0.1	90.5	Lifou	11
88.5	Mare	0.45	90.5	Paita	0.1
89.0	Iles-des-Pins	0.04	90.5	Poya	0.3
89.0	Noumea-Mont Coffyn	2.2	91.0	Bourail	0.3
89.0	Ponerihouen	3	91.0	Houailou	2
89.0	Pouebo	0.01	91.0	Koumac	4
89.5	Iles-des-Pins	0.1	91.0	Mont-Dore	0.1
89.5	Ouvea	0.55	91.0	Thio	0.01
90.0	Canala	0.01	91.0	Yate	0.12
90.0	Hienghene	0.01	91.5	Lifou	0.1

RADIO FRANCE INTER
24h in French via satellite from Paris

FM	Location	kW	FM	Location	kW
92.0	Bouloupari	2.2	93.0	Noumea-Mont Coffyn	2.2
92.0	Mont-Dore	3	94.0	Kone	3.5
93.0	Ponerihouen	3	94.0	Noumea-Mont Koghis	5

RFO 24h common prgr feed in French via satellite from Paris

FM	Location	kW
91.5	Noumea-Mont Coffyn	2.2

Other FM Stations:

	FM MHz	Location	kW		FM MHz	Location	kW
1)	93.5	Noumea-Mont Coffyn	1.0	4)	99.0	Thio	0.2
2)	95.0	Dumbea [Noumea]	1.0	4)	100.0	Belep	0.02
3)	96.0	Belep	0.0	4)	100.0	Mont-Dore	0.75
3)	96.0	Bouloupari	0.8	4)	100.0	Kaala-Gomen	0.3
3)	96.0	Kaala-Gomen	0.3	4)	100.0	Mont-Dore	2.2
3)	96.0	Mont-Dore	2.2	4)	100.0	Touho	0.35
3)	96.0	Thio	0.4	4)	100.4	Noumea	1.5
3)	96.5	Ouvea	0.4	4)	101.0	Iles-des-Pins	0.01
3)	97.0	Iles-des-Pins	0.4	4)	101.0	Ponerihouen	0.75
3)	97.0	Ponerihouen	0.8	4)	101.0	Pouebo	0.35
3)	97.0	Pouebo	0.4	4)	101.5	Mare	0.6
3)	97.4	Noumea	1.5	3)	102.0	Canala	0.7
3)	97.5	Mare	0.6	3)	102.0	Dumbea [Noumea]	1.5
4)	98.0	Canala	0.7	3)	102.0	Hienghene	0.05
4)	98.0	Dumbea [Noumea]	5.3	3)	102.0	Kone	1
4)	98.0	Hienghene	0.05	3)	102.0	Poum	0.01
4)	98.0	Kone	1	3)	102.0	Yate	0.01
4)	98.0	Poum	0.01	4)	102.5	Lifou	1.5
4)	98.0	Yate	0.01	3)	103.0	Bourail	0.07
3)	98.5	Lifou	1.5	4)	103.0	Houailou	0.25
4)	99.0	Bourail	0.07	3)	103.0	Koumac	1.5
4)	99.0	Houailou	0.25	3)	103.0	Thio	0.2
4)	99.0	Koumac	1.5	4)	103.5	Ouvea	0.15

Stations, addresses and other information
1) NRJ, 41/43 rue Sebastopol, B.P G5 - 98848 Noumea Cedex ☎ +687 687279446 🖷 +687 687279447 **W:** www.nrj.nc (live streaming). 24h – **2) R. Oceane**, 1, avenue d'Auteuill Lotissement FSH Koutio - 98835. Dumbea ☎ +687 687410095 🖷 +687 687410099 **LP:** President, Dumbea Communications – Robert Lucas, Dir: Veronique Loisel **E:** oceane.fm@lagoon.nc **W:** www.oceanefm.net (live streaming) 24h – **3) R. Djiido**, 29, rue du Marechal Juin - 98880 Noumea ☎ + 687 687253515 🖷 +687 687272187 **W:** www.radiodjiido.nc (live streaming) **E:** radiodjiido@radiodjiido.nc **LP:** Thierry Kameremoin 24h – **4) R. Rythme Bleu**, B.P 578 - 98845 Noumea Cedex ☎ +687 687254500 🖷 +687 687284928 **E:** rrb@lagoon.nc. 24h

NEW ZEALAND

LT: UTC +12h (26 Sep 10-3 Apr 11, 25 Sep 11-1 Apr 12: +13h) – **Pop**: 4.4 million – **Pr.L:** English, Maori, Samoan – **E.C:** 50Hz, 230V – **ITU:** NZL

RADIO SPECTRUM MANAGEMENT GROUP
Ministry of Economic Development
🖷 P.O. Box 2847, Wellington 6140. ☎ +64 4 962 2603 NZ Freephone 0508 RSM INFO [0508 776 463] 🖷 +64 4 499 0797 **W:** www.rsm.govt.nz **E:** info@rsm.govt.nz **LP:** Mgr Radio Spectrum Policy & Planning: Brian Miller; Mgr Radio Spectrum Management: Sanjai Raj. The RSMG is the statutory authority responsible for radio spectrum administration.

BROADCASTING STANDARDS AUTHORITY
🖷 P.O. Box 9213, Wellington ☎ +64 4 382 9508 🖷 +64 4 382 9543 **W:** www.bsa.govt.nz **E:** info@bsa.govt.nz **LP:** Chief Exec: Dominic Sheehan. The BSA is a statutory authority that recommends codes of broadcasting practice, broadcasting standards and ethical conduct and administers a complaints procedure.

NEW ZEALAND ON AIR
🖷 P.O. Box 9744, Wellington 6141. ☎ +64 4 382 9524 🖷 +64 4 382 9546 **W:** www.nzonair.govt.nz **E:** info@nzonair.govt.nz **LP:** Chief Exec: Jane Wrightson, Mgr Community Broadcasting: Keith Collins. NZOA is the operational funding agency for **Radio New Zealand**, the **Community Access Radio** network of 12 stns, the **Radio Reading Service**, the **National Pacific Radio Trust, Samoan Capital Radio,** student radio [**bNet**] and TV and New Media.

RADIO NEW ZEALAND (Non-commercial, Pub)
🖷 P.O. Box 123, Wellington 6140 ☎ +64 4 474 1999 🖷 +64 4 474 1730 **W:** www.radionz.co.nz **LP:** Chief Exec: Peter Cavanagh; Infrastructure Mgr: Matthew Finn; Transmission Mgr: Gary Fowles.
Network Stations: RNZ National (N), RNZ Concert (C), RNZ AM Network (AM), RNZ International [see International broadcasting section] - for FM listings including non-major market locations see website. **Prgr:** 24h from Wellington studios except for RNZ AM Network which only broadcasts when Parliament is in session [relays commercial stn Southern Star at other times]. **N:** RNZ News bulletins

Network Stations MW

MW	kHz	kW	Net	MW	kHz	kW	Net
Wellington	567	50	N	Napier-Hastings	909	5.0	AM
Napier-Hastings	630	10.0	N	New Plymouth	918	2.5	N
Alexandra	639	2.0	N	Timaru	918	2.5	N
Tauranga	657	10.0	AM	Christchurch	963	10.0	AM
Wellington	657	50.0	AM	Kaikohe	981	2.0	N
Christchurch	675	10.0	N	Masterton	1071	2.5	N
Invercargill	720	10.0	N	Nelson	1116	2.5	N
Tokoroa	729	2.5	N	Queenstown	1134	2.0	N
Auckland	756	10.0	N	Hamilton	1143	2.5	N
Dunedin	810	10.0	N	Rotorua	1188	0.4	N
Tauranga	819	10.0	N	Gisborne	1314	2	N
Kaitaia	837	2.0	N	Invercargill	1314	5	AM
Whangarei	837	2.5	N	Palmerston North	1449	2.5	N
Auckland	882	10.0	AM	Westport	1458	2.5	N
Dunedin	900	10.0	AM	Hamilton	1494	2.5	AM

Network Stations FM
Major Radio Market

FM (MHz)	N	C	FM (MHz)	N	C
Auckland	101.4	92.6	New Plymouth	101.2	91.6
Christchurch	101.7	89.7/99.7	Palmerston North	101.8	89.0
Dunedin	101.4	92.6/99.0/99.4	Queenstown	101.6	98.4
Gisborne	101.3	97.3	Rotorua	101.5	90.3
Hamilton	101.0	91.4	Taupo	101.6/104.8	98.4
Invercargill	101.2	90.0	Tauranga	101.0	91.4
Kapiti Coast	101.5	98.3	Timaru	101.1	99.5
Napier-Hastings	101.5	91.1	Wellington	101.3/101.7	92.5/96.1
Nelson	101.6	91.2	Whangarei	101.2/104.4	100.4/105.2

COMMUNITY ACCESS RADIO
There are 12 independent stns affiliated to the **Association of Community Access Broadcasters [ACAB]** **W:** www.acab.org.nz **E:** info@acab.org.nz. Each stn serves local urban communities with a variety of ethnic language, cultural and community group prgrs. Limited sponsorship is allowed. BBC World Service is carried overnight on several stns. 24h. **F.PL:** extension of network to include Tauranga [1368 AM/LPFM 106.7] and Rangiora-North Canterbury [104.9FM].

MW	KHz	kW	Station, location
1)	783	10	Wellington Access R., Wellington
2)	999	1.5	Manawatu Access R., Palmerston North
3)	1206	0.5	Community Access R., Hamilton
4)	1431	2	R.Kidnappers, Napier-Hastings
5)	1575	2.5	Toroa R., Dunedin
FM	**MHz**	**kW**	**Station, location**
6)	104.8		Fresh FM, Nelson
7)	92.7		Arrow FM, Masterton
8)	96.9	3.5	Plains FM, Christchurch
9)	96.4		R.Southland, Invercargill
10)	104.6		Planet FM, Auckland
11)	104.7		Coast Access R., Kapiti Coast
12)	104.4	5	Access R. Taranaki, New Plymouth

Addresses & other information
1) P.O. Box 9073, Marion Square, Wellington 6141 ☎ +64 4 385 7210 🖷 +64 4 385 7212 **W:** www.accessradio.org.nz **E:** info@accessradio.org.nz **Mgr:** Phil O'Brien – **2)** P.O. Box 4666, Manawatu Mail Centre, Palmerston North 4442 ☎+64 6 357 9340 🖷 +64 6 357 9345 **W:** www.accessmanawatu.co.nz **E:** info@accessmanawatu.co.nz **Mgr:**

Fraser Greig **Prgr:** BBC WS 0900-2100 overnight daily – **3)** P.O. Box 110, Waikato Mail Centre, Hamilton 3240 ☎ +64 7 834 2170 🖹 +64 7 834 2174 **W:** www.communityradio.co.nz **E:** info@communityradio.co.nz **Mgr:** Phil Grey **Prgr:** BBC WS overnight Su-Fri **FM:** 106.7 LPFM [Hamilton & Tauranga] – **4)** P.O. Box 680, Hastings 4156 ☎ +64 6 878 8710 🖹 +64 6 871 0590 **W:** www.radiokidnappers.org.nz **E:** david@radiokidnappers.org.nz **Mgr:** David Teesdale **FM:** 104.7 – **5)** 301 Moray Place, Dunedin ☎ +64 3 471 6161 🖹 +64 3 471 6162 **W:** www.toroaradio.co.nz **E:** admin@toroaradio.co.nz **Prgr:** BBC WS overnight 1200-1800 daily **F.PL:** FM – **6)** c/o NMIT, Private Bag 19, Nelson 7042 ☎ +64 3 546 9891 🖹 +64 3 546 9892 **W:** www.freshfm.net **E:** nelson@freshfm.net **Mgr:** Mike Williams **FM Network:** 89.2 Blenheim/95.2 Takaka/104.8 Nelson City-Tasman – **7)** 92Queen Street, Masterton ☎+64 6 378 0255 **W:** www.arrowfm.co.nz **E:** quiver@arrowfm.co.nz **Mgr:** Michael Wilson – **8)** P.O. Box 22297, Christchurch ☎ +64 3 365 7997 🖹 +64 3 340 0967 **W:** www.plainsfm.org.nz **E:** info@plainsfm.org.nz **Mgr:** Nicki Reece **Prgr:** BBC WS overnight 1200-1800 daily – **9)** P.O. Box 1, Invercargill ☎ +64 3 218 9891 🖹 +64 3 214 1425 **W:** www.radiosouthland.org.nz **E:** southernmatt@xtra.co.nz **Mgr:** Matt Rutherford **Prgr:** BBC WS overnight 1200-1800 daily – **10)** P.O. Box 44215, Pt Chevalier, Auckland 1246 ☎ +64 9 815 8600 🖹 +64 9 815 8620 **W:** www.planetfm.org.nz **E:** info@planetfm.org.nz **Mgr:** Terri Byrne – **11)** P.O. Box 213, Waikanae ☎/🖹 +64 4 293 4838 **W:** www.coastaccessradio.org.nz **E:** accessradio.kapiti@xtra.co.nz **Mgr:** Graeme Joyes – **12)** PO Box 445, Taranaki Mail Center, New Plymouth 4340. ☎ 06 757 4039

TE MANGAI PAHO
✉ P.O. Box 10004, Wellington 6143. ☎ +64 4 915 0700 🖹 +64 4 915 0701 **W:** www.tmp.govt.nz **E:** radio@tmp.govt.nz **LP:** CEO: John Bishara, Mgr Radio Portfolio: Carl Goldsmith. **TMP** is the operational funding agency for the 21 independent commercial Maori Iwi Radio stns that operate 24h and often network programs overnight.

MAORI IWI RADIO (Comm.)
All Iwi stns and Maori TV are connected by PungaNet2 a broadband internet system for prgr and data sharing, monitoring and archiving. **W:** www.irirangi.net

	MW	KHz	kW	Station, location
1)		585	2	R.Ngati Porou, Ruatoria
2)		603	5	R.Waatea, Auckland
3)		765	2.5	R.Kahungunu, Napier-Hastings
4)		1161	5	Te Upoko o te Ika, Wellington
5)		1440	0.2	Moana AM, Tauranga
	FM	**MHz**	**kW**	**Station, location**
6)		89.0		Te Arawa FM, Rotorua
7)		89.8		Kia Ora FM, Palmerston North
8)		90.5		Tahu FM, Christchurch
9)		90.6		Raukawa FM, Tokoroa
10)		90.8		Tautoko FM, Mangamuka Bridge
11)		91.7		Turanga FM, Gisborne
12)		91.9		Maniapoto FM, Te Kuiti
13)		94.4		Te Hiku o te Ika, Kaitaia
14)		94.8		Te Korimako o, Taranaki, New Plymouth
15)		95.4		R.Tainui, Ngaruawahia
16)		96.9		Atiawa Toa FM, Lower Hutt
17)		97.6		Tuwharetoa FM, Turangi
18)		98.4		Sun FM, Whakatane
19)		99.1		Ngati Hine FM, Whangarei
20)		99.5		Nga Iwi FM, Paeroa
21)		100.0		Awa FM, Whanganui

Addresses & other information
1) P.O. Box 55, Ruatoria 4043 ☎ +64 6 864 8020 🖹 +64 6 864 8023 **W:** www.radiongatiporou.co.nz **E:** manager@radiongatiporou.co.nz **Mgr:** Rene Robati **FM Network:** 88.2 /89.3/90.1/93.3/98.1 – **2)** P.O. Box 43157, Favona, Mangere, Manukau 2153 ☎ +64 9 275 9070 🖹 +64 9 275 8060 **W:** www.waatea603am.co.nz **E:** info@waatea603am.co.nz **Mgr:** Willie Jackson – **3)** P.O. Box 2406, Hastings 4153 ☎ +64 6 872 8943 🖹 +64 6 876 4157 **W:** www.radiokahungunu.co.nz **E:** pat@radio-kahungunu.co.nz **Mgr:** Patricia Te Rangi **FM:** 94.3 – **4)** P.O. Box 11812, Manners Street, Wellington 6142 ☎ +64 4 801 5002 🖹 +64 4 801 5009 **E:** wena@teupoko.co.nz **Mgr:** Wena Tait – **5)** P. O. Box 382, Seventh Avenue, Tauranga 3140 ☎ +64 7 571 0009 🖹 +64 7 571 0007 **E:** charlie@moanaradio.co.nz **Mgr:** Charlie Tawhiao **FM:** 98.2 – **6)** P.O. Box 883, Rotorua 3040 ☎ +64 7 349 2959 **E:** rodger@tearawa.com **Mgr:** Rodger Cunningham – **7)** P.O. Box 1341, Palmerston NorthCentral, Palmerston North 4440 ☎ +64 6 353 1881 🖹 +64 6 353 1880 **E:** danielle@rangitaane.co.nz **Mgr:** Danielle Harris – **8)** P.O. Box 13469, Armagh, Christchurch 8141 ☎ +64 3 371 3905 🖹 +64 3 371 3901 **E:** blade_jones@ngaitahu.iwi.nz **Mgr:** Blade Jones **FM Network:** 90.5/91.1/95.0/99.6 – **9)** P.O. Box 842, Tokoroa 3444 ☎+64 7 886 0127 🖹 +64 7 886 0947 **E:** wendy@raukawafm.com **Mgr:** Wendy Biddle **FM Network:** 90.6/95.7 – **10)** Mangamuka Bridge RD2,

Okaihau 0476 ☎+64 9 401 8991 🖹 +64 9 401 9746 **E:** cyrilchapman@clear.net.nz **Mgr:** Cyril Chapman **FM Network:** 90.8/92.8/98.2 – **11)** P.O. Box 1224, Gisborne 4040 ☎ +64 6 868 6821 🖹 +64 6 868 1564 **W:** www.turangafm.co.nz **E:** fred@turangafm.maori.nz **Mgr:** Fred Maynard **FM Network:** 91.7/95.5/98.0 – **12)** P.O. Box 416, Te Kuiti 3941 ☎ +64 7 878 1160 🖹 +64 7 878 3002 **W:** www.maniapotofm.co.nz **E:** info@maniapotofm.co.nz **Mgr:** Jaqui Taituha **FM Network:** 91.9/92.7/96.5/99.6 – **13)** P.O. Box 458, Kaitaia 0441 ☎ +64 9 408 3944 🖹 +64 9 408 1061 **E:** wiremu@tehiku.co.nz **Mgr:** William Harrison – **14)** P.O. Box 4232, Taranaki Mail Centre, New Plymouth 4340 ☎ + 64 6 757 9055 🖹 +64 6 757 9093 **E:** tipene@tekorimako.co.nz **Mgr:** Tipene O'Brien – **15)** P.O. Box 208, Ngaruawahia 3742 ☎ +64 7 824 5650 🖹 +64 7 824 5659 **Mgr:** Trina Koroheke **FM Network:** 94.5/96.3/96.5 – **16)** P.O. Box 36111, Waiwhetu, Lower Hutt 5043 ☎ +64 4 569 7993 🖹 + 64 4 560 3278 **E:** wluke@atiawa.co.nz **Mgr:** Wirangi Luke **FM Network:** 94.9/96.9 – **17)** P.O. Box 198, Turangi 3353 ☎+64 7 386 0935 🖹 +64 7 386 0994 **E:** katipo@tuwharetoa.co.nz **Mgr:** Katipo Te Hiini **FM Network:** 90.4/92.6/97.6/100.6 – **18)** P.O. Box 2090, Kopeopeo, Whakatane **T:** +64 7 308 0403 🖹 +64 7 308 0150 **Mgr:** William Pryor **FM Network:** 96.9/98.4 – **19)** P.O. Box 1127, Whangarei 0110 ☎ +64 9 438 6115 🖹 +64 9 438 5767 **E:** mike@ngatihinefm.co.nz **Mgr:** Michael Kake **FM Network:** 96.4/99.5 – **20)** P.O. Box 135, Paeroa 3640 ☎+ 64 7 862 6247 🖹 + 64 7 862 6279 **E:** nifm@ngaiwifm.co.nz **Mgr:** Caroline Kara **FM Network:** 92.2/99.5 – **21)** P.O. Box 430, Whanganui 4540 ☎ + 64 6 347 1402 🖹 +64 6 347 2339 **E:** geoff@awafm.co.nz **Mgr:** Geoff Mariu **FM Network:** 91.0/93.5/100.

NATIONAL PACIFIC RADIO TRUST [PACIFIC MEDIA NETWORK, Com]
✉ P.O. Box 99582, Newmarket, Auckland. ☎ +64 9 361 6656 🖹 +64 9 361 3966 **LP:** CEO: Tom Etuata. Independent charitable trust funded by NZ On Air and Ministry for Culture & Heritage **Network station:** Niu FM "The Beat of the Pacific" 24h English & Pacific community languages **W:** www.niufm.com **E:** info@niufm.com **N:** Pacific R. News
Major Radio Market

Location	MHz	Location	MHz
Whangarei	103.6	Wellington	103.7
Auckland	103.8	Wellington	100.7
Hamilton	103.4	Dunedin	99.0
Rotorua	103.9	Christchurch	104.1
Taupo	104.0	Napier-Hastings	103.6
New Plymouth	103.6	Invercargill	103.6
Palmerston N.	103.4		

Local Station MW: 531pi, Auckland 531KHz 5kW. 24h Pacific community languages **W:** www.radio531pi.co.nz **E:** info@radio531pi.com **N:** Pacific R. News

RADIO BROADCASTERS ASSOCIATION
✉ P.O. Box 3762, Auckland. ☎ +64 9 378 0788 🖹 +64 9 378 8180 **W:** www.rba.co.nz **E:** david@rba.co.nz **LP:** Exec.Dir: David Innes. **RBA** represents the interests of the NZ commercial radio industry and sponsors The NZ Radio Awards.

THE RADIO BUREAU
✉ P.O. Box 8049, Symonds Street, Auckland. ☎+64 9 302 1122 **W:** www.trb.co.nz **E:** gills@trb.co.nz **LP:** GM: Gill Stewart. **TRB** is a joint venture of RadioWorks and TRN and acts as a singe radio marketing and sales resource for advertising agencies.

MAJOR COMMERCIAL NETWORKS
RADIOWORKS
Corporate: Level 2,239 Ponsonby Road, Ponsonby, Auckland 1011. **Postal:** P.O. Box 8880, Symonds Street, Auckland 1150. ☎ 64 9 928 9300 🖹 +64 9 373 4000 **W:** www.mediaworks.co.nz **LP:** Group MD: Sussan Turner, CEO RadioWorks: Belinda Mulgrew. **Prgrs:** 24h **Other Media:** TV3, TV4 **Owner:** Ironbridge Capital [Australia].
Network Stations: P.O. Box 47560, Ponsonby, Auckland 1144 ☎ +64 9 928 9000 🖹 +64 9 361 1677 **RadioLive:** P.O. Box 8880, Symonds Street, Auckland 1150 ☎ +64 9 928 9270 🖹 +64 9 360 0390. **George FM:** P.O. Box 47864, Ponsonby, Auckland 1144 **T:** +64 9 928 9150 🖹 +64 9 360 0044. **Prgr:** 24h from Auckland studios. N.Live bulletins.
Brands: The Edge **W:** www.theedge.co.nz, Kiwi **W:** www.kiwifm.co.nz, R.Live **W:** www.radiolive.co.nz, LiveSPORT **W:** www.radiolivesport.co.nz, The Rock **W:** www.therock.co.nz, Solid Gold FM **W:** www.solidgoldfm.co.nz, George FM **W:** www.georgefm.co.nz. For full FM listings including non-major market locations see individual brand websites.
Radio Trackside: shares LiveSPORT frequencies overnight and weekends with independent local and international racing commentaries. Owned by the Totaliser Agency Board [TAB]: P.O. Box 38899, Wellington Mail Centre, Lower Hutt 5045 **W:** www.radiotrackside.co.nz

Radio Live

MW	kHz	kW	MW	KHz	kW
Auckland	702	10	Tauranga	1107	1
Christchurch	738	5	Wellington	1233	2
Rotorua	1107	1	Napier-Hastings	1368	1

LiveSPORT

MW	kHz	kW	MW	KHz	kW
Napier-Hastings	549	1	Dunedin	1206	2
Wellington	711	5	Invercargill	1224	2
Palmerston North	828	2	Timaru	1242	1
Tauranga	873	1	Christchurch	1260	2
Hamilton	954	2	Auckland	1476	5
Nelson	990	1	Gisborne	1485	1
Ashburton	1071	1	Rotorua	1548	0.9

FM	1	2	3	4	5	6	7
Whangarei	94.0	-	90.8	92.4	90.0	-	-
Auckland	94.2	102.2	100.6	-	90.2	93.8	96.6
Hamilton	97.8	-	100.2	-	93.0	93.8	-
Tauranga	97.8	-	100.0	-	94.2	92.6	-
Rotorua	99.9	-	95.1	-	92.7	91.1	-
Taupo	88.8	-	99.2	91.2	94.4	100.0	-
Gisborne	99.7	-	94.9	-	94.1	96.5	-
Napier-Hastings	98.3	-	-	-	95.1	91.9	-
New Plymouth	94.0	-	89.2	97.2	95.6	98.0	-
Palmerston North	93.0	-	93.8	-	95.4	94.6	-
Kapiti Coast	97.5	-	99.1	-	91.9	94.3	-
Wellington	91.7	102.1	98.9	-	96.5	97.3	-
Nelson	88.8	-	96.0	-	94.4	98.4	-
Christchurch	88.9	102.5	99.3	-	93.7	92.9	-
Timaru	95.5	-	89.9	-	91.5	97.1	-
Dunedin	91.8	-	96.6	-	93.4	90.2	-
Queenstown	95.2	-	91.2	93.6	100.0	97.6	96.8
Invercargill	97.2	-	94.0	-	90.8	98.0	-

1= The Edge, **2**= Kiwi, **3**= R.Live, **4**=LiveSPORT, **5**= The Rock, **6**= Solid Gold, **7**= George

Local Stations:
Local breakfast shows and other daytime prgrs [some stns only]. Overnight and weekends increasingly networked from Auckland except for R.Dunedin. **Prgr:** 24h. **N:** R.Live bulletins. **Brands:** The Breeze **W:** www.thebreeze.co.nz More FM **W:** www.morefm.co.nz Mai FM **W:** www.maifm.co.nz R.Dunedin **W:** www.radiodunedin.co.nz

Local Stations:

MW	KHz	kW	Station, location
1)	531	2	More FM, Alexandra
2)	891	5	The Breeze, Wellington
3)	1305	2.5	R.Dunedin, Dunedin
22)	1359	1	More FM, Queenstown

	Location	The Breeze	More FM	Mai FM
4)	Northland	-	91.6	98.0
5)	Auckland	93.4	91.8	88.6
6)	Waikato	99.4	92.2	-
7)	Tauranga	95.8	93.4	96.6
8)	Mercury Bay	96.6	-	-
9)	Rotorua	91.9	95.9	105.5
10)	Taupo	-	93.6	-
11)	Gisborne	-	98.9/90.1	-
12)	Hawkes Bay	97.5	92.7	105.5
13)	Taranaki	92.4	93.2	-
14)	Wanganui	-	92.8	97.0
15)	Manawatu	98.6	92.2	97.0
16)	Wairarapa	99.8	89.3/105.5	-
17)	Horowhenua	-	-	95.1
18)	Kapiti	100.7	90.3	-
2)	Wellington	94.0/98.5	94.7/99.7	-
19)	Nelson	97.6	92.8	-
20)	Marlborough	96.1/97.3/98.5	92.9/94.5/96.3	-
21)	Christchurch	94.5	92.1	-
3)	Dunedin	98.2	97.4	-
1)	Central Otago	96.7	90.3/94.3	-
22)	Queenstown	99.2	92.0/99.4	-
23)	Southland	91.6	89.2	-

Addresses & other information
1) P.O. Box 143, Alexandra 9340 ☎ +64 3 901 6200 🖷 +64 3 448 6502 **-2)** P.O. Box 11441, Manners Street, Wellington 6142 ☎+64 4 915 1000 🖷 +64 4 915 1009 – **3)** P.O. Box 1957, Dunedin 9054 **R.Dunedin:** ☎ +64 3 477 6934 **Mgr:** Cindy Davies **E:** cdavies@radioworks.co.nz **FM** 99.8MHz**The Breeze/More FM:** ☎ +64 3 951 3600 🖷 +64 3 477 6874 – **4)** P.O. Box 100, Whangarei 0140 ☎ +64 9 928 9990 🖷 +64 9 438 2348 **-5)** **The Breeze/More FM:** P.O. Box 8880, Symonds Street, Auckland 1150 ☎+64 9 928 9300 🖷 +64 9 373 4000 **Mai FM:** P.O. Box 68886, Newton, Auckland ☎ +64 9 977 7800 🖷 +64 9 977 7801 – **6)** P.O. Box 19293, Hamilton 3244 ☎ +64 7 958 7050 🖷 +64 7 838 2893 – **7)** P.O. Box 13344, Tauranga 3141 ☎ +64 7 928 7300 🖷 +64 7 577 0294 – **8)** P.O. Box 16, Whitianga **v** +64 7 866 5696 🖷 +64 7 866 2553 – **9)** P.O. Box 92, Rotorua 3040 ☎ +64 7 921 7630 🖷 +64 7 348 3830 – **10)** P.O.

Box 393, Taupo 3351 ☎+64 7 906 7500 🖷 +64 7 378 2701 – **11)** P.O. Box 468, Gisborne 4040 ☎+64 6 986 3700 🖷 +64 6 869 0037 – **12)** P.O. Box 193, Hastings 4156 ☎ +64 6 974 6150 🖷 +64 6 876 5626 – **13)** P.O. Box 869, Taranaki Mail Centre, New Plymouth **v** +64 6 968 6200 🖷 +64 6 757 5020 – **14)** P.O. Box 928, Wanganui 4540 ☎ +64 6 965 6300 🖷 +64 6 345 5592 – **15)** P.O. Box 446, Palmerston North Central, Palmerston North 4440 ☎ +64 6 952 6420 🖷 +64 6 356 1317 – **16)** P.O. Box 881, Masterton ☎+64 6 370 2548 🖷 +64 6 378 8877 – **17)** P.O. Box 603, Levin ☎+64 6 368 2827 🖷 +64 6 368 0415 – **18)** P.O. Box 132, Paraparaumu 5254 ☎ +64 4 903 0400 🖷 +64 4 297 2999 – **19)** P.O. Box 907, Nelson 7040 ☎+64 3 546 9670 🖷 +64 3 546 9427 – **20)** P.O. Box 930, Blenheim ☎+64 3 579 0393 – **21) The Breeze:** Private Bag 4750, Christchurch 8140 ☎ +64 3 961 3102 🖷 +64 3 366 5301 **More FM:** P.O. Box 25209, Victoria Street, Christchurch 8144 ☎+64 3 961 3322 🖷 +64 3 377 1993 – **22)** P.O. Box 224, Queenstown ☎+64 3 901 0810 🖷 +64 3 442 7799 – **23)** P.O. Box 1740, Invercargill ☎ +64 3 948 3900 🖷 +64 3 218 8015

Associated Radioworks FM Stations
These stns operate as independent local brands. Overnights networked from More FM.**Prgr:** 24h. **N:** R.Live bulletins.
1) Coromandel FM, Paeroa 89.0MHz, **2)** Times FM, Auckland (Orewa) 89.9MHz, **3)** Big River R., Balclutha 92.9MHz

Addresses & Other Information
1) PO Box 962, Thames 3540. ☎+64 7 868 6063 🖷 +64 7 868 6681 **Mgr:** Warren Male **W:** www.coromandelfm.co.nz **FM Network:** 89 .0/89.1/89.9/90.3/93.8/93.9/94.0/96.2/97.5 – **2)** PO Box 755, Orewa 0931. ☎ +64 9 928 9940 🖷 +64 9 427 0251 **GM:** Anna McGovern **E:** amcgovern@mediaworks.co.nz**W:** www.timesfm.co.nz **FM Network:** 89.9/93.9 – **3)** 1ˢᵗ Fl. PO John Street, Balclutha ☎+64 3 418 1969 **W:** www.bigriverradio.webs.com **FM Network:** 92.9/93.7

RHEMA BROADCASTING GROUP

🖳 **Corporate:** 53 Upper Queen Street, Auckland. 🖳 **Postal:** Private Bag 92636, Symonds Street, Auckland 1150. ☎ +64 9 307 1251 🖷 +64 9 309 6888 **W:** www.rbg.co.nz **L.P:** Chief Exec: John Fabrin. **Other Media:** Shine TV, UCB International **Owner:** NZ charity.
Network Stations. Prgr: 24h from Auckland studios. Southern Star also broadcasts on RNZ AM Network transmitters when Parliament is not in session. **N:** IRN bulletins.
Brands: Life FM **W:** www.lifefm.co.nz, NZ's Rhema **W:** www.rhema. co.nz, Southern Star **W:** www.sstar.co.nz, The Word/Bible Radio **W:** www.bibleradio.co.nz. For full FM listings including non-major market locations see individual brand websites.

NZ's Rhema

MW	kHz	kW	MW	KHz	kW
Tauranga	540	5	Dunedin	621	2
New Plymouth	540	3	Gisborne	684	5
Kaitaia	549	3	Nelson	801	5
Wanganui	594	2	Hamilton	855	2
Timaru	594	2	Wellington	972	5
Christchurch	612	2	Auckland	1251	5
Whangarei	621	2	Invercargill	1404	5

Southern Star

MW	kHz	kW	MW	KHz	kW
Tauranga	657	10	Christchurch	963	10
Wellington	657	50	Timaru	981	2.5
Auckland	882	10	Invercargill	1314	5
Dunedin	900	10	Hamilton	1494	3
Napier-Hastings	909	5			

Word/Bible Radio

MW	kHz	kW	MW	KHz	kW
Hamilton	576	2.5	Dunedin	1377	3
Invercargill	1026	2.5			

FM: NZ's Rhema: Palmerston North 91.4, Rotorua 93.5, Queenstown 94.4 Taupo 95.2, Napier-Hastings 99.1 – **Life FM** Kapiti Coast 96.7, Gisborne 100.5 – **Southern Star** Gisborne 92.5, Napier-Hastings 93.5 Nelson 93.6, Timaru 93.9, Dunedin 94.2, Hamilton 94.6, Tauranga 94.6, Palmerston North 96.2, Whangarei 98.8, Wellington 99.2, New Plymouth 99.6, Auckland 99.8, Invercargill 100.0

THE RADIO NETWORK [TRN]

🖳 **Corporate:** 54 Cook Street, Auckland. 🖳 **Postal:** Private Bag 92198, Auckland Mail Centre, Auckland 1142. ☎ +64 9 373 0000 🖷 +64 9 367 4802. **W:** www.radionetwork.co.nz **L.P:** CEO: John McElhinney, Dir. of Engineering: Norm Collinson. **Other Media:** APN News & Media newspapers and magazines **Owner:** The Australian Radio Network.
Network Stations. Prgr: 24h from Auckland studios **N:** NewstalkZB bulletins. **Brands:** Coast **W:** www.thecoast.net.nz Flava **W:** www. flava.co.nz Radio Hauraki **W:** www.hauraki.co.nz NewstalkZB **W:** www.newstalkzb.co.nz Radio Sport **W:** www.radiosport.co.nz EasyMix **W:** www.easymix.co.nz ZM **W:** www.zmonline.com For full FM listings including non-major market locations see individual brand websites.

Radio Sport

MW	kHz	kW	MW	KHz	kW
Nelson	549	1	Westport	1287	2
Invercargill	558	5	Auckland	1332	10
Dunedin	693	5	Rotorua	1350	1
Ashburton	702	1	Levin	1377	2
Whangarei	729	3	Timaru	1494	2.5
New Plymouth	774	5	Wellington	1503	5
Hamilton	792	5	Christchurch	1503	2.5
Wanganui	1062	1	Tauranga	1521	1
Palmerston North	1089	2.5	Blenheim	1539	1
Napier-Hastings	1125	1			

NewstalkZB

MW	kHz	kW	MW	KHz	kW
Rotorua	747	0.4	Auckland	1080	10
Masterton	846	2	Christchurch#	1098	5
Invercargill	864	10	Timaru	1152	2
Ashburton	873	1	Wanganui	1197	2
Palmerston N.	927	2	Kaikohe	1215	2
Gisborne	945	2	Hawera	1278	3
Tauranga	1008	10	Napier-Hastings	1278	2
Kaitaia	1026	2	Hamilton	1296	2.5
Whangarei	1026	2	Nelsoni	1341	2
Wellington#	1035	20	Oamaru	1395	2
Dunedin	1044	10	Tokoroa	1413	2
New Plymouth	1053	2			

Radio Hauraki

MW	kHz	kW	MW	KHz	kW
Christchurch	1017	2.5	Dunedin	1125	1

Coast

MW	kHz	kW	MW	KHz	kW
Whangare	900	2.5	Palmerston North	1548	1
Dunedin	954	1	Hawera	1557	2
New Plymouth	1359	2.5	Christchurch	1593	2.5
Napier-Hastings	1530	1			

FM	1	2	3	4	5	6	7
Whangarei		106.0	93.2				94.8
Whangarei			93.6				
Auckland	105.4	95.8	99.0	89.4		98.2	91.0
Hamilton			96.2	97.0			89.4
Tauranga	100.6	90.2	91.0			99.0	89.8
Rotorua	96.7	89.5	94.3				98.3
Taupo			92.8	96.0	90.4		
Napier-Hastings		96.7	99.9			90.3	95.9
Napier-Hastings							99.7
New Plymouth	92.6		90.8				90.6
Palmerston North			105.8				90.6
Kapiti Coast	95.9			89.5			91.1
Wellington	99.4		93.3		89.4	93.7	90.9
Nelson	100.8		90.4				96.8
Christchurch		106.5					90.9/91.3
Timaru							96.3
Dunedin			106.2				95.8
Queenstown			89.6/90.6/95.1				
Invercargill	92.4		93.2				95.6

1=Coast, 2=Flava, 3=R.Hauraki, 4=NewstalkZB, 5=R.Sport, 6=EasyMix, 7=ZM

Local StationClassic Hits

MW	kHz	kW	MW	KHz	kW
1) Takaka	1269	0.4	2) Picton	1584	0.4

Local Station Classic Hits FM

FM	MHz	FM	MHz	FM	MHz
23) Dunedin	89.4	18) Greymouth	90.9	2) Blenheim	96.9
11) Hawkes Bay	89.5	18) Greymouth	91.1	3) Northland	97.2
1) Nelson	89.6	20) Ashburton	92.5	18) Greymouth	97.3
13) Wanganui	89.6	16) Kapiti	92.7	4) Auckland	97.4
12) Taranaki	90.0	18) Greymouth	93.1	7) Rotorua	97.5
15) Masterton	90.1	21) Timaru	94.7	3) Northland	97.6
17) Wellington	90.1	6) Tauranga	95.0	19) Christchurch	97.7
24) Central Otago	90.4	3) Northland	95.6	14) Manawatu	98.4
25) Southland	90.4	24) Central Otago	96.2	22) Oamaru	98.4
18) Greymouth	90.5	8) Tokoroa	96.4	5) Waikato	98.6
12) Taranaki	90.7	19) Christchurch	96.5	21) Timaru	98.7
7) Rotorua	90.8	3) Northland	96.8	25) Southland	98.8
10) Gisborne	90.9	9) Taupo	96.8	24) Central Otago	99.9

Addresses & other information

1) P.O. Box 43, Nelson 7043 ☎ +64 3 548 1064 🖷 +64 3 546 2580 **GM:** Mike McElhinney **E:** mikemcelhinney@radionetwork.co.nz – 2) P.O. Box 225, Blenheim ☎+64 3 579 2969 🖷 +64 3 578 0981 **GM:** Thelma Sowman **E:** thelmasowman@radionetwork.co.nz – 3) P. O. Box 845, Whangarei ☎ +64 9 430 4950 🖷 +64 9 430 4968 **GM:** Murray Madden **E:** murraymadden@radionetwork.co.nz -4) Private Bag 92198, Auckland ☎ +64 9 373 0000 🖷 +64 9 367 4802 **GM:** Grant Lee **E:** grantlee@radionetwork.co.nz – 5) P.O. Box 489, Hamilton ☎+64 7 858 0700 🖷 +64 7 858 0730 **GM:** Craig Hobbs **E:** craighobbs@radionetwork.co.nz – 6) P.O. Box 642, Tauranga ☎+64 7 578 9139 🖷 +64 7 577 8522 **GM:** Andrew Love **E:** andrewlove@radionetwork.co.nz – 7) P.O. Box 1147, Rotorua ☎

+64 7 348 9089 🖷 +64 7 349 5527 **GM:** Aaron Gillions **E:** aarongillions@radionetwork.co.nz – 8) P.O. Box 272, Tokoroa ☎ +64 7 668 9431 🖷 +64 7 886 8391 **GM:** Jenny Shattock **E:** jennyshattock@radionetwork.co.nz – 9) P.O. Box 967, Taupo ☎ +64 7 376 0550 🖷 +64 7 378 0030 **GM:** Brian Jennings **E:** brianjennings@radionetwork.co.nz – 10) P.O. Box 1040, Gisborne ☎+64 6 867 2139 🖷 +64 6 867 8309 **GM:** Darryl Monteith **E:** darrylmonteith@radionetwork.co.nz – 11) P.O. Box 241, Napier ☎ +64 6 833 8400 🖷 +64 4 833 8421 **GM:** Rebecca Johnson **E:** rebeccajohnson@radionetwork.co.nz – 12) P.O. Box 141, New Plymouth ☎ +64 6 759 2460 🖷 +64 6 759 2440 **GM:** Richard Williams **E:** richardwilliams@radionetwork.co.nz – 13) P.O. Box 632, Wanganui ☎ +64 6 348 1176 🖷 +64 6 345 6402 **SM:** Megan Wishnowsky **E:** meganwishnowsky@radionetwork.co.nz – 14) P.O. Box 1045, Palmerston North ☎ +64 6 350 3550 🖷 +64 6 350 3580 **GM:** Richard Williams **E:** richardwilliams@radionetwork.co.nz – 15) P.O. Box 220, Masterton ☎+64 6 370 5014 🖷 +64 6 370 8460 **GM:** Phil Conn **E:** philconn@radionetwork.co.nz – 16) P.O. Box 596, Paraparaumu ☎+64 4 382 6677 🖷 +64 4 385 4210 **GM:** Rhys Nimmo **E:** rhysnimmo@radionetwork.co.nz – 17) P.O. Box 300, Wellington ☎+64 4 802 4710 🖷 +64 4 385 4210 **GM:** Rhys Nimmo **E:** rhysnimmo@radionetwork.co.nz – #NewstalkZB 1035MW/90.1FM carries local breakfast show - 18) P.O. Box 378, Greymouth ☎+64 3 768 7068 🖷 +64 3 768 7067 **GM:** Mike McElhinney **E:** mikemcelhinney@radionetwork.co.nz – 19) P.O. Box 1484, Christchurch ☎ +64 3 379 9600 🖷 +64 3 363 3510 **GM:** Andrew Britt **E:** andrewbritt@radionetwork.co.nz – #NewstalkZB 1098MW carries local breakfast show - 20) P.O. Box 465, Ashburton ☎+64 3 307 8927 🖷 +64 3 307 8930 **GM:** Kerry Treymane **E:** kerrytrymane@radionetwork.co.nz – 21) P.O. Box 275, Timaru ☎ +64 3 684 8152 🖷 +64 3 688 6733 – 22) P.O. Box 426, Oamaru ☎ +64 3 433 1090 🖷 +64 3 433 1087 **GM:** Gary Watling **E:** garywatling@radionetwork.co.nz – 23) P.O. Box 888, Dunedin ☎+64 3 474 8400 🖷 +64 3 474 8422 **GM:** Lee Piper **E:** leepiper@radionetwork.co.nz – 24) Level 1, 11 Earl Street, Church Lane, Queenstown ☎ +64 3 441 2784 🖷 +64 3 441 2785 **SM:** Craig McKenzie **E:** craigmckenzie@radionetwork.co.nz – 25) P.O. Box 802, Invercargill ☎+64 3 211 1500 🖷 +64 3 211 1532 **SM:** Nick Jeffrey **E:** nickjeffrey@radionetwork.co.nz

Associated TRN FM Stations

These stns operate as independent local brands. Overnights are networked from Classic Hits. **Prgr:** 24h. **N:** NewstalkZB bulletins.
1) R.Clutha 91.3MHz 1.5kW The Lane, Balclutha 9230. ☎ +64 3 418 2884. **Mgr:** Lorraine McLean **Prgr:** local 1900-2400 then relays Hokonui Gold **W:** www.hokonuigold.co.nz– **2) Hokonui Gold:** Hokonui Gold 'Southland's Hits & Memories' 94.8/95.2MHz, P.O. Box 292, Gore 9700 ☎+64 3 208 9325 🖷 +64 3 208 9326 **W:** www.hokonuigold.co.nz **Prgr:** local 1700-2200 then relays Classic Hits Southland.

INDEPENDENT STATIONS

MW	KHz	kW	Station, location
1)	729	0.1	Burn729am, Ranfurly
2)	756	0.8	Puketapu R., Palmerston
3)	783	10	Samoan Capital R., Wellington
4)	810	2	BBC World Service NZ, Auckland
5)	936	1	New Supremo, Auckland
6)	990	1	Apna 990 Auckland
7)	1179	5	R.Ake ,Auckland
8)	1242	2	1XXOne Double X, Whakatane
8)	1242	0.1	1XXOne Double X, Murupara
9)	1368	0.8/0.1	1XT Village R., Tauranga
10)	1386	10	R.Tarana, Auckland
11)	1413	1	3XP R. Ferrymead, Christchurch
12)	1440	1	Goldrush R., Lawrence
13)	1593	5	R.Samoa, Auckland
14)	1602	2.5	R.Reading Service, Levin

FM	MHz	kW	Station, location
17)	89.2		Kaitaia Community R, Kaitaia
19)	89.4		The Rhythm, Waihi Beach
21)	89.8		Ski FM, Ohakune
22)	90.3		Vision FM, Tokoroa
23)	90.3		Coast FM, Westport
8)	90.5		1XX One Double X, Whakatane
24)	90.6		R.Chinese Auckland
25)	90.8		Waihi Community R, Waihi
26)	90.8		Thames Valley FM, Thames
27)	91.0		Radio 1*, Dunedin
28)	92.2		R.Wanaka, Wanaka
8)	92.9		Kiwi FM, Te Puke
8)	93.0		Bayrock, Wanaka
31)	93.1		Port FM, Timaru
32)	93.5		Central FM, Waipukurau
34)	95.0		95bfm* ,Auckland
35)	95.8	3	Real Good Life FM, Auckland
68)	96.0		Next FM, Queenstown
36)	96.2		Big River R., Dargaville
37)	96.4		Gold FM, Waihi

FM	MHz	kW	Station, location
8)	97.7	2	Bayrock, Whakatane
38)	98.1		Raglan Community R., Raglan
39)	98.5	1.5	RDU*, Christchurch
40)	99.1		Peak FM, Taumarunui
42)	99.8		Sea Breeze FM, Himatangi Beach
43)	100.3		Blue FM, Kaikoura
33)	100.4		Cruise FM, Tokoroa
45)	100.7	4	Bay FM, Napier-Hastings
46)	100.9	5	F.PL, Christchurch
47)	104.9		Compass FM, Christchurch
50)	105.2	1.5	Cruize 105.2, New Plymouth
51)	105.2		Labotomy Pants FM, Hamilton
52)	105.2	4	Country R., Invercargill
54)	105.3	2	X105, Wellington
55)	105.4	8	BOP 105.4 FM, Tauranga
12)	105.6	1.25	Goldrush R., Queenstown
57)	105.6	5	F.PL, Nelson
59)	105.6		R.Waipu, Waipu
60)	105.7	10	Pulzar FM, Christchurch
61)	105.8		Wellsford Country R., Wellsford
63)	106.2	15	Big FM Auckland (inactive)
65)	106.4	2	Timeless Taupo FM, Taupo
66)	106.4		Heads FM, Mangawhai

Addresses & other information

1) 3 Charlemont St East, Ranfurly 9332 – **2)** 118 Ronaldsay St, Palmerston 9430 **W:** www.puketapuradio.com **Prgr:** Day: local community radio – Night: relay UK based Planet Caroline online stream – **3)** PO Box 6647, Marion Square, Wellington 6141 **Prgr:** Samoan "Siufofaga O Le Laumua" Mon-Fri 1900-0100 – **4)** Auckland Radio Trust, PO Box 800, Shortland Street, Auckland 1001 **W:** www.worldservice.co.nz **Prgr:** BBC World Service satellite relay, RNZI Dateline Pacific and local advertising – **5)** PO Box 12743, Penrose, Auckland 1642 **W:** www.chinesevoice.co.nz **Prgr:** Chinese [Mandarin] – **6)** Level 3, 362 Great North Rd, Henderson, Waitakere 0612 **W:** www.apna990.co.nz**Prgr:** Hindi – **7)** c/o Spoke Communications, PO Box 52148, Kingsland, Auckland 1352 **Prgr:** Contemporary Urban Maori – **8)**Radio Bay of Plenty Ltd, PO Box 383, Whakatane 3158 **Brands:** 1XX **W:** www.1xx.co.nz Kiwi FM [146 Jellicoe St, Te Puke 3119] Bayrock **W:** www.bayrock.co.nz **FM Network:** 97.7/99.3 and 93.0 Wanaka – **9)** PO Box 597, Seventh Avenue, Tauranga 3140 **FM:** LPFM 106.7 relays Community Access R. [Hamilton 1206AM] 0400-2200 daily. **F.PL:** fulltime Community Access R. – **10)** PO Box 5956, Wellesley Street, Auckland 1141 **W:** www.tarana.co.nz **Prgr:** Hindi – **11)** Box 19090, Woolston, Christchurch 8241 **W:** www.radioferrymead.co.nz**Prgr:** Fri 0300-Mon 1200 including automated prgrs, Statutory Holidays & Christmas New Year period – **12)** 24 Ross Place, Lawrence 9532 **Brands:** Goldrush 1440 **W:** www.goldrush1440.com Rush99 **W:** www.rush99.net.ms Pink FM [LPFM] **W:** www.pinkfm.net.ms Independent Network News **W:** www.nzinn.com – **13)** PO Box 200105, Papatoetoe Central, Manukau 2156 **W:** www.radiosamoa.co.nz **Prgr:** Samoan – **14)** PO Box 360, Levin 5500 **W:** www.radioreading.org.nz **Prgr:** Mon-Fri 2000-1800, Sat 1200-0500, Sun 0000-0800 – RNZ National at other times – **17)** PO Box 690, Kaitaia – **19)** 7 The Crescent, Waihi Beach 3611 – **21)** PO Box 201, Ohakune **FM Network:** 89.5/89.8 /93.4/94.4/94.6/96.6/106.2/106.3 **W:** www.934skifm.co.nz – **22)** 9 Glendevon Pl, Tokoroa – **23)** PO Box 249, Westport 7866 **W:** www.westportnews.co.nz **FM Network:** 90.3 and others – **24)** PO Box 82343, Highland Park, Auckland **W:** www.chineseradio.co.nz **Prgr:** Chinese – **25)** PO Box 260, Waihi 3641 – **26)** 529 Pollen Street, Thames – **27)** PO Box 1436, Dunedin – **28)** 42 Norrie Ave, Raglan 3225 **W:** www.raglanradio.com – **29)** PO Box 31244, Christchurch 8444 **W:** www.rdu.org.nz – **40)** PO Box 37, Raetihi 4632 **W:** www.twitter.com/peakfm958 **FM Network:** 92.7/99 and others – **42)** 42 Muapoko St, Himatangi Beach **W:** www.seabreezefm.com**Prgr:** Fri night, Sat-Sun only – **43)** Scarborough Street, Kaikoura – **45)** Heretaunga St, West, Hastings – **46)** c/o David Moore, 12 Walters Rd, Marshland, Christchurch – **47)** PO Box 27, Rangiora 7440 "The Voices of North Canterbury" **F.PL:** fulltime Community Access R. – **50)** 603 Devon St East, New Plymouth **W:** www.facebook.com/pages/new-plymouth/cruize-fm-1052 – **51)** PO Box 1183, Hamilton – **52)** 145 Islington St, Invercargill **W:** www.countryradionetwork.com/invercargill **FM Network:** 105.2 and LPFM – **54)** Level 1, 15 Walter St, Wellington **W:** www.x105.co.nz – **55)** Cnr Devonport Rd & Elizabeth St, Tauranga 3110 **W:** www.bopfm.co.nz – **57)** Wild Tomato Media, 243 Trafalgar

St, Nelson – **59)** Wave FM Ltd, 7 Finch St, Marsden Cove, One Tree Point, Ruakaka 0118 – **60)** PO Box 13209, Christchurch 8141 **W:** www.pulzarfm.co.nz (studio destroyed in earthquake September 2010) – **61)** 32 Rusty Brook Rd, Wellsford – **63)** Currently inactive but r. to return to air in 2011 under new ownership – **65)** Field & Sport Ltd, 23 Scannell Rd, Taupo – **66)** PO Box 180, Mangawhai 0540 – **68)** c/o Stuart Campbell, Neplusultra St, Cromwell 9191.
NB: + 16 FM stations below 1kW

Notes: 1) Over 250 FM stns are changing frequency between mid-2010 and mid-2011 and as final allocations are ongoing at time of print the information here is subject to change. Transmitter power changes are also taking place as necessary so technical information is also subject to change. **2)** All commercial FM licences are being renewed for 20 years in 2011 and licence holders must bid for continued rights to operate on current frequencies. Considerable changes are expected during 2011 **3)** Stns marked * are bNet student radio stns which is a loose prgr and advertising sales network **4)** An estimated 500 or more LPFM [1w or less] stns broadcast on 87.6-88.3 and 106.7-107.7 throughout the country. A regularly updated guide to these authorized but otherwise unlicenced stns is maintained by the Radio Heritage Foundation **W:** www.radioheritage.net

NICARAGUA

L.T: UTC -6h — **Pop:** 5.8 million — **Pr.L:** Spanish — **E.C:** 60Hz, 120V — **ITU:** NCG — **Intl. dialling code:** +505

TELCOR – INSTITUTO NICARAGÜENSE DE TELECOMUNICACIONES Y CORREOS
✉ Ave. Bolívar Esquina Diagonal a la Cancillería, Managua
☎ +505 2222 7350 **W:** www.telcor.gob.ni

ASOCIACION NICARAGUENSE DE RADIODIFUSION (ANIR)
✉ c/o R. Ya, Frente a la Universidad Centroamericana, Managua

CAMARA NICARAGUENSE DE RADIODIFUSION
✉ c/o R. Corporación, Cd. Jardín Q-20, Av. Ponciano Lombillo, Managua

Hrs of tr usually 24h – see list for variations. Call YN–

MW	Call	kHz	kW	Station, location & h. of tr.
MA01)	A3OW	540	25	R. Corporación, Managua: 0950-0505
CH01)	A2RQ	570	5	R. Veritas 5-70, Chinandega: 1030-0250
MA02)	A3LP	580	10	R. 5-80, Managua: 1030-0000 (Sat 1200-, Sun 1100-)
MA03)	A3MD	600	10	La Nueva R. Ya, Managua: 1000-0600 (SS 24h)
MA04)	N	620	50	R. Nicaragua, Managua: 1000-0400
MA04)	A4LR	640	10	La Mera Mera , Managua
MT01)	A6RS	650	5	R. Muzun, Matagalpa: 1100-0100
GR01)	RD	650	10/8	R. Diriangén "La Super D", Granada: 0950-2300
MA25)		*660	5	R. Máxima, Managua
ZE01)	RC	*670		R. Caribe, Pto Cabezas
MA05)	AM	680	10/2	R. La Primerísima, Managua: 1045-0500 (SS -2400)
MT02)	RH	690	10/5	R. Hermanos, Matagalpa: 1000-0400
MA06)	A3RC	720	25	R. Católica, Managua: 1000-0430
MA07)	A3LS	740	50	R. Sandino "La S Grande", Managua: 1000-0400
MA08)	A3RO	800	10	R. 800, Managua: 0800-0500
MA09)	FAOL	820	20	R. Ondas de Luz, Managua: 1000-0400
MA10)	A3NT	840	5	R. Noticias, Managua: 1030-0200
CT01)	CD	870	10	R. Centro, Juigalpa: 1100-0200
MA11)	A3EP	880	10	R. El Pensamiento, Managua: 1100-0300
MA12)	A3RT	900	5	R. Tiempo, Managua: 1050-0400
JI01)		910	5	R. Jinoteca, Jinoteca
MA13)	W	920	10	R. Mundial, Managua: 1000-0400
RS01)	ACTH	960	2.5	LV del Trópico Húmedo, San Carlos: 1000-0300
MA13)	A3NO	980	1	R. Redención Internac., Managua
MA15)	FF	1000	10	R. Mil, Managua: 1200-0400
NS02)	FAVP	1010	5	R. LV del Pinar, Ocotal: 1100-0400
JI02)	VJ	1040	2	LV de Jinoteca, Jinoteca
MS02)	LL	1050	3	R. Masaya, Masaya
ZE03)		*1060		LV del Atlántico, Bluefields
MA16)	A3LC	1080	10	R. 15 de Septiembre, Managua: 1100-0100
ES01)	HAAL	*1090	5	R. Alma Latina, Estelí: 1100-0400
LE01)	F2MT	*1110	1	R. Momotombo, La Paz Centro
MA17)	A3CP	1120	5	R. CEPAD "El Arco Iris del Amor", Managua: 1100-0100
NS03)		1130	0.5	Voz Evangélica de Jalapa, Jalapa
LE02)	A2RD	*1150	5	R. Darío, León
ES02)	HM	*1160	1	R. Satélite, Estelí
MA18)	A3AC	1200		1200 La Radio, Managua
MA19)	A3RA	1220	1	R. América, Managua: 1200-0400

MW Call	kHz	kW	Station, location & h. of tr.
AS01) MNG	1230	5	R. Manantial, Nueva Guinea: 1000-0300
MA20) A3RR	1240	5	R. Vida Managua
ES03) CR	1250	2.5	Cad. Radial Samaritano, Condega: 1000-0400
MT03) RA	1270	3	R. Amistad, Matagalpa
MA21) A2CC	1300	1	Canal 130 AM, Managua: 1200-2330
CH02) SC	1310	10/1	R. San Cristóbal, Chinandega: 1000-0200
MT04) A6RM	1330	5	R. Matagalpa, Matagalpa: 1100-0500
MA22) OS	*1340	1	R. Ondas Sonoras, Managua
MD01) AARS	*1370	1	R. Fronteras, Somoto
MA23) A3MA	*1400	10	R. María, Managua
LE03) RA	1410	3/1	La Estación de la Amistad, León: 1000-0200
ES04) AARL	1430	5	R. Liberación "La Tayacana", Estelí: 1100-0300
MA24 A3MR	1440	25	R. Maranatha, Managua: 1000-0500
BO01) RY	1470	1	R. Yarrince, Boaco
MA01) PT	*1500	1	R. Minuto, Managua
CA02 A4TS	1530	0.5	LV de Sta Teresa, Sta Teresa: 1400-2200

* = inactive, (r) = repeater, v = varying fq

Addresses and other information:

AS00) ATLANTICO SUR
AS01) TELCOR, 1½ c este, Nueva Guinea
B000) BOACO
B001) Casa del Finquero, 20 vrs al este, Boaco
CA00) CARAZO
CA02) Entrada II Calle, ½ c abajo, Sta Teresa
CH00) CHINANDEGA
CH01) Frente Iglesia de Guadalupe (or: Ap. 12), Chinandega – **CH02)** Club Eden, 2½ c. Al Sur, (or: Ap. 59), Chinandega
CT00) CHONTALES
CT01) Caracoles negros,Juigalpa **W:** http://radiocentro870.com
ES00) ESTELÍ
ES01) Esquina Norte de Hospital Adb, 2 c al norte, 3½ c este, Estelí– **ES02)** Esquina Sur-Oeste de la Escuela Nexo, 25 vrs al Río, Estelí – **ES03)** Instituto Bíblico Samaritado, Calle Principal, Condega – **ES04)** Shell, 1 c al norte, Estelí **W:** www.radioliberacion.com
GR00) GRANADA
GR01) Cuerpo de Bomberos 1c al N 1c al E No. 110, Granada ☎+505 2552 2040
JI00) JINOTEGA
JI01) Escuela Gabriela Mistral, ½ al Norte, Avenida Ernesto Rosales, Jinotega. ☎ +505 2782 2247 **W:** www.radiojinotega.hostoi.com – **JI02)** Cine Betty, 2½ c al norte, Jinotega
LE00) LEÓN
LE01) Del Pto del Mct, 1 c abajo, ½ c norte, León – **LE02)** Residencial Posada del Sol, Casa N° 93, León 1000-0300 – **LE03)** Unan 1½ c al norte, León FM 91.9
MA00) MANAGUA
MA01) Cd. Jardín Q-20, Av. Ponciano Lombillo (Apartado Postal 2442), Managua **E:** rc540@radio-corporacion.com **W:** www.radio-corporacion.com – **MA02)** Reparto El Carmen,Costado Oesyr frl Parque, Managua ☎+505 2268 2580 **W:** http://la580.com – **MA03)** Frente a la Universidad Centroamericana, Managua **W:** www.nuevaya.com.ni **E:** info@nuevaya.com.ni – **MA04)** Villa Fontana, Contiguo a TELCOR, Managua ☎+ 505 2277 2330 **W:** www.nicaragua620.com – **MA05)** Apartado Postal 4003 (or Barrio Bolonia, de Tica bus, 100 metros al sur, 100 metros al este), Managua **W:** www.radioprimerisima. com **E:** info@radiolaprimerisima.com – **MA06)** Altamira D'Este 621, Managua ☎+505 2278 0836 **W:** www.radiocatolica.org – **MA07)** Paseo Tiscapa (or: Ap. 4776), Managua **W:** www.lasandino.com.ni – **MA08)** Semaforos de Lozelsa 1c al lago, ½ abajo, Managua **W:** http://radio800.net – **MA09)** Costado Sur del Hospital Bautista N° 945, Managua ☎+505 2222 2250 – **MA10)** Ciudad Jardín, Casa N-10, Managua – **MA11)** Distribuidora Vicky, 4C Al lago, Casa 73, Managua ☎+505 278 1633 **W:** www.radioelpensamiento.com – **MA12)** Los Robles Gimn Atlas, 1c al E. 20 vs Al S N° 217, Managua ☎+505 2278 2540 – **MA13)** Reparto Miraflores, Rest. Munich 4c Al lago 1 c al Oe, Managua ☎+505 2266 6767 – **MA14)** Calle Edgar Lang, Managua – **MA15)** Urb. de Puntaldía, Managua ☎+505 2270 0096 – **MA16)** Altamira de Este contiguo Embajada de Taiwán, Managua ☎+505 2278 4040 **W:** www.radio15deseptiembre.com – **MA17)** Apartado 3091, Managua ☎+505 2248 2888] Managua – **MA18)** Managua – **MA19)** Foto Castillo 1 c Al sur ½ c, Villa Don Bosco E-182, Managua ☎+505 2244 2068 – **MA20)** Carret. Vieja a León, km 10 ¾, 500 m al N 200 al Oe, Managua ☎+505 2265 4919 – **MA21)** Carretera a Masaya, Km 12 ¾, 450 Metros al este, Managua ☎+505 2279 9491 – **MA22)** Bo La Cruz, Cine Blanca, 5c al N, ½ c al E, Casa 1112, Managua – **MA23)** De la Iglesia San Francisco 150 metros al este, Bolonia, Managua **E:** rmaria@cablenet.com.ni – **MA24)** Rotonda Metrocentro 1 c al sur, ½ C Abajo, Casa 41, Managua ☎+505 2278 5235 **W:** http://radiomaranatha.fm **E:** maranatha@cablenet.com.ni – **MA25)** Managua
MD00) MADRIZ
MD01) Somoto
MS00) MASAYA
MS01) Carr. a Managua km 24½, Masaya – **MS02)** Teatro Masaya, 1½ c al Oeste, N° 135, Masaya
MT00) MATAGALPA
MT01) Frente al Catedral, Matagalpa – **MT02)** Bo. Liberación Igl. Catedral. 1c al N 25 vs al Oe, Matagalpa ☎+505 2772 2964 FM: 92.3 **W:** www.radiohermanos.com – **MT03)** Detras de la Iglesia San José, Matagalpa – **MT04)** Rep. Brenes ½c al N, Matagalpa
NS00) NUEVA SEGOVIA
NS02) Perroquita Asunción, 1 cal norte, Ocotal **W:** www.radiolavozdel-pinar.com **FM:** 100.9MHz Stereo Mogotón, 101.7MHz R. Sí. – **NS03)** Jalapa. 102.1MHz
RS00) RIO SAN JUAN
RS01) Costado Norte de la Iglesia Católica, San Carlos
ZE00) ZELAYA
ZE01) Barrio 19 de Julio, Pto Cabezas – **ZE03)** Frente al Palacio Municipal, Barrio Beholdsen, Bluefields

FM in Managua (MHz): 89.1 Exitos – 89.5 R. Visión – 89.9 Tropicálida – 90.5 Nicaragua – 90.9 Estéreo Unión Radio –91.3 Futura – 91.7 La Primerísima – 92.1 Estación X – 92.7 Advent Estéreo – 93.1 La Buenísima – 93.5 Alfa Radio – 93.9 La Tigre – 94.3 Ondas de Luz – 94.7 Mujer – 95.1 La Pachanguera –95.5 Amor – 95.9 Estéreo Ritmo – 96.3 La Gran Cadena – 96.7 Furia Magic – 97.1 Estéreo Mía – 97.5 Corporación – 97.9 Salsa 98 – 98.3 R. Vida – 98.7 Romántica – 99.1 Stereo La Grande – 99.5 Universidad – 99.9 María – 100.3 La Musical – 100.7 Disney – 101.1 Güegüense – 101.5 Juvenil – 101.9 Uno – 102.3 Universidad – 102.7 Magic – 103.1 Bautista – 103.5 Maranatha – 103.9 Joya FM – 104.3 Estrella del Mar – 104.7 Hit – 105.1 Mi Preferida – 105.5 Rock FM – 105.9 Rica – 106.3 Galaxia, La Picosa – 106.7 Eco Romántico – 107.1 Sol – 107.5 Sandino – 107.9 Restauración

NIGER

LT: UTC +1h —— **Pop:** 14 million — **Pr.L:** French, Hausa, Zarma, Tamashek, Fulfulde, Arabic etc. — **E.C:** 50Hz, 220V — **ITU:** NGR

CONSEIL SUPÉRIEUR DE LA COMMUNICATION (CSC)
✉ Plateau I, Niamey ☎+227 722356 🖷 +227 20 722667 **L:P:** Chmn: Daouda Diallo. Vice Chmn: Hamidou Kô.

LA VOIX DU SAHEL – OFFICE DE RADIODIFFUSION-TÉLÉVISION DU NIGER (ORTN, Gov.)
✉ Maison de la Radio, B.P. 361, Niamey ☎+227 20 722272 🖷 +227 20 722548 **W:** telesahel.org **E:** ortny@ortn-niger.com **L:P:** DG: Amadou Harouna Yayé. Dir. Voix du Sahel: Mahaman Chamsou Maïgary. Tech. Secr: Mrs. Diaffra Fadimou Moumouni. Tech. Dir: Maraka Laouali.
MW: Niamey, Goudel 1125kHz 20kW.
SW: Niamey, Goudel 9705kHz 100kW (running at 40kW, irreg.).
FM (MHz): Maradi 88.4, Doutchi 89.7, Niamey 91.3, Zinder 91.3 2.5kW, Diffa 92.0, B. Konni 96.2, Madaoua 97.2, Tillaberi 99.0 10kW, Dosso 99.8, Tahoua 100.0, Agadez 106.8. All 1kW if not given otherwise. In addition 16 txs under 1kW.
D.Prgr in French/ethnic: 0500-2300 (Sun -2200). **N. in French:** 0545, 1200, 1900. **IS:** Local flute. **Ann:** F:"Ici la Voix du Sahel", A: "Idha'at al-Jumhuriya al-Niger, Sawt as-Sahel min Niamey".

Other stations:
R. Anfani FM: Niamey/Zinder/Maradi/Diffa 100MHz 1.5kW. Also rel. DW & VOA. **E:** anfani@intnet.ne – **Dounia FM,** Niamey: 89.0MHz 3kW. **E:** radioteledounianiger@yahoo.fr – **Espoir FM,** Niamey: 101MHz **W:** espoirfm@iniger.ne – **La Voix de l'Hemicycle,** Niamey: 95.1MHz –**Radio & Musique,** Niamey: 104.5MHz 1kW. Also rel. BBC African Sce. **E:** retm@intnet.ne – **R. Saraounia,** Niamey: 102.1MHz – **Sahara FM,** Agadez: fq. not known – **Tambara FM,** Niamey: 107MHz 0.5kW – **Ténéré FM,** Niamey: 98.0MHz 1kW.
R. Rurale stations on FM in Agadez, Bankilaré, Diffa, Dosso, Gaya, Maradi, Niamey, Tahoua, Tillaberi, Zinder.
Africa No.1, Niamey: 103.0MHz (see main entry under Gabon).
BBC African Sce, Niamey: 100.4MHz.
CRI: Niamey 106.0MHz, Agadez/Maradi/Zinder: fq not known.
RFI Afrique: Niamey/Maradi/Tahoua/Zinder 96.2MHz

NIGERIA

L.T: UTC +1h — **Pop:** 150 million — **Pr.L:** English, Yoruba, Hausa, Igbo — **E.C:** 50Hz, 230V — **ITU:** NIG

NATIONAL BROADCASTING COMMISSION (NBC)
✉ Road 14, Badagry Rd, Gwarinpa, Abuja ☎+234 1 2647867 **W:** www.nbc.gov.ng **L:P:** DG: Yomi Bolarinwa **L:P:** DG: Engr Bolarinwa.

FEDERAL RADIO CORPORATION OF NIGERIA (Gov.)
☞ Radio House, Herbert Macauley Way, Area 10, PMB 452, Garki, Abuja, Federal Capital Territory +234 9 2341103 🖹 +234 9 2346486 **W:** radionigeriaonline.com **LP:** DG: Barrister Yusuf Nuhu. Dir. Eng. Sces: Ibrahim Abdullahi.
1) FRCN Lagos, Broadcasting House, P.M.B. 12504, Ikoyi, Lagos, Lagos State. +234 1 2690301-5. L.P: Exec. Dir: Prince Atilade Atoyebi. R. One in English. **NB:** Nigerian N. from Lagos or Abuja at 0600, 1500 & 2100 is relayed by all FRCN stations and most state stations. Ann: "This is R. Nigeria, Lagos". Metro FM in English: 0500-2300 on 97.6MHz 20kW. Bond FM in Pidgin/English/Yoruba/Hausa/Igbo on 92.9MHz 20kW: 0430-2300 – **2) FRCN Abuja**, Broadcasting House, Gwangwalada, P.M.B. 71, Abuja, Federal Capital Territory +234 9 8821040 L.P: Exec. Dir: Shuaibu Ibrahim. D.Prgr: 0430-2305 (-2130 on 7275kHz) in English/Hausa/Igbo/Yoruba and others. Local N. in English 0500, 1700. Ann: "This is R. Nigeria, Abuja". F.PI: additional national channel for local languages – **3) FRCN Enugu**, Broadcasting House, Onitsha Rd, P.M.B. 1051, Enugu, Enugu State +234 42 254400 🖹 + 234 42 254173 L.P: Exec. Dir: Eddy Agwuegbo. 0430-2315 in English/Igbo/Tiv/Efik/Izon. F.PI: 100kW on 828kHz – **4) FRCN Ibadan**, Broadcasting House, Oba Adebimpe Rd, P.M.B. 5003, Dugbe, Ibadan, Oyo State +234 2 2414093 🖹 + 234 2 2413930 **W:** radionigeriaibadan.net **E:** info@radionigeriaibadan.net L.P: Exec. Dir: Princess Banke Ademola. D.Prgr: 0430-2305 in English/Yoruba/Edo/Igala/Urhobo. Ann: "R. Nigeria Ibadan, Station with distinction". – **5) FRCN Kaduna**, No. 7 Yakubu Gowon Way, P.O.Box 250, Kaduna, Kaduna State +234 62 235390 🖹 + 234 62 245392 L.P: Zonal Dir: Ladan Salihu, Chief Tech. Officer: Shehu A. Muhammad. Ch. 1 in Hausa: 0430-2300 on 594/6090kHz. Ch. 2 in English/Hausa/Fulfulde/Kanuri/Nupe: 0430-2300 on 1107/4770kHz. English N: 10500, 100, 1600, 1700, 2000. Ch. 3 in English: 0500-2300 on 96.1MHz. Karama FM in Hausa: 92.1MHz. Ann: "This is R. Nigeria, Kaduna".

MW		kHz	kW	MW		kHz	kW
4)	Alaho	567	50	3)	Enugu	828	25
4)	Moniya	576	25	2)	Gwagwalada	909	50
5)	Jaji	594	200	5)	Jaji	1107	25
4)	Ibadan	657	100				

SW		kHz	kW	SW		kHz	kW
5)	Kaduna	4770	50	5)	Kaduna	6090	50
3)	Enugu	*6025	10	2)	Abuja	**7275	100

*irregular/inactive. **also rep. on 7350khz for morning transmission.
FM (MHz): **1)** 92.9/97.6 **2)** 93.5 **3)** 92.85 **4)** 93.4 **5)** 92.1/96.1.
Further federal FM stations (MHz): Abakaliki 101.5, Abeokuta 94.5, Akure 102.5, Asaba 104.5, Awka 102.5, Bauchi 98.5, Benin 101.5, Benue (Makurdi)103.5, Bida (Minna) 100.5, Birnin-Kebbi 103.5, Maiduguri (Borno) 102.5, Calabar 99.5, Damaturu 104.5, Dutse 100.5, Ado-Ekiti 100.5, Gombe 103.5, Gusau 102.5, Kano 103.5, Kastina 104.5, Ilesha 95.5, Ilorin 103.3, Jalingo 100.5, Lafia 102.5, Lokoja 101.5, Osogbo 93.5, Owerri 100.5, Port-Harcourt 98.5, Sokoto 101.5, Umuahia103.5, Uyo 104, Yenogoa 101.5, Yola 101.5.
Aso FM: Abuja: 93.5MHz. **W:** www.asoradioonline.com

STATE RADIO AND OTHER STATIONS:

MW		kHz	kW	MW		kHz	kW
19)	Akure	*531	50	30)	Damaturu	801	20
17)	Sokoto	540	50	27)	Zuru	801	1
14)	Tukun Tawa	*549	25	14)	Azare	*846	10
37)	Ado	549	25	8)	Kafanchan	*882	25
23)	Calabar	*558	50	34)	Osu	*891	
12)	Owerri	567	50	18)	Abeokuta	*900	25
40)	Gusau	567		43)	Yola	917	50
6)	Abakaliki	*585	50	10)	Makurdi	918	50
9)	Maiduguri	603	50	27)	Birnin Kebbi	945	10
18)	Abeokuta	603	25	25)	Katsina	972	25
16)	Ilorin	612	50	35)	Otite	972	10
28)	Akwa	*621	20	20)	Ikeja	990	50
8)	Katabu	638	25	21)	Bauchi	990	50
7)	Benin City	*666	50	12)	Kontagora	1008	5
15)	Ojeowode	*675	25	34)	Iree	1008	10
35)	Ochaja	*693	10	33)	Dutse	1026	25
30)	Damaturu	*684	50	28)	Onitsha	*1062	10
31)	Wukari	*702	25	23)	Ugaga	1134	20
13)	Owerri	*720	50	12)	Bida	1143	10
14)	Jogana	729	50	24)	Jos	*1224	50
42)	Kaduna	747	60	31)	Jalingo	1269	10
9)	Damagum	*756	25	34)	Iwo	*1359	10
15)	Ibadan	756	100	8)	Zaria	*1359	50
12)	Minna	*756	50	35)	Esgbe	*1395	10
31)	Wukari	*774	10	26)	Abak	*1395	10
35)	Okene	*783	50	39)	Gombe	*1404	1
15)	Gambari	*792	25	11)	Yola	1440	10

NB: many transmitters are irregular or inactive, marked *.

FM (MHz): **6)** 96.1 **7)** 95.8 **8)** 90.8 **9)** 95.3 **10)** 95.0 **11)** 95.8 **12)** 91.2 **13)** 94.4 **14)** 89.3 **15)** 98.5 **16)** 99.9 **17)** 96.4 **18)** 91.4 **19)** 96.5 **20)** 107.5 **21)** 94.6 **22)** 99.1 **23)** 92.7 **24)** 90.5 **26)** 90.5 **28)** 88.5 **29)** 88.6/97.9 **31)** 90.6 **32)** 88.1 **34)** 89.5 **35)** 94.0 **36)** 97.1 **37)** 91.5 **38)** 97.3 **39)** 96.8 **41)** 98.1
State Radio information:
6) Enugu State Broadc. Sce (ESBS), Broadcasting House, Independence Layout, P.M.B. 01600, Enugu, Enugu State. Prgr 1 on MW: 0430-2300 in English/others. Prgr. 2 on FM ("Sunrise 96"): 0500-2100 – **7)** Edo State Broadc. Sce, P.M.B. 1012, Aduwawa, Benin City, Edo State. 0400-2305 in English + 12 local languages – **8)** Kaduna State Media Corp, Wurno Close, P.M.B. 2013, Kaduna, Kaduna State. 0430-2315 in English/Hausa – **9)** Borno Radio & TV Corp., P.M.B. 1020, Broadcasting House, Along Shehu Laminu Way, Maiduguri, Borno State **E:** brtvnews@yahoo.com 0400-2305 in English/Hausa/Kanuri/Marghi/Suwa/Babur-Bura – **10)** R. Benue, P.M.B. 102202, Makurdi, Benue State. Prgr. 1: 0430-2305 in English/others. Prgr. 2 on FM: 0500-2105. Ann: "This is R. Benue, Makurdi" – **11)** Adamawa Broadc. Corp. (ABC), P.M.B. 2123, Yola, Adamawa State. 0430-2300 (MW -1800) in English/Hausa + 6 Nigerian languages. Ann: "This is GBC Yola, your No. 1 Radio Station" – **12)** Niger State Media Corp. (Crystal R.), Radio House, Ibrahim Babangida St, P.M.B. 88, Minna, Niger State **E:** radioniger@yahoo.com 0430-2130 in English/others – **13)** Imo Broadc. Corp, Ebu Rd, P.O. Box 329, Owerri, Imo State. Prgr. 1: 0425-2305 on MW. Prgr. 2: 0440-2305 on FM. English: 0430-0630, 1100-1830, 2100-2300 (Sat/Sun 0100), other times Igbo – **14)** Kano State BC, 1 Ibrahim Taiwo Rd, Gidan Bello Dandago, P.M.B. 3014, Kano, Kano State. **W:** radiokanoonline.com Prgr 1 on MW: 0430-2320. Prgr. 2: on FM: 0550-2320 in English/Hausa. Ann: "Radio Kano" – **15)** Broadc. Corp. of Oyo State, P.M.B. 1, Akodi Post Office, Ibadan, Oyo State. Prgr. 1: 0400-2200 in English/Yoruba. Prgr 2: on FM: 0700-2100. Ann: "R. O-y-o" – **16)** Kwara State Broadc. Corp, Akpata Yakuba, P.M.B. 1345, Ilorin, Kwara State. 0400-2305 in English/others – **17)** Sokoto State BC, Moliba Adamawa Rd, Tudua Wada, P.M.B. 2156, Sokoto, Sokoto State. 0430-2305 in English/Hausa. Ann: "Rima Radio" – **18)** Ogun State BC, Ibara Housing Estate, P.M.B. 2084, Abeokuta, Ogun State. OGBC1 on MW, OGBC2 on FM: 0400-2400 in English/Yoruba – **19)** Ondo State Radio Corp, Broadcasting House, Oba-Ile, P.M.B. 709, Akure, Ondo State. 0400-2300 in English/others – **20)** Lagos State Broadc. Corp, Obafemi Awolowo Way, P.M.B. 21035, Ikeja, Lagos State. 0430-0005 in English/Yoruba. Ann: "NBC" – **21)** Bauchi Radio Corp, Broadc. House, Ahmadu Bello Way, P.M.B. 0133, Bauchi, Bauchi State. Prgr 1: 0430-2300 on MW, Prgr. 2: 0500-1700 (F.PI: 24h.) on FM in English/others – **22)** Rivers State Broadc. Corp, 4 Degema St, P.M.B. 5170, Port Harcourt, Rivers State. Prgr. 1: 0450-2310 on MW, Prgr. 2: 0450-2310 on FM in English/others – **23)** Cross River State Broadc. Corp. (CRBC), No. 8 IBB Way, P.M.B. 1035, Calabar, Cross River State **E:** crbc@skannet.com 0430-2315 in English/others – **24)** Plateau Radio & TV Corp. (PRTVC), 5 Joseph Gomwalk Rd, P.M.B. 2043, Jos, Plateau State. Ch. 1 on MW: 0500-2300, Ch. 2 on MW: 0500-2300 in English/others. Ann: "This is Radio Plateau 1 AM", "This is Radio Plateau 2, 90.5 FM Stereo" – **25)** Katsina State Radio & TV Sces (KSRTV), Former SDP State Headquarters, Batsari Rd, P.M.B. 2163, Katsina, Katsina State. 0430-2300 in English/others. Ann: "This is Katsina State R." – **26)** Akwa Ibom Broadc. Corp, 205 Aka Rd, P.M.B. 1122, Uyo, Akwa Ibom State. 0500-2300 in English/others – **27)** R. Kebbi, km 9 Kalgo Rd, Birnin Kebbi, Kebbi State. 0500-2300 in English/others – **28)** Anambra Broadc. Sce (ABS), off Arroma Junction, P.M.B. 5070, Awka, Anambra State. 0500-2300 in English/Igbo – **29)** Delta State Broadc. Sce, Broadc. House P.M.B. 5032, Asaba, Delta State. 0500-2300 in English/others – **30)** Yobe Broadc. Corp, km 6 Gujba Rd, P.M.B. 1044, Damaturu, Yobe State. 0500-2300 in English/others – **31)** Taraba State Broadc. Sces, Broadc. House, adjacent Gen. Sani Abacha State Secretariat, P.M.B. 1038, Jalingo, Taraba State. 0500-2300 in English/others – **32)** Broadc. Corp. of Abia State (BCA), Broadc. House, Government Station Layout, B.M.P. 7276, Umuahia, Abia State. **W:** www.bcanigeria.com **E:** bcaniger@bcnigeria.com 0500-2300 in English/others – **33)** Jigawa Broadc. Corp, Broadc. House, Kiyawa Rd, P.M.B. 7032, Dutse, Jigawa State. 0500-2205 – **34)** Osun State Broadc. Corp, Studio 1, Ita-Akogun St, P.M.B. 4425, Osogbo, Osun State. 0500-2300 in English/others – **35)** Kogi State Broadc. Corp, 1 Danladi Zakari Rd, P.M.B. 1095 GRA, Lokoja, Kogi State. 0500-2300 in English/others. Ann: "R. Kogi" – **36)** Nasawara Broadc. Sce (NBS), Tudun K. Nasarawa auri, Makurdi Rd, P.M.B. 97, Lafia, Nasarawa State. 24h in English/others – **37)** Broadc. Sce of Ekiti State, Old Ado Ekiti Local Government Secretariat, Okeyinmi, P.M.B. 5343, Ado, Ekiti State – **38)** Bayelsa State Broadc. Corp, P.M.B. 56, Ekeki, Yenagoa, Bayelsa State. "Glory FM" in English/others – **39)** Gombe State Broadc. Sce, Buhari Estate Rd, GRA, Gombe, Gombe State. 0500-2300 English/others – **40)** Zamfara State R, Mall. Yahaya Secretariat, Off Zaria Road, P.M.B. 01007, Gusau, Zamfara State – **41)** Ebonyi Broadcasting Service (EBBS), Ministry of Information Building, Government House Annex, Abakaliki, Ebonyi State.

Other stations:
42) Nagarta R, Nagarta Communications Complex, Katabu, Mararraban, Jos, P.O. Box 574, Kaduna. **W:** nagartaradio.tripod.com **E:** nagartaradio@yahoo.com **43) R. Gotel,** P.O. Box 5759, Modire (After Yola Bridge), Off Yola-Mubi Expressway Jimeta-Yola, Adamawa State. **E:** radiogotel@yahoo.com D.Prgr: 0500-2305.

Brilla FM, Eleganza 634, Adeyemo Alakija House, Victoria Island, Lagos: 88.9MHz – **Choice FM,** 103.5MHz – **Cool FM,** PMB 10096, Victoria Island, Lagos. FM: Abuja/Lagos 96.9MHz. **W:** www.coolfm.us **E:** info@coolfm.us – **Cosmo FM,** Plot 18, Pocket Estate, Independence Layout, Enugu: 105.5MHz – **Eko-FM,** Lagos: 89.75MHz – **Freedom R,** Plot 33, Sarki Dikko, Off Ibrahim Sani Abacha Rd, Gyadi-Gyadi, Kano: 99.5MHz – **Independent R,** Benin City. – **Ray Power 1:** Lagos/Abuja 100.5MHz. **Ray Power 2:** Lagos/Kano 106.5MHz + relays in other towns (Incl. rel. of BBC African Sce in English/Hausa) – **Rhythm FM,** 17A Commercial Ave, Yaba, Lagos: 93.7MHz – **Rhythm 94.7,** Hilltop, Karu, Abuja: 94.7MHz.

EXTERNAL SCE: Voice of Nigeria: see International Radio section

NIUE

L.T: UTC -11h — **Pop:** 1,398 — **Pr.L:** Niuean, English — **E.C:** 50Hz, 230V — **ITU:** NIU

BROADCASTING CORPORATION OF NIUE (BCN)
P.O. Box 68, Alofi, Niue Island ☎ +683 4026 🖷 +683 4217
E: gm.bcn@mail.gov.nu
L.P: Chmn: Hunukitama. GM: Patrick Lino CE: Trevor Tiakia
FM: 91.0MHz 0.5kW, 102.0MHz 0.1kW
D.Prgr: 1730-2000, 2230-0030, 0500-0830. **N:** on the h includes RNZI bulletins
Ann: "This is Radio Sunshine". **V.** by letter
Other Stations
1) Oka Rock FM, Alofi 107.9MHz, 2) Hakupu Community R., 3) Liku Community R.
Addresses and other information
1) OKA-KOA Shop, Commercial Center, Alofi **D.Prgr:** 24h **-2)** local solar-powered village radio **-3)** local solar-powered village radio

NORFOLK ISLAND (Australia)

L.T: UTC +11½h — **Pop:** 2,141 — **Pr.L:** English, Pitcairn Norfolk — **E.C:** 50Hz, 220V — **ITU:** NFK

NORFOLK ISLAND BROADCASTING SCE. (Gov.)
New Cascade Road [PO Box 456], Norfolk Island 2899, Australia ☎ +672 3 22137 🖷 +672 3 23298 **E:** news@radio.gov.nf **W:** http://vl2ni.nf **L.P:** Broadc. Mgr: George Smith
Radio Norfolk VL2NI: MW: 1566kHz 0.1kW **FM:** 89.9MHz 0.25kW
D.Prgr: Local prgr: 1930-0530 M-F, 1930-0230 Sat, 1930-0130 Sun. Operates as local community stn. Rel. RNZ Nat. Sce overnight on MW.

Other stations:

AM	kHz	kW	Station				
5)	1611	0.4	Vision FM (inactive)				
FM	**MHz**	**kW**	**Station**	**FM**	**MHz**	**kW**	**Station**
1)	91.9	0.25	2ABCRN	3)	95.9	0.25	2ABCRR
2)	93.9	0.25	2ABCFM	4)	98.9	0.25	2JJJ

Addresses and other information
1) 24h satellite relay ABC Radio National – **2)** 24h satellite relay ABC Classic FM – **3)** 24h satellite relay ABC Regional Radio – **4)** 24h satellite relay ABC JJJ Radio [r.98.2] – **5)** Locked Bag 3, Springwood QLD 4127 (currently inactive)

NORTHERN MARIANA ISLANDS (USA Commonwealth)

L.T: UTC +10h — **Pop:** 88,662 — **Pr.L:** English, Chamorro, Carolinian, Filipino — **E.C:** 60Hz, 110V — **ITU:** MRA

FEDERAL COMMUNICATIONS COMMISSION (FCC)
see USA for details

MW	kHz	kW	Station	MW	kHz	kW	Station
1)	1080	5	KCNM	2)	1440	3/0.5	KKMP
FM	**MHz**	**kW**	**Station**	**FM**	**MHz**	**kW**	**Station**
3)	88.1	1.8	KRNM	2)	92.1	0.01	K221EF
3)	89.1	0.25	KRNM	4)	97.9	6.5	KRSI
6)	89.9	1.8	KORU	4)	99.5	6.5	KPXP
3)	91.5	0.06	KMOP	5)	100.3	1.1	KWAW

FM	MHz	kW	Station	FM	MHz	kW	Station
1)	101.1	4.1	KNUT	1)	103.9	3.2	KZMI

Addresses and other information
1) Choice Broadcasting Company LLC, 543A N Marine Dr, Tamuning GU 96913 T: +671 4780104 F: +671 647 7480 **Prgr:** 24h – **2) Blue Continent Communications Inc.** PO Box 500815, Saipan 96950 T: +670 989 7673 **W:** www.myepacific.com **F.PL:** CP AM/FM 'Island Heat' under construction – **3) Marianas Educational Media Services.** 1st Floor, J. Perez Bldg, 138 Seaton Blvd, Hagatina, GU 96910-5136 **D.Prgr:** 24h relay KPRG Guam with NPR, Public Radio International and 6h daily BBC World Service. **F.PL:** KMOP 91.5 CP under construction – **4) Sorensen Pacific Broadcasting Inc.** GRC Building, Middle Road #305, Garapan, Saipan MP 96910 ☎ +1 670 2357996 🖷+1 670 2357998 **L.P:** SM: Tina Palacios **W:** www.myspace.com/power-99radio, KRSI The Rock www.radiopacific.com **D.Prgr:** 24h – **5) Magic 100FM.** 1st Fl, Naru Building, Susupe, Saipan 96950 ☎ +1 670 2345929 🖷 +1 670 2342262 **W:** www.magic100radio.com **E:** kwaw100.3@magic100radio.com **D.Prgr:** 24h – **6) Good News Broadcasting Corp.** 290 Chalan Palasyo, Agana Heights, GU 96910. **Prgr:** Religious

NORWAY

L.T: UTC +1h (27 Mar-30 Oct: +2h) — **Pop:** 4.9 million — **Pr.L:** Norwegian — **EC:** 50Hz, 230V — **ITU:** NOR

POST OG TELETILSYNET
Norwegian Post and Telecommunications Authority
PB 93, NO-4791 Lillesand ☎+47 22824600 🖷 +47 22824640 **W:** www.npt.no
NORKRING (Transmission provider)
Telenor Broadcast, Snarøyveien 30, NO-1331 Fornebu ☎+47 67892000 🖷 +47 67893611 **W:** www.norkring.no

NRK - NORSK RIKSKRINGKASTING AS (Pub.)
NO-0340 Oslo ☎+47 23047000 🖷+47 23047575 **Inf.Dpt:** +47 81565900 **E:** info@nrk.no **W:** www.nrk.no
L.P: DG: Hans-Tore Bjerkaas

LW/MW	kHz	kW	LW/MW	kHz	kW
Ingøy	153b	100	Røst	675f	20
Vigra	630e	100			

FM	P1	P2	P3	R.Norge+	kW
Alta	89.7b	94.6þ	91.3	101.0	3.5
Bagn	91.7h	95.3	88.0	102.1*	35/7*
Bangsberget	90.4h				4.1
Bergen	89.1d	94.8	99.0	102.5	46
Bjerkreim	94.2i	98.7	91.8	101.0*	60/6*
Bokn	93.5i	97.3	91.1	90.3	120
Bremanger	93.6j	98.1	91.3	103.2*	46/9.2*
Dikkevikfjell	87.8b	93.4þ	95.0		1.4
Førde	92.8j	98.7	97.1	102.0*	12/2.4*
Gamlemsvet	91.9e	96.3	90.0	102.8	50
Gausta	89.5m	96.4	98.7	103.1*	55/5.5*
	101.1a				6.7
Greipstad	88.8k	92.5	97.0	100.1	57.5
Grong	91.9g	96.6	88.9	102.0*	95/9.5*
Gulen	88.0j	94.5	97.6	101.4*	39/7.8*
Hadsel	92.4i	99.3	94.5	101.4*	30/6*
Halden	94.8p	89.1	101.5		72.5
Hammerfest	96.6b	87.7þ	93.6	102.8*	24/4.2*
Hasvik	90.1b	90.0þ	94.9		2
Hemnes	88.5f	99.8	96.1	104.2*	36/7.2*
Hestmannen	97.0f	90.7	93.5		1.2
Hovdefjell	87.8k	93.7	96.0	103.6*	25/5*
Hvitingen	91.7o				2.76
Iskuras	88.7b	96.1þ	92.0		2.2
Jetta	95.9h	95.5	91.1	101.6*	85/8.5*
Kappfjell	95.5f	99.4	93.4		1.3
Karasjok	87.9b	94.7þ			1.5
Kautokeino	90.3b	93.8þ	99.2		35
Kistefjell	91.8n	95.7þ	99.8	103.1*	44/4.4*
Kongsberg	91.3a	95.5	97.8*	102.5	60/30*
Kongsvinger	89.8c	93.9	96.1	107.2*	33/6.6*
	98.9q				11
Kopparen	88.3l	94.5	96.0	102.4*	40/8*
Lyngdal	97.6k	88.3	95.0	102.0*	50/10*
Lyngen	93.3n	97.5þ			4.2
Lønahorgi	93.3d	88.3	96.7	100.6*	48/4.8*
Melhus	92.4l	97.2	99.1	101.1*	60/2.4*
Mosvik	90.9g	93.4	93.4		33
Narvik	88.8f	98.9þ	91.1	101.1*	90/9*
Nordfjordeid	89.4j	99.3	92.3	101.2*	12/2.4
Nordhue	87.6c	97.1*	92.5	106.5	70/60*
Nordkapp	89.2b	95.4þ	98.2	102.5*	15/3*

FM	P1	P2	P3	R.Norge+	kW
Oslo	88.7q	100.0	93.5	103.9	90
Reinsfjell	89.1e	95.1	90.7	100.2*	24/2.4*
Salten	93.3f	95.5	89.8	100.4*	48/9.8*
Skien	88.2m	92.3	100.4	105.2	80
	90.3o				7.25
Sogndal	91.5j	95.1	98.7	103.9*	25/2.4*
Sprinklerfjell	94.1o				84
Steigen	90.3f	97.8þ	93.9	102.1*	102/106/107/7*
Stord	96.0d	99.6	92.6	101.8*	60/6*
Store Jekkir	99.9b	97.3þ	90.9	103.3	1.2/1.16/1.13/1.2
Tana	92.5b	97.0þ	91.1		24
Trolltind	88.2n	94.0þ	90.5	101.7*	50/5*
Tron	98.3c	88.6	94.3	102.5*	24/4.8*
Varanger	88.1b	91.8þ	100.2	105.8*	30/6*
Vega	89.3f	95.2	98.2	102.8*	55/5.5*
Andalsnes		99.9			2.2

+ more than 1800 lp txs less than 1kW +) Private comm. stn (see below) þ) carries Sámi Radio a-q) refers to reg. prgrs listed below.

FM	AN	AK	MP3	FM	AN	AK	MP3
Alta	90.8	93.4	96.3	Oslo	93.0	91.9	97.0
Bangsberget	88.4	89.3	93.2	Porsgrunn	95.8	97.4	90.8
Bergen	93.8	98.2	95.4	Stavanger	93.0	99.3	96.6
Bodø	94.8	90.9	97.2	Tromsø	89.8	94.4	96.8
Bokn	95.6	89.4	92.2	Trondheim	94.9	96.3	92.7
Fredrikstad	90.7	-	87.6	Vadsø	90.7	88.8	92.2
Kristiansand	94.0	98.6	96.7				

Transmitter powers 40W-3kW, typically 50-200W

P1: ✉ NO-7005 Trondheim ☎+47 73881400 🖷 +47 73881809 24h on FM and DAB. **N:** MF on the h. also 0530, 0630, 0730, 1130, 1530, 1630. Sat on the h. also 0630, 1130. Sun on the h. also 0430. Regional Prgrs: see below.
P2: ✉ NO-0340 Oslo ☎+47 23047297 🖷 +47 23047480 24h cultural prgr. on FM and DAB. **N:** MF on the h 0500-0000 except 1800, 1900 2000 and 2200. Sat 0530, 0630, 0730, 1130, 1630. Sat: on the h. 0500-0000 except 0900, 1200, 1800, 1900, 2000. Also 0630, 1130, 1530. Sun: on the h. 0500-2300 except 1200, 1300, 1600, 1800, 1900. Also 1530.
P3: ✉ NO-7005 Trondheim ☎+47 73881600 🖷 +47 73881609 24h youth prgr on FM and DAB. Rly P1 2300-0500. **N:** MF on the h 0500-2300 except 1600, 1800, 2000, 2200, also 0530, 0630, 0730, 0830. Sat: on the h. except 1600, 1800, 2000, 2100, 2200. Sun: on the h. except 2200.
Sámi NRK Sámiradio (special prgrs in Lappish): ✉ PB 183, NO-9730 Karasjok ☎+47 78469200 🖷 +47 78469223
D.Prgr: P1: Sun 2030-2100 (in Norwegian). P2: MF 1230-1300. Additional prgrs on P2 in northern Norway (marked þ in the frequency table) plus 90.1MHz (122W) in Oslo: MF 0600-0800, 1300-1630, (Fri also)1200-1230, 1630-1700), Sat/sun 1700-1800
AK: NRK Alltid Klassisk: ✉ NO-0340 Oslo ☎+47 23047882 🖷 +47 23048575. 24h FM and DAB. Classical music channel. Some rly P2.
AN: NRK Alltid Nyheter): ✉ NO-0340 Oslo ☎+47 23047000 🖷 +47 23045141 24h rolling news sce on FM and DAB. Rly BBC World Service most of the day Sat/Sun and 2100-0500 weekdays.
MP3: NRK MPETRE: ✉ NO-7005 Trondheim ☎+47 73881600 🖷 +47 73881609. 24h teenager channel based on techno/dance music on FM and DAB. Some rly of P3.

NRK REGIONAL SERVICES:
On P1. D.Prgr: MF 0503-0530, 0540-0600, 0603-0630, 0640-0650, 0654-0700, 0705-0718, 0725-0730, 0737-0800, 0803-0805, 0903-0905, 1003-1050, 1054-1059, 1103-1130, 1145-1159, 1203-1205, 1303-1305, 1403-1405, 1503-1530, 1533-1600, 1603-1630, 1703-1705. Sat: 0705-0707, 0803-0805, 0903-0905, 1003-1008, 1103-1108.
a) NRK Buskerud, PB 733, NO-3003 Drammen: 91.3/101.1MHz Some shared prgr with Telemark and Vestfold, as NRK Østafjells. – **b)** NRK Troms og Finnmark, PB 613, NO-9811 Vadsø: **153kHz**, 87.9/98.1/88.7/ 89.2/99.790.1//90.3/92.5/96.6/99.9MHz Some shared prgr with Troms and Nordland. – **c+h)** NRK Hedmark og Oppland, PB 174, NO-2601 Lillehammer: 90.4/91.7/ 95.9MHz and NO-2418 Elverum: 87.6/89.8/98.3MHz Separate news hourly, remaining prgrs shared – **d)** NRK Hordaland, PB 7777, NO-5020 Bergen: 89.1/93.3/ 96.0MHz – **e)** NRK Møre og Romsdal, NO-6025 Ålesund: **630kHz**, 89.1/91.9MHz – **f)** NRK Nordland, NO-8038 Bodø: **675kHz**, 88.5/88.8/89.3/90.3/92.4/93 .3/95.5/97.0MHz Some shared prgr with Finnmark and Troms. – **g+l)** NRK Trøndelag, NO-7005 Trondheim: 88.3/92.4MHz and Løshalla 15, NO-7712 Steinkjer: 90.9/ 91.9MHz Some separate newsbulletins, remaining prgrs shared– **i)** NRK Rogaland, PB 614, NO-4090 Hafrsfjord: 93.5/94.2MHz – **j)** NRK Sogn og Fjordane, PB 100, NO-6801 Førde: 88.0/89.4/91.5/ 92.8/93.6MHz – **k)** NRK Sørlandet, PB 413, NO-4664 Kristiansand: 87.8/88.8/97.6MHz– **m)** NRK Telemark, PB 284, NO-3901

Porsgrunn: 88.2/89.5MHz Some shared prgr with Buskerud and Vestfold, as NRK Østafjells. – **n)** NRK Troms og Finnmark, NO-9291 Tromsø: 88.2/91.8/ 93.3MHz Some shared prgr with Finnmark and Nordland. – **o)** NRK Vestfold, PB 120, NO-3101 Tønsberg: 90.3/91.7/94.1MHz Some shared prgr with Buskerud and Telemark, as NRK Østafjells. – **p)** NRK Østfold, PB 33, NO-1629 Gamle Fredrikstad: 94.8MHz – **q)** NRK Østlandssendingen, PB 4555 Nydalen, NO-0421 Oslo: 88.7/98.9MHz
V: by DX-Listeners' Club. Rec. e-mail reports to: nrk_qsl@dxlc.com
Ann: 1st Prgr: "P1". 2nd Prgr: "P2". 3rd Prgr: "Petre". Lappish: "Datlae Sámeradio, Kárássjagás"

DAB: One DAB-multiplex (12D) covering parts of Norway. Includes all NRK-channels, P4, and NRK Alltid Folkemusikk.
Regional DAB-multiplex (12C) for southeastern Norway. Includes NRK Østfold, Østlandssendinga, Buskerud, Vestfold, Telemark, NRK Stortinget, NRK P1 Oslofjord and Met. Oslofjord. In addition, regional multiplex 12B Rogaland, 12C Trøndelag and 13E Troms.

RADIO NORGE (Comm.)
✉ P.O.Box 144, NO-1601 Fredrikstad ☎+47 07270 🖷 +47 69707601
W: www.radionorge.com **LP:** MD: Bente Klemetsdal
FM: See freq table above. D.Prgr: 24h **N:** M-F: on the h. 0500-2300, also 0630, 0730, 1430, 1530. Sat: on the h. 0800-2300. Sun: on the h.0800-2300

P4 – RADIO HELE NORGE (Comm.)
✉ PB P4, NO-2626 Lillehammer ☎+47 61248444 🖷 +47 61248445
W: www.p4.no **LP:** MD: Kalle Lisberg
FM Main freqs (MHz): Bjerkreim: 88.5 1kW, Kistefjell: 89.4 4kW, Steigen: 91.5 10kW, Vega: 94.1 2kW, Ski: 94.5 1kW, Nordhue: 95.0 5kW, Horta: 95.6 1.1kW, Hammerfest: 96.8 7kW, Hadsel: 97.3 5kW, Førde: 98.4 4kW, Hemnes: 98.6 5kW, Sogndal: 99.9 5kW, Varanger: 102.8 8kW, Bokn: 102.8 120kW, Mosvik: 103.8 33kW, Greipstad: 104.9 38kW, Tron: 105.8 6kW, Halden: 106.1 72.5kW + 102 txs below 1kW
D.Prgr: 24h on FM and DAB. **N:** M-F: on the h 0500-2300, also 0530, 0730, 1430, 1530, 1630. Sat-Sun: on the h 0700-2300
LOCAL FM STATIONS
Around 300 low power FM commercial stns are in operation, some sharing freqs. Many of them organised through Lokalradioforbundet
W: www.lokalradioforbundet.no.
RADIO 1 NORGE (Comm.)
✉ PB 1102 Sentrum, NO-0104 Oslo ☎+47 22023300
🖷 +47 22952202 **W:** www.radio1.no
Network of stns in Oslo, Bergen, Trondheim and Stavanger.
NRJ NORGE (Comm.)
✉ Trondheimsveien 184, NO-0570 Oslo ☎+47 22797500
🖷 +47 22797501 **W:** www.nrj.no
Netw. of stns. in Oslo, Bergen, Trondheim, Stavanger and Romerike.
JÆRRADIOEN (Comm.)
✉ PB 10, NO-4301 Sandnes ☎+47 51979200 🖷 +47 51979256 **W:** www.jaerradioen.no
Major shareholder in 23 stns around the country with netw. prgrs

Internet: Most stns provide webstreams and/or oedemand audio sces

AFRTS (U.S. Mil.)
FM: Lifjell 101.5MHz (Stavanger). D.Prgr: 24h rly AFN Europe.
Satellite: Most NRK-channels, radio and TV (incl NRK regional TV), Radio Norge and P4 are available through satellite.

SVALBARD (SPITSBERGEN) (Norwegian Territory)
L.T: UTC +1h (27 Mar-30 Oct: UTC +2h) — **Pr.L:** Norwegian — **E.C:** 50Hz, 230V

MW: Longyearbyen: 1485kHz 1kW
D.Prgr: 24h Rly NRK P1 (incl. regional prgrs NRK Troms)

FM	P1	P2	P3	R.Norge	kW
Isfjord Radio	89.7	93.6	97.3		0.05
Longyearbyen	88.8	94.5	98.3	104.0*	0.045/0.013*
Ny-Alesund	91.3	94.8			0.12
Svea	89.1	92.0			0.025

OMAN

L.T: UTC +4h — **Pop:** 3.4 million — **Pr.L:** Arabic — **E.C:** 50Hz, 240V — **ITU:** OMA

MINISTRY OF INFORMATION
✉ Omanet Team, Oman Electronic Network, P.O.Box 600, 113 Muscat ☎+968 24603222 🖷 +968 24693770 **W:** www.omanet.om
E: omanet@omantel.net.om
L.P: Minister: Hamad bin Mohammed Al Rashdi

RADIO SULTANATE OF OMAN (RSO, Gov.)
✉ Ministry of Information, P.O. Box 600, 113 Muscat ☎+968 24 601538 📠+968 24 602831 **W:** www.oman-tv.gov.om **L.P:** DG Radio: Nasser Al-Sybani. DG Eng.: Mohd Salim Al Marhouby. Dir. Freq: Salim Al-Nomani. Dir. Trs: Saif Al-Rashedi.

MW	kHz	kW	MW	kHz	kW
Haima	576	100	Seeb	1242	200
Salahah	738	100			

F.PI: Renovation of MW network by 2010 with stations in Bideya 558kHz, Barka 1242kHz and Bahla 1278kHz.

FM	MHz	FM	MHz	FM	MHz
Sumail	87.8	Al-Badiaah	92.0	Qahwi Mts	97.0
Al-Hajer	87.8	Maskan	92.0	Al-Buraimi	97.1
Ja'alan Bani	88.1	Wadi Rajmaa	92.0	Sur	97.2
Al-Ghubrah	88.2	Qumiraa	92.2	Madha	97.4
Riyam	88.5	Tahwat	92.3	Ajran	97.5
Nizwa Main	88.5	Al-Khafdi	92.5	Sidab	97.8
Al-Ghashb	88.5	Al-Qhudhift	92.5	Al-Aabla	97.8
Sh'm	88.6	Wadi Mahram	92.6	Al-Taher	99.1
Al-Rustaq	88.7	Barkha	92.8	Ibri	99.1
Fanja	88.7	Al-Gafat	93.0	Tibat	99.2
Al-Saaidi	89.0	Siyyah	93.1	Mazraa al-Hadri	99.4
Heibi	89.0	Al Fi	93.1	City	100.5
Qurum	89.0	Al-Shurah	93.2	Al-Qalaat	100.5
Sicq	89.2	Ibra	93.2	Wadi Dimah	100.6
Hatta Mts	89.3	Tiwi	93.5	Lima	100.6
Al-Awabi	89.3	Bi Thaat	93.8	Amirat	101.2
Quriyat	89.5	Kumzar	94.0	Hulm	101.2
Al-Ansab	89.5	Al-Ain	94.1	Shinas	101.5
Butain	89.8	Mazrah	94.1	Maqal	101.7
Al-Hamra	89.9	M'zara	94.2	Qantab	101.8
Ruwi Murtafat	90.4	Bidbid	94.4	Al-Jinah	102.2
Salalah	90.4	Sabt	94.4	Khur al-Juramah	103.3
Wad	90.5	Sital	94.5	Qadi	103.4
Samad	90.7	Sultan Qaboos	94.6	Saham Main	103.7
Yamit	90.8	Wadi Bani Omar	94.7	Daba	103.8
Tawi al-nawa	90.9	Mazbar	95.3	Al-Jafnen	103.9
Tanuf	91.2	Wadi Sahtan	95.3	Al-Wasat	104.0
Dhank	91.2	Al-Hujirat	95.8	Rawtha Lima	104.0
Wadi Dima	91.3	Saifah	96.1	Al-Bustan	104.2
Thumrait	91.3	Yakka	96.1	Al-Filaj	105.7
Badaa	91.5	Yanqul	96.4	Thouyan	107.2
Hawraa Mts	91.5	Harat Mts	96.5	Bahla	107.3
Khasab	91.5	Al-Waqba	96.7	Fl'm	107.4
Barkah	91.7				

F.PI: 32 new FM transmitters by the end of 2009.
General Arabic prgr: 24h on MW & FM. **N:** 0300, 0700 (Fri 0830), 1300, 1600, 1700, 1900, 2000. **Al-Shabab (Youth) channel** on: Muscat/Salalah/Thamrat/Taqa/Murbat/Sadah 100MHz, Al-Batina region 91.7/101.5/ 103.6MHz, A'Dakhlia region/Al-Wusta/Al A'Sharqia 98.8MHz, Dhank 91.2MHz, Dhalkot/Rakhuot 94.5MHz. **R. Oman FM in English:** 0300-1800 on Muscat/Salalah 90.4MHz, Thumrait 91.3MHz. **Quran prgr.** in Muscat 93.2MHz and Salalah 96.7MHz.
Ann: A: "Idha'atul Saltanat al-Oman min Muscat." E: "This is the English Service of Radio Sultanate of Oman from Muscat".
External service on shortwave: see International Radio section.
BBC relay station: MW (702kHz 1500-2300 & 1413kHz 0030-0400, 1300-2100) & SW: for details see International Broadcasting section.

Other stations:
Al Wisal FM, Muscat: 96.5MHz. **W:** wisal.fm
Hala FM, Muscat: 102.7MHz. **W:** halafm.com
Hi FM, Muscat: 95.95MHz. **W:** hifmradio.com

PAKISTAN

L.T: UTC +5h — **Pop:** 177 million — **Pr.L:** Urdu, Punjabi, Sindhi, Pushto, Balochi, English — **E.C:** 50Hz, 230V — **ITU:** PAK

PAKISTAN ELECTRONIC MEDIA REGULATORY AUTHORITY (PEMRA)
✉ Green Trust Tower, 6th Floor F-6, Jinnah Ave, Blue Area, Islamabad **W:** pemra.gov.pk **E:** info@pemra.gov.pk **L.P:** Chairman: Mian Muhammad Javed.

PAKISTAN BROADCASTING CORPORATION (PBC, Gov.)
RADIO PAKISTAN
✉ Broadcasting House, Constitution Avenue, Islamabad 44000 ☎+92 51 9214278 📠+92 51 9223827 **W:** radio.gov.pk **E:** info@radio. gov.pk **L.P:** DG: Ghulam Murtaza Solangi. Dir. News: Haroon Abbassi

MW	kHz	kW	R	H of tr
Peshawar	540	300	N	0045-0405, 0600-1805
Khuzdar	567	150	B	1155-1808 NBS
Islamabad(Faqeerabad)	585	500	F	0045-0605, 0800-1900
Karachi (Landhi)	612	10	S	0215-0700, 0900-1645
Lahore-I	630	100	F	0200-1900
Karachi (Landhi)	639	100	S	0200-0400, 1300-1800 NBS
Quetta-I (Pishin)	756	150	B	0045-0805, 1000-1810
Karachi	828	100	S	0045-0405, 0600-1900
Hyderabad-I	1008	120	S	0045-0405, 0600-1808
Multan	1035	120	P	0045-0405, 0600-1808
Lahore-III	1080	50	P	1230-1705 NBS
Hyderabad-II	1098	10	S	1230-1705 NBS
Quetta (city)	1134	100	B	0200-0400, 1300-1800 NBS
Rawalpindi	1152	100	P	0200-0400, 1300-1800 NBS
Peshawar	1170	100	N	0200-0400, 1300-1800 NBS
Loralai	1251	10	B	1145-1615 NBS
Lahore II	1332	100	P	0200-1900 NBS
Bahawalpur	1341	10	P	0850-1808 NBS
Zhob	1449	10	B	1155-1600 NBS
Faisalabad	1476	10	P	0045-0820
Gilgit	1512	10	N	1000-1700
Skardu	1557	10	N	1000-1700
Turbat	1584	0.25	B	1300-1810 NBS
Sibi	1584	0.25	B	0755-1108
Chitral	1584	0.25	N	1050-1515
Abbottabad	1602	0.25	N	0845-1415
SW	kHz	kW		H. of tr
Islamabad	7470	100		0545-0715 (BS)

FM93: on 93.0MHz in Abbottabad, Bannu, Bhitshah, Chitral, Dera Ismail Khan, Gwadar, Hyderabad, Islamabad, Karachi, Khairpur, Kohat, Lahore, Larkana, Mianwali, Mlthi, Multan and Sargodha. Powers 2/3 kW.
Planet FM94: Islamabad 94.0MHz (English music channel).
FM101: information and entertainment channel on 101.0MHz in Daska Sialkot, Faisalabad, Hyderabad, Kalarkahar, Karachi, Lahore, Larkana, Muree, Peshawar and Quetta. Powers 2kW exc. Karachi 5kW. **W:** fm101.gov.pk
In addition, in Rawalpindi regional prgr. is carried on 93.5MHz and national prgr. on 93.0MHz.
(BS)= Balti & Sheena sces. (NBS)= National Broadcasting Service.
R=Region: N=No. We. Frontier Province & Northern tribal areas, F=Federal District of Islamabad, P=Punjab, S=Sindh, B=Balochistan.
D.Prgr: as above in Urdu, English and regional languages. Local IDs are usually heard at sign on/off. **NBS:** National Broadcasting Channel from Islamabad: 0200-1800 in Urdu & English. **N. in English:** 0300, 0500, 0800, 1100, 1300, 1600. 585/612/630/729/7561080/1098kHz carry Voice of Quran 0200-0700.
F.PI: **MW:** 1000kW to Lahore. 1000/500kW at Umarkot on 558kHz. 400kW for Peshawar 540kHz. 150kW to Quetta. 100 kW to Chamar, Gwadar, Hyderabad, Larkana, Multan, Muzaffarabad, Paranichar, Turbat and Dera Ismail Khan 1404kHz. 10kW to Abbottabad. **SW:** 1x100kW to be added at Islamabad (Rewat) and Karachi. **FM93:** is being extended to Umerkot, Sukkur, Nawabshah, Jacobabad, Dadu, Badin, Mirpurkhas, Sanghar and Thatta. In total the network will have 47 stns. Prgrs will be 40% English/Urdu and 60% regional.

EXTERNAL SERVICE: Radio Pakistan
See International Broadcasting section.

AZAD KASHMIR RADIO (AKR, Gov.)
✉ Broadcasting House, Muzaffarabad (AJK) 13100, via Pakistan.
L.P: Dep. Controller (Eng): Muhammad Iqbal.
MW: Muzaffarabad 792kHz 150kW, Mirpur 936kHz 100kW (r. 930kHz)
SW: Rawalpindi 4790kHz 10kW.
Islamabad (Rewat) 3975/4790/7265kHz 100kW (inactive).
FM: 93.0MHz
D.Prgr: Muzaffarabad channel: 0045-0445 & 1000-1810 on 792kHz Mirpur: 0045-0515 & 1100-1810. Rawalpindi-III channel (from Islamabad): 3975kHz (wi)/4790kHz (su) 10kW: 0045-0215 & 1445-1810, 10kW: 0230-0430, 1335-1430. 7265kHz 100kW: 0900-1215. Prgrs of all channels also contain R. Pakistan relays.
Ann: 792kHz: "Yeh Azad Kashmir Radio Muzaffarabad Hay". 936kHz: "Mediumwave na-sau-chattis (936) kHz par. Yeh Azad Kashmir Radio hay". SW: "Yeh Azad Kashmir Radio Trarkhal Hay".
IS: "Azad Kashmir" anthem at open and close.

Independent FM stations:
Apna Karachi 107: Karachi 107.0MHz. **W:** apnakarachi107.fm – **City FM 89:** Karachi/Lahore/Islamabad/Faisalabad 89.0MHz. **W:** www.cityfm89.com – **FM 100 Pakistan:** Islamabad/Karachi/Lahore 100.0MHz 5kW. **W:** fm100pakistan.com – **FM Sunrise Pakistan:** Jhelum 95.0MHz, Sardogha/Sahiwal 96.0MHz, Islamabad 97.0MHz.

W: fmsunrise.com – **Hot FM 105,** Karachi: 105.0MHz. **W:** hotfm.
com.pk – **Hum FM,** Islamabad/Karachi/lahore/Sukkur: 106.2MHz.
W: hum.fm – **Josh FM,** Karachi: 99.0MHz – **KUST FM 99,** Kohat:
98.0MHz. Web: radio.kust.edu.pk – **Mast FM 103,** Faisalabad/
Karachi/Lahore/Multan: 103.0MHz. **W:** mastfm103.com.pk – **Power
FM,** Abbottabad/Islamabad: 99.0MHz. **W:** power99.com.pk – **Punjab
Univ. FM,** Lahore: 104.6MHz 0.1kW **W:** pu.edu.pk/news/fm_sched-
ule2007.asp – **Radioactive 96,** Karachi: www.radioactive96.fm
– **R. Awaz:** 10 sites in Punjab on 105.0MHz **W:** radioawaz.com.
pk – **R. Buraq:** Abbottabad/Mardan/Peshawar/Sialkot: 104.0MHz.
W: radio-buraq.com – **R. One,** Gwadar/Islamabad/Karachi/Lahore:
91.0MHz. **W:** fm91.com.pk – **Super FM,** Bahawalnagar: 90.0MHz. **W:**
superfmnetwork.com – **VO Kashmir,** Muzaffarabad: 105.0MHz. **W:**
www.vokfm105.com – **Zab FM,** Islamabad/Karachi/Larkana: 106.6MHz
0.5kW **W:** zabfm.org
F.PI: Ewaz FM: Kharian/Haripur/Mandi Bahuddin 97.0/98.0MHz. **W:**
ewaz.com.pk

PALAU (USA associated)

L.T: UTC +9h — **Pop:** 20,796 — **Pr.L:** Palauan, English — **E.C:** 60Hz,
115/230V — **ITU:** PLW

T8AA BROADCASTING STATION (Gov)
✉ Box 279, Koror State, Republic of Palau 96940 ☎ +680 4882417
🖷 +680 4881932 **L.P:** SM: Salustiano Albert. **E:** ecoparadise@
palaugov.net
MW: Voice of Palau T8AA 1584kHz 5kW: 1900-1300 **N:** includes
R. Australia via satellite
FM: Eco-Paradise FM/BBC 87.9MHz: 24h relay BBC via satellite

Other Stations:
FM	MHz	kW	Station
4)	88.3	0.6	PWFM
1)	88.9		WPKR Island Rhythm
2)	89.5	0.5	WWFM
5)	91.5		R.Australia
2)	98.5	0.5	KDFM Pinoy FM
3)	102.5	0.75	KRST-FM

Addresses & other information
1) PO Box 2000, Koror 96940 ☎+680 4881359 **E:** rudimch@palaunet.
com **D.Prgr:** 24h – **2)** Diaz Broadcasting Co ✉ PO Box 1327, Koror
96940 ☎+680 4884848 🖷 +688 5874420 **L.P:** GM Alfonzo Diaz, Mgr
KDFM: Imelda Aban. WWFM English & Palaun, KDFM Filipino, classical, jazz, easy listening format. **W:** http://www.brouhaha.net/palau/
wwfm.html **E:** wwfm@palaunet.com **D.Prgr:** 24h – **3)** World Harvest
Radio, PO Box 12, South Bend IN 46614, USA ☎ (574) 2918200 🖷
(574) 2919043 **E:** whr@lesea.com **W:** www.whr.org **Prgr:** relays
English language religious prgrs from SW stn T8WH – **4)** Koror. 24h
contemporary music. **L.P:** Salvador Tellames – **5)** 24h satellite relay.
NB: Calls beginning with K and W are unofficial as Palau regulates
its own broadcasting spectrum [no longer the American FCC] but such
pseudo-calls are more familiar to locals compared to official ones
using the T8 prefix allocated to Palau.

EXTERNAL SERVICES: T8WH (Rlg.)
See International Broadcasting section.

PANAMA

L.T: UTC -5h — **Pop:** 3.3 million — **Pr.L:** Spanish — **E.C:** 60Hz, 110V
— **ITU:** PNR

AUTORIDAD NACIONAL DE SERVICIOS PUBLICOS
✉ Vía España, Edificio Office Park, Ciudad de Panamá (Apartado
Postal 0816-01235, Zona 5, Panamá ☎ + 507-508 4500
W: www.asep.gob.pa

ASOCIACION PANAMEÑA DE RADIODIFUSIÓN
✉ Ap. 55-1326, Panamá
Call HO–, * = inactive, (r) = repeater v = varying fq.
Hr of tr usually 24h.

MW	Call	kHz	kW	Station, location, hr. of tr.
PA01)	U23	540	10	R. Mía Chiriqui, David (r: 650)
PA02)	PU	540	10	R. Lider, Panamá: 1000-0400
PA03)	H2	560	3	RPC Radio, Colón (r: 610)
PA04)	S	570	5	R. Soberana, Panamá: 1000-0300
PA03)	H4	580	10	RPC Radio, David (r: 610)
PA03)	H3	590	5	RPC Radio, Chitré (r: 610)
PA03)	HM	610	10	RPC Radio, Panamá
HE01)	J35	630	2	R. Provincias, Chitré

MW	Call	kHz	kW	Station, location, hr. of tr.
CN01)	K22	640	2.5	R.CPR, Colón
PA14)		640	2.5	R. Panamá, La Palma
PA01)	S22	650	5	R. Mía "Cadena Nacional", Panamá
PA03)	F33	660	1	RPC Radio, Bocas del Toro (r: 610)
PA09)		660	5	La Nueva Exitosa, Sabana Grande (r: 930)
PA06)	LY	670	5	R. Hogar, Panamá
DA01)		680	5	Voz Sin Fronteras, Metetí: 1000-2400
PA26)	F32	680	5	Mujer AM, David
VE01)	R43	690	10	R. Veraguas, Santiago: 1000-0300
PA18)		690	5	R. Evangelio Vivo, Panamá
PA08)	Q51	710	10	KW Continente, Panamá
BT01)	B52	710	5	Ondas del Caribe, Bocas del Toro: 1000-0400
HE04)	B50	720	10	R. República, Chitré: 1100-0300
PA38)		730	3	Asamblea Nacional, Fort Sherman, Colón
CH03)	N26	740	5	R. Cristal, David
PA37)	R44	740	2.6	La Exitosa de Chorrera, La Chorrera
HE07)		750	5	R. La Inolvidable, Chitré: 1000-2300
PA38)		760	3	Asamblea Nacional, Bocas del Toro
PA10)	XO	760	5	LV del Istmo, Panamá: 1145-0300
HE05)	L83	*770	10	R. Nacional Herrera, Chitré: 1100-0100
CH04)	B55	780	10	R. Chiriquí, David: 1100-2300
PA07)		780	5	R. Recuerdo, Panamá
PA14)		790	6	R. Panamá, Santiago
CE02)		800	3	Tropical 800, Los Santos: 1030-0300
PA11)	G	810	1	R. 10, Panamá
CH05)	F28	820	3	R. Ritmo Chiriquí, David: 1100-0400
LS01)	R56	830	5	R. Península, Macaracas: 1000-0200
PA12)	L80	840	10	R. Nacional, Panamá
PA09)	T61	850	5	La Exitosa de Chiriquí, David (r: 930)
PA09)		*850	1	La Exitosa de Colón, Colón
HE03)	L55	860	10	R. Reforma, Chitré: 1030-0400
PA13)	HO	v870	5.5	R. Libre, Panamá
CN02)	B51	880	1	R. Visión Panamá, Colón
PA14)		880	2.5	R. Panamá, Bocas del Toro
PA14)		880	2.5	R. Panamá, Chiriquí
HE02)	Q62	890	5	R. Ritmo Stereo, Chitré.
PA14)	HA	*900	10	R. Panamá, Panamá
CH12)	L81	*910	10	R. Nacional, David
CN03)	L85	*910	10	R. Nacional, Colón
PA12)		*910	10	R. Nacional, Panamá
PA01)	S56	920	5	R. Mía Centrales, Los Santos (r: 650)
PA09)	R46	930	10	La Nueva Exitosa, Panamá
CH06)	K85	930	2	Mi Preferida Estéreo, Pto Armuelles: 1000- (SS 1200-) 0400
PA38)		940	5	Asamblea Nacional, Darien
CE04)	L84	*950	3	R. Nacional, Panamá
PA14)		*950	2.5	R. Panamá, Las Mercedes, Colón
PA05)		960	1	R. Capital, Panamá
CH07)	M33	960	1	CHT Stereo Digital, David
VE02)	S97	970	3	Ondas Centrales, Santiago: W 1000-0300, SS 1300-2400
PA38)		990	5	Asamblea Nacional, Chiriquí
PA15)		990	5	R. Impacto, Panamá: (Prgr: Filadelfia R.)
CE01)	K36	1000	10	R. Poderosa "La Fuerte", Aguadulce: 1000-0500
BT03)	L86	*1010	3	R. Nacional, Bocas del Toro
PA16)		1020	5	R. Ancón, Panamá: 1000-0400
LS02)	J2	1040	2.5	Ondas del Canajagua, Las Tablas
PA35)		1040	3	LV del Mamoni, Panamá
PA39)	J60	1060	3.5	R. LV de Panamá "La Auténtica", Panamá: 1100-0300
CE02)		1070	3	R. Estéreo Mi Favorita, Penonomé
LS04)		*1070	10	R. Nacional, Los Santos
PA07)	J24	1080	5	R. Mundo Internacional, Panamá
PA12)		*1090	10	R. Nacional, La Peña
PA08)	M92	1100	5	R. Sabrosa, Panamá
PA19)	M21	1120	5	R. Sonora, Panamá: 1030-0300
CE03)	U80	1130	2.5	R. Sensación, Aguadulce
PA20)	B49	1140	5	R. Panamericana, Panamá: 1100-0500
CH08)	C20	1160	5	Ondas Chiricanas, David
PA17)	WK	1160	10	R. Metrópolis, Panamá
VE03)	U	1180	10	AM Original, Santiago: 1100-0200
PA21)		1180	10	China Visión Panamá, Panamá
PA11)	E91	1210	1	R. Diez, Panamá
PA38)		1220	5	Asamblea Nacional, Veraguas
CH09)	M56	1240	3	Faro de David, David
PA17)		1240	1	Comunicación de Masas, Panamá
CE05)	LY	1250	5	R. Hogar, Penonomé: 0955-0300
PA22)	J22	1270	3	R. Tipy Q, Panamá
CH13)	S23	1290	3	R. Única, Chiriqui
CH13)		1290	5	R. Unica, Panama
CH13)		1290	5.5	R. Unica, Los Santos
CH10)	I417	1300	5	R. Baha'ís, Boca del Monte

MW	Call	kHz	kW	Station, location, hr. of tr.
PA24)		1310	12	R. María, Panamá
PA25)		1330	5	LV Poderosa, Panamá
LS05)		1340	2.5	R. Tipikal, Las Tablas
PA36)	Z38	1350	5	BBN R., Panamá
CH11)	B64	1370	1	R. Sitrachilco, Pto Armuelles: 1000-0200
PA26)		1380	10	Mujer AM, Panamá: 1130-0400
PA07)		1390	5	R. Mundo Internacional, Colón
PA27)	T40	1400	10	Digital Radio Luz, La Chorrera
LS03)	H779	1410	5	R. Mensabé, Las Tablas: 1000-0300
PA28)		1430	7.5	R. Kids, Panamá
PA29)		1450	5	R. Melodía, Panamá: 1000-0330
BT02)	D42	1460	0.5	LV de Almirante, Bocas del Toro: 1400-0400
PA30)		1470	5	R. La Primerísima, Panamá
PA38)		1490	3	Asamblea Nacional, Cocle
PA31)	A95	1510	5	Hosanna R., Panamá
PA32)		1530	10	R. Avivamiento, Panamá
PA33		1560	10	R. Adventista de Panamá, Panamá: 1000-0300
PA34)		1580	1	Hosanna Oeste, Panamá

Province abbreviations: (Provincias) BT = Bocas del Toro, CE = Coclé, CH = Chiriquí, CN = Colón, DA = Darién, HE = Herrera, LS = Los Santos, PA = Panamá, VE = Veraguas. **N.B:** These abbreviations are not recognized by the Post Office. Letters should carry the full name.

Addresses and other information:
BT00) BOCAS DEL TORO
BT01) Finca 13, Empalme, Changuinola ☎+507 758 6087 – **FM:** 90.1MHz – **BT02)** Calle 6 y Av. N. Almirante – **BT03)** Av. Central, Bocas del Toro.
CE00) COCLÉ
CE01) Ap. 090 (or: Vía al Puerto), Aguadulce **E:** r.poderosa@cwpanamá.net ☎ +507 997 4156 🖷 +507 997 4157 – **FM:** 99.9MHz – **CE02)** Av. Juan Demóstenes Arosemena, Galerías Aro, Penonomé **E:** darfer@cwpanama.net ☎+507 997 7167 🖷+507 997 1386. – **FM:** 91.7MHz – **CE03)** Calle Pablo Arosemena, Pueblo Nuevo, Aguadulce ☎/🖷+507 997 6280 **W:** www.plateadas.com/sensacion – **FM:** 103.7MHz– **CE04)** Villa Inter-americana, Penonomé. – **CE05)** La Esperanza, Penonomé ☎+507 997 8929 🖷+507 997 7340
CH00) CHIRIQUÍ
CH03) Ap. 540 (or: Av. 8 y Calle A Norte, Barrio Bolívar), David ☎+507 774 3852 - **FM:** 98.1MHz – **CH04)** Ap. 43 (or: Calle Central), David **E:** rguerra@chiriqui.com ☎+507 775 2822 - **FM:** 101.7MHz – **CH05)** Av. D.Noreste, Medio Oeste, David ☎+507 774 3352 🖷+507 774 0512 - **FM:** 93.1MHz– **CH06)** Ap. 44 (or: Barriada San José), Puerto Armuelles ☎+507 770 7408 - **FM:** 105.3MHz – **CH07)** Calle Central, David ☎+507 774 0755 🖷+507 774 8081 - **FM:** 107.9MHz – **CH08)** Ap. 172 (or: Calle Elisandro Calvo, Doleguita), David ☎+507 775 2742 - **FM:** 100.1MHz – **CH09)** Av.Estudiante, Calle 4 final, Edif.Hermanos Pinzón, David – **CH10)** Ap. 1187, David 1045-2300 🖷+507 726 5004 – **CH11)** Principal, Barriada Santa Fe, Puerto Armuelles ☎+507 770 744 🖷+507 770 7217 – **CH12)** Av. Primera Oeste, David. **CH13)** Zahita SA, Calle 45, Bella Vista, Edif. El Conquistador, Panamá. **E:** radiometropolis@hotmail.com
CN00) COLÓN
CN01) Calle 2da detras de Panamá All Brown, Colón ☎+507 441 1300 - **FM:** 101.5MHz, 103.5MHz – **CN02)** Calle 1 Paseo Washington, Colón ☎+507 433 1035 - **FM:** 105.3MHz – **CN03)** Av. Bolívar, Calle 9, Colón
DA00) DARIÉN
DA01) Calle Principal de Metetí, Metetí. (or: Ap. 87-0871 Panamá 7) ☎/🖷+507 299 6346 **E:** vozst@cwp.net.pa - **FM:** 100.1MHz
HE00) HERRERA
HE01) Ap. 423 (or: Urb.Las Mercedes), Chitré ☎+507 996 4127 🖷+507 996 2668 **W:** www.radioprovincias630.com – **HE02)** Paseo Enrique Geenzier, Chitré ☎+507 996 2061 – **FM:** 97.5 – **HE03)** Ap. 194, Chitré ☎+507 996 4223 - **FM:** 98.5MHz – **HE04)** Ap. 191, Chitré ☎+507 994 4627 – **FM:** 103.3MHz – **HE05)** Av. Pérez, Chitré – **HE07)** Ap. 375 (or: Calle Francisco Audia), Chitré. ☎+507 996 5302
LS00) LOS SANTOS
LS01) Calle Central, Macaracas - **FM:** 93.7MHz – **LS02)** Ap. 10 (or: Av. Belisario Porras, final), Las Tablas **E:** rocavi@cwp.net.pa ☎+507 994 6674 🖷+507 994 8133 – **LS03)** Ap. 20 (or: Av. Agustín Cano Castillero), Las Tablas ☎+507 994 6606 🖷 +507 994 8477. – **LS04)** Los Santos. – **LS05)** Los Cerritos, Las Tablas
PA00) PANAMÁ
PA01) Ap. 5117, Panamá 5. **W:** www.radiomiapanama.com **E:** rmia@sinfo.net ☎+507 227 2700 🖷 +507 227 2523 – **PA02)** Parque Lefebre, Edif.Sta Elena Torre II, Ofc.11, Panamá ☎ +507 222 0500 **E:** radio540@sinfo.net **W:** www.sinfo.net/radio540 – **PA03)** Ap. 0827-00116 (or: Avenida 12 de Octubre), Panamá. ☎+507 390 6700 **W:** www.rpcradio.com **E:** rpcradio@medcom.com.pa – **PA04)** Ap. 6-2323, El Dorado (or: Calle 63B, Casa N° 2), Panamá ☎+507 236 1940 **W:** www.radio-

soberana.com – **PA05)** Edif. Orion, P8, Via España frente al PIEX, Panamá ☎+507 263 0183 – **PA06)** Ap. 102 (or: San Francisco de la Caleta, via Cincuentario y Av. José Matilde Pérez), Panamá 9-A ☎ +507 270 0141 🖷 +507 270 0145 **E:** rhogas@cableonda.net **W:** www.sinfo.net/rhogar **PA07)** Radio Hit SA, Calle 50 y 77 San Francisco 35, (Apartado 0819-0391) Panamá **W:** http://estereobahia.com **PA08)** Ap. 87-1324, Panamá 7 (or: Vía Argentina, Edif. Carillón, Panamá) **W:** www.kwcontinente.com **E:** kwcontinente@cableonda.net ☎+507 223 8846 🖷+507 264 6230. – **PA09)** Ap. 7462, Panamá 5 **W:** www.sinfo.net/exitosa ☎ +507 225 2052 🖷 +507 225 7252 – **PA10)** Ap. 6-1192, (or: 66 Oeste N° 641) El Dorado, Panamá **E:** atrd@panama.c-com.net ☎ +507 229 3989 🖷 +507 261 3366. – **PA11)** La Gloria 31-B Bethania, Panamá – **PA12)** Ap. 4950 (or: Edif.Dorchester, P5), Panamá 5 ☎+507 269 6594 🖷+507 269 5910 – **PA13)** Edif. Dorchester, Vía España, Panamá 4. ☎+507 264 5239 – **PA14)** Calle 54 Obarrio, Edificio Plaza Globus, 2do piso, Panamá ☎+507 263 0121/263 0078 **W:** www.radiopanama.com.pa – **PA15)** Río Abajo, Calle 13, Panamá ☎+507 221 0110. Filadelfia Radio: Filadelfia Eglesia, Plaza El Conquistador, Local 61, Vía Domingo Diaz, Panamá. – **PA16)** Calle Cuba y Calle 37, Panamá. 🖷+507 225 0025 – **PA17)** Vía España y Calle 45, Edif. El Conquistador PB, Panamá **W:** www.sinfo.net/radiometropolis/ **E:** radiometropolis@hotmail.com 🖷+507 212 0112– **PA18)** Condomino Dorado N° 2, Ofc. 10A, Vía Ricardo J. Alfaro, Panamá **PA19)** Ap. 87-1165 (or: Calle 63 Oeste N° E-21, Urb. Los Angeles), Panamá 7 ☎/🖷 +507 236 3065 – **PA20)** Ap. 6956 (or: Vía José Agustín Arango), Panamá 5 – **PA21)** Sun Tower Mall, Av. Ricardo J.Alfaro, Panamá ☎+507 236 5363 🖷+507 236 9810 – **PA22)** Calle 45, Edif. Conquistador, Bella Vista, P2, Panamá ☎+507 269 4226. – **PA24)** Ap. 6- 4509, El Dorado, (or Urb. Los Angeles, Av. Los Periodistas, Casa D-5, 0819-5581 El Dorado, Panamá) **W:** www.radiomariapanama.org **E:** info.pan@radiomaria.org ☎+507 261 0449 🖷+507 261 1535 – **PA25)** Iglesia Internacional del Evagélico Cuadrangular, Los Andes N° 2, Panamá – **PA26)** Calle 51 51 Av Manuel María Icaza, Edif. Torre Cosmos, Panamá **PA27)** Ap. 473, La Chorrera – **PA28)** Río Abajo, Panamá – **PA29)** Ap. 87-3541 (or: Vía Fernández de Córdoba, Jardin Cosita Buena), Panamá 7 **E:** lizbethcardenas@hotmail.com ☎+507 229 1504 🖷+507 229 4850 – **PA30)** Edif.La Marqueta, Vía Porras, Panamá **E:** rlprimerisima@latinol.com ☎ +507 226 6476. **W:** www.sinfo.net/laprimerisima – **PA31)** Ap. 6-8229 (or: Calle Erick del Valle y Vía Argentina, Edif. Vicky 2), El Dorado, Panamá **W:** www.hosanna.pma.org/sradio.htm – **PA32)** Av. Ernesto T. Lefevre, Panamá – **PA33)** Ap. 3244, Panamá 3 (or: Carrasquilla, Calle 2da N° 39, Panamá) ☎ +507 214 6430 🖷 +507 263 4255 **W:** www.tagnet.org/radioadvpma **E:** celes@usa.com - Ids: "Radio Adventista - La Voz de la Esperanza" – **PA34)** Sky Phone S.A., Cerro Peñon, Panamá. – **PA35)** Av. José Agustin Arango N° 5013, Panamá. – **PA36)** Ap. 0860-00356, (or: Vía Cincuentenario final, a 300 metros del McDonald's de Río Abajo, después del Edificio La Reina) Panamá. **PA37)** Av. Las Américas, La Chorrera. – **PA38)** – Asamblea Nacional, Panamá (txs broade. the sessions of the National Assembly only) – **PA39)** Vía España y Calle 45, Edif. El Conquistador, PB, Panamá **E:** vozdepanama@hotmail.com
VE00) VERAGUAS
VE01) Ap. 48 (or: Calle 9 y vía Panamericana), Santiago ☎/🖷+507 958 7060 **E:** radioveraguas@cerco.net - **FM:** 102.1MHz – **VE02)** Ap. 131, Santiago. – **VE03)** Ap. 286 (or: Calle 10), Santiago ☎ +507 998 1655 🖷 +507 998 5580

FM in Panamá City (MHz): 88.1 Diez – 88.5 85.5 FM Stereo – 88.9 RCM Noticias – 89.3 Cool FM – 89.9 Estéreo 89 – 90.5 Super Q – 90.9 RPC Radio – 91.3 Los 40 Principales – 91.7 La Nueva 91.7 – 92.1 Super Estación – 92.5 La KY – 92.9 YXY – 93.5 Metrópolis – 93.9 Una voz cristiana en tu casa – 94.5 R. Panamá – 94.9 Hosanna Capital 94.9 – 95.3 La Nueva Exitosa – 95.9 KW Continente – 96.3 Estéreo Selecta – 96.7 Mía – 97.1 Caliente Panamá – 97.5 WAO 97½ – 97.9 Mix – 98.3 La Mega – 98.9 Ultra Estéreo – 99.3 La 99 – 99.7 Tropic Q – 100.1 Antena 8 – 100.5 Fabulosa – 101.1 Estéreo Azul – 101.5 Economía – 101.9 Nacional/SERTV – 102.1 Lo Nuestro – 102.5 FM Corazón – 103.1 Blast – 103.5 Quiubo Estéreo – 103.9 Mil – 104.3 Amor – 104.7 La Tipik – 105.1 Estéreo Vida – 105.7 La Nueva Bahía – 106.1 Nuevo Sol – 106.7 Rock & Pop – 107.3 Omega Stereo – 107.9 Estéreo Universidad.

PAPUA NEW GUINEA

L.T: UTC +10h — **Pop:** 6,057,263 — **Pr.L:** English, Tok Pisin, Motu + 30 ethnic langs — **E.C:** 50Hz, 240V — **ITU:** PNG

NATIONAL BROADCASTING CORPORATION (Gov)
🖃 P.O. Box 1359, Boroko N.C.D ☎ +675 325 5233 🖷 +675 325 6296 **E:** info@nbc.com.pg **W:** www.nbc.com.pg
L.P: MD Joseph Ealedona, Exec. Dir Kundu Netw.: Kathleen Sakias. **Networks:** Karai (National: English/Tok Pisin/Motu), Kundu (Provincial: English/Tok Pisin/Motu).

Karai Network (Voice of Papua New Guinea):

MW	kHz	kW			
Pt Moresby	585	10			
SW	**kHz**	**kW**	**SW**	**kHz**	**kW**
Pt Moresby P2K4	4890	100	Pt Moresby P2T9	9675	100
FM	**MHz**	**kW**	**FM**	**MHz**	**kW**
Pt Moresby†	90.7	1	Rabaul	103.3	0.3

D.Prgr: 1900-1400 daily **N:** on the h 1900-1400 **Format:** nationwide music and development prgrs including primary school broadcasts. **V:** card or letter, send reports direct to Port Moresby.
†=Tr. by 5) R. Central of Kundu network

Kundu Network (Provincial):

MW	kHz	kw	Station, location
1)	585	10	R. West Sepik "Maus Bilong Sandau", Vanimo
14)	675	10	R. East Sepik "Maus Bilong Sepi", Wewak
10)	810	10	R. Morobe "Maus Bilong Kund", Lae
9)	810	10	R. East New Britain "Maus Bilong Tavuvu", Rabaul
13)	864	10	R. Madang "Maus Bilong Garamut", Madang
2)	900	10	R. West New Britain "Singaut Bilong Tavur", Kimbe
6)	900	10	R. Eastern Highlands "Karai Bilong Kumul", Goroka
20)	1107	10	R. Milne Bay "Maus Bilong Chauka", Alotau

SW	kHz	kW	Station, location
4)	2410	10	R. Enga, Wabag
1)	3205	10	R. West Sepik "Maus Bilong Sandaun", Vanimo
10)	3220	10	R. Morobe "Maus Bilong Kundu", Lae
2)	3235	10	R. West New Britain "Singaut Bilong Tavur", Kimbe
7)	3245	10	R. Gulf "Voice of the Seagull", Kerema
13)	3260	10	R. Madang "Maus Bilong Garamut", Madang
17)	3275	10	R. Southern Highlands, Mendi
5)	3290	10	R. Central "Voice of the Conch-Shell" Port Moresby
18)	3305	10	R. Western "Voice of the Sunset", Daru
11)	3315	10	R. Manus "Maus Bilong Chauka", Lorengau
19)	3325	10	R. Buka "Maus Bilong Sankamap", Buka
14)	3335	10	R. East Sepik "Maus Bilong Sepik", Wewak
12)	3335	10	R. Northern "Voice of the People of Oro", Popondetta
15)	3355	10	R. Simbu "Karai Bilong Mumbu", Kundiawa
20)	3365	10	R. Milne Bay "Voice of Kula", Alotau
9)	3385	10	R. East New Britain "Maus Bilong Tavuvur", Rabaul
3)	3905	10	R. New Ireland "Singaut Bilong Drongo", Kavieng

NB: Some Kundu SW stns operate irr. and may be silent for extended periods because of funding or local problems.

FM	MHz	kW	Station, location
13)	90.4		R. Madang "Singaut Bilong Garamut", Madang
20)	90.5		R. Milne Bay "Voice of Kula", Alotau
15)	90.7		R. Simbu "Karai Bilong Mumbu", Kundiawa
1)	90.7		R. West Sepik "Maus Bilong Sandaun", Vanimo
14)	90.7		R. East Sepik "Maus Bilong Sepik", Wewak
3)	90.9	1	R. New Ireland "Singaut Bilong Drongo", Kavieng
7)	91.1	0.3	R. Gulf "Voice of the Seagull", Kerema
18)	91.5		R. Western "Voice of the Sunset", Daru
8)	91.5	0.1	R. Western Highlands "Eagle FM", Mount Hagen
6)	91.9	0.1	R. Eastern Highlands "Kam Gud FM", Goroka
6)	91.9		R. Eastern Highlands "Kam Gud FM", Kainantu
5)	95.5		R.Central "Voice of the Conch-Shell", Port Moresby
9)	98.0	1	R. East New Britain "Maus Bilong Tavuvur", Kinabot
19)	100.1		R. Buka "Maus Bilong Sankamap", Buka
2)	100.3	0.1	R. West New Britain "Singaut Bilong Tavur", Kimbe
11)	100.3	0.1	R. Manus "Maus Bilong Chauka", Manus
12)	100.3	0.1	R. Northern "Voice of the People of Oro", Popondetta
9)	101.3	0.3	R. East New Britain "Maus Bilong Tavuvur", Palmalmal
9)	104.3	0.2	R. East New Britain "Maus Bilong Tavuvur", Malmaluan
10)	105.0	1	R. Morobe "Kundu FM", Lae

D.Prgr: 2000-2200, 0800-1200v. **Format:** Local prgrs now with fewer music and listener request prgrs and more development prgrs (health, education, law and order) and sport. Stns may also carry some Karai Network prgrs. **V:** card or letter, email; send reports direct to stn.
F.PL NBC 2010-2015 Plan proposes converting existing MW and SW stns to FM only with satellite prgr delivery. R.Central 95.5 FM was the first FM outlet with satellite delivery from March 2010.

Addresses and other information:
1) P.O Box 37, Vanimo ☎ +675 857 1144/1149 🖷 +675 8571305 **–2)** P.O.Box 412, Kimbe ☎ +675 983 5600/5185/5010 🖷 +675 9835600 **–3)** P.O.Box 477, Kavieng ☎ +675 9842077 🖷 +675 984 2191 **– 4)** P.O.Box 300, Wabag ☎ +675 5471013 🖷 +675 5471069 (currently reported inactive after local tribal fighting destroyed the tower system in July 2007) **– 5)** P.O Box 1359, Boroko N.C.D ☎ +675 3217155 🖷 +675 3217110 **– 6)** P.O Box 311, Goroka ☎ +675 732 1618/1733/1607 🖷 +675 7321533 **– 7)** P.O.Box 36, Kerema ☎ +675 6481076 🖷 +675 6481103 **– 8)** P.O.Box 311, Mount Hagen ☎ +675 5421000 🖷 +675 5421001 **– 9)** P.O.Box 393, Rabaul ☎ +675 982 8966/67/68/69/70 🖷 +675 9828971 **–10)** P.O.Box 1262, Lae ☎ +675 472 1311/7520/4209 🖷 +675 4726423 **– 11)** P.O Box 505, Lorengau ☎ +675 4709079 🖷 +675

4709079 **– 12)** P.O.Box 137, Popondetta ☎ +675 329 7037/38 🖷 +675 3297362 **– 13)** P.O Box 2036, Jomba ☎ +675 852 2415/2301/2360 🖷 +675 8522360 **–14)** P.O Box 65, Wewak ☎ +675 856 2316/2398 🖷 +675 8562405 **– 15)** P.O.Box 228, Kundiawa ☎ +675 7351012 🖷 +675 7351012 **– 17)** P.O.Box 104, Mendi ☎ +675 549 1017/1020 🖷 +675 5491017 **– 18)** P.O.Box 23, Daru ☎ +675 645 9234/9151 🖷 +675 6459319 **– 19)** P.O.Box 35 Buka, Autonomous Province of Bougainville ☎ +675 9739911 🖷 +675 9739912 **– 20)** P.O.Box 111, Alotau ☎ +675 641 1028/1334 🖷 +675 6411028.

Other stations:

SW	Station, Location	kHz	kW
38)	R.Fly, Kiunga	3915	1
38)	R.Fly, Kiunga	5960	1
22)	Wantok R. Light, Pt Moresby	7325	1

FM		MHz	kW	FM		MHz	kW
23)	Gobo	88.0		33)	Bereina	99.5	
21	Vunapope	88.1		39)	Pt Moresby	99.5	
21)	Bereina	88.5		34)	Goroka	100.2	
24)	Porgera	88.7	0.1	34)	Kavieng	100.2	
26)	Kotubu	89.3	0.02	34)	Pt Moresby	100.3	1
27)	Pt Moresby	89.9	1	34)	Lae	100.3	
26)	Samberigi	90.9	0.02	34)	Lorengau	100.3	
21)	Rabaul	91.3	0.5	34)	Tabubil	100.3	
21)	Malmaluan	91.3		34)	Mount Hagen	100.4	
21)	Vanimo	91.5	0.3	34)	Popondetta	100.5	
26)	Kikori	92.3	0.5	34)	Buka	100.8	
24)	Mount Hagen	92.5	0.5	34)	Kimbe	100.8	
21)	Aitape	92.9	0.3	34)	Madang	100.8	
23)	Pt Moresby	93.1	1	34)	Wewak	100.8	
23)	Mount Hagen	93.5	0.3	34)	Kundiawa	101.0	
23)	Alotau	93.7	0.3	34)	Rama	101.0	
23)	Lae	93.7	0.3	34)	Rabaul	101.1	
23)	Madang	93.7	0.3	35)	Pt Moresby	101.9	
23)	Goroka	93.9	0.3	34)	Lae	102.1	
22)	Pt Moresby	93.9	1	21)	Pt Moresby	103.5	0.5
23)	Rabaul	93.9	0.3	34)	Tabubil	103.8	
21)	Kerema	94.3		25)	Pt Moresby	105.1	
29)	Morobe	94.7	0.5	12)	Lorengau	105.1	
23)	Kavieng	95.0	0.01	34)	Goroka	105.9	0.3
37)	Buka	95.3		33)	Ialibu	105.9	0.3
38)	Kiunga	95.3		22)	Kainantu	105.9	0.3
23)	Goroka	96.1	0.3	34)	Lae	105.9	0.3
23)	Lae	96.3	0.3	22)	Mendi	105.9	0.3
23)	Madang	96.5	0.3	34)	Mount Hagen	105.9	0.5
23)	Lorengau	96.5	0.02	22)	Rabaul	105.9	0.5
23)	Pt Moresby	96.5	1	22)	Wabag	105.9	0.5
23)	Rabaul	96.5	0.3	22)	Wewak	105.9	0.5
23)	Alotau	96.7	0.3	34)	Buka	105.9	0.5
23)	Kimbe	96.9	0.3	22)	Kimbe	105.9	0.5
23)	Mount Hagen	96.9	0.3	35)	Pt Moresby	106.7	
30)	Gobo	98.0		16)	Lae	106.9	0.1
23)	Kavieng	98.0	0.01	34)	Alotau	107.0	
30)	Moro	98.0		34)	Dimodimo	107.1	
31)	Wewak	98.0	1	34)	Kainguma	107.1	
21A)	Mount Hagen	98.1	0.3	34)	Waterholes	107.1	
32)	Pt Moresby	98.5	0.02	34)	Horeatoa	107.5	
23)	Misima Island	98.9		34)	Boregoro	107.7	

Addresses and other information
21) Voice of Blessed Peter ToRot Radio. A network of locally programmed Catholic community radio stations formerly part of the Catholic Radio Network and/or relays programs from Radio Maria PNG Inc: PO Box 8719, Boroko, Port Moresby. T: +675 325 9178 **Email:** info.png@radiomaria.org **GM:** Fr. Martin We-en – **21A)** Catholic local community radio, Mt.Hagen. **ID: Trinity FM - F.PL:** R.Gabriel, Sandaun – **22) Papua New Guinea Christian Broadcasting Network [Wantok R. Light]** P.O Box 1273, Port Moresby N.C.D ☎ +675 3262933 🖷 +675 3261104 **W:** www.wantokradio.org **E:** admin@ wantokradio.org **D.Prgr:** 24h via satellite. **Relay:** R.Bianmi, Western Province **F.PI:** 60+ licensed and repeater FM stations throughout the country and at prisons. Supported by HCJB and Evangelical Bible Missions through Life Radio Ministries, Griffin, GA USA **QSL:** reports can be sent to qsl@wantokradio.org – **23) PNG FM [Nau FM & Yumi FM]**, P.O. Box 744, Port Moresby, N.C.D ☎ +675 3201996 🖷 +675 3201995 **E:** aau@naufm.com.pg **W:** www.cfl.com.fj/radiopng **GM:** Adrian Au, **PrgrDir - Nau FM:** Turner Arifeae, **PrgrDir - Yumi FM:** Rosemary Botong **FM Networks: Nau FM** — English [urban westernised youth market] **Yumi FM** – Pidgin [local music and adult contemporary music] **D.Prgr:** 24h – **24) Piam Town R.**, Porgera Mine, Porgera – **26) CDI FM**, CDI Foundation Trust Fund, Allotment 16, Section 55, Lokua Ave, Boroko, P.O. Box 383, Port Moresby N.C.D ☎ +675 325 0706/1759 **W:** www.cdi.org.pg **E:** pr@ cdi.org.pg **D.Prgr:** 24h – **27) FM Central**, P.O Box 333, Port Moresby

N.C.D ☎ +675 3210533 – **28) KBBC Krai Bilong Baibel [Baibel FM]**, Bible Broadcasting Network, Mount Hagen – **29) FM Morobe [Positive R.]**, P.O Box 3310, Lae ☎ +675 4791477 – **30) Red FM**, Gobo – **31) Laif FM**, UCB Pacific Partners, PO Box 556, Wewak ☎ +675 4561400 **F.PI:** FM stations in Goroka, Madang and Port Moresby Associated with UCB International, Auckland, NZ – **32) R UPNG FM98.5**, University of Papua New Guinea, P.O. Box 320, University, Port Moresby N.C.D ☎ +675 3267191 – **33) Ere'Ere**, Bereina – **34) Kalang FM100**, Broadcast House, Five Mile, Boroko P.O Box 1534, Boroko N.C.D ☎ +675 325 6843 ▤ +675 325 1747 **L.P:** MD Peter Tariasi Associated with PNG Telikom who took over the service from NBC because of non-payment of electricity charges – **35) R. Australia**, 24h satellite relay – **36) BBC World Service**, 24h satellite relay **-37** New Dawn FM, Buka Island, Autonomous Province of Bougainville T: +675 973 9319 F: +675 973 9285 **Web:** www.bougainville.typepad.com/newdawn **E:** tambolema@daltron.com.pg **SM:** Aloysius Laufa – **38) R.Fly**, Ok Tedi Mining Company, Dakon Street, Tabubil. PO Box 1, Tabubil, Western Province ☎ +675 6493924 ▤ +675 6493023 **L.P:** Team Leader: Michael Miise. **QSL:** cards are reported to be available shortly and reply postage appreciated. Contact: Jobby Paiva **E:** jobby.paiva@oktedi.com **D.Prgr** 24h includes automated prgr overnight – **39)** Rait FM, PO Box 1106, Boroko, Port Moresby ☎ +675 323 8524 **SM:** Shane Amean **E:** shane@chm.com. pg **W:** www.chmsupersound.com

PARAGUAY

L.T: UTC -4h (13 Oct 10-10 Apr 11, 2 Oct 11-8 Apr 12: -3h) — **Pop:** 7 million — **Pr.L:** Spanish, Guaraní — **E.C:** 50Hz, 220V — **ITU:** PRG — **Int. dialling code:** +595

COMISIÓN NACIONAL DE TELECOMUNICACIONES (CONATEL)
Offices: Avenida Yegros 437, Asunción ☎ 21 440 020 **W:** www.conatel.gov.py **L.P:** Pres: Jorge Seall

MW	Call	kHz	kW	Station, location, h. of tr
AP1)	ZP16	550	20/10	R.Parque, Ciudad del Este: 0800-0100
AM1)	ZP15	570	1	R. LV del Amambay, Pedro Juan Caballero: 0930- (Sun 1000-) 0100
MI3)	ZP39	570	1	R. San Roque, Ayolas: 0900-0200, Sat: 0900-0000, Sun: 1000-2200
SP1)	ZP32	590	5	R. Ycuámandyyú, Villa de S. Pedro: 0900-0200
BO1)	ZP30	610	10/1	LV del Chaco Paraguayo, Filadelfia: 0900-0230
SP2)	ZP40	v620	10/1	R. Nasaindý, San Estanislao: (r. v619-616.6) 0900-0230
CG1)	ZP19	v640	8	R. Caaguazú, Coronel Oviedo (r. v640-645): 0900-0500
CA1)	ZP4	650	50	R. Uno, Asunción: 24h
AP2)	ZP26	660	6/12	R. Itapirú, Cd. del Este: 0800-0100
CA2)	ZP11	680	12	R. Caritas, Asunción: 24h
NE1)	ZP12	700	1	R. Carlos Antonio López, Pilar: (rel. R. Nal 920): 0900-0200
PH1)	ZP17	720	25	R. Pai Puku, Teniente Irala Férnandez 0900-0100 Sun; 1000-0030
CA3)	ZP7	730	30	R. Cardinal, Asunción: 24h
CP1)	ZP38	740	1/0.5	Hechizo AM, Caazapà
CA4)	ZP42	750	5	LV de la Policía Nal, Asunción: 0900-0200
CA5)	ZP70	780	50	R. Primero de Marzo, Asunción: 24h
CN1)	ZP27	800	5/3	R. Mbaracayá, Salto del Guairá: 0900-0300
GU1)	ZP6	v840	5	R. Guairá, Villarrica: (v835-840): 0900-2400
CR1)	ZP28	v860	25	LV de la Cordillera, Caacupé: 0900 (Sat/Sun 1000)-0400
CE1)	ZP33	v890	5/0.5	R. Tres de Febrero, Itá: (v885-890): 0900-1000
CA6)	ZP1	920	10/100	R. Nal. del Paraguay, Asunción: 0800-2400
CA7)	ZP9	970	80	R. 9-70 , Asunción: 0800-0400
AM2)	ZP31	980	5	R. Mburucuyá, Pedro Juan Caballero: 0900-0130
CE2)	ZP36	1000	5/0.5	R. Mil, San Antonio: 24h
CA8)	ZP14	1020	25/10	R. Nandutí, Asunción: 24h
MI1)	ZP43	1040	5	R. Arapisandú, San Ignacio: 0900-0200
NE2)	ZP13	v1060	3	R. Boquerón, Alberdi: (r. 1060.5): 1000-0300
CE3)	ZP25	1080	10	R. Nanawa, Luque: 1100 (Sun 0900)-2330
AM3)	ZP71	1100	5	R. Nú Verá, Capitán Bado: 0900-0200
CE4)	ZP24	1120	10	R. Nuevo Mundo, San Lorenzo
GU2)	ZP22	1140	5/2	R. Panambí Verá, Villarrica: 0900-0200
CA9)	ZP72	1160	15	R. Antena Dos, Asunción
CG2)	ZP52	1180	5/1	R. Coronel Oviedo – RCO-AM, Coronel Oviedo: 0900-0100
AP3)	ZP45	1190	5	LV de la Libertad, Henendarias
CE5)	ZP44	1200	10	R. Libre, Fernando de la Mora
CA10)	ZP3	1250	5	R. Asuncion: 0900-0400
CG3)	ZP21	1230	3/0.5	Fenix AM, Fortaleza del Este

MW	Call	kHz	kW	Station, location, h. of tr
GU3)	ZP34	1260	1	R. Panambi Vera, Villarrica
AP4)	ZP53	1280	10	R. LV del Este, Cd. del Este
CA11)	ZP10	1300	5	R. Fe y Alegria, Asunción
CA12)	ZP4	1330	10	R. Chaco Boreal, Asunción: 24h
CO1)	ZP37	1360	1	R. Yby Ya'u, Ybu Ya'u: 0900-0300
CO2)	ZP8	1380	1	R. Concepción, «LV del Norte»: 0930-0100
MI2)	ZP35	1410	2	R.Mangore, S. Juan Bautista: (n. 1430): 1000-0130
CO3)	ZP42	1420	5	R. Güyrá Campana, Horqueta (r. 1417)
CO4)	ZP29	1450	5	R. Vallemi, Vallemi
AM4)	ZP23	1480	5	R. Dos Fronteras, Bella Vista Norte: 1000-0100
CE6)	ZP20	1480	5	R. América, Nemby: 24h

SW	Call	kHz	kW	Name, location and h of tr
BO1)		6884	0.1	LV del Chaco Paraguayo, Chaco: (USB feeder): 0900-0230
PH1)		6890	0.1	R. Pa´i Puku, Tte. Irala Fernández (USB feeder): 0900-0100, Sun1000-0030
CA6)	ZPA1*9735		100	R. Nal. del Paraguay, Asunción
CE7)		12000	0.025	R.Licemil, Ypané

° = on-air stn name not confirmed, * = inactive, v = varying freq.

Addresses and other information:
AM00 (AMAMBAY)
AM1) Mcal López 336, Pedro Juan Caballero 36 72210 **W:** www.amambayfm.com **E:** administracion@amambayfm.com **FM:** 100.5MHz – **AM2)** Villa María Victoria, Fracción San Jorge, Pedro Juan Caballero **W:** www.nanduti.com.py/mburucuya.asp – **AM3)** Estrella c/4 de Enero, capitán Bado, Amambay ☎+595 37 262.– **AM4)** Calle Iturbe 146, Bella Vista Norte **FM:** 92.5MHz

AP00 (ALTO PARANA)
AP1) Ciudad del Este – **AP2)** Av Coronel Sánchez 3800, Cd del Este **W:** www.radioitapiru.com – **FM:** 96.1MHz – **AP3)** Juan E.O´Leary 152, 1a piso, Oficina 5, Hernandarias **E:** lavozdelalibertad@telesurf.com.py – **AP4)** Avenida San Blás No 353, Ciudad del Este ☎61 512 583 **E:** lavozam@hotmail.com

BO00 (BOQUERÓN)
BO1) Av Trebol 137E, Filadelfia, Chaco (✉ Casilla 984, Asunción) **W:** www.zp30.com.py ☎ 491 32031/32330 ▤ 491 32501 Operated by Mennonite Mission. **German** (20%) & 4 ethnic langs

CA00 (CAPITAL)
CA1) Av Mariscal López 2948 c/MacArthur, Asunción ☎ 2160 3400 **W:** www.radiouno.com.py **E:** info@radiouno.com.py– **CA2)** Kubischek 661 y Azara, (✉ Cas 1313), Asunción **W:** www. uca.edu.py/caritas **E:** caritas@caritas.com.py ☎521 213 570 ▤521 204 161 – **CA3)** Calles Comendador Nicolás Bó y Guaranies 1334, Lambaré (✉ Cas 247, Asunción) **W:** www.cardinal.com.py – **FM:** 92.3MHz – **CA4)** Comandancia de la Policia Nacional, El Paraguayo Independiente c/Chile, Asunción ☎ 21 492 515.– **CA5)** Av Perón y Concepción Prieto Yegros (✉ Cas 1456), Asunción ☎21 300 389 **W:** www.780am. com.py – **CA6)** Av Blas Garay 241e/Yegros e Iturbe, Asunción ☎21 4500776 ▤21 372 233 **W:** www.rnp920am.com **E:** info@rnp920.com – **CA7)** Av Rodriguez de Francia 34, Asunción ☎21 450 283 **W:** www. radio970am.com.py **E:** info@radio970am.com.py – **CA8)** Choferes del Chaco 1194, Asunción ☎21 604 308 ▤21 606074 **W:** www.nanduti. com.py **E:** publicidad@holdingderadio.com.py – **CA9)** Estados Unidos 2019, Asunción – **CA10)** Capitán Lombardo 174 y Av Artigas, Asunción ☎21 282 661 **W:** www.radioasuncion.com.py **E:** radioasuncion@cmm. com.py – **CA11)** O´Leary 1847, Asunción ☎21 390 576 ▤21 390 534 **W:** www.feyalegria.org.py **E:** oficinanacional@feyalegria.org.py – **CA12)** Alejo Garcia 2589 con Rio de la Plata, Asuncion ☎ 21 425 589 **Web:** www.chacoboreal.com.py **E:** info@chacoboreal.com.py

CE00 (CENTRAL)
CE1) Av Enrique Doldán Ibieta y Presidente Franco, Itá ☎24 32543 **Guaraní:** 0900-1000, 1330-1430, 1800-1845.– **CE2)** Av 25 de Mayo 1160 e/Brasil y Constitución, Central de Llamadas, Asuncion ☎21 209 000 **W:** www.radio1000.com.py **E:** info@radio1000.com.py – **CE3)** Av General Aquino 9999 y José Bonifacio, Luque ☎644-330 **W:** www.nanawa.infoluque.com – **CE4)** Coronel Romero y de las Residentes, San Lorenzo (Asunción: General Díaz 488, Ofic 34.) ☎21 582424 ▤21 586258 **Guaraní:** 0900-1000, 1700-1800 – **CE5)** Av Zavalas-Cué 1615, Fernando de la Mora ☎21 513 887 ▤ 21 509 876 **W:** www.fundacionlibre.org.py **E::** info@fundacionlibre.org.py – reception reports to: Dr Benjamín Fernández Bogado – **CE6)** Cas 2220, Asunción **W:** www.radioiglesia.com **E:** radioamerica@lycos.com ☎ 21 960 228 ▤ 21 963 149 – **CE7)** Liceo Militar "Acosta Nu", Ypané

CG00 (CAAGUAZÚ)
CG1) Ruta Mcal Estigarribia Km 131, Coronel Oviedo ☎▤521 200 370 **W:** www.radiomas.com.py – **FM:** 102.3MHz – **CG2)** Av Mariscal Estigarribia 304 casi Yrendague, Coronel Oviedo **E:** rco@radiocoroneloviedo.com.py ▤ 521 202579 - **Guaraní:** 50 % of the program-

ming 0900-0100 - **FM:** 91.9MHz FM del Sol – **CG3)** Fortaleza del Este

CN00 (CANINDEYU)
CN1) Defensa Nacional y Av Paraguay, Salto del Guairá ☎462 42 350

CO00 (CONCEPCIÓN)
CO1) Ruta V, Ybu Ya'u ☎39 210 250. 0900-0300. – **CO2)** Panchito López 241 entre Prof.Cabral y Screiber, (Cas 78), Concepción ☎ 31 42254 🖹 31 40919 **E:** radioconcepcion@infonorte.com.py – Prgrs in **Spanish & Guaraní.** 0930-0100. – **CO3)** José Luís Arbues c/Ruta 5, Horqueta ☎32 222 364 – **CO4)** Zona Urbana, Vallemi ☎351 230 329

CP00 (Caazapá)
CP1) Caazapá

CR00 (CORDILLERA)
CR1) Dr Venancio Pino y 3ra Proyectada, Caacupé ☎511 42363

GU00 (GUAIRA)
GU1) Pte Franco 788 y Alejo Garcia, Villarrica ☎🖹 54 142 130 **W:** www.fmguaira.com **E:** administracion@fmguaira.com **Guaraní:** 2100-2300.0900-2400 **FM:** 103.5MHz – **GU2)** General Caballero 650, Villarrica ☎541 42230 **W:** www.grupopanambi.com **E:** info@ grupopanambi.com - **FM:** 94.7MHz – **GÜ3)** Angostura y Olimpio, Bo Ybarotu, Villarrica ☎541 42229 **W:** www.grupopanambi.com **E:** info@ grupopanambi.com

MI00 (MISIONES)
MI1) Av Mariscal López y Capitan del Puerto, San Ignacio (Misiones) ☎82 232 374– **MI2)** Coronel Alfredo A Ramos esq San Juan, San Juan Bautista, Misiones ☎81 212 306 – **MI3)** Mil Viivendas, Ayola ☎7222 2433 🖹7222 2324 **W:** www.radiosanroque.org **E:** ventas@ radiosanroque.org

NE00 (ÑEEMBUCÚ)
NE1) Alberdi 998, Pilar ☎86 32254 **Web:** rnp920am.com **E:** zp12@ uninet.com.py – **NE2)** Av Mariscal López 379, Alberdi

PH00 (PRESIDENTE HAYES)
PH1) Km 389 de la Ruta Transchaco, Teniente Irala Fernández, Chaco ☎424 270 349 **W:** www.radiopaipuku.org.py **E:** radiopaipuku@ chaconet.com.py or rppuku@telesurf.com.py

SP00 (SAN PEDRO)
SP1) Ruta 11 Juana M de Lara, Villa de San Pedro **Web:** www.desde-paraguay.com/ycuamandyyu – **SP2)** Mariscal López y B Caballero, San Estanislao, San Pedro ☎43 4342 2095 🖹43 4342 0292 **W:** www. radionasaindy.com.py **E:** AM620@pla.net.py

FM in Metro Asunción (MHz): 87.9 LV de San Juan María Viané – 88.3 R.Nemby – 88.9 R.Sembrador – 89.1 R.Conquistador –90.1 FM Trinidad – 90.7 Ysapy – 89.1 La Estacion – 90.1 FM Trinidad – 90.7 Ysapy – 91.1 La Estacion – 91.5 R. Top Milenium – 13) 92.3 Cardinal Romance – 92.7 R.Fernando de la Mora – 93.1 FM Florida – 93.3 R. Luque – 93.5 R.Rebelde – 93.9 FM Universal – 94.3 RRGS – 94.7 R.Azul y Oro. – 21) 95.1 – 25) 95.5 Rock & Pop – 95.9 FM Amor – 96.5 R.Disney – **CA5:** Chaco 96.9 «R Mariscal Estigarríbia» – **CA5:** 97.1 FM Latina – 97.9 R.Nuevo Tiempo – 98.5 Yacyretá – 99.1 R.City – 100.1 R. Canal 100 – 100.5 FM Arpa –100.9 Monte Carlo – 101.3 R.SanPablo – 101.5 R.Universitaria – 102.1 R.Obedira – 102.5 R.Tropicana – 102.7 Aspen Classic FM – 103.1 FM Popular – 103.5 FM SSP – 103.7 R.Lambaré – 104.1 R.Planeta – 105.1 R.Venus– 105.5 R.Continental – 106.1 FM Paraguay – 106.9 R.Urbana – 107.3 R.Maria – 25) 107.7 FM Concert.

PERU

L.T: UTC -5h — **Pop:** 29.5 million— **Pr.L:** Spanish, Quechua, Aymara — **E.C:** 60Hz, 220V — **ITU:** PRU — **Int. dialling code: +**51

MINISTERIO DE TRANSPORTES, COMUNICACIONES, VIVIENDA Y CONSTRUCCION
Dirección General de Telecomunicaciones.
🖃 Av. 28 de Julio 800, Lima 1 ☎ 1433 7800, 1433 1212, 1433 0570 **W:** www.mtc.gob.pe **E:** dgt@mtcgob.pe
L.P: Dir. de Telecomunicaciones: Ing. Carlos A. Romero Sanjinés. Dir. Freq. Div: Ing. José Villa Gamboa

ASOCIACION DE RADIO Y TELEVISION DEL PERU (ARTV)
🖃 Jr. Manuel Corpancho 208, Santa Beatriz, Lima ☎ 1433 3908, 1433 3953 **L.P:** Pres: Humberto Maldonado Balbín. Dir: Daniel Linares Bazan

INSTITUTO NACIONAL DE COMUNICACION SOCIAL
🖃 Jr. de la Unión 264, Lima **L.P:** Hernán Valdizán C. Dir. Bc: Sra. Clarisa P. de Olivera

UNION DE RADIOEMISORAS DE PROVINCIAS DEL PERU (UNRAP)
🖃 Mariano Carranza 754, Santa Beatriz, Lima 1

° = on-air stn name not confirmed, * = inactive, v = varying freq.

MW	Call	kHz	kW	Station, location, H of tr.
LI01)	OBX4E	540	1	R. Inca del Perú, Lima: 24h
LL01)	OCX2D	540	1	R. San Antonio, Trujillo
LI02)	OBZ4L	v560	2	R. Oriente, Lima
LA01)	OBX1H	560		Radiomar, Chiclayo
LA02)	OAU1M	570	1	R. Univ. Nal. Pedro Ruiz Gallo, Lambayeque
LL51)	OCU2B	570		(Ruiz Marciano Esgardo), Huamachuco (F.P.I.)
CJ01)	OAX2E	580	10	R. Marañón, Jaén: 1000-0400 (Sat: 1000-0200 Sun: 100-1800
LL02)	OCY2L	580	1	R. El Sol, La Esperanza
LI03)	OAX4M	580		R. Maria, Lima: 1100-0500
AQ02)	OCX6V	590	1	R. Catedral, Miraflores, Arequipa
LL03)	OBX2B	590	1	R. Star, Trujillo: 24h
LI04)	OBZ4W	600	10	R. Cora, Lima: 24h
MO01)	OCX6D	600		R. Cultural, Ilabaya
TA01)	OAX6S	600		R. Cultura Toquepala, Ilabaya
CJ32)	OCY2I	610	5	R. Santa Monica, Chota: 1100-0100
PI02)	OBU1E	610	1	R. Santa Rosa, Sullana
AQ01)	OCX6B	620		R. Maria, Arequipa
CU60)	OAR7H	620		Apostol de Yanaoca, Yanaoca
LL04)	OAX2M	620	0.4	R. Chepen, Chepen
LI05)	OBU4B	620	10	R. Ovación, San Isidro
CJ04)	OAU2R	634		R. Cajamarca, Cajamarca:(n.f.: 1420): 0800-0200 (Sun. 0300)
LA03)	OAU1Y	640	1	R. La Luz, José Leonardo Ortiz
LI06)	OAZ4K	640	10	R. Del Pacifico, Lima: 1030-0430
PU01)	OBX7B	640	10	R. Onda Azul, Puno. 0900-0300
LL05)	OAX2N	650	1	R. Regional del Norte, Trujillo: 1000-0500
JU02)	OCX4L	660	1	R. Chinchaycocha, Junin
LA04)	OCX1U	660	5	R. J.H.C., Chiclayo: 1000-0500
LI07)	OCX4R	660	10	R. La Inolvidable, Lima: 1100-0700
PU02)	OAX7H	670	10	R. Nacional del Perú, Puno
CJ03)	OCY2Y	680	5	R. San Luis, Jaén
IC01)	OAX5E	680	1	R. Emisora del Pacifico, Ica: 1100-0600
LL06)	OBX2L	680	0.5	R. Amauta, Chócope. 1100-0500
LI08)	OBX4A	680	20	R. Tigre, San Isidro
PU40)		690		R.Manco Capac, Région Puno
LA07)	OCX1T	693	1.5	R. Horizonte, Chiclayo (n.f.: 770)
CU04)	OBU7K	700	1	R. La Salle, Urubamba
JU01)	OBU4J	700	1	R. La Luz, Huancayo
LL07)	OCY2H	700	1	R. Sausal Superior, Sausal
LI09)	OBZ4H	700	1	R.Integrida, San Miguel
PI04)	OBX1U	700	10	R. Cutivalú "LV del Desierto", Castilla: 1030-0100 Sun.1100-0100
PI49)	OCU1B	700		(Walson Javier Ruiz Zelada), Sechura (F.P.I.)
AQ04)	OAU6L	710	1	R. Amor, Arequipa
IC02)	OBX5Q	710	5	R. Programas del Perú, Ica: 1100-0600 (r.: 730)
MD01)	OCX7I	*710	10	R. Nacional del Peru, Puerto Maldonado
PU41)		710		R. Surupana, Azángaro
CU05)	OBU7D	720	1	R. Alegria, Wanchaq
JU04)	OAU4E	720	10	R. Sideral, La Oroya
LL08)	OAX2J	*720	25	R. Nacional del Perú, Trujillo
LA06)	OAU1Q	720	1	R. Frecuencia Oceánica, Lambayeque
CJ05)	OBU2Q	730	5	R. Maria, Cajamarca
LI10)	OAX4G	730	50	R. Programas del Peru RPP, San Isidro:24h
PI05)	OAX1D	730	10	R. del Pacifico, Piura: 1100-0500
AQ05)	OAX6C	740	10	R. Continental, Arequipa: 0900-0300
CU06)	OBU7C	740	1	Red Latino, Cusco
LL09)	OCX2X	740	1	R. El Puerto, Pascamoy
PA01)	OCX4X	750	5	R. Altura, Cerro de Pasco: 1000-0400
LL01)	OBX2K	760	0.5	R. Andino, Otuzco
LI11)	0BZ4X	760	10	R. Mar Plus, Chorillos: 24h
AQ06)	OBX6H	770	5	Radiomar, Uchumayo
LA07)	OCX1T	770	1.5	R. Horizonte, Chiclayo (r.: 693)
LO01)	OAX8M	*770	5	LV de la Selva, Iquitos: 0950-0300, Sun: 1100-1700
PU03)	OAU7D	770	2.5	R. LV del Allinccapac, Macusani
CJ07)	OBU2N	780	1	R. Coremarca, Bambamarca
LI12)	OAX4X	780	3	R. Victoria, Lima
PU04)	OAZ7S	780	10	R. Nuevo Tiempo, Juliaca: 1000-0500
TU01)	OAX1K	780	10	R. Nacional del Perú, Tumbes
CU07)	OAZ7H	790	5	R. La Luz, Cusco
LL11)	OAX2I	790	10	R. Programas del Perú, Trujillo
AQ07)	OBX6A	800	0.3	R. Porteña, Arequipa
IC03)	OBX5B	*800	1	R. Sur, Ica
JU05)	OBU4D	800	1	R. Vida, Huancayo
LI13)	OAU4H	800	0.5	R. La Luz, Huaral
PI06)	OCX1P	800	1	Telecom del Norte, Piura
AY19)	OBU5E	810	1	(Ivan Bendezu Vargas), Huamanga (F.P.I.)
LL12)	OAU2G	810	1	R. Apocali, Trujillo
PU05)	OAX7V	810	10	R. Programas del Perú, Juliaca: 24h
CJ43)	OBX2J	820	0.5	R. Nuevo Continene, Cajamarca (r: 1560)
LI14)	OAX4O	820	20	R. Libertad, Lima: 1000-0900
CU08)	OAZ7U	830	1	R. Inti Raimi, Santiago: 24h
CJ62)	OCU2M	830		R.Ebenezer, Bambamarca

MW	Call	kHz	kW	Station, location, H of tr.
HV01)	OAX3Y	830	1	R. La Selva, Rupa-Rupa (n.f.: 1280)
JU06)	OAU4C	830	1	CPN Radio, El Tambo (rel.: CPN 1470)
LL13)	OCX2Y	830	1	CPN Radio, Trujillo (rel.: CPN 1470)
TA02)	OAX6D	830	10	R. Nacional del Perú, Tacna
AN01)	OAX3S	840	1	R. Casma, Casma: 100-0300
AQ01)	OBX6Y	840	1	R. Azul, Arequipa: 24h
CJ08)	OAU2E	840	1	R. Nuevo Continente, San Ignacio (r.: 1575): 1200-2300
CU58)	OCU7I	840		R. Santa Cruz, Kunturkanki
CJ57)		850		R. San Juan, El Tambo
LI15)	OAX4A	850	40	R. Nacional del Perú, Lima: 24h
TA11)	OAU6S	*850		R. Nacional del Peru, Tarata: 1100-0100
PU35)	OBU7Z	850		R. Pachamama, Puno: 0845-0300 (Sat/Sun): 0845-2300
AY01)	OAU5Q	860	5	R. Educativo Macedonia, Ayacucho
CJ09)	OAU2J	860		CPN Radio, Cajamarca (rel.: CPN 1470)
PI08)	OCX1M	860	3	R. Nuevo Norte, Sullana
CU10)	OCX7R	870	1	R. Mundo, Wanchaq: 0900-1000
JU34)	OCX4D	870	2.5	R. Huancayo, Huancayo: 0900-0500
LA08)	OBX1F	870	10	R. Programas del Perú, Chiclayo: 24h
PU06)	OAU7O	870	5	R. Libertad, Puno
PI09)	OAU1G	870	1	R. San Pedro Chanel, Sullana
LL15)	OAX2P	880	1	R. Sintonia, Trujillo: 0900-0300
LI16)	OBZ4N	880	10	R. Union, Lima: 24h
CU56)	OCU7C	890		R. Laramani, Espinar
PU07)	OBX7S	890	1	R. Bahá'í del Lago Titicaca, Chiucuito: 0900-0200
CJ10)	OAU2N	890	1	R. Panorama, Cajamarca: 1000-0400
AQ11)	OBX6K	900	1	R. Nevada, Uchumayo: 24h
CJ11)	OAU2Q	900	1	R. Nor Oriental, Jaén
HU02)	OAX3E	900	1	R. Ribereña, Aucayuc: 1000-0500
LA09)	OCX1X	900	1	R. Sensacional, Ferreñafe: 1000-0300
LI17)	OBX4X	900	10	R. Felicidad, Lima: 1000-0300
AY02)	OAU5M	910	1	R. Estacion Wari, Ayacucho
PU08)	OAU7G	910	1	R. Vision del Altiplano, Juliaca
CU12)	OCX7M	920	1	R. Programas del Perú, Cusco: 24h
IC04)	OCX5C	920	0.1	R. Stelar, Chinca Alta
JU36)		920		R. Campesina, Juli: 0900-0300
LL16)	OBX2S	920	1	R. Ollantay, Virú
PI10)	OBX1J	920	10	R. Programas del Peru, Piura: 24h
SM01)	OAX9V	920	1	R. Marginal, Tocache
AQ12)	OBX6T	930	2.5	R. Yaravi, Arequipa: 0830-500
LL17)	OCX2V	930	1	R. Inti, Chepén
LI18)	OAX4E	930	5	R. Moderna "R.Papa", Lima: 24h
PU09)	OBU7T	930		R. Cadena Colca, Juliaca
CJ58)	OBX2G	940	1	R. CRN, Cutervo
CU13)	OBX7L	940	1.5	R. Willkamayu, Wanchaq: 24h
JU08)	OBU4E	940	1	R. Comericial, Jauja
PI47)	OBU1Y	940	1	R. Studio Satelite, Tambo Grande
AN02)	OBX3S	950	1	R. Programas del Perú, Chimbote
SM02)	OBX9L	950	1	R. Estación Láser, Rioja
AQ13)	OBX6S	*960	12	R. Hispania, Mariano Melgar
CU14)	OBU7P	960		R. Concierto Santa Monica, Espinar
JU09)	OCY4V	960	1	R. Manantial, Huancayo
LA11)	OBX1Y	v960	3	R. WSP, Chiclay:(r: on 958): 0900-0400
LI19)	OAX4D	960	10	R. Panamericana, Lima: 24h
LO02)	OBX8H	960	1	R. Diez, Iquitos
CJ13)	OAU2K	970	1	R. Lider del Norte, Cajamarca
CU15)	OAU7A	970	1	R. Tropicana, Wanchaq
IC05)	OBX5A	970	1	R. Comericial Sonora, Ica: 1100-0500
PI11)	OBX1V	970	1	R. La Capullana, Sullana: 1100-0100
PU11)	OBU7B	970	1	R. Qollasuyo, Juliaca
AQ14)	OAU6F	980	1	R. Universidad, Arequipa: 1000-0100
AY14)	OAX5K	980		R. LV de Huamanga, Huamanga
JU10)	OBU4H	980	1	R. OBU4H, Huancayo
LA12)	OAU1N	980	1	Radio y Sonido, Lambayque
AN04)	OBX3L	990		R. Peruana, Chimbote
CJ14)	OBX2M	990	0.5	R. Contumaza, Contumaza: 1000-0400
LI20)	OBX4J	990	12	R. Latina, Miraflores: 24h
LO03)	OBX8E	990	1	R. Programas del Perú, Iquitos
PA07)	OCU4A	990		(Pasco R&TV Oro E.I.R.L.) Huayllay (F.P.I.)
PI12)	OBU1C	990		R. OBU1C, Sechuar
TA04)	OAX6K	990	10	R. Continental, Tacna
AQ15)	OBX6R	1000	1	R. Edesa, Paucarpata
CJ41)	OAU2P	1000	2	R. Bambamarca, Bambamarca(n.f.: 1530)
CU16)	OAZ7P	1000	2	R. Prensa al Dia, Cusco
HV01)	OBX5W	1000	1	R. Lircay, Lircay
HU03)	OBX3V	1000		R. Huanuco
LA13)	OAU1P	1000	1	R. San Jose, Lambayque
AM10)	OBX9T			R. Ministerio Mundial, Utcubamba
AP01)	OAU5G	1010	1	R. Amistad, Abancay
CJ15)	OBX2P	*1010	1.5	R. San Francisco, Cajamarca
HU11)		1010		R. Bella, Tingo Maria
LI54)	OAX4U	1010	10	R. Cielo, Lima: 24h
PI13)	OBU1L	1010		LV de las Huaringas, Huancabamba: 1045-0200
TU02)	OBZ1C	1010	1	R. Sonora, Tumbes
AN06)	OBX3U	*1020	1	R. Nacional, Chimbote
LA38)	OBX1G	1020	5	R. Heroica, Chiclayo
CU17)	OBU7O	1020	1	R. Informes, Sicuani
JU11)	OBU4F	1020	1	R. Cristo Vive, Huancayo
PI14)	OBU1D	1020	1	R. La Luz, Piura
TA05)	OAU6J	1020	1	R. Internacional, Tacna
AQ16)	OCX6L	1020	1	R. Cumbia, Arequipa
CU18)	OCX7O	1030	1	R. HG-AM, Cusco
LL19)	OAX2U	1030	3	R. Imperio, Huamachuco
PU12)	OAX7N	1030	1	R. LV del Altiplano, Puno: 1000-0400
PU38)		1030		R. Sillustani, Atuncolla
CJ16)	OBX2O	1040	1	R. Nor Oriente, Jaén
CU19)	OAU7H	1040	1	Multimedio Sistena de R., Espinar
IC06)	OBX5U	1040	1	R. Andina del Pacifico, Ica
LI21)	OBX4O	1040	10	R. Metropolitiana, Miraflores
PI15)	OAZ1D	1040	1	R. Vecinal, Piura
AQ17)	OBX6D	1050	1	R. Bethel, Arequipa
CJ56)		1050		R. Campesina, Cajamarca
JU12)	OBZ4J	1050	1	R. Bolognesi, Huancayo
LL20)	OCX2B	1050	1	R. Maria, Chepen
LO04)	OBX8F	1050	1	CPN Radio, Iquitos (rel.: CPN 1470)
PU13)	OAZ7Q	1050	1	R. San Augistín, Juliaca
TU03)	OAU1C	1058	5	CPN R., Tumbes (rel.: CPN 1470) (n.f.: 1060)
CJ17)	OCY2O	v1060	5	R. Sudamerica, Cutervo: 1130-0300 (Sun: 1200-2300)
CU20)	OAU7U	1060		R. 1060, Cusco
LI22)	OCY4D	1060	1	R. Exito, Lima
MO02)	OAU6R	1060		R. La Luz, Ilo
PI41)	OBU1F	1060		R. 1060. Piura
TU03)	OAU1C	1060	5	CPN R., Tumbes (rel.: CPN 1470)(r.: 1058)
AQ18)	OAU6K	1070	1	R. Trinidad, Arequipa: 1000-0200
IC07)	OAX5A	1070	0.2	R. San Juan, San Juan de Marcona
JU13)	OBX4G	1070	1	R. Visión, San Ramón
LA14)	OAU1J	1070	1	R. Vida, Chiclayo
SM03)	OBX9J	1070	3	R. Andes, Tarapoto
CJ18)	OAU2L	1080	1	R. Nueva Vida, Cajamarca
CU21)	OAX7S	1080	2.2	R. Salkantay, Cusco: 0900-0400 (Sun: 1000-0200)
LI23)	OAU4I	1080	10	R. La Luz, Lima //1200, 1340: 24h
MO03)	OCX6X	1080	1	R. Futura, Ilo
PI16)	OBX1D	v1080	1.5	R. San Miguel, Piura: 0900-0600
PU14)	OCU7O	1080		R. 1080, Puno
AQ19)	OBX6X	1090	1	R. Amistad, Arequipa
AY05)	OAU5F	1090	1	R. Inti Andina, Aucara
CJ19)	OBX2A	1090	1	R. Cajabamba, Cajabamba: 0900-0300
JU14)	OCY4G	1100	1	Sonorama Radio, Huancayo: 1100-0300
LA15)	OBX1L	*1100	1	R. Star, Chiclayo
LI25)	OCX4S	1100		R. Imperial "LV de la Provincia", Cañete: 0900-0300
LL46)	OCU2E	1100	1	R. 1000, Julcan
PU15)	OBX7Z	1100	1	R. LTC, Juliaca
CJ20)	OCX2U	1110	1	R. Jaén, Jaén: 1030-0600
CU22)	OCX7T	1110	5	R. Comer, Cusco: 0900-0100 (Sun: 0900-1600)
LI24)	OAZ4W	1110	1	R. Antarki, Lima
MO04)	OCX6P	1110	1	R. Austral, Ilo
PI17)	OCX1R	1110	0.5	R. Centro Popular, La Union: 1000-0500
AQ20)	OCX6U	1120	1	R. Municipal, Cerro Colorado
AY06)	OAU5H	1120	1	R. Dif. Sonora Comunal - R. Quispillaccta, Ayacucho: 0900-1400. 2100-0100
CJ61)	OAM2F	1120		R. Dif. Cristiana RP, Chota
HV03)	OAU5W	1120		R. Huayllahuara
LL21)	OBX2I	1120	1.5	R. Dinamica, Trujillo:0900-2400
UC09)	OBX8R	1120		(Zosimo Rios Perez), Coronel Portillo (F.P.I.)
AP02)	OAU5A	1130	1	R. Armonia, Abancay (r.: 1587): 1030-0300
CJ21)	OAX2V	1130	1.2	R. Los Andes (RPP), Cjamarca
JU15)	OAZ4S	1130	1	R. Chanchamayo, Chanchamayo: 1000-0300
LI27)	OAX4N	1130	2.6	R. Bacán, Lince: 24h
PU16)	OAU7B	1130	1	R. San Gabriel, Juliaca
AN07)	OAX3R	1140	1	R. Bahia, Chimbote: 1000-0200 (Sun: 2200)
AQ21)	OAX6L	1140	1	R. Concordia, Arequipa (rel.: CPN 1470)
IC02)	OAX5W	1140	0.5	R. Chinchaysuyo, Chinca Alta: 1100-0600
JU16)	OCY4C	1140	1	R. Programas del Perú, Pilcomayo
LA16)	OAU1T	1140		R. Fraternal, Ferreñafe
LL48)	OCU2D	1140		Chami Radio, Otuzco: 1000-0200
PI18)	OBX1W	1140	1	R. Piura, Piura
CJ22)	OCY2E	1150	0.5	R. Chasqui Llacta, San Marcos: 1030-2200
CU23)	OCX7Q	1150	2.5	R. Universal, Santa Monica: 1000-0400
LO06)	OAX8D	*1150	10	R. Loreto, Iquitos
PA02)	OBU4K	1150	5	R. Mineria, Cerro de Pasco
PU17)	OAU7K	1150	2.5	R. Lider, Juliaca: 0900-0200
PI44)		1153		R. Ayabaca , Ayabaca (r.)
AY07)	OBX5O	1160	1	R. Huanta 2000, Huanta (r 1390): 1030-0130
CJ23)	OAU2T	1160	1	R. siglo 21, Chota
LL22)	OAX2C	1160	0.3	R. Libertad Mundo, Trujillo: 1100-0400
LA17)	OCX1S	1160	1	Radiales Nor Oriental del Marañon, Chiclayo
LI56)	OAX4C	1160	5	Onda Cero, Lima: 24h

MW	Call	kHz	kW	Station, location, H of tr.
MD02)	OCX7Z	1160	1	R. del Sur, Puerto Maldonado
MO05)	OBX6G*	1160	1	R. Nacional del Perú, Moquegua
AN08)	OAZ3K	*1170	1	R. Nor Peruana Chimbote
AQ22)	OBX6L	1170	10	R. Programas del Perú, Arequipa
CJ24)	OAU2M	1170	1	R. Layzon, Cajamarca
CU24)	OBU7F	1170	1	R. Bethel, Cusco
IC19)	OAU5V	1170		R. Horizonte La Voz del Agro, Pueblo Nuevo
JU17)	OCX4Y	1170	1	R. COSAT, Satipo: 1100-0200
L007)	OBX8M	1170	1	R. La Luz, Iquitos
PI19)	OCX1B	1170	1	R. San Juan, Talara
PU18)	OCX7Y	1170	0.5	R. Constelación, Puno
CJ25)	OBU2E	1180	1	R. La Luz, Jaen
JU18)	OCY4Z	1180	1	R. Libertad "R.RLJ", Junin
LL23)	OCX2A	1180	1	R. Americana, Quiruvilca
PI20)	OAZ1C	1180	1	R. Chulucanas, Chulucanas
AN09)	OBX3D	1190	5	R. Ancash, Huaraz
AQ23)	OCX6G	1190	1.5	R. Alas Peruanas
LA18)	OAX1E	1190	10	R. Em. del Pacifico, Chiclayo
CU25)	OAX7B	1190	2	R. Tawantinsuyo, Cusco: 1000-0300
AP03)	OBX5X	1200	1	R. Comercial, Abancay
CJ26)	OAU2A	1200	1	Frecuencia Pedagogica, Cajamarca
JU19)	OAU4G	1200	3	R. Andes, Huancayo
LI60)	OAX4B	1200	3	Cadena Radio 1200, Lima
PI45)	OAU1Q	1200	1	R. San Andres, Tambogrande
PU19)	OCX7S	1200	1	R. Cultura, Juliaca
TA06)	OAU6P	1200	1	R. La Luz, Tacna
UC01)	OBX8N	1200		R. La Luz, Pucallpa
CU26)	OAX7M	1210	1	R. Quillabamba, Quillabamba: 1000-0300
CU55)	OCU7B	1210	1	R. Santo Tomas
HU12)	OBX3X	1210		(Flor Ramirez de Villamonte), Huanuco (F.Pl.)
JU20)	OCY4T	1210	1	R. Galaxia, Satipo
LL24)	OAX2Q	1210	1	R. Universo, Trujillo: 24h
L005)	OAX8A	1210	5	R. Nacional del Perú, Iquitos: 1100-0500
PI21)		1210	1	R. Municipalidad, Ayabaca
AQ24)	OAX6X	1220	4	R. Melodia, Arequipa: 24h
CU02)	OAU7N	1220	1	R. Univ.Nal.San Antonio Abad, Cusco
IC09)	OAU5N	1220	1	R. La Luz, Ica
IC20)	OBU5I	1220		(Carlos Bautista Gutierrez), Pisco (F.Pl.)
LA19)	OCX1X	1220	3	R. Libertad, Chiclayo: 0900-0400
JU21)	OBZ4Y	1230	1	R. Selecciones,Tarma: 1100-0200 (Sun: 1800)
LL25)	OAX2T	1230	1	R. Albújar, Guadalupe: 1200-0400
MD03)	OBX7J	1230	0.5	R. Madre de Dios, Puerto Maldonado: 1000(Sun: 1100)-0200
PA03)	OBX4Z	*1230	1	R. LV de Oxapampa, Oxapampa
PU20)	OAU7V	1230		R. Frecuencia Amistad, Juliaca
AM01)	OBX9O	1240		R. Bagua Grande
AN10)	OAU3C	1240	1	R. La Luz, Chimbote
AQ25)	OAU6D	1240	5	R. Lider, Arequipa (r. 1236)
CU27)	OBX7M	1240	1	R. Túpac Amaru, Sicuani: 1000-0100
IC10)	OAU5U	1240	1	R. Eco, Ica
JU22)	OAU4V	1240	10	R. Cumbre, Chilca(R.Maria): 24h
LL26)	OAU2Y	1240	1	R. Bolivar, Bolivar
PI22)	OCX1C	1240	1	R. Sechura, Sechura
CJ27)	OAU2V	1250	1	HGV, Santa Cruz
CU28)	OBX7A	1250	3	R. Solar, Cusco: 1000-0500
LI30)	OAX4L	1250	5	R. Miraflores, Miraflores 24h
PI23)	OBZ1B	1250	1	R.B.N.S., Talara Alta
SM07)	OAX9K	1250	1	R. Comercial, San Martin
UC02)	OAX8P	1250	1	R. Pucallpa, Pucallpa: 1030-0500
PI43)		1250		R. Centinela, Huancabamba (r)
AN11)	OAU3G	1260	1	R. Periodico, Chimbote
AQ26)	OBX6D	1260	1	R. Mundial, Arequipa
AQ51)		1260		R. Mahanaim, Mollendo
HU04)	OAU3F	1260	1	R. La Luz, Huanuco
LA20)	OCX1O	1260	1	R. Nor Puruana, Chiclayo
LA21)	OAZ1A	1260	1	R. Ferrañafe, Ferrañafe
LL49)	OBX2C	1260		Telesistema Peruano, Otuzco
AY08)	OBX5S	1260	0.3	R. Nacional del Perú, Ayacucho
CU30)	OAU7S	1270	2	R. Horizonte, Cusco
JU23)	OBZ4T	1270	0.4	R. La Merced, Chanchamayo: 1100-1900, 2200-1200 (Sun 1200-1900)
LL28)	OCX2Z	1270	1	R. Estacion Latina, Cepén
LI31)	OAZ4H	1270	0.4	R. Huacho, Huacho
L008)	OAX8T	1270	1	R. Eco, Iquitos: 0900-0400
PI24)	OAU1S	1270	1	R. Nor Paita, Paita
LA22)	OAU1R	v1276	1	R. Gotas del Oro, Urrunaga(n.f.: 1280): 0900-0500
AN12)	OBX3C	*1280	1	R. Alopesa, Chimbote
AQ27)	OBX6P	1280	0.5	R. Fénix, Camaná
AQ28)	OCX6B	r1280	1	R. Trebol, Mariano Melgar (nf. 1290)
CJ28)	OBX2F	1280	1	R. Moderna, Cajamarca: 1000-0400 (Sun: 0300)
CU29)	OAU7K	1280	1	R. Stereo Nevada, Espinar
CU61)	OCU7F	1280	1	R. Ministerio Mundial, Sicuani
HU01)	OAX3Y	1280	1	R. La Selva, Rupa-Rupa (r: 830)
LA22)	OAU1R	1280	1	R. Gotas del Oro, Urrunaga (r: 1276)
AM02)	OAX9N	*1290	10	R. Nor Oriental, Bagua Grande
AQ28)	OCX6B	1290	1	R. Trebol, Mariano Melgar (r. 1280)
JU24)	OBU4S	1290		R. 1290, La Oroya
LI32)	OBU4Q	1290	1	S & RD, Hualmay
PI42)		1290		R. Satelite, La Union
PU21)	OAX7X	1290	0.3	R. Juliaca, Juliaca: 0900-0300 (nf. on 1300)
TU04)	OCX1Q	1290	1	R. Programas del Perú, Tumbes: 24h
AN13)	OAX3X	1300	0.5	R. Huascarán, Independencia: 1100-0300
CJ29)	OAU2I	1300	1	R. Paraiso, Cajabamba
CU31)	OAX7P	1300	5	R. Misercordia, Cusco: 1200-2130
JU25)	OAZ4B	v1300	1	R. Andina, Huancayo: 0900-1400 (Sun: 1000-0300)
LA23)	OAU1U	1300	1	R. Frecuencia Lider, Morrop
LI33)	OAU4S	1300	5	R. Comas, Comas: 1000-0500
PU21)	OAX7X	1300	0.3	R. Juliaca, Juliaca(r. 1290)
SM08)	OBX9P	*1300		R. La Luz, Tarapoto
TA07)	OBX6P	1300	0.4	R. Comercial Latina, Tacna (rel.: RPP 730)
UC03)	OAZ8B	1300		R. Nuevo Mundo, Pucallpa
AQ29)	OAU6N	1310	1	R. MCV, Paucarpata
AQ50)		1310		R. Libertad, Arequipa
CJ30)	OBX2D	1310	1	R. Chota, Chota: 1100-0300
LI34)	OBX4L	1310	1	R. Irvisa, Huacho
L009)	OBX8L	1310	1	R. Vision Amazonia (R.MIVIA), Iquitos
AN14)	OAX3U	*1320	1	R. Miramar, Chimbote
JU26)	OBU4T	1320	1	R. Corporacion, Huancayo
LA24)	OBU1S	1320	1	R. Frecuencia Popular, Olmos
LI55)	OAX4I	1320	1	R. La Cronica, Lima//R. Nacional 850
PU22)	OAU7W	1320	3	R. Peru, Juliaca: -0300
AQ30)	OVX6E	1330	1	R. Ondas del Misti, Mariano Melgar
AY09)	OAU5L	1330	0.5	R. Bethel, Huamanga
CU32)	OCX7K	1330	1	R. San Miguel, Wanchaq: 0900-0300 (Sun:1100-0200)
LA25)	OAU1A	1330	1	R. Dos Mil, Chiclayo: 1000-0400
AN15)	OAU3E	1340	1	Assn. Iglesia de Dios, Chimbote
AQ31)	OAU6T	1340	1	R. Comercial, Mollendo
CJ31)	OAU2S	1340	1	R. Shalom, Cajamarca
IC11)	OAX5D	1340	0.5	R. Chinca, Chinca Alta
JU27)	OAU4N	1340	1	R. Jauja, Jauja:_0900-0300
LI35)	OAQ4Q	1340	5.5	R. Alegria, Pucasana // 1080,1200
PU23)	OBU7V	1340		R. Sudamericana, Juliaca
AY17)	OBU5O	1350	1	(Walter Huayanay Quispe), Huamanga (F.Pl.)
CU33)	OBU7E	1350	1	R. Lider, Cusco
LA31)		1350	1	R. Vision, Chiclayo: 0845-0500
MO06)	OBX6F	1350	1	R. Ilo, Ilo: 24h
UC04)	OBX8D	1350	1	R. Super, Pucallpa: 1030-0500
HU05)	OAX3N	v1352	1	R. Ondas del Huallaga, Huanuco: 0930 (Sun:1100)-0300 (n.f.: 1350)
AN16)	OAU3A	1360	0.2	R. Intercontinental, Yungay
AQ32)	OCX6T	1360	1	R. Luza, Paucarpata
CU34)	OAX7R	v1360	2.5	R. Sicuani, Sicuani(r: 1362): 0900-0300
IC12)	OBZ5Z	1360	1	R. Cruz del Sur, Palpa: 1030-0300
JU28)	OAU4O	1360	1	R. Hecaburt, Tarma
LL31)	OBX2N	1360	1	R. Super Uno, Santiago de Cao
LI58)	OAX4Y	1360		R. Nueva Q-FM, Lima
PI25)	OBZ1A	1360	1	R. del Norte, Sullana
PU24)	OUA7L	1360	2.5	R. Continente, Juliaca
AP05)	OCX5A	1370	1	Inti Radio, Abanacy
CJ47)	OBU2U	1370	1	R. Satelite, Santa Cruz
CU35)	OAU1W	1370	1	R. Chiclayo, Chiclayo
LA27)	OAZ7J	1370	1	R. Santa Monica, Wanchaq: 0900-0300
LI37)	OAZ40	1370	0.5	R. Tres de Octubre (R.Cosmos), Huacho
MO07)	OAX6T	1370	1	R. Moquegua, Moquegua: 0930-0500
SM09)	OBX9A	1370		R. Palmera: 1100-0500
AQ33)	OAX6O	1380	2.5	R. San Martin, Arequipa: 1000-0130
CJ33)	OA2XW	v1380	1	R. Atahualpa, Cajamarca: 1100-0500
JU29)	OBU4L	1380	1	R. Chilca, Huancayo
LI38)	OCY4U	1380	1	R. Nuevo Tiempo, Lima
LL52)		1380		R. Fraternidad, Trujillo
PI26)	OBZ1D	1380	1	R. Bellavista, Bellavista
HU06)	OBX3I	v1382	1	R. Pilco Mozo, Huanuco (n.f.: 1380)
AQ34)		1390		Difusora Neptuno, Mollendo
AY07)	OBX5O	1390	1	R. Huanta 2000, Huanta(n.f.:1160)
CU36)	OAU7T	1390	1	R. Enlace, Kunturkanki: 1000-0300
LL32)	OAU2Z	1390	1	R. La Luz, Trujillo
LA28)	OAU1V	1390	1	R. Tropical, Morrope
PU25)	OCX7U	1390	1	R. Cultura, Yunguyo
AQ35)	OAX6J	1400	5	R. Landa, Arequipa
CA34)	OAU2H	1400	1	R. OAU2H, Cajamarca
CU37)	OAX7I	1400	1	R. La Hora, Cuzco: 1000-0200 (Sun: 1100-1700)
IC13)	OCX5B	1400	1	R. Interandina, Pisco: 24h
JU30)	OBX4H	1400	1	R. Luz, Tarma
LI39)	OBX4W	1400	2.5	R. Callao Super, Lima
PI27)	OCX1A	1400	1	R. MDY, Talara Alta: 1000-0600
CJ64)	OCU2Q	1410		(Maria Vasquez Bazan), San Marcos (F.Pl.)
LL33)	OAX2Y	1410	1	R. Heróica, Trujillo
LA29)	OBU1G	1410		R. Olmos, Olomos

MW	Call	kHz	kW	Station, location, H of tr.
LI40)	OBZ4V	1410	1	R. Universal, Santa Maria
LI57)	OBZ4C	1410	1	R. Bethel, Huacho
PU26)	OBU7A	1410	1	R. La Luz, Juliaca (R.Grégor)
TU05)	OBU1H	1410	1	R. La Luz, Tumbes
UC05)	OBX8I	1410		Dif. Comercial, Pucallpa
CJ04)	OAX2R	1420		R. Cajamarca, Cajamarca: (r.:634)
CJ38)	OBX2V	1420	5	R. Ilucan, Cutervo: (r.:1476): 1100-0300 (Sun: 1030-2300)
CU38)	OBU7L	1420		R. OBU7L, Yanaoca
LI41)	OBZ4G	1420	1	R. San Isidro, Lima
LO10)	OAZ8Z	1420	1	R. Oriente, Yurimaguas: 1000-0200
PI28)	OCX1H	1426	0.2	R. San Jose, La Union: 0900-0100 (n.f. 1420)
AM03)	OBX9H	1430	1	R. Utcubamba, Bagua Grande
AN17)	OBX3E	1430	1	R. Huarmey, Huarmey
AN18)	OAZ3H	*1430	1	R. Chavin, Chimbote
CU39)	OAZ7M	1430	1	CPN Radio, Cusco (r: CPN 1470)
JU31)	OAZ4V	1430	0.5	R. Universal, El Tambo: 1100-0500
LA39)		1430		R. Nueva Juventud, Tucumé
LL34)	OBX2T	1430	1	R. Santa Bárbara, Ascope
LI42)	OCU4L	1430		(Maria del Pilar Ruiz Coneja Pineda), San Vicente de Cañete (F.PI.)
PU27)	OBU7U	1430	1	R. Red, Andina
TA08)	OAU6M	1430	1	R. Lider, Tacna
AN19)	OAZ3O	1440	1	R. LV de Pomabamba, Pomabamba
AQ16)	OAX6R	1440		R. Santa Monica, Arequipa
CJ35)	OAU2O	1440	1	R. Frecuencia VH, Celendin
CU40)	OAU7M	1440	1	R. Regional, Sicuani
LA30)	OBX1T	1440	2	R. Cooperativa Tumán, Chiclayo: 1100-0300
LI43)	OAX4K	1440	1	R. Imperial 2, Lima (n.f.: 1440)
CJ59)		1450		R. Super Nueva Sencacion, Chirinos
CJ36)	OAU2W	1450	1	R. San Miguel, Cajamarca: 09000400
CU41)	OCX7W	1450	1	R. Santa Rosa, Santa Ana
JU45)	OBU4Y	1450		Rdif. E.G.C S.R.L., Huancayo
LL35)	OCX2J	1450	1.5	R. San Juan, Trujillo: 0900-0600
LI44)	OBX4K	1450	1	R. Fortaleza, Barranca
CJ36)	OAU2W	1458	1	R. San Miguel, San Miguel: (n.f.: 1450)
PI29)	OAX1V	1457		R. Sullana "LV de Chira", Sullana: (n.f 1460)
AQ37)	OBX6C	1460	1	R. Bahia, Mollendo
CJ37)	OBU2E	1460	1	R. Comercial, Jaén
CU42)	OBU7M	1460		R. OBU7M, Marcapata
IC14)	OAX5K	1460	2.5	R. Internacional, Pisco
JU32)	OCY4I	1460	0.5	R. Imperial, Junin
JU33)	OAZ4F	1460	1	R. La Oroya, La Oroya: 1000-0500
PI29)	OAX1V	1460	1	R. Sullana "LV de Chira", Sullana: (r. 1457)
PU28)	OAX7W	1460	10	R. El Sol de los Andes, Juliaca
AQ38)	OAU6E	1470	1	R. Victoria, Arequipa
CU43)	OAX7G	1470	1	R. Cusco, Cusco: 1000-0300 (Sun: 1100-0300)
LL37)	OCX2G	1470	1	R. Occidente, Quiruvilca
LI45)	OAU4B	1470	20	CPN Radio, Lima
TA09)	OAX6M	1470	0.8	R. Tacna, Tacna: 0900-0500
CU44)	AOZ7G	1475	1	R. Espinar, Yauri: (n.f.: 1480)
CJ38)	OBX2V	1476	5	R. Ilucan, Cutervo: (n.f.: 1420)
CJ44)	OBU2H	1480	0.5	R. San Lorenzo, Socota: (r.: 1584)
JU35)	OAU4A	1480	1	R. Laser, Santa Rosa de Sacco: 1100-2300
LI38)	OCX2C	1480	0.6	R. Comercial, Virú
LI46)	OCX4V	1480	1	R. K`ler, Paramonga
PI30)	OCX1L	1480	1	R. Supercontinental, Chulucanas: 1100-0500
AQ39)	OAX6Q	1490	1.3	R. Minuto, Cerro Colorado
AP10)	OBU5C	1490		Radiodifusora los Chankas, Andahuaylas
CU45)	OBU7I	1490	1	R. Chaski, Maras: 1000-2300. 0000-0200
IC15)	OAX5N	1490	0.3	R. Nazca, Nazca
LA31)	OAX1L	v1490	1	R. Vision, Chiclayo: 0845-0500
LO11)	OAX8F	1490	1	R. Nueva Atlantida, Iquitos: 0900(Sun: 1100) - 0500
PA04)	OBU4N	1490	1	R. La Luz, Cerro de Pasco
PU29)	OCX7P	1490	1	R. Emisora Frontera, Puno
CJ39)	OBU2J	1500	1	R. San Pablo, San Pablo
HU07)	OBX3J	1500	1	R. Luz y Sonido, Huanuco: 0900-0300
JU36)	OAU4W	1500	1	R. Wanka, Huancayo
LL39)	OBX2X	1500	1	R. Comercial, Trujillo: 0845-0300
LI47)	OBX4I	v1500	10	R. Santa Rosa, Lima
TA10)	OAU6B	1500	1	R. Bulevar, Tacna
AQ40)	OCX6Q	v1510	1	R. Las Vegas Arequipa: 24h
CJ40)	OCX2O	1510		R. Inca, Los Baños del Inca: 0900-0300 (r.: 1517)
CU47)	OBX7P	1510	1	R. El Sur, Wanchaq
IC16)	OAX5F	1510	1	R. LV Huamanga, Nazca:
JU37)	OCX4J	1510	1	R. Tarma, Tarma: 1000-0500
LL40)	OAU2U	1510	2.5	R. Virgin de la Alta Gracia, Huamachuco
LA32)	OBU1B	1510	1	R. OBU1B, Olmos
TU06)	OCX1V	1510	1	R. Tumbes, Tumbes
UC06)	OBX8K	1510	1	R. Centro de los Medios, Sepahua
CJ40)	OCX2O	1510	1	R. Inca, Los Baños del Inca: 0900-0300 (n.f.: 1510)
CU48)	OBU7X	1520		R. Fuentes Mollo, Espinar
LA26)	OAU1H	1520	1	R. Cristal, Chiclayo
PU30)	OAU7Y	1520	5	R. OAU7Y, Juliaca

MW	Call	kHz	kW	Station, location, H of tr.
SM11)	OAX9X	1520	1	R. Vision, Janjui: (r.: 1545)
AN21)	OBX3H	1530	1	CPN Radio, Chimbote: (rel.: CPN 1470)
CJ06)	OBX2R	1530	3	R. Oriental, Jaén
CJ41)	OAU2P	1530	2	R. Bambamarca, Bambamarca: 1000-0300
CU49)	OAZ7F	1530	0.5	Rdif. Espinar, Yauri (R.Confraternidad)
CU50)	OBX7N	1530	1	R. Ondas del Sur Oriente, Quillabamba
IC17)	OAU5R	1530	1	R. Universidad, San Juan Bautista
JU38)	OBZ4S	1530	1	R. 15-50, Huancayo: 1000-0600
LI49)	OBU4C	1530	10	R. Milenia, Lima
PI31)	OCX1Y	1530	1	R. Leomar, Bellavista
PU39)		1530		R. Capachica, Capachica
AQ41)	OAU6A	1540	1	R. Milenio Universal, Arequipa: 24h
CJ63)		1540	5	R.Turbomix, Cajamarca
CU51)	OCX7V	1540	1	R. Los Andes, Cusco
LL42)	OBU2A	1540	2	R. Mundial AM, Trujillo
LI50)	OBZ4U	*1540	1	R. Barranca, Barranca
PA05)	OBX4N	1540	0.3	R. Corporacion, Cerro de Pasco: 0900-0300
TU07)	OBX1B	1540	1	R. LV de la Frontera, Tumbes
SM11)	OAX9X	1545	1	R. Vision, Janjui: (n.f.: 1520)
AN22)	OAU3D	*1550	1	R. Cruz, Chimbote
AY10)	OBX5J	1550	1	R. San Cristobal, Carmen Alto (R.Maria)
CJ42)	OCX2R	1550	1	R. TV Chota, Chota
LA34)	OAX1D	1550	1	R. Superior, Monsefú: 1100-0400
LI51)	OBX4P	1550	1	R. Independencia, Lima
AQ42)	OCX6N	1550	1	R. La Union (R.La Luz), Arequipa
CJ43)	OBX2J	1560	0.5	R. Nuvo Continente, Cajamarca: 1000-0300 (n.f.: 820)
CU52)	OAZ7N	1560	1	R. Maria, Wanchaq
LO13)	OBX80	1560	1	R. Nuevo Mundo, Iquitos: 1100-0500
JU39)	OBU4G	1560	1	R. San Sebastian, Yauyos
PU32)	OAU7Z	1560	1.5	R. Carráviz, Juliaca
CJ44)	OBU2H	1564	0.5	R. San Lorenzo, Socota: (n.f.: 1480)
AN23)	OBX3N	1570	1	R. Chasqui, Yungay
AQ43)	OCX6I	1570	1	R. Willy, Uraca
CJ54)	OBU2L	1570		R. Colonial, Contumaza
CU62)	OCX1Y	1570		(Andrea Alejla Llanos Alca), Canchis (F.PI.)
HU08)	OBX3M	1570	1	R. San Martin, Huanuco
LI59)		1570		R. Bethel, Lima
LL47)	OCU2C	1570		Radiodifusora Julcan
PI32)	OCX1Z	1570	1	R. La Nueva Esperanza, Tambo Grande
LA36)	OAZ1F	1575	3/1	R. Naylamp, Lambayeque: (n.f.: 1580)
CJ08)	OAU2E	1575	1	R. Nuevo Continente, San Ignacio: 1200-2300 (n.f.: 840)
HV02)	OAU5J	1580	1	R. Virgen del Carmen, Huancavelica: 24h
JU40)	OAU4P	1580	1	R. San Juan, Tarma: 1030-0230 (Sun: 1100-2100)
LA36)	OBX1M	1580	3/1	R. Naylamp, Lambayeque: 1000-0200 (r.:1575)
SM12)		1580	1	R. Central, Bellavista: 1100-0300
CU53)	OBX7Q	1584	0.5	R. El Triunfo, Cusco: (n.f.:1580)
AR44)	OCX6S	1590	1	R. Mundo, Arequipa
AP02)	OAU5A	1587	1	R. Armonia, Abancay: 1030-0300 (n.f.1130)
LL43)	OBU2C	1590	1	Agro Radio, Trujillo
LI52)	OAZ4Z	1590	2	R. Agricultura "La Peruanísima" Lima
PU33)	OAU7C	1590	1	R. Huaynaroque, Juliaca
CJ45)	OCY2D	1600	1	R. Int., San Pablo: 1000-1700, 2100-0300
JU44)	OBU4R	1600	1	R. Nuevo Tiempo, Huancayo: 1000-0500
PI46)		1600		R. San Juan, Catacaos
AP09)		1610		R. Haquira, Haquira (r.)
AQ45)	OAU60	1610	0.5	R. Flor de los Andes, José Luis Bustamante y Rivero (R. El Sabor, Arequipa)
LL45)		1610		R. Carabamba, Julcán (r.)

SW	Call	kHz	kW	Station, location, h of tr.
HU09)	v3173			R. Municipal, Panao: 0900-1100, 2300-0200
HU07)	OAW3A	v3235	1	R. Luz y Sonido, Huánuco
HU05)	OAX3Q	v3330	5	R. Ondas del Huallaga, Huánuco: 1030-0200 (irr.)
AQ46)	OAW6B	3375	1	R. San Antonio, Callalli: (irr.)
AY07)	OAZ5B	v4747	0.5	R. Huanta 2000, Huanta: (n.f. 4755): 1100-0100
PI13)		*4750		R. San Francisco Solano, Sondor: (irr)
AY07)	OAZ5B	*4755	0.5	R. Huanta 2000, Huanta: (r: 4747)
JU37)	OCX4W	4775	0.5	R. Tarma, Tarma: 1100-0200
CJ47)	OAX2L	*4781	1	R. Satelite, Santa Cruz: 2300-0300
LA31)		4790	0.5	R. Vision, Chiclayo: 24h
LO01)	OAX8R	4824	10	LV de la Selva, Iquitos
CU34)	OAX7T	v4826	0.3	R. Sicuani "LV de Canchis", Sicuani: 1030-0300 (n.f.:4835)
CJ01)	OCX2E	4835	1	R. Marañon, Jaen: 1100-0200
AY17)	OAW5E	4850	1	R. Genesis, Huanta
CU37)	OAZ7A	4856	1	R. La Hora, Cusco: irr
HV02)	OAX5X	4887	1	R. Virgen del Carmen, Huancavelica: (nf: 4885), irr
UC07)	OAW8A	4940	1	R. San Antonio, Villa Atalaya: 2200-0030 (irr)
MD03)	OBX7I	4950	5	R. Madre de Dios, Puerto Maldonado: 1000(Sun: 1100)-0200
AY13)	OAX5S	4955	5	R. Cultural Amauta, Huanta: 1000-1400, 2100-0100
CU32)	OAZ7B	*4965	1	R. Santa Monica, Wanchaq

SW	Call	kHz	kW	Station, location, h of tr.
LI06)	OAZ4X	v4975	5	R. Pacifico, Lima
AN09)	OAZ3B	4992	5	R. Ancash, Huaraz: (n.f.: 4990)(irr.)
JU09)		4987		R. Manantial, Huancayo
CJ20)	OAX2S	v4996	2	R. Andina, Huancayo: (n.f.:4995)
PA01)	OBZ4B	5014	5	R. Altura, Cerro de Pasco
AM04)	OBX9K	v5020	5	R. Horizonte, Chachapoyas: 1100-0200
CU26)	OAX7Q	5025	5	R. Quillabamba, Quillabamba: 1000-0200
JU18)	OCY4Y	5039	1	R. Libertad de Junin, Junín: 1100-1400
PI13)	OAW1B	5059		LV de las Huarinjas, Huancabamba
CU50)		5120	1	R. Ondas del Sur Oriente, Quillabamba: 1000-0300
LL26)		5460		R. LV Bolivar, Bolivar: 2330-0130
LA24)		5485		R.Frecuencia Popular: 2330-0200 (r)
AM06)		5487		Reina de la Selva, Chachapoyas
AQ17	OAX6A	r5921		R. Bethel, Arequipa: 1045-0100
AQ24)	OBX6I	v5939	1	R. Melodia, Arequipa
LI12)	OAX4Q	v6020	3	R. Victoria, Lima
LI47)	OCY4H	v6047	10	R. Santa Rosa, Lima
CU23)	OAZ7Q	*6090		R. Universal, Santa Monica
LI16)	OBZ4O	v6115	10	R. Union, Lima: 24h
CU25)	OAX7C	6174	1	R. Tawantinsuyo, Cusco(nf: 6175)
LO10)	OAX8I	*6188	1	R. Oriente, Yurimaguas: 1000-0200 (n.f.: 6190)
CU43)	OAX7A	6195	1	R. Cusco, Cusco: (n.f.: 6195)
LI06)		9675	5	R. Pacifico, Lima: 1400-2300
LI12)	OCX4C	v9722	1	R. Victoria, Lima

Addresses and other information:
(Names of **departamentos** should be added to address info, main capitals excluded).

AM00 (AMAZONAS):
AM01) CP – Av Circunvalacion 1336, Bagua Grande – **AM02)** Simón Bolívar 433, Bagua Grande – **AM03)** Jr F Villareal 400, Utcubamba - **FM:** 96.9MHz – **AM04)** Jr Amazonas 1717 (Ap 69), Chachapoyas ☎4177 7793 📠4175 7004. **E:** rhorizonte@hotmail.com - **FM:** 99.9MHz – **AM05)** Jr Huayabamba 513, San Nicolas, Prov Rodríguez de Mendoza – **AM06)** Calle Ayacucho 944, Plaza Mayor, Chachapoyas ☎ 41 47 7203 📠 4147 7989 **W:** www.reinadelaselva.com - **E:** joreno@terra.com.pe - **FM:** 101.5MHz – **AM07)** Jr Amazonas 315, Aramango, Prov de Bagua – **AM10)** Calle Higos Urcos 651, Bagua Grande, Utcubamba
AN00 (ANCASH):
AN01) Av Nepeña Mza 8-C, Lote 3, Casma ☎4371 1266 - 1100-0300 - **FM:** 92.5MHz – **AN02)** Av Francisco Pizzarro, Chimbote - **FM:** 95.5MHz – **AN04)** Urb el Trapecio 2da etapa, MZ G, Lote 18, Chimbote - **FM:** 97.5MHz – **AN05)** Manzana 8 Lote Ind Primero de Mayo, Chimbote - **FM:** 105.1MHz – **AN07)** Pasaje los Jardines 129, Chimbote ☎4332 2391 – **AN08)** Pasaje Los Jardines 129, Chimbote - **FM:** 104.3MHz – **AN09)** Jr Francisco Araos 114 Independencia, Huaraz **W:** www.radioancash.org **E:** info@radioancash.org ☎4342 1381 4342 29992 - 24h - **FM:** 101.3MHz **AN10)** Av Enrique Meiggs 2013, Chimbote.- ☎ 43 805591 - **FM:** 89.9MHz – **AN11)** Calle Ramon Castilla MZ.J, Lote 11, Chimbote – **AN12)** Av Leoncio Prado 660, Chimbote - **FM:** 89.9MHz – **AN13)** Jr San Martin 655, Huaraz - **FM:** 104.5MHz – **AN14)** Av San pedro 246, Chimbote ☎4332 2279 - **FM:** 92.3MHz – **AN15)** Av Pardo 6788 ,1° de Mayo, Nuvo Chimbote 43-318-949 **W:** www.iglesia-dios. org **E:** pastorvalle@hotmail.com – **AN16)** Casero el Rayan, Yungay – **AN17)** Av Cabo Alberto Reyes 281, Huarmey - **FM:** 89.7MHz – **AN18)** Urb San Juan ZN 5, Chimbote - **FM:** 92.3MHz – **AN19)** Zona Cushuro, Pomabamba – **AN21)** Calle Aviación 298, Chimbote - **FM:** 103.5MHz – **AN22)** Jr Alfonso Ugarte 627 4° piso, Chimbote – **AN23)** Barrio Lucmapampa, Yungay – **AN24)** Plaza de Armas S/N, Chiquian.
AP00 (APURIMAC):
AP01) Av Seoane s/n, Abancay – **AP02)** Jr Cusco 319, Abancay - **FM:** 92.1MHz – **AP03)** Av Nuñez 401, Abancay – **AP05)** Av Seoane 337, Region Inca, Abancay 970 5246 **W:** www.corporacionsolar.com **E:** intradio@corporacionsolar.com - **FM:** 103.3 Solar FM – **AP06)** Jr Apurimac s/n, Chincheros – **AP07)** Jr Cuzco 206, Abancay - **FM:** 104.5MHz – **AP08)** Guillermo Cáceres Tresierra 381, Andahuaylas ☎8372 1511 - **FM:** 94.9MHz – **AP09)** Haquira, Dis de Haquira, Prov de Cotabambas **AP10)** Jr.Juan Antonio Trelles 278, Andahuaylas ☎83 421 208 **E:** ronalripa@yahoo.es - **FM:** 100.3MHz – **AP11)** Tres Cruces, Ocobamba, Provincia de Chincheros.
AQ00 (AREQUIPA):
AQ01) Calle San Juan de Dios 210, Zona Alto, Arequipa 95 9871 242 **W:** www.radioazularequipaperu.com **E:** radioazulamfm@hotmail.com - **FM:** 89.5MHz – **AQ02)** Av Goyeneche 818, Miraflores, Arequipa – **AQ03)** Lote 64-5-C Km 48 Panamericana Sur, La Joya – **AQ04)** Cl Dean Valdivia 418, Of.25, Urb Cercado, Arequipa ☎5480 7362 – **AQ05)** Av Independencia 56, Arequipa ☎5422 4109 - **FM:** 93.5MHz – **AQ06)** Cruce Ferroc, Puno Variante, Uchumayo – **AQ07)** Calle Alto de la Luna 334, Arequipa –**AQ09)** Calle Francia 120, Urb Satélite Chico-Paucarpata, Arequipa ☎54 424237 - **Quechua: Sun**0830-1030 - **FM:** 103.5MHz

– **AQ11)** Av Victor A Belaúnde C-8, Umacollo, Arequipa ☎5425 5888 📠 5425 1822 - **FM:** 97.1MHz – **AQ12)** Calle Junin s/n Porvenir, Miraflores, (Ap 17061), Arequipa ☎5426 3324 – Quechua: 2h per day **W:** www. radoyaravi.org.pe **E:** radioyaravi@terra.com - **FM:** 106.3MHz – **AQ13)** Calle Consuelo 404, Depto A, Arequipa ☎5421 9928 - **FM:** 98.5MHz in Majes, 102.1 in Tacna – **AQ14)** Av Independencia s/n 2° piso, Pabellón de la Cultura, Ciudad Universitaria (Cas 23), Arequipa.☎ 5428 7771 **W:** www.unsa.edu.pe **E:** radiouniversidad@unsa.edu.pe – **AQ15)** At 200 Millas La Pina, Paucarpata – **AQ16)** Av Independncia 905 – 2do piso, Arequipa ☎204 904 **W:** //santamonica.blogdiario.com - **FM:** 107.7MHz – **AQ17)** Av Union 225, Miraflores, Arequipa (Also in Lima: **LI59)**) **W:** www.bethelradio.fm **E:** betheltadio@bethelradio.fm – **AQ18)** Urb Chapi Chico, Mz A, Lt 4, Miraflores, Arequipa ☎ 5420 4847 **Quechua:** 1000-1200 – **AQ19)** Av Independencia 905, Arequipa ☎5420 1904 – **AQ20)** Calle Mariano Melgar 500, Cerro Colorado, Prov de Arequipa – **AQ21)** Av.La Paz 512, Arequipa ☎054 446053 - **FM:** 95.8MHz – **AQ22)** Av La Paz 511 "A", Of 312 – 3er piso, Arequipa ☎ 5428 7821 – **AQ23)** Av La Salle 124, José Luis Bustamente y Rivero, Urb Daniel Alcides Carrión G-14 ☎5443 1051 – **AQ24)** Calle San Camilo 501-A Cercado, Arequipa ☎5420 5811 5420 4420 **W:** www.radiomelodia.com.pe - **FM:** 104.3MHz – **AQ25)** Av Independencia 1819, Arequipa ☎5428 6438 – **AQ26)** Calle Pierola 209, Of 205, Arequipa – **AQ27)** Esq Av Lima y Bolognesi, Camaná – **AQ28)** Parque Azángaro 150, Miraflores, Arequipa – **AQ29)** Jr Benavides 405, Urb Selva Alegre, Arequipa – **AQ30)** Av República de Chile 123, Mariano Melgar, Prov de Arequipa – **AQ31)** Calle Cesar Vallejo 107, Mollendo – **AQ32)** Zona Rural Huayracpampa, Paucarpata, Prov de Arequipa – **AQ33)** Calle Deán Valdivia, Of-221 (56), Arequipa **E:** dominico@terra.com ☎5421 3301 📠5428 8229 - **FM:** 97.7MHz – **AQ34)** Calle Tupac Amaru A-18, Urb Miramar, Mollendo – **AQ35)** Sucre 409, Arequipa – **AQ37)** C.Baca Flor 410, (Cas 128), Mollendo ☎5453 2521 - **FM:** 101.5MHz – **AQ38)** Dean Valdivia 418, Piso 3, Cercado, Arequipa ☎5440 5480 – **AQ39)** Santo Domingo 113 , Galerias Gamesa Of 700, Arequipa, (Ap 2330) ☎5421 4997 **E:** radiominuto@terra.com.pe - **FM:** 99.9MHz – **AQ40)** Centro Comercial Independencia, Av Independencia 403-A, Ofic 433, 4° piso, Arequipa **W:** http://mollendovegas.com - **FM:** 99.3 MHz – **AQ41)** Av Brasil 612, Distrito de Alto Selva Alegre, Arequipa ☎ 5426 4319**Quechua:** 2000-2100 – **AQ42)** Alto Siguas s/n, Arequipa ☎5428 0414 – **AQ43)** Av Progreso 58, Corire, Uraca, Prov de Castilla - **FM:** 97.9MHz – **AQ44)** Calle Castilla 39, Urb Municipal, Arequipa – **AQ45)** Puerte Grau 122, Cercado de Arequipaa **E:** manuel_montes30@ hotmail.com – **AQ46)** Parroquia San Antonio de Padua, Plaza Principal s/n, Callalli, Prov de Caylloma **E:** rsan_antonio14@hotmail.com – **AQ47)** Calle Puno 820, Miraflores, Arequipa – **AQ49)** Sucre 409, Arequipa (Apartado Postal 105, Serpost, Cercado, Arequipa) **W:** www.radziovosdasalvacion.100megas.com – **AQ50)** Calle Trabada No 105 VI Centenario, Arequipa ☎54 202022 **W:** www.radiolibertadaqp. com **E:** radiolibertadaqp@hotmail.com – **AQ51)** Mollendo ☎54 284411
AY00 (AYACUCHO):
AY01) Av Los Rosales 199, Urb Jardin, Ayacucho - **FM:** 93.3MHz – **AY02)** Calle Nazareno 108H, Ayacucho **E:** macebu90@hotmail.com - **FM:** 95.3MHz – **AY05)** Plaza Mayor Felipe Guzman Poma, Aucara, Prov de Lucanas **E:** radioia1090@hotmail.com – **AY06)** Jr Chorro 274, Huamanga, Ayacucho **E:** aba-ay@wayna.rcp.net.pe ☎6483 6042 **Prgr** mainly in **Quechua** – **AY07)** Jr Gervasio Santillana 455, Huanta ☎6683 2105 - **FM:** 92.9MHz – **AY08)** Jr Piura s/n, Ayacucho - **FM:** 97.9MHz – **AY09)** Av Arica 105, San Juan Bautista, Huamanga **E:** ayacucho@bethelradio.net - **FM:** 93.9MHz – **AY10)** Local de Obispado, Carmen Alto, Prov de Huamanga - **FM:** 106.7MHz – **AY11)** Calle Nazareno 108, Ayacucho – **AY13)** Jr Cahuide 278, Huanta (☞Cas 24) ☎66 322153 **W:** www.turadioamauta.com **E:** radioamauta@ turadioamauta,com – **AY14)** Calle El Nazareno, 2do Pasaje 159, Cercado, Huamanga ☎66 318767 **W:** www.diariola-vozdehuamanga.com **E:** diariolavozdehuamanga@yahoo.com.ar- **FM:** 91.1MHz – **AY16)** Calle el Nazareno – 2do Pasaje no 159, Ayacucho. – **AY17)** Jr.Miguel Untiverso No. 431, Huanta. – **AY18)** Calle Manco Capac 157, Huamanga – **AY19)** Jr.Angel del Señor MZ C Lote 1A, Asociacion Los Mecanicos, Huamanga
CJ00 (CAJAMARCA):
CJ01) Jr Francisco de Orellana 343 (Apt 50), Jaén **W:** www.radio-maranon.org.pe **E:** correo@radiomaranon.org.pe ☎7673 1147 - **FM:** 96.1MHz – **CJ02)** Jr 24 de Junio 189, Huambos, Chota – **CJ03)** Km 5 Carretera Jaen-San Ignacio, Jaen - **FM:** 90.9MHz – **CJ04)** Jr Revilla Peréz 194, Cajamarca ☎7682 9067 - **FM:** 105.1MHz – **CJ05)** Predio Coliga, Cajamarca – **CJ06)** Av Mesones Muro 157, Jaén – **CJ07)** Jr 28 de Julio 712-716, Bambamarca **W:** www.radiocoremarca.com **E:** coremarca@radiocoremarca.com.pe ☎ 7635 3169 - **FM:** 101.1MHz – **CJ08)** Jr Villanueva Pinillos 330, San Ignacio ☎7471 6100 - **FM:** 96.3MHz – **CJ09)** Jr.Amazonas 725, Cajamarca - **FM:** 99.3MHz – **CJ10)** Jr.San Martin Cajamarca ☎4482 6251 📠7683 0238 – **CJ11)** Calle Zurumilla 1328, Jaén - **FM:** 94.9MHz – **CJ13)** Jr Huánuco 2367, Cajamarca ☎7682 3363 - **FM:** 90.3MHz – **CJ14)** Jr David León 601,

Contumazá – **CJ15)** Jr Dos de Mayo 271, Cajamarca ☎ 7636 9915 **W:** www.radiosanfranciscoperu.com.pe **E:** radiosanfrancisco@gmail.com - **FM:** 91.9MHz – **CJ16)** Calle Bolívar 1020, Jaén – **CJ17)** Jr Ramón Castilla 491, Cutervo ☎7673 7090 - **FM:** 97.7MHz – **CJ18)** Av. Via de Evitamiento S/N, Cajamarca **W:** www.radionuevavidacomunicaciones.com **E:** radiotvnc@radionuevavidacomunicaciones.com or radiotvnvc@hotmail.com – **CJ19)** Jr Lara s/n, Cajabamba **E:** radioca-jabamba@hotmail.com - - **FM:** 90.3MHz – **CJ20)** Jr Mariscal Castilla 439, Prov de Jaén - **FM:** 96.7MHz – **CJ21)** Av San Martin De Porres s/n, Cajamarca ☎7682 8566 - **FM:** 100.7MHz – **CJ22)** Jr Leoncio Prado 330, San Marcos ☎4485 8083 – **CJ23)** Av Inca Garcilazo de la Vega 473, Chota 35 1411 – **CJ24)** Jr Sullana 212, Cajamarca **W:** www.radiolayzon.com **E:** radiolayzon902@hotmail.com - **FM:** 90.5MHz – **CJ25)** Calle Santa Rosa 914 Pi 2, Sector Pueblo Nuevo, Jaén 76-318-974 - **FM:** 105.7MHz – **CJ26)** Av El Maestro 290, Cajamarca – **CJ27)** Jr Simon Bolivar 280, Santa Cruz – **CJ28)** Jr 2 de Mayo 484, Cajamarca – **CJ29)** Jr Silva 673, Cajabamba **E:** radioparaiso1300@hotmail.com – **CJ30)** Jr Santa Rosa No 674-680 (Ap 14), Chota ☎7635 1240 **W:** www.radiochota.com **W:** Jr Cajamarca s/n, Cajamarca – **CJ32)** Jr 30 de Agosto 641, Chota **W:** www.radiosantamonica.com/chota/ **E:** radiosantamonica@hotmail.com ☎7684 1477 📠7684 1132 - **FM:** 95.7MHz **CJ33)** Juan XXIII s/n (Plaza Bolognesi), Cajamarca **W:** www.yanacocha.com/pe/comunicandonos/myradio01.htm - **FM:** 105.1MHz – **CJ34)** Jr 5 Esquinos 563, Cajamarca ☎7682 3041 – **CJ35)** Jr Arica Cuadra s/n, Celendín ☎7655 5112 **W:** http//frecuenciavh.loquegustes.com **E:** frecuenciavh@hotmail.com - **FM:** 95.7MHz – **CJ36)** Jr. Alfonso Ugarte No 668, Cajamarca ☎7655 7075 **W::** www.radiosanmiguel-cajamarca.com**E:** sanmiguelradioradio@hotmail.com - **FM:** 101.1MHz – **CJ37)** Calle Libertad 430, Jaén – **CJ38)** Jr. Lima 290, Cutervo ☎7673 7010 📠7673 7269 **E:** radioilucan@hotmail.com (Sun: 1030-2300) - **FM:** 96.1MHz – **CJ39)** Av Bolognesi 851, San Pablo – **CJ40)** Jr Pachacutec 433, Los Baños del Inca ☎7680 1408 – **CJ41)** Jr Jorge Chávez 416, Bambamarca ☎4484 3260 📠4484 3078 - **FM:** 100.5MHz "Stereo Líder" – **CJ42)** Jr Micaela Bastidas 352, Chota – **CJ43)** Amazonas 655, Cajamarca - 1000-0300 – **CJ44)** Jr Castro Alfaro s/n, Socota, Cutervo, Prov de Cutervo – **CJ45)** Av Bolognesi 532, San Pablo, Prov de San Pablo - **FM:** 100.7MHz – **CJ47)** Jr Cutervo No 543, Santa Cruz ☎7684 4068– **CJ49)** Jr Bolognesi - 1300, Barrio La Almeda, Sector Los Delfines, Cajabamba, Prov de Cajabamba –**CJ53)** Dis de Faique – **CJ54)** Jr.Jose Galvez No 698, Contumaza – **CJ56)** Av Casanova 630, Cajamarca ☎76 346656 **W:** www.radiocampesinadecajamarca.com – **CJ57)** Juana Atalaya s/n Peru, El Tambo, Bambamarca 833 538 833 550 **W:** radiosanjuan.geoscopio.net **E:** rsjfcia1@hotmail.com – **CJ58)** Calle La Merced y Calle Comercio, Cutervo **E:** leo1551@hotmail.com – **CJ59)** Chirinos - **FM:** 99.3MHz – **CJ61)** 30 de Agost,, Chota – **CJ62)** Los Libertadores 250, Bambamarca, Hualgayoc – **CJ63)** Jr. Miguel Iglesia 483-489, Cajamarca ☎7636 6985 **W:** www.turbomix.com.pe - **FM:** 95.5MHz – **CJ64)** Jr.Leonico Prado 550, Pedrz Galvez, San Marcos

CU00 (CUSCO):
CU02) Ciudad Universitaria de Perayoc, Sotano del Pabellón "C", Cusco – **CU03)** Av.Ejercito 164, Cusco – **CU04)** Av.Charcahuaylla s/n, Distrito Maras, Prov Urubamba **E:** itdsalle@terra.com.pe - **FM:** 91.7MHz – **CU05)** Tres Cruces de Oro 205, Cusco – **CU06)** Jr Hipolito Unanue M4, Urb Industrial, Cusco – **CU07)** Calle Puputi K-3B, Cusco 84 805364 – **CU08)** Calle Inca 650, Santiago, Cusco 8422 8649 **E:** aqripinaf@hotmail.com - **Quechua:** 2 hrs: - 1100, 1500 – **CU10)** Cl Daniel A Carrioón 602, Cusco ☎8422 4371 – **CU12)** Calle "C" 13, Urb Jardin, Cusco - 24h – **CU13)** Av Infancia 527, Wanchaq ☎84 246391 **W:** www.radiow.tk **E:** radiorw@hotmail.com – **CU14)** Calle Nueve de Diciembre s/n, Espinar - **FM:** 103.9MHz – **CU15)** Asoc. Pro-Vivendi el Periodista Lt B-13, Wanchaq – **CU16)** Portal de Carnes 260, Cusco – **CU17)** Calle Sucre 107, Sicuani – **CU18)** Jr Ricardo Palma Mz.L-1, Urb Santa Monica, Cusco ☎8424 6201 - **FM:** 100.7MHz – **CU19)** Calle Anta s/n, Antanampa, Espinar - **FM:** 98.9MHz – **CU20)** Jr.Juan Espinoza Medrano P-13, Urb Rosas Pata, Cusco **E:** sernaquem@hotmail.com – **CU21)** Cl Triunfo 201, Cusco ☎84 23 6020 - **FM:** 92.7MHz – **CU22)** Lote E-11, Urb Bancopata, Cusco - **FM:** 100.1MHz – **CU23)** Jr Santos Chocano G-11, Urb Santa Monica, Cusco ☎8422 6765 📠8423 4494 **W:** www.radiouniversalcusco.com **E:** info@radiouniversalcusco.com - **FM:** 103.3MHz – **CU24)** Calle Meloc 417, Cusco – **CU25)** Av El Sol 830, Cusco ☎8422 8411 - - **FM:** 91.3MHz – **CU26)** Av Martin Pio Concha 339 Quillabamba **Quechua:** 1300-1430, 2100-0100 ☎84 281002 **W: http://**quillabanoticias.org/radioquillabamba **- FM:** 91.1MHz – **CU27)** Av Manuel Callo Cevallos 111, Cusco – **CU28)** Pasaje Constancia 102, Of 410, Wanchaq – **CU29)** Jr 22 de Febrero 104, Espinar ☎8430 1045 – **CU30)** Jr José Olaya M H-9, Urb Bancopata, Cusco ☎8425 2591 – **CU31)** Calle Sacsaywaman K-10, Urb Manuel Prado, Cusco - 1200-2130 - **FM:** 104.1MHz – **CU32)** Av. Huayna Capac 146, Wanchaq 8422 5160 – **CU33)** Calle Bélen 306, Cusco – **CU34)** Jr 2 de Mayo 206, Sicuani **W:** www.radiosicuani.org.pe **E:** radiosicuani@speedy.com.pe ☎ 8435 1136 - **Quechua:** 0930-1100, 2300-0300 0900-0300 - **FM:** 91.1MHz – **CU35)** Urb Marcavalle, P-20,

Wanchaq ☎8422 5357 📠8422 6555 0900-0300 - **FM:** 93.9MHz – **CU36)** Plaza de Armas s/n, El Descanso, Kunturkanki Canas, Prov. de Canas **E:** cpmaldonado@caritas.org.pe – **CU37)** Av Garcilazo 185, Cusco ☎8421 1371 – Rprts to Carlos Gamarra Moscoso, Av Garcilazo 411, Wanchác **E:** adalidcusco @hotmail.com – **CU38)** Av.Tupac Amaru s/n, Yanaoca, Prov de Canas – **CU39)** Jr Matara 526, Cusco - **FM:** 106.5MHz – **CU40)** Pasaje San Pablo 142, Sicuani - **FM:** 97.7MHz – **CU41)** Jr Independencia 143 Piso 2, Quillabamba – **CU42)** Plaza de Armas s/n, Marcapata, Provincia de Quispicanchi – **CU43)** Calle Saphi 601, Cusco ☎8423 5851 - **FM:** 90.1MHz – **CU44)** Av El Sol 230, Yauri, Prov de Espinar - **FM:** 103.1MHz – **CU45)** Alameda Pachacutec B-5, Urb Bancapata (Apt 713), Cusco. ☎8422 5052 – **CU47)** Huayna Capac 154, Wanchaq - **FM:** 100.1MHz – **CU48)** Av Panamericana 105, Yuari, Provincia de Espinar – **CU49)** Av Cusco s/n, Yauri, Provincia de Espinar – **CU50)** Jr Ricardo Palma 516, Quillabamba - **FM:** 96.5MHz – **CU51)** Cl Choquechaca 152, Cusco ☎8480 2444 – **CU52)** Calle Heladeros 220, Wanchaq - **FM:** 102.1MHz – **CU53)** Calle Siete Angelitos 715, San Blas, Cusco ☎8423 3101 8423 1881 - **FM :** 93.3MHz – **CU55)** Comunidad de Usañaje, Calle 28 de Junio No 507, Santo Tomas, Chumbivilcas – **CU56)** Av.San Martin No 305, Yauri, Espinar **E:** radiolaramani@latinmail.com – **CU57)** Urb Tambillo L-5, Urcos, Prov de Quispicanchi – **CU58)** Plaza de Armas s/n, Kunturkanki, Prov de Canas: radiosantacruz@peru.com - **FM:** 97.7MHz – **CU60)** Av.Arequipa s/n, Yanaoca – **CU61)** Calle Arequipa 590, Sicuani, Canchis – **CU62)** Av Arequipa s/n, Sicuani, Canchis

HV00 (HUANCAVELICA):
HV01) Puno 110, Lircay, Prov de Angaraes – **HV02)** Plaza Bolognesi 142, Cercado,, Huancavelica **W:** www.radiovirgendelcarmen.com **E:** jlopez_alvarado@hotmail.com ☎6745 1257 - 24h - SW:irr - **FM:** 99.1MHz – **HV03)** Calle Arequipa S/N, Huayllahuara.

HU00 (HUANUCO):
HU01) Av Raymondi 432, Rupa-Rupa ☎6256 2024 – **HU02)** Malecón Huallaga 1038, Aucayacu, José Crespo y Castillo – **HU03)** Jr Damasao Beraun 832 – Segundo Piso Oficina 4, Huanuco ☎6250 3049 – **HU04)** Jr.Leoncio Prado 163F, Huánuco ☎6251 6360 – **HU05)** Jr Leoncio Prado 723 (Cas 343), Huánuco ☎6251 1525 📠6251 2428 Prgr in **Sp & Quechua** - **FM:** 88.9MHz – **HU06)** Ruben Dario 128, Zona Cerro Paucarbamvilla, Distrito Amarilis, Huánuco ☎6251 2428 – **HU07)** Jr Dos de Mayo 1286, Of 308, Huánuco **E:** luzysonido@hotmail.com ☎6251 8500 📠6251 1985 **Quechua:** 1000-1200, 2300-0200 - **FM:** 105.7MHz – **HU08)** Jr Aguilar 744-746, Huánuco - **FM:** 100.1MHz – **HU09)** Jr Bolognesi 175, Distrito de Panao, Provincia de Pachitea (rept to Pablo Alfredo Albornoz Rojas, Jirón Tacna 385, Panao, Pachita - **FM:** 95.3MHz – **HU11)** Jr.Monzón, Cuadra 1, Tingo Maria – **HU12)** Jr. Abtao1163, Huanuco

IC00 (ICA):
IC01) Conde de Nieva 125, Ica – **IC02)** Av Ayacucho Esq Grau s/n, Ica ☎5623 1956 - **FM:** 105.3MHz – **IC03)** Av Conde de Nieva, Urb Luren, Ica - **FM:** 90.7MHz – **IC04)** Av San Martín 305, 2° piso, Chincha Alta, Prov de Chinca - **FM:** 89.7MHz – **IC05)** Calle Cajamarca 195, Ica - 1100-0500 - **FM:** 103.3MHz – **IC06)** Av Arenales 1370, Ica - **FM:** 98.7MHz – **IC07)** s/n Mz A21 Urb aa.Hh Tupac Amaru, San Juan de Marcona ☎5652 5268 – **IC08)** Jr Mauytua 189 (Ap 54), Chincha Alta, Prov de Chinca – **IC09)** Av.Arenales 579, Ica ☎5680 8460 – **IC10)** Calle L 204 – San Miguel, Ica.56-219-300 – **IC11)** Camino a San Juan, Chincha Alta, Prov Chincha – **IC12)** Los Portales de Escribanos 167, Plaza de Armas, Palpa **E:** radiocruzdelsur@hotmail.com - **FM:** 99.7MHz – **IC13)** Calle San Francisco, Pisco ☎5653 3150 - **FM:** 96.5MHz "Paracas" – **IC14)** Av 8 de Septiembre s/n (Cas 24), Pisco – **IC15)** San Martín 120, Urb Pencal, Nazca – **IC16)** Av los Incas 117-119 (57), Nazca – **IC17)** MZ 1, LT 1, Urb Los Viñedos, San Juan Bautista **E:** epcc@uosjb.edu.pe – **IC19)** Jr.Los Ángeles No 296 3er Piso, Publo Nuvo, Chincha 263-278 – **IC20)** Calle San Francisco 301, San Clemente, Pisco

JU00 (JUNIN):
JU01) Jr Lima 354, Edif Marakamí 7 piso, Huancayo – ☎ 6423 1935 - **FM:** 101.7MHz – **JU02)** Jr.Grau 642, Junin – **JU04)** Av.Tayacaja 324, Of 202, La Oroya – **JU05)** Av Jorge Chavez 851, Anexo Zaños Grande, El Tambo **E:** radiovida@hotmail.com – **JU06)** Calle Ancash 543, Of 208, Huancayo - **FM:** 103.1MHz – **JU07)** Av Las Palmeras 285,La oroya ☎6439 1595 – **JU08)** Jr Acolla 935, Jauja – **JU09)** Av Calmell del Solar 469-481, Sn Carlos Hye, Huancayo ☎6421 1312. **W:** www.somosmanantial.com **E:** manantialradio960@hotmail.com - **FM:** 94.9MHz – **JU10)** Jr.Huancas 251, A San Carlos, Huancayo – **JU11)** Esquina Prolongación Pachitea 136 y Pasaje Andaluz 106 4° piso, Huancayo – **JU12)** Jr Lima 400, 2° piso, Huancayo - **FM:** 92.9 – **JU13)** Calle Mercado 194, San Ramón – **JU14)** Calle Real 270, El Tambo, Huancayo ☎ 6424 5396 📠 6425 3921 - **FM:** 96.7MHz "R Futura" – **JU15)** Jr Tarma 545, La Merced ☎6453 1068.📠 6453 1304 - **FM :** 105.7MHz – **JU16)** Paseo La Breña 174 2 piso Of 202, Huancayo 6421 9990 - **FM:** 89.7MHz – **JU17)** Jr Manuel Prado 459, Satipo – **JU18)** Jr Cerro de Pasco 582, (Ap 2), Junín ☎64 344029 **W:** www.rlibertadjunin.com **E:** radiolibertadjunin@yahoo.es (SW-irr) - **FM:** 98.9MHz

– **JU19)** Av Ayacucho 7300, Huancayo – **FM:** 105.7MHz – **JU20)** Av Manuel Prado 239, Satipo – **JU21)** Jr Moquegua 648, Tarma – **JU22)** Jr Ancash 555, Chilca (Apt 245, Huancayo) **E:** radiocumbre fm@ hotmail.com ☎6421 8080 ▤6423 9189 - 24h – **FM:** 98.5MHz – **JU23)** Jr Junín 163, La Merced – **JU24)** Jr Dario Leon 198, 4° piso, La Oroya – **JU25)** Calle Real 175, Chilca, Huancayo ☎6423 1123 – **JU26)** Calle Real 1453, Huancayo – **JU27)** Jr Junin , 2° piso, Jauja ☎6436 2428 ▤6436 1850 – **JU28)** Jr Jauja 494, Tarma – **JU29)** Pasaje Andalz 175, Huancayo – **JU30)** Jr Moquegua 642, Tarma ☎6432 1864 – **JU31)** Av Jose Carlos Mariátegui 699, Urb Tambo, Huancayo ☎6424 1941 ▤6425 2840 - **FM:** 102.5MHz – **JU32)** Jr Cerro de Pasca 582, Junin – **JU33)** Marcavelle Block "F" 191,Santa Rosa de Sacco, La Oroya **E:** radiolaoroya@speedy.com.pe ☎6439 1401 ▤6439 1748 - **FM:** 100.1MHz – **JU34)** Calle Real 517, Of 403, Huancayo ☎6423 1831 **W:** www.ldwebstudios.com/radiohuancayo - **FM:** 104.3MHz – **JU35)** Av Arevaldo 484, Anexo Chuccus, Santa Rosa de Sacco, Prov de Yauli – **JU36)** Túpac Amaru sn, Alturaa Coliseo, Juli 51 554173 **E:** radio-campesinajuli@hotmail.com - **FM:** 107.1MHz – **JU37)** Jr Molino del Amo 167 (Cas.167), Tarma ☎6432 1510 ▤6432 1167 **E:** radiotarma@ terra.com.pe - **FM:** 99.3 & 101.7MHz "R Tropicana" in La Merced – **JU38)** Av Huancavelica 430, 2° piso (Ap 230), Huancayo ☎6423 3640 **W:** www.radio1550.com - **FM:** 88.9MHz – **JU39)** Jr Arequipa 572, Yauyos - **FM:** 98.3MHz – **JU40)** Jr Huancavelica 498, 2° piso, Tarma – **JU41)** Brasilia 200, Huancayo – **JU42)** Jr Tarma 551, La Merced ☎6453 1068 ▤6453 1304 – **FM:** 105.7MHz – **JU43)** H.Zevallos Gomez 231, La Oroya – **JU44)** Huancayo – **JU44)** Calle Principal de Lama, Pariahuanca, Huancayo – **JU45)** Av.Huancavlica N 439, Huancyo

LA00 (LAMBAYEQUE):
LA01) Km 4 de la Carretera Pimentel, Chiclayo – **FM:** 105.1MHz – **LA02)** Calle Juan XXXIII 391 (Ciudad Universitaria), Lambayeque **E:** universitariaradio@hotmail.com – **LA03)** Psje.Woyke Nr 179 3 piso Edifico Angelica, Chiclayo 74 237850 – **FM:** 88.5MHz – **LA04)** Juan Guglievan 984, Chiclayo – Quechua: **Sun**0600-1100 1000-0500 - **FM:** 107.7MHz – **LA06)** Alfonso Ugarte 505, Distrito San José, Prov de Lambayeque (Ap 67, Correo Central, Chiclayo) - **FM:** 103.3MHz – **LA07)** Jr Incanato 387, Altos, Distrito José Leonardo Ortiz, Chiclayo – **LA08)** Calle San José 462, Of 207, Chiclayo ☎7420 4786 - **FM:** 102.9MHz – **LA09)** Jr Nicanor Carmona 177, Ferreñafe – **LA10)** Calle Santa Cecilia s/n, Olmos, Prov Lambayeque – **LA11)** Av Pedro Ruiz 1123, 3° piso, Chiclayo - **FM:** 100.5MHz – **LA12)** Calle 28 de Julio 440, San José, Provincia de Lambayeque – **LA13)** Ca Antonio Monsalve Baca 204 P.J.Santa, Lambayeque ☎7428 3562 radiosanjose@hotmail.com **E:** radiosanjose@hotmail.com - **FM:** 97.7MHz – **LA14)** Calle Arica 1247, Chiclayo ☎ 7420 8523 – **LA15)** Av Saenz Peña 1046, Chiclayo - **FM:** 98.3MHz – **LA16)** Av Tupac Amaru 532, Ferreñafe – **LA17)** Calle San José 1084, Chiclayo – **LA18)** Calle F Villareal y A Arguedas s/n, Chiclayo – **FM:** 94.1MHz – **LA19)** Av Miguel Grau 350, Oficina ▤, Chiclayo ☎7423 6363 – **LA20)** Calle Las Violetas s/n, Chiclayo - **FM:** 94.9MHz – **LA21)** Calle Francisco Gonzales Burgán 717, Ferreñafe ☎7428 6351 – **LA22)** Calle 1 de Mayo 278, Urrunaga, Provincia de Chiclayo – **LA23)** Caserio Tranca Falupe, Morrope, Prov Lambayeque – **LA24)** Jr San Francisco 239, Olmos, Prov de Lambayeque – **LA25)** Nicolás de Pierola 335, Distrito de José Leonardo Ortiz, Pro de Chiclayo - **FM:** 107.7MHz – **LA26)** Empresa Capimag S.R.L., Calle Justicia 102, Urb. Túpac Amaru, Chiclayo – **LA27)** Av Balta 1144, Chiclayo **W:** http// es.geocities.com/radiochiclayofm/ – **LA28)** Av.Los Incas 946, Morrope, Prov de Lambayeque – **LA29)** Calle San José 148, Olmos, Prov de Lambayequea **E:** clori1009@yahoo.es – **LA30)** Av el Tren s/n, Chiclayo – **LA31)** Calle Juan Fanning N° 457, Urb San Juan, Chiclayo ☎7423 9889 **W:** www.radioperu.com **E:** informes@visionradioperu.com – 0845-0500 – **LA32)** Calle Tarata s/n, Olmos, Prov de Chiclayo – **LA34)** Av Mariscal Castilla 859, Monsefú, Prov de Chiclayo – **LA36)** Av Andrés Avelino Cáceres 800, Lambayeque ☎7428 3353 **E:** naylamp@llampal-lec.rep.net.pe ☎7428 3353 - **FM:** 96.1MHz – **LA38)** Juan Guelman de Cameros, Chiclayo **E:** lorryand@hotmail.com – **LA39)** Tucumé.

LI00 (LIMA):
LI01) Juan Vargas 147, Chorrillos, Lima ☎1438 8850 – **LI02)** Jr.Camana Pl6 615 of 605, Urb Cercado, Lima ☎347 0120 – **LI03)** Calle Mama Ocllo 2058, Lince, Lima **W:** www.radiomariaperu.org **E:** info.per@ radiomaria.org ☎1445 8217 ▤1446 7161 – 15 FM repeater stations – **LI04)** Av 28 de Julio 1004, Piso 4, Cercado, Lima ☎1424 7547 – **LI05)** Calle Miguel Dasso 144, Of 2A, San Isidro, Lima ☎ 1221 4107 **W:** www.ovacion.com **E:** ovacion@ovacion.com.pe – **LI06)** Av Guzman Blanco 465, 7° piso, (Ap 4236) Lima ☎1433 7879 ▤1433 3276 **W:** www.grupopacifico.org/radio.html **E:** administracion@gru-popapacifico.com – **LI07)** Justo Pastor Davila 197, Chorrillos ☎ 617 6600 251 3324 **W:** www.crpradio.com.pe **E:** crpradio@crpradio.com.pe – **LI08)** Av Manco Cápac 333, La Victoria, Lima 13 – **LI09)** Av La Marina 3099, San Miguel or Algarrobos 430. Urb. California, Trujillo. La Libertad ☎1578 2970 **W:** http://radiointegridad.blogspot.com **E:** R_Integridad@yahoo.com – **LI10)** Av Paseo de la República 3866,

2° piso, San Isidro ☎1215 0200 **W:** www.rpp.com.pe - 24h – **LI11)** Justo Pastor Dávila 197, Chorillos ☎1617 6600 ▤1251 3324 **W:** www. crpradio.com.pe **E:** crpradio@crpradio.com.pe – Signal downlinked to 39 repeaters - 24h – **LI12)** Avenida Arica 248, Breña, Lima ☎1424 5187 ▤1332 3094 **W:** www.ipda.com.pe **E:** sede_nacional@ipda.com. pe – **LI13)** Av Andrés Mármol Castellanos 230, Huaral ☎1335 5765 – **LI14)** Av Salaverry 1082, Jesús Maria, Lima **W:** radiolibertad.com. pe **E:** info@radiolibertad.com.pe ☎1266 0777 ▤1471 5319 – **LI15)** Av Petit Thouars 447, Santa Beatriz, Lima 1. Lima **W:** www.radionacional. com.pe ☎1433 8956 – **LI16)** Av José Pardo 138, Edifico Neptuno, Piso 16, Miraflores ☎1445 8549 ▤1445 8901 **W:** www.unionlaradio.com – **LI17)** Av Paseo de la Republica 3866, San Isidro **W:** www.felicidad. com.pe - 1000-0300 – **LI18)** Av Republica de Chile 295, Of. 1104, Urb. Santa Beatriz, Lima ☎1332 7779 **W:** www.modernaradiopapa.com **E:** informes@radiomoderna.com – **LI19)** Paseo Parodi 340, San Isidro **W:** www.radiopanamericana.com **E:** radio@panamericana.com.pe ☎1422 6787 ▤1422 1223 – Satellite signal downlinked by 60 FM repeaters - **FM:** 101.1MHz – **LI20)** Ignacio Merino 230, Santa Cruz, Miraflores – **LI31)** Av Andrés Mármol Castellanos 230, Huaral ☎1335 5765 – **LI14)** Av Salaverry 1082 – **LI20)** Ignacio Merino 230, Santa Cruz, Miraflores ☎1442 8810 **W:** www.radiolatina.com.pe **E:** contacto@radiolatina. com.pe – **LI21)** Calle Porta 130, Miraflores **W:** www.metropolitana-radioperuana.com **E:** Metropolitan@GrupoOnda.com – **LI22)** Julio C.Tello 152, Lince, Lima ☎1471 4291 **W:** www.radioexitoperu.com **E:** radioexicto@GrupoOnda.com – **LI23)** Jr Huancayo 288, Of 701, Lima. **W:** www.radioaluz.com ☎1433 4599 – **LI24)** Gerardo Unger N° 6347, San Martin de Porres ☎1537 3204 **W:** www.radioantarkiperu.com **E:** radioantarki@terra.com.pe – **LI25)** 2 de Mayo 573, Imperial ☎1284 8052 - 900-0300 – **LI27)** Jr.Bernardo Alcedo 375, Lince, Lima 14 **W:** www.radiobacan.com **E:** bacan@radiobacan.com ☎511 266 1856 ▤511 1471 3908 - 24h – **LI30)** Av Manco Cápac 495, Of 401, Miraflores ☎1444 1773 ▤1445 0126 – **LI31)** Jr Echenique 140, Huacho – **LI32)** Mz L, Lt 7, Urb La Esperanza, Hualmay – **LI33)** Av Estados Unidos 327, Urb Huaquillay, Comas **W:** www.radiocomas.com **E:** comasam@ radiocomas.com ☎1525 0094 ▤1525 0859 – **LI34)** Jr Atahualpa 148, 5° piso, Huacho ☎276 0557 – **LI37)** Av Grau 538, Huacho – **LI38)** Av Comandante Espinar N° 680, Miraflores, Lima ☎1610 7760 ▤1610 7761 **W:** www.nuevotiempo.org.pe **E:** radio@nuevotiempo. org.pe – **LI39)** Calle Juan de Carpio 140-144 2° piso, San Isidro, Lima 27 ☎1442 0482 ▤1442 1693 **W:** www.radiocallao.com **E:** radiocal-lao@hotmail.com – **LI40)** Ausejo Salas 153, Huacho ☎1323 1976 - 1000-0600 – **LI41)** Av Petit Thouars 1806, Lince, Lima – **LI42)** Mz C Lt 5, Asoc. Virgen de la Familia, Chilca, Cañete – **LI43)** Av Separadora Industrial s/n, Villa El Salvador ☎1291 3146 – **LI44)** Alfonso Ugarte 149, Barranca ☎1235 4238 – **LI45)** Cadena Peruana de Noticias, Gral Salaverry 152–156, Miraflores **E:** publicidad@gestion.com.pe **W:** www. cpnradio.com.pe ☎1447 1789 **NB:** CPN operates network of FM and AM stns all over the country – **LI46)** Av Recreo 317, Altos, Paramonga – **LI47)** Jr Carmaná 170, (Apt 4451 San Miguel), El Cercado, Lima **W:** www.radiosantarosa.com **E:** radiosantarosa@terra.com.pe ☎1427 7488 ▤1426 6587 **Quechua:** 1300 **English:** 0130.– **LI49)** Av.Arnaldo Márquez 1944 Jesús Maria, Lima 11 ☎1461 2222 ▤1461 7757 **W:** www.radiomilenia.com.pe **E:** milenia@radiomilenia.com.pe - 1000-0300 – **LI50)** Plaza de Armas 132, Barranca ☎1235 2301 – **LI51)** Jr. Yahuar Huaca Piso 3 106, Urb. Tahuantinsuyo, Independencia ☎1526 0469 **W:** www.radioindependenciadelperu.com **E:** wbaldeon@radio-independenciadelperu.com – **LI52)** Av Alfonso Ugarte 1428, Of 202, Breña (Cas.11-0625), Lima ☎1424 6677 **W:** www.laperuanisima.com **E:** laperunisima@yahoo.com - 1100-0400 – **LI54)** Jr. Almirante Guisse 1885, Lince (altura Cdra. 4 de Av. Canevaro / 18 de Av.Arenales ☎717 8485 **W:** www.radiocielo.pe – **LI55)**Av Petit Thouars 447, Santa Beatriz, Lima 1 **W:** http://www.radionacional.org/?option=ralacr ☎1433 1404 – **LI56)** Av. Paseo Parodi 340, San Isidro, Lima ☎1441 3050 **W:** www.ondacero.com.pe - **FM:** 98.1MHz – **LI57)** Huacho– **LI58)** CRP – Justo Pastor Davila 197, Chorrillos, Lima **W:** http//radionuavaqfm. com – **LI59)** Av. 28 de Julio 1781, La Victoria. Lima ☎1613 1701 **W:** www.bethelradio.fm **E:** bethaltadio@bethelradio.fm – **LI60)** Avda. Colonial 1467, Oficina 203 El Cercado, Lima ☎1425 4996 **W:** www. cadenaradio1200.com **E:** prensa@cadenaradio1200.com

LL00 (LA LIBERTAD):
LL01) Jr Trujillo 597, Otuzco ☎4443 6007 – **LL02)** Jr Benito Juarez 1753, La Esperanza ☎1438 – **LL03)** Calle Opalo 298 2° piso-A, Urb Sta.Ines, Trujillo ☎4420 2156 - **FM:** 98.3MHz – **LL04)** Calle Lima 599, Chepén – **LL05)** San Martín 472, Trujillo ☎4425 1792 – **LL06)** Av Los Incas s/n, Anexo Facala, Chócope, Prov de Ascope - **FM:** 102.5MHz – **LL07)** Calle Junín 23, Sausal, Prov de Ascope – **LL08)** Francisco Pizarro 532, Of 205, Trujillo - **FM:** 89.7MHz – **LL09)** Jr Ayacucho 65, Pacasmayo, Prov Pacasmayo – **LL10)** Jr Trujillo 597, Otuzco – **LL11)** Jr Marcelo Corne 224, Urb San Andrés, Trujillo ☎4429 4050 – **W:** www.radiolibertadmundo.com **FM:** 90.9MHz – **LL12)** Av V Belaunde MZ.L lote 15, Urb Santo Dominguito, Trujillo – **LL13)** Jr Francisco Pizarro 208, Of 302,.Trujillo ☎4424 2666 - **FM:** 107.5MHz –**LL15)** Av del Ejercito 717, Trujillo - 0900-0300 – **LL16)**

Alfonso Ugarte 222, Virú - **FM:** 102.MHz – **LL17)** Calle Trujillo 699-A, Chepén – **LL19)** Jr.Baltan 252, Huamachuco – **LL20)** Jr Atahualpa 795, Chepen ☎4456 2038 **W:** www.xaski.com/radiosansebastian/principal. htm - **FM:** 92.7MHz – **LL21)** Mz.B-1 2-A, Urb La Libertad, Trujillo ☎4421 7885 – **LL22)** Zepita 452, Trujillo ☎4424 9326 📠4425 2970 **W:** www. radiolibertadmundo.com **E:** contactanos@radiolibertadmundo.com — **LL23)** Jr Progreso 121, Quiruvilca – **LL24)** Bolívar 780 (Cas 1029), Trujillo ☎4423 3981 – **LL25)** Victoria 229, Guadalupe, Prov de Pacasmayo – **LL26)** Jr Caceres s/n, Bolívar, Prov de Bolívar ☎4423 0277 – **LL28)** Jr Progreso 759, Chepén - **FM:** 98.7MHz – **LL31)** Ca Camal 13, Cartavio, Prov.de Ascope ☎4443 2448 – **LL32)** Mercado La Hermelinda Puesto C 288 – seccion abarrotes, Trujillo ☎4480 3207 – **LL33)** Jr Gamarra 713, Of 405, Trujillo ☎4424 6211 – **LL34)** Ascope – **LL35)** Pasaje San Martin 300, Urb Alto Mochica, (Ap 352) Trujillo ☎4426 3592 – **LL37)** Jr Trujillo 281, Quiruvilca – **LL38)** Jr Grau s/n, Virú – **LL39)** Sebastian Barranca 469, Urb Chimú, Trujillo 0845-0300 – **LL40)** Pasaje Monseñor Damián Nicolau 108, Huamachuco **E:** radiolosandes@starmedia.com - **FM:** 103.1MHz – **LL41)** Calle Bolívar 130, Otuzco – **LL42)** Av Cacuro 459, 5° piso Of 901, Trujillo ☎4425 7431 - **FM:** 96.9MHz – **LL43)** Av Perú 608, Trujillo **E:** elparaisodedios@speedy.com.pe – **LL44)** Jr Zepita 450, Trujillo – **LL45)** Carabamba, Prov.Julcán – **LL46)** Calle Victor Julio Rossel 324, Julcan – **LL47)** Calle Progreso 218, Barrio Alto Coscomba, Julcan – **LL48)** Calle La Libertad 120, Otuzco 4443 6565 **W:** www.chamiradio. org.pe **E:** escribenos@chamiradio.org.pe – **LL49)** Calle San Antonio 880, 2-piso, Otuzco – **LL51)** Jr Balta 470, Huamachuco – **LL52)** Trujillo.

LO00 (LORETO):
LO01) Calle Abtao 255 (Ap 207), Iquitos ☎6526 5244 📠6526 5244 **E:** lavozdelaselva@xploratemex.com.pe - **FM:** 93.9MHz – **LO02)** Jr Elias Agurrie 857, Iquitos - **FM:** 104.5MHz – **LO03)** Prolong Perú s/n, Cuadra 19, Iquitos - **FM:** 98.9MHz – **LO04)** Jr Lima 821, Iquitos - **FM:** 90.1MHz – **LO05)** Av.Antonio Raymondi 331, Iquitos - **FM:** 101.3MHz – **LO06)** Arica 228, Iquitos ☎ 6523 3302 **E:** radioloreto@yahoo.es - **FM:** 103.5MHz – **LO07)** Plaza de Armas s/n, Iquitos – **LO08)** Jr Próspero 645 Iquitos (Cas 174) - **FM:** 105.9MHz – **LO09)** Av Abelardo Quiñones km 4.5 Carretera a Nauta, Iquitos – **LO10)** Calle Progreso 112-114, Yurimaguas, Alto Amazonas **W:** www.roriente.org **E:** oriente995@ yahoo.com ☎65 351611 - **FM:** 99.5MHz – **LO11)** Av Mariscal Cáceres 1037 (Ap 786), Iquitos (Reports to: Pablo Rojas Bordales, Jr Bermúdez 445), Iquitos ☎6523 4886 – **LO12)** Av Alfredo Vargas Guerra 440, Contama, Prov de Ucayali **E:** rmdpem@terra.com.pe – **LO13)** Av Miguel Grau 1029, Iquitos ☎22 4205

MD00 (MADRE DE DIOS):
MD01) Jr Guillermo Billingurst 406 PTO, Puerto Maldonado - **FM:** 101.3MHz – **MD02)** Nueva Plaza de Armas 200, Puerto Maldonado - **FM:** 105.500 – **MD03)** Jr Daniel A Carrion 387, Puerto Maldonado ☎8257 105000 - **FM:** 92.5MHz

MO00 (MOQUEGUA):
MO01) Edifico S-27 Barrio Azul Asiento Minero, Ilabaya, Prov de Jorge Basadre - **FM:** 101.3MHz – **MO02)** Jr Callao s/n, Ilo – **MO03** Alto Ilo, Sector Arenal G-6, Ilo - **FM:** 96.7MHz – **MO04)** Mz E lote 48 P, Joven J.F Kennedy, Ilo – **MO05)** Jr Tarapaca 260, Moquegua – **MO06)** Jr Moquegua 123, 5° piso, Ilo ☎5378 1313 - **FM:** 105.5MHz – **MO07)** Av Ayacucho 639 (Ap 22), Moquegua ☎53 761 542 - **FM:** 105.3MHz

PA00 (PASCO):
PA01) Plazuela Gamaniel Blanco, 127- 2°Nivel - Ninguno, Cerro de Pasco ☎ 63 422398 **W:** www.radiotvaltura.com - **FM:** 97.7MHz – **PA02)** Jr Puno s/n, Chaupimarca - **FM:** 102.5MHz – **PA03)** Jr Mullenbruck 468, Urb Cercado, Oxapampa ☎ 6376 2689 - **German:** - **FM:** 101.5MHz – **PA04)** Zona Denominada Principal, Chaupimarca – **PA05)** Jiron Huamachuco No 214, Cerro de Pasco ☎63 422 124 **W:** www.tvrcomunicaciones.com **PA06)** Jr.Huamachuco 221, Caupimarca – **PA07)** Jr. Daniel Alcids Carrion 250, Huayllay

PI00 (PIURA):
PI02) Calle Santa Ana 471, Urb Santa Rosa, Sullana –**PI04)** Jr San Ignacio de Loyola 300, Urb Miraflores, Piura **W:** www.radiocutivalu.org/ **E:** cutivalu@cipca.org.pe ☎7334 2802 ☎7334 2965 - **FM:** 100.5MHz – **PI05)** Ica 419, Of 206, Piura - 1100-0500 - **FM:** 92.1MHz – **PI06)** Santa Maria C., Piura - **FM:** 94.5MHz.– **PI08)** Ugarteche 490, Sullana - **FM:** 99.3MHz – **PI09)** Calle Santa Teresa Cdra 8 s/n, Urb Santa Rosa, Sullana **W:** www.chanel.edu.pe/serviciosl.php ☎7350 4760 📠7350 1382 – **PI10)** Calle Tacna 260 4 piso, (frente al banco de la nación), Piura ☎7330 3369 - **FM:** 103.3MHz – **PI11)** Calle San Martin N° 1041, Sullana ☎7350 3071 – **W:** //lacapullana.tripod.com.pe **E:** capullanaradio@latinmail.com.pe - **FM:** 95.7MHz – **PI12)** Jr Cesar Pinglo 345, Sechura – **PI13)** Parroquia San Miguel, Barrio El Altillo s/n, Huancabamba – Re to Correo Central, Huancabamba **E:** coylemareen@ speedy.com.pe – **PI14)** Av.Sánches Cerro 582 2do Piso, Piura ☎7330 4221 - **FM:** 107.9MHz – **PI15)** Calle Cusco 670, Piura ☎7332 3100 – **PI16)** Zona Industrial III Mz O, lote 10, Piura – **PI17)** Calle Unión 515, La Unión ☎7337 4106 – **PI18)** Zona Industrial Mz D Km 5, Carretera Piura, Sullana **E:** piuraradio@terra.com.pe - **FM:** 101.9MHz – **PI19)**

Urb Aproviser A2-1, Pariñas, Prov de Talara - **FM:** 98.1MHz – **PI20)** CI Lambayeque 1005, Chulucanas ☎7337 8627 – **PI21)** Jr Bolognesi Cuadra 1, Ayabaca **W:** www.muniayabaca.gob.pe ☎7347 1103 – **PI22)** Calle San Martín 354, Sechura – **PI23)** Calle 8 s/n, Talara Alta, Talara – **PI24)** Mz E 8, Fonavi I Etapa, Paita ☎ 7361 1885 - **FM:** 102.9MHz – **PI25)** Jr Leoncio Prado 425, Sullana – **PI26)** Madre de Dios 258, Sullana ☎7350 5026 – **PI27)** Jr Las Azucenas Urb El Milagro, Talara Alta, Talara. ☎7449 5431 – **PI28)** Calle Tumbes 641, La Unión – **PI29)** Jr Sucre 556, Sullana – **PI30)** Maria Prado de Bellido 500, Chulucanas, Prov Morropón ☎7337 8740 – **PI31)** Calle Cajamarca 485, Bellavista, Prov de Sullana - **FM:** 107.5MHz – **PI32)** Av Grau Cuadra 5, Cruceta San Lorenzo, Tambo Grande, Prov de Piura – – **PI34)** Calle San Miguel 207, Distrito de Sondor, Región Grau, Prov de Huancabamba - **FM:** 89.1MHz – **PI35)** Av Ramon Castilla 254, Huancabamba ☎7347 3369 – **PI37)** Av Quiles Escala s/n, Barrio San Francisco, Huancabamba – **PI38)** Av San Francisco Assisi s/n, Huarmaca, Prov de Huancabamba – **PI41)** AAHH.Alm Grau MZ C, Lt 21, Sector Coscomba, Piura – **PI42)** Calle Tumbes 729, La Union – **PI43)** Calle Grau 505, Centro, Huancabamba 56-219-300 – **PI44)** Calle Tomas E Velasquez s/n, Ayabaca.– **PI45)** Jr Lima 4ta cuadra, Tambo Grande **E:** jpaivalombardi@yahoo.com – **PI46)** Catacaos – **PI47)** Av Principal, Caserio La Peñita, Tambo Grande – **PI48)** Calle Alfonso Ugarte 118, Vice, Sechura – **PI49)** A 500m del Puente Virrila, Sechura

PU00 (PUNO):
PU01) Jr Conde Lemos 212, Puno ☎5135 1562.– **W:** www.radiondazul. com **E:** ondazul@radiondazul.com - 0900-0300 - Quechua & Aymara: 6h daily - **FM:** 7.35MHz "Stereo Azul" – **PU02)** Jr Arequipa 385, Puno – **PU03)** Plaza 28 de julio s/n, Macusani, Prov de Carabaya **E:** rvamacu@terra.com.pe - **FM:** 90.5MHz – **PU04)** Carretera a Arequipa, Km 6, Chulluaquiani (Casilla 344), Juliaca ☎5180 0601 **W:** www. nuevotiempo.pe **E:** nuevotiempojuliaca@hotmail.com – **PU05)** Calle San Román 116, Juliaca ☎5132 5357 - **FM:** 89.5MHz – **PU06)** Simon Bolivar 442, Puno – **PU07)** Av Panamericana 940, Chucuito (299, Puno), Prov de Puno **Aymara & Quechua** 0900-1400, 1900-0100 – **PU08)** Jr.Altiplano 206,Urb La Pampilla, Juliaca ☎5132 2313 - **FM:** 107.9MHz – **PU09)I)** Jr.Ramón Castilla 861 Cercado, Juliaca ☎502-319 – **PU11)** Hipolito Unanue 240, Juliaca - 107.9MHz – **PU12)** Jr Arequipa 845, 2do nivel, (Ap 130), Puno - 1000-0400 – **PU13)** Calle Mariano Pandía 166, 2° piso, Juliaca – **PU14)** Jr Lambayeque 520, Puno – **PU15)** Jr Unión 242, Juliaca **E:** radioltcj@latinmail.com ☎5132 2452 📠 5136 9450 - **FM:** 102.7MHz - **Quechua & Aymara:** 1000-1100, 1800-1900 – **PU16)** Av El Maestro 1140, Juliaca - **FM:** 88.9MHz – **PU17)** Jr Antonio (Sierra) No 178, prolongación Ramón Castilla, a dos cuadras del Cuartel Bolognesi, Juliaca 323-499.**W:** www.spinoza.bolhost.com **E:** radioliderjuliaca1150@yahoo.es – **PU18)** Jr Piura 167, Puno ☎5135 3680 - **Aymara & Quechua:** 0900-1100 – **PU19)** Jr Union 229, 3° piso, Juliaca – **PU20)** Jr Apurimac 1316, Barrio Manco Capac, Juliaca **E:** trebol34@mixmail.com – **PU21)** Ramon Castilla 949, Juliaca ☎5132 1372 **W:** www.radiojuliaca.com **E:** webmaster@radiojuliaca.com - **FM:** 90.9MHz – **PU22)** Jr Apurimac 644, Juliaca ☎5132 5733 – **PU23)** Jr 2 de Mayo 790, Juliaca – **PU24)** Jr Ricardo Palma 111, Juliaca ☎5132 7208 – **PU25)** Jr San Martin 134, Yunguyo – **PU26)** Jr.Moquegua 600, Juliaca **E:** asidersicuani@yahoo.es – **PU27)** Jr Chachani 220, Juliaca – **PU28)** Jr 2 de Mayo 258, Juliaca ☎5133 6103 - 0900-0200 Aymara & Quechua: 0900-1100, 1900-2100 - **FM:** 104.5MHz "El Sol de los Andes FM" – **PU29)** Av Titicaca 160 Int C, Puno – **PU30)** Av El Maestro 1140, Barrio Tupac Amaru, Juliaca – **PU32)** Jiron 2 de Mayo 149, Juliaca **E:** carraviz@gmail.com or carraviz@hotmail.com - Listeners correspondence to: Iván Tito Vizcarra Jiron José Galvez No 542, Barria Bellavista, Julica – **PU33)** Urbanización Rinconada Frente al Colegio, 3ra Etapa, Juliaca - **FM:** 107.3MHz – **PU34)** Jr 7 de Junio 580, Juliaca – **PU35)** Jr Acora N°222, Puno 366 222 **W:** www.pachamamaradio.org.pe **E:** relacionespublicas@pachamamaradio.org.pe – **PU36)** Túpac Amaru Altura Coliseo Juli s/n, Juli, Prov Chucuito ☎51 85 4173 **W:** webs.demasiado. com/radiocampesina.htm – **PU37)** Av El Sol N° 1384, Puno ☎365 888 – **PU38)** Distrito de Atuncolla – **PU39)** Distrito de Capachica – **PU40)** Region de Puno – **PU41)** Provincia de Azángaro.

SM00 (SAN MARTIN):
SM01) Jr San Martín 257, Tocache ☎4255 1031 – **SM02)** Jr Santo Toribio 1252, Rioja – **SM03)** Av Compagñón 410, Tarapoto – **SM07)** Jr.Martinez de Compagnion 442, piso 2, Tarapoto – **SM08)** Jr Bolognesi 180 Altos, Tarapoto – **SM09)** CI Bolognesi s/n, Bellavista ☎4254 4170 📠4254 4135 - **FM:** 95.1MHz – **SM11)** Peña Meza 467, Juanjuí – **SM12)** Jr Progreso 389, Bellavista ☎4254 4179 – **SM13)** Tocacha, Prov de Tocache - **FM:** 96.1MHz

TA00 (TACNA):
TA01) Edifico S-27 Barrio Azul Asiento Minero, Ilabaya - **FM:** 89.9MHz – **TA02)** Prolong Unanue 1041 (Cas 113), Tacna - **FM:** 99.9MHz – **TA03)** Av Varela 705, Tacna – **TA04)** Jr Sir Jones s/n (Cas 281), Tacna – **TA05)** Arias y Araguez 584, Tacna – **TA06)** Calle Candarave 645, Urb Bacigalupo, Tacna – **TA07)** Urb Jorge Chavez 9 (Ap 115), Tacna

- **FM:** 89.5MHz – **TA08)** Av Internacional 484, Tacna 5280 4100 - **FM:** 106.7MHz – **TA09)** Calle Aniceto Ibarra 436 (Cas 370), Tacna ☎5241 4871 📠5272 3745 **W:** www.radiotacnaladecana.com **E:** radio_tacna@hotmail.com - **FM:** 104.3MHz – **TA10)** Av San Martín de Porras 209, Natividad, Tacna **E:** radiobulevar@hotmail.com ☎5284 8537 – **TA11)** Jose Olaya s/n 2° piso, Tarata – **TA12)** Calle 2 Mayo 263, Tacna

TU00 (TUMBES):
TU01) Pza Alipio Rosales s/n, Tumbes – **FM:** 102.1MHz – **TU02)** Calle Tarapaca 163, Tumbes **TU03)** Av del Ejercito, cuadra 5, Tumbes - **FM:** 91.1MHz – **TU04)** Panamericana Norte Km 1321, Tumbes - **FM:** 100.5MHz – **TU05)** Calle Huáscar 502 3er Piso, Tumbes ☎7252 7002 - **FM:** 107.9MHz – **TU06)** Jr Bolívar 117, Tumbes ☎ 7252 3003 – **TU07)** Jr Piura 1010, Tumbes

UC00 (UCAYALI):
UC01) Calle Diego de Almagro s/n, Pucallpa – **UC02)** Jr Inmaculada 667, (Cas 263), Pucallpa ☎6157 8615 - **FM:** 89.1 – **UC03)** Av 9 de Diciembre 646, Pucallpa - **FM:** 102.5MHz – **UC04)** Jr Coronel Portillo 448-A, Pucallpa ☎6157 3876 📠6157 1540 - **FM :** 103.3MHz– **UC05)** Zona San Fernando, Callería, Pucallpa – **UC06)** Calle Padre Francisco Alvares s/n, Sephua, Prov de Atalaya – **FM:** 95.5MHz – **UC07)** Jr Iquitos 499, Villa Atalaya, Distrito de Raymondi, Prov de Atalaya ☎6146 1240 **E:** j541oen@hotmail.com or j541oen@yahoo.es - **FM:** 93.5MHz– **UC08)** Jr.Coronel Portillon No 460, Pucallapa. – **UC09)** Carretera Federico Basadre Km 37, Los Pinos, Distrito de Campoverde, Coronel Portillo

FM in Lima (MHz): 88.3 R.Magica – 88.9 R.Felicidad – 89.7 Emisoras Perúanos (RPP) – 90.5 CPN Radio –91.1 R.San Borja –91.9 Okey Radio – 92.5 R.Studio 92 – 93.1 R.Ritmo Romantica – 93.7 La Inplvidable – 94.3 R.Moda – 94.9 La Karibeña – 95.5 R.La Calle– 96.1 R.Pueblo Libre – 96.7 R.Capital – 97.3 R.Moda– 98.1 R.Onda Zero – 99.1 Doble Nueve – 100.1 R.Oasis – 101.1 R.Panamericana – **LI40)** 101.7 – **LI33)** 102.1 R.Oxigeno – 103.3 R.Unión **LI13)**104.1 –103.9 R.Nacional –104.7 Viva FM – 105.1 R.Fiesta – 106.3 Radio Mar – **LI58)** 107.1 R.Nueva Q – 107.7 R. Planeta

PHILIPPINES

L.T: UTC +8h — **Pop:** 100 million — **Pr.L:** Pilipino (Tagalog), English, Cebuano, Ilocano, Hiligaynon, Bicol — **E.C:** 60Hz, 220V — **ITU:** PHL

NATIONAL TELECOMMUNICATIONS COMMISSION (NTC) (Dept. of Transportation and Communications)
📮 NTC Bldg., BIR Road, East Triangle, Diliman, Quezon City 1104 ☎ +63 9254651 or 9267722 **W:** www.ntc.gov.ph
L.P: Commissioner: Gamaliel A. Cordoba. Deputy Commissioners: Douglas Michae; N. Mallillin, Jaime M. Fortes Jr. Chief Broadcast Sces Div: Alvin Bernard N. Blanco.

KAPISANAN NG MGA BRODKASTER NG PILIPINAS (KBP) (Assoc. of Broadcasters of the Philippines)
📮 6th Flr, LTA Bldg, 118 Perea Str, Legaspi Village, Makati C, 1226 NCR ☎ +63 (2) 815 1990/1/2 📠 +63 (2) 815 1993 **W:** www.kbp.org.ph
L.P: Chmn: Mr Ruperto S. Nicdao, Jr. Pres: Herman Z. Basbaño. Most stns are KBP members

PHILIPPINE FEDERATION OF CATHOLIC BROADCASTERS (Catholic Media Network, CMN)
📮 Unit 201 Sunrise Condominium, 226 Ortigas Ave, North Greenhills, San Juan, Manila 1503 NCR ☎ +63 (2) 7249850 **W:** www.catholicmedianetwork.org
L.P: Pres: Fr. Francis Lucas. Chmn: Bishop Bernardino Cortez. Exec. Dir.: Fr. James B. Reuter, S.J. (28 owned and affiliated stns on MW, 20 on FM)

PHILIPPINE BROADCASTING SERVICE (PBS, "Radyo ng Bayan") (Gov.)
📮 4/F Media Center Bldg., Visayas Ave., Del Monte, Quezon C, 1105 NCR ☎ +63 (2) 9203968 **W:** www.pbs.gov.ph
L.P: Dir. Gen: Mr John S. Manalili. Dep. Dir. Gen: Ms Monina S. Cespedes.
Manila stns: DZRB Radyo Balita (news sce) 738 kHz, DZSR Sports Radio 918 kHz, DZRM Radyo Magasin 1278 kHz, DWBR-FM Business Radio 104.3MHz
Regional MW stns: DWBT, San Antonio, Basco, 3900 Batanes. DWFB, Mariano Marcos State University Campus, Laoag C, 2900 Ilocos Norte. DWFR, Multipurpose Bldg, Provincial Capitol Compound, Bontoc, 2616 Mountain Province. DWLC, Perez Park, Lucena C, 4301 Quezon Province. DWPE, CSU Campus, Caritan Highway, Tuguegarao, 3500 Cagayan. DWRM, City Hall Compound, Puerto Princesa C, 5300 Palawan. DWRP, City Civic Center, Taal Ave, Naga C, 4400 Camarines Sur. DWRS, Poblacion, Tayug, 2445 Pangasinan. DXBN, City Hall

Compound, Brgy. Doongan, Butuan C, 8600 Agusan del Norte. DXIM, A. Velez Str, Cagayan de Oro C, 9000 Misamis Oriental. DXJS, Capitol Hills, Tandag, 8300 Surigao del Sur. DXJT, Brgy Maloro,Tangub C, 7214 Misamis Occidental. DXMR, Baliwasan Chico, Zamboanga C, 7000 Zamboanga del Sur. DXPT, Tubig Boh, Bongao, 7500 Tawi-Tawi. DXRG, Dugenio Str, Gingoog C, 9014 Misamis Oriental. DXRP, Door 5, PTA Complex, Magsaysay Park, 2nd District, Agdao, 8000 Davao C. DXSM, Camp Asturias, Jolo, 7400 Sulu. DXSO, Satellite Office, MSU Campus, Marawi C, 9700 Lanao del Sur.DYES, Capitol Compd, Borongan, 6800 Eastern Samar. DYLL, PNRC Youth Center Bldg, Bonifacio Drive, Iloilo C, 5000 Iloilo. DYMP, Govt Center, Candahug, Palo, Leyte. DYMR, CSCST Compound, Vicente Sotto, 6000 Cebu C. DYOG, Butel Building, Calbayog C, 6710 W. Samar. DYSL, Southern Leyte State University Compound, Sogod, 6606 Southern Leyte. DZAG, Don Mariano Marcos Memorial State University, Agoo, 2504 La Union. DZEQ, Polo Field, Pacdal Circle, Baguio C, 2600 Benguet. DZER, Boac, 4900 Marinduque. DZMQ, Tondaligan Beach, Dagupan C, 2400 Pangasinan. DZRK, Capitol Compound, Tabuk, 3800 Kalinga. DZVC Virac, State College Campus, 4800 Catanduanes.
See (PB) entries in the MW frequency list below for frequencies and powers. Regional stns usually relay news from Manila on the h., and also carry networked prgrs at times.
NB: A number of stns are operating irr or off the air for financial or other reasons.

	MW Call	kHz	kW	Net		MW Call	kHz	kW	Net
67)	DXGH	531	5	dz	58)	DZYI	711	5	ss
118)	DYDW	531	10		50)	DYOK	720	10	ak
47)	DZBR	531	5		7)	DZJO	720	5	
83)	DYRB	540	1		60)	DZSO	720	5	bo
54)	DZWT	540	10	cm	PB)	DWPE	729	10	
16)	DXHM	549	5	cm	84)	DXMY	729	5	
84)	DZXL	558	50		70)	DXOR	729	5	
67)	DXCH	567	5	dz	45)	DYEH	729	5	
PB)	DWRP	567	10		72)	DZGB	729	5	
73)	DXMF	576	10	bo	PB)	DZRB	738	60	
PB)	DYMR	576	10		62)	DXND	747	5	cm
PB)	DZMQ	576	5		84)	DYHB	747	10	
17)	DZHR	576	5	dz	50)	DZJC	747	10	ak
16)	DXCP	585	5	cm	PB)	DWRS	756	10	
16)	DYLL	585	1	su	9)	DWHL	756	1	
16)	DXDB	585	5	cm	42)	DWNW	756	5	
60)	DYWR	*594	10	bo	121)	DXBZ	756	10	
36)	DZBB	594	20	su	82)	DXJM	756	2	
10)	DZLL	603	10		83)	DXGS	765	5	
84)	DXPR	603	5		58)	DYAR	765	5	ss
22)	DZVV	603	5	bo	68)	DYPR	765	10	
75)	DWSP	612	5	dz	58)	DZYT	765	5	ss
84)	DYHP	612	10		41)	DWWW	774	25	
84)	DXDC	621	10	ag	PB)	DXSM	774	10	
PB)	DZVC	621	1		PB)	DXSO	774	10	
85)	DZTG	621	5		84)	DYRI	774	10	ag
13)	DYAG	630	5		94)	DXRA	783	10	
2)	DZMM	630	35		51)	DYME	783	5	
84)	DXKR	639	5		75)	DZNL	783	5	ak
85)	DZRL	639	1		95)	DWES	792	5	
67)	DWRH	648	10		38)	DWGV	792	5	
PB)	DWRM	648	10		PB)	DXBN	792	10	
84)	DXMB	648	5		73)	DXPD	792	5	bo
50)	DYRC	648	5	ak	66)	DYRR	792	5	
77)	DWRN	657	5		39)	DWFA	801	5	
130)	DXDD	657	5	cm	58)	DXBL	801	1	ss
PB)	DYES	657	1		58)	DXES	801	5	bo
84)	DYVR	657	5	ag	16)	DYKA	801	5	cm
98)	DZLU	657	1		35)	DYWC	801	5	cm
PB)	DXRP	666	10		60)	DZNC	801	10	bo
50)	DZRH	666	50/25		90)	DZRJ	810	10	
42)	DWLW	675	5		58)	DWAR	819	5	ss
103)	DXGD	675	1	cm	114)	DWMG	819	1	
85)	DYKC	675	5		101)	DXSC	819	1	
42)	DWGW	*684	1		53)	DXUM	819	10	ak
43)	DWJJ	684	1		84)	DYVL	819	10	ak
50)	DYEZ	684	10	ak	39)	DWZR	828	5	
33)	DZCV	684	5		84)	DXCC	828	10	ag
84)	DXBC	693	10	ag	46)	DYER	828	10/5	
85)	DXDX	693	5		134)	DZTC	828	1	
50)	DYKX	693	5	dz	87)	DXJS	837	10	
50)	DYPH	693	10	dz	58)	DXRE	837	5	ss
106)	DZPB	693	10		22)	DYFM	837	5	ss
31)	DZAS	702	50		30)	DZKE	*837	5	
84)	DXIC	711	10	ag	87)	DZRV	846	50	cm
58)	DXRD	711	5	ss	67)	DXGO	846	10	ak
29)	DYBR	*711	5		42)	DXWG	*855	1	
71)	DZLW	711	10		17)	DXZH	855	5	dz
60)	DZVR	711	5	bo	33)	DZGE	855	10	

MW	Call	kHz	kW	Net	MW	Call	kHz	kW	Net	MW	Call	kHz	kW	Net	MW	Call	kHz	kW	Net
58)	DWSI	864	5	ss	31)	DXKI	1062	5		36)	DYSI	1323	10	su	19)	DYAC	v1449	5	
128)	DYHH	864	10		28)	DZEC	1062	40		PB)	DZRK	1323	10		31)	DWRF	*1458	10	
58)	DZSP	864	5	ss	85)	DXKT	1071	5		58)	DWAY	1332	5	ss	97)	DYZZ	1458	10	
134)	DZWM	864	5	cm	110)	DYXT	1071	1		85)	DZKI	1332	1		120)	DZJV	1458	10	
58)	DXRB	873	5	ss	123)	DZSL	1071	1		91)	DXRL	*1341	10	su	87)	DWVR	1467	1	cm
58)	DXRT	873	5		16)	DWAM	*1080	1	cm	36)	DXXY	*1350	5	su	131)	DXVP	1467	5	cm
127)	DYUP	873	5		28)	DWIN	1080	5		PB)	DZER	*1350	5		92)	DWRB	1476	1	
1)	DZPA	873	5	cm	83)	DWRL	1080	5		48)	DZXQ	v1350	10		90)	DXRJ	1476	10	
33)	DZRC	873	5		85)	DXKS	1080	1		129)	DYSJ	1359	1		88)	DZYA	1476	1	
3)	DZWI	882	50		34)	DXRH	*1080	5		77)	DZYR	1359	5		67)	DYDH	1485	5	dz
62)	DXMS	882	10	cm	67)	DYBH	1080	5	dz	58)	DWTT	*1368	5		108)	DWSS	1494	10	
PB)	DXRG	882	1		111)	DXCM	1089	10	uk	85)	DXKO	1368	10		83)	DXOC	1494	5	
PB)	DYOG	882	1		39)	DYHR	*1089	1		85)	DZBS	1368	2.5		36)	DYBB	*1503	5	su
12)	DWHQ	*891	10		23)	DWAD	1098	10		15)	DZRA	1368	1		2)	DYAB	1512	10	
59)	DYSR	*891	10		58)	DXCL	1098	5	ss	85)	DXKP	1377	10		125)	DZAT	1512	10	
73)	DZGR	891	5		61)	DWDY	1107	10		55)	DXCR	1386	10		14)	DZME	1530	25	
63)	DWNE	900	5		91)	DXBB	1107	5		16)	DYYW	1386	5	cm	77)	DZYN	1539	5	
99)	DXIP	900	5		129)	DYIN	1107	5	bo	42)	DZTV	1386	25		36)	DZSD	1548	10	su
84)	DXRZ	900	5	ag	8)	DZOM	1107	1		17)	DYCH	1395	10	dz	16)	DYDM	1548	5	cm
22)	DYOW	900	5	bo	31)	DXAS	1116	5		132)	DZVT	1395	5	cm	43)	DZKV	*1548	5	
115)	DYLA	909	5		104)	DYTR	1116	10		58)	DXAQ	1404	-	ss	5)	DXID	1566	10	
91)	DYSP	909	5	su	113)	DZLB	1116	5		85)	DYKB	1404	1		PB)	DYMP	1566	7.8	
16)	DZEA	909	5		31)	DWAS	1125	5		91)	DWRA	1413	5	su	74)	DXRA	*1566	10	
PB)	DZSR	918	10		69)	DXGL	1125	5		33)	DYXW	1413	5		21)	DXJR	*1575	10	
122)	DWRS	927	10		91)	DXGN	1125	5	su	109)	DWBC	*1422	10		4)	DYAY	*1584	10	
64)	DXDA	927	5		22)	DZWN	1125	10	bo	18)	DXMU	1422	10		24)	DWBR	1584	1	
84)	DXMD	927	5	ag	26)	DWDD	1134	10		11)	DYZD	1422	5		124)	DXFM	1593	10	
103)	DXMM	927	5	cm	95)	DWJS	1134	5		119)	DZOR	1422	5		113)	DZUP	1602	1	
73)	DZLG	927	5	bo	111)	DXMV	1134	5	uk	89)	DYRS	1431	5		37)	DWGI	1638	0.6	
40)	DWIM	936	5		79)	DXOS	*1134			50)	DWDH	1440	10	dz	56)	DZBF	1674	1	
111)	DXDN	936	5	uk	77)	DYRM	v1134	1		100)	DXSI	1440	0.01						
PB)	DXIM	936	10		PB)	DWBT	1134	1		52)	DXSA	1449	5						
84)	DYCC	936	1		137)	DYAF	1143	10	cm										
85)	DYKW	936	1		31)	DZMR	1143	10											
82)	DZXT	936	1		51)	DWCL	1152	5											
PB)	DWFB	945	10		75)	DWCM	1161	10											
116)	DXDV	945	10		111)	DXDS	1161	1	uk										
58)	DXRO	945	5	ss	84)	DYKR	1161	5	ag										
4)	DYRO	*945	5		11)	DYRD	1161	5	cm										
PB)	DXJT	954	1		72)	DZMD	1161	5											
110)	DYMM	*954	5		PB)	DXMR	1170	10											
93)	DZAL	954	1		PB)	DYSL	1170	10											
20)	DZEM	954	40		65)	DZCA	*1170	10											
58)	DXYZ	963	5	ss	91)	DXYK	1179	5	su										
73)	DYMF	963	10	bo	60)	DYCX	1179	5											
136)	DZNS	963	5	cm	36)	DYSB	1179	5	su										
PB)	DWFR	972	5		86)	DZRS	1179	1											
45)	DWTI	972	5		60)	DXIF	1188	10	bo										
17)	DXKH	972	5	dz	22)	DXLX	1188	5	bo										
17)	DYSM	972	1	ak	2)	DYRV	*1188	1											
75)	DWMT	981	5	dz	82)	DZLT	1188	5											
22)	DXBR	981	10	bo	114)	DZXO	1188	5											
84)	DXDR	981	5	ag	98)	DWBA	1197	5											
88)	DXOW	981	10		31)	DXFE	1197	5											
42)	DYBQ	981	5		4)	DYRH	1197	5											
58)	DZRD	981	5	ss	6)	DWAN	1206	10											
107)	DZIQ	990	10/5		84)	DXRS	1206	5	ag										
91)	DXBM	990	5	su	118)	DYRF	1215	10	cm										
67)	DYTH	990	5	dz	50)	DWSR	1224	5	dz										
67)	DZMT	990	5	dz	28)	DXED	1224	10											
45)	DWMI	999	5		PB)	DZAG	1224	10											
84)	DXHP	999	1		87)	DWRV	1233	5	cm										
PB)	DXPT	999	1		31)	DYVS	1233	5											
91)	DYSS	999	5	su	32)	DWBL	1242	20											
PB)	DZEQ	999	5		105)	DXSY	1242	5											
16)	DWBS	1008	5	cm	27)	DXZB	1242	5											
102)	DWGO	1008	5		50)	DXPH	*1251	5											
85)	DXXX	1008	10		42)	DYRG	1251	1											
42)	DWDW	1017	10		72)	DZMS	1251	2.5											
PB)	DWLC	1017	10		49)	DWMC	1260	5											
44)	DXAM	1017	10		50)	DXRF	1260	5	dz										
138)	DXSN	1017	5	cm	97)	DYDD	1260	5											
4)	DYRP	*1017	10		28)	DZEL	1260	5											
73)	DXMC	1026	5	bo	91)	DWRC	1269	5	bo										
111)	DXMI	*1026	1		22)	DYWB	1269	10/5	bo										
58)	DZAR	1026	10		60)	DZVX	1269	5	bo										
88)	DYRL	1035	10		PB)	DZRM	1278	10											
111)	DYUM	*1035	5/2.5		91)	DXRC	*1287	5	su										
22)	DZWX	1035	5	bo	50)	DZZH	1287	5	dz										
88)	DXCO	1044	10		4)	DWLQ	*1296	5											
81)	DXLL	1044	5	uk	133)	DWPR	1296	5											
96)	DXML	1044	1		2)	DXAB	1296	5											
17)	DYMS	1044	5	ak	42)	DYJJ	1296	5											
60)	DZNG	1044	10	bo	28)	DYFX	*1305	10											
85)	DXKD	1053	10		25)	DWXI	1314	10											
112)	DYSA	1053	5	cm	57)	DXAD	v1322	5											

* = r. inactive

SW	Call	Location	kHz	kW	H of tr
PB)			6170v		2300v-1300v
PB)	DUR2	Marulas, Valenzuela	9580v	0.25	r. inactive

Alt. freq for 9580: 9620kHz (r. on 9619v). Operated by PBS, Philippine Broadc. Sce, these freqs. relay various PBS AM and FM sces. 6170kHz generally relays DZRM 1278kHz.

GENERAL NOTES:

Station identifications: Generally, stn IDs are given on the h. and half h. The English alphabet is used for the call letters, while the freq. is usually expressed in Spanish- or English-language numerals. Extensive stn details are included in sign on and sign off anns.

Callsign assignments: DU = Shortwave only; DW = Luzon; DX = Mindanao and Sulu; DY = Visayas and Palawan; DZ = Luzon.

Administrative divisions: Level 1: regions, 2: provinces, cities (C.), 3: municipalities, 4: barangays (brgy.). The National Capital Region (NCR) is also known as Metropolitan Manila or Metro-Manila.

NB: Cities may be referred to with or without "City", e.g. Baguio City or Baguio. Quezon City is always referred to by its full name.

Prgr. networks: ag=Radyo Agong – ak=Aksyon Radyo – bo=Bombo Radyo – cm=Catholic Media Network (CMN, see above and entry **16)** below) – dz=DZRH (key: 666kHz) – ss=Sonshine Radio (key: 1026kHz) – su=Super Radyo – uk=Radyo Ukay.

Web URLs for broadc. networks: FEBC: www.febc.org – Bombo Radyo: www.bomboradyo.com – R. Mindanao Netw: www.rmn.com. ph – Manila Broadc. Co: www.mbcradio.net – Sonshine Radio: www.sonshineradio.com – DZRH: dzrh.tripod.com

FM: A large number of FM stns are operating throughout the country.

Manila FM(MHz): – 88.3 DWCT-FM "Jam 88.3" (Raven Broadc. Corp.) – 89.1 DWAV "Wave 89.1" (Blockbuster Broadc. System) – 89.9 DWTM "Magic 89.9" (Quest Broadc. Inc.) – 50) 90.7 DZMB "Love Radio" – 91.5 DWKY "Energy FM" (Ultrasonic Broadc. Syst. Inc.) – 92.3 DWFM "U92" (Nation Broadc. Corp.) – 93.1 DWRX "Monster Radio" (Audiovisual Communicators Inc.) – 84) 93.9 DWKC "iFM" – 32) 94.7 DWLL "Mellow 947" – 28) 95.5 DWDM – 17) 96.3 DWRK "Easy Rock" – 36) 97.1 DWLS "Barangay LS" – 3) 97.9 DWQZ "97dot9 Home Radio" – 31) 98.7 DZFE "The Master's Touch" (FEBC) – 99.5 DWRT-FM "99.5 RT" (Real Radio Network Inc.) – 90) 100.3 DZRJ "RJ 100" – 67) 101.1 DWKYS-FM "Yes! FM" – 2) 101.9 DWRR "Tambayan 101.9" – 73) 102.7 DWSM "Star FM" – 103.5 DWKX "Max FM" (Advanced Media Broadc. Syst.) – PB) 104.3 DWBR "Business Radio" – 105.1 DWBM "Crossover" (Mareco Broadc. Network) – 90) 105.9 DWLA "RJ Underground Radio UR 105.9" – 106.7 DWET "Dream FM" (Associated Broadc. Co.) – 107.5 DWNU "NU 107" (Progressive Broadc. Corp.).

Addresses

For each entry the organisation or company name is followed by the call letters (in alphabetical order) and addresses of the stns licensed to the organisation. When contacting a stn, use Radio Station + the call

letters as stn name. In some cases the station may be operated by a different organisation than the licensee mentioned below.

PB) See separate listing for Philippine Broadc. Sce. Above. – **1)** Abra Community Btcg. Corp. DZPA Radyo Totoo, Blessed Arnold Janssen Communication Center, Zamora Str corner Rizal Str, Bangued, 2800 Abra – **2)** ABS-CBN Broadc. Corp (Radyo Patrol). DXAB, KM-4, Shrine Hills, Matina, 8000 Davao C. DYAB, ABS-CBN Broadc. Center, Jagobiao, Mandaue C, 6014 Cebu. DYRV, Catbalogan, 6700 Samar. DZMM, 15/F, Philcomcen Bldg, Ortigas Ave, Pasig C, NCR – **3)** Aliw Broadc. Corp. DWIZ, 5th Floor, Dominga Bldg, 2113 Pasong Tamo, Makati C, 1231 NCR – **4)** Allied Broadc. Center, Inc. DWLQ, Happy Valley Str, Ibabang Dupay, Lucena C, 4301 Quezon Province. DYAY, Manuel Quezon Str, 6000 Cebu C. DYRH, JTL Bldg, North Drive, Bacolod C, 6100 Negros Occidental. DYRO. Roxas C., Capiz. DYRP Radyo Tagring, General Luna Str, Iloilo C, 5000 Iloilo – **5)** Association of Islamic Dev't. Cooperative. DXID, Banale Dist, Pagadian C, 7016 Zamboanga del Sur – **6)** Metropolitan Manila Development Authority (MMDA). DWAN MMDA Traffic Radio 1206, MMDA Communications and Command Center, EDSA corner Orense Str, Guadalupe Nuevo, Makati C, NCR – **7)** Bayanihan Broadc. Corp. DZJO, Infanta, 4336 Quezon Province – **8)** Ben Viduya (OMARCO). DZOM, Calapan C, 5200 Mindoro Oriental – **9)** Beta Broadc. Syst. DWHL Radyo Apo, 8 Kessing Str, Olongapo C, 2200 Zambales – **10)** Bicol Broadc. Syst. DZLL, BBS Bldg, Balagtas Road, Magsaysay Ave, Naga C, 4400 Camarines Sur – **11)** Bohol Chronicle Radio Corp. DYRD, Dejaresco Bldg, Tagbilaran C, 6300 Bohol. DYZD, Brgy Tapon, Ubay, 6315 Bohol – **12)** Caceres Broadc. Corp. DWHQ Radyo Oragon, Diversion Rd., Tabuco, Naga C, 4400 Camarines Sur. – **13)** Cadiz Radio & TV Netw. DYAG, Cadiz C, 6121 Negros Occidental – **14)** Capitol Broadc. Center. DZME Radyo Uno, 5th Floor Victory Central Mall, Victory Liner Compound, 717 Rizal Avenue Extension, Monumento, Caloocan C, NCR – **15)** Catanduanes State College. DZRA, Virac, 4800 Catanduanes – **16)** Catholic Media Network (CMN). Most MW sts ID as Radyo Totoo. DWAM, Basilican Site, Batangas C, 4200 Batangas. DWBS Radio Veritas, 2/F Landco Business Park, Legaspi C, 4500 Albay. DXCP, Lagao, Gen. Santos C, 9500 South Cotabato. DXDB, Communications Media Center, San Isidro Cathedral Compound, Malaybalay C, 8700 Bukidnon. DXHM, Clergy House Compound, Madang, Mati C, Davao Oriental. DYDM, SJC Extension Campus, Mambajao, Maasin C, 6600 Southern Leyte. DYKA, St Joseph Bldg, San Jose de Buenavista, 5700 Antique. DYVW, Clergy House, Baybay Blvd, Borongan, 6800 Eastern Samar. DZEA, Brgy. Nalbo, Laoag C, 2900 Ilocos Norte– **17)** Cebu Broadc. Co. DXKH, Cagayan de Oro C, 9000 Misamis Oriental. DXZH, Zamboanga C, 7000 Zamboanga del Sur. DYCH, Tanke, Talisay C, 6045 Cebu. DYMS, San Bartolome Str, Catbalogan, 6700 Samar. DYSM, Brgy. Cawayan, Catarman, 6400 Northern Samar. DZHR, Tuguegarao, 3500 Cagayan. – **18)** Central Mindanao University. DXMU, Musuan, 8710 Bukidnon – **19)** Leyte State University. VISCA Compound, Bo Pangasudan, Baybay, 6521 Leyte – **20)** Christian Era Broadc. Sce. DZEM, Maligaya Bldg 2, 887 EDSA, Quezon C – **21)** COC Broadc. Netw. DXJR, Cagayan de Oro College, Max Suniel St, Carmen, Cagayan de Oro C, 9000 Misamis Oriental – **22)** Consolidated Broadc. Syst, Inc. DXBR, Bombo R. Broadc. Center, Arujville Subd, Brgy. Libertad, Butuan C, 8600 Agusan del Norte. DXES Bombo R. Broadc. Center, Amao Rd, Brgy. Bula, Gen. Santos C, 9500 South Cotabato. DXLX, Tambo, Brgy. Hinaplon, Iligan C, 9200 Lanao del Norte. DYFM, Sky City Tower, Mapa Str, Jaro, Iloilo C, 5000 Iloilo. DYOW, Bombo R. Broadc. Center, Arnaldo Blvd, Roxas C, 5800 Capiz. DYWB, Bombo Radyo Broadcast Center, Lacson Str, Mandalagan, Bacolod C, 6100 Negros Occidental. DZVN, Bombo R. Broadc. Center, Brgy. Tamag, Vigan, 2700 Ilocos Sur. DZWN, Bombo R. Broadc. Center, Maramba Bankers' Village, Bonuan Catacdang, 2400 Dagupan C, Pangasinan. DZWX, Bombo R. Broadc. Center, 87 Lourdes Subdivision Rd, Baguio C, 2600 Benguet – **23)** Crusaders Broadc. Syst. DWAD, 209 E. de la Paz Str, Mandaluyong C, 1550 NCR – **24)** Dawnbreaker's Foundation Inc. DWBR Radyo Baha'i, Bulac, Talavera, 3114 Nueva Ecija or P. O.Box 27, San José City 3121 – **25)** Delta Broadc. Syst. DWXI, Mathew Str, Multinational Village, Parañaque C, 1708 NCR – **26)** Dept. of National Defense. DWDD, Camp Aguinaldo, EDSA, Quezon C, 1110 NCR – **27)** DZXB/TV13 Cooperative, Inc. DXZB, Zamboanga C, 7000 Zamboanga del Sur – **28)** Eagle Broadc. Corp. DWIN, Bo. Lucao, Dagupan C, 2400 Pangasinan. DXED, Cabiguio Ave, Agdao, 8000 Davao C. DYFX, Tanke, Talisay C, 6045 Cebu. DZEC, R Aguila, Maligaya Bldg II, 887 EDSA, Quezon C. DZEL, Bo. Mayao, Lucena C, 4301 Quezon Province – **29)** East Visayan Broadc. DYBR, Sagcahan Rd, P.O. Box 80, Tacloban C, 6500 Leyte – **30)** Fairwaves Broadc. Netw. DZXE Radyo Tirador, Mira Hills, Vigan, 2700 Ilocos Sur – **31)** Far East Broadc. Co. DWAS, P.O. Box 78, Arimbay, Legaspi C, 4500 Albay. DWRF, P.O. Box 3222, Amungan, Iba, 2201 Zambales. DXAS, P.O. Box 349, Tugbungan, Zamboanga C, 7000 Zamboanga del Sur. DXFE, Circumferential Rd, Dona Vicente Village, 8000 Davao C. DXKI, P.O. Box 8004, Brgy Morales, Koronadal C, 9506 South Cotabato. DYVS, P.O. Box 393, Km. 7, Pahanocoy, Bacolod C, 6100 Negros Occidental. DZAS, 62 Karuhatan Rd, Karuhatan, Valenzuela C, 1441 NCR. DZMR, Maharlika Highway, Sefton

Village, Santiago City, 3311 Isabela – **32)** FBS Radio Netw. DWBL, Unit 908, Paragon Plaza, EDSA corner Reliance Str, Mandaluyong C, NCR – **33)** Filipinas Broadc. Netw. DYXW, Baruyan, San Jose, Tacloban C, 6500 Leyte. DZCV, Ugac Norte, Tuguegarao, 3500 Cagayan. DZGE Radyo Numero Uno, Nordia Resort, Baras, Canaman, Naga C, 4400 Camarines Sur. DZRC, Capt. Aquendes Drive, Legaspi C, 4500 Albay – **34)** First United Broadc. Corp. DXRH Radio Hermosa, Zamboanga C, 7000 Zamboanga del Sur – **35)** Franciscan Broadc. Corp. DYWC Radyo Bandilyo, Parish Compound, St Anthony of Padua Parish, Sibulan, Dumaguete C, 6201 Negros Oriental – **36)** GMA Netw, Inc. DZSD, Arellano St., Dagupan C, 2400 Pangasinan. DXXY, Dipolog C, 7100 Zamboanga del Norte. DYBB, 2nd Flr, Oscar R. Arcenas Bldg, Roxas Av, Roxas C, 5800 Capiz. DYSB, Bacolod C, 6100 Negros Occidental. DYSI, GMA Compound, MacArthur Drive, Jaro, Iloilo C, 5000 Iloilo. DZBB, GMA Netw. Center, EDSA corner Timog Ave, Diliman, 1103 Quezon C – **37)** Guzman Institute of Tech. DWGI, 509 Z.P. de Guzman, Quiapo, Manila, NCR – **38)** GV Broadc. Syst. DWGV Radyo Centro, Rizal Extension, Cut-Cut, Angeles C, 2009 Pampanga –**39)** Hypersonic Broadc. Center. DWFA, Maharlika Hwy, Sorsogon C, 4700 Sorsogon. DWZR Zoom Radio, Penaranda Str, Legaspi C, 4500 Albay. DYHR, Calbayog C, 6710 W. Samar – **40)** Insular Broadc. Syst. DWIM Radyo Mindoro, Brgy. Bayanihan, Calapan, 5200 Mindoro Oriental – **41)** Interactive Broadcast Media, Inc. DWWW, 23 E. Rodriguez Sr. Blvd, Quezon C – **42)** Intercontinental Broadc. Corp. DWDW, A.B. Fernandez Ave, Dagupan C, 2400 Pangasinan. DWGW Radyo Budyong, Penaranda Str, Legaspi C, 4500 Albay. DWLW, Brgy. Nangalisan, Laoag C, 2900 Ilocos Norte. DWNW Radyo Budyong, Mabolo Drive, Naga C, 4400 Camarines Sur. DXWG, Iligan C, 9200 Lanao del Norte. DYBQ Radyo Budyong Iloilo, Datu Puti Subdivision, ,Cubay, Jaro, Iloilo C, 5000 Iloilo. DYJJ, Roxas Ave, Roxas C, 5800 Capiz. DYRG Radyo Budyong, Roxas Ave Extension, Andagao, Kalibo, 5600 Aklan – DZTV Radyo Budyong, Quezon C., NCR – **43)** Kaissar Broadc. Netw. DWJJ Radyobisyon (Double J Ad Ventures), Celcor Compound, Bitas, Cabanatuan C, 3100 Nueva Ecija. DZKV, Lipa C, 4217 Batangas – **44)** Kalayaan Broadc. Syst. DXAM Radyo Rapido, Bug-ac, Matina, 8000 Davao C – **45)** Katigbak Enterprises (ConAmor Broadcasting Systems). DWMI, Calapan, 5200 Mindoro Oriental. DWTI, Broadcast Village, Ibabang Dupay, Lucena C, 4301 Quezon Province. DYEH, Puerto Princesa C, 5300 Palawan.– **46)** Puerto Princesa Broadc. Co. DYER Environment Radio, Puerto Princesa C, 5300 Palawan – **47)** Kumintang Broadc. Syst. DZBR R. Balisong, KBS Bldg, Capitol Hills, Batangas C, 4200 Batangas – **48)** Mabuhay Broadc. Syst., Inc. DZXQ Kaibigan ng Masa, Centerpoint Condominium, Dona Julia Vargas Ave., Ortigas Center, Pasig C, NCR – **49)** Magiliw Community Broadc. Co. DWMC, Tomana, Rosales, 2441 Pangasinan – **50)** Manila Broadc. Co. DWDH, Lucao District, Dagupan C, 2400 Pangasinan. DWSR, Lucena C, 4301 Quezon Province. DXPH, Prosperidad, 8500 Agusan del Sur. DXRF, Matina, 8000 Davao C. DYEZ, Wilrose Building, Burgos Str, Bacolod C, 6100 Negros Occidental. DYKX, Kalibo, 5600 Aklan. DYOK, Suite 301Carlos Uy Bldg, Diversion Rd, Manurriao, Iloilo C, 5000 Iloilo. DYPH, Puerto Princesa C, 5300 Palawan. DYVL, J. Romualdez corner Real Streets, Tacloban C, 6500 Leyte. DYRC, 3rd Floor, Cinco Centrum Building, Fuente Osmeña Blvd, 6000 Cebu C. DZJC, Brgy. 29, Rizal Street, St. Joseph District, Laoag C, 2900 Ilocos Norte. DZRH, MBC Bldg, Vicente Sotto Str, CCP Complex, Pasay C, 1300 NCR. DZZH, Cabit-an, Sorsogon C, 4700 Sorsogon. – **51)** Masbate Community Broadc. Co. DYCM, Bogo Amusement Complex, Taytayan, Bogo, 6010 Cebu. DYME, Tugbo Str, Masbate C, 5400 Masbate – **52)** Mindanao Broadc. Co. Inc. DXSA, Marawi C, 9700 Lanao del Sur – **53)** Univ. of Mindanao Broadcasting Netwk. DXUM, UMBN Broadcast Center, Multi-test Bldg, Ponciano Reyes St, 8000 Davao C – **54)** Mt. Province Broadc. Corp. DZWT, P.O. Box 156, Mount Beckel, La Trinidad, Baguio C, 2600 Benguet – **55)** Mt. View College. DXCR Voice of Hope, MVC, Valencia, 8709 Bukidnon – **56)** Municipality of Marikina. DZBF, R. Marikina, Second Floor, City Hall, Shoe Avenue, Marikina C, NCR – **57)** Muslim Mindanao Dev. Multi-Purpose Coop. DXAD Radio Ranao, Marcos Blvd, Saduc, Marawi C, 9700 Lanao del Sur – **58)** Swara Sug Media Corporation. DWAR, Laoag C, 2900 Ilocos Norte. DWAY, Cabanatuan C, 3100 Nueva Ecija. DWSI, North Eastern Foundation College, Santiago, 3311 Isabela. DWTT, Tarlac C, 2300 Tarlac. DXBL, Mangagoy, Bislig C, 8311 Surigao del Sur. DXAQ, Philippine-Japan Friendship Hwy, Catitipan, Davao C. DXCL, Cagayan de Oro C, 9000 Misamis Oriental. DXRB, Brgy Libertad, Butuan C, 8600 Agusan del Norte. DXRD, J.P Laurel Ave, Bajada, 8000 Davao C. DXRE, Lagao, Gen. Santos C, 9500 South Cotabato. DXRO, Don Roman Vilo Str, Cotabato C, 9600 Maguindanao. DXRT, Jolo, 7400 Sulu. DXYZ, San Jose Rd, Baliwasan, 7000 Zamboanga C. DYAR, 3ʳᵈ Fl. Astron Centre Bldg, Gorordo Ave, 6000 Cebu C. DZAR, Suite 3004, 30/F Jollibee Plaza Building, F Ortigas Jr Road, Ortigas Center, Pasig City, 1600 NCR. DZRD, Banuan Guesset, Dagupan C, 2400 Pangasinan. DZSP, San Pablo C, 4000 Laguna. DYZI, Calamagui 2nd, Ilagan, 3300 Isabela. DZYT, Cagayan Teachers College, Tuguegarao, 3500 Cagayan – **59)** National Council of Churches in the Philippines. DYSR, Camp Seasite Banilad, Dumaguete C, 6200 Negros Oriental – **60)**

Newsounds Broadc. Netw. DXIF, Bombo R. Broadc. Center, Corrales Ave, Cagayan de Oro C, 9000 Misamis Oriental. DYCX, San Jose de Buenavista, Antique. DYWR, Bombo R. Broadc. Center, Sto. Nino cor. Imelda Ave, Tacloban C, 6500 Leyte. DZNC, Bombo R. Broadc. Center, Barrio Menante II, Cauayan, 3305 Isabela. DZNG, Bombo R. Broadc. Center, Diversion Road, Brgy. Tabuko, Naga C, 4400 Camarines Sur. DZSO, Bombo Radyo Broadcast Center, Pennsylvania Ave, Parian, San Fernando C, 2500 La Union. DZVR, Bombo Radyo Broadcast Center, 48 A, Cabungaan Airport Ave, Laoag C, 2900 Ilocos Norte. DZVX, J. Pimentel Str, Daet, 4600 Camarines Norte – **61)** Northeastern Broadc. Sce. DWDY, Ground Floor, Isabela Hotel, Mirante Uno, Cauayan, 3305 Isabela – **62)** Notre Dame Broadc. Corp. DXMS, Sinsuat Ave cor. Rizal Ave, Cotabato C, 9600 Maguindanao. DXND, Daang Maharlika, Kidapawan C, 9400 North Cotabato – **63)** Nueva Ecija Provincial Gov. DWNE, Brgy Singalat, Palayan C, 3132 Nueva Ecija – **64)** Office of the Governor, Prov. of Agusan del Sur. DXDA Radyo Agusan, Patin-ay, Prosperidad, 8500 Agusan del Sur – **65)** Office of the Civil Defense. DZCA, Agham Rd. Science Garden, Pag-asa Planetarium, NCR – **66)** Ormoc Broadc. Co. DYRR, Bantigue, Ormoc C, 6541 Leyte – **67)** Pacific Broadc. Syst (subsidiary of Manila Broadc. Co.). DWRH, Santiago C, 3311 Isabela.. DXCH, Cotabato C, 9600 Maguindanao. DXGH, Purok Malakas, Lagao, Gen. Santos C, 9500 South Cotabato. DXGO, MBC Compound, Brgy. Duterte, R. Castillo Str, Agdao, 8000 Davao C. DYBH, Bacolod C, 6100 Negros Occidental. DYDH, Iloilo C, 5000 Iloilo. DYTH, Real Str, Tacloban C, 6500 Leyte. DZMT, Laoag C, 2900 Ilocos Norte – **68)** Palawan Broadc. Corp. DYPR, Rey Olivar Bldg., 61 Mabini St. Puerto Princesa C, 5300 Palawan. – **69)** PEC Broadc. Corp. DXGL, Butuan C, 8600 Agusan del Norte – **70)** Pedro N. Roa Broadc. DXOR, Don A. Velez Str, Cagayan de Oro C, 9000 Misamis Oriental – **71)** Peñafrancia Broadc. Corp. DZLW Radyo Isarog, Naga C, 4400 Camarines Sur – **72)** People's Broadc. Netw. DZGB, Mayona Building, Imperial Court Subdivision, Legaspi C, 4500 Albay. DZMD, Vinzons Ave, Daet, 4600 Camarines Norte. DZMS, Balobo Str, Sorsogon C, 4700 Sorsogon – **73)** People's Broadc. Sce. DXMC, Bombo R. Broadc. Center, Km 4 General Santos Drive, Koronadal C, 9506 South Cotabato. DXMF, Bombo R. Broadc. Center, San Pedro Str, 8000 Davao C. DXPD, Bombo R. Broadc. Center, North Diversion Road, Brgy. Banale, Pagadian C, 7016 Zamboanga del Sur. DYMF, 87-A. Borromeo Str, 6000 Cebu C. DZGR, Bombo R. Broadc. Center, Taft Str Extension, Brgy 5, Tuguegarao, 3500 Cagayan. DZLG, Bombo R. Broadc. Center, Tahao Road, Legaspi C, 4500 Albay. – **74)** Philippine Air Force. DZHH, Villamor Air Base, Pasay C, 1309 NCR – **75)** Philippine Broadc. Corp. DWCM, Caranglaan District, Dagupan C, 2400 Pangasinan. DWMT, Naga C, 4400 Camarines Sur. DWSP, Tuding, Itogon, nr Baguio City, Benguet. DZNL, Brgy. Pagdalagan, San Fernando C, 2500 La Union. – **77)** Philippine Radio Corp. DWRN, Manipit Rd, Queborac Bagumbayan, Naga C, 4400 Camarines Sur. DYRM, Bo. Calindangan, Dumaguete C, 6200 Negros Oriental. DZYM Radyo Asenso, Puerto Gallenero, Pag-asa, San Jose, 5100 Mindoro Occidental. DZYR, Catbangen, San Fernando C, 2500 La Union – **79)** Public Affairs Sce, Armed Forces of the Philippines. DXOS, Basilan Island, Basilan – **81)** R.T. Broadc. Specialistns Philippines. DXLL, Campaner Str, Zamboanga C, 7000 Zamboanga del Sur – **82)** Radio Corp. of the Philippines. DXJM, J & M Bldg, Villakananga, Butuan C, 8600 Agusan del Norte. DZLT, Bo. Ibabang Dupay, Lucena C, 4301 Quezon Province. DZXT, MacArthur H-way, Tarlac C, 2300 Tarlac – **83)** DWRL Radio, Inc (subsidiary of 82 above). DWRL, Purok 5, Rawis, Legaspi C, 4500 Albay. DXGS Radyo Asenso, NLSA Rd, Lagao, Gen. Santos C, 9500 South Cotabato. DXOC, Manabay, Catadman, Ozamis C, 7200 Misamis Occidental. DYRB, C. Padilla St., 6000 Cebu C – **84)** Radio Mindanao Netw. DXBC, Montilla Blvd, Butuan C, 8600 Agusan del Norte. DXCC, Canoy Bldg., Don Apolinar Velez Str, Cagayan de Oro C, 9000 Misamis Oriental. DXDC, San Francisco Bldg, cor. Anda & Bonifacio Stns, 8000 Davao C. DXDR, Bo. Mario Turno, Dipolog C, 7100 Zamboanga del Norte. DXHP, Flomencia Bldg. P. Castillo Mangagoy, Bislig C, 8311 Surigao del Sur. DXIC, Pafs Mejia Bldg, Roxas Str. cor Aguinaldo Str, Iligan C, 9200 Lanao del Norte. DXKR, Gen. Santos Drive, Koronadal C, 9506 South Cotabato. DXMB, Fortich Str, Malaybalay C, 8700 Bukidnon. DXMD, Bo. Obrero National Highway, Gen. Santos C, 9500 South Cotabato. DXMY, Esteros, RH 10, Cotabato C, 9600 Maguindanao. DXPR, Mercedes Str, San Jose Dist, Pagadian C, 7016 Zamboanga del Sur. DXRS, Km. 1 Rizal Str, Surigao C, 8400 Surigao del Norte. DXRZ, Zamaveco Bldg, Pilar Str, Zamboanga C, 7000 Zamboanga del Sur. DYCC, Brgy. Obrero, Calbayog C, 6710 W. Samar. DYHB, 4th Flt, SSS Bldg. Lacson Str, Bacolod C, 6100 Negros Occidental. DYHP, 2nd Flr, Gold Palace Bldg, 168 Osmeña Blvd, 6000 Cebu C. DYKR, C. Laserna Str, Kalibo, 5600 Aklan. DYRI, St Anne Bldg, Luna Str, La Paz, Iloilo C, 5000 Iloilo. DYVR, Punta, Tabuc, Roxas C, 5800 Capiz. DZXL, 4/F Guadelupe Commerical Complex, Guadelupe Nuevo, Makati C, 1200 NCR – **85)** Radio Philippines Netw. DXDX Radyo Ronda, Acharon Blvd, Gen. Santos C, 9500 South Cotabato. DXKD, Gonzales corner Lopez Jaena Str, Biasong, Dipolog C, 7100 Zamboanga del Norte. DXKO Radyo Ronda, Gusa, National Hwy, Cagayan de Oro C, 9000 Misamis Oriental. DXKP Radyo Ronda, Araulio Str, Brgy Datoc, Pagadian C, 7016 Zamboanga del

Sur. DXKS, Capitol Rd, Surigao C, 8400 Surigao del Norte. DXKT Radyo Ronda, Marfori Heights, 8000 Davao C. DXXX Radyo Ronda, Brgy Tugbungan, 7000 Zamboanga C. DYKB Radyo Ronda, Bo. Sumag, Bacolod C, 6100 Negros Occidental. DYKC, Maguikay, Mandaue C, 6014 Cebu. DYKM Radyo Ronda, Cagamayan, Binalbagan, 6107 Negros Occidental. DZBS Radyo Ronda, Agrix Supermarket cor. Magsaysay Ave. & Bakawkan, Baguio C, 2600 Benguet. DZKI Radyo Ronda, San Agustin, Iriga C, 4431 Camarines Sur. DZRL Radyo Ronda, Bo. Kawayan, Batac, 2906 Ilocos Norte. DZTG, 46 Rizal Str, Tuguegarao, 3500 Cagayan – **86)** Radio Sorsogon Netw, Inc. DZRS, Don Luis Lee Bldg, Plaza Bonifacio, Sorsogon C, 4700 Sorsogon – **87)** Radio Veritas Global Broadc. Syst. DWRV, Maharlika Highway, Bayombong, 3700 Nueva Vizcaya. DWVR, San Jose C, 3121 Nueva Ecija. DZRV, R. Veritas, 20/F The Centerpoint Bldg 1, 162 West Ave corner EDSA, Ortigas Center, Pasig C, 1600 NCR – **88)** Radyo Pilipino Corp (Radyo Asenso). DXCO, Atco Bldg, Capistrano & Gomez Str, Cagayan de Oro C, 9000 Misamis Oriental. DXOW, Mapa, 8000 Davao C. DYRL, Camaroli Av, Lupit Subd, Bacolod C, 6100 Negros Occidental. DZYA 2/F Tanglao Bldg, Balibago, Angeles C, 2009 Pampanga. – **89)** Ragde, Vicente & Sons. DYRS, Ragde Comp, Corner M. Endrinda Str and Broce Str, San Carlos C, 6127 Negros Occidental – **90)** Rajah Broadc. Netw (Radyo Bandido). DXRJ, RJ Clubhouse, Sta. Filomena, Iligan C, 9200 Lanao del Norte. DZRJ, Ventures Bldg 1, Gen. Luna Str, Makati C, NCR – **91)** Republic Broadc. Syst. (owned by GMA Network Inc.) DWRA, Baguio C, 2600 Benguet. DWRC, San Nicolas, 2901 Ilocos Norte. DXBB, Gen. Santos C, 9500 South Cotabato. DXBM, Cotabato C, 9600 Maguindanao. DXGM, Shrine Hills, Matina, 8000 Davao C. DXRC, Zamboanga C, 7000 Zamboanga del Sur. DXRL, 3/F Carisma Bldg., General Santos Drive, Koronadal C, 9506 South Cotabato. DXYK, Butuan C, 8600 Agusan del Norte. DYSP, Solid Rd, Brgy San Manuel, Puerto Princesa C, 5300 Palawan. DYSS, GMA Network Center, Nivel Hills, Apas, 6000 Cebu C – **92)** Ribbon Broadc. Netw. DWRB, 5/F, LCC Bldg, Lipa C, 4217 Batangas – **93)** Rinconada Broadc. Corp. DZAL Radyo Rinconada, UNEP Compound, San Roque, Iriga C, 4431 Camarines Sur – **94)** RMC Broadc. Co, Inc (Rizal Memorial Colleges). DXRA Radyo Arangkada, A. Pichon St., 8000 Davao C – **95)** Rolin Broadc. Enterprises. DWES, Narra, 5303 Palawan. DWJS, Roxas, 5308 Palawan – **96)** Rural Electrification Corp. DXML, MacArthur Hwy, Digos C, 8002 Davao del Sur – **97)** Siam Broadc. Netw. Corp. (Bantay Radyo). DYDD, Lapu-Lapu C, 6015 Cebu. DYZZ, Guihulongan, 6214 Negros Oriental – **98)** Satellite Broadc. Corp. DWBA, Bangued, 2800 Abra. DZLU, National College of Technology Campus, Barangay 1, San Fernando C, 2500 La Union – **99)** Southern Broadc. Netw. DXIP Bantay Radyo, 3/F Lachmi Shopping Mall, Bolton Street, 8000 Davao C – **100)** Southern Institute of Tech. DXSI, Cagayan de Oro C, 9000 Misamis Oriental – **101)** Southern Philippines Mass Comm. DXSC, Camp Navarro, Calarian, 7000 Zamboanga C – **102)** Subic Broadc. Corp. DWGO Gabay ng Olangapo, 1 Kasarinlan Rd, Olongapo, 2200 Zambales – **103)** Sulu Tawi-Tawi Broadc. Foundation. DXGD Radio for Peace, Bongao, 7500 Tawi-Tawi. DXMM Radyo Totoo, Gandasuli Str, Jolo, 7400 Sulu – **104)** Tagbilaran Broadc. Corp. DYTR, CPG Ave, Dampas, Tagbilaran C, 6300 Bohol – **105)** Times Broadc. Corp. DXSY, Mariano Marcos, Ozamis C, 7200 Misamis Occidental – **106)** Tirad Pass R/TV Broadc. Netw. DZTP Radyo Tirad Pass, San Nicolas, Candon, 2710 Ilocos Sur – **107)** Trans-Radio Broadc. Corp. (operated by Philippine Daily Inquirer). DZIQ Radyo Inquirer, 2/F Media Resources Plaza, Pasong Tirad cor. Mola Str, Brgy La Paz, Makati C, NCR – **108)** Supreme Broadc. Systems. DWSS, Paragon Plaza, EDSA, Mandaluyong C, NCR – **109)** United Broadc. Netw. DWBC, Bo. Upong del Norte, Quezon C, 1110 NCR – **110)** Universal Broadc. Syst (owned by Radio Mindanao Network). DYMM, Sunshine Village, Esperos Str, Tacloban C, 6500 Leyte. DYXT, Luna Str, Tagbilaran C, 6300 Bohol – **111)** University of Mindanao Broadc. Netwk (UMBN). DXCM, UM School Compound, Cotabato C, 9600 Maguindanao. DXDN, UM Tagum School Compound, Tagum C, 8100 Davao del Norte. DXDS, Digos C, 8002 Davao del Sur. DXMI, Iligan C, 9200 Lanao del Norte. DXMV, Mt. Kitangcad Cor. Kanlaon Street, Valencia, 8709 Bukidnon. DYUM, Ormoc C, 6541 Leyte – **112)** University of San Agustin. DYSA Radyo San Agustin, 2/F Univ. of S. Agustin, Gen. Luna Str, Iloilo C, 5901 Iloilo – **113)** University of the Philippines. DZLB, UP Los Banos College, 4031 Laguna. DZUP, Media Center, College of Mass Comunications, UP Campus Diliman, R, Magasay Ave corner Apacible Str, Quezon C, 1104 NCR – **114)** Vanguard Radio Netw (Radio Vanguard). DWMG, Solano, 3709 Nueva Vizcaya. DZXO, Ground Floor Diego Building, Maharlika Highway, Cabanatuan C, 3100 Nueva Ecija – **115)** Visayas Mindanao Confederation of Trade Unions. DYLA, Alu-Vimcontu Welfare Center, Pier Area, 6000 Cebu C – **116)** Vismin Radio & TV Broadc. Net. DXDV, Baan, Butuan C, 8600 Agusan del Norte – **118)** Word Broadc. Corp. DYDW Radio Diwa, Burayan, San José, Tacloban C, 6500 Leyte. DYRF Radyo Fuerza, Univ. of San Carlos, Pelaez Str, 6000 Cebu C – **119)** Zambales Btcg. & Devt. Corp. DZOR Radyo Olangapo, 1683 Rizal Ave, Olongapo C, 2200 Zambales – **120)** ZOE Broadc. Netw. DZJV, 140 Brgy Parian, Calamba, 4027 Laguna – **121)** Baganian Broadc. Corp. DXBZ Radyo Bagting, Bana Str, Sta Maria District, Pagadian C., 7015

Zamboanga del Sur – **122)** Solidnorth Broadcasting System. DWRS Commando Radio, Tamag, Vigan, 2700 Ilocos Sur – **123)** S.O.L. Telebroadcasting Station. DZSL, Purok 2, Talisay, Camarines Norte – **124)** Ranao Radio & TV Broadcast System Corp. DXFM, Pangarungan Village, Marawi C, 9700 Lanao del Sur – **125)** End Time Mission (Pentecostal Missionary Church of Christ 4th Watch). DZAT, Purok Rosal, Bo. Silangan Mayao, Lucena C., 4301 Quezon Province – **127)** University of the Philippines in the Visayas. DYUP UPV Radio, Miagao, Iliolo – **128)** Philippine Air Force, DYHH Radyo ng Hukbong Himpapawid, Bogo, 6010 Cebu – **129)** Inter-Island Broadc. Corp. (IBC). DYIN, Bombo Radyo Broadcast Center, Oyo Torong Str, cor. J. Magno. Str, Kalibo, 5600 Aklan. DYSJ, San Jose de Buenavista, Antique – **130)** Dan-ag sa Dakbayan Broadc. Corp. DXDD Radyo Kampana, New DXDD Bldg, Rizal. Str, Ozamis C, 7200 Misamis Occidental – **131)** Roman Catholic Archdiocese of Zamboanga Broadc. Network (RCA-ZBN). DXVP Radyo Verdadero, Sacred Heart Center, R.T. Lim Bvd, Zamboanga C, 7000 Zamboanga del Sur – **132)** Apostolic Vicariate of San Jose de Mindoro. DZVT Radyo Totoo, Labangan Poblacion, San Jose, 5100 Mindoro Occidental – **133)** Multipoint Broadc. Netwk. DWPR Power Radio, A.B. Fernandez Ave, Bolosan District, Dagupan C, 2400 Pangasinan – **134)** Alaminos City Broadc. Corporation. DZWM Radyo Totoo, St Joseph Cathedral Compound, Alaminos, 2404 Pangasinan – **135)** Government of Tarlac Province. DZTC, MacArthur Hwy, Tarlac C, 2300 Tarlac – **136)** Archdiocese of Nueva Segovia. DZNS Radyo Totoo, Brgy Pantay Fatima, Vigan, 2700 Ilocos Sur – **137)** Diocese of Bacolod. DYAF Radyo Veritas Bacolod, Rizal Str corner San Juan Str, Brgy. 11, Bacolod C, 6100 Negros Occidental – **138)** Silangan Broadcasting Corporation. DXSN Radyo Magbalantay, 55 Magallanes Str, Surigao C, 8400 Surigao del Norte.

EXTERNAL SERVICES: Radyo Pilipinas, Radio Veritas Asia, FEBC International Service, VOA/IBB
see International Broadcasting section

POLAND

L.T: UTC +1h (27 Mar-30 Oct: +2h) — **Pop:** 38.5 million — **Pr.L:** Polish — **E.C:** 50Hz, 230V — **ITU:** POL

KRAJOWA RADA RADIOFONII I TELEWIZJI (KRRiT)
(National Broadcasting Council)
✉ Skwer Ksiedza Kardynala Stefana Wyszynskiego Prymasa Polski 9, 01-015 Warszawa ☎ +48 225973000 🖹 +48 225973180 **E:** krrit@krrit.gov.pl **W:** www.krrit.gov.pl **L.P:** Pres: Jan Dworak
NB. KRRiT is the regulatory authority for broadcasting.

POLSKIE RADIO S.A. (PR) (Pub)
✉ al. Niepodleglosci 77/85, 00-977 Warszawa ☎ +48 226459212 🖹 +48 226453993 **E:** public.relations@polskieradio.pl **W:** www.polskieradio.pl; www.prsa.com.pl (corporate) **L.P:** Chmn: Jaroslaw Hasinski

LW	kHz	kW	Prgr
Solec Kujawski	225	1000	PR1

FM (MHz)	PR1	PR2	PR3	PR4	kW
Bialogard (Slawoborze)	106.0	98.2	101.5	-	10/2x15
Bialystok (Cieszynska)	-	106.4	-	91.1	1/0.1
Bialystok (Krynice)	92.3	-	96.0	-	30
Bogatynia (G.Wysoka)	92.8	-	-	-	1
Bydgoszcz (Foton)	-	-	-	96.2	1
Bydgoszcz (Trzeciewiec)	106.6	97.6	102.1	98.9	60/2x120/2
Czestochowa (Wreczyca)	87.5	90.6	91.7	-	10/2x60
Elblag (Jagodnik)	-	102.3	-	101.2	5/0.25
Gdansk (Chwaszczyno)	95.7	-	99.9	-	120
Gdansk	-	89.5	-	93.4	1/0.1
Gdynia (Oksywie)	-	97.2	-	-	1
Gizycko (Milki)	97.1	92.6	94.4	-	6/2x10
Gorzów Wlkp. (Janice)	-	-	-	105.4	1
Ilawa (Kisielice)	94.8	102.7	-	104.8	10/10/5
Jelenia Góra (Sniezne Kotly)	92.5	-	94.0	-	10
Kalisz (Mikstat)	100.0	95.6	102.5	94.2	10
Katowice (Kosztowy)	97.9	105.6	99.7	-	60
Kielce (Swiety Krzyz)	92.3	-	96.2	-	60
Kielce	-	102.7	-	87.6	1/0.1
Klodzko (Czarna Góra)	97.6	-	89.2	-	10
Klodzko	-	92.4	-	-	2
Konin	87.7	95.0	103.3	-	30/1/30
Koszalin (Gologóra)	107.9	93.8	97.4	-	60
Kraków (Choragwica)	89.4	-	99.4	-	60
Kraków (Krzemionki)	104.9	-	-	97.2	1/0.4
Krosno (Sucha Góra)	88.0	-	92.0	-	120
Krynica (G.Jaworzyna)	106.4	-	-	98.4	1
Lebork (Skórowo Nowe)	100.5	88.2	107.5	-	10
Legnica	-	105.3	-	-	2
Lezajsk (Giedlarowa)	96.8	-	98.9	-	10
Lobez (Toporzyk)	-	-	-	100.6	3

FM (MHz)	PR1	PR2	PR3	PR4	kW
Lódz	107.8	91.4	103.8	107.3	30/2x10/1.5
Lowicz	101.6	-	-	-	10
Lubaczów (Boble)	100.0	88.4	96.0	-	10
Luban (Nowa Karczma)	99.0	-	91.5	-	60/30
Lublin (Piaski)	90.8	-	104.2	-	30/90
Nowy Tomysl (Bolewice)	-	107.7	-	-	10
Olsztyn (Pieczewo)	93.0	93.7	99.1	97.9	30/2/120/0.1
Opole (Chrzelice)	88.3	94.5	90.3	-	60/10/60
Ostroleka (Lawy)	106.7	96.3	98.5	-	10/5/10
Pila (Staszyce)	-	102.5	-	-	10
Plock (Rachocin)	92.2	98.1	96.1	-	60/2.5/60
Poznan (Srem)	92.3	96.4	-	-	120
Przasnysz	105.9	107.1	-	-	10
Przemysl (Tatarska Góra)	87.8	94.1	99.6	-	5/1/5
Przysucha (Kozlowiec)	92.0	104.8	-	-	10
Rabka (G.Lubon Wielki)	93.4	90.4	-	-	5
Radom	-	-	-	104.6	1
Radom (Wacyn)	-	100.3	-	97.5	1
Ryki	105.1	88.7	-	-	10
Rzeszów (Baranówka)	-	105.8	-	91.5	1/0.1
Siedlce (Losice)	88.3	-	90.5	-	30
Slupsk	104.3	-	-	106.8	2.8/5
Suwalki (G.Krzemianucha)	105.5	92.0	96.6	-	20/30/30
Swieradów-Zdrój	-	93.2	-	90.5	10/1
Swinoujscie (Chrobrego)	107.7	-	-	-	10
Szczawnica (G.Prehyba)	88.0	-	94.7	-	10/5
Szczecin (Kolowo)	100.3	-	102.3	-	60
Szczecin (Warszewo)	-	96.3	-	88.4	1
Tarnów (G.Sw. Marcina)	-	-	-	99.9	2.5
Tarnów (Lichwin)	91.1	88.6	-	-	10
Wagrowiec (Golancz)	101.3	-	-	-	3
Walbrzych (G.Chelmiec)	-	87.9	99.8	94.3	5/5/0.5
Walcz (Rusinowo)	101.9	-	90.9	-	30
Warszawa (PKiN)	92.4	104.9	99.1	92.0	0.1/2.5/0.1/0.2
Warszawa (Raszyn)	102.4	-	98.8	-	120
Wisla (G.Skrzyczne)	91.5	-	100.8	-	10
Wloclawek (Szpetal Górny)	-	93.9	-	-	1
Wlodawa (Zolnierzy)	-	102.5	-	-	10
Wloszczowa (Dobromierz)	88.9	-	-	-	5
Wroclaw (G.Sleza)	98.8	-	100.2	-	120
Wroclaw	-	87.7	-	107.5	10/5
Zagan (Wichów)	91.2	104.7	87.8	-	30
Zakopane (G.Gubalówka)	92.8	90.9	98.2	-	10/0.3/10
Zamosc (Feliksówka)	105.7	-	-	95.3	10/1
Zamosc (Tarnawatka)	-	87.6	91.3	-	30
Zielona Góra (Jemiolów)	105.0	89.9	94.1	-	60
Zielona Góra (Ptasia)	-	-	-	104.0	2

NB: Sites with only txs below 1kW not listed.
D.Prgr: PR1 (Jedynka): 24h. – **PR2 (Dwójka):** 24h – **PR3 (Trójka):** 24h. – **PR4 (Czwórka):** 24h. – **External Service:** see Int Radio section.

PR REGIONAL STATIONS (Pub)
D.Prgr: All stations broadcast 24h. **PR R.Bialystok:** ul. Swierkowa 1, 15-328 Bialystok. **E:** radiobia@radio.bialystok.pl. On (MHz) 87.9 (Lomza 0.2kW), 89.4 (Bialowieza 0.1kW), 98.6 (G.Krzemianucha 30kW), 99.4 (Krynice 30kW), 104.1 (Mikstat 0.1kW). – **PR R.Dla Ciebe (RDC):** ul. Mysliwiecka 3/5/7, 00-977 Warszawa. **E:** radio@rdc.pl. On (MHz) 87.6 (Ostrów Mazowiecka 1kW), 89.1 (Wacyn 5kW), 100.8 (Ostroleka 0.25kW), 101.0 (Warszawa PKiN 10kW), 101.9 (Rachocin 60kW), 103.4 (Losice 120kW). – **PR R.Gdansk:** ul. Grunwaldzka 18, 80-006 Gdansk. **E:** poczta@radio.gdansk.pl. On (MHz) 91.1 (Skórowo Nowe 10kW), 102.0 (Slupsk 1kW), 103.7 (Chwaszczyno 120kW), 106.0 (Janków 1kW), 107.0 (Bytów 10kW). – **PR R.Katowice:** ul. Ligonia 29, 40-953 Katowice. **E:** sekretariat@radio.katowice.pl. On (MHz) 89.3 (Zabrze 0.5kW), 97.0 (Racibórz 1kW), 98.4 (Wreczyca 60kW), 101.2 (Bytków 1kW), 102.2 (Kosztowy 60kW), 103.0 (G.Skrzyczne 10kW). – **PR R.Kielce:** ul. Radiowa 4, 25-317 Kielce. **E:** radio@radio.kielce.com.pl. On (MHz) 90.4 (Kielce 0.25kW), 101.4 (Swiety Krzyz 120kW). – **PR R.Koszalin:** ul. Pilsudskiego 43-49, 75-502 Koszalin. **E:** radio@radio.koszalin.pl. On (MHz) 88.1 (Rusinowo 3kW), 89.7 (Toporzyk 0.1kW), 91.0 (Kolobrzeg 0.1kW), 92.5 (Slawoborze 15kW), 95.3 (Slupsk 1kW)*, 98.0 (G.Chelmska 0.1kW), 103.1 (Gologóra 60kW). *) incl. prgrs from Slupsk studio. – **PR R.Kraków:** ul. Slowackiego 22, 30-007 Kraków. **E:** radio@radio-krakow.pl. On (MHz) 87.6 (G. Lubon Wielki 5kW), 90.0 (G.Prehyba 10kW), 97.4 (Gorlice 2kW), 98.8 (Andrychów 1kW), 100.0 (G.Gubalówka 10kW), 101.0 (G. Sw. Marcina 10kW), 101.6 (Choragwica 60kW), 102.1 (G.Jaworzyna 1kW). – **PR R.Lódz:** ul. Narutowicza 130, 90-146 Lódz. **E:** studio@radiolodz.pl. On (MHz) 96.7 (Sieradz 0.5kW), 99.2 (Lódz 30kW), 103.0 (Wieruszów 1kW). – **PR R.Lublin:** ul. Obronców Pokoju 2, 20-030 Lublin. **E:** poczta@radio.lublin.pl. On (MHz) 89.9 (Lublin 0.1kW), 93.1 (Biala Podlaska 5kW), 102.2 (Piaski 90kW), 103.1 (Ryki 10kW), 104.2 (Feliksówka 30kW). – **PR R.Merkury:** ul. Berwinskiego 5, 60-765 Poznan. **E:** office@radio-merkury.pl. On (MHz) 91.1 (Mikstat 10kW), 91.9 (Zólwieniec 30kW), 98.3 (Golancz 10kW), 100.9 (Srem 120kW), 102.4 (Bolewice 3kW), 103.6 (Rusinowo 60kW). Local substation: **PR MC Radio:**

E: redakcja@mcradio.pl. On (MHz) 102.7 (Piatkowo 2kW). – **PR R.Olsztyn:** ul. Radiowa 24, 10-206 Olsztyn. **E:** radio@ro.com.pl. On (MHz) 99.6 (Milki 10kW), 103.2 (Pieczewo 120kW), 103.4 (Jagodnik 0.5kW). – **PR R.Opole** ul. Strzelców Bytomskich 8, 45-084 Opole. **E:** pro_fm@radio.opole. pl. On (MHz) 88.0 (Brzeg 1kW), 89.1 (Olesno 1kW), 92.6 (Paczków 1kW), 94.8 (Glubczyce 1kW), 96.3 (Kluczbork 30kW), 101.2 (Opole 1kW), 103.2 (Chrzelice 60kW), 105.1 (Strzelce Opolskie 1kW), 107.7 (Namyslów 1kW). – **PR R.PiK:** ul. Gdanska 48-50, 85-006 Bydgoszcz. **E:** radio@radiopik.pl. On (MHz) 100.1 (Trzeciewiec 120kW), 100.3 (Wloclawek 1kW), 106.9 (Brodnica 10kW). – **PR R.Rzeszów:** ul. Zamkowa 3, 35-032 Rzeszów. **E:** radiorz@radio.rzeszow. pl. On (MHz) 90.3 (Machów 1kW), 90.5 (Sucha Góra 120kW), 96.4 (Mielec 1kW), 99.2 (G.Jawor 5kW), 102.0 (Tatarska Góra 10kW), 102.9 (Giedlarowa 30kW), 103.7 (Boble 10kW), 106.7 (Magdalenka 2kW). – **PR R.Szczecin:** al. Wojska Polskiego 73, 70-481 Szczecin. **E:** sekretariat@radio.szczecin.pl. On (MHz) 92.0 (Kolowo 60kW), 98.7 (Slawoborze 10kW), 106.3 (Chrobrego 10kW). Substation: **PR R.Szczecin.FM:** E: sekretariat@szczecin.fm. On (MHz) 94.4 (Warszewo 0.5kW). – **PR R.Wroclaw:** ul. Karkonoska 8-10, 53-015 Wroclaw. **E:** sekretariatzarzadu@prw.pl. On (MHz) 89.0 (G.Wysoka 1kW), 95.5 (G.Chelmiec 5kW), 96.0 (Czarna Góra 10kW), 96.7 (Sniezne Kotly 10kW), 98.0 (G.Parkowa 0.1kW), 102.3 (G.Sleza 120kW), 103.6 (Nowa Karczma 60kW). Local substation: **PR R.RAM:** E: ram@prw.pl. On (MHz) 89.8 (Zórawina 6kW). – **PR R.Zachód:** ul. Kukulcza 1, 65-472 Zielona Góra. **E:** radio@zachod.pl. On (MHz) 103.0 (Jemiolów 1kW), 106.0 (Wichów 30kW). Local substations of R.Zachód: **PR R.Zielona Góra: E:** rzg@rzg.pl. On (MHz) 97.1 (Zielona Góra 1kW); **PR RMG FM:** ul. Warszawska 131, 66-400 Gorzów Wlkp. **E:** rmg@zachod.pl. On (MHz) 95.9 (Janice 1kW).

OTHER STATIONS

MW	kHz	kW	Location	Station
9)	531	0.8	Zywiec	Twoje R. Zywiec
9A)	531	0.8	Wlodawa	Twoje R. Wlodawa*
9)	963	0.5	Brzesko	Twoje R. Brzesko
9B)	963	0.1	Lipsko	Twoje R. Lipsko*
9C)	963	0.8	Lubaczów	Twoje R. Lubaczów*
9D)	963	0.1	Lubliniec	Twoje R. Lubliniec*
9E)	1062	0.8	Cmolas	Twoje R. Cmolas*
9F)	1062	0.5	Jaroslaw	Twoje R. Jaroslaw*
9G)	1062	0.8	Pulawy	Twoje R. Pulawy*
9)	1062	0.8	Skarzysko	Twoje R. Skarzysko
9)	1332	0.8	Pinczów	Twoje R. Pinczów
9H)	1404	0.8	Chojnice	Twoje R. Chojnice Plus*
9)	1485	0.8	Bilgoraj	Twoje R. Bilgoraj
9)	1485	0.8	Bielsko-Biala	Twoje R. Bielsko-Biala
9)	1485	0.8	Gorlice	Twoje R. Gorlice
9)	1485	0.8	Kielce	Twoje R. Kielce
9)	1485	1	Przemysl	Twoje R. Przemysl
9)	1485	0.8	Walcz	Twoje R. Walcz
9I)	1584	0.1	Andrychów	Twoje R. Andrychów*
9)	1584	0.5	Busko-Zdrój	Twoje R. Busko-Zdrój
9)	1584	0.8	Krosno	Twoje R. Krosno
9)	1584	0.8	Nowy Sacz	Twoje R. Nowy Sacz
9)	1584	0.8	Tarnobrzeg	Twoje R. Tarnobrzeg
9J)	1584	0.1	Ozorków	Twoje R. Ozorków*
9)	1584	0.8	Slupsk	Twoje R. Slupsk
9)	1584	0.8	Chelm	Twoje R. Chelm
9)	1602	0.8	Cieszyn	Twoje R. Cieszyn
9K)	1602	0.1	Ilza	Twoje R. Ilza*
9)	1602	0.8	Sanok	Twoje R. Sanok
9)	1602	0.8	Kraków	Twoje R. Kraków

*) incl. own prgrs. All stns relay the Twoje R. Gminne feed (alt. a regional stn of Polskie Radio)

FM	MHz	kW	Location	Station
3)	87.7	3	Miedzyzdroje	R. Maryja
3)	87.8	1	Biala Podlaska	R. Maryja
1A)	87.8	1	Kraków	R. RMF Classic
17)	87.9	25	Lublin	R. eR
3)	87.9	10	Lódz	R. Maryja
2)	88.0	2	Wagrowiec	R. ZET
1)	88.2	120	Kielce	R. RMF FM
3)	88.2	1	Ostrów Wlkp.	R. Maryja
2)	88.3	60	Zielona Góra	R. ZET
3)	88.3	1	Kutno	R. Maryja
3)	88.4	10	Bielsko-Biala	R. Maryja
8)	88.4	5	Poznan	R. Zlote Przeboje 88.4FM
3)	88.5	10	Slupsk	R. Maryja
61)	88.6	1	Skierniewice	R. RSC
2)	88.7	1	Koszalin	R. ZET
3)	88.7	1	Wagrowiec	R. Maryja
5A)	88.8	1	Lomza	R. Eska Lomza
16)	88.8	2	Lodz	Studenckie Radio Zak
3)	88.9	2	Gdansk	R. Maryja
3)	88.9	120	Wroclaw	R. Maryja
4)	88.9	15	Szczecin	R. Plus Szczecin
3)	89.0	1	Warszawa	R. Maryja
39)	89.2	1	Bialystok	R. Jard

FM	MHz	kW	Location	Station
1)	89.3	60	Koszalin	R. RMF FM
1)	89.3	30	Lublin	R. RMF FM
2)	89.4	60	Luban	R. ZET
3)	89.4	2	Stargard Szczec.	R. Maryja
4)	89.5	10	Gniezno	R. Plus Gniezno
5A)	89.5	1	Sanok	R. Eska Rzeszów
4)	89.6	1	Opole	R. Plus Opole
8)	89.8	1	Szczecin	R. Zlote Przeboje 89.8FM
3)	89.8	1	Mielec	R. Maryja
30)	89.8	1	Poznan	R. Emaus
1A)	89.9	1	Czestochowa	R. RMF Classic
19)	90.0	1	Rybnik	R. 90
7)	90.1	2	Lodz	R. WAWA Lodz
13)	90.1	10	Zamosc	Katolickie R. Zamosc
47)	90.1	1	Koscierzyna	R. Kaszëbë
63)	90.2	1	Bielsko-Biala	R. Aniol Beskidów
43)	90.2	1	Kolobrzeg	R. Kolobrzeg
3)	90.2	2	Kamiensk	R. Maryja
3)	90.3	1	Zielona Góra	R. Maryja
8)	90.4	1	Wroclaw	R. Zlote Przeboje 90.4FM
3)	90.6	5	Kraków	R. Maryja
4)	90.7	5	Radom	R. Plus Radom
5A)	90.7	2	Gdynia	R. Eska Trójmiasto
4)	90.7	2	Gryfice	R. Plus Gryfice
35)	90.8	1	Inowroclaw	R. GRA
14)	90.9	5	Walbrzych	Muzyczne R.
1)	91.0	120	Warszawa	R. RMF FM
8)	91.2	1	Katowice	R. Zlote Przeboje 91.2FM
1)	91.3	10	Lobez	R. RMF FM
38)	91.4	1	Pelplin	R. Glos
1B)	91.5	1	Slupsk	R. RMF Maxxx Pomorze
1)	91.5	15	Ostroleka	R. RMF FM
2)	91.6	10	Ryki	R. ZET
4)	91.7	1	Zielona Góra	R. Plus Zielona Góra
2)	91.8	10	Swinoujscie	R. ZET
3)	91.8	1	Ciechanów	R. Maryja
1)	91.9	30	Siedlce	R. RMF FM
2A)	92.0	1	Gdansk	R. Chilli ZET
59)	92.0	2.5	Wroclaw	R. Rodzina
8)	92.1	1	Bydgoszcz	R. Zlote Przeboje 92.1FM
2)	92.1	1	Wlodawa	R. ZET
2)	92.2	10	Opole	R. ZET
55)	92.3	1	Laziska Gorna	R. Express FM
47)	92.3	2	Gdansk	R. Kaszëbë
8)	92.5	1	Kraków	R. Zlote Przeboje 92.5FM
7)	92.6	1	Krosno	R. WAWA Rzeszów
2)	92.6	120	Lodz	R. ZET
42)	92.6	1	Sepólno Kraj.	R. Weekend
3)	92.7	10	Lebork	R. Maryja
8)	92.8	1	Opole	R. Zlote Przeboje 92.8FM
49)	92.9	1	Tomaszów Maz.	R. FaMa Tomaszów
2)	92.9	10	Gryfice	R. ZET
1)	92.9	10	Wroclaw	R. RMF FM
1)	93.0	60	Katowice	R. RMF FM
5A)	93.0	10	Poznan	R. Eska Poznan
3)	93.1	1	Krynica	R. Maryja
56)	93.2	1	Goleniów	R. Pólnoc
1)	93.3	120	Bydgoszcz	R. RMF FM
33)	93.3	1	Warszawa	R. VOX FM
25)	93.4	2	Gliwice	R. CCM
1)	93.5	10	Lódz	R. RMF FM
2)	93.6	120	Wroclaw	R. ZET
9)	93.7	1	Kraków	R. Planeta FM
1)	93.8	60	Luban	R. RMF FM
5A)	93.8	10	Gorzów Wlkp.	R. Eska Gorzów
3)	93.8	1	Kutno	R. Victoria
54)	93.9	1	Kedzierzyn-Kozle	R. Park FM
4)	94.0	1	Konskie	R. Plus Radom
24)	94.0	1	Warszawa	Antyradio 94 FM
27)	94.1	1	Elblag	R. El
12)	94.1	1	Jaslo	VIA - Kat. R. Rzeszów
38)	94.2	1	Kartuszy	R. Glos
1)	94.3	60	Plock	R. RMF FM
3)	94.4	1	Racibórz	R. Maryja
8)	94.4	1	Zary	R. Zlote Przeboje 98.1FM
3)	94.4	1	Tarnobrzeg	R. Maryja
5)	94.4	5	Bydgoszcz	R. Eska Bydgoszcz
23)	94.5	1	Grodzisk Maz.	R. Bogoria
3)	94.5	1	Ustrzyki Dolne	R. Maryja
1)	94.6	120	Poznan	R. RMF FM
5A)	94.6	1.5	Gdansk	R. Eska Trójmiasto
36)	94.7	10	Czestochowa	R. Fiat
47)	94.7	1	Rawa Maz.	R. Victoria
59)	94.8	2	Strzelin	R. Rodzina
1)	94.8	30	Zagan	R. RMF FM

FM	MHz	kW	Location	Station
4)	94.9	1	Jelenia Góra	R. Plus Legnica
31)	94.9	1	Sochaczew	R. Fama
2)	95.0	20	Lezajsk	R. ZET
3)	95.0	3	Szczecinek	R. Maryja
1)	95.1	1.6	Suwalki	R. RMF FM
64)	95.1	1	Zabrze	R. Planeta 95.1
2)	95.2	60	Szczecin	R. ZET
6)	95.2	1	Gdynia	R. TOK FM
3)	95.2	1	Swieradów-Zdrój	R. Maryja
3)	95.2	1	Sieradz	R. Maryja
1)	95.3	60	Olsztyn	R. RMF FM
1)	95.3	60	Opole	R. RMF FM
1)	95.4	10	Tarnów	R. RMF FM
3)	95.4	5	Skierniewice	R. Maryja
3)	95.4	1	Gniezno	R. Maryja
5B)	95.5	1	Katowice	R. Eska Rock
2)	95.6	120	Bydgoszcz	R. ZET
8)	95.6	1	Lublin	R. Zlote Przeboje 95.6FM
2)	95.7	10	Wisla	R. ZET
8)	95.7	1	Rzeszów	R. Zlote Przeboje 95.7FM
5B)	95.7	1	Szczecin	R. Eska Rock
3)	95.8	1	Hrubieszów	R. Maryja
1B)	95.8	1	Warszawa	R. RMF MAXXX Warszawa
5A)	95.9	1	Koszalin	R. Eska Koszalin
57)	95.9	1	Olsztyn	R. UWM FM
53)	96.0	6	Lódz	R. Parada
1)	96.0	60	Kraków	R. RMF FM
1B)	96.0	1	Olesnica	R. RMF MAXXX Olesnica
1)	96.1	1	Gorzów	R. RMF FM
1)	96.1	1	Legnica	R. RMF FM
4)	96.2	2	Zabrze	R. Plus Slask
1)	96.4	15	Bialogard	R. RMF FM
1B)	96.4	1	Gdansk	R. RMF MAXXX Trójmiasto
4)	96.5	10	Warszawa	R. Plus Warszawa
3)	96.5	10	Zamosc	R. Maryja
1)	96.6	30	Walcz	R. RMF FM
2)	96.6	10	Lebork	R. ZET
7)	96.7	3	Torun	R. WAWA Torun
1B)	96.7	2	Kraków	R. RMF MAXXX Kraków
52)	96.7	1	Skierniewice	R. Victoria
3)	96.7	7.5	Ilawa	R. Maryja
5A)	96.9	1	Szczecin	R. Eska Szczecin
2)	97.0	30	Poznan	R. ZET
3)	97.0	1	Lublin	R. Maryja
3)	97.0	2	Ciechanowiec	R. Maryja
1)	97.1	12	Wloszczowa	R. RMF FM
2)	97.2	1	Walbrzych	R. ZET
2)	97.3	60	Plock	R. ZET
5A)	97.3	1	Zamosc	R. Eska Zamosc
6)	97.4	1	Katowice	R. TOK FM
2)	97.5	30	Zagan	R. ZET
5A)	97.7	1	Kraków	R. Eska Kraków
2)	97.8	2.5	Szczawnica	R. ZET
2)	97.9	60	Walcz	R. ZET
1)	98.0	10	Kalisz	R. RMF FM
22)	98.1	60	Bialystok	R. Racja
5A)	98.1	2	Tarnów	R. Eska Tarnów
52)	98.1	1	Mszczonów	R. Victoria
34)	98.2	1	Przemysl	R. Fara
1)	98.4	120	Gdansk	R. RMF FM
28)	98.5	1	Leszno	R. Elka
63)	98.5	1	Zywiec	R. Aniol Beskidów
51)	98.2	2	Lodz	R. Niepokalanów
2)	98.6	1	Nysa	R. ZET
2)	98.7	10	Ilawa	R. ZET
3)	98.8	10	Gorzów Wlkp.	R. Maryja
1)	98.9	30	Konin	R. RMF FM
47)	98.9	2	Reda	R. Kaszëbë
5A)	99.0	1	Szczecinek	R. Eska Szczecinek
22)	99.2	2	Biala Podlaska	R. Racja
42)	99.3	1	Chojnice	R. Weekend
64)	99.4	11	Poznan	R. Planeta Poznan
3)	99.5	1	Lipiany	R. Maryja
1)	99.5	1	Kluczbork	R. RMF FM
4)	99.5	1	Slupsk	R. Plus Koszalin
9)	99.6	1	Konin	R. Planeta Konin
1B)	99.7	1	Koszalin	R. RMF MAXXX Pomorze
21)	99.9	1	Zywiec	R. Bielsko
3)	100.0	5	Zielona Góra	R. Maryja
8)	100.1	1	Warszawa	R. Zlote Przeboje 100.1FM
1)	100.1	120	Krosno	R. RMF FM
66)	100.2	2	Chorzów	R. Fest
3)	100.2	1	Gizycko	R. Maryja
1)	100.2	120	Bialystok	R. RMF FM
48)	100.3	1	Racibórz	R. Vanessa
3)	100.3	1	Bogatynia	R. Maryja
3)	100.4	1	Jelenia Góra	R. Maryja
3)	100.4	10	Ostróda	R. Maryja
3)	100.4	10	Ostrów Maz.	R. Maryja
4)	100.4	5	Lódz	R. Plus Lodz
59)	100.4	2	Nowa Ruda	R. Rodzina
3)	100.4	1	Nysa	R. Maryja
41)	100.5	1	Kraków	Radiofonia
3)	100.6	10	Torun	R. Maryja
40)	100.6	60	Czestochowa	R. Jasna Góra
3)	100.6	10	Krosno	R. Maryja
3)	100.6	10	Glogów	R. Maryja
3)	100.6	5	Parczew	R. Maryja
4)	100.7	10	Gorzów Wlkp.	R. Plus Gorzów
3)	100.7	5	Rabka	R. Maryja
2)	100.7	2	Zamosc	R. ZET
1)	100.8	10	Jelenia Góra	R. RMF FM
32)	100.8	1	Kielce	R. FaMa Kielce
2)	100.9	10	Slupsk	R. RMF FM
3)	100.9	1	Wloclawek	R. Maryja
2A)	101.0	1	Kraków	R. Chilli ZET
5A)	101.1	2	Kalisz	R. Eska Kalisz/Ostrów
1B)	101.1	2	Radomsko	R. RMF MAXXX Kielce/Rad.
1B)	101.1	5	Walbrzych	R. RMF MAXXX Walbrych
1)	101.1	30	Solina	R. RMF FM
3)	101.1	10	Zlotów	R. Maryja
1B)	101.2	1	Bytów	R. RMF MAXXX Pomorze
58)	101.2	1	Nowy Sacz	R. RDN Malopolska
1)	101.2	10	Swinoujscie	R. RMF FM
3)	101.2	10	Zagan	R. Maryja
8)	101.3	1	Pabianice	R. Zlote Przeboje 101.3FM
24)	101.3	1	Kraków	Antyradio 101.3 FM
3)	101.4	10	Lomza	R. Maryja
3)	101.4	10	Czersk	R. Maryja
2)	101.4	30	Suwalki	R. ZET
5B)	101.5	1	Wroclaw	R. Eska Rock
3)	101.6	10	Pisz	R. Maryja
1)	101.6	10	Klodzko	R. RMF FM
3)	101.6	1	Szczecin	R. Maryja
2A)	101.6	1	Poznan	R. Chilli ZET
62)	101.7	1	Kepno	R. Sud
46)	101.7	120	Siedlce	Katolickie R. Podlasie
4)	101.7	120	Gdansk	R. Plus Gdansk
1)	101.8	60	Lezajsk	R. RMF FM
1)	101.8	10	Zakopane	R. RMF FM
3)	102.0	10	Bielsk Podl.	R. Maryja
1)	102.0	10	Gizycko	R. RMF FM
5A)	102.0	1	Leszno	R. Eska Leszno
3)	102.3	10	Lubaczów	R. Maryja
3)	102.4	1	Kartuzy	R. Maryja
1B)	102.6	1	Czestochowa	R. RMF MAXXX Czestochowa
4)	102.6	20	Polkowice	R. Plus Legnica
3)	102.6	10	Tarnów	R. Maryja
2)	102.6	1	Bydgoszcz	R. Plus Bydgoszcz
18)	102.6	1	Elk	R. 5 Elk
4)	102.6	1	Koszalin	R. Plus Koszalin
51)	102.7	1	Skierniewice	R. Niepokalanów
4)	102.7	5	Rabka	R. Plus Podhale
37)	102.7	1	Bialystok	R. Orthodoxia
2)	102.8	3	Ostroleka	R. ZET
3)	102.8	1	Chelm	R. Maryja
3)	102.8	5	Kluczbork	R. Maryja
2)	102.8	5	Katowice	R. ZET
4)	102.8	1	Swieradów-Zdrój	R. Plus Legnica
1)	102.9	5	Walbrzych	R. RMF FM
9)	102.9	1	Slupca	R. Planeta Slupca
6)	102.9	1	Kraków	R. TOK FM
3)	102.9	10	Gryfice	R. Maryja
1B)	102.9	1	Lebork	R. RMF Maxx Pomorze
2)	102.9	12	Przysucha	R. ZET
8)	103.0	2	Gdansk	R. Zlote Przeboje 103.0i99.2
44)	103.0	1	Warszawa	R. Kolor 103 FM
2)	103.1	30	Solina	R. ZET
60)	103.1	1	Kalisz	R. Rodzina / R. Maryja
1)	103.2	10	Szczawnica	R. RMF FM
68)	103.3	1	Bialystok	R. 103i3 FM
2)	103.4	60	Czestochowa	R. ZET
3)	103.4	5	Lebork	R. RMF FM
1)	103.4	10	Przemysl	R. RMF FM
3)	103.5	1	Trzcinsko-Zdrój	R. Maryja
52)	103.5	5	Lowicz	R. Victoria
10)	103.5	1	Bydgoszcz	R. Roxy Bydgoszcz
58)	103.6	30	Tarnów	R. RDN Malopolska
50)	103.6	10	Lomza	R. Nadzieja
5A)	103.6	1	Lublin	R. Eska Lublin

FM	MHz	kW	Location	Station
3)	103.7	3	Katowice	R. Maryja
67)	103.7	1	Wroclaw	R. PiN
2)	103.8	10	Klodzko	R. ZET
10)	103.8	1	Kraków	R. Roxy Kraków
12)	103.8	10	Rzeszów	VIA - Kat. R. Rzeszów
39)	103.9	2	Bialystok	R. Jard 2
45)	103.9	1	Ciechanów	Kat. R. Ciechanów
2)	104.0	10	Swiecie	R. Maryja
2)	104.0	10	Gizycko	R. ZET
2)	104.1	60	Kraków	R. ZET
1B)	104.1	5	Pila	R. RMF MAXXX Pila
3)	104.2	10	Elblag	R. Maryja
3)	104.2	10	Bialogard	R. ZET
2)	104.2	10	Jelenia Góra	R. ZET
45)	104.3	1	Plock	Kat. R. Plock
4)	104.3	1	Lipiany	R. Plus Lipiany
3)	104.4	1	Stalowa Wola	R. Maryja
3)	104.4	10	Kalisz	R. ZET
3)	104.5	10	Wlodawa	R. Maryja
3)	104.5	5	Wielen	R. Maryja
34)	104.5	1	Krosno	R. Fara
2)	104.6	1	Hrubieszów	R. ZET
1B)	104.6	1	Krynica	R. RMF MAXXX Nowy Sacz
3)	104.6	2	Opole	R. Maryja
5A)	104.6	1	Torun	R. Eska Torun
3)	104.7	120	Bialystok	R. Maryja
3)	104.7	10	Lobez	R. Maryja
15)	104.7	2	Sieradz	Nasze R.
1)	104.7	3	Rabka	R. RMF FM
5A)	104.9	60	Wroclaw	R. Eska Wroclaw
5A)	104.9	10	Krosno	R. Eska Rzeszów
11)	104.9	1	Chelm	Bon Ton R.
1)	104.9	1	Koszalin	R. RMF FM
24)	105.0	1	Bielsko-Biala	Antyradio 106.4 FM
2)	105.0	120	Gdansk	R. ZET
3)	105.1	1	Przemysl	R. Maryja
3)	105.1	30	Konin	R. Maryja
3)	105.1	10	Elk	R. Maryja
20)	105.2	1	Zakopane	R. Alex
2)	105.2	5	Wielun	R. Maryja
2)	105.3	60	Kielce	R. ZET
2)	105.3	30	Koszalin	R. ZET
3)	105.3	1	Plonsk	R. Maryja
2)	105.4	30	Siedlce	R. ZET
10)	105.4	1	Poznan	R. Roxy Poznan
7)	105.5	1	Wroclaw	R. WAWA Wroclaw
3)	105.6	1	Kalisz	R. Maryja
5B)	105.6	1	Gdynia	R. Eska Rock
5A)	105.6	1	Pila	R. Eska Pila
5A)	105.6	3.2	Warszawa	R. Eska Warszawa
2)	105.7	20	Olsztyn	R. ZET
14)	105.8	5	Jelenia Góra	Muzyczne R.
1)	105.9	60	Czestochowa	R. RMF FM
46)	106.0	1	Zelechów	Katolickie R. Podlasie
10)	106.1	10	Wroclaw	R. Roxy Wroclaw
4)	106.1	10	Kraków	R. Plus Kraków
35)	106.1	5	Bydgoszcz	R. Gra
3)	106.2	10	Lidzbark Warm.	R. Maryja
1)	106.2	1	Warszawa	R. Warszawa
8)	106.2	1	Jelenia Góra	R. Zlote Przeboje 106.2FM
30)	106.2	1.5	Poznan	R. Emaus
3)	106.3	60	Plock	R. Maryja
3)	106.3	5	Klodzko	R. Maryja
3)	106.3	10	Zakopane	R. ZET
3)	106.3	20	Lezajsk	R. Maryja
1)	106.4	60	Zielona Góra	R. RMF FM
26)	106.4	10	Kalisz	R. Centrum Kalisz
24)	106.4	1	Zabrze	Antyradio 106,4 FM
1)	106.5	10	Elk	R. RMF FM
1B)	106.5	5	Kielce	R. RMF Maxx Kielce/Rad.
5A)	106.5	1	Zlocieniec	R. Eska Szczecinek
10)	106.6	1	Opole	R. Roxy Opole
21)	106.7	1	Bielsko-Biala	R. Bielsko
1B)	106.7	3	Gdynia	R. RMF Maxx Trójmiasto
1)	106.7	60	Szczecin	R. RMF FM
14)	106.7	5	Swieradów-Zd.	Muzyczne R.
3)	106.8	120	Poznan	R. Maryja
5A)	106.8	1	Bochnia	R. Eska Malopolska
5A)	106.9	10	Radom	R. Eska Radom
3)	107.0	5	Czestochowa	R. Maryja
64)	107.0	1	Gizycko	R. Planeta Gizycko
5B)	107.0	1.6	Kraków	R. Eska Rock
2)	107.0	120	Lublin	R. ZET
2)	107.1	30	Konin	R. ZET
3)	107.2	120	Kielce	R. Maryja
59)	107.2	2	Bystrzyca Klodzka	R. Rodzina
2)	107.3	120	Bialystok	R. ZET
1)	107.4	10	Ilawa	R. RMF FM
2)	107.4	30	Krosno	R. ZET
3)	107.4	2.5	Walbrzych	R. Maryja
5B)	107.4	1	Poznan	R. Eska Rock Poznan
3)	107.4	1	Koszalin	R. Maryja
2)	107.5	10	Warszawa	R. ZET
29)	107.6	60	Katowice	R. eM
3)	107.6	15	Zamosc	R. RMF FM
3)	107.7	10	Siedlce	R. Maryja
3)	107.7	20	Olsztyn	R. Maryja
2)	107.8	10	Tarnów	R. ZET
4)	107.9	1	Kielce	R. Plus Kielce
2)	107.9	20	Suwalki	R. Maryja
3)	107.9	10	Ryki	R. Maryja
4)	107.9	10	Opole	R. Plus Opole
2)	107.9	10	Przemysl	R. ZET

NB: Txs below 1kW not listed.

Addresses & other information:
1,1A,1B) al. Waszyngtona 1, 30-204 Kraków. **E:** redakja@rmf.fm – **2,2A)** ul. Zurawia 8, 00-503 Warszawa. **E:** radiozet@radiozet.pl – **3)** ul. Zwirki i Wigury 80, 87-100 Torun. **E:** radio@radiomaryja.pl – **4)** ul. Zarawia 8, 00-503 Warszawa (°). – **5A,B)** ul. Senatorska 13/15, 00-075 Warszawa. **E:** skrzynka@radioeska.com.pl (°). – **6)** ul. Czerska 14, 00-732 Warszawa. **E:** tokfm@tokfm.com.pl – **7)** ul. Senatorska 12, 00-082 Warszawa. **E:** radio@ wawa.com.pl (°) – **8)** ul. Czerska 14, 00-732 Warszawa. – **9)** ul. Fatimska 13A, 31-831 Kraków. Affiliates with own prgrs: **9A)** al. Pilsudskiego 10, 22-200 Wlodawa. **E:** tr_wlodawa@wp.pl. 0600-0800 (MF), 0900-1000 (SS), 1600-1700 (SS), 1500-1700 (MF). **9B)** ul. Ilzecka 6a, 27-300 Lipsko. E: lipsko@radiogminne.com.pl. 0600-0800, 1600-1800. **9C)** ul. M. Konopnickiej 9, 37-600 Lubaczów. **E:** radio.lubaczow@gmail.com. 0800-0900 (MF), 0900-1000 (Sat), 0900-1100 (Sun), 1600-1700 (MF), 1900-2000 (Wed). **9D)** 42-700 Lubliniec. **E:** redakcja@radiolubliniec.pl 0600-0700 (MF), 1600-1700 (MF). **9E)** ul. Szkolna 2, 36-105 Cmolas. **E:** radiocmolas@o2.pl. 0600-0800 (W), 1000-1400 (Sun), 1600-1800 (W). **9F)** ul. Czarnieckiego 16, 37-500 Jaroslaw. 1500-1600 (MF). **9G)** ul. Mickiewicza 2a, 24-100 Pulawy. **E:** silver7@neostrada.pl. 0600-0700 (MF), 1700-1800 (W), 1800-2100. **9H)** ul. Huberta Wagnera 1, 89-606 Chojnice. E: radio@chojnice24.pl. 0600-0800, 1600-1800. **9I)** ul. Krakowska 74, 34-120 Andrychów. **E:** radio@ um.andrychow.pl 0700-0800, 1100-1200, 1600-1700, 1700-1800 (Wed), 1900-2000 (Thu), 2000-2100. **9J)** 95-035 Ozorków. 0805-0900, 1600-1700. **9K)** 97-100 Ilza. **E:** biurowanda@pro.onet.pl. 0600-0700, 1500-1600. – **10)** ul. Czerska 14, 00-732 Warszawa. – **11)** ul. Wojslawicka 7, 22-100 Chelm. – **12)** ul. Zamkowa 4, 35-032 Rzeszów. – **13)** ul. Hetmana J. Zamoyskiego 1, 22-400 Zamosc. – **14)** ul. Ks. K. Wyszynskiego 45, 58-500 Jelenia Góra. – **15)** ul. Rynek 14, 98-200 Sieradz. – **16)** III Dom Studenta, al. Politechniki 7, 93-590 Lódz. – **17)** ul. Jana Pawla II 11, 20-535 Lublin. – **18)** ul. Bulwarowa 5, 16-400 Suwalki. – **19)** Os. Dabrówki 1b, 44-286 Wodzislaw Sl. – **20)** ul. Smrekowa 26A, 34-500 Zakopane. – **21)** ul. Olszówka 62, 43-300 Bielsko-Biala. – **22)** ul. Ciepla 1/7, 15-472 Bialystok. – **23)** ul. Kilinskiego 14, 05-825 Grodzisk Mazowiecki. – **24)** al. Komisji Edukacji Narodowej 93, 02-777 Warszawa. **25)** ul. Jana Pawla II 2, 44-100 Gliwice. – **26)** ul. Lazienna 6, 62-800 Kalisz. **27)** ul. 1-go Maja 2, 82-300 Elblag. – **28)** ul. Spóldzielcza 6, 64-100 Leszno. **29)** ul. Jordana 39, 40-953 Katowice. – **30)** ul. Zielona 2, 61-851 Poznan. – **31)** ul. Narutowicza 1/1, 96-500 Sochaczew. – **32)** ul. Piotrkowska 12/522, 25-510 Kielce. – **33)** Senatorska 13/15, 00-075 Warszawa. – **34)** ul. Katedralny 4, 37-700 Przemysl. – **35)** ul. Chrobrego 75, 88-100 Inowroclaw. **36)** al. Najswietszej Marii Panny 54, 42-200 Czestochowa. – **37)** ul. Antoniuk Fabryczny 13, 15-762 Bialystok. – **38)** ul. Biskupa Dominika 11, 83-130 Pelplin. – **39)** ul. Rzemieslnicza 4A, 15-703 Bialystok. – **40)** ul. O. Augustyna Kordeckiego 2, 42-225 Czestochowa. – **41)** ul. Rostafinskiego 8, 30-072 Kraków. – **42)** ul. Jana Pawla II 1B, 89-804 Chojnice. **43)** ul. Janusza Korczaka 2, 78-100 Kolobrzeg. – **44)** ul. Narbutta 41/43, 02-536 Warszawa. – **45)** ul. Tumska 3, 09-400 Plock. – **46)** ul. Pilsudskiego 62, 08-110 Siedlce. – **47)** al. Zeromskiego 32, 84-120 Wladyslawowo. – **48)** ul. Batorego 5, 47-400 Raciborz. – **49)** ul. Smugowa 1/11, 97-200 Tomaszów Mazowiecki. – **50)** ul. Sadowa 3, 18-400 Lomza. – **51)** ul. Zakroczymska 1, 00-225 Warszawa. – **52)** ul. Seminaryjna 6A, 99-400 Lowicz. – **53)** ul. Pilsudskiego 141, 92-318 Lódz. – **54)** ul. Piastowska 1, 47-200 Kedzierzyn-Kozle. – **55)** ul. Pilsudskiego 12, 43-100 Tychy. – **56)** ul. Mieszka I-go 30, 75-132 Koszalin. – **57)** ul. Kanafojskiego 1/14, 10-724 Olsztyn. – **58)** ul. Bema 14, 33-100 Tarnów. – **59)** ul. Katedralna 13, 50-328 Wroclaw. – **60)** ul. Widok 80/82, 62-810 Kalisz. – **61)** ul. Wita Stwosza 2/4, 96-100 Skierniewice. – **62)** ul. Jankowy 55, 63-600 Kepno. – **63)** ul. Sw. Jana Chrzciciela 14, 43-346 Bielsko-Biala. – **64)** ul. Zurawia 8, 00-503 Warszawa. **65)** ul. Floriańska 3, 03-707 Warszawa. – **66)** ul. Jana Pawla II 2, 44-100 Gliwice. – **67)** plac Inwalidów 10, 01-552 Warszawa. – **68)** ul Ks. A.Abramowicza 1A, 15-872 Bialystok. (°) Addresses of local outlets not listed

PORTUGAL

L.T: UTC (27 Mar-30 Oct: +1h) — **Pop:** 10.7 million — **Pr.L:** Portuguese — **E.C:** 50Hz, 220V — **ITU:** POR

ANACOM – Autoridade Nacional de Comunicações.
✉ HQ: Avenida José Malhoa, 12, 1099-017 Lisboa ☎+351 21 721 10 00 🖷 +351 21 721 10 01 **E:** info@anacom.pt
W: www.anacom.pt
Gov body responsible for licensing & monitoring radio & TV txs

APR – Associação Portuguesa de Radiodifusão (Assoc. of Portuguese Broadcasters)
✉ Av. das Descobertas, 17, 1400-091 Lisboa ☎+35 121 301 69 99, 301 54 53/9 🖷 +351 21 301 65 36 **E:** apr@apradiofusao.pt
W: www.apradiofusao.pt

RTP-Rádio e Televisão de Portugal, SGPS (Pub)
✉ Av. Marechal Gomes da Costa, 37, 1849-030 Lisboa ☎ +351 21 382 00 00 (or 00 98) **W:** www.rtp.pt **E:** (check www.rtp.pt)
LP: Chmn: Manuel Guilherme de Oliveira da Costa
Antena 1/Antena 2/Antena 3/RDP Africa/RDP Açores/RDP Madeira/RDP Internacional:
☎ +351 21 382 00 00. Antena 1 🖷 +351-21-382 00 70 🖷+351-21-382 00 05, Antena 2 ☎+351-21-382 02 82 🖷 +351-21-382 01 99, Antena 3 +351-21-382 02 02 🖷 +351-21-382 00 17, RDP África 🖷 +351-21-382 02 12 🖷 +351-21 382 00 81. News Dept ☎+351-21 382 00 02 🖷 +351-21-382 01 83.
LP: Chmn. bd of Dirs: Guilherme Costa, TD: Francisco Mascarenhas, Dir. of Antena 1/2/3:Rui Fernandes Pêgo, Dir. of RDP África: Jorge Oliveira Gonçalves, Reg. Dir (RDP Norte, Porto): José Alberto Lemos, Reg. Dir (RDP Centro, Coimbra): José Manuel Portugal, Reg. Dir (RDP Sul, Faro): Feliciano Estêvão, Reg. Dir. Açores & Madeira: see respective country entries, Dir. RDPI: Jorge Oliveira Gonçalves. Technical info. & support may be obtained from: gabinete.tecnologias@rtp.pt
Ann: "Antena 1, a rádio que liga Portugal", "Antena 2, a rádio clássica"

MW: Antena 1	kHz	kW	Antena 1	kHz	kW
Chaves	630	2	Castelo Branco	720	10
Miranda do Douro	630	10	Elvas	720	10
Montemor-o Velho	630	10	Faro	720	10
Bragança	666	10	Guarda	720	10
Castanheira do Ribatejo*	666	10	Miramar (Porto)	720	10
Covilhã	666	10	Mirandela	720	10
Valença	666	10	Lamego	756	2
Vila Real	666	10	Portalegre	1287	2
Viseu	666	10			

*) north of Lisbon; also known as "CEN" (Centro Emissor Nacional)

FM (MHz)	Ant. 1	Ant. 2	Ant. 3	kW
Alcoutim S	88.9	91.5	101.9	0.2
Arestal (Aveiro) C	106.7	95.2		0.5
Bornes N	92.8	91.1	102.1	10
Braga (Sameiro) N	91.3	88.0	103.0	10/2/10
Bragança N	96.4	98.2	104.2	10
Castelo Branco C	89.9	94.9	104.3	0.5
Coimbra C (a)			94.9	4
Elvas (VªBoim)	103.8	93.2	101.6	5.4
Faro (S.Miguel) (b) S	97.6	93.4	100.7	10
Gardunha C	96.4	93.9	101.3	10
Grândola C	99.2	90.6	103.6	10
Gravia C	104.5	106.8	107.9	0.1
Guarda C	94.7	88.4	100.6	6.4
Janas (Sintra)	96.9	96.0	103.8	0.2
Leiria C	98.7	104.2	106.4	1
Lisboa (Banática)	99.4	88.9		1
Lisboa (Monsanto) (c)	95.7	94.4	100.3	36/36/32
Lousã C	87.9	89.3	102.2	34/34/39
Manteigas B	104.8	91.6	100.3	0.5
Marão N	95.2	99.8	101.5	9
Marofa C	97.2	93.4	104.6	20/20/10
Mendro	87.7	91.1	102.4	20/20/44
Mértola C	90.9	92.2	100.1	0.4
Minhéu N	94.9	88.0	104.7	10
Miranda do Douro N	90.3	95.7	98.9	0.05
Moledo N	102.9	88.0	92.3	0.5
Monchique (Fóia) S	88.9	91.5	101.9	25
Montargil	93.6	99.6	105.0	3
Montejunto	98.3	88.0	105.2	10
Muro N	98.3	94.6	102.0	10
Paredes de Coura N	102.9		92.3	0.1
Portalegre	97.9	92.9	102.8	10
Porto (Mte. da Virgem)N	96.7	92.5	100.4	44/44/50
Santarém	98.8			0.4
S. Domingos N	87.9	89.3	103.7	0.2
Serra de Ossa	88.4	95.0	102.1	2/0.5/0.5

FM (MHz)	Ant. 1	Ant. 2	Ant. 3	kW
Tróia (Setúbal)	106.7	99.7	107.9	0.07
Valença N	98.2	89.6	104.0	10
Viseu C	88.2	97.5	101.8	0.5/0.5/0.7

RDP África: a) 103.4MHz 1kW**, b)** 99.1MHz 1kW**, c)** 101.5MHz 4kW
D.Prgrs: 24h. 1=Antena 1 (general pt rgrs, sport), 2=Antena 2 (serious music, culture), 3=Antena 3 (pop/rock music), RDP Africa (general prgrs aimed at the Portuguese-speaking African community . **C, N, S**: carry Antena 1 reg. prgrs. M-F 1300-1400 on VHF-FM only, while Antena 1 on MW continues with a nationwide prgr

RDP Reg. Centres, mainland:
N=RDP Norte: Rua Cândido dos Reis, 74, 4050-151 Porto ☎+351 22 339 99 00 🖷+351 22 339 99 02. Dir. José Alberto Lemos
C=RDP Centro: Rua Dr. José Alberto Reis, 74, 3000-232 Coimbra ☎+351 23 979 89 00 🖷 +351 239 72 42 53 Dir. José Manuel Portugal **E:** rdpcentro@rdp.pt
S=RDP Sul: Campo Senhora da Saúde, 8001-904 Faro ☎ +351 289 89 68 69 🖷+351 289 80 21 92 **E:** rdpsul@rdp.pt. Dir. Feliciano Estêvão
RDP Reg. Centres in the autonomous regions of Açores and Madeira: see respective entries.
RDP abroad: txs in Cape Verde, Guinea-Bissau, São Tomé & Príncipe, Mozambique, all relaying RDP África, in Timor, relaying RDPi and airing a local prgr, and in Bosnia for the Portuguese peace-keeping force: this tx operated by the military (see respective country entries).
RDP África: ✉ Av. Marechal Gomes da Costa, 37, 1849-030 Lisboa ☎ +351 21 382 00 00 🖷 +351 21 382 00 81 **LP:** Dir: Jorge Oliveira Gonçalves **E:** rdpafrica@rtp.pt

DAB: 28 T-DAB txs on 225.648MHz, ch. 12, block B, covering mainland's western & southern areas & carrying Antenas 1, 2, 3, RDP África and RDP Internacional.
Web Radio: Audio feeds at www.rtp.pt Antenas 1, 2 & 3 and RDP-África, RDP-Madeira and RDP-Açores . RTP web-only radiso: "Rádio Lusitânia", "Rádio Vivace", "Rádio República & "Antena 1 Vida"selectable in http://multimedia.rtp.pt/index.php?aud=1.
SATELLITE: Europe, N. Africa & Mid. East: Hot Bird 7A (13° E), Transponder 111 (10.723GHz), Ku Band, H. Pol., FEC: ¾, SyR: 29.9 Ms/s, RDPi: PID 1230 (stereo), SID 4630. RDP Antena 1: PID 1235, SID 4635, 24h. **Africa:** Intelsat 907 (27.5° W), Transponder 22/22, C Band, Right Circ. Pol. , Symbol Rate 10.850 MSps (Mega Symbols per second). Frequency 3841 MHz; prgrs: RDP África: audio PID d 412, RDPi: audio PID d 413, Antena 1: audio PID d 411.**Asia & Oceania:** Asiasat 2 (digital) (100.5°E), "European Bouquet", Transponder 10B (4GHz), C Band, Horiz. Pol. on 28.125 Ms/s, FEC ¾. RDP Internacional stereo on the audio ch. 704, RDP Antena 1 aired on the audio ch. 705 on ch.. **N.America & Hawaii:** AMC-4 (digital) (101° W), Transponder 22, Frequency 12169 MHz, Ku Band, H. Pol., SyR 3.003 Ms/s, FEC ¾. RDP Internacional audio 2 PID: 36 (stereo). **The Americas**: Intelsat 805 (55.5° W), Transponder 16 (4080MHz), C Band, V. Pol., SyR 4340 Ms/s, FEC ¾. RDP Internacional: PID 35. **S.America**: Intelsat IA8 (89°W), transponder 709, 11882.3Mhz, ku Band, Horiz. Pol., Sy.Rate 4883, FEC ½. RDP Internacional PID1230.

EXTERNAL SERVICES:
RDP Internacional see International Broadcasting section
Pro-Funk GmBH owns & operates Deutsche Welle relay stn Sines. see International Broadcasting Section

RÁDIO COMERCIAL, S.A. (Priv., comm.)
Owned by Media Capital Rádio – Radiofonia e Publicidade, S.A.
W: http://mcr.clix.pt/index.asp
✉ Rua Sampaio e Pina, 24-26, 1099-044 Lisboa ☎ +351 21 382 15 00 🖷 +351 21 382 15 89 **E:** info@radiocomercial.clix.pt, Northern office: Rua Tenente Valadim, 181, 4100 Porto ☎ +351 22 605 75 00
W: www.radiocomercial.clix.pt
LP: Dir. Gen. MCR Rádios: Jordi Jordà, Dir. Eng. Dept: Pinto Ventura

MW	kHz	kW	MW	kHz	kW
Avanca	783	*100	Belmonte (Benavente)	1035	100

*) currently inactive but possibly to be reactivated
D.Prgr: 24h **Ann** on MW: Rádio Clube Português

FM	MHz	kW		MHz	kW
Fóia (Monchique)	88.1	10	Grândola	96.8	10
Lamego	88.7	10	Monsanto (Lisboa)	97.4	44
Minhéu (Vila Real)	88.9	10	Monte da Virgem (Porto)	97.7	44
Lousã	90.8	44	Gardunha	98.2	10
Bornes (Chaves)	91.9	10	São Mamede	98.9	10
Mendro (Évora)	92.0	50	Valença	99.0	10
Bragança	93.9	10	Braga	99.2	10
Guarda	96.1	10	Montejunto	99.8	10
São Miguel (Faro)	96.1	10			

D.Prgr: 24h **Ann on FM:** Rádio Comercial. **Format:** music stn

Local FM stns in the same group:
M80 : (see Southern network) – **Rádio Clube Português** (**W**: radioclube.clix.pt)**:** txs in Lisboa 96.6 MHz 5 kW, Valongo 105.8 1.5 kW, Cantanhede (near Coimbra) 103.0MHz 2 kW, Santarém 97.7MHz 2kW, Aveiro 94.4 MHz 2 kW, Manteigas (near Guarda) 104.4 MHz 0.5 kW & Sabugal (near Guarda) 96.8 MHz 0.5 kW (see also Rádio Comercial MW). A new station, "**Star FM**"is planned to replace RCP on most (if not all) frequencies. – **Best Rock FM** (**W**: www.bestrock.clix.pt): txs in Moita (near Lisboa) 101.1MHz 1.5 kW, Matosinhos (near Porto) 89.5MHz 1.5 kW. – **Cidade FM** (**W**: www.cidadefm.clix.pt): txs in Lisboa 91.6MHz 5 kW, Porto 107.2MHz 0.5 kW, Redondo (Alentejo province) 97.2MHz 0.5 kW, Alcanena (Santarém) 99.3MHz 2kW, Penacova (Coimbra) & Loulé (Algarve prov.) 99.7MHz 1 / 2 kW , Vale de Cambra (near Aveiro) 101.0 MHz 0.5 kW, Viseu 102.8 MHz 2 kW, Amares (Braga) 104.4MHz 1 kW, Montijo (Lisboa region) 106.2MHz 1 kW – **Mix FM** (**W**: www.mix.clix.pt): tx ln Barreiro (Lisboa region) 103.0MHz 2kW – **Romântica FM** (**W**: www.romantica.clix.pt): txs in Amadora (Lisboa reg.) 107.2MHz 1.5 kW, Maia (near Porto) 100.8MHz 1.5 kW & Figueiró dos Vinhos (near Coimbra) 92.8MHz 1 kW – **MFM** (**W**: .mfm962.com) tx in Barreiro (Lisboa reg.) 96.2MHz 2kW

RÁDIO RENASCENÇA – Emissora Católica Portuguesa (Rlg/Comm)
🖃 Rua Ivens, 14, 1249-108 Lisboa ☎ +351 21 323 92 00 🖷 +351 21 323 92 20 **E:** mail@rr.pt **PR**: **E:** rp@rr.pt **W:** www.rr.pt and www.rfm.pt and www.radiosim.pt
LP: News Dir.: Dr. Francisco Sarsfield Cabral, PD (for R.Renascença): Dr Nelson Ribeiro, PD (for RFM): Dr. António Mendes, PD (for Rádio Sim): Dina Isabel

MW: Rádio Sim	kHz	kW	MW: Rádio Sim	kHz	kW
Braga	576	10	Coimbra	981	10
Muge †	594	100	Guarda	981	1
Vila Moura	891	1	Chaves	1251	1
Évora	927	1	Valongo	1251	10
Seixal ‡	963	10	Castelo Branco	1251	1
Bragança	981	1	Viseu	1251	10
Vila Real	981	1			

† usually 60-80kW; 2x10kW stand by units ‡ 1x1kW standby unit

FM (MHz)	RR	RFM	R. Sim	kW
Aveiro	102.5	97.4		0.2
Arrábida	105.8	89.9		10/12
Bornes	89.6	101.1		10
Braga		89.7	101.1	10/1
Bragança	105.7	99.5		10
Elvas (L)			102.3	0.1
Elvas/V.ª Boim (L)		107.1	99.8	1
Fóia	98.6	104.9		12
Gardunha	103.4	99.5		10
Guarda	90.2	104.0		10
Lamego	98.6	106.2		12
Leiria		107.7	95.1	1/3
Lisboa	103.4	93.2		50
Lousã	106.0	91.7		50/56
Marofa	94.2	103.0		16/10
Mendro	96.5	100.9		50
Minhéu	89.8	102.6		10
Montejunto	90.2	106.8		10
Muro	103.4	90.4		10/20
S. Mamede	95.3	101.1		10
São Miguel	103.8	89.6		10
Serra de Ossa	98.5	89.7		0.5/2
Sintra	105.0	106.6		0.3
Valença	100.0	95.4		10
Valongo		106.2		1
Monte da Virgem	93.7	104.1		50
Viseu		99.4	103.6	1
Vouzela		95.0		0.4

L =includes local prgrs. **D.Prgr.**: 24h. **Ann:** "Renascença – música e informação dia-a-dia.", " RFM. – só grandes músicas"
Stn IDs during local prgrs: Elvas "Rádio Sim de Elvas"
Some progs are simulcast on RR Canal 1 & R. Sim.
Local VHF-FM stns of the R. Renascença group:
Mega FM (**W:** www.mega.fm): txs in Lisboa 92.4MHz 5kW, Sintra (west of Lisboa) 88.0MHz 1 kW, Coimbra 90.0MHz 5kW, Aveiro 105.6MHz 1kW, Gondomar (Porto region) 90.6MHz 2kW and Braga 92.9 MHz 2kW.
Local VHF-FM stns relaying Rádio Sim:
R. Sim Foz do Ave - Vila do Conde (near Porto): 88.6MHz 2kW ; **R. Sim Rio Maior** - Rio Maior (near Santarém): 92.6MHz 1kW, 99.5MHz 0.050kW ; **R. Sim - Pal** - Palmela (near Lisboa): 102.2 MHz 1 kW; **R. Sim Alentejo** – Portel (near Évora): 97.5 MHz 0.5 kW

Regional FM networks:
Northern network
RADIO NOTÍCIAS, PRODUÇÕES E PUBLICIDADE, S.A.
🖃 Edifício "Altejo", Rua 3 da Matinha. 3º piso, sala 301, 1900-823 Lisboa ☎ +351 21 861 25 00 🖷 +351 21 861 25 07/8 **LP:** Chmn.: Joaquim Oliveira, TD: Jaime Silva, Dir tx network: José de Sousa TSF Northern office & Radiopress HQ: 🖃 Rua Gonçalo Cristóvão, 195, 4017-001 Porto ☎ +351 22 206 28 00 🖷 +351 22 206 28 03 **E:** tsf@tsf.pt **W:** www.tsf.pt
Station: TSF (priv., comm.)

FM	MHz	kW	FM	MHz	kW
Pena (Vouzela)	102.5	0.1	Guarda	106.6	10
Bornes	103.2	10	Minhéu	106.7	10
Gardunha	105.1	10	Braga	106.9	10
Valongo (Porto)	105.3	50	Bragança	107.0	10
Marofa	105.4	10	Lousã	107.4	50
Valença	105.7	10	Marão	107.6	10
Muro	106.5	10			

NB: the whole netw. relays key stn TSF 89.5MHz 5 kW Lisboa (cf. local FM stns list).TSF broadcasts 24 hours/ day via **R. Jovem** 105.4MHz 2kW Évora and **R. Santa Maria** 90.9MHz 2kW Faro (Algarve). TSF is also relayed most of the day via **Rádio Caldas** 103.1 MHz 1 kW Caldas da Rainha; in the Açores via **R.Comercial dos Açores-TSF** 99.4MHz 3kW Ponta Delgada (São Miguel isl.) and in Madeira via **R.Notícias-TSF** 100.0MHz 2kW Funchal. **Format:** mainly news. **D.Prgr:** 24h.

Southern network
LICENSEE: RÁDIO REGIONAL DE LISBOA, S.A.
(Owned by Média Capital Rádio) 🖃 see R.Comercial. **W:** m80.clix.pt
Station: M80 Rádio (priv., comm.)

FM	MHz	kW	FM	MHz	kW
Porto #	90.0	5	Monsanto (Lisboa)	104.3	50
Leiria#	93.0	2	São Miguel (Faro)	106.1	10
Penalva do Castelo#	95.6	0.5	Mendro (Évora)	106.4	50
Montejunto	96.4	10	Portalegre	106.7	10
Vila Real #	97.4	1	Fóia (Monchique)	107.1	10
Coimbra #	98.4	5	Grândola	107.5	10
Fafe #	103.8	1			

) txs of associated local stns.
Format: music oldies from 1970s-90s. **N:** every h on the h. **D.Prgr:** 24h. **Ann:** "M80 Rádio"

NFM Global, Lda.
Station: Rádio NFM (Priv., comm.)
🖃 Rua do Salgueirô, 69, 4585-208 Gandra PRD ☎ +351 224 155 350 🖷 +351 224 155 359 **E:** geral@nfm.pt **W:** radio.nfm.pt
FM: 89.2 MHz 2 kW Amarante (near Porto), Espinho (near Porto) 88.4 MHz 1 kW, Bombarral 94.8 MHz 1 kW, Aljezur (Algarve) 102.9 MHz 0.5 kW & Ponte de Sor (near Portalegre) 96.0 MHz 2 kW + 105.6 MHz 0.050 kW. All frequencies carry also local prgs. **D.Prgr:** 24h.

Local Stations (all priv. & comm.)
RADIO ALTITUDE
🖃 Rua Batalha Reis, 6300 Guarda ☎ +351 271 22 19 95 🖷 +351 271 22 14 92 **E:** altitude@altitude.fm **W:** www.altitude.fm **LP:** Dir: Rui Isidro
MW: 1584kHz 1kW 24h. (inactive, but planned to be reactivated)
FM: 90.9MHz 2kW 24h. **Ann:** "Altitude FM"

Other FM stations:

FM	MHz	kW	Station, location
62)	88.2	2	Ultra FM, Vila Franca de Xira
51)	88.5	2	R. Guadalupe, Serpa
54)	88.6	2	R. Jornal de Setúbal, Setúbal
48)	89.0	2	R. Mar, Póvoa de Varzim
63)	89.1	2	R. Lezíria, Vila Franca de Xira
40)	89.3	2	R. Praia, São Teotónio, Odemira
1)	89.7	2	R. Antena Livre, Abrantes
33)	90.4	5	R. Europa-Lisboa
59)	90.5	2	R. Cidade de Tomar, Tomar
60)	90.8	2	R. Geice, Viana do Castelo
12)	91.4	2	R. Iris FM, Benavente
55)	91.8	2	Algarve FM, Silves
43)	91.8	2	R. Club de Penafiel, Penafiel
9)	91.9	2	R. Local de Barcelos, Barcelos
15)	92.0	2	R. Beira Interior, Castelo Branco
37)	92.0	2	R. Amália, Loures
24)	92.1	2	R. Maiorca, Figueira da Foz
7)	92.7	2	ERA FM-Emissora Reg. de Amarante, Amarante
36)	92.8	2	Horizonte Tejo, Bobadela
15)	92.9	2	R. Clube do Minho
4)	93.9	2	R. Mirasado, Alcácer do Sal
65)	94.0	2	R. Cidade Hoje, Vila Nova de Famalicão

FM	MHz	kW	Station, location
30)	94.0	2	R. Douro Sul, Lamego
32)	94.0	2	R. 94 FM, Leiria
21)	94.1	2	Diana FM, Évora
20)	94.5	2	R. Despertar, Estremoz
17)	94.7	2	Voz do Sorraia, Coruche
45)	94.8	5	R. Festival, Porto
57)	94.8	2	R. Gilão, Tavira
66)	95.5	2	Gaia FM, Vila Nova de Gaia
27)	95.8	2	R. Fundação, Guimarães
47)	96.1	2	R. Onda Viva, Póvoa de Varzim
8)	96.5	2	R. Aveiro FM, Aveiro
2)	96.7	2	R. Tágide, Abrantes
56)	96.9	2	R. Horizonte-Algarve, Tavira
29)	97.0	2	R. Clube de Lamego, Lamego
5)	97.8	2	RADAR, Almada
28)	98.0	2	R. Santiago, Guimarães
58)	98.0	2	R. Hertz, Tomar
14)	98.1	3	R. Marginal, Cascais
46)	98.9	2	R. Nova, Porto
53)	98.9	2	R. Azul, Setúbal
23)	99.1	2	R. Clube Foz do Mondego, Figueira da Foz
3)	99.3	2	R. Soberania, Águeda
42)	100.1	2	R. Nova Era, Terra Verde, Paredes (rel 67)
44)	100.5	2	R. Portalegre, Portalegre
52)	100.6	2	R. Amália de Setúbal, Setúbal (rel R. Amália 92.0)
6)	100.8	2	R. Capital, Almada
17)	100.8	2	R. Urbana, Castelo Branco
31)	101.3	2	R. Liz, Leiria
39)	101.3	2	RNA-R.Nova Antena, Montemor-o-Novo
67)	101.3	2	R. Nova Era, Vila Nova de Gaia
10)	101.4	2	R. Pax, Beja
49)	101.7	2	R. Pernes, Santarém
16)	101.8	2	R. Juventude, Castelo Branco
35)	101.9	2	Estação Orbital, Loures (Sacavém)
25)	102.7	2	R. Clube de Gondomar, Gondomar (rel. 6)
50)	102.7	2	Antena Miróbriga, Santiago do Cacém
40)	103.1	2	Total FM, Loulé
22)	103.2	2	R. Telefonia do Alentejo, Évora
41)	103.7	2	R. Canção Nova, Ourém
12)	104.5	2	R. Voz da Planície, Beja
61)	104.6	2	R. Linear, Vila do Conde
64)	105.0	2	Digital FM, Vila Nova de Famalicão
68)	105.5	2	RCI-R. Club do Interior, Viseu
26)	105.8	2	R. "F", Guarda
13)	106.0	2	R. Antena Minho, Braga
69)	106.4	2	R. "Noar", Viseu
38)	107.1	2	R. Voz de Mangualde, Mangualde
18)	107.9	2	R. Universidade de Coimbra, Coimbra

Some stns use 0.05 kW repeaters + over 200 stns of less than 2kW

Addresses and other information:

Many stns have websites and may also have webcasting; most, if not all, have an email address.

Country codes must be added to tel. & fax nos

1) Rua General Humberto Delgado, Edifício Mira Rio, 2200-117 Abrantes ☎ 241 360170/1 🖷 241 360179 E: info@antenalivre.pt W: www.antenalivre.pt – **2)** Rua das Laranjeiras, s/n - Apartado 3 - 2206-906 Tramagal ☎ 241 897192 🖷 241 897859 E: radio.tagide@clix.pt – **3)** Rua José Sucena, 120-3º, 3750-157 Águeda ☎ 234 602133 🖷 241-624334 E: radiosoberania@mail.telepac.pt – **4)** Rua da Fábrica, Convento dos Frades, 7580-122 Alcácer do Sal ☎ 265 622981 🖷 265 622479 E: mirasado@mail.telepac.pt W: w3.mirasado.pt – **5)** Rua Viriato, 25-7º D, S.Sebastião da Pedreira, 1050-234 Lisboa ☎ 21 0105790 / 21 0105770 🖷 21 0105769 E: geral@radarlisboa.fm W: www.radarlisboa.fm – **6)** Rua Viriato, 25-4º D, S.Sebastião da Pedreira, 1050-234 Lisboa ☎ 21 0105760 🖷 21 2740781 E: geral@radiocapital. fm W: www.radiocapital.fm – **7)** Edifício Santa Luzia, 4600 Amarante ☎ 255 420480/9 🖷 255 425317 W: www.erafm927.com E: erafm@iol. pt – **8)** Av. Dr. Lourenço Peixinho, 15-1º G, 3800 Aveiro ☎ 234 424601 🖷 234 428965 E: regional@mail.telepac.pt – **9)** Centro Comercial Bolívar, lojas 45-49, 4750-180 Barcelos ☎ 253 823530 /1 🖷 253 181265 E: geral@radiobarcelos.com – **10)** Rua de Angola, Torre C-11º, 7800 Beja ☎ 284 325 011 🖷 284 326312 E: radio@radiopax.pt W: www.radiopax.pt – **11)** Rua da Misericórdia, 4, 7800-285 Beja ☎ 284 311 330 🖷 284 321446 E: radio@vozdaplanicie.pt W: www.vozdaplanicie.pt – **12)** Rua dos Operários Agrícolas, 2135 Samora Correia, ☎ 263 654840 🖷 263 655567 E: irisfm@mail.telepac.pt W: www.irisfm.pt – **13)** Praceta do Magistério, 36, 4700-222 Braga ☎ 253 309560 🖷 253 309569 E: armindo.veloso@antena-minho.pt, info@antena-minho.pt W: www. antena-minho.pt – **14)** Rua Viriato nº 25 6º-105-234 Lisboa ☎21 0105721 🖷21 3156555 E: net@marginal.fm, geral@marginal.fm W: www.marginal.fm – **15)** Avª 1º de Maio, 39, 3º Dtº / Apartado 178, 6000-909 Castelo Branco ☎ 272 321 050 🖷 272 320 488 E: radio.

interior@netvisao.pt – **16)** Rua Prof. Hugo correia Pardal, Edifício Plátano-loja "A", 6000-267 Castelo Branco ☎ 272 341758 🖷 272 347660 E: radio.juventude@netvisao.pt – **17)** Rua Cadetes de Toledo, lote 5- 1.º E, 6000-156 Castelo Branco ☎272 347676 🖷 272 328670 E: geral@urbana.fm W: www.radiourbana.pt – **18)** Apartado 1178 3001-501 Coimbra ☎ 239 410 410 & 239 410 426 🖷239 835 446 E: info@ ruc.pt, tecnica@ruc.pt W: www.ruc.pt – **19)** Rua do Couço, 29-r/c frt, 2100 Coruche ☎ 243 617436/0 🖷 243 617100 E: radiovozsorraia@ sapo.pt – **20)** Rua Bento de Jesus Caraça Bloco C 1º Andar - Apartado 76 , 7100-104 Estremoz ☎ 268 339454 🖷 268 339456 E: geral@ radiodespertar.net – **21)** MARÉ, EE08, 7000-500 Évora ☎ 266 700333 🖷 266 700555 E: geral@diana.fm com,dianaev@mail.pt – **22)** Estrada de Arraiolos, Aos Arcos da Cartuxo, Edifício Piçarra e Companhia, Lda. , 7000-862 Évora ☎ 266 730415 🖷 266 730416 E: telefonia@diariodosul.com.pt, radiotelefonia-doalentejo@hotmail.com W: www.diariodosul.com.pt – **23)** Rua dos Bombeiros Voluntários, 37-1.º, 3080-133 Figueira da Foz ☎ 233 040620 🖷 233 428134 E: fozdomondego.secretariado@gmail.com W: www.rcfm.web.pt – **24)** Edifício - CC Foz Center, Rua da República n.º 202 lj 36, 3080-036 Figueira da Foz ☎ 233 930500 🖷 233 930499 E: geral@radiomaiorcafm.com W: www.radiomaiorcafm.com – **25)** (see nr. 6) ☎ 21 2740765 🖷 21 2740781 – **26)** Rua Soeiro Viegas, 2-B, 6300-758 Guarda ☎ 271 221468 🖷 271 224407 E: radiof@radiof.com W: www.radiof.com – **27)** Rua Arqueólogo Mário Cardozo, 253-C, Guimarães Ed. Guimarães Palace 411, Apartado 358, 4800-116 Guimarães ☎ 253 420520/2/5/6 🖷 253 420529 E: geral@radiofunda-cao.net W: www.radiofundacao.net – **28)** Praceta de Santiago, 31, 4800 Guimarães ☎ 253 421700 🖷 253 421709 E: santiago@guimara-esdigital.com W: www.santiago.fm – **29)** Urbanização da Úrtigosa, Bloco 6, 5100-183 Lamego ☎ 254 609300/1 🖷 254 609309 E: geral@ rclamego.pt W: www.rclamego.pt – **30)** Rua Fausto Guedes Teixeira, Bloco 1, 5100 Lamego ☎ 254 611551/2 🖷 254 611551 E: radiodouro-sul@gmail.com W: www.radiodourosul.com – **31)** Urbanização Quinta do Amparo, lote 4, R/C – Esq. /Quinta da Matinha / Apartado 525, 2415-583 Leiria ☎ 244 817 707 🖷 244 813951 E: radio@lizfm.pt, davidsantos@lizfm.pt W: www.lizfm.pt – **32)** Av. dos Combatentes da Grande Guerra, Edifício Liz – 10º / Apartado 1113, 2400-122 Leiria, ☎ 244 860090 🖷 244 860098 E: geral@radio94fm.pt W: www.radio94fm. pt – **33)** Rua Latino Coelho, 50 - 1º, 1050-137 Lisboa ☎ 21 3510580 🖷 21 3510598 E: programas@radioeuropa.fm W: www.radioeuropa.fm Relays RFI in French at certain times daily – **34)** Sítio do Troto, 8135-030 Almancil ☎ 289 397666 🖷 289 397110 E: totalfm@totalfm.pt W: www.totalfm.pt – **35)** Travessa do Olival, 6, 2685 Sacavém ☎ 21 9401019 & 21 9427750 🖷 21 9427757 E: orbital@orbital.pt, estacaoor-bital@ mail.telepac.pt – **36)** Rua da Boa Vista, lote 20 - r/c, 2685-027 Bobadela Loures ☎ 21 9559215 & 21 9553113 🖷 21 9558465 E: geral@horizontefm.pt W: www.horizontefm.pt – **37)** Rua Viriato, nº 25. 3º Esquerdo, 1050-234 Lisboa ☎ 21 010 57 40 🖷21 315 57 69 E: geral@amalia.fm W: www.amalia.fm – **38)** Av. D. Henrique, Bloco 4-B-r/c, 3531 Mangualde ☎ 232 612363 🖷 232 611490 E: radiomangualde@netvisao.pt W: www.radiomangualde.pt – **39)** Rua Francisco José Mareco, lote 28, 7050-241 Montemor-o-Novo ☎ 266 086418 🖷 266 086 518 E: radionovaantena@gmail.com W: www.rna-montemor.net – **40)** Rua Frei Joaquim de Loulé, 25, 8100-579 Loulé ☎289 414000 🖷289 412400 – **41)** Estrada da Batalha, Edifício Canção Nova/ Apartado 199, 2496-908 Fátima ☎249 530600 🖷249 530609 – **42)** (see nr. 67) – **43)** Rua Alfredo Pereira, 14-2º / Apartado 14, 4564-909 Penafiel ☎ 255 710040 🖷 255 710049 E: info.mail@radioclube-penafiel.pt W: www.radioclube-penafiel.pt – **44)** Av. de Santo António, 22 , Edifício Régio 1, Atelier "A" e "B", 7300-074 Portalegre ☎ 245 300550 🖷 245 331630 E: geral@radioportalegre.pt W: www.radiopor-talegre.pt – **45)** Rua da Alegria, 582 - 9 esqº/frt , 4000-037 Porto ☎ 22 5370177 & 22 5370723 E: festival@radiofestival.pt W: www.radiofes-tival.pt – **46)** Rua João de Deus Barros, 265, Foz do Douro, 4100 Porto ☎ 22 6101390/5 🖷 22 6101420 E: nova@radionova.pt W: www. radionova.pt – **47)** Praça dos Combatentes, 15, 4990-439 Póvoa de Varzim ☎ 252 613686/888/878 🖷252 613898 E: radioondaviva@sapo. pt, geral@radioondaviva.pt W: www.radioondaviva.pt – **48)** Rua Comendador Francisco Alves Quintas, 835 - Pavilhão A 4490-489 Póvoa de Varzim ☎ 252 690140 🖷 252 690149 E: radiomar@radiomar.com W: www.radiomar.com – **49)** Rua da Fé, 1, 2035 Pernes & R. Pedro de Santarém, 10-3º dtº, 2000 Santarém ☎ 243 332922 & 243 332004 🖷 243 332 998 E: pernesradio@gmail.com W: www.radiopernes.pt – **50)** Rua Condes de Avillez, 19-21, Apartado 45, 7540-909 Santiago do Cacém ☎ 269 829920/8 E: antenamirobriga@antenamirobriga.com W: www.mirobriga.pt – **51)** (see nr. 40) – **52)** Rua Nossa Srª do Amparo, 15 - 3º "A", 2900-144 Setúbal ☎ 265 236445 🖷 265 236444 W: www.amalia.fm – **53) & 54)** Rua Dr. António Rodrigues Manito, 58 - r/c "B", 2900-061 Setúbal ☎ 265 089052 & 265 573601 🖷 265 089053 & 265 573639 E: radioazul98.9@sapo.pt, radioazul_setubal@hotmail.com, radiojornal@sapo.pt – **55)** Rua da Boa Vista, 5

- Enxerim, 8300-025 Silves , ☎ 282 440100 🖷 282 440102 **E:** algarve-fm@netvisao.pt **W:** www.algarvefm.pt **– 56)** Quinta do Rosal, Sítio de São Pedro, E.N. 125 / Apartado 252 , 8800-902 Tavira ☎ 281 380 240 🖷 281 380 249 **E:** horizontealgarve@gmail.com **W:** www.radiohorizonte.com **– 57)** Largo Santana, 1, 8800-902 Tavira ☎ 281 320240 🖷 281 325523 **E:** radiogilao@net.vodafone.pt **W:** www.radiogilao.sdv.pt **– 58)** Rua Centro Republicano, 135, Apartado 133, 2300-909 Tomar ☎ 249 323100/20 🖷 249 316995 **E:** radiohertz@radiohertz.pt **W:** www.radiohertz.pt **– 59)** Travessa da Cascalheira, n.º 27, 2301 Tomar ☎ 249 310010 🖷 249 310016 **E:** radio@cidadetomar.pt **W:** www.radio.cidadetomar.pt **– 60)** Praça 1º de Maio, 6-traseiras, 4900-534 Viana do Castelo ☎ 258 800400 🖷 258 800409 **E:** geral@radiogeice.com **W:** www.radiogeice.com **– 61)** Rua das Donas, 3, 4480-910 Vila do Conde ☎ 252 642426/7/8/9 🖷 252 642303 **E:** radiolinear@clix.pt, radiolinear@gmail.com **W:** www.radiolinear.pt **– 62)** Rua Fausto Nunes Dias, n.º5, 2600-145 Vila Franca de Xira ☎ 263 287590 🖷 263 287599 **E:** geral@ultrafm.pt **W:** www.ultrafm.pt **– 63)** Largo Marquês de Pombal, 2-7º,2600-222 Vila Franca de Xira ☎ 263 3286000 🖷 263 3286007 **E:** radioleziria@gmail.com **W:** www.radio89fm.net **– 64)** Rua 8 de Dezembro, 214, 4760 Vila Nova de Famalicão ☎ 252 308143/7 🖷 252 308144/9 **E:** comercial@opiniaopublica.pt, jfernandes@opiniaopublica.pt **W:** www.digitalfm.pt **– 65)** Edifício Vilarminda, loja 204 / Apartado 218, 4764-976 Vila Nova de Famalicão ☎ 252 301781/2 🖷 252 301789 **E:** geral@cidadehoje.pt **W:** www.cidadehoje.pt **– 66)** Rua Raimundo de Carvalho, 242 - Sala 1, 4430-185 Vila Nova de Gaia ☎ 22 3770840 🖷 22 3770849 **E:** geral@gaiafm.com **W:** www.gaiafm.com **– 67)** Rua das Camélias, 134-B, 4430-038 Vila Nova de Gaia ☎ 22 3770180 🖷 22 3759675 **E:** geral@radionovaera.pt **W:** www.radionovaera.pt **– 68)** Rua do Comércio 58, 3º Andar 3500-110 Viseu ☎ 232 423411/2 🖷 213 519134 **E:** info@rci.pt **W:** www.rci.pt **– 69)** Rua 5 de Outubro, 87-2º, 3500 Viseu ☎ 232 480030/2 🖷 232 425432 **E:** info@radionoar.pt comercial@radionoar.pt **W:** www.radionoar.pt

Military Stations:
CINCSOUTHLANT-Commander-in-Chief South Atlantic Area/ **CINCIBERLANT**-Commander-in-Chief Iberian Atlantic Area: ✉ 2780 OEIRAS ☎. PR: +351 214 404106 – **FM** 88.4MHz 0.1kW (inactive) **D. Prgr:** AFN in English

PUERTO RICO (USA Commonwealth)

L.T: UTC -4h — **Pop:** 4 million — **Pr.L:** Spanish, English — **E.C:** 60Hz, 120V — **ITU:** PTR

FEDERAL COMMUNICATIONS COMMISSION (FCC)
see USA for details

BROADCASTERS ASSOCIATION OF PUERTO RICO
✉ Suite 212, Cobians Plaza, 1607 Ave. Ponce de León, San Juan 00926
Most stns broadcast in Spanish only °=English or mainly English
d=directional antenna * = inactive
Hrs of tr usually 24h – see address section for variations.

MW Call	kHz	kW	Station, location.
1) WPAB	550	5	WPAB 550 Ponce, Ponce
2) WKAQ	580	10	R. Reloj, San Juan
3) WYEL	600	5	R. 600, Mayagüez
4) WEXS	610	1/0.25	X-61, Patillas
5) WUNO	630	5	NotiUno, San Juan
6) WAPA	680	10	Cadena WAPA "Guapa", San Juan
WAPA	680	0.4	Arecibo (synchr. WAPA)
7) WKJB	710	10/0.75	KJB "Radio Isla", Mayagüez
8) WIAC	740	10	Boricua 740, San Juan
8) WIAC	740	0.5/0.1	Boricua 740, Ponce (synchr)
9) WORA	760	5	NotiUno, Mayagüez
10) WKVM	810	50	R. Paz 810 AM, San Juan
11) WXEW	840	5/1	NotiUno/R. Victoria, Yabucoa
12) WABA	850	5/1	Waba "La Grande", Aguadilla
13) WQBS	870	5	La Gran Cadena QBS, San Juan
14) WYKO	880	1/0.5	La Poderosa 880, Sabana Grande
15) WFAB	890	0.25	La Nave 890, Ceiba
16) WPRP	910	4.4	NotiUno, Ponce
17) WYAC	930	2.5	Boricua, Cabo Rojo
18) WIPR	940	10	940 AM, San Juan
19) WCHQ	*960	1/1.7	La Radio Que Te Bendice, Quebradillas
20) WPRA	990	0.91	La Primera, Mayagüez
21) WOQI	1020	1/0.28	R. Coquí/La Señal de la Montaña, Adjuntas
22) WOSO°	1030	10	Total R. El Oso, San Juan
23) WZNA	1040	9/0.25	Zona 1040, Moca
24) WCGB	1060	5/0.5	La Roca, Rock R. Netw., Juana Díaz
25) WMIA	1070	0.5/2.5	R. Arecibo del Norte, Arecibo
26) WLEY	1080	0.25	R. Isla 1080, Cayey
27) WSOL	1090	0.25/0.7	La Nueva Sol 1090, San Germán

MW Call	kHz	kW	Station, location.
28) WVJP	1110	2.5/0.5	R. Caguas, Caguas
29) WMSW	1120	2.6/5	R. Once, Hatillo
30) WOIZ	1130	0.2/0.7	R. Antillas, Guayanilla
31) WQII	1140	10	Once Q, San Juan
32) WBQN	1160	5/2.5	Super Borinquén, Barceloneta-Manatí
33) WLEO	1170	0.25	R. Leo, Ponce
34) WBMJ	°1190	10/5	WBMJ Rock R. Netw. "La Roca", San Juan
35) WGDL	1200	0.25/1	La mejor AM, Lares
36) WHOY	1210	5	La Señal Activa de PR, Salinas
37) WNIK	1230	1	Única R., Arecibo
38) WALO	1240	1	R. Oriental/Cad. R. Puerto Rico,Humacao
39) WJIT	1250	0.25/1	R. Hit, Sabana
40) WISO	1260	2.5	R. Wiso/Cadena WAPA, Ponce
40) WISO	1260	2.5/0.9	R. Wiso/Cadena WAPA, Aguadilla (synchr)
40) WISO	1260	5/1.8	R. Wiso/Cadena WAPA, Mayagüez (synchr)
41) WCMN	1280	5/1	NotiUno 1280, Arecibo
42) WTIL	1300	1	La Voz Romántica, Mayagüez
43) WSKN	1320	5/2.3	R. Isla 1320, San Juan
44) WENA	1330	2/1.4	La Buena del Sur, Yauco
45) WWNA	1340	0.95	R. Una 1340, Aguadilla
46) WEGA	1350	2.5	Nueva Victoria, Vega Baja
47) WIVV	°1370	5/1	WIVV Rock Radio Network "La Roca", Vieques Isl. (r: 1190kHz)
49) WOLA	1380	1	Prócer, Voz de la Montaña, Barranquitas
50) WISA	1390	1	R. Puerto Rico 1390, Isabela
51) WIDA	1400	1	R. Vida AM, Carolina.
52) WRSS	1410	1	R. Progreso, San Sebastián
2) WUKQ	1420	1	R. Reloj, Ponce: rel R. Reloj 580kHz
54) WNEL	1430	5	NotiUno/R. Tiempo, Caguas
55) WCPR	1450	1	R. Coamo, Coamo
56) WRRE	1460	0.5/0.3	Sonido Santidad, Juncos
57) WLRP	1460	0.5	R. Raíces, San Sebastián
58) WKCK	1470	2.4/2.5	R. Cumbre, Orocovis
59) WMDD	1480	5	R. Tropical/Sonido 14-80, Fajardo
60) WDEP	1490	5/1	R. Isla, Ponce
61) WMNT	1500	1/0.25	R. Atenas, Manatí
62) WBSG	*1510	1	R. Voz, Lajas
63) WVOZ	1520	25	R. Voz/La Voz Boricua, San Juan
64) WUPR	1530	1/0.25	Exitos 15-30, Utuado
65) WIBS	1540	1d	R. Voz/R. Caribe, Guayama
66) WKFE	1550	0.25	La Isla/R. Café Dinámica Yauco
67) WRSJ	1560	5/0.75	La Bachatera del Norte, Bayamón
68) WPPC	1570	1/0.1	R. Felicidad, Peñuelas
69) WEKO	1580	5/2.5	R. Voz 1580, Morovis
70) WXRF	1590	1	R. Voz, Guayama
71) WLUZ	1600	5	R. Luz/Romántica 1600, Bayamón
72) WGIT	*1660	10/1	R. Voz/Gigante 16-60, Canóvanas

Addresses and other information:
1) Box 7243, Ponce 00732-7243 ☎+1 787 8405550 🖷+1 7878407077 **W:** www.wpabradio.com/index.html **E:** algabsas@wpabradio.com - **FM:** WOQI 93.3MHz, WIOC 105 1MHz, WOYE 94 1MHz, Mayagüez **– 2)** Box 364668, San Juan 00936-4668 ☎+1 7877585800 🖷+1 7877565220 **W:** www.univision.com - **FM:** KQ-105 La Primera 104.7MHz **– 3)** Box 1370, Mayagüez 00681-1370 - 0830-0400 **FM:** WAEL-FM 93.3MHz, Maricao **– 4)** Box 640, Patillas 00723-0640 **– 5)** Box 363222, San Juan 00936-3222 **W:** www.notiuno.com - **FM:** WFID 97.7MHz, Río Piedras **– 6)** Urb Baldrich, 134 Domenech Ave, Hato Rey 00918-3502 – 0900-0400 **W:** www.wapradio.com **– 7)** Box 1293, Mayagüez 00709-1293 - 0915-0400 **W:** www.radioisla1320.com - **FM:** WKJB-FM 99.1MHz **– 8)** Box 9023916, San Juan 00902-3916 **W:** www.boricua740.com - **FM:** 102.5MHz **– 9)** Box 43, Mayagüez 00681-0043 (or: P.O.Box 363222, San Juan, PR 00936) **W:** www.notiuno.com **– 10)** Urb Roosevelt, 415 Calle Carbonell, Hato Rey 00918-2866 - 0900-0500 - **FM:** WORO 92.5MHz, Corozal **– 11)** Box 100, Yabucoa 00767 - 0830-0300 **W:** www.victoria840.com **– 12)** 6 Calle Munoz Rivera St., Aguadilla 00603-5154 (or: P.O.Box 188, Aguadilla, PR 00605) **W:** www.waba.850.com **– 13)** Calle Bori 1508, Urb Autonsanti, San Juan 00927 - 0900-0400 **– 14)** Calle Dr. Felix Tio 34, Sabana Grande, PR 00637. **– 15)** P.O.Box 318, Río Blanco, PR 00744. **W:** www.radiounidadcristiana.com **– 16)** Box 7771, Ponce 00732-7771 **– 17)** Box 681, Cabo Rojo 00623-0681 - **FM:** WMIO 102.3MHz **– 18)** Box 190909, Hato Rey 00918-0909 ☎+1 787 7660555 🖷+1 787 2507694 **W:** www.tutv.puertorico.pr - **FM:** 91.3MHz, Allegro **– 19)** P.O.Box 846, Aguada, PR 00602. **W:** www.zona1040am.com **– 20)** Box 1293, Mayagüez 00681-1293 - 0900-0400 **W:** www.wpra990.com **– 21)** Box 704, Adjuntas 00601-0704 - 1000-0200 ☎+1 787 829 2010 🖷+1 787 829 4842 **– 22)** Box 9023940, San Juan 00902-3940 **W:** www.woso.com **– 23)** Box 846, Aguada, PR 00602-0846 **W:** www.zona1040.com **– 24)** Box 248, Juana Díaz 00795-0248 (or P.O.Box 367000, San Juan, PR 00936) - 0900-0300 - **W:** www.therockradio.org **– 25)** Box 1055, Arecibo 00613-1055 - 0925-0400 (Sun-0200) **– 26)** Box 1186, Cayey 00737-1186 (or 100 Gran Bulevar Paseo #403A, San Juan, PR

00926) **W:** www.radioisla1320.com – **27)** Box 5000, Suite 442, San Germán 00683-0442 - 0930(Sun 1000-) -0400 **W:** www.radiosol.com – **28)** Box 207, Caguas 00726-0207 - **FM:** 103.3MHz, Criolla – **29)** 550 Calle Truncado, Hatillo 00659-2712 (or P.O.Box 140961, Arecibo, PR 00614) - 1000-0200 **W:** www.radioonce.com – **30)** Box 561130, Guayanilla 00656-1130 - 0900-0200 ☎+1 787 835 1130 🖹+1 787 835 3130 **E:** radioantillas@yahoo.com **W:** www.radioantillas.4t.com – **31)** Cobian's Plaza, Santurce 00909-1820 (or: Box 193779, San Juan, PR 00919) - 1000-0400 – **32)** Box 1625 (or: Calle 16 H-6 Urb. Flamboyán), Manatí 00674-1625 - 1100-0200 ☎+1 787 854 2450 🖹+1 787 854 3738 – **33)** Box 7213, Ponce 00732-7213 - **FM:** WZAR 101.9MHz – **34)** Box 367000, San Juan 00936-7000 (or: Av Ponce de León N° 1409, P4, Santurce 00907) (// 48) ☎+1 787 724 1190 🖹+1 787 722 5395 **W:** www.therockradio.org **E:** radio@therockradio.org – **35)** Box 872, Lares, PR 00669 – **W:** www.wgdl1200am.com – **36)** Box 1148, Salinas 00751-1148 - 0900-0330 ☎+1 787 824 3420 🖹+1 787 824 8054 **E:** whoyam@coqui.net – **37)** Box 141526, Arecibo 00614 – **W:** www.unicaradio1230.com – -0200 **FM:** 106.5MHz – **38)** Box 1240 (or: P.O.Box 9230), Humacao 00792 **W:** www.waloradio.com – **39)** Box 316, Coamo 00769-0316 (or: P.O.Box 878, Vega Alta, PR 00692) – **40)** Box 7251, Ponce 00732-7251 (or: 155 San Antonio St., Floral Park, Hato Rey, PR 00917) - 1000-0300 – **41)** Box 436, Arecibo 00613-0436 - **FM:** 107.3MHz – **42)** Box 1360, Mayagüez 00681-1360 – **43)** Box 363222, San Juan 00936-3222 **W:** www.radioisla1320.com – **44)** Box 1330, Yauco 00698-1330 **W:** www.labuena1330.com – **45)** Box 7, Moca 00676-0007 - 0900-0400 **W:** http://radiouna1340.com – **46)** Box 1488, Vega Baja 00694-1488 - 1000-0200 – **48)** Address, **Web** and **E:** As 34) – **49)** Box 669-A, Barranquitas, PR 00794 - 0900-0200 – **50)** Box 750, Isabela 00662-0750 **W:** www.wisa1390.com **E:** wisa@prtc.net **FM:** WKSA 101.5MHz – **51)** Box 188, Carolina 00986-0188 **W:** www.cade-naradiovida.com - **FM:** 90.5MHz – **52)** Box 1410, San Sebastián 00685-1410 – **54)** Box 487, Caguas 00726-0487 ☎+1 787 744 3131 🖹+1 787 743 0252 **E:** buzoncadena@hotmail.com - **FM:** WPRM 98.5MHz, San Juan – **55)** Box 1863, Coamo 00769-1863 ☎+1 787 825 7061 🖹+1 787 825 1905 **W:** www.coamomall.com or coamomallradio/ – **56)** Box 1460, Las Piedras, PR 00771-1460 ☎+1 787 561 1460 🖹+1 787 734 2595 **W:** www.sonidosantidad.com – **57)** Box 1670, San Sebastián 00685-1670 - 0900-0400 – **58)** 10 Calle Pedro Arroyo, Orocovis 00720-2202 (or: P.O.Box 1210, Orocovis, PR 00720) **W:** www.cumbre1470.com 0900-0200 – **59)** Box 948, Fajardo 00738-0948 **W:** www.el1480.com **FM:** WDOY 96.5MHz – **60)** Box 7213, Ponce 00732-7213 **W:** www.radioisla1320.com – **61)** Box 6, Manatí 00674-0006 - 1030-0200 **W:** www.radioatenas.com **E:** info@radioatenas.com – **62)** Box 593, Lajas 00667-0593 (has requested permission to move to San Germán) – **63)** Calle Bori 1554, San Juan 00927-6113 - **FM:** 107.7MHz Carolina – **64)** Box 868, Utuado 00641-0868 – **W:** www.wupr.com 1000-0400 – **65)** Box 1540, Guayama 00785-1540 – **66)** Box 324, Yauco 00698-0324 (or: 100 Gran Bulevar Paseo #403A, San Juan, PR 00926) **W:** www.radio-isla1320.com – **67)** Box 4036, Carolina 00984-4036 (or: Calle Bori 1554, San Juan, PR 00927) – **68)** Box 9064, Ponce 00732-9064 - 1300-2300 **W:** www.wppc1570am.com **E:** wppcam@prtc.net – **69)** Calle Bori 1554, San Juan, PR 00927 – **70)** Calle Bori 1554, San Juan, PR 00927 – **71)** Box 9394, Santurce 00908-9394 - 1100-0500 ☎+1 787 729 1600 🖹+1 787 723 8685 – **72)** Calle Bori 1554, San Juan, PR 00927.

FM in San Juan (mHz): 89.7 WRTU University of San Juan – 91.3 WIPR-FM – 93.7 WZNT – 98.5 WPRM-FM – 99.1 WPRM-FM – 99.9 WIOA – 102.5 WIAC-FM – 104.7 WKAQ-FM – 105.7 WCAD

AFRTS

🖃 Naval Media Center, 2713 Mitscher Road SW, Washington, DC 20373-5819, USA **W:** www.afrts.osd.mil/afnonradio
MW: AFCN 1200kHz 0.25kW, Roosevelt Roads
SW: AFCN 7507kHz USB, Roosevelt Roads
FM: AFCN 101.5MHz 1kW, Roosevelt Roads. 93.1MHz 0.25kW, Sabana Seca. 90.5MHz 0.20kW, Aguadilla. 91.1MHz 0.20kW

L.T: UTC +3h — **Pop:** 850,000 — **Pr.L:** Arabic — **E.C:** 50Hz, 240V — **ITU:** QAT

SUPREME COUNCIL OF INFORMATION & COMMUNICATION TECHNOLOGY
W: ict.gov.qa

QATAR RADIO & TV CORPORATION (QRTC, Gov.)
🖃 P.O. Box 1414, Doha ☎+974 4894444 🖹 +974 4882888 **W:** www.qatarradio.net **E:** info@qatarradio.net **L.P:** Vice-Chairman for Radio & TV Corp: Abdul Rahman Saif Al-Mahmadi. Dir. of Broadc: Mubarak Jaham Al-Kawari.

MW	kHz	kW	Prgr.	H. of tr.
Al Arish	675	600	A	24h
Al Arish	*954	1500	A	24h

Arabic Prgr. (A): 24h. On **FM:** Al-Jumailiya 90.8MHz 40kW & 92.6MHz 20kW, Umm Said 93.4MHz, Al Kohr 97.6MHz 10kW, Markhiyah 102.0MHz 10kW, Al-Khaisah 102.6/103.4MHz, Al Ruwais 104.0.MHz. *954kHz carries Al-Jazeera TV sound. **Foreign prgr:** Doha 97.5MHz 10kW: English: 0300-1000, 1300-1600, 1900-2200. French: 1000-1300, Urdu 1600-1900. **Sowt al-Khaleej** (Voice of the Gulf): Doha 100.8MHz; Arabic language music sce. Al-Jazeera English TV audio: Doha 101.7MHz. **Ann:** "Idha'at Qatar min al-Doha".

Other stations:
Emarat FM, Manama: 92.3MHz. See UAE – **Middle East BC,** Markhiya 92.0MHz – **QF R.,** Doha: 91.4MHz **W:** qfradio.org.qa – **AFN Qatar,** Al Udeid Airbase 107.5MHz – **BBC World Sce,** Doha: 107.4MHz – **BFBS,** Al Udeid Airbase: 103.0/105.0MHz – **Deutsche Welle,** Doha: 94.4MHz – **Monte Carlo Doualiya,** Doha: 93.4MHz 1kW – **RFI,** Doha, 107.7MHz – **R. Sawa,** Doha: 92.6MHz 2kW

L.T: UTC +4h — **Pop:** 800,000 — **Pr.L:** French — **E.C:** 50Hz, 220V — **ITU:** REU

RADIO RÉUNION (Gov)
🖃 RFO Réunion, 1 rue Jean-Chatel, FR-97716 St. Denis Messag Cédex 9 ☎+262 262406767 🖹 +262 262216484 **W:** reunion.rfo.fr **L.P:** Directrice Regional: Dominique Richard.
MW: St. Pierre 666kHz 20kW, St. André 1224kHz 10kW.

FM	MHz	FM	MHz
Saint-André	87.9	Vincendo	90.6
Tampon/Gr. Coude	88.5	Saint-Gilles	90.7
Saint-Louis	89.0	Manapany	90.7
Saint-Denis/Maïdo	89.2	Salazie	90.8
Dextor	89.6	Saint-Paul	92.6
Plaine des Palmistes	89.6	Saint-Philippe	92.6
Sainte-Suzanne	90.0	Cilaos	93.5
Piton Hyacinthe	90.0	Saint-Anne	96.5
Le Platte	90.4		

D.Prgr: 24h. During nighttime 2300-0100 relay of RFI.
Ann: "Société Nationale de Radio-Télévision Française d'Outre Mer, Station de la Réunion".**IS:** "Séga & Maloya" (Réunion Folklore).

RADIO ARC-EN-CIEL (Rlg.)
🖃 BP 362, 18 bis rue Montreuil, FR-97467 St-Denis
☎+262 262201082 **W:** 7afm.com/7areligion/7aradio_v2.htm
E: arcenciel@7afm.com

FM	MHz	FM	MHz
St.Joseph	89.3	St. Giles	95.6
Le Pont	89.6	Montagne	103.4
St. Denis	91.3	St. Benoit	105.2
St. Pierre	94.7	St. Leu	106.3

Other stations:
R. Vie, B.P. 772, 97476 Saint Denis Cedex. FM: 105.5MHz 24h.
BBC World Sce in English rel. by R. Kanal: West Réunion 97.0MHz, East Réunion 98.1MHz and South Réunion 98.4MHz.
France Inter: 98.8MHz, 99.6MHz + 7 relay stns.
France Culture: 95.7MHz, 96.7MHz
R Kanal Ocean Indien: West 96.6MHz 2kW, East 98.1MHz 0.5kW, 100.4MHz 0.3kW, 102.6MHz 3kW, South 93.0MHz 1kW

L.T: UTC +2h (27 Mar-30 Oct: +3h) — **Pop:** 22.3 million — **Pr.L:** Romanian, Hungarian, German — **E.C:** 50Hz, 230V — **ITU:** ROU

CONSILIUL NATIONAL AL AUDIOVIZUALULUI (CNA)
🖃 Bd. Libertatii nr. 14, sector 5, 050706 Bucuresti ☎ +40 21 3055356 🖹 +40 21 3055354 **E:** cna@cna.ro **W:** www.cna.ro **L.P:** Pres: Rasvan Popescu
NB. CNA is the regulatory authority for broadcasting.

SOCIETATEA ROMÂNA DE RADIODIFUZIUNE (Pub)
🖃 Str. Berthelot nr. 60-64, 010105 Bucuresti ☎ +40 21 3031777 🖹 +40 21 3031726 **E:** comunicare@radioromania.ro **W:** www.radioromania.ro; www.srr.ro **L.P:** Pres: András István Demeter

LW/MW	kHz	kW	Prgr	MW	kHz	kW	Prgr
Brasov (Bod)	153	200	AS	Târgu Jiu²	558	400	1
Petrosani	531	15	1	Brasov (Bod)	567	50	1
Urziceni	531	15	AS	Satu Mare	567	50	1

MW	kHz	kW	Prgr	MW	kHz	kW	Prgr
Botosani	603	50	1	Bacau (Galbeni)	1179	400	1
Bucuresti (a)	603	30	AS	Resita (Vascau)	1179	10	1
Oradea	603	50	1	Brasov (Bod)[2]	1197	15	R/MG/L
Drobeta-T. Severin	603	15	R	Constanta (d)	1314	50	AS
Timisoara (b)	630	400	R	Craiova	1314	15	R
Voinesti	630	50	AS	Timisoara	1314	25	AS/MG
Sighetul M.	711	50	1	Târgu Mures[2]	1323	15	R/MG
Baia Mare	720	10	1	Galati	1332	50	1
Nufarul	720	15	1	Sighetul M.	1404	50	R/MG/L
Sinaia	720	15	1	Sibiu	1404	15	1
Lugoj (Boldur)[3]	756	400	1	Olanesti	1422	10	1
Bucuresti (c)	855	400	1	Constanta (d)	1458	100	1
Cluj (Jucu)[2]	909	200	R/MG	Nufarul	1530	15	R
Timisoara	909	50	1	Radauti	1530	15	1
Constanta (d)	909	25	R	Miercurea Ciuc[2]	1593	15	R/MG
Miercurea Ciuc	945	15	1	Ion Corvin	1593	15	1
Iasi (Uricani)[2]	1053	400	R	Oradea	1593	15	R
Cluj (Jucu)[1]	1152	400	1	Sibiu	1593	10	R/MG

(a) Herastrau (b) Ortisoara (c) Tâncâbesti (d) Valul lui Traian
R=Regional prgrs, L=Local prgrs (substudios)

NB. Txs are on the air 24h (except [1]=0300-2200, [2]=0400-2200, [3]=0600-2200), relaying R.România Actualitati at nighttime.

FM (MHz)	RR1	RR2	RR4	kW
Alexandria	91.8	89.7	-	10
Arad (Siria)	103.8	106.8	-	10
Bacau (Turn)	98.8	101.8	-	5
Baia Mare (Mogosa)	102.5	100.1	-	60
Bârlad	103.9	102.8	-	-
Bechet (Dabuleni)	99.1	-	-	2
Bihor (Stei)	91.0	105.8	-	60
Bistrita (Heniu)	103.9	101.3	-	4
Botosani (Săveni)	106.0	100.8	-	10
Brasov	102.5	105.0	-	2
Bucuresti (Herastrau)	105.3	101.3	104.8	10/10/2
Buzau (Dealul Istrita)	107.0	103.7	-	14
Calafat (Plenita)	90.2	101.1	-	-
Calarasi (Baneasa)	106.6	89.1	-	30
Câmpulung M. (Rarau)	96.0	98.7	-	30
Cluj-Napoca (Feleac)	88.8	101.0	-	30
Comanesti (Laposi)	104.7	101.4	-	30
Constanta (Techirghiol)	102.7	-	-	60
Craiova	88.7	-	-	10
Deva (Magura Boiu)	103.4	105.0	-	4
Drobeta - T. Severin (Balota)	91.4	105.8	-	30
Faget	89.8	-	-	5
Focsani (Magura Odob.)	102.5	101.0	-	60
Galati (Vacareni)	106.4	101.6	-	60
Gheorgheni (Suseni)	103.4	106.8	-	60
Giurghiu	104.6	102.6	-	2
Husi	101.7	-	-	2
Iasi (Pietrarie)	101.1	103.1	-	60
Lehliu Gara	106.2	-	-	1
Mahmudia	100.5	102.0	-	2
Mangalia	-	92.7	-	2
Moldova Noua	105.1	-	-	2
Negresti-Oas	89.4	-	-	2
Novaci (Cerbu)	92.9	89.5	-	100
Oradea	104.1	96.1	-	10/60
Petrosani (Parâng)	88.1	90.6	-	10
Piatra Neamt (Pietricica)	103.6	100.3	-	35
Ploesti (Costila)	102.2	104.1	97.6	100
Resita (Semenic)	102.5	-	-	100
Rm. Vâlcea (Cozia)	103.4	102.5	-	30
Sibiu (Paltinis)	101.8	103.7	-	60
Sighetul Marmatiei	106.2	-	-	2
Slobozia	96.3	-	-	2
Suceava (Mihoveni)	99.6	101.6	-	30
Sulina	100.9	-	-	2
Târgu Mures	93.6	104.9	-	10
Timisoara (Urseni)	106.4	100.7	-	15
Toplita	101.0	-	-	2
Tulcea	99.4	105.4	-	-
Tulcea (Topolog)	105.0	103.0	-	-
Turnu Magurele	105.0	-	-	30
Varatec	91.2	100.8	-	10
Vaslui	106.1	102.4	-	2
Vatra Dornei	100.7	107.7	-	2
Zalau (Mezes)	88.1	105.0	-	10

NB: Txs below 1kW not listed.

D.Prgr: RR1 (R. România Actualitati): 24h. – **RR2 (R. România Cultural):** 24h. – **RR4 (R. România Muzical "George Enescu"):** 24h. – **Antena Satelor (AS):** 0400-2000. – **"Program Maghiar-German" (MG)** for Hungarian & German minorities: W 1200-

1300 German & 1300-1400 Hungarian; Sun 0800-0820 Hungarian & 0820-0830 German. **E:** rgermana@radio.rornet.ro (German), laci@srr.ro (Hungarian).

EXTERNAL SERVICE R. Romania Int: see Int Radio section.

R. ROMÂNIA REGIONAL (Pub)
✉ Str. Berthelot nr. 60-64, 010105 Bucuresti ☎ +40 21 3031469 🖷 +40 21 1860 **E:** romaniaregional@ssr.ro **W:** www.romaniaregional.ro
NB: R. România Regional is the dept for regional broadcasting of SRR. All reg. stns rel. news from national netws at certain times.
R. Bucuresti: Str. Berthlot nr. 60-64, 010105 Bucuresti. **E:** radiobucuresti@srr.ro. Local service **"Antena Burestilor"** 24h on 98.3MHz (Bucuresti 10kW). – **R. Cluj:** Str. Donath nr. 160, 400293 Cluj-Napoca. **E:** office@radiocluj.ro. On 95.6MHz (Feleac 6kW): 24h; on (kHz) 909 (Cluj), 1404 (Sighetul M.), 1593kHz (Oradea & Sibiu): 0400-2000. Hungarian ("Kolozsvári Rádió") 0600-0800 (W); 0715-0730 (Sun); 1300 (Sun 1200)-1600, 1800-2000 (Mon), 1900-2000 (Tue/Wed). Also in Ukrainian. Local sub-studios own prgrs (otherwise rel. R. Cluj): a) **"Antena Sibiului"** Str. Brutarilor nr. 3, 550251 Sibiu. **E:** antena.sibiului@gmail.ro. On 95.4MHz (Sibiu): 0900-1200, 1400-1700; b) **"R. Sighet"** Str. Plevnei nr. 8, 435500 Sighetul Marmatiei. **E:** radiosighet@yahoo.com. On 1404kHz: 0430-0600, 1600-1800; incl. Hungarian and Ukrainian. – **R. Constanta:** Vila nr. 1, 900001 Mamaia. **E:** secretariat@radioconstanta.ro. Prgr 1 ("R.Constanta FM") on (MHz) 100.1 (Techirghiol 60kW), 105.4 (Tulcea). Not on the air 1 June - 1 Sep (txs carry R. Vacanta). Prgr 2: on (kHz) 909 (Valul lui Traian), 1530 (Mahmudia): 0400-2200. Incl. prgrs in Armenian, Aromanian, Greek, Russian, Tatar, Turkish. **"R. Vacanta"** (1 June - 1 Sep): on 100.1/105.4MHz: 24h. Multilingual tourist service (Romanian, English, French, German, Russian). **E:** radio.vacanta@rdsct.ro. – **R. Iasi:** Str. Catargi Lascar nr. 44, 700107 Iasi. **E:** secretariat@radioiasi.ro. On (MHz) 90.8 (Rarau 60kW), 96.3MHz (Pietrarie 10kW): 0400-2200; on 1053kHz: 0400-2000. Incl. prgrs in Ukrainian. – **R. Oltenia Craiova:** Str. Stirbei Voda nr. 3, 200352 Craiova. **E:** office@radiocraiova.ro. On (kHz) 603 (Drobeta-Turnu Severin), 1314 (Craiova) & (MHz) 102.9 (Craiova 1.3kW), 105.0 (Cerbu 10kW): 0400-2200. – **R. Resita:** Str. Maior Petru nr. 71, 320111 Resita. **E:** secretariat@radio-resita.ro. On 105.6MHz (Semenic 100kW): 0400-2000. Incl. prgrs in Croatian, Hungarian, German, Romany, Serbian, Slovak, Ukrainian. – **R. Târgu Mures:** Bd. 1 Decembrie 1918 nr. 109, 540445 Târgu Mures. **E:** office@radiomures.ro. On (MHz) 98.4 (Toplita 2kW), 98.9 (Suseni 60kW), 102.9 (Târgu Mures 10kW): 0400-2000; on (kHz) 1197 (Brasov) 1323 (Târgu Mures), 1593 (Miercurea Ciuc): 0400-2200. Hungarian ("Marosvásárhelyi Rádió"): MF 0900-1600, Sat 0600-0900 & 1200-1600, Sun 0800-1600. In German ("R. Neumarkt") on MW only: 0830-0930 (Sun), 1900-2000 (W, also via Sibiu 1593kHz). Local sub-studio with own prgr **"Antena Brasovului"** (other times rel. R.Târgu Mures): Bd. Eroilor nr. 29, 507246 Brasov. **E:** antenabv@rdsbv.ro. On 1197kHz: 0600-0700 (MF), 0700-0800 (Sun), 0900-1100 (Sat), 1600-1700. – **R. Timisoara:** Str. Pestalozzi nr. 14A, 300115 Timisoara. **E:** secretariat@radiotimisoara.ro. On 630kHz & (MHz) 103.8 (Faget 5kW), 105.9MHz (Urseni): 0400-2200. For ethnic minorities: on MW only: German ("R. Temeswar"): 1100-1200, Hungarian ("Temesvári Rádió"): 1200-1300, Serbian: 1300-1400; on MW+FM on Sun: 1400-1430 Bulgarian, 1430-1500 Czech, 1500-1600 Slovak, 1600-1625 Ukrainian, 1625-1700 Romany.

OTHER STATIONS

MW	kHz	kW	Location	Station
3)	1485	1	Botosani	R. Vocea Sperantei
3)	1485	1	Iasi	R. Vocea Sperantei
3)	1485	1	Medias	R. Vocea Sperantei
3)	1485	1	Oradea	R. Vocea Sperantei
3)	1584	1	Craiova	R. Vocea Sperantei
3)	1584	1	Radauti	R. Vocea Sperantei
3)	1584	1	Sighetul M.	R. Vocea Sperantei
3)	1584	1	Tecuci	R. Vocea Sperantei
3)	1584	1	Zalau	R. Vocea Sperantei
3)	1602	1	Bistrita	R. Vocea Sperantei
3)	1602	1	Piatra Neamt	R. Vocea Sperantei

FM	MHz	kW	Location	Station
14)	88.0	30	Bârlad	Kiss FM
2A)	88.4	10	Varatec	Info Pro
2A)	88.5	100	Semenic	Info Pro
5)	88.9	10	Mahmudia	Romantic FM
4)	89.0	30	Laposi	Guerrilla FM
2A)	89.1	10	Târgu Mures	Info Pro
10)	89.7	22	Suseni	R. Maria
2A)	90.0	60	Magura Odobesti	Info Pro
5)	90.6	30	Topolog	Romantic FM
10)	90.7	30	Turnu Magurele	National FM
13)	90.9	10	Mahmudia	National FM
14)	91.3	30	Dealul Istrita	Kiss FM
2A)	91.3	10	Urseni	Info Pro

FM	MHz	kW	Location	Station
2A)	91.4	60	Suseni	Info Pro
A)	91.7	30	Feleac	RFI relay
2A)	92.0	60	Vacareni	Info Pro
14)	92.0	10	Bistrita	Kiss FM
2A)	92.4	60	Paltinis	Info Pro
10)	92.5	10	Mezec	R. Maria
6)	93.0	10	Mahmudia	Realiatea FM
9)	93.0	6	Suseni	Trinitas FM
13)	93.4	30	Heniu	National FM
2A)	93.9	30	Baneasa	Info Pro
2A)	94.3	30	Cozia	Info Pro
2B)	94.9	60	Paltinis	Pro FM
9)	95.3	100	Costila	R. Trinitas
16)	95.5	10	Calafat	R. Galaxy
2A)	95.7	30	Bârlad	Info Pro
2A)	95.9	30	Magura Boiu	Info Pro
2A)	96.2	60	Techirghiol	Info Pro
8)	96.3	5	Cozia	R. Galaxy
9)	96.5	60	Acareni	R. Trinitas
5)	96.6	60	Paltinis	Romantic FM
2A)	96.7	10	Saveni	Info Pro
11)	96.9	60	Mogosa	R. Impact
4)	97.2	30	Turnu Magurele	Guerrilla FM
4)	97.8	60	Techirghiol	Guerrilla FM
2A)	97.9	10	Bucuresti	Info Pro
13)	98.1	10	Varatec	National FM
2A)	98.5	10	Parâng	Info Pro
2A)	98.5	30	Laposi	Info Pro
12)	98.5	10	Calafat	Logos Calafat
14)	98.5	30	Baneasa	Kiss FM
5)	98.5	60	Mogosa	Romantic FM
2A)	99.2	100	Pietrarie	Info Pro
2A)	99.2	30	Cozia	Kiss FM
14)	99.6	100	Costila	Info Pro
2A)	100.7	100	Carbu	Info Pro
13)	100.9	5	Faget	National FM
2A)	101.6	10	Mezes	Info Pro
1)	102.7	5	Mezes	Europa FM
1)	103.4	9.5	Galati	Europa FM
2A)	103.4	30	Mihoveni	Info Pro
14)	104.1	10	Parâng	Kiss FM
1)	104.4	5	Timisoara	Europa FM
2A)	104.5	30	Feleac	Info Pro
2A)	104.8	30	Rarau	Info Pro
2A)	105.3	60	Siria	Info Pro
2A)	105.3	60	Mogosa	Info Pro
1)	105.5	7.6	Suceava	Europa FM
2A)	105.5	30	Craiova	Info Pro
1)	105.8	7.6	Focsani	Europa FM
13)	105.9	10	Parâng	National FM
1)	106.1	5.2	Constanta	Europa FM
1)	106.3	5.1	Comanesti	Europa FM
1)	106.5	5.1	Iasi	Europa FM
1)	106.6	11.7	Feleac	Europa FM
1)	106.7	23.4	Bucuresti	Europa FM
1)	107.1	5.1	Târgu Jiu	Europa FM
1)	107.4	5	Tulcea	Europa FM
1)	107.5	5	Resita	Europa FM
2A)	107.6	60	Oradea	Info Pro
2A)	107.8	30	Heniu	Info Pro
2A)	107.9	60	Bigor (Curcubata)	Info Pro
7)	107.9	10	Tulcea	R. 21
2A)	107.9	30	Balota	Info Pro

NB: Txs below 5kW not listed.

Addresses & other information:
1) Str. Horia Macelariu nr. 36-28, sector 1, 013932 Bucuresti. **E:** europafm@europafm.ro – **2A,B)** Bd. Pache Protopopescu nr. 109, 021409 Bucuresti. **E:** 1A) infopro@infopro.ro; 1B) profm@profm.ro – **3)** Str. Erou Iancu Nicolae nr. 38-38A, 077190 Voluntari. **E:** rvs@rvs.ro – **4)** Piata Natiunilor Unite nr. 3-5, sector 4, 040012 Bucuresti. **E:** statmajor@radioguerilla.ro – **5)** Bd. Ficusului 44A, 013975 Bucuresti. **E:** stiri@romanticfm.ro – **6)** Piata Presei Libere nr. 1, sector 1, 013701 Bucuresti. **E:** office@realitatea.net – **7)** Str. Horia Macelariu nr. 36-28, sector 1, 013932 Bucuresti. **E:** radio21@radio21.ro – **8)** Bd. Carol nr. 43, 318688 Drobeta-Turnu Severin. E: office@radiogalaxy.ro – **9)** str. Cuza-Voda 51, 700038 Iasi. **E:** radio@trinitas.ro – **10)** Str. Barsei 18, 410423 Oradea. **E:** info.rom@radiomaria.org – **11)** Str. Gheorghe Doja nr. 181A, 720147 Suceava. **E:** office@radioimpactfm.ro – **12)** Str. Mitropolit Firmilian nr. 3, 200381 Craiova. – **13)** str. Fabricii nr. 46B, sector 6, 060823 Bucuresti. **E:** radio@nationalfm.ro – **14)** Splaiul Independentei nr. 202A, sector 6, 060022 Bucuresti. **E:** kissfm@kissfm.ro – **A)** Rel. RFI (France).

DAB (Trial): Bucuresti ch12A (223.936MHz). **Operator:** Radiocom

L.T: Moscow: UTC +3h (27 Mar-30 Oct: +4h); See VGTRK Reg. sces section for other zones — **Pop:** 140 million — **Pr.L:** Russian, numerous minority languages — **E.C:** 50Hz, 220V — **ITU:** RUS

FEDERALNAYA SLUZHBA PO NADZORU V SFERE SVYAZI, INFORMATSIONNYKH TEKHNOLOGIY I MASSOVYKH KOMMUNIKATSII (ROSKOMNADZOR)
⌧ 109074 Moskva, Kitaygorodskiy proyezd 7 ☎ +7 495 9876800 ▤ +7 4959876801 **E:** rsoc_in@rsoc.ru **W:** www.rsoc.ru
L.P: Head: Sergey K. Sitnikov **NB:** Licensing body for broadcasting.

VSEROSSIYSKAYA GOSUDARSTVENNAYA TELEVIZIONNAYA I RADIOVESHCHATELNAYA KOMPANIYA (VGTRK) (Gov)
⌧ 125040 Moskva, Yamskogo polya 5-ya ul. 19/21 ☎ +7 495 2514050▤ ☎ +7 495 2142347 **E:** vgtrk@vgtrk.com **W:** www.vgtrk.com
L.P: GD: Oleg Dobrodeyev
VGTRK is the national state broadcasting company of the Russian Federation and the holding for the federal radio networks R.Rossii, R.Mayak, R.Rossii.Kultura, Radio Yunost, Vesti FM, and the regional state broadcasting companies (see Regional Services section): **RADIO ROSSII** ⌧ 125040 Moskva, Yamskogo polya 5-ya ul. 19/21 ☎ +7 495 9506989 ▤ +7 495 2145366. **E:** mail@radiorus.ru **W:** www.radio-rus.ru – **RADIO MAYAK** ⌧ 125040 Moskva, Yamskogo polya 5-ya ul. 19/21 ☎ +7 495 9594207 ▤ +7 495 9558561 ☎ +7 495 9594198 **E:** pr@radiomayak.ru **W:** www.radiomayak.ru – **RADIO YUNOST (YuFM)** ⌧ 115326 Moskva, ul. Pyatnitskaya 25 ☎ +7 495 9506024 ▤ +7 495 9594198 **E:** info@radiounost.ru **W:** www.radiounost.ru – **RADIO ROSSII.KULTURA** ⌧ 125040 Moskva, Yamskogo polya 5-ya ul. 19/21 ☎ +7 495 6335558 ▤ +7 495 633557 **E:** reklama@cultradio.ru **W:** www.cultradio.ru – **VESTI FM** ⌧ 115162 Moskva, ul. Shabolovka 37 ☎ +7 495 2349768 **E:** vesti-fm@vgtrk.com **W:** radio.vesti.ru.

VGTRK HOME SERVICES (Gov)
Abbreviations: RR = R. Rossii (D-1 to D-4 refer to the timeshifted editions Dubl 1 - 4, cf. D.Prgr); RM = R. Mayak; Reg = Regional prgrs; Prgrs not produced by VGTRK: F = Foreign Service prgrs by Golos Rossii (GR) and/or rel. of foreign broadcasters; Rg = Region (decoding table see Reg. services chapter); Geographical location: E = European part of Russia, S = Siberia, FE = Russia's Far East. **NB.** Tr times subject to changes

LW/MW

Rg	Location	kHz	kW	Hour of tr	Prgr
KH	Komsomolsk-na-A., FE	153	1200	2000-1600	RR-D2, Reg
KN	Norilsk, S	162	150	2200-1800	RR-D3, Reg
KA	Bolshakovo, E	171	150	0200-2200	RR
NS	Oyash, S	171	250	2200-1800	RR-D3, Reg
RS	Yakutsk, FE	171	150	2200-1800	RR-D3, Reg
ZB	Kruchina, S	180	150	2100-1600	RM
KM	Yelizovo, FE	180	150	1800-1400	RR-D3, Reg
AM	Konstantinogradovka, FE	189	1200	2000-1600	RR-D2, Reg
MO	Kurovskaya, E	198	150	0230-2200	RM
SP	Olgino, S	198	150	0300-2200	RM
AM	Tynda, FE	207	150	2100-1600	RM
YV	Birobidzhan, FE	216	30	2000-1600	RR-D2, Reg
KN	Krasnoyarsk, S	216	150	2200-1800	RR-D3, Reg
KY	Surgut, S	225	1000	0000-2000	RR-D4, Reg
IR	Angarsk, S	234	500	2200-1800	RR-D3, Reg
MA	Arman, FE	234	150	1800-1400	RR-D1, Reg
TS	Kazan, E	252	150	0200-2200	RR, Reg
ZB	Kruchina, S	261	150	2100-1600	RR-D2, Reg
MO	Taldom, E	261	250	0200-2200	RR
BU	Selenginsk, S	279	150	2200-1800	RR-D3, Reg
SV	Yekaterinburg, S	279	150	0000-2000	RR-D4, Reg
RA	Gorno-Altaysk, S	279	50	2200-1800	RR-D3, Reg
SL	Yu-Sakhalinsk, FE	279	1000	1900-1500	RR-D1, Reg
CV	Cheboksary, E	531	30	0300-2200	RM, Reg
OB	Orenburg, E	540	50	0100-2000	RM
MO	Noginsk, E	549	75	0300-2200	RM
AM	Svobodnyy, FE	549	150	2100-1600	RM
RS	Yakutsk, FE	549	50	2100-1600	RM
RO	Novocherkassk, E	549	50	0300-2200	RM
KO	Syktyvkar, E	549	150	0300-2200	RM
KA	Kaliningrad, E	549	50	0300-2200	RM
MA	Magadan, FE	549	25	1900-1500	RM
SP	Krasnyy Bor, E	549	600	0300-2200	RM
PM	Tavrichanka, FE	549	150	1900-1500	RM
VG	Volgograd, E	567	1000	0200-2200	RR, Reg
RT	Kyzyl, S	567	150	2200-1800	RR-D3, Reg
NS	Oyash, S	576	500	0000-1900	RM
KH	Khabarovsk, FE	576	150	2000-1500	RM
KM	Yelizovo, FE	576	150	1800-1300	RM

Rg	Location	kHz	kW	Hour of tr	Prgr
IR	Angarsk, S	576	250	2200-1700	RM
PR	Sylva, E	585	150	0000-2000	RR-D4, Reg
KY	Surgut, S	594	1000	0100-2000	RM
KN	Krasnoyarsk, S	594	150	2300-1800	RM
UD	Izhevsk, E	594	40	0200-2200	RR
AM	Skovorodino, FE	603	30	2100-1600	RM
AM	Belogorsk, FE	603	30	2100-1600	RM
KY	Norilsk, S	612	25	2300-1800	RM
KT	Pedaselga, E	612	150	0300-2200	RM
KH	Khabarovsk, FE	621	50	2000-1600	RR-D2, Reg
KO	Syktyvkar, E	621	50	0200-2200	RR, Reg
DA	Makhachkala, E	621	50	0200-2200	RR, Reg
DA	Kochubey, E	621	5	0200-2200	RR, Reg
OM	Omsk, S	639	75	0000-2000	RR-D4, Reg
MU	Murmansk, E	657	150	0200-2200	RM
KH	Komsomolsk-na-A., FE	666	150	2000-1500	RM
KD	Sochi, E	666	25	0300-2200	RM
BA	Yazykovo, E	693	150	0000-2000	RR-D4
IR	Bratsk, S	702	7	2200-1700	RM
NE	Naryan-Mar, E	711	7	0200-2200	RR, Reg
KH	Nikolayevsk, FE	711	5	2000-1600	RR-D2, Reg
SL	Vestochka, E	§ 720	1000	1900-1100[a]	RM, F
KM	Palana, FE	738	25	1800-1400	RR-D1, Reg
CB	Chelyabinsk, S	738	40	0000-2000	RR-D4, Reg
KT	Pedaselga, E	765	150	0200-2200	RR, Reg
VN	Voronezh, E	774	30	0300-2200	RM
SL	Aleksandrovsk, FE	792	50	1800-1400	RR-D1, Reg
RK	Abakan, S	792	25	2200-1800	RR-D3, Reg
VG	Volgograd, E	810	500	0300-2200	RM
SV	Yekaterinburg, S	810	50	0100-2000	RM
PM	Razdolnoye, FE	810	150	2200-1800	RR-D3, Reg
RT	Kyzyl, S	828	150	2300-1800	RM
KX	Elista, E	846	42	0200-2200	RR, Reg
MO	Noginsk, E	‡ 846	150	various	Reg
PZ	Kamenka, E	855	50	0200-2200	RR
RS	Yakutsk, S	864	25	2000-1600	Reg
MO	Lesnoy, E	873	250	0200-2200	RR
SP	Olgino, S	873	75	0200-2200	RR
SA	Samara, E	873	100	0200-2200	RR, Reg
KA	Kaliningrad, E	873	50	0300-2200	RR
ST	Stavropol, E	882	7	0300-2200	RR, Reg
TY	Tyumen, S	891	5	0100-2200	RM
ME	Sovetskiy, E	900	20	0300-2200	RM
AR	Arkhangelsk, E	918	150	0200-2200	RR, Reg
DA	Makhachkala, E	918	50	0300-2200	RM, Reg
KG	Shumikha, S	918	5	0100-2200	RM
KG	Makushino, S	918	5	0100-2200	RM
KG	Shadrinsk, S	918	7	0100-2200	RM
OB	Matveyevka, E	936	5	0000-2000	RR-D4, Reg
RO	Novocherkassk, E	945	40	0200-2200	RR, Reg
BU	Zakamensk, S	963	25	2200-1800	RR-D3
BU	Guzinoozersk, S	963	1	2200-1800	RR-D3
CB	Yuryuzan, S	990	1	0100-2000	RM
KD	Tuapse, E	1008	5	0200-2200	RR
AR	Porog, E	1026	5	0300-2200	RM
AR	Urdoma, E	1026	5	0300-2200	RM
AR	Nyandoma, E	1026	5	0300-2200	RM
KM	Ust-Kamchatsk, FE	1062	1	1800-1300	RM
AM	Zeya, FE	1071	7	2100-1600	RM
MD	Kovylkino, E	1080	100	0200-2200	RR, Reg
KD	Tbilisskaya, E	§ 1089	1200	0200-2200[a]	RR, F
KM	Tilichiki, FE	1089	5	1800-1400	RR-D1, Reg
VO	Nikolsk, E	1098	5	0200-2200	RR, Reg
VO	Chagoda, E	1098	7	0200-2200	RR, Reg
KD	Sochi, E	1116	30	0200-2200	RR, Reg
MD	Kovylkino, E	1134	25	0300-2200	RM
BR	Shvedchiki, E	1134	7	0300-2200	RM
RO	Volgodonsk, E	1134	5	0200-2200	RR, Reg
RO	Salsk, E	1134	5	0200-2200	RR, Reg
MU	Murmansk, E	1134	75	0300-2200	RM
RO	Veshenskaya, E	1134	5	0200-2200	RR, Reg
IR	Tayshet, S	1143	7	2200-1700	RM
SA	Samara, E	1143	100	0300-2200	RM
KA	Bolshakovo, E	1143	150	0300-2400[a]	RM, F
KH	Komsomolsk-na-A., FE	1152	50	2000-1400	RR-D2, Reg
SR	Balakovo, E	1197	5	0200-2200	RR, Reg
SR	Balashov, E	1197	5	0200-2200	RR, Reg
SR	Yershov, E	1197	5	0200-2200	RR, Reg
ZB	Mogocha, S	1197	0.2	2000-1600	RR-D2, Reg
ZB	Ulety, S	1197	0.2	2000-1600	RR-D2, Reg
ZB	Nerchinsk, S	1197	0.2	2000-1600	RR-D2, Reg
ZB	Chernyshevsk, S	1197	0.2	2000-1600	RR-D2, Reg
AR	Plesetsk, E	1206	5	0300-2200	RM
ST	Neftekumsk, E	1251	1	0300-2200	RM
ST	Letnyaya Stavka, E	1251	5	0300-2200	RM
KC	Cherkessk, E	1251	7	0200-2200	RR, Reg
KC	Urup, E	1251	1	0200-2200	RR, Reg
BU	Bagdarin, S	1278	5	2200-1800	RR-D3, Reg
BU	Barguzin, S	1278	25	2200-1800	RR-D3, Reg
BU	Severobaykalsk, S	1278	7	2200-1800	RR-D3, Reg
CC	Groznyy	1278	50	0200-2200	RR, Reg
BU	Kyakhta, S	1287	5	2200-1800	RR-D3, Reg
SV	Serov, S	1305	7	0100-2000	RM
IR	Ust-Kut, S	1305	7	2200-1700	RM
OB	Pleshanovo, E	1314	1	0000-2000	RR-D4, Reg
RA	Ust-Kan, S	1350	5	2200-1800	RR-D1, Reg
RA	Ust-Ulagan, S	1350	5	2200-1800	RR-D1, Reg
PR	Perm, E	1359	50	0100-2000	RM
RA	Shebalino, S	1359	1	2300-1800	RM
RA	Onguday, S	1359	5	2300-1800	RM
RA	Choya, S	1359	1	2300-1800	RM
IR	Ust-Ilimsk, S	1359	7	2200-1700	RM
SL	Okha, FE	1377	5	1900-1400	RM
OB	Buguruslan, E	1395	5	0000-2000	RR-D4, Reg
RS	Sangar, FE	1413	5	2000-1600	RR-D2
ZB	Kuanda, S	1422	5	0000-2000	RR-D4
RA	Kosh-Agach, S	1440	5	2200-1800	RR-D3, Reg
RA	Ust-Koksa, S	1440	5	2200-1800	RR-D3, Reg
RA	Turachak, S	1440	5	2200-1800	RR-D3, Reg
BR	Unecha, E	1449	7	0300-2200	RM
MU	Monchegorsk, E	1449	42	0300-2200	RM
MU	Ostrovnoy, E	1449	7	0300-2200	RM
MU	Umba, E	1449	7	0300-2200	RM
MU	Kirovsk, E	1449	5	0300-2200	RM
MU	Kandalaksha, E	1449	1	0300-2200	RM
MU	Nikel, E	1449	1	0300-2200	RM
PR	Kudymkar, E	1458	7	0000-2000	RR-D4, Reg
RA	Onguday, S	1476	20	2200-1800	RR-D3, Reg
RS	Olekminsk, FE	1485	0.2	2000-1600	RR-D2
TY	Tyumen, S	1485	1	0000-2000	RR-D4, Reg
RS	Solnechnyy, FE	1485	2	2100-1600	RM
KM	Kamenskoye, FE	1485	1	1800-1400	RR-D1, Reg
YN	Krasnoselkup, S	1485	1	0000-2000	RR-D4
YN	Gazsale, S	1485	1	0000-2000	RR-D4
RS	Batagay, FE	1485	1	2000-1600	RR-D2
IR	Magistralnyy, S	1503	1	2200-1800	RR-D3, Reg
YN	Salekhard, S	1503	5	0100-2000	RM
KN	Boguchany, S	1521	5	2200-1800	RR-D3
MU	Zapolyarnyy, E	1521	7	0300-2200	RM
ZB	Krasnyy Chikoy, S	1530	5	2100-1600	RM
SL	Aleksandrovsk, FE	1548	10	1900-1400	RM
BU	Taksimo, S	1584	1	2200-1800	RR-D3, Reg
DA	Khunzakh, E	1584	1	0200-2200	RR
KM	Klyuchi, FE	1584	1	1800-1400	RR-D1, Reg
KM	Tigil, FE	1584	1	1800-1400	RR-D1, Reg
RS	Belk.Gora, FE	1584	0.2	2000-1600	RR-D2
RS	Khandyra, FE	1584	0.2	2000-1600	RR-D2
BU	Novo-Ilinsk, S	1602	1	2200-1800	RR-D3, Reg
BU	Ust-Barguzin, S	1602	1	2200-1800	RR-D3, Reg
KH	Chumikan, FE	1602	0.2	2000-1600	RR-D2

SW:

Rg	Location	kHz	kW	H of tr[o]	Su/Wi	Prgr
MO	Taldom, E	5905	250	1620-1800	Wi	RR
KM	Petropavlovsk-K., FE	5930	100	1800-1400	Su	RR-D1, Reg
MU	Monchegorsk, E	5930	50	0200-2200		RR, Reg†
MA	Arman, FE	5940	100	1800-1400		RR-D1, Reg
KM	Petropavlovsk-K., FE	6075	100	1800-1400	Wi	RR-D1, Reg
KN	Krasnoyarsk, S	6085	10	2200-1800		RR-D3, Reg
RT	Kyzyl, S	6100	1	2200-1800		RR-D3, Reg
RS	Yakutsk, FE	6150	5	2000-1600		RR-D2, Reg
AR	Arkhangelsk, E	6160	40	0200-2200		RR, Reg
BU	Selenginsk, S	6195	50	2200-1800		RR-D3, Reg
RS	Yakutsk, FE	7140	5	2000-1600		RR-D2, Reg
RS	Yakutsk, FE	7200	250	2000-1600		RR-D2, Reg
RS	Yakutsk, FE	7230	100	2000-1600		RR-D2, Reg
MO	Taldom, E	7310	250	1320-1600	Wi	RR
MA	Arman, FE	7320	100	1800-1400		RR-D1, Reg
MO	Taldom, E	7420	250	1820-2200	Su	RR
MO	Taldom, E	9410	250	1420-1800	Su	RR
MO	Taldom, E	9840	250	0500-0800	Wi	RR
MO	Taldom, E	12070	250	0500-0900	Su	RR
MO	Taldom, E	12075	250	0820-1300	Wi	RR
MO	Taldom, E	13665	250	0200-2200		RR

Keys: [a] Times may vary between broadcasting seasons depending on the schedule of GR; ‡) tx shared with commercial services (see "Other stations" section); §) tx shared with Foreign Service trs and/or relays (see Int Radio section); [o] for compatibilty reasons, the summer schedules are given in winter UTC: substract 1h during summer; Su) summer (27 Mar-30 Oct), Wi) winter (to 26 Mar/fr 31 Oct); †) reg prgr may be cancelled **FM:** All nat. (federal) channels are broadc. over a large netw. of FM txs.
D.Prgr: RR (R. Rossii): 24h. 0200-2200 via txs in European Russia,

and in four time-shifted editions in other parts of the country: "Dubl 4" (RR-D4) via txs in the region between Volga and the Urals: 0000-2000, "Dubl 3" (RR-D3) for We.Siberia: 2200-1800, "Dubl 2" (RR-D2) for Ea.Siberia: 2000-1600, "Dubl 1" (RR-D1) for the Russian Far East: 1800-1400. NB: These editions are not announced on-air, and some prgrs may be carried simultaneously in several editions. – **RM (R. Mayak), Mayak FM, R. Rossii.Kultura, Vesti FM, YuFM:** 24h.

VGTRK REGIONAL SERVICES (Gov)

The regional VGTRK branches (in form of reg. state broadcasting companies - GTRK: gosudarstvennaya teleradiokompaniya) provide reg. services in each of the regions of the Russian Federation. They transmit reg. news-blocks on txs shared with federal services, usually R. Rossii. Some GTRKs have also separate local or reg. outlets on own frequencies.

Ann: The stations typically identify with the name of their prgr (as listed below), or in the form "Radio Rossii - (name of regional capital)".

NB: Due to space limitations, schedules are listed only for regional stns that are also broadcasting on SW.

Local Time: "[+3]" - differences in hours to UTC (winter).

AD) Respublika Adygeya [+3]: GTRK "Adygeya", 385000 Maykop, ul. Zhukovskogo 24. **E:** trkra@radnet.ru **Reg:** On (MHz) Guzeripl 68.00, Maykop 69.08, Khamyshki 70.70, Koshekhabl 71.93, Takhtamukay 73.76 in Russian, Adyghian. Apart from own reg. prgrs, txs also relay GTRK "Kuban", Krasnodar (KD). Local channel "Adygeya+" on Maykop 67.88MHz.

AK) Altayskiy kray [+6]: GTRK "Altay", 656045 Barnaul, Zmeinogorskiy trakt 27a. **E:** altai@gtrk.ttb.ru **Reg:** On Gorno-Altaysk [RA] 279kHz + (MHz) Pavlovsk 66.53, Gornyak 66.59, Slavgorod 66.59, Mikhaylovka 66.59, Tselinnoye 66.74, Blagoveshchenka 66.95, Mamontovo 67.16, Ust-Kalmanka 67.85, Zmeinogorsk 68.15, B.Istok 68.27, Pankrushikha 68.36, Barnaul 68.60, Zarinsk 69.53, Rubtsovsk 69.68, Kamen-na-Obi 70.31, Shipunovo 70.35, Biysk 70.40. Local channel "Heart FM" on Barnaul 69.80/105.9MHz.

AM) Amurskaya oblast [+9]: GTRK "Amur", 675000 Blagoveshchensk, per. Svyatitelya Innokentiya 15. **E:** gtrkamurinformrv@tsl.ru **Reg:** On Konstantinogradovka 189kHz + (MHz) Skovorodino 67.22, Belogorsk 67.82, Zeya 68.24, Progress 68.36, Shimanovsk 67.52, Svobodnyy 69.92, Tynda 70.64, Blagoveshchensk 72.86.

AR) Arkhangelskaya oblast [+3]: GTRK "Pomorye", 163061 Arkhangelsk, ul. Popova 2. **E:** agtrk@pomorie.ru **Reg:** "R. Pomorye" on Arkhangelsk 918/6160kHz + (MHz) Arkhangelsk 66.08, Urdoma 66.38, Nyandoma 67.31, Karpogory 67.76, Plesetsk 69.23, Vazhskiy 69.56, Pogost 69.92, Pogor 70.19 (also via txs in region NE): 0410-0500 (MF), 0710-0800 (SS), 0910-1000 (SS), 1010-1100 (MF), 1510-1600 (MF).

AS) Astrakhanskaya oblast [+3]: GTRK "Lotos", 414000 Astrakhan, ul. Molodoy Gvardii 17. **E:** tvlotos@astranet.ru **Reg:** "R. Lotos" on (MHz) Astrakhan 66.02, Chernyy Yar 69.98, Tambovka 70.16.

BA) Resp. Bashkortostan [+5]: GTRK "Bashkortostan", 450076 Ufa, ul. Gafuri 9/1. **E:** gtrk@bashtv.ru **Reg:** "R. Bashkortostana" on (MHz) Neftekamsk 66.47, Mesyagutovo 66.86, Salavat 67.04, Baymak 67.16, Ufa 68.24, Oktyabrskiy 70.28, Belebey 70.61, Abzanovo 70.82, Burayevo 71.90, Beloretsk 72.05, Bakaly 103.7 in Bashkir, Russian. Local channel "R. Yuldash" on Ufa 105.5MHz & network in Bashkir, Tatar; local channel "R. Sputnik FM"on Ufa 107.0MHz & network in Russian.

BE) Belgorodskaya oblast [+3]: GTRK "Belgorod", 308000 Belgorod, pr. Slavy 60. **E:** trcblg@belgtts.ru **Reg:** On (MHz) Stroitel 66.17, Selivanovo 66.80, Rakitnoye 68.39, Biryuch 69.29, Belgorod 70.16, Volokonovka 70.49, Ivnya 70.64, Prokhorovka 70.76, Borisovka 71.03, Staryy Oskol 71.09, Borisovka 71.30, Alexeyevka 71.78, Chernyanka 72.02, Valuyki 73.64, Krasnoye 73.67.

BR) Bryanskaya oblast [+3]: GTRK "Bryansk", 241033 Bryansk, ul. Stanke Dimitrova 77. **E:** radio@br-tvr.ru **Reg:** On (MHz) Navlya 67.37, Bryansk 67.58, Shvedchiki 70.04, Unecha 70.55, Pocheb 71.54, Trubchevsk 73.94.

BU) Respublika Buryatiya [+8]: GTRK "Buryatiya", 670000 Ulan-Ude, ul. Erbanova 7. **E:** bgtrk@bgtrk.ru **Reg:** On Selenginsk 279/6195kHz + (MHz) Severobaykalsk 66.30, Zakamensk 66.68, Ulan-Ude 69.74, Kyakhta 70.16, Sukhara & Tankhoy & Vydryno 101.0, Baykalo-Kudara 102.0, Kabansk 104.0 in Russian, Buryat: 0110-0200 (SS), 0310-0325 (MF), 0510-0600 (MF), 1110-1200 (MF), 2210-2300 (Sun-Thu), 2310-2400 (Sun-Thu).

CB) Chelyabinskaya oblast [+5]: GTRK "Yuzhnyy Ural", 454000 Chelyabinsk, ul. Ordzhonikidze 54b. **E:** radio@cheltv.ru **Reg:** "R. Yuzhnyy Ural" on Chelyabinsk 738kHz + (MHz) Kartaly 66.65, Kyshtym 67.13, Yuryuzan 67.25, Stepnoye 68.36, Novoburino 70.82, Chelyabinsk 71.18, Zlatoust 71.69, Magnitogorsk 71.81. Local channel "Studiya 1"on (MHz) Miass 65.90, Yuryuzan 68.03, Zlatoust 72.47, Chelyabinsk 106.8.

CC) Chechenskaya respublika [+3]: GTRK "Vaynakh", 364000 Groznyy, ul. pr. Pobedy 22. **E:** gtrkvaynah@mail.ru **Reg:** On Groznyy 1287kHz + (MHz) Goragorskiy 72.44, Naurskaya 101.2, Shelkovskaya 103.1, Groznyy 103.6 in Russian, Chechen.

CK) Chukotskiy avt. okrug [+11]: GTRK "Chukotka", 686710 Anadyr, ul. Lenina 18. **E:** gtrk@anadyr.ru **Reg:** On (MHz) 101.6 (several txs), 102.8

(several txs), 104.7 (Anadyr & others) in Russian, Chukchi, Eskimo. Txs also relay GTRK "Magadan", Magadan MA. [NB: CK is subordinated to region MA].

CV) Chuvashskaya respublika - Chuvashiya [+3]: GTRK "Chuvashiya", 428003 Cheboksary, ul. Nikolayeva 4. **E:** chradio@tvr.chtts.ru **Reg:** "R. Chuvashii" on (MHz) Tsivilsk 67.04, Ibresi 70.85, Yadrin 71.12 in Russian, Chuvash. Separate prgrs in Chuvash on Cheboksary 531kHz.

DA) Respublika Dagestan [+3]: GTRK "Dagestan", 367020 Makhachkala, ul. Amet-Khana, r-n DSK. **E:** gtrk_dagestan@mail.ru **Reg:** Radiokanal "Chirag" on Makhachkala & Kochubey 621kHz + (MHz) Makhachkala 68.87, Kochubey 68.99, Gergebil 69.95 in Avar, Chechen, Kumyk, Lakk, Nogay, Rutul, Russian, Tsakhur. On Makhachkala 918kHz in Agul, Azeri, Dargwa, Lezgi, Russian, Tabasaran,Tatar.

IN) Ingushskaya respublika [+3]: GTRK "Ingushetiya", 366720 Nazran, pr. Bazorkina 72. **E:** tvi2002@mail.ru **Reg:** On Nazran 72.02MHz in Russian, Ingush.

IR) Irkutskaya oblast [+8]: GTRK "Irkutsk", 664003 Irkutsk, ul. Gorkogo 15. **E:** igtrk@irmail.ru **Reg:** On Angarsk 234kHz + (MHz) Zhmurovo 66.32, Zheleznogorsk 66.56, Ulkan 66.62, Tulun 66.74, Ust-Ilimsk 66.80, Chuna 67.10, Tayshet 69.80, Nizhneudinsk 70.04, Bratsk 70.28, Irkutsk 70.31, Ust-Kut 70.64, Zima 72.14, Ust-Ordynskiy 107.8. Incl. Ust-Ordynskiy okrug prgr in Buryat, Russian.

IV) Ivanovskaya oblast [+3]: GTRK "Ivteleradio", 153647 Ivanovo, ul. Teatralnaya 31. **E:** adm@itrk.ru **Reg:** On (MHz) Rodniki 70.13, Ivanovo 71.21, Furmanov 73.64.

KA) Kaliningradskaya oblast [+2]: GTRK "Kaliningrad", 236016 Kaliningrad, ul. Klinicheskaya 19. **E:** tv-rv@baltnet.ru **Reg:** "R. Yantar" on (MHz) Veselovka 65.90, Kaliningrad 66.02, Bolshakovo 70.19.

KB) Kabardino-Balkarskaya respublika [+3]: GTRK "Kabardino-Balkariya", 360000 Nalchik, pr. Lenina 3. **E:** vestikbr@mail.ru **Reg:** On (MHz) Samarkovo 66.62, Nalchik 70.52, Bulundu 73.70 in Kabardino-Circassian, Balkar, Russian.

KC) Karachayevo-Cherkesskaya respublika [+3]: GTRK "Karachayevo-Cherkesiya", 357100 Cherkessk, ul. Krasnoarmeyskaya 51. **E:** gtrk_kchr@rambler.ru **Reg:** On Cherkessk & Urup 1251kHz + (MHz) Krasnogorskaya 66.98, Karachayevsk 67.34, Adyge-Khabl 68.66, Zelenchuskaya 68.81, Kavkazkiy 68.21, Pregradnaya 69.80, Ispravnaya 69.83, Ust-Dzheguta 69.89, Teberda 69.98, Dombay 70.25, V.Mara 71.06, Storozhevaya 71.72, Kurdzhinovo 71.81, Khabez 71.84, Cherkessk 72.11, Urup 72.44 in Russian, Abaza, Circassian, Karachay, Nogay.

KD) Krasnodarskiy kray [+3]: GTRK "Kuban", 350038 Krasnodar, ul. Radio 5. **E:** owl@kubantv.ru **Reg:** "Radiostantsiya Kuban" on Sochi 1116kHz + (MHz) Tbilisskaya 66.20, Novorossiysk 67.97, Gelendzhik 68.30, Kanevskaya 68.36, Armavir 68.57, Temryuk 70.22, Krasnyy Kut & Psebay 70.43, Tuapse 70.46, Adler 70.58, Eysk 71.15, Glubskaya 71.21, Lazarevskoye 71.54, Krasnodar 71.81, Arkhipo-Osipovka 71.93, Sochi 71.93, Apsheronsk 72.20, Belaya Glina 73.25, Kushchevskaya 73.49, Primorsko-Aktarsk 73.70 (also via txs in region AD). - Territorialnoye otdeleniye GTRK "Kuban", 354000 Sochi, ul. Teatralnaya 11a. **E:** tv@sochi.ru. **Reg:** On Sochi 71.93MHz (tx shared with GTRK "Kuban").

KE) Kemerovskaya oblast [+6]: GTRK "Kuzbass", 650099 Kemorovo, ul. Krasnoarmeyskaya 137a. **E:** director@gtrk.kuzbass.net **Reg:** On (MHz) Yurga 66.11, Novokuznetsk 66.20, Kemerovo 66.56, Klyuchevaya 67.04, Leninsk-Kuznetskiy 69.71, Tashtagol 69.80, Anzhevo-Sudzhensk 70.40, Mezhdurechensk 70.64. Local channel "Kuzbass FM" on (MHz) Topki 65.93, Myski 69.38, Osinniki 68.42, Kemerovo 91.0, Klychevaya 101.2, Leninsk-Kuznetskiy 101.3, Anzhero-Sudzhensk 101.6, Mezhdurechensk 101.8, Guryevsk 102.5, Novokuznetsk 103.0, Yurga 103.6, Belovo 104.5.

KG) Kurganskaya oblast [+5]: GTRK "Kurgan", 640018 Kurgan, ul. Sovetskaya 105. **E:** headgtrk@acmetelecom.ru **Reg:** On (MHz) Shumikha 66.89, Makushino 68.48, Shadrinsk 69.23, Shatrovo 71.18, Kurgan 71.87.

KH) Khabarovskiy kray [+10]: GTRK "Dalnevostochnaya", 682632 Khabarovsk, ul. Lenina 4. **E:** main@dvtrk.khv.ru **Reg:** On (kHz) Komsomolsk-na-Amure 153/1152, Khabarovsk 621, Nikoleyevsk 711 + (MHz) Ayan 68.00, Chumikan 68.12, Komsomolsk 68.72, Okhotsk 69.32, Glebovo 69.47, Vyazemskiy 69.83, Bikin 69.92, Chegdomyn 70.16, Khabarovsk 72.80 (also via txs in region YV). Local channel "R. 101.8" on Khabarovsk 101.8MHz. - GTRK "Komsomolsk-na-Amure", 681000 Komsomolsk-na-Amure, ul. Molododvargeyskaya 7. **Reg:** On Komsomolsk-na-Amure 68.72MHz (tx shared with GTRK "Dalnevostochnaya").

KL) Kaluzhskaya oblast [+3]: GTRK "Kaluga", 248021 Kaluga, Pole Svobody 40a. **E:** gtrk@kaluga.ru **Reg:** On (MHz) Kaluga 66.23, Lyudinovo 67.91, Sukhinichi 69.62, Baryatino 70.79, Mosalsk 71.39, Spassk-Demensk 72.05, Zhizda 72.92, Khvastovichi 73.07, Yukhnov 104.9.

KM) Kamchatskiy kray [+11]: GTRK "Kamchatka", 683000 Petropavlovsk-Kamchatskiy, ul. Sovetskaya 62. **E:** gtrk@kamchatka.tv **Reg:** On (kHz) Yelizovo 180/5930(su)/6075(wi), Palana 738, Tilichiki 1089, Klyuchi & Tigil 1584 + Petropavlovsk 69.68MHz: 0010-0030 (MF), 0210-0230 (MF), 0610-0630 (MF), 0910-1000 (MF), 2010-2030 (Sun-Thu), 2110-2200 (Sun-Thu). Incl. prgr "Kamchatka rybatskaya" (for fishermen): Thu 2110-2140. Local channel on Petropavlovsk 103.5MHz. – Territorialnoye otdeleniye GTRK

"Kamchatka", 684620 Palana, ul. Obukhova 4. **E**: palana_tv@palana.ru **Reg**: On (kHz) Palana 738, Tilichiki 1089, Klyuchi & Tigil 1584 in Russian, Koryak.

KN) Krasnoyarskiy kray [+7]: GTRK "Krasnoyarsk", 660028 Krasnoyarsk, ul. Mechnikova 44a. **E**: referent@kgtrk.krsn.ru **Reg**: On (kHz) Norilsk 162*, Krasnoyarsk 216/6085 + (MHz) Dikson 66.13, Dudinka 66.25, Balakhta 66.32, B.Uluy 67.40, Krasnoyarsk 68.09, Solyanka 68.84, Novomikhaylovka 69.23, Uzhur 69.56, Norilsk 69.68*, Shira 70.16, Achinsk 70.52, Tyukhtet 71.42, Yeniseysk 72.74: 0020-0100 (MF), 0210-0300 (SS), 0610-0700 (MF), 1010-1100 (MF). *) Also carries subregional prgrs: Territorialnoye otdeleniye GTRK "Krasnoyarsk", 663370 Tura, ul. 50 let Oktyabrya 28. **E**: heglen@tura.evenkya.ru **Reg**: In Russian, Evenki on Tura 101.5MHz. — GTRK "Norilsk", 663300 Norilsk, nab. Urvantseva 10. **E**: secr@norilsk-tv.ru **Reg**: On Norilsk 69.68MHz. Territorialnoye otdeleniye GTRK "Norilsk", 647000 Dudinka, ul. Gorogog 15. **E**: gtrktaimyr@dudinka.krasnet.ru **Reg**: On Norilsk 162kHz in Russian, Nenets.

KO) Resp. Komi [+3]: GTRK "Komi Gor", 167610 Syktyvkar, Oktyabrskiy pr. 164. **E**: komigor@online.ru. **Reg**: On Syktyvkar 621kHz + (MHz) Yarashyu 65.90, Priuralsk 66.35, Ukhta 66.44, Vorkuta 66.60, Syktyvkar 66.80, Pechora 66.92, Ust-Tsilma 67.10, Sludka 67.16, Kadzherom 68.30, N.Odes 68.36, Troitsko-Pechorsk 68.60, Shoshka 68.72, Aykino 68.78, Kartayel 68.93, Shchelyabozh 68.93, Trakt 68.96, Meshchura 69.02, Mordino 69.05, Inta 69.08, Krasnobor 69.11, Okunevo 69.20, Usogorsk 69.56, Koygorodok 69.74, Myyeldino 69.83, Ust-Kulom 70.16, Petrun 70.28, Vetyu 70.40, Voyvozh 70.64, Vuktyl 71.06, Kuratovo 71.09, B.Pyssa 71.15, Usinsk 73.31, Vekshor 73.31 in Russian, Komi.

KS) Kostromskaya oblast [+3]: GTRK "Kostroma", 156005 Kostroma, ul. Nikitskaya 10. **E**: radio@gtrk.kmtn.ru **Reg**: On (MHz) Bogovarovo 66.20, Vokhma 66.98, Kostroma 69.86, Galich 66.74, Sharya 67.10, Ostrovskoye 72.26, Pavino 73.10, Chukhloma 73.64, Pyshchug 73.88.

KT) Respublika Kareliya [+3]: GTRK "Kareliya", 185002 Petrozavodsk, ul. Pirogova 2. **E**: gtrk@petrozavodsk.rfn.ru **Reg**: On Pedaselga 765kHz + (MHz) Nadvoitsy 66.29, Sortavala 67.13, Naystenyarvi 69.80, Loukhi 70.07, Kostomuksha 70.28, Petrozavodsk 70.52, Muyezerskiy 72.17, Medvezhyegorsk 72.47 in Russian, Finnish, Karelian, Vepsian.

KU) Kurskaya oblast [+3]: GTRK "Kursk", 305016 Kursk, ul. Sovetskaya 32. **E**: gtrk@kursk.rfn.ru **Reg**: On (MHz) Lgov 66.83, Kursk 69.71, Kshenskiy 72.41

KV) Kirovskaya oblast [+3]: GTRK "Vyatka", 610002 Kirov, ul. Uritskogo 34. **E**: tv@gtrk-vyatka.ru **Reg**: On (MHz) Vyatskiye Polyarny 66.35, Kirov 66.92, Kirs 66.86, Sovietsk 67.07, Klyuchi 67.91, Pinyug 70.55, Shmelevo 70.73, Urzhum 71.06, Omutninsk 71.33, Sanchursk 73.28.

KX) Respublika Kalmykiya [+3]: GTRK "Kalmykiy", 358000 Elista, ul. M. Gorkogo 34. **E**: kalmykiagtrk@mail.ru **Reg**: On Elista 846kHz + (MHz) Sadovoye 66.95, Elista 67.28, Utta 68.24, Ulan-Kholl 69.59 in Russian, Kalmyk.

KY) Khanty-Mansiyskiy avt. okrug [+5]: GTRK "Yugoriya", 626200 Khanty-Mansiysk, ul. Mira 7. **E**: gtrk@wsmail.ru **Reg**: "R. Yugry" on Surgut 225kHz + (MHz) Khanty-Mansiysk 66.08, Beloyarskiy 67.22, Langepas 67.28, Surgut 68.84, Nyagan 70.82, Beryozovo 71.42, Kogalym 71.30, Yugorsk 71.78, Nizhnevartovsk 72.56 in Russian, Khanti, Mansi. Txs also relay GTRK "Region-Tyumen", TY. [NB: KY is subordinated to region TY]

LI) Lipetskaya oblast [+3]: GTRK "Lipetsk", 398050 Lipetsk, pl. Plekhanova 1. **E**: regiontv@lipetsk.rfn.ru **Reg**: On (MHz) Lipetsk 66.53, Izmalkovo 73.79, Terbuny 101.9, Dankov 102.4, Dobrinka 102.7, Ploty 102.6, Chaplygin 103.3, Lev Tolstoy 103.8, Usman 104.0, Volovo 104.4.

MA) Magadanskaya oblast [+11]: GTRK "Magadan", 685024 Magadan, ul. Kommuny 8/12. **E**: center@magtrk.ru **Reg**: On (kHz) Arman 234/5940/7320 + (MHz) Susuman 70.16, Evensk 100.5, Myandzha & Palatka 101.0, Ola 101.3, Omchak & Ust-Omchug 102.5: 0210-0300, 2010-2100. Local channel on Magadan 105.0MHz.

MD) Respublika Mordoviya [+3]: GTRK "Mordoviya", 430000 Saransk, ul. Dokuchayeva 29. **E**: radiomordovii@mail.ru **Reg**: On Kovylkino 1080kHz + (MHz) Tengushevo 66.35, Saransk 66.68, Dubenki 67.28, B.Ignatovo 67.34, Ruzayevka 67.46, Yavas 67.67, Umet 68.33, B.Berezniki 68.42, Torbeyevo 68.69, Chamzinka 68.75, Kovylkino 69.14, Ardatov 69.53, St.Shaygovo 69.65, Insar 71.03, Romodanovo 71.12, Atyuryevo 71.33 in Russian, Mordvin.

ME) Respublika Mariy El [+3]: GTRK "Mariy-El", 424014 Yoshkar-Ola, ul. Osipenko 50. **E**: tv@tv.mari.ru **Reg**: On Sovietskiy 900kHz + (MHz) Yoshkar-Ola 70.34, Sovetskiy 71.21, Kozmodemyansk 72.20, Zvenigovo 73.16 in Russian, Mari.

MO) Moskva (Federal City) & Moskovskaya oblast [+3]: TRK "RTV-Podmoskovye", 123007 Moskva, ul. 1-ya Magistralnaya 14. **E**: info@rtvp.ru **Reg**: On Noginsk 846kHz + (MHz) Chekhov 65.90, Moskva 66.44, Domodedovo 67.61, Serebryanyye Prudy 67.67, Kolomna 67.70, Ruza 67.73, Kashira 68.54, Lukhovitsy 69.98, Vereya 70.34, Klin 70.43, Zosimova Pustyn 70.49, Odintsovo 70.52, Pravdinskiy 70.61, Ramenskoye 70.76, Solechnogorsk 70.82, Naro-Fominsk 70.94, Rogovo 71.69, Krasnogorsk 71.78, Ryazanovskiy 71.99, Lotoshino 72.77, Istra & Sergiyev Posad 91.5, Taldom & Yegoryevsk 100.2, Stupino 100.3, Istra 100.7, Sergiyev Posad 100.7, Podolsk 100.8, Mozhaysk & Ozery 101.6, Orekhovo-Zuyevo

102.1, Shatury 102.7, Dmitrov & Voskressensk 102.8, Serpukhov 104.0, Pavlovskiy Posad 104.5, Zaraysk 107.0, Noginsk 107.2.

MU) Murmanskaya oblast [+3]: GTRK "Murman", 183032 Murmansk, per. Rusanova 7. **E**: radio@tvmurman.com **Reg**: On (kHz) Murmansk 657*, Monchegorsk 5930* + (MHz) Murmansk 67.22, Kandalaksha 67.70, Revda 67.94, Alakurtti 69.50, Teriberka, Umba & Zapolyarnyy 69.74, Ostrovnoy 69.83, Kaneva & Sosnovka 70.01, Tumannyy 70.19, Kirovsk & Prirechnyy 70.34, Lovozero 70.73, Krasnoshchelye 72.38: 0410-0500 (MF), 0710-0800 (W), 0910-1000 (MF), 1510-1600 (W). *) reg prgrs may not always be carried.

NE) Nenetskiy avt. okrug [+3]: Territorialnoye otdeleniye GTRK "Pomorye", 164700 Naryan-Mar, ul. Smidovicha 19. **E**: zapolyarie@mail.ru **Reg**: On Naryan-Mar 711kHz + 66.20MHz in Russian, Nenets. Apart from own reg. prgrs, txs also relay prgrs from GTRK "Pomorye", Arkhangelsk AR. [NB: NE is subordinated to region AR]

NN) Nizhegorodskaya oblast [+3]: GTRK "Nizhniy Novgorod", 603060 Nizhniy Novgorod, ul. Belinskogo 9a. **E**: radio@nnov.rfn.ru **Reg**: On (MHz) Sergach 67.16, Pavlovo 67.85, N.Novgorod 67.94, Kovernino 69.53, Shakhunya 69.59, Arzamas 69.95, Lukoyanov 70.52, Krasnyye Baki 70.64, Vyksa 71.09, Kstovo 73.97.

NO) Novgorodskaya oblast [+3]: GTRK "Slaviya", 173620 Velikiy Novgorod, ul. B.Moskovskaya 106. **E**: radio@slavia.natm.ru **Reg**: "R. Slaviya" on (MHz) Borovichi 69.02, Proletariy 71.39, Zaluchye 71.93, Pestovo 100.0.

NS) Novosibirskaya oblast [+6]: GTRK "Novosibirsk", 630048 Novosibirsk, ul. Rimskogo-Korsakova 9. **E**: gtrkn@nsktv.ru **Reg**: On (MHz) Chulym 66.35, Vengerovo 66.35, Severnoye 66.50, Ust-Tarka 66.62, Suzun 66.68, Ordynskoye 66.74, Kargat 66.89, Kyshtovka 66.89, Chistoozernoye 67.04, Dovolnoye 67.28, Proletarskiy 67.52, Novosibirsk 67.88, Bagan 68.36, Ubinskoye 68.63, Kyshovka 68.84, Karasuk 68.93, Osinovskiy 68.93, Zdvinsk 68.96, Maslyanino 69.05, Kuybyshev 69.68, Kolsk 68.93, Krasnozerskoye 69.95, Cherepanovo 70.10, Moshkovo 70.22, Tatarsk 71.60, Kisilevka 71.66, Beloye 72.08 + txs below 0.1kW.

OB) Orenburgskaya oblast [+5]: GTRK "Orenburg", 460024 Orenburg, per. Televizionnyy 3. **E**: gtrc@orenburg.rfn.ru **Reg**: On (MHz) Orenburg 66.02, Buzuluk 66.62, Orsk 66.92, Yasnyy 69.71, Kuvandyk 70.04, Uralskoye 101.9, Sorochinsk 102.0, Bikkulovo 102.1, Zhdanovka 102.6, Svetlyy 102.9, Pervomayskiy 103.0, Kvarkeno & Saraktash 103.1, Saraktash 103.3, Donetskoye 103.4, Izobilnoye 103.5, Ilek 103.6, Sovosergiyevka 103.9, Sol-lletsk 104.6, Abdulino 105.1, Pleshanovo 105.2, Akbulak 105.5, Sharlyk 106.4, Alekseyevo 106.5, Tyulgan 106.6, Ponomarevka 106.9, Severnoye 107.0, Aleksandrovka 107.8 in Russian, Chuvash, Tatar.

OL) Orlovskaya oblast [+3]: GTRK "Oryol", 302028 Oryol, ul. 7 Noyabrya 43. **E**: post@ogtrk.oryol.ru **Reg**: On (MHz) Livny 67.19, Oryol 70.31.

OM) Omskaya oblast [+6]: GTRK "Irtysh", 644050 Omsk, pr. Mira 2. **E**: gtrk@rtr-omsk.ru **Reg**: On Omsk 639kHz + (MHz) Isilkul 66.50, Ust-Ishim 67.04, Nazyvayevsk 67.28, Tara 68.39, Khutora 70.43, Cherlak 71.06.

PM) Primorskiy kray [+10] GTRK "Vladivostok", 690091 Vladivostok, ul. Uborevicha 20a. **E**: ptr@ptr-vlad.ru **Reg**: On Razdolnoye 810kHz + (MHz) Nakhodka 66.74, Olga 67.58, Permskoye 67.79, Arsenyev 68.60, Kavalerovo 69.20, Dalnerechensk 69.32, Dalnegorsk 70.04, Novozhatkovo 70.64, Plastun & Terney 71.00, Vladivostok 71.84, Chkalovka 72.08, Kamenka & Krasnoperechenskiy & Mikhaylovka & Serafimovka 101.5, Rudnyy 101.9, Moryak Rybolov & Veselyy Yar 102.0, Agzu 103.5.

PR) Permskiy kray [+5]: GTRK "Perm", 614070 Perm, ul. Tekhnicheskaya 7. **E**: main@t7.ru **Reg**: "R. Permskogo kraya" on (kHz) Sylva 584, Kudymkar 1458 + (MHz) Perm 66.02, Kungur 66.65, Barda 67.10, Kudymkar 67.19*, Gayny 67.34*, Ust-Chernaya 68.84*, Ilyinskiy 68.93, Uinskoye 69.38, Karagay 69.53, Chusovoy 70.67, Krasnovishersk 71.33, Berezniki 71.87, Kosa 73.10*. – *) Also carries prgrs by Territorialnoye otdeleniye GTRK "Perm", 617240 Kudymkar, ul. Volodarskogo 18. **E**: kudtv@mail.ru. **Reg**: in Russian, Komi.

PS) Pskovskaya oblast [+3]: GTRK "Pskov", 180000 Pskov, ul. Nekrasova 50. **E**: oblradio@svs.ru **Reg**: On (MHz) Pskov 66.05, Trutnevo 67.34, Novosokolniki 67.94, Dedovichi 69.86, Glubokoye 70.01.

PZ) Penzenskaya oblast [+3]: GTRK "Penza", ul. Lermontova 39, 440602 Penza. **E**: gtrk@penza-trv.ru **Reg**: On (MHz) Pachelma 66.80, Sosnovoborsk 68.51, Meshcherskoye 66.84, Blagodatka 69.08, Penza 70.67, Narovchat 70.88, Gorodishche 100.4, Lunino 101.0, Neverkino 101.1, Issa 102.2, Nikolsk 106.1, M. Serdova 106.4.

RA) Respublika Altay [+7]: GTRK "Gornyy Altay", 659700 Gorno-Altaysk, ul. Choros-Gurkina 38. **E**: info@gtrk.gorny.ru **Reg**: On (kHz) Gorno-Altaysk 279, Ust-Kan & Ust-Ulagan 1350, Kosh-Agach & Turachak & Ust-Koksa 1440, Unguday 1476 + (MHz) Tashanta & Aktash 67.10, Gorno-Altaysk 67.22, Onguday 71.66, Tiyakhty 71.66 in Russian, Altai.

RK) Respublika Khakasiya [+7]: GTRK "Khakasiya", 662000 Abakan, ul. Vyatkina 12. **E**: vgtrk2003@mail.ru **Reg**: "R. Khakasiya" on Abakan 792kHz + (MHz) Abaza 66.02, Abakan 66.89, Kopyevo 68.00, Shira 70.16, Tashtyp 71.00, Askiz 72.59, Cheremushki 73.01, Sorsk 73.19, Sarala 73.52, Beya 73.61 in Russian, Khakass.

RO) Rostovskaya oblast [+3]: GTRK "Don-TR", 344101 Rostov-na-Donu, ul. 1-ya Barrikadnaya 18. **E**: dontr2@dontr.ru **Reg**: On (kHz) Salsk & Volgodonsk

& Veshenskaya 1134 + (MHz) Salsk 66.86, Morozovsk 67.07, Volgodonsk 70.13, Kamensk 70.28, Veshenskaya 70.67, Rostov-na-Donu 72.95

RS) Respublika Sakha (Yakutiya) [+9]: GTRK "Sakha", 677007 Yakutsk, ul. Ordzhonikidze 48. **E:** gtrksakha@yandex.ru **Reg:** On (kHz) Yakutsk 171/6150/7140/7200/7230 + (MHz) Neryungri 66.68, Aldan 69.38, Yakutsk 70.40 in Yakut, Russian: 0310-0500 (MF), 0910-1300 (MF), 2110-0100 (Sun-Thu), 2300-0600 (Fri/Sat). Local Channel "Kuyaar Duoraana" in Yakut on Yakutsk 864kHz + (MHz) Tit-Ary 102.0, Pokrovsk 102.3, various sites 102.5, Yakutsk 107.1.

RT) Respublika Tyva [+7]: GTRK "Tyva", 667003 Kyzyl, ul. Gornaya 31. **E:** tv@tuva.ru **Reg:** On Kyzyl 567/6100kHz + (MHz) Kyzyl 67.10, Shagonar 70.64 in Russian, Tuvinian: 0710-0700 (MF), 0910-0900 (MF).

RY) Ryazanskaya oblast [+3]: GTRK "Oka", 390006 Ryazan, ul. Skomoroshinskaya 20. **E:** okatv@org.etr.ru **Reg:** On (MHz) Kadom 68.03, Ryazan 69.32, Yermish 69.35, Mosolovo 71.66, Lesnoye-Konobeyevo 72.35

SA) Samarskaya oblast [+3]: GTRK "Samara", 443011 Samara, ul. Sovetskoy Armii 205. **E:** gtrk7@tvsamara.ru **Reg:** On Samara 873kHz + (MHz) Sergeyevsk 66.71, Lhvorostyanka 66.98, Zhigulevsk 67.31, Chelno-Vershiny 67.43, Pokhvistnevo 69.05, Kamyshla 69.47, Samara 70.31, Yelkhovka 70.76, B. Glushitsa 71.15, Shentala 71.33, Syzran 73.10, Oktyabrsk 100.4, Kinel-Cherkassy 107.7.

SL) Sakhalinskaya oblast [+10]: GTRK "Sakhalin", ul. Komsomolskaya 209, 693000 Yuzhno-Sakhalinsk. **E:** sakhtv@gtrk.sakhalin.su **Reg:** On (kHz) Vestochka 279, Aleksandrovsk-Sakhalinskiy 792 + (MHz) Smirnykh 68.69, Okha 69.20, Aleksandrovsk-Sakhalinskiy 69.44, Gornozavodsk 69.50, Nogliki 69.56, Poronaysk 69.92, Tomari 70.16, Uglegorsk 70.40, Shebunino 71.63, Korsakov 72.29, Nevelsk 101.7, Dolinsk 102.0, Kholmsk 104.8, Yuzhno-Sakhalinsk 106.0, Korsakov 107.6. Incl. prgr in Korean.

SM) Smolenskaya oblast [+3]: GTRK "Smolensk", 214025 Smolensk, ul. Nakhimova 1. **E:** rukovodstvo@smolgtrk.rfn.ru **Reg:** On (MHz) Smolensk 68.54, Smogiri 68.96, Vyazma 69.20, Roslavl 70.91.

SO) Respublika Severnaya Osetiya - Alaniya [+3]: GTRK "Alaniya", 362007 Vladikavkaz, Osetinskaya gorka 2. **E:** radio@osetia.ru **Reg:** On (MHz) Mozdok 71.78, Vladikavkaz 72.20 in Russian, Ossetic. Local station: "R. Ir" on 103.5MHz.

SP) Sankt Peterburg (Federal City) & Leningradskaya oblast [+3]: GTRK "Sankt-Peterburg", 197022 St.Peterburg, nab. reki Karpovki 43. **E:** info@rtr.spb.ru **Reg:** On St.Peterburg 873kHz + (MHz) Tikhvin 66.14, St.Peterburg 66.30, Kingisepp 67.67, Podporozhye 69.95, Yefimovskiy 72.05.

SR) Saratovskaya oblast [+3]: GTRK "Saratov", 410004 Saratov, 2-ya Sadovaya ul. 7. **E:** top@gtrk.renet.ru **Reg:** On Balakovo & Balashov & Yershov 1197kHz + (MHz) Perelyub 66.44, Yershov 66.48, Aleksandrov Gay 69.68, Balashov 70.16, Balakovo 70.52, Saratov 71.09.

ST) Stavropolskiy kray [+3]: GTRK "Stavropolye", 355000 Stavropol, ul. Artema 35a. **E:** referent@stavropolye.tv **Reg:** "R. Stavropol" on Stavropol 882kHz + (MHz) Ipatovo 66.77, Pyatigorsk 68.96, Stavropol 69.53, Neftekumsk 70.01. Regional channel "R. Rus" on (MHz) Stavropol 68.72/104.3MHz, Nevinnomyssk 67.67, Ipatovo 68.39, Kurskaya 68.72, Pyatigorsk 69.74/102.5 (partly), Blagodarnyy 71.78, Neftekumsk 71.90. Local channel "Pyataya vershina" on Pyatigorsk 69.74/102.5MHz.

SV) Sverdlovskaya oblast [+5]: GTRK "Ural", 620026 Yekaterinburg, ul. Lunacharskogo 212. **E:** radio@sgtrk.ru **Reg:** "R. Urala" on Yekaterinburg 279kHz + (MHz) Talitsa 65.93, Alapayevsk 66.50, Zaykovo 66.83, Nizhniye Sergi 67.01, Ivdel 67.76, Rezh 69.17, Bisert 69.20, Baranchinskiy 69.29, Serov 69.65, Andronovo 70.16, Afanasyevskoye & Azanka 70.43, Yekaterinburg 71.06, Pyshma 71.24, Kamyshlov 72.53.

TA) Tambovskaya oblast [+3]: GTRK "Tambov", 392720 Tambov, ul. Michurinskaya 8a. **E:** tgtrkbgt@tmb.ru **Reg:** On (MHz) Tambov 71.00, Tokaryovka 103.3, Staroyuryevo 103.6.

TL) Tulskaya oblast [+3]: GTRK "Tula", 300600 Tula, Staronikitskaya ul. 1. **E:** info@tula.rfn.ru **Reg:** On (MHz) Efremov 66.92, Tula 71.15, Novomoskovsk 72.35, Suvorov 102.4.

TO) Tomskaya oblast [+6]: GTRK "Tomsk", 634050 Tomsk, ul. Pushkina 19. **E:** adm@tvtomsk.ru **Reg:** "R. Tomsk" on Oyash (NS) 171kHz + (MHz) Aleksandrovskoye 66.02, Kozhevnikovo 67.01, Tomsk 67.22, Krivosheino 67.31, Malinovka 68.39, Parabel 68.45, Kolpashevo 68.87, Strezhevoy 66.78, St.Yuvala 68.15, Strezhevoy 68.51, Teguldet 68.60, Koplashevo 68.87, Podgornoye 69.20, Belyy Yar 69.56, Bakchar 70.01, Kargasok 71.42, Asino 71.57, Molchanovo 71.84, Krasnyy Yar 72.17, Chilino 72.20, Zyryanskoye 73.01, Volodino 73.40.

TS) Respublika Tatarstan [+3]: GTRK "Tatarstan", 420015 Kazan, ul. M. Gorkogo 15. **E:** secret@trrttv.ru **Reg:** "R.Tatarstana"/ "Tatarstan radiosi" on Kazan 252kHz + (MHz) sovkhoz im. Kirova 67.28, Kazan 68.48, Naberezhnyye Chelny 67.01, Bavly 68.45, Leninogorsk 68.63, Tetyushi 69.95, Bilyarsk 70.13, Aktanysh 70.22, Nurlat 70.46, Buinsk 70.61, Cheremshan 72.14, Nizhnekamsk 72.29, A.Saplyk 72.29, Almetyevsk 103.9, Bazarnyye Mataki 107.8 in Tatar, Russian.

TV) Tverskaya oblast [+3]: GTRK "Tver", 170000 Tver, ul. Vagzhanova 9. **E:** gtrktver@tvcom.ru **Reg:** On (MHz) Kashin 66.95, Andriyanovo 68.48, Selizharovo 69.68, Maksatikha 71.24, Kimry 71.48.

TY) Tyumenskaya oblast [+5]: GTRK "Region-Tyumen", 625013 Tyumen, ul.

Permyakova 6. **E:** gtrk@region-tyumen.ru **Reg:** On (MHz) Gagarino 66.89, Sladkovo 68.33, Masali 68.96, Yurginskoye 69.11, Uvat 71.42, Tyumen 71.66, Tobolsk 71.90, Shabanovo 72.17, Uporovo 73.19, Zavodoukovsk 104.1, Baykalovo 105.5 (also via txs in region KY).

UD) Udmurtskaya respublika [+3]: GTRK "Udmurtiya", 426004 Izhevsk, ul. Komunarov 216. **E:** adm@udmtv.ru **Reg:** On (MHz) Izhevsk 68.06, Balezino 70.94, Yar 72.11, Karakulino 98.4, Debyesy 101.0, Vavozh 102.7, Kambarka & Yukamenskoye 103.6, Mozhga 104.0, Valamaz 105.0, Syumsi 107.2 in Russian, Udmurt, Tatar, Mari.

UL) Ulyanovskaya oblast [+3]: GTRK "Volga", 432030 Ulyanovsk, ul. Simbirskaya 5. **E:** volga@mv.ru **Reg:** On (MHz) Novospasskoye 67.07, Dimitrovgrad 67.19, Veshkayma 70.40, Ulyanovsk 71.00, Inza 73.43, Surskoye 73.58.

VG) Volgogradskaya oblast [+3]: GTRK "Volgograd-TRV", 400066 Volgograd, ul. Mira 9. **E:** trk@volgograd-trv.ru **Reg:** On Volgograd 567kHz + (MHz) Mikhaylovka 66.83, Elton 66.95, Kamyshin 69.14, Chilekovo 69.44, Loboykovo & Uspenka 70.40, Volgograd 70.43, Yelan 70.49, Kletskiy 70.94, Surovikino 71.00.

VL) Vladimirskaya oblast [+3]: GTRK "Vladimir", 600000 Vladimir, ul. Bol. Moskovskaya 62. **E:** adm@vladtv.ru **Reg:** On (MHz) Murom 66.32, Petushki 66.70, Aleksandrov 67.47, Suzdal 68.72, Gorokhovets 69.08, Sudogda 69.47, Melenki 70.25, Vyazniki 70.28, Kovrov 70.67, Sobinka 70.61, Kolchugino 73.55, Gus-Khrustalnyy 103.1.

VN) Voronezhskaya oblast [+3]: GTRK "Voronezh", 394625 Voronezh, ul. Karl Marksa 114. **E:** tv@vgtv.vrn.ru **Reg:** On (MHz) Bobrov 67.04, Borisoglebsk 70.82, Boguchar 71.90, Voronezh 72.11.

VO) Vologodskaya oblast [+3]: GTRK "Vologda", 160000 Vologda, ul. Predtecheskaya 32. **E:** gtrk_vologda@pochta.ru **Reg:** On Chagoda & Nikolsk 1098kHz + (MHz) Cherepovets 66.38, Sludno 66.77, Vitegra & Yakutino 66.86, Ozerki 66.86, Syamzha 67.88, Nyuksenitsa 68.03, Vologda 69.05, Verkhovazhye 69.08, Totma 69.71, Lipin Bor 69.65, Kurilovo 70.07, Kharovsk 102.2, Vozhega 102.5.

YA) Yaroslavskaya oblast [+3]: GTRK "Yaroslaviya", 150014 Yaroslavl, ul. Bogdanovicha 20. **E:** gtrk@nordnet.ru **Reg:** On (MHz) Dubki 68.66, Volga 70.88, Lyubim 103.6, Danilov 104.3.

YN) Yamalo-Nenetskiy avt. okrug [+5]: GTRK "Yamal", 626600 Salekhard, ul. Lambinykh 3. **E:** gtrk@gtrk-yamal.ru **Reg:** On Salekhard 603kHz + (MHz) Muzhi 68.90, Nadym 71.78, Salekhard 71.99/100.6. Txs also relay GTRK "Region-Tyumen", Tyumen TY. [NB: YN is subordinated to region TY]

YV) Yevreyskaya avt. oblast [+10]: GTRK "Bira", 679016 Birobidzhan, ul. Oktyabrskaya 15. **E:** gtrkbira@biratv.rfn.ru **Reg:** On Birobidzhan 216kHz + (MHz) Nikolayevka 66.02, Obluchye 66.32, Smidovich 66.80, Birobidzhan 67.88, Khingansk 68.33, Birakan 68.60, Pashkovo 69.89, Bidzhan 70.07, Leninskoye 73.22, Kuldur 73.64 in Russian, Yiddish. Txs also relay GTRK "Dalnevostochnaya", Khabarovsk (KH).

ZB) Zabaykalskiy kray [+9]: GTRK "Chita", 672090 Chita, ul. Kostyushko-Grigorovicha 27. **E:** chrtv@chita.rfn.ru **Reg:** On (kHz) Kruchina 261, Mogocha & Nerchinsk & Ulety & Chernyshevsk 1197 + (MHz) Petrovsk-Zabaykalskiy 66.14, Chita 66.32, Khilok 68.09, Khada-Bulak 69.56, Kholbon 69.80, Orlovskiy 70.07, Krasnokamensk 70.67 in Russian, Buryat.
— Territorialnoye otdeleniye GTRK "Chita", ul. Bazara Rinchino 7, 687000 Aginskoye. **E:** abgtrk_agi@aginsk.chita.ru **Reg:** On Orlovskiy 70.07MHz in Russian, Buryat.

Special shortwave prgrs produced by regional GTRKs GTRK "Adygeya", GTRK "Kabardino-Balkariya", GTRK "Tatarstan" ("Na volne Tatarstana"): see Int Radio section.

RADIO KAVKAZ / CHECHNYA SVOBODNAYA (Gov)

✉ 115326 Moskva, ul. Pyatnitskaya 25 **E:** info@chechnyafree.ru **W:** www.chechnyafree.ru
Txs: Tbilisskaya 171kHz 1200kW, Groznyy 657kHz 50kW
D.Prgr: **R. Kavkaz** 24h (on LW: 0300-2100), exc. **Chechnya Svobodnaya** 1305-1330.
NB. Service for listeners in the Russian Caucasus, produced by the state broadcasting company Golos Rossii, with participation of the Ministry of Culture and Massmedia.

RADIO ORFEY (Gov)

✉ 115326 Moskva, ul. Pyatnitskaya 25 ☎ +7 495 6335401 🖷 +7 495 6901412 **E:** rgmc@muzcentrum.ru **W:** www.muzcentrum.ru
LP: GD: Irina Gerasimova
Txs: Olgino 1125kHz 150kW, Volgograd 1161kHz 75kW & FM network.
D.Prgr: 24h (on MW/FM: 0300-2100).

EXTERNAL SERVICE (Voice of Russia): see Int Radio section.

OTHER STATIONS

MW kHz	kW	Location	Station
1) 270	150	Oyash, S	R. Slovo
16) 531	5	Yuzhno-Sakhalinsk, FE	Avtoradio

MW	kHz	kW	Location	Station
A)	612	20	Kurkino, E	Relay Services
B)	666	10	Yekaterinburg, S	BBC relay
5)	675	40	Razdolnoye, FE	R. Radonezh
5)	684	10	St.Peterburg, E	R. Radonezh
D)	693	10	Kurkino, E	DW relay
G)	738	5	Kurkino, E	WRN relay
7)	765	5-20	Khabarovsk, FE[1]	R. Vostok Rossii
11)	783	75	Vladivostok, FE	R. Lemma
12)	810	10	Krasnoyarsk, S	Avtoradio-Krasnoyarsk
F)	810	20	Kurkino, E	VOA relay
8)	828	10	St.Peterburg, E	Radiogazeta Slovo
5) †	846	150	Noginsk, E	R. Radonezh
10)	864	25	Blagoveshchensk, FE	R. Shanson
E)	1044	20	Kurkino, E	RFE-RL relay
6)	1053	20	Krasnoyarsk, S	R. Shanson
4)	1053	10	St.Peterburg, E	R. Mariya
3)	1089	50	Krasnyy Bor, E	R. Teos
3)	1134	20	Kurkino, E	R. Teos
C/D)	1188	10	St.Peterburg, E	DW/RFI relay
3)	1188	10	Khabarovsk, FE	R. Teos
9)	1233	20	Yelizovo, FE	Yumor FM
15)	1242	2	Angarsk, S	Novyy den - Angarsk
B)	1260	10	Kurkino, E	BBC relay
B)	1260	10	St.Peterburg, E	BBC relay
7)	1269	7	Komsomolsk-na-A., FE	R. Vostok Rossii
17)	1269	5	Ussuriysk, FE	R. Ussuri
13)	1332	5	Irkutsk, S	Dobryye pesni
18)	1440	10	St.Peterburg, E	R. Zvezda
2)	1503	*20	Kurkino, E	R. Tsentr
14)	1602	2	Zhelezhnogorsk-Ilimskiy, S	Ekho Moskvy

† time-shared: used by regional gov radio at other times (see VGRTK tx table);*) 10kW at night; ¹) sync. netw. of txs in various towns

Addresses & other information:
1) 630011 Novosibirsk, ul. Kirova 3. E: rslovo@mail.ru – 2) 109012 Moskva, ul. Nikolskaya 7. E:1503am@radiocenter.net. Rel. religious prgrs. – 3) 190000 St.Peterburg, P.O.Box 110. E: radio@teos.org.ru – 4) 190068 St.Peterburg, P.O.Box 732. E: pr@radiomaria.ru – 5) 113326 Moskva, ul. Pyatnitskaya 25. E: radonezh@radonezh.ru – 6) 660021 Krasnoyarsk. E: news@7fm.ru. Rel. R.Shanson (Moskva). – 7) 680000 Khabarovsk, ul. Lenina 4. E: radio@erussia.khv.ru – 8) 197022 St.Peterburg, P.O.Box 122. Also rel. Pravoslavnoye R. (St.Peterburg). – 9) 683024 Petropavlovsk-Kamchatskiy, ul. Lukashevskogo 5. Rel. Yumor FM (Moskva). – 10) Rel. R.Shanson (Moskva). – 11) 690000 Vladivostok, ul. Strelnikova 3a. E: lemma@vtc.ru – 12) 660022 Krasnoyarsk, ul. Partizana Zheleznyaka 15. E: rozin@avtoradio1.g-service.ru – 13) 664003 Irkutsk, ul. Litvinova 20. E: altair_irk09@mail.ru. Rel. Dobryye pesni (Moskva). – 14) 665651 Zhelezhnogorsk-Ilimskiy, ul. Yangelya 6. Rel. Ekho Moskvy (Moskva). – 15) 665827 Angarsk, 11 mikrorayon 7/7a. E: surikovaav@angarsk-adm.ru – 16) 6693023 Yuzhno-Sakhalinsk, ul. Komsomolskaya 213A. E: office@astv.ru – 17) 692525 Ussuriysk, ul. Kirova 28. – 18) 119160 Moskva, Kolymazhnyy per. 14. E: radio@tvzvezda.ru – A) Various relay services incl. Golos Rossii (for listeners in Moscow). – B) Rel. BBC (UK). – C) Rel. RFI (France). – D) Rel. DW (Germany). – E) Rel. RFE-RL (USA). – F) Rel. VOA (USA). – G) Rel. WRN (UK).

FM	MHz	kW	Location	Station
14)	66.08	1	Barnaul	R. Shanson
1)	66.23	1	Nab.Chelny	Avtoradio
144)	66.29	1	Biysk	R. RIF
14)	66.35	4	Smolensk	R. Shanson
56)	66.68	4	Ufa	R. Ashkadar
36)	66.74	4	Chelyabinsk	Intervolna
14)	66.83	4	Pskov	R. Shanson
16)	66.86	1	Yuzhno-Sakhalinsk	Retro FM
168)	67.04	1	Tyumen	La Femme R.
167)	67.10	1	Solikamsk	Soyuz FM
147)	67.16	4	Cherepovets	R. Transmit
14)	67.46	4	Ufa	R. Shanson
4)	67.46	1	Kholmsk	Evropa Plus
52)	67.46	1	Yekaterinburg	R. Ekspress
155)	67.58	4	Pskov	Sedmoye Nebo
163)	67.70	4	Magnitogorsk	Seven Skies
24)	67.70	1	Ulyanovsk	NRJ
120)	67.79	1	Nab.Chelny	R. Kunel
1)	68.00	1	Moskva	Avtoradio
2)	68.00	1	Ulan-Ude	Russkoye R.
4)	68.09	1	Penza	Evropa Plus
9A)	68.21	1	Murmansk	Glavnoye R.
133)	68.39	2	Yekaterinburg	R. Si
12)	68.51	1	Samara	Dobryye pesni
132)	68.57	2	N.Novgorod	R. Obraz
16)	68.60	4	Omsk	Retro FM
95B)	68.66	1	Sankt-Peterburg	R. Vanya
173)	68.90	1	Cheboksary	Otkrytoye Radio
3)	69.11	1	Barnaul	Ekho Moskvy
13)	69.26	1	Moskva	RSN

FM	MHz	kW	Location	Station
2)	69.26	4	Shuyda	Russkoye R.
36)	69.41	1	Chebarkul	Intervolna
3)	69.51	2	Shelekhov	Ekho Moskvy
42)	69.68	1	Ufa	R. Pervyy kanal
4)	69.71	1	Nab.Chelny	Evropa Plus
58)	69.74	1	Ulyanovsk	Russkiy prospekt
38)	69.80	1	Tomsk	R. Dacha
172)	70.04	4	Perm	Vzrosloye R.
14)	70.13	1	Komsomolsk-na-A.	R. Shanson
146)	70.19	1	Moskva	R. Ultra
159)	70.34	1	Ryazan	R. Zvezda
8)	70.55	1	Omsk	Nashe R.
35)	70.58	4	Bobrov	Kanal Melodiya - Voronezh
15)	70.70	1	Chelyabinsk	DFM
16)	70.85	1	Kirov	Retro FM
157)	70.88	2	Novosibirsk	R. Yuniton
20)	71.00	1	Kostroma	R. Alla
16)	71.03	1	Magnitogorsk	Retro FM
3)	71.06	1	Abakan	Ekho Moskvy
35)	71.12	4	Boguchar	Kanal Melodiya - Voronezh
14)	71.18	1	Kursk	R. Shanson
130)	71.21	4	Barancha	Kanal Voskreseniye
82)	71.24	15	Sankt-Peterburg	R. Baltika
2)	71.30	2.5	Moskva	Russkoye R.
35)	71.39	1	Voronezh	Kanal Melodiya - Voronezh
13)	71.45	1	Nalchik	RSN
16)	71.57	1	Barnaul	Retro FM
14)	71.72	4	Kaluga	R. Shanson
2)	71.72	4	Balezino	Russkoye R.
1)	71.84	1	Kirov	Avtoradio
3)	71.93	1	Surgut	Ekho Moskvy
81)	72.11	1	Kaliningrad	R. Baltik plus
8)	72.14	5	Sankt-Peterburg	Nashe R.
159)	72.26	1	Bryansk	R. Zvezda
159)	72.26	1	Yaroslavl	R. Zvezda
59)	72.29	1	Chita	Populyarnoye R.
9A)	72.38	4	Borisoglebsk	Glavnoye R.
1)	72.41	1	N.Novgorod	Avtoradio
2)	72.41	1	Stavropol	Russkoye R.
14)	72.44	1	Kirov	R. Shanson
156)	72.44	1	Perm	R. 7 Hot
3)	72.44	2	Tyumen	Ekho Moskvy
166)	72.50	1	Krasnoyarsk	RuFM
2)	72.56	4	Tambov	Nashe R.
15)	72.71	4	Tula	DFM
159)	72.80	4	Lipetsk	R. Zvezda
130)	72.83	4	Yekaterinburg	Kanal Voskreseniye
2)	72.83	1.5	Samara	Russkoye R.
12)	73.16	1	Novosibirsk	Dobryye pesni
2)	73.19	1	Kirov	Russkoye R.
14)	73.25	1	Kaluga	Retro FM
14)	73.28	1	Biysk	R. Shanson
1)	73.37	1	Nizhnevartovsk	Avtoradio
9A)	73.43	4	Borisoglebsk	Glavnoye R.
8)	73.52	1	Chelyabinsk	Nashe R.
55)	73.58	1	Novosibirsk	Gorodskaya volna
9A)	73.58	4	Boguchar	Glavnoye R.
131)	73.61	1	Samara	R. Samara-Maximum
17)	73.64	1	Stavropol	Serebryanyy dozhd
18)	73.76	4	Bobrov	Yumor FM
80)	73.76	1	Angarsk	R. Avtos
10)	73.79	1	Biysk	R. 7
11)	73.82	10	Sankt-Peterburg	R. Maksimum
3)	73.82	2.5	Moskva	Ekho Moskvy
137)	73.88	1	Saransk	Vaygel
16)	73.88	1	Staryy Oskol	Retro FM
1)	73.91	1	Vladimir	Avtoradio
15)	73.94	1	Dubki	DFM
3)	73.94	1	Yaroslavl	Ekho Moskvy
4)	73.94	1	Omsk	Evropa Plus
105)	73.97	1	Kirov	R. Mariya
3)	73.97	1	Chelyabinsk	Ekho Moskvy
19)	87.5	2.5	Sankt-Peterburg	Dorozhnoye R.
78)	87.5	5	Moskva	Business FM
23)	87.6	1	Perm	Detskoye R.
1)	87.7	1	Blagoveshchensk	Avtoradio
16)	87.9	1	Khabarovsk	Retro FM
20)	87.9	1	Saratov	R. Alla
31)	87.9	5	Moskva	City FM
16)	88.0	1	Sankt-Peterburg	Retro FM
2)	88.0	1	Tolyatti	Russkoye R.
99)	88.0	1	Perm	Bolid FM
19)	88.1	1	Omsk	Dorozhnoye R.
13)	88.2	1	Surgut	RSN
1)	88.3	1	Vladivostok	Avtoradio
15)	88.3	1	Yekaterinburg	DFM

FM	MHz	kW	Location	Station
16)	88.3	1	Moskva	Retro FM
17)	88.3	1	Kazan	Serebryanyy dozhd
23)	88.3	1	Saratov	Detskoye R.
7)	88.3	1	Krasnodar	Militseyskaya volna
1)	88.4	10	Sankt-Peterburg	Avtoradio
1)	88.5	1	Komsomolsk	Avtoradio
1)	88.7	1	Khabarovsk	Avtoradio
14)	88.7	1	Blagoveshchensk	R. Shanson
126)	88.7	1	Saratov	R. Rekord
18)	88.7	5	Moskva	Yumor FM
23)	88.7	1	Krasnodar	Detskoye R.
95B)	88.7	1	Vyborg	R. Vanya
8)	88.8	5	Vladivostok	Nashe R.
18)	88.9	1	Sankt-Peterburg	Yumor FM
93)	89.1	1	Moskva	R. Jazz
23)	89.3	1	Kazan	Detskoye R.
4)	89.4	1	Perm	Evropa Plus
46)	89.5	1	Moskva	Megapolis FM
2)	89.6	1	Khabarovsk	Russkoye R.
174)	89.7	4	Sankt-Peterburg	R. Zenit
33)	89.9	1	Moskva	Keks FM
95A)	89.9	4	Vyborg	Hit R.
12)	89.9	1	Yuzhno-Sakhalinsk	Dobryye pesni
23)	90.0	1	Ulyanovsk	Detskoye R.
89)	90.1	5	Sankt-Peterburg	R. Ermitazh
140)	90.2	1	Yekaterinburg	R. SK
1)	90.3	5	Moskva	Avtoradio
20)	90.6	1	Krasnodar	R. Alla
7)	90.6	1	Samara	Militseyskaya volna
95A)	90.6	5	Sankt-Peterburg	Hit R.
1)	90.7	1	Perm	Avtoradio
12)	90.7	1	Kazan	Dobryye pesni
77)	90.7	1	Lipetsk	Lipetsk FM
11)	90.8	2	Yekaterinburg	R. Maksimum
164)	90.8	1	Moskva	Relaks FM
10)	91.0	1	Samara	R. 7
38)	91.0	1	Saratov	R. Dacha
17)	91.1	1	Lipetsk	Serebryanyy dozhd
19)	91.1	1	Kazan	Dorozhnoye R.
3)	91.1	1	Ufa	Ekho Moskvy
33)	91.1	10	Sankt-Peterburg	Keks FM
20)	91.2	1	Rostov-na-Donu	R. Alla
3)	91.2	5	Moskva	Ekho Moskvy
3)	91.2	1	Perm	Ekho Moskvy
10)	91.3	5	Vladivostok	R. 7
4)	91.3	1	Tolyatti	Evropa Plus
1)	91.4	1	Shatura	Avtoradio
20)	91.4	1	Ulyanovsk	R. Alla
23)	91.4	1	Izhevsk	Detskoye R.
3)	91.4	1	Yekaterinburg	Ekho Moskvy
15)	91.5	1	Ufa	DFM
3)	91.5	5	Sankt-Peterburg	Ekho Moskvy
17)	91.5	1	Kemerovo	Serebryanyy dozhd
9B)	92.0	5	Moskva	Govorit Moskva
20)	92.3	1	Kazan	R. Alla
38)	92.4	1	Moskva	R. Dacha
162)	92.8	1	Moskva	R. Karnaval
142)	93.2	5	Moskva	R. Sport
65)	93.6	5	Moskva	Kommersant FM
9A)	93.6	5	Kaliningrad	Glavnoye R.
18)	94.0	1	Kaliningrad	Yumor FM
63)	94.0	5	Moskva	Pioner FM
12)	94.4	5	Moskva	Dobryye pesni
152)	94.8	5	Moskva	Moya semya
143)	95.2	5	Moskva	Rock FM
16)	95.5	1	Kaliningrad	Retro FM
159)	95.6	10	Moskva	R. Zvezda
18)	95.6	1	N.Novgorod	Yumor FM
18)	95.7	1	Samara	Yumor FM
23)	95.7	1	Volgograd	Detskoye R.
151)	95.8	1	Gusev	R. na Vostoke
20)	95.9	1	Yekaterinburg	R. Alla
1)	96.0	1	Tolyatti	Avtoradio
19)	96.0	5	Moskva	Dorozhnoye R.
29)	96.0	1	N.Novgorod	Dinamit N.Novgorod
12)	96.2	2	Novosibirsk	Dobryye pesni
20)	96.2	1	Izhevsk	R. Alla
141)	96.3	1	Samara	Kot FM
2)	96.3	2	Kaliningrad	Russkoye R.
12)	96.4	1	N.Novgorod	Dobryye pesni
16)	96.4	1	Chelyabinsk	Retro FM
69)	96.4	5	Moskva	XFM
20)	96.8	1	N.Novgorod	R. Alla
23)	96.8	5	Moskva	Detskoye R.
23)	96.8	1	Chelyabinsk	Detskoye R.
16)	97.0	1	Novosibirsk	Retro FM
23)	97.2	1	Krasnoyarsk	Detskoye R.
57)	97.2	5	Moskva	R. Komsomolskaya pravda
2)	97.4	1	Chernyakhovsk	Russkoye R.
158)	98.0	10	Moskva	98 FM - 98 khitov
138)	98.4	5	Moskva	Svezheye R.
16)	98.6	1	Samara	Retro FM
1)	98.7	1	Novosibirsk	Avtoradio
33)	98.7	1	Chelyabinsk	Keks FM
16)	98.7	1	Krasnoyarsk	Retro FM
20)	98.8	10	Moskva	R. Alla
20)	98.8	1	Volgograd	R. Alla
136)	98.9	1	Yekaterinburg	Uralskiy ekspress
1)	99.1	2	Chelyabinsk	Avtoradio
10)	99.1	1	Krasnoyarsk	R. 7
18)	99.1	1	Voronezh	Yumor FM
23)	99.1	1	N.Novgorod	Detskoye R.
24)	99.1	1	Novosibirsk	NRJ
3)	99.1	1	Samara	Ekho Moskvy
57)	99.3	1	Tver	R. Komsomolskaya pravda
16)	99.4	5	Perm	Retro FM
78)	99.4	2	Yekaterinburg	Business FM
18)	99.5	1	Novosibirsk	Yumor FM
3)	99.5	1	Chelyabinsk	Ekho Moskvy
153)	99.6	5	Moskva	Finam FM
19)	99.6	1	Luban	Dorozhnoye R.
19)	99.6	1	Oryol	Dorozhnoye R.
4)	99.9	1	Samara	Evropa Plus
10)	100.0	1	N.Novgorod	R. 7
150)	100.0	1	Volgograd	R. Vedo
16)	100.0	1	Yekaterinburg	Retro FM
19)	100.0	4	Mosolovo	Dorozhnoye R.
2)	100.0	1	Oryol	Russkoye R.
32)	100.0	1	Chelyabinsk	Absolutnoye R.
8)	100.0	1	Perm	Nashe R.
1)	100.1	1	Kaliningrad	Avtoradio
110)	100.1	1	Izhevsk	R. Moya Udmurtia
17)	100.1	5	Moskva	Serebryanyy dozhd
18)	100.1	1	Magnitogorsk	Yumor FM
2)	100.1	1	Kuznetsk	Russkoye R.
5)	100.1	1	Rostov-na-Donu	Hit FM
6)	100.1	1	Tolyatti	Love R.
2)	100.2	1	Zheleznogorsk	Russkoye R.
4)	100.2	1	Almetyevsk	Evropa Plus
4)	100.2	1	Vologda	Evropa Plus
1)	100.3	5	Krasnoyarsk	Avtoradio
14)	100.3	1	Kostroma	R. Shanson
2)	100.3	2	Samara	Russkoye R.
4)	100.3	1	Bryansk	Evropa Plus
4)	100.3	1	Syktyvkar	Evropa Plus
4)	100.3	1	Voronezh	Evropa Plus
17)	100.4	4	N.Novgorod	Serebryanyy dozhd
2)	100.4	1	Chelyabinsk	Russkoye R.
4)	100.4	1	Petrozavodsk	Evropa Plus
8)	100.4	2	Yekaterinburg	Nashe R.
81)	100.4	1	Sovetsk	R. Baltik plus
133)	100.5	1	Nizhniy Tagil	R. Si
14)	100.5	1	Magnitogorsk	R. Shanson
16)	100.5	1	Tula	Retro FM
171)	100.5	1	Kazan	Tatar Radiosi
2)	100.5	1	Izhevsk	Russkoye R.
26)	100.5	5	Moskva	Best FM
4)	100.5	1	Kovrov	Evropa Plus
4)	100.5	10	Sankt-Peterburg	Evropa Plus
88)	100.5	4	Novokuznetsk	Apeks R.
15)	100.6	1	Chernyakhovsk	DFM
18)	100.6	1	Saratov	Yumor FM
2)	100.6	1	Tver	Russkoye R.
4)	100.6	4	Saransk	Evropa Plus
85)	100.6	1	Volgograd	Evropa Plus
23)	100.6	1	Tyumen	R. City
10)	100.7	1	Barnaul	Detskoye R.
126)	100.7	1	Stavropol	R. 7
157)	100.7	3	Rostov-na-Donu	R. Rekord
18)	100.7	1	Novosibirsk	R. Yuniton
38)	100.7	1	Ryazan	Yumor FM
4)	100.7	1	Kursk	R. Dacha
4)	100.7	1	Unecha	Evropa Plus
5)	100.7	1	Ussuriysk	Evropa Plus
92)	100.7	5	Perm	Hit R.
10)	100.7	1	Makhachkala	R. Priboy
14)	100.8	1	Orenburg	R. 7
159)	100.8	1	Mezhdurechensk	R. Shanson
18)	100.8	1	Chelyabinsk	R. Zvezda
19)	100.8	1	Vladimir	Yumor FM
2)	100.8	1	Krasnoyarsk	Dorozhnoye R.
2)	100.8	4	Tbilisskaya	Russkoye R.

FM	MHz	kW	Location	Station
1)	100.9	1	Kushchevskaya	Avtoradio
10)	100.9	1	Irkutsk	R. 7
117)	100.9	4	Mosolovo	R. OK
121)	100.9	10	Sankt-Peterburg	Piter FM
14)	100.9	1	Nizhniy Tagil	R. Shanson
16)	100.9	5	N.Novgorod	Retro FM
17)	100.9	1	Tuapse	Serebryanyy dozhd
19)	100.9	1	Lipetsk	Dorozhnoye R.
39)	100.9	1	Kaliningrad	R. Monte-Karlo
4)	100.9	1	Sterlitamak	Evropa Plus
4)	100.9	1	Oryol	Evropa Plus
40)	100.9	5	Moskva	R. Klassik
71)	100.9	1	Belgorod	R. 31
88)	100.9	2	Kemerovo	Apeks R.
14)	101.0	1	Samara	R. Shanson
14)	101.0	1	Tyumen	R. Shanson
17)	101.0	1	Magnitogorsk	Serebryanyy dozhd
1)	101.1	1	Sochi	Avtoradio
1)	101.1	1	Yakutsk	Avtoradio
10)	101.1	1	Makhachkala	R. 7
10)	101.1	1	Perm	R. 7
10)	101.1	1	Voronezh	R. 7
3)	101.1	1	Volgograd	Ekho Moskvy
4)	101.1	1	Murmansk	Evropa Plus
5)	101.1	1	Yeysk	Hit FM
7)	101.1	1	Novokuznetsk	Militseyskaya volna
12)	101.2	1	Kursk	Dobryye pesni
14)	101.2	1	Tolyatti	R. Shanson
15)	101.2	10	Moskva	DFM
16)	101.2	1	Krasnodar	Retro FM
18)	101.2	1	Chelyabinsk	Yumor FM
39)	101.2	1	Pechora	R. Monte-Karlo
4)	101.2	5	Yekaterinburg	Evropa Plus
14)	101.3	1	Shadrinsk	R. Shanson
16)	101.3	1	Vladimir	Retro FM
18)	101.3	1	Izhevsk	Yumor FM
2)	101.3	1	Buzuluk	Russkoye R.
2)	101.3	1	Saransk	Russkoye R.
3)	101.3	1	Orenburg	Ekho Moskvy
4)	101.3	1	Lipetsk	Evropa Plus
6)	101.3	1	Krasnoyarsk	Love R.
6)	101.3	1	Pyatigorsk	Love R.
8)	101.3	1	Kaliningrad	Nashe R.
1)	101.4	1	Rossosh	Avtoradio
1)	101.4	1	Sterlitamak	Avtoradio
1)	101.4	1	Tver	Avtoradio
100)	101.4	1	Shabanovo	R. Krasnaya Armiya
13)	101.4	1	Berezniki	RSN
165)	101.4	10	Sankt-Peterburg	Eldoradio
19)	101.4	1	Tula	Dorozhnoye R.
19)	101.4	1	Stavropol	Dorozhnoye R.
2C)	101.4	5	Angarsk	Hit FM
5)	101.4	1	N.Novgorod	Hit FM
5)	101.4	1	Tuapase	Hit FM
55)	101.4	2	Novosibirsk	Gorodskaya volna
7)	101.4	1	Oryol	Militseyskaya volna
9A)	101.4	1	Kostroma	Glavnoye R.
1)	101.5	1	Bryansk	Avtoradio
1)	101.5	1	Nab.Chelny	Avtoradio
115)	101.5	1	Perm	R. Nostalzhi
12)	101.5	1	Murmansk	Dobryye pesni
126)	101.5	5	Samara	R. Rekord
14)	101.5	2	Saratov	R. Shanson
16)	101.5	1	Novokuznetsk	Retro FM
169B)	101.5	1	Volgograd	Volgograd FM
19)	101.5	1	Ryazan	Dorozhnoye R.
2)	101.5	1	Kanevskaya	Russkoye R.
2)	101.5	1	Yurga	Russkoye R.
4)	101.5	1	Nizhniy Tagil	Evropa Plus
51)	101.5	1	Malaya Vishera	MV Diapazon
6)	101.5	1	Makhachkala	Love R.
1)	101.6	1	Arkhangelsk	Avtoradio
129)	101.6	1	Rostov-na-Donu	R. Rostova
13)	101.6	1	Saraktash	RSN
14)	101.6	2	Khanty-Mansiysk	R. Shanson
4)	101.6	1	Chelyabinsk	Evropa Plus
42)	101.6	1	Ufa	R. Pervyy Kanal
90)	101.6	2.5	Voronezh	Muz FM
1)	101.7	1	Astrakhan	Avtoradio
1)	101.7	1	Borisoglebsk	Avtoradio
13)	101.7	1	Abakan	RSN
14)	101.7	1	Krasnoyarsk	R. Shanson
149)	101.7	5	Vladivostok	R. VBC
2)	101.7	1	Komsomolsk-na-A.	Russkoye R.
21A)	101.7	5	Kingisepp	R. Gardarika
1)	101.7	1	Ulyanovsk	Evropa Plus
8)	101.7	5	Moskva	Nashe R.
9A)	101.7	1	Yaroslavl	Glavnoye R.
13)	101.8	1	Buzuluk	RSN
15)	101.8	1	Kemerovo	DFM
16)	101.8	1	Izhevsk	Retro FM
16)	101.8	1	Orenburg	Retro FM
16)	101.8	4	V.Novgorod	Retro FM
2)	101.8	1	Krasnodar	Russkoye R.
2)	101.8	1	Penza	Russkoye R.
2)	101.8	1	Stavropol	Russkoye R.
2)	101.8	1	Syktyvkar	Russkoye R.
4)	101.8	1	Tyumen	Evropa Plus
4)	101.8	1	Tver	Evropa Plus
50)	101.8	1	Oryol	Love Music
78)	101.8	1	Kaliningrad	Business FM
1)	101.9	1	Biysk	Avtoradio
1)	101.9	5	N.Novgorod	Avtoradio
10)	101.9	5	Surgut	R. 7
126)	101.9	2	Kazan	R. Rekord
14)	101.9	1	Barnaul	R. Shanson
17)	101.9	5	Sochi	Serebryanyy dozhd
19)	101.9	1	Ukhta	Dorozhnoye R.
2)	101.9	1	Makhachkala	Russkoye R.
4)	101.9	1	Buguruslan	Evropa Plus
4)	101.9	1	Nab.Chelny	Evropa Plus
4)	101.9	1	Novorossiysk	Evropa Plus
4)	101.9	1	Omsk	Evropa Plus
5)	101.9	1	Tulun	Hit FM
8)	101.9	5	Tula	Nashe R.
1)	102.0	1	Boguchar	Avtoradio
1)	102.0	1	Ryazan	Avtoradio
126)	102.0	1	Murmansk	R. Rekord
127)	102.0	10	Sankt-Peterburg	R. Roks
13)	102.0	1	Novotroitsk	RSN
14)	102.0	1	Achinsk	R. Shanson
15)	102.0	1	Yakutsk	DFM
17)	102.0	1	Kirovsk	Serebryanyy dozhd
17)	102.0	1	Kirovsk	Serebryanyy dozhd
18)	102.0	1	Yekaterinburg	Yumor FM
19)	102.0	1	Bryansk	Dorozhnoye R.
4)	102.0	1	Chita	Evropa Plus
4)	102.0	1	Smolensk	Evropa Plus
4)	102.0	1	Vladikavkaz	Evropa Plus
47)	102.0	1	Cheboksary	MFM
5)	102.0	1	Saransk	Hit FM
54)	102.0	4	Volgograd	Novaya volna
9A)	102.0	5	Perm	Glavnoye R.
1)	102.1	1	Lipetsk	Avtoradio
1)	102.1	1	Saratov	Avtoradio
10)	102.1	1	Kursk	R. 7
14)	102.1	1	Bratsk	R. Shanson
14)	102.1	1	Nizhnevartovsk	R. Shanson
16)	102.1	1	Verkhneuralsk	Retro FM
2)	102.1	1	Tomsk	Russkoye R.
38)	102.1	1	Samara	R. Dacha
39)	102.1	5	Moskva	R. Monte-Karlo
4)	102.1	1	Pskov	Evropa Plus
5)	102.1	1	Irkutsk	Hit FM
68)	102.1	1	Ulyanovsk	R. 2x2
133)	102.2	1	Kamensk-Uralskiy	R. Si
17)	102.2	1	Yaroslavl	Serebryanyy dozhd
19)	102.2	1	Volkhov	Dorozhnoye R.
2)	102.2	1	Abakan	Russkoye R.
2)	102.2	1	Astrakhan	Russkoye R.
2)	102.2	1	Belgorod	Russkoye R.
4)	102.2	1.5	Kirov	Evropa Plus
4)	102.2	1	Krasnodar	Evropa Plus
4)	102.2	1	Nalchik	Evropa Plus
1)	102.3	1	Armavir	Avtoradio
16)	102.3	1	Usolye-Sibirskoye	Retro FM
2)	102.3	1	Khabarovsk	Russkoye R.
2)	102.3	1	Neftekamsk	Russkoye R.
33)	102.3	1	Voronezh	Keks FM
4)	102.3	1	Ulan-Ude	Evropa Plus
6)	102.3	1	Syktyvkar	Love R.
79)	102.3	1	Tolyatti	R. Avgust
83)	102.3	1	Kolomna	R. Blago
11)	102.4	1	N.Novgorod	R. Maksimum
126)	102.4	1	Izhevsk	R. Rekord
15)	102.4	1	Barnaul	DFM
15)	102.4	1	Vladimir	DFM
16)	102.4	1	Kazan	Retro FM
17)	102.4	5	Nazran	Serebryanyy dozhd
177)	102.4	1	Novorossiysk	Roks FM
2)	102.4	1	Ukhta	Russkoye R.
21B)	102.4	10	Sankt-Peterburg	R. Metro

FM	MHz	kW	Location	Station	FM	MHz	kW	Location	Station
22)	102.4	1	Yeysk	Pervoye R.	13)	103.1	1	Kuvandyk	RSN
4)	102.4	1	Makhachkala	Evropa Plus	16)	103.1	1	Stavropol	Retro FM
4)	102.4	1	Nakhodka	Evropa Plus	16)	103.1	1	Lipetsk	Retro FM
4)	102.4	1	Tikhoretsk	Evropa Plus	17)	103.1	1	Omutinsk	Serebryanyy dozhd
6)	102.4	4	Tambov	Love R.	175)	103.1	1	Kaluga	Nika FM
74)	102.4	1	Mikhaylovka	R. Aprel	18)	103.1	3	Irkutsk	Yumor FM
9A)	102.4	1	Chelyabinsk	Glavnoye R.	18)	103.1	1	Biysk	Yumor FM
1)	102.5	1	Murmansk	Avtoradio	2)	103.1	1	Sochi	Russkoye R.
10)	102.5	1	Biysk	R. 7	21A)	103.1	5	Vyborg	R. Gardarika
104)	102.5	1	Ufa	R. Manhattan	4)	103.1	1	Staryy Oskol	Evropa Plus
123)	102.5	5	Moskva	Pervoye Popularnoye R.	6)	103.1	1	Khabarovsk	Love R.
14)	102.5	1	Sochi	R. Shanson	7)	103.1	1	Apatity	Militseyskaya volna
16)	102.5	1	Tomsk	Retro FM	94A)	103.1	1	Yakutsk	R. Viktoriya
167)	102.5	1	Lysva	Soyus FM	1)	103.2	1	Krasnodar	Avtoradio
17)	102.5	1	Ryazan	Serebryanyy dozhd	1)	103.2	1	V.Novgorod	Avtoradio
177)	102.5	1	Psebay	Roks FM	11)	103.2	5	Perm	R. Maksimum
19)	102.5	1	Kotkozero	Dorozhnoye R.	12)	103.2	1	Astrakhan	Dobryye pesni
2)	102.5	1	Omsk	Russkoye R.	12)	103.2	1	Tolyatti	Dobryye pesni
2)	102.5	5	Surgut	Russkoye R.	14)	103.2	1	Vladivostok	R. Shanson
2)	102.5	1	Tyumen	Russkoye R.	14)	103.2	1	Yekaterinburg	R. Shanson
2)	102.5	1	Magnitogorsk	Russkoye R.	18)	103.2	1	Nab.Chelny	Yumor FM
4)	102.5	1	Cheboksary	Evropa Plus	18)	103.2	1	Yelabuga	Yumor FM
4)	102.5	1	Yakutsk	Evropa Plus	2)	103.2	1	Pyatigorsk	Russkoye R.
4)	102.5	1	Yuzhno-Sakhalinsk	Evropa Plus	4)	103.2	3	Novosibirsk	Evropa Plus
76)	102.5	1	Kumertau	R. Aris	4)	103.2	1	Ryazan	Evropa Plus
96)	102.5	1	Yekaterinburg	Dzhem FM	5)	103.2	1	Belgorod	Hit FM
1)	102.6	1	Murom	Avtoradio	5)	103.2	1	Kurgan	Hit FM
10)	102.6	1	Novokuznetsk	R. 7	1)	103.3	2	Surgut	Avtoradio
10)	102.6	2	Saratov	R. 7	14)	103.3	1	Kemerovo	R. Shanson
10)	102.6	1	Shabanovo	R. 7	149)	103.3	1	Nakhodka	R. VBC
134)	102.6	1	Chita	R. Sibir	15)	103.3	1	Berezniki	DFM
14)	102.6	2	Novosibirsk	R. Shanson	15)	103.3	1	Murom	DFM
16)	102.6	1	Pskov	Retro FM	18)	103.3	1	Azov	Yumor FM
19)	102.6	1	Tikhvin	Dorozhnoye R.	2)	103.3	1	Balezino	Russkoye R.
2)	102.6	1	Bryansk	Russkoye R.	2)	103.3	1	Orsk	Russkoye R.
2)	102.6	1	Yaroslavl	Russkoye R.	22)	103.3	1	Krasnaya Polyana	Pervoye R.
4)	102.6	1	Kaluga	Evropa Plus	22)	103.3	4	Tbilisskaya	Pervoye R.
1)	102.7	1	Zheleznogorsk	Avtoradio	5)	103.3	1	Yaroslavl	Hit FM
102)	102.7	1	Vladivostok	R. Lemma	59)	103.3	1	Chita	Populyarnoye R.
161)	102.7	1	Tver	R. Pilot	61)	103.3	1	Kazan	Puls R.
172)	102.7	5	Perm	Vzrosloye R.	7)	103.3	1	Tula	Militseyskaya volna
2)	102.7	1	Kurgan	Russkoye R.	1)	103.4	1	Kirov	Avtoradio
22)	102.7	5	Krasnodar	Pervoye R.	1)	103.4	1	Obninsk	Avtoradio
4)	102.7	1	Astrakhan	Evropa Plus	1)	103.4	1	Voronezh	Avtoradio
5)	102.7	1	Syktyvkar	Hit FM	125)	103.4	5	N.Novgorod	R. Randevu
95B)	102.7	2	Tolyatti	R. Vanya	15)	103.4	10	Sankt-Peterburg	DFM
97)	102.7	1	V.Novgorod	R. 53	16)	103.4	1	Kovdor	Retro FM
1)	102.8	5	Krasnoyarsk	Avtoradio	19)	103.4	1	Arkhangelsk	Dorozhnoye R.
11)	102.8	10	Sankt-Peterburg	R. Maksimum	2)	103.4	1	Vladimir	Russkoye R.
14)	102.8	1	Voronezh	R. Shanson	22)	103.4	1	Tuapse	Pervoye R.
20)	102.8	1	Nab.Chelny	R. Alla	4)	103.4	4	Tambov	Evropa Plus
27)	102.8	1	Kazan	Bim R.	4)	103.4	1	Tomsk	Evropa Plus
3)	102.8	1	Vladikavkaz	Ekho Moskvy	1)	103.5	1	Chernyakhovsk	Avtoradio
4)	102.8	1	Berezniki	Evropa Plus	1)	103.5	1	Vladikavkaz	Avtoradio
4)	102.8	1	Kemerovo	Evropa Plus	16)	103.5	1	Petrozavodsk	Retro FM
4)	102.8	1	Orsk	Evropa Plus	16)	103.5	1	Cheboksary	Retro FM
48)	102.8	1	Penza	Most R.	19)	103.5	1	Ulyanovsk	Dorozhnoye R.
86)	102.8	1	Chaykovskiy	R. Dixi	2)	103.5	1	Smolensk	Russkoye R.
1)	102.9	1	Tambov	Avtoradio	28)	103.5	1	Bryansk	Bit R.
105)	102.9	1	Kirov	R. Mariya	4)	103.5	1	Saratov	Evropa Plus
15)	102.9	1	Samara	DFM	70)	103.5	1	Omsk	R. 3
177)	102.9	4	Armavir	Roks FM	8)	103.5	1	Chelyabinsk	Nashe R.
18)	102.9	1	Barnaul	Yumor FM	107)	103.6	5	Samara	R. Megapolis
19)	102.9	1	Podporozhye	Dorozhnoye R.	109)	103.6	5	Perm	Nashe pesni
2)	102.9	5	N.Novgorod	Russkoye R.	16)	103.6	1	Serov	Retro FM
36)	102.9	1	Chelyabinsk	Intervolna	19)	103.6	1	Volgograd	Dorozhnoye R.
4)	102.9	1	Vladimir	Evropa Plus	2)	103.6	4	Nab.Chelny	Russkoye R.
6)	102.9	1	Kaliningrad	Love R.	21A)	103.6	1	Priozersk	R. Gardarika
8)	102.9	1	Bratsk	Nashe R.	4)	103.6	1	Belgorod	Evropa Plus
116)	103.0	1	Noyabrsk	R. Noyabrsk	4)	103.6	1	Gus-Khrustalnyy	Evropa Plus
12)	103.0	1	Saratov	Dobryye pesni	4)	103.6	4	Stavropol	Evropa Plus
128)	103.0	1	Ufa	R. Roksana	5)	103.6	1	Magnitogorsk	Hit FM
14)	103.0	5	Moskva	R. Shanson	98)	103.6	1	Yakutsk	Kiin R.
15)	103.0	1	Magnitogorsk	DFM	10)	103.7	3	Krasnodar	R. 7
159)	103.0	2	Komsomolsk-na-A.	R. Zvezda	10)	103.7	1	Rostov-na-Donu	R. 7
180)	103.0	1	Nizhniy Tagil	Eko R.	10)	103.7	1	Saransk	R. 7
19)	103.0	1	Kirishi	Dorozhnoye R.	101)	103.7	1	Kursk	R. Kurs
2)	103.0	1	Ulyanovsk	Russkoye R.	11)	103.7	10	Moskva	R. Maksimum
3)	103.0	1	Bugulma	Ekho Moskvy	133)	103.7	5	Yekaterinburg	R. Si
4)	103.0	1	Izhevsk	Evropa Plus	135)	103.7	1	Abakan	R. Sibir
5)	103.0	2	Orenburg	Hit FM	14)	103.7	1	Astrakhan	R. Shanson
7)	103.0	1	Omsk	Militseyskaya volna	16)	103.7	1	Vladivostok	Retro FM
8)	103.0	1	Murmansk	Nashe R.	2)	103.7	1	Biysk	Russkoye R.
1)	103.1	1	Petrozavodsk	Avtoradio	2)	103.7	1	Ust-Ilimsk	Russkoye R.
1)	103.1	1	Volgograd	Avtoradio	4)	103.7	1	Orenburg	Evropa Plus
10)	103.1	1	Tyumen	R. 7	4)	103.7	5	V.Novgorod	Evropa Plus

FM	MHz	kW	Location	Station	FM	MHz	kW	Location	Station
44)	103.7	1	Khabarovsk	R. Vostok Rossii	4)	104.4	1	Beloretsk	Evropa Plus
5)	103.7	1	Orsk	Hit FM	4)	104.4	1	Novokuznetsk	Evropa Plus
1)	103.8	1	Syzran	Avtoradio	4)	104.4	5	Sochi	Evropa Plus
10)	103.8	1	Syktyvkar	R. 7	6)	104.4	1	Sovetsk	Love R.
15)	103.8	4	Leninsk-Kuznetskiy	DFM	7)	104.4	1	Murom	Militseyskaya volna
15)	103.8	1	Sterlitamak	DFM	1)	104.5	1	Petropavlovsk-K.	Avtoradio
15)	103.8	1	Tomsk	DFM	118)	104.5	2	Chelyabinsk	R. Olimp
175)	103.8	1	Sukhinichi	Nika FM	14)	104.5	1	Nakhodka	R. Shanson
178)	103.8	1	Yoshkar-Ola	Puls R.	14)	104.5	1	Yaroslavl	R. Shanson
19)	103.8	1	Yaroslavl	Dorozhnoye R.	145)	104.5	1	Yekaterinburg	Nashe pesni
2)	103.8	4	Arkhangelsk	Russkoye R.	15)	104.5	1	Vladikavkaz	DFM
2)	103.8	1	Kyzyl	Russkoye R.	18)	104.5	5	Volgograd	Yumor FM
4)	103.8	5	Irkutsk	Evropa Plus	2)	104.5	1	Ufa	Russkoye R.
4)	103.8	1	Krasnoyarsk	Evropa Plus	22)	104.5	4	Kanevskaya	Pervoye R.
4)	103.8	1	Penza	Evropa Plus	38)	104.5	2	N.Novgorod	R. Dacha
4)	103.8	1	Pyatigorsk	Evropa Plus	4)	104.5	1	Kaliningrad	Evropa Plus
6)	103.8	1	Voronezh	Love R.	4)	104.5	1	Yoshkar-Ola	Evropa Plus
8)	103.8	1	Izhevsk	Nashe R.	5)	104.5	1	Ukhta	Hit FM
87)	103.8	1	Sovetsk	R. Ekspress	60)	104.5	1	Murmansk	Power Hit R.
1)	103.9	1	Barnaul	Avtoradio	68)	104.5	1	Dmitrovgrad	R. 2x2
134)	103.9	1	Omsk	R. Sibir	72)	104.5	1	Izhevsk	R. Adam
15)	103.9	1	Chusovoy	DFM	8)	104.5	1	N.Novgorod	Nashe R.
15)	103.9	2	Novosibirsk	DFM	94B)	104.5	1	Yakutsk	R. Viktoriya-Sakha
2)	103.9	1	Kirov	Russkoye R.	95A)	104.5	1	Podporozhye	Hit R.
4)	103.9	5	N.Novgorod	Evropa Plus	10)	104.6	1	Kislovodsk	R. 7
4)	103.9	1	Ozyorsk	Evropa Plus	100)	104.6	1	Tyumen	R. Krasnaya Armiya
68)	103.9	5	Veshkaima	R. 2x2	134)	104.6	1	Tomsk	R. Sibir
90)	103.9	1	Tambov	Muz FM	147)	104.6	1	Cherepovets	R. Transmit
1)	104.0	1	Kamensk-Uralskiy	Avtoradio	15)	104.6	1	Rostov-na-Donu	DFM
11)	104.0	1	Novokuznetsk	R. Maksimum	16)	104.6	5	Irkutsk	Retro FM
126)	104.0	1	Tolyatti	R. Rekord	19)	104.6	1	Krasnoyarsk	Dorozhnoye R.
14)	104.0	1	Kazan	R. Shanson	2)	104.6	1	Vyborg	Russkoye R.
16)	104.0	1	Murmansk	Retro FM	24)	104.6	1	Kursk	NRJ
16)	104.0	1	Ufa	Retro FM	6)	104.6	2.5	Lipetsk	Love R.
16)	104.0	1	Volgograd	Retro FM	95A)	104.6	1	Volkov	Hit R.
19)	104.0	1	Yakutsk	Dorozhnoye R.	1)	104.7	1	Magnitogorsk	Avtoradio
2)	104.0	4	Maykop	Russkoye R.	1)	104.7	1	Nizhnevartovsk	Avtoradio
4)	104.0	1	Nizhnevartovsk	Evropa Plus	10)	104.7	5	Moskva	R. 7
53)	104.0	1	Novorossiysk	Novaya Rossiya	126)	104.7	1	Perm	R. Rekord
8)	104.0	10	Sankt-Peterburg	Nashe R.	15)	104.7	1	Kazan	DFM
1)	104.1	1	Rostov-na-Donu	Avtoradio	2)	104.7	5	Petrozavodsk	Russkoye R.
14)	104.1	1	Borisoglebsk	R. Shanson	38)	104.7	5	Vladivostok	R. Dacha
2)	104.1	2	Khanty-Mansiysk	Russkoye R.	6)	104.7	1	Belgorod	Love R.
24)	104.1	1	Ryazan	NRJ	8)	104.7	1	Arkhangelsk	Nashe R.
2A)	104.1	1	Chelyabinsk	Russkoye R.	8)	104.7	1	Krasnodar	Nashe R.
3)	104.1	1	Orsk	Ekho Moskvy	95A)	104.7	1	Luga	Hit R.
35)	104.1	1	Buguruslan	Kanal Melodiya - Voronezh	1)	104.8	1	Vladimir	Avtoradio
38)	104.1	1	Yekaterinburg	R. Dacha	1)	104.8	1	Samara	Avtoradio
4)	104.1	1	Bratsk	Evropa Plus	12)	104.8	1	Penza	Dobryye pesni
4)	104.1	1	Kursk	Evropa Plus	12)	104.8	2	Voronezh	Dobryye pesni
5)	104.1	1	Tryokhgornyy	Hit FM	14)	104.8	1	Makhachkala	R. Shanson
6)	104.1	1	Krasnoturinsk	Love R.	14)	104.8	1	Tver	R. Shanson
73)	104.1	1	Perm	R. Alfa	16)	104.8	1	Sterlitamak	Retro FM
10)	104.2	1	Ulyanovsk	R. 7	17)	104.8	1	Saratov	Serebryanyy dozhd
12)	104.2	1	Krasnodar	Dobryye pesni	2)	104.8	1	Kemerovo	Russkoye R.
14)	104.2	1	Surgut	R. Shanson	2)	104.8	1	Kostroma	Russkoye R.
147)	104.2	2	Babyevo	R. Transmit	24)	104.8	1	Nab.Chelny	NRJ
147)	104.2	1	Totma	R. Transmit	5)	104.8	1	Oryol	Hit FM
19)	104.2	1	Kingisepp	Dorozhnoye R.	6)	104.8	1	Monchegorsk	Love R.
2)	104.2	1	Magnitogorsk	Russkoye R.	82)	104.8	5	Sankt-Peterburg	R. Baltika
24)	104.2	1	Belgorod	NRJ	9A)	104.8	1	Smolensk	Glavnoye R.
24)	104.2	10	Moskva	NRJ	1)	104.9	1	Noyabrsk	Avtoradio
4)	104.2	1	Abakan	Evropa Plus	16)	104.9	1	Kirov	Retro FM
4)	104.2	5	Vladivostok	Evropa Plus	177)	104.9	1	Yeysk	Roks FM
6)	104.2	1	Cheboksary	Love R.	179)	104.9	1	Chelyabinsk	L-Radio
6)	104.2	5	Irkutsk	Love R.	2)	104.9	1	Vologda	Russkoye R.
7)	104.2	1	Pervomayskiy	Militseyskaya volna	22)	104.9	4	Novorossiysk	Pervoye R.
1)	104.3	1	Oryol	Avtoradio	4)	104.9	1	Tula	Evropa Plus
131)	104.3	2	Samara	R. Samara-Maximum	49)	104.9	1	Vladikavkaz	MSS
15)	104.3	2	Orenburg	DFM	6)	104.9	1	N.Novgorod	R. Vesna
16)	104.3	1	Saratov	Retro FM	68)	104.9	2	Novospasskoye	R. 2x2
19)	104.3	1	Khabarovsk	Dorozhnoye R.	1)	105.0	1	Novokuznetsk	Avtoradio
2)	104.3	1	Sterlitamak	Russkoye R.	1)	105.0	1	Sovetsk	Avtoradio
2)	104.3	1	Voronezh	Russkoye R.	10)	105.0	1	Ryazan	R. 7
57)	104.3	1	Vladimir	R. Komsomolskaya pravda	122)	105.0	2	Yekaterinburg	R. Pilot
7)	104.3	1	Kemerovo	Militseyskaya volna	14)	105.0	1	Tuapse	R. Shanson
1)	104.4	5	Tula	Avtoradio	154)	105.0	1	Neryungri	R. Voyazh
11)	104.4	1	Tuymazy	R. Maksimum	16)	105.0	1	Nakhodka	Retro FM
14)	104.4	5	Sankt-Peterburg	R. Shanson	18)	105.0	1	Astrakhan	Yumor FM
147)	104.4	1	Nikolsk	R. Transmit	24)	105.0	1	Ulan-Ude	NRJ
147)	104.4	1	Vologda	R. Transmit	3)	105.0	1	Omsk	Ekho Moskvy
16)	104.4	1	Tambov	Retro FM	3)	105.0	1	Tomsk	Ekho Moskvy
16)	104.4	1	Barnaul	Retro FM	3)	105.0	1	Ukhta	Ekho Moskvy
2)	104.4	1	Chusovoy	Russkoye R.	5)	105.0	1	Ufa	Hit FM
2)	104.4	1	Kyshtym	Russkoye R.	1)	105.1	1	Stavropol	Avtoradio
2)	104.4	1	Noyabrsk	Russkoye R.	14)	105.1	1	Rostov-na-Donu	R. Shanson

FM	MHz	kW	Location	Station	FM	MHz	kW	Location	Station
16)	105.1	1	Tyumen	Retro FM	7)	105.9	1	N.Novgorod	Militseyskaya volna
169A)	105.1	1	Volgograd	R. Sputnik	148)	106.0	1	Ussuriysk	R. Ussuri
2)	105.1	1	Berezniki	Russkoye R.	15)	106.0	1	Krasnodar	DFM
21A)	105.1	1	Luga	R. Gardarika	19)	106.0	1	Murmansk	Dorozhnoye R.
21A)	105.1	1	Tikhvin	R. Gardarika	2)	106.0	1	Balakovo	Russkoye R.
37)	105.1	1	Perm	Kama FM	25)	106.0	1	Irkutsk	AS FM
4)	105.1	1	Yaroslavl	Evropa Plus	4)	106.0	1	Magnitogorsk	Evropa Plus
9A)	105.1	1	Lipetsk	Glavnoye R.	4)	106.0	1	Ufa	Evropa Plus
1)	105.2	1	Chita	Avtoradio	4)	106.0	1	Vyborg	Evropa Plus
1)	105.2	2	Krasnoyarsk	Avtoradio	5)	106.0	1	Nab.Chelny	Hit FM
12)	105.2	1	Syktyvkar	Dobryye pesni	1)	106.1	1	Izhevsk	Avtoradio
160)	105.2	5	Moskva	Next FM	1)	106.1	1	Velikiye Luki	Avtoradio
177)	105.2	5	Krasnodar	Roks FM	126)	106.1	1	Podporozhye	R. Rekord
2)	105.2	2	Novosibirsk	Russkoye R.	19)	106.1	2	Samara	Dorozhnoye R.
3)	105.2	1	Makhachkala	Ekho Moskvy	2)	106.1	1	Yakutsk	Russkoye R.
5)	105.2	2	Megion	Hit R.	7)	106.1	1	Tomsk	Militseyskaya volna
6)	105.2	1	Magnitogorsk	Love R.	95A)	106.1	1	Kirshi	Hit R.
8)	105.2	1	Sochi	Nashe R.	10)	106.2	1	Lipetsk	R. 7
81)	105.2	1	Kaliningrad	R. Baltik plus	12)	106.2	1	Tobolsk	R. 7
1)	105.3	1	Kazan	Avtoradio	16)	106.2	1	Pyatigorsk	Retro FM
1)	105.3	1	Kemerovo	Avtoradio	17)	106.2	4	Omsk	Serebryanyy dozhd
14)	105.3	1	Orenburg	R. Shanson	19)	106.2	1	Kostroma	Dorozhnoye R.
15)	105.3	5	Vladivostok	DFM	19)	106.2	1	Kursk	Dorozhnoye R.
16)	105.3	1	Voronezh	Retro FM	2)	106.2	5	Perm	Russkoye R.
19)	105.3	1	Izhevsk	Dorozhnoye R.	4)	106.2	10	Moskva	Evropa Plus
2)	105.3	1	Saratov	Russkoye R.	4)	106.2	1	Murom	Evropa Plus
2)	105.3	1	Tula	Russkoye R.	5)	106.2	1	Kingisepp	Hit FM
4)	105.3	1	Kostroma	Evropa Plus	5)	106.2	4	Yekaterinburg	Hit FM
6)	105.3	5	Sankt-Peterburg	Love R.	6)	106.2	2	Novosibirsk	Love R.
1)	105.4	1	Tomsk	Avtoradio	6)	106.2	1	Ulyanovsk	Love R.
119)	105.4	1	Arkhangelsk	Mega FM	95A)	106.2	5	Kingisepp	Hit R.
177)	105.4	1	Pavlovskaya	Roks FM	10)	106.3	1	Tver	R. 7
2)	105.4	1	Kursk	Russkoye R.	126)	106.3	5	Sankt-Peterburg	R. Rekord
2)	105.4	1	Ryazan	Russkoye R.	15)	106.3	1	Saransk	DFM
20)	105.4	1	Samara	R. Alla	17)	106.3	1	Chelyabinsk	Serebryanyy dozhd
22)	105.4	4	Armavir	Pervoye R.	19)	106.3	1	Saratov	Dorozhnoye R.
1)	105.5	1	Kuznetsk	Avtoradio	2)	106.3	1	Nizhnevartovsk	Russkoye R.
12)	105.5	1	Oryol	Dobryye pesni	11)	106.4	1	Tula	R. Maksimum
147)	105.5	1	Belozersk	R. Transmit	11)	106.4	1	Surgut	R. Maksimum
2)	105.5	1	Murmansk	Russkoye R.	113)	106.4	1	Kostroma	Serebryanaya ladya
2)	105.5	1	Nalchik	Russkoye R.	17)	106.4	1	Barnaul	Serebryanyy dozhd
2)	105.5	1	Novokuznetsk	Russkoye R.	19)	106.4	1	Tambov	Dorozhnoye R.
2)	105.5	1	Staryy Oskol	Russkoye R.	2)	106.4	10	Irkutsk	Russkoye R.
6)	105.5	1	Tver	Love R.	41)	106.4	5	Vladivostok	Vladivostok FM
62)	105.5	1	Yuzhno-Sakhalinsk	R. 105.5	80)	106.4	10	Angarsk	R. Avtos
7)	105.5	1	Tuapse	Militseyskaya volna	95A)	106.4	1	Tikhvin	Hit R.
82)	105.5	5	Vyborg	R. Baltika	9A)	106.4	4	N.Novgorod	Glavnoye R.
1)	105.6	1	Yaroslavl	Avtoradio	1)	106.5	1	Ufa	Avtoradio
14)	105.6	5	Irkutsk	R. Shanson	16)	106.5	1	Bryansk	Retro FM
171)	105.6	1	Bogatye Saby	Tatar Radiosi	19)	106.5	4	Luga	Dorozhnoye R.
176)	105.6	4	Tyumen	Dipol FM	3)	106.5	1	Yaroslavl	Ekho Moskvy
2)	105.6	1	Volgograd	Russkoye R.	4)	106.5	1	Neftekamsk	Evropa Plus
22)	105.6	1	Primorsko-Akhtarsk	Pervoye R.	43)	106.5	1	Yakutsk	Lena R.
3)	105.6	1	Lipetsk	Ekho Moskvy	75)	106.5	1	Tyumen	R. Apriori
4)	105.6	1	Khabarovsk	Evropa Plus	9A)	106.5	1	Murmansk	Glavnoye R.
7)	105.6	1	Saransk	Militseyskaya volna	6)	106.6	5	Moskva	Love R.
10)	105.7	2	Novosibirsk	R. 7	6)	106.6	3	Samara	Love R.
16)	105.7	4	Omsk	Retro FM	1)	106.7	1	Kursk	Avtoradio
170)	105.7	1	Yakutsk	STV Radio	10)	106.7	4	Rodniki	R. 7
19)	105.7	1	V.Novgorod	Dorozhnoye R.	106)	106.7	1	Penza	Zolotoye FM
2)	105.7	10	Moskva	Russkoye R.	149)	106.7	1	Ussuriysk	R. VBC
2)	105.7	1	Yekaterinburg	Russkoye R.	19)	106.7	1	Kirov	Dorozhnoye R.
22)	105.7	1	Sochi	Pervoye R.	19)	106.7	1	Tver	Dorozhnoye R.
38)	105.7	1	Rostov-na-Donu	R. Dacha	19)	106.7	5	Vyborg	Dorozhnoye R.
39)	105.7	5	Vladivostok	R. Monte-Karlo	2)	106.7	1	Kingisepp	Russkoye R.
5)	105.7	1	Tolyatti	Hit FM	22)	106.7	4	Temryuk	Pervoye R.
5)	105.7	1	Izhevsk	Hit FM	38)	106.7	2	Novosibirsk	R. Dacha
1)	105.8	1	Kostroma	Avtoradio	5)	106.7	1	Listvyanka	Hit FM
10)	105.8	1	Vladimir	R. 7	7)	106.7	1	Bugulma	Militseyskaya volna
13)	105.8	1	Buguruslan	RSN	9A)	106.7	1	Ryazan	Glavnoye R.
15)	105.8	1	Pyatigorsk	DFM	1)	106.8	1	Omsk	Avtoradio
159)	105.8	1	Orenburg	R. Zvezda	14)	106.8	1	Krasnodar	R. Shanson
2)	105.8	1	Krasnoyarsk	Russkoye R.	19)	106.8	5	Belgorod	Dorozhnoye R.
3)	105.8	1	Kazan	Ekho Moskvy	19)	106.8	2	Voronezh	Dorozhnoye R.
3)	105.8	1	Saratov	Ekho Moskvy	4)	106.8	1	Kazan	Evropa Plus
6)	105.8	4	Tula	Love R.	4)	106.8	1	Murom	Evropa Plus
1)	105.9	1	Bugulma	Avtoradio	6)	106.8	1	Saratov	Love R.
14)	105.9	1	Chelyabinsk	R. Shanson	9A)	106.8	1	Stavropol	Glavnoye R.
14)	105.9	1	Ryazan	R. Shanson	10)	106.9	1	Sochi	R. 7
16)	105.9	1	Vladikavkaz	Retro FM	10)	106.9	4	Tolyatti	R. 7
167)	105.9	4	Suksun	Soyuz FM	111)	106.9	1	Sovetsk	Tilzitskaya volna
17)	105.9	1	Zverevo	Serebryanyy dozhd	112)	106.9	1	Murmansk	Bolshoye R.
19)	105.9	1	Kaliningrad	Dorozhnoye R.	124)	106.9	1	Vologda	R. Premyer
2)	105.9	1	Tambov	Russkoye R.	14)	106.9	1	N.Novgorod	R. Shanson
39)	105.9	10	Sankt-Peterburg	R. Monte-Karlo	3)	106.9	5	Tula	Ekho Moskvy
4)	105.9	1	Surgut	Evropa Plus	9A)	106.9	1	Vladimir	Glavnoye R.

FM	MHz	kW	Location	Station
1)	107.0	1	Belovo	Avtoradio
13)	107.0	1	Moskva	RSN
15)	107.0	1	Izhevsk	DFM
177)	107.0	4	Kanevskaya	Roks FM
2)	107.0	1	Vladivostok	Russkoye R.
2A)	107.0	1	Cheboksary	Russkoye R.
64)	107.0	1	Yekaterinburg	R. 107 FM
1)	107.1	1	Irkutsk	Avtoradio
108)	107.1	4	Rodniki	R. Most
14)	107.1	1	Pyatigorsk	R. Shanson
155)	107.1	1	Pskov	Sedmoye Nebo
6)	107.1	1	Tomsk	Love R.
117)	107.2	1	Ryazan	R. OK
19)	107.2	1	Petrozavodsk	Dorozhnoye R.
19)	107.2	1	Smolensk	Dorozhnoye R.
2)	107.2	1	Murom	Russkoye R.
2)	107.2	1	Orenburg	Russkoye R.
23)	107.2	1	Samara	Detskoye R.
45)	107.2	1	Saransk	MC Radio
84)	107.2	3	Voronezh	R. Borneo
126)	107.3	1	Vyborg	R. Rekord
126)	107.3	1	Luga	R. Rekord
15)	107.3	1	Chelyabinsk	DFM
2)	107.3	1	Kazan	Russkoye R.
5)	107.3	1	Kemerovo	Hit FM
103)	107.4	1	Tyumen	R. Yurga
114)	107.4	1	Sochi	Maks FM
14)	107.4	1	Yeysk	R. Shanson
2)	107.4	1	Rossosh	Russkoye R.
3)	107.4	5	N.Novgorod	Ekho Moskvy
34)	107.4	1	Tambov	Global FM
5)	107.4	5	Moskva	Hit FM
67)	107.4	1	Arkhangelsk	R. 29.ru - Region 29
71)	107.4	1	Staryy Oskol	R. 31
78)	107.4	10	Sankt-Peterburg	Business FM
8)	107.4	5	Novorossiysk	Nashe R.
9A)	107.4	1	Tolyatti	Glavnoye R.
14)	107.5	1	Tula	R. Shanson
15)	107.5	1	Nab.Chelny	DFM
15)	107.5	1	Nizhnekamsk	DFM
2)	107.5	1	Kovrov	Russkoye R.
2)	107.5	1	Ulan-Ude	Russkoye R.
3)	107.5	1	Penza	Ekho Moskvy
78)	107.5	1	Ufa	Business FM
9A)	107.5	1	Karachayevsk	Glavnoye R.
16)	107.6	1	Yakutsk	Retro FM
30)	107.6	2	Bryansk	Chistyye klyuchi
38)	107.6	1	Voronezh	R. Dacha
4)	107.6	1	Baymak	Evropa Plus
1)	107.7	1	Belgorod	Avtoradio
126)	107.7	2	Novosibirsk	R. Rekord
66)	107.7	1	Krasnodar	R. 107
139)	107.8	1	N.Novgorod	R. Privolzhye
2)	107.8	5	Sankt-Peterburg	Russkoye R.
6)	107.8	1	Kazan	Love R.
7)	107.8	5	Moskva	Militseyskaya volna
91)	107.8	1	Chelyabinsk	R. Narodnyy Khit
14)	107.9	1	Vladimir	R. Shanson
16)	107.9	1	Sochi	Retro FM
2)	107.9	1	Vladikavkaz	Russkoye R.
3)	107.9	1	Tolyatti	Ekho Moskvy

NB: Txs below 1kW not listed.

Addresses and other information:

1) 127083 Moskva, ul. 8 Marta 8. – **2A,B)** 123298 Moskva, 3-ya Khoroshevskaya ul. 12. – **3)** 119992 Moskva, ul. Novyy Arbat 11. – **4)** 109004 Moskva, ul. Stanislavskogo 21. – **5)** 123298 Moskva, 3-ya Khoroshevskaya ul. 12. – **6)** 127299 Moskva, ul. Bolshaya Akademicheskaya 5a. – **7)** 109180 Moskva, 3-y Golutinskiy per. 8/10. – **8)** Novorizhskoye shosse, 10km od MKAD, Kompleks "Riga Land", Moskva. – **9)** 115184 Moskva, B. Tatarskaya ul. 34. – **10)** 109004 Moskva, ul. Stanislavskogo 21. – **11)** 123298 Moskva, 3-ya Khoroshevskaya ul. 12. – **12)** 123298 Moskva, 3-ya Khoroshevskaya ul. 12. – **13)** 123298 Moskva, 3-ya Khoroshevskaya ul. 12. – **14)** 119049 Moskva, ul. Shabolovka 10. – **15)** 123298 Moskva, 3-ya Khoroshevskaya ul. 12. – **16)** 109004 Moskva, ul. Stanislavskogo 21. – **17)** 125083 Moskva, Petrovskogo-Razumovskaya aleya 12a. – **18)** 127427 Moskva, ul. Ak. Korolyova 19. – **19)** 199406 St.Peterburg, ul. Shevchenko 28. – **20)** 125040 Moskva, ul. Nizhnyaya Maslovka 9. – **21A,B)** 192007 St.Peterburg, Ligovskiy pr. 174. – **22)** 350038 Krasnodar, ul. Korolenko 2/1. – **23)** 129272 Moskva, ul. Trifonovskaya 57. – **24)** 127427 Moskva, ul. Korolyova 19. – **25)** 666036 Shelekhov, 4-y mikrorayon 32b. – **26)** 123056 Moskva., Gruzinskaya B. ul. 60. – **27)** 420021 Kazan, ul. Tukaya 91. – **28)** 241021 Bryansk, ul. Dimitrova 79. – **29)** 603022 N.Novgorod, Okskiy syezd 8. – **30)** 241021 Bryansk, ul. Dimitrova 79. – **31)** 129272 Moskva, Trifonovskaya 57. – **32)** 454091 Chelyabinsk, P.O.Box 13391. – **33)** 109004 Moskva, ul. Stanislavskogo 21. – **34)** 392000 Tambov, ul. Olega Koshevogo 14. – **35)** 394000 Voronezh. – **36)** 454091 Chelyabinsk, ul. Ordzhonikidze 81. – **37)** 614000 Perm, ul. Gazety "Zvezda" 26b. – **38)** 127299 Moskva, ul. B. Akademicheskaya 5a. – **39)** 123298 Moskva, 3-ya Khoroshevskaya ul. 12. – **40)** 125190 Moskva, Leningradskiy pr. 80. – **41)** 690091 Vladivostok, ul. Pologaya 66. – **42)** 450075 Ufa, ul. Blyukhera 15. – **43)** 677001 Yakutsk, ul. Bestuzheva-Marlinskogo 9/3a. – **44)** 680000 Khabarovsk, ul. Lenina 4. – **45)** 430000 Saransk, ul. Proletarskaya 39-43a. – **46)** 109028 Moskva, M.Trekhsvyatitelskiy per. 2/5. – **47)** 428028 Cheboksary, Traktostroitelei pr. 101. – **48)** 440026 Penza, ul. Lermontova 39. – **49)** 362021 Vladikavkaz, ul. Nikolayeva 84. – **50)** 302020 Oryol, Naugorskoye shosse 40. – **51)** 174260 Malaya Vishera, ul. Moskovskaya 21. – **52)** 620144 Yekaterinburg, ul. Khoryakova 104. – **53)** 353900 Novorossiysk, proyezd Skoblikova 10. – **54)** 400131 Volgograd, ul. Komsomolskaya 8. – **55)** 630087 Novosibirsk, ul. Nemirovicha-Danchenko 122. – **56)** 450076 Ufa, ul. Gafuri 9. – **57)** 127993 Moskva, Staryy Petrovsko-Razumovskiy proyezd 1/23. – **58)** 432602 Ulyanovsk, ul. Engelsa 21. – **59)** 672010 Chita, ul. Amurskaya 36. – **60)** 183038 Murmansk, ul. Yegorova 14. – **61)** 420066 Kazan, ul. Dekabristov 2. – **62)** 693023 Yuzhno-Sakhalinsk, ul. Komsomolskaya 213a. – **63)** 125130 Moskva, ul. Priorova 18. – **64)** 620075 Yekaterinburg, pr. Lenina 50l. – **65)** 125080 Moskva, ul. Vrubelya 4. – **66)** 350000 Krasnodar, ul. Gimnazicheskaya 51. – **67)** 163002 Arkhangelsk, pr. Novgorodskiy 32. – **68)** 432071 Ulyanovsk, ul. Narimanova 75. – **69)** 125040 Moskva, Leningradskiy pr. 30. – **70)** 644010 Omsk, ul. Dekabristov 130. – **71)** 308000 Belgorod, pr. Slavy 61. – **72)** 426069 Izhevsk, ul. Pesochnaya 11. – **73)** 614060 Perm, ul. Turgeneva 33a. – **74)** 403323 Mikhaylovka, ul. Kommuny 129-1. – **75)** 625013 Tyumen, ul. Tekstilnaya 1. – **76)** 453300 Kumertau, PKiO im. Gagarina. – **77)** 398050 Lipetsk, ul. Plekhanova 1. – **78)** 127287 Moskva, 2-ya Khutorskaya ul. 38a. – **79)** 445010 Tolyatti, ul. Sovetskaya 74a. – **80)** 665831 Angarsk, 6-y mikrorayon 5. – **81)** 236000 Kaliningrad, pr. Mira 87. – **82)** 197022 St.Peterburg, Kamennoostrovskiy pr. 67. – **83)** 140400 Kolomna, ul. Shilova 9. – **84)** 394071 Voronezh, ul. 20-letnaya Oktyabrya 66. – **85)** 625013 Tyumen, ul. Permyakova 7. – **86)** 617760 Chaykovskiy, ul. Lenina 6-3. – **87)** 238750 Sovetsk, ul. A.Nevskogo 6. – **88)** 650036 Novokuznetsk, pr. Lenina 90/4. – **89)** 194044 St.Peterburg, per. Krapivnyy 5-206. – **90)** 394000 Voronezh. – **91)** 454091 Chelyabinsk, ul. Vorovskogo 71. – **92)** 367000 Makhachkala, pr. Akushinskogo 13a. – **93)** 125190 Moskva, Leningradski pr. 80. – **94A,B)** 677000 Yakutsk, ul. Oktyabrskaya 16/2. – **95A,B)** 199406 St.Peterburg, ul. Shevchenko 27. – **96)** 620075 Yekaterinburg, pr. Lenina 41. – **97)** 173020 N.Novgorod, B.Moskovskaya ul. 106. – **98)** 677000 Yakutsk, ul. Kirova 17/3-3. – **99)** 614068 Perm, ul. Kuybysheva 37. – **100)** 625019 Tyumen, ul. Respubliki 211a. – **101)** 305004 Kursk, ul. Dimitrova 76. – **102)** 690000 Vladivostok, ul. Strelnikova 3a. – **103)** 625019 Tyumen, ul. Respubliki 53. – **104)** 450058 Ufa, pr. Oktyabrya 128/3. – **105)** 610000 Kiro, Oktyabrskiy pr. 120. – **106)** 440034 Penza, ul. Markina 1. – **107)** 443070 Samara, ul. Aerodromnaya 13. – **108)** 153025 Ivanovo, 5-y Severniy per. 18. – **109)** 614077 Perm, ul. Uralskaya 119. – **110)** 426069 Izhevsk, ul. Pesochnaya 9. – **111)** 238750 Sovetsk, ul. Goncharova 12. – **112)** 183038 Murmansk, ul. Lenina 68. – **113)** 156961 Kostroma, pl. Konstitutsii 1. – **114)** 354000 Sochi, ul. Severnaya 12. – **115)** 614000 Perm, ul. Lenina 50. – **116)** 626726 Noyabrsk, ul. Lenina 47-412. – **117)** 390023 Ryazan, ul. Tsilkovskogo 20. – **118)** 454090 Chelyabinsk, ul. Tsvillinga 46a. – **119)** 163000 Arkhangelsk, pr. Troitskiy 52. – **120)** 423827 Nab.Chelny, bul. Yunykh Lenintsev 9. – **121)** 195299 St.Peterburg, ul. Cherkasova 14. – **122)** 620102 Yekaterinburg, ul. Posadskaya 79. – **123)** 123242 Moskva, Novinskiy bul. 31. – **124)** 160035 Vologda, ul. Kozlenskaya 35. – **125)** 603123 N.Novgorod, ul. Semashko 37. – **126)** 198303 St.Peterburg, ul. Stachek 105. – **127)** 197376 St.Peterburg, Aptekarskiy pr. 6/4. – **128)** 450075 Ufa, ul. Blyukhera 15. – **129)** 344082 Rostov-na-Donu, Bolshaya Sadova ul. 10. – **130)** 620086 Yekaterinburg, ul. Repina 4. – **131)** 443010 Samara, ul. Nekrasovskaya 62. – **132)** 603068 N.Novgorod, Yarmarochnyy proyezd 10. – **133)** ul. Lenina 41, 620014 Yekaterinburg. – **134)** 634003 Tomsk, per. Marinskiy 8. – **135)** 655017 Abakan, ul. Vyatkina 12. – **136)** 454091 Yekaterinburg, ul. Marksa 38. – **137)** 30000 Saransk, ul. Goncharov 39. – **138)** 109004 Moskva, ul. Stanislavskogo 21. – **139)** 603600 N.Novgorod, ul. Belinskogo 9a. – **140)** 620109 Yekaterinburg, ul. Repina 15. – **141)** 443011 Samara, ul. Sovetskoy Armii 245e. – **142)** 115184 Moskva, B. Tatarskaya ul. 35. – **143)** 125568 Moskva, ul. Mitinskaya 49. – **144)** 659322 Biysk, ul. Dekabristov 10. – **145)** 620014 Yekaterinburg, pr. Lenina 41. – **146)** 105082 Moskva, ul. Baumanskaya 15. – **147)** 162610 Cherepovets, ul. Lenina 151. – **148)** 692525 Ussuriysk, ul. Kirova 28. – **149)** 690091 Vladivostok, ul. Uborevicha 20a. – **150)** 400131 Volgograd, ul. Kommunisticheskaya 6. – **151)** 238051 Gusev, u. Shkolnaya 11. – **152)** 117218 Moskva, Krzhizhanovskogo 29. – **153)** 1015054 Moskva, B. Strochenovskiy per. 22/25. – **154)** 678960 Neryungri, per. Lenina 6. – **155)** 180000 Pskov, ul. Lenina 6a. – **156)** 614000 Perm, ul. Lenina 50. – **157)** 630087 Novosibirsk, ul. Nemirovicha-Danchenko 122. – **158)** 103287 Moskva, Khutorskaya 2-ya ul. 38a. – **159)** 119160 Moskva, Kolymazhnyy per. 14. – **160)** 103000 Moskva, ul. Kazakova 16. – **161)** 170000 Tver, ul. Mednikovskaya 55/25. – **162)** 123022 Moskva, ul. 1905 goda 10. – **163)** 455026 Magnitogorsk, ul. Gagarina 35. – **164)** 129272 Moskva, ul. Trifonovskaya 57. – **165)** 197376 St.Peterburg, ul. Prof.Popova

47. – **166)** 660118 Krasnoyarsk, ul. Mate Zalki 24. – **167)** 614000 Perm, ul. Bolshevistskaya 84a. – **168)** 625019 Tyumen, ul. Respublika 211a. – **169A,B)** 400117 Volgograd, bul. 30-let Pobedy 74a. – **170)** 677027 Yakutsk, ul. Oktyabrskaya 10/1. – **171)** 420094 Kazan, ul. Golubzatnikova 20a. – **172)** 614077 Perm, bul. Gagarina 80. – **173)** 428000 Cheboksary, pr. Lenina 35. – **174)** 197101 St.Peterburg, ul. Kronverskaya 23. – **175)** 248021 Kaluga, ul. Moskovskaya 189. – **176)** 625000 Tyumen, ul. Geologorazvedchikov 28a. – **177)** 350000 Krasnodar, ul. Levanevskogo 57. – **178)** 424000 Yoshkar-Ola, ul. Sovetskaya 138. – **179)** 454091 Chelyabinsk, ul. Ordzhonikidze 41. – **180)** 622001 Nizhniy Tagil, pr. Lenina 4.

Radio via DTT: see TV section.

Foreign Service/Int relays on MW: (txs operated by Russian TV and Radio Broadcasting Network): Angarsk, S 1080kHz 1000kW, Belogorsk, FE 585kHz 1200kW; Bolshakovo, E 1143kHz 150kW, 1215 kHz 1200kW; Chita, S 801kHz 1200kW, Komsomolsk-na-Amure, FE 630kHz 500kW; Oyash, S 1026kHz 250kW; St. Peterburg, S 1494kHz 600kW; Tbilisskaya, E 1089/1170 kHz 1200kW; Vladivostok, FE 648kHz 1000kW, 1251 600kW; Yuzhnyy-Sakhalinsk, FE 720kHz 1000kW. See International Radio section

RWANDA

L.T: UTC +2h — **Pop:** 11 million — **Pr.L:** Kinyarwanda, Swahili, French, English — **E.C:** 50Hz, 230V — **ITU:** RRW

RWANDA UTILITIES REGULATORY AGENCY (RURA)
B.P. 6929, Kigali ☎+250 252584562 ☒ +250 252584563 **W:** www.rura.gov.rw **E:** info@rura.gov.rw

RADIODIFFUSION DE LA REPUBLIQUE RWANDAISE (Gov)
B.P. 83, Kigali ☎+250 76665/76180 ☒+250 25276182 **W:** www.orinfor.gov.rw/radio **E:** radiorwanda@yahoo.com **L.P:** Dir. Broadc: Mweusi Karake. Dir. Prgrs: Paul Ndamage. Ag. Ch.Editor: Willy Rukundo. Tech. Dir: Charles Nahayo.
SW: Kigali: 6055kHz 50kW. Rel. of channels I/II 0300-2100.
FM: Channel I (MHz): 89.8 Kinaira 0.5kW, 93.5 Nyarupfubire 0.5kW, 95.1 Mugogo/Rushaki 0.5kW, 97.6 Karongi 1kW, 100.7 Mt. Jali 5kW, 103.2 Byumba 0.5kW, 103.9 Butare 0.5kW + 8 trs under 0.5kW.
Channel II: Kigali 90.7MHz 5kW (irregular).
D.Prgr in Kinyarwanda/Swahili/French/English: 24h. **English**: Thurs 2000-2045. **N. in English:** 0515, 1830.
Ann: F: "Vous écoutez Radio Rwanda émettant de Kigali".

Other stations:
City R, Kigali: 88.3MHz 1kW – **Contact FM:** Kigali 89.7MHz 0.5kW – **Flash FM,** Kigali: 89.2MHz 0.5kW – **R. Communautaire:** Cyangugu 92.9MHz 30W, Karongi 96.5MHz 0.5kW, Butare 100.4MHz 0.5kW, Gisenyi 104.7MHz 0.1kW – **R. Izuba,** Kibungo: 100.0MHz 0.25kW – **R. Maria Rwanda:** Gitarama 88.6MHz 1kW, Kigali 97.3MHz 2kW, Karongi 99.8MHz. **E:** radiomariar@yahoo.fr – **R. 10 FM:** Mont Jali 87.6MHz 0.5kW **W:** www.danslevent.com/rwanda **E:** contact@danslevent.com – **R. Salus** (University R.), Butare: 97.0MHz 1kW, Kigali 101.9MHz 1kW. D. Prgr: 0300-2100. **W:** www. nur. ac.rw **E:** salusradio@yahoo.com – **R. Sana uRwanda** (Rlg.): Kigali 98.0MHz. – **R. Umucyo,** Kigali: 102.8MHz 0.1kW.
AWR: 106.4MHz 0.3kW – **BBC African Sce:** Karongi 93.3MHz, Kigali 93.9MHz, Butare 106.1MHz 3kW – **RFI Afrique:** Kigali 92.1MHz in F/E/Swahili – **Voice of America:** Kigali 104.3MHz 2kW.
Deutsche Welle: Kigali 96MHz 2kW. **Shortwave relay station:** see International radio section

SABA (Netherlands)

L.T: UTC -4h — **Pop:** 1,400 — **Pr.L:** Dutch (official), English — **E.C:** 60Hz, 110V — **ITU:** ATN

QFM, The Voice of Saba
The Bottom, Saba ☎ +599 416 3213 ☒ +599 416 3308 Owner: Max Nicholson
FM: 93.9MHz 1kW

SAMOA

L.T: UTC -11h (26 Sep 10-3 Apr 11: -10h) DST 2011-2012 subject to confirmation — **Pop:** 219,998 — **Pr.L:** Samoan, English — **E.C:** 50Hz, 230/410V — **ITU:** SMO

NATIONAL RADIO 2AP (Gov)
Ministry of Communications and Information Technology, Government of Samoa, Level 1, CA & CT Plaza, Savalolo, Apia, Samoa

☎ +685 26177 ☒ +685 24671 **W:** www.mcit.gov.ws **E:** mcit@mcit. gov.ws Studio ☎ +685 24790 ☒ +685 24789
L.P: CEO: Tua'imalo Asamu Ah Sam, Senior Programmer: Vaasiliega Lupati Lagaia, Principal Technician: Clement Warren
MW: Apia 540kHz 10kW, 747kHz 10kW (both running at 8kW)
Samoan/English: 1700 (Sun 2200)-1000 on 540kHz **Edu:** brdcsts on some weekdays for Samoa and Tokelau 1930-2030 on 747kHz **World N:** 1800W, 1900W, 1930W. **Local N:** 1630W, 1730W, 1830W, 0730W
Ann: E: "National Radio 2AP" or "Voice of the Nation' **IS:** Gong
V. by card or letter

Other Stations:

FM		MHz	kW	FM	MHz	kW
1)	Talofa FM	88.5	0.25	5) Showers of Blessings	97.5	0.5
8)	Mai FM 89.1	89.1	1	1) Magik FM	98.1	0.3
2)	Aiga Fesilafa'I R.	90.5		1) Talofa FM	99.9	0.25
3)	R. Graceland	90.9	1	1) K-Lite FM	101.1	0.25
1)	Talofa FM	91.5	0.25	6) R. Australia	102.0	
3)	R. Graceland	94.9		4) R. Laufou	103.1	2
4)	R. Laufou	95.1	0.25	7) Calvary Chapel R.	106.1	
1)	Star FM	96.1	0.25	3) R. Graceland	106.9	

Location: All Apia (Upolu) and surrounding area exc. R. Graceland 94.9MHz: and Talofa FM 91.5MHz.

Addresses and other information
1) Radio Polynesia Ltd, P.O.Box 762, Apia ☎+685 25148 ☒ +685 25147 **Brands:** Magik FM - "Samoa's #1 Hit Music Station' (English) ☎ Studio +685 33981, Talofa FM – "100% Local" (Samoan) ☎ Studio +685 33999, K-Lite FM "Memories are Good" (English) ☎ +685 33101 **N:** RNZI throughout the day, Star FM "Absolute Music Variety" (English) ☎ +685 33961 **W:** www.fmradio.ws **E:** corey@fmradio.ws **L.P:** CEO Corey Keil **D.Prgr:** 24h – **2)** Catholic Archdiocese of Samoa, P.O.Box 532, Apia **Prgr:** Samoan religious – **3)** Graceland Broadcasting Network, Levili, Apia ☎+685 20107 ☒+685 25487 **E:** gbn@post.com **D.Prgr:** Sunday 1800-1100, satellite relay from TBN – **4)** Youth For Christ Mission, Mulinu'u, Apia **Ann:** "Laufou o le Talalelei" **Prgr:** Samoan religious – **5)** Sogi, Apia. **Prgr:** local Samoan religious plus satellite relay – **6)** 24h satellite relay from Melbourne – **7)** 24h satellite relay **Prgr:** English religious – **8)** Samoa Quality Broadcasters, Mulin'u, Apia. CEO: Galuemalemana Ms Faresea Matafeo. **D.Prgr** 24h (English & Samoan)

SAMOA (AMERICAN)

L.T: UTC -11h — **Pop:** 65,628 — **Pr.L:** Samoan, English — **E.C:** 60Hz, 120V — **ITU:** SMA

FEDERAL COMMUNICATIONS COMMISSION (FCC)
see USA for details

MW	Call	kHz	kW	MW	Call	kHz	kW
1) Tafuna	KJAL*	585	5	2) Leone	KKHJ*	900	5
2) Leone	WVUV*	648	10				

FM	Call	MHz	kW	FM	Call	MHz	kW
6) Tafuna	New*	88.1	1.5	4) Ili'Ili	KULA-LP	95.1	0.1
7) Pago Pago	New*	88.9	0.3	4) W.District	KULA-LP	97.1	0.01
8) Nu'uuli	KMOA*	89.7	1.5	4) C.District	KULA-LP	99.1	0.01
9) Mapusaga	KPPO*	90.5	0.75	4) E.District	KULA-LP*	102.5	0.02
2) Pago Pago	KSBS-FM	92.1	15	2) Fagaitua	WVUV-AM	103.1	1.3
2) Pago Pago	KKHJ-FM	93.1	1.1	5) Pago Pago	KNWJ-FM	104.1	0.01
2) Pavaiai	KKHJ	93.7	0.01	5) Leone	KNWJ-FM	104.7	0.3

NB: stns marked * are currently silent.

Addresses and other information:
1) Asia Pacific Media Ministries, KJAL-AM, PO Box 1138, Pago Pago, American Samoa 96799. L.P: Vickie Haleck, SM. **ID:** "For You and Your Family" **NB:** uses 580kHz when on air. **F.PL:** CP 630kHz 5kW – **2)** South Seas Broadcasting Inc, PO Box 6758, Pago Pago, American Samoa 96799. L.P: Joey Cummings, GM ☎+1 684 633 4493 **W:** KKHJ-FM: www.khjradio.com **ID:** '93KHJ', **F.PL:** KKHJ CP 900kHz 5kW. **W:** WVUV-FM: www.wvuv.com **ID:** V103 'The People's Station' in Polynesian and Samoan, D.Prgr: 24h **F.PL:** WVUV CP 720kHz 5/2kW – **3)** Samoa Broadcasting System, PO Box 793, Pago Pago, American Samoa 96799-0793 ☎+1 684 633 7000 ☒+1 684 633 5727 **W:** www. ksbsfm92.com **E:** info@ksbsfm92.com **L.P:** Esther Prescott, GM **ID:** 'Island 92 - The Station That Belongs to You' **News:** hourly bulletins from RNZI, Radio Australia, BBC, NPR, VOA. D.Prgr: 24h – **4)** Pacific Islands Bible School, PO Box 1268, Pago Pago, American Samoa 96799 – **5)** Showers of Blessings Radio, PO Box 997777, Pago Pago, American Samoa 96799 ☎+1 684 699 8123. ☒+1 684 699 8126 **W:** www.fm104.org **E:** info@fm104.org – **6)** Leone Church of Christ, PO Box 5093, Pago Pago, American Samoa 96799 –**7)** Marianas Educational Media Services, 125 Tun Jesus Crisostomo St #302, Tamuning, GU 96913 – **8)** Horizon Christian Fellowship, 5331 Mt

Alifan Dr, San Diego CA 91111 – **9)** Second Samoan Congregational Church of Long Beach, 655 Cedar Ave, Long Beach CA 90802-1222

SAN MARINO

L.T. UTC +1h (27 Mar- 30 Oct: +2h) — **Pop.** 30.000 — **Pr.L:** Italian — **ITU:** SMR

SAN MARINO RTV (Gov)
✉ Viale J.F.Kennedy 13, RSM-47890.Repubblica di San Marino ☎ +378 0549 882000 🗎 +378 0549 882840 **E:** radio@sanmarinortv.sm **W:** www.sanmarinortv.sm **L.P:** Dir. .Mrs Carmen Lasorella Prgr.Dir.: G.Cesetti: T.D Fabio Pelliccioni
FM: 102.7MHz 30kW **D.Prgr:** 24h
San Marino Classic, E: classic@sanmarinoerv.sm **FM:** 103.2MHz 30kW **D.Prgr:** 24h Also carries govt meetings, live service. Prgr.Dir. S. Coveri **F.PI.** no plans to start on MW, assigned freq.711kHz
V: by QSL-card. Rpts to **E:** ufficiotecnico@sanmarinortv.sm

RADIO INTERNATIONAL SAN MARINO (Comm)
✉ Europa Radiodiffusione S.r.l. Strada Rovereta 42 RSM-47891 Falciano ☎ +378 0549 909905 🗎 +378 0549 941580 **E:** info@radioin-ternational.sm **Wb:** www.radiointernatinal.sm
FM: 94.25MHz 1kW **D.Prgr:** 24h

SÃO TOMÉ E PRÍNCIPE

L.T. UTC — **Pop:** 210,000 — **Pr.L:** Portuguese, Crioulo — **E.C:** 50Hz, 220V — **ITU:** STP

RÁDIO NACIONAL DE SÃO TOMÉ E PRÍNCIPE (RNSTP, Gov)
✉ Avenida Marginal 12 de Julho, C.P. 44, São Tomé ☎+239 222875 🗎 +239 223293 **E:** Rnstp04@cstome.net **L.P:** Dir: Artur Meneses de Pinho. CE: Felisberto Garcia.
MW: Pinheira 945kHz 20kW **FM:** 89.7/95.4/99.3MHz.
D.Prgr: 24h in Portuguese. **N:** 0700, 1300, 1630, 1930.
Ann: "Aqui São Tomé, Capital da República Democrática de S. Tomé e Príncipe, transmite a Rádio Nacional". **IS:** one note gong, guitar.

RDP África: São Tomé 92.8MHz 3kW, Príncipe 101.9MHz 70W.
RFI Afrique: 102.8MHz in French/Portuguese.
VOA: São José 105.5MHz 0.2kW in English/Portuguese.
VOA relay station: MW 1530kHz 600kW 0300-0630, 1600-2200 & SW. For further details see International Radio section under USA

SAUDI ARABIA

L.T. UTC +3h — **Pop:** 29 million — **Pr.L:** Arabic — **E.C:** 60Hz, 127/220V — **ITU:** ARS

MINISTRY OF CULTURE & INFORMATION (MOCI)
✉ Nasseriya Str, Riyadh 11161 ☎+966 1 4014440 🗎 +966 1 402 3570. **W:** moci.gov.sa
L.P: Dep. Min. of Eng. Affairs: Dr. Riyadh Najm.

BROADCASTING SERVICE OF THE KINGDOM OF SAUDI ARABIA (BSKSA, Gov.)
✉ P.O. Box 61718, Riyadh 11575 ☎+966 1 4425170 🗎 +966 1 4041692 **W:** www.saudiradio.net **E:** saudi-radio@moci.gov.sa

MW	kHz	kW	H of tr & Prgr.
Bisha	531	10	24h (Q)
Ar-Rass	549	10	0300-2300 (G)
Gizan	549	1	0300-2300 (G)
Qurayyat	549	20	24h (G)
Rafha	549	20	24h (G)
Jeddah	558	50	24h (G)
Abha	567	20	24h (Q)
Afif	567	15	24h (Q)
Gizan	576	20	24h (Q)
Riyadh	585	1200	0100-2300 (G)
Al-Hufuf	594	10	0300-2300 (G)
Duba	594	2000	0300-1500 (G)
Makkah	594	50	24h (G/P)
Al-Aflaj	612	15	24h (Q)
Hail	612	5	24h (Q)
Gizan	630	20	0300-2200 (G)
Najran	630	10	24h (Q)
Jeddah (Khumra)	648	2000	0300-2300 (G)
Rafha	657	20	24h (Q)
Abha	675	5	0300-2300 (G)
Afif	675	20	0300-2300 (G)

MW	kHz	kW	H of tr & Prgr.
Jeddah	684	50	24h (2)
Riyadh	684	10	24h (2)
Bisha	702	10	0300-2200 (2)
Duba	702	40	24h (2)
Najran	747	10	0300-2300 (G)
Al-Aflaj	765	20	24h (Q)
Al-Hufuf	765	10	24h (Q)
Qurayyat	765	20	24h (Q)
Ras al-Zawr	783	100	24h (2)
Jeddah	792	50	24h (Q)
Abha	810	20	0300-2200 (2)
Ras al-Zawr	855	10	24h (Q)
Ar-Rass	873	10	24h (Q)
Dammam	882	100	24h (Q)
Qurayyat II	900	1000	24h (G)
Al-Hufuf	927	20	24h (G)
Makkah	936	50	24h (Q)
Riyadh	936	50	24h (Q)
Hail	945	5	0300-2300 (G)
Madinah	981	20	24h (Q)
Duba	999	20	24h (Q)
Madinah	1017	20	24h (P)
Yanbu al-Bahr	1035	20	24h (2)
Bisha	1071	50	24h (G)
Najran	1080	10	0300-2200 (2)
Qurayyat	1089	20	24h (2)
Dammam	1098	100	24h (2)
Madinah	1116	20	24h (Q)
Madinah	1215	20	0300-2300 (G)
Dammam	1260	500	0300-2300 (G)
Riyadh	1422	20	0600-2100 (F)
Ras al-Zawr	1440	1600	24h (G)
Hafar Al-Batin	1467	50	24h (Q)
Jeddah (Khumra)	1512	1000	1300-0300 (Q)
Duba	1521	2000	1500-0300 (G)

FM(MHz)	G	2	Q	M/F
Aflaj	93.3	96.5	99.8	
Al-Baha	98.0	88.4	91.5	
Ar-Rass	96.1	102.9	99.4	
Arafat	92.2	94.0	90.8	
Arar	94.1	97.4	88.4	
Buraydah	89.3	95.6	93.2	
Dammam	92.8/93.8	94.7	90.0	103.6
Duba	89.5	95.8	92.6	
Jeddah	92.0/99.5	93.0	89.9	96.2
Jizan	95.1	88.8	91.9	
Jubail	107.7	105.6	95.3	
Kharj	93.0	96.0	90.0	
Mecca	94.7	98.0	91.5	
Medina	90.5	93.6	96.8	
Riyadh	91.2	94.4	100.0	97.7
Taif	96.5	93.3	99.8	106.9
Yanbu	97.4	93.6	90.9	

+ numerous low power stations under 10kW for local coverage.

General Prgr. (G) in Arabic: 24h incl. Call of Islam 0100-0300 & 1500-1700. Only part of A prgr trs operate 24h – **Second (2) Prgr in Arabic:** 24h, on most MW fqs 0300-2200. – **Quran (Q) prgr:** 24h incl. Call of Islam 0100-0300. – **Foreign Language (F) prgrs** (from either Riyadh or Jeddah studios): **English:** 0600-0800, 1000-1300, 1600-2100. **French:** 0800-1000, 1400-1600. – **Music (M) prgr:** 24h. – **Pilgrimage (P) Enlightment Radio:** 24h during two months of the "Haj" season (Dec-Feb) in Arabic/English/French/Persian/Turkish/Hausa/Indonesian/Urdu on MW and on FM 94.2/101.0MHz in Mina/Arafat/Muzdalifah.
Ann: General prgr: "Idha'at il-Mamlaka al-Arabiya as-Saudiya(, al-barnamig al-aam,) min ar-Riyadh". 2nd prgr: "Idha'at il-Mamlaka al-Arabiya as-Saudiya, al-barnamig at-thani min Jeddah". Call of Islam: "Idha'at Nidaa Al-Islam min Makka al-Mukaram". **E:** "This is Radio Riyadh/Jeddah".
IS: 'Ud' (oriental lute). Opens and closes with National Anthem.

EXTERNAL SERVICE: Saudi Radio: see International Radio section.

SAUDI ARAMCO RADIO (Serving the staff of Saudi Aramco Co.)
✉ P.O.Box 5000, Dhahran 31311 ☎+966 3 8723046 🗎 + 966 3 8726444 **W:** www.saudiaramco.com **E:** webmaster@aramco.com.sa
L.P: Supervisor, Radio & Media Services: Mohammad A. Atani.
Studio 1 (pop, rock and country music): Udhailiyah 88.8MHz, Dhahran 91.4MHz, Safaniya/Tanajib/Haradh 103.8MHz – **Studio 2** (easy listening, jazz and classical music): Udhailiyah 91.9MHz, Dhahran 101.4MHz, Safaniya/Tanajib/Haradh 107.9MHz. **D.Prgr:** 24h in English.

American Forces Network: 93.7/100.7/103.9/107.8MHz.

Panorama FM: Dammam 91.9MHz, Riyadh 96MHz, Tabuk 101.7MHz, Jeddah 102MHz, Madinah 102.3MHz, Buraidah (Al Qassim) 103.3MHz, Abha/Taif 104MHz. See main entry under UAE
New FM licenses granted to companies: Alf Alf, Ghayat Al-Ibdah, Rotana, Electronic Resources and Shams R.

SENEGAL

L.T: UTC — **Pop:** 14 million — **Pr.L:** French, Wolof, Mandinga, Soninké, Pular, others — **E.C:** 50Hz, 230V — **ITU:** SEN

CONSEIL NATIONAL DE REGULATION DE L'AUDIOVISUEL (CNRA)
✉ 15ème étage, Immeuble Fahd, Blvd Djily Mbaye, B.P. 50059, Dakar RP ☎+221 33 8499120 🖷 +221 33 8234785 **W:** www.cnra-sn.org **L.P:** Chairperson: Nancy Ngom Ndiaye.

RADIODIFFUSION TÉLÉVISION SÉNÉGALAISE (Gov.)
✉ Triangle Sud x Avenue El-Hadj Malick SY, B.P. 1765, Dakar ☎+221 33 8491212 🖷 +221 33 8223490 **W:** www.rts.sn **E:** rts@rts.sn
L.P: DG: Babacar Diagne. Dir. Radio: Oumar Seck. Dir. New Tech. & Development: Papa Abdou Diallo.

FM	N	I	R	M	kW
Bakel	95.9	107.3			5
Dakar	95.7	92.5	94.5	95.2	10
Diourbel	97.6	96.6	101.1		2
Fatick	95.7		92.8		0.5/2
Goudiry	106.0		91.1		0.5
Kaolack	103.0	107.0	97.9		5
Kédougou	94.6	97.7	100.0		2
Kolda	100.0	102.2	92.2		2
Koungheul	89.7		107.0		1
Linguère	92.1	89.0			5
Louga	95.0	101.8	88.7		5/2
Matam	95.6	89.1	100.6		2
Ndioum		98.4	92.7		5
Ourossogui	96.5	89.1	105.3		5/2
Podor			100.6		0.25
Richard Toll			89.6		0.1
Saint-Louis	91.9	90.1	96.3		10/5
Tambacounda	102.0	88.1	92.0		5
Thiès	96.9	94.9	100.6		5
Touba			99.2		0.25
Vélingara	99.0	89.1	92.2		2
Ziguinchor	95.2	100.2	98.9		5

N=Chaîne Nationale: 24h in French, Wolof and other national languages – **I=R. Sénégal Internationale:** 24h in French, Arabic, Portuguese and other languages – **M=RTS Mag FM:** 24h in French – **R=Chaîne Régionale:** 0600-2400, regional programming for 9 to 18 hours a day depending on station.
Ann: N: "Radiodiffusion Télévision Sénégalaise émettant de Dakar". I: "Radio Sénégal Internationale". **IS:** Melody on "Cora" (local harp).

SUD FM SEN RADIO
✉ Immeuble Fahd, Bld. Djily Mbaye x rue Macodou Ndiaye (5ème étage), Dakar +221 33 82255991 🖷 +221 33 8220250 **W:** sudfm.net

FM	MHz	kW	FM	MHz	kW
Sédhiou	88.0		Ziguinchor	95.6	2
Nioro-du-Rip	88.3	3	Bakel	97.5	1
Diourbel	91.1	2	Kédougou	97.9	6
Pikine	91.7	5	Matam	98.4	2
Banjul (GMB)	92.1		Dakar	98.5	3
St.-Louis	93.2	2	Tambac.	98.5	1.5
Kaolack	94.6	2	Thiès	102.2	0.2
Kolda	95.4	0.4	M'Bour	108.0	1

WEST AFRICA DEMOCRACY RADIO (WADR)
✉ P.O. Box 16650, Sacré-Coeur1, N°8408, Dakar-Fann +221 33 8691569 🖷 +221 33 8647009 **W:** www.wadr.org **E:** wadr@wadr.org
L.P: Essoh Honoré, Radio Production Officer.
FM: Dakar 94.9MHz 5kW. Also via community radios in Guinea, Liberia and Sierra Leone. **F.PI:** txs in Chad and Cameroon.

Other stations:
Afia FM, Dakar: 93.0MHz 5kW – **Aprosor FM,** Dakar: 104.4MHz 2kW – **Bamtaare Dowri FM,** Vélingara: 92.5MHz 10kW – **CMC:** Koungheul 92.1MHz 2kW, Kaolack 103.2MHz 2kW. **W:** cmcsenegal. org – **Convergence FM,** Dakar: 103.9MHz. **W:** convergencefm. com – **Diamono FM,** Dakar 100.8MHz 10kW – **Energie FM,** Dakar: 106.1MHz 1kW – **Express An-Nour FM:** Vélingara 87.7MHz 10kW, Bignona 88.3MHz 2kW, Kolda 91.9MHz 5kW, Kaolack 92.7MHz 2kW, Tambacounda 94.0MHz 2kW, Bakel 100.8MHz 2kW, Dakar 101.0MHz 5kW, Koungheul 102.1MHz 2kW, Kédougou 106.4MHz 10kW, Thiès 106.9MHz 5kW – **FM Téranga,** St. Louis 99.7/103.2MHz 2kW – **Jappo FM,** Dakar: 94.0MHz **W:** jappofm.net – **Jiida FM,** Dakar: 105.3MHz 8kW – **Jokkoo FM,** Dakar: 87.7MHz 5kW – **Lamp Fall FM (R. Xarnu Bi):** Touba 97.4MHz, Dakar 101.7MHz. **W:** alazhartouba.com/LampFallFm.htm – **Love FM,** Dakar: 107.3MHz 10kW – **Medina Baye FM,** Kaolack 90.1MHz 5kW – **Ocean FM,** St. Louis 88.9MHz, Kaolack 93.1MHz 5kW, Kébémer 93.9MHz, Dakar 98.7MHz 5kW, Ziguinchor 106.0MHz. **W:** oceanfm.sn – **R. Al-Hamdoudilah,** Dakar: 91.0MHz 5kW – **R. Connexion sans Frontières,** Dakar: 90.0MHz 5kW – **R. Djako,** Nioro-du-Rip: 88.7MHz 2kW – **R. Dunyaa:** Dakar 88.9MHz 3kW, Bignon/Diourbel 90.9MHz 4kW, Kébémer 91.5MHz 2kW, Louga 91.8MHz 2kW, Ziguinchor 92.4MHz 10kW, Bakel/Thiès 93.7MHz 5kW, S'dhiou 95.3MHz 2kW, Kédougou 98.6MHz 5kW, Kolda 98.7MHz 2kW, Dahra 102.8MHz 2kW, Kaolack 105.0MHz 3kW, Richard Toll 106.1MHz 2kW, St-Louis 106.3MHz – **R. Fass FM,** Thiès: 96.5MHz – **R. Futurs Medias:** Thiès 90.2/102.5MHz 5kW, Kaolack 93.9MHz 2kW, Dakar 94.0/99.4MHz 8/2kW. **W:** seneweb.com/news/elections2007/radios. php?radio=rfm – **R. Municipale de Dakar:** 99.5MHz 5kW. **W:** www. rmdfm.com – **R. Nostalgie:** Dakar 90.3MHz 3kW. **W:** nostalgie.sn – **R. Oxyjeunes,** Pikine: 103,4MHz 8kW. **E:** oxyjeunea@caramail.com – **Santé FM,** Dakar: 96.6MHz 10kW – **Sénégal Info:** Kolda 88.1MHz 2kW, Tambacounda 90.3MHz 5kW, Kaolack 91.9MHz 5kW, Ngor 97.8MHz 2kW, Dakar 98.0MHz 5kW, Ziguinchor 103.4MHz 5kW, Thiès 105.1MHz 2kW – **Sept FM:** Louga 93.2MHz 2kW, Dakar 97.3MHz 10kW, Thiès 98.5MHz – **Sine Saloum FM,** Kaolack: 96.4MHz 5kW – **Soxna FM,** Dakar: 99.9MHz 10kW – **Témoin FM,** Dakar: 107.0MHz – **Wal Fadjri FM:** Thiès 91.3MHz 2kW, Dakar 93.6/99.0MHz 5/10kW, Pikine 96.3MHz 2kW, Louga 98.3MHz 2kW, Tambacounda 99.3MHz 2kW, Kaolack 101.4MHz 2kW, Ziguinchor 102.4MHz **W:** walf.sn/radio – **Zik FM,** Dakar: 89.7MHz 5kW. **W:** zikfm.et
Community radio:
Afia FM, Dakar: 93.0MHz 5kW – Biyen FM, Thiès 89.5MHz 2kW – Ferlo FM, Dahra: 94.0MHz 2kW – Gaynaako FM, Podor 99.4MHz 0.5kW. Web: gaynaakofm.org – Jéeri FM, Keur Momar Sarr 97.0MHz 0.3kW – Maanoore FM, Dakar 89.4MHz 5kW – R. Penc Mi, Fissel: 90.6MHz 1kW. Email: pencmi_fm@yahoo.fr – R. Tim-Timol, Matam: 91.8MHz 1kW.

Africa No. 1: Dakar 102.0MHz 10kW. (see main entry under Gabon).
BBC African Sce: Dakar 105.6MHz 10kW. In French.
RFI Afrique: Ziguinchor 87.6MHz 5kW, Tambacounda 88.9MHz 2kW, Kaolack 91.5MHz 5kW, Dakar 92.0MHz 10kW, St-Louis 99.7MHz 5kW, Thiès 100.2MHz 5kW

SERBIA

L.T: UTC +1h (27 Mar-30 Oct: +2h) — **Pop:** 7.379.339 — **Pr.L:** Serbian — **E.C:** 50Hz, 220V — **ITU:** SRB

UDRUZENJE RADIOTELEVIZIJE SRBIJE d.O.O.
✉ Beogradska 70, 11000 Beograd ☎ +381 11 433718, 434688 🖷 +381 11 434023, 437280 **E:** yrtcoord@eunet.rs
L.P: MD: Ms. Vjera Nikolic.
Udruzenje Radiotelevizije Srbije comprises Radio-televisija Srbije and Voice of Serbia

RADIO-TELEVIZIJA SRBIJE
✉ Takovska 10, 11000 Beograd ☎ +381 11 3211000 **E:** kontaktcentar@rts.rs **W:** www.rts.rs
L.P: DG: Aleksandar Tijanic
R. Beograd: Hilendarska 2, 11000 Beograd ☎ +381 11 3248888 **L.P:** Dir: Slobodan Divjak Web: www.radiobeograd.co.rs

MW	kHz	kW	MW	kHz	kW
Bosilegrad	675	1d	Aleksinac	684	30d
Negotin	693	1d	Vranje	1296	10d
Nis	711	10d	Jagodina	1440	5d
Medvedja	765	1d	Crna Trava	1485	1d
Kladovo	999	1d	Tutin	1485	1a
Beograd 2/3	1008	1	Beograd 202	1503	10
Novi Pazar	1062	1a	Sjenica	1602	1d

a) Beograd1+ reg. prgr d) rel. Beograd 1

FM (MHz)	I	II/III	202	kW
Avala	95.3	97.6	104.0	75/75/130
Bajina Basta	93.5			2
Beograd	88.3			2
Besna Kobila	91.7	95.3	105.6	25/25/40
Bitovik	91.7			
Crni Vrh	89.7	99.3	101.0	25
Crveni Cot	94.5	96.5	101.8	75/75.130
Deli Jovan	87.7	94.9	98.9	25/25/40
Jastrebac	96.9	89.3	103.5	100

FM (MHz)	I	II/III	202	kW
Kopaonik	90.9	93.7	102.1	50/50/100
Ljubovija	91.5			
Maljen	104.5	96.3		25
Nis	99.5			3.5
Ovcar	88.1	90.1	101.6	25
Pirot	92.1			1
Rudnik	91.5			2
Subotica	88.9	101.1	98.5	50/50/0.3
Tornik	90.6	97.5	100.2	15
Trgovista	90.1			
Tupiznica	92.5	96.1	100.4	25
Vrsac	95.7			30

Additional low power local stns not mentioned.
R. Beograd 1: 24h. **N:** W 0303, 0330, 0400, 0430, 0500, 0540, 0700, 0800, 0900, 1000, 1100, 1200, 1300, 1400, 1600, 1700, 1830, 2100, 2200, 2300; Sun 0430, 0500, 0530, 0600, 0700, 0800, 0900, 1000, 1100, 1200, 1400, 1600, 1830, 2000, 2200, 2300. – **R. Beograd 2:** W 0400-1900 (Sun 0600-1900). **N:** W 1130, 1230, 1330, 1500, 1600, 1850. Sun 0630, 0730, 0930, 1055, 1130, 1330, 1730, 1850 – **R. Beograd 3:** 1900-2300 (Serious prgr.) – **R. Beograd 202:** 0400-2400 on 1503kHz, 104.8MHz + FM 202 (0000-0400 rel. R. Beograd 1) – **Stereorama:** 0700-1900SS on FM 202. Other times rel. Beograd 202

Local stations

MW/FM	kHz	kW	MHz	MW/FM	kHz	kW	MHz
Arandjelovac			98.9	Novi Pazar	1062	1	90.0
Bor			93.2	Pirot			95.8
Cacak			92.8	Pozarevac			90.1
Jagodina			97.3	Priboj	1485	1	87.6
Kladovo	1458	1	89.9	Prijepolje			95.9
Kragujevac			106.8	Smed. Palanka			88.3
Kraljevo			106.6	Smederevo			96.1
Krusevac			92.2	Soko Banja			90.5
Lazarevac			89.3	Uzice			92.0
Leskovac			99.0	Valjevo	1368	6	88.6
Loznica			107.4	Vranje	531	1	96.5
Majdanpek			96.7	Vrnjacka Banja			93.2
Mladenovac			90.8	Zajecar			98.1

FM stations with national coverage:

FM	MHz	FM	MHz	FM	MHz
Index	88.9	Radio S	94.9	Fokus	101.4
B92	92.5	Roadstar	98.5		

All stns have relays

Local FM stations in Beograd:

FM	MHz	FM	MHz	FM	MHz
Pingvin	90.9	Sport FM	100.4	Nostalgie	105.2
Pink	91.3	Studio B	100.8	TDI	91.8
MIP	93.7	Sport FM	100.4	City	106.3
TRI	95.8	MFM	102.2	TOP FM	106.8
Naxi FM	96.9	Novosti	104.7	Beta Plus	107.9

There are numerous local FM stns

VOJVODINA (Autonomous Province)

RADIO TELEVIZIJA NOVI SAD
Ignjata Pavlasa 3, 21000 Novi Sad ☎ +381 21 425588 📠 +381 21 423348 **E:** veroslava.pop@rtv.rs **W:** www.rtv.rs

MW	kHz	kW	Notes
Orlovat	1107	50	r. Beograd 1
Srbobran	1269	10	in Serbian
Novi Sad	1485	0.5	in Serbian

FM (MHz)	I	II	III	kW
Novi Sad	87.7	90.5	100.0	5
Subotica	99.3	92.5		15
Vrsac	99.6	91.7		3

I) in Serbian, II) in Hungarian, III) prgrs for national minorities.
R. Novi Sad 1: 24h in Serbian **R. Novi Sad 2:** 0400-2305 in Hungarian.

Local stations

FM	MHz	FM	MHz	FM	MHz
Apatin	98.7	Kovin	88.5	Stara Pazova	91.5
Backa Palanka	99.1	Odzaci	89.7	Subotica	91.5
Backa Topola	97.8	Pancevo	92.1	Temerin	93.5
Backi Petrovac	91.4	Ruma	102.7	Vrbas	95.5
Beocin	97.8	Sid	89.1	Vrsac	98.1
Indjija	96.0	Sombor	90.9	Zrenjanin	103.6
Kovacica	93.2	Srbobran	102.6		

There are numerous low-power local FM stns

EXTERNAL SERVICE: Voice of Serbia
See International Broadcasting section

SEYCHELLES

L.T: UTC +4h — **Pop:** 90,000 — **Pr.L:** Creole, English, French — **E.C:** 50Hz, 240V — **ITU:** SEY

MINISTRY OF INFORMATION TECHNOLOGY & COMMUNICATION (MITC)
Telecom Division, P. O. Box 1389, Oceangate House, Room 16, Victoria, Mahé ☎+248 382039 📠 +248 225325 **E:** telecom@seychelles.sc **L.P:** Dr. George Ah-Thew, Dir.

SEYCHELLES BROADCASTING CORPORATION (SBC, Pub.)
P.O. Box 321, Hermitage, Mahé ☎+248 289600 📠 +248 225641 **W:** www.sbc.sc **E:** sbcradtv@seychelles.sc
L.P: MD: Mr. Ibrahim Afif. Prgr. Mgr.(Radio) Ms. Thelma Pool. CE: Mr. Joyvani Chetty. Marketing & PR Mgr: Mrs. Jacqueline Moustache.
MW: Victoria 1368kHz 10kW.
FM: Anse Soleil 93.6MHz 0.25kW, Fairyland 93MHz 0.25kW, St.Louis 93.6MHz 1kW, Praslin 100.8MHz 0.03kW.
D.Prgr: **MW** (spoken word): MF 0200-0930 & 1200-1800, SS 0200-1800. **N: English:** 0300, 0600, 0900, 1500. **French:** 0330, 0700, 1300, 1700. **Creole:** 0230, 0500, 0800, 1600.
FM: Paradise FM (musical prgr.): 24h.
Ann: E: "This is SBC Radio". F: "Ici la Radio SBC". C: "Isi Radyo SBC"
IS: Instrumental music.

RFI Afrique: St.Louis 103.8MHz 1kW, Anse Soleil 102.8MHz 0.25kW.
BBC African Sce: St.Louis 106.2MHz 0.5kW.
BBC Indian Ocean relay station: see International Radio section

---

SIERRA LEONE

L.T: UTC — **Pop:** 6.5 million — **Pr.L:** English, Krio, Limba, Mende, Temne, others — **E.C:** 50Hz, 230V — **ITU:** SRL

INDEPENDENT MEDIA COMMISSION (IMC)
Kissy House, 54 Siaka Stevens Street, Freetown ☎+232 22 221835 **W:** www.imc-sl.org **E:** info@imc-sl.org
L.P: Chairperson: Mrs. Bernadette Cole.

SIERRA LEONE BROADCASTING CORPORATION (SLBC, Pub.)
New England, Freetown ☎+232 22 241919 📠 +232 22 240922
L.P: DG: Elvis Gbanabom Hallowell. Dir. Eng. Sces: A.K.Sheriff.
FM: Freetown 100.0MHz 4kW, Bo 96.5MHz 2kW, Kenema 93.5MHz 2kW, Kono 90.2MHz 1kW, Makeni 88.0MHz 1kW & 4 trs under 1kW.
D.Prgr: "Power FM": 0558-2400.

COTTON TREE NEWS (CTN, by Hirondelle Foundation)
FM: via R. Mount Aureol, Fourah Bay College, Freetown: 107.3MHz. Also relayed by many stations listed below.
For shortwave relays see Sierra Leone in COTB section.

Other stations:
Believers Broadcasting Network (BBN) (Rlg.), Freetown: 93.0MHz 2kW. **W:** www.bbn-sl.org – **Capital R:** Freetown 104.9MHz 4kW, Bo 102.3MHz 50W. L.P: CE: Dave Stanley **W:** capitalradio.sl – **Eastern R:** Kenema 101.9MHz, Kono 96.5MHz – **Kiss FM,** Bo: 104MHz. (Also rel. VOA) – **R. Bintumani,** Kabala: 93.7MHz – **R. Bontico,** Bonthe: 96.9MHz – **R. Galaxy,** Mahera: 106.1MHz – **R. Gbafth,** Mile 91: 91.0MHz – **R. Kolenten,** Kambia: 92.4MHz – **R. Mankneh,** Makeni: 95.1MHz – **R. Moa,** Kailahun: 105.5MHz – **R. Modcar,** Moyamba: 94.8MHz – **R. Maria,** Makeni 101.1MHz 0.5kW – **R. Numbura,** Bumbuna: 102.5MHz – **R. One,** Freetown: 103.7MHz – **R. Wanjei,** Pujehun: 101.1MHz – **R. Viascity,** Waterloo: 100.6MHz – **Skyy R,** Freetown: 106.6MHz – **Unity R,** Freetown: 98.4MHz – **Voice of Islam,** Freetown: 102.0MHz – **VO the Handicapped,** Freetown: 96.2MHz (mostly rel. BBC) – **VO the Peninsula,** Tombo: 96.0MHz – **VO Women,** Mattru Jong: 88.5MHz.
BBC African Sce: Freetown 94.3MHz 8kW, Bo 94.5MHz 120W, Kenema 95.3MHz 60W.
RFI Afrique: Freetown 89.9MHz in French/English.
VOA, Freetown: 102.4MHz

SINGAPORE

L.T: UTC +8h — **Pop:** 5.0 million — **Pr.L:** English, Chinese, Malay, Tamil — **E.C:** 50Hz, 230V — **ITU:** SNG — **Int. dialling code:** +65

MEDIA DEVELOPMENT AUTHORITY OF SINGAPORE
(Government statutory board)
140 Hill Street, MITA Building #04-01, Singapore 179369 ☎ +65

6837 9973 📠 +65 6336 8023 **W:** www.mda.gov.sg
LP: Chmn: Dr Tan Chin Nam. CEO: Dr Christopher Chia

MEDIACORP RADIO SINGAPORE PTE LTD (Comm.)
📧 Caldecott Broadcast Centre, Andrew Road, Singapore 299939 ☎ +65 6333 3888 📠 +65 6256 9533 **W:** www.mediacorpradio.sg
L.P: Chmn: Mr. Teo Ming Kian. CEO: Mr. Lucas Chow. Deputy CEO News, Radio & Print: Mr. Shaun Seow. MD, MediaCorp Radio: Ms Florence Lian. MD MediaCorp Technologies: Mr. Mock Pak Lum.
Stations: FM tx centre at Bukit Batok.

	FM MHz	Network	Format	Lang.	H of tr
1)	89.7	Ria 89.7FM	CHR	Malay	24h
2)	90.5	Gold 90FM	Gold	English	24h
3)	92.4	Symphony 92FM	Classical	English	24h
4)	93.3	Y.E.S. 93.3FM	AC	Chinese	24h
5)	93.8	938LIVE	N./Info	English	24h
6)	94.2	Warna 94.2FM	Full sce.	Malay	24h
7)	95.0	Class 95FM	AC	English	24h
8)	95.8	Capital 95.8FM *	N./Info	Chinese	24h
9)	96.3	96.3XFM	Int'l	***	24h
10)	96.8	Oli 96.8FM	Full sce.	Tamil	24h
11)	97.2	Love 97.2FM **	Easy	Chinese	24h
12)	98.7	987FM	CHR	English	24h
13)	99.5	Lush 99.5FM	Urban	English	24h

*) in Chinese: "Chengshi Pindao". **) in Chinese: "Zui'ai Pindao". ***)
D.Prgr: Mon-Fri 2250-0100 "Hello Singapore" in Japanese (2300 rel. NHK news). Mon-Fri 0100-0300 rel. R. France Int. in French. Mon-Fri 0700-0900 rel. Deutsche Welle in German. Daily 0900-1200 "Masti 96.3" in Hindi. Daily 1200-1400 " in Korean. At other times carries music interludes.
F.PI: Will move to a new broadcast centre at Bukit Batok in 2011.
Ownership: MediaCorp is wholly owned by Temasek Holdings, an investment company of the Government of Singapore.

DAB: Single frequency networks from two tx sites using 190.640MHz, block 7B, and 192.352MHz, block 7C.

SAFRA RADIO (Comm.)
Operated by the Singapore Armed Forces Reservists' Ass.
📧 Tower B #12-04, Defence Technology Towers, 5 Depot Rd, Singapore 109681 or Bukit Merah Central PO Box 1315, Singapore 911599 ☎ +65 6373 1924 📠 +65 6278 3039 **W:** power98.com.sg or www.883jia.com.sg
L.P: News Dir.: Low Mei Mei.
883JiaFM: 88.3MHz, 24h in Chinese
Power98FM: 98.0MHz, 24h in English

SPH UNIONWORKS PTE LTD (Comm.)
Joint venture of NTUC Media Co-operative (National Trade Unions Congress) & SPH MediaWorks (Singapore Press Holdings)
📧 1000 Toa Payoh North, News Centre Podium Block Level 3, Singapore 318994 ☎ +65 6319 1900 📠 +65 6319 1099 **W:** www.sphuww.com.sg
L.P: Gen. Mgr: Ms Goh Wee Wang. Prgr Dir Radio 91.3: Mr Jamie R. Meldrum. VP Radio 100.3: Ms Anna Lim
Radio 91.3 FM: 24h in English **Radio 100.3 FM:** 24h in Chinese

BBC SINGAPORE 88.9 FM
24h rel. of BBCWS in English. The FM tx at Bukit Batok is operated by MediaCorp Technologies.

BBC FAR EASTERN RELAY STATION (VT Communications Ltd)
📧 51 Turut Track, Singapore 718930 ☎ +65 6793 7511
See International Broadcasting section

SLOVAKIA

L.T: UTC +1h (27 Mar-30 Oct: +2h) — **Pop:** 5 million — **Pr.L:** Slovak — **E.C:** 50Hz, 230V — **ITU:** SVK

SLOVENSKY ROZHLAS (SLOVAK RADIO)
📧 Mytna 1, 817 55 Bratislava ☎ + 421 2 57273111
📠 + 421 2 57273559 **W:** www.slovakradio.sk **E:** sro@slovakradio.sk **Radio FM. W:** www.radiofm.sk **E:** info@radiofm.sk **L.P:** DG: Miloslava Zemková. PD: Lubos Machaj. Mus. Dir.: Rudolf Pepucha. CE: Robert Oravec

MW	kHz	kW	Prgr.
Kosice	702	5	S5 (daytime) + S3 (nighttime)
Nitra (Jarok)	1098	10	S5 (daytime) + S3 (nighttime)

FM (MHz)	S1	S2	S3	S4	kW
Banská Bystrica	90.1	101.5	102.0	105.4	100/100/0.1/2
Banská Stiavnica	99.0		102.6		20/20

FM (MHz)	S1	S2	S3	S4	kW
Bardejov	93.5	89.3	88.8	101.7	10/1/10/10
Borsky Mikulás			95.6	102.8	1
Bratislava	96.6	99.3	104.4	89.3	100/10/10/10
Cadca				91.8	0.5
Dolny Kubin				91.7	1
Kosice	96.6	100.3			100/35
Kosice (city)			96.2	101.2	0.5/1
Lucenec	103.6	88.2		98.0	10/2/10
Martin				91.8	0.5
Modry Kamen	90.9	88.5	103.1	98.3	10
Námestovo	102.4	100.4	88.7		10
Nitra	91.2	102.2			10/10
N. Mesto n.V.	103.2	100.7			10
Nové Zámky			94.6	102.8	1
Poprad	92.2	96.9		104.3	30
Presov			106.7	101.5	0.5
Rim. Sobota		95.0			1
Roznava	97.3	88.6	90.0	105.9	1/1/1/1
Ruzomberok	103.8	100.6	104.6	102.1	5
Snina	91.2		102.2	107.6	10
Stará Lubovna	89.1	102.3	96.1		10
Sturovo	96.3	91.7	106.2	103.7	10
Trebisov		89.2	106.7	101.3	10
Trencín	95.9		97.8	101.2	10
Trstená				91.9	10
Zilina	103.5	100.1	97.2	91.9	20/20/30/1
Zvolen		99.8	89.0		0.5/1

Addresses and other information:
S1 = Radio Slovensko: 24h (national prgr news). **S2** = Radio Regina: 24h (regional prgrs + prgrs for national minorities in Hungarian, Ukrainian, Ruthenian, German, Czech, Polish and Gypsy/Roma + relays of Radio Slovensko – S1). **S2 BA** = Radio Regina Bratislava, 📧 Mytna 1, 817 55 Bratislava 15. Mon-Fri 0335-2400, Sat 0500-2100, Sun 0700-2100. **S2 BB** = Radio Regina Banská Bystrica, 📧 L. Sáru 1, 975 68 Banská Bystrica. Mon-Fri 0335-1730, Sat 0500-2030, Sun 0800-1930. **S2 KE** = Radio Regina Kosice, 📧 Masarykova 7, 041 61 Kosice. Mon-Fri 0400-2030, Sat 0500-2030, Sun 0500-1600. **S3** = Radio Devín: 24h (cultural prgr) on FM and 1700-0500 on 702 and 1098kHz. **S4** = Radio FM: 24h (rock, pop and alternative music). N: on the h. **S5** = Radio Patria – production of prgrs for national minorities in Hungarian, Ukrainian, Ruthenian, German, Czech, Polish, Gypsy/Roma relayed on S2 and S5 txs 📧 Slovensky Rozhlas, HRNEV, Moyzesova 7, 040 01 Kosice **E:** nev@slovakradio.sk Prgrs for minorities (S5): Hungarian 0500-1700 on 702 and 1098kHz. Prgrs for national minorities (S5) on Radio Regina (S2): Mon+Wed+Fri+Sat+Sun 1700-1800, Tue+Thu 1700-1800, 1900-2000

EXTERNAL SERVICE: Radio Slovakia International
See International Broadcasting section

MAJOR PRIVATE STATIONS/NETWORKS:
ASOCIÁCIA NEZÁVISLYCH ROZHLASOVYCH STANIC (Association of Independent Radio Stations)
📧 Stúrova 9, 811 02 Bratislava ☎ +421 2 5296 2370
FUN RADIO (Comm.)
📧 Leskova 5, 815 25 Bratislava ☎ +421 2 52494601
📠 +421 2 52495535 **W:** www.funradio.sk
JEMNÉ MELODIE (Comm.)
📧 Dr. Vladimíra Clementisa 10, 815 25 Bratislava ☎ +421 2 48484811 📠 +421 2 52492701 **W:** www.jemnemelodie.sk
FM: see list below **D.Prgr:** 24h
RADIO VIVA (Comm.)
📧 Salviová 1, 830 00 Bratislava ☎ +421 2 48255500 **W:** www.radioviva.sk **FM:** see list below **D.Prgr:** 24h
RADIO EXPRES (Comm.)
📧 Lamacská cesta 1, 841 04 Bratislava ☎ +421 2 59308900 📠 +421 2 59308991 **W:** www.expres.sk **FM:** see list below **D.Prgr:** 24h
EUROPA 2 (Comm.)
📧 Seberíniho 1, 821 03 Bratislava ☎ +421 2 48224201 **W:** www.europa2.sk **FM:** see list below **D.Prgr:** 24h
RADIO LUMEN (Relig.)
📧 Kapitulská 2, 974 01 Banská Bystrica ☎ +421 48 4710800 📠 +421 48 4710840 **W:** www.lumen.sk **FM:** see list below **D.Prgr:** 24h
RADIO HEY! (Comm.)
📧 Jelsová 11, 831 01 Bratislava ☎ +421 2 59303030
📠 +421 2 54777777 **W:** www.radiohey.sk

Private Commercial FM Stations:

FM	MHz	kW	Station
Kosice	87.7	80	Fun R.
Banská Bystrica	87.7	10	R. Jemné melodie
Nové Mesto nad Váhom	88.0	8.5	R. Jemné melodie
Hlohovec	88.4	2	R. Expres

FM	MHz	kW	Station
Ruzomberok	88.4	1	R. Expres
Snina	88.5	10	R. Viva
Nitra	88.8	10	R. Hey!
Zvolen	89.0	1	R. Expres
Trencín	89.1	10	Fun R.
Ruzomberok	89.2	1	Fun R.
Rimavská Sobota	89.3	1	R. Expres
Ruzomberok	89.7	1	R. Lumen
Bratislava	89.7		R. Sity
Banská Bystrica	90.5	2	R. One
Trstená	90.7	1	R. Viva
Trnava	90.8	10	R. Jemné melodie
Presov	90.8	2	R. Kiss
Zilina	90.8	1	R. Hit FM
Moldava nad Bodvou	91.0	1	R. Expres
Ruzomberok	91.1	1	R. Viva
Lucenec	91.6	10	Fun R.
Dubnica nad Váhom	92.2	1	Fun R.
Stropkov	92.4	1	R. Viva
Levoca	92.6	1	R. Kiss
Zvolen	92.6	1	R. Expres
Nové Zámky	92.7	1	R. Expres
Ruzomberok	92.8	1	R. Viva
Brezno	93.0	1	R. Expres
Snina	93.2	1	R. Lumen
Banská Stiavnica	93.3	1	R. Lumen
Bratislava	93.8	6	R. Lumen
Lehota pod Vtácnikom	93.9	1	R. Beta
Bratislava	94.3	90	Fun R.
Zilina	94.5	13	R. Zet
Banská Bystrica	94.7	1	R. Rock
Liptovsky Mikulás	95.0	1	Fun R.
Nitra	95.2	10	R. Europa 2
Levoca	95.3	1	R. Expres
Bardejov	95.6	1	R. Viva
Roznava	95.7	1	R. Expres
Snina	95.9	10	R. Kiss
Lucenec	96.0	5	R. Viva
Cadca	96.1	1	R. Frontinus
Martin	96.2	2	R. Frontinus
Partizánske	96.4	1	R. Hit FM
Trstená	96.5	2	R. Expres
Michalovce	97.0	5	R. Kiss
Banská Bystrica	97.6	100	R. Hey!
Stropkov	97.8	1	R. Kiss
Handlová	98.1	1	R. Lumen
Bardejov	98.2	1	R. Expres
Nové Mesto nad Váhom	98.5	8.8	R. Europa 2
Kosice	98.6	100	R. Jemné melodie
Nové Zámky	98.7	1	R. Max
Ruzomberok	98.8	5	R. Hey!
Michalovce	99.0	1	R. Viva
Bardejov	99.1	1	R. Lumen
Zilina	99.2	25	Fun R.
Sturovo	99.4	1	R. Expres
Svit	99.5	1	R. Expres
Cadca	99.6		R. Viva
Bratislava	100.3		R. Hey!
Poprad	100.9	30	R. Europa 2
Lucenec	101.1	5	R. Expres
Roznava	101.4	1	R. Viva
Bratislava	101.8	90	R. Viva
Rimavská Sobota	102.4		R. Hey!
Trencín	102.5	1	R. Expres
Poprad	102.5	1	Fun R.
Zilina	102.8	1	R. Europa 2
Bardejov	102.8	1	Fun R.
Roznava	102.8	1	Fun R.
Strbské Pleso	102.9	2	R. Lumen
Kosice	102.9	1	Fun R.
Michalovce	103.3	2	R. Lumen
Presov	103.7	8	R. Europa 2
Banská Bystrica	104.0	95	Fun R.
Presov	104.1	1	R. Kiss
Prievidza	104.5	1	R. Hit FM
Bratislava	104.8	50	R. Europa 2
Poprad	104.8	1	R. Viva
Martin	104.9	2	R. Viva
Banská Stiavnica	105.1	10	R. Viva
Povazská Bystrica	105.2	1	R. Expres
Trencín	105.5	10	R. Viva
Stará Lubovna	105.7	1	R. Expres
Námestovo	105.8	10	R. Lumen
Presov	105.8	1	R. Viva
Banská Bystrica	106.0	50	R. Europa 2

FM	MHz	kW	Station
Kosice	106.2	20	R. Expres
Lucenec	106.3	3	R. Lumen
Roznava	106.3	1	R. Lumen
Modry Kamen	106.5	1	R. Expres
Bratislava	106.6	10	R.Jemné melodie
Banská Bystrica	106.6	1	R. Viva
Zilina	106.9	3	R.Jemné melodie
Dobsiná	107.0	1	R. Viva
Bardejov	107.1	10	R. Europa 2
Levice	107.1	4	Fun R.
Poprad	107.3		R. Hey!
Bratislava	107.6	10	R. Expres
Stará Lubovna	107.7	1	R. Jemné melodie

+ more than 60 relays of less than 1kW

SLOVENIA

L.T: UTC +1h (27 Mar-30 Oct: +2h)— **Pop:** 2 million— **Pr.L:** Slovenian — **E.C:** 50Hz, 220V — **ITU:** SVN

SLOVENIAN BROADCASTING COUNCIL (ATRP)
P.O. Box 418, 1000 Ljubljana ☎+386 1 5836385 ▤+386 1 5111101 **W:** www.srdf.si www.apek.si **E:** info.box@apek.si

RADIOTELEVIZIJA SLOVENIJA (Pub.)
Kolodvorska ulica 2, SI-1550 Ljubljana ☎+386 1 4752151 ▤ +386 1 4752150 **W:** rtvslo.si **E:** webmaster@rtvslo.si
LP: DG: Marko Filli. Dir. Radio: Vinko Vasle. Dir. Tech: Matej Zunkodic

MW	kHz	kW	Prgr.	MW	kHz	kW	Prgr.
Beli Kriz	549	15	1/K	Domzale	918	300	1
Nemcavci	558	15	1/MMR	B. Kriz	1170	15	C/RSI

C=R. Capodistria in Italian **RSI**=R. Slovenija International, **K**=R. Koper in Slovenian, **MMR**=Muravideki Magyar R. in Hungarian

FM (MHz)	Slo 1	Slo 2	Slo 3	Reg.	kW
Beli Kriz	92.0	94.1	96.1	104.3k/97.7c	5
				102.0si	1
Blejska Dobrava				100.4si	1
Boc				90.4m	2
Koper	92.2			104.1k	1
Krim	88.0	93.5	96.5		5
Krvavec	91.8	98.9	102.0		100
Kuk	90.8	87.8	96.4	100.6k	5
Kum	94.1	99.9	103.9		30
Ljubljana-Šance				100.8si	1
Nanos	92.9	95.3	105.7	88.6k	50/50/50/25
				103.1c	100
Pec	100.1	104.0	106.0		5
Pecarovci				87.6h	5
Plešivec	90.0	92.4	101.4		10
Pohorje	88.5	96.9	105.3	93.1m/102.8si	5/3/2/3/8
Skalnica				100.3k	2
Tinjan	89.3	98.9	98.1	107.6k/103.6c	6/6/6/6/5
				94.6si	0.2
Trdinov Vrh	90.9	97.6	100.6		7.5/10/11
Trstelj	92.6	94.3	102.2	96.7si	5

Reg. stns: c=R. Capodistria in Italian, h=MMR in Hungarian, k=R. Koper, m=R. Maribor, si=R. Slovenija Int.

R. Slovenija 1 "Prvi program": 24h. **N. in E & German:** 2130 – R. **Slovenija 2 "Val 202":** 0500 -2300. Other times relay R. Slovenija Int.- Pop + entertainment – **R. Slovenija 3 "Program ARS":** 24h. Serious music, educational – **R. SI, R. Slovenija International,** Ilichova ulica 33, SI-2000 Maribor. **E:**radio.si@rtvslo.si; Music and entertainment channel 24h on **FM** ("si") and **MW** 1170kHz 2300-0500.

RADIO KOPER – CAPODISTRIA (Pub.)
PO Box 117, SI-6000 Koper-Capodistria ☎+386 (5) 6685050 ▤ +386 (5) 6684500 (Slovenian Dept.) ☎+386 (5)6685440 (Italian Dept.) **W:** www.radiocapodistria.net **E:** radio.koper@rtvslo.si; radio.capodistria@rtvslo.si
R. Koper in Slovenian: 0500-2300 on 549kHz + FM ("k"). Other times rel. Slovenija 1 – **R. Capodistria in Italian:** 0500-2300 on 1170kHz + FM ("c"). 2300-0500 rel. R. Slovenija International.

RADIO MARIBOR (Pub.)
Ilichova ulica 33, SI-2000 Maribor ☎+386 2 4201555 **E:** radio.maribor@rtvslo.si **FM:** ("m").

MURAVIDEKI MAGYAR RADIO (Pub.)
Kranjceva ulica 10, SI-9220 Lendava +386 2 4299700 ▤+386 2 4299712 **W:** www.rtvslo.si/mmr/ **E:** mmr.studio@rtvslo.si

MW 558kHz + **FM** 87.6MHz. **D.Prgr:** 0445-2300. At other times rel. Slovenia 1 on MW and R. SI on FM.

OTHER STATIONS

MW	kHz	kW	Station	Location
A)	594	0.6	Primorski Val/R. Odmev	Cerkno
B)	648	10	R. Murski Val	Nemcavci

A) Platiševa ul. 39, SI-5282 Cerkno. **E:** info@radio-odmev.net **W:** www.radio-cerkno.si; www.primorskival.si **D.Prgr:** 1500-1900 R. Odmev. 1900-1500 Primorski Val (a joint programme of Alpski Val and R. Odmev)

B) Ul. Arhitekta Novaka 13, SI-9000 Murska Sobota **W:** www.radiomurskival.si **E:** murski.val@siol.net **D.Prgr:** 24h. A joint night programme of Koroski R., Murski val, R. Celje, R. Goldi, R. Kranj, R. Kum, R. Ptuj, R. Robin, R. Slovenske Gorice, R. Sora, R. Triglav and R. Univox is broadcast.

	FM	MHz	kW	Station	Location
1)		87.6	1	R. Europa 05	Ljubljana-Šance
2)		87.8	1	R. Salomon	Blejska Dobrava
7)		87.9	1	R. Hit	Hrvatini
3)		88.3	1	R. 1 Portoroz	Malija
3)		88.4	2	R. 1 Krvavec	Krvavec
6)		89.3	1	R. Študent	Ljubljana-Šance
18)		89.8	1	R. Ptuj	Majsperg
6)		90.0	2	R. Maxi	Ljutomer
3)		90.1	1	R. 1 Primorska	Nova Gorica
7)		90.2	1	R. Hit	Vrhnika
4)		90.6	1	R. 1 Orion	Krim
8)		91.7	1	R. Capris	Markovec
9)		92.6	1	R. Ljubljana	Ljubljana-Šance
10)		93.7	2	Stajerski Val	Boc
26)		93.8	1	R. Center	Markovec
40)		94.6	5	Murski Val	Pecarovci
43)		94.6	1	R. Sraka	Trdinov Vrh
41)		94.9	1	R. Veseljak	Ljubljana-Šance
41)		95.1	2	R. Celje	Boc
7)		95.6	3	R. Hit	Dobeno
12)		95.9	1	R. MARŠ	Maribor
13)		96.0	1	R. Triglav	Ravni Valvazor
14)		97.2	1	Koroški R.	Plešivec
3)		97.3	1	R. 1 Primorska	Hrvatini
3)		97.4	1	R. 1 Štajerska	Ljubcina
15)		97.3	1	R. Kranj	Smarjetna Gora
16)		98.1	1	R. Kum	Kum
17)		98.2	1	R. 94	Postojna
18)		98.2	1	R. Ptuj	Ptuj
3)		99.1	5	R. 1 Primorska	Trstelj
20)		99.5	1	R. Robin	Nova Gorica
21)		99.5	1	R. City	Ljubljana-Šance
3)		99.6	5	R. 1 Primorska	Koper
20)		100.0	1	R. Robin	Trstelj
21)		100.2	5	R. Aktual	Krim
22)		100.2	1	Net FM	Maribor
23)		100.6	1	R. City	Maribor
23)		100.8	1	R. City	Topolšica
21)		101.2	1	R. Aktual	Ljubljana-Šance
19)		101.3	1	R. Pohorje	Maribor
2)		101.6	1	R. Salomon	Ljubljana-Šance
24)		101.8	1	R. Rogla	Konjiska Gora
25)		102.1	4	R. Viva	M.Sobota/Bogojina
26)		102.4	1	R. Center	Ljubljana-Šance
35)		102.4	5	R. Ekspres	Hrvatini
35)		102.8	1	R. Ekspres	Portoroz/Šentanje
28)		103.0	5	Studio D	Trdinov Vrh
29)		103.2	1	R. Alfa	Rahtelov Vrh
31)		103.3	1	R. Dur	Ljubljana-Šance
26)		103.7	1	R. Center	Maribor
42)		103.7	1	Primorski Val/R. Odmev	Javornik
17)		104.1	1	R. 94	Ilirska Bistrica
30)		104.5	100	R. Ognjišče	Krvavec
7)		104.5	5	R. Hit	Trstelj
21)		104.8	2	R. Aktual	Boc
26)		104.9	1	R. Center	Nova Gorica
3)		105.0	1	R. 1 Dolenjska	Krsko
8)		105.1	1	R. Capris	Slavnik
32)		105.1	1	Primorski Val/Alpski Val	Kobariški Stol
33)		105.2	1	R. Laser	Rahtelov Vrh
34)		105.2	1	R. Antena	Ljubljana-Šance
40)		105.7	2	Murski Val	Zlatolicje
30)		105.9	5	R. Ognjišče	Kum
26)		106.4	6	R. Center	Tinjan
35)		106.4	4	R. Ekspres	Krim
36)		106.6	15	R. Krka	Trdinov Vrh
37)		107.0	1	Moj R.	Topolsica
7)		107.0	1	R. Hit	Markovec

	FM	MHz	kW	Station	Location
38)		107.1	1	R. 94 / 1TR	Rovte
30)		107.3	1	R. Ognjišče	Boc
30)		107.5	2	R. Ognjišče	Skalnica
8)		107.9	1	R. Capris	Portoroz
3)		107.9	1	R. 1 Stajerska	Maribor
3)		107.9	1	R. 1 107.9	Ljubljana-Šance

NB: Txs below 1kW not mentioned.

Addresses: 1) Leskoškova 9E, 1000 Ljubljana. **W:** www.radioeuropa05.si – **2)** Cesta 24.junija 23, 1532 Ljubljana **W:** www.radiosalomon. si – **3)** Stegne 11B, 1000 Ljubljana **W:** www.radio1.si – **4)** Pozarnice 78H, Vnanje Gorice, 1351 Brezovica **W:** www.r-orion.com – **5)** Cesta 27. aprila 31, 1000 Ljubljana **W:** www.radiostudent.si – **6)** Prešernova ul. 3, 9240 Ljutomer **W:** www.radiomaxi.com – **7)** Ljubljanska Cesta 36, 1230 Domzale **W:** www.radiohit.si – **8)** ul. 15.maja 10B, 6000 Koper **W:** www.radiocapris.com – **9)** Dunajska 270, Ljubljana **W:** www.radioljubljana.si – **10)** Aškercev trg 21, 3240 Šmarje pri Jelšah **W:** www.radio-stajerski-val.si – **11)** Cesta 24. junija 23, 1532 Ljubljana **W:** www.radioveseljak.si – **12)** Gosposvedska cesta 83, 2000 Maribor **W:** www.radiomars.si – **13)** Trg Toneta Curfaja 4, 4270 Jesenice **W:** www.radiotriglav.si – **14)** Meškova 21, 2380 Slovenj Gradec **W:** www.koroski-radio.si – **15)** Stritarjeva 6, 4000 Kranj **W:** www.radio-kranj.si – **16)** Trg Svobode 11A, 1420 Trbovlje **W:** www.radio94.si – **17)** Kazarje 10, 6230 Postojna **W:** www.radio-tednik.si – **18)** Raiceva 6, 2250 Ptuj **W:** www.radio-tednik.si – **19)** Partizanska cesta 24, 2000 Maribor **W:** www.radiopohorje.si – **20)** ul. Tolminskih Puntarjev 12, p.p. 22, 5000 Nova Gorica **W:** www.robin.si – **21)** Cesta 24. junija 23, 1532 Ljubljana **W:** www.radioaktual.si – **22)** Loška cesta 13, 2000 Maribor **W:** www.radionet.si – **23)** Slovenska ul. 35, 2000 Maribor **W:** www.radiocity.si – **24)** Škalska 7, 3210 Slovenske Konjice **W:** www.radiorogla.si – **25)** Bakovska ul. 2, 9000 Murska Sobota **W:** www.radioviva.si – **26)** Zelezna cesta 14, 1000 Ljubljana **W:** www.radiocenter.si – **28)** Seidlova 29, 8000 Novo Mesto **W:** www.studiod.si – **29)** Ronkova 4, 2380 Slovenj Gradec **W:** www.radio-alfa.si – **30)** Trg Brolo Št 11, 6000 Koper. **W:** radio.ognjisce.si – **31)** Zerjalova 8, 1210 Ljubljana-Šentvid **W:** www.radiodur.si – **32)** Poljubinj 89F, p.p.46, 5220 Tolmin **W:** www.alpskival.net; www.primorskival.si. **D.Prgr:** 1500-1900 Alpski Val. 1900-1500 Primorski Val (a joint programme of Alpski Val and R. Odmev) – **33)** Legen 101A, 2380 Slovenj Gradec **W:** www.laserr.si – **34)** Cesta na Brdo 27, 1000 Ljubljana **W:** www.radioantena.si – **35)** Stegne 21C, 1000 Ljubljana **W:** www.radioekspres.si – **36)** Ljubljanska Cesta 26, 8000 Novo Mesto. **W:** www.radiokrka.com – **37)** Kidriceva 2B, 3320 Velenje **W:** www.mojradio.com – **38)** Trzaška 148, 1370 Logatec – **40)** Arhitekta Novaka 13, 9000 Murska Sobota **W:** www.radiomurskival.si – **41)** Prešernova 19, 3000 Celje **W:** www.radiocelje.si – **42)** Platiševa ul. 39, 5282 Cerkno **W:** www.radio-cerkno.si. D.Prgr: 1500-1900 R. Odmev. 1900-1500 Primorski Val – **43)** Valancicevo 17, 8000 Novo Mesto. **W:** www.radiosraka.com

DAB: Krvavec ch12B. Programmes: Slo1/Slo2/Slo3/RSI

L.T: UTC +11h — **Pop:** 595,613 — **Pr.L:** Pidgin, English — **E.C:** 50Hz, 240V — **ITU:** SLM

SOLOMON ISLANDS BROADCASTING CORP. (Statutory Authority, Comm.)

Honiara: P.O. Box 645, Honiara ☎ +677 20051 ▯ +677 23159. Wantok FM ☎ +677 29600 ▯ +677 29600 Gizo: P. O. Box 78, Gizo, Western Province ☎ +677 60160 Lata: P. O. Box 46, Lata, Santa Cruz ☎ +677 53047 **L.P:** GM: Cornelius Rathamana.
W: www.sibconline.cm.sb **E:** sibcnews@solomon.com.sb

MW: R. Happy Isles [Voice of the Nation], Honiara 1035kHz 10kW (r. 6kW), R. Happy Lagoon, Gizo 945kHz [inactive] 10kW, R. Temotu, Lata 1386kHz [inactive] 5kW

SW: 5020kHz & 9545kHz 10kW. Relays R.Happy Isles **Schedule:** Oct-Mar 2100-1800 9545 Apr-Sep 0700-0000, 1800-2200 5020 [NB: times are often approximate depending on local requirements].

FM: Wantok FM, Honiara 96.3MHz, R.Happy Lagoon, Gizo 96.3MHz.

D.Prgr: local 1900-1130, BBC 1300-1900

N. in English: 2000, 2200 (R. Australia), 0130W (local), 0200 (R. Australia), 0500W (local), 0600 (BBC), 0730 (local), 1000 (R. Australia), 1100 (local).

Ann: "This is the SIBC, Radio Happy Isles". **IS:** Drum and Bamboo Pipes. **V.** by QSL-card (Send rpts to Honiara address, international mail often takes more than a month to reach the station. IRC's required).

Other Stations:

	FM	MHz	kW	Station
1)	Honiara	88.3		Gud Nius R
7)	Kia	89.5	0.1	R.Kia
7)	Tutuba	89.5	0.1	R.Tutuba

FM	MHz	kW	Station
6) Tetere	89.9	0.1	R.Bosco FM
7) Buala	91.1	0.1	R.Buala
7) Susubona	91.1	0.1	R.Susubona
7) Sigana	92.5	0.1	R.Sigana
7) Kolotubi	92.5	0.1	R.Kolotubi
7) Leleghia	94.3	0.1	R.Leleghia
7) Samasodu	94.3	0.1	R.Samasodu
2) Honiara	97.7		Paoa FM
3) Honiara	100.0		ZFM 100
2) Honiara	101.7		Paoa FM
4) Honiara	105.6		BBC
5) Honiara	107.0		R. Australia

Addresses and other information
1) UCB Pacific Partners, P.O. Box 1415, Honiara **W:** www.pacificpartners.org **E:** solomons@pacificpartners.org **F.PI:** FM relay at Gizo. – **2)** Star FM [Paoa FM] PO Box R331, Panatina Plaza, Prince Philip Hwy, Honiara ☎ +677 38984 🖷 +677 38980 **W:** www.solomonstarnews.com . **E:** paoafm@solomon.com.sb. **L.P:** GM Joel Lamani, News Editor: Uriel Matangani **3)** P.O. Box 100, Honiara ☎ +677 21100 🖷 +677 21100 **L.P:** Sammy 'Sharzy' Saeni. **E:** zfm@solomon.com.sb. – **4)** 24h satellite relay – **5)** 24h satellite relay **-6)** Catholic Communications Solomons, PO Box 647, Honiara T: +677 22125 F: +677 36333 **E:** ambrose@don-bosco.org.sb **L.P:** Father Ambrose Pereira. **F.PI:** Pupuraka FM, Visale **-7)** Isabel Province Community FM Network, c/o Office of the Premier, PO Box 4, Buala, Isabel Province. **D.Prgr:** 0600-1100 UTC daily. **N:** SIBC local news 0730. **Note:** Each stn is solar powered using 0.03kW actual power, and operates independently for each village area. Local commercials and family messages are also broadcast.

SOMALIA

L.T: UTC +3h — **Pop:** 10 million — **Pr.L:** Somali, Rahanwein (Maay), Arabic, English — **E.C:** 50Hz, 220V — **ITU:** SOM

RADIO MOGADISHU ("Voice of the Republic of Somalia", controlled by the Transitional Federal Government). **W:** radiomuqdisho.net **FM:** 90.0MHz.
RADIO AL-FURQAAN W: radioalfurqaan.com **E:** radioalfurqaan@yahoo.com **FM:** 106.5MHz.
RADIO BANADIR W: radiobanadir.com **E:** info@radiobanadir.com **FM:** 103.0MHz.
RADIO HORN AFRIK
W: www.hornafriknews.com **E:** feedback@hornafrik.com
FM: Mogadishu 99.9MHz, Baidoa/Marka 88.8MHz, Beled Weyne/ Kismaayo 99.9MHz.
RADIO SHABELE W: shabelle.net **E:** radio@shabelle.com
FM: Marka 92.0MHz, Mogadishu 101.5MHz, Baidoa: fq. unknown. Also rel. BBC.
HOLY KORAN RADIO (IQK) FM: Mogadishu 102.5MHz.
RADIO SIMBA W: simbanews.com **E:** simba@simbanews.com **FM:** Mogadishu 95.0MHz 1kW.
RADIO ANDALUS FM: Kismayo 88.8MHz, Elbur 100.5MHz, Baidoa&Jowhar (fqs unknown).
VOICE OF DEMOCRACY FM: Mogadishu: 93.5MHz.
VOICE OF PEACE W: codkanabadda.com **E:** codkanabadda@gmail.com **FM:** Galkayo 88.8MHz, Bosaso 95.0MHz, Mogadishu 98.8MHz.
STAR FM, Mogadishu: 97.0MHz. See main entry under Kenya.
BAR-KULAN RADIO (a joint project by African Union and United Nations)
W: bar-kulan.com **L.P:** David Smith, Dir. Farah Lamaane, Editor.
FM: Mogadishu 98.0MHz 3.5kW. D.Prgr in Somali: 0300-1900.
F.PI: FM relays in Garowe, Bosaaso and Galkayo.

SOMALILAND
(self-declared autonomous state in northwest Somalia)

RADIO HARGEISA
🖃 Ex-Indian Club, Tima-Cadde, near Main St, Hargeisa **E:** radiohargeisa@yahoo.com **L.P:** Dir: Muhammad Said Muhummad.
Hargeisa: **MW:** 693kHz. **SW:** 7145kHz 25kW (irreg.). **FM:** 98.2MHz.
D.Prgr in Somali: 1200/1500-1900.
Ann: "Halkani wa Radio Hargeysa, codka jamhuriyada Somaliland".

BBC World Sce: Hargeisa 89.0MHz.
Voice of America: Hargeisa 88.0MHz.

PUNTLAND
(self-declared autonomous state in northeast Somalia)

RADIO GALKAYO
🖃 Galkayo **W:** www.radiogaalkacyo.com **E:** Radiogaalkacyo@yahoo.

co.uk **L.P:** Dir: Hasan Muhammad Jama. Prgr. Mgr: Mohamod Yasin Issak. Editor: Abdullahi A. Dalab.
FM: Galkayo 88.2MHz, Garowe 89.0MHz.
D.Prgr in Somali: 0400-0700, 0800-2000.

RADIO HAGE SOMALIA
🖃 Galkayo. **E:** radiohagesom@gmail.com **L.P:** Dir: Abdikarim Nur Muhamud. Prgr. Prod.: Abdikani Ahmed Mahamed. Ed: Nur Ahmed Ali
SW: 6915kHz 0.5kW 0930-1130. **F.PI:** SW 3980kHz & FM 99.9MHz.

SOMALI BROADCASTING CORPORATION (SBC)
🖃 SBC Building, Airport Rd, Bosaso **E:** sbc@sbconline.net
FM: Bosaso 89.0, Qardho 88.7, Garowe 89.2.
RADIO MIDNIMO
E: radiomidnimo@hotmail.com **L.P:** MD: Abdishakur Mire Adam. **FM:** Bosaso 97.5MHz. **N. in Somali:** 1030, 1700.
RADIO DALJIR
W: radiodaljir.com **E:** daljir@radiodaljir.com
FM: Galkayo 103,0MHz, Bosaso/Garowe 87.9MHz, Burtinle 88.8MHz. Also r. VOA.
R. Garowe: 89.6MHz 1kW. **W:** www.radiogarowe
R. Hikma, Galkayo: freq. not known.
R. Horseed Media, Bosaso: 89.2MHz 1kW. **W:** horseedmedia.net
R. Mandeq, Balad Hawo: freq. not known.
R. Nugaal, Garowe: 95.0MHz
BBC World Sce (Arabic/English/Somali): Galkayo 96.6MHz

SOUTH AFRICA

L.T: UTC +2h — **Pop:** 50 million — **Pr.L:** English, Afrikaans, isiNdebele, isiXhosa, isiZulu, Sepedi, Sesotho, Setswana, siSwati, Tshivenda, Xitsonga — **E.C:** 50Hz, 230V — **ITU:** AFS — **Int. dialling code:** +27

SOUTH AFRICAN BROADCASTING CORPORATION (SABC) Pub
🖃 Private Bag X1, Auckland Park 2006 ☎ +27 11 714 9111 🖷 +27 11 714 9744 **W:** www.sabc.co.za **Regional offices:** PO Box 2551, Cape Town 8000 – PO Box 1588, Durban 4000 – PO Box 563, Bloemfontein 9300 – PO Box 1040, Port Elizabeth 6000 – PO Box 395, Polokwane 0700 – PO Box 2724, Nelspruit 1200 – PO Box 1008, Kimberley 8300 – Private Bag X2158, Mafikeng 2735.
L.P: Group CEO: Solly Mokoetle. Head of Regions: Charlotte Mampane. Head, PBS: Thami Ntenteni. Head Audience Services: Anton Heunis.
NB: All txs belong to SENTECH (the common carrier for broadcasting in South Africa), 🖃 Private Bag X06, Honeydew 2040
HOME SERVICES (Comm.)
MW:

Location	Station	kHz	kW
Meyerton	R.Metro FM	576	50
Komga	Umhlobo Wenene FM	846	100
Welgedacht	Ligwalagwala FM	1287	2

COUNTRYWIDE FM (Comm.)

Limpopo FM (MHz)	R. Sonder Grense	SAfm	R. 2000	5 FM	R. Metro
Blouberg	102.3	105.9	-	-	-
Hoedspruit	102.0	105.6	98.5	-	-
Louis Trichardt	100.7	104.3	97.2	-	-
Modimolle	102.9	106.5	-	-	-
Mokopane	101.4	105.0	97.9	91.4	106.7
Thabazimbi	101.9	105.5	98.4	-	-
Tzaneen	102.6	106.2	107.7	-	-
NW Province FM (MHz)	**R. Sonder Grense**	**SAfm**	**R. 2000**	**5 FM**	**R. Metro**
Christiana	103.6	107.2	-	-	-
Enzelsberg	101.6	105.2	-	-	-
Groot Marico	102.3	105.9	-	-	-
Klerksdorp	101.2	104.8	97.7	-	-
Piet Plessis	102.8	106.4	-	-	-
Pomfret	101.1	104.7	-	-	-
Rustenburg	100.7	104.3	97.2	-	-
Schweizer-Reneke	103.1	106.7	99.6	-	-
Zeerust	102.6	106.2	99.1	-	-
Gauteng FM (MHz)	**R. Sonder Grense**	**SAfm**	**R. 2000**	**5 FM**	**R. Metro**
Heidelberg	100.8	104.4	97.3	-	-
Helderkruin	-	-	-	104.0	-
Johannesburg	101.5	105.1	99.7	98.0	96.4
Menlo Park	102.1	105.7	98.6	-	-
Pretoria	101.0	104.6	97.5	89.9	92.4
Sunnyside	-	-	-	103.6	-
Welverdiend	102.0	105.6	98.5	107.3	-

Mpumalanga FM (MHz)	R. Sonder Grense	SAfm	R. 2000	5 FM	R. Metro
Carolina	103.0	106.6	-	-	-
Davel	103.5	107.1	100.0	90.4	-
Dullstroom	100.8	104.4	-	-	-
Lydenburg	102.8	106.4	-	-	-
eMalahleni	101.8	105.4	98.3	97.0	100.3
Nelspruit	102.5	106.1	99.0	91.1	-
Piet Retief	102.1	105.7	-	-	-
Sabie	104.2	107.9	-	-	-
Volksrust	102.6	106.2	-	-	-

Northern Cape FM (MHz)	R. Sonder Grense	SAfm	R. 2000	5 FM	R. Metro
Alexander Bay	102.2#	105.8#	98.7	92.2	-
Calvinia	101.5	105.1	-	-	-
Carnarvon	102.5	106.1	-	-	-
Colesberg	103.8	107.5	-	-	-
De Aar	102.0	105.6	-	-	-
Douglas	102.9	106.5	-	-	-
Faans Grove	103.0	106.6	-	-	-
Garies	100.7#	104.3#	-	-	-
Kimberley	101.0	104.6	97.5	91.0	-
Kuruman Hills	102.4	106.0	-	-	-
Pofadder	102.8	106.4	-	-	-
Prieska	100.8	104.4	-	-	-
Springbok	101.6	105.2	-	-	-
Upington	101.7	105.3	-	-	-
Victoria West	101.1	104.7	-	-	-
Williston	103.2		-	-	-

Free State FM (MHz)	R. Sonder Grense	SAfm	R. 2000	5 FM	R. Metro
Bethlehem	101.9	105.5	98.4	-	-
Bloemfontein	103.0	106.6	99.5	91.6	98.1
Boesmanskop	101.2	104.8	-	-	-
Ficksburg	103.7	107.3	-	-	-
Kroonstad	103.4	107.0	99.9	93.4	-
Ladybrand	102.1	105.7	-	-	-
Petrus Steyn	102.3	105.9	98.8	-	-
Senekal	101.1	104.7	97.6	-	-
Springfontein	102.6	106.2	99.1	-	-
Theunissen	102.5	106.1	99.0	92.5	-
Witsieshoek	101.3	104.9	-	-	-

Kwazulu Natal FM (MHz)	R. Sonder Grense	SAfm	R. 2000	5 FM	R. Metro
Donnybrook	102.7	106.3	99.2	-	-
Durban	100.8	104.4	97.3	89.9	93.0
Durban North	102.5	106.1	99.0	103.8	107.9
Eshowe	103.4	107.0	99.9	-	90.3
Glencoe	103.1	106.7	99.6	-	-
Greytown	101.7	105.3	98.2	-	-
Kokstad	101.0	104.6	-	-	-
Ladysmith	101.0	104.6	97.5	-	-
Matatiele	101.5	105.1	-	-	-
Mooi River	102.2	105.8	98.7	-	-
Nongoma	102.9	106.5	99.4	-	89.8
Pietermaritzburg	101.4	105.0	97.9	100.3	-
Port Shepstone	101.3	104.9	97.8	-	-
The Bluff	102.0	105.6	98.5	107.4	-
Ubombo	102.4	106.0	98.9	-	-
Vryheid	101.2	104.8	97.7	-	-

Western Cape FM (MHz)	R. Sonder Grense	SAfm	R. 2000	5 FM	R. Metro
Beaufort West	100.7@	104.3@	-	-	-
Constantiaberg	102.1	105.7	98.6	89.0	-
Ceres	103.7	107.3	-	-	-
Franschhoek	100.7	104.3	97.2	-	-
George	101.7	105.3	98.2	91.7	-
Grabouw	101.7	105.3	-	-	-
Hermanus	100.8	104.4	97.3	-	-
Hex River	102.0	105.6	-	-	-
Hout Bay	100.9	104.5	97.4	87.8	-
Kleinmond	104.2	107.9	-	-	-
Knysna	102.2	105.8	98.7	92.2	-
Ladysmith	101.4	105.0	-	-	-
Matjiesfontein	102.8	106.4	-	-	-
Montagu	104.2	107.9	-	-	-
Napier	102.4	106.0	-	-	-
Oudtshoorn	102.6	106.2	99.1	92.6	-
Paarl	101.6	105.2	98.1	88.5	-
Piketberg	101.1	104.7	97.6	-	-
Plettenberg	100.8	104.4	-	-	-
Riversdale	100.9	104.5	-	-	-
Sea Point	103.5	107.1	100.0	90.4	91.7
Simonstown	100.7	104.3	97.2	87.6	-
Stellenbosch	100.9	104.5	97.4	87.8	-
Table Mountain	102.6	106.2	99.1	89.9	88.6

FM (MHz)	RSG	SAfm	R. 2000	5 FM	R. Metro
Tygerberg	103.0	106.6	99.5	88.2	93.0
Uniondale	103.4	107.0	-	-	-
Vanrhynsdorp	103.4	107.0	-	-	-
Villiersdorp	103.3	106.9	99.8	-	-

Eastern Cape FM (MHz)	R. Sonder Grense	SAfm	R. 2000	5 FM	R. Metro
Aliwal North	101.7	105.3	-	-	-
Andrieskraal	103.2	105.8	-	-	-
Barkly East	100.9	104.5	-	-	-
Bedford	100.8	104.4	-	-	-
Burgersdorp	103.9	107.6	-	-	-
Butterworth	101.1	104.7	97.6	-	-
Cala	103.4	107.0	-	-	-
Cradock	102.7	106.3	-	-	-
East London	101.6	105.2	98.1	88.5	107.7
Elliot	101.4	105.0	-	-	-
Graaff-Reinet	103.3	106.9	-	-	-
Grahamstown	103.5	107.1	100.0	90.4	-
Hankey	101.0	104.6	-	-	-
Kareedouw	102.0	106.5	-	-	-
King Williams Tn	103.0	106.6	-	-	-
Mount Ayliff	103.2	106.8	99.7	-	-
Noupoort	101.4	105.0	-	-	-
Patensie	101.5	105.0	-	-	-
Parsons Hill (PE)	101.0	104.6	95.7	-	87.9
Paul Sauer Dam	103.6	107.2	-	-	-
Port Elizabeth	102.3	105.9	98.8	89.2	100.5
Port St.Johns	103.7	107.3	100.2	-	-
Queenstown	102.2	105.8	98.7	-	-
Suurberg	101.8	105.4	-	-	-
Ugie	102.6	106.2	-	-	-
Umtata	102.0	105.6	98.5	-	-
Willowmore	101.2	104.8	-	-	-

= mono – no RDS, @ = mono RDS

NATIONAL SW SERVICES: Meyerton (G.C: 26S35 028E08): 4 x 100kW txs + 1 standby tx

kHz	Sce.	H of tr	kHz	Sce.	H of tr
3320	RSG	1600-0500	9650	RSG	0800-1600
7285	RSG	0500-0800			

RSG = R. Sonder Grense (tr for Northern Cape region)

SABC PUBLIC BROADCASTING SERVICES (PBS) (Comm.)
Ikwekwezi FM: (isiNdebele): ✉ P.O.Box 11982, Hatfield 0028 **W:** www.ikwekwezifm.co.za ☎ +27 12 431 5301 📠 +27 12 431 5312 - **FM(MHz):** Middelburg 91.8/Kwamhlanga 93.8/Menlo Park 93.6MHz/Johannesburg 106.3/Davel 94.5 + 4 relays – **Lesedi FM:** (Sesotho): ✉ Private Bag X20707, Bloemfontein 9300 **W:** www.lesedifm.co.za ☎ +27 51 503 3091 📠 +27 51 503 3270 - **FM(MHz):** Bloemfontein 89.9/Johannesburg 88.4/Kroonstad 90.3MHz + 15 relays. – **Ligwalagwala FM:** (siSwati): ✉ Private Bag X11301, Nelspruit 1200 **W:** www.ligwalagwalafm.co.za ☎ +27 13 759 6600 📠 +27 13 755 3865 – **FM(MHz):** Nelspruit 92.5/Pretoria 89.3MHz + 10 FM relays and on MW 1287kHz – **Lotus FM:** ✉ Private Bag X1337, Durban 4000 **W:** www.lotusfm.co.za ☎ +27 31 362 5445 📠 +27 31 362 5202 - **FM(MHz):** Durban 87.7/Johannesburg 106.8/Pretoria 100.1 + 8 relays.– **Motsweding FM:** (Setswana): ✉ Private Bag X2158, Mmabatho 2735 **W:** www.motswedingfm.co.za ☎ +27 18 389 7104 📠 +27 18 389 7326 - **FM(MHz):** Mmabatho 88.7/Johannesburg 89.6/Pretoria 91.0/Rustenburg 87.6MHz + 24 relays.– **Munghana Lonene FM:** (Xitsonga): ✉ PO Box 395, Polokwane 0700 **W:** www.munghanalonenefm.co.za ☎ +27 15 290 0262 📠 +27 15 290 0171 - **FM(MHz):** Johannesburg 103.2/Pretoria 95.6/Tzaneen 92.6/Nelspruit 89.4MHz + 4 relays.– **Phalaphala FM:** (Tshivenda): ✉ PO Box 395, Polokwane 0700 **W:** www.phalaphalafm.co.za ☎ +27 15 290 0032 📠 +27 15 290 0170 - **FM(MHz):** Johannesburg 107.8/Tzaneen 99.1MHz + 6 relays.– **Radio Sonder Grense** (National service in Afrikaans): ✉ PO Box 91312, Auckland Park 2006 **W:** www.rsg.co.za ☎ +27 11 714 2702 📠 +27 11 714 3472 – On FM and SW as above.– **Radio X-K** (in !Xu and Khwe languages for Khoi San communities in N.Cape): ✉ PO Box 1008, Kimberley 8300 ☎ +27 53 831 8131 📠 +27 53 831 8127 - **FM(MHz):** Schmidtsdrift 99.4.– **R.2000:** ✉ Private Bag X1, Auckland Park 2006 **W:** www.radio2000.co.za ☎ +27 11 714 4085 📠 +27 11 714 4085 - **FM:** (as above).– **SAfm** (National service in English): ✉ PO Box 91162, Auckland Park 2006 **W:** www.safm.co.za ☎ +27 11 714 4442 📠 +27 11 714 4585 - **FM:** (as above).– **Thobela FM:** (Sepedi): ✉ PO Box 395, Polokwane 0700 **W:** www.thobelafm.co.za ☎ +27 15 290 0264 📠 +27 15 290 0172 - **FM(MHz):** Pretoria 87.9/Tzaneen 89.5MHz + 10 relays.– **TrueFM:** (isiXhosa & English): ✉ Private Bag X0037, Bhisho 5605 **W:** www.trufm.co.za ☎ +27 40 635 0117 📠 +27 40 635 0125 - **FM(MHz):** East London 104.1/Bhisho 100.3 + 2 relays.– **Ukhozi FM:** (isiZulu): ✉ PO Box 1588, Durban

4000 **W**: www.ukhozifm.co.za ☎ +27 31 362 5403 🖹 +27 31 362 5203 - **FM(MHz)**: Durban 90.8, 92.0, 92.5/Johannesburg 91.5/Pretoria 102.4MHz + 21 rel..– **Umhlobo Wenene FM**: (isiXhosa): 🖃 PO Box 1040, Port Elizabeth 6000 **W**: www.uwfm.co.za ☎ +27 41 391 1911 🖹 +27 41 373 2702 - **FM(MHz)**: Port Elizabeth 92.3/King Williams Town 93.0/Durban 96.2/Johannesburg 93.2/Cape Town 92.1MHz + 48 FM relays and on MW 846kHz.

SABC COMMERCIAL BROADCASTING SERVICES (CBS) (Comm.)
Good Hope FM:🖃 PO Box 2551, Cape Town 8000 **W**: www.good-hopefm.co.za ☎ +27 21 430 8276 🖹 +27 21 434 3392 - **FM**: Cape Town 95.3MHz +7 relays – **R.Metro FM:** 🖃 PO Box 91136, Auckland Park 2006 **W**: www.metrofm.co.za ☎ +27 11 714 2658 🖹 +27 11 714 4166 - **FM**: (as above), also on MW 576kHz – **5 FM:** 🖃 PO Box 91555, Auckland Park 2006 **W**: www.5fm.co.za ☎ +27 11 714 2905 🖹 +27 11 714 5714 – **FM**: (as above).

EXTERNAL SERVICE: Channel Africa
See International Broadcasting section

INDEPENDENT COMMUNICATIONS AUTHORITY OF SOUTH AFRICA (ICASA)
🖃 Private Bag X10002, Sandton 2146 ☎ +27 11 321 8200 🖹 +27 11 448 1870 **E**: info@icasa.org.za **W**: www.icasa.org.za
The ICASA is the regulator of telecommunications and the broadc. sectors. It issues licences for commercial and community stns.

PRIVATE STATIONS (Comm.)

	MW kHz	kW	Station	Location
1)	567	25	Cape Talk	Klipheuwel (Cape Town)

	FM MHz	kW	Station	Location
2)	89.0	-	M-Power FM	Piet Retief
2)	89.7	-	M-Power FM	Sabie
3)	89.8	-	North West FM	Rustenburg
4)	89.9	-	Capricorn FM	Thohoyandou
3)	91.8	-	North West FM	Mmabatho
3)	91.9	-	North West FM	Taung
5)	92.7	-	Talk R. 702	Johannesburg
3)	93.5	-	North West FM	Zeerust
6)	93.7	5	OFM	Sasolburg
7)	93.9	10	KFM	Beaufort West
7)	93.9	3	KFM	Garies
8)	93.9	6	R. Jacaranda	Rustenburg
8)	93.9	15	R. Jacaranda	Louis Trichardt
9)	94.0	5	Algoa FM	Bedford
8)	94.0	10	R. Jacaranda	Dullstroom
6)	94.0	9	OFM	Prieska
10)	94.0	25	East Coast R.	Durban
11)	94.0	0.1	Highveld Stereo	Heidelberg
7)	94.0	0.1	KFM	Hermanus
7)	94.1	13	KFM	Riversdale
8)	94.1	33	R. Jacaranda	Pretoria
9)	94.2	0.1	Algoa FM	Parson's Hill
10)	94.2	0.1	East Coast R.	Ladysmith
6)	94.2	-	OFM	Barkley West
6)	94.2	-	OFM	Boshof
6)	94.2	10	OFM	Kimberley
6)	94.3	-	OFM	Ventersburg
6)	94.3	12	OFM	Senekal
7)	94.3	10	KFM	Piketberg
2)	94.3	-	M-Power FM	Nelspruit
10)	94.4	10	East Coast R.	Vryheid
6)	94.4	10	OFM	Klerksdorp
10)	94.5	10	East Coast R.	Port Shepstone
7)	94.5	1.3	KFM	Tygerberg
7)	94.6	2.5	KFM	Ladismith
8)	94.6	10	R. Jacaranda	Mokopane
10)	94.6	0.3	East Coast R.	Pietermaritzburg
9)	94.6	10	Algoa FM	Noupoort
9)	94.6	-	Algoa FM	Colesberg
9)	94.6	-	Algoa FM	Middelburg
12)	94.7	-	Gagasi FM	North Coast
10)	94.7	12	East Coast R.	Matatiele
11)	94.7	38	Highveld Stereo	Johannesburg
7)	94.7	10	KFM	Calvinia
7)	94.8	17	KFM	Springbok
2)	94.8	-	M-Power FM	Carolina
8)	94.8	0.3	R. Jacaranda	Enzelberg
9)	94.8	10	Algoa FM	East London
7)	94.9	10	KFM	George
7)	94.9	-	KFM	Grabow
6)	94.9	8	OFM	Upington
10)	94.9	10	East Coast R.	Greytown

	FM MHz	kW	Station	Location
9)	94.9	10	Algoa FM	Aliwal North
9)	94.9	-	Algoa FM	Barkley East
9)	95.0	-	Algoa FM	Somerset East
9)	95.0	11	Algoa FM	Port Elizabeth
8)	95.0	11	R. Jacaranda	eMalahleni
6)	95.1	10	OFM	Bethlehem
8)	95.1	11	R. Jacaranda	Thabazimbi
8)	95.2	18	R. Jacaranda	Hoedspruit
7)	95.2	-	KFM	Hex River
11)	95.2	20	Highveld Stereo	Welverdiend
10)	95.2	0.1	East Coast R.	The Bluff
8)	95.3	9	R. Jacaranda	Piet Retief
6)	95.3	10	OFM	Ladybrand
9)	95.4	12	Algoa FM	Queenstown
10)	95.4	10	East Coast R.	Mooi River
7)	95.4	-	KFM	Alexander Bay
7)	95.4	0.2	KFM	Knysna
8)	95.5	0.1	R. Jacaranda	Groot Marico
8)	95.5	0.2	R. Jacaranda	Blouberg
6)	95.5	11	OFM	Petrus Steyn
9)	95.5	-	Algoa FM	Jeffreys Bay
9)	95.5	-	Algoa FM	Uitenhage
9)	95.5	16	Algoa FM	Port Elizabeth
10)	95.6	15	East Coast R.	Ubombo
6)	95.6	11	OFM	Kuruman Hills
7)	95.6	3	KFM	Napier
10)	95.7	6	East Coast R.	Durban North
8)	95.7	12	R. Jacaranda	Nelspruit
6)	95.7	10	OFM	Theunissen
6)	95.7	-	OFM	Welkom
7)	95.7	10	KFM	Carnarvon
7)	95.8	9	KFM	Oudtshoorn
8)	95.8	10	R. Jacaranda	Volksrust
8)	95.8	11	R. Jacaranda	Zeerust
8)	95.8	12	R. Jacaranda	Tzaneen
6)	95.8	-	OFM	Colesberg
6)	95.8	-	OFM	Springfontein
10)	95.9	10	East Coast R.	Donnybrook
9)	95.9	12	Algoa FM	Cradock
13)	95.9	35	Kaya FM	Johannesburg
4)	96.0	-	Capricorn FM	Mokopane
6)	96.0	10	KFM	Matjiesfontein
7)	96.0	5	KFM	Pofadder
6)	96.1	9	OFM	Douglas
8)	96.1	0.2	R. Jacaranda	Modimolle
10)	96.1	10	East Coast R.	Nongoma
9)	96.1	-	Algoa FM	Plattenberg Bay
9)	96.1	-	Algoa FM	Cape St. Francis
9)	96.2	10	Algoa FM	King Williams Town
9)	96.2	-	Algoa FM	Bisho
6)	96.2	10	OFM	Bloemfontein
8)	96.2	9	R. Jacaranda	Carolina
10)	96.3	10	East Coast R.	Glencoe
6)	96.3	-	OFM	Warrenton
6)	96.3	-	OFM	Vryburg
6)	96.3	10	OFM	Schweitzer-Reineke
9)	96.5	-	Algoa FM	Pearson
9)	96.5	-	Algoa FM	Aberdeen
9)	96.5	10	Algoa FM	Graaf-Reinet
7)	96.5	10	KFM	Villiersdorp
7)	96.6	17	KFM	Vanrhynsdorp
10)	96.6	10	East Coast R.	Eshowe
6)	96.6	10	OFM	Kroonstad
9)	96.7	-	Algoa FM	Port Alfred
9)	96.7	10	Algoa FM	Grahamstown
8)	96.7	10	R. Jacaranda	Davel
6)	96.8	5,5	OFM	Christiana
10)	96.9	0.1	East Coast R.	Newcastle
7)	96.9	20	KFM	Ceres
6)	96.9	-	OFM	Ficksburg
3)	97.0	-	North West FM	Klerksdorp
6)	97.1	-	OFM	Potchefstroom
7)	97.1	-	KFM	Kleinmond
7)	97.1	-	KFM	Montagu
3)	97.3	-	North West FM	Schweitzer-Reineke
4)	97.6	-	Capricorn FM	Tzaneen
4)	98.0	-	Capricorn FM	Hoedspruit
12)	98.5	0.3	Gagasi FM	Pietermaritzburg
14)	99.2	35	Y-FM	Johannesburg
12)	99.5	25	Gagasi FM	Durban
12)	100.1	6	Gagasi FM	Durban North
2)	101.6	-	M-Power FM	Dullstroom
15)	102.7	0.1	Heart 104.9	Paarl

FM	MHz	kW	Station	Location
16)	102.7	35	Classic FM	Johannesburg
12)	103.5	-	Gagasi FM	Port Shepstone
15)	104.9	1.3	Heart 104.9	Tygerberg
4)	105.4	-	Capricorn FM	Makhado
2)	105.8	-	M-Power FM	Standerton
2)	106.4	-	M-Power FM	eMalahleni

Addresses and other information:
1) ✉ Private Bag X567, Vlaeberg 8018 ☎ +27 21 446 4700 🖷 +27 21 446 4800 **LP:** Colleen Louw **W:** www.capetalk.co.za **E:** feedback@capetalk.co.za – **2)** ✉ PO Box 361, Nelspruit, 1200 ☎ +27 13 757 9700 🖷 +27 13 757 0248 **LP:** Bondo Ntuli **W:** www.mpowerfm.co.za **E:** info@mpowerfm.co.za – **3)** ✉ Postnet Suite 215, Private Bag X3172, Rustenburg 0300 ☎ +27 14 597 3915 🖷 +27 14 597 3345 **LP:** Shadrack Menyatswe **W:** www.northwestfm.co.za **E:** shadrack.menyatswe@gmail.com – **4)** ✉ Postnet Suite 93, Private Bag X9676, Polokwane, 0700 ☎ +27 15 291 0815 🖷 +27 15 291 0822 **LP:** Simphiwe Mdlalose **W:** www.capricornfm.co.za **E:** info@capricornfm.co.za – **5)** ✉ PO Box 5572, Rivonia 2128. ☎ +27 11 506 3702 🖷 +27 (11) 506 3663 **W:** www.702.co.za **E:** comment@702.co.za – **6)** ✉ PO Box 7117, Bloemfontein 9300 ☎ +27 51 505 0900 🖷 +27 51 505 0905 **LP:** Gary Stroebel **W:** www.ofm.co.za **E:** gary@ofm.co.za – **7)** ✉ Private Bag X945, Cape Town, 8000 ☎ +27 21 446 4700 🖷 +27 21 446 4800 **LP:** Colleen Louw. **W:** www.kfm.co.za **E:** colleen@primedia.co.za – **8)** ✉ PO Box 11961, Centurion 0046 ☎ +27 12 673 9100 🖷 +27 12 657 0105 **LP:** Alan Khan **W:** www.jacarandafm.com **E:** bridget@jacarandafm.com – **9)** ✉ PO Box 5973, Walmer, 6065 ☎ +27 41 505 9497 🖷 +27 41 583 5555 **LP:** Dave Tiltmann. **W:** www.algoafm.co.za **E:** dave.t@algoafm.co.za – **10)** ✉ P.O. Box 25095, Gateway, Umhlanga Rocks 4321 ☎ +27 31 570 9495 🖷 +27 86 679 4951. **LP:** Trish Taylor. **W:** www.ecr.co.za **E:** trish@ecr.co.za – **11)** ✉ PO Box 3438, Rivonia 2128 ☎ +27 11 506 3947 🖷 +27 11 506 3393 **LP:** Tery Volkwyn **W:** www.highveld.co.za **E:** comments@highveld.co.za – **12)** ✉ PO Box 4995, Durban 4001.☎ +27 31 580 5300 🖷 +27 31 566 3403 **LP:** Pearl Sokhulu **W:** www.gagasi995.co.za **E:** management@gagasi995.co.za – **13)** ✉ PO Box 434, Newtown, 2113 **W:** www.kayafm.co.za ☎ +27 11 634 9500 🖷 +27 11 634 9574 **LP:** Charlene Deacon **W:** www.kayafm.co.za **E:** pr@kayafm.co.za – **14)** ✉ cor.Albury Rd. & Dunkeld Cresc., South West Blocks, Dunkeld West Ext.8, Sandton 2196 ☎ +27 11 772 0800 🖷 +27 11 280 0421 **LP:** Kanthan Pillay **W:** www.yworld.co.za **E:** kanthan@yfm.co.za – **15)** ✉ PO Box 211, Greenpoint 8051 ☎ +27 21 406 8900 🖷 +27 21 406 8940 **LP:** Gavin Meiring **W:** www.1049.fm **E:** nikki@1049.fm – **16)** ✉ PO Box 782, Auckland Park 2006. ☎ +27 11 403 1027 🖷 +27 11 408 5451 **LP:** Mike Ford **W:** www.classicfm.co.za **E:** info@classicfm.co.za

COMMUNITY STATIONS

Numerous licences issued by ICASA with about 100 stns currently on the air, mainly on FM. The license term is usually 4 years.

MW	kHz	kW	Station	Location
1)	657	50	R. Pulpit/R.Kansel	Meyerton
2)	1422	1	Hellenic Radio	Bedfordview
3)	1485	1	R. Today	Honeydew
4)	1548	10	R. Islam	Lenasia
5)	1584	0.25	R. 1584	Pretoria

Addresses and other information:
1) PO Box 3436, Pretoria 0001 **W:** www.radiokansel.co.za or www.radiopulpit.co.za ☎ +27 12 334 1200 🖷 +27 12 334 1400 Relig. prgrs in English and Afrikaans, 24h.– **2)** PO Box 4077, Edenvale 1610 ☎ +27 11 453 3794 🖷 +27 11 453 3778 Prgrs in Greek and English 24h. – **3)** PO Box 2820, Parklands 2121 **W:** www.1485.org.za ☎ +27 11 880 0329 🖷 +27 86 601 2950 **E:** info@1485.org.za Prgrs in English for over 50s 24h. – **4)** PO Box 2580, Lenasia 1820 **W:** www.radioislam.co.za ☎ +27 11 854 7022 🖷 +27 11 854 7044 Prgrs in English, 24h. – **5)** Institute for Islamic Services, PO Box 46001, Belle Ombre 0142 **W:** www.islam.co.za/1584 **E:** info@islamservices.org ☎ +27 12 374 1584 🖷 +27 12 374 2448 Prgrs in English, Arabic, Afrikaans, Sesotho, 24h.
FM: Approx. 90 low-power stns operating on permanent licences

SPAIN

L.T: UTC +1h (27 Mar-30 Oct: +2h) — **Pop:** 40.5 million — **Pr.L:** Castilian, Catalan, Galician, Basque — **E.C:** 50Hz, 230V — **ITU:** E

MINISTERIO DE FOMENTO
Secretaría General de Comunicaciones
✉ Alcalá 50, Palacio de Comunicaciones, 28071 Madrid

RADIO NACIONAL DE ESPAÑA (RNE) (Pub)
✉ Casa de la Radio, Prado del Rey, 28223 Pozuelo de Alarcón
☎ +34 91 346 2030 🖷 +34 91 346 1769 **W:** www.rne.es
RNE1 (R. Nacional) and RNE5 (R. 5 Todo Noticias)

MW	kHz	kW	Net	Rg	Location
AS01	531	20	RNE5TN	AS	Oviedo °
AN02	531	10	RNE5TN	AN	Córdoba
GA02	531	10	RNE5TN	GA	Pontevedra
NA01	531	10	RNE5TN	NA	Pamplona °
VA01	558	50	RNE5TN	VA	València °
GA01	558	20	RNE5TN	GA	A Coruña °
EU02	558	20	RNE5TN	EU	Donosti-San Sebastián
MU01	567	50	RNE5TN	MU	Murcia °
AN03	567	5	RNE5TN	AN	Marbella
CA01	576	100	RNE5TN	CA	Barcelona °
MA01	585	600	RNE1	MA	Madrid °
AN01	603	50	RNE5TN	AN	Sevilla °
CL02	603	10	RNE5TN	CL	Palencia
CA02	612	10	RNE1	CA	Lleida (relay)
EU01	612	10	RNE1	EU	Vitoria-Gasteiz °
AN04	621	10	RNE1	AN	Jaén
BA01	621	10	RNE1	BA	Palma de Mallorca °
CL03	621	10	RNE1	CL	Avila (relay)
GA01	639	300	RNE1	GA	A Coruña °
AR01	639	50	RNE1	AR	Zaragoza °
EU03	639	50	RNE1	EU	Bilbo-Bilbao
AN05	639	20	RNE1	AN	Almería (relay)
CM03	639	10	RNE1	CM	Albacete (relay)
EX02	648	10	RNE1	EX	Badajoz (relay)
MA01	657	50	RNE5TN	MA	Madrid °
AN01	684	600	RNE1	AN	Sevilla °
AS01	693	5	RNE1	AS	Boal (rel. of Oviedo)
CM01	693	20	RNE1	CM	Toledo °
CA03	693	10	RNE1	CA	Tarragona (rel.of Barcelona)
AS01	729	100	RNE1	AS	Oviedo °
AN06	729	20	RNE1	AN	Málaga
RI01	729	20	RNE1	RI	Logroño °
CL01	729	10	RNE1	CL	Valladolid °
CM04	729	10	RNE1	CM	Cuenca
VA02	729	10	RNE1	VA	Alacant-Alicante (relay)
CA01	738	600	RNE1	CA	Barcelona °
AN07	747	10	RNE5TN	AN	Cádiz
VA01	774	100	RNE1	VA	València °
EX01	774	60	RNE1	EX	Cáceres °
EU02	774	50	RNE1	EU	Donosti-San Sebastián
GA03	774	20	RNE1	GA	Ourense (relay)
AN08	774	10	RNE1	AN	Granada (relay)
AN09	774	10	RNE1	AN	La Línea
CL04	774	10	RNE1	CL	León (relay)
CL05	774	10	RNE1	CL	Soria (relay)
CM05	801	25	RNE1	CM	Ciudad Real (relay)
GA04	801	20	RNE1	GA	Lugo (relay)
CA04	801	10	RNE1	CA	Girona (relay)
CL06	801	10	RNE1	CL	Burgos
CL07	801	10	RNE1	CL	Zamora
VA03	801	10	RNE1	VA	Castelló (relay)
MU01	855	300	RNE1	MU	Murcia °
CT01	855	10	RNE1	CT	Santander °
CA03	855	20	RNE1	CA	Tarragona (relay)
GA02	855	20	RNE1	GA	Pontevedra (relay)
AN10	855	10	RNE1	AN	Huelva (relay)
AR02	855	10	RNE1	AR	Teruel (relay)
CL08	855	10	RNE1	CL	Ponferrada (relay)
CL09	855	10	RNE1	CL	Salamanca (relay)
NA01	855	10	RNE1	NA	Pamplona-Iruñea °
AN03	855	5	RNE1	AN	Marbella (relay)
CM02	864	10	RNE1	CM	Socuellamos (rel.of Toledo)
BA01	909	10	RNE5TN	BA	Palma de Mallorca °
AR01	936	20	RNE5TN	AR	Zaragoza °
CL01	936	10	RNE5TN	CL	Valladolid °
VA02	936	10	RNE5TN	VA	Alacant-Alicante
AN11	972	5	RNE1	AN	Cabra (rel.of Sevilla)
GA05	972	2	RNE1	GA	Monforte de Lemos (relay)
AN08	1017	10	RNE5TN	AN	Granada
CL06	1017	10	RNE5TN	CL	Burgos
AN05	1098	25	RNE5TN	AN	Almería
GA04	1098	10	RNE5TN	GA	Lugo
CL03	1098	10	RNE5TN	CL	Avila
AN10	1098	5	RNE5TN	AN	Huelva
RI01	1107	25	RNE5TN	RI	Logroño °
CT01	1107	20	RNE5TN	CT	Santander °
EX01	1107	20	RNE5TN	EX	Cáceres °
AR02	1107	10	RNE5TN	AR	Teruel
CL08	1107	10	RNE5TN	CL	Ponferrada (rel.of León)
CL05	1125	10	RNE5TN	CL	Soria
CM01	1125	10	RNE5TN	CM	Toledo °
EU01	1125	10	RNE5TN	EU	Vitoria-Gasteiz °
VA03	1125	10	RNE5TN	VA	Castelló
EX02	1125	5	RNE5TN	EX	Badajoz
AN06	1152	20	RNE5TN	AN	Málaga

MW	kHz	kW	Net	Rg	Location
CA02)	1152	10	RNE5TN	CA	Lleida
CL07)	1152	10	RNE5TN	CL	Zamora
CM03)	1152	10	RNE5TN	CM	Albacete
MU02)	1152	10	RNE5TN	MU	Cartagena
GA03)	1305	25	RNE5TN	GA	Ourense
CM05)	1305	20	RNE5TN	CM	Ciudad Real
EU03)	1305	20	RNE5TN	EU	Bilbo-Bilbao
CL04)	1305	10	RNE5TN	CL	León
CM04)	1314	20	RNE5TN	CM	Cuenca
CA03)	1314	10	RNE5TN	CA	Tarragona
CL09)	1314	10	RNE5TN	CL	Salamanca
MA01)	1359	600		MA	Madrid °° (irr.)
GA06)	1413	20	RNE5TN	GA	Vigo
AN04)	1413	10	RNE5TN	AN	Jaén
CA04)	1413	5	RNE5TN	CA	Girona
AN09)	1503	5	RNE5TN	AN	La Linea (rel.of Cádiz)
GA05)	1503	2	RNE5TN	GA	Monforte de Lemos (rel. Lugo)

°= regional key stn °°= night-time transmission, 2000-0600 UTC approx. DRM tests relay RNE1.

FM	Location	RNE1	RNE2	RNE3	RNE4	RNE5	kW
Andalucía							
AN08)	Baza	92.6	97.3	87.8			5
AN02)	Cabra	95.1	89.5	103.8		88.0	1
AN02)	Córdoba					99.8	2
AN08)	Granada	104.2	96.4	94.4		98.5	1
AN01)	Guadalcanal		90.6				5
AN07)	Jerez	103.5	94.5	96.7	106.3		10
AN02)	Lagar de la Cruz	92.2	97.5	98.6			10
AN06)	Málaga		99.2	104.0		92.5	1
AN03)	Marbella					87.6	1
AN06)	Mijas	106.6	98.1	99.8		88.0	10
AN08)	Parapanda	103.0	91.1	93.9			5
AN05)	Pechina	100.9	92.4	94.9		106.7	5
AN10)	Punta Umbria	95.2	92.6	99.0		88.8	5
AN06)	Ronda	106.1	99.3	91.6		102.3	1
AN04)	Sierra Almadén	105.4	90.0	96.0			10
AN08)	Sierra Lújar	96.7	90.4	94.2			5
AN07)	Tajo	105.0	94.0	103.1			5
AN01)	Valencina	91.2	93.7	98.8		90.0	5
Aragon							
AR02)	Alcañiz	89.5				99.3	1
AR03)	Arguis	100.9	94.4	103.7		92.8	5
AR03)	Barbastro	89.6	97.4	105.1		100.2	1
AR01)	Caspe	90.2	99.0			103.7	1
AR01)	Ejea de los C.	94.8	98.9	106.4		91.2	
AR03)	Fraga	95.0	96.3	102.2		98.8	1
AR01)	Inogés	89.4	92.4	99.7		105.0	5
AR03)	Jaca	103.7	94.4	100.3		98.7	1
AR02)	Javalambre		90.0	93.9			1
AR01)	La Muela	94.5	90.9	96.3		103.6	10
AR02)	Montalban	90.5	92.7	96.4		105.1	
AR02)	Peracense	88.3	98.1	100.6		106.1	
AR02)	Teruel	104.7	89.2	94.5		95.6	1
Asturias							
AS01)	Avilés	100.0	87.9	95.6		102.9	1
AS01)	Boal	93.2	97.8	88.2		90.5	1
AS01)	Cangas Narcea	97.2	99.0	87.7		90.9	1
AS01)	Cangas Onis	88.8	92.5	104.0		100.3	1
AS01)	Gamoniteiro	102.5	92.2	94.4		104.4	10
AS02)	Gijón	99.2	98.5	102.0		89.9	5
AS01)	Ibías	95.8	98.7	102.9		105.1	1
AS01)	Llanes	106.1				97.3	1
AS01)	Los Oscos	89.7	104.0	105.7		96.1	1
AS01)	Luarca	96.8	93.8	100.3			1
AS01)	Mieres					101.8	1
AS01)	Oviedo	89.4	96.0	90.3		99.6	1
AS01)	Peñamelleras	93.9	96.4	100.7		104.6	
AS01)	San Martín	88.3	96.7	100.2		93.3	1
Baleares							
BA01)	Alfabia	90.1	87.9	92.3		104.5	10
BA01)	Ibiza	101.6	104.0	105.7		94.9	1
BA01)	Menorca	94.6	97.1	105.8		100.4	1
BA01)	Pollensa	93.2	95.4	97.4		99.7	1
Cantabria							
CT03)	Embalse Ebro	89.0	94.0	98.2		101.9	1
CT01)	Liérganes	96.9	93.0	102.9		105.0	10
CT02)	Torrelavega	99.5	97.9	103.4		89.4	1
Catalunya							
CA02)	Alpicat	94.6	89.2	97.8	87.9		10
CA02)	Baquéira	92.2	87.7	89.0	93.3		1
CA02)	Bossost	94.4	100.5	105.2	102.3		1
CA01)	Collserola	88.3	93.0	98.6	100.8	99.0	20
CA01)	Collsuspina	92.2	97.9	103.1	104.7		1
CA01)	Igualada	89.4	90.9	105.1	106.9		1
CA03)	Monte Caro	104.3	96.6	99.6	90.7		5

FM	Location	RNE1	RNE2	RNE3	RNE4	RNE5	kW
CA01)	Montserrat	94.3	99.0		103.8	98.8	2
CA03)	Musara	106.5	91.5	94.5	88.8	94.0	5
CA01)	Sant Pere Ribes	92.7	95.2	97.5	106.3	101.3	1
CA04)	Rocacorba	93.3	91.1	95.9	106.2	94.0	5
CA02)	Soriguera	99.9	103.6	106.4	90.6	97.2	1
CA03)	Ulldecona	95.0					5
CA02)	Viella	90.0	96.2	104.4	102.6		1
Castilla-León							
CL06)	Aranda Duero	90.0	92.7	101.6		106.2	1
CL03)	Arenas Pedro	102.4	90.3				1
CL03)	Avila	87.6	92.0	97.8		102.4	1
CL09)	Béjar	99.9	101.6	104.7			1
CL07)	Benavente	87.8	91.3	97.9		100.2	1
CL05)	Burgo Osma	96.1	98.4	88.7		102.8	1
CL06)	Burgos	93.6	90.3	91.2		106.6	1
CL04)	Castropodame	103.3	93.0	99.9		105.9	5
CL02)	Cervera	88.6	94.8	97.3		100.4	1
CL09)	El Cabaco	102.9	92.4	95.4			5
CL02)	Guardo	89.8	105.6			104.0	1
CL04)	León	97.1	91.1	89.3		102.2	1
CL02)	Palencia	91.8	101.0	97.6		88.0	1
CL06)	Pancorbo	89.7	92.0	101.7		104.5	1
CL07)	Pbla Samabria	93.6	103.5	100.3		91.9	1
CL09)	Salamanca	94.5	88.1	91.4		102.2	1
CL10)	Segovia	97.0				91.5	1
CL05)	Soria	89.7	91.5	94.3		104.7	2
CL01)	Valladolid	97.3	93.1	92.2		95.1	5
CL04)	Villablino	98.1	89.0	91.4		99.4	1
CL06)	Villadiego		102.3	103.3			1
CL04)	Villafranca	90.9	89.7	97.5		104.1	1
CL07)	Zamora	101.8	96.7	98.5		88.8	5
Castilla La Mancha							
CM03)	Almansa	91.2	98.6	95.6		94.4	1
CM03)	Chincilla	91.8	93.6	99.0		106.3	5
CM05)	Ciudad Real	95.7	92.8	94.1		88.8	1
CM04)	Cuenca	105.6	93.0	92.0		96.1	1
CM06)	Guadalajara	103.7	93.5	96.9		102.1	1
CM05)	La Mancha	101.0	89.8	94.5		106.8	10
CM05)	Puertollano	93.1	99.1	91.8		101.8	5
CM05)	Socuéllamos	94.0					1
CM01)	Talavera	97.8	105.5	94.7		89.4	5
CM01)	Toledo	102.0	103.9	106.4		99.9	1
CM05)	Valdepeñas	92.6	95.5	97.3		102.1	
Euskadi							
EU03)	Archanda	100.7	90.6	99.2		96.3	5
EU02)	Azcoitia	88.7	104.9	106.9			1
EU02)	Beasain	100.2	98.4	94.9			1
EU02)	Eibar	92.9	98.7	95.9			1
EU02)	Jaizquibel	104.7	90.0	92.1			10
EU02)	Monte Igueldo	87.6	99.5	98.9		93.3	1
EU03)	Oiz	106.4	105.3	102.1			5
EU01)	San León					93.3	1
EU03)	Sollube	105.9	93.9	95.4			5
EU02)	Tolosa	101.9	98.8	96.0			1
EU01)	Vitoria-Gasteiz	92.5	96.9	99.5		89.4	1
Extremadura							
EX02)	Badajoz	94.9	90.1	92.2		106.0	1
EX01)	Cáceres	95.1	101.7	93.7		88.2	1
EX02)	Mérida					101.3	1
EX01)	Montánchez	105.3	97.7	99.3			5
EX01)	Plasencia	88.6		99.3		104.4	1
Galicia							
GA03)	Barco	94.7	96.4	100.3		104.6	
GA02)	Domayo	90.1	92.1	97.4			5
GA04)	Monforte					88.8	1
GA03)	Monte Meda	102.8	91.2	94.3		106.8	5
GA01)	Monte Xalo	100.4	91.6	94.5		95.8	10
GA03)	Ourense	100.6	97.2	99.4		95.1	5
GA04)	Páramo	101.7	88.2	99.6		92.8	5
GA04)	Piedrafita	89.4	92.6	105.3		95.2	
GA02)	Pontevedra			88.3			1
GA07)	Santiago	103.1	98.1	99.0		93.7	5
GA03)	Verin	90.7	98.4	106.4		94.1	1
GA06)	Vigo					96.0	5
GA04)	Xistral	89.5	96.3	104.2		106.6	1
Madrid							
MA01)	Navacerrada	104.9	98.8	95.8			30
MA01)	Torrespaña	88.2	96.5	93.2		90.3	10
Murcia							
MU01)	Carrascoy	101.7	98.2	96.0		92.1	5
MU02)	Cartagena	102.9	94.5	97.5		103.5	1
MU01)	Jumilla	89.1	93.1	100.1			1
MU01)	Yecla	88.8	93.4	103.7			1
Navarra							
NA01)	Estella	89.0	101.2	100.5		90.9	1

FM	Location	RNE1	RNE2	RNE3	RNE4	RNE5	kW
NA01)	Gorramendi	88.3	99.0	100.6		95.3	1
NA01)	Ibañeta	89.6	93.8	103.4		101.9	1
NA01)	Isaba	90.3	95.1	103.0		91.8	1
NA01)	Leire	88.9	101.0	99.6		90.5	1
NA01)	Lesaka	90.6	94.8	97.0		102.2	1
NA01)	Monreal	106.1	97.5	93.0		95.7	5
NA01)	Pamplona	104.8	97.1	102.3		103.7	
NA01)	San Miguel	96.7	100.0	90.3		102.7	1
NA01)	Tudela	100.9	102.2	91.3		88.3	1
La Rioja							
RI01)	Logroño	95.4	88.1	89.9		97.2	1
RI01)	Moncalvillo	102.0	88.5	94.6		103.3	40
RI01)	Monte Yerga	87.6	106.8	96.5		105.4	1
Comunitat Valenciana							
VA02)	Aitana	104.8	88.6	99.7			10
VA02)	Alcoi	95.8	92.3	91.1		105.9	1
VA02)	Alicante	105.2	99.4	97.1		103.6	
VA03)	Benicasim	89.3	90.3	92.8		95.5	5
VA02)	Benidorm	87.6	97.8	102.1			
VA02)	Elda	93.9	88.1	97.6			1
VA01)	Monduber	97.4	99.3	100.1			5
VA01)	Monte Picayo	89.8	106.6	95.1		88.2	10
VA01)	Ontinyent	100.7	96.7	102.4			1
VA02)	Santa Pola	92.5	100.1	94.3		104.2	5
VA02)	Santa Pola					105.8	5
VA01)	Utiel	98.1	96.6	89.1		87.9	1
VA02)	Villena	90.7	97.1	101.1			1

RNE1 R. Nacional: (MW and FM): 24h. **N:** On the h. Regional prgrs from key stns of each region: Mon-Fri 0625-0630, 0650-0700, 0850-0900 (//R5TN), 1208-1300, 1400-1415, 1845-1900 (only Canarias, Galicia, La Rioja, Murcia & Navarra), Sat & Sun 1230 –1300 (//R5TN), Local: 0800-0815 – **RNE2 R.Clásica:** (FM): Classical music & cultural prgrs: 24h. – **RNE3:** (FM): Young people's music prgr: 24h. – **RNE4:** (FM): Regional network in Catalunya: 24h. in Catalán. – **RNE5TN:** (MW and FM): 'Todo Noticias'-All news: 0800-2300 (2300-0800 //R. Nacional). 0625-0630 //RN, 0850-0900 Regional //RN, 0845-0900 Local, Sat-Sun 1400-1415 Regional

Addresses for RNE regional key stns:
AN Andalucia: Edif.RTVE, Parque del Alamillo, 41092 Sevilla – **AR Aragón:** José Luís Albareda 1-3, 50004 Zaragoza – **AS Asturias:** Calle San Esteban de las Cruces 92, 33195 Oviedo – **BA Balears:** Aragó 26, 07006 Palma de Mallorca – **CA Catalunya:** C. Roc Boronat 127, 08018 Barcelona – **CL Castilla y León:** García Morato 27-29, 47007 Valladolid – **CM Castilla La Mancha:** Paseo de San Cristóbal s/n, 45002 Toledo – **CT Cantabria:** Polígono de Raos s/n, 39609 Camargo (Santander) – **EU Euskadi:** Plaza de Simón Bolívar 13, 01003 Vitoria-Gasteiz – **EX Extremadura:** Av. Ruta de la Plata 10, 10001 Cáceres – **GA Galicia:** Paseo Méndez Nuñez 12, 15006 A Coruña – **MA Madrid:** Casa de la Radio, Prado del Rey, 28223 Pozuelo de Alarcón – **MU Murcia:** La Olma 27-29, 30005 Murcia – **NA Navarra:** Emilio Arrieta 8, P8, 31002 Pamplona-Iruñea – **RI La Rioja:** Vara de Rey 42, 26002 Logroño – **VA Comunitat Valenciana:** Av.Colóm 13, 46004 València.

OTHER STATIONS
Only stns with MW broadcasts and FM networks are listed. A number of other stns are heard irr. There are approx. 2,300 FM stns
NATIONAL NETWORKS
(COPE) CADENA DE ONDAS POPULARES ESPANOLAS (Comm)
✉ Alfonso XI N° 4, 28014 Madrid ☎ +34 91-3090000 🖷 +34 91-5317517 **W:** www.cope.es.
FM stns ID as "Cadena 100". Local and regional programming on AM stns: Mon-Fri 0624, 0650, 0724, generally at xx27 and xx57 0757-1900, 1930, 2030, 2130, 2230, 2355, 2357. Sat xx27 and xx57 0757-1400, 1557, 1657, 1810, 1840, 1910, 1940, 2040, 2257, 2330. Sun xx27 and xx57 0957-1300, xx10 and xx40 1510-1840, 2157, 2330.
(D) CADENA DIAL (Comm)
✉ Gran Vía 32, 28013 Madrid ☎ +34 91-3470880 🖷 +34 91-5211753. **W:** .www.cadenadial.com Belongs to SER.
(EFM) EUROPA FM (Comm)
✉ Bueso Pineda 7, 28043 Madrid ☎+34 91 4134361 🖷+34 91 4137175 **W:** www.europafm.com
(ES) ESRADIO (Comm)
✉ C/ Juan Esplandiú 13, 28007 Madrid. **W:** www.esradio.fm
(IE) INTERECONOMIA (Comm)
✉ Paseo de la Castellana 36-38, 20846 Madrid. Prgrs: Intereconomía and Interpop **W:** www.intereconomia.com and www.interpop.es
(KFM) KISS FM (Comm) W: www.kissfm.com
(LFM) LOCA FM (Comm)
✉ C/ Enrique Larreta 12 bajo izq., 28036 Madrid **W:** www.locafm.com

(MFM) MAXIMA FM (Comm)
✉ C/ Gran Vía 32, 7ª planta, 28013 Madrid **W:** www.maxima.fm
(M80) M-80 RADIO (Comm)
✉ Gran Vía 32, 28013 Madrid ☎ +34 91-3470798 🖷 +34 91-5228693 **W:** www.m80radio.com Belongs to SER
(OCR) ONDA CERO RADIO (Comm)
✉ Calle de Ortega y Gasset 22-24. 28006 Madrid ☎ +34 91-5386300 🖷 +34 91-5386323 **W:** www.ondacero.es **Prgrs:** Onda Cero Radio; Onda Cero Melodía. Local and regional programming on AM stations: Mon-Fri 0527, 0555, 0620, 0655, 0720, 0827, 0855, 0927, 0955, 1025, 1130, 1400, 1527, 1557, 1627, 1657, 1727, 1757, 1800, 2030, 2130, 2230, 2257, 2340, 0030. Sat-Sun 0527, 0727, 0757, 0827, 0857, 0927, 0957, 1027, 1227, 1457, 1850, 2255, 2320, 2340. Times vary.
(PR) Punto Radio (Comm) W: www.puntoradio.com
(RA) RADIO TELEVISION AMISTAD (Rlg)
✉ Apartado 269, 08211 Castellar del Vallés (Barcelona). **W:** www.rtvamistad.com
(R&G) R&G ROCK&GOL (Comm)
✉ Diputació 238, 08013 Barcelona **W:** www.rockandgol.net
(RKM) RKM RADIO (Rlg)
✉ Carretera de Ajalvir a Daganzo km 1.7, Ajalvir, 28864 Madrid. **W:** www.rkmradio.com
(RM) RADIO MARIA (Rlg)
✉ Paseo de Lanceros 2 (Centro Comercial), Planta 1ª, 28024 Madrid. **W:** www.radiomaria.es
(RMA) RADIO MARCA (Comm)
✉ Avenida de San Luís 25, 28033 Madrid. **W:** www.marca.com/multimedia/radiomarca/index.html
(RO) Radiolé (Comm) W: www.radiole.com
(SER) SOCIEDAD ESPANOLA DE RADIODIFUSION (Comm)
✉ Gran Vía 32, 28013 Madrid ☎ +34 91-3470700 🖷 +34 91-3470709 **W:** www.cadenaser.com
FM stns ID as "Los 40 Principales, Cadena 40". Local and regional programming on AM stns: Mon-Fri 0550, 0620, 0650, 0720, 0827, 0855, 0930, 0957, 1003, 1030, 1057, 1120, 1410, 1530, 1630, 1720, 1925, 2157, 2255, 2330, 0000, 0030, 0159, 0259. Sat-Sun 0750, 0855, 0955, 1055, 1105 and xx23 or xx53 in the evening. Times vary.

REGIONAL NETWORKS
(AR) ARAGON RADIO
✉ María Zambrano 2, 50018 Zaragoza. **W:** www.aragonradio.es
(COM) COM RADIO CATALUNYA ONA MITJANA
✉ Travessera de les Corts 131-159, Recinte Martenitat, Pavello Cambo, 08028 Barcelona **W:** www.comradio.com
(CR) CORPORACIO CATALANA DE MITJANS AUDIOVISUALS
✉ Av. Diagonal 614-616, 08021 Barcelona ☎ +34 93-3069200 🖷 +34 93-3069201 **W:** www.catradio.cat **Prgrs:** Catalunya Ràdio; Catalunya Informació; Catalunya Música; iCat FM.
(CER) CANAL EXTREMADURA RADIO
✉ Av. de las Américas 1, 1°, 06800 Mérida (Badajoz). **W:** http://radio.canalextremadura.es/
(CSR) CANAL SUR RADIO
✉ Carr.San Juan de Aznalfarache km 1.300, 41920 Sevilla ☎ +34 95-5607600 🖷 +34 95-5607845 **W:** www.canalsur.es
Prgrs: Canal Fiesta: (Int. music and Spanish pop and rock music); Canal Sur Radio: (Andalucian and Spanish music, news and sports); Radio Andalucía Información.
(EI) EUSKA IRRATI TELEBISTA – RADIO TELEVISIÓN VASCA
✉ EiTB Donostia: Paseo Miramon 172, 20014 Donostia-San Sebastián ☎ +34 943 0116 00 🖷 +34 943 01 19 95.
✉ EiTB Bilbao: Capuchinos de Basurto 2, 48013 Bilbo-Bilbao ☎ +34 94656 30 00 🖷 +34 94 656 30 95.
✉ EiTB Vitoria: Domingo Martinez de Aragón, 5-7 bajo, 01006 Vitoria-Gasteiz ☎+34 945 01 25 00 🖷 +34 945 01 26 95.
✉ EiTB Iruña: Calle Tomás Caballero 2, 31005 Pamplona-Iruña. ☎ +34 948 01 22 00. 🖷 +34 948 15 34 85.
Prgrs: Euskadi Irratia (AM + FM), R. Euskadi (AM + FM), R. Vitoria (AM + FM), EiTB Música (FM), Gaztea Irratia (FM) **W:** www.eitb.com
(GR) GRUP FLAIX (Comm)
✉ Passeig de Gràcia 55, 08007 Barcelona. Prgrs: Flaix FM and Flaixbac **W:** www.flaixfm.cat and www.radioflaixbac.com
(IB3) !B3 RADIO ✉ C/ Manuel Azaña 7-A, 07006 Palma de Mallorca. **W:** http://ib3noticies.com/portada-radio
(+R) + RADIO
✉ Avda. Barón de Carcer 48, 46001 València. **W:** www.masradio.com
(OC) ONA CATALANA (Comm) W: www.ona-fm.cat
(OM) ONDA MADRID
✉ Paseo del Príncipe 3, 28223 Pozuelo de Alarcón (Madrid). **W:** www.ondamadrid.es
(ORM) ONDA REGIONAL MURCIA
✉ Avda. Libertad 6, 30009 Murcia. **W:** www.orm.es

(RAC) RAC (Comm)
Av. Diagonal 477, Planta 15, 08036 Barcelona. Prgrs: RAC1 and RAC 105. **W:** www.rac1.cat and www.rac105.cat/
(R9) RADIO AUTONOMIA VALENCIANA
☒ Av. Blasco Ibañez 136, 46002 València ☎ +34 96-3721011 🖹 +34 96-3728513 Prgrs: Sí Radio and Radio Nou **W:** www.siradio.com and www.radionou.com
(RCM) RADIO CASTILLA-LA MANCHA
☒ Edificio RTVCM, C/ Río Alberche s/n, Polígono Santa María de Benquerencia, 45007 Toledo. **W:** www.rtvsm.es
(RCL) RADIO CASTILLA Y LEÓN
Edificio Promecal Burgos, Avenida Castilla y León s/n, 09006 Burgos. **W:** www.rtvcyl.es
(RE) RADIO ESTEL (RIg)
☒ Comtes de Bell-lloc 67-69, 08014 Barcelona **W:** www.radioestel.com
(RG) RADIO GALEGA – COMPAÑÍA DE RADIO TELEVISION DE GALICIA ☒ Casa da Radio, Edificio de Usos Múltiples San Marcos, 15820 Santiago de Compostela ☎ +34 981-562323 🖹 +34 981-561150 **W:** www.crtvg.es
(RP) RADIO POPULAR - HERRI IRRATIA
☒ Plaza Segrado Corazón 5, 48011 Bilbo-Bilbao **W:** www.radio-popular.com ☒ Garibai 19, 2004 Donostia-San Sebastián **W:** www.loyolamedia.com
(RPA) RADIO DEL PRINCIPADO DE ASTURIAS
☒ Camino de las Clarisas 263, 33203 Gijón.
(RTT) RADIO TELE TAXI (Comm)
☒ C/ Sant Carles 40, 08922 Sta Coloma de Gramenet (Barcelona) **W:** www.radioteletaxi.com
(RV) RADIO VOZ
☒ Av. De la Prensa 84-85, Arteixo, 15142 A Coruña **W:** www.radiovoz.com

MW	kHz	kW	Net	Rg	Station, location	FM (MHz)
CA07)	540	50	OCR	CA	Onda Cero Catal., Barcelona	89.8
CA08)	666	50	SER	CA	R. Barcelona, Barcelona	93.9
MU05)	711	25	COPE	MU	COPE, Murcia	89.7
EU07)	756	10	EI	EU	R. Euskadi, Bilbo-Bilbao	91.7
CA09)	783	50	R&G	CA	COPE Miramar/Rock & Gol, Barcelona	
AN15)	792	50	SER	AN	R. Sevilla, Sevilla	97.1
MA05)	810	20	SER	MA	R. Madrid, Madrid	93.9
EU08)	819	10	EI	EU	R. Euskadi, Donosti-San Seb	96.5
CA13)	828	5	KFM	CA	Kiss FM Catalunya, Terrassa	95.5
AN16)	837	10	COPE	AN	COPE, Sevilla	99.6
CL15)	837	10	COPE	CL	COPE, Burgos	95.5
BA05)	837	5	COPE	BA	COPE, Eivissa	89.1
GA09)	837	5	COPE	GA	COPE, El Ferrol	88.7
AR05)	873	25	SER	AR	R. Zaragoza, Zaragoza	95.3
GA10)	873	10	SER	GA	R. Galicia, Stgo de Comp.	90.6
CA10)	882	50	COM	CA	COM Ràdio, Barcelona	91.0
VA07)	882	10	COPE	VA	COPE, Alacant-Alicante	95.6
AN17)	882	5	COPE	AN	COPE, Málaga	89.4
AS05)	882	5	COPE	AS	COPE, Gijón	103.6
CL16)	882	5	COPE	CL	COPE, Valladolid	88.5
EU10)	900	10	RP	EU	R. Popular, Bilbo-Bilbao	97.8
AN18)	900	5	COPE	AN	COPE, Granada	88.2
EX05)	900	5	COPE	EX	COPE Alta Extremadura, Cáceres	88.8
EX05)	900	5	COPE	EX	COPE Alta Extremadura, Plasencia (rel.)	
GA11)	900	5	COPE	GA	COPE, Vigo	87.8
MA06)	918	20	IE	MA	R. Intercontinental, Madrid	95.1
MA07)	954	50	OCR	MA	Onda Cero R., Madrid	97.2
EU09)	963	10	EI	EU	R. Euskadi, Vitoria-Gasteiz	90.9
EU11)	990	10	SER	EU	R. Bilbao, Bilbo-Bilbao	89.5
AN19)	990	5	SER	AN	R. Cádiz, Cádiz	89.4
MA08)	999	50	COPE	MA	COPE, Madrid	99.5
CA11)	1008	10	SER	CA	R. Girona, Girona	88.1
EX06)	1008	5	SER	EX	R. Extremadura, Badajoz	96.9
VA08)	1008	5	SER	VA	R. Alacant, Alacant-Alicante	91.0
CA12)	1026	10	SER	CA	R. Reus, Reus	101.4
AS06)	1026	5	SER	AS	R. Asturias, Oviedo	97.5
GA12)	1026	5	SER	GA	R. Vigo, Vigo	99.4
AN20)	1026	5	SER	AN	R. Jaén, Jaén	96.9
AN21)	1026	5	SER	AN	R. Jerez, J. de la Frontera	97.8
CL17)	1026	5	SER	CL	R. Salamanca, Salamanca	96.9
EU12)	1044	10	SER	EU	R. San Sebastián, Donosti-S Se	97.2
CL18)	1044	5	SER	CL	R. Valladolid, Valladolid	90.9
AR06)	1053	25	COPE	AR	COPE, Zaragoza	88.5
VA09)	1053	5	COPE	VA	COPE, Vila-Real	91.7
AN22)	1080	10	SER	AN	R. Granada, Granada	95.4
AR07)	1080	10	SER	AR	R. Huesca, Huesca	96.9
BA06)	1080	5	SER	BA	R. Mallorca, P. de Mallorca	94.1
CM11)	1080	5	OCR	CM	Onda Cero R., Toledo	100.8
GA13)	1080	5	SER	GA	R. Coruña, A Coruña	91.0
CM12)	1116	5	SER	CM	R. Albacete, Albacete	89.6
GA14)	1116	5	SER	GA	R. Pontevedra, Pontevedra	89.1

MW	kHz	kW	Net	Rg	Station, location	FM (MHz)
CL19)	1134	10	COPE	CL	COPE, Salamanca	90.0
AN23)	1134	5	COPE	AN	COPE, Jerez de la Frontera	92.4
BA07)	1134	5	COPE	BA	COPE, Ciutadella	89.6
CL20)	1134	5	COPE	CL	COPE, Astorga	87.6
CM13)	1134	5	COPE	CM	COPE, Puertollano	97.5
NA05)	1134	5	COPE	NA	COPE, Pamplona-Iruñea	87.9
AN24)	1143	5	COPE	AN	COPE, Jaén	88.8
AS05)	1143	5	COPE	AS	COPE, Oviedo	92.8
CA14)	1143	5	COPE	CA	COPE, Reus	89.7
GA15)	1143	5	COPE	GA	COPE, Ourense	92.4
VA10)	1179	50	SER	VA	R. València, València	94.2
RI05)	1179	2	SER	RI	R. Rioja, Logroño	91.7
EU09)	1197	10	EI	EU	Euskadi Irratia, Vitoria-Gasteiz	95.0
AN25)	1215	5	COPE	AN	COPE, Córdoba	87.6
CL21)	1215	5	COPE	CL	COPE, León	97.7
CT05)	1215	5	COPE	CT	COPE Cantabria, Santander	88.4
MU06)	1215	5	COPE	MU	COPE, Lorca	89.2
EU13)	1224	10	COPE	EU	COPE, Donostia-San Sebastián	88.5
AN26)	1224	5	COPE	AN	COPE, Huelva	91.9
AN27)	1224	5	COPE	AN	COPE, Almería	97.1
BA08)	1224	5	COPE	BA	COPE, Palma de Mallorca	97.6
CA15)	1224	5	COPE	CA	COPE, Lleida	96.0
CM14)	1224	5	COPE	CM	COPE, Albacete	95.4
GA16)	1224	5	COPE	GA	COPE, Lugo	90.0
AN28)	1260	5	SER	AN	R. Algeciras, Algeciras	95.7
MU07)	1260	5	SER	MU	R. Murcia, Murcia	91.3
CA16)	1269	10	COPE	CA	COPE Comarques Gironines, Figueres	89.4
CL15)	1269	10	COPE	CM	COPE, Ciudad Real	93.6
CL22)	1269	5	COPE	CL	COPE, Zamora	94.9
EX07)	1269	5	COPE	EX	COPE, Badajoz	89.1
CA17)	1287	10	SER	CA	R. Lleida, Lleida	92.6
CL23)	1287	5	SER	CL	R. Castilla, Burgos	89.1
GA17)	1287	5	SER	GA	R. Lugo, Lugo	91.8
VA11)	1296	50	COPE	VA	COPE, València	99.0
AN29)	1341	10	OCR	AN	Onda Cero R., Almería	93.8
CL24)	1341	5	SER	CL	R. León, León	88.2
CM16)	1341	5	OCR	CM	Onda Cero R., Ciudad Real	92.1
EU07)	1386	50	EI	EU	Euskadi Irratia, Bilbo-Bilbao	88.9
EU08)	1476	50	EI	EU	Euskadi Irratia, Donosti-San S.	94.4
CL25)	1485	10	SER	CL	R. Zamora, Zamora	89.8
CT06)	1485	10	SER	CT	R. Santander, Santander	90.9
AN30)	1485	6	OCR	AN	Onda Cero R., Antequera	96.3
CA18)	1485	5	PR	CA	Onda Rambla, Punto R., Vilanova	96.3
VA12)	1485	5	SER	VA	R. Alcoi, Alcoi	96.3
VA13)	1521	5	SER	VA	R. Castelló, Castelló	94.8
VA14)	1539	6	SER	VA	R. Elche - R. Elx, Elx	94.8
VA15)	1539	5	SER	VA	R. Manresa, Manresa	91.7
NA06)	1575	10	SER	NA	R. Pamplona, Pamplona	92.2
AN31)	1575	5	SER	AN	R. Córdoba, Córdoba	96.6
GA18)	1584	5	SER	GA	R. Ourense, Ourense	87.6
VA15)	1584	5	SER	VA	R. Gandia, Gandia	96.5
EU09)	1602	25	EI	EU	R. Vitoria, Vitoria-Gasteiz	104.1
AN32)	1602	5	SER	AN	R. Linares, Linares	94.9
CL26)	1602	5	SER	CL	R. Segovia, Segovia	93.6
MU08)	1602	5	SER	MU	R. Cartagena, Cartagena	102.3
VA16)	1602	5	SER	VA	R. Ontinyent, Ontinyent	95.3

* = inactive

FM	MHz	kW	Net	Rg	Station, location
AN71)	87.5	5		AN	R. Pinomar, Alhaurín de la Torre
MA21)	87.5	15		MA	Evolution FM, Madrid
AN41)	87.6	5	CSR	AN	Canal Sur R., Santo Pitar
CA25)	87.6	20	CR	CA	Catalunya Informació, Soriguera
AN35)	87.7	6	OCR	AN	Kiss FM, Jerez de Frontera
CA34)	87.7	20	RAC	CA	RAC1, Barcelona
AN36)	87.9	5	SER	AN	R. Morón, Morón de la Frontera
CA25)	87.9	5	CR	CA	iCAT FM, Collsuspina
CA37)	88.0	25	CR	CA	iCAT FM, La Mussara
EU25)	88.0	5		EU	R. Nervión, Bilbao
AN37)	88.3	5	CSR	AN	Canal Fiesta R., Algeciras
CA25)	88.4	5	CR	CA	Catalunya R., Montcaro
CA25)	88.6	20	CR	CA	Catalunya Música, Soriguera
CA35)	88.7	10		CA	R. RM, Barcelona
AN70)	88.8	40		AN	Play FM, Sevilla
CA38)	88.9	25	CR	CA	iCAT FM, Rocacorba
EU07)	88.9	20	EI	EU	Euskadi Irratia, Bilbo-Bilbao
MU17)	88.9	5		MU	Solo R., Murcia
MA05)	89.0	20	M80	MA	M80 Madrid, Madrid
CA07)	89.1	8	OCR	CA	R. Salut, Barcelona
CA32)	89.2	5	RTT	AR	R. Tele Taxi, Zaragoza
EU25)	89.2	5		EU	Élite FM, Bilbo
EU26)	89.2	5		EU	Segura Irratia, Segura
GA29)	89.2	5	RMA	GA	R. Marca, A Coruña
VA21)	89.2	5	OCR	VA	Kiss FM, Alacant-Alicante
NA11)	89.3	6	OCR	NA	Kiss FM, Pamplona
GA28)	89.4	6		GA	Ondas de Vida, Vigo

FM	MHz	kW	Net	Rg	Station, location
MU14)	89.4	6	COPE	MU	COPE, Cartagena
AN42)	89.5	20	CSR	AN	Canal Fiesta R., Huelva
AN62)	89.5	14		AN	Antena Sevilla R., Sevilla
BA11)	89.5	8	KFM	BA	Kiss FM, Palma de Mallorca
EX11)	89.5	6	KFM	EX	Kiss FM, Cáceres
AR13)	89.7	40	PR	AR	Punto R., Zaragoza
AN43)	89.8	8	CSR	AN	R. Andalucía, Granada
CA07)	89.8	10	OCR	CA	Onda Rambla, Barcelona
AS03)	89.8	8	RPA	AS	RPA, Gijón
CA36)	89.9	10	PR	CA	Onda Rambla Punto R., Girona
VA08)	90.0	8	M80	VA	M80 R., Alicante
AN38)	90.1	8	OCR	AN	Kiss FM, Málaga
VA30)	90.1	5		VA	Mi R., Valencia
CA25)	90.2	10	CR	CA	iCAT FM, Sant Celoni
AN35)	90.3	6	OCR	AN	Onda Cero R., Jerez de la Frontera
AN55)	90.3	10	CSR	AN	Canal Fiesta R., Córdoba
EU27)	90.3	5	LFM	EU	Loca FM, Bilbao
EX12)	90.4	5	OCR	EX	Onda Cero Melodía, Mérida
AN57)	90.5	70	CSR	AN	R. Andalucía, Almería
CA08)	90.5	8	M80	CA	M80 Barcelona, Barcelona
CL31)	90.5	5	SER	CL	R. Miranda, Miranda de Ebro
AN39)	90.8	5	SER	AN	R. Puerto, El Puerto de S.M.
AN38)	90.8	10	OCR	AN	Onda Cero R., Málaga
AN59)	90.8	64	CSR	AN	R. Andalucía, Sevilla
GA29)	90.8	8	RG	GA	R. Galega Música, Vigo
EU09)	90.9	20	EI	EU	R. Euskadi, Vitoria
MA11)	91.0	100	EFM	MA	Europa FM, Madrid
AS06)	91.1	6	D	AS	Dial Asturias, Oviedo
RI02)	91.1	6	COPE	RI	COPE Rioja, Logroño
EU07)	91.2	20	EI	EU	Euskadi Gaztea, Bilbo-Bilbao
CM08)	91.3	5	RCM	CM	R. Castilla-La Mancha, Guadalajara
AR04	91.4	40	KFM	AR	Kiss FM, Zaragoza
EX13)	91.4	10	SER	EX	Cadena 40, Plasencia
AN40)	91.4	8	EFM	AN	Europa FM, Córdoba
CL27)	91.5	8	KFM	CL	Kiss FM, Astorga
EU21)	91.5	8	OCR	EU	Kiss FM, Donosti-San Sebastián
EX05)	91.6	8	COPE	EX	COPE, Cáceres
VA17)	91.6	10	EFM	VA	Europa FM, València
EU07)	91.7	20	EI	EU	R. Euskadi, Bilbo-Bilbao
MA05)	91.7	100	D	MA	Dial Madrid, Madrid
AN41)	91.7	5	CSR	AN	Canal Sur R., Málaga
CL40)	91.7	5	KFM	CL	Kiss FM, Salamanca
CL41)	91.7	5	ES	CL	esRadio, Valladolid
VA08)	91.7	8	SER	VA	R. Alicante, Alicante
CA25)	91.9	15	CR	CA	Catalunya Música, Alpicat
CA34)	91.9	10	RAC	CA	RAC105, Girona
CM08)	91.9	5	RCM	CM	R. Castilla-La Mancha, Toledo
CT11)	91.9	6	OCR	CT	Onda Cero R., Santander
AR05)	92.0	40	SER	AR	Máxima FM, Zaragoza
CA25)	92.0	100	CR	CA	Catalunya Informació, Collserola
VA29)	92.0	10	PR	VA	LP Punto R., València
CA25)	92.2	10	CR	CA	Catalunya Música, Falsel
MA05)	92.4	14	SER	MA	Radiolé, Madrid
CA25)	92.5	100	CR	CA	iCAT FM, Colserola
CM09)	92.5	5	PR	CM	Punto R., Cuenca
EX08)	92.6	6	SER	EX	Cadena 40, Cáceres
CA32)	92.6	5	RTT	CA	R. Tele Taxi, València
GA21)	92.6	8	RV	GA	R. Voz, A Coruña
CM21)	92.7	6	KFM	CM	Kiss FM, Albacete
AN22)	92.8	8	D	AN	Cadena Dial, Granada
VA18)	92.8	6	EFM	VA	Europa FM, Elx-Elche
CL34)	92.9	6	OCR	CL	Punto R., Burgos
AN43)	93.1	5	CSR	AN	Canal Fiesta R., Baza
AN72)	93.1	8	D	AN	Cadena Dial, Málaga
EU04)	93.1	8		EU	T. Gorbea, Vitoria-Gasteiz
GA22)	93.1	5	OCR	GA	Onda Cero R., Pontevedra
VA31)	93.1	5		VA	R. Inter FM, València
AN73)	93.2	5	PR	AN	Runto R., Sevilla
VA08)	93.2	8	D	VA	Cadena Dial, Alicante
CA25)	93.3	5	CR	BA	Catalunya R., Alfabia
GA13)	93.4	8	SER	GA	R. Coruña 2, A Coruña
CA30)	93.4	5		CA	R. Segre, LLeida
AR05)	93.5	8	SER	AR	R. Zaragoza 2, Zaragoza
CA25)	93.5	20	OCR	CA	Onda Cero R., Barcelona
AN47)	93.8	5	EFM	AN	Europa FM, Almería
AN72)	93.8	5	RO	AN	Radiolé, Málaga
AN42)	94.0	20	CSR	AN	Canal Sur R., Huelva
GA23)	94.0	6	OCR	GA	Kiss FM, Vigo
GA07)	94.1	20	RG	GA	R. Galega, Santiago de Compostela
AN12)	94.2	5	PR	AN	Punto R., Jaén
CT03)	94.2	6	RMA	CT	R. Marca, Santander
AN55)	94.3	10	CSR	AN	Canal Sur R., Córdoba
CL23)	94.3	6	D	CL	Cadena Dial, Burgos
CA11)	94.4	10	SER	CA	R. Girona, Rocacorba
CL39)	94.4	5	SER	CL	Europa FM, Valladolid
EU08)	94.4	20	EI	EU	Euskadi Irratia,Donosti-S Sebastián
EX08)	94.4	6	SER	EX	SER, Cáceres
CA25)	94.5	5	CR	CA	Catalunya Informació, Collsuspina
AN43)	94.6	5	CSR	AN	Canal Fiesta R., Loja
EU07)	94.7	20	EI	EU	Euskadi Gaztea, Bilbo-Bilbao
AN15)	94.8	40	M80	AN	M80 Sevilla, Sevilla
AN41)	94.9	60	CSR	AN	R. Andalucía Información, Málaga
CA07)	94.9	20	EFM	CA	Europa FM, Barcelona
CM10)	95.0	6		CM	R. Surco, Albacete
EU09)	95.0	20	EI	EU	Euskadi Irratia, Vitoria-Gasteiz
AN43)	95.1	5	CSR	AN	Canal Sur R., Granada
CA11)	95.1	10	M80	CA	M80, Girona
MA17)	95.1	100		MA	R. Intereconomia, Madrid
AS08)	95.2	6	OCR	AS	Onda Cero R., Oviedo
CA31)	95.3	5	OCR	CA	Onda Cero R., Reus
AS03)	95.4	10	RPA	AS	RPA, Boal
AN44)	95.4	8	OCR	AN	Kiss FM, Cádiz
AN52)	95.6	8	OCR	AN	Kiss FM, Córdoba
BA02)	95.8	5		BA	Insel R., Mallorca
CA28)	95.8	20	KFM	CA	Kiss FM, Barcelona
AN45)	95.9	40	OCR	AN	Onda Cero R., Sevilla
CA32)	95.9	5	RTT	VA	R. Tele Taxi, Alicante
CA08)	96.0	20	RO	CA	Radiolé, Barcelona
MA)	96.0	20	LFM	MA	Loca FM, Madrid
RI03)	96.0	6	KFM	RI	Kiss FM, Logroño
EU04)	96.1	20	EI	EU	Euskadi Gaztea, Zaldiaran
VA10)	96.1	10	M80	VA	M80 Valencia, València
AN46)	96.2	20	CSR	AN	Canal Sur R., Cádiz
CL32)	96.2	6	OCR	CL	Europa FM, Salamanca
AN60)	96.2	6	D	AN	Dial Almería, Almería
VA22)	96.2	10	R9	VA	R. Nou, Elda
AN69)	96.3	5	EFM	AN	Europa FM, Antequera
MU03)	96.3	5		MU	AMC R. 5, Cartagena
CM24)	96.4	6	M80	CM	M80 Albacete, Albacete
EU08)	96.5	20	EI	EU	R. Euskadi, Donosti-San Sebastián
CA25)	96.5	10	CR	CA	iCAT FM, Montserrat
AS09)	96.5	8	SER	AS	R. Gijón, Gijón
VA22)	96.5	5	R9	VA	R. Nou, Alacant-Alicante
CL35)	96.5	5	OCR	CL	Kiss FM, León
AN43)	96.6	5	CSR	AN	Canal Sur R., Loja
GA08)	96.6	6	RG	GA	R. Galega Música, Friol
CA25)	96.7	25	CR	CA	Catalunya Música, Rocacorba
MU11)	96.7	6	KFM	MU	Kiss FM, Cartegena
AN22)	96.8	5	RO	AN	Radiolé, Granada
CM08)	96.8	10	RCM	CM	R. Castilla-La Mancha, Herencia
VA23)	96.9	8	KFM	VA	Kiss FM, València
CA08)	96.9	5	SER	CA	R. Barcelona 2, Barcelona
MA16)	96.9	8	RM	MA	R. María España, Madrid
CA25)	97.0	5	CR	BA	Catalunya Informació, Alfábia
EX08)	97.0	6	D	EX	Cadena Dial, Cáceres
AN43)	97.1	30	CSR	AN	Canal Fiesta R., Granada
AR05)	97.1	40	D	AR	Dial Zaragoza, Zaragoza
CL23)	97.1	6	SER	CL	R. Castilla, Burgos
MU04)	97.1	5		MU	OM R., Cartagena
MA07)	97.2	100		MA	Top Madrid, Madrid
AN42)	97.3	30	CSR	AN	R. Andalucía Información, Huelva
CA25)	97.3	10	CR	CA	Catalunya R., Montserrat
EU05)	97.3	8	COPE	EU	Cadena 100, Vitoria-Gasteiz
EU06)	97.3	5		EU	Fórmula Hit, Portugalete
VA05)	97.3	8		VA	Onda Melodía, Alicante
CA16)	97.4	10	COPE	CA	Cadena 100, Figueres
AR06)	97.5	40	COPE	AR	COPE, Zaragoza
MA28)	97.5	10		MA	Energy FM, Madrid
GA13)	97.6	8	M80	GA	M80, A Coruña
MA18)	97.6	8	OCR	MA	Onda Cero R., Alcalá de Henares
CL32)	97.6	6	OCR	CL	Onda Cero R., Salamanca
CA32)	97.7	20	RTT	CA	R. Tele Taxi, Barcelona
VA11)	97.7	20		VA	97 Punto 7, València
AN48)	97.9	25	CSR	AN	Canal Fiesta R., Jaén
AR06)	97.9	40	COPE	AR	Cadena 100, Zaragoza
AN13)	98.0	30	COPE	AN	COPE, Alanís
BA 03)	98.0	50		BA	Ona Mallorca, Palma de Mallorca
EU14)	98.0	8	COPE	EU	Cadena 100, Donosti-San Sebastián
EU15)	98.0	8		EU	R. Álava, Vitoria-Gasteiz
MA13)	98.0	100	OCR	MA	Onda Cero R., Madrid
AN49)	98.1	5	SER	AN	R. Huelva, Huelva
CL18)	98.1	8	M80	CL	M80, Valladolid
EX03)	98.1	5	CER	EX	Canal Extremadura R., Zafra
VA07)	98.1	10	COPE	VA	R. Rock & Gol, Alicante
RI04)	98.2	6	PR	RI	Punto R., Logroño
CA25)	98.3	10	CR	CA	Catalunya Informació, Montserrat
CM23)	98.3	5	KFM	CM	Kiss FM, Toledo
NA02)	98.3	5		NA	98.3 R., Pamplona
CA25)	98.4	5	CR	BA	iCAT FM, Alfábia
VA10)	98.4	20	D	VA	Dial Mediterraneo, València

FM	MHz	kW	Net	Rg	Station, location
CA25	98.5	10	CR	CA	Catalunya Informació, Montcaro
CA40	98.5	10		CA	Ona FM, Girona
CT11	98.5	6	OCR	CT	Kiss FM, Santander
CM22	98.5	5	OCR	CM	Onda Cero R., Talavera de la Reina
AR05	98.6	40	M80	AR	M80 Zaragoza, Zaragoza
CL36	98.6	6		CL	R. Arlanzón, Burgos
CM08	98.6	5	RCM	CM	R. Castilla-La Mancha, Valdepeñas
EU11	98.8	5	M80	EU	M80, Bilbao
MU12	98.8	6		MU	Solo R. /Cartagena R., Cartagena
BA16	98.8	5		BA	Ultima Hora Punto R., Palma de Mallorca
CA11	98.9	10	D	CA	Cadena Dial, Rocacorba
VA26	98.9	5	COPE	VA	COPE, Benidorm
CL42	99.1	5	KFM	CL	Kiss FM, Zamora
EU17	99.1	5		EU	Fórmula Hit, Vitoria-Gasteiz
MA03	99.1	10	ES	MA	esRadio, Madrid
CL17	99.3	6	D	CL	Cadena Dial, Salamanca
MU13	99.3	8	OCR	MU	Kiss FM, Murcia
AN74	99.4	30	CSR	AN	R. Andalucía Información, Cádiz
AN55	99.4	60	CSR	AN	R. Andalucía Información, Córdoba
AR12	99.4	40	OCR	AR	Onda Cero R., Zaragoza
CA08	99.4	10	D	CA	Dial Barcelona, Barcelona
CL33	99.4	8	OCR	CL	Kiss FM, Valladolid
AN51	99.5	8	EFM	AN	Europa FM, Granada
CA05	99.6	10	GR	CA	Flaix FM, Girona
CA25	99.7	5	CR	CA	Catalunya R., Collsuspina
EU16	99.8	20	EI	EU	Herri-Iratia, Loyola
MA04	99.8	10		MA	R. Sol XXI, Madrid
RI05	99.8	6	M80	RI	M80, Logroño
AN19	99.9	8	D	AN	Dial Bahía, Cádiz
BA13	99.9	8		BA	R. Balear Ciutat, Palma de Mallorca
AS03	100.0	8	RPA	AS	RPA, Los Oscos
CA09	100.0	20	COPE	CA	Cadena 100, Barcelona
CA34	100.1	5	RAC	CA	RAC1, Rocacorba
EU07	100.1	20	EI	EU	EITB Irratia, Bilbo-Bilbao
CA25	100.2	35	CR	CA	Catalunya R., La Mussara
GA08	100.2	5	RG	GA	R. Galega Música, Xistral
AN45	100.3	40	OCR	AN	Kiss FM, Sevilla
CM12	100.3	5	SER	CM	SER Albacete
MA19	100.4	10		MA	Onda Joven Getafe, Getafe
AN54	100.4	6	SER	AN	R. Málaga, Málaga
CL18	100.4	5	D	CL	Cadena Dial, Valladolid
EU22	100.4	5	SER	EU	Cadena 40, Vitoria-Gasteiz
MA09	100.4	5		MA	R. Círculo, Madrid
VA10	100.4	20	SER	VA	R. Valencia 2, Valencia
AR12	100.5	10	EFM	AR	Europa FM, Zaragoza
AS03	100.5	20	RPA	AS	RPA, Gijón
CA25	100.5	5	CR	CA	Catalunya Música, Collsuspina
CM08	100.5	5	RCM	CM	R. Castilla-La Mancha, Guadalajara
CM23	100.5	5		CM	R. Santa María, Toledo
GA12	100.6	8	SER	GA	R. Vigo 2, Vigo
AN48	100.6	30	CSR	AN	Canal Sur R., Jaén
BA16	100.6	5	GR	BA	Flaix FM, Mallorca
GA08	100.6	5	RG	GA	R. Galega Música, Monte Páramo
CA25	100.7	80	CR	CA	Catalunya R., Alpicat
CA05	100.7	10	GR	CA	Flaixbac FM, Rocacorba
MA08	100.7	20	COPE	MA	COPE, Madrid
AN55	100.8	5	CSR	AN	Canal Sur R., Córdoba
EX03	100.8	6	CER	EX	Canal Extremadura R., Badajoz
VA19	100.8	6	COPE	VA	Cadena 100, Elx-Elche
GA25	100.9	30	RG	GA	R. Galega, Xesteiras
BA11	101.0	5	OCR	BA	Onda Cero, Mallorca
CA34	101.0	5	RAC	CA	RAC1, Mont Caro
EU05	101.0	8	COPE	EU	COPE, Vitoria
AN54	101.1	10	M80	AN	M80 Málaga, Málaga
CT16	101.1	6	M80	CT	M80, Santander
VA23	101.2	20	OCR	VA	Onda Cero R., València
AN56	101.2	5	OCR	AN	Onda Cero R., Huelva
GA12	101.2	5	M80	GA	M80, Vigo
VA22	101.2	6	R9	VA	R. Nou, Benidorm
MA14	101.3	100		MA	Onda Madrid, Madrid
AN55	101.3	60	CSR	AN	Canal Fiesta R., Córdoba
AN43	101.3	5	CSR	AN	R. Andalucía Información, Loja
AS03	101.4	10	RPA	AS	RPA, Avilés
RI02	101.4	6	COPE	RI	Cadena 100, Logroño
AN15	101.5	40	SER	AN	Radiolé, Sevilla
CA25	101.5	10	CR	CA	Catalunya Música, Collserola
EU18	101.5	24	OCR	EU	Onda Cero R., Bilbo-Bilbao
AN53	101.6	5	EFM	AN	Europa FM, Mijas
CM08	101.6	5	RCM	CM	R. Castilla-La Mancha, Cuenca
CA38	101.7	25	CR	CA	Catalunya Informació, Rocacorba
AN67	101.8	6	SER	AN	Cadena 40, Almería
AS04	101.8	5		AS	R. Langreo, Sama de Langreo
CA06	101.8	5	GR	CA	Flaix FM, Tarragona
AN41	101.9	5	CSR	AN	Canal Fiesta R., Archidona
AN59	101.9	60	CSR	AN	Canal Fiesta R., Sevilla
EU09	101.9	8	EI	EU	Euskadi Irratia, Amurrio
GA30	101.9	6	RMA	GA	R. Marca, Vigo
AN14	102.0	8	COPE	AN	COPE, Cádiz
CA09	102.0	20	COPE	CA	COPE, Barcelona
CA25	102.2	25	CR	CA	Catalunya R., Rocacorba
AN42	102.2	30	CSR	AN	Canal Fiesta R., Huelva
VA24	102.2	100	R9	VA	R. Nou, València
GA25	102.3	40	RG	GA	R. Galega, Domaio
BA06	102.3	5	M80	BA	M80 Mallorca, Palma de Mallorca
AN68	102.4	40	D	AN	Dial Sevilla, Sevilla
CA25	102.4	10	CR	CA	Catalunya Música, Montserrat
CT06	102.4	6	SER	CT	SER, Liérganes
EU28	102.4	6	OCR	EU	Onda Cero R., Vitoria
AN57	102.5	20	CSR	AN	Canal Fiesta R., Almería
CA25	102.5	10	CR	CA	Catalunya Música, Montcaro
CM08	102.5	5	RCM	CM	R. Castilla-La Mancha, Ciudad Real
EU29	102.5	8	OCR	EU	Onda Cero R., Donosti-San Sebastián
AN22	102.5	8	D	AN	Dial Granada, Granada
EX03	102.6	60	CER	EX	Canal Extremadura R., Montánchez
EU23	102.6	15	COPE	EU	Bizkaia Irratia, Bilbo-Bilbao
MA13	102.7	100	CR	MA	Kiss FM, Madrid
GA26	102.7	8	OCR	GA	OndaCeroR.,Acoruña
AN72	102.8	8	SER	AN	Cadena 40, Mijas
CA25	102.8	100	CR	CA	Catalunya R., Collserola
CL43	102.8	8	PR	CL	Punto R., Valladolid
BA15	102.8	5	SER	BA	R. Ibiza, Ibiza
VA32	102.8	90	R9	VA	Sí R., Benicassim
AN71	103.1	5		AN	R. Pinomar, Málaga
VA22	103.0	120	R9	VA	R. Nou, Aitana
EU24	103.0	8		EU	R. Egüín, Hernani
AN15	103.2	28	SER	AN	R. Sevilla 2, Sevilla
AR05	103.2	10	RO	AR	Radiolé, Zaragoza
EU07	103.2	20	EI	EU	R. Euskadi, Bilbo-Bilbao
MA22	103.2	5		MA	R. Gladys Palmera, Madrid
VA33	103.2	10	EFM	VA	Europa FM, València
CA29	103.4	10	RE	CA	R. Estel, Rocacorba
CL37	103.4	6	OCR	CL	Punto R., Salamanca
CM08	103.4	5	RCM	CM	R. Castilla-La Mancha, Puertollano
EU09	103.4	20	EI	EU	EITB Musika, Vitoria-Gasteiz
AN43	103.5	5	CSR	AN	Canal Sur R., Sierra de Lújar
CA08	103.5	20	SER	CA	Ona Catalana, Barcelona
EU08	103.5	20	EI	EU	Euskadi Gaztea,Donosti-S. Sebastián
VA24	103.5	10	R9	VA	R. Nou, Gandía
CA28	103.5	5		CA	Ona Música, Barcelona
AN55	103.6	30	CSR	AN	Canal Sur R., Córdoba
EU30	103.7	24	PR	EU	Punto R., Bilbo-Bilbao
GA08	103.7	70	RG	GA	R.Galega, Monte Páramo
VA24	103.7	90	R9	VA	R. Nou, Castelló
GA30	103.8	6	RV	GA	R. Voz, Vigo
AN59	103.9	30	CSR	AN	Canal Fiesta R., Sevilla
GA25	103.9	40	RG	GA	R. Galega, Bailadora
MA23	103.9	30		MA	Zeta R., Madrid
MU07	103.9	8	D	MU	Dial Murcia, Murcia
CM08	104.0	10	RCM	CM	R. Castilla-La Mancha, Chinchilla
EX03	104.0	5	CER	EX	Canal Extremadura R., Cáceres
AN24	104.1	6	COPE	AN	R. Rock & Gol, Córdoba
AN60	104.1	6	SER	AN	Kiss FM, Almería
AN43	104.2	6	CSR	AN	Canal Sur R., Granada
AS11	104.2	5		AS	R. Amistad, Gijón
CA08	104.2	20	SER	CA	Máxima FM, Barcelona
GA31	104.2	5	RG	GA	R. Galega Música, Santiago de Compostela
MA05	104.3	37	MFM	MA	Máxima FM, Madrid
EU07	104.4	20	EI	EU	Euskadi Irratia, Bilbo-Bilbao
AN42	104.5	30	CSR	AN	Canal Sur R., Huelva
CA25	104.5	25	CR	CA	Catalunya Informació, La Mussara
AN41	104.6	30	CSR	AN	Canal Sur R., Málaga
AN59	104.6	60	CSR	AN	R. Andalucía Información, Sevilla
MU16	104.6	6	ORM	MU	Onda Regional Murcia, Cartagena
VA20	104.6	10		VA	Pulsa, Elx-Elche
GA12	104.7	6	MFM	GA	Máxima FM, Vigo
AN50	104.8	30	CSR	AN	Canal Sur R., Jerez de la Frontera
AR14	104.8	10	RKM	AR	RKM, Zaragoza
EU31	104.8	5	RKM	EU	RKM, Vitoria-Gasteiz
GA25	104.8	40	RG	GA	R. Galega, Monte Meda
AN57	104.8	20	CSR	AN	Canal Sur R., Almería
AN43	104.9	30	CSR	AN	Canal Sur R., Granada
EU09	104.9	8	EI	EU	Euskadi Gaztea, Amurrio
BA04	105.0	5	GR	BA	Flaix FM, Ibiza-Eivissa
CA34	105.0	20	RAC	CA	RAC105, Barcelona
AN59	105.1	30	CSR	AN	Canal Sur R., Sevilla
CA32	105.1	10	RTT	CA	R. Tele Taxi, Rocacorba
MA24	105.1	5		MA	Energy R., Madrid
CL39	105.2	8	OCR	CL	Onda Cero R., Valladolid
MU16	105.3	60	ORM	MU	Onda Regional, Murcia
AS03	105.4	5	RPA	AS	RPA, Oviedo

FM	MHz	kW	Net	Rg	Station, location
CA25	105.4	20	CR	CA	Catalunya Música, La Mussara
MA05	105.4	100	SER	MA	R. Madrid 2, Madrid
AN75	105.5	10		AN	Spectrum FM, Mijas
AN55	105.5	6	CSR	AN	Canal Sur 1, Cabra
CL38	105.5	6	EFM	CL	Kiss FM, Burgos
VA22	105.5	10	R9	VA	Sí R., Sierra del Cid
EU32	105.6	8	PR	EU	Punto R., Vitoria-Gasteiz
CA41	105.7	20	GR	CA	Flaix FM, Barcelona
MA25	105.7	5		MA	Rockservatorio FM, Madrid
AR12	105.8	40	OCR	AR	Kiss FM, Zaragoza
AN41	105.8	30	CSR	AN	Canal Fiesta R., Málaga
BA17	105.8	5	RKM	BA	RKM, Alfábia
MA14	106.0	30		MA	Onda Madrid R., Madrid
AN55	106.1	6	CSR	AN	R. Andalucía Información, Cabra
BA12	106.1	8	D	BA	Cadena Dial, Palma de Mallorca
CA41	106.1	20	GR	CA	Faixbac FM, Barcelona
GA27	106.1	8	RV	GA	R. Voz, Santiago de Compostela
EU33	106.2	8	PR	EU	Punto R., Donosti-San Sebastián
MA26	106.3	10	PR	MA	Punto R., Madrid
AS03	106.4	60	RPA	AS	RPA, Gamonitéiro
CA42	106.4	10	LFM	CA	Loca FM, Barcelona
VA28	106.5	5	OCR	VA	Onda Cero Alicante, Alicante
CA29	106.6	20	RE	CA	R. Estel, Barcelona
MA27	106.6	5		MA	R. Corazón Tropical, Madrid
EU34	106.7	8	AFM	EU	Fiss FM, Vitoria-Gasteiz
BA03	106.8	100	IB3	BA	IB3, Alfábia
CA04	106.8	10	EFM	CA	Europa FM, Rocacorba
AN76	106.9	50	RMA	AN	R. Marca, Sevilla
CA43	106.9	5		CA	R. Kanal Musical, Barcelona
MU18	106.9	8	PR	MU	Punto R., Murcia
AN65	107.1	5		AN	R. Guadalete/Fórmula Hit, Bornos
MA20	107.2	15		MA	Fiesta FM, Madrid
CA44	107.4	20		CA	Grama R., Barcelona

NB: stns less than 5kW omitted
Addresses and other information:

AN00) ANDALUCIA
AN01) Edif.RTVE, Parque del Alamillo, 41092 Sevilla. – **AN02)** Góngora 3, 14002 Córdoba. – **AN03)** Av.Ricardo Soriano 11, 29600 Marbella. – **AN04)** Av.de Granada 57, P1, 23001 Jaén. – **AN05)** Hermanos Machado 3, 04004 Almería. – **AN06)** Av.de la Aurora 40, 29006 Málaga. – **AN07)** Av.de Andalucía 67, 11007 Cádiz. – **AN08)** Plaza Carretas 5, 18009 Granada. – **AN09)** Real 24, 11300 La Línea de la Concepción. – **AN10)** La Fuente 4, 21004 Huelva. – **AN11)** Cervantes 11, 14940 Cabra. – **AN12)** C/ Bartolomé 32 bajo, 23001 Jaén. – **AN13)** C/ Triana 8, 41380 Alanis. – **AN14)** C/ Algeciras 1, 2º modulo 8, 11011 Cádiz. – **AN15)** Rafael González Abreu 6, 41001 Sevilla E: radiosevilla@cadenaser.com – **AN16)** Rioja 4, 41001 Sevilla. E: sevilla@cadenacope.net – **AN17)** Linaje 2, 29001 Málaga E: malaga@cadenacope.net – **AN18)** Gran Vía de Colón 28, 18001 Granada E: granada@cadenacope.net – **AN19)** Paseo Marítimo 1, Edif.Reina Victoria, 11010 Cádiz E: sercadiz@cadenaser.com – **AN20)** Obispo Aguilar 1, 23001 Jaén. W: www.radiojaen.es – **AN21)** Guadalete 12, 11403 Jerez de la Frontera. W: www.radiojerez.com – **AN22)** Santa Paula 2 (or: Ap.158), 18001 Granada W: www.radiogranada.es E: radiogranada@radiogranada.es – **AN23)** San Agustín 11 (or: Ap.364), 11403 Jerez de la Frontera E: jerez@cadenacope.net – **AN24)** Federico Mendizábal 10, 23001 Jaén E: jaen@cadenacope.net – **AN25)** Plaza Cardenal Toledo 4, 14001 Córdoba E: cordoba@cadenacope.net – **AN26)** José María Amoz 2, 21001 Huelva E: huelva@cadenacope.net – **AN27)** Padre Luque 11, 04001 Almería E: almeria@cadenacope.net – **AN28)** General Castaños 2, 11201 Algeciras E: radioalgeciras@unionradio.es – **AN29)** Av.Federico García Lorca 105, 04005 Almería. – **AN30)** San Agustín 4, 29200 Antequera. – **AN31)** García Lovera 3, 14002 Córdoba. W: www.radiocordoba.com –**AN32)** Plaza Ramón y Cajal 8, 23700 Linares E: radiolinares@unionradio.es – **AN35)** Gaitan 10, 11402 Jerez de la Frontera. – **AN36)** Paseo Nuevo 60-2, 41530 Morón de la Frontera. – **AN37)** Plaza de España 15, 2º y 3º, 11006 Cádiz. – **AN38)** Peregrinos 3, 29002 Málaga. – **AN39)** Misericordia 10, 11500 Puerto de Santa María. – **AN40)** Manuel Ruiz Maya 8, 14004 Córdoba. – **AN41)** Carr.de Cádiz 307, Av.Velázquez, 29004 Málaga. – **AN42)** Plaza de San Pedro 3 y 4, 21001 Huelva. – **AN43)** Urb. Bola de Oro, C/ Laguna de Aguas Verdes 11, 18008 Granada. – **AN44)** C/ Dr. Manuel Ruíz Maya 8, 5º, 11004 Cádiz. – **AN45)** Pabellón Once, Isla de Cartuja, 41092 Sevilla. – **AN46)** Plaza de España 15, 11006 Cádiz. – **AN47)** Avenida Federico García Lorca 105, 04005 Almería. – **AN48)** Av.del Ejército Español 6, 23007 Jaén. – **AN49)** Mendez Nuñez 15-5-6, 21001 Huelva. – **AN50)** Corredera 53, 11402 Jerez de la Frontera. – **AN51)** Recogidas 37, 18005 Granada. – **AN52)** Barroso 4-2, 14003 Córdoba. – **AN53)** C/ Ramón Gómez de la Serna 22, 29600 Marbella. – **AN54)** Palestina 1, 29007 Málaga. – **AN55)** Glorieta de Guadalhorce s/n, "Antigua Estación de RENFE", 14008 Córdoba. – **AN56)** Arquitecto Pérez Carasa 14-16, 21001 Huelva. – **AN57)** Centro Residencial Oliveros, C/ Maestro

Serrano 9, 2º-B, 04004 Almería. –**AN59)** Edificio Canal Sur, Av. José Gálvez 1, 41092 Isla de la Cartuja (Sevilla). – **AN60)** Av.Cabo de Gata 2, 04007 Almería. – **AN61)** Placentines 2, 41004 Sevilla. – **AN62)** Av.de la Borbolla 47, 41013 Sevilla. – **AN65)** San Jerónimo 7, 11640 Bornos. –**AN67)** Av.Mediterráneo s/n, De Laura 2, 04001 Almería. – **AN68)** Rafael Gonzáles Abreu 6, 41001 Sevilla. – **AN69)** C/ San Agustín 4, 29200 Antequera. –**AN70)** Sevilla W: www.playfm.es. – **AN71)** CL. Coín Parcela 1273, 29130 Alhaurín de la Torre (Málaga). – **AN72)** C/ Dr. Manuel Domínguez "Ed. Bulevar 2", 29001 Málaga. – **AN73)** Edificio de Oficinas del Estadio Olímpico, Isla de la Cartuja, 41092 Sevilla. – **AN74)** C/ Capinteros de Ribera 2, 11002 Cádiz. – **AN75)** Av. Valle Inclán 12, of. 3, La Campana, Nueva Andalucía, 29660 Marbella. – **AN76)** Av. República Argentina 25-9º-B, 41011 Sevilla.

AR00) ARAGON
AR01) José Luís Albareda 1-3, 50004 Zaragoza. – **AR02)** Nueva 1, 44001 Teruel. – **AR03)** José Gil Caves 1, 22005 Huesca. **AR04)** Zaragoza. – **AR05)** Paseo de la Constitución 21, 50001 Zaragoza E: radiozaragoza@unionradio.es – **AR06)** Paseo de Sagasta 50 (or: Ap.42), 50006 Zaragoza. E: programas.zaragoza@cadenacope.net – **AR07)** Calle Alcalde Cardedera 1, 22080 Huesca. W: www.radiohuesca.com – **AR11)** Coso 46, 50004 Zaragoza. – **AR12)** Paseo Echegaray y Caballero 76, 50003 Zaragoza. – **AR13)** Vocento, C/ Ramón y Cajal, 50015 Zaragoza. – **AR14)** REMAR, Av. Cataluña, 50003 Zaragoza.

AS00) ASTURIAS
AS01) C/ San Esteban de las Cruces 92, 33195 Oviedo. – **AS02)** Plaza del Instituto 3, 33201 Gijón. – **AS03)** Camino de las Clarisas 263, 33203 Gijón. – **AS04)** C/ Dorado 5, 2º-A, 33900 Sama de Langreo. – **AS05)** Carr.de la Costa 87 (or: Ap.235), 33205 Gijón. – **AS06)** Asturias 19, Bajo, 33004 Oviedo. W: www. radioasturias.com – **AS07)** Prado Picón 16, 33008 Oviedo. W: www.copeasturias.com E: c-oviedo@arra-kis.es – **AS08)** C/ Cervantes 27, 5º,,33003 Oviedo. – **AS09)** Jovellanos 1, 33202 Gijón. – **AS11)** Gijón.

BA00) BALEARES
BA01) Aragó 26, 07006 Palma de Mallorca. – **BA02) Paseo Marítimo 26, 07014 Palma de Mallorca.** – BA03) **C/ Font i Monteros 21, 07003 Palma de Mallorca. BA04)** Carrer Antoni Jaume 8 bajo, 07800 Ibiza-Eivissa **BA05)** Felip II Nº 28, 07800 Eivissa E: ibiza@cadenacope.net **BA06)** Rector Bertomeu Martorell 35, Son Xigala, 07013 Palma de Mallorca E: informativos.mallorca@cadenaser.com – **BA07)** Av.Negrete 3, 07760 Ciutadela W: www.telyse.net/cope-menorca E: menorca@cadenaser.com – **BA08)** Av.Jaume III Nº 18, 07012 Palma de Mallorca E: mallorca@cadenacope.net – **BA09)** C/ Manuel Azaña 7-A, 07006 Palma de Mallorca. – **BA11)** Forners 7, Edif. Once, 07002 Palma de Mallorca. – **BA12)** Passeig Mallorca 32, 07012 Palma de Mallorca. – **BA13)** Menacor 171, 07007 Palma de Mallorca. – **BA15)** Avenida Sant Jordi s/n, 07800 Figueretes (Ibiza). – **BA16)** C/ Gremi Selleters i Basters 14, "Polígon Son Castelló", 07009 Palma de Mallorca. – **BA17)** Mallorca.

CA00) CATALUNYA
CA01) C. Roc Boronat 127, 08018 Barcelona. – **CA02)** Carrer Lluis Companys 1, 25003 Lleida. – **CA03)** Rambla Nova 23, 43003 Tarragona. – **CA04)** Gran Vía Jaume I Nº 60, 17001 Girona. E-mail: emisora.girne@rtve.es – **CA05)** Av. Jaume I 76, 17002 Girona. – **CA06)** Plaça del Pati 2, entlo, 43800 Valls – **CA07)** La Rambla 88-94, 08002 Barcelona E: ondacero@ ondacero.es – **CA08)** Casp 6, 08010 Barcelona. E: radio-barcelona@unionradio.es – **CA09)** Diputació 238, 08013 Barcelona W: www.fm/copebarcelona E: barcelona@cadenacope.net – **CA10)** Travessera de les Corts 131-159, Recinte Martenitat, Pavello Cambo, 08028 Barcelona W: www.comradio.com – **CA11)** Placa Josep Pla 2, 17001 Girona E: radiogirona@unionradio.es – **CA12)** Tomàs Bergadà 3, 43204 Reus E: radioreus@unionradio.es – **CA13)** Carrer de Aragón 390-394, 2a planta, 08013 Barcelona – **CA14)** Lluvera 54-56, 43204 Reus E: reus@cadenacope.net – **CA15)** Acadèmia 14, 25002 Lleida. W: www.copelleida.com E: lerida@cadenacope.net – **CA16)** Sèquia 3, 17001 Girona E: girona@cadenacope.net – **CA17)** Vila Antònia 5, 25007 Lleida E: lleida@cadenaser.com – **CA18)** Relay Onda Rambla Barcelona (Local adress: Rambla Nova 69, 43003 Tarragona) – **CA19)** Calle Nou 47, 08240 Manresa E: informatius@els40.com – **CA25)** Av Diagonal 614-616, 08021 Barcelona. E: info@catradio.cat – W: www. catradio.cat – **CA27)** Bulidor s/n, Polígnon Industrial 1, 08960 St Just D (Barcelona). – **CA28)** Aragó 390-394, P2, 08013 Barcelona. E: onamusica@onacataluna.com – **CA29)** C/ Comtes de Bell-lloc 67-69, 08014 Barcelona E: radioestel@radioestel.com – **CA30)** Del Riu 6, 25007 Lleida. – **CA31)** Av.Sant Jordi 25, L3 baixos, 43201 Reus. – **CA32)** C/ Sant Carles 40, 08922 Sta Coloma de Gramenet (Barcelona). – **CA34)** Av. Diagonal 477, 15º, 08006 Barcelona. – **CA35)** Camí Ral 551, 1º, 08302 Mataró. – **CA36)** Rambla de la Llibertat 6, 17004 Girona. – **CA37)** C/ Ramón y Cajal 36, 3º, 43001 Tarragona. – **CA38)** Carretera de Barcelona 33, 4º, 17001 Girona. – **CA39)** Rambla d'Arago 43, 1º, 25003 Lleida.- **CA40)** Granvia de Jaume I 29-2º, 17001 Girona. – **CA41)** Paseo de Gràcia 55-57, 9º, 08007 Barcelona. – **CA42)** Barcelona. – **CA43)**

Granvía de les Corts Catalanes 645, 2º-1, 08007 Barcelona. – **CA44)** C/ Josép Alselm Clavé 5-7, 08921 Santa Colomada Gramanet.

CL00) CASTILLA Y LEÓN
CL01) García Morato 27-29, 47007 Valladolid. – **CL02)** Becerro de Bengoa 9, 34002 Palencia. – **CL03)** Santa Clara 2, 05001 Avila. – **CL04)** Ordoño II N° 28, 24001 León. – **CL05)** Campo 5, 42001 Soria. – **CL06)** Calle Barrio Gimeno 11, 09004 Burgos. – **CL07)** Av.de Requejo 21, 49012 Zamora. – **CL08)** Ave María 11, (or Apartado de Correos 105, 24480 Ponferrada) 24400 Ponferrada. – **CL09)** Plaza de Colón 4, 37001 Salamanca. – **CL10)** Paseo Ezequiel Gonzales 24, 40002 Segovia. – **CL15)** Av.del Cid 8, 09005 Burgos **E:** informativos.burgos@cadenacope.net – **CL16)** Duque de la Victoria 23, 47001 Valladolid **E:** direccion.valladolid@cadenacope.net – **CL17)** C/ Veracruz 2 bajo, 37008 Salamanca **W:** www.radiosalamanca.com – **CL18)** C/ La Estación 3, 47004 Valladolid **E:** radiovalladolid@cadenaser.com – **CL19)** Sol Oriente 11-15, 37002 Salamanca **E:** salamanca@cadenacope.net – **CL20)** Hermanos La Salle 2, 24700 Astorga **E:** astorga@cadenacope.net – **CL21)** Lope de Vega 1, 24002 León **W:** www.copeleon.com **E:** leon@cadenacope.net – **CL22)** Plaza Fernández Duró 3 (or: Ap.42), 49001 Zamora **E:** zamora@cadenacope.net – **CL23)** Venerables 8, 09005 Burgos **E:** radiocastilla.redaccion@unionradio.es – **CL24)** Villafranca 6, 24001 León **W:** www.radioleon.com **E:** radioleon@radioleon.com – **CL25)** Calle Santa Ana 6, 49006 Zamora **E:** radioz@teleline.es – **CL26)** Plaza Cirilo Rodríguez 2, 40001 Segovia **W:** www.radiosegovia.com – **CL27)** Astorga. – **CL31)** Vitoria 24, 09200 Miranda de Ebro. – **CL32)** Bermejeros 14, 37001 Salamanca. – **CL33)** Rastrojo 5, 47014 Valladolid. – **CL34)** Vitoria 29, 09004 Burgos. – **CL35)** Julio del Campo 4-6, 24002 León. – **CL36)** Plaza de los Vadillos 5, 09005 Burgos. – **CL37)** Aliso 2 bajo, 37004 Salamanca. – **CL38)** 09004 Burgos. – **CL39)** Edif.Promecal, c/los Astros s/n, 47009 Valladolid. **CL40)** Salamanca. – **CL41)** Valladolid. – **CL42)** Zamora. – **CL43)** C/ Manuel Canesi Acevedo 1, 47406 Valladolid.

CM00) CASTILLA-LA MANCHA
CM01) Paseo de San Cristóbal s/n, 45002 Toledo. – **CM02)** Ramiro Ledesma 8, 13630 Socuéllamos. – **CM03)** Nuestra. Sra. De Araceli 1, Edif.Las Torres, 02002 Albacete. – **CM04)** Radio Nacional de España 2 (or: Ap.18), 16003 Cuenca. – **CM05)** Ronda del Carmen s/n (or: Ap.150), 13002 Ciudad Real **E:** emisora.cr.rne@rtve.es – **CM06)** Plaza de Consejo, Centro Cívico, 19001 Guadalajara. – **CM07)** Ronda del Canillo 35, 45600 Talavera de la Reina. – **CM08)** Polígono Santa María de Benquerencia, C/ Río Alberche s/n, 45007 Toledo. – **CM09)** C/ Carretería 10, 1º-B, 16002 Cuenca. **CM10)** Travesía Scripreste Gutiérrez 2, 3º, 02600 Villarobledo. – **CM12)** Avenida de la Estación 5, 02001 Albacete. **E:** radioalbacete@unionradio.es – **CM13)** Alejandro Prieto 2, 13500 Puertollano **E:** puertollano@cadenacope.net – **CM14)** Tesifonte Gallego 9, 02002 Albacete **E:** albacete@cadenacope.net – **CM15)** Pasaje San Isidro 3, 13001 Ciudad Real **E:** ciudadreal@cadenacope.net – **CM21)** Av.de la Estación 5, 02001 Albacete. – **CM22)** Av. del Principe 25, 45600 Talavera de la Reina. – **CM23)** Calle Trinidad 12, 45002 Toledo. **E:** rtvdiocesana@planalfa.es – **CM24)** 02001 Albacete.

CT00) CANTABRIA
CT01) Polígono de Raos s/n, 39609 Camargo (Santander) – **CT02)** Av.del Besaya 1 (or: Ap.46), 39300 Torrelavega – **CT03)** C/ José María Pereda 23, 39100 Santa Cruz de Bezana. – **CT05)** Rualasal 5, 39001 Santander **E:** santander@cadenaser.com – **CT06)** Pasaje de la Peña 2, Edif.Simeon 11, 39008 Santander **W:** www.radiosantander.com **E:** informativos@radiosantander.com – **CT11)** Fernandez de Isla 14B, 39008 Santander.

EU00) EUSKADI
EU01) Plaza de Simón Bolívar 13, 01003 Vitoria-Gasteiz – **EU02)** Paseo de los Fueros 2, 20006 Donosti-San Sebastián **E:** emisora.ss.rne@the.es – **EU03)** Licenciado Poza 55, 48013 Bilbo-Bilbao. – C/ Polorínviejo 4, 01003 Vitoria-Gasteiz. – **EU04)** C/ Domingo Martínez de Aragón 5-9, 01006 Vitoria-Gasteiz. – **EU05)** C/ San Antonio 2 bajo, 01005 Vitoria-Gasteiz. – **EU06)** Grupo Alonso Allende 21, Lonja izquierda, 48920 Portugalete. – **EU07)** Capuchinos de Basurto 2, 48013 Bilbo-Bilbao – **EU08)** Miramón 172, 20004 Donostia-San Sebastián **W:** www.eitb.com/euskara/ – **EU09)** C/ Domingo Martínez de Aragón 5-7 bajo, 01006 Vitoria-Gasteiz **W:** www.eitb.com/radiovitoria – **EU10)** Alameda Mazarredo 47, 48009 Bilbo-Bilbao **W:** www.radiopopular.com – **EU11)** C/ Epalza 8, 48007 Bilbo-Bilbao **E:** radiobilbao@unionradio.es – **EU12)** Paseo Portuetxe 51, Edificio ACB, 20018 Donostia-San Sebastián **E:** radiosansebastian@cadenaser.com – **EU13)** Miracruz 9, 20001 Donostia-San Sebastián **W:** www.herri-irratia.com – **EU14)** C/ Miracruz 9, 20001 Donosti-San Sebastián. – **EU15)** Portal de Villarreal 6, 01002 Vitoria-Gasteiz. – **EU16)** Loiola Santutegia, 20730 Azpeitia – **EU17)** Vitoria-Gasteiz – **EU21)** Av.de la Libertad 17, 20004 -Donosti-San Sebastián – **EU22)** General Alava 10-6 Depto 9, 01005 Vitoria-Gasteiz – **EU23)** Fontecha y Salazar 9-5, 48007 Bilbo-Bilbao – **EU24)** Eziago Poligonoa 10B, 20120 Hernani. – **EU25)** C. Hurtado de Amezaga 27, 48008 Bilbo-Bilbao. – **EU26)** C/ Esteban Zurbano 20, 20214 Segura

(Guipúzcoa). – **EU27)** Gordóniz 44, 12º, 48002 Bilbao. – **EU28)** C/ San Prudencio 8-A,, 5º, 01005 Vitoria-Gasteiz. – **EU29)** Paseo Federico García Lorca 10, 4º, Puerta 1-2, 20014 Donosti-San Sebastián. – **EU30)** Ribera de Elorrieta 7, 48015 Bilbo-Bilbao. – **EU31)** C/ José Lejarreta 11, 01003 Vitoria-Gasteiz. – **EU32)** C/ Capelamendi 1, Polígono de Betoño, 01013 Vitoria-Gasteiz. – **EU33)** Parque Empresarial Zuazu, Ed. Ulía 8, 20018 Donosti-San Sebastián. – **EU34)** Vitoria-Gasteiz.

EX00) EXTREMADURA
EX01) Av.Ruta de la Plata 10, 10001 Cáceres – **EX02)** Plaza de España 5, 06002 Badajoz. – **EX03)** Avenida de las Américas 1, 1º, 06800 Mérida. – **EX05)** Sánchez Herrero 2, 10002 Cáceres **E:** caceres@cadenacope.net – **EX06)** Ramón Albarrán 2, 06002 Badajoz **E:** radioextremadura@unionradio.es – **EX07)** Menacho 12, 06001 Badajoz **E:** badajoz@cadenacope.net – **EX08)** C/ Profesor Rodríguez Moñino 1, 8º-A, 10003 Cáceres – **EX11)** Av.de España 9-6, 10004 Cáceres – **EX12)** Av.de Portugal s/n, Ctro Comercial El Foro, 06800 Mérida – **EX13)** Santa Isabel 4, 10600 Plasencia – **EX16)** Luis Alvarez Lancero 8, 10001 Cáceres.

GA00) GALICIA
GA01) Paseo Méndes Nuñez 12, (or: Ap.199), 15006 A Coruña – **GA02)** Lepanto 7, 36001 Pontevedra – **GA03)** Rua de Progreso 115 (or: Ap.268), 32003 Ourense – **GA04)** Ourense 59-63 (or: Ap.73), 27004 Lugo – **GA05)** Plaza de España 4, 27400 Monforte de Lemos – **GA06)** Av.García Barbón 36, 36201 Vigo – **GA07)** San Marcos s/n, Edif.TVE, 15780 Santiago de Compostela. – **GA08)** Rúa Pascual Veiga 12-14 baixo dereita, 27002 Lugo. – **GA09)** Plaza de España 5-6, 15403 El Ferrol **E:** ferrol@cadenacope.net – **GA10)** San Pedro de Mezonzo 3 (or: Ap 469), 15701 Santiago de Compostela **E:** radiogalicia@unionradio.es – **GA11)** Principe 57, 36202 Vigo **E:** vigo@cadenacope.net – **GA12)** Areal 6-8, 36201 Vigo **W:** www.radiovigo.es – **GA13)** Plaza de Ourense 3, 15004 A Coruña **W:** www.radiocoruna.es – **GA14)** Castelao 3 B, 36001 Pontevedra **E:** ser@radiopontevedra.com – **GA15)** Rua de Progreso 89, 32003 Ourense **E:** orense@cadenacope.net – **GA16)** Rua de Valiño s/n, 27002 Lugo **E:** lugo@cadenacope.net – **GA17)** Plaza de Santo Domingo 3, 27001 Lugo **W:** www.radiolugo.com – **GA18)** Rua do Paseo 30 (or: Ap.1017), 32003 Ourense **W:** www.radioorense.com **E:** cadenaser@radioorense.com – **GA21)** Concepción Arenal 11-13, 15006 A Coruña – **GA22)** Salvador Moreno 30, 36001 Pontevedra – **GA23)** Av.García Barbón 104, 36201 Vigo – **GA25)** Casa de la Radio, San Marcos, 15820 Santiago de Compostela – **GA26)** C/ Marcial del Adalid 8, 15005 A Coruña – **GA27)** Salguiriños de Arriba 44, bajo, 15890 Santiago de Compostela. – **GA28)** Apartado de Correos 3114, 36208 Vigo.- **GA29)** Rúa Benito Corbal 14, 2º, 36001 Pontevedra – **GA30)** Av. García Barbón 28, 36201 Vigo. – **GA31)** Rúa Costa Rica 6, 7º, 15005 A Coruña.

MA00) MADRID
MA01) Casa de la Radio, Prado del Rey, 28223 Pozuelo de Alarcón – **MA02)** C/ Enrique Larreta 12, Madrid. – **MA03)** C/ Juan Esplandiú 13, 8º, 28007 Madrid. – **MA04)** C/ San Bernardo 20, 3º, Centro, 28015 Madrid – **MA05)** Gran Vía 32, 28013 Madrid **E:** redaccion@cadenaser.com – **MA06)** Modesto Lafuente 42, 28003 Madrid **W:** www.radiointer.com **E:** radiointer@radiointer.com – **MA07)** Fuerteventura 12, 28703 San Sebastián de los Reyes – **MA08)** Alfonso XI N° 4, 28014 Madrid **W:** www.cope.es **E:** programas.madrid@cadenacope.net – **MA09)** Círculo de Bellas Artes, C/ Alcalá 42, 5 planta, 28014 Madrid.- **MA11)** Bueso Pineda 7, 28043 Madrid – **MA13)** Calle de Ortega y Gasset 22-24, 28006 Madrid – **MA14)** Pso del Principe 3, Cd.de la Imagem, 28223 Pozuelo de Alarcón **W:** www.telemadrid.com – **MA16)** Av.de los Arqueros s/n, 28024 Madrid **E:** radiomaria@arsenet.com – **MA17)** Paseo de la Castellana 36-38, 28046 Madrid **W:** www.interecomia.com – **MA18)** Sta Clara 7, 28801 Alcalá de Henares – **MA19)** Casa de la Juventud, Guadalajara 1, 28901 Getafe – **MA20)** C/ Juan Español 47, local bajo, 28026 Madrid. – **MA21)** C. Francisco Silvela 122 bajo, 28002 Madrid. – **MA22)** Madrid. – **MA23)** Madrid. – **MA24)** C/ Orense 18, piso 8, of. 9, 28020 Madrid. – **MA25)** Orense 70, 10º- izq., 28020 Madrid. – **MA26)** Paseo Recoletos 21, 1º-derecha, 28004 Madrid. – **MA27)** C/ Esteban Collante 9, 28019 Madrid. – **MA28)** Madrid.

MU00) MURCIA
MU01) La Olma 27-29, 30005 Murcia **E:** emisora.mu.rne@rtve.es – **MU02)** Paseo Alfonso XIII N° 51, 30203 Cartagena. – **MU03)** C/ Bucarest 29, 30391 Cartagena. **MU04)** C/ Carmen Conde 46, 1º, 30203 Cartagena. – **MU05)** Arco de Santo Domingo 2-3, Edif.Fontanar, 30001 Murcia **E:** murcia@cadenacope.net – **MU06)** Av.Juan Carlos I N° 63, 30800 Lorca **E:** lorca@cadenacope.net – **MU07)** Calle Radio Murcia 4, 30001 Murcia **E:** radiomurcia@unionradio.es – **MU08)** Real 70, 30201 Cartagena **E:** informativos.cartagena@cadenaser.com – **MU11)** Edif.Mediterráneo, Puerta Murcia 11, 30201 Cartagena – **MU12)** C/ Carmen 521, 1º-A, 30201 Cartagena – **MU13)** Madre de Dios 15, 30004 Murcia – **MU14)** Mayor 31, 30280 Cartagena – **MU16)** Av.Libertad 6, bajo, 30009 Murcia. – **MU17)** Pza de los Apóstoles 7, 30011 Murcia. – **MU18)** Ed. delPeriódico La Verdad, Camino Viejo de Monteagudo

s/n, 30160 Murcia.

NA00) NAVARRA
NA01) R1: Emilio Arrieta 8, P8, 31002 Pamplona-Iruñea; R5: Aoiz 17, 31004 Pamplona-Iruñea. – Ed. Ciencias Sociales, Universidad de Navarra, Campus Universitario s/n, 31080 Pamplona.-Iruñea – **NA05)** Amaya 2-B, 31002 Pamplona-Iruñea **E:** pamplona@cadenacope.net – **NA06)** Polígono Plazaola, Manzana F, 2° A, 31195 Aizoain (or Apartado de Correos 71, 31080 Pamplona) **E:** informativosnavarra@cadenaser.com – **NA11)** Plaza del Castillo 43, 31001 Pamplona-Iruñea – **NA12)** Cortes de Navarra 1, 31002 Pamplona-Iruñea.

RI00) LA RIOJA
RI01) Vara de Rey 42, (or: Ap.247), 26002 Logroño – **RI02)** Residencia Universitaria Francisco Jordán, Av. Madre de Dios 17, 26001 Logroño - **RI03)** C/ Estambrera 36, 1°, 26006 Logroño. – **RI04)** C/ General Vara del Rey 74, 26002 Logroño. – **RI05)** Av.de Portugal 12 (or: Ap.149), 26001 Logroño **W:** www.radiorioja.com – **RI11)** 26001 Logroño – **RI12)** Miguel Villanueva 2, Ofc.5, 26001 Logroño.

VA00) CUMUNITAT VALENCIANA
VA01) Av Colóm 13, 46004 València – **VA02)** Angel Lozano 18, 03001 Alacant-Alicante – **VA03)** Passeig de la Ribalta 5, 12001 Castelló – **VA04)** Juan Carlos I 37, 03202 Elx. - **VA05)** Plaça dels Sports 7-8, Ed. Sabater, 03590 Altea – **VA07)** Rambla de Méndez Nuñez 45, 03002 Alacant-Alicante **E:** alicante@cadenacope.net – **VA08)** Calderón de la Barca 26, 03004 Alacant-Alicante **E:** alicante@cadenaser.com – **VA09)** Av.Francisco Tàrrega 69, 12540 Vila-Real **E:** castellon@cadenacope. net – **VA10)** Don Juan de Austria 3, 46002 València **E:** valencia@cadenaser.com – **VA11)** Passatge Dr.Sierra 2, 46004 València **W:** www.cope.es/valencia – **VA12)** Doctor Sempere 16B y C, Bajos, 03803 Alcoi **E:** radioalcoy@radioalcoy.com – **VA13)** Moyano 5, 12002 Castelló **W:** www.radiocastellon.comdc – **VA14)** Dr.Caro 43, 03201 Elx **W:** www.rtvelche.com – **VA15)** Calle Loreto 32, 46700 Gandia **E:** ser@radiogandia.net – **VA16)** Ereta 2A (or: Ap.84), 46870 Ontinyent **W:** www.radioontinyent.com – **VA 17)** C/ Hort dels Frares 12, 46600 Alzira. – **VA18)** C/ Doctor Caro 18 entresuelo derecha, 03201 Elx. – **VA19)** C/ La Pira 10, 03201 Elx. **VA20)** C/ Almorida 2, 4° derecha, 03201 Elx. – **VA21)** Av.Maissonnave 19-21, 03003 Alacant-Alicante – **VA22)** C/ Segura 19, 03004 Alacant-Alicante – **VA23)** C/ San Vicente 16, entreplanta 1°, 46001 València – **VA24)** Av.Blasco Ibañez 136, 46022 València – **VA26)** Vía Emilia Ortuño 5, 03500 Benidorm – **VA28)** Paseo Explanada de España 26, 03001 Alicante – **VA29)** C/ Els Gremis 1, 46014 València. **VA30)** C.C. Alfafar, Pl. Alquería de la Culla 4, 1401, Alfafar, 46910 Valencia. – **VA31)** C/ Lebón 7, 46023 València. – **VA32)** Av. Blasco Ibáñez 134, 46022 València. – **VA33)** Av. Aragón 30, 46031 València.

AMERICAN FORCES RADIO & TV SERVICE (Mil.)
ZFM 92.1MHz, Morón de la Frontera ⬚ Base Aerea USAF, 41530 Morón de la Frontera – **FM102 Navy** 102.5MHz, Rota. ⬚ Base Naval, 11520 Rota. **D.Prgr:** All stns 24h

SRI LANKA

LT: UTC +5½h — **Pop:** 19 million — **Pr.L:** Sinhala, Tamil, English — **E.C:** 50Hz, 230V — **ITU:** CLN

SRI LANKA BROADCASTING CORPORATION (Pub)
⬚ P.O. Box 574, Independence Square, Colombo 7 ☎+94 11 2697491 🖷 +94 11 2691568 **E:** ddge@slbc.lk **W:** www.slbc.lk
LP: Chairman Hudson Samarasinghe, DG: Samantha Weliweriya, Dep. DG (Eng.) H. M. Jackson, Dep. DG (Finance): N P W Perera, Dir. News: M. I. A. Jayaranthbe, Dir. Eng. (Studios): C. N. D. Siriwardene.

MW: Irattaperiyakulam 855kHz 20kW. Wanni Regional service.

FM (MHz)	A	B	C	D	E	F
Colombo†	98.3	93.3	101.3	105.6	95.6	87.6
Deniyaya	99.6	89.3	104.8	92.8	90.9	102.6
Haputale	90.3	96.4	102.0	107.9	98.4	92.2
Hunasgiriya	102.0	107.3	98.8	94.2	96.4	92.2
Karagahatenna	107.6	92.7	102.4	104.5	99.6	95.0
Radella	97.0	106.9	103.5	105.6	100.2	94.4
Yatiyantota	90.3	99.6	94.2	104.8	96.4	92.2
Palali				92.2	–	

† = also on **F** at 91.2MHz

A = Sinhala National Sce 2300-1600., **B** = Sinhala Commercial Sce 24h, **C** = Tamil National Sce 2300-1715, **D** = Tamil Commercial Sce 2300-1700, **E** = English Commercial Sce 0000-1700, **F** = City FM 24h, **Sports Sce:** operates irregularly as required using freqs of Tamil National Sce and also Palali 102.0.
Vidula (Children's channel): Yatiyantota 102.6MHz, 0000-1630 in Sinhala, Tamil and English.

Regional FM services: Akkaraipattu: Haputale 102.0MHz – **Anuradhapura:** Karaghatenna 90.6MHz – **Batticaloa:** Karaghatenna 97.0MHz **Jaffna:** 102.0 MHz – **Kandy:** Radella 89.7MHz, Hunasgiriya 89.3MHz – **Kurunegala:** Karaghatenna 99.6MHz – **Matara:** Haputale 105.4MHz, Deniyaya 92.8MHz 5 kW. **F.PI:** Jaffna to be a nat. sce. Regional Sce operates 2300-0230 & 1000-1530.

Ann: A: "Me Sri Lanka Guwan Viduli Sansthave Welanda Sevaya". B: "Me Sri Lanka Guwan Viduli Sansthava Swadeshiya Sevaya". C: "Illangar Oliparappu Koothuthapanam Tamil Sevai". E: "This is the Sri Lanka Broadcasting Corporation"

F.PI: Installation of 50kW MW transmitter in Puttalam on 1125 kHz to tr. to southern India and to strengthen coverage to northern Sri Lanka.

EXTERNAL SERVICE: SLBC see International Broadcasting section

MAJOR COMMERCIAL NETWORKS

ASSET RADIO BROADCASTING Ltd
⬚ Gregories Rd., Colombo 7 ☎+94 11 2671166 🖷 +94 11 2671167
Neth FM in Sinhala: Gongala 93.9MHz, Colombo 95.0MHz, Kandy 100.4MHz, Laggala 105.4MHz, Nayabedda 105.9MHz

COLOMBO COMMUNICATIONS (PVT) Ltd
⬚ 2/9 2nd Floor, Liberty Plaza, 250 R.A. de Mel Mw., Colombo 3 ☎+94 11 2577924-7, 2330718-9 🖷 +94 11 2577929
E! FM: Kandy 93.2MHz, Clombo 100.4MHz, Gongala 104.5MHz
RAN FM: Nayabedda 91.5MHz, Gongala 95.0MHz, Gammaduwa 101.3MHz, Colombo 102.2MHz, Ratnapura 104.5MHz
Shree FM: Gongala/Ratnapura 93.2MHz, Hunasgiriya 95.8MHz, East 98.8MHz, Colombo 99.0MHz, Magalkanda 99.3MHz

HIRU MEDIA NETWORKS
Gold FM: Gongala 94.7MHz, Colombo 99.9MHz, Hunasgiriya 99.9MHz, Gammaduwa 102.7MHz, Nuwara Eliya 104.2MHz
Hiru FM: Hunasgiriya 94.7MHz, Gammaduwa 94.7MHz, Nuwara Eliya 95.3MHz, Colombo 96.7MHz, Gongala 96.7MHz, Uva 107.0MHz
Sooriyan FM: Jaffna 93.0MHz, Gammaduwa 97.3MHz, Nuwara Eliya 97.9MHz, Colombo 103.2MHz, Ruhuna 103.2MHz, Uva 103.4MHz

INDEPENDENT TELEVISION NETWORK
⬚ Wickramasinghepura, Battaramulla 10120 ☎+94 11 2774424 🖷+94 11 2774591 **E:** itn@slt.lk **W:** www.itn.lk **L.P:** Chmn.: Rosmand Senarathna, Gen Mgr: W Wijesinghe
Lakhande: Karagahatenna 87.9MHz, Nayabedda 88.5MHz, Deniyaya 97.6MHz
Prime Radio: Kandy 95.5MHz, Galle 99.0MHz, Colombo 104.5MHz
Vasantham: Colombo 97.3MHz, Karagahatenna 97.6MHz

MBC NETWORKS (PVT) Ltd
⬚ 109 2nd Floor Collettes Bldg., Rt. Hon. D.S. Senanayake Mw., Colombo 8 ☎ +94 11 2689234-6.
Shakthi FM: ⬚ 7 Braybrooke Place, Colombo. Nayabedda 91.2MHz, Hunasgiriya 91.5MHz, Nuwara Eliya 103.8MHz, Colombo 105.1MHz, Gammaduwa 105.1MHz, Gongala 105.1MHz
Sirasa FM: ⬚ PO Box 25, Araliya Uyana, Pannipitiya. Nuwara Eliya 88.8MHz, Gammaduwa 101.7MHz, Hunasgiriya 106.2MHz, Gongala 106.2MHz, Colombo 106.5MHz, Nayabedda 106.5MHz
Y FM: ⬚ 7 Braybrooke Place, Colombo. Kandy 91.2MHz, Colombo 92.6MHz, Gammaduwa 99.1MHz, Gongala 101.3MHz, Nayabedda 101.3MHz
Yes FM: ⬚ as MBC Networks. Kandy 88.2MHz, Gongala 88.2MHz, Colombo 89.5MHz, Nuwara Eliya 101.0MHz.

TNL RADIO NETWORK (Telshan Networks Ltd)
⬚ 58 Tower Bldg., Station Rd., Colombo 4 ☎+94 11 2584107
L.P: Chmn. & MD: Shan Wickremesinghe, Comm. Dir: Ms. Ishini Wickremesingohe
Rhythm FM: Kandy 87.6MHz, Karagahatenna 87.6MHz, Gongala 95.6MHz, Colombo 100.7MHz
Eisira: ⬚ Dampe, Piliyandala. Piliyandala 88.0MHz, Nuwara Eliya 88.0MHz, Deniyaya 89.0MHz, Karagahatenna 89.0MHz, Kandy 93.7MHz, Colombo 93.9MHz, Ratnapura 97.2MHz
Lite 89.2: Colombo 89.2MHz, Nuwara Eliya 90.0MHz, Gongala 92.5MHz, Kandy 98.2MHz
TNL 101.7 FM: ⬚ 52 5th Lane, Colombo 3 ☎+94 11 2575000. Gongala 87.9MHz, Kandy 92.5MHz, Colombo 101.7MHz

Other stations:
Trans World Radio India MW: 882kHz 400kW and **Family Radio**

MW: 873kHz and 400kW. Broadcasts for Sri Lanka and southern India. See International section for details.
DEUTSCHE WELLE RELAY STATION and **IBB RELAY STATION** see International Broadcasting section

ST BARTHÉLEMY (France)

LT: UTC -4h — **Pop:** 7,500 — **Pr.L:** French, Creole, English — **E.C:** 50Hz, 230V — **ITU:** BLM

RADIO GUADELOUPE (Pub)
c/o Morne Bernard-Destrellan, B.P. 180, 97122 Baie-Mahault, Guadeloupe. ☎+590 590939696 📠 +590 590939682.
FM: 88.6MHz 0.3kW

RADIO SAINT-BARTH
BP 1113, 97014 St Barthélemy. ☎+590 590 27 74 74 📠 +590 590 27 74 10. **L.P:** Président Clemenceau Magras
FM: 98.7MHz 0.3kW, 100.7MHz 0.3kW, 103.7MHz 0.3kW.

R. France Internationale: via R. St. Barth 100.7MHz

ST EUSTATIUS (Netherlands)

LT: UTC -4h — **Pop:** 2,500 — **Pr.L:** Dutch (official), English — **E.C:** 60Hz, 110V — **ITU:** ATN

RADIO STATIA
The Mall, Korthalsweg, Oranjestad ☎ +599 318 2722
MW: Call: PJE-3, 1120kHz 1kW
FM: 92.3MHz

ST HELENA (UK)

LT: UTC — **Pop:** 4,000 — **Pr. L:** English — **E.C:** 50Hz, 240V — **ITU:** SHN

RADIO ST HELENA
Pouncey's, St Pauls, St Helena Island, So. Atlantic Ocean STHL 1ZZ ☎+290 4669 📠 +290 4542 **E:** radio.sthelena@cwimail.sh
W: www.news.co.sh **L.P:** SM: Gary Walters

MW: 1548kHz 1kW
SW: 'RSH Day' on 11092.5kHz 1kW in USB. (see DX groups for day)
D.Prgr: 24h. Relays BBC World Service 1200-1500 MF, 1300-1900 Sat, 22.00-0700 daily, 0700-1800 Sun **Ann:** "Radio St. Helena" **V.** by QSL-card, Rp. (3 IRCs). Email Rpt. not accepted. Rec not returned

SAINT FM
St Helena Media Productions Ltd, Jamestown, St Helena, So. Atlantic Ocean STHL 1ZZ. ☎+290 2660/2488 **E:** fm@cwimail.sh
W: www.saint.fm (& Webcasting) **L.P:** SM: Mike Olsson
FM (MHz): 93.1 0.25kW, 95.1 0.03kW, 100.7 0.03kW
D.Prgr: 24h. **Ann:** "Saint FM". **V.** by QSL-card, Rp. (2 IRCs). Email Rpt. accepted. Rec not returned

ST KITTS & NEVIS

LT: UTC -4h — **Pop:** 50,000 — **Pr.L:** English — **E.C:** 60Hz, 220V — **ITU:** SCN

ZIZ RADIO (Gov. Comm.)
P.O. Box 331, Springfield, Basseterre, St. Kitts ☎ +1 869 465 2621 📠 +1 869 466 2159 **L.P:** GM: Winston McMahon
Email: info@zizonline.com **W:** www.zizonline.com
MW: Radio ZIZ 555kHz 10kW. Rel. BBC World Service 0400-0800.
FM: Radio ZIZ 95.9/96.1/96.9MHz – **Big Wave 96.7 FM** 96.7MHz

RADIO PARADISE (Rlg.)
Bath Plains, P.O. Box 508, Charlestown, Nevis ☎ +1 869 469 1994 📠 +1 869 469 1642 (**Addr in USA:** P.O. Box A, Santa Ana, CA 92711 ☎ +1 (714) 832 2950 📠 +1 (714) 730 0661) **W:** www.radioparadiseonline.com **L.P:** Local mgr: Andre Gilbert
MW: St. Kitts 820kHz 50kW
D.Prgr: 24h. Local prgrs: 0900-2300; sometimes later. Other times: relays audio feed from TBN satellite TV.
Ann: "This is R. Paradise, 820 on your AM dial, broadc. from St. Kitts-Nevis"

VOICE OF NEVIS (Comm.)
Bath Plains, P.O. Box 195, Charlestown, Nevis ☎ +1 869 469 1616/1700 📠 +1 869 469 5329 **W:** www.vonradio.com
L.P: GM: Evered Herbert
MW: 860kHz 10kW **D.Prgr:** 24h **Ann:** "This is Von Radio on 860 AM"

Other stations:
CHOICE FM, Rams, Stoney Grove, Charlestown, Nevis ☎ +1 869 469 5300 **W:** www.choicefm1053.com **FM:** 105.3MHz – **DOMINION RADIO**, PO Box 513, Basseterre ☎ +1 869 465 1597 **W:** www.dominionradiskn.com **FM:** 91.5MHz. Format: Rlg. – **FREEDOM FM**, The Cable Bldg., Suite 2, Cayon St., Basseterre. **L.P.:** Clement Juni Liburd. **FM:** 106.5MHz – **GEM RADIO, FM:** 93.1MHz (rel. Trinidad) – **GOODWILL RADIO** P.O. Box 98, Lodge Village, St. Kitts ☎ +1 869 465 7795 📠 +1 869 465 9556 **Email:** info@goodwillfm.com **W:** www.goodwillradiskn.com **L.P:** SM: Denis Nilsson. CE: Nigel Brown. **FM:** 103.3MHz (Nevis) & 104.5MHz (St. Kitts). Format: Rlg. – **KYSS FM - The Love FM**, 51A Stadium View, Sandy Point ☎ +1 869 466 5978 📠 +1 869 466 1746 **E:** info@kyssonline.com **W:** www.kyssonline.com **FM:** 102.5MHz – **PRAISE FM**, Hamilton Estate, Charlestown, Nevis ☎ +1 869 667 0351 **W:** www.praisefm993.site40.net **L.P.:** Steve Huggins. **FM:** 99.3MHz Format: Gospel – **RADIO ONE**, St. Kitts & Nevis Broadcasting Corp., Bakers Corner, Basseterre ☎ +1 869 466 0941 **FM:** 94.1MHz – **RADIO ST. KITTS NEVIS**, Reef Broadcasting, #79 Castle Coakley, Christiansted, VI 99820, USA ☎ +1 401 573 1620. **FM:** 90.7MHz (relays WAXJ 103.5, US Virgin Isl.) – **SUGAR CITY ROC FM**, Unit C24, The Sands ☎ +1 869 466 1113 **W:** www.sugarcityrock.com **FM:** 90.3MHz – **VIBZ, FM:** 98.3MHz (reported on 97.9) (rel. Antigua) – **WINN FM**, Unit C24, Newtown Bay Rd, Basseterre, St. Kitts ☎ +1 869 466 9586 📠 +1 869 466 7904 **E:** info@winnfm.com **W:** www.winnfm.com **L.P:** SM: Clive Bacchus. Dir: Charles Wilkin. **FM:** 98.9MHz

ST LUCIA

LT: UTC -4h — **Pop:** 174,000 — **Pr.L:** English, Creole — **E.C:** 50Hz, 220V — **ITU:** LCA

RADIO ST. LUCIA COMPANY LTD. (Gov. Comm.)
Morne Fortune, P.O. Box 660, Castries ☎ +1 758 452 2337/9 📠 +1 758 453 1578 **E:** rsl@candw.lc **W:** www.rslonline.com
L.P: MD: Mary Polius . CEN: Othneil Robinson. Prgr. Coordinator: Garfield Alexander. Dir. News: Shelton Daniel
FM: 97.3MHz 3kW, 97.7MHz 0.5kW. **D.Prgr:** 24h.

Other stations:
CARIBBEAN HARMONY. FM: 107.9MHz. – **CATHOLIC TV BROADCASTING SERVICE**, Micoud Str, Castries ☎ +1 758 452 7050. **FM:** 87.75MHz (TV sound of EWTN, USA) – **HIT RADIO MUSIC POWER**, Reduit, Gros Islet ☎ +1 758 452 0048 **W:** www.hitradiomusicpower.com **FM:** 94.1/96.5/96.7MHz (+ rel. in Antigua & Dominica) – **HOT FM**, Chef Harry Drive, The Morne, Castries ☎ +1 758 452 6040 📠 +1 758 452 1462 **W:** www.caribbeanhotfm.com **L.P:** Mgr: Patrick Smith **FM:** 96.1/105.3MHz – **JOY FM**, Castries **W:** www.joyfmstlu.com **FM:** 90.1/96.9MHz Format: Rlg. – **PRAYZ FM RADIO**, Sir John Compton Highway, Sans Solicis, PO Box CP6141, Castries ☎ +1 758 452 1022 **W:** www.prayzfm.org **FM:** 98.5MHz. Format: Rlg – **RADIO CARIBBEAN INTERNATIONAL**, 11 Mongiraud St., P.O.Box 121, Castries ☎ +1 758 452 2636 📠 +1 758 452 2669 **W:** www.rcistlucia.com **L.P:** GM: Pet Gibson. SM: Peter Ephraim. **FM:** 99.1/101.1MHz (+ rel. in Dominica) – **RADIO 100 HELEN FM**, P.O. Box 621, Castries ☎ +1 758 451 7260 📠 +1 758 453 1737 **W:** www.htsstlucia.com/radio_100.htm. **FM:** 100.1/100.3/103.5MHz – **RFI**, Soufriere. **FM:** 102.1MHz. Format: Community radio – **RHYTHM FM**, P.O. Box 584, Castries ☎ +1 758 450 9494 📠 +1 758 451 6217 **W:** www.rhythmfm.net **L.P:** MD Dwayne Mendes. **FM:** 95.5/99.5MHz – **RIZZEN 102FM**, Castries ☎ +1 758 451 3057 📠 +1 758 451 3011 **W:** www.rizzen102.com **FM:** 99.7/102.5/102.9MHz. Format: Rlg. – **THE WAVE**, American Drywall Building, Vide Boutielle, P.O. Box CP 5631, Castries ☎ +1 758 451 6400 📠 +1 758 452 2633 **W:** www.thewavestlucia.com **FM:** 93.7/94.5MHz

ST MAARTEN (Netherlands)

LT: UTC -4h — **Pop:** 33,100 — **Pr.L:** Dutch (official), English — **E.C:** 60Hz, 120V — **ITU:** ATN

BUREAU TELECOMMUNICATION AND POST
C.W.G. Buncamper Road, Harbour View Lot 3 / Unit D, P.O.Box 5054; St. Maarten ☎ +1 721 5 425557 📠 +1 721 5424817

MW Call	kHz	kW	Station, location
1) PJD-2	1300	1	The Voice of St. Maarten, Philipsburg

FM	MHz	kW	Station, location
4)	91.9		Island 92, Simpson Bay
5)	94.7		Mix 94.7, Philipsburg
6)	96.3		Oasis 96.3, Philipsburg
8)	98.1		Pearl FM, Philipsburg
3)	99.9		Radio Soualiga, Choice FM

FM	MHz	kW	Station, location
6)	101.1		Laser 101
1)	102.7	3.5	PJD3 The Voice of St. Maarten, Philipsburg
7)	105.5		Tropixx 105.5
2)	107.9	1	Gem Radio Network, Philipsburg

Addresses and other information
1) "The Voice of St. Maarten" Plaza 21 Shopping Centre, P.O. Box 366, Philipsburg, St. Maarten ☎ +1 721 55422580, 5422764 🖷 +1 721 55424905, 5425531 GM/Dir: Donald R. Hughes. MW: 0930-0600(Mon 0400). **N**: 1030, 1700, 0000. FM: 0930-0400. **N**: 1200, 1800, 0000. – **2)** Relays prgrs from Trinidad.– **3)** W.J.A. Nisbeth Road 23, PO Box 1029, Philipsburg, St Maarten ☎ +1 721 5422049. 🖷 +1 721 5425791 **W:** www.sxmradio.com/choicefm.html – **4)** Caribe Broadcasting System, 2nd floor, Federal Express Building, Simpson Bay , St. Maarten ☎ +1 721 5443377 **E:** info@island92.com **W:** www.island92.com – **5)** Lighthouse Broadcasting Network, 5 Brooks Towers, Suites 5A and 5B, Philipsburg ☎+1 721 5425773 🖷 +1 721 5425778 – **6)** A.T. Illidge Road 106, Suite 2, Philipsburg ☎+1 721 5432200/2201 🖷 +1 721 5432229 **E:** marketing@philbroad.com **W:** www.laser101.fm – **7)** ☎ +1 721 5437960 🖷 +1 721 5811055 – **8)** Fort Belair Road 3, Philipsburg ☎ +1 721 5430462 **E:** pearlstudio@caribserve.net **W:** www.pearlfmradio.com

ST MARTIN (France)

LT: UTC -4h — **Pop:** 36,000 — **Pr. L:** French, Creole, English, Dutch — **E.C:** 50Hz, 230V — **ITU:** MAF

RADIO GUADELOUPE (Pub)
🖳 Quartier Bellevue-Marigot, 97100 Saint Martin ☎+590 590291716 **FM:** St. Martin 88.9MHz 0.3kW

RCI - RADIO CARAÏBES INTERNATIONAL (Comm.)
🖳 B.P. 173, Marigot, F-97150 Saint Martin. ☎+590 590875406 🖷 +590 590878887 **RCI FM:** 105.0MHz 0.3kW **RCI2 FM:** 102.1MHz + relays in Guadeloupe & Martinique.

RADIO SAINT MARTIN (Comm.)
🖳 Port de Marigot, 97150 Saint Martin **L.P.:** Mgr: H. Cocks. **FM:** 95.3MHz **D.Prgr:** 1000-0500(Sun 0400) in French & English exc. Spanish: 2000-2100W.

RADIO VOIX CHRETIENNES DE ST. MARTIN (Rlg.)
🖳 B.P. 103, Marigot, F-97150 Saint Martin. ☎+590 590873159 **L.P:** Mgr: Father Cornelius Charles **FM:** 106MHz 0.25kW. **D.Prgr:** 0845-0530 in English & French.

RADIO CALYPSO
🖳 10, rue du Général de Gaule, 97150 Saint Martin ☎+590 590 522222 🖷 +590 590 52 22 23 **W:** www.radiocalypso.net **E:** calypsopub@powerantilles.com **FM:** 102.1MHz 1kW

ST PIERRE ET MIQUELON (France)

LT: UTC -3h (13 Mar-6 Nov: -2h) — **Pop:** 7,000 — **Pr.L:** French — **E.C:** 50Hz, 220V — **ITU:** SPM

RFO ST PIERRE ET MIQUELON
🖳 B.P. 4227-97500 St. Pierre et Miquelon. ☎+508 508413824. **L.P:** Dir: Joseph Edern. Dir. Tec: Daniel Beugin. Head of N: Jacques Barret. **FM:** St. Pierre 97.9MHz 10W, 99.9MHz 0.5kW, Miquelon 98.9MHz 50W.
D.Prgr: 0930-0230. **Rel. France-Inter:** 0230-0930. **N (local):** 1000, 1530, 2200, (**Rel. France-Inter:**) 1100, 1200, 1300, 1400, 1800, 1900 **Ann:** "Ici RFO, Station de Saint-Pierre et Miquelon"
IS: La Marseillaise **V.** by QSL-card. Rec. acc.

R. Atlantique 🖳 B.P. 1282-97500 ☎+508 508412493 **W:** www.cheznoo.net/radioatlantique **FM:** 102.1MHz (also rel. R. France Int.)

ST VINCENT & THE GRENADINES

LT: UTC -4h — **Pop:** 100,000 — **Pr.L:** English — **E.C:** 50Hz, 230V — **ITU:** VCT

NATIONAL BROADCASTING CORPORATION RADIO ST. VINCENT AND THE GRENADINES – NBCSVG (Gov. Comm.)
🖳 Richmond Hill, PO. Box 705, Kingstown ☎ +1 784 457 1111 🖷 +1 784 456 2749 **W:** www.nbcsvg.com **E:** nbcsvgadmin@nbcsvg.com

L.P: Chmn: Kenneth Browne. GM: Corlita Ollivierre. Dep. GM: Raphael King. PM: Juanita Francois. Tech Dir: Lynford Byron. N Ed: Lesley De Bique.
MW: 700kHz 10kW **FM:** 89.7MHz 1kW, 90.7MHz 1kW, 107.5MHz 1kW **D.Prgr:** 24h. Relays BBC 0400-0930. **N:** 1130, 1630, 2230 – Su 1230 only. **Ann:** "NBC Radio".

Other stations:
BEQUIA COMMUNITY HIGH SCHOOL, PO Box 75BQ, Bequia. FM: 89.3MHz – **CCR - CROSS COUNTRY RADIO**, 50 Vigie Highway, PO Box 1000, Kingstown ☎ +1 784 458 5555 🖷 +1 784 456 4117. **W:** ccrradiosvg.com. L.P.: SM Carlos Meloni – **FM:** 88.5/104.3MHz. Format: Rlg – **EZEE RADIO**, PO Box 617, Kingstown ☎ +1 784 456 1078. - **FM:** 91.1/100.5/102.7MHz. Format: Easy listening – **GARIFUNA RADIO**, Learning Resource Center, Sandy Bay Community. **W:** garifunaradio. webs.com – **FM:** 89.1MHz – **GEM RADIO**. Format: Community radio – **HITZ-FM**, St Vincent Broadcasting Corp., Dorsetshire Hill, P.O. Box 617, Kingstown ☎ +1 784 456 1078 🖷 +1 784 456 1015. **W:** www.svgbc.com/hitz_fm.htm - **FM:** 91.5/103.7MHz. Format: Urban Caribbean – **HOT 97FM**, 1 Melville Street, PO Box 1716, Kingstown ☎ +1 784 452 9797 🖷 +1 784 456 2462 **W:** www.hot97svg.com - **FM:** 93.1/97.1MHz. Format: Urban Caribbean – **HYPE 101.5 ASYLUM RADIO**. FM: 101.5MHz – **JEM RADIO**, Hopewell Rd, PO Box 1419, Kingstown ☎ +1 784 451 3827 - **FM:** 89.1MHz. Format: Rlg. – **NICE RADIO**, BDS Company Ltd., Dorsetshire Hill, PO Box 324, Kingstown ☎ +1 784 456 1013 🖷 +1 784 456 5556. **W:** www.niceradio.org. L.P.: Mgr: Douglas Defreitas – **FM:** 90.3/96.7/101.3MHz – **PRAISE FM**, Sion Hill, P.O.Box 443, Kingstown ☎ +1 784 456 1057 🖷 +1 784 456 1696. **W:** www.praisefmsvg.com. L.P.: PD Donny Daniel - **FM:** 95.7/105.7MHz. Format: Rlg – **STAR FM**, Murray's Rd, McKies Hill, PO Box 1651, Kingstown ☎ +1 784 453 7827 🖷 +1 784 485 7827. **W:** www.star983fm.com - **FM:** 98.3/104.7MHz - **TOTAL FM**, 6 Mckies Hill, PO Box 360, Kingstown ☎ +1 784 457 1234 - **FM:** 100.5MHz – **WEFM**, Lower Questelles, PO Box 1346, Kingstown ☎ +1 784 457 9992 🖷 +1 784 457 7123. **E:** wefm@vincysurf.com. **W:** 999wefm.com. L.P.: MD Julius Williams - **FM:** 99.9MHz.

SUDAN

LT: UTC +3h — **Pop:** 42 million — **Pr.L:** Arabic, Nubian, Ta Bedawie, south Sudanic languages — **E.C:** 50Hz, 240V — **ITU:** SDN

MINISTRY OF INFORMATION & COMMUNICATION
W: www.sudannow.net **L.P:** Minister: Al-Zahawi Ibrahim Malek

SUDAN RADIO & TV CORPORATION - SUDAN RADIO(Gov.)
🖳 P.O. Box 1094, Mulazmin, Omdurman ☎+249 87 572956 🖷 +249 87 556006 **Regional** 🖳 P.O. Box 126, Juba. **W:** sudanradio.info srtc.gov.sd **E:** info@sudanradio.info info@srtc.gov.sd
L.P: Dir: Mr. Mutasim Fadul. DG Eng. & Tech. Sces: Abbas Sidig.

MW	kHz	kW	Prgr	MW	kHz	kW	Prgr
Nyala	540	50	R	Wadi Halfa	873	5	R
El Obeid	639	10	R	Malakal	909	5	R
Kassala	666	10	R	Al-Foula	945	5	R
Juba	693	100	R	Khartoum	963	100	H/S
Khartoum	747	10	K	Al-Damazin	1026	5	R
Port Sudan	747	5	R	Wou	1071	5	R
Omdurman	765	50	G	Reiba	1296	600	G
Al-Fashir	801	5	R	Al-Gadarif	1485	5	R
Dongola	819	10	R	Kadogli	1602	5	R
Wad Madani	873	10	R				

SW: Khartoum (Al-Aitahab) 7200kHz 100kW 0200-2100.
FM: in Khartoum: 90.0MHz Prgr S, 95.0MHz Prgr G, 98.0MHz Prgr F, 100.0MHz Prgr N, 105.0MHz Prgr H.

Prgrs: G=General Prgr (incl. **National Unity R**. 1000-1200) in Arabic: 24h (exc. 765kHz 0000-1000 and 963 kHz 1000-0600, exc. R. Peace transmission). **N:** 0400, 0700, 1300, 1600, 1900, 2000. **H=Holy Quran R:** 24h. **K=Khartoum State R:** 0300-0700, 1300-1900. **N=Omdurman FM100. S=R. As-Salam** (Peace) on 963kHz. **T=Two Niles Radio:** 24h. Incl. relays of Deutsche Welle in English. **F=Foreign languages prgr.** in English/French. **R=Regional stations.** Not all the regional stations are confirmed active and some are known to be off the air or on low power because of maintenance problems. Powers listed are the nominal ones. When on air, regional stations carry a mixture of local prgrs and relays of the General Service. Gezira R, Wad Madani: web: gbc.info.sd .**South Sudan R**, Juba 693kHz rep. with news in English between 1530-1445.
Ann: "Huna Omdurman, Idha'atu-l-Gumhuriya as-Sudan"
IS: Sudanese music.

RADIO PEACE (Rlg.)
✉ Global Endeavor, P.O. Box 905, Spotsylvania, VA 22553, USA
W: globalendeavor.org **E:** info@edmedia.org **LP:** Peter Stover, Mgr.
SW: 4740kHz 1kW (Inactive at time of editing, relocating to Juba.)
D.Prgr: Arabic, English and Sudanic langs: 4740kHz: Mon-Fri 0230-0415, 1600-1820v **Ann:** E: "This is Radio Peace"

MIRAYA FM (Joint project by UNMIS and Hirondelle Foundation)
✉ P.O. Box 69, Plot 16-17,Block 25, New Bridge St, Manshiya, Khartoum ☎+249 1 87087777 📠 +249 1 87089465 **W:** mirayafm.
org **E:** mirayasudan@mirayafm.org **LP:** Editor-in-Chief: Jean-Claude Labreque. TD: Sonam Tobgyal.
FM: Juba/Malakal/Rumbek/Wau 5kW, Torit/Yambio/Maridi/Yei/Bor 1kW, all on 101.0MHz. 24h in Arabic, Juba Arabic and English.
F.PI: txs in Khartoum and Darfur, also on 101.0MHz.
Relays on shortwave: see International Radio section.

SUDAN RADIO SERVICE (A project by Educational Development Center)
✉ P.O. Box 425, Plot 48 ,Block I Korok, Juba ☎ +254 20 3870906 📠 +254 20 3876520 **W:** sudanradio.org **E:** info@sudanradio.org
L.P: Chief of Party: Jon Newstrom.
FM: Juba 98.6MHz 1kW.
Relays on shortwave: see International Radio section.

Other stations:
Al-Rabaa FM, Khartoum 94.0MHz. **W:** alrabaafm.com – **Bentiu FM,** Bentiu 99.0MHz – **Capital FM,** Juba: 101.0MHz – **Channel 4,** Khartoum: 94.0MHz. For main entry see UAE – **Internews** stations: **Naath FM,** Leer 88.0MHz 1kW, **Nhomlaau FM,** Malualkon 88.0MHz 1kW, **R. Al-Mujtama,** Kurmuk 88.0MHz 500W, **VO Community,** Kauda: 88.0/107.0MHz 300W. **F.PI:** Turalei 1kW – **Khartoum FM,** Khartoum: 89.0MHz. **W:** kfm89.net – **Mango FM,** Khartoum 96.0MHz. **W:** mango96.com Also rel. BBC – **OUS R,** Khartoum: 89.5MHz. **W:** ousmedia.com – **R. Rumbek,** Rumbek: 98.0MHz – **Spirit FM,** Yei: freq. not known – **Sports FM,** Khartoum: 104.0MHz. **W:** sportsfm. com – **Sudan Catholic R. Netw.** (rlg.): **Bakhita R,** Juba: 91.0MHz. **VO Love,** Malakal: 93.6MHz. **E:** scr.network@gmail.com **F.PI:** 6 more FM stations in Southern Sudan and Nuba Mountains – **Voice of Life,** Kakwa & Madi: 100.9MHz.
BBC World Sce, Juba: 88.2MHz (English), 90.0MHz (Arabic).
DW/Monte-Carlo Doualiya: Khartoum 93.0MHz, Juba 90.4MHz.
R. France Int: Juba 90.4MHz.
R. Sawa, Khartoum: 97.5MHz.
F.PI: UNAMID R. in Darfur on 101 MHz. Web: unamid.unmissions.com

L.T: UTC -3h — **Pop:** 500,000 — **Pr.L:** Dutch, English, Sranang Tongo, Sarnami Hindi, Javanese. — **E.C:** 60Hz, 110/115/127/220V — **ITU:** SUR — **Int. dialling code:** 597

TELECOMMUNICATE BEDRIJF SURINAME (TELESUR) (Gov)
✉ P.O. Box 1839, Paramaribo ☎ 474242/473944 📠 404800 **W:** www.
telesur.sr **E:** telesur@sr.net

STICHTING RADIO-OMROEP SURINAME (SRS)
✉ P.O. Box 271, Paramaribo ☎ 498115 📠 498116 **E:** radiosrs@sr.net
MW: 725kHz 5kW (inactive)
FM: Paramaribo 96.3MHz 1kW, 93.1MHz 0.1kW, Coronie, Wageningen, Nickerie, Mesago all 94.7MHz 0.1kW, Albina 105.7MHz 0.1kW
D.Prgr: 0730-0400 (Sat 0430, Sun 0300). **Sarnami Hindi:** 0730-0830 on all freqs, W 1900-2030 & 2315-2400 on 600kHz + 93.1MHz.
Javanese: 0830-0930. Other times in Dutch. **N. in Dutch:** W 1000, 1100, 1200, 1730, 2000, 2345, 0100; Sun 1600, 1900, 0000 **Ann:** "Dit is de Stichting Radio Omroep Suriname, de SRS in Paramaribo" **V.** by QSL-folder or letter. Rec. acc.

PRIVATE COMMERCIAL STATIONS:

MW	kHz	kW	Name and Location
1)	600	10	R. Paramaribo, Paramaribo, r. "not heard"
2)	820	1	R. Apintie, Paramaribo, r. "not heard"
3)	914	3	R. Nickerie, Nieuw Nickerie, r. "not heard"

SW: R. Apintie, Paramaribo 4990kHz 1kW (inactive)
NB: only FM stns reported active
FM: Paramaribo: 1) R. Paramaribo (Rapar) 89.7MHz 4kW, ACME Broadc. Netw. 92.1MHz 1.5kW – **2)** R. Apintie 97.1MHz 1kW – **4)** Radika NV 98.3MHz 1kW – **5)** R. Sangeet Mala 96.3MHz 1kW – **6)** R. 10 88.1MHz 2kW – **7)** Ampies Broadc. Corp. 101.7MHz 2.5kW, R. Koyeba 104.9MHz 2kW, R. Garuda 105.7MHz 3.5kW – **8)** R. Shalom 94.5MHz

1kW, R. Portjajab 95.3MHz 1.5kW – **9)** R. Zon 107.5MHz 1kW +5 other stns below 1kW. **Other areas:** 25 stns below 1kW

Addresses and other information
1) P.O. Box 975, Paramaribo ☎ 499995 24h. **Sarnami Hindi:** W 0845-0930, 1500-1600; Mon-Thurs 2100-0100; Fri 2200-0200; Sat 2200-0300; Sun 0845-1330, 1930-2100. **Javanese:** W 0800-0845, 1900-2100. Other times in Dutch. **N. in Dutch:** 1830 Ann: «Dit is R.P. Internationaal, the hot one», or «dit is R. Paramaribo, RAPAR N.V. op de 500 meter AM Band en 89.7 FM» – **2)** P.O. Box 595, Paramaribo ☎ +597 400450 📠 400684 **W:** www.apintie.sr **E:** apintie@sr.net 0730-0400 (Sun 0300). **Sarnami Hindi:** W 1900-2030, Sun 1800-1900. **Javanese:** W 2030-2100, Sun 1900-1930. Other times in Dutch. **N. in Dutch:** 1730, 2200 Ann: «U luistert naar R. Apintie op AM-FM stereo en special voor het binnenland op de kortegolf in de zestig meter band». **IS:** The beat of the Apintie drum **V.** by letter. – **3)** R. Nickerie, Waterloostraat 3, Nieuw Nickerie ☎ 231462. - 0900-0030 **Javanese:** 2100-2230. Other times in Sarnami Hindi **IS:** «Surinam hamara pyara desh» sung by G. Kallasing. – **4)** R. Radika, Indira Ghandiweg 165, Paramaribo ☎ +597 482800, 482910 - 0800-0300 in Sarnami Hindi plus some Dutch. – **5)** R. Sangeet Mala, Indira Gandhiweg 40, Paramaribo ☎ 482390 - 0800-0400 (Sat 0500) in Sarnami Hindi. **N. in Dutch:** 0830, 1300. **N. in Hindi:** 0915, 2300. – **6)** R. 10, Letitia Vriesdelaan 5, Paramaribo ☎ 410881 📠 410885 **E:** radio10@cq-link. sr – **7)** Ampies Broadc. Corp., Maystraat 57, Paramaribo ☎ 464609 📠 464680 **W:** www.abcsuriname.com **E:** info@abcsuriname.com – **8)** R. Shalom, Malebatrumstraat 10-12 BV, Paramaribo ☎ 422630 📠 422737 **E:** shalom@sr.net – **9)** R. Zon, Burenstraat 60, Paramaribo ☎ 475261 📠 420233

L.T: UTC +2h — **Pop:** 1.1 million — **Pr.L:** English, Siswati — **E.C:** 50Hz, 230V — **ITU:** SWZ

SWAZILAND POSTS & TELECOMMUNICATIONS CORPORATION (SPTC)
✉ Phutfumani Bldg, Warner St, P.O Box 125, Mbabane ☎+268 2405 2000 📠 +268 24052020 **W:** www.sptc.co.sz **E:** info@sptc.co.sz
L.P: MD: Nathi Dlamini.

SWAZILAND BROADCASTING AND INFORMATION SERVICES (Gov.)
✉ Corner Gwamile & Msakato Str, P.O. Box 338, Mbabane H100 ☎+268 24042761 📠 +268 24042774 **W:** www.gov.sz/home.
asp?pid=65 **E:** sbisnews@africaonline.co.sz
L.P: Dir: Percy Simelane. Asst. Principal prgrs Officer: Phesheya Dube. A/Prgr. Coordinator: Austin Dlamini. A/Tr. Engineer: Christopher Motsa.
FM: 88.5/91.6/93.6/105.2MHz 10kW + 4 low power relays.
English Sce: D.Prgr 0255-1800 on FM 91.6/93.6MHz. **N:** 0400, 0500, 1600. **Siswati Sce: D.Prgr** 0255-2100 on FM88.5/105.2MHz.
Ann: E: "This is the English sce. of Radio Swaziland". Siswati: "Lona ngu Mawakato waka Ngwane".
IS: at s/on, Cilongo (Swazi instrument). English Sce: cock crow, fanfare, spoken ID, instrumental theme. **F.PI:** 100kW tx on 954kHz.

TRANS WORLD RADIO - VOICE OF THE CHURCH
✉ P.O. Box 4544, Corner Martin & Tenbergen St,Manzini ☎+268 25054845 📠 +268 25054809 **L.P:** Nat. Dir: Nelson Vilakati, Adm: Tryphinah Dlamini, PM: Abel Vilakati.
FM: 95/97/101MHz.
TWR Africa: MW Mpangela Ranch 1170kHz 50kW 1600-2030 & SW. For further details see International Radio section.

F.PI: private radio stations.

L.T: UTC +1h (27 Mar-30 Oct: +2h) — **Pop:** 9.2 million — **Pr.L:** Swedish — **E.C:** 50Hz, 230V — **ITU:** S

TERACOM AB
Responsible for distribution of prgrs. produced by Sveriges Radio (Swedish Broadcasting Corporation) and by most of the commercial radio stns and community radio associations.
HQ: (✉ Box 1366, SE-17227 Sundbyberg ☎ +46 8 55542000 📠 +46 8 55542001 **W:** www.teracom.se **L.P:** MD: Stephan Guiance

PTS
PTS (Post-och telestyrelsen) is the authority that supervises activities in radio, telecom and datacom.

✉ Box 5398, SE-10249 Stockholm ☎ +46 8 6785500 **W:** www.pts.se
E: pts@pts.se **LP:** Director General .Göran Marby

SVERIGES RADIO AB
(Swedish Broadcasting Corporation) (Public)
HQ: Sveriges Radio AB,
Radiohuset, Oxenstiernsgatan 20, Stockholm (✉SE-10510 Stockholm)
☎+ 46 8 7845000 📠+ 46 8 7841500 **W:** www.sr.se **E:** lyssnarservice@sr.se **LP:** MD Mats Svegfors.

FM (MHz)	1	2	3	4	kW
24) Arvidsjaur	89.4	94.2	97.1	100.6	60
20) Bollnäs	88.4	91.7	96.0	103.8	60
19) Borlänge	89.4	93.0	97.7	101.3	60
25) Borås	88.5	94.6	97.9	102.9	10
14) Bäckefors	92.7	96.8	99.1	102.2	60
6) Emmaboda	93.0	96.7	99.7	101.8	60
7) Emmaboda				95.6	60
5) Finnveden	90.1	94.2	99.9	103.4	30
24) Gällivare	88.3	94.9	98.5	100.9	60
20) Gävle	88.1	97.4	99.8	102	60
13) Göteborg	89.3	96.3	99.4	101.9	60
12) Halmstad	87.7	91.2	95.4	97.3	60
10) Halmstad				102.6	3
11) Helsingborg	89.8	95.7	98.4	103.2	6
20) Hudiksvall	87.6	90.2	93.7	100.7	60
10) Hörby	88.8	92.4	97.0	101.4	60
11) Hörby				89.5	5
24) Kalix	91.3	93.6	97.9	102.2	60
9) Karlshamn	90.3	93.4	98.3	100.4	15
9) Karlskrona	89.1	95.0	97.7	100.7	10
17) Karlstad	90.5	94.2	96.5	103.5	15
24) Kiruna	89.1	92.7	96.4	102.7	60
4) Kisa	90.5	92.5	96.9	103.6	30
23) Lycksele	92.9	95.4	98.7	103.3	60
11) Malmö	87.9	93.3	98.0	102.0	6
11) Malmö*				100.6	6
19) Mora	92.2	96.7	99.0	101.0	60
4) Motala	91.1	94.0	98.2	101.2	20
3) Norrköping	90.0	93.5	98.7	102.3	60
4) Norrköping				94.8	60
5) Nässjö	89.6	92.1	99.0	102.1	60
24) Pajala	90.8	93.0	95.9	100.2	60
23) Skellefteå	93.8	96.3	100.0	103.9	60
16) Skövde	88.9	95.1	97.5	100.3	60
21) Sollefteå	87.9	93.5	98.1	101.2	60
1) Stockholm	92.4	96.2	99.3	103.3	60
1) Stockholm**		89.6			0.9
1) Stockholm***			93.8		0.9
23) Storuman	87.6	91.2	99.0	102.5	60
21) Sundsvall	92.7	96.9	99.2	102.8	60
16) Sunne	90.9	94.5	98.5	101.8	60
22) Sveg	90.6	94.9	97.9	102.2	60
22) Tåsjö	89.9	94.7	97.5	100.8	60
23) Tåsjö				88.2	60
2) Uppsala	90.3	93.3	96.6	102.5	20
12) Varberg	90.4	93.6	98.8	103.8	10
8) Visby	87.6	94.1	97.2	100.2	60
6) Vislanda	88.0	90.6	94.7	101.0	20
23) Vännäs	88.5	92.1	95.8	103.6	60
7) Västervik	88.3	91.8	96.0	102.7	60
15) Västerås	90.7	95.8	98.0	100.5	60
21) Änge	93.2	95.6	99.6	103.1	60
22) Änge				94.5	60
24) Älvsbyn	90.6	94.5	99.4	102.9	60
17) Örebro	87.9	91.5	99.6	102.8	60
21) Örnsköldsvik	90.8	94.4	9787	100.1	60
22) Östersund	87.9	91.5	94.0	100.4	60
2) Östhammar	89.1	92.8	95.5	101.6	60
24) Överkalix	88.9	91.7	99.0	103.2	15

+ 360 low power txs. A comprehensive list of all stns is available on **W:** www.teracom.se
*) Broadcasts "Din Gata", a local P3-prgr only for the Malmö-area.
**) P6 International. Prgrs for minorities and immigrants plus classical music. 24h. 2200-0200 and 0300-0600 relay of BBC World Service.
***) SR Metropol. Music for young people, 24 h (additional frequencies in greater Stockholm area: 97.6, 102.9 MHz).

First Prgr. W: www.sr.se/p1 (news & spoken word): MF 0429-0045,SS0455-0030. News:MF 0430, 0500, 0530, 0600, 0630, 0700, 0800, 0900, 1000, 1130, 1300, 1400, 1445, 1500, 1545, 1645, 1800, 1900, 2000, 2100, 2200, 2300,2400. **Wrp** (incl. forecast for Swedish waters): 0455, 0555, 0655, 11155, 1455, 2050. **Rel. of 1st programme on SW:** See Foreign Sce. schedule.

Second Prgr. W: www.sr.se/p2 24h: Classical music and jazz prgrs, Sami, Finnish and prgrs for immigrants.
Third Prgr. W: www.sr.se/p3: light music, entertainment, current affairs, news for listeners under 40s: 24h.
Fourth Prgr. W: www.sr.se/p4 24h: Regional network 0511 - 1700. 1700-0500 relay of P4 network from Stockholm (frequencies as above, addresses given below).
P4 Radio Stockholm. W: www.sr.se/stockholm 103.3 MHz Local prgrs for Stockholm area. 0503-2303. 2303-0503 Relay of main P4 prgr.
Ann: Nat. Prgr. "Sveriges Radio" and the service e.g. "Sveriges Radio P1"

Digital Radio (DAB): Malmö 2kW, Göteborg 2kW, Stockholm/Nacka 4.0kW, Enköping 2.0kW, Uppsala 2.0kW, Södertälje 2.0kW, Älvsbyn 3.8kW Single frequency network on 225.648MHz, block 12B with SR Atlas, , SR Klassiskt, SR Minnen, SR P1, , P7 Sisuradio, P3 Star and SR knattekanal.

Addresses of Regional Centres
1) SR Stockholm, Pipersgatan 45, 107 80 Stockholm. – **2)** SR Uppland, Box 1552, 751 45 Uppsala. – **3)** SR Sörmland, Box 641, 631 08 Eskilstuna. – **4)** SR Östergötland, Box 500, 601 07 Norrköping. – **5)** SR Jönköping, 551 92 Jönköping. – **6)** SR Kronoberg, Box 62, 351 03 Växjö. – **7)** SR Kalmar, 391 83 Kalmar. – **8)** SR Gotland, Box 1324, 621 24 Visby. – **9)** SR Blekinge, Box 305, 371 25 Karlskrona. – **10)** SR Kristianstad, Box 505, 291 25 Kristianstad. – **11)** SR Malmöhus, 211 01 Malmö. – **12)** SR Halland, Box 133, 301 04 Halmstad. – **13)** SR Göteborg, Pumpvägen 2, 405 13 Göteborg. – **14)** SR Väst, Box 654, 451 24 Uddevalla. – **15)** SR Västmanland, Box 850, 721 22 Västerås. – **16)** SR Skaraborg, 541 24 Skövde. – **17)** SR Värmland, Box 98, 651 03 Karlstad. – **18)** SR Örebro, Västra Bangatan 15, 701 80 Örebro. – **19)** SR Dalarna, Box 123, 791 23 Falun. – **20)** SR Gävleborg, Box 6702, 801 74 Gävle. – **21)** SR Västernorrland, 851 79 Sundsvall. – **22)** SR Jämtland, Lingonvägen 7 B, 831 62 Östersund. – **23)** SR Västerbotten, Mariehemsvägen 4, 906 15 Umeå. – **24)** SR Norrbotten, Nygatan 3, 971 71 Luleå. **25)** SR Sjuhärad, Box 27, 503 05 Borås.
Sameradion, Föreningsgatan 15, 981 23 Kiruna. Responsible for programs in Sami **E:** avvudanboddu@sr.se.
Sisuradio, Box 703222, 107 23 Stockholm
SR Metropol, Pipersgatan 45, SE-107 80 Stockholm **E:** metropol@sr.se
Further information about email addresses, tel numbers and prgrs can be found at **W:** www.sr.se

SVERIGES UTBILDNINGSRADIO AB (Pub)
(Swedish Educational Broadcasting Company)
✉ UR, Tulegatan 7, 113 95 Stockholm ☎ + 46 8 7840000
W: www.ur.se **E:** kundtjanst@ur.se. **LP:** MD:Erik Fichtelius **FM:** See Swedish Radio
Local Commercial Radio:

Location	MHz	kW	Net	Location	MHz	kW	Net
Skellefteå	92.4	1.5	C	Borlänge	105.5		
Vimmerby	100.0	1	C	Sundsvall	105.5	1	A
Södertälje	100.8	3	C	Borås	105.5	1	A
Falun	103.2	3	A	Stockholm	105.5	2	C
Östersund	104.0	3	C	Luleå	105.6	2	C
Färjestaden	104.0	3	C	Nyköping	105.7	4	A
Uddevalla	104.2	1	A	Uddevalla	105.7	2	C
Oskarström	104.2	1	A	Växjö	105.8	1	A
Umeå	104.2	5	A	Göteborg	105.9	3	A
Stockholm	104.3	1	A	Helsingborg	106.0	1	A
Växjö	104.3	2	C	Jönköping	106.0	5	C
Karlstad	104.4	10	C	Lund	106.1	4	A
Visby	104.4	6	C	Västerås	106.1	3	A
Eskilstuna	104.5	10	A	Visby	106.1	6	A
Finnveden	104.6	1	A	Borlänge	106.3	1	B
Mora	104.6	1	C	Stockholm	106.3	1	B
Örebro	104.7	2	A	Örebro	106.3	2	A
Örnsköldsvik	104.8	10	A	Karlshamn	106.3	1	C
Gävle	104.9	1	A	Skövde	106.4	3	A
Trollhättan	105.0	3	C	Karlskrona	106.4	1	C
Stockholm	105.1	1	B	Uppsala	106.5	4	D
Hudiksvall	105.1	3	A	Varberg	106.5	1	A
Jönköping	105.1	4	A	Norrköping	106.5	3	A
Stockholm	105.1	1.3	B	Sundsvall	106.6	5	A
Malmö	105.2	1	B	Gävle	106.7	1	A
Gällivare	105.2	10	A	Malmö	106.7	3	A
Göteborg	105.3	1	B	Mora	106.8	1	A
Uppsala	105.3	2	C	Gnosjö	106.8	3	C
Skellefteå	105.4	2	A	Norrköping	106.9	3	A
Färjestaden	105.4	2	A	Luleå	106.9	3	A
Karlstad	105.4	5	A	Västerås	106.9	1	A
Kalmae	105.4	1	A	Linköping	106.9	2	A

Location	MHz	kW	Net	Location	MHz	kW	Net
Malmö	107.0	3	A	Göteborg	107.3	3	A
Hudiksvall	107.0	3	C	Kristianstad	107.3	1	C
Kiruna	107.0	3	A	Eskilstuna	107.3	3	A
Umeå	107.0	3	A	Helsingborg	107.6	1	C
Borås	107.1	3	C	Skövde	107.6	5	C
Örnsköldsvik	107.1	5	C	Nyköping	107.7	4	C
Halmstad	107.2	1	C				

+ 30 stns below 1kW. **NB:** These stns belong to networks. There are a number of commercial local stns on FM. A comprehensive list can be found on **W:** www.teracom.se

Addresses and other information:
A) MIX MEGAPOL, 115 78 Stockholm **W:** www.mixmegapol.com **E:** info@mixmegapol.com
B) NRJ, P.O. Box17115, 10462 Stockholm ☎ +46 8 6589800 **W:** www.nrj.se **E:** info@nrj.se
C) RIX FM, Box 17820, 118 94 Stockholm ☎ +46 8 56272000 🖷+46 8 56272082 **W:** www.rixfm.com **E:** rix@rixfm.com
D) RADIO GULD, 85172 Sundsvall ☎ +46 60 197080 **W:** www.guld.nu **E:** guld@st.nu

Private stations in the Stockholm area
FM(MHz): 88.0 City 1, 96.3 City 2, 97.3 Solna-Sundbyberg, 101.1 Radio Sydost, 101.9 Star FM, 104.3 Mix Megapol, 104.7 Lugna Favoriter, 105.1 Energy, 105.5 Rix FM, 105.9 The Voice, 106.3 Bandit, 106.7 Rockklassiker, 107.1 Vinyl, 107.5 Studio

COMMUNITY STATIONS
Närradio is open for any non-commercial organization, whose main activity is other than broadcasting. The organization may obtain a permit for community radio broadcasting by PTS. The txs are made available at a nominal fee and built and operated by Teracom AB. The txs have powers of 10-400W and the target is the local community. There are more than 200 txs in operation. Frequency range: 88-108MHz. A few Närradio stns also broadcast commercial prgrs. A comprehensive list of Närradio-stns is at **W:** www.teracom.se

Närradio in greater Stockholm area:
FM(MHz):, 88.2 Radio Sigtuna, 88.9 Sydväst, 90.5 MRS, 91.4 Tyresö, 91.6 BMU, 94.2 Järfälla, 94.5 Hit FM, 94.6 Sollentuna, 95.3 City 2, 97.3 Solna/Sbg, 97.8 Lidingö, 98.3 Radio Nord, 08.5 Haninge, 99.9 Nacka, 101.4 Radio Viking, 103.7 Radio Österåker and 107.8 Radio Roslagen.

SWITZERLAND

L.T: UTC +1h (27 Mar-30 Oct: +2h) — **Pop:** 7.6 million — **Pr.L:** German, Swiss-German dialects, French, Italian, Rumantsch — **E.C:** 50Hz, 230V — **ITU:** SUI

SRG SSR IDÉE SUISSE / SSR - Société Suisse de Radiodiffusion et Télévision (Pub.)
📧 Giacomettistrasse 1, Postfach 570, CH-3000 Bern 31 ☎ +41 31 3509111 🖷+41 31 3509256 **W:** srgssrideesuisse.ch **L.P:** Pres: Jean Bernard Münch. DG: Armin Walpen. DG adj: Daniel Eckmann.

GERMAN LANGUAGE NETWORK
Schweizer Radio DRS (SR DRS)
📧 Radiodirektion SR DRS, Novarastrasse 2, CH-4002 Basel ☎ +41 61 365 3411 🖷 +41 61 365 3483 **W:** drs.ch **E:** radiofon@srdrs.ch
L.P: Dir: Iso Rechsteiner.

FM (MHz)	DRS 1	DRS 2	DRS 3	RR	kW
Bantiger	88.2	93.2	99.3		5
Biel-Magglingen	-	99.7			1
Castel S.Pietro	98.8				10
Celerina	91.9	100.3	106.3	89.1	0.75
Chalavornaire	93.2				1
Chasseral	103.0		105.3		20
Feschel	88.2	90.3	101.5		1.5
Froburg	96.0	98.7	91.3		1.2
Gebidem	89.4	93.9	103.9		2.4
Glarus	91.4				1
Haute-Nendaz	92.0				2.5
Leucel	88.1				6
Castel S. Pietro	98.8				10
Niederhorn	93.6	97.2	105.8		4.5
Paudo	96.9				2.2
Pfaender	96.3	97.7	107.5		4
Ravoire	102.1				1.9
Rigi	90.9	96.6	103.8		30
Salève	87.8				1
San Salvatore	96.3				17
Säntis	101.5	95.4	105.6		55

FM (MHz)	DRS 1	DRS 2	DRS 3	RR	kW
Schaltenrein	90.7				1
Solothurn	89.7	98.0			1.5
St. Chrischona	90.6	99.0	103.6		33
Tarasp	101.3	103.9	95.1	98.7	0.25
Uetliberg	94.6	99.6	105.8		1.6
Valzeina	93.8	102.5	104.3	90.3	0.75

+ numerous trs less than 1kW. – RR=Rumantsch (see below).
1st Prgr(DRS 1): 24h. **Local Broadc. in German:** MF 0532-0534,MF 0615-0617, MF 0632-0637,MF 0712-0714, Mon-Sat 1103-1113, MF + Sun1630-1700, Sat 1630-1640 – **2nd Prgr(DRS 2):** 24h. – **3rd Prgr(DRS 3):** 24h – DRS Virus: Satellite Hot Bird 12399 GHz, DAB and cable 24h – **DRS 4 News:** Satellite Hot Bird 12399 GHz, DAB and cable 24h. All prgrs also via Internet.

RUMANTSCH LANGUAGE NETWORK
Radio e Televisiun Rumantscha (RTR)
📧 Theaterweg 1, CH-7002 Chur ☎+41 81 2557575 🖷 +41 81 2557500 **W:** www.rtr.ch **L.P:** Dir: Mariano Tschuor.
FM: see RR netw. (above) **D.Prgr:** 0500-2300, 2300-0500 rel. DRS1.

FRENCH LANGUAGE NETWORK
Radio Télévision Suisse (RTS) - Radio Suisse Romande (RSR)
📧 20 Quai Ernest-Ansermet, CP 234, CH-1211 Geneve 8 ☎+41 22 708 2020 🖷 +41 22 708 9800 **W:** rsr.ch **L.P:** MD: Gilles Marchand.

FM (MHz)	RSR1	RRS2	RSR3	RSR4	kW
Anzere				106.5	0.5
Bantiger	95.1				5
Castel S.Pietro	87.8				10
Chasseral	102.3	100.3	104.2		20
Chamossaire	105.1/98.1	95.1	88.6		1
Chaux de Fo.	92.3	96.3	103.4		1.4
Dôle, La	91.2	100.1	105.6		19
Feschel	91.4	96.1	107.4		1.5
Gebidem	90.8				2.4
Gibloux	91.0	92.5	88.6		1.0
Hte Nendaz	94.4	96.5	106.0		2.5
Les Ordons	94.2	99.6	104.8		9
Leucel	102.6	96.2	98.5		6
Martigny				90.8	0.6
Montmagny	92.3	92.0	89.1		4
Mt. Pélerin	91.6	101.5	90.6		2
Nendaz				97.5	0.5
Paudo	105.3				2.8
Premier	94.7	100.8	104.7		10
Ravoire	93.2	106.9	100.5		1.9
Salève	94.9	100.7	104.4	90.8	1
San Salvatore	104.0				17
Säntis	99.9				55
Schaltenrein	96.9				1
Vallée de Joux	99.5	87.6	101.4		1
Vercorin				98.5	0.5

+ many txs less than 1kW.
RSR1 (La Première): 24h. **N:** On the h. (exc. 2100, 2200). Also 0530, 0630, 1130, 2130 – **RSR2 (Espace 2):** 24h. **N:** 0500, 0600, 0700, 0800, 1200(M-F), 1600, 1800, 2100, 2300 – **RSR3 (Couleur 3):** 24h non-stop music and N – **RSR4 (Option Musique):** 24h.

ITALIAN LANGUAGE NETWORK
Radiotelevisione Svizzera (RSI)
📧 Via Canevascini, Casella postale 235, CH-6903 Lugano-Besso ☎+41 91 803 5111 🖷 +41 91 803 5355 **W:** rsi.ch **E:** info@rsi.ch
L.P: Dir: Dino Balestra.

FM (MHz)	RSI I	RSI 2	RSI 3	kW
Celerina	104.3			2
Chasseral	107.3			20
Gebidem	96.7			2.4
Leucel	97.8			6
Castel S. Pietro	88.8	98.8	104.5	10
Paudo	89.4	93.5	107.4	2.8
Rigi	106.2			30
Salève	97.1			2
S. Salvatore	88.1	91.5	106.0	17
Valzeina	95.8			1.5

+ numerous txs less than 1kW.
RSI1 (Rete 1): 24h. **N:** hourly (not 1200, 1800) + 1130, 1730. – **RSI2 (Rete 2):** 24h. **N:** 0530, 0630, 0730, 0830, 0930, 1030, 1130, 1200, 1330, 1430, 1530, 1630, 1800, 2200, 2300. Classical music and cultural prgrs – **RSI3 (Rete 3):** 24h. **N:** 0530, 0630, 0730, 0830, 0930, 1330, 1530, 1800, 2120, 2300.

WORLD RADIO SWITZERLAND
📧 Passage de la Radio 2, CH-1205 Geneva ☎+41 22 7087444 🖷 +41

22 7087454 **W:** www.worldradio.ch **E:** management@worldradio.ch
FM; Salève 101.7MHz 2kW. **D.Prgr** in English: 24h.

Private stations, main networks (FM, MHz):
R. Central: 89.3 Matt, 89.4 Steinerberg, 89.4 Wangen, 89.6 Menzingen,
92.1 Weesen, 93.6   Stans, 97.7 Haslen, 97.7 Spiringen, 98.8
Seelisberg, 99.0 Giswil/Sarnen, 99.2 Küssnacht, 99.2 Zug, 100.1
Sonnenberg, 101.8 Andermatt, 101.8 Feusisberg, 101.8 Mattgratt, 102.6
Ingenbohl, 103.0 Attinghausen. **Web:** www.radiocentral.ch
R. Grischa: 87.9 Avers, 89.3 Lohn, 89.4 Safien, 89.6 Ruschein, 89.8
Lenzerheide, 94.5 Sils i.D. Crap, 95.0 Sedrun, 95.1 Feldis, 97.2 Davos,
98.5 Arosa, 99.2 Malix, 99.7 Gotschnagrat, 102.3 Bergün, 102.4 Vals,
102.7 Splügen, 106.2 Mon, 107.0 Curaglia, 107.0 Valzeina, 107.4
Morissen. **Web:** www.radiogrischa.ch
R. Pilatus: 87.6 Kussnacht/Schupfheim/Zug, 92.0 Kriens, 92.8
Dagmersellen, 93.0 Sursee, 95.7 Menzingen, 95.8   Beckenried/
Ennetbürgen, 100.7 Willisau, 103.4 Igenbohl/Sarnen, 104.4 Engelberg/
Ennetmoos/Giswil/Höchweidwald, 104.6 Seelisberg, 104.9 Sonnenberg,
106.5 Werthenstein. **Web:** www.radiopilatus.ch
R. Sunshine: 88.0 Rooterberg/Zug, 88.8 Einsielden/Horgen/
Widen/Willisau, 90.0 Igenbohl/Kriens, 93.4 Kriens/Menzingen, 93.6
Dagmersellen, 94.2 Giswil/Mattgratt/Sarnen/Sursee, 99.6 Seelisberg/
Lopper/Werthenstein. **Web:** www.sunshine.ch
R. Top: 87.9 Gossau, 88.2 Oberwinterthur, 88.5 St. Gallen, 90.0 St.
Gallen, 93.1   Wil, 95.0 Feusisberg, 97.7 Strichboden, 98.9 Fluringen,
99.1 Ebnatt Kappel, 99.4 Niederhasli, 99.5 Kreuzlingen, 100.2 Reutenen
Homburg, 101.9 Elgg, 103.3 Schaffhausen, 103.5 Winterthur, 104.5
Zürich. **Web:** www.toponline.ch/radiotop.shtml
RJB: 90.0 Bienne, 90.3Sonceboz, 90.4 Plagne, 91.8 Lovresse, 95.5
Cormoret/Moutier, 95.9 Court Frête, 96.3 Gerolfingen, 97.5 Orvin, 101.2
Prêles/Tavannes, 102.0 Tramelan, 103.4 Delémont, 104.5 Péry, 104.9
Gals, **Web:** www.rjb.ch
RTN: 87.9 Bôle, 90.2 Monts de Verrières, 93.2 Les Brenets, 93.4 Noiraigue,
94.2 Fleurier/Haut-Geneveys, 95.8 Marin, 97.5 Dombresson Clémesin,
97.5 La Brévine/Villiers, 98.2 Locle, 98.2 Montmagny/Neuchatel, 101.0
Brot, 101.7 Chaux-de-Fonds, 107.6 Yverdon. **W:** www.rtn.ch
NB: Tx powers vary from a few W to 5 kW. 40 more stns in operation.

DAB: Palette DeutschSchweiz: channel 12C 227.360MHz. Prgrs:
DRS1, DRS2, DRS3, DRS Musigwälle, DRS Virus, R. Rumantsch,
Swiss Pop, Swiss Classic, Swiss Jazz, RSR La Premiere+, RSI Rete
Uno+ & DRS 4 News WRS + – **Palette Suisse Romande:** channel
12A 225.648MHz. Prgrs: RSR La Première, RSR Espace 2, RSR Couleur
3, RSR Option Musique, Swiss Pop. Swiss Classic, Swiss Jazz, R.
Rumantsch, DRS 1+, RSI Rete Uno+, DRS Musigwälle+ & World R.
Switzerland – **Palette Svizzera Italiana:** channel 12 A 223.936 MHz.
Prgrs: Rete Uno, Rete Due, Rete Tre, DRS Musigwälle, R. Rumantsch,
Option Musique , Swiss Classic, Swiss Pop, Swiss Jazz & World R.
Switzerland – **Palette Rumantsch:** channel 12 D 229.072 MHz. Prgrs:
DRS1+ DRS2, DRS3+, DRS Musigwälle, DRS Virus+, R. Rumantsch,
Swiss Pop+, Swiss Classic+, Swiss Jazz+, RSR La Premiere+, RSI Rete
Uno+ ,DRS 4 News, WRS+ – **Palette Privat SMC Deutschschweiz:**
channel 7 D 194.064 MHz. Prgrs: Backstageradio+, Lifechannel+, Open
Broadcast+, Radio Basel+, Radio Inside+ ,Energy+, Eviva+ ,Top Two+,
Swissmountainholyday Radio+, R. Maria+, Option Musique +, Rete
Tre+, DRS 4 News+, CH Classic+

SYRIA

L.T: UTC +2h (25 Mar-1 Nov: +3h) — **Pop:** 20 million — **Pr.L:** Arabic
— **E.C:** 50Hz, 220V — **ITU:** SYR

MINISTRY OF INFORMATION
⌨ Mezzeh Autostrad, Dar al Ba'th Building, Damascus ☎+963 11
6664681 🖷 +963 11 6664681 **W:** www.moi.gov.sy **E:** info@moi.gov.sy
L.P: Talib Qadi Amin, Asst. Minister.

SYRIAN RADIO AND TV (Gov.)
⌨ Radio & TV Directorate, Ommayad Square, Damascus ☎+963 11
2720700 🖷 +963 11 2234930 **W:** www.rtv.gov.sy **L.P:** DG: Fayez Al
Sayegh. Dir. Eng: Adnan Salhah. Dir. Radio: Mahmoud Al Joma'at.

MW	kHz	kW	Prgr	MW	kHz	kW	Prgr
Adra	567	300	1	Al-Hassake	918	200	1
Homs	594	100	2/Y	Homs	936	100	1
Sabboura	666	600	2/M	Deir ez-Zor	954	50	2
Sarakeb	747	100	1	Tartus	1071	100	*
Tartus	783	300	1/E	Al-Hassake	1125	100	2
Deir ez-Zor	v828	200	1	Sarakeb	1314	50	A/E/Y
Adra	873	50	2	*relays R. Al-Nour, LBN			

FM	1	2/M	Y	kW
Afrin		90.3	96.6	
Al-Hassake	89.9		99.5	
Aleppo	96.1	99.4	89.9	10
Bloudan	93.5		91.9	
Damascus	95.0	98.3	88.7	
Deir ez-Zor			87.8	150
Maalaqa			99.0	
Nabi Saleh	89.0			
Raqqah	103.7		93.7	
Slenfe	94.9	91.7	88.6	150
Sweida	92.6	87.8	100.9	150

General Prgr (1): 24h. Incl. **VO Armed Forces:** 1530-1600 and **R.
Palestine** prgr: 1700-1730.
Voice of the People (2): 0400-2200.
Voice of Youth (Y) 0400-2400. **Aleppo local prgr (A):** 1300-1600.
Musical prgr. (M) in English/French: 1800-2000. **E:** External Service
is relayed by 2nd network. **NB:** the segments and networks carried on
each frequency may vary.
Ann: 1: "Idha'at Dimashq". 2: "Huna Idha'at Sowt as-Sha'ab min
Dimashq". Y: "Huna Sowt as-Shabab". R. Palestine: "Idha'at Falasteen
min Dimashq". Aleppo: "Idha'at al-Halab".

EXTERNAL SERVICE: R. Damascus: see International Radio section.

Other stations:
Al Madina FM: Slenfe 100.5MHz, Aleppo/Damascus 101.5MHz.
W: www.almadinafm.com – **Arabesque FM:** Aleppo/Damascus
102.3MHz, Slenfe 106.9MHz. **W:** arabesque.fm – **Farah FM:** Aleppo/
Damascus 97.3MHz. **W:** www.farah.fm – **R. Gecko (UN):** Camp
Faouar, Golan: 103.8MHz. **W:** www.radio-gecko.com **E:** office@
radio-gecko.com – **Melody FM:** Aleppo/Damascus 99.5MHz. **W:**
melodysyria.com – **Mix FM:** Damascus: 105.7MHz. **W:** mixfm-
syria.com – **Ninar FM:** Aleppo 88.3MHz, Slenfe 89.6MHz, Damascus
93.8MHz. **W:** ninarweb.com – **Rotana Style FM:** Slenfe 103.3MHz,
Aleppo/Damascus 105.0MHz. **W:** www.rotanastyle.com – **R. Fann:**
Aleppo 89.0MHz, Damascus 89.2MHz, Slenfe 106.1MHz. **W:** fann-
fm.com – **Sawt el-Ghad,** Damascus: 99.9MHz. **W:** sawtelghad.com
– **Shahba FM,** Aleppo 94.0MHz. **W:** www.shahbafm.com – **Sham
FM:** Damascus 92.3MHz, Aleppo 95.3MHz, Slenfe 101.8MHz. **W:**
www.shamfm.fm – **Syria Al-Ghad FM:** Damascus 104.2MHz, Aleppo
104.4MHz, Slenfe 107.4MHz. **W:** syriaalghad.com

TAIWAN (Rep. of China)

L.T: UTC +8h — **Pop:** 23 million — **Pr.L:** Mandarin(Chinese), Amoy,
Hakka — **E.C:** 60Hz, 110V — **ITU:** CHN (**WRTH:** TWN)

BROADCASTING CORPORATION OF CHINA (BCC)
(Priv. Comm.)
⌨ 375 Sungchiang Rd, Chungshan Ward, Taipei 104 ☎ + 886 2 2501
9688 🖷 + 886 2 2501 8834 **W:** www.bcc.com.tw **E:** pr@bcc.com.tw
L.P: Chairman: Kuang Hsiang-Hsia

Call: BE followed by the callsign below

	MW Call	Location	kHz	kW	Netw.
10)	D65	Ilan	630	10	N
1)	D34	Taipei	657	20	N
6)	D92	Tainan	711	10	N
4)	D58	Taichung	720	10	C
1)	D57	Taipei	747	10	H
4)	D28	Taitung	819	10	C
4)	D43	Taichung	837	10	C
9)	D27	Hualien	855	10	N
7)	D25	Kaohsiung	864	10	N
2)	G77	Hsinchu	882	10	N
6)	D24	Tainan	891	10	L
7)	D79	Kaohsiung	909	10	C
1)	D55	Taipei	954	20	C
8)	D88	Taitung	1008	10	C
2)	D53	Hsinchu	1017	10	C
5)	D26	Chia-i	1035	10	C
4)	D23	Taichung	1062	10	L
11)	D72	Yuli	1116	3.5	N
12)	D68	Puli	1152	1	C
10)	D86	Ilan	1161	10	C
3)	D89	Miaoli	1161	10	C
9)	D32	Hualien	1188	10	C
5)	D52	Kaohsiung	1224	10	C
6)	D47	Tainan	1296	10	N
5)	D63	Chia-i	1350	10	N
11)	D74	Yuli	1386	3.5	C

MW	Call	Location	kHz	kW	Netw.
3)	D54	Miaoli	1413	10	N
12)	D67	Puli	1413	1	N

N=News Netw, C=Country Netw, H=Hakka Ch, L=local

FM	Location	P	F	M	kW
1)	Taipei	103.3	105.9	96.3	35/10
3)	Huoyenshan	102.9	101.5	96.1	10/0.1
4)	Taichung	102.1	106.9	96.3	35/10
5)	Chentoshan	103.1	104.3	96.1	10/0.1
7)	Kaohsiung	103.3	105.9	96.3	35/10
8)	Taitung	102.1	106.9	96.3	5/2.5
9)	Hualien	102.1	106.9	96.3	5/2.5
10)	Ilan	102.1	102.9	96.1	5/2.5
11)	Yuli*	103.3	105.7		2.5/1
12)	Puli*	107.3			0.1
13)	Chinmen*	96.3			1

P=Pop Netw, F=Formosa Netw, M=Music Netw. *) relay station.

D.Prgr: News Network: 24h in Mandarin. – **Country Network:** 24h in Amoy. – **Hakka Channel:** 24h in Hakka. – **Pop Network:** 24h in Mandarin. – **Formosa Network:** 24h in Amoy. – **Music Network (i radio):** 24h in Mandarin.

Addresses of local stations:
2) 3, 9th Flr, 55 Tungkuang Rd, Hsinchu 300. – **3)** 78, Lane 1008, Chungshan Rd, Kaomiao Li, Miaoli 360. – **4)** 35th Flr, 758 Chungming So. Rd, Taichung 402. – **5)** 121 Wufeng So. Rd, Chia-i 600. – **6)** 5, 19th Flr, 248, Sec. 2, Yunghua Rd, Anping, Tainan 708. – **7)** 1, 24th Flr, 91 Chungshan 2nd Rd, Chienchen, Kaohsiung 806. – **8)** 23, Lane 52, Kuilin No. Rd, Taitung 950. – **9)** 25 Shuiyuan Str, Hualien 970. – **10)** 8 Kuchie Rd, Chuangwei Village, Ilan 263. – **11)** Yuli (relay st.) – **12)** Puli (relay st.) – **13)** Chinmen (relay stn), relays News Netw: W2300-0100, W0800-1000, Sun1300-1400.

Ann: Mandarin: "Chungkuo Kuangpo Kungssu" or "Chungkuo Kuangpo Kungssu, (location) Kuangpo Tientai", Amoy: "Tiyon Gok Kon Po Kon Sih, (location) Kon Po Den Tai"

EXTERNAL SERVICES: Radio Taiwan International
see International Broadcasting section

HAN SHENG BROADCATING CORPORATION (Gov)
(operated by General Political Warfare Bureau, Ministry of National Defense)
✉ B, 5th Flr, 3, Sec. 1, Hsin-i Rd, Chungcheng, Taipei 100 ☎ + 886 2 2321 5191 ▤ + 886 2 2396 2657 **W:** www.voh.com.tw

MW	Call	Location	kHz	kW
1)	C22	Taipei	684	10
1)	C25	Taoyuan	693	10
1g)	C32	Tainan	693	10
1d)	C33	Hualien	792	10
1f)	C38	Penghu	846	10
1)		Taoyuan	936	5
1)	C31	Yunlin	1089	10
1)	C22	Taipei	1116	10
1)	C30	Ilan	1116	10
1f)	C44	Penghu	1269	10
1b)	C27	Taichung	1287	10
1e)	C36	Tsoying	1332	10
1d)	C40	Hualien	1359	5

FM	Call	Location	MHz	kW
1g)	C28	Tainan	101.3	35
1)	C26	Miaoli	104.5	35
1d)	C35	Hualien	104.5	3
1)	C39	Taitung	105.3	3
1)	C24	Taipei	106.5	35
1)		Ilan*	106.5	
1d)	C34	Hualien	107.3	35
1c)	C37	Kaohsiung	107.3	3
1h)		Chinmen*	107.3	

*) relay station

D.Prgr: MW: 2100-1600, FM 24h. **Rel. RTI "Sound of Taiwan" domestic sce:** Indonesian prgr: D1300-1400 on MW, MF2105-2200 on MW, Thai prgr: MF0420-0500 on MW & FM, D1200-1300 on MW, 2300-2400 on Fri & Sat on MW & FM, Vietnamese prgr: D2200-2300 on MW (+FM on Fri & Sat), D1400-1500 on MW – **1a)** Kuo Kuang Broadc. St, 122, Sec. 1, Chungching So. Rd, Taipei 100. – **1b)** 178 Chenhsing Rd, Taichung 401. – **1d)** 643 Chungcheng Rd, Hualien 970. – **1e)** 40 Mingte New Village, Tsoying, Kaohsiung 801. – **1f)** Chukuang Ying, Makong, Penghu 880. – **1g)** 139 Fuhsing Rd, Yongkang City, Tainan 710. – 1h) Chinmen.

BROADCASTS TO MAINLAND:
VOICE OF KUANGHUA (Gov)
✉ P.O.Box 1700, Taipei ☎ + 886 2 2603 0429 ▤ + 886 2 2603 0433 **W:** www.khmusic.com.tw

MW location	kHz	kW	Location	kHz	kW
Hsinfeng	711	250	Kuanyin	846	250
Kuanyin	801	250	Hsinfeng	981	250

D.Prgr. 24h

SW Location	kHz	kW
Kuanyin	9745	250

D.Prgr. 0745-0005
Ann: "Hansheng Kuangpo Tientai, Kuanghua chih Sheng."

FU HSING BROADCATING CORPORATION (Gov)
(operated by Military Information Bureau, Ministry of National Defense)
✉ 5, Lane 280, Sec. 5, Chungshan No. Rd, Taipei 111 ☎ + 886 2 2882 3450 ▤ + 886 2 2881 8218 **W:** www.fhbs.com.tw

MW	Call	Location	kHz	kW
1)	H7	Taipei 1	558	1
1)	H2	Taipei 2	594	10
2)	H38	Taichung 2	594	5
3)	H44	Kaohsiung 1	594	10
3)	H56	Kaohsiung 2	846	10
1)	H3	Taipei 1	909	10
1)	H5	Taipei2	1089	5
2)	H34	Taichung 2	1089	10

SW	Call	Location	kHz	kW
1)		Kuanyin	9410	10
1)		Kuanyin	9774	10
1)		Kuanyin	15375	10

FM	Call	Location	MHz	kW
2)	I44	Taichung 1	107.8	10

D.Prgr: 1st Netw. on 558/909kHz, 2nd Netw. on 594/1089kHz, both 24h. 3rd Netw. on 9410/9774/15375kHz 2300-0100, 0400-0600, 0800-1000, 1100-1300 for China Mainland.

Local Stations: 2) 81 Chungtai Rd, Chunshe Li, Nantun, Taichung 408. 1st Netw. on 107.8MHz, 2nd Netw. on 594/1089kHz. – **3)** 819 Chengching Rd, Niaosung Village, Kaohsiung 833. 1st Netw. on 594kHz, 2nd Netw. on 846kHz.
Ann: "Fu Hsing Kuangpo Tientai, (location) Tai".

OTHER PUBLIC & COMMERCIAL STATIONS (Call: BE.)

MW	Call	Station	Location	kHz	kW
6a)		Taiwan	Tahsi	621	1
6c)		Taiwan	Sungling	630	10
1b)	V59	Cheng Sheng	Taichung 2	657	10
1c)		Cheng Sheng	Peikang	675	1
8b)	P24	Ching Cha	Taichung	702	10
11)	E43	Shih Hsin	Taipei	729	0.5
31)	L2	Yuyeh	Penghu	738	100
12)		Sheng Li	Makung	756	1
6b)	V94	Taiwan	Taichung	774	10
13)	V88	Hsien Sheng	Taoyuan	774	10
12)	V56	Sheng Li	Tainan 1	774	1
14)	V79	Keelung	Keelung	792	10
7)		Chien kuo	Hsinhua	801	10
15)	V54	Kuo Sheng	Changhua	810	10
1)	V35	Cheng Sheng	Taipei	819	5
8d)	P28	Ching Cha	Kaohsiung	819	10
1a)	V72	Cheng Sheng	Chia-i	855	5
17)	V24	Min Pen	Taipei 2	855	1
18)	V52	Chung Sheng	Taichung	864	10
19a)		Feng Ming	Penghu	882	1
7)	V85	Chien Kuo	Hsinying	954	10
6c)	V84	Taiwan	Chunghsing	963	10
19)	V68	Feng Ming	Kaohsiung 2	981	10
1b)	V58	Cheng Sheng	Taichung 1	990	10
8e)	P38	Ching Cha	Ilan	990	1
8f)	P34	Ching Cha	Hualien	990	10
22)	V92	Tien Nan	Taipei	999	1
1f)	V60	Cheng Sheng	Kaohsiung	1008	5
21a)		Tien Sheng	Yuanli	1026	1
23)	V51	Chung Hua	Sanchung	1026	1
3)		Cheng Kung	Kaohsiung	1044	10
20)	V64	Yen Sheng	Hualien 1	1044	5
1d)	V82	Cheng Sheng	Ilan	1062	5
24)	V74	Min Li	Pingtung	1062	5
25)	V96	Tien Sheng	Tainan	1071	1
16)	G28	Kaohsiung	Kaohsiung	1089	10
8a)	P26	Ching Cha	Hsinchu	1116	5
8d)	P25	Ching Cha	Kaohsiung	1116	5
1c)	V36	Cheng Sheng	Yunlin	1125	5
8g)	P40	Ching Cha	Taitung	1125	1
26)	G26	Taipei	Taipei	1134	10
31)	L3	Taiwan Yuyeh	Penghu	1143	100
28)	V70	Hua Sheng	Taipei 1	1152	5
19)	V67	Feng Ming	Kaohsiung 1	1161	1
6a)		Taiwan	Kuanhsi	1170	1

MW Call		Station	Location	kHz	kW
15a)		Kuo Sheng	Erhlin	1179	2.5
6)	V46	Taiwan	Taipei 2	1188	10
12)	V57	Sheng Li	Tainan 2	1188	1
6a)	V62	Taiwan	Hsinchu	1206	10
21a)		Tien Sheng	Pengshan	1215	1
28)	V71	Hua Sheng	Taipei 2	1224	1
23)		Chung Hua	Juifang	1233	1
20)		Yen Sheng	Hualien 2	1242	1
1a)		Cheng Sheng	Taipao	1260	1
8)	P22	Ching Cha	Taipei	1260	10
1e)	V37	Cheng Sheng	Taitung	1269	1
24)		Min Li	Fangliao	1287	1
17)	V23	Min Pen	Taipei 1	1296	1
8c)	P32	Ching Cha	Tainan	1314	5
21)	V76	Tien Sheng	Chunan	1314	10
6)	V45	Taiwan	Taipei 1	1323	1
6c)		Taiwan	Puli	1332	1
23)	V50	Chung Hua	Sanchung 1	1350	2.5
27)		Chin Hsi	Kaohsiung	1368	1
1f)		Cheng Sheng	Tafa	1395	10
30)	V78	Yi Shih	Keelung	1404	10
7)		Chien Kuo	Kuanyin	1422	5
4)	E32	Chiao Yu	Taipei	1494	10
4a)	E34	Chiao Yu	Changhua	1494	5
8a)		Ching Cha	Hsinchu	1512	10
31a)		Yuyeh	Ilan	1593	3

FM Call	MHz	kW	FM Call	MHz	kW	FM	Call	MHz	kW
11)	88.1		38)	97.5	3	5c)		100.8	
4f)	88.9	1	67) N74	97.5	1	8f)	P35	101.3	2
4l)	88.9	1	39)	97.7	3	8g)	P39	101.3	1
61)	90.5		81)	97.7	3	8e)		101.3	1
83)	91.3		61a)	97.9	3	4)	E33	101.7	30
4n)	91.5		80)	97.9	3	4b)	E36	101.7	30
63)	91.7		4i)	98.1	3	75)		102.3	3
32)	92.1	3	40) M23	98.1	3	49)	M27	102.5	3
10)	92.3	3	38b)	98.3	3	64)		102.5	3
33a)	92.3	3	41)	98.3	3	4d)	E38	102.9	5
38a)	92.3	3	47a)	98.3	3	4a)	E35	103.5	30
77)	92.5	3	82)	98.5		4g)	E40	103.5	10
69)	92.5	3	74)	98.5	3	4c)	E37	103.7	5
33)	92.7	3	62)	98.7	3	4h)	E41	103.9	3
66)	92.9	3	72)	98.7	3	32a)		103.9	3
8d)	P42	93.1	25	41a) M31	98.9	3	1)	M22	104.1	3
26) G25	93.1	13	4k)	99.1	4	8)	P29	104.9	35
55)	93.3	3	78)	99.1	3	8c)	P31	104.9	4
56)	93.5	3	43)	99.1	3	8d)	P32	104.9	25
34)	93.7	3	4m)	99.3	3	8e)	P30	104.9	35
57)	93.7	3	44)	99.3	3	8b)		105.1	35
8)	P41	94.3	10	63)	99.3	3	4k)		105.3	4
8f)	P44	94.3	5	M24	99.5	3	50)		105.5	3
8g)	P45	94.3	5	65)	99.5	3	70)		105.5	3
16) G29	94.3	21	76) N77	99.5	3	71)		105.7	3
8b)	P43	94.5	30	46)	99.7	3	51)	M29	106.1	3
35)	96.7	3	47b)	99.7	3	8g)	P29	106.1	2
58)	96.9	3	47) M30	99.9	3	73)		106.5	3
59)	96.9	3	4e)	100.1	1	79)		106.7	3
36)	97.1	3	5b)	100.1	30	52)	M28	106.9	3
47c)	97.1	3	4f) E39	100.3	1	61b)		107.1	3
68) N61	97.1	3	4d)	100.5		29)		107.3	3
4c)	97.3	3	48) M26	100.7	3	53)	M25	107.7	3
37)	97.3	3	5) M3	100.7	30	4j)		107.7	3
60)	97.3	3	5a)	100.7	30	54)		107.7	3

NB: + more than 70 low-powered community FM stns

Addresses and other information:
1) Cheng Sheng Broadc. Corp., 7th Flr, 1, Lane 66, Sec. 1, Chungching So. Rd, Taipei 10045. 24h (exc. Sun 1800-2100) on 819kHz, 24h on 104.1MHz – **1a)** Erhlin, Kangwei Li, Taipao City, Chia-i 612. 24h. – **1b)** 760, Sec. 2, Chunghsing Rd, Tali City, Taichung 412. 1st prgr on 990kHz, 2nd prgr on 657kHz, both 24h. – **1c)** 10 Shuiyuan Rd, Huwei Town, Yunlin 632. – **1d)** 45 Chienchun Rd, Ilan 260. 24h. – **1e)** 21, Lane 380, Hsinsheng Rd, Taitung 950. 24h (Sat 2120-Sun 1500). – **1f)** 838 Chengching Rd, Niaosung Village, Kaohsiung 833. Kaohsiung St. on 1008kHz, Tafa St. on 1395kHz, both 24h **W:** www.csbc.com.tw – **3)** Chengkung Broadc. St, 63 Chunglin 3rd Rd, Kaohsiung. 24h (exc. Sun 1600-2100). – **4)** Chiao Yu Broadc. System – National Education Radio, 41 Nanhai Rd, Taipei 10066. On MW, FM both 24h. **Rel. BBC-WS** 1500-1600 & 2200-2300, **Rel. RFI** 1600-1700 and Chinese 0930-1030. **Ann:** "Chiao Yu chih Sheng, Chiao Yu Kuangpo Tientai". – **4a)** 5-1 Hukang Rd, Changhua 500. – **4b)** 380 Kuangtung 3rd Rd, Kaohsiung 80656. – **4c)** 457 Tunghsing Rd, Hualien 970. 1st prgr on 103.7MHz, 2nd prgr on 97.3MHz. – **4d)** 135, Ma Hengheng Rd, Taitung 95047. 1st prgr on 102.9MHz, 2nd prgr on 100.5MHz – **4e)** Keelung (relay st.). – **4f)** Yuli (relay st.) 1st prgr on 100.3MHz, 2nd prgr on 88.9MHz. – **4g)** Ilan (relay st.). – **4h)** Miaoli (relay st.). – **4i)** Nantou (relay

st.). – **4j)** Chia-i (relay st.). – **4k)** Penghu (relay st.) 1st prgr on 99.1MHz, 2nd prgr on 105.3MHz. – **4l)** Chinmen (relay st.). – **4m)** Hengchun (relay st.). – **4n)** Matzu (relay st.). **W:** www.ner.gov.tw – **5)** International Community Radio Taipei (ICRT), 5 19th Flr, 107, Sec. 1, Chungshan Rd, Hsinchuang 24250. 24h in English. **Rel. BBC News** 2300-2330. **F.PI:** The power output of 100.7MHz will be decreased. A new relay stn will be installed. – **5a)** Kaohsiung. – **5b)** Taichung – **5c)** Chia-i **W:** www.icrt.com.tw – **6)** Taiwan Broadc. Co, 4th Flr, 87 Shuiyuan Rd, Chungcheng, Taipei 100. 1st prgr on 1323kHz, 24h. Thai prgr: Sat 2300-0100, Vietnamese prgr: Sun 0500-0700, Indonesian prgr: Sun 0900-1000, Fri 2300-2400, Tagalog prgr: Sun 1000-1200. 2nd prgr on 1188kHz, 24h. Vietnamese prgr: Sun 1100-1200. **W:** www.taiwanradio.com.tw – **6a)** Lane 506, Kaofeng Rd, Hsinchu 300. 24h. – **6b)** 2309 Tzuli Tuan, Taya Village, Taichung. 24h. – **6c)** 258-1 Fentsao Rd, Tsaotun Town, Nantou 542. 24h. – **7)** Chien Kuo Broadc. St, 78 Chienkuo Rd, Hsinying City, Tainan 730. 24h. – **8)** Chingcha Broadc. St (Police Radio Station), 17 Kuangchou Str, Chungcheng, Taipei 10066. Four networks in 24h: AM Nat. Evergreen Netw. (ANE), AM Reg. Public Security Traffic Netw. (ARS), FM Nat. Public Security Traffic Netw. (FNS) and FM Reg. Public Security Traffic Netw. (FRS). ANE on 1260kHz, FRS on 94.3MHz – **8a)** 1-1 Chiahsing Rd, Chupei City, Hsinchu 302. ARS on 1512kHz. – **8b)** 99 Po-ai Str, Nantun, Taichung 408. ANE on 702kHz, FNS on 105.1MHz, FRS on 94.5MHz – **8c)** 85-21, Nanshih, Nanshih Li, Matou Town, Tainan 721. ARS on 1314kHz, FNS on 104.9MHz – **8d)** 455 Po-ai 4th Rd, Tsoying Ward, Kaohsiung 813. ANE on 1116kHz, FNS on 104.9MHz, FRS on 93.1MHz. – **8e)** 48 Sec. 5, Chungsan Rd, Ilan 26054. ANE on 990kHz, FNS on 104.9MHz, FRS on 101.3MHz. – **8f)** 21-2 Fuchien Rd, Hualien 970. , ANE on 990kHz, FNS on 101.3MHz, FRS on 94.3MHz. – **8g)** 2nd Flr, 100, Hsinchan, Taitung 950. , ANE on 1125kHz, FNS on 101.3, 106.5MHz, FRS on 94.3MHz. **W:** www.prs.gov.tw –**10)** Chin Sheng Broadc. St, 25th Flr, 206 Kuanghua 1st Rd, Lingya, Kaohsiung 802. 24h. – **11)** Shih Hsin Broadc. St, 1, Lane 17, Sec. 1, Mushan Rd, Wenshan Ward, Taipei 116. AM: 2255(Fri 2300)-1605(Sun 1305), FM: 2255-1600 (Sun1305). **W:** www.shrs.shu.edu.tw – **12)** Shengli chih Sheng (Voice of Victory) Broadc. Co, 22, Sec. 1, Chienkang Rd, Tainan 700. 1st Prgr. on 774kHz, 24h. 2nd Prgr. on 1188kHz, 24h. Makung St. on 756kHz, 24h. **Ann:** "Tainan Sheng Li chih Sheng Kuangpo Tientai". **W:** www.e-go.org.tw/victor/ – **13)** Hsien Sheng Broadc. Co, 1, 16th Flr, Lane 505, Chungshan Rd, Taoyuan 330. 24h.– **14)** Keelung Broadc. St, 12th Flr, 13 Chungsu Rd, Keelung 200. 24h. – **15)** Kuo Sheng Broadc. Co, 35 Wenchuan Rd, Pakuashan, Changhua 500. 24h. – **15a)** 2 Taiping Rd, Erhlin Town, Changhua 526.– **16)** Kaohsiung Broadc. St, 90 Hsinchiang Rd, Kushan, Kaohsiung 804 (operated by Kaohsiung City Council). Two prgr. on 1089kHz, 94.3MHz, both 2150-1600. **W:** www.kbs.gov.tw – **17)** Min Pen Broadc. Co, 6th Flr, 325, Sec. 3, Huanho So. Rd, Taipei 108. 1st Prgr on 1296kHz, 2nd Prgr on 855kHz, both 24h **W:** www.mingpen.com.tw – **18)** Chung Sheng Broadc. Co, 134 Kuangfu Rd, Taichung 400.2055-1800. **W:** www.864.com.tw/ – **19)** Feng Ming Broadc. Co, 492 Chiuju 2nd Rd, Sanmin Ward, Kaohsiung 807. 1st Prgr on 1161kHz, 2nd Prgr on 981kHz, both 24h **W:** www.fengmin.com.tw – **19a)** Chentieh Hsien, Li 38, Makung, Penghu.– **20)** Yen Sheng Broadc. St, 31, Sec. 1, Nanpin Rd, Tungchang, Chi-an Village, Hualien 973. On 1044, 1242kHz, both 2055(Sat 2200)-1800(Sun 1600).– **21)** T'ien Sheng Broadc. St, 285 Kungyi Rd, Chunan Town, Miaoli 350. 24h.– **21a)** 8, Kozhuang, Chungshue Rd, Yuanli Town, Miaoli 358. Yuanli St. on 1026kHz, Pengshan St. on 1215kHz, both 24h (exc. Sun 1600-2200). – **22)** Tien Nan Broadc. St, 1st Flr. 29-1, Sec. 2, Hangchou So. Rd, Taipei 106. Golden Ch. on 999kHz. 24h– **23)** Chung Hua (China) Broadc. Co, 6th Flr, 238 Hopien No. Str, Sanchung City, Taipei 241. 1st Prgr on 1350kHz, 2nd prgr on 1026kHz, both 24h **Ann:** "Chung Hua Kuangpo Tientai Ti I/Erh Tai". Juifang Relay St, on 1233kHz, 24h **W:** www.e-go.org.tw/chbc – **24)** Min Li Broadc. St, 57-20 Minsheng Rd, Pingtung 900. 24h.– **25)** Tien Sheng Broadc. St, 11, 15th Flr, 149, Sec. 1, Linsen Rd, Tainan 701. 24h (exc. Sun 1555-2155) **W:** www.am1071.com.tw. – **26)** Taipei Broadc. St, 4th Flr, 62-2, Sec. 3, Chungshan No. Rd, Taipei 10452 (operated by Taipei City Council). AM "Ho Hi Yan" Ch. on 1134kHz, 2300-1500. FM "City Info" Ch. on 93.1MHz, 24h. **Rel: BBC-WS**: MF1400-1500, MF2200-2300. **W:** www.radio.taipei.gov.tw – **27)** Chin Hsi Broadc. Co, 2nd Flr, 461 Wenfu Rd, Tsoying, Kaohsiung 813. 24h **W:** www.am1368. com.tw – **28)** Hua Sheng Broadc. Co, 18 Huasheng Str, Shihlin Ward, Taipei 111. 1st Prgr on 1152kHz, 2nd Prgr on 1224kHz, both 24h **W:** www. hsradio.com.tw – **29)** Lan Yang FM Broadc. St, 12th Flr, 186, Sec. 3, Chungcheng Rd, Wuchie Village, Ilan. 24h **W:** www.frs.gov.tw – **30)** Yi Shih Broadc. St, 75 Paisan Str, Chitu Ward, Keelung 206. 24h **Ann:** "Keelung Yi Shih Kuangpo Tientai". **W:** /yishih.ehosting.com.tw/ – **31)** Yuyeh Broadc. St (Fishery Radio Station), 5 Yukang No. 2nd Rd, Kaohsiung 806. 24h Weather rpt. at every h. Program is also relayed on Wed exc final Wed in month at 0800-0830 & 0830-0900 on 15290kHz of RTI, 0900-0930 & 0930-1000 on 11550kHz of RTI. **Ann:** "Hi-giap Kong-po'-tian-tai" **W:** www.frs.gov.tw – **31a)** Ilan (relay st.). – **32)** Fei Tieh (UFO) Broadc. Co (UFO Netw), 25th Flr, 102, Sec. 2, Lossufou Rd, Taipei 100. 24h **W:** www.uforadio.com.tw – **32a)** Nan Taiwan chih Sheng (Voice of South Taiwan), 20th Flr, 12 Po-ai 3th Rd, Tsoying, Kaohsiung 813. UFO Netw: Taichung, Changhua, Nantou

district 89.9MHz, Miaoli 91.3MHz, Yunlin & Chia-i district 90.5MHz, Ilan 89.9MHz, Penghu 89.7MHz, Hualien & Taitung 91.3MHz – **33)** Yachou (Asia) Broadc. St (Asia FM Netw), 2, 22nd Flr, 102 Chungping Rd, Taoyuan 330. 24h **W:** www.asiafm.com.tw – **33a)** Ya Tai Broadc. St, 1, 15th Flr, 307, Tapei Rd, Hsinchu. – **34)** Sheng Tu Broadc. Co, 233 Fentsao Rd, Tsaotun Town, Nantou 542. 24h **W:** www.fm937.com.tw – **35)** Huan Yu Broadc. Co (Uni Radio), 3, 6th Flr, 675, Sec. 1, Chingkuo Rd, Hsinchu 300. 24h **W:** www.turc967.com.tw – **36)** Ilan chih Sheng (Voice of Ilan) Chung Shan Broadc. Co, 12th Flr, 289-3 Kungcheng Rd, Lotung Town, Ilan 265. 24h **W:** www.super971.com.tw – **37)** Green Peace Broadc. St, 1, 14th Flr, 97, Sec. 4, Chunghsing Rd, Sanchung City, Taipei 241. 24h **W:** www. greenpeace.com.tw – **38)** Kuai Le (Happy) Broadc. St, 3, 8th Flr, 63 Santo 4th Rd, Lingya, Kaohsiung 802. 24h (exc. Sun 1800-2200). – **38a)** Chia-i (Chia Le Broadc St), 1, 16th Flr, 193, Hsiaoya Rd, Chia-i 600. – **38b)** Hualien (Huan Le Broadc St), 1, Lane 120, Tunghsing 2nd Str, Minhsiang Li, Hualien 970. ET FM (Tung Sen) Netw: Taichung 89.5MHz, Taipei 89.3MHz, Pingtung 91.3MHz – **39)** Hao Chia Ting Broadc. Co (Family 977 Broadc. Network), 37th Flr, 789 Chungming So. Rd, Taichung 402. 24h **W:** www.family977.com.tw – **40)** Taiwan Chuan Min Broadc. St (News 98), 1, 25th Flr, 100, Sec. 2, Lossufu Rd, Taipei 100. 24h – **41)** Kang Tu Broadc. St (Best Radio), 1, 34th Flr, 80 Mintsu 1st Rd, Kaohsiung 807. 24h **W:** www.bestradio.com.tw – **41a)** Chin Yue Broadc. St (Best 989), 6th Flr, 88, Sec. 2, Chunghsiao East Rd, Chungcheng, Taipei 100. 24h Haoshih (Best) Netw: Taichung 90.3MHz, Pingtung 89.3MHz, Hualien 93.5MHz – **43)** Ta Chien Broadc. St (Super 99.1), 8th Flr, 83 Hsuehshih Rd, Taichung 404 **W:** www.superfm99-1.com.tw – **44)** Hsin Sheng FM Broadc. St, 1, 19th Flr, 37 Chianchung 1st Rd, Hsinchu 300 **W:** www.ss-radio.com.tw – **45)** Shen Nong (Farmer Radio) Broadc. Co, 10th Flr, 234 Peiping Rd, Huwei Town, Yunlin 632.24h **W:** www.fm995.com.tw – **46)** Taipei Ai Yue Broadc. Co, 7th Flr, 47 Tunghsing Rd, Hsin-i, Taipei 110. 24h **W:** www.prtmusic.com. tw – **47)** Ta Chung Broadc. Co (Kiss Radio), 2, 34th Flr, 6 Minchuan 2nd Rd, Kaohsiung 806. 24h **W:** www.kiss.com.tw – **47a)** Ta Miaoli FM Broadc. St, 3, 16th Flr, 1 Chanchien, Shangmiao Li, Miaoli 360. – **47b)** Nantou Broadc. St, 1A, 37th Flr, 760 Chungming So. Rd, Taichung 402. – **47c)** Tainan chih Yin Broadc. St, 18th Flr, 1-119 Chunghua Rd, Yongkang City, Tainan 710 – **48)** Taichung Broadc.Co, 21st Flr, 345, Sec. 1, Chungkang Rd, Taichung 403. 24h **W:** www.lucky7.com.tw – **49)** Ku Tu Broadc. Co, 1, 15th Flr, 77, Sec. 2, Chunghua East Rd, Tainan 701. 24h **W:** www.fm1025. com.tw – **50)** Tung Shan He FM Broadc. St, 13th Flr, 162-1, Sec. 3, Chunching Rd, Lotung Town, Ilan 265. – **51)** Chuan Kuo Broadc. Co, 1, 10th Flr, 1-18, Sec. 2, Chungkang Rd, Taichung 407. 24h **W:** www.taichungnet. com.tw – **52)** Taoyuan Broadc. St (TBC Radio), 9th Flr, 859, Sec. 1, Chunghua Rd, Chungli City, Taoyuan 320. 24h **W:** www.tbcradio.com.tw – **53)** Tung Taiwan Broadc. St, 55 Chunghsing Rd, Hualien 970. 2200-1600. Wuho Relay St also on 107.7MHz – **54)** Taipei chih Yin (Sound of Taipei) Broadc. Co, B, 1, 10th Flr, 15-1, Sec. 1, Hanchou So. Rd, Taipei 100. 24h. Relay St: Taichung 91.5MHz, Kaohsiung 90.1MHz **W:** www.hitfm. com – **55)** Yun Chia Broadc. St, 9th Flr, 617 Chungshan Rd, Chia-i 600. 24h **W:** www.fm933.com.tw – **56)** Hsin Kechia Broadc. St, 1, 16th Flr, 411 Huannan Rd, Pingchen City, Taoyuan 324. 24h – **57)** Pao Tao Kechia Broadc. St, 2, 17th Flr, 91, Sec. 2, Lossufu Rd, Taipei 106. 24h Taipei & Ilan 93.7MHz, Taoyuan 98.7MHz, Hsinchu & Miaoli 102.5MHz **W:** www. formosahakka.org.tw – **58)** Tien Tien (Sky) Broadc. St, 42nd Flr, 760 Chungming So. Rd, Taichung 402. 24h **W:** tw.myblog.yahoo.com/ sky9692004 – **59)** Chu Jen (Boss) Broadc. St, 16th Flr, 121-8 Tachang 2nd Rd, Kaohsiung 807. 24h – **60)** Ai Yu chih Sheng Broadc. St, 7, Lane 828, Sec. 3, Chinma Rd, Changhua 500. 24h **W:** tw.myblog.yahoo.com/fm973-fm973 – **61)** Kaohsiung Hsiakang chih Sheng Broadc. St, 1st Flr, 11 Teshun Rd, Kaohsiung. **W:** www.smileradio.com.tw – **61a)** Tainan Kaihsuan Broadc. St, 2, 21th Flr, 425 Chunghua Rd, Yungkang City, Tainan 710 – **61b)** Chia-i Huanchiu Broadc. St, 1, 19th Flr, 25 Pingtien, Chianghsi Village, Fanliu, Chia-i – Other Smile Netw: Hsinchu on 90.3, Taichung 105.5, Pingtung on 90.9/91.3/92.5MHz – **62)** Mei Jih Broadc. Co (Sakura Radio), 1, 7th Flr, 1-67 Wuchuan Rd, Taichung 403. 24h **W:** www.fm987. com.tw – **63)** Chin Ma chih Sheng Broadc. St, 8, 10th Flr, 80, Sec. 2, Kuangfu Rd, Sanchung City, Taipei. 2100-1430. – **64)** Pei Tai chih Sheng Broadc. St, 16th Flr, 5, Lane 2, Shen-aokang Rd, Keelung 201. 2300(Sun 0300)-1600.– **65)** Lan Yu Broadc. St, 147, Yujen, Hongtou, Lanyu Village, Taitung 95241.24h **W:** www.lanan.org.tw/radeo.htm – **66)** Cheng Shih Broadc. St, 28th Flr, 758 Chungming So. Rd, Taichung 402. **W:** www. goldfm.com.tw – Other Gold FM Netw st: Taipei 90.1MHz – **67)** IC chih Yin, IC Broadc. Co. Ltd., 2, 11th Flr, 287, Sec. 2, Kuangfu Rd, Hsinchu 30071. 24h **W:** www.ic975.com.tw – **68)** Ta Han chih Yin (Voice of Hakka) Broadc. St, 1-1 Hsintung Rd, Toufen Town, Miaoli 351. 24h **W:** www. fm971.com.tw – **69)** Chuan Chiu Tung Broadc. St, 2, 17th Flr, 143, Sec. 2, Yuanhuan No. Rd, Fengyuan City, Taichung 420. – **70)** Huanhsi chih Sheng (Happy Radio) Broadc. St, 37th Flr, 760 Chungming So. Rd, So. Ward, Taichung. 24h **W:** www.happy1055.com.tw – **71)** Tzumei (Sister Radio) Broadc. St, 4th Flr, 32, Lane 416, Sec. 1, Linsen Rd, Huwei Town, Yunlin 632.24h **W:** www.sister-radio.com.tw – **72)** Ching Chun Broadc. St, 15-2, 53, Sec. 2, Lin'an Rd, No. Ward, Tainan 704. – **73)** Chih Nan Broadc. St, 6,

21st Flr, 3 Tzuchiang 3rd Rd, Lingya, Kaohsiung 802. – **74)** Pao Tao Hsin Sheng (Super FM 98.5) Broadc. St, 1, 2nd Flr, 3, Sec. 1, Tunhua So. Rd, Taipei 105. 24h **W:** www.superfm98-5.com.tw – **75)** Hung Sheng Broadc. St, 6th Flr, 30 Chinyang Str, Miaoli 360. 24h – **76)** Tung Fang Broadc. St, 13rd Flr, 168, Sec. 3, Chunching Rd, Lotung Town, Ilan 265. 24h – **77)** Ta Pao Sang Broadc. St, 83 Lienhang Rd, Taitung 950. 24h – **78)** Yang Kuang Broadc. St, 6, 21st Flr, 3, Tzuchiang 3rd Rd, Lingya, Kaohsiung 802. – **79)** Kao Ping Hsi Broadc. St, 17th Flr, 161 Chiuta Rd, Chiuchu, Tashu Village, Kaohsiung 840. – **80)** Ka Ma Lan Broadc. St, 36, Lane 175, Sec. 3, Chungcheng Rd, Wuchie Village, Ilan 268. – **81)** Taiwan Sheng Yin Broadc. St, Taipei. 24h – **82)** Feifanyin Broadc. St (Libra Radio), 40, Lane 40, Sec. 2, Shuangshih Rd. No. Ward, Taichung 40455. 24h **W:** www. libraradio.com.tw – **83)** Taipei Tuhui Hsiuhsien Yinyueh Tai (Taipei City Leisure Music St)(Bravo FM 91.3), 4, 3rd Flr, 148, Sec. 4, Chunghsiao East Rd, Taan Ward, Taipei, 10627, 24h **W:** www.bravo913.com.tw

DAB: Pao Tao Hsin Sheng (Super FM 98.5) organized 3 prgrs of themselves, Super FM 99.1 and Radio Taiwan International.
F.PI: 12 FM stns using freqs over 104.1MHz will shift to new freqs between 92.1 and 104.0MHz

L.T: UTC +5h — **Pop:** 7.3 million — **Pr.L:** Tajik, Russian — **E.C:** 50Hz, 220V — **ITU:** TJK

KUMITAI TELEVIZION VE RADIOI
(State Committee for Radio & TV)
✉ k. Sheroz 31, 734025 Dushanbe ☎ +992 37 2276569 🖷 +992 37 2213495 **L.P:** Chmn: Asadullo Rakhmonov
NB: In addition to being a state broadcaster, the committee is also responsible for issuing licenses to private radio stations.

LW/MW	kHz	kW	Prgr	MW	kHz	kW	Prgr
Dushanbe (a)	252	150	TR1	Khujand	819	15	TR1
Dushanbe	549	40	TR2	Orzu	1161	40	TR2
Orzu	702	150	TR1	(a) Yangiyul			

NB: Unknown freqs: TR1 tx in Panjakent; TR2 tx in Khorugh (1kW).

SW	kHz	kW	Prgr
Dushanbe (a)	4765	50	TR1	(a) Yangiyul

FM (MHz)	TR1	TR2	TR3	kW
Ayvani		107.8	-	-
Dushanbe	70.64	72.20	-	17
Dushanbe		102.2	106.5	-
Khujand	72.56	69.80	106.5	17/17/-
Khorugh	102.0	104.0	103.0	-
Panj		100.3	-	-
Qurghonteppa	67.88	66.32	-	17

+ low power txs.
D.Prgr: TR1 (Radioi Tojikiston) 24h on FM; on LW/MW/SW: 2300-2000. N. Russian: 0400-0430, 1000-1030. N. Uzbek: 1030-1100. – **TR2 (Sadoi Dushanbe)** 0100-1900 in Tajik, Russian. – **TR3 (Radioi Farhang)** 0100-1900.
External Service (Ovozi Tojik): see International Radio section.

OTHER STATIONS

MW	kHz	kW	Location	Station
A)	1323	7	Dushanbe	VOR relay
A)	1503	7	Dushanbe	VOR relay

FM	MHz	kW	Location	Station
4)	101.5	0.2	Kulob	R. Mavji ozod
2)	102.4	-	Qurghonteppa	R. Vatan
7A)	103.0	-	Dushanbe	R. Rusii Oriyono
7B)	103.0	-	Khujand	R. Imruz
B)	103.7	-	Dushanbe	R. Zvezda
1A)	103.7	4	Khujand	R. Tiroz
5)	104.0	-	Dushanbe	AFM
3A)	104.5	0.3	Dushanbe	R. Aziya FM
4)	105.9	-	Isfara	R. Mavji ozod
2)	106.0	0.1	Dushanbe	R. Vatan
6)	106.1	-	Istaravshan	R. AVIS-Plus
1B)	106.7	-	Khujand	R. Tiroz Plus
3B)	107.0	0.1	Dushanbe	R. Aziya Plyus
7B)	107.0	-	Dushanbe	R. Imruz
8)	107.5	1	Ghafurov	R. Jahonoro

Addresses & other information:
1A,B) k. Tiroz 9, 735700 Khujand. **E:** radio@tiroz.org – **2)** pr. S.Sherozi 16, 734018 Dushanbe. **E:** info@vatan.tj – **3A,B)** k. Bokhtar 35/1, 734002 Dushanbe. **E:** radio@asiaplus.tj – **4)** k. Umari Hazom 18, 735330 Vose. – **5)** pr. S.Ayni 27/17, 734000 Dushanbe. **E:** radio@afm.tj – **6)** k. A.Mirrajabov 10, 735610 Istaravshan. **E:** avis@avis.tj – **7A,B)** pr. Rudaki 95/1, 734001 Dushanbe. **E:** info@orionomedia.tj. 7A) rel. Russkoye R. (Russia) – **8)** k. Lenin 22, Ghafurov. – **A)** Rel. Voice of Russia (Russia). – **B)** Rel. R.Zvezda (Russia).

Int relays on MW: (txs operated by Teleradiokom) Dushanbe (Yangiyul) 1143kHz 150kW, 1251kHz 100kW; Orzu 648/801kHz 1000kW, 972kHz 500kW & operated on behalf of IBB (USA): 972kHz 800kW. See Int Radio section

TANZANIA

L.T: UTC +3h — **Pop:** 41 million — **Pr.L:** Swahili, English — **E.C:** 50Hz, 230V— **ITU:** TZA

TANZANIA COMMUNICATIONS REGULATORY AUTHORITY (TCRA)

✉ Mawasiliano House, Plot 304, Ali Hassan Mwinyi/Nkomo Rd, P.O Box 474, Dar es Salaam ☎+255 22 2118947 🖷 +255 22 2116664 **W:** www.tcra.go.tz **E:** dg@tcra.go.tz
L.P: DG: John Nkoma. Dir. Broadc. Affairs: Habbi Gunze.

TANZANIA BROADCASTING CORPORATION(TBC,Gov.)

✉ P. O. Box 9191, Dir. of Broadc., Nyerere/Mandera Rd, Dar es Salaam ☎+255 51 860760/6 🖷 +255 51 865577 **E:** radio@ud.co.tz
L.P: DG: Dunstan Tido Mhando. Dir. Radio: Ms. Edah Sanga. TD: Harold Limo. Dir. News: Ms. Susan Mungi. Dir. PR: Ngalimecha Ngayoma

MW	kHz	kW	MW	kHz	kW
Dodoma	603	100/10	Kigoma	711	100/10
Mbeya	621	50/10	Mwanza	720	50/10
Nachingwea	648	100/10	Songea	990	100/10
Kunduchi+	657	2 x 50	Arusha	1215	50/10

+) near Dar es Salaam

FM: Arusha 91.6MHz, Dar es Salaam 89.9/92.35MHz, Dodoma 87.7MHz, Kigoma 88.4MHz, Lindi 93.5MHz, Mbeya/Masasi/Nachingwea 92.3MHz, Mwanza 89.2MHz, Songea 98.7MHz I: Dar es Salaam 94.6MHz.
TBC Taifa in Swahili on MW: 0200-2100. **TBC FM in Swahili:** 24h. MW & FM channels may opt out at times to carry regional prgrs. **I= International Sce in English:** 94.6MHz 0200-1915.

PRIVATE STATIONS

Clouds FM, P.O. Box 31513, Dar es Salaam. FM: Dar es Salaam 88.4MHz 2kW, Arusha 98.6MHz, Mwanza 99.4MHz. 0150 – **R. Five**, P.O. Box 11843, Arusha: 105.7MHz. **E:** impala@cybernet.co.tz – **R. Free Africa,** P.O. Box 1732, Post Road, Mwanza. **W:** radiofreeafricatz.com **E:** info@radiofreeafricatz.com Swahili Sce: MW/FM: Mwanza 1377kHz 50kW, Shinyanga 98.6MHz, Mbeya 99.3MHz, Arusha 89MHz, Mwanza 89.8MHz (also rel. BBC/VOA/RTD/DW). English sce: **Kiss FM**: Mbeya 88.2MHz, Mwanza 88.7MHz, Arusha 89.9MHz, Dar es Salaam 89MHz, Shinyanga 96.4MHz – **R. Imaan** (IRlg.): Morogoro 96.0MHz – **R. Kheri** (Rlg.) Dar es Salaam 104.1MHz – **R. Kwizera**, P.O. Box 154, Ngara Field Office, Ngara. **W:** www.jrs.net/countries/eaf.php?lang=en **E:** eastern.africa@jrs.net - **FM:** 97.9MHz (in Swahili, also rel. RFI Afrique in English) – **R. One**, P.O. Box 4374, Dar es Salaam. **W:** www.ippmedia. com/ipp/radio **E:** itv@ipp.co.tz MW: Moshi 1323kHz 10kW (Swahili), Dar es Salaam 1440kHz 10kW (English). FM (all 5kW): Dar es Salaam 89.5MHz, Mwanza 102.5MHz, Dodoma 100.8MHz, Arusha 95.3MHz – **East Africa R**, P.O.Box 4374, Dar es Salaam: 87.8MHz (also relayed on FM in Kampala, Uganda, & Nairobi, Kenya). **W:** www.eastafricafm. com **E:** admin@eastafricafm.com – **R. Maria Tanzania**, P.O. Box 34573, Dar es Salaam. FM: Songea 89.1MHz, Iringa 90.4MHz, Mbeya 91.9MHz, Morogoro 102.0MHz, Unguja/Pemba 103.5MHz, Mwanza 106.0MHz, Arusha 106,7MHz. Arusha 2kW, Pemba 0.5kW, others 1kW. **W:** www.radiomariatanzania.co.tz – **R. Sauti ya Injili** (Rlg.), Lutheran R. Centre, P.O. Box 777, Moshi **W:** www.elct.org/TechServ/Radio **E:** Redio@elct.org FM: Kidia 92.2MHz, Tanga 96.0MHz, Arusha 96.2MHz, Rombo 96.4MHz, Morogoro 99.6MHz, Same 100.4MHz, Usambara 102.6MHz, Kibaya 102.9MHz. F.PI: FM trs in Bukoba and Iringa and a low-power SW transmitter – **R. Sauti ya Quran** (Rlg.) Dar es Salaam: 102.0MHz – **Tanzanite R. FM**, Arusha: 96.1MHz – **R. Tumaini**: P.O. Box 9916, Dar es Salaam. FM: Dar es Salaam 96.3MHz, Kibahe 91.4MHz. **R.Tumaini 2:** Dar es Salaam: 105.9MHz.

ZANZIBAR & PEMBA
(autonomous islands)

VOICE OF TANZANIA - ZANZIBAR (Gov.)

✉ P.O. Box 1178, Zanzibar, Tanzania ☎+255 54 31088/9 **E:** Karumehouse@tvz.co.tv **L.P:** Dir. of Broadc: Yussuf Omar Chunda. CE: Ali Aboud Talib.
MW: Chumbuni 585kHz 10kW (F.PI: 50kW).
SW: Dole: 6015/11735kHz 50kW (irregular/inactive).
FM: Unguja 97.4MHz, Pemba 90.5MHz.
D.Prgr in Swahili: 0300-0600 on 6015kHz, 0900-2100 on 585kHz

1400v-2100 11735kHz. FM ("Spice FM"): 0300-2100. **N:** Local bulletins at 0400, 1200, 1600, 1800, 1900. Rel. R.Tanzania from Dar es Salaam at 1000, 1700. **In English:** irr. 1800. **Ann:** "Hii ni Sauti Tanzania, Zanzibar". **English:** "Voice of Tanzania, Zanzibar."

Other stations:
Al-Noor FM (Rlg.), Zanzibar: 92.6MHz 2kW. **W:** alnoorcharity.org
R. Maria Tanzania (Rlg.) Pemba: 103.5MHz.
R. Tumaini 2, Zanzibar: 105.9MHz.
BBC African Sce: Zanzibar 94.1MHz, Pemba 93.5MHz.
RFI Afrique: Dar es Salaam 94.6MHz in F/E/Swahili

THAILAND

L.T: UTC +7h — **Pop:** 66.5 million — **Pr.L:** Thai — **E.C:** 50Hz, 220V — **ITU:** THA

NATIONAL TELECOMMUNICATIONS COMMISSION (NTC)

✉ **Public Relations Bureau of the NTC:** 87 Phahonyotin Rd. Soi 8, Phayatai Bangkok 10400 ☎ +66 2271-3511 🖷 +66 2290-5240 **W:** eng.ntc.or.th
L.P: Chmn: Prof. Prasit Prapinmongkolkarn,
The NTC controls administrative, legal, technical and programming aspects of broadcasting in Thailand pending the establishment of the National Broadcasting and Telecommunications Commission to supervise the broadcasting industry.

MW STATIONS:

	kHz	kW	Province +)		kHz	kW	Province +)
RT)	531	25	Maha Sarakham	RT)	837	10	Pathum Thani#
39)	540	5	Bangkok	RT)	846	10	Phetchabun
RT)	549	100	Lampang (E)	2)	855	5	Prachin Buri
RT)	549	10	Mukdahan	RT)	864	10	Tak
RT)	558	50	Songkhla (E)	RT)	864	10	Si Sa Ket
RT)	558	10	Kanchanaburi	RT)	864	10	Phatthalung
24)	567	5	Chaiyaphum	16)	873	5	Bangkok
17)	576	5	Bangkok	RT)	891	1000	Sara Buri#
7)	585	5	Phrae	RT)	909	10	Loei
32)	585	5	Chumphon	RT)	909	25	Surin
9)	594	5	Bangkok	30)	918	10	Chiang Mai
32)	603	5	Khon Kaen	RT)	918	100	Nakhon Pathom #
10)	612	5	Lop Buri	RT)	927	20	Chanthaburi (E)
25)	612	5	Chiang Mai	RT)	927	10	Nong Khai
RT)	621	100	Khon Kaen (E)	8)	936	10	Pattani
1)	630	5	Bangkok	RT)	936	50	N. Sawan (E)
RT)	639	10	Lamphun	12)	945	10	Bangkok
RT)	639	10	N. Si Thammarat	6)	945	10	Kalasin
RT)	648	25	Khon Kaen	12)	954	10	Phitsanulok
24)	657	5	Bangkok	12)	954	10	Chanthaburi
7)	666	5	Tak	RT)	963	25	Krabi (E)
6)	666	5	Surin	18)	963	10	Bangkok
29)	675	5	Bangkok	34)	972	10	Phetchabun
8)	684	5	N. Si Thammarat	38)	981	10	Pathum Thani#
39)	684	5	Udon Thani	RT)	981	25	Mae Hong Son
11)	693	5	Saraburi	RT)	981	20	N. Phanom
19)	711	5	Chiang Mai	RT)	981	25	Yala
RT)	711	20	U. Ratchathani (E)	RT)	990	10	N. Ratchasima
RT)	711	5	Lop Buri	7)	999	10	Chiang Rai
RT)	720	10	Krabi	33)	999	10	Bangkok
29)	720	5	Chon Buri	32)	1008	10	N. Ratchasima
RT)	729	25	N. Ratchasima	12)	1017	10	Prachuap KK
32)	738	5	Chiang Mai	RT)	1026	50	Phitsanulok
32)	738	5	Songkhla	RT)	1026	10	Yala
37)	747	5	Bangkok	31)	1035	10	Bangkok
6)	747	5	Udon Thani	35)	1044	10	Khon Kaen
34)	756	5	Narathiwat	8)	1044	10	N. Si Thammarat
35)	756	5	Surin	4)	1053	10	Lampang
31)	765	5	Lampang	1)	1053	10	Bangkok
12)	765	5	Lop Buri	12)	1062	10	Udon Thani
18)	774	5	Rayong	RT)	1062	10	Phuket
36)	774	5	Udon Thani	40)	1071	10	Bangkok
RT)	783	10	Ranong	31)	1071	10	Tak
7)	783	5	Kamphaeng Phet	32)	1080	10	Chiang Rai
19)	792	20	Bangkok	32)	1080	10	N. Sawan
3)	801	5	N. Sawan	32)	1080	10	Yala
12)	801	5	Chiang Rai	31)*1089	10	Bangkok	
12)	801	5	U. Ratchathani	12)	1089	10	Udon Thani
RT)	810	20	Nong Khai	30)	1098	10	Songkhla
RT)	810	10	Kanchanaburi	RT)	1098	10	Tak
RT)	810	10	Trang	25)	1107	10	Samut Sakhon#
RT)	819	10	Pathum Thani#	6)	1107	10	Khon Kaen
32)	828	5	N. Si Thammarat	7)	1116	10	Phitsanulok
7)	828	5	Sukhothai	RT)	1116	10	Phangnga
34)	837	5	Sakon Nakhon	RT)	1125	25	Chanthaburi

kHz	kW	Province +)	kHz	kW	Province +)
RT) 1134	10	Lampang	4)*1350	10	Lampang
6) 1134	10	N. Ratchasima	32) 1350	10	Trang
27) 1143	10	Bangkok	33) 1350	10	Bangkok
37) 1152	10	Chiang Mai	6) 1359	10	Sakhon Nakhon
37) 1152	10	Khon Kaen	RT) 1368	25	Nan
20) 1161	20	Bangkok	RT) 1368	10	Buri Ram
29) 1161	10	U. Ratchathani	12) 1368	10	N. Pathom
29) 1170	10	Chanthaburi	22) 1377	10	Phitsanulok
29) 1170	10	Phitsanulok	RT) 1377	10	Chumphon
36) 1179	10	Bangkok	13) 1386	10	Bangkok
34) 1179	10	Chiang Rai	34) 1395	10	Chiang Rai
35) 1188	10	Sakon Nakhon	RT) 1404	25	Songkhla
7) 1188	10	Phitsanulok	24) 1404	10	Yasothon
5) 1188	10	Sa Kaeo	5) 1404	10	Suphan Buri
26) 1197	10	Lop Buri	RT) 1422	10	Amnat Charoen
RT) 1206	10	Satun	33) 1422	10	Bangkok
5) 1206	10	Prachuap KK	30) 1422	10	Phitsanulok
35) 1215	10	Phrae	12) 1431	10	N. Ratchasima
6) 1215	10	U. Ratchathani	29) 1431	5	Songkhla
RT) 1215	50	Surat Thani	6) 1440	10	N. Phanom
12) 1224	10	Chiang Rai	32) 1440	10	Samut Sakhon
12) 1224	10	N. Sawan	7) 1449	10	Phichit
12) 1233	10	Bangkok	10) 1449	10	Chumphon
32) 1233	10	Udon Thani	24) 1458	10	Si Sa Ket
7)*1242	10	Lampang	29) 1458	10	Phuket
7) 1242	10	Phetchabun	RT) 1467	100	Pathum Thani (E) #
RT) 1242	50	Surat Thani (E)	RT) 1476	50	Lamphun
24) 1251	10	Roi Et	27) 1494	10	Bangkok
12) 1251	5	Bangkok	24) 1503	10	Surat Thani
RT) 1260	25	Chiang Mai	RT) 1512	10	Yasothon (F.P.I.)
25) 1269	10	Songkhla	35) 1512	10	Phayao
15) 1269	10	Bangkok	12) 1512	10	Songkhla
28) 1287	10	Samut Prakan#	34)*1512	10	Uthai Thani
7)*1287	10	Uttaradit	34) 1521	10	Bangkok
32) 1287	10	U. Ratchathani	32) 1530	10	Uttaradit
RT) 1296	10	Pattani	5) 1530	10	Chanthaburi
39) 1305	10	Bangkok	23) 1539	10	Kanchanaburi
25) 1314	10	Khon Kaen	11) 1557	10	Phetchabun
12) 1323	10	Chiang Mai	RT) 1557	10	Trat
12) 1323	10	Surat Thani	—	15751000	Ayutthaya
14) 1332	10	Bangkok	RT)*1584	1	Loei
12) 1332	10	Maha Sarakham	8)*1584	1	Phatthalung
RT) 1341	20	Loei	31)*1593	10	Buri Ram
RT) 1341	25	U. Ratchathani	RT) 1593	10	Ratchaburi
RT) 1341	10	Phangnga			

*) r. inactive. +) N=Nakhon, U=Ubon, KK=Khiri Khan, RT=R. Thailand. #) Bangkok area.

NB: Mobile 1kW units for Army use with no advertisements have been registered for 747, 1242, 1485, 1584, and 1602kHz

BANGKOK FM(MHz): (exc. R. Thailand and community radio stns):
87.5 Sathaanii Witthayu Ratthasapha (Parliament R. St.) – 88.5 Sor. Thor. Ror. 1, "Luukthung Thailand Network" – 89.0 Yaan Kraw "Chill FM" – 89.5 Rajamangala University of Technology "Sweet FM" – 90.0 Phon Neung Ror. Or. – 90.5 Wor. Phor. Thor. – 91.0 Sor. Wor. Phor. – 91.5 Yaan Kraw, "Hot FM" – 92.0 Wor. Sor. Sor. – 93.0 Sor. Thor. Ror. 1, "Cool 93 Fahrenheit" – 94.0 Thor. Thor. Bor. (Sathaanii Witthayu Thorathat Kongthap Bok, Army TV Station), "EFM" – 94.5 Jor. Sor., "Luukthung FM" – 95.0 Or. Sor. Mor. Thor "Luukthung Mahaanakhon" – 96.0 Ror. Dor, "Sport R." – 96.5 Or. Sor. Mor. Thor, "Modern Radio FM96.5" – 97.5 Or. Sor. Mor. Thor, "Modern Radio Seed 97.5 FM" – 98.0 Phon Neung Ror. Or. – 98.5 Neung Por. Nor, "Good FM" – 99.0 Or. Sor. Mor. Thor, "Active R." – 99.5 Sathaanii Witthayu 9-1-9, "Traffic Radio Society (TRS)" – 100.0 Jor. Sor. Roi – 100.5 Or. Sor. Mor. Thor, "Modern Radio FM100.5" – 101.0 Sathaanii Witthayu Kong Banchaakaan Thahaan Suungsut (Supreme Command HQ), "101 Radio Report One" – 101.5 Sathaanii Witthayu Chulaa or "Witthayu Chulaa" or "CU FM" – 102.0 Khor. Sor. Thor. – 102.5 Thor. Or, "Get 102.5" – 103.0 Jor. Sor – 103.5 Thor. Thor. Bor, "FM One" – 104.0 Or. Sor. – 104.5 Phon Por. Thor. Or. (Kong Phon Thahaan Peun Yai Tosue Akart Yaan, Anti-Aircraft Artillery Division), "Fat Radio" – 105.5 or Sor. Mor. Thor "Eazy FM" – 106.0 Sor. Thor. Ror. 1 – 106.5 Neung Por. Nor, "Green Wave" – 107.0 Or. Sor. Mor. Thor, "Met 107".
OTHER FM STATIONS: A large number of FM stns belonging to R. Thailand or other operators are on air throughout Thailand, including approx. 6,000 community radio stns in Bangkok and the provinces with max. permitted tx power 30W.

English language prgrs in Bangkok (freqs carrying substantial content in English): 918kHz R.Thailand Network 6 (English news & features, western light & classical music) – 88.0 Radio Thailand "Wave FM" – 95.5MHz "Virgin Hitz" – 107.0MHz "Met 107" (Or Sor Mor Thor).

GOVERNMENT PUBLIC RELATIONS DEPT. (Gov.)
Soi Aree Samphan, Rama VI Road, Bangkok 10400
☎ +66 2618-2323 📠 +66 2618-2364/2399 **W:** http://thailand.prd. go.th (general info in English). This body operates the NBT radio & TV services (R.Thailand & Television Thailand).**L.P:** DG: Mr Phachern Khampho.

THE NATIONAL BROADCASTING SERVICES OF THAILAND (NBT) – RADIO THAILAND (Sathaanii Witthayu Krachaisiang Haeng Pratheet Thai, Sor. Wor. Thor.) (Gov.)
236 Vibhavadi Rangsit Superhighway, Din Daeng, Huay Khwang, Bangkok 10320 ☎ +66 2277-8181 📠 +66 2277-8182 & 2277-5881 **W:** http://nbt.prd.go.th
L.P: Exec. Dir. R. Thailand: Mrs Kuntalee Buasuwan. Head World Sce: Mrs Amporn Samosorn
D.Prgr: (SWT=Sathaanii Witthayu Krachaisiang Haeng Pratheet Thai)
Prgr. 1 (Khreungkhai Thii Neung or SWT Pheua Khwamruu Sara Lae Borihan Saa Tarana): 2200-1700 on 891kHz in Bangkok (tx site: Nong Khae, Sara Buri), and in full or in part on many RT regional AM stns. **N:** On the h –**Prgr 2: News and local service** (SWT Pheua Khaosaan Lae Borikaan Thongthin): 2200-2000 on 819kHz (tx site: Rangsit, Pathum Thani), also in part on many RT regional MW stns. – **R. Thailand Network 6:** 2200-1700 **English & Thai** exc. **Malay:** 0600-0730, **Chinese:** 0730-0900, **Burmese:** 0910-1000, **Lao:** 1010-1100, **Khmer:** 1105-1200, on 918kHz (tx site: Salaya, Nakhon Pathom), also in part on some RT regional FM stns inc. Chiang Mai 98MHz, Phuket 90.5MHz, Samui 96.75MHz, Songkhla 102.25MHz – **Educational Prgr. 2** (SWT Raikarn Song Pheua Karn Seuksaa): 2300-1700 on 837kHz (tx site: Bang Phun, Pathum Thani). **N:** rel. Prgr. 1. Mostly carries educational prgrs from Ramkhanghaeng University at other times – **Bangkok FM prgrs:** 88.0MHz "Wave FM." (10kW), 92.5MHz (10kW), 93.5MHz "Digital R. HD One" (10kW), 95.5MHz "Virgin Hitz" (10 kW), 97.0MHz (10kW) "Bangkok Today R.", 105.0MHz (10kW) prgr for children and families (SWT Pheua Dek Lae Khropkhrua). **Selected Reg. Stations:** 49 Prachasamphan Rd., Tambon Chang Khlan, Muang Dist., **Chiang Mai** 50100 **FM:** 93.25 & 98.0MHz; 1476 kHz: Prgr in Thai and minority langs for hill tribes 2200-1600. Kasikon Thungsang Rd, Muang Dist., **Khon Kaen** 40000 **FM:** 98.5 & 99.5MHz. Soi Sathaban Ratchaphat Phuket, Thepkasatri Rd, Tambon Ratsada, Muang Dist., **Phuket** 83000 **FM:** 90.5 & 96.75MHz. 439 Mu 2, Songkhla - Ko Yo Road, Tambon Phawong, Muang Dist., **Songkhla** 90100 **FM:** 89.5, 90.5 & 102.25MHz
Addresses of other regional stations: Most stns can be reached by quoting "Sathaanii Witthayu Sor. Wor. Thor." or "Radio Thailand" and the location given in the freq. list, followed by the phrase "Muang District" and finally the city, which is generally the same as the location. Exceptions are the following: **Lamphun:** the MW txs are in Lamphun, but the studios are in Chiang Mai. – **Loei** 909kHz: located in Dansai district. – **Nong Khai** 927kHz: located in Bung Kan district. – **Phangnga:** The addr. for 1116kHz is Takua Pa District, Phangnga 82110. – **Saraburi:** Studios in Bangkok. The addr. for the 1000kW tx on 891kHz is Rim Klong Hog Wa, Mu 4, Nong Rong, Nong Khae, Saraburi 18140. – **Tak:** The addr. for 1098kHz is 14 Asia Hwy, Mae Sot District, Tak 63110. – **Yala:** The addr. for 1026kHz is Betong District, Yala 95110, for 981kHz Raman District, Yala 95140

EXTERNAL SERVICE: Radio Thailand
see International Broadcasting section.

NATIONAL EDUCATION RADIO (Sathaanii Witthayu Krachaisiang Haeng Pratheet Thai Pheua Kaan Seuksaa, Sor. Wor. Sor.)
Soi Aree Samphan, Rama VI Rd, Samsen, Phaya Thai, Bangkok 10400
☎ +66 2271-3448 📠 +66 2245-7083 **W:** http://edu.prd.go.th
D.Prgr: continuous 0000 on 1467kHz in the Bangkok area and nationwide on all "E" txs. Regional stns carry own prgrs and relay Bangkok.

OTHER STATIONS:
GENERAL NOTES: News: Stns are generally required to relay N. from R. Thailand at 0000 & 1200 daily, each 30 mins, and to relay time signal and national anthem at 0000 and 1100, if on the air at those times.
Station IDs: Both short names, e.g. Wor. Por. Tho, and long names may serve as stn identifications, usually preceded by "Thiinii" ("This is"), "Thiinii Sathaanii Witthayu (Krachaisiang)" ("This is R. St.") or "Khun kamlang rap fang" ("You are listening to"). Changwat=province. Amphoe=district (dt.) prgrs are often supplied by separate production companies. The Thai name for Bangkok is 'Krung Thep' or 'Krung Thep Mahanakhon'. **Thai numerals:** 0 = suun, 1 = neung (et), 2 = song, 3 = saam, 4 = sii, 5 = haa, 6 = hok, 7 = jet, 8 = paet, 9 = kao, 10 = sip, 20 = sip, 100 = roi, 1000 = phan; thii = number, jut = decimal point.

RT) Radio Thailand see separate entry above – **1) Mor. Thor. Bor. Sip Et** (Monthon Thahaan Bok Thii Sip Et, 11th Military Circle). ✉ 145 Rama V Rd, Dusit Region, Bangkok 10300. **Ann:** "Suan Mitsakawan". Prgrs produced by KCS Radio – **2) Mor. Thor. Bor. Sip Song** (Monthon Thahaan Bok Thii Sip Song, "Siang Khai Chakkrapong", 12th Military Circle, "Voice of Chakkrapong Camp"). ✉ Chakkrapong Camp, Dong Phra Ram, Prachin Buri 25000 – **3) Mor. Thor. Bor. Thii Saam Sip Et** (Monthon Thahaan Bok Thii Saam Sip Et, 31st Military Circle). ✉ Jiraprawat Camp, Nakhon Sawan 60000 – **4) Mor. Thor. Bor. Saam Sip Song** (Monthon Thahaan Bok Thii Saam Sip Song, 32nd Military Circle). ✉ Headquarters of the 32nd Army Area, Surasak Montri Camp, Phahonyothin Rd, Phichai, Lampang 52000 –. **5) Thor. Phor. Neung** (Kongthap Phaak Thii Neung, 1st Army Area). HQ: ✉ Headquarters of the 1st Army Area, Suan Mitsakawan, Rajchadamnern Nok Ave, Dusit Region, Bangkok 10300. **Regional stns:** 9 Mu 4, Bang Kacha, Chanthaburi 22000 – Phairirayodet Camp, Suwansri Rd, Tha Kasem, Sa Kaeo 27000 – Kao Kuat, Kraw Plub Pla, Ratchaburi 70000 – Ban Sam Liam, Mu 4, Don Pho Thong, Suphan Buri 72000 – **6) Thor. Phor. Song** (Kongthap Phaak Thii Song, 2nd Army Area). HQ: ✉ Suranari Camp, Ratchadamnoen Rd, Nong Phailom, Nakhon Ratchasima 30000. **Regional stns:** Aphai Rd, Nai Muang, Kalasin 46000 – Si Phatcharin Camp, Sila, Khon Kaen 40000 – Phra Yot Muang Khwang Camp, Nakhon Phanom-Sakon Nakhon Rd, Khurukhu, Nakhon Phanom 48000 – Krit Siwara Camp, That Naveng, Sakon Nakhon 47000 – Wirawatyothin Camp, Phakdichumphon Rd, Nok Muang, Surin 32000 – Sapphasiti Prasong Camp, Warin Chamrap District, Ubon Ratchathani 34190 – Yutthasin Prasit Camp, Non Sung Rd, Udon Thani 41330 – **7) Thor. Phor. Saam** (Kongthap Phaak Thii Saam, 3rd Army Area). ✉ Headquarters of the 3rd Army Area, Somdet Phra Ekathosarot Camp, Aranyik, Phitsanulok 65000. **Regional stns:** Mengrai Maharat Camp, Chiang Rai 57000 – 236/5 Mu 3, Nakhon Sawan - Kamphaeng Phet Rd, Nakhon Chum, Kamphaeng Phet 62000 – Khalang Nakhon Camp, Nong Krating, Lampang 52000 – Phokun Pha Muang Camp, 166/1 Mu 1, Wat Pa, Lom Sak District, Phetchabun 67110 – 104/1 Mu 5, Ban Krot Ngam, Ban Na, Wachirabarami District, Phichit 66140 – Ban Mai, Ratsadon Uthit Rd, Nai Wiang, Phrae 54000 – Bypass Road, Pak Khwae, Sukhothai 64000 – Charot Withithong Rd, Nam Ruem, Tak 63000 – Phichai Dap Hak Camp, 109 Mu 8, Uttaradit 53000 – **8) Thor. Phor. Sii** (Kongthap Phaak Thii Sii, 4th Army Area). HQ: ✉ Wachirahwud Camp, Ratchadamnoen-Pak Nun Rd, Nakhon Si Thammarat 80000. **Regional stns:** Aphai Borirak Rd, Chumphon, King-Ampoe Si Nakharin, Phatthalung 93000 – Senanarong Camp, Kho Hong, Hat Yai District, Songkhla 90110 – Ban Na San District, Surat Thani 84120 – Charoen Pradit Rd, Rusamilae, Pattani 94000 – **9) Thor. Phor. Saam** (Kong Phon Thahaan Peun Yai Tosue Akart Yaan, Anti-Aircraft Artillery Division), ✉ Kiak Kay Junction, Thahaan Road, Bangsue, Dusit Region, Bangkok 10300 – **10) Wor. Sor. Por.** (Witthayu Suun Karn Thahaan Peun Yai, Artillery Centre R. St.). ✉ 301 Phahonyothin Camp, Artillery Centre, Khao Phra Ngam, Lop Buri 15160. **Regional st:** Khet Udomsak Camp, Wang Mai, Chumphon 86000 – **11) Siang Adison** (Suun Kaan Thahaan Maa, Cavalry Centre, "Voice of Adison"). ✉ Saraburi Cavalry Centre, Adison Camp, Mitraphap Rd, Pak Phrieo, Saraburi 18000. **Regional st:** Saraburi-Lom Sak Rd, Nong Khwai, Lom Sak District, Phetchabun 67110 – **12) Thor. Or.** (Thahaan Akart, Royal Thai Airforce). ✉ Tor. Or. 01, 1233kHz, Don Muang: 171 Mu 2, Phahonyothin Rd, Khlong Thanon, Sai Mai, Bangkok 10220. Tor. Or. 01, 945kHz, Min Buri: 74 Mu 2, Nimit Mai, Sai Kong Tin, Min Buri, Bangkok 10510. Tor. Or. 06, 1251kHz: The Empress Hotel, 1091/343 Phetchaburi Tat Mai Road, Charurat, Makassan, Ratcha Thewi, Bangkok 10400. **Regional stns:** Thor. Or. 02: 301 Wing 2, 1st Air Division, Khao Phra Ngam Rd, Lop Buri 15160 – Thor. Or. 03: Wing 1, Mu 3, Nong Phai Lom, Nakhon Ratchasima 30000 – Thor. Or. 04: 305 Mu 4, Wing 4, 3rd Air Division, Takhli District, Nakhon Sawan 60140 – Thor. Or. 05: Wing 53, 4th Air Division, Ko Lak, Prachuap Khiri Khan 77000 – Thor. Or. 7: Surat Thani Airport Entrance, Huatoey, Phunphin District, Surat Thani 84130 – Thor. Or. 08: 38 Mu 14, Ban Nongphai, Chayangkun Rd, Khamyai, Ubon Ratchathani 34000 – Thor. Or. 09: 549 Mu 9, Wing 23, Thahaan Rd, Makkhaeng, Udon Thani 41000 – Thor. Or. 10: Wing 46, 3rd Air Division, Yaek Khok Matum, Phitsanulok - Wangthong Rd, Nai Muang, Phitsanulok 65000 – Thor. Or. 11: 99 Mu 8, Wing 56, Khok Muang, Khlong Hoykhong District, Songkhla 90110 – Thor. Or. 12: Flying Training School, Malaimaen Rd, Kratip, Kamphaeng Saen District, Nakhon Pathom 73180 – Thor. Or. 13: 90 Mu 3, Suthep, Chiang Mai 50200 – Thor. Or. 14: Wapiprathum Rd, Wangnang, Maha Sarakham 44000 – Thor. Or. 15, 141 Mu 1, Buasali, Mae Lao District, Chiang Rai 57250. 1st prgr: 801kHz. 2nd prgr: 1224kHz – Thor. Or. 16: 1049 Tha Chalaep Rd, Talat, Chanthaburi 22000 – **13) Sathaanii Witthayu Pheua Kaan Kaset** (Agricultural R. St.). ✉ Agricultural Radio Section, Phahonyothin Road, Lat Yao, Bang Khen Region, Bangkok 10900 – **14) Sor. Sor.** (Sathaanii Witthayu Amphon Sathaan, Phraratchawang Dusit, Amphon Sathan Throne Radio Station). ✉ Dusit Palace, Ratchawithi Rd, Chitralada, Dusit Region, Bangkok 10303 **D.Prgr:** Tu-Sa 0330-0500, 0900-1200, Su 0230-0500, M

silent – **15) Kho. Sor. Thor. Bor.** (Kromkarn Khon Song Thahaan Bok, Army Transportation Dept.). ✉ Army Transportation Broadcasting Station, Transport School Compound, Thahaan Road, Dusit Region, Bangkok 10300 – **16) Wor. Kor. Thor. Mor.** (Sathaanii Witthayu Krung Theep Mahaanakhon, Bangkok Radio Station). ✉ 192 Sarasin Rd, Lumphini Park, Pathum Wan Region, Bangkok 10330 – **17) Tor. Chor. Dor.** (Tamruat Trawen Chaidaen, Border Patrol Police). ✉ Bang Khen Police Dept. Club, Vibhavadi-Rangsit Rd, Bang Khen Bangkok 10210 – **18) Phon Mor. Song** (Sathaanii Witthayu Kong Phan Thahaan Maa Thii Song, 2nd Cavalry Division), ✉ Samsen Rd, Bang Krabeu, Dusit Region, Bangkok 10300 Bangkok. **Regional st:** Rayong-Ban Khai Rd, Nam Khok, Rayong 21000 – **19) Wor. Phor. Thor.** (Witthayu Kromkarn Phalang Ngan Thahaan, Defence Energy Dept. R. St.). ✉ New Building, Sukhumvit 24, Phra Khanong, Bangkok 10250. **Regional st:** 141/3 Mu 4, Don Kaeo Rd, Chotana, Mae Rim District, Chiang Mai 50180 – **20) Wor. Sor. Sor.** (Witthayu Seuksa, Educational Radio). ✉ Educational Technology Centre, Si Ayutthaya Rd, Ratcha Thewi, Bangkok 10400 – **22) Wor. Phon Sii** (Witthayu Kong Phon Thii Sii, 4th Infantry Division). ✉ Headquarters of the 4th Infantry Division, Somdet Phra Naresuan Maharat Camp, Phitsanulok 65000 – **23) Phon Ror. Kao** (Kong Phon Thahaan Raap Thii Kao, 9th Infantry Division), ✉ Surasi Camp, Kanchanaburi 71190 – **24) Jor. Sor.** (Krom Jaye Thahaan Suesarn, Army Signals Department). ✉ Jor. Sor. 1, Rama V Rd, Saphan Daeng, Bangsue, Dusit Region, Bangkok 10300. **Regional stns:** Jor. Sor. 2, Tharathibodi Rd, Thakham, Phunphin District, Surat Thani 84130 – Jor. Sor. 3, Prasert Songkhram Camp, Kongphon Si Rd, Nua Muang, Roi Et 45000 – Jor. Sor. 4, 104 Thetsaban 1 Rd, Nai Muang, Yasothon 35000 – Jor. Sor. 5, 5 Mu 2 Ban Lao, Ban Lao, Chaiyaphum 36000 – Jor. Sor. 6, 1543/23 Srisumang Rd, Muang Tai, Si Sa Ket 33000 – **25) Mor. Kor.** (Mahawitthayalai Kasetsart, Kasetsart University). HQ: ✉ 50 Phahonyothin Rd, Bang Khen, Chatuchak, Bangkok 10900. Bangkok. Tr. located at Nongkhaem in Samut Sakhon province. **Regional stns:** 301/1 Mu 5, Paphai, Sansai District, Chiang Mai 50210 – 86/8 Maliwan Rd, Muang Kao, Sitan, Khon Kaen 40000 – 424 Mu 3, Kanchanawanit Rd, Phawong, Songkhla 90100 – **26) Jor. Sor. Lor.** (Jangwat Thahaan Bok Lop Buri, Lop Buri Army Province). ✉ 13th Military Circle, Narai Maharat Rd, Lop Buri 15000 – **27) Or. Sor. Mor. Thor.** (Ongkarn Suesarn Muanchon Haeng Pratheet Thai, Mass Communications Org. of Thailand, MCOT "Modern Radio"). ✉ 63/1 Rama IX Rd, Huay Khwang, Bangkok 10320 – **28) Sor. Or. Thor.** (Sathaanii Witthayu Krom Utiniyom Witthayaa, Meteorological Department R. St.), ✉ 4353 Sukhumvit Rd, Bangna, Bangkok 10260 – **29) Sor. Thor.** (Siang Chaak Thahaan Reua, Voice of the Navy). ✉ Sor. Thor. Ror. 2: Phutianan Stadium, Phra Khanong, Bangna District, Bangkok 10260. **Regional stns:** Sor. Thor. Ror. 3: 99/1 Mu 1, Phuket 83000 – Sor. Thor. Ror. 4: 9/9 Thetsaban-Phatthana Rd, Wat Mai, Chanthaburi 22000 – Sor. Thor. Ror. 5: 652 Mu 2, Sattahip District, Chon Buri 20180 – Sor. Thor. Ror. 6: Songkhla Naval Station, Thale Luang Rd, Bo Yang, Songkhla 90000 – Sor. Thor. Ror. 7: Mae Klang River Operation Unit, Nakhon Phanom 48000 – Sor. Thor. Ror. 8: Ban Khlong Mek, Tha Chang, Phrom Phiram District, Phitsanulok 65150 – Sor. Thor. Ror. 9: Ban Thung Sawang, Ubon-Takan Rd, Rai Noi, Ubon Ratchathani 34000 – **30) Sor. Wor. Phor.** (Sathaanii Witthayu Phitaksantirat, Police R. St.). ✉ Radio Broadcasting Section, 2nd Communication Division, Directorate of Police Communications, Police Department, Bang Khen Region, Bangkok 10900. **Regional stns:** 40 Mu 1, Chotana Rd, Maesa, Mae Rim District, Chiang Mai 50180 – Sor. Wor. Phor. 2, Suranarai Rd, Cho Ho, Nakhon Ratchasima 30310 – Sor. Wor. Phor. 3, Banphru, Hat Yai District, Songkhla 90250 – Sor. Wor. Phor. 4, Pracha Uthit Rd, Nai Muang, Phitsanulok – **31) Neung. Por. Nor.** (Krom Praisanii Thoralek, Post & Telegraph Dept.). ✉ Chaengwattana-Thungsonghong Rd, Don Muang, Bangkok 10210. 1035kHz=Phaak Phiset, 1089kHz=Mor. Sor. Thor. **Regional stns:** Bypass Rd, Nai Muang, Buri Ram 31000 – 219 Mu 4, Lampang-Hang Chat Rd, Pong Yang Khok, Hang Chat District, Lampang 52190 – 2/7 T. Nong Luang, Mahat Thai Bamrung Rd, Nong Luang, Tak 63000 – Ban Nong Bu, Rop Muang Rd, Samphrao, Udon Thani 41000– **32) Wor. Por. Tho.** (Witthayu Prachaam Thin, Local R, Communications Division, Signals Dept, Royal Thai Army). ✉ Wor. Por. Tho. 8: Kamphaeng Phet Akkharayothin Camp, Suan Luang, Krathum Baen District, Samut Sakhon 74110. **Regional stns:** Wor. Por. Tho. 2: Kawila Camp, Kongsai, Wat Ket, Chiang Mai 50000 – Wor. Por. Tho. 3: 001 Na Khai Suranari, Phanibut Rd, Pho Klang, Nakhon Ratchasima 30000 – Wor. Por. Tho. 4: Thep Sattri Si Sunthon Camp, Kabang, Thung Song District, Nakhon Si Thammarat 80310 – Wor. Por. Tho. 5: 5 Kanchanawanit Rd, Hat Yai District, Songkhla 90110 – Wor. Por. Tho. 6: Sapphasiti Prasong Camp, Warin Chamrap District, Ubon Ratchathani 34190 – Wor. Por. Tho. 7: Phrai Prachak Sinlaprakhom, Thahaan Rd, Mak Khaeng, Udon Thani 41000 –Wor. Por. Tho. 9: Chiraprawat Camp, Na Khu Chiraprawat Rd, Nakhon Sawan 60000 – Wor. Por. Tho. 10: Mengrai Maharat Barracks, Chiang Rai 57000 – Wor. Por. Tho. 12: 140 Kasikonthungsang Rd, Sila, Khon Kaen 40000 – Wor. Por. Tho. 14: Phichai Dap Hak Camp, 13/7 Prachanimit Rd, Tha It,

NATIONAL RADIO

Uttaradit 53000 – Wor. Por. Tho. 15, Khet Udomsak Camp, Wang Mai, Chumphon 86190– Wor. Por. Tho. 16: 35 Sukayang Rd, Sateng, Yala 95000 – Wor. Por. Tho. 17: Trang-Palian Rd, Ban Khuan, Trang 92000 – **33) Phon Neung Ror. Or.** (Kong Phon Thii Neung Raksaa Phra Ong, 1st Infantry Division, Royal Guard). Phitsanulok Rd, Dusit Region, Bangkok 10300. 999kHz=Phaak Phiset, 1350kHz=Phaak Pokkati. – **34) Nor. Thor. Phor.** (Nuai Bannachakaan Thahaan Phatthanaa, Armed Forces Development Command AFDC, Supreme Command HQ). Sathaanii Witthayu 919, Phitsanulok Rd, Dusit Region, Bangkok 10300. **Regional Stns:** Sathaanii Witthayu 914, Suan Sak Kieo Tap Yong, Ban Pong 00, Mae Chan, Mae Chan District, Chiang Rai 57110. 1395kHz in Thai, 1179kHz prgr in Thai and minority langs for hilltribes. – Sathaanii Witthayu 934, Mu 2 Ban Khao Kiw, Uthai-Thapthan Rd, Sakaekrang, Uthai Thani 61000 – Sathaanii Witthayu 912, 13 Chan Uthit Rd, Bang Nak, Narathiwat 96000 in Thai and Malay – Sathaanii Witthayu 921, 114 Mu 1, Na Saeng, Lom Kao District, Phetchabun 67120 – Sathaanii Witthayu 909, Ban Rung Phatthana, Sakon Nakhon-Nakhon Phanom Rd, That Naweng, Sakon Nakhon 47000 – **35) Kor. Wor. Sor.** (Kitkarn Witthayu Krachaisiang, Radio & TV Division, Army Signals Dept). HQ: Radio Broadcasting & Television Division, Signals Department, Royal Thai Army, Rama V Rd, Saphan Daeng, Bangsue, Dusit Region, Bangkok 10300. **Regional stns:** Kor. Wor. Sor. 1, Surin-Prasat Rd, Nok Muang, Surin 32000 – Kor. Wor. Sor. 2, Yantarakit Sokon Rd, Sung Men District, Phrae 54130 – Kor. Wor. Sor. 3, 1879 Mu 14, That Choeng Chum, Sakon Nakhon 47000 – Kor. Wor. Sor. 4, 383 Super Highway, Ban Dom, Phayao 56000 – Kor. Wor. Sor. 5, 252 Mitraphap Rd, Ban Phai District, Khon Kaen 40110 –**36) Sor. Sor. Sor.** (Siang Sam Yot, Crime Suppression Division, Royal Thai Police). Section 1, Superintendency 2, Command Division, Crime Suppression Division, Phahonyothin Rd, Bangkok 10900. **Regional st:** 195 Mu 8, Udon-Nong Samrong Rd, Mumon, Udon Thani 41000 – **37) Ror. Dor.** (Kromkarn Raksaa Dindaen, Territorial Defence Dept.). HQ: 2 Charoen Krung Rd, Suan Chaochet, Phra Nakhon Region, Bangkok 10200. **Regional stns:** Nong Ho, Chotana Rd, Chang Peuak, Chiang Mai 50000 – Sri Phatcharin Camp, Raat Khaneung Rd, Nai Muang, Khon Kaen 40000 – **38) Mahaawitthayalai Thammasat** (Thammasat University), Faculty of Journalism and Mass Communications, Thammasat University, Prachan Rd, Phra Nakhon Region, Bangkok 10200 – **39) Yaan Kraw** (4th Cavalry Battalion, Armoured Unit, Royal Guard). HQ: Military Armoured Car School, 1156 Samsen Road, Bangkabrue, Dusit Region, Bangkok 10300. **Regional st:** Mitraphap Rd, Nong Bua Udon Thani 41000 – **40) Sathaanii Witthayu Rattasapha** (Parliament R. Station). Parliament House, Uthong Nai Rd, Dusit Region, Bangkok 10300.

International Relays: Radio Saranrom, Radio Liberty, Voice of America, Radio Farda, BBC – see International radio section

TIMOR-LESTE

LT: UTC +9h — **Pop:** 1.1 million — **Pr.L:** Tetun, Portuguese, Indonesian — **E.C:** 50 Hz, 220V — **ITU:** TLS

AUTORIDADE REGULADORA DAS COMUNICAÇÕES (ARCOM)
 Secretária de Estado dos Transportes, Equipamentos e Comunicações, Ministério das Infra-Estruturas, Av. Bispo de Medeiros, Díli ☎ +670 3339343 +670 3339339 **W:** www.arcomtl.com

RÁDIO E TELEVISÃO DE TIMOR-LESTE (RTTL) (Pub.)
 Edifício da Rádio e Televisão, Rua Mercado Municipal, Díli ☎ +670 7231152 **W:** www.rttlep.com **L.P:** Acting Dir: Julio Manuel Correia. RTTL administers Rádio Timor-Leste (RTL) and TV Timor-Leste (TVTL).

RÁDIO TIMOR-LESTE (RTL) (Pub.)
, Rua Mercado Municipal, Díli ☎ +670 3321827 or +670 7231158 **W:** www.rttlep.com **E:** radiotimorleste@gmail.com **L.P:** Dir: Rosário Maia Martins

MW (kHz)	kW	Location		
684	1	Díli (alt. freq. 680)		

FM (MHz)	kW	Location	FM (MHz)	kW	Location
88.7	0.5	Maliana	95.0	1	Baucau
90.1		Ermera	96.3		Ainaro
90.9	0.3	Aileu	96.3	1	Same
91.7	4	Díli	97.1	0.5	Lospalos
92.1	0.3	Oecussi	97.6		Díli (Antena 2)
93.1	0.3	Suai	98.5	0.3	Viqueque
94.5	0.3	Manatuto	99.5		Liquisa

D.Prgr in Tetun, Portuguese and Indonesian: 2045-1200. Regional stn hours may be limited due to restricted electricity supply.
N. in Tetun: 2200, 0800. **N. in Portuguese:** 2300, 0700. **N. in Indonesian:** 0000, 1000. All N. Mon-Fri only.

TIMOR-LESTE ASSOCIATION OF COMMUNITY RADIO STATIONS (Asosiasaun Radio Komunidade Timor-Leste - ARKTL)
 c/o Timor-Leste Media Development Centre, Rua Sebastião, Colmera, Díli ☎ +670 3324475 **W:** http://arktl.wordpress.com ARKTL's main role is as advocate for all community and independent stns.

COMMUNITY RADIO CENTRE (Centru Radio Comunidade) (CRC)
 CNE Building, Rua Bispo de Medeiros, Area of Quintal Boot, PO Box 160, Santa Cruz, Díli ☎ +670 3331227 or +670 7237890 **W:** www.crc-tl.org **E:** admin@crc-tl.org **L.P:** Mgr: Luis Evaristo dos Santos. CRC supports stations 8, 10, 11, 12, 13, 15, 17, 19 and 21.

COMMUNITY AND INDEPENDENT STATIONS

MW	kHz	kW	Station
1)	1404	2.5/5	R. Timor Kmanek, Díli

FM	MHz	kW	Station
2)	88.8	0.15	Direito FM, Díli
3)	89.5	0.5	Voz FM, Díli
4)	89.7		R. Comunidade Maubisse Mau-Loko, Maubisse
5)	90.0		R. Akademika, Universidade Nal de Timor-Leste, Díli
6)	91.2		R. Lalenok Ba Ema Hotu, Díli
7)	91.7	0.2	R. Comunidade Maliana, Maliana
8)	92.3	0.1	R. Comunidade Café Ermera, Gleno
9)	92.3	0.1	R. Comunidade Tokodede, Liquisa
10)	93.3	0.1	R. Comunidade Atoni Lifau, Oecussi
11)	94.5	0.1	R. Comunidade Cova Taroman, Suai
12)	94.7	0.3	Centru R. Comunidade (CRC) R, Díli
13)	95.1	0.1	R. Comunidade 1912 Dom Boaventura, Same
14)	95.8		Liberdade FM, Díli
15)	96.1	0.1	R. Comunidade Ili Uai, Manatuto
16)	97.0		R. Suara Timor Lorosae, Díli
17)	97.1	0.1	R. Comunidade Rai Husar, Aileu
18)	97.9	0.8	R. Povo Viqueque, Viqueque
19)	98.1	0.1	R. Comunidade Lian Tatamailau, Ainaro
1)	98.5	0.1	R. Timor Kmanek, Díli
20)	99.5	0.16	R. Rakambia, Díli
21)	99.9	0.1	R. Comunidade Lian Matebian, Baucau
22)	100.1	0.3	R. Comunidade Lospalos, Lospalos
23)	100.5	0.1	R. Lorico Lian, Díli
24)	102.0		R. Klibur, Díli
25)	102.5	0.15	R. Koulimai, Bukoli
26)	104.0		R. Sapientia, Díli
27)	107.9		R. Fini Lorosae, Baucau

RELAYS OF INTERNATIONAL STATIONS:
RDP Internacional, Díli 105.3MHz 1 KW – **RDP Antena 1**, Díli 103.1MHz – **BBC World Service**, Díli 95.3vMHz – **R. Australia**, Díli 106.4MHz.

TOGO

LT: UTC — **Pop:** 6 million — **Pr.L:** French, Ewé, Kabyè, Kotokoli, Mina — **E.C:** 50Hz, 127(Lomé)/220V — **ITU:** TGO

HAUTE AUTORITÉ DE L'AUDIOVISUEL ET DE LA COMMUNICATION (HAAC)
 Lomé. **L.P:** Pres: Philippe Evegno, Vice Pres: Wiyao Dadja Pouwi

RADIODIFFUSION-TÉLÉVISION TOGOLAISE (Gov.)
 B.P. 434, Lomé ☎ +228 2212493 +228 2213673 **W:** www.radiolome.tg **E:** radiolome@radiolome.tg **L.P:** Dir: Bawa Semedo. CE: Dodzi Soares.
FM: Agou 88.3MHz, Alédjo 92.7MHz, Dapaong 88.3MHz, Badou 99.3MHz, Lomé 99.5MHz.
D.Prgr: French/Ethnic: 24h. **English:** 1940. **Ann:** "Radio Lomé". **IS:** Soft tempo chime.

RADIO KARA (Regional station)
 B.P. 21, Kara ☎ +228 606060 **L.P:** Dir: Kao Pérézi. CE: Tete Anani
FM: Kara 91.5MHz, Dapaong 91.9MHz, Agou 94.5MHz, Alédjo 99.3MHz, Lomé 101.5MHz.
D.Prgr: 0525-0905, 1200-1435, 1625-2105. **Ann:** "Radiodiffusion Kara".

Other stations:
R. Avenir, 76 Blvd. de la Kara, Quartier Doumassessé, B.P. 20183, Lomé: 104.3MHz. – **R. Carré-Jeune**, Quartier Adidogomé, B.P. 2550, Lomé: 103.1MHz. – **R. de L'Evangile**, Lomé: 100.3MHz. – **R. Delta Santé**, Aneho: 106.1MHz. (Also rel. RFI) – **R. Evangile Jésus Vous Aime**, Bretelle de Klimamé, B.P. 2313, Lomé. **FM:** Lomé 100.2MHz, Agou 104.1MHz – **R. Maria Togo**, n°155 de la rue 158, Hédzranawoé,

B.P. 30162, Lomé **W:** www.radiomaria.org **E:** rmariatg@ids.tg **FM:** Dapaong 88.5MHz 0.25kW, Lomé 98.8MHz, Kara 101.5MHz, Kpalimé/Sokodé 104.5MHz – **R. Missionnaire,** Quartier Tomdé Kara, B.P. 170, Kara: 106.3MHz. **E:** emc_kara@yahoo.com – **R. Nana FM,** Angle Rues Tanou et Djossi, B.P. 6035, Lomé: 95.5MHz. **E:** petdog2@yahoo.fr – **Océan FM,** Aneho: 93.1MHz – **R. Rurale:** Pagouda 88.9MHz, Notsè 100.1MHz, Dapaong 102.5MHz – **Sport FM,** Tokoin Habitat, B.P. 8675, Lomé: 91.9MHz **W:** www.radiosportfm.com – **R. Tropik FM,** Quartier Wuiti, B.P. 2276, Lomé: 93.1MHz. **E:** tropikfm@nomade.fr – **Zephyr FM,** B.P. 20017, Lomé. **W:** www.zephyr.tg **E:** zephyr@zephyr.tg **FM:** Lomé 92.3MHz, Kara 95.5MHz, Atakpamé 102.9MHz – **R. Zion,** Adidogomé, B.P. 13853, Lomé. **FM:** Lomé 94.3MHz, Kpalimé 102.5MHz.
Africa No 1: Lomé 102MHz.
BBC African Sce in English/French: Lomé 97.5MHz
RFI Afrique: Lomé 91.5MHz, Aledjo 95.9MHz, Agou 98.3MHz

TOKELAU (New Zealand)

L.T: UTC -10h — **Pop:** 1,416 — **Pr.L:** Tokelauan, English —**E.C:** 50Hz, 240V — **ITU:** TOK

TOKELAU COMMUNITY RADIO
Office for the Council for the Ongoing Government, Tokelau Office, Apia, Samoa
LP: Acting Mgr: Aleki Silao **Prgr:** local community news and information, educational talks, weather reports and contemporary and traditional music. Each community sttn operates independently from separate studios.

FM	MHz	kW	Station
Atafu Atoll	107.5	0.005	R.Atafu FM
Fakaofo Atoll	107.5	0.005	R.Fakaofo FM
Nukunonu Atoll	107.5	0.005	R.Nukunonu FM

TONGA

L.T: UTC +13h — **Pop:** 120,898 — **Pr.L:** Tongan, English — **E.C:** 50Hz, 240V — **ITU:** TON

TONGA BROADCASTING COMMISSION
(Independent Statutory Board, part-comm.)
P.O. Box 36, Nuku'alofa ☎ +676 23295, 23555, 23556, 📠 +676 24417 Fangatongo, Neiafu, Vava'u T: +676 70827, 70843 **W:** www.tonga-broadcasting.com **E:** contactus@tonga-broadcasting.com
L.P: GM: Mrs. 'Elenoa 'Amanaki. Tech Mgr: Sioeli Maka Tohi
MW: A3Z 1017kHz 10kW **FM:** 90.0MHz 0.1kW (Kool FM) + relay at Fangatongo, Neiafu, Vava'u
R. Tonga on 1017kHz: **D.Prgr:** 1900-1100. **N. in English:** 1800 (BBC), 1900 (ABC), 0000 (ABC or RNZI), 0700 (local), 0715 (ABC) **Ann:** "Radio Tonga" **Kool FM** on 90.0MHz **D.Prgr:** 24h **Ann:** "Kool 90 FM", "The Call of the Friendly Isles"

Other Stations:

FM	MHz	kW	Station
9) Vaipoa	88.0		Niuatoputapu Min. of Inf. & Communications R.
8) Nuku'alofa	88.1		FM 88.1
1) Nuku'alofa	88.6		R.Nuku'alofa
6) Neiafu,Vava"u	88.6		FM Peau Vava'u
2) Neiafu,Vava'u	89.0		Letio Faka-Kalistiana 89FM
3) Nuku'alofa	89.1		Tonga R.Magic 89.1.
7) Neiafu, Vava'u	89.3		PIG FM1
4) Nuku'alofa	93.1	0.2	Letio Faka-Kalistiane 93FM
3) Neiafu,Vava'u	101.1		Vava'u R.
5) Nuku'alofa	103.0		R.Australia

Addresses and other information
1) Nuku'alofa. **ID:** '88.6 FM' or '88FM' ☎ +676 24901 Txt: +676 76511 **W:** www.bebo.com/radionukualofa **D.Prgr:** 24h – **2)** UCB Pacific Partners, P.O.Box 478, Nuku'alofa ☎ +676 27327 **W:** www.pacificpartners.org **E:** tonga@pacificpartners.org. **L.P:** Mgr Loni Akolo. **D.Prgr:** 24h **Call:** A3R – **3)** 13 Vaha'akolo Road, Kaipongipongi, Nuku'alofa ☎ +676 25891 📠 +676 25600 **W:** www.tongaradio.com **E:** a3rv@tongaradio.com, magic@tongaradio.com **L.P:** Mgr Phillip Vea. **Call:** A3V – **4)** UCB Pacific Partners, P.O.Box 95, Neiafu, Vava'u. ☎ +676 70223 **L.P:** Mgr Willy Florian **D.Prgr:** 24h – **5)** 24h satellite relay – **6)** ☎ +676 71128/7129 – **7) ID:** "FM1" ☎ +676 71479 – **8)** BroadCom-TMN Group, Ngeieia, Nuku'alofa. **L.P:** Dir: Maka Tahi, C.Ops Mgr: Siaosi Lavaka. **D.Prgr:** 24h in Tongan & English [RNZI]. **F.PL:** nationwide FM coverage to other islands – **9)** Old Catholic Priest's residence, Vaipoa, Niuatoputapu, Northern Tonga. **Call:** A3NTT. Low solar power community radio established May 2010 as part of earthquake and tsunami relief operations.

TRINIDAD & TOBAGO

L.T: UTC -4h — **Pop:** 1.4 million — **Pr.L:** English — **E.C:** 60Hz, 115V — **ITU:** TRD

CARIBBEAN NEW MEDIA GROUP LTD. (Gov. Comm.)
11a Maraval Rd, Port of Spain ☎ +1 868 622 4141
FM: Talk City 91.1MHz – **Next** 99.1MHz – **Sweet** 100.1MHz

TRINIDAD BROADCASTING COMPANY (Comm.)
Second Floor, Guardian Building, 22-24 St. Vincent St., P.O. Box 716, Port of Spain ☎ +1 868 623 9202/3/6/7 📠 +1 868 625 1782 **W:** www.radiotrinidad.com & www.aakashvaniradio.com **L.P:** GM: Steve Dipnarine.
MW: Inspirational 7-30 AM: 730kHz 20kW, 24h **N:** On the h 0600-1900
FM: The Best Mix: 95.1MHz – **The Vibe City 105:** 105.1/105.5MHz – **Sangeet:** 106.1MHz – **Aakash Vani:** 106.5MHz

GEM RADIO NETWORK (Comm.)
3 A Queens Park West, Port of Spain ☎ +1 868 625 8426 📠 +1 868 624 3234 **L.P:** Regional GM: Cheryl Chambers
Hott 93: 93.1MHz (Tobago), 93.5MHz (Port of Spain)

HCU COMMUNICATIONS GROUP (Comm.)
1 Mulchan Seuchan Rd, Endeavour, Chaguanas ☎ +1 868 665 3630 📠 +1 868 672 1059 **E:** bollywood@homeviewtnt.com **L.P:** Chmn: Mohan Jaikaran. CEO: Marcel Mahabir. SM: Joy Mahabir
FM: Shakti Hot Like Pepper: 97.5MHz – **Bollywood Masala** 101.1MHz 25kW. Format: East Indian

C.L. COMMUNICATIONS GROUP (Comm.)
Level 4, Long Circular Mall, Long Circular Road, St. James. Radio 90.5, Suite 5, Valpark Shopping Plaza, Valsayn ☎ +1 868 622 4124, Radio 90.5: 645 8083 103FM: 628 9222 📠 +1 868 622 6693 **E:** radio90fm@homeviewtnt.com, radio97@wow.net, 103fm@homeviewtnt.com, radio104@tstt.net.tt
FM: Radio 90.5: 90.5MHz (East Indian Stn) – **Music Radio 9-7:** 97.1/97.9MHz – **103FM:** 103.1MHz (East Indian stn) – **Heartbeat 103.5:** 103.5MHz (For woman) – **Ebony 104:** 104.1MHz

CITADEL LIMITED (Comm.)
20 Rust St., St. Claire, Port of Spain & 47 Tragarete Rd, Newton, Port of Spain ☎ +1 868 622 9292 📠 +1 868 628 0251 **W:** www.i955fm.com & www.hitz107fm.com
L.P: CEO: Louis Lee Sing. VP Prgr: Anthony Lee. VP Marketing: Ian Lee Sing. VP Fin: Charlene Quamina-Vincent
FM: i95.5 FM: 95.5/95.7MHz Format: News/talk/current affairs – **Red 96.7:** 96.7MHz Format: Urban – **Hitz 107.1:** 107.1MHz. Format: AC.

Other Stations:
BBC. FM: 98.7MHz – **City FM,** 88-90 Abercromby St., Port of Spain ☎ +1 868 627 6937 **W:** www.city94fm.com.**L.P:** MD Brian Knight. **FM:** 94.1MHz – **Heritage R.,** 2-12 Hilda Lazzari Terrace, San Fernando ☎ +1 868 657 5153 📠 +1 868 653 9248 **W:** heritageradiott.com **FM:** 101.7MHz – **Isaac 98.1,** 115A Woodford Street, Newtown, Port of Spain ☎ +1 868 628 0904 **W:** www.isaac981.com **L.P:** CEO Margaret Elcock. **FM:** 98.1MHz Format: Rlg – **Life R., FM:** 99.5MHz – **More FM. W:** www.1077musicforlife.com **FM:** 107.7MHz – **Music For Life,** 177 Tragarete Road, Woodbrook ☎ +1 868 628 1047 **W:** www.morefmtrinidad.com **FM:** 104.7MHz – **Power 102,** 90 Abercromby St., Port of Spain ☎ +1 868 627 6937 **W:** www.power102fm.com **L.P:** MD Ingrid Isaac. **FM:** 102.1/102.5MHz – **Pulse 91.5 FM,** Bolan Amar Building, Pole Carew Street, Woodbrook ☎ +1 868 628 0827. **FM:** 91.5MHz – **R. Jaagriti,** Corner Pasea Main Road Ext and Churchill Roosevelt Highway, Tunapuna ☎ +1 868 663 8743 📠 +1 868 663 8691 **W:** www.jaagriti.com **L.P:** MD: Sat Maharaj. **FM:** 102.7MHz. Format: Rlg. (hindu) – **R. Tambrin,** 3 Picton Street, Scarborough, Tobago ☎ +1 868 639 3437 📠 +1 868 660 7357 **E:** tambrin@tstt.net.tt **W:** www.tambrintobago.com **L.P:** GM: George Leacock. SM: Garth James. **FM:** 92.7MHz – **R. Toco,** Galera Road, Toco ☎ +1 868 670 0068 **L.P:** CEO: Michael Als. **FM:** 106.7MHz 1.2kW. Community Radio for NE Trinidad. Relays VoA 0200-1200 – **R. Trinbago,** V.L. Communications Ltd., Victoria Park Suites, 14-17 Park Street, Port of Spain ☎ +1 868 627 8340/9484 **FM:** 94.7MHz – **Sidewalk R.,** Curepe Mall, Curepe ☎ +1 868 663 8691 **FM:** 92.1MHz – **Soca 91.9,** 56 Maraval Road, Port of Spain ☎ +1 868 628 3469/4351 **W:** www.919socafm.com **FM:** 91.9MHz – **WACK FM,** 129c Coffee Street, San Fernando ☎ +1 868 652 9774 **W:** www.wackradio901fm.com **L.P:** CEO: Kenny Phillips. **FM:**

90.1MHz – **W.E.F.M.**, 153 Tragarete Road, Port of Spain ☎ +1 868 628 9696 🖷 +1 868 622 9387 **E:** info@96.1wefm.com **W:** www.96wefm.com L.P: CEO: Kenny Phillip. MD: Diane Phillips. **FM:** 96.1MHz Format: Urban Caribbean – **WMJX FM**, 9 Long Circular Road, St. James ☎ +1 868 628 9561 **W:** www.wmjxfm.com/cms2/ L.P.: GM: Keith Cadet. **FM:** 100.5MHz. Format: Smooth Jazz.

TRISTAN DA CUNHA (UK)

LT: UTC — **Pop:** 275 — **Pr.L:** English — **E.C:** 50Hz, 220V — **ITU:** TRC

TRISTAN BROADCASTING SERVICE (Gov.)
🖃 The Administrator, Tristan da Cunha, So. Atlantic via Cape Town, South Africa. **E:** tristan.radio@yahoo.co.uk **L.P:** Head of Telecommunications: Andy Repetto.
FM: Tristan Radio 93.5MHz 25W **D.Prgr:** Mon/Wed/Fri 1630-1800, Sun 1000-1200

Other stations:
Atlantic FM 93.5MHz, **Saint FM** 95.0MHz (rebroadcast from St Helena), and **BFBS 1**

TUNISIA

LT: UTC +1h — **Pop:** 10.5 million — **Pr.L:** Arabic — **E.C:** 50Hz, 115/220V — **ITU:** TUN

OFFICE NATIONAL DE LA TÉLÉDIFFUSION (ONT)
🖃 Cité Ennasim I, Borjel, B.P. 399, 1080 Tunis Cedex ☎+216 71801177 🖷+216 71781927 **W:** www.telediffusion.net.tn **E:** ont@telediffusion.net.tn

RADIO TUNISIENNE (Gov.)
🖃 71 Ave. de la Liberté, TN-1002 Tunis ☎+216 71847300 🖷+216 71785146 **W:** rtci.tn **E:** ittisal@ertt.nat.tn
L.P: DG Radio: Chaouki Aloui. Dir. Tech. Radio: Moncef Fathallah.

MW	kHz	kW		Prgr.	Times
Gafsa	585	350	N	0400-2400	
Tunis	630	300	N	24h	
Mednine	684	10	N	0400-2400	
Remada	882	1	N	0400-2400	
Tunis	963	100	I/C	0400(Mon 0900)-2300	
FM (MHz)	**N**	**C**	**I**	**Y**	**kW**
Ain Draham	90.3		93.4	96.6	6
Biadha		105.4	101.8	95.0	50
Gabes		93.3			1
Ghraba (Sfax)	93.0	103.0	99.5	93.0	60
Gorrâa	89.1		95.4	89.1	4
Harkoussia				92.5	
Kasserine			99.2	89.6	49
Kchabta	102.6		93.8	97.0	
Kef Errand	89.8			99.4	
Remada	103.4	99.9	93.4	90.3	80
Souk Jomaa				91.3	
Trozza		87.7		90.8	
Tunis	105.3		98.2	88.6	1
Zaghouan	94.3		94.3	96.5	20
Zarzis			90.7	93.9	72

National Channel (N) in Arabic**:** 24h. **N.** on the h. – **Cultural channel (C)** in Arabic: 1100-2300 on 963kHz & FM. – **R. Tunis Chaîne Internationale (RTCI) (I):** French/others: 0500-2300, 963kHz 0500-1100. – **R. Jeunes** (Youth R.) **(Y):** 24h.

Regional stations : **R. Gafsa,** Avenue Habib Bourguiba, 2100 Gafsa: 88.3MHz, 89.MHz 40kW, Biadha 91.8MHz, Chambi 92.7MHz 50kW, 93.5MHz – **R. Le Kef,** Rue Mongi Slim, 7100 Le Kef. FM: Souk-Jomaa 92.2MHz 50kW, 90.0MHz, Aïn Draham 90.3MHz, Ghardimaou 94.1MHz, Sidi Youssef 95.8Mhz, Sidi Salem 96.2MHz, Le Kef 96.8MHz, Nefta 99.6MHz, Nebeur 100.1MHz, Goraa 102.2MHz, 103.1MHz, 106.7MHz – **R. Monastir,** Rue Farhat Hached, 5019 Monastir. FM: Harkoussia 95.7MHz, Trozza 97.3MHz 5kW, Sousse 99.0MHz, Zaghouan 104.7MHz, Monastir 106.1MHz – **R. Sfax,** Route de Menzel Chaker Road, 3058 Sfax. FM: Djerba 89.0MHz, Ksour-Essaf 100.2MHz, Trozza 100.8MHz, Sfax 105.2MHz – **R. Tataouine,** Cité 7 Novembre, 3263 Tataouine: 87.6MHz 70kW, Techout 89.5MHz, 92.2MHz, 94,3MHz 80kW, 96.6Mhz 80kW, 102.6MHz.

Ann: National Channel: "Huna Tunis, Idha'atu-l-Wataniya at-Tunisiya". Cultural Channel: "Huna idha'at-Tunis at-thakafiya". F: "Ici Radio Tunisie Internationale"
Relays for abroad on shortwave: see International Radio section.

Other stations:
R. Mosaïque FM, Immeuble Montplaisir, Tunis. FM: Hammamet 88.9MHz, Sidi Bou Said 90.3MHz 0.1kW, Nabul 92.9MHz, Tunis 94.9MHz 5kW. D.Prgr: 24h in French and English. **W:** mosaiquefm.net
R. Jawhara FM, Kairouan 89.4MHz, Sousse 102.5/107.3MHz 1kW. 0500-2400. **W:** jawharafm.net
Shems FM: Monastir 90.6MHz, Sousse 93.7MHz, Bizerta 95.7MHz, Sfax 96.2MHz, Tunis 101.7MHz, Cap Bon 106.5MHz. **W:** shemsfm.net
Zitouna FM: El Ghraba 89.9MHz 60kW, Nefta 91.4MHz 1kW, Gorraa 92.2MHz 3kW, Trozza 94.0MHz 5kW, Tataouine 94.4MHz 1kW, Ksour-Essaf 96.9MHz, Zaghouan 97.6MHz 80kW, Souk-Jomaa 97.8MHz, Biadha 98.3MHz 35 kW, Bizerte 99.1MHz 5kW, Gabes 99.8MHz 1kW, Ain Draham 100.4MHz 5kW, Zarzis 100.7MHz 63kW, Tozeur 102.3MHz 1kW, Nabeul 102.9MHz 7kW, Remada 103.4MHz 80kW, Sidi Bou Said 106.9MHz. **W:** www.zitounafm.net

TURKEY

LT: UTC +2h (27 Mar-30 Oct: +3h) — **Pop:** 73 million — **Pr.L:** Turkish — **E.C:** 50Hz, 230V — **ITU:** TUR

SUPREME BOARD OF RADIO AND TELEVISION(RTÜK)
🖃 Bilkent Plaza B2 Blok, 06530 Bilkent/Ankara ☎+90 312 2975000 **W:** www.rtuk.gov.tr **E:** rtuk@rtuk.gov.tr **L.P:** Pres: Davut Dursun.

TÜRKIYE RADYO-TELEVIZYON KURUMU (TRT) (Turkish Radio-Television Corporation)
🖃 TRT Genel Mudurlugu, TRT OR-AN, Yerleskesi, 06540 Ankara ☎+90 312 4904300 🖷 +90 312 4912817
W: www.trt.net.tr **E:** aktifhat@trt.net.tr
Regional addr: TRT Ankara Radyosu Müdürlüğü, Atatürk Bulvari, N:o 39 Sihhiye, Ankara – TRT Istanbul Müdürlüğü, Harbiye, Istanbul – TRT Izmir Müdürlüğü, 1420 Sokak No:87/A Kahramanlar, Izmir – TRT Çukurova Müdürlüğü,Inonu Bulvari 30 Ocak Mah. No:48 Liman B Kapisi Karsisi, 33130 Mersin – TRT Antalya Müdürlüğü, Memur Evleri Tonguc Caddesi 19, 07050 Antalya – TRT Trabzon Müdürlüğü, Adnan Kahveci Bulvari 70, 61200 Sogutlu-Akcabat/Trabzon – TRT Diyarbakir Müdürlüğü, Istasyon Caddesi N:o 1, 21100 Diyarbakir
L.P: DG: Ibrahim Sahin. Dep. DG (Eng.): Alaettin Korkmaz. Dep. DG (Bc. Sces): Zeynel Koç. Dir. Radio Dept:: Ömer Altan Bahadir. Transmitters Dept: Erol Büyükkaya. Radio Dept: Senol Goka.

LW & MW	kHz	kW	N	MW	kHz	kW	N
Mersin**	630	300	1/R	Izmir	927	200	1
Çatalca*	702	600	VOT	Trabzon	954	300	R
Antalya	891	300	1/R	Diyarbakir	1062	300	K

*) Istanbul, **Çukurova regional. 1=TRT1, 4=TRT4, R=Reg, K=Kurdish., VOT=VO Turkey Turkish Sce 24h.

FM: Location	TRT1	TRT2	TRT3	TRT4	Türkü	TRT6	Nagmek	kW
Adana	96.7	92.5	89.2	105.1				30
Adiyaman	88.8	94.4	90.8	103.3	90.8			30
Afyon	97.0	93.0		94.0				5/30
Agri	88.2	92.2	95.2	101.8	95.2			30
Amasya	94.7	93.9	99.6	107.3				30
Ankara-Cankaya	93.3	88.0	91.2	98.6				30
Ankara-Yenimehalle	107.8	100.3	102.8	103.7	98.6	102.8		30
Antalya	88.4	95.6	91.6	100.6			92.1	30/1
Antalya-Alanya	90.3	92.7		94.4				5
Antalya-Kas	98.8	93.6						1/5
Aydin-Kusadasi	98.7	90.2	93.5	103.5			96.6	5/30
Balikesir	92.6	98.4	96.4					30
Balikesir-Ayvalik	88.4	90.4	95.4	103.5			101.1	5
Bingöl	99.2	97.2	90.2	105.0	99.6			30
Bitlis	98.0	94.2	90.6	100.0				5
Bozkurt	88.3	95.8		91.8	103.5			30
Bursa	99.6	95.0	97.6					30
Bursa-Gundoglu	87.9	98.9	91.1	103.5				30/5
Çanakkale	93.0	89.5	97.0					30
Çankin-Eldivan	88.4	91.6	98.8	100.8	98.8			30
Çankiri-Ilgaz	93.2	98.4	90.0		90.0			30
Çorum	105.7	103.7	101.2					5
Cizre	101.9			103.8		94.5		5
Denizli	95.2	93.2	90.0	101.5			101.0	30/5
Diyarbakir	98.4	95.5		107.3		88.4		30
Edirne (Uzunköprü)	97.9	89.0	91.0					30
Elazig-Baskil	94.8	92.8	89.6	102.2				30
Erzincan	88.0	93.2	91.2					30
Erzurum	90.8	98.8	92.7	102.6	96.8			30/5
Erzurum-Oltu	95.6	93.6		98.1	89.1			5
Eskisehir	89.0	96.8	94.4	101.3				30
Eskisehir-Sivrihisar	90.2	98.4	96.2	104.4				30
Gaziantep	92.0	97.6	95.2	101.9				30
Hatay	93.6	91.2	100.0	103.5				30/5

FM: Location	TRT1	TRT2	TRT3	TRT4	Türkü	TRT6	Nagmek	W
Isparta	94.0	89.6	99.2		102.4		99.2	30
Istanbul	95.6	91.4	88.2	103.4			101.6	100
Izmir	94.7	91.2	88.0	100.5				100
Izmir-Karaburun	90.8	93.8	99.1		101.6		88.7	5
Izmit-Kocaeli	90.5	96.0	93.6	100.5				30
Kahramanm.-Elbistan	91.4	96.4			94.0			30
Karaman	96.4	98.6		106.5	90.8			30
Kars	100.8	89.5		91.3	103.3			30
Kastamonu		91.5		101.9				30/5
Kayseri	89.4	97.2	99.2	93.3				30
Kilis	94.4	90.8		98.3			88.8	5
Kirklareri-Demirköy	94.5	90.0	103.6	92.0				30
Kirsehir	97.6	92.0		102.5	88.8			30
Konya-Tuzlukcu	89.2	95.8		101.0	92.4			30
Kütahya	92.0	95.4	92.1	88.1				30
Mardin-Nusaybin	96.8	93.4		91.0				30
Mersin-Silifke	98.3	88.8	94.4	100.8	102.2			30/5
Mugla-Bodrum	94.6	99.3	89.4	100.3				5
Mugla-Datca	95.8	107.1	102.9	92.6			96.6	30
Mugla-Fethiye	97.7	94.5	89.3	93.7				5
Mugla-Köycegiz	105.1	99.8	95.4					30
Mugla-Marmaris	98.2	90.9	95.0	101.0				5
Mugla-Yatagan	88.8	92.0		102.2			96.6	30
Mus	90.9	98.7		105.0		90.2		5
Nevsehir-Avanos	99.6	95.0	93.7	102.5	88.8			5
Nigde	90.0	95.6		105.7	93.2			30
Ordu-Persembe	99.9	95.6		97.6				30
Samsun	95.2	92.8	90.8	96.8				30
Sanliurfa	98.1	102.5						5/30
Sanliurfa-Suruc	97.1	100.3		105.5				30
Siirt-Kurtalan	99.6	105.6	103.6	101.6		103.6		30
Sirnak	101.6	97.7				103.8		30
Sivas	93.6	98.3	90.4	100.9				30
Trabzon	88.8	95.0	92.0	97.0				30
Tunceli	89.2	92.4		101.1	106.9			30
Usak	105.5	101.9		95.7				30
Van	94.8	89.3	98.2	100.3		102.3		30/5
Van-Özalp	93.2	91.2		101.2	97.6			5
Yozgat	98.0	96.0	89.8					5
Zonguldak	88.8	97.2	99.2	93.4				5/30

+400 stations 1kW or less.

1= Radyo Bir (spoken word): 24h on FM. MW: 0400-0800 on 630/702/891kHz. In Turkish exc. in Bosnian (Mon) Arabic (Tues), Kurdish (Wed/Fri) and Circassian (Thu): 0410-0445 – **2 = (TRT-FM)** (popular music): 24h in Turkish – **3)= Radyo Üc** (classical music): 24h in Turkish exc. N. in English/French/German (3 min's each): 0503, 0803, 1003, 1303, 1503, 1803. Tourist Prgrs (3 min's each on Saturdays): English: 1715, French: 1515, German: 2015 – **4= Radyo Dört** (art & folk music): 24h on FM, 0700-1600 on 927kHz – **TRT Türkü** (Turkish folk music channel): 24h – **6= Radyo Sese** in Kurdish: 0400-1500 on 1062kHz & 24h on FM – **TRT Nagme** (Turkish Art Music Channel): 24h – **Regional prgrs** (Bölgesel R): On 630/891/954kHz: 0700-1600 & on FM: 0400-1600 – **Armenian Sce.** on 106.3MHz – **TRT Ankara Radyosu** on105.6MHz. **Ann:** TRT-1: "Burasi TRT Radyo Bir", TRT-4: "Burasi TRT Radyo Dört". Reg.: e.g. Antalya: "Burasi TRT Antalya Radyosu."

EXTERNAL SERVICE: Voice of Turkey
See International Radio section.

Other stations; main networks:

FM (MHz)	1)	2)	3)	4)	5)	6)	7)	8)	9)	10)
Abant		104.5								104.8
Adana	105.4	89.6	102.9	103.8	96.0		106.3	90.8	101.9	92.0
Adiyaman					92.0		97.0	100.5		
Afyon			90.4		95.1	104.6	88.8	94.9		
Agri					100.4		88.8	96.2		
Akhisar			93.0			106.2		105.2		
Aksaray			93.0		90.5	97.2	92.0	105.0		
Aksehir						104.5	97.0	99.5		
Alanya		100.0	89.3	98.0		89.9	88.8	93.5		
Amasya							104.0	101.6		
Ankara	102.4	88.8	100.0	105.3	97.2	94.5	89.8	90.0	107.4	90.8
Antalya	90.2	89.7	100.0	89.3	102.6	90.9	101.2	97.6	95.3	94.2
Ardahan					95.5	98.0	88.8	99.0		
Artvin							99.0	100.4		
Aydin	100.2		92.3		95.8		94.5	91.0		
Ayvalik			98.3		93.6					105.2
Balikesir	98.7	88.8	100.0	90.7	93.5	94.3	97.2	106.8	88.5	100.8
Bandirma			89.3		107.7	106.2	94.4	107.0		
Bartin					95.5	98.4	98.0	100.0		
Batman							88.8	99.1		
Bayburt			91.5			98.0	95.0			
Bilecik							92.4	102.0		

FM (MHz)	1)	2)	3)	4)	5)	6)	7)	8)	9)	10)
Boyabat								88.8	99.0	
Bafra						101.0	105.5	96.0		
Bingol								96.0	103.3	
Bitlis						94.5		88.8	101.0	
Bodrum	104.8	103.5	100.0	90.3	92.4					89.0
Bolu	92.0		100.0	94.2		97.1	89.2	90.0	107.6	
Boyabat								88.8	99.0	
Bucak								100.0		
Burdur			92.0			88.0	88.8	104.5		
Bursa	92.0	89.8	100.0	89.2	97.2	101.6	104.6	88.8	104.5	90.8
Ceyhan							107.7	90.6	97.8	
Cizre						88.1	93.0			
Çanakkale			89.3			99.5	105.3	88.8	101.0	
Çankiri						90.1		104.0	92.8	
Çerkezkoy	101.9							100.2		
Çesme			89.6	100.0	89.3	97.2				93.3
Çorlu	100.3		100.0				101.6	91.3		
Çorum						91.3	97.7	89.5		
Demirci							104.6	97.4		
Denizli	92.0	88.0	100.0	89.3	96.0	107.7	88.8	96.8	90.8	
Develi								99.0		
Didim		103.5								
Dinar							88.8	96.5		
Diyarbakir	92.0	89.7	100.3	92.3		90.4	98.0	88.7	101.0	97.5
Dogubeyazit					100.5					
Düzce	92.0					107.0	88.8	105.0	96.3	
Edirne			101.3	102.6	90.5	103.0	104.3	91.3	98.4	
Edremit					90.7	92.0	96.0			
Elazig						104.0	92.4	99.4	94.1	
Erzincan						89.5	88.8	93.8		
Erzurum	91.8	90.4	100.7	89.3		94.6	91.5	98.0	94.0	
Eskisehir	88.6	100.0	100.5	89.3		92.6	106.3	91.8	106.6	
Fethiye	102.0		100.3	96.7	96.4		99.0	99.7	98.5	93.2
Gaziantep	107.5	92.4	100.7	102.5	103.7	96.6	107.0	88.8	103.0	99.3
Gazipasa							97.5			
Gerede	104.0					94.5	102.7	88.8	107.4	104.8
Giresun						91.1		89.6	93.0	
Gölcük										90.8
Gümüshane						92.0		91.0	102.0	
Hakkari						98.0		88.8		
Hatay						88.5	106.7	105.0	101.0	
Igdir						95.5		88.8	95.5	
Iskenderun			101.0				94.6	100.6	95.5	
Isparta				91.8		96.0		88.8	100.0	
Istanbul	92.0	89.8	100.0	89.2	97.2	94.1	104.6	88.8	107.6	90.8
Izmir	96.2	89.6	100.0	89.3	97.2	96.7	101.3	95.2	96.9	90.8
K.Maras			89.0					92.5	94.0	
Karaman								88.8	103.9	
Kars						90.6		88.8	102.7	
Kayseri	93.9	105.0	100.0	88.7	98.9		100.2	92.5	88.5	96.2
Kirikkale						105.5	89.8	88.8	105.0	
Kirklareli								88.8	107.6	
Kirsehir								92.5	94.4	
Kocaeli	92.0	89.8	100.0	89.2		90.7	104.6	88.8	107.6	
Konya	95.1	89.9		89.3	101.7	92.9	112.6	102.0	93.4	
Kusadasi	104.5		100.0	105.8	92.7		88.6			
Kütahya	93.5	94.1		91.3		92.6	107.0	88.4	93.8	90.8
Ladik	92.0									
Malatya	92.0	89.9	103.3	97.8		91.7	104.6	105.2	105.5	94.5
Manisa			89.1			101.3	99.1	103.3	105.4	
Mardin							96.5	97.0		
Marmaris	109.6	105.5	100.0	89.3	96.1		97.9			
Mersin	98.1	88.1	105.0	90.3		94.5	103.0	106.8	105.3	
Milas							91.0	96.9		
Mugla	103.0		100.3	89.3		104.6	93.1	102.6	90.1	
Mus							88.8	102.0		
Nevsehir						96.3	90.7	102.0		
Nigde						94.5	99.9	103.0		
Ordu							102.0	92.5		
Osmaniye	105.4	89.6				98.4	99.9	95.3	101.7	
Rize				94.6			106.5	101.0		
Sakarya	101.3		100.7	101.8	91.2	104.6	102.6	107.8	97.8	
Samsun	99.7	106.2	100.0	88.0		94.1	92.0	103.0	96.3	
Saraykoy						107.4				
Siirt							88.8	91.0		
Sinop						94.3	88.8	99.0		
Sivas	96.2		102.2				91.0	100.1		
Sivrihisar						97.5	102.2	94.9		
Soke							96.9			
Sanliurfa	92.5					92.0	103.0	93.5	105.4	
Tekirdag	96.2		90.3	96.0		104.6	104.4	105.4	91.8	
Tokat						104.0	96.5			
Trabzon	92.2	100.4		89.3		94.6	93.0	92.5	102.8	90.3
Tunceli						94.5		100.0		
Usak						106.0		105.0	99.0	

FM (MHz)

	1)	2)	3)	4)	5)	6)	7)	8)	9)	10)
Van							104.6	92.0	96.0	
Yalova		100.0				94.1				
Yozgat							102.0	94.2	100.4	
Zile							98.0	95.0	103.5	
Zonguldak	92.7			90.0			104.6	100.0	107.0	91.0

Contact details:
1) Kral FM ✉ Mehmet Akif Mah, Inönü Cad. Star Sk. No :2 Ikitelli (Merkez-Dr. Eminpasa Sokak No :20/3 Cagaloglu), Istanbul ☎+90 212 4489060 🖷 +90 212 4489158 **W:** www.kralfm.com.tr – **2) Show R.** ✉ Büyükdere Caddesi No 163, Zincirlikuyu, Istanbul ☎+90 212 2851260 🖷 +90 212 2851297 **W:** www.showradyo.com.tr – **3) Power FM** ✉ Power Media Center, Ali Riza Gürcan Cad. N. 27, 34173 Merter, Istanbul ☎+90 212 4490900 🖷 +90 216 4816364 **W:** www.powerfm.com.tr **E:** powerteknik@powerfm.com.tr – **4) Alem FM** ✉ Davutpasa Cad. Merkez Efende Mah. No 34, Topkapi, Zeytinburnu, Istanbul ☎+90 212 2305858 **W:** www.alemfm.com **E:** info@alemfm.com – **5) Metro FM** ✉ Mehmet Akif Mah. Basin Ekspres Yolu Star Sok. No 2, Ikitelli, Istanbul ☎+90 212 4489860 🖷 +90 212 4489365 **W:** www.metrofm.com.tr – **6) Polis Radyosu** ✉ Emniet Genel Müdürlügü, Haberlesma Dairesi, Baskanligi Radyo TV ve Foto Film, Sube Md. Necatibey caddes No. 105, Anittepe-Ankara ☎+90 312 4123000 **W:** www.polisradyosu.net **E:** polisradyosu@egm.gov.tr – **7) Radyo 7** ✉ Otakçilar Cad No: 60, Eyüp, Istanbul ☎+90 212 5675454 🖷 +90 212 5677797 **W:** www.radyo7.com **E:** radyo7@radyo7.com – **8) Burc FM** ✉ Ferah Mahallesi Resatbey Sokak No:12, 34692 Buyukcamlica, Istanbul ☎+90 216 4433176 **W:** burcfm.com.tr **E:** burcfm@burcfm.com.tr – **9) Akra FM** ✉ Bulgurlu Mahallesi Duhanci Haci Mahmut Sokak No:35, Kucukcamlica, Istanbul ☎+90 216 3252265 🖷 +90 216 3277633 **W:** akradyo.net **E:** akradyo@akradyo.net – **10) Super FM** ✉ Basin Ekspres Yolu Ikitelli, Istanbul ☎+90 212 368 6210 **W:** superfm.com.tr **E:** superfm@superfm.com.tr
In addition there are about 30 national, 100 regional and 1000 local stations in operation on FM.

AFN INCIRLIK AIR BASE BROADCASTING STN (Mil.)
☎+90 322 3166421 **W:** www.afrts.osd.mil **E:** 39abw.pa@incirlik. af.mil **MW:** 1590kHz 5W. **FM:** 107.1MHz on cable

TURKMENISTAN

LT: UTC +5h — **Pop:** 4.8 million — **Pr.L:** Turkmen — **E.C:** 50Hz, 220V — **ITU:** TKM

TÜRKMEN RADIOSY (Gov)
✉ Magtymguly köçesi 89, 744000 Asgabat ☎ +993 12 351515 🖷 +993 12 394470

LW/MW	kHz	kW	Net	MW	kHz	kW	Net
Asgabat	*279	150	1	Türkmenabat	927	50	1
Asgabat	576	150	2	Serhetabat	1080	5	1
Asgabat	*675	150	1	Asgabat	1125	20	1
Türkmenbasy	675	10	2	Syrtagta	1233	40	1
Ekarça	720	1	1	Dasoguz	1233	5	1
Etrek	720	1	1	Türkmenbasy	1476	10	1

*) status uncertain

SW	kHz	kW	Net	SW	kHz	kW	Net
Asgabat	**4930	50	2	Asgabat	5015	20	1

**) inactive

FM (MHz)	Net 1	Net 2	kW	FM	Net 1	Net 2	kW
Asgabat	71.12	69.68	4	Dasoguz	69.32	67.22	4
Asgabat	103.2	104.4	10	Dasoguz	100.7	103.0	1
Baharly	...	70.64	4	Magdanly	104.2	106.7	4
Balkanabat	70.28	72.02	4	Tejen	70.52	72.14	4
Balkanabat	100.4	101.9	1	Türkmenabat	66.95	68.77	4
Baýramaly	...	70.27	4	Türkmenbasy	69.23	67.19	4
Boldumsaz	105.6	...	4	Türkmenbasy	100.2	101.7	10

...) unknown freq.

D.Prgr: Net 1 (Watan): 24h. On SW: 2100-1900. – **Net 2 (Çar tarapdan/Miras): Çar tarapdan:** 0100-0400, 0700-0900, 1400-1700; **Miras:** 0400-0700, 0900-1400, 1700-2300. – **Net 3 (Owaz):** 24h in Turkmen, Russian, English. On Asgabat 101.3MHz (nationwide FM network is under construction)

TURKS & CAICOS ISLANDS (UK)

LT: UTC -5h (13 Mar-6 Nov: -4h) — **Pop:** 40,000 — **Pr.L:** English — **E.C:** 60Hz, 110/220/440V — **ITU:** TCA

RADIO TURKS & CAICOS (Gov. Comm.)
✉ P.O. Box 69, Grand Turk ☎ +1 649 946 2010 🖷 +1 649 946 1600 **E:** teamrtc@rtc107fm.com **W:** www.rtc107fm.com **L.P:** MD: Yasmin Blues. PM: Audley Astwood

FM: 101.9MHz (Grand Turk/Salt Cay & South Caicos), 103.9MHz (North & Middle Caicos), 105.9MHz & 107.7MHz (Providenciales)
D.Prgr: 24h. Local prgr: 1100-0300; at other times relays country satellite stn. On 105.9: "RTC" (Official Government News Radio).
Ann: "This is Radio Turks & Caicos on Grand Turk, Turks & Caicos Islands"

RADIO VISION CRISTIANA INTERNACIONAL (Rlg.)
✉ North End, So. Caicos ☎ +1 649 946 6601 **W:** www.radiovision.net
L.P: Mgr: Bob Rodríguez. CEN: Peter Polano
MW: So. Caicos 530kHz
D.Prgr. in Spanish: rel. WWRV 1330, NY, USA
Ann: "R. Visión Cristiana Internacional, transmitiendo para todo el área del Caribe, Sudamérica y la parte sur de los Estados Unidos."

CARIBBEAN CHRISTIAN RADIO (Rlg.)
Operating under licence from West Indies Broadcasting Ltd.
✉ P.O. Box 200, Grand Turk (✉ **in USA:** PO Box 220, Roswell, NM 88201 ☎ +1 575 622 0658) **W:** www.superpower1020.com
L.P: GM: Reo Stubbs. SM and CEN: Jerry Kiefer
MW: Grand Turk 1020kHz 20kW (inactive)
D.Prgr in English: 1100-0600 **Ann:** "Super Power 1020".

WIV FM RADIO (Comm.)
✉ WIV Building, Leeward Highway, Box 324, Providenciales ☎ +1 888 628 9391 **W:** www.power925fm.com **L.P:** Kenny Caughlin
FM (MHz): 90.5/92.5/93.9/99.9/101.5/102.5 MHz
PraiseHim FM: 90.5MHz (Gospel) – **Island FM:** 93.9MHz (Island music) – **KISS FM:** 102.5MHz (Light rock) – **Power 92.5 FM:** 92.5MHz (Hit music)

Other stations:
KIST (Rlg.), Providenciales: 106.3MHz. Grand Turk: 94.9MHz 2kW. Format: Gospel – **Life R. - ZIBF**, Communication Network, Basden Hill, So. Caicos. **FM: Life R. 1:**105.5MHz 0.6kW, **Life R. 2:** 107.1MHz – **Tropical Reggae Breeze**, The Bight, Providenciales: 105.5MHz – **WDDR R.**, Box 262, Providenciales: 88.7MHz (0.25kW) – **ZVIC (Victory In Christ)** (Rlg.), Providenciales: 96.7MHz

TUVALU

LT: UTC +12h — **Pop:** 12,373 — **Pr.L:** Tuvaluan, English —**E.C:** 50Hz, 240V (Funafuti only) — **ITU:** TUV

TUVALU MEDIA CORPORATION (Gov.)
✉ Private Mail Bag, Vaiaku, Funafuti ☎ +688 20139 🖷 + 688 20732
L.P: GM Melali Taape. Head of Tech. Sces: John Sammons
Web: www.tuvalu-news.tv
FM: Radio Tuvalu 100.1MHz 0.02kW
D.Prgr: 1830-2000, 2325-0100, 0625-10000 daily. **N. in English:** 1910, 0710 **Ann:** E: "This is Radio Tuvalu" **V.** by letter.
BBC World Service via satellite at other times: 2000-2325, 0100-0625, 1000-1830

UGANDA

LT: UTC +3h — **Pop:** 33 million — **Pr.L:** Luganda, Swahili, English — **E.C:** 50Hz, 240V — **ITU:** UGA

UGANDA COMMUNICATIONS COMMISSION (UCC)
✉ 12th Floor, Communications House, Plot 1, Colville Street, P. O. Box 7376, Kampala ☎+256 41 4339000 🖷 +256 41 4348832 **W:** www.ucc. co.ug **E:** ucc@ucc.co.ug

UBC RADIO (Pub.)
✉ P.O. Box 2038, Plot 17-19, Nile Ave, Kampala ☎+256 41 4257256 🖷 + 256 41 4257252 **W:** www.ubconline.co.ug **E:** ubc@ubconline. co.ug **L.P:** Chmn: Chris B. Katuramu. Mg. Dir: Musinguzi Mugasa. Commissioners: Radio Broadc: Jack Turyamwijuka. Ag. Contr. of Prgrs (Radio): Charles Byekwaso. Ag. Principal Eng. (Radio): Yona Hamala.

MW: Network is being rebuilt. Freqs rep. inactive, except Mityana 576 kHz rep. carrying Star FM and Kabale 999kHz for Red channel.

SW	kHz	kW	Ch.	Times
Kampala	4976	10	Red	0200-0600, 1300-2105
Kampala	7195	10	Red	0600-1300

FM (MHz)	Blue	Red	Butebo	Star FM	Magic
Fort Portal	98.8				
Jinja				95.7	
Kabale	93.7				
Kampala	105.7	98.0	107.3	87.5	87.5
Lira		100.0			

FM (MHz)	Blue	Red	Butebo	Star FM	Magic
Masaka	99.5				96.9
Mbale			96.9		
Mbarara	97.4				
Masindi	105.0				
Soroti				96.7	

Red Channel in English, Swahili, 5 Northern languages: MF 0200-2105, SS 0345-2105 – **Blue Channel** in 6 ethnic langauges: MF 0300-2105, SS 0345-2105 – **Butebo Channel** in 10 ethnic languages – **Star FM** in Luganda. **Magic FM** in English
Ann: E: "UBC Radio". **IS:** local xylophone.

DUNAMIS SHORTWAVE (Rlg.)
(a joint project by High Adventure Canada and Dunamis FM).
✉ P.O. Box 4260, Kampala. **W:** www.biblevoice.org **E:** dunamis4750@hotmail.com **L.P:** CE: David Firth.
SW: Mukono 4750kHz 1kW. **D.Prgr:** in English/Ethnic 1500-1900v.

Other stations:
African R, Kampala: 104.5MHz – **All Karamoja FM,** Moroto: 94.7MHz – **Arua One FM,** Arua: 88.7MHz 2kW – **Bamboo FM,** Jinja: 107.6MHz – **Basoga Bainho,** Jinja: 87.7MHz – **Beat FM,** Kampala: 96.3MHz – **Buddu BS,** Masaka: 98.8MHz – **Bukedde FM:** Kampala 100.5MHz, Masaka 106.8MHz. **W:** bukedde.co.ug – **Bunuoryo BS,** Masindi: 98.2MHz – **Busiro FM,** Kakiri: 107.5MHz – **Busoga FM,** Jinja: 96MHz – **Campus FM,** Kampala: 106.6MHz – **Capital FM:** Kampala 91.3MHz, Mbale 90.9MHz, Mbarara 88.7MHz – **City FM,** Kampala: 98.1MHz – **Continental FM,** Kumi: 94.7MHz – **Dembe FM,** Kampala: 90.4MHz – **Dunamis FM,** Kampala: 103MHz – **East Africa R,** Kampala: 99MHz (cf. Tanzania) – **Eastern Voice,** Bugiri: 102.3MHz – **Elgon FM,** Kapchorwa 89.2MHz – **Etop R,** Soroti: 99.4MHz – **Eye FM,** Iganga: 98.8MHz – **Impact & Alpha FM:** Mbale 98.5MHz, Masaka 101.5MHz 1kW, Kampala 102.1MHz 4kW. **W:** www.victoryuganda. org – **Kibaale Community R:** 91.7MHz – **Kiira FM,** Jinja: 88.6MHz – **Liberty FM,** Hoima 89.0MHz **Maranatha FM,** Jinja: 104.7MHz – **Mbale FM:** 90.1MHz – **Mega FM,** Gulu: 102.1MHz – **Nile BS,** Jinja: 89.4MHz – **Open Gate FM,** Mbale: 103.2MHz – **Power FM,** Kampala: 104.1MHz – **Prime R,** Kampala: 91.9MHz – **R. Apac,** Apac 92.9MHz 0.4kW, Odokomit 106.5MHz 0.1kW **W:** www.interconnection. org/radioapac – **R. KFM,** Kampala: 93.3MHz **W:** www.monitor.co.ug – **R. Kitara,** Masindi: 101.8MHz – **R. Kyoga Veritas,** Soroti: 91.5MHz 1kW. **E:** socadido@yahoo.co.uk – **R. Lira,** Lira: 95.3MHz – **R. Mama,** Kampala: 101.7MHz. **W:** interconnection.org/umwa/community_radio. html – **R. Maria Uganda,** Masaka 94MHz 40W, Mbale 101.8MHz, Kampala 103.7MHz 40W, Fort Portal 104.6MHz, Mbarara 105.4MHz. **E:** info.uga@radiomaria.org – **R. One,** Kampala: 90.0MHz – **R. Pacis,** Arua: 90.9/94.5MHz 1kW. **W:** radiopacis.org – **R. Paidha,** Nebbi: 87.8MHz – **R. Rukungiri:** 96.9MHz – **R. Sapientia,** Kampala: 94.4MHz 5kW. **W:** radiosapientia.com – **R. Simba,** Kampala: 97.3MHz **W:** www.simba.fm – **R. Skynet,** Mityana: 96.9MHz – **R. Two,** Kampala: 87.9MHz – **R. Unity,** Lira: 97.7MHz – **R. Wa,** Lira: 89.8MHz – **R. West FM:** Mbarara 102.2MHz, Tooro 91.0MHz, Kabale & Masak – **Rhino FM,** Lira: 96.1MHz – **Rock Mamba FM,** Tororo 106,8MHz – **Sanyu FM,** Kampala 88.2MHz – **Spirit FM,** Mukono: 96.6MHz – **Ssuubi FM,** Kampala 104.9MHz. **W:** ssuubifm.net – **Star FM,** Kampala: 100.0MHz – **Step FM,** Mbale: 99.8MHz – **Super FM,** Kampala: 88.5MHz – **Vision Voice FM,** Kampala 94.8MHz. **W:** newvision.co.ug – **VO Africa:** Kampala 92.3MHz – **VO Kigezi,** Kabale 89.5MHz – **VO Teso,** Soroti: 88.4MHz – **Top R,** Kampala: 89.6MHz – **VO Toro,** Kampala 100.5MHz, Fort Portal 101MHz, Mbarara 95MHz, Mubende 97.5MHz – **Touch FM,** Kampala: 95.9MHz 1kW. **W:** www.touch.fm

BBC African Sce: Kampala 101.3MHz, Mbale/Mbarara 107.3MHz in English/Swahili/Kinyarwanda.
RFI Afrique: Kampala 93.7MHz in F/E/Swahili

UKRAINE

LT: UTC +2h (27 Mar-30 Oct: +3h) — **Pop:** 45.7 million — **Pr.L:** Ukrainian, Russian — **E.C:** 50Hz, 220V — **ITU:** UKR

NATSIONALNA RADA UKRAINI ZA PYTAN TELEBACHENNIA I RADIOMOVLENNIA
(National Council for TV and Radio Broadcasting)
✉ vul. Prorizna 2, 01601 Kyiv ☎ +380 44 2786832 📠 +380 44 2787490 **E:** pressa@nrada.gov.ua **W:** www.nrada.gov.ua
L.P: Chmn: Volodymyr Manzhosov
NB: The Council is the regulatory authority for broadcasting.

NATSIONALNA RADIOKOMPANIA UKRAINY (Gov)
✉ vul. Khreschatyk 26, 01001 Kyiv ☎ +380 44 2396103 📠 +380 44 2791170 **E:** euroradiodep@nrcu.gov.ua **W:** www.nrcu.gov.ua
L.P: Pres: Taras Avrakhov

LW/MW	kHz	kW	Prgr°	MW	kHz	kW	Prgr°
Kyiv (a)	207	500	UR1	Dnipropetrovsk	873	10	DN
Kyiv (a)	*549	150	UR2	Lviv (a)	936	600	UR1
Lviv (b)	*549	75	UR2	Starobilsk	936	3	UR1,LU
Mykolaiv (c)	*549	150	UR2	Mykolaiv (c)	972	500	UR1
Vinnytsia	*549	50	UR2	Luhansk	*1134	5	UR1
Oktiabrske	*648	150	UR1	Dokuchaievsk	*1242	40	UR3
Chernivtsi	657	25	UR3/F	Oktiabrske	*1242	50	UR3
Uzhhorod	675	25	UR1	Dokuchaievsk	1359	50	DO
Dokuchaievsk	711	40	UR1,DO	Chernivtsi	1377	50	UR1
Lutsk	810	5	UR1,VO	Mukachevo	1377	7	MY
Novodnistrovsk	819	7	CV	Mykolaiv (c)[1]	1431	1000	UR3
Kharkiv (d)	837	150	UR1	Vinnytsia	*1530	30	UR1
Chernivtsi	837	30	CV	Putyla	1557	1	UR1
Vinnytsia	873	7	VI	Verkhovyna	1584	1	UR1,IF

(a) Brovary (b) Krasne (c) Luch (d) Taranivka F=External Service
*) Inactive at editorial deadline (tx usage is subject to change frequently) °) for reg. outputs, see "Regional Stations".[1]=1700-2200.

FM: UR1 (Rg=Region; txs also carry reg. prgrs, cf. below)

Rg	Location	MHz	kW	Rg	Location	MHz	kW
SU	Shostka	65.93	4	DN	Nikopol	69.38	1
VO	Kovel	66.02	4	DO	Kramatorsk	69.41	1
PO	Krasnohorivka	66.08	4	CH	Chernihiv	69.47	4
CH	Bakhmach	66.38	1	ZK	Uzhhorod	69.53	4
SU	Bilopillia	66.50	4	CV	Novodnistrovsk	69.59	1
RI	Antopil	66.53	4	DO	Donetsk	69.77	4
OD	Kamianske	66.59	4	MY	Mykolaiv	69.80	4
KR	Oktiabrske	66.68	4	ZH	Olevsk	69.80	4
CH	Kholmy	66.71	4	TE	Lozova	69.83	4
KR	Krasnoperekopsk	66.80	4	MY	Pervomaisk	69.92	4
KH	Kirovohrad	66.98	4	ZK	Khust	70.04	4
LV	Lviv	67.04	4	VI	Bershad	70.10	4
KA	Kharkiv	67.13	4	ZK	Rakhiv	70.19	2
KR	Sevastopol	67.25	4	OD	Odesa	70.52	4
OD	Kotovsk	67.25	4	ZP	Zaporizhia	70.73	4
DO	Mariupol	67.34	4	KM	Kulchiivtsi	70.76	4
KM	Khmelnytskyi	67.70	4	CH	Pryluky	71.00	4
CK	Buky	67.88	4	IF	Ivano-Frankivsk	71.24	4
LU	Rovenky	68.08	4	DN	Kryvyi Rih	71.63	1
LU	Starobilsk	68.08	4	KR	Kerch	71.66	4
DN	Dnipropetrovsk	68.36	4	VI	Vinnytsia	71.69	4
KY	Kyiv	68.51	15	KE	Kherson	71.90	4
ZP	Berdiansk	68.57	4	ZH	Andriivka	71.90	4
ZP	Melitopol	68.72	4	KA	Izium	72.08	4
LU	Luhansk	68.75	2	CK	Cherkasy	72.20	4
SU	Trostianets	68.75	4	KR	Sovietskyi	72.20	4
OD	Zhovten	68.99	4	OD	Izmail	72.53	1
ZP	Komysh-Zoria	68.99	4	ZH	Olevsk	100.2	1
KE	Vasylivka	69.23	4	ZH	Ovruch	104.2	1
CV	Chernivtsi	69.26	1	VI	Bershad	105.0	1

NB: Txs below 1kW not listed.
UR2: Kyiv 105.0MHz (5kW) – **UR3:** Kyiv 72.86MHz (5kW), Lutsk 101.9MHz (1kW), Liubeshiv 105.2 (1kW), Horokhiv 106.5 (1kW) & txs below 1kW
D.Prgr: UR1 (Persha prohrama): 0330 (SS 0400)-2300. – **UR2 (Promin):** 24h. – **UR3 (R. Kultura):** 0330 (SS 0400)-2300.
External Service (R. Ukraine Int): see Int Radio section.

REGIONAL STATIONS (Gov)
D.Prgr: (on UR1 FM freqs: generally 0445-0500 (MF), 0410-0430 (SS), 0610-0630, 1040-1100 (Sat), 1340-1400 (MF), 1345-1400 (Sun), 1545-1600 (Sat), 1610-1700, 1800-1830; times may vary for some stations. Most regional broadc. companies also run separate local/regional channels, see below. Obl. = oblast (region); ODTRK = oblasne derzhavna teleradiokompania (regional state b'casting company).
CH) Chernihivska obl: Chernihivska ODTRK, pr. Zhovtnevoi revolyutsii 62, 14001 Chernihiv. **E:** info@trk.cn.ua. Reg. prgr "R. Siver-Tsentr" via UR1 txs. – **CK)** Cherkaska obl: Cherkaska ODTRK, vul. B.Vishnevetskoho 35/1, 18001 Cherkasy. **E:** tvros@icu.net.ua. Reg prgr "Ros R." via UR1 txs. Local channel "R. 101 Dalmatin" (vul. Khreschatyk 195, 18002 Cherkasy, **E:** fm101@ukr.net): 24h on 101.0MHz. – **CV)** Chernivetska obl: Chernivetska ODTRK, vul. Holovna 91, 58001 Chernivtsi. **E:** dtrkb-buk@chv.ukrpack.net. Reg. prgr "R. Bukovyna" via UR1 txs. Reg. channel "R. Bukovyna" 0600-2000 on (kHz) 819 (Novodnistrovsk), 837 (Chernivtsi) + 67.97MHz. – **DN)** Dnipropetrovska obl: Dnipropetrovska ODTRK, vul. Televiziina 3, 49010 Dnipropetrovsk. **E:** dneprtrk@a-teleport.com. Reg. prgr via UR1 txs. Local channel "R. Mryia": 0400-2200 on 873kHz. – **DO)** Donetska obl: Donetska ODTRK, vul. Kuibysheva 61, 83016 Donetsk. **E:** beta@k61.donetsk.ua. Reg. prgr "R. Donechyna" via UR1 txs. Reg. channel "R. Tsentr:" 0500-1700 on 71.75MHz (Donetsk) & 1700-2000 on 1359kHz (Dokuchaievsk). – **IF)** Ivano-Frankivska obl: Ivano-Frankivska ODTRK, vul. Sichovykh striltsiv 30-a, 76000 Ivano-Frankivsk. **E:** odtrk@itc.if.ua. Reg. prgr via

UR1 FM txs & 1584kHz (Verkhovyna). – **KA)** Kharkivska obl: Kharkivska ODTRK, vul. Chernyshevskoho 22, 310000 Kharkiv. **E:** oblradio@vostok.net. Reg. prgr "Hovoryt Kharkiv" on FM. Local channel "R. Kharkiv" on 69.20MHz: 0600-1400. – **KE)** Khersonska obl: Khersonska ODTRK "Skifia", vul. Perekopska 10, 73000 Kherson. **E:** program@skifiya.ks.ua. Reg. prgr "R. Dnipro" via UR1 txs. Local channel "R. Tavria" on 100.6MHz: 0400-2215. – **KH)** Kirovohradska obl: Kirovohradska ODTRK, pl. Kirova 1, 25022 Kirovohrad. **E:** teleradio@host.kr.ua. Reg. prgr "R. Skifia-Tsentr" via UR1 txs. Reg. channel "R. Skifia-Tsentr" on (MHz) 100.7 (Znamianka), 101.2 (Smoline), 106.7 (Kirovohrad), 70.22 (Svitlovodsk), 73.34 (Onufriivka): 24h. – **KM)** Khmelnytska ODTRK "Podillia-Tsentr", vul. Volodymyrska 92, 29000 Khmelnytskyi. **E:** tvradio@rp.km.ua. Reg. prgr "Hovoryt Khmelnytskyi" via UR1 txs. Local channel "R. Podillia-Tsentr": 24h on 104.6MHz. – **KR)** Respublika Krym (Crimea): DTRK "Krym", vul. Studentska 14, 95610 Simferopol. **E:** tv@tv.crimea.com. Reg. prgr "R. Krym" via UR1 txs in Ukrainian, Russian, Krymo-Tatar. – **KY)** Kyivska obl: Kyivska DTRK, vul. Khreschatyk 5v, 01001 Kyiv. **E:** program@skifiya.ks.ua. Reg. prgr "Holos Kyieva" via UR1 txs. Reg. channel "Studia Maidan" on 72.08 (Kyiv), 106.8 (Bila Tserkva): 0400-1900. – **LU)** Luhanska obl: Luhanska ODTRK, vul. Demokhina 25, 91000 Luhansk. **E:** lgtrk@lep.lg.ua. Reg. prgr via UR1 txs. Local channel "R. Puls FM": 24h on 103.6MHz. – **LV)** Lvivska obl: Lvivska ODTRK, vul. Kniazia Romana 6, 79005 Lviv. **E:** ltv@litech.lviv.ua. Reg. prgr "Lvivske R." via UR1 txs. – **MY)** Mykolaivska obl: Mykolaivska ODTRK, pr. Lenina 24-b, 54029 Mykolaiv. **E:** ogtrk@mksat.net. Reg. prgr "Buzka khvylia" in Ukrainian, Russian via UR1 txs. Local channel "R. Mykolaiv": 0500-2000 on 1377kHz + 72.91MHz. – **OD)** Odeska obl: Odeska ODTRK, vul Troitska 43-b, 270011 Odesa. **E:** odtrk@farlep.net. Reg. prgr "Chornomorskyi maiak" via UR1 txs in Ukrainian & Russian, incl. Romanian Sun 1610-1700, Bulgarian/Gagauz Sun 1800-1830. – **PO)** Poltavska obl: Poltavska ODTRK "Ltava", vul. Rozy Liuksemburg 1, 36000 Poltava. **E:** admin@ltava.poltava.org. Reg. prgr "R. Ltava" via UR1 txs. Reg. channel "R. Vasha khvylia" on (MHz) 66.86 (Krasnohorivka), 101.8 (Poltava), 105.4 (Kremenchuk), 106.3 (Krasnohorivka): 0500-2300. – **RI)** Rivnenska obl: Rivnenska ODTRK, vul. Kotliarevskoho 20-a, 33028 Rivne. **E:** rodtrk@ukrwest.net. Reg. prgr "R. Krai" via UR1 txs. – **SU)** Sumska obl: Sumska ODTRK, vul. Petropavlovska 125, 244021 Sumy. **E:** glavredtv@strc.sumy.ua. Reg. prgr "Hovoryt Sumy" via UR1 txs. – **TE)** Ternopilska obl: Ternopilska ODTRK, bul. Tarasa Shevchenka 17, 46021 Ternopil. **E:** admin@odtrk.te.ua. Reg. prgr via UR1 txs. Local channel "R. Lad" on 71.03MHz. – **VI)** Vinnytska obl: Vinnytska ODTRK, vul. Teatralna 15, 21100 Vinnytsia. mail@vodtrk.com.ua. Reg prgr "Hovoryt Vinnytsia" via UR1 txs. Local channel "R. Khvylia" on 873kHz: 0500-1900. – **VO)** Volynska obl: Volynska ODTRK, vul. Horkoho 12, 43000 Lutsk. **E:** odtrk@fk.lutsk.ua. Reg. prgr via UR1 txs. Reg. channel "R. Lutsk" on (MHz) 68.48 (Kovel), 107.3 (Lutsk): 0430-2200. – **ZH)** Zhytomyrska obl: Zhytomyrska ODTRK, vul. Teatralna 7, 10014 Zhytomyr. **E:** tvradio@zt.ukrpack.net. Reg. prgr via UR1 txs. Reg. channel "Zhytomyrska khvylia" on (MHz) 71.12 (Andriivka), 103.4 (Zhytomyr): 0500-2300. – **ZK)** Zakarpatska obl: Zakarpatska ODTRK, Kyivska nab. 18, 88018 Uzhhorod. **E:** petryk@karpaty.uzhgorod.ua. Reg. prgr via UR1 txs. Reg. channel "R. Tysa" on (MHz) 102.2 (Rakhiv), 103.0 (Uzhhorod), 104.3 (Velykyi Bereznyi), 106.6 (Mizhhiria), 106.8 (Khust): 0400-2200. – **ZP)** Zaporizka obl: Zaporizka ODTRK, vul. Matrosova 24, 69057 Zaporizhia. **E:** zdtrk@zp.ukrtel.net. Reg. prgr "R. Zaporizhia" in Ukrainian, Russian via UR1 txs. Reg. channel "Z FM" on (MHz) 103.4 (Dniprorudne), 103.7 (Zaporizhia), 103.9 (Tokmak): 0400-2300.

OTHER STATIONS

MW	kHz	kW	Location	Station
A)	594	50	Kyiv (a)	BBC relay
A)	612	10	Kharkiv (b)	BBC relay
74)	765	75	Petrivka	R. Maiak

SW	kHz	kW	Location	Station
75)	11980	0.3	Zaporizhia	Dniprovska khvylia

FM	MHz	kW	Location	Station
50)	67.70	1	Kyiv	R. ROKS
73)	69.68	1	Kyiv	R. Maria
30)	87.5	1	Odesa	FM1-Pershe R.
11)	87.9	1	Odesa	Era FM
23)	88.0	1	Simferopol	L'Radio
1)	88.1	1	Dnipropetrovsk	R. Alla
14)	88.5	1	Odesa	R. 5 - Retro FM
29)	88.5	1	Dnipropetrovsk	Music R.
11)	88.6	1	Lviv	Era FM
70)	89.7	1	Odesa	R. Pivdenna Stolitsia
14)	90.1	1	Simferopol	R. 5 - Retro FM
50)	90.2	1	Odesa	R. ROKS
7)	90.3	1	Sumy	Nashe R.
14)	90.4	1	Mariupol	R. 5 - Retro FM
17)	90.8	1	Luhansk	Stilne R. - Perets FM
11)	90.9	1	Sumy	Era FM
26)	90.9	1	Dnipropetrovsk	MFM
4)	91.0	1	Odesa	GALA-Radio
10)	91.1	1	Kryvyi Rih	Avtoradio-Ukraina

FM	MHz	kW	Location	Station
11)	91.1	1	Simferopol	Era FM
8)	91.1	1	Lviv	R. Sharmanka
14)	91.3	1	Sumy	R. 5 - Retro FM
8)	91.3	1	Vinnytsia	R. Sharmanka
1)	91.4	1	Odesa	R. Alla
14)	91.5	1	Berdiansk	R. 5 - Retro FM
23)	91.5	1	Lviv	L'Radio
12)	91.6	1	Kryvyi Rih	Lux FM
14)	92.4	1	Kyiv	R. 5 - Retro FM
3A)	92.8	1	Kyiv	Europa Plus Ukraina
22)	94.2	1	Kyiv	R. Renesans
23)	95.2	1.5	Kyiv	L'Radio
68)	95.6	2	Kyiv	Dzhem FM
11)	96.0	2	Kyiv	Era FM
12)	96.1	1	Donetsk	Lux FM
5)	96.4	1	Kyiv	Hit FM Ukraina
49)	96.8	2	Kyiv	DJ FM
41)	98.0	1	Kyiv	R. Kyiv
16)	98.5	1	Kyiv	Russkoye R. Ukraina
34)	99.0	2	Kyiv	R. Nostalgie 99FM
14)	99.1	1	Melitopol	R. 5 - Retro FM
14)	99.1	1	Poltava	R. 5 - Retro FM
1)	99.3	1	Vinnytsia	R. Alla
1)	99.4	1	Kyiv	R. Alla
46)	99.4	1	Kherson	R. Sofia
8)	99.4	1	Donetsk	R. Sharmanka
10)	100.0	1	Poltava	Avtoradio-Ukraina
17)	100.0	1	Donetsk	Stilne R. - Perets FM
4)	100.0	5	Kyiv	GALA-Radio
4)	100.2	4	Kryvyi Rih	GALA-Radio
10)	100.3	1	Vinnytsia	Avtoradio-Ukraina
4)	100.3	1	Zaporizhia	GALA-Radio
10)	100.4	1	Odesa	Avtoradio-Ukraina
14)	100.4	1	Ivano-Frankivsk	R. 5 - Retro FM
8)	100.4	1	Luhansk	R. Sharmanka
1)	100.5	1	Kharkiv	R. Alla
12)	100.5	2	Dnipropetrovsk	Lux FM
18)	100.5	2	Kyiv	Narodne R.
25)	100.5	1	Donetsk	Mega-Radio
10)	100.6	1	Chernihiv	Avtoradio-Ukraina
13)	100.6	1	Simferopol	R. Melodia
53)	100.6	2	Korosten	R. Rekord FM
7)	100.7	1	Rivne	Nashe R.
24)	100.8	1	Lviv	Lvivska khvylia
5)	100.8	1	Mariupol	Hit FM Ukraina
61)	100.8	1	Zaporizhia	R. Univers/R. ROKS
14)	100.9	1	Kremenchuk	R. 5 - Retro FM
14)	100.9	1	Lutsk	R. 5 - Retro FM
14)	100.9	1	Vinnytsia	R. 5 - Retro FM
11)	101.0	1	Kramatorsk	Era FM
15)	101.0	1	Kryvyi Rih	R. Shanson
5)	101.0	1	Odesa	Hit FM Ukraina
54)	101.0	1	Sokal	R. Sokal
13)	101.1	3	Kyiv	R. Melodia
13)	101.1	1	Luhansk	R. Melodia
16)	101.1	1	Dnipropetrovsk	Russkoye R. Ukraina
3A)	101.1	1	Kharkiv	Europa Plus Ukraina
11)	101.2	1	Melitopol	Era FM
12)	101.2	1	Khmelnytskyi	Lux FM
5)	101.2	1	Donetsk	Hit FM Ukraina
6)	101.2	1	Kherson	Kiss FM
17)	101.3	1	Poltava	Stilne R. - Perets FM
13)	101.4	1	Rivne	R. Melodia
3A)	101.4	1	Kerch	Europa Plus Ukraina
57)	101.4	1	Sumy	R. SveSweet/MFM
13)	101.5	1	Dnipropetrovsk	R. Melodia
14)	101.5	1	Kirovohrad	R. 5 - Retro FM
14)	101.5	1	Ternopil	R. 5 - Retro FM
16)	101.5	1	Krasnohorivka	Russkoye R. Ukraina
17)	101.5	1	Kharkiv	Stilne R. - Perets FM
29)	101.5	3	Kyiv	Music R.
1)	101.6	1	Cherkasy	R. Alla
16)	101.6	1	Mykolaiv	Russkoye R. Ukraina
16)	101.7	1	Khmelnytskyi	Russkoye R. Ukraina
16)	101.7	1	Zhytomyr	Russkoye R. Ukraina
5)	101.7	1	Lviv	Hit FM Ukraina
50)	101.7	1	Kramatorsk	R. ROKS
6)	101.7	1	Mariupol	Kiss FM
7)	101.7	1	Simferopol	Nashe R.
10)	101.8	1	Luhansk	Avtoradio-Ukraina
6)	101.8	1	Odesa	Kiss FM
62)	101.8	1	Zaporizhia	R. Velykyi Luh
15)	101.9	1	Kyiv	R. Shanson
17)	101.9	1	Kherson	Stilne R. - Perets FM
55)	101.9	2	Khust	R. Zakarpattia FM
35)	102.0	1	Sevastopol	R. Bryz

FM	MHz	kW	Location	Station
5)	102.0	2	Dnipropetrovsk	Hit FM Ukraina
67)	102.0	1	Pryluky	R. Galaktika Plus
14)	102.1	1	Khmelnytskyi	R. 5 - Retro FM
6)	102.1	1	Mykolaiv	Kiss FM
69)	102.1	1	Donetsk	Klasne R.
37)	102.2	2	Odesa	R. Feel
26)	102.3	1	Luhansk	MFM
43)	102.3	1	Simferopol	Trans-M-Radio
5)	102.3	1	Poltava	Hit FM Ukraina
12)	102.4	1	Chernivtsi	Lux FM
4)	102.4	1	Kholmets	GALA-Radio
6)	102.4	1	Kharkiv	Kiss FM
13)	102.5	1	Khmelnytskyi	R. Melodia
14)	102.5	4	Lviv	R. 5 - Retro FM
36)	102.5	1	Tokmak	R. Nostalzi Zaporizhia
3A)	102.5	1	Dnipropetrovsk	Europa Plus Ukraina
5)	102.5	1	Kherson	Hit FM Ukraina
9)	102.5	2	Kyiv	Prosto R.
13)	102.6	1	Donetsk	R. Melodia
14)	102.6	1	Lubyn	R. 5 - Retro FM
13)	102.7	1	Kovel	R. Melodia
7)	102.7	4	Kryvyi Rih	Nashe R.
7)	102.7	1	Zhytomyr	Nashe R.
13)	102.8	1	Kamianets-Pod.	R. Melodia
2)	102.8	1	Mariupol	Best FM
65)	102.8	1	Pryluky	R. Planeta
7)	102.8	1.2	Mykolaiv	Nashe R.
16)	102.9	1	Luhansk	Russkoye R. Ukraina
7)	102.9	1	Cherkasy	Nashe R.
7)	102.9	3	Dnipropetrovsk	Nashe R.
13)	103.0	1	Kharkiv	R. Melodia
14)	103.0	1	Rivne	R. 5 - Retro FM
8)	103.0	1	Ivano-Frankivsk	R. Sharmanka
12)	103.1	5	Kyiv	Lux FM
3A)	103.1	1	Kherson	Europa Plus Ukraina
6)	103.1	1	Zaporizhia	Kiss FM
7)	103.1	1	Khmelnytskyi	Nashe R.
1)	103.2	1	Kryvyi Rih	R. Alla
13)	103.2	1	Melitopol	R. Melodia
18)	103.2	1	Odesa	Narodne R.
11)	103.3	1	Cherkasy	Era FM
22)	103.3	1	Dnipropetrovsk	R. Renesans
5)	103.4	1	Sumy	Hit FM Ukraina
15)	103.5	1	Kremenchuk	R. Shanson
59)	103.5	1	Ternopil	R. TON
50)	103.6	2	Kyiv	R. ROKS
6)	103.6	1	Khmelnytskyi	Kiss FM
8)	103.6	2	Kryvyi Rih	R. Sharmanka
11)	103.7	1	Kherson	Era FM
16)	103.7	1	Cherkasy	Russkoye R. Ukraina
40)	103.7	1	Simferopol	R. Pilot
5)	103.7	1	Rivne	Hit FM Ukraina
63)	103.7	1	Vinnytsia	R. Takt
12)	103.8	1	Ivano-Frankivsk	Lux FM
16)	103.8	1	Kirovohrad	Russkoye R. Ukraina
17)	103.8	1	Bila Tserkva	Stilne R. - Perets FM
27)	103.8	5	Odesa	Moye Radio
1)	103.9	1	Lviv	R. Alla
11)	103.9	1	Zhytomyr	Era FM
14)	103.9	1	Evpatoria	R. 5 - Retro FM
14)	103.9	1	Kovel	R. 5 - Retro FM
16)	103.9	1	Kremenchuk	Russkoye R. Ukraina
69)	103.9	1	Kramatorsk	Klasne R.
2)	104.0	2	Kharkiv	Best FM
3A)	104.0	1	Mariupol	Europa Plus Ukraina
5)	104.0	1	Luhansk	Hit FM Ukraina
8)	104.0	1	Dnipropetrovsk	R. Sharmanka
8)	104.0	4	Kyiv	R. Sharmanka
16)	104.1	1	Vinnytsia	Russkoye R. Ukraina
17)	104.1	1	Mykolaiv	Stilne R. - Perets FM
26)	104.1	1	Donetsk	MFM
13)	104.2	1	Starobilsk	R. Melodia
36)	104.2	1	Polohy	R. Nostalzhi (R. Slavia)
17)	104.3	1	Simferopol	Stilne R. - Perets FM
20)	104.3	2	Odesa	Armianske R.
6)	104.3	1	Lviv	Kiss FM
7)	104.3	1	Chernihiv	Nashe R.
13)	104.4	1	Kherson	R. Melodia
71)	104.4	1	Novovolynsk	R. Nova
12)	104.5	1	Ternopil	Lux FM
16)	104.5	2	Sevastopol	Russkoye R. Ukraina
23)	104.5	1	Cherkasy	L'Radio
5)	104.5	1	Zhytomyr	R. Klub 104.5FM
7)	104.5	5	Kharkiv	Nashe R.
8)	104.5	1	Zaporizhia	R. Sharmanka
11)	104.6	1	Kirovohrad	Era FM
32)	104.6	2	Kyiv	NRJ
9)	104.6	1	Mykolaiv	Prosto R.
12)	104.7	1	Lviv	Lux FM
13)	104.7	1	Kryvyi Rih	R. Melodia
16)	104.7	1	Donetsk	Russkoye R. Ukraina
16)	104.7	1	Melitopol	Russkoye R. Ukraina
5)	104.7	1	Chernihiv	Hit FM Ukraina
10)	104.8	1	Dnipropetrovsk	Avtoradio-Ukraina
16)	104.8	1	Kherson	Russkoye R. Ukraina
33)	104.8	1	Simferopol	R. Asol
3A)	104.8	1	Luhansk	Europa Plus Ukraina
7)	104.8	1	Lutsk	Nashe R.
16)	104.9	1	Odesa	Russkoye R. Ukraina
18)	104.9	1	Zhytomyr	Narodne R.
13)	105.0	1	Poltava	R. Melodia
14)	105.0	1	Chernivtsi	R. 5 - Retro FM
13)	105.1	1	Mykolaiv	R. Melodia
13)	105.1	1	Sumy	R. Melodia
15)	105.1	1	Zaporizhia	R. Shanson
6)	105.1	4	Donetsk	Kiss FM
12)	105.2	1	Uzhhorod	Lux FM
14)	105.2	1	Krasnohorivka	R. 5 - Retro FM
6)	105.2	1	Kryvyi Rih	Kiss FM
13)	105.3	1	Dubrovytsia	R. Melodia
14)	105.3	1	Shostka	R. 5 - Retro FM
15)	105.3	1	Dnipropetrovsk	R. Shanson
16)	105.3	1	Mariupol	Russkoye R. Ukraina
5)	105.3	1	Kirovohrad	Hit FM Ukraina
56)	105.3	1	Sieverodonetsk	R. STV
9)	105.3	3	Odesa	Prosto R.
10)	105.4	1	Lviv	Avtoradio-Ukraina
15)	105.4	1	Simferopol	R. Shanson
66)	105.4	1	Khmelnytskyi	OK FM
13)	105.5	1	Lutsk	R. Melodia
17)	105.5	2	Kyiv	Stilne R. - Perets FM
2)	105.5	1	Donetsk	Best FM
52)	105.5	1	Luhansk	R. Ekho
14)	105.6	1	Zhytomyr	R. 5 - Retro FM
16)	105.6	1	Sumy	Russkoye R. Ukraina
17)	105.6	1	Sevastopol	Stilne R. - Perets FM
5)	105.6	1	Krasnohorivka	Hit FM Ukraina
5)	105.6	1	Ternopil	Hit FM Ukraina
7)	105.6	1	Zaporizhia	Nashe R.
14)	105.7	1	Kamianets-Pod.	R. 5 - Retro FM
17)	105.7	1	Kramatorsk	Stilne R. - Perets FM
8)	105.7	1	Kharkiv	R. Sharmanka
13)	105.8	1	Kirovohrad	R. Melodia
16)	105.8	1	Yalta	Russkoye R. Ukraina
17)	105.8	1	Mariupol	Stilne R. - Perets FM
19)	105.8	2	Dnipropetrovsk	Avtoradio-Dnipro
50)	105.8	1	Poltava	R. ROKS
16)	105.9	1	Kryvyi Rih	Russkoye R. Ukraina
10)	106.0	1	Donetsk	Avtoradio-Ukraina
21)	106.0	1	Berdiansk	Azovska khvylia 106.4FM
39)	106.0	2	Kyiv	R. Dynamo
48)	106.0	1	Odesa	R. Odesa-Mama
7)	106.0	1	Lviv	Nashe R.
10)	106.1	1	Kharkiv	Avtoradio-Ukraina
12)	106.1	1	Cherkasy	Lux FM
7)	106.1	1	Luhansk	Nashe R.
8)	106.1	1	Ternopil	R. Sharmanka
47)	106.1	2	Zhytomyr	Z Radio 106.2FM
14)	106.2	1	Bila Tserkva	R. 5 - Retro FM
17)	106.2	1	Kirovohrad	Stilne R. - Perets FM
3A)	106.2	1	Zaporizhia	Europa Plus Ukraina
7)	106.2	1	Kherson	Nashe R.
28)	106.2	1	Kotovsk	R. Sanna
36)	106.3	1	Melitopol	R. Nostalzhi Zaporizhia
10)	106.4	1	Ivano-Frankivsk	Avtoradio-Ukraina
15)	106.4	1	Donetsk	R. Shanson
3A)	106.4	1	Yalta	Europa Plus Ukraina
60)	106.4	2	Rivne	R. Trek
7)	106.4	1	Vinnytsia	Nashe R.
4)	106.5	1	Luhansk	GALA-Radio
6)	106.5	2	Kyiv	Kiss FM
7)	106.5	1	Mariupol	Nashe R.
15)	106.6	1	Kharkiv	R. Shanson
38)	106.6	2	Odesa	R. Glas
42)	106.6	1	Simferopol	R. Lider
5)	106.6	1	Zaporizhia	Hit FM Ukraina
31)	106.6	1	Chernivtsi	Blysk FM
14)	106.7	1	Kherson	R. 5 - Retro FM
45)	106.7	1	Lviv	R. Nezalezhnist
15)	106.8	1	Chernihiv	R. Shanson
15)	106.8	1	Poltava	R. Shanson
3A)	106.8	5	Donetsk	Europa Plus Ukraina

FM	MHz	kW	Location	Station
51)	106.8	2	Kolomyia	R. Siaivo
6)	106.8	3	Dnipropetrovsk	Kiss FM
12)	106.9	1	Lutsk	Lux FM
5)	106.9	1	Kryvyi Rih	Hit FM Ukraina
6)	106.9	1	Luhansk	Kiss FM
11)	107.0	1	Kharkiv	Era FM
14)	107.0	1	Zaporizhia	R. 5 - Retro FM
3B)	107.0	2	Kyiv	Europa FM
5)	107.0	1	Sevastopol	Hit FM Ukraina
6)	107.0	1	Kramatorsk	Kiss FM
72)	107.0	1	Sumy	Diva R.
28)	107.0	1	Odesa	R. Sanna
12)	107.1	1	Mykolaiv	Lux FM
13)	107.1	1	Zarichne	R. Melodia
50)	107.1	1	Khmelnytskyi	R. ROKS
28)	107.1	1	Izmail	R. Sanna
8)	107.1	1	Cherkasy	R. Sharmanka
13)	107.2	1	Lviv	R. Melodia
14)	107.2	1	Kholmets	R. 5 - Retro FM
16)	107.2	1	Chernihiv	Russkoye R. Ukraina
4)	107.2	1	Donetsk	GALA-Radio
11)	107.3	1	Luhansk	Era FM
44)	107.3	1	Dnipropetrovsk	R. Mix
10)	107.4	2.5	Kyiv	Avtoradio-Ukraina
11)	107.4	1	Kryvyi Rih	Era FM
3A)	107.4	1	Odesa	Europa Plus Ukraina
4)	107.4	1	Kharkiv	GALA-Radio
17)	107.5	1	Dubrovytsia	Stilne R. - Perets FM
36)	107.5	1	Zaporizhia	R. Nostalzhi Zaporizhia
50)	107.6	1	Kherson	R. ROKS
7)	107.6	1	Donetsk	Nashe R.
12)	107.7	1	Zhytomyr	Lux FM
27)	107.7	1	Chernihiv	MFM
58)	107.7	1	Dnipropetrovsk	107.7FM - Dushevne R.
11)	107.8	1	Mykolaiv	Era FM
11)	107.8	1	Vinnytsia	Era FM
4)	107.8	1	Simferopol	GALA-Radio
10)	107.9	1	Sumy	Avtoradio-Ukraina
13)	107.9	1	Zaporizhia	R. Melodia
7)	107.9	1	Kirovohrad	Nashe R.
7)	107.9	5	Kyiv	Nashe R.
7)	107.9	2	Odesa	Nashe R.

NB: Txs less than 1kW not listed.

Addresses & other information:
1) vul. V.Arnautska 15, 65012 Odesa. – **2)** pr-t Tychyny 2, 02098 Kyiv. – **3A,B)** vul. Frunze 39, 04080 Kyiv. – **4)** vul. Saksahanskoho 91, 01032 Kyiv. – **5)** vul. V.Khvoiky 15/15, 04655 Kyiv. – **6)** vul. V.Khvoiky 15/15, 04655 Kyiv. – **7)** vul. Otto Shmidta 6, 04107 Kyiv. – **8)** b-r Shevchenka 54/1, 01032 Kyiv. – **9)** vul. Tereshkovoi 15, 65078 Odesa. – **10)** vul. Frunze 39, 04080 Kyiv. – **11)** b-r Verkhovnoy Rady 20, 02100 Kyiv. – **12)** vul. Volodymyrska 61/11, 01033 Kyiv. **13)** vul. Moskovskyi 9b, 04273 Kyiv. – **14)** vul. Frunze 39, 04080 Kyiv. – **15)** b-r Shevchenka 54/1, 01032 Kyiv. – **16)** vul. V.Khvoiky 15/15, 04655 Kyiv. – **17)** vul. Dovshenka 14, 03057 Kyiv. – **18)** b-r Lesi Ukrainki 34, 01133 Kyiv. – **19)** pr-t Kirova 111b, 49054 Dnipropetrovsk. – **20)** vul. Saslavskoho 12/14, 65007 Odesa. – **21)** Melitopolska shose 20, 71100 Berdiansk. – **22)** b-r Shevchenka 54/1, 01032 Kyiv. – **23)** b-r Verkhovnoy Rady 20, 02100 Kyiv. – **24)** vul. Hutsulska 9a, 79008 Lviv. – **25)** pr-t Illicha 100, 83052 Donetsk. – **26)** vul. Artema 145a, 83015 Donetsk. – **27)** vul. Artyleriyska 1, 65039 Odesa. – **28)** vul. Ak.Borobiova 1, 65031 Odesa. – **29)** vul. Frunze 39, 04080 Kyiv. –**30)** vul. Artyleriyska 1, 65039 Odesa. – **31)** vul. Eminesku 2, 58000 Chernivtsi. – **32)** vul. Otto Shmidta 6, 04107 Kyiv. – **33)** vul. Radio 4, 95038 Simferopol. – **34)** vul. Saperno-Slobodska 25, 03039 Kyiv. – **35)** vul. 4-a Bastionna 5, 90011 Sevastopol. – **36)** vul. Sedova 12, 69044 Zaporizhia. – **37)** vul. Troiitska 50, 65045 Odesa. – **38)** vul. Kanatna 83, 65107 Odesa. – **39)** vul. Frunze 39, 04080 Kyiv. – **40)** Simferopol – **41)** vul. Dehtiarivska 37, 03680 Kyiv. – **42)** vul. Balaklavska 68, 95048 Simferopol. – **43)** vul. Marshala Zhukova 17, 95035 Sevastopol. – **44)** vul. Suchkova 2a, 49010 Novomoskovsk. – **45)** vul. Kn. Romana 12, 79005 Lviv. – **46)** vul. Lavrenova 25-2, 73000 Kherson. – **47)** vul. Yana Hamarnyka 6a, 10000 Zhytomyr. – **48)** vul. Saslavskoho 10/12, 65004 Odesa. – **49)** b-r Shevchenka 54/1, 01032 Kyiv. – **50)** vul. V.Khoiky 15/15, 04655 Kyiv. – **51)** vul. Sichovykh Striltsiv 23, 78200 Kolomyia. – **52)** vul. Demokhina 27/62, 91016 Luhansk. – **53)** vul. Peremohy 51-8, 10002 Zhytomyr. –**54)** vul.Sheptytskoho 105, 80000 Sokal. – **55)** vul. Magritycha 6, 89600 Mukachevo. –**56)** vul. Haharina 93, 93400 Severodonetsk. – **57)** vul. Kirova 25, 40030 Sumy. – **58)** Dnipropetrovsk. – **59)** b-r Shevchenka 5, 46001 Ternopil. – **60)** vul. 16-ho Lypnia 38, 33000 Rivne. – **61)** vul. Zaliznichna 24, 69063 Zaporizhia. – **62)** vul. Matrosova 8a, 69057 Zaporizhia. – **63)** vul. Soborna 59, 21000 Vinnytsia. – **64)** vul. Kyiivska 1, 10014 Zhytomyr. – **65)** vul. Pyriatynska 129, 17500 Pryluky. – **66)** pr-t Miru 69, 29015 Khmelnytskyi. – **67)** vul. Pyriatynska 129, 17500 Pryluky. – **68)** vul. Tymoshenka 2L, 04212 Kyiv. – **69)** vul. Artema 131a, 83015 Donetsk. – **70)** vul. Uspenska 77, 65011 Odesa. – **71)** pr-t Druzhby

27, 45400 Novovolynsk. – **72)** vul. Kharkivska 5, 40024 Sumy. – **73)** vul. Sribnokilska 8, 01001 Kyiv. – **74)** vul. Fontanska doroha 33, 65009 Odesa. – **75)** PPRK "Aleks", vul. 8-ho Marta 48, 69068 Zaporizhia. **E:** radiodh@ rambler.ru. Rel. UR1 (alt. regional prgr from Zaporizhia): Sat/Sun 0600-0800 & 0900-1100. On the air irregular. – **A)** Rel. BBC (UK)

UNITED ARAB EMIRATES

LT: UTC +4h — **Pop:** 5 million — **Pr.L:** Arabic — **E.C:** 50Hz, 220V — **ITU:** UAE

NATIONAL MEDIA COUNCIL
✉ P.O. Box 17, Abu Dhabi ☎+971 2 4453000 🖷 +971 2 4452504 **W:** www.uaeinteract.com **L.P:** Chmn: Saqr Ghubash Saeed Ghubash.

ABU DHABI MEDIA COMPANY (Pub.)
✉ 4th St, Sector 18, Zone 1, Abu Dhabi ☎+971 2 4144000 🖷 +971 2 4144001 **W:** admedia.ae **L.P:** CEO: Ahmed Ali Mohamed Al Bloushi. Dir. Radio: Jaber Obaid. Head R&TV Eng: Mahmood Al-Redha.

MW	kHz	kW	Prgr.	Times
Al-Dhabbiya	657	100	AR/A	24h
Maqtaa	810	50	A	24h
Al-Ain	828	1	A	24h

A=Abu Dhabi FM in Arabic: 1400-2000 on 657 kHz. MW txs relay also Quran prgr and Emarat FM at times. On FM(MHz): Abu Dhabi 90.0, Dubai 98.4, Ras al-Khaimah 89.7, Fujairah 106.0, Habshan 100.1, Liwa 103.7, Jabal Al-Dhanna 97.3.
AN=Asianet R. in Urdu: 2000-1400.
Holy Quran R: 24h on 810kHz & FM (MHz) Dubai 88.2, Al-Ain 88.6, Abu Dhabi 98.1MHz, Fujairah 95.6, Ras al-Khaimah 105.2.
Emarat FM: 24h of FM (MHz): Jabal al-Dhanna 92.4, Abu Dhabi 95.8, Al-Ain 94.9, Liwa 95.6, Dubai 97.1, Ras al-Khaimah 88.5, Habshan 98.4, Fujairah 103.9.
R. 1: Abu Dhabi 100.5MHz 1kW, Dubai 104.1MHz 10kW. **R. 2:** Abu Dhabi 106.0MHz 1kW, Dubai 93.3MHz 10W.
Star FM: Abu Dhabi 92.4MHz, Dubai 99.9MHz.
Sound of Music: Abu Dhabi 90.0MHz, Al Ain 99.6MHz, Liwa 103.7MHz, Dubai 99.9MHz, northern Emirates 89.7/103.9/106.0MHz.
Abu Dhabi Classic FM: Abu Dhabi 91.6MHz, Al Ain 105.2MHz.
Ann: A: "Imarat FM", "Shabakat-u Abu Dhabiy".

RADIO ASIA
✉ Dolphin Recording Studio, P.O. Box 31876, Dubai ☎+971 4 3491011 🖷 +971 4 3421387 **L.P:** GM: Brij Bhalla. PD: Vettoor G. Sreedharan. **W:** www.radioasiauae.com **E:** admin@radioasia.ae
MW: Ras al Khaimah 1269kHz 200kW. **D.Prgr:** 24h in Malayalam (on MW also segments of Tamil and Telugu). **FM: Oxygen:** 94.7MHz 20kW. **Ann:** "12-69 AM Radio Asia".

RAS AL-KHAIMAH BROADCASTING STATION
✉ RAK Media, P.O. Box 141, Ras al Khaimah ☎+971 7 851151 🖷 +971 7 353441
MW: 1152kHz 200kW. **FM:** 95.3MHz.
D.Prgr: 0200-1600 in Arabic, 1600-0200 in Bengali/Kannada/Marathi/ Filipino/Indonesian. **Ann: A:** "Idha'atu-I-Imarat min R'as al-Khaimah".

UMM AL QUWAIN BROADCASTING STATION
✉ Shamal Media Services, P.O. Box 1106, Umm al Quwain ☎+971 6 5657106 🖷 +971 6 5651806
UAQ FM: 97.8MHz in Arabic. **W:** uaqfm.com
Hum FM: 106.2MHz in Hindi/Urdu. **W:** humfm.com
Holy Quran R: MW 846kHz 20kW. **D.Prgr:** 0200-1900 in Arabic.

CHANNEL 4 RADIO NETWORK
✉ P.O Box 55137, Pyramid Centre, Office No 202A, 2nd Floor, Near Al Nasr Cinema, Karama, Dubai ☎+971 4 3374445 🖷 +971 4 3377360 **L.P:** MD: Mohammad Murad. Prgr. Controller: Peter Gowers.
Stations: all in Ajman. **R. 4 FM:** 89.1MHz in Hindi/English. **W:** radio4fm.com – **Channel 4 FM:** 104.8MHz 1kW in English. **W:** channel4fm. com – **Al Rabea FM:** 107.8MHz 1kW in Arabic. **W:** 1078fm.com

FUJAIRAH MEDIA
✉ Fujairah ☎+971-9-2244100 🖷 +971-9-2244101 **W:** fujairahmedia. com **FM:** Al Rufaisa Mountains 98.4MHz in Arabic – **Spice FM,** 105.4MHz. **W:** www.radiospicefm.com

ARABIAN RADIO NETWORK (ARN)
✉ P.O. Box 502255, CNN Bldg, Media City 103, Dubai ☎+971 4 3912000 🖷 +971 4 3912007 **W:** arnonline.com **E:** pr@arnonline.com **L.P:** MD: Abdel Latif Al Sayegh. COO: Steve Smith.

FM: Dubai 92: 92.0MHz 5kW in English. **W:** dubai92.com – **Noor Dubai:** 93.9MHz 5kW in Arabic. **W:** noordubai.com – **Pulse 95.3:** 95.3MHz 5kW in South Asian languages – **Hit FM:** 96.7MHz 5kW in Tamil. **W:** hit967.com – **Al-Arabiya:** 99.0MHz 10kW in Arabic – **Al-Khaleejiya:** 100.9MHz 5kW in Arabic – **City FM:** 101.6MHz 5kW in Hindi. **W:** city1016.ae – **Dubai Eye:** 103.8MHz 5kW in English **W:** www.dubaieye1038.com – **Virgin R:** Dubai 104.4MHz 30kW. **W:** virginradiodubai.com

MBC FM & PANORAMA FM

P.O. Box 75335, MBC Building, Media City, Dubai ☎+971 4 3919713 +971 4 3916683 **W:** mbc.net **E:** contactus@mbc.ae
L.P: Dir: Hassan Muawad.
FM: Dubai freq. not known, Abu Dhabi 98.1MHz. Transmitters also in Bahrain, Iraq, Jordan, Kuwait, Qatar, Saudi Arabia, Sudan and Palestine West Bank.

BBC World Sce: Abu Dhabi 90.3MHz.
BBG - R. Sawa: MW: Al-Dhabbiya 1170kHz 800kW 24h. FM: Dubai 90.5MHz 5kW, Abu Dhabi 98.7MHz 10kW. For more details see Interational Radio section (USA).
BBG - R. Aap Ki Dunyaa: Al-Dhabbiya 1539kHz 600kW 1400-0200.
BBG - R. Farda: Al-Dhabbiya 1575kHz 800kW 24h. For further details see International radio section under USA.
Monte Carlo Doualiya: Dubai 95.3MHz

UNITED KINGDOM

L.T: UTC (27 Mar-30 Oct: +1h) — **Pop:** 62 million — **Pr.L:** English Welsh— **E.C:** 50Hz, 230V — **ITU:** G — **Int.Dialling Code:** +44

CROWN DEPENDENCIES
NB: The Channel Islands and the Isle of Man are dependencies of the British Crown and are not part of the United Kingdom. They are included here for editorial convenience.

BRITISH BROADCASTING CORPORATION (Pub)
The BBC is an independent body created by Royal Charter and operates under licence. Broadcasting House, Portland Place, London W1A 1AA. ☎ +44 20 7580 4468 **W:** www.bbc.co.uk **LP:** Chairman: Sir Michael Lyons; DG: Mark Thompson; Deputy DG: Mark Byford; Dir Audio & Music: Tim Davie; Dir Vision: Jana Bennett; Dir BBC People: Lucy Adams; Dir Future Media & Technology: Erik Huggers; CEO: Caroline Thomson; CEO BBC Worldwide: John Smith

LW/MW:

Radio 4	kHz	kW	Radio 4	kHz	kW
Burghead	198	50	Crystal Palace	720	0.8
Droitwich	198	500	Redruth	756	2
Westerglen	198	50	Enniskillen	774	1
Newcastle	603	2	Plymouth	774	1
Londonderry	720	0.25	Redmoss	1449	2
Lisnagarvey	720	10	Carlisle	1485	1

Radio 5 Live	kHz	kW	Radio 5 Live	kHz	kW
Barrow	693	1	Clevedon	909	50
Bexhill	693	1	Exeter	909	1
Brighton	693	1	Fareham	909	1
Burghead	693	25	Lisnagarvey	909	10
Droitwich	693	150	Londonderry	909	1
Enniskillen	693	1	Bournemouth	909	0.25
Folkestone	693	1	Moorside Edge	909	200
Postwick	693	10	Redruth	909	2
Redmoss	693	1	Westerglen	909	50
Stagshaw	693	50	Whitehaven	909	1
Start Point	693	50	Tywyn	990	1
Brookmans Park	909	150			

England, Isle of Man, Channel Is FM (all stereo)

FM	R1	R2	R3	R4	kW
Barnstaple	98.1	88.5	90.7	92.9	1
Beacon Hill	98.4	88.7	90.9	93.1	1
Belmont	98.3	88.8	90.9	93.1	16
Bilsdale	98.6	89.0	91.2	93.4	5
Bow Brickhill	98.2	88.6	90.8	93.0	10
Bristol	98.9	89.3	91.5	93.7	1.3
Chatton	99.7	90.1	92.3	94.5	5.6
Crystal Palace	98.5	88.8	91.0	93.2	4
Douglas (I.O.M.)	98.0	88.4	90.6	92.8	11
Guildford	97.7	88.1	90.3	92.5	3
Caversham	99.4	89.8	92.0	94.2	1
Holme Moss	98.9	89.3	91.5	93.7	250
Keighley	98.5	88.9	91.1	93.3	1
Les Platons (C.I.)	97.1	89.6	91.1	94.8	16
Manningtree	97.7	88.1	90.3	92.5	5
Morecambe Bay	99.6	90.0	92.2	94.4	10
North Hessary Tor	97.7	88.1	90.3	92.5	160

FM	R1	R2	R3	R4	kW
Oxford	99.1	89.5	91.7	93.9	46
Pendle Forest	97.8	90.2	92.6	94.6	1
Peterborough	99.7	90.1	92.3	94.5	40
Pontop Pike	98.1	88.5	90.7	92.9	134
Redruth	99.3	89.7	91.9	94.1	25
Ridge Hill	98.2	88.6	90.8	93.0	10
Rowridge	98.2	88.5	90.7	92.9	250
Sandale	97.7	88.1	90.3	92.5	250
Stanton Moor	99.4	89.8	92.0	94.2	1.2
Sutton Coldfield	97.9	88.3	90.5	92.7	250
Swingate (Dover)	99.5	90.0	92.4	94.4	11
Tacolneston	99.3	89.7	91.9	94.1	250
Winter Hill	98.2	88.6	90.8	93.0	4
Woolmoor	99.6	90.2	92.2	94.4	5
Wrotham	*98.8	89.1	91.3	93.5	125*/250

+ 74 low power txs less than 1kW

STATIONS: Radio 1: New music genres for youth audience 24h **N:** on the half h at peak times. Newsbeat M-F 1245, 1745 – **Radio 2:** Popular and specialist music, entertainment: 24h **N:** on the h – **Radio 3:** Classical music, jazz, world music, arts: 24h **N:** 0700, 0800, 0900(SS), 1300, 1700(MF), 1800(MF) – **Radio 4:** News, documentaries, drama, entertainment, and cricket on LW/MW in season: 0520-0100; relays BBCWS 0100-0520. **N:** 0530, then on the h (not 1000 Sun, 1100 Sun, 1500 Sat) – **Radio 5 Live:** News & sport: 24h **N:** on the h and half h – **Radio 6 Music:** New and archive music: 24h (digital only) – **Radio 7:** Comedy, drama, books, children's prgrs: 24h (digital only)

BBC LOCAL RADIO

MW Station		Location	kHz	kW
1)	Three Counties R.	Luton	630	0.2
6)	R. Cornwall	Redruth	630	2
6)	R. Cornwall	Bodmin	657	0.5
37)	R. York	Fulford	666	0.5
11)	Essex	Manningtree	729	0.2
15)	Hereford & Worcester	Worcester	738	0.037
8)	R. Cumbria	Carlisle	756	1
11)	Essex	Chelmsford	765	0.5
18)	R. Kent	Littlebourne	774	0.7
20)	R. Leeds	Farnley	774	0.5
8)	R. Devon	Barnstaple	801	2
8)	R. Cumbria	Barrow	837	1
19)	R. Lancashire	Preston	855	1
26)	R. Norfolk	Postwick	855	1.5
26)	R. Norfolk	West Lynn	873	0.3
10)	R. Devon	Exeter	990	1
33)	R. Solent	Fareham	999	1
4)	R. Cambridgeshire	Chesterton Fen	1026	0.5
17)	Jersey	Trinity	1026	1
31)	R. Sheffield	Sheffield	1035	1
9)	R. Derby	Burnaston Lane	1116	1
14)	Guernsey	Rohais	1116	0.5
30)	Sussex	Bexhill	1161	1
1)	Three Counties R.	Bedford	1161	0.1
37)	R. York	Scarborough	1260	0.5
36)	Wiltshire	Lacock	1332	0.4
33)	R. Solent for Dorset	Bournemouth	1359	0.85
30a)	Surrey	Duxhurst	1368	0.5
36)	Wiltshire	Swindon	1368	0.1
22)	R. Lincolnshire	Lincoln	1368	2
12)	R. Gloucestershire	Berkeley Heath	1413	0.5
12)	R. Gloucestershire	Bourton-on-the-Water	1413	0.5
8)	R. Cumbria	Whitehaven	1458	0.5
10)	R. Devon	Torquay	1458	2
25)	Newcastle	Wrekenton	1458	2
16)	R. Humberside	Hull	1485	2
30)	Sussex	Brighton	1485	1
23)	R Merseyside	Wallasey	1485	2
34)	R. Stoke	Sideway	1503	1
11)	Essex	Southend-on-Sea	1530	0.15
3)	R. Bristol	Mangotsfield	1548	5
19)	R. Lancashire	Oxcliffe	1557	0.25
3)	Somerset & R. Bristol	Taunton	1566	1
28)	R. Nottingham	Clipstone	1584	1
15)	Hereford & Worcester	Woofferton	1584	0.3
18)	R. Kent	Rusthall	1602	0.25
FM	**Station**	**Location**	**MHz**	**kW**
31)	R. Sheffield	Sheffield	88.6	0.3
17)	Jersey	Les Platons	88.8	3.8
1)	Three Counties R.	Epping Green	90.4	0.1
20)	R. Leeds	Holme Moss	92.4	5.6
14)	Guernsey	Les Touillets	93.2	1
34)	R. Stoke	Alsagers Bank	94.6	6.1
2)	R. Berkshire	Henley	94.6	0.25
15)	Hereford & Worcester	Ridge Hill	94.7	1

FM	Station	Location	MHz	kW
1)	Three Counties R	Aylesbury	94.7	0.2
31)	R. Sheffield	Chesterfield	94.7	0.4
7)	Coventry & Warwickshire	Meriden	94.8	2.2
10)	R. Devon	Huntshaw Cross	94.8	0.675
24)	London	Crystal Palace	94.9	4
3)	R. Bristol	Dundry Lane	94.9	0.5
22)	R. Lincolnshire	Belmont	94.9	8
30)	Sussex	Newhaven	95.0	0.1
5)	Tees	Bilsdale	95.0	10
12)	R. Gloucestershire	Stroud	95.0	0.1
13)	R Manchester	Holme Moss	95.1	5.6
28)	R. Nottingham	Newark	95.1	0.2
26)	R. Norfolk	Stoke Holy Cross	95.1	1
6)	R. Cornwall	Caradon Hill	95.2	4.3
8)	R. Cumbria	Kendal	95.2	0.1
29)	Oxford	Beckley	95.2	5.8
11)	Essex	South Benfleet	95.3	1.2
30)	Sussex	Brighton	95.3	1.2
9)	R. Derby	Stanton Moor	95.3	1.2
20)	R. Leeds	Luddenden	95.3	0.083
2)	R. Berkshire	Windsor	95.4	1
25)	Newcastle	Pontop Pike	95.4	10
1)	Three Counties R.	Sandy Heath	95.5	1
3)	Somerset & R. Bristol	Mendip	95.5	9
19)	R. Lancashire	Hameldon Hill	95.5	1.6
28)	R. Nottingham	Mansfield	95.5	1
35)	R. Suffolk	Lowestoft	95.5	2
37)	R. York	Olivers Mount	95.5	0.25
26)	R. Norfolk	West Runton	95.6	2.0
8)	R. Cumbria	Sandale	95.6	15
38)	WM (West Midlands)	Sutton Coldfield	95.6	1
4)	R. Cambridgeshire	Peterborough	95.7	5.1
10)	R. Devon	Plymouth	95.7	1
23)	R. Merseyside	Allerton Park	95.8	4
5)	Tees	Whitby	95.8	0.1
10)	R. Devon	Exeter	95.8	0.4
35)	R. Suffolk	Aldeburgh	95.9	2
16)	R. Humberside	High Hunsley	95.9	9.6
4)	R. Cambridgeshire	Cambridge	96.0	1
9)	R. Derby	Buxton	96.0	1.5
32)	R. Shropshire	The Wrekin	96.0	4.8
25)	Newcastle	Chatton	96.0	5.6
8)	R. Cumbria	Morecambe Bay	96.1	3.2
33)	R. Solent	Rowridge	96.1	1
18)	R. Kent	Wrotham	96.7	8.7
18)	R. Kent	Folkestone	97.6	0.1
1)	Three Counties R.	High Wycombe	98.0	0.2
20)	R. Leeds	Keighley	102.7	0.05
10)	R. Devon	North Hessary Tor	103.4	15
11)	Essex	Great Braxted	103.5	12
36)	Wiltshire	Salisbury	103.5	1
36)	Wiltshire	Swindon	103.6	0.5
27)	R. Northampton	Geddington	103.6	0.8
7)	Coventry & Warwickshire	Lark Stoke	103.7	1.4
37)	R. York	Acklam Wold	103.7	2
25)	Newcastle	Hexham	103.7	0.1
1)	Three Counties R.	Zouches Farm	103.8	0.5
28)	R. Nottingham	Mapperley Ridge	103.8	1
33)	R. Solent for Dorset	Bincombe Hill	103.8	0.5
6)	R. Cornwall	Redruth	103.9	18
19)	R. Lancashire	Winter Hill	103.9	2
20)	R. Leeds	Beecroft Hill	103.9	0.1
35)	R. Suffolk	Manningtree	103.9	5
15)	Hereford & Worcester	Great Malvern	104.0	2
30a)	Surrey	Reigate	104.0	3.8
8)	R. Cumbria	Whitehaven	104.1	1
31)	R. Sheffield	Holme Moss	104.1	4.4
34)	R. Stoke	Stafford	104.1	0.075
2)	R. Berkshire	Hannington	104.1	3
18)	R. Kent	Swingate	104.2	10
27)	R. Northampton	Northampton	104.2	4
36)	Wiltshire	Naish Hill	104.3	0.6
10)	R. Devon	Beacon Hill	104.3	1
37)	R. York	Woolmoor	104.3	0.5
2)	R. Berkshire	Reading	104.4	1
26)	R. Norfolk	Great Massingham	104.4	4.2
15)	Hereford & Worcester	Redditch	104.4	0.1
30)	Sussex	Heathfield	104.5	10
1)	Three Counties R.	Bow Brickhill	104.5	2.2
9)	R. Derby	Drum Hill	104.5	5.4
19)	R. Lancashire	Lancaster	104.5	2
13)	R. Manchester	Saddleworth	104.6	0.1
15)	Hereford & Worcester	Kidderminster	104.6	0.5
30a)	Surrey	Guildford	104.6	3
3)	R. Bristol	Bath	104.6	0.082

FM	Station	Location	MHz	kW
35)	R. Suffolk	Great Barton	104.6	2
12)	R. Gloucestershire	Churchdown Hill	104.7	2
30)	Sussex	Burton Down	104.8	2
21)	R. Leicester	Copt Oak	104.9	8
36)	Wiltshire	Marlborough	104.9	0.1

+ 16 low power txs less than 0.1kW

Addresses

1) 1 Hastings St, Luton LU1 5XL ☎1582 637400 **E:** 3cr@bbc.co.uk – **2)** Caversham Park, Peppard Road, Reading RG4 8TZ ☎118 9464200 **E:** radio.berkshire.news@bbc.co.uk – **3)** PO Box 194, Bristol BS99 7QT ☎117 9741111 **E:** radio.bristol@bbc.co.uk; BBC Somerset, Broadcasting House, Park Street, Taunton TA1 4DA ☎1823 323956 **E:** somerset@bbc.co.uk – **4)** PO Box 96, 104 Hills Road, Cambridge CB2 1LD ☎1223 259696 **E:** cambs@bbc.co.uk – **5)** PO Box 95FM, Broadcasting House, Newport Road, Middlesbrough TSI 5DG ☎1642 225211 **E:** tees@bbc.co.uk – **6)** Phoenix Wharf, Truro TR1 1UA ☎1872 275421 **E:** radio.cornwall@bbc.co.uk – **7)** Priory Place , Coventry CV1 5SQ ☎24 76551000 **E:** coventry@bbc.co.uk – **8)** Annetwell Street, Carlisle CA3 8BB ☎1228 592444 **E:** radio.cumbria@bbc.co.uk – **9)** PO Box 104.5, Derby DE1 3HL ☎1332 361111 **E:** radio.derby@bbc.co.uk – **10)** Broadcasting House, Seymour Road, Mannamead, Plymouth PL3 5YQ ☎1752 260323 **E:** radio.devon@bbc.co.uk – **11)** 198 New London Road, PO Box 765, Chelmsford CM2 9XB ☎1245 616000 **E:** essex@bbc.co.uk – **12)** London Road, Gloucester GL1 1SW ☎1452 308585 **E:** radio.gloucestershire@bbc.co.uk – **13)** PO Box 951, Oxford Road, Manchester M60 1SD ☎161 200 2000 **E:** radiomanchester@bbc.co.uk – **14)** Bulwer Ave, St Sampson, Guernsey GY2 4LA ☎ 1481 200600 **E:** bbcguernsey@bbc.co.uk – **15)** Hylton Road, Worcester WR2 5WW ☎1905 748485 **E:** bbchw@bbc.co.uk and 43 Broad Street, Hereford HR4 9HH ☎1432 355252 – **16)** Queens Court, Queens Gardens, Hull HU1 3RH ☎1482 323232 **E:** radio.humberside@bbc.co.uk – **17)** 18 Parade Road, St. Helier, Jersey JE2 3PL ☎1534 870000 **E:** radiojersey@bbc.co.uk – **18)** The Great Hall, Mount Pleasant Road, Tunbridge Wells TN1 1QQ ☎1892 670000 **E:** radio.kent@bbc.co.uk – **19)** 20-26 Darwen Street, Blackburn BB2 2EA ☎1254 262411 **E:** radio.lancashire@bbc.co.uk – **20)** 2 St Peters Square, Leeds LS9 8AH ☎113 244 2131 **E:** leeds@bbc.co.uk – **21)** 9 St Nicholas Place, Leicester LE1 5LB ☎116 251 6688 **E:** radioleicester@bbc.co.uk – **22)** PO Box 219, Lincoln LN1 3XY ☎1522 511411 **E:** radio.lincolnshire@bbc.co.uk – **23)** PO Box 958, Liverpool L69 1ZJ ☎151 708 5500 **E:** radio.merseyside@bbc.co.uk – **24)** Egton Wing, Broadcasting House, Portland Place, London W1A 1AA ☎20 7224 2424 **E:** yourlondon@bbc.co.uk – **25)** Broadcasting Centre, Barrack Road, Newcastle-Upon-Tyne NE99 1RN ☎191 222 4141 **E:** tyne@bbc.co.uk – **26)** The Forum, Millennium Plain, Norwich NR2 1BH ☎1603 617411 **E:** radio.norfolk@bbc.co.uk – **27)** Broadcasting House, Abington Street, Northampton NN1 2BH ☎1604 239100 **E:** radio.northampton@bbc.co.uk – **28)** London Road, Nottingham NG2 4UU ☎115 955 0500 **E:** radio.nottingham@bbc.co.uk – **29)** 269 Banbury Road, Oxford OX2 7DW ☎8459 311444 **E:** oxford@bbc.co.uk – **30)** Broadcasting House, 40-42 Queen's Road, Brighton BN1 3XB **E:** sussex@bbc.co.uk – **30a)** Broadcasting Centre, Guildford GU2 5AP ☎1483 306306 **E:** surrey@bbc.co.uk – **31)** 54 Shoreham Street, Sheffield S1 4RS ☎114 2731177 **E:** radio.sheffield@bbc.co.uk – **32)** 2-4 Boscobel Drive, Shrewsbury SY1 3TT ☎1743 248484 **E:** radio.shropshire@bbc.co.uk – **33)** Broadcasting House, 10 Havelock Road, Southampton SO14 7PW ☎23 8063 2811 **E:** radio.solent@bbc.co.uk – **34)** Cheapside, Hanley, Stoke-on-Trent ST1 1JJ ☎01782 00080 **E:** radio.stoke@bbc.co.uk – **35)** Broadcasting House, St. Matthew's Street, Ipswich IP1 3EP ☎1473 250000 **E:** suffolk@bbc.co.uk – **36)** 56-58 Prospect Place, Swindon SN1 3RW ☎ 1793 513626 **E:** wiltshire@bbc.co.uk – **37)** 20 Bootham Row, York YO30 7BR ☎1904 641 351 **E:** radio.york@bbc.co.uk – **38)** The Mailbox, Birmingham B1 1RF ☎121 567 6767 **E:** radio.wm@bbc.co.uk.
D.Prgr: Stns generally carry local prgrs from 0600 to 1800/1900, regional prgrs until 2400/0100, then BBC Radio 5 Live overnight.

BBC SCOTLAND

⌨40 Pacific Quay, Glasgow G51 1DA ☎ 141 422 6000
W: www.bbc.co.uk/radioscotland
MW: R. Scotland: Burghead 810kHz 100kW, Westerglen 810kHz 100kW, Redmoss 810kHz 5kW, Dumfries 585kHz 2kW

FM stereo	1FM	R2	R3	R4	RS/L	kW
Ashkirk	98.7	89.1	91.3	103.9	93.5f	50
Ben Gullipen	98.3	88.7	90.9	104.9	93.1	1
Black Hill	99.5	89.9	92.1	*95.8	94.3	250/200*
Bressay	97.9	88.3	90.5	94.9	92.7ac	43
Clettraval	97.7	88.1	90.3	95.1	92.5d	2
Daliburgh	98.9	89.3	91.5	95.9	93.7d	1
Darvel	99.1	89.5	91.7	104.3	93.9	10
Durris	99.0	89.4	91.6	95.9	93.8a	2.1
Eitshal	99.4	89.8	92.0	95.1	94.2d	5

FM stereo	1FM	R2	R3	R4	RS/L	kW
Forfar	97.9	88.3	90.5	94.9	92.7	17
Fort William	98.9	89.3	91.5	95.9	93.7d	3
Glengorm	99.1	89.5	91.7	96.1	93.9d	5
Keelylang Hill	98.9	89.3	91.5	96.0	93.7ab	41
Kirkton Mailer	98.6	89.0	91.2	94.6	93.4	1
Meldrum	98.3	88.7	90.9	95.3	93.1a	150
Melvaig	98.7	89.1	91.3	95.7	93.5d	50
Oban	98.5	88.9	91.1	95.3	93.3d	3.6
Rosemarkie	99.2	89.6	91.8	103.6	94.0d	20
Rumster Forest	99.7	90.1	92.3	95.6	94.5d	10
Sandale	97.7	88.1	90.3	92.5	94.7e	250
Skriaig	98.1	88.5	90.7	94.8	92.9d	30
So. Knapdale	98.9	89.3	91.5	95.6	93.7	2.2

+ 32 low power txs less than 1kW
RS/L=R. Scotland + local news – a) RS: Aberdeen – b) RS: Orkney – c) RS: Shetland – d) RS: Inverness – e) RS: Dumfries – f) RS: Selkirk.
D.Prgr: FM: M-F 0600-0030, SaSu: 0600-2400. Relays BBC Radio 5 Live overnight. **MW:** As FM SaSu as FM
Local Services (FM only). Freqs as above.
a) Beechgrove Terrace, Aberdeen AB15 5ZT. M-F: 0654, 0750, 0958, 1158, 1254, 1558, 1754. **W:** www.bbc.co.uk/northeastscotland – b) Castle Str, Kirkwall, Orkney KW15 1DF: M-F 0730-0800, Fri 1810-1900 – c) Pitt Lane, Lerwick, Shetland ZE1 0DW: M-F 1730-1800, Fri 1810-1900 – d) 7 Culduthel Rd, Inverness IV2 4AD M-F: 0654, 0750, 0958, 1158, 1254, 1558, 1754 – e) Elmbank, Lovers Walk, Dumfries DG1 1NZ: M-F: 0654, 0750, 0958, 1158, 1254, 1558, 1754. **W:** www.bbc.co.uk/southscotland – f) Ettrick Riverside, Dunsdale Rd, Selkirk TD7 5EB M-F: 0654, 0750, 0958, 1158, 1254, 1558, 1754.

RADIO NAN GAIDHEAL
✉ 52 Church Street, Stornoway HS1 2LS ☎ 1851 705000 🖷 1851 704633 **W:** www.bbc.co.uk/radionangaidheal
MW: Redmoss 990kHz 1kW(a)

FM	MHz	kW	FM	MHz	kW
Glengorm	103.5	5	Meldrum	104.2	150
Clettraval	103.7	2	Eitshal	104.3	2
So. Knapdale	103.7	2.2	Kirkton Mailer	104.5	1
Forfar	103.7	1	Rumster Forest	104.5	10
Melvaig	103.9	50	Oban	104.6	3.6
Craigkelly	104.1	5	Black Hill	104.7	10
Daliburgh	104.2	1	Skriaig	104.7	30
Fort William	104.2	3	Rosemarkie	104.9	20

+ 16 low power txs less than 1kW
D.Prgr: Own prgrs in Gaelic and relays of BBC R. Scotland. (a) rel local news for Aberdeen in English:0958 and 1254.

BBC WALES
✉ Broadcasting House, Llantrisant Rd, Llandaff, Cardiff CF5 2YQ ☎ 29 2032 2000 🖷 29 2055 5960
E: radiowales@bbc.co.uk **W:** www.bbc.co.uk/wales

FM stereo	R1	R2	R3	R4	kW
Blaenplwyf	98.3	88.7	90.9	104.0	250
Carmel	98.0	88.4	90.6	92.8	2.5
Haverfordwest	98.9	89.3	91.5	104.9	20
Kilvey Hill	99.1	89.5	91.7	94.6	1
Llanddona	99.4	89.8	92.0	103.6	21
Llandrindod Wells	98.7	89.1	91.3	103.8	2.8
Llangollen	98.5	88.9	91.1	93.3	15.6
Wenvoe	99.5	89.9	92.1	94.3	250

FM stereo	R. Wales	R.Cymru	kW
Blaenplwyf	95.3	93.1	250/120
Carmel	95.1	104.6	3/3.2
Haverfordwest	95.9	93.7	20
Kilvey Hill	93.9	104.2	1
Llanddona	94.8	94.2	21/10
Llandrindod Wells		93.5	2.8
Llangollen		104.3	15.6
Wenallt	103.9		2
Wenvoe		96.8	250

+ 42 low power txs less than 1kW
MW R. Wales: Forden 882kHz 1kW, Llandrindod Wells 1125kHz 1kW, Penmon 882kHz 10kW, Tywyn 882kHz 5kW, Washford 882kHz 100kW, Wrexham 657kHz 2kW.
D.Prgr: R Wales: 0600(SaSu0500)-0100. Relays BBC World Service overnight. **R Cymru:** 0500-0100. Relays BBC Radio 5 overnight.

BBC NORTHERN IRELAND
✉ Broadcasting House, 25-27 Ormeau Avenue, Belfast BT2 8HQ ☎ 28 9033 8000 🖷 28 9032 6453
W: www.bbc.co.uk/northernireland
MW: Enniskillen 873kHz 1kW, Lisnagarvey 1341kHz 100kW

FM stereo	R1	R2	R3	R4	R.Ulster	kW
Brougher Mountain	99.0	89.4	91.6	95.6	93.8	9.8

FM stereo	R1	R2	R3	R4	R.Ulster	kW
Camlough	98.3	88.7	90.9	104.6	93.1	4
Divis	99.7	90.1	92.3	96.0	94.5	250/125
Limavady	99.2	89.6	91.8	94.0	95.4	3.4
Londonderry	98.3	88.7	90.9	94.9	93.1h	31/10

+ 5 low power txs of less than 1kW – h) **R. Foyle** (see below).
R. Ulster: Enniskillen 873kHz 1kW, Lisnagarvey 1341kHz 100kW.
D.Prgr: MF 0630-0000(SaSu-0700-0200). Other times rel. BBC R5.

BBC RADIO FOYLE
✉ 8 Northland Rd, Londonderry BT48 7GD ☎ 28 7137 8600 **E:** radio.foyle@bbc.co.uk
MW: Londonderry 792kHz 1kW **FM:** 93.1MHz 31kW.
D.Prgr: 24h.Own prgs. and relay BBC R. Ulster. Local N. M-F hourly 0800-1700, Sa 1300.

BBC ASIAN NETWORK
✉ The Mailbox, Birmingham B1 1RF ☎ 121 567 6767 **E:** asian.network@bbc.co.uk **W:** www.bbc.co.uk/asiannetwork

MW	kHz	kW	MW	kHz	kW
Sedgley	828	0.2	Gunthorpe	1449	0.15
Freemen's Common	837	0.5	Langley Mill	1458	5

D.Prgr: 0500-0100, relay BBC Radio 5 Live overnight.
BBC Asian Network relays

MW	kHz	kW	MW	kHz	kW
R. Leeds	774a	0.7	R. Derby	1116c	1
R. Sheffield	1035b	1			

Key: a) Mon-Fri 1900-0100, b) Mon-Fri 1600-0100, c) Mon-Fri 1900-0100, SS 1800-0030. Local Asian programming is also carried on many BBC local radio stns at various times.
V: BBC does not QSL national domestic sces. BBC local stns V. by letter.

ARQIVA
✉ Crawley Court, Winchester SO21 2QA ☎1962 923434
W: www.arqiva.com. Operates most BBC domestic and many commercial radio tx sites. Formerly National Grid Wireless.

EXTERNAL SERVICE: BBC World Service
See International Broadcasting section.

OFFICE OF COMMUNICATIONS (Ofcom)
(Regulatory Authority)
✉ Riverside House, 2A Southwark Bridge Road, London SE1 9HA ☎ 20 7981 3000 🖷 20 7981 3333 **W:** www.ofcom.org.uk **L.P:** CEO: Ed Richards. Ch Operating Officer: Jill Ainscough; Partner (Strategy): Peter Phillips, Partner (Ext Affairs): Christopher Woolard, Partner (Content): Stuart Purvis.

RADIO CENTRE
✉ 77 Shaftesbury Ave, London W1D 5DU ☎ 20 7306 7800 🖷 20 7306 7801 **W:** www.radiocentre.org - Radio Centre represents commercial radio to Government, Ofcom, Copyright Societies and other organizations concerned with radio.

DIGITAL RADIO (DAB): DAB trs are on Band 3. **BBC Digital Radio:** BBC Radios 1, 1Xtra, 2, 3, 4, 5 Live, 5 Live Sports Extra, 6 Music, 7, BBC Asian Network, BBC World Service and regional srvcs in Scotland, Wales and N.Ireland are in a single frequency network on 225.648MHz. **Digital One:** ✉ 33-34 Alfreds Place, London WC1E 7DP ☎+44 (207) 299 8670 **W:** www.ukdigitalradio.com. Progr includes: Absolute Radio, Absolute 80s, Amazing Radio, BFBS Radio, Classic FM, talkSPORT, Planet Rock, Premier Christian Radio, Smooth Radio, UCB UK. **All prgrs** are in a single frequency network for England on 222.064MHz, block 11D, and for Scotland on 223.936MHz. **Local multiplexes: Arqiva:** Ayr. **Bauer,** Central Lancashire, Humberside, Leeds, Liverpool, South Yorkshire, Teesside, Tyne & Wear. **CE Digital,** Greater London, Birmingham, Manchester. **Digital Radio Group,** Greater London. **MXR,** North-East England, North-West England, South Wales/Severn Estuary, West Midlands, Yorkshire. **MuxCo** Chester & Wrexham*, Derbyshire*, Gloucestershire*, Lincolnshire*, Hereford & Worcester*, Mid & W Wales*, N Wales*, Northamptonshire*, Surrey, & N Sussex*, Somerset*, N Yorkshire*, Oxford* (*planned). **Now Digital,** Bournemouth, Bristol & Bath, Cambridge, Cardiff & Newport, Coventry, Exeter & Torbay, Kent, Leicester, Norwich, Nottingham, Peterborough, Reading & Basingstoke, So. Hampshire, Southend & Chelmsford, Sussex Coast, Swindon & West Wiltshire, Wolverhampton/Shrewsbury & Telford. **Score Digital,** Dundee & Perth, Edinburgh, Glasgow, Inverness, Northern Ireland. **South West Digital Radio,** Plymouth & Cornwall. **Switchdigital,** Aberdeen, Central Scotland, Greater London. **UTV-Bauer** Bradford & Huddersfield, Stoke-on-Trent, Swansea. **3G:** Isle of Man. **DRM:** Tests of local DRM have taken place in London on 25.7 and 26MHz using up to 1kW

NATIONAL COMMERCIAL STATIONS:

CLASSIC FM

30 Leicester Square, London WC2H 7LA
☎ 20 7343 9000 📠 20 7344 2700 **W:** www.classicfm.co.uk

FM	MHz	kW	FM	MHz	kW
Cumbria	99.9	250	Blaen Plwyf	101.1	10
No.Hessary Tor	100.0	160	Holme Moss	101.1	250
Angus	100.1	10.3	Darvel	101.3	8
Sutton Coldfield	100.1	250	Oxford	101.3	46
Bath	100.2	0.2	Swansea	101.3	1
Douglas I.O.M.	100.2	1	Bristol	101.4	0.2
Bradford	100.3	0.5	Inverness	101.4	11
Pontop Pike	100.3	130	Tacolneston	101.5	250
Rowridge	100.3	250	Redruth	101.5	10
Milton Keynes	100.4	1.9	Gt. Ormes Head	101.6	2.5
Ridge Hill	100.4	5	Bilsdale	101.6	2
Belmont	100.5	6.4	Leeds	101.6	0.5
Londonderry	100.5	31	Black Hill	101.7	250
Meldrum	100.5	150	Sheffield	101.7	0.5
Presely	100.5	7.13	Wenvoe	101.7	250
Crystal Palace	100.6	2	Dover	101.8	5.2
Arfon	100.7	18.75	Morecambe Bay	101.8	6.4
Swindon	100.8	0.72	Reading	101.8	0.5
Selkirk	100.9	10	Brighton	101.9	0.4
Wrotham	100.9	250	Divis	101.9	250
Fenham	101.0	0.05	Peterborough	101.9	35

D.Prgr: 24h **N:** on the h

TALKSPORT

18 Hatfields, London SE1 8DJ ☎ 20 7959 7800 📠 20 7959 7808
W: www.talksport.net

MW	kHz	kW	MW	kHz	kW
Bournemouth	1053	1	Brookmans Park	1089	400
Brighton	1053	2	Dartford Tunnel	1089	0
Droitwich	1053	500	Lisnagarvey	1089	13
Dumfries	1053	10	Moorside Edge	1089	400
Londonderry	1053	1	Redmoss	1089	2
Plymouth	1053	1	Redruth	1089	2
Postwick	1053	18	Washford	1089	80
Stockton	1053	1	Westerglen	1089	125
Tonbridge	1053	4	Boston	1107	1
Dundee	1053	1	Fareham	1107	1
Exeter	1053	1	Lydd	1107	2
Hull	1053	1	Reigate/Crawley	1107	1
Inverness	1053	1	Torbay	1107	1
Clipstone	1071	1	Wallasey	1107	1
Newcastle	1071	1			

D.Prgr: 24h **N:** on the h

ABSOLUTE RADIO

1 Golden Square, London W1F 9DJ ☎ 20 7434 1215 📠 20 7434 1197 **W:** www.absoluteradio.co.uk

MW	kHz	kW	MW	kHz	kW
Bournemouth	1197	0.3	Plymouth	1215	1
Brighton	1197	1	Redmoss	1215	2.3
Cambridge	1197	0.2	Redruth	1215	2
Torbay	1197	1	Washford	1215	100
Trowell	1197	1	Westerglen	1215	100
Wallasey	1197	0.4	Wrekenton	1215	2
Gloucester	1197	0.3	Kings Heath	1233	1
Hoo (Kent)	1197	2	Manningtree	1233	1
Oxford	1197	0.3	Reading	1233	0.2
Brookmans Park	1215	125	Sheffield	1233	0.3
Dartford Tunnel	1215	0.004	Swindon	1233	0.1
Droitwich	1215	105	Boston	1242	2
Fareham	1215	1	Dundee	1242	1
Hull	1215	0.3	Sideway	1242	1
Lisnagarvey	1215	16	Stockton	1242	1
Moorside Edge	1215	200	Guildford	1260	1
Norwich	1215	1	Lydd	1260	1

D.Prgr: 24h (rock & contemporary music). **N:** on the h
FM: London 105.8MHz 4kW

GOLD

30 Leicester Square, London WC2H 7LA ☎ 20 7766 6000 📠 20 7766 6100 **W:** www.mygoldmusic.co.uk

MW	kHz	kW	Location	MW	kHz	kW	Location
40)	603	0.4	Littlebourne	87)	936	0.18	Naish Hill
76)	774	0.14	Gloucester	41)	945	0.7	Bexhill
72)	792	0.28	Bedford	69)	945	0.2	Derby
43)	828	0.27	Bournemouth	64)	990	0.09	Wolverhampton
72)	828	0.2	Luton	69)	999	0.25	Nottingham

MW	kHz	kW	Location	MW	kHz	kW	Location
64)	1017	0.63	Shrewsbury	88)	1332	0.6	Peterborough
34)	1152	3	Birmingham	42)	1359	0.2	Cardiff
104)	1152	0.83	Norwich	92)	1359	0.28	Chelmsford
82)	1152	0.32	Plymouth	84)	1359	0.27	Coventry
87)	1161	0.16	Swindon	89)	1431	0.14	Reading
104)	1170	0.28	Ipswich	92)	1431	0.35	S'thend-on-Sea
43)	1170	0.12	Portsmouth	45)	1458	5	Manchester
40)	1242	0.32	Maidstone	89)	1485	1	Newbury
104)	1251	0.76	Bury St.Eds.	41)	1521	0.64	Crawley
86)	1260	1.6	Bristol	33)	1548	97.5	London
100)	1260	0.64	Wrexham	72)	1557	0.76	Northampton
42)	1305	0.2	Newport	43)	1557	0.5	Southampton
41)	1323	0.5	Brighton				

(Numbers refer to local studio address - see list below). **D.Prgr:** 24h.

HEART

30 Leicester Square, London WC2H 7LA ☎ 20 7766 6222 **W:** www.heart.co.uk

FM	kHz	kW	Location	FM	kHz	kW	Location
40)	95.9	0.27	Thanet	82)	101.2	1.15	Salcombe
40)	96.1	0.2	Ashford	92)	101.7	0.1	Harlow
92)	96.1	0.5	Colchester	82)	101.9	0.5	Ivybridge
82)	96.2	2.5	N Devon	41)	102.0	0.2	Hastings
86)	96.3	2	Bristol	87)	102.2	0.5	W Wiltshire
92)	96.3	1	Southend	43)	102.3	2	Bournemouth
100)	96.3	1.25	Llandudno	41)	102.4	8.2	Eastbourne
82)	96.4	1.6	Torbay	76)	102.4	2	Gloucester
104)	96.4	2	Bury St.Edm'ds	104)	102.4	3.3	Norwich
72)	96.6	4	Northampton	41)	102.6	2	Chelmsford
72)	96.6	0.5	Watford	89)	102.6	9	Oxford
43)	96.7	0.5	Winchester	86)	102.6	4	Mendip
72)	96.9	0.9	Bedford	41)	102.7	3.6	Reigate
41)	96.9	0.1	Newhaven	88)	102.7	4	Peterborough
40)	97.0	0.5	Dover	40)	102.8	1	Dunkirk
82)	97.0	1	Exeter	89)	102.9	3.4	Hannington
82)	97.0	2	Plymouth	100)	103.0	5	Caernarfon
89)	97.0	1	Reading	88)	103.0	1	Cambridge
86)	97.1	0.2	W.Somerset	82)	103.0	1	Stockland Hl
104)	97.1	3.4	Ipswich	76)	103.0	0.1	Stroud
100)	97.1	1	Wirral	86)	103.0	0.1	Weston-S-Mare
87)	97.2	0.7	Swindon	40)	103.1	4	Maidstone
82)	97.3	0.1	Ilfracombe	72)	103.3	2	Milton Keynes
82)	97.4	0.3	Banbury	100)	103.4	1.4	Wrexham
43)	97.5	0.85	Portsmouth	89)	103.4	0.1	Henley
72)	97.6	1	Luton	41)	103.5	1	Brighton
82)	100.5	0.3	Totnes	38)	106.0	8	Copt Oak
82)	100.8	0.1	Dartmouth	33)	106.2	4	London

(Numbers refer to local studio address - see list below)

COMMERCIAL RADIO STATIONS

MW	kHz	kW	Station or Slogan	Location
-	531	0.001	occasional RSLs	
175)	558	1	Spectrum R.	London
129)	756	0.63	R. Maldwyn, Magic 756	Newtown
56)	828	0.12	Magic 828	Leeds
128)	855	0.15	Sunshine R.	Ludlow
126)	936	1	Fresh R.	Hawes
1)	963	0.95	Buzz Asia	E. London
93)	963	0.2	Asian Sound R.	Haslingden
1)	972	1	Buzz Asia	W. London
53)	990	0.25	Magic AM	Doncaster
60)	999	0.8	Magic 999	Preston
153)	1026	1.7	Downtown R.	Belfast
155)	1035	0.78	Northsound Two	Aberdeen
160)	1035	0.32	West Sound AM	Ayr
18)	1035	2.5	Kismat R.	London
161)	1107	1.5	Moray Firth R.	Inverness
-	1134	0.001	LPAMs	
33)	1152	23.5	LBC	London
158)	1152	3.6	Clyde 2	Glasgow
45)	1152	1.5	Magic 1152	Manchester
61)	1152	1.8	Magic 1152	Newcastle
54)	1161	0.35	Magic 1161	Hull
156)	1161	1.4	Tay AM	Dundee
127)	1170	0.58	Swansea Sound	Swansea
59)	1170	0.32	Magic 1170	Stockton
181)	1170	0.2	Signal Two	Stoke-on-Trent
-	1251	0.001	LPAMs	
151)	1260	0.29	Sabras R.	Leicester
106)	1278	0.43	Pulse 2	Bradford
-	1278	0.001	LPAMs/RSLs	
-	1287	0.001	LPAMs	
148)	1296	10	R. XL	Birmingham

MW	kHz	kW	Station or Slogan	Location
53)	1305	0.15	Magic AM	Barnsley
98)	1305	0.5	Premier Christian R.	Epsom
98)	1305	0.5	Premier Christian R.	Chingford
98)	1332	1	Premier Christian R.	London
-	1350	0.001	LPAMs	
93)	1377	0.08	Asian Sound R.	Ashton Moss
24)	1386	0.003	R. JCom	Leeds
-	1386	0.001	LPAMs	
-	1404	0.001	LPAMs	
98)	1413	0.5	Premier Christian R.	Heathrow
98)	1413	0.5	Premier Christian R.	Dartford Marshes
126)	1413	0.1	Fresh R.	Skipton
126)	1413	0.04	Fresh R.	Richmond
126)	1431	0.01	Fresh R.	Ilkley & Settle
-	1431	0.001	LPAMs	
-	1449	0.001	LPAMs	
167)	1458	125	Sunrise R.	London
234)	1521	0.01	Flame CCR	Wirral
106)	1530	0.74	Pulse 2	Huddersfield
109)	1530	0.05	Celtic Music R.	Glasgow
53)	1548	0.74	Magic AM	Sheffield
52)	1548	1	Magic 1548	Liverpool
159)	1548	2.2	Forth 2	Edinburgh
195)	1566	0.8	County Sound R.	Guildford
-	1575	0.001	LPAMs/RSLs	
156)	1584	0.21	Tay AM	Perth
11)	1584	0.2	London Turkish R.	London
83)	1602	0.07	Desi R.	Southall
-	1602	0.001	occasional RSLs	

FM	MHz	kW	Name or Slogan	Location
-	87.7	-	RSLs/LPFMs	
-	87.9	-	RSLs	
91)	95.2	0.2	Kingdom FM	Dunfermline
33)	95.8	4	Capital FM	London
91)	96.1	0.5	Kingdom FM	Glenrothes
2)	96.1	0.2	Rother FM	Rotherham
124)	96.2	4	SIBC	Shetland
63)	96.2	0.2	KMFM	Tonbridge
99)	96.2	1	Mix 96	Aylesbury
77)	96.2	2.6	North Norfolk R.	Stody
188)	96.2	0.1	The Revolution	Oldham
145)	96.2	0.625	Yorkshire Coast R.	Scarborough
169)	96.2	0.1	Touch R.	Coventry
69)	96.2	1	Trent FM*	Nottingham
56)	96.3	2.5	R. Aire	Leeds
176)	96.3	0.2	Rock R.	Paisley
36)	96.4	0.2	Real R. - NE	Hexham
153)	96.4	2	Downtown R.	Limavady
34)	96.4	10	BRMB	Birmingham
185)	96.4	1.5	The Wave	Swansea
195)	96.4	3	The Eagle	Guildford
156)	96.4	0.8	Tay FM	Perth
181)	96.4	0.25	Signal One	Congleton
163)	96.4	3	CFM	Carlisle
197)	96.4	0.1	KMFM	Folkestone
31)	96.4	0.1	Compass FM	Grimsby
179)	96.5	0.1	R. Wave	Blackpool
157)	96.5	0.12	West Sound	Stranraer
153)	96.6	8.2	Downtown R.	Brougher Mountain
115)	96.6	0.1	Midwest R.	Blandford
26)	96.6	0.2	RNA FM	Arbroath
59)	96.6	8.9	TFM	Bilsdale
102)	96.6	0.4	R. Ceredigion	Lampeter
142)	96.6	0.45	Nevis R.	Fort William
161)	96.6	0.45	Moray Firth R	Cairngorm
25)	96.6	0.4	Spirit FM	Chichester
160)	96.7	2.2	West FM	Ayr
52)	96.7	8.2	R. City	Liverpool
144)	96.7	0.55	City Beat	Belfast
196)	96.7	3	KL.FM	King's Lynn
171)	96.7	0.2	Ashbourne R.	Ashbourne
70)	96.7	0.1	Wyvern FM	Kidderminster
161c)	96.7	0.1	Moray Firth R./Kinnaird	Fraserburgh
162)	96.8	5	R. Borders	Selkirk
125)	96.8	0.5	Lochbroom FM	Polbain
54)	96.9	9.4	Viking FM	Hull
155)	96.9	11	Northsound One	Aberdeen
181)	96.9	0.2	Signal 1	Stafford
170)	96.9	3.2	The Bay	Morecambe Bay
142)	97.0	0.25	Nevis R.	Glencoe
157)	97.0	1	West Sound	Dumfries
84)	97.0	1.8	Mercia FM	Coventry
61)	97.1	10	Metro R.	Newcastle
121)	97.1	0.3	NECR	Braemar
121)	97.1	0.1	NECR	Turriff

FM	MHz	kW	Name or Slogan	Location
133)	97.1	0.275	Kestrel FM	Haslemere
153)	97.1	0.08	Downtown R.	Larne
102)	97.1	3	R. Carmarthenshire	Carmel
64)	97.2	2	Beacon R.	Wolverhampton
78)	97.2	0.2	Kiss 101	Bristol
16)	97.2	0.31	Causeway Coast R.	Coleraine
118)	97.2	0.625	Wessex FM	Dorchester/Weymouth
147)	97.2	1	Stray FM	Harrogate
33)	97.3	4	LBC	London
159)	97.3	9.8	Forth 1	Edinburgh
161)	97.4	6.25	Moray Firth R.	Inverness
60)	97.4	2	Rock FM	Preston/Blackpool
154)	97.4	3.2	Cool FM	Belfast
42)	97.4	0.5	Red Dragon FM*	Newport
53)	97.4	0.4	Hallam FM	Sheffield
102)	97.4	0.4	R. Ceredigion	Penwaun
115)	97.4	0.125	Midwest R.	Shaftesbury
134)	97.4	0.24	The Beach	Southwold
36)	97.5	9	Smooth R - NE	Tyne & Wear
106)	97.5	0.5	The Pulse	Bradford
28)	97.5	0.36	Heartland FM	Pitlochry
160)	97.5	0.15	West FM	Girvan
66)	97.5	0.2	Scarlet FM	Llanelli
70)	97.6	0.8	Wyvern FM	Hereford
159)	97.6	0.1	Forth 1	Edinburgh
45)	97.7	1	XFM	Manchester
184)	99.8	0.75	2BR	Burnley
68)	99.8	0.75	KCFM	Hull
50)	99.9	0.5	R. Norwich	Stoke Holy Cross
58)	100.0	4	Kiss 100	London
101)	100.1	0.2	Lakeland R.	Kendal
12)	100.2	2	Dream 100 FM	Colchester
122)	100.3	20	Real R - Scotland	Black Hill
37)	100.4	5	Smooth R.	Winter Hill
8)	100.5	1	Five FM	Newry
36)	100.7	8.5	Real R. - NE	Bilsdale
47)	100.7	11	Heart FM	Sutton Coldfield
101)	100.8	0.12	Lakeland R.	Windermere
200)	100.8	0.13	Total Star	Porlock
78)	101.0	40	Kiss 101	Mendip
122)	101.1	10	Real R – Scotland	Edinburgh
8)	101.1	0.4	Five FM	Newry
178)	101.2	1.65	Waves R.	Peterhead
57)	101.2	6.26	Q101.2 West	Brougher Mountain
36)	101.2	0.2	Smooth R.-NE	Newton
55)	101.4	0.2	Smooth R.	Drum Hill
133)	101.6	0.1	Kestrel FM	Alton
63)	101.6	0.4	KMFM	Wrotham
36)	101.8	8.5	Real R. - NE	Burnhope
133)	101.8	0.11	Kestrel FM	Petersfield
121)	101.9	0.22	NECR	Tullich
105)	102.0	1	Town FM	Ipswich
80)	102.0	0.45	Dearne FM	Barnsley
130)	102.0	0.1	Peak FM	Matlock
45)	102.0	0.5	Galaxy*	Manchester
146)	102.0	1.25	Spire FM	Salisbury
133)	102.0	0.1	Kestrel FM	Alton
169)	102.0	2.6	Touch R.	Stratford upon Avon
177)	102.0	0.2	Wave 102	Dundee
121)	102.1	1.25	NECR	Inverurie
193)	102.1	1.2	Bay Radio	Swansea
159)	102.2	0.5	Forth 1	Penicuik
15)	102.2	4	Smooth R.	Croydon
112)	102.2	6.4	Lincs FM	Belmont
194)	102.2	2.5	Pirate FM	Caradon Hill
47)	102.2	1	Galaxy*	Birmingham
163)	102.2	0.815	CFM	Workington
125)	102.2	0.7	Lochbroom FM	Ullapool
158)	102.3	0.6	Clyde 1	Rothesay
25)	102.3	0.5	Spirit FM	Littlehampton
153)	102.3	0.5	Downtown R.	Ballymena
142)	102.3	0.8	Nevis R.	Fort William
20)	102.4	0.625	Touch R.	Tamworth
200)	102.4	4	Total Star	Minehead
153)	102.4	10	Downtown R.	Londonderry
158)	102.4	0.6	Clyde 1	Rosneath
145)	102.4	0.1	Yorkshire Coast R.	Bridlington
182)	102.4	0.1	Wish FM	Wigan
142)	102.4	0.8	Nevis R.	Glenachulish
158)	102.5	15	Clyde 1	Glasgow
106)	102.5	2	The Pulse	Halifax
161a)	102.5	1.2	Moray Firth R./Caithness	Thurso
102)	102.5	20	R. Pembrokeshire	Haverfordwest
163)	102.5	0.1	CFM	Penrith
181)	102.6	4	Signal 1	Stoke-on-Trent

FM	MHz	kW	Name or Slogan	Location
61)	102.6	0.125	Metro R.	Alnwick
121)	102.6	0.3	NECR	Kildrummy
62)	102.7	9	Cuillin FM	Isle of Skye
150)	102.8	0.5	Star R. NE	Burnhope
96)	102.8	0.9	Ram FM *	Derby
70)	102.8	1	Wyvern FM	Worcester
156)	102.8	5	Tay FM	Dundee
194)	102.8	10	Pirate FM	Redruth
161d)	102.8	1	Moray Firth R./Keith R.	Keith
53)	102.9	0.45	Hallam FM	Barnsley
19)	102.9	3.14	Q102.9 FM	Londonderry
51)	103.0	4	Key 103	Manchester
143)	103.0	4	Isles FM	Stornoway
155)	103.0	0.174	Northsound One.	Peterhead
174)	103.0	0.1	Your R.	Dumbarton
157)	103.0	0.7	West Sound	Kirkcudbright
64)	103.1	2.7	Beacon R.	Shrewsbury
14)	103.1	0.5	Central FM	Stirling
153)	103.1	1.8	Downtown R.	Newry
77)	103.2	0.25	North Norfolk R. (2 txs)	N. Norfolk
42)	103.2	2	Red Dragon FM*	Cardiff
43)	103.2	2	Galaxy*	Southampton
32)	103.2	0.1	Mansfield 103.2	Mansfield
170)	103.2	0.1	The Bay	Kendal
168)	103.2	0.4	Sunrise FM	Bradford
150)	103.2	0.4	Star R - NE	Darlington
61)	103.2	0.12	Metro R.	Hexham
121)	103.2	0.3	NECR	Colpy
74)	103.3	0.17	High Peak R.	Buxworth
74)	103.3	0.1	High Peak R.	Hope Valley
114)	103.3	0.05	London Greek R.	London
102)	103.3	5.8	R. Ceredigion	Blaen Plwyf
158)	103.3	0.1	Clyde 1	Rosneath
6)	103.3	0.4	Oban FM	Oban
122)	103.3	0.5	Real R.-Scotland	Penicuik
53)	103.4	1.6	Hallam FM	Doncaster
153)	103.4	0.2	Downtown R.	Newcastle
162)	103.4	0.5	R. Borders	Eyemouth
39)	103.4	0.16	Sun FM	Sunderland
163)	103.4	0.4	CFM	Whitehaven
134)	103.4	2	The Beach	Lowestoft
150)	103.5	0.2	Star R - NE	Northallerton
21)	103.7	4	Channel 103 FM	Jersey
141)	104.7	2.5	Minster FM	York
5)	104.7	1.25	Island FM	Guernsey
186)	104.9	0.64	Imagine FM	Stockport
33)	104.9	2.9	XFM	London
49)	105.1	3.1	Galaxy Yorkshire*	Emley Moor
85)	105.1	2.5	Atlantic FM	E Cornwall
95)	105.1		Southend R.	Southend-on-Sea
35)	105.2	30	Smooth R.	Glasgow
73)	105.2	11	Kerrang!	Sutton Coldfield
164)	105.2	10	Wave 105	Solent
123)	105.2	3	Real R. - Wales	Carmel
48)	105.3	8.4	Galaxy NE*	Burnhope
37)	105.4	5	Real R. - NE	Winter Hill
81)	105.4	5	Leicester Sound*	Billesdon
91)	105.4	0.1	Kingdom FM	Fife
58)	105.4	4	Magic 105.4	Croydon
123)	105.4	5	Real R. - Wales	Cardiff
111)	105.5	1.6	Palm FM	Torbay
94)	105.6	1	Kiss 105-108	Cambridge
49)	105.6	0.5	Galaxy Yorkshire*	Bradford
49)	105.6	0.25	Galaxy Yorkshire*	Sheffield
79)	105.6	0.25	Midwest R.	Yeovil
63)	105.6	0.25	KMFM	Maidstone
29)	105.6	0.1	Newbury Sound	Newbury
44)	105.7	10	Galaxy Scotland*	Edinburgh
22)	105.7	11	Smooth R.	Sutton Coldfield
123)	105.7	9.4	Real R. - Wales	Presely
10)	105.8	1.9	U105	Belfast
48)	105.8	0.2	Galaxy NE*	Hexham
164)	105.8	0.625	Wave 105	Poole
49)	105.9	9.6	Galaxy Yorkshire*	Hull
123)	105.9	1	Real R. - Wales	Newport
52)	105.9	7.5	City Talk	Liverpool
107)	106.0	4	The Coast 106	Solent
140)	106.0	0.6	Six FM	Cookstown
197)	106.0	0.1	KMFM	Canterbury
123)	106.0	1/0.2	Real R.Wales (2 txs)	Swansea/Carmarthen
75)	106.0	0.6	Two Lochs R.	Gairloch
94)	106.1	4	Kiss 105-108	Stoke Holy Cross
44)	106.1	20	Galaxy Scotland*	Glasgow
37)	106.1	1	Rock R.	Manchester
131)	106.2	3.12	Real R. Yorks	Emley Moor
123)	106.2	0.5	Real R.- Wales	Fishguard
103)	106.2	1.5	Sunshine R.	Hereford
91)	106.3	0.15	Kingdom FM	Fife
30)	106.3	0.9	Bridge FM	Bridgend
27)	106.3	0.2	Dee 106.3	Chester
94)	106.4	20	Kiss 105-108	Mendelsham
48)	106.4	8.9	Galaxy NE*	Bilsdale
121)	106.4	0.3	NECR	Cock Bridge
191)	106.4	0.4	Bright	Haywards Heath
74)	106.4	0.25	High Peak R.(2 txs)	Buxton/Glossop
117)	106.4	0.1	Andover Sound	Andover
33)	106.5	0.5	Argyll FM	Campbeltown
138)	106.5	0.5	The Severn	Shrewsbury
187)	106.5	1	Jack FM	Bristol
132)	106.5	0.1	Central R.	Preston
107)	106.6	0.3	The Coast	Poole
199)	106.6	0.25	Time 106.6	Slough
55)	106.6	10.8	Smooth R.	Waltham
79)	106.6	0.25	Midwest R.	Chard
75)	106.6	2	Two Lochs R.	Loch Ewe
25)	106.6	0.4	Spirit FM	Midhurst
46)	106.6	0.25	Perth FM	Perth
23)	106.7	0.1	Jack FM	Stevenage
160)	106.7	0.6	West FM	Rothesay
173)	106.7	0.62	R Plymouth	Ft Staddon
193)	106.8	4	Nation R.	Cardiff
191)	106.8	0.1	Bright 106.4	Lewes
150)	106.8	0.2	Star R. - NE	Durham
197)	106.8	0.1	KMFM	Dover
136)	106.8	0.2	Connect FM	Peterborough
80)	106.8	0.5	Ridings FM	Wakefield
65)	106.8	0.3	Jack FM	Oxford
180)	106.8	19	Original FM	Aberdeen
23)	106.9	0.28	Jack FM	Hertford
152)	106.9	0.15	Silk FM	Macclesfield
174)	106.9	0.1	Your R.	Helensburgh
85)	107.0	11	Atlantic FM	W Cornwall
184)	107.0	0.5	The Bee	Blackburn
113)	107.0	0.1	Isle of Wight R.	Chillerton Down
4)	107.0	0.1	Oak FM	Loughborough
90)	107.0	0.2	Reading 107	Reading
135)	107.0	0.62	Seven FM	Ballymena
103)	107.0	1	Sunshine R.	Monmouth
33)	107.1	0.1	Choice FM	N. London
108)	107.1	0.1	Trax FM	Doncaster
172)	107.1	0.14	Speysound R	Aviemore
198)	107.1	0.12	Star R.	Ely
33)	107.1	0.625	Argyll FM	Ballygroggan
169)	107.1	0.1	Rugby FM	Rugby
138)	107.1	0.1	The Severn	Oswestry
140)	107.2	0.25	Six FM	Dungannon
165)	107.2	0.2	Breeze 107	Winchester
182)	107.2	0.18	Wire FM	Warrington
110)	107.2	0.2	Rutland FM	Oakham
137)	107.2	0.2	Juice	Brighton
9)	107.2	0.2	KMFM	Thanet
192)	107.2	0.66	Star R.	Bristol
138)	107.2	0.2	The Wyre	Kidderminster
119)	107.3	1	Exeter FM	Exeter
169)	107.3	0.2	Touch R.	Warwick
193)	107.3	1.25	Nation R.	Swansea
136)	107.4	0.2	Connect FM	Kettering
200)	107.4	0.1	Total Star	Bridgwater
130)	107.4	0.2	Peak FM	Chesterfield
138)	107.4	0.1	The Severn FM	Telford
182)	107.4	0.18	Tower FM	Bolton
165)	107.4	0.2	Breeze 107	Portsmouth
102)	107.5	0.1	R. Pembrokeshire	Fishguard
200)	107.5	0.1	Total Star	Cheltenham
190)	107.5	0.34	Time FM	Romford
7)	107.5	0.15	Sovereign FM	Eastbourne
200)	107.5	0.1	Total Star	Warminster
17)	107.6	0.5	Juice FM	Liverpool
116)	107.6	0.1	Kestrel FM	Basingstoke
116)	107.6	1	Fire R.	Bournemouth
131)	107.6	0.2	Real R. - Yorks	Bradford
67)	107.6	0.2	Banbury Sound	Banbury
197)	107.6	0.5	KMFM	Ashford, Kent
36)	107.7	5	Smooth R.-NE	Eston Nab
200)	107.7	0.2	Total Star	Swindon
120)	107.7	0.1	Radio Nova	Weston Super Mare
95)	107.7	0.1	Chelmsford R.	Chelmsford
94)	107.7	0.2	Kiss 105-108	Peterborough
131)	107.7	0.2	Real R. - Yorks	Sheffield
183)	107.7	0.1	Splash FM	Worthing

FM MHz	kW	Name or Slogan	Location
3) 107.7	0.17	The Wolf	Wolverhampton
33) 107.7	0.5	Argyll FM	South Knapdale
149) 107.8	0.1	Arrow FM	Hastings
139) 107.8	0.8	R. Jackie	SW London
165) 107.8	0.9	Breeze 107	Southampton
189) 107.9	0.2	L107	Hamilton
4) 107.9	0.2	Oak FM	Hinckley
108) 107.9	0.2	Trax FM	Worksop
198) 107.9	0.1	Star R.	Cambridge
200) 107.9	0.1	Total Star	Bath
166) 107.9	0.2	Glide FM	Oxford
132) 107.9	0.2	Dune FM	Southport
63) 107.9	0.2	KMFM	Medway

+approx 50 relays of less than 0.1kW *Capital FM from Jan. 2011
H. of tr: Most stns operate 24h Some stns carry automated prgrs outside peak hours

MAJOR COMMERCIAL RADIO GROUPS:
BAUER RADIO Ltd ▣ Mappin House, 4 Winsley Str, London W1W 8HF ☎ 207 182 8000 **W:** www.bauermedia.co.uk
GLOBAL RADIO ▣ 30 Leicester Square, London WC2H 7LA ☎ 20 7766 6000 ▤ 20 7766 6111 **W:** www.thisisglobal.com/radio
GMG RADIO Ltd ▣ Laser House, Waterfront Quay, Salford M50 3XW ☎ 161 886 8800 **W:** www.gmgradio.com
TINDLE RADIO Ltd ▣ Radio House, Orion Court, Gt Blakenham IP5 0LW ☎ 1473 836100 ▤ 1473 836136 **W:** www.tindleradio.com
UKRD GROUP Ltd ▣ Carn Brea Studios, Barncoose Ind. Est.,, Redruth TR15 3RQ ☎ 1209 310435 ▤ 1209 310406 **W:** www.ukrd.com
UTV RADIO (GB) Ltd ▣ 18 Hatfields, London SE1 8DJ ☎ 20 7959 7900 **W:** www.utvradio.com

Addresses & other information
1) Radio House, Bridge Rd, Southall UB2 4AT ☎ 20 8574 6666 **W:** www.buzzasia.com – **2)** Aspen Court, Bessemer Way, Rotherham S60 1FB ☎ 1709 369991 **W:** www.rotherfm.co.uk – **3)** 2nd Floor, Mander House, Wolverhampton WV1 3NB ☎ 1902 571070 **W:** www.thewolf.co.uk – **4)** 3 Martins Court, Telford Way, Coalville LE67 3HD ☎ 1530 278200 **W:** www.oak-fm.co.uk – **5)** 12 Westerbrook, St Sampsons, Guernsey GY2 4QQ ☎ 1481 242000 **W:** www.islandfm.com – **6)** 132 George Street, Oban PA34 5NT ☎ 1631 570057 **W:** www.obanfm.co.uk – **7)** 14 St Mary's Walk, Hailsham BN27 1AF ☎ 1323 442700 **W:** www.sovereignfm.com – **8)** Win Business Park, Canal Quay, Newry BT35 6PH ☎ 28 3083 5550 **W:** www.fivefm.co.uk – **9)** 183 Northdown Rd, Cliftonville, Margate, CT9 2TA ☎ 1843 220222 **W:** www.kmfm.co.uk – **10)** Haveland House, Ormeau Road, Belfast BT7 1EB ☎ 28 9033 2102 **W:** www.u105.com – **11)** 185b High Road, Wood Green, London N22 6BA ☎ 20 8881 0606 **W:** www.londonturkishradio.org – **12)** Northgate House, St Peters Street, Colchester CO1 1HT ☎ 1206 764466 **W:** www.dream100.com – **13)** 27-29 Longrow, Campbeltown PA28 6ER ☎ 1586 551800 **W:** www.argyllfm.co.uk – **14)** 201-203 High Street, Falkirk FK1 1DU ☎ 1324 611654 **W:** www.centralfm.co.uk – **15)** 26-27 Castlereagh Street, London W1H 5DL ☎ 20 7706 4100 **W:** www.smoothradio.co.uk – **16)** 24 Cloyfin Road, Coleraine BT52 2NU ☎ 28 7035 9100 **W:** www.q972.fm – **17)** 27 Fleet Street, Liverpool L1 4AR ☎ 151 707 3107 **W:** www.juicefm.com – **18)** Radio House, Bridge Rd, Southall UB2 4AT ☎ 20 8574 6666 **W:** www.kismatradio.com – **19)** Riverview Suite, 87 Rosdowney Road, Londonderry BT47 5SU ☎ 2871 344449 **W:** www.q102.fm – **20)** 5-6 Aldergate, Tamworth, B79 7DJ ☎ 1827 318000 **W:** www.touchradio.co.uk – **21)** 6 Tunnell Street, St Helier, Jersey JE2 4LU ☎ 1534 888103 **W:** www.channel103.com – **22)** 3rd Floor, Crown House, Beaufort Court, 123 Hagley Road, Birmingham B16 8LD ☎ 121 452 1057 **W:** www.smoothradio.co.uk – **23)** The Pumphouse, Knebworth Park SG3 6HQ ☎ 1438 810900 **W:** www.jack106.com – **24)** MAZCC, 311 Stonegate Rd, Leeds LS17 6AZ ☎ 113 218 5836 **W:** www.radioj-com.com – **25)** 9/10 Dukes Court, Bognor Road, Chichester PO19 8FX ☎ 1243 773600 **W:** www.spiritfm.net – **26)** Arbroath Infirmary, Rosemount Road, Arbroath DD11 2AT ☎ 1241 879660 **W:** www.radionorthangus.co.uk – **27)** 2 Chantry Court, Chester CH1 4QN ☎ 1244 391000 **W:** www.dee1063.com – **28)** 9 Alba Place, Pitlochry PH16 5BH ☎ 1796 474040 **W:** www.heartlandfm.co.uk – **29)** 42 Bone Lane, Newbury RG14 5SD ☎ 1635 841600 **W:** www.newburysound.co.uk – **30)** PO Box 1063, Bridgend CF35 6WY ☎ 845 890 4000 **W:** www.bridge.fm – **31)** 26a Wellowgate, Grimsby DN32 0RA ☎ 1472 346666 **W:** www.compassfm.co.uk – **32)** The Media Suite, Brunts Business Centre, Samuel Brunts Way, Mansfield NG18 2AH ☎ 1623 646666 **W:** www.mansfield103.co.uk – **33)** 30 Leicester Square, London WC2H 7LA ☎ 20 7766 6000 **W:** www.capitalfm.com Choice: www.choice-fm.co.uk LBC: www.lbc.co.uk XFM: www.xfm.co.uk – **34)** 9 Brindleyplace, 4 Oozells Square, Birmingham B1 2DJ ☎ 121 566 5200 **W:** www.brmb.co.uk – **35)** Unit 1130, Glasgow Business Park, G69 6GA ☎ 141 781 1011 **W:** www.smoothradio.co.uk – **36)** Marquis Court, Team Valley Trading Estate,

Gateshead NE11 0RU ☎ 191 440 7500 **W:** www.realradio.co.uk www.smoothradio.co.uk – **37)** Laser House, Waterfront Quay, Salford M50 3XW ☎ 161 886 8800 **W:** www.realradio.co.uk; www.smoothradio.co.uk; www.rockradiomanchester.co.uk– **38)** City Link, Nottingham NG2 4NG ☎ 115 910 6105 – **39)** PO Box 1034, Sunderland SR5 2YL ☎ 191 548 1034 **W:** www.sun-fm.co.uk – **40)** Radio House, John Wilson Business Park, Whitstable CT5 3QX ☎ 1227 772004 – **41)** Radio House, Franklin Rd., Brighton BN41 1AF ☎ 1273 430111 – **42)** Radio House, Atlantic Wharf, Cardiff CF10 4DJ ☎ 29 2094 2900 **W:** www.reddragonfm.co.uk – **43)** Apple Ind. Estate, Whittle Ave, Fareham PO15 5SX ☎ 1489 587600 **W:** www.galaxysouthcoast.co.uk – **44)** 4 Winds Pavilion, Pacific Quay, Glasgow G51 1EB ☎ 141 566 6106 **W:** www.galaxyscotland.co.uk – **45)** 4 Exchange Quay, Salford M5 3EE ☎ 161 662 4700 **W:** www.galaxymanchester.co.uk; www.xfmmanchester.co.uk – **46)** Moncreiffe Business Centre, Perth PH2 8DG ☎ 1738 634109 **W:** www/perthfm.co.uk – **47)** 1 The Square, 111 Broad Street, Birmingham B15 1AS ☎ 121 695 0000 **W:** www.galaxybirmingham.co.uk – **48)** Kingfisher Way, Silverlink Business Park, Wallsend NE28 9NX ☎ 191 444 2500 **W:** www.galaxynortheast.co.uk – **49)** 2a Joseph's Well, Hanover Walk, Leeds LS3 1AB ☎ 113 308 5100 **W:** www.galax-yyorkshire.co.uk – **50)** 29 Yarmouth Rd, Norwich NR7 0EE ☎ 845 365 6999 **W:** www.999radionorwich.com – **51)** Castle Quay, Castlefield, Manchester M15 4PR ☎ 161 288 5000 **W:** www.key103.co.uk www.manchestersmagic.co.uk – **52)** St Johns Beacon, 1 Houghton Street, Liverpool L1 1RL ☎ 151 472 6800 **W:** www.magic1548.co.uk, www.radiocity.co.uk – **53)** 900 Herries Road, Sheffield S6 1RH ☎ 114 2091000 **W:** www.magicam.co.uk, www.hallamfm.co.uk – **54)** Commercial Road, Hull HU1 2SG ☎ 1482 325141 **W:** Magic FM: www.magic1161.co.uk Viking FM: www.vikingfm.co.uk – **55)** 2 Alder Court, Rennie Hogg Road, Nottingham NG2 1RX ☎ 115 986 1066 **W:** www.smoothradio.co.uk – **56)** PO Box 2000, 51 Burley Road, Leeds LS3 1LR ☎ 113 283 5500 **W:** www.magicaire.co.uk www.magic828.co.uk – **57)** 42A Market Street, Omagh BT78 1EN ☎ 28 6632 0777 **W:** www.q101west.fm – **58)** Mappin House, 4 Winsley Street, London W1W 8HF ☎ 20 7182 8000 **W:** Magic: www.magic.fm Kiss: www.totalkiss.com – **59)** Radio House, Yales Crescent, Thornaby, Stockton-on-Tees TS17 6AA ☎ 1642 888222 **W:** www.tfmradio.co.uk www.magic1170.co.uk – **60)** PO Box 974, Preston PR1 1XS ☎ 1772 477700 **W:** Rock FM: www.rockfm.co.uk Magic 999: www.magic999.com – **61)** 55 Degrees North, Pilgrim St., Newcastle upon Tyne NE1 6BF ☎ 191 230 6100 **W:** www.magic1152.co.uk, www.metroradio.co.uk – **62)** Stormyhill Road, Portree IV51 9DY ☎ 1478 611797– **W:** www.cuillinfm.co.uk – **63)** Medway House, Ginsbury Close, Strood, Rochester ME2 4DU ☎ 1634 711079 **W:** www.kmfm.co.uk – **64)** 267 Tettenhall Road, Wolverhampton WV6 0DQ ☎ 1902 461300 **W:** www.beaconradio.co.uk – **65)** 270 Woodstock Rd, Oxford OX2 7NW ☎ 1865 315980 **W:** www.jackfmoxford.com – **66)** PO Box 971, Llanelli SA15 1YH ☎ 845 8907000 **W:** www.radiocar-marthenshire.co.uk – **67)** Unit 9A Manor Park, Banbury OX16 3TB ☎ 1295 661070 **W:** www.banburysound.co.uk – **68)** Planet House, 2 Woodshouse St. Hull HU9 1NU ☎ 1482 333999 **W:** www.kcfm.co.uk – **69)** Chapel Quarter, Maid Marian Way, Nottingham NG1 6JR ☎ 115 873 1500 **W:** www.trentfm.co.uk – **70)** Kirkham House, John Comyn Drive, Worcester WR3 7NS ☎ 1905 545510 **W:** www.wyvernfm.co.uk – **71)** 5 Abbey Court, Fraser Road, Bedford, MK44 3WH ☎ 1234 235010 – **72)** Chiltern Road, Dunstable LU6 1HQ ☎ 1582 676200 – **73)** 20 Lionel Street, Birmingham B3 1AQ ☎ 845 0531052 **W:** www.kerrangradio.co.uk – **74)** Smithbrook Close, High Peak SK23 0QD ☎ 1298 813144 **W:** www.highpeakradio.co.uk – **75)** Mansegate, Gairloch IV21 2LR ☎ 870 712106 **W:** www.2lr.co.uk – **76)** Bridge Studios, Eastgate Centre, Gloucester GL1 1SS ☎ 1452 572400 – **77)** Breck Farm, Stody, Norfolk NR24 2ER ☎ 1263 860808 **W:** www.north-norfolkradio.com – **78)** 26 Passage Street, Bristol BS1 1SE ☎ 117 9010101 **W:** www.totalkiss.com – **79)** 72 Middle Street, Yeovil BA20 1DJ ☎ 1935 848488 **W:** www.midwestradio.co.uk – **80)** Unit 7 Network Centre, Zenith Park, Whaley Road, Barnsley S75 1HT ☎ 1226 321733 **W:** www.dearnefm.co.uk www.ridingsfm.co.uk – **81)** 6 Dominus Way, Merdian Business Park, Leicester LE19 1RP ☎ 116 256 1300 **W:** www.leicestersound.co.uk – **82)** Hawthorn Hse., Exeter Business Park, Exeter EX1 3QS ☎ 1392 444444 – **83)** Panjabi Centre, 30 Sussex Road, Southall UB2 5EG ☎ 20 8564 9591 **W:** www.desiradio.org.uk – **84)** Hertford Place, Coventry CV1 3TT ☎ 2476 868200 **W:** www.merciafm.co.uk – **85)** 10 Wheal Kitty Workshops, St Agnes TR5 0RD ☎ 1872 554400 **W:** www.atlantic.fm **W:** www.ten17.co.uk – **86)** 1 Passage Str., Bristol BS2 0JF ☎ 117 984 3200 – **87)** Chiseldon House, Stonehill Green, Westlea, Swindon SN5 7HB ☎ 1793 842600 – **88)** PO Box 225, Queensgate Centre, Peterborough PE1 1XJ ☎ 1733 460460 – **89)** The Chase, Calcot, , Reading RG31 7RB ☎ 118 9454400 – **90)** Radio House, Madejski Stadium, Reading RG2 0FN ☎ 118 986 2555 **W:** www.reading107fm.com - **91)** Haig House, Haig Business Park, Markinch, Fife KY7 6AQ ☎ 1592 753753 **W:** www.kingdomfm.co.uk – **92)** 31 Glebe Road, Chelmsford CM1 1QG ☎ 1254 524500 – **93)** Nynex Bldg, Shield Drive,

Manchester M28 5PR ☎ 161 288 1000 **W:** asiansoundradio.co.uk – **94)** Reflection House, Olding Road, Bury St Edmunds IP33 3TA ☎ 1284 715300 **W:** www.totalkiss.com – **95)** Icon Bldg., Western Esplanade, Southend-on-Sea SS1 1EE ☎ 1702 455070 **W:** www.southendradio. com www.chelmsfordradio.co.uk – **96)** 35-36 Irongate, Derby DE1 3GA ☎ 1332 324000 **W:** www.ramfm.co.uk – **97)** The Stanley Centre, Kelvin Way, Crawley RH10 9SE ☎ 1293 636000 **W:** www.mercuryfm.co.uk – **98)** 22 Chapter Street, London SW1P 4NP ☎ 20 7316 1300 **W:** www. premier.org.uk – **99)** Friars Square Studios, 11 Bourbon Street, Aylesbury HP20 2PZ ☎ 1296 399396 **W:** www.mix96.co.uk – **100)** The Studios, Mold Rd, Wrexham LL11 4AF ☎ 1978 752200 – **101)** Lakeland Food Park, Plumgarths, Crook Road, Kendal LA8 8QJ ☎1539 737 380 **W:** www.lakelandradio.co.uk – **102)** Unit 14, Old School Estate, Station Road, Narbeth SA67 7DU ☎1834 869384 **W:** www.radiopembrokeshire.com www.ceredigionradio.co.uk – **103)** Suite 5, Penn House, Broad Str. Hereford HR4 9AP ☎ 1432 360246 **W:** www.sunshineradio. co.uk – **104)** St George's Plain, 47-49 Colegate, Norwich NR3 1DB ☎ 1603 630621– **105)** Radio House, Orion Court, Gt Blakenham, Ipswich IP6 0LW ☎ 473 836102 **W:** www.town102.com – **106)** Forster Square, Bradford BD1 5NE ☎ 1274 203040 **W:** www.pulse.co.uk www.pulsegold.co.uk – **107)** Roman Landing, Kingsway, Southampton SO14 1BN ☎ 23 8030 4100 **W:** www.thecoast106.com – **108)** 5 Sidlings Court, Doncaster DN4 5NU ☎ 1302 341166 **W:** www.traxfm.co.uk – **109)** 153 Queen Str., Glasgow G1 3BJ ☎ 141 548 3397 **W:** www.celticmusicradio.net – **110)** 40 Melton Road, Oakham LE15 6AY ☎ 1572 757868 **W:** www.rutlandradio.co.uk – **111)** Marble Court, Lymington Rd, Torquay TQ1 4FB ☎ 1803 321055 **W:** www.palm.fm – **112)** Witham Park, Waterside South, Lincoln LN5 7JN ☎ 1522 549900 **W:** www.lincsfm. co.uk – **113)** Dodnor Park, Newport, Isle of Wight PO30 5XE ☎ 1983 822557 **W:** www.iwradio.co.uk – **114)** LGR house, 437 High Road, London N12 0AP ☎ 20 8349 6950 **W:** www.lgr.co.uk – **115)** Longmead Studios, Shaftesbury SP7 8QQ ☎ 1747 855711 **W:** www.midwestradio. co.uk – **116)** 307 Holdenhurst Rd, Bournemouth BH8 8BX ☎ 1202 443600 **W:** www.thenewfireradio.com – **117)** 3 Eastgate House, Andover SP10 1EP ☎ 11264 336000 **W:** www.andoversound.com – **118)** 18 Trinity Street, Dorchester DT1 1DJ ☎ 1305 250333 **W:** www. wessexfm.co.uk – **119)** 6a Cranmere Court, Lustleigh Close, Exeter EX2 8PW ☎ 1392 823557– **120)** 11 Beaconsfield Rd, Weston-super-Mare BS23 1YE ☎ 1934 624455– **121)** The Shed, School Rd, Kintore, Inverurie AB51 0UX ☎ 1467 632909 **W:** www.necrfm.co.uk – **122)** Parkway Court, Baillieston, Glasgow G69 6GA ☎ 141 781 1011 **W:** www.realradio.co.uk – **123)** Ty-Nant Court, Cardiff CF15 8LW ☎ 2920 315100 **W:** www.realradioco.uk – **124)** Market Street, Lerwick, Shetland ZE1 0JN ☎ 1595 695299 **W:** www.sibc.co.uk – **125)** Mill Street Industrial Estate, Ullapool IV26 2UN ☎ 1854 613131 **W:** www.lochbroomfm.co.uk – **126)** The Watermill, Broughton Farm, Skipton BD23 3AG ☎ 845 2242052 **W:** www.freshradio.co.uk – **127)** PO Box 1170, Swansea SA4 3AB ☎ 1792 511964 **W:** www.swanseasound.co.uk – **128)** Unit 11, Burway Trading Estate, Ludlow SY8 1EN ☎ 1584 877755 **W:** www.sunshineradio.co.uk – **129)** The Studios, The Park, Newtown SY16 2NZ ☎ 1686 623555 **W:** www.magic756.net – **130)** Radio House, Foxwood Road, Chesterfield S41 9RF ☎ 1246 269138 **W:** www.peakfm. net – **131)** 1 Sterling Court, Tingley, Wakefield WF3 1EL ☎113 238 1114 **W:** www.realradio.co.uk – **132)** The Power Station, Victoria Way, Southport PR8 1RR ☎ 1704 502500 **W:** www.dunefm.co.uk www.central1065.co.uk – **133)** Paddington House, Festival Place, Basingstoke RG21 7LJ ☎ 1256 694000 **W:** www.kestrelfm.com – **134)** PO Box 103.4, Lowestoft NR32 2TL ☎ 845 3451035 **W:** www.thebeach.co.uk – **135)** Woodside Rd Industrial Estate, Ballymena BT42 4PT ☎ 28 2564 8777 **W:** www.sevenfm.co.uk – **136)** 5 Church Street, Peterborough PE1 1XB ☎ 1733 898106 **W:** www.connectfm.com – **137)** 170 North Street, Brighton BN1 1EA ☎ 1273 387107 **W:** www.juicebrighton.com – **138)** Shropshire Newspapaers, Waterloo Road, Ketley, Telford TF1 5HU ☎ 1952 280011 **W:** www.thesevern.co.uk weww.thewyre.com – **139)** 110-112 Tolworth Broadway, Surbiton KT6 7JD ☎ 20 8288 1300 **W:** www. radiojackie.com – **140)** 2c Park Avenue, Cookstown BT80 8AH ☎28 8675 8696 **W:** www.sixfm.co.uk – **141)** 1 Chessingham Park, Dunnington, York YO19 5SE ☎ 1904 488888 **W:** www.minsterfm.co.uk – **142)** Ben Nevis Estate, Claggan, Fort William PH33 6PR ☎ 1397 700007 **W:** www.nevisradio.co.uk – **143)** PO Box 333, Stornoway, Isle of Lewis HS1 2PU ☎ 1851 703333 **W:** www.isles.fm – **144)** Arena Bldg., 85 Ormeau Rd, Belfast BT7 1SH ☎ 28 9023 4967 **W:** www.citybeat967.co.uk – **145)** PO Box 962, Scarborough YO11 3ZP ☎ 1723 581700 **W:** www. yorkshirecoastradio.com – **146)** City Hall Studios, Malthouse Lane, Salisbury SP2 7QQ ☎ 1722 416644 **W:** www.spirefm.co.uk – **147)** The Hamlet, Hornbeam Park Avenue, Harrogate HG2 8RE ☎ 1423 522972 **W:** www.strayfm.com – **148)** KMS House, Bradford Street, Birmingham B12 0JD ☎ 121 753 5353 **W:** www.radioxl.net – **149)** Priory Meadow Centre, Queen Square, Hastings TN34 1PJ ☎ 1424 461177 **W:** www. arrowfm.co.uk – **150)** Radio House, 11 Woodland Rd, Darlington DL3 7BJ ☎ 1325 341801 **W:** www.thisisstar.com – **151)** Radio House, 63

Melton Road, Leicester LE4 6PN ☎ 116 261 0666 **W:** www.sabrasradio.com – **152)** 140 Moss Lane, Macclesfield SK11 7YT ☎ 1625 268000 **W:** www.silkfm.com – **153)** Newtownards, Co Down BT23 4ES ☎ 28 9181 5555 **W:** www.downtown.co.uk – **154)** PO Box 974, Belfast BT1 1RT ☎ 28 9181 7181 **W:** www.coolfm.co.uk – **155)** Abbottswell Road, West Tullos, Aberdeen AB12 3AJ ☎ 1224 337000 **W:** www.northsound1.co.uk – **156)** 6 North Isla Street, Dundee DD3 7JQ ☎ 01382 200800 **W:** www.radiotay.co.uk – **157)** Unit 40, The Loreburne Centre, High Street, Dumfries DG1 2BD ☎ 1387 250999 **W:** www.westsoundradio.com – **158)** Clydebank Business Park, Clydebank, Glasgow G81 2RX ☎ 141 565 2200 **W:** www.clyde1.com – **159)** Forth House, Forth Street, Edinburgh EH1 3LE ☎ 131 556 9255 **W:** www.forthone.com – **160)** Radio House, 54a Holmston Road, Ayr KA7 3BE ☎ 1292 283662 **W:** www.westsound.co.uk www.westfm.co.uk – **161)** Scorguie Place, Inverness IV3 8UJ ☎ 1463 224433 **W:** www.mfr.co.uk – **161a)** Neil Gunn Drive, Thurso KW14 7QU ☎01847 890000 **W:** www.caithnessfm. co.uk –**161c)** Old Thomas Walker Hospital, Charlotte Street, Fraserburgh AB43 9LS ☎ 01346 512010 – **161d)** 59a Land Street, Keith AB55 5AN ☎ 01542 886080 **W:** www.keithcommunityradio.com – **162)** Tweeside Park, Galashiels TD1 3TD ☎ 1896 759444 **W:** www.radioborders.com – **163)** PO Box 964, Carlisle CA1 3NG ☎ 1228 818964 **W:** www.cfmradio.com – **164)** 5 Manor Court, Barnes Wallis Road, Segensworth East, Fareham PO15 5TH ☎ 1489 481050 **W:** www.wave105.com – **165)** St Marys Stadium, Southampton SO14 5FP ☎ 23 8063 5151 **W:** www. thebreeze107.com – **166)** 270 Woodstock Rd, Oxford OX2 7NW ☎ 1865 315980 **W:** www.glidefm.co.uk – **167)** Radio House, Bridge Rd, Southall UB2 4AT ☎ 20 8574 6666 **W:** www.sunriseradio.com – **168)** 55 Leeds Rd, Bradford BD1 5AF ☎ 1274 735043 **W:** www.sunriseradio.fm – **169)** Holly Farm Business Park, Honily, Kenilworth CV8 1NP ☎ 1926 485600 **W:** www.touchradio.co.uk www.rugbyfm.co.uk – **170)** PO Box 969, 24 St George's Quay, Lancaster LA1 3LD ☎ 1524 848747 **W:** www.thebay. co.uk – **171)** St Monicas House, Windmill Lane, Ashbourne DE6 1EY ☎ 1335 346967 **W:** www.ashbourneradio.co.uk – **172)** Plot 4A, Dalfaber Industrial Estate, Aviemore PH22 1ST ☎ 1479 811888 **W:** www.speysoundradio.com – **173)** 3 Crescent Ave. Mews, Plymouth PL1 3AP ☎ 1752 389532 **W:** www.radioplymouth.com – **174)** Unit 3, 80 Castlegreen Str., Dumbarton G82 1JB ☎ 1389 742855 **W:** www.yourradio.com – **175)** 4 Inchgate Place, London SW8 3NS ☎ 20 7627 4433 **W:** www. spectrumradio.net – **176)** Unit 1130, Parkway Court, Glasgow G69 6GA ☎ 141 781 0963 **W:** www.rockradiofm.co.uk – **177)** 8 South Tay Street, Dundee DD1 1PA ☎ 1382 901000 **W:** www.wave102.co.uk – **178)** 7 Blackhouse Circle, Blackhouse Industrial Estate, Peterhead AB42 1BW ☎ 1779 491012 **W:** www.wavesfm.co.uk – **179)** Mowbray Drive, Blackpool FY3 7JR ☎ 1253 650300 **W:** www.wave965.com – **180)** Craigshaw Rd, Aberdeen AB12 3AR ☎ 1224 294860 **W:** www.originalfm.com – **181)** 67-73 Stoke Road, Stoke-on-Trent ST4 2SR ☎ 1782 441 300 **W:** www.signal1.co.uk – **182)** Orrell Lodge, Orrell Road, Wigan WN5 8HJ ☎ 1942 761024 **W:** www.wishfm.net www.towerfm.co.uk – **183)** Guildbourne Centre, Worthing BN11 1LZ ☎ 1903 210772 **W:** www.splashfm.com – **184)** 2 Petre Court, Petre Rd, Accrington BB5 5HH ☎ 1282 690000 **W:** www.2br.co.uk www.thebee. co.uk – **185)** PO Box 964, Swansea SA4 3AB ☎ 1792 511964 **W:** www. thewave.co.uk – **186)** Regent House, Heaton Lane, Stockport SK4 1BX ☎ 161 609 1400 ▤ **W:** www.imaginefm.net – **187)** County Gates, Ashton Rd, Bristol BS3 2JH ☎117 966 1065 **W:** www.jackbristol.com – **188)** PO Box 962, Oldham OL1 31JF ☎ 161 621 6500 **W:** www.revolutiononline.co.uk – **189)** 69 Bothwell Rd, Hamilton ML3 ODW ☎ 1698 303420 **W:** www.L107.com – **190)** Lambourne House, 7 Western Road, Romford RM1 3LD ☎ 1708 731643 **W:** www.timefm.com – **191)** Unit 34 Market Place Shopping Centre, Burgess Hill RH15 9NP ☎ 1444 248127 **W:** www.bright1064.com – **192)** County Gates, Ashton Rd, Bristol BS3 2JH ☎ 117 966 1065 **W:** www.starbristol.co.uk – **193)** Newby House, Neath Abbey Industrial Estate, Neath SA10 7DR ☎ 845 025 1000 **W:** www.swanseabayradio.com www.nationwales.com – **194)** Carn Brea Studios, 102 Wilson Way, Redruth TR15 3XX ☎ 1209 314400 **W:** www.piratefm.co.uk – **195)** Dolphin House, North Street, Guildford GU1 4AA ☎ 1483 300964 **W:** www.countysound.co.uk. 964eagle.co.uk – **196)** 18 Blackfriars Street, Kings Lynn PE30 1NN ☎ 1553 772777 **W:** www.klfmradio.co.uk – **197)** 34-36 North Street, Ashford TN24 8JR ☎ 1233 623232 **W:** www.kmfm.co.uk – **198)** 20 Mercers Row, Cambridge CB5 8HY ☎ 1223 305107 **W:** www.star107. co.uk – **199)** addr as stn 167) ☎ 845 194 1066 **W:** www.time1066.com – **200)** 8 Manchester Park, Tewksbury Rd, Cheltenham GL51 9EJ ☎ 1242 699555 **W:** www.totalstar.co.uk

MANX RADIO (Comm.)
▢ Douglas Head, Douglas, Isle of Man IM1 5BW ☎ 1624 682600 ▤ 1624 682604 **W:** www.manxradio.com **L.P:** MD: Anthony Pugh.
MW: 1368kHz Foxdale 20kW **FM:** 89.0MHz Snaefell 4kW / 97.2MHz Carnane 11kW / 103.7MHz Jurby 4kW
D.Prgr: 24h Separate prgrs on MW: MF 0730-0830, Sun 2000-2200

ENERGY FM (Comm.)
100 Market Street, Douglas, Isle of Man IM1 2PH ☎ 1624 611936
📠 1624 664699 **E:** mail@energyfm.net **W:** www.energyfm.net **FM:** 91.2MHz Snaefell 1.2kW, 93.4MHz Jurby 2kW, 98.6MHz Carnane 2kW (+ relays on 98.4/102.4/105.2) **D.Prgr:** 24h

3FM (Comm.)
45 Victoria Street, Douglas, Isle of Man IM1 3RS ☎ 1624 616333
📠 1624 614333 **W:** www.three.fm
FM: 104.2MHz (Ramsey & Port St Mary), 105MHz (Carnane 2kW), 106.6MHz (Snaefell), 106.2MHz (Peel) **D.Prgr:** 24h

BRITISH FORCES BROADCASTING SCE.
(a division of Services Sound & Vision Corp.)
SSVC, Narcot Lane, Gerrards Cross SL9 8TN ☎ 1494 878354 📠 1494 878552 **E:** adminofficer@bfbs.com **W:** www.bfbs.com
LP: Contr BFBS Radio: Charles Foster
MW: BFBS Gurkha Radio in Nepali on 1134kHz (Bramcote, Catterick & Sandhurst), 1251kHz (York), 1278kHz (Folkestone-main studio & Stafford), 1287kHz (Blandford, Brecon, Maidstone, Chippenham).
FM: BFBS on 100.6MHz (Lisburn) 101MHz (Belfast), 106.5MHz (Antrim), 107.5MHz (Ballykinlar) **DAB:** Digital One. See under Afghanistan, Ascension Island, Belgium, Belize, Bosnia, Brunei, Canada, Cyprus, Falkland Islands, Germany, Gibraltar, Nepal, Netherlands, Qatar for other sces.

GARRISON FM
Shute Road, Catterick Garrison DL9 4AF ☎ 1748 830050 **E:** hq@garrisonfm.com **W:** wwwgarrisonfm.com **LP:** MD: Mark Page. **MW:** Low power sce on 1287kHz (Catterick, Bassingbourn, Glencourse, Wattisham) **FM:** 98.5MHz (Edinburgh), 102.5MHz (Aldershot), 106.8MHz (Salisbury), 106.9MHz (Catterick), 107MHz (Colchester)

Restricted Service Licences (RSL) Licences are granted for low power special event stns operating for up to 28 days (occ. longer) usually on FM (occ. on MW)
LPAM (Low Power AM stations) There are currently about 70 stns on the air with txs of 0.001kW e.r.p. Freqs used: 1134, 1251, 1278, 1287, 1350, 1386, 1404, 1431, 1449, 1575kHz
LPFM (Low Power VHF/FM stations) There are currently about 18 stns on the air, most on 87.7MHz, with txs of typically 50mW

Community Radio Small-scale, low-power, non-profit community radio sces to serve a particular neighbourhood. Most on FM with 25W. 185 stns on air as of October 2010. For updated listing of short-term RSLs, long-term RSLs (LPAMs) and Community Radio see Radio Broadcast Licensing at **W:** www.ofcom.org.uk/radio
Community Audio Distribution Systems (CADS) licence-exempt service for religious and community events using 27MHz Citizens Band.

UNITED STATES OF AMERICA

L.T: See World Time Table (DST where applicable: 13 Mar-6 Nov) — **Pop:** 307 million — **Pr.L:** English — **E.C:** 60Hz, 110V — **ITU:** USA

FEDERAL COMMUNICATIONS COMMISSION (FCC)
Govt. licensing agency for broadcast stations
445 12th Street, SW, Washington, D.C. 20554 ☎ +1 202 418-0190
📠 +1 202 418-0232 **W:** www.fcc.gov **E:** fccinfo@fcc.gov
LP: Chmn: Julius Genachowski. Commissioners: Michael J. Copps, Robert M. McDowell, Mignon Clyburn, Meredith Attwell Baker
The FCC is an independent federal agency composed of commissioners appointed by the President with the consent of the Senate. One of its major activities is the general regulation of broadcasting, visual as well as aural. This regulation may be divided into three phases: 1) The allocation of spectrum space to the different types of broadcast services 2) consideration of applications to build and operate individual stations 3) regulation of their operations. Broadcasting is handled by the FCC Mass Media Bureau.

Call letter assignments: International agreement provides for the identification of the country of a radio stn by the first letter or first two letters of the stn's assigned call signal and for this purpose apportions the alphabet among different nations. USA nations use the initial letters K, N and W exclusively, and part of the A series. For broadcast stns, calls beginning with K are assigned to stns west of the Mississippi River, incl. Guam and No. Marianas Is, while W is assigned to broadcast stns east of the Mississippi, incl. Puerto Rico. Calls consist of four letters, to which FM or TV may be added, with a hyphen, for FM or TV stns.
A few exceptions with stns east of the Mississippi using a "K" callsign and stns west of the Mississippi using a "W" callsign will be noted. These are old callsigns that were assigned before the geographical division was introduced and are retained by special permission. Similarly some very old callsigns using only three letters may be retained by the stns that once were assigned these callsigns.

Stations: More than 14,500 stns are operating (MW & FM). As of June 2010 there were 4,786 licensed AM stns.

MAJOR NETWORKS PROVIDING AM STATION PROGRAMMING

CBS RADIO
1515 Broadway, New York, NY 10036 ☎ +1 212 846-3939 **W:** www.cbsradio.com **LP:** Pres. & CEO: Dan Mason
CBS Radio is a division of CBS Corporation and operates 130 radio stns. The overall mix of each radio stn's programming is designed to fit the stns specific format and serve its local community. In addition, CBS Radio owns the CBS Radio Network, with sales and affiliate relations handled by Westwood One, providing hourly newscasts to more than 1000 news and news/talk formatted stns.

CITADEL MEDIA, INC
13725 Montfort Drive, Dallas, TX 75240 ☎ +1 972-991-9200 📠 261 Madison Avenue, New York, NY 10016 ☎ +1 212-735-1700 **W:** www.citadelmedianetworks.com **LP:** Pres.: James Robinson.
Citadel Media (which changed its name from ABC Radio Networks in April 2009) is owned by Citadel Broadcasting. Citadel Broadcasting owns 58 AM stns. Citadel Media has more than 4,400 affiliate radio stations. Prgrs and services include ABC News Radio (distributed on behalf of Disney), The Mark Levin Show, The Michael Baisden Show, The Huckabee Report and a variety of 24/7 music formats.

CLEAR CHANNEL RADIO
200 East Basse Road, San Antonio, TX 78209 ☎ +1 210 822-2828 **W:** www.clearchannel.com **LP:** Pres/CEO, Clear Channel R: John Hogan
Clear Channel operates over 800 radio stns in the USA. In addition, Clear Channel's Premiere Radio Network syndicates more than 90 prgrs to more than 5,000 radio affiliates, including Rush Limbaugh, Jim Rome, Ryan Seacrest, Steve Harvey, Delilah, Jim Rome and Fox Sports Radio.

CNN RADIO
Box 105573, Atlanta, GA 30348. ☎ +1 404 827-1500 **W:** www.cnnradio.com **LP:** Turner Chmn. & CEO: Philip Kent.
Cable News Network is owned by Turner Broadcasting System, Inc, a Time Warner Company. CNN Radio provides news, sports, business and feature reports. CNN and Westwood One are partnered together to distribute CNN Radio prgrs to radio stns across the U.S. Under the agreement, Westwood One syndicates the CNN Radio Network, the most widely distributed radio news network in North America.

CUMULUS MEDIA, INC
3280 Peachtree Road, NW Suite 2300, Atlanta, GA 30305. ☎ +1 404 949-0700 **W:** www.cumulus.com **LP:** Chmn, Pres & CEO: Lewis W Dickey, Jr.
Cumulus Media, Inc. operates over 350 radio stations in the USA.

DISNEY & ESPN MEDIA NETWORKS, INC
Walt Disney Comp., 500 S. Buena Vista St., Burbank, CA 91521-9722 ☎ +1-818-560-1000 **W:** disney.go.com **LP:** Co-Chmn, Disney Media Networks and Pres, ESPN, Inc. & ABC Sports: George W. Bodenheimer.
Disney Media Networks comprise a vast array of broadcast, cable, radio, publishing and Internet businesses. Key radio areas include Disney-ABC Television Group (which manages Radio Disney Network) and ESPN Inc.

FOX NEWS RADIO
Fox News Radio, 1211 Ave. of the Americas, 18th Floor, New York, NY 10036 ☎ +1 212-301-5439 **W:** www.foxnewsradio.com
Fox News Radio provides radio stns with hourly newscasts at both top and bottom of the hour. Newscasts are anchored by Fox News Channel correspondents and provided 24 hours a day, 7 days a week. The service also provides radio stns with radio-anchored breaking news coverage of crisis events both nationally and internationally. Additionally, Fox News Radio syndicates talk prgrs to affiliates across the country.

IRN/USA RADIO NETWORK
USA Radio Network, Inc, PO Box 383230, Germantown, TN 38183.

☎ +1 800 325-0919 **W:** www.irnusaradio.com **LP:** Pres.: Larry Bates. USA Radio Network, inc. and Information Radio Network merged in Febraury 2008 to form a new company called IRN/USA Radio Network with a combined affiliate base of almost 2500 radio stns.

NATIONAL PUBLIC RADIO (NPR) (non-comm.)

✉ 635 Massachusetts Ave, NW, Washington, D.C. 20001 ☎ +1 202 513-2000 **W:** www.npr.org **LP:** Pres. & CEO: Vivian Schiller.

NPR is a producer and distributor of non-commercial news, talk, and enter-tainment prgrs. A privately supported, not-for-profit membership organization, NPR serves an audience of 27.5 million Americans each week in partnership with more than 860 independently operated, non-commercial public radio sttns. Each NPR Member Station serves local listeners with a distinctive combination of national and local prgrs.

TRITON RADIO NETWORKS/DIAL-GLOBAL

Triton Radio Networks: ✉ 220 West 42nd Street 3rd Floor, New York, N.Y. 10036, ☎ +1 212 419-2900, 📠 +1 212 896-5341 **W:** www.tritonradionetworks.com

Dial-Global: ✉ 11812 San Vicente Blvd., 3rd floor, Suite 350, Los Angeles, CA 90049 ☎ +1 310 820-8666 **W:** www.dial-global.com **LP:** Pres/CEO Triton Radio Networks: Spencer Brown; Pres/ CEO Dial Global: Ken Williams/David Landau

Dial Global, a division of Triton Radio Networks, is owned by the Triton Media Group, LLC. Dial Global owns the assets of Transtar Radio Networks, Jones Radio Networks and Waitt Radio Networks. Dial Global Programming produces and syndicates approximately 100 music prgrs and prep services in a variety of formats to more than 6,000 radio stns nationwide.

WESTWOOD ONE, INC.

✉ 40 West 57th Street, 5th floor, New York, NY 10019 ☎ +1 212 614-2000 **W:** www.westwoodone.com **LP:** Pres: Roderick M. Sherwood III.

Westwood One is the largest independent provider of network radio programming and the largest provider of traffic information in the U.S. Westwood One serves more than 5,000 radio stns and provides over 150 news, sports, music, talk and entertainment prgrs, features and live events to numerous media partners. Through its Metro Networks division, Westwood provides traffic reporting and local news, sports and weather to over 2,500 radio stns.

OTHER NATIONAL ORGANIZATIONS:

NATIONAL ASSOCIATION OF BROADCASTERS

✉ 1771 N Str, NW, Washington, DC 20036 ☎ +1 202 429-5300 **W:** www.nab.org **LP:** Pres/CEO: Gordon H. Smith.

The NAB is a trade association that advocates on behalf of more than 8,300 free, local radio and television stns and also broadcast networks before Congress, the Federal Communications Commission and the Courts.

NATIONAL ASSOCIATION OF SHORTWAVE BROADCASTERS, INC.

✉ 10400 NW 240th Str, Okeechobee, FL 34972 ☎ +1 863 763-0281 📠 +1 863 763-1034 **W:** www.shortwave.org **LP:** Pres: Jeff White. Vice Pres: W. Glen Tapley.

The NASB represents the interests of FCC-licensed broadcasters in the private sector of the US International Shortwave Broadcast community

COMMERCIAL STATIONS

Especially in multi-station markets, radio stns concentrate their prgrs to appeal to a given segment of the population or a given listening taste. Many stns devote their entire broadcast day to news and/or talk prgrs. Others specialize in hit music (adult contemporary, top 40), country music, oldies (e.g. hits of the fifties and sixties), big bands/ standards, black (urban contemporary, jazz, rhythm & blues), religious services and inspirational music, classical music, ethnic prgrs.

Today satellites are widely used for the distribution of prgrs. Numerous such networks are in operation. Stns making extensive use of network prgrs may have only one local identification per hour, usually on top of the hour. The former "clear channel" stns today have a protected area extending to 700 miles. Outside this area, the frequencies are also used by other stns. A few stns have been granted temporary licenses for increased powers to combat interference from neighbouring countries. Many daytime stns may now operate after local sunset using low or very low powers.

With the large decrease in AM listening in favour of FM, an increasing number of AM stns go off the air for a longer or shorter period due to economical difficulties. The latest development is that stns on the so-called regional channels, previously limited to 5kW power, may now

apply for up to 50kW, limited only by the required protection of other stns. Relaxed ownership rules have allowed groups of co-owned stns to form in larger markets with the group stns often broadcasting from a common studio address.

Digital broadcasting in the AM band is being tested by a number of stns, using the IBOC system. According to this system, digital signals are emitted on both sides of the stns AM signal, so that both AM and digital receivers can recover the st audio. AM listeners may experience the IBOC signal as an increased noise level on the chs adjacent to the nominal channel of the emitting stn.

MEDIUMWAVE
Explanations

Call: Station call letters. All stns are required to announce their actual call letters and city of licence once per hour as close as possible to the top of the hour.

Ant: Type of licence and use of directional antenna. The symbols mean as follows: **U** means Unlimited Time operation, i.e. up to 24 hours a day. **U1** without directional antenna, **U2** with directional antenna at night, **U3** with directional antenna at all hours, same pattern, day and night, **U4** with directional antenna at all hours, different patterns day and night, **U5** with directional antenna daytime, non-directional at night, **U6** with directional antenna at night and during critical hours, **U7** with different directional patterns for day, critical hours and night, **U8** as U7 but non-directional day, **U9** means directional day and night (different patterns), but non-directional during critical hours (usually on reduced power), **U10** means directional during critical hours only, **U11** means separate patterns for daytime and critical hours, nondirectional nights. **D** is daytime operation (between local sunrise and local sunset). The symbols mean as follows: **D** is daytime operation (between local sunrise and local sunset), **D1** without directional antenna, **D2** with directional antenna during critical hours only, **D3** with directional antenna, **D4** with directional antenna, differenct patterns during critical and non-critical hours, **D5** with directional antenna except during critical hours. **L** is limited time, and means a st. West of the dominant st. can operate from as early as sunrise at the dominant sts location; A st. East of the dominant st. can operate as late as the dominant sts sunset. Number indicates directional pattern as under "U" above.

D: Daytime power in kW. **N:** Nighttime power in kW.

City of License and Sta: City and State that the license has been issued to. **N.B:** Hawaii and Alaska are listed under separate country headings.

Scope: Due to the large number of stns in operation, the following list has been limited to stns operating at 2.5kW or more during the daytime (and in some cases where power has been increased to 2.5kW or more at night).

STATES: AL Alabama, AR Arkansas, AZ Arizona, CA California, CO Colorado, CT Connecticut, DE Delaware, FL Florida, GA Georgia, IA Iowa, ID Idaho, IL Illinois, IN Indiana, KS Kansas, KY Kentucky, LA Louisiana, MA Massachusetts, MD Maryland, ME Maine, MI Michigan, MN Minnesota, MO Missouri, MS Mississippi, MT Montana, NC North Carolina, ND North Dakota, NE Nebraska, NH New Hampshire, NJ New Jersey, NM New Mexico, NV Nevada, NY New York, OH Ohio, OK Oklahoma, OR Oregon, PA Pennsylvania, RI Rhode Island, SC South Carolina, SD South Dakota, TN Tennessee, TX Texas, UT Utah, VA Virginia, VT Vermont, WA Washington, WI Wisconsin, WV West Virginia, WY Wyoming.

MW	Call	kHz	Ant.	D	N	Sta	City of License
1)	KRXA	540	U4	10	0.5	CA	Carmel Valley
2)	KVIP	540	U1	2.5	0.01	CA	Redding
3)	WFLF	540	U4	50	46	FL	Pine Hills
4)	WDAK	540	U1	4	0.03	GA	Columbus
5)	KWMT	540	U3	5	0.17	IA	Fort Dodge
6)	KMLB	540	U4	5	1	LA	Monroe
9)	WETC	540	U4	4	0.5	NC	Wendell-Zebulon
7)	KNMX	540	U3	5	0.02	NM	Las Vegas
8)	WLIE	540	U4	2.5	0.22	NY	Islip
10)	WWCS	540	U4	5	0.5	PA	Canonsburg
12)	KTZN	550	U1	5	5	AK	Anchorage
11)	WASG	550	U1	10	0.14	AL	Atmore
13)	KFYI	550	U1	5	1	AZ	Phoenix
14)	KUZZ	550	U4	5	5	CA	Bakersfield
15)	KRAI	550	U2	5	0.5	CO	Craig
16)	WAYR	550	U3	5	0.06	FL	Orange Park
17)	WDUN	550	U2	10	2.5	GA	Gainesville
19)	KFRM	550	U3	5	0.11	KS	Salina
20)	KTRS	550	U2	5	5	MO	Saint Louis
21)	KBOW	550	U2	5	1	MT	Butte
23)	KFYR	550	U2	5	5	ND	Bismarck

MW	Call	kHz	Ant.	D	N	Sta	City of License
22)	WGR	550	U2	5	5	NY	Buffalo
24)	WKRC	550	U4	5	1	OH	Cincinnati
25)	KOAC	550	U4	5	5	OR	Corvallis
27)	KCRS	550	U4	5	1	TX	Midland
26)	KTSA	550	U2	5	5	TX	San Antonio
29)	WSVA	550	U2	5	1	VA	Harrisonburg
28)	WDEV	550	U4	5	1	VT	Waterbury
30)	KARI	550	U4	5	2.5	WA	Blaine
31)	WSAU	550	U4	15	20	WI	Wausau
32)	WOOF	560	U1	5	0.11	AL	Dothan
33)	KSFO	560	U2	5	5	CA	San Francisco
34)	KLZ	560	U3	5	5	CO	Denver
35)	WQAM	560	U1	5	1	FL	Miami
36)	WIND	560	U4	5	5	IL	Chicago
37)	WMIK	560	U1	2.5	0.08	KY	Middlesboro
40)	WHYN	560	U3	5	1	MA	Springfield
39)	WFRB	560	U4	5	0.05	MD	Frostburg
38)	WGAN	560	U4	5	5	ME	Portland
41)	WEBC	560	U4	5	5	MN	Duluth
42)	KWTO	560	U4	5	4	MO	Springfield
43)	KMON	560	U2	5	5	MT	Great Falls
44)	WFIL	560	U4	5	5	PA	Philadelphia
45)	WVOC	560	U4	5	5	SC	Columbia
47)	WNSR	560	U4	4.5	0.07	TN	Brentwood
46)	WHBQ	560	U4	5	1	TN	Memphis
48)	KLVI	560	U2	5	5	TX	Beaumont
49)	KPQ	560	U2	5	5	WA	Wenatchee
50)	WJLS	560	U2	4.5	0.47	WV	Beckley
51)	WAAX	570	U2	5	0.5	AL	Gadsden
53)	KCFJ	570	U1	5	0.04	CA	Alturas
52)	KLAC	570	U2	5	5	CA	Los Angeles
54)	WTBN	570	U4	5	5	FL	Pinellas Park
55)	WTNT	570	U2	5	1	MD	Bethesda
59)	WWNC	570	U1	5	5	NC	Asheville
56)	KSNM	570	U1	5	0.15	NM	Las Cruces
57)	WMCA	570	U3	5	5	NY	New York
58)	WSYR	570	U4	5	5	NY	Syracuse
60)	WKBN	570	U2	5	5	OH	Youngstown
61)	WNAX	570	U2	5	5	SD	Yankton
62)	KLIF	570	U4	5	5	TX	Dallas
63)	KACP	570	U4	5	5	UT	Salt Lake City
64)	KVI	570	U1	5	5	WA	Seattle
65)	KRSA	580	U3	5	5	AK	Petersburg
66)	KSAZ	580	U2	5	0.39	AZ	Marana
67)	KMJ	580	U3	50	50	CA	Fresno
68)	KUBC	580	U2	5	1	CO	Montrose
69)	WDBO	580	U2	5	5	FL	Orlando
70)	WGAC	580	U2	5	0.84	GA	Augusta
71)	KIDO	580	U2	5	5	ID	Nampa
72)	WILL	580	U3	5	0.1	IL	Urbana
73)	WIBW	580	U2	5	5	KS	Topeka
74)	KJMJ	580	U2	5	1	LA	Alexandria
75)	WTAG	580	U4	5	5	MA	Worcester
76)	WTCM	580	U4	50	1.1	MI	Traverse City
77)	WKSK	580	U1	5	0.03	NC	West Jefferson
78)	WHP	580	U2	5	5	PA	Harrisburg
80)	WKTY	580	U4	5	0.74	WI	La Crosse
79)	WCHS	580	U2	5	5	WV	Charleston
81)	KHAR	590	U1	5	5	AK	Anchorage
82)	KZHS	590	U1	5	0.06	AR	Hot Springs
84)	KTIE	590	U4	2.5	0.96	CA	San Bernardino
83)	KTHO	590	U2	5	0.5	CA	South Lake Tahoe
85)	WDIZ	590	U2	1.7	2.5	FL	Panama City
86)	WDWD	590	U2	12	4.5	GA	Atlanta
88)	KID	590	U2	5	1	ID	Idaho Falls
89)	WVLK	590	U4	5	1	KY	Lexington
90)	WEZE	590	U3	5	5	MA	Boston
91)	WJMS	590	U2	5	1	MI	Ironwood
92)	WKZO	590	U4	5	5	MI	Kalamazoo
95)	WGTM	590	U4	5	5	NC	Wilson
93)	KXSP	590	U1	5	5	NE	Omaha
94)	WROW	590	U4	5	1	NY	Albany
96)	KUGN	590	U2	5	5	OR	Eugene
97)	WARM	590	U3	5	5	PA	Scranton
98)	KLBJ	590	U2	5	1	TX	Austin
99)	KSUB	590	U2	5	1	UT	Cedar City
100)	KQNT	590	U1	5	5	WA	Spokane
101)	KOGO	600	U4	5	5	CA	San Diego
102)	KCOL	600	U4	5	0.5	CO	Wellington
103)	WBWL	600	U2	5	5	FL	Jacksonville
104)	WMT	600	U2	5	5	IA	Cedar Rapids
105)	WKYH	600	U1	5	0.04	KY	Paintsville
107)	WCAO	600	U3	5	5	MD	Baltimore
106)	WFST	600	U1	5	0.12	ME	Caribou
108)	KGEZ	600	U4	5	1	MT	Kalispell
109)	WSJS	600	U4	5	5	NC	Winston-Salem

MW	Call	kHz	Ant.	D	N	Sta	City of License
110)	KSJB	600	U3	5	5	ND	Jamestown
111)	WREC	600	U2	5	5	TN	Memphis
112)	KROD	600	U2	5	5	TX	El Paso
113)	KTBB	600	U4	5	2.5	TX	Tyler
114)	WAGG	610	U2	5	1	AL	Birmingham
115)	KAVL	610	U4	4.9	4	CA	Lancaster
116)	KEAR	610	U1	5	5	CA	San Francisco
117)	KVLE	610	U1	5	0.21	CO	Vail
118)	WIOD	610	U4	5	5	FL	Miami
119)	KDAL	610	U2	5	5	MN	Duluth
120)	KCSP	610	U1	5	5	MO	Kansas City
123)	WFNZ	610	U4	5	1	NC	Charlotte
121)	WGIR	610	U2	5	1	NH	Manchester
122)	KNML	610	U2	5	5	NM	Albuquerque
124)	WTVN	610	U2	5	5	OH	Columbus
125)	KRTA	610	U4	2.5	5	OR	Medford
126)	WIP	610	U3	5	5	PA	Philadelphia
127)	KILT	610	U4	5	5	TX	Houston
128)	KVNU	610	U2	10	1	UT	Logan
129)	WVBE	610	U4	5	1	VA	Roanoke
130)	KONA	610	U4	5	5	WA	Kennewick
132)	KGTL	620	U1	5	5	AK	Homer
131)	WJHX	620	U1	5	0.09	AL	Lexington
133)	KTAR	620	U2	5	5	AZ	Phoenix
134)	KJOL	620	U1	5	0.07	CO	Grand Junction
135)	WDAE	620	U2	5.6	5.5	FL	Saint Petersburg
136)	WTRP	620	U1	2.5	0.12	GA	La Grange
138)	WZON	620	U2	5	5	ME	Bangor
139)	WJDX	620	U2	5	1	MS	Jackson
142)	WDNC	620	U4	5	1	NC	Durham
140)	WSNR	620	U4	3	7.6	NJ	Jersey City
141)	WHEN	620	U2	5	1	NY	Syracuse
143)	KPOJ	620	U2	25	10	OR	Portland
144)	WKHB	620	U1	5.5	0.05	PA	Irwin
145)	WGCV	620	U1	2.5	0.12	SC	Cayce
146)	WRJZ	620	U2	5	5	TN	Knoxville
147)	KMKI	620	U2	5	4.5	TX	Plano
148)	WVMT	620	U4	5	5	VT	Burlington
150)	WTMJ	620	U4	50	10	WI	Milwaukee
149)	WWNR	620	U1	5	0.02	WV	Beckley
151)	KJNO	630	U1	5	1	AK	Juneau
152)	KIAM	630	U1	10	3.1	AK	Nenana
153)	KHOW	630	U4	5	5	CO	Denver
154)	WMAL	630	U4	10	5	DC	Washington

MW	Call	kHz	Ant.	D	N	Sta	City of License
155)	WBMQ	630	U1	4.8	0.04	GA	Savannah
156)	KFXD	630	U4	5	5	ID	Boise
157)	WLAP	630	U4	5	1	KY	Lexington
158)	KJSL	630	U4	5	5	MO	Saint Louis
159)	KPLY	630	U2	5	1	NV	Reno
160)	KWRO	630	U1	5	0.04	OR	Coquille
161)	WPRO	630	U2	5	5	RI	Providence
162)	KSLR	630	U4	5	4.3	TX	San Antonio
163)	KCIS	630	U4	5	2.5	WA	Edmonds
164)	KYUK	640	U1	10	10	AK	Bethel
165)	KFI	640	U4	50	50	CA	Los Angeles
166)	WMEN	640	U2	7.5	0.46	FL	Royal Palm Beach
167)	WGST	640	U4	50	1	GA	Atlanta
168)	WOI	640	U2	5	1	IA	Ames
169)	KTIB	640	U4	5	1	LA	Thibodaux
170)	WNNZ	640	U4	50	1	MA	Westfield
171)	KGVW	640	U4	10	1	MT	Belgrade
173)	WFNC	640	U1	10	1	NC	Fayetteville
172)	WWJZ	640	U4	50	0.95	NJ	Mount Holly
174)	WHLO	640	U4	5	0.5	OH	Akron
175)	WWLS	640	U4	5	1	OK	Moore
177)	WXSM	640	U2	10	0.81	TN	Blountville
176)	WCRV	640	U2	50	0.48	TN	Collierville
178)	KENI	650	U1	50	50	AK	Anchorage
179)	KSTE	650	U4	21.4	0.92	CA	Rancho Cordova
181)	WNMT	650	U2	10	1	MN	Nashwauk
182)	WSM	650	U1	50	50	TN	Nashville
183)	KMTI	650	U4	10	0.9	UT	Manti
184)	KGAB	650	U2	8.5	0.5	WY	Orchard Valley
186)	KFAR	660	U1	10	10	AK	Fairbanks
185)	WXQW	660	U2	10	0.85	AL	Fairhope
187)	KTNN	660	U2	50	50	AZ	Window Rock
188)	KWVE	660	U4	8	6	CA	Oildale
189)	WBHR	660	U4	10	0.5	MN	Sauk Rapids
191)	KEYZ	660	U4	5	5	ND	Williston
190)	WFAN	660	U1	50	50	NY	New York
192)	KXOR	660	U1	10	0.07	OR	Junction City
193)	WLFJ	660	D1	50		SC	Greenville
194)	KSKY	660	U4	20	0.7	TX	Balch Springs
195)	KAPS	660	U4	10	1	WA	Mount Vernon
197)	KDLG	670	U1	10	10	AK	Dillingham
196)	WYLS	670	D1	4.8		AL	York
198)	KWXI	670	D1	5		AR	Glenwood
199)	KIRN	670	U3	5	3	CA	Simi Valley
200)	KLTT	670	U4	50	1.4	CO	Commerce City
201)	WWFE	670	U4	50	1	FL	Miami
203)	KBOI	670	U2	50	50	ID	Boise
204)	WSCR	670	U4	50	50	IL	Chicago
205)	KMZQ	670	U4	30	0.6	NV	Las Vegas
206)	WIEZ	670	D1	5.4		PA	Lewistown
207)	WMTY	670	D1	2.5		TN	Farragut
208)	WRJR	670	U5	20	0.003	VA	Claremont
209)	KBRW	680	U1	10	10	AK	Barrow
210)	KNBR	680	U1	50	50	CA	San Francisco
211)	WCNN	680	U4	50	10	GA	North Atlanta
213)	WRKO	680	U4	50	50	MA	Boston
212)	WCBM	680	U4	50	20	MD	Baltimore
214)	WNZKn	680	N2	0	2.5	MI	Dearborn Heights
215)	WDBC	680	U4	10	1	MI	Escanaba
216)	KFEQ	680	U4	5	5	MO	Saint Joseph
217)	KKGR	680	D1	5		MT	East Helena
219)	WPTF	680	U2	50	50	NC	Raleigh
218)	WINR	680	U4	5	0.5	NY	Binghamton
220)	WMFS	680	U2	10	5	TN	Memphis
221)	KKYX	680	U2	50	10	TX	San Antonio
222)	KOMW	680	D1	5		WA	Omak
224)	WOGO	680	U4	2.5	0.5	WI	Hallie
223)	WKAZ	680	U4	10	0.22	WV	Charleston
225)	WJOX	690	U2	50	0.5	AL	Birmingham
226)	WADS	690	D3	3.2		CT	Ansonia
227)	WOKV	690	U2	50	25	FL	Jacksonville
229)	KGGF	690	U4	10	5	KS	Coffeyville
230)	WIST	690	U4	10	5	LA	New Orleans
231)	WNZK	690	D3	2.5		MI	Dearborn Heights
232)	KTSM	690	U4	10	10	TX	El Paso
233)	WZAP	690	U1	10	0.01	VA	Bristol
234)	WELD	690	U1	3	0.01	WV	Fisher
235)	KBYR	700	U1	10	10	AK	Anchorage
236)	KMBX	700	U1	2.5	0.7	CA	Soledad
238)	WTUB	700	D1	2.5		MA	Orange-Athol
237)	WDMV	700	D3	5		MD	Walkersville
239)	WLW	700	U1	50	50	OH	Cincinnati
240)	KGRV	700	U1	23	0.47	OR	Winston
241)	KSEV	700	U4	5	1	TX	Tomball
242)	KALL	700	U4	50	10	UT	North Salt Lake City
243)	KXLX	700	U2	10	0.6	WA	Airway Heights

MW	Call	kHz	Ant.	D	N	Sta	City of License
244)	KBMB	710	U4	22	3.9	AZ	Black Canyon City
245)	KFIA	710	U4	25	1	CA	Carmichael
246)	KSPN	710	U2	50	10	CA	Los Angeles
247)	KNUS	710	U3	5	5	CO	Denver
248)	WAQI	710	U4	50	50	FL	Miami
249)	WUFF	710	D1	2.5		GA	Eastman
250)	WEKC	710	L1	4.2		KY	Williamsburg
251)	KEEL	710	U4	50	5	LA	Shreveport
252)	KCMO	710	U4	10	5	MO	Kansas City
254)	WEGG	710	D1	2.5		NC	Rose Hill
255)	KXMR	710	U7	50	4	ND	Bismarck
253)	WOR	710	U4	50	50	NY	New York
256)	KGNC	710	U4	10	10	TX	Amarillo
257)	WFNR	710	D3	10		VA	Blacksburg
258)	KIRO	710	U2	50	50	WA	Seattle
259)	WDSM	710	U2	10	5	WI	Superior
260)	KOTZ	720	U1	10	10	AK	Kotzebue
261)	WRZN	720	U2	10	0.25	FL	Hernando
262)	WVCC	720	D1	7.97		GA	Hogansville
264)	WGN	720	U1	50	50	IL	Chicago
265)	WGCR	720	D1	25		NC	Pisgah Forest
266)	KDWN	720	U2	50	50	NV	Las Vegas
267)	KFIR	720	U1	5	0.18	OR	Sweet Home
268)	KSAH	720	U4	10	0.89	TX	Universal City
269)	WSTT	730	U1	5	0.02	GA	Thomasville
270)	KBSU	730	U4	15	0.5	ID	Boise
271)	WMTC	730	U3	5	0.05	KY	Vancleve
272)	WACE	730	U1	5	0.008	MA	Chicopee
273)	KURL	730	U1	5	0.23	MT	Billings
274)	WZGV	730	U1	10	0.19	NC	Cramerton
275)	WPIT	730	U1	5	0.02	PA	Pittsburgh
276)	WXTR	730	U1	8	0.02	VA	Alexandria
277)	WMSP	740	U4	10	0.23	AL	Montgomery
278)	KBRT	740	U3	10	0.11	CA	Avalon
279)	KCBS	740	U4	50	50	CA	San Francisco
280)	KVOR	740	U4	3.3	1.5	CO	Colorado Springs
281)	WSBR	740	U3	2.5	0.94	FL	Boca Raton
282)	WYGM	740	U4	50	50	FL	Orlando
283)	WNOP	740	U4	2.5	0.03	KY	Newport
285)	WPAQ	740	U1	10	0.007	NC	Mount Airy
286)	KVOX	740	U7	50	0.94	ND	Fargo
284)	WNYH	740	U3	5	0.04	NY	Huntington
287)	KRMG	740	U4	50	25	OK	Tulsa
288)	KTRH	740	U4	50	50	TX	Houston
289)	WDGY	740	D3	2.5		WI	Hudson
290)	KFQD	750	U1	50	50	AK	Anchorage
291)	WSB	750	U1	50	50	GA	Atlanta
292)	WNDZ	750	D3	15		IN	Portage
293)	KBNN	750	D1	5		MO	Lebanon
294)	KERR	750	U2	50	1	MT	Polson
295)	KMMJ	750	L3	10.5		NE	Grand Island
296)	KHWG	750	U1	10	0.25	NV	Fallon
297)	KXL	750	U4	50	20	OR	Portland
298)	KAMA	750	U4	10	1	TX	El Paso
299)	KOAL	750	U2	10	6.8	UT	Price
300)	KMTL	760	D1	10		AR	Sherwood
301)	KFMB	760	U2	5	50	CA	San Diego
302)	KKZN	760	U4	50	1	CO	Thornton
304)	WLCC	760	U4	10	1	FL	Brandon
303)	WEFL	760	U4	3	1.5	FL	Tequesta
306)	KCCV	760	U4	6	0.2	KS	Overland Park
307)	WVNE	760	U1	25		MA	Leicester
308)	WJR	760	U1	50	50	MI	Detroit
310)	WCIS	760	D1	3.5		NC	Morganton
309)	WCHP	760	U4	35	0.01	NY	Champlain
311)	KTKR	760	U4	50	1	TX	San Antonio
313)	KCHU	770	U4	9.7	9.7	AK	Valdez
312)	WVNN	770	U2	7	0.25	AL	Athens
314)	KCBC	770	U4	50	4.1	CA	Manteca
315)	WWCN	770	U4	10	0.63	FL	North Fort Myers
316)	KUOM	770	D1	5		MN	Minneapolis
317)	KATL	770	U2	10	1	MT	Miles City
321)	WLWL	770	D1	5		NC	Rockingham
318)	KKOB	770	U2	50	50	NM	Albuquerque
320)	WABC	770	U1	50	50	NY	New York
319)	WTOR	770	D3	13		NY	Youngstown
322)	KAAM	770	U4	10	1	TX	Garland
323)	WYRV	770	D1	5		VA	Cedar Bluff
324)	KTTH	770	U4	50	5	WA	Seattle
326)	KNOM	780	U1	25	14	AK	Nome
325)	WZZX	780	D1	5		AL	Lineville
327)	KAZM	780	U2	5	0.25	AZ	Sedona
328)	WBBM	780	U1	50	50	IL	Chicago
329)	WXME	780	U1	5	0.06	ME	Monticello
330)	WTME	780	U1	10	0.01	ME	Rumford
331)	WIIN	780	D1	5		MS	Ridgeland

MW	Call	kHz	Ant.	D	N	Sta	City of License	MW	Call	kHz	Ant.	D	N	Sta	City of License
333)	WCKB	780	U1	7	0.001	NC	Dunn	421)	KICY	850	U10	50	50	AK	Nome
334)	WWOL	780	D1	10		NC	Forest City	420)	WXJC	850	U4	50	1	AL	Birmingham
332)	KKOH	780	U2	50	50	NV	Reno	422)	KOA	850	U1	50	50	CO	Denver
335)	WAVA	780	D1	12		VA	Arlington	423)	WREF	850	D1	2.5		CT	Ridgefield
337)	KCAM	790	U1	5	5	AK	Glennallen	424)	WRUF	850	U2	5	5	FL	Gainesville
336)	WTSK	790	U1	5	0.03	AL	Tuscaloosa	425)	WFTL	850	U4	5	1	FL	West Palm Beach
339)	KURM	790	U2	5	0.5	AR	Rogers	427)	WAIT	850	D3	2.5		IL	Crystal Lake
338)	KNST	790	U4	5	0.5	AZ	Tucson	428)	WEEI	850	U4	50	50	MA	Boston
342)	KFPT	790	U4	5	2.5	CA	Clovis	429)	WWJC	850	D1	10		MN	Duluth
340)	KWSW	790	U1	5	0.11	CA	Eureka	431)	KFUO	850	L1	5	5	MO	Clayton
341)	KABC	790	U2	5	5	CA	Los Angeles	430)	WQST	850	D3	10		MS	Forest
344)	WLBE	790	U2	5	1	FL	Leesburg	432)	WKIX	850	U2	10	5	NC	Raleigh
343)	WAXY	790	U4	5	5	FL	South Miami	433)	WKNR	850	U4	50	4.7	OH	Cleveland
345)	WQXI	790	U2	28	1	GA	Atlanta	434)	WKGE	850	U3	10	10	PA	Johnstown
347)	KXXX	790	U1	5	0.02	KS	Colby	435)	WKVL	850	D3	50		TN	Knoxville
348)	WKRD	790	U4	5	1	KY	Louisville	436)	KJON	850	D3	5		TX	Carrollton
349)	WSGW	790	U4	5	1	MI	Saginaw	437)	KEYH	850	U4	10	0.18	TX	Houston
350)	KGHL	790	U1	5	1.8	MT	Billings	438)	WTAR	850	U4	50	25	VA	Norfolk
351)	WBLO	790	U1	2.5	0.02	NC	Thomasville	439)	KHHO	850	U4	10	1	WA	Tacoma
352)	KFGO	790	U2	5	5	ND	Fargo	440)	KTRB	860	U2	50	50	CA	San Francisco
353)	WAEB	790	U4	3.6	1.5	PA	Allentown	441)	WGUL	860	U4	5	1.5	FL	Dunedin
354)	WPRV	790	U2	5	5	RI	Providence	443)	WAEC	860	U12	5	0.5	GA	Atlanta
356)	WETB	790	U1	5	0.07	TN	Johnson City	442)	WDMG	860	U2	5	5	GA	Douglas
355)	WMC	790	U2	5	5	TN	Memphis	444)	KKOW	860	U2	10	5	KS	Pittsburg
358)	KBME	790	U4	5	5	TX	Houston	446)	WSBS	860	U1	2.7	0.004	MA	Great Barrington
357)	KFYO	790	U4	5	1	TX	Lubbock	445)	WBGR	860	U4	2.5	0.06	MD	Baltimore
359)	WNIS	790	U3	5	5	VA	Norfolk	447)	KPAM	860	U2	50	15	OR	Troutdale
360)	KGMI	790	U2	5	1	WA	Bellingham	448)	WWDB	860	D3	10		PA	Philadelphia
361)	KJRB	790	U4	5	3.8	WA	Spokane	449)	KONO	860	U2	5	0.9	TX	San Antonio
362)	WAYY	790	U2	5	5	WI	Eau Claire	450)	KKAT	860	U1	10	0.19	UT	Salt Lake City
363)	KINY	800	U1	10	7.6	AK	Juneau	451)	WOAY	860	U1	10	0.01	WV	Oak Hill
364)	KBRV	800	U1	5	0.02	ID	Soda Springs	453)	KSKO	870	U1	10	10	AK	McGrath
365)	WNNW	800	U1	3	0.24	MA	Lawrence	452)	WQRX	870	D1	10		AL	Valley Head
366)	WVAL	800	U4	2.6	0.85	MN	Sauk Rapids	454)	KRLA	870	U4	50	3	CA	Glendale
367)	WTMR	800	U4	5	0.5	NJ	Camden	455)	WWL	870	U3	50	50	LA	New Orleans
368)	KQCV	800	U4	2.5	1	OK	Oklahoma City	456)	WLVP	870	U4	10	1	ME	Gorham
369)	WSVS	800	U1	10	0.27	VA	Crewe	457)	WKAR	870	D3	10		MI	East Lansing
371)	WDUX	800	U3	5	0.5	WI	Waupaca	458)	KPRM	870	U2	25	1	MN	Park Rapids
370)	WVHU	800	U1	5	0.18	WV	Huntington	461)	WTCG	870	D1	5		NC	Mount Holly
372)	WCKA	810	U4	50	0.5	AL	Jacksonville	459)	KLSQ	870	U2	5	0.43	NV	Whitney
373)	KGO	810	U3	50	50	CA	San Francisco	460)	WHCU	870	U2	5	1	NY	Ithaca
374)	WEUS	810	U4	10	0.4	FL	Orlovista	462)	WPWT	870	D1	10		TN	Colonial Heights
375)	WTHV	810	D1	2.5		GA	Hahira	463)	KFLD	870	U1	10	0.25	WA	Pasco
376)	WEKG	810	D1	5		KY	Jackson	464)	KLRG	880	U2	50	0.22	AR	Sheridan
377)	WMJH	810	D1	3.6		MI	Rockford	465)	KKMC	880	U4	10	10	CA	Gonzales
379)	WHB	810	U2	50		MO	Kansas City	466)	WZAB	880	U4	4	5	FL	Sweetwater
378)	WSJC	810	U2	50	0.5	MS	Magee	467)	WBKZ	880	D1	5		GA	Jefferson
380)	KSWV	810	U1	5	0.01	NM	Santa Fe	468)	KJJR	880	U1	10	0.5	MT	Whitefish
381)	WGY	810	U1	50	50	NY	Schenectady	472)	WPEK	880	D1	5		NC	Fairview
382)	WQIZ	810	D1	5		SC	Saint George	469)	KRVN	880	U2	50	50	NE	Lexington
383)	KBHB	810	U1	25	0.06	SD	Sturgis	470)	KHAC	880	U1	10	0.43	NM	Tse Bonito
384)	WMGC	810	U1	5	0.006	TN	Murfreesboro	471)	WCBS	880	U1	50	50	NY	New York
385)	WPIN	810	D1	4.2		VA	Dublin	473)	WRFD	880	D1	23		OH	Columbus
386)	KTBI	810	D1	50		WA	Ephrata	474)	KWIP	880	U1	5	1	OR	Dallas
387)	KCBF	820	U1	10	10	AK	Fairbanks	475)	WMDB	880	U1	2.5	0.002	TN	Nashville
388)	WWBA	820	U4	50	1	FL	Largo	476)	KJOJ	880	U4	10	1	TX	Conroe
389)	WCPT	820	U2	5	1.5	IL	Willow Springs	477)	KIXI	880	U4	50	10	WA	Mercer Island-Seattle
390)	WWFD	820	U2	4.3	0.43	MD	Frederick	478)	WMEQ	880	U2	10	0.21	WI	Menomonie
391)	WWLZ	820	U2	4.1	0.85	NY	Horseheads	480)	KBBI	890	U1	10	10	AK	Homer
392)	WNYC	820	U4	10	1	NY	New York	479)	WYAM	890	D1	2.5		AL	Hartselle
393)	WOSU	820	U2	5	0.79	OH	Columbus	481)	KLFF	890	U2	5	5	CA	Arroyo Grande
394)	WBAP	820	U1	50	50	TX	Fort Worth	482)	WJTP	890	D1	5		GA	Lithia Springs
395)	KUTR	820	U8	5	2.5	UT	Taylorsville	483)	KDJQ	890	U2	50	0.25	ID	Meridian
396)	WGGM	820	U4	10	1	VA	Chester	484)	WLS	890	U1	50	50	IL	Chicago
397)	KGNW	820	U4	50	5	WA	Burien-Seattle	485)	WAMG	890	U4	25	6	MA	Dedham
398)	KFLT	830	U2	50	1	AZ	Tucson	486)	WHJA	890	D1	10		MS	Laurel
399)	KNCO	830	U2	5	5	CA	Grass Valley	487)	WBAJ	890	D1	50		SC	Blythewood
400)	KLAA	830	U2	50	20	CA	Orange	489)	KVOZ	890	U2	10	1	TX	Del Mar Hills
402)	WFNO	830	U4	5	0.75	LA	Norco	488)	KTXV	890	U4	20	0.25	TX	Mabank
403)	WCRN	830	U4	50	50	MA	Worcester	490)	KDXU	890	U2	10	10	UT	Saint George
404)	WCCO	830	U1	50	50	MN	Minneapolis	491)	WKNV	890	D3	10		VA	Fairlawn
405)	KOTC	830	D1	10		MO	Kennett	492)	KZPA	890	U1	5	5	AK	Fort Yukon
406)	WTRU	830	U4	50	10	NC	Kernersville	493)	WJWL	900	U4	10.5	1.08	DE	Georgetown
407)	WEEU	830	U4	20	6	PA	Reading	494)	WMOP	900	U1	2.7	0.02	FL	Ocala
408)	KUYO	830	D1	25		WY	Evansville	495)	WJLG	900	U1	4.35	0.15	GA	Savannah
409)	WBHY	840	D1	10		AL	Mobile	497)	WLSI	900	U1	3.5	0.12	KY	Pikeville
410)	KMPH	840	U4	5	5	CA	Modesto	498)	KTIS	900	U4	50	0.5	MN	Minneapolis
411)	WHGH	840	D1	10		GA	Thomasville	499)	WYCV	900	U1	2.5	0.25	NC	Granite Falls
412)	WHAS	840	U1	50	50	KY	Louisville	500)	WCPA	900	U4	2.5	0.5	PA	Clearfield
413)	KWDF	840	D1	8		LA	Ball	501)	WKDA	900	U1	5	0.1	TN	Lebanon
414)	KTIC	840	D1	5		NE	West Point	502)	KREH	900	U1	5	0.01	TX	Pecan Grove
415)	KXNT	840	U4	50	25	NV	North Las Vegas	503)	WKDW	900	U1	2.5	0.12	VA	Staunton
416)	WCEO	840	D3	50		SC	Columbia	504)	KIYU	910	U1	5	5	AK	Galena
417)	KVJY	840	U4	5	1	TX	Pharr	506)	KLCN	910	U1	5	0.08	AR	Blytheville
418)	WKTR	840	D3	8.2		VA	Earlysville	505)	KGME	910	U2	5	5	AZ	Phoenix
419)	KMAX	840	U1	10	0.28	WA	Colfax	509)	KECR	910	U4	5	5	CA	El Cajon

MW	Call	kHz	Ant.	D	N	Sta	City of License
508)	KNEW	910	U2	20	5	CA	Oakland
507)	KOXR	910	U3	5	1	CA	Oxnard
510)	KPOF	910	U1	5	1	CO	Denver
511)	WLAT	910	U2	5	5	CT	New Britain
512)	WTWD	910	U3	5	5	FL	Plant City
513)	WRFV	910	U2	5	5	GA	Valdosta
514)	WSUI	910	U2	5	4	IA	Iowa City
515)	WAEI	910	U2	5	5	ME	Bangor
516)	WFDF	910	U4	50	25	MI	Farmington Hills
517)	WALT	910	U1	5	1	MS	Meridian
519)	WSRP	910	U2	5	5	NC	Jacksonville
520)	KCJB	910	U4	5	1	ND	Minot
518)	KBIM	910	U2	5	0.5	NM	Roswell
521)	WLTP	910	U5	5	0.04	OH	Marietta
522)	WAVL	910	U4	5	0.06	PA	Apollo
523)	WSBA	910	U3	5	1	PA	York
524)	WOLI	910	U4	3.6	0.89	SC	Spartanburg
525)	WJCW	910	U2	5	1	TN	Johnson City
526)	WEPG	910	U1	5	0.09	TN	South Pittsburg
527)	KRIO	910	U1	5	5	TX	McAllen
528)	KWDZ	910	U4	5	1	UT	Salt Lake City
529)	WRNL	910	U2	5	1.5	VA	Richmond
530)	KKSN	910	U4	5	5	WA	Vancouver
531)	WHSM	910	U1	5	0.07	WI	Hayward
532)	KSRM	920	U1	5	5	AK	Soldotna
533)	KARN	920	U2	5	5	AR	Little Rock
535)	KVIN	920	U4	0.5	2.5	CA	Ceres
534)	KPSI	920	U4	5	1	CA	Palm Springs
536)	KLMR	920	U2	5	0.5	CO	Lamar
537)	WDMC	920	U4	5	1	FL	Melbourne
538)	WGKA	920	U1	14	0.49	GA	Atlanta
540)	KYFR	920	U4	5	2.5	IA	Shenandoah
539)	WBAA	920	U2	5	1	IN	West Lafayette
541)	WTCW	920	U1	4.2	0.04	KY	Whitesburg
542)	KDHL	920	U4	5	5	MN	Faribault
547)	WPCM	920	U2	5	0.05	NC	Burlington-Graham
543)	KBAD	920	U2	5	0.5	NV	Las Vegas
544)	KIHM	920	U4	4.6	0.85	NV	Reno
545)	WGHQ	920	U4	5	0.07	NY	Kingston
546)	WIRD	920	U1	5	0.08	NY	Lake Placid
548)	WHJJ	920	U2	5	5	RI	Providence
549)	KKLS	920	U3	5	0.11	SD	Rapid City
550)	KYST	920	U4	5	1	TX	Texas City
551)	KVEL	920	U4	5	1	UT	Vernal
552)	WURA	920	U4	7	0.97	VA	Quantico
554)	KGTK	920	U2	3	0.007	WA	Olympia
553)	KXLY	920	U1	20	5	WA	Spokane
556)	WOKY	920	U4	5	1	WI	Milwaukee
555)	WMMN	920	U2	5	0.2	WV	Fairmont
559)	KTKN	930	U1	5	1	AK	Ketchikan
560)	KNSA	930	U1	2.5	2.5	AK	Unalakleet
557)	WEZZ	930	U2	5	0.04	AL	Monroeville
558)	WGAD	930	U2	5	0.5	AL	Rainbow City
561)	KAPR	930	U1	2.5	0.07	AZ	Douglas
562)	KAFF	930	U1	5	0.03	AZ	Flagstaff
563)	KHJ	930	U2	5	5	CA	Los Angeles
565)	KIUP	930	U1	5	0.1	CO	Durango
564)	KRKY	930	U1	4.5	0.12	CO	Granby
566)	WFXJ	930	U2	5	5	FL	Jacksonville
567)	WLSS	930	U4	5	3	FL	Sarasota
568)	WMGR	930	U2	5	0.5	GA	Bainbridge
569)	KSEI	930	U2	5	5	ID	Pocatello
570)	WTAD	930	U2	5	1	IL	Quincy
571)	WAUR	930	U4	2.5	4.2	IL	Sandwich
572)	WKCT	930	U2	5	0.5	KY	Bowling Green
573)	WFMD	930	U4	5	2.5	MD	Frederick
574)	WBCK	930	U1	5	1	MI	Battle Creek
575)	KKIN	930	U1	2.5	0.36	MN	Aitkin
577)	KWOC	930	U2	5	0.5	MO	Poplar Bluff
576)	WSFZ	930	U2	3.8	3.1	MS	Jackson
578)	KMPT	930	U2	5	1	MT	East Missoula
584)	WYFQ	930	U2	5	1	NC	Charlotte
583)	WDLX	930	U2	5	1	NC	Washington
579)	KOGA	930	U4	5	0.5	NE	Ogallala
580)	WGIN	930	U2	5	5	NH	Rochester
581)	WPAT	930	U2	5	5	NJ	Paterson
582)	WBEN	930	U2	5	5	NY	Buffalo
585)	WKY	930	U2	5	5	OK	Oklahoma City
586)	KAGI	930	U1	5	0.12	OR	Grants Pass
587)	KSDN	930	U4	5	1	SD	Aberdeen
588)	WSEV	930	U1	5	0.14	TN	Sevierville
589)	KLUP	930	U2	5	1	TX	Terrell Hills
590)	WLLL	930	U1	9	0.04	VA	Lynchburg
591)	KYAK	930	U1	10	0.12	WA	Yakima
593)	WLBL	930	U1	5	0.07	WI	Auburndale
592)	WRVC	930	U2	5	1	WV	Huntington

MW	Call	kHz	Ant.	D	N	Sta	City of License
594)	KROE	930	U1	5	0.11	WY	Sheridan
595)	KYNO	940	U4	50	50	CA	Fresno
596)	WINZ	940	U2	50	10	FL	Miami
597)	WMAC	940	U2	50	10	GA	Macon
600)	KPSZ	940	U4	10	5	IA	Des Moines
599)	WMIX	940	U4	5	1.5	IL	Mount Vernon
601)	WYLD	940	U4	10	0.5	LA	New Orleans
602)	WIDG	940	U1	5	0.004	MI	Saint Ignace
603)	WCPC	940	U4	50	0.25	MS	Houston
605)	WKYK	940	U2	4.6	0.25	NC	Burnsville
604)	KVSH	940	U1	5	0.01	NE	Valentine
606)	KICE	940	U4	10	0.06	OR	Bend
607)	WECO	940	U1	5	0.01	TN	Wartburg
609)	KIXZ	940	U4	5	1	TX	Amarillo
608)	KTFS	940	U1	2.5	0.01	TX	Texarkana
610)	KOBY	940	U1	10	0.03	UT	Cedar City
612)	WNRG	940	U1	5	0.01	VA	Grundy
611)	WKGM	940	U2	10	3.1	VA	Smithfield
613)	KXJK	950	U1	5	0.08	AR	Forrest City
614)	KAHI	950	U4	5	5	CA	Auburn
615)	KRWZ	950	U3	5	5	CO	Denver
616)	WTLN	950	U2	12	5	FL	Orlando
617)	WGTA	950	U1	5	0.11	GA	Summerville
618)	WGOV	950	U1	3.5	0.06	GA	Valdosta
623)	KOEL	950	U4	5	0.5	IA	Oelwein
620)	KNJY	950	U4	3.5	0.03	ID	Boise
619)	KOZE	950	U4	5	1	ID	Lewiston
621)	WNTD	950	U2	1	5	IL	Chicago
622)	WXLW	950	U3	5	0.11	IN	Indianapolis
625)	WROL	950	U3	5	0.09	MA	Boston
624)	WCTN	950	U3	2.5	0.03	MD	Potomac-Cabin John
626)	WWJ	950	U4	50	50	MI	Detroit
628)	KWOS	950	U1	5	0.5	MO	Jefferson City
627)	WHSY	950	U1	5	0.06	MS	Hattiesburg
629)	KMTX	950	U2	5	5	MT	Helena
630)	KNFT	950	U1	5	0.22	NM	Bayard
631)	KDCE	950	U1	4.2	0.08	NM	Espanola
632)	WIBX	950	U3	5	5	NY	Utica
633)	KTBR	950	U1	3.4	0.02	OR	Roseburg
634)	WPEN	950	U2	25	21	PA	Philadelphia
635)	WJKB	950	U4	10	6	SC	Moncks Corner
636)	WORD	950	U1	5	5	SC	Spartanburg
637)	WAKM	950	U1	5	0.08	TN	Franklin
638)	KPRC	950	U2	5	5	TX	Houston
639)	KJTV	950	U4	5	0.5	TX	Lubbock
640)	WXGI	950	U4	3.9	0.04	VA	Richmond
641)	KJR	950	U4	50	50	WA	Seattle
642)	WBES	950	U2	5	1	WV	Charleston
644)	WERC	960	U2	5	5	AL	Birmingham
643)	WLPR	960	U1	6	0.03	AL	Prichard
646)	KCGS	960	U1	5	0.04	AR	Marshall
645)	KKNT	960	U4	5	5	AZ	Phoenix
647)	KIXW	960	U1	5	0.02	CA	Apple Valley
648)	KKGN	960	U3	5	5	CA	Oakland
649)	WELI	960	U2	5	5	CT	New Haven
650)	WJYZ	960	U4	5	0.39	GA	Albany
651)	WRFC	960	U2	5	2.5	GA	Athens
653)	KMA	960	U2	5	5	IA	Shenandoah
652)	WSBT	960	U4	5	5	IN	South Bend
654)	WPRT	960	U1	3.8	0.01	KY	Prestonsburg
656)	WFGL	960	U4	2.5	1	MA	Fitchburg
655)	WTGM	960	U4	5	5	MD	Salisbury
657)	WHAK	960	U1	5	0.13	MI	Rogers City
658)	KLTF	960	U1	5	0.03	MN	Little Falls
659)	KZIM	960	U2	5	0.5	MO	Cape Girardeau
660)	KFLN	960	U1	5	0.09	MT	Baker
664)	WCRU	960	U4	10	0.5	NC	Dallas
665)	WRNS	960	U2	5	1	NC	Kinston
661)	KNEB	960	U4	5	0.35	NE	Scottsbluff
662)	KNDN	960	U1	5	0.16	NM	Farmington
663)	WEAV	960	U4	5	5	NY	Plattsburgh
666)	KLAD	960	U2	5	5	OR	Klamath Falls
668)	WHYL	960	U1	5	0.02	PA	Carlisle
667)	WATS	960	U1	5	0.05	PA	Sayre
669)	KGKL	960	U2	5	1	TX	San Angelo
670)	KOVO	960	U2	5	1	UT	Provo
671)	WFIR	960	U2	5	5	VA	Roanoke
672)	KALE	960	U2	5	1	WA	Richland
675)	KFBX	970	U1	10	10	AK	Fairbanks
674)	WERH	970	D1	5		AL	Hamilton
673)	WTBF	970	U1	5	0.04	AL	Troy
676)	KVWM	970	U1	5	0.19	AZ	Show Low
677)	KHTY	970	U4	1	5	CA	Bakersfield
678)	KNWZ	970	U1	5	1	CA	Coachella
679)	KFEL	970	U1	3.2	0.18	CO	Pueblo
680)	WFLA	970	U4	25	11	FL	Tampa

MW	Call	kHz	Ant.	D	N	Sta	City of License
681)	WNIV	970	U1	5	0.03	GA	Atlanta
682)	WVOP	970	U1	4	0.06	GA	Vidalia
683)	KFTA	970	U2	2.5	0.9	ID	Rupert
684)	WFSR	970	U1	5	0.02	KY	Harlan
685)	WGTK	970	U4	5	5	KY	Louisville
686)	WZAN	970	U2	5	5	ME	Portland
687)	WZAM	970	U1	5	0.06	MI	Ishpeming
688)	KQAQ	970	U3	5	0.5	MN	Austin
689)	KBUL	970	U2	5	5	MT	Billings
694)	WYSE	970	U1	5	0.03	NC	Canton
695)	WDAY	970	U2	5	5	ND	Fargo
690)	KJLT	970	U1	5	0.05	NE	North Platte
692)	WNYM	970	U4	50	50	NJ	Hackensack
691)	KNUU	970	U4	5	0.5	NV	Paradise
693)	WNED	970	U3	5	5	NY	Buffalo
696)	WFUN	970	U1	5	1	OH	Ashtabula
697)	KCFO	970	U2	2.5	1	OK	Tulsa
698)	KXFD	970	U2	5	5	OR	Portland
699)	WBGG	970	U4	5	5	PA	Pittsburgh
700)	WJMX	970	U2	10	3	SC	Florence
701)	WKCI	970	U2	5	1	VA	Waynesboro
702)	KTTO	970	U2	5	1	WA	Spokane
703)	WHA	970	U1	5	0.05	WI	Madison
704)	KCAB	980	U1	5	0.03	AR	Dardanelle
706)	KINS	980	U2	5	0.5	CA	Eureka
705)	KFWB	980	U1	5	5	CA	Los Angeles
707)	KDBV	980	U4	10	10	CA	Salinas
708)	WTEM	980	U4	50	5	DC	Washington
711)	WDVH	980	U1	5	0.16	FL	Gainesville
709)	WRNE	980	U2	4	1	FL	Gulf Breeze
710)	WHSR	980	U4	5	2.2	FL	Pompano Beach
712)	WPGA	980	U1	2.6	0.08	GA	Perry
713)	KSPZ	980	U4	5	1	ID	Ammon
714)	KOKA	980	U1	5	0.07	LA	Shreveport
715)	WCAP	980	U4	5	5	MA	Lowell
716)	KKMS	980	U3	5	5	MN	Richfield
718)	KMBZ	980	U2	5	5	MO	Kansas City
717)	WAKK	980	U1	5	0.15	MS	McComb
721)	WAAV	980	U2	5	5	NC	Leland
719)	KVLV	980	D1	5		NV	Fallon
720)	WOFX	980	U2	5	5	NY	Troy
722)	WONE	980	U2	5	5	OH	Dayton
723)	WILK	980	U2	5	5	PA	Wilkes-Barre
724)	WBZK	980	U2	3	0.16	SC	York
725)	KDSJ	980	U2	5	1	SD	Deadwood
726)	WYFN	980	U2	5	5	TN	Nashville
727)	KRTX	980	U2	1	4	TX	Rosenberg-Richmond
728)	KSVC	980	U2	10	1	UT	Richfield
729)	WFHG	980	U2	5	1	VA	Bristol
730)	KBBO	980	U4	5	0.5	WA	Selah
732)	WCUB	980	U4	5	5	WI	Two Rivers
731)	WHAW	980	U1	25	0.04	WV	Lost Creek
733)	KTKT	990	U4	10	1	AZ	Tucson
734)	KATD	990	U4	10	5	CA	Pittsburg
735)	KTMS	990	U5	5	0.5	CA	Santa Barbara
736)	KRKS	990	U2	6.6	0.39	CO	Denver
737)	WXCT	990	U4	2.5	0.08	CT	Southington
739)	WMYM	990	U4	5	5	FL	Miami
738)	WDYZ	990	U4	50	14	FL	Orlando
741)	WDEO	990	U4	9.2	0.25	MI	Ypsilanti
742)	KRMO	990	U1	2.5	0.04	MO	Cassville
744)	WEEB	990	U1	10	0.02	NC	Southern Pines
743)	WDCX	990	U4	5	2.5	NY	Rochester
745)	WNTP	990	U4	50	10	PA	Philadelphia
746)	WNTW	990	U3	10	0.1	PA	Somerset
747)	WALE	990	U4	50	5	RI	Greenville
749)	WNML	990	U2	10	10	TN	Knoxville
748)	KWAM	990	U4	10	0.45	TN	Memphis
750)	KFCD	990	U4	7	0.92	TX	Farmersville
751)	WNRV	990	U1	5	0.01	VA	Narrows-Pearisburg
752)	KCEO	1000	U4	2.5	0.25	CA	Vista
753)	WYBT	1000	D1	5		FL	Blountstown
754)	WMVP	1000	U4	50	50	IL	Chicago
755)	WXTN	1000	D1	5		MS	Benton
756)	KKIM	1000	U1	10	0.03	NM	Albuquerque
757)	WLNL	1000	D1	5		NY	Horseheads
758)	KTOK	1000	U4	5	5	OK	Oklahoma City
759)	KXRB	1000	U4	10	0.1	SD	Sioux Falls
760)	WMUF	1000	D4	5		TN	Paris
761)	KOMO	1000	U2	50	50	WA	Seattle
762)	WCOC	1010	U1	5	0.04	AL	Dora
763)	KXXT	1010	U5	15	0.25	AZ	Tolleson
766)	KCHJ	1010	U4	5	1	CA	Delano
765)	KIQI	1010	U3	10	0.5	CA	San Francisco
764)	KXPS	1010	U4	3.6	0.4	CA	Thousand Palms
767)	KSIR	1010	U3	25	0.28	CO	Brush
768)	WJXL	1010	U4	50	30	FL	Jacksonville Beach
769)	WQYK	1010	U4	50	5	FL	Seffner
770)	WGUN	1010	U1	50	0.07	GA	Atlanta
772)	KXEN	1010	U4	50	0.5	MO	Saint Louis
771)	WMOX	1010	U4	10	1	MS	Meridian
775)	WFGW	1010	U7	50	0.5	NC	Black Mountain
773)	WCNL	1010	U1	10	0.03	NH	Newport
774)	WINS	1010	U3	50	50	NY	New York
776)	KOOR	1010	D1	4.5		OR	Milwaukie
777)	WHIN	1010	U1	5	0.04	TN	Gallatin
779)	KTNZ	1010	U4	5	0.5	TX	Amarillo
780)	KLAT	1010	U4	5	3.6	TX	Houston
778)	KBBW	1010	U4	10	2.5	TX	Waco
781)	KIHU	1010	U13	50	0.19	UT	Tooele
782)	WPMH	1010	U3	5	0.44	VA	Portsmouth
783)	KOAN	1020	U2	10	10	AK	Eagle River
784)	KTNQ	1020	U4	50	50	CA	Los Angeles
785)	WHDD	1020	D1	2.5		CT	Sharon
786)	WURN	1020	U4	8.9	0.98	FL	Kendall
787)	WSBX	1020	D1	10		GA	Ochlocknee
788)	KMMQ	1020	U4	50	1.4	NE	Plattsmouth
789)	KCKN	1020	U4	50	50	NM	Roswell
790)	KDKA	1020	U1	50	50	PA	Pittsburgh
791)	WRIX	1020	D1	10		SC	Homeland Park
792)	KDYK	1020	U5	4	0.4	WA	Union Gap
794)	KFAY	1030	U4	10	1	AR	Farmington
793)	KVOI	1030	U1	1		AZ	Cortaro
795)	KJDJ	1030	U1	2.5	0.7	CA	San Luis Obispo
796)	WONQ	1030	U2	45	1.7	FL	Oviedo
797)	WEBS	1030	U1	5	0.003	GA	Calhoun
798)	WNVR	1030	U7	10	0.12	IL	Vernon Hills
799)	KBUF	1030	U2	2.5	1.2	KS	Holcomb
801)	WBZ	1030	U3	50	50	MA	Boston
800)	WWGB	1030	D3	50		MD	Indian Head
802)	WUFL	1030	D3	5		MI	Sterling Heights
803)	WCTS	1030	U4	50	1	MN	Maplewood
804)	KCWJ	1030	U4	5	0.5	MO	Blue Springs
805)	WDRU	1030	D3	50		NC	Creedmoor
806)	WNOW	1030	D3	9.4		NC	Mint Hill
807)	KDUN	1030	U4	50	0.63	OR	Reedsport
808)	WGSF	1030	U12	50	1	TN	Memphis
809)	KCTA	1030	L1	50		TX	Corpus Christi
810)	KMAS	1030	U1	10	1	WA	Shelton
811)	WBGS	1030	D5	10		WV	Point Pleasant
812)	KTWO	1030	U2	50	50	WY	Casper
813)	KCBR	1040	D1	15		CO	Monument
815)	WLVJ	1040	U4	25	1.1	FL	Boynton Beach
814)	WHBO	1040	U2	3.6	0.42	FL	Pinellas Park
816)	WPBS	1040	D1	50		GA	Conyers
818)	WHO	1040	U1	50	50	IA	Des Moines
821)	WSGH	1040	U4	9.1	0.18	NC	Lewisville
819)	WNJE	1040	U7	15	1.5	NJ	Flemington
820)	WYSL	1040	U7	20	0.5	NY	Avon
822)	WJTB	1040	D1	5		OH	North Ridgeville
823)	WZSK	1040	D1	10		PA	Everett
824)	WKTI	1040	D1	10		TN	Powell
825)	KGGR	1040	D1	3.3		TX	Dallas
827)	KJPG	1050	U5	10	0.007	CA	Frazier Park
826)	KTCT	1050	U4	50	10	CA	San Mateo
829)	WJSB	1050	D1	3.1		FL	Crestview
828)	WROS	1050	U4	5	0.01	FL	Jacksonville
830)	WFAM	1050	U1	5	0.08	GA	Augusta
831)	WBQH	1050	U1	3.5	0.04	MD	Silver Spring
832)	WTKA	1050	U4	10	0.5	MI	Ann Arbor
833)	KLOH	1050	U4	9.38	0.43	MN	Pipestone
834)	KMTA	1050	U1	10	0.13	MT	Miles City
837)	WFSC	1050	U1	5	0.15	NC	Franklin
835)	WSEN	1050	U3	2.5	0.01	NY	Baldwinsville
836)	WEPN	1050	U3	50	50	NY	New York
838)	KORE	1050	U1	5	0.14	OR	Springfield-Eugene
839)	WHSC	1050	U3	5	0.47	SC	Conway
841)	WBRG	1050	U1	4	0.09	VA	Lynchburg
840)	WVXX	1050	U3	5	0.35	VA	Norfolk
843)	KEYF	1050	U1	5	0.26	WA	Dishman
842)	KBLE	1050	U1	5	0.44	WA	Seattle
844)	WADC	1050	U1	5	0.14	WV	Parkersburg
845)	KDUS	1060	U2	5	0.5	AZ	Tempe
846)	KTNS	1060	U1	5	0.02	CA	Oakhurst
847)	KRCN	1060	U1	50	0.11	CO	Longmont
848)	WIXC	1060	U1	50	5	FL	Titusville
849)	WKNG	1060	D1	50		GA	Tallapoosa
851)	KBGN	1060	D1	10		ID	Caldwell
852)	WMCL	1060	U3	2.5	0.002	IL	McLeansboro
853)	WLNO	1060	U4	50	5	LA	New Orleans
854)	WBIX	1060	U7	40	2.5	MA	Natick
855)	WHFB	1060	U1	5	0.001	MI	Benton Harbor

MW	Call	kHz	Ant.	D	N	Sta	City of License
857)	WXNC	1060	D1	4		NC	Monroe
856)	KKVV	1060	U1	5	0.04	NV	Las Vegas
858)	WILB	1060	D3	5		OH	Canton
859)	KYW	1060	U3	50	50	PA	Philadelphia
860)	KGFX	1060	U4	10	1	SD	Pierre
861)	KXPL	1060	D1	10		TX	El Paso
862)	KIJN	1060	D3	10		TX	Farwell
863)	KDYL	1060	U1	10	0.14	UT	South Salt Lake
864)	WAPI	1070	U2	50	5	AL	Birmingham
865)	KNX	1070	U1	50	50	CA	Los Angeles
866)	WNVY	1070	U1	15	0.02	FL	Cantonment
867)	WFRF	1070	D1	10		FL	Tallahassee
868)	WFNI	1070	U4	50	10	IN	Indianapolis
869)	KLIO	1070	U2	10	1	KS	Wichita
870)	KVKK	1070	U2	10	5	MN	Verndale
871)	KHMO	1070	U4	5	1	MO	Hannibal
872)	KATQ	1070	U1	5	0.05	MT	Plentywood
875)	WNCT	1070	U4	50	10	NC	Greenville
873)	WTWK	1070	D1	5		NY	Plattsburgh
874)	WSCP	1070	D1	2.5		NY	Sandy Creek-Pulaski
876)	WKOK	1070	U2	10	1	PA	Sunbury
877)	WCSZ	1070	U4	50	1.5	SC	Sans Souci
879)	WFLI	1070	U4	50	2.5	TN	Lookout Mountain
878)	WDIA	1070	U4	50	5	TN	Memphis
880)	KNTH	1070	U4	10	5	TX	Houston
881)	KWEL	1070	D1	2.5		TX	Midland
882)	WINA	1070	U2	5	5	VA	Charlottesville
884)	WTSO	1070	U4	10	5	WI	Madison
883)	WIWS	1070	D1	10		WV	Beckley
886)	KUDO	1080	U1	10	10	AK	Anchorage
885)	WKAC	1080	D1	5		AL	Athens
887)	KSCO	1080	U2	10	5	CA	Santa Cruz
888)	WTIC	1080	U2	50	50	CT	Hartford
889)	WHIM	1080	U4	50	10	FL	Coral Gables
890)	WHOO	1080	U7	19	0.19	FL	Kissimmee
891)	WFTD	1080	D4	50		GA	Marietta
893)	KVNI	1080	U2	10	1	ID	Coeur d' Alene
894)	WNWI	1080	U2	3	2.6	IL	Oak Lawn
895)	WKJK	1080	U4	10	1	KY	Louisville
896)	WKGX	1080	D1	5		NC	Lenoir
897)	KFXX	1080	U4	50	10	OR	Portland
898)	WWNL	1080	D4	50		PA	Pittsburgh
899)	WALD	1080	D1	9		SC	Johnsonville
900)	KRLD	1080	U2	50	50	TX	Dallas
901)	KSLL	1080	D1	10		UT	Price
902)	KAAY	1090	U2	50	50	AR	Little Rock
903)	KNCR	1090	D1	10		CA	Fortuna
904)	KMXA	1090	U4	50	0.5	CO	Aurora
905)	WFCV	1090	D4	2.5		IN	Fort Wayne
907)	WILD	1090	D1	4.8		MA	Boston
906)	WBAL	1090	U2	50	50	MD	Baltimore
908)	KEXS	1090	D4	8		MO	Excelsior Springs
909)	KBOZ	1090	U2	5	5	MT	Bozeman
910)	WTSB	1090	D1	9		NC	Selma
911)	KLWJ	1090	D1	2.5		OR	Umatilla
912)	WCZZ	1090	D1	5		SC	Greenwood
913)	WHGG	1090	D1	10		TN	Kingsport
914)	KVOP	1090	U4	5	0.5	TX	Plainview
915)	KPTK	1090	U4	50	50	WA	Seattle
916)	WAQE	1090	D1	5		WI	Rice Lake
917)	KFNX	1100	U4	50	1	AZ	Cave Creek
918)	KAFY	1100	U2	4.2	0.8	CA	Bakersfield
919)	KFAX	1100	U3	50	50	CA	San Francisco
920)	KNZZ	1100	U12	50	10	CO	Grand Junction
921)	WWWE	1100	D1	5		GA	Hapeville
922)	WCGA	1100	D1	10		GA	Woodbine
923)	WZFG	1100	U2	50	0.44	MN	Dilworth
924)	KKLL	1100	D1	5		MO	Webb City
925)	KWWN	1100	U4	22	2	NV	Las Vegas
926)	WHLI	1100	D3	10		NY	Hempstead
927)	WTAM	1100	U1	50	50	OH	Cleveland
928)	KDRY	1100	U2	11	1	TX	Alamo Heights
929)	WTWN	1100	D1	5		VT	Wells River
930)	WISS	1100	D1	2.5		WI	Berlin
932)	KAGV	1110	U1	10	10	AK	Big Lake
931)	WTOF	1110	D1	10		AL	Bay Minette
933)	KGFL	1110	D1	5		AR	Clinton
934)	KDIS	1110	U4	50	20	CA	Pasadena
935)	KLIB	1110	U4	5	0.5	CA	Roseville
936)	WTIS	1110	D1	5		FL	Tampa
938)	WMBI	1110	D1	4.2		IL	Chicago
939)	WUPE	1110	D3	5		MA	Pittsfield
940)	WJML	1110	D3	10	0.01	MI	Petoskey
944)	WBT	1110	U2	50	50	NC	Charlotte
941)	KFAB	1110	U2	50	50	NE	Omaha
942)	WCCM	1110	D3	5		NH	Salem
943)	KYKK	1110	D1	5		NM	Humble City
945)	WGNZ	1110	D4	2.5		OH	Fairborn
946)	KBND	1110	U12	10	5	OR	Bend
947)	WNAP	1110	D5	4.8		PA	Norristown
948)	WPMZ	1110	D3	5		RI	East Providence
949)	WSLV	1110	D1	2.5		TN	Ardmore
950)	KTEK	1110	D3	2.5		TX	Alvin
951)	KVTT	1110	D3	20		TX	Mineral Wells
952)	WYRM	1110	D3	50		VA	Norfolk
953)	KZSJ	1120	U1	5	0.15	CA	San Martin
954)	WUST	1120	D1	20		DC	Washington
955)	WBNW	1120	U4	5	1	MA	Concord
957)	KMOX	1120	U1	50	50	MO	Saint Louis
956)	WTWZ	1120	D1	10		MS	Clinton
958)	WSME	1120	D1	6		NC	Camp Lejeune
959)	KPNW	1120	U3	50	50	OR	Eugene
960)	KANN	1120	U4	10	1	UT	Roy
961)	WACQ	1130	D1	25		AL	Carrville
963)	KRDU	1130	U4	5	6.2	CA	Dinuba
962)	KSDO	1130	U4	10	10	CA	San Diego
964)	WWBF	1130	U2	2.5	0.5	FL	Bartow
965)	WLBA	1130	D1	10		GA	Gainesville
966)	KWKH	1130	U4	50	50	LA	Shreveport
967)	WDFN	1130	U4	50	10	MI	Detroit
968)	KFAN	1130	U4	50	25	MN	Minneapolis
970)	WPYB	1130	D1	6.5		NC	Benson
971)	KBMR	1130	U1	10	0.02	ND	Bismarck
969)	WBBR	1130	U2	50	50	NY	New York
972)	KTRP	1130	U4	25	0.49	OR	Mount Angel
973)	WEAF	1130	U1	5	0.007	SC	Camden
974)	KTMR	1130	D3	25		TX	Converse
975)	WISN	1130	U4	50	10	WI	Milwaukee
977)	KSLD	1140	U1	10	10	AK	Soldotna
976)	WBXR	1140	D4	15		AL	Hazel Green
978)	KLTK	1140	D1	5		AR	Centerton
980)	KNWQ	1140	U4	5	2.5	CA	Palm Springs
979)	KHTK	1140	U4	50	50	CA	Sacramento
981)	WQBA	1140	U4	50	10	FL	Miami
982)	WRMQ	1140	U2	5	0.008	FL	Orlando
983)	KGEM	1140	U2	10	10	ID	Boise
984)	WVEL	1140	D1	5		IL	Pekin
985)	WVHF	1140	D3	5		MI	Kentwood
987)	KCXL	1140	U1	4	0.006	MO	Liberty
986)	WSAO	1140	D1	5		MS	Senatobia
988)	KYDZ	1140	U2	10	2.5	NV	North Las Vegas
989)	WCJW	1140	D3	2.5		NY	Warsaw
990)	KSOO	1140	U2	10	5	SD	Sioux Falls
991)	KHFX	1140	U4	5	0.71	TX	Cleburne
992)	KYOK	1140	D3	5		TX	Conroe
993)	WRVA	1140	U3	50	50	VA	Richmond
994)	WXLZ	1140	D1	2.5		VA	Saint Paul
995)	KZMQ	1140	D1	10		WY	Greybull
996)	WJRD	1150	U2	20	1	AL	Tuscaloosa
997)	KCKY	1150	U4	5	1	AZ	Coolidge
998)	KTLK	1150	U4	50	44	CA	Los Angeles
999)	KNRV	1150	U4	10	1	CO	Englewood
1000)	WMRD	1150	U1	2.5	0.04	CT	Middletown
1001)	WDEL	1150	U4	5	5	DE	Wilmington
1002)	WTMP	1150	U4	10	0.5	FL	Egypt Lake
1003)	WJEM	1150	U3	5	0.1	GA	Valdosta
1005)	KWKY	1150	U4	2.5	1	IA	Des Moines
1004)	WGGH	1150	U4	5	0.04	IL	Marion
1006)	KSAL	1150	U2	5	5	KS	Salina
1007)	WMST	1150	U1	2.5	0.05	KY	Mount Sterling
1008)	WJBO	1150	U4	15	5	LA	Baton Rouge
1009)	WWDJ	1150	U4	5	5	MA	Boston
1010)	KSEN	1150	U4	10	5	MT	Shelby
1013)	WGBR	1150	U4	5	0.8	NC	Goldsboro
1011)	KDEF	1150	U2	5	0.5	NM	Albuquerque
1012)	WUTI	1150	U4	5	1	NY	Utica
1014)	WCUE	1150	U3	5	0.5	OH	Cuyahoga Falls
1015)	KAGO	1150	U2	5	1	OR	Klamath Falls
1016)	KLPM	1150	U1	5	0.01	OR	Portland
1017)	WLLI	1150	U1	5	0.03	PA	Huntingdon
1018)	WAVO	1150	U1	5	0.05	SC	Rock Hill
1019)	KIMM	1150	U2	5	0.5	SD	Rapid City
1020)	WGOW	1150	U2	5	1	TN	Chattanooga
1021)	WCRK	1150	U2	5	0.5	TN	Morristown
1022)	KHRO	1150	U1	5	0.38	TX	El Paso
1023)	WNLR	1150	U1	2.5	0.03	VA	Churchville
1024)	KQQQ	1150	U1	10	0.02	WA	Pullman
1025)	KKNW	1150	U2	10	6	WA	Seattle
1027)	WEAQ	1150	U1	5	0.04	WI	Chippewa Falls
1028)	WHBY	1150	U4	20	25	WI	Kimberly
1026)	WELC	1150	D1	5		WV	Welch
1029)	WEWC	1160	U5	5	0.25	FL	Callahan

MW	Call	kHz	Ant.	D	N	Sta	City of License
1030)	WIWA	1160	U4	2.5	0.5	FL	Saint Cloud
1031)	WCFO	1160	U5	50	0.16	GA	East Point
1032)	WYLL	1160	U4	50	50	IL	Chicago
1034)	WQRT	1160	U4	5	0.99	KY	Florence
1033)	WKCM	1160	U2	2.5	1	KY	Hawesville
1036)	WMET	1160	U4	50	1.5	MD	Gaithersburg
1035)	WSKW	1160	U1	10	0.73	ME	Skowhegan
1037)	KCTO	1160	U4	5	0.23	MO	Cleveland
1043)	WTEL	1160	U1	5	0.25	NC	Red Springs
1042)	WJFJ	1160	U2	10	0.5	NC	Tryon
1038)	WOBM	1160	U4	5	8.9	NJ	Lakewood Township
1039)	WVNJ	1160	U4	20	2.5	NJ	Oakland
1040)	WABY	1160	U1	5	0.57	NY	Mechanicville
1041)	WPIE	1160	U4	5	0.31	NY	Trumansburg
1045)	WCCS	1160	U4	10	1	PA	Homer City
1044)	WBYN	1160	U4	4	1	PA	Lehighton
1046)	WCRT	1160	U2	50	1	TN	Donelson
1048)	KVCE	1160	U4	35	1	TX	Highland Park
1047)	KRDY	1160	U4	10	1	TX	San Antonio
1049)	KSL	1160	U1	50	50	UT	Salt Lake City
1050)	WODY	1160	U2	5	0.25	VA	Fieldale
1052)	KJNP	1170	U1	50	21	AK	North Pole
1051)	WACV	1170	U4	10	1	AL	Montgomery
1053)	KCBQ	1170	U4	50	2.9	CA	San Diego
1054)	KLOK	1170	U4	50	5	CA	San Jose
1055)	WAVS	1170	U2	5	0.25	FL	Davie
1056)	WLBH	1170	D3	5		IL	Mattoon
1057)	KOWZ	1170	U1	2.5	0.005	MN	Waseca
1059)	WCXN	1170	D1	7.7		NC	Claremont
1058)	WCLN	1170	D1	5		NC	Clinton
1060)	KFAQ	1170	U2	50	50	OK	Tulsa
1061)	WQVA	1170	D1	10		SC	Lexington
1062)	KPUG	1170	U2	10	5	WA	Bellingham
1063)	WWVA	1170	U2	50	50	WV	Wheeling
1065)	WPLX	1180	U7	5	0.02	AR	Turrell
1064)	KYET	1180	U1	10	0.25	AZ	Williams
1066)	KERN	1180	U4	50	10	CA	Wasco - Greenacres
1067)	VOA	1180	U3	100	100	FL	Marathon
1068)	WZQZ	1180	D1	5		GA	Trion
1069)	WXLA	1180	D4	10		MI	Dimondale
1070)	KYES	1180	U7	50	5	MN	Rockville
1071)	WJNT	1180	U12	50	0.5	MS	Pearl
1072)	KOFI	1180	U2	50	10	MT	Kalispell
1075)	WMYT	1180	D3	10		NC	Carolina Beach
1073)	KOIL	1180	U4	25	1	NE	Bellevue
1074)	WHAM	1180	U1	50	50	NY	Rochester
1076)	WFGN	1180	D1	2.5		SC	Gaffney
1077)	WVLZ	1180	D1	10		TN	Knoxville
1078)	KGOL	1180	U4	50	1	TX	Humble
1079)	KLAY	1180	U2	5	1	WA	Lakewood
1080)	WEUV	1190	D1	2.5		AL	Moulton
1082)	KREB	1190	D1	5		AR	Bentonville-Bella Vista
1081)	KNUV	1190	U4	5	0.25	AZ	Tolleson
1083)	KXMX	1190	U4	20	1.3	CA	Anaheim
1084)	KVCU	1190	U1	6.8	0.11	CO	Boulder
1085)	WAMT	1190	U2	4.7	0.23	FL	Pine Castle-Sky Lake
1086)	WAFS	1190	D1	25		GA	Atlanta
1087)	WOWO	1190	U2	50	9.8	IN	Fort Wayne
1088)	WBIS	1190	D3	10		MD	Annapolis
1089)	KKOJ	1190	D3	5		MN	Jackson
1091)	KQQZ	1190	U4	10	0.02	MO	Desoto
1092)	KPHN	1190	U2	5	0.5	MO	Kansas City
1090)	WMEJ	1190	D1	5		MS	Bay Saint Louis
1095)	WIXE	1190	U1	5	0.07	NC	Monroe
1093)	KXKS	1190	U1	10	0.02	NM	Albuquerque
1094)	WLIB	1190	U4	10	30	NY	New York
1096)	KEX	1190	U2	50	50	OR	Portland
1097)	WSDQ	1190	D1	5		TN	Dunlap
1098)	KFXR	1190	U4	50	5	TX	Dallas
1100)	WNWC	1190	U7	4.8	0.02	WI	Sun Prairie
1099)	WVUS	1190	U1	4.5	0.02	WV	Grafton
1101)	KYAA	1190	U2	25	10	CA	Soquel
1102)	WPTK	1200	U4	10	1	FL	Pine Island Center
1103)	WRTO	1200	U4	20	4.5	IL	Chicago
1104)	WXKS	1200	U4	50	50	MA	Newton
1105)	WCHB	1200	U4	50	15	MI	Taylor
1106)	WXIT	1200	D1	10		NC	Blowing Rock
1107)	WSML	1200	U1		1	NC	Graham
1108)	KFNW	1200	U4	50	13	ND	West Fargo
1110)	WRKK	1200	U4	10	0.25	PA	Hughesville
1109)	WKST	1200	U2	5	1	PA	New Castle
1111)	WMIR	1200	U1	6.5	0.01	SC	Atlantic Beach
1112)	WJES	1200	U1	10	0.004	SC	Saluda
1113)	WAMB	1200	D1	50		TN	Nashville
1114)	WOAI	1200	U1	50	50	TX	San Antonio
1115)	WAGE	1200	U2	5	1	VA	Leesburg
1116)	WTXK	1210	U1	10	0.003	AL	Pike Road
1117)	KEVT	1210	U2	10	1	AZ	Sahuarita
1118)	KEBR	1210	U5	5	0.5	CA	Rocklin
1119)	KPRZ	1210	U4	20	10	CA	San Marcos
1120)	WNMA	1210	U4	47	2.5	FL	Miami Springs
1121)	WDGR	1210	D1	10		GA	Dahlonega
1122)	WILY	1210	U11	10	0.003	IL	Centralia
1123)	WSKR	1210	U2	10	1	LA	Denham Springs
1124)	WJNL	1210	D1	50		MI	Kingsley
1125)	KGYN	1210	U2	10	10	OK	Guymon
1127)	WPHT	1210	U1	50	50	PA	Philadelphia
1126)	WANB	1210	D1	5		PA	Waynesburg
1128)	KOKK	1210	U4	5	0.9	SD	Huron
1130)	WMPS	1210	U2	10	0.25	TN	Bartlett
1129)	WSBI	1210	D1	10		TN	Static
1131)	KUBR	1210	U4	10	5	TX	San Juan
1132)	KUNF	1210	U1	10	0.25	UT	Washington
1133)	KTBK	1210	U4	27.5	10	WA	Auburn-Federal Way
1135)	KRSV	1210	U1	5	0.25	WY	Afton
1134)	KHAT	1210	U2	10	1	WY	Laramie
1136)	KDOW	1220	U1	5	0.14	CA	Palo Alto
1137)	WSLM	1220	U3	5	0.08	IN	Salem
1138)	KLBB	1220	U1	5	0.25	MN	Stillwater
1141)	WDYT	1220	U5	25	0.1	NC	Kings Mountain
1140)	WENC	1220	U1	5	0.15	NC	Whiteville
1139)	WGNY	1220	U4	10	0.18	NY	Newburgh
1142)	WHKW	1220	U3	50	50	OH	Cleveland
1143)	WFAX	1220	U1	5	0.04	VA	Falls Church
1145)	WZOB	1250	U1	5	0.12	AL	Fort Payne
1144)	WAPZ	1250	U1	5	0.08	AL	Wetumpka
1146)	KHIL	1250	U1	5	0.19	AZ	Willcox
1148)	KZER	1250	U4	2.5	1	CA	Santa Barbara
1147)	KLLK	1250	U4	5	2.5	CA	Willits
1149)	WHNZ	1250	U4	25	5.9	FL	Tampa
1150)	KYYS	1250	U4	25	3.7	KS	Kansas City
1151)	WARE	1250	U4	5	2.5	MA	Ware
1152)	WNEM	1250	U4	5	1.1	MI	Bridgeport
1153)	KBRF	1250	U2	5	2.2	MN	Fergus Falls
1154)	WHNY	1250	U2	5	1	MS	McComb
1155)	KIKC	1250	U1	5	0.13	MT	Forsyth
1159)	WGHB	1250	U4	5	2.5	NC	Farmville
1158)	WBRM	1250	U1	5	0.06	NC	Marion
1156)	WGAM	1250	U4	5	5	NH	Manchester
1157)	WMTR	1250	U4	5	7	NJ	Morristown
1160)	WLEM	1250	U1	2.5	0.03	PA	Emporium
1161)	WEAE	1250	U2	5	5	PA	Pittsburgh
1162)	WTMA	1250	U1	5	1	SC	Charleston
1164)	KDEI	1250	U2	5	1	TX	Port Arthur
1163)	KZDC	1250	U4	25	2	TX	San Antonio
1165)	KNEU	1250	U1	5	0.12	UT	Roosevelt
1166)	WDVA	1250	U2	5	5	VA	Danville
1167)	WKDL	1250	U3	5	0.03	VA	Warrenton
1168)	KWSU	1250	U1	5	5	WA	Pullman
1169)	KKDZ	1250	U2	5	5	WA	Seattle
1171)	WSSP	1250	U4	5	5	WI	Milwaukee
1170)	WYKM	1250	D1	5		WV	Rupert
1172)	WYDE	1260	U1	5	0.04	AL	Birmingham
1173)	KGIL	1260	U4	20	7.5	CA	Beverly Hills
1174)	KSFB	1260	U1	5	1	CA	San Francisco
1175)	WWRC	1260	U1	5	5	DC	Washington
1176)	WFTW	1260	U2	2.5	0.13	FL	Fort Walton Beach
1177)	WSUA	1260	U4	50	20	FL	Miami
1178)	WUFE	1260	D1	5		GA	Baxley
1179)	WTJH	1260	U1	5	0.03	GA	East Point
1184)	KTIA	1260	U3	5	0.03	IA	Boone
1181)	KBLY	1260	U1	5	0.06	ID	Idaho Falls
1180)	KWEI	1260	U1	5	0.06	ID	Weiser
1182)	WSDZ	1260	U4	20	5	IL	Belleville
1183)	WNDE	1260	U2	5	5	IN	Indianapolis
1185)	KBRH	1260	U1	5	0.12	LA	Baton Rouge
1186)	WMKI	1260	U2	5	5	MA	Boston
1187)	WPNW	1260	U4	10	1	MI	Zeeland
1188)	KSGF	1260	U4	50	5	MO	Springfield
1192)	WKXB	1260	U4	5	0.5	NC	Asheboro
1189)	WFJS	1260	U2	5	2.5	NJ	Trenton
1190)	KTRC	1260	U1	5	1	NM	Santa Fe
1191)	WSKO	1260	U2	5	5	NY	Syracuse
1193)	WWMK	1260	U4	10	5	OH	Cleveland
1194)	WNXT	1260	U1	5	1	OH	Portsmouth
1196)	WRIE	1260	U1	5	5	PA	Erie
1195)	WPHB	1260	U1	5	0.03	PA	Philipsburg
1198)	WPJF	1260	U1	5	0.01	SC	Greenville
1197)	WHYM	1260	U1	5	0.05	SC	Lake City
1199)	KWYR	1260	U1	5	0.14	SD	Winner
1200)	WNOO	1260	U1	5	0.02	TN	Chattanooga
1201)	WDKN	1260	U1	5	0.01	TN	Dickson

MW	Call	kHz	Ant.	D	N	Sta	City of License
1202)	KSML	1260	U1	4.5	0.07	TX	Diboll
1203)	WCHV	1260	U4		2.5	VA	Charlottesville
1204)	WWVT	1260	U1	5	0.02	VA	Christiansburg
1205)	WXCE	1260	U4	5	5	WI	Amery
1206)	KPOW	1260	U2	5	1	WY	Powell
1207)	WIJD	1270	U1	5	0.1	AL	Prichard
1208)	KDJI	1270	U1	5	0.13	AZ	Holbrook
1210)	KFUT	1270	U4	5	0.75	CA	Thousand Palms
1209)	KJUG	1270	U2	5	1	CA	Tulare
1211)	WRLZ	1270	U4	25	5	FL	Eatonville
1212)	WNOG	1270	U4	5	5	FL	Naples
1213)	WNLS	1270	U2	5	5	FL	Tallahassee
1214)	WSHE	1270	U1	5	0.18	GA	Columbus
1215)	WJJC	1270	U1	5	0.17	GA	Commerce
1217)	KPDA	1270	U1	5	1	ID	Twin Falls
1218)	WKBF	1270	U2	5	5	IL	Rock Island
1219)	WCMR	1270	U4	5	1	IN	Elkhart
1220)	KSCB	1270	U1	5	0.02	KS	Liberal
1222)	WSPR	1270	U4	5	5	MA	Springfield
1221)	WCBC	1270	U4	5	1	MD	Cumberland
1223)	WMKT	1270	U2	27	5	MI	Charlevoix
1224)	WXYT	1270	U4	50	50	MI	Detroit
1226)	WWWI	1270	U2	5	5	MN	Baxter
1225)	KWEB	1270	U4	5	1	MN	Rochester
1232)	WCGC	1270	U4	10	0.5	NC	Belmont
1231)	WMPM	1270	U1	5	0.14	NC	Smithfield
1228)	WTSN	1270	U4	5	5	NH	Dover
1227)	KBZZ	1270	U2	13	5	NV	Sparks
1229)	WHLD	1270	U3	5	1	NY	Niagara Falls
1230)	WDLA	1270	U1	5	0.08	NY	Walton
1233)	KRVT	1270	U4	5	1	OK	Claremore
1234)	KAJO	1270	U1	10	0.04	OR	Grants Pass
1235)	WLBR	1270	U4	5	1	PA	Lebanon
1236)	WHGS	1270	U1	10	0.21	SC	Hampton
1237)	KNWC	1270	U4	5	2.3	SD	Sioux Falls
1238)	WLIK	1270	U2	5	0.5	TN	Newport
1239)	KFLC	1270	U4	50	5	TX	Fort Worth
1240)	WHEO	1270	D1	5		VA	Stuart
1241)	KBAM	1270	U1	5	0.08	WA	Longview
1242)	KIML	1270	U2	5	1	WY	Gillette
1243)	WMXB	1280	U2	5	0.5	AL	Tuscaloosa
1244)	KXEG	1280	U1	2.5	0.04	AZ	Phoenix
1245)	KXTK	1280	U4	10	2.5	CA	Arroyo Grande
1246)	KBNO	1280	U4	5	5	CO	Denver
1247)	WDSP	1280	U1	5	0.04	FL	De Funiak Springs
1248)	WIBB	1280	U1	5	0.09	GA	Macon
1249)	WGBF	1280	U2	5	1	IN	Evansville
1250)	WODT	1280	U3	5	5	LA	New Orleans
1252)	WPKZ	1280	U4	5	1	MA	Fitchburg
1251)	WFAU	1280	U2	5	5	ME	Gardiner
1253)	WWTC	1280	U2	5	5	MN	Minneapolis
1254)	KVXR	1280	U4	5	1	MN	Moorhead
1259)	WYAL	1280	D1	5		NC	Scotland Neck
1256)	KRZE	1280	U1	5	0.1	NM	Farmington
1255)	KDOX	1280	U1	5	0.02	NV	Henderson
1258)	WADO	1280	U4	50	7.2	NY	New York
1257)	WHTK	1280	U2	5	5	NY	Rochester
1260)	KRVM	1280	U4	5	1.5	OR	Eugene
1261)	WHVR	1280	U4	5	0.5	PA	Hanover
1262)	WJST	1280	U2	4.9	1	PA	New Castle
1264)	WANS	1280	U2	5	1	SC	Anderson
1263)	WJAY	1280	U1	4.2	0.27	SC	Mullins
1265)	WMCP	1280	U4	5	0.5	TN	Columbia
1266)	KZNS	1280	U4	50	0.67	UT	Salt Lake City
1267)	WYVE	1280	U1	2.5	0.16	VA	Wytheville
1269)	KPTQ	1280	U4	5	0.12	WA	Spokane
1268)	KIT	1280	U1	5	1	WA	Yakima
1270)	WNAM	1280	U4	50	5	WI	Neenah-Menasha
1271)	WOPP	1290	U4	2.5	0.5	AL	Opp
1273)	KDMS	1290	U1	2.5	0.1	AR	El Dorado
1272)	KUOA	1290	U1	5	0.03	AR	Siloam Springs
1276)	KPAY	1290	U2	5	5	CA	Chico
1274)	KAZA	1290	U4	5	0.08	CA	Gilroy
1275)	KKDD	1290	U4	5	5	CA	San Bernardino
1277)	WWTX	1290	U1	2.5	0.03	DE	Wilmington
1279)	WCFI	1290	U2	5	1	FL	Ocala
1278)	WJNO	1290	U4	10	4.9	FL	West Palm Beach
1281)	WCHK	1290	U4	10	0.5	GA	Canton
1280)	WTKS	1290	U2	5	5	GA	Savannah
1282)	KOUU	1290	U5	50	0.02	ID	Pocatello
1283)	WIRL	1290	U4	5	5	IL	Peoria
1284)	KMMM	1290	U4	5	0.5	KS	Pratt
1285)	WCBL	1290	U1	5	0.05	KY	Benton
1286)	WKLB	1290	U1	5	0.03	KY	Manchester
1287)	WNBN	1290	U1	2.5	0.09	MS	Meridian
1288)	KGVO	1290	U2	5	1	MT	Missoula
1293)	WHKY	1290	U4	50	1	NC	Hickory
1292)	WJCV	1290	U1	5	0.04	NC	Jacksonville
1289)	KKAR	1290	U2	5	5	NE	Omaha
1290)	WKBK	1290	U3	5	5	NH	Keene
1291)	WNBF	1290	U2	9.3	5	NY	Binghamton
1294)	WHIO	1290	U2	5	5	OH	Dayton
1295)	KUMA	1290	U2	5	5	OR	Pendleton
1296)	WFBG	1290	U2	5	1	PA	Altoona
1297)	WRNI	1290	U4	10	10	RI	Providence
1298)	WATO	1290	U4	5	0.5	TN	Oak Ridge
1299)	KIVY	1290	U1	2.5	0.17	TX	Crockett
1301)	KRGE	1290	U2	5	5	TX	Weslaco
1300)	KWFS	1290	U1	5	0.07	TX	Wichita Falls
1302)	WDZY	1290	U1	25	0.04	VA	Colonial Heights
1305)	WMCS	1290	U4	5	5	WI	Greenfield
1304)	WKLJ	1290	U1	5	0.05	WI	Sparta
1303)	WVOW	1290	U2	5	1	WV	Logan
1306)	KOWB	1290	U4	5	1	WY	Laramie
1307)	WKXM	1300	U1	5	0.03	AL	Winfield
1308)	KWCK	1300	D1	5		AR	Searcy
1311)	KWRU	1300	U2	5	1	CA	Fresno
1310)	KPMO	1300	U1	5	0.07	CA	Mendocino
1309)	KAZN	1300	U4	23	1	CA	Pasadena
1312)	KCSF	1300	U1	5	1	CO	Colorado Springs
1315)	WMEL	1300	U4	5	1	FL	Cocoa Beach
1314)	WFFG	1300	U3	2.5	2.5	FL	Marathon
1313)	WQBN	1300	U5	5	0.16	FL	Temple Terrace
1316)	WMTM	1300	U1	5	0.006	GA	Moultrie
1319)	KGLO	1300	U4	5	5	IA	Mason City
1317)	KLER	1300	U2	5	1	ID	Orofino
1318)	WRDZ	1300	U4	4.5	4	IL	La Grange
1320)	WLXG	1300	U2	2.5	1	KY	Lexington
1322)	WIBR	1300	U4	5	1	LA	Baton Rouge
1321)	KSYB	1300	U1	5	0.03	LA	Shreveport
1323)	WJZ	1300	U4	5	5	MD	Baltimore
1324)	WOOD	1300	U3	20	20	MI	Grand Rapids
1325)	WOAD	1300	U1	5	1	MS	Jackson
1332)	WSYD	1300	U2	5	1	NC	Mount Airy
1326)	KBRL	1300	U3	5	0.13	NE	McCook
1328)	WPNH	1300	U1	5	0.08	NH	Plymouth
1329)	WIMG	1300	U4	3.2	1.3	NJ	Ewing
1327)	KCMY	1300	U2	5	0.5	NV	Carson City
1330)	WKXL	1300	U1	5	2.5	NY	Lancaster
1331)	WGDJ	1300	U4	10	8	NY	Rensselaer
1333)	WJMO	1300	U3	5	5	OH	Cleveland
1334)	KAKC	1300	U4	5	1	OK	Tulsa
1335)	KAPL	1300	U2	20	5	OR	Phoenix
1336)	WKZN	1300	U4	5	0.5	PA	West Hazleton
1337)	KOLY	1300	U1	5	0.11	SD	Mobridge
1338)	WMTN	1300	U1	5	0.09	TN	Morristown
1339)	WNQM	1300	U2	50	5	TN	Nashville
1340)	KVET	1300	U4	5	1	TX	Austin
1341)	WKCY	1300	U1	6.4	0.005	VA	Harrisonburg
1342)	KKOL	1300	U4	50	47	WA	Seattle
1343)	WCLG	1300	U1	2.5	0.04	WV	Morgantown
1345)	WHEP	1310	U1	2.5	0.04	AL	Foley
1344)	WJUS	1310	U1	5	0.03	AL	Marion
1346)	KIHP	1310	U1	5	0.5	AZ	Mesa
1348)	KIQQ	1310	U1	5	0.11	CA	Barstow
1347)	KMKY	1310	U3	5	5	CA	Oakland
1349)	KFKA	1310	U2	5	1	CO	Greeley
1350)	WICH	1310	U4	5	5	CT	Norwich
1351)	WYND	1310	U1	10.4	0.11	FL	Deland
1352)	WAUC	1310	U3	5	0.5	FL	Wauchula
1354)	WPBC	1310	U1	2.5	0.03	GA	Decatur
1353)	WOKA	1310	U1	3.9	0.03	GA	Douglas
1355)	KLIX	1310	U2	5	2.5	ID	Twin Falls
1356)	WTLC	1310	U2	5	1	IN	Indianapolis
1357)	WDOC	1310	U1	5	0.02	KY	Prestonsburg
1358)	KMBS	1310	U1	5	0.04	LA	West Monroe
1360)	WORC	1310	U4	5	1	MA	Worcester
1359)	WLOB	1310	U4	5	5	ME	Portland
1362)	WDTW	1310	U4	5	5	MI	Dearborn
1361)	WCCW	1310	U4	15	7.5	MI	Traverse City
1363)	KGLB	1310	U4	2.5	0.27	MN	Glencoe
1364)	KZRG	1310	U4	5	1	MO	Joplin
1365)	KEIN	1310	U1	5	1	MT	Great Falls
1372)	WISE	1310	U2	5	1	NC	Asheville
1371)	WGSP	1310	U1	5	0.24	NC	Charlotte
1370)	WTIK	1310	U4	5	1	NC	Durham
1373)	KNOX	1310	U2	5	5	ND	Grand Forks
1366)	WADB	1310	U4	2.5	1	NJ	Asbury Park
1367)	KKNS	1310	U2	5	0.5	NM	Corrales
1369)	WRVP	1310	U3	5	0.03	NY	Mount Kisco
1368)	WTLB	1310	U4	5	0.5	NY	Utica
1374)	KNPT	1310	U2	5	1	OR	Newport

MW	Call	kHz	Ant.	D	N	Sta	City of License
1375)	WBFD	1310	U1	2.5	0.08	PA	Bedford
1376)	WNAE	1310	U1	5	0.09	PA	Warren
1377)	WDKD	1310	U1	5	0.06	SC	Kingstree
1378)	WDOD	1310	U2	5	5	TN	Chattanooga
1379)	WDXI	1310	U2	5	1	TN	Jackson
1381)	KTCK	1310	U4	9	5	TX	Dallas
1380)	KAHL	1310	U3	5	0.28	TX	San Antonio
1383)	WDCT	1310	U4	5	0.5	VA	Fairfax
1382)	WGH	1310	U4	20	5	VA	Newport News
1384)	KZXR	1310	U1	5	0.06	WA	Prosser
1386)	WIBA	1310	U2	5	5	WI	Madison
1385)	WSLW	1310	D1	5		WV	White Sulphur Springs
1387)	WENN	1320	U2	5	0.11	AL	Birmingham
1388)	KWHN	1320	U2	5	5	AR	Fort Smith
1389)	KCTC	1320	U4	5	5	CA	Sacramento
1390)	WATR	1320	U4	5	1	CT	Waterbury
1393)	WLQY	1320	U4	5	5	FL	Hollywood
1391)	WBOB	1320	U2	50	5	FL	Jacksonville
1392)	WDDV	1320	U4	5	1	FL	Venice
1394)	WHIE	1320	U1	5	0.08	GA	Griffin
1395)	KNCB	1320	U1	5	0.05	LA	Vivian
1396)	WARL	1320	U4	5	5	MA	Attleboro
1397)	WILS	1320	U4	25	1.9	MI	Lansing
1398)	WDMJ	1320	U1	5	0.13	MI	Marquette
1399)	KOZY	1320	U2	5	5	MN	Grand Rapids
1401)	KSIV	1320	U2	4.6	0.27	MO	Clayton
1400)	WRJW	1320	U1	5	0.07	MS	Picayune
1405)	WCOG	1320	U4	5	5	NC	Greensboro
1406)	WKRK	1320	U1	5	0.06	NC	Murphy
1407)	KHRT	1320	U1	2.5	0.31	ND	Minot
1402)	KOLT	1320	U2	5	1	NE	Scottsbluff
1403)	WDER	1320	U4	10	1	NH	Derry
1404)	WHHO	1320	U1	5	0.02	NY	Hornell
1408)	WJAS	1320	U2	5	5	PA	Pittsburgh
1409)	WISW	1320	U2	5	2.5	SC	Columbia
1410)	KELO	1320	U2	5	5	SD	Sioux Falls
1411)	WGOC	1320	U2	5	0.5	TN	Kingsport
1412)	WMSR	1320	U1	5	0.07	TN	Manchester
1413)	KXYZ	1320	U2	5	5	TX	Houston
1414)	KFNZ	1320	U3	5	5	UT	Salt Lake City
1415)	WVNZ	1320	U4	5	0.008	VA	Richmond
1416)	KXRO	1320	U1	5	1	WA	Aberdeen
1417)	WFHR	1320	U2	5	0.5	WI	Wisconsin Rapids
1419)	KXLJ	1330	U1	10	3	AK	Juneau
1418)	WZCT	1330	U1	5	0.03	AL	Scottsboro
1420)	KJLL	1330	U2	2	5	AZ	South Tucson
1422)	KWKW	1330	U2	5	5	CA	Los Angeles
1421)	KLBS	1330	U2	0.42	5	CA	Los Banos
1423)	WJNX	1330	U4	5	1	FL	Fort Pierce
1425)	WEBY	1330	U5	25	0.07	FL	Milton
1424)	WCVC	1330	D1	5		FL	Tallahassee
1426)	WMLT	1330	U2	5	0.5	GA	Dublin
1429)	KWLO	1330	U4	5	5	IA	Waterloo
1427)	WKTA	1330	U4	5	0.11	IL	Evanston
1428)	WVHI	1330	U2	5	1	IN	Evansville
1430)	KNSS	1330	U2	5	5	KS	Wichita
1431)	WKDP	1330	U4	5	0.01	KY	Corbin
1432)	KVOL	1330	U2	5	1	LA	Lafayette
1434)	WRCA	1330	U4	25	17	MA	Watertown
1433)	WJSS	1330	U2	5	0.5	MD	Havre de Grace
1435)	WTRX	1330	U4	5	1	MI	Flint
1436)	WLOL	1330	U4	9.7	5.1	MN	Minneapolis
1437)	KGAK	1330	U2	5	1	NM	Gallup
1438)	WWRV	1330	U4	10	5	NY	New York
1439)	WEBO	1330	U1	5	0.03	NY	Owego
1440)	KKPZ	1330	U3	5	5	OR	Portland
1441)	WFNN	1330	U4	5	5	PA	Erie
1442)	WBHV	1330	U3	5	0.03	PA	Somerset
1444)	WPJS	1330	U1	3.2	0.02	SC	Conway
1443)	WYRD	1330	U2	5	5	SC	Greenville
1445)	KCKM	1330	U2	5	1	TX	Monahans
1446)	WBTM	1330	U2	5	1	VA	Danville
1448)	WITM	1330	U1	5	0.03	VA	Marion
1447)	WESR	1330	U1	5	0.05	VA	Onley-Onancock
1449)	KMBI	1330	D1	5		WA	Spokane
1450)	WHBL	1330	U4	5	1	WI	Sheboygan
1451)	KOVE	1330	U1	5	0.25	WY	Lander
1452)	WJBY	1350	U2	5	1	AL	Gadsden
1453)	KZTD	1350	U1	2.5	0.07	AR	Cabot
1454)	KTDD	1350	U4	5	0.6	CA	San Bernardino
1455)	KSRO	1350	U2	5	5	CA	Santa Rosa
1456)	KDZA	1350	U1	5	0.28	CO	Pueblo
1457)	WINY	1350	U1	5	0.07	CT	Putnam
1458)	WFNS	1350	U3	2.5	0.11	GA	Blackshear
1459)	WRWR	1350	U2	15	0.5	GA	Warner Robins
1462)	KRNT	1350	U2	5	5	IA	Des Moines
1460)	KTIK	1350	U2	5	0.6	ID	Nampa
1461)	WIOU	1350	U4	5	1	IN	Kokomo
1463)	WWWL	1350	U2	5	5	LA	New Orleans
1467)	WQNX	1350	U3	2.5	0.02	NC	Aberdeen
1468)	WZGM	1350	U1	10	0.05	NC	Black Mountain
1464)	WEZS	1350	U1	5	0.11	NH	Laconia
1465)	WHWH	1350	U4	5	5	NJ	Princeton
1466)	KABQ	1350	U2	5	0.5	NM	Albuquerque
1469)	WARF	1350	U3	5	5	OH	Akron
1470)	WOYK	1350	U2	5	1	PA	York
1471)	KCOX	1350	U1	5	0.03	TX	Jasper
1472)	KCOR	1350	U2	5	5	TX	San Antonio
1473)	WBLT	1350	U1	5	0.04	VA	Bedford
1475)	WNVA	1350	U1	5	0.03	VA	Norton
1474)	WGPL	1350	U4	5	5	VA	Portsmouth
1476)	KRLC	1350	U2	5	1	WA	Clarkston-Lewiston
1478)	WIXI	1360	U1	12	0.04	AL	Jasper
1477)	WMOB	1360	U3	5	0.21	AL	Mobile
1479)	KPXQ	1360	U2	50	1	AZ	Glendale
1481)	KFIV	1360	U4	4	0.95	CA	Modesto
1480)	KLSD	1360	U1	5	1	CA	San Diego
1482)	KHNC	1360	U4	10	1	CO	Johnstown
1483)	WDRC	1360	U2	5	5	CT	Hartford
1484)	WHNR	1360	U4	5	2.5	FL	Cypress Gardens
1485)	WCGL	1360	U1	5	0.08	FL	Jacksonville
1486)	WKAT	1360	U1	5	1	FL	North Miami
1487)	KSCJ	1360	U1	5	5	IA	Sioux City
1488)	WKMI	1360	U2	5	1	MI	Kalamazoo
1489)	KKBJ	1360	U2	5	2.5	MN	Bemidji
1493)	WCHL	1360	U2	5	1	NC	Chapel Hill
1490)	WNJC	1360	U4	5	0.8	NJ	Washington Twnshp
1491)	KBUY	1360	U1	9	0.2	NM	Ruidoso
1492)	WYOS	1360	U4	5	0.5	NY	Binghamton
1494)	WSAI	1360	U2	5	5	OH	Cincinnati
1495)	WWOW	1360	U1	5	0.03	OH	Conneaut
1497)	KOHU	1360	U2	4.3	0.5	OR	Hermiston
1496)	KUIK	1360	U2	5	5	OR	Hillsboro
1499)	WMNY	1360	U2	5	1	PA	McKeesport
1498)	WPPA	1360	U4	5	0.5	PA	Pottsville
1500)	WELP	1360	U1	5	0.03	SC	Easley
1502)	KDJW	1360	U4	6	0.32	TX	Amarillo
1503)	KWWJ	1360	U4	5	1	TX	Baytown
1501)	KMNY	1360	U4	50	0.89	TX	Hurst
1504)	WWWJ	1360	U1	5	0.03	VA	Galax
1505)	WHBG	1360	U1	5	0.009	VA	Harrisonburg
1506)	KKMO	1360	U1	5	5	WA	Tacoma
1508)	WTAQ	1360	U4	10	5	WI	Green Bay
1507)	WMOV	1360	D1	5		WV	Ravenswood
1509)	KRKK	1360	U2	5	1	WY	Rock Springs
1511)	KWRM	1370	U4	5	2.5	CA	Corona
1510)	KRAC	1370	U4	5	0.5	CA	Quincy
1512)	KZSF	1370	U3	5	5	CA	San Jose
1514)	WOCA	1370	U1	5	0.03	FL	Ocala
1513)	WCOA	1370	U2	5	5	FL	Pensacola
1515)	WLOP	1370	U1	5	0.03	GA	Jesup
1518)	KDTH	1370	U2	5	5	IA	Dubuque
1517)	WGCL	1370	U4	5	0.5	IN	Bloomington
1519)	KGNO	1370	U1	5	0.23	KS	Dodge City
1520)	WGOH	1370	U1	5	0.02	KY	Grayson
1522)	WVIE	1370	U4	50	7.7	MD	Pikesville
1521)	WDEA	1370	U4	5	5	ME	Ellsworth
1523)	WLJW	1370	U4	5	1	MI	Cadillac
1524)	KXTL	1370	U1	5	5	MT	Butte
1530)	WLLN	1370	U3	5	0.04	NC	Lillington
1529)	WGIV	1370	U1	16	0.04	NC	Pineville
1528)	WTAB	1370	U1	5	0.1	NC	Tabor City
1531)	KWTL	1370	U1	12	0.27	ND	Grand Forks
1525)	WFEA	1370	U4	5	5	NH	Manchester
1526)	WRWD	1370	D1	5		NY	Ellenville
1527)	WXXI	1370	U1	5	5	NY	Rochester
1532)	WSPD	1370	U2	5	5	OH	Toledo
1533)	WKMC	1370	U3	5	0.03	PA	Roaring Spring
1534)	WDEF	1370	U2	5	5	TN	Chattanooga
1535)	KJCE	1370	U4	5	0.5	TX	Rollingwood
1536)	KSOP	1370	U2	5	0.5	UT	South Salt Lake
1538)	WHEE	1370	D1	5		VA	Martinsville
1537)	WSHV	1370	U1	4.2	0.04	VA	South Hill
1541)	WCCN	1370	U1	5	0.04	WI	Neillsville
1540)	WVMR	1370	D1	5		WV	Frost
1539)	WVLY	1370	U1	5	0.02	WV	Moundsville
1542)	WVSA	1380	U1	5	0.03	AL	Vernon
1544)	KDXE	1380	U4	5	2.5	AR	North Little Rock
1543)	KLPZ	1380	U2	2.5	0.05	AZ	Parker
1545)	KTKZ	1380	U4	5	5	CA	Sacramento
1546)	WFNW	1380	U4	5	0.5	CT	Naugatuck
1547)	WELE	1380	U4	5	2.5	FL	Ormond Beach

MW	Call	kHz	Ant.	D	N	Sta	City of License
1548)	WWMI	1380	U2	5	5	FL	Saint Petersburg
1549)	WAOK	1380	U2	5	4.2	GA	Atlanta
1550)	WTJK	1380	U2	5	5	IL	South Beloit
1551)	WKJG	1380	U4	5	5	IN	Fort Wayne
1552)	KCNW	1380	U1	2.5	0.02	KS	Fairway
1553)	WMJR	1380	U1	5	0.03	KY	Nicholasville
1554)	WPYR	1380	U4	5	0.06	LA	Baton Rouge
1555)	WPHM	1380	U4	5	5	MI	Port Huron
1556)	KLIZ	1380	U2	5	5	MN	Brainerd
1557)	KSLG	1380	U4	5	1	MO	Saint Louis
1560)	WKJV	1380	U2	25	1	NC	Asheville
1561)	WTOB	1380	U4	5	2.5	NC	Winston-Salem
1558)	WABH	1380	U4	10	0.45	NY	Bath
1559)	WKDM	1380	U3	5	5	NY	New York
1562)	KMUS	1380	U4	7	0.25	OK	Sperry
1563)	KSRV	1380	U2	5	1	OR	Ontario
1564)	WNRI	1380	U1	2.5	0.01	RI	Woonsocket
1565)	WNRR	1380	U1	4	0.07	SC	North Augusta
1566)	KOTA	1380	U2	5	5	SD	Rapid City
1567)	WHEW	1380	D1	2.8		TN	Franklin
1568)	WLRM	1380	U4	2.5	1	TN	Millington
1569)	KHEY	1380	U1	5	0.5	TX	El Paso
1570)	KWMF	1380	U5	4	0.16	TX	Pleasanton
1572)	WBTK	1380	U2	5	5	VA	Richmond
1571)	WSYB	1380	U2	5	1	VT	Rutland
1573)	KRKO	1380	U2	5	5	WA	Everett
1574)	WOTE	1380	U3	3.9	1.8	WI	Clintonville
1575)	WHMA	1390	U2	5	1	AL	Anniston
1576)	KFFK	1390	U1	5	0.03	AR	Rogers
1577)	KLTX	1390	U4	5	3.6	CA	Long Beach
1578)	KLOC	1390	U4	5	5	CA	Turlock
1579)	KGNU	1390	U4	5	0.13	CO	Denver
1580)	WAJD	1390	U1	5	0.05	FL	Gainesville
1581)	WGRB	1390	U4	5	5	IL	Chicago
1582)	WZQQ	1390	D1	5		KY	Hazard
1584)	WPLM	1390	U4	5	5	MA	Plymouth
1583)	WEGP	1390	U4	25	10	ME	Presque Isle
1585)	WLCM	1390	U5	0.97	4.5	MI	Holt
1586)	KXSS	1390	U4	2.5	1	MN	Waite Park
1589)	KJPW	1390	U1	5	0.11	MO	Waynesville
1587)	WROA	1390	U4	5	5	MS	Gulfport
1588)	WMER	1390	U1	5	0.1	MS	Meridian
1594)	WEED	1390	U1	5	0.03	NC	Rocky Mount
1595)	KRRZ	1390	U1	5	1	ND	Minot
1591)	KENN	1390	U2	5	1.3	NM	Farmington
1590)	KHOB	1390	U2	5	0.5	NM	Hobbs
1592)	WEOK	1390	U3	5	0.1	NY	Poughkeepsie
1593)	WFBL	1390	U4	5	5	NY	Syracuse
1597)	WMPO	1390	U1	5	0.12	OH	Middleport-Pomeroy
1596)	WNIO	1390	U2	9.5	4.8	OH	Youngstown
1598)	KWOD	1390	U1	5	0.69	OR	Salem
1599)	WLAN	1390	U4	5	1	PA	Lancaster
1600)	WSPO	1390	U2	5	5	SC	Charleston
1602)	WYXI	1390	U1	2.5	0.06	TN	Athens
1601)	WTJS	1390	U2	5	1	TN	Jackson
1603)	KLGN	1390	U2	5	0.5	UT	Logan
1605)	WZHF	1390	U4	5	5	VA	Arlington
1606)	WKPA	1390	U1	4.7	0.03	VA	Lynchburg
1604)	WCAT	1390	U2	5	5	VT	Burlington
1607)	KJOX	1390	U4	5	0.39	WA	Yakima
1608)	WRIG	1390	U4	10	7.2	WI	Schofield
1610)	WNGL	1410	U2	5	5	AL	Mobile
1609)	WIQR	1410	U2	5	1	AL	Prattville
1612)	KMYC	1410	U4	5	1	CA	Marysville
1611)	KCAL	1410	U2	5	4	CA	Redlands
1613)	WPOP	1410	U2	5	5	CT	Hartford
1614)	WDOV	1410	U4	5	5	DE	Dover
1616)	WMYR	1410	U2	5	5	FL	Fort Myers
1617)	WRHB	1410	U1	5	0.09	FL	Leesburg
1615)	WHBT	1410	U1	5	0.01	FL	Tallahassee
1618)	WKKP	1410	U1	2.5	0.05	GA	McDonough
1619)	KKLO	1410	U4	5	0.5	KS	Leavenworth
1620)	KGSO	1410	U1	5	1	KS	Wichita
1621)	WHLN	1410	U1	5	0.04	KY	Harlan
1625)	WRJD	1410	U4	5	0.29	NC	Durham
1622)	KOOQ	1410	U2	5	0.5	NE	North Platte
1623)	WELM	1410	U2	5	1	NY	Elmira
1624)	WNER	1410	U1	3.5	0.05	NY	Watertown
1626)	WING	1410	U4	5	5	OH	Dayton
1627)	KBNP	1410	U1	5	0.009	OR	Portland
1628)	WLSH	1410	D3	5		PA	Lansford
1629)	KQV	1410	U4	5	5	PA	Pittsburgh
1630)	WRIS	1410	U1	5	0.07	VA	Roanoke
1632)	WIZM	1410	U1	5	5	WI	La Crosse
1631)	WSCW	1410	D1	5		WV	South Charleston
1633)	KWYO	1410	U1	5	0.35	WY	Sheridan
1634)	WACT	1420	U1	5	0.1	AL	Tuscaloosa
1636)	KBHS	1420	U1	5	0.08	AR	Hot Springs
1635)	KMOG	1420	U2	2.5	0.5	AZ	Payson
1637)	KSTN	1420	U4	5	1	CA	Stockton
1638)	WLIS	1420	U4	5	0.5	CT	Old Saybrook
1640)	WDJA	1420	U4	5	0.5	FL	Delray Beach
1639)	WBRD	1420	U4	2.5	1	FL	Palmetto
1641)	WRCG	1420	U1	5	0.07	GA	Columbus
1642)	WKWN	1420	U1	2.5	0.11	GA	Trenton
1646)	WOC	1420	U4	5	5	IA	Davenport
1644)	KIGO	1420	U1	32	0.01	ID	Saint Anthony
1645)	WIMS	1420	U1	5	5	IN	Michigan City
1647)	WVJS	1420	U4	5	1	KY	Owensboro
1648)	WBSM	1420	U3	5	1	MA	New Bedford
1649)	KTOE	1420	U2	5	5	MN	Mankato
1650)	WIGG	1420	U1	5	0.07	MS	Wiggins
1651)	WASR	1420	U1	5	0.13	NH	Wolfeboro
1653)	WACK	1420	U4	5	0.5	NY	Newark
1652)	WLNA	1420	U4	5	1	NY	Peekskill
1654)	WHK	1420	U2	5	5	OH	Cleveland
1656)	WCOJ	1420	U1	5	5	PA	Coatesville
1655)	WCED	1420	U2	5	0.5	PA	Du Bois
1657)	WEMB	1420	U1	5	0.02	TN	Erwin
1658)	WKCW	1420	U1	22	0.06	VA	Warrenton
1660)	KITI	1420	U4	5	5	WA	Centralia-Chehalis
1659)	KUJ	1420	U1	5	0.9	WA	Walla Walla
1661)	WTCR	1420	U2	5	0.5	WV	Kenova
1662)	WFHK	1430	D1	5		AL	Pell City
1665)	KFIG	1430	U3	5	5	CA	Fresno
1663)	KMRB	1430	U4	50	9.8	CA	San Gabriel
1664)	KVVN	1430	U4	1	2.5	CA	Santa Clara
1666)	KEZW	1430	U2	10	5	CO	Aurora
1668)	WTMN	1430	U1	10	0.04	FL	Gainesville
1667)	WOIR	1430	U2	5	0.5	FL	Homestead
1669)	WLKF	1430	U1	5	1	FL	Lakeland
1670)	WLTG	1430	U4	5	5	FL	Panama City
1671)	WGFS	1430	U1	3.9	0.21	GA	Covington
1672)	WDAL	1430	U1	2.5	0.07	GA	Dalton
1673)	WATB	1430	U5	50	0.17	GA	Decatur
1674)	WXNT	1430	U2	5	5	IN	Indianapolis
1677)	WPNI	1430	U4	5	0.01	MA	Amherst
1676)	WKOX	1430	U2	5	1	MA	Everett
1675)	WNAV	1430	U2	5	1	MD	Annapolis
1678)	WION	1430	U2	4.7	0.33	MI	Ionia
1679)	KZQZ	1430	U4	5	5	MO	Saint Louis
1684)	WDEX	1430	U4	2.5	2.5	NC	Monroe
1685)	WMNC	1430	U1	2.7	0.04	NC	Morganton
1686)	WDJS	1430	U4	10	5	NC	Mount Olive
1680)	KRGI	1430	U2	5	1	NE	Grand Island
1681)	WNSW	1430	U2	5	1	NJ	Newark
1682)	KCRX	1430	U2	5	1	NM	Roswell
1683)	WENE	1430	U2	5	5	NY	Endicott
1687)	KTBZ	1430	U4	25	5	OK	Tulsa
1688)	KYKN	1430	U2	5	5	OR	Keizer
1689)	WVAM	1430	U2	5	1	PA	Altoona
1690)	WBLR	1430	U1	5	0.14	SC	Batesburg
1692)	WOWW	1430	U2	2.5	2.5	TN	Germantown
1691)	WPLN	1430	U2	15	1	TN	Madison
1694)	KEES	1430	U2	5	1	TX	Gladewater
1693)	KCOH	1430	U4	5	1	TX	Houston
1695)	KLO	1430	U4	10	5	UT	Ogden
1696)	WDIC	1430	D1	5		VA	Clinchco
1698)	KCLK	1430	U1	5	1	WA	Asotin
1697)	KBRC	1430	U2	5	1	WA	Mount Vernon
1699)	WLWI	1440	U1	5	5	AL	Montgomery
1701)	KTUV	1440	U2	5	0.24	AR	Little Rock
1700)	KAZG	1440	U1	5	0.05	AZ	Scottsdale
1703)	KVON	1440	U4	5	1	CA	Napa
1702)	KUHL	1440	U2	5	1	CA	Santa Maria
1704)	KRDZ	1440	U1	5	0.21	CO	Wray
1705)	WWCL	1440	U1	5	1	FL	Lehigh Acres
1706)	WPRD	1440	U2	5	1	FL	Winter Park
1707)	WGMI	1440	U1	2.5	0.06	GA	Bremen
1708)	WGIG	1440	U2	5	1	GA	Brunswick
1709)	KPTO	1440	U2	2.5	0.35	ID	Pocatello
1710)	WGEM	1440	U4	5	1	IL	Quincy
1711)	WROK	1440	U5	5	0.27	IL	Rockford
1712)	KMAJ	1440	U3	5	1	KS	Topeka
1714)	WVEI	1440	U2	5	5	MA	Worcester
1713)	WRED	1440	U2	5	5	ME	Westbrook
1715)	WMAX	1440	U2	5	2.5	MI	Bay City
1716)	KDIZ	1440	U2	5	0.5	MN	Golden Valley
1717)	WRBE	1440	D1	5		MS	Lucedale
1720)	WBLA	1440	U1	5	0.19	NC	Elizabethtown
1719)	WLXN	1440	U2	5	1	NC	Lexington
1718)	WFNY	1440	U2	5	0.5	NY	Gloversville

MW	Call	kHz	Ant.	D	N	Sta	City of License
1721)	WHKZ	1440	U4	5	5	OH	Warren
1722)	KMED	1440	U1	5	1	OR	Medford
1723)	KODL	1440	U2	5	1	OR	The Dalles
1725)	WCDL	1440	U1	5	0.03	PA	Carbondale
1724)	WNPV	1440	U4	2.5	0.5	PA	Lansdale
1726)	WGVL	1440	U2	5	5	SC	Greenville
1727)	WZYX	1440	U1	5	0.06	TN	Cowan
1729)	KPUR	1440	U2	5	1	TX	Amarillo
1728)	KETX	1440	U1	5	0.09	TX	Livingston
1730)	KTNO	1440	U4	50	0.35	TX	University Park
1731)	WKLV	1440	U1	5	0.07	VA	Blackstone
1734)	WNFL	1440	U4	5	0.5	WI	Green Bay
1733)	WHIS	1440	U1	5	0.5	WV	Bluefield
1732)	WAJR	1440	U4	5	0.5	WV	Morgantown
1735)	WHCJ	1460	U2	5	0.5	AL	Cullman
1736)	WHAL	1460	U1	4	0.14	AL	Phenix City
1737)	KTYM	1460	U4	5	0.5	CA	Inglewood
1738)	KION	1460	U3	10	10	CA	Salinas
1739)	KZNT	1460	U2	5	0.5	CO	Colorado Springs
1742)	WZEP	1460	U1	10	0.18	FL	Defuniak Springs
1741)	WNPL	1460	U4	7	2	FL	Golden Gate
1740)	WQOP	1460	U2	15	5	FL	Jacksonville
1743)	WXEM	1460	U1	5	0.19	GA	Buford
1746)	KXNO	1460	U2	5	5	IA	Des Moines
1745)	WKAM	1460	U2	2.5	0.5	IN	Goshen
1747)	WEKB	1460	U1	5	0.11	KY	Elkhorn City
1748)	WXOK	1460	U1	4.7	0.29	LA	Port Allen
1749)	WXBR	1460	U2	5	1	MA	Brockton
1750)	WBRN	1460	U2	5	2.5	MI	Big Rapids
1751)	KKAQ	1460	U1	2.5	0.15	MN	Thief River Falls
1752)	KHOJ	1460	U3	5	0.08	MO	Saint Charles
1758)	WEWO	1460	U4	5	5	NC	Laurinburg
1759)	WHBK	1460	U1	5	0.13	NC	Marshall
1760)	KLTC	1460	U2	5	5	ND	Dickinson
1753)	KXPN	1460	U1	5	0.05	NE	Kearney
1755)	WIFI	1460	U3	5	0.5	NJ	Florence
1754)	KENO	1460	U4	10	0.62	NV	Las Vegas
1756)	WDDY	1460	U2	5	5	NY	Albany
1757)	WHIC	1460	U2	3.7	5	NY	Rochester
1761)	WBNS	1460	U2	5	1	OH	Columbus
1762)	WTKT	1460	U2	5	4.2	PA	Harrisburg
1763)	WGMF	1460	U4	5	1	PA	Tunkhannock
1765)	KCLE	1460	U4	5	0.7	TX	Burleson
1764)	KBRZ	1460	U1	5	0.12	TX	Missouri City
1767)	WKDV	1460	U4	5	5	VA	Manassas
1766)	WRAD	1460	U4	5	0.5	VA	Radford
1768)	KARR	1460	U4	5	2.5	WA	Kirkland
1769)	KUTI	1460	U2	5	3.7	WA	Yakima
1770)	WBUC	1460	U1	5.5	0.02	WV	Buckhannon
1771)	KNXN	1470	U1	2.5	0.03	AZ	Sierra Vista
1772)	KUTY	1470	U4	5	5	CA	Palmdale
1773)	KIID	1470	U4	5	1	CA	Sacramento
1774)	WMMW	1470	U4	2.5	2.5	CT	Meriden
1775)	WMGG	1470	U1	5	0.5	FL	Dunedin
1776)	WWNN	1470	U4	50	2.5	FL	Pompano Beach
1777)	WRGA	1470	U2	5	5	GA	Rome
1779)	KWSL	1470	U4	5	5	IA	Sioux City
1778)	WMBD	1470	U4	5	5	IL	Peoria
1780)	WBFC	1470	U1	2.5	0.02	KY	Stanton
1781)	KLCL	1470	U1	5	0.5	LA	Lake Charles
1784)	WAZN	1470	U4	1.4	3.4	MA	Watertown
1783)	WJDY	1470	U3	5	0.04	MD	Salisbury
1782)	WLAM	1470	U3	5	5	ME	Lewiston
1785)	WFNT	1470	U4	5	1	MI	Flint
1786)	KMNQ	1470	U4	5	5	MN	Brooklyn Park
1787)	WNAU	1470	U2	2.5	0.5	MS	New Albany
1791)	WWBG	1470	U4	10	5	NC	Greensboro
1790)	WJPI	1470	D1	5		NC	Plymouth
1789)	WTOE	1470	U1	5	0.1	NC	Spruce Pine
1788)	WNYY	1470	U2	5	1	NY	Ithaca
1792)	WSAN	1470	U2	5	5	PA	Allentown
1793)	WQXL	1470	U1	11	0.1	SC	Columbia
1794)	WVOL	1470	U4	5	1	TN	Berry Hill
1795)	KYYW	1470	U2	5	1	TX	Abilene
1796)	KWRD	1470	D1	5		TX	Henderson
1798)	WBTX	1470	D1	5		VA	Broadway-Timberville
1797)	WTZE	1470	D1	5		VA	Tazewell
1800)	KELA	1470	U1	5	1	WA	Centralia-Chehalis
1799)	KBSN	1470	U4	5	1	WA	Moses Lake
1802)	WBKV	1470	U4	2.5	2.5	WI	West Bend
1801)	WRWB	1470	U1	5	0.07	WV	Huntington
1804)	WQOH	1480	U1	5	0.02	AL	Irondale
1803)	WABB	1480	U2	5	4.4	AL	Mobile
1806)	KTHS	1480	U1	5	0.06	AR	Berryville
1805)	KPHX	1480	U2	5	0.5	AZ	Phoenix
1808)	KGOE	1480	U1	5	1	CA	Eureka
1809)	KYOS	1480	U2	5	5	CA	Merced
1807)	KVNR	1480	U4	5	5	CA	Santa Ana
1810)	WKGC	1480	U1	5	0.03	FL	Southport
1812)	WYZE	1480	U1	5	0.04	GA	Atlanta
1811)	WGUS	1480	U2	5	5	GA	Augusta
1813)	KRXR	1480	U1	5	0.09	ID	Gooding
1814)	WPFR	1480	U1	5	1	IN	Terre Haute
1815)	KQAM	1480	U4	5	1	KS	Wichita
1816)	WGCK	1480	D1	5		KY	Neon
1817)	WSAR	1480	U3	5	5	MA	Fall River
1818)	WGVU	1480	U2	2	5	MI	Kentwood
1819)	WSDS	1480	U4	0.75	3.8	MI	Salem Township
1820)	KKCQ	1480	U1	5	0.09	MN	Fosston
1824)	WGFY	1480	U4	4.4	5	NC	Charlotte
1825)	WPFJ	1480	U1	5	0.01	NC	Franklin
1822)	WLEA	1480	U1	2.5	0.01	NY	Hornell
1821)	WZRC	1480	U4	5	5	NY	New York
1823)	WADR	1480	D1	5		NY	Remsen
1826)	WHBC	1480	U4	15	5	OH	Canton
1827)	WDJO	1480	U4	4.5	0.3	OH	Cincinnati
1828)	WUBA	1480	U4	5	1	PA	Philadelphia
1829)	WBBP	1480	U1	5	0.04	TN	Memphis
1830)	KNIT	1480	U4	5	1.9	TX	Dallas
1832)	KLVL	1480	U4	5	0.5	TX	Pasadena
1831)	KCHL	1480	U3	2.5	0.09	TX	San Antonio
1836)	WPWC	1480	U4	5	0.5	VA	Dumfries-Triangle
1834)	WTOX	1480	U4	6.3	1.5	VA	Glen Allen
1835)	WTOY	1480	U1	5	0.02	VA	Salem
1833)	WCFR	1480	U1	5	0.02	VT	Springfield
1837)	KBMS	1480	U2	1	2.5	WA	Vancouver
1838)	WLMV	1480	U4	5	5	WI	Madison
1839)	KSJX	1500	U4	10	5	CA	San Jose
1840)	WFIF	1500	D3	5		CT	Milford
1841)	WFED	1500	U4	50	50	DC	Washington
1842)	WDPC	1500	D4	5		GA	Dallas
1844)	WBRI	1500	D3	5		IN	Indianapolis
1845)	WLQV	1500	U4	50	10	MI	Detroit
1846)	KSTP	1500	U2	50	50	MN	Saint Paul
1847)	KFNN	1510	U3	22	0.1	AZ	Mesa
1849)	KIRV	1510	D3	10		CA	Fresno
1848)	KSPA	1510	U4	10	1	CA	Ontario
1850)	KPIG	1510	U4	8	2.4	CA	Piedmont
1851)	KCKK	1510	U4	10	25	CO	Littleton
1852)	WWBC	1510	D4	50		FL	Cocoa
1853)	WWZN	1510	U7	50	50	MA	Boston
1854)	WJKN	1510	D3	5		MI	Jackson
1856)	KCTE	1510	D3	10		MO	Independence
1855)	KMRF	1510	D4	5		MO	Marshfield
1857)	WFAI	1510	D3	2.5		NJ	Salem
1859)	WWSM	1510	D3	5		PA	Annville-Cleona
1858)	WPGR	1510	U4	5	0.001	PA	Monroeville
1860)	KMSD	1510	U1	5	0.01	SD	Milbank
1861)	WLAC	1510	U2	50	50	TN	Nashville
1862)	KBED	1510	D3	5		TX	Nederland
1863)	KLLB	1510	D1	10		UT	West Jordan
1864)	KGA	1510	U4	50	15	WA	Spokane
1865)	WRRD	1510	D4	23		WI	Waukesha
1867)	KMPG	1520	D4	5		CA	Hollister
1866)	KVTA	1520	U4	10	1	CA	Port Hueneme
1868)	WBZW	1520	U4	5	0.35	FL	Apopka
1869)	WEXY	1520	U2	3.5	0.25	FL	Wilton Manors
1870)	WDCY	1520	D1	2.5		GA	Douglasville
1871)	WHOW	1520	D1	5		IL	Clinton
1872)	WLGC	1520	D1	5		KY	Greenup
1873)	KFXZ	1520	U6	10	0.5	LA	Lafayette
1875)	WIZZ	1520	D3	10		MA	Greenfield
1874)	WTRI	1520	D3	17		MD	Brunswick
1876)	KOLM	1520	U8	10	0.8	MN	Rochester
1877)	KRHW	1520	U7	5	1.6	MO	Sikeston
1880)	WDSL	1520	D1	5		NC	Mocksville
1879)	WARR	1520	D1	5		NC	Warrenton
1878)	WWKB	1520	U3	50	50	NY	Buffalo
1881)	KOKC	1520	U2	50	50	OK	Oklahoma City
1882)	KGDD	1520	U3	50	15	OR	Oregon City
1883)	KYND	1520	D5	3		TX	Cypress
1884)	KVDW	1530	D1	2.5		AR	England
1885)	KFBK	1530	U4	50	50	CA	Sacramento
1886)	KCMN	1530	U1	15	0.01	CO	Colorado Springs
1887)	WDJZ	1530	D3	5		CT	Bridgeport
1888)	WYMM	1530	D3	50		FL	Jacksonville
1889)	WTTI	1530	D4	10		GA	Dalton
1890)	WLCO	1530	D3	5		MI	Lapeer
1891)	KQSP	1530	U3	8.6	0.01	MN	Shakopee
1892)	WRPM	1530	D1	10		MS	Poplarville
1893)	WLLQ	1530	D3	10		NC	Chapel Hill
1894)	WCKY	1530	U2	50	50	OH	Cincinnati

MW	Call	kHz	Ant.	D	N	Sta	City of License
1895)	KXTD	1530	D3	5		OK	Wagoner
1899)	KZNX	1530	U7	10	0.22	TX	Creedmoor
1897)	KGBT	1530	U8	50	10	TX	Harlingen
1898)	KLBW	1530	D1	2.5		TX	New Boston
1896)	KCLR	1530	D1	5		TX	Ralls
1900)	KASA	1540	U3	10	0.01	AZ	Phoenix
1901)	KMPC	1540	U4	50	37	CA	Los Angeles
1903)	KXEL	1540	U2	50	50	IA	Waterloo
1904)	WACA	1540	D1	5		MD	Wheaton
1907)	WYNC	1540	D1	2.5		NC	Yanceyville
1905)	WXEX	1540	U1	5	0.003	NH	Exeter
1906)	WDCD	1540	U3	50	50	NY	Albany
1908)	WNWR	1540	D3	50		PA	Philadelphia
1909)	WECZ	1540	D1	5		PA	Punxsutawney
1910)	WTBI	1540	D1	10		SC	Pickens
1912)	KGBC	1540	U4	2.5	0.25	TX	Galveston
1911)	KEDA	1540	U4	5	1	TX	San Antonio
1913)	KZMP	1540	U4	32	0.75	TX	University Park
1914)	WREJ	1540	U4	10	0.007	VA	Richmond
1915)	KXPA	1540	U2	5	5	WA	Bellevue
1916)	WLOR	1550	U4	50	0.04	AL	Huntsville
1917)	KUAZ	1550	D1	50		AZ	Tucson
1918)	KWRN	1550	U2	5	0.5	CA	Apple Valley
1919)	KXEX	1550	U4	5	2.5	CA	Fresno
1920)	KFRC	1550	U4	10	10	CA	San Francisco
1921)	WDZK	1550	U4	5	2.4	CT	Bloomfield
1924)	WNZF	1550	U2	8.7	0.25	FL	Bunnell
1922)	WRHC	1550	U4	10	0.5	FL	Coral Gables
1923)	WAMA	1550	U1	10	0.13	FL	Tampa
1926)	WTHB	1550	U1	5	0.01	GA	Augusta
1925)	WAZX	1550	U5	50	0.01	GA	Smyrna
1927)	WKTF	1550	U1	10	0.02	GA	Vienna
1928)	WPFC	1550	U1	5	0.04	LA	Baton Rouge
1929)	WNTN	1550	U1	10	0.003	MA	Newton
1931)	KAPE	1550	U3	5	0.04	MO	Cape Girardeau
1930)	KESJ	1550	U2	5	5	MO	Saint Joseph
1932)	KLFJ	1550	U1	5	0.02	MO	Springfield
1934)	KIVA	1550	U1	10	0.02	NM	Albuquerque
1933)	KXTO	1550	U1	2.5	0.09	NV	Reno
1935)	KYAL	1550	U3	2.5	0.04	OK	Sapulpa
1936)	WITK	1550	U4	10	0.5	PA	Pittston
1937)	WBSC	1550	U2	10	5	SC	Bennettsville
1938)	WIGN	1550	U1	35	0.006	TN	Bristol
1939)	WQZQ	1550	U1	2.5	0.01	TN	Clarksville
1940)	KMRI	1550	U1	10	0.34	UT	West Valley City
1942)	WKBA	1550	D3	10		VA	Vinton
1941)	WVAB	1550	U1	5	0.009	VA	Virginia Beach
1944)	KRPI	1550	U4	50	10	WA	Ferndale
1943)	KKAD	1550	U2	50	12	WA	Vancouver
1946)	WHIT	1550	D3	5		WI	Madison
1945)	WMRE	1550	U1	5	0.006	WV	Charles Town
1947)	WCMA	1560	D1	50		AL	Daleville
1948)	KNZR	1560	U2	25	10	CA	Bakersfield
1949)	WINV	1560	D1	5		FL	Beverly Hills
1950)	WINT	1560	D1	5		FL	Melbourne
1951)	KLNG	1560	D1	10		IA	Council Bluffs
1952)	WPAD	1560	U7	10	1	KY	Paducah
1953)	WNWN	1560	D3	4.1		MI	Portage
1954)	WQEW	1560	U4	50	50	NY	New York
1955)	WCNW	1560	D3	5		OH	Fairfield
1956)	WAGL	1560	D3	50		SC	Lancaster
1957)	KKAA	1560	U4	10	10	SD	Aberdeen
1958)	KGOW	1560	U4	50	15	TX	Bellaire
1959)	KTXZ	1560	U4	2.5	2.5	TX	West Lake Hills
1960)	WSBV	1560	D1	2.5		VA	South Boston
1961)	KVAN	1560	U4	10	0.7	WA	Burbank
1962)	KZIZ	1560	U2	5	0.9	WA	Pacific
1963)	WCRL	1570	U1	2.5	0.06	AL	Oneonta
1965)	KCVR	1570	U4	5	0.5	CA	Lodi
1966)	KPRO	1570	U3	5	0.19	CA	Riverside
1964)	KTGE	1570	U4	5	5	CA	Salinas
1967)	KPIO	1570	U1	7	0.01	CO	Loveland
1969)	WTWB	1570	U1	5	0.01	FL	Auburndale
1968)	WVOJ	1570	U1	10	0.03	FL	Fernandina Beach
1970)	WIGO	1570	U1	5	0.05	GA	Morrow
1972)	WFRL	1570	U3	5	0.5	IL	Freeport
1974)	WNSH	1570	U1	30	0.08	MA	Beverly
1973)	WNST	1570	U1	5	0.23	MD	Towson
1975)	KYCR	1570	U1	3.8	0.23	MN	Golden Valley
1976)	KAKK	1570	U1	9.5	0.25	MN	Walker
1978)	KBCV	1570	U4	5	3	MO	Hollister
1977)	WIZK	1570	D1	3.2		MS	Bay Springs
1980)	WNCA	1570	U1	5	0.28	NC	Siler City
1981)	WECU	1570	U1	3.8	0.02	NC	Winterville
1979)	WFLR	1570	U1	5	0.44	NY	Dundee
1983)	WPGM	1570	U1	2.5	0.22	PA	Danville
1982)	WISP	1570	U4	5	0.9	PA	Doylestown
1985)	WNKX	1570	U1	5	0.07	TN	Centerville
1984)	WCLE	1570	U1	5	0.08	TN	Cleveland
1986)	WYTI	1570	U1	2.5	0.22	VA	Rocky Mount
1987)	WLKD	1570	U1	5	0.5	WI	Minocqua
1988)	WVOK	1580	U1	2.5	0.02	AL	Oxford
1989)	KMIK	1580	U2	50	50	AZ	Tempe
1990)	KBLA	1580	U4	50	50	CA	Santa Monica
1991)	KREL	1580	U1	10	0.14	CO	Colorado Springs
1994)	WNTF	1580	D3	10		FL	Bithlo
1992)	WTCL	1580	D1	10		FL	Chattahoochee
1993)	WSRF	1580	U4	10	5	FL	Fort Lauderdale
1995)	WGVN	1580	U3	10	0.04	KY	Georgetown
1996)	WHFS	1580	U4	50	0.27	MD	Morningside
1997)	WPMP	1580	U4	5	0.05	MS	Pascagoula-Moss Pt
1998)	WLIM	1580	U8	10	0.5	NY	Patchogue
1999)	WVKO	1580	U4	3.2	0.29	OH	Columbus
2000)	WDAB	1580	U1	5	0.001	SC	Travelers Rest
2001)	WNPZ	1580	D1	5		TN	Knoxville
2002)	WLIJ	1580	U1	5	0.01	TN	Shelbyville
2003)	WTTN	1580	U5	5	0.004	WI	Columbus
2004)	WVNA	1590	U2	5	1	AL	Tuscumbia
2006)	KBJT	1590	U1	4.7	0.03	AR	Fordyce
2005)	KYNG	1590	U1	2.5	0.05	AR	Springdale
2008)	KLIV	1590	U1	5	5	CA	San Jose
2007)	KUNX	1590	U4	5	5	CA	Ventura
2010)	WPSL	1590	U1	5	0.06	FL	Port Saint Lucie
2009)	WRXB	1590	U4	5	1	FL	Saint Pete Beach
2013)	WALG	1590	U4	5	1	GA	Albany
2012)	WQCH	1590	D1	5		GA	Lafayette
2011)	WXRS	1590	U1	2.5	0.02	GA	Swainsboro
2015)	WAIK	1590	U3	5	0.05	IL	Galesburg
2016)	WNTS	1590	U4	5	0.5	IN	Beech Grove
2017)	KVGB	1590	U2	5	5	KS	Great Bend
2018)	WTVB	1590	U2	5	1	MI	Coldwater
2019)	KCNN	1590	U4	5	1	MN	East Grand Forks
2020)	WZRX	1590	U2	5	1	MS	Jackson
2026)	WCSL	1590	U1	10	0.03	NC	Cherryville
2025)	WHPY	1590	U3	5	0.02	NC	Clayton
2021)	KTCH	1590	U3	2.5	0.04	NE	Wayne
2023)	WSMN	1590	U1	5	5	NH	Nashua
2022)	KQLO	1590	U1	5	0.06	NV	Sun Valley
2024)	WGGO	1590	U1	5	0.01	NY	Salamanca
2027)	WAKR	1590	U2	5	5	OH	Akron
2028)	KTIL	1590	U2	5	1	OR	Tillamook
2029)	WHGT	1590	U2	5	1	PA	Chambersburg
2031)	WPWA	1590	U2	2.5	1	PA	Chester
2030)	WPSN	1590	U1	2.5	0.01	PA	Honesdale
2032)	WARV	1590	U4	5	5	RI	Warwick
2033)	WKTP	1590	U3	5	5	TN	Jonesborough
2035)	KGAS	1590	U1	2.5	0.12	TX	Carthage
2036)	KELP	1590	U4	5	0.8	TX	El Paso
2034)	KMIC	1590	U2	5	5	TX	Houston
2037)	WFTH	1590	U1	5	0.01	VA	Richmond
2038)	KLFE	1590	U2	5	5	WA	Seattle
2039)	WIXK	1590	U1	5	0.09	WI	New Richmond
2040)	WHIY	1600	U1	5	0.5	AL	Huntsville
2041)	WXVI	1600	U1	5	1	AL	Montgomery
2042)	KNWA	1600	U1	5	0.05	AR	Bellefonte
2043)	KGST	1600	U2	5	5	CA	Fresno
2045)	KAHZ	1600	U2	5	5	CA	Pomona
2044)	KUBA	1600	U2	5	2.5	CA	Yuba City
2046)	KEPN	1600	U2	5	5	CO	Lakewood
2047)	WAMS	1600	U1	5	5	DE	Dover
2048)	WZNZ	1600	U1	5	0.08	FL	Atlantic Beach
2049)	WHTY	1600	U4	5	4.7	FL	Riviera Beach
2050)	WAOS	1600	U1	20	0.06	GA	Austell
2051)	KGYM	1600	U2	5	5	IA	Cedar Rapids
2052)	KLEB	1600	U4	5	0.25	LA	Golden Meadow
2053)	WUNR	1600	U3	20	20	MA	Brookline
2054)	WHNP	1600	D1	2.5		MA	East Longmeadow
2055)	WAAM	1600	U4	5	5	MI	Ann Arbor
2056)	KPNP	1600	U3	5	5	MN	Watertown
2057)	KATZ	1600	U2	5	3.5	MO	Saint Louis
2061)	WIDU	1600	U4	5	0.14	NC	Fayetteville
2058)	KRKE	1600	U1	10	0.17	NM	Albuquerque
2060)	WEHH	1600	U4	5	0.17	NY	Elmira Heights
2059)	WWRL	1600	U4	25	5	NY	New York
2062)	KOPB	1600	U2	5	1	OR	Eugene
2063)	WAYC	1600	U1	2.7	0.01	PA	Bedford
2064)	WKZK	1600	U1	4	0.02	SC	North Augusta
2065)	WATX	1600	U1	2.5	0.02	TN	Algood
2066)	WMQM	1600	U1	50	0.03	TN	Lakeland
2067)	KRVA	1600	U4	25	0.93	TX	Cockrell Hill
2068)	KOKE	1600	U4	5	0.7	TX	Pflugerville

MW	Call	kHz	Ant.	D	N	Sta	City of License
2069)	KTUB	1600	U2	5		1	UT	Centerville
2070)	WCPK	1600	U1	4.2	0.02	VA	Chesapeake
2071)	WXMY	1600	D1	5		VA	Saltville
2072)	KVRI	1600	U4	50		10	WA	Blaine
2075)	WRPN	1600	U4	5		5	WI	Ripon
2074)	WZZW	1600	U1	5	0.02	WV	Milton
2073)	WKKX	1600	U1	5	0.03	WV	Wheeling
2076)	KSMH	1620	U1	10		1	CA	West Sacramento
2077)	WNRP	1620	U1	10		1	FL	Gulf Breeze
2078)	WDND	1620	U1	10		1	IN	South Bend
2079)	KOZN	1620	U1	10		1	NE	Bellevue
2080)	WTAW	1620	U1	10		1	TX	College Station
2081)	KYIZ	1620	U1	10		1	WA	Renton
2082)	WRDW	1630	U1	10		1	GA	Augusta
2083)	KCJJ	1630	U1	10		1	IA	Iowa City
2084)	KKGM	1630	U1	10		1	TX	Fort Worth
2085)	KRND	1630	U1	10		1	WY	Fox Farm
2086)	KDIA	1640	U2	10		10	CA	Vallejo
2087)	WTNI	1640	U1	10		1	MS	Biloxi
2088)	KFXY	1640	U4	10		1	OK	Enid
2089)	KDZR	1640	U1	10		1	OR	Lake Oswego
2090)	KBJA	1640	U1	10		1	UT	Sandy
2091)	WKSH	1640	U1	10		1	WI	Sussex
2092)	KYHN	1650	U1	10		1	AR	Fort Smith
2093)	KFOX	1650	U1	10	0.49	CA	Torrance
2094)	KBJD	1650	U1	10		1	CO	Denver
2095)	KCNZ	1650	U1	10		1	IA	Cedar Falls
2096)	KSVE	1650	U1	8.5	0.85	TX	El Paso
2097)	WHKT	1650	U1	10		1	VA	Portsmouth
2098)	KTIQ	1660	U1	10		1	CA	Merced
2099)	WCNZ	1660	U1	10		1	FL	Marco Island
2100)	KXTR	1660	U1	10		1	KS	Kansas City
2101)	WQLR	1660	U1	10		1	MI	Kalamazoo
2103)	WBCN	1660	U1	10		1	NC	Charlotte
2104)	KQWB	1660	U1	10		1	ND	West Fargo
2102)	WWRU	1660	U4	10		10	NJ	Jersey City
2105)	KRZI	1660	U1	10		1	TX	Waco
2106)	KXOL	1660	U1	10		1	UT	Brigham City
2108)	KHPY	1670	U4	10		9	CA	Moreno Valley
2107)	KNRO	1670	U1	10		1	CA	Redding
2109)	WFSM	1670	U1	10		1	GA	Dry Branch
2110)	WTDY	1670	U1	10		1	WI	Madison
2111)	KGED	1680	U1	10		1	CA	Fresno
2112)	WOKB	1680	U1	10		1	FL	Winter Garden
2113)	KRJO	1680	U1	10		1	LA	Monroe
2114)	WPRR	1680	U1	10	0.68	MI	Ada
2115)	WTTM	1680	U1	10		1	NJ	Lindenwold
2116)	KNTS	1680	U1	10		1	WA	Seattle
2117)	KFSG	1690	U1	10		1	CA	Roseville
2118)	KDDZ	1690	U1	10		1	CO	Arvada
2119)	WMLB	1690	U1	10		1	GA	Avondale Estates
2120)	WVON	1690	U1	10		1	IL	Berwyn
2121)	WPTX	1690	U1	10		1	MD	Lexington Park
2122)	WEUP	1700	U1	10		1	AL	Huntsville
2123)	WJCC	1700	U1	10		1	FL	Miami Springs
2124)	KBGG	1700	U1	10		1	IA	Des Moines
2125)	KVNS	1700	U1	8.8	0.88	TX	Brownsville
2126)	KKLF	1700	U1	10		1	TX	Richardson

Addresses

1) 495 Elder Ave #7, Sand City, CA 93955-3547 – **2)** 1139 Hartnell Ave., Redding, CA 96002-2113 – **3)** 2500 Maitland Center Pkwy #401, Maitland, FL 32751- 4122 – **4)** 1501 13th Ave, Columbus, GA 31901-1908 – **5)** 200 N 10th St, Fort Dodge, IA 50501-3925 – **6)** 1109 Hudson Lane, Monroe, LA 71201-6003 – **7)** 304 South Grand Ave, Las Vegas, NM 87701-3873 – **8)** 2395 Ocean Ave #3, Ronkonkoma, NY 11779-5670 – **9)** 3305 Durham Drive #111, Raleigh, NC 27603-3579 – **10)** 400 Ardmore Drive, Pittsburgh, PA 15221-3019 – **11)** 2070 North Palofox St, Pensacola, FL 32501-2145 – **12)** 800 East Dimond Blvd #3-370, Anchorage, AK 99515-2058 – **13)** 4686 E Van Buren St #300, Phoenix, AZ 85008-6967 – **14)** 3223 Sillect Ave., Bakersfield, CA 93308-6329 – **15)** 1111 W. Victory Way, Craig, CO 81625-2950 – **16)** 2500 Russell Rd., Green Cove Springs, FL 32043-9492 – **17)** 1102 Thompson Bridge Rd, Gainesville, GA 30501-1706 – **19)** 1815 Meadowlark Rd, Clay Center, KS 67432-8201 – **20)** 638 West Port Plaza, Saint Louis, MO 63146-3106 – **21)** 660 Dewey Blvd, Butte, MT 59701-2318 – **22)** 500 Corporate Parkway #200, Buffalo, NY 14226-1263 – **23)** 3500 E. Rosser Ave, Bismarck, ND 58501-3398 – **24)** 8044 Montgomery Rd #650, Cincinnati, OH 45236-2959 – **25)** 7140 SW Macadam Ave, Portland, OR 97219-3013 – **26)** 4050 Eisenhauer Rd, San Antonio, TX 78218-3409 – **27)** 1330 East 8th St #207, Odessa, TX 79761-4731 – **28)** 9 Stowe St, Waterbury, VT 05670-1820 – **29)** 1820 Heritage Center Way, Harrisonburg, VA 22801-8451 – **30)** 4840 Lincoln Rd, Blaine, WA 98230-9602 or PO Box 75150 RPO White Rock, White Rock, BC V4B 5L3 – **31)** 557 Scott St,

Wausau, WI 54403-4829 – **32)** 2518 Columbia Hwy, Dothan, AL 36303-5402 – **33)** 900 Front St, San Francisco, CA 94111-1427 – **34)** 2821 S Parker Rd #1205, Aurora, CO 80014-2708 – **35)** 194 NW 187th St, Miami, FL 33169-4050 – **36)** 25 NW Point Blvd #400, Elk Grove, IL 60007-1030 – **37)** PO Box 608, Middlesboro, KY 40965-0608 – **38)** 420 Western Ave, South Portland, ME 04106-1704 – **39)** 242 Finzel Rd, Frostburg, MD 21532-4009 – **40)** 1331 Main St, Springfield, MA 01103-1669 – **41)** 14 East Central Entrance, Duluth, MN 55811-5508 – **42)** 3000 Chestnut Expressway, Springfield, MO 65802-2528 – **43)** 20 3rd St N #231, Great Falls, MT 59401-3188 – **44)** 117 Ridge Pike, Lafayette Hill, PA 19444-1900 – **45)** 316 Greystone Blvd, Columbia, SC 29210-8007 – **46)** 6080 Mount Moriah Rd Ext., Memphis, TN 38115-2698 – **47)** 1815 Division St #110, Nashville, TN 37203-2753 – **48)** 2885 Interstate 10 E, Beaumont, TX 77702-1001 – **49)** 231 N. Wenatchee Ave, Wenatchee, WA 98801-2009 – **50)** 102 N. Kanawha St, Beckley, WV 25801-4715 – **51)** 304 South 4th St, Gadsden, AL 35901-5213 – **52)** 3400 Olive Ave #550, Burbank, CA 91505-5544 – **53)** PO Box 880, Alturas, CA 96101-0580 – **54)** 5211 West Laurel St, Tampa, FL 33607-1736 – **55)** 1801 Rockville Pike, Rockville, MD 20852-1633 – **56)** 1355 California Ave, Las Cruces, NM 88001-4130 – **57)** 777 Terrace Ave #16, Hasbrouck Heights, NJ 07604-3100 – **58)** 500 Plum St #100, Syracuse, NY 13204-1427 – **59)** 13 Summerlin Rd, Asheville, NC 28806-2800 – **60)** 7461 South Ave, Youngstown, OH 44512-5789 – **61)** WNAX Bldg - 1609 East Hwy 50, Yankton, SD 57078-6406 – **62)** 3500 Maple Ave. #1600, Dallas, TX 75219-3945 – **63)** 2801 Decker Lake Drive, West Valley City, UT 84119-2330 – **64)** 140 4th Ave North #340, Seattle, WA 98109-4932 – **65)** 11 North 12th St, Petersburg, AK 99833-0650 – **66)** 1011 North Craycroft Rd #400, Tucson, AZ 85711-7313 – **67)** 1071 West Shaw St, Fresno, CA 93771-3702 – **68)** 106 Rose Lane, Montrose, CO 81401-3823 – **69)** 4192 North John Young Pkwy, Orlando, FL 32804-2696 – **70)** 4051 Jimmie Dyess Pkwy, Augusta, GA 30909-9469 – **71)** 827 Park Blvd #201, Boise, ID 83712-7782 – **72)** Campbell Hall - 300 N Goodwin Ave, Urbana, IL 61801-2316 – **73)** 1200 SW Executive Drive, Topeka, KS 66615-3850 – **74)** 601 Washington St, Alexandria, LA 71301-8028 – **75)** 96 Stereo Lane, Paxton, MA 01612-1376 – **76)** 314 E. Front St, Traverse City, MI 49684-2528 – **77)** 240 Radio Rd, West Jefferson, NC 28694-ND – **78)** 600 Corporate Cir #100, Harrisburg, PA 17110-9787 – **79)** 1111 Virginia St East, Charleston, WV 25301-2406 – **80)** 201 State St, La Crosse, WI 54601-3246 – **81)** 301 Arctic Slope Ave #200, Anchorage, AK 99518-3035 – **82)** 208 Buena Vista Rd, Hot Springs, AR 71913-8208 – **83)** 2520 Lake Tahoe Blvd #E, South Lake Tahoe, CA 96150-7726 – **84)** 701 N. Brand Blvd #550, Glendale, CA 91203-1235 – **85)** 1834 Lisenby Ave., Panama City, FL 32405-3713 – **86)** 900 Circle 75 Pkwy SE #1320, Atlanta, GA 30339-3095 – **88)** 1406 Commerce Way, Idaho Falls, ID 83401-1233 – **89)** 300 W. Vine St 3rd Flr, Lexington, KY 40507-1807 – **90)** 500 Victory Rd #2, Quincy, MA 02171-3132 – **91)** 222 S. Lawrence St, Ironwood, MI 49938-2524 – **92)** 4200 W. Main St, Kalamazoo, MI 49006-2766 – **93)** 5030 North 72nd St, Omaha, NE 68134-2363 – **94)** 6 Johnson Rd, Latham, NY 12110-5638 – **95)** 4002 NC Hwy 42 West, Wilson, NC 27893-7774 – **96)** 1200 Executive Pkwy #440, Eugene, OR 97401-2169 – **97)** 600 Baltimore Drive, Wilkes-Barre, PA 18702-7901 – **98)** 8309 North Interstate 35, Austin, TX 78753-5771 – **99)** 750 Ridgeview Drive #204, Saint George, UT 84770-2665 – **100)** 808 East Sprague Ave, Spokane, WA 99202-2126 – **101)** 9660 Granite Ridge Drive, San Diego, CA 92123-2657 – **102)** 4270 Byrd Drive, Loveland, CO 80538-7074 – **103)** 10245 Centurion Pkwy North #109, Jacksonville, FL 32256-0569 – **104)** 600 Old Marion Rd NE, Cedar Rapids, IA 52402-2152 – **105)** 330 2nd Ave, Paintsville, KY 41240-1034 – **106)** 670 Sweden St, Caribou, ME 04736-3419 – **107)** 711 West 40th St #350, Baltimore, MD 21211-2190 – **108)** 2995 US Highway 93 South, Kalispell, MT 59901-8640 – **109)** 875 W. 5th St, Winston-Salem, NC 27101-2505 – **110)** 2400 8th Ave SW, Jamestown, ND 58401-6623 – **111)** 2650 Thousand Oaks Blvd #4100, Memphis, TN 38118-2451 – **112)** 4180 North Mesa St, El Paso, TX 79902-1420 – **113)** 1001 E. Southeast Loop 323 #455, Tyler, TX 75701-9600 – **114)** 950 22nd St North #1000, Birmingham, AL 35203-5312 – **115)** 352 "E" Ave #K4, Lancaster, CA 93535-4505 – **116)** 260 Hegenberger Rd, Oakland, CA 94621-1491 – **117)** 614 Kimbark St, Longmont, CO 80501-4911 – **118)** 7601 Riviera Blvd, Miramar, FL 33023-6574 – **119)** 715 East Central Entrance, Duluth, MN 55811-5596 – **120)** 7000 Squibb Rd, Mission, KS 66202-3233 – **121)** 195 McGregor St #810, Manchester, NH 03102-3755 – **122)** 500 4th St NW, Albuquerque, NM 87102-5324 – **123)** 1520 South Blvd #300, Charlotte, NC 28203-3701 – **124)** 2323 West 5th Ave #200, Columbus, OH 43204-4988 – **125)** 511 Rossanley Drive, Medford, OR 97501-1771 – **126)** 2 Bala Plaza 7th Flr, Bala Cynwyd, PA 19004-1501 – **127)** 24 East Greenway Plaza #1900, Houston, TX 77046-2428 – **128)** 810 West 200 North, Logan, UT 84321-3726 – **129)** 3934 Electric Rd, Roanoke, VA 24018-4513 – **130)** 2823 W. Lewis St, Pasco, WA 99301-6700 – **131)** 50 Highway 26, Alabaster, AL 35007-4865 – **132)** PO Box 109, Homer, AK 99603-0109 – **133)** 7740 N. 16th St, Phoenix, AZ 85020-4479 – **134)** 1360 E. Sherwood Drive, Grand Junction, CO 81501-7546 – **135)** 4002 W. Gandy Blvd, Tampa, FL 33611-

3410 – **136)** 806 New Franklin Rd., La Grange, GA 30240-1859 – **138)** 861 Broadway, Bangor, ME 04401-2916 – **139)** 1375 Beasley Rd, Jackson, MS 39206-2018 – **140)** 2508 Coney Island Ave 2nd Flr, Brooklyn, NY 11223-5026 – **141)** 500 Plum St #100, Syracuse, NY 13204-1427 – **142)** 3100 Highwoods Blvd #140, Raleigh, NC 27604-1065 – **143)** 4949 SW. Macadam Ave, Portland, OR 97201-3912 – **144)** 1918 Lincoln Hwy, North Versailles, PA 15137-2706 – **145)** 2440 Millwood Ave, Columbia, SC 29205-1128 or PO Box 2355, West Columbia, SC 29171-2355 – **146)** 1621 E. Magnolia Ave, Knoxville, TN 37917-7825 – **147)** 2221 E. Lamar Blvd #300, Arlington, TX 76006-7419 – **148)** 118 Malletts Bay Ave, Colchester, VT 05446-2009 – **149)** 306 S Kanawha St, Beckley, WV 25801-5619 – **150)** 720 E. Capitol Drive, Milwaukee, WI 53212-1308 – **151)** 3161 Channel Dr. #2, Juneau, AK 99801-7815 – **152)** PO Box 474, Nenana, AK 99760-0474 – **153)** 4695 S. Monaco St, Denver, CO 80237-3403 – **154)** 4400 Jenifer St NW #400, Washington, DC 20015-2183 – **155)** 214 Television Circle, Savannah, GA 31406-4519 – **156)** 827 Park Blvd #201, Boise, ID 83712-7782 – **157)** 2601 Nicholasville Rd, Lexington, KY 40503-3307 – **158)** 10845 Olive Blvd #160, Saint Louis, MO 63141-7792 – **159)** 2900 Sutro St, Reno, NV 89512-1616 – **160)** 320 Central Ave #519, Coos Bay, OR 97420-2272 – **161)** 1502 Wampanoag Trail, Riverside, RI 02915-1075 – **162)** 9601 McAllister Freeway #1200, San Antonio, TX 78216-4686 – **163)** 19319 Fremont Ave N, Shoreline, WA 98133-3800 – **164)** PO Box 468, Bethel, AK 99559-0468 – **165)** 3400 Olive Ave #550, Burbank, CA 91505-5544 – **166)** 2100 Park Central Blvd #100, Pompano Beach, FL 33064-2219 – **167)** 1819 Peachtree Rd. NE #700, Atlanta, GA 30309-1849 – **168)** 204 Communications Bldg - Iowa State Univ, Ames, IA 50011-0001 – **169)** 108 Green St, Thibodaux, LA 70301-3144 – **170)** 131 County Circle, Amherst, MA 01003-9257 – **171)** 2050 Amsterdam Rd, Belgrade, MT 59714-8957 – **172)** 501 Office Center Drive #190, Fort Washington, PA 19034-3268 – **173)** 1009 Drayton Rd, Fayetteville, NC 28303-3887 – **174)** 7755 Freedom Ave NW, North Canton, OH 44720-6905 – **175)** 4045 NW 64th St #600, Oklahoma City, OK 73116-2615 – **176)** 6401 Poplar Ave #640, Memphis, TN 38119-4808 – **177)** 162 Free Hill Rd, Gray, TN 37615-3144 – **178)** 800 E. Dimond Blvd #3-370, Anchorage, AK 99515-2058 – **179)** 1440 Ethan Way #200, Sacramento, CA 95825-2214 – **181)** 807 West 37th St, Hibbing, MN 55746-2856 – **182)** 2804 Opryland Drive, Nashville, TN 37214-1209 – **183)** 1600 West 500 North, Manti, UT 84642-5503 – **184)** 1912 Capitol Ave #300, Cheyenne, WY 82001-3659 – **185)** 2800 Dauphin St #104, Mobile, AL 36606-2400 – **186)** 819 1st Ave #A, Fairbanks, AK 99701-4449 – **187)** PO Box 2569, Window Rock, AZ 86515-2569 – **188)** 3000 W MacArthur Blvd #500, Santa Ana, CA 92704-7497 – **189)** 1010 2nd St. North , Sauk Rapids, MN 56379-2527 – **190)** 345 Hudson St Fl11, New York, NY 10014-4502 – **191)** 410 E. 6th St, Williston, ND 58801-5552 o – **192)** 895 Country Club Rd #A200, Eugene, OR 97401-6015 – **193)** 2420 Wade Hampton Blvd, Greenville, SC 29615-1107 – **194)** 6400 North Belt Line Rd #110, Irving, TX 75063-6065 – **195)** 2029 Freeway Drive, Mount Vernon, WA 98273-5470 – **196)** 11474 US Hwy 11, York, AL 36925-9764 – **197)** 565 Seward Highway, Dillingham, AK 99576-ND – **198)** 108 Highway 70 East #11, Glenwood, AR 71943-8800 – **199)** 3301 Barham Blvd #300, Los Angeles, CA 90068-1477 – **200)** 2821 S Parker Rd #1205, Aurora, CO 80014-2708 – **201)** 330 SW 27th Ave #207, Miami, FL 33135-2957 – **203)** 1419 W. Bannock St, Boise, ID 83702-5234 – **204)** 180 North Stetson St #1000, Chicago, IL 60601-6822 – **205)** 3999 Las Vegas Blvd S #K, Las Vegas, NV 89119-1097 – **206)** 12 E. Market St, Lewistown, PA 17044-2123 – **207)** 517 Watt Rd, Knoxville, TN 37922-1110 – **208)** 2202 Mt. Jolliff Rd, Chesapeake, VA 23321-1416 – **209)** PO Box 109, Barrow, AK 99723-0109 – **210)** 55 Hawthorne St. #1100, San Francisco, CA 94105-3932 – **211)** 3535 Piedmont Court NE #14-1200, Atlanta, GA 30305-4608 – **212)** 1726 Reisterstown Rd #117, Pikesville, MD 21208-2986 – **213)** 20 Guest St 3rd Flr, Brighton, MA 02135-2040 – **214)** Tower 14-21700 Northwestern Hwy #1190, Southfield, MI 48075-4923 – **215)** 604 Ludington St, Escanaba, MI 49829-3830 – **216)** 4104 Country Lane, Saint Joseph, MO 64506-4921 o – **217)** 1400 11th Ave #3, Helena, MT 59601-7996 – **218)** 320 N. Jensen Rd, Vestal, NY 13850-2111 – **219)** 3012 Highwoods Blvd #200, Raleigh, NC 27604-1031 – **220)** 1835 Moriah Woods Blvd, Memphis, TN 38117-7122 – **221)** 8122 Datapoint Drive #600, San Antonio, TX 78229-3446 – **222)** 320 Emery Drive, Omak, WA 98841-9237 – **223)** 1111 Virginia St East, Charleston, WV 25301-2406 – **224)** 2396 Hallie Rd, Chippewa Falls, WI 54729-7519 – **225)** 244 Goodwin Crest Dr. #300, Birmingham, AL 35209-3700 – **226)** 261 Portsea St, New Haven, CT 06519-2104 – **227)** 8000 Belfort Parkway #100, Jacksonville, FL 32256-6971 – **229)** 306 W 8th St, Coffeyville, KS 67337-5829 – **230)** 1218 Decatur St #B, New Orleans, LA 70116-2608 – **231)** Tower 14-21700 Northwestern Hwy #1190, Southfield, MI 48075-4923 – **232)** 4045 N. Mesa St, El Paso, TX 79902-1526 – **233)** 11373 Wallace Pike, Bristol, VA 24202-2743 – **234)** 126 Kessel Rd, Fisher, WV 26818-4012 – **235)** 1399 West 34th Ave #202, Anchorage, AK 99503-3659 – **236)** 67 Garden Court, Monterey, CA 93940-5302 – **237)** 5028 Wisconsin Ave NW #304, Washington, DC 20016-4118 – **238)**

362 Green St, Gardner, MA 01440-1348 – **239)** 8044 Montgomery Rd #650, Cincinnati, OH 45236-2959 – **240)** 196 Main St, Winston, OR 97496-ND – **241)** 11451 Katy Freeway #125, Houston, TX 77079-2004 – **242)** 1903 W Research Way, Salt Lake City, UT 84119-5684 – **243)** 500 West Boone Ave, Spokane, WA 99201-2404 – **244)** 501 North 44th St #425, Phoenix, AZ 85008-6587 – **245)** 1425 River Park Dr. #520, Sacramento, CA 95815-4524 – **246)** 800 W Olympic Blvd #A-200, Los Angeles, CA 90015-1360 – **247)** 3131 S. Vaughn Way #601, Aurora, CO 80014-3516 – **248)** 800 S. Douglas Rd #111, Coral Gables, FL 33134-3187 – **249)** 855 College St, Eastman, GA 31023-6771 – **250)** 402 Main St, Williamsburg, KY 40769-1126 – **251)** 6341 West Port Ave, Shreveport, LA 71129-2415 – **252)** 5800 Foxridge Drive, Mission, KS 66205-2333 – **253)** 111 Broadway 3rd Flr, New York, NY 10006-1992 – **254)** 3228 South US Hwy 117, Rose Hill, NC 28458-8498 – **255)** 3500 E. Rosser Ave, Bismarck, ND 58501-3398 – **256)** 3505 Olsen Blvd #117, Amarillo, TX 79109-3096 – **257)** 7080 Lee Highway, Fairlawn, VA 24141-8416 – **258)** 1820 Eastlake Ave East, Seattle, WA 98102-3711 – **259)** 715 East Central Entrance, Duluth, MN 55811-5596 – **260)** PO Box 78, Kotzebue, AK 99752-0078 – **261)** 3988 N Roscoe Rd, Hernando, FL 34442-3141 – **262)** 154 Boone Drive, Newman, GA 30263-2801 – **264)** 435 N. Michigan Ave, Chicago, IL 60611-4076 – **265)** 1455 East Tropicana Ave #800, Las Vegas, NV 89119-8326 – **266)** 3400 New Hendersonville Rd, Pisgah Forest, NC 28768-8614 – **267)** 28041 Pleasant Valley Rd, Sweet Home, OR 97386-9599 – **268)** 4050 Eisenhauer Rd, San Antonio, TX 78218-3409 – **269)** 2194 US Hwy 319 South, Thomasville, GA 31792-1417 – **270)** 1910 University Drive, Boise, ID 83725-0399 – **271)** 1036 Highway 541, Jackson, KY 41339-9434 – **272)** 326 Chicopee St, Chicopee, MA 01013-1797 or PO Box 1, Springfield, MA 01101-0001 – **273)** 636 Haugen St, Billings, MT 59101-5671 – **274)** 1366 Startown Rd, Lincolntown, NC 28152-5033 – **275)** 7 Parkway Center - 875 Greentree Rd #625, Pittsburgh, PA 15220-3508 – **276)** 1801 Rockville Pike #405, Rockville, MD 20852-5604 – **277)** 1 Commerce St #300, Montgomery, AL 36104-3542 – **278)** 3183 Airway Ave #D, Costa Mesa, CA 92626-4611 – **279)** 865 Battery St, San Francisco, CA 94111-1503 – **280)** 6805 Corporate Drive #130, Colorado Springs, CO 80919-5903 – **281)** 6699 N. Federal Hwy #200, Boca Raton, FL 33487-1671 – **282)** 2500 Maitland Center Pkwy #401, Maitland, FL 32751-4122 – **283)** 5440 Moeller Ave, Cincinnati, OH 45212-1211 – **284)** 100-25 Queens Blvd #1CC, Forest Hills, NY 11375-2417 – **285)** 2147 Springs Rd, Mount Airy, NC 27030-2447 – **286)** 1020 South 25th St, Fargo, ND 58103-2312 – **287)** 7136 S. Yale Ave. #500, Tulsa, OK 74136-6325 – **288)** 2000 West Loop South #300, Houston, TX 77027-3510 – **289)** 300 St Croix Trail South, Lakeland, MN 55043-ND or PO Box 25130, Saint Paul, MN 55125-0130 – **290)** 301 Arctic Slope Ave #200, Anchorage, AK 99518-3035 – **291)** 1601 W. Peachtree St. NE, Atlanta, GA 30309-2663 – **292)** 5625 North Milwaukee Ave, Chicago, IL 60646-6221 – **293)** 18553 Gentry Rd, Lebanon, MO 65536-5748 – **294)** 36581 N. Reservoir Rd, Polson, MT 59860-8677 – **295)** 128 South 4th St, O'Neill, NE 68763-1814 – **296)** 1050 W Williams Ave, Fallon, NV 89406-2634 – **297)** 0234 S.W. Bancroft St, Portland, OR 97201-4237 – **298)** 2211 E. Missiouri Ave # S-300, El Paso, TX 79903-3831 – **299)** 1899 Carbonville Rd, Helper, UT 84526-ND or PO Box 875, Price, UT 84501-0875 – **300)** 301 Brookswood St #208, North Little Rock, AR 72120-4200 – **301)** 7677 Engineer Rd, San Diego, CA 92111-1582 – **302)** 4695 S. Monaco St, Denver, CO 80237-3408 – **303)** 2090 Palm Beach Lake Blvd #701, West Palm Beach, FL 33409-6508 – **304)** 3514 W Arch St #200, Tampa, FL 33607-4901 – **306)** 10550 Barkley St, Overland Park, KS 66212-1824 – **307)** 70 James St #140, Worcester, MA 01603-1038 – **308)** 3011 W. Grand Blvd #800, Detroit, MI 48202-3086 – **309)** 137 Rapids Rd , Champlain, NY 12919-4945 – **310)** 2828 NC 126, Morganton, NC 28655-8264 – **311)** 6222 West Interstate 10, San Antonio, TX 78201-2097 – **312)** 1717 US Hwy 72 E, Athens, AL 35611-4413 – **313)** 128 Pioneer Drive, Valdez, AK 99686-ND – **314)** 10948 Cleveland Ave, Oakdale, CA 95361-9709 – **315)** 20125 S. Tamiami Trail, Estero, FL 33928-2117 – **316)** 330 21st Ave. South #610, Minneapolis, MN 55455-4550 – **317)** 818 Main St, Miles City, MT 59301-3221 – **318)** 500 4th St NW, Albuquerque, NM 87102-2102 – **319)** 904 Center St, Lewiston, NY 14092-1737 or 600 The East Mall, Suite #400, Toronto, ON M9B 4B1 – **320)** 2 Penn Plaza #1700, New York, NY 10121-1701 – **321)** 275 River Rd, Rockingham, NC 28380-1536 – **322)** 3201 Royalty Row, Irving, TX 75062-4961 – **323)** 504 Middle Creek Rd, Cedar Bluff, VA 24609-ND – **324)** 1820 Eastlake Ave East, Seattle, WA 98102-3711 – **325)** 801 Noble St #30, Anniston, AL 36201-5698 – **326)** PO Box 988, Nome, AK 99762-0988 – **327)** 3400 W Highway 89a, Sedona, AZ 86336-4914 – **328)** 180 North Stetson St #1100, Chicago, IL 60601-6723 – **329)** 274 Britton Rd, Monticello, ME 04760-3110 – **330)** 243 Main St, Norway, ME 04268-5914 – **331)** 265 Highpoint Drive, Ridgeland, MS 39157-6018 – **332)** 595 E. Plumb Lane, Reno, NV 89502-3503 – **333)** 17336 US Highway 421 South, Dunn, NC 28334-5580 – **334)** 1381 W. Main St, Forest City, NC 28043-2525 – **335)** 1901 North Moore St #200, Arlington, VA 22209-1746 – **336)** 142 Skyland Blvd. East, Tuscaloosa, AL 35405-4096 – **337)**

PO Box 249, Glennallen, AK 99588-0249 – **338)** 3202 North Oracle Rd, Tucson, AZ 85705-3820 – **339)** 113 East New Hope Rd, Rogers, AR 72758-6058 – **340)** 1101 Marsh Rd, Eureka, CA 95501-1574 – **341)** 3321 S. La Cienega Blvd, Los Angeles, CA 90016-3114 – **342)** 351 W Cromwell Ave #108, Fresno, CA 93711-6115 – **343)** 20450 NW 2nd Ave, Miami, FL 33169-2505 – **344)** 32900 Radio Rd, Leesburg, FL 34788-3903 – **345)** 3350 Peachtree Rd. NE #1610, Atlanta, GA 30326-1040 – **347)** 1065 S. Range Ave, Colby, KS 67701-3505 – **348)** 4000 Radio Drive #1, Louisville, KY 40218-4568 – **349)** 1795 Tittabawassee Rd, Saginaw, MI 48604-9431 – **350)** 222 N. 32nd St - 10th Floor, Billings, MT 59101-1911 – **351)** 1607 Country Club Drive, High Point, NC 27626-4559 – **352)** 1020 S. 25th St, Fargo, ND 58103-3212 – **353)** 1541 Alta Drive #400, Whitehall, PA 18052-5632 – **354)** 1502 Wampanoag Trail, Riverside, RI 02915-1075 – **355)** 1835 Moriah Woods Blvd, Memphis, TN 38117-7122 – **356)** 231 Brandonwood Drive, Johnson City, TN 37604-2156 – **357)** 4413 82nd St #300, Lubbock, TX 79424-3395 – **358)** 2000 West Loop South #300, Houston, TX 77027-3510 – **359)** 500 Dominion Tower - 999 Waterside Drive, Norfolk, VA 23510-3300 – **361)** 2219 Yew St Rd, Bellingham, WA 98229-8898 – **361)** 1601 E. 57th Ave, Spokane, WA 99223-6623 – **362)** 944 Harlem St, Altoona, WI 54707-1127 – **363)** 1107 W 8th St #2, Juneau, AK 99801-1896 – **364)** 213 East 2nd St, Soda Springs, ID 83276-1411 – **365)** 462 Merrimack St, Methuen, MA 01844-5804 – **366)** 1010 2nd St. North, Sauk Rapids, MN 56379-2527 – **367)** 2775 Mt Ephraim Ave, Camden, NJ 08104-3295 – **368)** 1919 North Broadway Ave, Oklahoma City, OK 73103-4499 – **369)** 1032 Melody Lane, Crewe, VA 23930-ND – **370)** 134 4th Ave, Huntington, WV 25701-1253 or 9801 Radio Park Rd, Catlettsburg, KY 41129-8824 – **371)** 200 Tower Rd, Waupaca, WI 54981-1699 – **372)** 188 John Turner BRdcast Blvd., Jacksonville, AL 36265-6659 or PO Box 8, Anniston, AL 36202-0008 – **373)** 900 Front St, San Francisco, CA 94111-1450 – **374)** 999 Douglas Ave #3318, Altamonte Springs, FL 32714-5213 – **375)** 2352 Jaycee Shack Rd, Valdosta, GA 31602-6475 – **376)** 1501 Hargis Lane, Jackson, KY 41339-1102 – **377)** 2422 Burton St SE, Grand Rapids, MI 49546-4806 – **378)** 130 Radio Station Drive, Magee, MS 39111-4399 – **379)** 6721 West 121st St, Overland Park, KS 66209-2003 – **380)** 102 Taos St, Santa Fe, NM 87505-3832 – **381)** 1203 Troy-Schenectady Rd #201, Latham, NY 12205-5579 – **382)** 2 Beeco Rd, Greer, SC 29650-1004 – **383)** 1162 Junction Ave, Sturgis, SD 57785-2149 – **384)** 2514 Eugenia Ave, Nashville, TN 37211-2117 – **385)** 145 Jackson St NE, Blacksburg, VA 24060-3931 – **386)** 55 Alder St NW #3 , Ephrata, WA 98823-1663 – **387)** 819 1st Ave #A, Fairbanks, AK 99701-4449 – **388)** 4300 West Cypress St #1040, Tampa, FL 33607-4185 – **389)** 6012 S Pulaski Rd, Chicago, IL 60629-4538 – **390)** 3400 Idaho Ave NW #200, Washington, DC 20016-3000 – **391)** 2205 College Ave, Elmira, NY 14903-1201 – **392)** 160 Varick St, New York, NY 10013-1220 – **393)** 2400 Olentangy River Rd, Columbus, OH 43210-1027 – **394)** 2221 E. Lamar Blvd #300, Arlington, TX 76006-7419 – **395)** 3701 Harrison Rd, Ogden, UT 84403-2059 – **396)** 4301 W. Hundred Rd, Chester, VA 23831-1737 – **397)** 2201 6th Ave #1500, Seattle, WA 98121-1840 – **398)** 7355 N. Orcale Rd. #102, Tucson, AZ 85704-6353 – **399)** 1255 E. Main St. #A, Grass Valley, CA 95945-5711 – **400)** 2000 E Gene Autry Way, Anaheim, CA 92806-6143 – **402)** 3841 Veterans Memorial Blvd. #201, Metairie, LA 70002-5624 – **403)** 82 Franklin St, Worcester, MA 01608-1982 – **404)** 625 2nd Ave South #200, Minneapolis, MN 55402-1961 – **405)** 932 County Rd 448, Poplar Bluff, MO 63901-9018 – **406)** 4405 Providence Lane #D, Winston-Salem, NC 27106-3226 – **407)** 34 North 4th St, Reading, PA 19601-3996 – **408)** 1423 S. Beverly St, Casper, WY 82609-4131 – **409)** 6530 Spanish Fort Blvd #B, Spanish Fort, AL 36527-5014 or PO Box 1328, Mobile, AL 36633-1328 – **410)** 1192 Norwegian Ave, Modesto, CA 95350-3643 – **411)** 221 Pall Bearer Rd, Thomasville, GA 31792-1101 – **412)** 4000 Radio Drive #1, Louisville, KY 40218-4568 – **413)** 3735 Rigolette Rd., Pineville, LA 71360-7365 – **414)** 1011 N. Lincoln St. , West Point, NE 68788-1003 – **415)** 6655 W. Sahara Ave #D110, Las Vegas, NV 89146-0846 – **416)** 4801 East Independence Blvd #815, Charlotte, NC 28212-5490 – **417)** 1201 North Jackson Ave #900, McAllen, TX 78501-5764 – **418)** PO Box 7111, Charlottesville, VA 22906-7111 – **419)** 1114 North Almon St, Moscow, ID 83843-8507 – **420)** 120 Summit Pkwy #200, Birmingham, AL 35209-4741 – **421)** PO Box 820, Nome, AK 99762-0820 – **422)** 4695 S. Monaco St, Denver, CO 80237-3403 – **423)** 198 Main St, Danbury, CT 06810-6662 – **424)** 3200 Weimer Hall, Gainesville, FL 32611 – **425)** 2100 Park Central Blvd #100, Pompano Beach, FL 33064-2219 – **427)** 5625 N. Milwaukee Ave, Chicago, IL 60646-6221 – **428)** 20 Guest St 3rd Flr, Brighton, MA 02135-2040 – **429)** 1120 E. McCuen St, Duluth, MN 55808-2199 – **430)** 18844 Highway 80, Forest, MS 39074-4410 – **431)** 85 Founders Lane, Saint Louis, MO 63105-3085 – **432)** 4601 Six Forks Rd #520, Raleigh, NC 27609-5210 – **433)** 1301 E 9th St #252, Cleveland, OH 44114-1800 – **434)** 104 South Center St #400, Ebensburg, PA 15931-1656 – **435)** 517 N Watt Rd, Knoxville, TN 37922-1110 or PO Box 760, Louisville, TN 37777-0760 – **436)** 8828 N Stemmons Fwy #106, Dallas, TX 75247-3720 – **437)** 3000 Bering Drive, Houston, TX 77057-5708 – **438)** 500 Dominion Tower - 999 Waterside Drive, Norfolk, VA 23510-3300 – **439)** 351 Elliott Ave West #300, Seattle, WA 98119-4150 – **440)** 300 Broadway #8, San Francisco, CA 94133-4545 – **441)** 5211 W Laurel St, Tampa, FL 33607-1736 – **442)** 601 West Roanoke Drive, Fitzgerald, GA 31750-3633 – **443)** 1465 Northside Dr. NW #218, Atlanta, GA 30318-4239 – **444)** 1162 East Hwy 126, Pittsburg, KS 66762-8712 – **445)** 305 Washington Ave 4th Flr, Towson, MD 21204-4748 – **446)** 425 Stockbridge Rd, Great Barrington, MA 01230-1233 – **447)** 6605 SE Lake Rd, Portland, OR 97222-2161 – **448)** 555 East City Ave #330, Bala Cynwyd, PA 19004-1137 – **449)** 8122 Datapoint Dr. #600, San Antonio, TX 78229-3446 – **450)** 434 Bearcat Drive, Salt Lake City, UT 84115-2520 – **451)** 240 Central Ave, Oak Hill, WV 25901-3006 – **452)** 2278 Wortham Lane, Grovetown, GA 30813-5103 – **453)** 870 Airport Rd , McGrath, AK 99627-ND – **454)** 701 N. Brand Blvd #550, Glendale, CA 91203-1235 – **455)** 400 Poydras St #900, New Orleans, LA 70130-3738 – **456)** 447 Congress St #3B, Portland, ME 04101-3505 – **457)** 283 Comm. Arts Bldg. - M.S.U., East Lansing, MI 48824-1212 – **458)** PO Box 49, Park Rapids, MN 56470-0049 – **459)** 6767 W Tropicana Ave #102, Las Vegas, NV 89103-4755 – **460)** 1751 Hanshaw Rd., Ithaca, NY 14850-9105 – **461)** PO Box 1149, Clayton, NC 30525-1149 – **462)** 340 Martin Luther King Blvd, Bristol, TN 37620-3996 – **463)** 2621 West A St, Pasco, WA 99301-4702 – **464)** 10000 Warden Rd., North Little Rock, AR 72120-3656 – **465)** 30 East San Joaquin St #105, Salinas, CA 93901-2946 – **466)** 2828 W Flagler St, Miami, FL 33135-1337 – **467)** 1186 West BRd St, Athens, GA 30606-3050 – **468)** 2432 US Hwy 2 East, Kalispell, MT 59901-2310 or PO Box 5409, Kalispell, MT 59903-5409 – **469)** 1007 Plum Creek Parkway , Lexington, NE 68850-2621 – **470)** PO Box 9090, Window Rock, AZ 86515-9090 – **471)** 524 West 57th St, New York, NY 10019-2924 – **472)** 13 Summerlin Rd, Asheville, NC 28806-2600 – **473)** 8101 N. High St #360, Columbus, OH 43235-1442 – **474)** 1405 E. Ellendale Ave, Dallas, OR 97338-1709 – **475)** 209 10th Ave South #342, Nashville, TN 37203-0758 – **476)** 3000 Bering Drive, Houston, TX 77057-5708 – **477)** 3650 131st SE #550, Bellevue, WA 98006-1334 – **478)** 619 Cameron St, Eau Claire, WI 54703-4700 – **479)** 1301 Central Pkwy SW, Decatur, AL 35601-4817 – **480)** 3913 Kachemak Way, Homer, AK 99603-7618 – **481)** 560 Higuera St #G, San Luis Obispo, CA 93401-3850 – **482)** 2800 Shallowford Rd NE, Atlanta, GA 30341-5217 – **483)** 1050 Clover Drive, Boise, ID 83703-5714 – **484)** 190 N. State St, Chicago, IL 60601-3398 – **485)** 122 Green St #2, Worcester, MA 01604-4138 – **486)** 6555 US Hwy 98 #8, Hattiesburg, MS 39402-8699 – **487)** 241 Riverchase Way #A, Lexington, SC 29072-9470 – **488)** 10613 Bellaire Blvd #900, Houston, TX 77072-5221 – **489)** 4501 North McColl Rd, McAllen, TX 78504-2431 – **490)** 750 Ridgeview Drive #204, Saint George, UT 84770-2665 – **491)** 145 Jackson St NE, Blacksburg, VA 24060-3931 – **492)** 1936 East Third Ave , Fort Yukon, AK 99740-0050 – **493)** 233 NE Front St, Milford, DE 19963-1431 – **494)** 2320 NE 2nd St #5, Ocala, FL 34470-6992 – **495)** 214 Television Circle, Savannah, GA 31406-4519 – **497)** 1240 Radio Drive, Pikeville, KY 41501-4779 – **498)** 3003 Snelling Ave North, Saint Paul, MN 55113-1599 – **499)** 398 South Main St, Granite Falls, NC 28630-8535 – **500)** 801 E DuBois Ave, DuBois, PA 15801-3643 – **501)** 2514 Eugenia Ave, Nashville, TN 37211-2117 – **502)** 10613 Bellaire Blvd #900, Houston, TX 77072-5221 – **503)** 207 University Blvd #200, Harrisonburg, VA 22801-3752 – **504)** PO Box 165, Galena, AK 99741-0165 – **505)** 4686 E Van Buren St #300, Phoenix, AZ 85008-6967 – **506)** 125 South 2nd St, Blytheville, AR 72315-3413 – **507)** 200 South A St #400, Oxnard, CA 93030-5717 – **508)** 340 Townsend St #4, San Francisco, CA 94107-1698 – **509)** 11865 Moreno Ave, Lakeside, CA 92040-1110 – **510)** 3455 W. 83rd Ave. Westminster, Denver, CO 80030-4005 – **511)** 135 Burnside Ave, East Hartford, CT 06108-3466 – **512)** 5211 West Laurel St, Tampa, FL 33607-1736 – **513)** 3765 N. John Young Parkway, Orlando, FL 32804-3213 – **514)** 710 S. Clinton St Bldg, Iowa City, IA 52242-4214 – **515)** 184 Target Industrial Circle #207, Bangor, ME 04401-5718 – **516)** 1000 Town Center #2810, Southfield, MI 48075-1183 – **517)** 3436 Highway 35 N, Meridian, MS 39301-1509 – **518)** 1301 North Main St, Roswell, NM 88201-5013 – **519)** 1223 W. New Bern Rd, Kinston, NC 28504-4713 – **520)** 1000 20th Ave SW, Minot, ND 58701-6447 – **521)** 6006 Grand Central Ave, Parkersburg, WV 26105-9125 or PO Box 5559, Vienna, WV 26105-5559 – **522)** 120 Beale Rd, Sarver, PA 16055-9403 – **523)** 5989 Susquehanna Plaza Drive, Hellam, PA 17406-8910 – **524)** 225 S Pleasantburg Dr #3B, Greenville, SC 29607-2533 – **525)** 162 Freehill Rd, Gray, TN 37615-3144 – **526)** 105 N Ash Ave, South Pittsburg, TN 37380-1565 – **527)** 4300 S. US Highway 281, Edinburg, TX 78539-9650 – **528)** 2801 Decker Lake Drive #100, West Valley City, UT 84119-2330 – **529)** 3245 Basie Rd, Richmond, VA 23228-3404 – **530)** 5110 SE Stark St, Portland, OR 97215-1751 – **531)** 16880 West US Highway 63, Hayward, WI 54843-7186 – **532)** 40960 Kalifornsky Beach Rd, Kenai, AK 99611-6445 – **533)** 700 Wellington Hills Rd, Little Rock, AR 72211-2026 – **534)** 2100 E Tahquitz Canyon Way, Palm Springs, CA 92262-7046 – **535)** 961 N. Emerald Ave #A, Modesto, CA 95351-1556 – **536)** 7350 US Hwy 50, Lamar, CO 81052-9563 – **537)** 1800 Turtle Mound Rd, Melbourne, FL 32934-8105 – **538)** 2970 Peachtree Rd. NW #700, Atlanta, GA 30305-

4919 – **539)** 712 3rd St, West Lafayette, IN 47907-2005 – **540)** 290 Hegenberger Rd, Oakland, CA 94621-1436 – **541)** PO Box 228, Mayking, KY 41837-0228 – **542)** 601 Central Ave North, Faribault, MN 55021-1307 – **543)** 8755 W Flamingo Rd, Las Vegas, NV 89147-8667 – **544)** 3256 Penryn Rd #100, Loomis, CA 95650-8052 – **545)** 715 Route 52, Beacon, NY 12508-1047 – **546)** 159 Santanoni Ave, Saranac Lake, NY 12983-2478 – **547)** 1109 Tower Drive, Burlington, NC 27215-4425 – **548)** 75 Airport St, Providence, RI 02905-4722 – **549)** 660 Flormann St #100, Rapid City, SD 57701-4679 – **550)** 7322 Southwest Frwy #500, Houston, TX 77074-2084 – **551)** 2495 N. Vernal Ave, Vernal, UT 84078-ND – **552)** 17700 Van Buren Drive, Dumfries, VA 22025-2036 – **553)** 500 W. Boone Ave, Spokane, WA 99201-2497 – **554)** 1700 SE Mile Hill Drive #201A, Port Orchard, WA 98366-3553 – **555)** 450 Leonard Ave Extension, Fairmont, WV 26554-3878 – **556)** 12100 W. Howard Ave, Greenfield, WI 53228-1851 – **557)** 873 South Alabama Ave, Monroeville, AL 36460-2507 – **558)** 750 Walnut St, Gadsden, AL 35901-4139 – **559)** 526 Stedman St., Ketchikan, AK 99901-6629 – **560)** 565 Seward Highway, Dillingham, AK 99576-ND – **561)** 3222 South Richey Ave, Tucson, AZ 85713-5453 – **562)** 1117 West Route 66, Flagstaff, AZ 86001-6213 – **563)** 1845 W Empire Ave, Burbank, CA 91504-3402 – **564)** PO Box 7069, Breckenridge, CO 80424-7069 – **565)** 190 Turner Dr #G, Durango, CO 81303-8231 – **566)** 11700 Central Pkwy, Jacksonville, FL 32224-2600 – **567)** 5211 West Laurel St, Tampa, FL 33607-1736 – **568)** 203 West Shotwell St, Bainbridge, GA 39819-3903 – **569)** 544 N. Arthur Ave, Pocatello, ID 83204-3002 – **570)** 329 Maine St, Quincy, IL 62301-3928 – **571)** 1496 Bellevue St #202, Green Bay, WI 54311-4205 – **572)** 804 College St, Bowling Green, KY 42101-2133 – **573)** 5966 Grove Hill Rd, Frederick, MD 21703-6012 – **574)** 390 Golden Ave, Battle Creek, MI 49015-4598 – **575)** 37208 US Hwy 169, Aitkin, MN 56431-4195 – **576)** 574 Hwy 51 N #F, Ridgeland, MS 39157-2607 – **577)** 1015 W. Pine St, Poplar Bluff, MO 63901-4839 – **578)** 3250 South Reserve St #200, Missoula, MT 59801-8236 – **579)** 113 W 4th St, Ogallala, NE 69153-2508 – **580)** 815 Lafayette Rd, Portsmouth, NH 03801-5406 – **581)** 449 Broadway Fl2, New York, NY 10013-2549 – **582)** 500 Corporate Parkway #200, Buffalo, NY 14226-1263 – **583)** 525 Evans St, Greenville, NC 27858-2311 – **584)** 11530 Carmel Commons Blvd, Charlotte, NC 28226-3976 – **585)** 4045 NW 64th St #600, Oklahoma City, OK 73116-2615 – **586)** 1250 Siskiyou Blvd., Ashland, OR 97520-5010 – **587)** 3304 S Highway 281, Aberdeen, SD 57401-8792 – **588)** 430 State Highway 165 #C, Branson, MO 65616-3541 – **589)** 9601 McAllister Freeway #1200, San Antonio, TX 78216-4686 – **590)** 105 Whitehall Rd, Lynchburg, VA 24501-6706 – **591)** PO Box 31000, Spokane, WA 99223-3016 – **592)** 401 11th St #200, Huntington, WV 25701-2235 – **593)** 821 University Ave, Madison, WI 53706-1412 – **594)** 1726 KROE Lane, Sheridan, WY 82801-9681 – **595)** 1415 Fulton St, Fresno, CA 93721-1609 – **596)** 7601 Riviera Blvd, Miramar, FL 33023-6574 – **597)** 544 Mulberry St #500, Macon, GA 31201-8258 – **599)** 3501 Broadway St, Mount Vernon, IL 62864-2202 – **600)** 1416 Locust St, Des Moines, IA 50309-3014 – **601)** 929 Howard Ave, New Orleans, LA 70113-1148 – **602)** 7119 W M-68, Indian River, MI 49749-9472 – **603)** 1189 North Jackson St, Houston, MS 38851-8273 – **604)** 126 W. 3rd St, Valentine, NE 69201-1826 – **605)** 401 Saw Mill Hollow Rd, Burnsville, NC 28714-9789 – **606)** 345 SW Cyber Drive #100, Bend, OR 97702-1045 – **607)** 305 N. Church St, Wartburg, TN 37887-3164 o – **608)** 615 Olive St, Texarkana, TX 75501-5512 – **609)** 6214 West 34th Ave, Amarillo, TX 79109-4006 – **610)** 1105 N Iron Springs Rd, Cedar City, UT 84720-6526 or P O Box 819, Cedar City, UT 84721-0819 – **611)** 13379 Great Springs Rd, Smithfield, VA 23430-6930 – **612)** 1011 Radio Drive, Grundy, VA 24614-6157 – **613)** 501 E. Broadway St, Forrest City, AR 72335-3801 – **614)** 985 Lincoln Way #103, Auburn, CA 95603-5255 – **615)** 7800 East Orchard Rd #400, Greenwood Village, CO 80111-2599 – **616)** 1188 Lake View Rd, Altamonte Springs, FL 32714-2713 – **617)** 1800 Lake Park Drive #99, Smyrna, GA 30080-7689 – **618)** 2973 US Hwy 84 West, Valdosta, GA 31601 – **619)** 2560 Snake River Ave, Lewiston, ID 83501-9685 – **620)** 624 3rd St South, Nampa, ID 83651-3840 – **621)** 541 North Fairbanks Court #1260, Chicago, IL 60611-3319 – **622)** 645 Industrial Drive, Franklin, IN 46131-9617 or PO Box 47307, Indianapolis, IN 46247-0307 – **623)** 501 Sycamore St #300, Waterloo, IA 50703-4651 – **624)** 7825 Tuckerman Ln #217, Potomac, MD 20854-3241 – **625)** 500 Victory Rd #2, Quincy, MA 02171-3132 – **626)** 26495 American Drive, Southfield, MI 48034-6114 – **627)** 63 Braswell Rd, Hattiesburg, MS 39401-9730 – **628)** 3109 South 10 Mile Drive, Jefferson City, MO 65109-1012 – **629)** 516 Fuller Ave, Helena, MT 59601-3301 – **630)** 1560 North Corbin St, Silver City, NM 88061-6526 – **631)** 403 W. Pueblo Drive, Espaola, NM 87532-2530 – **632)** 9418 State Route 49, Marcy, NY 13403-2342 – **633)** 1250 Siskiyou Blvd, Ashland, OR 97520-5010 – **634)** 1 Bala Plaza #424, Bala Cynwyd, PA 19004-1403 – **635)** 60 Markfield Drive #4, Charleston, SC 29407-7907 – **636)** 25 Garlington Rd, Greenville, SC 29615-4613 – **637)** 222 Mallory Station Rd., Franklin, TN 37067-0201 – **638)** 2000 West Loop South #300, Houston, TX 77027-3510 – **639)** 9800 University Ave, Lubbock, TX 79423-5302 – **640)** 701 German School Rd., Richmond, VA

23225-5357 – **641)** 351 Elliott Ave West #300, Seattle, WA 98119-4150 – **642)** 817 Suncrest Place, Charleston, WV 25303-2302 – **643)** 6530 Spanish Fort Blvd #B, Spanish Fort, AL 36527-5014 or PO Box 1328, Mobile, AL 36633-1328 – **644)** 600 Beacon Parkway W #400, Birmingham, AL 35209-3118 – **645)** 2425 E. Camelback Rd #570, Phoenix, AZ 85016-4250 – **646)** 260 Battle St, Marshall, AR 72650-9440 – **647)** 12370 Hesperia Rd #16, Victorville, CA 92392-5808 – **648)** 340 Townsend St #4, San Francisco, CA 94107-1698 – **649)** 495 Benham St, Hamden, CT 06514-2009 – **650)** 809 S. Westover Blvd, Albany, GA 31707-4953 – **651)** 1010 Tower Place, Bogart, GA 30622-3052 – **652)** 1301 E Douglas Rd, Mishawaka, IN 46545-1732 – **653)** 209 N. Elm St, Shenandoah, IA 51601-1139 – **654)** 1240 Radio Drive, Pikeville, KY 41501-4779 – **655)** 351 Tilghman Rd, Salisbury, MD 21804-1920 – **656)** 356 BRd St, Fitchburg, MA 01420-3030 – **657)** 1491 M-32 West, Alpena, MI 49707-8194 – **658)** 16405 Haven Rd, Little Falls, MN 56345-6400 – **659)** 324 Broadway St, Cape Girardeau, MO 63701-7331 – **660)** 3600 Highway 7 North, Baker, MT 59313-ND – **661)** 1928 East Portal Place, Scottsbluff, NE 69361-2727 – **662)** 1515 W. Main St, Farmington, NM 87401-3896 – **663)** 1500 Hegeman Ave, Colchester, VT 05446-3116 – **664)** 4405 Providence Lane #D, Winston-Salem, NC 27106-3226 – **665)** 1361 Colony Drive, New Bern, NC 28562-4129 – **666)** 404 Main St, Klamath Falls, OR 97601-6021 – **667)** 204 Desmond St, Sayre, PA 18840-2004 – **668)** 1703 Walnut Bottom Rd, Carlisle, PA 17013-9151 – **669)** 1301 South Abe St, San Angelo, TX 76903-7245 – **670)** 515 South 700 East #1C, Salt Lake City, UT 84102-2802 – **671)** 3934 Electric Rd, Roanoke, VA 24018-4513 – **672)** 830 N. Columbia Center Blvd #B2, Kennewick, WA 99336-7756 – **673)** 67 W Court Square, Troy, AL 36081-2611 – **674)** 1597 Military St South, Hamilton, AL 35570-5026 – **675)** 546 9th Ave, Fairbanks, AK 99701-4902 – **676)** 1838 Commerce Drive #A, Lakeside, AZ 85929-7007 – **677)** 1100 Mohawk St #280, Bakersfield, CA 93309-7417 – **678)** 1321 North Gene Autry Trail, Palm Springs, CA 92262-5473 – **679)** 201 N. Industrial Park Rd., Excelsior Springs, MO 64024-1736 – **680)** 4002 West Gandy Blvd #A, Tampa, FL 33611-3410 – **681)** 2970 Peachtree Rd. NW #700, Atlanta, GA 30305-4919 – **682)** 1501 Mount Vernon Rd, Vidalia, GA 30474-3031 – **683)** 120 South 300 West, Rupert, ID 83350-9667 – **684)** 125 S. Main St, Harlan, KY 40831-2109 – **685)** 9960 Corporate Campus Drive #3600, Louisville, KY 40223-4070 – **686)** 420 Western Ave, South Portland, ME 04106-1704 – **687)** 121 North Front St, Marquette, MI 49855-4300 – **688)** 109 East Clark St, Albert Lea, MN 56007-2420 – **689)** 27 N 27th St, Billings, MT 59101-2357 – **690)** 201 S. Bailey Ave, North Platte, NE 69101-5406 o – **691)** 1455 E Tropicana Ave #550, Las Vegas, NV 89119-6592 – **692)** 777 Terrace Ave #602, Hasbrouck Heights, NJ 07604-3113 – **693)** 140 Lower Terrace, Buffalo, NY 14202-4303 – **694)** 1190 Patton Ave, Asheville, NC 28806-2706 – **695)** 301 8th St South, Fargo, ND 58103-1826 – **696)** 3226 Jefferson Rd, Ashtabula, OH 44004-9112 – **697)** 5800 East Skelly Drive #150, Tulsa, OK 74135-6416 – **698)** 2040 SW 1st Ave, Portland, OR 97201-5302 – **699)** 200 Fleet St 4th Flr, Pittsburgh, PA 15220-2910 – **700)** 181 E. Evans St. #311, Florence, SC 29506-2512 – **701)** 207 University Blvd #200, Harrisonburg, VA 22801-3752 – **702)** PO Box 2482, Kirkland, WA 98083-2482 – **703)** 821 University Ave, Madison, WI 53706-1412 – **704)** 2705 E. Parkway Drive, Russellville, AR 72802-2006 – **705)** 5670 Wilshire Blvd #200, Los Angeles, CA 90036-5611 – **706)** 1101 Marsh Rd., Eureka, CA 95501-1574 – **707)** 229 Pajaro St #205, Salinas, CA 93901-3499 – **708)** 1801 Rockville Pike, Rockville, MD 20852-1633 – **709)** 312 East Nine Mile Rd #29D, Pensacola, FL 32514-1475 – **710)** 6699 N Federal Hwy #200, Boca Raton, FL 33487-1660 – **711)** 100 NW 76th Drive #2, Gainesville, FL 32607-6659 – **712)** 1691 Forsyth St, Macon, GA 31201-1407 – **713)** 854 Lindsay Blvd., Idaho Falls, ID 83402-1820 – **714)** 208 North Thomas Drive, Shreveport, LA 71107-6520 – **715)** 243 Central St, Lowell, MA 01852-2214 – **716)** 2110 Cliff Rd, Eagan, MN 55122-2347 – **717)** 206 North Front St, McComb, MS 39648-3916 – **718)** 7000 Squibb Rd, Mission, KS 66202-3233 – **719)** 1155 Gummow Drive, Fallon, NV 89406-9453 – **720)** 1203 Troy-Schenectady Rd #201, Latham, NY 12110-1046 – **721)** 3233 Burnt Mill Rd #4, Wilmington, NC 28403-2655 – **722)** 101 Pine St, Dayton, OH 45402-2925 – **723)** 305 Hwy 315, Pittston, PA 18640-3987 – **724)** 400 Pineview St, Rock Hill, SC 29730-3444 – **725)** 745 Main St, Deadwood, SD 57732-1015 – **726)** 11530 Carmel Commons Blvd, Charlotte, NC 28226-3976 – **727)** 912 Curtis Ave, Pasadena, TX 77502-2402 – **728)** 390 East Annabella Rd, Richfield, UT 84701-2692 – **729)** 901 E. Valley Drive, Bristol, VA 24201-4903 – **730)** 1200 Chesterly Drive #160, Yakima, WA 98902-7345 – **731)** 300 Harrison Ave, Weston, WV 26452-2100 – **732)** 1915 Mirro Drive, Manitowoc, WI 54220-6715 – **733)** 3871 North Commerce Drive, Tucson, AZ 85705-2983 – **734)** 145 Natoma St, San Francisco, CA 94105-3734 – **735)** 414 E. Cota St, Santa Barbara, CA 93101-1624 – **736)** 3131 S. Vaughn Way #601, Aurora, CO 80014-3516 – **737)** 440 Old Turnpike Rd, Plantsville, CT 06479-1678 – **738)** 610 Sycamore St #220, Celebration, FL 34747-4996 – **739)** 2150 W 68th St #202, Hialeah, FL 33016-1802 – **741)** 24 Frank Lloyd Wright Drive, Ann Arbor, MI 48105-9755 – **742)** 1569 N. Central Ave, Monett, MO 65708-

1104 – **743)** 2494 Browncroft Blvd, Rochester, NY 14625-1410 – **744)** 1650 Midland Rd, Southern Pines, NC 28387-2111 o – **745)** 117 Ridge Pike, Lafayette Hill, PA 19444-1900 – **746)** 109 Plaza Drive #2, Johnstown, PA 15905-1212 – **747)** 1185 North Main St, Providence, RI 02904-1824 – **748)** 5495 Murray Rd, Memphis, TN 38119-3703 – **749)** 4711 Old Kingston Pike, Knoxville, TN 37919-5207 – **750)** 12900 Preston Rd #100, Dallas, TX 75230-1312 – **751)** 1848 Clay St SE, Roanoke, VA 24013-2614 – **752)** 1835 Aston Drive, Carlsbad, CA 92008-7310 – **753)** 20872 NE Kelley Ave, Blountstown, FL 32424-1115 – **754)** 190 N. State St, Chicago, IL 60601-3302 – **755)** PO Box 369, Lexington, MS 39095-0369 – **756)** 4125 Carlisle Blvd NE, Albuquerque, NM 87107-4848 – **757)** 3134 Lake Rd, Horseheads, NY 14845-3103 – **758)** 1900 NW Expressway St #1000, Oklahoma City, OK 73118-1854 – **759)** 5100 South Tennis Lane, Sioux Falls, SD 57108-2212 – **760)** 110 India Rd, Paris, TN 38242-7565 – **761)** 140 4th Ave North #340, Seattle, WA 98109-4932 – **762)** 6475 Highway 78 #73, Cordova, AL 35550-4101 – **763)** 2800 North 44th St #100, Phoenix, AZ 85008-1560 – **764)** 1321 North Gene Autry Trail, Palm Springs, CA 92262-5473 – **765)** 145 Natoma St, San Francisco, CA 94105-3734 – **766)** 5100 Commerce Drive, Bakersfield, CA 93309-0684 – **767)** 220 State St, Fort Morgan, CO 80701-2116 – **768)** 9090 Hogan Rd, Jacksonville, FL 32216-4648 – **769)** 9721 Executive Circle Drive North #200, Saint Petersburg, FL 33702-2439 – **770)** 2901 Mountain Industrial Blvd., Tucker, GA 30084-3011 – **771)** 451 Highway 11 & 80, Meridian, MS 39301-2779 – **772)** 5615 Pershing Ave #12, Saint Louis, MO 63112-1757 – **773)** 103 Hanover St, Newport, NH 03766-1098 or PO Box 2295, New London, NH 03257-2295 – **774)** 345 Hudson St FI11, New York, NY 10014-4502 – **775)** 3 Porters Cove Rd, Asheville, NC 28805-2834 or PO Box 159, Black Mountain, NC 28711-0159 – **776)** 5110 SE Stark St, Portland, OR 97215-1751 – **777)** 1625 Hwy 109 North, Gallatin, TN 37066-8135 – **778)** 1019 Washington Ave, Waco, TX 76701-1256 – **779)** 3639 Wolffin Ave, Amarillo, TX 79102-2119 – **780)** 5100 Southwest Freeway, Houston, TX 77056-7308 – **781)** 3256 Penryn Rd #100, Loomis, CA 95650-8052 – **782)** 2202 Jolliff Rd, Chesapeake, VA 23321-1416 – **783)** 4700 Business Park Blvd #E-44A, Anchorage, AK 99503-7176 – **784)** 655 N. Central Ave #2500, Glendale, CA 91203-1447 – **785)** 67 Main St, Sharon, CT 06069-2018 – **786)** 2828 Coral Way #110, Coarl Gables, FL 33145-3214 – **787)** 131 Doe Run Circle, Thomasville, GA 31757-0923 – **788)** 5011 Capitol Ave, Omaha, NE 68132-2921 – **789)** 1700 La Luz Rd, Roswell, NM 88201-ND – **790)** 420 Ft Duquesne Blvd #100, Pittsburgh, PA 15222-1416 – **791)** 100 E. Shockley Ferry Rd, Anderson, SC 29624-3746 – **792)** 706 Butterfield Rd, Yakima, WA 98901-2021 – **793)** 3222 S. Richey Ave., Tucson, AZ 85704-7738 – **794)** 4209 N Frontage Rd, Fayetteville, AR 72703-5002 – **795)** 121 W Alvin Ave, Santa Maria, CA 93458-3002 – **796)** 1355 E Altamonte Drive, Altamonte Springs, FL 32701-5011 – **797)** 427 S. Wall Street, Calhoun, GA 30701-2431 – **798)** 3656 W Belmont Ave, Chicago, IL 60618-5328 – **799)** 1402 E. Kansas Ave., Garden City, KS 67846-5806 – **800)** 5210 Auth Rd #500, Suitland, MD 20746-4354 – **801)** 1170 Soldiers Field Rd, Boston, MA 02134-1004 – **802)** 42669 Garfield Rd. #328, Clinton Township, MI 48038-5024 or PO Box 1030, Sterling Heights, MI 48311-1030 – **803)** 900 Forestview Lane North, Plymouth, MN 55441-5934 – **804)** 18920 E Valley View Pkwy #C, Independence, MO 64055-7020 – **805)** 4405 Providence Lane #D, Winston-Salem, NC 27106-3226 – **806)** 4321 Stuart Andrew Blvd #E, Charlotte, NC 28217-1588 – **807)** 136 North 7th St, Reedsport, OR 97467-1503 – **808)** 3654 Park Ave, Memphis, TN 38111-5626 – **809)** 1602 South Brownlee Blvd, Corpus Christi, TX 78404-3134 – **810)** 210 West Cota St, Shelton, WA 98584-2264 – **811)** 303 8th St, Point Pleasant, WV 25550-1209 – **812)** 150 Nichols Ave, Casper, WY 82601-1816 – **813)** 5050 Edison Ave. #218, Colorado Springs, CO 80915-3450 – **814)** 4300 West Cypress St #1040, Tampa, FL 33607-4185 – **815)** 2100 Park Central Blvd #100, Pompano Beach, FL 33064-2219 – **816)** 1885 Beaver Ridge Cir NW, Norcross, GA 30071-3847 – **818)** 2141 Grand Ave, Des Moines, IA 50312-5303 – **819)** Two Penn Plaza 17th Floor, New York, NY 10121-0101 – **820)** 5620 S. Lima Rd, Avon, NY 14414-9791 – **821)** 3025 Waughtown St #G, Winston-Salem, NC 27107-1679 – **822)** 105 Lake Ave, Elyria, OH 44035-5013 – **823)** 151 E. 1st Ave, Everett, PA 15537-1351 – **824)** 1533 Amherst Rd, Knoxville, TN 37909-1204 – **825)** 5787 South Hampton Rd #285, Dallas, TX 75232-2290 – **826)** 55 Hawthorne St. #1100, San Francisco, CA 94105-3914 – **827)** 3256 Penryn Rd #100, Loomis, CA 95650-8052 – **828)** 5590 Rio Grande Ave., Jacksonville, FL 32254-1354 – **829)** 506 West 1st Ave, Crestview, FL 32536-2420 – **830)** 552 Laney-Walker Extension, Augusta, GA 30901-3014 – **831)** 3400 Idaho Ave NW #200, Washington, DC 20016-3000 – **832)** 1100 Victors Way #100, Ann Arbor, MI 48108-5220 – **833)** 608 State Highway 30, Pipestone, MN 56164-1458 – **834)** 508 Main St, Miles City, MT 59301-3047 – **835)** 8456 Smokey Hollow Rd, Baldwinsville, NY 13027-8222 – **836)** Two Penn Plaza 17th Floor, New York, NY 10121-0101 – **837)** 180 Radio Hill Rd, Franklin, NC 28734-6927 – **838)** 2080 Laura St, Springfield, OR 97477-2197 – **839)** 11640 Highway 17 Bypass, Murrells Inlet, SC 29576-9332 – **840)** 700 Monticello Ave #301, Norfolk, VA 23510-2538 – **841)**

539 Ragland Rd, Madison Heights, VA 24572-ND – **842)** PO Box 2482, Kirkland, WA 98083-2482 – **843)** 1601 E. 57th Ave, Spokane, WA 99223-6623 – **844)** 5 Rosemar Circle, Parkersburg, WV 26104-1203 – **845)** 1900 W. Carmen St, Guadalupe, AZ 85283-2559 – **846)** 40356 Oak Park Way, Oakhurst, CA 93612-8872 – **847)** 614 Kimbark St, Longmont, CO 80501-4911 – **848)** 4300 W Cypress St #1040, Tampa, FL 33607-4185 – **849)** 102 Parkwood Circle, Carrollton, GA 30117-8353 – **851)** 3303 E. Chicago St, Caldwell, ID 83605-6904 – **852)** RR1 Box 46A, McLeansboro, IL 62859-9701 – **853)** 401 Whitney Ave #160, Gretna, LA 70056-2573 – **854)** 100 Mt. Wayte Ave, Framingham, MA 01701-5705 – **855)** 2100 Fairplain Ave, Benton Harbor, MI 49022-6828 – **856)** 3185 S. Highland Drive #13, Las Vegas, NV 89109-1029 – **857)** 4801 East Independence Blvd #803, Charlotte, NC 28212-5497 – **858)** 4365 Fulton Drive NW, Canton, OH 44718-2823 – **859)** 400 Market St, Philadelphia, PA 19106-2513 – **860)** 214 W. Pleasant Drive, Pierre, SD 57501-2472 – **861)** 2211 E. Missiouri Ave #N-300, El Paso, TX 79903-3807 – **862)** 205 9th St, Farwell, TX 79325-ND – **863)** 3606 South 500 West, Salt Lake City, UT 84115-4208 – **864)** 244 Goodwin Crest Drive #300, Birmingham, AL 35209-3700 – **865)** 5670 Wilshire Blvd #200, Los Angeles, CA 90036-5611 – **866)** 2070 N. Palafox St, Pensacola, FL 32501-2145 – **867)** 4015 North Monroe St, Tallahassee, FL 32303-2139 – **868)** 40 Monument Circle #400, Indianapolis, IN 46204-3014 – **869)** 4200 N. Old Lawrence Rd, Wichita, KS 67219-3211 – **870)** 11 Bryant Ave SE, Wadena, MN 56482-1543 – **871)** 119 N. 3rd St, Hannibal, MO 63401-0711 – **872)** 112 East 3rd Ave, Plentywood, MT 59254-2223 – **873)** 372 S. Dorset St, South Burlington, VT 05403-6363 – **874)** 235 Walton St, Syracuse, NY 13202-1533 – **875)** 2929 Radio Station Rd, Greenville, NC 27834-0864 – **876)** 1227 County Line Rd, Sellinsgrove, PA 17870-8188 or PO Box 1070, Sunbury, PA 17801-0870 – **877)** 200 N. Highway 25 Bypass, Greenville, SC 29617-1108 – **878)** 2650 Thousand Oaks Blvd #4100, Memphis, TN 38118-2451 – **879)** 621 O'Grady Drive, Chattanooga, TN 37419-1305 – **880)** 6161 Savoy Drive #1200, Houston, TX 77036-3363 – **881)** 310 West Wall St #104, Midland, TX 79701-5123 – **882)** 1140 Rose Hill Drive, Charlottesville, VA 22903-5128 – **883)** 306 S Kanawha St, Beckley, WV 25801-5619 – **884)** 2651 S. Fish Hatchery Rd, Fitchburg, WI 53711-5410 or PO Box 99, Madison, WI 53701-0099 – **885)** 19245 Hwy 127, Athens, AL 35614-6805 – **886)** 4700 Business Park Blvd #E-44A, Anchorage, AK 99503-7176 – **887)** 2300 Portola Drive, Santa Cruz, CA 95062-4203 – **888)** 10 Executive Drive, Farmington, CT 06032-2841 – **889)** 2828 W Flagler St, Miami, FL 33135-1337 – **890)** 1160 S. Semoran Blvd #A, Orlando, FL 32807-1461 – **891)** 3405 Duluth Park Lane, Duluth, GA 30096-3259 – **893)** 504 E. Sherman Ave, Coeur d' Alene, ID 83814-2731 – **894)** 934 West 138th St, Riverdale, IL 60827-1673 – **895)** 4000 Radio Drive #1, Louisville, KY 40218-1568 – **896)** 827 Fairview Drive SW, Lenoir, NC 28645-6023 – **897)** 0700 SW Bancroft St, Portland, OR 97239-4226 – **898)** 5316 William Flynn Hwy #3N, Gibsonia, PA 15044-9646 – **899)** 375 Hendersonville Highway, Walterboro, SC 29488-4511 – **900)** 4131 North Central Expressway, Dallas, TX 75204-2102 – **901)** 63 East Main St, Price, UT 84501-3031 – **902)** 700 Wellington Hills Rd, Little Rock, AR 72211-2026 – **903)** 2200 Smith Lane, Fortuna, CA 95540-2771 – **904)** 777 Grant St, Denver, CO 80203-3501 – **905)** 3737 Lake Ave, Fort Wayne, IN 46805-5554 – **906)** 3800 Hooper Ave, Baltimore, MD 21211-1313 – **907)** 500 Victory Rd, Quincy, MA 02171-3139 – **908)** 201 Industrial Park Rd, Excelsior Springs, MO 64024-1736 – **909)** 5445 Johnson Rd, Bozeman, MT 59718-8333 – **910)** PO Box 90, Smithfield, NC 27577-0090 – **911)** 80898 Powerline Rd, Umatilla, OR 97882-9309 – **912)** 210 Montague Ave, Greenwood, SC 29649-1935 – **913)** 340 Martin Luther King Blvd, Bristol, TN 37620-2313 – **914)** 3218 Quincy St, Plainview, TX 79072-1906 – **915)** 1000 Dexter Ave North #100, Seattle, WA 98109-3582 – **916)** 1859 21st Ave, Rice Lake, WI 54868-9502 – **917)** 2001 N. 3rd St #102, Phoenix, AZ 85004-1439 – **918)** 4043 Geer Rd, Hughson, CA 95326-9715 – **919)** 39138 Fremont Blvd 3rd Flr, Fremont, CA 94538-1305 – **920)** 1360 E. Sherwood Dr, Grand Junction, CO 81501-7546 – **921)** 1465 Northside Dr. NW #218, Atlanta, GA 30318-4220 – **922)** 714 Narrow Way, Saint Simons Island, GA 31522-9712 – **923)** 64 Broadway N, Fargo, ND 58102-4934 – **924)** 1411 Locust St, Saint Louis, MO 63103-2332 – **925)** 8755 W Flamingo Rd, Las Vegas, NV 89147-8667 – **926)** 234 Airport Plaza Blvd #5, Farmingdale, NY 11735-3938 – **927)** 6200 Oak Tree Blvd 4th Flr, Independence, OH 44131-2510 – **928)** 16414 San Pedro Ave #575, San Antonio, TX 78232-2277 – **929)** 1047 Route 302, Wells River, VT 05081-9742 – **930)** 112 North Pearl St, Berlin, WI 54923-1570 – **931)** 2500 Battleship Pkwy, Mobile, AL 36602-8003 – **932)** 4723 King David St, Houston Pake, AK 99694-ND – **933)** 360 Main St, Clinton, AR 72031-6622 – **934)** 3800 West Alameda Ave, Burbank, CA 91505-4300 – **935)** 3463 Ramona Ave. #15, Sacramento, CA 95826-3827 – **936)** 311 112th Ave. NE, St Petersburg, FL 33716-3394 – **938)** 820 N.LaSalle St, Chicago, IL 60610-3214 – **939)** 211 Jason St, Pittsfield, MA 01201-5998 – **940)** 2175 Click Rd, Petoskey, MI 49770-8818 – **941)** 5010 Underwood Ave, Omaha, NE 68132-2297 – **942)** 462 Merrimack St, Methune, MA 01844-5804 – **943)** 1423 West Bender Blvd, Hobbs, NM 88240-9252 –

944) 1 Julian Price Place, Charlotte, NC 28208-5211 – **945)** 8010 North Main St, Dayton, OH 45405-2249 – **946)** 63088 NE 18th St #200, Bend, OR 97701-7102 – **947)** 2311 Old Arch Rd, Norristown, PA 19401-2013 – **948)** 1270 Mineral Spring Ave, North Providence, RI 02904-4637 – **949)** 26321 Stateline Rd West, Ardmore, TN 38449-3083 – **950)** 6161 Savoy Drive #1140, Houston, TX 77036-3323 – **951)** 11061 Shady Trail, Dallas, TX 75229-5603 – **952)** 700 Monticello Ave #305, Norfolk, VA 23510-2517 – **953)** 2670 S. White Rd #165, San Jose, CA 95148-2083 – **954)** 2131 Crimmins Lane, Falls Church, VA 22043-1962 – **955)** 144 Gould St #155, Needham Heights, MA 02494-2338 – **956)** 4611 Terry Rd #C, Jackson, MS 39212-5646 – **957)** 1 South Memorial Drive #600, Saint Louis, MO 63102-2498 – **958)** 410 New Bridge St #3B, Jacksonville, NC 28540-4759 – **959)** 1500 Valley River Drive #350, Eugene, OR 97401-2163 – **960)** 2201 S. 6th St, Las Vegas, NV 89104-2999 – **961)** 320 Barnett Blvd, Tallassee, AL 36078-1506 – **962)** 136 South Oak Knoll Ave #300, Pasadena, CA 91101-2624 – **963)** 83 East Shaw Ave #150, Fresno, CA 93710-7622 – **964)** 1130 Radio Rd, Bartow, FL 33830-7600 – **965)** 5815 Westside Rd, Austell, GA 30106-3179 – **966)** 6341 West Port Ave, Shreveport, LA 71129-2415 – **967)** 27675 Halsted Rd, Farmington Hills, MI 48331-3511 – **968)** 1600 Utica Ave South #440, Minneapolis, MN 55416-1480 – **969)** 731 Lexington Ave, New York, NY 10022-1331 – **970)** 2234 Hodges Chapel Rd, Benson, NC 27504 – **971)** 3500 E. Rosser Ave, Bismarck, ND 58501-3398 – **972)** 9134 Silverwood Court, Granite Bay, CA 95746-7239 – **973)** 2440 Millwood Ave, Columbia, SC 29205-1128 or PO Box 2355, West Columbia, SC 29171-2355 – **974)** 3050 Post Oak Blvd #1688, Houston, TX 77056-6527 – **975)** 12100 W. Howard Ave, Milwaukee, WI 53228-1851 – **976)** 2926 Huntsville Hwy #D, Fayetteville, TN 37334-6687 – **977)** 40960 Kalifornsky Beach Rd, Kenai, AK 99611-6445 – **978)** 1504 W Persimmon St, Rogers, AR 72756-3350 – **979)** 5244 Madison Ave, Sacramento, CA 95841-3004 – **980)** 1321 North Gene Autry Trail, Palm Springs, CA 92262-5473 – **981)** 800 S. Douglas Rd #111, Coral Gables, FL 33134-3187 – **982)** 1355 E Altamonte Drive, Altamonte Springs, FL 32701-5011 – **983)** PO Box 714, Boise, ID 83701-0714 – **984)** 120 Eaton St, Peoria, IL 61603-4217 – **985)** 1919 Eastern Ave SE, Grand Rapids, MI 49507-2721 – **986)** 15963 Highway 4 East, Senatobia, MS 38668-5786 – **987)** 310 S. La Frenz Rd, Liberty, MO 64068-7944 – **988)** 6655 W. Sahara Ave #D110, Las Vegas, NV 89146-0846 – **989)** 3258 Merchant Rd, Warsaw, NY 14569-9320 – **990)** 5100 South Tennis Lane, Sioux Falls, SD 57108-2212 – **991)** 1302 N Shepherd Dr, Houston, TX 77008-3752 – **992)** 300 Bryant Rd, Conroe, TX 77303-1796 – **993)** 3245 Basie Rd, Richmond, VA 23228-3404 – **994)** PO Box 1299, Lebanon, VA 24266-1299 – **995)** 1949 Mountain View Drive, Cody, WY 82414-4932 or PO Box 1210, Cody, WY 82414-1210 – **996)** 5455 Jug Factory Rd, Tuscaloosa, AL 35405-4213 – **997)** 1445 West Baseline Rd, Phoenix, AZ 85041-7010 – **998)** 3400 Olive Ave #550, Burbank, CA 91505-5544 – **999)** 1582 S Parker Rd #204, Denver, CO 80231-2716 – **1000)** 777 River Rd, Middletown, CT 06457-3922 – **1001)** 2727 Shipley Rd, Wilmington, DE 19810-3299 – **1002)** 407 N Howard Ave #200, Tampa, FL 33606-1575 – **1003)** 2325 Hwy 84 East, Valdosta, GA 31601-2410 – **1004)** 1801 East Main St, Marion, IL 62959-5115 – **1005)** 6626 Dubuque Trail, Norwalk, IA 50211-9645 – **1006)** 131 N. Santa Fe Ave, Salina, KS 67401-2615 – **1007)** 22 West Main St, Mount Sterling, KY 40353-1314 – **1008)** 5555 Hilton Ave. #500, Baton Rouge, LA 70808-2564 – **1009)** 500 Victory Rd #2, Quincy, MA 02171-3132 – **1010)** 830 Oilfield Ave, Shelby, MT 59474-1641 – **1011)** 10424 Edith Blvd NE, Albuquerque, NM 87113-2408 – **1012)** 135 White Bridge Rd, Middletown, NY 10940-7319 – **1013)** 2581 US Hwy 70 West, Goldsboro, NC 27530-9553 – **1014)** 290 Hegenberger Rd, Oakland, CA 94621-1436 – **1015)** 404 Main St, Klamath Falls, OR 97601-6021 – **1016)** 5110 SE Stark St, Portland, OR 97215-1751 – **1017)** 1 Forever Drive, Holidaysburg, PA 16648-3029 – **1018)** 5732 N. Tryon St, Charlotte, NC 28213-6802 – **1019)** 11 Main St, Rapid City, SD 57701-2831 – **1020)** 821 Pineville Rd, Chattanooga, TN 37405-2633 – **1021)** 510 West Economy Rd, Morristown, TN 37814-3223 – **1022)** 5426 North Mesa St, El Paso, TX 79912-5421 – **1023)** 35 Eagle Rock Lane, Churchville, VA 24421 – **1024)** 801 Old Wawawai Rd., Pullman, WA 99163-9002 – **1025)** 3650 131st Ave #550, Bellevue, WA 98006-1334 – **1026)** 18385 Coal Heritage Rd, Welch, WV 24801-9773 – **1027)** 944 Harlem St, Altoona, WI 54720-1127 – **1028)** 2800 East College Ave, Appleton, WI 54915-3255 – **1029)** 8384 Baymeadows Rd #1, Jacksonville, FL 32256-7486 – **1030)** 4365 Kennedy Ave, Cordova, FL 32812-8214 – **1031)** 1100 Spring St #610, Atlanta, GA 30309-2828 – **1032)** 25 NW Point Blvd #400, Elk Grove Village, IL 60007-1030 – **1033)** 1115 Tamarack Rd #500, Owensboro, KY 42301-6988 – **1034)** 635 West 7th St #400, Cincinnati, OH 45203-1549 – **1035)** 208 Middle Rd, Skowhegan, ME 04976-5023 – **1036)** 8121 Georgia Ave #806, Silver Spring, MD 20910-4945 – **1037)** 310 S. La Frenz Rd, Liberty, MO 64068-7944 – **1038)** 1015 Atlantic City Blvd, Bayville, NJ 08721-3541 or PO Box 927, Toms River, NJ 08754-0927 – **1039)** 1086 Teaneck Rd #4F, Teaneck, NJ 07666-4858 – **1040)** 100 Saratoga Village Blvd #21, Malta, NY 12020-3703 – **1041)** 1705 Lake St, Elmira, NY 14901-1220 – **1042)** 60 Courthouse St, Columbus, NC 28722

– **1043)** PO Box 711, Red Springs, NC 28377-0711 – **1044)** 619 Alexander Rd, Princeton, NJ 08540-6000 – **1045)** 840 Philadelphia St, Indiana, PA 15701-3922 – **1046)** 15 Century Blvd #101, Nashville, TN 37214-3692 – **1047)** 84 NE Loop 410 #143, San Antonio, TX 78216-5835 – **1048)** 11451 Katy Freeway #125, Houston, TX 77079-2004 – **1049)** 55 North 300 West, Salt Lake City, UT 84180-1109 – **1050)** 1675 Grandview Rd, Martinsville, VA 24112-2319 – **1051)** 4101 Wall St #A, Montgomery, AL 36106-3656 – **1052)** 2501 Mission Rd, North Pole, AK 99705-6361 – **1053)** 9255 Towne Centre Drive #535, San Diego, CA 92121-3038 – **1054)** 2905 King St, San Jose, CA 95122-1518 – **1055)** 6360 SW 41st Place, Davie, FL 33314-3412 – **1056)** PO Box 1848, Mattoon, IL 61938-1848 – **1057)** 255 Cedardale Drive SE, Owatonna, MN 55060-4425 – **1058)** 118 E. Main St, Clinton, NC 28328-4029 – **1059)** 785 US Highway 70 SW, Hickory, NC 28602-5096 – **1060)** 4590 E. 29th St, Tulsa, OK 74114-6208 – **1061)** 7707 St Andrews Rd, Irmo, SC 29063-2835 – **1062)** 2219 Yew St Rd, Bellingham, WA 98229-8855 – **1063)** 1015 Main St, Wheeling, WV 26003-2782 – **1064)** 812 East Beale St, Kingman, AZ 86401-5925 – **1065)** 6655 Poplar Ave #200, Germantown, TN 38138-0643 – **1066)** 1400 Easton Drive #144, Bakersfield, CA 93309-9404 – **1067)** VOA - 330 Independence Ave. SW, Washington, DC 20547-0003 – **1068)** 10143 Commerce St, Summerville, GA 30747-1356 – **1069)** 600 Cavanaugh Rd, Lansing, MI 48910-5299 – **1070)** 1310 2nd St NW #A, Sauk Rapids, MN 56379-2532 – **1071)** 731 S. Pear Orchard Rd #27, Ridgeland, MS 39157-4839 – **1072)** 317 First Ave. East, Kalispell, MT 59901-9601 – **1073)** 5011 Capitol Ave, Omaha, NE 68132-2921 – **1074)** 100 Chestnut St #1700, Rochester, NY 14604-2418 – **1075)** 201 North Front St #805, Wilmington, NC 28401-5089 – **1076)** 470 Leadmine Rd, Gaffney, SC 29340-4037 – **1077)** 802 S Central Ave, Knoxville, TN 37902-1207 – **1078)** 6161 Savoy Drive #1140, Houston, TX 77036-3323 – **1079)** 10025 Lakewood Dr SW #B, Tacoma, WA 98499-3897 – **1080)** 2609 Jordan Lane NW, Huntsville, AL 35816-1030 – **1081)** 4041 N Central Ave #1000, Phoenix, AZ 85012-3310 – **1082)** 1780 West Holly St, Fayetteville, AR 72703-1307 – **1083)** 701 N. Brand Blvd #550, Glendale, CA 91203-1235 – **1084)** UMC Campus Box 207, Boulder, CO 80309-1001 – **1085)** 1160 S Semoran Blvd, Orlando, FL 32807-1461 – **1086)** 2970 Peachtree Rd. NW #700, Atlanta, GA 30305-4919 – **1087)** 2915 Maples Rd., Fort Wayne, IN 46816-3199 – **1088)** 2131 Crimmins Lane, Falls Church, VA 22043-1962 – **1089)** 71991 US Hwy 71 South, Jackson, MS 56143-ND – **1090)** 1190 Hollywood Blvd, Bay Saint Louis, MS 39520-1662 – **1091)** 6500 West Main St #315, Belleville, IL 62223-3700 – **1092)** 1100 Main St #1950, Kansas City, MO 64105-5173 – **1093)** 2000 Randolph Rd SE #103, Albuquerque, NM 87106-2146 – **1094)** 3 Park Ave FL 41, New York, NY 10016-5902 – **1095)** 1700 Buena Vista Drive, Monroe, NC 28112-6306 – **1096)** 4949 S.W. Macadam Ave, Portland, OR 97239-3997 – **1097)** 105 Ash Ave, South Pittsburg, TN 37380-1513 – **1098)** 14001 Dallas Pkwy #300, Dallas, TX 75240-7369 – **1099)** Pruntytown Pike, RR5, Box 28, Grafton, WV 26354-9680 – **1100)** 5606 Medical Circle, Madison, WI 53719-1232 – **1101)** 651 Cannery Row #1, Monterey, CA 93940-1050 – **1102)** 2824 Palm Beach Blvd, Fort Myers, FL 33916-1503 – **1103)** 625 N. Michigan Ave #300, Chicago, IL 60611-3163 – **1104)** 10 Cabot Rd #302, Medford, MA 02155-5173 – **1105)** 3250 Franklin St, Detroit, MI 48207-4219 – **1106)** 738 Blowing Rock Rd, Boone, NC 28607-4840 – **1107)** 875 W. 5th St, Winston-Salem, NC 27101-2505 – **1108)** 5702 52nd Ave South, Fargo, ND 58104-5605 – **1109)** 219 Savannah-Gardner Rd, New Castle, PA 16101-5546 – **1110)** 1559 W. 4th St, Williamsport, PA 17701-5650 – **1111)** 4337 Big Barn Drive, Little River, SC 29566-6802 – **1112)** 637 East Durst St, Greenwood, SC 29649-2713 – **1113)** 1617 Lebanon Rd, Nashville, TN 37210-3217 – **1114)** 6222 West Interstate 10, San Antonio, TX 78201-2097 – **1115)** 2131 Crimmins Lane, Falls Church, VA 22043-1962 – **1116)** 1359 Carmichael Way, Montgomery, AL 36106-3629 – **1117)** 2955 E Broadway Blvd, Tucson, AZ 85716-5311 – **1118)** 4135 Northgate Blvd, Sacramento, CA 95834-1226 – **1119)** 9255 Towne Centre Dr. #535, San Diego, CA 92121-3038 – **1120)** 7250 NW 58th St, Miami, FL 33166-3719 – **1121)** 4325 Steve Reynolds Blvd, Norcross, GA 30093-3362 – **1122)** 300 S. Poplar St, Centralia, IL 62801-3900 – **1123)** 5555 Hilton Ave #500, Baton Rouge, LA 70808-2564 – **1124)** 2175 Click Rd, Petoskey, MI 49770-8818 – **1125)** 2300 North Lelia St, Guymon, OK 73942-2840 – **1126)** 369 Tower Rd, Waynesburg, PA 15370-3663 – **1127)** 2 Bala Plaza 7th Flr, Bala Cynwyd, PA 19004-1501 – **1128)** 1726 Dakota Ave South, Huron, SD 57350-4024 – **1129)** 1079 East Trinity Lane, Nashville, TN 37207-3043 – **1130)** 6080 Mount Moriah Rd Extension, Memphis, TN 38115-2645 – **1131)** 4501 North McColl Rd, McAllen, TX 78504-2431 – **1132)** 750 Ridgeview Drive #204, Saint George, UT 84770-2665 – **1133)** 1400 West Main St, Auburn, WA 98001-5230 – **1134)** 300 South 2nd St #204, Laramie, WY 82070-3650 – **1135)** 10399 State Hwy 238, Afton, WY 83110-ND – **1136)** 39138 Fremont Blvd 3rd Flr, Fremont, CA 94538-1305 – **1137)** 1308 East Hwy 56, Salem, IN 47167-9690 – **1138)** 104 Main St North, Stillwater, MN 55082-5076 – **1139)** 661 Little Britain Rd, Newburgh, NY 12553-6150 – **1140)** 108 Radio Station Rd, Whiteville, NC 28472-4906 – **1141)** 201 South College St #100, Charlotte, NC

28244-0010 – **1142)** 4 Summit Park Drive #150, Independence, OH 44131-6921 – **1143)** 161 Hillwood Ave #B, Falls Church, VA 22046-2983 – **1144)** 2821 U.S. Highway 231, Wetumpka, AL 36093-1222 – **1145)** PO Box 680748, Fort Payne, AL 35968-1608 – **1146)** 900 Patte Rd, Willcox, AZ 85643-3408 – **1147)** 140 North Main, Lakeport, CA 95453-4815 – **1148)** 200 South "A" St #400, Oxnard, CA 93030-5717 – **1149)** 4002 W. Gandy Blvd, Tampa, FL 33611-3410 – **1150)** 813 S 7th St, Kansas City, KS 66105-2003 – **1151)** 3 Converse St #101, Palmer, MA 01069-1538 – **1152)** 107 North Franklin St, Saginaw, MI 48607-1263 – **1153)** 728 Western Ave, Fergus Falls, MN 56537-1095 – **1154)** 63 Braswell Rd, Hattiesburg, MS 39401-9750 – **1155)** 210 W Front St, Forsyth, MT 59327-ND – **1156)** 149 Main St #210, Nashua, NH 03060-2725 – **1157)** 55 Horsehill Rd, Cedar Knolls, NJ 07927-2003 – **1158)** 147 N. Garden St, Marion, NC 28752-3709 – **1159)** 525 Evans St, Greenville, NC 27858-2311 – **1160)** 241 W 4th St, Emporium, PA 15834-1047 – **1161)** 400 Ardmore Blvd, Pittsburgh, PA 15221-3019 – **1162)** 4230 Faber Place Drive #100, North Charleston, SC 29405-8512 – **1163)** 4050 Eisenhauer Rd, San Antonio, TX 78218-3409 – **1164)** 601 Washington St, Alexandria, LA 71301-8028 – **1165)** RR2 Box 2384, Roosevelt, UT 84066-9523 – **1166)** 1 Radio Lane, Danville, VA 24541-5235 – **1167)** 9540 Godwin Drive, Manassas, VA 20110-4165 – **1168)** Murrow Comm Cntr - WSU, Pullman, WA 99163-ND – **1169)** 200 1st Ave West #104, Seattle, WA 98119-4291 – **1170)** 714 Nicholas St, Rupert, WV 25984 – **1171)** 11800 W. Grange Ave, Hales Corners, WI 53130-1099 – **1172)** 120 Summit Pkwy #200, Birmingham, AL 35209-4719 – **1173)** 1500 Cotner Ave, Los Angeles, CA 90025-3303 – **1174)** 3256 Penryn Rd #100, Loomis, CA 95650-8052 – **1175)** 1901 N Moore St #200, Arlington, VA 22209-1706 – **1176)** 225 Hollywood Blvd NW, Fort Walton Beach, FL 32548-4725 – **1177)** 2100 Coral Way, Coral Gables, FL 33145-2635 – **1178)** 4005 Golden Isle West, Baxley, GA 31513-7972 – **1179)** 3079 Campbellton Rd SW #104, Atlanta, GA 30311-5400 – **1180)** 1156 N Orchard St, Boise, ID 83706-2234 – **1181)** PO Box 699, Blackfoot, ID 83221-0699 – **1182)** 1978 Interbelt Business Center Drive, Saint Louis, MO 63114-5760 – **1183)** 6161 Fall Creek Rd, Indianapolis, IN 46220-5032 – **1184)** 900 8th St, Boone, IA 50036-2920 – **1185)** 2825 Government St, Baton Rouge, LA 70806-5412 – **1186)** 309 Waverly Oaks Rd #103, Waltham, MA 02452-8403 – **1187)** 425 Centerstone Ct #1, Zeeland, MI 49464-2249 – **1188)** 2330 W Grand St, Springfield, MO 65802-4900 – **1189)** PO Box 7509, Trenton, NJ 08628-0509 – **1190)** 2502 Camino Entrada #C, Santa Fe, NM 87507-4911 – **1191)** 1064 James St, Syracuse, NY 13203-2704 – **1192)** 1119 Eastview Drive, Asheboro, NC 27203-4576 – **1193)** 175 Ken Mar Industrial Pkwy, BRdview Heights, OH 44147-2950 – **1194)** 604 Chillicothe St #405, Portsmouth, OH 45662-4024 – **1195)** 1884 Port Matilda Hwy, Philipsburg, PA 16866-3128 – **1196)** 471 Robison Rd West, Erie, PA 16509-5425 – **1197)** 51 Commerce St, Sumter, SC 29150-5014 – **1198)** 20 Grand Ave #C, Greenville, SC 29607-2161 – **1199)** 346 South Main St, Winner, SD 57580-1832 – **1200)** 1108 Hendricks St, Chattanooga, TN 37406-3159 – **1201)** 106 E. College St, Dickson, TN 37055-1828 – **1202)** 121 South Cotton Square, Lufkin, TX 75904-2933 – **1203)** 1150 Pepsi Place #300, Charlottesville, VA 22901-2865 – **1204)** 3520 Kingsbury Circle, Roanoke, VA 24014-1356 – **1205)** 328 100th St, Amery, WI 54001-4024 – **1206)** 912 Lane 11, Powell, WY 82435-9222 – **1207)** 273 Azalea Rd #A03, Mobile, AL 36609-1970 – **1208)** 1838 Commerce Drive #A, Lakeside, AZ 85929-7007 – **1209)** 1401 West Caldwell Ave, Visalia, CA 93277-7725 – **1210)** 1321 North Gene Autry Trail, Palm Springs, CA 92262-5473 – **1211)** 6106 Hoffner Ave, Orlando, FL 32822-4906 – **1212)** 2824 Palm Beach Blvd, Ft Myers, FL 33916-1503 – **1213)** 325 John Knox Rd. #G, Tallahassee, FL 32303-4161 – **1214)** 1501 13th Ave, Columbus, GA 31901-1908 – **1215)** 1801 North Elm St, Commerce, GA 30529-2347 – **1216)** 21361 Highway 30, Twin Falls, ID 83301-0197 – **1217)** 1035 Lincoln Rd #205, Bettendorf, IA 52722-4149 – **1218)** 25802 County Rd 26, Elkhart, IN 46517-9132 – **1219)** 1410 N. Western Ave, Liberal, KS 67901-2212 – **1220)** 35 Baltimore St, Cumberland, MD 21502-3024 – **1221)** 34 Sylvan St, West Springfield, MA 01089-3444 – **1222)** 2095 South US Highway 131, Petoskey, MI 49770-9216 – **1223)** 31555 W 14 Mile Rd #102, Farmington Hills, MI 48334-1286 – **1224)** 1530 Greenview Drive SW #200, Rochester, MN 55902-1080 – **1225)** 305 West Washington St, Brainerd, MN 56401-2923 – **1226)** 961 Matley Lane #120, Reno, NV 89502-2119 – **1227)** 101 Back Rd, Dover, NH 03820-5012 – **1228)** 1420 Main St #2A, Buffalo, NY 14209-1733 – **1229)** 34 Chestnut St, Oneonta, NY 13820-2466 – **1230)** PO Box 57, Smithfield, NC 27577-0057 – **1231)** 5732 N. Tryon St, Charlotte, NC 28607-4835 – **1232)** 2530 E 71st St #C, Tulsa, OK 74136-5577 – **1233)** 888 Rogue River Highway, Grants Pass, OR 97527-5209 – **1234)** 440 Rebecca Rd, Lebanon, PA 17046-1705 – **1235)** 1816 Savannah Hwy, Hampton, SC 29924-6545 – **1236)** 6300 South Tallgrass Ave, Sioux Falls, SD 57108-8184 – **1237)** 640 W. Hwy 25/70, Newport, TN 37821-8068 – **1238)** 7700 Carpenter Freeway Fl2, Dallas, TX 75247-4829 – **1239)** 3824 Wayside Rd, Stuart, VA 24171-2506 – **1240)** 1130 14th Ave, Longview, WA 98632-3017 – **1241)** 2810 Southern Drive, Gillette, WY 82718-9369 – **1242)** 601 Greensboro Ave #507, Tuscaloosa, AL 35401-1795 – **1243)** 2800 North 44th St #100, Phoenix, AZ 85008-1559 – **1244)** 880 Via Esteban #C, San Luis Obispo, CA 93420-2462 – **1245)** 600 Grant St #600, Denver, CO 80203-3540 – **1246)** 6163 S 2nd St, De Funiak Springs, FL 32435-2709 – **1247)** 7080 Industrial Way, Macon, GA 31206-7538 – **1248)** 117 SE 5th St, Evansville, IN 47708-1639 – **1249)** 929 Howard Ave, New Orleans, LA 70113-1148 – **1250)** 150 Whitten Rd, Augusta, ME 04330-6021 – **1251)** 762 Water St, Fitchburg, MA 01420-6481 – **1252)** 2110 Cliff Rd, Saint Paul, MN 55122-2347 – **1253)** 216 Belmont Rd, Grand Forks, ND 58201-4620 – **1254)** 73 Spectrum Blvd, Las Vegas, NV 89101-4838 – **1255)** 204 E Broadway, Farmington, NM 87401-6418 – **1256)** 100 Chestnut St #1700, Rochester, NY 14604-2418 – **1257)** 485 Madison Ave, New York, NY 10022-5803 – **1258)** 25539 NC Hwy 125, Scotland Neck, NC 27874-ND – **1259)** PMB 2371574 1574 Coburg Rd. #237, Eugene, OR 97401-4802 – **1260)** 275 Radio Rd, Hanover, PA 17331-1140 – **1261)** 219 Savannah-Gardner Rd, New Castle, PA 16101-5546 – **1262)** 3004 East Highway 76, Mullins, SC 29574-7396 or PO Box 1020, Marion, SC 29571-1020 – **1263)** 100 E. Shockley Ferry Rd, Anderson, SC 29624-3746 – **1264)** 886 Mt Olivet Rd, Columbia, TN 38401-8031 – **1265)** 515 South 700 East #1C, Salt Lake City, UT 84102-2802 – **1266)** 110 West Spiller Ave, Wytheville, VA 24382-1953 – **1267)** 4010 Summitview Ave, Yakima, WA 98908-2966 – **1268)** 808 East Sprague Ave, Spokane, WA 99202-2126 – **1269)** 491 South Washburn St #400, Oshkosh, WI 54904-6733 – **1270)** 1101 Cameron Rd., Opp, AL 36467-2407 – **1271)** 2250 West Sunset Ave #3, Springdale, AR 72762-5187 – **1272)** 1904 W. Hillsboro St, El Dorado, AR 71730-6806 – **1273)** 1820 Cochrane Rd, Morgan Hill, CA 95037-9029 – **1274)** 2030 Iowa Ave #A, Riverside, CA 92507-7412 – **1275)** 2654 Cramer Lane, Chico, CA 95928-8838 – **1276)** 920 West Basin Rd #400, New Castle, DE 19720-1013 – **1277)** 3071 Continental Drive, West Palm Beach, FL 33407-3274 – **1278)** 3621 NW 10th St., Ocala, FL 34475-4541 – **1279)** 245 Alfred St, Savannah, GA 31408-3205 – **1280)** 3235 Satellite Blvd #230, Duluth, GA 30096-8688 – **1281)** 436 N. Main St, Pocatello, ID 83204-3018 – **1282)** 331 Fulton St #1200, Peoria, IL 61602-1475 – **1283)** 30129 East US Hwy 54, Pratt, KS 67124-8304 – **1284)** 1039 Eggners Ferry Rd, Benton, KY 42025-8070 – **1285)** PO Box 448, Manchester, KY 40962-0448 – **1286)** 266 23rd St, Meridian, MS 39301-1728 – **1287)** 3250 South Reserve St #200, Missoula, MT 59801-8236 – **1288)** 5011 Capitol Ave, Omaha, NE 68132-2921 – **1289)** 69 Stanhope Ave, Keene, NH 03431-1577 – **1290)** 59 Court St #100, Binghamton, NY 13901-3293 – **1291)** 907 Lejeune Blvd, Jacksonville, NC 28540-5916 – **1292)** 526 Main Ave SE, Hickory, NC 28602-1103 – **1293)** 1414 Wilmington Ave, Dayton, OH 45420-1568 – **1294)** 2003 NW 56th St, Pendleton, OR 97801-4593 – **1295)** 1 Forever Drive, Holidaysburg, PA 16648-3029 – **1296)** I Union Station, Providence, RI 02903-1758 – **1297)** 517 N Watt Rd, Knoxville, TN 37922-1110 – **1298)** 102 South 5th St, Crockett, TX 75835-2037 – **1299)** 2525 Kell Blvd #200, Wichita Falls, TX 76308-1008 – **1300)** 2720 Highway 83, Weslaco, TX 78596-1225 – **1301)** 413 Stuart Circle #110, Richmond, VA 23220-3754 – **1302)** 204 Main St #201, Logan, WV 25601-3943 – **1303)** 113 W. Oak St, Sparta, WI 54656-1712 – **1304)** 6800 Gogel Hope Rd, Menomonee Falls, WI 53051-4441 – **1305)** 3525 Soldier Springs Rd, Laramie, WY 82070-8903 – **1306)** 655 Fairview Rd, Winfield, AL 35594-4755 – **1307)** 111 N Spring St, Searcy, AR 72143-7712 – **1308)** 747 E. Green St #400, Pasadena, CA 91101-2148 – **1309)** 1250 Siskiyou Blvd, Ashland, OR 97520-5010 – **1310)** 145 Natoma St Fl4, San Francisco, CA 94105-3734 – **1311)** 6805 Corporate Drive #130, Colorado Springs, CO 80919-1977 – **1312)** 5203 North Armenia Ave, Tampa, FL 33603-1407 – **1313)** 1 Boot Key, Marathon, FL 33050 – **1314)** 2355 Pluckebaum Rd, Cocoa, FL 32926-5179 – **1315)** 100 WMTM Rd, Moultrie, GA 31788-4104 – **1316)** N72W12922 Good Hope Rd, Orofino, ID 83544-9629 – **1317)** 3110 Upper Fords Creek Rd, Orofino, ID 83544-9629 – **1318)** 401 North Michigan Ave #2010, Chicago, IL 60611-4206 – **1319)** 341 South Yorktown Pike, Mason City, IA 50401-4533 – **1320)** 401 West Main St #301, Lexington, KY 40507-1646 – **1321)** 1526 Corporate Drive, Shreveport, LA 71107-6338 – **1322)** 650 Wooddale Blvd, Baton Rouge, LA 70806-2980 – **1323)** 1423 Clarkview Rd, Baltimore, MD 21209-2134 – **1324)** 77 Monroe Center NW #1000, Grand Rapids, MI 49503-2912 – **1325)** 731 S. Pear Orchard Rd #27, Ridgeland, MS 39157-4839 – **1326)** 1811 West O St, McCook, NE 69001-4264 – **1327)** 1960 Idaho St, Carson City, NV 89701-5324 – **1328)** 110 Babbitt Rd, Franklin, NH 03235-2105 – **1329)** 1842 S. BRd St, Trenton, NJ 08610-6002 – **1330)** 5426 William St, Lancaster, NY 14086-9320 – **1331)** 51 S Pearl St, Albany, NY 12207-1500 – **1332)** 2147 Springs Rd, Mount Airy, NC 27030 – **1333)** 2510 St Clair Ave NE, Cleveland, OH 44114-4013 – **1334)** 2625 South Memorial Drive #A, Tulsa, OK 74129-2623 – **1335)** 7590 Highway 238, Jacksonville, OR 97530-9728 – **1336)** 305 Hwy 315, Pittston, PA 18460-3987 – **1337)** 118 E. 3rd St East, Mobridge, SD 57601-2511 – **1338)** 510 W Economy Rd, Morristown, TN 37814-3223 o – **1339)** 1300 WWCR Ave, Nashville, TN 37218-3800 – **1340)** 3601 South Congress Ave #F, Austin, TX 78704-7280 – **1341)** 207 University Blvd #200, Harrisonburg, VA 22801-3752 – **1342)** 2201 6th Ave #1500, Seattle, WA 98121-1840 – **1343)** 343 High St, Morgantown, WV 26505-5515 – **1344)**

16 Martin Luther King St, Selma, AL 36703-3109 – **1345)** PO Box 1747, Foley, AL 36536-1747 – **1346)** 3256 Penryn Rd #100, Loomis, CA 95650-8052 – **1347)** 963 Industrial Rd #I, San Carlos, CA 94070-4146 – **1348)** 650 S E St #H, San Bernadino, CA 92408-1946 – **1349)** 820 11th Ave, Greeley, CO 80631-3246 – **1350)** PO Box 551, Norwich, CT 06360-0551 – **1351)** 316 E. Taylor Rd., Deland, FL 32724-7817 – **1352)** 1310 S. Florida Ave., Wauchula, FL 33873-9479 – **1353)** 1310 Walker St West, Douglas, GA 31533-7952 – **1354)** 2400 Pleasant Hill Rd #200, Duluth, GA 30096-1705 – **1355)** 415 Park Ave, Twin Falls, ID 83301-7752 – **1356)** 21 East St Joseph St, Indianapolis, IN 46204-1025 – **1357)** 95 Jackson St, Prestonsburg, KY 41653-1010 – **1358)** 1133 OLE Hwy 15, West Monroe, LA 71291-1726 – **1359)** 779 Warren Ave, Portland, ME 04103-1176 – **1360)** 122 Green St #2, Worcester, MA 01604-4138 – **1361)** 300 East Front St #450, Traverse City, MI 49684-5720 – **1362)** 27675 Halsted Rd, Farmington Hills, MI 48331-3511 – **1363)** 5300 Edina Industrial Blvd #200, Minneapolis, MN 55439-2922 – **1364)** 2702 East 32nd St, Joplin, MO 64804-4307 – **1365)** 3313 15th St #F, Great Falls, MT 59405-ND – **1366)** 2401 State Route 66, Ocean, NJ 07712-3962 – **1367)** 1606 Central Ave SE #104, Albuquerque, NM 87106-4478 – **1368)** 39 Kellogg Rd, New Hartford, NY 13413-2849 – **1369)** 419 Broadway, Paterson, NJ 07501-2104 – **1370)** 707 Leon St, Durham, NC 27704-4125 – **1371)** 4801 East Independence Blvd #815, Charlotte, NC 28212-5497 – **1372)** 1190 Patton Ave, Asheville, NC 28806-2706 – **1373)** 1185 9th St NE, Grand Forks, ND 58201-5569 – **1374)** 906 SW Alder St, Newport, OR 97365-4712 – **1375)** 134 East Pitt St, Bedford, PA 15522-1311 – **1376)** 310 2nd Ave, Warren, PA 16365-2407 – **1377)** 51 Commerce St, Sumter, SC 29150-5014 – **1378)** 2615 South BRd St, Chattanooga, TN 37408-3100 – **1379)** 1 Radio Park Drive, Jackson, TN 38305-4124 – **1380)** 8023 Vantage Drive #840, San Antonio, TX 78230-4771 – **1381)** 3500 Maple Ave. #1310, Dallas, TX 75219-3931 – **1382)** 5589 Greenwich Rd #200, Virginia Beach, VA 23462-6565 – **1383)** 3231 Old Lee Highway, Fairfax, VA 22030-1501 – **1384)** 152101 W. County Rd 12, Prosser, WA 99350-7265 – **1385)** 276 Seneca Trail, Ronceverte, WV 24970-1343 – **1386)** 2651 S. Fish Hatchery Rd., Madison, WI 53711-5400 – **1387)** 950 22nd St North #1000, Birmingham, AL 35203-5312 – **1388)** 311 Lexington Ave, Fort Smith, AR 72901-3842 – **1389)** 5345 Madison Ave, Sacramento, CA 95841-3141 – **1390)** 79 Baldwin Ave, Waterbury, CT 06706-1854 – **1391)** 2360 St Johns Bluff Rd S #2, Jacksonville, FL 32246-2310 – **1392)** 1779 Independence Blvd, Sarasota, FL 34234-2106 – **1393)** 10800 Biscayne Boulevard #810, Miami, FL 33161-7402 – **1394)** 1000 Memorial Drive, Griffin, GA 30223-4446 – **1395)** 17525 Highway 1, Vivian, LA 71082-9526 – **1396)** 127 Dorrance St, Providence, RI 02903-2828 – **1397)** 600 Cavanaugh Rd, Lansing, MI 48910-5299 – **1398)** 1009 West Ridge St #A, Marquette, MI 49855-3963 – **1399)** 507 SE 11th St, Grand Rapids, MN 55744-3950 – **1400)** 2438 Highway 43 South, Picayune, MS 39466-7486 – **1401)** 1750 S Brentwood Blvd #811, Saint Louis, MO 63144-1344 – **1402)** 2002 Char Ave, Scottsbluff, NE 69361-2255 – **1403)** 8 Lawrence Rd, Derry, NH 03038-4191 – **1404)** 1484 Beech St, Hornell, NY 14843-9404 – **1405)** 3404 West Wendover Ave #H, Greensboro, NC 27407-1524 – **1406)** 427 Hill St, Murphy, NC 28906-3509 – **1407)** 3600 County Rd 19 South, Minot, ND 58701-ND – **1408)** 900 Parish St, 3rd FLR, Pittsburgh, PA 15220-3407 – **1409)** 1801 Charleston Hwy #J, Cayce, SC 29033-2019 or PO Box 5106, Columbia, SC 29250-0626 – **1410)** 500 S. Phillips Ave, Sioux Falls, SD 57104-6825 – **1411)** 162 Free Hill Rd, Gray, TN 37615-3144 – **1412)** 1030 Oakdale St, Manchester, TN 37355-5618 – **1413)** 1782 W Sam Houston Pkwy North, Houston, TX 77043-2723 – **1414)** 434 Bearcat Drive, Salt Lake City, UT 84115-2520 – **1415)** 306 West BRd St, Richmond, VA 23220-4219 – **1416)** 1308 Coolidge Rd, Aberdeen, WA 98520-6317 – **1417)** 645 25th Ave North, Wisconsin Rapids, WI 54495-3294 – **1418)** 1111 East Willow St, Scottsboro, AL 35768-2210 – **1419)** 1105 West 9th St, Juneau, AK 99801-1811 – **1420)** 4433 E Broadway Blvd #210, Tucson, AZ 85711-3536 – **1421)** 401 Pacheco Blvd., Los Banos, CA 93635-4277 – **1422)** 3301 Barham Blvd #201, Los Angeles, CA 90068-1477 – **1423)** 4100 Metzger Rd, Fort Pierce, FL 34947-1712 – **1424)** 201 West Park Ave #11, Tallahassee, FL 32301-7760 – **1425)** 7179 Printers Alley, Milton, FL 32583-5347 – **1426)** 807 Bellevue Ave, Dublin, GA 31021-4847 – **1427)** 4320 Dundee Rd, Northbrook, IL 60062-1703 – **1428)** 2207 E Morgan Ave #J, Evansville, IN 47711-4355 – **1429)** 514 Jefferson St, Waterloo, IA 50701-5422 – **1430)** 2120 N. Woodlawn St #352, Wichita, KS 67208-1881 – **1431)** 821 Adams Rd, Corbin, KY 40701-4708 – **1432)** 3225 Ambassadore Caffery Pkwy, Lafayette, LA 70506-7214 – **1433)** 1605 Level Rd, Havre de Grace, MD 21078-1727 – **1434)** 552 Massachusetts Ave #201, Cambridge, MA 02139-4088 – **1435)** 4511 Miller Rd #G, Flint, MI 48507-1107 – **1436)** 919 Lilac Drive North, Golden Valley, MN 55422-4615 – **1437)** 401 E. Coal Ave, Gallup, NM 87301-6099 – **1438)** 419 Broadway, Paterson, NJ 07501-2104 – **1439)** 57 North Ave, Owego, NY 13827-1392 – **1440)** 9700 SE Eastview Drive, Happy Valley, OR 97086-6975 – **1441)** 1 Boston Store Place, Erie, PA 16501-2312 –

1442) 970 Tripoli St, Johnstown, PA 15902-1119 – **1443)** 25 Garlington Rd, Greenville, SC 29615-4613 – **1444)** 1516 4th Ave #B, Conway, SC 29526-5032 – **1445)** 1200 S Stockton Ave, Monahans, TX 79756-4060 – **1446)** 710 Grove St, Danville, VA 24541-1704 – **1447)** 22479 Front St, Accomac, VA 23301-1641 or PO Box 460, Onley, VA 23418-0460 – **1448)** 2065 Highway 16, Marion, VA 24354-4047 – **1449)** 5408 S. Freya St, Spokane, WA 99223-7114 – **1450)** 2100 Washington Ave, Sheboygan, WI 53081-7042 – **1451)** 1530 Main St, Lander, WY 82520-2658 – **1452)** 750 Walnut St, Gadsden, AL 35901-4139 – **1453)** 308 S Broadway St, Cabot, AR 72201-2324 or PO Box 2275, Little Rock, AR 72203-2275 – **1454)** 2030 Iowa Ave #A, Riverside, CA 92507-7412 – **1455)** 1410 Neotomas Ave. #200, Santa Rosa, CA 95405-7533 – **1456)** 106 West 24th St, Pueblo, CO 81003-2408 – **1457)** 45 Pomfret St, Putnam, CT 06260-1827 – **1458)** 7515 Blythe Island Hwy, Brunswick, GA 31523-6261 – **1459)** 1350 Radio Loop, Warner Robins, GA 31088-3626 – **1460)** 1419 W. Bannock St, Boise, ID 83702-5234 – **1461)** 671 East County Rd 400 South, Kokomo, IN 46902-8101 – **1462)** 1416 Locust St., Des Moines, IA 50309-3014 – **1463)** 400 Poydras St #900, New Orleans, LA 70130-3738 – **1464)** 277 Union Ave, Laconia, NH 03246-3114 – **1465)** 449 Broadway Fl2, New York, NJ 10013-2549 – **1466)** 5411 Jefferson St NE #100, Albuquerque, NM 87109-3485 – **1467)** PO Box 1350, Aberdeen, NC 28315-1350 – **1468)** 101 West St #2, Black Mountain, NC 28711-3161 – **1469)** 7755 Freedom Ave NW, North Canton, OH 44720-6905 – **1470)** 1051 Dairy Lane, Elizabethtown, PA 17022-9547 – **1471)** 1408 E Gibson St, Jasper, TX 75951-6123 – **1472)** 1777 N.E. Loop 410 #400, San Antonio, TX 78217-5217 – **1473)** 1035 Avalon Drive, Forest, VA 24551-2970 – **1474)** 645 Church St #400, Norfolk, VA 23510-1712 – **1475)** 214 Walnut Drive SE, Wise, VA 24293-ND – **1476)** 805 Stewart Ave, Lewiston, ID 83501-4709 – **1477)** 2500 Battleship Pkwy, Mobile, AL 36602-8003 – **1478)** 409 9th Ave, Jasper, AL 35501-3731 – **1479)** 2425 East Camelback Rd #570, Phoenix, AZ 85016-4250 – **1480)** 9660 Granite Ridge Drive, San Diego, CA 92123-2657 – **1481)** 2121 Lancey Drive, Modesto, CA 95355-3000 – **1482)** 2 S Parish Ave, Johnstown, CO 80534-7800 – **1483)** 869 Blue Hills Ave, Bloomfield, CT 06002-3710 – **1484)** 1505 Dundee Rd, Winter Haven, FL 33884-1013 – **1485)** 3890 Dunn Ave #804, Jacksonville, FL 32218-6429 – **1486)** 2828 W Flagler St, Miami, FL 33135-1337 – **1487)** 2000 Indian Hills Drive, Sioux City, IA 51104-1602 – **1488)** 4154 Jennings Drive, Kalamazoo, MI 49048-1087 – **1489)** 2115 Washington Ave S, Bemidji, MN 56601-8918 – **1490)** The Tower Commons Office Complex # 302, 1374 Highway 41-47, Sewell, NJ 08080 – **1491)** 1096 Mechem Drive #G3, Ruidoso, NM 88345-7057 – **1492)** 59 Court St #100, Binghamton, NY 13901-3293 – **1493)** 88 Vilcom Center #130, Chapel Hill, NC 27514-1660 – **1494)** 8044 Montgomery Rd #650, Cincinnati, OH 45236-2959 – **1495)** 229 BRd St, Conneaut, OH 44030-2616 – **1496)** 3355 NE Cornell Rd, Hillsboro, OR 97124-5018 – **1497)** 80404 Cooney Lane, Hermiston, OR 97838-6613 – **1498)** 212 S. Centre St, Pottsville, PA 17901-3532 – **1499)** 900 Parish St, Pittsburgh, PA 15220-3425 – **1500)** 100 Cross Hill Way, Easley, SC 29640-8854 – **1501)** 5801 Marvin D Love Freeway #409, Dallas, TX 75237-2319 – **1502)** 701 S Pierce St #101, Amarillo, TX 79101-2428 – **1503)** 4638 Decker Drive, Baytown, TX 77520-1418 – **1504)** 325 Poplar Knob Rd, Galax, VA 24333-4106 – **1505)** 1820 Heritage Center Way, Harrisonburg, VA 22801-8451 – **1506)** 2201 6th Ave #1500, Seattle, WA 98121-1840 – **1507)** 527 Gibbs St, Ravenswood, WV 26164-1011 – **1508)** 1420 Bellevue St, Green Bay, WI 54311-5649 – **1509)** 2717 Yellowstone Rd, Rock Springs, WY 82901-2813 – **1510)** 395 Main St, Quincy, CA 95971-9121 – **1511)** 210 Radio Rd, Corona, CA 92879-1722 – **1512)** 3031 Tisch Way #3 Plaza West, San Jose, CA 95128-2530 – **1513)** 6565 N W St #270, Pensacola, FL 32505-1797 – **1514)** 1515 E. Silver Springs Blvd #134, Ocala, FL 34470-6830 – **1515)** 2420 Waycross Highway, Jesup, GA 31545-2332 o – **1517)** 120 West 7th St #400, Bloomington, IN 47404-3869 – **1518)** 346 West 8th St, Dubuque, IA 52001-4649 – **1519)** 2601 Central Ave #C, Dodge City, KS 67801-6212 – **1520)** PO Box 487, Grayson, KY 41143-0487 – **1521)** 49 Acme Rd, Brewer, ME 04412-1545 – **1522)** 1726 Reisterstown Rd #117, Pikesville, MD 21208-2986 – **1523)** 1101 South Cass St, Traverse City, MI 49684-3235 – **1524)** 750 Dewey Blvd #1, Butte, MT 59701-3200 – **1525)** 500 North Commercial St, Manchester, NH 03101-1151 – **1526)** 20 Tucker Drive, Poughkeepsie, NY 12603-1644 – **1527)** 280 State St, Rochester, NY 14614-1033 – **1528)** PO Box 127, Tabor City, NC 28463-0127 – **1529)** 9349 China Grove Church Rd, Pineville, NC 28134-8531 – **1530)** 910 E McNeill St, Lillington, NC 27546-7483 – **1531)** 216 Belmont Rd, Grand Forks, ND 58201-4620 – **1532)** 125 S. Superior St, Toledo, OH 43602-1790 – **1533)** 2513 6th Ave, Altoona, PA 16602-2129 – **1534)** 2615 BRd St, Chattanooga, TN 37408-3100 – **1535)** 4301 Westbank Drive #301, Austin, TX 78746-4400 – **1536)** 1285 West 2320 South, Salt Lake City, UT 84119-1448 – **1537)** 26256 Highway Forty Seven, South Hill, VA 23970-ND – **1538)** 1129 Chatham Heights, Martinsville, VA 24112-2149 – **1539)** 1143 Main St, Wheeling, WV 26003-2722 –

1540) RR1 Box 139, Dunmore, WV 24934-9712 – **1541)** 1201 E. Division St, Neillsville, WI 54456-2123 – **1542)** PO Box 630, Vernon, AL 35592-0630 – **1543)** 816 West 6th St, Parker, AZ 85344-4599 – **1544)** 203 Beale St #204, Memphis, TN 38103-3727 – **1545)** 1425 River Park Drive #520, Sacramento, CA 95815-4524 – **1546)** 182 Grand St #215, Waterbury, CT 06702-1914 – **1547)** 432 S. Nova Rd, Ormond Beach, FL 32174-6121 – **1548)** 11300 4th St N, Saint Petersburg, FL 33716-2918 – **1549)** 400 Colony Square NE #800, Atlanta, GA 30361-6318 – **1550)** 1 Parker Place #485, Janesville, WI 53545-4078 – **1551)** 2915 Maples Rd., Fort Wayne, IN 46816-3199 – **1552)** 4535 Metropolitan Ave, Kansas City, KS 66106-2599 – **1553)** 195 Moore Drive, Lexington, KY 40503-2918 – **1554)** 8230 Summa Ave, Baton Rouge, LA 70809-3421 – **1555)** 808 Huron Ave, Port Huron, MI 48060-3705 – **1556)** 13225 Dogwood Drive, Baxter, MN 56425-8669 – **1557)** 22 Morgan St, Saint Louis, MO 63102-2558 – **1558)** 7035 E Washington St Ext, Bath, NY 14810-ND – **1559)** 449 Broadway 2nd Floor, New York, NY 10013-2549 – **1560)** 70 Adams Hill Rd, Asheville, NC 28806-3841 – **1561)** 3025 Waughtown St #G, Winston-Salem, NC 27107-1679 – **1562)** 8321 East 61st #202, Tulsa, OK 74133-1911 – **1563)** 1725 North Oregon St, Ontario, OR 97914-1541 – **1564)** 786 Diamond Hill Rd, Woonsocket, RI 02895-1499 – **1565)** 2743 Perimeter Pkwy #200, Augusta, GA 30909-6487 – **1566)** 518 St Joseph St, Rapid City, SD 57701-2717 o – **1567)** 1811 Carters Creek Pike, Franklin, TN 37064-6823 – **1568)** 6960 Bucknell Rd, Millington, TN 38053-7502 – **1569)** 4045 N. Mesa St, El Paso, TX 79902-1526 – **1570)** 3308 Broadway #401, San Antonio, TX 78209-6550 – **1571)** 67 Merchants Row, Rutland, VT 05701-5910 – **1572)** 3600 West BRd St #696, Richmond, VA 23230-4916 – **1573)** 2707 Colby Ave #1380, Everett, WA 98201-3568 – **1574)** 1456 East Green Bay St, Shawano, WI 54166-2258 – **1575)** 801 Noble St #30, Anniston, AL 36201-5698 – **1576)** 1780 West Holly St, Fayetteville, AR 72703-1307 – **1577)** 136 South Oak Knoll Ave #300, Pasadena, CA 91101-2624 – **1578)** 4043 Geer Rd, Hughson, CA 95326-9715 – **1579)** 4700 Walnut St, Boulder, CO 80301-2548 – **1580)** 7120 SW 24th Ave., Gainesville, FL 32607-3705 – **1581)** 233 N Michigan Ave #2800, Chicago, IL 60601-5519 – **1582)** 516 Main St, Hazard, KY 41701-1775 – **1583)** 28 Houlton Rd, Presque Isle, ME 04769-5206 – **1584)** 17 Columbus Rd, Plymouth, MA 02360-4810 – **1585)** 1613 Lawrence Hwy, Charlotte, MI 48813-8844 – **1586)** 640 Lincoln Ave SE, Saint Cloud, MN 56304-1024 – **1587)** 10250 Lorraine Rd, Gulfport, MS 39503-6005 – **1588)** 1106 18th Ave, Meridian, MS 39301-4101 – **1589)** 313 Old Route 66, Saint Robert, MO 65584-ND or PO Box D, Waynesville, MO 65583-0480 – **1590)** 3301 North Bensing Rd, Hobbs, NM 88240-8803 – **1591)** 212 W Apache St, Farmington, NM 87401-6235 – **1592)** 2 Pendell Rd, Poughkeepsie, NY 12601-1500 – **1593)** 8456 Smokey Hollow Rd, Baldwinsville, NY 13027-8222 – **1594)** 115 North Church St, Rocky Mount, NC 27804-5402 – **1595)** 1000 20th Ave SW, Minot, ND 58701-6447 – **1596)** 7461 South Ave, Youngstown, OH 44512-5789 – **1597)** 39540 Bradbury Rd, Middleport, OH 45760-9703 – **1598)** 5110 SE Stark St, Portland, OR 97215-1751 – **1599)** 1685 Crown Ave, Lancaster, PA 17601-6310 – **1600)** 2294 Clements Ferry Rd, Charleston, SC 29492-7729 – **1601)** 122 Radio Rd, Jackson, TN 38301-3465 – **1602)** 104 Cherry St, Athens, TN 37303-ND – **1603)** 810 West 200 North, Logan, UT 84321-3726 – **1604)** 372 Dorset St, South Burlington, VT 05403-6212 – **1605)** 13321 New Hampshire Ave #207, Silver Springs, MD 20904-3450 – **1606)** 2043 10th St NE, Roanoke, VA 24012-5309 – **1607)** 1200 Chesterly Drive #160, Yakima, WA 98902-7345 – **1608)** 557 Scott St, Wausau, WI 54403-4829 – **1609)** 921 East Main St, Prattville, AL 36066-5621 – **1610)** 366 S Section St, Fairhope, AL 36532-ND – **1611)** 1950 S Sunwest Lane #302, San Bernadino, CA 92408-3227 – **1612)** 1605 Simpson Lane, Marysville, CA 95901-9747 – **1613)** 10 Columbus Blvd #24, Hartford, CT 06106-1973 – **1614)** 1575 McKee Rd #206, Dover, DE 19904-1382 – **1615)** 3411 W Tharpe St, Tallahassee, FL 32303-1139 – **1616)** 5043 Tamiami Trail East, Naples, FL 34113-4127 – **1617)** 203 N 3rd St, Leesburg, FL 34748-5105 – **1618)** 940 Brownlee Rd, Jackson, GA 30233-2418 – **1619)** 1411 Locust St, Saint Louis, MO 63103-2332 – **1620)** 1632 South Maize Rd, Wichita, KS 67209-3912 – **1621)** 100 Eversole St #1, Harlan, KY 40831-2346 – **1622)** 1301 E. 4th St, North Platte, NE 69101-4302 – **1623)** 1705 Lake St, Elmira, NY 14901-1299 – **1624)** 134 Mullin St, Watertown, NY 13601-3616 – **1625)** 707 Leon St, Durham, NC 27704-4125 – **1626)** 717 E. David Rd, Dayton, OH 45429-5218 – **1627)** 278 SW Arthur St, Portland, OR 97201-4745 – **1628)** 2147 Market St, Nesquehoning, PA 18240-1422 or P. O. Box D, Lansford, PA 18232-0801 – **1629)** Centre City Towers - 650 Smithfield St #520, Pittsburgh, PA 15222-3913 – **1630)** 219 Luckett St. NW, Roanoke, VA 24017-6812 – **1631)** 100 Kanawha Terrace, Saint Albans, WV 25177-2771 – **1632)** 201 State St, La Crosse, WI 54601-3246 – **1633)** 1716 KROE Lane, Sheridan, WY 82801-9681 – **1634)** 3900 11th Ave, Tuscaloosa, AL 35401-7056 – **1635)** 500 E Tyler Pkwy, Payson, AZ 85541-3276 – **1636)** 208 Buena Vista Rd., Hot Springs, AR 71913-8208 – **1637)** 2171 Ralph Ave, Stockton, CA 95206-3699 – **1638)** 777 River Rd, Middletown, CT 06457-3922 – **1639)** 3912 US Highway 301 North,

Ellenton, FL 34222-2333 – **1640)** 2710 W. Atlantic Ave., Delray Beach, FL 33445-4431 – **1641)** 1820 Wynnton Rd, Columbus, GA 31906-2930 – **1642)** 12544 North Main St, Trenton, GA 30752-2227 – **1644)** PO Box 84, Jerome, ID 83338-0084 – **1645)** 720 Franklin St, Michigan City, IN 46360-3506 – **1646)** 3535 E. Kimberly Rd, Davenport, IA 52807-2583 – **1647)** 1115 Tamarack Rd #500, Owensboro, KY 42301-6988 – **1648)** 22 Sconticut Neck Rd, Fairhaven, MA 02719-1930 – **1649)** 59346 Madison Ave, Mankato, MN 56001-8518 – **1650)** 959 Magnolia Drive North, Wiggins, MS 39577-3630 – **1651)** 73 Varney Rd, Wolfeboro, NH 03894-4351 – **1652)** 715 Route 52, Beacon, NY 12508-1047 – **1653)** 187 Vienna Rd., Newark, NY 14513-9124 – **1654)** 4 Summit Park Drive #150, Independence, OH 44131-6921 – **1655)** 12 West Long Ave, Du Bois, PA 15801-2100 – **1656)** 40 Rickert Rd, Doylestown, PA 18901-2326 – **1657)** 101 Riverview Rd, Erwin, TN 37650-8722 – **1658)** 9540 Godwin Drive, Manassas, VA 20110-4165 – **1659)** 45 Campbell Rd, Walla Walla, WA 99362-9597 – **1660)** 1133 Kresky Ave, Centralia, WA 98531-3789 – **1661)** 134 4th Ave, Huntington, WV 25701-1220 or 9801 Radio Park Rd, Catlettsburg, KY 41129-8824 – **1662)** 22 Cogswell Ave, Pell City, AL 35125-2438 – **1663)** 747 E. Green St #400, Pasadena, CA 91101-2148 – **1664)** 342 Day St, San Francisco, CA 94131-2313 – **1665)** 351 W Cromwell Ave #108, Fresno, CA 93711-6115 – **1666)** 4700 S Syracuse St #1050, Denver, CO 80237-2713 – **1667)** 13085 SW 133rd Ct, Miami, FL 33186-5850 – **1668)** 100 NW 76th Drive #2, Gainesville, FL 32607-6659 – **1669)** 404 West Lime St, Lakeland, FL 33815-4651 – **1670)** 3100 E 15th St, Panama City, FL 32405-7421 – **1671)** 1151 Hendricks St SW, Covington, GA 30014-6601 – **1672)** 613 Silver Circle, Dalton, GA 30721-4551 – **1673)** 3589 North Decatur Rd, Scottdale, GA 30079-1867 – **1674)** 9245 N. Meridian St #300, Indianapolis, IN 46260-1832 – **1675)** 236 Admiral Drive, Annapolis, MD 21401-3123 – **1676)** 10 Cabot Rd #302, Medford, MA 02155-5173 – **1677)** 100 William T Morrissey Blvd, Dorchester, MA 02125-3300 – **1678)** 1150 Haynor Rd, Ionia, MI 48846-8522 – **1679)** 6500 West Main St #315, Belleville, IL 62223-3700 – **1680)** 3205 West North Front St, Grand Island, NE 68803-4024 – **1681)** 651 Marshall St, Elizabeth, NJ 07206-1214 – **1682)** 200 West 1st St, Roswell, NM 88203-4668 – **1683)** 320 N. Jensen Rd, Vestel, NY 13850-2111 – **1684)** 3901 Weddington Rd, Monroe, NC 28110-9513 – **1685)** 1103 North Green St, Morganton, NC 28655-9003 – **1686)** 990 North Center St Extension, Mount Olive, NC 28365-2704 – **1687)** 2625 South Memorial Drive, Tulsa, OK 74129-2600 – **1688)** 4205 Cherry Ave NE, Keizer, OR 97303-4856 or PO Box 1430, Salem, OR 97308-1430 – **1689)** 1 Forever Drive, Holidaysburg, PA 16648-3029 – **1690)** 2278 Wortham Lane, Grovetown, GA 30813-5103 – **1691)** 630 Mainstream Drive, Nashville, TN 37228-1204 – **1692)** 6080 Mount Moriah Rd Extension, Memphis, TN 38115-2698 – **1693)** 5011 Almeda Rd, Houston, TX 77004-5996 – **1694)** 1001 East Southeast Loop 323 #455, Tyler, TX 75701-9600 – **1695)** 257 E 200 S #400, Salt Lake City, UT 84111-2073 – **1696)** 2298 Rose Ridge, Clintwood, VA 24228-7738 – **1697)** 2029 Freeway Drive, Mount Vernon, WA 98273-5470 – **1698)** 403 Capital St, Lewiston, ID 83501-1815 – **1699)** 1 Commerce St #300, Montgomery, AL 36104-3549 – **1700)** 4343 East Camelback Rd #200, Phoenix, AZ 85018-8306 – **1701)** 723 W. Daisy L Gatson Bates Drive, Little Rock, AR 72202-3704 – **1702)** 1101 S Broadway, Santa Maria, CA 93463-2689 – **1703)** 1124 Foster Rd, Napa, CA 94558-6520 – **1704)** 32992 US Highway 34, Wray, CO 80758-9161 – **1705)** 7573 NW 1st St, Lehigh Acres, FL 33972-9303 or PO Box 50580, Fort Myers, FL 33994-0580 – **1706)** 222 Hazard St, Orlando, FL 32804-3030 – **1707)** 613 Tallapoosa St W, Bremen, GA 30110-1838 – **1708)** 3833 US Highway 82, Brunswick, GA 31523-7735 – **1709)** c/o Klien Assc 299 S Main St #1300, Salt Lake City, UT 84111 – **1710)** 513 Hampshire St, Quincy, IL 62301-2928 – **1711)** 3901 Brendenwood Rd., Rockford, IL 61107-2200 – **1712)** 825 S. Kansas Ave #100, Topeka, KS 66612-1233 – **1713)** 779 Warren Avene, Portland, ME 04103-1007 – **1714)** 20 Guest St 3rd Flr, Brighton, MA 02135-2040 – **1715)** 24 Frank Lloyd Wright Drive, Ann Arbor, MI 48105-9755 – **1716)** 2000 Elm Sreet SE, Minneapolis, MN 55414-2531 – **1717)** 3276 Highway 198 West, Lucedale, MS 39452-7947 – **1718)** 101 South Main St, Gloversville, NY 12078-3820 – **1719)** 200 Radio Drive, Lexington, NC 27292-8010 – **1720)** 512 Peanut Rd, Elizabethtown, NC 28337-8811 – **1721)** 4 Summit Park Drive #150, Independence, OH 44131-6921 – **1722)** 3624 Avion Drive, Medford, OR 97504-4011 – **1723)** 404 East 2nd St, The Dalles, OR 97058-2412 – **1724)** 1210 Snyder Rd, Lansdale, PA 19446-4614 – **1725)** 1049 N Sekol Ave, Carbondale, PA 18504-1098 – **1726)** 105 N Spring St #113, Greenville, SC 29601-2859 – **1727)** 540 Cumberland St West, Cowan, TN 37318-3115 – **1728)** 115 Radio Rd, Livingston, TX 77351-7702 – **1729)** 301 South Polk St #100, Amarillo, TX 79101-1404 – **1730)** 5787 South Hampton Rd #340, Dallas, TX 75232-6335 or QSL's to CE Mortenson BRdcasting Co 3270 Blazer Pkwy Suite 101, Lexington, KY 40509-1847 – **1731)** 950 Kenbridge Rd, Blackstone, VA 23824-3105 – **1732)** 1251 Earl L. Core Rd, Morgantown, WV 26505-5881 – **1733)** 900 Bluefield Ave, Bluefield, WV 24701-2760 – **1734)** 1420 Bellevue St, Green Bay, WI 54311-5649 – **1735)** 1707

Warnke Rd NW, Cullman, AL 35055-2231 – **1736)** 1501 13th Ave, Columbus, GA 31901-1908 – **1737)** 6803 West Blvd, Inglewood, CA 90302-1895 – **1738)** 903 N. Main St, Salinas, CA 93906-3912 – **1739)** 7150 Campus Drive #150, Colorado Springs, CO 80920-3157 – **1740)** 1611 Atlantic Blvd, Atlantic Beach, FL 32233-2516 – **1741)** 2824 Palm Beach Blvd, Fort Myers, FL 33916-1503 – **1742)** 449 North 12th St, Defuniak Springs, FL 32433-0411 – **1743)** 5815 Westside Rd, Austell, GA 30106-3179 o – **1745)** 930 East Lincoln Ave, Goshen, IN 46528-3504 – **1746)** 2141 Grand Ave, Des Moines, IA 50312-5303 – **1747)** 1240 Radio Drive, Pikeville, KY 41501-4779 – **1748)** 650 Wooddale Blvd, Baton Rouge, LA 70806-2930 – **1749)** 60 Main St, Brockton, MA 02301-4040 – **1750)** 18720 16 Mile Rd, Big Rapids, MI 49307-9303 – **1751)** 1433 Main Ave North, Thief River Falls, MN 56701-1141 – **1752)** 4424 Hampton Ave, Saint Louis, MO 63109-2232 – **1753)** 403 E. 25th St, Kearney, NE 68847-5515 – **1754)** 8755 West Flamingo Rd, Las Vegas, NV 89147-8667 – **1755)** The Tower Commons Office Complex # 302, 1374 Highway 41-47, Sewell, NJ 08080 – **1756)** 52 Corporate Circle #K, Albany, NY 12203-5176 – **1757)** 6325 Sheridan Drive, Williamsville, NY 14221-4801 – **1758)** 1338 Bragg Blvd, Fayetteville, NC 28301-4202 – **1759)** 1055 Skyway Drive, Marshall, NC 28753-3809 – **1760)** 11291 39th St SW, Dickinson, ND 58601-9206 – **1761)** 605 South Front St #300, Columbus, OH 43215-5626 – **1762)** 600 Corporate Cir #100, Harrisburg, PA 17110-9787 – **1763)** PO Box 701, Tunkhannock, PA 18657-0701 – **1764)** 10614 Rockley Rd, Houston, TX 77099-3514 – **1765)** 919 N. Main St, Cleburne, TX 76033-3853 – **1766)** 7080 Lee Highway, Fairlawn, VA 24141-8416 – **1767)** 9540 Godwin Drive, Manassas, VA 20110-4165 – **1768)** 290 Hegenberger Rd, Oakland, CA 94621-1436 – **1769)** 4010 Summitview Ave, Yakima, WA 98908-2966 – **1770)** 1 Washington Ave, Elkins, WV 26241-3160 – **1771)** 3222 S Richey Ave, Tucson, AZ 85713-5453 – **1772)** 570 E. Ave Q9, Palmdale, CA 93550-2354 – **1773)** 8265 Sierra College Blvd #312, Roseville, CA 95661-9403 – **1774)** 869 Blue Hills Ave, Bloomfield, CT 06002-3789 – **1775)** 4300 West Cypress St #1040, Tampa, FL 33607-4185 – **1776)** 6699 N. Federal Hwy #200, Boca Raton, FL 33487-1671 – **1777)** 20 John Davenport Drive NW, Rome, GA 30165-2536 – **1778)** 331 Fulton St #1200, Peoria, IL 61602-1475 – **1779)** 1113 Nebraska St, Sioux City, IA 51105-1438 – **1780)** 2401 Paint Creek Rd, Stanton, KY 40380-9272 – **1781)** 900 North Lake Shore Drive, Lake Charles, LA 70601-2120 – **1782)** 447 Congress St #3B, Portland, ME 04101-3505 – **1783)** 351 Tilghman Rd, Salisbury, MD 21804-1920 – **1784)** 500 West Cummings Park #2600, Woburn, MA 01801-6503 – **1785)** 3338 E. Bristol Rd, Burton, MI 48529-1408 – **1786)** 1516 E Lake St #200, Minneapolis, MN 55407-1997 – **1787)** 240 Moss Hill Drive, New Albany, MS 38652-3400 – **1788)** 1751 Hanshaw Rd, Ithaca, NY 14850-9105 – **1789)** 401 Saw Mill Hollow Rd, Burnsville, NC 28714-9789 – **1790)** 126 E Water St, Plymouth, NC 27962-1330 – **1791)** 3025 Waughtown St #G, Winston-Salem, NC 27107-1679 – **1792)** 1541 Alta Drive #400, Whitehall, PA 18052-5622 – **1793)** 2440 Millwood Ave, Columbia, SC 29205-1128 or PO Box 2355, West Columbia, SC 29171-2355 – **1794)** 1320 Brick Church Pike, Nashville, TN 37207-5038 – **1795)** 3911 South 1st St, Abilene, TX 79605-1639 – **1796)** 1101 Kilgore Drive, Henderson, TX 75652-5129 – **1797)** 900 Bluefield Ave, Bluefield, WV 24701-2760 – **1798)** 166 North Main St, Broadway, VA 22815-9702 – **1799)** 2241 W. Main St, Moses Lake, WA 98837-2826 – **1800)** 1635 S. Gold St, Centralia, WA 98531-8997 – **1801)** 703 3rd Ave, Huntington, WV 25701-1421 – **1802)** 2410 S. Main St #A, West Bend, WI 53095-5270 – **1803)** 1551 Springhill Ave #A, Mobile, AL 36604-3283 – **1804)** 40 Park Raod #B, Pleasant Grove, AL 35127-1910 – **1805)** 824 E. Washington St, Phoenix, AZ 85034-1088 – **1806)** 1 Radio Drive, Berryville, AR 72616-ND – **1807)** 13749 Beach Blvd, Westminster, CA 92683-3204 – **1808)** 5640 S. Broadway St, Eureka, CA 95503-6997 – **1809)** 1020 West Main St, Merced, CA 95340-4521 – **1810)** 5230 W. Highway 98, Panama City, FL 32401-1058 – **1811)** 4051 Jimmie Dyess Pkwy, Augusta, GA 30909-9469 – **1812)** 1111 Boulevard SE, Atlanta, GA 30312-3895 – **1813)** 501 S Lincoln Ave, Jerome, ID 83338-3026 – **1814)** 18889 N 2350th St, Dennison, IL 62423-2523 – **1815)** 1632 South Maize Rd, Wichita, KS 67209-3912 – **1816)** 486 Lakeside Dr, Jenkins, KY 41537-8917 – **1817)** 1 Home St, Somerset, MA 02720-5229 – **1818)** 301 Fulton St West, Grand Rapids, MI 49404-6492 – **1819)** 580 West Clark Rd., Ypsilanti, MI 48198-3488 – **1820)** 35006 US Highway 2 East, Fosston, MN 56542-9268 – **1821)** 449 Broadway 2nd Flr, New York, NY 10013-2549 – **1822)** 5942 County Route 64, Hornell, NY 14843-9730 – **1823)** 185 Genesee St, Utica, NY 13501-2102 – **1824)** 1100 South Tryon St #210, Charlotte, NC 28203-4297 – **1825)** 185 Franklin Plaza Drive, Franklin, NC 28734-3249 – **1826)** 550 Market Ave South, Canton, OH 44702-2103 – **1827)** 4445 Lake Forest Dr #420, Cincinnati, OH 45242-3733 – **1828)** 111 Presidential Blvd #100, Bala Cynwyd, PA 19004-1009 – **1829)** 369 East GE Patterson Ave, Memphis, TN 38126-3301 – **1830)** 771 Helena St, Dallas, TX 75217-5129 – **1831)** 1211 West Hein Rd, San Antonio, TX 78220-3301 – **1832)** 10614 Rockley Rd, Houston, TX 77099-3514 – **1833)** 18 Park St, Springfield, VT 05156-3023 – **1834)**

306 West BRd St, Richmond, VA 23220-4219 – **1835)** 504 23rd St NW, Roanoke, VA 24017-5414 – **1836)** 14416 Jefferson Davis Hwy #20, Woodbridge, VA 22191-2890 – **1837)** 601 Main St #400, Vancouver, WA 98660-3404 – **1838)** 730 Ray O Vac Lane, Madison, WI 53711-2472 – **1839)** 519 Wooster Ave, San Jose, CA 95116-ND or 145 Natoma St, San Francisco, CA 94105-3734 – **1840)** 90 Kay Ave, Milford, CT 06460-5495 – **1841)** 3400 Idaho Ave NW #200, Washington, DC 20016-3000 – **1842)** 8451 S. Cherokee Blvd. #B, Douglasville, GA 30134-8520 – **1844)** 4802 E 62nd St, Indianapolis, IN 46220-5296 – **1845)** 2 Radio Plaza St, Ferndale, MI 48220-ND – **1846)** 3415 University Ave West, Saint Paul, MN 55114-2099 – **1847)** 4800 N. Central Ave. Phoenix, Mesa, AZ 85012-1722 – **1848)** 8729 9th St #110, Rancho Cucamonga, CA 91730-4312 – **1849)** 3401 Holland Ave, Fresno, CA 93722-4197 – **1850)** 28 2nd St #501, San Francisco, CA 94105-3461 – **1851)** 11203 E Peakview Ave, Greenwood Village, CO 80111-6811 – **1852)** 1150 W. King St., Cocoa, FL 32922-8618 – **1853)** 308 Victory Rd, Quincy, MA 02171-3129 – **1854)** 106 E Main St, Spring Arbor, MI 49283-9701 – **1855)** 1411 Locust St, Saint Louis, MO 63103-2332 – **1856)** 6721 West 121st St, Overland Park, KS 66209-2003 – **1857)** 704 North King St #604, Wilmington, DE 19801-3535 – **1858)** 151 Penns Grove Lane, Latrobe, PA 15650-3745 – **1859)** 621 Cumberland St #4, Lebanon, PA 17042-8500 – **1860)** 15096 South Dakota Highway 15, Milbank, SD 57252-5954 – **1861)** 55 Music Square West, Nashville, TN 37203-3207 – **1862)** 755 South 11th St #102, Beaumont, TX 77701-3723 – **1863)** 868 East 5900 South, Murray, UT 84107-7650 – **1864)** 1601 E. 57th Ave, Spokane, WA 99223-6623 – **1865)** 310 W. Wisconsin Ave #100, Milwaukee, WI 53203-2224 – **1866)** 2284 S. Victoria Ave #2-G, Ventura, CA 93003-6626 – **1867)** PO Box 369, Hollister, CA 95024-0369 – **1868)** 1188 Lake View Rd, Altamonte Springs, FL 32714-2713 – **1869)** 7250 NW 58th St, Miami, FL 33166-3719 – **1870)** 8451 S. Cherokee Blvd. #B, Douglasville, GA 30134-8520 – **1871)** 2980 US Highway 51, Clinton, IL 61727-9479 – **1872)** 1401 Winchester Ave, Ashland, KY 41101-7555 – **1873)** 3225 Ambassadore Caffery Pkwy, Lafayette, LA 70506-7214 – **1874)** 10 Radio Lane, Brunswick, MD 21788-1645 – **1875)** 369 Shelburne Rd, Greenfield, MA 01301-9653 – **1876)** 122 4th Ave SW, Rochester, MN 55902-3339 – **1877)** 125 S. Kingshighway St, Sikeston, MO 63801-2943 – **1878)** 500 Corporate Parkway #200, Buffalo, NY 14226-1265 – **1879)** 824 US Hwy 158 West Bypass, Warrenton, NC 27589-9796 – **1880)** 1117 Radio Rd, Statesville, NC 28677-3399 – **1881)** 400 E. Britton Rd, Oklahoma City, OK 73114-7507 – **1882)** 5110 SE Stark St, Portland, OR 97215-1751 – **1883)** 16620 Cypress Rosehill Rd, Cypress, TX 77429-1424 or PO Box 19886, Houston, TX 77224-9886 – **1884)** 204 Bucky Beaver St, England, AR 72076-4907 – **1885)** 1440 Ethan Way #200, Sacramento, CA 95825-2214 – **1886)** 5050 Edison Ave. #218, Colorado Springs, CO 80915-3540 – **1887)** 177 State St, Bridgeport, CT 06604-4872 – **1888)** 5900 Picketville Rd, Jacksonville, FL 32254-1172 – **1889)** PO Box 216, Dalton, GA 30722-0216 – **1890)** 3338 East Bristol Rd, Burton, MI 48529-1408 – **1891)** 1107 Hazeltine Blvd #301, Chaska, MN 55318-1065 – **1892)** 103 Progress Rd, Poplarville, MS 39470-3388 – **1893)** 3025 Waughtown St #G, Winston-Salem, NC 27107-1634 – **1894)** 8044 Montgomery Rd #650, Cincinnati, OH 45236-2959 – **1895)** 5807 S. Garnett St #F, Tulsa, OK 74146-6824 – **1896)** 4501 North McColl Rd, McAllen, TX 78504-2431 – **1897)** 200 South 10th #600, McAllen, TX 78501-4869 – **1898)** 1190 Daniels Chapel Rd, New Boston, TX 75570-ND – **1899)** 912 S Capital of Texas Hwy #400, Austin, TX 78746-6176 – **1900)** 1445 W. Baseline Rd, Phoenix, AZ 85041-7010 – **1901)** 3700 Wilshire Blvd #600, Los Angeles, CA 90010-3013 – **1903)** 514 Jefferson St, Waterloo, IA 50701-5422 – **1904)** 11141 Georgia Ave #310, Wheaton, MD 20902-4658 – **1905)** 11 Downing Ct, Exeter, NH 03833-1903 – **1906)** 4243 Albany St, Albany, NY 12205-4609 – **1907)** 545 Fire Tower Rd, Yanceyville, NC 27379-ND – **1908)** 200 Monument Rd #6, Bala Cynwyd, PA 19004-1726 – **1909)** 904 North Main St, Punxsutawney, PA 15767-2641 – **1910)** 3931 Whitehorse Rd, Greenville, SC 29611-5599 – **1911)** 510 S. Flores St, San Antonio, TX 78204-1289 – **1912)** 1302 N Shepherd Dr, Houston, TX 77008-3752 – **1913)** 400 Las Colinas Blvd E #1033, Irving, TX 75039-5599 – **1914)** 306 West BRd St, Richmond, VA 23220-4219 – **1915)** 114 Lakeside Ave, Seattle, WA 98122-6542 – **1916)** 1550 The Boardwalk #1, Huntsville, AL 35816-ND – **1917)** Univ of Arizona, Tucson, AZ 85721-0067 – **1918)** 15165 7th St #D, Victorville, CA 92392-3816 – **1919)** 139 West Olive Ave, Fresno, CA 93728-3035 – **1920)** 865 Battery St, San Francisco, CA 94111-1503 – **1921)** 160 Chapel Rd #101, Manchester, CT 06042-1625 – **1922)** 330 SW 27th Ave #207, Miami, FL 33135-2957 – **1923)** 4201 N Armenia Ave, Tampa, FL 33607-6446 – **1924)** 2405 E Moody Blvd #402, Bunnell, FL 32110-5994 – **1925)** 3338 Peachtree Rd NE #3005, Atlanta, GA 30326-1471 – **1926)** 411 Radio Station Rd, North Augusta, SC 29841-9411 – **1927)** 7120 US Highway 41, Vienna, GA 31092-4605 – **1928)** 6943 Titian Drive, Baton Rouge, LA 70806-2767 – **1929)** 143 Rumford Ave, Auburndale, MA 02466-1311 – **1930)** 4104 Country Lane, Saint Joseph, MO 64506-4921 – **1931)** 901 S. Kingshighway, Cape Girardeau, MO 63703-8003 – **1932)**

430-C State Highway 165 South, Branson, MO 65616-3541 – **1933)** 2580 Wrondel Way, Reno, NV 89502-3702 – **1934)** 1213 San Pedro Drive NE, Albuquerque, NM 87110-6725 – **1935)** 2448 East 81st St #5500, Tulsa, OK 74137-4201 – **1936)** 944 Exeter Ave, Exeter, PA 18643 – **1937)** 226 Radio Rd, Bennettsville, SC 29512-6183 – **1938)** 101 Lee St, Bristol, VA 24201-4355 – **1939)** 1824 Murfreesboro Pike, Nashville, TN 37217-3208 – **1940)** 314 S Redwood Rd, Salt Lake City, UT 84104-3536 – **1941)** 2202 Jolliff Rd, Chesapeake, VA 23321-1416 – **1942)** 2043 10th St NE, Roanoke, VA 24012-5309 – **1943)** 6605 SE Lake Rd, Portland, OR 97222-2161 – **1944)** 5538 Imhoff Rd, Ferndale, WA 98248-9177 – **1945)** 510 Pegasus Court, Winchester, VA 22602-4596 – **1946)** 730 Ray O Vac Lane, Madison, WI 53711-2472 – **1947)** RR1 Box 189, Daleville, AL 36322-9747 – **1948)** 3561 Pegasus Drive #107, Bakersfield, CA 93308-0658 – **1949)** 4554 S Suncoast Blvd, Homosassa, FL 34446-1103 – **1950)** 1800 W. Hibiscus Blvd #138, Melbourne, FL 32901-2624 – **1951)** 120 South 35th St #2, Council Bluffs, IA 51501-3203 – **1952)** 6000 Bristol Drive, Paducah, KY 42003-9213 – **1953)** 4200 W. Main St, Kalamazoo, MI 49006-2766 – **1954)** 2 Penn Plaza 17th Floor, New York, NY 10121-0101 – **1955)** 8686 Michael Lane, Fairfield, OH 45014-3096 – **1956)** 101 S. Woodland Drive, Lancaster, SC 29720-2244 – **1957)** 3980 S. Dakota St, Aberdeen, SD 57401-8585 – **1958)** 5353 West Alabama #415, Houston, TX 77056 – **1959)** 9434 Parkfield Drive, Austin, TX 78758-6227 – **1960)** PO Box 778, South Boston, VA 24592-0778 – **1961)** 216 N 11th Ave, Pasco, WA 99301-5461 – **1962)** 2600 S. Jackson St, Seattle, WA 98144-2499 – **1963)** 908 2nd Ave East, Oneonta, AL 35121-2506 – **1964)** 548 E Alisal St, Salinas, CA 93905-2760 – **1965)** 6820 Pacific Ave. #3A, Stockton, CA 95207-2604 – **1966)** 7351 Lincoln Ave, Riverside, CA 92504-4618 – **1967)** 201 N. Industrial Park Rd., Excelsior Springs, MO 64024-1786 – **1968)** 8384 Beavermeadows Rd #1, Jacksonville, FL 32256-7486 – **1969)** 127 Glenn Rd, Auburndale, FL 33823-2401 – **1970)** 2424 Old Rex Morrow Rd, Ellenwood, GA 30294-3901 – **1972)** 834 N Tower Rd, Freeport, IL 61032-8650 – **1973)** 1550 Hart Rd, Towson, MD 21286-1697 – **1974)** PO Box 2442, South Hamilton, MA 01982-0442 – **1975)** 2110 Cliff Rd, Eagan, MN 55122-2347 – **1976)** PO Box 49, Park Rapids, MN 56470-0049 – **1977)** 150 Bay Ave, Laurel, MS 39440-4510 or PO Box 548, Bay Springs, MS 39422-0548 – **1978)** 1111 S Glenstone Ave #3-102,, Springfield, MO 65804 – **1979)** 3568 Lenox Rd, Geneva, NY 14456-2058 – **1980)** 17890 US Hwy 64 West, Siler City, NC 27344-1631 – **1981)** PO Box 1534, Greenville, NC 27835-1534 – **1982)** 40 Rickert Rd, Doylestown, PA 18901-2326 – **1983)** 8 East Market St, Danville, PA 17821-2917 – **1984)** 1860 Executive Park NW #E, Cleveland, TN 37312-2743 – **1985)** 150 Highway 50, Centerville, TN 37033-5996 – **1986)** 275 Glenwood Drive, Rocky Mount, VA 24151-2136 – **1987)** 3616 State Highway 47, Rhinelander, WI 54501-8819 – **1988)** 1215 Church St, Oxford, AL 36203-1639 – **1989)** 4602 E University Drive #150, Phoenix, AZ 85034-7423 – **1990)** 747 E. Green St #400, Pasadena, CA 91101-2148 – **1991)** 614 Kimbark St, Longmont, CO 80501-4911 – **1992)** PO Box 300, Greensboro, FL 32330-0300 – **1993)** 1510 NE 162nd St, North Miami Beach, FL 33162-4716 – **1994)** 3765 N. John Young Parkway, Orlando, FL 32804-3213 – **1995)** 2601 Nicholasville Rd, Lexington, KY 40503-3307 – **1996)** 4200 Parliament Place #300, Lanham, MD 20706-1881 – **1997)** 5115 Telephone Rd, Pascagoula, MS 39567-1130 – **1998)** 41 Pennsylvania Ave, Medford, NY 11763-3717 – **1999)** 4673 Winterset Drive, Columbus, OH 43220-8113 – **2000)** 830 Old Buncombe Rd, Travelers Rest, SC 29690-9467 – **2001)** 2330 Merchant Drive, Knoxville, TN 37912-5136 – **2002)** 236 Woodland Drive, Shelbyville, TN 37160-6759 – **2003)** 100 Stoddart St, Beaver Dam, WI 53916-1306 – **2004)** 509 North Main St, Tuscumbia, AL 35674-2048 – **2005)** 4209 N Frontage Rd, Fayetteville, AR 72703-5002 – **2006)** 303 N. Spring St, Fordyce, AR 71742-3317 – **2007)** 2284 Victoria Ave. #2-G, Ventura, CA 93003-6626 – **2008)** 750 Story Rd, San Jose, CA 95122-2604 – **2009)** 3551 42nd Ave S #B106, Saint Petersburg, FL 33711-4369 – **2010)** 4100 Metzger Rd, Fort Pierce, FL 34947-1712 – **2011)** 2 Radio Loop, Swainsboro, GA 30401-5673 – **2012)** PO Box 746, Lafayette, GA 30728-0746 – **2013)** 1104 W. BRd Avene, Albany, GA 31707-4340 – **2014)** 2100 Lee St, Evanston, IL 60202-1539 – **2015)** 55 Public Square, Monmouth, IL 61462-1755 – **2016)** 3745 W Washington St, Indianapolis, IN 46241-1503 – **2017)** 1200 Baker Ave, Great Bend, KS 67530-4523 o – **2018)** 182 N Angola Rd, Coldwater, MI 49036-9554 – **2019)** 1185 9th St NE, Thompson, ND 58278-9343 or PO Box 13638, Grand Forks, ND 58208-3638 – **2020)** 1375 Beasley Rd, Jackson, MS 39206-2018 – **2021)** 85592 574th Ave, Wayne, NE 68787-7043 – **2022)** 2450 Wrondell Way #G, Reno, NV 89502-3767 – **2023)** 149 Main St #210, Nashua, NH 03060-2725 – **2024)** 231 N. Union St, Olean, NY 14760-2663 – **2025)** 911 W. Main St, Clayton, NC 27520-1620 – **2026)** 1366 Startown Rd, Lincolnton, NC 28092-8038 – **2027)** 1795 West Market St, Akron, OH 44313-7001 – **2028)** 170 3rd St, Tillamook, OR 97141-9489 – **2029)** 16221 National Pike, Hagerstown, MD 21740-2150 – **2030)** 575 Grove St, Honesdale, PA 18431-1041 – **2031)** 12 Kent Rd, Aston, PA 19014-1498 – **2032)** 19 Luther Ave, Warwick, RI 02886-4615

– **2033)** 222 Commerce St, Kingsport, TN 37660-4319 – **2034)** 3120 Southwest Freeway #610, Houston, TX 77098-4521 – **2035)** 215 South Market St, Carthage, TX 75633-2623 – **2036)** 6900 Commerce Ave, El Paso, TX 79915-1102 – **2037)** 227 E. Belt Blvd, Richmond, VA 23224-1205 – **2038)** 2201 6th Ave #1500, Seattle, WA 98121-1840 – **2039)** 125 East 3rd St, New Richmond, WI 54017-1800 – **2040)** 2609 Jordan Lane NW, Huntsville, AL 35816-1030 – **2041)** 912 South Perry St, Montgomery, AL 36104-5002 – **2042)** 600 South Pine St, Harrison, AR 72601-5828 – **2043)** 1110 E Olive Ave, Fresno, CA 93728-3535 – **2044)** 1479 Sanborn Rd, Yuba City, CA 95993-6042 – **2045)** 747 E. Green St #400, Pasadena, CA 91101-2148 – **2046)** 7800 East Orchard Rd #400, Greenwood Village, CO 80111-2599 – **2047)** 1205 Walker Rd, Dover, DE 19904 – **2048)** 1611 Atlantic Blvd, Atlantic Beach, FL 32233-2516 – **2049)** 824 US Highway 1 #260, North Palm Beach, FL 33408-3876 – **2050)** 5815 Westside Rd SW, Austell, GA 30106-3179 – **2051)** 1110 26th Ave SW, Cedar Rapids, IA 52404-3430 – **2052)** 11603 Highway 308, Larose, LA 70373 – **2053)** 60 Temple Place #200, Boston, MA 02111-1324 – **2054)** 15 Hampton Ave, Northampton, MA 01060-3809 – **2055)** 4230 Packard St, Ann Arbor, MI 48108-1597 – **2056)** 6500 Brooklyn Blvd, Brooklyn Center, MN 55429-1754 – **2057)** 1001 Highlands Plaza Drive West #100, Saint Louis, MO 63110-1339 – **2058)** 1213 San Pedro Drive NE, Albuquerque, NM 87110-6725 – **2059)** 333 7th Ave #1401, New York, NY 10001-5021 – **2060)** 1705 Lake St, Elmira, NY 14901-1299 – **2061)** 1338 Bragg Blvd, Fayetteville, NC 28301-4202 – **2062)** 7140 SW Macadam Ave, Portland, OR 97219-3013 – **2063)** 134 East Pitt St, Bedford, PA 15522-1311 – **2064)** 2 Milledge Rd, Augusta, GA 30904-3063 – **2065)** 259 S. Willow Ave #A, Cookeville, TN 38501-3140 – **2066)** 3704 Whittier Rd, Memphis, TN 38108-2649 – **2067)** 9780 Walnut St #405, Dallas, TX 75243-2356 – **2068)** 9434 Parkfield Drive, Austin, TX 78758-6227 – **2069)** 2722 South Redwood Rd #1, Salt Lake City, UT 84119-8410 – **2070)** 645 Church St #400, Norfolk, VA 23510-1712 – **2071)** 188 Valley Rd, Saltville, VA 24370-ND – **2072)** 4840 Lincoln Rd, Blaine, WA 98230-9602 or PO Box 75150 RPO White Rock, White Rock, BC V4B 5L3 – **2073)** 1201 Main St, Wheeling, WV 26003-2844 – **2074)** 134 4th Ave, Huntington, WV 25701-1253 or 9801 Radio Park Rd, Catlettsburg, KY 41129-8824 – **2075)** 112 Watson St, Ripon, WI 54971-1327 o – **2076)** 3256 Penryn Rd #100, Loomis, CA 95650-8052 – **2077)** 7251 Plantation Rd, Pensacola, FL 32504-6334 – **2078)** 3371 W. Cleveland Rd Ext. #310, South Bend, IN 46628-9780 – **2079)** 5011 Capitol Ave, Omaha, NE 68132-2921 – **2080)** 2700 Earl Rudder Freeway South #5000, College Station, TX 77845-5011 or PO Box 3248, Bryan, TX 77805-3248 – **2081)** 2600 S. Jackson St, Seattle, WA 98144-2499 – **2082)** 4051 Jimmie Dyess Pkwy, Augusta, GA 30909-9469 – **2083)** 845 Quarry Rd #120, Iowa City, IA 52241-2212 – **2084)** 5787 South Hampton Rd #108, Dallas, TX 75232-6377 or QSL's to CE Mortenson BRdcasting Co 3270 Blazer Pkwy Suite 101, Lexington, KY 40509-1847 – **2085)** 415 E 3rd St, Cheyenne, WY 82007-1479 – **2086)** 3260 Blume Drive #520 Plaza II, Richmond, CA 94806-5715 – **2087)** 1909 East Pass Rd #D11, Gulfport, MS 39507-3778 – **2088)** 316 E. Willow Rd, Enid, OK 73701-1514 – **2089)** 3030 SW Moody Ave #210, Portland, OR 97201-4868 – **2090)** 1762 S Main St, Salt Lake City, UT 84115-1912 – **2091)** W223 N3251 Shady Lane, Pewaukee, WI 53072-4194 – **2092)** 311 Lexington Ave, Fort Smith, AR 72901-3842 – **2093)** 4525 Wilshire Blvd 3rd Flr, Los Angeles, CA 90010-3845 – **2094)** 3131 S. Vaughn Way #601, Aurora, CO 80014-3516 – **2095)** 721 Shirley St, Cedar Falls, IA 50613-1513 – **2096)** 5426 North Mesa St, El Paso, TX 79912-5442 – **2097)** 2202 Mt. Jolliff Rd, Chesapeake, VA 23321-1416 – **2098)** 1020 West Main St, Merced, CA 95340-4521 – **2099)** 5043 Tamiami Trail East, Naples, FL 34113-4127 – **2100)** 7000 Squibb Rd, Mission, KS 66202-3233 – **2101)** 4200 W. Main St, Kalamazoo, MI 49006-2749 – **2102)** 449 Broadway Fl2, New York, NY 10013-2549 – **2103)** 1520 South Blvd #300, Charlotte, NC 28203-3701 – **2104)** 2720 South 7th Ave SW, Fargo, ND 58103-8710 – **2105)** 220 S. 2nd St #2B2, Waco, TX 76701-2250 – **2106)** 80 S Redwood Rd #211, North Salt Lake, UT 84054-2920 – **2107)** 3360 Alta Mesa Drive, Redding, CA 96002-2831 – **2108)** 2636 N Ontario St, Burbank, CA 91504-2514 – **2109)** 7080 Industrial Way, Macon, GA 31216-7538 – **2110)** 730 Ray O Vac Lane, Madison, WI 53711-2472 – **2111)** 139 W. Olive Ave, Fresno, CA 93728-3035 – **2112)** 3765 N. John Young Parkway, Orlando, FL 32804-3213 – **2113)** 1109 Hudson Lane, Monroe, LA 71201-6003 – **2114)** 3777 44th St SE, Grand Rapids, MI 49512-3945 – **2115)** 449 Broadway Fl2, New York, NY 10013-2549 – **2116)** 2201 6th Ave #1500, Seattle, WA 98121-1840 – **2117)** 3463 Ramona Ave #15, Sacramento, CA 95826-3827 – **2118)** 12136 Bayaud Ave #125, Lakewood, CO 80228-2115 – **2119)** 1100 Spring St #610, Atlanta, GA 30309-2828 – **2120)** 1000 East 87th St, Chicago, IL 60619-6397 – **2121)** 28095 Three Notch Rd #2B, Mechanicsville, MD 20659-3373 – **2122)** 2609 Jordan Lane NW, Huntsville, AL 35816-1030 – **2123)** 7250 NW 58th St, Miami, FL 33166-3719 – **2124)** 4143 109th St, Urbandale, IA 50322-7925 – **2125)** 901 East Pike Blvd, Weslaco, TX 78596-4937 – **2126)** 3500 Maple Ave. #1600, Dallas, TX 75219-3945

FM STATIONS IN MAJOR METROPOLITAN AREAS

FM Callsign	MHz	Location	kW
Atlanta Area			
W201CC	88.1	Buford	0.019
WIVL	88.3	Jasper	0.2
WRAS	88.5	Atlanta	100
WKEU-FM	88.9	The Rock	5
WMSL	88.9	Athens	20
W205CI	88.9	Tallapoosa	0.01
WBCX	89.1	Gainesville	0.84
WRFG	89.3	Atlanta	65
WYFW	89.5	Winder	6
W209CD	89.7	Buford	0.007
W209CG	89.7	Tallapoosa	0.01
WABE	90.1	Atlanta	96
W213BE	90.5	Snellville	0.01
W214AS	90.7	Waleska	0.01
WMVV	90.7	Griffin	18
WUWG	90.7	Carrollton	0.43
WREK	91.1	Atlanta	40
WWEV-FM	91.5	Cumming	8.9
WMVW	91.7	Peachtree City	13
WCLK	91.9	Atlanta	6
WBTR-FM	92.1	Carrollton	0.58
W221AZ	92.1	Lilburn	0.027
W221CG	92.1	Kennesaw	0.038
W222AF	92.3	Marietta	0.01
WEKS	92.5	Zebulon	12
WZGC	92.9	Atlanta	66
WVFJ-FM	93.3	Manchester	27
W229AG	93.7	Sandy Plains	0.25
WSTR	94.1	Smyrna	100
WUBL	94.9	Atlanta	100
WBTS	95.5	Doraville	40
WKLS	96.1	Atlanta	100
W242AU	96.3	Athens-Clarke County	0.018
WWLG	96.7	Peachtree City	2.15
WSRV	97.1	Gainesville	100
WUMJ	97.5	Fayetteville	7.9
W250BC	97.9	Riverdale	0.25
WSB-FM	98.5	Atlanta	100
WPPI	98.9	Tallapoosa	1.85
WWWQ	99.7	Atlanta	100
W261BG	100.1	Morrow	0.005
WNNX	100.5	College Park	12.5
W265BD	100.9	Woodstock	0.007
WKHX-FM	101.5	Marietta	100
WLKQ-FM	102.3	Buford	4.2
WPZE	102.5	Mableton	3
W275BK	102.9	Decatur	0.16
WVEE	103.3	Atlanta	100
WALR-FM	104.1	Greenville	100
WFSH-FM	104.7	Athens	24
WHLB-LP	104.9	Cartersville	0.1
WBZY	105.3	Bowdon	61
WWVA-FM	105.7	Canton	20
WHLE-LP	106.3	Atlanta	0.1
WYAY	106.7	Gainesville	77
WTSH-FM	107.1	Rockmart	45
WAMJ	107.5	Roswell	45
WHTA	107.9	Hampton	27
WPCG-LP	107.9	Canton	0.1
Baltimore Area			
WMUC-FM	88.1	College Park	0.01
WYPR	88.1	Baltimore	15.5
WAMU	88.5	Washington	50
WEAA	88.9	Baltimore	12.5
WPFW	89.3	Washington	50
WTMD	89.7	Towson	10
WCSP-FM	90.1	Washington	36
WKHS	90.5	Worton	17.5
WETA	90.9	Washington	5
W215BY	90.9	Church Hill	0.027
WHFC	91.1	Bel Air	1.1
WBJC	91.5	Baltimore	50
WERQ-FM	92.3	Baltimore	37
WPOC	93.1	Baltimore	19.5
WD2XAB	93.5	Columbia	2
WKYS	93.9	Washington	25
W231BG	94.1	Sunnyburn	0.025
WIAD	94.7	Bethesda	50
WRBS-FM	95.1	Baltimore	50
WPGC-FM	95.5	Morningside	50
WWIN-FM	95.9	Glen Burnie	3
W241AO	96.1	Wye Mills	0.01
WHUR-FM	96.3	Washington	16.5
WCEI-FM	96.7	Easton	12.5

FM Callsign	MHz	Location	kW
WASH	97.1	Washington	17.5
W248AO	97.5	Baltimore	0.008
WIYY	97.9	Baltimore	13.5
W252BR	98.3	Edgemere	0.005
WMZQ-FM	98.7	Washington	4.2
WLZL	99.1	Annapolis	45
WIHT	99.5	Washington	50
W260BM	99.9	Annapolis	0.003
W260BV	99.9	Aberdeen	0.01
W261CD	100.1	Baltimore	0.002
WBIG-FM	100.3	Washington	40
WZBA	100.7	Westminster	25
WWDC	101.1	Washington	25
WLIF	101.9	Baltimore	13.5
WMMJ	102.3	Bethesda	2.9
WQSR	102.7	Baltimore	50
WRNR-FM	103.1	Grasonville	6
WTOP-FM	103.5	Washington	44
WXCY	103.7	Havre De Grace	33
WPRS-FM	104.1	Waldorf	0.45
WZFT	104.3	Baltimore	29
W284BE	104.7	Havre De Grace	0.01
W285EJ	104.9	White Marsh	0.01
WAVA-FM	105.1	Arlington	19.5
WJZ-FM	105.7	Catonsville	50
W291BA	106.1	Baltimore	0.004
WWMX	106.5	Baltimore	8.3
WRQX	107.3	Washington	21.5
WMVK-LP	107.3	Perryville	0.008
WFSI	107.9	Annapolis	50
Boston Area			
WMBR	88.1	Cambridge	0.72
WRPS	88.3	Rockland	0.105
WIQH	88.3	Concord	0.1
WBMT	88.3	Boxford	0.66
WERS	88.9	Boston	4
WHAB	89.1	Acton	0.008
WGBH	89.7	Boston	100
WZBC	90.3	Newton	1
WBUR-FM	90.9	Boston	40
WSHL-FM	91.3	Easton	0.1
WDJM-FM	91.3	Framingham	0.1
WZLY	91.5	Wellesley	0.007
WUML	91.5	Lowell	1.4
WMLN-FM	91.5	Milton	0.17
WMFO	91.5	Medford	0.125
WMWM	91.7	Salem	0.13
WAVM	91.7	Maynard	0.016
WUMB-FM	91.9	Boston	0.66
W221CH	92.1	Lawrence	0.25
WBOS	92.9	Brookline	18.5
WMKK	93.7	Lawrence	42
WJMN	94.5	Boston	9.2
WHRB	95.3	Cambridge	3
W242AA	96.3	Beacon Hill	0.005
WTKK	96.9	Boston	22.5
WYAJ	97.7	Sudbury	0.004
WKAF	97.7	Brockton	2.7
WBZ-FM	98.5	Boston	9
WCRB	99.5	Lowell	37
WHHB	99.9	Holliston	0.017
WBRS	100.1	Waltham	0.025
WZLX	100.7	Boston	21.5
WFNX	101.7	Lynn	1.7
WKLB-FM	102.5	Waltham	14
WODS	103.3	Boston	21
W279BQ	103.7	Gloucester	0.02
WBMX	104.1	Boston	21
WRBB	104.9	Boston	0.019
WBOQ	104.9	Gloucester	3.2
WROR-FM	105.7	Framingham	23
WMJX	106.7	Boston	21.5
WXKS-FM	107.9	Medford	20.5
Charlotte Area			
WPIR	88.1	Hickory	26.5
WMTG-LP	88.1	Mount Gilead	0.03
W201DI	88.1	Monroe	0.17
WGWG	88.3	Boiling Springs	50
W202BW	88.3	Harrisburg	0.01
W205BY	88.9	Lexington	0.01
WNSC-FM	88.9	Rock Hill	100
WRYN	89.1	Hickory	0.85
WRFE	89.3	Chesterfield	33
WDAV	89.9	Davidson	100
WRBK	90.3	Richburg	7.5
WFAE	90.7	Charlotte	100

FM Callsign	MHz	Location	kW
W216BI	91.1	Lexington	0.04
WYFG	91.1	Gaffney	100
W217AX	91.3	Harrisburg	0.01
WFBK	91.5	Fort Mill	0.1
WSGE	91.7	Dallas	7.5
W219CH	91.7	Lowrys	0.01
W220DL	91.9	Statesville	0.04
WRCM	91.9	Wingate	30
WQNC	92.7	Harrisburg	6
W224BN	92.7	Hickory	0.04
WSEQ-LP	92.9	Hudson	0.1
W225BD	92.9	Statesville	0.25
WRHJ-LP	93.1	Rock Hill	0.04
WOGR-FM	93.3	Salisbury	0.1
WYFQ-FM	93.5	Wadesboro	8.7
WWLV	94.1	Lexington	43
W232AX	94.3	Rock Hill	0.05
WNKS	95.1	Charlotte	100
WXRC	95.7	Hickory	100
WIBT	96.1	Shelby	100
W243BY	96.5	Charlotte	0.25
WKKT	96.9	Statesville	100
WPEG	97.9	Concord	95
WEHB-LP	98.3	Wadesboro	0.02
W252BU	98.3	Dallas	0.25
W253BA	98.5	Indian Trail	0.02
W254AZ	98.7	Belmont	0.09
WVOY-LP	98.9	Jefferson	0.06
W256BP	99.1	Charlotte	0.01
WLRZ-LP	99.3	Hickory	0.02
WBT-FM	99.3	Chester	7.7
WRFX	99.7	Kannapolis	84
W261AP	100.1	Kings Mountain	0.01
W262BM	100.3	Charlotte	0.01
WPZS	100.9	Indian Trail	6
WWGT-LP	100.9	Lincolnton	0.1
W267AG	101.3	Salisbury	0.04
W267AM	101.3	Mocksville	0.03
WBAV-FM	101.9	Gastonia	100
WGSP-FM	102.3	Pageland	2.55
WLYT	102.9	Hickory	31
W277CB	103.3	Charlotte	0.25
WSOC-FM	103.7	Charlotte	100
W282BP	104.3	Charlotte	0.25
W282AX	104.3	Gaffney	0.25
WKQC	104.7	Charlotte	100
WNOW-FM	105.3	Gaffney	51
W289BO	105.7	Marion	0.15
WOLS	106.1	Waxhaw	21
WEND	106.5	Salisbury	84
WRHM	107.1	Lancaster	2.4
WLNK	107.9	Charlotte	100
Chicago Area			
WSSD	88.1	Chicago	0.01
WLTL	88.1	La Grange	0.18
WCRX	88.1	Chicago	0.1
WNTH	88.1	Winnetka	0.1
WETN	88.1	Wheaton	0.25
WAES	88.1	Lincolnshire	0.15
WBMF	88.1	Crete	0.09
WLRA	88.1	Lockport	0.14
WZRD	88.3	Chicago	0.1
WXAV	88.3	Chicago	0.15
WDGC-FM	88.3	Downers Grove	0.25
WAWF	88.3	Kankakee	1.25
WHCM	88.3	Palatine	0.1
WCLR	88.3	Arlington Heights	1
WDSO	88.3	Chesterton	0.4
WHFH	88.5	Flossmoor	1.5
WHPK-FM	88.5	Chicago	0.1
WHSD	88.5	Hinsdale	0.125
WGBK	88.5	Glenview	0.185
WSEH	88.5	South Elgin	0.67
W203AJ	88.5	Michigan City	0.013
WLUW	88.7	Chicago	0.1
WCSF	88.7	Joliet	0.1
WEGN	88.7	Kankakee	3
WSRI	88.7	Sugar Grove	0.6
WRSE	88.7	Elmhurst	0.32
WGVE-FM	88.7	Gary	2.1
WARG	88.9	Summit	0.5
WRRG	88.9	River Grove	0.1
WEPS	88.9	Elgin	0.74
WIIT	88.9	Chicago	0.017
W205AZ	88.9	Kankakee	0.01
W205CC	88.9	Dekalb	0.01

FM Callsign	MHz	Location	kW
WOTW	88.9	Monee	0.1
WMXM	88.9	Lake Forest	0.295
WONC	89.1	Naperville	1.5
W206BL	89.1	Mount Prospect	0.12
WLPR-FM	89.1	Lowell	2.4
WNUR-FM	89.3	Evanston	7.2
W207BI	89.3	University Park	0.013
WKKC	89.3	Chicago	0.28
WBEW	89.5	Chesterton	4
WONU	89.7	Kankakee	35
WHLP	89.9	Hanna	8
WMBI-FM	90.1	Chicago	100
WMTH	90.5	Park Ridge	0.008
WRTE	90.5	Chicago	0.073
WRTW	90.5	Crown Point	3.1
WDCB	90.9	Glen Ellyn	5
WKCC	91.1	Kankakee	2.6
WGTD	91.1	Kenosha	3.2
W217BM	91.3	Elgin	0.01
WBEZ	91.5	Chicago	8.3
W219CD	91.7	Elgin	0.01
WJCH	91.9	Joliet	50
W221BY	92.1	Elgin	0.015
WPWX	92.3	Hammond	50
WKIF	92.7	Kankakee	3
WCPT-FM	92.7	Arlington Heights	1.8
WXRT-FM	93.1	Chicago	14
WVIX	93.5	Joliet	6
WLIT-FM	93.9	Chicago	6
W232BL	94.3	Joliet	0.005
WJKL	94.3	Glendale Heights	3.5
WLS-FM	94.7	Chicago	20.5
WVLI	95.1	Kankakee	2.3
WVUR-FM	95.1	Valparaiso	0.036
W236BD	95.1	Michigan City	0.013
WIIL	95.1	Union Grove	50
WNUA	95.5	Chicago	8.3
WERV-FM	95.9	Aurora	3
WEFM	95.9	Michigan City	3
WBBM-FM	96.3	Chicago	19
W244BO	96.7	Park Ridge	0.01
WSSR	96.7	Joliet	3.1
WCOE	96.7	La Porte	3
WWDV	96.9	Zion	50
WDRV	97.1	Chicago	8.3
W248BB	97.5	Hillside-Chicago	0.003
WLUP-FM	97.9	Chicago	6
WCCQ	98.3	Crest Hill	3
WVLP-LP	98.3	Valparaiso	0.1
WFMT	98.7	Chicago	6
WUSN	99.5	Chicago	24.2
WCPQ	99.9	Park Forest	50
WILV	100.3	Chicago	8.3
WRXQ	100.7	Coal City	2.45
W264BF	100.7	Englewood	0.002
WKQX	101.1	Chicago	8.3
WLGS-LP	101.5	Lake Villa	0.1
WTMX	101.9	Skokie	4.2
WXLC	102.3	Waukegan	3
WYCA	102.3	Crete	1.05
W272BZ	102.3	Portage	0.027
WVAZ	102.7	Oak Park	35
WCSJ-FM	103.1	Morris	6
WVIV-FM	103.1	Highland Park	6
W276BM	103.1	Park Forest	0.019
WKSC-FM	103.5	Chicago	17
WWYW	103.9	Dundee	2.55
WXRD	103.9	Crown Point	3
WJMK	104.3	Chicago	24.2
WCFL	104.7	Morris	50
WOJO	105.1	Evanston	8.4
WYKT	105.5	Wilmington	1.3
WZSR	105.5	Woodstock	1.6
WLJE	105.5	Valparaiso	1.25
WCFS-FM	105.9	Elmwood Park	25.1
WSRB	106.3	Lansing	4.1
W292DJ	106.3	Lake Bluff	0.011
WPPN	106.7	Des Plaines	50
W294BA	106.7	Valparaiso	0.055
W295AC	106.9	La Porte	0.013
WSPY-FM	107.1	Plano	1.5
WZVN	107.1	Lowell	1.3
WGCI-FM	107.5	Chicago	3.7
WLEY-FM	107.9	Aurora	21
Dallas & Fort Worth Area			
KNTU	88.1	Denton	100

FM Callsign	MHz	Location	kW
KJRN	88.3	Keene	23
KEOM	88.5	Mesquite	61
KTCU-FM	88.7	Fort Worth	10
KSQX	89.1	Springtown	3
KNON	89.3	Dallas	55
KVRK	89.7	Sanger	14
KERA	90.1	Dallas	100
KTXG	90.5	Greenville	100
KCBI	90.9	Dallas	100
K218EB	91.5	Greenville	0.25
KKXT	91.7	Dallas	100
KXEZ	92.1	Farmersville	1.95
K221FM	92.1	Irving	0.1
KZPS	92.5	Dallas	100
KLIF-FM	93.3	Haltom City	50
K227AZ	93.3	Godley	0.1
KIKT	93.5	Greenville	1.8
KLNO	94.1	Fort Worth	100
KSOC	94.5	Gainesville	100
KLTY	94.9	Arlington	100
KHYI	95.3	Howe	15
K240DS	95.9	Garland	0.115
KSCS	96.3	Fort Worth	100
KEGL	97.1	Fort Worth	100
K248BC	97.5	Dallas	0.075
KLAK	97.5	Tom Bean	32
KBFB	97.9	Dallas	100
K252EB	98.3	Cleburne	0.01
KLUV	98.7	Dallas	100
KFZO	99.1	Denton	1.2
KPLX	99.5	Fort Worth	100
KJKK	100.3	Dallas	100
KWRD-FM	100.7	Highland Village	100
WRR	101.1	Dallas	100
KYLP-LP	101.5	Greenville	0.1
KDGE	102.1	Ft. Worth-Dallas	100
K273BJ	102.5	Dallas	0.25
KDMX	102.9	Dallas	100
KESN	103.3	Allen	100
KVIL	103.7	Highland Pk-Dallas	100
KTDK	104.1	Sanger	7
KKDA-FM	104.5	Dallas	100
KZMP-FM	104.9	Pilot Point	42
KRLD-FM	105.3	Dallas	100
KHKS	106.1	Denton	100
KRVF	106.9	Kerens	21.5
K295BF	106.9	Greenville	0.25
KDXX	107.1	Benbrook	0.2
KMVK	107.5	Fort Worth	53
KESS-FM	107.9	Lewisville	100
Denver Area			
KVOD	88.1	Lakewood	1.2
KGNU-FM	88.5	Boulder	4
KDAB	88.9	Central City	0.1
K206DB	89.1	Fort Collins	0.01
KUVO	89.3	Denver	22.5
KILE-FM	89.5	Woodland Park	0.1
KCFR-FM	90.1	Denver	50
KGQD	90.3	Fraser	0.4
K213EG	90.5	Littleton	0.003
KGUD	90.7	Longmont	0.1
KENC	90.7	Estes Park	0.2
K215ER	90.9	Loveland	0.006
KLDV	91.1	Morrison	100
K218DT	91.5	Lake George	0.009
K219LF	91.7	Idaho Springs	0.026
K220IY	91.9	Lafayette	0.115
K220JN	91.9	Granby	0.115
KJMN	92.1	Castle Rock	42
KWOF	92.5	Broomfield	57
KTCL	93.3	Wheat Ridge	71
K228EM	93.5	Woodland Park	0.015
K229AC	93.7	Ward	0.028
KRKU	93.9	Loveland	0.3
K231AA	94.1	Boulder	0.205
K232EF	94.3	Estes Park	0.25
KRKS-FM	94.7	Lafayette	100
K235BT	94.9	Masonville	0.065
K237CY	95.3	Ft. Collins	0.01
KPTT	95.7	Denver	100
KXPK	96.5	Evergreen	100
KBCO	97.3	Boulder	100
K248AS	97.5	Woodland Park	0.095
K251AB	98.1	Longmont	0.25
KYGO-FM	98.5	Denver	100
KQMT	99.5	Denver	100

FM Callsign	MHz	Location	kW
K258BE	99.5	Estes Park	0.013
K260AL	99.9	Arvada	0.205
KIMN	100.3	Denver	100
KOSI	101.1	Denver	100
KTNI-FM	101.5	Strasburg	97
K269CL	101.7	Evergreen	0.035
K269AE	101.7	Boulder	0.103
KAMV-LP	101.9	Brighton	0.1
K270AL	101.9	Granby	0.01
KKHI	101.9	Centennial	9.5
KRKY-FM	102.1	Estes Park	6
KCUV	102.3	Greenwood Vlg	1
K273BX	102.5	Pinecliffe	0.01
K274BW	102.7	Berthoud	0.25
KRFX	103.5	Denver	100
KKFN	104.3	Longmont	100
KREV-LP	104.7	Estes Park	0.1
KELS-LP	104.7	Greeley	0.061
KXKL-FM	105.1	Denver	100
K288EX	105.5	Lakewood	0.028
KALC	105.9	Denver	100
K292FM	106.3	Denver	0.012
KBPI	106.7	Denver	100
KDHT-FM	107.1	Bennett	97
KQKS	107.5	Lakewood	100
NB: Synchronized translators not listed.			
Detroit Area			
WBFH	88.1	Bloomfield Hills	0.36
WHPR-FM	88.1	Highland Park	0.011
WDTR	88.1	Monroe	1.2
WSDP	88.1	Plymouth	0.3
WNFA	88.3	Port Huron	1.3
WCBN-FM	88.3	Ann Arbor	0.2
WXOU	88.3	Auburn Hills	0.11
WAKL	88.9	Flint	0.38
W205BH	88.9	Port Huron	0.025
WPHS	89.1	Warren	0.1
WEMU	89.1	Ypsilanti	16
WWKM	89.1	Imlay City	1.5
W206BI	89.1	Hamtramck	0.01
WHFR	89.3	Dearborn	0.27
WBLD	89.3	Orchard Lake	0.015
WAHS	89.5	Auburn Heights	0.1
WOVI	89.5	Novi	0.1
W208BB	89.5	Royal Oak	0.045
WTAC	89.7	Burton	25
WRCJ-FM	90.9	Detroit	42
WFUM-FM	91.1	Flint	17.5
WSGR-FM	91.3	Port Huron	0.12
WGTE-FM	91.3	Toledo	13.5
WUOM	91.7	Ann Arbor	110
WMXD	92.3	Detroit	45
WDRQ	93.1	Detroit	26.5
WHMI-FM	93.5	Howell	5.2
W230BI	93.9	Oxford	0.125
W232BH	94.3	Holly	0.007
WCSX	94.7	Birmingham	14
WFBE	95.1	Flint	34
WKQI	95.5	Detroit	100
WDVD	96.3	Detroit	21
WXYT-FM	97.1	Detroit	50
W248AQ	97.5	Harrison	0.001
WYDM	97.5	Monroe	0.049
WJLB	97.9	Detroit	50
WTWR-FM	98.3	Luna Pier	3.4
W252BX	98.3	Detroit	0.17
WDZH	98.7	Detroit	50
W256AY	99.1	Detroit	0.04
WYCD	99.5	Detroit	17.5
WKKO	99.9	Toledo	50
WNIC	100.3	Dearborn	50
WRIF	101.1	Detroit	11
WRVF	101.5	Toledo	33
WDET-FM	101.9	Detroit	48
W272CA	102.3	Detroit	0.038
WHTD	102.7	Mount Clemens	50
WWWW-FM	102.9	Ann Arbor	49
WQUS	103.1	Lapeer	2.6
WMUZ	103.5	Detroit	50
WOMC	104.3	Detroit	190
WMRP-LP	104.7	Mundy Township	0.1
W284BQ	104.7	Warren	0.05
WIOT	104.7	Toledo	50
WMGC-FM	105.1	Detroit	50
WWCK-FM	105.5	Flint	25
W288BK	105.5	Rochester Hills	0.012

FM Callsign	MHz	Location	kW
W288BT	105.5	St. Clair	0.008
WWWM-FM	105.5	Sylvania	4.3
WDMK	105.9	Detroit	20
W292DK	106.3	Westland	0.01
WDTW-FM	106.7	Detroit	61
WSAQ	107.1	Port Huron	6
WQKL	107.1	Ann Arbor	3
WGPR	107.5	Detroit	50
WCRZ	107.9	Flint	50
WMLZ-LP	107.9	Temperance	0.1

NB: +Several stns from Windsor, Ontario (see listings under Canada for details).

Houston-Galveston Area

FM Callsign	MHz	Location	kW
K201EU	88.1	Katy	0.25
K201FA	88.1	Freeport	0.05
K201DZ	88.1	Port Bolivar	0.12
KFTG	88.1	Pasadena	0.7
KAFR	88.3	Conroe	100
KUHF	88.7	Houston	100
KSBJ	89.3	Humble	100
K208DG	89.5	Galveston	0.25
KACC	89.7	Alvin	5.6
K210DF	89.9	Lake Jackson	0.25
KPFT	90.1	Houston	100
KJIC	90.5	Santa Fe	36
KCPC	90.7	Sealy	1.7
KTSU	90.9	Houston	18.5
KYBJ	91.1	Lake Jackson	5
K217DJ	91.3	Kemah	0.183
K217DP	91.3	Barker	0.248
KPVU	91.3	Prairie View	31
K218EJ	91.5	Galveston	0.25
KTRU	91.7	Houston	50
K221CV	92.1	Sealy	0.077
KROI	92.1	Seabrook	22
KKBQ-FM	92.9	Pasadena	100
KQBU-FM	93.3	Port Arthur	100
K227BD	93.3	Freeport	0.2
KKRW	93.7	Houston	100
KTBZ-FM	94.5	Houston	100
K236AR	95.1	Angleton	0.041
KKHH	95.7	Houston	100
KHMX	96.5	Houston	100
KTHT	97.1	Cleveland	0.8
KFNC	97.5	Beaumont	8
KBXX	97.9	Houston	100
KTJM	98.5	Port Arthur	100
KODA	99.1	Houston	100
KVST	99.7	Willis	2.95
KSHN	99.9	Liberty	26.5
K260BM	99.9	Sugar Land	0.099
KILT-FM	100.3	Houston	100
KKHT-FM	100.7	Winnie	100
KLOL	101.1	Houston	100
KSTB	101.5	Crystal Beach	6
KMJQ	102.1	Houston	100
KLTN	102.9	Houston	100
KHJK	103.7	La Porte	100
KRBE	104.1	Houston	100
KAMA-FM	104.9	Deer Park	15
KTWL	105.3	Hempstead	9.2
KPTY	105.3	Crystal Beach	6
KORG-LP	105.3	Cleveland	0.072
KHCB-FM	105.7	Houston	100
KOVE-FM	106.5	Galveston	100
K294BH	106.7	Simonton	0.25
KHPT	106.9	Conroe	100
KGLK	107.5	Lake Jackson	100
KQLC-LP	107.9	Brookshire	0.063
KQQK	107.9	Beaumont	100

Los Angeles Area

FM Callsign	MHz	Location	kW
K201AR	88.1	Banning	0.009
KKJZ	88.1	Long Beach	30
K201CD	88.1	Victorville	0.01
KUCR	88.3	Riverside	0.15
KCLU-FM	88.3	Thousand Oaks	3.2
KSBR	88.5	Mission Viejo	0.6
KCSN	88.5	Northridge	0.37
KISL	88.7	Avalon	0.2
KSPC	88.7	Claremont	3
K205DZ	88.9	Devore	0.01
KUCI	88.9	Irvine	0.2
K205EP	88.9	La Canada	0.01
KTLW	88.9	Lancaster	5.8
KXLU	88.9	Los Angeles	2.9
K205BH	88.9	Victorville	0.009

FM Callsign	MHz	Location	kW
K206AA	89.1	Laguna Beach	0.04
KCRU	89.1	Oxnard	0.85
KUOR-FM	89.1	Redlands	0.035
KPCC	89.3	Pasadena	0.6
K208AM	89.5	Newport Beach	0.075
K209CN	89.7	Gorman	0.01
KSGN	89.7	Riverside	3
KCRW	89.9	Santa Monica	6.9
KBPK	90.1	Buena Park	0.019
K211EY	90.1	Palmdale	0.01
K211DK	90.1	Santa Ana	0.01
KSAK	90.1	Walnut	0.004
K212FA	90.3	Temple City	0.01
KPFK	90.7	Los Angeles	110
K216EM	91.1	Arcadia	0.01
K216FM	91.1	Pacoima	0.02
K216FA	91.1	Quartz Hill	0.01
KUSC	91.5	Los Angeles	39
KVCR	91.9	San Bernardino	3.8
K220FR	91.9	Simi Valley	0.01
KHHT	92.3	Los Angeles	43
K224DK	92.7	Fontana	0.005
KJLL-FM	92.7	Fountain Valley	0.69
KHJL	92.7	Thousand Oaks	3.1
KCBS-FM	93.1	Los Angeles	28.5
KDEY-FM	93.5	Ontario	5
KDAY	93.5	Redondo Beach	4.2
KXOS	93.9	Los Angeles	35
KEBN	94.3	Garden Grove	6
KJVA-LP	94.3	San Bernardino	0.1
KBUA	94.3	San Fernando	3
KTWV	94.7	Los Angeles	58
K236AW	95.1	Lancaster	0.01
KFRG	95.1	San Bernardino	50
KBBY-FM	95.1	Ventura	12.5
KLOS	95.5	Los Angeles	63
KOCP	95.9	Camarillo	1
KFSH-FM	95.9	La Mirada	6
KRQB	96.1	San Jacinto	1.4
KXOL-FM	96.3	Los Angeles	22
KCAL-FM	96.7	Redlands	1.75
KWIZ	96.7	Santa Ana	6
KLJR-FM	96.7	Santa Paula	0.28
K246AR	97.1	Green Valley Lake	0.01
KAMP-FM	97.1	Los Angeles	7
KLYY	97.5	Riverside	72
KLAX-FM	97.9	East Los Angeles	33
K251AH	98.1	Grand Terrace	0.008
K252EI	98.3	Rialto	0.03
KRCV	98.3	West Covina	6
KYSR	98.7	Los Angeles	75
KGGI	99.1	Riverside	2.55
KKLA-FM	99.5	Los Angeles	10
KOLA	99.9	San Bernardino	29.5
K261AB	100.1	Newhall	0.007
KSWD	100.3	Los Angeles	15
KAEH	100.9	Beaumont	5.3
KRTH	101.1	Los Angeles	54
KORM-LP	101.5	Corona	0.1
KWVS-LP	101.5	Malibu	0.1
KOCI-LP	101.5	Newport Beach	0.042
KSCA	101.9	Glendale	4.8
KJLH	102.3	Compton	5.6
KIIS-FM	102.7	Los Angeles	7.2
K276EF	103.1	Muscoy	0.01
KDLE	103.1	Newport Beach	0.3
KDLD	103.1	Santa Monica	3.7
KOST	103.5	Los Angeles	12.5
KRCD	103.9	Inglewood	4.1
KCXX	103.9	Lake Arrowhead	0.18
K280DT	103.9	Thousand Oaks	0.005
KBIG-FM	104.3	Los Angeles	65
KCAQ	104.7	Oxnard	20
KQIE	104.7	Redlands	1.35
KKGO	105.1	Los Angeles	35
KGIC-LP	105.5	Corona	0.059
KXRS	105.5	Hemet	6
KBUE	105.5	Long Beach	3.9
KVVS	105.5	Rosamond	6
KPWR	105.9	Los Angeles	25
KGMX	106.3	Lancaster	3
KALI-FM	106.3	Santa Ana	6
K292CR	106.3	Simi Valley	0.004
KROQ-FM	106.7	Pasadena	6.5
K295AI	106.9	Muscoy	0.007
KSSE	107.1	Arcadia	6

FM Callsign	MHz	Location	kW
KLVE-FM	107.5	Los Angeles	32
KWVE-FM	107.9	San Clemente	0.53
KNJR-LP	107.9	Thousand Oaks	0.088

NB: Synchronized translators not listed.

Miami-Fort Lauderdale Area

FM Callsign	MHz	Location	kW
WRGP	88.1	Homestead	0.165
WGNK	88.3	Pennsuco	6
WKPX	88.5	Sunrise	3
WMFL	88.5	Florida City	7.7
WDNA	88.9	Miami	7.4
WRMB	89.3	Boynton Beach	100
WKCP	89.7	Miami	100
WAFG	90.3	Fort Lauderdale	3
WVUM	90.5	Coral Gables	5.9
WLFE	90.9	Cutler Bay	100
WLRN-FM	91.3	Miami	47
W220DU	91.9 (F.PI 91.7)	W. Deerfield Bch.	0.08
WMKL	91.9	Hammocks	25
WCMQ-FM	92.3	Hialeah	31
WHDR	93.1	Miami	100
W228BY	93.5	Allapattah	0.12
W228BV	93.5	Fort Lauderdale	0.01
WMIA-FM	93.9	Miami Beach	100
W233AP	94.5	Oakland Park	0.027
WMGE	94.9	Miami Beach	100
W237CP	95.3	Miami	0.07
W237BD	95.3	Pompano Beach	0.08
W237CI	95.3	Miami Beach	0.25
WXDJ	95.7	N. Miami Beach	40
WPOW	96.5	Miami	100
WKEZ-FM	96.9	Tavernier	25
W245BC	96.9	Lauderdale Lakes	0.067
W245BF	96.9	North Miami	0.1
WFLC	97.3	Miami	100
WRMF	97.9	Palm Beach	100
WURM	98.3	Goulds	100
WEDR	99.1	Miami	100
W258BQ	99.5	Homestead	0.25
WKIS	99.9	Boca Raton	100
W262AN	100.3	Tamarac	0.055
WHYI-FM	100.7	Fort Lauderdale	100
WLYF	101.5	Miami	100
WKLG	102.1	Rock Harbor	100
WMXJ	102.7	Pompano Beach	100
WMIB	103.5	Fort Lauderdale	100
WEAT-FM	104.3	West Palm Beach	100
WORZ-LP	104.3	Key Largo	0.1
WHQT	105.1	Coral Gables	100
WWWK	105.5	Islamorada	50
WBGG-FM	105.9	Fort Lauderdale	100
WRAZ-FM	106.3	Leisure City	50
WRMA	106.7	Fort Lauderdale	100
W296BP	107.1	Coral Gables	0.25
WAMR-FM	107.5	Miami	95

Minneapolis-St Paul Area

FM Callsign	MHz	Location	kW
KRLX	88.1	Northfield	0.1
WAJC	88.1	Newport	1.2
KJGT	88.3	Waconia	11
KBEM-FM	88.5	Minneapolis	2.9
W204DS	88.7	Glencoe	0.115
WUSG-LP	88.7	Cambridge	0.1
WRFW	88.7	River Falls	3
KCMP	89.3	Northfield	100
KPCS	89.7	Princeton	40
KMOJ	89.9	Minneapolis	6.2
KFAI	90.3	Minneapolis	0.9
KMKL	90.3	North Branch	15
K214DF	90.7	Golden Valley	0.01
K215DU	90.9	Hutchinson	0.23
KNOW-FM	91.1	Minneapolis	100
K218DK	91.5	Bloomington	0.216
WMCN	91.7	St. Paul	0.005
K220JP	91.9	Minneapolis	0.013
W220DO	91.9	North Branch	0.038
W221BS	92.1	Waite	0.055
K221ES	92.1	Albertville	0.25
KQRS-FM	92.5	Golden Valley	100
W225AP	92.9	St. Paul	0.17
W227BF	93.3	Shoreview	0.01
KXXR	93.7	Minneapolis	100
KSTP-FM	94.5	St. Paul	100
KNOF	95.3	St. Paul	6
KBEK	95.5	Mora	25
KRDS-FM	95.5	New Prague	6
WPCA-LP	95.7	Amery	0.05
W239AM	95.7	Hudson	0.25

FM Callsign	MHz	Location	kW
WDMO	95.7	Baldwin	4
WLKX-FM	95.9	Forest Lake	3
KQCL	95.9	Faribault	3
KHTC	96.3	Edina	19
KTCZ-FM	97.1	Minneapolis	100
K249ED	97.7	Albertville	0.17
KTIS-FM	98.5	Minneapolis	100
KSJN	99.5	Minneapolis	100
K260BA	99.9	Coon Rapids	0.17
KTLK-FM	100.3	Minneapolis	100
W264BR	100.7	Falcon Heights	0.01
KDWB-FM	101.3	Richfield	100
KEEY-FM	102.1	St. Paul	100
K273BH	102.5	Fridley	0.041
W273BX	102.5	Mora	0.25
WLTE	102.9	Minneapolis	100
K277AS	103.3	Big Lake	0.01
K279AZ	103.7	Cottage Grove	0.25
KZJK	104.1	St. Louis Park	100
K283BG	104.5	Minneapolis	0.099
WGVX	105.1	Lakeville	2.6
WGVY	105.3	Cambridge	25
WGVZ	105.7	Eden Prarie	3.8
KWNG	105.9	Red Wing	12
KLCI	106.1	Elk River	9.1
WEVR-FM	106.3	River Falls	6
KDXL	106.5	St. Louis Park	0.008
K293BA	106.5	Elko	0.196
KUOM-FM	106.5	St. Louis Park	0.006
K294AM	106.7	West St. Paul	0.17
KTMY	107.1	Coon Rapids	22
KBGY	107.5	Faribault	48
KQQL	107.9	Anoka	100

New York Area

FM Callsign	MHz	Location	kW
WCWP	88.1	Brookville	0.1
WXBA	88.1	Brentwood	0.18
WBGO	88.3	Newark	4.5
WEDW-FM	88.5	Stamford	2
WKWZ	88.5	Syosset	0.125
WPOB	88.5	Plainview	0.125
WPSC-FM	88.7	Wayne	0.2
WRHU	88.7	Hempstead	0.47
WMNJ	88.9	Madison	0.008
WSIA	88.9	Staten Island	0.011
WFRS	88.9	Smithtown	1.5
WFDU	89.1	Teaneck	0.55
WNYU-FM	89.1	New York	8.3
WDDM	89.3	Hazlet	0.01
WGSS	89.3	Copiague	0.035
WSOU	89.5	South Orange	2.4
W208AU	89.5	Massapequa	0.236
WOBH	89.7	Lindenhurst	1.775
WKCR-FM	89.9	New York	1.35
W211AI	90.1	Stamford	0.25
WUSB	90.1	Stony Brook	3.6
WKNJ-FM	90.3	Union Township	0.009
WMSC	90.3	Upper Montclair	0.001
WHCR-FM	90.3	New York	0.008
WHPC	90.3	Garden City	0.5
WKRB	90.3	Brooklyn	0.01
WFUV	90.7	New York	47
WFMU	91.1	East Orange	1.25
W217AF	91.3	Huntington Station	0.25
WNYE	91.5	New York	2
W220AA	91.9	Parlin	0.006
WSHR	91.9	Lake Ronkonkoma	6
WXRK	92.3	New York	18
WQBU-FM	92.7	Garden City	2
WPAT-FM	93.1	Paterson	22
WVIP	93.5	New Rochelle	2.95
W228BI	93.5	Smithtown	0.055
WNYC-FM	93.9	New York	6
WMJC	94.3	Smithtown	2.6
WFME	94.7	Newark	37.2
W235BB	94.9	Hauppauge	0.01
WPLJ	95.5	New York	19
WXNY-FM	96.3	New York	26
WCTZ	96.7	Port Chester	3.3
W245BA	96.9	Manorville	0.01
WQHT	97.1	New York	29.5
WALK-FM	97.5	Patchogue	39
WSKQ-FM	97.9	New York	12.5
WKJY	98.3	Hempstead	3
WRKS	98.7	New York	6
WBAI	99.5	New York	4.3
WHTZ	100.3	Newark	13

FM Callsign	MHz	Location	kW
W264BT	100.7	Edison	0.008
WCBS-FM	101.1	New York	16.8
W268AN	101.5	Plainview	0.01
WRXP	101.9	New York	29.5
WBAB	102.3	Babylon	6
WWFS	102.7	New York	50
W276AQ	103.1	Fort Lee	0.035
W276AV	103.1	Stamford	0.003
WBZO	103.1	Bay Shore	3
WKTU	103.5	Lake Success	17
WFAS-FM	103.9	Bronxville	1.3
WAXQ	104.3	New York	17
W283BA	104.5	Selden	0.01
WDDM	104.7	Hazlet	0.005
WWPR-FM	105.1	New York	17
W289AD	105.7	Selden	0.25
WQXR-FM	105.9	Newark	1.59
WBLI	106.1	Patchogue	49
WLTW	106.7	New York	17
W296BT	107.1	Brooklyn	0.019
WBLS	107.5	New York	4.2
WMNJ	107.9	Madison	0.008

NB: Synchronized translators not listed.

Philadelphia Area

FM Callsign	MHz	Location	kW
WNJS-FM	88.1	Berlin	0.02
WPEB	88.1	Philadelphia	0.001
WXPN	88.5	Philadelphia	5
WYBF	89.1	Radnor Township	0.7
WXVU	89.1	Villanova	0.1
WRDV	89.3	Warminster	1
WKVP	89.5	Cherry Hill	2
WDNR	89.5	Chester	0.008
WGLS-FM	89.7	Glassboro	0.75
WRTI	90.1	Philadelphia	12.5
WHYY-FM	90.9	Philadelphia	13.5
WDBK	91.5	Blackwood	0.1
WSRN-FM	91.5	Swarthmore	0.11
WMPH	91.7	Wilmington	0.1
WKDU	91.7	Philadelphia	0.8
WLBS	91.7	Bristol	0.1
WCUR	91.7	West Chester	0.1
WKDU	91.7	Philadelphia	0.033
WXTU	92.5	Philadelphia	15
WMMR	93.3	Philadelphia	25
WSTW	93.7	Wilmington	47.1
WYSP	94.1	Philadelphia	15
WPST	94.5	Trenton	50
WRSD	94.9	Folsom	0.014
W235AP	94.9	Radnor	0.002
W236AF	95.1	Burlington	0.12
WBEN-FM	95.7	Philadelphia	11
WRDW-FM	96.5	Philadelphia	9.6
W245AG	96.9	Gladwyne	0.009
W246AQ	97.1	Collingswood	0.01
W246AR	97.1	Bensalem	0.074
WZZE	97.3	Glen Mills	0.018
WPEN-FM	97.5	Burlington	26
WOGL	98.1	Philadelphia	10
WUSL	98.9	Philadelphia	32
WJBR-FM	99.5	Wilmington	50
WHHS	99.9	Havertown	0.009
WPHI-FM	100.3	Media	33
W264BH	100.7	Mount Holly	0.016
WBEB	101.1	Philadelphia	14
WIOQ	102.1	Philadelphia	32
WMGK	102.9	Philadelphia	43
W278AK	103.5	Village Green	0.08
WPPZ-FM	103.9	Jenkintown	0.37
WRFF	104.5	Philadelphia	11.5
WDAS-FM	105.3	Philadelphia	42
W289AZ	105.7	Trenton	0.01
WISX	106.1	Philadelphia	22.5
WKDN	106.9	Camden	38
W297AD	107.3	Philadelphia	0.02
WRNB	107.9	Pennsauken	0.78

Pittsburg Area

FM Callsign	MHz	Location	kW
WRWJ	88.1	Murrysville	1
WWOM	88.1	Ellwood City	0.45
WRCT	88.3	Pittsburgh	1.75
WQED-FM	89.3	Pittsburgh	28
WDUQ	90.5	Pittsburgh	26
WYEP-FM	91.3	Pittsburgh	18
WNJR	91.7	Washington	0.95
WCAL	91.9	California	3
WKPL	92.1	Ellwood City	2.5
WPTS-FM	92.1	Pittsburgh	0.016

FM Callsign	MHz	Location	kW
WKPL	92.1	Ellwood City	2.6
WLTJ	92.9	Pittsburgh	43
KDKA-FM	93.7	Pittsburgh	41
W231BM	94.1	Clairton	0.01
WSW-FM	94.5	Pittsburgh	50
WJPA-FM	95.3	Washington	2.15
WKST-FM	96.1	Pittsburgh	49
WRRK	96.9	Braddock	45
W248AR	97.5	Monroeville	0.01
W249BD	97.7	West View	0.19
W250AU	97.9	Jeannette	0.038
WPKV	98.3	Duquesne	4.5
W257CD	99.3	Pittsburgh	0.01
WSHH	99.7	Pittsburgh	15.5
W261AX	100.1	Weirton	0.075
WZPT	100.7	New Kensington	14.5
WORD-FM	101.5	Pittsburgh	43
WDVE	102.5	Pittsburgh	55
WOGI	104.3	East Liverpool	15
WPGB	104.7	Pittsburgh	13
W288BO	105.5	Pittsburgh	0.01
WXDX-FM	105.9	Pittsburgh	15.5
WAOB-FM	106.7	Beaver Falls	37
WDSY-FM	107.9	Pittsburgh	36

Phoenix Area

FM Callsign	MHz	Location	kW
KNAI	88.3	Phoenix	22.5
KPHF	88.3	Phoenix	22.5
K204DR	88.7	Laveen	0.011
K205CI	88.9	Phoenix	0.01
KLVK	89.1	Fountain Hills	2.5
KBAQ	89.5	Phoenix	12.5
K209DV	89.7	Scottsdale	0.01
K210DY	89.9	Black Canyon City	0.25
KZAI	89.9	Superior	45
KFLR-FM	90.3	Phoenix	100
K214DN	90.7	Surprise	0.01
K216FO	91.1	Guadalupe	0.011
KJZZ	91.5	Phoenix	100
K219DZ	91.7	Rio Verde	0.01
KTAR-FM	92.3	Glendale	100
K224CJ	92.7	Phoenix	0.01
KDKB	93.3	Mesa	100
KOOL-FM	94.5	Phoenix	100
KYOT-FM	95.5	Phoenix	100
K240DC	95.9	Buckeye	0.06
K240BD	95.9	Ft. Mcdowell	0.25
KSWG	96.3	Wickenburg	6.4
KMXP	96.9	Phoenix	100
K247BH	97.3	Goodyear	0.04
KUPD	97.9	Tempe	100
KPKX	98.7	Phoenix	100
K257CD	99.3	Phoenix	0.018
K259AS	99.7	Star Valley	0.055
KESZ	99.9	Phoenix	100
KSLX-FM	100.7	Scottsdale	100
KZON	101.5	Phoenix	100
K270BA	101.9	Wickenburg	0.115
KNIX-FM	102.5	Phoenix	100
K276BZ	103.1	Sunflower	0.02
KLNZ	103.5	Glendale	62
KEXX	103.9	Gilbert	100
KZZP	104.7	Mesa	100
KLVA	105.5	Casa Grande	50
KHOT-FM	105.9	Paradise Valley	36
KHOT-FM1	105.9	Glendale	7
KOMR	106.3	Sun City	23
KKMR	106.5	Arizona City	8.6
KWSS-LP	106.7	Scottsdale	0.1
KDVA	106.9	Buckeye	6
KVVA-FM	107.1	Apache Junction	23.5
KMLE	107.9	Chandler	100

Portland Area

FM Callsign	MHz	Location	kW
KBVM	88.3	Portland	3.5
KMUZ	88.5	Turner	0.03
KTFH	88.7	Lees Camp	0.1
KLVP	88.7	Sandy	3.7
KMHD	89.1	Gresham	7.9
KPFR	89.5	Pine Grove	7
KJVH	89.5	Longview	0.1
KQAC	89.9	Portland	3.7
KWBX	90.3	Salem	0.14
KZRI	90.3	Welches	0.24
KSLC	90.3	Mcminnville	0.30
KLWO	90.3	Longview	0.4
KBOO	90.7	Portland	26.5
K216EH	91.1	Colton	0.01

FM Callsign	MHz	Location	kW
KZME	91.1	Brightwood	0.38
KOPB-FM	91.5	Portland	73
K220IN	91.9	Portland	0.01
KGON	92.3	Portland	100
K224DL	92.7	Portland	0.02
K224CP	92.7	Hazel Dell	0.01
K225BF	92.9	Turner	0.02
KRYP	93.1	Gladstone	1.6
KKJC-LP	93.5	Mcminnville	0.1
K228EU	93.5	Vancouver	0.10
KPDQ-FM	93.9	Portland	52
K231AM	94.1	Woodland	0.12
KZPT	94.3	Government Camp	3.4
KLYK	94.5	Kelso	3
KNRK	94.7	Camas	6.3
KXTG	95.5	Portland	100
K240CZ	95.9	Tigard	0.02
K240DA	95.9	Stevenson	0.09
KKJC-LP	96.3	Mcminnville	0.1
K242AF	96.3	Portland	0.03
K242AB	96.3	Salem	0.25
KWLZ-FM	96.3	West Linn	2.9
KQRZ-LP	96.3	Hillsboro	0.1
KQSO-LP	96.3	Newberg	0.001
KYCH-FM	97.1	Portland	100
K248BS	97.5	Newberg	0.003
KRRC	97.9	Portland	0.01
KNRQ-FM	97.9	Tualatin	3.7
K250AE	97.9	Longview	0.25
KPPK	98.3	Rainier	1.6
KUPL-FM	98.7	Portland	37
KWJJ-FM	99.5	Portland	52
KKRZ	100.3	Portland	100
KUFO-FM	101.1	Portland	100
KQRZ-LP	101.5	Hillsboro	0.1
K268BN	101.5	Eufaula	0.25
KINK	101.9	Portland	100
K272EL	102.3	Portland	0.03
K273AJ	102.5	Elwood	0.01
K273AI	102.5	Ariel	0.01
K274AR	102.7	Gresham	0.01
KQSO-LP	102.9	Newberg	0.001
KKCW	103.3	Beaverton	100
KXPC-FM	103.7	Lebanon	0.6
KFIS	104.1	Scappoose	7
K283BL	104.5	Beaverton	0.01
K284BM	104.7	Longview	0.04
KRSK	105.1	Salem	100
K288FT	105.5	Portland	0.05
KUKN	105.5	Longview	0.7
KFBW	105.9	Vancouver	22.5
KLTH	106.7	Lake Oswego	100
K296FT	107.1	West Haven	0.03
KXJM	107.5	Banks	71

NB: Synchronized translators not listed.

San Diego Area

FM Callsign	MHz	Location	kW
KSDS	88.3	San Diego	22
KSDW	88.9	Temecula	1.15
K206AC	89.1	San Diego	0.004
KPBS-FM	89.5	San Diego	26
K210CL	89.9	Lemon Grove	0.001
KOPA	91.3	Pala	0.1
KSOQ-FM	92.1	Escondido	0.58
KHTS-FM	93.3	El Cajon	50
K229BO	93.7	Rancho Bernardo	0.01
KMYI	94.1	San Diego	77
KMYT	94.5	Temecula	0.54
KBZT	94.9	San Diego	26.5
KUSS	95.7	Carlsbad	28
KSIQ	96.1	Campo	25
KYXY	96.5	San Diego	26.5
K245AI	96.9	San Pasqual	0.01
KSON	97.3	San Diego	50
KIFM	98.1	San Diego	26.5
K252BF	98.3	Temecula	0.003
K253AD	98.5	Oceanside	0.01
KLVJ	100.1	Julian	0.11
KFMB-FM	100.7	San Diego	30
KGB-FM	101.5	San Diego	50
KPRI	102.1	Encinitas	30
KLQV	102.9	San Diego	30
KTMQ	103.3	Temecula	1.25
KSCF	103.7	San Diego	26.5
KIOZ	105.3	San Diego	26
KLNV	106.5	San Diego	50
KSSD	107.1	Fallbrook	3

FM Callsign	MHz	Location	kW
KHHS-LP	107.5	San Diego	0.038
KRLY-LP	107.9	Alpine	0.002
KWVE-FM	107.9	San Clemente	0.8

NB: +Several stns from Tijuana (see Mexico for details).

San Francisco Area

FM Callsign	MHz	Location	kW
KCRH	89.9	Hayward	0.018
K210EH	89.9	Bolinas	0.01
KZSU	90.1	Stanford	0.5
K211EZ	90.1	Livermore	0.01
K212BJ	90.3	Dublin	0.2
KUSF	90.3	San Francisco	2.85
KSJS	90.5	San Jose	1.5
KVHS	90.5	Concord	0.41
KWMR	90.5	Pt. Reyes Station	0.235
KALX	90.7	Berkeley	0.5
K214CS	90.7	Sonoma	0.004
KAIS	90.7	Tracy	0.21
K208DO	90.9	Napa	0.01
K216AX	91.1	Laurel	0.01
KCSM	91.1	San Mateo	11.5
K216FV	91.1	Concord	0.01
KSVY	91.3	Sonoma	8
KKUP	91.5	Cupertino	0.2
KRVH	91.5	Rio Vista	0.05
KASK	91.5	Fairfield	0.075
KALW	91.7	San Francisco	1.9
K220BV	91.9	San Jose	0.01
K220JB	91.9	Leisure Town	0.01
K221DQ	92.1	Petaluma	0.01
KKDV	92.1	Walnut Creek	3
KSJO	92.3	San Jose	32
KREV	92.7	Alameda	3.6
KRZZ	93.3	San Francisco	50
K228DM	93.5	Vacaville	0.01
KXZM	93.7	Felton	0.028
KJZY	93.7	Sebastopol	6
KPFA	94.1	Berkeley	59
KBAY	94.5	Gilroy	45
KYLD	94.9	San Francisco	30
KRTY	95.3	Los Gatos	0.87
KUIC	95.3	Vacaville	0.49
K238AF	95.5	Santa Rosa	0.24
KBWF	95.7	San Francisco	6.9
K240CD	95.9	Soquel	0.028
KSQQ	96.1	Morgan Hill	4.7
KOIT-FM	96.5	San Francisco	24
KLLC	97.3	San Francisco	82
K249DJ	97.7	San Pablo	0.01
KFFG	97.7	Los Altos	3.3
KISQ	98.1	San Francisco	75
KUFX	98.5	San Jose	10
KSOL	98.9	San Francisco	6.1
KSQL	99.1	Santa Cruz	1.1
K257BE	99.3	Los Gatos.	0.01
KVYN	99.3	St. Helena	6
KMVQ-FM	99.7	San Francisco	45
KZST	100.1	Santa Rosa	6
KBRG	100.3	San Jose	14.5
KVVZ	100.7	San Rafael	6
K264AQ	100.7	Mountain View	0.002
K265CV	100.9	Fremont	0.008
K265DI	100.9	Sausalito	0.08
K265CY	100.9	San Jose	0.007
KIOI	101.3	San Francisco	125
KKIQ	101.7	Livermore	4.5
KXFX	101.7	Santa Rosa	2.2
K269DF	101.7	Daly City	0.013
KDFC-FM	102.1	San Francisco	33
KBLX-FM	102.9	Berkeley	6.6
K276DT	103.1	Santa Rosa	0.25
K276EK	103.1	Vacaville	0.01
KSCU	103.3	Santa Clara	0.03
KKSF	103.7	San Francisco	10
KHKK	104.1	Modesto	50
K281BB	104.1	Vacaville	0.009
KFOG	104.5	San Francisco	13.5
KMHX	104.9	Rohnert Park	6.6
KCNL	104.9	Sunnyvale	6
KITS	105.3	San Francisco	16.5
KVVF	105.7	Santa Clara	50
K289AS	105.7	Cotati	0.01
KMEL	106.1	San Francisco	69
KSHC-LP	106.5	St. Helena	0.002
KEZR	106.5	San Jose	42
KFRC-FM	106.9	San Francisco	80
KLVS	107.3	Stockton	8.1

FM Callsign	MHz	Location	kW
KOWS-LP	107.3	Occidental	0.003
K298AZ	107.5	Santa Rosa	0.017
KSAN	107.7	San Mateo	8.9
K300AO	107.9	Santa Rosa	0.01

NB: Synchronized translators not listed.

Seattle-Tacoma Area

FM Callsign	MHz	Location	kW
K201EM	88.1	Olympia	0.14
K201AB	88.1	West Seattle	0.12
K201EN	88.1	Everett	0.005
K201EB	88.1	Mount Vernon	0.008
K201ET	88.1	Port Townsend	0.002
K201EX	88.1	Greenwater	0.005
KPLU-FM	88.5	Tacoma	68
KMIH	88.9	Mercer Island	0.03
K205DF	88.9	Enumclaw	0.002
K206CJ	89.1	Issaquah	0.003
K206DL	89.1	Granite Falls	0.005
K206DO	89.1	Cape George	0.002
K206DM	89.1	Bremerton	0.011
K207AZ	89.3	Gig Harbor	0.033
KAOS	89.3	Olympia	1.25
K207AP	89.3	Sumner	0.025
KNHC	89.5	Seattle	8.5
KWFJ	89.7	Roy	1
KGRG-FM	89.9	Auburn	0.25
KGHP	89.9	Gig Harbor	1.35
KASB	89.9	Bellevue	0.06
KUPS	90.1	Tacoma	0.1
KPLI	90.1	Olympia	0.1
KEXP-FM	90.3	Seattle	4.7
KSER	90.7	Everett	5.8
KVTI	90.9	Tacoma	51
KBCS	91.3	Bellevue	8
KSQM	91.5	Sequim	0.7
KSVR	91.7	Mount Vernon	0.17
KXOT	91.7	Tacoma	23
K220HD	91.9	Fall City	0.004
K221FJ	92.1	Tacoma	0.15
KQMV	92.5	Bellevue	60
K225AX	92.9	White Center	0.085
KUBE	93.3	Seattle	100
K229BL	93.7	Gig Harbor	0.058
KMPS-FM	94.1	Seattle	73
K233BU	94.5	White Center	0.062
KUOW-FM	94.9	Seattle	100
KJR-FM	95.7	Seattle	100
KXXO	96.1	Olympia	65
KJAQ	96.5	Seattle	53
KGY-FM	96.9	Mccleary	11
KIRO-FM	97.3	Tacoma	55
KOMO-FM	97.7	Oakville	69
K249DX	97.7	Redmond	0.075
KING-FM	98.1	Seattle	68
KWJZ	98.9	Seattle	68
KDDS-FM	99.3	Elma	64
K258BJ	99.5	Everett	0.005
KISW	99.9	Seattle	68
KKWF	100.7	Seattle	68
KPLZ-FM	101.5	Seattle	100
KZOK-FM	102.5	Seattle	73
K277AE	103.3	Seattle	0.25
KMTT	103.7	Tacoma	68
K281AS	104.1	Tacoma	0.092
KBSG-LP	104.5	Fall City	0.032
KMCQ	104.5	Covington	7.1
K284AT	104.7	Everett	0.041
K285FJ	104.9	Brinnon	0.013
KCMS	105.3	Edmonds	54
K289AK	105.7	Orting	0.009
KBKS-FM	106.1	Tacoma	73
K293AY	106.5	Enumclaw	0.001
KRWM	106.9	Bremerton	49
KNDD	107.7	Seattle	68

St Louis Area

FM Callsign	MHz	Location	kW
KDHX	88.1	St. Louis	42
WSIE	88.7	Edwardsville	50
W206AN	89.1	Carlinville	0.08
KCLC	89.1	St. Charles	50
KTBJ	89.3	Festus	25
WGRN	89.5	Greenville	0.3
WARW	89.5	Dorsey	1.5
KCFV	89.5	Ferguson	0.1
KNLH	89.5	Cedar Hill	0.068
WCBW-FM	89.7	East St. Louis	0.25
KGNX	89.7	Ballwin	0.12
WLCA	89.9	Godfrey	1.5

FM Callsign	MHz	Location	kW
KGNV	89.9	Washington	1
KGNA-FM	89.9	Arnold	0.15
W211AD	90.1	Granite City	0.06
WTSG	90.1	Carlinville	5
KRHS	90.1	Overland	0.014
KWUR	90.3	Clayton	0.009
KWMU	90.7	St. Louis	100
WIBI	91.1	Carlinville	50
KSIV-FM	91.5	St. Louis	85
K220HT	91.9	St. Louis	0.099
WIL-FM	92.3	St. Louis	100
W224BJ	92.7	Carlyle	0.17
W226BC	93.1	Brighton	0.019
KBDZ	93.1	Perryville	50
KSD	93.7	St. Louis	74
KSHE	94.7	Crestwood	100
K236AZ	95.1	Gray Summit	0.015
WFUN-FM	95.5	Bethalto	24.5
WOLG	95.9	Carlinville	6
KIHT	96.3	St. Louis	100
WCXO	96.7	Carlyle	2.1
KFTK	97.1	Florissant	100
WDLJ	97.5	Breese	2.5
KHZR	97.7	Potosi	26.5
KYKY	98.1	St. Louis	90
KLJY	99.1	Clayton	100
WZJM-LP	99.9	Freeburg	0.1
KFAV	99.9	Warrenton	10.5
KDJR	100.1	De Soto	2
WSDD	100.3	Alton	50
KFNS-FM	100.7	Troy	6
WXOS	101.1	East St. Louis	100
K268BF	101.5	Bellefontaine	0.099
WGEL	101.7	Greenville	6
KLPW-FM	101.7	Elsberry	3.1
KEZK-FM	102.5	St. Louis	100
KLOU	103.3	St. Louis	90
W279AO	103.7	Mascoutah	0.01
W280DR	103.9	Greenville	0.25
WHHL	104.1	Hazelwood	50
KSLQ-FM	104.5	Washington	3
WNSV	104.7	Nashville	3.4
KMJM-FM	104.9	Columbia	7.8
W286AJ	105.1	Jerseyville	0.12
K286BG	105.1	Washington	0.016
WAOX	105.3	Staunton	6
KPNT	105.7	St. Genevieve	100
WSMI-FM	106.1	Litchfield	50
WARH	106.5	Granite City	90
KSLZ	107.7	St. Louis	100

Tampa-St. Petersburg Area

FM Callsign	MHz	Location	kW
WJIS	88.1	Bradenton	100
W202CB	88.3	Bayonet Point	0.027
WMNF	88.5	Tampa	7
WYFE	88.9	Tarpon Springs	60
WSMR	89.1	Sarasota	50
W207BU	89.3	Bayonet Point	0.01
WKFA	89.3	St. Catherine	3.9
WUSF	89.7	Tampa	100
WJUF	90.1	Inverness	21
WLVF-FM	90.3	Haines City	0.75
WBVM	90.5	Tampa	77
WKES	91.1	Lakeland	100
WCIE	91.5	New Port Richey	16.5
WFTI-FM	91.7	St. Petersburg	3
WYFO	91.9	Lakeland	25
WHGN	91.9	Crystal River	41
WLTQ-FM	92.1	Venice	11.5
W221CE	92.1	Wesley Chapel So.	0.027
WYUU	92.5	Safety Harbor	50
WFLZ-FM	93.3	Tampa	100
WLLD	94.1	Lakeland	100
W233AV	94.5	Gulfport	0.01
WWRM	94.9	Tampa	100
W237CW	95.3	Pinellas Park	0.014
W237DI	95.3	West Tampa	0.115
WBTP	95.7	Clearwater	100
WAPQ-LP	95.9	Avon Park	0.1
WTMP-FM	96.1	Dade City	2.8
WLAS-LP	96.1	Bartow	0.1
W242AK	96.3	Lakeland	0.055
W243AK	96.5	Bradenton	0.055
WQRD-LP	96.5	Gibsonton	0.1
WSLR-LP	96.5	Sarasota	0.023
WWVD-LP	96.5	East Tampa	0.1
W244BE	96.7	Brandon	0.08

FM Callsign	MHz	Location	kW
W244BJ	96.7	Frostproof	0.019
WEKJ-LP	96.7	Chassahowitzka	0.1
WZPH-LP	96.7	Dade City	0.1
WSUN-FM	97.1	Holiday	22
W247AF	97.3	Sarasota	0.08
WPCV	97.5	Winter Haven	100
WXTB	97.9	Clearwater	100
WWRZ	98.3	Fort Meade	27
W254AI	98.7	Auburndale	0.055
WSJT	98.7	Holmes Beach	50
W255CC	98.9	Sarasota	0.12
WQYK-FM	99.5	St. Petersburg	100
WXJB-FM	99.9	Homosassa	2.3
WMTX	100.7	Tampa	100
W266AI	101.1	Chassahowitzka	0.17
WPOI	101.5	St. Petersburg	100
WHPT	102.5	Sarasota	100
W274AX	102.7	Groveland	0.17
WKJO-LP	102.7	Brooksville	0.086
W274BB	102.7	Haines City	0.01
W275AX	102.9	Fort Meade	0.027
WFUS	103.5	Gulfport	100
W280DW	103.9	Brandon	0.099
W280EA	103.9	Ruskin	0.099
WKZM	104.3	Sarasota	6
W283AM	104.5	Arcadia	0.027
WRBQ-FM	104.7	Tampa	100
WZSP	105.3	Nocatee	4.1
WDUV	105.5	New Port Richey	47
WTZB	105.9	Englewood	25
W291AG	106.1	Highland City	0.027
WJQB	106.3	Spring Hill	25
WCTQ	106.5	Sarasota	13
WZZS	106.9	Zolfo Springs	5
W295BH	106.9	Sarasota	0.055
WXGL	107.3	St. Petersburg	100
W299AU	107.7	Zolfo Springs	0.019
WSRZ-FM	107.9	Coral Cove	47
WWMA-LP	107.9	Avon Park	0.1
Washington DC Area			
WMUC-FM	88.1	College Park	0.01
WYPR	88.1	Baltimore	15.5
WAMU	88.5	Washington	50
WPFW	89.3	Washington	50
W209BY	89.7	Woodbridge	0.008
WCSP-FM	90.1	Washington	36
WETA	90.9	Washington	75
WGTS	91.9	Takoma Park	27
WERQ-FM	92.3	Baltimore	37
WWXT	92.7	Prince Frederick	2.85
WPOC	93.1	Baltimore	19.5
WD2XAB	93.5	Columbia	2
WKYS	93.9	Washington	24.5
WIAD	94.7	Bethesda	50
WRBS-FM	95.1	Baltimore	50
WPGC-FM	95.5	Morningside	50
WWIN-FM	95.9	Glen Burnie	3
WHUR-FM	96.3	Washington	16.5
WASH	97.1	Washington	17.5
W248AO	97.5	Baltimore	0.008
W249BE	97.7	Alexandria	0.01
WSMD-FM	98.3	Mechanicsville	3
WMZQ-FM	98.7	Washington	50
WLZL	99.1	Annapolis	45
WIHT	99.5	Washington	50
W260BM	99.9	Annapolis	0.003
WBIG-FM	100.3	Washington	40
WWDC	101.1	Washington	25
WMJS-LP	102.1	Prince Frederick	0.082
WMMJ	102.3	Bethesda	2.9
W275BO	102.9	Chantilly	0.013
WTOP-FM	103.5	Washington	44
WPRS-FM	104.1	Waldorf	20
WZFT	104.3	Baltimore	20
W284BC	104.7	Waldorf	0.01
WAVA-FM	105.1	Arlington	34
W288BS	105.5	Reston	0.25
WVRX	105.9	Woodbridge	40
W291BA	106.1	Baltimore	0.004
WWMX	106.5	Baltimore	8.3
WJFK-FM	106.7	Manassas	40
WRQX	107.3	Washington	21.5
WFSI	107.9	Annapolis	50

ARMED FORCES RADIO & TELEVISION SERVICE

AFRTS Broadcast Center, 1363 Z Street, Bldg. 2730, March Air Reserve Base, CA 92518-2017 ☎ +1 909 413-2236 **W:** www.afrts. osd.mil or www.afrts.dodmedia.osd.mil

The AFRTS Broadcast Center, located at March Air Reserve Base near Riverside, California, is the sole prgr source for military radio and television outlets overseas. These outlets serve American service men and women, Department of Defense (DoD) civilians, and their families stationed in over 150 countries around the world where English language broadcast service is unavailable or inadequate. Known as AFRTS-BC, the Broadcast Center is responsible for reflecting an accurate cross-section of what is widely available to stateside audiences of the American radio and television industry. The global AFRTS radio and television network service is called AFN, the American Forces Network.

STATIONS: Details of the AFRTS on-air broadcast services are listed under the countries concerned. AFRTS SW services are listed in the International Radio section.

ARMY BROADCASTING SERVICE (ABS)

601 North Fairfax Street, Suite 340, Alexandria, VA 22314-2040. **E:** abs@afn.army.mil or myafn@dodmedia.osd.mil

LP: Commander, ABS: Col David R. Apt

ABS broadcasts for soldiers, civilians and their families serving overseas. The ABS international networks and stns broadcast American radio and television to United States Army soldiers, civilians and their families serving across the globe. ABS manages three Armed Forces Radio & Television Service broadcast networks and two independent broadcast stns. These stns broadcast satellite news, entertainment and information to the majority of United States Army soldiers, civilians and their families stationed overseas.

STATIONS: American Forces Network Europe is the largest ABS broadcast outlet and serves the United States military community in Western Europe, Northern Africa, the Mediterranean and the Balkans - American Forces Korea Network serves the United States military community on the Korean Peninsula - AFN Kwajalein is located in the Marshall Islands - AFN Honduras is located on the Soto Cano Air Base, Honduras. See the various countries for details.

INTERNATIONAL BROADCASTING

Government-operated, private and religious stns on SW are listed in the International Broadcasting section. Some stns of the latter categories also target a domestic audience.

URUGUAY

L.T: UTC -3h (3 Oct 10 – 13 Mar 11, 2 Oct 11– 12 Mar 12: -2h) — **Pop:** 3.5 million — **Pr.L:** Spanish — **E.C:** 50Hz, 220V — **ITU:** URG **Int. dialling code:** +598

DIRECCION NACIONAL DE TELECOMUNICACIONES

Ministerio de Industria, Energia y Mineria.
Av. Uruguay 988 (Casilla de Correo 927), 11100 MontevideoEdifico ciudadela Sarandi 690 D, 2° entrepiso ☎ 2 915 0856 **E:** info@dinatel. miem.gub.uy **W:** www.dinatel.gub.uy **L.P:** Gustavo Gómez

UNIDAD REGULADORA DE SERVICIOS DE COMUNICACIONES (URSEC)

Av. Uruguay 988 (Casilla de Correo 927), 11100 Montevideo ☎ 2 902 8082 2 2 900 5708 **E:** radiodifusion@ursec.gub.uy **W:** www.ursec. gub.uy **L.P:** Presidente: Sr. Ing. Gabriel Lombide

ASOCIACION NACIONAL DE BROADCASTERS URUGUAYOS (ANDEBU)

Carlos Quijano 1264, 11100 Montevideo ☎ 2 902 1525, 908 0037 ▤ 2 902 1540 **E:** andebu@adinet.com.uy **W:** www.andebu.com.uy

COOPERATIVA DE RADIO EMISORAS DEL INTERIOR (CORI)

Av. 18 de Julio 948, Oficina 603, 11000 Montevideo ▤ 2 902 9047 **W:** http://www.cori.com.uy **E:** coriamfm@adinet.com.uy

RADIOS AM DEL INTERIOR (RAMI)

Nueva York 1618, 11800 Montevideo ▤ 2 9047279
W: www.ramiradiosdelinterior.com **E:** rami@adinet.com.uy

RED ORO

Rio Negro 1337, Esc. 209, 11100 Montevideo ☎ 2 903-1678 ▤ 2 900-3916 **E:** redoro@adinet.com.uy

SERVICIO OFICIAL DE DIFUSIÓN, RADIOTELEVISIÓN Y ESPECTÁCULOS (S.O.D.R.E.) (Gov)
Sarandí 450, 11000 Montevideo ☎ 2 9155378 **W**: www.sodre.gub.uy
E: dirradio@sodre.gub.uy
L.P.: Director: Sergio Sacomani. Tchn. Director: José Cuello

MW	Call	kHz	kW	Station, location, h. of tr.
CO01)	CW1	550	25/10	R. Colonia, Colonia: 24h
MO01)	CX58	580	5	R. Clarín, Montevideo: 24h
MO02)	CX4	610	50	R. Rural, Montevideo: 0900-0400
MO03)	CX6	650	50/25	S.O.D.R.E. "R. Clásica", Montevideo: 24h
RN01)	CW68	680	1/0.7	R. Young, Young: 0900-0300
MO04)	CX8	690	10/15	R. Sarandí, Montevideo: 24h
MO05)	CX10	730	5/2.5	R. Continente, Montevideo: 24h
SA01)	CW27	740	5/1	R. Tabaré, Salto: 0900-0300
MO06)	CX12	770	5	R. Oriental, Montevideo: 24h
MO07)	CX14	810	50/25	R. El Espectador, Montevideo: 0800-0500
SA02)	CW23*	820	1/0.5	R. Cultural, Salto
MO08)	CX16	850	50	R. Carve, Montevideo: 0825-0300
MO09)	CX18	890	5	R. Sport 890, Montevideo: 24h
AR1)	CW17	900	3/1	R. Frontera, Artigas
MO10)	CX20	930	50/25	R. Monte Carlo, "la Super R.", Montevideo: 24h
DU01)	CW96	960	3/1	R. Yí, Durazno: 1000-0200
MO11)	CX22	970	20/5	R. Universal, Montevideo: 1030-0100
MO12)	CX24	1010	20	R. Nuevotiempo, Montevideo: 0900-0400
SA03)	CW102	1020	0.1	R. Libertadores, Salto: 0700-0300
MO03)	CX26	1050	25	S.O.D.R.E. "R.Uruguay", Montevideo: F.pl: 24h
MO14)	CX28	1090	15	R. Imparcial, Montevideo: 24h
TA01)	CX111	1110	2/1	R. Paso de los Toros, Paso de los Toros: 1100-0200
SA04)	CW31	1120	10/2	R. Salto, Salto: 0900-0300
MO15)	CX30	1130	20/5	R. Nacional, Montevideo: 24h
TT01)	CW116	1160	1	R. Agraria del Uruguay, Cerro Chato: 0800-0100
MO16)	CX32	1170	10	Radiomundo, Montevideo: 1100-0300
AR02)	CW118	1180	10	LV de Artigas, Artigas: 0900-0300
FL01)	CW33	1200	2	La Nueva Radio, Florida: 24h
SO02)	CW121	1210	2	Difusora Soriano, Mercedes: 24h
MA01)	CV121	1210	2.5/0.25	R. RBC, Piriápolis: 24h
TT02)	CW121	1210	1	R. El Libertador, Villa Vergara
RI05)	CX122	1220	1	R. Reconquista, Rivera: 1100-0400
PA01)	CW35	1240	5/1	R. Paysandú, Paysandú: 0900-0400
MO17)	CX36	1250	10	R. Centenario, Montevideo:24h
AR03)	CW125	1250	5	R. Bella Unión, Bella Unión: 0900-0300
RO01)	CW37	1260	3/1	Dif. Rochense, Rocha: 0900-0300
AR04)	CV127	1270	4/2	R. Cuareim, Artigas: 0900-0300
TA02)	CX128	1280	3/1	R. Tacuarembó, Tacuarembó: 0845-0300
MO03)	CX38	1290	10	S.O.D.R.E. "Em. del Sur", Montevideo: F.Pl: 24h
PA02)	CW39	1320	1/0.5	R. LV de Paysandú, Paysandú: 0900-0300
RO02)	CW132	1320	2	R. Fortaleza, Rocha: 1000-0200
MO19)	CW40	1330	5	R. Fénix, Montevideo: 1000-0600
CL01)	CW53	1340	10	LV de Melo, Melo: 0800-0300
CL02)	CW136	1360	1/0.25	R. Río Branco, Río Branco: 1055-0200
SJ01)	CW41	1360	2.5/0.5	Radio 41, San José: 24h
MO20)	CX42	1370	10/2.5	Em. Ciudad de Montevideo: 1100-0300
RI01)	CV137A	1370	0.5/0.25	R. Real, Minas de Corrales: 0930-0130
RN02)	CW137	1370	0.25/0.1	R. San Javier, San Javier: 24h
TT03)	CW45	v1390	7.5/3	Dif. Treinta y Tres, Treinta y Tres: 0800-0300
TA03)	CW140	1400	25	R. Zorrilla de San Martín, Tacuarembó: 0900-0300
MO21)	CW44	1410	10/5	AM Libre, Montevideo: 24h
SA05)	CW141	1410	1/0.25	R. Turística, Salto: 0900-0300
LA01)	CW43	1420	5	R. Lavalleja, Minas: 0900-0300
PA03)	CV142	1420	1	R. Felicidad, Paysandú: 0830-0130
DU02)	CW25	1430	5/1	R. Durazno, Durazno: 0830-0300
RI02)	CX144	1440	3/0.5	R. Rivera, Rivera: 0830 (Su: 1000)-0300
MO22)	CX46	1450	10/5	R. América, Montevideo: (0930-0630)
SA06)	CW145	1450	2/0.5	R. Arapey, Salto: 24h
CO02)	CX146	1460	3/1	R. Carmelo, Carmelo: 0930-0030 (Su: 1000-0100)
LA02)	CW146	1460	0.25	R. José Battle y Ordóñez, José Battle y Ordóñez
CA01)	CX147	1470	15	R. Cristal del Uruguay,Las Piedras: 24h
CL03)	CW147	1470	1/0.5	R. Maria, Melo: 24h
RO04)	CW148	1480	3/0.8	R. Universo, Castillos: 0900-0300
RI03)	CW43B	1480	5/1.5	R. Internacional, Rivera: 0800-0300
RN03)	CW148	1480	1/0.7	Difusora Río Negro, Young:0900-0300
AR05)	CV149	1490	1/0.25	R. del Centro, Baltasar Brum: 0900-0100
CO03)	CV149	1490	5/4	R. del Oeste, Nueva Helvecia: 0930-0300
RN04)	CX151	1510	1/0.25	R. Rincón, Fray Bentos:0915-0230
MA02)	CW57	1510	2/0.5	R. San Carlos, San Carlos: 0800-0400
TA04)	CW151	1510	0.5	R. Ibirapitá, San Gregorio de Polanco:1000-0200
CL04)	CX152	1520	2	R. Acuarela, Melo: 0900-0300
SO03)	CV152	1520	0.1	R. Paz, "la nueva Radio", Guichón1000-0030
CO4)	CW153	1530	0.25	Em. Cono Sur, Nueva Palmira: 0900-0300
PA04)	CW154	1540	0.1	R. Charrúa, Paysandú: 1000-0300
TT04)	CX154	1540	0.25	R. Patria, Treinta y Tres: 0800-0300
CO05)	CV154	1540	1	R. Centro, Cardona: 0900-0200
SO04)	CV155	1550	1/0.25	R. Agraciada, Mercedes: 24h
DU03)	CX155	1550	2	R. Sarandí del Yí, Sarandí del Yí: 1030-0130
MA03)	CW51	1560	3/0.5	R. Maldonado, Maldonado: 24h
FO01)	CX156	1560	2/0.5	Dif. Americana, Trinidad: 0930-0130
RI04)	CV156	1560	1/0.25	R. Vichadero: 1000-0200
CA02)	CX157	1570	2/0.5	R. Canelones: 0930-0230
AR06)	CW157A	1570	0.25/0.1	Em. Celeste, Tomás Gomensoro: 0900-0300
LA03)	CW54	1580	0.5	Emisoras del Este, Minas: 0800-0200
SO05)	CW158	1580	1/0.5	R. San Salvador, Dolores: 0900 (Sat&Sun 0800)-0300
RO05)	CW159	1590	1/0.25	R. Regional "La Nueva Radio", Lascano
CO06)	CX159	1590	10	R. Real, Colonia: 1000-0300
CA03)	CV160	1600	2	R. Continental, Pando: 0915-0300
RN05)	CX160	1600	1	R. Litoral, Fray Bentos:0900-0300

SW	Call	kHz	kW	Station, location
TA05)		v5900	0.03	Em. Chaná, Tacuarembó: 1200-0200
MO20)	CXA142A	*6010	10	Em. Ciudad de Montevideo
MO09)	CXA61	6045	0.3	R. Sport 890, Montevideo: (USB) rel. MO03)
RO04)	CWA148	6055	0.3	R. Universo, Castillos
AR02)	CXA3	*6075	1	LV de Artigas, Artigas: irr.
MO18)	CXA4	6125	0.35	S.O.D.R.E, Montevideo: irr (rel. MO03)
MO10)	CXA20	*6140	1	R. Montecarlo. Montevideo
DU03)	CWA155	*6154	2	Banda Oriental, Sarandí del Yi
MO20)	CXA142A	*9650	10	Em. Ciudad de Montevideo
MO10)	CXA72	*9595	1	R. Montecarlo Carlo, Montevideo
MO18)	CXA4	*9620	0.35	S.O.D.R.E., Montevideo
MO06)	CXA7	*11735	1	R. Oriental, Montevideo

° = on-air stn name not confirmed, * = inactive, v = varying freq.

Addresses and other information
AR00) ARTIGAS
AR01) Av Lecueder 815, 55000 Artigas ☎4772 1230 - **FM**: 88.3MHz "Frontera FM" – **AR02)** Av Lecueder 483, 55000 Artigas ☎4772 2447 📠4772 4744 **W**: www.radioartigas.com.uy **E**: cx118@radioartigas.com.uy - **FM**: 90.7MHz "Amatista FM" – 105.5MHz "Norte FM" – **AR03)**Enrique Ferreira 1550, 55100 Bella Unión ☎4779 2058 📠4772 4744 **W**: www.radiobellaunion.com.uy **E**: radiobellaunion@gmail.com - **FM**: 105.5MHz "Stereo Norte FM" – **AR04)** Av Lecueder 167, 55000 Artigas ☎4772 2867 **W**: www.radiocuareim.com.uy **E**: racua@adinet.com.uy –**AR05)** Batlle y Ordóñez y 25 de Agosto, 55001 Baltasar Brum, Artigas ☎4776 2109 **W**: www.radiodelcentro.com **E**: radiodelcentro_95@hotmail.com – **AR06)**18 de Julio y 19 de Abril, 55002 Tomás Gomensoro ☎4777 2157

CA00) CANELONES
CA01) Av Artigas 781, 90200 Las Piedras, Canelones ☎4236 44775 📠4236 44814 **W**: www.radiocristaldeluruguay.com **E**: cx147cristal@hotmail.com or info@radiocristaldeluruguay.com – **CA02)** J.T González 434, 90000 Canelones ☎4332 2589 📠4332 2040 **E**: cx157@adinet.com.uy – **FM**: 101.1MHz – **CA03)** Av Artigas 977, 91000 Pando ☎4229 22512 📠4229 24440 **Web.**: www.radiocontinental.com.uy **E**: gerencia@radiocontinental.com.uy

CL00) CERRO LARGO
CL01) Remigio Castellanos 721, 37000 Melo ☎4642 2105 📠4642 3226 **W**: www.lavozdemelo.com **E**: lavozdemelo@yahoo.com – **CL02)** Virrey Arredondo 986, 37100 Río Branco ☎4675 2009 **E**: am1360@adinet.com.uy – **CL03)** Treinta y Tres 949, 37100 Melo, Depto de Cerro Largo ☎4642 2387 **W**: www.radiomaria.org.uy **E**: info.ury@radiomaria.org – **CL04)** José Pedro Varela 750, Melo ☎4642 2051 📠4642 1264 **E**: acuarelaradio@yahoo,com or correo@radioacuarela

CO00) COLONIA
CO01) Rivadavia 383, 70000 Colonia ☎4522 2006 📠4522 2961 **W**: radiocolonia.com.uy **E**: cw1@adinet.com.uy - **FM**: 93.5MHz "FM Mágica" – **CO02)**19 de Abril 444, 70100 Carmelo ☎4542 3558 📠4542 2520 **W**: www.radiocarmelo.com **E**: radiocarmelo@adinet.com.uy - 0930-0030 – **CO03)** Calle Berna 1375, 70201 Nueva Helvecia ☎4554 4409 📠4554 4217 **W**: www.corporacionro.com **E**: deloeste@adinet.com.uy - **FM**: 90.7MHz "Reflejos" – **CO04)** Chile 1162 y Gral Artigas, 70101 Nueva Palmira, Depto de Colonia ☎📠4544 6053 **E**: radio_conosur@hotmail.com – **CO05)** Boulevard Cardona s/n y Rivera, 75.200 Cardona ☎4536 9315 – **CO06)** Av Gral Flores 472, 70000 Colonia ☎4522 2030 **W**: www.radioreal.com.uy **E**: radioreal@adinet.com.uy

DU00) DURAZNO
DU01) Zorrilla de San Martín 875, 97000 Durazno ☎4362 2701 📠4362 3297 **E**: multimyi@adinet.com.uy – **DU02)** Bv Gral Fructuoso 501, 97000 Durazno ☎4362 2015 📠4362 2058 **W**: www.radiodurazno.com **E**: am1430@adinet.com.uy, info@radiodurazno.com - **FM**: 95.1MHz "Radio City" – **DU03)** Calle Sarandí del Yi 428, 97100 Sarandí del Yí ☎📠4367 9155 **E**: norasan@adinet.com.uy - **FM**: 89.5MHz "Scala FM"

FL00) FLORIDA
FL01) Antonio Ma Fernández 800, 94000 Florida ☎4352 2026 **W:** www.cw33florida.com.uy **E:** cw33@adinet.com.uy - **FM:** 88.7MHz "Claridad"
F000) FLORES
F001) 25 de Agosto 724, 85000 Trinidad ☎4364 2229 📠364 37550
LA00) LAVALLEJA
LA01) José E Rodó 530, 30000 Minas ☎📠4442 2304 – **LA02)** Camino Nacional s/n, 30200 José Batlle y Ordóñez ☎📠4266 2132 – **LA03)** Treinta y Tres 632, 30000 Minas ☎4442 3092 📠442 8714 **E:** federalfm@adinet.com.uy - **FM:** 107.3MHz "Federal FM".
MA00) MALDONADO
MA01) Chacabuco y Moreno, 20200 Piriápolis ☎📠4432 2771 **W:** www.radiorbc.com.uy **E:** rbc1210@adinet com.uy – **MA02)** Calle Sarandí 775, entre 18 de Julio y Treinta y Tres, 20400 San Carlos ☎4266 9162 📠4426 69575 **W:** www.radiosancarlos.8k.com **E:** radiosancarlos@ adinet.com.uy or rsc@adinet.com.uy – **MA03)** Zelmar Michelini 819, 20000 Maldonado ☎4422 3872 📠4422 2555 **E:** am1560@adinet.com. uy - **N:** every ½h - **FM:** 103.5MHz "FM Punta del Este"
M000) MONTEVIDEO
M001) Av 18 de Julio 1516, P.9, Esc 7, 11200 Montevideo ☎240 06877 📠240 15841 **W:** www.radioclarin.com **E:** clarinam580@adinet. com.uy – **M002)** Joaquin Suarez 3409, 11700 Montevideo ☎233 60610 **W:** www.cx4radiorural.com **E:** ruralcx4@adinet.com.uy – **M003)** Sarandí 430, 11000 Montevideo **W:** www.sodre.gub.uy ☎291 5387 📠291 61933 - CX6: - Clásica: - 1000-0300 - CX26: – Uruguay - 0900-0300 - CX38 Uruguayan music Media Prgr: SS1400-1500, 0200-0300 "Radioactividades" on 1050kHz Re to Cas, 7011, 11000 Montevideo. **E:** radioactividades@sodre.gub.uy – **M004)** Enriqueta Compte y Riquet 1250, 11800 Montevideo ☎220 82612 📠220 36906 **W:** www.sarandi. com.uy **E:** direccion@sarandi890.com.uy – **M005)** Germán Barbato 1472, 11100 Montevideo ☎902 4038 📠290 24038 **E:** cx10@adinet. com.uy – **M006)** Cerrito 475, 11000 Montevideo **W:** www.oriental.com. uy **E:** admin@oriental.com.uy ☎291 61130 – **M007)** Río Branco 1481, 11100 Montevideo **W:** www.espectador. com **E:** am810@espectador.com.uy – **M008)** Mercedes 973, 11100 Montevideo ☎202 6162 📠290 20126 **W:** www.carve850.com.uy **E:** carve@sadrep.com.uy – **M009)** Enriqueta Compte y Rique 1250, 11800 Montevideo ☎220 4163 📠220 3786 **W:** www.sport890.com.uy **E:** sport890@sport890.com.uy – Rprts to: fgopar34@gmail.com – **M010)** Av 18 de Julio 1224, 11100 Montevideo ☎290 14433 **W:** www.radiomontecarlo.com.uy **E:** cx20@radiomontecarlo.com.uy – **M011)** Av 18 de Julio, 1220, 3er piso, 11100 Montevideo ☎290 26022 📠290 26050 **W:** www.22universal.com **E:** info@22universal.com – **M012)** Mercedes 973, 11100 Montevideo ☎902 6712 📠902 9110 **E:** prensa@portalx. com.uy – **M014)** Av del Libertador Brig Gral Lavalleja 1708, ap 101, Edificio Carioca, 11800 Montevideo ☎292 41514 📠292 42323 **E:** radioimparcial@netgate.com.uy – **M015)** Plaza Independencia 846, EP, 11100 Montevideo ☎290 25640 **W:** www.radionacional.com.uy **E:** la30@radionacional.com.uy – **M016)** Rambla Armenia 1647, Montevideo ☎262 89626 📠262 89627 – **M017)** Av 18 de Julio 1357, Oficina 202, 11200 Montevideo ☎290 30302 📠290 30307 **E:** radio36@gmail.com **W:** www.radio36.com.uy – **M019)** Canelones 1969, 11200 Montevideo ☎📠240 83292 **W:** www.cx40radiofenix.com **E:** radiofenix@adinet.com.uy – – **M020)** Arenal Grande 2093, 11800 Montevideo ☎292 40142 📠292 40700 **W:** www.emisoraciudaddemontevideo.com **E:** CX42@emisoraciudaddemontevideo.com.uy – **M021)** Garibaldi 2579, Montevideo **W:** www.1410amlibre.com.uy **E:** 1410amlibre@1410amlibre.com.uy ☎248 73565 – **M022)** Emilio Frugoni 1312, Montevideo ☎240 90094 📠240 89314 **E:** americaam@redfacil.com.uy
PA00) PAYSANDÚ
PA01) Av España 1629, 60000 Paysandú ☎4722 3617 📠4722 2954 **W:** www.radiopaysandu.com.uy – **PA02)** 18 de Julio 614, 60000 Paysandú ☎4722 2267 📠4722 4970 **E:** cw39@adinet.com.uy – **PA03)** 33 Orientales 946,1° piso, 60000 Paysandú ☎4722 4020 📠4722 4020 **W:** www.paysandu.com/radiofelicidad **E:** radiofelicidad@adinet.com. uy – **PA04)** Tte. Cnl Francisco Bicuda y Ruta 3 Gral Artigas, 60000 Paysandú ☎📠4722 4856 **E:** cw154@adinet.com.uy
RI00) RIVERA
RI01) Dr Dávison s/n, 40002 Minas de Corrales ☎4658 2073 – **RI02)** Dr Gabriel Anolles 441, 40000 Rivera ☎📠4622 3230 - **E:** larivera@adinet. com.uy or radiorivera@gmail.com – **RI03)** Av Sarandí 792, 40000 Rivera ☎4622 3259 📠4622 3422 **E:** internac@gmail.com - - **FM:** 94.5MHz – **RI04)** Bulevar Artigas casi Rivera, 40003 Vichadero ☎4654 2018 **E:** radiosamfm@hotmail.com – **RI05)** Francisco Acuña de Figueroa 887, 40000 Rivera ☎4622 5893 **W:**www.multimediadelnorte.com/reconquista **E:** reconquista1220@hotmail.com – **FM:** 90.6MHz
RN00) RIO NEGRO
RN01) Rincón 1689, 65100 Young 📠4567 2071 **E:** am680@adinet. com.uy – **RN02)** 27 de Julio casi Basilio Lubkov, San Javier ☎4569 2005 📠4569 2089 **E:** radiosanjavier@hotmail.com – **RN03)** Rincón

1811, 65100 Young ☎4567 5143 **E:** imagenfm@adinet.com.uy - **FM:** 89.1MHz "Imágen FM" – **RN04)** 25 de Mayo 3164 al 3168, 65000 Fray Bentos ☎4562 2022 📠4562 2653 **W:** www.agenda.org.uy/radiorincon **E:** rinconprensa@adinet.com.uy or prensa.rincon@gmail.com – **RN05)** 18 de Julio y 25 de Agosto, 65000 Fray Bentos ☎4562 3100 & 4562 3100 📠4562 3528 **W:** www.radiolitoral.com.uy **E:** litoral@adinet.com.uy
RO00) ROCHA
RO01) Ramirez 127, 27000 Rocha ☎4472 2250 📠4472 2650 **W:** www. difusorarochense.com.uy **E:** direccion@difusorarochense.com.uy – **FM:** 91.5MHz & 106.3MHz – **RO02)** Zorrilla de S Martin 200, 27000 Rocha ☎4472 2460 📠4472 3973 – **RO04)** 18 de Julio 1322, 27200 Castillos ☎4475 8054 📠4475 8755 **W:** www.universoam.com **E:** am1480@adi-net.com.uy (radio) grupouniverso@adinet.com.uy (dir.) – **RO05)** Nicolás Corbo 1152, 27300 Lascano. **W:** www.lanuevaradiolascano.com **E:** lanuevaradio@adinet.com.uy ☎4456 9280 & 4456 4380
SA00) SALTO
SA01) Uruguay 1416, 50000 Salto ☎4734 0298 📠4733 3222 **W:** www. saltouruguay.com/radiotabre **E:** amtabare@adinet.com.uy, radiotabare. hotmail.com – **SA02)** Lavalleja 48, 50000 Salto ☎📠4732 4330 - **FM:** 106.5MHz "Emisora del Éxodo." – **SA03)** Uruguay 1416, 50000 Salto ☎4733 3222 **W:** http://amlibertadores.com **E:** libertadores@mundonet.com.uy – **SA04)** Brasil 715, 50000 Salto ☎4733 2615 📠4733 3414 **W:** www.agenda.org.uy/radiosalto **E:** cw31salto@adinet.com.uy & radiosalto@adinet.com.uy – **FM:** 88.3MHz "Emisora del Lago" – **SA05)** Calle Artigas 1014, 50000 Salto ☎4732 7759 📠4732 6264 **W:** http://agenda.org.uy/radioturistica – **SA06)** Artigas 101, 50000 Salto ☎4732 6264 **W:** http://radioarapey.agenda.org.uy **E:** radioarapey@hotmail.com
SJ00) SAN JOSÉ
SJ01 Evaristo Ciganda 511, 80000 San José ☎📠4342 6444 **W:** www. radio41.com.uy **E:** radionoticias41@adinet.com.ut
S000) SORIANO
S002) De Castro y Careaga 568, 75000 Mercedes ☎4532 3430 📠4532 2977 **W:** www.difusorasoriano.com.ar **E:** difusorasoriano@adinet.com. uy - **FM:** 89.3MHz "Em del Hum" – **S003)**Luis Alberto de Herrera 346, 60008 Guichón, Depto de Paysandú ☎4742 2053 📠4742 2297 **W:** www.pazlanuevaradio.net **E:** lanuevapaz@adinet.com.uy – **S004)** Colón 86, 75000 Mercedes ☎4532 8536 (Adm.), 4532 8538 (AM Studio) **W:** http://agenda.org.uy/radioagraciada **E:** radioagraciada@ adinet.com.uy - **FM:** 100.3MHz "Galicia" – **S005)** Av Asencio 1695, 75100 Dolores, Depto de Soriano ☎4534 2110 📠4534 2691 **W:** www.radiosansalvador.com **E:** gerencia@radiosansalvador.com.uy or radiosan@adinet.com.uy - **FM:** 89.7MHz "Skorpio"
TA00) TACUAREMBÓ
TA01) 18 de Julio 743, 45100 Paso de los Toros ☎📠4664 2333 **W:** www.pasodelostoros.com **E:** am1110@adinet.com.uy or radiopasodelostoros@pasodelostoros.com - **FM:** 91.9MHz "Toros FM" – **TA02)** 18 de Julio 112, 45000 Tacuarembó ☎4263 20214 📠4263 2495 **W:** radiotacuarembo.com **E:** radiotacuaremboam@radiotacuarembo.com - **FM:** 92.5MHz, 104.5MHz – **TA03)** 18 de Julio 302, 45000 Tacuarembó ☎4632 2605 📠4622 2779 – **W:** www.radiozorrilla.com **E:** zsm@adinet.com.uy - **FM:** 88.9MHz "Em de la Música" – **TA04)** Gral Artigas 193, 42500 San Gregorio de Polanco, Tacuarembó ☎4639 4547 📠4639 2495 – **TA05)** Sr. Omar Lima, R.Zorrilla de San Martin, 18 de Julio 304, 45000 Tacuarembó
TT00) TREINTA Y TRES
TT01) Juan Muñoz s/n, 30204 Cerro Chato, Depto Treinta y Tres ☎4466 2200 📠4466 2225 **E:** radioagraria@adinet.com.uy Rpts. to: cx2ua@hotmail.com – **TT02)** Marcelo Barreto s/n, Villa Vergara, 33000 Treinta y Tres ☎4458 2917 **W:** www.ellibertador.com.uy **E:** ellibertador@adinet.com.uy – **TT03)** Pablo Zufriátegui 1076, 33000 Treinta y Tres ☎4452 22476 📠4452 2340 **W:** difusoratreintaytres.com **E:** cw45@adinet.com.uy – **TT04)** Juan A.Lavalleja 1530, 33000 Treinta y Tres ☎4452 3532 📠452 2423 **W:** www.radiopatria.com.uy **E:** radiopatria@adinet.com.uy

FM in Montevideo (MHz): all stns 10-100kW
89.1 Uni-Radio (LP stn) – 90.3 FM Oldies – 91.1 R.Futura – **6)** 91.9 R.Disney – 92.5 Urbana FM – **19)** 93.9 Océano – **M003)** 94.7 Emisora del Sur (SODRE) – 95.5 Em. Del Plata – 96.3 Alfa FM – **M003)** 97.1 Babel (SODRE) – 97.9 M24 – 98.7 Diamante FM – 99.5 Em. del Sol – 100.3 Aire FM – 101.3 LV de la Esperanza – 101.9 Azul FM – 103.7 Latina FM – **8)** 104.3 Radiocero – 105.9 Galaxia FM – 106.7 La Ley FM. In the rest of the country there are 172 FM outlets. There are 38 Community FM low-power authorized stns in the country.

UZBEKISTAN

L.T: UTC +5h — **Pop:** 27.6 million — **Pr.L:** Uzbek, Russian — **E.C:** 50Hz, 220V — **ITU:** UZB

O'ZBEKISTON ALOQA VA AXBOROTLASHTIRISH

AGENTLIGI (OzAAA)
(Communications and Information Agency of Uzbekistan)
✉ A. Navoiy ko'chasi, 28A, 100011 Toshkent ☎ +998 71 2384107 🖷
+998 71 2398782 **E:** info@aci.uz **W:** www.aci.uz
L.P: DG: Xakim A. Muxitdinov
NB. OzAAA is the licensing authority for broadcasting.

O'ZBEKISTON MILLY TELERADIOKOMPANIYASI (Gov)
✉ A. Navoiy ko'chasi 69, 100011 Toshkent ☎ +998 71 2141250 🖷
+998 71 2411332. **E:** info@mtrk.uz **W:** www.mtrk.uz
🖳 Radio studios: Xorazm ko'chasi 49, 100047 Toshkent
L.P: Pres: Alisher Xadjayev

FM (MHz)	UZR1*	UZR2	kW	FM	UZR1*	UZR2	kW
Andijon		105.2	1	Samarqand	105.2i	101.9	0.1
Buxoro	102.0c	103.9	2	Termiz		103.1k	104.6	-
Navoiy	106.6g	105.8	2	Toshkent		103.1	104.0	4
Qarshi	102.3e	103.1	1	Urganch		103.5l	101.5	4

NB. Selected district capitals shown *) incl. reg prgrs (see below)
...) freq not known
D.Prgr: UZR1 (O'zbekiston): 0000-2000 in Uzbek. – **UZR2 (Yoshlar):**
0000-2000 in Uzbek, Russian. – **Local Station: Kanali Toshkent** on
Toshkent 87.9MHz 4kW.

O'ZBEKISTON MTRK REGIONAL STATIONS (Gov) (via UZR1 txs)
a) Andijon TRK: Istiqlol ko'chasi 9, 170120 Andijon. – **b) Buxoro
TRK:** Eshanov ko'chasi 20, 200120 Buxoro. – **c) Farg'ona TRK:**
150100 Farg'ona. – **d) Jizzax TRK:** Rashidov maydon, 130100 Jizzax.
– **e) Qashqadaryo TRK:** 180100 Qarshi. – **f) Qoraqalpog'iston TRK:**
Dustnazarov ko'chasi 20, 230100 Nukus. – **g) Navoiy TRK:** Xalklar
Do'stligi ko'chasi 32, 210100 Navoiy. – **h) Namangan TRK:** Holhanov
ko'chasi 1, 160136 Namangan. – **i) Samarqand viloyati TRK:** 140100
Samarqand. – **j) Sirdaryo TRK:** 120100 Guliston. – **k) Surxandaryo
TRK:** 190100 Termiz. – **l) Xorazm TRK:** 220100 Urganch.

OTHER STATIONS
FM	MHz	kW	Location	Station
8)	88.4	1	Toshkent	Navro'z FM
7)	90.0	1	Toshkent	A'lo FM
1B)	100.5	2	Toshkent	Oriat FM
3)	101.0	1	Toshkent[1]	O'zbegim taronasi
2)	101.5	1	Toshkent	R. Grand
5)	102.0	2	Toshkent	R. Hamroh
6)	102.7	1	Toshkent[1]	Vodiy sadosi
9A)	103.5	1	Toshkent	R. Poytaxt
9A)	104.5	1	Samarqand	R. Poytaxt
4)	105.0	4	Toshkent	R. Terra
10)	105.8	1	Toshkent	Zamin FM
1A)	106.5	4	Toshkent	Oriat Dono
6)	106.9	1	Angren[1]	Vodiy sadosi
9B)	107.2	1	Toshkent	R. Poytaxt-Inform
10)	107.3	1	Andijon	Zamin FM
10)	107.4	1	Buxoro	Zamin FM

[1] +txs in other towns on same freq. (synchr. network)
NB: Txs below 1kW not listed.
Addresses & other information:
1A,B) Xorazm ko'chasi 6 1, 100000 Toshkent. 1A) in Uzbek, **E:** mail@
oriatdono.uz; 1B) in Russian, **E:** mail@oriatfm.uz – **2)** Bunyodkor ko'chasi
15, 100043 Toshkent. **E:** radio@grand.uz – **3)** Shaxrisabz ko'chasi 16a,
100000 Toshkent. **E:** ut101@mail.ru – **4)** Toshkent. – **5)** M.Riyoziy
ko'chasi 30b, 100007 Toshkent. **E:** hamroh@mail.ru – **6)** Mirobod
ko'chasi 39/1A, 100000 Toshkent. **E:** mtrkh@intal.uz – **7)** Toshkent.
– **8)** Mukumiy ko'chasi 178, 100096 Toshkent. **E:** navruz88.4@uzpak.uz
– **9A,B)** Movaraunnahr ko'chasi 14, 100000 Tashkent. **E:** radio1072@
rambler.ru – **10)** Xamid Olimjon square 13A, 100000 Toshkent

VANUATU

L.T. UTC + 11h — **Pop:** 218,519 — **Pr.L:** Bislama, English, French
— **E.C:** 50Hz, 230V — **ITU:** VUT

VANUATU BROADCASTING AND TELEVISION CORPORATION (VBTC)
🖳PMB 049, Port Vila. ☎ +678 22999 🖷 +678 22026
MW: Emten Lagoon (Port Vila) 1125kHz 10kW, Santo (Luganville)
1179kHz 10kW
SW: Emten Lagoon (Port Vila) 3945kHz 10kW 5055kHz 10kW 7260kHz
10kW **Schedule:** A10 – 3945 0700-2000, 5055 0700-1800; B10 – 3945
0700-2000, 7260 0700-1800. Relays MW.
FM: Paradise FM, Luganville 98.1MHz 0.2kW, **Tudei FM,** Port Vila
98.0MHz 0.25kW
D.Prgr: 1125/1179 1900-1115 (Sun 1000)

Other Stations:
FM	MHz	kW	Station
6) Aniwa Is	89.0	0.1	CREST FM
7) Siviri	89.0		Taleva 89 FM
1) Port Vila	90.0	0.3	Laef FM
2) Port Vila	99.0	0.25	BBC World Service
2) Luganville	99.0	0.25	BBC World Service
3) Port Vila	100.0	0.2	France-Inter
4) Port Vila	102.0		China R. International
4) Luganville	102.0		China R. International
9) Port Vila	103.0	0.2	R. Australia
6) Isangel	104.0	0.1	CREST FM
4) Lakatoro	106.0		China R. International
5) Port Vila	107.0	0.3	Capital FM107
8) Saratamata	-	-	-

Addresses and other information:
1) Vanuatu Christian Broadcasting Network [VCBN], P.O.Box 674, Port
Vila ☎ +678 26408 **E:** nenes@vanuatu.com.vu **W:** www.vcbn.org
L.P: CEO: Jenny Joy James **F.PL:** Laef FM Lenakel [Tanna Isl] – **2)** 24h
satellite relay – **3)** 24h satellite relay in French – **4)** 24h satellite relay
in English – **5)** Top Flr, Laguna Bldg, Port Vila. ☎ +678 23847 **E:** sales@
fm107vanuatu.com **W:** www.fm107vanuatu.com **L.P:** Mgr Sera Cakau
ID: "Capital FM107" **D.Prgr:** 24h – **6)** CREST Radio Society of Tafea,
Isangel Station, Tanna Island ☎ +678 68054 **E:** gudfella@vanuatu.
com.vu **L.P:** Mgr: David Kiel. **Network:** 3 community FM stations in
Tafea Province – **7)** community FM station, Siviri, North Efate – **8) F.PL:**
new community FM station, Penama Province – **9)** 24h satellite relay in
Bislama, French and English.

VATICAN CITY STATE

L.T: UTC +1h (27 Mar- 30 Oct: +2h) — **Pop:** 900 — **E.C:** 50Hz, 220V
— **ITU:** CVA

RADIO VATICANA (Rlg.)
🖳 Vatican Radio, 00120 Vatican City ☎ +39 06 6988 3551 **Int. Rel:**
☎ +39 06 6988 3551 🖷 +39 06 6988 4565
Email: promo@vatiradio.va **W:** www.radiovaticana.org
L.P: DG: Rev. Federico Lombardi S.J.; TD: Sandro Piervenanzi; CE:
Maurizio Venuti; Head of Int. Rel: Giacomo Ghisani; Vatican Radio
Museum, guided visiting tour c/o Palazzo Pio XII, Piazza Pia 3,
E: contact museo_rv@vatircanadio.org **W:** www.radiovaticana.org/
museo_tecnico/it/index.asp
MW: 585kHz 5kW, 1260kHz 5kW, 1530kHz 150kW, 1611kHz 20kW
(AM, also DRM test, 15-100kW) **FM:** 93.3/103.8/105.0MHz 10kW
Progr: Europa Programma 1 1530kHz (0330-2310)/ 93.3MHz 24h;
Europa Programma 2 1260kHz (0130-2130); 1611kHz (0520-0030);
One-o-Five Live 105.0MHz 24h (Multil.) (Relay Studio A 2330-0530);
585kHz 24h (Classical Music & rely One-o-Five Live105); Radio
Vaticana International Service Roma area 103.8MHz 24h **W:** www.
radiovaticana.org/it1/sched_eur1.asp **W:** www.radiovaticana.org/it1/
sched_eur2.asp
ANN: Before all transmissions: Latin: "Laudetur Jesus Christus"
(Praised be Jesus Christ), repeated in the language of the broadcast,
then stn identification – **IS:** "Christus Vincit" – **V.** by QSL-card.

EXTERNAL SERVICE: Vatican Radio see Int. Broadcasting section

VENEZUELA

L.T: UTC -4½h — **Pop:** 26.4 million — **Pr.L:** Spanish — **E.C:** 60Hz,
120V — **ITU:** VEN — **Intl. dialling code:** +58

COMISION NACIONAL DE TELECOMUNICACIONES (CONATEL)
🖳 Avenida Veracruz con Calle Cali, Edificio CONATEL, Urb. Las
Mercedes, Caracas 1060 🖷 212 993 8801 **W:** www.conatel.gob.ve **E:**
conatel@conatel.gob.ve

CAMARA VENEZOLANA DE LA INDUSTRIA DE RADIODIFUSION
🖳 Ap. 3955, Caracas 1060 ☎ +58 212 2634855, 2634528 🖷 +58
212 2614783
In 2009, the Venezuelan authorities announced the closure of some 240
AM and FM stns for failure to comply with licensing requirements. At
time of editing, these measures have still not been carried out. Those
known to have closed are marked with an *.
Hrs of tr usually 24h – see address section for variations

MW	Call	kHz	kW	Station, location
AM01)		540	10	LV de Manapiare, San Juan de Manapiare

MW	Call	kHz	kW	Station, location
ZU01)	OY	540	50/25	R. Perijá, La Villa del Rosario
DC01)	KE	550	50	YVKE Mundial, Caracas
DC02)	RH	560	50	R. Nal. "RNV", Cd. Guayana (r: 630)
TA01)	PJ	560	20/10	R. Éxitos "Latina 5-60", Rubio
AR09)	LX	570	100	R. Rumbos, Villa de Cura
DC02)		*580		R. Nacional "RNV", Maturín (r: 630)
ZU02)	MJ	580	50/10	LV de la Fe, Maracaibo
DC04)	KL	590	20	R. Continente, Caracas
BA01)	SW	600	15	R. Alto Llano, Sta Bárbara de Barinas
SU01)	QB	*600	10	R. Sucre, Cumaná
AN01)	XY	610	10	R. Centro 6-10, Cantaura
LA01)	SE	610	10	R. Cristal, Barquisimeto
AP01)	ZC	620	50/25	R. Fe y Alegría Los Llanos,Guasdualito
ZU03)	NO	620	10	R. Libertad, Cabimas
DC02)	KA	630	50/25	R. Nal "Canal Informativo", Caracas
AN02)	QO	640	30	Actualidad 640, Puerto La Cruz
LA02)	MU	640	10/5	R. Carora, Carora
AR01)	LH	650	50/20	Aragüeña 650, Maracay
AN03)	QZ	660	10	R. Anaco, Anaco
DC02)		*660		R. Nacional, El Callao (r: 630)
FA01)	NA	660	10	Ondas de los Médanos "Tu R. Popular", Coro
DC03)	LL	670	100	R. Rumbos, Caracas
BA02)	ZJ	680	10	R. Llanera "R. 1400", Barinas
SU02)	QR	v680	10	R. Continente Cumaná, Cumaná
LA03)	MR	690	50/20	R. Barquisimeto, Barquisimeto
BO01)	PQ	700	5/2	R. Sur, Puerto Ordaz
ZU04)	MH	700	10	R. Popular, Maracaibo
DC05)	KY	710	50/20	R. Capital, Caracas
AP02)	XE	720	10	R. Elorza, Elorza
NE01)	QE	720	50	R. Venezuela "Oriente", Porlamar
LA04)	MT	730	10	R. Universo, Barquisimeto
TA02)	OO	*730	10	R. Frontera, San Antonio del Táchira
BO02)	NQ	740	50	R. Caroni "Q-FM", Puerto Ordaz
ZU05)	NC	740	10	CNB 740 La Zuliana, Maracaibo
DC06)	KS	750	100	RCR 750 "Radio Caracas", Caracas
AN04)	QQ	760	10	R. Puerto La Cruz/La Doble Q, Pto. La Cruz
TR01)	SO	760	10	Simpatía 760 AM, Trujillo
DC02)	KK	770	50/20	R. Nacional, Valencia (r: 630)
FA02)	MN	780	10	R. Coro, Coro
TA03)	OD	780	50/20	Ecos del Torbes, San Cristóbal
DC02)		*790		R. Nacional, Cd.Bolívar (r: 630)
DC07)	KC	790	10	R. Venezuela 7-90, Caracas
LA05)	XM	790	50	R. Minuto "La Barquisimetana", Barquisimeto
CA01)	LP	810	50	Super Radio 810, Valencia
BO03)	SH	820	50	R. Guayana, Upatá
FA03)	XG	*820	25/10	R. Guadalupana, Coro
TA04)	KU	820	10	R. Altura 820, La Grita
DC08)	LT	830	25	R. Nueva Sensación, Caracas
LA04)	MY	840	10	8-40 AM, Barquisimeto
MO01)	UZ	840	10/5	Guarapiche 8-40 "La Primera", Maturín
CA02)	RV	850	10	RV-850, Valencia
ZU06)	ZC	850	10	R. Fe y Alegría, Maracaibo
GU01)	YE	860	20/10	Enlace 8-60, Valle de la Pascua
TA05)	OL	860	10	Mundial 8-60, San Cristóbal
AN06)	RU	870	10	Unión R. Deportes 870, Puerto La Cruz
LA11)	MP	870	10	Unión R. Notícias, Barquisimeto
BO04)	YM	880	10	R. Venezuela, Puerto Ordaz
DC02)	KV	880	10	RNV 880 Canal Musical, Caracas
FA04)		880	10	R. Paraguaná (Continente), Punto Fijo
AN07)	VO	890	10	R. Oriente, El Tigre
CA03)	LW	890	10	R. América, Valencia
ZU07)	MD	900	25	R. Venezuela "Mara Ritmo", Maracaibo
DC10)	RQ	910	50/20	R. Q 910 "Tu AM Center", Caracas
CO01)	QU	920	10/5	R. San Carlos, San Carlos
NE02)	QX	920	20	R. Nueva Esparta, Porlamar
AR02)	LJ	930	10	R. Maracay, Maracay
AN08)	LU	940	10	R. Fe y Alegría El Tigre, El Tigre
BA03)	ZR	940	15	R. Continental, Barinas
FA05)	NN	940	10	R. Punto Fijo, Punto Fijo
DC11)	KG	950	50	YVKG AM Popular, Caracas
MO02)	RB	960	50/20	R. Monagas, Maturín
PO01)		960	25	R. Venezuela "Llanera 9-60", Acarigua
TA06)	SS	960	10	R. San Sebastián, San Cristóbal
AN09)		*970	10	Mundial 970, Barcelona
AR03)	LR	970	10	R. Continente 970 Maracay, Maracay
TR02)	SD	970	15	R. Turismo, Valera
AN10)	QM	980	10	Unión R. Notícias "LV de El Tigre", El Tigre
DC02)		*980		R. Nacional, Maracaibo (r: 630)
DC12)	RT	990	20	R. Tropical "99-0", Caracas
LA06)	TA	990	10	R. Venezuela "Tricolor", Barquisimeto
CA04)	NM	1000	10	R. Caribeña Mil AM (Continente), Morón

MW	Call	kHz	kW	Station, location
TA11)	OA	1000	10	R. Táchira, San Cristóbal
AR04)	PC	1010	10	R. Aragua, Cagua
BO05)	QF	1010	10	R. Venezuela, Cd.Bolívar
NE03)	RS	1020	10	R. Mundial Margarita, La Asunción
YA01)	TW	1020	25	R. Alegría, Chivacoa
ZU08)	MX	1020	50/10	R. Continente/R. Calendario, Maracaibo
MI01)	TD	1030	25/10	R. Valles del Tuy, Ocumare del Tuy
PO02)	QY	1030	20	R. Onda 1030, Guanare
CA05)	LB	1040	20	LV de Carabobo, Valencia
ME01)	ON	1040	20/10	Mundial Los Andes, Mérida
DC02)	KZ	1050	25	RNV "R. Educativa", Caracas
DC02)	PO	1050	20	R. Nacional, Cabudare
GU02)	LN	1060	10	R. Guárico 1060 AM, S. Juan de los Morros
TA07)	OE	1060	10	Unión R. Noticias, San Cristóbal
AP03)		1070	10	Superior 1070 Biruaca, San Fernando de Apure
PO03)		1070	25	Contacto 1070, Ospino
TA08)	PX	1070	5	R. El Sol, La Fría
ZU09)	MA	1070	10	Mundial Zulia, Maracaibo
AN11)	QJ	1080	10	R. Barcelona, Barcelona
AR05)	NR	1080	10	R. Venezuela, Maracay
DC13)	SZ	1090	20	Unión R. Notícias1090, Caracas
YA02)	PB	1090	10	R. Yaracuy "Operadora 1090 AM", S. Felipe
ZU10)	TG	1090	3	Melódica 1090, Machiques
BO06)	SV	1100	10	R. Angostura, Cd.Bolívar
ME02)	OP	1100	10	R. Fe y Alegría, Tovar
CA06)	RX	1110	10	Unión R. Notícias, Valencia
SU03)	QT	1110	10	R. Venezuela Carúpano, Carúpano
AP04)	SK	1120	20/10	R. Dif.del Sur, San Fernando de Apure
MO03)	XZ	1120	5	R. República "La Estación Feliz", Maturín
ZU11)	MF	1120	10	Ondas del Lago "Super Ondas", Maracaibo
AM02)	PY*	°1130	15	R. Amazonas, Puerto Ayacucho
LA07)	KQ	1130	10	R. Popular, Barquisimeto
DC16)	RL	1130	20/10	R. Ideal, Maiquetía
NE04)		*1140	10	R. Porlamar "LV del Caribe", Porlamar
BO07)	QD	1150	10	Ecos del Orinoco, Cd.Bolívar
FA06)	MV	1150	10	R. Venezuela "Ondas del Caribe", Punto Fijo
ME03)	OK	1160	1	R. Universidad, Mérida
MI02)	RR	1160	20/10	R. Notícias RN 1160, Guarenas
PO04)	QV	*1170	20/10	R. Acarigua, Acarigua
VA01)	KW	1170	10	R. Bolivariana "R. 1070", Maiquetía
AR06)	LQ	1180	10	LV de la Victoria "Super Suave 11-80", La Victoria
MO04)	OR	1180	10	R. Maturín, Maturín
ZU12)	NJ	1180	10	R. Petrolera, Cd.Ojeda
BA04)	RE	1190	20/10	R. Barinas 1190 AM Estéreo, Barinas
BO08)	PF	1190	20/10	Ondas de Libertad, San Félix
TA09)	ZD	1190	10	R. Dif. Cult. del Táchira "Paz Vital 11-90", San.Cristóbal
DC14)	OZ	1200	10	R. Tiempo, Caracas
MO05)	SF	1200	10	R. Dimensión, Caripito
ZU13)	NH	1200	1	Ondas del Escalante, Sta Bárbara del Zulia
AN12)	ZT	1210	10	R. Anzoátegui, Barcelona
AP05)	RD	1220	10	LV de Apure, San Fernando de Apure
CA07)	VM	1220	10/5	R. Venezuela, Valencia
ZU14)	ZO	1220	20/10	R. Aeropuerto 1220, Maracaibo
MI03)	NT	*1230	10	R. Barlovento, Caucagua
TR03)	OH	1230	10	R. Valera, Valera
BO09)	PZ	1250	20/10	Latina 12-50, Pto Ordaz
ZU15)	ML	1250	1	R. Cabimas, Cabimas
DC15)	RM	1260	10	BBN R., Caracas
YA03)	RY	1260	10	R. Horizonte, Nirgua
DA01)	TR	*1270	5	R. Tucupita, Tucupita
DC02)		*1270		R. Nacional, Ureña (r: 630)
ME04)	OU	1270	10	R. Ondas Panamericanas, El Vigía
GU03)	QS	1280	10/5	R. Zaraza (Continente), Zaraza
TR04)	OF	1280	10	R. Trujillo, Trujillo
CA08)	LF	1290	10	R. Puerto Cabello, Puerto Cabello
DC10)	KH	1300	10/8	R. Recuerdos 1300, Caracas
ZU16)	NS	1300	10	Deportes Unión R., Maracaibo
DC02)	SM	1300	10	R. Nacional "RNV", Barcelona (r: 630)
DC02)	SL	1310	1	R. Naciona "RNV", Guri (r: 630)
TR05)	TS	1310	5	R. Andina "Sonido 13-10", Isnotú
AR07)	WP	1320	10/5	R. Apolo, Turmero
LA08)	SG	1320	5	R. Colonial, El Tocuyo
DC02)		1330		R. Nacional "RNV", La Paragua (r: 630)
GU04)	OY	1330	5	R. Los Llanos, Calabozo
ZU17)	TU	*1330	10	R. Regional, Cd.Ojeda
DC17)	NE	1340	10	R. Uno, Caracas
AN13)	NT	1350	5	R. Eclipse "R. Guanipa",El Tigrito
FA07)	TJ	1350	5	R. Falcón, Puerto Cumarebo

MW	Call	kHz	kW	Station, location
MI04)	TZ	1360	5	R. YVTZ, Charallave
YA01)	TW	1360	5	R. Alegría, Chivacoa
ZU18)	TI	1360	10	R. Internacional, Maracaibo
GU05)	OQ	1370	5	Unión R. Notícias/R. La Pascua, Valle de la Pascua
ME05)	JI	1370	10	R. Continente Cumbre, Mérida
PO05)	SV	1370	5	R. Portuguesa, Araure
BO10)	ME	1380	5	R. Revelación, Cd.Bolivar
CA09)	NG	1380	10	Ondas del Mar, Puerto Cabello
ZU19)	TL	1380	10	R. Triunfo 13-80, Caja Seca
DC18)	ZA	1390	20	R. Fe y Alegría, Caracas
LA09)	TT	1390	10	R. Terepaima, Cabudare
ZU20)	ZO	1390	10	R. Lumen , Maracaibo (Prgr: R. Católica Mundial)
GU06)	NF	1400	1	R. Sabana, El Sombrero
PO06)	ST	1410	5	R. Turén, Turén
TR01)	SP	1410	10	R. Simpatía, Valera
DC21)		1420	5	R. Sintonía, Caracas
LA10)	RW	1420	10/5	R. Cardenal, Carora
ZU21)	NZ	*1420	5	R. Marabina 1420, Maracaibo
AN14)	TP	1430	25	R. Bahía, Puerto La Cruz
BO11)	TM	1430	10/5	R. Caicara, Caicara del Orinoco
CA10)	NB	1430	10	R. Satélite 14-30, Guacara
GU07)	RF	1440	5	R. Orituco, Altagracia del Orituco
PO07)	ZI	1440	10	R. Estelar 14-40, Guanare
TA10)	TY	1440	1	R. Sucesos, Táriba
BO12)	XC	1450	10/5	R. Mega Visión, San Felix
VA02)	KJ	1450	10/8	R. María, Caracas
ZU22)	ZQ	1450	10	Informativa 14-50, Los Puertos de Altagracia
TR07)	RJ	1460	5	R. Jardín, Boconó
CA11)	LW	1470	10	CNB La Valenciana 14-70, Valencia
SU04)	SY	1470	10	Unión R. Notícias/R. Vibración, Carúpano
FA08)		1480		R. Cumarebo, Cumarebo
DC19)	XD	1490	10	La Dinámica, Caracas
ME06)	SQ	1490	1	R. Mérida 14-90, Mérida
ZU11)	RP	1490	10/5	R. El Sol, Maracaibo
AR08)		*1500		R. Galaxia, San Mateo
SU05)	RZ	1500	10/5	R. 2000 AM, Cumaná
CA12)		1510	20	Informativa "LV del Centro", Güigüe
MI05)	IC	*1520	25	R. Bonita La Guapa, Guatire
YA04)	NP	1530	10	R. San Felipe el Fuerte, San Felipe
MI06)	MW	*1550	10	R. Metropolitana, Los Teques
ZU24)	XO	1550	10/5	R. Impacto La Poderosa, Cd. Ojeda
ME07)	LZ	1560	10/5	R. Dif. Andina, Mérida
GU08)	YV	1580	10/5	R. Venezolana, Calabozo
SU06)	TK	*1580	10/5	Manzanares 15-80, Cumaná: (r: RQ-910 "Tu AM Center")
ZU25)	YO	1580	10	R Celestial, San Francisco, Maracaibo
DC20)	UD	1590	10	R. Deporte 15-90, Caracas

Call YV–,° = also on SW, * = inactive, (r) = repeater, v = varying fq.

SW	Call	kHz	kW	Name and h of tr
AM02)	PA	*4940	1	R. Amazonas, Pto. Ayacucho:1000-0400

Other stations
CNB - CIRCUITO NACIONAL BELFORT
⬚ Quinta CNB, Av.Los Naranjos, La Florida, Caracas **W:** www.cnb.com.ve.
CIRCUITO AM CENTER
⬚ CentroComercialConcresa, Nivel 1, Circuito Center, Prados del Este, Caracas 1080, Edo.Miranda ☎ +58 212 976-2013 **E:** feloespinosa@cantv.et.
CIRCUITO RADIAL ALFA OMEGA
⬚ Calle 25, Con Calle 67, Sector El Paraíso, frente Al Colegio La Epifanía, Maracaibo, Edo.Zulia ☎ +58 261 783-2524
CIRCUITO POPULAR
⬚ Boulevard de Sabana Grande, Torre Provincial, P10, Sabana Grande, Caracas 1050 ☎ +58 212 762 5052
CIRCUITO RADIO CARACAS RADIO
⬚ Av.Páez, Quinta RCR, El Paraíso, Caracas 1021 ☎ +58 212 481-3590
CIRCUITO RADIAL CONTINENTE
⬚ Calle La Joya, Edif.Cosmos, PH, Chacao, Caracas 1060, Edo. Miranda ☎ +58 212 267-3132 🖷 +58 212 267-1223 **W:** http:// radiocontinente.tripod.om.ve **E:** produccion@radiocontinente. zzn.com
CIRCUITO RADIO VENEZUELA
⬚ Av.Rómulo Gallegos, Edif.KLM, P12, Ofcs CyD, Los Palos Grandes, Caracas 1062, Edo.Miranda ☎ +58 212 286-8492 **W:** www. radiovenezuela.com.ve **E:** radiovenezuela@hotmail.com
CIRCUITO SATELITAL RUMBOS
⬚ Av.Francisco de Miranda, Multicentro Empresarial del Este, Edif.

Libertador, Núcleo A, P7, Chacao, Caracas 1060, Edo.Miranda ☎ +58 212 263-3236 🖷 212 263-2212 **E:** radiorumbos@ip-net.work.net
CORPORACIÓN REGIONAL BRADCASTING
⬚ Calle 74, Entre Av. 3Dy3E, Edif.Televisa, Sector La Lago, Maracaibo 4002, Edo. Zulia ☎ +58 261 792-9217
GRUPO RADIAL DE ORIENTE
⬚ Urb.Tricentenaria, Centro ComercialTricentenaria, P2, Ofcs 03y09, Barcelona 6001, Edo.Anzoátegui ☎ +58 281 277-1743
🖷 281 277-1776 **E:** radioanzoategui@hotmail.com
RADIO CADENA MUNDIAL
⬚ Calle Nueva York, Edif.Manzanillo, P2, Las Mercedes, Caracas 1060, Edo.Miranda ☎ 212 993-9391 **E:** prensayvke@cantv.net.
UNION RADIO
⬚ Av.Mohedano, Entre Calle Los Granados y 1ª transversal, Edif. Splendor, La Castellana, Caracas 1060, Edo.Miranda ☎ +58 212 263-5133 **W:** www.unionradio.com.ve.

State abbreviations: AM = Amazonas, AN = Anzoátegui, AP = Apure, AR = Aragua, BA = Barinas, BO = Bolívar, CA = Carabobo, CO = Cojedes, DA = Delta Amacuro, DC = Distrito Capital, FA = Falcón, GU = Guárico, LA = Lara, ME = Mérida, MI = Miranda, MO = Monagas, NE = Nueva Esparta, PO = Portuguesa, SU = Sucre, TA = Táchira, TR = Trujillo, VA = Vargas, YA = Yaracuy, ZU = Zulia.
N.B: These abbreviations are not officially recognized by the Venezuelan Post Office. Letters should therefore carry the full name.

Addresses and other information:
AM00) AMAZONAS
AM01) San Juan de Manapiare ☎ +58 248 978-0249. – **AM02)** Av. Bolívar 4 c/c Av. La Guardia, Puerto Ayacucho 7101 - 0830-0400 ☎ +58 248 521-4892 🖷 +58 248 214-769 (Reception reports to: Jorge García Rangel, Calle Roma, Qta Costa Rica N° A-16, Urb.Alto Barinas, Barinas 5201, Barinas, Venezuela).
AN00) ANZOÁTEGUI
AN01) Av Hospital cruce con Calle Freites, Edif.Radio Centro, Cantura 6007 ☎ +58 282 455-1414 – **AN02)** Av 5 de Julio, Edif Los Angeles, Sotanos 1y2, Puerto La Cruz 6023 ☎ +58 281 265-1953 **W:** www. unionradio.com.ve/ – **AN03)** Calle Cajigal cruce con Av.Nueva Esparta N° 39, Edif.Radio City, planta baja, Anaco 6003 - 0900-0400 ☎ +58 282 125-2055– **AN04)** Calle Arismendi N° 20, Edif.Radio Puerto La Cruz, PB, Puerto La Cruz 6023 - 0955-0300 ☎ +58 281 265-3512 – **AN06)** Av.Municipal,Torre Porteñas, Mezzanina, Ofc 2-4, Puerto La Cruz 6023 ☎ +58 281 267-0870 **W:** www.cnb.com.ve – **AN07)** Calle Guayana, Centro Comercial Bleu Hill, P1, Local 4, El Tigre 6034 ☎ +58 283 235-0902 **E:** Oriente89cero@terra.com.ve – **AN08)** Av.Simon Rodríguez con 8va Calle Norte, Complejo Cultural Simón Rodríguez, El Tigre 6034 - 0900-0300 ☎ +58 283 231-6330 **W:** www.feyalegria.com/ Venezuela **E:** feyalegria@cantv.net o irfaeltigre@ cantv.net – **AN09)** Av.Intercomunal "Andrés Bello", Centro Comercial Géminis, P3, Local 9, Barcelona 6001 ☎ +58 281 276-2986 – **AN10)** Av Francisco de Miranda N° 196, Al lado del Banco Provincial, El Tigre 6034 - 1000-0300 ☎+58 283 235-2801 – **AN11)** Av Miranda cruce con Av.San Carlos, Edif.Radio Barcelona, P2, Barcelona 6001 ☎ +58 281 277-1080 – **AN12)** Urb Tricentenaria, Centro Comercial Tricente-naria, P2, Ofc 3y9, Barcelona 6001 ☎ +58 281 277-1743 **E:** radioanzoategui@ hotmail.com – **AN13)** Av.Intercomunal El Tigre El Tigrito, Detrás de Elite Motors, Casa Amarilla,El Tigrito 6035 ☎ +58 283 255-0556 **E:** eclipse@telcel.net.ve – **AN14)** Av Municipal, Torre Pelicano, P8, Apto 8-4, Puerto La Cruz 6023 ☎ +58 281 269-9522
AP00) APURE
AP01) Carr Nacional, Vía Elorza La Arenosa, Edif.Fe y Alegría, Guasdualito 5063 - 0900-0400 ☎ +58 278 332-0233 **W:** www.fey-alegria org/ve **E:** rcepeda@cantv.net o feyalegria@cantv.net – **AP02)** Calle 9 con Cra 4, Municipio Rómulo Gallegos, Elorza 7007 ☎ +58 240 929-1051 – **AP03)** Av.Fuerzas Armadas, Edif.Superior, P1, San Fernando de Apure 7001 0930-0400 ☎ +58 247 341-0070 – **AP04)** Calle Carlos Rodríguez Rincones, Gobernación del Estado Apure, San Fernando de Apure 7001 ☎ +58 247 341-1114 – **AP05)** Av Miranda, Edif.Don António Cestari, San Fernando de Apure 7001 ☎ +58 247 341-1768
AR00) ARAGUA
AR01) Calle Coromoto, Urb.Calicanto, Torre Capitolio, P9, Ofc.B-9, Urb. Calicanto, Maracay 2101 - 0900-0500 ☎ +58 243 246-2809 **W:** www unionradio.com.ve **E:** rb650@telcel.net.ve – **AR02)** Calle Boyacá, Edif Centro, P9, Ofc 1, Maracay 2101 - 1000-0600 ☎ +58 243 246-5591 – **AR03)** Av Miranda Oeste N° 149, Entre Carabobo y Pinhincha, Edif. Canaobre, PH, Maracay 2101 - 0900-0400 ☎ +58 243 246-4560 **W:** http://radiocontinente.tripod.com.ve/ **E:** produccion@radiocontinente. zzn.com – **AR04)** Calle Boyacá Nte N° 9, Edif Radio Aragua, Ofc.1y2, Cagua 2122 - 0900-0400 ☎ +58 244 395-4123 – **AR05)** Urb.Calicanto, Calle Coromoto, Norte 6, Detrás de la Maestranza Cesar Girón, Maracay 2101 ☎ +58 243 245-5643 **W:** www radiovenezuela.com.

ve **E:** radiovenezuela@hotmail.com – **AR06)** Edif Veliz, Calle Aldao, frente a la Plaza Rivas, La Victoria 2126 **W:** www.cnb.com.ve/ – **AR07)** Av.Bermúdez, Torre Apolo, PB, entre Mariño y Bolívar, Turmero 2115 ☎ +58 244 6630-2928 – **AR08)** San Mateo – **AR09)** Calle Páez, N° 138, Detrás del Teatro de La Opera, Maracay 2126 ☎ +58 243 232-9835

BA00) BARINAS
BA01) Cra 3 N° 7-39, entre Calles 7 y 8, Santa Bárbara de Barinas 5210 - 0900-0500 ☎ +58 278 221-1133 – **BA02)** Av Sucre Quinta Claret N° 17-46, Barinas 5201 - 0900-0500 ☎ +58 273 552-7976 – **BA03)** Av Marqués del Pumar, Edif.Radio Continental, Barinas 5201 ☎ +58 273 552-4021 – **BA04)** Av Sucre con Av Agustín Codazzi, Edif Circuito Sensacional Barinas, Barinas 5201 ☎ +58 273 552-24.17 **W:** www. radiovenezuela.com.ve **E:** radiovenezuela@hotmail.com

B000) BOLÍVAR
B001) Av.Guasipati, Edif Piarde, PH, Puerto Ordaz 8015 ☎ +58 286 922-99.08 **E:** sur700am@yahoo.com or – **B002)** Urb.Altavista Calle Caura, Edif Los Bancos, P4, Puerto Ordaz 8015 ☎ +58 286 961-8411 **E:** cesargonzales1284@hotmail.com – **B003)** Av.Raúl Leoni, Edif. Antonelli, PB, Upatá 8026 - 0900-0500 ☎ +58 288 221-1457 **E:** rodriguayana@hotmail.com – **B004)** Av.Venezuela, Centro Comercial Venezuela, Local 14-15, P1, Urb Villa Colombia, Puerto Ordaz 8015 ☎ +58 286 923-7051 **W:** www.radiovenezuela com.ve **E:** radiovenezuela@ hotmail.com – **B005)** Calle Dalla Costa, Alto N° 5, Cd Bolívar 8001 - 0900-0400 ☎ +58 285 632-3743 **W:** www.radiovenezuela.com.ve **E:** radiovenezuela@hotmail.com – **B006)** Final Paseo Heres, Edif.Tovar, P2, Cd Bolívar 8001 ☎ +58 285 652-2202 – **B007)** Paseo Meneses, Centro Comercial Meneses, PA, Locales 11 y 12, Cd Bolívar 8001 ☎ +58 285 632-0413 – **B008)** Calle México, Parcela El Roble, Detrás de la Estación de Servicio Volfo, Sector La Antena, San Félix 8024 - 0900-0300 ☎ +58 286 932-2366 **E:** ondasdelibertad@cantv.net – **B009)** Calle El Tocuyo, Centro Comercial Plaza, P2, Pto Ordaz 8015 - 0900-0400 ☎ +58 286 922-1346 – **B010)** Av 19 de Abril, Edif La Disinca, P.B., Cd Bolívar 8001 ☎ 285 632-6564 (Offices and studios in Cd.Bolívar, tx site Cantaura, Anzoátegui) **W:** http://revelacion1380am.com – **B011)** Calle Constitución N° 78, Caicara del Orinocco 7107 ☎ +58 284 666.74.30 – **B012)** Av.Della Costa, Edif.Flor Motors, PB, San Felix 8024 ☎ +58 286 932-2662

CA00) CARABOBO
CA01) Av.Girardot con Calle Montes de Oca, Edif.Normal, Piso PH, Valencia 2001 - 0900-0500 ☎ +58 241 857-1910 – **CA02)** Av Bolívar Norte, Edif Felpo, P7, Ofc 3-3, Valencia 2001 ☎ +58 241 857-5850 – **CA03)** Calle Girardot, Entre Urdaneta y Boyacá N° 98-28, Valencia 2001 - 0900-0400 ☎ +58 241 857-4868 **E:** tomas@cantv.net – **CA04)** Carr.Panamericana, Edif.Radio Mil, Morón 2004 - 0900-0400 ☎ +58 242 372-0283 – **CA05)** Av Rosarito, Torre Trebol, P1, Ofc 13, Urb Lomas del Este, Valencia 2001 - 0900-0400 ☎ +58 241 857-4111 **E:** mbranger@ mixmail com – **CA06)** Av.Bolívar Norte, Torre Banavén, P12, Ofc.12-9, Valencia 2001 ☎ +58 241 821-1213 **W:** : www.unionradio.com.ve – **CA07)** Av.Rotaria, Edif.El Parque, PB Local 2, Urb.Lomas del Este, Valencia 2001 ☎ +58 241 857-4864 **W:** www.radiovenezuela.com **E:** radiovenezuela@hotmail.com – **CA08)** Av.Marina, Edif.Diproca, PB, Local 3, Puerto Cabello 2024 ☎ +58 242 361-3103 **W:** www.unionradio.com.ve **W:** http://diproca.com/rpc **E:** radiorpc@ telcel.net.ve – **CA09)** Av Bolívar, Edif Sabatino, P1, Urb.Rancho Grande, Puerto Cabello 2024 - 0900-0400 ☎ +58 242 362-3569 **W:** www.cnb. com.ve/ **E:** ondasdelmar1380am@gmail.com – **CA10)** Final de la Calle Jacinto con Calles Ricaurte y Girardot, Edif.Radio Satélite, Guacara 2015 - 0930-0400 ☎ +58 245 564-0979 **E:** mbranger@mixmail.com – **CA11)** Av.Montes de Oca, Edif.Don Pelayo, P12, Valencia 2001 ☎ +58 241 857-4270 **W:** www.latina.com.ve/ **E:** latina@unete.com.ve – **FM:** 99.1MHz – **CA12)** Av Miranda,Edif Padre Cecilio Ávila, PH, Güigüe 2010 ☎ +58 245 541-1819.

C000) COJEDES
C001) Av.Sucre, Edif.General Manuel Manrique, P3, Local 46, San Carlos 2201 - 0955-0400 ☎ +58 285 433-0051 **W:** www.unionradio. com.ve **E:** radiosancarlos@cantv.net

DA00) DELTA AMACURO
DA01) Calle Petión cruse con Calle La Paz, Tucupita 6401 - 0900-0500 ☎ +58 287 721-0022.

DC00) DISTRITO CAPITAL
DC01) Calle Nueva York Cruce con Av.Rio de Janeiro, Edif YVKE Mundial, P1, Las Mercedes, Caracas 1060, Edo.Miranda ☎ +58 212 993-6242 – ▤ +58 212 993 2267 **E:** prensayvke @cantv.net – **DC02)** Final Calle Las Marías, Edif.Radio Nacional de Venezuela, entre Chapellín y Country Club, La Frorida, Caracas 1050, Edo Miranda - 1000-0400 ☎ +58 212 730-6022 **W:** www.rnv.gov.ve **E:** rnv@rnv.gov.ve – **DC03)** Av.Francisco de Miranda, Multicentro Empresarial del Este, Edif.Libertador, P7, Núcleo A Chacao, Caracas 1060, Edo Miranda ☎ +58 212 263-3236 **W:** www.tycom.com.ve/rumbos **E:** radiorumbos@ip-net.work.net – **DC04)** Calle La Joya, Edif Cosmos PH, Chacao, Caracas 1060, Edo Miranda ☎ +58 212 267-3132 **W:** www.radiocontinente.da.ru/ or http://radiocon-

tinente.tripod.com.ve **E:** produccion@radiocontinente.zzn.com – **DC05)** Av.Francisco de Miranda, Centro Comercial Los Ruices, P3, Los Ruices, Caracas 1071, Edo Miranda - 1000-0600 ☎ +58 212 238-1630 – **DC06)** Av José A Paez, Quinta RCR, El Paraiso, Caracas 1021, Distrito Capital ☎ +58 212 481-3590 **W:** www.radionet.com.ve **E:** jnestares@etheron. net – **DC07)** Av.Rómulo Gallegos, Edif.KLM, P12, Ofc CyD, Los Palos Grandes, Caracas 1062, Edo Miranda ☎ +58 212 286-84.92 **W:** www. radiovenezuela.com.ve **E:** radiovenezuela@hotmail.com – **DC08)** Av Santiago de Chile, Quinta Radio Sensación, Los Caobos, Caracas 1050 Disatrito Capital - 0900-0500 ☎ +58 212 793-6458 **E:** radiosensacion@ cantv.net – **DC10)** Centro Comercial Concresa, Nivel 1, Circuito Center, Prados del Este, Caracas 1080, Edo.Miranda ☎ +58 212 976.20.13 **E:** feloespinosa@cantv.net – **DC11)** Edif Provincial, Boulevard de Sabana Grande (frente al Gran Café)Caracas 1050, Distrito Capital - 1000-0400 ☎ +58 212 761 0411, 761 4853 **W:** www.cadenastereo.com. ve **E:** balfaro@cadenastereo.com.ve – **DC12)** Puente Nuevo a Puerto Escondido, Edif.Torre del Oeste, P1, El Silencio, Caracas 1010, Distrito Capital (or: Ap.3674, Caracas 1010-A) ☎ +58 212 482-1111 **E:** radio-tropical@cantv.net – **DC13)** Av.Mohedano, Entre Calle Los Granados y 1ª transversal, Edif.Splendor, La Castellana, Caracas 1060, Edo.Miranda ☎ +58 212 263-5133 **W:** www.unionradio.com.ve – **DC14)** Av Los Mangos N° 49, Qta.Radio Tiempo, La Florida, Caracas 1050-A, Edo. Miranda ☎ +58 212 730-3889 **E:** radiotiempo@telcel.net.ve – **DC15)** Av Los Mangos con Av.Valencia Parpacén, Qta. Marisabel (BBN), La Florida, Caracas 1050, Edo.Miranda ☎ +58 212 731-8161 **W:** www. bbnradio.org – **DC16)** Centro Comercial Uslar, P15, Ofc 152, Montalbán, Caracas 1021, Distrito Capital ☎ +58 212 443-6680 **E:** radioideal@ cantvc.net – **DC17)** Edif Mundial, Av Tamanaco, El Rosal, Caracas 1060 – **DC18)** Calle 3B, Edif.C-207, P2, (detrás del McDonald's), La Urbina, Caracas 1070, Edo.Miranda ☎ +58 212 242 2919 **W:** www.feyalegria. org/Venezuela/ **E:** radiofya@ cantv.net or radiofyanacional@cantv. net – **DC19)** Av.Boulevard Brasil N° 74, de Santa Ana a Providencia La Pastora, Caracas 1010, Distrito Capital ☎ +58 212 862-6415 **E:**dinamica@cantv.net – **DC20)** Av Circunvalación del Sol, Centro Profesional Sta Paula, Torre A, P5 Ofc 51, Caracas 1061, Edo.Miranda ☎ +58 212 985-2907 **W:** radiodeporte.8m.com **E:** radiodeporte@cantv. net – **DC21)** Calle La Joya, Torre Cosmos, P9, Ofc 9A, Chacao, Caracas 1060, Edo.Miranda (or: Centro Comercial El Pichacho, P8, San António de los Altos 1204) ☎ +58 212 264-0782

FA00) FALCÓN
FA01) Calle Bolívar, Edif.Don Cosme, P2, Coro 4101 - 0900-0400 ☎ +58 268 253-2095 **E:** olm660@cantv.net – **FA02)** Calle Bolívar, Edif. Don Cosme, P2, Coro 4101 ☎ +58 268 253-1448 **E:** radiocoro780am @cantv.net – **FA03)** Calle Palmasola cruce Con Calle Federación, Edif.Arquidiocesano, Coro 4101 - 0900-0400 ☎ +58 268 251-5455 **E:** rguadalu@funflc.org.ve –**FA04)** Urb Los Caciques, Calle Falcón, Qta.Paraguaná, Punto Fijo 4102 ☎ +58 269 246-5194 – **FA05)** Calle Talavera, entre Calles Comercio y Arismendi, Edif.Radio Punto Fijo, Punto Fijo 4102 - 0900-0500 ☎ +58 269 245-1232 – **FA06)** Av.Ecuador, Entre Calles Comercio y Arismendi, Punto Fijo 4102 ☎ +58 269 245-6565 – **FA07)** Av.Bolívar, Edif.Colonial Planta Baja, Urb.Alta Vista, Puerto Cumarebo 4167 ☎ +58 268 747-1437 – **FA08)** Centro Ciudad Comercial Tamanaco (CCCT), Torre B, P7, Ofc 704, Chuao, Caracas 1060, Edo Miranda ☎ +58 212 959-0075

GU00) GUÁRICO
GU01) Av RómuloGallegos, Edif.Flor de Pascua, Loc 2, Valle de la Pascua 2307 ☎ +58 235 341-2860 **E:** radioenlace860@ cantv.net – **GU02)** Av Principal La Moreras, Edif.Ghersy N° 28, San Juan de los Morros 2301 - 1030-0330 ☎ +58 246 431-6494 **E:** feloespinosa@cantv.net – **GU03)** Calle Concordia, Qta Puerto Arturo N° 35, Zaraza 2332 - 1000-0300 ☎ +58 238 762-1032 **W:** www.circuitoz.com or htttp://radiocontinente. tripod.com.ve **E:** circuitoz@hotmail com – **GU04)** Cra 12, Altos del Teatro Paez, Frente a La Bomba, Calabozo 2312 - 0900-0300 ☎ +58 246 871-3639 **E:** rlosllanos@telcel.net.ve – **GU05)** Av 5 de Julio N° 20, Valle de la Pascua 2307 - 1000-0400 ☎ +58 235 341-3255 **W:** www. unionradio.com.ve – **GU06)** Calle Alegría, Qta.Galia, El Sombrero 2319 - 1000-0200 ☎ +58 246 616-3111 – **GU07)** Calle Andrés Eloy Blanco, Altagracia de Orituco 2320 – **GU08)** Cra 12, Entre Calles 3 y 4 N° 3-57, Calabozo 2312 - 1000-0200 ☎ +58 246 871-1776.

LA00) LARA
LA01) Av Venezuela con Calles 13 y 14, Edif.Radio Cristal, Barquisimeto 3001 ☎ +58 251 252-3610 – **LA02)** Calle Sucre Entre Cras 7 y 8, La Casita, Carora 3040 - 1000-0400 ☎ +58 251 421-6142 **W:** www. angelfire.com/la/carora/ **E:** radiocarora@cantv.net or antena@cantv. net – **FM (MHz):** 100.5 – **LA03)** Calle 4 con Cra 3, Qta.Técnica, Urb. del Este, Barquisimeto 3002 ☎ +58 251 255-1690 **E:** feloespinosa@ cantv.net – **LA04)** Av Venezuela con Calles 32 y 33, Edif.Don Martín, P4, Apto 4-A, Barquisimeto 3001 ☎ +58 251 232-8145 **W:** http:// radiocontinente,.tripod.com.ve **E:** radiouniverso@cantv.net – **LA05)** Av.Pedro León Torres, Centro Comercial Venrol, locales 29 y 30, Barquisimeto 3001 ☎ +58 251 446-2494 **E:** radiominuto@cantv.net

– **LA06)** Av Vargas, Cra 16, Edif Tricolor, Barquisimeto 3001 ☎ +58 251 231-2311 **W:** radiovenezuela.com.ve **E:** radiovenezuela@ hotmail.com – **LA07)** Calle 29, Entre Calles 18 y 19, Casa N° 18-74, Barquisimeto 3001 - 0900-0400 ☎ +58 251 232-5386 **E:** radiopopular@ cantv.net – **LA08)** Calle 10 cruce con Calle 9, Casa S/N, El Tocuyo 3018 ☎ +58 253 663-1040 – **LA09)** Av.Libertador, altos de la Farmacia San Rafael, Cabudare 3023 ☎ +58 251 261-1884 – **LA10)** Av Bolívar, Edif Guillermo, Locales 2 y 3, Carora 3040 - 1000-0400 ☎ +58 252 421-3292 – **LA11)** Av Los Leones, Centro Empresarial Caracas, P5, Ofc 5-2, Barquisimeto 3002 - 1000-0400 ☎ +58 251 255-0981 **W:** www unionradio.com.ve

ME00) MÉRIDA
ME01) Calle 44 N° 3-57, Diagonal al Colegio de Médicos, Mérida 5101 ☎ +58 274 263-0921 – **ME02)** Cra 4 N° 6-46, Frente a la Plaza Bolívar, Tovar 5143 - 0900-0300 ☎ +58 275 873-3574 – **ME03)** Av Gonzalo Pico, Bajando por La Facultad de Ingeniería, Qta.Radio Universidad, Mérida 5101 ☎ +58 274 252-1931 – **ME04)** Av Bolívar, Esquina Calle 11, N° 10-87, El Vigía 5145 ☎ +58 275 881-4140 – **ME05)** Av.Andrés Bello, Centro Comercial Las Tapias, P3, Ofc.40-41, Mérida 5101 - 1000-0400 ☎ +58 274 266-1355 ☐ +58 274 266-4712 **W:** http://radiocontinente.tripod.com.ve **E:** radiocumbrecrc@hotmail.com – **ME06)** Av 3, Esquina con Calle 22, Mérida 5101 ☎+58 274 263-0466 **E:** feloespinosa@cantv.net – **ME07)** Av Urdaneta, Edif La Huaca, PH, Mérida 5101 ☎ +58 274 263-1560 **W:** www.latinmail.com

MI00) MIRANDA
MI01) Calle Urdaneta N° 29, Edif Radio Valles del Tuy, P1, Ocumare del Tuy 1209 - 0930-0400 ☎ +58 239 225-0832 **E:** rvdt1030am@telcel. net ve – **MI02)** Edif Electricidad De Caracas, Semi Sotano, Frente a La Plaza Bolívar, Guarenas 1220 ☎ +58 212 362-1160 **E:** rvdt1030am@ telcel.net.ve – **MI03)** Calle Real Pantoja, al lado del Estadium De Barlovento, Caucagua 1246 ☎ +58 234 662-1045 – **MI04)** Final Av.Tosta Gracía, Resd.Boal, Mezz.2, Charallave 1200 - 1000-0300 ☎ +58 239 248-9984 **E:** yvtz1360@cantv – **MI05)** Av Concepción con Ricaurta N° 16, Guatire 1221 - 0930-0400 ☎ +58 212 341-4885 **E:** rbonita@cantv.net – **MI06)** Calle Rivas, Edif Centro Empresarial, Torre Chocolate, P7, Ofc.7-A y 7-B, Los Teques 1201 ☎ +58 212 364-8513 **W:** www.radiometropolitana.com **E:** metropolitana@hotmail.com

MO00) MONAGAS
MO01) Cra 5 N° 33, Antigua Calle Boyacá, Maturín 6201 ☎ +58 291 641-1083 – **MO02)** Av Bolívar,Edif.Radio Monagas, P1, Maturín 6201 ☎ +58 291 641-4495 **FM (MHz):** 93.5 – **MO03)** Calle Monagas, Edif Isnotú, PB, Maturín 6201 ☎ +58 291 641-3282 **E:** rdsn94 @cantv.net – **MO04)** Calle Sucre, Edif.Radio Maturín, PB, Maturín 6201 - 0900-0400 ☎ +58 291 641-1439 **E:** radiomaturin@cantv.net – **MO05)** Av Bolívar, Edif.Radio Dimensión, PB, Caripito 6211 - 1000-0300 ☎ +58 291 772-1082 **E:** mrosque@cantv.net

NE00) NUEVA ESPARTA
NE01) Calle La Marina, Edif.Sta Rita, Nivel 3, Porlamar 6301 ☎ +58 295 261-6656 **W:** www.radiovenezuela.com.ve **E:** radiovenezuela@ hotmail.com – **NE02)** Av Miranda, Edif.Best, P2, Porlamar 6301 - 1000-0400 ☎ +58 295 261-2857 – **NE03)** Calle Girardot, Urb.Cocheima, Edif Doña Teresa, P3, La Asunción 6311 - 1000-0500 ☎ +58 295 242-1383 – **NE04)** Av 4 de Mayo, Centro Comercial Real, Local 2, Porlamar 6301 - 1000-0400 ☎ +58 295 263-2626.

PO00) PORTUGUESA
PO01) Av 36-A, cruce con Av.Las Lagrimas, Qta.Acarigua, frente al Parque Mario Nerio, Acaragua 3301 ☎ +58 255 621-3182 **W:** www. radiovenezuela.com.ve **E:** radiovenezuela@hotmail.com – **PO02)** Cra 9, Esq Calle 15, Edif D'Zonno, P3, Apto 8, Guanare 3310 - 0900-0600 ☎ +58 257 251-0396 – **PO03)** Intercepción de la Autopista José António Páez con Carr.Nacional, Ospino 3319 - 1000-0400 ☎ +58 256 328-2077 – **PO04)** Av 35 con Calle 29, Edif Tricolor, Acarigua 3301 ☎ +58 255 621-2067 **W:** publiworldnet.com/radioacarigua **E:** radioacarigua@cantv.net – **PO05)** Av 24 entre Calles 5 y 6, Araure 3303 - 1000-0400 – **PO06)** Av Peñalver con Calle 31, Edif Los Andes, PB, Turén 3308 - 0900-0400 – **PO07)** Av Los Próceres, Urb.Francisco de Miranda, Edif.Radial, Guanare 3310 - 0900-0400 ☎ +58 257 251-5886 ☐ +58 257 253-5476 **E:** rotelpa@cantv.net

SU00) SUCRE
SU01) Av.Perimetral, Edif.Libertad, P2, Cumaná 6101 ☎ +58 293 433-2201 – **SU02)** Av.Gran Mariscal Sucre N° 30, Cumaná 6101 - 1000-0500 ☎ +58 293 451-2090 **W:** http://radiocontinente.tripod.com.ve/ – **SU03)** Av.Juncal, Edif.Siglo XX, P1, Locales B1 y B-2, Carúpano 6124 ☎ +58 294 332-1402 - 0900-0400 **W:** www.radiovenezuela.com.ve **E:** radiovenezuela@hotmail.com – **SU04)** Av Independencia 141, Edif Plaza, PB, Carúpano 6124 ☎ +58 294 331-0214 **W:** www.unionradio. com.ve **E:** solar1015fmvibracion@cantv.net – **SU05)** Av Santa Rosa 18, Sector La Copita, frente a la Iglesia Santa Rosa de Lima, Cumaná 6101 - 1000-0400 ☎ +58 293 432-2037 **E:** radio2000@cantv.net – **SU06)** Av Miranda, Qta.Tere, Cumaná 6101 ☎ +58 293 433-4432 **E:** feloespinosa@cantv.net

TA00) TACHIRA
TA01) Av 19 N° 13-61, Rubio 5030 - 1000-0400 ☎ +58 276 762-2263 – **TA02)** Av.1 de Mayo (Cra 3), Edif.Centro Civico, P7, San Antonio del Táchira 5007 ☎ +58 276 771-2083 – **TA03)** Calle 9 N° 8-16, San Cristóbal 5001 - 0900-0400 (SS – 0600) ☎ +58 276 341-4189 **E:** ecos1947@cantv.net – **TA04)** Av Fco de Cáceres 9-88, Qta Delia Mercedes, La Grita 5022 - 1000-0400 ☎ +58 277 881-2583 – **TA05)** Av Las Lomas, Edif.Primo Centro, Locales 3-12 y 3-13, San Cristóbal 5001 - 0900-0500 ☎ +58 276 341-7524 **E:** mundial860@cantv.net – **TA06)** Av 18 de Abril, Qta.Circuito Lider, San Cristóbal 5001 - 1000-0500 ☎ +58 276 347-4594 – **TA07)** Pasaje Acueducto N° 24-60, Barrio Obrero, San Cristóbal 5001 ☎ +58 276 355-0801 **W:** www.unionradio.com.ve/ – **TA08)** Calle 2, Edif Illinois, P2, La Fria 5020 ☎ +58 277 541-1667 – **TA09)** Av 19 de Abril con Av 8, La Concordia, San Cristóbal 5001 - 1000-0400 ☎ +58 276 344-6536 – **TA10)** Cra 4 N° 1-35, Táriba 5017 0950-0400 ☎ +58 276 394-0758 –**TA11)** Cra 9 cruce con Calle 9, Edif.El Ciclón, P4, San Cristóbal 5001 - 1000-0400 ☎ +58 276 356-7444

TR00) TRUJILLO
TR01) Av 11, entre Calles 12y13 N° 12-56, Valera 3101 ☎ +58 271 225-7835 **E:** simpatia@cantv.net – **TR02)** Av Bolívar con Calle 15, Edif. Grasso, P1, Valera 3101 - 0900-0400 ☎ +58 271 234-3755 – **TR03)** Av 10 entre Calles 9y10, Edif.Radio Valera, Local 9-31, Valera 3101 - 0900-0400 ☎ +58 271 225-3978 **W:** www.envalera.com/radiovalera **E:** radiovalera@envalera.com – **TR04)** Calle Independencia N° 10-11, Trujillo 3102 ☎ +58 272 236-3080 **W:** www.cnb.com.ve – **TR05)** Calle Iglesia, José Gregorio Hernández, Isnotu 3109 - 0900-0500 – **TR07)** CalleBolívar, Plaza la Alameda, Edif.Radio Jardín, Boconó 3103 ☎ +58 272 652-5574 **E:** radiojardin@cantv.net

VA00) VARGAS
VA01) Av Soublette, Edif Las Américas B, P16, Maiquetía 1161 - 1000-0400 ☎+58 212 332-5401 – **VA02)** 3era Norte Av Guaicaipuro, Quinta Mirna, Caracas ☎ radiomariavenezuela@hotmail.com

YA00) YARACUY
YA01) Av 10, Entre Calles 7y8, Edif.Alegría, Chivacoa 3202 ☎ +58 251 883-1692 – **YA02)** Prolongación 5ta Av.Urb.Andrés Eloy Blanco, Sector la Aduana, San Felipe 3201 ☎ +58 254 231-5055 ☐ +58 254 234.3453 **E:** radioyaracuym@cantv.net – **YA03)** Urb Las Tunitas, Av 4, entre Calles 4 y 5, Nirgua 3205 - 1000-0200 ☎ +58 254 572-2192 – **YA04)** Av.Cartagena, entre Calles 19 y 20, Edif.Radio San Felipe, San Felipe 3201 ☎ +58 254 231-5046.

ZU00) ZULIA
ZU01) Calle Central, Edif.Radio Perijá, P2, La Villa del Rosario 4047 - 0900-0400 ☎ +58 263 451-1158 – **ZU02)** Calle 64 Esq.Av 3e, Edif. La Voz de la Fe, Sector Don Bosco, Maracaibo 4002 (or: P.O.Box 459, Maracaibo 4002-A) ☎ +58 261 792-3712 **W:** www.aciprensa.com/ radio/ vene.htm – **ZU03)** Av El Muelle N° 1, Edif.Radio Libertad, frente a la Plaza Bolívar, Cabimas 4013 - 0900-0400 ☎ +58 264 241-1057 **W:** www.cnb.com.ve **E:** orbitafm@telcel.net.ve – **ZU04)** Av 11 N° 87-46, Edif 95.5, PB, Sector Veritas, Maracaibo 4002 - 1000-0400 ☎ +58 261 798-5674 – **ZU05)** Av 25 con Calle Paraíso N° 24-88, Maracaibo 4005 - 0900-0400 ☎ +58 261 759-1182 **W:** www.cnb.com.ve **E:** maracaibo@ cnb.com.ve – **ZU06)** Av 3-E N° 63-50, Sector Don Bosco, Maracaibo 4002 - 0900-0400 ☎ +58 261 791-0237 **W:** www.feyalegria.org/ve **E:** rfya850@cantv.net or irfamaracaibo@cantv.net or glombardi@cantv. net – **ZU07)** Calle 67 cruce con Av 27, detrás del Colegio La Epifanía, Sector Santa María, Maracaibo 4005 ☎ +58 261 752-8383 **W:** www. radiovenezuela.com.ve **E:** radiovenezuela@ hotmail.com – **ZU08)** Av Edif.Radio Calendario, Sector Grano de Oro, frente al Stadium Alejandro Borges, Maracaibo 4005 ☎ +58 261 783-4673 **W:** www.radioconti- nente.da.ru or http://radiocontinente.tripod.com.ve **E:** produccion@ radiocontinente.zzn.com – **ZU09)** Edif Radio Zulia, Av 23 con Calle 79 1 de Mayo,Maracaibo 4005 ☎ +58 261 752-9614 – **ZU10)** Av Gral Trias, Machiques 4021, Distrito Perijá - 0900-0500 – **ZU11)** Calle 74, Entre Av 3Dy3E, Edif.Televisa, Sector La Lago, Maracaibo 4002 ☎ +58 261 792-9217 – **ZU12)** Calle Manrique, Edif.Raquel, PB, Cd.Ojeda 4019 - 0900-0700 ☎ +58 265 631-3707 – **ZU13)** Av 5 N° 2-21, Sta Bárbara del Zulia 5148 - 1000-0300 ☎ +58 275 555-2400 – **ZU14)** Av 3H, Edif.Plaza, Local 2, Sierra Maestra, Maracaibo 4005 ☎ +58 261 792-1330 – **ZU15)** Calle 74, Entre Av 3Dy3E, Edif Televisa, Sector La Lago, Maracaibo 4002 (or: Av Andrés Bello, Edif Ambrosio, Cabimas 4013) (Office in Maracaibo, tx site Cabimas) - 1000-0500 ☎ +58 261 792-9217 – **ZU16)** Av 8 Esq Calle 73, N° 72-75, Edif Radiolandia, Sector Santa Rita, Maracaibo 4020 ☎ +58 261 797-1539 – **ZU17)** Av.Intercomunal Cabimas-Cd.Ojeda, con Calle La Planta, Cd.Ojeda 4019 - 0900-0500 ☎ +58 265 662-5093 – **ZU18)** Calle 27 con Av.12 N° 12-10, Edif.Camsa, Maracaibo 4002 (or: Calle Bello Lago, Santa Cruz de Mara 4005) (Office in Maracaibo, tx site Santa Cruz de Mara) ☎ +58 262 867-9400 – **ZU19)** Av El Terminal, Centro Comercial Nuevo Mundo, Local 2, Caja Seca 3156 - 0900-0400 ☎ +58 271 772-1087 – **ZU20)** Iglesia de María en Pentecostés, Urb.San Jacinto, primera entrada, El Moján, Maracaibo 4005 1030-0500 ☎ +58 261 757 3779 **E:** lumen

2000vzla@cantv.net or spev@cantv.net – **ZU21)** Edif.R.Marabina, Av 25 con Calle 67 N° 24-88, Sector Paraiso, Maracaibo 4002 – **ZU22)** Sector La Salina, Los Puertos de Altagracia 4036 - 0900-0500 ☎ +58 261 791-1712 – **ZU24)** Cra N Con Av.51, Zona Industrial, Cd.Ojeda 4019 ☎ +58 261 751-7540 – **ZU25)** Calle 13, Edif.La Linda, Sector Sierra Maestra, Municipio San Francisco, Maracaibo 4008 ☎ +58 261 735-4892

FM in Caracas (MHz): 88.1 Imagen – 88.9 Romántica – 89.7 X FM – 90.3 Unión Noti– 91.1 Nacional (classical) – 91.9 Avila – 92.9 Tu FM – 94.1 Hot 94 – 94.9 Alternativa – 95.5 Jazz – 96.3 Estrella – 96.9 X FM – 97.7 Em.Cultural – 99.1 Mágica – 99.9 Éxitos –100.7 Ateneo – 101.5 Kys – 101.9 Tiuna – *102.3 CNB –.7 Original –103.3 Sonorama – 103.9 R. Activa/Canal Juvenil (RNV) – 104.5 Rumbera – 105.3 Planeta – 105.9 Sonera – 106.5 Fiesta 106 – 106.9 Playa 107 – 107.3 La Mega Estación107 – 107.9 Onda

VIETNAM

L.T: UTC +7h — **Pop:** 89.6 million — **Pr.L:** Vietnamese, ethnic — **E.C:** 50Hz, 127/230V — **ITU:** VTN

DÀI TIÉNG NÓI VIÊT NAM
(VOV, RADIO THE VOICE OF VIETNAM) (Gov.)
✉ 58 Quan Su Str, Hanoi ☎ +84 (48) 254953 📠 +84 48255765
W: www.vov.org.vn **E:** qhqt.vov@hn.vnn.vn
L.P: DG: Mr. Vu Van Hien. Deputy DG (Int. Rel): Ms. Hoang Minh Nguyet. Deputy DG (Technical): Mr. Doan Viet Trung.

MW	kHz	Net	kW	Station, location, h of tr
1)	549	2	700	Hung Yen, (Site: My Hao): 2145-1700
1)	v558	2	50	Ho Chi Minh C., Quan Tre: 2145-1700
12)	576	2,P	50	Khanh Hoa, Nha Trang: 2145-1700
1)	594	1	50	Danang, (Site: An Hai): 2145-1700
2)	610	H	100	Ho Chi Minh City: 2130-1600
1)	630	1	200	Quang Binh, Dong Hoi: 2145-1700
1)	648	1	50	Binh Dinh, Quy Nhon, (Site: An Nhon): 2145-1700
14)	650	P	3	Bac Giang, Ha Bac: 1030-1100 (r. inactive)
1)	v657	1	50	Ho Chi Minh C., Quan Tre: 2145-1700
1)	666	1	50	Khanh Hoa, Nha Trang: 2145-1700
1)	675	1	700	Hung Yen, (Site: My Hao): 2145-1700
1)	690	1	55	Dac Lac, Buon Me Thuot: inactive
17)	702	2,O,D	50	Danang, (Site: An Hai): 2145-1700*
1)	711	1	500	Can Tho, Thoi Long: 2145-1700
1)	729	2	200	Quang Binh, Dong Hoi: 2145-1700
60)	740	2,P	50	Binh Dinh, Quy Nhon, (Site: An Nhon): 2145-1700
1)	v747	4		Ho Chi Minh C., Quan Tre: 2200-1330*
62)	750	P	7.5	Thai Nguyen: 2200-1530*
20)	756	P	10	Long An, Tan An: 2200-1100*
21)	765	P		Thai Binh: 0300-1250* (r. inactive)
22)	v782	P	15	Nghe An, Vinh: ?2200-?* (r. inactive)
1)	783	2	500	Can Tho, Thoi Long: 2145-1700
47)	819	4		Dac Lac, Buon Me Thuot: 2200-1600*
24)	828	P	50	Son La: 2200-1400*
26)	837	P	10	Can Tho: 2200-1200*
27)	846	P	10	Thanh Hoa: 0400-1030*
1)	873	3,4	500	Can Tho, Thoi Long: 2200-1600*
30)	891	1,P	10	Lam Dong, Da Lat: 2200-1600*
31)	v898	P	10	Ha Tinh: 2200-1200*
33)	900	1,P	10	Kon Tum: 2200-1600*
35)	909	P	10	Ca Mau: 2200-1330*
38)	v930	P	5	Ben Tre: 2155-1315*
39)	954	P	10	Vinh Long: 2200-1145*
41)	972	1,P	10	Quang Ngai: 2200-1600*
61)	981	1,P	10	Gia Lai. Pleiku
1)	981	1,4	10	Son La: 2200-1400*
36)	999	P	10	Lang Son
44)	1035	P		Quang Ninh, Halong: 2200-?* (r. inactive)
63)	1035	P	10	Hoa Binh: 2215-1100*
45)	v1050	P		Tra Vinh: 2230-1145*
46)	1089	P	10	Kien Giang, Rach Gia: 2200-1200*
37)	1089	P	10	Cao Bang
48)	1098	1,P	10	Binh Thuan (Phan Thiet): 2215-1100*
19)	1098	P	10	Thua Thien Hue, Hue: 2145-1310*
49)	1098	P	10	Lao Cai: 2200-1400*
50)	1116	P	1	Bac Lieu: 2200-1130*
13)	1125	P	10	Tay Ninh: 2200-1200*
53)	1170	P	10	An Giang, Long Xuyen: 2200-1330*
1)	1242	E	2000	Can Tho, Thoi Long: 1100-1600+
56)	1242	P	10	Soc Trang: 2200-?*

*) Split schedule, see Regional stns below for details. +) Recently r. on reduced power.

SW	kHz	Net	kW	Location, h. of tr
24)	v4740	P	1	Son La: inactive
1)	5925	2	50	Xuan Mai: 2145-1700
1)	5975	1	50	Hanoi: 2145-1700
1)	6020	4	20	Buon Me Thuot: 2200-1600
1)	6165	4	50	Xuan Mai: 2200-2300, 2330-2400, 1130-1400
28)	v6442	P	1	Dien Bien: r. inactive
1)	7210	1	20	Buon Me Thuot: 2145-1700
1)	7435	1	100	Son Tay: 2255-1200+
1)	9530	1	50	Xuan Mai: 2145-1700
1)	9635	1	100	Son Tay: 2145-1500+
1)	9850	4	50	Xuan Mai: 0400-0600
1)	9875	2	50	Hanoi: 0150v-1000
1)	11720	1	100	Son Tay: 2145-1200+

SW Stations: Hanoi 50kW (Me Tri, G.C: 105.47E 21.01N). Xuan Mai (also known as CK2) 15/50kW (GC: 105.33E 20.43N). Buon Me Thuot 2x20kW (G.C: 108.03E 12.41N). Son Tay (see International Broadcasting section). +) transmission for Gulf of Tonkin area.
Netw.: 1/2/3: Voice of Vietnam 1st/2nd/3rd national prgr – **4:** Voice of Vietnam minorities network – **D:** Radio & TV Danang – **E:** Voice of Vietnam external sces – **H:** Voice of the People of Ho Chi Minh City – **P:** Provincial sce – **Q:** Radio & TV Quang Nam (Hoi An, Quang Nam Province).
FM: Most services are also carried by numerous FM stns. Details for Hanoi and Ho Chi Minh City are shown here, others in the address section where known.

FM(MHz) Local	kW	VOV1	kW	VOV2	kW	VOV3	kW	VOV5	kW
1) Hanoi	90.0	100.0m		96.5		102.7t		105.5m	
2) HCMC	99.9	20	-99.1s			104.5q	10	105.7q	

1) ✉ 5 Huynh Thuc Khang, Dong Da District, Hanoi. m) Me Tri. q) Quan Tre. t) High power tx at Tam Dao in Vinh Phuc province. s) Sai Gon FM, operated by HTV (HCMC TV),
Prgrs. from Hanoi
VOV1, news & current affairs: 2145-1700. **FM:** Hanoi & Quang Nam 100.0MHz, Ha Giang 103.2MHz, Son La/Dien Bien Phu 104.3MHz, and also relayed in part by many regional sts. **N:** 2200, 2300, 0100, 0300, 0500, 1100, 1300. Some freqs relay VOV3 1700-2145.
VOV2, economic, social, cultural & education prgrs: 2145-1700. **FM:** Hanoi 96.5MHz, Buon Me Thuot 102.7MHz. **N:** 2200, 2300, 0200, 0400, 0600, 1100.
VOV3, news & music prgrs: 24h on **FM** (Quang Binh 105.1MHz, Vinh 98.0, Hue 106.1MHz, Danang 106.0MHz (5kW), Quang Nam 102.5MHz, Qui Nhon 103.1, Nha Trang 106.5MHz, Phu Yen 96.0, Dac Nong 96.6,Tay Ninh 101.0MHz and as above). Also on 873kHz 0000-0200, 1330-1600. Inc. **Xone FM:** music prgrs for young people, MF 2300-0100, 0900-1600, SS 0600-1600.
VOV4, ethnic language prgrs: Bana, Ede, Giarai, Hmong, K'Hor (Koho), Sedang, Thai, M'Nong: 2200-1600 on 819 & 6020kHz + FM for Central Highlands (incl relays of VOV1 in Vietnamese). **Dao, H'Mong (Ho Mong), Thai:** 2200-2300, 0000-0030, 1145-1400 on 981, 6165kHz, 0400-0530 on 981 & 9850kHz for Northern Vietnam. **Khmer (Kho Me), Cham, Vietnamese:** 2200-2400, 0200-0600, 0730-1330 on 747v & 873kHz for Ho Chi Minh City and the Mekong Delta.
VOV5, prgrs for foreigners: Hanoi 105.5MHz, Ho Chi Minh City 105.7MHz. **Cambodian:** 0800-0830. **Cantonese:** 0000-0030. **Chinese:** 0400-0430, 1100-1130. **English:** 0030-0130, 0500-0600, 0900-1030, 1200-1300, 1400-1500, 1600-1730. **French:** 0130-0230, 0600-0700, 1300-1330. **Indonesian:** 0730-0800. **Japanese:** 0430-0500, 1330-1400. **Lao:** 0700-0730. **Russian:** 0230-0300, 0830-0900. **Spanish:** 1030-1100. **Thai:** 1130-1200. **Vietnamese:** 0300-0400, 1500-1600.
VOV Traffic Channel: 2230-1800. **FM:** Hanoi 91.1MHz, Ho Chi Minh C. 91.0MHz.
Ann: "Dây là Tiéng Nói Viêt Nam, phát thanh tù Hà Nôi, thu dô nuóc Công Hòa Xá Hôi Chu Nghia Viêt Nam"". Khmer: "Thini Vithayu Samlang Vietnam".

Regional stations
General remarks: Schedules are partly based on monitoring. Further local txs are carried by many stns, especially in the 2200-2400, 0330-0630 and 0930-1400 periods, details unk. A few provincial stns also operate longer h. in daytime on Sun. mornings, especially in the Mekong Delta area. Many provincial stns also relay news from Hanoi, especially at 2300-2330, 0500-0545/0600 and 1100-1130/1145 if they are on air at those times. Relays of Hanoi as a rule are not included in the schedules below. Stns broadc. only on FM have been omitted.
Ann: Provincial services, usually identify as "Radio & TV (name of province)", in Vietnamese: "Dài Phàt Thanh Truyên Hình (name)".

Addresses and other information:
1) National freqs. See above for details – 2) 3 Nguyen Dinh Chieu,

Dist. 1, Ho Chi Minh City. H: 2130-1600 - **FM**: 99.9MHz. Districts: Binh Chanh (An Lac) 94.3MHz, Cu Chi 106.5MHz, Hoc Mon 93.6MHz, Nha Be 96.5MHz – **10)** Group 8, Tran Phu Ward, Ha Giang Town. P: r. 0300-0330v, 1230 – **11)** National Road 13, Phu Hoa Ward, Thu Dau Mot Town. Binh Duong Province. - **FM**: 92.5MHz – **12)** 70 Tran Phu, Nha Trang. P: ?2230-2200, 0430-0500, 1030-1100 – **13)** 188 Ward 3, 30/4 Rd, Tay Ninh. 2200-2400 (Sun 0230), 0400-0500, 1000-1200. - **FM**: 92.7MHz (P) – **14)** Nguyen Van Cu, Bac Giang Town – **17) Q:** Tran Phu Road, Tan Thanh Ward, Tam Ky Town, Quang Nam: 2220-2245, 0400-0430, 1145-1215. - **FM**: 97.6MHz (P), Hoi An 91.2MHz. **D:** 19 Le Loi, Da Nang. 2245-2300, 0430-0445, 1215-1315. - **FM**: 94.5 (P) – **18)** Dong Khoi Road, Group 3, Tam Hoa Ward, Bien Hoa City, Dong Nai Province. P: 2200-2300, 0430-0530, 1130-1230. - **FM**: 92.5MHz – **19)** 14 Ngo Quyen / 2A Tran Cao Van, Hue City. P: 2200-2300, 0400-0500, 0955-1100, 1200-1315. - **FM**: 93.0MHz (P) – **20)** 15/21 - 1st National Road, Tan An Town. P: 2200-0030, 0430-?, 1000-1100/1210 – **21)** 2 Le Loi St, Binh Town. P: 0300-0500, ? -1250 – **22)** 01 Nguyen Thi Minh Khai, Vinh City, Nghe An Province. P: 2230-2300, 0330-0500, 1030-1100 - **FM**: 103.1MHz (P), 98.0MHz (3) – **24)** Group 12, Quyet Thang Ward, Son La Town. P: 2200-2400, 0400-0600, 1200-1400 – **26)** 213 30 Thang 4 St, Can Tho City. P: 2200-2400, 0400-0600, 0900-1100 - **FM**: 97.3MHz (P, C) – **27)** 8 Hac Thanh St, Thanh Hoa City. P: 2200-2300, 0250-0600, 0930-1045 – **28)** 279 Muong Thanh Ward, Dien Bien Phu Town. P: 2200-2400, 0400-0600, 1200-1400 – **30)** 12 Tran Hung Dao, Da Lat City, Lam Dong Province. P: 2330-0030, 0400-0630, 0900-1100, 1230-1330 – **31)** 28 - Phan Dinh Phung, Ha Tinh Town. P: 2200-2330, 0400-0600, 1000-1400 - **FM**: 97MHz – **32)** Ngo Quyen Ward, Vinh Yen Town. - **FM**: 89.7MHz– **33)** 258A Phan Dinh Phung St, Kon Tum. P: 2215-2300, 2330-2400, 0345-0500, 1015-1100, 1145-1215 – **35)** 413 Nguyen Trai, Ward 9, Ca Mau Town. P: 2200-2400, 0400-0500, 1000-1330 – **36)** 9 Hoang Van Thu Rd, Lang Son Town. P: 1030-1100, 1130-1200 - **FM**: 88.2/101.0MHz – **37)** Be Van Dan Rd, Cao Bang Town. P: 0300-0500, 1200-1400, also VOV1 – **38)** Km 98/1 Tran Quoc Tuan, Ward 4, Ben Tre Province. P: 2155-2300, 0430-0545, 1200-1315. **FM:** Co Lach 88.5MHz – **39)** 50 Pham Thai Buong, Ward 4, Vinh Long Town, Vinh Long Province. P: 2200-2350(Sat 0215, Sun 0645), 0400-0645, 0900-1145. - **FM:** 90.2MHz. Districts (MHz): Vinh Long Township 93.5, Binh Minh 96.0, Long Ho 98.0, Mang Thit 95.5, Tam Binh 99.7, Tra On 93.2 – **41)** 165 Hung Vuong St, Quang Ngai Town. 1: 2200-2230, 2330-0030, 0500-0600, 1100-1600. P: 2230-2300, 0400-0500, 1030-1100 - **FM**: 89.6MHz (P) – **44)**1 Le Loi Road, Yet Kien Ward, Ha Long City, Quang Ninh Province. P: 2300, 0400 (occ. 0430)-0500, 1030-?- **FM:** 97.5MHz – **45)** 18 A Le Loi St, Tra Vinh Town. prgrs in Vietnamese/ Khmer: 2230-0015, 0430-0530, 1000-1145 - **FM:** 96.6MHz– **46)** Dong Da, Vinh Lac, Rach Gia Town. prgrs in Vietnamese, Khmer: 2230-2330(Sun 0230), 0400-0530, 0900-1145 – **47)** 1 Nguyen Chi Thanh, Buon Me Thuot, Dak Lak Province– **48)** Nguyen Tat Thanh Road, Phan Thiet City, Binh Thuan Province. P: 2215-2300, 0415-0500, 1000-1100 – **49)** 200 Hoang Lien Road, Coc Leu Ward, Lao Cai. P: 2200-2300, 0330-0430, 0945-1100, 1145-1400. - **FM**: 95.2MHz (P/1) – **50)** Tra Kha, Highway 1A, Ward 8, Bac Lieu. P: 2200-2300, 0955-1130. - **FM**: 93.8MHz – **53)** 45/1 Tran Hung Dao, Long Xuyen City, An Giang Province. P: 2200-0015 (Sun0145), 0330-0600, 0900-1330 in Vietnamese/Khmer – **FM**: Districts: Chau Doc 92.1MHz, An Phu 96.0MHz, Tinh Bien 107.1MHz – **54)** 32 To Hieu Road, Ha Dong Town, Ha Tay Province. - **FM**: 96.0MHz (P) – **56)** 357/1, Le Hong Phong, Soc Trang. P: 2200-2400(Sun 0205), 0400-0600, 0900-? – **60)** 181 Le Hong Phong Rd, Quy Nhon City. P: 2230-2300, 0430-0500, 1145-1230. - **FM:** 93.2MHz – **61)** 2 Hung Vuong Rd, Pleiku. 2215-2300, 0415-0500, 1015-1100. **FM:** 93.7MHz – **62)** 2200-2300, 0300-0530, 1230-1530 in Chinese/Vietnamese

VIRGIN ISLANDS (AMERICAN)

L.T: UTC -4h — **Pop:** 115,000 — **Pr.L:** English, Spanish, Creole — **E.C:** 60Hz, 110V — **ITU:** VIR

FEDERAL COMMUNICATIONS COMMISSION (FCC)
see USA for details

MW	Call	kHz	kW	MW	Call	kHz	kW
1)	WSTX	970	5/1	4)	WRRA	1290	0.5/0.3
2)	WVWI	1000	5/1	5)	WSTA	1340	1
3)	WGOD	1090	0.25	4)	WDHP	1620	10/1

FM	Call	MHz	kW	FM	Call	MHz	kW
6)	WIVH	90.1	1	11)	WMYP	98.3	1.9
7)	WXZT	90.9	10	12)	WVIQ	99.5	32
8)	WVVI	93.5	9.6	1)	WSTX	100.3	50
9)	WJKC	95.1	50	2)	WWKS	101.3	50
10)	WIVI	96.1	2.4	13)	WEVI	101.7	0.9
3)	WGOD	97.9	50	14)	WIUJ	102.9	1.5

FM	Call	MHz	kW	FM	Call	MHz	kW
4)	WAXJ	103.5	6	2)	WWJZ	105.3	30
15)	WZIN	104.3	44	17)	WVGN	107.3	1.7
16)	WMNG	104.9	6				

Addresses and other information
1) Fort Louise Augusta, St. Croix 00822 *or* P.O. Box 25680, Christiansted, St. Croix 0824-1580 ☎ +1 340 773-0390 Stns: WSTX-AM: News/sport/talk. WSTX-FM: reggae – **2)** Gark LLC, Box 8209, St. Thomas 00801 ☎ +1 340 776 5260 Stns: KISS 101.3 FM (A/C), 105 JAMZ (CHR/rythmic on 105.3) & Radio One (News/talk/sport – on 1000kHz) – **3)** 3AGN, Box 5012, Charlotte Amalie, St. Thomas 00803-5012 ☎ +1 340 774 4498 **W:** www.3abn.org Stn: '3ABN Radio' (Rlg.) – **4)** Reef Broadcasting Inc., 79A Castle Oakley, PO Box 755, Christiansted, St. Croix 00820 ☎ +1 340 772 1290 **W:** www.reefbroadcasting.com Stns: WRRA (Island Pop), WAXJ: News/talk & WDHP: Local Information/Music – **5)** Box 1340, Charlotte Amalie, St. Thomas 00801 ☎ +1 344 774 1340 🖳 +1 340 776 1316 **W:** www.wsta.com **Ann:** "Lucky 13" – **6)** Gospel Media Institute, Christiansted, St. Croix ☎ +1 340 778 2852 **Ann:** "West Indies Voice Of Hope" (Rlg.) – **7)** Christiansted, St, Croix – **8)** P.O. Box 25868, Christiansted, St. Croix 00824-1868 ☎ +1 340 773 5935 **Ann:** Caribbean Country – **9)** Isle 95, 5020 Anchors Way, Box 4084, Christiansted, St. Croix 00824 ☎ +1 340 773 0995 **W:** www.isle95.com **Ann:** "Isle 95" (Urban/reggae) – **10)** Box 2179, Charlotte Amalie, St. Thomas 00803 ☎ +1 340 776 9696 **Ann:** Hitz 96 – **11)** DBA J&J Broadcasters, PO Box 142755, Arecibo, 00614 PR. **Ann:** "Latino 98.3"(Spanish/Tropical/Variety) – **12)** JKC Communications, PO Box 4084, Christiansted, St. Croix 00822 ☎ +1 340 773 1180. **Ann:** "Sunny 99.5"(A/C) – **13)** Frontline Missions Int'l, PO Box 892, Christiansted 00821 – ☎ +1 340 719 9384 **W:** www.wevifm.org **Ann:** "Power 101.7 FM" (Rlg.) – **14)** Virgin Isl. Youth Development Radio, P.O.Box 2477, Charlotte Amalia, St. Thomas 00803 ☎ +1 340 776 1029 **W:** www.wiuj.com (Public/Educational) – **15)** PO Box 306117, Charlotte Amalie, St. Thomas VI00803 ☎ +1 340 776 1043 **Ann:** "The Buzz" (Active Rock) – **16)** Choice Communications Corp., Christiansted, St. Croix 00820 ☎ +1 340 713 9666 **Ann:** "Moongoose 104.9 FM" (Classic hits) – **17)** LKK Group Corp., PO Box 4084, Christiansted, St. Croix 00822 ☎ +1 800 275 6437 **W:** www.wvgn.org (Public).

VIRGIN ISLANDS (UK)

L.T: UTC -4h — **Pop:** 24,000 — **Pr.L:** English — **E.C:** 60Hz, 110V — **ITU:** VRG

VIRGIN ISLANDS BROADCASTING LTD. (Comm.)
🖳 P.O. Box 78, Road Town, Tortola, BVI ☎ +1 284 494 2250/2430/6994 🖳 +1 284 494 1139 **E:** zbvi@caribsurf.com **W:** www.zbviradio.com & www.zbvi.vi **L.P:** MD: Meritt Herbert. GM: Harvey Herbert. Ops Mgr: Mrs. Sandra Warrican. Production: Iris Jones
MW: ZBVI 780kHz 10kW
D.Prgr: MF 0930-0200, Sat 1000-0300, Sun 1100-0200 **N:** W on the h exc. 1000, 1700, 1900, 0100. Sun 1600, 2000. **Local N:** 1100, 2230. **Ann:** "This is ZBVI Radio from Tortola"

Other statsions:
ZJKC ISLE 95 FM: 90.9MHz 1kW (relay WJKC St.Croix USVI) – **ZKING**, Christian Broadcasting Network, PO Box 2993, Horse Parh, Road Town, Tortola ☎ +1 284 494 4600 **FM:** 104.9MHz – **ZROD**, P.O. Box 992, Road Town, Tortola, BVI ☎ +1 284 494 1037 🖳 +1 284 494 4564 **W:** www.zrodfm.com **L.P:** GM: Rodney Herbert. **FM:** 103.7MHz – **ZVCR**, P.O.Box 43, Road Town, Tortola BVI ☎ +1 284 494 7305 **W:** www.zvcr1069fm.com **FM:** 106.9MHz.

WALLIS & FUTUNA (France)

L.T: UTC +12h — **Pop:** 13,000 — **Pr.L:** French, Wallisian — **E.C:** 50Hz, 220V — **ITU:** WAL

RFO RADIO WALLIS ET FUTUNA
🖳 B.P.102, Pointe Matala, 98600 Mata-Utu, Iles de Wallis et Futuna (par Nouméa, Nouvelle-Calédonie) ☎ +681 681722020 🖳 +681 681722346 **W:** www.wallisfutuna.rfo.fr (live streaming) **L.P:** Dir. Gerard Christian Hoarau
FM(MHz): Sigave 90.6/Aio 91.0/Sigave 98.0/ Uvea 100.0/Uvea 101.0
D.Prgr: 24h **Ann:** "Bonjour, vous écoutez Radio Wallis et Futuna"
V. by letter

France Inter: 🖳 Mata-Utu, Uvea
FM: 103.0MHz **D.Prgr:** 24h satellite relay Paris in French

YEMEN

L.T: UTC +3h — **Pop:** 24 million — **Pr.L:** Arabic — **E.C:** 50Hz, 220/230V — **ITU:** YEM

MINISTRY OF INFORMATION
P.O. Box 19560, Al-Zubairy St, San'a ☎+967 1 215116/7/8 +967 1 207716 **W:** www.yemeninfo.gov.ye **E:** yemen-info@y.net.ye
L.P: Hassan al-Lawzi, Minister.

YEMEN RADIO & TV CORPORATION (YRTC) (Gov)
Tech. Dept, 26 September St, PO Box 2371, San'a ☎+967 1 282060 282053 **W:** www.yemenrt.com www.adenradio.net **E:** info@yemenradio.net, adenradio@yemen.net.ye
L.P: DG: Dr Abdullah Ali Al-Zalab. Tech. Dir: Mohammed H. Bather. Dir. Eng: Ali Ahmed Altashi. Head Tr. Station: Ismail Hussein Al-Nono.

MW	kHz	kW	Netw.	Times
Unknown location	602		G	-2300
San'a	711	200	G	0300-1900
Mukalla	765	50	G/L	1500-2300
Al-Hiswah	792	100	2	0300-0800,1100-2300
San'a	837	30	Y	0300-2300
Taiz	891		L	0300-2100
Hudaydah	909	750	G	1500-1900
San'a	v1008	600	G	1400-2100
Taiz	1071	30	G	0300-2300
Hudaydah	1125	50	L	0300-2300
Al-Hiswah	1188	100	G	0300-2300

FM: Ad-Dali 96.7MHz 5kW, Ad-Damigh 99.9MHz 5kW, Aden 99.0/102.5MHz (G/2), Al-Ashmur 92.6MHz 5kW, Hudaydah 90.4MHz, Mukalla 89.5/95.4MHz, Riam 92.4MHz 5kW, San'a 88.1/92.5MHz (G/2), 89.9/96.5MHz (Y).
G=General prgr. from San'a: 0300-2215. **English:** 1800-1900.
2=Second prgr. from Aden: 0300-0800, 1100-2130 (Fri 0255-2130). **English:** 1600-1630. **French:** 1705-1725.
Y=Youth prgr. on 837kHz and 89.9/96.5MHz + 3 other trs.
L=Local prgr; times vary by station, between 0600-2100, also rel. General prgr. Mukalla R. frequencies are shared by Sayun R.
IS: Flute. **Ann:** "Idha'atu-l-Jumhuriyah al-Yamaniyah min (town: San'a/Adan)". E: "This is Republic of Yemen Radio broadcasting from (town: San'a/Aden)".
Relays for abroad on shortwave: see International radio section

ZAMBIA

L.T: UTC +2h — **Pop:** 12 million — **Pr.L:** English, Bemba, Lozi, Lunda, Nyanja, Tonga, Chichewa, others — **E.C:** 50Hz, 230V — **ITU:** ZMB

COMMUNICATIONS AUTHORITY OF ZAMBIA (CAZ)
Plot 3141, Lumumba Rd, P.O. Box 36871, Lusaka ☎+260 21 241236/246702/246557/240463 +260 21 246701
W: www.caz.gov.zm **E:** info@caz.gov.zm

ZAMBIA NATIONAL BROADCASTING CORPORATION (ZNBC) (Pub)
P.O. Box 50015, Mass Media Complex, Alick Nkhata Rd, Lusaka 10101 ☎+260 21 254989 +260 21 254013 **W:** www.znbc.co.zm
L.P: Acting DG: Juliana Mwila. Dir. Tech. Sces: Edward H. Mwanza. Dir. Prgrs: Maxwell N'gandu. Regional Contr. (North): Reuben Kajokoto. Public Rel. Mgr: Mirriam Mtonga.

SW	kHz	kW	Sce.	Times
Lusaka	5915	100	R1	0245-2205
Lusaka	6165	100	R2	0245-2205

NB: Both transmitters rep. inactive.

FM(MHz)	R1	R2	R4	kW
Chipata	93.1	96.3		1
Kabwe			92.2	0.5
Kapirimoshi	97.8	88.2		1
Kasama	88.0	92.3		1
Kitwe	98.7	95.4	92.2	3/2
Livingstone			95.5	1/0.5
Lusaka	102.6	95.8	88.2	
Mansa	88.6	91.7		1
Mongu	95.1	91.9		1
Ndola			94.5	0.1
Senkobo	89.3	97.1		1
Solwezi	95.1	91.6		1

R. One in 7 Zambian languages: 0245-2205. **N. in English** (rel. R2): 0500, (Mon-Sat) 0600, 1115, 1800 — **R. Two in English:** 0245-2205.

N: 0400, 0500, 0600W, 0800W, 1000W, 1115MF, 1400MF, 1600W, 1800, 2000, 2100 — **R. Four (music channel) in English:** 0240-2205. Also rel. VOA.
F.PI: upgrading of FM network.
Ann: E: "This is Radio Two of ZNBC broadcasting from Lusaka". Chichewa: "Kuno ndi ku Zambia National Broadcasting Corporation wa Lusaka." **IS:** "Call of the Fish Eagle".

Other stations:
Breeze FM, Chipata: 99.6MHz. **W:** www.breezefm.com — **Chikuni Community R,** Monze: 91.8MHz 0.5kW. **W:** www.chikuniradio.org – **Choice FM,** Lusaka: 107.8MHz. **E:** choice@microlink.zm — **Five FM,** Lusaka: 105.1MHz — **Flava FM,** Kitwe: 96.4MHz. **W:** www.flavafm.co.zm – **Horn FM,** Lusaka: 94.2MHz — **R. Icengelo,** Kitwe: 89.1MHz – **R. Liambayi,** Mongu 101.9MHz — **R. Maria Zambia,** Chipata. FM: Kanjala 90MHz 0.3kW. **E:** info.zam@radiomar.org F.PI: another tr in Lusaka — **R. Musi O Tunya,** Livingstone: 106MHz — **R. Phoenix,** 12th Floor, ZIMCO House, Cairo Rd, Private Bag E702, Lusaka. **W:** www.radiophoenix.co.zm FM: Lusaka 89.5MHz, Kabwe 100MHz, Kitwe 100.5MHz, Chingola 104MHz, Kapiri/Mposhi 104.5MHz, Ndola/ Luanshya 107.6MHz – **R. Q-FM,** 15th floor Indeco House, P.O. Box 30896, Lusaka **W:** www.qfmradio.com FM: Lusaka 93.2MHz, Kabwe 96.7MHz, Choma 89.8MHz, Kitwe 90.0MHz, Mumbwa 89MHz, Namwala 90.6MHz — **Sky FM,** P.O. Box 31165, Plot 55, Luwato Rd, Roma, Lusaka. FM: Choma 88.8MHz, Zimba 93.8MHz, Monze 95.1MHz, Livingstone 102.4MHz, Lusaka 104MHz. **E:** skyfmbcast@ zamtel.zm — **Yatsani R,** P.O. Box 320147, Bauleni, Lusaka: 99.1MHz 2kW. **W:** www.yatsani.com
BBC African Sce: Lusaka 98.0MHz 2kW, Kitwe 102.8MHz 2kW.
RFI Afrique: Lusaka 100.4MHz 2kW, Kitwe 92.8MHz 1kW.
Christian Voice: FM: Lusaka 106.2MHz, Ndola 99.3MHz, Kitwe 105.8MHz. For SW see International Radio section

ZIMBABWE

L.T: UTC +2h — **Pop:** 11 million — **Pr.L:** English, Shona, Ndebele, Chewa — **E.C:** 50Hz, 220V — **ITU:** ZWE

BROADCASTING AUTHORITY OF ZIMBABWE (BAZ)
Block A, Emerald Park 30, The Chase, P.O. Box MP 843, Mt. Pleasant, Harare ☎+263 4 333032/48 +263 4 333041 **E:** baz@ comone.co.zw **L.P:** Acting Chairperson: Pikirai Deketeke.

ZIMBABWE BROADCASTING HOLDINGS (Gov)
P.O. Box HG 444, Broadcasting Centre, Pockets Hill, Highlands, Harare ☎+263 4 498610, SW station: ☎+263 4 22104 +263 4 498613 **W:** zbc.co.zw **L.P:** DG: Henry Muradzikwa. CEO: Happison Muchechetere. Head Radio & TV: Abigail Mvududu.

SW	kHz	kW	Prgr	Times	
Gweru	3396	100	2	1630-0530	
Gweru	6045	100	2	0530-1630	
FM (MHz)	**R1**	**R2**	**R3**	**R4**	**kW**
Beithbridge	-	98.1	-	105.2	-
Bulawayo	90.0	96.3	99.6	103.1	10
Chiredzi	93.3	95.5	98.8	102.3	-
Chivhu	93.3	96.5	103.3	106.8	-
Gokwe	-	96.8	89.6	103.5	-
Gwanda	105.8	95.4	98.7	102.2	-
Gweru	90.7	93.9	97.2	100.7	5
Harare	92.8	96.0	99.3	102.8	10
Hwange	91.5	98.2	94.7	-	-
Kadoma	88.5	94.8	98.1	101.6	10
Karoi	99.9	96.6	93.4	90.3	-
Kenmur	90.4	93.5	96.7	-	-
Lowveld	101.1	88.0	91.1	94.3	-
Masvingo	106.5	92.9	99.4	102.9	-
Mount Darwin	-	95.2	99.2	102.0	-
Mutare	105.3	89.1	98.7	105.8	3
Mutorashanga	104.7	94.3	91.1	101.1	-
Nyanga	105.5	91.7	94.9	101.7	-
Sabi/Chipinge	94.5	97.8	101.3	-	-
Victoria Falls	92.9	96.1	99.4	-	-

F.PI: community radio stations: Sunshine R. in Harare and Skies R. in Bulawayo.
1) S-FM: mainly in English: 24h — **2) R. Zimbabwe:** in Shona/ Ndebele/English: 24h — **3) Power FM:** youth programme in English: 24h. **N:** on the h. — **4) National FM:** in 14 minority languages: 24h.
Ann: In addition to programme names: "ZBC".

EXTERNAL SERVICE: Voice of Zimbabwe; see Int. radio section

INTERNATIONAL RADIO

Section Contents

Initial entries for each letter, see Main Index for full details.

NB: The copy deadline for this section was 10 November 2010

Features & Reviews

National Radio

International Radio

Frequency Lists

Terrestrial Television

Reference

NB: Country abbreviation codes are shown after the country name. The three-letter codes after each frequency are transmitter site codes. These, and the Area/Country codes in the Area column, can all be decoded by referring to the tables in the Reference Section. Where a frequency has an asterisk etc. after it, see the '**KEY**' section at the end of the schedule entry. The txs listed are the installed base for each transmission site and do not reflect any details of the txs being coupled, run at reduced power or remaining unused. Where **Webcast** is shown, the letter(s) after indicate the service(s) available: D=Audio on demand; L=Live audio; P=Podcast.

AFGHANISTAN (AFG)

RADIO AFGHANISTAN (Gov)
⌨ See National Radio section.
MW: [KAB] Kabul, Pol-e Charkhi: 1296kHz 400kW; [KHO] Khost, Tani: 612kHz 200kW (both operated by IBB).
Notes: Only MW infrastructure used by IBB listed.

ALASKA (ALS)

KNLS INTERNATIONAL (Rlg)
⌨ P.O. Box 473, Anchor Point, AK 99556, USA.
☎ +1 907 2352326.
E: knls@aol.com **W:** www.knls.org (English); www.knls.net (Russian); www.smzg.org (Mandarin).
Webcast: D
⌨ 605 Bradley Court, Franklin, TN 37067, USA. (Studio, WCBC)
☎ +1 615 3718707. 🖷 +1 615 3718791.
L.P: Pres, WCBC: Charles Caudill; Snr Producer (English): Rob Scobey; Chief Engineer: Kevin K. Chambers; Frequency Coordinator: Jeff Jaworski.
SW: [NLS] Anchor Point, AK: 2 x 100kW (1 in service, 1 awaiting possible decommission).
kHz: *6120, 9615, 9655, 9680*

Winter Schedule 2010/2011

English	Days	Area	kHz
1000-1100	daily	EAs	9615nls
1200-1300	daily	EAs	9615nls
1500-1600	daily	RUS	9655nls
Mandarin	**Days**	**Area**	**kHz**
0800-0900	daily	CHN	9655nls
0900-1000	daily	CHN	9655nls
1100-1200	daily	CHN	9615nls
1300-1400	daily	CHN	9680nls
1400-1500	daily	CHN	9615nls
Russian	**Days**	**Area**	**kHz**
1600-1800	daily	RUS	6120nls

Ann: English: "Broadcasting from the top of the world, this is KNLS, Anchor Point, Alaska, USA".
V: QSL-card. Rec. acc. (For email rpt, the subject should be: Reception Report)
Notes: On air since 23 Jul 1983. Owned by World Christian Broadcasting Inc. (WCBC). See USA for corporate details.

ALBANIA (ALB)

RADIO TIRANA (Pub)
⌨ Rruga Ismail Qemali 11, Tirana, Albania.
☎ +355 4 2223650. 🖷 +355 4 2223650.
E: radiotirana-english@hotmail.com; radiotirana@rtsh.al
W: www.rtsh.al
Webcast: L
L.P: Dir: Martin Leka; Dir, Technical: Agron Aranitasi; Head of RTV Monitoring: Mrs. Drita Çiço (drita.cico@yahoo.com).
MW: [FLA] Fllaka: 1215/1395/1458kHz 500kW.
SW: [SHI] Shijak: 2 x 100kW.
SAT: Eutelsat W2.
kHz: *1395, 1458, 5970, 6000, 6040, 6100, 6130, 7390, 7425, 7435, 7465, 7530, 9895, 11635, 13640*

Winter Schedule 2010/2011

Albanian	Days	Area	kHz
0000-0130	daily	NAm	6130shi, 7425shi
0730-0900	daily	Eu	1458fla, 7390shi
0900-1000	daily	Eu	1395fla, 7390shi
1500-1630	daily	Eu	1458fla
2130-2300	daily	Eu	5970shi
2130-2300	daily	NAm	7435shi
English	**Days**	**Area**	**kHz**
0130-0145	.twtfss	NAm	6130shi
0245-0300	.twtfss	NAm	6130shi
0330-0400	.twtfss	NAm	6100shi
0430-0500	.twtfss	NAm	6100shi
1530-1600	mtwtfs.	NAm	13640shi
1945-2000	mtwtfs.	NAm	11635shi
1945-2000	mtwtfs.	Eu	7465shi
2100-2130	mtwtfs.	NAm	9895shi
2100-2130	mtwtfs.	Eu	7530shi
French	**Days**	**Area**	**kHz**
1830-1900	mtwtfs.	Eu	7465shi
2000-2030	mtwtfs.	Eu	7465shi
German	**Days**	**Area**	**kHz**
1900-1930	mtwtfs.	Eu	1458fla
2030-2100	mtwtfs.	Eu	7465shi
Greek	**Days**	**Area**	**kHz**
1645-1700	mtwtfs.	Eu	1458fla
Italian	**Days**	**Area**	**kHz**
1800-1830	mtwtfs.	Eu	6000shi
2000-2030	mtwtfs.	Eu	6000shi
Serbian	**Days**	**Area**	**kHz**
1900-1915	mtwtfs.	Eu	6040shi
2115-2130	mtwtfs.	Eu	1458fla
Turkish	**Days**	**Area**	**kHz**
1930-2000	mtwtfs.	ME	1458fla

Ann: Albanian: "Radio Tirana per Bashkatdhetaret"; English: "This is Radio Tirana"; French: "Ici Tirana"; German: "Hier ist Radio Tirana"; Greek: "Sas milun ta Tirana"; Italian: "Parla Tirana"; Serbian: "Govori Tirana"; Turkish: "Burasi Tiran Radyosu".
V: QSL-card.
Notes: Radio Tirana is the External Sce of the public broadcaster Albanian Radio & TV (Radiotelevizioni Shqiptar).

CHINA RADIO INTERNATIONAL RELAY
L.P: Mgr: Han Shuhe.
SW: [CER] Cërrik, Shtermen: 6 x 150kW.
Notes: The Cërrik transmitting stn is owned by Radiotelevizioni Shqiptar (RTSH), and was leased to China Radio International for 15 years in 2003.

ALGERIA (ALG)

RADIO ALGÉRIENNE (Gov)
⌨ 21 Boulevard des Martyrs, 16000 Algiers, Algeria.
☎ +213 21483790. 🖷 +213 21230823.
E: technique@algerian-radio.dz **W:** www.radioalgerie.dz
Webcast: L
SAT: Atlantic Bird 3, Badr 6, Galaxy 25, Hot Bird 6, Nilesat 102, NSS 7.
kHz: *5865, 7295, 7455, 9390*

Winter Schedule 2010/2011

Arabic	Days	Area	kHz
0400-0600	daily	NAf,CAf	5865iss
0500-0600	daily	NAf,CAf	7295iss
0600-0700	daily	NAf,CAf	5865iss*, 7295iss**
1800-2000	daily	NAf,CAf	9390iss
1900-2200	daily	NAf,CAf	7455iss
2100-2300	daily	NAf,CAf	5865iss

Key: * Nov-Feb; ** Mar.
Ann: Arabic: "Huna Al-Djazair".
Notes: Relays of Home Sce 'Koran' prgr.

ANGOLA (AGL)

ANGOLAN NATIONAL RADIO (Gov)
⌨ C.P. 1329, Luanda, Angola.
☎ +244 22 2323172. 🖷 +244 22 2324647.
E: rna@rna.ao **W:** www.rna.ao
Webcast: L
L.P: DG: Eduardo Magalhães; Dir, Technical: Ale Fernandes.
MW: [MUL] Luanda, Mulenvos: 945kHz 25kW.
SW: [MUL] Luanda, Mulenvos: 1 x 15kW.

FM/DAB: FM: 101.4MHz (Luanda, 4kW).
kHz: *945, 7217*

Winter Schedule 2010/2011

English	Days	Area	kHz
2200-2300	daily	Af	945mul, 7217mul*

French	Days	Area	kHz
2100-2200	daily	Af	945mul, 7217mul*

Key: * Running on low power.
Ann: English: "This is Luanda, the International Service of the Angolan National Radio"; Portuguese: "Rádio Nacional de Angola".
V: QSL-card.

ANGUILLA (AIA)

CARIBBEAN BEACON (UNIVERSITY NETWORK RELAY)
⌨ See "University Network", under USA.
E: beacon@anguillanet.com
✉ P.O. Box 690, Anguilla, British West Indies.
☎ +1 809 4974340. 🖷 +1 809 4974311.
L.P: Chief Engineer: Kevon Mooney; Frequency Mgr: George McClintock.
MW: [AIA] The Valley: 690kHz 10kW, 1610kHz 25kW (run at 8/10kW).
SW: [AIA] The Valley: 1 x 100kW.
FM/DAB: FM: 100.1MHz (The Valley, 35kW).
V: QSL-card.
Notes: Carries programming from Melissa Scott's "University Network" (see main entry under USA for schedules).

ARGENTINA (ARG)

RADIODIFUSIÓN ARGENTINA AL EXTERIOR (RAE) (Pub)
⌨ Casilla de Correo 555, Correo Central, C1000WAF Buenos Aires, Argentina.
☎ +54 11 43256368. 🖷 +54 11 43259433.
E: rae@radionacional.gov.ar; argentinainternationalradio@gmail.com
W: www.radionacional.gov.ar
Webcast: L
L.P: Dir: Luis María Barassi.
SW: [BUE] Buenos Aires, General Pacheco: 2 x 50, 1 x 100kW.
kHz: *6060, 11710, 15345*

Winter Schedule 2010/2011

English	Days	Area	kHz
0200-0300	.twtfs.	Am	11710bue
1800-1900	mtwtf..	Eu	15345bue
French	**Days**	**Area**	**kHz**
0300-0400	.twtfs.	Am	11710bue
2000-2100	mtwtf..	Eu	15345bue
German	**Days**	**Area**	**kHz**
1700-1800	mtwtf..	Eu	15345bue
2100-2200	mtwtf..	Eu	15345bue
Italian	**Days**	**Area**	**kHz**
1900-2000	mtwtf..	Eu	15345bue
Japanese	**Days**	**Area**	**kHz**
0100-0200	.twtfs.	As	11710bue
1000-1100	mtwtf..	SAm	6060bue
1000-1100	mtwtf..	As	11710bue
Portuguese	**Days**	**Area**	**kHz**
0000-0100	mtwtf..	Am	11710bue
1100-1200	mtwtf..	Am	6060bue, 11710bue
Spanish	**Days**	**Area**	**kHz**
1200-1400	mtwtf..	Am	11710bue
1800-0300	.....s	Am	6060bue*, 15345bue*
2000-0230	.....s.	Am	6060bue*
2000-0230	.....s.	Eu	15345bue*
2200-2400	mtwtf..	Eu	15345bue
2200-2400	mtwtf..	Am	6060bue, 11710bue

Key: * Relay of RNA 870kHz.
Ann: English: "This is RAE, the International Service of the Argentine Radio", "RAE, Buenos Aires".
V: QSL-card.
Notes: RAE is the External Sce of the national public-service broadcaster Servicio Oficial de Radiodifusión (SOR). General Pacheco site

---

may close in the near future, due to possible environmental issues.

ARMENIA (ARM)

PUBLIC RADIO OF ARMENIA (FOREIGN SERVICE) (Pub)
⌨ A. Manoogian Street 5, 0025 Yerevan, Armenia.
☎ +374 10 558010. 🖷 +374 10 551513.
E: ak@arradio.am **W:** www.int.armradio.am
Webcast: L (www.armradio.am/live)
L.P: Dir, Foreign Service: Amasia Hovhannisyan.
MW/SW: Leased from Radio CJSC.
SAT: Eutelsat W7, Hot Bird 6.
kHz: *1314, 1395, 4810*

Winter Schedule 2010/2011

Arabic	Days	Area	kHz
1900-1930	daily	ME	4810yer
Assyrian	**Days**	**Area**	**kHz**
1530-1545	daily	ME	1395yer, 4810yer
Azeri	**Days**	**Area**	**kHz**
1345-1400	daily	ME	1314erv, 4810yer
1400-1415	mtwtf..	ME	1314erv, 4810yer
Farsi	**Days**	**Area**	**kHz**
1430-1500	daily	ME	1314erv, 4810yer
Kurdish	**Days**	**Area**	**kHz**
1315-1345	daily	ME	1314erv, 4810yer
1600-1630	daily	ME	1395yer
Turkish	**Days**	**Area**	**kHz**
1400-1415	.....ss	ME	1314erv, 4810yer
1415-1430	mtwtf..	ME	1314erv, 4810yer
Yezidi	**Days**	**Area**	**kHz**
1500-1530	daily	ME	1395yer, 4810yer

Ann: Arabic: "Huna Idha'at Jumhuriyat al-Yermaniyah min Yerevan".
V: QSL-card. Rp. (1 IRC)
Notes: Broadcasts may be changed or cancelled without notice. In some languages the prgrs identify as "Voice of Armenia".

RADIO CJSC (Tx Operator)
⌨ Noratus 3333, Armenia.
☎ +374 264 62640. 🖷 +374 264 30440.
E: info@radio-int.am **W:** www.radio-int.am
L.P: Dir: Gagik Aloyan.
MW: [ERV] Gavar, Noratus: 864/1350/1377kHz 1000kW; [YER] Yerevan, Arinj: 1395kHz 150kW.
SW: [ERV] Gavar, Noratus: 3 x 100, 3 x 1000kW; [YER] Yerevan, Arinj: 1 x 100kW.
Notes: Radio CJSC is the operator of high power transmitting centres in Armenia.

ASCENSION ISLAND (ASC)

BBC ATLANTIC RELAY STATION
⌨ English Bay, Ascension Island.
☎ +247 4458. 🖷 +247 6135.
E: niki.roy@atlantis.co.ac
L.P: Transmitter Engineer: Nicola Nicholls.
SW: [ASC] English Bay: 6 x 250kW.
V: QSL-letter. (For direct report)
Notes: Owned by the BBC and operated by Babcock International Group PLC (see under United Kingdom).

AUSTRALIA (AUS)

RADIO AUSTRALIA (Pub)
⌨ P.O. Box 428G, Melbourne, VIC 3001, Australia.
☎ +61 3 96261500. 🖷 +61 3 96261899.
E: english@ra.abc.net.au **W:** www.radioaustralia.net.au
Webcast: D/L/P
L.P: CEO: Dr Michael McCluskey; Senior Editor: Hanh Tran; Transmission Mgr: Nigel Holmes.
SW: Leased from Broadcast Australia.
SAT: Intelsat 5/10, NSS 6.
kHz: *5955, 5995, 6020, 6080, 7240, 9475, 9500, 9560, 9580, 9590, 9630, 9660, 9710, 9965, 11550, 11650, 11660, 11695, 11745, 11760, 11880, 11945, 12080, 13590, 13630, 13690, 15160, 15230, 15240,*

15290, 15350, 15415, 15515, 15560, 17585, 17715, 17750, 17795, 17840, 17845, 21725

Winter Schedule 2010/2011

Burmese	Days	Area	kHz
0100-0130	daily	SEA	17585hbn
2300-2330	daily	SEA	5955dha
Chinese	**Days**	**Area**	**kHz**
1300-1430	daily	EAs	9475shp, 9965hbn, 11660shp, 11760tsh
1600-1630	daily	EAs	9965hbn
English	**Days**	**Area**	**kHz**
0000-0200	daily	Pac	17715shp
0000-0800	daily	Pac	15240shp
0200-0500	daily	Pac	15515shp, 21725shp
0430-0500	daily	As	15415shp
0500-0800	daily	Pac	15160shp
0500-0900	daily	As	13630shp
0530-0600	daily	As	15415shp
0600-0630	.....ss	As	15290tnn
0600-0630	.....ss	SEA	15415shp
0630-0700	daily	As	15415shp
0700-0900	daily	Pac	9710shp
0700-1300	daily	As	9475shp
0700-1300	daily	Pac	11945shp
0800-0900	daily	Pac	5995brn
0800-1400	daily	Pac	9580shp
0800-1600	daily	Pac	9590shp
1100-1200	daily	Pac	5995brn, 12080brn+
1100-1300	daily	SEA	9965hbn
1100-1400	daily	Pac	6020shp, 9560shp
1200-1400	daily	Pac	5995brn+
1400-1700	daily	Pac	7240shp
1400-1800	daily	Pac	5995shp
1400-2000	daily	Pac	6080shp
1430-1700	daily	As	11660shp
1430-1900	daily	As	9475shp
1600-2000	daily	Pac	9710shp
1700-2000	daily	Pac	9580shp
1700-2100	daily	Pac	11880shp
1800-2000	daily	Pac	7240shp
1900-2200	daily	As	9500shp
2000-2100	.....ss	Pac	6080shp, 7240shp, 12080brn
2000-2200	daily	Pac	11650shp, 11660shp
2100-2200	daily	As,Pac	11695shp
2100-2200	daily	Pac	9660brn, 12080brn
2100-2300	daily	Pac	13630shp, 15515shp
2200-2300	daily	Pac	9660brn*, 12080brn*
2200-2330	daily	As	15240tnn
2200-2400	daily	Pac	15230shp, 15560shp
2200-2400	daily	SEA	13590hbn
2300-0200	daily	Pac	17795shp
2300-0700	daily	Pac	13690shp
2300-0800	daily	Pac	9660brn
2300-0900	daily	Pac	12080brn
2330-0400	daily	As	15415shp
2330-0700	daily	As	11750shp
Indonesian	**Days**	**Area**	**kHz**
0000-0030	daily	SEA	17840hbn
0000-0030	mtwtf..	SEA	13350tnn
0400-0430	daily	SEA	11550tnn, 15415shp, 17840hbn
0500-0530	daily	SEA	15415shp, 17845hbn
0500-0530	mtwtf..	SEA	11745tnn
0600-0630	mtwtf..	SEA	15290tnn, 15415shp
2200-2330	daily	As	15415shp
2200-2330	daily	As,Pac	11695shp
2200-2330	daily	SEA	9630dha, 11550tnn
Tok Pisin	**Days**	**Area**	**kHz**
0900-1100	daily	Pac	5995brn, 6020shp, 9710shp, 12080brn
2000-2100	mtwtf..	Pac	6080shp, 7240shp, 12080brn

Key: + DRM; * BBC World Service relay.
Ann: English: "This is Radio Australia broadcasting from studios in Melbourne, Victoria".
IS: "Waltzing Matilda" prior to opening, on all freqs. Foreign language broadcasts start with the laugh of the Kookaburra.
V: QSL-card.
Notes: Radio Australia is the External Sce of the public service Australian Broadcasting Corporation (ABC).

HCJB GLOBAL VOICE AUSTRALIA (Rlg)
✉ P.O. Box 291, Kilsyth, VIC 3137, Australia.
☎ +61 3 87208000. 📠 +61 3 87208020.
E: office@hcjb.org.au **W:** www.hcjb.org.au
Webcast: P
✉ 579 Packsaddle Rd, Kununurra, WA 6743. (Transmitter Site)
L.P: CEO: Dale Stagg; Frequency Mgr: Ian Williams; Transmitter Site Mgr: Peter Michalke.
SW: [KNX] Kununurra: 2 x 100kW (a further 2 x 100kW are on order).
kHz: *11750, 15340, 15400, 15525*

Winter Schedule 2010/2011

Amoy	Days	Area	kHz
0000-0030	mtwtf..	EAs	15525knx
1130-1200	mtwtf..	EAs	15400knx
Bhojpuri	**Days**	**Area**	**kHz**
0115-0130	...t...	As	15400knx
1315-1330	...t...	As	15340knx
Chhattisgarhi	**Days**	**Area**	**kHz**
0230-0245	.....ss	As	15400knx
1430-1445	.....ss	As	15340knx
English	**Days**	**Area**	**kHz**
0245-0300	daily	As,SEA	15400knx
0730-0930	daily	As	11750knx
1230-1300	mtwtf.s	EAs	15400knx
1300-1330	daily	EAs	15400knx
1445-1530	daily	As	15340knx
Gujarati	**Days**	**Area**	**kHz**
0115-0130	m......	As	15400knx
1315-1330	m......	As	15340knx
Hindi	**Days**	**Area**	**kHz**
0200-0230	daily	As	15400knx
Hindi/Rawang	**Days**	**Area**	**kHz**
1330-1400	daily	As	15340knx
Hmar	**Days**	**Area**	**kHz**
0230-0245	....f..	As	15400knx
1430-1445	....f..	As	15340knx
Indonesian	**Days**	**Area**	**kHz**
0000-0030	mtwtfs.	SEA	15400knx
1145-1200	daily	SEA	15340knx
1200-1230	mtwtfs.	SEA	15340knx
2345-2400	daily	SEA	15400knx
Japanese	**Days**	**Area**	**kHz**
2230-2300	....fs.	EAs	15525knx
Kurukh	**Days**	**Area**	**kHz**
0230-0245	m..t...	As	15400knx
1430-1445	m..t...	As	15340knx
Malay	**Days**	**Area**	**kHz**
1200-1230	......s	SEA	15340knx
Malayalam	**Days**	**Area**	**kHz**
0115-0130	......s	SEA	15400knx
1315-1330	......s	SEA	15340knx
Mandarin	**Days**	**Area**	**kHz**
0000-0030	.....ss	EAs	15525knx
1000-1130	daily	EAs	15400knx
1130-1200	.....ss	EAs	15400knx
1200-1230	daily	EAs	15400knx
1330-1400	daily	EAs	15400knx
2200-2230	daily	EAs	15525knx
2230-2300	mtwt..s	EAs	15525knx
2300-2400	daily	EAs	15525knx
Marathi	**Days**	**Area**	**kHz**
0115-0130	..w....	As	15400knx
1315-1330	..w....	As	15340knx
Marwari	**Days**	**Area**	**kHz**
0230-0245	.t.....	As	15400knx
1430-1445	.t.....	As	15340knx
Nepali	**Days**	**Area**	**kHz**
0100-0115	daily	As	15400knx
1300-1315	daily	As	15340knx
Punjabi	**Days**	**Area**	**kHz**
0115-0130	....f..	As	15400knx

Punjabi	Days	Area	kHz
1315-1330	...f..	As	15340knx
Rawang	**Days**	**Area**	**kHz**
0000-0030	......s	SEA	15400knx
0030-0100	daily	SEA	15400knx
Tamil	**Days**	**Area**	**kHz**
0115-0130	.....s.	As	15400knx
1315-1330	.....s.	As	15340knx
Telugu	**Days**	**Area**	**kHz**
0230-0245	..w....	As	15400knx
1430-1445	..w....	As	15340knx
Urdu	**Days**	**Area**	**kHz**
0115-0130	.t....	As	15400knx
0130-0200	daily	As	15400knx
1315-1330	.t.....	As	15340knx
1400-1430	daily	As	15340knx

Ann: English: "You are listening to HCJB Global Voice, Melbourne, Australia".
V: QSL-card.
Notes: HCJB Global branch and transmitting stn; for corporate details see under USA. Txs are to be moved to new site near the existing facility and is expected to be operational in early/mid 2011.

BROADCAST AUSTRALIA (Tx Operator)
P.O. Box 1212, Crows Nest, NSW 1585, Australia.
☎ +61 2 8113 4666. 🖷 +61 2 8113 4646.
E: info@broadcastaustralia.com.au
W: www.broadcastaustralia.com.au
Level 10, Tower A, 799 Pacific Highway, Chatswood, NSW 2067, Australia. (HQ)
LP: Chmn: Gerry Moriarty; Group CEO: Graeme Barclay.
SW: [BRN] Brandon: 3 x 10kW; [SHP] Shepparton: 7 x 100kW.
Notes: Broadcast Australia is the national transmitter network operator and is owned by Canada Pension Plan Investment Board (CPPIB). Has purchased 2 x100kW DRM-ready transmitters for use at Shepparton and Tenant Creek.

AUSTRIA (AUT)

RADIO Ö1 INTERNATIONAL (ORF) (Pub)
Argentinierstrasse 30a, A-1040 Wien, Austria.
☎ +43 1 5010116060. 🖷 +43 1 5010116066.
E: roi.service@orf.at **W:** oe1.orf.at
Webcast: D/L/P
LP: DG: Dr. Monika Lindner; Dir, Technical: Peter Moosmann.
SW: Uses txs provided by ORS.
SAT: Astra 1H.
kHz: *6155, 7325, 9840, 17855*

Winter Schedule 2010/2011

German	Days	Area	kHz
0000-0100	mtwtfs.	NAm,CAm	7325mos
0100-0130	mtwtfs.	SAm	9840mos
0600-0710	.....ss	Eu	6155mos
0600-0715	mtwtf..	Eu	6155mos
1300-1330	mtwtfs.	AUS,NZL	17855mos

Ann: English: "Radio Austria International".
IS: Blue Danube waltz.
V: QSL-card. Rec. acc.
Notes: Relays of ORF's domestic service "Ö1 Inforadio". Includes news in English/French at 0710-0715, Mon-Fri.

TWR EUROPE (Rlg)
Postfach 141, A-1235 Wien, Austria.
☎ +43 1 863120. 🖷 +43 1 8631220.
E: twre@twr.org **W:** www.twreurope.org
Other European branches: P.O. Box 176, 3780 BD Voorthuizen, The Netherlands; P.O. Box 12, 820 02 Bratislava 22, Slovakia.
SAT: Eurobird 1.
kHz: *864, 999, 1035, 1233, 1350, 1377, 1395, 1413, 1467, 1548, 5915, 6105, 6120, 7220, 7300, 9430, 9440, 9495, 9685, 9800, 11695*

Winter Schedule 2010/2011

Arabic	Days	Area	kHz
0300-0330	mtwtf..	ME,NAf	1233cgr
2000-2015	daily	ME	1377erv
2015-2030	.....s.	Eu	1395fla
2015-2030	m.wtfss	ME	1377erv
2025-2101	mt....s	ME,NAf	1233cgr
2025-2116	.....s.	ME,NAf	1233cgr

Arabic	Days	Area	kHz
2025-2216	..wtf..	ME,NAf	1233cgr
2030-2100	daily	ME	1377erv
2101-2116	.....s	ME,NAf	1233cgr
2101-2116	.t....	ME,NAf	1233cgr
2101-2116	m......	ME,NAf	1233cgr
2115-2130	...t..	NAf	1467rou
2116-2131	.....s.	ME,NAf	1233cgr
2116-2216	mt....s	ME,NAf	1233cgr
2130-2145	...t..	NAf	1467rou
2130-2145	...f..	NAf	1467rou
2130-2215	.....s	NAf	1467rou
2130-2245	.....s	NAf	1467rou
2130-2300	mtw....	NAf	1467rou
2131-2216	.....s.	ME,NAf	1233cgr
2145-2200	...t..	NAf	1467rou
2145-2300	...f..	NAf	1467rou
2200-2300	...t..	NAf	1467rou
2215-2230	.....s.	NAf	1467rou
2230-2245	.....s.	NAf	1467rou
2245-2300	.....s.	NAf	1467rou
2245-2300	.....s	NAf	1467rou
Armenian	**Days**	**Area**	**kHz**
1629-1659	mtwtf..	ME	9685wer, 11695wer
Assyrian	**Days**	**Area**	**kHz**
2015-2030	.t....	ME	1377erv
Belarusian	**Days**	**Area**	**kHz**
1500-1530	m......	Eu	7300wer, 9495mos
2030-2100	m......	Eu	999kch
Bosnian	**Days**	**Area**	**kHz**
2045-2100	.....s	Eu	1395fla
2100-2130	.....s	Eu	1395fla
Bulgarian	**Days**	**Area**	**kHz**
1800-1830	daily	Eu	1548kch
Croatian	**Days**	**Area**	**kHz**
2030-2045	....ss	Eu	1395fla
2030-2100	mtwtf..	Eu	1395fla
2100-2130	.....s.	Eu	1395fla
Czech	**Days**	**Area**	**kHz**
0700-0715	mt.tf..	Eu	5915mco, 7220mco
2015-2030	.....s.	Eu	1467rou
2015-2045	mtwtf.s	Eu	1467rou
English	**Days**	**Area**	**kHz**
0745-0850	.....s	Eu	6105nau, 9800mco
0800-0850	mtwtf..	Eu	6105nau, 9800mco
0815-0850	.....s.	Eu	6105nau, 9800mco
1445-1500	daily	CAs	1467bis
1840-1855	daily	ME	864erv
2300-2315	mtwtfs.	Eu	1467rou
2300-2345	.....s	Eu	1467rou
Farsi	**Days**	**Area**	**kHz**
1815-1830	mtwt.ss	ME	1377erv
1830-2000	daily	ME	1377erv
French	**Days**	**Area**	**kHz**
2030-2100	.....s.	NAf	1467rou
Hebrew	**Days**	**Area**	**kHz**
2000-2030	daily	ME	1350erv
Hungarian	**Days**	**Area**	**kHz**
0930-1000	daily	Eu	9430mco
1925-2000	daily	Eu	1395fla
Kabyle	**Days**	**Area**	**kHz**
2045-2115	mtwtf..	NAf	1467rou
2045-2130	......s	NAf	1467rou
2100-2130	.....s.	NAf	1467rou
Kalderash	**Days**	**Area**	**kHz**
1915-1945	mtwtf..	Eu	1548kch
Karakalpak	**Days**	**Area**	**kHz**
1755-1810	....ss	CAs	864erv
Kazakh	**Days**	**Area**	**kHz**
1500-1515	daily	CAs	1467bis
1715-1725	daily	CAs	1467bis
Kurmanji	**Days**	**Area**	**kHz**
1915-1930	daily	ME	1350erv
Kyrgyz	**Days**	**Area**	**kHz**
1545-1615	daily	CAs	1467bis
1700-1715	mt.....	CAs	1467bis
1745-1755	daily	CAs	1467bis

No.Caucasus languages	Days	Area	kHz
1825-1840	daily	RUS	864erv
Polish	**Days**	**Area**	**kHz**
0645-0700	daily	Eu	5915mco, 7220mco
2000-2015	.....s.	Eu	1395fla
2000-2030	mtwtf.s	Eu	1395fla
Qashqai	**Days**	**Area**	**kHz**
1815-1830	....f..	ME	1377erv
Romani (Balkan)	**Days**	**Area**	**kHz**
1830-1845	daily	Eu	1548kch
Romani (Vlax)	**Days**	**Area**	**kHz**
1915-1945	mtwtf..	Eu	1548kch
Romanian	**Days**	**Area**	**kHz**
1100-1130	.....s.	Eu	9440mco
1629-1659	.....s.	Eu	6120wer
1845-1915	daily	Eu	1548kch
1915-1945	.....ss	Eu	1548kch
Russian	**Days**	**Area**	**kHz**
0300-0500	daily	RUS	1035ttu
1500-1530	.twtf..	RUS	7300wer, 9495mos
1500-1600	.....ss	RUS	7300wer, 9495mos
1515-1530	daily	CAs	1467bis
1530-1545	mtwtf..	CAs	1467bis
1530-1545	.....ss	CAs	1467bis
1645-1700	daily	CAs	1467bis
1700-1800	m.w.f..	RUS	1035ttu
1755-1810	mtwtf..	CAs	864erv
1810-1825	daily	CAs	864erv
1900-2100	daily	RUS	1035ttu
2000-2030	....f..	CAs	1350erv
2000-2030	m......	RUS	999kch
2000-2100	.t....s	RUS	999kch
2015-2045	....s.	RUS	999kch
2030-2100	..wtf..	RUS	999kch
Serbian	**Days**	**Area**	**kHz**
0900-0930	.....s.	Eu	9430mco
1945-2000	daily	Eu	1548kch
2100-2130	mtwtf..	Eu	1395fla
Slovenian	**Days**	**Area**	**kHz**
2045-2100	.....s.	Eu	1395fla
Sorani	**Days**	**Area**	**kHz**
1855-1910	daily	ME	864erv
Sous/Tachelhit	**Days**	**Area**	**kHz**
2115-2130	....f..	NAf	1467rou
Tajik	**Days**	**Area**	**kHz**
1700-1715	..wtfss	CAs	1467bis
Tamazight	**Days**	**Area**	**kHz**
2115-2130	m.w....	NAf	1467rou
Tarifit	**Days**	**Area**	**kHz**
2115-2130	.t.....	NAf	1467rou
Turkmen	**Days**	**Area**	**kHz**
1725-1740	daily	CAs	864erv
Ukrainian	**Days**	**Area**	**kHz**
1830-1900	daily	Eu	1413kch
2000-2015	....s.	Eu	999kch
2000-2030	..wtf..	Eu	999kch
2045-2100	.....s.	Eu	999kch
Uyghur	**Days**	**Area**	**kHz**
1615-1645	daily	CAs	1467bis
Uzbek	**Days**	**Area**	**kHz**
1715-1745	daily	CAs	1467bis
1800-1815	daily	CAs	1377erv

Ann: English: "This is Trans World Radio. The following programme is in the ... language"; "This is TWR UK".
V: QSL-card.
Notes: TWR branch. Owned by TWR. For corporate details, see under USA.

ÖSTERREICHISCHE RUNDFUNKSENDER GMBH & CO KG (ORS) (Tx Operator)
✉ Würzburggasse 309, A-1136 Wien, Austria.
☎ +43 1 870400. 📠 +43 1 8704012773.
E: office@ors.at **W:** www.ors.at
L.P: CEO: Michael Wagenhofer, MD: Norbert Grill.
SW: [MOS] Moosbrunn: 4 x 100, 2 x 500kW.

Notes: ORS is the national transmitter network operator.

BAHRAIN (BHR)

RADIO BAHRAIN (Gov)
📺 See National Radio section.
Webcast: L (English Sce)
L.P: Dir, Technical Affairs: Abdulla Al-Baloushi.
SW: [ABH] Abu Hayan: 2 x 60kW; 1 x unknown low power (SSB).
SAT: Arabsat 3A.
kHz: 6010, 9745

Winter Schedule 2010/2011
Arabic	Days	Area	kHz
0000-2400	daily	ME	9745abh
English	**Days**	**Area**	**kHz**
0000-2400	daily	ME	6010abh

Ann: Arabic: "Idhaat al-Bahrain".
IS: Local composition, played on guitar and violin.
V: QSL-letter.
Notes: Relays of Home Sce prgrs.

BANGLADESH (BGD)

BANGLADESH BETAR (Pub)
✉ 121 Kazi Nazrul Islam Avenue, Shah Bagh, Dhaka-1000, Bangladesh.
☎ +880 2 8618119. (Ext. Sce) 📠 +880 2 8612012.
E: ts-betar@bdonline.com; betar.external@yahoo.com (Dir, Ext Sce) **W:** www.betar.org.bd
Webcast: D/L
☎ +880 2 8651083. (DG) 📠 +880 2 8612021. (DG)
L.P: DG: AKD Shamim Chowdhury; Deputy DG (Prog): A. S. M. S. Apel Mahmood; Dir, External Sce: Setab Uddin Ahmed.
SW: [DKA] Dhaka, Khabirpur: 2 x 250kW (1 transmitter in use).
Ann: English: "This is the External Service of Bangladesh Betar".
IS: Local composition, played on violin and tanpura.
V: QSL-card. (Rpt to Senior Engineer (Research Wing). Email rpt to: rrc@dhaka.net)
Notes: External service of the national public broadcaster Bangladesh Betar, which began broadcasting on 1 Jan 1972. No SW broadcasts at time of publication, but is expected to return in late 2011.

BELARUS (BLR)

BELARUSKAJE RADYJO (Gov)
📺 See National Radio section.
Webcast: L
kHz: 1170, 7255

Winter Schedule 2010/2011
Belarusian	Days	Area	kHz
0500-0800	daily	RUS	1170sas, 7255mns
1600-1800	daily	RUS	1170sas, 7255mns

V: QSL-card.
Notes: SW frequencies for domestic coverage: see National Radio section.

RADIO STATION BELARUS (Gov)
✉ Cyrvonaja Street 4, 220807 Minsk, Belarus.
☎ +375 17 2395852. 📠 +375 17 2848574.
E: radiostation-belarus@tvr.by **W:** www.radiobelarus.tvr.by
Webcast: L
L.P: Dir: Navum Halpiarovic; Head, Foreign Language Dept: Vjacaslaú Lakcjušyn.
MW/SW: Leased from Belaruski Radyjotelevizijny Peredajucy Centr.
FM/DAB: See National Radio section.
kHz: 1170, 6155, 7360, 7390

Winter Schedule 2010/2011
Belarusian	Days	Area	kHz
1200-1500	daily	Eu	7360mns, 7390mns
English	**Days**	**Area**	**kHz**
2100-2300	mtwtfs.	Eu	1170sas, 6155mns, 7360mns, 7390mns
2120-2300	......s	Eu	1170sas, 6155mns, 7360mns, 7390mns
French	**Days**	**Area**	**kHz**
2040-2100	......s	Eu	1170sas, 6155mns, 7360mns, 7390mns

German	Days	Area	kHz
1900-2040	......s	Eu	1170sas, 6155mns, 7360mns, 7390mns
1900-2100	mtwtfs.	Eu	1170sas, 6155mns, 7360mns, 7390mns

Polish	Days	Area	kHz
1700-1900	daily	Eu	7360mns, 7390mns
1805-1900	daily	Eu	6155mns

Russian	Days	Area	kHz
1500-1700	daily	Eu	7360mns, 7390mns
2300-2400	daily	Eu	6155mns, 7360mns, 7390mns

Spanish	Days	Area	kHz
2100-2120	......s	Eu	1170sas, 6155mns, 7360mns, 7390mns

Ann: Belarusian: "Havoryc Radyjostancyja Belarus"; English: "This is Radio Station Belarus", "You are listening to Radio Station Belarus"; German: "Hier ist die Radiostation Belarus".
V: QSL-card.
Notes: Radio Station Belarus is the External Sce of the National State Radio-TV Company of Belarus.

BELARUSKAJE RADIOTELEVIZIJNY PEREDAJUCY CENTR (Tx Operator)
✉ vul. Engelsa 22, 220030 Minsk, Belarus.
☎ +375 17 2270845. 🖷 +375 17 2271084.
E: inbox@brtpc.by **W:** www.brtpc.by
L.P: Dir: Andrej Karaim.
MW: [SAS] Sasnovy: 1170kHz 1000kW (run at 700kW).
SW: [MNS] Minsk, Kalodziščy: 1 x 75, 1 x 150, 1 x 250kW.
Notes: Belaruskaje Radyjotelevizijny Peredajucy Centr, a subsidiary of the Ministry of Post & Telecommunications, is the national transmitter network operator.

BELGIUM (BEL)

RADIO VLAANDEREN INTERNATIONAAL (RVI) (Pub)
✉ P.O. Box 26, B-1043 Bruxelles, Belgium.
☎ +32 2 7413807. 🖷 +32 2 7416295.
E: info@rvi.be **W:** www.rvi.be
Webcast: L
L.P: Net Mgr: Jan Kudde.
MW: Leased from Norkring België.
SAT: Astra 1H/4A, Hot Bird 8, Intelsat 7, Optus D2, SatMex 6, SES 1, Thaicom 5.
kHz: 927

	Winter Schedule 2010/2011		
Dutch	Days	Area	kHz
0500-2205	daily	Eu	927wol

Ann: Dutch: "Dit is Radio Vlaanderen Internationaal". (or ID for domestic channel)
IS: Opening: "Tussen Maas en Schelde".
V: QSL-card.
Notes: RVi is the External Sce of the public broadcaster Vlaamse Radio en Televisie (VRT). It consist of two channels (R. Vlaanderen & R. Vlaanderen Info); the MW frequency carries R. Vlaanderen Info (mainly a relay of domestic VRT channel Radio 1).

RTBF INTERNATIONAL (Pub)
✉ Local 3P09, 52 Bd Reyers, B-1044 Bruxelles, Belgium.
☎ +32 2 7374014. 🖷 +32 2 7373032.
E: rtbfi@rtbf.be **W:** www.rtbfi.be
Webcast: D/L/P
L.P: Dir: Gerard Delacroix.
MW: [WAV] Wavre: 612kHz 300kW.
FM/DAB: FM: 99.2MHz (Kinshasa).
SAT: Atlantic Bird 3.
kHz: 621

	Winter Schedule 2010/2011		
French	Days	Area	kHz
0400-2315	mtwtf..	WEu	621wav
0500-2315	......s	WEu	621wav
0500-2400	.....s.	WEu	621wav

V: Email only. Does not send QSL-cards via regular mail.
Notes: Transmissions are relays of RTBF Home Sce prgrs.

RTR RADIO TRAUMLAND (Comm)
✉ P.O. Box 15, B-4730 Raeren, Belgium.
☎ +32 87 301722.

E: rtr-radio@skynet.be **W:** www.rtr-radio.de
✉ Postfach 1142, D-52157 Roetgen, Germany.
kHz: 6180

	Winter Schedule 2010/2011		
German	Days	Area	kHz
1300-1500	......s	Eu	6180wer

V: QSL-card. Rp. (1 Euro, 1 USD or 1 IRC)
Notes: RTR Radio Traumland promotes holiday destinations and hotels for a German speaking audience.

TDPRADIO
✉ P.O. Box 1, B-2310 Rijkevorsel, Belgium.
☎ +32 33 147800. 🖷 +32 33 141212.
E: daniel@tdpradio.com; ludo@tdpradio.com
W: www.tdpradio.com
Webcast: L
L.P: PM: Daniël Versmissen; Technical Mgr: Ludo Maes.
kHz: 6015, 15755

	Winter Schedule 2010/2011		
English	Days	Area	kHz
0700-0800	m......	Eu	6015iss+
0800-0900	.t.....	Eu	6015iss+
0900-1000	..w....	Eu	6015iss+
1000-1100	...t...	Eu	6015iss+
1100-1200	....f..	Eu	6015iss+
1200-1300	.....s.	Eu	6015iss+
1300-1400	......s	Eu	6015iss+
2100-2200	daily	NAm	15755bon+

Key: + DRM.
V: QSL-card.
Notes: TDPradio broadcasts dance music mixes by Belgian Club DJs.

TDP (Broker)
✉ c/o Ludo Maes, P.O. Box 1, B-2310 Rijkevorsel, Belgium.
☎ +32 33 147800. 🖷 +32 33 141212.
E: info@transmitter.org **W:** www.airtime.be
L.P: Owner: Ludo Maes.
V: QSL-card (For brokered transmissions). Rp.
Notes: TDP brokers air time for prgrs with political or religious background.

NORKRING BELGIË (Tx Operator)
✉ Jules Bordetlaan 160, B-1140 Evere, Belgium.
☎ +32 2 3639900. 🖷 +32 2 7454537.
E: info@norkring.be **W:** www.norkring.be
L.P: CEO: Bart Bosmans.
MW: [WOL] Wolvertem: 1512kHz 300kW.
Notes: Norkring België, a subsidiary of Norkring (Norway), is a national transmitter operator.

BENIN (BEN)

TWR RELAY STATION
✉ B.P. 1039, Parakou, Benin.
L.P: Chief Engineer: Paul Cox.
MW: [PAR] Parakou: 1566kHz 100kW.
SW: [PAR] Parakou: 1 x 100kW (Planned).
Notes: Owned by TWR. For corporate details, see under USA. For schedule, see TWR Africa, under South Africa. SW transmitting station planned - licensing, funding and construction permitting.

BONAIRE (ATN)

TWR BONAIRE (Rlg)
✉ P.O. Box 388, Bonaire.
☎ +599 7178800. 🖷 +599 7178808.
E: 800am@twr.org **W:** www.twr.org/americas.html
MW: [TWB] Bonaire, Belnem: 800kHz 100kW.
FM/DAB: FM: 89.5MHz (Bonaire).
kHz: 800

	Winter Schedule 2010/2011		
Baniwa	Days	Area	kHz
0845-0900	.....s.	CAm,SAm	800twb
English	Days	Area	kHz
0200-0400	daily	CAm,SAm	800twb
Kreyol	Days	Area	kHz
0000-0030	daily	CAm,SAm	800twb
Macuxi	Days	Area	kHz
0845-0900	......s	CAm,SAm	800twb

Portuguese	Days	Area	kHz
0745-0800	daily	CAm,SAm	800twb
0845-0900	mtwtf..	CAm,SAm	800twb
2300-2400	daily	CAm,SAm	800twb
Spanish	**Days**	**Area**	**kHz**
0000-0200	daily	CAm,SAm	800twb
0900-1215	daily	CAm,SAm	800twb
1600-1800	daily	CAm,SAm	800twb
2100-2300	daily	CAm,SAm	800twb

Ann: English: "This is the international sound of the Caribbean, Trans World Radio, Bonaire".
V: QSL-card.
Notes: TWR branch. Owned by TWR. For corporate details, see under USA.

RADIO NETHERLANDS WORLDWIDE RELAY STATION
✉ P.O. Box 45, Kralendijk, Bonaire.
SW: [BON] Bonaire, Tolo: 3 x 250kW.
V: QSL-card. (For direct rpt)
Notes: Owned by Radio Netherlands Worldwide.

BOSNIA AND HERZEGOVINA (BIH)

INTERNATIONAL RADIO SERBIA RELAY STATION
✉ KTCB, 76300 Bijeljina, Bosnia & Herzegovina.
SW: [BIJ] Bijeljina, Jabanuša: 2 x 500kW.
Notes: Owned and operated by International Radio Serbia.

BOTSWANA (BOT)

IBB RELAY STATION BOTSWANA
✉ IBB Transmitting Station, Private Bag 38, Selebi-Phikwe, Botswana.
☎ +267 2610932. 🖷 +267 2610185.
LP: SM: Thomas Powell.
MW: [SEL] Selebi-Phikwe: 909kHz 600kW (Backup: 50kW).
SW: [BOT] Moepeng Hill: 4 x 100kW.

BRAZIL (B)

EMPRESA BRASIL DE COMUNICAÇÃO (EBC) (Gov)
✉ SCRN 702 / 3 Block B, Brasilia DF CEP 70323-900, Brasil.
☎ +55 61 37995200.
E: cao@radiomec.com.br **W:** www.ebc.com.br
LP: Pres: Ima Célia Guimarães Vieira.
SW: [BRA] Brasilia, Rodeador Park: 1 x 250kW.
V: QSL-card.
Notes: Leases air time to other broadcasters, such as CRI.

BULGARIA (BUL)

RADIO BULGARIA (Pub)
✉ P.O. Box 900, 1000 Sofia, Bulgaria.
☎ +359 2 9336633. 🖷 +359 2 8650560.
E: radiobulgaria@bnr.bg **W:** www.bnr.bg
Webcast: D
🏢 4 Dragan Tsankov Blvd., 1040 Sofia, Bulgaria. (Studio)
☎ +359 2 9336 505. (English Sce)
LP: PD: Valentin Stoyanov; Frequency Mgr: Ivo Ivanov.
MW/SW: Leased from NURTS Bulgaria.
kHz: 747, 1224, 5900, 6200, 7300, 7400, 9400, 9700, 9800, 11700, 11900, 15700

Winter Schedule 2010/2011

Albanian	Days	Area	kHz
0630-0700	mtwtf..	Eu	747pet, 1224vdn
0700-0800	.....ss	Eu	747pet, 1224vdn
1700-1730	daily	Eu	747pet, 1224vdn
2000-2100	daily	Eu	747pet, 1224vdn
Bulgarian	**Days**	**Area**	**kHz**
0100-0200	daily	NAm	5900pld, 7400pld
0100-0200	daily	SAm	6200pld, 7300pld
0500-0600	.....ss	Eu	747pet, 1224vdn, 5900pld, 5900sof, 7400pld, 7400sof
0500-0800	....f..	Eu	9400sof*,+
0530-0600	mtwtf..	Eu	747pet, 1224vdn, 5900pld, 5900sof, 7400pld, 7400sof

Bulgarian	Days	Area	kHz
0700-1000	.....ss	Eu	11900sof*,+
1000-1300	mtwt...	Eu	11900sof*,+
1400-1500	daily	Eu	5900sof, 7400sof, 11700pld, 15700pld
1600-1700	daily	Eu	747pet, 1224vdn, 5900sof, 7400sof
1600-1700	daily	ME,CAs	5900pld, 7400pld
1600-1700	daily	SAf	15700pld
1900-2000	daily	Eu	1224vdn, 1224vdn, 9700sof+
1900-2000	daily	Eu,ME	5900pld
2200-2300	daily	Eu	5900pld, 7300pld
English	**Days**	**Area**	**kHz**
0000-0100	daily	NAm	5900pld, 7400pld
0300-0400	daily	NAm	5900pld, 7400pld
0730-0800	daily	Eu	5900pld, 7400pld
1000-1030	.....ss	Eu	11900sof+
1830-1900	daily	Eu	6200pld, 7400pld, 9700sof+
2200-2300	daily	Eu	6200pld, 7400pld
French	**Days**	**Area**	**kHz**
0200-0300	daily	NAm	5900pld, 7400pld
0700-0730	daily	Eu	5900pld, 7400pld
1800-1830	daily	Eu	6200pld, 7400pld, 9700sof+
2100-2200	daily	Eu	6200pld, 7400pld
German	**Days**	**Area**	**kHz**
0630-0700	daily	Eu	5900pld, 7400pld
1730-1800	daily	Eu	6200pld, 7400pld, 9700sof+
2000-2100	daily	Eu	6200pld, 7400pld
Greek	**Days**	**Area**	**kHz**
0600-0630	mtwtf..	Eu	747pet, 1224vdn
0600-0700	.....ss	Eu	747pet, 1224vdn
1730-1800	daily	Eu	747pet, 1224vdn
2100-2200	daily	Eu	747pet, 1224vdn
Russian	**Days**	**Area**	**kHz**
0400-0500	daily	CAs	5900pld, 7400pld
0400-0500	daily	Eu	5900sof, 7400sof
0600-0630	daily	Eu	5900sof, 7400sof
1500-1600	daily	CAs	5900pld, 7400pld
1500-1600	daily	Eu	5900sof, 7400sof
1630-1700	daily	Eu	9800sof+
1700-1730	daily	Eu	5900sof, 7400sof
1900-2000	daily	Eu	6200sof, 7400sof
Serbian	**Days**	**Area**	**kHz**
0700-0730	mtwtf..	Eu	747pet, 1224vdn
0800-0900	.....ss	Eu	747pet, 1224vdn
1800-1830	daily	Eu	747pet, 1224vdn
2200-2300	daily	Eu	747pet, 1224vdn
Spanish	**Days**	**Area**	**kHz**
0000-0100	daily	SAm	6200pld, 7300pld
0200-0300	daily	CAm	6200pld
0200-0300	daily	SAm	6200pld, 7300pld
0700-0730	daily	Eu	6200pld, 7300pld
1730-1800	daily	Eu	5900pld, 9400pld
2100-2200	daily	Eu	5900pld, 7300pld
Turkish	**Days**	**Area**	**kHz**
0600-0630	daily	ME	6200pld, 7300pld
1830-1900	daily	ME	747pet, 1224vdn, 5900pld

Key: + DRM; * Relay of domestic service (Horizont).
Ann: Albanian: "Ju flet Radio Bulgaria"; Bulgarian: "Tuk e Radio Bâlgarija"; English: "This is Radio Bulgaria"; French: "Ici Radio Bulgarie"; German: "Hier spricht Radio Bulgarien"; Greek: "Akute ti Vulgariki Radiofonia"; Russian: "V efire Radio Balgaria"; Serbian: "Radio Bugarska"; Spanish: "Ésta es Radio Bulgaria"; Turkish: "Burasi Bulgaristan Radyosu".
IS: Opening: The first music phrase from "Bulgarian Suite" for orchestra, by Pancho Vladiguerov.
V: QSL-card. Rp (1 IRC). Rec acc.
Notes: Radio Bulgaria is the External Sce of the public-service Bulgarian National Radio (Balgarsko Natsionalno Radio).

SPACELINE LTD (Broker)
P.O. Box 812, 1000 Sofia, Bulgaria.
+359 2 9652270. +359 73 832269.
E: broadcast@bitex.com
L.P: GM: Dimitar Todorov.
Notes: Spaceline Ltd brokers air time for the SW facilities of the transmitter operator NURTS, in Sofia and Plovdiv.

NURTS BULGARIA (Tx Operator)
bul. Peyo K. Yavorov 2, 1164 Sofia, Bulgaria.
+359 2 8613700. +359 2 9515550.
E: central.office@btc.bg **W:** www.btc.bg
L.P: Dir: Svetoslav Tanev Zhelev.
MW: [PET] Petrich, Novo Konomladi: 747kHz 500kW; [VDN] Vidin, Vodna: 1224kHz 500kW.
SW: [PLD] Plovdiv, Padarsko: 3 x 250, 2 x 500kW; [SOF] Sofia, Kostinbrod: 1 x 15, 4 x 50, 3 x 100, 1 x 250kW.
Notes: NURTS is the Bulgarian national transmitter operator, a joint-venture of Vivacom (Bulgaria) and Mancelord Ltd (Cyprus).

CANADA (CAN)

RADIO CANADA INTERNATIONAL (RCI) (Pub)
1400, boulevard René-Lévesque Est, Montréal, Québec, Canada H2L 2M2.
+1 514 5977500. +1 514 5977760.
E: info@rcinet.ca **W:** www.rcinet.ca
Webcast: D/L/P
L.P: Mgr: Jean Larin; French and English prgrs: Elzbieta Olechowska; Foreign language prgrs: Roger Tetrault; Engineering: Jaques Bouliane; Audience Relations: Bill Westenhaver; Schedules: Steve Lemay.
SW: [SAC] Sackville, NB: 3 x 100, 6 x 250kW.
SAT: Hot Bird 6.
kHz: 5965, 5995, 6025, 6100, 6160, 7265, 7325, 9490, 9555, 9560, 9565, 9570, 9610, 9635, 9670, 9740, 9755, 9770, 9785, 9800, 9830, 9880, 11785, 11845, 11865, 11935, 11975, 11990, 12015, 13650, 13700, 15305, 15365, 17765, 17790

Winter Schedule 2010/2011

Arabic	Days	Area	kHz
0300-0400	daily	ME	6025smg
0400-0500	daily	ME	5995smg, 7265skn
1205-1305	daily	NAm	7325sac
2000-2100	daily	ME,NAf	11865sac, 13650sac
2005-2105	daily	NAm	9610sac

English	Days	Area	kHz
0000-0100	daily	SEA	9880kun
0005-0105	.twtfs.	NAm	9755sac
0105-0205	daily	NAm	9755sac
1500-1600	daily	SAs	9635xia, 11975uru
1605-1805	daily	NAm	9610sac, 9800sac+
1800-1900	daily	NAf,WAf	17790sac
1800-1900	daily	NAf,EAf	11845smg
1800-1900	daily	CAf,NAf	15365sac
1800-1900	daily	EAf,ME	9740kas
2200-2300	daily	NAm	9800sac+

French	Days	Area	kHz
1805-2005	daily	NAm	9610sac, 9800sac+
1900-2000	daily	Af	9670skn, 9770kas, 11845skn, 13650sac, 15305sac, 17790sac
2100-2200	daily	NAf	11845sac
2100-2200	daily	WAf	15365sac
2105-2205	daily	NAm	6100sac
2300-2330	daily	FE	6160kim

Mandarin	Days	Area	kHz
0000-0100	daily	FE	9565kim, 11785kim, 12015pht
0205-0305	daily	NAm	9755sac
1100-1200	daily	FE	9490pht, 9570kim
1405-1505	daily	NAm	9610sac
1500-1600	daily	FE	5965yam, 9560yam
2200-2300	daily	FE	6160kim
2205-2305	daily	NAm	6100sac

Portuguese	Days	Area	kHz
0005-0105	m.....s	NAm	9755sac
2000-2100	....fss	SAm	15305sac, 17765sac
2100-2200	....fss	SAm	15305sac, 17765sac
2200-2300	....fss	SAm	11990sac, 15305sac

Russian	Days	Area	kHz
1505-1605	daily	NAm	9610sac
1600-1630	daily	RUS,CAs	9830smg, 11935smg
1700-1730	daily	RUS,CAs	9555rmp, 11935wof

Spanish	Days	Area	kHz
0000-0100	daily	SAm	11990sac, 13700sac
0100-0200	daily	CAm,NAm	6100sac
0200-0300	daily	CAm,NAm	9800sac
0305-0405	daily	NAm	9755sac
1305-1405	daily	NAm	7325sac
2300-2400	daily	CAm,SAm	9785sac, 11990sac
2305-2400	daily	NAm	6100sac

Key: + DRM.
Ann: English: "This is Radio Canada International"; French: "Ici Radio Canada Internationale".
IS: First bar of Canadian National Anthem.
V: QSL-card.
Notes: Radio Canada International is the External Sce of the public service Canadian Broadcasting Corporation (CBC).

BIBLE VOICE BROADCASTING (BVB) (Rlg)
P.O. Box 425, Station E, Toronto, Ontario, Canada M6H 4E3.
+1 800 550 4670.
E: mail@biblevoice.org **W:** www.biblevoice.org
Webcast: D/P
kHz: 5950, 6030, 6110, 6130, 6145, 7220, 7355, 7365, 7410, 7425, 7485, 9345, 9390, 9460, 9510, 9925, 11875, 11895, 11915, 12035, 13635, 13670, 13810, 17545

Winter Schedule 2010/2011

Adja	Days	Area	kHz
1945-2000	.....s.	WAf	9510wer

Amharic	Days	Area	kHz
1630-1700	.t.....	EAf	13810iss
1630-1830	...t...	EAf	13810iss
1730-1800	.t.....	EAf	13810iss

Arabic	Days	Area	kHz
0430-0500	.t.t..	ME	7410wer
0500-0515	....f..	ME	7410wer
0900-1000	....f..	NAf	17545wer
1700-1720	mt.tf..	ME	11915wer
1700-1735	..w....	ME	11915wer
1715-1730	....f..	ME	7355wer
1715-1800	m.w....	ME	7355wer
1830-1900	daily	ME	6130wer
2045-2115	mt.tfss	Af	6145nau

Cantonese	Days	Area	kHz
1405-1420	.twtf.	EAs	9345alm

Chinese	Days	Area	kHz
1405-1435	mt.....	EAs	9345alm

Dari	Days	Area	kHz
0230-0330	daily	WAs	7410wer
1600-1630	daily	WAs	11895wer

Dinka	Days	Area	kHz
1700-1730	daily	EAf	11875wer

Dzongkha	Days	Area	kHz
1400-1415	...t...	As	7485dsb

English	Days	Area	kHz
0030-0100	....fss	ME	5950wer
0800-0830	......s	Eu	7220wer
0800-0845	.....s.	Eu	7220wer
1345-1415	......s	As	13635iss
1400-1500	......s	As	13635iss
1405-1430	......ss	EAs	9345alm
1430-1500	......s.	As	13635iss
1500-1515	......s	As	12035wer
1515-1545	.....s.	As	13670wer
1530-1600	..t....	As	13670wer
1630-1915	......s	ME	9460nau
1645-1700	m.w....	ME	9460nau
1645-1715	....f..	ME	9460nau
1645-1720	.t.....	ME	9460nau
1645-1745	...t...	ME	9460nau

English	Days	Area	kHz
1645-1830	.....s.	ME	9460nau
1800-1815	.....s.	ME	7365wer
1800-1830	.....s	ME	6110wer
1800-1900	.....s.	ME	6110wer
1900-2000	......s	Eu	6030wer
1915-1945	.....s.	Eu	6030wer

Farsi	Days	Area	kHz
1630-1830	daily	ME	9925wer
1800-1830	m.w.f..	ME	7365wer
1800-1900	.t.t..	ME	7365wer
1830-1900	......s	ME	7365wer

French	Days	Area	kHz
1930-1945	.....s.	WAf	9510wer
2045-2115	..w....	Af	6145nau

Fur	Days	Area	kHz
1730-1745	...f..	EAf	11875wer

Hebrew	Days	Area	kHz
1800-1900	.t.....	ME	9460nau

Hindi	Days	Area	kHz
0030-0100	mtwt...	ME	5950wer
1415-1500	....f..	As	7485dsb

Japanese	Days	Area	kHz
1230-1245	......s	EAs	9390dsb
1245-1300	......s	EAs	9390dsb

Nepali	Days	Area	kHz
1415-1430	...t...	As	7485dsb

Nuer	Days	Area	kHz
1630-1700	daily	EAf	11875wer

Oromo	Days	Area	kHz
1600-1630	m..tf..	EAf	13810iss

Russian	Days	Area	kHz
1900-1915	...f...	Eu	6030wer
1900-1930	.t.....	Eu	6030wer

Somali	Days	Area	kHz
1800-1830	...fss	EAf	13810iss

Spanish	Days	Area	kHz
1800-1830	......s	WAf	7425wer

Swahili	Days	Area	kHz
1830-1845	......s	WAf	9510iss

Tigrinya	Days	Area	kHz
1700-1730	mtw.f..	EAf	13810iss

Ukrainian	Days	Area	kHz
1900-1915	...t...	Eu	6030wer

Urdu	Days	Area	kHz
1530-1600	..w.f..	As	13670wer

V: QSL-card.
Notes: Joint radio mission of Bible Voice (UK) and High Adventure Gospel Communication Ministries (Canada).

ÉGLISE DU CHRIST (Rlg)
✉ 2500-2510, rue Charland, Montréal, Québec, Canada H1Z 1C5.
☎ +1 514 3876163. 🖷 +1 514 3871153.
E: egliseduchrist@videotron.ca **W:** www.eglise-du-christ.org
kHz: 15265

Winter Schedule 2010/2011

French	Days	Area	kHz
1500-1530	...t...	NAf	15265skn

CHILE (CHL)

CVC LA VOZ (Rlg)
✉ Casilla 395 Talagante, Santiago, Chile.
☎ +56 2 8557046. 🖷 +56 2 8557053.
E: ondacorta@cvclavoz.cl **W:** www.cvclavoz.cl (Shortwave)
W: www.cvclavoz.com
L.P: Senior Engineer: Gisela Vergara.
SW: [SGO] Santiago, Calera de Tango: 8 x 100kW.
kHz: 9635, 9780, 17680, 17860

Winter Schedule 2010/2011

Spanish	Days	Area	kHz
1100-1200	daily	CAm,SAm	9780sgo
1100-2200	daily	SAm	9635sgo
1200-0100	daily	SAm	17680sgo
1800-2000	daily	SAm	17860sgo+

Spanish	Days	Area	kHz
2200-0100	daily	SAm	9635sgo

Key: + DRM
V: QSL-card.
Notes: Owned by Voz Cristiana S.A., Emisora Christian Vision (Chile) Ltda., a branch of Christian Vision, United Kingdom. For corporate details, see United Kingdom under "CVC International".

CHINA (CHN)

CHINA RADIO INTERNATIONAL (CRI) (Gov)
✉ 16a, Shijingshan Rd, Beijing 100040, P.R. China.
☎ +86 10 68891001. 🖷 +86 10 68891582.
E: crieng@cri.com.cn
W: www.chinabroadcast.cn; www.cri.cn
Webcast: D/L
W: www.crienglish.com (English)
L.P: GD: Wang Gengnian; CE: Zhou Yi; Dir, English Sce: Yang Lei.
MW: See SARFT for tx information.
SW: See SARFT for tx information.
SAT: Apstar 6, Chinasat 6B/9, Intelsat 1R/7/8/9/10, Superbird C.
kHz: *603, 684, 702, 963, 1017, 1044, 1080, 1170, 1188, 1215, 1269, 1296, 1323, 1341, 1422, 1440, 1458, 1521, 5905, 5910, 5915, 5955, 5960, 5965, 5970, 5975, 5980, 5985, 5990, 6005, 6010, 6020, 6025, 6040, 6055, 6060, 6065, 6070, 6075, 6080, 6090, 6095, 6100, 6105, 6110, 6115, 6135, 6140, 6145, 6150, 6160, 6165, 6175, 6180, 6185, 6190, 7205, 7210, 7215, 7220, 7225, 7230, 7235, 7245, 7250, 7255, 7260, 7265, 7275, 7285, 7290, 7295, 7300, 7305, 7310, 7315, 7320, 7325, 7330, 7335, 7340, 7345, 7350, 7360, 7365, 7380, 7385, 7390, 7395, 7400, 7405, 7410, 7415, 7420, 7425, 7430, 7435, 7440, 7445, 9410, 9415, 9420, 9425, 9430, 9435, 9440, 9450, 9455, 9460, 9470, 9490, 9515, 9525, 9535, 9540, 9550, 9555, 9560, 9565, 9570, 9580, 9585, 9590, 9600, 9610, 9615, 9620, 9640, 9645, 9655, 9665, 9670, 9675, 9685, 9690, 9695, 9700, 9705, 9710, 9720, 9730, 9740, 9745, 9760, 9765, 9770, 9785, 9790, 9795, 9800, 9825, 9855, 9860, 9865, 9870, 9880, 11600, 11610, 11620, 11640, 11650, 11665, 11680, 11690, 11695, 11700, 11710, 11725, 11730, 11750, 11760, 11770, 11780, 11785, 11790, 11795, 11820, 11845, 11855, 11860, 11870, 11875, 11880, 11885, 11900, 11920, 11935, 11945, 11955, 11970, 11975, 11980, 11990, 12015, 12070, 12110, 13580, 13590, 13600, 13610, 13620, 13640, 13645, 13650, 13655, 13665, 13670, 13675, 13680, 13700, 13710, 13715, 13720, 13730, 13735, 13740, 13750, 13780, 13790, 13850, 13855, 15110, 15120, 15125, 15130, 15135, 15140, 15145, 15160, 15170, 15190, 15205, 15210, 15220, 15225, 15230, 15245, 15250, 15270, 15335, 15340, 15350, 15425, 15435, 15440, 15445, 15465, 15525, 15565, 15600, 15620, 15625, 15665, 17485, 17490, 17495, 17500, 17505, 17540, 17560, 17570, 17630, 17640, 17650, 17670, 17680, 17690, 17710, 17720, 17725, 17735, 17740, 17750, 17830, 17855*

Winter Schedule 2010/2011

Albanian	Days	Area	kHz
1600-1700	daily	Eu	1215fla
1900-2000	daily	Eu	6020szg, 7315kas
2100-2130	daily	Eu	6145iss

Amoy	Days	Area	kHz
0100-0200	daily	SEA	9460kun, 9550kun, 9610kun, 9860bei, 11945kun, 11980kun, 15425xia, 17495bei
0200-0300	daily	SEA	15425xia, 17495bei
1400-1500	daily	SEA	9655kun, 11650kun

Arabic	Days	Area	kHz
0500-0700	daily	Af,ME	17485uru
0500-0700	daily	ME	9590cer
0500-0700	daily	NAf	5985cer, 7210cer
1600-1700	daily	CAf,EAf	15125bko
1600-1700	daily	ME	7205spb
1600-1800	daily	NAf,ME	7300kas, 9555cer, 11725cer
1830-1930	daily	CAf,EAf	11640bko
2000-2200	daily	NAf	6185cer, 7215cer
2000-2200	daily	Af,ME	6100xia

Bengali	Days	Area	kHz
0200-0300	daily	SAs	9655kun, 11640kun
1300-1400	daily	SAs	9600bji
1300-1500	daily	SAs	9490kun, 11610kun
1400-1500	daily	SAs	1269xuw

Bengali	Days	Area	kHz
1500-1600	daily	SAs	9610kun, 9690kun

Bulgarian	Days	Area	kHz
1100-1200	daily	Eu	7220cer
1700-1800	daily	Eu	1458fla
1830-1900	daily	Eu	6020szg, 7265uru, 9695jin
2030-2100	daily	Eu	7320kun, 9720uru

Burmese	Days	Area	kHz
1100-1200	daily	SEA	1188kun, 9880kun
1300-1400	daily	SEA	1188kun, 9880bji
1300-1500	daily	SEA	11780kun

Cantonese	Days	Area	kHz
0000-0100	daily	SEA	11820xia, 17495bei
0400-0500	daily	EAs	15160jin
0400-0500	daily	NAm	9790hab
0400-0600	daily	EAs	13655xia
0500-0600	daily	EAs	15170jin
0700-0800	daily	EAs	11640jin, 13610xia
1000-1100	daily	Pac	15440kun, 17670kun
1100-1200	daily	SEA	603dof, 9590kun, 9645bei
1100-1200	daily	Pac	9540bei, 13580kun
1200-1300	daily	NAm	9560sac, 9570hab
1700-1800	daily	EAf	7220kas, 7325uru
1900-2000	daily	Eu	7215szg, 9770kas
2300-2400	daily	SEA	6140kun, 7325kun, 9425jin, 9460kun, 11935kun

Chaozhou	Days	Area	kHz
0700-0800	daily	SEA	13730xia, 15145xia
1100-1200	daily	SEA	9440kun, 11875kun
1800-1900	daily	Eu	6010uru, 7285xia

Croatian	Days	Area	kHz
1700-1800	daily	Eu	7335bei, 9610kas
2100-2200	daily	Eu	6135bei, 7225bei

Czech	Days	Area	kHz
1100-1200	daily	Eu	15225kas, 17570kas
1900-1930	daily	Eu	7325bei
1900-2000	daily	Eu	7415uru
1930-2000	daily	Eu	7305iss
2000-2100	daily	Eu	963por
2230-2330	daily	Eu	1458fla

English	Days	Area	kHz
0000-0100	daily	EAs	9425bei
0000-0100	daily	NAm	6005sac
0000-0100	daily	SAs	6180kas, 7425kas
0000-0200	daily	SEA	11650bei, 11885xia
0000-0200	daily	NAm	6020cer, 9570cer
0100-0200	daily	SAs	6075kas, 6175kas, 9420kas
0100-0200	daily	Eu	9410kas
0100-0200	daily	NAm	6005sac, 6080sac, 9580hab
0200-0300	daily	SAs	13640kas
0200-0400	daily	SAs	11785kas
0300-0400	daily	NAm	9690nob, 9790hab
0300-0400	daily	SAs	15110kas
0300-0500	daily	EAs	9460bei, 13620xia, 15120bei
0300-0600	daily	NAm	6190sac
0400-0600	daily	CAs	17725xia, 17855bei
0500-0600	daily	NAm	5960sac
0500-0600	daily	ME,NAf	7220cer
0500-0700	daily	Af,ME	17505kas
0500-0900	daily	SAs	11880kas, 15465kas, 17540kas
0500-1100	daily	SAs	15350kas
0600-0700	daily	ME	11770kas, 15145kas
0600-0700	daily	NAf	11750cer
0600-0700	daily	NAm	6115sac
0600-0800	daily	SEA	13645xia, 17710bei
0700-0800	daily	Af,ME	15125kas
0700-0900	daily	Eu	1215fla, 11785cer
0700-1300	daily	Eu	17490kas
0800-0900	daily	Af,ME	15625kas

English	Days	Area	kHz
0800-1000	daily	EAs	9415xia
0900-1000	daily	Eu	15270kas, 17570uru
0900-1000	daily	SAs	17750kas
0900-1100	daily	Pac	15210kun, 17690jin
1000-1100	daily	SAs	15190uru
1000-1100	daily	EAs	5955xia
1000-1100	daily	CAs	7215xia, 11640bei
1000-1200	daily	SEA	13590bei, 13720xia
1100-1200	daily	SAs	9570kas, 11795kas, 13645kas
1100-1200	daily	NAm	5960sac
1100-1300	daily	Eu	13665cer
1100-1300	daily	SAs	11650uru
1100-1300	daily	SEA	1269xuw
1100-1600	daily	EAs	5955bei
1200-1300	daily	SEA	684dof, 1188kun, 9600kun, 9645bei, 9730kun
1200-1300	daily	CAs	11690xia
1200-1300	daily	SAs	7250kas, 9460kas, 12015kas
1200-1300	daily	Pac	9760kun
1200-1300	daily	Eu	13790uru
1200-1400	daily	SEA	1341hdu, 11980kun
1200-1400	daily	Pac	11760kun
1300-1400	daily	Pac	11900kun
1300-1400	daily	SEA	9730bei
1300-1400	daily	SAs	7300kas, 9655kas
1300-1400	daily	NAm	9570hab, 11885sac
1300-1400	daily	Eu	13670kas
1300-1500	daily	CAs	9765xia
1300-1500	daily	NAm	15230sac
1300-1600	daily	SEA	9870xia
1400-1500	daily	NAm	13675sac
1400-1500	daily	SAs	7300uru, 9460uru
1400-1500	daily	ME	11665uru
1400-1500	daily	Eu	9700kas, 9795uru
1400-1600	daily	SEA	1188kun
1400-1600	daily	WAf,CAf	17630bko
1400-1600	daily	NAm	13740hab
1500-1600	daily	NAf,ME	6095kas, 9720uru
1500-1600	daily	SAs	7405uru, 9785jin
1500-1600	daily	SEA	7325bei
1500-1700	daily	Eu	9435kas, 9525kas
1500-1800	daily	SAs	1323uru
1600-1700	daily	SEA	6060kun
1600-1700	daily	SAs	7235kas
1600-1700	daily	ME	6100kas, 9600jin
1600-1700	daily	NAf,ME	7420uru
1600-1800	daily	Eu	7255kas
1600-1800	daily	SEA	1080xuw
1600-1800	daily	SAf	7435jin, 9570bei
1700-1800	daily	Eu	7205bei
1700-1800	daily	ME	6165bei
1700-1800	daily	SAs	6140kas, 7410kas, 7425kas
1700-1800	daily	SEA	6090kun, 7420kun
1700-1900	daily	Eu	6100bei
1800-1900	daily	Eu	7405bei
1900-2100	daily	Af,ME	7295kas, 9440kun
2000-2100	daily	CAf,SAf	5985bei
2000-2130	daily	CAf,SAf	11640bko
2000-2200	daily	Eu	5960cer, 7285cer, 7415kas, 9600kas
2100-2200	daily	SAf	7205xia
2100-2200	daily	CAf,SAf	7405bei
2200-2300	daily	EAs	5915bei
2200-2400	daily	Eu	1440mrn
2300-0100	daily	SEA	11790xia
2300-0200	daily	Eu	7350kas
2300-2400	daily	CAm	5990hab
2300-2400	daily	EAs	6145bei
2300-2400	daily	NAm	6040sac, 11970sac
2300-2400	daily	SAs	5915kas, 7415kas
2300-2400	daily	SEA	9610kun

Esperanto	Days	Area	kHz
1100-1200	daily	EAs	6100uru, 7210uru
1300-1400	daily	SEA	9440nnn, 11650bei
1700-1800	daily	Eu	1215fla, 7245xia
1930-2030	daily	Eu	7265uru, 9745uru
2200-2300	daily	SAm	7315kas, 9860kas

Farsi	Days	Area	kHz
1500-1530	daily	ME	6165uru, 9600kas, 9765kun
1800-1900	daily	ME	7295bei, 7325bei

Filipino	Days	Area	kHz
1130-1200	daily	SEA	1341hdu
1430-1500	daily	SEA	1341hdu

French	Days	Area	kHz
0600-0800	daily	Eu	15220uru
0800-1300	daily	Eu	702cdm
1200-1400	daily	Eu	15205kas
1300-1400	daily	Eu	13710kas
1400-1600	daily	WAf,Eu	11920cer, 13670cer
1600-1800	daily	Eu	7350kas
1800-2000	daily	NAf,WAf	6055cer, 7385cer
1800-2000	daily	Eu	5970cer, 7360cer
1800-2300	daily	Eu	702cdm
1830-2030	daily	WAf	9645kun
1830-2230	daily	WAf	7350uru
2030-2130	daily	Eu	7215sam
2030-2230	daily	Eu	6115bei
2100-2200	daily	Eu	1440mrn
2130-2230	daily	WAf	11975bko

German	Days	Area	kHz
0600-0800	daily	Eu	15245uru, 17720kas
1600-1800	daily	Eu	5970cer, 7380cer
1800-2000	daily	Eu	6160xia, 7395kas, 9615uru
1900-2100	daily	Eu	1440mrn
2100-2300	daily	Eu	963por

Hakka	Days	Area	kHz
0000-0100	daily	SEA	9460kun, 9550kun, 9610kun, 9860jin
0400-0500	daily	SAs	15350kas, 17540kas
0400-0500	daily	SEA	17505xia, 17710bei
1600-1700	daily	EAf	6090xia
1600-1700	daily	SAs	7325uru

Hausa	Days	Area	kHz
0800-0900	daily	WAf	7295bko
1630-1730	daily	WAf	9620kas, 9670kun
1730-1830	daily	WAf	9450kas, 9685kun
1800-1830	daily	WAf	11640bko

Hindi	Days	Area	kHz
0300-0400	daily	SAs	11640kas, 13720kas, 15210kas, 15350kas
1300-1400	daily	SAs	1269xuw, 1422kas, 7265uru, 9450kas
1500-1600	daily	SAs	7225uru, 7265kas
1500-1700	daily	SAs	1269xuw
1600-1700	daily	SAs	1422kas, 5915kas, 7395kun

Hungarian	Days	Area	kHz
1000-1100	daily	Eu	15220uru, 17570kas
1900-1930	daily	Eu	7440xia, 9560uru
2000-2100	daily	Eu	1458fla
2030-2100	daily	Eu	7390jin, 9585kas
2130-2200	daily	Eu	6145iss, 7250uru

Indonesian	Days	Area	kHz
0830-0930	daily	SEA	15135kun, 17735kun
1030-1130	daily	SEA	11700kun, 15135kun
1330-1430	daily	SEA	11955kun, 15135kun

Italian	Days	Area	kHz
0600-0700	daily	Eu	15620kas
1500-1800	daily	Eu	702cdm
1800-1900	daily	Eu	1458fla
1800-1900	daily	WEu	7340kas, 7435jin
2030-2130	daily	WEu	7265uru, 7310kas

Japanese	Days	Area	kHz
1000-1100	daily	EAs	9440xia
1000-1400	daily	EAs	7325jin
1100-1300	daily	EAs	7260xia
1100-1600	daily	EAs	1044hnl
1300-1400	daily	EAs	7215xia
1400-1500	daily	EAs	7210xia, 7410jin
1500-1600	daily	EAs	5980xia, 7220jin
2200-2300	daily	EAs	5985xia, 7440bei
2300-2400	daily	EAs	9435xia, 9695jin

Khmer	Days	Area	kHz
1030-1130	daily	SEA	684dof, 15160nnn, 17680kun
1200-1300	daily	SEA	9440kun, 11680nnn
1400-1500	daily	SEA	6055nnn, 9600nnn
2300-0100	daily	SEA	9765nnn, 11990nnn

Korean	Days	Area	kHz
1100-1400	daily	EAs	1323hdn
1100-1500	daily	EAs	1017cah, 5965xia
2100-2200	daily	EAs	7290xia
2100-2300	daily	EAs	1323hdn
2200-2300	daily	EAs	7210xia

Lao	Days	Area	kHz
1230-1330	daily	SEA	7360kun, 9785kun
1430-1530	daily	SEA	1080xuw, 7360kun, 9675kun

Malay	Days	Area	kHz
0930-1030	daily	SEA	15135kun, 17680kun
1230-1330	daily	SEA	11955kun, 15600kun

Mandarin	Days	Area	kHz
0000-0100	daily	EAs	11780jin, 11900bei
0000-0100	daily	NAm	6040sac
0000-0100	daily	SEA	9435kun, 11845xia, 11975xia
0000-0200	daily	SEA	13580bei
0000-0400	daily	EAs	13655xia
0100-0200	daily	SEA	11640xia
0100-0200	daily	SAs	7250uru, 7300kas
0100-0400	daily	EAs	15160jin
0200-0300	daily	NAm	9580hab, 9690nob
0200-0300	daily	SAm	7330kas, 11695bei
0200-0300	daily	SAs	9825kas
0200-0400	daily	NAm	6020cer, 9570cer
0300-0400	daily	SAs	9450kas, 17540kas
0300-0600	daily	EAs	15130bei
0400-0500	daily	SAs	13640kas, 15170kas
0500-0600	daily	SAs	15140kas
0500-0700	daily	EAs	13620xia, 15120bei
0600-0700	daily	EAs	13655xia, 15170jin
0600-0800	daily	SEA	13750kun, 17740xia
0600-1300	daily	Eu	17650kas
0700-0900	daily	Eu	11855cer
0700-0900	daily	EAs	17830kas
0800-0900	daily	EAs	9880bei, 11640jin, 13610xia
0800-1000	daily	CAs	17560xia
0800-1000	daily	RUS	15565xia
0900-1000	daily	EAs	7430jin, 9440xia
0900-1000	daily	Pac	15440kun, 17670kun
0900-1100	daily	EAs	5965bei
0900-1100	daily	SAs	15525uru, 17500kas
0900-1100	daily	SEA	13850bei, 15250kun, 15340xia
0900-1200	daily	SEA	11980kun
1000-1100	daily	EAs	6020bei, 7255xia
1100-1200	daily	EAs	7435bei
1100-1200	daily	Pac	11620bei, 15440kun
1100-1200	daily	SAs	9515kas
1200-1300	daily	EAs	7395bei
1200-1300	daily	ME	15110uru
1200-1300	daily	SAs	7205kas, 9655kas
1200-1300	daily	SEA	11640xia
1200-1400	daily	NAf,ME	9665kas, 11790kas
1200-1400	daily	SAs	9540kun
1200-1400	daily	SEA	7435nnn, 9855bei
1300-1400	daily	EAs	7205bei
1300-1400	daily	SEA	7215xia
1300-1400	daily	Eu	13855kas

Mandarin	Days	Area	kHz
1300-1400	daily	ME	13650uru
1300-1500	daily	Eu	702cdm
1400-1500	daily	EAs	7400bei
1400-1500	daily	Eu	9430kas, 11785kas
1400-1500	daily	ME	11610uru
1400-1500	daily	SAs	7445kas, 9730kas
1400-1500	daily	SEA	6040xia, 7410bei
1400-1700	daily	EAs	1323hdn
1500-1600	daily	SAs	7235kas, 9560kas
1500-1600	daily	SEA	5910bei, 9455kun
1500-1600	daily	Eu	9705kas, 9740kas
1500-1600	daily	EAs	7255bei
1500-1600	daily	NAm	13675sac
1600-1700	daily	NAm	17735sac
1730-1830	daily	Eu	6150szg, 7445uru
1730-1830	daily	NAf,ME	7275uru, 7315kun, 9695kun
2000-2100	daily	Af,ME	7245kas, 9865kun
2000-2100	daily	CAf,SAf	7305xia
2000-2100	daily	Eu	7335szg, 7440bei
2200-2300	daily	SEA	6100kun, 6140kun, 7220kun, 7325kun, 7440kun
2200-2300	daily	Af,ME	7220kun, 7405uru
2200-2300	daily	EAs	7305jin
2200-2300	daily	ME	5975kas
2200-2300	daily	SAf	7430jin
2230-2400	daily	NAf	11975bko
2300-2400	daily	Eu	7300uru
2300-2400	daily	SEA	7425kun
2300-2400	daily	WAf	7295bko
2300-2400	daily	EAs	9555bei

Mongolian	Days	Area	kHz
0000-0100	daily	EAs	9470xia, 11875bei
1100-1200	daily	EAs	7400huh, 9450uru
1200-1300	daily	EAs	1323uru, 5915huh, 5990huh
1300-1400	daily	EAs	6100uru, 7285bei
1400-1500	daily	EAs	5915huh, 5990huh
2300-2400	daily	EAs	6185xia, 7205xia

Nepali	Days	Area	kHz
0130-0230	daily	SAs	11860kun
0130-0330	daily	SAs	13780kun
0230-0330	daily	SAs	11730kun
1400-1500	daily	SAs	7220kun, 7435kun
1500-1600	daily	SAs	7215kun, 9535kun

Pashto	Days	Area	kHz
0200-0230	daily	WAs	6065kas, 7350kas, 15435xia
1500-1600	daily	ME	7435kun, 9620kas
1530-1600	daily	WAs	6165uru

Polish	Days	Area	kHz
1900-2000	daily	Eu	963por
2000-2100	daily	Eu	6020szg, 6145iss, 7405uru
2130-2230	daily	Eu	1458fla

Portuguese	Days	Area	kHz
0000-0100	daily	SAm	9435kas
1900-2000	daily	CAf,SAf	5985bei, 7365bei, 7405xia, 9535bji
1900-2000	daily	Eu	7335jin, 9730kas
1930-2000	daily	CAf,SAf	11640bko
2200-2300	daily	Eu	6175cer, 7260uru
2200-2300	daily	SAm	9410kas, 9685kas
2300-0100	daily	SAm	6100bei
2300-2400	daily	SAm	13650hab

Romanian	Days	Area	kHz
0900-1000	daily	Eu	7285cer, 9460cer
1800-1900	daily	Eu	1215fla
1900-1930	daily	Eu	7305iss
1900-2000	daily	Eu	6145uru
1930-2000	daily	Eu	7440xia

Russian	Days	Area	kHz
0100-0200	daily	RUS	5905kas, 13600xia
0200-0400	daily	RUS	5915kas

Russian	Days	Area	kHz
0200-0500	daily	RUS	17640xia
0300-0400	daily	RUS	11710uru, 15435xia, 17710jin
0300-0500	daily	Eu	963por
0400-0500	daily	RUS	5905kas
0400-0600	daily	RUS	15445kas, 15665kas
0800-1000	daily	RUS	15335kas, 15665kas
1000-1100	daily	RUS	7400huh
1000-1200	daily	RUS	5915huh, 7290szg
1000-1600	daily	EAs	963hdn
1100-1200	daily	EAs	1323uru
1100-1200	daily	RUS	6080bei
1100-1600	daily	EAs	1323hei
1100-2000	daily	CAs	1521uru
1200-1300	daily	RUS	6100bei, 7215xia, 7410szg, 9590szg, 9685uru
1200-1400	daily	RUS	5905kas
1300-1400	daily	RUS	5915huh, 5990huh, 7255szg, 9870xia
1300-1500	daily	EAs	1323uru
1400-1500	daily	RUS	7330xia, 9450szg
1400-1600	daily	RUS	6005kas
1500-1600	daily	RUS	5915huh, 5965bei, 5990huh, 6025xia, 6105szg, 6180uru
1600-1700	daily	RUS	6025kas, 7215szg
1600-1800	daily	RUS	6040uru, 7265uru
1700-1800	daily	RUS	7410bji
1700-1900	daily	Eu	963por
1700-1900	daily	RUS	6070xia
1800-1900	daily	RUS	7210uru, 7255szg, 9535iss
1900-2000	daily	RUS	6100bei, 6110xia, 7245bji
2000-2100	daily	RUS	7255bji, 9525bei
2300-2400	daily	RUS	5990huh, 7405huh

Serbian	Days	Area	kHz
1200-1300	daily	Eu	7345cer
2000-2030	daily	Eu	7315uru, 7390xia, 9585kas
2100-2130	daily	Eu	7325xia, 7425jin, 7440kun
2200-2300	daily	Eu	1215fla

Sinhala	Days	Area	kHz
1400-1500	daily	SAs	7265kas, 9610jin
2330-0030	daily	SAs	6100kun, 7260kas

Spanish	Days	Area	kHz
0000-0100	daily	CAm	5990hab
0000-0100	daily	SAm	15120hab
0100-0300	daily	SAm	9590kas, 9710kas
0300-0400	daily	SAm	9665bra
0600-0800	daily	Eu	15135kas
2100-2300	daily	Eu	6020szg, 9640kas
2200-2300	daily	LAm	13700sac
2200-2300	daily	SAm	9490bei
2200-2400	daily	Eu	7210cer, 7250uru
2300-0100	daily	SAm	9590kas, 9800kas
2300-2400	daily	Eu	6175cer

Swahili	Days	Area	kHz
1600-1700	daily	EAf	7320xia
1600-1800	daily	EAf	5985bei
1700-1800	daily	EAf	7400xia

Tagalog	Days	Area	kHz
1130-1200	daily	SEA	7410jin, 12070xia
1130-1230	daily	SEA	12110kun
1200-1230	daily	SEA	9720xia
1430-1500	daily	SEA	7325bei, 12110kun

Tamil	Days	Area	kHz
0200-0300	daily	SAs	11870kas, 13715kas
0300-0400	daily	SAs	13600kun, 13735kas
1400-1500	daily	SAs	9570kas, 9665kas
1500-1600	daily	SAs	9730kas, 13600kas

Thai	Days	Area	kHz
1130-1230	daily	SEA	1080xuw, 7360kun, 9785kun

Thai	Days	Area	kHz
1330-1430	daily	SEA	1080xuw, 7360kun, 9785kun

Turkish	Days	Area	kHz
1500-1600	daily	ME	7230cer, 9565cer
1600-1700	daily	ME	6165uru, 7325kun
1900-2000	daily	ME	7255kun, 9655kun
1930-2000	daily	ME	1170arm, 7225msk

Urdu	Days	Area	kHz
0100-0200	daily	SAs	6020kas, 7360kas
0200-0300	daily	SAs	6020kas, 7290kas
1400-1600	daily	SAs	1422kas, 6075kas, 7285kas

Vietnamese	Days	Area	kHz
0000-0100	daily	SEA	11770bei, 13680xia
0400-0500	daily	SEA	11650kun
0400-0600	daily	SEA	11740xia
0500-0600	daily	SEA	11640kun
1100-1200	daily	SEA	11990xia
1100-1300	daily	SEA	11600bji
1100-1500	daily	SEA	1296kun
1100-1600	daily	SEA	9550bei
1300-1400	daily	SEA	9685xia
1300-1600	daily	SEA	603dof
1400-1500	daily	SEA	9685bji
1400-1600	daily	SEA	684dof
1600-1700	daily	SEA	6010bei, 7360kun
2300-0100	daily	SEA	603dof
2300-2400	daily	SEA	7220xia, 9415bei

Key: * CRI News Radio.

Ann: Arabic: "Idha'at as-Sin ad-Duwaliyah"; English: "This is China Radio International, broadcasting from Beijing"; German: "Hier ist Radio China International"; Indonesian: "Inilah Radio CRI, China Radio International"; Japanese: "Kochirawa Pekin Hoso, Chugoku Kokusai Hosokyoku desu"; Korean: "Jungguk gukje bangsonggugimnida"; Malay: "Inilah Radio Antarabangsa China, dalam bahasa Melayu"; Mandarin: "Zhongguo guoji guangbo diantai"; Mongolian: "Hyatadyn Olon Ulsyn Radio"; Russian: "Govorit Meždunarodnoye Radio Kitaya"; Spanish: "Esta es Radio Internacional de China"; Swahili: "Hii ni Radio China kimataifa"; Vietnamese: "Day la dai phatthanh quoc te Trung quoc".

IS: First bars of the National Anthem.

V: QSL-card.

Notes: Founded on 3 Dec 1941. China Radio International is the External Sce produced under the roof of the State Administration of Radio, Film and Television of the P.R. of China (SARFT). Some MW programme content provided by Radio86 (Futuvision), see under Finland for schedule.

VOICE OF GUANGXI BEIBU GULF, BEIBU BAY RADIO (Gov)

⌨ 75 Minzu Dadao, Nanning, Guangxi 530022, P.R.China.
☎ +86 771 5802999. 🖷 +86 771 5802555.
W: www.bbrmedia.com; www.bbrmedia.com/english
Webcast: L
SW: See SARFT for tx information.
kHz: 5050, 9820

Winter Schedule 2010/2011			
Cantonese	**Days**	**Area**	**kHz**
1000-1100	daily	SEA	5050nnn, 9820nnn†
Mandarin	**Days**	**Area**	**kHz**
1330-1500	daily	SEA	5050nnn, 9820nnn†
2300-0100	daily	SEA	5050nnn, 9820nnn†
Thai	**Days**	**Area**	**kHz**
1100-1130	mtwtf..	SEA	5050nnn, 9820nnn†
Vietnamese	**Days**	**Area**	**kHz**
1100-1130	.....ss	SEA	5050nnn, 9820nnn†
1130-1330	daily	SEA	5050nnn, 9820nnn†
1500-1600	daily	SEA	5050nnn, 9820nnn†

Key: † Irregular.

Ann: English: "This is Guangxi Beibu Bay Radio", "BBR"; Mandarin: "Guangxi Bei-bu Wan zhi sheng"; Vietnamese: "Tieng noi vinh bac phong guang tay".

V: QSL-letter.

Notes: Beibu Bay Radio is a joint project of Guanxi People's Broadcasting Station and China Radio International.

YUNNAN PEOPLE'S BROADCASTING STATION - THE VOICE OF SHANGRI-LA (Gov)

⌨ Voice of Shangri-La, 182 Renmin Xi Lu, Kunming, Yunnan 650031, P.R.China.
☎ +86 871 5310211. 🖷 +86 871 5361744.
E: admin@ynradio.net **W:** ynradio.net
SW: See SARFT for tx information.
SAT: Chinasat 9.
kHz: *900, 6035*

Winter Schedule 2010/2011			
Burmese	**Days**	**Area**	**kHz**
0200-0300	daily	BRM	900deh
0700-0800	daily	BRM	900deh
1400-1500	daily	BRM	900deh
Chinese	**Days**	**Area**	**kHz**
0000-0030	daily	SEA	6035kun
0100-0130	daily	SEA	6035kun
1000-1030	daily	SEA	6035kun
1100-1130	daily	SEA	6035kun
1300-1330	daily	SEA	6035kun
1400-1430	daily	SEA	6035kun
2300-2330	daily	SEA	6035kun
Vietnamese	**Days**	**Area**	**kHz**
0030-0100	daily	SEA	6035kun
1030-1100	daily	SEA	6035kun
1130-1300	daily	SEA	6035kun
1330-1400	daily	SEA	6035kun
1430-1500	daily	SEA	6035kun
2230-2300	daily	SEA	6035kun
2330-2400	daily	SEA	6035kun

Ann: English: "This is the Voice of Shangri-La brought to you by Yunnan Radio"; Mandarin: "Xianggelila zhi sheng".

V: QSL-letter.

STATE ADMINISTRATION OF RADIO, FILM AND TV (SARFT) (Tx Operator)

⌨ 2 Fuxingmenwai Street, Xicheng District, Beijing 100866, P.R.China.
☎ +86 10 66093114. 🖷 +86 10 86092437.
E: sarft@chinasarft.gov.cn **W:** www.sarft.gov.cn
L.P: Minister: Wang Taihua.

MW: [CAH] Changchun, Jilin: 1017kHz 500kW; [DEH] Luxi, Dehong, Yunnan: 900kHz 100kW; [DOF] Dongfang, Hainan: 603/684kHz 600kW; [HDN] Huadian, Jilin: 963/1323kHz 600kW; [HDU] Huadu, Guangzhou, Guangdong: 1341kHz 300kW; [HEI] Shuangyashan, Heilongjiang: 1323 kHz 200kW; [HNL] Henglin, Changzhou, Jiangsu: 1044kHz 600kW; [KAS] Kashi, Saibagh, Xinjiang: 1422kHz 600kW; [KUN] Kunming, Anning, Yunnan: 1188/1296kHz 300kW; [URU] Ürümqi, Hutubi, Xinjiang: 1323/1521kHz 500kW; [XUW] Xuanwei, Yunnan: 1080/1269kHz 600kW.

SW: [BEI] Beijing, Doudian: 150/500kW; [BJI] Baoji, Shaanxi, Xinjie: 150kW; [HUH] Hohhot, Bikeqi, Nei Menggu: 4 x 100kW; [JIN] Jinhua, Lanxi, Zhejiang: 2 x 100, 3 x 500kW; [KAS] Kashi, Saibagh, Xinjiang: 2 x 100, 8 x 500kW; [KUN] Kunming, Anning, Yunnan: 1 x 50; 2 x 100, 5 x 500kW; [NNN] Nanning, Guangxi: 2 x 15; 2 x 100kW; [SZG] Shijiazhuang, Nanpozhuang, Hebei: 2 x 500kW; [URU] Ürümqi, Hutubi, Xinjiang: 9 x 100, 8 x 500kW; [XIA] Xi'an, Xianyang, Shaanxi: 150/500kW.

Notes: SARFT in an executive branch under the State Council of the Peoples Republic of China.

COSTA RICA (CTR)

RADIO EXTERIOR DE ESPAÑA RELAY STATION

⌨ Radio Nacional de España, c/o Delegación en Costa Rica, Apartado 677-2010, Zapote, San José, Costa Rica.
☎ +506 2904620. 🖷 +506 2329340.
L.P: Engineer: Luis García.
SW: [CRI] Cariari de Pococí: 3 x 100kW.

CROATIA (HRV)

VOICE OF CROATIA (GLAS HRVATSKE) (Pub)

⌨ Prisavlje 3, 10000 Zagreb, Croatia.
☎ +385 1 6342602. 🖷 +385 1 6343305.
E: voiceofcroatia@hrt.hr; hrt@hrt.hr **W:** www.hrt.hr
Webcast: D/L
L.P: GM, HRT: Josip Popovac.

SW: Uses txs provided by Odašiljaci i veze d.o.o.
SAT: Hot Bird 6, Optus D2, SES 1.
kHz: *3985, 6165, 7370, 7375, 17860*

Winter Schedule 2010/2011

Croatian/Various	Days	Area	kHz
0000-0400	daily	NAm	7375wer
0200-0600	daily	NAm	7375wer
0600-0755	daily	Eu	6165dea
0700-1100	daily	AUS,NZL	17860sng
0800-1455	daily	Eu	7370dea
1500-2125	daily	Eu	6165dea
2130-0555	daily	Eu	3985dea
2300-0400	daily	SAm	7375wer

Ann: Croatian: "Hrvatske Radio, kratki val", "Glas Hrvatske"; English: "This is Croatian Radio, you are listening to the Voice of Croatia"; Spanish: "La Voz Croacia".
IS: Tune to Dubrovnik's poem "Lovely, Dear, Sweet Liberty", played on celeste.
V: QSL-card.
Notes: A service of the public broadcaster Hrvatska Radiotelevizija (HRT) for listeners abroad. "Glas Hrvatske" consists of Home Sce prgrs in Croatian (part relays HR1), plus newscasts in English 2315-2330, 0300-0315, 0700-0605, 1100-1005, 1700-1715, 1805-1815; Spanish 2330-2345, 0330-0345, 1300-1305; Italian 1500-1520 (HRT R.Rijeka); Hungarian 1830-1840 (HRT R.Osijek) & German 1900-1905.

ODAŠILJACI I VEZE D.O.O. (OIV) (Tx Operator)
✉ ul. grada Vukovara 269d, 10000 Zagreb, Croatia.
☎ +385 1 6186000. 🖷 +385 1 6186100.
E: oiv@oiv.hr **W:** www.oiv.hr
L.P: Chmn: Denis Nikola Kulišic.
SW: [DEA] Deanovec: 2 x 10, 1 x 100kW.
Notes: Odašiljaci i veze is the national transmitter network operator.

CUBA (CUB)

RADIO HABANA CUBA (RHC) (Gov)
✉ Apartado 6240, La Habana 10600, Cuba.
☎ +53 7 877 5524. 🖷 +53 7 8776531.
E: inforhc@enet.cu **W:** www.radiohc.cu
Webcast: L (media.enet.cu/radiohabanacuba; media.enet.cu/radiohabanacubaingles)
L.P: DG: Isidro Betancourt Silva; Editor in Chief: Pedro Otero Cabañas; Dir, Technical: Rene Martinez Villavicencio; Head of Correspondence Dept: Rosario Lafita Fernández; Frequency Mgr: Prof. Arnaldo Coro Antich; Engineer: Luis Pruna Amer.
SW: Transmitters operated by Radiocuba.
SAT: Hispasat 1D.
kHz: *5040, 6000, 6010, 6050, 6060, 6095, 6120, 6140, 6150, 9640, 9770, 9820, 11690, 11730, 11760, 11770, 12010, 12040, 13680, 13750, 13780, 15120, 15360, 15370, 15390, 17750*

Winter Schedule 2010/2011

Arabic	Days	Area	kHz
2030-2100	daily	Eu	11770hab

Creole	Days	Area	kHz
0100-0130	daily	Car	5040hab
2300-2330	daily	SAm	15370hab

English	Days	Area	kHz
0000-0100	daily	Car	5040hab
0100-0500	daily	NAm	6000hab, 6050hab
0500-0700	daily	Am	6150hab
0500-0700	daily	NAm	6010hab, 6060hab
2000-2100	daily	Am	11760hab

Esperanto	Days	Area	kHz
0700-0730	......s	NAm	6010hab
1500-1530	......s	Am	11760hab
2100-2130	......s	CAm	11760hab
2230-2300	......s	SAm	15370hab

French	Days	Area	kHz
0130-0200	daily	Car	5040hab
1930-2000	daily	Eu	11770hab
2100-2130	mtwtfs.	Am	11760hab
2230-2300	mtwtfs.	SAm	15370hab

Portuguese	Days	Area	kHz
2000-2030	daily	Eu	11770hab
2300-2400	daily	SAm	15390hab
2330-2400	daily	SAm	15370hab

Quecha	Days	Area	kHz
0000-0030	daily	SAm	15370hab

Spanish	Days	Area	kHz
0000-0500	daily	Am	11760hab
0000-0500	daily	NAm	6060hab
0000-0500	daily	SAm	9770hab, 15390hab
0000-1100	daily	Car	6120hab
0200-1100	daily	Car	5040hab
0700-1100	daily	SAm	6060hab
0700-1100	daily	NAm	6050hab
0700-1300	daily	Am	6150hab
1100-0500	daily	Am	6140hab
1100-1300	daily	NAm	6000hab, 6095hab
1100-1500	......s	Am	11760hab
1100-1500	daily	SAm	15120hab, 15360hab
1100-2000	mtwtfs.	Am	11760hab
1100-2300	daily	SAm	15390hab
1100-2400	daily	CAm	12040hab
1300-1500	daily	NAm	13680hab, 13780hab
1300-2330	daily	CAm	11730hab
1400-1800	......s	Car	11690hab**
1400-1800	......s	NAm	13750hab**
1400-1800	......s	SAm	15370hab**, 17750hab**
1400-1800	......s	CAm	13680hab**
1500-2000	daily	Car	11690hab
1530-2000	......s	Am	11760hab
2030-2400	daily	Car	9820hab
2100-2400	daily	Car	5040hab
2100-2400	daily	Eu	11770hab
2300-0500	daily	SAm	12010hab
2330-0100	mtwtf..	NAm	6000hab*, 9640hab*

Key: * "Mesa Redonda" prgr; ** "Aló Presidente" prgr.
Ann: English: "This is Radio Havana, Cuba".
V: QSL-card and letter. (Email to: radiohc@enet.cu)
Notes: Radio Habana Cuba is the External Sce of the state-owned Instituto Cubano de Radio y Television (ICRT). Frequencies and schedule are highly variable.

RADIOCUBA (Tx Operator)
✉ Habana No 406, e/ Obispo y Obrapía, Habana Vieja, Ciudad de La Habana, Cuba.
☎ +53 7 8607181. 🖷 +53 7 8603107.
E: dirgeneral@radiocuba.cu
L.P: DG: Justo Moreno García.
SW: [HAB] La Habana, three sites: Bauta, Corralillo (G.C. 22N57 082W33): 1 x 50, 6 x 100kW; Bejucal, Casualidad (G.C. 22N52 082W20): 3 x 50, 1 x 100kW; Quivicán, San Felipe (G.C. 22N50 082W18): 5 x 250kW. The status of the Bejucal site is uncertain.
Customers: Radio Habana Cuba, Radio Rebelde, Radio Nacional de Venezuela, China Radio International.
Notes: Radiocuba, a state operated company that forms part of the Ministry of Information and Communications, is the national transmitter network operator.

CYPRUS (CYP)

CYPRUS BROADCASTING CORPORATION (CYBC) (Pub)
✉ P.O. Box 24824, 1397 Nicosia, Cyprus.
☎ +357 22862000. 🖷 +357 22314050.
E: rik@cybc.com.cy **W:** www.cybc.com.cy
Webcast: L
L.P: DG/CEO: Themis Themistocleous.
SW: Uses txs provided by BBC.
SAT: Apstar 2R, Hellas Sat 2, Optus D2.
kHz: *6180, 7210, 9760*

Winter Schedule 2010/2011

Greek	Days	Area	kHz
2215-2245	....fss	Eu	6180cyp, 7210cyp, 9760cyp

Ann: Greek: "Radiofonikon Idryma Kyprou".
IS: "Avkoritssa" (guitar).
V: QSL-card. Rec. acc.
Notes: The transmissions are relays of CBC's Home Sce prgrs.

BBC MIDDLE EAST RELAY STATION
✉ P.O. Box 54912, Limassol, Cyprus.

☎ +357 24332511. 📄 +357 24332595.
L.P: Senior Engineer: S. Welch.
MW: [CYP] Zygi: 1323kHz 200kW; [ZAK] Zakaki (located at Lady's Mile in the Akrotiri Sovereign Base Area): 639/720kHz 500kW.
SW: [CYP] Zygi: 2 x 250, 8 x 300kW.
V: QSL-card. (For direct report)
Notes: Owned by the BBC and operated by Babcock International Group plc (see under United Kingdom).

MONTE CARLO RADIODIFFUSION RELAY STATION
📄 Cape Gkreko, Cyprus.
MW: [CGR] Cape Gkreko: 990kHz 600kW (leased by IBB), 1233kHz 1200kW.
Notes: Transmitting station owned by Monte Carlo Radiodiffusion, see Monaco for corporate details.

CYPRUS - Northern

RADIO BAYRAK INTERNATIONAL (Gov)
📄 Bayrak Radio Television Corporation, P.O. Box 417, Lefkosa, via Mersin 10, Turkey.
☎ +90 392 2255555. 📄 +90 392 2254581.
E: brt@brtk.net **W:** www.brtk.net
Webcast: L
L.P: Chmn, BRTC: Yilmaz Baskaya; Head of R.Bayrak Int: Ülfet Kortmaz.
SW: [ISK] Yeni Iskele: 1 x 25kW.
FM/DAB: FM: 87.8MHz (Sinandagi, 10kW); 105.0MHz (Selvilitepe, 5kW).
SAT: Turksat 3A.
kHz: 6150

Winter Schedule 2010/2011
Greek/English/

Various	Days	Area	kHz
0000-2400	daily	Eu,ME,NAf	6150iskt, *

Key: † Irregular; * Running on lower power.
Ann: English: "This is Bayrak International, the Voice of the Turkish Republic of Northern Cyprus".
V: QSL-letter.
Notes: Foreign language channel of the state broadcaster Bayrak Radyo Televizyon Kurumu (BRTK). News broadcasts in the following languages/times:- English: 1200, 1700; Greek: 1230, 1730; Arabic, French, German & Russian: Mon-Wed and Fri, 1215-1240; Headlines in English/Greek: Mon-Fri, 1000 and 1400.

CZECH REPUBLIC (CZE)

RADIO PRAGUE (Pub)
📄 Vinohradská 12, 120 99 Praha 2, Czech Republic.
☎ +420 2 21552933. 📄 +420 2 21552903.
E: cr@radio.cz **W:** www.radio.cz
Webcast: D/L/P
L.P: Dir: Miroslav Krupicka; Editor in Chief: Gerald Schubert.
SW: Leased from Ceské Radiokomunikace A.S. and foreign relays.
SAT: Astra 3A (Also on WRN via Eurobird 1, Galaxy 25, Hot Bird 6).
kHz: 5930, 5980, 5995, 6055, 7345, 7355, 7410, 7420, 7435, 9790, 9855, 9880, 11600, 21745

Winter Schedule 2010/2011

Czech	Days	Area	kHz
0030-0100	daily	SAm	9790lit
0230-0300	daily	NAm	7410lit
0330-0400	daily	NAm	7345lit
0930-1000	daily	SEu	11600lit
1030-1100	daily	WAf	21745lit
1200-1230	daily	NEu	9880lit
1330-1400	daily	WEu	6055lit
1630-1700	daily	EAf,ME	7435lit
2030-2100	daily	WEu,SEu	5930lit
2200-2230	daily	SEu,SAm	5930lit

English	Days	Area	kHz
0100-0130	daily	NAm	7410lit
0200-0230	daily	NAm	7410lit
0400-0430	daily	NAm	7345lit
0430-0500	daily	EAf,ME	9855lit
0800-0830	daily	NEu	7345lit
1000-1030	daily	WAf	21745lit
1130-1200	daily	NEu	9880lit
1400-1430	daily	SAs	11600lit
1700-1730	daily	NEu	5930lit
1800-1830	daily	NEu	5930lit
2100-2130	daily	NEu	5930lit
2230-2300	daily	CAf,WAf	7355lit
2330-2400	daily	NAm	5930lit

French	Days	Area	kHz
0700-0730	daily	WEu	5930lit
0830-0900	daily	SEu,WAf	11600lit
1430-1500	daily	SEu,WAf	11600lit
1730-1800	daily	WEu	5930lit
1930-2000	daily	WEu,SEu	5930lit
2300-2330	daily	NAm	5930lit

German	Days	Area	kHz
0730-0800	daily	WEu	5930lit
1100-1130	daily	CEu	7345lit
1300-1330	daily	CEu	6055lit
1600-1630	daily	WEu	5930lit

Russian	Days	Area	kHz
0500-0530	daily	EEu	5980lit
1230-1300	daily	EEu	6055lit
1530-1600	daily	EEu	7420lit
1830-1900	daily	EEu	5995lit

Spanish	Days	Area	kHz
0000-0030	daily	SAm	9790lit
0130-0200	daily	CAm,NAm	7410lit
0300-0330	daily	SAm	7345lit
0900-0930	daily	SEu	11600lit
1500-1530	daily	SEu	11600lit
1900-1930	daily	SEu	5930lit
2000-2030	daily	SEu	5930lit
2130-2200	daily	SEu,SAm	5930lit

Ann: English: "Welcome to Radio Prague, the External Service of Czech Radio".
IS: Fanfare from Dvorák's 9th Symphony ("From the New World"), played on French horn.
V: QSL-card. Rec. acc. (Online form at: www.radio.cz/en/report)
Notes: Radio Prague (Ceský Rozhlas 7) is the External Sce of the public service Czech Radio (Ceský Rozhlas).

CESKÉ RADIOKOMUNIKACE A.S. (Tx Operator)
📄 U Nákladového nádrazí 3144, 130 00 Praha 3, Czech Republic.
☎ +420 242411111. 📄 +420 242417595.
E: spolecnost@radiokomunikace.cz
W: www.radiokomunikace.cz
📄 Mahlerovy sady 1, 130 00 Praha 3, Czech Republic. (Broadcasting Division)
☎ +420 242418901. 📄 +420 242418858.
L.P: CEO: Jane Hannah.
SW: [LIT] Litomyšl: 3 x 100kW.
Notes: Ceské Radiokomunikace A.S. is the national transmitter network owner.

DJIBOUTI (DJI)

IBB RELAY STATION DJIBOUTI
📄 IBB Transmitting Station, Djibouti.
MW: [DJI] Djibouti, Dorale: 1431kHz 600kW.

ECUADOR (EQA)

HCJB, LA VOZ DE LOS ANDES (Rlg)
📄 Villalengua OE2-52 y Av. 10 de Agosto, Quito, Pichincha, Ecuador 1717691.
☎ +593 2 2266808. 📄 +593 2 2267263.
E: vozandes@hcjb.org.ec **W:** www.radiohcjb.org
Webcast: L/P
L.P: Frequency Mgr: Douglas Weber.
SW: [QUI] Quito, Mount Pichincha: 1 x 10, 1 x 100kW (planned).
SAT: ALAS-HCJB.
kHz: 1251, 3955, 9835, 11920

Winter Schedule 2010/2011

Chechen	Days	Area	kHz
1600-1630	......s	RUS	3955sit

Dari	Days	Area	kHz
1530-1545	daily	CAs	1251dsb
German (High)	**Days**	**Area**	**kHz**
1700-1730	daily	Eu	3955sit
2330-2400	daily	SAm	9835sgo
German (Low)	**Days**	**Area**	**kHz**
1630-1700	daily	Eu	3955sit
2300-2330	daily	SAm	9835sgo
Hazaragi	**Days**	**Area**	**kHz**
1545-1600	m.....s	CAs	1251dsb
Kulina	**Days**	**Area**	**kHz**
2245-2300	daily	SAm	11920sgo
Portuguese	**Days**	**Area**	**kHz**
2300-0045	daily	SAm	11920sgo
Russian	**Days**	**Area**	**kHz**
1530-1600	......s	RUS	3955sit
Turkmen	**Days**	**Area**	**kHz**
1600-1615	daily	CAs	1251dsb
Uzbek	**Days**	**Area**	**kHz**
1545-1600	.twtfs.	CAs	1251dsb

Ann: Russian: "Golos na Andite".
V: QSL-card. Rp (1 IRC). Rec. acc., but cannot be returned.
Notes: HCJB Global branch and transmitting stn; for corporate details see under USA. Broadcasts in Spanish, Quichua, Waorani and Cofán will continue on 6050kHz and will be assumed by Vozandes Media. See National Radio section, under Ecuador.

EGYPT (EGY)

RADIO CAIRO (Gov)
✉ P.O. Box 1186, 11511 Cairo, Egypt.
☎ +20 2 25789461. 🖷 +20 2 25789491.
E: freqmeg@yahoo.com; enginfo@ertu.org (ERTU Engineering) **W:** www.ertu.org
Webcast: L (live.sis.gov.eg/live)
LP: Pres, ERTU: Osama El Sheikh; Chmn, Broadcasting Sector: Intesar Shalaby; Chmn, Engineering Sector: Hamdy Mounir.
SW: [ABS] Abis: 8 x 250, 1 x 500kW; [ABZ] Abu Zaabal: 13 x 100, 1 x 250, 4 x 500kW.
kHz: 6270, 9250, 9280, 9295, 9305, 9745, 9855, 9900, 9915, 9990, 11510, 11590, 11740, 11750, 12170, 13580, 15040, 15060, 15065, 15080, 15285, 15710, 17510, 17810, 17870

Winter Schedule 2010/2011

Afar	Days	Area	kHz
1600-1700	daily	EAf	15285abz
Albanian	**Days**	**Area**	**kHz**
1500-1600	daily	Eu	13580abz
Amharic	**Days**	**Area**	**kHz**
1730-1900	daily	EAf	15285abz
Arabic	**Days**	**Area**	**kHz**
0030-0430	daily	NAm	11590abz
0700-1100	daily	WAf	17510abz*
1015-1215	daily	WAs	15060abz
1300-1600	daily	WAf	15080abs
1700-2300	daily	SDN	9250abz
1900-0030	daily	CAf,EAf	9295abz**
1900-0700	daily	Eu,NAm	9305abs*
2000-2200	daily	Pac	9855abz
2330-0045	daily	SAm	9250abs, 9900abz
Dari	**Days**	**Area**	**kHz**
1300-1400	daily	WAs	15065abz
English	**Days**	**Area**	**kHz**
0200-0330	daily	NAm	6270abz
1215-1330	daily	SAs	17870abz
1600-1800	daily	CAf,SAf	12170abz
1900-2030	daily	WAf	11510abz
2115-2245	daily	Eu	6270abz
2300-0030	daily	NAm	11590abz
Farsi	**Days**	**Area**	**kHz**
1330-1530	daily	WAs	15040abz
French	**Days**	**Area**	**kHz**
2000-2115	daily	Eu	6270abz
2030-2230	daily	WAf	9280abs
German	**Days**	**Area**	**kHz**
1900-2000	daily	Eu	6270abz

Hausa	Days	Area	kHz
1800-2100	daily	WAf	9990abs
Indonesian	**Days**	**Area**	**kHz**
1230-1400	daily	SEA	15710abs
Italian	**Days**	**Area**	**kHz**
1800-1900	daily	Eu	6270abz
Pashto	**Days**	**Area**	**kHz**
1400-1600	daily	WAs	15065abz
Portuguese	**Days**	**Area**	**kHz**
2215-2330	daily	SAm	9900abz
Russian	**Days**	**Area**	**kHz**
1900-2000	daily	Eu	9280abs
Somali	**Days**	**Area**	**kHz**
1700-1730	daily	EAf	15285abz
Spanish	**Days**	**Area**	**kHz**
0045-0200	daily	LAm	9900abz, 9915abs
0045-0200	daily	NAm	6270abz
Swahili	**Days**	**Area**	**kHz**
0400-0600	daily	EAf,CAf	9745abz, 11740abs
1530-1730	daily	EAf,CAf	17810abz
Turkish	**Days**	**Area**	**kHz**
1700-1900	daily	ME	9280abs
Urdu	**Days**	**Area**	**kHz**
1600-1800	daily	SAs	6270abz
Uzbek	**Days**	**Area**	**kHz**
1500-1600	daily	CAs	11750abz

Key: * General prgr; ** "Voice of the Arabs" prgr. Note: These programmes are 1 hour earlier from May-Sept.
Ann: English: "You are tuned to Radio Cairo"; Arabic: "Sout al-Arab, min al Qahira", "Sowt-il Afrikiy min al-Qahira".
V: QSL-card.
Notes: Radio Cairo is the External Sce of the Egyptian Radio & TV Union (ERTU).

EQUATORIAL GUINEA (GNE)

RADIO AFRICA (Rlg)
✉ P.O. Box 3741, Cantonment, Accra, Ghana.
E: radioafrica@myway.com
W: www.radioafricanetwork.com; radiopanam.com/africa.htm
Webcast: Planned.
✉ 7011 Koll Center Pkwy Ste 250, Pleasanton, CA 94566-3253 USA. (PABC)
SW: [BAT] Bata: 1 x 50kW (R. Nacional de Guinea Ecuatorial site).
kHz: 15190

Winter Schedule 2010/2011

English	Days	Area	kHz
0800-1300	daily	Af	15190bat*
1700-2300	daily	Af	15190bat*

Key: * Broadcast times dependent on airtime bookings.
Ann: English: "Radio Africa".
V: QSL-card. (Rpt to US address)
Notes: Run by Pan American Broadcasting Inc. (see USA for corporate details). Actual transmission time varies acc. to bookings and therefore broadcasts appear to be irregular. The tx is leased from Radio Nacional de Guinea Ecuatorial (see National Radio section).

ESTONIA (EST)

TARTU PERERAADIO (Rlg)
✉ Annemõisa 8, 50708 Tartu, Estonia.
☎ +372 7488458. 🖷 +372 7488458.
E: pereraadio@pereraadio.ee **W:** www.pereraadio.ee
Webcast: D/P
L.P: Dir: Paavo Pihlak.
MW: [TTU] Tartu, Kavastu: 1035kHz 200kW.
Notes: Among others, rebroadcasts TWR prgrs for listeners abroad, see TWR schedule.

ETHIOPIA (ETH)

RADIO ETHIOPIA (Gov)
✉ P.O. Box 654, Addis Ababa, Ethiopia.
☎ +251 11 551011. 🖷 +251 11 552263.
E: etv2@ethionet.et **W:** www.erta.gov.et
L.P: Head: Moges Taffese; PD: Melesse Edea Beyi.
SW: [GJW] Geja Jewe: 3 x 100kW.
kHz: 7165, 9560

Winter Schedule 2010/2011

Afar	Days	Area	kHz
1300-1400	daily	EAf,ME	7165gjw†, 9560gjw±

Arabic	Days	Area	kHz
1400-1500	daily	EAf,ME	7165gjw†, 9560gjw±

English	Days	Area	kHz
1600-1700	daily	EAf,ME	7165gjw†, 9560gjw±

French	Days	Area	kHz
1700-1800	daily	ERI	7165gjw†
1700-1800	daily	EAf,ME	9560gjw±

Somali	Days	Area	kHz
1200-1300	daily	EAf,ME	7165gjw†, 9560gjw±

Key: ± Variable frequency; † Irregular.
Ann: English: "You're tuned to the External Service of Radio Ethiopia".
V: QSL-card.
Notes: External Sce of the national state broadcaster Radio Ethiopia.

FINLAND (FIN)

RADIO86
⌨ Pinninkatu 55, 33100 Tampere, Finland.
☎ +358 3 4108 9035. 📠 +358 3 4108 9001.
E: editors@radio86.com **W:** www.radio86.com
W: www.radio86.co.uk (English)
L.P: MD, Futuvision Media Ltd: Zhao Yinong; Editor in Chief: Jutta Valkeinen; Editors: Geni Raitisoja, Stina Björkell.
MW: Leased from Digita Oy.
kHz: 963

Winter Schedule 2010/2011

Estonian	Days	Area	kHz
0500-0600	daily	Eu	963por

Lithuanian	Days	Area	kHz
0600-0700	daily	Eu	963por

Notes: Radio86 is a subsidiary of FutuVision Media Ltd. (Finland), and a partner of China Radio International (CRI). Among other activities, the Finnish studio of Radio86 produces prgrs for CRI in Baltic languages. FutuVision Media Ltd. also arranges the relays of CRI prgrs via the Pori MW transmitter of Digita Oy (Finland), details of which are shown in the CRI schedule (see under China).

SCANDINAVIAN WEEKEND RADIO (SWR)
⌨ P.O. Box 99, FI-34801 Virrat, Finland.
☎ +358 400 995559. 📠 +358 3 4755776.
E: info@swradio.net **W:** www.swradio.net
L.P: Chief Editor: Esa Saunamäki; QSL Mgr: Alpo Heinonen.
MW: [VIR] Virrat, Liedenpohja: 1602kHz 0.1kW.
SW: [VIR] Virrat, Liedenpohja: 1 x 0.05, 1 x 0.1kW.
kHz: 1602, 5980, 6170, 11690, 11720

Winter Schedule 2010/2011

English/Finnish	Days	Area	kHz
0600-0900	.....s.	Eu	5980vir
0800-1400	.....s.	Eu	11720vir
0900-1500	.....s.	Eu	6170vir
1400-1700	.....s.	Eu	11690vir
1500-1900	.....s.	Eu	5980vir
1700-1900	.....s.	Eu	11720vir
1900-2200	.....s.	Eu	6170vir, 11690vir
2200-0600	....fs.	Eu	6170vir
2200-2200	....fs.	Eu	1602vir
2200-2300	....f..	Eu	11720vir
2300-0800	....fs.	Eu	11690vir

Key: * 1st day of month.
Ann: English: "You are listening to Scandinavian Weekend Radio".
V: QSL-card. Rp. (2 IRC/2 USD), (Rpt form available on website)
Notes: SWR is a licensed radio station run by volunteers, promoting the radio hobby. On air since 1st July 2000. Broadcasts every first Friday/Saturday of month and Christmas day.

DIGITA OY (Tx Operator)
⌨ P.O. Box 135 (Jämsänkatu 2), 00521 Helsinki, Finland.
☎ +358 20 411711.
E: communications@digita.fi **W:** www.digita.fi
L.P: Dir: Dr. Pauli Heikkilä.
MW: [POR] Pori: 963kHz 600kW.

FRANCE (F)

MONTE CARLO DOUALIYA (Gov)
⌨ Maison de la Radio, 116 Avenue du Président Kennedy, F-75220 Paris Cedex 16, France.
☎ +33 1 56401717. 📠 +33 1 56401700.
W: www.france24.com/ar
Webcast: D/L/P
L.P: CEO: Alain de Pouzilhac; DG: Genevieve Goëtzinger.
MW: Leased from Monte Carlo Radiodiffusion.
FM/DAB: FM: Throughout the Middle East and North Africa.
SAT: Astra 1H, Badr 6, Eurobird 2, Eutelsat W3A, Galaxy 19, Nilesat 101.
kHz: 1233

Winter Schedule 2010/2011

Arabic	Days	Area	kHz
0400-2020	daily	NAf,ME	1233cgr

Ann: Arabic: "Monte Carlo Doualiya".
V: QSL-card.
Notes: Part of RFI Group since 1996.

RADIO FRANCE INTERNATIONALE (RFI) (Gov)
⌨ Maison de la Radio, 116 Avenue du Président-Kennedy, F-75762 Paris Cedex 16, France.
☎ +33 1 56401212. 📠 +33 1 42303071.
E: english.service@rfi.fr **W:** www.rfi.fr
Webcast: L/P
⌨ B.P. 9516, F-75016 Paris Cedex 16, France.
W: www.rfi.ro
L.P: Pres/Chmn: Alain de Pouzilhac; CEO: Christine Ockrent; GM: Geneviève Goëtzinger; Dir, Communications: Françoise Hollmann.
MW/SW: Leased from TDF & foreign relays.
SAT: AB4, Afristar, Astra1H, Asiasat 3, AsiaStar, Anik F1/F1R, Atlantic Bird 3, Badr 6, Echostar 3, Eutelsat W2/W3/W4, Galaxy 3C/19, Hispasat 1C, Hot Bird 8, Intelsat 701/903/907, IS7/10, NIMIQ 1, NSS7, Optus D2, Sirius, Solidaridad 2.
kHz: 603, 684, 1098, 1296, 1503, 3965, 5900, 5905, 5925, 5995, 6145, 6175, 7205, 7215, 7220, 7315, 7325, 7340, 7350, 7375, 7380, 7425, 9565, 9765, 9790, 9795, 9800, 9805, 9825, 9835, 9955, 11605, 11665, 11670, 11690, 11700, 11705, 11725, 11790, 11830, 11860, 11875, 11995, 12015, 13640, 13680, 13695, 13750, 15160, 15170, 15300, 15315, 15360, 15515, 15530, 15605, 15680, 17610, 17620, 17630, 17660, 17850, 21580, 21620, 21690

Winter Schedule 2010/2011

Chinese	Days	Area	kHz
0930-1030	daily	EAs	5900irk, 7325tnn, 11875tnn
2200-2300	daily	EAs	1098kou, 7350nvs
2200-2400	daily	EAs	603luk, 11665tnn
2300-2400	daily	EAs	9955tnn

English	Days	Area	kHz
0400-0430	mtwtf..	EAf	7315iss, 9805iss
0500-0530	mtwtf..	EAf	9805iss, 11995iss
0600-0630	mtwtf..	EAf	13680iss, 15160iss
0600-0630	mtwtf..	CAf,WAf	7315iss*, 9765iss**
0700-0730	daily	WAf,CAf	11725iss*, 15605iss**
1200-1230	daily	EAf	21620iss
1600-1700	daily	Af	15605iss

French	Days	Area	kHz
0400-0500	daily	EAf	7215iss
0400-0700	daily	Af	9790iss
0500-0600	daily	CAf	7340iss*, 11700iss**
0500-0700	daily	CAf	11605mey
0600-0700	daily	CAf	11700iss*, 15300iss**
0600-0700	daily	NAf,WAf	5925iss*, 7340iss**
0700-0800	daily	NAf,WAf	9790iss*, 11700iss
0700-0800	daily	CAf	15170mey
0700-1000	daily	CAf	17850iss
0700-1000	daily	NAf,WAf	13695iss
0700-1800	daily	NAf,CAf	15300iss
0800-1600	daily	WAf	17620iss
1100-1130	daily	SEA	15680tnn
1130-1200	daily	NAm	13640guf
1130-1200	daily	Atl	6175iss
1130-1200	daily	CAm	17610iss
1200-1300	daily	CAf	17660mey

French	Days	Area	kHz
1200-1330	daily	CAf	21580iss
1300-1400	daily	SEA	684dof
1600-1700	daily	CAf,WAf	17850iss
1600-1700	daily	SEA	1296kun
1700-1800	daily	NAf,WAf	13695iss
1700-1800	daily	CAf,WAf	11995iss*
1700-2000	daily	CAf	11705iss
1700-2000	daily	WAf	21690guf
1800-1900	daily	NAf,WAf	11995iss
1800-1900	daily	WAf,CAf	9790iss*, 15300iss**
1800-2000	daily	NAf,WAf	13695iss**
1900-2000	daily	WAf	6175iss*
1900-2000	daily	Eu	3965iss+
1900-2100	daily	WAf,CAf	9790iss
2000-2200	daily	WAf	6175iss
2000-2200	daily	CAf,WAf	7205iss
2200-1800	daily	Eu	3965iss+

Hausa	Days	Area	kHz
0600-0630	daily	WAf,CAf	6145iss*, 7220iss, 9805iss**, 11690iss**
0700-0730	daily	CAf	13750iss*
0700-0730	daily	WAf	9805iss*, 11830iss
0700-0730	daily	WAf,CAf	15315iss**
1600-1700	daily	WAf,CAf	15315iss

Khmer	Days	Area	kHz
1200-1300	daily	SEA	1503fan

Persian	Days	Area	kHz
1430-1500	daily	ME	15360iss, 17850iss
1700-1800	daily	ME	9795iss

Portuguese	Days	Area	kHz
0600-0700	daily	CAf	11830mey
1700-1800	daily	Af	12015iss*, 15530iss

Russian	Days	Area	kHz
1400-1430	daily	Eu	11860iss*, 15605iss, 17850iss**
1600-1630	daily	Eu	9800iss, 11670iss
1900-2000	daily	Eu	5905iss, 7425iss

Spanish	Days	Area	kHz
0100-0130	daily	CAm	5995guf
1000-1030	daily	CAm	7375guf, 9825guf
1200-1230	daily	CAm	15515guf
2100-2130	daily	CAm	17630guf

Swahili	Days	Area	kHz
0430-0500	daily	EAf,CAf	7340mey*, 9835mey**
0530-0600	daily	EAf,CAf	11790mey
1500-1600	.....s.	CAf,EAf	12015mey

Vietnamese	Days	Area	kHz
1400-1500	daily	SEA	7380tnn
1500-1600	daily	SEA	1296kun, 9565tsh

Key: + DRM; * to Feb 28; ** from 1 Mar.
Ann: French: "Ici Paris, Radio France Internationale".
V: QSL-card.
Notes: RFI is the External Sce of the public broadcaster Radio France. RFI is a subsidiary of the holding company "Audiovisuel Exterieur de la France (AEF)". For RFI Arabic prgrs, see under Monte Carlo Doualiya.

GOLOS PRAVOSLAVIYA
(LA VOIX DE L'ORTHODOXIE) (Rlg)
BP 20-416, F-75366 Paris Cedex 08, France.
☎ +33 1 49770366. 🖷 +33 1 43534066.
E: voix.orthodoxie@wanadoo.fr
W: www.voixorthodoxie.org; www.russie.net/orthodoxie/vo
Webcast: L (via R. Grad Petrov, www.grad-petrov.ru)
nab. Leyt.Shmidta 39, 199034 St.Petersburg, Russia. (Studio in Russia)
☎ +7 812 3232867. 🖷 +7 812 3232867.
L.P: Gen Sec: Michel Woolff.
kHz: 7515

	Winter Schedule 2010/2011		
Russian	**Days**	**Area**	**kHz**
1530-1600	.t.f..	RUS	7515alm

V: QSL-card.
Notes: Russian-orthodox prgr, launched in 2000.

TÉLÉDIFFUSION DE FRANCE S.A. (TDF) (Tx Operator)
106 Avenue Marx Dormoy, 92541 Montrouge Cedex, France.

☎ +33 1 49651000. 🖷 +33 1 46574850.
W: www.tdf.fr
10 rue d'Oradour-sur-Glane, F-75732 Paris Cedex 15, France. (Radio Division)
☎ +33 1 55951553. 🖷 +33 1 55952137.
L.P: GD: Philippe Levrier; Dir, Communications: Laure Frugier.
SW: [ISS] Issoudun: 21 x 500kW.
V: QSL-card. (For RFI and other broadcaster relays via ISS)
Notes: TDF is the national French transmitter network operator, with shortwave transmitting facilities in France and French Guiana.

FRENCH GUIANA (GUF)

TÉLÉDIFFUSION DE FRANCE RELAY STATION
Direction TDF Outre-Mer, BP 7024, 97307 Cayenne Cedex, French Guiana.
☎ +594 594350550.
E: fabrice.esnay@tdf.fr
SW: [GUF] Cayenne, Montsinéry: 1 x 250, 4 x 500kW.
V: QSL-card.

GABON (GAB)

AFRICA NO.1 (Comm)
33 Rue du Faubourg Saint Antoine, F-75011 Paris.
☎ +33 1 55075801. 🖷 +33 1 55079748.
E: guykalenda@africa1.com **W:** www.africa1.com
Webcast: L/P
BP 1, Libreville, Gabon.
☎ +241 04030255. 🖷 +241 742133.
L.P: Dir, Admin/ Finance: Cecile Hatchy; Coordinator/Producer: Guy Kaleda.
SW: [GAB] Moanda, Moyabi: 3 x 500kW.
FM/DAB: FM Relays in some African countries.
SAT: Atlantic Bird 3, NSS 7.
kHz: 9580

	Winter Schedule 2010/2011		
French	**Days**	**Area**	**kHz**
0500-2300	daily	Af	9580gab

Ann: French: "Africa Numéro Un".
V: QSL-card.
Notes: Africa 1 is produced by Africa Média S.A.; a subsidiary of the Libyan state broadcaster LJBC.

GERMANY (D)

DEUTSCHE WELLE (DW) (Pub)
D-53110 Bonn, Germany.
☎ +49 228 4290. 🖷 +49 228 4293000.
E: info@dw-world.de **W:** www.dw-world.de
Webcast: D/L/P
Kurt-Schumacher-Str. 3, 53113 Bonn, Germany. (Studio)
L.P: DG: Erik Bettermann; Head, DW-Radio/dw-world.de: Christian Gramsch; Head, DW-TV: Christoph Lanz; MD, Distribution: Guido Baumhauer.
SW: Via leased foreign relays.
SAT: AMC 1, AsiaSat 3S, Atlantic Bird 3, Badr 4, Express AM1, Hot Bird 8, Intelsat 7/9, Nilesat 102, Nimiq 4, Superbird C2.
kHz: 693, 999, 1188, 1548, 3995, 5830, 5845, 5905, 5915, 5925, 5945, 5965, 6030, 6050, 6075, 6090, 6130, 6155, 6180, 6225, 7240, 7280, 7285, 7300, 7380, 7395, 7405, 7410, 7425, 7470, 9380, 9430, 9445, 9450, 9485, 9510, 9535, 9545, 9560, 9635, 9655, 9690, 9715, 9720, 9735, 9755, 9785, 9800, 9855, 9865, 11605, 11645, 11665, 11690, 11695, 11720, 11725, 11830, 11855, 11865, 11875, 11945, 12005, 12010, 12025, 12035, 12045, 12055, 12065, 12070, 13590, 13625, 13735, 13780, 13810, 15205, 15275, 15410, 15440, 15600, 15620, 15640, 17520, 17700, 17710, 17770, 17800, 17860, 21550, 21560, 21780

	Winter Schedule 2010/2011		
Amharic	**Days**	**Area**	**kHz**
1400-1500	daily	ETH	11645kig, 15640trm
Bengali	**Days**	**Area**	**kHz**
1530-1600	daily	SAs	1548trm
Chinese	**Days**	**Area**	**kHz**
1300-1330	daily	CHN	12010sng
1300-1400	daily	CHN	6225alm, 9380dsb, 11945trm, 13735trm
2300-2400	daily	CHN	6090kim, 9865sng, 11830ppk

Dari	Days	Area	kHz
0830-0900	daily	AFG	15640dha, 17710trm
1330-1400	daily	AFG	12065trm, 17860rmp
English	**Days**	**Area**	**kHz**
0000-0100	daily	EAs,SEA	9445trm, 9785trm, 11855sng
0200-0300	daily	EAs	15205trm+
0300-0400	daily	SAs	1548trm, 11695trm
0400-0500	daily	Af	5905sin, 5945sin, 6180kig, 9450kig, 15600trm
0500-0530	daily	Af	6130sin, 6155sin, 6180kig, 9755kig, 12045kig
0600-0630	daily	WAf	5945wof, 7240sin, 15205kig
0900-1000	daily	EAs	17710trm, 21780trm
1600-1700	daily	SAs	1548trm, 5845nak+, 5965trm, 9560trm, 13590trm+
1700-1800	daily	SAs	1548trm+, 5845nak+
1900-1930	daily	Af	9735kig, 13780trm, 15275sin
2000-2100	daily	Af	9735kig, 13780trm, 15275trm
2000-2100	daily	WAf	9690wof
2100-2200	daily	WAf	7280sin, 9545trm, 11865kig, 13780trm
Farsi	**Days**	**Area**	**kHz**
1830-1930	daily	IRN	5925kch, 7470smf
French	**Days**	**Area**	**kHz**
1200-1300	daily	Af	15275kig, 15440kig, 17520rmp, 17800kig, 21550sin
1700-1800	daily	Af	9535kig, 12035wof, 13625trm, 15275wof, 17800sin
German	**Days**	**Area**	**kHz**
0000-0200	daily	SAs	1548trm, 6225kch*, 7285trm, 7395rmp**
0000-0200	daily	CAm	9655kig, 11665asc, 12025kig
0000-0800	daily	Eu	6075sin
0200-0400	daily	Eu,ME	6075rmp
0400-0500	daily	Eu	13780dha**
0400-0600	daily	Af	6075wof, 13780arm*, 17800trm
0500-0600	daily	Eu	6075skn
0500-0800	daily	Eu	3995skn
0600-0700	daily	NAf	7410wof
0600-0800	daily	Eu	6075wof
0600-0800	daily	Eu,Af	9545wof***
0600-0800	daily	SAf	12005kig
0600-0800	daily	WAf	15275kig
0800-1000	daily	AUS,NZL	9450bon
0800-1000	daily	SEu,ME	9545skn
0800-1100	daily	SEu,ME	13780skn
0800-1200	daily	Eu	6075skn
0800-1200	daily	SEA,AUS	17520trm
1000-1100	daily	Car,SAm	9865hri
1000-1200	daily	CAm,NAm	5905bon
1000-1200	daily	SEA	21780trm
1100-1200	daily	SAm	17770asc
1100-1400	daily	SEu,ME	13780sin
1200-1400	daily	SAs	15640trm
1200-1400	daily	SEA	21780rmp
1200-1430	daily	SAs	1548trm
1200-2000	daily	Eu	6075wof
1400-1600	daily	SEu,ME	13780trm, 15275kig, 17800sin
1600-1700	daily	Af	12070kig
1600-1800	daily	Af	12055kig
1600-1800	daily	EAf	9545wof
1600-1800	daily	SEu,ME	13780rmp
1600-2200	daily	Eu	3995skn

German	Days	Area	kHz
1700-1800	daily	Af	12070trm
1800-1900	daily	Af	13780trm
1800-2000	daily	Af	9545rmp, 11725kig, 12070wof, 15640sin
2000-2100	daily	Eu	12070kig
2000-2200	daily	AUS,SEA	9510trm, 11605kig
2000-2400	daily	Eu	6075sin
2200-2400	daily	SEA	11875dha
2200-2400	daily	CAm	12025kig
2200-2400	daily	SAm	11865sin, 15640hri
2300-2400	daily	SEA	6050trm
Hausa	**Days**	**Area**	**kHz**
0630-0700	daily	WAf	7240sin, 12045kig, 15205kig
1300-1400	daily	WAf	15275kig, 17800kig, 21560sin
1800-1900	daily	WAf	9430kig, 11690kig, 15275sin
Indonesian	**Days**	**Area**	**kHz**
1200-1300	daily	INS	9655trm, 15620trm
2200-2300	daily	INS	7380trm, 9720kig, 11605mdc
Pashto	**Days**	**Area**	**kHz**
0800-0830	daily	AFG	15640dha, 17710trm
1400-1430	daily	AFG	12065trm, 17860rmp
Portuguese	**Days**	**Area**	**kHz**
0530-0600	daily	Af	9800kig, 12045kig
1930-2000	daily	Af	9735kig, 13780trm, 15275sin
Russian	**Days**	**Area**	**kHz**
0100-0200	daily	FER	7405rmp, 15640trm, 17700trm
0200-0300	daily	RUS,CAs	5905sin
0200-0400	daily	RUS,CAs	15640trm
0300-0400	daily	RUS,CAs	5905rmp, 9800sin
0400-0500	daily	RUS,CAs	13810kig, 17700trm
0500-0600	daily	RUS,Eu	17700trm
0500-0630	daily	RUS,Eu	5915rmp, 15640kig
0600-0630	daily	RUS,Eu	999kch*
1500-1600	daily	RUS	11720rmp
1500-1700	daily	RUS	13625wof
1500-1800	daily	RUS	9715wof, 15640sin
1600-1700	daily	RUS	999kch*, 11865rmp
1700-1800	daily	RUS	11605kig, 11645trm
1800-2000	daily	RUS	5830tac*, 9635trm, 9715sin, 11605kig
1900-2000	daily	RUS	999kch*
2000-2100	daily	RUS	6180rmp, 7425trm
Russian/ Belarusian	**Days**	**Area**	**kHz**
2000-2030	mtwtf..	BLR	6030rmp
Russian/German	**Days**	**Area**	**kHz**
0000-1700	daily	RUS	1188spb*
0000-2400	daily	RUS	693msk*
2000-2400	daily	RUS	1188spb*
Swahili	**Days**	**Area**	**kHz**
0300-0400	daily	EAf	5925kig, 6180kig, 9485asc, 9855sin, 15600trm
1000-1100	daily	EAf	9800kig, 12045kig, 15410kig
1500-1600	daily	EAf	7300kig, 9800kig, 11645kig
Urdu	**Days**	**Area**	**kHz**
1430-1500	daily	SAs	1548trm, 12065trm
1700-1730	daily	SAs	15620mos

Key: + DRM; * to 31 Dec; ** from 1 Jan; *** from 1 Feb. For joint DW/BBC DRM broadcasts, see "DRM International Transmissions".
Ann: English: "This is Deutsche Welle Radio"; German: "Hier ist die Deutsche Welle"; Russian: "Nemetskaya Volna".
V: QSL-card. (Rpt to DW Customer Service)
Notes: Deutsche Welle is a public service broadcaster established to serve German nationals abroad as well as foreign listeners around the world. Transmissions in some Asian languages are jammed. Relayed

via local FM/MW stations in Africa, Asia and Europe.

RADIO 700 (Comm)
✉ Kuchenheimer Str. 155, D-53881 Euskirchen, Germany.
☎ +49 2251 921300. 🖷 +49 2251 921303.
E: info@radio700.eu
W: www.radio700.de; www.radio700.eu
Webcast: L
L.P: Project Coordinator: Bernd Frinken.
SW: [KLL] Kall, Krekel: 1 x 1kW.
FM/DAB: FM: 90.1MHz (Elsenborn, 1 kW); 101.7MHz (Sankt Vith, 1 kW), both in Belgium, near German border.
kHz: 6005

	Winter Schedule 2010/2011		
German	**Days**	**Area**	**kHz**
0800-1800	daily	Eu	6005kll

Notes: Produced by Funkhaus Euskirchen e.V.

EVANGELISCHE MISSIONS-GEMEINDEN (Rlg)
✉ Jahnstrasse 9, D-89182 Bernstadt, Germany.
☎ +49 7348 948026. 🖷 +49 7348 948027.
kHz: 6055, 9605, 11840

	Winter Schedule 2010/2011		
German	**Days**	**Area**	**kHz**
1130-1200	.....ss	Eu	6055wer
Russian	**Days**	**Area**	**kHz**
1200-1230	.....s.	RUS	11840nau
1600-1630	.....s.	RUS,Eu	9605wer

V: QSL-card.
Notes: Rebroadcasts prgrs of various Protestant missions.

LUTHERISCHE STUNDE (Rlg)
✉ Postfach 1162, 27363 Sottrum, Germany.
☎ +49 4264 2436. 🖷 +49 4264 2437.
E: info@lutherischestunde.de **W:** www.lutherischestunde.de
Webcast: D/L
L.P: CEO: Manfred R. Weingarten.
kHz: 1215, 1323, 7310

	Winter Schedule 2010/2011		
German	**Days**	**Area**	**kHz**
1845-1900	..w....	Eu	1215klg, 1323wbr, 7310msk

V: QSL-card. (Email to: p.schmid@lutherischestunde.de)

MISSIONSWERK HEUKELBACH (Rlg)
✉ Sülemickerstraße 15, D-51700 Bergneustadt, Germany.
☎ +49 2261 9450. 🖷 +49 2261 94537.
E: info@missionswerk-heukelbach.de
W: www.missionswerk-heukelbach.de
Webcast: D/L (live on: rtl1440.com)
kHz: 630, 693, 1215, 1323, 1431, 1440, 7310

	Winter Schedule 2010/2011		
German	**Days**	**Area**	**kHz**
0415-0430	mtwtf..	Eu	1440mrn
0445-0500	.....s.	Eu	1440mrn
0500-0515	......s	Eu	1440mrn
0615-0630	......s	Eu	1440mrn
1845-1900	daily	Eu	1440mrn
1945-2000	mt.tfss	Eu	630klu, 693bln, 1215klg, 1323wbr, 1431dsd, 7310msk

V: QSL-card.

RADIO SANTEC (Rlg)
✉ Postfach 5643, D-97006 Würzburg, Germany.
☎ +49 931 3903264. 🖷 +49 931 3903195.
E: info@radio-santec.com **W:** www.radio-santec.com
Webcast: D/L
✉ Marienstrasse 1, D-97070 Würzburg, Germany. (Studio)
kHz: 630, 693, 1215, 1323, 1431, 7310

	Winter Schedule 2010/2011		
German	**Days**	**Area**	**kHz**
1200-1300	.....s.	Eu	1323wbr
1600-1700	..t...	Eu	630klu, 693bln, 1431dsd
1800-1900	......s	Eu	1215klg, 1323wbr, 7310msk

Ann: English: "You are listening to Radio Santec"; German: "Hier

hören Radio Santec".
V: QSL-card.
Notes: Radio ministry of the Universal Life Church (Universelles Leben).

HAMBURGER LOKALRADIO
✉ c/o Kulturzentrum Lola, Lohbrügger Landstrasse 8, D-21031 Hamburg, Germany.
☎ +49 40 72692422. 🖷 +49 40 72692423.
E: redaktion@hamburger-lokalradio.de **W:** www.hhlr.de
L.P: Editor-in-Chief: Michael Kittner.
kHz: 6045

	Winter Schedule 2010/2011		
German	**Days**	**Area**	**kHz**
1000-1100	......s	Eu	6045wer

Key: * 1st Sunday of month.
Notes: Hamburger Lokalradio is a non-commercial community radio station.

MV BALTIC RADIO
✉ Seestrasse 17, D-19089 Göhren, Germany.
☎ +49 3861 301380. 🖷 +49 3861 3029720.
E: info@mvbalticradio.de **W:** www.mvbalticradio.de
L.P: Producer: Roland Rohde.
kHz: 6140

	Winter Schedule 2010/2011		
German	**Days**	**Area**	**kHz**
1000-1100	......s	Eu	6140wer

Ann: German: "Hier ist MV Baltic Radio. Wir sendet auf 6140 kHz in Europaband".
V: QSL-card.
Notes: Produced by R&R Medienservice. Provides relays for other prgr producers on the 3rd and 4th Sunday of the month.

RADIO ÖÖMRANG
✉ Tanenwai 24, D-25946 Nebel-Westerheide, Germany.
☎ +49 4682 2688. 🖷 +49 4682 2262.
E: familie-koelzow@t-online.de
L.P: Producer: Arjan Kölzow.
Ann: English: "The Free Voice of Frisian People, welcome to our broadcast".
Notes: Radio Öömrang ("Radio Amrum") is a once-a-year broadcast in the Frisian language, produced by the radio amateur Arjan Kölzow on the island of Amrum in North Germany. First broadcast in February 2006.

IBB RELAY STATIONS GERMANY
✉ IBB Transmitting Station Lampertheim, Postfach 1145, D- 68601 Lampertheim, Germany.
L.P: SM: James Lambert.
SW: [BIB] Biblis: 11 x 100kW; [LAM] Lampertheim: 9 x 100kW.

MEDIA BROADCAST GMBH (Tx Operator)
✉ Joseph-Schumpeter-Allee 17, 53227 Bonn, Germany.
☎ +49 228 55055022. 🖷 +49 228 55055019.
E: info@media-broadcast.com
W: www.media-broadcast.com
✉ Raimundstrasse 48-54, 60431 Frankfurt, Germany. (Int Sales)
☎ +49 69 959565510. 🖷 +49 69 959565519.
L.P: Chmn: Patrick Babin; CEO: Helmut Egenbauer.
MW: [BLN] Berlin, Zehlendorf: 693kHz 250kW; [DSD] Dresden, Wilsdruff: 1431kHz 150/250kW; [KLU] Königslutter: 630kHz 16/100kW; [WBR] Wachenbrunn: 1323kHz 150/1000kW.
SW: [NAU] Nauen: 2 x 100 (plus 1 x 100kW awaiting installation), 4 x 500kW; [WER] Wertachtal: 2 x 100, 14 x 500kW.
V: QSL-card. (For relayed stns, Email rpts: qsl-shortwave@media-broadcast.com).
Notes: Media Broadcast GmbH, part of the TDF Group (France), is a major transmitter network operator in Germany and owns the SW transmitting centres in Nauen and Wertachtal.

GREECE (GRC)

RADIOFONIKOS STATHMOS MAKEDONIAS (ERT3) (Pub)
✉ Aggelaki 2, 546 36 Thessaloniki, Greece. (Admin)
☎ +30 2310299400. 🖷 +30 2310299550.
E: eupro@ert3.gr **W:** www.ert3.gr
Webcast: L
✉ Aggelaki 14, 546 36 Thessaloniki, Greece. (Radio)

L.P: Chmn: Tassos Spiliopoulos; GM: Thalia Ioannidou-Karathanassi; Dir, Radio: Xanthos Chytas.
SW: Via ERT shortwave station in Kalochori-Pantichi (for tx details, see Voice of Greece (ERA5)).
SAT: Hellas Sat 2.
kHz: 7450, 9935

Winter Schedule 2010/2011

Greek	Days	Area	kHz
1100-1650	daily	Eu	9935avl
1700-2250	daily	Eu	7450avl

Ann: Greek: "Edo Thessaloniki, Radiofonikos Stathmos Makedonias, Trito Programma, Vrahea".
V: QSL-card.
Notes: ERT3 ("Radio Station Macedonia") is a regional station of the public broadcaster ERT, transmitting on MW & FM in the Macedonia region in N.Greece (see National Radio section) and on SW for Greek listeners in Europe.

VOICE OF GREECE (ERA5) (Pub)
✉ Messogion 432, 15342 Aghia Paraskevi Attikis, Athens, Greece.
☎ +30 2106066310. 📠 +30 2106066309.
E: era5@ert.gr **W:** www.voiceofgreece.gr
Webcast: D/L
L.P: Dir: Zinovia Sirivli; Head, Foreign Language Broadcast Dept: Sihanis Zilber; Head, Greeks Living Abroad Dept: Ageliki Barka; Frequency Mgr: Demetri Vafeas.
SW: [AVL] Vathy (Avlida municipality), Kalochori-Pantichi: 2 x 100kW, 1 x 250kW. One 100kW tx is used for the ERT3 prgr (see Radiofonikos Stathmos Makedonias).
FM/DAB: FM: 107.0MHz (Athens).
SAT: Anik F3, Apstar 2R, EchoStar 15, Eutelsat W6, Hellas Sat 2, Hot Bird 6, Intelsat 8/10-02, NSS 806, Optus D2.
kHz: 7450, 7475, 9420, 11645, 12105, 15630, 15650

Winter Schedule 2010/2011

English	Days	Area	kHz
0600-0700	daily	Eu	11645avl*
0900-1000	......s	Eu	15630avl
0900-1000	......s	Eu,As,Am	9420avl

French	Days	Area	kHz
0700-0800	daily	Eu	11645avl**

German	Days	Area	kHz
0900-0930	m.wtfss	Eu	11645avl

Greek	Days	Area	kHz
0300-0550	daily	ME,SAs,Pac	7450avl
0700-0800	daily	Eu	15630avl
0800-0900	m.wtfss	Eu,As,Am	9420avl
0800-0900	m.wtfss	Eu	15630avl
0900-1000	m.wtfs.	Eu,As,Am	9420avl
0900-1000	m.wtfs.	Eu	15630avl
1100-1200	m.wtfss	Eu,Af,SAm	15650avl
1100-1200	m.wtfss	Eu,As,Am	9420avl
1200-0800	daily	Eu,As,Am	9420avl
1200-1550	daily	Eu,Af,SAm	15650avl
1600-1950	daily	Eu,Af,SAm	15630avl
2000-0650	daily	Eu,Am	7475avl
2300-0250	daily	ME,SAs,Pac	12105avl

Russian	Days	Area	kHz
0930-1000	m.wtfss	Eu	11645avl

Spanish	Days	Area	kHz
0800-0900	daily	Eu	11645avl

Key: * BBC relay Mon-Fri; ** RFI relay Mon-Fri.
Ann: Greek: "ERA pente, Foni tis Elladas".
IS: The opening notes of the Greek folk song "Tsopanakos imouna" (Once I was a Shepherd Boy), played on flute and sheep bells.
V: QSL-card. (Also accepts email rpt to: apodimos_era5@ert.gr)
Notes: The Voice of Greece (ERA5) is the External Sce of the public broadcaster Elliniki Radiofonia Teleorassi (ERT). Some prgrs are relays of ERT's domestic multilingual channel "R.Filia".

GUAM (GUM)

ADVENTIST WORLD RADIO GUAM (KSDA) (Rlg)
✉ P.O. Box 8990, Agat, Guam 96928.
☎ +1 671 5652289. 📠 +1 671 5652983.
E: aproffice@awr.org **W:** www.awr.org
L.P: SM: Victor Shepherd; Chief Engineer: Brook Powers.

SW: [SDA] Agat, Facpi Point: 4 x 100kW. (+ 1 x 100kW used as maintenance backup)
kHz: 5985, 9585, 9625, 9655, 9720, 9790, 9800, 9810, 9920, 11655, 11685, 11690, 11695, 11700, 11730, 11770, 11825, 11850, 11855, 11870, 11895, 11935, 11940, 11955, 11965, 11995, 12025, 12035, 12105, 15255, 15260, 15320, 15370, 15495, 15660, 17635, 17880

Winter Schedule 2010/2011

Assamese	Days	Area	kHz
1330-1400	..w...s	IND	15660sda

Bengali	Days	Area	kHz
1300-1330	daily	As	15660sda

Burmese	Days	Area	kHz
0000-0030	daily	As	17635sda
1430-1500	daily	As	11770sda

Cebuano	Days	Area	kHz
1030-1100	....fs.	SEA	11870sda

Chin	Days	Area	kHz
1400-1430	daily	As	11940sda

English	Days	Area	kHz
1330-1400	.t...s	SEA	11935sda
1330-1400	mt...s.	As	15660sda
1500-1530	daily	IND	12025sda
1600-1630	daily	IND	9585sda, 11690sda
1630-1700	daily	IND	9790sda
2230-2300	daily	SEA	15320sda

Hindi	Days	Area	kHz
1530-1600	daily	IND	12105sda

Hmong	Days	Area	kHz
1330-1400	...tf..	SEA	15660sda

Ilocano	Days	Area	kHz
1030-1100	......s	SEA	11870sda

Ilonggo	Days	Area	kHz
1030-1100	..wt...	SEA	11870sda

Indonesian	Days	Area	kHz
1100-1130	daily	SEA	15495sda
2200-2230	daily	SEA	11965sda

Javanese	Days	Area	kHz
1130-1200	m.w.f.	SEA	15260sda
2200-2230	m.w.f..	SEA	11850sda

Kannada	Days	Area	kHz
1530-1600	daily	IND	11690sda

Karen	Days	Area	kHz
0030-0100	daily	SEA	17635sda
1430-1500	mtwtf.s	SEA	11940sda

Khmer	Days	Area	kHz
1300-1330	...tfss	SEA	11935sda

Korean	Days	Area	kHz
1200-1300	daily	EAs	9800sda
2100-2200	daily	EAs	5985sda

Lao	Days	Area	kHz
1330-1400	...t.s.	SEA	11935sda

Malay	Days	Area	kHz
1300-1330	mtw....	SEA	11935sda

Malayalam	Days	Area	kHz
1530-1600	daily	IND	11955sda

Mandarin	Days	Area	kHz
0000-0200	daily	CHN	17880sda
0000-0200	daily	EAs	12035sda
0100-0200	daily	CHN	17635sda
1000-1100	daily	CHN	11995sda, 15260sda
1100-1200	daily	CHN	11825sda, 12035sda
1100-1200	daily	EAs	11730sda
1200-1300	daily	CHN	11825sda, 12035sda
1200-1300	daily	EAs	11855sda
1300-1400	daily	CHN	9920sda
1400-1500	daily	CHN	9810sda, 12035sda
2100-2200	daily	CHN	9720sda
2100-2200	daily	EAs	9625sda
2200-2300	daily	CHN	11895sda
2200-2300	daily	EAs	11685sda
2300-2400	daily	CHN	15370sda
2300-2400	daily	EAs	11700sda

Marathi	Days	Area	kHz
1530-1600	daily	IND	15495sda

Mizo	Days	Area	kHz
1500-1530	daily	IND	11695sda
Mongolian	**Days**	**Area**	**kHz**
1030-1100	daily	EAs	11730sda
Russian	**Days**	**Area**	**kHz**
0300-0330	daily	RUS	17635sda
1330-1400	daily	RUS	9655sda
Sinhala	**Days**	**Area**	**kHz**
1400-1430	daily	SEA	15255sda
Sundanese	**Days**	**Area**	**kHz**
1130-1200	.t.t.s.	SEA	15260sda
2200-2230	.t.t.s.	SEA	11850sda
Tagalog	**Days**	**Area**	**kHz**
1030-1100	mt.....	SEA	11870sda
Tamil	**Days**	**Area**	**kHz**
1500-1530	daily	IND	11685sda
Telugu	**Days**	**Area**	**kHz**
1500-1530	daily	IND	11655sda
Thai	**Days**	**Area**	**kHz**
1330-1400	m.w.f.	SEA	11935sda
Urdu	**Days**	**Area**	**kHz**
1600-1630	daily	IND	9790sda
Vietnamese	**Days**	**Area**	**kHz**
2300-2400	daily	VTN	15320sda

Ann: English: "From the beautiful island of Guam in the West Pacific, this is Adventist World Radio, the Voice of Hope".
V: QSL-card. (Rp not required, but rpt must be sent to: AWR Asia/Pacific Region, Ruko Palm Spring, Blok A-4, # 6-8, Batam Center, 29461, Batam, Indonesia)
Notes: AWR branch and transmitting stn; for corporate headquarters, see under USA.

TWR ASIA (KTWR) (Rlg)
✉ P.O. Box 8780, Agat, Guam 96928.
☎ +1 671 8288637. 🖷 +1 671 8288636.
E: ktwrfreq@guam.twr.org **W:** www.ktwr.net
Webcast: D
✉ Block 750C, Chai Chee Road #02-16/17, Technopark @ Chai Chee, Singapore 469003.
☎ +65 6444 8661. 🖷 +65 6444 3053.
E: info@twr.asia **W:** www.twr.asia
SW: [TWR] Merizo: 5 x 100kW (one tx used as backup).
kHz: 9345, 9370, 9585, 9910, 9920, 9975, 11590, 11620, 11680, 11840, 11870, 12075, 12105, 13765, 15170, 15200

Winter Schedule 2010/2011

Assamese	Days	Area	kHz
1330-1400	mtwtf..	As	12075twr
Balinese	**Days**	**Area**	**kHz**
0900-0915	mt..fss	INS	15200twr
Bengali	**Days**	**Area**	**kHz**
1315-1330	daily	As	11870twr
Burmese	**Days**	**Area**	**kHz**
1200-1245	mtwt...	BRM	13765twr
1200-1300	....fss	BRM	13765twr
Cantonese	**Days**	**Area**	**kHz**
1330-1400	mtwtf..	EAs	9975twr
English	**Days**	**Area**	**kHz**
0820-0900	mtwtf.s	Pac	15170twr
0830-0910	mtwtfs.	Pac	11840twr
1400-1425	m..t...	SEA	9975twr
1400-1435	.tw.fss	SEA	9975twr
Hui	**Days**	**Area**	**kHz**
1230-1300	.....ss	EAs	9370twr
Indonesian	**Days**	**Area**	**kHz**
0945-1030	daily	INS	15200twr
Karen (S'gaw)	**Days**	**Area**	**kHz**
1300-1330	daily	BRM	9585twr
Kokborok	**Days**	**Area**	**kHz**
1230-1300	mtwtf..	As	11870twr
1245-1300	......s	As	11870twr
Korean	**Days**	**Area**	**kHz**
1345-1430	......s	EAs	11620twr
1345-1445	.....s.	EAs	11620twr
1345-1500	mtwtf..	EAs	11620twr

Madura	Days	Area	kHz
0915-0945	daily	INS	15200twr
Mandarin	**Days**	**Area**	**kHz**
0930-1100	daily	EAs	12105twr
1015-1100	mtwtf..	EAs	11590twr
1100-1200	daily	EAs	11680twr
1100-1230	mtwtf..	EAs	9910twr
1130-1200	daily	EAs	9975twr
1230-1330	mtwtf.s	EAs	9975twr
1415-1500	mtwtf..	EAs	9975twr
1500-1600	daily	EAs	9345twr
Manipuri	**Days**	**Area**	**kHz**
1345-1400	......s	As	12075twr
Nosu Yi	**Days**	**Area**	**kHz**
1200-1215	daily	EAs	9975twr
Santhali	**Days**	**Area**	**kHz**
1300-1315	daily	As	11870twr
1330-1345	......s	As	12075twr
Sundanese	**Days**	**Area**	**kHz**
1030-1100	daily	INS	15200twr
Torajanese	**Days**	**Area**	**kHz**
0900-0915	..wt...	INS	15200twr
Vietnamese	**Days**	**Area**	**kHz**
1100-1130	daily	VTN	11840twr
1400-1415	mtwtf..	VTN	9920twr
1400-1500	.....ss	VTN	9920twr

Ann: English: "This is your Station for Inspiration, KTWR, Agana".
IS: "We've a story to tell the Nations", played on an organ.
V: QSL-card. Rp. (3 IRCs)
Notes: Owned by TWR Inc (USA). For corporate details, see under USA.

INDIA (IND)

ALL INDIA RADIO (AIR) (Pub)
✉ External Services Division, P.O. Box 500, New Delhi-110001, India.
☎ +91 11 23715411. 🖷 +91 11 23710057.
E: airlive@air.org.in **W:** www.allindiaradio.org
Webcast: D/L
✉ Akashvani Bhavan, 1 Sansad Marg, New Delhi-110001, India. (Studio)
E: gosesdair@yahoo.co.in **W:** www.newsonair.com (News)
LP: Prasar Bharati Corp: Chmn: Shri Arun Bhatnagar; CEO: Shri B.S Lalli; DG, Ext Sces: J.K. Das; Dep Dir: Nayyer Sadruddin.
MW: [JAL] Jalandhar: 702kHz 300kW; [KKT] Chinsurah: 594/1134kHz 1000kW; [TUT] Tuticorin: 1053kHz 200kW.
SW: [ALG] Aligarh: 4 x 250kW; [BGL] Bengaluru, Doddaballapur: 6 x 500kW; [DEL] Delhi, two sites: Khampur (G.C. 28N49 077E07): 7 x 250kW; Kingsway (G.C. 28N43 077E12): 3 x 50, 2 x 100kW; [GKP] Gorakhpur: 1 x 50kW; [GUW] Guwahati: 50/200kW; [MUM] Mumbai: 1 x 100kW; [PAN] Panaji: 2 x 250kW.
SAT: Insat 4B.
kHz: 702, 1053, 3945, 4860, 5990, 6045, 6055, 6155, 6165, 6180, 6280, 7250, 7270, 7305, 7340, 7370, 7400, 7410, 7420, 7550, 9415, 9445, 9575, 9595, 9620, 9635, 9690, 9810, 9835, 9905, 9910, 9950, 11585, 11620, 11645, 11710, 11715, 11730, 11735, 11740, 11775, 11840, 11850, 11935, 11985, 12025, 13605, 13695, 13710, 13770, 13795, 15040, 15050, 15075, 15140, 15175, 15185, 15235, 15260, 15770, 15795, 17510, 17670, 17705, 17740, 17800, 17810, 17845, 17860, 17875, 17890

Winter Schedule 2010/2011

Arabic	Days	Area	kHz
0430-0530	daily	ME	11730del, 15770alg, 17845del
1730-1945	daily	ME	6180bgl, 9905alg, 11585del
Baluchi	**Days**	**Area**	**kHz**
1500-1600	daily	SAs	6165del, 7340mum, 9620alg, 11585del
Bengali	**Days**	**Area**	**kHz**
0300-0430	daily	SAs	7420guw
0800-1100	daily	SAs	7420guw
1445-1515	daily	SAs	7420guw
1600-1730	daily	SAs	7420guw
Burmese	**Days**	**Area**	**kHz**
1215-1315	daily	SEA	11620bgl, 11710del, 15040del

Chinese	Days	Area	kHz
1145-1315	daily	EAs	11840del, 15795bgl, 17705bgl

Dari	Days	Area	kHz
0300-0345	daily	WAs	9835del, 9910alg, 11735bgl
1315-1415	daily	WAs	7410del, 9910del

English	Days	Area	kHz
1000-1100	daily	As,Pac	13710bgl
1000-1100	daily	EAs	15235del, 17800bgl
1000-1100	daily	Pac	17510del, 17895bgl
1000-1100	daily	SAs	1053tut, 7270cni, 15260del
1330-1500	daily	SEA	9690bgl, 11620del, 13710bgl
1745-1945	daily	EAf	7400del, 9415del, 11935mum
1745-1945	daily	NAf,WAf	7410del, 9445del, 13605bgl
1745-1945	daily	Eu	6280del, 7550del, 9950del+, 11620bgl
2045-2230	daily	Eu	6280del, 7550del, 9445del, 9950del+
2045-2230	daily	Eu,Pac	11620bgl
2045-2230	daily	Pac	11715pan
2245-0045	daily	EAs	11645del, 13605bgl
2245-0045	daily	SEA	6055del, 7305bgl

Farsi	Days	Area	kHz
0400-0430	daily	ME	11730del, 15770alg, 17845del
1615-1730	daily	ME	9905alg, 11585del

French	Days	Area	kHz
1945-2030	daily	NAf,WAf	6180bgl, 7410del, 13605bgl

Gujarati	Days	Area	kHz
0415-0430	daily	EAf	15075bgl, 15185del
1515-1600	daily	EAf	11620bgl, 15175pan

Hindi	Days	Area	kHz
0315-0415	daily	EAf	15075bgl, 15185del
0315-0415	daily	ME	11840del, 13695bgl, 15075bgl
0430-0530	daily	EAf	15075bgl, 15185del
1615-1730	daily	EAf	9950del, 15075del, 17670bgl
1615-1730	daily	ME	12025pan, 13770bgl
1945-2030	daily	Eu	7550bgl
1945-2045	daily	Eu	6280bgl, 9950del+, 11620bgl
2300-2400	daily	SEA	11740pan, 13795bgl

Indonesian	Days	Area	kHz
0845-0945	daily	SEA	15770alg, 17510del, 17875bgl

Kannada	Days	Area	kHz
0215-0300	daily	ME	11985bgl, 15075bgl

Malayalam	Days	Area	kHz
1730-1830	daily	ME	12025pan

Nepali	Days	Area	kHz
0130-0230	daily	SAs	3945gkp, 7420guw, 9810del, 11715del
0700-0800	daily	SAs	7250gkp, 7420guw, 9595del, 11850del
1330-1430	daily	SAs	3945gkp, 4860del, 7420guw, 11775pan

Pashto	Days	Area	kHz
0215-0300	daily	WAs	9835del, 9910alg, 11735bgl
1415-1530	daily	WAs	7410del, 9910del

Punjabi	Days	Area	kHz
0800-0830	daily	SAs	702jal
1230-1430	daily	SAs	702jal

Russian	Days	Area	kHz
1615-1715	daily	Eu	9595del, 11620bgl, 15140del

Saraiki	Days	Area	kHz
1130-1200	daily	SAs	702jal

Sindhi	Days	Area	kHz
0100-0200	daily	SAs	5990del, 7370del, 9635alg
1230-1500	daily	SAs	6165del, 7340mum, 9620alg, 11585del

Sinhala	Days	Area	kHz
0045-0115	daily	SAs	1053tut, 7270cni, 11740pan, 11985del
1300-1500	daily	SAs	1053tut, 7270cni, 15050del

Swahili	Days	Area	kHz
1515-1615	daily	EAf	9950del, 13605alg, 17670bgl

Tamil	Days	Area	kHz
0000-0045	daily	SAs	1053tut, 7270cni, 9835del, 11740pan, 11985del
0000-0045	daily	SEA	11740pan, 13795bgl
0115-0330	daily	SAs	1053tut
1100-1300	daily	SAs	1053tut
1115-1215	daily	SAs	7270cni, 15050del, 17860del
1115-1215	daily	SEA	13710bgl, 15770alg, 17810pan
1500-1530	daily	SAs	1053tut

Telugu	Days	Area	kHz
1215-1245	daily	SEA	13710bgl, 15770alg, 17810pan

Thai	Days	Area	kHz
1115-1200	daily	SEA	15235pan, 17740del

Tibetan	Days	Area	kHz
1215-1330	daily	SAs	7420guw, 9575del, 11775pan

Urdu	Days	Area	kHz
0015-0430	daily	SAs	702jal, 6155bgl, 7340mum, 9595del, 11620alg
0830-1130	daily	SAs	702jal, 7250gkp, 7340mum, 9595del, 11620del
1430-1735	daily	SAs	3945gkp
1430-1930	daily	SAs	702jal, 4860del, 6045del

Key: + DRM.

Ann: Dari: "Inja Delhi"; English: "This is the General Overseas Service of All India Radio"; Hindi: "Yeh Akashvani ki videsh prasaran sewa hai"; Nepali: "Yo All India Radio ho"; Sinhala: "Me All India Radio videshiya sevayai"; Tamil: "Idi Akashvani videsh sewai".

V: QSL-card. (Rpt to the Director of Spectrum Management, All India Radio, Room No.204, Akashvani Bhavan, New Delhi-110001, India. Email rpt to: spectrum-manager@air.org.in - online reception report form available at www.allindiaradio.org/recepfdk.html)

Notes: External Sce of the national public broadcaster Prasar Bharati Corporation. Began broadcasting in 1936. AIR is planning to install the following DRM-ready transmitters: 2 x 100kW at Dehli Kingsway, 2 x 250kW at Aligargh and 1 x 500kW at Bengalaru.

ATHMEEYA YATRA RADIO (Rlg)
✉ P.O. Box 12, Manjadi Junction P.O., Tiruvalla-5, Kerala 689 105, India.
☎ +91 469 2630654.
E: info@athmeeyayathra.org; info@ayasia.org
W: www.athmeeyayathra.org/radio.php; www.ayasia.org
Webcast: D
L.P: Pres: Dr K.P. Yohannan.
kHz: 6140, 6160, 7215, 7240, 9820, 11645, 12005, 15285

Winter Schedule 2010/2011

Adi	Days	Area	kHz
2345-2400	.....ss	As	6160dha

Ao	Days	Area	kHz
1400-1415	.....s.	As	12005wer

Assamese	Days	Area	kHz
2345-2400	mt......	As	6160dha

Awadhi	Days	Area	kHz
1600-1615	mtw....	As	11645wer

Bagheli	Days	Area	kHz
1545-1600	..wtf..	As	11645wer

Bagri	Days	Area	kHz
0030-0045	mt.....	As	7215wer

Banjara	Days	Area	kHz
1415-1430	.....ss	As	15285wer

Bantawa	Days	Area	kHz
2330-2345	......s	As	7240wer

Bengali	Days	Area	kHz
1515-1530	...fss	As	12005wer
2330-2345	mtwt..s	As	6160dha

Bhili	Days	Area	kHz
1400-1415	..t...	As	15285wer

Bhojpuri	Days	Area	kHz
1445-1500	.....ss	As	15285wer

Bodo	Days	Area	kHz
0000-0015	mt.....	As	7240wer

Bondo	Days	Area	kHz
1400-1415	..w...	As	15285wer

Bundelkhandi	Days	Area	kHz
1545-1600	mt.....	As	11645wer

Burmese	Days	Area	kHz
2345-2400	.twt...	As	7240wer

Chakma	Days	Area	kHz
0015-0030	mt.....	As	7240wer

Chhattisgarhi	Days	Area	kHz
1530-1545	mt.....	As	11645wer

Chin	Days	Area	kHz
2345-2400	m.....s	As	7240wer

Chin (Thado/Kuki)	Days	Area	kHz
1345-1400	.....ss	As	12005wer

Dari	Days	Area	kHz
1545-1615	.....s.	As	11645wer
1615-1630	mtw...s	As	9820dha

Deori	Days	Area	kHz
1345-1400	...tf..	As	12005wer

Deshiya	Days	Area	kHz
1345-1400	..wt...	As	15285wer

Divehi	Days	Area	kHz
1600-1615	..tfs.	As	9820dha

Dogri	Days	Area	kHz
0045-0100	...tf..	As	7215wer

Dzongkha	Days	Area	kHz
1430-1445	mtw....	As	12005wer

Gamit	Days	Area	kHz
1415-1430	..wt...	As	15285wer

Garhwali	Days	Area	kHz
0030-0045	.....ss	As	7215wer

Garo	Days	Area	kHz
1415-1430	.....ss	As	12005wer

Gojri	Days	Area	kHz
1230-1245	m......	As	15285wer

Gondi	Days	Area	kHz
1400-1415	...f..	As	15285wer

Gujarati	Days	Area	kHz
0000-0015	m.....s	As	6140dha
1245-1300	...tf..	As	15285wer

Gurung	Days	Area	kHz
2330-2345	.t.....	As	7240wer

Halam	Days	Area	kHz
1400-1415	mt.....	As	12005wer

Haryanvi	Days	Area	kHz
1430-1445	.....ss	As	15285wer

Hindi	Days	Area	kHz
0115-0130	.twtfs.	As	7215wer
1315-1330	.....ss	As	15285wer
1445-1500	mtwtf..	As	15285wer
1615-1630	daily	As	11645wer

Ho	Days	Area	kHz
0015-0030	....f..	As	7240wer

Kangri	Days	Area	kHz
0045-0100	.....ss	As	7215wer

Kannada	Days	Area	kHz
0030-0045	mtwtf..	As	6140dha

Karbi	Days	Area	kHz
1400-1415	...tf..	As	12005wer

Karbi	Days	Area	kHz
2345-2400	....fs.	As	7240wer

Kashmiri	Days	Area	kHz
1230-1245	..wt...	As	15285wer

Kaubru	Days	Area	kHz
1345-1400	..w...	As	12005wer

Khandesi	Days	Area	kHz
1315-1330	..w...	As	15285wer

Khariya	Days	Area	kHz
0030-0045	..w...	As	7215wer

Khasi	Days	Area	kHz
1430-1445	.....ss	As	12005wer

Khurukh	Days	Area	kHz
0000-0015	..wt...	As	7240wer

Kokborok	Days	Area	kHz
1345-1400	mt.....	As	12005wer
2330-2345	....fs.	As	6160dha

Konkani	Days	Area	kHz
0000-0015	.tw...	As	6140dha

Konyak	Days	Area	kHz
1430-1445	...tf..	As	12005wer

Kotwalia	Days	Area	kHz
1430-1445	....f..	As	15285wer

Koya	Days	Area	kHz
1245-1300	.....ss	As	15285wer

Kui	Days	Area	kHz
1415-1430	mt.....	As	15285wer

Kukna	Days	Area	kHz
1415-1430	....f..	As	15285wer

Kupiya	Days	Area	kHz
1400-1415	......s	As	15285wer

Ladakhi	Days	Area	kHz
1230-1245	.t.....	As	15285wer

Lepcha	Days	Area	kHz
1500-1515	.tw...	As	12005wer

Limbu	Days	Area	kHz
2330-2345	....fs.	As	7240wer

Magahi	Days	Area	kHz
0045-0100	..w...	As	7215wer
1500-1515	...tf..	As	12005wer

Magar (Eastern)	Days	Area	kHz
2330-2345	m......	As	7240wer

Magar (Kham)	Days	Area	kHz
1445-1500	.....ss	As	12005wer

Maithili	Days	Area	kHz
1530-1545	..wtf..	As	11645wer

Malayalam	Days	Area	kHz
0015-0030	daily	As	6140dha
1600-1615	mtw...s	As	9820dha

Malto	Days	Area	kHz
1330-1345	..t...	As	15285wer

Malto (FMPB)	Days	Area	kHz
1330-1345	...f..	As	15285wer

Marathi	Days	Area	kHz
0100-0115	...fss	As	7215wer
1315-1330	...tf..	As	15285wer

Marwari	Days	Area	kHz
0100-0115	m......	As	7215wer
1345-1400	mt.....	As	15285wer

Meitei	Days	Area	kHz
1330-1345	mtw....	As	15285wer
1500-1515	.....ss	As	12005wer

Mising	Days	Area	kHz
0000-0015	...fss	As	7240wer
1415-1430	..wtf..	As	12005wer

Mouchi	Days	Area	kHz
1430-1445	..t...	As	15285wer

Mundari	Days	Area	kHz
1515-1530	..wt...	As	12005wer

Nepali	Days	Area	kHz
1300-1315	daily	As	15285wer

Netakani	Days	Area	kHz
1345-1400	......s	As	15285wer

Newari	Days	Area	kHz
1445-1500	...tf..	As	12005wer
Nocte	**Days**	**Area**	**kHz**
1330-1345	mt.....	As	12005wer
2345-2400	..wt...	As	6160dha
Oriya	**Days**	**Area**	**kHz**
1230-1245	...fss	As	15285wer
Pashto	**Days**	**Area**	**kHz**
1545-1615	.....s	As	11645wer
1615-1630	...tfs.	As	9820dha
Punjabi	**Days**	**Area**	**kHz**
0100-0115	..wt...	As	7215wer
1245-1300	mtw....	As	15285wer
Rajasthani	**Days**	**Area**	**kHz**
0045-0100	m......	As	7215wer
0045-0115	.t.....	As	7215wer
Rengma	**Days**	**Area**	**kHz**
1400-1415	..w....	As	12005wer
Rongmei	**Days**	**Area**	**kHz**
1330-1345	.....s	As	12005wer
Sadri	**Days**	**Area**	**kHz**
1600-1615	...tf..	As	11645wer
Sambalpuri	**Days**	**Area**	**kHz**
1430-1445	mt.....	As	15285wer
Santhali	**Days**	**Area**	**kHz**
0015-0030	..wt...	As	7240wer
1315-1330	m......	As	15285wer
1415-1430	mt.....	As	12005wer
Sarchopa	**Days**	**Area**	**kHz**
1445-1500	mtw....	As	12005wer
Sherpa	**Days**	**Area**	**kHz**
1500-1515	m......	As	12005wer
2330-2345	..wt...	As	7240wer
Sindhi	**Days**	**Area**	**kHz**
1530-1545	.....ss	As	11645wer
Sinhala	**Days**	**Area**	**kHz**
0000-0015	...tfs.	As	6140dha
2345-2400	.....f.	As	6160dha
Soura	**Days**	**Area**	**kHz**
1400-1415	mt.....	As	15285wer
Sumi	**Days**	**Area**	**kHz**
1400-1415	.....s	As	12005wer
Tamang	**Days**	**Area**	**kHz**
1515-1530	mt.....	As	12005wer
Tamil	**Days**	**Area**	**kHz**
0100-0130	daily	As	6140dha
Tangkhul	**Days**	**Area**	**kHz**
1330-1345	...fs.	As	12005wer
Telugu	**Days**	**Area**	**kHz**
0045-0100	daily	As	6140dha
Tharu	**Days**	**Area**	**kHz**
1330-1345	..wt...	As	12005wer
Tibetan (Amdo)	**Days**	**Area**	**kHz**
0015-0030	.....s.	As	7240wer
1330-1345	.....s.	As	15285wer
Tibetan (Lhasa)	**Days**	**Area**	**kHz**
0015-0030	.....s	As	7240wer
1330-1345	.....s	As	15285wer
Tulu	**Days**	**Area**	**kHz**
0030-0045	....ss	As	6140dha
Urdu	**Days**	**Area**	**kHz**
1345-1400	....fs.	As	15285wer
Vadari	**Days**	**Area**	**kHz**
1315-1330	.t.....	As	15285wer
Varti	**Days**	**Area**	**kHz**
1430-1445	..w....	As	15285wer
Vasavi	**Days**	**Area**	**kHz**
0030-0045	...tf..	As	7215wer
Yerukala	**Days**	**Area**	**kHz**
1400-1415	.....s.	As	15285wer

Ann: Malayalam: "Athmeeya Yathra".
V: QSL-card.
Notes: Formerly known as "Gospel For Asia (GFA)". Owned by Believers Church. Also known as "AY Radio" or "Athmik Yatra (Spiritual Journey)".

CVC THE VOICE ASIA (Rlg)
✉ P.O. Box 1, Kangra, Pin Code 176001, Himachal Pradesh, India.
E: sav@thevoiceasia.com; enquiry@cvc.tv
W: www.thevoiceasia.com
Webcast: L/P
✉ P.O. Box 2, Ludhiana, Pin Code 141008, Punjab, India.
LP: Dir: Danny Choranji; Operations Mgr: Narinder Choranji.
SAT: Hot Bird 6.
kHz: *6260, 9500, 9975, 11805*

Winter Schedule 2010/2011			
Hindi	Days	Area	kHz
0000-0400	daily	IND	6260tac
0100-0400	daily	IND	9975tac
0400-1100	daily	IND	11805tac
1100-1400	daily	IND	9500tac
1400-2000	daily	IND	6260tac

V: QSL-card. Rp.
Notes: Owned by CVC International Ltd, a branch of Christian Vision, United Kingdom. On air since 1 Jun 2002. For corporate details, see UK under "CVC International".

TWR INDIA (Rlg)
✉ P.O. Box 4310, Delhi 110019, India.
☎ +91 11 26515790. 🖷 +91 11 6868049.
E: info@twr.in **W:** www.twr.in; radio882.com
Webcast: D/L
kHz: *882, 5920, 6115, 7295, 7320, 11965, 17650*

Winter Schedule 2010/2011			
Awadhi	Days	Area	kHz
1400-1415	.t.....	As	7320nvs
Banjara	**Days**	**Area**	**kHz**
1215-1230	mt.....	As	882put
1215-1245	..wtfss	As	882put
Bengali	**Days**	**Area**	**kHz**
0030-0045	mtwtf..	As	11965irk
1315-1330	.....s.	As	5920irk
2230-2245	m......	As	882put
2230-2300	.twtfss	As	882put
2245-2300	m......	As	882put
Bhili	**Days**	**Area**	**kHz**
1315-1345	mtwtf..	As	882put
1500-1515	.....ss	As	5920irk
Bhojpuri	**Days**	**Area**	**kHz**
0045-0115	mtwtf..	As	11965irk
Bodo	**Days**	**Area**	**kHz**
1330-1345	daily	As	17650wer
Brajbhasha	**Days**	**Area**	**kHz**
1345-1400	..w....	As	7320nvs
Bundeli	**Days**	**Area**	**kHz**
1345-1400	.....s.	As	5920irk
Chhattisgarhi	**Days**	**Area**	**kHz**
1245-1315	mtwtf..	As	882put
1300-1315	....ss	As	882put
1600-1615	....ss	As	882put
Chodhri	**Days**	**Area**	**kHz**
1445-1500	....ss	As	5920irk
Dari	**Days**	**Area**	**kHz**
1615-1630	.....s.	As	7295sam
Deccani	**Days**	**Area**	**kHz**
1730-1800	mtwtf..	As	882put
1745-1800	....ss	As	882put
Dhodiya	**Days**	**Area**	**kHz**
1515-1530	..wt...	As	5920irk
Dogri	**Days**	**Area**	**kHz**
1315-1330	mtwtf..	As	7320nvs
Dzongkha	**Days**	**Area**	**kHz**
0115-0130	mt....s	As	11965irk
English	**Days**	**Area**	**kHz**
1200-1215	daily	As	882put
1415-1430	...tf..	As	882put
1500-1515	mtwt..	As	882put
1615-1630	mtw....	As	882put
1615-1645	.....s	As	882put
1645-1700	.....s.	As	882put
1645-1715	.....s	As	882put

English	Days	Area	kHz
1700-1715	.....s.	As	882put
1700-1730	mtwtf..	As	882put

Gamit	Days	Area	kHz
1500-1515	mtw....	As	5920irk

Garhwali	Days	Area	kHz
1330-1345	.....s	As	5920irk
1330-1345	.....s.	As	5920irk
1415-1430	mtwtf..	As	7320nvs

Gondi	Days	Area	kHz
1545-1600	.....ss	As	882put
1545-1615	mtwtf..	As	882put

Gujarati	Days	Area	kHz
1315-1345	.....s.	As	882put
1345-1415	......s	As	882put
1400-1415	.....s.	As	882put
2300-2330	.twtfs.	As	882put

Haryanvi	Days	Area	kHz
1400-1415	..w....	As	7320nvs

Hindi	Days	Area	kHz
0030-0115	......s	As	11965irk
1315-1330	.....ss	As	7320nvs
1330-1345	....f..	As	5920irk
1330-1345	daily	As	7320nvs
1345-1400	.....ss	As	7320nvs
1400-1415	.....ss	As	7320nvs
1415-1430	.....ss	As	7320nvs
1430-1445	daily	As	7320nvs
1515-1530	mtwtfs.	As	7320nvs
1530-1545	mtwtf..	As	7320nvs

Ho	Days	Area	kHz
1300-1315	.....s.	As	5920irk

Kannada	Days	Area	kHz
0045-0115	mtwtf.s	As	882put
0115-0130	....s.	As	882put
1500-1515	....f..	As	882put

Kashmiri	Days	Area	kHz
1515-1530	......s	As	7320nvs

Khariya	Days	Area	kHz
1400-1415	......s	As	5920irk

Konkani	Days	Area	kHz
1615-1630	...tf..	As	882put
1630-1700	mtwtf..	As	882put

Kotwalia	Days	Area	kHz
2315-2330	......s	As	882put

Koya	Days	Area	kHz
1230-1245	mt.....	As	882put
1530-1545	......s	As	882put

Kui	Days	Area	kHz
1245-1300	.....s.	As	5920irk

Kukna	Days	Area	kHz
2300-2315	.t...s	As	882put

Kumaoni	Days	Area	kHz
1300-1315	......s	As	5920irk

Kurukh	Days	Area	kHz
1345-1400	......s	As	5920irk
1415-1430	...tfs.	As	5920irk

Kutchi	Days	Area	kHz
2315-2330	m......	As	882put

Maghai	Days	Area	kHz
1415-1430	m....s	As	5920irk

Maithili	Days	Area	kHz
1330-1345	mt.....	As	5920irk
1345-1415	mtwtf..	As	5920irk

Malayalam	Days	Area	kHz
0030-0045	daily	As	882put
0100-0115	.....s.	As	882put
1615-1630	.....s.	As	882put
2330-2400	daily	As	882put

Marathi	Days	Area	kHz
1415-1430	.t.....	As	882put
1430-1500	mtwtfs.	As	882put
1445-1500	......s	As	882put

Marwari	Days	Area	kHz
1315-1330	mt....s	As	5920irk

Mewari	Days	Area	kHz
1315-1330	...tf..	As	5920irk

Mouchi	Days	Area	kHz
1515-1530	mt.....	As	5920irk

Mundari	Days	Area	kHz
1415-1430	.tw....	As	5920irk

Nepali	Days	Area	kHz
0045-0115	.....s.	As	11965irk

Oriya	Days	Area	kHz
1400-1415	.....s.	As	5920irk
1415-1430	......s	As	882put
1500-1515	......s	As	882put
1500-1530	.....s.	As	882put
1515-1530	......s	As	882put
1515-1545	mtwtf..	As	882put

Pashto	Days	Area	kHz
1600-1615	.....s.	As	7295sam
1600-1630	mtwtf..	As	7295sam

Punjabi	Days	Area	kHz
1315-1345	..w....	As	5920irk
1445-1515	daily	As	7320nvs

Sadri	Days	Area	kHz
1430-1445	.....ss	As	5920irk

Santhali	Days	Area	kHz
1245-1300	......s	As	5920irk

Sindhi	Days	Area	kHz
1430-1500	mtwtf..	As	5920irk

Tamil	Days	Area	kHz
0000-0030	daily	As	882put
1715-1745	.....ss	As	882put

Telugu	Days	Area	kHz
1345-1400	.....s.	As	882put
1345-1415	mtwtf..	As	882put
1415-1430	m...s.	As	882put
1430-1445	......s	As	882put
1530-1545	.....s.	As	882put

Tibetan	Days	Area	kHz
1330-1345	..t...	As	5920irk

Tulu	Days	Area	kHz
0045-0100	......s	As	882put
1630-1645	.....s.	As	882put

Urdu	Days	Area	kHz
1500-1530	daily	As	6115sam

Varli	Days	Area	kHz
1415-1430	..w....	As	882put

Vasavi	Days	Area	kHz
1500-1515	...tf..	As	5920irk

V: QSL-card. (Online form for reception reports)
Notes: TWR branch. Owned by TWR. For corporate details, see under USA.

INDONESIA (INS)

VOICE OF INDONESIA (VOI) (Pub)
✉ P.O. Box 1157, Jakarta, 10110, Indonesia.
☎ +62 21 3456811. 🖷 +62 21 3500990.
E: english@voi.co.id **W:** www.voi.co.id
Webcast: D/L
L.P: Dir: Sutrisno Santoso.
SW: [JAK] Jakarta, Cimanggis: 2 x 50, 3 x 100, 9 x 250kW.
FM/DAB: FM: 89.0MHz (Jakarta).
SAT: Telkom 1.
kHz: 9526

Winter Schedule 2010/2011

Arabic	Days	Area	kHz
1600-1700	daily	Eu,NAf,ME	9526jak*

Chinese	Days	Area	kHz
1100-1200	daily	As,Pac	9526jak*

English	Days	Area	kHz
1000-1100	daily	As,Pac	9526jak*
1300-1400	daily	As,Pac	9526jak*
1500-1600	daily	As,Pac	9526jak*
1900-2000	daily	Eu,NAf,ME	9526jak*

French	Days	Area	kHz
2000-2100	daily	Eu,NAf,ME	9526jak*

German	Days	Area	kHz
1800-1900	daily	Eu,NAf,ME	9526jak*
Indonesian	**Days**	**Area**	**kHz**
1400-1500	daily	As,Pac	9526jak*
Japanese	**Days**	**Area**	**kHz**
1200-1300	daily	As,Pac	9526jak*
Spanish	**Days**	**Area**	**kHz**
1700-1800	daily	Eu,NAf,ME	9526jak*

Key: * Schedule and operation variable.
Ann: English: "This is the Voice of Indonesia, in Jakarta"; Spanish: "La Voz de Indonesia en Jakarta".
V: QSL-card.
Notes: The Voice of Indonesia is the External Sce of the state broadcaster Radio Republik Indonesia.

IRAN (IRN)

VOICE OF THE ISLAMIC REPUBLIC OF IRAN (VOIRI) (Gov)

✉ P.O. Box 19395-6767, Tehran, Iran.
☎ +98 21 22013687; +98 21 22162731. 🖷 +98 21 22162775.
E: englishradio@irib.ir; bm@irib.ir
W: www.irib.ir/worldservice
Webcast: D/L
☎ +98 21 22013720.
LP: Pres: Dr Ali Larijani; Dir. Int. Rel: M.Safdari.
MW: [AHW] Ahwaz, Bandar-e Mahshar: 576/1080kHz 750kW; [BNB] Bonab: 639kHz 400kW; [BNT] Bandar-e Torkaman: 1449kHz 400kW; [CHB] Chabahar: 765kHz 1000kW; [KER] Kerman: 1224kHz 600kW; [KIA] Bandar-e Kiashahr: 702kHz 500kW; [MHD] Mahidasht: 720kHz 750kW; [QSH] Qasr-e Shirin: 612/1161kHz 600kW; [TYB] Tayebad: 720kHz 400kW; [ZAB] Zabol: 1098kHz 100kW.
SW: [AHW] Ahwaz, Bandar-e Mahshar: 2 x 250kW; [KAM] Tehran, Kamalabad: 10 x 100, 3 x 250, 1 x 350, 12 x 500kW; [SIR] Sirjan: 10 x 500kW; [ZAH] Zahedan: 2 x 500kW.
SAT: AsiaSat 3S, Badr 5/6, Hot Bird 8, Galaxy 19, Intelsat 10/902, Nilesat 102, Telstar 12.
kHz: 576, 612, 639, 702, 720, 765, 1080, 1098, 1161, 1224, 1386, 1449, 3955, 3965, 3985, 5890, 5915, 5920, 5940, 5945, 5950, 5955, 5980, 5995, 6010, 6015, 6030, 6035, 6055, 6065, 6070, 6090, 6095, 6100, 6110, 6115, 6120, 6145, 6155, 6170, 6175, 6185, 6200, 6205, 7200, 7205, 7215, 7225, 7230, 7240, 7250, 7265, 7285, 7295, 7305, 7315, 7320, 7325, 7340, 7345, 7350, 7355, 7360, 7370, 7380, 7420, 7925, 9510, 9520, 9540, 9585, 9610, 9675, 9685, 9690, 9710, 9715, 9740, 9770, 9790, 9820, 9830, 9850, 9865, 9885, 9895, 9900, 9905, 9915, 9940, 11600, 11640, 11655, 11670, 11685, 11695, 11715, 11775, 11860, 11925, 11935, 11965, 11990, 12025, 13600, 13620, 13650, 13680, 13710, 13725, 13740, 13760, 13790, 13800, 13810, 15085, 15150, 15220, 15235, 15260, 15390, 15400, 15440, 15460, 15515, 15550, 15560, 17630, 17660, 17680, 17690, 17780, 17810, 21640

Winter Schedule 2010/2011

Albanian	Days	Area	kHz
0630-0730	daily	Eu	13810kam, 15235sir
1830-1930	daily	Eu	6100sir, 7285kam
2030-2130	daily	Eu	6100sir, 9740kam
Arabic	**Days**	**Area**	**kHz**
0000-2400	daily	ME	1224ker
0130-0330	daily	ME	612qsh
0230-0330	daily	ME	1161qsh
0230-0530	daily	ME	7350kam, 9895zah
0230-1630	daily	ME	576ahw
0330-0430	daily	ME	5915kam*, 7925sir*
0430-1530	daily	ME	612qsh
0530-0830	daily	ME	13790kam
0530-1430	daily	ME	13800zah
0530-1630	daily	NAf,ME	15550sir
0830-1030	daily	NAf,ME	9885sir
1030-1430	daily	ME	13790kam
1430-1730	daily	ME	9830kam
1630-0330	daily	ME	1080ahw
1630-0530	daily	NAf,ME	6065sir
1630-2130	daily	ME	1161qsh
1730-1300	daily	ME	765chb
Armenian	**Days**	**Area**	**kHz**
0300-0330	daily	ME	5915sir, 7295sir
0930-1000	daily	Cau	9690kam, 15220sir

Armenian	Days	Area	kHz
1630-1730	daily	Cau	6185sir, 7230sir
Azeri	**Days**	**Area**	**kHz**
0330-0530	daily	Cau	702kia
0330-0530	daily	ME	6200sir
1430-1700	daily	ME	702kia, 6200sir
Bengali	**Days**	**Area**	**kHz**
0030-0130	daily	SAs	765chb, 5915kam, 6100kam
0830-0930	daily	ME	13680kam
1430-1530	daily	SAs	7380kam, 11600kam
Bosnian	**Days**	**Area**	**kHz**
0530-0630	daily	Eu	13760sir, 15235kam
1730-1830	daily	Eu	7200kam, 7295sir
2130-2230	daily	Eu	5950sir, 9710kam
Chinese	**Days**	**Area**	**kHz**
1200-1300	daily	EAs	9900kam, 11670sir, 13650sir, 15150sir
2330-0030	daily	EAs	5945sir, 7325sir, 9710kam
Dari	**Days**	**Area**	**kHz**
0300-0630	daily	ME	1098zab, 9885kam, 11935ahw
0300-0630	daily	WAs	720tyb
0830-1200	daily	ME	11670kam
0830-1430	daily	ME	13725ahw
0830-1500	daily	ME	720tyb, 1098zab
1200-1500	daily	ME	9940kam
English	**Days**	**Area**	**kHz**
0130-0230	daily	NAm	6120sir**, 7250kam**
1030-1130	daily	ME	702kia
1030-1130	daily	SAs	15460kam, 17630kam
1030-1130	daily	WAs	765chb
1530-1630	daily	SAs	9915sir, 11655kam
1930-2030	daily	Eu	6010kam, 6115sit, 7320sir
1930-2030	daily	SAf	11695sir, 11860kam
French	**Days**	**Area**	**kHz**
0630-0730	daily	Eu,NAf	13600kam, 15560kam
1830-1930	daily	Eu	5980kam, 6115sit, 7380sir
1830-1930	daily	WAf	11775kam
Georgian	**Days**	**Area**	**kHz**
1700-1800	daily	ME	702kia
German	**Days**	**Area**	**kHz**
0730-0830	daily	Eu	15085kam, 17690sir
1730-1830	daily	Eu	3955sit, 6205sir, 7380kam
Hausa	**Days**	**Area**	**kHz**
0600-0700	daily	WAf	17810sir
1830-1930	daily	WAf	9715kam, 11965sir
Hebrew	**Days**	**Area**	**kHz**
0430-0500	daily	ME	9820kam, 11925sir
1200-1230	daily	ME	13740sir, 15390kam
Hindi	**Days**	**Area**	**kHz**
0230-0300	daily	SAs	7340sir, 9510sir
1430-1530	daily	SAs	7370kam, 9585sir
Indonesian	**Days**	**Area**	**kHz**
1230-1330	daily	SEA	15515kam, 17690sir
2230-2330	daily	SEA	7315kam, 9675sir
Italian	**Days**	**Area**	**kHz**
0630-0730	daily	Eu	9770sit, 13620kam, 15085kam
1930-2000	daily	Eu	5890kam, 7380kam
Japanese	**Days**	**Area**	**kHz**
1330-1430	daily	EAs	9585kam, 9905sir
2100-2200	daily	EAs	5995sir, 6145sir
Kazakh	**Days**	**Area**	**kHz**
0130-0230	daily	CAs	7205sir, 7265sir
1530-1630	daily	CAs	9540kam, 9850sir
Kurdish	**Days**	**Area**	**kHz**
0330-0430	daily	ME	612qsh, 639bnb
0430-0530	daily	ME	6170kam, 9610sir
1330-1630	daily	ME	639bnb, 5920kam

Pashto	Days	Area	kHz
0230-0330	daily	WAs	765chb, 6095kam, 6155sir
0730-0830	daily	WAs	1098zab, 11990sir, 15440ahw
1230-1330	daily	WAs	765chb, 7225sir, 9520kam
1430-1530	daily	WAs	5890sir
1630-1730	daily	WAs	6015sir, 7345ahw
Russian	**Days**	**Area**	**kHz**
0300-0330	daily	RUS	702kia, 7370kam, 9510sir
0500-0530	daily	RUS	12025kam, 13680sir, 17680sir, 17780sir
1430-1530	daily	RUS	1449bnt, 7345kam, 7420sit, 9610sir, 9685ahw
1700-1800	daily	RUS	3965kam, 6090ahw
1800-1900	daily	RUS	1386sit, 6035sir, 7305kam
1930-2030	daily	RUS	702kia, 3985kam, 7205sir
Spanish	**Days**	**Area**	**kHz**
0030-0230	daily	LAm	7240kam
0030-0330	daily	LAm	6110kam
0530-0630	daily	Eu	13710kam, 15400sir
2030-2130	daily	Eu	5950kam, 6055sit, 7200sir
Swahili	**Days**	**Area**	**kHz**
0400-0500	daily	EAf	13680kam, 15260sir
0830-0930	daily	EAf	17660sir, 21640sir
1730-1830	daily	EAf	11715sir
1730-1830	daily	EAf,ME	9830kam
1800-1830	daily	ME	702kia
Tajik	**Days**	**Area**	**kHz**
0100-0230	daily	CAs	720tyb, 5955sir, 7355kam
1600-1730	daily	CAs	720tyb, 5955sir, 7200kam
Turkish	**Days**	**Area**	**kHz**
0430-0600	daily	ME	9865kam, 11640kam
1600-1730	daily	ME	6175kam, 7315kam
1830-1930	daily	ME	639bnb, 702kia
Turkmen	**Days**	**Area**	**kHz**
0000-0500	daily	CAs	1449bnt
1330-1430	daily	CAs	1449bnt
1530-1830	daily	CAs	1449bnt
Urdu	**Days**	**Area**	**kHz**
0130-0230	daily	ME	1098zab
0130-0230	daily	SAs	765chb, 3965zah, 6030kam, 6185ahw
1300-1430	daily	ME	11685kam
1300-1430	daily	SAs	5940sir, 9790kam
1330-1430	daily	SAs	765chb
1530-1730	daily	SAs	765chb, 5890kam
Uzbek	**Days**	**Area**	**kHz**
0230-0300	daily	CAs	6175kam, 7360sir
1500-1600	daily	SAs	6070kam, 7215sir

Key: * "VO Palestine" prgr, also 1830-1930, during Arabic prgr; ** "VO Justice" prgr.
Ann: Arabic: "Huna Tahran - Sawt al Jumhuriya al Islamiya fi Iran"; English: "This is the Voice of the Islamic Republic of Iran", V.O.J. prgr: "This is the Voice of Justice"; Farsi: "Inja Tehran ast, sedaye jomhuriye eslamiye Iran"; French: "Ici Tehran, la Voix de la République Islamique de l'Iran"; Russian: "Govorit Tegeran, Golos Islamskoy Respubliki Iran".
IS: "Love's Rainfall", by Nasser Cheshmazar.
V: QSL-card. Rec. acc.
Notes: The Voice of the Islamic Republic of Iran is the External Sce of the state broadcaster IRIB. The prgr "Voice of Justice" is aimed at listeners in the USA. The "Voice of the Palestinian Islamic Revolution" prgr targets listeners in the territories under Palestinian Authority. External Sce in Farsi/Persian is known as "Seda-ye Ashena" (www.sedayeashna.ir) and is not available via terrestrial broadcasting methods.

IRELAND (IRL)

RTÉ RADIO WORLDWIDE (Pub)
✉ RTÉ, Donnybrook, Dublin 4, Ireland.
☎ +353 1 208 3111. 🖷 +353 1 208 3080.
E: hearus@rte.ie **W:** www.rte.ie/radio/worldwide.html
Webcast: D/L/P
L.P: DG: Cathal Goan; MD, Radio: Clare Duignan; Dir, Comms: Bride Rosney.
SAT: (via WRN English) Afristar 1, AsiaSat 2, Eurobird 1, Eutelsat 1/W7, Hot Bird 6, Intelsat 7/10, Galaxy 19, Nimiq 1, XM3/4, Sirius FM 5, Superbird C2.
kHz: *6225*

Winter Schedule 2010/2011

English/ Irish Gaelic	Days	Area	kHz
1930-2030	daily	Af	6225mey

Notes: RTÉ is Ireland's national Public Service broadcaster.

ISRAEL (ISR)

KOL ISRAEL (VOICE OF ISRAEL) (Pub)
✉ P.O. Box 1082, Jerusalem 91010, Israel.
☎ +972 2 6248715. 🖷 +972 2 5302327.
E: radio-int@iba.org.il; reception@iba.org.il
W: www.intkolisrael.com; www.radis.org (Persian/Farsi)
Webcast: D/L
L.P: Dir: Shmuel Ben-Zvi.
SW: Leased from Bezeq.
SAT: AMOS 2.
kHz: *9985, 11595*

Winter Schedule 2010/2011

Persian	Days	Area	kHz
1500-1600	daily	ME	9985isr*, 11595isr**
1600-1630	mtwt..s	ME	9985isr*, 11595isr**

Key: * Alternative Frequency 13850kHz; ** Alternative Frequency 15760kHz.
Ann: Farsi/Persian: "Inja Sedaye Israel".
V: Kol Israel does not verify reception reports, rpt should be sent to Bezeq.
Notes: Kol Israel is the External Sce of the Israel Broadcasting Authority (IBA).

GALEI TZAHAL
✉ See National Radio section.
Webcast: L
SW: [LOD] Lod: 2 x 10kW.
kHz: *6973, 15785*

Winter Schedule 2010/2011

Hebrew	Days	Area	kHz
0000-2400	daily	Eu	6973lod, 15785lod

Notes: Relay of domestic channel.

BEZEQ (THE ISRAEL TELECOMMUNICATIONS CORP. LTD.) (Tx Operator)
✉ P.O. Box 1088, Jerusalem 91010, Israel.
☎ +972 3 6800029. 🖷 +972 3 6800030.
E: mosheor@bezeq.com **W:** www.bezeq.com
L.P: CEO: Jacob Gelbard.
SW: [ISR] Yavne: 1 x 100, 4 x 300, 1 x 500kW.
Notes: Established in 1984. Bezeq is the national transmitter network operator in Israel.

ITALY (I)

RAI INTERNATIONAL (Pub)
✉ P.O. Box 320, I-00100 Roma, Italy.
☎ +39 06 33542526. 🖷 +39 06 33170767.
E: raiinternational@rai.it **W:** www.rai-international.rai.it
Webcast: L
L.P: Dir: Roberto Morrione.
MW: Leased from RAIWAY.
SAT: AsiaSat 5, EchoStar 1/2/3/4/6/9, Eutelsat W4, Galaxy 3C, Hot Bird 6/9, Intelsat 7/10, Optus C1, PAS2/6B/9.
kHz: *567, 657, 900, 1107*

Winter Schedule 2010/2011

Arabic	Days	Area	kHz
1430-1445	daily	Med	567cls

Italian/English/

French	Days	Area	kHz
2330-0500	daily	Eu	657nap*, 900mil*, 1107rom*

Key: * "Notturno Italiano": Music and cultural prgr with English and French news from 0000-0400, at approx 5 mins past the hour.
Ann: Italian:"RAI International - Notturno Italiano".
V: QSL-card.
Notes: RAI is conducting DRM Tests on MW (carrying relays of various Italian networks, including "Notturno Italliano").

IRRS-SHORTWAVE (NEXUS-IBA)

☞ P.O. Box 11028, I-20110 Milano, Italy.
☎ +39 02 2666971. 🗎 +39 02 70638151.
Email: info@nexus.org
Web: www.nexus.org (General); www.egradio.org (Religious relays)
Webcast: L (mp3.nexus.org)
L.P: Pres: Alfredo E. Cotroneo.
SW: Leased from Towercom (Slovakia).
kHz: 6090, 7385, 9510, 15710

Winter Schedule 2010/2011

English	Days	Area	kHz
0900-1000	......s.	Eu,ME,NAf	9510rso
1030-1300	......s	Eu,ME,NAf	9510rso
1900-2000	daily	Eu,ME,NAf	6090rso

English/Arabic	Days	Area	kHz
0300-0600	daily	Af	7385rso
1400-1700	daily	Af	15710rso

Ann: English: "This is IRRS, the Italian Radio Relay Service".
IS: S/on: Triumphal Scene from Aida (Verdi); S/off: Prisoners' Chorus (Verdi).
V: QSL-card. Rp. (Rpt by email to: reports@nexus.org)
Notes: NEXUS-IBA is a provider of relay services for prgr producers and broadcasters, via Internet (24/7) and leased tx facilities in Slovakia.

RAIWAY (Tx Operator)

☞ Via Teulada 66, I-00195 Roma, Italy.
☎ +39 06 33177022. 🗎 +39 06 33177022.
E: raiway@rai.it **W:** www.raiway.rai.it
L.P: Dir: Mirco Palmieri.
MW: [CLS] Caltanissetta: 567kHz 20kW; [MIL] Milano: 900kHz 600kW; [NAP] Napoli: 657kHz 120kW; [ROM] Roma: 846kHz 50kW, 1107kHz 10kW.

JAPAN (J)

RADIO JAPAN (NHK WORLD) (Pub)

☞ NHK World, Radio Japan, NHK, 2-1, Jinnan 2-chome, Shibuya-ku, Tokyo. 150-8001, Japan.
☎ +81 3 34651111. 🗎 +81 3 34811350.
E: nhkworld@nhk.jp
W: www.nhk.or.jp/nhkworld/index.html (English); www.nhk.or.jp/nhkworld/japanese/info/select.html (Japanese)
Webcast: D/L/P
L.P: DG, NHK: Makato Harada; Pres, NHK: Shigeo Fukuchi; DG, Broadcasting: Hidemi Hyuga.
SW: Leased from KDDI & foreign relays.
SAT: Hot Bird 6, Intelsat 8/9/10.
kHz: 738, 1350, 5955, 5960, 5975, 5980, 6035, 6075, 6085, 6090, 6110, 6120, 6130, 6140, 6145, 6155, 6160, 6165, 6185, 6190, 6195, 6200, 7225, 7395, 9540, 9560, 9575, 9585, 9605, 9620, 9625, 9670, 9695, 9750, 9760, 9770, 9790, 9795, 9825, 9835, 9840, 9850, 9875, 11655, 11665, 11715, 11740, 11760, 11780, 11815, 11860, 11905, 11910, 11935, 11945, 12045, 13640, 13650, 13840, 15195, 15205, 15290, 15325, 17560, 17605, 17690, 17735, 17810, 21560

Winter Schedule 2010/2011

Arabic	Days	Area	kHz
0400-0430	daily	ME,NAf	6035erv
0700-0730	daily	ME,NAf	11905iss
2115-2145	daily	ME	1350erv*

Bengali	Days	Area	kHz
1300-1345	daily	SAs	11860tac

Burmese	Days	Area	kHz
1030-1100	daily	SEA	11740sng
1430-1500	daily	SEA	11740sng
2340-2400	daily	SEA	13650yam

Chinese	Days	Area	kHz
0900-0930	daily	As	6090yam
1200-1230	daily	As	6090yam
1300-1330	daily	As	6190yam
1300-1330	daily	SEA	11740sng
1430-1500	daily	As	6190yam
1530-1600	daily	As	6190yam
1600-1630	daily	SEA	9540yam
2230-2250	daily	As	9560yam
2240-2300	daily	SEA	13650yam
2340-2400	daily	SEA	15195yam, 17810yam

English	Days	Area	kHz
0500-0530	daily	Eu	5975rmp
0500-0530	daily	NAm	6110sac
0500-0530	daily	SAf	9770iss
0500-0530	daily	SAs	15205tac
0500-0530	daily	SEA	17810yam
1000-1030	daily	SEA	9605yam
1000-1030	daily	Pac	9625yam, 9840yam
1000-1030	daily	SAs	11780tac
1100-1130	....f..	WEu	9760wof+
1200-1230	daily	Eu	9790wer
1200-1230	daily	NAm	6120sac
1200-1230	daily	Pac	9625yam
1200-1230	daily	SEA	9695yam
1300-1330	daily	SAs	9875yam
1400-1430	daily	CAf	21560iss
1400-1430	daily	SAs	9875yam
1400-1430	daily	SEA	5955yam

French	Days	Area	kHz
0530-0600	daily	CAf,WAf	9850wer, 13840mdc
1230-1300	daily	WAf	17690mdc

Hindi	Days	Area	kHz
1345-1430	daily	SAs	9585tac

Indonesian	Days	Area	kHz
0945-1030	daily	SEA	6140sng
1315-1400	daily	SEA	5955yam
2310-2340	daily	SEA	17810yam

Japanese	Days	Area	kHz
0200-0300	daily	SEA	11860yam
0200-0400	daily	SAm	11935bon
0200-0500	daily	As	15195yam
0200-0500	daily	CAm	5960sac
0200-0500	daily	ME,NAf	17560yam
0200-0500	daily	SAs	15325yam
0200-0500	daily	SEA	17810yam
0700-0800	daily	EAs	6145yam, 6165yam
0700-1700	daily	As	9750yam
0800-0900	daily	SAm	9825yam
0800-1000	daily	SEA	11740sng
0800-1000	daily	WAf	15290iss
0900-1000	daily	SAm	9795sac
1000-1700	daily	SEA	11815yam
1300-1500	daily	CAm	11655sac
1500-1700	daily	CAf	17735iss
1500-1700	daily	SAs	12045yam
1700-1900	daily	SEA	7225yam
1700-1900	daily	As	6035yam
1700-1900	daily	ME,NAf	9575dha
1700-1900	daily	SAf	11945iss
1700-1900	daily	SAm	9835yam
1900-2200	daily	ME,NAf	9670yam
2000-2100	daily	Pac	9625yam
2000-2200	daily	As	6085yam
2000-2400	daily	As	11910yam
2100-2200	daily	Pac	13640yam
2100-2200	daily	SEA	6075yam
2200-2300	daily	ME,NAf	9620wer
2200-2400	daily	SAm	17605bon
2200-2400	daily	SEA	11665yam

Korean	Days	Area	kHz
0915-0945	daily	EAs	6160yam
1130-1200	daily	EAs	6090yam
1230-1300	daily	EAs	6190yam
1400-1430	daily	EAs	6190yam

Korean	Days	Area	kHz
1500-1530	daily	EAs	6190yam
2210-2230	daily	EAs	9560yam

Persian	Days	Area	kHz
0330-0400	daily	ME,NAf	6155iss
1430-1500	daily	ME,NAf	12045wer**

Portuguese	Days	Area	kHz
0230-0300	daily	SAm	6145sgo
0930-1000	daily	SAm	6145sgo

Russian	Days	Area	kHz
0330-0400	daily	Eu	6130wer
0430-0500	daily	Eu	738msk, 5980sit
0530-0600	daily	EAs	11715yam, 11760yam
0800-0830	daily	EAs	6145yam, 6165yam
1130-1200	....f..	WEu	9760wof+
1130-1200	daily	EAs	6185yam
1330-1400	daily	EAs	6190yam
1700-1730	daily	Eu	738msk

Spanish	Days	Area	kHz
0400-0430	daily	SAm	6195bon
0500-0530	daily	CAm	6195bon
1000-1030	daily	CAm	6120sac
1000-1030	daily	SAm	6195bon

Swahili	Days	Area	kHz
0315-0400	daily	EAf	7395mdc

Thai	Days	Area	kHz
1130-1200	daily	SEA	11740sng
1230-1300	daily	SEA	9695yam
2300-2320	daily	SEA	13650yam

Urdu	Days	Area	kHz
1430-1515	daily	SAs	6200tac

Vietnamese	Days	Area	kHz
1100-1130	daily	SEA	9695yam
1230-1300	daily	SEA	11740sng
2320-2340	daily	SEA	13650yam

Key: + DRM; * Relayed in the West Bank on 89.3MHz FM (Jenin) and 107.2MHz FM (Ramallah); ** Relayed on 88.0MHz FM (Kabul and Herat).
Ann: English: "This is NHK World, Radio Japan, in Tokyo"; Indonesian: "Inilah Radio Jepang, NHK World, siaran bahasa Indonesia"; Japanese: "Kochirawa NHK Warudo, Rajio Nippon, NHK no kokusaihoso desu"; Korean: "Yeogineun NHK World, Radio Ilbonimnda"; Mandarin: "Zheli shi riben guoji guangbo diantai, NHK huanqiu guangbowang".
IS: Melody "Kazoe Uta".
V: QSL-card.
Notes: Radio Japan is the External Sce of the public broadcaster NHK. The Japanese programmes include relays of NHK domestic Radio 1. Indonesian is broadcast daily from 1406-1451 on 89.2MHz FM in Jakarta, plus 34 other cities in Indonesia. Bengali is broadcast daily from 1500-1545 on 97.6MHz FM in Dhaka, plus 6 other cities in Bangladesh.

KDDI CORPORATION (Tx Operator)
Garden Air Tower, 10-10, Iidabashi 3-chome, Chiyoda-ku, Tokyo 102-8460, Japan.
☎ +81 3 33470077. 🖷 +81 3 33475845.
W: www.kddi.com
L.P: Pres/Chmn: Tadashi Onodera.
SW: [YAM] Koga, Yamata, Ibaraki prefecture: 4 x 100, 7 x 300kW.
Notes: KDDI Corporation is a major national telecommunications provider.

JORDAN (JOR)

RADIO JORDAN (Gov)
P.O. Box 909, Amman, Jordan.
☎ +962 6 4757410. 🖷 +962 6 4207862.
E: eng@jrtv.gov.jo; feedback@jrtv.gov.jo **W:** www.jrtv.jo
Webcast: L
L.P: Dir, Foreign Sces: Haytham Al-Etoom.
SW: [AKA] Al Karanah: 2 x 500kW.
kHz: 9830, 11960, 15290

Winter Schedule 2010/2011

Arabic	Days	Area	kHz
0500-0600	daily	Eu	11960aka*
1130-1230	daily	Eu,NAf	15290aka*
1845-2115	daily	Eu	9830akat

Key: * Times variable; † Irregular.
Ann: Arabic: "Huna Amman, Idha'atu-l-mamlaka al-urduniyya al-hashimiyya"; English: "This is Radio Jordan, broadcasting from Amman", "Radio Jordan, 96.3 FM".
V: QSL-card. Rec. acc.
Notes: Relays of Home Sce prgrs.

KAZAKHSTAN (KAZ)

KAZTELERADIO (Tx Operator)
pr. Al-Farabi 118,050040 Almaty, Kazakhstan.
☎ +7 727 2717501. 🖷 +7 727 2717630.
E: kazteleradio@mail.online.kz **W:** www.kazteleradio.kz
L.P: Pres: Erlan S. Bayjanov.
SW: [ALM] Almati, Two sites: Almati, Beserke (G.C. 43N30 077E00): 9 x 100kW; Qaraturiq, Tolqin (G.C. 43N39 078E56): 4 x 1000kW.
Notes: Kazteleradio is the national transmitter network owner in Kazakhstan.

KOREA, (D.P.R.) (KRE)

VOICE OF KOREA (VOK) (Gov)
Pyongyang, Democratic People's Republic of Korea.
☎ +850 2 3816035. 🖷 +850 2 3814416.
MW/SW: Uses txs provided by the Ministry of Telecommunications.
SAT: Thaicom 3.
kHz: 621, 3250, 3560, 4405, 6070, 6185, 6285, 7210, 7220, 7235, 7570, 7580, 9325, 9335, 9345, 9650, 9730, 9850, 9975, 9990, 11535, 11545, 11710, 11735, 11910, 12015, 13650, 13760, 15100, 15180, 15245

Winter Schedule 2010/2011

Arabic	Days	Area	kHz
1500-1600	daily	ME,NAf	9990kuj, 11545kuj
1700-1800	daily	ME,NAf	9990kuj, 11545kuj

Chinese	Days	Area	kHz
0000-0100	daily	EAs	3560kuj
0000-0100	daily	SEA	13650kuj, 15100kuj
0200-0300	daily	EAs	4405kuj, 7220kuj, 9345kuj, 9730kuj
0300-0400	daily	EAs	3560kuj
0300-0400	daily	SEA	13650kuj, 15100kuj
0800-0900	daily	EAs	3560kuj, 7220kuj, 9345kuj
1100-1200	daily	EAs	3560kuj, 7220kuj, 9345kuj
1300-1400	daily	SEA	6185kuj, 9850kuj
2100-2300	daily	EAs	7235kuj, 9345kuj, 9975kuj, 11535kuj

English	Days	Area	kHz
0100-0200	daily	EAs	4405kuj, 7220kuj, 9345kuj, 9730kuj
0100-0200	daily	LAm	11735kuj, 13760kuj, 15180kuj
0200-0300	daily	EAs	3560kuj
0200-0300	daily	SEA	13650kuj, 15100kuj
0300-0400	daily	EAs	4405kuj, 7220kuj, 9345kuj, 9730kuj
1000-1100	daily	LAm	6285kuj, 9335kuj
1000-1100	daily	SEA	6185kuj, 9850kuj
1300-1400	daily	NAm	9335kuj, 11710kuj
1300-1400	daily	Eu	3560kuj
1300-1400	daily	Eu	7570kuj, 12015kuj
1500-1600	daily	NAm	9335kuj, 11710kuj
1500-1600	daily	Eu	7570kuj, 12015kuj
1500-1600	daily	EAs	3560kuj
1600-1700	daily	ME,NAf	9990kuj, 11545kuj
1800-1900	daily	EAs	3560kuj
1800-1900	daily	Eu	7570kuj, 12015kuj
1900-2000	daily	ME,NAf	9975kuj, 11535kuj
1900-2000	daily	SAf	7210kuj, 11910kuj
2100-2200	daily	EAs	3560kuj
2100-2200	daily	Eu	7570kuj, 12015kuj

French	Days	Area	kHz
0100-0200	daily	EAs	3560kuj

French	Days	Area	kHz
0100-0200	daily	SEA	13650kuj, 15100kuj
0300-0400	daily	LAm	11735kuj, 13760kuj, 15180kuj
1100-1200	daily	LAm	6285kuj, 9335kuj
1100-1200	daily	SEA	6185kuj, 9850kuj
1400-1500	daily	EAs	3560kuj
1400-1500	daily	Eu	7570kuj, 12015kuj
1400-1500	daily	NAm	9335kuj, 11710kuj
1600-1700	daily	NAm	9335kuj, 11710kuj
1600-1700	daily	EAs	3560kuj
1600-1700	daily	Eu	7570kuj, 12015kuj
1800-1900	daily	ME,NAf	9975kuj, 11535kuj
1800-1900	daily	SAf	7210kuj, 11910kuj
2000-2100	daily	EAs	3560kuj
2000-2100	daily	Eu	7570kuj, 12015kuj

German	Days	Area	kHz
1600-1700	daily	EAs	4405kuj
1600-1700	daily	Eu	6285kuj, 9325kuj
1800-2000	daily	EAs	4405kuj
1800-2000	daily	Eu	6285kuj, 9325kuj

Japanese	Days	Area	kHz
0700-1250	daily	EAs	621chj, 3250pyo, 4405kuj, 7580kuj, 9650kuj
0900-1250	daily	EAs	6070kng
2100-2350	daily	EAs	621chj, 3250pyo, 4405kuj, 7580kuj, 9650kuj

Korean	Days	Area	kHz
0000-0050	daily	EAs	4405kuj**, 7220kuj**, 9345kuj**, 9730kuj**
0700-0750	daily	EAs	3560kuj**, 7220kuj**, 9345kuj**
0900-0950	daily	EAs	3560kuj*, 7220kuj*, 9345kuj*, 9975kuj**, 11735kuj**
0900-0950	daily	Eu	13760kuj**, 15245kuj**
1000-1050	daily	EAs	3560kuj**, 7220kuj**, 9345kuj**
1200-1250	daily	EAs	3560kuj*, 7220kuj**, 9345kuj**
1200-1250	daily	LAm	6285kuj*, 9335kuj*
1200-1250	daily	SEA	6185kuj*, 9850kuj*
1300-1350	daily	Eu	6285kuj*, 9325kuj**
1300-1350	daily	EAs	4405kuj**
1400-1450	daily	SEA	6185kuj*, 9850kuj*
1700-1750	daily	EAs	3560kuj*
1700-1750	daily	Eu	7570kuj*, 12015kuj*
1700-1750	daily	NAm	9335kuj*, 11710kuj*
2000-2050	daily	EAs	4405kuj*
2000-2050	daily	Eu	6285kuj*, 9325kuj*
2000-2050	daily	ME,NAf	9975kuj*, 11535kuj*
2000-2050	daily	SAf	7210kuj*, 11910kuj*
2300-2350	daily	EAs	3560kuj*, 7235kuj*, 9345kuj*, 9975kuj*, 11535kuj*
2300-2350	daily	Eu	7570kuj*, 12015kuj*

Russian	Days	Area	kHz
0700-0900	daily	EAs	9975kuj, 11735kuj
0700-0900	daily	Eu	13760kuj, 15245kuj
1400-1600	daily	Eu	6285kuj, 9325kuj
1400-1600	daily	EAs	4405kuj
1700-1800	daily	EAs	4405kuj
1700-1800	daily	Eu	6285kuj, 9325kuj

Spanish	Days	Area	kHz
0000-0100	daily	LAm	11735kuj, 13760kuj, 15180kuj
0200-0300	daily	LAm	11735kuj, 13760kuj, 15180kuj
1900-2000	daily	EAs	3560kuj
1900-2000	daily	Eu	7570kuj, 12015kuj
2200-2300	daily	Eu	7570kuj, 12015kuj
2200-2300	daily	EAs	3560kuj

Key: * Produced by KCBS; ** Produced by Pyongyang BS.

Ann: Arabic: "Huna Sowt al Koriya"; English: "This is the Voice of Korea"; French: "La Voix de la Corée"; German: "Hier ist die Stimme Koreas"; Japanese: "Choson no koe hoso desu"; Korean: "Joson Jung-ang Pangsong-imnida", "Pyongyang Pangsong-imnida"; Mandarin: "Chaoxian zhi sheng guangbo diantai"; Russian: "Govorit Golos Korei"; Spanish: "Aqui la Voz de Corea".
IS: Song of General Kim Il Sung. Opening music: National Anthem.
V: QSL-card.
Notes: The Voice of Korea is the External Sce of the Radio & TV Broadcasting Committee of the Democratic People's Republic of Korea.

PYONGYANG BROADCASTING STATION (PYONGYANG PANGSONG)

Pyongyang, Democratic People's Republic of Korea.
kHz: 621, 657, 684, 729, 801, 855, 3250, 3320, 6250, 6400

Winter Schedule 2010/2011

Korean	Days	Area	kHz
0000-0700	daily	EAs	621chj, 3250pyo
1300-1800	daily	EAs	621chj
1300-2030	daily	EAs	3250pyo
2100-1800	daily	EAs	684sag
2100-1900	daily	EAs	801hwd, 3320pyo, 6250pyo, 6400kng
2100-2030	daily	EAs	657kan, 729sep, 855swo

Ann: Korean: "Pyongyang Pangsong-imnida".
IS: Song of General Kim Il Sung. Opening & closing music: National Anthem.

MINISTRY OF POSTS & TELECOMMUNICATIONS (Tx Operator)

Oesong-dong, Central District, Pyongyang, Democratic People's Republic of Korea.
☎ +850 2 3813180. 🖷 +850 2 3814418.
L.P: Dir: Sung Su Ri.
MW: [CHJ] Chongjin: 621kHz 500kW; [HWD] Hwadae: 801kHz 500kW; [KAN] Kangnam: 657kHz 1500kW; [SAG] Samgo: 684kHz 250kW; [SEP] Sepo: 729kHz 50kW; [SWO] Sangwon: 855kHz 500kW.
SW: [KNG] Kanggye: 5 x 200kW; [KUJ] Kujang: 5 x 200kW; [PYO] Pyongyang: 10 x 200kW; Unknown location: 4 x 250kW.
Notes: The Ministry of Posts and Telecommunications owns and operates the transmitter network in the Democratic People's Republic of Korea.

KOREA, Rep. of (KOR)

KBS WORLD RADIO (Pub)

International Broadcasting, Korean Broadcasting System, 18 Yeouido-dong, Yeongdeungpo-gu, Seoul 150-790, Republic of Korea.
☎ +82 2 7813885. (English) 🖷 +82 2 7813694.
E: rki@kbs.co.kr; english@kbs.co.kr **W:** world.kbs.co.kr
Webcast: D/L/P
L.P: Pres: Kim In-Kyu; Exec. Producer, World Radio: Paek Seung Yeop.
MW: [DAN] Dangjin (HLCA): 972kHz 1500kW; [KIM] Gimje (HLSR): 1170kHz 500kW.
SW: [KIM] Gimje: 8 x 100, 3 x 250kW; [HWA] 3 x 100kW.
SAT: AsiaSat 2, Eurobird 1, Galaxy 25, Hot Bird 6, Intelsat 10, Nilesat 101.
kHz: 738, 972, 1170, 1440, 3955, 6045, 6065, 6095, 6155, 7235, 7275, 9430, 9515, 9560, 9565, 9570, 9580, 9640, 9650, 9705, 9760, 9770, 9805, 11795, 11810, 15160

Winter Schedule 2010/2011

Arabic	Days	Area	kHz
2000-2100	daily	ME,Af	9430sin
English	**Days**	**Area**	**kHz**
0000-0100	daily	WEu	1440mrn***
0200-0300	daily	SAm	9580kim
0800-0900	daily	SEA	9570kim
1100-1130	.....s.	Eu	9760wof+
1200-1300	daily	NAm	9650sac
1300-1400	daily	SEA	9570kim
1600-1700	daily	Eu	9515kim
1600-1700	daily	SEA	9640kim
1800-1900	daily	Eu	7275kim
2200-2230	daily	Eu	3955skn

French	Days	Area	kHz
2100-2200	daily	Eu	3955skn

German	Days	Area	kHz
0700-0730	daily	WEu	1440mrn+, 6095jun+
2000-2100	daily	Eu	3955skn

Indonesian	Days	Area	kHz
1200-1300	daily	SEA	9570kim
1400-1500	daily	SEA	9570kim
2200-2300	daily	SEA	9805kim

Japanese	Days	Area	kHz
0100-0200	daily	SAm	9580kim
0200-0300	daily	SAm	11810kim
0800-0900	daily	FE	6155kim, 7275kim
1000-1100	daily	FE	9805kim
1100-1300	daily	FE	1170kim

Korean	Days	Area	kHz
0000-0400	daily	FE	1170kim**
0300-0400	daily	SAm	11810kim
0400-2400	daily	FE	972dan*
0700-0800	daily	Eu	6045skn
0900-1000	daily	ME,Af,Eu	15160kim
0900-1100	daily	FE	7275kim
0900-1100	daily	SEA	9570kim
1000-1100	daily	FE	1170kim
1200-1300	daily	FE	7275kim
1400-1500	daily	NAm	9650sac
1400-2400	daily	FE	1170kim**
1600-1800	daily	Eu	7275kim
1600-1800	daily	ME,Af,Eu	9705kim
1700-1900	daily	Eu	9515kim

Mandarin	Days	Area	kHz
1130-1230	daily	CHN	6065kim
1130-1230	daily	SEA	9770kim
1300-1400	daily	FE	1170kim, 7275kim
2200-2300	daily	CHN	7275kim
2300-2400	daily	SEA	9805kim

Russian	Days	Area	kHz
1730-1800	daily	Eu	738msk
1800-1900	daily	Eu	7235rmp

Spanish	Days	Area	kHz
0100-0200	daily	SAm	11810kim
0200-0230	daily	NAm	9560sac
0600-0700	daily	Eu	6045sac
1100-1200	daily	SAm	11795sac

Vietnamese	Days	Area	kHz
0100-0200	daily	SEA	9565kim
1030-1100	daily	SEA	9770kim
1500-1600	daily	SEA	9640kim

Key: + DRM; * KBS Global Korean Network 1st programme; ** KBS Global Korean Network 2nd programme; *** Simulcast on www.rtl1440.com.

Ann: Arabic: "Huna KBS World Radio min Si'ul"; English: "This is KBS World Radio, the overseas service of the Korean Broadcasting System, coming to you from Seoul, the capital of the Republic of Korea"; German: "Hier ist KBS World Radio aus Seoul, der Auslandssender der Republik Korea"; Indonesian: "Inilah siaran bahasa Indonesia, KBS World Radio, yang dipancarkan langsung dari ibu kota Republik Korea, Seoul"; Japanese: "Kochirawa Kankoku Souru kara okurishiteimasu KBS no rajio kokusai hoso, KBS warudo rajio desu"; Korean: "Yeogineun Daehan Minguk Seoul-eseo bonaedeurineun KBS World Radio urimal bangsong-imnida"; Mandarin: "Zheli shi Hanguo guoji guangbo diantai, zai Dahanminguo shoudu Shou'er wei nin boyin"; Spanish: "Esto es KBS World Radio, emitiendo desde Seúl, Republica de Corea."; Vietnamese: "Day la chuong trinh phat thanh tieng Viet cua dai KBS World Radio phat thanh tu Seoul Han quoc".

IS: Korean children's song "Dar-a Dar-a Balgeun Dar-a (Oh, Bright Moon)", played on a glockenspiel. Original music "Dawn" composed by Kim Hee Jyo, with KBS symphony orchestra.

V: QSL-card.

Notes: KBS World Radio is the External Sce of the public broadcaster Korean Broadcasting System (KBS). KBS Global Korean Network Programmes are services for ethnic Koreans living outside of the Republic of Korea. Indonesian is broadcast daily from 1200-1300 on 102.6MHz in Jakarta (Indonesia).

FEBC KOREA (Rlg)
✉ P.O. Box 88, Seoul 121-707, Republic of Korea.

☎ +82 2 3200114. 🖷 +82 2 3200229.
E: febcadm@febc.net
W: english.febc.net (English); www.febc.net (Korean)
Webcast: L
LP: Chmn: Dr Billy Kim; Pres: Eun Gi Kim.
MW: [JEJ] Jeju (HLAZ): 1566kHz 250kW; [SEO] Seoul, Incheon (HLKX): 1188kHz 100kW.
kHz: *1188, 1566*

Winter Schedule 2010/2011

English	Days	Area	kHz
1100-1230	daily	SEA	1188seo

Japanese	Days	Area	kHz
1230-1345	daily	J	1566jej

Korean	Days	Area	kHz
1600-1100	daily	SEA	1188seo
1800-1100	daily	SEA	1566jej

Mandarin	Days	Area	kHz
1100-1230	daily	CHN	1566jej
1345-1730	daily	CHN	1566jej
1500-1600	daily	CHN	1188seo
1800-1900	daily	CHN	1566jej

Russian	Days	Area	kHz
1730-1800	daily	RUS	1566jej

Ann: English: "It's 8 o'clock and time for daily English segment on HLKX 1188 on your AM radio dial"; Japanese: "Kochirawa kirisuto-kyo hosokyoku FEBC desu"; Korean: "Jungpa Cheonbaek-palsip-pal (1188) kHz, Pyojun FM Paeng-nyuk-jeom-gu (106.9) MHz, Areumdaun Chanyanggwa Guwon-eui Gibbeun Sosigeul Jeonhaneun Geukdong Bangsong-imnida.", "Yeogineun Daehan Minguk Jeju Geukdong Bangsong-imnida"; Mandarin: "HLKX. Zheli shi zhongpo 1188 (yao yao ba ba) qianhe, Yiyou Diantai di 2 (er) dai.", "HLAZ. Zheli shi zhongbo 1566 (yao wu liu liu) qianhe, Yiyou Diantai di 1 (yi) dai".

V: QSL-card.

Notes: Owned by Far East Broadcasting Company (FEBC), see USA for corporate details.

KUWAIT (KWT)

RADIO KUWAIT (Gov)
✉ P.O. Box 193, Safat 13002, Kuwait.
☎ +965 22423773. 🖷 +965 22456660.
E: info@media.gov.kw
W: www.moinfo.gov.kw; www.media.gov.kw
LP: Assistant Under-Secretary for Radio Affairs, Ministry of Information: Khalid Al-Enezi; Dir, Frequency Management: Hani al-Naqi.
SW: [KBD] Kuwait, Kabd: 5 x 500kW.
SAT: AsiaSat 5, Badr 4, Hispasat 1C, Hot Bird 9, Nilesat 101, NSS 7.
kHz: *5960, 6050, 7250, 9750, 11630, 13650, 15515, 15540, 17550, 21540*

Winter Schedule 2010/2011

Arabic	Days	Area	kHz
0200-0900	daily	ME	5960kbd
0500-0900	daily	EAs	15515kbd
0930-1600	daily	WAf,CAf	11630kbd*
1000-1500	daily	Eu,NAm	21540kbd
1100-1600	daily	NAf	9750kbd
1600-2100	daily	ME	6050kbd
1700-2000	daily	NAm	13650kbd
2000-2400	daily	NAm	17550kbd‡

English	Days	Area	kHz
1800-2100	daily	Eu,NAm	15540kbd‡

Persian	Days	Area	kHz
0800-1000	daily	ME	7250kbd

Key: * Occasionally carries Quran prgr; ‡ inactive at time of publication.
Ann: Arabic: "Huna al-Kuwait".
V: QSL-folder. Rec. acc.
Notes: Arabic prgrs are relays of Home Sce networks.

IBB RELAY STATION KUWAIT
📡 IBB Transmitting Station, c/o US Embassy, P.O. Box 77, Safat 13001, Kuwait City, Kuwait.
LP: SM: Walter Patterson.
MW: [KWT] Kuwait, Umm Al-Rimam: 1548kHz 600kW, 1593kHz 150kW.
SW: [KWT] Kuwait, Umm Al-Rimam: 4 x 250kW.
V: QSL-card.

KYRGYZSTAN (KGZ)

SHORTWAVE RELAY SERVICE (Rlg)
⌨ Bishkek, Kyrgyzstan.
L.P: Dir, Technical: Timur Karimov.
SW: Leased from Kyrgyztelekom.
kHz: 5130

Winter Schedule 2010/2011

Various	Days	Area	kHz
1600-1800	daily	CAs	5130bis†

Key: † Irregularly varying schedule. Alternative frequency 6030kHz.
Notes: Rebroadcasts religious programming in Central Asian languages. Activity and schedule varies acc. to airtime bookings.

KYRGYZTELECOM (Tx Operator)
⌨ Chui avenue 96, 720000 Bishkek, Kyrgyzstan.
☎ +996 312 681616. 🖷 +996 312 662424.
E: info@kt.kg **W:** www.kt.kg
L.P: Pres: Damir A. Jumaev.
MW: [BIS] Bishkek, Krasnaya Rechka: 1467kHz 150kW.
SW: [BIS] Bishkek, Krasnaya Rechka: 1 x 100kW.
Notes: Kyrgyztelecom is the national tx operator.

LAOS (LAO)

LAO NATIONAL RADIO (Gov)
⌨ See National Radio section.
Webcast: D
SW: [VIE] Vientiane: 1 x 10kW.
FM/DAB: FM: 97.25MHz (Vientiane, 2kW).
kHz: 7145

Winter Schedule 2010/2011

English	Days	Area	kHz
0600-0630	daily	SEA	7145vie‡
1330-1400	daily	SEA	7145vie‡
French	**Days**	**Area**	**kHz**
0530-0600	daily	SEA	7145vie‡
1300-1330	daily	SEA	7145vie‡
Khmer	**Days**	**Area**	**kHz**
0000-0030	daily	SEA	7145vie‡
1230-1300	daily	SEA	7145vie‡
Thai	**Days**	**Area**	**kHz**
1130-1200	daily	SEA	7145vie‡
Vietnamese	**Days**	**Area**	**kHz**
1200-1230	daily	SEA	7145vie‡
2330-2400	daily	SEA	7145vie‡

Key: ‡ Inactive at time of publication.
IS: National Anthem.
V: QSL-card. No IRCs required.

LIBYA (LBY)

VOICE OF AFRICA (Gov)
⌨ General Centre of Broadcasting, P.O. Box 4677, Tripoli, Libya.
☎ +218 21 4449206. 🖷 +218 21 4446875.
E: info@ljbc.net; info@voiceofafrica.com.ly
W: www.ljbc.net; www.voiceofafrica.com.ly
Webcast: L
MW: [TRI] Tripoli: 1251kHz 400kW.
SW: [SAB] Sabrata: 2 x 500kW.
SAT: AsiaSat 3S, Atlantic Bird 3, Hispasat 1C.
kHz: 1251, 11860, 11965, 15215, 15660, 17725, 21695

Winter Schedule 2010/2011

Arabic	Days	Area	kHz
1700-0400	daily	NAf,Eu	1251tri†
English	**Days**	**Area**	**kHz**
1400-1600	daily	CAf,EAf	17725sab, 21695sab
French	**Days**	**Area**	**kHz**
1600-1700	daily	NAf,CAf	15660sab, 17725sab
1700-1800	daily	NAf,CAf	11965sab, 15215sab
Hausa	**Days**	**Area**	**kHz**
1800-1900	daily	WAf	15215sab
1800-2000	daily	WAf	11965sab
1900-2000	daily	WAf	11860sab
Swahili	**Days**	**Area**	**kHz**
1200-1400	daily	CAf,EAf	17725sab, 21695sab

Key: † Irregular.
Ann: Arabic: "Idha'at Sowt Afrikiya min al-jamahiriya al Ozma";

English: "This is the Voice of Africa from the Great Jamahiriya".
V: QSL-folder.
Notes: External Sce of the Libyan Jamahirya Broadcasting Corp. (LJBC). LJBC also owns the station "Africa No 1" (see Gabon).

LITHUANIA (LTU)

RADIO BALTIC WAVES INTERNATIONAL (RBWI)
⌨ Švitrigailos g. 11a-211, LT-03228 Vilnius, Lithuania.
☎ +370 5 2652532. 🖷 +370 5 2652532.
E: radio@balticwaves.cjb.net
L.P: Dir: Rolandas Stirblys; Project Coordinator: Rimantas Pleikys.
MW/SW: Leased from LRTC.
V: QSL-card. (For relayed stns)
Notes: RBWI markets air time on MW relay facilities in Lithuania for foreign broadcasters.

ZILIONIS RADIO TV CONSULTING (Broker)
⌨ P.O. Box 3000, LT-02003 Vilnius 13, Lithuania.
☎ +370 6 8576840. 🖷 +370 5 2652532.
E: consult@zilionis.com **W:** www.zilionis.com/airtime
L.P: Dir: Sigitas Zilionis.

LIETUVOS RADIJO IR TELEVIZIJOS CENTRAS (LRTC) (Tx Operator)
⌨ Sausio 13-osios g. 10, LT-04347 Vilnius, Lithuania.
☎ +370 5 2525300. 🖷 +370 5 2525325.
E: info@lrtc.net **W:** www.lrtc.net
L.P: GD: Gediminas Stirbys.
MW: [SIT] Kaunas, Sitkunai: 1386kHz 150/500kW; [VLN] Vilnius: 612kHz 100kW.
SW: [SIT] Kaunas, Sitkunai: 1 x 100kW.
Notes: LRTC is the national transmitter network operator.

LUXEMBOURG (LUX)

RTL RADIO (Comm)
⌨ 45, boulevard Pierre Frieden, L-1543 Luxembourg.
☎ +352 42 1422175. 🖷 +352 42 1422756.
E: webmaster@rtlgroup.com; fabien.culot@rtlgroup.com (RTL Group) **W:** www.rtlgroup.com
Webcast: L (radio.rtl.lu/wma.asx)
L.P: CEO: Alain Berwick.
MW/SW: Uses txs provided by Broadcasting Center Europe.
SAT: Astra 1H/KR.
kHz: 236, 1440, 6095

Winter Schedule 2010/2011

French	Days	Area	kHz
0000-2400	daily	F	236bdw*
German	**Days**	**Area**	**kHz**
0645-0700	daily	D	6095jun+
0730-0800	daily	D	1440mrn+, 6095jun+

Key: + DRM; * relay of RTL France.
Ann: German: "RTL Radio".
Notes: The RTL Group is Europe's largest TV, radio and production company and is majority-owned by Bertelsmann AG (Germany).

BROADCASTING CENTER EUROPE (BCE) (Tx Operator)
⌨ 45, boulevard Pierre Frieden, L-1543 Luxembourg.
☎ +352 24 806605. 🖷 +352 24 806609.
E: contact@bce.lu **W:** www.bce.lu
Webcast: L
L.P: CEO: Alain Flammang; Head, Radio/TV transmissions: Eugène Muller.
LW: [BDW] Beidweiler: 234kHz 2 x 1000kW.
MW: [MRN] Marnach: 1440kHz 1200kW.
SW: [JUN] Junglinster: 2 x 250kW (used exclusively in DRM mode).
V: QSL-card. (Online report form available at www.bce.lu/transmission/lwsw/report)
Notes: BCE was founded in January 2000, as a result of the merger of various technical entities from the RTL Group.

MACEDONIA (MKD)

RADIO MAKEDONIJA (Pub)
⌨ Blvd. "Goce Delcev" bb, 1000 Skopje, FYR Macedonia.
☎ +389 2 321193. 🖷 +389 2 311821.

E: radiomakedonija@mr.com.mk **W:** www.mrt.com.mk
LP: Dir, MRT: Grigori Popovski.
MW: Leased from Makedonska Radiodifuzija.
kHz: 810

Winter Schedule 2010/2011

Albanian	Days	Area	kHz
2000-2030	mtwtf..	Eu	810sko

Bulgarian	Days	Area	kHz
1900-1930	mtwtf..	Eu	810sko

Greek	Days	Area	kHz
1930-2000	mtwtf..	Eu	810sko

Macedonian	Days	Area	kHz
1830-1900	mtwtf..	Eu	810sko
1900-2100	.....s.	Eu	810sko
2100-0200	mtwtf..	Eu	810sko

Serbian	Days	Area	kHz
2030-2100	mtwtf..	Eu	810sko

Ann: Macedonian: "Radio Makedonija".
V: QSL-letter.
Notes: External Sce of the public service broadcaster Makedonska Radio Televizija (MRT).

MAKEDONSKA RADIODIFUZIJA (Tx Operator)
✉ blvd. "Goce Delcev" bb, 1000 Skopje, FYR Macedonia.
☎ +389 2 3297100. 📠 +389 2 3225520.
E: info@jpmrd.gov.mk **W:** www.jpmrd.gov.mk
MW: [SKO] Skopje, Sveti Nikole: 810kHz 1200kW.
V: QSL-letter.
Notes: Makedonska Radiodifuzija is the national transmitter network provider.

MADAGASCAR (MDG)

MADAGASCAR WORLD VOICE (Rlg)
✉ 605 Bradley Court, Franklin, TN 37067, USA. (Headquarters)
☎ +1 615 3718707. (HQ)
W: www.yas3na.com (Arabic)
SW: [MHJ] Mahajanga: 3 x 100kW. (Under construction)
Notes: Owned by World Christian Broadcasting, Inc. (USA).

RADIO NETHERLANDS WORLDWIDE RELAY STATION
✉ P.O. Box 404, Antananarivo, Madagascar.
SW: [MDC] Talata Volonondry: 1 x 50, 1 x 250, 2 x 300kW.
V: QSL-card. (For direct rpt)

MALAYSIA (MLA)

VOICE OF MALAYSIA (VOM) (Pub)
✉ P.O. Box 11272, 50740 Kuala Lumpur, Malaysia.
☎ +60 3 22887826. 📠 +60 3 22847594.
E: suaramalaysia@rtm.gov.my; vom@rtm.gov.my
W: www.vom.com.my
Webcast: L
MW: [TUA] Kota Kinabalu: 1475kHz 700kW.
SW: [KAJ] Kajang: 9 x 100, 2 x 500kW.
SAT: Measat 3.
kHz: 1475, 6050, 6175, 7235, 9750, 11885, 15295

Winter Schedule 2010/2011

Chinese	Days	Area	kHz
1000-1200	daily	SEA,Pac	7235kaj+, 11885kaj+, 15295kaj

English	Days	Area	kHz
0600-0800	daily	SEA,Pac	6175kaj, 9750kaj, 15295kaj
0800-1000	daily	SEA,Pac	6175kaj*, 9750kaj*, 15295kaj*

Malay	Days	Area	kHz
0400-0600	daily	SEA,Pac	6175kaj, 9750kaj, 15295kaj
1200-1400	daily	SEA	15295kaj
1400-1600	daily	SEA	6050kaj*, 6175kaj*, 9750kaj*

Tagalog	Days	Area	kHz
1100-1330	daily	SEA	1475tua

Key: + DRM (tests, schedule variable); * "Voice of Islam" prgr.

Ann: English: "This is the Voice of Malaysia"; Malay: "Inilah suara Malaysia".
IS: First bar of National Anthem "Negara Ku" (Chimes).
V: QSL-card. Rec. acc.
Notes: The Voice of Malaysia is the External Sce of the public service Radio Televisyen Malaysia (RTM). Frequencies and schedule variable.

MALI (MLI)

CHINA RADIO INTERNATIONAL RELAY
SW: [BKO] Bamako, Kati: 2 x 100kW.
V: QSL-card. (Rpt to CRI, in China)
Notes: The shortwave facilities are leased to CRI by Radiodiffusion-Télevision du Mali.

MOLDOVA

Transnistria

✉ ul. Pravdy 31, MD-3300 Tiraspol, Moldova.
☎ +373 533 77758.
E: radiopmr@inbox.ru **W:** radiopmr.org
Webcast: L
LP: Dir: Anatoliy A. Kirsa; Editor-in-Chief: Yekaterina N. Poshelyuk.
MW/SW: Leased from Pridnestrovskiy Radioteletsentr.
kHz: 999, 6240

Winter Schedule 2010/2011

English	Days	Area	kHz
1830-1900	daily	Eu	6240kch
2030-2100	daily	Eu	6240kch
2230-2300	daily	Eu	6240kch

French	Days	Area	kHz
1900-1930	daily	Eu	6240kch
2100-2130	daily	Eu	6240kch
2300-2330	daily	Eu	6240kch

German	Days	Area	kHz
1930-2000	daily	Eu	6240kch
2130-2200	daily	Eu	6240kch
2330-2400	daily	Eu	6240kch

Russian	Days	Area	kHz
0300-0500	mtwtf..	Eu	999kch
1800-1830	daily	Eu	6240kch
2000-2030	daily	Eu	6240kch
2200-2230	daily	Eu	6240kch

Ann: English: "Here is Tiraspol, the capital of the Pridnestrovyan Moldavian Republic", "Radio PMR"; German: "Hier ist Tiraspol"; Russian: "Vy slushaete programmu Pridnestrovya Radio Pridnestrovsko-Moldavskoy Respubliki".
V: QSL-letter.
Notes: Produced by the state broadcaster of the self-proclaimed Pridnestrovian Moldavian Republic, in Eastern Moldova.

PRIDNESTROVSKIY RADIOTELETSENTR (Tx Operator)
✉ MD-4006 Maiac, Pridnestrovian Moldavian Republic, Moldova.
☎ +373 210 66500.
E: prtc@idknet.com
LP: DG: Vitaliy Kucherenko.
MW: [KCH] Grigoriopol, Maiac: 999/1413/1548kHz 1000kW.
SW: [KCH] Grigoriopol, Maiac: 5 x 1000kW.
Notes: Pridnestrovskiy Radioteletsentr (owned by RTRN, Russia) provides high power MW & SW transmitting facilities.

MONACO (MCO)

RADIO MONACO (Comm)
✉ 7 rue du Gabian, Gildo Pastor Center, 98000 Monaco.
☎ +377 97 700621. 📠 +377 97 985051.
E: info@radio-monaco.com **W:** radio-monaco.com
kHz: 4368, 8728, 13146, 17260

Winter Schedule 2010/2011

French	Days	Area	kHz
1200-1203	mtwtf..	Atl,Med	4368mco*, 8728mco*, 13146mco*, 17260mco*

Key: * USB.
Ann: English: "Radio Monaco".
V: QSL-email.

Notes: News/weather relays for seamen, via Monaco Radio/Naya Radio utility station. Irregular.

MONTE CARLO RADIODIFFUSION (MCR)
(Tx Operator)
📧 10-12 quai Antoine 1er, MC-98000 Monte Carlo, Monaco.
☎ +377 97974700. 🖷 +377 97974707.
E: mcradiodiffusion@mcr.mc **W:** www.mcr.mc
L.P: DG: Jean-Pierre Margossian.
LW: [ROU] Roumoules (France): 216kHz 2000kW.
MW: [CDM] Col de la Madone (France): 702kHz 2 x 600kW; [ROU] Roumoules (France): 1467kHz 1000kW.
SW: [MCO] Fontbonne, Mont Agel (France): 2 x 100, 1 x 500kW.
Notes: MCR (a subsidiary of Télédiffusion de France) is the national transmitter network owner in Monaco and also maintains high power transmitting centres in France and Cyprus.

MONGOLIA (MNG)

VOICE OF MONGOLIA (Pub)
📧 P.O. Box 365, Ulaanbaatar 13, Mongolia.
☎ +976 11 327900. 🖷 +976 11 323096.
E: densmaa9@yahoo.com **W:** www.vom.mn; www.mnb.mn
Webcast: D
L.P: Dir, Foreign Sce: Mrs Narantuya B; Mail Editor: Mrs Densmaa Zorigt.
MW/SW: Leased from MRTBN.
kHz: *990, 9665, 12085*

Winter Schedule 2010/2011

English	Days	Area	kHz
1030-1100	daily	As	12085uba
1530-1600	daily	As	9665uba
Japanese	**Days**	**Area**	**kHz**
0900-0930	daily	As	12085uba
1500-1530	daily	As	9665uba
Mandarin	**Days**	**Area**	**kHz**
1000-1030	daily	As	990uba, 12085uba
1430-1500	daily	As	990uba, 9665uba
Mongolian	**Days**	**Area**	**kHz**
0930-1000	daily	As	990uba, 12085uba
1400-1430	daily	As	990uba, 9665uba

Ann: English: "This is the Voice of Mongolia"; Mongolian: "Ulaanbaataraas yarij baina".
V: QSL-card. Rec. acc. (non returnable), Rp (2 IRCs or 1 USD) appreciated.
Notes: The Voice of Mongolia is the External Sce of the Mongolian National Radio & TV.

MONGOLIAN RADIO AND TELEVISION BROADCASTING NETWORK (MRTBN) (Tx Operator)
📧 P.O. Box 1126, Ulaanbaatar, Mongolia.
☎ +976 21 248810. 🖷 +976 21 248812.
E: rnts@mongol.net **W:** rtbn.gov.mn
L.P: Dir: T. Gantömör.
MW: [CHO] Choybalsan: 1350kHz 500kW; [UBA] Ulaanbaatar, Hönhör: 990kHz 500kW.
SW: [UBA] Ulaanbaatar, Hönhör: 3 x 50, 1 x 100, 1 x 250, 1 x 500kW.
Notes: MRTBN is the national transmitter operator in Mongolia.

MOROCCO (MRC)

RADIO MAROCAINE (Gov)
📧 BP 1042, 10000 Rabat, Morocco.
☎ +212 37 766880. 🖷 +212 37 766888.
W: www.snrt.ma
Webcast: L (Live streams of HS channels)
L.P: Dir, Radio: Mohamed El Boukili; Dir, Foreign Broadcasting: Abdelouahad Belghiti Alaoui; Dir, Foreign Rel: Abdelkader Bouazza.
SAT: Arabsat 2B/2D/3A, Astra 1E, Eutelsat W2, Hot Bird 6.
kHz: *15341, 15345*

Winter Schedule 2010/2011

Arabic	Days	Area	kHz
0900-1500	daily	Af	15341nad±
1500-2200	daily	Af	15345nad

Key: ± Variable frequency.
Ann: Arabic: "Huna Ribat, idha'atu-l-mamlaka al Maghribiyya".
V: QSL-card. Rec. acc. Rp.
Notes: Relays of Home Sce prgrs for listeners abroad.

RADIO MÉDITERRANÉE INTERNATIONALE (MEDI 1) (Comm)
📧 3/5 Rue M'Sallah, 90000 Tanger, Morocco.
☎ +212 5 39936363. 🖷 +212 5 39935755.
E: medi1@medi1.com **W:** www.medi1.com
Webcast: D/L
L.P: DG: M. Pierre Casalta.
LW: [NAD] Nador: 171kHz 2000kW.
SW: [NAD] Nador: 2 x 250kW.
FM/DAB: FM: Transmitters in Morocco, France and Belgium.
SAT: Astra 1H.
kHz: *171, 9575*

Winter Schedule 2010/2011

Arabic/French	Days	Area	kHz
0000-2400	daily	Af,ME,Eu	171nad, 9575nad

Ann: Arabic: "Mahataat Medi an"; French: "Ici Medi 1, Radio Méditerranée Internationale".
V: QSL-card.
Notes: Medi 1 is a Moroccan-French joint venture.

NETHERLANDS (HOL)

RADIO NETHERLANDS WORLDWIDE (RNW)
📧 P.O. Box 222, 1200 JG Hilversum, Netherlands.
☎ +31 35 6724211. (Switchboard, open 24 hours) 🖷 +31 35 6724239. (English language service)
E: letters@rnw.nl **W:** www.radionetherlands.nl
Webcast: D/L/P
☎ +31 35 6724784. (For journalists/media professionals)
L.P: Chmn, Supervisory Board: Dr B.R. Bot; GD: Jan C. Hoek; Editor in Chief: Rik R. Rensen; Frequency Mgr: Jan Peter Werkman.
SAT: AMC4, Astra 1L, Hot Bird 8, Intelsat 7, Optus D2, SatMex 6, Sirius 4, Thaicom 5.
kHz: *1296, 5860, 5935, 5955, 6020, 6035, 6040, 6100, 6120, 6145, 6165, 6195, 7285, 7360, 7425, 9720, 9750, 9795, 9810, 9830, 9865, 9895, 11615, 11655, 11935, 12065, 12075, 12080, 13700, 15255, 15280, 15315, 15540, 15595, 15750, 17505, 17605*

Winter Schedule 2010/2011

Dutch	Days	Area	kHz
0000-0030	daily	LAm	6145bon
0100-0130	daily	NAm	6195bon
0300-0330	daily	NAm	6100bon
0330-0400	daily	SAm	6195bon
0400-0430	daily	NAm	6165bon
0600-0630	daily	NAm	6165bon
0600-0630	daily	Pac	9865bon
0600-0700	daily	Eu,ME	9830wer
0600-0700	daily	Eu,WAf	6120nau
0600-0800	daily	Eu	5955sin, 9895kch
0700-0900	daily	Eu	6120smg
0800-0830	daily	SEA,Pac	15750sai
0800-0900	daily	Eu	5955wer, 9895nau
0800-0900	daily	Eu,NAf	11935smg
0900-1100	mtwtf.s	Eu	5955wer
0900-1100	mtwtf..	Eu	1296orf, 6035iss, 6120wer
0900-1600	......s	Eu	9895smg
0930-1000	mtwtfs.	SAm	6020bon
1100-1130	daily	EAs	7360pht
1100-1600	......s	Eu	5955wer
1200-1230	daily	EAs	12065pht
1230-1300	daily	SEA	9795tin
1300-1330	daily	SAs	12065pht, 15255dha
1300-1330	daily	SEA,Pac	12080pht
1600-1700	daily	Eu,ME	9750iss
1600-1700	daily	Eu	1296orf, 5955rmp
1600-1800	daily	Eu,NAf	9895wer
1700-1730	daily	SAf	7285mdc
1700-1800	daily	Eu	5955kch
1730-1800	daily	EAf,ME	11615mdc, 11655mdc
2100-2130	daily	NAf,WAf	6040smg
2100-2130	daily	SAm	17605bon
2100-2130	daily	WAf	13700bon
2200-2230	daily	NAf,WAf	6040smg
2200-2230	daily	SAm	15315bon, 15540bon

English	Days	Area	kHz
1000-1100	daily	SEA	9720pht, 12065pht
1400-1500	daily	SAs	12080pht
1400-1600	daily	SAs	15595mdc
1500-1600	daily	SAs	12080pht
1800-1900	daily	SAf	6020mdc
1800-2100	daily	CAf,WAf	11655mdc
1900-2000	daily	EAf,ME	9895trm
1900-2000	daily	WAf	11615mey
1900-2100	daily	CAf,EAf	7425mdc
2000-2100	daily	WAf	5935sin
Indonesian	**Days**	**Area**	**kHz**
1100-1200	daily	SEA	9795sng, 17505dha
1200-1300	daily	SEA	12075sng
2200-2300	daily	SEA	5860pht, 15280sai
2300-2400	daily	SEA	6120sng
Spanish	**Days**	**Area**	**kHz**
0000-0400	daily	LAm	6165bon
1100-1130	daily	CAm,NAm	6165bon
1130-1230	daily	SAm	6165bon
1200-1230	daily	CAm	9810bon

Ann: English: "This is Radio Netherlands, the Dutch International Service"; Indonesian: "Inilah Radio Nederland di Hilversum dengan siaran dalam bahasa Indonesia"; Spanish: "Transmite R. Nederland desde la ciudad de Hilversum en Holanda".
V: QSL-card.
Notes: RNW is a government funded public broadcaster serving Dutch citizens abroad, listeners in the former Dutch colonies and an international audience around the world.

NEW ZEALAND (NZL)

RADIO NEW ZEALAND INTERNATIONAL (RNZI) (Pub)
✉ P.O. Box 123, Wellington, New Zealand.
☎ +64 4 4741437. 🖷 +64 4 4741433.
E: info@rnzi.com **W:** www.rnzi.com
Webcast: D/L/P
LP: CEO & Chief Editor: Peter Cavanagh; Technical Mgr: Adrian Sainsbury; Transmission Engineer: Andy Anderson.
SW: [RAN] Rangitaiki: 2 x 100kW.
kHz: 5950, 7440, 9765, 9870, 9890, 11675, 11725, 13660, 13730, 15720, 17675

Winter Schedule 2010/2011			
English	**Days**	**Area**	**kHz**
0500-0700	daily	Pac	11725ran*, 13730ran*, +
0700-0800	daily	Pac	11675ran*, +
0700-1100	daily	Pac	9765ran*
0800-1200	daily	Pac	9870ran*, +
1100-1300	daily	Pac	13660ran*
1300-1550	daily	Pac	5950ran*
1550-1650	daily	Pac	5950ran*, +, 7440ran*
1650-1750	daily	Pac	9765ran*, 9890ran*, +
1750-1850	daily	Pac	11675ran*, +
1750-2150	daily	Pac	11725ran*
1850-1950	daily	Pac	15720ran*, +
1950-2050	daily	Pac	17675ran*, +
2050-2150	daily	Pac	15720ran*, +
2150-0500	daily	Pac	15720ran*, 17675ran*, +

Key: + DRM; * Schedule variable.
Ann: English: "This is Radio New Zealand International, the Voice of the Pacific"; Maori: "Te reo irirangi o Aotearoa, o te Moana-nui-a-Kiwa".
V: QSL-card. Rp (2 IRC or 2 USD). Rec. not accepted.
Notes: RNZI is the External Sce of the public broadcaster Radio New Zealand. The SW transmissions are in English, with news in various Pacific languages. A new 300kW DRM capable transmitter, increased programming and languages, plus local FM relays in major Pacific urban centers are planned. 1st & 3rd Thursdays of each month are site maintenance days and transmissions may be reduced or there may be test transmissions during this period.

NIGERIA (NIG)

VOICE OF NIGERIA (VON) (Gov)
✉ Broadcasting House, Ikoyi, P.M.B. 40003, Falomo, Lagos, Nigeria.
☎ +234 1 2693078. 🖷 +234 1 2691944.
E: info@voiceofnigeria.org **W:** www.voiceofnigeria.org

Webcast: D/P
✉ 6th and 7th Floors, Radio House, Herbert Macaulay Way Garki, Abuja, Nigeria. (Abuja Office)
LP: Chmn: Stella Effah Attoe; DG: Alhaji Abubakar Jijiwa; PD: Kabir Mohammed Ahmed.
SW: [AJA] Abuja, Lugbe: 3 x 250kW (under construction, due to be become active in the near future, with a total of 5 x 250kW transmitters); [IKO] Ikorodu: 3 x 250kW.
kHz: 7255, 9690, 11770, 15120

Winter Schedule 2010/2011			
Arabic	**Days**	**Area**	**kHz**
1730-1800	daily	NAf,ME	15120iko
English	**Days**	**Area**	**kHz**
0455-0700	daily	Eu	15120iko
0900-1500	daily	WAf	9690iko
1500-1600	daily	WAf	15120iko
1800-1900	daily	Eu,Af	15120iko
1900-2000	daily	WAf	7255iko
French	**Days**	**Area**	**kHz**
0700-0800	daily	Eu	15120iko
2000-2100	daily	WAf	7255iko
Fulfulde	**Days**	**Area**	**kHz**
2100-2200	daily	WAf	7255iko
Hausa	**Days**	**Area**	**kHz**
0800-0900	daily	WAf	9690iko
2200-2300	daily	WAf	7255iko
Igbo	**Days**	**Area**	**kHz**
1700-1730	daily	WAf	9690iko
Swahili	**Days**	**Area**	**kHz**
1600-1630	daily	EAf	11770iko
Yoruba	**Days**	**Area**	**kHz**
1630-1700	daily	WAf	9690iko

Ann: English: "You're listening to the Voice of Nigeria, Lagos".
IS: As Home Sce. Also bells playing the first bars of the National Anthem, 15 minutes before the commencement of each transmission block.
V: QSL-card.
Notes: The Voice of Nigeria is the External Sce of the Federal Radio Corporation of Nigera (FRCN).

NORTHERN MARIANA IS (MRA)

FEBC SAIPAN (KFBS) (Rlg)
✉ P.O. Box 500209, Saipan, MP 96950-0209, USA.
☎ +1 670 3229088. 🖷 +1 670 3223060.
E: saipan@febc.org **W:** febc.org
LP: Dir: Robert Springer.
SW: [FBS] Saipan, Marpi: 4 x 100kW.
kHz: 11580, 11650, 11850, 11990, 12090, 15580

Winter Schedule 2010/2011			
Bashkir	**Days**	**Area**	**kHz**
1245-1300	...f..	RUS	11650fbs
Chuvash	**Days**	**Area**	**kHz**
1245-1300	..t...	RUS	11650fbs
Gorontalo	**Days**	**Area**	**kHz**
0830-0900	...tfss	SEA	11850fbs
Indonesian	**Days**	**Area**	**kHz**
1030-1130	daily	SEA	15580fbs
Javanese	**Days**	**Area**	**kHz**
1000-1030	daily	SEA	15580fbs
Karakalpak	**Days**	**Area**	**kHz**
1345-1400	....fs.	RUS	11650fbs
Kazakh	**Days**	**Area**	**kHz**
1330-1330	daily	CAs	11650fbs
1345-1400	m.....s	CAs	11650fbs
Komi	**Days**	**Area**	**kHz**
1230-1245	.....s.	RUS	11650fbs
Kyrgyz	**Days**	**Area**	**kHz**
1330-1345	daily	CAs	11650fbs
Madura	**Days**	**Area**	**kHz**
0830-0900	mtw....	SEA	11850fbs
Mandarin	**Days**	**Area**	**kHz**
1300-1400	daily	EAs	11580fbs
Mari	**Days**	**Area**	**kHz**
1230-1245	..w....	RUS	11650fbs

Mongolian	Days	Area	kHz
0930-1000	daily	EAs	11990fbs
Osetian	**Days**	**Area**	**kHz**
1345-1400	...t...	RUS	11650fbs
Russian	**Days**	**Area**	**kHz**
1100-1230	daily	RUS	11650fbs
Tatar	**Days**	**Area**	**kHz**
1230-1245	...f...	RUS	11650fbs
1345-1400	.t.....	RUS	11650fbs
Udmurt	**Days**	**Area**	**kHz**
1230-1300	.....s	RUS	11650fbs
1245-1300	.....s.	RUS	11650fbs
Uzbek	**Days**	**Area**	**kHz**
1345-1400	..w....	CAs	11650fbs
Vietnamese	**Days**	**Area**	**kHz**
2230-2300	daily	SEA	12090fbs

V: QSL-card. Rp. (2 IRCs), Rec. acc.
Notes: Owned by Far East Broadcasting Company (FEBC), see USA for corporate details.

IBB "ROBERT E. KAMOSA" TRANSMITTING STATION
✉ IBB "Robert E. Kamosa" Transmitting Station, P.O. Box 504969, Saipan, MP 96950, USA.
☎ +1 670 2331624. 🖷 +1 670 2331614.
L.P: SM: Gary Shirk; Transmitter Plant Supervisor, Tinian: Dan Harnett.
SW: [SAI] Saipan, Agingan Point: 3 x 100kW; [TIN] Tinian: 2 x 250, 6 x 500kW.
Notes: IBB owned transmitting station. Maintained & operated by Rome Research Corporation (RRC).

NORWAY (NOR)

NORKRING (Tx Operator)
✉ Telenor Broadcast, N-1331 Fornebu, Norway.
☎ +47 67892000. 🖷 +47 67893611.
E: norkring@telenor.no **W:** www.norkring.no
L.P: CEO: Torbjørn Ødegård Teigen.
SW: [KVI] Kvitsøy: 2 x 500kW.

OMAN (OMA)

RADIO SULTANATE OF OMAN (Gov)
✉ P.O. Box 600, 113 Muscat, Oman.
☎ +968 24 603888. 🖷 +968 24 604629.
E: feedback_rd@oman-radio.gov.om
W: www.oman-tv.gov.om
Webcast: L
L.P: Dir, Foreign Sce: Shakar Al-Araimi; Dir, Engineering: Mohd Salim Al Morhouby; Frequency Mgr: Salim Al-Nomani.
SW: [SEB] Seeb: 1 x 100kW; [THU] Thumrait: 1 x 100kW.
SAT: Arabsat 3A, AsiaSat 3S, Hispasat 1C, Hot Bird 4, Nilesat 101, NSS 7, Galaxy 25.
kHz: 15140, 15355, 17590

Winter Schedule 2010/2011			
Arabic	**Days**	**Area**	**kHz**
0200-0300	daily	EAf	15355thu
0400-0600	daily	EAf	17590thu
1500-2200	daily	Eu,ME	15140thu
English	**Days**	**Area**	**kHz**
0300-0400	daily	EAf	15355thu
1400-1500	daily	Eu,ME	15140thu

Ann: Arabic: "Idha'atu Saltanat Oman min Muscat"; English: "Radio Sultanate of Oman".
V: QSL-folder.
Notes: Relays of Home Sce programmes in Arabic and English.

BBC EASTERN RELAY STATION
✉ P.O. Box 40, 422 Al Ashkarah, Oman.
E: rebers@omantel.net.om
L.P: Resident Engineer: Dave Battey.
MW: [SLA] A'Seela: 702/1413kHz 800kW.
SW: [SLA] A'Seela: 3 x 250kW.
V: QSL-card. (For direct report)
Notes: Owned by the BBC and operated by Babcock International Group PLC (see under United Kingdom).

PAKISTAN (PAK)

RADIO PAKISTAN (Gov)
✉ Broadcasting House, Constitution Avenue, Islamabad 4400, Pakistan.
☎ +92 51 9210689. (News Room) 🖷 +92 51 9222432.
E: info@radio.gov.pk; infonews@radio.gov.pk
W: www.radio.gov.pk
L.P: DG, PBC: Murtaza Solangi; GD: S Auwar Mehmood; Dir, Technical: Muhammad Iqbel; Dir, Overseas Liaison: Muhammad Sharif Shad.
SW: [ISL] Islamabad, Rawat: 5 x 100, 2 x 250kW.
kHz: 6235, 7470, 7510, 7530, 9340, 9345, 9670, 11510, 11570, 11575, 11580, 11880, 15100, 15490, 15540, 15620, 17700, 17830

Winter Schedule 2010/2011			
Balti	**Days**	**Area**	**kHz**
0445-0530	daily	SAs	7470isl
Bengali	**Days**	**Area**	**kHz**
0900-1000	daily	SAs	9345isl, 15620isl
Chinese	**Days**	**Area**	**kHz**
1200-1300	daily	EAs	9670isl, 11510isl
Dari	**Days**	**Area**	**kHz**
1445-1545	daily	WAs	6235isl
English	**Days**	**Area**	**kHz**
1100-1105	daily	Eu	15100isl, 17700isl
1600-1610	daily	ME,Af	7510isl, 11575isl
Farsi	**Days**	**Area**	**kHz**
1700-1800	daily	ME	6235isl, 7470isl
Gujarati	**Days**	**Area**	**kHz**
1145-1215	daily	SAs	9340isl, 11570isl
Hindi	**Days**	**Area**	**kHz**
1045-1145	daily	SAs	9340isl, 11570isl
Nepali	**Days**	**Area**	**kHz**
1000-1030	daily	SAs	9345isl, 11570isl
Pashto	**Days**	**Area**	**kHz**
1345-1445	daily	WAs	6235isl
Sheena	**Days**	**Area**	**kHz**
0530-0615	daily	SAs	7470isl
Sinhala	**Days**	**Area**	**kHz**
1230-1300	daily	SAs	11880isl, 15540isl
Tamil	**Days**	**Area**	**kHz**
1300-1330	daily	SAs	11880isl, 15540isl
Urdu	**Days**	**Area**	**kHz**
0045-0215	daily	SEA	11580isl, 15490isl
0500-0700	daily	ME	15100isl, 17830isl
0830-1100	daily	Eu	15100isl, 17700isl
1330-1530	daily	ME	7530isl, 11575isl
1700-1900	daily	Eu	7530isl, 9340isl

Ann: English: "This is Radio Pakistan"; Urdu: "Ye Radio Pakistan hai".
V: QSL-card.
Notes: Some transmissions are irregular. FPI: new transmitter site at Landhi Karachi, planned to be operational in next two years.

PALAU (PLW)

T8WH - WORLD HARVEST RADIO INTERNATIONAL (Rlg)
✉ P.O. Box 66, Koror, Republic of Palau PW96940.
SW: [HBN] Babeldaob, Medorm: 4 x 100kW.
SAT: Galaxy 16.
kHz: 9930, 9955, 9965, 13745, 15680

Winter Schedule 2010/2011			
English/Various	**Days**	**Area**	**kHz**
1400-1430	daily	EAs,SEA	9955hbn, 9965hbn
1430-1500	daily	EAs,SEA	9930hbn
1600-1300	daily	EAs,SEA	9930hbn
1630-1900	daily	EAs,SEA	9965hbn
1800-2200	daily	EAs,SEA	9955hbn
2200-1400	daily	EAs,SEA	9965hbn
English/			
Vietnamese	**Days**	**Area**	**kHz**
0030-0400	daily	EAs,SEA	15680hbn
0430-0500	daily	EAs,SEA	15680hbn
0530-1300	daily	EAs,SEA	15680hbn

Mandarin/ Vietnamese	Days	Area	kHz
1300-1400	daily	EAs,SEA	13745hbn

V: QSL-card.
Notes: Transmitting station owned by World Harvest Radio International, see USA for corporate details. Former callsigns: T8BZ, KHBN. Registered freqs shown, actual usage depends on airtime bookings.

PHILIPPINES (PHL)

RADIO PILIPINAS OVERSEAS (Gov)
✉ 4th Floor, PIA Bldg, Visayas Ave, Quezon City, Manila 1100, Philippines.
☎ +63 2 9242267. 🖷 +63 2 9242745.
E: radyo_pilipinas_overseas@yahoo.com
W: www.pbs.gov.ph
Webcast: L (www.pia.gov.ph/pbsradio.asp?fi=dzrp)
L.P: SM: Magtanggol Rodriguez.
SW: Uses facilities provided by IBB.
kHz: 11730, 11880, 11890, 15190, 15285, 17770

Winter Schedule 2010/2011

English	Days	Area	kHz
0200-0330	daily	ME	11880pht, 15285pht, 17770pht

Filipino	Days	Area	kHz
1730-1930	daily	ME	11730pht, 11890pht, 15190pht

Ann: English: "This is Radio Pilipinas, the Overseas Service of the Philippines Broadcasting Service, PBS. Radio Pilipinas is reaching you from Manila, Philippines", "Radio Pilipinas Overseas Service, the Voice of the Philippines".
V: QSL-card. Rp (2 IRCs). Rec. acc.
Notes: Radio Pilipinas Overseas is the External Sce of the state-owned Philippine Broadcasting Service - Bureau of Broadcast Services (PBS-BBS). Official callsign: DZRP.

FEBC PHILIPPINES (Rlg)
✉ 62 Karuhatan Road, 1441 Valenzuela City 0560, Philippines.
☎ +63 2 2925603; +63 2 2921152. 🖷 +63 2 2925790.
E: info@febcintl.org **W:** febc.ph
Webcast: D/L
L.P: Pres: Dan Andrew S. Cura.
SW: [BOC] Bocaue, Bulacan: 3 x 50, 1 x 100kW; [IBA] Iba, Zambales: 2 x 100kW.
kHz: 5990, 7410, 7460, 7480, 7505, 9400, 9405, 9430, 9435, 9445, 9465, 9625, 9730, 9795, 9920, 9940, 11820, 11835, 12055, 12070, 12095, 12120, 15320, 15330, 15380, 15435, 15450, 15455, 15525, 15600

Winter Schedule 2010/2011

Achang	Days	Area	kHz
1230-1245	mtw.fs.	As	12095boc

Akha	Days	Area	kHz
1215-1230	daily	As	11835boc

Bahnar	Days	Area	kHz
1145-1200	..wtfs.	As	9920boc

Banjar	Days	Area	kHz
0830-0900	m.w.f..	As	11820boc, 15320boc

Batak Toba	Days	Area	kHz
1000-1030	daily	As	15450boc

Bru	Days	Area	kHz
1345-1400	...tfs.	As	7480iba

Bugis	Days	Area	kHz
0930-1000	daily	As	15380boc

Burmese	Days	Area	kHz
1330-1430	daily	As	12120boc
2330-0100	daily	As	15600boc

Cantonese	Days	Area	kHz
1600-1630	daily	As	7505iba

Cham	Days	Area	kHz
1330-1345	mt....s	As	7480iba

Chin (Asho)	Days	Area	kHz
0100-0115	mtw...s	As	15600boc

Chin (Daai)	Days	Area	kHz
1245-1300	daily	As	11835boc

Chin (Khumi)	Days	Area	kHz
0100-0115	...tfs.	As	15600boc

Chin (Mro)	Days	Area	kHz
0130-0200	daily	As	15435boc

Chin (Thado)	Days	Area	kHz
0100-0115	daily	As	15435boc

Chinese	Days	Area	kHz
0030-0200	daily	As	15455boc
0600-0700	daily	As	15450iba
0700-0900	daily	As	15525boc
0900-1400	daily	As	9400iba
0900-1630	daily	As	9430boc
1400-1600	daily	As	7505iba
2230-0030	daily	As	9405boc
2300-0100	daily	As	12070iba

Chrau	Days	Area	kHz
1130-1145	mt....s	As	9920boc

Chru	Days	Area	kHz
1130-1145	..wtfs.	As	9920boc

Hmong (Black)	Days	Area	kHz
1100-1130	daily	As	12095boc

Hmong (Blue)	Days	Area	kHz
1300-1330	.....ss	As	9625boc
2300-2330	.....ss	As	9730boc

Hmong (White)	Days	Area	kHz
1300-1330	mtwtf..	As	9625boc
2300-2330	mtwtf..	As	9730boc

Hre	Days	Area	kHz
1200-1215	daily	As	9920boc

Hui	Days	Area	kHz
1400-1430	daily	As	7460boc

Indonesian	Days	Area	kHz
0900-0930	daily	As	15450boc
2230-2300	.tw.f..	As	9435boc
2300-2330	mtwtf.s	As	9435boc

Iu Mien	Days	Area	kHz
1200-1230	daily	As	12095boc
1230-1300	...t....	As	12095boc
2230-2300	daily	As	5990boc

Jarai	Days	Area	kHz
1100-1130	...tfs.	As	9920boc

Javanese	Days	Area	kHz
1030-1100	daily	As	15380boc

Jeh	Days	Area	kHz
1145-1200	mt....s	As	9920boc

Jingpho	Days	Area	kHz
1145-1200	daily	As	11835boc

Karen (Pa'o)	Days	Area	kHz
0115-0130	...tfs.	As	15600boc
1100-1115	daily	As	15330boc

Katu	Days	Area	kHz
1330-1345	..wtfs.	As	7480iba

Khmer	Days	Area	kHz
1200-1300	daily	As	7410iba
2300-2400	daily	As	9445iba

Khmu	Days	Area	kHz
0000-0015	daily	As	9795iba
1330-1400	daily	As	9625boc

Koho	Days	Area	kHz
1230-1300	mtw...s	As	9920boc

Lahu	Days	Area	kHz
0015-0045	daily	As	12055boc

Lao	Days	Area	kHz
1130-1200	daily	As	12095boc
2330-2400	daily	As	9795boc

Lisu	Days	Area	kHz
1300-1330	daily	As	12120boc

Makassarese	Days	Area	kHz
0900-0930	daily	As	15380boc

Malay	Days	Area	kHz
2230-2330	.....s.	As	9435boc

Mandarin	Days	Area	kHz
1430-1500	daily	As	7460boc

Maru	Days	Area	kHz
0130-0145	mtw...s	As	15600boc

Meitei	Days	Area	kHz
0115-0130	daily	As	15435boc
Minangkabau	**Days**	**Area**	**kHz**
0930-1000	daily	As	15450boc
Mon	**Days**	**Area**	**kHz**
1115-1145	daily	As	15330boc
2300-2330	daily	As	9795boc
Muong	**Days**	**Area**	**kHz**
1300-1315	mtw...s	As	7480iba
1315-1330	mt....s	As	7480iba
Naga	**Days**	**Area**	**kHz**
1230-1245	daily	As	11835boc
Nung	**Days**	**Area**	**kHz**
1345-1400	mtw...s	As	7480iba
Palaung (Pale)	**Days**	**Area**	**kHz**
2330-2345	daily	As	12055boc
Rade	**Days**	**Area**	**kHz**
1100-1130	mtw...s	As	9920boc
Rawang	**Days**	**Area**	**kHz**
1200-1215	daily	As	11835boc
Roglai	**Days**	**Area**	**kHz**
1215-1230	mtw...s	As	9920boc
Sedang	**Days**	**Area**	**kHz**
1300-1315	...tfs.	As	7480iba
Shan	**Days**	**Area**	**kHz**
0000-0045	daily	As	15435boc
Stieng Bulo	**Days**	**Area**	**kHz**
1315-1330	..wtfs.	As	7480iba
Sundanese	**Days**	**Area**	**kHz**
1000-1030	daily	As	15380boc
Tai (Dam)	**Days**	**Area**	**kHz**
1215-1230	...tfs.	As	9920boc
Tai (Lu)	**Days**	**Area**	**kHz**
1030-1100	daily	As	12095boc
2345-0015	daily	As	12055boc
Tai (Nua)	**Days**	**Area**	**kHz**
0045-0100	daily	As	15435boc
Tibetan (Khams)	**Days**	**Area**	**kHz**
1300-1330	daily	As	9465boc
Uyghur	**Days**	**Area**	**kHz**
1500-1530	daily	As	9940boc
Wa	**Days**	**Area**	**kHz**
0045-0100	daily	As	12055boc
Zaiwa	**Days**	**Area**	**kHz**
0130-0145	...tfs.	As	15600boc
Zhuang	**Days**	**Area**	**kHz**
1330-1400	daily	As	9465boc

Ann: English: "This is FEBC Radio, broadcasting from Manila, Philippines".
V: QSL-card. Rp. preferred (3 IRCs)
Notes: Owned by Far East Broadcasting Company (FEBC), see USA for corporate details. Callsign: DZAS.

RADIO VERITAS ASIA (Rlg)
P.O. Box 2642, Quezon City, Manila 1166, Philippines.
☎ +63 2 9390011. 📠 +63 2 9390011.
E: rvaprogram@rveritas-asia.org **W:** www.rveritas-asia.org
Webcast: D/L
L.P: GM: Fr. Roberto (Bobby) M. Ebisa; PD: Rev. Msgr. Pietro Nguyen Van Tai; Dir, Technical: Engr. Alex Movilla.
SW: [PUG] Palauig: 1 x 50, 2 x 250kW.
kHz: 6115, 9515, 9520, 9570, 9615, 9645, 9670, 9720, 9865, 11710, 11730, 11850, 11870, 11935, 11945, 15225, 15280, 15350, 15435, 15450, 15520, 15530, 17830, 17860

Winter Schedule 2010/2011

Bengali	Days	Area	kHz
0030-0100	daily	As	11945pug
1400-1430	daily	As	11870pug
Burmese	**Days**	**Area**	**kHz**
1130-1200	daily	SEA	15450pug
2330-2400	daily	SEA	9720pug
Chin	**Days**	**Area**	**kHz**
1430-1500	daily	SEA	9520pug
Chin (Zomi)	**Days**	**Area**	**kHz**
0130-0200	daily	SEA	15520pug

Hindi	Days	Area	kHz
0030-0100	daily	As	11710pug
1330-1400	daily	As	11870pug
Hmong	**Days**	**Area**	**kHz**
1200-1230	daily	SEA	11935pug
Kachin	**Days**	**Area**	**kHz**
1230-1200	daily	SEA	15225pug
2330-2400	daily	SEA	9645pug
Karen	**Days**	**Area**	**kHz**
0000-0030	daily	SEA	11935pug
1200-1230	daily	SEA	15225pug
Khmer	**Days**	**Area**	**kHz**
0130-0200	daily	SEA	15280pug
1000-1030	daily	SEA	11850pug
Mandarin	**Days**	**Area**	**kHz**
1000-1200	daily	CHN	9615pug
2100-2300	daily	CHN	6115pug
Russian	**Days**	**Area**	**kHz**
0200-0300	daily	RUS	17830pug
1500-1600	daily	RUS	9570pug
Sinhala	**Days**	**Area**	**kHz**
0000-0030	daily	SEA	9865pug, 11730pug
1330-1400	daily	SEA	9520pug
Tagalog	**Days**	**Area**	**kHz**
1500-1600	daily	SEA	15350pug
2300-2330	daily	SEA	9720pug
Tamil	**Days**	**Area**	**kHz**
0030-0100	daily	As	11935pug
1400-1430	daily	As	9520pug
Telugu	**Days**	**Area**	**kHz**
0100-0130	daily	As	15530pug
1430-1500	daily	As	9515pug
Urdu	**Days**	**Area**	**kHz**
0100-0130	daily	As	15280pug, 17860pug
1430-1500	daily	As	15435pug
Vietnamese	**Days**	**Area**	**kHz**
0130-0130	daily	SEA	15530pug
1030-1130	daily	SEA	11850pug
1300-1330	daily	SEA	11850pug
2330-2400	daily	SEA	9670pug

Ann: English: "This is Radio Veritas Asia, broadcasting from Quezon City, Philippines".
V: QSL-card.
Notes: Catholic station, on air since 11 April 1969. Owned by the "Philippine Radio Educational and Information Center" (PREIC), composed of Filipino bishops and professionals.

IBB RELAY STATIONS PHILIPPINES
Station Manager, IBB Philippines Transmitting Station, PSC 500 Box 28, DPO AP 96515-1000, USA.
☎ +63 45 9820254. 📠 +63 45 9821402.
IBB Transmitting Station (Poro Point), San Fernando, La Unión, Philippines.
☎ +63 72 8882747. 📠 +63 72 8885133.
L.P: SM: Dennis G. Brewer; Poro Pt. Transmitter Plant Supervisor: Sheldon Daitch.
MW: [PHP] Poro Point (DWVA): 1170kHz 1000kW.
SW: [PHT] Tinang: 3 x 50, 12 x 250kW.

POLAND (POL)

POLISH RADIO (EXTERNAL SERVICE) (Pub)
P.O. Box 46, 00-977 Warszawa, Poland.
☎ +48 22 6453302. 📠 +48 22 6453952.
E: zagranica@polskieradio.pl
W: www.polskieradio.pl/zagranica; www.thenews.pl/radio (English)
Webcast: L/P
L.P: Dir: Marek Cajzner.
SAT: AsiaSat 3S, Galaxy 25, Hot Bird 9, PAS 4.
kHz: 3975, 5895, 5920, 5980, 5990, 6000, 6040, 6050, 6100, 6135, 7265, 9460, 9470, 9490, 9580, 9650, 9850, 11785, 11860, 11905, 13835, 13850, 15175, 15245, 17670, 17715

Winter Schedule 2010/2011

Belarusian	Days	Area	kHz
1430-1530	daily	Eu	15245wof
1630-1800	daily	Eu	6050skn

English	Days	Area	kHz
1300-1400	daily	Eu	9460mos, 11860wof
1800-1900	daily	Eu	5895kvi+, 9650dha

German	Days	Area	kHz
1230-1300	daily	Eu	9470wof, 9850wof
1630-1700	daily	Eu	6100wof
2030-2100	daily	Eu	3975skn+, 6000dha

Hebrew	Days	Area	kHz
1900-1930	daily	ME	7265rmp

Polish	Days	Area	kHz
1130-1200	daily	Eu	11785mos, 15175wof
2200-2300	daily	Eu	5980dha, 5990skn

Russian	Days	Area	kHz
1200-1230	daily	RUS	17670wof, 17715wof
1400-1430	daily	RUS	13850rmp, 15245wof
1530-1600	daily	RUS	9580wof
1900-1930	daily	RUS	5920wof
2000-2030	daily	RUS	6135wof

Ukrainian	Days	Area	kHz
1430-1630	daily	UKR	11905wof
1600-1630	daily	UKR	13835wof
1930-2000	daily	UKR	9490rmp
1930-2030	daily	UKR	6040wof

Key: + DRM.
Ann: English: "This is the External Service of Polish Radio"; German: "Hier ist der Auslandsdienst des Polnischen Rundfunks"; Polish: "Polskie Radio dla zagranicy".
V: QSL-card.
Notes: External Sce of the public broadcaster Polskie Radio.

PORTUGAL (POR)

RDP INTERNACIONAL
✎ Av. Marechal Gomes da Costa, 37 - bloco B - 2°, 1849-030 Lisboa, Portugal.
☎ +351 21 7947000. 🖷 +351 21 7947570.
E: rdp.internacional@rtp.pt **W:** www.rtp.pt
Webcast: D/L/P
L.P: Dir: Jorge Oliveira Gonçalves; Dir, Engineering Technology: Teresa Beatriz Abreu.
SW: [LIS] São Gabriel: 4 x 300kW. Also uses facilities of Pro-Funk GmbH in Sines, see DW relay station.
SAT: AsiaSat 5, AMC4, Galaxy 28, Hot Bird 8, Intelsat 805/907.
kHz: 7285, 7345, 7360, 9455, 9795, 9815, 9855, 9860, 11635, 11655, 11665, 11885, 11960, 12020, 12040, 13720, 15465, 15520, 15555, 15560, 15690, 17745, 17820, 17840, 21655

Winter Schedule 2010/2011

Portuguese	Days	Area	kHz
0000-0300	.twtfs.	SAm	9855lis, 11655lis
0000-0300	.twtfs.	NAm	9455lis
0600-0700	mtwtf..	Eu	7345lis
0700-1300	mtwtf..	Eu	9815lis
0745-0900	mtwtf..	Eu	7360sin
0800-1055	.....ss	WAf,SAm	15555lis
0800-1100	.....ss	Af	15520lis, 17745lis, 17840lis
0800-1200	.....ss	Eu	12020lis
0930-1100	.....ss	Eu	9815sin+
1100-1300	mtwtf..	WAf,SAm	21655lis
1100-1300	mtwtf..	Af	17745lis
1100-1300	.....ss	WAf,SAm	21655lis
1200-1500	.....ss	Eu	11885lis
1300-1700	.....ss	NAm	15560lis
1300-1700	mtwtf..	NAm	15560lis*
1400-1600	mtwtf..	ME,SAs	15690lis
1500-1700	.....ss	Af	15520lis
1500-1700	.....ss	Eu	11635lis
1700-1900	daily	Eu	9860lis
1700-1900	.....ss	NAm	17820lis
1700-1900	mtwtf..	NAm	17820lis*
1700-2000	mtwtf..	Af	13720lis
1700-2000	mtwtf..	WAf,SAm	15465lis
1700-2100	.....ss	Af	13720lis
1700-2100	.....ss	WAf,SAm	15465lis
1800-2000	daily	Eu	9795lis
1900-2100	.....ss	NAm	12040lis

Portuguese	Days	Area	kHz
1900-2400	mtwtf..	NAm	12040lis*
2000-2300	daily	Eu	9795lis*
2000-2400	daily	Af	11665lis*
2000-2400	daily	WAf,SAm	11960lis*
2100-2400	.....ss	NAm	12040lis*
2300-2400	daily	Eu	7285lis*

Key: + DRM; * During special, usually sports, events only.
Ann: Portuguese: "RDP Internacional - Rádio Portugal".
IS: Opens with tune on Portuguese guitar, followed by station ID, National Anthem and then frequency announcement. Closes with frequency announcement.
V: QSL-card. Rec. acc. (Email reports to: isabel.venes@rtp.pt)

DEUTSCHE WELLE RELAY STATION
✎ Pro-Funk GmbH, Monte Mudo, 7520-065 Sines, Portugal.
☎ +351 269870280. 🖷 +351 269870290.
E: profunk@mail.telepac.pt
SW: [SIN] Sines: 3 x 250kW (DRM capable).

ROMANIA (ROU)

RADIO ROMANIA INTERNATIONAL (RRI) (Pub)
✎ P.O. Box 1-111, 014700 Bucuresti, Romania.
☎ +40 21 3031357; +40 21 3031465. 🖷 +40 21 2232613.
E: rri@rri.ro **W:** www.rri.ro
Webcast: L/P
L.P: Deputy DG,SRR & Head of RRI: Doru Vasile Ionescu.
MW/SW: Leased from Radiocom.
SAT: Eutelsat W2, Galaxy 25, Hot Bird 6, Optus D2.
kHz: 5875, 5900, 5910, 5915, 5950, 5955, 5975, 5990, 5995, 6010, 6015, 6020, 6025, 6030, 6065, 6100, 6110, 6115, 6125, 6130, 6145, 6175, 6180, 6200, 7210, 7220, 7300, 7305, 7310, 7315, 7325, 7345, 7350, 7370, 7380, 7415, 7430, 7435, 7445, 7450, 9525, 9535, 9610, 9620, 9635, 9655, 9660, 9665, 9690, 9765, 9790, 9805, 9875, 11710, 11730, 11790, 11825, 11870, 11895, 11905, 11940, 11960, 11970, 15150, 15155, 15160, 15170, 15255, 15260, 15290, 15330, 15370, 15380, 15430, 15460, 15540, 17745, 17765, 17775, 17780, 17800, 17870, 21600

Winter Schedule 2010/2011

Arabic	Days	Area	kHz
0730-0800	daily	NAf,ME	11710tig, 11905gal, 15155gal, 15330tig
1500-1600	daily	NAf,ME	9655gal, 11730gal, 15290tig, 17540tig

English	Days	Area	kHz
0100-0200	daily	NAm	6145gal, 7325gal
0400-0500	daily	ME	9690gal, 11895gal
0400-0500	daily	NAm	6130tig, 7305tig
0630-0700	daily	AUS,NZL	17780gal, 21600tig
0630-0700	daily	Eu	6020gal+, 7370tig
1200-1300	daily	Af	15430gal, 17765gal
1200-1300	daily	Eu	11970tig, 15460tig
1800-1830	daily	Eu	6020kvi+
1800-1900	daily	Eu	6065tig+, 7415tig
2130-2200	daily	Eu	6030gal, 7380gal
2130-2200	daily	NAm	6115tig, 7310tig
2300-2400	daily	Af	5915tig, 7300tig
2300-2400	daily	Eu	6015gal, 7220gal

French	Days	Area	kHz
0200-0300	daily	NAm	5975gal, 7325gal
0600-0630	daily	Eu	6100gal+, 7370gal
0600-0630	daily	Af	9690tig, 11790tig
1000-1100	......s	Eu	15260gal, 17870gal
1100-1200	daily	Af	17870tig
1100-1200	daily	Eu	15150tig, 15255gal, 17800gal
1700-1800	daily	Eu	7370tig, 9690tig
1800-1900	daily	Eu	7350gal
2000-2100	daily	Eu	7380gal
2100-2130	daily	Eu	6030gal+, 7370gal+

German	Days	Area	kHz
0700-0730	daily	Eu	7210gal, 9450tig+
1300-1400	daily	Eu	11970tig, 15460tig
1700-1730	daily	Eu	5875kvi+
1900-2000	daily	Eu	7370tig, 9805tig+

Italian	Days	Area	kHz
1500-1530	daily	Eu	9875tig*
1700-1730	daily	Eu	7415tig*
1900-1930	daily	Eu	6180tig+*

Macedonian	Days	Area	kHz
1530-1600	daily	Eu	6125tig*
1730-1800	daily	Eu	6015tig*
1930-2000	daily	Eu	6110tig*

Mandarin	Days	Area	kHz
0500-0530	daily	EAs	15160tig, 17870tig+
1400-1430	daily	EAs	5900tig, 9660tig

Romanian	Days	Area	kHz
0100-0300	daily	NAm	5910tig, 7345tig
0500-0600	daily	Eu	6145gal, 7220gal
0800-0900	......s	Af	11730tig, 15370tig
0800-0900	......s	ME	15430gal, 17775gal
0900-1000	......s	Af	15380tig, 15430tig, 17745gal, 17775tig
1000-1100	......s	Af	15380tig, 17780tig
1300-1400	daily	Eu	9610tig
1300-1500	daily	Eu	11940gal, 15170gal
1600-1700	daily	Eu	9655gal, 11870gal
1700-1800	daily	Af	5995gal, 7435gal
1800-2100	daily	Eu	5990gal
1900-2000	daily	Eu	7430gal

Russian	Days	Area	kHz
0530-0600	daily	RUS	6175tig+, 7210tig
1430-1500	daily	RUS	9535tig, 11870tig
1600-1700	daily	RUS	6030tig+, 7445tig

Serbian	Days	Area	kHz
1630-1700	daily	Eu	6025tig*
1830-1900	daily	Eu	5955tig*
2030-2100	daily	Eu	6010tig*

Spanish	Days	Area	kHz
0000-0100	daily	SAm	9665tig, 11960tig
0000-0100	daily	SAm,CAm	7315gal, 9525gal
0300-0400	daily	SAm	7325gal, 9635tig, 9765gal, 11825tig
2000-2100	daily	Af	7430tig
2000-2100	daily	Eu	9620tig
2200-2300	daily	SAm	7380tig, 9790tig

Ukrainian	Days	Area	kHz
1600-1630	daily	Eu	6130tig*
1800-1830	daily	Eu	6200tig*
2000-2030	daily	Eu	5950tig*

Key: + DRM; * via Saftica site.
Ann: English: "You are tuned to Radio Romania International, broadcasting from Bucharest".
V: QSL-card. (Online reception report form available)
Notes: Radio Romania International is the External Sce of the public broadcaster "Radio Romania". Romanian language prgrs includes relays of Home Sce networks. Transmissions may be shortened, or cancelled, at times.

RADIOCOM (Tx Operator)
🖃 Bd. Libertatii 14, sector 5, 050706 Bucuresti, Romania.
☎ +40 21 5003131. 🖷 +40 21 3073131.
E: relatiipublice@radiocom.ro **W:** www.radiocom.ro
L.P: DG: Gabriel Grecu.
SW: [GAL] Bacau, Galbeni: 2 x 300kW; [TIG] Bucuresti, two sites: Tiganesti (G.C. 44N45 026E06): 3 x 300kW; Saftica (G.C. 44N38 026E05): 1 x 100kW.
Notes: Radiocom is the national transmitter network owner.

RUSSIA (RUS)

GTRK "ADYGEYA" (Gov)
🖃 ul. Zhukovskogo 24, 385000 Maykop, Russia.
☎ +7 87722 23542. 🖷 +7 87722 203039.
E: adigradio@mail.ru
L.P: Dir: Azamat Borus.
SW: Leased from RTRN.
kHz: 6005

Winter Schedule 2010/2011

Adyghian	Days	Area	kHz
1800-1900	m..f..	ME	6005arm*

Adyghian	Days	Area	kHz
1900-2000	......s	ME	6005arm

Key: * Mondays also in Arabic and Turkish.
Notes: Special prgr for Adygey expatriates in the Middle East, produced by the regional state broadcasting company GTRK "Adygeya". Prgrs for domestic audience: see National Radio section.

GTRK "KABARDINO-BALKARIYA" (Gov)
🖃 ul. Nogmova 38, 360000 Nalchik, Russia.
☎ +7 8662 775861. 🖷 +7 8662 774024.
E: vestikbr@mail.ru **W:** www.gtrk-kbr.ru
L.P: Dir: Lyudmila B. Kazancheva.
SW: Leased from RTRN.
kHz: 6005

Winter Schedule 2010/2011

Kabardino-Circassian	Days	Area	kHz
1830-1900	..w...s	ME	6005arm

Karachay-Balkar	Days	Area	kHz
1830-1900	...t...	ME	6005arm

Notes: Special prgr for Balkar and Kabardin expatriates in the Middle East, produced by the regional state broadcasting company GTRK "Kabardino-Balkariya". Prgrs for domestic audience: See National Radio section.

NA VOLNE TATARSTANA (TATARSTAN DULKYNYNDA) (Gov)
🖃 GTRK "Tatarstan", ul. Sh. Usmanova 9, 420095 Kazan, Russia.
☎ +7 843 5547170. 🖷 +7 843 5543201.
E: secret@trttv.ru **W:** trt-tv.ru
L.P: Dir, GTRK "Tatarstan": Ayrat M. Sibagatullin.
SW: Leased from RTRN.
kHz: 9860, 11610, 15105

Winter Schedule 2010/2011

Tatar/Russian	Days	Area	kHz
0510-0600	daily	RUS	15105sam
0710-0800	daily	RUS,CAs	9860sam
0910-1000	daily	RUS,Eu	11610sam

Ann: Russian: "V efire programma na volne Tatarstana"; Tatar: "Efirda Tatarstan dulkininda programmasi".
V: QSL-card. (2 IRCs, 1 USD or 1 Euro from QSL-manager Ildus Ibatullin, P.O. Box 134, 420136 Kazan, Russia).
Notes: "Na volne Tatarstana" is produced by GTRK "Tatarstan" for ethnic Tatars living outside of Tatarstan. Also relayed locally in Tatarstan on LW and FM.

RADIO ROSSII (Gov)
🖃 See National Radio section.
Webcast: L
SW: Leased from RTRN.
kHz: 1296, 5905, 7310, 9840, 12075

Winter Schedule 2010/2011

Russian	Days	Area	kHz
0500-0800	daily	Eu,Atl	9840msk
0825-1300	daily	Eu,Atl	12075msk
1325-1600	daily	Eu,Atl	7310msk
1625-2200	daily	Eu,Atl	5905msk
2200-1800	daily	CAs	1296dsb

VOICE OF RUSSIA - MEZHDUNARODNOYE RUSSKOYE RADIO (MRR) (Gov)
Webcast: D/L
kHz: 630, 693, 801, 1026, 1170, 1314, 1431, 7225, 7250

Winter Schedule 2010/2011

Russian	Days	Area	kHz
0000-0300	daily	CAs	1026nvs
0000-0400	daily	ME,CAs,As	7225sam
0300-0900	daily	CAs,ME	801dsb
0400-0500	daily	CAm	7250arm
0400-1300	daily	ME,CAs	1314erv
0500-0600	daily	ME	1170arm
0500-1000	daily	Eu	630klu, 693bln, 1431dsd

Ann: Russian: "Radiokompaniya Golos Rossii predstavlyayet programmu Mezhdunarodnogo Russkogo Radio".

VOICE OF RUSSIA (VOR) (Gov)

✉ Pyatnitskaya 25, 115326 Moscow, Russia.
☎ +7 495 9506331. 🖷 +7 495 9512017.
E: letters@ruvr.ru **W:** www.ruvr.ru
Webcast: D/L/P
L.P: Chmn: Andrey Bystritskiy.
MW/SW: Leased from RTRN.
SAT: Eutelsat W4, Express AM1/11, Yamal 201.
kHz: 585, 603, 612, 630, 648, 693, 720, 801, 972, 999, 1026, 1080, 1089, 1143, 1170, 1215, 1251, 1269, 1314, 1323, 1377, 1413, 1431, 1494, 1503, 1548, 4975, 5900, 5905, 5920, 5935, 5940, 5945, 5965, 5975, 5985, 6000, 6005, 6020, 6030, 6040, 6060, 6065, 6085, 6090, 6105, 6120, 6130, 6135, 6140, 6145, 6155, 6170, 6180, 6240, 7205, 7210, 7215, 7220, 7230, 7240, 7250, 7260, 7270, 7280, 7290, 7295, 7300, 7305, 7310, 7320, 7325, 7330, 7345, 7350, 7400, 7430, 7440, 9470, 9475, 9480, 9660, 9675, 9695, 9720, 9800, 9820, 9840, 9855, 9865, 9875, 9880, 9885, 9900, 9965, 11600, 11605, 11630, 11635, 11655, 11660, 11985, 12025, 12030, 12035, 12055, 12060, 12070, 13735, 15240, 15450, 15735, 17650, 17665, 17805

Winter Schedule 2010/2011

Arabic	Days	Area	kHz
1600-1700	daily	ME	1314erv, 5945nvs, 7215spb
1600-1900	daily	Af	9480msk
1600-2100	daily	Af	5920spb
1700-1800	daily	ME	7305msk
1700-2000	daily	Af,ME	7400dsb
1700-2000	daily	ME	9820dsb
1800-1900	daily	ME	5935spb, 6060spb
1800-2000	daily	ME	6020arm
1800-2100	daily	ME	1314erv, 7345dsb
1900-2100	daily	ME	5965irk
1900-2100	daily	Af,ME	5975nvs
2300-0000	daily	ME	648dsb
2300-2400	daily	ME	1314erv, 1377erv

Dari/Pashto	Days	Area	kHz
1300-1500	daily	ME	9900sam
1300-1500	daily	WAs	648dsb, 801dsb, 972dsb, 4975dsb, 15450arm

English	Days	Area	kHz
0000-0400	daily	NAm,CAm	7250arm
0000-0500	daily	NAm,CAm	6240kch
0300-0400	daily	NAm	13735vld
0300-0400	daily	NAm,SAm	7440lvi
0300-0500	daily	NAm	12030ppk
0400-0500	daily	NAm	13735vld
0400-0600	daily	As	15735kna+
0500-0700	daily	NAm	9840ppk, 9855vld
0600-0900	daily	Eu	1323wbr
0700-0900	daily	Eu	11635msk+
0700-1100	daily	AUS,NZL	17805irk
0800-1000	daily	As	1251dsb
0800-1100	daily	AUS,NZL	17665kna
0800-1100	daily	SEA,AUS,NZL	17650dsb
1000-1200	daily	SEA	7205tch
1200-1300	daily	As	7340irk+, 7350tch, 9695sam, 11660dsb
1300-1500	daily	SEA	7205tch
1400-1500	daily	As	11660dsb, 12055msk
1500-1600	daily	SEA	7260vld, 9660xia
1500-1600	daily	As	1251dsb
1500-1600	daily	Eu	5905msk+, 9675msk+
1500-2000	daily	As,ME	4975dsb
1600-1700	daily	As	972dsb, 7305nvs, 11630dsb
1600-1700	daily	Eu	6130msk
1600-1700	daily	SEA	7330ppk
1600-1800	daily	ME,Af	9470msk
1600-1900	daily	SEA	9880ppk
1700-1800	daily	As	1269xuw
1700-1900	daily	As	1251dsb, 7240ppk
1700-1900	daily	SEA	7330ppk
1800-1900	daily	Af	7270erv
1800-1900	daily	ME	7305msk
1800-2200	daily	Eu	7330msk

English	Days	Area	kHz
1900-2000	daily	Af,Eu	12060msk
1900-2000	daily	ME	5985msk
2000-2400	daily	Eu	1215klg
2100-2200	daily	Eu	7290msk
2200-2300	daily	Eu	7300msk
2300-2400	daily	NAm,CAm	7250arm

Farsi	Days	Area	kHz
1600-1800	daily	ME	648dsb, 1377erv, 5935spb, 6020arm, 7345dsb
1700-1800	daily	ME	7205spb

French	Days	Area	kHz
1700-1800	daily	Eu,Af	7330msk
1700-1800	daily	Eu	6145klg+, 9675msk+
1700-1900	daily	Af	7295tch
1700-2100	daily	Af	11985erv
1700-2200	daily	Af,Eu	11600msk
1700-2200	daily	Eu,Af	6130msk
1800-1900	daily	Eu	12060msk
1900-2200	daily	Eu	6120klg
2000-2200	daily	Eu	1323wbr, 6105klg+
2000-2200	daily	Af,Eu	12060msk

German	Days	Area	kHz
1000-1100	daily	Eu	9720klg, 11655lvi
1000-1300	daily	Eu	630klu, 693bln, 1323wbr, 1431dsd
1100-1300	daily	Eu	7325klg+
1600-1700	daily	Eu	9675msk+
1600-1800	daily	Eu	630klu, 1431dsd, 7220sam
1600-2000	daily	Eu	693bln, 1215klg, 1323wbr
1800-1900	daily	Eu	1143klg, 5940sam
1800-2000	daily	Eu	7310sam
1900-2000	daily	Eu	630klu, 1431dsd

Hindi	Days	Area	kHz
1300-1400	daily	As	7340irk+, 7350tch, 9885dsb, 11630dsb, 11660dsb, 12055msk
1500-1600	daily	As	972dsb, 5900sam, 7305nvs, 7340irk+, 7350tch, 9885dsb, 11630dsb

Italian	Days	Area	kHz
1800-1900	daily	Eu	5975klg, 6040klg+, 6145klg+, 7230msk, 7320spb
2230-2330	daily	Eu	1548kch

Japanese	Days	Area	kHz
1200-1300	daily	As	6180ppk
1200-1400	daily	As	630kna, 720iuj, 6085irk

Kurdish	Days	Area	kHz
1700-1800	daily	ME	1314erv, 5945nvs, 6005arm, 7270msk

Mandarin	Days	Area	kHz
1100-1300	daily	As	801tch, 1080irk, 5965vld
1100-1500	daily	As	585blg, 648vld, 1251vld, 6170khb
1200-1300	daily	As	7330khb
1400-1500	daily	As	801tch, 1080irk, 5940nvs, 5965vld

Mongolian	Days	Area	kHz
1300-1400	mtwtfs.	As	801tch, 1080irk, 5940nvs, 5965vld

Portuguese	Days	Area	kHz
0000-0100	daily	SAm	6135dsb, 9865sam, 9965erv
2200-2300	daily	Eu	5920dsb, 5940sam, 6090arm, 6120klg, 7340nvs
2300-2400	daily	SAm	9965erv, 11605guf

Russian	Days	Area	kHz
0000-0300	daily	ME	1314erv

Russian	Days	Area	kHz
0000-0400	daily	CAm	7220kch
0000-0400	daily	CAm,SAm	7430erv
0200-0300	daily	CAs,ME	801dsb
0200-0300	daily	SAm	6065spb
0200-0400	daily	As	15240ppk, 15735kna+
0200-0400	daily	CAs	972dsb, 1503dsb
0200-0500	daily	CAs,ME	648dsb
0300-0400	daily	CAs,ME	7305arm
0300-0700	daily	CAs	12070msk
0400-0800	daily	Eu	1548kch
0400-1600	daily	ME	1377erv
0500-0700	daily	ME	1089arm
0600-0800	daily	AUS,NZL	17650dsb
0600-1300	daily	CAs	972dsb
0700-0900	daily	CAs,ME	648dsb
0700-1600	daily	Eu	999kch
0900-1000	daily	CAs,ME	801dsb
0900-1100	daily	Eu	7325klg+
0900-1400	daily	Eu	1215klg
1100-1200	daily	AUS,NZL	17650dsb
1100-1300	daily	CAs,ME	1323dsb
1200-1300	daily	CAs,ME	648dsb, 801dsb
1200-1600	daily	CAs	12025sam
1200-1700	daily	Eu	5940sam
1300-1500	daily	As	7330khb
1300-1500	daily	Eu	9675msk+
1300-1500	daily	SEA	7260vld
1300-1600	daily	Eu	612vln, 630klu, 693bln, 1323wbr, 1431dsd, 7325msk+
1300-1600	daily	ME	1314erv, 7205msk
1300-1700	daily	ME	1503dsb
1300-1700	daily	SEA,AUS,NZL	9800irk
1300-1700	daily	CAs,As,SEA	9840msk
1300-1800	daily	Eu	1143klg
1400-1500	daily	As	1251dsb
1400-1500	daily	Eu	5905msk+
1400-1600	daily	CAs	5945nvs
1400-1600	daily	Eu	1170sas
1400-1600	daily	ME	1089arm, 9470msk
1400-2400	daily	CAs	1026nvs
1500-1600	daily	CAs,ME	648dsb
1500-1600	daily	ME	7215spb
1500-1700	daily	ME	6140msk
1500-1800	daily	CAs,ME	801dsb
1600-1700	daily	SEA	7260vld
1600-1700	daily	As	1251dsb, 5900sam, 7240ppk
1600-2100	daily	ME	1089arm
1600-2200	daily	Eu	1494spb
1700-1800	daily	Eu	7230msk
1700-1900	daily	Eu	999kch
1700-1900	daily	Eu,ME	5985msk
1700-1900	daily	ME	1170arm, 7215spb
1700-2000	daily	CAs	12035sam
1800-1900	daily	Eu	630klu, 1431dsd, 7290msk
1800-2000	daily	Eu	1170sas
1800-2300	daily	CAs,ME	648dsb
1900-2000	daily	Eu	6155sam
1900-2000	daily	ME	7325sam
1900-2000	daily	CAs,ME	1323dsb, 7305msk
1900-2100	daily	Eu	1413kch, 5940sam, 7230msk
1900-2200	daily	CAs,ME	801dsb
1900-2200	daily	Eu	1143klg
1900-2200	daily	ME	1503dsb
1900-2300	daily	CAs	1143dsb
2000-2100	daily	RUS	612msk
2000-2100	daily	Eu	7290msk
2000-2200	daily	ME,CAs	7270nvs
2000-2300	daily	Eu	630klu, 693bln, 1431dsd
2100-2200	daily	Eu	7300msk

Russian	Days	Area	kHz
2100-2300	daily	ME	1314erv
2100-2400	daily	Eu	999kch
2100-2400	daily	ME	1170arm
2200-2300	daily	Eu	1323wbr
2200-2300	daily	RUS	612msk
2300-2400	daily	ME	5935arm

Serbian	Days	Area	kHz
1600-1800	daily	Eu	1548kch, 5975klg, 6000msk, 6040klg+, 7320spb
2100-2230	daily	Eu	1548kch, 6030sam

Spanish	Days	Area	kHz
0100-0200	daily	SAm	6065spb
0100-0300	daily	SAm	9865sam, 9875guf
0100-0500	daily	CAm	9965erv
0100-0600	daily	CAm	7280arm
0100-0600	daily	SAm	6135dsb, 7210msk
0200-0600	daily	SAm	9475dsb
0300-0400	daily	SAm	6065spb
0300-0600	daily	CAm	7335guf
2100-2200	daily	Eu	5920dsb, 5940sam, 6090arm, 7340nvs

Turkish	Days	Area	kHz
1500-1700	daily	ME	1170arm, 5985spb, 6005arm, 7270msk

Urdu	Days	Area	kHz
1400-1500	daily	As	5900sam, 7340irk+, 7350tch, 9885dsb, 11630dsb

Vietnamese	Days	Area	kHz
1200-1300	daily	As	603dof, 7205tch, 7260vld

Key: + DRM.

Ann: English: "This is Moscow, you are tuned to the World Service of the Voice of Russia"; French: "Vous ecoutez la Voix de la Russie".

V: QSL-card.

Notes: The Voice of Russia (VOR) is the External Sce produced as a state owned, but independent, venture and is not associated with the All-Russian State Radio and Television Broadcasting Company (VGTRK).

YEVANGELSKIYE CHTENIYA (Rlg)

2-y Raushskiy per. 4, 115035 Moskva, Russia.

☎ +7 485 9515793.

E: e-c-r@pravoverie.ru **W:** e-c-r.pravoverie.ru

Webcast: D

LP: Chief Editor: Theodore V. Kalinin.

kHz: *612, 1089*

Winter Schedule 2010/2011			
Russian	Days	Area	kHz
1600-1700	mt.t..s	RUS	612msk
2100-2200	daily	RUS	612msk
2100-2200	daily	ME	1089arm

RADIOAGENCY-M (Broker)

Novokhoroshevskiy proyezd 18, 123308 Moskva, Russia.

☎ +7 499 1919161. 📠 +7 499 1918591.

E: abat@radioagency.ru

LP: Dir: Anatoliy Batyushkin.

V: QSL-card. (For brokered stns)

Notes: Radioagency-M brokers air time for high power medium and shortwave txs owned by RTRN in Russia.

RUSSIAN TELEVISION AND RADIO BROADCASTING NETWORK (RTRN) (Tx Operator)

ul. Ak.Korolyova 13, 129515 Moskva, Russia.

☎ +7 495 2175141. 📠 +7 495 2175228.

E: glavred@rtrn.ru **W:** www.rtrs.ru; www.rtrn.ru

LP: GD: Gennadiy Sklyar.

MW: [ARM] Krasnodar, Tbilisskaya: 1089/1170kHz 1200kW; [BLG] Belogorsk, Konstantinogradovka: 585kHz 1200kW; [EKB] Yekaterinburg: 666kHz 10kW; [IRK] Irkutsk, Angarsk: 1080kHz 1000kW; [IUJ] Yuzhno-Sakhalinsk, Vestochka: 720kHz 1000kW; [KLG] Kaliningrad, Bolshakovo: 1143kHz 150kW, 1215kHz 1200kW; [KNA] Komsomolsk-na-Amure: 630kHz 500kW; [MSK] Moskva, Kurkino: 612/1503kHz 20kW, 693/1260kHz 10kW; [NVS] Novosibirsk, Oyash:

1026kHz 250kW; [SPB] St. Peterburg, Krasnyy Bor: 1188/1260kHz 10kW, 1494kHz 600kW; [TCH] Chita, Kruchina: 801kHz 1200kW; [VLD] Vladivostok, Razdolnoye: 648/1251kHz 1000kW.

SW: [ARM] Krasnodar, Tbilisskaya: 8 x 100, 1 x 250, 4 x 1000kW; [IRK] Irkutsk, Angarsk: 2 x 100, 4 x 250, 2 x 2000kW; [KHB] Khabarovsk: 7 x 100, 4 x 120kW; [KLG] Kaliningrad, Bolshakovo: 9 x 80kW; [KNA] Komsomolsk-na-Amure: 4 x 100; 1 x 200, 2 x 250kW; [MSK] Moskva, three sites: Kurovskaya (G.C. 55N35 039E08): 2 x 80, 8 x 100, 1 x 150, 6 x 250kW; Taldom (G.C. 56N45 037E37): 12 x 100, 4 x 250, 3 x 1000kW; Lesnoy (G.C. 56N04 037E57): 1 x 150, 15 x 250kW; [NVS] Novosibirsk, two sites: Novosibirsk (G.C. 54N55 082E51): 23 x 100kW; Oyash (G.C. 55N30 083E41): 3 x 1000kW; [PPK] Petropavlovsk, Yelizovo: 4 x 100, 2 x 250kW; [SAM] Samara: 8 x 250kW; [SPB] St. Peterburg, Krasnyy Bor: 18 x 200kW; [TCH] Chita, Kruchina: 1 x 100, 2 x 250, 2 x 1000kW; [VLD] Vladivostok, Razdolnoye: 2 x 100, 2 x 120, 2 x 200, 3 x 250, 2 x 1000kW.

Notes: RTRN is the national transmitter network operator in Russia.

RWANDA (RRW)

DEUTSCHE WELLE RELAY STATION
⌕ Kigali, Rwanda.
SW: [KIG] Kigali, Kinyinya: 4 x 250kW.

SÃO TOMÉ E PRÍNCIPE (STP)

IBB RELAY STATION SÃO TOMÉ
⌕ IBB Transmitting Station, CP 522, São Tomé, São Tomé e Príncipe.
☎ +239 223406. 🖷 +239 223406.
L.P: Acting SM: Shannon White.
MW: [SAO] Pinheira: 1530kHz 600kW.
SW: [SAO] Pinheira: 5 x 100kW.
V: QSL-email. Does not have own QSL-cards.

SAUDI ARABIA (ARS)

BROADCASTING SERVICE OF THE KINGDOM OF SAUDI ARABIA (BSKSA) (Gov)
⌕ P.O. Box 61718, Riyadh-11575, Saudi Arabia.
☎ +966 1 4425170. 🖷 +966 1 4041692.
E: eng@saudiradio.net.sa (English Service)
W: www.saudiradio.net
Webcast: L
L.P: GM: Suleiman Al-Kalifa; Dir, Engineering & Frequency Management: Suleiman Al-Samnan.
SW: [JED] Jeddah: 6 x 50, 2 x 100kW; [RIY] Riyadh: 4 x 350, 8 x 500kW.
SAT: AsiaSat 5, Badr 4, Galaxy 19, Hispasat 1C, Hot Bird 8/9, Nilesat 102, NSS 7.
kHz: *7240, 9555, 9580, 9675, 9715, 9870, 9885, 11785, 11820, 11855, 11915, 11930, 11935, 13710, 13775, 15120, 15170, 15205, 15225, 15250, 15285, 15380, 15435, 15490, 15560, 15570, 17615, 17625, 17660, 17705, 17730, 17740, 17785, 17805, 17895, 21505, 21670*

Winter Schedule 2010/2011

Arabic	Days	Area	kHz
0300-0600	daily	ME,EAf	9580jed**,†
0300-0600	daily	Eu,CAs,WAs	15170riy*
0300-0800	daily	CAs,EAs	17895riy*
0300-1000	daily	ME	9715riy*
0600-0900	daily	Eu	17740riy***
0600-0900	daily	ME	15380riy*
0600-0900	daily	NAf	17730riy***
0600-1700	daily	EAf,ME	11855jed**,†
0900-1200	daily	EAs,SEA	17570riy*
0900-1200	daily	SAs,SEA	17615riy*
0900-1200	daily	NAf	17805riy***
0900-1200	daily	Eu	15490riy***
0900-1200	daily	ME	11935riy*
1000-1700	daily	ME	11785riy*
1200-1400	daily	ME	15380riy*
1200-1400	daily	SAs,SEA	17625riy*
1200-1500	daily	NAf	17895riy*, 21505riy***
1200-1500	daily	Eu	17705riy***
1400-1600	daily	SAf	17615riy*
1500-1800	daily	Eu	15435riy***,~

Arabic	Days	Area	kHz
1500-1800	daily	NAf	13710riy*, 15225riy***,~
1600-1800	daily	Eu	15205riy*
1600-1800	daily	WAf,CAf	17560riy*
1700-2200	daily	EAf,ME	9580jed**,†
1800-2300	daily	WAf,CAf	11930riy*
1800-2300	daily	Eu	9870riy***, 11820riy*
1800-2300	daily	NAf	9555riy***, 11915riy*
Bengali	**Days**	**Area**	**kHz**
1200-1500	daily	SAs	15120riy
English	**Days**	**Area**	**kHz**
1000-1230	daily	WAf,CAf	15250riy
French	**Days**	**Area**	**kHz**
0800-1000	daily	WAf	17785riy
1400-1800	daily	WAf	17660riy
Indonesian	**Days**	**Area**	**kHz**
0900-1200	daily	SEA	21670riy
Persian	**Days**	**Area**	**kHz**
1500-1800	daily	ME	7240riy
Swahili	**Days**	**Area**	**kHz**
0400-0700	daily	EAf	15285riy
Tajik/Turkmen/ Uyghur/Uzbek	**Days**	**Area**	**kHz**
1500-1800	daily	CAs	9885riy
Turkish	**Days**	**Area**	**kHz**
1800-2100	daily	ME	9675riy
Urdu	**Days**	**Area**	**kHz**
1200-1500	daily	SAs	13775riy

Key: * Quran prgr; ** 2nd prgr; *** General prgr; ~ "Call of Islam" prgr, 1500-1700.
Ann: Arabic (General Prgr): "Idha'at al mamlaka alarabiya al saudiyah min al Riyadh"; English: "Radio Riyadh", "Broadcasting Service of the Kingdom of Saudi Arabia".
IS: 'Ud' (Oriental Lute). Opens and closes with National Anthem.
V: No longer issues QSL-cards.
Notes: The SW transmissions in Arabic are relays of Home Sce prgrs. Plans to install 4 x 250kW DRM-ready transmitters at Jeddah.

SERBIA (SRB)

INTERNATIONAL RADIO SERBIA (Gov)
⌕ P.O. Box 200, 11000 Beograd, Serbia.
☎ +381 11 3244455. 🖷 +381 11 3232014.
E: radioju@sbb.rs **W:** www.glassrbije.org
Webcast: D/L
L.P: Dir: Milorad Vujovic; Deputy Dir: Igor Mladenovic; Deputy Editor in Chief: Kiril Panov.
SW: [BEO] Beograd, Stubline: 1 x 10kW & via relay station in Bosnia-Herzegovina.
SAT: Eutelsat Sesat 1.
kHz: 6100, 6190, 7230

Winter Schedule 2010/2011

English	Days	Area	kHz
0130-0200	mtwtfs.	NAm,Eu	6190bij
1930-2000	daily	Eu	6100bij
2200-2230	daily	Eu	6100bij
French	**Days**	**Area**	**kHz**
2130-2200	daily	Eu	6100bij
German	**Days**	**Area**	**kHz**
2100-2130	mtwtf.s	Eu	6100bij
Russian	**Days**	**Area**	**kHz**
1900-1930	daily	RUS	6100bij
Serbian	**Days**	**Area**	**kHz**
0100-0130	mtwtfs.	NAm,Eu	6190bij
0100-0200	......s	NAm,Eu	6190bij
0200-0230	..w....	NAm,Eu	6190bij
2030-2100	mtwtf.s	Eu	6100bij
2030-2130	.....s.	Eu	6100bij
2230-2300	daily	AUS	7230bij
Spanish	**Days**	**Area**	**kHz**
2000-2030	daily	Eu	6100bij

Ann: English: "This is the Engish Service of International Radio Serbia"; Serbian: "Medjunarodni Radio Srbija"; Spanish: "Esta es Radio Serbia".
V: QSL-card. Rec. acc.

SEYCHELLES (SEY)

BBC INDIAN OCEAN RELAY STATION
P.O. Box 448, Victoria, Mahé, Seychelles.
☎ +248 78496. 🖷 +248 78500.
L.P: Resident Engineer: Barrie Elding.
SW: [SEY] Mahé: 2 x 250kW.
V: QSL-card. (For direct report)
Notes: Owned by the BBC and operated by Babcock International Group PLC (see under United Kingdom).

SINGAPORE (SNG)

BBC FAR EAST RELAY STATION
51 Turut Track, Singapore 718930.
☎ +65 67937511. 🖷 +65 67937834.
SW: [SNG] Singapore: 4 x 100, 5 x 250kW.
V: QSL-card. (For direct report)
Notes: Owned by the BBC and operated by Babcock International Group PLC (see under United Kingdom).

SLOVAKIA (SVK)

RADIO SLOVAKIA INTERNATIONAL (Pub)
Mýtna 1, P.O. Box 55, 817 55 Bratislava 15, Slovak Republic.
☎ +421 2 57273734. 🖷 +421 2 52496282.
E: drahoslava.valocka@slovakradio.sk; englishsection@slovakradio.sk **W:** www.rsi.sk
Webcast: D/P
L.P: Dir: Mária Mikušová; Chief Editor: Ladislav Kubiš.
SW: Leased from Towercom A.S.
SAT: Astra 3A.
kHz: 5915, 6010, 6040, 6055, 6080, 6190, 7240, 7345, 9440, 9445, 9460, 9485, 9540, 11600, 11610, 13625, 13715, 15460

Winter Schedule 2010/2011

English	Days	Area	kHz
0100-0130	daily	SAm	9440rso
0100-0130	daily	NAm	6040rso
0700-0730	daily	As,AUS	13715rso, 15460rso
1730-1800	daily	WEu	5915rso, 6010rso
1930-2000	daily	WEu	5915rso, 7345rso
French	**Days**	**Area**	**kHz**
0200-0230	daily	NAm	6040rso
0200-0230	daily	SAm	9440rso
1800-1830	daily	WEu	5915rso, 6055rso
2030-2100	daily	WEu	5915rso, 7345rso
German	**Days**	**Area**	**kHz**
0800-0830	daily	WEu	5915rso, 6055rso
1430-1500	daily	WEu	6055rso, 7345rso
1700-1730	daily	WEu	5915rso, 6010rso
1900-1930	daily	WEu	5915rso, 7345rso
Russian	**Days**	**Area**	**kHz**
1400-1430	daily	EEu,As	9540rso, 13625rso
1600-1630	daily	EEu,As	6190rso, 7240rso
1830-1900	daily	EEu,As	5915rso, 9485rso
Slovak	**Days**	**Area**	**kHz**
0130-0200	daily	NAm	6040rso
0130-0200	daily	SAm	9440rso
0730-0800	daily	As,AUS	13715rso, 15460rso
1630-1700	daily	WEu	5915rso, 6055rso
2000-2030	daily	WEu	5915rso, 7345rso
Spanish	**Days**	**Area**	**kHz**
0230-0300	daily	SAm	6080rso, 9440rso
1530-1600	daily	WEu	9445rso, 11600rso
2100-2130	daily	WEu	9460rso
2100-2130	daily	SAm	11610rso

Ann: English: "You are listening to Radio Slovakia International".
V: QSL-card.
Notes: Began broadcasting in 1993. Radio Slovakia International is the External Sce of the public-service "Slovak Radio (Slovenský Rozhlas)".

TOWERCOM A.S. (Tx Operator)
Cesta na Kamzik 14, 810 05 Bratislava, Slovak Republic.
☎ +421 2 49220111. 🖷 +421 2 44461042.
E: info@towercom.sk **W:** www.towercom.sk

L.P: Chmn: Roman Fischer.
SW: [RSO] Rimavská Sobota: 4 x 250kW (run at lower power).
Notes: Towercom is the operator of transmitter networks in the Slovak Republic.

SOUTH AFRICA (AFS)

CHANNEL AFRICA (Pub)
P.O. Box 91313, Auckland Park 2006, South Africa.
☎ +27 11 7142255. 🖷 +27 11 7142072.
E: matemm@channelafrica.org; dawetimj@sabc.co.za (Scheds)
W: www.channelafrica.org
Webcast: D/L/P
L.P: GM: David Moloto; Head of Web/Technical: Maurice Mate; Prgr Mgr: Lungi Daweti.
SW: Leased from Sentech.
SAT: PAS 10.
kHz: 3345, 6120, 7230, 9625, 15235, 15255, 17770

Winter Schedule 2010/2011

English	Days	Area	kHz
0300-0400	daily	EAf	6120mey
0300-0500	daily	SAf	3345mey
0400-0700	daily	SAf	7230mey
0600-0700	daily	WAf	15255mey
0700-1200	daily	SAf	9625mey
1500-1600	daily	SAf	9625mey
1700-1800	daily	WAf	15235mey
French	**Days**	**Area**	**kHz**
1600-1700	daily	WAf	15235mey
Lozi	**Days**	**Area**	**kHz**
1300-1400	daily	SAf	9625mey
Nyanja	**Days**	**Area**	**kHz**
1200-1300	daily	SAf	9625mey
Portuguese	**Days**	**Area**	**kHz**
1400-1500	daily	SAf	9625mey
Swahili	**Days**	**Area**	**kHz**
1500-1600	daily	EAf	17770mey

Ann: English: "You're listening to Channel Africa coming to you from Johannesburg"; "You are listening to Channel Africa, the voice of the African Renaissance, broadcasting live from Johannesburg, South Africa".
IS: Birds chirping and native melody.
V: QSL-card.
Notes: Channel Africa is the External Sce of the public-service South African Broadcasting Corporation (SABC).

CVC 1AFRICA RADIO (Rlg)
P.O. Box 3933, Tygervalley 7536, South Africa. (Studio)
☎ +27 21 9506900. 🖷 +27 21 9419261.
E: 1africa@cvc.tv; radio@1africa.tv **W:** www.1africa.tv
Webcast: L/P
L.P: SM: Charles Maboshe; Transmitter Engineer: John Kawele.
SW: For tx details see "CVC Radio Christian Voice", under Zambia.
SAT: PAS 7/10.
kHz: 9430, 9505, 13590

Winter Schedule 2010/2011

English	Days	Area	kHz
0400-0600	daily	Af	9430lus
0600-2000	daily	Af	13590lus
2000-2200	daily	Af	9505lus

Ann: English: "Number 1 Africa - CVC"; "One Life, One Way, One Africa".
V: QSL-letter.
Notes: Produced in South African studios. Owned by South African based CVC Media, a branch of Christian Vision, United Kingdom. For corporate details, see United Kingdom under "CVC International".

TWR AFRICA (Rlg)
P.O. Box 4232, 1620, Kempton Park, South Africa. (Postal)
☎ +27 11 9742885. 🖷 +27 11 9749960.
E: info@twr.org.za **W:** www.twrafrica.org
Webcast: D/L/P
San Croy Business Park, Die Agora Road, Croydon 1619, South Africa. (Physical Address)
W: www.twr.org.za (TWR South Africa)
SAT: PAS 7.

kHz: *1170, 1566, 3200, 3240, 4760, 4775, 5995, 6025, 6120, 6130, 7315, 9475, 9500, 9525, 9635, 15360*

Winter Schedule 2010/2011

Amharic	Days	Area	kHz
1700-1715	.....s	EAf	9500man
1700-1730	mt..fs.	EAf	9500man
1730-1800	.....s.	EAf	9500man
Bemba	**Days**	**Area**	**kHz**
1615-1630	.t.....	SAf	6130man
Chewa	**Days**	**Area**	**kHz**
0400-0430	daily	SAf	5995man
0430-0500	.....ss	SAf	5995man
ChiChewa	**Days**	**Area**	**kHz**
1600-1615	mtwtf.	SAf	6130man
1615-1630	m.....s	SAf	6130man
Chokwe	**Days**	**Area**	**kHz**
1820-1835	daily	SAf	6130man
Dendi/Songhai	**Days**	**Area**	**kHz**
2010-2025	daily	WAf	1566par
English	**Days**	**Area**	**kHz**
0255-0325	.....s	SAf	3200man
0255-0400	daily	WAf	1566par
0430-0500	daily	WAf	1566par
0430-0500	mtwtf..	SAf	3200man
0430-0800	mtwtf..	SAf	4775man
0500-0800	.....ss	SAf	4775man
0500-0800	daily	CAf	9500man
0502-0800	daily	SAf	6120man
0530-0600	daily	WAf	1566par
1425-1455	daily	SAf	6025man
1525-1555	.....ss	SAf	6025man
1700-2000	daily	SAf	3200man
1700-2030	.....s.	SAf	3200man
1700-2105	daily	SAf	1170man
1740-1825	daily	WAf	1566par
1800-1900	daily	SAf	9500man
Ewe/Twi	**Days**	**Area**	**kHz**
0500-0530	daily	WAf	1566par
Fiote	**Days**	**Area**	**kHz**
1905-1920	...f..	SAf	6130man
Fongbe	**Days**	**Area**	**kHz**
1855-1910	daily	WAf	1566par
French	**Days**	**Area**	**kHz**
1440-1525	.....s.	SAf	9635man
1935-1950	daily	CAf	9525man
1950-2020	.....s.	CAf	9525man
2040-2245	daily	WAf	1566par
Fulfulde/Pulaar	**Days**	**Area**	**kHz**
1925-2010	daily	WAf	1566par
German	**Days**	**Area**	**kHz**
0400-0430	mtwtf..	SAf	3200man, 4775man
0400-0500	.....ss	SAf	3200man, 4775man
Hadiya	**Days**	**Area**	**kHz**
1645-1700	...fs.	EAf	9500man
Hausa	**Days**	**Area**	**kHz**
0400-0430	daily	WAf	1566par
1700-1740	daily	WAf	1566par
Igbo/Moore	**Days**	**Area**	**kHz**
2025-2040	daily	WAf	1566par
Kambaata	**Days**	**Area**	**kHz**
1630-1645	...fs.	EAf	9500man
Kanuri	**Days**	**Area**	**kHz**
1910-1925	daily	WAf	1566par
KiKongo	**Days**	**Area**	**kHz**
1850-1905	.twtf.s	SAf	6130man
Kimbundu	**Days**	**Area**	**kHz**
1950-2005	daily	SAf	6130man
Kuanyama	**Days**	**Area**	**kHz**
1905-1920	......s	SAf	6130man
Lingala	**Days**	**Area**	**kHz**
1905-1935	daily	CAf	9525man
Lomwe	**Days**	**Area**	**kHz**
0342-0357	daily	SAf	4775man
1510-1555	daily	SAf	7315man

Luchazi	Days	Area	kHz
1905-1920	..w....	SAf	6130man
Lunyaneka	**Days**	**Area**	**kHz**
1905-1920	.....s.	SAf	6130man
Luvale	**Days**	**Area**	**kHz**
1850-1905	m......	SAf	6130man
1905-1920	...t...	SAf	6130man
Makhuwa	**Days**	**Area**	**kHz**
1355-1425	.....s.	SAf	7315man
1455-1510	daily	SAf	7315man
Malagasy	**Days**	**Area**	**kHz**
1455-1525	mtwtf.s	SAf	9635man
Ndau	**Days**	**Area**	**kHz**
0325-0340	daily	SAf	3240man
1600-1630	.....s.	SAf	4760man
1615-1645	......s	SAf	4760man
1645-1700	daily	SAf	4760man
Ndebele	**Days**	**Area**	**kHz**
0255-0310	.....s.	SAf	3200man
0255-0325	mtwtf..	SAf	3200man
1455-1525	daily	SAf	6025man
1525-1555	mtwtf..	SAf	6025man
Oromo	**Days**	**Area**	**kHz**
1630-1730	..wt...	EAf	9500man
1715-1745	.....s	EAf	9500man
1730-1800	mtwtf..	EAf	9500man
Oromo/Borana	**Days**	**Area**	**kHz**
1645-1700	mt...s	EAf	9500man
Portuguese	**Days**	**Area**	**kHz**
1355-1425	.....s	SAf	7315man
1425-1455	daily	SAf	7315man
1630-1645	m..t...	SAf	4760man
1850-1905	.....s.	SAf	6130man
1905-1920	mt.....	SAf	6130man
1920-1950	daily	SAf	6130man
2005-2020	......s	SAf	6130man
Shangaan	**Days**	**Area**	**kHz**
1545-1615	.....s	SAf	4760man
1630-1645	.tw.f..	SAf	4760man
Shona	**Days**	**Area**	**kHz**
0255-0325	daily	SAf	3240man
1555-1625	daily	SAf	6025man
Swahili	**Days**	**Area**	**kHz**
1700-1745	daily	EAf	9475man
1745-1815	.....ss	EAf	9475man
Tshwa	**Days**	**Area**	**kHz**
1600-1630	mtwtf..	SAf	4760man
Umbundu	**Days**	**Area**	**kHz**
1750-1820	mtwtf..	SAf	6130man
1835-1850	daily	SAf	6130man
Urdu	**Days**	**Area**	**kHz**
1400-1415	daily	As	15360man
Yoruba	**Days**	**Area**	**kHz**
1825-1855	daily	WAf	1566par
Zulu	**Days**	**Area**	**kHz**
1630-1700	daily	SAf	1170man

Ann: English: "You are listening to TWR Broadcasting from Manzini, Swaziland".

IS: Last bar of "We've a story to tell the Nations", played on hand bells.

V: QSL-folder. Rp. (IRCs appreciated, 3 IRCs for airmail reply)

Notes: Owned by TWR Inc, USA. For corporate details, see under USA. TWR Africa owns transmitting stations in Benin and Swaziland.

AMATEUR RADIO MIRROR INTERNATIONAL

✉ P.O. Box 90438, Garsfontein 0042, South Africa.
☎ +27 11 6752393. 🖷 +27 11 6752793.
E: armi@sarl.org.za
W: www.sarl.org.za/public/armi/armi.asp
Webcast: L
SW: Leased from Sentech.
kHz: *3215, 7205, 17860*

	Winter Schedule 2010/2011		
English	**Days**	**Area**	**kHz**
0800-0900	......s	SAf	7205mey

English	Days	Area	kHz
0800-0900	......s	EAf	17860mey
2005-2105	m......	SAf	3215mey

V: QSL-card.
Notes: ARMI is a weekly prgr about amateur radio, shortwave listening and electronics, produced by the South African Radio League.

SENTECH PTY. (Tx Operator)
✉ Private Bag X06, Honeydew 2040, South Africa.
☎ +27 11 4388883. 🖷 +27 11 6917107.
E: support@sentechsa.com **W:** www.sentech.co.za
L.P: Chmn: Quraysh Patel; Acting CEO: Ms Beverley Ngwenya.
SW: [MEY] Meyerton, Tygerberg: 10 x 100, 4 x 250, 3 x 500kW.
V: QSL-card. (For relayed stations)
Notes: Sentech Pty. is the owner of the transmitter networks in South Africa.

SPAIN (E)

RADIO EXTERIOR DE ESPAÑA (REE) (Pub)
✉ Casa de la Radio, Avenida de la Radio y la Televisión, 4 Pozuelo de Alarcón 28223, Madrid, Spain.
☎ +34 91 3461034. 🖷 +34 91 3461815.
E: ree@rtve.es; secretariatecnica.ree@rtve.es
W: www.rtve.es/radio/radio-exterior
Webcast: D/L/P
☎ +34 91 3461149.
L.P: Dir: Josefina Benéitez; Foreign Language prgrs: Jose J. Amorena Zabalza.
SW: [NOB] Noblejas: 6 x 250kW.
SAT: Asiasat 5, EchoStar9/Galaxy 23, Hot Bird 8, Hispasat 1C, NSS 7.
kHz: 3350, 5965, 5970, 6055, 6125, 7265, 7275, 9535, 9570, 9590, 9605, 9620, 9630, 9665, 9675, 9690, 9765, 9780, 11625, 11680, 11755, 11765, 11780, 11815, 11895, 11910, 11940, 12030, 12035, 13720, 15110, 15125, 15170, 15325, 15385, 15585, 17595, 17715, 17755, 17770, 17850, 21540, 21570, 21610

Winter Schedule 2010/2011

Arabic	Days	Area	kHz
1700-1900	daily	ME	11765nob
1900-2100	mtwtf..	ME	12030nob
1900-2100	mtwtf..	NAf	7265nob
2000-2200	.....ss	NAf	7265nob

English	Days	Area	kHz
0000-0100	daily	NAm	5970nob*
1900-2000	mtwtf..	Af	9605nob
1900-2000	mtwtf..	Eu	9665nob
2200-2300	.....ss	Eu	6125nob

French	Days	Area	kHz
1800-1900	mtwtf..	Eu	9665nob
1900-2000	......s	ME	12030nob
1900-2000	.....s.	Af	9590nob
2000-2100	mtwtf..	ME	9605nob
2000-2100	mtwtf..	Af	9570nob
2300-2400	.....ss	Eu	5970nob
2300-2400	daily	NAm	6055nob

Ladino	Days	Area	kHz
0115-0145	.t.....	SAm	11780nob
0415-0445	.t.....	NAm	9690nob
1425-1455	m......	ME	15325nob

Portuguese	Days	Area	kHz
1830-1900	mtwtf..	NAm	17850cri
1830-1900	mtwtf..	Af	17755nob
1830-1900	mtwtf..	SAm	15125cri, 17715nob
1830-1900	mtwtf..	Eu	7275nob
1830-1900	mtwtf..	CAm	9765cri
1830-1900	mtwtf..	LAm	17595nob
2100-2200	mtwtf..	SAm	11680nob

Russian	Days	Area	kHz
1700-1730	mtwtf..	Eu	11755nob

Spanish	Days	Area	kHz
0000-0200	daily	NAm	9630cri+
0000-0400	daily	SAm	9765cri
0100-0600	daily	NAm	6055nob
0200-0600	daily	CAm	3350cri
0200-0600	mtwtf..	NAm	9675cri
0400-0800	daily	SAm	5965cri
0500-0600	mtwtf..	Eu	12035nob

Spanish	Days	Area	kHz
0500-0700	daily	ME	11895nob
0500-0900	daily	Eu	9780nob+
0600-0900	daily	Eu	12035nob
0700-0900	daily	Pac	17770nob
0800-1300	daily	Eu	13720nob
0900-1500	daily	Af	21540nob
0900-1700	daily	ME	21610nob
0900-1700	daily	Eu	15585nob
1200-1400	daily	SEA	11910bei
1200-1500	mtwtf..	CAm	9765cri
1200-1500	mtwtf..	SAm	11815cri
1200-1500	mtwtf.s	NAm	15170cri
1200-2300	......s	CAm	9765cri
1200-2300	......s	SAm	15125cri
1300-1400	....ss	Eu	13720nob
1300-1700	daily	SAm	21570nob
1300-1830	mtwtf..	LAm	17595nob
1300-1900	...ss	Am	17595nob
1500-1700	mtwtfs.	Af	15385nob
1500-2200	......s	Af	17755nob
1500-2300	......s	NAm	17850cri
1600-2300	.....s.	CAm	9765cri
1600-2300	.....s.	NAm	17850cri
1600-2300	.....s.	SAm	15125cri
1700-1830	daily	SAm	17715nob
1700-1830	mtwtf..	Eu	7275nob
1700-1830	mtwtf..	Af	17755nob
1700-2200	.....s.	Af	17755nob
1700-2200	...ss	Eu	9665nob
1700-2300	...ss	Eu	7275nob
1800-1830	mtwtf..	CAm	9765cri
1800-1830	mtwtf..	NAm	17850cri
1800-1830	mtwtf..	SAm	15125cri
1900-2000	mtwtf..	NAm	17850cri
1900-2000	mtwtf..	SAm	15125cri
1900-2000	mtwtf..	CAm	9765cri
1900-2300	daily	NAm	15110nob
1900-2300	mtwtf..	Eu	7275nob
1900-2300	...ss	LAm	11940nob
2200-2300	...ss	Af	11625nob
2200-2300	daily	Af	7265nob
2300-0200	daily	SAm	11680nob
2300-0500	daily	SAm	9620nob
2300-0500	daily	LAm	6125nob
2300-0500	daily	NAm,CAm	9535nob

Key: + DRM; * Alternative frequency 6055kHz. Spanish languge broadcasts include the following segments:- Basque: Mon-Fri 1330-1400; Catalan & Galician: Mon-Fri 2345-2355, repeated next morning at 0505-0515 and multilingual (Arabic, English, French, Portuguese and Russian) news at 1530-1545.
Ann: Arabic: "Idha'atu Isbania al-Jariyia"; English: "This is Radio Exterior de Espana, broadcasting from Madrid"; French: "Radio Extérieure d'Espagne"; German: "Hier ist der Spanische Auslandssender Radio Exterior de España"; Spanish: "Radio Exterior de España".
V: Does not verify.
Notes: REE is the External Sce of the public broadcaster Radio Nacional de España.

SRI LANKA (CLN)

SRI LANKA BROADCASTING CORPORATION (SLBC)
✉ P.O. Box 574, Colombo 7, Sri Lanka.
☎ +94 11 2697491. 🖷 +94 11 2691568.
E: chmnslbc@slbc.lk (Chairman); ddge@slbc.lk (DG, Engineering)
W: www.slbc.lk
Webcast: L
✉ Independence Square, Colombo 7, Sri Lanka. (Studio)
☎ +94 26 2222097.
L.P: Chmn: Hudson Samarasinghe; DG: Samantha Weliweriya; Deputy DG, Engineering: H.M Jacson.
MW: [PUT] Puttalam: 873/882kHz 400kW, 1125kHz 50kW (expected to be in service late 2010). 873kHz used for SLBC relays; 882kHz leased to TWR (TWR schedule, see under India); 1125kHz leased to WRN.
SW: [EKA] Colombo, Ekala: 10 x 10, 3 x 35, 2 x 100, 2 x 300kW.
kHz: 6005, 7190, 9770, 11750, 11905, 15745

Winter Schedule 2010/2011

English	Days	Area	kHz
0130-0300	mtwtfs.	SAs	6005eka, 9770eka, 15745eka
0300-0500	......s	SAs	6005eka, 9770eka, 15745eka

English/Hindi	Days	Area	kHz
1530-1630	daily	SAs	11905trm

Hindi	Days	Area	kHz
0020-0230	daily	SAs	7190eka, 11905eka

Malayalam	Days	Area	kHz
1000-1100	daily	As	7190eka, 11905eka

Sinhala	Days	Area	kHz
1630-1900	daily	ME	11750trm

Tamil	Days	Area	kHz
1100-1215	daily	As	7190eka, 11905eka

Telugu	Days	Area	kHz
0930-1000	daily	As	7190eka, 11905eka

Ann: English: "This is the Sri Lanka Broadcasting Corporation", "Radio Ceylon calling out to India".
IS: Melody on drums.
V: QSL-card. Rp.

DEUTSCHE WELLE RELAY STATION
✉ 92/2, D.S. Senanayake Mawatha, Colombo 8, Sri Lanka. (Correspondence)
☎ +94 11 2699449. 🖷 +94 11 2699450.
L.P: Resident Engineer: R. Groschkus.
MW: [TRM] Trincomalee: 1548kHz 400kW.
SW: [TRM] Trincomalee: 1 x 250, 2 x 300kW (plus 1 reserve transmitter).
V: QSL-card. (Rpt to DW Bonn, not to relay stn)
Notes: Schedule see under Deutsche Welle (Germany).

IBB RELAY STATION SRI LANKA
✉ IBB Transmitting Station, P.O. Box 14, Negombo, Sri Lanka.
☎ +94 32 2255931. 🖷 +94 32 2255822.
L.P: SM: Glenn Britt.
SW: [IRA] Iranawila: 3 x 250, 4 x 500kW.

SWAZILAND (SWZ)

TWR RELAY STATION
✉ P.O. Box 64, Manzini, Swaziland.
☎ +268 5052781. 🖷 +268 5055333.
L.P: Chief Engineer: Steve Stavropoulos.
MW: [MAN] Manzini, Mpangela Ranch: 1170kHz 50kW.
SW: [MAN] Manzini, Mpangela Ranch: 1 x 50, 3 x 100kW.
Notes: Owned by TWR. For corporate details, see under USA. For schedule, see TWR Africa, under South Africa.

SWEDEN (S)

IBRA RADIO (Rlg)
✉ SE-141 99 Stockholm, Sweden.
☎ +46 8 6089680. 🖷 +46 8 6089650.
E: ibra@ibra.se; info@ibra.se
W: www.ibra.org (English); www.ibra.se (Swedish)
✉ Regulatorvägen 11, 141 49 Huddinge, Sweden.
L.P: Dir: Gösta Åkerlund; PR: Maria Levander.
SAT: Hot Bird 6.
kHz: *5900, 5910, 7320, 7445, 9420, 9635, 9945, 11740, 11785, 12045*

Winter Schedule 2010/2011

Arabic	Days	Area	kHz
1700-1800	daily	ME	12045wof
1800-1830	daily	ME	5910sam
1800-1900	daily	ME	9420rmp
1800-1930	daily	Af	9635skn
1930-1945	daily	Af	9635skn
1945-2015	daily	Af	9635skn
Chinese	Days	Area	kHz
1200-1230	daily	EAs	7320nvs
1200-1300	daily	EAs	5900ppk
Fon	Days	Area	kHz
2000-2015	mt.....	WAf	7445rmp
Jula	Days	Area	kHz
1945-2000	mt..fss	WAf	7445rmp

Malinke	Days	Area	kHz
1945-2000	..wt...	WAf	7445rmp
Mandarin	Days	Area	kHz
1100-1200	daily	EAs	9945hbn
Moore	Days	Area	kHz
1930-1945	mtw...s	WAf	7445rmp
Somali	Days	Area	kHz
1730-1800	daily	EAf	11740mey
Swahili	Days	Area	kHz
1730-1800	daily	Af	11785skn
Tamajeq	Days	Area	kHz
1930-1945	...tfs.	WAf	7445rmp
Zarma	Days	Area	kHz
2000-2015	..wtf..	WAf	7445rmp

V: QSL-card.
Notes: IBRA Radio (part of IBRA Media), is the radio ministry of the Swedish Pentecostal Movement. On air since July 1955.

SWITZERLAND (SUI)

RADIO FREUNDES-DIENST (Rlg)
✉ Freundes-Dienst International, Buhaldenstrasse 30, CH-5023 Biberstein, Switzerland.
☎ +41 62 8272727. 🖷 +41 62 8393003.
E: info@freundesdienst.org **W:** www.freundesdienst.org
L.P: Dir: Joseph Schmid, Samuel J. Schmid.
kHz: *1440*

Winter Schedule 2010/2011

German	Days	Area	kHz
0415-0430	......s	Eu	1440mrn
0430-0445	mtwtfs.	Eu	1440mrn
1830-1845	daily	Eu	1440mrn

V: QSL-card.
Notes: Freundes-Dienst International (Friends Service International) is a Swiss government recognized non-profit foundation. Began radio broadcasts in 1959.

RADIO RÉVEIL (Rlg)
✉ Chemins Chapons-des-Prés 4, CH-2022 Bevaix (NE), Switzerland.
☎ +41 32 8470610. 🖷 +41 32 8462547.
E: contact@paroles.ch **W:** www.paroles.ch
Webcast: L (www.eclair6.com)/P
kHz: *9760*

Winter Schedule 2010/2011

French	Days	Area	kHz
1830-1845	.t.t...	CAf	9760iss

Ann: French: "Radio Réveil Paroles de Vie".
V: QSL-letter.

STIMME DES TROSTES (Rlg)
✉ Missionswerk Arche, Rosenbüelstrasse 48, CH-9642 Ebnat-Kappel, Switzerland.
☎ +41 71 992 25 00. 🖷 +41 71 992 25 55.
E: info@missionswerk-arche.ch
W: www.missionswerk-arche.ch
L.P: Secretary: Herbert Skutzik.
kHz: *1440, 6055*

Winter Schedule 2010/2011

German	Days	Area	kHz
0415-0430	.....s.	Eu	1440mrn
1200-1215	.....s	Eu	6055wer*
1815-1830	.....s.	Eu	1440mrn*

Key: * Every second week.
Notes: Produced by Missionshaus Arche.

SYRIA (SYR)

RADIO DAMASCUS (Gov)
✉ P.O. Box 4702, Damascus, Syria.
☎ +963 11 720700. 🖷 +963 11 2234336.
E: radiodamascusenglish@yahoo.com
W: www.radio-damascus.net
Webcast: D/P (Podcasts are at: radiodamascusenglish.podomatic.com)
W: www.syriaonline.sy/radio.php
L.P: GD: Khudr Omran; Dir, Engineering: M.Bara; Dir, Public Rel: Mrs. Awafet Haffar.

MW: [HMS] Homs, Saraqeb: 594kHz 100kW; [TAR] Tartus: 783kHz 300kW.
SW: [ADR] Adra: 2 x 500kW.
SAT: AsiaSat 2, Badr 4, Hot Bird 3, Nilesat 101.
kHz: *594, 783, 9330, 12085*

Winter Schedule 2010/2011

Arabic/French/

English	Days	Area	kHz
1900-2000	daily	ME	594hms
2100-2200	daily	Eu,NAm,Pac	9330adr, 12085adr‡

French	Days	Area	kHz
1900-2000	daily	Eu,NAm	9330adr, 12085adr‡

German	Days	Area	kHz
1800-1900	daily	Eu	9330adr, 12085adr‡

Hebrew	Days	Area	kHz
1600-1830	daily	ME	783tar

Russian	Days	Area	kHz
1700-1800	daily	Eu	9330adr, 12085adr‡
1830-1900	daily	ME	783tar

Spanish	Days	Area	kHz
2200-2300	daily	Eu,LAm	9330adr, 12085adr‡

Turkish	Days	Area	kHz
1600-1700	daily	ME	9330adr, 12085adr‡

Key: ‡ Inactive at time of publication.
Ann: Arabic: "Idha'atu-l-jumhuriyati-l'arabiyya as-suriyya min dimashq"; English: "You are listening to Radio Damascus, the External Service of the Syrian Broadcasting System", "Welcome to the Broadcasting Service of the Syrian Arab Republic calling from Damascus"; French: "Ici Damas"; Hebrew: "Kol Damasek".
IS: Guitar.
V: QSL-card.
Notes: Radio Damascus is the External Sce of the state broadcaster Organisme de la Radio-TV Arabe Syrienne (ORTAS).

TAIWAN (Rep. of China) (TWN)

RADIO TAIWAN INTERNATIONAL (RTI) (Gov)
✉ 55 Pei'an Road, Taipei 10462, Taiwan.
☎ +886 2 28856168. 📠 +886 2 28862382.
E: rti@rti.org.tw **W:** www.rti.org.tw
Webcast: D/L
✉ Post Box 4914, P.O. Safdarjung Enclave, New Delhi 110029 India. (RTI India)
L.P: Chmn: Kuang Hsiang Hsia; Pres: Wang Tan-Ping; VP: Hsiao Hsu-Tsen.
MW: [FAN] Fangliao: 1503kHz 600kW; [KOU] Kouhu: 1098/1557kHz 300kW; [LUK] Lukang: 603kHz 500kW, 1008kHz 300kW; [MIN] Minhsiung: 1206/1422kHz 100kW; [TAI] Taiwan (see note below).
SW: [HUW] Huwei: 4 x 100, 1 x 300kW; [KOU] Kouhu: 3 x 100kW; [MIN] Minhsiung: 1 x 50kW; [PAO] Paochung: 5 x 100kW; [TAI] Taiwan (see note below); [TNN] Tainan: 4 x 250kW; [TSH] Tanshui: 3 x 300kW.
FM/DAB: DAB: 220.064MHz.
SAT: SatMex 5.
kHz: *603, 1008, 1098, 1206, 1422, 1503, 1557, 3955, 3965, 3985, 5950, 6075, 6085, 6105, 6120, 6145, 6150, 6875, 6890, 7365, 7380, 7385, 7445, 7555, 7570, 9365, 9625, 9660, 9680, 9685, 9735, 9780, 9840, 11520, 11550, 11605, 11625, 11635, 11640, 11655, 11665, 11710, 11715, 11765, 11850, 11875, 11885, 11915, 11985, 11995, 12055, 15225, 15265, 15270, 15290, 15320, 15440, 15465, 15690*

Winter Schedule 2010/2011

Amoy	Days	Area	kHz
0100-0200	daily	EAs	1422min
0500-0600	daily	EAs	1008luk, 1422min
0900-1000	daily	EAs	1422min
1000-1100	daily	EAs	15465pao
1200-1300	daily	EAs	1206min
1200-1300	daily	SEA	11715tnn
1300-1400	daily	EAs	11625pao

Cantonese	Days	Area	kHz
0200-0230	daily	SEA	15440yfr
0400-0430	daily	SEA	15320pao
0900-1000	daily	EAs	15465pao
1000-1030	daily	SEA	11625pao, 15270pao
1200-1230	daily	EAs	6105kou, 11915tnn

Cantonese	Days	Area	kHz
1500-1530	daily	SEA	11550tnn
1500-1600	daily	SEA	7380pao

English	Days	Area	kHz
0100-0200	daily	SAs	11875tnn
0200-0300	daily	NAm	5950yfr, 9680yfr
0230-0300	daily	EAs	1422min
0300-0400	daily	NAm	6875yfr
0300-0400	daily	SEA	15320pao
0500-0600	daily	NAm	6875yfr
1100-1200	daily	SEA	7445pao, 11715tnn
1600-1700	daily	SAs	11550tnn, 12055iss
1700-1800	daily	Eu	11850iss*, 15690iss**
1800-1900	daily	Eu	3965iss

French	Days	Area	kHz
1900-2000	daily	Af	9365iss*, 11875iss**
1900-2000	daily	Eu	3985skn

German	Days	Area	kHz
1900-2000	daily	Eu	3955skn
2100-2200	daily	Eu	3965iss

Hakka	Days	Area	kHz
0200-0230	daily	EAs	1422min
0230-0300	daily	SEA	15440yfr
0430-0500	daily	SEA	15320pao
1030-1100	daily	SEA	11625pao, 15270pao
1230-1300	daily	EAs	6105kou, 11915tnn
1530-1600	daily	SEA	11550tnn

Indonesian	Days	Area	kHz
0300-0500	daily	EAs	1422min
1000-1100	daily	SEA	11520pao, 11550tnn
1200-1300	daily	EAs	1422min
1200-1300	daily	SEA	11625pao
1400-1500	daily	SEA	11875tnn

Japanese	Days	Area	kHz
0800-0900	daily	EAs	11605tnn
1100-1200	daily	EAs	9735tnn
1300-1400	daily	EAs	9735tnn

Mandarin	Days	Area	kHz
0000-0100	daily	EAs	1422min
0300-0400	daily	NAm	5950yfr
0300-0800	daily	EAs	1557min
0400-0500	daily	EAs	1008luk
0400-0500	daily	NAm	6875yfr
0400-0600	daily	EAs	11640kou, 11885huw
0400-0600	daily	SEA	15290tnn
1000-1100	daily	EAs	1422min, 6105kou
1000-1200	daily	EAs	1503fan
1000-1400	daily	EAs	6150kou, 9780kou
1000-1500	daily	EAs	6085huw
1000-1700	daily	EAs	603luk, 1008luk, 7385kou, 11665tsh
1100-1200	daily	SEA	11625pao
1100-1300	daily	EAs	11710tnn
1100-1700	daily	EAs	9680huw
1300-1330	daily	EAs	1503fan
1300-1400	daily	SEA	15265tnn
1300-1500	daily	SEA	7445pao
1300-1700	daily	EAs	1098kou
1400-1800	daily	EAs	6075kou, 6145pao
1500-1700	daily	EAs	7365tsh
1600-1700	daily	EAs	1503fan
2200-2400	daily	SEA	11635pao
2200-2400	daily	EAs	6105tsh, 6150kou, 11710tnn, 11885huw
2300-0300	daily	EAs	9660kou
2300-2400	daily	EAs	9685kou

Russian	Days	Area	kHz
1100-1200	daily	EAs	11985huw
1400-1500	daily	CAs	15225iss
1700-1800	daily	Eu	6120iss*, 9840iss**

Spanish	Days	Area	kHz
0200-0300	daily	SAm	7570yfr, 11995guf
0400-0500	daily	CAm	6890yfr

Spanish	Days	Area	kHz
0600-0700	daily	CAm	6875yfr
2000-2100	daily	Eu	3965iss
2300-2400	daily	SAm	11885yfr
Thai	**Days**	**Area**	**kHz**
1300-1500	daily	EAs	1422min
1400-1500	daily	SEA	11635pao
1500-1600	daily	SEA	1503fan, 7555pao
2200-2300	daily	SEA	1503fan
2200-2400	daily	SEA	7445pao
2300-2400	daily	EAs	1422min
Vietnamese	**Days**	**Area**	**kHz**
0900-1000	daily	SEA	15270pao
1100-1200	daily	EAs	1422min
1200-1300	daily	SEA	11765tnn
1300-1400	daily	EAs	1206min
1400-1500	daily	SEA	9625tnn
2330-0030	daily	SEA	11655tnn

Key: * To 28 Feb; ** from 1 Mar.
Ann: English: "This is Radio Taiwan International"; Indonesian: "Inilah Radio Taiwan Internasional"; Japanese: "Kochirawa Taiwan Kokusai Hoso, RTI, Chukaminkoku Chuohosokyoku no nihongobangumi desu"; Mandarin: "Cheli shih Chungyang Kuangpo Tientai, Taiwan chih Yin".
V: QSL-card. Rec. acc.
Notes: Formed in 1998, when the Broadcasting System of the Ministry of Defense was joined with the international section of the Broadcasting Corporation of China. Schedule includes some CBS networks. [TAI] is used as a general term for Taiwanese sites when the exact site is unknown. Programmes to mainland China are jammed by "China National Radio (CNR)" 1st programme transmissions.

TAJIKISTAN (TJK)

VOICE OF TAJIK (OVOZI TOJIK) (Gov)
Sheroz St. 31, 734025 Dushanbe, Tajikistan.
☎ +992 37 2277417. ⊟ +992 37 2211198.
MW/SW: Leased from Teleradiokom.
kHz: 1143, 7245

Winter Schedule 2010/2011

Arabic	Days	Area	kHz
1200-1300	daily	ME	1143dsb, 7245dsb
Dari	**Days**	**Area**	**kHz**
0600-0800	daily	WAs	1143dsb, 7245dsb
English	**Days**	**Area**	**kHz**
1300-1400	daily	WAs,ME	1143dsb, 7245dsb
Farsi	**Days**	**Area**	**kHz**
0400-0600	daily	ME	1143dsb, 7245dsb
1600-1800	daily	ME	1143dsb, 7245dsb
Hindi	**Days**	**Area**	**kHz**
1100-1200	daily	As	1143dsb, 7245dsb
Russian	**Days**	**Area**	**kHz**
0800-1000	daily	CAs	1143dsb, 7245dsb
Tajik	**Days**	**Area**	**kHz**
0200-0400	daily	CAs	1143dsb, 7245dsb
1400-1600	daily	CAs	1143dsb, 7245dsb
Uzbek	**Days**	**Area**	**kHz**
1000-1100	daily	CAs	1143dsb, 7245dsb

V: QSL-letter.
Notes: External Sce under the roof of the State Committee for TV and Radio Broadcasting.

TELERADIOKOM (Tx Operator)
Internatsionalnaya Street 85, 734001 Dushanbe, Tajikistan.
☎ +992 37 2210912. ⊟ +992 37 2217974.
E: info@teleradiocom.tj **W:** www.teleradiocom.tj
L.P: DG: Suhrob Aliyev.
MW: [DSB] Two sites: Dushanbe, Yangiyul (G.C: 38N29 068E48): 1143kHz 150kW, 1251kHz 100kW; Orzu (G.C: 37N31 068E48): 648/801kHz 1000kW, 972kHz 500kW (also other freqs may be used); operated on behalf of IBB (USA): 972kHz 800kW.
SW: [DSB] Dushanbe, Yangiyul (G.C: 38N29 068E48): 1 x 50, 5 x 100kW; Orzu (G.C: 37N31 068E48): 2 x 1000kW; operated on behalf of IBB (USA): 1 x 250kW.
Notes: Teleradiokom, a subsidiary of the Telecommunications Ministry, is the national transmitter network owner.

THAILAND (THA)

RADIO SARANROM (Gov)
Sri Ayudhya Road, Bangkok 10400, Thailand.
☎ +66 28435095. ⊟ +66 26435093.
W: www.mfa.go.th/web/151.php
L.P: Dir, Broadcasting Division: Narumit Hinshiranan.
MW: Uses facilities provided by IBB.
kHz: 1575

Winter Schedule 2010/2011

Thai	Days	Area	kHz
1030-1100	daily	SEA	1575bph
1100-1130	mtwtf..	SEA	1575bph
1200-1230	mtwtf..	SEA	1575bph
1500-1530	daily	SEA	1575bph
2230-2400	mtwt..s	SEA	1575bph

V: QSL-card.
Notes: Service for Thai's living in South East Asia, funded by the Ministry of Foreign Affairs.

RADIO THAILAND WORLD SERVICE (HSK9) (Gov)
Public Relations Department, Royal Thai Government, 236 Vibhavadi Rangsit Road, Ding Daeng, Bangkok 10400, Thailand.
☎ +66 2 6919917. ⊟ +66 2 2776139.
E: english@hsk9.org; feedback@hsk9.org **W:** www.hsk9.org
Webcast: L
SW: Txs provided by IBB.
SAT: Thaicom 3.
kHz: 7235, 7255, 7465, 7570, 9535, 9720, 9725, 11730, 11870, 12040, 13745, 15275

Winter Schedule 2010/2011

Burmese	Days	Area	kHz
1145-1200	daily	SEA	7235udo
English	**Days**	**Area**	**kHz**
0000-0030	daily	NAm	13745udo
0030-0100	daily	NAm	13745udo
0200-0230	daily	NAm	15275udo
0530-0600	daily	Eu	11730udo
1230-1300	daily	As,Pac	9720udo
1400-1430	daily	As,Pac	9725udo
1900-2000	daily	Eu	7570udo
2030-2045	daily	Eu	9535udo
German	**Days**	**Area**	**kHz**
2000-2015	daily	Eu	9535udo
Japanese	**Days**	**Area**	**kHz**
1300-1315	daily	EAs	7465udo
Khmer	**Days**	**Area**	**kHz**
1115-1130	daily	SEA	7255udo
Lao	**Days**	**Area**	**kHz**
1130-1145	daily	SEA	7235udo
Malay	**Days**	**Area**	**kHz**
1200-1215	daily	SEA	11870udo
Mandarin	**Days**	**Area**	**kHz**
1315-1330	daily	CHN	7465udo
Thai	**Days**	**Area**	**kHz**
0100-0200	daily	NAm	13745udo
0230-0330	daily	NAm	15275udo
1000-1100	daily	SEA	12040udo
1330-1400	daily	EAs	7465udo
1800-1900	daily	Eu	7570udo
2045-2115	daily	Eu	9535udo
Vietnamese	**Days**	**Area**	**kHz**
1100-1115	daily	SEA	7255udo

Ann: English: "This is HSK9, Radio Thailand's World Service broadcasting from the Public Relations Department in Bangkok".
IS: Gongs and chimes.
V: QSL-card.
Notes: Radio Thailand World Service is the External Sce of the Thai Government Public Relations Department.

BBC ASIA RELAY STATION
P.O. Box 20, Muang, Nakhon Sawan 60000, Thailand.
☎ +66 56227275. ⊟ +66 56227277.
SW: [NAK] Nakhon Sawan: 4 x 250kW.
V: QSL-card. (For direct report)
Notes: Owned by the BBC and operated by Babcock International Group PLC (see under United Kingdom).

IBB RELAY STATIONS THAILAND
☞ IBB Transmitting Station, Rangsit-Bangpoon Road, Bangkok, Thailand.
☎ +66 25815191.
☞ IBB Transmitting Station (Udon Thani), P.O. Box 99, Amphur Muang, Udon Thani 41000, Thailand.
L.P: SM: Jack A. Fisher.
MW: [BPH] Ban Phachi, Rasom: 1575kHz 1000kW.
SW: [UDO] Udon Thani (Udorn), Ban Dung: 7 x 500kW.

TUNISIA (TUN)

RADIO TUNISIENNE (Gov)
☞ See National Radio section.
Webcast: D/L/P
SW: Uses txs provided by ONT.
SAT: Atlantic Bird 1/4A, Badr 6, Galaxy 19, Hot Bird 8.
kHz: 7225, 7275, 7335, 7345, 9725, 12005

Winter Schedule 2010/2011

Arabic	Days	Area	kHz
0200-0510	daily	NAf,ME	9725sfa, 12005sfa
0400-0625	daily	Eu	7275sfa
0600-0810	daily	NAf	7335sfa
1600-2000	daily	NAf,ME	9725sfa, 12005sfa
1700-2110	daily	NAf	7225sfa
1900-2310	daily	NAf	7345sfa

Ann: Arabic: "Huna Tunis, Idha'at al-wataniya at-Tunisiya".
V: QSL-card. (Rpt in Arabic or French to L'Office National de la Télédiffusion).

OFFICE NATIONAL DE LA TÉLÉDIFFUSION (ONT)
(Tx Operator)
☞ Cité Ennassim I - Borjel, BP 399, Tunis 1080, Tunisia.
☎ +216 71801177. ▤ +216 71781927.
E: ont@telediffusion.net.tn **W:** www.telediffusion.net.tn
L.P: Pres:/DG: Saiid Aljane.
SW: [SFA] Sfax, Sidi Mansour: 3 x 100, 2 x 500kW.
Notes: ONT is the national transmitter network owner.

TURKEY (TUR)

VOICE OF TURKEY (VOT) (Pub)
☞ P.O. Box 333, Yenisehir, Ankara 06443, Turkey.
☎ +90 312 4633271. ▤ +90 312 4633355.
E: tsr@trt.net.tr **W:** www.trt-world.com
Webcast: D/P
☞ TRT/Oran Sitesi A Blok No: 427, Ankara 06109, Turkey. (Studio)
☎ +90 312 4633372. (English Desk)
L.P: Dir, Foreign Broadcasting: Osman Dilekçi.
SW: [CAK] Çakirlar: 3 x 250, 2 x 500kW; [EMR] Emirler: 5 x 500kW.
SAT: Galaxy 19, Hot Bird 8, Optus D2, Thaicom 5, Türksat 2A/3A.
kHz: 5960, 5965, 5970, 5980, 6000, 6050, 6120, 6185, 7205, 7240, 7245, 9410, 9495, 9530, 9610, 9650, 9655, 9665, 9700, 9785, 9820, 9840, 11620, 11680, 11735, 11795, 11815, 11835, 11925, 11955, 11965, 11985, 12035, 13625, 13640, 15200, 15245, 15350, 15360, 15480, 17715, 17755

Winter Schedule 2010/2011

Arabic	Days	Area	kHz
1000-1100	daily	NAf	11955cak
1000-1100	daily	ME	15245emr
1500-1600	daily	NAf	15200emr
1500-1600	daily	NAf,ME	9665cak
Azeri	**Days**	**Area**	**kHz**
0800-0900	daily	Cau,ME	11835cak
1630-1730	daily	ME	5965emr
Bulgarian	**Days**	**Area**	**kHz**
1200-1230	daily	Eu	7245cak
Chinese	**Days**	**Area**	**kHz**
1200-1300	daily	EAs	17715cak
Dari	**Days**	**Area**	**kHz**
1600-1630	daily	WAs	11680cak
1700-1730	daily	WAs	11680cak
English	**Days**	**Area**	**kHz**
0400-0500	daily	Eu,NAm	9655emr
0400-0500	daily	ME	7240cak
1330-1430	daily	Eu	12035cak
1330-1430	daily	As,Pac	11735emr
1930-2030	daily	Eu	6050emr

English	Days	Area	kHz
2130-2230	daily	As,Pac	9610cak
2300-2400	daily	Eu,NAm	5960emr
Farsi	**Days**	**Area**	**kHz**
0930-1100	daily	ME	11795emr
1600-1700	daily	ME	9530emr
French	**Days**	**Area**	**kHz**
2030-2130	daily	Af	6050emr
2030-2130	daily	Eu	5970emr
Georgian	**Days**	**Area**	**kHz**
1100-1200	daily	Cau	9840cak
German	**Days**	**Area**	**kHz**
1230-1330	daily	Eu	17755emr
1830-1930	daily	Eu	7205cak
Italian	**Days**	**Area**	**kHz**
1500-1530	daily	Eu	6185cak
Kazakh	**Days**	**Area**	**kHz**
1430-1500	daily	CAs	9785emr
Pashto	**Days**	**Area**	**kHz**
1630-1700	daily	WAs	11680cak
Russian	**Days**	**Area**	**kHz**
1400-1500	daily	RUS	9410cak
Spanish	**Days**	**Area**	**kHz**
0200-0300	daily	Am	9650cak
0200-0300	daily	SAm	9410emr
1730-1830	daily	Eu	9495cak
Tatar	**Days**	**Area**	**kHz**
1100-1130	daily	RUS	15360cak
Turkish	**Days**	**Area**	**kHz**
0100-0300	daily	Eu	6000emr
0500-0700	daily	ME	9820cak
0500-0700	daily	Eu	9700emr
0700-1000	daily	ME	11925cak
0700-1400	daily	ME	15480cak
0700-1400	daily	Eu	15350emr
1400-1700	daily	Eu	11815cak
1700-2200	daily	Eu	5980cak
1700-2200	daily	NAf,ME	6120emr
Turkmen	**Days**	**Area**	**kHz**
1300-1330	daily	CAs	11965cak
Urdu	**Days**	**Area**	**kHz**
1300-1400	daily	SAs	11985emr
Uyghur	**Days**	**Area**	**kHz**
0300-0400	daily	CAs	13640cak
1500-1600	daily	CAs	11620emr
Uzbek	**Days**	**Area**	**kHz**
1130-1200	daily	CAs	13625cak
1700-1730	daily	CAs	11680cak

Ann: English: "This is the Voice of Turkey"; German: "Hier ist die Stimme der Türkei"; Spanish: "Esta es La Voz de Turquia"; Turkish: "Burasi Türkiye'nin Sesi Radyosu".
V: QSL-card.
Notes: The Voice of Turkey is the External Sce of the public service Turkish Radio-TV Corporation, TRT (Türkiye Radyo-Televizyon Kurumu).

UKRAINE (UKR)

RADIO UKRAINE INTERNATIONAL (RUI) (Gov)
☞ vul. Kreschatyk 26, 01001 Kyiv, Ukraine.
☎ +380 44 2782534. ▤ +380 44 2287356.
E: vsru@nrcu.gov.ua **W:** www.nrcu.gov.ua
Webcast: D/L
L.P: Dir, RUI WS: Alexander Wild; Chief Editor, English Section: Zhanna Mescherska.
MW/SW: Leased from Concern RRT.
SAT: Sirius 4.
kHz: 657, 6030, 6140, 7435, 7440, 9410

Winter Schedule 2010/2011

English	Days	Area	kHz
0100-0200	daily	NAm	7440lvi
0800-0900	daily	Eu	9410smf
2000-2100	daily	Eu	6030khr
2300-2400	daily	NAm	7440lvi
German	**Days**	**Area**	**kHz**
1800-1900	daily	Eu	6030khr
2100-2200	daily	Eu	6140khr

Romanian	Days	Area	kHz
1800-1830	daily	Eu	657crn
2030-2100	daily	Eu	657crn
2200-2230	daily	Eu	657crn

Ukrainian	Days	Area	kHz
0000-0100	daily	NAm	7440lvi
0200-0300	daily	NAm	7440lvi
0900-1100	daily	Eu	9410smf
1500-1800	daily	Eu	7435khr
1900-2000	daily	Eu	6030khr

Ann: English: "This is Radio Ukraine International"; Ukrainian: "Hovorit Kyiv, Vsesvitnia sluzhba Radio Ukrayiny".
V: QSL-card.
Notes: Radio Ukraine International is the External Sce of the state broadcaster "Natsionalna Radiokompaniia Ukraini (NRKU)". May cease SW broadcasts as of 1st January 2011.

CONCERN RRT (Tx Operator)
⌑ vul. Dorohozhytska 10, 04112 Kyiv, Ukraine.
☎ +380 44 2262260. 🖷 +380 44 4408722.
E: rrt@rrt.ua **W:** www.rrt.ua
L.P: GD: Anatolyi M. Antonenko.
MW: [CRN] Chernivtsi: 657kHz 30kW; [KHR] Kharkiv, Taranivka: 612kHz 10kW; [KYV] Kyiv, Brovary: 594kHz 50kW.
SW: [KHR] Kharkiv, Taranivka: 4 x 100kW; [KYV] Kyiv, Brovary: 4 x 100kW; [LVI] Lviv, Krasne: 2 x 1000kW; [SMF] Mykolaiv, Luch: 1 x 100, 2 x 250, 2 x 1000, 2 x 1200kW.
Notes: Concern RRT is the national transmitter network operator.

UNITED ARAB EMIRATES (UAE)

ABU DHABI MEDIA CO. (Tx Operator)
⌑ 4th St, sector 18, Abu Dhabi, United Arab Emirates.
☎ +971 2 4144000. 🖷 +971 2 4144001.
E: Via website. **W:** www.admedia.ae
L.P: Chmn: H.E Mohamed Khalaf Al-Mazrouei; CEO: Edward Borgerding; Exec. Dir, Radio/TV: Karim Sarkis.
MW: [DHA] Dhabbaya: 1170/1539/1575kHz 800kW.
SW: [DHA] Dhabbaya: 4 x 500kW.
Notes: Abu Dhabi Media is the state broadcaster and transmitter operator. Under special appointment, the Dhabbaya transmitting station is operated and maintained by Babcock International Group PLC (UK) until 2011.

UNITED KINGDOM (G)

BBC WORLD SERVICE (Pub)
⌑ Bush House, Strand, London, WC2B 4PH, United Kingdom.
☎ +44 20 72403456. 🖷 +44 20 75571258.
E: worldservice.letters@bbc.co.uk
W: www.bbc.co.uk/worldservice
Webcast: D/L/P
L.P: Dir: Peter Horrocks; Head of Asia-Pacific Region: Behrouz Afagh; Head of Americas-Europe: Nikki Clarke; Head of Africa & Middle East Region: Jerry Timmins.
MW/SW: Uses txs provided by Babcock International Group plc (formerly VT Communications) & foreign relays.
SAT: Afristar, Astra 1H/2B, Badr 4 (ArabSat), Hot Bird 8, Intelsat 10-02, NileSat 102, Telstar 10/18.
kHz: 198, 594, 612, 639, 648, 666, 675, 702, 720, 1251, 1260, 1296, 1323, 1413, 1503, 3255, 3380, 3915, 3955, 5790, 5845, 5865, 5875, 5890, 5905, 5910, 5920, 5940, 5945, 5955, 5965, 5970, 5975, 5985, 5990, 6005, 6020, 6030, 6040, 6055, 6065, 6080, 6085, 6090, 6095, 6100, 6110, 6115, 6135, 6140, 6145, 6155, 6165, 6170, 6175, 6180, 6190, 6195, 7205, 7225, 7235, 7255, 7260, 7265, 7270, 7305, 7315, 7325, 7335, 7350, 7355, 7360, 7375, 7380, 7385, 7390, 7395, 7400, 7405, 7410, 7415, 7425, 7435, 7445, 7465, 7490, 7505, 7540, 7600, 9410, 9440, 9450, 9460, 9470, 9480, 9485, 9495, 9505, 9510, 9580, 9585, 9595, 9605, 9610, 9615, 9650, 9685, 9730, 9740, 9750, 9810, 9815, 9860, 9915, 11620, 11680, 11685, 11710, 11730, 11740, 11750, 11760, 11770, 11785, 11795, 11800, 11805, 11810, 11820, 11830, 11850, 11860, 11890, 11895, 11915, 11920, 11925, 11950, 11955, 11965, 11970, 11995, 12005, 12015, 12035, 12065, 12095, 13590, 13660, 13675, 13725, 13750, 13775, 13790, 13810, 13820, 13825, 13845, 13865, 15105, 15150, 15180, 15285, 15310, 15335, 15360, 15400, 15420, 15490, 15510, 15575, 15745, 15750, 15790, 17640, 17685, 17720, 17760, 17780, 17790, 17820, 17825, 17830, 17870, 17885, 21470, 21595, 21630

Winter Schedule 2010/2011

Arabic	Days	Area	kHz
0300-0400	daily	NAf	6040skn
0300-0400	daily	ME	5790skn, 6055cyp
0300-0500	daily	ME	5875cyp, 7390cyp
0300-2200	daily	ME	720zak
0330-2300	daily	ME	639zak
0400-0500	daily	ME	13660sla
0400-0500	daily	NAf	5790skn
0400-0600	daily	ME	9915cyp
0400-0700	daily	NAf	6110skn, 7325rmp
0500-0600	daily	ME	15790sla
0500-0800	daily	ME	11820cyp
0500-1600	daily	ME	5905cyp
0600-0700	daily	NAf	11680cyp
0600-0800	daily	ME	13660cyp, 15790cyp
0700-0800	daily	NAf	11680rmp
0800-1000	daily	NAf	13660rmp, 15180cyp
1400-1600	daily	ME	12095cyp, 15790cyp
1400-1700	daily	ME	5875cyp
1500-1600	daily	NAf	13660skn
1500-1700	daily	NAf	12095rmp
1500-2100	daily	ME	702sla
1600-1800	daily	ME	9915mos
1600-1800	daily	NAf	9915cyp
1600-2100	daily	ME	7375cyp
1700-2100	daily	NAf	6110rmp
1700-2100	daily	ME	6195sla
1700-2100	daily	ME,NAf	5790skn
1800-2100	daily	NAf	9915cyp

Azeri	Days	Area	kHz
0300-0315	mtwtf..	CAs	6085rmp, 7335dha, 7410mos
1600-1630	daily	CAs	6090cyp, 9615skn, 12005rmp
1700-1730	daily	CAs	6030cyp, 7465rmp, 9750rmp

Bengali	Days	Area	kHz
0030-0100	daily	SAs	6065nak, 9510sng, 11750sng
0130-0200	daily	SAs	9510nak, 11995sng
1330-1400	daily	SAs	5845nak, 9510sla, 11850sng
1630-1700	daily	SAs	5865nak, 7270sng, 9650sng

Burmese	Days	Area	kHz
0000-0030	daily	SEA	5875nak, 9510sng, 11750sng
0200-0230	daily	SEA	7380nak, 9480nak, 11995sng
1345-1430	daily	SEA	7205sng, 9585sng, 11685sng

Dari	Days	Area	kHz
0030-0100	daily	WAs	1413sla, 5875cyp, 7435rmp
0130-0200	daily	WAs	5875cyp, 6195cyp, 7435rmp
0200-0230	daily	CAs	1251dsb
0230-0300	daily	WAs	6140sla, 6195cyp, 7435cyp
0930-1000	daily	CAs	1251dsb
0930-1030	daily	WAs	13820sla, 17720nak
1400-1500	daily	CAs	1251dsb
1400-1500	daily	WAs	6195skn, 7405sla
1400-1500	daily	WEu	11950cyp
1700-1730	daily	WAs	5865nak, 5875cyp, 9505sng
1800-1900	daily	CAs	1413sla, 7505nak, 9505cyp
1800-1900	daily	WAs	1251dsb, 5865nak

English	Days	Area	kHz
0000-0100	daily	FE	5970sla
0000-0100	daily	SAs	6195sng, 9410nak, 12095sng, 13725nak
0000-0100	daily	WAs	7360cyp
0000-0200	daily	FE	15360nak
0000-0200	daily	SEA	15335sng

English	Days	Area	kHz
0000-2400	daily	WEu	648orf
0000-2400	daily	CHN	675hkg
0100-0200	daily	SAs	5970sla, 11750sng, 12095nak
0100-0200	daily	SEA	17685sng
0100-0300	daily	SAs	15310nak
0100-0400	daily	CAs	5940cyp
0100-0520	daily	Eu	198dro
0200-0230	daily	SAs	1413sla
0200-0300	daily	SAs	12095sla
0200-0300	daily	EAf	7445sey
0200-0300	daily	ME	5875cyp
0200-0730	mtwtf..	ME	1323cyp
0200-2300	.....ss	ME	1323cyp
0300-0330	daily	ME	639zak
0300-0400	daily	SAf	7445asc
0300-0400	daily	WAf	6145mey
0300-0400	daily	ME	1413sla
0300-0400	daily	Eu	5940skn
0300-0400	daily	EAf	6100cyp, 9460sey
0300-0400	daily	CAs	12095sla
0300-0600	daily	SAf	3255mey
0300-0600	daily	SAs	15310sla
0300-0600	daily	WAs	9410cyp
0300-0600	daily	CAf	7255asc
0300-0700	daily	SAs	17790nak
0300-2200	daily	SAf	6190mey
0330-0600	daily	EAf	11860sey
0400-0500	daily	WAf	9460mey
0400-0500	daily	EAf	12035cyp
0400-0600	daily	CAs	15360sla
0400-0700	daily	WAf	6005asc
0500-0530	......s	EAf	15420sey
0500-0600	mtwtf..	EAf	15420sey, 15420sey
0500-0700	daily	NAf	17640cyp
0500-0700	daily	RUS	5875rmp, 12095cyp
0500-0700	daily	CEu	1296orf+
0500-0700	daily	WAf	11770mey
0500-0800	daily	Eu	3955skn
0600-0700	daily	WAf	9410asc, 12015asc
0600-0700	daily	WAs	11760cyp
0600-0800	.....ss	EAf	15420sey
0600-1400	daily	SAs	15310nak
0600-1600	daily	SAf	9860mey
0700-0800	daily	RUS	5875mos+
0700-0800	daily	WAf	11770asc, 13820asc, 17830mey
0700-1000	daily	WAf	15400asc
0700-1300	daily	SAs	17790sla
0700-1300	daily	WAs	15575cyp
0700-1400	daily	ME	11760sla
0800-0900	daily	Eu	5875wof
0800-0900	daily	WEu	9610sin+
0800-1000	daily	CAf	17830asc
0800-1300	daily	EAf	17640sey
0800-1400	daily	SAf	21470sey
0900-1000	daily	FE	17760nak
0900-1100	daily	SEA	6195nak, 6195sng, 15285sng
0900-1200	daily	FE	11895nak
0900-1600	daily	SEA	9740sng
0900-2300	mtwtf..	ME	1323cyp
1000-1100	.....ss	CAf	17830asc
1000-1100	.....ss	WAf	15400asc
1000-1300	daily	FE	9605nak
1100-1130	daily	WAf	15400asc
1100-1600	daily	SEA	6195sng
1100-1800	daily	CAf	17830asc
1200-1600	daily	FE	5875nak
1215-1300	mtwtf..	Car	9410hri, 11860guf
1300-1400	daily	CAs	11805cyp
1300-1400	daily	SAs	1413sla
1300-1400	daily	EAf	15420sey
1300-1400	daily	Eu	15575skn
1300-1500	daily	SAs	9410sla
1300-1700	daily	SAf	17640asc
1400-1500	daily	CAs	9915cyp
1400-1500	daily	WAs	11760cyp
1400-1600	daily	SAs	5845nak, 13590trm+
1400-1700	daily	SAs	5975sng
1400-1700	daily	EAf	15420cyp
1500-1530	daily	EAf	9410sey, 11860sey, 15105mey
1500-1600	daily	WAs	9485cyp
1500-1600	daily	SAs	7395sla
1500-1700	daily	Eu,ME	11830rmp
1500-2100	daily	WAf	15400asc
1530-1700	.....s.	EAf	9410sey, 11860sey, 15105mey
1600-1700	daily	SAs	7355sla
1600-1800	daily	WAf	13790asc
1600-1800	daily	SAs	9740sng
1600-2200	daily	SAf	3255mey
1615-1700	......s	EAf	9410sey, 11860sey, 15105mey
1630-1700	mtwtf..	EAf	9410sey
1700-1746	daily	EAf	9410sey, 11860sey
1700-1800	.....ss	CAs	1251dsb
1700-1800	daily	SAs	5975nak
1700-1830	daily	SAs	5975sla
1700-1900	daily	EAf	15420mey
1700-2100	daily	NAf	12095cyp
1730-1800	daily	SAs	1413sla
1800-1830	daily	WAs	7260cyp
1800-1830	daily	SAs	7355nak
1800-2000	daily	CAs	5955sla
1800-2000	daily	RUS	5875rmp, 7225cyp
1800-2000	daily	WAs	5945cyp
1800-2100	daily	NAf	9615skn
1800-2100	daily	WAf	11810asc
1830-2100	daily	EAf	6005sey, 9410sey
1900-2100	daily	ME	1413sla
2100-2200	daily	SAf	5910sey
2100-2200	daily	SEA	6195nak
2100-2200	daily	WAf	7465mey
2100-2300	daily	FE	5965sla
2100-2300	daily	NAf	9915asc
2100-2300	daily	WAf	12095asc
2100-2400	daily	FE	5875nak
2100-2400	daily	SEA	3915sng
2200-0200	daily	SEA	9740sng
2200-2300	daily	FE	6135vld
2200-2300	daily	WAf	5910mey
2200-2400	daily	SEA	6195sng
2300-2400	daily	SEA	11955sng
2300-2400	daily	FE	6135sla, 7385nak
2330-2400	daily	FE	6170kim

English/Russian	Days	Area	kHz
0200-2100	daily	RUS	666ekb, 1260msk, 1260spb

English/Russian/Ukrainian	Days	Area	kHz
0000-2400	mtwtf..	UKR	612khr
0400-2000	mtwtf..	UKR	594kyv
1600-2100	.....s.	UKR	612khr

Farsi	Days	Area	kHz
0230-0300	daily	ME	1413sla
0230-0330	daily	ME	7400wof
0230-0330	daily	CAs	1251dsb
0230-0430	daily	ME	5985rmp, 6165cyp
0330-0400	.....ss	CAs	1251dsb
0330-0430	daily	ME	7435cyp
1600-1700	daily	WEu	6195skn
1600-1700	daily	ME	1413sla, 6155sla, 6195cyp, 9810skn
1630-1700	daily	ME	6090cyp, 6115cyp, 9505sng

French	Days	Area	kHz
0430-0500	daily	WAf	6135asc
0430-0500	daily	EAf	15490sey
0430-0500	daily	CAf	7415asc
0600-0630	daily	WAf	6135asc, 7305asc

French	Days	Area	kHz
0600-0630	daily	NAf	6055skn, 7350skn
0700-0730	daily	CAf	15490mey
0700-0730	daily	WAf	12095asc
1200-1230	daily	CAf	21630asc
1200-1230	daily	NAf	15420sin
1200-1230	daily	WAf	17780asc
1800-1830	daily	SAf	7465mey
1800-1830	daily	NAf	9605rmp
1800-1830	daily	CAf	11860asc
1800-1830	daily	WAf	17640asc, 17885asc

Hausa	Days	Area	kHz
0530-0600	daily	WAf	5975rmp, 6135asc, 7305asc
0630-0700	daily	WAf	7255rmp, 9440asc, 11800asc
1400-1430	daily	WAf	15105asc, 17780asc, 21630sin
1430-1700	.....s.	WAf	17780rmp
1930-2000	daily	WAf	11890asc, 15105asc, 17885asc
2000-2030	...f..	WAf	11890skn, 15105mey, 17885asc

Hindi	Days	Area	kHz
0100-0130	daily	SAs	1413sla, 6065sla, 6085arm, 9510nak, 11995sng, 13675nak
0230-0300	daily	SAs	11760sla, 15510nak, 17760sng, 17825nak
1400-1500	daily	SAs	1413sla, 5865dsb, 6030nak, 7395nak, 9505sla, 11620cyp
1700-1730	daily	SAs	1413sla, 6155sng, 7260sla, 9685sng, 11740nak

Indonesian	Days	Area	kHz
1100-1130	mtwtf..	SEA	7390sng, 9510nak, 11920sng
2200-2300	mtwt..s	SEA	5955nak, 6080sng, 7235sng, 9730sng

Kinyarwanda/			
Kirundi	Days	Area	kHz
0500-0600	.....s.	EAf	11925mey, 15420sey
0530-0600	......s	EAf	11925mey, 15420sey
1630-1700	mtwtf..	EAf	11860sey, 15105mey
1830-1900	mtwtf..	EAf	7425mey, 9815cyp, 11860cyp

Kyrgyz	Days	Area	kHz
1300-1330	daily	CAs	12095sla, 13845cyp, 15180cyp

Mandarin	Days	Area	kHz
1300-1530	mtwtf..	FE	9605sng, 11890sng
1300-1530	mtwtf..	CHN	6095kim, 7540nak
2200-2300	mtwt..s	CHN	5890nak, 7325sng, 9470sng
2200-2330	mtwt..s	CHN	6020sla, 6170kim, 7490nak

Nepali	Days	Area	kHz
1500-1530	daily	SAs	5865dsb, 9595sla, 11915sng

Pashto	Days	Area	kHz
0100-0130	daily	WAs	5875cyp, 6195cyp, 7435rmp
0200-0230	daily	WAs	6140sla, 6195cyp, 7435cyp
0300-0330	daily	WAs	6140sla, 6195rmp, 7435cyp
0745-0800	daily	WAs	15750sla, 17820nak
0830-0930	daily	WAs	13820sla, 17720nak
1030-1130	daily	WAs	13820sla, 17720nak
1045-1055	daily	WAs	17820dha
1045-1100	daily	WAs	11750sng
1345-1355	daily	WAs	12095cyp, 13810cyp, 15750dha
1500-1600	daily	WAs	7360cyp, 9810cyp
1500-1600	daily	WEu	6195skn

Pashto	Days	Area	kHz
1730-1800	daily	WAs	5865nak, 5875cyp, 9505sng

Portuguese	Days	Area	kHz
0430-0530	mtwtf..	SAf	3380mey, 6145mey, 7305asc
2030-2100	mtwtf..	SAf	3380mey, 6135mey, 7260mey, 11860asc
2030-2100	mtwtf..	WAf	5875rmp, 7415skn

Russian	Days	Area	kHz
0300-0330	mtwtf..	RUS	5965wof, 6110rmp, 7265cyp
0330-0400	mtwtf..	CAs	1251dsb
0500-0600	mtwtf..	RUS	9450cyp
0500-0700	mtwtf..	RUS	5790rmp, 7425rmp, 11970cyp
0600-0700	mtwtf..	RUS	13775cyp
1400-1500	mtwtf..	RUS	13750rmp, 15150rmp
1500-1600	mtwtf..	RUS	9915rmp, 11730rmp
1500-1800	mtwtf..	RUS	5920cyp, 7325cyp
1600-1800	mtwtf..	RUS	5990rmp, 7425rmp
1700-1800	mtwtf..	CAs	1251dsb
1700-1830	.....ss	RUS	5920cyp, 5990rmp, 7325cyp, 7425rmp

Sinhala	Days	Area	kHz
1630-1700	daily	SAs	5955dha, 7600nak, 9615sla

Somali	Days	Area	kHz
0400-0430	daily	EAf	9815cyp, 12015dha
1100-1130	daily	EAf	17780cyp, 21595cyp
1400-1500	daily	EAf	11860sey, 15510sey, 17870cyp
1500-1700	.....s.	EAf	13825cyp, 15510cyp
1800-1830	daily	EAf	6005sey, 9410sey, 9815cyp

Spanish	Days	Area	kHz
1200-1215	mtwtf..	Car	9410hri, 11860guf

Swahili	Days	Area	kHz
0300-0330	daily	EAf	5975asc, 9610sey, 11720mey
0400-0430	daily	EAf	7415asc, 11720mey, 11785sey
1430-1746	.....s.	EAf	15745mey
1530-1615	......s	EAf	9410sey, 11860sey, 15105mey
1530-1630	mtwtf..	EAf	9410sey, 11860sey, 15105mey
1746-1800	daily	EAf	7465mey, 9410sey, 11860sey

Tajik	Days	Area	kHz
1000-1030	daily	CAs	1251dsb
1830-1900	....f..	CAs	1251dsb

Tamil	Days	Area	kHz
1545-1615	daily	SAs	6135sla, 7600nak, 9605sla, 11965skn

Urdu	Days	Area	kHz
0130-0200	daily	SAs	1413sla, 6065sla, 6085arm, 13675nak
0130-0200	daily	WAs	7375cyp
0300-0330	daily	SAs	7325dha, 15510nvs, 17760nak, 17825nak
0730-0745	daily	SAs	15750sla, 17820nak
1030-1045	daily	SAs	11750sng, 17820dha
1330-1345	daily	SAs	12095cyp, 13810cyp, 15750dha
1355-1400	daily	SAs	12095cyp, 13810cyp, 15750dha
1500-1545	daily	SAs	7600nak
1500-1600	daily	SAs	1413sla, 6175sla, 9505nak, 12065rmp

Uzbek	Days	Area	kHz
1300-1330	daily	CAs	11730cyp, 13865cyp, 17790rmp
1300-1400	daily	CAs	1251dsb
1600-1630	daily	CAs	6115cyp, 6180arm, 9495nak, 11795wof
1630-1700	daily	CAs	1251dsb

Vietnamese	Days	Area	kHz
1430-1445	daily	SEA	1503fan, 6135sng, 7315sng, 9580sng

Key: + DRM. For joint BBC/DW DRM broadcasts, see "DRM International Transmissions".
Ann: English: "This is London, you are listening to the World Service of the BBC".
V: Does not verify reception reports.
Notes: BBC World Sce prgrs in English and other languages are relayed by local stns in many countries. Transmissions in some Asian languages are jammed.

CVC INTERNATIONAL (Rlg)
The Pavilion, Manor Drive, Coleshill, West Midlands, B46 1DL, United Kingdom.
☎ +44 1675 435500. 🖷 +44 1675 435501.
E: admin@cvuk.org **W:** www.christianvision.com
L.P: Dir, Int Broadcasting: Andrew Flynn.
V: QSL-card.
Notes: CVC International is a service of Christian Vision Ltd. Christian Vision owns transmitting stations in Australia (CVC Asia Pacific), Chile (CVC La Voz) and Zambia (Radio Christian Voice).

FEBA RADIO (Rlg)
Ivy Arch Road, Worthing, Sussex, BN14 8BX, United Kingdom.
☎ +44 1903 237281. 🖷 +44 1903 205294.
E: info@feba.org.uk; angela@feba.org.uk
W: www.feba.org.uk
E: lifechange@feba.org.uk
W: www.febaradio.net (Schedules etc.)
L.P: Chief Executive: John Bartlett.
kHz: 6125, 6140, 6180, 7230, 7235, 7315, 7485, 7510, 9400, 9550, 9595, 9650, 9850, 9900, 11875, 11985, 12045, 15205, 15215, 15220, 15250

Winter Schedule 2010/2011

Amharic	Days	Area	kHz
1600-1630	daily	EAf	11875kig
1600-1630	...tfss	EAf	9900erv
1630-1700	...tfs	EAf	9850dha
1630-1700	daily	EAf	9900erv

Arabic	Days	Area	kHz
0800-0830	daily	ME	15220mos
1900-1930	daily	ME	7235wer
1900-2030	daily	ME	9550kig

Bengali	Days	Area	kHz
0030-0045	...tfs	As	7485tac

Bengali (Rural)	Days	Area	kHz
0015-0030	daily	As	7485tac
1500-1530	daily	As	7485tac

Dari	Days	Area	kHz
0230-0300	daily	WAs	6125dha
1500-1530	daily	WAs	9400erv

English	Days	Area	kHz
1400-1430	......s	As	12045dha

French	Days	Area	kHz
1830-1845	daily	CAF,WAf	15250asc

Guragena	Days	Area	kHz
1600-1630	mtw....	EAf	9900erv

Hassinya/Pulaar	Days	Area	kHz
2145-2215	mt.tf..	WAf	11985asc

Hindi	Days	Area	kHz
0030-0045	..w...s	As	7485tac
0045-0100	daily	As	7485tac

Kashmiri	Days	Area	kHz
1445-1500	..wtfs.	As	9650dha

Malayalam	Days	Area	kHz
1400-1415	mtwtfs.	As	12045dha

Oromo	Days	Area	kHz
1700-1730	daily	EAf	9595kig

Pashto	Days	Area	kHz
0200-0230	daily	WAs	6125dha

Silte	Days	Area	kHz
1730-1800	daily	EAf	7510erv

Somali	Days	Area	kHz
1700-1730	daily	EAf	6180dha

Telugu	Days	Area	kHz
0130-0200	...tf.s	As	6140dha

Tibetan	Days	Area	kHz
1200-1230	daily	As	15215dha

Tigrinya	Days	Area	kHz
1630-1700	mtw...s	EAf	9850dha
1730-1757	daily	EAf	9595kig

Urdu	Days	Area	kHz
0200-0215	mtwtfs.	WAs	7315dha
0200-0230	......s	WAs	7315dha
0800-0830	daily	WAs	15205dha
1400-1430	daily	WAs	7230nvs
1430-1445	daily	As	9650dha

Various (Minority langs)	Days	Area	kHz
0030-0045	mt.....	As	7485tac
0130-0200	mtw..s.	As	6140dha
0215-0230	mtwtfs.	WAs	7315dha
0300-0315	daily	WAs	6125dha
1415-1430	mtwtfs.	As	12045dha
1430-1500	daily	WAs	7230nvs
1445-1500	mt....s	As	9650dha

V: Does not verify reception reports.
Notes: A division of Far East Broadcasting Company (FEBC), see USA for corporate details.

WRN BROADCAST
Wyvil Court, 10 Wyvil Road, London, SW8 2TG, United Kingdom.
☎ +44 20 78969000. 🖷 +44 20 78969007.
E: contactus@wrn.org **W:** www.wrn.org
L.P: MD: David Treadway; Dir, Technical: Tim Ashburner; Dir, Development: Jeff Cohen; Dir: Karl Miosga; Head of Operations: Michael Ward.
SAT: EuroBird, Galaxy 19, Hot Bird 6, Thaicom 3.
Notes: Previously known as "World Radio Network (WRN)". WRN Broadcast is a multichannel 24h news and information network via satellite and Internet, carrying prgrs from major world broadcasters. WRN Broadcast feeds can be heard on numerous AM/FM radio stations and cable around the globe. WRN Broadcast also brokers shortwave transmitter air time.

BABCOCK INTERNATIONAL GROUP PLC (Tx Operator)
33 Wigmore Street, London, W1 1QX, United Kingdom. (Corp HQ)
☎ +44 20 7355 5300. 🖷 +44 20 7355 5360.
E: Via website.
W: www.babcock-online.co.uk; www.babcock.co.uk (Corporate)
Blue Fin Building, 110 Southwark Street, London, SE1 0TA, United Kingdom. (Media Management Centre)
☎ +44 20 79690000. 🖷 +44 20 73555360.
L.P: CEO: Peter Rogers.
MW: [ORF] Orfordness: 648/1296kHz 500kW.
SW: [RMP] Rampisham: 10 x 500kW; [SKN] Skelton: 11 x 250, 6 x 300kW; [WOF] Woofferton: 6 x 250, 4 x 300kW.
Notes: In July 2010, Babcock International Group PLC acquired VT Group PLC, the owner and operator of the medium and shortwave transmitting centres in the UK that are used by the BBC World Service. With the take-over, it will also operate the British overseas relay stations under a management contract.

UNITED STATES OF AMERICA (USA)

AMERICAN FORCES RADIO AND TELEVISION SERVICE (AFRTS) (Gov)
NMC Det AFRTS-DMC, 23755 Z St., Bldg. 2730, Riverside, CA 92518-2017, USA.
☎ +1 951 4132236. 🖷 +1 951 4132457.
E: afrtops1@hq.afis.osd.mil; technologist@dodmedia.osd.mil **W:** afrts.dodmedia.osd.mil (Gen); myafn.dodmedia.osd.mil/shortwave. aspx (Shortwave)
601 N. Fairfax Street, Room 360, Alexandria, VA 22314, USA. (HQ/Engineering)
L.P: Dir: Melvin W. Russell.
SW: [BAR] Barrigada, Guam; [DGA] Diego Garcia, BIO; [KEW] Saddlebunch Keys, FL.
SAT: AMC 1, Galaxy 25, Hot Bird 4/6, Intelsat 10-02/701/906, NSS 5/6/7.
kHz: 4319, 5446, 5765, 7812, 12133, 12759, 13362

Winter Schedule 2010/2011

English	Days	Area	kHz
0000-2400	daily	Atl	5446kew*, 7812kew*, 12133kew*
0400-1500	daily	IOc	12759dga
0800-2000	daily	Pac	5765bar
1500-0400	daily	IOc	4319dga
2000-0800	daily	Pac	13362bar

Key: All in USB; * 0.5kHz higher than shown frequency.
Ann: English: "You're listening to AFN", "This is National Public Radio".
V: QSL-card.
Notes: AFRTS is an activity of the American Forces Information Service (AFIS) and a DoD Field Activity under the Assistant Secretary of Defence (Public Affairs).

BBG - AFIA DARFUR RADIO (Gov)
E: info@afiadarfur.com **W:** afiadarfur.com
kHz: 5885, 7275, 9380, 9780, 9805, 9815, 9845, 11615

Winter Schedule 2010/2011

Arabic	Days	Area	kHz
0300-0330	daily	SDN	5885smg, 7275sao, 9845ira
1800-1830	daily	SDN	9380ira, 9805wer, 11615smg
1900-1930	daily	SDN	9780smg, 9805sao, 9815wer

Ann: Arabic: "Afia Darfur".
Notes: BBG funded station for listeners in the Darfur region of Sudan and also in Eastern Chad, launched on 29 Sept 2008.

BBG - RADIO FARDA (Gov)
1201 Connecticut Avenue NW, Washington, D.C. 20036, USA.
☎ +1 202 8287220. 📠 +1 202 8287235.
E: comment@radiofarda.com **W:** www.radiofarda.com
Webcast: L/P
Vinohradská 159A, 100 00 Prague 10, Czech Republic. (HQ/Studio)
☎ +420 221 124113. 📠 +420 221 122622.
L.P: Dir: Andres Ilves; Dep Dir: Hossein Aryan.
SAT: AsiaSat 3D, Badr 4, Hot Bird 8, Intelsat 907, Telstar 12.
kHz: 1575, 5810, 5850, 5860, 6115, 7520, 7580, 9340, 9430, 9520, 9550, 9760, 9785, 9850, 11690, 11750, 12015, 13615, 13680, 15410, 15535, 15690, 17815, 17840, 21715

Winter Schedule 2010/2011

Farsi	Days	Area	kHz
0000-0100	daily	IRN	6115kwt
0000-0300	daily	IRN	5860ira
0000-2400	daily	IRN	1575dha
0100-0300	daily	IRN	6115bib
0200-0230	daily	IRN	9430lam
0230-0400	daily	IRN	9430bib
0230-0500	daily	IRN	9550lam
0230-1300	daily	IRN	15690ira
0300-1400	daily	IRN	5860kwt
0400-0500	daily	IRN	9430wer
0400-1100	daily	IRN	13615ira
0430-0500	daily	IRN	12015ira
0500-0600	daily	IRN	12015ira
0500-1200	daily	IRN	9520bib
0600-0700	daily	IRN	17840wer
0600-0930	daily	IRN	15535ira
0700-0900	daily	IRN	9760lam
0700-0930	daily	IRN	21715udo
0700-1030	daily	IRN	17840ira
0900-1100	daily	IRN	17815lam
0930-1200	daily	IRN	11690lam
0930-1230	daily	IRN	21715ira
1030-1200	daily	IRN	17840ira
1100-1400	daily	IRN	13615lam, 15410bib
1200-1230	daily	IRN	17840lam
1230-1600	daily	IRN	13680wer
1300-1400	daily	IRN	11750bib, 15690ira
1400-1500	daily	IRN	11750bib
1400-1600	daily	IRN	13615lam, 15410skn
1600-1700	daily	IRN	7520udo

Farsi	Days	Area	kHz
1600-1730	daily	IRN	13615lam
1600-2300	daily	IRN	7580ira
1700-1800	daily	IRN	9785lam
1700-2100	daily	IRN	7520udo
1800-1900	daily	IRN	9850wer
1900-2000	daily	IRN	9850lam
1900-2130	daily	IRN	9340ira
1930-2230	daily	IRN	5850ira
2100-2400	daily	IRN	7520ira
2230-2400	daily	IRN	5810kwt

Ann: Farsi: "Radyo Farda".
V: QSL-card.
Notes: BBG funded station for listeners in Iran, launched in December 2002. 24h on satellite & FM. Transmissions on medium wave are jammed.

BBG - RADIO FREE AFGHANISTAN (Gov)
1201 Connecticut Avenue NW, Washington, D.C. 20036, USA.
☎ +1 202 4576900. 📠 +1 202 4576992.
E: afghan@rferl.org
W: www.azadiradio.org
Webcast: D/L
Vinohradská 159A, 100 00 Prague 10, Czech Republic. (HQ/Studio)
☎ +420 2 21122370. 📠 +420 2 21123245.
W: pa.azadiradio.org (Pashto); da.azadiradio.org (Dari)
L.P: Dir: Hashem Mohmand.
SAT: AsiaSat 3S, Hot Bird 8.
kHz: 1296, 9335, 9990, 12140, 15335, 17530, 19010

Winter Schedule 2010/2011

Dari	Days	Area	kHz
0330-0430	daily	AFG	1296kab, 9335kwt, 12140kwt, 15335udo
0530-0630	daily	AFG	1296kab, 12140kwt, 17530ira, 19010kwt
0730-0830	daily	AFG	1296kab, 12140kwt, 17530kwt, 19010kwt
0930-1030	daily	AFG	1296kab, 12140kwt, 17530kwt, 19010kwt
1130-1230	daily	AFG	1296kab, 9335kwt, 9990udo, 12140kwt
1330-1430	daily	AFG	1296kab, 9335kwt, 12140kwt

Pashto	Days	Area	kHz
0230-0330	daily	AFG	1296kab, 9335kwt, 12140kwt, 15335udo
0430-0530	daily	AFG	1296kab, 12140kwt, 15335udo, 17530kwt
0630-0700	daily	AFG	17530ira
0630-0730	daily	AFG	1296kab, 12140kwt, 19010kwt
0700-0730	daily	AFG	17530ira
0830-0930	daily	AFG	1296kab, 12140kwt, 17530kwt, 19010kwt
1030-1130	daily	AFG	1296kab, 9990udo, 12140kwt, 19010kwt
1230-1330	daily	AFG	1296kab, 9335kwt, 9990udo, 12140kwt

Ann: Dari: "Inja Radyoi Azadi"; Pashto: "Da Azadi Radyo".
V: QSL-card.
Notes: BBG funded station for listeners in Afghanistan, launched in January 2001. Produced in the RFE/RL studios in Prague, Czech Republic.

BBG - RADIO FREE ASIA (RFA) (Gov)
2025 M Street NW, Suite 300, Washington, D.C. 20036, USA.
☎ +1 202 5304900. 📠 +1 202 5307794.
E: contact@rfa.org; info@rfa.org **W:** www.rfa.org
Webcast: D/L/P
L.P: Chmn: Victor Ashe; Pres: Libby Liu; Dir, Production: A.J. Janitschek; Dir, Comms/Ext Relations: John A. Estrella.
SAT: NSS 12, Thaicom 5.
kHz: 648, 1098, 1350, 1359, 1503, 5780, 5790, 5810, 5850, 5855, 5860, 6005, 6010, 6095, 7210, 7285, 7355, 7385, 7415, 7445, 7460, 7470, 7480, 7495, 7515, 7540, 7550, 7570, 9325, 9355, 9385, 9435, 9455, 9480, 9645, 9670, 9690, 9725, 9780, 9790, 9815, 9825, 9835, 9875, 9905, 9990, 11585, 11590, 11605, 11695, 11740, 11775, 11790,

11795, 11850, 11880, 11900, 11945, 11965, 11975, 11980, 12010, 12080, 12105, 12115, 12130, 13580, 13625, 13670, 13710, 13725, 13745, 13810, 15120, 15135, 15140, 15150, 15160, 15220, 15375, 15550, 15665, 15690, 15700, 17515, 17615, 17715, 17730, 17750, 17880, 21490, 21540, 21695

Winter Schedule 2010/2011

Burmese	Days	Area	kHz
0030-0130	daily	BRM	12115ira, 13710tin, 15700tin
1230-1330	daily	BRM	7515tin, 11795tin, 12105ira
1330-1400	daily	BRM	7515tin, 11795ira
1330-1430	daily	BRM	12105ira
1400-1430	daily	BRM	11795kwt
1630-1730	daily	BRM	7570tin

Cantonese	Days	Area	kHz
1400-1500	daily	CHN	5810tin, 7470tin
2200-2300	daily	CHN	9780tin, 11740kwt, 11775tin

Khmer	Days	Area	kHz
1230-1330	daily	CBG	13810ira, 15160tin
2230-2330	daily	CBG	5790ira, 9355ira, 11850tin

Korean	Days	Area	kHz
1500-1700	daily	KRE	5860tin, 7210irk, 9385tin
1500-1900	daily	KRE	648vld
1700-1900	daily	KRE	5860tin, 9385ira
2100-2200	daily	KRE	648vld, 1350cho, 5860tin, 7460uba, 9385tin

Lao	Days	Area	kHz
0000-0100	daily	LAO	9815sai, 15690tin
1100-1130	daily	LAO	15120ira
1100-1200	daily	LAO	9325ira
1130-1200	daily	LAO	15120sai

Mandarin	Days	Area	kHz
0300-0700	daily	CHN	11980dsb, 13710tin, 15150tin, 15665sai, 17615sai, 17880tin, 21540tin
1500-1600	daily	CHN	9790sai
1500-1700	daily	CHN	13725tin
1500-1800	daily	CHN	9905hbn
1500-1900	daily	CHN	7445tin
1500-2200	daily	CHN	5810tin, 11945dsb
1600-1900	daily	CHN	7415tin
1600-2200	daily	CHN	9455sai
1700-1900	daily	CHN	13670tin
1700-2200	daily	CHN	9355sai
1800-2000	daily	CHN	7385tai, 9905sai, 11790sai
1900-2100	daily	CHN	5860tin
1900-2200	daily	CHN	1098tai, 1098tai, 6095tin, 9875hbn
2000-2100	daily	CHN	11900sai
2000-2200	daily	CHN	7355tai, 7495tin
2100-2200	daily	CHN	11900tin
2300-2400	daily	CHN	7540dsb, 9825tin, 11775tin, 11975sai, 13745kwt, 15550tin

Tibetan	Days	Area	kHz
0100-0300	daily	CHN	7470kwt, 9670dsb, 11695dha, 15220tin, 17730uba
0600-0700	daily	CHN	17515dsb, 17715kwt, 21490tin, 21695dha
1000-1100	daily	CHN	9690sit, 15140lam, 17750kwt
1100-1200	daily	CHN	15375dha
1100-1400	daily	CHN	7470uba, 9435dsb, 11590kwt
1200-1400	daily	CHN	13625tin, 15375dsb
1500-1600	daily	CHN	5780dsb, 7470kwt, 11585kwt, 11880dha

Tibetan	Days	Area	kHz
2200-2300	daily	CHN	6005tin, 9835lam
2200-2400	daily	CHN	7470dsb
2300-2400	daily	CHN	6010dha, 7550kwt, 9875sit

Uyghur	Days	Area	kHz
0100-0200	daily	CHN	7480dsb, 9480sit, 9645dha, 9690dha, 12010tin
1600-1700	daily	CHN	7285dsb, 7470ira, 9725dha, 12080sai

Vietnamese	Days	Area	kHz
1400-1430	daily	VTN	1503tai, 1503tai
1400-1500	daily	VTN	5855tin, 7515tin, 9990sai, 11605tai, 12130ira, 13580ira
2300-2330	daily	VTN	1359tai
2300-2400	daily	VTN	1359tai
2330-0030	daily	VTN	5850ira, 11605tai, 11965tin, 15135tin
2330-2400	daily	VTN	1359tai

Ann: At the start of the transmission period on each frequency in English: "This is Radio Free Asia. The following program is in ...".
V: QSL-card. (Rpt to 'Reception Reports', Radio Free Asia, 2025 M. Street NW, Washington, DC 20036, USA. **E:** qsl@rfa.org or www.techweb.rfa.org)
Notes: BBG funded station, launched in September 1996 and aimed at listeners in East & South East Asia. Transmissions are jammed in parts of the target area. Burmese language prgr includes segments in Arakanese, Chin Kachin, Karen, Karenni, Mon and Shan.

BBG - RADIO FREE EUROPE/RADIO LIBERTY (RFE/RL) (Gov)

🖃 1201 Connecticut Avenue NW, Washington, D.C. 20036, USA. (Corporate Office)
☎ +1 202 4576900. 🖷 +1 202 4576992.
E: webteam@rferl.org **W:** www.rferl.org
Webcast: D/L (www.rferl.org/Howtolisten.aspx)
🖃 Vinohradská 159A, 100 00 Prague 10, Czech Republic. (HQ/Studio)
☎ +420 2 21121111. 🖷 +420 2 21123013.
L.P: Pres: Jeffrey Gedmin; Dir, Broadcasting: Michele DuBach; Dir, Communications: Donald Jensen.
SAT: AsiaSat 2, Hot Bird 3, Intelsat 907, NSS 703.
kHz: *612, 864, 5820, 5840, 5885, 5895, 5910, 5925, 5940, 5955, 5990, 6000, 6015, 6055, 6060, 6105, 6120, 6135, 6150, 7215, 7220, 7235, 7260, 7270, 7275, 7285, 7290, 7295, 7305, 7395, 7480, 7550, 9310, 9360, 9405, 9430, 9445, 9465, 9485, 9520, 9525, 9535, 9560, 9570, 9595, 9625, 9640, 9680, 9695, 9715, 9760, 9780, 9790, 9805, 9835, 9850, 11605, 11715, 11730, 11790, 11795, 11805, 11870, 11955, 11980, 11985, 12015, 12025, 13755, 15130, 15230, 15250, 15265, 15285, 15590, 17770*

Winter Schedule 2010/2011

Avar	Days	Area	kHz
0400-0420	daily	Cau	5885kwt, 15230ira
1600-1620	daily	Cau	11605skn, 11730lam

Azeri	Days	Area	kHz
1600-1700	daily	Cau	7480ira, 9485nau

Belarusian	Days	Area	kHz
0400-0500	daily	BLR	6105wer
0400-0600	daily	BLR	612vln, 6120wer
0500-0600	daily	BLR	6105lam
1600-1700	daily	BLR	7220wer
1600-1800	daily	BLR	9520wof
1600-2200	daily	BLR	612vln
1700-1800	daily	BLR	7220bib
1800-2000	daily	BLR	6150lam, 9570lam
2000-2200	daily	BLR	5840kwt, 7220wer

Chechen	Days	Area	kHz
0420-0440	daily	Cau	5885kwt, 15230ira
1620-1640	daily	Cau	11605skn, 11730lam

Circassian	Days	Area	kHz
0440-0500	daily	Cau	5885kwt, 15230ira
1640-1700	daily	Cau	11605skn, 11730lam

Kazakh	Days	Area	kHz
0100-0200	daily	CAs	7235lam, 9790udo
1300-1400	daily	CAs	9445kwt, 15265lam

Kyrgyz	Days	Area	kHz
1200-1230	daily	CAs	11955pht, 13755ira, 15265kwt
1500-1530	daily	CAs	9445lam, 11790wer

Moldovan (Romanian)	Days	Area	kHz
0500-0530	mtwtf..	MDA	5955bib
1600-1630	.....ss	MDA	5910bib
1700-1730	mtwtf..	MDA	6135bib
1900-1930	mtwtf..	MDA	6135bib

Russian	Days	Area	kHz
0400-0500	daily	RUS	6015lam
0400-0600	daily	RUS	5925bib
0400-0700	daily	RUS	9760lam, 17770udo
0500-0600	daily	RUS	9535kwt
0600-0700	daily	RUS	9535bib
0600-0800	daily	RUS	15250pht
0700-0800	daily	RUS	7290lam, 12015kwt, 15285udo
0900-1100	daily	RUS	7220pht, 9360tin, 15130bib
1300-1400	daily	RUS	9715sai, 9850lam
1300-1500	daily	RUS	15130lam
1400-1500	daily	RUS	7305lam, 9715bib
1500-1600	daily	RUS	7270udo, 11870wof, 11985lam
1500-1700	daily	RUS	11805bib
1600-1700	daily	RUS	12015lam, 12025lam
1700-1800	daily	RUS	9405bib, 9640bib, 9805lam, 12025lam
1800-1900	daily	Cau	9525lam*, 9780bib*
1800-1900	daily	RUS	6105bib, 9560bib, 9625lam
1800-2000	daily	RUS	9405wof
1900-2000	daily	CAs	5990lam
1900-2000	daily	RUS	9430bib
2000-2100	daily	RUS	6150lam, 9465lam
2000-2200	daily	RUS	5895udo
2100-2200	daily	RUS	6105pht, 7395kwt

Tajik	Days	Area	kHz
0100-0400	daily	CAs	11795udo
0100-0300	daily	CAs	7275kwt
0300-0400	daily	CAs	9520bib
1400-1500	daily	CAs	7215kwt
1400-1700	daily	CAs	9695lam
1500-1700	daily	CAs	7260udo

Tatar/Bashkir	Days	Area	kHz
0400-0500	daily	RUS	5940lam, 7285lam
0600-0700	daily	RUS	11730kwt
1600-1700	daily	RUS	9310pht, 11980wer
2000-2100	daily	RUS	5990lam

Turkmen	Days	Area	kHz
0200-0300	daily	CAs	864erv, 7295lam
0200-0400	daily	CAs	12015udo
0300-0400	daily	CAs	6000lam
1400-1500	daily	CAs	9445bib
1400-1600	daily	CAs	6055udo
1500-1600	daily	CAs	9835lam
1530-1600	daily	CAs	864erv
1600-1800	daily	CAs	5820kwt, 6060lam

Uzbek	Days	Area	kHz
0200-0300	daily	CAs	9680ira, 12025udo
0200-0400	daily	CAs	15590udo
0300-0400	daily	CAs	9680lam, 12025kwt
1400-1500	daily	CAs	9595wer, 11715lam, 12015wer
1500-1530	daily	CAs	864erv
1600-1700	daily	WAs	7550kwt
1600-1700	daily	CAs	9625bib, 9760lam

Key: * Special service for Caucasus: "Ekho Kavkaza" ("Echo of the Caucasus"). See www.ekhokavkaza.com
Ann: R. Liberty: Belarusian: "Havoryc Radyjo Svaboda"; Kazakh: "Azattyq Radiosinan sövlep turmiz"; Kyrgyz: "Azattiq Radioyosinan söylöbüz"; Russian: "Govorit Radio Svoboda"; Tajik: "Injo Radioi Ozodi"; Tatar/Bashkir: "Azatliq Radiosi söyli"; Turkmen: "Gepleýär

Azatlyk Radiosy"; Ukrainian: "Hovorit Radio Svoboda"; Uzbek: "Ozodlik Radiosidan gapiramiz".
V: QSL-card.
Notes: BBG funded station for listeners in Eastern Europe and the successor states to the former USSR. On air since July 1950.

BBG - RADIO FREE IRAQ (Gov)
⌖ 1201 Connecticut Avenue NW, Washington, D.C. 20036, USA.
☎ +1 202 4576900. 🗎 +1 202 4576992.
E: iraq@rferl.org **W:** www.iraqhurr.org
Webcast: D/L/P
⌖ Vinohradská 159A, 100 00 Prague 10, Czech Republic. (HQ/Studio)
☎ +420 2 21121111. 🗎 +420 2 21123013.
L.P: Dir: Sergei Danilochkin.
SAT: AsiaSat 3S, Hot Bird 8.
kHz: *1593*

	Winter Schedule 2010/2011		
Arabic	**Days**	**Area**	**kHz**
0200-0700	daily	IRQ	1593kwt
1500-1530	daily	IRQ	1593kwt
1830-2000	daily	IRQ	1593kwt
2100-2300	daily	IRQ	1593kwt

Ann: Arabic: "Idha'at al-Iraq al-khar min Prag".
V: QSL-card.
Notes: BBG funded station for listeners in Iraq, launched in October 1998. Produced in the RFE/RL studios in Prague, Czech Republic.

BBG - RADIO MARTÍ (Gov)
⌖ P.O. Box 521868, 2200 NW 72 Avenida, Miami, FL 33152-9998, USA.
☎ +1 305 4377000. 🗎 +1 305 4377016.
E: info@martinoticias.com **W:** www.martinoticias.com
Webcast: L
L.P: Dir: Carlos A García-Pérez.
FM/DAB: FM: 102.5 MHz (Florida Coast, WPIK, 50kW).
SAT: Hispasat 1C, NSS 806.
kHz: *1180, 5745, 5980, 6030, 7365, 7405, 9565, 9825, 11930, 13820, 15330*

	Winter Schedule 2010/2011		
Spanish	**Days**	**Area**	**kHz**
0000-0100	daily	CUB	9825grv
0000-0400	daily	CUB	7365grv
0000-2400	daily	CUB	1180mth
0100-0300	daily	CUB	9825sac
0300-0400	daily	CUB	7405grv
0400-0500	.twtfss	CUB	6030grv, 7365grv
0400-0700	.twtfss	CUB	7405grv
0500-1000	.twtfss	CUB	6030grv
0700-1000	.twtfss	CUB	5980grv
1000-1200	daily	CUB	6030grv
1000-1300	daily	CUB	5980grv
1100-1400	daily	CUB	5745grv
1200-1400	daily	CUB	7405grv
1300-2200	daily	CUB	11930grv
1400-2000	daily	CUB	15330grv
1400-2200	daily	CUB	13820grv
2000-2200	daily	CUB	9565sac
2200-0400	daily	CUB	6030grv
2200-2400	daily	CUB	7405grv, 9565grv

Ann: Spanish: "Radio Martí, retransmitiendo para Cuba desde Miami, Estados Unidos de America".
V: QSL-card.
Notes: BBG funded station for listeners in Cuba, launched in May 1985. Jammed.

BBG - RADIO MASHAAL (Gov)
⌖ 1201 Connecticut Avenue NW, Washington, D.C. 20036, USA.
☎ +1 202 4576900. 🗎 +1 202 4576992.
W: www.mashaalradio.org
Webcast: L/D/P (www.rferl.org/howtolisten/PK/ondemand.html and www.mashaalradio.org/audio/ondemand)
⌖ Vinohradská 159A, 100 00 Prague 10, Czech Republic. (HQ/Studio)
☎ +420 2 21121111.
W: www.mashaalradio.com
L.P: Acting Dir: Amanullah Ghilzai; Engineer: Abdul Rahim.

FM/DAB: FM: 100.5MHz (Gardaiz, Khost and Kunar in Afghanistan, 1100-1300 UTC daily).
SAT: Hot Bird 8.
kHz: *621, 9360, 12130, 13580, 15715, 15750*

Winter Schedule 2010/2011

Pashto	Days	Area	kHz
0400-0900	daily	AFG	15715wer
0400-1000	daily	AFG	13580udo
0400-1300	daily	AFG	12130ira
0400-1300	daily	AFG	621kho
0900-1100	daily	AFG	15715ira
1000-1100	daily	AFG	15750udo
1100-1300	daily	AFG	9360udo, 13580ira

Ann: Pashto: "Daa Mashaal Radyo".
Notes: BBG funded service of RFE/FL for Pashto speaking listeners in the Pakistan/Afghanistan border region. Radio Mashaal was launched on 15 January 2010 in order to counter the growing number of Islamic extremist radio stations in the region.

BBG - RADIO SAWA (Gov)
7600 Boston Boulevard, Springfield, VA 22153, USA.
☎ +1 703 6885200. 🖷 +1 703 6885255.
E: comments@radiosawa.com **W:** www.radiosawa.com
Webcast: D/L
L.P: Dir: Gary Thatcher.
SAT: Badr 4, Hot Bird 8, Intelsat 907, Nilesat 101, NSS12.
kHz: *990, 1170, 1431, 1548*

Winter Schedule 2010/2011

Arabic	Days	Area	kHz
0000-2400	daily	ME	1170dha
0000-2400	daily	IRQ	1548kwt
0000-2400	daily	EGY	990cgr
1645-0400	daily	SDN	1431dji

Ann: Arabic: "Radio Sawa".
V: QSL-card.
Notes: BBG funded station for young Arab listeners in the Middle East & North Africa, launched on 23 March 2002. 24h on satellite & FM.

BBG - VOA ASHNA RADIO (Gov)
Room 3200, 330 Independence Avenue SW, Washington, D.C. 20237, USA.
☎ +1 202 6193136 (Dari); +1 202 0327619. (Pashto)
🖷 +1 202 3825193 (Dari); +1 202 2125260. (Pashto)
E: dari@voanews.com (Dari); pashto@voanews.com (Pashto) **W:** www.voanews.com/dari; www.voanews.com/pashto
Webcast: D/L/P
L.P: Dir: Beth Mendelson.
FM/DAB: FM: 100.5MHz (throughout Afghanistan).
kHz: *1296, 5780, 5925, 7405, 7560, 9335, 9445, 9770, 9975, 11840, 12140*

Winter Schedule 2010/2011

Dari	Days	Area	kHz
0130-0230	daily	AFG	1296kab, 7560ira, 9335kwt
1530-1630	daily	AFG	1296kab, 9770wer, 9975kwt, 12140kwt
1730-1830	daily	AFG	1296kab, 5780kwt, 7560udo, 9445udo
1930-2030	daily	AFG	1296kab, 5780kwt, 7560kwt

English	Days	Area	kHz
2030-0030	daily	AFG	1296kab*, 7405kwt*

Pashto	Days	Area	kHz
0030-0130	daily	AFG	1296kab, 5925ira, 7560kwt
1430-1530	daily	AFG	1296kab, 9335kwt, 11840ira, 12140kwt
1630-1700	daily	AFG	9770wer
1630-1730	daily	AFG	1296kab, 9975kwt, 12140kwt
1700-1730	daily	AFG	9770wer
1830-1930	daily	AFG	1296kab, 5780kwt, 7560udo

Key: * Includes news in Farsi at 0000-0005.
Ann: Dari: "In Radyoi Ashna"; Pashto: "Da VOA Ashna Radyo".
V: QSL-card.

Notes: BBG funded service for listeners in Afghanistan, launched April 2004. Produced in the VOA studios.

BBG - VOA DEEWA RADIO (Gov)
330 Independence Avenue SW, Washington, D.C. 20237, USA.
☎ +1 202 2050403. 🖷 +1 202 3825218.
E: deewaradio@voanews.com
W: www.voadeewaradio.com; www.voanews.com/deewa/news
Webcast: D/L
FM/DAB: FM: 100.5MHz (various locations in Afghanistan).
kHz: *621, 5835, 7455, 7495, 9370, 9380, 9565, 11895*

Winter Schedule 2010/2011

Pashto	Days	Area	kHz
0100-0200	daily	AFG	9370ira, 9380udo
0100-0400	daily	AFG	621kho, 11895ira
0200-0300	daily	AFG	9380udo
0200-0400	daily	AFG	9370ira
0300-0400	daily	AFG	9380udo
1300-1400	daily	AFG	7455ira, 7495udo, 9370ira, 9565ira
1300-1900	daily	AFG	621kho
1400-1500	daily	AFG	7495udo, 9370udo, 9565wer
1400-1600	daily	AFG	7455ira
1500-1530	daily	AFG	9370udo
1500-1600	daily	AFG	7495udo
1500-1800	daily	AFG	5835ira
1530-1600	daily	AFG	9370udo
1600-1800	daily	AFG	9370udo
1600-1900	daily	AFG	7455ira, 7495udo
1800-1900	daily	AFG	5835kwt, 9370ira

Ann: Pashto: "Deewa Radio".
Notes: BBG funded service for Pashto speaking listeners in the Afghanistan-Pakistan border area. Launched 29 September 2006. Produced in the VOA studios.

BBG - VOA RADIO AAP KI DUNYAA (Gov)
330 Independence Avenue SW, Washington, D.C. 20237, USA.
☎ +1 202 6191933.
E: urdu@voanews.com **W:** www.voanews.com/urdu/news
Webcast: D/L/P
L.P: Head of Urdu Service: Dr. Brian Q. Silver.
kHz: *972, 1539, 7480, 9520, 11675, 12020*

Winter Schedule 2010/2011

Urdu	Days	Area	kHz
0100-0200	daily	SAs	9520ira*, 12020udo*
1400-0200	daily	SAs	972dsb*, 1539dha*
1400-1500	daily	SAs	7480ira*, 11675ira*

Key: * Includes 5 minutes of news in English at 1900, 2000, 2100 and 2200. Also relays Deewa Radio at 2005-2100 & 0005-0100.
Ann: Urdu: "Radyo Aap Ki Dunyaa".
V: QSL-card.
Notes: BBG funded station for listeners in Pakistan, launched May 2004. Produced in the VOA studios. Between 1900-0100 a trilingual service is carried, consisting of VOA news in English, VOA R. Aap Ki Dunyaa in Urdu and VOA Deewa R. in Pashto.

BBG - VOA STUDIO 7 (Gov)
Voice of America, Africa Division, 330 Independence Avenue SW, Washington, D.C. 20237, USA.
☎ +1 202 2059942. (Then select #11) 🖷 +1 202 2034230.
E: studio7@voanews.com
W: www.voanews.com/zimbabwe/news
Webcast: D/P
kHz: *909, 4930, 12080, 15775*

Winter Schedule 2010/2011

English	Days	Area	kHz
1720-1740	.....ss	ZWE	909bot, 4930bot, 12080mdc, 15775sao
1730-1800	mtwt...	ZWE	909bot, 4930bot, 12080mdc, 15775sao

English/Ndebele/ Shona	Days	Area	kHz
1800-1830	....f..	ZWE	909bot, 4930bot, 12080sao, 15775sao
1830-1900	mtwtf..	ZWE	909bot, 15775sao

Ndebele	Days	Area	kHz
1730-1800	....f..	ZWE	909bot, 4930bot, 12080mdc, 15775sao
1740-1800	.....ss	ZWE	909bot, 4930bot, 12080mdc, 15775sao
1800-1830	mtwt...	ZWE	909bot, 4930bot, 12080sao, 15775sao
1830-1900	mtwtf..	ZWE	12080sao

Shona	Days	Area	kHz
1700-1720	.....ss	ZWE	909bot, 4930bot, 12080mdc, 15775sao
1700-1730	mtwtf..	ZWE	909bot, 4930bot, 12080mdc, 15775sao

Ann: English: "You are listening to Studio 7, for Zimbabwe, from the Voice of America".
V: QSL-card.
Notes: BBG funded station for listeners in Zimbabwe, launched in April 2003. Produced in the VOA studios.

BBG - VOICE OF AMERICA (VOA) (Gov)
✉ 330 Independence Avenue SW, Washington, D.C. 20237, USA.
☎ +1 202 2034959. (Public Relations) 🖷 +1 202 2034960.
E: askvoa@voanews.com **W:** www.voanews.com
Webcast: D/L/P
L.P: Dir: Danforth Austin; Acting Chief of Staff: Barbara Brady; Associate Dir, Operations: Mark L. Prahl; Exec Editor: Steve Redisch.
SAT: AsiaSat 3S, Eutelsat W7, Hot Bird 8, Intelsat 907, NSS 12/806, Telstar 12.
kHz: 648, 909, 1170, 1188, 1431, 1530, 1575, 1593, 4930, 4940, 4960, 5830, 5835, 5850, 5890, 5930, 5945, 5955, 5960, 5975, 5980, 6020, 6035, 6040, 6045, 6060, 6080, 6105, 6135, 6140, 6170, 6180, 7205, 7220, 7230, 7235, 7255, 7260, 7265, 7280, 7295, 7315, 7325, 7340, 7390, 7405, 7420, 7425, 7430, 7440, 7460, 7465, 7480, 7495, 7520, 7525, 7530, 7560, 7570, 7575, 9310, 9315, 9320, 9325, 9355, 9360, 9395, 9415, 9420, 9430, 9435, 9445, 9480, 9485, 9490, 9495, 9505, 9520, 9530, 9540, 9545, 9555, 9565, 9570, 9585, 9620, 9640, 9645, 9670, 9680, 9690, 9705, 9715, 9725, 9745, 9755, 9760, 9775, 9780, 9785, 9810, 9815, 9825, 9845, 9855, 9860, 9875, 9885, 9930, 9945, 9960, 11560, 11635, 11655, 11675, 11695, 11705, 11720, 11740, 11750, 11765, 11775, 11780, 11805, 11820, 11825, 11840, 11850, 11855, 11885, 11905, 11920, 11925, 11930, 11965, 12005, 12010, 12045, 12055, 12070, 12080, 12120, 12150, 13580, 13600, 13635, 13640, 13650, 13710, 13715, 13740, 13765, 15115, 15165, 15185, 15205, 15225, 15265, 15290, 15385, 15460, 15540, 15550, 15580, 15590, 15620, 15670, 15730, 17645, 17650, 17715, 17740, 17780, 17850, 17860, 17895, 21570, 21580

Winter Schedule 2010/2011

Albanian	Days	Area	kHz
0600-0630	daily	Eu	6035bib
1700-1730	daily	Eu	7235bib
1930-2000	daily	Eu	11740sao

Amharic	Days	Area	kHz
1800-1900	daily	EAf	9320ira, 9485nau, 9860kwt, 11675wer, 11905mey

Azeri	Days	Area	kHz
1830-1900	daily	Cau	7315bib, 9445bib, 9495wer

Bengali	Days	Area	kHz
1600-1700	daily	SAs	1575bph, 7405udo, 11850pht

Burmese	Days	Area	kHz
0000-0030	daily	SEA	1575bph
0130-0300	daily	SEA	12120ira, 15115pht, 17780pht
1130-1230	daily	SEA	11965pht, 15550pht, 17850ira
1430-1500	daily	SEA	1575bph, 12120kwt
1430-1530	daily	SEA	9325pht, 11965pht
1500-1530	daily	SEA	12120pht
1530-1600	daily	SEA	1575bph
1530-1630	daily	SEA	9355pht, 11560pht
2300-0030	daily	SEA	7430pht, 12120pht
2300-2330	daily	SEA	9325ira
2330-0030	daily	SEA	9325pht

Cantonese	Days	Area	kHz
1300-1400	daily	EAs	7390tin, 9705sai

Cantonese	Days	Area	kHz
1300-1500	daily	EAs	1170php
1400-1500	daily	EAs	7390pht, 9705sai

Croatian	Days	Area	kHz
0530-0600	daily	Eu	6035bib
1930-1945	daily	Eu	6135bib, 7235kwt

English	Days	Area	kHz
0030-0100	daily	SEA	1575bph, 9715udo, 11695pht, 15205pht
0030-0100	daily	EAs,SAs,SEA	15290pht
0030-0100	daily	SEA,Pac	15185pht
0030-0100	daily	ME	1593kwt, 6170kwt
0030-0100	daily	EAs	9490pht, 12005pht
0030-0100	daily	SAs,SEA	9325pht
0100-0130	daily	ME	1593kwt
0100-0200	daily	ME	7325ira
0100-0200	daily	SAs	9435udo, 11705udo
0130-0200	.twtfs.	LAm	5960grv, 7465grv
0130-0200	.twtfs.	ME	1593kwt
0300-0430	daily	Af	1530sao, 9885bot
0300-0500	daily	Af	15580bot
0300-0600	daily	Af	909bot, 4930bot, 6080sao
0400-0500	daily	Af	4960sao
0430-0600	daily	Af	9885bot
0500-0600	daily	EAf,ME	15580bot
0600-0700	daily	Af	909bot, 1530sao, 6080sao, 9885mey
0600-0700	daily	EAf,SAf	15580bot
1100-1130	.....ss	SEA	1575bph
1130-1200	.....ss	SEA	1575bph
1200-1300	daily	EAs	1170php, 9640pht, 11750pht
1200-1300	daily	EAs,SEA	7575ira
1200-1300	daily	SEA,Pac	11705udo
1300-1400	.....ss	SEA,Pac	11705udo
1300-1400	.....ss	SAs,SEA	9760pht
1300-1400	.....ss	EAs,SEA	7575pht
1300-1400	.....ss	EAs	9640pht
1400-1500	mtwtf..	SAs	7575udo, 12150ira
1400-1500	mtwtf..	SAs,SEA	9760udo
1400-1500	daily	Af	15580mey, 17650sao, 17715bot
1400-1530	daily	Af	6080bot
1400-1700	daily	Af	4930bot
1500-1600	daily	SAs,SEA	7520pht, 7520pht, 9930ira
1500-1600	daily	Af	15580sao, 17715bot, 17895sao
1500-1600	daily	EAs	6140udo, 9760pht, 9945ira
1500-1600	daily	ME	11765lam, 12055lam
1500-1600	daily	SAs	7575udo, 12150ira
1530-1700	daily	Af	6080mey
1600-1700	mtwtf..	EAs	1170php
1600-1700	daily	Af	909bot, 1530sao, 9395ira, 13600sao, 15460lam, 15580sao, 17895bot
1630-1700	mtwtf..	Af	9785nau*, 11905wer*, 13635wer*
1700-1730	daily	EAf	15580grv
1700-1800	daily	Af	6080sao, 17895bon
1700-1800	daily	ME	13635lam
1730-1800	daily	EAf	15580bot
1800-1830	.....ss	Af	909bot
1800-1830	.....ss	SAf	4930bot
1800-1830	daily	Af	13635ira, 15580bot
1800-2000	daily	Af	6080sao
1830-1900	.....ss	Af	909bot
1830-1900	daily	Af	13635ira
1830-1900	.....ss	SAf	909bot
1830-1930	daily	Af	15580bot
1830-2100	daily	Af	4930bot
1900-2000	daily	ME	7480ira, 9585bib

English

English	Days	Area	kHz
1900-2030	daily	Af	4940sao
1900-2100	daily	Af	909bot
1930-2000	daily	Af	15580bon
2000-2100	daily	Af	1530sao, 15580bon
2000-2100	mtwtf..	ME	9420bib, 9490lam
2000-2200	daily	Af	6080sao
2030-2100	.....ss	Af	4940sao
2100-2200	daily	Af	1530sao, 15580bot
2200-2300	mtwt..s	EAs	7220pht, 7570pht
2200-2300	mtwt..s	SAs,SEA	5835pht
2200-2300	mtwt..s	SEA,Pac	9490sai
2200-2300	mtwt..s	CAs	7425kwt
2230-2300	daily	EAs	5850pht, 7230udo, 9570ira
2230-2400	....fs.	SEA	1575bph
2300-0030	daily	ME	1593kwt
2300-2400	daily	EAs	6180udo, 7220pht, 7460pht, 7570pht, 11655pht, 11840pht
2300-2400	daily	SAs,SEA	7480pht
2300-2400	daily	SEA	5830pht
2300-2400	daily	SEA,Pac	9490sai
2330-2400	daily	SEA	13640tin

Farsi

Farsi	Days	Area	kHz
0230-0330	daily	IRN	9495wer
0230-0330	daily	ME	7205wer, 9745ira
1530-1630	daily	ME	11775bib
1530-1700	daily	ME	11705bib
1530-1730	daily	ME	9320udo
1530-1830	daily	ME	1593kwt
1630-1730	daily	ME	9540wer
1700-1830	daily	ME	9760bib
1730-1830	daily	ME	9680nau, 9825wer
1800-1900	daily	ME	648dsb
1830-1930	daily	ME	9680nau, 9825wer
1900-1930	daily	ME	5850ira, 5850ira

French

French	Days	Area	kHz
0530-0600	mtwtf..	Af	1530sao
0530-0630	mtwtf..	Af	4960sao, 6020bot, 7265mdc, 9480bot, 9505sao
1830-1900	daily	Af	1530sao, 15225bon, 15620bot
1900-2000	daily	Af	1530sao, 15225bon
2000-2030	daily	Af	9780sao, 9815bot, 12080bot, 15225bon, 15620sao
2030-2100	.....ss	Af	6040sao, 9775sao, 9815bot, 12080bot, 15225bot
2100-2130	mtwtf..	Af	9435skn, 9680mdc, 9780sao, 9815bot

Georgian

Georgian	Days	Area	kHz
1600-1700	daily	Cau	7390wer
1600-1800	daily	Cau	11840bib
1700-1800	daily	Cau	9310udo

Hausa

Hausa	Days	Area	kHz
0500-0530	daily	WAf	1530sao, 4960sao, 6040sao, 9780bot
0700-0730	daily	WAf	4960sao, 12070sao, 15620bot
1500-1530	daily	WAf	9780sao, 11705sao, 15620bot
2030-2100	mtwtf..	WAf	4940sao, 6035sao, 9780mey, 11705nau, 11885smg

Indonesian

Indonesian	Days	Area	kHz
1130-1230	daily	SEA	7255udo, 9725pht, 15165pht
1400-1500	...tfs.	SEA	9360pht, 11635sai
2200-0030	daily	SEA	9620udo, 15205pht
2200-2400	daily	SEA	11805pht

Khmer

Khmer	Days	Area	kHz
1330-1430	daily	SEA	1575bph, 9325pht, 11965pht
2200-2230	daily	SEA	1575bph, 6060pht, 7260pht, 13640tin

Kinyarwanda/ Kirundi

Kinyarwanda/ Kirundi	Days	Area	kHz
0330-0400	daily	CAf	9540sao
0330-0430	daily	CAf	7340bot, 11750sao
0400-0430	daily	CAf	9540sao
1600-1630	.....s.	CAf	11750mey, 12010sao, 15730bot

Korean

Korean	Days	Area	kHz
1200-1300	daily	EAs	9555sai
1200-1500	daily	EAs	1188seo, 5890tin, 7235tin
1300-1500	daily	EAs	9555pht
1900-2100	daily	EAs	648vld, 5835pht, 6060udo, 7420udo

Kurdish

Kurdish	Days	Area	kHz
0500-0600	daily	ME	5945bib, 9430nau, 9690lam
1400-1500	daily	ME	1593kwt, 11805bib, 13710wer, 15265bot
1700-1800	daily	ME	7480ira, 11820lam, 11855bib
2000-2100	daily	ME	1593kwt

Lao

Lao	Days	Area	kHz
1230-1300	daily	SEA	1575bph, 9810sai, 11930pht

Mandarin

Mandarin	Days	Area	kHz
0000-0100	daily	EAs	7495udo, 9545pht, 11925pht
0000-0300	daily	EAs	15385pht, 17645pht, 21580tin
0100-0200	daily	EAs	7495pht, 9545pht
0100-0300	daily	EAs	11925pht
0700-1000	daily	EAs	11855udo
0700-1100	daily	EAs	9845pht, 13650tin, 13765udo
0700-1200	daily	EAs	15670udo
0800-1100	daily	EAs	11965udo
0800-1200	...tfs.	EAs	11720dsb
0800-1200	..w....	EAs	11720pht
0900-1100	daily	EAs	9855tin
1000-1400	daily	EAs	9530pht
1100-1200	daily	EAs	9825sai
1100-1300	daily	EAs	12045pht
1100-1500	daily	EAs	9785pht
1200-1300	daily	EAs	6040udo
1200-1400	daily	EAs	11635udo
1200-1500	daily	EAs	9825pht
1300-1400	daily	EAs	6040udo, 12045sai
1300-1500	daily	EAs	7295nvs
1400-1500	daily	EAs	6040pht, 6105dsb, 7525pht
2200-2300	daily	EAs	6045udo, 7440udo, 9545pht, 9755udo, 9875udo, 11925pht

Oromo

Oromo	Days	Area	kHz
1730-1800	mtwtf..	EAf	9320ira, 9485nau, 9860kwt, 11675bot, 11905wer

Portuguese

Portuguese	Days	Area	kHz
1000-1030	.....ss	SAf	11825mdc, 17850ira
1700-1730	daily	Af	15670sao
1700-1800	daily	Af	1530sao, 9395bot, 17740grv
1730-1800	daily	Af	15670sao
1800-1830	mtwtf..	Af	1530sao, 15670bot, 17740grv

Somali

Somali	Days	Area	kHz
0330-0400	daily	EAf	5975lam, 11780ira, 15620ira
1300-1400	daily	EAf	13580lam, 15620mdc
1600-1630	daily	EAf	1431dji, 15620bot
1600-1730	daily	EAf	13580ira

Somali	Days	Area	kHz
1630-1700	daily	EAf	15620wer
1700-1730	daily	EAf	15620ira
1730-1800	daily	EAf	13580ira, 15620sao

Spanish	Days	Area	kHz
0000-0100	daily	LAm	5890grv, 9725grv, 9885grv
0100-0200	.twtfs.	LAm	5890grv, 9725grv, 9885grv
1230-1300	mtwtf	LAm	9885grv, 13715grv, 15590grv
1300-1400	daily	LAm	9885grv, 13715grv, 15590grv

Swahili	Days	Area	kHz
1630-1700	daily	EAf	9565bot, 13740bot, 15730sao

Tibetan	Days	Area	kHz
0000-0100	daily	SAs	5980ira, 7255udo, 9645udo
0300-0400	daily	SAs	15540pht
0300-0600	daily	SAs	17860udo, 21570pht
0400-0600	daily	SAs	15540udo
1400-1500	daily	SAs	7255kwt, 7280pht, 9315udo, 9670lam
1600-1700	daily	SAs	7530pht, 7560udo, 11920pht

Tigrinya	Days	Area	kHz
1900-1930	mtwtf..	EAf	9320kwt, 9485nau, 9860smg, 11675ira, 11905dha

Uzbek	Days	Area	kHz
1500-1530	daily	CAs	5930wer, 6105kwt, 9960tin
1500-1530	daily	WAs	9415pht

Vietnamese	Days	Area	kHz
1300-1330	daily	SEA	1575bph, 9325pht, 11695pht
1500-1600	daily	SEA	1170php, 5955pht, 9520ira, 9725pht
2230-2330	daily	SEA	6060pht, 13640tin

Key: * Special prgr for Sudan.

Ann: At the start and end of the transmission period on each frequency, English: "This is the Voice of America, Washington D.C., signing on/off". Before all foreign language programs: "This is the Voice of America. The following program is in... (language)".

V: QSL-card. (Email to: letters@voa.gov)

Notes: Launched in 1942, under the roof of the U.S. Foreign Information Service (FIS). From 1953-1994, financed by the U.S. Information Agency (USIA). BBG funded since April 1994. Some transmissions in Asian languages are jammed. Some programs in Portuguese, directed to Angola, are from the "Vision Angola/VOA Multipress" service.

BROADCASTING BOARD OF GOVERNORS (BBG) (Gov)
▣ 330 Independence Avenue SW, Washington, D.C. 20237, USA.
☎ +1 202 2034400. 🖷 +1 202 2034585.
E: publicaffairs@bbg.gov **W:** www.bbg.gov
L.P: Chmn: Walter Isaacson; Exec Dir: Jeffrey N. Trimble; CFO: Maryjean Buhler.
Notes: On 1 October 1999, the Broadcasting Board of Governors (BBG) became the independent, autonomous entity responsible for all U.S. government and government sponsored, non-military, international broadcasting.

INTERNATIONAL BROADCASTING BUREAU (IBB) (Gov)
▣ 330 Independence Avenue SW, Washington, D.C. 20237, USA.
☎ +1 202 4017000. 🖷 +1 202 6191241.
E: pubaff@ibb.gov **W:** www.ibb.gov
▣ 3919 VOA Site B Road, Grimesland, NC 27837, USA. (Greenville Transmitting Site)
L.P: Dir: Richard M Lobo; Acting Deputy Dir: Danforth Austin; Dir, Engineering/Technical Sces: Andre V. Mendes; Chief of Staff: Maria Skiba Lennon; SM, Greenville Transmitter site: David Strawman.
MW: [MTH] Marathon Key, FL: 1180kHz 1 x 100kW.
SW: [GRV] Greenville, NC: 6 x 250 (including 3 x 500kW units running at 250kW), 2 x 500kW (Due to close in 2011).
V: QSL-card.
Notes: Under the supervision of the Broadcasting Board of Governors

(BBG), the International Broadcasting Bureau (IBB) provides the administrative and engineering support for U.S. government funded non-military international broadcast services. The IBB Office of Engineering and Technical Services manages, operates, and maintains a network of domestic and overseas transmitting stations in Botswana, Djibouti, Germany, Kuwait, Philippines, Northern Mariana Islands, São Tomé, Sri Lanka, Thailand and USA.

THE DISCO PALACE (Comm)
▣ c/o Alyx & Yeyi, LLC, 5201 Blue Lagoon Drive, 8th Floor, Miami, FL 33126, USA.
☎ +1 305 5728070. 🖷 +1 305 5728674.
E: info@thediscopalace.com **W:** www.thediscopalace.com
Webcast: L
kHz: 6015, 15755

Winter Schedule 2010/2011

English	Days	Area	kHz
1400-1500	daily	Eu	6015iss+
2000-2100	daily	NAm	15755bon+

Key: + DRM.
Ann: English: "Feel the Music- The Disco Palace!".
Notes: 24hr Internet radio station that began broadcasting on SW (DRM only) in February 2010. Plays music from the 1970 and 80s disco era. Technical services are provided by TDP.

ADVENTIST WORLD RADIO (AWR) (Rlg)
▣ 12501 Old Columbia Pike, Silver Spring, ML 20904-6600, USA.
☎ +1 301 6806304. 🖷 +1 301 6806303.
E: info@awr.org **W:** www.awr.org
Webcast: D/L/P
▣ Milbanke Court, Milbanke Way, Bracknell, Berkshire RG12 1RP, United Kingdom. (Europe/Africa Regions)
☎ +44 1344 401401. 🖷 +44 1344 401419.
L.P: Pres: Elder Dowell W. Chow; Regional Dir, Africa: Ray Allen; Regional Dir, Americas: Dowell W. Chow; Regional Dir, Asia/Pacific: Jonathan Wagiran; Regional Dir, Europe: Tihomir Zestic; Frequency Manager: Claudius Dedio.
SAT: Hot Bird 6.
kHz: 3215, 3345, 5970, 5975, 6045, 6090, 6100, 7315, 7370, 9505, 9515, 9535, 9595, 9605, 9610, 9625, 9770, 9805, 9830, 11675, 11725, 11750, 11755, 11760, 11775, 11795, 11800, 11830, 11895, 11925, 11955, 11975, 12010, 13755, 15155, 15240, 15440, 15445, 15495, 17575, 17605, 17670

Winter Schedule 2010/2011

Acholi	Days	Area	kHz
1800-1830	.....s.	EAf	9515mos

Afar	Days	Area	kHz
1430-1500	mtwtf.s	EAf	17605mos

Amharic	Days	Area	kHz
0330-0400	daily	EAf	7370wer

Arabic	Days	Area	kHz
0700-0800	daily	NAf	11975wer
1830-1900	daily	NAf	9605mos
1900-1930	daily	NAf	11760wer
1900-2000	daily	As	9535nau
1900-2000	daily	ME	15155mey
1900-2100	daily	NAf,ME	11800mey
2000-2100	daily	ME	15155mey

Arabic (Juba)	Days	Area	kHz
1800-1830	.t.....	EAf	9515mos

Bari	Days	Area	kHz
1800-1830	m......	EAf	9515mos

Bengali	Days	Area	kHz
1230-1300	daily	As	15495nau

Bulgarian	Days	Area	kHz
0400-0430	daily	Eu	5975wer
1600-1630	daily	Eu	6100wer

Dinka	Days	Area	kHz
1800-1830	...t...	EAf	9515mos

Dyula	Days	Area	kHz
2000-2030	daily	WAf	9770mos

English	Days	Area	kHz
1200-1230	daily	As	15495nau
1530-1600	daily	As	11675wer
1800-1830	daily	SAf	3215mey, 3345mey
1830-1900	daily	EAf	11830mey
2100-2130	daily	WAf	9830mos

English (Colloquial)	Days	Area	kHz
1800-1830	..w....	EAf	9515mos
Farsi	**Days**	**Area**	**kHz**
0330-0430	daily	ME	6090mos
1630-1730	daily	ME	9830mos
French	**Days**	**Area**	**kHz**
0430-0500	daily	NAf	6045mos
0800-0830	daily	NAf	12010wer
1930-2000	daily	CAf	9625mos
2000-2030	daily	CAf	11755mey
2000-2030	daily	NAf	9805wer
2030-2100	daily	WAf	9805mos
Fulfulde	**Days**	**Area**	**kHz**
1900-1930	daily	WAf	15240mey
Hausa	**Days**	**Area**	**kHz**
1900-1930	daily	WAf	9625mos
Hindi	**Days**	**Area**	**kHz**
1530-1600	daily	IND	11895wer
Ibo	**Days**	**Area**	**kHz**
1930-2000	daily	CAf	11750mey
Italian	**Days**	**Area**	**kHz**
1000-1100	......s	Eu	9610nau
Kabyle	**Days**	**Area**	**kHz**
0800-0830	daily	NAf	11975wer
1730-1800	daily	NAf	9595wer
Malagasy	**Days**	**Area**	**kHz**
0230-0330	daily	SAf	3215mdc
1430-1528	daily	SAf	3215mdc
Mandarin	**Days**	**Area**	**kHz**
1300-1330	mtwtf..	WAf	13755nau
1330-1500	daily	CHN	11725nau
Masai	**Days**	**Area**	**kHz**
1730-1800	daily	EAf,CAf	11925mey
Moro	**Days**	**Area**	**kHz**
1800-1830	......s	EAf	9515mos
Nepali	**Days**	**Area**	**kHz**
1500-1530	daily	As	11675wer
Oromo	**Days**	**Area**	**kHz**
0300-0330	daily	EAf	7370wer
1730-1800	daily	EAf	11795wer
Punjabi	**Days**	**Area**	**kHz**
0230-0300	daily	PAK	5970mos
1500-1530	daily	IND	11955wer
Somali	**Days**	**Area**	**kHz**
1630-1700	daily	EAf	17575iss
Swahili	**Days**	**Area**	**kHz**
1700-1730	daily	EAf,CAf	11925mey
Tachelhit	**Days**	**Area**	**kHz**
0830-0900	daily	NAf	12010wer
1930-2000	daily	NAf	11760wer
Tigrinya	**Days**	**Area**	**kHz**
0300-0330	daily	EAf	7315wer
Turkish	**Days**	**Area**	**kHz**
1500-1530	daily	ME	11775mos
Urdu	**Days**	**Area**	**kHz**
0200-0230	daily	PAK	5970mos
1400-1430	daily	PAK	15440mos
1600-1630	daily	PAK	9505mos
Uyghur	**Days**	**Area**	**kHz**
1300-1330	.....ss	CHN	13755nau
Vietnamese	**Days**	**Area**	**kHz**
0100-0200	.....s.	VTN	15445tai
1300-1400	daily	VTN	17670mdc
Yoruba	**Days**	**Area**	**kHz**
2030-2100	daily	WAf	11755mey
Zande	**Days**	**Area**	**kHz**
1800-1830	....f..	EAf	9515mos

Ann: English: "You're listening to Adventist World Radio, the Voice of Hope"; French: "Ici la Radio Mondiale Adventiste, la Voix de l'Esperance"; German: "Sie hören Adventist World Radio, die Stimme der Hoffnung"; Italian: "La Voce della Speranza".
IS: Various arrangements of the melody "Lift Up the Trumpet".
V: QSL-card.

Notes: Owned by Adventist Broadcasting Service, Inc. AWR owns the transmitting station KSDA in Guam.

CHRISTIAN SCIENCE SENTINEL (Rlg)
⌂ 1 Norway Street C04-10, Boston, MA 02115-3195, USA.
☎ +1 617 4502893. 🖶 +1 617 4502893.
E: herald@csps.com **W:** www.sentinelradio.com
Webcast: D
kHz: 5960, 6055

Winter Schedule 2010/2011

German	Days	Area	kHz
1000-1100	......s	Eu	6055wer
Russian	**Days**	**Area**	**kHz**
1900-2000	.....s.	Eu	5960wer

Notes: Produced by the Christian Science Publishing Society (The First Church of Christ, Scientist).

ETERNAL GOOD NEWS (Rlg)
⌂ 400 E. Wilshire Boulevard, Oklahoma City, OK 73105, USA.
☎ +1 405 3591235; +1 405 3400877.
E: eternalgoodnews@sbcglobal.net
W: www.oldpaths.net/works/radio/wilshire
Webcast: D
LP: Producer/Preacher: Brother Germaine Lockwood.
kHz: 15525

Winter Schedule 2010/2011

English	Days	Area	kHz
1130-1145	.....f..	SAs	15525dha

V: QSL-letter. (Rpt to speaker, Mr.George Bryan. Email rpt to: gabry@juno.com)
Notes: Produced by the Wilshire Church of Christ in Oklahoma City, USA.

FAMILY RADIO WORLDWIDE (Rlg)
⌂ 290 Hegenberger Rd., Oakland, CA 94621, USA.
☎ +1 510 5686200. 🖶 +1 510 6337983.
E: international@familyradio.com; info@familyradio.org
W: www.familyradio.com
Webcast: D/L
⌂ 10400 NW 240th Street, Okeechobee, FL 34972, USA (Studio)
☎ +1 863 7630281. 🖶 +1 863 7631034.
LP: Pres: Harold Camping; International Mgr: David Hoff.
SW: [YFR] Okeechobee, FL: 2 x 50, 10 x 100kW & foreign relays.
SAT: Astra 2B, Atlantic Bird 4A, Hot Bird 6, Intelsat 10.
kHz: 747, 863, 873, 1197, 1359, 1503, 1557, 3230, 3955, 3975, 5745, 5820, 5825, 5900, 5930, 5950, 5960, 5985, 5995, 6000, 6005, 6010, 6020, 6045, 6050, 6070, 6075, 6085, 6090, 6100, 6105, 6115, 6120, 6140, 6150, 6220, 6225, 6230, 6240, 6280, 6875, 6885, 6890, 6915, 7220, 7230, 7240, 7260, 7265, 7295, 7300, 7305, 7310, 7340, 7360, 7385, 7395, 7455, 7460, 7490, 7510, 7520, 7550, 7560, 7565, 7570, 7590, 7600, 7730, 9310, 9320, 9355, 9405, 9430, 9440, 9450, 9455, 9460, 9465, 9480, 9485, 9495, 9500, 9505, 9515, 9525, 9530, 9535, 9540, 9545, 9555, 9575, 9590, 9595, 9605, 9630, 9660, 9680, 9685, 9690, 9695, 9705, 9715, 9720, 9770, 9800, 9840, 9850, 9855, 9885, 9895, 9920, 9930, 9945, 9955, 9960, 9985, 11520, 11530, 11535, 11550, 11560, 11565, 11570, 11580, 11610, 11615, 11630, 11665, 11690, 11700, 11720, 11725, 11730, 11740, 11820, 11825, 11830, 11855, 11865, 11875, 11885, 11895, 11935, 11955, 11970, 11975, 11995, 12015, 13605, 13615, 13655, 13660, 13695, 13700, 13740, 13820, 15115, 15130, 15195, 15210, 15250, 15315, 15325, 15355, 15400, 15440, 15520, 15565, 15795, 17505, 17535, 17540, 17545, 17555, 17575, 17660, 17690, 17735, 17760, 17810, 18930, 18980, 21455, 21680, 21745, 21840

Winter Schedule 2010/2011

Amharic	Days	Area	kHz
1600-1700	daily	EAf	11955iss
1700-1800	daily	EAf	6045dha
Arabic	**Days**	**Area**	**kHz**
0500-0600	daily	ME,NAf	7520yfr, 11580yfr
0700-0800	daily	ME,NAf	9985yfr
1600-1645	daily	ME,NAf	15250yfr
1600-1700	daily	ME,NAf	9430wer, 11995wer
1700-1800	daily	ME,NAf	9530skn, 9850wer, 11690wer
1800-1900	daily	ME,NAf	7220wer, 9660skn, 9840wer
1900-2000	daily	ME,NAf	5745wer, 9500wer
2000-2100	daily	ME,NAf	9515nau, 17690yfr

Arabic

	Days	Area	kHz
2100-2200	daily	ME,NAf	6010wer, 11665yfr
2100-2300	daily	ME,NAf	5960nau
2200-2245	daily	ME,NAf	15115yfr

Assamese

	Days	Area	kHz
1400-1500	daily	IND	9440arm

Bengali

	Days	Area	kHz
1300-1500	daily	As	13820nau

Bulgarian

	Days	Area	kHz
1800-1900	daily	Eu	7600erv

Burmese

	Days	Area	kHz
1100-1200	daily	As	6220huw
1200-1300	daily	As	11570huw
1300-1400	daily	As	7560alm
1800-1900	daily	As	1503fan

Cantonese

	Days	Area	kHz
0800-0900	daily	EAs,SEA	1557kou

Cebuano

	Days	Area	kHz
1200-1300	daily	SEA	5900ppk

Czech

	Days	Area	kHz
1800-1900	daily	Eu	6090rmp

English

	Days	Area	kHz
0000-0100	daily	CAm,SAm	7360yfr, 11720yfr, 11730yfr
0000-0100	daily	NAm	5950yfr, 6085yfr
0000-0200	daily	NAm	15440yfr
0000-0500	daily	NAm	9505yfr
0100-0200	daily	CAm,SAm	6100yfr
0100-0500	daily	NAm	7455yfr
0200-0300	daily	NAm	9525yfr
0200-0300	daily	CAm,SAm	5930yfr, 6885yfr, 6890yfr
0300-0400	daily	CAm,SAm	9930yfr, 9985yfr
0300-0500	daily	SAf	1197mas
0400-0500	daily	NAm	9715yfr
0400-0600	daily	NAm	5950yfr
0400-0700	daily	NAm	9680yfr
0600-0700	daily	Eu	11530yfr
0600-0700	daily	CAm,SAm	6000yfr
0600-0700	daily	Af	9985yfr
0600-0800	daily	Eu	5745yfr
0700-0800	daily	CAm,SAm	9495yfr
0700-0900	daily	Af	11580yfr
0700-1000	daily	NAm	6875yfr
0700-1400	daily	NAm	7455yfr
0900-1100	daily	SEA	9465pao
1000-1100	daily	EAs	9460irk, 9460nvs
1000-1400	daily	NAm	6890yfr
1100-1200	daily	CAm,SAm	6000yfr, 11725yfr, 11830yfr
1100-1200	daily	EAs	7300irk
1100-1200	daily	NAm	6875yfr
1200-1300	daily	CAm,SAm	11530yfr, 17545yfr
1200-1400	daily	NAm	11970yfr
1300-1400	daily	NAm	11830yfr
1300-1400	daily	SEA	6075tch, 9310alm, 11520pao
1300-1500	daily	SEA	11560huw
1300-1600	daily	NAm	11855yfr
1400-1500	daily	EAs	5995ppk, 5995ppk, 6115ppk, 6115ppk
1400-1500	daily	SEA	6070tch
1400-1500	daily	NAm	13695yfr
1400-1500	daily	IND	9485irk
1400-1700	daily	EAs,SEA	1557kou
1400-1700	daily	NAm	11565yfr, 17760yfr
1500-1600	daily	CAm,SAm	15210yfr
1500-1600	daily	IND	6280tsh, 9495dha, 12015dha, 21840asc
1500-1600	daily	NAm	15795yfr
1600-1700	daily	CAm,SAm	6085yfr
1600-1700	daily	IND	11740dha
1600-1700	daily	Af	17540yfr, 17690yfr
1600-1700	daily	RUS	1503msk
1600-1800	daily	Eu	18980yfr

English (continued)

	Days	Area	kHz
1600-1900	daily	SAf	1197mas
1600-2000	daily	NAm	13695yfr
1700-1800	daily	Af	7230mdc, 7385mdc, 21680yfr
1700-1800	daily	NAm	15795yfr
1700-2000	daily	EAs	1359fan
1700-2000	daily	SEA	1359fan
1700-2200	daily	NAm	17555yfr
1800-1900	daily	Af	6045mey, 7240skn, 9895dha, 11665wer
1800-2000	daily	Af	7395mdc
1800-2200	daily	Af	15115yfr
1800-2200	daily	Eu	6915yfr
1800-2200	daily	NAm	17535yfr
1900-2000	daily	Af	3230mey, 6020mdc, 9480yfr, 9705yfr, 9885dha
1900-2000	daily	CAm,SAm	6085yfr
1900-2000	daily	Eu	15565yfr
1900-2200	daily	Af	9480nau
1900-2200	daily	EAs	1557kou
2000-2100	daily	CAm,SAm	17555yfr
2000-2100	daily	EAs	1503fan
2000-2100	daily	Eu	5745yfr, 9850arm
2000-2100	daily	RUS	1503msk
2000-2100	daily	SEA	1503fan
2000-2100	daily	Af	11615asc, 15520asc
2000-2200	daily	Af	15195asc
2000-2200	daily	Eu	7510alm
2000-2200	daily	SAf	1197mas
2100-2400	daily	NAm	5950yfr
2200-2300	daily	Af	17690yfr
2200-2400	daily	NAm	11740yfr, 15440yfr
2300-2400	daily	CAm,SAm	9430yfr, 15400yfr

Farsi

	Days	Area	kHz
1600-1700	daily	ME	11885nau
1700-1800	daily	ME	6105nau

French

	Days	Area	kHz
0000-0100	daily	CAm	15400yfr
0500-0600	daily	Eu,Af	9985yfr, 11530yfr
0600-0700	daily	Eu,Af	7520yfr, 11580yfr
0800-0900	daily	Eu,Af	9985yfr
1000-1100	daily	CAm	9680yfr, 11740yfr
1200-1300	daily	NAm	13695yfr
1300-1400	daily	CAm	11740yfr
1600-1700	daily	NAm	11855yfr
1700-1800	daily	Eu,Af	6225mey, 15115yfr
1800-1900	daily	Eu,Af	15565yfr, 17690yfr
1830-1930	daily	Eu,Af	17660asc
1900-2000	daily	Eu,Af	9695wer, 15795yfr, 17690yfr
2000-2100	daily	Eu,Af	7590alm, 9595wer
2100-2200	daily	Eu,Af	7305wer
2100-2200	daily	CAm	17575yfr
2200-2300	daily	Eu,Af	9355yfr
2300-2400	daily	NAm	6085yfr

German

	Days	Area	kHz
0500-0600	daily	Eu	7730yfr
0700-0800	daily	Eu	11530yfr
1700-1800	daily	Eu	17760yfr
1800-1900	daily	Eu	7490erv, 15795yfr, 21455yfr
1900-2000	daily	Eu	7490erv
2000-2100	daily	Eu	11565yfr

Greek

	Days	Area	kHz
1800-1900	daily	Eu	7240skn

Gujarati

	Days	Area	kHz
1500-1600	daily	IND	9800nau, 11610wer

Hausa

	Days	Area	kHz
1800-1900	daily	Af	9535nau
1900-2000	daily	Af	9685dha

Hindi

	Days	Area	kHz
1400-1500	daily	IND	15520dha
1400-1600	daily	IND	13700nau

Hindi	Days	Area	kHz
1600-1700	daily	IND	6280tsh, 9405wer
Hungarian	**Days**	**Area**	**kHz**
1800-1900	daily	Eu	3975wer
Igbo	**Days**	**Area**	**kHz**
1800-1900	daily	Af	11875asc
Indonesian	**Days**	**Area**	**kHz**
0000-0100	daily	SEA	11865pao
1100-1200	daily	SEA	11550tai
1200-1300	daily	SEA	11520pao
1200-1400	daily	SEA	9485irk
1400-1500	daily	SEA	1359fan
2300-2400	daily	SEA	1359fan
Italian	**Days**	**Area**	**kHz**
0600-0700	daily	Eu	9355yfr
1700-1800	daily	Eu	18930yfr
1800-1900	daily	Eu	17760yfr
1900-2000	daily	Eu	6000msk
Japanese	**Days**	**Area**	**kHz**
1000-1100	daily	J	7265nvs
Kannada	**Days**	**Area**	**kHz**
1300-1400	daily	As	17735dha
1500-1600	daily	As	13655wer
Khmer	**Days**	**Area**	**kHz**
1200-1300	daily	SEA	17505dha
Kikongo	**Days**	**Area**	**kHz**
1900-2000	daily	Af	11955nau
Kinyarwanda/			
Kirundi	**Days**	**Area**	**kHz**
1800-1900	daily	Af	9770mey
Kituba	**Days**	**Area**	**kHz**
1800-1900	daily	Af	9595mey
Korean	**Days**	**Area**	**kHz**
0800-0900	daily	EAs	11895tai
1100-1200	daily	EAs	9460irk
1200-1300	daily	EAs	6005kna
Kurdish	**Days**	**Area**	**kHz**
1700-1800	daily	ME	9630wer
Lingala	**Days**	**Area**	**kHz**
1900-2000	daily	Af	13740dha
Ilocano	**Days**	**Area**	**kHz**
1100-1200	daily	SEA	1359fan, 5900irk
Malagasy	**Days**	**Area**	**kHz**
1600-1700	daily	Af	6225mey
Malay	**Days**	**Area**	**kHz**
0230-0300	daily	SEA	873put
0900-1030	daily	SEA	873put
1400-1500	daily	SEA	15315wer
Mandarin	**Days**	**Area**	**kHz**
0000-0200	daily	EAs,SEA	1503fan
0500-0700	daily	NAm	5985yfr
0500-1000	daily	EAs,SEA	1503fan
0900-1000	daily	EAs	11565tai
0900-1100	daily	EAs	9545tai, 9945tai
0900-1400	daily	EAs,SEA	1557kou
1000-1100	daily	EAs	9920tai
1100-1200	daily	EAs	9720nvs
1100-1400	daily	EAs	5995ppk, 6115ppk
1100-1600	daily	EAs	6240pao
1200-1300	daily	EAs	11535tai
1300-1400	daily	EAs,SEA	747min
1300-1400	daily	NAm	13695yfr
1500-1600	daily	NAm	11830yfr
1700-1900	daily	EAs,SEA	1557kou
2000-2100	daily	EAs,SEA	1359fan
2200-0300	daily	EAs,SEA	1557kou
2200-2400	daily	EAs	6230pao
2300-2400	daily	EAs	9540tai
Marathi	**Days**	**Area**	**kHz**
1400-1500	daily	IND	9855dha
1500-1600	daily	IND	5825erv
Nepali	**Days**	**Area**	**kHz**
1400-1500	daily	As	5825tac

Oriya	Days	Area	kHz
1400-1500	daily	IND	15325wer
Oromo	**Days**	**Area**	**kHz**
1600-1700	daily	EAf	13660nau
Pashto	**Days**	**Area**	**kHz**
1500-1600	daily	WAs	7550erv
Polish	**Days**	**Area**	**kHz**
0700-0800	daily	Eu	7730yfr
1800-1900	daily	Eu	5820erv, 7590erv
2000-2100	daily	Eu	11665yfr
Portuguese	**Days**	**Area**	**kHz**
0000-0100	daily	B	9430yfr, 9690yfr, 11885yfr
0100-0200	daily	B	7520yfr, 9930yfr, 11825yfr
0200-0300	daily	B	7520yfr
0300-0400	daily	B	7520yfr, 7730yfr
0400-0500	daily	Af	11580yfr
0700-0800	daily	Eu	9355yfr
0800-1000	daily	B	9680yfr
0800-1100	daily	B	6105yfr, 9605yfr
0900-1100	daily	B	9575yfr
1000-1100	daily	B	6105yfr
1200-1300	daily	B	11830yfr
1300-1400	daily	B	11530yfr
1400-1500	daily	B	15210yfr
1500-1600	daily	B	15355yfr
1700-1800	daily	Af	17690yfr
1700-2000	daily	B	17575yfr
1900-2000	daily	Af	3955mey, 6100mey
2100-2145	daily	Eu	11565yfr
2100-2200	daily	Af	17690yfr
2200-2400	daily	B	7360guf, 9690yfr, 17575yfr
2300-2400	daily	B	11720yfr
Punjabi	**Days**	**Area**	**kHz**
1400-1600	daily	As	6150arm
1600-1700	daily	As	6070arm
Romanian	**Days**	**Area**	**kHz**
0600-0700	daily	Eu	7730yfr
1800-1900	daily	Eu	6050wer
2000-2100	daily	Eu	9355yfr
Russian	**Days**	**Area**	**kHz**
0400-0500	daily	RUS	7520yfr
0400-0600	daily	RUS	1503msk
1200-1300	daily	RUS	9320dsb
1500-1700	daily	RUS	9955tnn
1600-1700	daily	RUS	21745hbn
1600-1800	daily	RUS	21745yfr
1700-1900	daily	RUS	6140wer
1800-2000	daily	RUS	1503msk
1900-2000	daily	RUS	18930yfr
Serbian	**Days**	**Area**	**kHz**
1900-2000	daily	Eu	3975wer
Sesotho	**Days**	**Area**	**kHz**
1800-1900	daily	Af	12015iss
Setswana	**Days**	**Area**	**kHz**
1800-1900	daily	Af	13660iss
Shona	**Days**	**Area**	**kHz**
1700-1800	daily	Af	17505asc
Sindhi	**Days**	**Area**	**kHz**
1400-1500	daily	As	13655wer
Somali	**Days**	**Area**	**kHz**
1700-1800	daily	Af	11665rmp
Spanish	**Days**	**Area**	**kHz**
0000-0100	daily	CAm,NAm	9715yfr, 11855yfr
0000-0100	daily	CAm,SAm	5985guf, 13615yfr
0000-0200	daily	CAm,SAm	5985yfr
0000-0500	daily	CAm,SAm	9355yfr
0100-0200	daily	CAm,NAm	5950yfr, 6890yfr, 9525yfr
0100-0200	daily	CAm,SAm	7570yfr, 11885yfr
0200-0300	daily	CAm,NAm	9930yfr
0200-0300	daily	CAm,SAm	9985yfr, 11825yfr

Spanish	Days	Area	kHz
0300-0400	daily	CAm,NAm	6890yfr, 9525yfr, 9680yfr
0300-0500	daily	CAm,SAm	5985yfr
0400-0500	daily	CAm,SAm	9930yfr
0400-0500	daily	CAm,SAm	7730yfr, 9985yfr
0500-0600	daily	CAm,SAm	5745yfr
0500-0600	daily	CAm,SAm	6000yfr
0500-0600	daily	Eu	9355yfr
0500-0700	daily	CAm,NAm	9495yfr
0500-1300	daily	CAm,NAm	9715yfr
0600-0700	daily	CAm,NAm	5950yfr
0700-0800	daily	Eu	7520yfr
0700-0800	daily	CAm,NAm	9680yfr
0700-1100	daily	CAm,SAm	6000yfr
0800-1000	daily	CAm,NAm	9495yfr
0800-1000	daily	CAm,SAm	5745yfr, 11740yfr
0800-1400	daily	CAm,NAm	9555yfr
0900-1000	daily	CAm,NAm	6890yfr
1000-1600	daily	CAm,NAm	6085yfr
1100-1200	daily	CAm,NAm	9575yfr
1100-1300	daily	CAm,NAm	9605yfr
1100-1300	daily	CAm,SAm	11740yfr
1200-1600	daily	CAm,NAm	11725yfr
1200-1600	daily	CAm,SAm	13615yfr
1300-1400	daily	CAm,NAm	15355yfr
1300-2000	daily	CAm,NAm	15130yfr
1400-1500	daily	CAm,NAm	11830yfr
1400-1500	daily	CAm,NAm	15355yfr
1400-1600	daily	CAm,SAm	11740yfr, 17555yfr
1500-1600	daily	CAm,NAm	13695yfr
1600-1700	daily	Eu	18930yfr
1700-1800	daily	CAm,SAm	17535yfr
1700-1900	daily	CAm,SAm	6085yfr
1800-1900	daily	Eu	6120nau, 18930yfr
2000-2100	daily	CAm,NAm	13695yfr
2000-2300	daily	CAm,SAm	5985yfr
2000-2400	daily	CAm,NAm	11855yfr, 15130yfr
2100-2200	daily	Eu	9355yfr
2100-2300	daily	CAm,NAm	11700yfr
2200-2300	daily	CAm,SAm	9465yfr, 11580yfr, 11665yfr
2300-2400	daily	CAm,SAm	5985yfr, 9355yfr, 9495yfr, 13615yfr

Swahili	Days	Area	kHz
1600-1700	daily	Af	9590mdc
1700-1800	daily	Af	11975iss
1900-2000	daily	Af	9660mey

Tagalog	Days	Area	kHz
1000-1100	daily	SEA	1359fan
1100-1200	daily	SEA	11520pao
1200-1300	daily	SEA	1359fan, 9310alm
2200-2300	daily	SEA	1359fan

Tamil	Days	Area	kHz
0130-0230	daily	As	863put
1030-1130	daily	As	873put
1400-1500	daily	As	17810dha
1500-1600	daily	As	11935nau

Telugu	Days	Area	kHz
1300-1400	daily	As	17810dha

Thai	Days	Area	kHz
1200-1300	daily	SEA	9450nvs
1900-2000	daily	SEA	1503fan

Turkish	Days	Area	kHz
1700-1800	daily	ME	9430skn
1800-1900	daily	ME	9885rmp

Urdu	Days	Area	kHz
1400-1600	daily	As	7565kch
1600-1700	daily	As	7295nvs, 7590erv

Uzbek	Days	Area	kHz
1400-1500	daily	CAs	13605wer

Vietnamese	Days	Area	kHz
0000-0100	daily	SEA	11630pao
1000-1100	daily	SEA	9455tai

Vietnamese	Days	Area	kHz
1200-1300	daily	SEA	7340irk, 7460pao
1300-1400	daily	SEA	7260tai, 7310irk, 9960tai
1300-1500	daily	SEA	1503fan
1600-1700	daily	SEA	1359fan
1700-1800	daily	SEA	1503fan
2100-2200	daily	SEA	1359fan
2300-2400	daily	SEA	1503fan

Xhosa	Days	Area	kHz
1800-1900	daily	Af	11820wer

Yoruba	Days	Area	kHz
1900-2000	daily	Af	11665asc

Zulu	Days	Area	kHz
1900-2000	daily	Af	11820wer

Ann: English: "You are listening to Family Radio, the Sound of the New Life", "This is your Family Radio, International Broadcast Station WYFR, Okeechobee, Florida, the United States of America"; German: "Dies ist Ihr Familienradio, die internationale Radiostation WYFR, in Okeechobee, Florida, Vereinigte Staaten von Amerika".
V: QSL-card. (No tapes accepted)
Notes: Began International SW broadcasts in 1973. Owned by Family Stations, Inc. The call Letters WYFR are used only for transmissions from the Okeechobee, FL facility.

FAR EAST BROADCASTING COMPANY (FEBC) (Rlg)
✉ P.O. Box 1, La Mirada, CA 90637-0001, USA.
☎ +1 562 9474651. 🖷 +1 562 9430160.
E: Via website. **W:** www.febc.org
Webcast: D
✉ FEBC International Ltd, 30 Lorong Ampas, #07-01 Skywaves Industrial Bldg., Singapore 328783.
☎ +65 6392 3154. 🖷 +65 6392 3156.
E: info@febcintl.org **W:** www.febcintl.org
LP: Chmn: Dr Doug Pennoyer; Pres: Gregg J. Harris.
Notes: FEBC maintains more than 35 recording studios in various countries producing religious programming for listeners in the Far East. FEBC owns transmitting stations in South Korea, the Philippines and Northern Mariana Islands.

HCJB GLOBAL VOICE (Rlg)
✉ P.O. Box 39800, Colorado Springs, CO 80949-9800, USA.
☎ +1 719 5909800. 🖷 +1 719 5909801.
E: info@hcjb.org **W:** www.hcjb.org
LP: Pres, HCJB Global: Wayne Pederson; SM: John E. Beck; PD: Alex Saks; Frequency Mgr: Douglas Weber.
V: QSL-card.
Notes: Owned by World Radio Missionary Fellowship, Inc. HCJB owns transmitting stations in Australia and Ecuador.

HMONG WORLD CHRISTIAN RADIO (Rlg)
✉ P.O. Box 600427, St. Paul, MN 55105, USA.
☎ +1 651 3034386.
E: giatoulee@comcast.net; voiceofhope@comcast.net
W: www.hwcr.us
Webcast: D/L
✉ P.O. Box 132, Cottage Grove, St. Paul, MN 55106, USA.
LP: Pres: Rev. Gia Tou Lee.
kHz: 9540

Winter Schedule 2010/2011			
Hmong	Days	Area	kHz
1300-1330	.....s.	NAm	9540hri*

Key: * Via World Harvest Radio.
Ann: English: "Hmong World Christian Radio".
Notes: Founded August 2005. Religious prgr for Hmong speakers in China and South East Asia.

KJES RADIO (Rlg)
✉ Our Lord's Ranch, 230 High Valley Rd., Vado, NM 88072-7221, USA.
☎ +1 505 2332090. 🖷 +1 505 2333019.
E: kjesroots@gmail.com
LP: Pres: Fr Rick Thomas; GM: Michael Reuter.
SW: [JES] Vado, NM: 1 x 5 (standby), 1 x 50kW.
kHz: 7555, 11715, 15385

Winter Schedule 2010/2011			
English/Spanish	Days	Area	kHz
0200-0330	daily	NAm	7555jes

English/Spanish	Days	Area	kHz
1400-1700	daily	NAm,CAm	11715jes
1900-2100	daily	CAm	15385jes

Ann: English: "This is KJES Radio, broadcasting from the Lord's Ranch".
V: QSL-card. Rp.
Notes: Catholic station. Transmissions are part of a rehabilitation programme for young people. KJES ("King Jesus Eternal Savior") has been licensed since November 1992.

KVOH - LA VOZ DE RESTAURACIÓN (Rlg)
🖃 P.O. Box 56320, Los Angeles, CA 90056, USA.
☎ +1 323 7662454. 📠 +1 323 7662458.
E: kvoh@restauracion.com **W:** www.restauracion.com
Webcast: L
🖃 4409 W. Adams blvd., Los Angeles, CA 90007, USA. (Studio)
LP: Administration: Rene A. Hernandez Harris.
SW: [VOH] Rancho Simi, CA: 2 x 50kW.
kHz: *9975, 17775*

		Winter Schedule 2010/2011	
Spanish	Days	Area	kHz
0100-0800	daily	CAm	9975voh
1300-1500	daily	CAm	9975voh
1500-0100	daily	CAm	17775voh

Ann: English: "This is KVOH, La Voz de Restauración broadcasting"; Spanish: "Ésta es KVOH, La Voz de Restauración".
V: QSL-card.
Notes: Owned by Iclesias de Restauración Inc.

PAN AMERICAN BROADCASTING (Rlg)
🖃 Suite 250, 7011 Koll Center Parkway, Pleasanton CA 94566-3253 USA.
☎ +1 925 4629800. 📠 +1 925 4629808.
E: info@panambc.com **W:** www.radiopanam.com
Webcast: D
LP: Pres: Gene Bernald.
kHz: *6040, 13645*

		Winter Schedule 2010/2011	
Arabic	Days	Area	kHz
1930-2015	......s	NAf	6040wer
1930-2030	......s	NAf	6040wer
English	Days	Area	kHz
1400-1430	......s	IND	13645wer
1415-1430	mtwtfs.	IND	13645wer
1430-1445	......s	IND	13645iss

V: Online form. (For broadcasts from the 'Radio Africa' service only)
Notes: Pan American Broadcasting, Inc rebroadcasts prgrs of various religious production studios, via leased air time through international tx providers and via its service 'Radio Africa' (see Equatorial Guinea).

RADIO PAYAM-E DOOST (Rlg)
🖃 P.O. Box 765, Great Falls, VA 22066, USA.
☎ +1 703 6718888. 📠 +1 301 2926947.
E: payam@bahairadio.org **W:** www.bahairadio.org
Webcast: D/L
SAT: Galaxy 19, Hot Bird 8.
kHz: *7460, 7480*

		Winter Schedule 2010/2011	
Farsi	Days	Area	kHz
0230-0315	daily	ME	7460kch
1800-1845	daily	ME	7480kch

Ann: Farsi: "Payam-e Doost".
Notes: Payam-e Doost ("Message from a friend") is a satellite radio station run by members of the Baha'i Faith in the USA. Regular relays on shortwave started 21 April 2001 and on satellite from May 2002. Jammed.

SUAB XAA MOO ZOO (Rlg)
🖃 c/o Christian & Missionary Alliance, P.O. Box 35000, Colorado Springs, CO 80935-3500, USA.
E: suabxaamoozoo@yahoo.com
W: www.suabxaamoozoo.com
Webcast: D
kHz: *7530*

		Winter Schedule 2010/2011	
Hmong	Days	Area	kHz
2230-2300	daily	SEA	7530tai

THE OVERCOMER MINISTRY (Rlg)
🖃 P.O. Box 691, Walterboro, SC 29488, USA.
☎ +1 843 5384202. 📠 +1 843 5384202.
E: brotherstair@overcomerministry.org
W: www.overcomerministry.org
Webcast: L
LP: Owner: Ralph G. Stair.
SAT: ABS 5, Galaxy 19, Hot Bird 6, Optus D2, Thaicom 5.
kHz: *5945, 6065, 9860, 13810, 17485*

		Winter Schedule 2010/2011	
English	Days	Area	kHz
1400-1500	daily	Eu,NAf	13810nau
1500-1600	daily	Af	17485wer
1900-2000	daily	Eu,NAf	6065mos, 9860wer
1900-2000	daily	Eu	5945wer

Ann: English: "You have been listening to the International Broadcast - The Overcomer".
V: QSL-card. (Rpt to: overcomer@overcomerministry.com)
Notes: Owned by Faith Cathedral Fellowship, Inc.

TWR (Rlg)
🖃 P.O. Box 8700, Cary, NC 27512, USA.
☎ +1 919 4603700. 📠 +1 919 4603702.
E: info2@twr.org **W:** www.twr.org
W: Regional branches: www.twrafrica.org (Africa); www.twreurope.org (Europe); www.twr.in (India); www.twr.asia; www.ktwr.net (Asia)
LP: Chmn: Dr Thomas J. Lowell; Pres/CEO: Lauren Libby.
V: QSL-card.
Notes: TWR owns transmitting facilities in Benin, Guam, Bonaire and Swaziland.

UNIVERSITY NETWORK (DR GENE SCOTT) (Rlg)
🖃 P.O. Box 1, Los Angeles, CA 90053-0001, USA.
☎ +1 818 2408151.
W: www.pastormelissascott.com; www.drgenescott.com
Webcast: L
SAT: Galaxy 19.
kHz: *6090, 11775*

		Winter Schedule 2010/2011	
English	Days	Area	kHz
1000-2200	daily	NAm	11775aia
2200-1000	daily	NAm	6090aia

V: Does not verify reception reports.
Notes: Transmits prgrs produced by Pastor Melissa Scott and recordings of the late Dr. Gene Scott.

WEWN - EWTN SHORTWAVE RADIO (Rlg)
🖃 5817 Old Leeds Rd., Irondale, AL 35210-2164, USA.
☎ +1 205 2712900. 📠 +1 205 2712926.
E: radio@ewtn.com **W:** www.ewtn.com
Webcast: D/L/P
🖃 EWTN, PO Box 157, Station A, Etobicoke, Ontario, Canada M9C 4V2.
LP: Pres, EWTN: Michael P. Warsaw; GM, Radio: Frank Leurck; SM: Richard Jones; Frequency Mgr: Joseph A. Dentici; Affiliate Engineering Mgr: Glen Tapley.
SW: [EWN] Vandiver, AL: 4 x 500kW.
SAT: Eurobird 1, Galaxy 11, PAS 3R/8/10.
kHz: *5810, 7555, 9390, 11520, 11550, 11870, 12050, 13830, 15610*

		Winter Schedule 2010/2011	
English	Days	Area	kHz
0000-0600	daily	ME	11520ewn
0600-0900	daily	Af	11520ewn
0900-1200	daily	SEA	9390ewn
1200-1900	daily	Eu	15610ewn
1900-2400	daily	Af	15610ewn
Spanish	Days	Area	kHz
0000-0500	daily	CAm,NAm	5810ewn
0000-1000	daily	SAm	11870ewn
0500-1300	daily	NAm	7555ewn
1000-1700	daily	SAm	12050ewn
1300-1800	daily	CAm,NAm	11550ewn
1300-2400	daily	SAm	13830ewn
1800-2400	daily	CAm,NAm	12050ewn

Ann: English: "This is WEWN, Global Catholic Radio, Birmingham, Alabama, USA".
V: QSL-card. Rp (3 IRCs).

Notes: Owned by the Eternal Word TV Network, Inc. Began broadcasting in December 1992. 3 transmitters in use, 4th transmitter used as a maintenance backup.

WHRI - WORLD HARVEST RADIO INTERNATIONAL (Rlg)

✉ P.O. Box 12, South Bend, IN 46624, USA.
☎ +1 574 2918200. 🖷 +1 574 2919043.
E: whr@lesea.com **W:** www.whr.org
Webcast: L
✉ LeSEA Broadcasting, 61300 S Ironwood Rd, South Bend, IN 46614, USA.
W: www.lesea.com
L.P: GM: Peter Sumrall; Dir, Engineering: Larry Vehorn.
SW: [HRI] Cypress Creek, SC: 1 x 100, 3 x 500kW & [HBN] T8WH (see Palau).
SAT: Galaxy 16.
kHz: 5875, 5920, 7315, 7335, 7385, 7465, 7520, 7555, 7570, 7590, 9410, 9470, 9490, 9495, 9505, 9540, 9595, 9615, 9640, 9840, 9895, 11565, 15180, 15665, 15680, 17520, 17540, 21630

Winter Schedule 2010/2011

English	Days	Area	kHz
0000-0200	mtwtf.s	NAm,Eu	7590hri*
0000-1300	daily	NAm	5875hri
0100-1200	daily	NAm	7315hri
0200-0300	mtwtf.s	NAm,Eu	7385hri*
0300-0400	mtwtf.s	NAm,Eu	7590hri*
0400-0500	....ss	Eu,Af	9640hri
0400-1000	mtwtf..	NAm,Eu	7465hri
0500-0600	......s	EAs,SEA	11565hri
0600-0700	.....ss	Eu,Af	9615hri
0700-1100	......s	EAs,SEA	11565hri
0800-0900	mtwtf..	EAs,SEA	11565hri
1000-1200	.....ss	NAm	7520hri
1200-1300	daily	NAm	7315hri, 7385hri
1200-1400	daily	Eu,Af	15665hri
1300-2200	daily	NAm	9840hri
1400-1500	daily	Eu,Af	17540hri
1500-1600	mtwtfs.	Af	21630hri
1500-2300	daily	NAm	15180hri
1600-1900	daily	Af	21630hri
1900-2000	daily	Af	17520hri
2000-2100	mtw...s	NAm,Eu	7570hri*
2000-2100	......s	NAm,Eu	9895hri
2000-2200	...tfs.	NAm,Eu	15665hri
2100-2200	daily	NAm,Eu	9490hri**
2100-2300	mtwtf.s	NAm,Eu	7555hri*
2200-2300	daily	Af	9615hri
2200-2400	mtwtf.s	NAm,Eu	9505hri****
2300-0100	daily	NAm	9470hri
2300-2400	.....ss	NAm,Eu	7335hri***
English/Various	**Days**	**Area**	**kHz**
1300-1500	daily	NAm	9540hri
1500-1600	......s	NAm,Eu	15680hri
Spanish/English	**Days**	**Area**	**kHz**
0300-0400	.....ss	SAm	7385hri
0400-0500	mtwtf..	SAm	7385hri
0500-0600	....s.	SAm	7385hri
0600-0700	.....ss	SAm	7385hri
0700-0800	mtwtf..	SAm	7385hri
0800-0900	....s.	SAm	7385hri
0900-1200	mtwtf..	SAm	7385hri
1200-1300	.....ss	CAm	9410hri
1300-1400	.....ss	SAm	9495hri
1400-1900	....s.	SAm	15665hri
1400-1900	daily	SAm	9495hri
1900-2300	daily	SAm	9595hri
2300-0200	mtwtf..	SAm	7385hri
2300-1300	mtwtf..	SAm	5920hri

Key: * To 6 Feb; ** from 7 Feb; *** to 6 Mar; **** from 7 Mar.
Ann: English: "This is World Harvest Radio International".
V: QSL-card. (Online reception report form)
Notes: World Harvest Radio International is a shortwave radio network owned by LeSEA Broadcasting, Inc. Registered freqs shown, actual usage depends on airtime bookings. On air since 25 Dec 1985.

WINB (Rlg)

✉ P.O. Box 88, Red Lion, PA 17356, USA.
☎ +1 717 2445360. 🖷 +1 717 2460363.
E: info@winb.com **W:** www.winb.com
Webcast: L
L.P: Sales/Frequency Mgr: Hans Johnson.
SW: [INB] Red Lion, PA: 1 x 50kW.
kHz: 9265, 13570

Winter Schedule 2010/2011

English	Days	Area	kHz
1000-1500	daily	CAm	9265inb**
1100-1600	daily	CAm	9265inb*
1500-2100	daily	CAm	13570inb**
1600-2200	daily	CAm	13570inb*
2100-0300	daily	CAm	9265inb**
2200-0400	daily	CAm	9265inb*

Key: * To 12 Mar; ** from 13 Mar.
Ann: English: "This is WINB, Red Lion, Pennsylvania in the United States of America".
V: QSL-card.
Notes: Owned by World International Broadcasters, Inc. Operational since October 1962.

WJHR RADIO INTERNATIONAL (Rlg)

✉ 5920 Oak Manor Drive, Milton, FL 32570, USA.
E: wjhr@usa.com
W: calvaryscall.org/Media3.html (Mt Calvary Baptist Church, WJHR Page)
L.P: Owner: George Scott Mock.
SW: [JHR] Milton, FL: 1 x 50kW.
kHz: 15550

Winter Schedule 2010/2011

English	Days	Area	kHz
1400-2200	..w....	NAm	15550jhr*

Key: * SSB.
Ann: English: "WJHR Radio International, Milton, Florida".
Notes: WJHR (John Hill Radio) Started test transmissions in November 2009.

WORLD CHRISTIAN BROADCASTING INC. (Rlg)

✉ 605 Bradley Court, Franklin, TN 37067, USA.
☎ +1 615 3718707.
E: info@wrorldchristian.org **W:** www.worldchristian.org
L.P: Pres: Charles H. Caudill; Vice President, Dev: Andy Baker; Dir, Engineering: Kevin Chambers.
Notes: World Christian Broadcasting, Inc owns the shortwave station KNLS (see under Alaska) and is preparing to set up a SW relay station in Southern Africa (see under Madagascar).

WRNO WORLDWIDE (Rlg)

✉ P.O. Box 895, Fort Worth, TX 76101, USA.
☎ +1 817 8509990. 🖷 +1 817 8509994.
E: wrnoradio@mailup.net **W:** www.wrnoradio.com
Webcast: L
L.P: Chmn: Robert E. Mawire.
SW: [RNO] New Orleans, LA: 1 x 50kW.
kHz: 7505

Winter Schedule 2010/2011

English	Days	Area	kHz
0200-0500	daily	NAm,CAm	7505rno

Ann: English: "From New Orleans, Louisiana, you're listening to WRNO Worldwide broadcasting from the United States of America".
IS: "When the Saints go marching in".
V: QSL-card. Rp. (2 IRCs).
Notes: Owned by Good News World Outreach. Registered frequencies shown, actual schedule varies acc. to airtime bookings.

WTJC - FUNDAMENTAL BROADCASTING NETWORK (FBN) (Rlg)

✉ 520 Roberts Road, Newport, NC 28570, USA.
☎ +1 252 2234600. 🖷 +1 252 2232201.
E: fbn@fbnradio.com **W:** www.fbnradio.com
Webcast: L
✉ 123 Grace Baptist Ln., Newport, NC 28570, USA. (Home)
L.P: GM, FBN: Michael Ebron; Missionary/Engineer: David Robinson.
SW: [TJC] Newport, NC: 1 x 50kW.
kHz: 9370

Winter Schedule 2010/2011

English	Days	Area	kHz
0000-2400	daily	NAm	9370tjc

V: QSL-card. Rp. (1 USD or IRC. Rpts to Mrs Robinson, Grace Missionary Church, 520 Roberts Road, Newport, NC 28570 USA)
Notes: Owned by Paxson Communications Corporation. FBN includes the shortwave station WTJC ("Working Till Jesus Comes"), which has been on air since Oct 1999.

WTWW (Rlg)

✉ 1784 West Northfield Blvd, #305 Murfreesboro, TN 37129-1702, USA.
E: george@wtww.us **W:** wtww.us
Webcast: L
LP: Owner/GM: George McClintock; SM: Dan Dixon.
SW: [TWW] Lebanon, TN: 1 x 50, 2 x 100kW.
kHz: 5080, 5755, 9479, 9480, 9990

Winter Schedule 2010/2011

English	Days	Area	kHz
1100-2200	daily	NAm,Eu	9990tww***
1100-2200	daily	NAm	9480tww***
1200-2300	daily	NAm,Eu	9990tww**
1200-2300	daily	NAm	9479tww**
1300-2200	daily	NAm,Eu	9990tww*
1300-2200	daily	NAm	9480tww*
2200-1100	daily	NAm,Eu	5080tww***, 5755tww***
2200-1300	daily	NAm,Eu	5080tww*, 5755tww*
2300-1200	daily	NAm,Eu	5080tww**, 5755tww**

Key: * To 31 Jan; ** 1 Feb-12 Mar; *** from 13 Mar.
Ann: English: "This is WTWW, Lebanon, Tennessee, USA".
V: QSL-card.
Notes: WTWW ("We Transmit World Wide") is owned by Leap of Faith, Inc and has been on air since Feb 2010 (tests during Jan 2010). Registered freqs shown, actual usage may vary. The tx on the assigned freq of 9480kHz is, as of Nov 2010, being operated on 9479kHz.

WWCR - WORLDWIDE CHRISTIAN RADIO (Rlg)

✉ 1300 WWCR Avenue, Nashville, TN 37218, USA.
☎ +1 615 2551300. 🖷 +1 615 2551311.
E: wwcr@wwcr.com **W:** www.wwcr.com
Webcast: L
LP: GM: Eric Westenberger; Ops Mgr: Brady Murray; Frequency Mgr: Dr Jerry Plummer.
SW: [WCR] Nashville, TN: 4 x 100kW.
kHz: 3215, 4840, 5070, 5890, 5935, 7465, 7490, 9350, 9980, 9985, 12160, 13845, 15825

Winter Schedule 2010/2011

English	Days	Area	kHz
0000-1200	daily	NAm,Eu	4840wcr**
0000-1200	daily	Af	5935wcr**
0100-0900	daily	NAm,Eu,Af	3215wcr**
0100-1100	daily	NAm,Af	5890wcr**
0100-1300	daily	Af	5935wcr*
0100-1300	daily	NAm,Eu	4840wcr*
0200-1200	daily	NAm,Eu,Af	3215wcr*
0200-1200	daily	NAm,Af	5890wcr*
0900-1100	daily	NAm,Eu,Af	9985wcr**
1000-1200	daily	NAm,Eu,Af	9985wcr*
1100-0100	daily	NAm,Af	9980wcr**
1100-2000	daily	NAm,Eu,Af	15825wcr**
1200-0200	daily	NAm,Af	9980wcr*
1200-1600	daily	Af	7490wcr**
1200-2100	daily	NAm,Eu,Af	15825wcr*
1200-2400	daily	NAm,Eu	13845wcr**
1300-0100	daily	Af	13845wcr*
1300-1600	daily	NAm,Eu	7490wcr*
1600-2000	daily	Af	12160wcr**
1600-2100	daily	NAm,Eu	12160wcr*
2000-2200	daily	Af	9350wcr**
2000-2400	daily	NAm	7465wcr**
2100-2300	daily	NAm	7465wcr*
2100-2300	daily	Af	9350wcr*
2200-0100	daily	NAm,Eu,Af	7465wcr**
2200-2400	daily	Af	5070wcr**

English	Days	Area	kHz
2300-0100	daily	Af	5070wcr*
2300-0200	daily	NAm,Eu,Af	7465wcr*, 7490wcr**

Key: * To 12 Mar; ** From 13 Mar.
Ann: English: "This is World Wide Christian Radio-WWCR, Nashville, Tennessee, USA".
V: QSL-card. Rp. preferred (1 IRC). Rec. acc.
Notes: Owned by WNQM, Inc. Registered freqencies shown, actual usage may vary.

WWRB (Rlg)

✉ c/o Airline Transport Communications Inc., Listener Services, P.O. Box 7, Manchester, TN 37349-0007, USA.
☎ +1 931 7286063; +1 931 7286087. 🖷 +1 931 7286087.
E: Via website. **W:** www.wwrb.org
Webcast: L
✉ 6755 Shady Grove Road, Morrison, TN 37355, USA. (Studio)
LP: Owner & CE: Dave Frantz.
SW: [WRB] Manchester, TN: 4 x 100kW (+ 1 x 100kW reserve).
kHz: 3185, 3215, 5050, 5745, 9385

Winter Schedule 2010/2011

English	Days	Area	kHz
1200-2300	daily	NAm,CAm	9385wrb
2200-0100	daily	NAm	3215wrb
2200-0400	daily	NAf	5745wrb
2200-1300	daily	NAm,CAm	5050wrb
2300-1200	daily	NAm	3185wrb

V: QSL-card. (Email rpts not accepted)
Notes: A subsidiary of Airline Transport Communications Inc. Registered freqs shown, actual usage varies.

WBCQ - THE PLANET

✉ 274 Britton Road, Monticello, ME 04760, USA.
☎ +1 207 5389180.
E: wbcq@wbcq.com **W:** www.wbcq.com
Webcast: D/L
LP: Owner: Allan Weiner.
SW: [BCQ] Monticello, ME: 4 x 50kW.
kHz: 5110, 7415, 9330, 15420

Winter Schedule 2010/2011

English	Days	Area	kHz
0000-2400	daily	NAm,CAm	5110bcq*, 7415bcq*
1200-0600	daily	NAm,CAm	9330bcq*
1200-2300	daily	NAm,CAm	15420bcq*

Key: * USB/AM.
Ann: English: "This is WBCQ, Monticello, Maine, the United States of America. The Planet".
V: QSL-card. Rp. (SASE)
Notes: Owned by A. Weiner/Becker Broadcast Systems, Inc. Leases air time to religious and other prgr producers. Registered frequencies shown, actual schedule varies acc. to airtime bookings. On air since 8 September 1998.

WRMI - RADIO MIAMI INTERNATIONAL

✉ 175 Fontainebleau Blvd., Suite 1N4, Miami, FL 33172 USA.
☎ +1 305 5599764. 🖷 +1 305 5598186.
E: info@wrmi.net **W:** www.wrmi.net
Webcast: L
LP: GM: Jeff White; Dir, Technical: Jose Raul Mena.
SW: [RMI] Hialeah, FL: 1 x 5 (standby), 1 x 50kW.
kHz: 9955

Winter Schedule 2010/2011

Various	Days	Area	kHz
0000-1500	daily	CAm,SAm	9955rmi
1500-1700	daily	NAm	9955rmi
1700-2400	daily	CAm,SAm	9955rmi

Ann: English: "This is WRMI, Radio Miami International".
V: QSL-card.
Notes: Owned by Radio Miami International, Inc. On air since June 1994. WRMI provides air time for prgrs by various production companies (mainly in English and Spanish) and rebroadcasts international radio stations. Some prgrs aimed at a Cuban audience are jammed.

DT HOLDINGS LLC (Broker)

✉ 300 East 75th Street, Suite 50, New York, NY 10021, USA.
☎ +1 917 5392494. 🖷 +1 208 4603547.
E: d.robinson@dtholdings.com; sales@dtholdings.com

W: www.dtholdings.com
L.P: MD: Daniel Robinson.
Notes: DT Holdings brokers air time for transmitter facilities in Tajikistan and Uzbekistan.

UZBEKISTAN (UZB)

RADIOALOQA, RADIOESHITTIRISH VA TELEVIDENIYE MARKAZI (RRTM) (Tx Operator)
📧 Amir Timur Street 109a, 100084 Toshkent, Uzbekistan.
☎ +998 71 2356516. 📠 +998 71 2344517.
E: info@crrt.uz **W:** www.crrt.uz
L.P: GD: Shomansur Sh. Abidxodjayev.
SW: [TAC] Toshkent: 19 x 100kW.
Notes: RRTM, a division of the State Communications and Information Agency of Uzbekistan, is the national transmitter network operator in Uzbekistan.

VATICAN CITY STATE (CVA)

VATICAN RADIO (Rlg)
📧 Piazza Pia 3, I-00120 Vatican City.
☎ +39 06 69883945. 📠 +39 06 69883463.
E: sedoc@vatiradio.va **W:** www.radiovaticana.org
Webcast: D/L/P
L.P: GD: Fr Federico Lombardi; PD: Fr Andrzej Koprowski; Dir, Technical: Sandro Piervenanzi; Int Rel: Giacomo Ghisani.
MW: [SMG] Santa Maria di Galeria: 1530kHz 600kW (run at greatly reduced power), 1611kHz 50kW; [VAT] Vatican City: 585/1260kHz 5kW.
SW: [SMG] Santa Maria di Galeria: 4 x 100, 5 x 500kW; [VAT] Vatican City: 1 x 10, 1 x 80kW.
SAT: Hot Bird 8, Intelsat 904/907, VSAT 1.
kHz: 585, 1260, 1530, 1611, 4005, 5885, 5895, 5900, 5910, 5965, 5985, 6020, 6040, 6060, 6185, 7250, 7290, 7305, 7335, 7355, 7360, 7365, 7370, 7385, 7395, 7435, 7585, 9580, 9585, 9600, 9610, 9635, 9645, 9660, 9695, 9755, 9800, 9850, 11625, 11715, 11740, 11850, 11910, 13765, 15235, 15460, 15595, 17765, 21680

Winter Schedule 2010/2011

Albanian	Days	Area	kHz
0620-0700	daily	Eu	1260vat, 1611smg
2000-2020	daily	Eu	1260vat, 1611smg, 6185smg, 7355smg

Amharic	Days	Area	kHz
0400-0415	daily	Af	7360smg, 9660mdc, 9660mdc
1630-1645	daily	Af	11625smg, 13765smg

Arabic	Days	Area	kHz
0500-0530	daily	Eu	9645smg
0500-0530	daily	ME	11715smg
0500-0600	daily	Eu	1260vat
0745-0805	mtwtf..	Eu	5965smg, 7250smg, 9645smg
0745-0805	mtwtfs.	Af	15595smg
1630-1700	daily	NAf	7290smg
1630-1700	daily	Af	9635smg
1630-1700	daily	Eu	1260vat
2140-2200	daily	Af	7250smg
2140-2200	daily	Eu	1611smg+, 4005vat, 5885smg

Armenian	Days	Area	kHz
0310-0330	daily	Eu	1260vat, 6185smg, 9645smg
1650-1710	daily	Eu	1611smg, 7365smg*, 9585smg
1650-1710	daily	ME	11715smg

Belarusian	Days	Area	kHz
0420-0440	daily	Eu	1260vat, 6185smg, 7335smg
1800-1820	daily	Eu	1260vat, 1611smg, 6185smg*, 7365smg, 9585smg**

Bulgarian	Days	Area	kHz
0540-0600	daily	Eu	1611smg, 6185smg, 7335smg
1920-1940	daily	Eu	1260vat, 1611smg, 6185smg, 7365smg

Croatian	Days	Area	kHz
0350-0410	daily	Eu	4005vat
1750-1810	daily	Eu	4005vat, 5885smg, 7250smg

Czech	Days	Area	kHz
0410-0425	daily	Eu	4005vat, 5965smg
1830-1845	daily	Eu	4005vat, 5885smg, 7250smg

English	Days	Area	kHz
0140-0200	daily	As	5895dsb, 7335smg**, 7335tac*, 9580tac**
0250-0310	daily	CAm,SAm	7305smg
0250-0320	daily	NAm	6040sac
0300-0320	mtwtfs.	As	15460pug
0300-0330	daily	Af	7360smg, 9660mdc, 9660mdc
0500-0530	daily	Af	7360smg, 9660mdc, 11625smg
0600-0620	daily	Eu	7250smg
0600-0620	mtwtfs.	Eu	4005vat
0600-0630	mtwtf..	Eu	5965smg
0600-0630	daily	Eu	1530smg
0630-0700	daily	Af	7360smg, 9660smg, 11625smg
0730-0745	mtwtf..	Eu	585vat, 1530smg, 1611smg, 5965smg, 7250smg, 9645smg
0730-0745	mtwtfs.	Af	15595smg
0730-0745	mtwtfs.	Eu	4005vat, 11740smg
1100-1130	.....s	Eu	7250smg
1530-1550	mtwtf..	As	7585dsb, 11850smg, 13765smg
1530-1600	.....s.	Af	13765smg
1530-1600	.....s.	As	7585dsb, 11850smg
1715-1730	daily	NAf	7290smg
1715-1730	daily	Eu	585vat, 4005vat, 5885smg, 7250smg, 9645smg
1730-1800	daily	Af	9755smg, 11625smg, 13765smg
2000-2030	daily	Af	7365smg, 9755smg, 11625smg
2045-2130	daily	Eu	9800sac+
2050-2120	daily	Af	7250smg
2050-2120	daily	Eu	1530smg, 4005vat, 5885smg
2300-2345	daily	NAm	7370smg+
2330-2400	daily	Eu	1611smg+

English (Mass)	Days	Area	kHz
1130-1300	....f..	Af	15595smg, 17765smg

Esperanto	Days	Area	kHz
2020-2030	..wt...	Eu	1260vat, 1611smg, 6185smg, 7355smg
2020-2030	......s	Eu	1530smg, 4005vat, 5885smg, 7250smg
2020-2030	......s	Af	7435smg
2250-2315	......s	Eu	4005vat

Finnish	Days	Area	kHz
0600-0620	m....s.	Eu	1260vat, 1611smg, 6185smg, 7335smg
1940-2000	....f.s	Eu	1260vat, 1611smg, 6185smg, 7355smg

French	Days	Area	kHz
0230-0250	daily	CAm,SAm	7305smg
0230-0250	daily	NAm	6040bon, 6040sac
0230-0300	daily	Af	7360smg
0430-0500	daily	Af	7360smg, 9660mdc, 9660mdc
0540-0600	daily	Eu	5965smg, 7250smg
0540-0600	mtwtfs.	Eu	4005vat
0600-0628	daily	Af	7360smg, 9660smg, 11625smg
0715-0730	mtwtfs.	Eu	4005vat, 11740smg
0715-0730	mtwtf..	Eu	585vat, 1530smg, 1611smg, 5965smg, 7250smg, 9645smg

French	Days	Area	kHz
0715-0730	mtwtfs.	Af	15595smg
1200-1215	daily	Eu	5965smg
1200-1215	mtwtf..	Eu	585vat, 1611smg+
1700-1715	daily	Eu	585vat, 4005vat, 5885smg, 7250smg, 9645smg
1700-1715	daily	NAf	7290smg
1700-1728	daily	Af	11625smg, 13765smg
2030-2050	daily	Af	7250smg, 11625smg
2030-2050	daily	Eu	1530smg, 4005vat, 5885smg
2030-2100	daily	Af	7365smg, 9755smg

German	Days	Area	kHz
0520-0540	daily	Eu	4005vat, 5965smg
1500-1515	daily	Eu	5885smg, 6060smg+, 7250smg, 9645smg
1920-1940	daily	Eu	4005vat, 5885smg, 7250smg
2310-2330	daily	Eu	1611smg+

Hindi	Days	Area	kHz
0040-0100	daily	As	5895dsb, 7335smg**, 7335tac*, 9580tac**
0200-0220	daily	As	15460pug
1430-1450	daily	As	7585dsb, 11850smg, 13765smg

Hungarian	Days	Area	kHz
0440-0500	daily	Eu	4005vat, 5965smg
1810-1830	daily	Eu	4005vat, 5885smg, 7250smg

Italian	Days	Area	kHz
0000-0030	daily	Eu	1611smg+
0620-0630	daily	Eu	4005vat, 7250smg
0700-0715	mtwtf..	Eu	585vat, 1530smg, 1611smg, 5965smg, 7250smg, 9645smg
0700-0715	mtwtfs.	Af	15595smg
0700-0715	mtwtfs.	Eu	4005vat, 11740smg
0915-1015	......s	Eu	1611smg+
0930-1015	......s	Eu	585vat
1100-1115	mtwtfs.	Eu	5965smg
1100-1130	mtwtf..	Eu	585vat, 1611smg
1300-1330	daily	Af	15595smg, 21680smg
1300-1330	daily	Eu	585vat, 1611smg+, 5965smg, 7250smg, 9645smg
1530-1600	daily	Eu	5885smg, 7250smg
1530-1600	...f..	Eu	9645smg
1630-1700	daily	Eu	585vat, 5885smg, 7250smg, 9645smg
1830-1940	daily	Eu	585vat
2000-2020	......s	Eu	4005vat
2000-2020	daily	Af	7250smg, 7435smg
2000-2020	daily	Eu	1530smg, 5885smg
2000-2030	mtwtfs.	Eu	4005vat
2200-2220	daily	Eu	585vat, 1611smg+
2200-2250	......s	Eu	4005vat
2200-2310	daily	Eu	5885smg
2200-2310	mtwtfs.	Eu	4005vat
2245-2310	daily	Eu	585vat, 1611smg+

Italian (Mass)	Days	Area	kHz
0830-0930	......s	Eu	585vat
0845-0905	......s	Af	15595smg

Latin (Angelus)	Days	Area	kHz
1100-1130	......s	Af	9645smg, 15595smg, 17765smg
1100-1130	......s	Eu	585vat, 1530smg, 1611smg+, 11740smg
1100-1130	......s	SAm	21680smg
1100-1200	......s	Eu	5965smg

Latin (Compline)	Days	Area	kHz
2220-2245	daily	Eu	585vat, 1611smg+

Latin (Mass)	Days	Area	kHz
0630-0700	daily	Af	15595smg
0630-0700	daily	Eu	585vat, 1530smg, 11740smg

Latin (Mass)	Days	Area	kHz
0630-0700	mtwtfs.	Eu	4005vat, 5965smg, 7250smg, 9645smg
0630-0710	......s	Af	9645smg
0630-0710	......s	Eu	4005vat, 5965smg, 7250smg

Latin (Rosary)	Days	Area	kHz
1940-2000	daily	Af	7250smg, 7435smg, 9755smg, 11625smg
1940-2000	daily	Eu	585vat, 1530smg, 4005vat, 5885smg
1940-2010	daily	Eu	7365smg

Latin (Vespers)	Days	Area	kHz
1600-1630	daily	Eu	5885smg, 7250smg, 9645smg

Latvian	Days	Area	kHz
0500-0520	daily	Eu	6185smg, 7335smg
1840-1900	daily	Eu	1260vat, 1611smg, 6185smg*, 7365smg, 9585smg**

Lithuanian	Days	Area	kHz
0440-0500	daily	Eu	1260vat, 6185smg, 7335smg
1820-1840	daily	Eu	1260vat, 1611smg, 6185smg*, 7365smg, 9585smg**

Liturgy	Days	Area	kHz
0715-0845	......s	Eu	1611smg
0830-0930	......s	Eu	7250smg
0930-1050	......s	Eu	11740smg
0930-1100	......s	Af	15595smg, 17765smg

Malayalam	Days	Area	kHz
0120-0140	daily	As	5895dsb, 7335smg**, 7335tac*, 9580tac**
0240-0300	daily	As	15460pug
1510-1530	daily	As	7585dsb, 11850smg, 13765smg

Mandarin	Days	Area	kHz
1230-1315	daily	EAs	5985nvs, 6020pug
1230-1315	daily	As	15235smg
2200-2245	daily	EAs	7395smg, 9600tin
2200-2245	daily	As	5900smg

Music	Days	Area	kHz
1050-1100	......s	Eu	7250smg, 11740smg
1530-1600	mtwt.s.	Eu	9645smg
2020-2030	mt..fs.	Eu	6185smg
2020-2030	mtwtfs.	Af	7435smg
2020-2030	mtwtfs.	Eu	5885smg

Norwegian	Days	Area	kHz
0600-0620	.t.....	Eu	1260vat, 1611smg, 6185smg, 7335smg
1940-2000	m......	Eu	1260vat, 1611smg, 6185smg, 7355smg

Papal Audience	Days	Area	kHz
0900-1015	..w....	Eu.	5965smg
0915-1015	..w....	Eu	585vat, 1611smg+

Polish	Days	Area	kHz
0500-0520	daily	Eu	4005vat, 5965smg
1515-1530	daily	Eu	5885smg, 6060smg+, 7250smg, 9645smg
1900-1920	daily	Eu	4005vat, 5885smg, 7250smg

Portuguese	Days	Area	kHz
0030-0057	daily	CAm,SAm	7305smg
0030-0057	daily	SAm	9610sac
0030-0100	daily	Eu	1260vat
0530-0600	daily	Af	7360smg, 9660smg, 11625smg
0900-0930	mtwtf..	Eu	1260vat
1000-1030	mtwtf..	Eu	1260vat
1000-1030	mtwtfs.	SAm	21680smg
1415-1430	daily	Af	9645smg
1415-1430	daily	Eu	1260vat, 7250smg
1500-1600	...t...	Eu	1260vat
1600-1630	daily	Eu	1260vat

Portuguese	Days	Area	kHz
1800-1858	daily	Af	9755smg, 11625smg, 13765smg

Romanian	Days	Area	kHz
0520-0540	daily	Eu	1611smg, 6185smg, 7335smg
0710-0830	......s	Eu	7250smg, 9645smg
1900-1920	daily	Eu	1260vat, 1611smg, 6185smg, 7365smg

Russian	Days	Area	kHz
0330-0400	daily	Eu	1260vat, 6185smg
0330-0400	daily	RUS	7335smg, 9645smg
0905-0930	......s	RUS	15595smg
0905-1050	......s	RUS	17765smg
1330-1400	daily	Eu	1260vat
1330-1400	daily	RUS	5900tac, 9695smg
1710-1730	daily	RUS	7365smg
1710-1738	daily	RUS	9585smg, 11715smg
1710-1740	daily	Eu	1611smg, 6185smg*
2100-2130	daily	Eu	1260vat
2100-2130	daily	RUS	5910smg, 7385smg

Slovak	Days	Area	kHz
0425-0440	daily	Eu	4005vat, 5965smg
1845-1900	daily	Eu	4005vat, 5885smg, 7250smg

Slovenian	Days	Area	kHz
0330-0350	daily	Eu	4005vat
1730-1750	daily	Eu	4005vat, 5885smg, 7250smg

Somali	Days	Area	kHz
0345-0400	......s	Af	7360mdc, 9660mdc, 9660mdc
1615-1628	.....s.	Af	11625smg, 13765mdc

Spanish	Days	Area	kHz
0100-0145	daily	CAm,SAm	7305smg
0100-0145	daily	Eu	1260vat
0100-0145	daily	SAm	9610bon, 9610sac, 11910smg
0145-0230	daily	CAm,SAm	7305smg
0145-0230	daily	SAm	9610bon, 9610sac, 11910smg
0320-0400	daily	CAm,SAm	7305smg
0320-0400	daily	NAm	6040sac
1130-1200	mtwtf..	Eu	1260vat
1130-1200	mtwtf..	SAm	21680smg
1400-1415	daily	Af	9645smg
1400-1415	daily	Eu	1260vat, 7250smg
1500-1600	m...f..	Eu	1260vat
1730-1800	daily	Eu	1260vat
1900-1930	.....s.	Af	9755smg, 11625smg
2120-2140	daily	Af	7250smg
2120-2140	daily	Eu	1611smg+, 4005vat, 5885smg

Swahili	Days	Area	kHz
0330-0345	......s	Af	9660mdc
0330-0345	daily	Af	7360mdc, 9660mdc
0330-0400	mtwtfs.	Af	9660mdc
0345-0400	mtwtfs.	Af	7360mdc, 9660mdc
1600-1615	daily	Af	11625smg, 13765mdc
1615-1628	mtwtf..	Af	11625smg, 13765mdc

Swedish	Days	Area	kHz
0600-0620	..wtf.s	Eu	1260vat, 1611smg, 6185smg, 7335smg
1940-2000	.twt.s.	Eu	1260vat, 1611smg, 6185smg, 7355smg

Tagalog	Days	Area	kHz
2020-2100	......s	Eu	1260vat, 1611smg

Tamil	Days	Area	kHz
0100-0120	daily	As	5895dsb, 7335smg**, 7335tac*, 9580tac**
0220-0240	daily	As	15460pug
1450-1510	daily	As	7585dsb, 11850smg, 13765smg

Tigrinya	Days	Area	kHz
0415-0430	daily	Af	7360smg, 9660mdc

Tigrinya	Days	Area	kHz
1645-1700	daily	Af	11625smg, 13765smg

Ukrainian	Days	Area	kHz
0400-0420	daily	Eu	1260vat, 6185smg, 7335smg
0715-0845	......s	Eu	9850smg, 11740smg
1740-1800	daily	Eu	1611smg, 7365smg, 9585smg**
1740-1800	daily	UKR	6185smg*

Urdu	Days	Area	kHz
0025-0040	m..t...	As	5895dsb, 7335smg**, 7335tac*, 9580tac**
1415-1430	..w...s	As	11850smg, 13765smg

Vietnamese	Days	Area	kHz
1315-1400	daily	As	13765smg, 15235smg
2315-2400	daily	SEA	7395smg, 9600smg

Key: + DRM; * to 5 Mar; ** from 6 Mar. Some prgrs are also broadcast on 'Holy' days in addition to their usual schedule.
Ann: Before all transmissions: Latin: "Laudetur Jesus Christus" (Praised be Jesus Christ), repeated in the language of the broadcast, then station identification. English: "This is the English program of Vatican Radio".
V: QSL-card.
Notes: On air since 12 Feb 1931.

VENEZUELA (VEN)

RADIO NACIONAL DE VENEZUELA - ANTENA INTERNACIONAL (Gov)
✉ Final calle Las Marías, El Pedregal de Chapellin, Caracas D.F., Zona Postal 1050, Venezuela.
☎ +58 212 7306666. 📠 +58 212 7311457.
E: canalinternacionalrnv@gmail.com; canalinternacional@rnv.gov.ve **W:** www.rnv.gov.ve
✉ Cesar Mendez, M., 705NW 111 C.T., Apt 8, Miami, FL 33172, USA. (Reception Reports)
L.P: Dir: Helena Salcedo.
SW: [CLZ] Calabozo (SW transmitting centre under construction). Currently leases airtime from Radiocuba (see Cuba).
kHz: *6060, 6180, 11670, 11680, 11690, 11705, 13680, 13750, 15250, 15290, 15370, 17705, 17750*

Winter Schedule 2010/2011			
Spanish	Days	Area	kHz
1400-1800	......s	Am	11690hab*, 13680hab*, 13750hab*, 15370hab*, 17750hab*

Spanish/English	Days	Area	kHz
1000-1100	daily	Am	6180hab
1100-1200	daily	Am	6060hab
1200-1300	daily	Am	11705hab
1500-1600	daily	Am	11680hab
1900-2000	daily	Am	15290hab
2000-2100	daily	Am	17705hab
2200-2300	daily	Am	11670hab
2300-2400	daily	Am	13680hab, 15250hab

Key: * "Aló Presidente" prgr.
Ann: Spanish: "Vd. escucha Radio Nacional de Venezuela, Antena Internacional", "Radio Nacional de Venezuela transmitiendo desde Caracas", "Canal Internacional de radio Nacional de Venezuela".
V: QSL-card. (Email reports to: aherrera@enet.cu - only for transmissions via Cuba).
Notes: Transmissions include irregular segments in English.

VIETNAM (VTN)

VOICE OF VIETNAM (OVERSEAS SERVICE) (VOV) (Gov)
✉ 58 Quan Su Street, Hanoi, Vietnam.
☎ +84 4 39344231. 📠 +84 4 39344230..
E: englishsection@vov.org.vn
W: www.vovnews.vn; www.vov.org.vn
Webcast: L
L.P: Chmn: Prof Dr Vu Van Hien; Deputy DG, Ext Relations: Dao Duy Hua; Deputy DG, Technical: Doan Viet Trung.
MW: [OMO] Can Tho, Ô Môn: 1242kHz 2000kW.
SW: [HAN] Hanoi, Me Tri: 2 x 50kW; [VNI] Son Tay: 11 x 100kW & via leased foreign relays.
FM/DAB: FM: 105.5MHz (Hanoi); 105.7MHz (Ho Chi Minh City).
SAT: Vinasat 1.

kHz: *1242, 3985, 5955, 5970, 6175, 7220, 7280, 7285, 7370, 9550, 9730, 9840, 12000, 12020*

Winter Schedule 2010/2011

Chinese	Days	Area	kHz
1100-1130	daily	SEA	1242omo
1100-1130	daily	As	7220vni, 12000vni
1200-1230	daily	As	7220vni, 12000vni
1300-1330	daily	As	7220vni, 12000vni
2230-2300	daily	As	9840vni, 12020vni

English	Days	Area	kHz
0100-0130	daily	NAm	6175sac
0230-0300	daily	NAm	6175sac
0330-0400	daily	NAm	6175sac
1000-1030	daily	As	9840vni, 12020vni
1100-1130	daily	As	7285han
1130-1200	daily	As	9840vni, 12020vni
1230-1300	daily	As	9840vni, 12020vni
1330-1400	daily	As	9840vni, 12020vni
1500-1530	daily	As	7285han, 9840vni, 12020vni
1600-1630	daily	Eu	7280vni, 9730vni
1600-1630	daily	ME	7220vni, 9550vni
1800-1830	daily	Eu	5955mos
1900-1930	daily	Eu	7280vni, 9730vni
2030-2100	daily	Eu	7280vni, 9730vni
2030-2100	daily	ME	7220vni, 9550vni
2330-2400	daily	As	9840vni, 12020vni

French	Days	Area	kHz
1200-1230	daily	As	7285han
1630-1700	daily	ME	7220vni, 9550vni
1830-1900	daily	Eu	7280vni, 9730vni
1930-2000	daily	Eu	5955mos, 7280vni, 9730vni
2100-2130	daily	Eu	7280vni, 9730vni
2100-2130	daily	ME	7220vni, 9550vni

German	Days	Area	kHz
2030-2100	daily	Eu	3985skn
2100-2130	daily	Eu	3985skn

Indonesian	Days	Area	kHz
1030-1100	daily	As	9840vni, 12020vni
1200-1230	daily	SEA	1242omo
1300-1330	daily	As	9840vni, 12020vni
1430-1500	daily	As	9840vni, 12020vni
1430-1500	daily	SEA	1242omo
2300-2330	daily	As	9840vni, 12020vni

Japanese	Days	Area	kHz
1100-1130	daily	As	9840vni, 12020vni
1200-1230	daily	As	9840vni, 12020vni
1400-1430	daily	As	9840vni, 12020vni
2200-2230	daily	As	9840vni, 12020vni

Khmer	Days	Area	kHz
1030-1100	daily	As	7285han
1230-1300	daily	SEA	1242omo
1230-1300	daily	As	7285han
2230-2300	daily	As	7285han

Lao	Days	Area	kHz
1330-1430	daily	SEA	1242omo
1330-1430	daily	As	7285han
2300-2400	daily	As	7285han

Russian	Days	Area	kHz
1130-1200	daily	As	7220vni, 12000vni
1230-1300	daily	As	7220vni, 12000vni
1630-1700	daily	Eu	7280vni, 9730vni
2000-2030	daily	Eu	5970wof, 7280vni, 9730vni

Spanish	Days	Area	kHz
0300-0330	daily	NAm	6175sac
0400-0430	daily	NAm	6175sac
1800-1830	daily	Eu	7280vni, 9730vni

Thai	Days	Area	kHz
1130-1200	daily	As	7285han
1130-1200	daily	SEA	1242omo
1300-1330	daily	As	7285han
1300-1330	daily	SEA	1242omo
1430-1500	daily	As	7285han
1530-1600	daily	As	7285han
2200-2230	daily	As	7285han

Vietnamese	Days	Area	kHz
0000-0100	daily	As	7285han
0130-0230	daily	NAm	6175sac
0430-0530	daily	NAm	6175sac
1500-1600	daily	ME	7220vni, 9550vni
1500-1600	daily	SEA	1242omo
1700-1800	daily	Eu	7280vni, 9730vni
1830-1930	daily	Eu	5955mos
2130-2230	daily	Eu	7370wof

Ann: English: "You are listening to Radio The Voice of Vietnam".
V: QSL-card.
Notes: The External Sce (VOV5) of the national broadcaster Voice of Vietnam. Schedule and frequency usage variable. Chinese prgr consist of Mandarin with some segments in Cantonese.

YEMEN (YEM)

REPUBLIC OF YEMEN RADIO (Gov)
⌨ See National Radio section.
SW: [SAN] Sana'a: 1 x 50, 1 x 100kW.
kHz: *6135, 9780*

Winter Schedule 2010/2011

Arabic	Days	Area	kHz
0300-0600	daily	ME	9780san†
0500-1500	daily	ME	6135san
1200-1800	daily	ME	9780san†
1900-2200	daily	ME	9780san†

English	Days	Area	kHz
1800-1900	daily	ME	9780san†

Key: † Irregular.

ZAMBIA (ZMB)

CVC RADIO CHRISTIAN VOICE (Rlg)
⌨ Private Bag E606, Lusaka, Zambia.
☎ +260 1 273191. 🖷 +260 1 279183.
E: voicefm@zamnet.zm **W:** www.voiceafrica.net
L.P: Acting SM: Mwiza Sinyangwe; Head of Transmissions: Edward Phiri.
SW: [LUS] Lusaka, Makeni Ranch: 2 x 100kW.
kHz: *4965, 6065*

Winter Schedule 2010/2011

English	Days	Area	kHz
0500-1700	daily	Af	6065lus
1700-2200	daily	Af	4965lus

Ann: English: "Radio Christian Voice".
Notes: Owned by South African based CVC Media, a branch of Christian Vision (for corporate details see CVC International, under United Kingdom). Service for Southern Africa, produced in CVC studios in Lusaka.

ZIMBABWE (ZWE)

VOICE OF ZIMBABWE (Gov)
⌨ 24, 7th St, Gweru, Zimbabwe.
☎ +263 4 498610. 🖷 +263 4 498613.
E: voiceof_zimbabwe@yahoo.com **W:** www.zbc.co.zw
Webcast: P
L.P: GM: Happison Muchechetere.
SW: [GWE] Gweru, Guinea Fowl: 1 x 100kW.
kHz: *4828, 5975*

Winter Schedule 2010/2011

English/Ndebele/ Shona	Days	Area	kHz
0430-1530	daily	Af	5975gwe†
1530-0430	daily	Af	4828gwe†

Key: † Irregular.
Ann: English: "This is the Voice of Zimbabwe".
Notes: Produced by the state-controlled "Zimbabwe Broadcasting Holdings (ZBH)". Established in 2007, but became active in July 2010.

CLANDESTINE AND OTHER TARGET BROADCASTS

Clandestine Broadcasts (Clan) are politically-motivated broadcasts produced by groups opposed to the government of the target country. **Other Target Broadcasts** are produced by non-governmental or governmental organisations and targetted at zones of regional or local conflict. Most COTBs are transmitted via the facilities of international transmitter operators.

Target: CAMEROON (CME)

SAWTU LINJIILA (VOICE OF THE GOSPEL) (Rlg)
✉ Centre Multimedia Chretien, B.P. 02, Ngaoundéré, Cameroon.
E: administration@sawtulinjiila.org; sawtulinjiila@yahoo.fr **W:** www.sawtulinjiila.org/sawtu; www.lutheranworld.org/lwf
LP: Dir: Rev Yaya Bournang.
kHz: *9800*

	Winter Schedule 2010/2011		
Fulfulde	**Days**	**Area**	**kHz**
1830-1900	daily	CAf	9800wer

Ann: Fulfulde: "Sawtu Linjiila".
Notes: On air, in various forms, since Nov 1966 and most recently (from March 2008) via Media Broadcast GMBH (see International Radio section under "Germany").

Target: CHINA (CHN)

IMG (Comm)
✉ McCormack House, Hogarth Business Park, Burlington Lane, Chiswick, London, W4 2TH, United Kingdom. (International HQ)
☎ +44 208 2335300. 🖷 +44 208 2335301.
E: web.feedback@imgworld.com **W:** www.imgworld.com
✉ 767 5th Avenue, New York, NY 10153, USA. (USA HQ)
LP: Chmn/CEO: Theodore J. Forstmann; Pres, IMG Int: Ian Todd.
kHz: *9300*

	Winter Schedule 2010/2011		
Mandarin	**Days**	**Area**	**kHz**
1245-1900	.....ss	CHN	9300tac

Notes: IMG is a sports, entertainment and media enterprise. IMG is currently providing reports/commentaries of British Premier league matches to China on SW, via the services of WRN Broadcast (see International Radio section, under "United Kingdom").

TAKAI SHINLINGDE SOUCHI (KEY TO OPEN THE SPIRIT) (Rlg)
✉ 766 Dongping Road, Taiping City, Taichung County 411, Taiwan.
☎ +886 4 2278 2091.
E: ikutao@ms47.hinet.net
W: ikt.webu.com.tw; www.ikttv.org
kHz: *1098, 7460*

	Winter Schedule 2010/2011		
Mandarin	**Days**	**Area**	**kHz**
1100-1157	daily	CHN	7460huw
1200-1300	daily	CHN	1098kou

Notes: On air since 1 May 2010. Programme sponsored by four organizations, including the I-Kuan Tao (IKT) Chong Cheng Fund Association, and promotes Taoism.

MINGHUI RADIO
✉ P.O. Box 250759, New York, NY 10025, USA.
E: editor@minghui.org **W:** www.mhradio.org
Webcast: D
SAT: Eutelsat W5.
kHz: *6030*

	Winter Schedule 2010/2011	
Chinese	**Days**	**Area**	**kHz**
1300-1400	daily	CHN	6030tsh

Ann: Mandarin: "Zhe shi Minghui Guangbo Diantai".
Notes: Falun Gong-affiliated and sister station to "Sound of Hope Radio". On air since 30 Dec 2005. Broadcasts 24hrs a day to Asia via Satellite.

SOUND OF HOPE RADIO INTERNATIONAL
✉ 6-4, Lane 84, GuóTài St, North District, Taichung 404, Taiwan.
E: contact@soundofhope.org; allenz@soundofhope.org
W: www.soundofhope.org
Webcast: D/L/P
✉ P.O. Box 70456, Sunnyvale, CA 94086, USA; 1010 Corporation Way, Palto Alto, CA 94303, USA.
☎ +1 866 432 7764. 🖷 +1 415 2765861.

E: 9ping@soundofhope.org
W: sohnetwork.com; asia-cast.com/shortwave-broadcasts (English)
LP: Dir: Allen Zeng.
kHz: *6280, 7105, 7280, 7310, 7525, 7560, 7565, 7590, 9150, 9380, 9450, 9540, 9635, 9990, 10500, 10970, 11100, 11250, 11760, 11765, 13100, 13680, 13800, 13970, 14400, 14700, 14970, 15140, 15750, 15850, 15970, 16100, 16700, 17650, 17920, 18180*

	Winter Schedule 2010/2011		
Chinese	**Days**	**Area**	**kHz**
0000-2400	daily	CHN	9150tai±, 9380tai±, 10500tai±, 10970tai±, 11100tai±, 11250tai±, 13100tai±, 13800tai±, 13970tai±, 14700tai±, 14970tai±, 15140tai±, 15850tai±, 15970tai±, 16100tai±, 17920tai±
0900-1100	.....ss	CHN	9540tsh, 11760tsh
1100-1300	daily	CHN	7280tsh
1200-1230	daily	CHN	15750dsb±
1300-1400	daily	CHN	7310tsh
1400-1430	daily	CHN	9990dsb±
1400-1600	daily	CHN	9450pao
1500-1530	daily	CHN	7525dsb±
1530-1600	daily	CHN	7590erv±
1600-1630	daily	CHN	7565dsb±
1600-1700	daily	CHN	11765tsh
2000-1300	daily	CHN	13680tai±
2200-1100	daily	CHN	18180tai±
2200-1700	daily	CHN	14400tai±, 16700tai±, 17650tai±
2200-2300	daily	CHN	7105tsh, 9635tsh
2200-2400	....fs.	CHN	6280tsh
2200-2400	daily	CHN	7560dsb±
2300-2400	daily	CHN	7310tsh

Key: ± Variable frequency.
Ann: Mandarin: "Xiwang zhi sheng guoji guangbo diantai".
V: QSL-card.
Notes: Established in June 2003. Falun Gong-related "Sound of Hope Radio International" is the shortwave programme of Sound of Hope Radio Network Inc. It is a global provider of Chinese language news and cultural programming for the Chinese community in over 20 major cities in USA, Canada, Australia, Germany, Sweden, Denmark, Taiwan and mainland China. SW broadcasts are subject to 'Firedragon',aka 'Firedrake', Chinese music jammer. At the top of the hour the jammer is off air for 15 minutes. Also reported on 1053kHz at approx 1700-1900. Relays 'Falun Dafa Radio' broadcasts.

VOICE OF CHINA (Clan)
✉ 2261 Morello Avenue, Suite A, Pleasant Hill, CA 94523, USA.
☎ +1 510 6872354. 🖷 +1 510 6877396.
E: info@china21century.org **W:** www.china21century.org
LP: Exec Producer: Hu Juying (Lily Hu).
kHz: *7270*

	Winter Schedule 2010/2011		
Mandarin	**Days**	**Area**	**kHz**
1400-1500	daily	CHN	7270tai
2300-2400	daily	CHN	7270tai

Ann: Mandarin: "Zhongguo zhi yin".
V: QSL-card.
Notes: On air since April 1991. Produced by the Foundation for China in the 21st Century, a U.S.-based non-profit organisation. Jammed.

VOICE OF TIBET (Clan)
✉ Voice of Tibet Foundation, Kirkegata 5, 0153 Oslo, Norway. (Administration)
☎ +47 22111209.
E: editor@vot.org; oystalme@gmail.com **W:** www.vot.org
Webcast: D
✉ Narthang Building, Gangchen Kyishong, Dharamsala-176215 H.P., India. (Main Editorial Office)

☎ +91 1892 228179. 🖷 +91 1892 224913.
L.P: Project Mgr: Øystein Alme; Project Coordinator: Chophel Norbu; Editor in Chief: Karma Yeshi.
SAT: PAS 10.
kHz: *15430, 15521, 15523, 15527, 15542, 15548, 15568, 15572, 15578, 15582, 17560*

Winter Schedule 2010/2011

Mandarin	Days	Area	kHz
1200-1210	daily	CHN	15523dsb
1210-1230	daily	CHN	15542dsb
1330-1342	daily	CHN	15521dsb
1342-1400	daily	CHN	15527dsb

Tibetan	Days	Area	kHz
1230-1235	daily	CHN	15548dsb
1235-1307	daily	CHN	15568dsb
1307-1330	daily	CHN	15572dsb
1330-1400	daily	IND,NPL	15430dha
1330-1407	daily	CHN	15578dsb
1400-1430	daily	IND,NPL	17560mdc
1407-1430	daily	CHN	15582dsb

Ann: Mandarin: "Zheli shi Nuowei Xizang zhi Sheng Guangbo Diantai huayu jiemu"; Tibetan: "'Di nor we bod kyi rlung 'phrin khang yin".
V: QSL-card.
Notes: On air since July 1996. Licensed radio station in Norway, run by the Voice of Tibet Foundation. Established by the organisations Worldview Rights, the Norwegian Human Rights House and the Norwegian Tibet Committee. Jammed. Frequencies are often changed.

Target: CONGO (Dem. Rep. of) (COD)

RADIO OKAPI
🖳 See National Radio section.
Webcast: D
kHz: *11690*

Winter Schedule 2010/2011

French/Various	Days	Area	kHz
0400-0500	daily	COD	11690mey

Ann: French: "Radio Okapi".
V: QSL-card. (1 USD)

Target: CUBA (CUB)

WRMI - RADIO MIAMI INTERNATIONAL
🖳 See International Radio section, under "USA".
SW: See International Radio section, under "USA".
Notes: WRMI relays a number of anti-Government broadcasts to Cuba, in Spanish, from various programmme producers. Usually on 9955kHz.For programme details, see www.wrmi.net/schedule.php

RADIO REPÚBLICA (Clan)
🖳 P.O. Box 110235, Hialeah, FL 33011, USA.
☎ +1 305 2794416.
E: radiorepublica@gmail.com
W: www.radiorepublica.org
Webcast: L
L.P: Asst to Prgr Coordinator: Maria A. Lima.
kHz: *5954, 9490, 9965*

Winter Schedule 2010/2011

Spanish	Days	Area	kHz
0000-0300	mtwtf..	CUB	9490sac
1200-1500	daily	CUB	9965†,*
2100-0400	daily	CUB	5954†,*

Key: † Irregular; * Via ELCOR tx in Guápiles, Costa Rica.
Ann: Spanish: "Esta es Radio República. La voz del Directorio Democrático Cubano", "Radio República, la voz de Cuba libre".
Notes: On air since August 2005. Produced by Directorio Democrático Cubano. Jammed. Also on additional freqs, which may include 5910, 6100, 6135, 6155, 9735kHz & others. Freqs/schedules are changed without notice to escape from jamming.

Target: DJIBOUTI (DJI)

LA VOIX DE DJIBOUTI (LVD)
☎ +33 1 76660452 (France); +1 613 4821744 (USA).
E: info@lavoixdedjibouti.com
W: www.lavoixdedjibouti.com
Webcast: D
kHz: *21525*

Winter Schedule 2010/2011

French/Somali/

Various	Days	Area	kHz
1200-1300	...t...	DJI	21525sam

Ann: French: "Vouz ecoutez la Voix de Djibouti"; Somali: Halkan wa Idha'adda ... Radio Odki Djibouti".
V: Email.
Notes: On SW since 7 January 2010.

Target: ERITREA (ERI)

DIMTSI WEGAHTA (VOICE OF THE DAWN) (Clan)
🖳 Based in Mekelle, Ethiopia.
W: www.sallina.com
Webcast: D
L.P: SM: Mr Hagos.
kHz: *918*

Winter Schedule 2010/2011

Arabic/Tigrinya	Days	Area	kHz
0300-0600	daily	ERI	918mek
1400-2100	daily	ERI	918mek

Ann: Tigrinya: "Dimtsi Wegahta".
Notes: Previously reported as "Eastern Radio" when operating from Sudan. Run by an Eritrean opposition organisation.

EPDP RADIO / VOICE OF ERITREAN PEOPLE (Clan)
🖳 c/o Eriträisches Zentrum, Neue Mainzer Str. 24, D-60311 Frankfurt, Germany.
☎ +49 69 24248583. 🖷 +49 69 24248637.
E: comments@harnnet.org
W: www.harnnet.org/index.php/epdp-radio
Webcast: D
kHz: *11775*

Winter Schedule 2010/2011

Various	Days	Area	kHz
1700-1800	...t...	ERI	11775nau

Ann: Arabic: "Idha'at Sawt al-Hurriya, Sawt u-Hesb as-Sha'ab Demokratiya Eritriyah"; Tigriniya: "Ezi Dimtsi Nharnet yaw. Dimtsi Nharnet, Dimtsi Hesbi Demokrasi Ertran".
V: QSL-letter.
Notes: On air since March 2000. Produced by the Eritrean People's Democratric Party (EPDP).

VOICE OF ASENA (Clan)
E: aseye.asena@googlemail.com; aseye@asena-online.com **W:** www.assenna.com
Webcast: D
L.P: Dir: Amanuel Eyasu.
kHz: *9605*

Winter Schedule 2010/2011

Tigrinya	Days	Area	kHz
1730-1800	m...f..	ETH	9605sam

Ann: Tigrinya: "Ezi dimtsi Asena Eyu".
Notes: Started broadcasts on 16 Feb 2009.

VOICE OF DEMOCRATIC ALLIANCE (Clan)
🖳 c/o Eritrean Democratic Alliance P.O. Box 13043, Khartoum, Sudan.
E: erit_alliance_2008@yahoo.com
W: www.erit-alliance.com (EDA)
kHz: *7165, 9560*

Winter Schedule 2010/2011

Afar	Days	Area	kHz
1530-1600	.t.t.s.	EAf	7165gjw†, 9560gjw±

Arabic	Days	Area	kHz
1500-1530	m.w.f.s	EAf	7165gjw†, 9560gjw±

Kunama	Days	Area	kHz
1530-1600	m.w.f..	EAf	7165gjw†, 9560gjw±

Tigrinya	Days	Area	kHz
1500-1530	.t.t.s.	EAf	7165gjw±, 9560gjw±
1530-1600	......s	EAf	7165gjw†, 9560gjw±

Key: ± Variable frequency; † Irregular.
Ann: Arabic: "Sawt al-Tahalufa al-Dimuqrati".
Notes: On air since April 2005. Produced by the Eritrean Democratic Alliance (EDA). Via txs of the Ethiopian state broadcaster 'Radio Ethiopia' (see International Radio section, under "Ethiopia").

VOICE OF ERITREA (Clan)
W: www.harnnet.org
Webcast: D
kHz: *7165, 9560*

Winter Schedule 2010/2011

Tigrinya	Days	Area	kHz
0400-0430	.t.t.s.	ERI	7165gjw†, 9560gjw±
1800-1830	.t.t.s.	ERI	7165gjw†, 9560gjw±

Key: ± Variable frequency; † Irregular.
Ann: Tigrinya: "Dimtsi Ertrai".
Notes: Produced by Eritrean People's Democratic Party (EPDP) (was produced by the now defunct EPM, which merged with EDP and EPP to form EPDP).

VOICE OF MESELNA DELINA (Clan)
🖃 17326 Edwards Road, Suite A230, Cerittos, CA 90703, USA.
☎ +1 562 9262424. 🖷 +1 562 9262423.
E: admin@meseley.net **W:** meseley.net
Webcast: D (Archived broadcasts only)
☎ +27 12 4404749. (South Africa)
kHz: *9605*

Winter Schedule 2010/2011

Tigrinya	Days	Area	kHz
1730-1800	.t.t.s.	ERI	9605sam

Ann: Tigrinya: "Ezi Dimtsi Meselna Delina".
Notes: On air since January 2005. Produced by Tesfa Delina Foundation, Inc.

VOICE OF PEACE AND DEMOCRACY OF ERITREA (Clan)
🖃 c/o Radio Ethiopia, P.O. Box 1020, Addis Ababa, Ethiopia.
kHz: *7165, 9560*

Winter Schedule 2010/2011

Tigrinya	Days	Area	kHz
0400-0500	m.w.f.	ERI	7165gjw†, 9560gjw±
1800-1830	m.w.f.	ERI	7165gjw†, 9560gjw±

Key: ± Variable frequency; † Irregular.
Ann: Tigrinya: "Yeh Radio Demtsi Selaman Demokratia Ertrai".
Notes: On air since February 1999. Via txs of the Ethiopian state broadcaster 'Radio Ethiopia' (see International Radio section, under "Ethiopia").

Target: ETHIOPIA (ETH)

EOTC HOLY SYNOD RADIO (Rlg)
🖃 P.O. Box 7097, Los Angeles, CA 90007, USA.
☎ +1 408 2306629.
E: radioteam@eotcholysynod.org; eotcholysynod@eotcholysynod.org
W: www.eotcholysynod.org
Webcast: D
Notes: On the air since July 2007. Produced by the Holy Synod of the Ethiopian Orthodox Tewahedo Church. No SW transmissions at time of publication.

ETHIOPIA ADERA DIMTSE RADIO (Clan)
E: ecadf@ecadforum.com; info@ecadforum.com
W: www.ecadforum.com/YE_ETHIOPIA_ADERA_DIMTSE.html
Webcast: D
Ann: Amharic: "Yeh ye Ethiopia Adera Dimtse Radio nay".
Notes: No SW transmissions at time of publication.

GINBOT 7 DIMTS RADIO (Clan)
🖃 P.O. Box 4916, 2003EX, Haarlem, The Netherlands.
☎ +44 203 2869661.
E: g7radio@ginbot7.org; info@ginbot7.org
W: www.ginbot7.org
Webcast: D
🖃 8647 Richmond Highway, 652, Alexandria, VA 22309, USA; P.O. Box 56281, London, N4 9BH, United Kingdom.

SW: Broadcast via txs of "The Voice of the Broad Masses of Eritrea" in Asmara, Eritrea (see National Radio section, under "Eritrea").
kHz: *837, 5945, 6170, 7175, 7185*

Winter Schedule 2010/2011

Amharic	Days	Area	kHz
0600-0630	.t.t.s.	ETH	837asm, 5945asm±, 6170asm±, 7175asm±, 7185asm±
1700-1730	.t.t.s.	ETH	837asm, 5945asm±, 6170asm±, 7175asm±, 7185asm±

Key: ± Variable frequency.
Ann: Amharic: "Yeh Ginbots Sabat Dimtse now".
Notes: Operated by "Ginbot 7: Movement for Justice, Freedom and Democracy". On air since September 2008.

MELEKET ETHIOPIA RADIO (Clan)
E: radio@meleketethiopia.org
W: www.meleketethiopia.org
Webcast: D
Ann: Amharic: "Yeh Meleket Ethiopia Radio".
Notes: On air since end of October 2009. No SW transmissions at time of publication.

RADIO BILAL (Clan)
🖃 Bilal Communication Inc., 4324 Georgia Avenue NW, Washington, D.C. 20011, USA.
☎ +1 202 2391485.
E: bilalradio@yahoo.com **W:** www.radio-bilal.com/radio
Webcast: D
kHz: *9345*

Winter Schedule 2010/2011

Amharic	Days	Area	kHz
1800-1830	daily	ETH	9345sam

Ann: Amharic: "Yeh Radio Bilal".

RADIO DEMOCRACY (Clan)
🖃 P.O. Box 8141, Silver Spring, MD 20910, USA.
☎ +1 301 5784465. 🖷 +1 202 4498363.
E: questions@eprp-ihapa.com **W:** www.eprp-ihapa.com
Webcast: D
🖃 P.O .Box 88675, Los Angeles, CA 90009, USA.
kHz: *21555*

Winter Schedule 2010/2011

Amharic	Days	Area	kHz
0900-1000	......s	ETH	21555sam

Ann: Amharic: "Yeh Democracia".
Notes: On air since end of October 2009. Produced by Ethiopian People's Revolutionary Party - EPRP (Democratic).

RADIO XORIYO (Clan)
🖃 P.O. Box 27618, Toronto, Ontario, Canada, M3A 3B8.
E: raadioxoriyo@yahoo.com **W:** www.ogaden.com
Webcast: D
Ann: Somali: "Ku soo dhawaada Radio Xoriyo codkii ummadda Ogadeniya".
V: QSL-email.
Notes: On air since May 2000. Produced by the Ogaden National Liberation Front (ONLF). No transmissions on shortwave at time of publication.

VOICE OF ETHIOPIAN UNITY (Clan)
🖃 c/o Finote Democracy, P.O. Box 73337, Washington, D.C. 20056, USA.
🖷 +1 202 2917645.
E: efdpu@finote.org **W:** www.finote.org
Webcast: D
kHz: *11830*

Winter Schedule 2010/2011

Amharic	Days	Area	kHz
1700-1800	..w...s	ETH	11830nau

Ann: Amharic: "Yih Finote Demokrasi ye Ethiopia andinet dimtsi now".
V: QSL-letter.
Notes: On the air since January 2000. Produced by Finote Democracy, a group of Ethiopians in Europe and the USA. Linked with the Ethiopian Peoples Revolutionary Party.

VOICE OF OMORO LIBERATION FRONT (Clan)
kHz: 11760

Winter Schedule 2010/2011
Oromo	Days	Area	kHz
1600-1630	...t.s	ETH	11760wer

VOICE OF OROMIYA LIBERATION (Clan)
Webcast: D
W: www.oromoliberationfront.org/sbo.html
Ann: Amharic: "Yeh ye Oromonne sana dimtse now".
Notes: No SW transmissions at time of publication.

VOICE OF OROMO LIBERATION (Clan)
✉ Postfach 510620, D-13366 Berlin, Germany.
☎ +49 30 4943372. 🖷 +49 30 4943372.
E: sbo13366@aol.com
W: www.oromoliberationfront.org/sbo.html
Webcast: D
L.P: Secretary, SBO Committee: Taye Teferra.
kHz: 11810

Winter Schedule 2010/2011
Amharic	Days	Area	kHz
1730-1800	..w....	ETH	11810iss
Oromo	Days	Area	kHz
1700-1730	..w....	ETH	11810iss
1700-1800	......s	ETH	11810iss

Ann: Amharic: "Radio Bilisummaa Oromoo"; Oromo: "Kun Sagalee Bilisummaa Oromoo".
V: QSL-letter.
Notes: On air since July 1988 (transmitting from outside of Ethiopian territory since 1996). Produced by the Oromo Liberation Front (OLF). Also carried via "Voice of the Broad Masses of Eritrea" transmitters (see National Radio section, under "Eritrea").

Target: GAMBIA (GMB)

BAATI REWMI RADIO (VOICE OF THE NATION)
✉ c/o STGDP, P.O. Box 48321, Doraville, GA 30362, USA.
E: matarss1@yahoo.com; banka.manneh@savethegambia.org (STGDP) W: www.savethegambia.org
Webcast: D
L.P: Pres, STGDP: Banka Manneh.
Ann: Engish: "Save the Gambia Democracy project brings you Baati Rewmi".
V: QSL-email.
Notes: On the air since 22 May 2010. Produced by the Save The Gambia Democracy Project (STDGP). Brokered by Jeff White (WRMI). Known previously (2005/6) as "Voices from the Diaspora". No SW broadcasts at time of publication.

Target: INDIA (IND)

VOICE OF JAMMU & KASHMIR FREEDOM MOVEMENT (Clan)
✉ P.O. Box 102, Muzaffarabad 13100, Pakistan.
L.P: PM: Rehan J.
kHz: 3995

Winter Schedule 2010/2011
Kashmiri/Various	Days	Area	kHz
0300-0400	daily	IND	3995isl†
1300-1445	daily	IND	3995isl†

Key: † Irregular.
Ann: Urdu: "Ye Sadaye Hurriyat Jammu Kashmir hai".
V: QSL-letter. No rp.
Notes: On air since 1999. Produced by the Jammu and Kashmir Freedom Movement (JKFM). Presumed to use trs of the Pakistan BC in Rewat (Islamabad). Languages used in addition to Kashmiri are Urdu, Balti, Gojri, English and Hindi.

Target: IRAN (IRN)

RADIO RAHAYE IRAN (Clan)
✉ c/o Oghab-e-Iran, 5319 University Dr. # 310, Irvine, CA 92612, USA.
☎ +1 949 9395227. 🖷 +1 949 6442122.
E: eagleofiran@gmail.com
W: www.eagleofiran.com (Oghab-e-Iran)
Webcast: L
kHz: 5825

Winter Schedule 2010/2011
Persian	Days	Area	kHz
1630-1730	m.w.f..	IRN	5825erv

Ann: Farsi: "In ja Los Angeles, Radiae Rahaie Eran, barnamae mahsuse guruhee nizamii siyasii Ughabi Eran".
Notes: Produced by the US-based opposition group Eagle of Iran (Oghab-e-Iran). On SW since October 2010.

RADIO VOICE OF KURDISTAN (Clan)
✉ c/o Kurdistan Democratic Party of Iran, Storgården 50, SE-58644 Linköping, Sweden.
☎ +964 770 1597268. (Sulaimaniya, Iraq)
E: info@radiokurdistan.net
kHz: 3930

Winter Schedule 2010/2011
Farsi	Days	Area	kHz
0300-0330	daily	IRN	3930-±
1400-1430	daily	IRN	3930-±
Kurdish	Days	Area	kHz
0200-0300	daily	IRN	3930-±
1300-1400	daily	IRN	3930-±

Key: ± Variable frequency.
Ann: Farsi: "Im Radyo Sedaye Kordestane"; Kurdish: "Era Radyo Dengi Kurdistana".

VOICE OF IRANIAN KURDISTAN (Clan)
✉ Reportedly based in Salah Al-Din, Iraqi Kurdistan.
☎ +871 762280073. (Intelsat) 🖷 +871 762280074. (Intelsat)
E: info@rdkiran.com; pdkicanada@pdki.org
W: www.rdkiran.com
Webcast: D
✉ c/o AFK, BP 102, F-75623 Paris Cedex 13, France (Correspondence); PDKI Canada Bureau, P.O. Box 29010, London, ON, Canada N6G 2V3.
SW: Transmitter reported to be in the Salah Al Din area of Northern Iraq.
kHz: 4870

Winter Schedule 2010/2011
Farsi	Days	Area	kHz
0430-0500	daily	IRN	4870-±
1430-1500	daily	IRN	4870-±
Kurdish	Days	Area	kHz
0300-0430	daily	IRN	4870-±
1330-1430	daily	IRN	4870-±

Key: ± Variable frequency.
Ann: Farsi: "In Seda-ye Kordestan-e Iran"; Kurdish: "Erê Dengê Kurdistana Îranê".
V: QSL-email.
Notes: On air 1973-1975, and again since 1980. Produced by the Democratic Party of Iranian Kurdistan (PDKI). Jammed.

Target: IRAQ (IRQ)

DENGÊ MEZOPOTAMYA (VOICE OF MESOPOTAMIA)
✉ Roj NV, Fabriekstraat 6, B-9470 Denderleeuw, Belgium.
☎ +32 53 648827. 🖷 +32 53 680779.
E: info@denge-mezopotamya.com
W: www.denge-mezopotamya.com
Webcast: L
L.P: SM: Zerdest Peri.
kHz: 7540, 11530

Winter Schedule 2010/2011
Kurdish	Days	Area	kHz
0500-1500	daily	ME	11530smf
1500-2100	daily	ME	7540smf

Ann: Kurdish: "Radyoya Dengê Mezopotamya".
Notes: On air since May 2001. Station is licensed in Belgium and is produced by the Kurdish-Belgian media production company Roj NV.

Target: KOREA, (D.P.R.) (KRE)

VOICE OF FREEDOM (VOICE OF MARTYRS) (Rlg)
P.O. Box 92, Yangcheon, Yangcheon-gu, Seoul, Republic of Korea.
☎ +82 2 2699 0976. 🖷 +82 2 2699 0978.
W: www.fnkvof.com
kHz: *6240*

Winter Schedule 2010/2011

Korean	Days	Area	kHz
1600-1700	daily	KRE	6240tac

Ann: Korean: "Daeham Minguk Seoul-eseo bonae deurineun Jayu-e Sori, Bukjoseon Seongyo Bangsong-imnida".
Notes: On air since 31 October 2009 as successor to the "Voice of Free Radio", which was on air from 8 March 2008 to 30 September 2009.

VOICE OF WILDERNESS (Rlg)
P.O. Box 8, Yeongdong, Seoul, 135-660, Republic of Korea.
☎ +82 2 7968846. 🖷 +82 2 7968846.
E: main@cornerstone.or.kr; info@cornerstonelive.net
W: www.cornerstone.or.kr; cornerstonelive.net
Webcast: D
Cornerstone Ministries Int., P.O. Box 4002, Tustin, CA 92781, USA.
☎ +1 714 4840042. 🖷 +1 714 4840046.
E: info@cornerstoneusa.org (USA)
W: www.cornerstoneusa.org (under construction)
kHz: *1566, 9390*

Winter Schedule 2010/2011

Korean	Days	Area	kHz
1300-1400	daily	KRE	9390dsb*
2000-2100	.....s.	KRE	1566jej**

Key: * via BVB; ** via HLAZ.
Ann: Korean: "Gwangya-e Sori Bangsong-imnida".
Notes: Produced by Cornerstone Ministries International. Previously known as "North Korea Missionary Broadcast".

FURUSATO NO KAZE/ILBON-E BARAM
Policy Planning Division, Headquarters for the Abduction Issue, Cabinet Secretariat, 1-6-1 Nagata-cho, Chiyoda-ku, Tokyo 100 8968, Japan.
☎ +81 3 52532111. 🖷 +81 3 35813781.
E: info@rachi.go.jp **W:** www.rachi.go.jp/jp/shisei/radio
Webcast: D
kHz: *9780, 9950, 9965, 9975*

Winter Schedule 2010/2011

Japanese	Days	Area	kHz
1330-1357	daily	KRE	9950tai
1430-1500	daily	KRE	9950hbn
1600-1630	daily	KRE	9780tnn

Korean	Days	Area	kHz
1300-1330	daily	KRE	9950tai
1500-1530	daily	KRE	9975hbn
1530-1600	daily	KRE	9965hbn

Ann: Japanese: "Furusato no Kaze"; Korean "Ilbon-e Baram".
V: QSL-letter.
Notes: On air since July 2007. "Ilbon-e-Baram" broadcasts in Korean only and is sometimes referred to as "Nippon no Kaze". "Furusato no Kaze" broadcasts in Japanese only.

RADIO FREE CHOSUN
(121-821) 3F, 384-20 Mangwon-dong, Mapo-gu, Seoul, Republic of Korea.
☎ +82 2 3330577. 🖷 +82 2 3330577.
E: rfchosun@rfchosun.org
W: www.rfchosun.org; rfchosun.org
Webcast: D
L.P: Pres: Han Ki Hong; Programme Mgr: An Kyoung Hee.
kHz: *6225, 7505, 11560*

Winter Schedule 2010/2011

Korean	Days	Area	kHz
1200-1300	daily	KRE	11560erv
1500-1600	daily	KRE	6225dsb
2000-2100	daily	KRE	7505tac

Ann: Korean: "Inmin-e sori, jeongni-e hamseong, Jayu Joseon Bangsong-imnida".

Notes: Launched on 5 December 2005. Operated by the Network for North Korean Democracy and Human Rights.

SHIOKAZE (SEA BREEZE)
c/o COMJAN, 2-3-8-401 Koraku Bunkyo Ward, Tokyo 112-0004, Japan.
☎ +81 3 56845058. 🖷 +81 3 56845059.
E: chosakai@circus.ocn.ne.jp
W: www.chosa-kai.jp/SWR.html
Webcast: D (very limited archive)
L.P: Producer/Editor: Tatsuru Murao (also Director of COMJAN).
kHz: *5955, 5985*

Winter Schedule 2010/2011

English/Korean/ Japanese/ Mandarin	Days	Area	kHz
1400-1430	daily	KRE	5985yam
2030-2100	daily	KRE	5955yam

Ann: English: "This is the Shiokaze, Sea Breeze, from Tokyo, Japan"; Japanese: "JSR, Kochirawa Shiokaze desu"; Korean: "Yeogineun Shiokaze, Badatbaramimnida"; Mandarin: "Zheshi Shiokaze, Chaofeng Bosong".
V: QSL-card. (Issued if a 10 USD money order donation is sent together with report)
Notes: Operational since 20 October 2005, by the Investigation Commission on Missing Japanese, Probably Related to North Korea (COMJAN). Aimed at reaching Japanese citizens believed to have been abducted to North Korea. Jammed. Frequencies subject to change without notice.

VOICE OF FREEDOM
1, 3-ga, Yongsan-dong, Yongsan-gu, Seoul, Republic of Korea.
☎ +82 2 7484662.
FM/DAB: FM: 103.1MHz, 107.3MHz (Republic of Korea, exact location unknown).
Ann: Korean: "Yeogineun Jayu-eui Sori Bang-imnida".
Notes: Station re-activated on 24 May 2010, after 6 years off air. Reported to be administered by the Republic of Korea Government Defence Department. In the future may also use MW, in addition to FM. Operates to the following daily schedule, in Korean, targeting the Democratic People's Republic of Korea: 0300-0400, 0900-1300, 1500-1900 and 2100-0100 on both 103.1MHz and 107.3MHz FM.

ECHO OF HOPE (VOH) (Clan)
Based in, and transmitted from, Republic of Korea.
SW: Relayed via KBS World Radio transmitters (See under Republic of Korea).
kHz: *3985, 6003, 6348*

Winter Schedule 2010/2011

Korean	Days	Area	kHz
0254-0854	.....ss	KRE	3985hwa, 6003hwa, 6348hwa
0854-1900	daily	KRE	3985hwa, 6003hwa, 6348hwa
1900-2303	mtwt..s	KRE	3985hwa, 6003hwa, 6348hwa

Ann: Korean: "Huimang-e meari pangsong-imnida, VOH".
Notes: Until June 1973 was broadcasting under the name "The Voice of Reunification". Claims to be run by the General Union of Overseas Compatriots, but is operated by the South Korean National Intelligence Service. Jammed. "VOH" is an abbreviation of "Voice of Hope".

FREE NORTH KOREA RADIO (Clan)
P.O. Box 92, Mok-dong, Yangcheon-gu, Seoul 158-600, Republic of Korea.
☎ +82 2 26990977. 🖷 +82 2 26990978.
E: mini6915@hanmail.net **W:** www.fnkradio.com
Webcast: D (various audio/video archives)
L.P: Dir: Kim Seong Min.
kHz: *7505, 7530*

Winter Schedule 2010/2011

Korean	Days	Area	kHz
1200-1400	daily	KRE	7505dsb
1900-2100	daily	KRE	7530erv

Ann: Korean: "Jigeumbuteo Daehan Minguk sudo Seoul-eso bonae-neun Jayu Bukhan Bangsong-eul sijakhagesseumnida".
Notes: Founded on 20 April, 2004 by North Korean defectors. On air since 7 December 2005.

NORTH KOREA REFORM RADIO (Clan)

290-96, Sindang 6-dong, Jung-gu, Seoul, 100-824, Republic of Korea.
☎ +82 1 22426512. 🖷 +81 2 64426512.
E: nkreform@naver.net **W:** www.nkreform.net; www.nkreform.com
Webcast: D
kHz: *7590*

Winter Schedule 2010/2011

Korean	Days	Area	kHz
1500-1600	daily	KRE	7590tac

Ann: Korean: "Inmini baraneun saeroun sesang-ul hamgge ggumg-guneun Joseon Gyaehyeok Bangsong-imnida".
Notes: On the air since December 2007.

OPEN RADIO FOR NORTH KOREA (Clan)

P.O. Box 158, Mapo, Seoul, 121-600, Republic of Korea.
☎ +82 50 54707470; +82 10 71512785. 🖷 +82 50 54717470.
E: nkradio@nkradio.com
W: www.nkradio.org; english.nkradio.org
Webcast: D
3901 Fair Ridge Dr. Faifax, VA 22033, USA.
☎ +1 202 2462571.
L.P: Exec Dir: Tae Kung Ha.
FM/DAB: FM: 92.3MHz (Republic of Korea, exact site and power unknown).
kHz: *774, 7480, 7560*

Winter Schedule 2010/2011

Korean	Days	Area	kHz
1400-1500	daily	KRE,CHN	7560tac
1900-2000	daily	KRE	774mbc
2100-2200	daily	KRE,CHN	7480erv

Ann: Korean: "Jayue gwangjang, huimang-e sori, Yeollin Bukhan Bangsong-imnida".
V: QSL-card.
Notes: Active since 7 December 2005. The station broadcasts various programmes aimed at North Korea and NE China, and is produced in Republic of Korea. Also broadcasts on FM in Korean to KRE, with the following schedule: 1900-2000 daily on 92.3MHz

VOICE OF THE PEOPLE (Clan)

Based in, and transmitted from, Republic of Korea.
kHz: *3480, 3912, 4450, 6518, 6600*

Winter Schedule 2010/2011

Korean	Days	Area	kHz
0500-2305	daily	KRE	3480goy, 3912goy, 4450goy, 6518goy, 6600goy

Ann: Korean: "Inmin-e sori pangsong-imnida".
Notes: On air since June 1985. Claims to be run by the Korean Workers Union, but is operated by the South Korean National Intelligence Service. Jammed.

Target: KOREA, Rep. of (KOR)

PYONGYANG BRANCH OF THE ANTI-IMPERIALIST NATIONAL DEMOCRATIC FRONT (PYONGYANG BROADCASTING STATION RELAY)

c/o AINDF Mission, Munsu-dong, Taedonggang District, Pyongyang, Democratic People's Republic of Korea.
☎ +850 2 3814505.
E: aindf@celery.ocn.ne.jp (AINDF)
W: aindf.dyndns.org (AINDF)
c/o AINDF, Grenier Osawa 107, 40 Nando-machi, Shinjuku-ku, Tokyo 162-0837, Japan.
☎ +81 3 52610331. 🖷 +81 3 52610332.
E: ndfsk@celery.ocn.ne.jp **W:** ndfsk.dyndns.org
kHz: *1053, 3480, 4450, 4557*

Winter Schedule 2010/2011

Korean	Days	Area	kHz
0755-1405	daily	KOR	1053hju, 3480won, 4450pyo, 4557hju
2155-0400	daily	KOR	1053hju, 3480won, 4450pyo, 4557hju

Ann: Korean: "Yeogineun Banje Minjok Minju Jeonseon Pyongyang Jibuimnida" (at s/on and s/off).
IS: "We Are One".
V: QSL-letter. (Rpt to Japan address)

Notes: Broadcasting service claimed to be provided by the Pyongyang Mission of the [South Korean] Anti-imperialist National Democratic Front (AINDF). Believed to be a North Korean government operation. Jammed.

Target: LAOS (LAO)

MOJ THEM RADIO

P.O. Box 75666, Saint Paul, MN 55175-0666, USA.
☎ +1 651 2302422
E: hmoob@mojthem.com **W:** www.mojthem.com
Webcast: D
Notes: No SW transmissions at time of publication.

Target: LEBANON (LBN)

RADIO MASHREQ (Clan)

P.O. Box 52341, 4062 Limassol, Cyprus. (Correspondence)
E: almachrek@gmail.com
Suite 115, 5505 Connecticut Avenue N.W., Washington, D.C. 20015-2601, USA.
☎ +1 202 4780261. 🖷 +1 202 4780261.
MW: [MEA] Metula: 756kHz 50kW.
FM/DAB: FM: 99.1MHz.
kHz: *756*

Winter Schedule 2010/2011

Arabic	Days	Area	kHz
0500-2200	daily	LBN	756mea

Ann: Arabic: "Al-idha'at al-Mashriqiyah".
V: QSL-email.
Notes: On air since January 2001 (tests since December 2000). Presumed to be run by the Israeli Secret Service. Officially produced by the news agency Carmelnews.

Target: MYANMAR (BRM)

DEMOCRATIC VOICE OF BURMA (Clan)

P.O. Box 6720, St Olavs Pass, N-0130 Oslo, Norway.
☎ +47 22868486. 🖷 +47 22868471.
E: acn@dvb.no **W:** www.dvb.no
Webcast: D (Available only on: burmese.dvb.no)
L.P: Exec Dir/Chief Editor: Aye Chan Naing.
kHz: *7440, 7510, 11515, 17790*

Winter Schedule 2010/2011

Burmese	Days	Area	kHz
1430-1530	daily	BRM	11515erv, 17790mdc
2330-0030	daily	BRM	7440wer, 7510erv

Ann: Burmese: "Democratic Myanmar a-Than".
V: QSL-card.
Notes: On the air since July 1992. Founded by the exile organisation National Coalition Government of the Union of Burma. DVB holds a Norwegian broadcasting license. Broadcasts are in Burmese, with 15 mins per transmission allocated to the Arakan, Chin, Karen, Karenni, Kayan, Kachin, Mon and Shan languages.

Target: PAKISTAN (PAK)

RADIO SEDAYEE KASHMIR (Clan)

c/o All India Radio (AIR), Akashvani Bhavan, Sansad Marg, New Delhi-110001, India.
kHz: *4870, 6100*

Winter Schedule 2010/2011

Dogri	Days	Area	kHz
0310-0330	daily	PAK	4870del
0810-0830	daily	PAK	6100del
1510-1530	daily	PAK	4870del

Kashmiri	Days	Area	kHz
0230-0310	daily	PAK	4870del
0730-0810	daily	PAK	6100del
1430-1510	daily	PAK	4870del

Ann: Urdu: "Ye Radio Sedayee Kashmir".
Notes: On air since early 2003. Radio Sedayee Kashmir is a prgr representing the views of the Indian government in the dispute with Pakistan over Kashmir.

Target: SIERRA LEONE (SRL)

COTTON TREE NEWS
▣ CTN, Fourah Bay College, Mass Communication Department, Mount Aureole Suburb, Freetown, Sierra Leone.
☎ +232 33 596737; +232 76 413201.
E: info@hirondelle.org **W:** www.cottontreenews.org
Webcast: D
LP: Project Co-ordinator: Graeme Loten; Editor in Chief: Joshua P Nicol; Admin Officer: Zainab Kamara.
FM/DAB: FM: 103.0; 107.3MHz (Freetown).
kHz: *11875*

Winter Schedule 2010/2011

English/Krio/
Limba/Mende/

Temne	Days	Area	kHz
0730-0800	daily	SRL	11875rmp

Ann: English: "CTN"; "This is CTN".
Notes: Produced by Fondation Hirondelle, Media for Peace and Dignity, in partnership with Fourah Bay College at the University of Sierra Leone, and the United Nations Integrated Peacebuilding Office in Sierra Leone (UNIPSIL). CTN is funded by the European Union, Ireland and Germany.

Target: SOMALIA (SOM)

BAR-KULAN RADIO
▣ Studio located in Nairobi, Kenya.
E: barkulanradio@gmail.com **W:** www.bar-kulan.com
Webcast: D/L
LP: Dir: David Smith.
kHz: *9960, 15750*

Winter Schedule 2010/2011

Somali	Days	Area	kHz
0500-0600	daily	SOM	15750dha
1600-1700	daily	SOM	9960mey

Ann: Somali: "Radio Bar-Kulan".
V: QSL-card.
Notes: Operational since March 2010. UN-aided and in support of the UN and African Union Mission in Somalia (AMISOM). Run by the same management team that set up "Radio Okapi".

IRIN RADIO
▣ P.O. Box 30218, 00100 Nairobi, Kenya.
☎ +254 733 860082. ▤ +254 20 622129.
E: feedback@irinnews.org; ben@irinnews.org
W: www.irinnews.org/radio.aspx
Webcast: D/P
☎ +1 917 3679228. (IRIN New York Office)
LP: Dir: Ben Parker.
kHz: *17680*

Winter Schedule 2010/2011

Somali	Days	Area	kHz
0830-0930	daily	EAf	17680dha

Ann: Somali: "Halkalee Walanta Somalia Radio IRIN".
Notes: Produced by the UN Office for the Coordination of Humanitarian Affairs. On shortwave from 1 April 2008. IRIN is an abbreviation of 'Integrated Regional Information Network'.

RADIO DAMAL
W: under construction, URL had not been announced at editorial deadline.
kHz: *11740, 11970, 15700*

Winter Schedule 2010/2011

Somali	Days	Area	kHz
0400-0700	daily	SOM	15700dha
1830-1930	daily	SOM	11740wof
1930-2030	daily	SOM	11970dha

Notes: On air since 4 Nov 2010.

RADIO HORYAAL (Clan)
▣ P.O. Box 51045, Scarborough, Ontario, Canada M1L 4T2.
E: radio@horyaal.net **W:** www.horyaal.net
Webcast: D/P
LP: PM: Ahmed H Nur.
Ann: Somali: "Halkanee, Waa Radio Horyaal odka'shad ku Somaliland".

Notes: On air, intermittently, since March 2005. No SW transmissions at time of publication.

Target: SUDAN (SDN)

MIRAYA FM
▣ See National Radio Section.
Webcast: D/L/P (www.mirayafm.org)
SAT: Nilesat 101.
kHz: *9670, 15710*

Winter Schedule 2010/2011

Arabic/English	Days	Area	kHz
0300-0600	daily	SDN	9670rso
1400-1700	daily	SDN	15710rso

Ann: English: "Radio Miraya".

RADIO DABANGA
▣ c/o Press Now, Witte Kruislaan, 55, 1217 AM Hilversum, The Netherlands.
☎ +31 35 6254350. ▤ +31 35 6254310.
E: radiodabanga@yahoo.com **W:** www.radiodabanga.com
Webcast: D/P
W: www.radiodabanga.org
LP: Network Mgr: Leon Willems.
kHz: *7315, 11515, 11615, 13590, 13740*

Winter Schedule 2010/2011

Arabic (Darfuri)	Days	Area	kHz
0430-0600	daily	EAf	7315iss, 13590dha
1530-1630	daily	EAf	13740wer
1530-1730	daily	EAf	11515mdc
1630-1730	daily	EAf	11615wer

Ann: Darfuri Arabic: "Radio Dabanga".
Notes: On air since 15 November 2008. Radio Dabanga is part of the Radio Darfur Network, a project of Press Now.

SUDAN RADIO SERVICE (SRS)
▣ See National Radio Section.
Webcast: D
kHz: *9590, 9840, 11770, 11785, 11805, 13720, 17700, 17745*

Winter Schedule 2010/2011

Arabic	Days	Area	kHz
1600-1630	mtwt.ss	SDN	11770mey, 11785mey
1600-1630	daily	SDN	17700asc
Arabic (Juba)	**Days**	**Area**	**kHz**
0400-0500	daily	SDN	11805dha, 13720dha
0500-0530	daily	SDN	13720dha, 13720dha
1500-1700	daily	SDN	17745sin, 17745sin
Bari	**Days**	**Area**	**kHz**
1700-1730	..t...	SDN	9590dha, 9840dha
Dinka	**Days**	**Area**	**kHz**
1700-1730	m......	SDN	9590dha, 9840dha
English	**Days**	**Area**	**kHz**
0530-0600	daily	SDN	13720dha, 13720dha
1730-1800	daily	SDN	9590dha, 9840dha
Fur	**Days**	**Area**	**kHz**
1630-1700	m..t...	SDN	11770mey, 11785mey, 17700asc
Maasalit	**Days**	**Area**	**kHz**
1630-1700	..w...s	SDN	11770mey, 11785mey, 17700asc
Moru/Nuer	**Days**	**Area**	**kHz**
1700-1730	..w....	SDN	9590dha, 9840dha
Shilluk	**Days**	**Area**	**kHz**
1700-1730	....f..	SDN	9590dha, 9840dha
Zagawa	**Days**	**Area**	**kHz**
1630-1700	.t..s.	SDN	11770mey, 11785mey, 17700asc
Zande	**Days**	**Area**	**kHz**
1700-1730	.t.....	SDN	9590dha, 9840dha

Ann: English: "You are listening to the Sudan Radio Service on ... kHz".
V: QSL-letter.
Notes: On air since July 2003. SRS is a project of Education Development Center, Inc. (USA). It is funded by the U.S. Agency for International Development (USAID).

Target: UGANDA (UGA)

RADIO Y'ABAGANDA (Rlg)
E: rediyoyabaganda@gmail.com; info@ababaka.com
W: www.ababaka.com
Webcast: D
kHz: *17725*

Winter Schedule 2010/2011

Luganda	Days	Area	kHz
1700-1730	.....s.	UGA	17725iss

Ann: English: "Radio Y'Abaganda", "Baganda Radio".
Notes: Internet radio station, also known as "Baganda Radio". On SW since 27 March 2010. Run by a Bagandan group known as 'Abaganda Ba Katonda - Brothers and Sisters of God'.

Target: VIETNAM (VTN)

NEW HORIZON RADIO (Rlg)
P.O. Box 18031, Spokane, WA 99228-0031, USA.
☎ +1 509 4897017.
E: lienlac@radiochantroimoi.com
W: www.radiochantroimoi.com
Webcast: D/P
P.O. Box 48, Hishi Yodogawa, Osaka 555, Japan. (Correspondence)
kHz: *1503*

Winter Schedule 2010/2011

Vietnamese	Days	Area	kHz
1330-1430	daily	VTN	1503fan

Ann: Vietnamese: "Dài là dài phat thanh Chân Trời Mói".
V: QSL-card.
Notes: Reported to have been on the air since 1993. The prgr promotes religious freedom in Vietnam.

RADIO HOA-MAI (Clan)
P.O. Box 842064, Houston, TX 77284-2064, USA.
☟ +1 281 2608266. (VNPP)
E: radio@hoamai.org W: radiohoamai.us
Webcast: D
L.P: Prgr Host: Nguyen Cong Bang.
kHz: *9930*

Winter Schedule 2010/2011

Vietnamese	Days	Area	kHz
1300-1330	.t.t...	VTN	9930hbn

Notes: On air since April 2005. Produced by Hoa-Mai Club.

Target: WEST BANK & GAZA (PSE)

RADIO AL-QUDS
P.O. Box 5092, Damascus, Syria.
L.P: Dir: Abu Shadi.
MW: 702kHz Exact location/power unknown.
FM/DAB: FM: 107.4MHz (location unknown, 24h)
kHz: *702*

Winter Schedule 2010/2011

Arabic	Days	Area	kHz
0600-2300	daily	PSE	702

Ann: Arabic: "Idha'at Al-Quds".
V: QSL-card.
Notes: On air since January 1988. Operated by the Popular Front for the Liberation of Palestine - General Command (PFLP-GC), with approval by Syrian authorities. The txs are located in South Western Syria.

Target: WESTERN SAHARA (AOE)

RADIO NACIONAL DE LA R.A.S.D.
Polisario Front, BP 10, El Mouradia, Algiers, Algeria.
☎ +213 49 923525.
E: rasdradio@yahoo.es; sario@saharaoccidental.info
W: web.jet.es/rasd/radionacional
Webcast: L
MW: [RBN] Rabouni: 1550kHz 50kW (presumed).
SW: [RBN] Rabouni: 1 x 20kW.

kHz: *1550, 6297*

Winter Schedule 2010/2011

Arabic	Days	Area	kHz	
0600-0800	daily	NAf	1550rbn*,	6297rbn±
1700-2300	daily	NAf	1550rbn*,	6297rbn±
Spanish	**Days**	**Area**	**kHz**	
2300-2330	daily	NAf	1550rbn*,	6297rbn±

Key: ± Variable frequency; * Alternative frequency 700kHz.
Ann: Arabic: "Huna el-estudiohaay al-markaziya al-wataniya, Sowt al-sha'ab a-Sahraui al-mukafa"; Spanish: "Ésta es la Radio Nacional de la República Arabe Saharaui Democrática".
V: QSL-letter.
Notes: On air since 28 Dec 1975. Operated by the Polisario Front.

Target: ZIMBABWE (ZWE)

ZIMBABWE COMMUNITY RADIO
☎ +263 913 103262.
E: editor@zicora.com W: www.zicora.com
Webcast: L/P
kHz: *4895*

Winter Schedule 2010/2011

English/Ndebele/ Shona	Days	Area	kHz
1755-1855	daily	ZWE	4895mey

Ann: English: "Zimbabwe Community Radio, tell it like it is".

RADIO VOICE OF THE PEOPLE (Clan)
P.O. Box 5750, Harare, Zimbabwe.
☎ +263 91 308052. ☟ +263 4 707123; +263 4 706988.
E: voxpop@ecoweb.co.zw W: www.radiovop.com
Webcast: L
L.P: MD: John Masuku.
kHz: *11610*

Winter Schedule 2010/2011

Various	Days	Area	kHz
0400-0500	daily	ZMB	11610mdc

Ann: English: "This is Radio VOP, Zimbabwe's Alternative Voice", "You are tuned to Radio Voice of the People".
V: QSL-letter.
Notes: On air since June 2000. Funded by the Soros Foundation, run by former staff of the Zimbabwe Broadcasting Corporation (ZBC). The prgr "Voice of the People" strives to broadcast ideas and information for the general development of the country, socially, politically and culturally. Jammed.

SW RADIO AFRICA (Clan)
P.O. Box 243, Borehamwood, Herts, WD6 4WA, United Kingdom.
☎ +44 20 83871441. ☟ +44 20 83871416.
E: mail@swradioafrica.com W: www.swradioafrica.com
Webcast: D/L/P
L.P: Founder: Ms Gerry Jackson; Tech Mgr: Keith Farquharson.
kHz: *4880*

Winter Schedule 2010/2011

English/Ndebele/ Shona	Days	Area	kHz
1700-1900	daily	ZWE	4880mey

Ann: English: "SW Radio Africa, Zimbabwe's Independent Voice".
V: QSL-card.
Notes: On air since December 2001. Produced by a London based group of Zimbabwean exiles. Jammed.

FREQUENCY LISTS

Section Contents

Features & Reviews

National Radio

International Radio

Frequency Lists

(For country codes and transmitter codes, please see the decode tables in the Reference section)

Please note that the North America MW listing has been removed in order to increase the number of MW stations in the main USA listing

Terrestrial Television

Reference

EUROPE, AFRICA, NEAR & MIDDLE EAST

Abbreviations: AFN=American Forces Network, BBCWS=British Broadcasting Corporation World Service, COPE= Cadena de Ondas Populares Espanolas, DRM=Digital Radio Mondiale, LPAMs=Low Power AM station, OCR=Onda Cero Radio, RFI=Radio France International, RSL=Restricted Service Licences, TWR=Trans World Radio, VOA=Voice of America.

kHz	kW	Ctry	Station, location
153	2000/1000	ALG	R. Algérienne 1, Béchar
	500/250	D	Deutschlandfunk, Donebach
	100	NOR	NRK P1/Troms og Finnmark, Ingøy
	200	ROU	Antena Satelor, Brasov
162	2000/1000	F	France Inter, Allouis
171	1600	MRC	Medi 1, Nador
	1200	RUS	GR Prgr. Kavkaz/Checnya Sv, Tbilisskaya
	150	RUS	R. Rossii, Bolshakovo
177	500	D	Deutschlandr. Kultur, Zehlendorf (Oranienbg.)
183	2000	D	Europe 1, Felsberg (Saarlouis)
189	300	ISL	RUV Rás 2, Gufuskálar
198	2000/1000	ALG	R. Algérienne 1, Ouargla
	50	G	BBC R4, Burghead
	500	G	BBC R4, Droitwich
	50	G	BBC R4, Westerglen
	150	RUS	R. Mayak, Kurovskaya (Moskva)
	150	RUS	R. Mayak, Olgino (Sankt-Peterburg)
207	500/250	D	Deutschlandfunk, Aholming
	100	ISL	RUV Rás 1, Eidar
	400	MRC	SNRT National Netw, Azilal
	250	UKR	UR1, Kyiv
216	900	F	RMC Info, Roumoules
225	1000	POL	PR 1, Solec Kujawski
234	2000	LUX	RTL, Beidweiler
243	0.2	DNK	DR6, Kalundborg (DRM, inactive)
252	1500/750	ALG	R. Algérienne 3, Tipaza
	300	IRL	RTE Radio 1, Summerhill (Clarkstown)
	150	RUS	R. Rossii/Reg, Kazan
261	75	BUL	Horizont, Sofia Vakarel
	250	RUS	R. Rossii, Taldom (Moskva)
270	650	CZE	CRo 1, Uherské Hradisté
279	500	BLR	BR 1, Sasnovy
531	600	ALG	R. Algérienne 1, F'kirina (Aïn Beïda)
	10	ARS	BSKSA Quran prgr, Bisha
	50	BOT	R. Botswana, Maun
	10	E	RNE5 TN, Cordoba
	20	E	RNE5 TN, Oviedo
	10	E	RNE5 TN, Pamplona
	10	E	RNE5 TN, Pontevedra
	100/50	FRO	Kringvarp Føroya Útvarpið, Akraberg
	0.001	G	RSLs
	500	IRN	IRIB R. Iran, Azarshahr
	600	IRN	IRIB R. Iran, Iranshahr
	100	ISR	KI Reshet Alef, Yavne
	50	NIG	Ondo State R. Corp, Akure
	0.8	POL	Twoje R, Wlodawa
	0.8	POL	Twoje R, Zywiec
	15	ROU	Antena Satelor, Urziceni
	15	ROU	R. România Actualitati, Petrosani
	30	RUS	R. Mayak/Reg, Cheboksary
	1	SRB	R. Vranje, Vranje
540	50	E	OCR Catalunya, Barcelona
	2000/1000	HNG	Kossuth R, Solt
	200	IRN	IRIB R. Iran, Mashhad
	100	KEN	KBC Swahili Sce, Voi
	600	KWT	R. Kuwait Main prgr, Kabd
	50	MLI	ORTM, Bamako
	300	MRC	SNRT National Netw./R, Tahadart (Tanger)
	10	MWI	MBC R. 1, Mangochi
	50	NIG	Sokoto State BC, Sokoto
	50	RUS	R. Mayak, Orenburg
	50	SDN	SRTC Regional, Nyala
549	600	ALG	R. Algérienne 1, Sidi Hamadouche
	51	ARS	BSKSA General prgr, Qurayyat +3 stns
	70	AZE	Azärbaycan R, Gäncä
	200	D	Deutschlandfunk, Nordkirchen/Thurnau
	20	GAB	RTG 2, Oyem
	400	IRN	IRIB R. Iran, Sirjan

kHz	kW	Ctry	Station, location
	25	NIG	Broadc. Sce of the Ekiti State, Ado
	25	NIG	Kano State BC, Tukun Tawa
	10	RKS	RTK Radio Kosova, Prishtinë
	50	RUS	R. Mayak, Kaliningrad
	600	RUS	R. Mayak, Krasnyy Bor (St. P.)
	75	RUS	R. Mayak, Noginsk (Moskva)
	50	RUS	R. Mayak, Novocherkassk
	150	RUS	R. Mayak, Syktyvkar
	15	SVN	R. Koper, Beli Kriz
558	10	ALG	R. Algérienne 1/R. Ouargla, Touggourt
	50	ARS	BSKSA General prgr, Jeddah
	50	BOT	R. Botswana, Muchenje
	10	CYP	Cy BC 1, Paphos
	20	E	RNE5 TN, A Coruña
	20	E	RNE5 TN, San Sebastián
	50	E	RNE5 TN, Valencia
	100	EGY	ERTU Educ. prgr, Cairo Abu Zaabal
	1	G	Spectrum Radio, London
	1000	IRN	IRIB R. Farhang, Gheslagh
	25	KEN	KBC Western Sce, Kapsimotwa
	10	MWI	MBC R. 1, Karonga
	50	NIG	Cross River State BC, Calabar
	400	ROU	R. România Actualitati, Tirgu Jiu
	10	SVN	MMR / R. Slovenija 1, Nemcavci
567	25	AFS	Cape Talk, Cape Town
	20	ARS	BSKSA Quran prgr, Afif/Abha
	5	E	RNE5 TN, Marbella
	50	E	RNE5 TN, Murcia
	60	I	RAI Radiouno/Reg, Bologna
	20	I	RAI Radiouno/Reg, Caltanissetta
	50	KEN	KBC Swahili Sce, Garissa
	50	NIG	FRCN Ibadan, Alaho
	50	NIG	Imo BC, Owerri
	-	NIG	Zamfara State R, Gusau
	100	ROU	R. România Actualitati, Brasov/Satu Mare
	1000	RUS	R. Rossii/Reg, Volgograd
	300	SYR	Syrian R. 1, Damascus Adra
576	50	AFS	SABC Metro FM, Meyerton
	400/200	ALG	R. Algérienne R. Béchar
	20	ARS	BSKSA Quran prgr, Gizan
	25	CNR	RNE R. Nacional, Las Palmas
	95	D	SWR Cont.Ra, Mühlacker
	100	E	RNE5 TN, Barcelona
	750	IRN	IRIB R. Iran/VOIRI, Mahshahr
	25	NIG	FRCN Ibadan, Moniya
	100	OMA	R. Sultanate of Oman, Haima
	10	POR	R. Sim, Braga
	100	UGA	UBC Star FM, Mityana
584	5	AFG	R. Badakhshan, Faizabad (inactive)
585	1200	ARS	BSKSA General prgr, Riyadh
	5	CVA	Vatican R, Vatican City
	600	E	RNE R. Nacional, Madrid
	5	F	FIP, Paris
	2	G	BBC R. Scotland, Dumfries
	600	IRN	IRIB R. Quran, Tehran
	50	NIG	Enugu State BC, Abakaliki
	150	RUS	R. Rossii/Reg, Perm
	350	TUN	ERTT National prgr, Gafsa
	10	TZA	Vo Tanzania Zanzibar, Chumbuni
594	5	AFG	R. Faryab, Maimana
	2000	ARS	BSKSA General prgr, Duba + 1 stn
	50	ARS	BSKSA General/Pilgrimage prgr, Makkah
	100	ETH	R. Ethiopia, Bahir Dar
	400	IRN	IRIB Regional, Shiraz (Dehnow)
	20	IRQ	Radio Al-Nas, Baghdad
	30	MWI	MBC R. 1, Lilongwe
	200	NIG	FRCN Kaduna, Jaji
	70	POR	R. Sim, Muge

kHz	kW	Ctry	Station, location
	40	RUS	R. Rossii, Izhevsk
	0.6	SVN	R.Odmev, Cerkno
	100	SYR	Syrian R. 2nd/Youth prgr, Homs
	50	UKR	BBC, Kyiv
595	50	MRC	SNRT A/R, Oujda (alt. on 594kHz)
602		YEM	YRTC General prgr, unknown location
603	100	CYP	CyBC 3, Nicosia
	20	D	Oldiestar R, Zehlendorf (Oranienburg)
	10	E	RNE5 TN, Palencia
	50	E	RNE5 TN, Sevilla
	50	EGY	ERTU Koran prgr, Sohag
	300	F	France Info, Lyon
	2	G	BBC R. 4, Newcastle
	0.4	G	Gold, Littlebourne
	100	IRN	IRIB R. Iran, Zahedan + 1 stn
	20	IRQ	IMN Republic of Iraq R, Mosul
	10	MDR	RDP Madeira, Pico do Areeiro (inactive)
	50	NIG	Borno R. & TV Corp, Maiduguri
	25	NIG	Ogun State BC, Abeokuta
	30	ROU	Antena Satelor, Bucuresti
	100	ROU	R. România Actualitati, Botosani/Oradea
	15	ROU	R. România Actualitati, Drobeta-T. Severin
	100/10	TZA	R. Tanzania, Dodoma
610	0.5	IRQ	R. Kull al-Iraq, Nasiriya
612	20	ARS	BSKSA Quran prgr, Al-Aflaj/Hail
	100	BHR	R. Bahrain Quran prgr, Manama
	100	BIH	BH Radio 1, Sarajevo
	10	E	RNE R. Nacional, Lleida
	10	E	RNE R. Nacional, Vitoria
	600	IRN	VOIRI, Qasr-e-Shirin
	200	JOR	R. Jordan Main prgr, Amman
	100	KEN	KBC Swahili Sce, Ngong
	100	LTU	Relays via R. Baltic Waves, Vilnius
	300	MRC	SNRT National Netw./R, Sebaa-Aioun
	50	NIG	Kwara State BC, Ilorin
	150	RUS	R. Mayak, Pedaselga (Petrozavodsk)
	10	UKR	BBC, Kharkiv
621	200	AFG	BBG Deewa R./R. Mashal, Khost
	300	BEL	RTBF International, Wavre
	100	BOT	R. Botswana, Selebi-Phikwe
	300/10	CNR	RNE R. Nacional, Santa Cruz de Tenerife
	10	E	RNE R. Nacional, Avila
	10	E	RNE R. Nacional, Jaén
	10	E	RNE R. Nacional, Palma de Mallorca
	1000	EGY	ERTU VO Arabs, Al-Mansura Batra
	50	IRN	IRIB Regional, Bandar Abbas
	50	IRN	IRIB Regional, Birjand
	150	MDA	R. PMR, Grigoriopol
	20	NIG	Anambra BS, Akwa
	5	RUS	R. Rossii/Reg, Kochubey
	50	RUS	R. Rossii/Reg, Makhachkala
	50	RUS	R. Rossii/Reg, Syktyvkar
	50/10	TZA	R. Tanzania, Mbeya
630	20	ARS	BSKSA 2nd prgr, Gizan
	10	ARS	BSKSA Quran prgr, Najran
	100/16	D	VOR, Scheppau (Braunschweig)
	2	G	BBC R. Cornwall, Redruth
	0.2	G	BBC Three Counties R, Luton
	100	KWT	R. Kuwait Quran prgr, Kuwait city
	75	MDG	RNM 1, Antananarivo (rep. inactive)
	100	NOR	NRK P1/Møre og Romsdal Reg, Vigra
	14	POR	Antena 1, Montemor-o-Velho + 2 stns
	50	ROU	Antena Satelor, Voinesti
	400	ROU	R. Timisoara/R.R. Act, Ortisoara
	300	TUN	ERTT National prgr, Tunis Djedeida
	300	TUR	TRT 1/Çukurova Reg, Mersin Kazanli
638	25	NIG	Kaduna State Media Corp, Katabu
639	500	CYP	BBC Arabic Sce, Zakaki (Ladies Mile)
	1530	CZE	CRo 2/CRo 6, Liblice (Praha) + Ostrava
	300	E	RNE R. Nacional, A Coruña
	10	E	RNE R. Nacional, Albacete
	20	E	RNE R. Nacional, Almeria
	50	E	RNE R. Nacional, Bilbao
	50	E	RNE R. Nacional, Zaragoza
	400	IRN	VOIRI, Bonab
	50	KEN	KBC English/Northern Sce, Garissa
	100	LSO	LNBS R. Lesotho, Lancer's Gap

kHz	kW	Ctry	Station, location
	1	MKD	R. Stip
	10	SDN	SRTC Regional, El Obeid
	50	UGA	UBC R, Kampala (inactive)
648	50	ALG	R. Algérienne, Aïn Beïda (F.P.I.)
	2000	ARS	BSKSA General prgr, Jeddah Khumra
	50	BOT	R. Botswana, Mopipi
	50	E	RNE R. Nacional, Badajoz
	500	G	BBC World Sce, Orfordness
	50	GMB	GRTS, Bonto
	50	IRN	IRIB Regional, Shahr-e-Kord
	300	LBY	LJBC VO Africa, Tobruk (inactive)
	10	SVN	R. Murski Val, Nemcavci
	100/10	TZA	R. Tanzania, Nachingwea
657	50	AFS	R. Pulpit / R. Kansel, Meyerton
	20	ARS	BSKSA Quran prgr, Rafha
	50	E	RNE5 TN, Madrid
	0.5	G	BBC R. Cornwall, Bodmin
	2	G	BBC R. Wales, Wrexham
	120	I	RAI Radiouno/International, Napoli
	25	I	RAI Radiouno/Reg, Bolzano
	55	I	RAI Radiouno/Reg, Pisa (Coltano)
	100	IRN	IRIB Regional, Zahedan
	1	IRQ	IMN Republic of Iraq R, Kirkuk
	200	ISR	KI Reshet Bet, Yavne
	100	NIG	FRCN Ibadan, Ibadan
	50	RUS	GR Prgr Kavkaz/Chechya Sv, Groznyy
	150	RUS	R. Rossii/Reg, Murmansk
	2x50	TZA	R. Tanzania, Dar-es-Salaam
	100	UAE	Asianet R./Abu Dhabi FM, Al-Dhabbiya
	25	UKR	UR3 Kultura/RUI, Chernivtsi
666	10	ALG	R. Algérienne 1/R. Tindouf
	150	D	SWR Cont.Ra, Rohrdorf
	50	E	SER R. Barcelona, Barcelona
	0.5	G	R. York, Fulford
	100	GRC	ERA R. Filia/Kosmos, Athens Megara
	50	IRN	IRIB R. Iran, Shushtar
	50	NIG	Edo State BS, Benin City
	52	POR	RTP Antena 1, 6 stns
	20	REU	R. Réunion, St. Pierre
	25	RUS	R. Mayak, Sochi
	10	SDN	SRTC Regional, Kassala
	50	SYR	Syrian R. 2/M, Damascus Adra
675	25	ARS	BSKSA General prgr, Abha/Afif
	120	HOL	R. Maria Nederland, Lopik
	50	IRN	IRIB Regional, Hamadan
	1	IRQ	IMN Republic of Iraq R, Baghdad
	50	KEN	KBC Swahili Sce, Marsabit
	100	LBY	LJBC Main prgr, Benghazi
	50	MWI	MBC R. 1, Ekwendeni
	25	NIG	R. Oyo, Ojeowode
	20	NOR	NRK P1/Nordland Reg, Røst
	600	QAT	Qatar RTC, Al Arish
	1	SRB	RTS Beograd 1, Bosilegrad
	25	UKR	UR1, Uzhhorod
684	60	ARS	BSKSA 2nd prgr, Jeddah/Riyadh
	600	E	RNE R. Nacional, Sevilla
	100	ETH	R. Ethiopia, Metu
	100	IRN	IRIB Regional, Gaem (Mashhad)
	10	MAU	MBC R. Maurice, Malherbes
	50	NIG	Yobe BC, Damaturu
	10	RUS	R. Radonezh, Sankt-Petersburg
	10	SRB	RTS Beograd 1, Aleksinac
	50	TUN	ERTT National prgr, Mednine
693	10	ALG	R. Algérienne 1/R. Adrar, Reggane
	5	ALG	RA 2, Aboudid (Ain el Hammam)
	20	ARS	BSKSA Quran prgr, Tabuk
	10	AZR	RDP Açores, Santa Barbara
	25	BOT	R. Botswana, Shakawe
	10	CYP	CyBC 1, Limassol
	250	D	VOR, Zehlendorf (Oranienburg)
	10	E	RNE R. Nacional, Tarragona
	20	E	RNE R. Nacional, Toledo
	50/1	G	BBC R. 5 Live, 10 stns
	150	G	BBC R. 5 Live, Droitwich
	30	I	RAI Milano, DRM test
	20	I	RAI Radiouno/Reg, Potenza
	100	IRN	IRIB Regional, Bandar Abbas

kHz	kW	Ctry	Station, location
	50	JOR	R. Jordan Main prgr, Ruweished
	10	NIG	Kogi State BC, Ochaja
	20	RUS	Deutsche Welle, Kurkino (Moskva)
	150	RUS	R. Rossii, Yazykovo (Ufa)
	100	SDN	SRTC Regional, Juba
	1	SOM	R. Hargeisa, Hargeisa
	1	SRB	RTS Beograd 1, Negotin
702	25	ALG	R. Algérienne 3/R. Laghouat
	50	ARS	BSKSA 2nd prgr, Bisha/Duba
	5	D	NDR Info Spezial, Flensburg
	10	EGY	ERTU Reg./Koran prgr, Asswan
	10	EGY	ERTU Reg./Koran/Sports, El Kharga
	200	F	China R. Int, Col de la Madone
	100	IRN	IRIB R. Iran, Bushehr
	500	IRN	VOIRI, Kiashahr
	100	KEN	KBC Swahili Sce, Meru
	25	NIG	Taraba State BS, Wukari
	800	OMA	BBC Arabic Service, A'Seela
	5	SVK	SR R. Patria/Devín, Kosice
	-	SYR	R. Al-Quds, SW Syria
	600	TUR	Voice of Turkey, Çatalca (Istanbul)
711	2	D	SWR Cont.Ra, Heilbronn (DRM tests)
	25	E	COPE Murcia
	100	EGY	ERTU Youth & Sports prgr, Tanta
	300	F	France Info/Bretagne Reg, Rennes
	200	IRN	IRIB Regional, Ahwaz
	50	LBY	LJBC Main prgr, Jeffren (inactive)
	300	MRC	SNRT National Netw./R, Laâyoune
	50	ROU	R. România Actualitati, Sighetul Marmatiei
	7	RUS	R. Rossii/Reg, Naryan-Mar
	10	SRB	RTS Beograd 1/R. Nish
	100/10	TZA	R. Tanzania, Kigoma
	40	UKR	UR1/Reg, Dokuchaievsk
	200	YEM	YRTC General prgr, San'a
720	10	CNR	RNE5 TN, Santa Cruz de Tenerife
	500	CYP	BBC Arabic Sce, Zakaki (Ladies Mile)
	85	D	WDR 2/VERA, Langenberg
	10/0.3	G	BBC R. 4, Lisnagarvey + 2 stns
	750	IRN	IRIB R. Iran, Mahidasht
	400	IRN	IRIB Regional/VOIRI, Tayebad
	50	NIG	Imo BC, Owerri
	60	POR	RTP Antena 1, 6 stns
	40	ROU	R. România Actualitati, Isaccea + 2 stns
	50/10	TZA	R. Tanzania, Mwanza
729	1	D	BR On3radio, Hof/Würzburg
	10	E	RNE R. Nacional, Alacant-Alicante
	10	E	RNE R. Nacional, Cuenca
	20	E	RNE R. Nacional, Logroño
	20	E	RNE R. Nacional, Málaga
	100	E	RNE R. Nacional, Oviedo
	10	E	RNE R. Nacional, Valladolid
	0.2	G	BBC Essex, Manningtree
	100	GRC	ERA NET, Athens Bogiati
	50	NIG	Kano State BC, Jogana
	100	UGA	UBC R, Palisa (inactive)
738	5	ALG	R. Algérienne 1/R. Illizi, In Amenas
	600	E	RNE R. Nacional, Barcelona
	0.04	G	BBC Hereford & W, Worcester
	50	IRN	IRIB Regional, Dayyer
	50	MOZ	Antena Nacional, Maputo
	100	OMA	R. Sultanate of Oman, Salalah
	5	RUS	WRN Relay, Kurkino (Moskva)
747	10	ARS	BSKSA General prgr, Najran
	10	BUL	Horizont/Turkish Sce, Shumen Salmanovo
	300	BUL	R. Bulgaria, Petrich
	25	CNR	RNE5 TN, Las Palmas
	10	E	RNE5 TN, Cádiz
	10	GMB	GRTS, Basse
	400/200	HOL	NPS Radio 5, Zeewolde
	150	IRN	IRIB R. Iran, Gonbad
	100	IRN	IRIB Regional, Kerman
	100	KEN	KBC English/Central Sce, Ngong
	60	NIG	Nagarta R, Kaduna
	10/5	SDN	SRTC Regional, Khartoum/Port Sudan
	100	SYR	Syrian R. 1, Sarakeb
756	100/200	D	DLF, Ravensburg/Scheppau
	10	E	R. Euskadi, Bilbao

kHz	kW	Ctry	Station, location
	10	EGY	ERTU Reg./Koran prgr, Qena
	2	G	BBC R 4, Redruth
	1	G	R. Cumbria, Carlisle
	0.6	G	R. Maldwyn, "Magic 756", Newtown
	100	IRN	IRIB Regional, Rasht
	-	IRQ	R. Dar as-Salam, Basra
	50	ISR	R. Mashreq/PAL-Voice, Metulla
	10	MWI	MBC R. 1, Blantyre
	25	NIG	Borno Radio & TV Corp, Damagum
	100	NIG	R. Oyo, Ibadan
	2	POR	RTP Antena 1, Lamego
	400	ROU	R. România Actualitati, Lugoj (Boldur)
765	50	ARS	BSKSA Quran prgr, 3 stns
	0.5	G	BBC Essex, Chelmsford
	10	GRC	ERA Regional, Ioannina
	100	IRN	IRIB Regional, Shahr-e-Kord
	1000	IRN	VOIRI/IRIB, Chabahar
	50	MOZ	EP de Nampula, Nampula
	150	RUS	R. Rossii/Reg, Petrozavodsk
	50	SDN	SRTC General prgr, Omdurman
	1	SRB	RTS Beograd 1, Medvedja
	70	UKR	R. Maiak, Odesa (Petrivka)
	50	YEM	YRTC Local/General prgr, Mukalla
774	50	AGL	EP de Benguela, Benguela
	2	BIH	R. Tuzla
	75	BUL	R. Varna
	5	D	WDR 2/VERA, Bonn
	40	E	RNE R. Nacional, 4 stns
	60	E	RNE R. Nacional, Cáceres
	20	E	RNE R. Nacional, Ourense
	50	E	RNE R. Nacional, San Sebastián
	100	E	RNE R. Nacional, Valencia
	1000	EGY	ERTU Middle East prgr, Alexandria Abis
	1	G	BBC R4, Enniskillen/Plymouth
	0.1	G	Gold, Gloucester
	0.7	G	R. Kent, Littlebourne
	0.5	G	R. Leeds/BBC Asian Network, Farnley
	100	IRN	IRIB Regional, Arak
	10	NIG	Taraba State BS, Wukari
	30	RUS	R. Mayak/Reg, Voronezh
783	5	ALG	R. Algérienne 1/R. Illizi, Djanet
	10	ALG	R. Algérienne 1/R. Souf, El Oued
	100	ARS	BSKSA 2nd prgr, Ras al-Zawr
	100	D	MDR Info, Wiederau (Leipzig)
	50	E	COPE Miramar/Rock & Gol, Barcelona
	150	IRN	IRIB Regional, Iranshahr
	50	MTN	R. Mauritanie, Nouakchott
	50	NIG	R. Kogi, Okene
	300	SYR	Syrian R. 1/Ext. Sce, Tartus
792	50	ARS	BSKSA Quran prgr, Jeddah
	1	BIH	R. Banovici
	2	D	NDR Info Spezial, Lingen
	50	E	SER R. Sevilla
	300	F	France Info, Limoges
	1	G	BBC R. Foyle, Londonderry
	0.3	G	Gold, Bedford
	100	GRC	ERA Spor, Thessaloniki Malgara
	100	IRN	IRIB Regional, Zanjan (Zanjan)
	20	LBY	LJBC Main prgr, Sirt (inactive)
	25	NIG	R. Oyo, Gambari
	100	YEM	YRTC 2nd prgr, Al-Hiswah
801	150	AZE	Azärbaycan R, Haciqabul (Pirsaat)
	100	BHR	R. Bahrain General prgr, Manama
	110	D	BR On3radio, München/Nürnberg
	40	E	RNE R. Nacional, 4 stns
	25	E	RNE R. Nacional, Ciudad Real
	20	E	RNE R. Nacional, Lugo
	100	ETH	VO Amhara State, Bahir Dar
	2	G	R. Devon, Barnstaple
	50	IRN	IRIB R. Iran, Kashmar
	200/2000	JOR	R. Jordan main prgr, Ajlun (inactive)
	1	NIG	R. Kebbi, Zuru
	20	NIG	Yobe BC, Damaturu
	5	SDN	SRTC Regional, Al-Fashir
810	20	ARS	BSKSA 2nd prgr, Abha
	20	E	SER R. Madrid
	100	G	BBC R. Scotland, Burghead

kHz	kW	Ctry	Station, location
	5	G	BBC R. Scotland, Redmoss
	100	G	BBC R. Scotland, Westerglen
	12	HNG	Magyar Katolikus R, Lakihegy
	100	IRN	IRIB R. Iran, Khorramabad
	5/2.5	IRQ	R. Um al-Qura, Baghdad
	1200	MKD	MR 1 & R. Makedonija, Sveti Nikole
	50	MOZ	EP de Gaza, Xai-Xai
	10	MWI	MBC R. 1, Bangula
	500	RUS	R. Mayak, Volgograd
	20	RUS	VOA relay, Kurkino (Moskva)
	50	UAE	Abu Dhabi FM, Maqtaa
	100	UGA	UBC R, Gulu (inactive)
819	10	E	R. Euskadi, Gasteiz-Vitoria
	1000	EGY	ERTU General prgr, Al-Mansura Batra
	1	F	Sud Radio, Toulouse
	20	I	RAI Radiouno/Reg, Trieste
	30	IRN	IRIB Regional, Sari
	10	IRQ	R. Al-Amal, Basra
	10	MAU	MBC R. Mauritius, Malherbes
	10	SDN	SRTC Regional, Dongola
	7	UKR	UR R. Bukovyna, Novodnistrovsk
828	20	ARS	BSKSA General prgr, Medinah
	1	AZR	RDP Açores, Monte das Cruzes
	20/5	D	NDR Info Spezial, Hannover
	10	D	SWR Cont.ra, Freiburg
	5	E	Kiss FM Catalunya, Terrassa
	100	ETH	R. Ethiopia, Arba Minch
	0.2	G	BBC Asian Network, Sedgley
	0.3	G	Gold, Bournemouth
	0.2	G	Gold, Luton
	0.1	G	Magic 828, Leeds
	20/5	HOL	R. 10 Gold, Heinenoord
	50	IRN	IRIB Regional, Tabas
	300	LBY	LJBC Main prgr, Sabha (inactive)
	25	NIG	FRCN Enugu
	10	RUS	Radiogazeta Slovo, Sankt-Peterburg
	200	SYR	Syrian R. 1, Deir-ez-Zor
	1	UAE	Abu Dhabi FM, Al-Ain
837	5	ALG	R. Algérienne 3, Béchar
	10	AZR	RDP Açores, Pico da Barrosa
	10	CNR	COPE, Las Palmas
	10	E	COPE, Burgos
	5	E	COPE, Eivissa
	5	E	COPE, El Ferrol
	10	E	COPE, Sevilla
	100	ERI	VO the Broad Masses 2, Asmara
	100	ETH	R. Oromiya, Robe (Bale)
	200	F	France Info, Nancy
	0.5	G	BBC Asian Netw, Freemen's Common
	1	G	R. Cumbria, Barrow
	300	IRN	IRIB Isfahan Reg, Habibabad
	7	UKR	UR R. Khvylia, Vinnytsia
	150	UKR	UR1, Kharkiv (Taranivka)
	30	YEM	YRTC 2nd/General prgr, San'a (vf)
840	20	TCD	RNT, N'djamena-Gredia
846	100	AFS	SABC Umhlobo Wenene FM, Komga
	20	ARS	BSKSA Quran prgr, Buraida
	25	I	RAI Radiouno/Reg, Roma (DRM, inactive)
	10	IRN	IRIB R. Iran, Mianeh
	20	IRQ	IMN Republic of Iraq R, Nasiriya (inactive)
	100	KEN	KBC Swahili Sce, Nyamninia
	10	NIG	Bauchi R. Corp, Azare
	150	RUS	R. Radonezh/Podmoskovya, Noginsk
	42	RUS	R. Rossii/Reg, Elista
	20	UAE	Holy Quran Radio, Umm al Qiwain
855	100	ARS	BSKSA Quran prgr, Ras al-Zawr
	10	D	DRadio Wissen, Berlin-Britz (DRM)
	55	E	RNE R. Nacional, 6 stns
	300	E	RNE R. Nacional, Murcia
	20	E	RNE R. Nacional, Pontevedra
	50	E	RNE R. Nacional, Santander
	20	E	RNE R. Nacional, Tarragona
	100	ETH	R. Ethiopia, Harar
	1	G	BBC R. Lancashire, Preston
	1.5	G	BBC R. Norfolk, Postwick
	0.2	G	Sunshine R, Ludlow
	10	JOR	R. Jordan Main prgr, Amman

kHz	kW	Ctry	Station, location
	400	ROU	R. România Actualitati, Tancabesti
	50	RUS	R. Rossii, Kamenka
864	1000	ARM	Relays, Gavar
	10	BUL	Horizont/Turkish Sce, Samuil
	75	BUL	R. Blagoevgrad, Blagoevgrad
	10	E	RNE R. Nacional, Socuellamos
	500	EGY	ERTU Koran prgr, Santah
	300	F	France Bleu Ile de France, Paris (Stereo)
	50	IRN	IRIB Regional, Qasr-e-Shirin
	10	IRQ	IMN Republic of Iraq R, Ramadi (inactive)
873	10	ALG	R. Algérienne 1/R. Ghardaïa
	10	ARS	BSKSA Quran prgr, Ar-Rass
	50	BOT	R. Botswana, Gantsi
	60	BUL	Hristo Botev/R. Stara Zagora
	150	D	AFN Power Network, Weisskirchen
	10	E	SER R. Galicia, Stgo de Compostela
	25	E	SER R. Zaragoza
	100	ETH	R. Ethiopia, Addis Ababa
	0.3	G	BBC R. Norfolk, West Lynn
	1	G	BBC R. Ulster, Enniskillen
	40	HNG	Magyar R. 4, Lakihegy/Pécs
	1	I	RAI Radiouno/Reg, Taranto
	50	IRN	IRIB Regional, Bojnurd
	100	IRN	IRIB Regional, Mazandaran prov.
	75	MDA	R. Moldova, Chisinau
	50	MOZ	EP de Sofala, Beira
	50	RUS	R. Rossii, Kaliningrad
	250	RUS	R. Rossii, Lesnoy (Moskva)
	75	RUS	R. Rossii, Olgino (Sankt-Peterburg)
	100	RUS	R. Rossii/Reg, Samara
	10/5	SDN	SRTC Reg, Wad Madani/Wadi Halfa
	50	SYR	Syrian R. 2, Damascus Adra
	10	UKR	UR R. Mryia, Dnipropetrovsk
882	100	ARS	BSKSA Quran prgr, Dammam
	20	CNR	COPE Tenerife, La Laguna
	20	D	MDR Info, Wachenbrunn
	50	E	COM R, Barcelona
	10	E	COPE, Alacant-Alicante
	5	E	COPE, Gijón
	5	E	COPE, Málaga
	5	E	COPE, Valladolid
	10	EGY	ERTU General prgr, Matruh
	10/5/1	G	BBC Wales, Penmon/Tywyn/Forden
	100	G	BBC Wales, Washford
	60	IRN	IRIB Regional, Mahabad
	5	IRQ	R. Dar as-Salam, Mosul
	10	ISR	KI Reshet Bet, She'ar-Yeshuv
	50	KEN	KBC Swahili Sce, Kitale
	5	MNE	R. Podgorica 1, Podgorica
	25	NIG	Capital Sound, Kafanchan
	7	RUS	R. Mayak/Reg, Stavropol
891	600/300	ALG	R. Algérienne 1, Ouled Fayet (Algér)
	30	AZE	Azärbaycan R, Baki
	100	ETH	R. Ethiopia, Dese
	20/5	HOL	R. 538, Hulsberg
	50	IRN	IRIB Regional, Yasuj/Dehdasht
	50	LSO	LNBS Ultimate FM, Lancer's Gap
	-	NIG	Osun State BC, Osu
	10	POR	R. Sim, Vilamoura
	300	TUR	TRT 1/Reg, Antalya
900	1000	ARS	BSKSA General prgr, Qurayyat II
	5	E	COPE Alta Extremadura, Plasencia (relay)
	10	E	COPE, Cáceres/Plasencia
	5	E	COPE, Granada
	5	E	COPE, Vigo
	10	E	R. Popular, Bilbao
	600	I	RAI Radiouno/International, Milano
	600	IRN	IRIB R. Iran, Tehran
	100	KEN	KBC English Sce, Marania
	25	NIG	Ogun State BC, Abeoukuta
	20	RUS	R. Mayak/Reg, Sovietskiy
909	10	AFG	R. Kunduz
	0.05	AFG	R. Paktia, Gardez
	10	ALG	R. Algérienne 1, Tamanrasset
	600	BOT	VOA, Mopeng Hill (Selebi-Phikwe)
	10	E	RNE5 TN, Palma de Mallorca
	50/1	G	BBC R.5 Live, 9 stns

kHz	kW	Ctry	Station, location
	150	G	BBC R.5 Live, Brookmans Park
	200	G	BBC R.5 Live,Moorside Edge
	50	IRN	IRIB R. Iran, Lar
	25	IRQ	IMN Republic of Iraq R, Basra
	20	LBY	LJBC Main prgr, Giaghboub (inactive)
	50	NIG	FRCN Abuja, Gwagwalada
	200	ROU	R. Cluj, Jucu
	25	ROU	R. Constanta, Valu lui Traian
	50	ROU	R. România Actualitati, Timisoara
	5	SDN	SRTC Regional, Malakal
	20	UGA	UBC R, Kampala (inactive)
	750	YEM	YRTC General prgr, Hudaydah
917	50	NIG	R. Gotel, Yola
918	10	CYP	CyBC 3, Paphos
	20	E	R. Intercontinental, Madrid
	10	EGY	ERTU Educ./Sports/Koran, Hurghada
	10	EGY	ERTU General prgr, Bawiti
	100	ETH	VO the Dawn, Mekele
	50	IRN	IRIB Regional, Jiroft
	50	NIG	R. Benue, Makurdi
	50	RUS	R. Mayak/Reg, Makhachkala
	150	RUS	R. Rossii/Reg, Arkhangelsk
	300	SVN	R. Slovenija 1, Domzale
	100	SYR	Syrian R. 1, Al-Hassake
927	10	ALG	R. Algérienne 1/R. Adrar, Timimoun
	20	ARS	BSKSA 2nd prgr, Al-Hufuf
	100	BEL	RVI Info, Wolvertem
	50	IRN	IRIB R. Iran, Dorud
	50	ISR	KI Reshet Bet, Acre (Akko)
	10	ISR	KI Reshet Bet, Eilat
	100	KEN	KBC Swahili Sce, Malindi
	1	POR	R. Sim, Évora
	200	TUR	TRT 4, Izmir
936	10	AFG	R. Zabul, Qalat
	100	ARS	BSKSA Quran prgr, Makkah/Riyadh
	8	BEN	ORTB R. Regionale, Parakou
	20	E	RNE5 TN, Alicante
	20	E	RNE5 TN, Valladolid
	20	E	RNE5 TN, Zaragoza
	10	EGY	ERTU General prgr, Salum
	100	EGY	ERTU Om Kalthoum prgr, Cairo
	1	G	Fresh Radio, Hawes
	0.2	G	Gold, Naish Hill
	10	I	RAI Radiouno/Reg, Trapani
	20	I	RAI Radiouno/Reg, Venezia
	100	IRN	IRIB Regional, Urumiyeh
	20	IRQ	R. As-Safir, Basra
	100	MRC	SNRT C/R, Agadir
	5	RUS	R. Rossii/Reg, Matveyevka
	100	SYR	Syrian R. 1, Homs
	600	UKR	UR1, Lviv Krasne
	3	UKR	UR1/Reg, Starobilsk
945	5	ARS	BSKSA General prgr, Hail
	25	BOT	R. Botswana, Mmathethe
	100	ERI	Vo the Broad Masses 1, Asmara
	300	F	France Bleu/Reg, Toulouse
	0.7	G	Gold, Bexhill
	0.2	G	Gold, Derby
	5	GRC	ERA Regional, Larissa
	100	IRN	IRIB Regional, Dehgolan
	50	ISR	Galei Tzahal, Yavne
	10	NIG	R. Kebbi, Birnin Kebbi
	15	ROU	R. România Actualitati, Miercurea Ciuc
	40	RUS	R. Rossii/Reg, Novocherkassk
	5	SDN	SRTC Regional, Al-Foula
	20	STP	R. Nacional, Pinheira
954	250	CZE	CRo 2/CRo 6, Dobrochov (Brno) + 2 stns
	50	E	Onda Cero R, Madrid
	3	ETH	R. Sidama, Yirgalem
	10	GRC	ERA Regional, Heraklion (inactive)
	100	KEN	KBC English Sce, Nyamninia
	1500	QAT	Qatar RTC (Al-Jazeera TV sound), Al Arish
	50	SYR	Syrian R. 2, Deir ez-Zor
	300	TUR	TRT Reg, Trabzon
963	45	BUL	Horizont, Sofia/Malko Tarnovo
	75	BUL	Horizont/R. Shumen
	50	BUL	Horizont/Turkish Sce, Kardzali

kHz	kW	Ctry	Station, location
	100	CYP	CyBC 1, Nicosia
	10	E	R. Euskadi, Donosti-San Sebastián
	600	FIN	CRI relay, Pori
	0.2	G	Asian Sound R, Haslingden
	1	G	Buzz Asia, East London
	50	IRN	IRIB R. Iran, Birjand
	20	KWT	R. Kuwait Multilingual/Main prgr, K. City
	50	MOZ	EP de Tete, Tete
	0.5	POL	Twoje R, Brzesko
	0.1	POL	Twoje R, Lipsko
	0.8	POL	Twoje R, Lubaczow
	0.1	POL	Twoje R, Lubliniec
	10	POR	R. Sim, Seixal
	100	SDN	SRTC Peace R./Koran prgr, Khartoum
	100	TUN	ERTT Int. & Cultural ch, Tunis Djedeida
972	50	BOT	R. Botswana, Sebele (inactive)
	100	D	NDR Info Spezial, Hamburg
	5	E	RNE R. Nacional, Cabra
	2	E	RNE R. Nacional, Monforte de Lemos
	100	ETH	R. Ethiopia, Robe (Bale)
	1	G	Buzz Asia, West London
	100	IRN	IRIB Regional, Ilam
	50	LBY	LJBC Main prgr, Sirt
	5	MRC	RNE R. Nacional, Mellilla
	25	NIG	Katsina State R, Katsina
	10	NIG	R. Kogi, Otite
	1200	UKR	UR1, Mykolaiv
981	100	ALG	R. Algérienne 2, Ouled Fayet (Algér)
	20	ARS	BSKSA Quran prgr, Madinah
	2	EGY	ERTU General prgr, Abu Simbel/Baris
	10	EGY	ERTU Reg./Koran prgr, Assiut
	200	GRC	ERA Spor, Athens Megara
	10	I	RAI Reg, Trieste
	100	IRN	IRIB R. Iran, Hamadan
	100	KEN	KBC English Sce, Voi
	13	POR	R. Sim, Coimbra + 3 stns
989	1	ETH	R. Ethiopia FS relay, Addis Ababa
990	50	AGL	EP do Bié, Kuito
	600	CYP	R. Sawa, Cape Greco
	100	D	Deutschlandradio Kultur, Berlin-Britz
	10	E	SER R. Bilbao
	5	E	SER R. Cádiz
	1	G	BBC R. 5 Live, Tywyn
	1	G	BBC R. Devon, Exeter
	1	G	BBC R. Nan Gaidheal, Redmoss
	0.1	G	Gold, Wolverhampton
	0.3	G	Magic AM, Doncaster
	400	IRN	IRIB R. Iran, Shiraz (Dehnow)
	50	NIG	Bauchi R. Corp, Bauchi
	2	NIG	Lagos State BC, Ikeja
	1	RUS	R. Mayak, Yuryuzan
	100/10	TZA	R. Tanzania, Songea
999	5	AFG	R. Helmand, Lashkar Ga (inactive)
	20	ARS	BSKSA Quran prgr, Duba
	50	E	COPE, Madrid
	0.3	G	Gold, Nottingham
	0.8	G	Magic 999, Preston
	1	G	R. Solent, Fareham
	20	I	RAI Radiouno/Reg, Perugia
	20	I	RAI Radiouno/Reg, Rimini
	50	I	RAI Radiouno/Reg, Torino (Volpiano)
	2	I	RAI Radiouno/Reg, Vibo Valentia
	50	IRN	IRIB R. Iran, Baneh
	20	IRQ	R. Bilad, Baghdad
	500	MDA	R. PMR/Relays, Grigoriopol
	5	MLT	R. Malta 1, Bizbizja
	1	SRB	RTS Beograd 1, Kladovo
	100	UGA	UBC R Red Channel, Kabale
1008	7	BLR	BR Kanal Kultura, Hrodna
	10	CNR	Punto Radio, Las Palmas
	5	E	SER R. Alacant-Alicante
	5	E	SER R. Extremadura, Badajoz
	10	E	SER R. Girona
	1	EGY	ERTU General prgr, El Farafra
	10	EGY	ERTU Reg. prgr, El Fayoum
	100	EGY	ERTU Reg./Hebrew prgr, El Arish
	100	GRC	ERA Regional, Kerkyra (Corfu)

kHz	kW	Ctry	Station, location	kHz	kW	Ctry	Station, location
	200	HOL	Groot Nieuws R, Flevoland		10	NIG	Enugu State BS, Onitsha
	100	IRN	IRIB Regional, Semnan		0.8	POL	Twoje R, Cmolas
	20	IRQ	Sowt al-Fadhila, Najaf		0.5	POL	Twoje R, Jaroslaw
	50	MOZ	EP de Maputo, Maputo		0.8	POL	Twoje R, Pulawy
	10	NIG	Niger State Media Corp, Kontagora		0.8	POL	Twoje R, Skarzysko
	10	NIG	Osun State BC, Iree		1	SRB	RTS Beograd 1/R. Novi Pazar
	5	RUS	R. Mayak, Tuapse		300	TUR	TRT Kurdish prgr, Diyarbakir
	1	SRB	RTS Beograd 2/3, Beograd	**1071**	5	ALG	R. Algérienne 1, Illizi
	600	YEM	YRTC General prgr, San'a		20	ARS	BSKSA General prgr, Bisha
1017	10	AFG	R. Ghazni		25	BOT	R. Botswana, Jwaneng
	10	ALG	R. Algérienne 1/R. Batna		100	EGY	ERTU Adults prgr, Cairo Abu Zaabal
	20	ARS	BSKSA Pilgrimage R, Madinah		1	G	TalkSport, Clipstone/Newcastle
	100	D	SWR Cont.Ra, Wolfsheim (Mainz)		100	IRN	IRIB R. Ma'aref, Qom (Alborz)
	10	E	RNE5 TN, Burgos		20	IRQ	IMN R. Babil, Hilla
	10	E	RNE5 TN. Granada		5	SDN	SRTC Regional, Wou
	0.6	G	Gold, Shrewsbury		100	SYR	R. Al-Nour (LBN) relay, Tartus
	100	IRN	IRIB R. Iran, Zahedan		30	YEM	YRTC General prgr, Taiz
	10	IRQ	IMN Republic of Iraq R, Karbala	**1080**	10	ARS	BSKSA 2nd prgr, Najran
	1	MDR	PEF, Santana (inactive)		10	E	Onda Cero R, Toledo
1026	10	ALG	RA 1/R. Ouargla, Hassi Messaoud		5	E	SER R. Coruña, A Coruña
	35	BLR	BR Kanal Kultura, Brest/Miadzel/Salihorsk		5	E	SER R. Granada
	5	E	SER R. Asturias, Oviedo		10	E	SER R. Huesca
	5	E	SER R. Jaén		5	E	SER R. Mallorca, Palma de M.
	5	E	SER R. Jerez, J. de la Frontera		20	EGY	ERTU General prgr, El Minya/Luxor
	10	E	SER R. Reus		3	ETH	R. Fana, Addis Ababa
	5	E	SER R. Salamanca		10	GRC	ERA Regional, Orestias
	5	E	SER R. Vigo		0.01W	HOL	Dream R, Limburg
	1	G	BBC Jersey, Trinity		750	IRN	IRIB Regional/VOIRI, Mahshahr
	0.5	G	BBC R. Cambrigeshire, Chesterton Fen		10	ISR	KI Arabic prgr, Yavne
	1.7	G	Downtown R, Belfast		10	LBY	LJBC Main prgr, Kufra (inactive)
	100	IRN	IRIB Regional, Tabriz		5	MRC	SNRT Quran prgr, Casablanca (rep. on 1089)
	50	MOZ	EP de Manica, Chimoio		100	RUS	R. Rossii/Reg, Kovylkino
	25	NIG	Jigawa BC, Dutse	**1088**	25	AGL	RNA Canal A, Mulenvos
	7	RUS	R. Mayak, Nyandoma + 2 stns	**1089**	10	ALG	R. Algérienne 1/R. Adrar
	5	SDN	SRTC Regional, Al-Damazin		20	ARS	BSKSA 2nd prgr, Qurayyat
	10	UGA	UBC R, Mbale (inactive)		400	G	TalkSport, Brookmans Park
1035	20	ARS	BSKSA 2nd prgr, Yanbu al-Bahr		400	G	TalkSport, Moorside Edge
	130	EST	R. Eli/TWR, Tartu		80/1	G	TalkSport, Washford + 4 stns
	10	ETH	R. Oromiya, Adama (Nazret)		125	G	TalkSport, Westerglen
	1	G	BBC R. Sheffield/Asian Netw, Sheffield		50	IRN	IRIB Regional, Biarjmand
	1	G	Kismat Radio, London		50	RUS	R. Teos, Krasnyy Bor (Sankt-Peterburg)
	0.8	G	Northsound Two, Aberdeen		1200	RUS	VOR, Tbilisskaya
	0.3	G	West Sound AM, Ayr	**1098**	100	ARS	BSKSA 2nd prgr, Dammam
	2	I	RAI Radiouno/Reg, Lecce		2x50	CYP	BRT Bayrak Radyo 1, Iskele (Trikomo)
	10	I	RAI Radiouno/Reg, Pescara		25	E	RNE5 TN, Almeria
	50	IRN	IRIB Regional, Yazd		10	E	RNE5 TN, Avila
	20	JOR	R. Jordan Main prgr, Amman		5	E	RNE5 TN, Huelva
	30	POR	R Club, Belmonte		20	E	RNE5 TN, Lugo
1044	7	AFG	R. Farah (inactive)		100	IRN	VOIRI, Zabol
	10	CYP	CyBC 3, Limassol		12	RUS	R. Rossii/Reg, Chagoda/Nikolsk
	20	D	MDR Info, Wilsdruff (Dresden)		10	SVK	SR R. Patria/Devín, Nitra
	10	E	SER R. San Sebastian, Donosti-S.S.	**1107**	400	AFG	RTV Afghanistan, Pol-e-Charkhi
	5	E	SER R. Valladolid		10	D	AFN Power Network, Kaiserslautern
	200	ETH	R. Ethiopia, Mekele		10	D	AFN The Eagle, Vilseck
	150	GRC	RS Makedonias 1, Thessaloniki Perea		20	E	RNE5 TN, Caceres
	50	IRN	IRIB R. Iran, Dehloran		20	E	RNE5 TN, Camargo
	-	IRQ	V. of the South, Basra		25	E	RNE5 TN, Logroño
	100	KEN	KBC English Sce, Malindi (inactive)		10	E	RNE5 TN, Ponferrada
	300	MRC	SNRT C, Sebaa-Aioun		10	E	RNE5 TN, Teruel
	20	RUS	R. Svoboda (RFERL), Kurkino (Moskva)		600	EGY	ERTU Reg./Palestine prgr, Al-M. Batra
1053	5	E	COPE, Vila Real		1.5	G	Moray Firth R, Inverness
	25	E	COPE, Zaragoza		2/0.5	G	TalkSport, 6 stns
	100	ETH	R. Oromiya, Nekemte (F.P.I.)		10	I	RAI Radiouno/International, Roma
	500	G	TalkSport, Droitwich + 12 stns		50	IRN	IRIB Regional, Sabzevar
	100	IRN	IRIB R. Iran, Khorramabad		100	KEN	KBC Swahili Sce, Maralal
	25	IRN	IRIB Regional, Saravan		1	MWI	MBC R. 1, Nkhota Kota
	3	IRQ	R. As-Salam, Baghdad (alt. 1030kHz)		25	NIG	FRCN Kaduna, Jaji
	100	LBY	LJBC Main prgr, Tripoli		50	SRB	RTN, rel. of Beograd 1, Orlovat
	400	ROU	R. Iasi, Uricani	**1116**	20	ARS	BSKSA 2nd prgr, Madinah
	10	RUS	R. Mariya, Sankt-Peterburg		50	DJI	RTD, Djibouti
1062	20/1	CZE	Country R, Praha		5	E	SER R. Albacete
	250	DNK	DR P4 news & weather, Kalundborg		5	E	SER R. Pontevedra
	10	I	RAI Radiouno/Reg, Ancona		0.5	G	BBC Guernsey, Rohais
	25	I	RAI Radiouno/Reg, Cagliari		1	G	BBC R Derby/Asian N, Burnaston Lane
	2	I	RAI Radiouno/Reg, Catania		5	HNG	1/Nyugat-Dunánt. Reg, Mosonmagyaróvár
	2	I	RAI Radiouno/Reg, Trento		15	HNG	MR1/Eszak-Magyarországi Reg, Miskolc
	50	IRN	IRIB R. Iran, Kerman		0.5	HOL	R. Bloemendaal, Bloemendaal

kHz	kW	Ctry	Station, location
	2	I	RAI Radiouno/Reg, Aosta
	2	I	RAI Radiouno/Reg, Bari
	20	I	RAI Radiouno/Reg, Cuneo
	10	I	RAI Radiouno/Reg, Palermo
	200	IRN	IRIB R. Iran, Ardekan (inactive)
	20	IRQ	R. Dar as-Salam, Baghdad
	30	RUS	R. Rossii/Reg, Sochi
1125	10	BEL	RTBF Vivacité, Wavre
	150/75	BLR	BR Kanal Kultura, Minsk
	5	E	RNE5 TN, Badajoz
	10	E	RNE5 TN, Castelló
	10	E	RNE5 TN, Gasteiz-Vitoria
	10	E	RNE5 TN, Soria
	10	E	RNE5 TN, Toledo
	1	G	R. Wales, Llandrindod Wells
	10	IRN	IRIB Regional, Nehbandan
	50	IRN	IRIB Regional, Qazvin
	500	LBY	LJBC Main prgr, El Beida
	1	MDR	RDP Madeira, Ponta do Pargo
	20	NGR	ORTN La Voix du Sahel, Niamey
	150	RUS	R. Orfey, Olgino (Sankt-Peterburg)
	100	SYR	Syrian R. 2, Al-Hassake
	50	YEM	YRTC Local prgr, Hudaydah
1134	10	AGL	EP do Bengo, Mulenvos
	5	E	COPE, Astorga
	5	E	COPE, Ciutadella
	5	E	COPE, Jerez de la Frontera
	5	E	COPE, Pamplona
	5	E	COPE, Puertollano
	10	E	COPE, Salamanca
	0.003	G	BFBS Gurkha R, 3 stns
	0.001	G	LPAMs
	600	HRV	HR Voice of Croatia, Zadar
	50	IRN	IRIB R. Iran, Bojnurd
	50	KEN	KBC English Sce, Kitale
	100	KWT	R. Kuwait Sports/Main prgr, Kabd
	20	NIG	Cross River State BC, Ugaga
	75	RUS	R. Mayak, Murmansk + 2 stns
	5	RUS	R. Rossii/Reg, Salsk + 2 stns
	20	RUS	R. Teos, Kurkino
	5	UKR	UR1, Luhansk
1143	0.3	D	AFN Bavaria, Bamberg/Schweinfurt
	12	D	AFN Power Network, Stuttgart + 2 stns
	1	D	AFN SHAPE, Mönchengladbach
	5	E	COPE, Jaén
	2	E	COPE, Ourense
	5	E	COPE, Oviedo
	2	E	COPE, Reus
	50	IRN	IRIB R. Iran, Yasuj
	10	NIG	Niger State Media Corp, Bida
	100	RUS	R. Mayak, Samara
	150	RUS	R. Mayak/VOR, Bolshakovo
1152	10	AGL	EP do Zaire, Mbanza Congo
	10	E	RNE5 TN, Albacete
	10	E	RNE5 TN, Cartagena
	10	E	RNE5 TN, Lleida
	20	E	RNE5 TN, Málaga
	10	E	RNE5 TN, Zamora
	4	G	Clyde 2, Glasgow
	3	G	Gold, Birmingham
	0.8	G	Gold, Norwich
	0.3	G	Gold, Plymouth
	24	G	LBC News, London
	1.5	G	Magic 1152, Manchester
	2	G	Magic 1152, Newcastle
	100	IRN	IRIB R. Iran, Tabriz (inactive)
	50	KEN	KBC Swahili Sce, Wajir
	400	ROU	R. România Actualitati, Cluj (Jucu)
	200	UAE	Ras Al-Khaimah BS, Ras al-Khaimah
1161	5	ALG	RA 1/R. Tamanrasset, In Salah
	20	BUL	Horizont/Turkish Sce, Dulovo/Targovishte
	100	EGY	ERTU Reg. prgr, Tanta
	1	G	BBC Sussex, Bexhill
	0.1	G	BBC Three Counties R, Bedford
	0.2	G	Gold, Swindon
	0.4	G	Magic 1161, Hull
	1	G	Tay AM, Dundee
	600	IRN	VOIRI, Qasr-e-Shirin

kHz	kW	Ctry	Station, location
	75	RUS	R. Orfey, Dubovka (Volgograd)
1170	25	AGL	EP do Huambo, Huambo
	700	BLR	R. Belarus/VOR relay, Sasnovy
	0.3	G	Gold, Ipswich
	0.1	G	Gold, Portsmouth
	0.3	G	Magic 1170, Stockton
	0.2	G	Signal Two, Stoke-on-Trent
	0.6	G	Swansea Sound, Swansea
	750	IRN	IRIB R. Iran, Abadan
	50	IRN	IRIB R. Iran, Damghan
	1200	RUS	VOR/Relays, Tbilisskaya
	15	SVN	R. Capodistria/R. Slovenija Int, Beli Kriz
	50	SWZ	TWR Mpangela Ranch (Manzini)
	800	UAE	R. Sawa, Al-Dhabbiya
1179	25	CNR	SER R. Club Tenerife, Santa Cruz
	10	D	SR Antenne Saar, Heusweiler
	2	E	SER R. Rioja, Logroño
	50	E	SER R. València
	10	EGY	ERTU General prgr, Qena
	50	GRC	RS Makedonias 2/ERA4, Thess. Malgara
	50	IRN	IRIB R. Iran, Chabahar
	50	IRN	IRIB Regional, Gonbad
	30	IRQ	R. Voice of Iraq, Baghdad
	50	MOZ	EP da Zambézia, Quelimane
	400	ROU	R. România Actualitati, Bacau (Galbeni)
	10	ROU	R. România Actualitati, Resita
1188	3	D	MDR Info, Reichenbach (Görlitz)
	10	EGY	ERTU General prgr, Ras Gharib
	400	HNG	Magyar R. 4, Marcali/Szolnok
	300	IRN	IRIB R. Payam, Tehran
	10	RUS	DW/RFI, Sankt-Peterburg
	100	YEM	YRTC General prgr, Al-Hiswah
1197	10	AGL	EP de Malange, Malange
	50	E	Euskadi Irratia, Gasteiz-Vitoria
	6	G	Absolute R, 9 stns
	50	IRN	IRIB Regional, Dasht
	50	IRN	IRIB Regional, Moghan
	1	IRQ	R. Dar as-Salam, Kirkuk
	100	LSO	Family R, Lancer´s Gap
	1	MKD	R. Kriva Palanka
	15	ROU	R.Târgu Mures/Ant. Brasovului, Brasov
	15	RUS	R. Rossii/Reg, Balakovo/Balashov/Ershov
1200	0.5	AFG	R. Day Kundi, Nili (inactive)
1206	300	F	France Info, Bordeaux
	50	IRN	IRIB Birjand regional
	-	IRQ	VO People of Kurdistan, Sulaimaniyah
	50	ISR	KI Arabic prgr, Acre (Akko)
	50	MOZ	EP de Inhambane, Inhambane
	1	ROD	Mauritius BC R. Rodrigues, Citronelle
	5	RUS	R. Mayak, Plesetsk
1215	500	ALB	CRI relay, Fllakë
	20	ARS	BSKSA General prgr, Madinah
	50	BOT	R. Botswana, Mahalapye
	5	E	COPE, Córdoba
	5	E	COPE, Léon
	5	E	COPE, Lorca
	5	E	COPE, Santander
	660	G	Absolute R, 14 stns
	50	IRN	IRIB Regional, Chalus
	-	IRQ	IMN Republic of Iraq R, Tikrit
	1200	RUS	VOR, Bolshakovo
	10	TZA	R. Tanzania, Arusha
1224	300	BUL	R. Bulgaria, Vidin
	5	E	COPE, Albacete
	5	E	COPE, Almería
	10	E	COPE, Donosti-San Sebastián
	5	E	COPE, Huelva
	5	E	COPE, Lleida
	5	E	COPE, Lugo
	5	E	COPE, Palma de Mallorca
		IRN	IRIB R. Iran, unknown location
	400	IRN	VOIRI, Kerman
	20	ISR	Galei Tzahal, Beersheba
	50	MOZ	EP de Cabo Delgado, Pemba
	50	NIG	Plateau RTV Corp, Jos
	10	REU	R. Réunion, St. Andre (F.PI.)
	30	RUS	R. Ura, Elista
1233	10	AGL	EP da Huila, Lubango

kHz	kW	Ctry	Station, location
	600	CYP	Monte-Carlo Doualiya/TWR, Cape Greco
	1.6	G	Absolute R, 5 stns
	50	IRN	IRIB Regional, Abadeh
	50	KEN	KBC English Sce, Marsabit
1242	150	F	France Info, Marseille
	4	G	Absolute R, 4 stns
	0.3	G	Gold, Maidstone
	50	IRN	IRIB R. Iran, Zanjan
	200	OMA	R. Sultanate of Oman, Seeb
1251	-	G	BFBS Gurkha Service, York
	0.8	G	Gold, Bury St. Edmunds
	0.001	G	LPAMs
	25	HNG	MR1/Eszag-Alföldi Reg, Nyíregyháza
	25	HNG	MR1/Nyugat-Dunántúli Reg, Szombathely
	10	HOL	NPS Radio 5, Hulsberg
	100	IRN	IRIB R. Iran, Kiashahr
	400	LBY	LJBC Main prgr./VO Africa, Tripoli
	22	POR	R. Sim, Valongo/Viseu + 2 stns
	6	RUS	R. Mayak, Letnaya Stavka/Neftekumsk
	7	RUS	R. Mayak/Reg, Cherkessk
1260	10	AGL	EP do Kuanza Norte, N'dalatando
	500	ARS	BSKSA General prgr, Dammam
	2	CVA	Vatican R, Vatican City
	5	E	SER R. Algeciras
	5	E	SER R. Murcia
	1.5	G	Absolute R, Lydd/Guildford
	0.5	G	BBC R. York, Scarborough
	2	G	Gold, Bristol
	0.6	G	Gold, Wrexham
	0.3	G	Sabras R, Leicester
	10	IRN	IRIB R. Iran, Khur
	50	MOZ	EP do Niassa, Lichinga
	10	RUS	BBC World Sce, Moskva/Sankt-Peterburg
1269	20	CNR	R. ECCA, Las Palmas
	300	D	Deutschlandfunk, Neumünster
	5	E	COPE, Badajoz
	10	E	COPE, Ciudad Real
	10	E	COPE, Figueres
	5	E	COPE, Zamora
	50	IRN	IRIB R. Iran, Khalkhal
	100	KWT	R. Kuwait Classical music, Kabd
	10	NIG	Taraba State BS, Jalingo
	10	SRB	RTN Serbian Sce, Srbobran
	200	UAE	R. Asia, Ras al-Khaimah
1278	25	AGL	EP de Cabinda, Tenda
	10	EGY	ERTU General prgr, Asswan
	300	F	France Bleu/Reg, Strasbourg
	0.002	G	BFBS Gurkha R, Folkestone/Stafford
	0.001	G	LPAMs/RSLs
	0.4	G	Pulse 2, Bradford
	10	GRC	ERA Regional, Florina
	100	IRN	IRIB R. Iran, Kermanshah
1287	2	AFS	SABC Ligwalagwala FM, Welgedacht
	5	ARS	BSKSA Quran prgr, Makkah
	5	E	SER R. Castilla, Burgos
	10	E	SER R. Lleida
	5	E	SER R. Lugo
	0.003	G	BFBS Gurkha R, 3 stations
	0.004	G	Garrison Radio, 4 stations
	0.001	G	LPAMs
	100	IRN	IRIB Regional, Lar
	2	POR	RTP Antena 1, Portalegre
	50	RUS	R. Rossii/Reg, Groznyy
1296	400	AFG	R. Free Afgh. / VOA, Pol-e-Charkhi
	10	AGL	EP do Uíge, Uíge
	75	BUL	Hristo Botev, Kardzhali
	50	E	COPE, Valencia
	35	G	BBC World Sce, Orfordness (DRM/relays)
	10	G	R. XL, Birmingham
	50	IRN	IRIB Regional, Qazvin
	10	IRN	IRIB Regional, Zabol
	600	SDN	SRTC General prgr, Reiba
	10	SRB	RTS Beograd 1, Vranje
1305	10	AFG	R. Kandahar
	1	ARS	BSKSA 2nd prgr, Taif
	20	E	RNE5 TN, Bilbao
	20	E	RNE5 TN, Ciudad Real
	10	E	RNE5 TN, León
	25	E	RNE5 TN, Ourense
	10	EGY	ERTU General prgr, Assiut
	0.2	G	Gold, Newport
	0.2	G	Magic AM, Barnsley
	0.5	G	Premier Christian R, Chingford/Epsom
	50	IRN	IRIB Regional, Bushehr
	50	ISR	Galei Tzahal, Rosh-Pina
	50	KEN	KBC Swahili/Eastern Sce, Wajir
	7	RUS	R. Mayak, Serov
1314	10	AGL	EP do Namibe, Namibe
	1000	ARM	PRA FS/VOR relay, Gavar
	20	E	RNE5 TN, Cuenca
	10	E	RNE5 TN, Salamanca
	10	E	RNE5 TN, Tarragona
	11	EGY	ERTU General prgr, Hurghada/1 st.
	1	EGY	ERTU Reg./Koran prgr, Abu Simbel
	10	GRC	ERA Regional, Tripolis
	2	I	RAI Radiouno/Reg, Matera
	50	IRN	IRIB R. Iran, Ardabil + 1 stn
	50	ROU	Antena Satelor, Constanta (Valu lui Traian)
	25	ROU	Antena Satelor, Timisoara
	15	ROU	R. Oltenia Craiova, Craiova
	1	RUS	R. Rossii/Reg, Pleshanovo
	50	SYR	Syrian R. 2/Aleppo R/Ext. Sce, Sarakeb
1323	200	CYP	BBC World Sce, Zyyi
	1000/150	D	VOR, Wachenbrunn
	0.5	G	Gold, Brighton
	50	IRN	IRIB Regional, Jolfa
	15	ROU	R. Târgu Mures
	10	TZA	R. One Swahili channel, Moshi
1332	50	CZE	CRo2/CRo6, Moravské Budejovice
	0.4	G	BBC R. Wiltshire, Lacock
	0.6	G	Gold, Peterborough
	1	G	Premier Christian R, London
	2	HOL	Hot R. Plus, Lopik
	300	IRN	IRIB Tehran City R, Tehran
	1	MDR	RDP Madeira, Monte Funchal
	0.8	POL	Twoje R, Pinczów
	50	ROU	R. România Actualitati, Galati
1341	10	E	Onda Cero R, Almeria
	5	E	Onda Cero R, Ciudad Real
	5	E	SER R. León
	100	EGY	ERTU Cult./Songs prgr, Cairo Abu Zaabal
	20	EGY	ERTU General prgr, Idfu/Siwa
	10	EGY	ERTU Koran/Educ./Sports prgr, Bawiti
	100	G	BBC R. Ulster, Lisnagarvey
	300	HNG	Magyar Katolikus R, Balatonsz./Szolnok
	10	IRN	IRIB R. Iran, Bam/Sirjan
	-	IRQ	Voice of Komal, Kirkuk
	10	KWT	R. Kuwait Quran/2nd prgr, Magwa
1350	1000	ARM	TWR/DW relays, Gavar
	50	BOT	R. Botswana, Tshabong
	10	EGY	ERTU General prgr, Quseir
	10	F	R. Orient, Nice (Fontbonne)
	0.001	G	LPAMs
	30	GEO	Abkhaz State Radio, Soxum
	5	HNG	MR1/Nyugat-Dunántúli Reg, Györ
1359	10	E	RNE R. Nacional, Madrid (DRM)
	100	ETH	VO Tigray Revolution, Addis Ababa
	0.8	G	BBC R. Solent, Bournemouth
	0.2	G	Gold, Cardiff
	0.3	G	Gold, Chelmsford
	0.3	G	Gold, Coventry
	50	IRN	IRIB Regional, Darab
	25	NIG	Capital Sound, Zaria
	10	NIG	Osun State BC, Iwo
	50	RUS	R. Mayak, Perm
	50	UKR	UR R. Tsentr, Dokuchaievsk
1368	10	EGY	ERTU General prgr, El Kharga
	1	EGY	ERTU Koran/Educ./Sports, El Farafra
	2	G	BBC R. Lincolnshire, Lincoln
	0.1	G	BBC R. Wiltshire, Swindon
	0.5	G	BBC Surrey, Duxhurst
	20	G	Manx Radio, Foxdale
	0.4	I	Challenger R, Villa Estense (tests)
	100	IRN	IRIB Regional, Gorgan
	20	ISR	Galei Tzahal, Shivta
	10	SEY	SBC Radio, Victoria

kHz	kW	Ctry	Station, location
	6	SRB	R. Valjevo
1377	1000	ARM	VOR/TWR relays, Gavar
	300	F	France Info, Lille
	0.1	G	Asian Sound R, Ashton Moss
	50	IRN	IRIB Regional, Chabahar
	50	IRN	IRIB Regional, Paveh
	50	TZA	R. Free Africa, Mwanza
	50	UKR	UR1, Chernivtsi
1386	5	AFG	RTA R. Paktin Voice, Zareh Sharan
	10	AGL	EP da Lunda Sul, Saurimo
	50	E	Euskadi Irratia, Bilbao
	1	EGY	ERTU General prgr, Barnis
	10	EGY	ERTU Reg./Koran prgr, Luxor
	0.001	G	LPAMs
	0.003	G	R Jcom, Leeds
	50	GUI	R. Rurale, Labé (inactive)
	100	KEN	KBC English/Northern Sce, Maralal
	500	LTU	Relays via R. Baltic Waves Int, Sitkunai
1394	4	MDG	RNM 1, Antananarivo
1395	500	ALB	R. Tirana/TWR relay, Fllakë
	150	ARM	Public R. of Armenia, Yerevan
	20	HOL	Big L International/KBC R, Trintelhaven
	50	IRN	IRIB R. Iran, Hajiabad
	0.5	IRQ	R. Shanasheel, Basra
	10	NIG	Akwa Ibom BC, Abak
	10	NIG	R. Kogi, Egbe
	5	RUS	R. Rossii/Reg, Buguruslan
1404	10	AGL	EP do Bié, Kuito
	20	F	France Bleu Frequenza Mora, Ajaccio
	45	F	France Info, Grenoble/Pau/Dijon
	20	F	France Info/Bretagne Reg, Brest
	0.001	G	LPAMs/RSLs
	100	GRC	ERA Regional, Komotini
	1	I	R. Luna, Dinazzano di Casalgrande
	10	IRN	IRIB R. Iran, Dasht
	0.1	IRQ	R. Ashur, Mosul (vf)
	20	LBY	LBJC Quran prgr, Tripoli (inactive)
	10	MWI	MBC R. 1, Chitipa
	10	NIG	Gombe State BS, Gombe
	0.8	POL	Twoje R. RCH Plus, Chojnice
	15	ROU	R. România Actualitati, Sibiu
	50	ROU	R.Cluj/R. Sighet, Sighetul Marmatiei
1413	1	D	SWR Cont.Ra, Ulm
	5	E	RNE5 TN, Girona
	10	E	RNE5 TN, Jaén
	20	E	RNE5 TN, Vigo
	1	G	BBC R. Gloucestershire, 2 stns
	0.14	G	Fresh R, Skipton/Richmond
	1	G	Premier Christian R, 2 stns
	10	IRN	IRIB Regional, Estahban
	500	MDA	VOR, Grigoriopol
	750	OMA	BBC World Sce, A'Seela
1422	1	AFS	Hellenic R, Bedfordview
	50	ALG	R. Algérienne C, Ouled Fayet (Algér)
	20	ARS	BSKSA Foreign prgr, Riyadh
	400	D	Deutschlandfunk, Heusweiler
	10	EGY	ERTU Koran/Educ./Sports, Ras Gharib
	10	EGY	ERTU Reg./Koran prgr, Salum
	100	IRN	IRIB R. Iran, Kermanshah (inactive)
	10	MWI	MBC R. 1, Matiya
	10	ROU	R. România Actualitati, Olanesti
1431	250/150	D	VOR/Relays, Wilsdruff (Dresden)
	600/300	DJI	R. Sawa, Djibouti (Dorale)
	0.02	G	Fresh Radio, Ilkley/Settle
	0.1	G	Gold, Reading
	0.4	G	Gold, Southend
	0.001	G	LPAMs
	0.35	GRC	1431 AM, Thessaloniki
	2	I	RAI Radiouno, Foggia
	200	IRN	IRIB R. Iran, Habibabad (Isfahan)
	50	IRN	IRIB Regional, Lamerd
1440	10	AFG	R. Nangarhar, Jalalabad
	10	AGL	EP da Lunda Norte, Dundo
	1600	ARS	BSKSA General prgr, Ras al-Zawr
	50	CAF	R. Centrafrique, Bangui
	1200/120	LUX	RTL R./Relays, Marnach (also DRM)
	10	NIG	Adamawa BC, Yola

kHz	kW	Ctry	Station, location
	10	RUS	R. Zvezda, Sankt-Peterburg
	5	SRB	RTS Beograd 1, Jagodina
	10	TZA	R. One English channel, Dar es Salaam
1449	0.2	G	BBC Asian Network, Gunthorpe
	2	G	BBC R.4, Redmoss
	0.001	G	LPAMs
	2	I	RAI Radiouno/Reg, Biella
	4	I	RAI Radiouno/Reg, Bolzano
	2	I	RAI Radiouno/Reg, Sondrio
	400	IRN	VOIRI, Bandar-e-Torkamen
	500	LBY	LJBC Main prgr, Al-Assah
	70	RUS	R. Mayak, Monchegorsk + 6 stns
1458	10	AGL	EP do Moxico, Luena
	500	ALB	R. Tirana/CRI relay, Fllakë
	10	BHR	R. Bahrain General prgr, Manama
	5	G	BBC Asian Network, Langley Mill
	2	G	BBC Newcastle, Wrekenton
	0.5	G	BBC R. Cumbria, Whitehaven
	2	G	BBC R. Devon, Torquay
	5	G	Gold, Manchester
	125	G	Sunrise Radio, London
	2	GIB	GBC R. Gibraltar, Wellington Front
	10	IRN	IRIB Regional, Birjand
	20	ISR	KI Reshet Alef, Eilat/She'ar-Yeshuv
	5	MYT	RFO Mayotte, Pamandzi
	100	ROU	R. România Actualitati, Constanta
	7	RUS	R. Rossii/Reg, Kudymkar
	1	SRB	R. Kladovo
1467	10	AGL	EP do Kuando-Kubango, Menongue
	50	ARS	BSKSA Quran prgr, Hafar Al-Batin
	40	F	R. Maria France, Col de la Madone
	1000	F	Trans World R, Roumoules
	100	IRN	IRIB Regional, Qom (Alborz)
1476	1	AZE	Azärbaycan R, Sixli
	50	E	Euskadi Irratia, Donosti-San Sebastián
	10	EGY	ERTU Reg./Koran prgr, El Minya
	20	IRN	IRIB Regional, Marivan
1485	1	AFS	R. Today, Honeydew
	10	AGL	EP do Kuanza Sul, Sumbe
	0.6	D	AFN Bavaria, Hohenfels/Garmisch-P.
	0.3	D	AFN Power Network, Ansbach
	0.5	D	Oldiestar, Berlin (DRM)
	6	E	Onda Cero R, Antequera
	5	E	Onda Rambla/Punto R, Vilanova
	5	E	SER R. Alcoi
	10	E	SER R. Santander
	10	E	SER R. Zamora
	10/1	EGY	ERTU Educ./Koran/Sports prgr, El Tur
	1	G	BBC R 4, Carlisle
	2	G	BBC R. Humberside, Hull
	2	G	BBC R. Merseyside, Wallasey
	1	G	BBC Sussex, Brighton
	1	G	Gold, Newbury
	1	GRC	ERA Regional, Patras
	1	GRC	ERA Regional, Volos
	0.2	HNG	Régió R, Mohács
	0.5	I	Broadcast Italia, Roma
	1	IRN	IRIB R. Iran, Damghan
	10	IRN	IRIB Regional, Abadan
	10	IRN	IRIB Regional, Jahrom
	10	IRN	IRIB Regional, Khoy
	10	JOR	R. Jordan Main prgr, Aqaba
	1	LBY	LJBC Quran prgr, Brach (inactive)
	1.25	LVA	R. Merkurs, Riga
	1	MRC	SER R. Melilla, Melilla
	1	NOR	NRK P1/Troms og F, Longyearbyen
	0.8	POL	Twoje R, Bielsko-Biala
	0.8	POL	Twoje R, Bilgoraj
	0.8	POL	Twoje R, Gorlice
	0.8	POL	Twoje R, Kielce
	1	POL	Twoje R, Przemysl
	0.8	POL	Twoje R, Walcz
	0.5	POL	Twoje R, Zakopane
	4	ROU	R. Vocea Sperantei, Bacau + 3 stns
	5	SDN	SRTC Regional, Al-Gadarif
	1	SRB	R. Priboj
	0.5	SRB	RTN Serbian Sce, Novi Sad

kHz	kW	Ctry	Station, location
	1	SRB	RTS Beograd 1, Crna Trava
	1	SRB	RTS Beograd 1/R. Tutin
1494	20	F	France Bleu Frequenza Mora, Bastia
	29	F	France Info, Clermont Ferrand + 2 stns
	100	GRC	ERA Regional, Rhodos
	20	IRN	IRIB R. Iran, Maku
	1000	JOR	R. Jordan Main prgr, Al Karanah (inactive)
	50	MDA	R. Moldova, Cahul/Edinet
	600	RUS	VOR, Krasnyy Bor (Sankt-Peterburg)
1500	0.1	AFG	R. Nuristan
	0.1	AFG	R. Samangan, Aybak (inactive)
	-	AFG	R. Sar-e-Pol
	6	AFG	RTA R. Badghis, Qalay-e-Naw
1503	10	AGL	EP de Benguela, Benguela
	0.1	AZR	AFN, Lajes
	1	BIH	R. 1503 Zavidovici
	5	E	RNE5 TN, La Linea
	2	E	RNE5 TN, Monforte de Lemos
	25	EGY	ERTU Reg. prgr, El Arish
	1	G	BBC R. Stoke, Sideway
	200	IRN	IRIB R. Iran, Bushehr
	20/10	RUS	R. Tsentr, Kurkino (Moskva)
	10	SRB	RTS Beograd 202, Beograd
1512	1000	ARS	BSKSA Quran prgr, Jeddah Khumra
	100	GRC	ERA Regional, Chania
	0.5	I	Onda Media Broadcast, San Pietro in Casale
	50	IRN	IRIB Regional, Ardabil
1521	2000	ARS	BSKSA General prgr, Duba
	10	BHR	R. Bahrain 2nd prgr, Manama
	5	E	SER R. Castelló
	0.01	G	Flame Christian & Community R, Wirral
	0.6	G	Gold, Reigate
	100	IRN	IRIB R. Farhang, Kiashahr (inactive)
	7	RUS	R. Mayak, Zapolyarnyy
1530	10	AGL	EP de Cabinda, Tenda
	75	CVA	Vatican R, S.M. di Galeria
	0.2	G	BBC Essex, Southend-on-Sea
	0.05	G	Celtic Music R, Glasgow
	0.7	G	Pulse 2, Huddersfield
	50	IRN	IRIB R. Iran, Yazd
	1	MDR	PEF, Funchal
	15	ROU	R. România Actualitati, Radauti
	15	ROU	R.Constanta, Mahmudia (Nufaru)
	600	STP	VOA, Pinheira
1539	700	D	Evangeliumsrundfunk, Mainflingen
	50	DJI	RTD, Djibouti
	6	E	SER R. Elche, Elx-Elche
	5	E	SER R. Manresa, Manresa
	50	IRN	IRIB Regional, Gorgan
	600	UAE	VOA R. Aap Ki Dunyaa, Al-Dhabbiya
1548	10	AFS	R. Islam, Lenasia
	5	G	BBC R. Bristol, Mangotsfield
	2	G	Forth 2, Edinburgh
	97	G	Gold, London
	1	G	Magic 1548, Liverpool
	0.7	G	Magic AM, Sheffield
	10	IRN	IRIB R. Iran, 2 st's
	10	IRN	IRIB Regional, Ferdows/Larijan
	300/600	KWT	R. Sawa, Kuwait
	500	MDA	VOR/TWR/CRI, Grigoriopol
	1	SHN	R. St. Helena, Jamestown
1550	0.1	AFG	R. Herat
	50	ALG	R. Nacional de la RASD, Rabouni
1557	300	F	France Info, Nice (Fontbonne)
	0.8	G	Gold, Northampton
	0.5	G	Gold, Southampton
	0.3	G	R. Lancashire, Oxcliffe
	0.05	HOL	Magic Jazz AM, Amsterdam
	50	IRN	IRIB R. Iran, Zabol
1566	100	BEN	TWR Africa, Parakou
	0.6	G	BBC Somerset, Taunton
	0.8	G	County Sound R, Guildford
	1	I	Challenger R, Villa Estense (DRM)
	100	IRN	IRIB R. Iran, Bandar Abbas
1575	10	AFG	R. Kunar, Asadabad
	10	E	SER R. Pamplona
	5	E	SER R.Córdoba
	10	EGY	ERTU Koran/Educ./Sports, Quseir

kHz	kW	Ctry	Station, location
	0.001	G	LPAMs/RSLs
	2	I	RAI Radiouno/Reg, Campobasso
	50	I	RAI Radiouno/Reg, Genova
	2	I	RAI Radiouno/Reg, Gorizia
	1	I	RAI Radiouno/Reg, Nuoro
	10	IRN	IRIB Regional, Qayen
	2	MAU	BBC WS, Bigara
	800	UAE	R. Farda, Al-Dhabbiya
1584	10	AFG	R. Balkh, Mazar-e-Sharif
	0.5	AFG	R. Ghor, Chaghcharan
	2	AFG	R. Nimroz, Zaranj
	0.25	AFS	R. 1584, Pretoria
	1	BHR	R. Bahrain English prgr, Manama
	1	BIH	R. Bosanski Petrovac
	5	E	SER R. Gandía, Gandia
	5	E	SER R. Ourense, Ourense
	10	EGY	ERTU Reg./Koran prgr, Idfu
	1	EGY	ERTU Reg./Koran/Sports prgr, Baris
	1	FIN	R Hami (occ. operation)
	0.3	G	BBC Hereford & Worcester, Woofferton
	1	G	BBC R. Nottingham, Clipstone
	0.2	G	London Turkish R, London
	0.2	G	Tay AM, Perth
	1	GRC	ERA Regional, Serres
	1	GRC	RS Amaliadas, Amalias
	0.1	HOL	R. Paradijs, Utrecht
	10	I	R. Studio X, Momigno
	1	I	R. Verona, Verona
	2	I	RAI Radiouno/Reg, Terni
	10	IRN	IRIB R. Iran, Maku
	5	MRC	SER Radiolé, Ceuta
	0.1	POL	Twoje R, Andrychów
	0.5	POL	Twoje R, Busko-Zdrój
	0.8	POL	Twoje R, Chelm
	0.8	POL	Twoje R, Krosno
	0.8	POL	Twoje R, Nowy Sacz
	0.1	POL	Twoje R, Ozorków
	0.8	POL	Twoje R, Slupsk
	0.8	POL	Twoje R, Tarnobrzeg
	5	ROU	R. Vocea Sperantei, Bistrita + 4 stns
	7	RUS	R. Rossii, Khunzakh
	1	UKR	UR1/Reg, Verkhovyna
1588	0.1	IRQ	R. Shrara, Baghdad
1590	0.005	TUR	AFN Incirlik (Adana)
1593	20	D	WDR Kiraka, Langenberg (DRM)
	10	EGY	ERTU Reg./Koran prgr, Matruh
	150	KWT	VOA/R. Free Iraq, Kuwait
	10	ROU	Antena Sibiului/R. Cluj, Sibiu
	15	ROU	R. Cluj, Oradea
	15	ROU	R. Constanta, Ion Corvin
	15	ROU	R. Târgu Mures, Miercurea Ciuc
1601	1	GRC	ERA Regional, Kavala
1602	10	AFG	R. Khost
	25	E	R. Vitoria, Vitoria
	5	E	SER R. Cartagena
	5	E	SER R. Linares
	5	E	SER R. Ontinyent
	5	E	SER R. Segovia
	10/1	EGY	ERTU Reg./Koran prgr, Nag Hamadi
	10	EGY	ERTU Reg./Koran/Sports, Siwa
	1	F	R. Orient, Nîmes
	0.4	FIN	Scandinavian Weekend R, Virrat
	0.3	G	BBC R. Kent, Rusthall
	0.1	G	Desi R, Southall
	0.001	G	RSLs
	1	GRC	ERA Regional, Kozani
	1	HOL	R. Seagull/Waddensee, Harlingen
	1	IRN	IRIB R. Iran, Bafq/Estahban/Kazerun
	1	IRN	IRIB Regional, Dezful
	1	IRN	IRIB Regional, Semnan
	0.5	POL	Twoje R, Cieszyn
	0.1	POL	Twoje R, Ilza
	0.5	POL	Twoje R, Krakow
	0.8	POL	Twoje R, Sanok
	2	ROU	R. Vocea Sperantei, Bistrita + 1 stn
	5	SDN	SRTC Regional, Kadogli
	1	SRB	RTS Beograd 1, Sjenica
1611	20/30	CVA	Vatican R, S.M. di Galeria (also DRM)

EAST ASIA & PACIFIC

Abbreviations peculiar to the E.Asia/Pacific section of MW freq. lists: AF = allocated freq. C. = City. PO = Present operation on. Proj. = Projected station. Rptr. = repeater. Trtr = translator.
Australia: The numeral preceding the call letters indicates the state: 2 = New South Wales. 3 = Victoria. 4 = Queensland. 5 = South Australia. 6 = Western Australia. 7 = Tasmania. 8 = Northern Territory. ACT = Australian Capital Territory. **China, P.R:** If several locations are listed for one frequency, the power listed applies to the first entry. For full details see country section. **Indonesia:** Only RRI stns included. For details of other stns see country section. **Philippines:** Province Abbreviations: Ag Nte = Agusan del Norte; Ag Sur = Agusan del Sur; Ant = Antique; Boh = Bohol; Bat = Batangas; Buk = Bukidnon; Bul = Bulacan; Cag = Cagayan; Cam Nte = Camarines Norte; Cam Sur = Camarines Sur; Dvo Nte = Davao del Norte; Dvo Sur = Davao del Sur; Isa = Isabela; I.Nte = Ilocos Norte; I.Sur = Ilocos Sur; LU = La Union; Lanao Nte = Lanao del Norte; Lanao Sur = Lanao del Sur; Mag = Maguindanao; Mas = Masbate; M Octal = Mindoro Occidental; Mind Or = Mindoro Oriental; Mis Occ = Misamis Occidental; Mis Or = Misamis Oriental; Mt Prov = Mountain Province; Neg Occ = Negros Occidental; Neg Or = Negros Oriental; Nva Viz = Nueva Vizcaya; Pam = Pampanga; Pang = Pangasinan; Que = Quezon; Riz = Rizal; S Cot = South Cotabato; S Leyte = Southern Leyte; S Sur = Surigao del Sur; Sor = Sorsogon; Tar = Tarlac; Z Nte = Zamboanga del Norte; Z Sur = Zamboanga del Sur; Zamb = Zambales. **Russia:** Regions in the Asian parts of Russia: Sib. = Siberia. FE = Far East.

kHz	kW	Ctry	Call	Station, location
153	1200	RUS		R. Rossii + Reg., Komsomolsk, FE
162	150	RUS		R. Rossii + Reg., Norilsk, Sib.
164	500	MNG		MRT (1), Ulaanbaatar
171	250	RUS		R. Rossii + Reg., Oyash, Sib.
	150	RUS		R. Rossii + Reg., Yakutsk, FE
180	150	RUS		R. Mayak, Kruchina, Sib.
	150	RUS		R. Rossii + Reg., Yelizovo, FE
189	1200	RUS		R. Rossii + Reg., Konstantinogradovka, FE
207	150	RUS		R. Mayak, Tynda, FE
209	75	MNG		MRT (1), Dalanzadgad/Choybalsan
	30	MNG		MRT (1), Ölgiy
216	30	RUS		R. Rossii + Reg., Birobidzhan, FE
	150	RUS		R. Rossii + Reg., Krasnoyarsk, Sib.
225	1000	RUS		R. Rossii + Reg., Surgut, Sib.
227	75	MNG		MRT (1), Altay
234	500	RUS		R. Rossii + Reg., Angarsk, Sib.
	500	RUS		R. Rossii + Reg., Arman, FE
252	150	TJK		TR (1), Yangiyul
261	150	RUS		R. Rossii + Reg., Kruchina, Sib.
270	150	RUS		R. Slovo, Novosibirsk, Sib.
279	50	RUS		R. Rossii + Reg., Gorno-Altaysk, Sib.
	150	RUS		R. Rossii + Reg., Selenginsk, Sib.
	150	RUS		R. Rossii + Reg., Yekaterinburg, Sib.
	500	RUS		R. Rossii + Reg., Yuzhno-Sakhalinsk, FE
	150	TKM		Turkmen Radio (1), Asgabat
531	10	AUS	6DL	ABC (RR), Dalwallinu
	0.5	AUS	5RTI	Adelaide (HPONS)
	5	AUS	4KZ	Innisfail
	5	AUS	2PM	Port Macquarie
	5	AUS	3GG	Warragul
	10	CHN		ZJ
	300	IND		AIR (A), Jodhpur
	10	J	JOQG	NHK (1), Morioka
	1	J		NHK (1), Nago
	5	NZL		531pi, Auckland
	2	NZL		More FM, Alexandra
	5	PHL	DZBR	Kumintang Bc. System, Batangas C., Bat.
	5	PHL	DXGH	Pacific Bc. System, Gen. Santos C, S Cot
	10	PHL	DYDW	Word Bc. Corp., Tacloban C., Leyte
	5	RUS		Avtoradio, Yuzhno-Sakhalinsk, FE
	25	THA		R. Thailand, Maha Sarakham
540	10	AUS	4QL	ABC (RR), Longreach
	5	AUS	7SD	Scottsdale
	50	CHN		CNR1
		CHN		NM; QH
	20	IND		AIR, Aizawl
	2/10	INS		RRI, Bandung
	1	J	JOSK	NHK (1), Kitakyushu
	1	J		NHK (1), Matsumoto
	5	J	JOMG	NHK (1), Miyazaki
	1	J		NHK (1), Nanao/Ishigaki
	5	J	JOJG	NHK (1), Yamagata
	10	KOR		KBS, Hongseong
	1	KOR		KBS, Jangsu/Jangheung/Jeomchon
	2.5	NZL		NZ's Rhema, New Plymouth
	5	NZL		NZ's Rhema, Tauranga
	300	PAK		PBC, Peshawar
	10	PHL	DYRB	DWRL Radio, Inc., Cebu C.

kHz	kW	Ctry	Call	Station, location
	10	PHL	DZWT	Mt. Province BC, Baguio C., Benguet
	50	RUS		R. Mayak, Orenburg
	8	SMO		National Radio 2AP, Apia
	5	THA		Yaan Kraw, Bangkok
549	50	AUS	2CR	ABC (RR), Orange
	25	CHN		EN; NM (2 stns)
	1200	CHN		FJ (CNR5)
	100	IND		AIR (A), Ranchi
	10	J	JOAP	NHK (1), Okinawa
	2.5	NZL		NZ's Rhema, Kaitaia
	1	NZL		R. Sport, Nelson
	1	NZL		R. Trackside/LiveSPORT, Napier-Hastings
	5	PHL	DXHM	Catholic Media Netw., Madong, Mind Or
	25-500	RUS		R. Mayak, FE , 4 stns (sync.)
	100	THA		R. Thailand, Lampang
	10	THA		R. Thailand, Mukdahan
	40	TJK		TR (2), Dushanbe
	700	VTN		Hung Yen (2), My Hao
550	5	HWA	KMVI	Wailuku, Maui
	50	AUS	6WA	ABC (RR), Wagin
558	5	AUS	4AM	Atherton (Mareeba)
	2	AUS	7BU	Burnie
	5	AUS	4GY	Gympie
	100	BGD		Bangladesh Betar, Khulna
	200	CHN		XJ; FJ; NM (2 stns); YN
	10	FJI		Fiji Bc. Comm. Ltd. (1), Suva
	100	IND		AIR (B), Mumbai
	20	J	JOCR	CRK, Kobe
	250	KOR	HLQH	KBS, Daegu
	5	NZL		R. Sport, Invercargill
	50	PHL	DZXL	Radio Mindanao Netw., Pasig C., NCR
	10	THA		R. Thailand, Kanchanaburi
	50	THA		R. Thailand, Songkhla
	1	TWN	BEH7	Fu Hsing BC (1), Taipei
	50	VTN		Ho Chi Min C. (2), Quan Tre (v)
565	10	MLA		RTM Labuan, Tenom
567	10	AUS	4JK	ABC (RR), Julia Creek
	0.1	AUS	6...	ABC (RR), W. A., 4 stns
	0.5	AUS	2BH	Broken Hill
	10	CHN		JS (CNR1)
	20	CHN		TJ; EN
	10	GUM	KGUM	Agana
	20	HKG		RTHK (3), Golden Hill
	300	IND		AIR, Dibrugarh
	100	J	JOIK	NHK (1), Sapporo
	100	KOR	HLKF	KBS, Jeonju
	200	LAO		Lao National Radio, Vientiane
	50	NZL		RNZ National, Wellington
	150	PAK		PBC, Khuzdar
	5	PHL	DXCH	Pacific Bc. System, Cotabato C., Mag.
	10	PHL	DWRP	Pilippine Bc. Sce., Naga C., Cam. Sur
	150	RUS		R. Rossii + Reg., Kyzyl, Sib.
	5	THA		Jor. Sor. 5, Chaiyaphum
570	1	HWA	KQNG	Lihu'e, Kauai
576	50	AUS	2RN	ABC (RN), Sydney
	200	BRM		MRTV, Yangon
	10	CHN		YN; ZJ(v); EN; FJ
	200	IND		AIR, Alappuzha

kHz	kW	Ctry	Call	Station, location
	1	J	JODG	NHK (1), Hamamatsu
	10	J	JOHG	NHK (1), Kagoshima
	40	KGZ		R. DDD, Osh
	5	KOR		AFNK, Munsan
	1	KOR		KBS, Suncheon
	100	NPL		R. Nepal, Surkhet
	2.5	NZL		The Word/Bible Radio, Hamilton
	5	PHL	DZHR	Cebu Bc. Co., Tuguegarao, Cag.
	10	PHL	DXMF	People's Bc. Sce., Davao C., Dvo Sur
	10	PHL	DYMR	Pilippine Bc. Sce., Cebu C
	5	PHL	DZMQ	Pilippine Bc. Sce., Dagupan C., Pang.
	150	RUS		R. Mayak, FE, 2 stns, (sync)
	250-500	RUS		R.Mayak, Sib., 2 stns, (sync)
	5	THA		Tor. Chor. Dor., Bangkok
	150	TKM		Turkmen Radio (2), Asgabat
	50	VTN		Khanh Hoa (2/P), Nha Trang
585	10	AUS	6PB	ABC (PNN), Perth
	10	AUS	7RN	ABC (RN), Hobart
	10	AUS	2WEB	W. Region Educ. Bc., Bourke
	50	CHN		JS + 12 stns
	200	CHN		Southeast BC, FJ
	300	IND		AIR (A), Nagpur
	50	INS		RRI, Surabaya
	10	J	JOPG	NHK (1), Kushiro
	20	LAO		Lao National Radio, Khantabouly
	2	NZL		R. Ngati Porou, Ruatoria
	500	PAK		PBC, Islamabad (Faqeerabad)
	5	PHL	DXCP	Catholic Media Netw., Gen. Santos C., S Cot
	1	PHL	DYLL	Pilippine Bc. Sce., Iloilo C.
	10	PNG		NBC, Port Moresby
	10	PNG		R. West Sepik, Vanimo
	1200	RUS		VoR, Belogorsk, FE
	5	SMA	KJAL	Asia Pac. Media Min., Tafuna (r. 580/inactive)
	5	THA		Thor. Phor. 3, Phrae
	5	THA		Wor. Por. Tho. 15, Chumphon
590	7.5	HWA	KSSK	Honolulu, Oahu
594	50	AUS	3WV	ABC (RR), Horsham
	200	BRM		MRTV, Nay Pyi Taw
	200	BRU		RTV Brunei, Tutong
	300	CHN		XZ; SD
	1000	IND		AIR (FS), Chinsurah
	300	J	JOAK	NHK (1), Tokyo
	10	KOR		KBS, Yeongju
	5	NZL		NZ's Rhema, Timaru
	2	NZL		NZ's Rhema, Wanangui
	5	PHL	DXDB	Catholic Media Netw., Malaybalay, Buk.
	20	PHL	DZBB	GMA Network, Inc., Quezon C., NCR
	150	RUS		R. Mayak, Krasnoyarsk, Sib.
	1000	RUS		R. Mayak, Surgut, Sib.
	5	THA		Phon. Por. Thor. Or., Bangkok
	10	TWN	BEH44	Fu Hsing BC (2), Kaohsiung
	5	TWN	BEH38	Fu Hsing BC (2), Taichung
	10	TWN	BEH2	Fu Hsing BC (2), Taipei
	50	VTN		Danang (1), An Hai
603	10	AUS	2RN	ABC (RN), Nowra
	10	AUS	4CH	ABC (RR), Charleville
	2	AUS	6PH	ABC (RR), Port Hedland
	200	CHN		EN + 38 stns
	600	CHN		HA (FS)
	200	IND		AIR, Ajmer
	5	J	JOOG	NHK (1), Obihiro
	5	J	JOKK	NHK (1), Okayama
	500	KOR	HLSA	KBS, Namyang (Seoul)
	5	NZL		R. Waatea, Auckland
	10	PHL	DZLL	Bicol Bc. System, Naga C., Cam Sur
	5	PHL	DZVV	Cons. Bc. Syst., Inc., Vigan, I. Sur
	5	PHL	DXPR	R. Mindanao Netw., Pagadian C, Z Sur
	30	RUS		R. Mayak, Belogorsk/Skovorodino, FE
	5	THA		Wor. Por. Tho. 12, Khon Kaen
	500	TWN		CBS (RTI), Lukang
610	100	VTN		Ho Chi Minh C. (H)
612	50	AUS	4QR	ABC (MS), Brisbane
	10	AUS	6RN	ABC (RN), Dalwallinu
	100	CHN		FJ; LN; SC
	200	IND		AIR (A), Bengaluru
	100	J	JOLK	NHK (1), Fukuoka
	150	KGZ		Kyrgyz R. (2), Bishkek
	2	NZL		NZ's Rhema, Christchurch
	10	PAK		PBC (2), Karachi (Landhi)
	5	PHL	DWSP	Philippine Bc. Corp., Itogon, Benguet
	10	PHL	DYHP	R. Mindanao Netw., Cebu C.
	25	RUS		R. Mayak, Norilsk, Sib.
	5	THA		Mor. Kor., Chiang Mai
	5	THA		Wor. Sor. Por., Lop Buri
620	5	HWA	KHNU	Hilo, Hawaii
	10	HWA	KHNU	Kalaoa, Hawaii (rptr)
	5	HWA	KHNU	Na'alehu, Hawaii (rptr)
621	50	AUS	3RN	ABC (RN), Melbourne
	2	AUS	6EL	Bunbury
	200	CHN		HL; QH; HB; SC; YN; SD
	20	HKG		RTHK (P), Golden Hill
	100	IND		AIR (A), Patna
	3	J	JOCG	NHK (1), Asahikawa
	1	J		NHK (1), Iida/Nobeoka
	1	J	JOOK	NHK (1), Kyoto
	10	KOR		KBS, Seogwipo
	10	KOR		KBS, Taebaek
	1	KOR		KBS, Yeongdong
	500	KRE		Pyongyang BS/VoK, Chongjin
	2	NZL		NZ's Rhema, Dunedin
	2	NZL		NZ's Rhema, Whangarei
	1	PHL	DZVC	Pilippine Bc. Sce., Virac, Catanduanes
	10	PHL	DXDC	R. Mindanao Netw., Davao C., Dvo Sur
	5	PHL	DZTG	R. Philippines Netw., Tuguegarao, Cag.
	50	RUS		R. Rossii + Reg., Khabarovsk, FE
	100	THA		R. Thailand, Khon Kaen
	1	TWN		Taiwan BC, Tahsi
630	10	AUS	2PB	ABC (PNN), Sydney
	0.4	AUS	7RN	ABC (RN), Queenstown
	5	AUS	6AL	ABC (RR), Albany
	50	AUS	4QN	ABC (RR), Townsville
	100	BGD		Bangladesh Betar (B), Dhaka
	200	CHN		JX; HEN (CNR2)
	2.5	CKH		R. Cook Is., Rarotonga
	10	GUM	KUAM	Agana
	100	IND		AIR, Thrissur
	50	INS		RRI, Makassar
	5	KOR		KBS, Inje
	10	KOR		KBS, Yeosu
	10	NZL		RNZ National, Napier-Hastings
	100	PAK		PBC (1), Lahore
	35	PHL	DZMM	ABS-CBN Bc. Corp., Quezon C., NCR
	5	PHL	DYAG	Cadiz R. And TV Netw., Cadiz C., Cam. Sur
	500	RUS		VoR, Komsomolsk, FE
	5	THA		Mor. Thor. Bor. 11, Bangkok
	10	TWN	BED65	BCC (N), Ilan
	10	TWN		Taiwan BC, Sungling
	200	VTN		Quang Binh (1), Dong Hoi
639	2	AUS	8RN	ABC (RN), Katherine
	1	AUS	4MS	ABC (RR), Mossman
	10	AUS	5CK	ABC (RR), Port Pirie
	5	AUS	2HC	Coff's Harbour
	200	CHN		BJ (CNR1)
	2	FJI		Fiji Bc. Comm. Ltd. (1), Lautoka
	100	IND		AIR, Kohima
	5	J	JOIP	NHK (1), Oita
	10	J	JOPB	NHK (2), Shizuoka
	5	J	JOWN	STV, Hakodate
	50	KOR	HLKC	KBS, Gaebong (Seoul)
	2	NZL		RNZ National, Alexandra
	100	PAK		PBC, Karachi (Landhi)
	5	PHL	DXKR	R. Mindanao Netw., Koronadal, Cot. Sur
	1	PHL	DZRL	R. Philippines Netw., Batac, I. Nte
	75	RUS		R. Rossii + Reg., Omsk, Sib.
	10	THA		R. Thailand, Lamphun
	20	THA		R. Thailand, N. Si Thammarat
640	10	LAO		Lao Nationale Radio, Vientiane
648	2	AUS	6GF	ABC (RR), Kalgoorlie
	10	AUS	2NU	ABC (RR), Tamworth
	150	CHN		GD + 6 stns
	200	IND		AIR (A), Indore
	10	J		AFN, Okinawa C.

kHz	kW	Ctry	Call	Station, location
	5	J	JOIG	NHK (1), Toyama
	1	KOR		KBS, Boseong
	100	NPL		R. Nepal, Dhankuta
	5	PHL	DYRC	Manila Bc. Co., Cebu C.
	10	PHL	DWRH	Pacific Bc. System, Santiago C., Isa.
	10	PHL	DWRM	Pilippine Bc. Sce., Pto. Princesa, Palawan
	5	PHL	DXMB	R. Mindanao Netw., Malaybalay, Buk.
	1000	RUS		VoR, Razdolnoye, FE
	25	THA		R. Thailand, Khon Kaen
	1000	TJK		Various relays, Orzu
	50	VTN		Binh Dinh (1), An Nhon
650	10	HWA	KRTR	Honolulu, Oahu
657	2	AUS	8RN	ABC (RN), Darwin
	10	AUS	2BY	ABC (RR), Byrock
	2	AUS	6--	Perth (HPONS)
	300	CHN		EN; JL; ZJ
	200	IND		AIR (A), Kolkata
	50	KOR	HLKM	KBS, Chuncheon
	1500	KRE		Pyongyang BS, Kangnam
	20	MLA		RTM, Gerik
	10	NZL		Parl. Bc/So. Star, Tauranga
	50	NZL		Parl. Bc/So. Star, Wellington
	5	PHL	DXDD	Dan-ag sa Dakbayan Bc. Corp., Ozamis C.
	1	PHL	DYES	Philippine Bc. Sce., Borongan, E. Samar
	5	PHL	DWRN	Philippine R. Corp., Naga C., Cam. Sur
	5	PHL	DYVR	R. Mindanao Netw., Roxas C., Capiz
	1	PHL	DZLU	Satellite Bc. Corp., S. Fernando C., LU
	5	THA		Jor. Sor. 1, Bangkok
	20	TWN	BED34	BCC (N), Taipei
	10	TWN	BEV59	Cheng Sheng BC (2), Taichung
	50	VTN		Ho Chi Minh C. (1), Quan Tre (v)
666	5	AUS	2CN	ABC (MS), Canberra
	2	AUS	4CC	Biloela (trtr)
	1	AUS	6LN	Carnarvon
	2	AUS	4LM	Mt. Isa
	200	CHN		QH + 9 stns
	600	CHN		VO Strait, FJ
	100	IND		AIR (B), New Delhi
	100	J	JOBK	NHK (1), Osaka
	20	NCL		R. Nouvelle Caledonie, Noumea
	50/25	PHL	DZRH	Manila Bc. Co., Makati C, NCR
	10	PHL	DXRP	Pilippine Bc. Sce., Davao C., Dvo Sur
	10	RUS		BBC, Yekaterinburg, Sib.
	150	RUS		R. Mayak, Komsomolsk, FE
	5	THA		Thor. Phor. 2, Surin
	5	THA		Thor. Phor. 3, Tak
	50	VTN		Khanh Hoa (1), Nha Trang
670	10	HWA	KPUA	Hilo, Hawaii
675	10	AUS	2CO	ABC (RR), Albury
	5	AUS	6BE	ABC (RR), Broome
	200	CHN		NM + 5 stns
	10	HKG		RTHK (6), Peng Chau
	20	IND		AIR, Bhadravathi
	20	IND		AIR, Chhattarpur
	100	IND		AIR, Itanagar
	5	J	JOVK	NHK (1), Hakodate
	5	J	JOUG	NHK (1), Yamaguchi
	10	KOR		KBS, Jeonju
	10	NZL		RNZ National, Christchurch
	5	PHL	DWLW	Intercontinental Bc. Corp., Laoag C., I. Nte
	5	PHL	DYKC	R. Philippines Netw., Mandaue, Cebu
	1	PHL	DXGD	Sulu Tawi-Tawi Bc. Found., Bongao, Tawi-Tawi
	10	PNG		R. East Sepik, Wewak
	40	RUS		R. Radonezh, Razdolnoye, FE
	5	THA		Sor. Thor. Ror. 2, Bangkok
	150	TKM		Turkmen R. (1), Asgabat
	10	TKM		Turkmen R. (2), Türkmenbasy
	1	TWN		Cheng Sheng BC, Peikang
	700	VTN		Hung Yen (1), My Hao
684	1	AUS	8RN	ABC (RN), Tennant Creek
	5	AUS	6BS	ABC (RR), Busselton
	10	AUS	2KP	ABC (RR), Kempsey
	1200	CHN		FJ (CNR6)
	600	CHN		HA (FS)
	50	CHN		HL + 8 stns
	10	FJI		Fiji Bc. Comm. Ltd. (1), Labassa

kHz	kW	Ctry	Call	Station, location
	200	IND		AIR (A), Kargil
	100	IND		AIR (A), Kozhikode
	100	IND		AIR, Port Blair
	5	J	JODF	IBC, Morioka
	1	J	JOLO	IBC, Ofunato
	5	J	JOAG	NHK (1), Nagasaki
	250	KRE		Pyongyang BS, Samgo
	100	NPL		R. Nepal, Pokhara
	5	NZL		NZ's Rhema, Gisborne
	5	PHL	DZCV	Filipinas Bc. Netw.,. Tuguegarao, Cag.
	5	PHL	DWJJ	Kaissar Bc. Netw., Cabanatuan, Nva. Ecija
	10	PHL	DYEZ	Manila Bc. Co., Bacolod, Neg. Occ.
	5	THA		Thor. Phor. 4, N. Si Thammarat
	5	THA		Yaan Kraw, Udon Thani
	1	TLS		R. Timor-Leste, Díli (Mt. Kutulau)
	10	TWN	BEC22	Han Sheng BC, Taipei
690	10	HWA	KHNR	Honolulu, Oahu
693	2	AUS	5SY	ABC (RR), Streaky Bay
	5	AUS	4KQ	Brisbane (Newstead)
	0.5	AUS	4LM	Cloncurry (trtr)
	5	AUS	3AW	Melbourne
	5	AUS	6WR	R. Stn. 6WR, Kununurra
	0.5	AUS	4KZ	Tully (trtr)
	1000	BGD		Bangladesh Betar (A), Dhaka
	400	BRM		MRTV, Nay Pyi Taw
	300	CHN		SN; HL
	500	J	JOAB	NHK (2), Tokyo
	5	NZL		R. Sport, Dunedin
	1	PHL	DYKH	Manila Bc. Co., Kalibo, Aklan
	10	PHL	DYPH	Manila Bc. Co., Pto. Princesa, Palawan
	10	PHL	DXBC	R. Mindanao Netw., Butuan, Ag. Nte
	1	PHL	DXDX	R. Philippines Netw., Gen. Santos C., S. Cot.
	10	PHL	DZTP	Tirad Pass R/TV Bc. Netw., Candon, I. Sur
	150	RUS		R. Rossii, Yazykovo, Sib,
	5	THA		Siang Adison, Saraburi
	10	TWN	BEC32	Han Sheng BC, Tainan
	10	TWN	BEC25	Han Sheng BC, Taoyuan
702	50	AUS	2BL	ABC (MS), Sydney
	10	AUS	6KP	ABC (RR), Karratha
		CHN		GD (CRI DS)
	200	CHN		JS + 8 stns
	200	IND		AIR (A) (FS), Jalandhar
	2/10	INS		RRI, Manokwari
	10	J	JOFB	NHK (2), Hiroshima
	10	J	JOKD	NHK (1), Kitami
	50	KRE		Korean Central BS (C/R), Chongjin
	10	NZL		R. Live, Auckland
	1	NZL		R. Sport, Ashburton
	50	PHL	DZAS	FEBC, Valenzuela, NCR
	7	RUS		R. Mayak, Bratsk, Sib.
	150	TJK		TR (1), Orzu
	10	TWN	BEP24	Ching Cha BS, Taichung
	50	VTN		Danang (2/O/D), An Hai
705	10	LAO		Lao National Radio, Luang Prabang
711	10	AUS	4QW	ABC (RR), Roma/St. George
	10	CHN		QH + 7 stns
	200	IND		AIR, Siliguri
	500	KOR	HLKA	KBS, Sorae (Seoul)
	5	NZL		Trackside/LiveSPORT, Wellington
	5	PHL	DZVR	Newsounds Bc. Netw., Laoag, I. Nte
	10	PHL	DZLW	Peñafrancia Bc. Corp., Naga C., Cam. Sur
	5	PHL	DXIC	R. Mindanao Netw., Iligan C., Lanao Nte
	5	PHL	DXRD	Swara Sug Media Corp., Davao C., Dvo. Sur
	5	PHL	DZYI	Swara Sug Media Corp., Ilagan, Isa.
	5	RUS		R. Rossii + Reg., Nikolayevsk, FE
	20	THA		R. Thailand, U. Ratchathani
	5	THA		Wor. Por. Thor., Chiang Mai
	5	THA		Wor. Sor. Por., Lop Buri
	10	TWN	BED92	BCC (C), Tainan
	250	TWN		VO Kuanghua, Hsinfeng
	500	VTN		Can Tho (1), Thoi Long
720	50	AUS	6WF	ABC (MS), Perth
	0.05	AUS	2RN	ABC (RN), Armidale
	4	AUS	4AT	ABC (RR), Atherton
	0.4	AUS	2ML	ABC (RR), Murwillumbah
	2	AUS	3MT	ABC (RR), Omeo

kHz	kW	Ctry	Call	Station, location
	200	CHN		BJ (CNR2)
	1	CHN		SC; AH
	5	HWA	KUAI	'Ele'ele, Kauai
	200	IND		AIR (A), Chennai
	10	INS		RRI, Ambon
	1	J	JOIL	KBC, Kitakyushu
	500	KRE		Korean Central BS (C/R), Wiwon (Kanggye)
	10	NZL		RNZ National, Invercargill
	5	PHL	DZJO	Bayanihan Bc. Corp., San Juan, NCR
	10	PHL	DYOK	Manila Bc. Co., Iloilo C
	5	PHL	DZSO	Newsounds Bc. Netw., San Fernando, LU
	1000	RUS		R. Mayak/VoR, Vestochka, FE
	10	THA		R. Thailand, Krabi
	5	THA		Sor. Thor. Ror. 5, Chon Buri
	1	TKM		Turkmen Radio (1), Ek-Arça/Etrek
	10	TWN	BED58	BCC (N), Taichung
729	50	AUS	5RN	ABC (RN), Adelaide
		BRM		MRTV, Yangon
	200	CHN		JX; EN
	100	IND		AIR (A), Guwahati
		INS		RRI, Nabire
	50	J	JOCK	NHK (1), Nagoya
	50	KRE		Pyongyang BS, Sepo
	5	NCL		R. Nouvelle Caledonie, Touho
	0.1	NZL		Burn 729AM, Ranfurly
	2.5	NZL		R. Sport, Whangarei
	2.5	NZL		RNZ National, Tokoroa
	5	PHL	DYEH	Katigbak Enterprises, Pto. Princesa, Palawan
	5	PHL	DXOR	Pedro N. Roa Bc., Cagayan de Oro C., Mis. Or.
	5	PHL	DZGB	People's Bc. Netw., Legaspi C., Albay
	10	PHL	DWPE	Pilippine Bc. Sce., Tuguegarao, Cag.
	5	PHL	DXMY	R. Mindanao Netw., Cotabato C., Mag.
	25	THA		R. Thailand, N. Ratchasima
	0.5	TWN	BEE43	Shih Hsin BS, Taipei
	200	VTN		Quang Binh (2), Dong Hoi
738	50	AUS	2NR	ABC (RR), Grafton
	5	AUS	6MJ	ABC (RR), Manjimup
	200	CHN		HN; JL; XJ; ZJ
	200	IND		AIR (A), Hyderabad
	1	J		KNB, Takaoka
	5	J	JOLR	KNB, Toyama
	10	J	JORR	RBC, Naha, Okinawa
	100	KOR	HLKG	KBS, Daegu
	10	MAC		R. Vilaverde
	5	NZL		R. Live, Christchurch
	20	OCE		R. Polynesie, Mahina
	60	PHL	DZRB	Pilippine Bc. Sce., Quezon C., NCR
	40	RUS		R. Rossii + Reg., Chelyabinsk, Sib.
	25	RUS		R. Rossii + Reg., Palana, FE
	5	THA		Wor. Por. Tho. 2, Chiang Mai
	5	THA		Wor. Por. Tho. 5, Songkhla
	100	TWN	BEL2	Taiwan Yuyeh, Penghu
740	50	VTN		Binh Dinh (2/P), An Nhon
747	10/5.5	AUS	7PB	ABC (PNN), Hobart
	0.2	AUS	8JB	ABC (RR), Jabiru
	10	AUS	4QS	ABC (RR), Toowoomba
	5	AUS	6SE	Esperance
	1	AUS	6FMS	Exmouth
	1	CHN		AH (CRI DS 1)
		CHN		BJ (CNR12)
	100	CHN		YN + 29 stns
	300	IND		AIR (A), Lucknow
	10	INS		RRI, Bengkulu
	500	J	JOIB	NHK (2), Sapporo
	20	KAZ		R. Liberty, Qaraghandy
	100	KOR	HLKH	KBS, Gwangju
	0.4	NZL		NewstalkZB, Rotorua
	10	PHL	DZJC	Manila Bc. Co., Laoag C, I. Nte
	5	PHL	DXND	Notre Dame Bc. Corp., Kidapawan, N. Cot.
	10	PHL	DYHB	R. Mindanao Netw., Bacolod, Neg. Occ.
	8	SMO		National Radio 2AP, Apia
	5	THA		Ror. Dor., Bangkok
	5	THA		Thor. Phor. 2, Udon Thani
	10	TWN	BED57	BCC (H), Taipei
		VTN		Ho Chi Minh City (4), Quan Tre (v)
750	1	CHN		SX

kHz	kW	Ctry	Call	Station, location
	7.5	VTN		Thai Nguyen (P)
756	10	AUS	3RN	ABC (RN), Wangaratta (rptr)
	2	AUS	2TR	ABC (RR), Taree
	2	AUS	6TZ	Margaret River
	150	CHN		HL (CNR1)
	100	IND		AIR, Jagdalpur
	2/10	INS		RRI, Purwokerto
	10	J	JOGK	NHK (1), Kumamoto
	100	KOR		KBS, Yeoju
	0.8	NZL		Puketapu R., Palmerston
	10	NZL		RNZ National, Auckland
	150	PAK		PBC (1), Quetta (Pishin)
	10	PHL	DXBZ	Baganian Bc. Corp., Pagadiani, Z Sur
	1	PHL	DWHL	Beta Bc. Syst., Olongapo C., Zamb.
	5	PHL	DWNW	Intercontinental Bc. Corp., Naga C., Cam. Sur
	10	PHL	DWRS	Pilippine Bc. Sce., Tayug, Pang.
	2	PHL	DXJM	R. Corp. of the Philippines, Butuan C., Ag. Nte
	5	THA		Kor. Wor. Sor. 1, Surin
	5	THA		Nor. Thor. Phor., Narathiwat
	1	TWN		Sheng Li, Makung
	10	VTN		Long An (P), Tan An
760	10	HWA	KGU	Honolulu, Oahu
765	5	AUS	2EC	Bega
	0.5	AUS	4GC	Hughenden (trtr)
	0.5	AUS	8HOT	Katherine (trtr)
	5	AUS	5CC	Port Lincoln
	0.1	AUS	6SAT	Tom Price/Paraburdoo. (trtr)
	10	CHN		EN + 5 stns
	600	CHN		FJ (CNR5)
	200	IND		AIR (A), Dharwad
	1	INS		RRI, Tual
	5	J	JOPF	KRY, Shunan
	5	J	JOJF	YBS, Kofu
	10	KOR	HLCQ	MBC, Daejeon
	50	KRE		Korean Central BS (C/R), Hyesan
	2.5	NZL		R. Kahungunu, Napier-Hastings
	5	PHL	DXGS	DWRL Radio, Inc., Gen. Santos C., S. Cot.
	10	PHL	DYPR	Palawan Bc. Corp., Pto. Princesa, Palawan
	5	PHL	DYAR	Swara Sug Media Corp., Cebu C.
	5	PHL	DZYT	Swara Sug Media Corp., Tuguegarao, Cag.
	5	RUS		R. Vostok Rossii, Bikin, FE (sync)
	20	RUS		R. Vostok Rossii, Khabarovsk, FE (sync)
	5	THA		Neung. Por. Nor., Lampang
	5	THA		Thor. Or. 2, Lop Buri
774	50	AUS	3LO	ABC (MS), Melbourne
	5	AUS	4TO	Townsville
		CHN		BJ (Bejing Foreign)
	100	CHN		HB; SX; LN; XJ
	2.5	FJI		Fiji Bc. Comm. Ltd. (2), Namara
	100	IND		AIR, Shimla
		INS		RRI, Fak-Fak
	500	J	JOUB	NHK (2), Akita
	10	KOR	HLAN	MBC, Chuncheon
	10	KOR	HLAJ	MBC, Jeju
	5	NZL		R. Sport, New Plymouth
	25	PHL	DWWW	Interactive Bc. Media, Inc., Quezon C., NCR
	1	PHL	DXSM	Philippine Bc. Sce., Jolo, Sulu
	10	PHL	DXSO	Pilippine Bc. Sce., Marawi C., Lanao Sur
	10	PHL	DYRI	R. Mindanao Netw., Iloilo C
	5	THA		Phon. Mor. 2, Rayong
	5	THA		Sor. Sor. Sor., Udon Thani
	10	TWN	BEV88	Hsien Sheng BC, Taoyuan
	1	TWN	BEV56	Sheng Li (1), Tainan
	10	TWN	BEV94	Taiwan BC, Taichung
783	2	AUS	8AL	ABC (RR), Alice Springs
	2	AUS	6VA	Albany
	100	CHN		EB(2 stns); GD
		CHN		VO Strait, FJ
	20	HKG		RTHK (5), Golden Hill
	20	IND		AIR (C), Chennai
	2/10	INS		RRI, Ende
	10	KOR		KBS, Yeongwol
	10	NZL		Access R. /Samoan Cap. R., Wellington
	5	PHL	DYME	Masbate Comm. Bc. Co., Masbate C.
	5	PHL	DZNL	Philippine Bc. Corp., San Fernando, LU
	10	PHL	DXRA	RMC Bc. Co., Inc., Davao C., Dvo Sur

kHz	kW	Ctry	Call	Station, location
	75	RUS		R. Lemma, Vladivostok, FE
	10	THA		R. Thailand, Ranong
	5	THA		Thor. Phor. 3, Kamphaeng Phet
	500	VTN		Can Tho (2), Thoi Long
790	5	HWA	KKON	Kealakekua, Hawaii
792	25	AUS	4RN	ABC (RN), Brisbane
	200	CHN		GX + 7 stns
	100	IND		AIR (A), Pune
	1	J		NHK (1), Takada/Naze
	1	J		NHK (1), Takayama/Enbetsu
	50	KOR	HLSQ	Seoul Bc. System, Goyang (Seoul)
	100	NPL		R. Nepal, Kathmandu
	5	NZL		R. Sport, Hamilton
	150	PAK		Azad Kashmir Radio, Muzaffarabad
	5	PHL	DWGV	GV Bc. System, Angeles C, Pampanga
	5	PHL	DYRR	Ormoc Bc. Co., Ormoc C., Leyte
	5	PHL	DXPD	People's Bc. Sce., Pagadian, Z. Sur
	10	PHL	DXBN	Pilippine Bc. Sce., Butuan, Ag. Nte
	5	PHL	DWES	Rolin Bc. Enterprises, Narra, Palawan
	25	RUS		R. Rossii + Reg., Abakan, Sib.
	50	RUS		R. Rossii + Reg., Aleksandrovsk-Sakh., FE
	20	THA		Wor. Por. Thor., Bangkok
	10	TWN	BEC33	Han Sheng BC, Hualien
	10	TWN	BEV79	Keelung BS, Keelung
801	2	AUS	4QY	ABC (RR), Cairns
	2	AUS	5RM	Berri
	5	AUS	2RF	Gosford (HPONS)
		CHN		AH (CRI DS2)
	10	CHN		EB + 23 stns
		CHN		SD (CRI DS)
	50	CHN		Zhujiang EBS, GD
	10	GUM	KTWG	TWR, Agana
	200	IND		AIR, Jabalpur
	1	INS		RRI, Medan
	10	INS		RRI, Semarang
	500	KRE		Pyongyang BS, Hwadae
	10	MLA		RTM Labuan, Kudat
	1.6	NZL		NZ's Rhema, Nelson
	5	PHL	DYKA	Catholic Media Netw., San José, Ant.
	5	PHL	DXES	Consolidated Bc. System, Inc., Gen. Santos C., S. Cot.
	5	PHL	DYWC	Franciscan Bc. Corp., Dumaguete, Neg. Or.
	5	PHL	DWFA	Hypersonic Bc. Center, Sorsogon C.
	10	PHL	DZNC	Newsounds Bc. Netw., Cauayan, Isa.
	1	PHL	DXBL	Swara Sug Bc. Media Corp., Bislig, Surigao S.
	1200	RUS		VoR, Chita (Atamanovka), Sib.
	5	THA		Mor. Thor. Bor. No. 31, N. Sawan
	5	THA		Thor. Or. 15, Chiang Rai
	5	THA		Thor. Or. 8, U. Ratchathani
	1000	TJK		Various relays, Orzu
	10	TWN		Chien kuo, Hsinhua
	250	TWN		VO Kuanghua, Kuanyin
810	20	AUS	6RN	ABC (RN), Perth
	10	AUS	2BA	ABC (RR), Bega
	200	CHN		ZJ + 6 stns
	1	FJI		Fiji Bc. Comm. Ltd. (2), Labassa
	300	IND		AIR (A), Rajkot
	7.5	INS		RRI, Merauke
	50	J		AFN, Tokyo
	20	KOR	HLCT	MBC, Daegu
	50	KRE		Korean Central BS (C/R), Kaesong
	10	NPL		R. Nepal, Dipayal
	5	NZL		BBC WS NZ, Auckland
	10	NZL		RNZ National, Dunedin
	10	PHL	DZRJ	Rajah Bc. Netw., Manila, NCR
	10	PNG		R. East New Britain, Rabaul
	10	PNG		R. Morobe, Lae
	10	RUS		Avtoradio-Krasnoyarsk, K., Sib.
	50	RUS		R. Mayak, Yekaterinburg, Sib.
	150	RUS		R. Rossii + Reg., Razdolnoye, FE
	10	THA		R. Thailand, Kanchanaburi
	20	THA		R. Thailand, Nong Khai
	10	THA		R. Thailand, Trang
	10	TWN	BEV54	Kuo Sheng, Changhua
819	10	AUS	2GL	ABC (RR), Glen Innes
	5	AUS	6KW	ABC (RR), Kununurra
	200	CHN		SX; XJ(3 stns)
	200	IND		AIR (A), New Delhi
	5	J	JONK	NHK (1), Nagano
	20	KOR	HLCN	MBC, Gwangju
	500	KRE		Korean Central BS (C), Pyongyang
	10	NZL		RNZ National, Tauranga
	10	PHL	DYVL	Manila Bc. Co., Tacloban, Leyte
	1	PHL	DXSC	So. Philippines Mass. Comm., Zamboanga C., Z. Sur
	5	PHL	DWAR	Swara Sug Media Corp., Laoag C, I. Nte
	10	PHL	DXUM	Univ. of Mindanao Bc. Netw, Davao C, Dvo Sur
	1	PHL	DWMG	Vanguard R. Netw., Solano, Nva Viz
	10	THA		R. Thailand, Pathum Thani
	15	TJK		TR (1), Khujand
	10	TWN	BED28	BCC (N), Taitung
	5	TWN	BEV35	Cheng Sheng, Taipei
	10	TWN	BEP28	Ching Cha, Kaohsiung
		VTN		Dac Lac (4), Buon me Tuhot
828	10	AUS	6GN	ABC (RR), Geraldton
	10	AUS	3GI	ABC (RR), Sale
	1	AUS	4GC	Charters Towers
	50	CHN		BJ + 6 stns
	20	IND		AIR (B), Panaji
	20	IND		AIR, Silchar
	300	J	JOBB	NHK (2), Osaka
	2	NZL		R. Trackside/LiveSPORT, Palmerston No.
	100	PAK		PBC (1), Karachi
	1	PHL	DZTC	Govt of Tarlac Prov., Tarlac C.
	5	PHL	DWZR	Hypersonic Bc. Center, Legaspi C., Albay
	10/5	PHL	DYER	Pto. Princesa Bc. Co., Pto. Princesa, Palawan
	10	PHL	DXCC	R. Mindanao Netw., Cagayan de Oro C., Mis. Or.
	150	RUS		R. Mayak, Kyzyl, Sib.
	5	THA		Thor. Phor. 3, Sukhothai
	5	THA		Wor. Por. Tho. 4, N. Si Thammarat
	50	VTN		Son La (P)
830	10	HWA	KHVH	Honolulu, Oahu
837	1	AUS	6ED	ABC (RR), Esperance
	10	AUS	4RK	ABC (RR), Rockhampton
	0.5	AUS	7XS	Queenstown
		CHN		CNR1
		CHN		FJ (CNR5)
	20	CHN		HL; EN; XJ; LN; FJ
	100	IND		AIR (A), Vijayawada
	1	J		NHK (1), Nayoro
	10	J	JOQK	NHK (1), Niigata
	50	KOR	HLKY	CBS, Seoul
	2	NZL		RNZ National, Kaitaia
	2.5	NZL		RNZ National, Whangarei
	10	PHL	DYFM	Consolidated Bc. System, Inc., Iloilo C.
	10	PHL	DXJS	Pilippine Bc. Sce., Tandag, S Sur
	5	PHL	DXRE	Swara Sug Media Corp., Gen. Santos C., S. Cotab.
	5	THA		Nor. Thor. Phor., Sakon Nakhon
	10	THA		R. Thailand, Pathum Thani
	10	TWN	BED43	BCC (C), Taichung
	10	VTN		Can Tho (P)
846	10	AUS	2RN	ABC (RN), Canberra
	2.5	AUS	6CA	ABC (RR), Carnarvon
	5	AUS	4EL	Cairns
	100	BGD		Bangladesh Betar, Rajshahi (Bogra)
	10	CHN		BJ (CRI DS3)
	100	CHN		EN + 27 stns
	200	IND		AIR (A), Ahmedabad
	5	J		NHK (1), Koriyama
	1	J		NHK (1), Uwajima/Hitoyoshi
	10	KIR		R. Kiribati, Bariki
	5	KOR		KBS, Yanggu
	10	KOR	HLAU	MBC, Ulsan
	2	NZL		NewstalkZB, Masterton
	50	PHL	DZRV	R. Veritas, Quezon C., NCR
	10	THA		R. Thailand, Phetchabun
	10	TWN	BEH56	Fu Hsing BC (2), Kaohsiung
	10	TWN	BEC38	Han Sheng BC, Penghu
	250	TWN		VO Kuanghua, Kuanyin
	10	VTN		Thanh Hoa (P)
850	5	HWA	KHLO	Hilo, Hawaii

kHz	kW	Ctry	Call	Station, location
855	10	AUS	4QO	ABC (RR), Eidsvold
	10	AUS	4QB	ABC (RR), Pialba
	2	AUS	3CR	Community R. Fed. Ltd., Collingwood
	50	CHN		YN (CNR2); XJ
	20	CLN		SLBC, Irattaperiyakulam
	2/10	INS		RRI, Mataram
	50	INS		RRI, Medan (P. Cermin)
	10	KOR	HLCX	MBC, Jeonju
	500	KRE		Pyongyang BS, Sangwon
	2	NZL		NZ's Rhema, Hamilton
	5	PHL	DXZH	Cebu Bc. Co., Zamboanga C., Z. Sur
	10	PHL	DZGE	Filipinas Bc. Netw.., Naga C., Cam. Sur
	10	PHL	DXGO	Pacific Bc. System, Davao C., Dvo Sur
	5	THA		Phon. Mor. Song. 12, Prachin Buri
	10	TWN	BED27	BCC (N), Hualien
	1	TWN	BEV72	Cheng Sheng, Chia-i
	1	TWN	BEV24	Min Pen (2), Taipei
864	2	AUS	7RPH	Bc. Sce. for Handicapped Inc., Hobart
	2	AUS	6AM	Northam
	2	AUS	4GR	Toowoomba
	50	CHN		AH; ZJ (2 stn's); EN; EB
	10	HKG		Hong Kong Comm. Bc. Co., Peng Chau
	100	IND		AIR, Shillong
	2/10	INS		RRI, Cirebon
	1	J	JOXN	CRT, Nasu
	5	J	JOPR	FBC, Fukui
	3	J	JOHE	HBC, Asahikawa
	1	J		HBC, Enbetsu
	3	J	JOQF	HBC, Muroran
	10	J	JOXR	ROK, Naha, Okinawa
	1	J	JOSO	SBC, Matsumoto
	100	KOR	HLKR	KBS, Gangneung
	10	NZL		NewstalkZB, Invercargill
	5	PHL	DZWM	Alamino City Bc. Corp., Alaminos, Pang.
	10	PHL	DYHH	Philippine Amn Force, Bogo, Cebu
	5	PHL	DZSP	Swara Sug Media Corp., San Pablo C., Laguna
	5	PHL	DWSI	Swara Sug Media Corp., Santiago, Isa.
	10	PNG		R. Madang
	25	RUS		R. Shanson, Blagoveshchensk, FE
	25	RUS		Yakutsk (Reg.), Sib.
	10	THA		R. Thailand, Phattahalung
	10	THA		R. Thailand, Si Sa Ket
	10	THA		R. Thailand, Tak
	10	TWN	BED25	BCC (N), Kaohsiung
	10	TWN	BEV52	Chung Sheng, Taichung
873	2	AUS	6DB	ABC (RR), Derby
	2	AUS	4--	Innisfail (HPONS)
	5	AUS	2GB	Sydney
	100	BGD		Bangladesh Betar, Chittagong
	200	CHN		China Huayi BC, Fuzhou
	200	CHN		HL + 7 stns
	400	CLN		SLBC/Family R, Puttalam
	300	IND		AIR (B), Jalandhar
	500	J	JOGB	NHK (2), Kumamoto
	250	KRE		Korean Central BS (C/R), Sinuiju
	1	NZL		Newstalk ZB, Ashburton
	1	NZL		R. Trackside/LiveSPORT, Tauranga
	5	PHL	DZPA	Abra Comm. Bc. Corp., Bangued, Abra
	5	PHL	DZRC	Filipinas Bc. Netw.., Legaspi C., Albay
	10	PHL	DXJS	Pilippine Bc. Sce., Tandag, S. Sur
	5	PHL	DXRB	Swara Sug Media Corp., Butuan C., Ag. Nte
	5	PHL	DXRT	Swara Sug Media Corp., Jolo, Sulu
	5	PHL	DYUP	Univ. of the Philippines, Miagao, Iloilo
	5	THA		Wor. Kor. Thor. Mor., Bangkok
	500	VTN		Can Tho (3/4), Thoi Long
880	10	HWA	KHCM	Honolulu, Oahu
882	5	AUS	4BH	Brisbane
	10	AUS	6PR	Perth
	2	AUS	3YB	Warrnambool
	100	CHN		FJ + 10 stns
	400	CLN		TWR, Puttalam
	300	IND		AIR, Imphal
		INS		RRI, Kendari
	10	J	JOPK	NHK (1), Shizuoka
	1	J		STV, Esashi
	3	J	JOWS	STV, Kushiro

kHz	kW	Ctry	Call	Station, location
	20	KOR	HLKI	KBS, Daejon
	250	KRE		Korean Central BS (C/R), Wonsan
	75	MNG		MRT (1), Mörön
	10	NZL		Parl. Bc. /So. Star, Auckland
	50	PHL	DWIZ	Aliw Bc. Corp., Navotas, NCR
	10	PHL	DXMS	Notre Dame Bc. Corp., Cotabato C., Mag.
	10	PHL	DYOG	Pilippine Bc. Sce., Calbayog, W. Samar
	1	PHL	DXRG	Pilippine Bc. Sce., Gingoog, Mis. Or.
	10	TWN	BEG77	BCC (N), Hsinchu
	1	TWN		Feng Ming, Penghu
891	50	AUS	5AN	ABC (MS), Adelaide
	5	AUS	4TAB	Townsville (HPONS)
	200	CHN		NX; LN; NM; XJ
	20	IND		AIR, Rampur
	10	INS		RRI Ternate
	10	INS		RRI, Malang
	20	J	JOHK	NHK (1), Sendai
	250	KOR	HLKB	KBS, Busan
	5	NZL		The Breeze, Wellington
	5	PHL	DZGR	People's Bc. Sce., Tuguegarao, Cag.
	1	RUS		R. Mayak + Reg., Tyumen, Sib.
	1000	THA		R. Thailand, Saraburi
	10	TWN	BED24	BCC (L), Tainan
	10	VTN		Lam Dong (1/P), Da Lat
898	10	VTN		Ha Tinh (P) (v)
900	2	AUS	8HA	Alice Springs
	2	AUS	6BY	Bridgetown
	2	AUS	7AD	Devonport
	5	AUS	2LM	Lismore
	5	AUS	2LT	Lithgow
		CHN		BJ (CRI DS4)
	10	CHN		QH (CNR2)
	100	CHN		YN + 32 stns
	5	HWA	KNUI	Kahului, Maui
	100	IND		AIR, Kadapa
	5	J	JOHF	BSS, Yonago
	5	J	JOHO	HBC, Hakodate
	5	J	JOZR	RKC, Kochi
	50	KOR	HLKV	MBC, Seoul
	2.5	NZL		Coast, Whangarei
	10	NZL		Parl. Bc. /So. Star, Dunedin
	5	PHL	DYOW	Consolidated Bc. System, Inc., Roxas, Capiz
	5	PHL	DWNE	Nueva Ecija Prov. Gov., Cabanatuan C., Nva Viz.
	5	PHL	DXRZ	R. Mindanao Netw., Zamboanga C., Z . Sur
	5	PHL	DXIP	Southern Bc. Netw., Davao C.
	10	PNG		R. Eastern Highlands, Goroka
	10	PNG		R. West New Britain, Kimbe
	10	VTN		Kon Tum (1/P)
909	100	CHN		FJ (CNR6)
	50	CHN		TJ + 5 stns
	100	IND		AIR, Gorakhpur
	5/10	INS		RRI, Sorong
	10	J	JOCB	NHK (2), Nagoya
	5	J	JOVX	STV, Abashiri
	10	KOR		KBS, Gumi
	5	NZL		Parl. Bc. /So. Star, Napier-Hastings
	5	PHL	DZEA	Catholic Media Netw., Laoag C, I. Nte
	5	PHL	DYSP	Republic Bc. System, Pto. Princesa, Palawan
	5	PHL	DYLA	Visayas Mindanao C. of TU, Cebu C
	10	THA		R. Thailand, Loei
	25	THA		R. Thailand, Surin
	10	TWN	BED79	BCC (C), Kaohsiung
	10	TWN	BEH3	Fu Hsing BC (1), Taipei
	10	VTN		Ca Mau (P)
918	2/2.5	AUS	4VL	Charleville
	2	AUS	2XL	Cooma
	2	AUS	6NA	Narrogin
	200	CBG		Nat. Radio of Kampuchea, Steung Meanchey
	200	CHN		SD
	300	IND		AIR, Suratgarh
	1	J	JOPN	KRY, Iwakuni
	1	J	JOPM	KRY, Shimonoseki
	1	J		YBC, Shinjo
	1	J		YBC, Tsuruoka/Yonezawa
	5	J	JOEF	YBC, Yamagata
	50	KOR		KBS, Yeoncheon

kHz	kW	Ctry	Call	Station, location
	2.5	NZL		RNZ National, New Plymouth
	2.5	NZL		RNZ National, Timaru
	10	PHL	DZSR	Pilippine Bc. Sce., Quezon C., NCR
	5/7	RUS		R. Mayak, Sib., 3stns, (synch)
	100	THA		R. Thailand, Nakhon Pathom
	10	THA		Sor. Wor. Phor. 1, Chiang Mai
927	5	AUS	4CC	Gladstone
	5	AUS	3UZ	Melbourne
		CHN		FJ (CNR6)
	200	CHN		GZ + 24 stns
	2.5	FJI		Fiji Bc. Comm. Ltd. (1), Sigatoka
	100	IND		AIR, Visakhapatnam
	25	INS		RRI, Pekanbaru
	5	J	JOFG	NHK (1), Fukui
	5	J	JOKG	NHK (1), Kofu
	1	J		NHK (1), Wakkanai/Tsuyama
	10	KOR		KBS, Buyeo
	1	KOR		KBS, Hongcheon/Hadong
	50	KRE		Korean Central BS (C/R), Hwangju (Sariwon)
	2	NZL		NewstalkZB, Palmerston No.
	5	PHL	DXDA	Office of Governor, San Francisco, Ag. Sur
	5	PHL	DZLG	People's Bc. Sce., Legaspi, Albay
	5	PHL	DXMD	R. Mindanao Netw., Gen. Santos C., S. Cot.
	10	PHL	DWRS	Solid North Bc., Vigan, I. Sur
	5	PHL	DXMM	Sulu Tawi-Tawi Bc. Found., Jolo,Sulu
	20	THA		R. Thailand, Chanthaburi
	10	THA		R. Thailand, Nong Khai
930		CHN		ZJ
	5	VTN		Ben Tre (P) (v)
936	5	AUS	6FX	(PBS), Perth
	10	AUS	7ZR	ABC (MS), Hobart
	10	AUS	4PB	ABC (PNN), Brisbane
	200	CHN		AH; NM
	100	IND		AIR (A), Tiruchirapalli
	5	J	JOTR	ABS, Akita
	1	J		MRT, 4 stns
	5	J	JONF	MRT, Miyazaki
	10	KOR		KBS, Changwon
	1	NZL		New Supremo, Auckland
	100	PAK		Azad Kashmir Radio, Mirpur (r. 930)
	5	PHL	DWIM	Insular Bc. System, Calapan, Mind. Or.
	5	PHL	DXIM	Philippine Bc. Sce., Cagayan de Oro C., Mis. Or.
	1	PHL	DZXT	R. Corp. of the Philippines, Tarlac C.
	1	PHL	DYCC	R. Mindanao Netw., Calbayog C, W. Samar
	1	PHL	DYKW	R. Philippines Netw., Binalgaban, Neg. Occ.
	5	PHL	DXDN	Univ. of Mindanao, Tagum C., Dvo Nte
	50	THA		R. Thailand, N. Sawan
	10	THA		Thor. Phor. 4, Pattani
	5	TWN		Han Sheng BC, Taoyuan
940	10	HWA	KKNE	Honolulu, Oahu
945	2	AUS	3UZ	Bendigo (HPONS)
	1	AUS	4HI	Dysart (trtr)
	50	CHN		HL (2 stns); SD; HB
	400	CHN		JL (CNR 1)
	100	IND		AIR, Sambalpur
	1	J		NHK (1), Fukue
	1	J	JOQP	NHK (1), Hikone
	3	J	JOIQ	NHK (1), Muroran
	5	J	JOXK	NHK (1), Tokushima
	10	KOR		KBS, Boeun
	2	NZL		NewstalkZB, Gisborne
	10	PHL	DWFB	Pilippine Bc. Sce., Laoag C., I. Nte
	5	PHL	DXRO	Swara Sug Media Corp., Cotabato C., Mag.
	10	PHL	DXDV	Vismin R. and TV Bc. Net, Butuan C., Ag. Nte
	10	SLM		R. Hapi Lagun (SIBC), Gizo
	10	THA		Thor. Or. 1, Bangkok
	10	THA		Thor. Phor. 2, Kalasin
954	0.35	AUS	4EL	Gordonvale (trtr)
	5	AUS	2UE	Sydney
	50	CHN		NM + 7 stns
	200	IND		AIR, Najibabad
	10	INS		RRI, Kendari
	100	J	JOKR	TBS Radio, Tokyo
	1	NZL		Coast, Dunedin
	2	NZL		R. Trackside/LiveSPORT, Hamilton
	40	PHL	DZEM	Christian Era Bc. Sce., Quezon City, NCR

kHz	kW	Ctry	Call	Station, location
	1	PHL	DXJT	Pilippine Bc. Sce., Tangub C, Mis Octal
	5	PHL	DZAL	Rinconada Bc. Corp., Iriga C., Cam Sur
	10	THA		Thor. Or. 10, Phitsanulok
	10	THA		Thor. Or. 16, Chantaburi
	20	TWN	BED55	BCC (C), Taipei
	10	TWN	BEV85	Chien Kuo, Hsinying
	10	VTN		Vinh Long (P)
963	2	AUS	6TZ	Bunbury
	5	AUS	2RG	Griffith
	5	AUS	5SE	Mt. Gambier
	5	AUS	4WK	Warwick
	20	BGD		Bangladesh Betar, Sylhet
	600	CHN		JL (CRI)
	50	CHN		LN; ZJ; HB; EB; XJ
	20	IND		AIR, Jalgaon
	2/10	INS		RRI, Jember
	5	J	JOTG	NHK (1), Aomori
	5	J	JOZK	NHK (1), Matsuyama
	1	J	JOSP	NHK (1), Saga
	1	J		NHK (1), Yonago/Hagi
	10	KOR	HLCR	KBS, Andong
	10	KOR	HLKS	KBS, Jeju
	10	NZL		Parl. Bc. /So. Star, Christchurch
	5	PHL	DZNS	Archdiocese of Nueva Segovia, Vigan, I. Sur
	10	PHL	DYMF	People's Bc. Sce., Cebu C.
	5	PHL	DXYZ	Swara Sug Media Corp., Zamboanga C., Z. Sur
	1	RUS		R. Rossii, Guzino-Ozersk, Sib.
	20	RUS		R. Rossii, Zakamensk, Sib.
	10	THA		Phon Mor. 2, Bangkok
	25	THA		R. Thailand, Krabi
	10	TWN	BEV84	Taiwan BC, Chunghsing
972	2	AUS	5PB	ABC (PNN), Adelaide
	0.3	AUS	2DU	Cobar (trtr)
	5	AUS	2MW	Murwillumbah
	150	CHN		EN; HL; XJ
	300	IND		AIR (A), Cuttack
	50	INS		RRI, Surakarta
	1500	KOR	HLCA	KBS, Dangjin
	5	NZL		NZ's Rhema, Wellington
	5	PHL	DXKH	Cebu Bc. Co., Cagayan de Oro C., Mis. Or.
	1	PHL	DYSM	Cebu Bc. Co., Catarman, N. Samar
	5	PHL	DWTI	Katigbak Enterprises, Lucena C, Que.
	5	PHL	DWFR	Pilippine Bc. Sce., Bontoc, Mt. Prov.
	10	THA		Nor. Thor. Phor., Phetchabun
	500/800	TJK		Variousrelays, Orzu
	10	VTN		Quang Ngai (1/P)
981	2	AUS	3HA	Hamilton
	2	AUS	6KG	Kalgoorlie
	5	AUS	2NM	Muswellbrook
	200	CHN		JL; JX (CNR1)
	100	IND		AIR, Raipur
	1	J		NHK (1), Kisofukushima/Sasebo
	2	NZL		RNZ National, Kaikohe
	2.5	NZL		Southern Star, Timaru
	10	PHL	DXBR	Consolidated Bc. System, Inc., Butuan C., Ag. Nte
	10	PHL	DYBQ	Intercontinental Bc. Corp., Iloilo C.
	5	PHL	DWMT	Philippine Bc. Corp., Naga C., Cam. Sur
	5	PHL	DXDR	R. Mindanao Netw., Dipolog, Z. Nte
	10	PHL	DXOW	Radyo Pilipino Corp., Davao C., Dvo Sur
	5	PHL	DZRD	Swara Sug Media Corp., Dagupan C., Pang.
	10	THA		Mahaawittayalai Thammasat, Pathum Thani
	25	THA		R. Thailand, Mae Hong Son
	20	THA		R. Thailand, Nakhon Phanom
	25	THA		R. Thailand, Yala
	1	TWN	BEV68	Feng Ming (2), Kaohsiung
	250	TWN		VO Kuanghua, Hsinfeng
	10	VTN		Gia Lai (1/P), Pleiku
	10	VTN		Son La (1/4)
990	0.5	AUS	3RN	ABC (RN), Albury-Wodonga
	0.5	AUS	8GO	ABC (RR), Gove
	5	AUS	6RPH	Perth
	5	AUS	4RO	Rockhampton
	100	CHN		SH; YN; EB
	5	HWA	KHBZ	Honolulu, Oahu
	300	IND		AIR (A), Jammu
	10	J	JORK	NHK (1), Kochi

kHz	kW	Ctry	Call	Station, location
	10	KOR	HLAP	MBC, Changwon
	500	MNG	MRT (FS)	Ulaanbaatar
	1	NZL		Apna 990, Auckland
	1	NZL		R. Trackside/LiveSPORT, Nelson
	5	PHL	DZMT	Pacific Bc. System, Laoag, I. Nte
	5	PHL	DYTH	Pacific Bc. System, Tacloban C., Leyte
	5	PHL	DXBM	Republic Bc. System, Cotabato C., Mag.
	10/5	PHL	DZIQ	Trans-Radio Bc. Corp., Makati C, NCR
	10	THA		Sor. Wor. Phor. 2, N. Ratchasima
	10	TWN	BEV58	Cheng Sheng (1), Taichung
	10	TWN	BEP34	Ching Cha, Hualien
	1	TWN	BEP38	Ching Cha, Ilan
999	2	AUS	2NB	ABC (RR), Broken Hill
	5	AUS	2ST	Nowra
	10	BGD		Bangladesh Betar, Thakurgaon
	200	CHN		LN + 7 stns
	10	FSM	V6AF	Independent Baptist Church, Pohnpei
	1	IND		AIR, Almora
	20	IND		AIR, Coimbatore
	1/150	INS		RRI, Jakarta
	1	J		NHK (1), Fukuyama/Hachinoe
	1	J		NHK (1), Nakamura
	10	KOR	HLCL	CBS, Gwangju
	250	KRE		Korean Central BS (C/R), Hamhung
	1.5	NZL		Manawatu Access R., Palmerston No
	5	PHL	DWMI	Katigbak Enterprises, Calapan, Mind. Or.
	5	PHL	DZEQ	Pilippine Bc. Sce., Baguio C., Benguet
	1	PHL	DXPT	Pilippine Bc. Sce., Bongao, Tawi-Tawi
	1	PHL	DXHP	R. Mindanao Netw., Bislig, S. Sur
	5	PHL	DYSS	Republic Bc. System, Cebu C.
	10	THA		Phon. Neung Ror. Or., Bangkok
	10	THA		Thor. Phor. 3, Chiang Rai
	1	TWN	BEV92	Tien Nan BS, Taipei
	10	VTN		Lang Son (P)
1008	10	AUS	4TAB	Brisbane (Ipswich)
	0.3	AUS	2TAB	Canberra (HPONS)
	5	AUS	7TAB	Launceston (HPONS)
		CHN		BJ (CNR DS3)
	50	CHN		TJ + 19 stns
	200	CHN		YN (CNR1)
	100	IND		AIR (B), Kolkata
		INS		RRI, Gorontalo
	10	INS		RRI, Madiun
	50	J	JONR	ABC, Osaka
	50	KOR		KBS, Gangneung
	10	NZL		NewstalkZB, Tauranga
	120	PAK		PBC (1), Hyderabad
	5	PHL	DWBS	Catholic Media Netw., Sto. Domingo, Albay
	10	PHL	DXXX	R. Philippines Netw., Zamboanga C., Z. Sur
	5	PHL	DWGO	Subic Bc. Corp., Olongapo C., Zamb.
	10	THA		Wor. Por. Tho 3, N. Ratchasima
	10	TWN	BED88	BCC (C), Taitung
	300	TWN		CBS (RTI), Lukang
	5	TWN	BEV60	Cheng Sheng, Kaohsiung
1017	0.5	AUS	6WH	ABC (RR), Wyndham
	1	AUS	6TAB	Bunbury (HPONS)
	5	AUS	2KY	Sydney
	10	CHN		EB; GD
	100	CHN		JL (CNR8/FS)
		CHN		ZJ (CNR1)
	20	IND		AIR (B), Chennai
	10	IND		AIR, New Delhi
	50	J	JOLB	NHK (2), Fukuoka
	10	KOR	HLAW	MBC, Andong
	2.5	NZL		R. Hauraki, Christchurch
	10	PHL	DWDW	Intercontinental Bc. Corp., Dagupan C., Pang.
	10	PHL	DXAM	Kalayaan Bc. System, Davao C., Dvo Sur
	10	PHL	DWLC	Pilippine Bc. Sce., Lucena C., Que.
	5	PHL	DXSN	Silangan Bc. Corp., Surigao C., S. Nte
	10	THA		Thor. Or. 5, Prachuap KK
	10	TON	A3Z	Tonga Bc. Comm., Nuku'alofa
	10	TWN	BED53	BCC (C), Hsinchu
1026	10	AUS	3PB	ABC (PNN), Melbourne
	5	AUS	4AA	Mackay
	2	AUS	6NW	Port Hedland
		CHN		FJ (CNR6)
	200	CHN		GZ; BJ + 4 stns
	20	IND		AIR (A), Allahabad
	5	INS		RRI, Serui
	1	KOR		KBS, Geochang/Hwacheon
	2	NZL		Newstalk ZB, Kaitaia/Whangarei
	2.5	NZL		The Word Bible Radio, Invercargill
	5	PHL	DXMC	People's Bc. Sce., Koronadal, S. Cot.
	10	PHL	DZAR	Swara Sug Media Corp., Quezon C, NCR
	250	RUS		VoR, Oyash, Sib.
	50	THA		R. Thailand, Phitsanulok
	10	THA		R. Thailand, Yala
	1	TWN	BEV51	Chung Hua (2), Sanchung
	1	TWN		Tien Sheng, Yuanli
1035	2	AUS	2EA	Wollongong (SBS)
	50	CHN		CNR1
	10	IND		AIR (B), Guwahati
	1/5	INS		RRI, Bandar Lampung
		INS		RRI, Palu
	1	J	JOHD	NHK (2), Takamatsu
	1	J	JOIC	NHK (2), Toyama
	1	J		NHK (2), Tsuruoka
	1	KOR	HLCP	KBS, Pohang
	20	NZL		Newstalk ZB, Wellington
	120	PAK		PBC, Multan
	5	PHL	DZWX	Consolidated Bc. System, Inc., Baguio C, Benguet
	10	PHL	DYRL	Radyo Pilipino Corp., Bacolod C., Neg. Occ.
	10	SLM		R. Happy Isles (SIBC), Honiara
	10	THA		Phaak Phiset, Bangkok
	10	TWN	BED26	BCC (C), Chia-i
	10	VTN		Hoa Binh (P)
1040	10	HWA	KLHT	Honolulu, Oahu
1044	1	AUS	6BR	ABC (RR), Bridgetown
	2	AUS	2UH	ABC (RR), Muswellbrook
	0.5	AUS	4WP	ABC (RR), Weipa
	2	AUS	5AU	Port Pirie
	600	CHN		JS (FS)
	10	CHN		XJ (2 stns); YN
	10	HKG		Metro Bc. Corp., Peng Chau
	100	IND		AIR (A), Mumbai
	10	INS		RRI Tahuna
	10	INS		RRI, Biak
	10	INS		RRI, Sibolga
	1	KOR		AFNK, Chuncheon
	10	KOR		KBS, Samcheok/Jecheon
	10	NZL		NewstalkZB, Dunedin
	10	PHL	DYMS	Cebu Bc. Co., Catbalogan, W. Samar
	10	PHL	DZNG	Newsounds Bc. Netw., Naga C., Cam. Sur
	5	PHL	DXLL	R. T. Bc. Specialists Phil., Zamboanga C., Z. Sur
	5	PHL	DXCO	Radyo Pilipino Corp., Cayagan de Oro, Mis. Or.
	1	PHL	DXML	Rural Electrification Corp., Digos, Dvo. Sur
	10	THA		Kor. Wor. Sor. 5, Khon Kaen
	10	THA		Thor. Phor. 4, N. Si Thammarat
	10	TWN	BEV98	Cheng Kung, Kaohsiung
	5	TWN	BEV64	Yen Sheng (1), Hualien
1050		CHN		ZJ
		VTN		Tra Vinh (P) (v)
1053	0.5	AUS	4--	Brisbane (HPONS)
	5	AUS	2CA	Canberra, ACT
	20	BGD		Bangladesh Betar, Rangpur
		CHN		BJ (CNR 10)
	50	CHN		LN + 13 stns
	200	IND		AIR (FS), Tuticorin
	20	IND		AIR, Leh
	10	INS		RRI, Jayapura
	50	J	JOAR	CBC, Nagoya
	500	KRE		Pyongyang BS, Haeju
	2	NZL		NewstalkZB, New Plymouth
	10	PHL	DXKD	R. Philippines Netw., Dipolog, Z. Nte
	5	PHL	DYSA	Univ. of San Agustin, Iloilo C.
	50	RUS		Reg., Orenburg
	10	THA		Mor. Thor. Bor. 11, Bangkok
	10	THA		Mor. Thor. Bor. 32, Lampang
1060	5	HWA	KIPA	Hilo, Hawaii
1062	2	AUS	5MV	ABC (RR), Renmark/Loxton
	2	AUS	4TI	ABC (RR), Thursday Isl.
	150	CHN		Zhujiang EBS, GD; HL

kHz	kW	Ctry	Call	Station, location
	10	IND		AIR, Pasighat
	50	KOR	HLKQ	KBS, Cheongju
	1	NZL		R. Sport, Wanganui
	40	PHL	DZEC	Eagle Bc. Corp., Quezon C., NRC
	5	PHL	DXKI	FEBC, Koronadal C., S. Cot.
	1	RUS		R. Mayak, Ust-Kamchatsk, FE
	10	THA		R. Thailand, Phuket
	10	THA		Thor. Or. 9, Udon Thani
	10	TWN	BED23	BCC (L), Taichung
	5	TWN	BEV82	Cheng Sheng, Ilan
	5	TWN	BEV74	Min Li, Pingtung
1071	2	AUS	6WB	Katanning
	2	AUS	4SB	Kingaroy
	5	AUS	3EL	Maryborough
	50	CHN		TJ + 8 stns
	1000	IND		AIR (FS), Rajkot
	20	J	JOFK	NHK (1), Hiroshima
	5	J	JOWM	STV, Obihiro
	100	KAZ		Shygyz Qazaqstan OTRK, Ösqemen
	1	NZL		R. Trackside/LiveSPORT, Ashburton
	2.5	NZL		RNZ National, Masterton
	5	PHL	DXKT	R. Philippines Netw., Davao C., Dvo Sur
	1	PHL	DZSL	S. O. L. Telebc. Station, Talisay, Cam Nte
	1	PHL	DYXT	Universal Bc. System, Tagbilaran C., Bohol
	7	RUS		R. Mayak, Zeya, FE
	10	THA		Neung. Por. Nor., Tak
	10	THA		SW Rattasapha, Bangkok
	1	TWN	BEV96	Tien Sheng, Tainan
1080	2	AUS	2MO	Gunnedah
	5	AUS	7TAB	Hobart (HPONS)
	2	AUS	6IX	Perth
	10	BGD		Bangladesh Betar, Rajshahi
	10	CHN		JS + 5 stns
	300	CHN		YN(+FS)
	5	HWA	KWAI	Honolulu, Oahu
	2/10	INS		RRI, Singaraja
	5	KOR		AFNK, Daegu
	10	KOR	HLAT	MBC, Yeosu
	1500	KRE		Korean Central BS (C/R), Haeju
	5	MRA	KCNM	Choice Bc. Comp., Chalan Kiya, Saipan
	10	NZL		NewstalkZB, Auckland
	50	PAK		PBC (3), Lahore
	5	PHL	DWRL	DWRL Radio, Inc., Legaspi C., Albay
	5	PHL	DWIN	Eagle Bc. Corp., Dagupan C., Pang.
	5	PHL	DYBH	Pacific Bc. System, Bacolod C., Neg Occ
	1	PHL	DXKS	R. Philippines Netw., Surigao, S. Nte
	5	PHL	DXRH	First United Bc. Corp., Zamboanga C., Z. Sur
	1000	RUS		VoR, Angarsk, Sib.
	10	THA		Wor. Por. Tho. 10, Chiang Rai
	10	THA		Wor. Por. Tho. 16, Yala
	10	THA		Wor. Por. Tho. 9, N. Sawan
	5	TKM		Turkmen Radio (1), Serhetabat
1089	5	AUS	3WM	Horsham
	5	AUS	2EL	Orange
	600	CHN		FJ (CNR6)
	200	CHN		LN; HN
	20	IND		AIR, Naushera
	20	IND		AIR, Udipi
	10	J	JOHB	NHK (2), Sendai
	10	KOR	HLCH	KBS, Chungju
	2.5	NZL		R. Sport, Palmerston No.
	10	PHL	DXCM	Univ. of Mindanao, Cotabato C., Mag.
	5	RUS		R. Rossii + Reg., Tilichiki, FE
	10	THA		Neung. Por. Nor, Udon Thani
	10	THA		Neung. Por. Nor., Bangkok
	10	TWN	BEH34	Fu Hsing BC (2), Taichung
	5	TWN	BEH5	Fu Hsing BC (2), Taipei
	10	TWN	BEG28	Kaohsiung, Kaohsiung
	10	TWN	BEC31	Han Sheng BC, Yunlin
	10	VTN		Cao Bang (P)
	10	VTN		Kien Giang (P), Rach Gia
1098	0.2	AUS	2RN	ABC (RN), Goulburn
	5	AUS	7LA	Launceston
	2	AUS	4LG	Longreach
	2	AUS	6MD	Merredin
	10	CHN		ZJ + 22 stns
	2/10	INS		RRI, Jambi
	2/10	INS		RRI, Sumenep
	1	J	JOMF	NBC, Sasebo
	5	J	JOGF	OBS, Oita
	5	J	JOWO	RFC, Koriyama
	1	J	JOSW	SBC, Iida
	5	J	JOSR	SBC, Nagano
		KAZ		BBC relay, Almati
	20	KOR	HLCJ	KBS, Jinju
	25	MHL	V7AB	R. Marshalls, Majuro
	5	NZL		NewstalkZB, Christchurch
	50	PAK		PBC (2), Hyderabad
	10	PHL	DWAD	Crusaders Bc. System, Mandaluyong C., NCR
	5	PHL	DXCL	Swara Sug Media Corp., Cagayan de Oro C., Mis. Or.
	10	THA		R. Thailand, Tak
	10	THA		Sor. Wor. Phor. 3, Songkhla
	300	TWN		CBS (RTI), Kouhu
	10	VTN		Binh Thuan (1/P), Phan Thiet
	10	VTN		Lao Cai (P)
	10	VTN		Thua Tien Hue (P), Hue
1107	5	AUS	2EA	Sydney (SBS)
	120	CHN		XJ + 7 stns
	20	IND		AIR, Gulbarga
	1/5	INS		RRI, Kupang
	1/10	INS		RRI, Yogyakarta
	1	J		MBC, Akune/Oguchi/Sendai
	20	J	JOCF	MBC, Kagoshima
	5	J	JOMR	MRO, Kanazawa
	1	J		MRO, Nanao
	10	KOR	HLAV	MBC, Pohang
	1	NZL		R. Live, Tauranga/Rotorua
	1	PHL	DZOM	Ben Viduya, Calapan C., Mind. Or.
	5	PHL	DYIN	Inter-Island Broadc. Corp., Kalibo, Aklan
	10	PHL	DWDY	Northeastern Bc. Sce., Cauayan, Isa.
	5	PHL	DXBB	Republic Bc. System, Gen. Santos C., S. Cot.
	10	PNG		R. Milne Bay, Alotau
	10	THA		Mor. Kor., Samut Sakhon
	10	THA		Thor. Phor. 2, Khon Kaen
1110	5	HWA	KAOI	Kihei, Maui
1116	6.3/17	AUS	4BC	Brisbane
	2	AUS	6MM	Mandurah
	5	AUS	3AK	Melbourne
	600	CHN		FJ (CNR5)
	120	CHN		HL (CNR2)
	200	CHN		SC; AH; HAN; SD
	300	IND		AIR (A), Srinagar
	5	J	JODR	BSN, Niigata
	5	J	JOAF	RNB, Matsuyama
	1	J	JOAL	RNB, Niihama
	1	J	JOAM	RNB, Uwajima
	2.5	NZL		RNZ National, Nelson
	5	PHL	DXAS	FEBC, Zamboanga C., Z. Sur
	10	PHL	DYTR	Tagbilaran Bc. Corp., Tagbilaran C, Bohol
	5	PHL	DZLB	Univ. of the Philippines, Los Banos, Laguna
	10	THA		R. Thailand, Phangnga
	10	THA		Thor. Phor. 3, Phitsanulok
	3.5	TWN	BED72	BCC (N), Yuli
	5	TWN	BEP26	Ching Cha, Hsinchu
	5	TWN	BEP25	Ching Cha, Kaohsiung
	10	TWN	BEC30	Han Sheng BC, Ilan
	10	TWN	BEC22	Han Sheng BC, Taipei
	1	VTN		Bac Lieu (P)
1125	5	AUS	5MU	Murray Bridge
	2	AUS	1RPH	Print-Hand. Radio of ACT Inc., Canberra
	50	CHN		HB(v); EB
	20	IND		AIR, Tezpur
	1	J		NHK (2), Hagi/Nayoro
	1	J	JOIZ	NHK (2), Muroran
	10	J	JOAD	NHK (2), Naha, Okinawa
	1	J	JOOC	NHK (2), Obihiro
	1	J		NHK (2), Takayama
	1	J	JOLC	NHK (2), Tottori
	0.5	NZL		R. Hauraki, Dunedin
	1	NZL		R. Sport, Napier-Hastings
	10	PHL	DZWN	Consolidated Bc. System, Inc., Dagupan C., Pang.

kHz	kW	Ctry	Call	Station, location
	5	PHL	DWAS	FEBC, Legaspi C., Albay
	10	PHL	DXGL	PEC Bc. Corp., Butuan C., Ag. Nte
	5	PHL	DXGM	Republic Bc. System, Davao C., DvoSur
	25	THA		R. Thailand, Chanthaburi
	20	TKM		Turkmen Radio (1), Asgabat
	5	TWN	BEV36	Cheng Sheng, Yunlin
	1	TWN	BEP40	Ching Cha, Taitung
	10	VTN		Tay Ninh (P)
	10	VUT		VBTC, Emten Lagoon
1130	1	HWA	KPHI	Honolulu, Oahu
1134	2	AUS	2AD	Armidale
	5	AUS	3CS	Colac
	2	AUS	6TZ	Collie(trtr)
	10	CHN		GD + 4 stns
	1200	CHN		QH (CNR1)
	1000	IND		AIR (N/FS), Chinsurah (Mogra)
	1/25	INS		RRI, Banjarmasin
	100	J	JOQR	NCB, Tokyo
	500	KOR	HLKC	KBS, Hwaseong
	2	NZL		RNZ National, Queenstown
	100	PAK		PBC, Quetta
	10	PHL	DWDD	Dept. of Nat. Defence, Quezon C., NCR
	1	PHL	DYRM	Philippine R. Corp., Dumaguete, Neg. Or. (v)
	1	PHL	DWBT	Philippine Bc. Sce., Basco, Batanes
	5	PHL	DWJS	Rolin Bc. Enterprises, Roxas, Palawan
	5	PHL	DXMV	Univ. of Mindanao, Valencia, Buk.
	10	THA		R. Thailand, Lampang
	10	THA		Thor. Phor. 2, N. Ratchasima
	10	TWN	BEG26	Taipei BS, Taipei
1143	5	AUS	4HI	Emerald
	2	AUS	2HD	Newcastle
	10	CHN		BJ (CNR8)
	100	CHN		EN + 31 stns
	20	IND		AIR, Ratnagiri
	20	IND		AIR, Rohtak
	20	J	JOBR	KBS, Kyoto
	10	NPL		R. Nepal, Bardibas
	2.5	NZL		RNZ National, Hamilton
	10	PHL	DYAF	Diocese of Bacolod, Bacolod, Neg. Occ.
	10	PHL	DZRM	FEBC, Santiago C, Isa
	7	RUS		R. Mayak, Tayshet, Sib.
	10	THA		Or. Sor. Mor. Thor., Bangkok
	150	TJK		TR (2)/FS/Various relays, Yangiyul
	100	TWN	BEL3	Taiwan Yuyeh, Penghu
1147		CHN		Jiaozuo EBS, HEN
1152	10	AUS	6PNN	ABC (PNN), Busselton
	10	AUS	6RN	ABC (RN), Manjimup
	2	AUS	2WG	Wagga Wagga
	150	CHN		HN; NM (2 stns); LN
	2.5	FJI		Fiji Bc. Comm. Ltd. (1), Rakiraki
		INS		RRI, Lhokseumawe
	10	J	JORB	NHK (2), Kochi
	10	J	JOPC	NHK (2), Kushiro
	10	KOR	HLCW	KBS, Wonju
	2	NZL		Newstalk ZB, Timaru
	100	PAK		PBC, Rawalpindi
	5	PHL	DYCM	Masbate Comm. Bc. Co., Bogo, Cebu
	50	RUS		R. Rossii + Reg., Komsomolsk, FE
	10	THA		Ror. Dor., Chiang Mai
	10	THA		Ror. Dor., Khon Kaen
	1	TWN	BED68	BCC (C), Puli
	5	TWN	BEV70	Hua Sheng (1), Taipei
1161	1	AUS	7FG	ABC (RR), Fingal
	10	AUS	5PA	ABC (RR), Naracoorte
	2	AUS	4FC	Maryborough
	10	BGD		Bangladesh Betar, Rangamati
		CHN		CNR1
	10	CHN		SD + 4 stns
	20	IND		AIR, Trivandrum
	0.25	KOR		AFNK, Uijeongbu
	20	KOR	HLKU	MBC, Busan
	5	NZL		Te Upoko o Te Ika, Wellington
	5	PHL	DYRD	Bohol Chronicle R. Corp., Tagbilaran C, Bohol
	5	PHL	DZMD	People's Bc. Netw., Daet, Cam. Nte
	10	PHL	DWCM	Philippine Bc. Corp., Dagupan C., Pang.
	5	PHL	DYKR	R. Mindanao Netw., Kalibo, Aklan
	1	PHL	DXDS	Univ. of Mindanao, Davao C., Dvo. Sur
	10	THA		Sor. Thor. Ror. 9, U. Ratchathani
	20	THA		Wor. Sor. Sor., Bangkok
	40	TJK		TR (2), Orzu
	10	TWN	BED86	BCC (C), Ilan
	10	TWN	BED89	BCC (C), Miaoli
	1	TWN	BEV67	Feng Ming (1), Kaohsiung
1170	5	AUS	2CH	Sydney
	10	BGD		Bangladesh Betar (C), Dhaka
	10	CHN		SD (3 stns) + 5 stns
	0.25	GUM		Managament Adv. Sce., Agana
	1	IND		AIR, Hyderabad (stand-by)
		INS		RRI, Semarang
	500	KOR	HLSR	KBS, Gimje
	100	PAK		PBS, Peshawar
	10	PHL	DYSL	FEBC, Sogod, S Leyte
	1000	PHL		IBB Relay Stn., Poro Pt, Luzon
	10	PHL	DXMR	Philippine Bc. Sce., Zamboanga C., Z. Sur
	10	THA		Sor. Thor. Ror. 4, Chantaburi
	10	THA		Sor. Thor. Ror. 8, Phitsanulok
	1	TWN		Taiwan BC, Kuanhsi
	10	VTN		An Giang (P), Long Xuyen
1179	5	AUS	3RPH	Ass. for the Blind, Melbourne
	100	CHN		HB + 3 stns
	20	IND		AIR, Rewa
	2/10	INS		RRI, Padang
	50	J	JOOR	MBS, Osaka
	5	NZL		R. Ake, Auckland
	5	PHL	DYCX	Newsounds Bc. Netw., S. J. de Buenavista, Antique
	1	PHL	DZRS	R. Sorsogon Netw., Inc., Sorsogon C.
	5	PHL	DXYK	Republic Bc. System, Butuan C., Ag. Nte
	5	PHL	DYSB	GMA Netw., Inc., Bacolod C., Neg. Occ
	10	THA		Nor. Thor. Phor, Chiang Rai
	10	THA		Sor. Sor. Sor., Bangkok
	2.5	TWN		Kuo Sheng BC, Erhlin
	10	VUT		VBTC, Santo
1180	1	HWA	KORL	Honolulu, Oahu
1188	2	AUS	6XM	ABC (RR), Exmouth
	2	AUS	2NZ	Inverell
	10	CHN		EB(2 stns); JL
	300	CHN		YN (FS)
	50	IND		AIR (C), Mumbai
	1	INS		RRI, Manado
	10	J	JOKP	NHK (1), Kitami
		KAZ		BBC relay, Sariagas
	100	KOR	HLKX	FEBC, Seoul
	0.4	NZL		RNZ National, Rotorua
	5	PHL	DXLX	Consolidated Bc. System, Inc., Iligan, Lanao Nte
	10	PHL	DXIF	Newsounds Bc. Netw., Cagayan de Oro C., Mis. Or.
	5	PHL	DZLT	R. Corp. of the Philippines, Lucena C, Que.
	5	PHL	DZXO	Vanguard R. Netw., Cabanatuan, Nva. Ecija
	5	RUS		R. Teos, Khabarovsk, FE
	10	THA		Kor. Wor. Sor. 3, Sakon Nakhon
	10	THA		Thor. Phor. 3, Phitsanulok
	10	THA		Thor. Phor. Neung, Sa Kaeo
	10	TWN	BED32	BCC (C), Hualien
	1	TWN	BEV57	Sheng Li (2), Tainan
	10	TWN	BEV46	Taiwan BC (2), Taipei
1197	2	AUS	5RPH	R. 5RPH, Adelaide
	0.5/1	AUS	4BI	Switch FM, Brisbane
	10	CHN		HL; SH; FJ(v); SD
	1	IND		AIR, Shillong (stand-by)
	20	IND		AIR, Tirunelveli
	10	INS		RRI, Palangkaraya
	5	J	JOYF	IBS, Mito
	1	J	JOFO	RKB, Kitakyushu
	1	J		RKC, Nakamura
	1	J		RKK, 3 stns
	10	J	JOBF	RKK, Kumamoto
	1	J		STV, 3 stns
	3	J	JOWL	STV, Asahikawa
	10	KAZ		BBC relay, Astana
	1	KOR		AFNK, Dongducheon
	10	MLA		RTM Labuan, Kudat (v)

kHz	kW	Ctry	Call	Station, location
	2	NZL		NewstalkZB, Wanganui
	5	PHL	DYRH	Allied Bc. Center, Inc., Bacolod, Neg. Occ.
	5	PHL	DXFE	FEBC, Davao C., Dvo Sur
	5	PHL	DWBA	Satellite Bc. Corp., Bangued, Abra
	0.2	RUS		R. Rossii + Reg., Sib., 4stns
	10	THA		Jor. Tor. Lor, Lop Buri
1206	5	AUS	2CC	Canberra, ACT
	5	AUS	2GF	Grafton
	2	AUS	6TAB	Perth (HPONS)
	150	CHN		JL + 9 stns
	200	IND		AIR, Bhawanipatna
	1	KOR		KBS, Jeongseon/Cheongsong
	0.5	NZL		Community Access R, Hamilton
	2	NZL		Trackside/LiveSPORT, Dunedin
	10	PHL	DWAN	MMDA Traffic Radio, Quezon C, NCR
	5	PHL	DXRS	R. Mindanao Netw., Surigao C, S. Nte
	10	THA		R. Thailand, Satun
	10	THA		Thor. Phor. Neung, Prachuap KK
	100	TWN		CBS (RTI), Minhsiung
	10	TWN	BEV62	Taiwan BC, Hsinchu
1210	1	HWA	KZOO	Honolulu, Oahu
1215	0.5	AUS	6NM	ABC (RR), Northam
	0.35	AUS	2TAB	Bowral (HPONS)
	0.25	AUS	4HI	Moranbah (trtr)
	50	CHN		GD (CNR7)
	20	CHN		LN (CNR2)
	10	CHN		XJ; HB; HL
	20	IND		AIR (N), New Delhi (Kingsway)
	20	IND		AIR, Pudducherri
	0.5/10	INS		RRI, Samarinda
	1	J	JOBW	KBS, Hikone
	2	J	JOBO	KBS, Maizuru
	10	KOR	HLAK	MBC, Jinju
	2	NZL		NewstalkZB, Kaikohe
	10	PHL	DYRF	Word Bc. Corp., Cebu C.
	10	THA		Kor. Wor. Sor. 2, Phrae
	50	THA		R. Thailand, Surat Thani
	10	THA		Thor. Phor. 2, U. Ratchathani
	1	TWN		Tien Sheng, Pengshan
1224	5	AUS	6RN	ABC (RN), Busselton
	5	AUS	3EA	Melbourne (SBS)
	5	AUS	2RPH	R. for the Print-Handicapped, Sydney
		CHN		FJ (CNR6)
	100	CHN		GX; JS
	20	IND		Srinagar (C)
	10	J	JOJK	NHK (1), Kanazawa
	20	KOR		KBS, Gwangju
	1	MHL		AFN, Kwajalein
	2	NZL		R. Trackside/LiveSPORT, Invercargill
	10	PHL	DXED	Eagle Bc. Corp., Davao C., Dvo Sur
	5	PHL	DWSR	Manila Bc. Co., Lucena, Que.
	10	PHL	DZAG	Pilipine Bc. Sce., Agoo, LU
	10	THA		Thor. Or. 15, Chiang Rai
	10	THA		Thor. Or. 4, N. Sawan
	10	TWN	BED52	BCC (C), Kaohsiung
	1	TWN	BEV71	Hua Sheng (2), Taipei
1233	10	AUS	2NC	ABC (MS), Newcastle
	120	CHN		XJ (2stns); HN; JS
	20	IND		AIR, Tura
	0.2/1/5	INS		RRI,Pontianak
	5	J	JOUR	NBC, Nagasaki
	5	J	JOGR	RAB, Aomori
	1	KOR		KBS, Pyeongchang
	2	NZL		R. Live, Wellington
	5	PHL	DYVS	FEBC, Bacolod, Neg. Occ.
	5	PHL	DWRV	R. Veritas, Bayombong, Nva Viz.
	20	RUS		Yumor FM, Yelizovo, FE
	10	THA		Thor. Or. 1, Bangkok
	10	THA		Wor. Por. Tho. 7, Udon Thani
	40	TKM		Turkmen Radio (1), Syrtagta
	1	TWN		Chung Hua, Juifang
1242	2	AUS	8TAB	Darwin (HPONS)
	2	AUS	5AU	Port Augusta
	5	AUS	3GV	Sale
	2	AUS	4AK	Toowoomba
	1	CHN		LN + 5 stns
	100	IND		AIR, Varanasi

kHz	kW	Ctry	Call	Station, location
	10	INS		RRI, Bogor
	100	J	JOLF	NBS, Tokyo
	10	KOR	HLSB	MBC, Wonju
	0.1	NZL		1XX – One Double X, Murupara
	2	NZL		1XX – One Double X, Whakatane
	1	NZL		R. Trackside/LiveSPORT, Timaru
	5	PHL	DXZB	DXZB/TV13 Coop., Inc., Zamboanga C., Z. Sur
	20	PHL	DWBL	FBS R. Netw., Pasig C., NCR
	5	PHL	DXSY	Times Bc. Corp., Ozamis C, Mis. Occ.
	2	RUS		Novyy den - Angarsk, Angarsk, Sib.
	50	THA		R. Thailand, Surat Thani
	10	THA		Thor. Phor. 3, Phetchabun
	1	TWN		Yen Sheng (2), Hualien
	2000	VTN		Can Tho (E), Thoi Long
1250	1	CHN		ZJ
1251	2	AUS	2DU	Dubbo
		CHN		BJ(CRI DS1)
	100	CHN		QH + 26 stns
	20	IND		AIR, Sangli
	10	INS		RRI, Banda Aceh
	10	KOR	HLKT	CBS, Daegu
	5	NZL		NZ's Rhema, Auckland
	10	PAK		PBC, Loralai
	1	PHL	DYRG	Intercontinental Bc. Corp., Kalibo, Aklan
	2.5	PHL	DZMS	People's Bc. Netw., Sorsogon C.
	600	RUS		VoR, Ussuriysk, FE
	10	THA		Jor. Sor. 3, Roi Et
	5	THA		Thor. Or. 6, Bangkok
	100	TJK		Various relays, Yangiyul
	10	TWN	BEC29	Han Sheng BC, Kaohsiung
1260	1	AUS	6KA	Karratha
	2	AUS	3SR	Shepparton
	2	AUS	4MW	Thursday Island
	1	CHN		XZ; LN; HN
	20	IND		AIR, Ambikapur
	20	J	JOIR	TBC, Sendai
	5	KOR	AFNK	Busan
	10	KOR		KBS, Namwon
	2	NZL		R. Trackside/LiveSPORT, Christchurch
	5	PHL	DZEL	Eagle Bc. Corp., Lucena C., Que.
	5	PHL	DWMC	Magiliw Comm. Bc. Co., Rosales, Pang.
	5	PHL	DXRF	Manila Bc. Co., Davao C., Dvo Sur
	10	PHL	DYDD	Siam Bc. Netw. Corp., Lapu-Lapu C., Cebu
	25	THA		R. Thailand, Chiang Rai
	1	TWN		Cheng Sheng, Taipao
	10	TWN	BEP22	Ching Cha, Taipei
1269	5	AUS	6RN	ABC (RN), Busselton
	5	AUS	2SM	Sydney
	200	CHN		FJ; SX; JL; JS
	600	CHN		YN (FS/VoR)
	20	IND		AIR, Agartala
	20	IND		AIR, Madurai
	1	J	JOFM	HBC, Esashi
	1	J	JOHW	HBC, Obihiro
	1	J		JRT, Ikeda
	5	J	JOJR	JRT, Tokushima
	1	KOR		KBS, Gurye
	10	KOR		KBS, Yangju
	2	NZL		Classic Hits, Takaka
	10/5	PHL	DYWB	Consolidated Bc. System, Inc., Bacolod, Neg. Occ
	5	PHL	DZVX	Newsounds Bc. Netw., Daet, Cam. Nte
	10	PHL	DWRC	Republic Bc. System, San Nicolas, I. Nte
	5	RUS		R. Ussuri, Ussuriysk, FE
		RUS		R. Vostok Rossii, Komsomolsk, FE
	10	THA		Kho. Sor. Thor. Bor., Bangkok
	10	THA		Mor. Kor, Songkhla
	1	TWN	BEV37	Cheng Sheng, Taitung
	10	TWN	BEC44	Han Sheng BC, Penghu
	50	UZB		UZR (2), Zarafshon
1270	5	HWA	KNDI	Honolulu, Oahu
1278	5	AUS	3EE	Melbourne
	100	CHN		EB; HL; JX
	10	IND		AIR (C), Lucknow
	50	J	JOFR	RKB, Fukuoka
	1	KOR		KBS, Hapcheon
	2.5	NZL		Newstalk ZB, Hawera
	2	NZL		Newstalk ZB, Napier-Hastings

kHz	kW	Ctry	Call	Station, location
	10	PHL	DZRM	Philippine Bc. Sce., Quezon C, NCR
	7	RUS		R. Rossii + Reg., Severobaykalsk, Sib.
	5	RUS		R. Rossii, Bagdarin, Sib.
	10	VTN		Soc Trang (P)
1287	2	AUS	2TM	Tamworth
	10	BGD		Bangladesh Betar, Barishal
		CHN		CNR1
	30	CHN		GD + 8 stns
	100	IND		AIR (A), Panaji
	20/25	INS		RRI, Palembang
	50	J	JOHR	HBC, Sapporo
	150	KGZ		KGR (1), Bishkek
	10	KOR	HLAX	MBC, Cheongju
	10	KOR	HLAF	MBC, Gangneung
	2	NZL		R. Sport, Westport
	5	PHL	DZZH	Manila Bc. Co., Sorsogon C.
	5	RUS		R. Rossii + Reg., Kyakhta, Sib.
	10	THA		Sor. Or. Thor., Samut Prakan (Bangkok)
	10	THA		Wor. Por. Tho. 6, U. Ratchathani
	10	TWN	BEC27	Han Sheng BC, Taichung
	1	TWN		Min Li, Fangliao
1296	10	AUS	6RN	ABC (RN), Wagin
	5	AUS	4RPH	Queensland R. for the Print-H', Brisbane
	25	CHN		SH + 5 stns
	300	CHN		YN (FS/RFI)
	10	IND		AIR, Darbhanga
	10	J	JOTK	NHK (1), Matsue
	2.5	NZL		NewstalkZB, Hamilton
	10	PHL	DXAB	ABS-CBN Bc. Corp., Davao C., Dvo Sur
	5	PHL	DYJJ	Intercontinental Bc. Corp., Roxas C., Capiz
	5	PHL	DWPR	Multipoint Broadc. Netwk., Dagupan C, Pang.
	10	THA		R. Thailand, Pattani
	10	TWN	BED47	BCC (N), Tainan
	1	TWN	BEV76	Min Pen (1), Taipei
1305	2	AUS	5RN	ABC (RN), Renmark/Loxton
		CHN		CNR2
		CHN		Jinan,SD
	20	IND		AIR, Parbhani
	10	KOR		KBS, Uljin
	2.5	NZL		R. Dunedin
	7	RUS		R. Mayak, Ust-Kut, Sib.
	10	THA		Yaan Kraw, Bangkok
1314	10	BGD		Bangladesh Betar, Cox's Bazar
	5	AUS	3BT	Ballarat
	5	AUS	2TAB	Wollongong (HPONS)
	150	CHN		CQ + 5 stns
	1	IND		AIR (B), Cuttack
	20	IND		AIR, Bhuj
	50	J	JOUF	OBC, Osaka
	10	KOR	HLCM	CBS, Jeonbuk
	5	NZL		Parl. Bc. /So. Star, Invercargill
	2	NZL		RNZ National, Gisborne
	10	PHL	DWXI	Delta Bc. System, Parañaque, NCR
	10	THA		Mor. Kor., Khon Kaen
	5	TWN	BEP32	Ching Cha, Tainan
	10	TWN	BEV76	Tien Sheng, Chunan
1320	5	HWA	KEWA	Ewa Beach, Oahu
1323	2	AUS	5DN	Adelaide
	0.4	AUS	1KIX	Canberra (HPONS)
	600	CHN		JL; HL (FS)
	500	CHN		XJ (FS)
	20	CHN		ZJ + 5 stns
	20	IND		AIR (C), Kolkata
	1	J	JOFP	NHK (1), Fukushima
	1	J		NHK (1), Yamada
	1	KOR		KBS, Yeonggwang/Ulleung
	10	PHL	DYSI	GMA Netw., Inc., Iloilo C.
	5	PHL	DXAD	Muslim Mindanao DMPC, Marawi, Lanao Sur (v)
	10	PHL	DZRK	Pilipine Bc. Sce., Tabuk, Kalinga
	10	THA		Thor. Or. 13, Chiang Mai
	10	THA		Thor. Or. 7, Surat Thani
	7	TJK		VoR relay, Dushanbe
	1	TWN	BEV45	Taiwan BC (1), Taipei
1332	5	AUS	4BU	Bundaberg
	2	AUS	3SH	Swan Hill
	100	CHN		EN; JL; FJ (2 stns); GS

kHz	kW	Ctry	Call	Station, location
	10	IND		AIR, Tezu
	10	INS		RRI, Jakarta
	50	J	JOSF	Tokai R., Nagoya
	10	KOR	HLAO	MBC, Chungju
	10	NZL		R. Sport, Auckland
	100	PAK		PBC (2), Lahore
	1	PHL	DZKI	R. Philippines Netw., Iriga C., Cam. Sur
	5	PHL	DWAY	Swara Sug Media Corp., Cabanatuan, Nva. Ecija
	5	RUS		Dobryye pesni, Irkutsk, Sib.
	10	THA		Or. Sor., Bangkok
	10	THA		Thor. Or. 14, Maha Sarakham
	10	TWN	BEC36	Han Sheng BC, Tsoying
	1	TWN		Taiwan BC, Puli
1341	5	AUS	3--	Geelong (HPONS)
	5	AUS	2TAB	Newcastle (HPONS)
	100	CHN		GD (CNR 1/FS)
	100	CHN		HL + 6 stns
	1	IND		AIR, Kohima
	1/5	INS		RRI, Tanjung Pinang
	1	J		NHK (1), Iwaki/Minamata
	25	KAZ		Qazaq Radiosi, Aqtaw
		KAZ		RFE/RL relay, Almati
	10	KOR		KBS, Gimpo
	2	NZL		NewstalkZB, Nelson
	10	PAK		PBC, Bahawalpur
	20	THA		R. Thailand, Loei
	10	THA		R. Thailand, Phangnga
	25	THA		R. Thailand, Ubon Ratchathani
1350	5	AUS	2LF	Young
	50	CHN		YN; NM; JX(v); LN
	1	FSM	V6A	Baptist Church, Weno, Chuuk (r. inactive)
	0.25	GUM		Managament Adv. Sce., Agana
	1	IND		AIR (C), Jalandhar
	20	IND		AIR, Kupwara
	10	INS		RRI, Tarakan
	20	J	JOER	RCC, Hiroshima
	10	KOR	HLAQ	MBC, Samcheok
	500	MNG		MR1/Relays, Choybalsan
	1	NZL		R. Sport, Rotorua
	10	PHL	DZXQ	Mabuhay Bc. System, Inc., Pasig C., NCR (v)
	10	RUS		R. Rossii + Reg., Ust-Kan/Ust-Ulagan, Sib.
	10	THA		Phon. Neung Ror. Or., Bangkok
	10	THA		Wor. Por. Tho. 17, Trang
	10	TWN	BED63	BCC (N), Chia-i
	2.5	TWN	BEV50	Chung Hua (1), Sanchung
1359	0.2	AUS	3--	Mildura (HPONS)
	0.25	AUS	4WK	Toowoomba (trtr)
		CHN		CNR1
	1	KOR		AFNK, Songtan
	2.5	NZL		Coast, New Plymouth
	1	NZL		More FM, Queenstown
	1	PHL	DYSJ	Inter-Island Broadc. Corp., S. J. de Buenavista, Antique
	1	PHL	DYSL	Philippine Bc. Sce., Sogod, S. Leyte fjernes?
	5	PHL	DZYR	Philippine R. Corp., S. Fernando, LU
	1	RUS		R. Mayak, Choya/Shebalino, Sib.
	5	RUS		R. Mayak, Onguday, Sib.
	7	RUS		R. Mayak, Ust-Ilimsk, Sib.
	10	THA		Thor. Phor. 2, Sakhon Nakhon
	5	TWN	BEC40	Han Sheng, Hualien
	250	TWN		RTI (WYFR), Fangliao
1368	2	AUS	2GN	Goulburn
	10	CHN		HL; FJ (v); HB
	20	IND		AIR (C), New Delhi
	1	J	JOTS	HBC, Wakkanai
	5	J	JOHP	NHK (1), Takamatsu
	1	J	JOLG	NHK (1), Tottori
	1	J		NHK (1), Tsuruoka
	1	KOR		KBS, Muju
	2	KRE		Korean Central BS (E), Pyongyang
	0.8/0.1	NZL	1XT	Village R., Tauranga
	1	NZL		R. Live, Napier-Hastings
	1	PHL	DZRA	Catanduanes State College, Virac, Catanduanes
	2.5	PHL	DZBS	R. Philippines Netw., Baguio C., Benguet
	10	PHL	DXKO	R. Philippines Netw., Cagayan de Oro C., Mis. Or.
	10	THA		R. Thailand, Buri Ram

kHz	kW	Ctry	Call	Station, location
	25	THA		R. Thailand, Nan
	10	THA		Thor. Or. 12, N. Pathom
	1	TWN		Chin Hsi, Kaohsiung
1370	6.2	HWA	KUPA	Pearl City, Oahu
1377	5	AUS	3MP	Melbourne
	200	CHN		China Huayi BC, Fuzhou, FJ
	600	CHN		HEN (CNR1)
	10	CHN		SD + 5 stns
	20	IND		AIR (B), Hyderabad
	10	INS		RRI, Tolitoli
	1	J		NHK (2), Hachinohe
	1	J	JOAC	NHK (2), Nagasaki
	5	J	JOUC	NHK (2), Yamaguchi
	2	NZL		R. Sport, Levin
	2.5	NZL		The Word/Bible R., Dunedin
	10	PHL	DXKP	R. Philippines Netw., Pagadian C., Z. Sur
	5	RUS		R. Mayak, Okha, FE
	10	THA		R. Thailand, Chumphon
	10	THA		Wor. Phon 4, Phitsanulok
1386	50	CHN		TJ + 5 stns
	20	IND		AIR, Gwalior
	10	J	JOHC	NHK (2), Kagoshima
	10	J	JOJB	NHK (2), Kanazawa
	10	J	JOQC	NHK (2), Morioka
	5	J	JOKB	NHK (2), Okayama
	10	KOR	HLAM	MBC, Mokpo
	2	KRE		Korean Central BS (E), Pyongyang
	10	NZL		R. Tarana, Auckland
	5	PHL	DYVW	Catholic Media Netw., Borongan, E. Samar
	25	PHL	DZTV	Intercontinental Bc. Corp., Quezon C, NCR
	10	PHL	DXCR	Mt. View College, Valencia, Buk.
	5	SLM		R. Temotu (SIBC), Lata, Sta. Cruz Is.
	10	THA		SW Pheua Kaan Kaset, Bangkok
	3.5	TWN	BED74	BCC (1), Yuli
1395	0.2	AUS	2LG	ABC (RR), Lithgow
	5	AUS	5AA	Adelaide
	50	CHN		AH; NM (2 stns); FJ
	20	IND		AIR, Bikaner
	1	INS		RRI, Wamena
	1	J	JOCE	CRK, Toyooka
	1	J	JOWE	RFC, Wakamatsu
	10	KOR		KBS, Cheorwon
	2	NZL		NewstalkZB, Levin
	5	PHL	DZVT	Apostolic Vicariate of S. J., San Jose, Min. Occ.
	10	PHL	DYCH	Cebu Bc. Co., Talisay C., Cebu
	10	THA		Nor. Thor. Phor., Chiang Rai
	10	TWN		Cheng Sheng, Tafa
1404	4	AUS	6TAB	Busselton (HPONS)
	2	AUS	2PK	Parkes/Forbes
	50	CHN		FJ; LN; HB; ZJ
	20	IND		AIR, Gangtok
	5	J	JOQL	HBC, Kushiro
	1	J	JOVO	SBS, Hamamatsu
	10	J	JOVR	SBS, Shizuoka
	1	KGZ		KGR (1), Cholpon-Ata
	7	KGZ		KGR (1), Haidarkan/Naryn
	20	KGZ		KGR (1), Jojomel
		KGZ		KGR (1), Orgochor
	10	KOR	HLKP	CBS, Busan
	5	NZL		NZ's Rhema, Invercargill
		PHL	DXAQ	End Time Mission, Lucena C., Que
	1	PHL	DYKB	R. Philippines Netw., Bacolod, Neg. Occ.
	10	THA		Jor. Sor. 4, Yasothon
	25	THA		R. Thailand, Songkhla
	10	THA		Thor. Phor. Neung, Suphan Buri
	2.5/5	TLS		R. Timor Kmanek, Díli
	10	TWN	BEV78	Yi Shih, Keelung
1413	5	AUS	2EA	Newcastle (SBS)
	0.5	AUS	3--	Shepparton (HPONS)
	10	BGD		Bangladesh Betar, Comilla
	5	CHN		XJ + 4 stns
	20	IND		AIR, Kota
	5	INS		RRI, Sungai Liat
	50	J	JOIF	KBC, Fukuoka
	0.3	NZL		3XP R. Ferrymead, Christchurch
	2	NZL		NewstalkZB, Tokoroa
	5	PHL	DYXW	Filipinas Bc. Netw.,. Tacloban C., Leyte

kHz	kW	Ctry	Call	Station, location
	5	PHL	DWRA	Republic Bc. System, Bauio C., Benguet
	5	RUS		R. Rossii, Sangar, FE
	10	TWN	BED54	BCC (N), Miaoli
	1	TWN	BED67	BCC (N), Puli
1420	5	HWA	KKEA	Honolulu, Oahu
1422	5	AUS	3XY	Melbourne (HPONS)
	1	AUS	4AM	Port Douglas (trtr)
	0.4	AUS	6GS	Wagin (HPONS)
	20	CHN		SH (2 stns); SX; SC
	600	CHN		XJ (CNR1/8/FS)
	0.5	CHR	6ABCRN	ABC Radio National relay, Phosphate Hill
	50	J	JORF	RF, Yokohama
	5	PHL	DYZD	Bohol Chronicle R. Corp., Ubay, Bohol
	5	PHL	DXMU	Central Mindanao Univ., Musuan, Buk.
	1	PHL	DZOR	Zambales Bc. & Dev. Corp., Olongapo C., Zamb.
	5	RUS		R. Rossii, Kuanda, Sib.
	10	THA		Phon. Neung Ror. Or., Bangkok
	10	THA		R. Thailand, Amnat Charoen
	10	THA		Sor. Wor. Phor. 4, Phitsanulok
	100	TWN		CBS (RTI), Minhsiung
	5	TWN		Chien Kuo, Kuanyin
1431	2	AUS	2RN	ABC (RN), Wollongong
	2	AUS	6TAB	Calgoorlie (HPONS)
	10	BGD		Bangladesh Betar, Bandorban
	10	CHN		EB + 6 stns
	1	J		BSS, Izumo
	1	J	JOHL	BSS, Tottori
	5	J	JOZF	GBS, Gifu
	1	J		NBC, Fukue
	1	J	JOWW	RFC, Iwaki
	5	J	JOVF	WBS, Wakayama
	2	NZL		R. Kidnappers, Napier-Hastings
	5	PHL	DYRS	Ragde, Vicente & Sons, San Carlos, Neg. Occ.
	5	THA		Sor. Thor. Ror. 6, Songkhla
	10	THA		Thor. Or. 3, N. Ratchasima
1440	2	AUS	1SBS	Canberra (SBS)
	50	CHN		NM (2 stns); GX; LN
	1	IND		AIR, Kurseong
	3	J		STV, Muroran
	50	J	JOWF	STV, Sapporo
	1	J		STV, Tomakomai
		KAZ		BBC relay, Qizilorda
	1-0.25	KOR		AFNK, 4 stns
	3/0.5	MRA	K...	Blue Continent Comm., Tamuning (u constr.)
	1	NZL		Goldrush R., Lawrence
	0.2	NZL		Moana AM, Tauranga
	10	PHL	DWDH	Manila Bc. Co., Dagupan C., Pang.
	0.01	PHL	DXSI	So. Inst. of Tech., Cagayan de Oro C., Mis. Or.
	5	RUS		R. Rossii + Reg., Turachak, Sib.
	5	RUS		R. Rossii + Reg., Ust-Koksha/Kosh-Agach, Sib.
	10	THA		Thor. Phor. 2, N. Phanom
	2	AUS	6TAB	Mandurah (HPONS)
1449	5	AUS	2MG	Mudgee
	20	CHN		JX; SD(2 stns); FJ
	10	FSM	V6AH	FSMBS R. Pohnpei, Kolonia
	1	IND		AIR, Kanpur
	5	J	JOQM	HBC, Abashiri
	1	J		RNC, Marugame
	5	J	JOKF	RNC, Takamatsu
	10	KOR	HLQB	KBS, Ulsan
	5	MLD		Raajje Radio, Malé
	2.5	NZL		RNZ National, Palmerston No.
	10	PAK		PBC, Zhob
	5	PHL	DYAC	Leyte State University, Baybay, Leyte (v)
	5	PHL	DXSA	Mindanao Bc. Co., Inc., Marawi C, Lanao Sur
	10	THA		Thor. Phor. 3, Phichit
1458	2	AUS	2PB	ABC (PNN), Newcastle
	200	CHN		NM; JS; LN
	20	IND		AIR, Barmer
	20	IND		AIR, Bhagalpur
	1	J		IBS, Sekijo
	1	J	JOYL	IBS, Tsuchiura
	1	J	JOUO	NBC, Saga
	1	J		RCC, Shobara
	1	J	JOWR	RFC, Fukushima

kHz	kW	Ctry	Call	Station, location
	1	KOR		KBS, Hamyang/Bonghwa
	2.5	NZL		RNZ National, Westport
	10	PHL	DYZZ	Siam Bc. Netw. Corp., Gihulngan, Neg. Occ.
	10	PHL	DZJV	ZOE Bc. Netw., Calamba, Laguna
	10	THA		Jor. Sor. 6, Si Sa Ket
	10	THA		Sor. Thor. Ror. 3, Phuket
1460	5	HWA	KRHA	Honolulu, Oahu
1467	2	AUS	3ML	Mildura
	17.5	CHN		JX; EB; SD
	2.5	FJI		Fiji Bc. Comm. Ltd. (2), Rakiraki
	100	IND		AIR, Jeypore
	1	J	JOVB	NHK (2), Hakodate
	1	J	JOMC	NHK (2), Miyazaki
	1	J	JONB	NHK (2), Nagano
	1	J	JOID	NHK (2), Oita
	1	J		NHK (2), Wakkanai
	150	KGZ		TWR relay, Bishkek
	50	KOR	HLKN	KBS, Mokpo
	1	PHL	DWVR	R. Veritas, San Jose C., Nva Ecija
	5	PHL	DXVP	RCA-ZBN, Zamboanga C., Z. Sur
	100	THA		R. Thailand, Pathum Thani
1476	1	AUS	5MG	ABC (RR), Mt. Gambier
	0.5	AUS	2--	Penrith (HPONS)
	2	AUS	4ZR	Roma
	200	CHN		HL (CNR2)
	10	CHN		QH + 8 stns
	1	IND		AIR (A), Jaipur
	1	J		NHK (2), Iida
	700	MLA		RTM Labuan, Tuaran (r1475)
	5	NZL		R. Trackside/LiveSPORT, Auckland
	10	PAK		PBC, Faisalabad
	1	PHL	DZYA	Radyo Pilipino Corp., Angeles C, Pamp.
	10	PHL	DXRJ	Rajah Bc. Netw., Iligan C., Lanao Nte
	1	PHL	DWRB	Ribbon Bc. Netw., Lipa C, Bat
	20	RUS		R. Rossii + Reg., Onguday, Sib.
	50	THA		R. Thailand, Lamphun
	10	TKM		Turkmen R. (1), Türkmenbasy
1485	0.1	AUS	2RN	ABC (RN), Wilcannia
	0.05-0.2	AUS		ABC (RR), 2 stns
	10	CHN		HB + 15 stns
	0.25	DGA		AFRTS, Diego Garcia
	1	IND		AIR, 10 stns
	1	J	JOPL	KRY, Hagi
	1	J	JOGO	RAB, Hachinohe
	1	KOR		KBS, Gongju/Goheung
	1	NZL		R. Trackside/LiveSPORT, Gisborne
	5	PHL	DYDH	Pacific Bc. System, Iloilo C.
	1/2	RUS		R. Mayak, FE, 2 stns (sync.)
	1	RUS		R. Rossii + Reg., Kamenskoye, FE
	1	RUS		R. Rossii + Reg., Tyumen, FE
	0.2-1	RUS		R. Rossii, FE, 2 stns (sync.)
	1	RUS		R. Rossii, Sib., 2 stns (sync.)
1494	2	AUS	2AY	Albury
	1	CHN		AH + 4 stns
	5	FSM	V6AI	FSMBS R. Yap, Colonia
	1	J	JOTL	HBC, Nayoro
	1	J		RSK, 5 stns
	10	J	JOYR	RSK, Okayama
	0.1	LHW		R. Lord Howe Island
	2.5	NZL		Parl. Bc. /So. Star, Hamilton
	2.5	NZL		R. Sport, Timaru
	5	PHL	DXOC	DWRL Radio, Inc., Ozamis C., Mis. Occ.
	10	PHL	DWSS	Supreme Bc. Systems DWSS, Pasig C., NCR
	10	THA		Or. Sor. Mor. Thor., Bangkok
	5	TWN	BEE34	Chiao Yu, Changhua
	10	TWN	BEE32	Chiao Yu, Taipei
1500	10	HWA	KUMU	Honolulu, Oahu
1503	5	AUS	3KND	(PBS), Port Melbourne
	5	AUS	2BS	Bathurst
	10	CHN		HN; AH(v); LN; ZJ
	1	FSM	V6AJ	FSMBS R. Kosrae, Tofol
	1	IND		AIR (B), Vijayawada
	10	J	JOUK	NHK (1), Akita
	1	J		NHK (1), Aso
	1	KOR		KBS, Gimcheon
	2.5	NZL		R. Sport, Christchurch
	5	NZL		R. Sport, Wellington
	5	PHL	DYBB	GMA Netw., Inc., Roxas C., Capiz
	5	RUS		R. Mayak, Salekhard, Sib.
	1	RUS		R. Rossii + Reg., Magistralnyy, Sib.
	10	THA		Jor. Sor. 2, Surat Thani
	7	TJK		VoR, Dushanbe
	600	TWN		CBS (RTI), Fangliao
1512	10	AUS	2RN	ABC (RN), Newcastle
	5	AUS	6BAY	Geraldton
	10	CHN		GS; NM; SD
	20	IND		AIR, Kokrajhar
		INS		RRI, Bukittinggi
	1	J		NHK (2), Koriyama/Matsumoto
	5	J	JOZB	NHK (2), Matsuyama
	0.25-0.1	KOR		AFNK,5 stns
	10	PAK		PBC, Gilgit
	10	PHL	DYAB	ABS-CBN Bc. Corp., Cebu C.
	10	PHL	DZAT	End Time Mission, Lucena C., Que
	10	THA		Kor. Wor. Sor. 4, Phayao
	10	THA		Thor. Or. 11, Songkhla
	10	TWN		Ching Cha, Hsinchu
1521	2	AUS	2QN	Deniliquin
	10	CHN		NM + 27 stns
	500	CHN		XJ (FS)
	1	IND		AIR, Aurangabad
	10	IND		AIR, Tawang
	1	J	JOTC	NHK (2), Aomori
	1	J	JOFC	NHK (2), Fukui
	1	J	JODC	NHK (2), Hamamatsu
	1	J		NHK (2), Ishigaki/Nakamura
	1	J	JOJC	NHK (2), Yamagata
	1	J		NHK (2), Yonago
	1	NZL		R. Sport, Tauranga
	5	RUS		R. Rossii, Boguchany, Sib.
	10	THA		Nor. Thor. Phor, Bangkok
1530	2	AUS	2VM	Moree
	50	CHN		ZJ; JL; SX
	0.25	GUM	KVOG	Unk. loc.
	20	IND		AIR, Agra
	1	J	JODO	BSN, Joetsu
	1	J	JOXF	CRT, Utsunomiya
	1	J	JOEO	RCC, Fukuyama
	1	J		RCC, Mihara
	5	KOR		AFNK, Seoul (Yongsan)
	1	NZL		Coast, Napier-Hastings
	10	PHL	DZME	Capitol Bc. Center, Quezon C., NCR
	5	RUS		R. Mayak, Krasnyy Chikoy, Sib.
	10	THA		Thor. Phor. Neung, Chanthaburi
	10	THA		Wor. Por. Tho. 14, Uttaradit
1539	10	AUS	5TAB	Adelaide (HPONS)
	1	AUS	2RF	Sydney (HPONS)
	10	CHN		CNR1
	1	KOR		KBS, Gosan
	1	NZL		R. Sport, Blenheim
	5	PHL	DZYM	Philippine R. Corp., San José, Mind. Occ.
	10	THA		Phon Ror. Kao, Kanchanaburi
1540	5	HWA	KREA	Honolulu, Oahu
1548	50	AUS	4QD	ABC (RR), Emerald
	200	CHN		SD; HN
	400	CLN		DW, Trincomalee
	1	NZL		Coast, Palmerston No
	0.9	NZL		R. Trackside/LiveSPORT, Rotorua
	5	PHL	DYDM	Catholic Media Netw., Maasin C., So. Leyte
	10	PHL	DZSD	GMA Netw., Inc., Dagupan C., Pang.
	10	RUS		R. Mayak + Reg., Aleksandrovsk-Sak., FE
1557	0.5	AUS	5TAB	Renmark/Loxton (HPONS)
	2	AUS	2RE	Taree
	25	CHN		EB (2 stns); LN
	2	NZL		Coast, Hawera
	10	PAK		PBC, Skardu
	10	THA		R. Thailand, Trat
	10	THA		Siang Adison, Phetchabun
	300	TWN		CBS (WYFR), Kouhu
1566	0.2	AUS	4GM	ABC (RR), Gympie
	5	AUS	3NE	Wangaratta
	10	CHN		EB; GS; SX; JL
	1000	IND		AIR (N), Nagpur (Buttibori)
	250/100	KOR	HLAZ	FEBC,Jeju

kHz	kW	Ctry	Call	Station, location
	0.1	NFK	VL2NI	Norfolk Isl.
	10	PHL	DXID	Ass. of Islamic Dev. Coop., Pagadian C., Z. Sur
	7.8	PHL	DYMP	Pilippine Bc. Sce., Cebu C
1570	15	HWA	KUAU	Ha'iku, Maui
1575	5	AUS	2RF	Wollongong (HPONS)
	2	CHN		LN; JL
	1	J		AFN, Iwakuni
	0.6	J		AFN, Misawa
	0.25	J		AFN, Sasebo
	2.5	NZL		Toroa R, Dunedin
	1000	THA		R. Saranrom/BBG, Ayutthaya
1584	0.05-0.1	AUS		ABC (RR), 3 stns
	0.2	AUS	4VL	Cunnamulla (trtr)
	0.2	AUS	2EC	Narooma (trtr)
	0.5	AUS	4CC	Rockhampton (trtr)
	10	CHN		SX (2 stns) + 6 stns
	0.1	HKG		RTHK (3), Chung Hom Kok
	1	IND		AIR, 10 stns
	1	KOR		KBS, Danyang/Geumsan
	1	KOR		KBS, Sancheong
	0.4	NZL		Classic Hits, Picton
	0.25	PAK		PBC, Turbat/Chitral/Sibi
	1	PHL	DWBR	Dawnbreaker's Found., Talavera, Nva Ecija
	5	PLW	T8AA	Voice of Palau, Malakal Island, Koror
	1/5	RUS		R. Mayak, Aykhal/Ust-Nera, FE
	1	RUS		R. Rossii + Reg., Klyuchi/Tigil, FE
	0.2	RUS		R. Rossii, B. Gora/Khandyra, Sib.
	1	RUS		R. Rossii, Taksimo, Sib.
1593	5	AUS	3RG	Melbourne (HPONS)
	0.2	AUS	2--	Murwillumbah (HPONS)
	10	CHN		HL (2 stns); XJ
	600	CHN		JS (CNR1)
	5	FSM	V6AK	FSMBS R. Chuuk, Weno
	10	IND		AIR (A), Bhopal
	10	J	JOTB	NHK (2), Matsue
	10	J	JOQB	NHK (2), Niigata
	2.5	NZL		Coast, Christchurch

kHz	kW	Ctry	Call	Station, location
	5	NZL		R. Samoa, Auckland
	10	PHL	DXFM	Ranao Radio & TV Bc. Sys. Corp, Marawi C., Lanao Sur
	50	RUS		R. Yunost, Sib.
	10	THA		R. Thailand, Ratchaburi
	3	TWN		Taiwan Yuyeh, Ilan
1602	0.05-0.25	AUS		ABC (RR), 3 stns
	1	CHN		Jiangsu N,JS
	1	IND		AIR, 11 stns
	1	J		NHK (2), 6 stns
	1	J	JOCC	NHK (2), Asahikawa
	1	J	JOFD	NHK (2), Fukushima
	1	J	JOSB	NHK (2), Kitakyushu
	1	J	JOKC	NHK (2), Kofu
	1	KOR		KBS, Sabuk
	2.5	NZL		R. Reading Service, Levin
	0.25	PAK		PBC, Abbottabad
	1	PHL	DZUP	Univ. of the Philippines, Quzon C, NCR
	1	RUS		Ekho Moskvy, Zeleznogorsk-Ilimskiy, Sib.
	0.2	RUS		R. Rossii, Chumikan, FE
	1	RUS		R. Rossii, Sib., 2 stns (sync.)
1611	0.001-0.4	AUS		17 stns (HPONS)
1613		KRE		Frontline Soldiers Radio (v)
1620	0.4	AUS		8 stns (HPONS)
1629	0.1-0.4	AUS		11 stns (HPONS)
1638	0.4	AUS		4 stns (HPONS)
	0.01	PHL	DWGI	Guzman Inst. Of Tech., Manila, NCR
1647	0.4	AUS		3 stns (HPONS)
1656	0.05-0.4	AUS		4 stns (HPONS)
1665	0.4	AUS	2MM	Sydney (HPONS)
1674	1	PHL	DZBF	Mun. of Marikina, Marikina C., NCR
1683	0.4	AUS	2---	Sydney (HPONS)
1692	0.4	AUS	4---	Nanango (HPONS)
1701	0.1	AUS	4---	Brisbane (HPONS)
	0.4	AUS	3VMV	Somerton (HPONS)
	0.4	AUS	2---	Sydney (HPONS)

NORTH AMERICA

The North American MW frequency listing has been removed from this edition in order to make space for over 1,000 extra MW stations and their addresses in the main USA listing. To find the stations previously listed here, please go to the entries for Alaska, Canada and United States of America in the National Radio section.

CENTRAL AMERICA, CARIBBEAN & MEXICO

Abbreviations: Broadc.=Broadcasting, Corp.=Corporation, Em=Emisora, LV=La Voz, Nal=Nacional, Nat=National, Sce=Service.
Call signs: Costa Rica TI_, Cuba CM_, Dominican Republic HI_, El Salvador YS_, Guatemala TG_, Honduras HR_, Mexico XE_, Nicaragua YN_, Panama HO_

kHz	kW	Ctry	Call	Station, location
530	10	CTR	RI	R. Sinfonola, Cartago
	10	CUB	BQ	R. Enciclopedia, HA
		TCA		R. Vision Cristiana, Caicos
540	1	CUB	BA	R. Rebelde, Sancti Spíritus, SS
	5	DOM	CM	R. ABC, Sto Domingo
	10	GRD		Klassic AM, G.B.N. R., St. George's
	0.02	GTM		R. Amistad, San Pedro de Laguna
		GTM		R. Cobán, Cobán
	0.1	MEX	SURF	540 AM, Tijuana
	5/2.5	MEX	HS	La Norteñita, Los Mochis

kHz	kW	Ctry	Call	Station, location
	20/2.5	MEX	WF	La Poderosa del Oriente/La Ke Buena AM 540, Ixtapaluca
	5	MEX	TX	La TX/La Ranchera de Paquime, Nuevo Casas Grandes
	5/1	MEX	MIT	LV de BalúnCanán, Comitán
	1	MEX	WA	W R., Monterrey
	150	MEX	WA	W R., San Luis Potosí
	25	NCG	A30W	R. Corporación, Managua
	10	PNR	PU	R. Lider, Panamá
	10	PNR	U23	R. Mía Chiriqui, David

kHz	kW	Ctry	Call	Station, location
	5	SLV	HV	La Estación de la Palabra, San Salvador
550	5	CTR	SCL	R. Santa Clara, Cd. Quesada
	10	CUB	BA	R. Rebelde, Guantánamo, GU
	1	CUB	BA	R. Rebelde, Manzanillo, GR
	30	CUB	BA	R. Rebelde, Pinar del Río, PR
	0.5	HND	XD	R. Manantial, Sta Rosa de Copán
	1	HND	XT	R. X, Tegucigalpa
		JMC		R. Jamaica, Kingston
	5/0.15	MEX	PL	La Super Estación, Cd. Cuauhtémoc
	1	MEX	ACD	Los 40 Principales, Acapulco
	1.5/0.25	MEX	HLL	Los40 Principales, Salina Cruz
	2.5/1	MEX	ZK	Poder 55, Tepatitlán
	2/0.35	MEX	QW	Q-W La Poderosa, Mérida
	2.5/0.15	MEX	TNC	R.Aztlán, Tepic
	5/0.25	MEX	KL	W R., Jalapa
	5	PTR	WPAB	WPAB 550 Ponce, Ponce
	2	SLV	FG	R. Cristo Te Llama, Sonsonate
555	10	SCN		Ziz R., St Kitts
560	30	CUB	BA	R. Rebelde, Moa, HO
	3	DOM	AA	R. Ritmos, Santiago
	10	GTM	RV	R. 560, Guatemala
	1	GTM		R. Quetzal, Malacatán
	1	HND	KL	R. Reloj, San Pedro Sula
	1	HND	VF	R. Valladolid, Comayagua
	5	HND	RZ	VRZ R. Juticalpa, Juticalpa
	10/1	MEX	MZA	Fórmula Melódica del Pacífico, Manzanillo
	5/1	MEX	XZ	Ke Buena, Zacatecas
	1.4/0.25	MEX	GIK	LaAcerera, Monclova
	0.75/0.5	MEX	OC	LaMejor, México
	5/1	MEX	QAA	La Poderosa, Chetumal
	10/0.1	MEX	SRD	La Tremenda, Santiago Papasquiaro
	2/0.25	MEX	IN	LV del Valle, Cintalapa
	1/0.5	MEX	YO	R. Lobo, Huatabampo
	3	PNR	H2	RPC R., Colón
570	5	CTR	ELR	R. Libertad, San José
	30	CUB	BD	R. Reloj, Santa Clara, VC
	10/5	DOM	MS	R. Crystal, Sto Domingo
	1	GTM	PA	R. Palmeras, Escuintla
		HND	LP	R. América, Tela
	1	HND	OX	R. El Triunfo, Choluteca
	2.5	MEX	ME	El Poder del Oriente, Valladolid
	1	MEX	TJ	La Mexicana, Torreón
	5/0.5	MEX	BJB	La Sabrosita, Monterrey
	0.5/0.25	MEX	UK	LaU-K/XEUK, Caborca
	10/1	MEX	VX	Mass R./La Grande de Tabasco, Villahermosa
	5/2.5	MEX	OA	O-A R. Mexicana, Oaxaca
	2/1.7	MEX	LQ	R. 5-70, Morelia
	0.25	MEX	TD	R. Red/La Z, Tecuala
	0.5	MEX	VJP	R. Xicotepec, Xicotepec de Juárez
	5	NCG	A2RQ	R. Veritas 5-70, Chinandega
	5	PNR	S	R. Soberana, Panamá
	10	SLV	KT	R. Exus, San Salvador
580	5	CUB	BA	R. Rebelde, Baracoa, GU
	10	CUB	BA	R. Rebelde, Mantua, PR
	5	DOM	FS	R. Montecristi, Montecristi
	5	GTM	Y	R. Progreso, Guatemala
	3	HND	ZQ	R. Noticias STC, Tegucigalpa
		HND	EO	Super Estrella de Occidente, Sta Rosa de Copán
		JMC		R. Jamaica, Kingston
	10/1	MEX	AV	Canal 58, Guadalajara
	1/0.25	MEX	HO	La Fuerza de la Palabra, Cd.Obregón
	1/0.25	MEX	UE	La Invasora, Tuxtla Gutiérrez
	1	MEX	HP	La Más Prendida, Cd.Victoria
	5/0.5	MEX	MU	La Rancherita del Aire, Piedras Negras
	1/0.25	MEX	YI	Mix FM, Cancún:
	5/0.7	MEX	FI	R. Mexicana, Chihuahua
	1/0.5	MEX	DZ	R. Ondas, Córdoba
	0.25	MEX	UAQ	R. Universidad, Querétaro
	10	NCG	A3LP	R. 5-80, Managua
	10	PNR	H4	RPC R., David
	10	PTR	WKAQ	R. Reloj, San Juan
590	5	CTR	RN	R. Nacional, San José
	30	CUB	BF	R. Musical, La Habana, CH
	10/5	DOM	DV	R. Santa María, La Vega
	5	GTM	RQ	R. Quiché, Sta Cruz del Quiché

kHz	kW	Ctry	Call	Station, location
		HND	LP3	R. América, San Pedro Sula
		HND	RE	R. Renacer, Catacamas
	10/1	MEX	CJU	La Explosiva 590, Puerto Vallarta
	1	MEX	BH	La Mejor, Hermosillo
	5/0.5	MEX	FD	La Mejor, Reynosa
	1	MEX	OM	R. Fórmula, Coatzacoalcos
	1	MEX	E	R. Fórmula, Primera Cadena, Durango
	2.5/10	MEX	PH	Sabrosita 590 , México
	5/1	MEX	ZZZ	Triple Z, Tapachula
	10/0.25	MEX	GTO	TuRecuerdo, León
	10	PNR	H3	RPC R., Chitré
600	150	CUB	BA	R. Rebelde, Urbano Noris, HO
		DOM		Celestial 600, Santo Domingo
		DOM	SD	R. Santo Domingo, El Seybo
	1	GTM	RC	R. Campesina, Escuintla
		HND	EK	R. Orion, La Ceiba
	1	MEX	TA	600 Solo Hits, Zitácuaro
	10/1	MEX	OCH	K'in R., Ococingo
	5/1	MEX	BB	La Comadre, Puros Éxitos, Acapulco
	5/1	MEX	HW	La Fiera Digital, Rosario
	5/1	MEX	CV	La Gran Compañía, Cd.Valles
	2.5	MEX	LAZ	La Mejor, Cd.Guzmán
	1/0.5	MEX	MN	La Regiomontaña, Monterrey
	20/1	MEX	Z	R. Fórmula, Segunda Cadena, Mérida
	1	MEX	DN	R. Noticias, Torreón
	10	NCG	A3MD	La Nueva R. Ya, Managua
	5	PTR	WYEL	R. 600, Mayagüez
	3	SLV	NK	Vox FM, San Salvador
610	15	CTR	RMV	R. María, San José
	1	CUB	BA	R. Rebelde, Bahía Honda, PR
	1	CUB	BD	R. Reloj, Trinidad, SS
	5	DOM	JR	R. Amanecer, Santiago
	1	DOM	SD	R. Santo Domingo, Pedernales:
	5	GTM	GA	R. Alianza, Guatemala
	10	HND	LP	R. América, Sta Rosa de Copán
	10	HND	LD	R. América,Tegucigalpa
	10	MEX	UM	Candela FM, Valladolid
	1/0.5	MEX	JA	Conexión 610/R. Fórmula, Jalapa
	5/0.1	MEX	EL	El Super Canal 610, Fresnillo
	1/0.5	MEX	GS	La GS/La Ley, Guasave
	1/0.5	MEX	KZ	La Poderosa, Tehuantepec
	5/0.5	MEX	BX	La Primera, Sabinas
	1	MEX	SAC	R. Lobo, Saltillo
	5/1	MEX	UF	Variadísima, Uruapan
	10	PNR	HM	RPC R., Panamá
	1/0.25	PTR	WEXS	X-61, Patillas
620	10	ATG		ABBS, St. John's
	25	CUB	BA	R. Rebelde, Colón, MA
	1	CUB	BA	R. Rebelde, Moa, HO
	10	DOM	SD	R. Santo Domingo, Sto Domingo
	5	GTM	PQ	R. 6-20, San Cristóbal
	1	HND	LP5	R. América, Comayagua
	1	HND	LP	R. América, Juticalpa
	1	HND	LP17	R. Continental, San Pedro Sula
		HND	LO	R. Litoral, Tocoa
	1/0.25	MEX	GH	La Lupe, Reynosa
	5/1	MEX	BU	La Norteñita, Chihuahua
	5	MEX	SS	La Tremenda, Ensenada
	1/0.5	MEX	CK	R. 6.20, Durango
	10	MEX	NK	R. 6-20, México
	2.5/1	MEX	HGR	R. Fórmula 620, Villahermosa
	2.5/0.5	MEX	WZ	R. Novedades, San Luis Potosí
	5/1	MEX	OO	W R., Tepic
	50	NCG	N	R. Nicaragua, Managua
630	5	CUB	BC	R. Progreso, Pinar del Río, PR
	1	DOM	SD	R. Santo Domingo, San Juan
		GTM	EL	R. Cultural Porvenir, Sta Elena
	1	HND	LP	R. América, Choluteca
	1	HND	LP7	R. América, La Ceiba
	1	HTI		Rdif. Jérémienne, Jérémie
	5	MEX	JR	Coral 630, Zihuatanejo
	1/0.25	MEX	FX	Doble X, Guaymas
	5/0.25	MEX	OPE	Exa, Mazatlán
	10	MEX	FB	F-B La Estación que da las noticias, Monterrey
	0.5	MEX	CCQ	Frecuencia Turquesa, Cancún
	10/0.5	MEX	JB	Jalisco R., Guadalajara

kHz	kW	Ctry	Call	Station, location
	10/0.75	MEX	FU	LaNueva Voz, Cosamaloapan
	1/0.15	MEX	ERO	R. Tamaulipas, Altamira
	2	PNR	J35	R. Provincias, Chitré
	5	PTR	WUNO	NotiUno, San Juan
	10	SLV	LN	R. Promesa, San Salvador
640	20	CTR	ALY	R. Rica, San José
	50	CUB	BC	R. Progreso, Guanabacoa, HA
	10	CUB	BC	R. Progreso, Las Tunas, LT
	5	CUB	BA	R. Rebelde, Las Mercedes, GR
	10	DOM	SD	R. Santo Domingo, Santiago
	40	GLP		R. Guadeloupe, Point-à-Pitre
	1	HND	UP	R. Centro, Tegucigalpa
		HND	JT	R. Jerusalen, Sta Bárbara
	5/1	MEX	HDL	Aro-AM, Huajuapán de León
	1/0.25	MEX	TAM	Ke Buena, Cd.Victoria
	10	MEX	NQ	N-Q La Superestación, Tulancingo:
	5	MEX	JUA	R. Recuerdo, Canal 640, Cd.Juárez
	5/1	MEX	YQ	R. Uno, Fresnillo
	10/1	MEX	HHI	R. Uno/La Número Uno, Hidalgo del Parral
	5/1	MEX	WM	Suprema 64, San Cristóbal de las Casas
	10	NCG	A4LR	La Mera Mera , Managua
	2.5	PNR		R. Panamá, La Palma
	2.5	PNR	K22	R.CPR, Colón
650	1	CUB	BA	R. Rebelde, Media Luna, GR
	1	CUB	BA	R. Rebelde, Stgo de Cuba, SC
	15/5	DOM	AT	R. Universal, Sto Domingo
	25	HND	VS	Nuestra Señora de La Esperanza, San Pedro Sula
	15	HND	LP	R. América, Danlí
	1	HND	VS	R. Católica Olancho, Olanchito
		HND	TA	R. Turquesa, Siguatepeque
	5	MEX	CHH	Capital Máxima, Chilpancingo
	1	MEX	IY	Espectacular, Río Verde
	1/0.5	MEX	VILL	La Comadre, Villahermosa
	10/2.5	MEX	EJ	La Z, Puerto Vallarta
	5/1	MEX	ZM	La Zamorana, Zamora
	5/0.2	MEX	PX	LV de Ángel/R. Fórmula, Puerto Ángel
	1/0.25	MEX	VSS	R. 13, Hermosillo
	1	MEX	VG	R. Fórmula, Primera Cadena, Mérida
	0.5d	MEX	RCG	R. Vida, Cd. Acuña
	5/1	MEX	TNT	W R., R. 65, Los Mochis
	10/8	NCG	RD	R. Diriangén, Granada
	5	NCG	A6RS	R. Muzun, Matagalpa
	5	PNR	S22	R. Mía "Cadena Nacional", Panamá
660	30	CUB	BC	R. Progreso, Santa Clara, VC
	3	DOM	AM	R. Visión Cristiana, Santiago:
	3	GTM	Q	La Voz de Quetzaltenango, Quetzaltenango
	3	HND	NN18	LV de Honduras, La Ceiba
		HND	KV	R. Betania, Choluteca
	5	HTI		R. Lumiere, P-au-P
	50/10	MEX	EY	6-60 La Consentida, Aguascalientes
	2.5/0.25	MEX	SJC	Cabo6-60, San José del Cabo
	5	MEX	AR	La Mexicana, Tampico
	10/1	MEX	FZ	Noti-R. 6-60, Monterrey
	5	MEX	ACB	R. 6-60/La Tremenda Número Uno, Cd. Delicias
	1/0.5	MEX	YG	R. 660/R. Fiesta Mexicana, Matías Romero
	1	MEX	CPR	R. Chan Santa Cruz-LV de los Mayas, Felipe Carillo Puerto
	50	MEX	DTL	R. Ciudadana, México
	1/0.5	MEX	WX	R. Mexicana, Durango
	5	NCG		R. Máxima, Managua
	5	PNR		La Nueva Exitosa, Sabana Grande
	1	PNR	F33	RPC R., Bocas del Toro
670	10	CTR	TNT	R. Managua, San José
	50	CUB	BA	R. Rebelde, Arroyo Arenas, CH
		CUB	BD	R. Reloj
	5	DOM	BS	R. Dial, San Pedro de Macorís
	1	DOM	SD	R. Santo Domingo, Barahona
	1	HND	NN20	LV de Honduras, Sta Rosa de Copán
	10	HND	N	LV de Honduras, Tegucigalpa
	1/0.1	MEX	OG	ABC R. 670, Querétaro
	5/0.5	MEX	OB	La Máquina Musical, Pichucalco
	5/1	MEX	IS	La Rancherita Consentida, Cd.Guzmán
	1	MEX	SIC	La Romántica, Córdoba
	5/1	MEX	LH	La Zeta 670, Acaponeta
	5/0.25	MEX	TOR	X-E-Tor R. Ranchito, Torreón

kHz	kW	Ctry	Call	Station, location
		NCG	RC	R. Caribe, Pto Cabezas
	5	PNR	LY	R. Hogar, Panamá
680	1	CUB	BC	R. Progreso, Cienfuegos, CI
	1	CUB	BCDB	R. Progreso, Stgo de Cuba, SC
	10	CUB	BA	R. Rebelde, Ciego de Ávila, CA
	3	DOM	JX	R. Zamba, San Ignacio de Sabaneta
	10	GTM	VP	R. Norte, Cobán
	1	HND	NN7	LV de Honduras, Danlí
	1	HND	NN10	LV de Honduras, Juticalpa
	10	HND	NN8	LV de Honduras, San Pedro Sula
	10	HND	NN2	LV de Honduras, Siguatepeque
	1	HND	NN9	LV de Honduras, Tela
	10	HND	NN11	LV de Honduras, Tocoa
	5	MEX	OAX	Aro AM, Oaxaca
	1/0.25	MEX	FO	Éxtasis Digital, Chihuahua
	.5/1	MEX	PY	Foro 6-80, Mérida
	5/2.5	MEX	CHG	Ke Buena, Chilpancingo:
	1/0.1	MEX	FJ	La Consentida/R. Fórmula, Teziutlán
	1/0.5	MEX	ORO	La Mera Jefa, Guasave
	1	MEX	SON	La Mexicana, Hermosillo
	5/3	MEX	KQ	La Mexicana, Tapachula
	10/3	MEX	LG	LG, La Grande, León
	10/2	NCG	AM	R. La Primerísima, Managua
	5	PNR	F32	Mujer AM, David
	5	PNR		Voz Sin Fronteras, Metetí
	0.4	PTR	WAPA	Arecibo
	10	PTR	WAPA	Cadena WAPA, San Juan
690	50	AIA		Caribbean Beacon, The Valley
	20	CUB	BC	R. Progreso, Jovellanos, MA
	10	DOM	AW	R. Guarachita "La Poderosa", Sto Domingo
	1	GTM	VB	R. Tamazulapa, Jutiapa
	1	HND	NN3	LV de Honduras, Choluteca
	2.5	MEX	AFA	Ke Buena, Coatzacoalcos
	100/5	MEX	N	La 69, México
	10/1	MEX	RG	La Deportiva 6-90/La R-G, Monterrey
	2/0.25	MEX	ST	La Invasora, Mazatlán
	2.5	MEX	XL	La Ley, Pátzcuaro
	5/1	MEX	CS	La Mejor, Manzanillo
	50/2	MEX	MA	M-A/La Madre de Todas, Fresnillo
	78/50	MEX	WW	W R. América, LV del Pueblo, Tijuana
	10/5	NCG	RH	R. Hermanos, Matagalpa
	5	PNR		R. Evangelio Vivo, Panamá
	10	PNR	R43	R. Veraguas, Santiago
700	10	CTR	JC	FCNradio.com, San José
	1	CUB	BQ	R. Enciclopedia, Guantánamo, GU
		CUB	BA	R. Progreso, Baracoa
	1	CUB	BA	R. Rebelde, Sancti Spíritus, SS
	0.6	DOM	DC	R. Mao, Mao, Valverde
	1	GTM	AJ	R. Inspiración, Escuintla
	15	GTM	HR	R. Mundial, Guatemala
		HND		LV de Honduras, Olanchito
	5	HND	KL	R. Reloj, Tegucigalpa
		JMC		R. Jamaica, Kingston
	2.5/0.1	MEX	VC	Ke Buena, Córdoba
	5	MEX	LX	La Ke Buena, Zitácuaro
	5/0.25	MEX	GD	La Poderosa, Hidalgo del Parral
	5d	MEX	ETCH	LV de los Tres Ríos, Etchojoa
	5	MEX	XPUJ	LV del Corazón de la Selva, X'pujil
	1	MEX	DKR	R. Red, Guadalajara
	2.5/0.5	MEX	RV	R. Villa, Villahermosa
	12	SLV		R. Mi Gente, San Miguel
	12	SLV		R. Mi Gente, San Salvador
	10	VCT		R. St Vincent and the Grenadines, Kingstown
710	30	CUB	BA	R. Rebelde, Camagüey, CM
	10	CUB	BA	R. Rebelde, Holguín, HO
	150	CUB	BA	R. Rebelde, La Julia, HA
	50	CUB	BA	R. Rebelde, Santa Clara, VC
		DOM	WP	Onda del Caribe, San Cristóbal
	1	GTM	XL	R. Tecún Umán, Quetzaltenango
	1	HND	NN13	LV de Honduras, Yoro
	3	HND	RH	LV de Occidente, Sta Rosa de Copán
	2.5	HND	KN	LV de Olancho, Catacamas
	2	HND	LK	R. Comayagua/LV Católica, Comayagua
	1	HND	UP3	R. Rock 'n Pop, San Pedro Sula
	1	MEX	MAR	Amor, Acapulco
	1	MEX	OLA	Huasteca, Tampico
	10	MEX	MP	Interferencia 7 Diez, Mexico

kHz	kW	Ctry	Call	Station, location
	5/0.25	MEX	BL	La Ke Buena, Culiacán
	5/0.5	MEX	RPO	La Ley 710, Oaxaca
	7/0.1	MEX	DP	La Ranchera de Cuauhtémoc, Cd.Cuauhtémoc
	5/0.25	MEX	LZ	La Reina, Torreón
	1	MEX	RL	La R-L de Colima, Colima
	1/0.25	MEX	PS	La Super Grupera, Guaymas
	5/0.25	MEX	YK	La Z, Mérida
	1/0.25	MEX	SMR	R. Fórmula, San Luis Potosí
	1	MEX	RK	R. Korita, Tepic
	.5/1	MEX	ON	R. Mexicana, Tuxtla Gutiérrez
	10	PNR	Q51	KW Continente, Panamá
	5	PNR	B52	Ondas del Caribe, Bocas del Toro
	10/0.75	PTR	WKJB	KJB"R. Isla", Mayagüez
720	1	CUB	BA	R. Rebelde, Cienfuegos, CI
	5	DOM	EF	R. Cayacoa, Higüey
	1.8	DOM	AQ	R. Norte, Santiago
	1	GTM	RO	R. Corona, Morales
	1	HND	NN3	R. Caribe, La Ceiba
		HND	ZN	R. San Lorenzo, San Lorenzo
	1	HTI		R. Lumière, Petite Riv.
		JMC		R. Jamaica, Kingston
	2d	MEX	CPQ	La Estrella Maya Que Habla, Felipe Carillo Puerto
	8/0.25	MEX	DE	La Kaliente, Saltillo
	1/0.5	MEX	VU	Magia, Mazatlán
	10/0.25	MEX	AVR	R.Fórmula, Primera Cad. Nal, Veracruz
	1/0.25	MEX	QZ	Ritmo 720/La Máquina Musical, San Juan de los Lagos
	25	NCG	A3RC	R. Católica, Managua
	10	PNR	B50	R. República, Chitré
	1	SLV	RA	Qué Buena, San Salvador
730	1	CTR		R. Pacífico, Puntarenas
	20	CTR	HB	Sin Fronteras, Desamparados
	10	CUB	BC	R. Progreso, Nueva Gerona, IJ
	10	DOM	Z	R. HIZ, Broadc. Nac., Sto Domingo
	10	GTM	N	R. Cultural, Guatemala
	0.25	HND	XG	R. Cadena Dial, Sta Bárbara
	1	HND	NN4	R. Exitos, Tegucigalpa
	100	MEX	X	Estadio W, México
	1/0.25	MEX	EBC	Ke Buena, Ensenada
	5/0.1	MEX	PQ	La 73/La Sabrosita, Cd.Muzquiz
	5/1	MEX	GDL	La Explosiva, Guadalajara
	10d	MEX	PET	LV de los Mayas, Peto
	10/1	MEX	LBC	R. La Giganta 730 AM, Loreto
	10	MEX	SOS	R. Uno, Agua Prieta
	10/5	MEX	VF	R. Villaflores, Villaflores
	50/1	MEX	HB	R. Viva Villa, Hidalgo del Parral
	3	PNR		Asamblea Nacional, Fort Sherman, Colón
	20	TRD		Inspirational 7-30 AM, TBC, Port of Spain
740	10	CUB	KO	R. Angulo, Sagua de Tanamo, HO
	20	CUB	BC	R. Progreso, Camagüey, CM
	1	HND	IH	7-40 La Super Grande, Juticalpa
		HND	VC	LV Evangélica, Olanchito
	1	HND	QQ	R. Intibuca, La Esperanza
	1	HND	TG2	R. Satélite, San Pedro Sula.
	1	HTI		R. Lumière, Pignon
	1	MEX	VAY	Amor, Puerto Vallarta
	10/1	MEX	KV	Exa FM, Villahermosa
	1	MEX	LTZ	Globo 740/R. Fórmula , Aguascalientes
	5/1	MEX	POR	La Explosiva/R. Fórmula, Putla de Guerrero
	2/0.25	MEX	GF	R. Fiesta, Gutiérrez Zamora
	10/1	MEX	QN	R. Fórmula 1ª Cadena, Torreón
	20/10	MEX	CAQ	R. Fórmula QR Cancún, Cancún:
	10/1	MEX	CW	R. Variedades, Los Mochis
	5/1	MEX	OF	Romántica, Celaya
	50	NCG	A3LS	R. Sandino, Managua
	2.6	PNR	R44	La Exitosa de Chorrera, La Chorrera
	5	PNR	N26	R. Cristal, David
	0.5/0.1	PTR	WIAC	Boricua 740, Ponce
	10	PTR	WIAC	Boricua 740, San Juan
750	10	CUB	BC	R. Progreso, Palmira, CI
	1	CUB	BC	R. Progreso, Trinidad, SS
	5	DOM	DB	R. Jesús es el Señor, Santiago
	1/0.1	MEX	RASA	Candela 750/Candela Pasión Grupera, San Luis Potosí
	10/1	MEX	URM	Fiesta Mexicana, Uruapán
	10/0.1	MEX	CORO	Ke Buena, Loma Bonita

kHz	kW	Ctry	Call	Station, location
	1/0.25	MEX	MG	La Ke Buena, Arriaga
	1/0.75	MEX	OH	La Pantera, Camargo
	5/0.25	MEX	KOK	La Poderosa, Acapulco
	10d	MEX	JMN	LV de los Cuatro Pueblos, Jesús María
	10/0.25	MEX	TI	R.Fiesta, La Más Picuda, Tempoal
	1/0.25	MEX	CSI	Vida 750, Culiacán
	5	PNR		R. La Inolvidable, Chitré
760	5	CTR	LX	R. Columbia, San José
	1	CUB	BD	R. Reloj, La Habana, HA
	10	CUB	BD	R. Reloj, Las Mercedes, GR
	5	DOM	CO	R. Cordillera, Sto Domingo
	5	GTM	HB	Nueva R. Super, Guatemala
	2.5	HND	XW	R. Comayagüela/Stereo Azul, Comayagüela
		HND	IJ	R. Jicatuyo, San José de Colinas
	2	HTI		R. Lumière, Cayes
	70/10	MEX	ABC	ABC R., México
	1/0.5	MEX	ES	Antena Musical 7-60, Chihuahua
	5/5	MEX	DGO	La Mejor, Durango
	2.5/0.5	MEX	YW	Mexicanísima, Mérida
	5/1	MEX	EB	R. Fiesta, Cd.Obregón
	5/1	MEX	ZZ	R. Gallito , Guadalajara
	5/0.1	MEX	NY	R. Geny, Nogales
	5/0.5	MEX	RA	R. Uno, San Cristóbal las Casas
	3	PNR		Asamblea Nacional, Bocas del Toro
	5	PNR	XO	LV del Istmo, Panamá
	5	PTR	WORA	NotiUno, Mayagüez
	5	SLV	KL	YSKL La Poderosa, San Miguel
	1	SLV	KL	YSKL La Poderosa, Sonsonate
		SLV	KL	YSKL La Poderosa, Zacateluca
770		CUB	BA	R. Rebelde
	5	DOM	MD	R. Activa, Santiago
	1	GTM	BX	R. Nueva Fraternidad, Quetzaltenango
	0.5	HND	MV	R. Aguán, Olanchito
	1	HND	RD	R. Majestad "LV del Guayape", Juticalpa
	10	HND	NN21	R. Norte, San Pedro Sula
		HND	PI	R. Sui Generis, Comayagua
		JMC		R. Jamaica, Kingston
	1d	MEX	MRO	Aro-AM, Matias Romero
	1d	MEX	HUA	Aro-AM, Sta Cruz Huatulco
	5/1.5	MEX	ML	La Ranchera, Apatzingán
	10/1	MEX	IH	La Unica, Fresnillo
	1/0.1	MEX	REV	Los 40 Principales, Los Mochis
	10	MEX	ANT	LV de las Huastecas, Tancanhuitz de los Santos
	2.5/1	MEX	ACH	R. Fórmula, Primera Cadena, Monterrey
	5/1	MEX	SUR	Tu Ritmo Musical, Chilapa
	5/0.5	MEX	QRV	Ultra 770, La R., Veracruz
	10	PNR	L83	R. Nacional Herrera, Chitré
	10	SLV	KL	YSKL La Poderosa, San Salvador
780	10	CTR	RA	R. América, San José
		CUB	BA	R. Rebelde
	0.5	DOM	BO	R. Constanza, Constanza
	1	GTM	CK	Sultana La Cristiana, Zacapa
	1	HND	SE	Alabanza Estéreo, Choluteca
		HND	ON	R. Sonora, La Ceiba
	10	HTI		Eben-Ezer, Mirebalais
	0.5	HTI		R. Lumière, Jérémie
	5/1	MEX	ZN	EXA FM, Celaya
	10/0.25	MEX	WGR	Exa FM, Monclova
	5/1	MEX	TS	La Máquina Musical, Tapachula
	5/1	MEX	SFT	La Triple T/La Caliente, San Fernando
	10d	MEX	GLO	LV de la Sierra Juárez, Guelato de Juárez:
	5/1	MEX	XY	LV del Balsas, Cd.Altamirano
	5/0.5	MEX	LD	R. Costa, Autlán
	2.3/0.25	MEX	MTS	R.Fórmula, Tampico
	10	PNR	B55	R. Chiriquí, David
	5	PNR		R. Recuerdo, Panamá
	1	SLV	KL	YSKL La Poderosa, Sta Ana
	1	SLV	KL	YSKL La Poderosa, Usulután
	10	VRG		ZBVI, Tortola
790	30	MEX	BD	R. Rubí, Pinar del Río, PR
	5	DOM	L	R. Centro, Sto Domingo
	3	GTM	O	R. Festival, Guatemala
	1	HND	FI	R. Feliz, Sta Bárbara
	3	HND	TG	R. Satélite, Tegucigalpa
	2.5/1	MEX	UP	Candela, Tizimín
	50/1	MEX	RC	Formato 21, México
	1/0.5	MEX	FE	Mi R. 790 - La Fiesta, Nuevo Laredo

kHz	kW	Ctry	Call	Station, location
	1/0.25	MEX	SU	R. 790/La Dinámica, Mexicali
	10/5	MEX	BI	R. B-I, La Estación que da las Notícias, Aguascalientes
	0.25d	MEX	GAJ	R. Fórmula, Primera Cadena, Guadalajara
	5/0.75	MEX	NT	R. La Paz/R. Fórmula, La Paz
	1/0.5	MEX	COV	R. Lobo, Poza Rica
	5/0.4	MEX	RPC	R. Ranchito, Chihuahua
	25/5	MEX	VA	R. Tabasco, La Em. del Hogar, Villahermosa
	1	MEX	GZ	W R., Torreón
	6	PNR		R. Panamá, Santiago
800	100	ATN	PJB	TWR, Kralendijk, Bonaire
	5	CTR	SD	R. La Gigante, San José
	1	CUB	BC	R. Progreso, Manzanillo, GR
	1	DOM	VM	R. Bonao, Bonao
	1	GTM	YZ	R. Rosa, Chiquimulilla
	1	HND	DL	R. Corporación, Comayagua
	3	HND	MA	R. Moderna, San Pedro Sula
		HND	GW	R. Patria, Catacamas
	1	HND	QN	R. Sonora, Danlí
		HND	MD	R. Yoro, Jocon
	0.5/0.25	MEX	SPN	ESPN800, Tijuana
	5/1	MEX	GX	Fiesta Mexicana , San Luis de la Paz
	1	MEX	QT	La Poderosa, Veracruz
	2/0.25	MEX	ZR	La Traviesa de Coahuila, Zaragoza
	10/2.5	MEX	DR	La Tremenda, Montemorelos
	5d	MEX	ZV	LV de la Montaña, Tlapa de Comonfort
	1/0.1	MEX	AN	R. Alegría, Ocotlán
	150	MEX	ROK	R. Cañon, Cd.Juárez
	5/1	MEX	UI	R. Comitán, Comitán
	10	NCG	A3RO	R. 800, Managua
	3	PNR		Tropical 800, Los Santos
	12	SLV	AX	R. María El Salvador, San Salvador
810	1	BAH		BCB, ZNS3, Nassau
	10	CUB	BC	R. Progreso, Guantánamo, GU
	5	DOM	AV	R. Salvación, Baní
		GTM		R. Circuito San Juan, San Juan
		GTM	END	R. Constelación, San Marcos
		GTM		R. Moapán, Sta Elena
	6	HND	VC	LV Evangélica, La Ceiba
	3	HND	LP24	R. Valle, Choluteca
	0.05	HTI		R. Atlantique, Gonaives
	1/0.5	MEX	IM	Fiesta Mexicana, Saltillo
	10/0.25	MEX	UX	LaLegendaria, Tepic
	50/1	MEX	FW	R. Estrella, Tampico
	5/1	MEX	ZC	R. Felicidad, Río Grande
	7/0.6	MEX	AGR	R. Fórmula, Primera Cadena, Acapulco
	5/1	MEX	HT	R. Huamantla, Huamantla
	0.1	MEX	IC	R. I-C, Campeche
	1/0.5	MEX	EMM	R. La Salmantina, Salamanca
	1d	MEX	SB	R. Mexicana/La S B, Santa Bárbara
	1/0.1	MEX	RI	R. Rey, Reynosa
	3/0.25	MEX	MAX	Radiomax, Tecomán
	.5/1.5	MEX	OE	Romántica, Tapachula
	2.5/0.25	MEX	RB	SolEstéreo, Cozumel
	5/0.25	MEX	RSV	Tribuna R., Cd.Obregón
	2/0.25	MEX	MQ	Yóol lik/R. Mayub, Mérida
	1	PNR	G	R. 10, Panamá
	50	PTR	WKVM	R. Paz 810 AM, San Juan
	1.5	SLV	DA	R. Imperial, Sonsonate
	2	SLV	FA	R. Lorenzana, San Vicente
820	2.5	CTR	GC	R. Centro AM, San José
	10	CUB	BE	R. Ciudad de la Habana, Santa Catalina, CH
	10	CUB	BC	R. Progreso, Ciego de Avila, CA
		CUB	BD	R. Reloj, Contramaestre
	2.5	DOM	AZ	R. Santiago, Santiago
	10	GTM	TO	R. Kyrios, Guatemala
	5	HND	LP16	R. Moderna, Tegucigalpa
	7/3	HND	KW	R. Sultana, Sta Rosa de Copán
	2.5/0.1	MEX	KG	Golden Hits/ABC R., Córdoba
	10/1	MEX	BA	La Consentida, Guadalajara
	10/1	MEX	BM	La Mera Mera, San Luis Potosí
	0.75d	MEX	ESC	R. Escárcega, Escárcega
	3.5/0.5	MEX	ABCA	R. Frontera, Mexicali
	1/0.25	MEX	UDO	R. Universidad de Occidente, Los Mochis
	1/0.5	MEX	YN	Romántica 8-20, Oaxaca
	1d	MEX	GRC	Soy Guerrero, Coyuca de Catalán
	10/0.5	MEX	DRD	W R., Durango

kHz	kW	Ctry	Call	Station, location
	20	NCG	FAOL	R. Ondas de Luz, Managua
	3	PNR	F28	R. Ritmo Chiriquí, David
	50	SCN		R. Paradise, Nevis
830	5	CUB	BD	R. Reloj, Holguín, HO
	10	DOM	JB	R. HIJB, Sto Domingo
	5	GTM	AV	R. Satélite, Mazatenango
	1	HND	JB	Cadena Radial Impacto, Comayagua
		HND	TB	R. Colón, Tocoa
	1	HND	VQ	R. Excelsior, Juticalpa
	1	HND	RU	R. Uno, San Pedro Sula
	5	MEX	DR	Digital 99, Guaymas
	3/0.25	MEX	LN	La Caliente 830 AM, Linares
	5	MEX	IK	La Norteñita 8-30AM/R. Fórmula, Piedras Negras
	6	MEX	TLX	La Poderosa/R. Tlaxiaco, Tlaxiaco
	5/1	MEX	VQ	La Superestación, La Grande de Sinaloa, Culiacán
	8d	MEX	PUR	LV de los P'urhepechas, Cheran
	1	MEX	DQ	R. Alegría/ LV Amiga de los Tuxtlas, San Andrés Tuxtla
	10/5	MEX	ITE	R. Capital, México
	5/1	MEX	ZQ	R. Futurama, Villahermosa
	10/0.5	MEX	LK	R. Mexicana, Zacatecas
	5	PNR	R56	R. Península, Macaracas
	5	SLV	PX	R. Pax, San Miguel
840	10	CUB	HW	Doblevé, Santa Clara, VC
	1	CUB	BQ	R. Enciclopedia, La Fé, IJ
	1	CUB	BC	R. Progreso, Las Tunas, LT
	1	CUB	KC	R. Revolución, Stgo de Cuba, SC
	1	DOM	AB	R. Isabel de Torres, Puerto Plata
		GTM		R. Idea 840, Jutiapa
	2.5	GTM		R. Luz, San Pedro Carchá
	1	HND	CR	Dif. Cristiana de R. "DCR", Choluteca
	10	HTI		R. 4VEH, Cap Haitien
	5/1	MEX	XXX	Fiesta Mexicana/Fiesta Digital, Tamazula
	1/0.1	MEX	FJ	La Consentida, Teziutlán
	2.5/0.1	MEX	PV	La Fiera Grupera, Papantla
	1d	MEX	MY	La Jefa, Cd.Mante
	10/2.5	MEX	IO	La Más Picuda, Tuxtla Gutiérrez
	5/1	MEX	FG	La Pachanga, Celaya
	1/0.25	MEX	TEY	R. Sensación, Tepic
	5	NCG	A3NT	R. Noticias, Managua
	10	PNR	L80	R. Nacional, Panamá
	5/1	PTR	WXEW	NotiUno/R. Victoria, Yabucoa
	10	SLV	FB	R. San Biblia, San Salvador
850	2	CTR	RDR	R. Cartago, Cartago
	1	CUB	BC	R. Progreso, Trinidad, SS
	1	CUB	BD	R. Reloj, Nueva Gerona, IJ
	5	DOM	UA	R. Clarín, Santiago
	5	DOM	GA	R. Guarocuya, Barahona
	10	GTM	X	R. Ciro, Guatemala
	0.5	HND	IF	R. Inspiración, La Entrada
	10	HND	UP	R. Televisión, Tegucigalpa
	3/1	MEX	MIA	850 Notícias, Información Que Sirve, Guadalajara
	10/1	MEX	TQ	La Q Orizabeña/R. Fórmula, Orizaba
	1	MEX	ZF	La Rancherita Contenta, Mexicali
	5/0.5	MEX	RTM	La Zeta, Macuspana
	1d	MEX	ZI	Maxistar, Zacapu
	1	MEX	JAQ	R. Felicidad, Jalpan
	1/0.2	MEX	US	R. Universidad de Sonora, Hermosillo
	5/1	MEX	M	Renacimiento 850 , Chihuahua
	5	PNR	T61	La Exitosa de Chiriquí, David
	1	PNR		La Exitosa de Colón, Colón
	5/1	PTR	WABA	Waba "La Grande", Aguadilla
860	10	ATN	PJZ-86	R. Curom, Willemstad, Curaçao
	10	CUB	BA	R. Rebelde, Arroyo Arenas, CH
	1	CUB	BD	R. Reloj, Baracoa, GU
	5	CUB	BD	R. Reloj, Colón, MA
	10	DOM	UA	R. Clarín, Sto Domingo
		HND	NZ	La Respuesta es la Cruz, Olanchito
	0.5	HND	LS	R. Dinorama, La Paz
		HND	BV	R. Piedra Blanca-LV de Nuestra Gente, Catacamas
	10	HND	BS	R. San Pedro, San Pedro Sula
	3	HTI		R. Men Kontre, Cayes
	5/0.5	MEX	RRF	860 AM, Mérida
	5/0.25	MEX	DB	Canal 86, Tonalá

kHz	kW	Ctry	Call	Station, location
	1/0.5	MEX	DU	D-U la que le gusta a Usted, Durango
	2.5	MEX	PLA	La Mexicana, Aguascalientes
	5/0.25	MEX	HX	La Mia, Ciudad Obregón
	10/7.5	MEX	MO	La Poderosa 860, Tijuana
	1/0.15	MEX	ZX	LV de Usumacinta, Tenosique
	1/0.25	MEX	NW	Máxima 103.3, Culiacán
	1	MEX	IW	R. 860, Uruapan
	5	MEX	CCN	R. Caribe, Cancún
	5/1	MEX	CTL	R. Chetumal, Chetumal
	1/0.25	MEX	TW	R. Fiesta, Tampico
	5/0.1	MEX	AL	R. Mundo/R. Fórmula, Manzanillo
	1/0.5	MEX	ZOL	R. Noticias 860, Cd.Juárez
	5/2	MEX	NL	R. Recuerdo, Monterrey
	4.5/10	MEX	UN	R. UNAM, México
	10	PNR	L55	R. Reforma, Chitré
	10	SCN		Voice of Nevis, Nevis
	1	SLV	RC	R. Tecana, Sta Ana
870	10	CTR	UCR	R. 870 UCR, San Pedro Montes de Oca
	1	CUB	BD	R. Reloj, Sancti Spíritus, SS
	4	DOM	VG	R. La Vega, La Vega
	0.5	GTM	L	R. Victoria, Mazatenango
	5	HND	H9	R. Nacional de Honduras, La Ceiba
	3	HND	H4	R. Nacional de Honduras, Nacaome
	5	HND	H10	R. Nacional de Honduras, Puerto Lempira
	1	HTI		R. Express, Jacmel
	1/0.5	MEX	AMO	AMO 870, Irapuato
	10d	MEX	TAR	LV de la Sierra Tarahumara, Guachochi
	1/0.1	MEX	LY	R. Fórmula, Morelia
	5/0.25	MEX	ACC	R. Fórmula/LV del Puerto, Puerto Escondido
	0.5	MEX	NG	R. Huauchinango, Huauchinango
	1/0.25	MEX	FIL	R. Notícias, Mazatlán
	1	MEX	GRO	Soy Guerrero, Chilpancingo
	10	NCG	CD	R. Centro, Juigalpa
	5.5	PNR	HO	R. Libre, Panamá
	5	PTR	WQBS	La Gran Cadena QBS, San Juan
	10	SLV	AR	R. Renacer, San Salvador
880	12.5	CUB	BC	R. Progreso, Pinar del Río, PR
	10	GTM	J	R. Nuevo Mundo, Guatemala
	5	HND	H5	R. Nal de Honduras, Sta Rosa de Copán
	10	HND	H	R. Nacional de Honduras, Tegucigalpa
	0.3	HTI		R. Independance, Gonaïves
	10/1	MEX	TC	880 AM, Torreón
	10/2	MEX	PNK	Canal 88/Superestación, Los Mochis
	10/0.5	MEX	QQQ	Ke Buena, Villahermosa
	10/1	MEX	YV	La Invasora, Córdoba
	5/1	MEX	EM	La M Mexicana, Río Verde
	2.5/1	MEX	IG	Los 40 Principales, Iguala
	20/1	MEX	AAA	R. 880/La Triple A, Guadalajara
	5/0.25	MEX	V	R. Fórmula, Primera Cad Nal, Chihuahua:
	10	NCG	A3EP	R. El Pensamiento, Managua
	2.5	PNR		R. Panamá, Bocas del Toro
	2.5	PNR		R. Panamá, Chiriquí
	1	PNR	B51	R. Visión Panamá, Colón
	1/0.5	PTR	WYKO	La Poderosa 880, Sabana Grande
	1	SLV	CD	R. Ritmo, Stgo de María
890	10	CTR	BAS	R. Heredia, Heredia
	25	CUB	BC	R. Progreso, Chambas, CA
	1	CUB	BC	R. Progreso, Santa Clara, VC
	3	DOM	OR	La Consentida, Mao, Valverde
	4/5	DOM	PJ	R. Continental, Sto Domingo
	1	GTM	HU	R. Escuintla, Escuintla
	1	HND	H2	R. Nacional de Honduras, Comayagua
	1	HND	H	R. Nacional de Honduras, Danlí
	1	HND	H6	R. Nacional de Honduras, El Paraíso
	3	HND	H7	R. Nacional de Honduras, Juticalpa
	5	HND	H8	R. Nacional de Honduras, Olanchito
	10	HND	H9	R. Nacional de Honduras, Siguatepeque
	10	HND	H3	R. Nal de Honduras, San Pedro Sula
	0.5	HTI		R. Trans Artibonite, Gonaïves
	1	HTI		Voix du Nord'est, Forte Liberte
	10/0.5	MEX	NZ	La Sinaloense, Culiacán
	5/0.5	MEX	AK	R. Consentida, Acámbaro
	1/0.25	MEX	BY	R. Fórmula, Tuxpán
	10/1	MEX	FRT	R. Frontera, Comitán
	1/0.25	MEX	PNA	R. Joya/R. Fórmula, Tepic
	5/1	MEX	PC	Sonido Estrella, Zacatecas
	5	PNR	Q62	R. Ritmo Stereo, Chitré.

kHz	kW	Ctry	Call	Station, location
	0.25	PTR	WFAB	La Nave 890, Ceiba
	3	SLV	LA	R. Renacimiento, Sta Ana
900	5	BRB		CBC R. 900 AM, St.Michael
	25	CUB	BC	R. Progreso, Cacocum, HO
		DOM	FK	R. Amanecer, Neiba
	5/1	DOM	EN	R. Puerto Plata, Puerto Plata
	1	GTM	MA	R. Amatique, Puerto Barrios
	1	HND	UP	R. Centro, Choluteca
	1	HND	UP6	R. Satélite, La Ceiba
	1	MEX	ED	La Líder 900 AM, Arneca
	5/1	MEX	DT	La Reina, Cd.Cuahtémoc
	10/2.5	MEX	OK	OK Notícias/R. Tráfico, Monterrey
	1/0.75	MEX	TAK	Radiorama Siglo XXI, Tapachula
	250	MEX	W	W R., México
	50/10	MEX	WB	W R., Veracruz
	5	NCG	A3RT	R. Tiempo, Managua
	10	PNR	HA	R. Panamá, Panamá
	2	SLV	QJ	R. Cristo Te Llama, San Salvador
910	5	CTR	UM	BBN, San José
	25	CUB	HA	R. Cadena Agramonte, Camagüey, CM
	10	CUB	BL	R. Metropolitana, La Lisa, CH
	5	CUB	BD	R. Reloj, Bolondrón, MA
	3	DOM	L81	R. 91 "La Grande", Bonao
	10	GTM	KL	R. Fe y Esperanza, Guatemala
	10	HND	VS	R. Católica "LV de Suyapa", Tegucigalpa
	2.5	HND	NM	R. Comunidad, Ocotepeque
	10/1	MEX	NAY	La Poderosa, Puerto Vallarta
	5/1	MEX	ACM	R. Exitos, Cárdenas
	5/0.1	MEX	ACN	R. Fórmula León/R. Uno, León:
	10/2.5	MEX	OL	R. Impacto, Teziutlán
	0.25	MEX	AO	R. Mexicana, Mexicali
	5	NCG		R. Jinotega, Jinotega
	3	PNR	L85	R. Nacional, Colón
	10	PNR		R. Nacional, Darién
	10	PNR	L81	R. Nacional, David
	4.4	PTR	WPRP	NotiUno, Ponce
920	1	CUB	BC	R. Progreso, Pilón, GR
	1	CUB	BD	R. Reloj, Moa, HO
	10	DOM	BA	R. 9-20 AM-Stereo "Power", Sto Domingo
	0.2	GTM	RS	R. Cultural, Escuintla
	5	HND	SK	R. Catacamas, Catacamas
		HND	VS	R. Católica, Tocoa
	1	HND	H11	R. Nacional de Honduras, Danlí
	1	HND	RM	R. Sistema, Comayagua
	1	HND	ZV	Una Voz que clama en el desierto, San Pedro Sula
	5/0.5	MEX	CQ	C-Q/La Ranchera de Culiacán, Culiacán
	5/1	MEX	RE	La Comadre, Puros Éxitos, Celaya:
	1/0.25	MEX	MJ	La Fronteriza del Aire, Piedras Negras
	10/0.5	MEX	VV	La Poderosa/R. Fórmula, Tuxtla Gutiérrez
	10	MEX	LE	La Preferida, Tampico
	1	MEX	ZAR	La Z, Puebla
	5/0.2	MEX	RCA	Planeta, Torreón
	5/1	MEX	HQ	R. Capital, Hermosillo
	1/0.15	MEX	PNX	R. Costa/Ke Buena, Santiago Pinotepa Nal
	5/2.5	MEX	LCM	R. La Mexicana, Cd.Lázaro Cárdenas
	1.5/0.5	MEX	TEB	R. Mar, Campeche
	10	MEX	LT	R. María, Tlaquepaque
	1/0.25	MEX	QD	R. Noticias 920, Chihuahua
	10	NCG	W	R. Mundial, Managua
	5	PNR	S56	R. Mía Centrales, Los Santos
930	5	CTR	RCR	R. Costa Rica, Guadalupe
	1	CUB	BD	R. Reloj, La Jaiba, MA
	1	CUB	BD	R. Reloj, Stgo de Cuba, SC
	25	CUB	IP	R. Surco, Ciego de Ávila, CA
	10	DOM	CK	Ondas del Yaque, Santiago
		HND	CQ	Cadena R. Samaritano, La Ceiba
		HND	LD	R. Estéreo Leed, Nacaome
	5	HTI		R. Cap Haitien, Cap Haitien
	0.1	HTI		R. Echo 2000, Val. de Jacmel
	1	MEX	ZU	La Explosiva, Zacapu
	2.5/0.2	MEX	UL	La Picosita, Mérida
	1/0.25	MEX	SHT	La Poderosa, Saltillo
	10/1	MEX	U	La U de Veracruz, Veracruz
	5d	MEX	TLA	LV de la Mixteca, Tlaxiaco
	1	MEX	TTT	Magia 930, Colima
	5/2.5	MEX	MK	M-K R. Mexicana, Huixtla

kHz	kW	Ctry	Call	Station, location
	2/1	MEX	CY	R. Diversión, Huejutla
	10/3	MEX	QS	Romance en R./R. Fórmula, Fresnillo
	10	PNR	R46	La Nueva Exitosa, Panamá
	2	PNR	K85	Mi Preferida Estéreo, Pto Armuelles
	2.5	PTR	WYAC	Boricua, Cabo Rojo
		SLV		R. Rey de Gloria, San Salvador
940	1	CUB	BC	R. Progreso, Sancti Spíritus, SS
	10	CUB	BD	R. Reloj, Central España, MA
	10	CUB	BD	R. Reloj, Holguín, HO
	3	DOM	AS	R. Metro, Montecristi
	5	GTM	TL	Eventos Católicos R., Guatemala
	1	GTM	TL	Eventos Católicos R., Sacatepeque
	1	HND	BO	R. Cadena Occidental, La Entrada
	1	HND	CR	R. Dif. Cristiana de R., Tegucigalpa
	0.25	HTI		R. St Marc, St marc
	0.2	HTI		Rdif. Jacmelienne, Jacmel
	1/0.1	MEX	MMM	940 AM Oldies/R. Fórmula, Mexicali
	1d	MEX	HE	La Melódica, Atotonilco
	1d	MEX	RKS	La Poderosa, Reynosa
	50	MEX	Q	La Q 9-40/Bésame 9-40, México
	15/0.5	MEX	YJ	Mix 9-40, Sabinas
	10/2.5	MEX	OL	R. Impacto, Teziutlán
	1d	MEX	RLA	R. Santa Rosalía, Santa Rosalía
	1/0.25	MEX	REC	W R., Villahermosa
	5	PNR		Asamblea Nacional, Darien
	10	PTR	WIPR	940 AM, San Juan
950	10	CUB	BD	R. Reloj, La Habana, HA
	1	CUB	BD	R. Reloj, Mayarí Arriba, SC
	10	DOM	IG	R. Popular, Sto Domingo
	1	GTM	AF	R. Indiana, Mazatenango
	1	HND	QL	Centro Radial Hondureño, Siguatepeque
		HND	QJ	R. Agalta, San Esteban
	6	HND		R. Choloma, Choloma
	1.5	HND	ZE	R. Cortés AM, Puerto Cortés
		HND	XI	R. El Camino, Olanchito
	1	MEX	CAA	La 950, Aguascalientes
	10/0.1	MEX	PB	La Grande/R. Amor, Hermosillo
	5/0.5	MEX	MEX	La Mexicana, Cd.Guzmán
	3/0.9	MEX	MAB	La Poderosa, Cd.del Carmen
	1/0.5	MEX	FA	La Poderosa, Chihuahua
	.5/1	MEX	ZE	La Poderosa, Santiago Ixcuintla
	10d	MEX	OJN	LV de la Chinantla, San Lúcas Ojitlán
	5/0.5	MEX	ORF	R. Exitos, Los Mochis
	20/5	MEX	KAM	R. Fórmula Californias, Tijuana
	5/1	MEX	ACA	R. Fórmula, Segunda Cadena, Acapulco
	10/1	MEX	CEL	R. Lobo Bajío, Celaya
	5/1	MEX	RN	R. Naranjera, Monterrey
	1/0.25	MEX	TUG	Radiorama Siglo XXI, Tuxtla Gutiérrez
	5/2	MEX	TO	Romántica, Tampico
	3	PNR	L84	R. Nacional, Penonomé
	2.5	PNR		R. Panamá, Las Mercedes, Colón
	1	SLV	HG	R. Cristo Te Llama, San Miguel
960	5	CTR	SD	R. Actual 960, San José
	1	CUB	BQ	R. Enciclopedia, Matanzas, MA
	0.25	CUB	BF	R. Musical, Ciego de Avila, CA
	1	CUB	BD	R. Reloj, Cienfuegos, CI
	10	CUB	BD	R. Reloj, Guantánamo, GU
	5/1	DOM	FF	LV del Atlántico, Puerto Plata
		HND	XB	R. Bautista Buenas Nuevas, Puerto Lempira
	1	HND	YF	R. Fergusón, Choluteca
	1	MEX	MM	960 Notícias, Morelia
	1	MEX	XC	ABC R. 960, Taxco
	1	MEX	CZ	ABC R., San Luis Potosí
	1/0.25	MEX	OZ	Amor, Jalapa
	5/1	MEX	TAP	Imperio/La Poderosa, Tapachula
	5/0.5	MEX	ROO	La Guadalupana, Chetumal
	5/1	MEX	K	La R. 9-60/La Estación Grande, Laredo
	10/2.5	MEX	HK	LV de Guadalajara, Guadalajara
	10/1	MEX	FAMA	R. Fama, Cd.Camargo
	1/0.5	MEX	GB	R. Fiesta, Coatzacoalcos
	1/0.5	MEX	IQ	R. Norteña, Cd.Obregón
	1/0.5	MEX	UQ	R. Variedades, Zihuatanejo
	0.5/0.1	MEX	KS	XEKS 960/LV del Tiempo, Saltillo
	2.5	NCG	ACTH	LV del Trópico Húmedo, San Carlos
	1	PNR	M33	CHT Stereo Digital, David
	1	PNR		R. Capital, Panamá
	1/1.7	PTR	WCHQ	La R. Que Te Bendice, Quebradillas
	0.5	SLV	TW	R. Centro, Sonsonate
970	5/1	DOM	CV	R. Barahona, Barahona
	6	DOM	VP	R. Olímpica, La Vega
	5	GTM	AX	R. Continental, Guatemala
		HND	KI	La Picosa, N. Ocotepeque
	2	HND	LY	R. Milenium, Tegucigalpa
	10/5	MEX	VT	970 AM Stereo, Villahermosa
	5/0.5	MEX	MH	Candela FM, Mérida
	1/0.5	MEX	ZAZ	De Mil Amores 9-70, Zacatecas
	10/1	MEX	VOX	Fiesta Mexicana, Mazatlán
	10/5	MEX	J	La J Mexicana, Cd.Juárez
	5/0.25	MEX	EZ	La Mejor, Caborca
	1/0.5	MEX	MF	La Mejor, Monclova
	1	MEX	BJ	R. 9-70, Cd. Victoria
	1/0.25	MEX	CJ	R. Apatzingán, Apatzingán
	50/4	MEX	RFR	R. Fórmula, Primera Cadena, México
	1	MEX	O	R. Gallito, Matamoros
	1/0.5	MEX	SW	R. Madera/La Mera Mera, Cd. Madera
	1	MEX	UG	R. Universidad de Guanajuato, Guanajuato
	3	PNR	S97	Ondas Centrales, Santiago
	5	SLV	MS	R. UTEC, San Salvador
	5/1	VIR	WSTX	WSTX-AM, St Croix
980	10	CTR	RC	R. Alajuela, Alajuela
	1	CUB	KV	La Voz del Níquel, Moa, HO
	5	CUB	B	R. COCO, Sapo, CH
	1	CUB	BD	R. Reloj, Bayamo, GR
	1	GTM	MQ	R. Retama, San Marcos
		HND	VC	LV Evangélica, Siguatepeque
	2	HND	ZC	R. Monumental, San Pedro Sula
	1	HND	AO	R. Tocoa, Tocoa
		HND	UI	Super 10, Catacamas
	5/0.2	MEX	LC	Dual Stereo, La Piedad
	1/0.25	MEX	KE	KE-98, Solo para tí, Navojoa
	1	MEX	JK	La Poderosa, Cd.Delicias
	2.5/0.5	MEX	FQ	LV de la Ciudad del Cobre, Cananea
	5/0.5	MEX	NR	R. 980, Nueva Rosita
	1	MEX	XT	R. Capital/Capital Máxima, Tepic
	1	MEX	FS	R. Matamoros, Código 9-80, Izúcar de Matamoros
	5	MEX	QO	R. Romance, Cosamaloapan
	5	MEX	TU	R. Tampico, Tampico
	1	NCG	A3NO	R. Redención Internac., Managua
990	1	CUB	AM	R. Guamá, San Luís, PR
	1	DOM	SA	R. Cibao HI-SA 9-90, Santiago
	1	GTM	AL	R. Perla de Oriente, Chiquimula
	3.5	HND	PR	R. Paz, Choluteca
	1	HND	OJ	R. Vida Cristiana, La Ceiba
	1	MEX	ATM	A Toda Máquina, Morelia
	1	MEX	IU	Cristal, Oaxaca
	5/0.25	MEX	HZ	HZ La Pura Sabrosura, La Paz
	1/0.1	MEX	BC	La Buena Onda, Cd.Guzmán
	10/1	MEX	TG	La Grande del Sureste, Tuxtla Gutiérrez
	50	MEX	T	La T Grande, Monterrey
	10/2.5	MEX	ID	R. Álamo, Álamo
	10/3	MEX	FP	R. Alegría, Xalpa
	5/0.25	MEX	ER	R. Lobo, Cd.Cuauhtémoc
	1.4/3	MEX	CL	Rockola 990, Mexicali
	20/5	MEX	PI	W R., Chilpancingo
	5	PNR		Asamblea Nacional, Chiriquí
	5	PNR		R. Impacto, Panamá
	0.91	PTR	WPRA	La Primera, Mayagüez
1000	1	CTR	MIL	100.7/Mil FM, San José
		CUB	CH	R. Cadena Habana
	1	CUB	NM	R. Granma, Manzanillo, GR
	1	CUB	BF	R. Musical, Camagüey, CM
	1	CUB	BF	R. Musical, Sancti Spíritus, SS
	5/1	DOM	HG	R. Beller, Dajabón
		GTM		R. Cultural y Educativa, Patzún
		GTM	)	R. Revelación y Verdad, Guatemala
	1	HND	XZ	R. Alfa, Tegucigalpa
	6	HND		R. Río de Piedras, Lempira
	10/1	MEX	TAC	Exa FM, Tapachula
	1	MEX	MMS	Ke Buena, Mazatlán
	1	MEX	FV	La Rancherita, Cd.Juárez
	5/0.25	MEX	MYL	Los 40 Principales, Mérida
	1	MEX	CSV	Máxima FM, Coatzacoales
	1/0.25	MEX	MIL	Planeta Mil, Los Mochis

kHz	kW	Ctry	Call	Station, location
	1/0.1	MEX	NLT	R. Fórmula/Laredo R., Nuevo Laredo
	50/20	MEX	OY	R. Mil, México
	1/0.5	MEX	HPC	R. Mil/R. Fórmula, Hidalgo del Parral
	1/0.5	MEX	RZ	W R., León
	10	NCG	FF	R. Mil, Managua
	10	PNR	K36	R. Poderosa, Aguadulce
	1	SLV	HH	Estación H, Sta Ana
	5/1	VIR	WVWI	KISS 101.3 FM, St Thomas
1010	1	CUB	BF	R. Musical, Holguín, HO
		DOM	JA	R. Comercial, Salcedo
		DOM	JA	R. Comercial, San Juan de la Maguana
	10	DOM	JA	R. Comercial, Sto Domingo
	1	GTM		R. Caribe, Izabal
	1	GTM	XI	R. Ixil, Nebaj
		HND	AE	R. Apaguiz, Danlí
	1	HND	CD	R. Constelación. Juticalpa
		HND	QN	R. Sonora, San Pedro Sula
		HND		R. Visión Cristiana, Tocoa
	2/0.5	MEX	DX	CBC R., Ensenada
	50/5	MEX	HL	Estadio W, Guadalajara
	5/0.5	MEX	FM	La Máquina Tropical, Veracruz
	0.5/0.25	MEX	KD	LaMejor, Cd.Acuña
	5/1	MEX	VK	La Poderosa 10-10 AM, Torreón
	5/0.5	MEX	LO	La X, Chihuahua
	5d	MEX	TUMI	LV Mazahua Otomi/LV de la Sierra Oriente, Tuxpán
	10/1	MEX	PA	Punto 10 R., Cholula
	1d	MEX	HGO	R. Hidalgo, Huejutla
	0.5/0.2	MEX	XN	R. Ures, Ures
	5/1	MEX	WS	Romántica, Culiacán
	5	NCG	FAVP	R. LV del Pinar, Ocotal
	3	PNR	L86	R. Nacional, Bocas del Toro
1020	5	CTR	TIC	LV de la Liberación, San José
	10	CUB	M	Cadena CMKS, Guantánamo, GU
		CUB	CH	R. Cadena Habana
	10	CUB	AM	R. Guamá, Bahía Honda, PR
		CUB	AM	R. Guamá, Los Palacios, PR
	5	CUB	BD	R. Reloj, Jorobo, LT
	10	DOM	TS	R. Enriquillo, Neyba
	5	GTM	CM	R. Frontera, Pajapita
		HND	PN	R. Visión Cristiana Roca de Salvación, Marcovia
	1/0.1	MEX	WO	97.7, Chetumal
	5/0.5	MEX	PR	Los 40 Principales, Poza Rica:
	1	MEX	KH	R. Centro, Querétaro
	1	MEX	PIC	R. Hits, Tepic
	5/1	MEX	OU	Sensación Estéreo, Huajuapan de León
	1	MEX	VE	W R., Colima
	5	PNR		R. Ancón, Panamá
	1/0.28	PTR	WOQI	R. Coquí/La Señal de la Montaña, Adjuntas
	5	SLV	CA	R. Internacional/La Máxima, San Salvador
	20	TCA		Caribbean Christian R., Grand Turk
1030	1	CUB	AM	R. Guamá, La Palma, PR
		CUB	BF	R. Musical
	5	DOM	DL	R. Novedades, Santiago
	10	GTM	UX	R. Panamericana, Guatemala
	1	HND	UP3	R. Rock 'n Pop, Tegucigalpa
	1	HND	RJ	R. Ticante, Ocotepeque
		HTI		R. Ginen, P-au-P
	20/2	MEX	LJ	La Ke Buena, Lagos de Moreno
	1/0.5	MEX	PAV	La Picosita, Tampico
	10	MEX	SSD	La Tremenda, Ensenada
	1/0.25	MEX	BCC	Los 40 Principales, Cd.del Carmen
	10/0.25	MEX	VFS	LVde la Frontera Sur, Las Margaritas
	5d	MEX	NKA	LV del Gran Pueblo, Felipe Carillo Puerto
	50/5	MEX	QR	R. Centro, México
	10/1	MEX	MPM	R. Fama, Los Mochis
	5/0.5	MEX	YC	R. Fórmula, Cd.Juárez
	1/0.5	MEX	TEKA	R. T-K, Juchitán
	5/1	MEX	IE	Stereo 1030 AM, Matehuala
	1/0.5	MEX	VP	W R., Acapulco
	10	PTR	WOSO	Total R. El Oso, San Juan
	1	SLV	RM	R. Frontera, Ahuachapán
1040	5	CTR	AC	R. Fides, San José
	2	CTR	HG	R. Nosara, Hojancha
	1	CUB	LL	R. Victoria, Puerto Padre, LT
	10	DOM	ON	CDN R., Sto Domingo
	1	GTM	JP	R. Oriental, Jalapa

kHz	kW	Ctry	Call	Station, location
	3	HND	NNY	Exitos, San Pedro Sula
		HND	VC	LV Evangélica, Danlí
		HND	VC	LV Evangélica, Juticalpa
	1	HND	MJ	R. Renovación, Comayagua
	2.5/1	MEX	GR	Imagen, Jalapa
	5/0.25	MEX	GYS	La Primera, Guaymas
	5/0.75	MEX	CH	R. Capital, Toluca
	1/0.25	MEX	SAG	R. Lobo, Irapuato
	10/1	MEX	BBB	R. Mujer, Guadalajara
	5/0.5	MEX	PLE	R. Palanque, Palenque
	5/0.25	MEX	HES	Radiorama Siglo XXI, Chihuahua
	2	NCG	VJ	LV de Jinotega, Jinotega
	3	PNR		LV del Mamoni, Panamá
	2.5	PNR	J2	Ondas del Canajagua, Las Tablas
	9/0.25	PTR	WZNA	Zona 1040, Moca
1050	10	CUB	LL	R. Victoria, Las Tunas, LT
	1.5	DOM	CB	R. Hispaniola, Santiago
	5/1	GTM	SL	LV de los Cuchumatanes, Huehuetenango
		HND	OK	R. Roatán, Roatán
	1	MEX	DC	1050 Notícias, Aguascalientes
	15	MEX	ZUM	ABC R., Chilpancingo
	5	MEX	RIO	La Poderosa, Ixtlán del Río
	1/0.5	MEX	IP	La Poderosa, Uruapán
	100	MEX	G	La Ranchera 1050 , Monterrey
	10/5	MEX	TAB	LV de Tabasco, Villahermosa
	10/1	MEX	BCS	R. Cultura Surcalifornia, La Paz
	35/2.5	MEX	QOO	R. Imagen, Cancún
	5d	MEX	JF	R. Max/R. Sensación, Tierra Blanca
	10	MEX	D	Radiorama Siglo 21/W R., Mexicali
	3	NCG	LL	R. Masaya, Masaya
1060	1	CTR	LL	R. Columbia, San Isidro del General
	5	CUB	M	Cadena CMKS, Baracoa, GU
		CUB	DL	R. 26, Matanzas
	1	CUB	L	R. Victoria, Amancio Rodríguez, LT
		DOM	AJ	R. Amanecer, San Pedro de Macorís
	1	DOM	XF	R. Azua, Azua
	10	GTM		R. Favorita, Guatemala
	2	HND	KT	La Catracha, Tegucigalpa
	0.5	HND	FA	R. Peña Blanca, Sta Barbara
	100/20	MEX	EP	R. Educación, México
	1	NCG		LV del Atlántico, Bluefields
	3.5	PNR	J60	R. LV de Panamá, Panamá
	5/0.5	PTR	WCGB	La Roca, Rock R. Netw., Juana Díaz
1070	10	CUB	M	Cadena CMKS, Guantánamo, GU
	1	CUB	AM	R. Guamá, Pinar del Río, PR
	5/1	DOM	BI	HIBI R. 1070, San Francisco de Macorís
	3/2	GTM	D	LV de Occidente, Quetzaltenango
	3	HND	GR	Cadena Guaymuras, El Paraíso
	1	HND	LE	R. Unica AM, San Pedro Sula
		HND	BB	R. Unidad Evangélica, Catacamas
	2.5	HND	QN	R. Sonora, Siguatepeque
	10/1	MEX	SP	10-70 R. Notícias, Guadalajara
	1/0.2	MEX	AGS	Digital 101.3,/Solo Exitos Acapulco:
	1/0.25	MEX	IT	Exa FM, Cd.del Carmen
	2.5	MEX	RPR	Extasis Digital, Tuxtla Gutiérrez
	5/0.25	MEX	EI	Ke Buena, San Luis Potosí
	1/0.1	MEX	MI	La Poderosa, Matehuala
	1/0.25	MEX	OBS	R. Fórmula, Cd.Obregón
	1/0.25	MEX	GY	R. Lobo, Tehuacán
	3	PNR		R. Estéreo Mi Favorita, Penonomé
	10	PNR		R. Nacional, Los Santos
	0.5/2.5	PTR	WMIA	R. Arecibo del Norte, Arecibo
	1	SLV	AN	LV de los Ausoles, Ahuachapán
1080	1	CTR	FC	Faro del Caribe, San José
	10	CUB	CH	R. Cadena Habana, Güines, MB
	1	DOM	MC	R. RPQ Sport, Sto Domingo
	1	GTM	LU	R. Novedad, Zacapa
		HND	IE	R. Evangélica Senda de Vida, Nacaome
	1	HND	ID	R. Miramar, Tela
	20	HTI		R. Nationale, P-au-P
	1/0.5	MEX	UU	La Mejor, Colima
	1/0.5	MEX	CN	Los 40 Principales, Irapuato
	5	MEX	AX	Magía, Oaxaca
	0.5/0.25	MEX	PAB	R.Celebridad, La Paz
	10/0.25	MEX	XK	R. Fórmula, Poza Rica:
	1/0.25	MEX	DY	R. Gallo, San LuisRíoColorado
	5/0.25	MEX	TUL	R. Mexiquense Valle de México, Tultitlán

kHz	kW	Ctry	Call	Station, location
	5d	MEX	JLV	Sistema Jaliscience, Puerto Vallarta
	10	NCG	A3LC	R. 15 de Septiembre, Managua
	5	PNR	J24	R. Mundo Internacional, Panamá
	0.25	PTR	WLEY	R. Isla 1080, Cayey
	6	SLV	ME	R. CRET, San Salvador
1090	1	CUB	KO	R. Angulo, Moa, HO
	1	CUB	CH	R. Cadena Habana, La Salud, CH
	1	CUB	AM	R. Guamá, Santa Lucia, PR
	3	DOM	JM	R. Amistad, Santiago
	1	HND	CQ	Cad. Radial Samaritano, Tegucigalpa
	1	HND	LB	R. La Mejor, Sta Rosa de Copán
	5/1	MEX	LB	La Buenísima, La Barca
	1/0.5	MEX	IL	La Comadre, Veracruz
	10/2.5	MEX	HR	La HR, al Servicio de Puebla/R. Fórmula, Puebla
	1d	MEX	WL	La Romántica, Nuevo Laredo
	5/0.5	MEX	AU	Milenio TV, Monterrey
	10	MEX	MCA	R. 1090, La Grande de las Huastecas, Pánuco
	2.5/1	MEX	XE	R. Grupo Fórmula Querétaro, Querétaro
	10/0.25	MEX	FC	XEFC1090, Mérida
	50	MEX	PRS	XX 1090 AM, Rosarito
	5	NCG	HAAL	R. Alma Latina, Estelí
	10	PNR		R. Nacional, La Peña
	0.25/0.7	PTR	WSOL	LaNueva Sol 1090, San Germán
	3	SLV	MG	R. 1090, Atiquizaya
	1	SLV		R. CRET, Sta Ana
	0.25	VIR	WGOD	3ABN, St Thomas
1100	5	CTR	SCR	R. Chorotega, Santa Cruz
	5	CTR	SBC	R. Guápiles, Guápiles
	1	CUB	KO	R. Angulo, Banes, HO
	1	CUB	CH	R. Cadena Habana, La Habana, CH
	1	DOM	PS	R. Comercial, Nagua
	1	DOM	RB	R. Jimaní, Jimaní
	1	DOM	MP	R. Ocoa, San José de Ocoa
	1	DOM	HD	R. Oriente, San Pedro de Macorís
	1	GTM	SR	R. Superior, Coatepeque
		HND	AJ	R. Antena 5, Catacamas
	1	HND	ND	R. Esperanza, La Esperanza
	1	HND	FQ	R. Máxima, Olanchito
	1	HND	VA	R. Tiempo/R. Fama, San Pedro Sula
	1/0.25	MEX	PO	Imagen, San Luis Potosí
	10/1	MEX	HTY	La Tremenda, Tlapacayan
	5	MEX	BV	R. Alegría, Moroleón
	1	MEX	BAC	R. Asunción/R. Sur California, Bahía Asunción
	5/0.5	MEX	TGO	R. Cañón, Tlaltenango
	4d	MEX	CAN	R. Mundo Maya Turquesa, Cancún
	1d	MEX	GRM	Soy Geurrero, Ometepec
	1/0.5	MEX	NAS	Única 1100 AM, Navojoa
	5	PNR	M92	R. Sabrosa, Panamá
	3	SLV	RF	R. Universidad Don Bosco, San Salvador
1110	10	CUB	KO	R. Angulo, Holguín, HO
		CUB	CH	R. Cadena Habana
	2.5	DOM	TC	R. Jarabacoa, Jarabacoa
	1/0.5	DOM	OS	R. Marién, Dajabón
	1	GTM	MK	R. Verapaz, Cobán
		HND	QN	R. Sonora, Choluteca
	1/0.2	MEX	PVJ	Ke Buena, Puerto Vallarta
	5/1	MEX	LEO	La Rancherita, León
	0.4d	MEX	TEO	La Señal de Oaxaca, Teotitlán de Flores Magon
	0.5d	MEX	TUX	La Señal de Oaxaca, Tuxtepec
	1/0.25	MEX	VS	Maxima 96, Hermosillo
	1d	MEX	OQ	Notigape 11-10/R. Fórmula, Reynosa
	0.25d	MEX	PU	Patronato Cultural Monclova, Monclova
	1/0.5	MEX	WR	R. Guadalupana, Cd.Juárez
	50	MEX	RED	R. Red, México
	1	NCG	F2MT	R. Momotombo, La Paz Centro
	2.5/0.5	PTR	WVJP	R. Caguas, Caguas
	2.5	SLV	CL	R. Horizonte, San Miguel
1120	1	CTR	ACE	R. Miel, Alajuela
	1	CUB	KO	R. Angulo, Mayarí, HO
	5	CUB	CH	R. Cadena Habana, Artemisa, AR
		CUB	BA	R. Rebelde
		DOM		R. Antillas, Barahona
		DOM	CN	R. Metro Hit, Samaná
	10	DOM	CN	R. Metro Hit, Sto Domingo
	0.5	GTM	C	R. Poderosa , Guatemala
	2	HND	TL	R. Fiesta, Tegucigalpa

kHz	kW	Ctry	Call	Station, location
		HND	VR	R. Marchala, Ocotepeque
	1/0.5	MEX	GV	11-20 Notícias, Querétaro
	1/0.1	MEX	POP	Fórmula 11-20 AM, Puebla
	5/0.5	MEX	TQE	La Morena 1230 AM, La Más Choca de Todas, Tenosique
	2/0.25	MEX	ZB	R. Oro/La Tremenda, Oaxaca
	1	MEX	TR	R. Panorámica, Cd.Valles
	1	MEX	RUY	R. Universidad, Mérida
	0.5	MEX	UNO	R. Uno La Popular , Guadalajara
	0.4/0.1	MEX	MX	Sonido 1120, Mexicali
	5	NCG	A3CP	R. CEPAD, Managua
	5	PNR	M21	R. Sonora, Panamá
	2.6/5	PTR	WMSW	R. Once, Hatillo
	3	SLV	LR	Una Voz que Clama en el Desierto, San Salvador
1130	5	CUB	KA	R. Angulo/Ecos del Sagua, Sagua de Tánamo, HO
	1	CUB	BQ	R. Enciclopedia, Santa Clara, VC
	10/1	DOM	RL	CDN R., Santiago
	1	GTM	VR	Em. Unidas LV de la Costa Sur, Retalhuleu
	1	HND	HP	R. Pinares, Siguatepeque
		HND		R. Pirata, Sonaguera
	5	HND	PL	R. Progreso, El Progreso
	1	HND	BT	R. San Francisco, San Francisco de la Paz
	10/5	MEX	TOL	11-30 Notícias, Toluca
	10/1	MEX	ZL	Capital 11-30, Jalapa
	1d	MEX	HN	Ke Buena/Mariachi Estéreo, Nogales
	1/0.25	MEX	MOS	La Invasora, Los Mochis
	10/2.5	MEX	YZ	La Poderosa, Aguascalientes
	1d	MEX	LUP	R. Lupita, Las Varas
	1/0.1	MEX	FN	R. Moderna, Uruapan:
	0.5	NCG		Voz Evangélica de Jalapa, Jalapa
	2.5	PNR	U80	R. Sensación, Aguadulce
	0.2/0.7	PTR	WOIZ	R. Antillas, Guayanilla
	1	SLV	LG	R. Chaparrastique, San Miguel
	1	SLV	AJ	R. Moderna, Sta Ana
1140	5	CTR	DKN	R. Nueva, Guápiles
	1	CUB	NL	R. Bayamo, Media Luna, GR
	5	CUB	CH	R. Cadena Habana, La Habana, CH
	1	CUB	HC	R. Camagüey, Camagüey, CM
		CUB	DP	R. Ciudad Bandera, Cárdenas, MA
	5	CUB	BQ	R. Enciclopedia, Loma de la Cruz, HA
		CUB	BF	R. Musical, Villa Clara, VC
	3	DOM	RA	R. Anacaona, San Juan de la Maguana
		HND	VC	LV Evangélica, Choluteca
	1	HND	UL	R. Pico Bonito1140 AM, La Ceiba
	1	MEX	TE	1140 Punto Digital, Tehuacán
	1d	MEX	PEC	Hidalgo R., San Bartolo Tutotepec
	5/0.5	MEX	LIA	La Tremenda, Morelia
	50	MEX	MR	M-R Deportes, Monterrey
	5/1	MEX	XF	R. Felicidad, León
	1/0.5	MEX	TEC	R. Tecpatán, Tecpatán
	5	PNR	B49	R. Panamericana, Panamá
	10	PTR	WQII	Once Q, San Juan
1150	10	CUB	NL	R. Bayamo, Entronque Bueycito, GR
	5	DOM	AS	Onda Musical, Sto Domingo
	10	GTM	T	R. Sonora, Guatemala
	5	HND	AV	Ondas del Ulúa, Sta Bárbara
		HND	LP12	R. Universal, Tegucigalpa
	50/10	MEX	JP	El Fonógrafo, México
	1.5/0.5	MEX	TVR	La Nueva Azul, Tuxpán
	5/0.3	MEX	SO	La Poderosa, Cd.Obregón
	10/1	MEX	XP	La Super Buena, Uruapan
	1/0.5	MEX	JS	R. Exitos/JS Digital, Hidalgo del Parral
	.5/1	MEX	BF	R. Extremo, San Pedro
	1	MEX	RM	R. Fórmula, Mexicali
	5/1	MEX	XM	R. Jerez, Jerez de García Salinas
	50/1	MEX	AD	R. Metrópoli, Guadalajara
	1/0.5	MEX	QUE	R. Querétaro
	10/0.15	MEX	UAS	R.Universidad/ R. UA Sinaloa, Culiacán
	5	NCG	A2RD	R. Darío, León
	1	SLV	CF	R. María Zona Oriental, San Miguel
1160	10	ATG		Caribbean R. Lighthouse, St. John's
	1	CTR	CA	R. Columbia, Puntarenas
		CUB	NL	R. Bayamo, Pilón, GR
	5	DOM	BE	Radiolandia, Santiago
	1	GTM	RI	R. Izabal, Morales

kHz	kW	Ctry	Call	Station, location
	0.5	HND	GF	R. El Paraíso, El Paraíso
	1	HND	VZ	R. Juan Pablo II, Siguatepeque
		HND	HZ	R. Liberación, Tocoa
		HND	BJ	R. Nueva Palestina, Nueva Palestina
		HND	FJ	R. País "LV del Valle de Sula", Progreso
	2.5	MEX	IW	Canal Stereo Juvenil, Aruapan
	10	MEX	QIN	LV del Valle, San Quintín
	5/0.25	MEX	BE	R. Perote, Perote
	1/0.1	MEX	GI	R. Reyna, La Gigante del Cuadrante, Tamazunchale
	2.5/0.5	MEX	VW	R. Sensación, Acámbaro
	1	NCG	HM	R. Satélite, Estelí
	5	PNR	C20	Ondas Chiricanas, David
	10	PNR	WK	R. Metrópolis, Panamá
	5/2.5	PTR	WBQN	Super Borinquén, Barceloneta-Manatí
	1	SLV	RG	R. Corporación, Sta Ana
1170	10	CUB	M	Cadena CMKS, Maisí, GU
		CUB	BA	R. Rebelde
		DOM	JS	Cadena Espacial, Azua
	5	GTM	RL	R. Cadena Landívar, Quetzaltenango
	2	HND	AF	R. Campeonísima, Choluteca
		HTI		R. Tropicale Internationale, Jérémie
	1/0.5	MEX	MDA	R. Ley 11-70, Monclova
	1	MEX	IB	La Primera, Caborca
	10/2.5	MEX	UVA	La Rancherita, Aguascalientes
	1/0.1	MEX	JTF	Prisma La Poderosa/Prisma Musical, Zacoalco de Torres
	.5/1	MEX	ZS	R. Hit/La Explosiva, Coatzacoalcos
	5/0.1	MEX	FEM	R. Manantial, Hermosillo
	10/2.5	MEX	CD	R. Oro, Puebla
	1/0.25	MEX	RLK	Super Stereo Miled, Atlacomulco
	5d	MEX	RT	Voz 1170/R. Formula, Reynosa
	0.25	PTR	WLEO	R. Leo, Ponce
		SLV	CR	R. Cristo Viene, San Miguel
	0.5	SLV	CB	R. Pentecostés, Sonsonate
1180	5	CTR	PJ	R. Victoria, Heredia
		CUB	MN	La Voz del Toa, Baracoa
	1	CUB	BA	R. Rebelde, Mayarí Arriba, SC
	50	CUB	BA	R. Rebelde, Villa María, CH
		CUB	BD	R. Reloj, Nueva Gerona, IJ
	10	DOM	BE	R. Mil, Sto Domingo
		GTM		R. 10, Guatemala
	1/0.8	HND	VS	R. Congolon, Gracias
	1	HND	AZ	R. La Tigre, Tegucigalpa
	0.5	MEX	AH	Ke Buena, Juchitán
	10/1	MEX	GN	La Gigante, Piedras Negras
	1/0.8	MEX	YA	La Picosa, Irapuato
	10/5	MEX	FR	R. Felicidad, Los Éxitos de Siempre, México
	10d	MEX	UBS	R. Universidad Autonoma de Baja California Sur, La Paz
	5/1.5	MEX	DCH	Romántica 11-80, Cd.Delicias
	10	PNR	U	AM Original, Santiago
	10	PNR		China Visión Panamá, Panamá
	5	SLV	VG	R. VEA, San Salvador
1190	1	CUB	DL	R. 26, La Caridad, MA
		CUB	JD	R. Coral, Guamá, SC
	1	CUB	GL	R. Sancti Spíritus, Trinidad, SS
	10	DOM	AG	Azul 11-90 Bachatisima, Santiago
	1	HND	GK	R. Brassabola, Minas de Oro
		HND		R. Ecológica de Olancho, Catacamas
		HND	ZQ	R. Notícias STC, El Progreso
	0.3	HTI		R. Grand Anse, Jérémie
	10/2.5	MEX	TOT	ABC R., Tampico
	0.25/0.1	MEX	MBC	CBCR., Mexicali
	10/0.1	MEX	CT	Contacto 11-90, Monterrey
	5/0.25	MEX	PP	La Comadre, Puros Exitos
	5	MEX	JPA	La Grande, Cuernavaca
	5/0.1	MEX	PZ	R. Norteña, Cd.Juárez
	5/1	MEX	SOL	R. Sol, la pura ley, Cd.Hidalgo
	2.5/1	MEX	XQ	R. Universidad, San Luis Potosí
	50/10	MEX	WK	W R./W Guadalajara, Guadalajara
	10/5	PTR	WBMJ	WBMJ Rock R. Netw, San Juan
1200	5	CTR	TQ	R. Cucú, San José
	10	CUB	KC	R. Revolución, Palma Soriano, SC
	1	CUB	GL	R. Sancti Spíritus, Sancti Spíritus, SS
	1	DOM	MR	R. Caracol, Azua
		DOM	AH	R. VEN - Voz Evangelica Nal, Sto Domingo
	12	GTM	RJ	R. Unción, Jutiapa
	1	HND	SI	R. Impacto, Tela
		HTI		Voix de la Paix, Port de Paix
	1	MEX	AGA	La Bonita, Aguascalientes
	1/0.25	MEX	YF	R. Fórmula Hermosillo, Hermosillo
	1d	MEX	PAS	R. Punta Abreojos, Punta Abreojos
	5	MEX	QJAL	R. Querétaro, Jalpan
	2.5	MEX	QY	Uno Más Uno R., Toluca
	1/0.25	MEX	WT	W R., Culiacán
	1/0.3	MEX	PW	W R., Poza Rica
		NCG	A3AC	1200 La R., Managua
	0.25/1	PTR	WGDL	La mejor AM, Lares
	1	SLV	KJ	R. Sirama, San Miguel
1210	1	CUB	J	R. Revolución, Chivirico, SC
	1	CUB	J	R. Revolución, Mayarí Arriba, SC
	10	CUB	GL	R. Sancti Spíritus, Sancti Spíritus, SS
	5	DOM	CJ	R. Merengue, San Francisco de Macorís
	10/5	GTM	MX	R. Miel, Guatemala
	1	HND	MY	LV Evangélica, La Entrada
	1	HND	RO	R. Capital, Comayagüela
	5d	MEX	COPA	LV de los Vientos, Copainalá
	5/1	MEX	PUE	Méxicana/R. Fórmula, Puebla
	10/0.25	MEX	BD	R.Centro, Jalapa
	5/1	MEX	VZ	R. La Veraz, Acayucan
	1	MEX	ITC	R. Tecnológico, Celaya
	1	PNR	E91	R. Diez, Panamá
	5	PTR	WHOY	La Señal Activa de PR, Salinas
	1	SLV	CG	R. América/R. La Paz, Zacatecoluca
1220	1	CTR	Q	R. Fe y Poder, Limón
	10	CUB	DL	R. 26, Central España, MA
	5	CUB	BY	R. Caribe, IJ
		DOM		R. HIN, Sto Domingo
	1	HND	OP	R. Costeña Ebenezer, San Pedro Sula
	3	HND	SD	R. Destellos de Luz, Sabá
		HND		R. Sintonía, Juticalpa
	1	HND	YS	R. Suari, Marcala
	1	HTI		Voix du Plateau Central, Hinche
	100	MEX	B	La B Grande, México
	2.5d	MEX	SAL	R. Universidad, Saltillo
	1	NCG	A3RA	R. América, Managua
	5	PNR		Asamblea Nacional, Veraguas
1230	3	CUB	DL	R. 26, Unión de Reyes, MA
	1	DOM	PM	R. Moca, Moca
		GTM		R. América, Cuyotenango
	1	GTM	AT	R. Atlántida, Puerto Barrios
	0.25	HND	CQ	R. Samaritano, San Marcos de Colón
	10	HND	QW	R. Tela, Tela
	1	HTI		Voix de L'ave Maria, Cap Haitien
	20/1	MEX	TVH	La Morena 1230, La Más Choca de Todas, Villahermosa
	1	MEX	IZ	R. Fórmula Cadena 3, Monterrey
	10/2	MEX	EX	R. Fórmula, Culiacán
	1/0.25	MEX	DKN	R. Fórmula, Segunda Cadena, Guadalajara
	1	MEX	LP	R. Pía, La Piedad
	1	MEX	TCP	W R., Tehuacan
	5	NCG	MNG	R. Manantial, Nueva Guinea
	1	PTR	WNIK	Única R., Arecibo
1240	1	BAH		BCB, ZNS2, Freeport
	1	CTR	WC	R. Corobici, Cañas
	10	CUB	DL	R. 26, Bolondrón, MA
	1	DOM		R. María, Santo Domingo
	1	DOM	AU	R. Vida, Puerto Plata
	5	GTM	K	R. Luz, Guatemala
	1	HND	ZC	R. Vanguardial, Tegucigalpa
	1	HND	VN	R. Venus, Sta Bárbara
	1	MEX	VM	Amor 107, Piedras Negras
	1	MEX	WG	Cambio 1240, Cd.Juárez
	1	MEX	BQ	FM 105, Guaymas
	.5/1	MEX	CE	Ke Buena, Oaxaca
	2.5/0.5	MEX	OV	La Picosa/R. Fórmula, Orizaba
	3	MEX	RD	R. Lobo, Pachuca
	1	MEX	SI	R. Positiva, Santiago Ixcuintla
	0.5/0.5	MEX	RPA	R. Ranchito, Morelia
	10/2.5	MEX	RO	R. Recuerdo, Aguascalientes
	1	MEX	BN	Radiola, Cd.Delicias
	2.5	MEX	LM	Romántica 12-40, Tuxtla Gutiérrez
	1	MEX	CG	Romántica, Nogales

kHz	kW	Ctry	Call	Station, location
	1/0.25	MEX	S	W R., Tampico
	5	NCG	A3RR	R. Vida Managua
	1	PNR		Comunicación de Masas, Panamá
	3	PNR	M56	Faro de David, David
	1	PTR	WALO	R. Oriental/Cad. R. Puerto Rico,Humacao
	0.5	SLV	MT	R. Metapán, Metapán
	1	SLV	QN	R. Norteña, San Miguel
1250	0.25	CUB	KS	R. Playitas, Imías, GU
	5	DOM	BC	LV del Progreso, San Francisco de Macorís
	5	DOM	RJ	R. Juventud, La Romana
	1	GTM		LV Cristiana, Totonicapán
	1	GTM	PY	R. Payakí, Esquipulas
	1	HND	YF	R. Cristiana 1250, Comayagua
	1	HND	DG	R. Oriental, Danlí
		HND	YL	R. Sonaguera, Sonaguera
		HND		Super R., San Pedro Sula
	5/1	MEX	JX	Cadena R. Uno, Grupo Fórmula, Querétaro
	10/1	MEX	DK	DK 12-50, Guadalajara
	1/0.5	MEX	SC	La Pantera/R. 1250, Sabinas
	1/0.5	MEX	DL	R. 13/DL/Fuerza de la Palabra, Hermosillo
	10	MEX	TF	R. Fórmula, Segunda Cadena, Veracruz
	5/0.25	MEX	AT	R. Imagen/Nueva Imagen, Hidalgo del Parral
	5/0.5	MEX	SJ	R. Saltillo, Saltillo
	5/0.5	MEX	ZT	R. Tribuna, Puebla
	1/0.25	MEX	TEJ	Sistema XEGEM "R. Mexiquense", Tejupilco
	2.5	NCG	CR	Cadena Radial Samaritano, Condega
	5	PNR	LY	R. Hogar, Penonomé
	0.25/1	PTR	WJIT	R. Hit, Sabana
1260	5	CTR	DIO	R. Emaús, San Vito de Coto Brus
	5	CUB	BF	R. Enciclopedia, Arroyo Arenas, CH
	1	DOM	T	R. Recuerdos, Sto Domingo
	1	HND	FP	R. Amistad, San Marcos de Colón
	1	MEX	QL	Catedral de la Música, Zamora
	1/0.25	MEX	R	Hits 12-60, Linares
	20/10	MEX	L	La 12-60 AM, México
	1/0.25	MEX	ZH	La Estación que se Escucha, Salamanca
	5/1	MEX	JY	La Mejor, El Grullo
	5/0.5	MEX	SA	La Mexicana, Culiacán
	1	MEX	TBV	La Poderosa, Tierra Blanca
	10d	MEX	JAM	LV de la Costa Chica, Santiago Jamiltepec
	1	MEX	MTV	R. Lobo de Mina, Minatitlán
	5/1	MEX	XR	R. Mensajera, Cd.Valles
	5/0.25	MEX	OG	R. Ranchito, Ojinaga
	1/0.25	MEX	MW	R. San Luis/Sonido Z, San Luis Río Colorado
	2.5/0.9	PTR	WISO	R. Wiso/Cadena WAPA, Aguadilla
	5/1.8	PTR	WISO	R. Wiso/Cadena WAPA, Mayagüez
	2.5	PTR	WISO	R. Wiso/Cadena WAPA, Ponce
	12	SLV	AA	R. Abba, San Salvador
1270	2.5	ABW		R. 1270 AM, Oranjestad
	1	CUB	BQ	R. Enciclopedia, Varadero, MA
	10	CUB	BD	R. Reloj, Camagüey, CM
	1	DOM	TA	R. Ambiente, Baní
	1.2	DOM	DA	R. Metro-Hit 12-70, Santiago
	2.5	GTM	CQ	R. Exclusiva, Guatemala
	1	HND	OF	Ecos del Celaque, Gracias
	1	HND	QN	R. Sonora, Tegucigalpa
	1	MEX	VHT	Bésame 12-70, Villahermosa
	0.5	MEX	AZ	Canal 1270, Zeta 13, Tijuana
	1/0.5	MEX	GL	Digital 12-70, Navojoa
	0.5/0.15	MEX	WN	ElFonógrafo del Recuerdo, Torreón
	10/0.15	MEX	RPL	LaPoderosa RPL, León
	3	MEX	QH	Milenium R., Ixmiquilpán
	1.5/0.5	MEX	HD	R. Universidad, Durango
	1/0.25	MEX	RRR	Romántica, Papantla
	2/0.5	MEX	RRT	Sport R., Cd.Madero
	3	NCG	RA	R. Amistad, Matagalpa
	3	PNR	J22	R. Tipy Q, Panamá
	1	SLV	QZ	R. W, San Miguel
1280	2	CTR	GV	Visión 1280, San José
	1	CUB	KW	R. Mambí, Stgo de Cuba, SC
		DOM	JH	Cadena Espacial, Sto Domingo
	2.5	GTM	VY	R. Zamaneb, Salamá
		HND		R. Armonía, Juticalpa
	1	HND	RF	R. Cadena de Notícias, San Pedro Sula
	1	HND	BN	R. San Miguel, Marcala
	1	HND	AM	R. Unción AM, Olanchito
		HTI		R. Transcaribbean International, Jean Rabel

kHz	kW	Ctry	Call	Station, location
	1/0.5	MEX	EG	ABC R., Puebla
	10/1	MEX	AW	A-W Notícias, Monterrey
	2.5/1	MEX	CAM	Kiss FM, Campeche
	2/1	MEX	AG	La Poderosa, Córdoba
	1/0.5	MEX	BW	Palabra Viva, Chihuahua
	0.5/0.25	MEX	BON	R.Fórmula, Tercera Cadena, Guadalajara
	.5/1.15	MEX	SQ	R. San Miguel. San Miguel de Allende
	1	MEX	TUT	R. Tamaulipas, Tula
	1/0.1	MEX	KY	Romántica 12-80, Huixtla
	5/1	PTR	WCMN	NotiUno 1280, Arecibo
	1	SLV	QV	R. CRET, Sta Ana
		SLV		R. Emaús, San Vicente
1290	1	CUB	E	CMHW, Rancho Veloz, VC
	5	CUB	BQ	R. Enciclopedia, La Habana, HA
	5	CUB	BC	R. Progreso, La Pastora, HA
	0.5	DOM	BD	R. Jánico, Santiago
		GTM		R. Miramundo, Zacapa
	1	HND	NN26	R. Choluteca, Choluteca
	1	HND	GS	R. HRGS/Bay Island Christian Netw., Utila
	1/0.5	MEX	IX	Enlace Digital 12-90/La Pantera, Sahuayo
	5/0.25	MEX	FAC	La Mera Mera, Salvatierra
	10/1	MEX	NX	R. Mujer, Mazatlán
	0.25d	MEX	TH	R. Palizada, Palizada
	10/1	MEX	DA	R. Trece, La Fuerza de la Palabra, México
	1/0.25	MEX	AP	Romántica 12-90, Cd.Obregón
	3	PNR	S23	R. Única, Chiriqui
	5.5	PNR		R. Unica, Los Santos
	5	PNR		R. Unica, Panama
	1	SLV	MA	R. Chalatenango, Chalatenango
	0.5/0.3	VIR	WRRA	WRRA, St Croix
1300	1	ATN	PJD-2	VO St Maarten, Philipsburg, St Maarten
	1	CTR	GL	La Fuente Musical, Cartago
	1	CUB	BQ	R. Enciclopedia, Las Tunas, LT
	1	DOM	KQ	R. R./La Doz de HIZ, Sto Domingo
	1	HND	IV	R. C.C.I., Tegucigalpa
	5	HND	LR	R. Sta Rosa, Sta Rosa de Copán
	1/0.25	MEX	JL	La 130, La Ley, Guamuchil
	1	MEX	KW	La Guadalupana, Morelia
	1	MEX	HU	La Que Manda, Martinez de la Torre
	10/0.75	MEX	XV	La Z, León
	50	MEX	P	R. 13/R. Centro, Cd.Juárez
	1/0.25	MEX	AWL	R. Jacala/Hidalgo R.,Jacala
	1/0.1	MEX	XW	W R., Nogales
	1	NCG	A2CC	Canal 130 AM, Managua
	5	PNR	I417	R. Baha'ís, Boca del Monte
	1	PTR	WTIL	La Voz Romántica, Mayagüez
		SLV	KG	R. Llanera, San Miguel
	6	SLV	LV	W-LV de la Verdad, San Salvador
1310	1	CUB	E	CMHW, Sagua La Grande, VC
	1	DOM	MH	R. Real, La Vega
	1	GTM	AN	R. LV de los Altos, Quetzaltenango
	2.5	HND	VC	LV Evangélica, San Pedro Sula
	1	HND	RL	R. Libertad, Marcala, La Paz
	5	HND	CM	R. Universidad de Agricultura, Catacamas
	5/0.25	MEX	VB	Digital 102.9, Monterrey
	.5/1	MEX	HV	H-V 1310, Veracruz
	5/0.25	MEX	AM	La M Grande, Matamoros
	1d	MEX	BTS	R. Bahía de Tortugas, Bahía de Tortugas
	1	MEX	C	R. Enciso, Tijuana
	5/1	MEX	HIT	R. Felicidad, Puebla
	1	MEX	LPZ	R. La Paz, La Paz
	1/0.1	MEX	FH	R. Plan de Agua Prieta, Agua Prieta
	1/0.25	MEX	RU	R. Vital, Chihuahua
	10/1	MEX	TIA	R. Vital, Guadalajara
	1d	MEX	GRT	Soy Guerrero, Taxco
	5	MEX	HY	Stereo Joya, la Música de tu Vida, Querétaro
	5	MRT		R. Martinique, Fort de France
	10/1	NCG	SC	R. San Cristóbal, Chinandega
	12	PNR		R. María, Panamá
	5	SLV	RV	R. Veritas, Stgo de María
1320	2.5	ABW		Voz di Aruba, Oranjestad
	1	CTR	LX	R. Columbia, San Carlos
	0.5	CUB	CW	R. Artemisa, Artemisa, AR
	1	CUB	BQ	R. Enciclopedia, Sancti Spíritus, SS
	1	CUB	BQ	R. Enciclopedia, Stgo de Cuba, SC
	1/0.5	DOM	BZ	R. Centro, San Juan de la Maguana
	1	GTM	ME	R. Quezada, Jutiapa

kHz	kW	Ctry	Call	Station, location
	1	HND	MG	R. Bahía "La Super Grande", La Ceiba
	10/0.1	MEX	CPN	1320 Notícias La Mexicana, Piedras Negras
	2.5/1	MEX	PAR	La Buena, Villahermosa
	2.5/0.25	MEX	JZ	LaCampera/R. Fórmula, Cd.Jimenez:
	20	MEX	NET	México
	1/0.5	MEX	NM	R. 1320, Estación sin fronteras, Aguascalientes
	0.5/0.25	MEX	SR	R.Cachanía, Santa Rosalia
	10/2	MEX	RJ	RJ 1320, La Ranchera de Mazatlán, Mazatlán
	10/1	MEX	NI	Romántica 13-20, Uruapán
	10/2	MEX	UH	X R., Tuxtepec
	5/2.3	PTR	WSKN	R. Isla 1320, San Juan
	1	SLV	AH	R. Emanuel, La Unión
1330	3	DOM	VC	R. Visión Cristiana, Sto Domingo
	5.5	GTM	MU	Unión R., Guatemala
		HND		R. Emisora Evangélica, Tegucigalpa
	1	HND	FL	R. Florida, La Entrada
	10	HTI		R. Haiti Inter, P-au-P
	5/0.9	MEX	AJ	1330 AM R., Saltillo
	10	MEX	MAC	La Poderosa, Manzanillo
	1/0.1	MEX	RP	La Tremenda, Cd.Madero
	0.5d	MEX	EV	R. Festival, Izúcar de Matamoros
	4/0.25	MEX	WQ	R. Triunfadora, Monclova
	5/1	MEX	BO	R. Variedades, Irapuato
	5	NCG	A6RM	R. Matagalpa, Matagalpa
	5	PNR		LV Poderosa, Panamá
	2/1.4	PTR	WENA	La Buena del Sur, Yauco
	5	SLV	HQ	R. Cristo Te Llama, San Salvador
1340	5	CTR	HR	R. Sideral, San Ramón
	0.25	CUB		AFRTS, Guantánamo Bay
	10	CUB	FL	R. Ciudad del Mar, Palmira, CI
	1	HND	CQ	Cadena Radial Samaritano, Comayagua
	10	HND	TQ	R. 1430/R. El Mundo, San Pedro Sula
		HND		Telecolor R, Catacamas
	1	MEX	AA	13-40 AM, Mexicali
	1	MEX	NV	91X La Experiencia, Monterrey
	1	MEX	APM	Candela, Apatzingán
	10/5	MEX	LU	Ke Buena Puebla, Cd.Serdán
	1	MEX	RPV	La Cotorra, Cd.Victoria
	1	MEX	QB	La Divertida/R. Fórmula, Tulancingo
	1d	MEX	QE	La Mera Mera, Escuinapa
	1	MEX	CR	La Zeta, Morelia
	1	MEX	BK	Mega ¡Sí pega!, Nuevo Laredo
	1	MEX	MT	Mi R. 13-40 - Nostalgia, Matamoros
	1	MEX	DH	R. Amistad, Cd.Acuña
	1/0.5	MEX	RCH	R. Exitos, Ojinaga
	1	MEX	OS	R. Mujer, Cd.Obregón
	5/1	MEX	DKT	R. Ranchito, Guadalajara
	1	MEX	CI	Romántica 13-40, Acapulco
	5	MEX	ASM	Romántica 13-40, Cuernavaca
	2	MEX	SL	Señal 13-40, Golden Music, San Luis Potosí
	1	NCG	OS	R. Ondas Sonoras, Managua
	2.5	PNR		R. Tipikal, Las Tablas
	0.95	PTR	WWNA	R. Una 1340, Aguadilla
	1	SLV	XW	R. Novedades, Usulután
	1	VIR	WSTA	WSTA, St Thomas
1350	1	CUB	FL	R. Ciudad del Mar, Aguada de Pasajeros, CI
	10	CUB	LM	R. Libertad, Puerto Padre, LT
	1	DOM	JD	Ondas del Yuna, Bonao
	1	DOM	PM	R. Rutas Musical, La Romana
	1	GTM	MC	R. Monja Blanca, Cobán
		HND	EL	R. Estelar, La Ceiba
	1	HND	JV	R. Henecan, San Lorenzo
	0.25	HTI		R. Dame Marie, Dame Marie
	5/1	MEX	CAH	La Popular 13-50/La Voz de Soconusco, Cacahoatán
	10d	MEX	CTZ	LV de la Sierra Norte, Cuetzalán
	1/0.25	MEX	ZD	Mi R. 1350, La Preferida, Camargo
	8	MEX	LBL	R. Centro, San Luis Río Colorado
	5/0.5	MEX	TB	R. Laguna, Torreón
	5/1	MEX	QK	Tropicalísima 13-50, México
	5	PNR	Z38	BBN R., Panamá
	2.5	PTR	WEGA	Nueva Victoria, Vega Baja
1360	5	CTR	DS	R. R. 1360, San José
	1	CUB	HA	R. Cad. Agramonte, Rodolfo Ramírez Esquível, CM
		DOM	XZ	R. Tropical, Sto Domingo
	10	GTM	LK	R. Tic Tac, Guatemala
	1	HND	BS	R. San Pedro, Tegucigalpa
	5	HND	BH	R. Sta Bárbara, Sta Bárbara
	1	MEX	KF	Canal 13-60, Iguala
	1/0.4	MEX	DI	La Nueva, Chihuahua
	5/0.5	MEX	UD	La U de Tuxtla/La Máquina Musical, Tuxtla Gutiérrez
	10d	MEX	ZON	LV de la Sierra de Zongolica, Zongolica
	1/0.25	MEX	Y	R. Fiesta Retro, Celaya
	5	SLV	FM	Super R., San Salvador
1370	1	CUB	HA	R. Cad. Agramonte/R. Nuevitas, Nuevitas, CM
	5	DOM	RP	R. Seybo, El Seybo
	1	GTM	AC	LV de Colomba, Colomba
	1	HND	SQ	R. El Shaddai R., Siguatepeque
	1	HND	ST	R. Fraternidad, San Pedro Sula
		HND	ZG	R. Guayapeña, Catacamas
	0.5	HTI		R. Citadelle, Cap Haitien
	1	HTI		Rdif. Cayenne, Cayes
	5/0.5	MEX	GNK	Fiesta Mexicana, Nuevo Laredo
	10/1	MEX	PJ	Frecuencia Deportiva, Guadalajara
	1	MEX	A	Ke Buena, Campeche
	1/0.25	MEX	RPU	La Z, Durango
	5	MEX	HF	R. Fórmula, Nogales
	10	MEX	MON	R. Fórmula, Segunda Cadena, Monterrey
	1/0.5	MEX	SV	R. Nicolaita, Morelia
	5/1.5	MEX	JE	R. Reyna, Dolores Hidalgo
	0.5	MEX	HG	Romántica, Mexicali
	1	NCG	AARS	R. Fronteras, Somoto
	1	PNR	B64	R. Sitrachilco, Pto Armuelles
	5/1	PTR	WIVV	WIVV Rock R. Network, Vieques Isl.
	1	SLV	KO	R Lluvias de Bendición, San Miguel
1380	1	CTR	MS	R. Guanacaste, Liberia
	10	CUB	HA	R. Cad. Agramonte, Central Brasil, CM
	1	DOM	SC	R. Nacional, Santiago
	1	GTM	EB	R. Momostenango Educativa, Momostenango
	1	HND	EJ	R. Monjaras, Choluteca
	0.5	HND	AH	R. Redención, Jutiapa
	5/1	MEX	GW	Mazz W, Cd.Victoria
	1/0.1	MEX	VD	R. Sensación, Allende
	50/5	MEX	CO	Romántica 13-80, México:
	1/0.5	MEX	RS	Romántica 1380, Torreón
	10/1	MEX	TP	Sensación FM, Jalapa
	10	PNR		Mujer AM, Panamá
	1	PTR	WOLA	Prócer, Voz de la Montaña, Barranquitas
1390	1	CUB	BT	R. Jaruco, Jaruco, HA
	1	DOM	AR	R. San Cristóbal
	1	HND	VC	LV Evangélica, Sta Rosa de Copán
	10/5	HND	VC	LV Evangélica, Tegucigalpa
	1	MEX	OR	La Papaya Tropicalísima, Reynosa
	5/1	MEX	XO	La Super Buena, Cd.Mante
	5/0.1	MEX	KT	La Súper Estación, Tecate
	10/2.5	MEX	TY	Los 40 Principales, Tecomán
	1/0.15	MEX	QC	LV de Pto Peñasco/La Reyna del Mar, Pto Peñasco
	1	MEX	CTA	R. Cauatla, Cuautla
	10/0.25	MEX	RW	R.Fórmula, León
	0.5d	MEX	ZG	R. Mezquital y Huasteca Hidalguense, Ixmiquilpán
	5/1	MEX	TL	R. Ola, Tuxpán
	5	PNR		R. Mundo Internacional, Colón
	1	PTR	WISA	R. Puerto Rico 1390, Isabela
		SLV		R. Fraternidad de Jesucristo, Chalchuapa
		SLV		R. Getsemani, La Unión
1400	1	CTR	GJ	R. Sinaí, San Isidro del General
	1	CUB	BF	R. Musical, Matanzas, MA
	1	DOM	AC	Ondas del Valle, La Vega
	5	GRD		Harbour Light of the Windwards, Carriacou
	1	GTM	RB	R. Porteña, Puerto Barrios
	1	HND	AU	R. Alegre, Sava Colón
	1	HND	YT	R. Estrella de Oro, San Pedro Sula
		HND	BO	R. Punto, Comayagua
		HND	UV	R. Universitaria, Catacamas
	1	MEX	VI	EXA FM 99.1, San Juan del Río
	1	MEX	AC	Ke Buena de Aguascalientes, Aguascalientes
	1	MEX	PF	La Efectiva/La Rancherita, Ensenada
	2.5/1	MEX	XI	La I de Ixtapan, Ixtapan de la Sal
	0.25	MEX	WU	La Poderosa, Matehuala
	1	MEX	KJ	Mariachi Stereo, Acapulco

kHz	kW	Ctry	Call	Station, location
	5/1	MEX	OJ	R. Horizonte/R. Fórmula, Cd.Lázaro Cárdenas
	5	MEX	SH	R. Sabinas, Cd.Sabinas
	0.25	MEX	AB	R. Santa Ana, Santa Ana
	5	MEX	I	R. Trece, Morelia
	1	MEX	UBJ	R. Universidad Benito Juárez, Oaxaca
	10	NCG	A3MA	R. María, Managua
	10	PNR	T40	Digital R. Luz, La Chorrera
	1	PTR	WIDA	R. Vida AM, Carolina.
	1	SLV	JI	LV del Litoral, Usulután
1410	1	CUB	HA	R. Cadena Agramonte, Sta Cruz, CM
	1	CUB	BQ	R. Enciclopedia, Pinar del Río, PR
	3/0.5	DOM	CH	R. 14-10, Barahona
	1/0.5	DOM	JJ	R. Grí-Grí, Río San Juan
	1	DOM	AE	R. Tricolor, Sto Domingo
	5	GTM	GH	R. Xelajú, Quetzaltenango
	3	HTI		Voix de Nord-ouest, Port de Paix
	2/1	MEX	ZHO	Aquamarina R., Zihuatanejo
	2.5/10	MEX	KB	Canal 14-10, Guadalajara
	1/0.25	MEX	AS	Ke Buena, Nuevo Laredo
	1/0.1	MEX	YD	La Grande de Madero, Torreón
	10/0.5	MEX	CF	La Mexicana, Los Mochis
	1/0.25	MEX	CUA	R. Universidad, Campeche
	2.5/1	MEX	BS	Sinfonola, La Mas Perrona, México
	5/0.5	MEX	IR	XEIR, La Señal Perfecta, Cd.Valles
	3/1	NCG	RA	La Estación de la Amistad, León
	5	PNR	H779	R. Mensabé, Las Tablas
	1	PTR	WRSS	R. Progreso, San Sebastián
1420	1	CTR	RPN	R. Pampa, Liberia
	15	DOM	FD	R. Oro, Cotuí
	1	GTM	RP	R. Capital, Guatemala
		HND	GB	R. Sabanagrande, Sabanagrande
	1	HND	SL	R. Stereo Actualidad, Trinidad
	0.5	HTI		R. Messie Continental, Dessalines
		MEX	PK	1R. Felicidad 14-20, Pachuca
	10/1	MEX	WE	La Estación Familiar, Irapuato
	5/1	MEX	H	La H, Antología Vallenata, Monterrey
	1d	MEX	KMX	La Super X, Sayula
	5/0.5	MEX	F	Línea Deportiva, Cd.Juárez
	2	MEX	XX	R. Mexicana/R. Fórmula 1420, Tijuana
	1d	MEX	AFQ	Romántica, Minatitlan
	1	MEX	EW	W1420/LV del Bajo Bravo, Matamoros
	1	MEX	WJ	WJ Fórmula, Tehuacán
	1	PTR	WUKO	R. Reloj, Ponce
1430	3	CTR	RDVC	R. San Carlos, Cd. Quesada
	10	CUB	JY	R. Surco/R. Amanecer, Primero de Enero, CA
	6	DOM	JC	R. Emanuel, Santiago
	1.2	GTM	AG	LV de Huehuetenango
		HND	QV	La Nueva Potencia, Olanchito
		HND		Ministerios Cristianos Fuente de Vida, Juticalpa
	1	HND	VM	R. Maranatha, La Paz
	1	HND	FO	R. Shekina, Puerto Cortés
	5/0.5	MEX	OX	Exa FM 106.5, Cd.Obregón
	1	MEX	COC	Inolvidable, Colima
	5/0.15	MEX	WD	La Grande de Ciudad Miguel Alemán, Cd. Miguel Alemán
	0.25	MEX	RAC	La Número Uno en Campeche/R. Fórmula, Campeche
	5/1	MEX	LL	Latido 14-30 AM, Veracruz
	5/1	MEX	TT	R. Tlaxcala, Tlaxcala
	5	NCG	AARL	R. Liberación, Estelí
	7.5	PNR		R. Kids, Panamá
	5	PTR	WNEL	NotiUno/R. Tiempo, Caguas
1440	10	CUB	JP	R. Surco, Ciego de Avila, CA
	5	DOM	AK	R. Impactante, Sto Domingo
	5	DOM	AD	R. San Juan, San Juan de la Maguana
	0.5	GTM	MS	R. Nacional, Mazatenango
	5	HND	RD	R. Belén, La Ceiba
	0.5	HND	RY	R. Ekklesia Int., San Marcos de Colón
	10/1	MEX	ABCJ	ABC R./Corazón , Guadalajara
	25/5	MEX	EST	Cambio 14-40, México
	1/0.15	MEX	VSD	La Señal del Progreso, Cd. Constitución
	25	NCG	A3MR	R. Maranatha, Managua
1450	1	CUB	CL	R. Güines, Güines, MB
	1	CUB	LN	R. Maboas, Amancio Rodríguez, LT
		DOM		R. Alfa y Omega, Sto Domingo
	10	DOM	AC	R. Util, Salcedo
	1	GTM	LG	R. Hosanna, Guatemala

kHz	kW	Ctry	Call	Station, location
	1	HND	BR	R. Cultural, La Entrada
		HND	GB	R. Sabanagrande, Tegucigalpa
	1	MEX	CM	Bonita, Cd.Mante
	1	MEX	BP	Bonita, Torreón
	5/1	MEX	JM	La Caliente, Monterrey
	2/1	MEX	RY	La Poderosa V del Sur, Arcelia
	5/1	MEX	RDO	La R. 14-50, Reynosa
	10/1	MEX	CU	La Rancherita, Los Mochis
	0.4d	MEX	PNO	La Señal de Oaxaca, Santiago Pinotepa Nal
	5/1	MEX	NA	R. Capital, Querétaro
	0.5	MEX	DJ	R. Clave, Magdalena
	1	MEX	GC	R. Impacto, Sahuayo
	1	MEX	KM	Mina/R. Fórmula, Minatitlán
	1	MEX	JD	R. Mundo 14-50, Poza Rica
	1/0.25	MEX	ARE	R. Pegüis/R. Lobo, Ojinaga
	10/1	MEX	CB	R. Ranchito, San Luis Río Colorado
	5	PNR		R. Melodía, Panamá
	1	PTR	WCPR	R. Coamo, Coamo
	1	SLV	KR	R. Restauración, San Miguel
1460	1	CTR	LX	R. Columbia, Ciudad Quesada
	1	CUB	HA	R. Cadena Agramonte, Sola, CM
	0.5	DOM	AN	R. Renacimiento, Hato Mayor del Rey
	2.5	GTM	RN	R. Petén, Flores
	0.5	HND	CX	LV de Patuca, Catacamas
	2.5	HND	GC	R. Conga, San Pedro Sula
		HND	FR	R. Firmamento, La Paz
	0.5	HND	KS	R. Ministerio Bautista, Yoro
	0.2	HTI		Voix du Nord, Cap Haitien
	1/0.1	MEX	JH	ABC R., Jalapa
	5/0.5	MEX	KC	Estéreo Exitos, Oaxaca
	0.5	PNR	D42	LV de Almirante, Bocas del Toro
	0.5	PTR	WLRP	R. Raíces, San Sebastián
	0.5/0.3	PTR	WRRE	Sonido Santidad, Juncos
1470	1	CUB	GE	R. Ciudad Bandera, Cárdenas, MA
	1	DOM	DE	LV de la Alabanza, San Francisco de Macorís
		DOM	CV	R. Barahona, Provincia Independencia
		DOM	CH	R. Vibra -La Deportiva, Barahona
		HND	XH	R. Globo Grupera, Nacaome
	1	MEX	IRG	La Campirana, Irapuato
	1/0.5	MEX	IND	LV Sierra Hidalguense, Tlanchinol
	10/0.25	MEX	HI	MiR. 1470, Puro Cañonazo, Ciudad Miguel Alemán
	5/1	MEX	CAV	Play 14-70, Tocando Tu Memoria, Durango
	1/0.1	MEX	ACE	R. Fórmula Mazatlán, Mazatlán
	50/5	MEX	AI	R. Fórmula, Tercera Cadena, México
	10/5	MEX	RCN	R. Hispana 14-70 San Diego y Tijuana, Tijuana
	2.5/0.5	MEX	BAL	R. Voz Maya de México, Bécal
	1	NCG	RY	R. Yarrince, Boaco
	5	PNR		R. La Primerísima, Panamá
	2.4/2.5	PTR	WKCK	R. Cumbre, Orocovis
1480	2	CTR	AW	R. El Sol, Puntarenas
	5	DOM	AH	R. Villa, Sto Domingo
	5	GTM	HB	R. Horizontes, Guatemala
	1	HND	EZ	LV de Misiones "R. MI", Comayagüela
	1	HND	WP	R. Soberanía, San Marcos, Ocotepeque
	2/1	MEX	ZJ	Ciudad 1480, Guadalajara
	1/0.5	MEX	HM	H-M R., Cd.Delicias
	1/0.1	MEX	XU	La Poderosa, Monclova
	2.5d	MEX	CARH	LV del Pueblo Hña-hñu, Cárdonal
	5/0.15	MEX	VIC	R. Tamaulipas, Cd.Victoria
	10/1	MEX	TKR	TKR Rancherita y Regional, Monterrey
	5/0.25	MEX	NS	Z14, Solo Exitos, Navojoa
	5	PTR	WMDD	R. Tropical/Sonido 14-80, Fajardo
1490	1	CUB	KN	R. Mayarí, Mayarí, HO
	3	DOM		R. Universal, Santiago:
	1	GTM	RE	R. Modelo, Retalhuleu
	1	HND	OM	R. Omega "Sonido Internacional", La Esperanza
		HND	OE	R. Pijol, Morazán
	1.2	HND	GO	R. Porteña, Puerto Cortés
	1	HND	AQ	La Caliente, Agua Prieta
	1/0.25	MEX	SK	La Super K/La Costeñita, Cd.Ruiz
	1	MEX	MS	R. Mexicana, Matamoros
	1	MEX	CJC	R. Net, Cd.Juárez
	1/0.25	MEX	FF	R. Norteña, Matehuala
	1	MEX	YT	R. Teocelo, Teocelo
	1	MEX	KN	R. Variedades, Huetamo
	5/1	MEX	GT	W R., Zamora

kHz	kW	Ctry	Call	Station, location
	3	PNR		Asamblea Nacional, Cocle
	5/1	PTR	WDEP	R. Isla, Ponce
1500	1	CTR	ASF	R. R.1500, Sarapiqui
	1	CUB	BQ	R. Enciclopedia, Holguín, HO
	0.5	DOM	PA	R. Higüey, Higüey
		DOM	RD	R. Juan Pablo Duarte, Elías Piña
		HND		R. MI-EL, Sabá
		HND	VP	R. Sion, La Ceiba
	1	HND	TX	R. Victoria, Choluteca
	0.4d	MEX	JQ	La Explosiva, Parras
	50	MEX	DF	R. Fórmula, Segunda Cadena, México
	1/0.5	MEX	FL	R. Santa Fe, Guanajuato
	1	NCG	PT	R. Minuto, Managua
	1/0.25	PTR	WMNT	R. Atenas, Manatí
	1	SLV	CS	R. Pentecostal, Usulután
1510	1	CUB	BQ	R. Enciclopedia, Moa, HO
	10/3	DOM	BL	R. Pueblo, Sto Domingo
	5	GTM	DX	R. Centroamericana del Amor, Guatemala
	1	HND	EM	R. Emanuel, Ocotepeque
	1	HND	PG	R. Gualcho, Tegucigalpa
	50d	MEX	QI	La Nueva R., Monterrey
	0.25	MEX	HUI	R. Huichapán, Huichapán
	5	PNR	A95	Hosanna R., Panamá
	1	PTR	WBSG	R. Voz, Lajas
1520	1	CTR	LX	R. Columbia, Cartago
	1	DOM	WJ	R. Samaná "R. 15-20", Samaná
	1	GTM	RS	R. Superior, Coatepeque
		GTM		R. Taysal, Sta Elena de la Cruz
	1	HND	CR	Dif. Cristiana de R. "DCR", San Pedro Sula
		HND	MQ	R. Manantial de Vida Eterna, Juticalpa
		HND	DF	R. Rios de Agua Viva, Siguatepeque
	1	HND	HJ	R. Santiago, Yoro
	5	MEX	JCC	La 1520, Cd.Juárez
	1d	MEX	VO	La Furia, San Rafael
	1d	MEX	VUC	La Norteñita, Allende
	1d	MEX	YP	Mazz 15-20, Cd. Mante
	1	MEX	EH	R. Exitos, San Luis Río Colorado
	2	MEX	ART	Señal 152, Jojutla
	1/0.25	MEX	ATL	Sist. XEGEM "R. Mexiquense", Atlacomulco
	25	PTR	WVOZ	R. Voz/La Voz Boricua, San Juan
1530	1	DOM	JN	R. 1530, Santiago
		HND		Super Q, Choluteca
	50/1	MEX	UR	Fiesta 15-30, México
	1	MEX	GQ	La Reyna de los Reyes, Los Reyes
	10/0.1	MEX	SD	Los 40 Principales, Silao
	0.5	NCG	A4TS	LV de Sta Teresa, Sta Teresa
	10	PNR		R. Avivamiento, Panamá
	1/0.25	PTR	WUPR	Exitos 15-30, Utuado
1540	8	BAH		BCB, ZNS1, Nassau
	1	CTR	CUB	Enlace R., Pavas
	1	CUB	ES	R. Sagua, Sagua La Grande, VC
	1	DOM	BUv	LV de La Romana, La Romana
	1	DOM	FP	R. Criolla Comercial, Sto Domingo
	1	GTM		R. Cultura y Deportes, Guatemala
		HND	VK	R. Nuevo Mundo, Tegucigalpa
	1/0.25	MEX	NC	La Auténtica 15-40, Celaya
	5	MEX	HOS	La Poderosa, Hermosillo
	2.5	MEX	RTP	La Poderosa/Impacto, San Martín Texmelucán
	1/0.5	MEX	STN	R. Red, Monterrey
	1d	PTR	WIBS	R. Voz/R. Caribe, Guayama
1550		CUB	BA	R. Rebelde, Nuevitas, CA
	1	HND	JO	R. Campeona, Comayagua
	1	HND	JX	R. Cristiana Nueva Vida, San Pedro Sula
		HND		R. Wuampu, Dulce Nombre de Culmi
	1	MEX	BG	CBC R., Tijuana
	5/0.25	MEX	NU	La Rancherita, Nuevo Laredo
	1	MEX	REL	R. Michoacán, Morelia
	10	MEX	RUV	R. Universidad Veracruzana, Jalapa
	0.25	PTR	WKFE	La Isla/R. Café Dinámica Yauco
	5	SLV		R. Sanidad Divina, San Salvador
1560	5	CTR	OAR	R. Nicoya, Nicoya
	1	CUB	BQ	R. Enciclopedia, Ciego de Avila, CA
	1/0.5	DOM	PZ	R. Pedernales, Pedernales
	1	DOM	GL	R. Única, Santiago
	1	DOM		R. Universidad UASD, Santo Domingo
		GTM		R. Inspiración, Quetzaltenango
		HND	FD	R. Mi Preferida, Choluteca
	1/0.25	MEX	MAS	Ke Buena, Salamanca
	5d	MEX	SE	LV de Campeche, Champotón
	50/10	MEX	INFO	México
	5/1	MEX	LAC	R. Azul/LV del Balsas, Cd.Lázaro Cárdenas
	20/0.15	MEX	CHZ	R.Lagarto, La Voz Viva de Chiapas, Chiapa de Corzo
	1d	MEX	JPV	R. Viva, Cd. Juárez
	10	PNR		R. Adventista de Panamá, Panamá
	5/0.75	PTR	WRSJ	R. Bachatera del Norte, Bayamón
1570		CUB	BQ	R. Enciclopedia, Las Tunas, LT
	10	GTM	VE	VEA, Guatemala
		HND	TF	Difusora Cristiana Torre Fuerte, Gracias
	2.5	HND	RF	R. Cad Nal de Noticias, Tegucigalpa
	100	MEX	RF	La Poderosa, Cd.Acuña
	1/0.1	PTR	WPPC	R. Felicidad, Peñuelas
1580	0.25	CTR	RCVT	LV de Talamanca, Talamanca
	1	CTR	LG	R. Casino, Siguirres, Limón
	0.25	CTR	RCC	R. Cultural de Corredores,
	0.25	CTR	RCLS	R. Cultural Los Santos
	0.25	CTR	RSCM	R. Cultural Maleku
	0.5	CTR	RCP	R. Cultural Pejibaye, Pérez Zeledón
	0.5	CTR	RCS	R. Cultural Santiago
	0.5	CTR	RCT	R. CulturalTilarán
	0.25	CTR	RCLC	R. Sistema Cultural de La Cruz
	0.25	CTR	RCL	R. Sistema Cultural de Los Chiles
	1	CUB	HA	R. Cad. Agramonte,Santa Cruz del Sur, CM
	10	DOM	AJ	R. Amanecer, Sto Domingo
	1	DOM	PK	R. Neiba, Neiba
		HND		R. La Voz Lenca, La Esperanza
	10	MEX	DM	DM Notícias, Hermosillo
	1/0.5	MEX	AF	La Temeraria 15-80, Celaya
	20	MEX	VAB	Stereo Miled, Valle del Bravo
	1/0.25	MEX	LI	Super 94.7, Chilpancingo
	1	PNR		Hosanna Oeste, Panamá
	5/2.5	PTR	WEKO	R. Voz 1580, Morovis
		SLV		R. Poder y Gloria, Santa Ana
1590	1.5	CTR	LGJ	R. 16, Grecia
		CUB	BQ	R. Progreso, Manzanillo, GR
	1	DOM	SF	R. Libertad, Santiago
	1	GTM	XC	R. Triunfadora, Chimaltenango
		HND	BX	R. Perla, El Progreso
		HND	ZL	R. Zol, Choluteca
	1/0.25	MEX	BZ	Extasis Digital, Cd.Delicias
	1/0.1	MEX	PT	La Nueva Misantla R., Misantla
	1	MEX	HC	R. Bahía, Ensenada
	20/10	MEX	VOZ	La Mexicana, México
	1	PTR	WXRF	R. Voz, Guayama
1600	2.5	CTR	RCCH	R. Cultural Chirripó
	0.25	CTR	RCBA	R. Cultural de Buenos Aires
	0.25	CTR	RCP	R. Cultural de Pital
	0.25	CTR	RCT	R. Cultural de Turrialba
	0.25	CTR	RCU	R. Cultural de Upala
	2.5	CTR	RCPV	R. Cultural Puerto Viejo
	0.5	CTR	RCSG	R. Cultural San Gabriel
	0.5	CTR	RPQ	R. Más, Pto Quepos
	1.5	CTR	MQ	R. Pococí, Guápiles
	2.5	CTR	CC	R. R. Cima, Pto Golfito
	0.25	CTR	RSCN	R. Sistema Cultural Nicoyano
	5	DOM	FG	R. Revelación en América, Sto Domingo
	1	HND	PC	R. Luz y Vida, San Luís
		HND	PQ	R. Poderosa, Tegucigalpa
	5	MEX	GEM	Sist. XEGEM "R. Mexiquense", Metepec
	1d	MEX	TPA	Soy Guerrero, Tlapa de Comonfort
	5	PTR	WLUZ	R. Luz/Romántica 1600, Bayamón
1610	30	AIA		The Caribbean Beacon, The Valley
		CUB	BD	R. Reloj
	0.25d	MEX	UACH	R. Chapingo, Chapingo
1620		CUB	BA	R. Rebelde
		DOM	SR	R. Taina/Planeta, San Pedro de Macorís
1630	10/1	VIR	WDHP	WDHP, St Croix
	10/1	MEX	UT	R. Universidad UABC, Mexicali
1640	1/0.5	DOM		R. Juventus Don Bosco, Sto Domingo
1650	5/3	DOM		RADECO, Santiago
1660	10/1	DOM		Fundación Lama, Sto Domingo
	10/1	PTR	WGIT	R. Voz/Gigante 16-60, Canóvanas
1680	1	DOM	SV	R. Senda 1680 AM, San Pedro de Macorís
1700	5/1	DOM		R. Eternidad, Sto Domingo
	10	MEX	PE	San Diego 1700, Tecate

SOUTH AMERICA
(excluding Brazil)

NB: Brazil has been excluded to save space – see country entry for frequencies

Abbreviations: Dif=Difusora, Em=Emisora, LV=La Voz, Nal=Nacional, SF=Santafé.

kHz	kW	Ctry	Call	Station, location
150	1	CLM	HKT71	Macheta
530		ARG		LV de las Madres, Buenos Aires
	1	EQA	DC1	R. Iris/530 AM, Quito
	15	FLK		Falkland I. Radio Service, Port Stanley
540	10/5	ARG	LU17	R. Golfo Nuevo, Pto. Madryn
	25/1	ARG	LRA14	R. Nal., Santa Fé
	5	ARG	LRA25	R. Nal.,Tartagal
		ARG		R.Italia, Villa Martelli
		BOL	LP77	Radiodifusora Victoria, La Paz
	1	CHL	CB54	R. Ignacio Serrano, Melipilla
	1	CHL	CD54	R. R.Ainil, Valdivia
	20	CLM	KA	R. Auténtica Básica, SF de Bogotá
	25	EQA	FA2	R. Tropicana, Guayaquil
	1	PRU	OBX4E	R. Inca del Perú, Lima
	1	PRU	OCX2D	R. San Antonio, Trujillo
	10	VEN		LV de Manapiare, San Juan de Manapiare
	50/25	VEN	OY	R. Perijá, La Villa del Rosario
550	2	CHL	CC55	R. Concepcion, Penco
	1	CHL	CD55	R. LV. de la Tierra, Angol
	50	CLM	HF	R. Nal., Marinilla
	50	CLM	ZQ	R. Nal., Neiva
	50	EQA	GM1	R. Reloj, Quito
	4	FLK		BFBS, Bush Rincon
	20/10	PRG	ZP16	R.Parque, Ciudad del Este
	25/10	URG	CW1	R. Colonia, Colonia
	50	VEN	KE	YVKE Mundial, Caracas
560	25/5	ARG	LV1	R. Colón, San Juan
	10/5	ARG	LT15	R. del Litoral, Concordia
	25/5	ARG	LRA13	R. Nal., Bahia Blanca
	25/1	ARG	LRA9	R. Nal., Esquel
	10/1	ARG	LRA16	R. Nal., La Quiaca
	15	BOL	LP03	R. El Mundo, La Paz
	25/10	CLM	PF	LV de la Pampa, Maicao
	10	CLM	GS	R. Nal., Tunja
	25	EQA	RN2	C. R. E. Satelital, Guayaquil
	10	GUY		NCN, Georgetown
	2	PRU	OBZ4L	R. Oriente, Lima
		PRU	OBX1H	Radiomar, Chiclayo
	20/10	VEN	PJ	R. Exitos "Latina 5-60", Rubio
	50	VEN	RH	R. Nal. "RNV", Cd. Guayana
570		ARG		R. Argentina, Buenos Aires
	50	CHL	CB57	R. Agricultura, Santiago
	100	CLM	ND	R. Nal de Colombia, SF de Bogotá
	10	EQA	CE1	R. El Sol, Quito
	1	PRG	ZP15	R. LV del Amambay, Pedro Juan Caballero
	1	PRG	ZP39	R.San Roque, Ayolas
		PRU	OCU2B	Huamachuco (F.PI.)
	1	PRU	OAU1M	R. Univ. Nal. Pedro Ruiz Gallo, Lambayeque
	100	VEN	LX	R. Rumbos, Villa de Cura
580	10/5	ARG	LU20	R. Chubut, Trelew
	25/5	ARG	LW1	R. Univ. Nal. de Córdoba, Córdoba
	10	BOL	LP01	R. Panamericana, La Paz
	50/10	CLM	HP	R. Nal., Cali
	10	EQA	PC2	R. Uno, Guayaquil
	1	PRU	OCY2L	R. El Sol, La Esperanza
	10	PRU	OAX2E	R. Marañón, Jaén
		PRU	OAX4M	R. Maria, Lima
	5	URG	CX58	R. Clarín, Montevideo
	50/10	VEN	MJ	LV de la Fe, Maracaibo
		VEN		R. Nacional "RNV", Maturín
590	25/5	ARG	LS4	R. Continental, Buenos Aires
	4	ARG	LV12	R. Independencia, San Miguel de Tucumán
	25/1	ARG	LRA30	R. Nal., San Carlos de Bariloche
	1	CHL	CA59	R. Horizonte, Antofagasta
	10	CHL	CD59	R. Patagonica, Punta Arenas
	1	CHL	CC59	R. Portales, Concepción
	50	CLM	CR	W R., Medellín
	10	EQA	SP1	R. Carrousel, Quito
	5	PRG	ZP32	R. Ycuámandyyú, Villa de S. Pedro
	1	PRU	OCX6V	R. Catedral, Miraflores, Arequipa
	20	VEN	KL	R. Continente, Caracas
600	20/5	ARG	LU5	R. Neuquén, Neuquén
	10	BOL	CH01	R. ACLO, Sucre
	1	BOL	LP35	Radioemisoras del Recobro, La Paz
	10	CHL	CB60	R. Monumental, Santiago
	1	CLM	Z95	LV de los Awas, Ricaurte el Diviso
	50	CLM	HJ	R. Libertad, Barranquilla
	50	EQA	XY2	R. Nal. del Ecuador, Guayaquil
	10	PRU	OBZ4W	R. Cora, Lima
		PRU	OAX6S	R. Cultura Toquepala, Ilabaya
		PRU	OCX6D	R. Cultural, Ilabaya
	1	PRU	OBX2B	R. Star, Trujillo
	10	SUR		R. Paramaribo, Paramaribo
	15	VEN	SW	R. Alto Llano, Sta Bárbara de Barinas
	10	VEN	QB	R. Sucre, Cumaná
610	5/1	ARG		La Buena R., Villa Lynch
	1	ARG	LRK201	R. Solidaridad, Añatuya
	5	CHL	CD61	R. Puerto Aysen, Puerto Aysén
	30	CLM	KL	La Cariñosa, SF de Bogotá
	50	CLM	D90	R. Nal., Uríbia
	10	EQA	MJ1	R. Caravana AM, Quito
	10/1	PRG	ZP30	LV del Chaco Paraguayo, Filadelfia
	5	PRU	OCY2I	R. Santa Monica, Chota
	1	PRU	OBU1E	R. Santa Rosa, Sullana
	50	URG	CX4	R. Rural, Montevideo
	10	VEN	XY	R. Centro 6-10, Cantaura
	10	VEN	SE	R. Cristal, Barquisimeto
620	25/5	ARG	LRA28	R. Nal., La Rioja
	25/5	ARG	LRA26	R. Nal., Resistencia
	25/7	ARG	LRA18	R. Nal., Río Turbio
	25/5	ARG	LT17	R. Provincia de Misiones, Posadas
	10/5	ARG	LV4	R. San Rafael, San Rafael
	10	BOL	LP02	R. San Gabriel, La Paz
	10	CHL	CC62	R. Bío-Bío, Concepción
	1	CHL	CA62	R. Norte Verde, Ovalle
	50/20	CLM	EL	Colmundo, Cali
	15	CLM	VP	Colmundo, Cartagena
	50	EQA	XY3	R. Nal. del Ecuador, Loja
	10/1	PRG	ZP40	R. Nasaindy, San Estanislao
		PRU	OAR7H	Apostol de Yanaoca, Yanaoca
	0.4	PRU	OAX2M	R. Chepen, Chepen
		PRU	OCX6B	R. Maria, Arequipa
	10	PRU	OBU4B	R. Ovación, San Isidro
	50/25	VEN	ZC	R. Fe y Alegría Los Llanos,Guasdualito
	10	VEN	NO	R. Libertad, Cabimas
630	10/5	ARG	LU4	R. Dif. Patagonia Argentina, Comodoro Rivadavia
	25/5	ARG	LS5	R. Rivadavia, Buenos Aires
	25/5	ARG	LW8	R. San Salvador de Jujuy
	10	CHL	CB63	R. Stela Maris, Valparaíso
	10	CLM	WC	LV del Guainía, Puerto Inírida
	10	CLM	FD	R. Manizales, Manizales
	10	EQA	HA2	Ondas Quevedeñas, Quevedo
	50/25	VEN	KA	R. Nal "Canal Informativo", Caracas
634		PRU	OAU2R	R. Cajamarca, Cajamarca
640	10/5	ARG	LU18	R. El Valle, "640 AM", General Roca
	25/5	ARG	LRA24	R. Nal., Río Grande
	10/5	ARG	LV15	R. Villa Mercedes
	10	CHL	CD64	R. Cooperativa AM, Temuco
	0.25	CHL	CC64	R. Portales, Curico
	10	CLM	BJ	RCN, Santa Marta
		EQA		R. Morena AM, Guayaquil
	50	EQA	XY1	R. Nal. del Ecuador, Quito
	8	PRG	ZP19	R. Caaguazú, Coronel Oviedo
	10	PRU	OAZ4K	R. Del Pacifico, Lima
	1	PRU	OAU1Y	R. La Luz, José Leonardo Ortiz

kHz	kW	Ctry	Call	Station, location
	10	PRU	OBX7B	R. Onda Azul, Puno
	30	VEN	QO	Actualidad 640, Puerto La Cruz
	10/5	VEN	MU	R. Carora, Carora
650		ARG		R.Reporter, Buenos Aires
	15	BOL	LP11	R. Dif. Integración, El Alto
	100	CLM	KH	RCN Antena 2, SF de Bogotá
	5	EQA	FD4	R. Visión Manta, Manta
	50	PRG	ZP4	R. Uno, Asunción
	1	PRU	OAX2N	R. Regional del Norte, Trujillo
	50/25	URG	CX6	S.O.D.R.E. "R. Clásica", Montevideo
	50/20	VEN	LH	Aragüeña 650, Maracay
660	1/5	ARG	LT41	R. LV del Sur Entrerriano, Gualeguaychú
		ARG		R. Popular, Claypole
	1	BOL	SC10	R. ABC, Santa Cruz
	50	CHL	CB66	R. UC, Santiago
	25	CLM	QS	Colmundo, Cúcuta
	20	CLM	JM	R. Auténtica, Cali
	30	EQA	LG2	R. Carrousel, Guayaquil
	6/12	PRG	ZP26	R. Itapirú, Cd. del Este
	1	PRU	OCX4L	R. Chinchaycocha, Junin
	5	PRU	OCX1U	R. J.H.C., Chiclayo
	10	PRU	OCX4R	R. La Inolvidable, Lima
	10	VEN	NA	Ondas de los Médanos, Coro
	10	VEN	QZ	R. Anaco, Anaco
		VEN		R. Nacional, El Callao
670		ARG		R. Antartida, San Justo
	25/5	ARG	LT4	R. Dif. Misiones, Posadas
	25/5	ARG	LRI209	R. Mar del Plata, Mar del Plata
		ARG		R. Maranata,Lomas del Mirador
	1	ARG	LRA52	R. Nal., Chos Malal
	25/5	ARG	LRA11	R. Nal., Comodoro Rivadavia
	10	CLM	R33	R. U.I.S., Bucaramanga
	50	CLM	PL	RCN Antena 2, Medellín
	12/5	EQA	FF1	R. Jesús del Gran Poder, Quito
	10	PRU	OAX7H	R. Nacional del Perú, Puno
	100	VEN	LL	R. Rumbos, Caracas
680	25	ARG	LT3	R. Cerealista, Rosario
		ARG		R. Magna, San Martin
	25/5	ARG	LV6	R. Nihuil, Mendoza
	25/5	ARG	LU12	R. Río Gallegos, Río Gallegos
	5	BOL	LP27	R. Andina, La Paz
	10	CHL	CC68	R. Cooperativa, Concepción
		CLM	ZO	R. Nal., Barranquilla (Sabanagrande)
	25/12	EQA	VP2	R. Atalaya, Guayaquil
	12	PRG	ZP11	R. Caritas, Asunción
	5	PRU	OAX5E	Emisora del Pacifico, Ica
	0.5	PRU	OBX2L	R. Amauta, Chócope
	5	PRU	OCY2Y	R. San Luis, Jaén
	20	PRU	OBX4A	R. Tigre, San Isidro
	1/0.7	URG	CW68	R. Young, Young
	10	VEN	QR	R. Continente Cumaná, Cumaná
	10	VEN	ZJ	R. Llerana "R. 1400", Barinas
690		ARG		AM Dakota, Buenos Aires
	10/3	ARG	LU19	R. LV de Comahue, Cipolletti
	25/5	ARG	LRA4	R. Nal., Salta
	10	CHL	CD69	R. Estrella del Mar, Ancud
	10	CHL	CB69	R. Santiago, Santiago
	1	CLM	Z73	LV Indígena de Uberaba, Apartado
	50/12	CLM	CZ	R. Recuerdos, SF de Bogotá
	50d	EQA	JB1	LV de los Andes, Quito
	5	EQA	FA4	Sucre Portoviejo, Portoviejo
		PRU		R. Manco Capac, Región Puno
	1/10	URG	CX8	R. Sarandí, Montevideo
	50/20	VEN	MR	R. Barquisimeto, Barquisimeto
693	1.5	PRU	OCX1T	R. Horizonte, Chiclayo
700	25/5	ARG	LV3	R. Córdoba, Córdoba
	5	CHL	CD70	R. Magallanes, Punta Arenas
	1	CHL	CA70	R. Nibsan, Copiapó
	1	CHL	CD70	R. Valdivia, Valdivia
	120	CLM	CX	W R., Cali
	50	EQA	RS2	Sucre Guayaquil, Guayaquil
	1	GUY		NCN, Linden
	1	PRG	ZP12	R. Carlos Antonio López, Pilar
	10	PRU	OBX1U	R. Cutivalú "LV del Desierto", Castilla
	1	PRU	OBU4J	R. La Luz, Huancayo
	1	PRU	OBU7K	R. La Salle, Urubamba
	1	PRU	OBZ4H	R. R.Integridad, San Miguel
	1	PRU	OCY2H	R. Sausal Superior, Sausal
		PRU	OCU1B	Sechura (F.Pl.)

kHz	kW	Ctry	Call	Station, location
	10	VEN	MH	R. Popular, Maracaibo
	5/2	VEN	PQ	R. Sur, Puerto Ordaz
	50	ARG	LRL202	R. Diez, Buenos Aires
710	25/5	ARG	LRA19	R. Nal., Pto. Iguazú
	25/1	ARG	LRA17	R. Nal., Zapala
	10	BOL	OR17	R. Pío XII, Siglo Veinte
	5	CLM	YD	R. La Paz, Paipa
	10	CLM	NX	R. Super, Medellín
	8	EQA	ER5	Escuelas Radiofónicas Populares, Riobamba
	1	PRU	OAU6L	R. Amor, Arequipa
	10	PRU	OCX7I	R. Nacional del Peru, Puerto Maldonado
	5	PRU	OBX5Q	R. Programas del Perú, Ica
		PRU		R. Surupana, Azángaro
	50/20	VEN	KY	R. Capital, Caracas
720	25/5	ARG	LV10	R. de Cuyo, Mendoza
	1	ARG	LRA59	R. Nal., Gobernador Gregores
	10	BOL	LP06	R. La Cruz del Sur, La Paz
	2.5	BOL	LP05	R. Yungas, Chulumani
	1	CHL	CA72	R. Portales, Iquique
	30	CLM	AN	Emisoras Unidas, Barranquilla
	50	CLM	ZX	R. Dif. Nal., Rionegro
	25	CLM	VO	Transmisora Quindío, Armenia
	10	EQA	GB4	LV de Portoviejo, Portoviejo
	5	EQA	MO3	R. Matovelle, Loja
	5	EQA	IC1	R. Municipal, Quito
	10	EQA	UE3	R. Única, Machala
	25	PRG	ZP17	R. Pai Puku, Teniente Irala Férnandez
	1	PRU	OBU7D	R. Alegria, Wanchaq
	1	PRU	OAU1Q	R. Frecuencia Oceánica, Lambayque
	25	PRU	OAX2J	R. Nacional del Perú, Trujillo
	10	PRU	OAU4E	R. Sideral, La Oroya
	10	VEN	XE	R. Elorza, Elorza
	50	VEN	QE	R. Venezuela "Oriente", Porlamar
725	5	SUR		SRS, Paramaribo
730		ARG		R. Guarani AM, San Justo
	10/1	ARG	LU23	R. Lago Argentino, El Calafate
	20/5	ARG	LRA27	R. Nal., Catamarca
	20/5	ARG	LRA3	R. Nal., Santa Rosa
		ARG		R. General Gümes "La Radio Mundial", Buenos Aires
	3	BOL	SC01	R. Mensaje, Montero
	1	CHL	CD73	R. Angelina, Los Angeles
	1	CHL	CD73B	R. Aysén, Pto. Aysén
	10	CHL	CB73	R. Cooperativa AM, Valparaíso
	100	CLM	CU	Cad. MelodíaRadio Lider, SF de Bogotá
	15	CLM	TJ	R. Uno, Montería
	10	EQA	MG2	R. Guayaquil, Guayaquil
	30	PRG	ZP7	R. Cardinal, Asunción
	10	PRU	OAX1D	R. del Pacifico, Piura
	5	PRU	OBU2Q	R. Maria, Cajamarca
	50	PRU	OAX4G	R. Programas del Peru RPP, San Isidro
	5/2.5	URG	CX10	R. Continente, Montevideo
	10	VEN	OO	R. Frontera, San Antonio del Táchira
	10	VEN	MT	R. Universo, Barquisimeto
740	25/5	ARG	LRH251	R. Chaco, Resistencia
	10/1	ARG	LRI200	R. Municipal
	1	ARG	LRA55	R. Nal., Alto Río Senguer
	10	CLM	HB	Ecos de Pasto, Pasto
	50	CLM	NS	R. Guatapurí, Valledupar
	10	EQA	SE4	R. Libertad, Quito
	10	EQA	GC1	R. Melodía "Canal 7-40", Quito
	1/0.5	PRG	ZP38	Hechizo AM, Caazapà
	10	PRU	OAX6C	R. Continental, Arequipa
	1	PRU	OCX2X	R. El Puerto, Pascamayo
	1	PRU	OBU7C	Red Latino, Cusco
	5/1	URG	CW27	R. Tabaré, Salto
	10	VEN	NC	CNB 740 La Zuliana, Maracaibo
	50	VEN	NQ	R. Caroni "Q-FM", Puerto Ordaz
750		ARG		R. AM 7-50, Lomas de Zamora
	100/10	ARG	LRA7	R. Nal., Córdoba
	50	CLM	DK	Caracol Colombia, Medellín
	5	CLM	LH	LV de Yopal, Yopal
	30	EQA	RC2	Caravana AM, Guayaquil
	5	PRG	ZP42	LV de la Policía Nal, Asunción
	5	PRU	OCX4X	R. Altura, Cerro de Pasco
	100	VEN	KS	RCR 750 "Radio Caracas", Caracas
760	25/5	ARG	LU6	Emisora. Atlántica, Mar del Plata
	5	BOL	CO02	R. Cosmos, Cochabamba
	50	BOL	LP07	R. Fides, La Paz

kHz	kW	Ctry	Call	Station, location
		BOL	CO56	R.Casachun Coca, Lauca Ñ
	50	CHL	CB76	R. Cooperativa, Santiago
	30/10	CLM	AJ	RCN, Barranquilla
	25	EQA	QR1	R. Quito "LV de la Capital", Quito
	10	GUY		NCN, Georgetown
	0.5	PRU	OBZ2K	R. Andino, Otuzco
	10	PRU	0BZ4X	R. Mar Plus, Chorillos
	10	VEN	QQ	R. Puerto La Cruz/La Doble Q, Pto. La Cruz
	10	VEN	SO	Simpatía 760 AM, Trujillo
770		ARG		Amplitud 770, Lomas del Mirador
	5/1	ARG		R. Cooperativa, Buenos Aires
	10	CHL	CD77	R. Agricultura, Temuco
	1	CHL	CD77	R. Cooperativa, Castro
	100	CLM	JX	RCN, SF de Bogotá
	25/12	EQA	MF2	R. El Telégrafo, Guayaquil
	5	PRU	OAX8M	LV de la Selva, Iquitos
	1.5	PRU	OCX1T	R. Horizonte, Chiclayo
	2.5	PRU	OAU7D	R. LV del Allinccapac, Macusani
	5	PRU	OBX6H	Radiomar, Uchumayo
	100/25	URG	CX12	R. Oriental, Montevideo
	50/20	VEN	KK	R. Nacional, Valencia
780	25/5	ARG	LV8	R. Libertador, Mendoza
	5	ARG	LRA12	R. Nal., Santo Tomé
	5/1	ARG	LRA10	R. Nal., Ushuaia
	5	ARG	LRF210	R. Tres, Trelew
	10	CHL	CD78	R. Sago AM, Osorno
	10	CLM	C21	Antena del Río, Barrancabermeja
	10	CLM	ZG	LV del Valle, Cali
	10/5	CLM	ZW	R. Almirante, Riohacha
	10/2	EQA	CM1	R. Colón AM, Quito
	50	PRU	ZP70	R. Primero de Marzo, Asunción
	1	PRU	OBU2N	R. Coremarca, Bambamarca
	10	PRU	OAX1K	R. Nacional del Perú, Tumbes
	10	PRU	OAZ7S	R. Nuevo Tiempo, Juliaca
	3	PRU	OAX4X	R. Victoria, Lima
	50/20	VEN	OD	Ecos del Torbes, San Cristóbal
	10	VEN	MN	R. Coro, Coro
790	5	ARG	LV19	R. Malargüe
	25/5	ARG	LR6	R. Mitre "AM 80," Buenos Aires
	25/5	ARG	LRA22	R. Nal, San Salvador de Jujuy
	1/0.25	ARG	LT46	R. Provincial, Bernardo de Irigoyen
	1	CLM	NC	Ecos del Combeima, Ibagué
	50	CLM	DC	R. Caracol, Medellín
	50	CLM	ZR	R. Nal., Villavicencio
	50	CLM	BU	R. Nal., Zambrano
		EQA		R. Paraíso, Maldonado
		EQA		Su Radio 790 AM, Otavalo
	5	PRU	OAZ7H	R. La Luz, Cusco
	10	PRU	OAX2I	R. Programas del Perú, Trujillo
	50	VEN	XM	R. Minuto, Barquisimeto
		VEN		R. Nacional, Cd.Bolívar
	10	VEN	KC	R. Venezuela 7-90, Caracas
800	1/0.25	ARG		R. AM 800 Wajzugun, San Martin de los Andes
	1/0.25	ARG	LT43	R. Mocoví, Charata
	1/0.25	ARG	LV23	R. Rio Atuel, General Alvear
	24/5	ARG	LU15	R. Viedma
	1	BOL	CH12	R. Churuquella, Sucre
	5	BOL	LP08	R. Libertad, La Paz
	0.25	BOL	LA09	R. Santa Clara, Sorata
	5/1	CHL	CB80	R. Maria, Viña del Mar
	100	CLM	BW	RCN, Bucaramanga
	25	EQA	ML2	K 800, Guayaquil
	5	EQA	FB1	R. Sensación 800, Quito
	5/3	PRG	ZP27	R. Mbaracayá, Salto del Guairá
	0.5	PRU	OAU4H	R. La Luz, Huaral
	0.3	PRU	OBX6A	R. Porteña, Arequipa
	1	PRU	OBX5B	R. Sur, Ica
	1	PRU	OBU4D	R. Vida, Huancayo
		PRU	OBU6D	Tacna (F.P.I.)
	1	PRU	OCX1P	Telecom del Norte, Piura
810		ARG		R. La Gauchita, Morón
	10/1	ARG		R. Mitre AM 810, Córdoba
	200	CLM	CY	Caracol Colombia, SF de Bogotá
	5	EQA	VT2	R. Atalaya, El Milagro
		EQA		Sucre Ambato, Ambato
		PRU	OBU5E	Huamanga (F.P.I.)
	1	PRU	OAU2G	R. Apocali, Trujillo
	10	PRU	OAX7V	R. Programas del Perú, Juliaca
	50/25	URG	CX14	R. El Espectador, Montevideo
	50	VEN	LP	Super Radio 810, Valencia
820	5/1	ARG	LRI208	Estacion 820, Lomas de Zamora
	1/0.25	ARG	LRK221	R. Ciudad Perico, Perico
	25/5	ARG	LRA8	R. Nal., Formosa
	5/1	ARG	LU24	R. Tres Arroyos- "LV del Pueblo"
	10	BOL	LP10	R. Altiplano, La Paz
	1	CHL	CD82	R. Concordia, La Unión
	1	CHL	CC82	R. Maria Inmaculada, Concepción
	0.25	CHL	CA82	R. Pampa, Pedro de Valdivia
	10/1	CHL	CA82A	R. Portales, La Serena
	10/5	CHL	CB82	Radioem. Carabineros de Chile, Santiago
	50	CLM	ED	Caracol Colombia, Cali
	10	CLM	AD	R. Vigía, Cartagena
	1	EQA	RF4	Canal Manabita, Portoviejo
	5	EQA	VI5	LV de Ingapirca, Cañar
	25	EQA	UP1	R. Unión, Quito
	20	PRU	OAX4O	R. Libertad, Lima
	0.5	PRU	OBX2J	R. Nuevo Continene, Cajamarca
	1	SUR		Apintie, Paramaribo
	1/0.5	URG	CW23	R. Cultural, Salto
	10	VEN	KU	R. Altura 820, La Grita
	25/10	VEN	XG	R. Guadalupana, Coro
	50	VEN	SH	R. Guayana, Upatá
830	5	ARG		R. del Pueblo, Buenos Aires
		ARG		R. Filadelfia, San Justo
	1/0.5	ARG	LT21	R. Municipal, Alvear
	0.25	ARG	LV18	R. Municipal, San Rafael
	25	ARG	LU14	R. Provincia de Santa Cruz, Río Gallegos
	10/5	ARG	LT8	R. Rosario, Rosario
	25	CLM	DM	R. Reloj, Medellín
	25	EQA	RM2	R. Huancavilca, Guayaquil
	4.5	EQA	RP5	R. Promoción, Riobamba
	1	PRU	OAU4C	CPN Radio, El Tambo (rel.
	1	PRU	OCX2Y	CPN Radio, Huancayo
	1	PRU	OAZ7U	R. Inti Raimi, Santiago
	1	PRU	OAX3Y	R. La Selva, Rupa-Rupa
	10	PRU	OAX6D	R. Nacional del Perú, Tacna
		PRU	OCU2M	R. Ebenezer, Bambamarca
	25	VEN	LT	R. Nueva Sensación, Caracas
840	25/5	ARG	LU2	R. Bahía Blanca, Bahía Blanca
	3	ARG		R. General Belgrano, Buenos Aires
	10/5	ARG	LT12	R. General Madariaga, Paso de los Libres
	25/5	ARG	LV9	R. Salta, Salta
	3	BOL	LP75	R. Atipiri, El Alto
	10	CHL	CB84	R. Portales, Valparaíso
	10	CHL	CD84	R. Santa María, Coyhaique
	30	CLM	KK	H J Doble K, Neiva
	30	CLM	BI	Ondas del Caribe, Santa Marta
	1	EQA	EM4	R. Costa Azul, Portoviejo
	50	EQA	PN1	R. Vigía, Quito
	5	PRG	ZP6	R. Guairá, Villarrica
	1	PRU	OBX6Y	R. Azul, Arequipa
	1	PRU	OAX3S	R. Casma, Casma
	1	PRU	OAU2E	R. Nuevo Continente, San Ignacio .
		PRU	OCU7I	R. Santa Cruz, Kunturkanki
	10	VEN	MY	8-40 AM, Barquisimeto
	10/5	VEN	UZ	Guarapiche 8-40 "La Primera", Maturín
850	10	ARG		LV de Amérca, San Miguel
	1	BOL	PO01	R. 21 de Diciembre, Mina Catavi
	5	BOL	SC03	R. María Auxiliadora, Montero
	50	CLM	KC	W R., SF de Bogotá
	20/12	EQA	VS2	R. San Francisco, Guayaquil
	40	PRU	OAX4A	R. Nacional del Perú, Lima
		PRU	OAU6S	R. Nacional del Peru, Tarata
		PRU	OBU7Z	R. Pachamama, Puno
		PRU		R. San Juan, El Tambo
	50	URG	CX16	R. Carve, Montevideo
	10	VEN	ZC	R. Fe y Alegría, Maracaibo
	10	VEN	RV	RV-850, Valencia
860		ARG		R. Digital, Lanus
	5/1	ARG		R. Municipal, Chilecito
	1	ARG	LRA56	R. Nal., Perito Moreno
	10	BOL	LP12	R. Nueva América, La Paz
		BOL	BE01	R. Paitití, Guayaramerín
	10	CHL	CC86	R. Inés de Suárez, Concepción
	12	CLM	NJ	LV del Cañaguate, Valledupar
	10	CLM	FP	Voces de Occidente, Buga
	10	EQA	PC1	R. Positiva, Quito
	25	PRG	ZP28	LV de la Cordillera, Caacupé

kHz	kW	Ctry	Call	Station, location
		PRU	OAU2J	CPN Radio, Cajamarca
	3	PRU	OCX1M	R. Nuevo Norte, Sullana
	5	PRU	OAU5Q	R. Educativo Macedonia, Ayacucho
	20/10	VEN	YE	Enlace 8-60, Valle de la Pascua
	10	VEN	OL	Mundial 8-60, San Cristóbal
870	100	ARG	LRA1	R. Nal., Buenos Aires
	0.6	BOL	CO31	LV del Campesino, Sipe Sipe
	5	CLM	GD	Em. Reina de Colombia, Chiquinquirá
	10	CLM	LA	LV del Tolima, Ibagué
	25	CLM	SB	R. Mar Caribe Int., Barranquilla
	5	CLM	ZH	Vida AM, Medellín
	20	EQA	NY2	R. Cristal, Guayaquil
	1	EQA	GS6	R. Píllaro, Píllaro
	2.5	PRU	OCX4D	R. Huancayo, Huancayo
	5	PRU	OAU7O	R. Libertad, Puno
	1	PRU	OCX7R	R. Mundo, Wanchaq
	10	PRU	OBX1F	R. Programas del Perú, Chiclayo
	1	PRU	OAU1G	R. San Pedro Chanel, Sullana
	10	VEN	RU	Unión R. Deportes 870, Puerto La Cruz
	10	VEN	MP	Unión R. Notícias, Barquisimeto
875		BOL	OR21	R. Eucaliptos, Eucaliptos
880	10	ARG	LU14	R. Provincia de Santa Cruz, Las Heras
		BOL	LP42	R. Inca, El Alto
	10	CHL	CB88	R. Colo Colo, Santiago
	20	CLM	GE	Caracol Colombia, Bucaramanga
	10	CLM	FH	R. Regional Independiente, Anserma
	50/40	EQA	RP1	R. Católica Nacional, Quito
	1	PRU	OAX2P	R. Sintonia, Trujillo
	10	PRU	OBZ4N	R. Union, Lima
	10	VEN		R. Paraguaná (Continente), Punto Fijo
	20/10	VEN	YM	R. Venezuela, Puerto Ordaz
	10	VEN	KV	RNV 880 Canal Musical, Caracas
890	25/5	ARG	LV11	Em. Santiago del Estero, Santiago del Estero
		ARG		R. Libre, San Justo
	25/5	ARG	LU33	R. Pampeana, Santa Rosa
	10	CHL	CC89	R. Interamericana, Concepción
	10	CHL	CA89	R. León XIII, Pozo Almonte
	20	CHL	CD89	R. Nal., Punta Arenas
	10	CLM	CE	R. Continental, SF de Bogotá
	0.25	CLM	HK093	R. Ecos de Soledad, Soledad
	20	CLM	PM	R. Galeón, Santa Marta
	1	EQA	TL5	Ondas del Chimborazo, Riobamba
	25/20	EQA	RS3	R. Superior, Machala
	5/0.5	PRG	ZP33	R. Tres de Febrero, Itá
	1	PRU	OBX7S	R. Bahá´í del Lago Titicaca, Chiucuito
		PRU	OCU7C	R. Laramani, Espinar
	1	PRU	OAU2N	R. Panorama, Cajamarca
	25	URG	CX18	R. Sport 890, Montevideo
	10	VEN	LW	R. América, Valencia
	10	VEN	VO	R. Oriente, El Tigre
900	1	ARG		R. Municipal, 25 de Mayo
	25/5	ARG	LT7	R. Provincia de Corrientes, Corrientes
	0.25	BOL	TA01	R. LV Nacional, Tarija
	5/0.1	BOL	LP36	R. Popular, La Paz
	1	CHL	CB90	Cablenoticias, Valparaíso
	1	CHL	CD90	R. LV de la Costa, Osorno
	1	CHL	CC90	R. Mayor, Chillán
	5	CHL	CA90	R. Universidad, Copiapó
	10	CLM	EY	LV de Cali, Cali
	15/5	CLM	DD	R. Super, Cúcuta
	5	EQA	OF4	R. Chone, Chone
	1	EQA	RR5	R. Reloj, Cuenca
	10	EQA	VA1	Sucre Quito, Quito
	10	PRU	OBX4X	R. Felicidad, Lima
	1	PRU	OBX6K	R. Nevada, Uchumayo
	1	PRU	OAU2Q	R. Nor Oriental, Jaén
		PRU	OAX3E	R. Ribereña, Aucaycu
	1	PRU	OCX1D	R. Sensacional, Ferreñafe
	3/1	URG	CW17	R. Frontera, Artigas
	25	VEN	MD	R. Venezuela "Mara Ritmo", Maracaibo
902	1	BOL	CO33	R. Central Misionera, Cochabamba
905	1.5	BOL	SC09	R. Norte, Montero
910	25/5	ARG	LR5	R. La Red, Buenos Aires
	50/5	ARG	LRA23	R. Nal.,San Juan
	15	CLM	S52	Colombia Estereo, Florencia
	10	CLM	DO	LV del Rio Grande, Medellín
	1	CLM	TT	Ondas del Porvenir, Samacá
	30	CLM	MY	RCN, San Andrés
	2	EQA	BO2	Nueva R. Colón,Guayaquil
	5	EQA	GE5	R. Mundial, Riobamba
	1	PRU	OAU5M	R. Estacion Wari, Ayacucho
	1	PRU	OAU7G	R. Vision del Altiplano, Juliaca
	50/20	VEN	RQ	R. Q 910 "Tu AM Center", Caracas
914	3	SUR		R. Nickerie, Nieuw Nickerie
920	3	BOL	CH11	R. Encuentro, Sucre
	1	BOL	LP65	R. San Andres de Topohoco, Topohoco
	1	CHL	CD92	R. 920, Temuco
	10	CLM	SJ	Colmundo, Ibagué
	30	CLM	AA	Em. Fuentes, Cartagena
	10	CLM	JN	Ondas del Mayo, Pasto
	10	EQA	RU3	CRO - Compañía Radiofónica Orense, Machala
	1	EQA	AB1	R. Democrácia, Quito
	10/100	PRG	ZP1	R. Nal. del Paraguay, Asunción
		PRU		R. Campesina, Juli
	1	PRU	OAX9V	R. Marginal, Tocache
	1	PRU	OBX2S	R. Ollantay, Virú
	1	PRU	OCX7M	R. Programas del Perú, Cusco
	10	PRU	OBX1J	R. Programas del Peru, Piura
	0.1	PRU	OCX5C	R. Stelar, Chinca Alta
	20	VEN	QX	R. Nueva Esparta, Porlamar
	10/5	VEN	QU	R. San Carlos, San Carlos
930		ARG		R. Nativa - "LV de Nuestra Gente", Ciudad Madero
	25/5	ARG	LV7	R. Tucumán, San Miguel de Tucumán
	5/1	ARG	LV28	R. Villa María, Villa María
	10	CHL	CB93	R. Nuevo Mundo, Santiago
	10	CHL	CD93	R. Reloncaví, Puerto Montt
	30	CLM	CS	LV de Bogotá, SF de Bogotá
	5	EQA	VI2	Canal Tropical, Guayaquil
	5	EQA	BA6	R. Ambato, Ambato
		PRU	OBU7T	R. Cadena Colca, Juliaca
	1	PRU	OCX2V	R. Inti, Chepén
	5	PRU	OAX4E	R. Moderna "R.Papa", Lima
	2.5	PRU	OBX6T	R. Yaravi, Arequipa
	50/25	URG	CX20	R. Monte Carlo, "la Super R.", Montevideo
	10	VEN	LJ	R. Maracay, Maracay
940	3/5	ARG	LRH200	R. Chajarí, Chajarí
	20/5	ARG	LRJ241	R. Dimensión, San Luís
		BOL	LP13	R. Metropolitana, La Paz
	1	BOL	CH13	R.Chuquisaca XXI, Sucre
	1	CHL	CA94	R. 9-40, Copiapó
		CHL	CC94	R. Armonia, Viña del Mar
	1	CHL	BC94	R. Valentín Letelier, Valparaíso
	10	CLM	GB	R. Calima, Cali
	25	CLM	TL	RCN, Cúcuta
		EQA		R. Austral del Ecuador, Cuenca
	5	EQA	BZ1	R. Dif. de la Casa de la Cultura Ecuatoriana, Quito
	1	PRU	OBU4E	R. Comericial, Jauja
	1	PRU	OBX2G	R. CRN, Cutervo
	1	PRU	OBU1Y	R. Studio Satelite, Tambo Grande
	1.5	PRU	OBX7L	R. Willkamayu, Wanchaq
	15	VEN	ZR	R. Continental, Barinas
	10	VEN	LU	R. Fe y Alegría El Tigre, El Tigre
	10	VEN	NN	R. Punto Fijo, Punto Fijo
941	0.8	BOL	CO49	R. San Lorenzo, Colcapirhua
950	25/5	ARG	LR3	R. Belgrano, Buenos Aires
	25/5	ARG	LT16	RSP - R. Sáenz Peña, Roque Saénz Peña
	3	BOL	CO01	R. Yurac Molino, Chimboata
	8	CLM	UJ	Armonias Boyacenses, Tunja
	15	CLM	FN	Caracol Colombia, Pereira
		EQA		Chaskis del Norte, Ibarra
	10	EQA	DE2	GRD R. Internacional, Guayaquil
	3	EQA	UE5	R. Colta, Colta
	1	PRU	OBX9L	R. Estacón Láser, Rioja
	1	PRU	OBX3S	R. Programas del Perú, Chimbote
	50	VEN	KG	YVKG AM Popular, Caracas
960	10/1	ARG	LRA6	R. Nal., Mendoza
	10/3	ARG	LU13	R. Necochea, Necochea
	1	BOL	PO02	R. Kollasuyo, Potosí
	10	BOL	SC04	R. Santa Cruz, Santa Cruz
	10	CHL	CB96	R. Carrera, Santiago
	10	CHL	CD96	R. Polar, Punta Arenas
	1	CLM	HX	Candela AM, Bucaramanga
	15	CLM	R31	Candela, San Andrés
	50	CLM	HN	Caracol Colombia, Magangué
	1	EQA	JX6	LV del Santuario, Baños

kHz	kW	Ctry	Call	Station, location
	1	EQA	NC1	R. Cosmopolita, Quito
	1	EQA	SA5	Sono Onda Internacional, Cuenca
		PRU	OBU7P	R. Concierto Santa Monica, Espinar
	1	PRU	OBX8H	R. Diez, Iquitos
	12	PRU	OBX6S	R. Hispania, Mariano Melgar
	1	PRU	OCY4V	R. Manantial, Huancayo
	10	PRU	OAX4D	R. Panamericana, Lima
	3	PRU	OBX1Y	R. WSP, Chiclayo
	3/1	URG	CW96	R. Yi, Durazno
	50/20	VEN	RB	R. Monagas, Maturín
	10	VEN	SS	R. San Sebastián, San Cristóbal
	25	VEN		R. Venezuela "Llanera 9-60", Acarigua
962	1	BOL	LP43	R. Huayna Potosí, Milluni
965	10	EQA	OT1	R. Católica Nal, Sto Domingo de los Colorados
970	25/5	ARG	LV2	R. General Paz "AM 970", Córdoba
		ARG		R. Genesis, Buenos Aires
	1/0.25	ARG	LT25	R. Guaraní, Curuzú Cuatiá
	1	CHL	CD97	R. Austral, Valdivia
	1	CHL	CA97	R. Calama, Calama
	1	CHL	CC97	R. Lautaro, Talca
	1	CHL	CD97A	R. Patagonia Chilena, Coyhaique
	30	CLM	VK	Armonias del Caquetá, Florencia
	0.25	CLM	HKX59	R. Quimbaya, Calarca
	10	CLM	CI	R. Super, SF de Bogotá
	10	CLM	ME	RCN Guajira, Maicao
	20	EQA	AW2	R. Católica Nal. del Ecuador, Guayaquil
	1	EQA	MB1	R. Imperio, Ibarra
	80	PRG	ZP9	R. 9-70 , Asunción
	1	PRU	OBX5A	R. Comericial Sonora, Ica
	1	PRU	OBX1V	R. La Capullana, Sullana
	1	PRU	OAU2K	R. Lider del Norte, Cajamarca
	1	PRU	OBU7B	R. Qollasuyo, Juliaca
	1	PRU	OAU7A	R. Tropicana, Wanchaq
	20/5	URG	CX22	R. Universal, Montevideo
	10	VEN		Mundial 970, Barcelona
	10	VEN	LR	R. Continente 970 Maracay, Maracay
	15	VEN	SD	R. Turismo, Valera
972	0.6	BOL	CO31	LV del Campesino, Sipe Sipe
980	3/1	ARG	LU37	R. General Pico "Radio37",
		ARG	LRG387	R. Luján AM, Valcheta
	5	ARG	LT39	R. Victoria, Victoria
		ARG		R.Regional, San Miguel
	5	BOL	CO04	R. Esperanza, Aiquile
	2.5	BOL	LP14	R. Mar Plus, La Paz
	5	CHL	CB98	R. Agricultura, Valparaíso
	1	CHL	CA98	R. Univ. de Tarapaca, Arica
	15	CLM	JV	La 980 Sensacional, Cúcuta
	100	CLM	ES	RCN, Cali
	5	EQA	CL3	R. Cariamanga, Cariamanga
	1	EQA	JI5	R. El Prado, Riobamba
	5	PRG	ZP31	R. Mburucuyá, Pedro Juan Caballero
		PRU	OAX5K	R. LV de Huamanga, Huamanga
	1	PRU	OBU4H	R. OBU4H, Huancayo
	1	PRU	OAU6F	R. Universidad, Arequipa
	1	PRU	OAU1N	Radio y Sonido, Lambayque
		VEN		R. Nacional, Maracaibo
	10	VEN	QM	Unión R. Notícias, El Tigre
990	25/5	ARG	LRH203	AM 990, Formosa
	1	ARG	LRJ201	R. Calingasta, Barreal
	25/5	ARG	LR4	R. Splendid AM 990, Buenos Aires
	1	CHL	CC99	R. El Roble, Parral
	5	CLM	HI	LV de Garagoa, Garagoa
	100	CLM	DB	RCN, Medellín
	15	EQA	EW2	Frecuencia Mil, Guayaquil
	25	EQA	GH1	R. Tarquí, Quito
		PRU	OCU4A	R. & TV Oro, Huayllay (F.P.I.)
	10	PRU	OAX6K	R. Continental, Tacna
	0.5	PRU	OBX2M	R. Contumaza, Contumaza
	12	PRU	OBX4J	R. Latina, Miraflores
	1	PRU	OBU1C	R. OBU1C, Sechuar
		PRU	OBX3L	R. Peruana, Chimbote
	1	PRU	OBX8E	R. Programas del Perú, Iquitos
	20	VEN	RT	R. Tropical "99-0", Caracas
	10	VEN	TA	R. Venezuela "Tricolor", Barquisimeto
1000	1/0.25	ARG	LT42	R. Del Iberá, Mercedes
	1/0.25	ARG	LU16	R. Río Negro, Villa Regina
		ARG		R.Sintonia, José C.Paz
	10	BOL	OR03	R. Bahá´í de Bolivia, Caracollo

kHz	kW	Ctry	Call	Station, location
	3	BOL	BE02	R. Dif. Trópico, Trinidad
	1	BOL	LP44	R. Mística, La Paz
	1	BOL	SC33	R. Piraí, Santa Cruz
	10	CHL	CB100	R. RRB, Santiago
	10	CLM	JG	R. Nal., Manizales
	50	CLM	ZP	R. Nal., Yopal
	0.8	CLM		R. Panamericana, Cajibío
	15	CLM	AQ	RCN, Cartagena
	1	EQA	NT3	Dinamita Mil AM, Catamayo
	5/0.5	PRG	ZP36	R. Mil, San Antonio
	2	PRU	OAU2P	R. Bambamarca, Bambamarca
	1	PRU	OBX6R	R. Edesa, Paucarpata
		PRU	OBX3V	R. Huanuco
	1	PRU	OBX5W	R. Lircay, Lircay
	2	PRU	OAZ7P	R. Prensa al Dia, Cusco
	1	PRU	OAU1P	R. San Jose, Lambayque
		PRU	OBX9T	R. Ministerio Mundial, Utcubamba
	10	VEN	NM	R. Caribeña Mil AM (Continente), Morón
	10	VEN	OA	R. Táchira, San Cristóbal
1010	1/0.25	ARG	LW2	R. Emis. Tartagal, Tartagal
	1	ARG	LRA28	R. Nacional, La Rioja // LRA28 - 620
		ARG		R. Onda Latina, Buenos Aires
	20/10	ARG	LV16	R. Rio Cuarto, Rio Cuarto
		BOL	CO03	R. LV de Sipe Sipe, Sipe Sipe
	15	CLM	JR	Caracol Colombia, Neiva
	10/5	CLM	BN	LV del Galeras, Pasto
	10	CLM	OP	Oxígeno R., Barranquilla
	15	CLM	ZD	R. Panzenú, Montería
	10	CLM	CN	R. Reloj/W R., SF de Bogotá
	10	CLM	IX	R. Yarima, Barrancabermeja
	3	EQA	RZ2	R. Amiga, Guayaquil
	2.5	EQA	RV5	R. Visión, Cuenca
	15	EQA	NR6	TSB R. Líder, Ambato
		PRU	OBU1L	LV de las Huaringas, Huancabamba
	1	PRU	OAU5G	R. Amistad, Abancay
		PRU		R. Bella, Tingo Maria
	10	PRU	OAX4U	R. Cielo, Lima
	1.5	PRU	OBX2P	R. San Francisco, Cajamarca
	1	PRU	OBZ1C	R. Sonora, Tumbes
	20	URG	CX24	R. Nuevotiempo, Montevideo
	10	VEN	PC	R. Aragua, Cagua
	10	VEN	QF	R. Venezuela, Cd.Bolívar
1011		CHL	CD101	R. Nielol, Temuco
1020	25/5	ARG	LRJ214	AM Mil 20 - La R. de la Gente, San Juan
	1	ARG	LRA58	R. Nacional, Río Mayo
	10/5	ARG	LT10	R. Univ. Nal. del Litoral, Santa Fé
	10	BOL	LP15	R. Illimani - R.Patria Nueva, La Paz
	5	CHL	CC102	R. Amiga, Talca
	10	CLM	DQ	Emisora Claridad, Medellín
	10	CLM	KS	LV del Llano, Villavicencio
	15	CLM	DZ	R. Primavera, Bucaramanga
	10	CLM	FT	R. Super, Ibagué
	10	CLM	FQ	RCN, Pereira
	3	EQA	GO3	Canal Estelar, Santa Rosa
	5/3	EQA	CR6	R. Surcos, Guaranda
	5	EQA	HR1	RTU (Radio y Televisión Unidas), Quito
	25/10	PRG	ZP14	R. Nandutí, Asunción
	1	PRU	OBU4F	R. Cristo Vive, Huancayo
	5	PRU	OBX1G	R. Heroica, Chiclayo
	1	PRU	OBU7O	R. Informes, Sicuani
	1	PRU	OAU6J	R. Internacional, Tacna
	1	PRU	OBU1D	R. La Luz, Piura
	1	PRU	OBX3U	R. Nacional, Chimbote
	0.1	URG	CW102	R. Libertadores, Salto
	25	VEN	TW	R. Alegría, Chivacoa
	50/10	VEN	MX	R. Continente/R. Calendario, Maracaibo
	10	VEN	RS	R. Mundial Margarita, La Asunción
1030	25/5	ARG	LS10	R. del Plata, Buenos Aires
		BOL	PO27	R. Em. Comunitaria Colquechacam ,
		BOL	CH19	R. Em.Comunitaria Mojocoya , Mojocoya
	3	BOL	CO48	R. Independencia, Ayopaya
	3	BOL	OR26	R. Orinaca, Orinaca
	3	BOL	BE20	R. Patria Nueva , Riberalta
		BOL	CO51	R. Totora, Totora
	1	CHL	CD103	R. Chiloé, Castro
	1	CHL	CD103A	R. Payne AM, Puerto Natales
	1	CHL	CB103	R. Progreso, Talagante
	15	CLM	RF	Ondas del Cesar, Aguachica
	5	CLM		Ondas del Vaupés, Mitú

kHz	kW	Ctry	Call	Station, location
	1	CLM	GX	R. Progreso de Córdoba, Lorica
	30	CLM	ER	RCN Antena 2, Cali
	10	CLM	DJ	RCN LV de los Libertadores, Duitama
	5	EQA	RF2	R. Punto 1030/Ecuantena, Guayaquill
	1	PRU	OCX6L	R. Cumbia, Arequipa
	1	PRU	OCX7O	R. HG-AM, Cusco
	3	PRU	OAX2U	R. Imperio, Huamachuco
	1	PRU	OAX7N	R. LV del Altiplano, Puno
		PRU		R. Sillustani, Atuncolla
	20	VEN	QY	R. Onda 1030, Guanare
	25/10	VEN	TD	R. Valles del Tuy, Ocumare del Tuy
1040		BOL	CH18	R. Em. Comunitaria 12 de Marzo, Tarabuco
		BOL	OR28	R. Em. Comunitaria Qaqachaca, Qaqachaca
		BOL	SC37	R. Em.Comunitaria Camiri, Camiri
		BOL	TA11	R. Em.Comunitaria Libertad, Villamontes
		BOL	SC38	R. Em.Comunitaria San Juan, San Julián
		BOL	SC30	R. San José, San José de Chiquitos
	0.25	BOL	CO20	R. Sipe Sipe, Quillacollo
	1	BOL	PO03	R. Villazón, Villazón
	1	CHL	CD104	R. Raíces, Curacautín
	15	CLM	UB	Colmundo, Pasto
	15	CLM	CJ	Colmundo, SF de Bogotá
	15	CLM	FM	LV de Armenia, Armenia
	15	CLM	BF	LV del Norte, Cúcuta
	10	CLM	SY	R. 1040/La Caucana 10-40, Popayán
	15	CLM	AI	R. Tropical, Barranquilla
	3	EQA	CW1	LV del Valle, Machachi
	3	EQA	GB6	R. Colosal, Ambato
	10/5	EQA	EV5	R. Splendit, Cuenca
	5	PRG	ZP43	R. Arapisandú, San Ignacio
	1	PRU	OAU7H	Multimedio Sistena de R., Espinar
	1	PRU	OBX5U	R. Andina del Pacifico, Ica
	10	PRU	OBX4O	R. Metropolitana, Miraflores
	1	PRU	OBX2O	R. Nor Oriente, Jaén
	1	PRU	OAZ1D	R. Vecinal, Piura
	20	VEN	LB	LV de Carabobo, Valencia
	20/10	VEN	ON	Mundial Los Andes, Mérida
1043	1	BOL	LP45	R. Bolivianísima, La Paz
1050	10/5	ARG		Concepto AM, Buenos Aires
	10	ARG	LV27	R. San Francisco, San Francisco
	5	BOL	SC06	R. El Mundo, Santa Cruz
		BOL	CO52	R. Em.Comunitaria , Independencia
		BOL	PO26	R. Em.Comunitaria Caizad , Caizad
	3	BOL	OR27	R. Sabaya Sabaya
	1	CHL	CD105	R. Armonía, Osorno
	15	CLM	BB	Caracol Colombia, Valledupar
	15	CLM	S62	Cusiana R., Yopal
	10	CLM	FZ	La Cariñosa del Centro, Antena 2, Espinal
	5	CLM	IO	LV de la Conquista, Granada
	10	CLM	LZ	LV del Cinaruco/Caracol, Arauca
	10	CLM	GU	R. Bucarica, Bucaramanga
	10	CLM	NG	R. Palmira, Palmira
	10	CLM	DR	R. Unica, Medellín
	5/3	EQA	IM1	LV de Imbabura, Ibarra
	5	EQA	RQ2	R. Águila, Guayaqui
	1	PRU	OBX8F	CPN Radio, Iquitos
	1	PRU	OBX6D	R. Bethel, Arequipa
	1	PRU	OBZ4J	R. Bolognesi, Huancayo
		PRU		R. Campesina, Cajamarca
	1	PRU	OCX2B	R. Maria, Chepen
	1	PRU	OAZ7Q	R. San Augustín, Juliaca
	25	URG	CX26	S.O.D.R.E. "R.Uruguay", Montevideo
	20	VEN	PO	R. Nacional, Cabudare
	25	VEN	KZ	RNV "R. Educativa", Caracas
1058	5	PRU	OAU1C	CPN R., Tumbes
1060		ARG		R. Las Naciones, Monte Grande
	10	BOL	LP38	R. Eco Loyola, La Paz
	0.5	BOL	SC07	R. LV de la Frontera, Pto. Suárez
	1.5	BOL	OR01	R. Noticias, Oruro
	50	CHL	CB106	R. Maria, Santiago
	1	CLM	YX	Caracoli, Sincelejo
	5	CLM	LY	R. Delfín, Riohacha
	10	CLM	MV	R. Furatena, Chiquinquirá
	1	CLM	MG	R. Litoral, Turbo
	15	CLM	OV	R. Surcolombiana, Neiva
	15	CLM	FJ	RCN Caldas, Manizales
	5	EQA	MG6	R. Ecos del Pueblo, Saquisilí
		EQA		R. Fiesta, Machala
		EQA		R. Richi, El Empalme

kHz	kW	Ctry	Call	Station, location
	0.05	GUF		RFO Guyane, St. Laurent du Maroní
	3	PRG	ZP13	R. Boquerón, Alberdi
	5	PRU	OAU1C	CPN R., Tumbes
		PRU	OAU7U	R. 1060, Cusco
		PRU	OBU1F	R. 1060. Piura
	1	PRU	OCY4D	R. Exito, Lima
		PRU	OAU6R	R. La Luz, Ilo
	5	PRU	OCY2O	R. Sudamerica, Cutervo
	10	VEN	LN	R. Guárico 1060 AM, S. Juan de los Morros
	10	VEN	OE	Unión R. Noticias, San Cristóbal
1070	25/5	ARG	LR1	R. El Mundo, Buenos Aires
	20	CLM	AH	Em. Atlántico, Barranquilla
	30	CLM	CG	R. Santa Fé, SF de Bogotá
	15	CLM	VR	R. Super, Popayán
	1	EQA	VP1	R. Libertad, Quito
	1	EQA	RS1	R. Lubakán, Santo Domingo de los Colorados
	5	EQA	CJ5	R. LV de Tomebamba, Cuenca
	10	GUF		RFO Guyane, Matoury
	3	PRU	OBX9J	R. Andes, Tarapoto
	0.2	PRU	OAX5A	R. San Juan, San Juan de Marcona
	1	PRU	OAU6K	R. Trinidad, Arequipa
	1	PRU	OAU1J	R. Vida, Chiclayo
	1	PRU	OBX4G	R. Visión, San Ramón
	25	VEN		Contacto 1070, Ospino
	10	VEN	MA	Mundial Zulia, Maracaibo
	5	VEN	PX	R. El Sol, La Fría
	10	VEN		Superior 1070 Biruaca, San Fernando de Apure
1075	0.5	BOL	SC05	R. Agricultura, Portachuelo(n.f.
1080		ARG		R. Claridad, Monte Grande
	25/5	ARG	LU3	R. del Sur, Bahía Blanca
	10/1	ARG		R. Departamento Minas, Andacollo
	25/5	ARG	LW4	R. Orán/R.Maria
		BOL	LP76	LV de la Mayoria, Caranavi
	1	BOL	CH02	R. Dif. Colosal, Sucre
	1	CHL	CD108	R. Los Confines, Angol
	1	CHL	CA108	R. Río Elqui, Vicuña
	10	CLM	AX	La 1080, Medellín
	10	CLM	AW	LV de Montería, Montería
	10	CLM	MH	Melodía AM, Floridablanca
	10	CLM	KT	R. Autentica/R. Macarena, Villavicencio
	10	CLM	JF	R. Eco, Cali
	15	CLM	JS	R. Pontoná, La Dorada
	1	EQA	AB4	R. Contacto, Manta
	10	EQA	BH6	R. Latacunga, Latacunga
	10	EQA	KD2	Sistema 2, Guayaquil
	10	PRG	ZP25	R. Nanawa, Luque
		PRU	OCU7O	R. 1080, Puno
	1	PRU	OCX6X	R. Futura, Ilo
	10	PRU	OAU4I	R. La Luz, Lima
	1	PRU	OAU2L	R. Nueva Vida, Cajamarca
	2.2	PRU	OAX7S	R. Salkantay, Cusco
	1.5	PRU	OBX1D	R. San Miguel, Piura
	10	VEN	QJ	R. Barcelona, Barcelona
	10	VEN	NR	R. Venezuela, Maracay
1090	0.5	ARG		Libertad AM 1100, Rosario
		ARG		R. Décadas, Hurlingham
		ARG		R. Nuestras Raíces, Valentín Alsina
	3	BOL	CO06	R. Cultura, Cochabamba
	15	CLM	BC	Caracol Colombia, Cúcuta
	10	CLM	IH	Caracol Colombia, Sogamoso
	10	CLM	JB	LV de los Pijaos, Guamo
	10	CLM	IA	Oxígeno R., Manizales
	10	CLM	IG	R. Autentica, Florencia
	5	CLM	OM	R. Bucanero, Cartagena
	5	EQA	VI1	R. Irfeyal, Quito
	1	PRU	OBX6X	R. Amistad, Arequipa
	1	PRU	OBX2A	R. Cajabamba, Cajabamba
	1	PRU	OAU5F	R. Inti Andina, Aucara
	15	URG	CX28	R. Imparcial, Montevideo
	3	VEN	TG	Melódica 1090, Machiques
	10	VEN	PB	R. Yaracuy "Operadora 1090 AM", S. Felipe
	20	VEN	SZ	Unión R. Notícias1090, Caracas
1100	1.3	ARG		R. Estilo, Glew
		BOL	LP29	R. Chaka, Pucarani
	4	BOL	LP04	R. Mundial, La Paz
	1	BOL	OR06	R. Universidad de Oruro
	10	CHL	CB110	R. Integridad, Viña del Mar
	1	CHL	CA110	R. La Portada, Antofagasta

kHz	kW	Ctry	Call	Station, location
	10	CLM	CN	BBN R., SF de Bogotá
	15	CLM	AT	Caracol Colombia, Barranquilla
	5	CLM	MK	Emisora Ideal, Planeta Rica
	5/1	CLM	GI	LV de Colombia, Socorro
	2	CLM	EF	LV del Vichada, Puerto Carreño
	15	CLM	YZ	R. Super, Neiva
	5	CLM	GQ	Transmisora Surandes, Andes
	5/2	EQA	GR6	R. Novedades, Latacunga
	1.5	EQA	LE7	R. Oriental, Tena
	5	PRG	ZP71	R. Nú Verá, Capitán Bado
		PRU	OCU2E	R. 1000, Julcan
		PRU	OCX4S	R. Imperial "LV de la Provincia", Cañete
	1	PRU	OBX7Z	R. LTC, Juliaca
	1	PRU	OBX1L	R. Star, Chiclayo
	1	PRU	OCY4G	Sonorama Radio, Huancayo
	10	VEN	SV	R. Angostura, Cd.Bolívar
	10	VEN	OP	R. Fe y Alegría, Tovar
1110	25/5	ARG	LS1	R. de la Ciudad"La Porteña", Buenos Aires
	10	CHL	CD111	R. La Frontera, Temuco
	1	CLM	PA	LV de las Islas, San Andrés
	5	CLM	GP	LV del Río Arauca, Arauca
	10	CLM	EW	Oxígeno 1110, Cali
	10	CLM	DI	R. Bolivariana, Medellín
	15	CLM	ZE	R. Piragua, Sincelejo
	10	CLM	JP	RCN, Villavicencio
	10	EQA	JR1	R. Clásica, Quito
	5	EQA	JC5	R. Ondas Azuayas, Cuenca
	5	EQA	RP6	R. Pelileo, Pelileo
	1	PRU	OAZ4W	R. Antarki, Lima
	1	PRU	OCX6P	R. Austral, Ilo
	0.5	PRU	OCX1R	R. Centro Popular, La Union
	5	PRU	OCX7T	R. Comer, Cusco
	1	PRU	OCX2U	R. Jaén, Jaén
	2/1	URG	CX111	R. Paso de los Toros, Paso de los Toros
	10	VEN	QT	R. Venezuela Carúpano, Carúpano
	10	VEN	RX	Unión N. Notícias, Valencia
1115	1	BOL	PO18	R. Difusoras Independencia, Atocha
1120		ARG		AM Tango, Buenos Aires
		ARG		Em. Santiago y Copla, Gregoria de Laferrere
	25/5	ARG	LV5	R. Sarmiento, San Juan
		ARG		R.Sudamericana, Victoria
	1	BOL	BE03	R. Estación El Dorado, Trinidad
		BOL	LP48	R. Revelacion El Gran Yo Soy, El Alto
		BOL	CO53	R.Porvenir, Tiquipaya
	5	CLM	Q92	Colombia Mía, Yopal, CS
	10	CLM	KQ	J.B.R., Tunja
	15	CLM	GH	Oxígeno R., Bucaramanga
	5	CLM	JC	R. Matecaña, Pereira
	10	CLM	TI	Vox Dei, Cúcuta
	2	EQA	EB1	Canal 1120, San Gabriel
	5	EQA	FV2	Estación Intercontinental, Guayaquil
	10	EQA	LE1	R. Dif. Marañon, Sto Domingo de los Colorados
	3	EQA	AS7	R. Variedades del Puyo, El Puyo
	10	PRG	ZP24	R. Nuevo Mundo, San Lorenzo
		PRU	OBX8R	Coronel Portillo (F.PI.)
	1	PRU	OAU5H	R. Dif. Sonora Comunal - R. Quispillaccta, Ayacucho
	1.5	PRU	OBX2I	R. Dinamica, Trujillo
		PRU	OAU5W	R. Huayllahuara
	1	PRU	OCX6U	R. Municipal, Cerro Colorado
		PRU	OAM2F	R.dif. Cristiana RP, Chota
	10/2	URG	CW31	R. Salto, Salto
	10	VEN	MF	Ondas del Lago "Super Ondas", Maracaibo
	20/10	VEN	SK	R. Dif.del Sur, San Fernando de Apure
	5	VEN	XZ	R. República "La Estación Feliz", Maturín
1125	0.5	BOL	SC08	R. Cruceña, Cotoca
	0.3	BOL	OR02	R. Em. Cooperativa Poopó, Poopó
1130	10	ARG		R. Cadena Vida, Buenos Aires
		ARG		R. Carisma, El Talar
	25/5	ARG	LRA21	R. Nal., Santiago del Estero
	10	CLM	AC	Em. Riomar, Barranquilla
	1	CLM	NN	Ondas del Río, Magangué
	15	CLM	QQ	Oxígeno R., Pasto
	15	CLM	VA	Vida AM Básica, SF de Bogotá
	5	EQA	PV6	R. Centro, Ambato
	5/3	EQA	RD1	R. Punto, Ibarra
		EQA		R. Sibimbe AM, Ventanas
		EQA		Romántica AM, Machala

kHz	kW	Ctry	Call	Station, location
	1	PRU	OAU5A	R. Armonia, Abancay
	2.6	PRU	OAX4N	R. Bacán, Lince
	1	PRU	OAZ4S	R. Chanchamayo, Chanchamayo
	1.2	PRU	OAX2V	R. Los Andes (RPP), Cjamarca
	1	PRU	OAU7B	R. San Gabriel, Juliaca
	20/5	URG	CX30	R. Nacional, Montevideo
	15	VEN	PY	R. Amazonas, Puerto Ayacucho
	20/10	VEN	RL	R. Ideal, Maiquetía
	10	VEN	KQ	R. Popular, Barquisimeto
1140		ARG		R. Independencia, Remedios de Escalada
	10/1	ARG	LU22	R. Tandil, Tandil
		ARG		R. La Luna, El Palomar
	2	BOL	LP49	R. Pico Verde, Chulumani
		BOL	CO45	R. San Isidro, Colami
	75	CLM	CB114	R. Nal., Santiago
	10	CLM	RW	Caracol Villavicencio, Villavicencio
	10	CLM	KO	R. Esperanza, Cartagena
	10	CLM	DL	R. Paisa, Medellín
	10	CLM	CL	R. Panamericana, Girardot
		CLM		R. Piendamo, Piendamo
	10	CLM	RN	RCN, Barbosa
	1	EQA	AZ5	R. Alfa Musical, Cuenca
	1.5	EQA	FB2	R. Cóndor, Guayaquil
	4	EQA	MF4	R. Rumbos, Portoviejo
	5	EQA	IR1	Raíz 11-40, Quito
	5/2	PRG	ZP22	R. Panambí Verá, Villarrica
		PRU	OCU2D	Chami Radio, Otuzco
	1	PRU	OAX3R	R. Bahia, Chimbote
	0.5	PRU	OAX5W	R. Chinchaysuyo, Chinca Alta
	1	PRU	OAX6L	R. Concordia, Arequipa
		PRU	OAU1T	R. Fraternal, Ferreñafe
	1.5	PRU	OBX1W	R. Piura, Piura
	1	PRU	OCY4C	R. Programas del Perú, Pilcomayo
		VEN		R. Porlamar "LV del Caribe", Porlamar
1143	1	BOL	SC31	R. Colonia, Yapacani
1145	1	BOL	LP17	R. Chuquiago Musical, La Paz
1150	10/5	ARG	LT9	R. Brigadier López, Santa Fé
	1	ARG	LRA51	R. Nal., Jáchal
	10	ARG	LRA2	R. Nal., Viedma
	5	ARG		R. Sagrada Familia, San Justo
	10	ARG	LRH202	R. Tupá Mbaé, Posadas
		BOL	OR22	R. 24 de Noviembre, Eucaliptos
	0.2	BOL	TA02	R. Chaco, Yacuíba
	0.5	BOL	OR07	R. El Cóndor, Oruro
	0.3	BOL	LP50	R. Guaqui, Guaqui
	15	CLM	FI	Caracol Colombia, Armenia
	1	CLM	GJ	JB R., Duitama
	1	CLM	TE	LV del Chocó, Quibdó
	10	CLM	BT	R. Catatumbo, Ocaña
	5	CLM	FP	RCN, Neiva
	10	EQA	GB5	LV de Riobamba, Riobamba
	10	EQA	AV3	R. Luz y Vida, Loja
	0.5	PRU	OCY2E	R. Chasqui Llacta, San Marcos
	2.5	PRU	OAU7K	R. Lider, Juliaca
	10	PRU	OAX8D	R. Loreto, Iquitos
	5	PRU	OBU4K	R. Mineria, Cerro de Pasco
	2.5	PRU	OCX7Q	R. Universal, Santa Monica
	10	VEN	QP	Ecos del Orinoco, Cd.Bolívar
	10	VEN	MV	R. Venezuela "Ondas del Caribe", Punto Fijo
1153		PRU		R. Ayabaca , Ayabaca
1160	5/10	ARG	LRH253	R. Cataratas, Pto. Iguazú
	10/2.5	ARG	LU32	R. Coronel Olavarría, Olavarría
		ARG		R. Excelsior, Monte Grande
	1	ARG	LRA57	R. Nal., El Bolsón
	5	BOL	SC11	R. Centenario, "La Nueva", Sta. Cruz
	10	BOL	LP33	R. Continental, La Paz
	1	BOL	CH03	R. Nuevo Mundo, Sucre
	3/1	BOL	CO08	R. RTC, Cochabamba
	1	CHL	CC116	R. Ancoa, Linares
	1	CHL	CD116A	R. Baha'i, Temuco
		CHL	CD116	R. El Espectador de America, La Serena
	10	CLM	S31	Colombia Mía, Barrancabermeja
	15	CLM	OC	Ecos de Colombia, SF de Bogotá
	5	CLM	AZ	Frecuencia Bolivariana "tu emisora", Montería
	15	CLM	AU	Ondas del Orteguaza, Florencia
		CLM		Ondas del Puerto, La Virginia
	10	CLM	BL	R. Aeropuerto, Barranquilla
	10	CLM	EC	R. San José de Cúcuta, Cúcuta
	10	CLM	EV	R. Unica, Cali

kHz	kW	Ctry	Call	Station, location
	5	CLM	ZV	RCN R. Las Lajas, Ipiales
		EQA		LV del Pueblo, Azoguez
	1	EQA	WD4	R. Cenit, Portoviejo
	1	EQA	UR6	R. Runatacuyaj, Latacunga
	2	EQA	VR3	R. Vía, Machala
	5	EQA	CP1	Super Auténtica, La Radio 11-60, Quito
	15	PRG	ZP72	R. Antena Dos, Asunción
	5	PRU	OAX4C	Onda Cero, Lima
	1	PRU	OCX7Z	R. del Sur, Puerto Maldonado
	1	PRU	OBX5O	R. Huanta 2000, Huanta
	0.3	PRU	OAX2C	R. Libertad Mundo, Trujillo
	1	PRU	OBX6G	R. Nacional del Perú, Moquegua
	1	PRU	OAU2T	R. siglo 21, Chota
	1	PRU	OCX1S	Radiales Nor Oriental del Marañón, Chiclayo
	1	URG	CW116	R. Agraria del Uruguay, Cerro Chato
	20/10	VEN	RR	R. Notícias RN 1160, Guarenas
	1	VEN	OK	R. Universidad, Mérida
1170	5	ARG		R. Mi País, Hurlingham
	25/3	ARG	LRA29	R. Nal., San Luis
	3	CHL	CD117	R. Natales, Puerto Natales
	10	CLM	NW	Caracol Colombia, Cartagena
	10	CLM	E74	Meridiano 70, Arauca
	10	CLM	PB	Ondas de Macondo, Valledupar
	10	CLM	BX	Ondas del Meta, Villavicencio
	10	CLM	KW	R. Nutibara, Medellín
	10	CLM	GA	R. Recuerdos, Tunja
	1	CLM	JE	RCN, Tuluá
	5	EQA	JV5	R. Central, Riobamba
	5	EQA	RV2	R. Filadelfia, Guayaquil
	5	EQA		R. Trébol AM, Zaruma
		PRU	OBU7F	R. Bethel, Cusco
	0.5	PRU	OCX7Y	R. Constelación, Puno
	1	PRU	OCX4Y	R. COSAT, Satipo
		PRU	OAU5V	R. Horizonte La Voz del Agro, Pueblo Nuevo
	1	PRU	OBX8M	R. La Luz, Iquitos
	1	PRU	OAU2M	R. Layzon, Cajamarca
	1	PRU	OAZ3K	R. Nor Peruana Chimbote
	10	PRU	OBX6L	R. Programas del Perú, Arequipa
	1	PRU	OCX1B	R. San Juan, Talara
	10	URG	CX32	Radiomundo, Montevideo
	20/10	VEN	QV	R. Acaricua, Acarigua
	10	VEN	KW	R. Bolivariana "R. 1070", Maiquetía
1180		ARG	LRI357	R. de la Sierra, Tandil
	5	BOL	OR04	R. Central, Oruro
	1	BOL	LP18	R. Emisora Ingavi, Viacha
		BOL	PO04	R. Kollasuyo, Potosí
		BOL	LP73	R. LV de Dios, Al Alto
	1	BOL	CO38	Radioem. 20 de Septiembre, Arbieto
	50	CHL	CB118	R. Portales, Santiago
	15	CLM	FX	Caracol Colombia, Manizales
		CLM		Em. Coorpurabá, Apartadó
	5	CLM	WA	LV del Guaviare, San José del Guaviare
	20	CLM	GK	R. Santander 2, Bucaramanga
	10/5	CLM	JT	RCN, Ibagué
		EQA		LV del Volante, Portoviejo
	12.5	EQA	LR1	Nueva Em. Central, Quito
	4	EQA	DP5	R. Cuenca, Cuenca
	1.2	EQA	R1	R. Familiar, Julio Andrade
	5/1	PRG	ZP52	R. Coronel Oviedo - RCO-AM, Coronel Oviedo
	1	PRU	OCX2A	R. Americana, Quiruvilca
	1	PRU	OAZ1C	R. Chulucanas, Chulucanas
	1	PRU	OBU2E	R. La Luz, Jaen
	1	PRU	OCY4Z	R. Libertad "R.RLJ", Junin
	10	URG	CX118	LV de Artigas, Artigas
	10	VEN	LQ	LV de la Victoria "Super Suave 11-80", La Victoria
	10	VEN	OR	R. Maturín, Maturín
	10	VEN	NJ	R. Petrolera, Cd.Ojeda
1190	25/5	ARG	LR9	R. América, Capital Federaf
	50	ARG	LRA15	R. Nal., San Miguel de Tucumán
	10	CLM	CT	LV de la Costa, Barranquilla
	15	CLM	EO	Ondas del Valle, Cartago
	1	CLM	KI	R. Barají, Sahagún
	10	CLM	CV	R. Cordillera, SF de Bogotá
	10	CLM	KG	R. Mira, Tumaco
	2	EQA	DE2	Estudio Universidad Católica, Guayaquil
	1	EQA	RF6	R. El Sol, Pujilí
	5	PRG	ZP45	LV de la Libertad, Henendarias
	1.5	PRU	OCX6G	R. Alas Peruanas

kHz	kW	Ctry	Call	Station, location
	5	PRU	OBX3D	R. Ancash, Huaraz
	10	PRU	OAX1E	R. Em. del Pacífico, Chiclayo
	2	PRU	OAX7B	R. Tawantinsuyo, Cusco
	20/10	VEN	PF	Ondas de Libertad, San Félix
	20/10	VEN	RE	R. Barinas 1190 AM Estéreo, Barinas
	10	VEN	ZD	R. Dif. Cult. del Táchira, San.Cristóbal
1195	1	BOL	CO09	R. Independencia, Quillacollo
		CLM		Ondas del Ranchería, Barrancas
1200	5/1.5	ARG	LT6	R. Goya, Goya
	1	ARG	LRA6	R. Nal. Mendoza, Valle de Uspallata
	0.25	BOL	CO10	R. 24 de Noviembre, Arani
		BOL	CH10	R. Mauro Nuñes, Villa Serrano
	5	BOL	SC12	R. Oriental, Santa Cruz
	10	CHL	CD120	R. Agricultura, Los Angeles
	10	CLM	CD	Em. Nueva Epoca, Fusagasugá
	10	CLM	LR	La Cariñosa, Antena2, Sogamoso
	10	CLM	BZ	Ondas del Riohacha, Riohacha
	15	CLM	IJ	R. 1200 "LV de la Raza", Medellín
	10	CLM	BV	R. Príncipe, Cartagena
	10	CLM	NF	R. Super, Cali
	5	EQA	RE2	LV del Trópico, Quevedo
	5	EQA	RM5	R. El Mercurio, Cuenca
	5	EQA	CS1	R. Super K, La Líder, Sangolquí
		EQA		R. U Cadena Sur, Sta Rosa
	10	PRG	ZP44	R. Libre, Fernando de la Mora
	3	PRU	OAX4B	Cadena Radio 1200, Lima
	1	PRU	OAU2A	Frecuencia Pedagogica, Cajamarca
	3	PRU	OAU4G	R. Andes, Huancayo
	1	PRU	OBX5X	R. Comercial, Abancay
	1	PRU	OCX7S	R. Cultura, Juliaca
		PRU	OBX8N	R. La Luz, Pucallpa
		PRU	OAU6P	R. La Luz, Tacna
	1	PRU	OAU1Q	R. San Andres, Tambogrande
	2	URG	CW33	La Nueva R., Florida
	1	VEN	NH	Ondas del Escalante, Sta Bárbara del Zulia
	10	VEN	SF	R. Dimensión, Caripito
	10	VEN	OZ	R. Tiempo, Caracas
1210		ARG	ZT	R. La Luz, Lomas de Mirador
	5/1	ARG	LRI229	R. Las Flores, Las Flores
		ARG		R. Mailín, Gregorio de Laferrere
	5	CHL	CD121	R. Armonia, Puerto Montt
		CHL	CA121	R. Universidad de Antofagasta, Antofagasta
	1	CHL	CC121	R. Universidad de Talca, Talca
		CHL	CB121	R. Valparaiso, Valparaiso
	10	CLM	BE	La Cariñosa, Antena 2, Cúcuta
	10	CLM	BQ	La Cariñosa, Pereira
	10	CLM	FR	R. Recuerdos, Neiva
	10	EQA	VC3	R. Centinela del Sur, Loja
	20	EQA	BJ2	R. El Mundo, Guayaquil
	3	EQA	JM6	R. Sira, Ambato
		PRU	OBX3X	Huanuco (F.P.I.)
	1	PRU	OCY4T	R. Galaxia, Satipo
	1	PRU		R. Municipalidad, Ayabaca
	5	PRU	OAX8A	R. Nacional del Perú, Iquitos
	1	PRU	OAX7M	R. Quillabamba, Quillabamba
		PRU	OCU7B	R. Santo Tomas
	1	PRU	OAX2Q	R. Universo, Trujillo
	2	URG	CX121	Difusora Soriano, Mercedes
	1	URG	CW121	R. El Libertador, Villa Vergara
	2.5/0.25	URG	CV121	R.RBC, Piriápolis
	10	VEN	ZT	R. Anzoátegui, Barcelona
1215		BOL	CO55	R. Tupuñan, El Paso
1220	5/1	ARG	LRL328	LV del Aire, Buenos Aires
		ARG		R. LRC - "La Radio de Chaco", Pres. Roque Sanez Peña
	1	ARG	LRI224	R. Onda Marina, Mar del Plata
	1	BOL	OR09	R. Batallón Topáter, Oruro
		BOL	CO11	R. El Cóndor, Arque
	1	BOL	LP19	R. Splendid, La Paz
	1	CHL	CA122	R. La Caribeña, La Serena
	10	CHL	CD122	R. Santa Maria de Guadalupe, Temuco
	10	CLM	KR	R. María, "LV Católica de su Hogar", SF de Bogotá
	15	CLM	FF	R. Reloj, Barranquilla
	10	CLM	NM	R. Viva Cultural Bolívar, Ipiales
	10	CLM	MT	RCN La R., San Gil
	10	CLM	AV	RCN, Montería
	3/5	EQA	EB6	Ecos de Bolívar, Guaranda
	10	EQA	AP1	R. Marañón, Quito

kHz	kW	Ctry	Call	Station, location
		PRU	OBU5I	Pisco (F.PI.)
	1	PRU	OAU5N	R. La Luz, Ica
	3	PRU	OCX1X	R. Libertad, Chiclayo
	4	PRU	OAX6X	R. Melodía, Arequipa
	1	PRU	OAU7N	R. Univ.Nal.San Antonio Abad, Cusco
	1	URG	CX122	R. Reconquista, Rivera
	10	VEN	RD	LV de Apure, San Fernando de Apure
	20/10	VEN	ZO	R. Aeropuerto 1220, Maracaibo
	10/5	VEN	VM	R. Venezuela, Valencia
1230		ARG		R. Claridad, Monte Grande
	25/5	ARG	LT2	R. Gen. San Martín "R.Dos", Rosario
		ARG		R. La Bendición, General Pico
	5/1	ARG	LW5	R. Libertador, General San Martin
		ARG		R. Litoral, Isidro Casanova
		ARG		R. Creativa, CA Buenos Aires
	15	CLM	GV	Colmundo, Bucaramanga
	10	CLM	IL	Minuto de Dios, Medellín
	10	CLM	BR	Oxígeno R., Tunja
	10	CLM	KL	R. Calidad "La Cariñosa", Cali
	1	CLM	TP	R. Colina, Girardot
	1	CLM	MJ	RCN Antena 2, Maicao
	3	EQA	RI1	CRI-Centro Radiofónico de Imbabura, Ibarra
	1	EQA	RL6	LV de Saquisilí y Libertador, Saquisilí
	15	EQA	FV2	R. Galáctica, Guayaquil
	3	EQA	MV5	R. Popular, Cuenca
	5	EQA	FG4	Sucre Esmeraldas, Esmeraldas
	3/0.5	PRG	ZP21	Fenix AM, Fortaleza del Este
	1	PRU	OAX2T	R. Albújar, Guadalupe
		PRU	OAU7V	R. Frecuencia Amistad, Juliaca
	1	PRU	OBX4Z	R. LV de Oxapampa, Oxapampa
	0.5	PRU	OBX7J	R. Madre de Dios, Puerto Maldonado
	1	PRU	OBZ4Y	R. Selecciones,Tarma
	10	VEN	NT	R. Barlovento, Caucagua
	10	VEN	OH	R. Valera, Valera
1240		ARG		R. Cadena Uno, Buenos Aires
		BOL	LP51	R. Achocalla, Achocalla
	2	BOL	TA03	R. Los Andes, Tarija
	1	BOL	CO12	R. San Miguel, Arani
	0.25	CHL	CA124	R. Principal Chuquicamata, Calama
	10	CHL	CB124	R. Universidad de Santiago, Santiago
	5	CLM	GN	R. Barrancabermeja, Barrancabermeja
	5	CLM	JA	R. Buenaventura, Buenaventura
	1	CLM	GO	R. Caribabare, Saravena
	10	CLM	FG	RCN, Calarcá
	5	EQA	RF3	R. Fenix, Zaruma
	1	EQA	PA1	R. Metropolitana, Quito
		EQA	LA5	R. Musical, Riobamba
		PRU	OBX9O	R. Bagua Grande
	1	PRU	OAU2Y	R. Bolivar, Bolivar
	10	PRU	OAU4V	R. Cumbre, Chilca (R.Maria)
		PRU	OAU5U	R. Eco, Ica
	1	PRU	OAU3C	R. La Luz, Chimbote
	5	PRU	OAU6D	R. Lider, Arequipa
	1	PRU	OCX1C	R. Sechura, Sechura
		PRU	OBX7M	R. Túpac Amaru, Sicuani
	5/1	URG	CW35	R. Paysandú, Paysandú
1250	1	ARG		R. Estirpe Nacional., San Justo
	1	BOL	SC14	R. Amboró, Santa Cruz
	0.1	BOL	PA01	R. Frontera, Cobija
	2.5	BOL	CH04	R. La Plata, Sucre
	0.5	BOL	CO13	R. Nacional, Cochabamba
	0.4	BOL	OR10	R. Oruro, Oruro
	0.5	BOL	SC13	R. Sararenda, Camiri
		BOL	PO20	R. Uncía, Uncia
	10	CHL	CD125	R. Armonía, Valdivia
	1	CHL	CA125	R. Santa Maria de Guadalupe, La Serena
	10	CLM	CA	Capital R., SF de Bogotá
	10	CLM	OK	Em. ABC, Barranquilla
	1	CLM	EM	LV de Corozal, Corozal
	15	CLM	HS	R. Reloj, Cúcuta
	10	CLM	FV	R. Viva/R. María, Pasto
	3	EQA	MY1	LV del Triunfo, Sto Domingo de los Colorados
	10	EQA	EM1	Ondas Carchenses, Tulcán
	10	EQA	HB2	R. Tricolor, Guayaquil
	5	PRG	ZP3	R. Asuncion
	1	PRU	OAU2V	HGV, Santa Cruz
		PRU	OAX9K	R. Comercial, San Martin
	5	PRU	OAX4L	R. Miraflores, Miraflores
	1	PRU	OAX8P	R. Pucallpa, Pucallpa
	3	PRU	OBX7A	R. Solar, Cusco
	1	PRU	OBZ1B	R.B.N.S., Talara Alta
	5	URG	CW125	R. Bella Unión, Bella Unión
	10	URG	CX36	R. Centenario, Montevideo
	20/10	VEN	PZ	Latina 12-50, Pto Ordaz
	1	VEN	ML	R. Cabimas, Cabimas
1255		BOL	LP51	R. Achocalla, Achocalla
1258		PRU		R. Centinela, Huancabamba
1260	10/5	ARG	LT14	R. General Urquiza, Paraná
	2	ARG		R. Oasis, Victoria
		ARG		R. Olivia, CA Buenos Aires
		ARG		R. Panamericana, CA Buenos Aires
		BOL	CO54	R. LV de la Esperanza, Quillacollo
	10	BOL	OR20	R. Nacional de Huanuni, Huanuni
	2	BOL	LP20	Radioemisora Unidas, La Paz
	2	CHL	CC126	R. Condell, Curicó
	10	CHL	CA126	R. Nal., Arica
	10	CHL	CD126	R. Santa Maria de Guadalupe, Punta Arenas
	5	CLM	NO	Bésame AM, Duitama
	5	CLM	DV	Caracol Colombia, Ibagué
	1	CLM	HU	Caracol Colombia, San Andrés
	5	CLM	LX	Minuto de Dios Eco Llanero, Villavicencio
	2	CLM	OU	Ondas del Amazonas, Leticia
	5	CLM	DA	R. Auténtica, Medellín
	5	CLM	ET	R. María, Cali
	5	CLM	TM	R. Sonar, Ocaña
	5	CLM	OH	RCN Cesar, Valledupar
	10	EQA	MO1	LV del Santuario del Quinche, Quito
	1	EQA	RB3	R. Benemérita, Sta Rosa
	3	EQA	RO6	R. Calidad, Ambato
	2	EQA	PB5	R. Contacto XG, Cuenca
	1	PRG	ZP34	R. Panambi Vera, Villarrica
	1	PRU	OAZ1A	R. Ferrañafe, Ferrañafe
	1	PRU	OAU3F	R. La Luz, Huanuco
		PRU		R. Mahanaim, Mollendo
	1	PRU	OBX6D	R. Mundial, Arequipa
	0.3	PRU	OBX5S	R. Nacional del Perú, Ayacucho
	1	PRU	OCX10	R. Nor Puruana, Chiclayo
		PRU	OAU3G	R. Periodico, Chimbote
		PRU	OBX2C	Telesistema Peruano, Otuzco
	3/1	URG	CW37	Dif. Rochense, Rocha
	10	VEN	RM	BBN R., Caracas
	10	VEN	RY	R. Horizonte, Nirgua
1265	0.4	BOL	PO05	R. Uncía, Uncía
1270	5	ARG	LRA20	R. Nal., Las Lomitas
	25/50	ARG	LS11	R. Provincia de Buenos Aires, La Plata
	1	BOL	LP21	R. Vanguardia, Colquiri
	10	CHL	CB127	R. Festival, Viña del Mar
	5	CLM	TX	Bésame AM, Bucaramanga
	5	CLM	IM	Colmundo, Pereira
	5	CLM	Q99	Colombia Mía, San José del Guaviare
	2	CLM	AR	La Cariñosa, Antena 2,Cartagena
	1	CLM	XQ	LV Amiga, Ubaté
	1.5	CLM	KJ	LV de Curumaní, Curumaní
	1	CLM	SV	LV de Orito, Orito
	5	CLM	BM	R. Internacional, Honda
	3	EQA	LD4	R. Junín, Junín
	15	EQA	UM2	R. Universal, Guayaqui
	1	PRU	OAX8T	R. Eco, Iquitos
	1	PRU	OCX2Z	R. Estacion Latina, Cepén
	2	PRU	OAU7S	R. Horizonte, Cusco
	0.4	PRU	OAZ4H	R. Huacho, Huacho
	0.4	PRU	OBZ4T	R. La Merced, Chanchamayo
	1	PRU	OAU1S	R. Nor Paita, Paita
	4/2	URG	CV127	R. Cuareim, Artigas
		VEN		R. Nacional, Ureña
	10	VEN	OU	R. Ondas Panamericanas, El Vigía
	5	VEN	TR	R. Tucupita, Tucupita
1275	0.5	BOL	SC15	R. Chané, Mineros
1276	1	PRU	OAU1R	R. Gotas del Oro, Urrunaga
1280		ARG		El Sonido de la Gente, Gregorio de Laferrere
		ARG		R. Eco Porteña, CA Buenos Aires
		ARG		R. Mística, Libertad
		ARG		R. Punto, Buenos Aires
	10/5	ARG	LU11	R. Trenque Lauquén, Tr. Lauquén
		BOL	LP68	R. Ondas del Titicaca, Huarina
	1	CHL	CC128	R. Arturo Prat Chacón AM, San Carlos
	10	CHL	CD128	R. del Sur, Osorno
	5	CLM	LR	Caracol Colombia, Pasto

kHz	kW	Ctry	Call	Station, location
	5	CLM	RP	Ecos de Tibú, Tibú
	5	CLM	HO	Impacto Popular, San Juan del Cesar
	1	CLM	NQ	LV del Río Suárez, Barbosa
	5	CLM	SO	R. Playa Mendoza, Barranquilla
	5	CLM	TK	R. Super, Caicedonia
	5	CLM	CM	R. Sur, Pitalito
	5	CLM	MB	R. Suroeste, Concordia
	5	CLM	KN	R. Única, SF de Bogotá
	1	EQA	IN4	LV del Sur de Manabí, Jipijapa
	1	EQA	NW5	R. Canal Tropical, Riobamba
	10	PRG	ZP53	R. LV del Este, Cd. del Este
		PRU	OBX3C	R. Alopesa, Chimbote
	0.5	PRU	OBX6P	R. Fénix, Camaná
	1	PRU	OAU1R	R. Gotas del Oro, Urrunaga
	1	PRU	OAX3Y	R. La Selva, Rupa-Rupa
		PRU	OCU7R	R. Ministerio Mundial, Sicuani
	1	PRU	OBX2F	R. Moderna, Cajamarca
		PRU	OAU7K	R. Stereo Nevada, Espinar
	1	PRU	OCX6B	R. Trebol, Mariano Melgar
	3/1	URG	CX128	R. Tacuarembó, Tacuarembó
	10	VEN	OF	R. Trujillo, Trujillo
	10/5	VEN	QS	R. Zaraza (Continente), Zaraza
1290	1	ARG	LRI371	R. Amanecer, Reconquista
		ARG		R. Interactiva, Ciudad Madero
	5/1	ARG	LRJ212	R. Murialdo, Villa Nueva de Guaymallén
		ARG		R. Provinciana, San Miguel
	1	BOL	OR12	Radiodifusoras Minería, Oruro
	0.25	CHL	CA129	R. Coya, María Elena
	0.25	CHL	CC129	R. Doce Noventa, Los Angeles
	5	CLM	SZ	Colombia Mía, Saravena, AR
	5	CLM	TH	LV de las Estrellas, Medellín
	5	CLM	NE	LV del Ariari, Granada
	5	CLM	EB	LV del Turismo, Santa Marta
	5	CLM	OI	R. Chacurí, Sampués
	5	CLM	MC	R. Viva 12-90, Cali
	5	CLM	KY	RCN, Girardot
	1	EQA	OF2	Canal Milagreño, El Milagro
	3	EQA	JA5	LV del Río Tarqui, Cuenca
	0.5	EQA	VM6	R. Once de Noviembre, Latacunga
	1	EQA	NS1	R. Popular, Atuntaqui
		PRU	OBU4S	R. 1290, La Oroya
	0.3	PRU	OAX7X	R. Juliaca, Juliaca
	10	PRU	OAX9N	R. Nor Oriental, Bagua Grande
	1	PRU	OCX1Q	R. Programas del Perú, Tumbes
		PRU		R. Satelite, La Union
	1	PRU	OCX6B	R. Trebol, Mariano Melgar
	1	PRU	OBU4Q	S & RD, Hualmay
	10	URG	CX38	S.O.D.R.E. "Em. del Sur", Montevideo
	10	VEN	LF	R. Puerto Cabello, Puerto Cabello
1300		ARG		Plus Radio, Lanús
		ARG		R. Identidad, Buenos Aires
	10/5	ARG	LRA5	R. Nal., Rosario
	5	BOL	BE18	R. Bandera Beniana, Trinidad
	0.3	BOL	PO08	R. Chichas, Siete Suyos
	1	BOL	SC16	R. Coronel Eduardo Avaroa, Sta. Cruz
	1	BOL	PO07	R. Fides, Potosí
	0.15	BOL	PO06	R. Juan XXIII, Uyuni
	2.5	BOL	CH05	R. Loyola, Sucre
	15/6	BOL	LP23	R. Sol, "Poder de Diós", La Paz
	5	CHL	CB130	R. Tierra, Santiago
	5	CLM	RB	CRB Cadena Radial Boyacense, Tunja
	5	CLM	OG	LV de las Antillas, Cartagena
	5	CLM	NB	Onda 5, Bucaramanga
	5	CLM	LD	Oxígeno R., Pereira
	5	CLM	IN	R. Eucha, Belalcázar
	5	CLM	EA	R. Lumbí, Mariquita
	5	CLM	UA	R. Sindamanoy, Mocoa
	5	EQA	DC2	R. Cenit, Guayaqui
	5	EQA	R1	R. Festival, Sto Domingo de los Colorados
		EQA		R. La Paz, Guaranda
	2/1	EQA	RS7	R. Sucumbios, Nueva Loja
	5	PRG	ZP10	R. Fe y Alegria, Asunción
	1	PRU	OAZ4B	R. Andina, Huancayo
	5	PRU	OAX4S	R. Comas, Comas
	0.4	PRU	OAX6P	R. Comercial Latina, Tacna
	1	PRU	OAU1U	R. Frecuencia Lider, Morrop
	0.5	PRU	OAX3O	R. Huascarán, Independencia
	0.3	PRU	OAX7X	R. Juliaca, Juliaca
		PRU	OBX9P	R. La Luz, Tarapoto
	5	PRU	OAX7P	R. Misercordia, Cusco
		PRU	OAZ8B	R. Nuevo Mundo, Pucallpa
	1	PRU	OAU2I	R. Paraiso, Cajabamba
	10	VEN	NS	Deportes Unión R., Maracaibo
	10/8	VEN	KH	R. Recuerdos 1300, Caracas
1310		ARG		Gesell Radio, Villa Gesell
		ARG		R. AM 13-10, Buenos Aires
	1	ARG		R. Dr. Gregorio Alvarez, Piedra del Aguila
		ARG		R. Imagen, Castelar
	1	ARG	LRA42	R. Nal., Gualeguaychú
	10	BOL	CO14	R. San Rafael, Cochabamba
	5	CLM	DG	Caracol Colombia, Monteria
	7	CLM	AK	LV de la Patria Celestial, Barranquilla
	5	CLM	WD	Micrófono Civico, Palermo
	5	CLM	JZ	R. Manantial, SF de Bogotá
	5	CLM	LM	R. Santa Bárbara
	5	CLM	TQ	R. Tasajero, Cúcuta
	5	CLM	IR	RCN Urabá, Apartadó
	0.5	EQA	AI5	Eco de los Andes, Cumanda
	1	EQA	CP3	LV de El Oro, Pasaje
	20	EQA	GB1	R. Nal. Espejo, Quito
	3	EQA	CI5	T. V. O., Biblián
	1	PRU	OBX2D	R. Chota, Chota
	1	PRU	OBX4L	R. Irvisa, Huacho
		PRU		R. Libertad, Arequipa
	1	PRU	OAU6N	R. MCV, Paucarpata
	1	PRU	OBX8L	R. Vision Amazonia, Iquitos
	5	VEN	TS	R. Andina "Sonido 13-10", Isnotú
	1	VEN	SL	R. Naciona "RNV", Guri
	10	VEN	SM	R. Nacional "RNV", Barcelona
1320	5/3	ARG	LU10	R. Azul, Azul
		ARG		R. Máster, Luján
		ARG		R. Mística, Libertad
	0.25	ARG	LV24	R. Río Tunuyán, Tunuyán
		ARG		R. S´Combro, José de Paz
		ARG		R. Area 1, Caseros
		ARG		R. Ciudad, Remedios de Escalada
		BOL	LP52	R. Panorama, Achocalla
	0.25	CHL	CA132	R. Estrella del Norte, Vallenar
	5	CHL	CD132	R. Lincoyan, Mulchén
	5	CLM	NO	La Cariñosa, Girardot
	5	CLM	MS	R. Fiesta, Barrancabarmeja
	5	CLM	HT	R. Guateque, Guateque
	10	CLM	QI	R. Leda Int., San Andrés
	1	CLM	NK	R. Luna, Palmira
	1	CLM	TA	R. María, Medellín
	5	CLM	LV	R. Onda Fantastica, Fundación
	10	EQA	JD6	R. Continental, Ambato
	3	EQA	FR2	R. Guayaquil, Babahoyo
	1	EQA	VO4	R. Stéreo Carrizal, Calceta
	1	PRU	OBU4T	R. Corporacion, Huancayo
		PRU	OBU1S	R. Frecuencia Popular, Olmos
		PRU	OAX4I	R. La Cronica, Lima//R. Nacional 850
	1	PRU	OAX3U	R. Miramar, Chimbote
	3	PRU	OAU7W	R. Peru, Juliaca
	2	URG	CW132	R. Fortaleza, Rocha
	1/0.5	URG	CW39	R. LV de Paysandú, Paysandú
	10/5	VEN	WP	R. Apolo, Turmero
	10	VEN	SG	R. Colonial, El Tocuyo
1330		ARG		R. Mailín, Gregorio de Laferrere
	3	BOL	OR08	R. América, Oruro
	1	BOL	TA04	R. Frontera, Yacuíba
	3	CHL	CB133	La Mexicana, Santiago
	3/1.5	CHL	CD133	R. Vicente Pérez Rosales, Puerto Montt
	0.25	CLM	HKR33	Alcadía de Salamina, Salamina
	5	CLM	FE	Antena 2, Pereira
	5	CLM	LS	Caracol Colombia, Popayán
	5	CLM	NR	La Caliente 13-30, San Gil
	1	CLM	MP	LV de Aguachica, Aguachica
	5	CLM	AP	R. Auténtica, Cartagena
	1	CLM	RD	R. Fénix de Oriente 1330 AM, El Peñol
	3	EQA	O1	GRC AM-Grupo Radial Carisma, El Angel
		EQA		Lomas Stereo 2000, Guayaquil
	5	EQA	RV3	Nacional El Oro, Machala
	2	EQA	LW5	R. Visión Cristiana, Cuenca
	3	EQA		R. Visión Cristiana, Quito
	10	PRG	ZP4	R. Chaco Boreal, Asunción
	0.5	PRU	OAU5L	R. Bethel, Huamanga
	1	PRU	OAU1A	R. Dos Mil, Chiclayo

kHz	kW	Ctry	Call	Station, location
	1	PRU	OVX6E	R. Ondas del Misti, Mariano Melgar
	1	PRU	OCX7K	R. San Miguel, Wanchaq
	5	URG	CX40	R. Fénix, Montevideo
	5	VEN	OY	R. Los Llanos, Calabozo
		VEN		R. Nacional "RNV", La Paragua
	10	VEN	TU	R. Regional, Cd.Ojeda
1340		ARG		AM Renacer, Moreno
	0.5	BOL	LP39	R. Copacabana, Copacabana
	1	BOL	SC17	R. Grigotá, Santa Cruz
	0.5	BOL	LP40	R. Jach'a Suyu, Corocoro
	0.35	BOL	LP22	R. San Francisco, Apolo
	10	CHL	CB134	R. Colo Colo, Valparaíso
	1	CHL	CC134	R. La Discusión, Chillán
	1	CHL	CD134	R. Panguipulli, Panguipulli
	0.5	CLM	VL	Brisas del Catatumbo, Tibú
	5	CLM	FB	Fiesta 13-40, SF de Bogotá
	5	CLM	KD	La Cariñosa/Antena 2, Neiva
	1	CLM	NP	R. Comunal, Nariño
	5	CLM	IS	R. El Sol, Buenaventura
	5	CLM	PY	R. Lemas, Cúcuta
	5	CLM	FA	R. Olímpica AM, Barranquilla
	1	CLM	NY	R. Unica, Bucaramanga
	5	CLM	HA	RCN Nariño, Pasto
	5	CLM	HY	RCN Sucre, Sincelejo
		EQA		LV de su amigo, Esmeraldas
	1	EQA		Ondas de Esperanza, Loja
	5	EQA	RT6	R. Paz y Bien, Ambato
		PRU	OAU3E	Assn. Iglesia de Dios, Chimbote
	5.5	PRU	OAQ4Q	R. Alegria, Pucasana
	0.5	PRU	OAX5D	R. Chinca, Chinca Alta
		PRU	OAU6T	R. Comercial, Mollendo
	1	PRU	OAU4N	R. Jauja, Jauja
	1	PRU	OAU2S	R. Shalom, Cajamarca
		PRU	OBU7V	R. Sudamericana, Juliaca
	10	URG	CW53	LV de Melo, Melo
	5	VEN	NE	R. Uno, Caracas
1350	25/5	ARG	LS6	R. Buenos Aires, Buenos Aires 24h
	5/1	ARG	LRJ747	R. Sucesos, Córdoba
	1	BOL	CH06	América Radiodifusión, Sucre
	2.5	BOL	CO05	R. Cochabamba, "CBA"
	2.5	BOL	SC18	R. Ichilo, Yapacaní
	1	CHL	CA135	R. Riquelme, Coquimbo
	0.25	CLM	HKZ98	Alcaldía de Caicedonia, Caicedonia
	5/1	CLM	HW	Em. Ecos del Río, Puerto Boyacá
	5	CLM	OC	La Cariñosa, Ant. 2, Santa Marta
	10	CLM	EN	R. Armonía, Cali
	1	CLM		R. Cultural 2001, Pailitas
	5	CLM	DS	R. Ondas de la Montaña, Medellín
	1	CLM	MN	R. Perijá, Codazzi
	5	CLM	HL	R. Reloj, Ibagué
	5	CLM	LO	RCN Antena 2/La Cariñosa, Caucasia
	2/1	EQA	SF5	LV de San Fernando, San Fernando
	3	EQA	VP2	Teleradio 13-50 AM Digital , Guayaquil
		PRU	OBU5O	Huamana (F.P.I.)
	1	PRU	OBX6F	R. Ilo, Ilo
	1	PRU	OBU7E	R. Lider, Cusco
		PRU	OBX8D	R. Super, Pucallpa
	1	PRU		R. Vision, Chiclayo
	5	VEN	ZZ	R. Eclipse "R. Guanipa",El Tigrito
	5	VEN	TJ	R. Falcón, Puerto Cumarebo
1352	1	PRU	OAX3N	R. Ondas del Huallaga, Huanuco
1355	0.25	BOL	CO15	R. Armonía, Cliza
1360		ARG		R. Nuestra Señora de Itatí - "R.Itatí", Morón
	5	BOL	LP16	Radiodifusoras Jiménez, El Alto
	5	CHL	CC136	R. Universidad del Bío Bío, Concepción
	1	CLM	RA	Eco 13-60 "La Superestación", Pereira
	10/5	CLM	PK	LV de Abejorral, Abejorral
	5	CLM	TU	Oxígeno R., Cartagena
	5	CLM	MI	R. Auténtica, Melgar
	1	CLM	KV	R. Láser, Zapatoca
	0.5	CLM		R. Segovia, Segovia
	3	EQA	MT	Oyambaro AM, Tumbaco
	1	EQA	RJ5	R. América, Riobamba
	5	EQA	HG3	R. Jerusalem AM, Machala
	1	PRG	ZP37	R. Yby Ya'u, Ybu Ya'u
	2.5	PRU	OUA7L	R. Continente, Juliaca
	1	PRU	OBZ5Z	R. Cruz del Sur, Palpa
	1	PRU	OBZ1A	R. del Norte, Sullana
	1	PRU	OAU4O	R. Hecaburt, Tarma

kHz	kW	Ctry	Call	Station, location
	0.2	PRU	OAU3A	R. Intercontinental, Yungay
	1	PRU	OCX6T	R. Luza, Paucarpata
	2.5	PRU	OAX7R	R. Sicuani, Sicuani
	1	PRU	OBX2N	R. Super Uno, Santiago de Cao
		PRU	OAX4Y	R.Nueva Q-FM, Lima
	2.5/0.5	URG	CW41	R. 41, San José
	1/0.25	URG	CW136	R. Río Branco, Río Branco
	5	VEN	TW	R. Alegría, Chivacoa
	10	VEN	TI	R. Internacional, Maracaibo
	5	VEN	TZ	R. YVTZ, Charallave
1370	5/3	ARG		AM-1370, Isidro Casanova
	1	ARG	LRA54	R. Nal., Ingeniero Jacobacci
	1	BOL	LP24	R. Agricultura, Achacachi
	0.15	BOL	CO16	R. Libertad, Cliza
	0.5	BOL	PO24	LV de Minero, Siglo XX
	5/3	BOL	OR11	Radiodifusoras Coral, Oruro
	1	CHL	CD137	R. Portales ,Temuco
	1	CLM	BD	La Nueva R. Guaimaral, Cúcuta
	5	CLM	BO	Minuto de Dios, Barranquilla
	5	CLM	KX	R. Mundial, SF de Bogotá
	1	CLM	NI	R. Sabana, Sincelejo
	1	CLM	JQ	RCN Antena 2, Zarzal
	10	CLM	EQ	RCN Cauca, Popayán
	2.5	CLM	NU	RCN, Rionegro
	2	EQA	JS1	Ecos Andinos, Pimampiro
	5	EQA	VO2	LV del Milagro, El Milagro
		EQA		R. El Rocio, Biblián
	2	EQA	RP7	R. Pastaza, El Puyo
	5	EQA	ER3	R. Progreso, Loja
	1	PRU	OCX5A	Inti Radio, Abanacy
	1	PRU	OAU1W	R. Chiclayo, Chiclayo
		PRU	OAX6T	R. Moquegua, Moquegua
		PRU	OBX9A	R. Palmera
	1	PRU	OAZ7J	R. Santa Monica, Wanchaq
		PRU	OBU2U	R. Satelite, Santa Cruz
	0.5	PRU	OAZ40	R. Tres de Octubre (R.Cosmos), Huacho
	10/2.5	URG	CX42	Em. Ciudad de Montevideo
	0.5/0.25	URG	CV137A	R.Real, Minas de Corrales
	0.25/0.1	URG	CV137	R.San Javier, San Javier
	10	VEN	JI	R. Continente Cumbre, Mérida
	5	VEN	SV	R. Portuguesa, Araure
	5	VEN	OQ	Unión R. Notícias/R. La Pascua, Valle de la Pascua
1380	5/1	ARG		LV del Sudeste, Necochea
		ARG		R. Buenas Nuevas, Merlo
		ARG		R. Los Toldos, Los Toldos
		ARG		R. Redentor, Claypole
	0.25	BOL	CO17	R. 16 de Noviembre, Sacaba
	1.5	BOL	CO34	R. Bandera Tricolor, Cochabamba
	0.5	BOL	TA06	R. Luis de Fuentes, Tarija
		BOL	LP66	R. Misericordia, El Alto
	50	CHL	CB138	R. Corporación, Santiago
	1	CLM	EJ	Armonías del Palmar, Palmira
	3	CLM	LG	LV de La Dorada, La Dorada
	3	CLM	JD	R. Nuestra Señora del Encuentro con Dios, Medellín
	5	CLM	ID	R. Potencia Latina, La Plata
	5	CLM	MM	R. Recuerdos, Valledupar
	5	CLM	EE	RCN, Tunja
	1	EQA	OA3	La Mejor, Balsas
	5	EQA	C1	R. Cristal, Quito
	5	EQA		R. Mera, Ambato
	1	PRG	ZP8	R. Concepción, "LV del Norte"
	1	PRU	OAX2W	R. Atahualpa, Cajamarca
	1	PRU	OBZ1D	R. Bellavista, Bellavista
		PRU	OBU4L	R. Chilca, Huancayo
		PRU		R. Fraternidad, Trujillo
	1	PRU	OCY4U	R. Nuevo Tiempo, Lima
	2.5	PRU	OAX6O	R. San Martin, Arequipa
	10	VEN	NG	Ondas del Mar, Puerto Cabello
	5	VEN	ME	R. Revelación, Cd.Bolivar
	10	VEN	TL	R. Triunfo 13-80, Caja Seca
1382	1	PRU	OBX3I	R. Pilco Mozo, Huanuco
1390		ARG		R. Ribera Sur, Ingeniero Budge
	10	ARG	LR11	R. Univ., La Plata
	0.25	BOL	PO11	R. LV Minera del Sud, Mina Telamayu
		BOL	CO50	R. Mancomunidad Andina, Cochabamba
	1	CLM	ZY	La Primera, Bucaramanga
	5	CLM	YW	R. Auténtica, Pacho

kHz	kW	Ctry	Call	Station, location
	5	CLM	FY	R. Avenida, Espinal
	0.1	CLM		R. Ciudad de Antioquia, Santa Fé de Antioquia
	5	CLM	FO	Red de los Andes, La Voz de Siempre, Manizales
	1	EQA	HE4	LV de Esmeraldas, Esmeraldas
	3	EQA	DN5	R. Atenas, Riobamba
	5	EQA	EA5	R. Tropicana, Cuenca
	1.5	EQA	IE1	R. Uno, Urcuquí
		PRU		Difusora Neptuno, Mollendo
	1	PRU	OCX7U	R. Cultura, Yunguyo
	1	PRU	OAU7T	R. Enlace, Kunturkanki
	1	PRU	OBX5O	R. Huanta 2000, Huanta
	1	PRU	OAU2Z	R. La Luz, Trujillo
	1	PRU	OAU1V	R. Tropical, Morrope
	7.5/3	URG	CW45	Dif. Treinta y Tres, Treinta y Tres
	20	VEN	ZA	R. Fe y Alegría, Caracas
	10	VEN	ZO	R. Lumen, Maracaibo
	10	VEN	TT	R. Terepaima, Cabudare
1400		ARG		R. AM 1400, Luján
	10/1	ARG	LRG202	R. Cumbre, Neuquén
		ARG		R. del Buen Ayre, Ituziangó
		ARG		R. Gama, Valentín Alsina
		BOL	CH20	R. Antena 2000, Sucre
		BOL	LP41	R. Comunidad, Patacamaya
	1	BOL	SC19	R. Libertador, Santa Cruz
	5	BOL	LP25	R. Nacional de Bolivia, La Paz
	0.25	CHL	CA140	R. Altisima, Iquique
	5	CHL	CD140	R. La Amistad, Los Angeles
	5	CHL	CD140A	R. Viento del Sur, Puerto Montt
	0.25	CLM	HKX22	Alcaldía de Majagual, Majagual
	0.25	CLM	HKZ25	Alcaldía de Ovejas, Ovejas
	0.25	CLM		Brisas del Sinú, Tierralta
	1	CLM	TY	Caracol Colombia, Vélez
	1	CLM	IT	Ecos del Atrato, Quibdó
	5	CLM	KM	Em. Mariana de Bogotá, SF de Bogotá
	5	CLM	HM	La Cariñosa de Armenia, Calarcá
	2.5	CLM	D31	LV de Cimitarra, Cimitarra
	1	CLM	WY	LV de los Samanes, Quilichao
	5	CLM	DF	LV de Niquel, Montelíbano
	1.5	CLM		LV de Samaniego, Samaniego
	0.45	CLM		R. Cañaveral, Morales
	5	CLM	JJ	R. Ipiales, Ipiales
	5	CLM	AS	R. Uno/RCN Antena 2, Barranquilla
	1	CLM	LL	RCN Antena 2, Santa Bárbara
	1	CLM	BK	Voz Grancolombia, Cúcuta
		EQA		Impacto 1400 AM, Latacunga
	10	EQA	FL2	R. Z Uno, Guayaquil
	2.5	PRU	OBX4W	R. Callao Super, Lima
	1	PRU	OCX5B	R. Interandina, Pisco
	1	PRU	OAX7I	R. La Hora, Cuzco
	0.5	PRU	OAX6J	R. Landa, Arequipa
	1	PRU	OBX4H	R. Luz, Tarma
	1	PRU	OCX1A	R. MDY, Talara Alta
	1	PRU	OAU2H	R. OAU2H, Cajamarca
	25	URG	CX140	R. Zorrilla de San Martín, Tacuarembó
	1	VEN	NF	R. Sabana, El Sombrero
1410	5/1	ARG	LRG203	R. Capital "Antena 10", Santa Rosa
	5/1	ARG		R. Folclorismo, José Léon Suárez
		ARG		R. Fundacion, Rafael Calzada
	0.25	BOL	OR14	R. Atlántida, Oruro
	0.25	BOL	SC20	R. Roboré, Roboré
	1	CHL	CD141	R. Loncoche, Loncoche
	3	CHL	CB141	R. Quinta Región, Valparaíso
	0.25	CLM	HKP86	Alcaldía de Chiquinquira, Chiquinquira
	5	CLM	DU	Em. Cultural Universidad de Antioquia, R. Universidad, Medellín
	2	CLM	P79	R. Evangélica, Uribia
	5	CLM	EI	R. Guadalajara, Buga
	1	CLM	HKP79	R. Universidad, Tunja
	5	CLM	FS	RCN, Honda
	1	EQA	FR4	LV de Quinindé, Quinindé
	1	EQA		Ondas Cisnerinas, Riobamba
	1	EQA	GC5	R. Centro Gualaceo, Gualaceo
	1	EQA	EC1	R. El Tiempo "Em.del Amor", Quito
	1	EQA	CQ2	R. Net AM, El Milagro
	2	PRG	ZP35	R. Mangore, S. Juan Bautista
		PRU	OBX8I	Dif.Comercial, Pucallpa
		PRU	OBZ4C	R. Bethel, Huacho
	1	PRU	OAX2Y	R. Heróica, Trujillo

kHz	kW	Ctry	Call	Station, location
		PRU	OBU7A	R. La Luz, Juliaca
	1	PRU	OBU1H	R. La Luz, Tumbes
		PRU	OBU1G	R. Olmos, Olomos
	1	PRU	OBZ4V	R. Universal, Santa Maria
		PRU	OCU2Q	San Marcos (F.PI.)
	10/5	URG	CX44	AM Libre, Montevideo
	1/0.25	URG	CW141	R. Turística, Salto
	10	VEN	SP	R. Simpatía, Valera
	5	VEN	ST	R. Turén, Turén
1420	1/0.25	ARG	LRI220	AM La Marea, Buenos Aires
		ARG		R. Génesis 2000, General Conesa
	1	BOL	CO18	R. Centro, Cochabamba
	1.5	BOL	TA05	R. Guadalquivir, Tarija
	1	BOL	CH15	R. Real Audiencia, Sucre
	1	CHL	CC142	R. Maule, Cauquenes
	1	CHL	CB142	R. Panamericana, Santiago
	5	CLM	BH	Caracol Colombia/R. Magdalena, Santa Marta
	1	CLM	D23	Ecos de Frontino, Frontino
	1	CLM	LE	La Cariñosa, Antena 2, Ibagué
	5	CLM	AP	R. Autentica, Cartagena
	1	CLM	SN	R. Lenguerque, Zapatoca
	5	CLM	HK	R. Recuerdos, Manizales
		EQA		Corazón AM, Machala
		EQA	VN7	LV del Napo, Tena
	1	EQA	MA6	R. Alternativa, Salcedo
	3	EQA	RN1	R. Bahá'í, Otavalo
	5	PRG	ZP42	R. Güyrá Campana, Horqueta
		PRU	OAU2R	R. Cajamarca, Cajamarca
	5	PRU	OBX2V	R. Ilucan, Cutervo
		PRU	OBU7L	R. OBU7L, Yanaoca
	1	PRU	OAZ8Z	R. Oriente, Yurimaguas
	1	PRU	OBZ4G	R. San Isidro, Lima
	1	URG	CX142	R. Felicidad, Paysandú
	5	URG	CW43	R. Lavalleja, Minas
	10/5	VEN	RW	R. Cardenal, Carora
	5	VEN	NZ	R. Marabina 1420, Maracaibo
	5	VEN		R. Sintonía, Caracas
1426	0.2	PRU	OCX1H	R. San Jose, La Union
1430		ARG		R. AM 1430 - "La Radio de los Cunumí Guazú"
	0.25	ARG	LRI223	R. Balcarce, Balcarce
		ARG		R. José de S. Martín, "La Pionera",El Jagüel
	1/0.25	ARG	LV26	R. Río Tercero, Río Tercero
	1/0.25	ARG	LT24	R. San Nicolás, San Nicolás
		ARG		R. Shekinah, Merlo
		ARG		R.Victoria, La Plata
	0.15	BOL	PO17	R. Centinela, Tupiza
	0.25	BOL	CO19	R. Nuestra Señora de Burgos, Mizque
	5	CLM	KU	1430 AM, SF de Bogotá
	1	CLM	IU	Armonías del Ingrumá, Riosucio
	5	CLM	PW	Colmundo, Barranquilla
	1	CLM	EG	LV de Belalcázar, Popayán
	0.5	CLM	G42	R. Alejandría, Alejandría
	2	CLM	BP	R. Cariongo, Pamplona
	0.25	CLM	HKX73	R. Ciudad de Pereira, Pereira
	0.25	CLM	X61	R. Dif. Cultural del Quindío, Armenia
	5	CLM	QX	R. Majagual, Corozal
	0.5	CLM	HKK38	R. Manantial, Sibundoy
	1	CLM	CK	R. Sensación, Yarumal
	2	CLM	MF	R. Venus, Puerto Berrío
	5	EQA	CV3	Ondas del Zamora, Canal Juvenil, Loja
	10	EQA	MB2	R. Federal, Virgen de Fátima
	3.5	EQA	GF1	R. Futura 14-30, Quito
	5	EQA	JC6	R. Guaranda, Guaranda
	1	PRU	OAZ7M	CPN Radio, Cusco
	1	PRU	OAZ3H	R. Chavin, Chimbote
	1	PRU	OBX3E	R. Huarmey, Huarmey
	1	PRU	OAU6M	R. Lider, Tacna
		PRU		R. Nueva Juventud, Tucumé
		PRU	OBU7U	R. Red, Andina
	1	PRU	OBX2T	R. Santa Bárbara, Ascope
	0.5	PRU	OAZ4V	R. Universal, El Tambo
	1	PRU	OBX9H	R. Utcubamba, Bagua Grande
		PRU	OCU4L	San Vicente de Cañete (F.PI.)
	5/1	URG	CW25	R. Durazno, Durazno
	25	VEN	TP	R. Bahía, Puerto La Cruz
	10/5	VEN	TM	R. Caicara, Caicara del Orinoco
	10	VEN	NB	R. Satélite 14-30, Guacara

kHz	kW	Ctry	Call	Station, location
1440	0.25	ARG	LU36	R. Coronel Suárez, Coronel Suárez
		ARG		R. Cristo Viene, Mar del Plata
	5/1	ARG	LRI221	R. General Obligado, Reconquista
		ARG		R. Impacto, Tapiales
	1/0.25	ARG	LV20	R. Laboulaye, Laboulaye
	1	ARG	LRA53	R. Nal., S. Martín de los Andes
	1	BOL	LP26	R. Batallón Colorados, La Paz
	0.25	BOL	CO42	R. Bolivia, Cochabamba
	0.5	BOL	SC32	R. Oriente, Camiri
	2/1	BOL	SC21	R. Yaguarí, Vallegrande
	1	CHL	CA144A	R. Agricultura, La Serena
	1	CHL	CC144	R. El Sembrador, Chillán
	1	CHL	CA144	R. Santa Maria de Guadalupe, Arica
	0.25	CLM	HKT58	Alcaldía de Ubala, Ubala
	5	CLM	NZ	Colmundo, Medellín
	5	CLM	BM	R. Internacional, Honda
	5	CLM	EK	R. Reloj, Tuluá
	1	CLM	IB	RCN Caquetá, Florencia
	5	CLM	GM	RCN, Sogamoso
		EQA		Mi Radio AM, Machala
	2.8	EQA	OV5	Ondas del Volante, Azogues
	3/5	EQA	AQ6	R. Fenix, Latacunga
	2.5	EQA	DY4	R. Iris, Esmeraldas
	5	EQA	DF1	R. Panorama, Ibarra
	2	PRU	OBX1T	R. Cooperativa Tumán, Chiclayo
	1	PRU	OAU2O	R. Frecuencia VH, Celendin
	1	PRU	OAX4K	R. Imperial 2, Lima
	1	PRU	OAZ3O	R. LV de Pomabamba, Pomabamba
	1	PRU	OAU7M	R. Regional, Sicuani
	J	PRU	OAX6R	R. Santa Monica, Arequipa
	3/0.5	URG	CX144	R. Rivera, Rivera
	10	VEN	ZI	R. Estelar 14-40, Guanare
	5	VEN	RF	R. Orituco, Altagracia del Orituco
	1	VEN	TY	R. Sucesos, Táriba
1445	0.5	BOL	SC22	Super Broadcasting Alborada, "SBA", Santa Cruz
	0.5	CLM		Em. R. Unión, La Palma
1450	5/1	ARG		R. El Sol, Buenos Aires
	5/1	ARG	LRI211	R. Las Cuarenta, San Juan
		BOL	OR23	R. Amanacer, Huari
		BOL	PA03	R. Amazonia, Cobija
	1	BOL	OR13	R. Em. Bolivia, Oruro
	1	BOL	SC24	R. Verde y Blanco, Santa Cruz
	5	CHL	CC145	R. Libertad, Curicó
	1	CHL	CD145	R. Sta Maria de Guadalupe, Puerto Varas
	1	CHL	CB145	R. Universidad Técnica, Valparaíso
	5	CLM	NL	La Cariñosa, Ant. 2, Manizales
	0.5	CLM		LV del Cauca, El Bordo
		CLM	TT	Ondas del Porvenir, Samacá
	5	CLM	BY	Oxígeno R., Flandes
	5	CLM	HH	R. Católica Metropolitana, Bucaramanga
	0.2	CLM		R. LV del Nordeste, Remedios
	1	CLM	MX	R. Mancomoján, Carmen de Bolívar
	1	CLM	E20	R. María, Urrao
	1	EQA	SC1	AS La Radio, Tabacundo
	10	EQA	SC5	R. Calidad, Riobamba
	1	EQA	DR	R. Minutera, Guayaquil
	1	EQA	SE2	R. Santa Elena, Santa Elena
	5	PRG	ZP29	R. Vallemi, Vallemi
	1	PRU	OBX4K	R. Fortaleza, Barranca
	1.5	PRU	OCX2J	R. San Juan, Trujillo
		PRU	OAU2W	R. San Miguel, Cajamarca
	1	PRU	OCX7W	R. Santa Rosa, Santa Ana
		PRU		R. Super Nueva Sencacion, Chirinos
		PRU	OBU4Y	Rdif. E.G.C S.R.L., Huancayo
	10/5	URG	CX46	R. América, Montevideo
	2/0.5	URG	CW145	R. Arapey, Salto
	10	VEN	ZQ	Informativa 14-50, Los Puertos de Altagracia
	10/8	VEN	KJ	R. María, Caracas
	10/5	VEN	XC	R. Mega Visión, San Felix
1455	0.5	BOL	CO39	R. Magnal, Capinota
1457	1	PRU	OAX1V	R. Sullana "LV de Chira", Sullana
1458	1	PRU	OAU2W	R. San Miguel, San Miguel
1460	1	ARG	LRK204	R. 21, Yerba Buena
		ARG		R. Contacto, San Antonio de Padua
		ARG		R. Jerusalén, Jerusalen
	0.25	ARG	LU30	R. Maipú, Maipú
	0.1	ARG	LU34	R. Pigüé, Pigüé
	1/0.25	ARG	LT29	R. Venado Tuerto, Venado Tuerto
		BOL	CO44	R. Morochata, Morochata
	10	CHL	CA146	R. Antofagasta, Antofagasta
	1	CHL	CC146	R. Armonía, Concepción
	1	CHL	CB146	R. Yungay, Santiago
	2.5	CLM	FL	Agustiniana Minuto de Dios, San Agustín
	0.25	CLM	HKY73	Alcaldía de San Andrés, San Andrés
	5	CLM	JW	Em. Nuevo Continente, SF de Bogotá
	1	CLM	MN	LV de Amalfi "La Primera", Amalfi
	1	CLM	E26	R. Capiro, La Ceja
	5	CLM	JH	R. Ciudad Milagro, Armenia
	5	CLM	TF	R. María, Turbo
	1	CLM	IW	R. Monumental, Cúcuta
	1	CLM	AL	R. Sincelejo, Sincelejo
	5	CLM	VH	R. Uno/RCN Antena 2, Barranquilla
	5	CLM	ZU	RCN Antena 2, Pasto
	5	EQA	AA7	LV de Gualaquiza, Gualaquiza
	5	EQA	IC6	R. Nuevos Horizontes, Latacunga
	1	PRU	OBX6C	R. Bahia, Mollendo
	1	PRU	OBU2E	R. Comercial, Jaén
	10	PRU	OAX7W	R. El Sol de los Andes, Juliaca
	0.5	PRU	OCY4I	R. Imperial, Junin
	2.5	PRU	OAX5K	R. Internacional, Pisco
	1	PRU	OAZ4F	R. La Oroya, La Oroya
		PRU	OBU7M	R. Marcapata
	1	PRU	OAX1V	R. Sullana "LV de Chira", Sullana
	3/1	URG	CX146	R. Carmelo, Carmelo
	0.25	URG	CV146	R. José Batlle y Ordoñez, José Batlle y Ordoñez
	5	VEN	RJ	R. Jardín, Boconó
1461		BOL	CO36	R. LV del Pueblo de Dios, Cochabamba
1470		ARG		Cadena 1470, Lanús
	0.25	ARG	LU26	R. Coronel Dorrego "La Dorrego", Coronel Dorrego
	1/0.25	ARG	LT20	R. Junín, Junín
		ARG		R. Mburucuya, José León Suarez
	1	ARG		R. Municipal, Luis Beltrán
	1/0.25	ARG	LT26	R. Nuevo Mundo, Colón
	1/0.25	ARG	LT28	R. Rafaela, Rafaela
	0.25	BOL	CH07	R. Cordech, Alcalá
	1	CHL	CB147	R. Romantica, San Antonio
	0.25	CLM	HKO96	Alcaldía de Baranoa, Baranoa
	5	CLM	PX	Colmundo, Cartagena
	0.25	CLM	JS20	Ecos de Palo Cabildo, Palo Cabildo
	5	CLM	TB	Ondas de Ibagué, Ibagué
	5	CLM	HQ	R. Futurama, Pacho
	5	CLM	NT	R. Huellas 1470, Cali
	5	CLM	IM	R. Popular, Medellín
	1	CLM	JIF	R. Tres Fronteras, Puerto Asís
	1	CLM	HJB63	R. Uno, Iza
	5	EQA	JC1	Ecos de Cayambe, Cayambe
	1.5	EQA	LD2	R. Ecos de Naranjito, Naranjito
	20	PRU	OAU4B	CPN Radio, Lima
	1	PRU	OAX7G	R. Cusco, Cusco
	1	PRU	OCX2G	R. Occidente, Quiruvilca
	0.8	PRU	OAX6M	R. Tacna, Tacna
	1	PRU	OAU6E	R. Victoria, Arequipa
	15	URG	CX147	R. Cristal del Uruguay,Las Piedras
	1/0.5	URG	CW147	R. Maria, Melo
	10	VEN	LW	CNB La Valenciana 14-70, Valencia
	10	VEN	SY	Unión R. Notícias/R. Vibración, Carúpano
1475		BOL	CO22	R. Tiraque, Tiraque
	1	PRU	AOZ7G	R. Espinar, Yauri
1476	5	PRU	OBX2V	R. Ilucan, Cutervo
1480	1	ARG	LU27	R. Dolores, Dolores
		ARG		R. Lider, Mariano Acosta
		ARG		R. Sensaciones, Tapiales
	0.1	BOL	PO12	Patrimonio Radiodifusión, Potosí
		BOL	LP58	R. Amor de Diós, La Paz
	1/0.8	BOL	CO32	R. Chiwalaki, Vacas
		BOL	CO40	R. Domingo Savio, Villa Independencia
	1/0.25	CHL	CA148	R. Amanecer, Ovalle
	1	CHL	CD148	R. General Baquedano, Valdivia
	1	CHL	CC148	R. La Amistad AM, Tomé
	0.25	CLM	HKR44	Alcaldía de Victoria, Victoria
	0.25	CLM		LV del Samán, Bochalema
	1	CLM	VB	R. Guayabal, Armero, Guayabal
	2.5	CLM	OD	R. Rodadero, Santa Marta
	1	CLM	TC	R. Sonsón, Sonsón
	5	CLM	FC	R. Unica, Pereira

kHz	kW	Ctry	Call	Station, location
	5	CLM	TZ	RCN Antena 2, Bucaramanga
	3	EQA	BS3	Oro Radio AM, Machala
	3	EQA	WP5	R. Atlántida, Alausí
	3	EQA	JV4	R. LV de Jipijapa, Jipijapa
	1	EQA	MC1	R. Municipal, Cotacachi
	5	EQA	CY6	R. Popular de la Maná, La Maná
	5	PRG	ZP20	R. América, Nemby
	1	PRG	ZP23	R. Dos Fronteras, Bella Vista Norte
	0.6	PRU	OCX2C	R. Comercial, Virú
	1	PRU	OCX4V	R. K'ler, Paramonga
	1	PRU	OAU4A	R. Laser, Santa Rosa de Sacco
	0.5	PRU	OBU2H	R. San Lorenzo, Socota
	1	PRU	OCX1L	R. Supercontinental, Chulucanas
	1/0.7	URG	CX148	Difusora Rio Negro, Young
	5/1.5	URG	CW43B	R. Internacional, Rivera
	3/0.8	URG	CW148	R. Universo, Castillos
		VEN		R. Cumarebo, Cumarebo
1485	1	BOL	C023	R. LV del Valle, Punata
1490	0.1	ARG	LU25	R. Carhué, Carhué
		ARG		R. Cielo Nuevo, Isidro Casanova
	1	ARG	LV22	R. Huinca Renancó, Huinca Renancó
		ARG		R. Vida, Mar del Plata
	0.25	BOL	SC23	R. Mairana, Mairana
	0.25	BOL	BE05	R. Moxos, San Ignacio de Moxos
	0.35	BOL	LP30	R. Pedro Domingo Murillo, Quime
	1	BOL	OR15	R. San José, San José, Oruro
	1	CHL	CA149	R. Chañaral, Chañaral
	1	CHL	CB149	R. El Canelo de Nos AM, San Bernardo
	5	CHL	CD149A	R. Malleco, Victoria
	0.2	CLM	J76	Alcaldía de El Peñon, El Peñon
	0.2	CLM	HKW24	Alcaldía de Guaitarilla, Guaitarilla
	5	CLM	BS	Em. Punto Cinco, SF de Bogotá
	1	CLM	JO	LV de San Marcos, San Marcos
	1	CLM	AG	R. Garzón, Garzón
	5	CLM	AY	R. Vida Nueva, Barranquilla
	5	CLM	ZB	Robles 14-90, La Nueva, Tuluá
	1	EQA	VY2	La R. Dinámica, Guayaquil
		EQA		Poderosa 14-90, Quito
	3	EQA	AI6	R. Moderna, Píllaro
	5	EQA	SM5	R. Santa María, Azogues
	2.5	EQA	AE4	R. Unión, Esmeraldas
	1	PRU	OBU7I	R. Chaski, Maras
	1	PRU	OCX7P	R. Emisora Frontera, Puno
	1	PRU	OBU4N	R. La Luz, Cerro de Pasco
	1.3	PRU	OAX6Q	R. Minuto, Cerro Colorado
	0.3	PRU	OAX5N	R. Nazca, Nazca
	1	PRU	OAX8F	R. Nueva Atlantida, Iquitos
	1	PRU	OAX1L	R. Vision, Chiclayo
		PRU	OBU5C	Radiodifusora los Chankas, Andahuaylas
	1/0.25	URG	CV149	R. del Centro, Baltasar Brum
	5/4	URG	CX149	R. del Oeste, Nueva Helvecia
	10	VEN	XD	La Dinámica, Caracas
	10/5	VEN	RP	R. El Sol, Maracaibo
	1	VEN	SQ	R. Mérida 14-90, Mérida
1495	2.5	BOL	CO21	El Mundo Radiodifusión, Sacaba
		BOL	CO40	R. Domingo Savio, Villa Independencia
1500	5/1	ARG	LRI214	R. Bonaerense, Lavalloi
	1/0.25	ARG		R. Municipal, Gral. Conesa
	0.25	ARG	LT34	R. Nuclear, Zárate
	5/1	BOL	LP31	R. Chuquisaca, El Alto
		BOL	CO47	R. Comuicacion Cristiana, Quillacollo
		BOL	LP71	R. Huaycheño, Puerto Acosta
	1	BOL	SC25	R. Sagrado Corazón, Mineros
	1	CHL	CC150	R. Centenario, San Javier
	1	CHL	CA150	R. Santa Maria de Guadalupe, Iquique
	1	CHL	CD150	R. Tierra del Fuego, Puerto Porvenir
	1	CHL	CB150	R. Trasandina, Los Andes
	5	CLM	UW	R. María, Manizales
	5	CLM	SH	R. Reloj, Moniquirá
	5	CLM	TW	R. Sumapaz, Fusagasugá
	5	CLM	LJ	Sonora, La Voz de la Red, Cali
	5	EQA	HG2	LV del Río Vinces, Vinces
	1	EQA	RO1	R. Otavalo, Otavalo
	5	EQA	AD4	R. Satélite, El Carmen
	1	PRU	OAU6B	R. Bulevar, Tacna
	1	PRU	OBX2X	R. Comercial, Trujillo
	1	PRU	OBX3J	R. Luz y Sonido, Huanuco
		PRU	OBU2J	R. San Pablo, San Pablo
	10	PRU	OBX4I	R. Santa Rosa, Lima
	1	PRU	OAU4W	R. Wanka, Huancayo
	10/5	VEN	RZ	R. 2000 AM, Cumaná
		VEN		R. Galaxia, San Mateo
1510		ARG		LV del Oeste,Libertad
		ARG		R. Alabanza, Guernica
		ARG		R. AM Líder, Martinez
	1/0.25	ARG	LRI253	R. Belgrano, Suardi
	1/0.25	ARG	LV21	R. Champaqui, Villa Dolores
		ARG		R. Urkupiña, Buenos Aires
	0.25	BOL	TA08	R. 27 de Diciembre, Villamontes
	0.05	CHL	CD151	R. La Trompeta de Dios, Loncoche
	1/0.5	CHL	CA151	R. Luís Alvarez Sierra, Illapel
	1	CHL	CC151	R. Rancagua, Rancagua
	0.25	CLM	HKZ94	Alcaldía de Buenaventura, Buenaventura
	0.25	CLM	HKZ98	Alcaldía de Caicedonia, Caicedonia
	0.25	CLM	HKZ93	Alcaldía de Versalles, Versalles
	1	CLM	HX	Candela AM, Bucaramanga
	1	CLM	HKY41	Colombia Mía, Barrancabermeja, SS
	5	CLM	D24	LV de La Unión, La Unión
	0.5	CLM		LV de los Cedros, Libanó
	1	CLM	A22	LV de San Luis, San Luis de Gaceno
	1	CLM	ZA	R. Cristal, Armenia
	0.5	EQA	HD2	Inst. Oceanográfico de la Armada, Guayaquil
	3	EQA	JV7	R. Ecos del Oriente, Lago Agrio
	5	EQA		R. Monumental, Quito
		EQA		R. Net, Ambato
	2	EQA	RC5	R. Punto C 1510 AM, Cañar
	1	EQA	RY6	R. Runacunapac Yachana, Simiátug
	10	EQA	UC3	R. Unión Calvense, Cariamanga
	1	PRU	OBX8K	R. Centro de los Medios, Sepahua
	1	PRU	OBX7P	R. El Sur, Wanchaq
		PRU	OCX2O	R. Inca, Los Baños del Inca
	1	PRU	OCX6Q	R. Las Vegas Arequipa
	1	PRU	OAX5F	R. LV Huamanga, Nazca
	1	PRU	OBU1B	R. OBU1B, Olmos
	1	PRU	OCX4J	R. Tarma, Tarma
	1	PRU	OCX1V	R. Tumbes, Tumbes
	2.5	PRU	OAU2U	R. Virgin de la Alta Gracia, Huamachuco
	0.5	URG	CW151	R. Ibirapitá, San Gregorio de Polanco
	1/0.25	URG	CX151	R. Rincón, Fray Bentos
	2/0.5	URG	CW57	R. San Carlos, San Carlos
	20	VEN		Informativa "LV del Centro", Güigüe
1517	1	PRU	OCX2O	R. Inca, Los Baños del Inca
1520		ARG		Cadena D, Monte Chingolo
		ARG		R. AM Fortaleza, Ezeiza
	5/1	ARG		R. Chascomús, Chascomús
	0.25	ARG	LT38	R. Gualeguay, Gualeguay
		ARG		R. Metropolitana "R.Metro", Ciudadela
	3	ARG		R. Norteña, Los Polvorines
		BOL	LP59	R. La Luz del Tiempo, El Alto
	0.25	BOL	LP32	R. LV del Cobre, Corocoro
		BOL	OR08	R. Melodía, Oruro
	1	BOL	SC26	R. Petrolera, Sta. Cruz
	0.1	CHL	CD152	R. Aníbal Pinto, Lautaro
	1	CHL	CB152	R. Integración, San Antonio
	1	CHL	CC152	R. Nueva Soberanía, Linares
	0.25	CLM	HKT20	Alcaldía de Montería, Montería
	0.1	CLM	HKW43	Alcaldía de Tangua, Tangua
	1	CLM	RL	Antena de los Andes, Santa Rosa de Cabal
	0.3	CLM		Brisas del Palmar, Caucasia
	0.25	CLM	T21	Colombia Mía, Tierralta, CO
	1	CLM	MZ	Ecos de la Sierra Flor, Sincelejo
	1	CLM	J98	Em. Una Voz de la Frontera, Puerto Santander
	5	CLM	LI	J-C R. Pasión Extrema, SF de Bogotá
	1	CLM	MA	LV de Suroeste, Jericó
	1	CLM	AM	R. Altamizal, Dolores
	0.5	CLM	HKS24	R. Cristalares Timbío, Timbío
	5	CLM	LQ	R. Minuto, Barranquilla
	1	CLM	HKW37	R. Universidad, Pasto
	1	CLM	V37	R.Pueblo Viejo, Zipacon
		CLM		Sonoradio 1520 AM, Viterbo
	2.5	EQA	RI5	LV de Guamote, Guamote
	1	EQA	RN2	LV de Naranjal, El Naranjal
	1	EQA	TI1	R. Ibarra, Ibarra
		PRU	OAU1H	R. Cristal, Chiclayo
		PRU	OBU7X	R. Fuentes Mollo, Espinar
	5	PRU	OAU7Y	R. OAU7Y, Juliaca
	1	PRU	OAX9X	R. Vision, Janjui
	2	URG	CX152	R. Acuarela, Melo

kHz	kW	Ctry	Call	Station, location
	0.1	URG	CV152	R. Paz, "la nueva Radio", Guichón
	25	VEN	IC	R. Bonita La Guapa, Guatire
1530	1	ARG	LRJ200	R. Centro Morteros, Morteros
		ARG		R. Esencia "LV del Litoral", San Miguel Oeste
		ARG		R. Eco Porteña, CA Buenos Aires
	1	BOL	CO24	R. Don Bosco, Kami
	0.5	BOL	BE07	R. Em. Ballivián, San Borja
	0.1	BOL	PO13	R. Horizonte, Huanuni
		BOL	LP69	R. Huaycheño, La Paz
	0.25	BOL	PO10	R. Litoral, Llica
	1	CHL	CB153	R. Nexo, Quillota
	1	CHL	CD153	R. Nuvo Mundo, Puerto Montt
	1	CHL	CC153	R. Portales (R. Corporación), Lota
	1	CHL	CA153	R. Portales, Copiapó
	0.25	CLM	HKN85	Alcaldía de Anza, Anza
	0.1	CLM	HKS58	Alcaldía de El Copey, El Copey
	0.25	CLM	HKN57	Alcaldía de San Juan de Uraba, San Juan de Uraba
	0.25	CLM	HKN79	Alcaldía de Uramita, Uramita
	0.25	CLM	V82	Alcaraván R., Puerto Lleras
	1	CLM	JB	Caracol Sevilla, Sevilla
	0.25	CLM	HKN65	Colombia Mía, Caucasia, AN
	1	CLM	HKR73	Ecos del Pacífico, Guapí
		CLM	HKS56	Fascinación AM, Becerril
	5	CLM	OZ	LV de la Prov. de Padilla, San Juan del Cesar
		CLM		R. Integración, Morales
	5	CLM	DN	Yeshu'a LV de Jesucristo, Medellín
	5	EQA	MP2	LV de la Península, La Libertad
	5	EQA	CC5	Ondas Cañaris AM, R. Universitaria Católica, Azogues
	1	EQA	MZ6	R. Deportes 15-30, Pelileo
	3	EQA	VP5	R. LV de Pallatanga, Pallatanga
	1	PRU	OBX3H	CPN Radio, Chimbote
	1	PRU	OBZ4S	R. 15-50, Huancayo
	2	PRU	OAU2P	R. Bambamarca, Bambamarca
		PRU		R. Capachica, Capachica
	1	PRU	OCX1Y	R. Leomar, Bellavista
	10	PRU	OBU4C	R. Milenia, Lima
		PRU	OBU7N	R. Ondas del Sur Oriente, Quillabamba
	3	PRU	OBX2R	R. Oriental, Jaén
		PRU	OAU5R	R. Universidad, San Juan Bautista
	5	PRU	OAZ7F	Rdif. Espinar, Yauri (R.Confraternidad)
	0.25	URG	CW153	Em. Cono Sur, Nueva Palmira
	10	VEN	NP	R. San Felipe el Fuerte, San Felipe
1540		ARG		AM Tango, Buenos Aires
	0.25	ARG	LU28	Cadena Uno, Gen.Madariaga
		ARG		R. Amanecer, CA Buenos Aires
		ARG		R. Cotidiana, Merlo
		ARG		R. Fuego, Longchamps
	0.25	ARG	LT35	R. Mon, Pergamino
	0.8	BOL	LP34	R. Sariri, Escoma
		BOL	CO46	R. Wiña Kalpachaj, Tarata
	1	CHL	CC154	R. Portales (R. Centra), Chillán
	1	CHL	CD154	R. San José de Alcudia, Río Bueno
	1	CHL	CB154	R. Sudamérica, Santiago
	0.25	CLM	HKP50	Alcaldía de Arjona, Arjona
	0.15	CLM	HKR80	Alcaldía de Sacama, Sacama
	1	CLM	HKZ52	Colombia Mía, Chaparral, TO
	1	CLM	B89	Em. Brisas del Río Chico, Belmira
	1	CLM	HD	LV del Petróleo, Barrancabermeja
	0.25	CLM		LV Dorada, Segovia
	2	CLM	RQ	R. Austral, Túquerres
	5	CLM	ZF	R. Cóndor, Manizales
	0.25	CLM		R. El Sur, San Vicente de Chucurí
	0.5	EQA	MH	Cotopaxi Digital, Latacunga
	0.25	EQA	VB7	LV del Upano, Macas
	1	EQA	DP1	R. Caracol, Quito
	3	EQA	FM2	R. Cristal de Ventanas, Babahoyo
		EQA		R. Flecha AM, Machala
	1	PRU	OBZ4U	R. Barranca, Barranca
	0.3	PRU	OBX4N	R. Corporacion, Cerro de Pasco
	1	PRU	OCX7V	R. Los Andes, Cusco
	1	PRU	OBX1B	R. LV de la Frontera, Tumbes
	1	PRU	OAU6A	R. Milenio Universal, Arequipa
	2	PRU	OBU2A	R. Mundial AM, Trujillo
	5	PRU		R. Turbomix, Cajamarca
	1	URG	CV154	R. Centro, Cardona
	0.1	URG	CW154	R. Charrúa, Paysandú

kHz	kW	Ctry	Call	Station, location
	0.25	URG	CX154	R. Patria, Treinta y Tres
1543		BOL	LP67	R. Bendita Trinidad Espirito de Dios, El Alto
1545		BOL	TA09	R. Emisoras Villamontes, Villamontes
	0.35	BOL	CO25	R. Mejillones, Tarata
	1	PRU	OAX9X	R. Vision, Janjui
1550	0.25	ARG	LT32	R. Chivilcoy, Chivilcoy
		ARG		R. Esperanza, Gregorio de Laferrere
	0.25	ARG	LT40	R. LV de la Paz, La Paz
		ARG		R. Popular, José León Suárez
	5/0.25	ARG	LT23	R. Regional, San Jenaro Norte
		ARG		R. Trompeta de Diós, Isidro Casanova
		ARG		R. Urkupiña, Buenos Aires
	10	BOL	LP28	R. Caranavi, Caranavi
	1	BOL	SC27	R. Tamengo, Pto. Quijarro
	1	CHL	CC155	R. Manuel Rodríguez, San Fernando
	1	CHL	CB155	R. Provincial AM, Putaendo
	0.25	CHL	CD155	R. Regional, Traiguén
	0.1	CLM	HKW53	Alcaldía de El Tablón, El Tablón
	0.1	CLM	HKW55	Alcaldía de Guachucal, Guachucal
	0.25	CLM	HKW50	Alcaldía de Mallama, Mallama
	1	CLM	HKW38	Colombia Mía, Pitalito, HU
	5	CLM	HKX29	Colombia Mía, Tibú, NS
	1	CLM	NC	Em. Revivir en Cristo, Cali
	5	CLM	UN	LV del Río Arma, Aguadas
	5	CLM	ZI	MCI R. 15-50, SF de Bogotá
	0.5	CLM		Ondas del Nechí, Campamento
	5	CLM	QD	R. Bésame, Armenia
	5	CLM	CB	R. El Sol, Barranquilla
	5	EQA	AD5	LV de Chaguarurco, Santa Isabel
	2	EQA	AD2	LV del Triunfo, El Triunfo
	2	EQA	EI6	R. Montalvo, Ambato
	1	PRU	OAU3D	R. Cruz, Chimbote
	1	PRU	OBX4P	R. Independencia, Lima
	1	PRU	OBX5J	R. San Cristobal, Carmen Alto (R.Maria)
	1	PRU	OAX1D	R. Superior, Monsefú
	1	PRU	OCX2R	R. TV Chota, Chota
	1/0.25	URG	CV155	R. Agraciada, Mercedes
	2	URG	CW155	R. Sarandí del Yí, Sarandí del Yí
	10/5	VEN	XO	R. Impacto La Poderosa, Cd. Ojeda
	10	VEN	MW	R. Metropolitana, Los Teques
1555	0.5	CLM		R. Parroquial, El Santuario
1560	0.5/0.25	ARG		AM 1560 "La R. de la Gente", Tandil
		ARG		R. Antena Lobos, Lobos
		ARG		R. Castañares, Ituzaingó
		ARG		R. Ebenezer, Ezeiza
	2.5/1.5	ARG	LT11	R. Gral. Francisco Ramírez, Villaguay
	0.25	ARG	LT33	R. Nueve de Julio, 9 de Julio
		ARG		R. Restauración, Llavallol
	0.5	BOL	CO26	1° de Octubre, Capinota
	1	BOL	OR19	R. Occidental, Oruro
		BOL	LP53	R. Tawantinsuyo, Taraco
	0.5	BOL	CO27	R. Urkupiña, Quillacollo
	1	CHL	CB156	R. Manantial, Talagante
	5/3	CHL	CA156	R. Parinacota, Putre
	1	CHL	CD156	R. Parque Nacional, Villarrica
	0.25	CLM	HKO35	Alcaldía de Cañasgordas, Cañasgordas
	0.25	CLM	HKW90	Alcaldía de Villavicencio, Villavicencio
	5	CLM	LP	La Cariñosa, Antena 2, Tuluá
	1	CLM	PZ	R. Codazzi, Codazzi
	0.5	CLM	HKS65	R. Tamalameque, Tamalameque
	5	CLM	CP	RCN Antena 2, Arbelaez
	1	CLM	XZ	Santa María de la Paz R., Medellín
	5	CLM	HE	Voces Rovirenses, Málaga
	1.5	EQA	ZD1	Ecos Culturales de Urcuquí, Urcuquí
	2	EQA	TR3	LV del Guabo, El Guabo
	2	EQA	CS2	R. Sideral, Daule
	1.5	PRU	OAU7Z	R. Carráviz, Juliaca
	1	PRU	OCX6N	R. La Union (R.La Luz), Arequipa
	1	PRU	OAZ7N	R. Maria, Wanchaq
		PRU	OBX80	R. Nuevo Mundo, Iquitos
	0.5	PRU	OBX2J	R. Nuvo Continente, Cajamarca
	1	PRU	OBU4G	R. San Sebastian, Yauyos
	2/0.5	URG	CX156	Dif. Americana, Trinidad
	3/0.5	URG	CW51	R. Maldonado, Maldonado
	1/0.25	URG	CV156	R. Vichadero
	10/5	VEN	LZ	R. Dif. Andina, Mérida
1564	0.5	PRU	OBU2H	R. San Lorenzo, Socota
1570	5/1	ARG	LRI223	Lomas de Zamora
	2	ARG		R. AM Rocha, La Plata

kHz	kW	Ctry	Call	Station, location
		ARG		R. La Morena de Itati, Grand Bourg
		ARG		R. Melody, Remedios de Escalada
	0.5	BOL	SC28	R. 1° de Mayo, 1° de Mayo
	0.25	CHL	CD157	R. Acuarela, Nueva Imperial
	7	CHL	CC157A	R. Familia de Talca, Talca
	1	CHL	CC157	R. Niebla, Rancagua
	0.25	CLM	HKX78	Alcaldía de Balboa, Balboa
	0.15	CLM	HKU42	Alcaldía de Cajica, Cajica
	0.25	CLM	HKQ83	Alcaldía de Maripi, Maripi
	0.25	CLM	HKQ82	Alcaldía de Sta María, Sta María
	0.25	CLM	HKP58	Alcaldía de Sta Rosa Sur, Sta Rosa Sur
	2	CLM		Arc. Armada de Colombia, Pto Leguizamo
	1	CLM	E96	Colombia Mía, Palmira, VA
		CLM		LV de Fomeque, Fomeque
	1	CLM	ZT	R. Auténtica, Manizales
	1	CLM	C22	R. Ciudad Dabeiba, Dabeiba
	1	CLM	TG	R. María, Machetá
	0.1	CLM	HKX80	R. Marsella, Marsella
	0.2	CLM	HKR66	R. Universidad de la Amazonia, Florencia
	0.5	CLM	HJR66	Timbiqui Estéreo, Timbiqui
	0.5	EQA		Ondas Quereñas, Quero
	1	EQA		R. LV Espíritu Santo de Dios, Manta
	10	EQA	PG1	R. Nucanchic, Maldonado
		PRU	OCU7L	Canchis (F.P.I.)
		PRU		R. Bethel, Lima
	1	PRU	OBX3N	R. Chasqui, Yungay
		PRU	OBU2L	R. Colonial, Contumaza
	1	PRU	OCX1Z	R. La Nueva Esperanza, Tambo Grande
	1	PRU	OBX3M	R. San Martin, Huanuco
	1	PRU	OCX6I	R. Willy, Uraca
		PRU	OCU2C	Radiodifusora Julcan
	0.25/0.1	URG	CW157A	Em.Celeste, Tomás Gomensoro
	2/0.5	URG	CX157	R. Canelones
1575	3/1	PRU	OAZ1F	R. Naylamp, Lambayeque
	1	PRU	OAU2E	R. Nuevo Continente, San Ignacio
1578	1	BOL	CO24	R. Don Bosco, Kami
1580	1	ARG		R. 26. de Julio, Longchamps
	0.25	ARG	LT36	R. Chacabuco, Chacabuco
	0.25	ARG	LT27	R. LV del Montiel, Villaguay
		ARG		R. Provincial de Sierra Colorada, Sierra Colorada
		ARG		R. Tradición, Isidro Casanova
		ARG		R. Tradición, San Martín
	1	BOL	SC29	R. Andrés Ibáñez, Santa Cruz
		BOL	LP62	R. El Fuego del Espíritu Santo, El Alto
	1	CHL	CC158	R. Colchagua, Santa Cruz
	0.5	CHL	CD158A	R. Continental, Collipulli
	0.25	CHL	CD158	R. Millaray, Cañete
	0.25	CLM	HKU42	Alcaldía de Cajica, Cajica
	0.1	CLM	HKW74	Alcaldía de Pupiales, Pupiales
	0.25	CLM	HKT34	Alcaldía de San Antero, San Antero
		CLM		Alcaldía de Yaguará, Yaguará
	5	CLM	RM	Caracol Colombia, Sincelejo
		CLM	TE	LV de Chocó, Quibdó
	1	CLM	LC	LV del Banco, El Banco
	0.15	CLM	HKS46	R. Alcaldía de Padilla, Padilla
	5	CLM	QZ	R. María, Barranquilla
	1	CLM	DE	R. Miraflores, Rovira
	5	CLM	SQ	R. Robledo/RCN Antena 2, Cartago
	1	CLM	KB	R. Zulima, Villa del Rosario
	5	CLM	QT	Sonríéle a Jesús R., SF de Bogotá
	1	EQA	LF1	Ecos de Orellana, Machach
	3	EQA	TP5	Ecos del Portete, Girón
	5	EQA	VA4	Estación de la Alegría, Esmeraldas
	0.25	EQA	AB3	Ondas de Paltas, Catacocha
	1	PRU		R. Central, Bellavista
	3/1	PRU	OBX1M	R. Naylamp, Lambayeque
	1	PRU	OAU4P	R. San Juan, Tarma
	1	PRU	OAU5J	R. Virgen del Carmen, Huancavelica
	0.5	URG	CW54	Emisoras del Este, Minas
	1/0.5	URG	CW158	R. San Salvador, Dolores
	10/5	VEN	TK	Manzanares 15-80, Cumaná
	10	VEN	YO	R Celestial, San Francisco, Maracaibo
	10/5	VEN	YV	R. Venezolana, Calabozo
1584	0.5	PRU	OBX7Q	R. El Triunfo, Cusco
1587	1	PRU	OAU5A	R. Armonia, Abancay
1590		ARG		R. Cristiana Adonal, Bánfield Oeste
		ARG		R. Guaviyú, Gregorio de Laferrere
		ARG		R. Olivera, General Rodriguez
	3	BOL	TA07	R. Bermejo, Bermejo
		BOL	SC34	R. Globo, La Guardia
		BOL	LP61	R. Kollasuyo Marka, Tiawanaku
	0.5	BOL	OR16	R. Producciones Pusisuyu, Oruro
	1	CHL	CB159	R. Aconcagua, San Felipe
	0.25	CHL	CC159	R. Rengo, Rengo
		CLM	HKS72	Alcaldía de La Gloria, La Gloria
	5	CLM	IP	BBN 15-90 R., Envigado
	1	CLM	QM	Ecos de la Miel, Samaná
	5	CLM	WB	Em Nuestra Sra del Socorro, Socorro
		CLM		Ondas del Rioseco, Rioseco
		CLM		R. Espacial, Andalucía
	1	EQA	RZ1	R. Mensaje, Cayambe
	1	EQA	QT6	R. Panamericana, Quero
	0.25	EQA	AS2	R. Record, La Libertad
	1	PRU	OBU2C	Agro Radio, Trujillo
	2	PRU	OAZ4Z	R. Agricultura "La Peruanísima" Lima
	1	PRU	OAU7C	R. Huaynaroque, Juliaca
	1	PRU	OCX6S	R. Mundo, Arequipa
	10	URG	CX159	R. Real, Colonia
	1/0.25	URG	CW159	R. Regional "La Nueva Radio", Lascano
	10	VEN	UD	R. Deporte 15-90, Caracas
1600	1.2	ARG		R. Armonia, José Ingenieros
		ARG		R. Metropolitana, Luís Guillón
	0.5	BOL	CO28	R. Continental, Punata
	1	BOL	LP63	R. La Voz del Espirito Santo, El Alto
	0.25	CHL	CC160	R. Llacolén, Concepción
	0.25	CHL	CB160	R. Nuevo Tiempo, Santiago
	0.25	CHL	CB160A	Radiocable, Viña del Mar
	0.15	CLM	HKZ79	Alcaldía de Cajamarca, Cajamarca
	0.25	CLM	HKO63	Alcaldía de Jardín, Jardín
	0.25	CLM	HKX83	Alcaldía de La Celia, Celia
	0.25	CLM	HKT39	Alcaldía de Valencia, Valencia
	0.15	CLM	HKZ77	Alcaldía de Venadillo, Venadillo
	5	CLM	HV	Armonías Zipaquireñas, Zipaquirá
	5	CLM	HKO72	Colombia Mía, Carepa, AN
		CLM	HKX84	Em. Mundial, Dosquebradas
	0.25	CLM		LV de Aranzazu
	0.25	CLM	HKR52	LV de Colina, Risaralda
	1	CLM		LV del Rosario, Junín
		CLM		R. Bello Horizonte, Pesca
		CLM		R. Fortaleza, Sogamoso
	0.25	CLM		R. Impacto Cristiano, Popayán
	0.25	CLM	F33	R. Restauración, Cali
		EQA		Ondas de Caluma, Caluma
		EQA		R. Ilusión 1600 AM, Puembo
	1	PRU	OCY2D	R. Internacional, San Pablo
		PRU	OBU4R	R. Nuevo Tiempo, Huancayo
		PRU		R. San Juan, Catacaos
	2	URG	CV160	R. Continental, Pando
	1	URG	CX160	R. Litoral, Fray Bentos
1610	0.5	ARG		R. Buenas Nuevas, Laboulaye
	0.2	ARG		R. Fósil, Rosario
		ARG		R. Guaviyú, Gregorio de Laferrere
	0.05	ARG		R. Luz del Mundo, Rafael Calzada
		CLM		Armonías de Occidente, Medellín
		CLM		R. Estelar, Santuario
		PRU		R. Carabamba, Julcán
	0.5	PRU	OAU6O	R. Flor de los Andes, José Luis Bustamente y Rivero
		PRU		R. Haquira, Haquira
1613	1	CLM		R. Ideal, Umbita
1620		ARG		R. AM 16-20, Mar del Plata
		ARG		R. Sión, Monte Grande
		ARG		R.Italia, Villa Martelli
1630		ARG		AM Restauración, Hurlingham
		ARG		R. AM Súpe Sport, Lomas de Zamora
1640		ARG		Hosanna AM 1640, Isidro Casanova
		ARG		R. Nueva Bolivia , Buenos Aires
1650	1/0.5	ARG	LRI227	Antares AM 1650 "La R. de la Familia", Pilar
		ARG		R.Fenix. Temperley
1660		ARG		R. Reivir, Isidro Casanova
		ARG		R.Esperanca, Virrey Del Pino
1670		ARG		R. Bethel, Banfield
1680		ARG		R.Hosanna Tropical, Ezeiza
1690	1/0.25	ARG		R. Apocalipsis II, San Justo
1700		ARG		R. Cristiana Príncipe con Dios, Banfield Oeste
1710		ARG		AM 1710 - R. Estudio ESBA, Buenos Aires

SHORTWAVE STATIONS OF THE WORLD
November 2010 - World Copyright WRTH Publications Ltd

For country and site codes, see relevant tables in reference section. Stations marked as 'dom' in the site column are domestic/national broadcasts. Stations marked with 'STF' in the site column are Standard Time/Frequency transmissions. The column 'N' indicates Notes. Symbols used in the 'N' column are '+', indicating DRM transmissions; '±' which indicates variable frequency; '†' for irregular transmissions and ‡ for frequencies that were inactive at time of publication.

kHz	N	kW	Ctry	Site	Station, location
2310		50	AUS	dom	Northern Territory SW Sce, Alice Springs
2325		50	AUS	dom	Northern Territory SW Sce, Tennant Creek
2350			KRE	dom	KCBS/Reg, Sariwon
2368		1	AUS	dom	R. Symban, Sydney NSW
2380		0.25	B	dom	R. Educadora, Limeira
2485		50	AUS	dom	Northern Territory SW Sce, Katherine
2500		10	CHN	STF	BPM, Kinshan
		5	HWA	STF	WWVH, Kauai
		2.5	USA	STF	WWV, Ft. Collins
2850			KRE	dom	KCBS, Pyongyang
2960	‡	0.3	INS	dom	RPDT2 Manggarai, Ruteng
3025	†		KRE	dom	Frontline Soldiers Radio
3173		1	PRU	dom	R. Municipal, Panao
3185		100	USA	WRB	WWRB
3200		50	SWZ	MAN	TWR Africa
3205	‡	10	PNG	dom	R. West Sepik, Vanimo
3215		100	AFS	MEY	Adventist World R. (AWR)
		100	AFS	MEY	Amateur R. Mirror Int.
		50	MDG	MDC	Adventist World R. (AWR)
		50	MDG	dom	R. Feon'ny Filazantsara, Talata Volonondry
		100	USA	WCR	WWCR
		100	USA	WRB	WWRB
3220			KRE	dom	KCBS/Reg, Hamhung
		10	PNG	dom	R. Morobe, Lae
3230		100	AFS	MEY	Family Radio Worldwide
3235		10	PNG	dom	R. West New Britain, Kimbe
		1	PRU	dom	R. Luz y Sonido, Huánuco
3240		50	SWZ	MAN	TWR Africa
3250		1	HND	dom	R. Luz y Vida, San Luís
		50	KRE	PYO	Pyongyang Broadcasting Stn.
		100	KRE	PYO	Voice of Korea
3255		100	AFS	MEY	BBC World Service
		1	B	dom	Rdif. 6 de Agosto, Xapuri
3260		10	PNG	dom	R. Madang, Madang
3275		10	PNG	dom	R. Southern Highlands, Mendi
3280		50	CHN	dom	VO Pujiang, Shanghai
		2	EQA	dom	LV del Napo, Tena
3289		10	MDG	dom	R. Nasionaly Malagasy, Ambohidrano
3290		10	GUY	dom	National Comms Netw, Georgetown
		10	PNG	dom	R. Central, Port Moresby (Boroko)
3305	‡	10	PNG	dom	R. Western, Daru
3310		10	BOL	dom	R. Mosoj Chaski, Cochabamba
3315		10	PNG	dom	R. Manus, Lorengau
3320		100	AFS	dom	SABC R. Sonder Grense, Meyerton
		50	KRE	PYO	Pyongyang Broadcasting Stn.
3325	†	2.5	B	dom	R. Mundial, São Paulo
		10	INS	dom	RRI, Palangkaraya
		10	PNG	dom	R. Buka (Bougainville), Buka
3330		10	CAN	STF	CHU, Ottawa
		5	PRU	dom	R. Ondas del Huallaga, Huánuco
3335		10	PNG	dom	R. East Sepik, Wewak
3340		2	HND	dom	R. Misiones Int, Comayagüela
3345		100	AFS	MEY	Adventist World R. (AWR)
		100	AFS	MEY	Channel Africa
		0.5	BOL	dom	R. Ayopaya, Independencia
		10	INS	dom	RRI, Ternate
	†	10	PNG	dom	R. Northern, Popondetta
3350		100	CTR	CRI	R. Exterior de España (REE)
			KRE	dom	KCBS/Reg, Pyongyang
3355	†	10	PNG	dom	R. Simbu, Kundiawa
3365		1	B	dom	R. Cultura, Araraquara
		10	PNG	dom	R. Milne Bay, Alotau
3375		1	B	dom	R. Municipal, São Gabriel da Cachoeira
	†	1	PRU	dom	R. San Antonio, Callalli
3380		100	AFS	MEY	BBC World Service
		1	EQA	dom	Centro Radiofonico de Imbabura, Ibarra
3385		10	PNG	dom	R. East New Britain, Rabaul
3390		1	CAF	dom	R. ICDI, Boali
3391		1	BOL	dom	R. Emisora, Camargo
3396		100	ZWE	dom	ZBC R. Zimbabwe, Gweru
3480		50	KOR	GOY	Voice of the People
		5	KRE	WON	KCBS Relay (AINDF)
3560		50	KRE	KUJ	Voice of Korea
3579	‡	0.5	INS	dom	RSPK Ngada
3810		1	EQA	STF	HD2IOA, Guayaquil
3815		0.2	GRL	dom	KNR, Tasiilaq
3900		5	CHN	dom	Hulun Buir, Hailar
3905		10	PNG	dom	R. New Ireland, Kavieng
3912		50	KOR	GOY	Voice of the People
3915		1	PNG	dom	R. Fly, Kiunga
		100	SNG	SNG	BBC World Service
			KRE	dom	KCBS/Reg, Hyesan
3925		10	J	dom	R. Nikkei 1, Sapporo
		50	J	dom	R. Nikkei 1, Tokyo
3930	±		IRQ	—	Radio Voice of Kurdistan
3940			KRE	dom	KCBS/Reg, Chongjin
3945		50	IND	dom	All India R, Gorakhpur
		50	IND	GKP	All India Radio (AIR)
		10	J	dom	R. Nikkei 2, Tokyo
		5	VUT	dom	R. Vanuatu, Port Vila
3950		50	CHN	dom	Xinjiang, Ürümqi
3955		100	AFS	MEY	Family Radio Worldwide
		100	G	SKN	BBC World Service
		250	G	SKN	KBS World Radio
		250	G	SKN	Radio Taiwan International
		100	LTU	SIT	HCJB La Voz de los Andes
		100	LTU	SIT	VO the Islamic Rep. of Iran
3957	‡	10	INS	dom	RRI, Palu
3960			KRE	dom	KCBS/Reg, Kanggye
		2.5	LBR	dom	Star R, Monrovia
3965	+	1	F	ISS	Radio France Int. (RFI)
		250	F	ISS	Radio Taiwan International
		500	IRN	KAM	VO the Islamic Rep. of Iran
		500	IRN	ZAH	VO the Islamic Rep. of Iran
3970			KRE	dom	KCBS/Reg, Wonsan
3975		250	D	WER	Family Radio Worldwide
	+	100	G	SKN	Polish Radio (Ext Sce)
	‡	100	PAK	dom	Azad Kashmir R, Islamabad
3976	†	10	INS	dom	RRI, Pontianak
3985		100	CHN	dom	CNR2 China Business R, Golmud
		250	G	SKN	Radio Taiwan International
		250	G	SKN	Voice of Vietnam (VOV)
		10	HRV	DEA	Voice of Croatia
		500	IRN	KAM	VO the Islamic Rep. of Iran
		100	KOR	HWA	Echo of Hope (VOH)
3987	‡	1	INS	dom	RRI, Manokwari
3990		15	CHN	dom	Gannan PBS, Hezuo
		50	CHN	dom	Xinjiang, Ürümqi
3995		250	G	SKN	Deutsche Welle
		5	INS	dom	RRI, Kendari
	‡	100	PAK	dom	Azad Kashmir R, Islamabad
	†	100	PAK	ISL	VO Jammu Kashmir Fr Mov.
4005		10	CVA	VAT	Vatican Radio
4010		100	KGZ	dom	KGR1, Bishkek
4050		100	KGZ	dom	R. Rossii relay, Bishkek
4052	‡	0.5	GTM	dom	R. Verdad, Chiquimula
4220	‡	15	CHN	dom	Qinghai, Xining
4319		3	BIO	DGA	AFRTS (AFN Feeder)
4330		50	CHN	dom	Xinjiang, Ürümqi

kHz	N	kW	Ctry	Site	Station, location
4368		10	F	MCO	Radio Monaco
4405		50	KRE	KUJ	Voice of Korea
4409		0.5	BOL	dom	R. Eco, Reyes
4413	‡	1	LAO	dom	R. Nationale Lao, Sam Neua
4450	‡	0.25	BOL	dom	R. Estación Frontera, Cobija
		50	KOR	GOY	Voice of the People
		50	KRE	PYO	KCBS Relay (AINDF)
4451		1	BOL	dom	R. Santa Ana, Santa Ana del Yacuma
4460		100	CHN	dom	CNR1 VO China, Beijing
4498			BOL	dom	R. Estambul, Guayaramerin
4500		50	CHN	dom	Xinjiang, Ürümqi
4556		0.12	BOL	dom	R. Hitachi, Guayaramerín
4557		15	KRE	HJU	KCBS Relay (AINDF)
4600		0.2	BOL	dom	R. Perla del Acre, Cobija
4605	†	1	INS	dom	RRI, Serui
4685		1	BOL	dom	R. Paitití, Guayaramerín
4700			BOL	dom	R. San Miguel, Riberalta
4717		1	BOL	dom	R. Yura, Yura
4732	†		BOL	dom	R. Universitaria, Cobija
4740	‡	1	SDN	dom	R. Peace, Juba
	‡	1	VTN	dom	VOV Provincial Sce, Son La
4747		0.5	PRU	dom	R. Huanta 2000, Huanta
4750		100	BGD	dom	Bangladesh Betar, Khabirpur
		100	CHN	dom	CNR1 VO China, Nanning/Hailar
		15	CHN	dom	Qinghai PBS, Xining
		20	INS	dom	RRI, Makassar
	†		PRU	dom	R. San Francisco Solano, Sondor
		1	UGA	dom	Dunamis SW, Mukono
4755		10	B	dom	R. Imaculada Conceição, Campo Grande
4760		10	IND	dom	All India R, Leh
		8	IND	dom	All India R, Port Blair
	‡	1	LBR	dom	R. ELWA, Monrovia
		25	SWZ	MAN	TWR Africa
4762			BOL	dom	R Chicha, Tocla
			BOL	dom	R. Guanay, Guanay
4765		50	TJK	dom	Tajik R. 1st prgr, Yangiyul
4770		50	NIG	dom	FRCN, Kaduna
4775		1	B	dom	R. Congonhas, Congonhas
		50	IND	dom	All India R, Imphal
		0.5	PRU	dom	R. Tarma, Tarma
		50	SWZ	MAN	TWR Africa
4780		50	DJI	dom	RTD, Djibouti
		1	GTM	dom	R. Cultural Coatán, San Sebastián Coatán
4781			BOL	dom	R.Tacana, Tumupasa
		3	EQA	dom	R. Oriental, Tena
4785	†	1	B	dom	R. Brasil 5000, Campinas
4790		10	INS	dom	RRI, Fak-Fak
	‡	100	PAK	dom	Azad Kashmir R, Islamabad
		10	PAK	dom	Azad Kashmir R, Rawalpindi
		0.5	PRU	dom	R. Visión, Lambayeque
4795		15	KGZ	dom	KGR1, Bishkek
4796			BOL	dom	R. Lípez, Uyuni
4800		100	CHN	dom	CNR1 VO China, Golmud
		1	GTM	dom	R. Buenas Nuevas, San Sebastián
		50	IND	dom	All India R, Hyderabad
		0.5	MEX	dom	XERTA R. Transcontinental, México
4805		10	B	dom	R. Dif. do Amazonas, Manaus
			PRU	dom	R. Rasuwilca, Ayacucho
4810		100	ARM	dom	Public R. of Armenia, Yerevan
		100	ARM	YER	Public Radio of Armenia (FS)
		50	IND	dom	All India R, Bhopal
4815		10	B	dom	R. Dif. Londrina
		1	EQA	dom	R. Buen Pastor, Saraguro
4819		5	HND	dom	LV Evangélica, Tegucigalpa
4820		100	CHN	dom	Xizang, Lhasa
		50	IND	dom	All India R, Kolkata
4824		10	PRU	dom	LV de la Selva, Iquitos
4825		10	B	dom	R. Canção Nova, Cachoeira Paulista
		5	B	dom	R. Educadora, Bragança
4826		0.3	PRU	dom	R. Sicuani LV de Canchis, Sicuani
4828	†	100	ZWE	GWE	Voice of Zimbabwe
4830		10	MNG	dom	Mongolian R. 1, Altay
4835		50	AUS	dom	Northern Territory SW Sce, Alice Springs
		0.5	BOL	dom	R. Virgen de los Remedios, Tupiza
		10	IND	dom	All India R, Gangtok
		1	PRU	dom	R. Marañón, Jaén
4840		50	IND	dom	All India R, Mumbai
		100	USA	WCR	WWCR
4845		10	B	dom	R. Cultura, Manaus
	†	1	B	dom	R. Meteorologia Paulista, Ibitinga
			BOL	dom	R. Norteño, Caranavi
		100	MTN	dom	R. Mauritanie, Nouakchott
4850	†	50	IND	dom	All India R, Kohima
4856		2	PRU	dom	R. La Hora, Cusco
4860		50	IND	dom	All India R, Delhi
		50	IND	DEL	All India Radio (AIR)
4865		5	B	dom	R. Alvorada, Londrina
		5	B	dom	R. Missões da Amazônia, Óbidos
		5	B	dom	R. Verdes Florestas, Cruzeiro do Sul
		5	BOL	dom	R. Logos, Santa Cruz de la Sierra
4870	±		IRQ	—	Voice of Iranian Kudistan
		5	EQA	dom	LV del Upano, Macas
		100	IND	DEL	Radio Sedayee Kashmir
		10	INS	dom	RRI, Wamena
4875	‡	10	B	dom	R. Dif. Roraima, Boa Vista
4876		10	BOL	dom	R. La Cruz del Sur, La Paz
4880		100	AFS	MEY	SW Radio Africa
		50	IND	dom	All India R, Lucknow
4885		2	B	dom	R. Clube do Pará, Belém
	‡	1	B	dom	R. Maria, Anápolis
		5	B	dom	Rdif. Acreana, Rio Branco
4895		100	AFS	MEY	Zimbabwe Community Radio
		5	B	dom	R. Novo Tempo, Campo Grande
		50	IND	dom	All India R, Kurseong
		10	MNG	dom	Mongolian R. 2, Mörön
4900		1	GUI	dom	R. Familia, Timbi-Madina
4905		1	B	dom	R. Anhanguera, Araguaína
		100	CHN	dom	Xizang, Lhasa
4909			EQA	dom	R. Chaskis, Otavalo
4910		50	AUS	dom	Northern Territory SW Sce, Tennant Creek
		50	IND	dom	All India R, Jaipur
4915		10	B	dom	R. Daquí, Goiânia
		25	B	dom	R. Dif. Macapá, Macapá
4919		12	EQA	dom	R. Quito LV de la Capital, Quito
4920		100	CHN	dom	Xizang, Lhasa
		50	IND	dom	All India R, Chennai
	†	1	INS	dom	RRI, Biak
4925		5	B	dom	R. Educação Rural, Tefé
	‡	10	INS	dom	RRI, Jambi
4930		100	BOT	BOT	BBG - VO America (VOA)
		100	BOT	BOT	BBG - VOA Studio 7
	‡	50	TKM	dom	Türkmen R. 2nd prgr, Asgabat
4935		1	B	dom	R. Capixaba, Vitória
4940		50	CHN	dom	VO Strait News Channel, Fuzhou
		50	IND	dom	All India R, Guwahati
	†	1	PRU	dom	R. San Antonio, Villa Atalaya
		100	STP	SAO	BBG - VO America (VOA)
		1	B	dom	R. Amazonas, Puerto Ayacucho
4950		25	AGL	dom	R. Nacional de Angola, Mulenvos
		50	CHN	dom	VO Pujiang, Shanghai
		50	IND	dom	All India R, Srinagar (Kashmir)
		5	PRU	dom	R. Madre de Dios, Puerto Maldonado
4955		5	PRU	dom	R. Cultural Amauta, Huanta
4958		3	BOL	dom	R. Trópico, Trinidad
4960	†	5	EQA	dom	R. Federación Shuar, Sucúa
		100	STP	SAO	BBG - VO America (VOA)
4965	‡	5	B	dom	R. Alvorada, Parintins
		50	IND	dom	All India R, Shimla
	†	1	PRU	dom	R. Santa Mónica, Wanchaq
		100	ZMB	LUS	CVC R. Christian Voice
4970		50	IND	dom	All India R, Shillong
4975		1	B	dom	R. Iguatemi (A Nossa Voz), Osasco
		10	CHN	dom	Fujian PBS News channel, Fuzhou
		5	PRU	dom	R. Pacífico, Lima
		100	TJK	DSB	Voice of Russia (VOR)
4976		10	UGA	dom	UBC R. Red channel, Kampala
4980		50	CHN	dom	Xinjiang, Ürümqi
4985		10	B	dom	R. Brasil Central, Goiânia
4987		1	PRU	dom	R. Manantial, Huancayo
4990		10	CHN	dom	Hunan PBS News channel, Xiangtan
		50	IND	dom	All India R, Itanagar
		1	SUR	dom	R. Apintie, Paramaribo

kHz	N	kW	Ctry	Site	Station, location
4992	‡	5	PRU	dom	R. Ancash, Huaraz
4996		5	RUS	STF	RWM, Moscow
5000		2	ARG	STF	LOL SHN, Buenos Aires
		20	CHN	STF	BPM, Kinshan
		10	HWA	STF	WWVH, Kauai
5000		10	USA	STF	WWV, Ft. Collins
	‡	2	VEN	STF	YVTO, Caracas
5005	†	50	GNE	dom	Rdif. de Guinea Ecuatorial, Bata
	†	5	NPL	dom	R. Nepal, Khumaltar
5010	†	1	DOM	dom	R. Pueblo, Santo Domingo
		50	IND	dom	All India R, Thiruvananthapuram
		10	MDG	dom	R. Nasionaly Malagasy, Ambohidrano
		100	MDG	dom	R. Nasionaly Malagasy, Ambohidrano
5014		1	PRU	dom	R. Altura, Cerro de Pasco
5015		50	IND	dom	All India R, Delhi
		20	TKM	dom	Türkmen R. 1st prgr, Asgabat
5020		5	PRU	dom	R. Horizonte, Chachapoyas
		10	SLM	dom	Solomon Islands BC, Honiara
5025		50	AUS	dom	Northern Territory SW Sce, Katherine
		10	BEN	dom	ORTB, Parakou
		50	CUB	dom	R. Rebelde, La Habana
		5	PRU	dom	R. Quillabamba, Quillabamba
5030	‡	100	BFA	dom	RTB, Ouagadougou
		10	MLA	dom	RTM Sarawak, Kuching
5035		10	B	dom	R. Aparecida, Aparecida
		1	CAF	dom	R. Centrafrique, Bangui
5039		1	PRU	dom	R. Libertad, Junín
5040		10	CHN	dom	Fujian PBS News channel, Fuzhou
		250	CUB	HAB	Radio Habana Cuba
		10	EQA	dom	LV del Upano, Macas
		50	IND	dom	All India R, Jeypore
5041		50	BRM	dom	R. Myanma, Yangon
5045		10	B	dom	R. Cultura do Pará, Belém
5050		15	CHN	NNN	Beibu Bay Radio
		10	IND	dom	All India R, Aizawl
		100	USA	WRB	WWRB
5055		1	B	dom	R. Difusora, Cáceres
	†	5	B	dom	R. Jornal A Crítica, Manaus
	‡	2	VUT	dom	R. Vanuatu, Port Vila
5059		1	PRU	dom	LV de las Huarinjas, Huancabamba
5060		50	CHN	dom	Xinjiang, Ürümqi
	±	10	ERI	dom	VO the Broad Masses 2nd prgr, Asmara
5066		1	COD	dom	R. Télé Candip, Bunia
5070		100	USA	WCR	WWCR
5075			CHN	dom	VO Pujiang, Shanghai
5080		100	USA	TWW	WTWW
5110		50	USA	BCQ	WBCQ
5120	†	1	PRU	dom	Ondas del Suroriente, Quillabamba
5130	†	100	KGZ	BIS	SW Relay Service
5240		100	CHN	dom	Xizang, Lhasa
5446		3	USA	KEW	AFRTS (AFN Feeder)
5460			PRU	dom	R. LV Bolivar, Bolivar
5470	‡	10	LBR	dom	R. Veritas, Monrovia
5486		1	PRU	dom	R. Frecuencia Popular, Olmos
5487		0.06	PRU	dom	La Reyna de la Selva, Chachapoyas
5580		0.25	BOL	dom	R. San José, San José de Chiquitos
5681			BOL	dom	R. San Rafael, Cochabamba
5745		250	USA	GRV	BBG - R. Martí
		100	USA	YFR	Family Radio Worldwide
		100	USA	WRB	WWRB
5755		100	USA	TWW	WTWW
5765		3	GUM	BAR	AFRTS (AFN Feeder)
5770		10	BRM	dom	Defence Forces BC, Taunggyi
5780		250	KWT	KWT	BBG - VOA Ashna Radio
		200	TJK	DSB	BBG - R. Free Asia (RFA)
5790		250	CLN	IRA	BBG - R. Free Asia (RFA)
		300	G	SKN	BBC World Service
		500	G	RMP	BBC World Service
5810		250	KWT	KWT	BBG - R. Farda
		250	MRA	TIN	BBG - R. Free Asia (RFA)
		250	USA	EWN	WEWN - EWTN Shortwave
5820		100	ARM	ERV	Family Radio Worldwide
		250	KWT	KWT	BBG - RFE/RL
5825		100	ARM	ERV	Family Radio Worldwide
		100	ARM	ERV	Radio Rahaye Iran
		100	UZB	TAC	Family Radio Worldwide

kHz	N	kW	Ctry	Site	Station, location
5830		250	PHL	PHT	BBG - VO America (VOA)
		250	UZB	TAC	Deutsche Welle
5835		250	CLN	IRA	BBG - VOA Deewa Radio
		250	KWT	KWT	BBG - VOA Deewa Radio
5835		250	PHL	PHT	BBG - VO America (VOA)
5840		250	KWT	KWT	BBG - RFE/RL
5845		100	THA	NAK	BBC World Service
		250	THA	NAK	BBC World Service
	+	100	THA	NAK	Deutsche Welle
5850		250	CLN	IRA	BBG - R. Farda
		250	CLN	IRA	BBG - R. Free Asia (RFA)
		250	CLN	IRA	BBG - VO America (VOA)
		250	PHL	PHT	BBG - VO America (VOA)
5855		500	MRA	TIN	BBG - R. Free Asia (RFA)
5860		50	CHN	dom	VO Jinling, Nanjing
		250	CLN	IRA	BBG - R. Farda
		250	KWT	KWT	BBG - R. Farda
		125	MRA	TIN	BBG - R. Free Asia (RFA)
		250	MRA	TIN	BBG - R. Free Asia (RFA)
		250	PHL	PHT	R. Netherlands Worldwide
5865		500	F	ISS	Radio Algeriennne
		250	THA	NAK	BBC World Service
		200	TJK	DSB	BBC World Service
5875	+	40	AUT	MOS	BBC World Service
		300	CYP	CYP	BBC World Service
		250	G	WOF	BBC World Service
		500	G	RMP	BBC World Service
	+	65	NOR	KVI	Radio Romania International
		250	THA	NAK	BBC World Service
		100	USA	HRI	WHRI - World Harvest R. Int.
5885		250	CVA	SMG	BBG - Afia Darfur Radio
		100	CVA	SMG	Vatican Radio
		250	KWT	KWT	BBG - RFE/RL
5890		500	IRN	KAM	VO the Islamic Rep. of Iran
		500	IRN	SIR	VO the Islamic Rep. of Iran
		250	MRA	TIN	BBG - VO America (VOA)
		250	THA	NAK	BBC World Service
		250	USA	GRV	BBG - VO America (VOA)
		100	USA	WCR	WWCR
5895	+	40	NOR	KVI	Polish Radio (Ext Sce)
		250	THA	UDO	BBG - RFE/RL
		250	TJK	DSB	Vatican Radio
5900		100	BUL	SOF	Radio Bulgaria
		170	BUL	PLD	Radio Bulgaria
		300	BUL	PLD	Radio Bulgaria
		500	CVA	SMG	Vatican Radio
		300	ROU	TIG	Radio Romania International
		250	RUS	IRK	Family Radio Worldwide
		250	RUS	PPK	Family Radio Worldwide
		200	RUS	PPK	IBRA Radio
		500	RUS	IRK	Radio France Int. (RFI)
		250	RUS	SAM	Voice of Russia (VOR)
		100	UZB	TAC	Vatican Radio
5905		250	ATN	BON	Deutsche Welle
		100	CHN	KAS	China Radio Int. (CRI)
		300	CYP	CYP	BBC World Service
		500	F	ISS	Radio France Int. (RFI)
		500	G	RMP	Deutsche Welle
		250	POR	SIN	Deutsche Welle
		250	RUS	MSK	Radio Rossii
	+	250	RUS	MSK	Voice of Russia (VOR)
5910		100	AFS	MEY	BBC World Service
		500	CHN	BEI	China Radio Int. (CRI)
			CLM	dom	Marfil Estéreo, Puerto Lleras
		500	CVA	SMG	Vatican Radio
		100	D	BIB	BBG - RFE/RL
		300	ROU	TIG	Radio Romania International
		250	RUS	SAM	IBRA Radio
		250	SEY	SEY	BBC World Service
5915		50	BRM	dom	Myanma R. Min & Educ. Sce, Naypyitaw
		50	BRM	dom	Myanma R. Padauk Myay prgr, Yangon
		100	CHN	HUH	China Radio Int. (CRI)
		100	CHN	KAS	China Radio Int. (CRI)
		150	CHN	BEI	China Radio Int. (CRI)
		100	F	MCO	TWR Europe
		500	G	RMP	Deutsche Welle

kHz	N	kW	Ctry	Site	Station, location
		500	IRN	KAM	VO the Islamic Rep. of Iran
		500	IRN	SIR	VO the Islamic Rep. of Iran
		300	ROU	TIG	Radio Romania International
5915		150	SVK	RSO	Radio Slovakia Int.
	‡	100	ZMB	dom	ZNBC R. One, Lusaka
5920		250	CYP	CYP	BBC World Service
		125	G	WOF	Polish Radio (Ext Sce)
		500	IRN	KAM	VO the Islamic Rep. of Iran
		250	RUS	IRK	TWR India
		200	RUS	SPB	Voice of Russia (VOR)
		1000	TJK	DSB	Voice of Russia (VOR)
		250	USA	HRI	WHRI - World Harvest R. Int.
5921		1	PRU	dom	R. Bethel, Arequipa
5925		50	CHN	dom	CNR5 VO Zhonghua, Beijing
		250	CLN	IRA	BBG - VOA Ashna Radio
		100	D	BIB	BBG - RFE/RL
		500	F	ISS	Radio France Int. (RFI)
		500	MDA	KCH	Deutsche Welle
		250	RRW	KIG	Deutsche Welle
		50	VTN	dom	VO Vietnam 2nd prgr, Xuan Mai
5927		1	BOL	dom	Radiodifusoras Minería, Oruro
5930		100	CZE	LIT	Radio Prague
		250	D	WER	BBG - VO America (VOA)
		50	RUS	dom	R. Rossii/Reg, Monchegorsk (Murmansk)
		100	RUS	dom	R. Rossii/Reg, Yelizovo
		100	USA	YFR	Family Radio Worldwide
5935		100	CHN	dom	Xizang, Lhasa
		250	POR	SIN	R. Netherlands Worldwide
		100	RUS	ARM	Voice of Russia (VOR)
		400	RUS	SPB	Voice of Russia (VOR)
		100	USA	WCR	WWCR
5939		1	PRU	dom	R. Melodía, Arequipa
5940		10	B	dom	Voz Missionária, Camboriú
		300	CYP	CYP	BBC World Service
		100	D	LAM	BBG - RFE/RL
		300	G	SKN	BBC World Service
		500	IRN	SIR	VO the Islamic Rep. of Iran
		100	RUS	dom	R. Rossii/Reg, Arman
		200	RUS	NVS	Voice of Russia (VOR)
		250	RUS	SAM	Voice of Russia (VOR)
5945		100	CHN	dom	CNR1 VO China, Beijing
		300	CYP	CYP	BBC World Service
		100	D	BIB	BBG - VO America (VOA)
		100	D	WER	The Overcomer Ministry
	±	10	ERI	ASM	Ginbot 7 Dimts Radio
		300	G	WOF	Deutsche Welle
		500	IRN	SIR	VO the Islamic Rep. of Iran
		250	POR	SIN	Deutsche Welle
		250	RUS	NVS	Voice of Russia (VOR)
5950		250	D	WER	Bible Voice Broadcasting (BVB)
		100	ETH	dom	VO Tigray Revolution, Geja
		500	IRN	KAM	VO the Islamic Rep. of Iran
		500	IRN	SIR	VO the Islamic Rep. of Iran
		100	NZL	RAN	R. New Zealand Int. (RNZI)
	+	50	NZL	RAN	R. New Zealand Int. (RNZI)
		100	ROU	TIG	Radio Romania International
		100	USA	YFR	Family Radio Worldwide
		100	USA	YFR	Radio Taiwan International
5952		5	BOL	dom	R. Pío XII, Siglo Veinte
5954	†		CTR	—	Radio República
5955		100	AUT	MOS	Voice of Vietnam (VOV)
		10	B	dom	R. Gazeta, São Paulo
		150	CHN	BEI	China Radio Int. (CRI)
		500	CHN	XIA	China Radio Int. (CRI)
		50	CHN	dom	CNR8 VO Minorities, Beijing
		100	D	BIB	BBG - RFE/RL
		500	D	WER	R. Netherlands Worldwide
		500	G	RMP	R. Netherlands Worldwide
		500	IRN	SIR	VO the Islamic Rep. of Iran
		300	J	YAM	Radio Japan (NHK World)
		100	J	YAM	Shiokaze
		250	MDA	KCH	R. Netherlands Worldwide
		250	OMA	SLA	BBC World Service
		250	PHL	PHT	BBG - VO America (VOA)
		250	POR	SIN	R. Netherlands Worldwide
		100	ROU	TIG	Radio Romania International

kHz	N	kW	Ctry	Site	Station, location
		250	THA	NAK	BBC World Service
		250	UAE	DHA	BBC World Service
5955		250	UAE	DHA	Radio Australia
5960		150	ALB	CER	China Radio Int. (CRI)
		250	CAN	SAC	China Radio Int. (CRI)
		250	CAN	SAC	Radio Japan (NHK World)
		50	CHN	dom	Xinjiang, Ürümqi
		100	D	WER	Christian Science Sentinel
		250	D	NAU	Family Radio Worldwide
		250	KWT	KBD	Radio Kuwait
		1	PNG	dom	R. Fly, Kiunga
		500	TUR	EMR	Voice of Turkey (VOT)
		250	USA	GRV	BBG - VO America (VOA)
5965	‡	8	B	dom	R. Transmundial, Santa Maria
			BOL	dom	R. Nacional, Huanuni
		150	CHN	BEI	China Radio Int. (CRI)
		500	CHN	BEI	China Radio Int. (CRI)
		500	CHN	XIA	China Radio Int. (CRI)
		250	CLN	TRM	Deutsche Welle
		100	CTR	CRI	R. Exterior de España (REE)
		100	CVA	SMG	Vatican Radio
		250	CVA	SMG	Vatican Radio
		250	G	WOF	BBC World Service
		300	J	YAM	Radio Canada Int. (RCI)
		100	MLA	dom	RTM Klasik Nasional FM, Kajang
		250	OMA	SLA	BBC World Service
		100	RUS	VLD	Voice of Russia (VOR)
		250	RUS	IRK	Voice of Russia (VOR)
		500	TUR	EMR	Voice of Turkey (VOT)
5970		150	ALB	CER	China Radio Int. (CRI)
		100	ALB	SHI	RadioTirana
		300	AUT	MOS	Adventist World R. (AWR)
		10	B	dom	R. Itatiaia, Belo Horizonte
		15	CHN	dom	Gannan PBS, Hezuo
		250	E	NOB	R. Exterior de España (REE)
		250	G	WOF	Voice of Vietnam (VOV)
		250	OMA	SLA	BBC World Service
		500	TUR	EMR	Voice of Turkey (VOT)
		100	UKR	dom	UR1, Kyiv
5975		250	ASC	ASC	BBC World Service
		100	CHN	KAS	China Radio Int. (CRI)
		100	CHN	dom	CNR8 VO Minorities, Beijing
		100	D	WER	Adventist World R. (AWR)
		100	D	LAM	BBG - VO America (VOA)
		500	G	RMP	BBC World Service
		500	G	RMP	Radio Japan (NHK World)
		250	OMA	SLA	BBC World Service
		300	ROU	GAL	Radio Romania International
		120	RUS	KLG	Voice of Russia (VOR)
		250	RUS	NVS	Voice of Russia (VOR)
		250	SNG	SNG	BBC World Service
		250	THA	NAK	BBC World Service
		50	VTN	dom	VO Vietnam 1st prgr, Hanoi
	†	100	ZWE	GWE	Voice of Zimbabwe
5980	‡	10	B	dom	R. Guarujá, Florianópolis
		500	CHN	XIA	China Radio Int. (CRI)
		250	CLN	IRA	BBG - VO America (VOA)
		100	CZE	LIT	Radio Prague
		0.1	FIN	VIR	Scandinavian Weekend R.
		500	IRN	KAM	VO the Islamic Rep. of Iran
		100	LTU	SIT	Radio Japan (NHK World)
		250	TUR	CAK	Voice of Turkey (VOT)
		250	UAE	DHA	Polish Radio (Ext Sce)
		250	USA	GRV	BBG - R. Martí
5984			BOL	dom	R. Cooperativa, Huanuni
5985		150	ALB	CER	China Radio Int. (CRI)
		50	BRM	dom	Myanma R, Naypyitaw
		500	CHN	BEI	China Radio Int. (CRI)
		500	CHN	XIA	China Radio Int. (CRI)
		500	G	RMP	BBC World Service
		500	GUF	GUF	Family Radio Worldwide
		100	GUM	SDA	AWR Guam (KSDA)
		100	J	YAM	Shiokaze
		100	RUS	NVS	Vatican Radio
		100	RUS	MSK	Voice of Russia (VOR)
		200	RUS	SPB	Voice of Russia (VOR)

kHz	N	kW	Ctry	Site	Station, location
		100	USA	YFR	Family Radio Worldwide
5990		250	B	dom	R. Senado, Brasília
		100	CHN	HUH	China Radio Int. (CRI)
		250	CUB	HAB	China Radio Int. (CRI)
		100	D	LAM	BBG - RFE/RL
	‡	100	ETH	dom	R. Ethiopia, Geja
		500	G	RMP	BBC World Service
		300	G	SKN	Polish Radio (Ext Sce)
		250	IND	DEL	All India Radio (AIR)
		100	PHL	BOC	FEBC Philippines
		300	ROU	GAL	Radio Romania International
5995		10	AUS	BRN	Radio Australia
		100	AUS	SHP	Radio Australia
	+	8	AUS	BRN	Radio Australia
		100	CVA	SMG	Radio Canada Int. (RCI)
		100	CZE	LIT	Radio Prague
		250	GUF	GUF	Radio France Int. (RFI)
		500	IRN	SIR	VO the Islamic Rep. of Iran
		50	MLI	dom	ORTM, Bamako (Kati)
		300	ROU	GAL	Radio Romania International
		250	RUS	PPK	Family Radio Worldwide
		100	SWZ	MAN	TWR Africa
5996		1	BOL	dom	R. Loyola, Sucre
6000		100	ALB	SHI	RadioTirana
		10	B	dom	R. Guaíba, Porto Alegre
		150	CHN	dom	CNR 2, Xianyang
		250	CUB	HAB	Radio Habana Cuba
		100	D	LAM	BBG - RFE/RL
		10	IND	dom	All India R, Leh
		500	RUS	MSK	Family Radio Worldwide
		250	RUS	MSK	Voice of Russia (VOR)
		500	TUR	EMR	Voice of Turkey (VOT)
		250	UAE	DHA	Polish Radio (Ext Sce)
		100	USA	YFR	Family Radio Worldwide
6003		100	KOR	HWA	Echo of Hope (VOH)
6005		250	ASC	ASC	BBC World Service
		250	CAN	SAC	China Radio Int. (CRI)
		100	CHN	KAS	China Radio Int. (CRI)
		10	CLN	EKA	Sri Lanka Broadcasting Corp.
		4	CME	dom	R. Buea, Buea
		1	D	KLL	Radio 700
		250	MRA	TIN	BBG - R. Free Asia (RFA)
		250	RUS	KNA	Family Radio Worldwide
		100	RUS	ARM	GTRK "Adygeya"
		100	RUS	ARM	GTRK "Kabardino-Balkariya"
		100	RUS	ARM	Voice of Russia (VOR)
		250	SEY	SEY	BBC World Service
6010		5	B	dom	R. Inconfidência, Belo Horizonte
		10	BHR	ABH	Radio Bahrain
		5	BLR	dom	Belaruskaje R. 1/Reg, Brest
	‡	1	CHL	dom	R. Parinacota, Putre
		500	CHN	BEI	China Radio Int. (CRI)
		500	CHN	URU	China Radio Int. (CRI)
		100	CHN	dom	CNR11, Xi'an
		5	CLM	dom	LV de tu Conciencia, Puerto Lleras
		250	CUB	HAB	Radio Habana Cuba
		250	D	WER	Family Radio Worldwide
		500	IRN	KAM	VO the Islamic Rep. of Iran
		1	MEX	dom	R. Mil Onda Corta, México
		100	ROU	TIG	Radio Romania International
		150	SVK	RSO	Radio Slovakia Int.
		500	UAE	DHA	BBG - R. Free Asia (RFA)
6015		100	CHN	dom	Xinjiang, Ürümqi
		100	D	LAM	BBG - RFE/RL
	+	60	F	ISS	TDPradio
	+	60	F	ISS	The Disco Palace
		500	IRN	SIR	VO the Islamic Rep. of Iran
		100	ROU	TIG	Radio Romania International
		300	ROU	GAL	Radio Romania International
	†	50	TZA	dom	VO Tanzania Zanzibar, Dole
6020		300	ALB	CER	China Radio Int. (CRI)
		250	ATN	BON	R. Netherlands Worldwide
		100	AUS	SHP	Radio Australia
		10	B	dom	R. Gaúcha, Porto Alegre
		100	BOT	BOT	BBG - VO America (VOA)
		100	CHN	KAS	China Radio Int. (CRI)
6020		500	CHN	BEI	China Radio Int. (CRI)
		500	CHN	SZG	China Radio Int. (CRI)
		50	IND	dom	All India R, Shimla
		50	MDG	MDC	Family Radio Worldwide
		250	MDG	MDC	R. Netherlands Worldwide
	+	65	NOR	KVI	Radio Romania International
		250	OMA	SLA	BBC World Service
		250	PHL	PUG	Vatican Radio
		3	PRU	dom	R. Victoria, Lima
	+	300	ROU	GAL	Radio Romania International
		100	RUS	ARM	Voice of Russia (VOR)
		20	VTN	dom	VO Vietnam, Buôn Mê Thuôt
6025		10	BOL	dom	R. Patria Nueva, La Paz
		100	CHN	KAS	China Radio Int. (CRI)
		500	CHN	XIA	China Radio Int. (CRI)
		100	CVA	SMG	Radio Canada Int. (RCI)
	†	1	DOM	dom	R. Amanecer Internacional, Santo Domingo
	‡	10	NIG	dom	FRCN, Enugu
		100	ROU	TIG	Radio Romania International
		100	SWZ	MAN	TWR Africa
6030		1	CAF	dom	R. ICDI, Boali
		0.1	CAN	dom	CKMX, Calgary
		100	CHN	dom	CNR1 VO China, Beijing
		300	CYP	CYP	BBC World Service
		125	D	WER	Bible Voice Broadcasting (BVB)
		100	ETH	dom	R. Oromiya, Geja
		500	G	RMP	Deutsche Welle
		50	IND	dom	All India R, Delhi
		500	IRN	KAM	VO the Islamic Rep. of Iran
	+	300	ROU	GAL	Radio Romania International
		300	ROU	GAL	Radio Romania International
	+	300	ROU	TIG	Radio Romania International
		250	RUS	SAM	Voice of Russia (VOR)
		250	THA	NAK	BBC World Service
		100	TWN	TSH	Minghui Radio
		100	UKR	KHR	Radio Ukraine Int. (RUI)
		250	USA	GRV	BBG - R. Martí
6035		100	ARM	ERV	Radio Japan (NHK World)
		30	BTN	dom	Bhutan Broadcasting Sce, Thimpu
		100	CHN	KUN	Yunnan PBS
		5	CLM	dom	LV del Guaviare, San José del Guaviare
		100	D	BIB	BBG - VO America (VOA)
		250	F	ISS	R. Netherlands Worldwide
		500	IRN	SIR	VO the Islamic Rep. of Iran
		300	J	YAM	Radio Japan (NHK World)
		100	STP	SAO	BBG - VO America (VOA)
6040		100	ALB	SHI	RadioTirana
		250	ATN	BON	Vatican Radio
		5	BLR	dom	Belaruskaje R. 1/Reg, Hrodna
		250	CAN	SAC	China Radio Int. (CRI)
		160	CAN	SAC	Vatican Radio
		500	CHN	URU	China Radio Int. (CRI)
		500	CHN	XIA	China Radio Int. (CRI)
		150	CHN	dom	CNR2 China Business R, Beijing
		50	CHN	dom	Nei Menggu, Hohhot
		250	CVA	SMG	R. Netherlands Worldwide
		250	D	WER	Pan American Broadcasting
		300	G	SKN	BBC World Service
		125	G	WOF	Polish Radio (Ext Sce)
		50	IND	dom	All India R, Jeypore
		250	PHL	PHT	BBG - VO America (VOA)
	+	15	RUS	KLG	Voice of Russia (VOR)
		100	STP	SAO	BBG - VO America (VOA)
		150	SVK	RSO	Radio Slovakia Int.
		250	THA	UDO	BBG - VO America (VOA)
6045		100	AFS	MEY	Family Radio Worldwide
		300	AUT	MOS	Adventist World R. (AWR)
		250	CAN	SAC	KBS World Radio
		100	D	WER	Hamburger Lokalradio
		300	G	SKN	KBS World Radio
		100	IND	DEL	All India Radio (AIR)
	†	0.45	MEX	dom	R. Universidad, San Luis Potosí
		250	THA	UDO	BBG - VO America (VOA)
		250	UAE	DHA	Family Radio Worldwide
		2	URG	dom	R. Sarandí, Montevideo
6045		100	ZWE	dom	ZBC R. Zimbabwe, Gweru

kHz	N	kW	Ctry	Site	Station, location
6047		10	PRU	dom	R. Santa Rosa, Lima
6050		100	CHN	dom	Xizang, Lhasa
		250	CLN	TRM	Deutsche Welle
		250	CUB	HAB	Radio Habana Cuba
		100	D	WER	Family Radio Worldwide
		10	EQA	dom	HCJB La Voz de los Andes, Pichincha
		300	G	SKN	Polish Radio (Ext Sce)
		500	KWT	KBD	Radio Kuwait
		100	MLA	dom	RTM, Kajang
		100	MLA	KAJ	Voice of Malaysia
		250	TUR	CAK	Voice of Turkey (VOT)
		500	TUR	EMR	Voice of Turkey (VOT)
6054		3	BOL	dom	R. Juan XXIII, San Ignacio de Velasco
6055		150	ALB	CER	China Radio Int. (CRI)
		100	CHN	NNN	China Radio Int. (CRI)
		250	CYP	CYP	BBC World Service
		100	CZE	LIT	Radio Prague
		100	D	WER	Christian Science Sentinel
		125	D	WER	Ev. Missions-Gemeinden
		250	D	WER	Stimme Des Trostes
		250	E	NOB	R. Exterior de España (REE)
		300	G	SKN	BBC World Service
		250	IND	DEL	All India Radio (AIR)
		50	J	dom	R. Nikkei 1, Tokyo
		100	LTU	SIT	VO the Islamic Rep. of Iran
		50	RRW	dom	Rdif. de la Republique Rwandaise, Kigali
		150	SVK	RSO	Radio Slovakia Int.
		250	THA	UDO	BBG - RFE/RL
6060		30	ARG	dom	R. Nacional, Buenos Aires
		50	ARG	BUE	Rdif. Argentina al Exterior
		10	B	dom	Super R. Deus é Amor, Curitiba
		150	CHN	KUN	China Radio Int. (CRI)
		15	CHN	dom	Sichuan PBS Life Channel, Xichang
		100	CUB	HAB	R. Nacional de Venezuela
		250	CUB	HAB	Radio Habana Cuba
+		125	CVA	SMG	Vatican Radio
		100	D	LAM	BBG - RFE/RL
		250	PHL	PHT	BBG - VO America (VOA)
		200	RUS	SPB	Voice of Russia (VOR)
		250	THA	UDO	BBG - VO America (VOA)
6065		300	AUT	MOS	The Overcomer Ministry
		100	CHN	KAS	China Radio Int. (CRI)
		150	CHN	dom	CNR2 China Business R, Beijing
†		50	IND	dom	All India R, Kohima
		500	IRN	SIR	VO the Islamic Rep. of Iran
		250	KOR	KIM	KBS World Radio
		250	OMA	SLA	BBC World Service
+		300	ROU	TIG	Radio Romania International
		800	RUS	SPB	Voice of Russia (VOR)
		250	THA	NAK	BBC World Service
		100	ZMB	LUS	CVC R. Christian Voice
6070		7.5	B	dom	R. Capital (Super R. Deus é Amor), RDJ
		5	BLR	dom	Belaruskaje R. 1/Reg, Brest
		1	CAN	dom	CFRX, Toronto
		500	CHN	XIA	China Radio Int. (CRI)
		500	IRN	KAM	VO the Islamic Rep. of Iran
		125	KRE	KNG	Voice of Korea
‡		2	LBR	dom	R. ELWA, Monrovia
		250	RUS	ARM	Family Radio Worldwide
		250	RUS	TCH	Family Radio Worldwide
6075			BOL	dom	R. Causachun Coca, Lauca Ñ
		100	CHN	KAS	China Radio Int. (CRI)
		15	CHN	dom	Yushu PBS, Qinghai
		250	G	SKN	Deutsche Welle
		300	G	SKN	Deutsche Welle
		300	G	WOF	Deutsche Welle
		500	G	RMP	Deutsche Welle
		300	J	YAM	Radio Japan (NHK World)
		250	POR	SIN	Deutsche Welle
		250	RUS	TCH	Family Radio Worldwide
		100	RUS	dom	R. Rossii/Reg, Yelizovo
		100	TWN	KOU	Radio Taiwan International
6080		100	AFS	MEY	BBG - VO America (VOA)
		100	AUS	SHP	Radio Australia
6080		5	B	dom	R. Daquí, Goiânia
		10	B	dom	R. Marumby, Curitiba
		150	BLR	dom	Belaruskaje R. 1, Minsk
		5	BOL	dom	R. San Gabriel, La Paz
		100	BOT	BOT	BBG - VO America (VOA)
		250	CAN	SAC	China Radio Int. (CRI)
		500	CHN	BEI	China Radio Int. (CRI)
		100	CHN	dom	CNR1 VO China, Golmud
		7	CHN	dom	Hulun Buir, Hailar
		250	SNG	SNG	BBC World Service
		100	STP	SAO	BBG - VO America (VOA)
		150	SVK	RSO	Radio Slovakia Int.
6085	+	10	D	dom	Bayern 5 Aktuell, München
		500	G	RMP	BBC World Service
		50	IND	dom	All India R, Delhi
		300	J	YAM	Radio Japan (NHK World)
		500	RUS	ARM	BBC World Service
		50	RUS	dom	R. Rossii/Reg, Krasnoyarsk
		100	RUS	IRK	Voice of Russia (VOR)
		300	TWN	HUW	Radio Taiwan International
		100	USA	YFR	Family Radio Worldwide
6090		100	AIA	AIA	University Network (DGS)
		300	AUT	MOS	Adventist World R. (AWR)
		10	B	dom	R. Bandeirantes, São Paulo
‡		10	CHL	dom	R. Esperanza, Temuco
		150	CHN	KUN	China Radio Int. (CRI)
		500	CHN	XIA	China Radio Int. (CRI)
		100	CHN	dom	CNR2 China Business R, Golmud
		250	CYP	CYP	BBC World Service
		100	ETH	dom	VO Amhara State, Geja
		500	G	RMP	Family Radio Worldwide
		250	IRN	AHW	VO the Islamic Rep. of Iran
		300	J	YAM	Radio Japan (NHK World)
		250	KOR	KIM	Deutsche Welle
‡		10	LBR	dom	R. Veritas, Monrovia
		50	NIG	dom	FRCN, Kaduna
		200	RUS	ARM	Voice of Russia (VOR)
		150	SVK	RSO	IRRS Shortwave
6095		500	CHN	KAS	China Radio Int. (CRI)
		250	CUB	HAB	Radio Habana Cuba
		500	IRN	KAM	VO the Islamic Rep. of Iran
		250	KOR	KIM	BBC World Service
+		50	LUX	JUN	KBS World Radio
+		60	LUX	JUN	RTL Radio
		500	MRA	TIN	BBG - R. Free Asia (RFA)
6100		100	AFS	MEY	Family Radio Worldwide
		100	ALB	SHI	RadioTirana
		250	ATN	BON	R. Netherlands Worldwide
		250	CAN	SAC	Radio Canada Int. (RCI)
		250	CHN	KAS	China Radio Int. (CRI)
		100	CHN	URU	China Radio Int. (CRI)
		150	CHN	KUN	China Radio Int. (CRI)
		500	CHN	BEI	China Radio Int. (CRI)
		500	CHN	URU	China Radio Int. (CRI)
		500	CHN	XIA	China Radio Int. (CRI)
		250	CYP	CYP	BBC World Service
		100	D	WER	Adventist World R. (AWR)
		125	G	WOF	Polish Radio (Ext Sce)
		100	IND	DEL	Radio Sedayee Kashmir
		500	IRN	KAM	VO the Islamic Rep. of Iran
		500	IRN	SIR	VO the Islamic Rep. of Iran
		125	KRE	dom	KCBS, Kanggye
+		300	ROU	GAL	Radio Romania International
		0.5	RUS	dom	R. Rossii/Reg, Kyzyl
		100	USA	YFR	Family Radio Worldwide
6105	‡	5	B	dom	R. Canção Nova, Cachoeira Paulista
		10	BOL	dom	R. Panamericana, La Paz
		500	CHN	SZG	China Radio Int. (CRI)
		100	D	BIB	BBG - RFE/RL
		100	D	LAM	BBG - RFE/RL
		250	D	WER	BBG - RFE/RL
		500	D	NAU	Family Radio Worldwide
		100	D	NAU	TWR Europe
		250	KWT	KWT	BBG - VO America (VOA)
		0.25	MEX	dom	Candela FM, Mérida
6105		250	PHL	PHT	BBG - RFE/RL
	+	15	RUS	KLG	Voice of Russia (VOR)

kHz	N	kW	Ctry	Site	Station, location
		250	TJK	DSB	BBG - VO America (VOA)
		100	TWN	KOU	Radio Taiwan International
		100	TWN	TSH	Radio Taiwan International
		100	USA	YFR	Family Radio Worldwide
6110		250	CAN	SAC	Radio Japan (NHK World)
		500	CHN	XIA	China Radio Int. (CRI)
		100	CHN	dom	Xizang-Tb, Lhasa
		125	D	WER	Bible Voice Broadcasting (BVB)
	†	100	ETH	dom	R. Fana, Geja
		250	G	RMP	BBC World Service
		300	G	SKN	BBC World Service
		500	G	RMP	BBC World Service
		50	IND	dom	All India R, Srinagar
		500	IRN	KAM	VO the Islamic Rep. of Iran
		100	ROU	TIG	Radio Romania International
6115		75	BLR	dom	Belaruskaje R. 1, Minsk
		250	CAN	SAC	China Radio Int. (CRI)
		500	CHN	BEI	China Radio Int. (CRI)
		50	CHN	dom	VO Strait Amoy channel, Fuzhou
		10	CLM	dom	LV del Llano, Villavicencio
	†	50	COG	dom	R. Congo, Brazzaville
		250	CYP	CYP	BBC World Service
		100	D	BIB	BBG - R. Farda
		50	J	dom	R. Nikkei 2, Tokyo
		250	KWT	KWT	BBG - R. Farda
		100	LTU	SIT	VO the Islamic Rep. of Iran
		250	PHL	PUG	Radio Veritas Asia
	†	10	PRU	dom	R. Union, Lima
		300	ROU	TIG	Radio Romania International
		200	RUS	PPK	Family Radio Worldwide
		250	RUS	SAM	TWR India
6120		250	AFS	MEY	Channel Africa
		100	ALS	NLS	KNLS International
	‡	10	B	dom	R. Globo (Super R. Deus é Amor), SP
		250	CAN	SAC	Radio Japan (NHK World)
		50	CHN	dom	Xinjiang, Ürümqi
		250	CUB	HAB	Radio Habana Cuba
		250	CVA	SMG	R. Netherlands Worldwide
		250	D	WER	BBG - RFE/RL
		250	D	NAU	Family Radio Worldwide
		500	D	NAU	R. Netherlands Worldwide
		500	D	WER	R. Netherlands Worldwide
		100	D	WER	TWR Europe
		500	F	ISS	Radio Taiwan International
		500	IRN	SIR	VO the Islamic Rep. of Iran
		120	RUS	KLG	Voice of Russia (VOR)
		250	SNG	SNG	R. Netherlands Worldwide
		50	SWZ	MAN	TWR Africa
		500	TUR	EMR	Voice of Turkey (VOT)
6125		100	CHN	dom	CNR1 VO China, Beijing
		100	CHN	dom	CNR1 VO China, Shijiazhuang
		250	E	NOB	R. Exterior de España (REE)
	†	10	INS	dom	RRI, Nabire
		100	ROU	TIG	Radio Romania International
		250	UAE	DHA	FEBA Radio
		0.35	URG	dom	SODRE R. Uruguay, Montevideo
6130		100	ALB	SHI	RadioTirana
		100	CHN	dom	Xizang-Tb, Lhasa
		125	D	WER	Bible Voice Broadcasting (BVB)
		250	D	WER	Radio Japan (NHK World)
		50	LAO	dom	R. Nationale Lao, Vientiane
		250	POR	SIN	Deutsche Welle
		100	ROU	TIG	Radio Romania International
		300	ROU	TIG	Radio Romania International
		200	RUS	MSK	Voice of Russia (VOR)
		100	SWZ	MAN	TWR Africa
		50	SWZ	MAN	TWR Africa
6135		250	AFS	MEY	BBC World Service
		250	ASC	ASC	BBC World Service
		25	B	dom	R. Aparecida, Aparecida
		10	BOL	dom	R. Santa Cruz, Santa Cruz
		500	CHN	BEI	China Radio Int. (CRI)
		100	D	BIB	BBG - RFE/RL
6135		100	D	BIB	BBG - VO America (VOA)
		125	G	WOF	Polish Radio (Ext Sce)
		30	MDG	dom	R. Nasionaly Malagasy, Ambohidrano
		250	OMA	SLA	BBC World Service
		250	RUS	VLD	BBC World Service
		100	SNG	SNG	BBC World Service
		1000	TJK	DSB	Voice of Russia (VOR)
		50	YEM	SAN	Republic of Yemen Radio
6140		100	CHN	KAS	China Radio Int. (CRI)
		100	CHN	KUN	China Radio Int. (CRI)
		100	CHN	dom	CNR8 VO Minorities, Beijing
	†	5	CLM	dom	Cadena Melodía R Lider, Bogotá
		250	CUB	HAB	Radio Habana Cuba
		250	D	WER	Family Radio Worldwide
		100	D	WER	MV Baltic Radio
		250	OMA	SLA	BBC World Service
		200	RUS	MSK	Voice of Russia (VOR)
		250	SNG	SNG	Radio Japan (NHK World)
		250	THA	UDO	BBG - VO America (VOA)
		250	UAE	DHA	Athmeeya Yatra Radio
		250	UAE	DHA	FEBA Radio
		100	UKR	KHR	Radio Ukraine Int. (RUI)
6145		100	AFS	MEY	BBC World Service
		250	AFS	MEY	BBC World Service
		250	ATN	BON	R. Netherlands Worldwide
		100	CHL	SGO	Radio Japan (NHK World)
		150	CHN	BEI	China Radio Int. (CRI)
		500	CHN	URU	China Radio Int. (CRI)
		15	CHN	dom	Qinghai PBS, Xining
		125	D	NAU	Bible Voice Broadcasting (BVB)
		250	F	ISS	China Radio Int. (CRI)
		500	F	ISS	China Radio Int. (CRI)
		500	F	ISS	Radio France Int. (RFI)
		500	IRN	SIR	VO the Islamic Rep. of Iran
		300	J	YAM	Radio Japan (NHK World)
		300	ROU	GAL	Radio Romania International
	+	15	RUS	KLG	Voice of Russia (VOR)
		100	TWN	PAO	Radio Taiwan International
6150		8	B	dom	R. Record, São Paulo
		500	CHN	SZG	China Radio Int. (CRI)
		250	CUB	HAB	Radio Habana Cuba
	†	12	CYP	ISK	Radio Bayrak International
		100	D	LAM	BBG - RFE/RL
		50	IND	dom	All India R, Itanagar
		300	RUS	ARM	Family Radio Worldwide
		100	TWN	KOU	Radio Taiwan International
6155		300	AUT	MOS	Radio Ö1 International
		250	BLR	MNS	Radio Station Belarus
		10	BOL	dom	R. Fides, La Paz
		150	CHN	dom	CNR2 China Business R, Beijing
		500	F	ISS	Radio Japan (NHK World)
		500	IND	BGL	All India Radio (AIR)
		500	IRN	SIR	VO the Islamic Rep. of Iran
		100	KOR	KIM	KBS World Radio
		250	OMA	SLA	BBC World Service
		250	POR	SIN	Deutsche Welle
		250	RUS	SAM	Voice of Russia (VOR)
		250	SNG	SNG	BBC World Service
6160	†	10	B	dom	R. Rio Mar, Manaus
		1	CAN	dom	CBC R. One, St. John's
		0.5	CAN	dom	CBC R. One, Vancouver
		250	CHN	XIA	China Radio Int. (CRI)
		300	J	YAM	Radio Japan (NHK World)
		100	KOR	KIM	Radio Canada Int. (RCI)
		40	RUS	dom	R. Rossii/Reg, Arkhangelsk
		250	UAE	DHA	Athmeeya Yatra Radio
6165		250	ATN	BON	R. Netherlands Worldwide
		1	BOL	dom	R. Logos, Santa Cruz de la Sierra
		500	CHN	BEI	China Radio Int. (CRI)
		500	CHN	URU	China Radio Int. (CRI)
		50	CHN	dom	CNR6 VO Shenzhou, Beijing
		250	CYP	CYP	BBC World Service
		10	HRV	DEA	Voice of Croatia
		250	IND	DEL	All India Radio (AIR)
		300	J	YAM	Radio Japan (NHK World)
6165		250	TCD	dom	ONR & TV du Tchad, N'Djamena
		50	VTN	dom	VO Vietnam Minorities prgr, Xuan Mai
	‡	100	ZMB	dom	ZNBC R. Two, Lusaka
6170	±	10	ERI	ASM	Ginbot 7 Dimts Radio

kHz	N	kW	Ctry	Site	Station, location
		0.1	FIN	VIR	Scandinavian Weekend R.
		500	IRN	KAM	VO the Islamic Rep. of Iran
		250	KOR	KIM	BBC World Service
		250	KWT	KWT	BBG - VO America (VOA)
		0.25	PHL	dom	Philippine BS, Marulas
		100	RUS	KHB	Voice of Russia (VOR)
6174		1	PRU	dom	R. Tawantinsuyo, Cusco
6175		150	ALB	CER	China Radio Int. (CRI)
		250	CAN	SAC	Voice of Vietnam (VOV)
		100	CHN	KAS	China Radio Int. (CRI)
		100	CHN	dom	CNR1 VO China, Beijing
		500	F	ISS	Radio France Int. (RFI)
		500	IRN	KAM	VO the Islamic Rep. of Iran
		100	MLA	KAJ	Voice of Malaysia
		250	OMA	SLA	BBC World Service
	+	300	ROU	TIG	Radio Romania International
6180		100	CHN	KAS	China Radio Int. (CRI)
		500	CHN	URU	China Radio Int. (CRI)
		100	CHN	dom	CNR1 VO China, Lingshi
		100	CHN	dom	CNR8 VO Minorities, Beijing
		100	CUB	HAB	R. Nacional de Venezuela
		250	CYP	CYP	Cyprus Broadcasting Corp.
		100	D	WER	RTR Radio Traumland
		500	G	RMP	Deutsche Welle
		500	IND	BGL	All India Radio (AIR)
	+	100	ROU	TIG	Radio Romania International
		250	RRW	KIG	Deutsche Welle
		200	RUS	ARM	BBC World Service
		250	RUS	PPK	Voice of Russia (VOR)
		250	THA	UDO	BBG - VO America (VOA)
		250	UAE	DHA	FEBA Radio
6185		150	ALB	CER	China Radio Int. (CRI)
		250	B	dom	R. Nacional da Amazônia, Brasília
		15	CHN	dom	China Huayi BC, Fuzhou
		500	CHN	XIA	China Radio Int. (CRI)
		100	CVA	SMG	Vatican Radio
		250	CVA	SMG	Vatican Radio
		500	CVA	SMG	Vatican Radio
		250	IRN	AHW	VO the Islamic Rep. of Iran
		500	IRN	SIR	VO the Islamic Rep. of Iran
		300	J	YAM	Radio Japan (NHK World)
		200	KRE	KUJ	Voice of Korea
		10	MEX	dom	R. Educación Onda Corta, México
		500	TUR	CAK	Voice of Turkey (VOT)
6190		100	AFS	MEY	BBC World Service
		250	BIH	BIJ	International Radio Serbia
		5	BLR	dom	Belaruskaje R. 1, Mahilioŭ
		250	CAN	SAC	China Radio Int. (CRI)
		100	CHN	dom	CNR2 China Business R, Golmud
		50	CHN	dom	Xinjiang, Ürümqi
		15	D	dom	Deutschlandfunk, Berlin-Britz
		50	IND	dom	All India R, Delhi
		300	J	YAM	Radio Japan (NHK World)
		150	SVK	RSO	Radio Slovakia Int.
6195		250	ATN	BON	R. Netherlands Worldwide
		250	ATN	BON	Radio Japan (NHK World)
		250	CYP	CYP	BBC World Service
		300	CYP	CYP	BBC World Service
		300	G	SKN	BBC World Service
		500	G	RMP	BBC World Service
		250	OMA	SLA	BBC World Service
		50	RUS	dom	R. Rossii/Reg, Selenginsk
		250	SNG	SNG	BBC World Service
		250	THA	NAK	BBC World Service
6196		1	PRU	dom	R. Cusco, Cusco
6200		100	BUL	SOF	Radio Bulgaria
		170	BUL	PLD	Radio Bulgaria
		100	CHN	dom	Xizang, Lhasa
		500	IRN	SIR	VO the Islamic Rep. of Iran
		100	ROU	TIG	Radio Romania International
		100	UZB	TAC	Radio Japan (NHK World)
6205		500	IRN	SIR	VO the Islamic Rep. of Iran
6210		1	COD	dom	R. Kahuzi, Bukavu
6214			BOL	dom	R. Amor de Dios, La Paz
6215		0.5	ARG	dom	R. Baluarte, Puerto Iguazú
6220		100	TWN	HUW	Family Radio Worldwide

kHz	N	kW	Ctry	Site	Station, location
6225		100	AFS	MEY	Family Radio Worldwide
		250	AFS	MEY	Family Radio Worldwide
		100	AFS	MEY	RTÉ Radio Worldwide
		500	KAZ	ALM	Deutsche Welle
		300	MDA	KCH	Deutsche Welle
		100	TJK	DSB	Radio Free Chosun
6230		100	TWN	PAO	Family Radio Worldwide
6235		100	PAK	ISL	Radio Pakistan
6240		500	MDA	KCH	Radio PMR
		500	MDA	KCH	Voice of Russia (VOR)
		100	TWN	PAO	Family Radio Worldwide
		100	UZB	TAC	Voice of Freedom
6250		20	GNE	dom	Rdif. de Guinea Ecuatorial, Malabo
		50	KRE	PYO	Pyongyang Broadcasting Stn.
6260		100	UZB	TAC	CVC The Voice Asia
6270		250	EGY	ABZ	Radio Cairo
6280		500	IND	BGL	All India Radio (AIR)
		300	TWN	TSH	Family Radio Worldwide
		300	TWN	TSH	Sound of Hope Radio Int.
6285		200	KRE	KUJ	Voice of Korea
6297	±	20	ALG	RBN	Radio Nacional de la RASD
6348		100	KOR	HWA	Echo of Hope (VOH)
6400		50	KRE	KNG	Pyongyang Broadcasting Stn.
6518		100	KOR	GOY	Voice of the People
6585			BOL	dom	R. Nueva Esperanza, El Alto
6600		50	KOR	GOY	Voice of the People
6700	†	1	AFG	dom	Peace R, Orgun
6875		100	USA	YFR	Family Radio Worldwide
		100	USA	YFR	Radio Taiwan International
		100	USA	YFR	Family Radio Worldwide
6885		100	USA	YFR	Family Radio Worldwide
6890	†	10	ETH	dom	R. Fana, Addis Ababa
		100	USA	YFR	Family Radio Worldwide
		100	USA	YFR	Radio Taiwan International
6915		0.5	SOM	dom	R. Hage Somalia, Galkayo
		100	USA	YFR	Family Radio Worldwide
6950		100	CHN	dom	CNR1 VO China, Shijiazhuang
6973		10	ISR	LOD	Galei Tzahal
7105		20	MDG	dom	R. Nasionaly Malagasy, Ambohidrano
		300	TWN	TSH	Sound of Hope Radio Int.
7110	‡	100	ETH	dom	R. Ethiopia, Geja
7120	±	10	ERI	dom	VO the Broad Masses 2nd prgr, Asmara
7125	†	50	GUI	dom	R. Guineé, Conakry
7130		10	MLA	dom	RTM Sarawak, Kuching
7145	‡	10	LAO	VIE	Lao National Radio
	‡	25	SOM	dom	R. Hargeisa, Hargeisa
7165	†	100	ETH	GJW	Radio Ethiopia
		100	ETH	GJW	VO Peace & Dem. of Eritrea
	±	100	ETH	GJW	Voice of Democratic Alliance
	†	100	ETH	GJW	Voice of Democratic Alliance
	†	100	ETH	GJW	Voice of Eritrea
7175	±	100	ERI	ASM	Ginbot 7 Dimts Radio
	±	100	ERI	dom	VO the Broad Masses 2nd prgr, Asmara
7185	±	10	ERI	ASM	Ginbot 7 Dimts Radio
7190		10	CLN	EKA	Sri Lanka Broadcasting Corp.
7195		10	UGA	dom	UBC R. Red channel, Kampala
7200		50	BRM	dom	Myanma R, Naypyitaw
		500	IRN	KAM	VO the Islamic Rep. of Iran
		500	IRN	SIR	VO the Islamic Rep. of Iran
		100	SDN	dom	SRTC, Khartoum
7205		100	AFS	MEY	Amateur R. Mirror Int.
		100	CHN	KAS	China Radio Int. (CRI)
		150	CHN	BEI	China Radio Int. (CRI)
		500	CHN	BEI	China Radio Int. (CRI)
		500	CHN	XIA	China Radio Int. (CRI)
		50	CHN	dom	Xinjiang, Ürümqi
		250	D	WER	BBG - VO America (VOA)
		500	F	ISS	Radio France Int. (RFI)
		500	IRN	SIR	VO the Islamic Rep. of Iran
		400	RUS	SPB	China Radio Int. (CRI)
		250	RUS	MSK	Voice of Russia (VOR)
7205		400	RUS	SPB	China Radio Int. (CRI)
		500	RUS	TCH	Voice of Russia (VOR)
		250	SNG	SNG	BBC World Service
		250	TUR	CAK	Voice of Turkey (VOT)
7210		150	ALB	CER	China Radio Int. (CRI)
		100	CHN	URU	China Radio Int. (CRI)

kHz	N	kW	Ctry	Site	Station, location
		500	CHN	URU	China Radio Int. (CRI)
		500	CHN	XIA	China Radio Int. (CRI)
		20	CHN	dom	Yunnan PBS Minority Sce, Kunming
		300	CYP	CYP	Cyprus Broadcasting Corp.
	±	100	ERI	dom	VO the Broad Masses 1st prgr, Asmara
	†	100	ETH	dom	R. Fana, Geja
		50	IND	dom	All India R, Kolkata
		200	KRE	KUJ	Voice of Korea
		300	ROU	TIG	Radio Romania International
		250	RUS	IRK	BBG - R. Free Asia (RFA)
		500	RUS	MSK	Voice of Russia (VOR)
		20	VTN	dom	VO Vietnam, Buôn Mê Thuôt
7215		150	ALB	CER	China Radio Int. (CRI)
		500	CHN	KUN	China Radio Int. (CRI)
		500	CHN	SZG	China Radio Int. (CRI)
		500	CHN	XIA	China Radio Int. (CRI)
		250	D	WER	Athmeeya Yatra Radio
		500	F	ISS	Radio France Int. (RFI)
		500	IRN	SIR	VO the Islamic Rep. of Iran
		250	KWT	KWT	BBG - RFE/RL
		500	RUS	SAM	China Radio Int. (CRI)
		200	RUS	SPB	Voice of Russia (VOR)
7217		15	AGL	MUL	Angolan National Radio
		15	AGL	dom	R. Nacional de Angola, Mulenvos
7220		150	ALB	CER	China Radio Int. (CRI)
	†	50	CAF	dom	R. Centrafrique, Bangui
		100	CHN	KAS	China Radio Int. (CRI)
		150	CHN	KUN	China Radio Int. (CRI)
		500	CHN	JIN	China Radio Int. (CRI)
		500	CHN	KUN	China Radio Int. (CRI)
		500	CHN	XIA	China Radio Int. (CRI)
		100	CHN	dom	CNR2 China Business R, Golmud
		100	D	BIB	BBG - RFE/RL
		250	D	WER	BBG - RFE/RL
		125	D	WER	Bible Voice Broadcasting (BVB)
		250	D	WER	Family Radio Worldwide
		500	F	ISS	Radio France Int. (RFI)
		100	F	MCO	TWR Europe
		200	KRE	KUJ	Voice of Korea
		500	MDA	KCH	Voice of Russia (VOR)
		250	PHL	PHT	BBG - RFE/RL
		250	PHL	PHT	BBG - VO America (VOA)
		300	ROU	GAL	Radio Romania International
		250	RUS	SAM	Voice of Russia (VOR)
		100	VTN	VNI	Voice of Vietnam (VOV)
7225		500	CHN	BEI	China Radio Int. (CRI)
		500	CHN	URU	China Radio Int. (CRI)
		100	CHN	dom	CNR8 VO Minorities, Beijing
		10	CHN	dom	Sichuan PBS City Life Sce, Chengdu
		300	CYP	CYP	BBC World Service
		500	IRN	SIR	VO the Islamic Rep. of Iran
		300	J	YAM	Radio Japan (NHK World)
		250	RUS	MSK	China Radio Int. (CRI)
		250	RUS	SAM	VOR - Mezhd. Russkoye Radio
		250	TUN	SFA	Radio Tunisienne
7230		100	AFS	MEY	Channel Africa
		150	ALB	CER	China Radio Int. (CRI)
	‡	100	BFA	OUA	RTB, Ouagadougou
		250	BIH	BIJ	International Radio Serbia
		100	CHN	dom	CNR1 VO China, Xi'an
		50	CHN	dom	Xinjiang, Ürümqi
		50	IND	dom	All India R, Kurseong
		500	IRN	SIR	VO the Islamic Rep. of Iran
		50	MDG	MDC	Family Radio Worldwide
		250	RUS	NVS	FEBA Radio
		100	RUS	dom	R. Rossii/Reg, Yakutsk
		250	RUS	MSK	Voice of Russia (VOR)
		250	THA	UDO	BBG - VO America (VOA)
7235		5	BLR	dom	Belaruskaje R. 1, Mahilioú
7235		100	CHN	KAS	China Radio Int. (CRI)
		100	D	LAM	BBG - RFE/RL
		100	D	BIB	BBG - VO America (VOA)
		250	D	WER	FEBA Radio
		500	G	RMP	KBS World Radio
		50	IND	dom	All India R, Delhi
		200	KRE	KUJ	Voice of Korea

kHz	N	kW	Ctry	Site	Station, location
		250	KWT	KWT	BBG - VO America (VOA)
+		100	MLA	KAJ	Voice of Malaysia
		250	MRA	TIN	BBG - VO America (VOA)
		100	SNG	SNG	BBC World Service
		250	THA	UDO	R. Thailand World Service
7240		500	ARS	RIY	BSKSA
		100	AUS	SHP	Radio Australia
		100	CHN	dom	Xizang, Lhasa
		250	D	WER	Athmeeya Yatra Radio
		300	G	SKN	Family Radio Worldwide
		50	IND	dom	All India R, Mumbai
		500	IRN	KAM	VO the Islamic Rep. of Iran
		250	POR	SIN	Deutsche Welle
		250	RUS	PPK	Voice of Russia (VOR)
		150	SVK	RSO	Radio Slovakia Int.
		250	TUR	CAK	Voice of Turkey (VOT)
7245		150	CHN	BJI	China Radio Int. (CRI)
		500	CHN	KAS	China Radio Int. (CRI)
		500	CHN	XIA	China Radio Int. (CRI)
		150	CHN	dom	CNR2 China Business R, Beijing
		100	MTN	dom	R. Mauritanie, Nouakchott
		100	TJK	DSB	Voice of Tajik (Ovozi Tojik)
		250	TUR	CAK	Voice of Turkey (VOT)
7250		100	CHN	KAS	China Radio Int. (CRI)
		500	CHN	URU	China Radio Int. (CRI)
		100	CHN	dom	CNR8 VO Minorities, Beijing
		100	CVA	SMG	Vatican Radio
		250	CVA	SMG	Vatican Radio
		50	IND	dom	All India R, Gorakhpur
		50	IND	GKP	All India Radio (AIR)
		500	IRN	KAM	VO the Islamic Rep. of Iran
		500	KWT	KBD	Radio Kuwait
		500	RUS	ARM	Voice of Russia (VOR)
		500	RUS	ARM	VOR - Mezhd. Russkoye Radio
7255		250	ASC	ASC	BBC World Service
		250	BLR	MNS	Belaruskaje Radyjo
		150	CHN	BEI	China Radio Int. (CRI)
		150	CHN	BJI	China Radio Int. (CRI)
		500	CHN	KAS	China Radio Int. (CRI)
		500	CHN	KUN	China Radio Int. (CRI)
		500	CHN	SZG	China Radio Int. (CRI)
		500	CHN	XIA	China Radio Int. (CRI)
		100	CHN	dom	Xizang, Lhasa
		500	G	RMP	BBC World Service
		250	IND	dom	All India R, Aligarh
		250	KWT	KWT	BBG - VO America (VOA)
		250	NIG	IKO	Voice of Nigeria
		250	THA	UDO	BBG - VO America (VOA)
		250	THA	UDO	R. Thailand World Service
7260		500	AFS	MEY	BBC World Service
		100	CHN	KAS	China Radio Int. (CRI)
		500	CHN	URU	China Radio Int. (CRI)
		500	CHN	XIA	China Radio Int. (CRI)
		50	CHN	dom	Xinjiang, Ürümqi
		250	CYP	CYP	BBC World Service
		250	MNG	dom	Mongolian R. 2, Ulaanbaatar
		50	MNG	dom	Mongolian R. 2, Ulaanbaatar
		250	OMA	SLA	BBC World Service
		250	PHL	PHT	BBG - VO America (VOA)
		500	RUS	VLD	Voice of Russia (VOR)
		250	THA	UDO	BBG - RFE/RL
		250	TWN	TAI	Family Radio Worldwide
	‡	10	VUT	dom	R. Vanuatu, Port Vila
7265		5	BLR	dom	Belaruskaje R. Kanal Kultura, Hrodna
		100	CHN	KAS	China Radio Int. (CRI)
		500	CHN	URU	China Radio Int. (CRI)
		250	CYP	CYP	BBC World Service
		250	E	NOB	R. Exterior de España (REE)
7265		500	G	RMP	Polish Radio (Ext Sce)
		300	G	SKN	Radio Canada Int. (RCI)
		500	IRN	SIR	VO the Islamic Rep. of Iran
		500	MDG	MDC	BBG - VO America (VOA)
	‡	100	PAK	dom	Azad Kashmir R, Islamabad
		250	RUS	NVS	Family Radio Worldwide
7270		250	ARM	ERV	Voice of Russia (VOR)
		50	CHN	dom	Nei Menggu, Hohhot

kHz	N	kW	Ctry	Site	Station, location
		100	IND	dom	All India R, Chennai
		100	IND	CNI	All India Radio (AIR)
		10	MLA	dom	RTM Sarawak, Kuching
		250	RUS	MSK	Voice of Russia (VOR)
		250	RUS	NVS	Voice of Russia (VOR)
		100	SNG	SNG	BBC World Service
		250	THA	UDO	BBG - RFE/RL
		100	TWN	TAI	Voice of China
7275		500	CHN	URU	China Radio Int. (CRI)
		100	CHN	dom	CNR1 VO China, Beijing
		50	CHN	dom	Xinjiang, Ürümqi
		250	E	NOB	R. Exterior de España (REE)
		250	KOR	KIM	KBS World Radio
		250	KWT	KWT	BBG - RFE/RL
	‡	100	NIG	dom	FRCN, Abuja
		100	STP	SAO	BBG - Afia Darfur Radio
		500	TUN	SFA	Radio Tunisienne
7280		5	BLR	dom	Belaruskaje R. 1/Reg, Hrodna
		50	CHN	dom	VO Strait Life Channel, Fuzhou
		50	IND	dom	All India R, Guwahati
		250	PHL	PHT	BBG - VO America (VOA)
		250	POR	SIN	Deutsche Welle
		500	RUS	ARM	Voice of Russia (VOR)
		300	TWN	TSH	Sound of Hope Radio Int.
		100	VTN	VNI	Voice of Vietnam (VOV)
7285		100	AFS	dom	SABC R. Sonder Grense, Meyerton
		150	ALB	CER	China Radio Int. (CRI)
		100	CHN	KAS	China Radio Int. (CRI)
		500	CHN	BEI	China Radio Int. (CRI)
		500	CHN	XIA	China Radio Int. (CRI)
		250	CLN	TRM	Deutsche Welle
		100	D	LAM	BBG - RFE/RL
		500	IRN	KAM	VO the Islamic Rep. of Iran
		250	MDG	MDC	R. Netherlands Worldwide
		50	MLI	dom	ORTM, Bamako (Kati)
		300	POR	LIS	RDP Internacional
		200	TJK	DSB	BBG - R. Free Asia (RFA)
		50	VTN	HAN	Voice of Vietnam (VOV)
7290		100	CHN	KAS	China Radio Int. (CRI)
		500	CHN	SZG	China Radio Int. (CRI)
		500	CHN	XIA	China Radio Int. (CRI)
		100	CHN	dom	CNR1 VO China, Beijing
		125	CVA	SMG	Vatican Radio
		100	D	LAM	BBG - RFE/RL
		50	IND	dom	All India R, Thiruvananthapuram
		10	INS	dom	RRI, Nabire
		100	RUS	MSK	Voice of Russia (VOR)
7295		500	CHN	BEI	China Radio Int. (CRI)
		500	CHN	KAS	China Radio Int. (CRI)
		50	CHN	dom	Xinjiang, Ürümqi
		100	D	LAM	BBG - RFE/RL
		500	F	ISS	Radio Algeriennne
		10	IND	dom	All India R, Aizawl
		500	IRN	SIR	VO the Islamic Rep. of Iran
	†	100	MLA	dom	RTM Traxx FM, Kajang
		100	MLI	BKO	China Radio Int. (CRI)
		200	RUS	NVS	BBG - VO America (VOA)
		250	RUS	NVS	Family Radio Worldwide
		250	RUS	SAM	TWR India
		500	RUS	TCH	Voice of Russia (VOR)
7300		170	BUL	PLD	Radio Bulgaria
		100	CHN	KAS	China Radio Int. (CRI)
		500	CHN	KAS	China Radio Int. (CRI)
		500	CHN	URU	China Radio Int. (CRI)
		100	D	WER	TWR Europe
		300	ROU	TIG	Radio Romania International
7300		250	RRW	KIG	Deutsche Welle
		100	RUS	IRK	Family Radio Worldwide
		250	RUS	MSK	Voice of Russia (VOR)
7305		250	ASC	ASC	BBC World Service
		500	CHN	JIN	China Radio Int. (CRI)
		500	CHN	XIA	China Radio Int. (CRI)
		100	CHN	dom	CNR1 VO China, Shijiazhuang
		125	CVA	SMG	Vatican Radio
		500	CVA	SMG	Vatican Radio
		100	D	LAM	BBG - RFE/RL
7310		500	D	WER	Family Radio Worldwide
		250	F	ISS	China Radio Int. (CRI)
		250	IND	BGL	All India Radio (AIR)
		500	IRN	KAM	VO the Islamic Rep. of Iran
		300	ROU	TIG	Radio Romania International
		200	RUS	ARM	Voice of Russia (VOR)
		250	RUS	MSK	Voice of Russia (VOR)
		250	RUS	NVS	Voice of Russia (VOR)
7310		500	CHN	KAS	China Radio Int. (CRI)
		50	CHN	dom	Xinjiang, Ürümqi
		300	ROU	TIG	Radio Romania International
		250	RUS	IRK	Family Radio Worldwide
		240	RUS	MSK	Lutherische Stunde
		240	RUS	MSK	Missionwerk Heukelbach
		250	RUS	MSK	Radio Rossii
		250	RUS	MSK	Radio Santec
		250	RUS	SAM	Voice of Russia (VOR)
		300	TWN	TSH	Sound of Hope Radio Int.
7315		500	CHN	KAS	China Radio Int. (CRI)
		500	CHN	KUN	China Radio Int. (CRI)
		500	CHN	URU	China Radio Int. (CRI)
		150	CHN	dom	CNR2 China Business R, Xi'an
		250	D	WER	Adventist World R. (AWR)
		100	D	BIB	BBG - VO America (VOA)
		5	F	ISS	Radio Dabanga
		500	F	ISS	Radio France Int. (RFI)
		50	IND	dom	All India R, Shillong
		500	IRN	KAM	VO the Islamic Rep. of Iran
		300	ROU	GAL	Radio Romania International
		250	SNG	SNG	BBC World Service
		50	SWZ	MAN	TWR Africa
		250	UAE	DHA	FEBA Radio
		250	USA	HRI	WHRI - World Harvest R. Int.
7320		500	CHN	KUN	China Radio Int. (CRI)
		500	CHN	XIA	China Radio Int. (CRI)
		500	IRN	SIR	VO the Islamic Rep. of Iran
		250	RUS	NVS	IBRA Radio
		100	RUS	dom	R. Rossii/Reg, Arman
		250	RUS	NVS	TWR India
		200	RUS	SPB	Voice of Russia (VOR)
		250	RUS	SPB	Voice of Russia (VOR)
7325		300	AUT	MOS	Radio Ö1 International
		250	CAN	SAC	Radio Canada Int. (RCI)
		150	CHN	KUN	China Radio Int. (CRI)
		500	CHN	BEI	China Radio Int. (CRI)
		500	CHN	JIN	China Radio Int. (CRI)
		500	CHN	KUN	China Radio Int. (CRI)
		500	CHN	URU	China Radio Int. (CRI)
		500	CHN	XIA	China Radio Int. (CRI)
		250	CLN	IRA	BBG - VO America (VOA)
		250	CYP	CYP	BBC World Service
		250	G	RMP	BBC World Service
		50	IND	dom	All India R, Jaipur
		500	IRN	SIR	VO the Islamic Rep. of Iran
		1	PNG	dom	Wantok R. Light, Port Moresby
		300	ROU	GAL	Radio Romania International
	+	15	RUS	KLG	Voice of Russia (VOR)
	+	250	RUS	MSK	Voice of Russia (VOR)
		250	RUS	SAM	Voice of Russia (VOR)
		250	SNG	SNG	BBC World Service
		100	TWN	TNN	Radio France Int. (RFI)
		500	UAE	DHA	BBC World Service
7330		500	CHN	KAS	China Radio Int. (CRI)
		500	CHN	XIA	China Radio Int. (CRI)
		100	RUS	KHB	Voice of Russia (VOR)
7330		250	RUS	MSK	Voice of Russia (VOR)
		250	RUS	PPK	Voice of Russia (VOR)
7335		500	CHN	BEI	China Radio Int. (CRI)
		500	CHN	JIN	China Radio Int. (CRI)
		500	CHN	SZG	China Radio Int. (CRI)
		100	CHN	dom	CNR2 China Business R, Xi'an
		100	CVA	SMG	Vatican Radio
		500	CVA	SMG	Vatican Radio
		250	GUF	GUF	Voice of Russia (VOR)
		50	IND	dom	All India R, Imphal
		500	TUN	SFA	Radio Tunisienne

kHz	N	kW	Ctry	Site	Station, location
		250	UAE	DHA	BBC World Service
		250	USA	HRI	WHRI - World Harvest R. Int.
		100	UZB	TAC	Vatican Radio
7340		500	AFS	MEY	Radio France Int. (RFI)
		100	BOT	BOT	BBG - VO America (VOA)
		500	CHN	KAS	China Radio Int. (CRI)
		100	CHN	dom	Xinjiang, Ürümqi
		500	F	ISS	Radio France Int. (RFI)
		100	IND	dom	All India R, Mumbai
		100	IND	MUM	All India Radio (AIR)
		500	IRN	SIR	VO the Islamic Rep. of Iran
		250	RUS	IRK	Family Radio Worldwide
	+	15	RUS	IRK	Voice of Russia (VOR)
		200	RUS	NVS	Voice of Russia (VOR)
7345		150	ALB	CER	China Radio Int. (CRI)
		100	CHN	dom	CNR1 VO China, Beijing
		100	CZE	LIT	Radio Prague
		200	CZE	LIT	Radio Prague
		250	IRN	AHW	VO the Islamic Rep. of Iran
		500	IRN	KAM	VO the Islamic Rep. of Iran
		300	POR	LIS	RDP Internacional
		300	ROU	TIG	Radio Romania International
		150	SVK	RSO	Radio Slovakia Int.
		100	TJK	DSB	Voice of Russia (VOR)
		100	TUN	SFA	Radio Tunisienne
7350		100	CHN	KAS	China Radio Int. (CRI)
		500	CHN	KAS	China Radio Int. (CRI)
		500	CHN	URU	China Radio Int. (CRI)
		100	CHN	dom	CNR11, Xi'an
		300	G	SKN	BBC World Service
		500	IRN	KAM	VO the Islamic Rep. of Iran
		100	NIG	dom	FRCN, Abuja
		300	ROU	GAL	Radio Romania International
		250	RUS	NVS	Radio France Int. (RFI)
		500	RUS	TCH	Voice of Russia (VOR)
7355		250	CVA	SMG	Vatican Radio
		100	CZE	LIT	Radio Prague
		100	D	WER	Bible Voice Broadcasting (BVB)
		500	IRN	KAM	VO the Islamic Rep. of Iran
		250	OMA	SLA	BBC World Service
		250	THA	NAK	BBC World Service
		100	TWN	TAI	BBG - R. Free Asia (RFA)
7360		150	ALB	CER	China Radio Int. (CRI)
		75	BLR	MNS	Radio Station Belarus
		100	CHN	KAS	China Radio Int. (CRI)
		100	CHN	KUN	China Radio Int. (CRI)
		100	CHN	dom	CNR11, Xi'an
		250	CVA	SMG	Vatican Radio
		500	CVA	SMG	Vatican Radio
		300	CYP	CYP	BBC World Service
		500	GUF	GUF	Family Radio Worldwide
		500	IRN	SIR	VO the Islamic Rep. of Iran
		250	MDG	MDC	Vatican Radio
		250	PHL	PHT	R. Netherlands Worldwide
		250	POR	SIN	RDP Internacional
		100	USA	YFR	Family Radio Worldwide
7365		500	CHN	BEI	China Radio Int. (CRI)
		250	CVA	SMG	Vatican Radio
		100	D	WER	Bible Voice Broadcasting (BVB)
		300	TWN	TSH	Radio Taiwan International
		250	USA	GRV	BBG - R. Martí
7370	+	125	CVA	SMG	Vatican Radio
		250	D	WER	Adventist World R. (AWR)
		250	G	WOF	Voice of Vietnam (VOV)
7370		10	HRV	DEA	Voice of Croatia
		100	IND	dom	All India R, Delhi
		50	IND	dom	All India R, Delhi
		100	IND	DEL	All India Radio (AIR)
		500	IRN	KAM	VO the Islamic Rep. of Iran
	+	300	ROU	GAL	Radio Romania International
		300	ROU	GAL	Radio Romania International
		300	ROU	TIG	Radio Romania International
7375		150	CHN	dom	CNR2 China Business R, Beijing
		300	CYP	CYP	BBC World Service
		100	D	WER	Voice of Croatia
		125	D	WER	Voice of Croatia

kHz	N	kW	Ctry	Site	Station, location
		250	GUF	GUF	Radio France Int. (RFI)
7380		150	ALB	CER	China Radio Int. (CRI)
		250	CLN	TRM	Deutsche Welle
		50	IND	dom	All India R, Chennai
		250	IRN	SIR	VO the Islamic Rep. of Iran
		500	IRN	KAM	VO the Islamic Rep. of Iran
		300	ROU	GAL	Radio Romania International
		300	ROU	TIG	Radio Romania International
		250	THA	NAK	BBC World Service
		100	TWN	TNN	Radio France Int. (RFI)
		100	TWN	PAO	Radio Taiwan International
7385		150	ALB	CER	China Radio Int. (CRI)
		100	CHN	dom	Xizang, Lhasa
		250	CVA	SMG	Vatican Radio
		50	MDG	MDC	Family Radio Worldwide
		150	SVK	RSO	IRRS Shortwave
		250	THA	NAK	BBC World Service
		100	TWN	TAI	BBG - R. Free Asia (RFA)
		100	TWN	KOU	Radio Taiwan International
		100	USA	HRI	WHRI - World Harvest R. Int.
		250	USA	HRI	WHRI - World Harvest R. Int.
7390		100	ALB	SHI	RadioTirana
		150	BLR	MNS	Radio Station Belarus
		500	CHN	JIN	China Radio Int. (CRI)
		500	CHN	XIA	China Radio Int. (CRI)
		300	CYP	CYP	BBC World Service
		250	D	WER	BBG - VO America (VOA)
		9	IND	dom	All India R, Port Blair
		250	MRA	TIN	BBG - VO America (VOA)
		250	PHL	PHT	BBG - VO America (VOA)
		100	SNG	SNG	BBC World Service
7395		150	CHN	BEI	China Radio Int. (CRI)
		500	CHN	KAS	China Radio Int. (CRI)
		500	CHN	KUN	China Radio Int. (CRI)
		250	CVA	SMG	Vatican Radio
		500	CVA	SMG	Vatican Radio
		250	G	RMP	Deutsche Welle
		250	KWT	KWT	BBG - RFE/RL
		50	MDG	MDC	Family Radio Worldwide
		250	MDG	MDC	Radio Japan (NHK World)
		250	OMA	SLA	BBC World Service
		250	THA	NAK	BBC World Service
7400		100	BUL	SOF	Radio Bulgaria
		170	BUL	PLD	Radio Bulgaria
		300	BUL	PLD	Radio Bulgaria
		100	CHN	HUH	China Radio Int. (CRI)
		150	CHN	BEI	China Radio Int. (CRI)
		500	CHN	XIA	China Radio Int. (CRI)
		300	G	WOF	BBC World Service
		250	IND	DEL	All India Radio (AIR)
		1000	TJK	DSB	Voice of Russia (VOR)
7405		100	CHN	HUH	China Radio Int. (CRI)
		500	CHN	BEI	China Radio Int. (CRI)
		500	CHN	URU	China Radio Int. (CRI)
		500	CHN	XIA	China Radio Int. (CRI)
		500	G	RMP	Deutsche Welle
		250	KWT	KWT	BBG - VOA Ashna Radio
		250	OMA	SLA	BBC World Service
		250	THA	UDO	BBG - VO America (VOA)
		250	USA	GRV	BBG - R. Martí
7410		300	AUT	MOS	BBC World Service
		100	CHN	KAS	China Radio Int. (CRI)
		150	CHN	BJI	China Radio Int. (CRI)
7410		500	CHN	BEI	China Radio Int. (CRI)
		500	CHN	JIN	China Radio Int. (CRI)
		500	CHN	SZG	China Radio Int. (CRI)
		100	CHN	dom	CNR8 VO Minorities, Beijing
		100	CZE	LIT	Radio Prague
		250	D	WER	Bible Voice Broadcasting (BVB)
		250	G	WOF	Deutsche Welle
		250	IND	DEL	All India Radio (AIR)
		50	PHL	IBA	FEBC Philippines
7415		250	ASC	ASC	BBC World Service
		100	CHN	KAS	China Radio Int. (CRI)
		500	CHN	KAS	China Radio Int. (CRI)
		500	CHN	URU	China Radio Int. (CRI)

kHz	N	kW	Ctry	Site	Station, location
		300	G	SKN	BBC World Service
		250	MRA	TIN	BBG - R. Free Asia (RFA)
		100	ROU	TIG	Radio Romania International
		300	ROU	TIG	Radio Romania International
		50	USA	BCQ	WBCQ
7420		150	CHN	KUN	China Radio Int. (CRI)
		500	CHN	URU	China Radio Int. (CRI)
		100	CHN	dom	Nei Menggu, Hohhot
		100	CZE	LIT	Radio Prague
		50	IND	dom	All India R, Guwahati
		50	IND	dom	All India R, Hyderabad
		50	IND	GUW	All India Radio (AIR)
		100	LTU	SIT	VO the Islamic Rep. of Iran
		250	THA	UDO	BBG - VO America (VOA)
7425		500	AFS	MEY	BBC World Service
		100	ALB	SHI	RadioTirana
		100	CHN	KAS	China Radio Int. (CRI)
		500	CHN	JIN	China Radio Int. (CRI)
		500	CHN	KUN	China Radio Int. (CRI)
		250	CLN	TRM	Deutsche Welle
		100	D	WER	Bible Voice Broadcasting (BVB)
		500	F	ISS	Radio France Int. (RFI)
		250	G	RMP	BBC World Service
		500	G	RMP	BBC World Service
		250	KWT	KWT	BBG - VO America (VOA)
		250	MDG	MDC	R. Netherlands Worldwide
7430		500	ARM	ERV	Voice of Russia (VOR)
		500	CHN	JIN	China Radio Int. (CRI)
		50	IND	dom	All India R, Bhopal
		250	PHL	PHT	BBG - VO America (VOA)
		300	ROU	GAL	Radio Romania International
		300	ROU	TIG	Radio Romania International
7435		100	ALB	SHI	RadioTirana
		100	CHN	NNN	China Radio Int. (CRI)
		150	CHN	BEI	China Radio Int. (CRI)
		500	CHN	JIN	China Radio Int. (CRI)
		500	CHN	KUN	China Radio Int. (CRI)
		100	CVA	SMG	Vatican Radio
		300	CYP	CYP	BBC World Service
		100	CZE	LIT	Radio Prague
		500	G	RMP	BBC World Service
		300	ROU	GAL	Radio Romania International
		100	UKR	KHR	Radio Ukraine Int. (RUI)
		50	VTN	dom	VO Vietnam 1st prgr, Hanoi
7440		100	CHN	KUN	China Radio Int. (CRI)
		150	CHN	BEI	China Radio Int. (CRI)
		500	CHN	BEI	China Radio Int. (CRI)
		500	CHN	KUN	China Radio Int. (CRI)
		500	CHN	XIA	China Radio Int. (CRI)
		125	D	WER	Democratic VO Burma
		50	IND	dom	All India R, Lucknow
		100	NZL	RAN	R. New Zealand Int. (RNZI)
		250	THA	UDO	BBG - VO America (VOA)
		500	UKR	LVI	Radio Ukraine Int. (RUI)
		500	UKR	LVI	Voice of Russia (VOR)
7445		250	ASC	ASC	BBC World Service
		100	CHN	KAS	China Radio Int. (CRI)
		500	CHN	URU	China Radio Int. (CRI)
		100	CHN	dom	CNR8 VO Minorities, Beijing
		160	G	RMP	IBRA Radio
		500	MRA	TIN	BBG - R. Free Asia (RFA)
		300	ROU	TIG	Radio Romania International
		250	SEY	SEY	BBC World Service
7445		100	TWN	PAO	Radio Taiwan International
7450		100	CHN	dom	Xizang, Lhasa
		100	GRC	AVL	RS Makedonias (ERT3)
		100	GRC	AVL	Voice of Greece (ERA5)
7455		250	CLN	IRA	BBG - VOA Deewa Radio
		500	F	ISS	Radio Algeriennne
		100	USA	YFR	Family Radio Worldwide
7460		500	MDA	KCH	Radio Payam-e Doost
		250	MNG	UBA	BBG - R. Free Asia (RFA)
		250	PHL	PHT	BBG - VO America (VOA)
		100	PHL	BOC	FEBC Philippines
		100	TWN	PAO	Family Radio Worldwide
		300	TWN	HUW	Takai Shinlingde Souchi

kHz	N	kW	Ctry	Site	Station, location
7465		100	AFS	MEY	BBC World Service
		250	AFS	MEY	BBC World Service
		500	AFS	MEY	BBC World Service
		100	ALB	SHI	RadioTirana
		500	G	RMP	BBC World Service
		250	THA	UDO	R. Thailand World Service
		250	USA	GRV	BBG - VO America (VOA)
		250	USA	HRI	WHRI - World Harvest R. Int.
		100	USA	WCR	WWCR
7470		250	CLN	IRA	BBG - R. Free Asia (RFA)
		250	KWT	KWT	BBG - R. Free Asia (RFA)
		250	MNG	UBA	BBG - R. Free Asia (RFA)
		250	MRA	TIN	BBG - R. Free Asia (RFA)
		100	PAK	dom	R. Pakistan, Islamabad
		100	PAK	ISL	Radio Pakistan
		250	TJK	DSB	BBG - R. Free Asia (RFA)
		250	UKR	SMF	Deutsche Welle
7475		100	GRC	AVL	Voice of Greece (ERA5)
7480		300	ARM	ERV	Open Radio for North Korea
		250	CLN	IRA	BBG - RFE/RL
		250	CLN	IRA	BBG - VO America (VOA)
		250	CLN	IRA	BBG - VOA R. Aap Ki Dunyaa
		500	MDA	KCH	Radio Payam-e Doost
		50	PHL	PHT	BBG - VO America (VOA)
		50	PHL	IBA	FEBC Philippines
		200	TJK	DSB	BBG - R. Free Asia (RFA)
7485		250	TJK	DSB	Bible Voice Broadcasting (BVB)
		100	UZB	TAC	FEBA Radio
7490		300	ARM	ERV	Family Radio Worldwide
		250	THA	NAK	BBC World Service
		100	USA	WCR	WWCR
7495		250	MRA	TIN	BBG - R. Free Asia (RFA)
		250	PHL	PHT	BBG - VO America (VOA)
		250	THA	UDO	BBG - VO America (VOA)
		250	THA	UDO	BBG - VOA Deewa Radio
7505		100	PHL	IBA	FEBC Philippines
		250	THA	NAK	BBC World Service
		100	TJK	DSB	Free North Korea Radio
		50	USA	RNO	WRNO Worldwide
		200	UZB	TAC	Radio Free Chosun
7510		300	ARM	ERV	Democratic VO Burma
		300	ARM	ERV	FEBA Radio
		200	KAZ	ALM	Family Radio Worldwide
		250	PAK	ISL	Radio Pakistan
		200	KAZ	ALM	Golos Pravoslaviya
7515		125	MRA	TIN	BBG - R. Free Asia (RFA)
7520		500	CLN	IRA	BBG - R. Farda
		250	PHL	PHT	BBG - VO America (VOA)
		250	THA	UDO	BBG - R. Farda
		100	USA	YFR	Family Radio Worldwide
		250	USA	HRI	WHRI - World Harvest R. Int.
7525		250	PHL	PHT	BBG - VO America (VOA)
±		100	TJK	DSB	Sound of Hope Radio Int.
7530		100	ALB	SHI	RadioTirana
		300	ARM	ERV	Free North Korea Radio
		250	PAK	ISL	Radio Pakistan
		250	PHL	PHT	BBG - VO America (VOA)
		100	TWN	TAI	Suab Xaa Moo Zoo
7540		250	THA	NAK	BBC World Service
		500	TJK	DSB	BBG - R. Free Asia (RFA)
		300	UKR	SMF	Dengê Mezopotamya
7550		300	ARM	ERV	Family Radio Worldwide
		250	IND	DEL	All India Radio (AIR)
7550		500	IND	BGL	All India Radio (AIR)
		250	KWT	KWT	BBG - R. Free Asia (RFA)
		250	KWT	KWT	BBG - RFE/RL
7555		100	TWN	PAO	Radio Taiwan International
		50	USA	JES	KJES Radio
		250	USA	EWN	WEWN - EWTN Shortwave
		250	USA	HRI	WHRI - World Harvest R. Int.
7560		250	CLN	IRA	BBG - VOA Ashna Radio
		200	KAZ	ALM	Family Radio Worldwide
		250	KWT	KWT	BBG - VOA Ashna Radio
		250	THA	UDO	BBG - VO America (VOA)
		250	THA	UDO	BBG - VOA Ashna Radio
±		100	TJK	DSB	Sound of Hope Radio Int.

kHz	N	kW	Ctry	Site	Station, location
		100	UZB	TAC	Open Radio for North Korea
7565		300	MDA	KCH	Family Radio Worldwide
	±	100	TJK	DSB	Sound of Hope Radio Int.
7570		200	KRE	KUJ	Voice of Korea
		125	MRA	TIN	BBG - R. Free Asia (RFA)
		250	PHL	PHT	BBG - VO America (VOA)
		250	THA	UDO	R. Thailand World Service
		100	USA	YFR	Family Radio Worldwide
		100	USA	YFR	Radio Taiwan International
		250	USA	HRI	WHRI - World Harvest R. Int.
7575		250	CLN	IRA	BBG - VO America (VOA)
		250	PHL	PHT	BBG - VO America (VOA)
		250	THA	UDO	BBG - VO America (VOA)
7580		250	CLN	IRA	BBG - R. Farda
		200	KRE	KUJ	Voice of Korea
7585		100	TJK	DSB	Vatican Radio
7590		100	ARM	ERV	Family Radio Worldwide
		300	ARM	ERV	Family Radio Worldwide
	±	100	ARM	ERV	Sound of Hope Radio Int.
		200	KAZ	ALM	Family Radio Worldwide
		250	USA	HRI	WHRI - World Harvest R. Int.
		100	UZB	TAC	North Korea Reform Radio
7600		300	ARM	ERV	Family Radio Worldwide
		250	THA	NAK	BBC World Service
7620		50	CHN	dom	CNR5 VO Zhonghua, Beijing
7645		100	USA	WCR	WWCR
7730		100	USA	YFR	Family Radio Worldwide
7812		3	USA	KEW	AFRTS (AFN Feeder)
7850		3	CAN	STF	CHU, Ottawa
7925		500	IRN	SIR	VO the Islamic Rep. of Iran
7935		100	CHN	dom	CNR8 VO Minorities, Lingshi
8728		10	F	MCO	Radio Monaco
9150	±	1	TWN	TAI	Sound of Hope Radio Int.
9170		50	CHN	dom	CNR6 VO Shenzhou, Beijing
9250		250	EGY	ABS	Radio Cairo
		250	EGY	ABZ	Radio Cairo
9265		50	USA	INB	WINB
9280		250	EGY	ABS	Radio Cairo
9295		100	EGY	ABZ	Radio Cairo
9300		100	UZB	TAC	IMG
9305		250	EGY	ABS	Radio Cairo
9310		200	KAZ	ALM	Family Radio Worldwide
		250	PHL	PHT	BBG - RFE/RL
		250	THA	UDO	BBG - VO America (VOA)
9315		250	THA	UDO	BBG - VO America (VOA)
9320		250	CLN	IRA	BBG - VO America (VOA)
		250	KWT	KWT	BBG - VO America (VOA)
		250	THA	UDO	BBG - VO America (VOA)
		100	TJK	DSB	Family Radio Worldwide
9325		250	CLN	IRA	BBG - R. Free Asia (RFA)
		250	CLN	IRA	BBG - VO America (VOA)
		200	KRE	KUJ	Voice of Korea
		250	PHL	PHT	BBG - VO America (VOA)
		50	PHL	PHT	BBG - VO America (VOA)
9330		500	SYR	ADR	Radio Damascus
		50	USA	BCQ	WBCQ
9335		200	KRE	KUJ	Voice of Korea
		250	KWT	KWT	BBG - R. Free Afghanistan
		250	KWT	KWT	BBG - VOA Ashna Radio
9340		250	CLN	IRA	BBG - R. Farda
		100	PAK	ISL	Radio Pakistan
		250	PAK	ISL	Radio Pakistan
9345		100	GUM	TWR	TWR Asia (KTWR)
9345		200	KAZ	ALM	Bible Voice Broadcasting (BVB)
		200	KRE	KUJ	Voice of Korea
		100	PAK	ISL	Radio Pakistan
		250	RUS	SAM	Radio Bilal
9350		100	USA	WCR	WWCR
9355		250	CLN	IRA	BBG - R. Free Asia (RFA)
		100	MRA	SAI	BBG - R. Free Asia (RFA)
		250	PHL	PHT	BBG - VO America (VOA)
		100	USA	YFR	Family Radio Worldwide
9360		250	MRA	TIN	BBG - RFE/RL
		250	PHL	PHT	BBG - VO America (VOA)
		250	THA	UDO	BBG - Radio Mashaal
9365		500	F	ISS	Radio Taiwan International

kHz	N	kW	Ctry	Site	Station, location
9370		250	CLN	IRA	BBG - VOA Deewa Radio
		100	GUM	TWR	TWR Asia (KTWR)
		250	PHL	PHT	BBG - VOA Deewa Radio
		250	THA	UDO	BBG - VOA Deewa Radio
		50	USA	TJC	WTJC
9380		250	CLN	IRA	BBG - Afia Darfur Radio
		250	THA	UDO	BBG - VOA Deewa Radio
		100	TJK	DSB	Deutsche Welle
	±	1	TWN	TAI	Sound of Hope Radio Int.
9385		250	CLN	IRA	BBG - R. Free Asia (RFA)
		250	MRA	TIN	BBG - R. Free Asia (RFA)
		100	USA	WRB	WWRB
9390		500	F	ISS	Radio Algeriennne
		200	TJK	DSB	Bible Voice Broadcasting (BVB)
		200	TJK	DSB	Voice of Wilderness
		250	USA	EWN	WEWN - EWTN Shortwave
9395		100	BOT	BOT	BBG - VO America (VOA)
		250	CLN	IRA	BBG - VO America (VOA)
9400		300	ARM	ERV	FEBA Radio
		170	BUL	PLD	Radio Bulgaria
	+	50	BUL	SOF	Radio Bulgaria
		100	PHL	IBA	FEBC Philippines
9405		100	D	BIB	BBG - RFE/RL
		500	D	WER	Family Radio Worldwide
		300	G	WOF	BBG - RFE/RL
		100	PHL	BOC	FEBC Philippines
9410		250	ASC	ASC	BBC World Service
		500	CHN	KAS	China Radio Int. (CRI)
		50	CHN	dom	CNR5 VO Zhonghua, Beijing
		300	CYP	CYP	BBC World Service
		250	OMA	SLA	BBC World Service
		250	SEY	SEY	BBC World Service
		250	THA	NAK	BBC World Service
		250	TUR	CAK	Voice of Turkey (VOT)
		500	TUR	EMR	Voice of Turkey (VOT)
		10	TWN	dom	Fu Hsing Broadcasting Station, Kuanyin
		250	UKR	SMF	Radio Ukraine Int. (RUI)
		250	USA	HRI	BBC World Service
		250	USA	HRI	WHRI - World Harvest R. Int.
9415		500	CHN	BEI	China Radio Int. (CRI)
		500	CHN	XIA	China Radio Int. (CRI)
		250	IND	DEL	All India Radio (AIR)
		250	PHL	PHT	BBG - VO America (VOA)
9420		100	CHN	KAS	China Radio Int. (CRI)
		100	CHN	dom	CNR1 VO China, Lingshi
		100	CHN	dom	CNR1 VO Minorities, Lingshi
		100	D	BIB	BBG - VO America (VOA)
		250	G	RMP	IBRA Radio
		170	GRC	AVL	Voice of Greece (ERA5)
9425		150	CHN	BEI	China Radio Int. (CRI)
		500	CHN	JIN	China Radio Int. (CRI)
		500	IND	dom	All India R, Bengaluru
9430		500	CHN	KAS	China Radio Int. (CRI)
		100	D	BIB	BBG - R. Farda
		100	D	LAM	BBG - R. Farda
		250	D	WER	BBG - R. Farda
		100	D	BIB	BBG - RFE/RL
		250	D	NAU	BBG - VO America (VOA)
		250	D	WER	Family Radio Worldwide
		100	F	MCO	TWR Europe
		300	G	SKN	Family Radio Worldwide
		100	PHL	BOC	FEBC Philippines
		250	POR	SIN	KBS World Radio
9430		250	RRW	KIG	Deutsche Welle
		100	USA	YFR	Family Radio Worldwide
		100	ZMB	LUS	CVC 1Africa Radio
9435		100	CHN	KUN	China Radio Int. (CRI)
		500	CHN	KAS	China Radio Int. (CRI)
		500	CHN	XIA	China Radio Int. (CRI)
		300	G	SKN	BBG - VO America (VOA)
		100	PHL	BOC	FEBC Philippines
		250	THA	UDO	BBG - VO America (VOA)
		200	TJK	DSB	BBG - R. Free Asia (RFA)
9440		250	ASC	ASC	BBC World Service
		100	CHN	NNN	China Radio Int. (CRI)
		150	CHN	KUN	China Radio Int. (CRI)

kHz	N	kW	Ctry	Site	Station, location
		500	CHN	KUN	China Radio Int. (CRI)
		500	CHN	XIA	China Radio Int. (CRI)
		100	F	MCO	TWR Europe
		300	RUS	ARM	Family Radio Worldwide
		100	STP	SAO	BBG - VO America (VOA)
		150	SVK	RSO	Radio Slovakia Int.
9445		250	CLN	TRM	Deutsche Welle
		100	D	BIB	BBG - RFE/RL
		100	D	LAM	BBG - RFE/RL
		100	D	BIB	BBG - VO America (VOA)
		500	IND	BGL	All India Radio (AIR)
		250	KWT	KWT	BBG - RFE/RL
		50	PHL	IBA	FEBC Philippines
		150	SVK	RSO	Radio Slovakia Int.
		250	THA	UDO	BBG - VOA Ashna Radio
9450		250	ATN	BON	Deutsche Welle
		100	CHN	KAS	China Radio Int. (CRI)
		100	CHN	URU	China Radio Int. (CRI)
		500	CHN	KAS	China Radio Int. (CRI)
		500	CHN	SZG	China Radio Int. (CRI)
		250	CYP	CYP	BBC World Service
	+	300	ROU	TIG	Radio Romania International
		250	RRW	KIG	Deutsche Welle
		250	RUS	NVS	Family Radio Worldwide
		100	TWN	PAO	Sound of Hope Radio Int.
9455		150	CHN	KUN	China Radio Int. (CRI)
		100	CHN	dom	CNR1 VO China, Lingshi
		100	CHN	dom	CNR8 VO Minorities, Lingshi
		100	MRA	SAI	BBG - R. Free Asia (RFA)
		300	POR	LIS	RDP Internacional
		100	TWN	TAI	Family Radio Worldwide
9460		100	AFS	MEY	BBC World Service
		150	ALB	CER	China Radio Int. (CRI)
		100	AUT	MOS	Polish Radio (Ext Sce)
		100	CHN	KAS	China Radio Int. (CRI)
		150	CHN	BEI	China Radio Int. (CRI)
		150	CHN	KUN	China Radio Int. (CRI)
		500	CHN	URU	China Radio Int. (CRI)
		100	D	NAU	Bible Voice Broadcasting (BVB)
		250	RUS	IRK	Family Radio Worldwide
		250	RUS	NVS	Family Radio Worldwide
		250	SEY	SEY	BBC World Service
		150	SVK	RSO	Radio Slovakia Int.
9465		100	D	LAM	BBG - RFE/RL
		100	PHL	BOC	FEBC Philippines
		100	TWN	PAO	Family Radio Worldwide
		100	USA	YFR	Family Radio Worldwide
9470		500	CHN	XIA	China Radio Int. (CRI)
		100	CHN	dom	Xinjiang, Ürümqi
		125	G	WOF	Polish Radio (Ext Sce)
		250	IND	dom	All India R, Aligarh
		500	RUS	MSK	Voice of Russia (VOR)
		250	SNG	SNG	BBC World Service
		100	USA	HRI	WHRI - World Harvest R. Int.
9475		100	AUS	SHP	Radio Australia
		100	SWZ	MAN	TWR Africa
		500	TJK	DSB	Voice of Russia (VOR)
9479		100	USA	TWW	WTWW
9480		100	CHN	dom	CNR11, Xi'an
		500	D	NAU	Family Radio Worldwide
		100	LTU	SIT	BBG - R. Free Asia (RFA)
		500	RUS	MSK	Voice of Russia (VOR)
9480		250	THA	NAK	BBC World Service
		100	USA	YFR	Family Radio Worldwide
		100	USA	TWW	WTWW
9485		250	ASC	ASC	Deutsche Welle
		250	CYP	CYP	BBC World Service
		250	D	NAU	BBG - RFE/RL
		250	D	NAU	BBG - VO America (VOA)
		500	RUS	IRK	Family Radio Worldwide
		150	SVK	RSO	Radio Slovakia Int.
9490		100	CAN	SAC	Radio República
		150	CHN	KUN	China Radio Int. (CRI)
		500	CHN	BEI	China Radio Int. (CRI)
		100	CHN	dom	Xizang, Lhasa
		100	D	LAM	BBG - VO America (VOA)

kHz	N	kW	Ctry	Site	Station, location
		250	G	RMP	Polish Radio (Ext Sce)
		100	MRA	SAI	BBG - VO America (VOA)
		250	PHL	PHT	BBG - VO America (VOA)
		250	PHL	PHT	Radio Canada Int. (RCI)
		250	USA	HRI	WHRI - World Harvest R. Int.
9495		100	AUT	MOS	TWR Europe
		250	D	WER	BBG - VO America (VOA)
	†	5	GEO	dom	Abkhaz State R, Soxum
		250	THA	NAK	BBC World Service
		250	TUR	CAK	Voice of Turkey (VOT)
		250	UAE	DHA	Family Radio Worldwide
		100	USA	YFR	Family Radio Worldwide
		250	USA	HRI	WHRI - World Harvest R. Int.
9500		100	AUS	SHP	Radio Australia
		100	CHN	dom	CNR1 VO China, Shijiazhuang
		250	D	WER	Family Radio Worldwide
		100	SWZ	MAN	TWR Africa
		100	UZB	TAC	CVC The Voice Asia
9505		300	AUT	MOS	Adventist World R. (AWR)
		8	B	dom	R. Record, São Paulo
		50	CHN	dom	VO Strait News Channel, Fuzhou
		250	CYP	CYP	BBC World Service
		250	OMA	SLA	BBC World Service
		250	SNG	SNG	BBC World Service
		100	STP	SAO	BBG - VO America (VOA)
		250	THA	NAK	BBC World Service
		100	USA	YFR	Family Radio Worldwide
		250	USA	HRI	WHRI - World Harvest R. Int.
		100	ZMB	LUS	CVC 1Africa Radio
9510		50	CHN	dom	Xinjiang-Mo, Ürümqi
		250	CLN	TRM	Deutsche Welle
		125	D	WER	Bible Voice Broadcasting (BVB)
		100	F	ISS	Bible Voice Broadcasting (BVB)
		500	IRN	SIR	VO the Islamic Rep. of Iran
		250	OMA	SLA	BBC World Service
		250	SNG	SNG	BBC World Service
		150	SVK	RSO	IRRS Shortwave
		250	THA	NAK	BBC World Service
9515		300	AUT	MOS	Adventist World R. (AWR)
		10	B	dom	R. Marumby/Novas de Paz, Curitiba
		100	CHN	KAS	China Radio Int. (CRI)
		150	CHN	dom	CNR2 China Business R, Beijing
		250	D	NAU	Family Radio Worldwide
		100	KOR	KIM	KBS World Radio
		100	KOR	KIM	KBS World Radio
		250	PHL	PUG	Radio Veritas Asia
9520		50	CHN	dom	Nei Menggu, Hohhot
		250	CLN	IRA	BBG - VO America (VOA)
		250	CLN	IRA	BBG - VOA R. Aap Ki Dunyaa
		100	D	BIB	BBG - R. Farda
		100	D	BIB	BBG - RFE/RL
		300	G	WOF	BBG - RFE/RL
		500	IRN	KAM	VO the Islamic Rep. of Iran
		250	PHL	PUG	Radio Veritas Asia
9525		500	CHN	BEI	China Radio Int. (CRI)
		500	CHN	KAS	China Radio Int. (CRI)
		100	D	LAM	BBG - RFE/RL
		300	ROU	GAL	Radio Romania International
		100	SWZ	MAN	TWR Africa
		100	USA	YFR	Family Radio Worldwide
9526		250	INS	JAK	Voice of Indonesia
9530	†	10	B	dom	R. Transmundial, Santa Maria
9530		100	CHN	dom	CNR11, Xi'an
		300	G	SKN	Family Radio Worldwide
		250	PHL	PHT	BBG - VO America (VOA)
		500	TUR	EMR	Voice of Turkey (VOT)
		50	VTN	dom	VO Vietnam 1st prgr, Hanoi
9535		150	CHN	BJI	China Radio Int. (CRI)
		500	CHN	KUN	China Radio Int. (CRI)
		100	D	NAU	Adventist World R. (AWR)
		100	D	BIB	BBG - RFE/RL
		500	D	NAU	Family Radio Worldwide
		250	E	NOB	R. Exterior de España (REE)
		500	F	ISS	China Radio Int. (CRI)
	†	5	GEO	dom	Abkhaz State R, Soxum
		250	KWT	KWT	BBG - RFE/RL

kHz	N	kW	Ctry	Site	Station, location
		300	ROU	TIG	Radio Romania International
		250	RRW	KIG	Deutsche Welle
		250	THA	UDO	R. Thailand World Service
9540		500	CHN	BEI	China Radio Int. (CRI)
		500	CHN	KUN	China Radio Int. (CRI)
		250	D	WER	BBG - VO America (VOA)
		500	IRN	KAM	VO the Islamic Rep. of Iran
		300	J	YAM	Radio Japan (NHK World)
		100	STP	SAO	BBG - VO America (VOA)
		150	SVK	RSO	Radio Slovakia Int.
		100	TWN	TAI	Family Radio Worldwide
		100	TWN	TSH	Sound of Hope Radio Int.
		100	USA	HRI	Hmong World Christian Radio
		100	USA	HRI	WHRI - World Harvest R. Int.
9545		250	CLN	TRM	Deutsche Welle
		250	G	WOF	Deutsche Welle
		300	G	SKN	Deutsche Welle
		500	G	RMP	Deutsche Welle
		250	PHL	PHT	BBG - VO America (VOA)
		10	SLM	dom	Solomon Islands BC, Honiara
		100	TWN	TAI	Family Radio Worldwide
9550	‡	10	B	dom	R. Boa Vontade, Porto Alegre
		150	CHN	KUN	China Radio Int. (CRI)
		500	CHN	BEI	China Radio Int. (CRI)
		100	D	LAM	BBG - R. Farda
		250	RRW	KIG	FEBA Radio
		100	VTN	VNI	Voice of Vietnam (VOV)
9555		150	ALB	CER	China Radio Int. (CRI)
		500	ARS	RIY	BSKSA
		150	CHN	BEI	China Radio Int. (CRI)
		500	G	RMP	Radio Canada Int. (RCI)
		100	MRA	SAI	BBG - VO America (VOA)
		250	PHL	PHT	BBG - VO America (VOA)
		100	USA	YFR	Family Radio Worldwide
9560		100	AUS	SHP	Radio Australia
		250	CAN	SAC	China Radio Int. (CRI)
		250	CAN	SAC	KBS World Radio
		100	CHN	KAS	China Radio Int. (CRI)
		500	CHN	URU	China Radio Int. (CRI)
		50	CHN	dom	Xinjiang, Ürümqi
		250	CLN	TRM	Deutsche Welle
		100	D	BIB	BBG - RFE/RL
	±	100	ETH	GJW	Radio Ethiopia
	±	100	ETH	GJW	VO Peace & Dem. of Eritrea
	±	100	ETH	GJW	Voice of Democratic Alliance
	±	100	ETH	GJW	Voice of Eritrea
		300	J	YAM	Radio Canada Int. (RCI)
		300	J	YAM	Radio Japan (NHK World)
9565		150	ALB	CER	China Radio Int. (CRI)
		20	B	dom	Super R. Deus é Amor, Curitiba
		100	BOT	BOT	BBG - VO America (VOA)
		100	CAN	SAC	BBG - R. Martí
		250	CLN	IRA	BBG - VOA Deewa Radio
		250	D	WER	BBG - VOA Deewa Radio
		250	KOR	KIM	KBS World Radio
		100	KOR	KIM	Radio Canada Int. (RCI)
		250	TWN	TSH	Radio France Int. (RFI)
		500	USA	GRV	BBG - R. Martí
9570		300	ALB	CER	China Radio Int. (CRI)
		100	CHN	KAS	China Radio Int. (CRI)
		100	CHN	BEI	China Radio Int. (CRI)
		100	CHN	dom	CNR2 China Business R, Golmud
9570		250	CLN	IRA	BBG - VO America (VOA)
		250	CUB	HAB	China Radio Int. (CRI)
		100	D	LAM	BBG - RFE/RL
		250	E	NOB	R. Exterior de España (REE)
		100	KOR	KIM	KBS World Radio
		100	KOR	KIM	Radio Canada Int. (RCI)
		250	PHL	PUG	Radio Veritas Asia
9575		50	IND	dom	All India R, Delhi
		50	IND	DEL	All India Radio (AIR)
		250	MRC	NAD	Radio Méditerranée Int.
		500	UAE	DHA	Radio Japan (NHK World)
		100	USA	YFR	Family Radio Worldwide
9580	†	50	ARS	JED	BSKSA
		100	AUS	SHP	Radio Australia
		100	CHN	dom	Xizang, Lhasa
		250	CUB	HAB	China Radio Int. (CRI)
		125	G	WOF	Polish Radio (Ext Sce)
		250	GAB	GAB	Africa No.1
		250	KOR	KIM	KBS World Radio
		100	SNG	SNG	BBC World Service
		100	UZB	TAC	Vatican Radio
9585		500	CHN	KAS	China Radio Int. (CRI)
		100	CVA	SMG	Vatican Radio
		250	CVA	SMG	Vatican Radio
		100	D	BIB	BBG - VO America (VOA)
		100	GUM	SDA	AWR Guam (KSDA)
		100	GUM	TWR	TWR Asia (KTWR)
		500	IRN	KAM	VO the Islamic Rep. of Iran
		500	IRN	SIR	VO the Islamic Rep. of Iran
		100	SNG	SNG	BBC World Service
		100	UZB	TAC	Radio Japan (NHK World)
9587	†	10	B	dom	Super R. Deus é Amor, São Paulo
9590		150	ALB	CER	China Radio Int. (CRI)
		100	AUS	SHP	Radio Australia
		150	CHN	KUN	China Radio Int. (CRI)
		500	CHN	KAS	China Radio Int. (CRI)
		500	CHN	SZG	China Radio Int. (CRI)
		250	E	NOB	R. Exterior de España (REE)
		250	MDG	MDC	Family Radio Worldwide
		250	UAE	DHA	Sudan Radio Service
9595		100	AFS	MEY	Family Radio Worldwide
		100	D	WER	Adventist World R. (AWR)
		250	D	WER	BBG - RFE/RL
		500	D	WER	Family Radio Worldwide
		250	IND	dom	All India R, Aligarh
		100	IND	DEL	All India Radio (AIR)
		250	IND	DEL	All India Radio (AIR)
		50	J	dom	R. Nikkei 1, Tokyo
		250	OMA	SLA	BBC World Service
		250	RRW	KIG	FEBA Radio
		250	USA	HRI	WHRI - World Harvest R. Int.
9599	†	10	MEX	dom	R. UNAM, México
9600		100	CHN	NNN	China Radio Int. (CRI)
		150	CHN	BJI	China Radio Int. (CRI)
		150	CHN	KUN	China Radio Int. (CRI)
		500	CHN	JIN	China Radio Int. (CRI)
		500	CHN	KAS	China Radio Int. (CRI)
		50	CHN	dom	Xinjiang, Ürümqi
		500	CVA		Vatican Radio
		250	MRA	TIN	Vatican Radio
9605		300	AUT	MOS	Adventist World R. (AWR)
		250	D	WER	Ev. Missions-Gemeinden
		250	E	NOB	R. Exterior de España (REE)
		250	G	RMP	BBC World Service
		300	J	YAM	Radio Japan (NHK World)
		250	OMA	SLA	BBC World Service
		250	RUS	SAM	Voice of Asena
		250	RUS	SAM	Voice of Meselna Delina
		250	SNG	SNG	BBC World Service
		250	THA	NAK	BBC World Service
		100	USA	YFR	Family Radio Worldwide
9610		250	ATN	BON	Vatican Radio
		250	CAN	SAC	Radio Canada Int. (RCI)
		160	CAN	SAC	Vatican Radio
		100	CHN	KUN	China Radio Int. (CRI)
		150	CHN	KUN	China Radio Int. (CRI)
9610		500	CHN	JIN	China Radio Int. (CRI)
		500	CHN	KAS	China Radio Int. (CRI)
		50	CHN	dom	CNR8 VO Minorities, Beijing
		100	D	NAU	Adventist World R. (AWR)
		500	IRN	SIR	VO the Islamic Rep. of Iran
	+	90	POR	SIN	BBC World Service
		100	ROU	TIG	Radio Romania International
		250	SEY	SEY	BBC World Service
		500	TUR	CAK	Voice of Turkey (VOT)
9615		100	ALS	NLS	KNLS International
		500	CHN	URU	China Radio Int. (CRI)
		300	G	SKN	BBC World Service
		250	OMA	SLA	BBC World Service
		250	PHL	PUG	Radio Veritas Asia

kHz	N	kW	Ctry	Site	Station, location
		250	USA	HRI	WHRI - World Harvest R. Int.
9620		100	CHN	KAS	China Radio Int. (CRI)
		500	CHN	KAS	China Radio Int. (CRI)
		150	CHN	dom	CNR2 China Business R, Beijing
		500	D	WER	Radio Japan (NHK World)
		250	E	NOB	R. Exterior de España (REE)
		250	IND	ALG	All India Radio (AIR)
		300	ROU	TIG	Radio Romania International
		250	THA	UDO	BBG - VO America (VOA)
9625		100	AFS	MEY	Channel Africa
		300	AUT	MOS	Adventist World R. (AWR)
		15	BOL	dom	R. Fides, La Paz
		100	CAN	dom	CBC North Quebec SW Sce, Sackville
		100	D	BIB	BBG - RFE/RL
		100	D	LAM	BBG - RFE/RL
		100	GUM	SDA	AWR Guam (KSDA)
		300	J	YAM	Radio Japan (NHK World)
		100	PHL	BOC	FEBC Philippines
		250	TWN	TNN	Radio Taiwan International
9630		10	B	dom	R. Aparecida, Aparecida
		100	CHN	dom	CNR1 VO China, Golmud
		100	CHN	dom	CNR1 VO China, Lingshi
		100	CHN	dom	CNR8 VO Minorities, Lingshi
	+	100	CTR	CRI	R. Exterior de España (REE)
		500	D	WER	Family Radio Worldwide
		250	UAE	DHA	Radio Australia
9635		50	CHL	SGO	CVC La Voz
		500	CHN	XIA	Radio Canada Int. (RCI)
		250	CLN	TRM	Deutsche Welle
		125	CVA	SMG	Vatican Radio
		300	G	SKN	IBRA Radio
		250	IND	ALG	All India Radio (AIR)
		50	MLI	dom	ORTM, Bamako (Kati)
		300	ROU	TIG	Radio Romania International
		100	SWZ	MAN	TWR Africa
		100	TWN	TSH	Sound of Hope Radio Int.
		50	VTN	dom	VO Vietnam 1st prgr, Hanoi
9640		500	CHN	KAS	China Radio Int. (CRI)
		250	CUB	HAB	Radio Habana Cuba
		100	D	BIB	BBG - RFE/RL
		100	KOR	KIM	KBS World Radio
		250	PHL	PHT	BBG - VO America (VOA)
		250	USA	HRI	WHRI - World Harvest R. Int.
9645		8	B	dom	R. Bandeirantes, São Paulo
		500	CHN	BEI	China Radio Int. (CRI)
		500	CHN	KUN	China Radio Int. (CRI)
		100	CHN	dom	CNR1 VO China, Beijing
		100	CHN	dom	CNR8 VO Minorities, Beijing
		100	CVA	SMG	Vatican Radio
		250	CVA	SMG	Vatican Radio
		250	PHL	PUG	Radio Veritas Asia
		250	THA	UDO	BBG - VO America (VOA)
		500	UAE	DHA	BBG - R. Free Asia (RFA)
9650		100	AFS	dom	SABC R. Sonder Grense, Meyerton
		250	CAN	SAC	KBS World Radio
		200	KRE	KUJ	Voice of Korea
		100	SNG	SNG	BBC World Service
		500	TUR	CAK	Voice of Turkey (VOT)
		250	UAE	DHA	FEBA Radio
		250	UAE	DHA	Polish Radio (Ext Sce)
9655		100	ALS	NLS	KNLS International
		100	CHN	KAS	China Radio Int. (CRI)
9655		100	CHN	KUN	China Radio Int. (CRI)
		150	CHN	KUN	China Radio Int. (CRI)
		500	CHN	KUN	China Radio Int. (CRI)
		100	CHN	dom	CNR1 VO China, Lingshi
		100	CHN	dom	CNR8 VO Minorities, Lingshi
		250	CLN	TRM	Deutsche Welle
		100	GUM	SDA	AWR Guam (KSDA)
		300	ROU	GAL	Radio Romania International
		250	RRW	KIG	Deutsche Welle
		500	TUR	EMR	Voice of Turkey (VOT)
9660		250	AFS	MEY	Family Radio Worldwide
		10	AUS	BRN	Radio Australia
		150	CHN	XIA	Voice of Russia (VOR)
		500	CVA	SMG	Vatican Radio
		300	G	SKN	Family Radio Worldwide
		250	MDG	MDC	Vatican Radio
		300	ROU	TIG	Radio Romania International
		100	TWN	KOU	Radio Taiwan International
9665		250	B	BRA	China Radio Int. (CRI)
		10	B	dom	Voz Missionária, Camboriú
		100	CHN	KAS	China Radio Int. (CRI)
		50	CHN	dom	CNR5 VO Zhonghua, Beijing
		250	E	NOB	R. Exterior de España (REE)
			KRE	dom	KCBS, Pyongyang
		250	MNG	UBA	Voice of Mongolia
		300	ROU	TIG	Radio Romania International
		250	TUR	CAK	Voice of Turkey (VOT)
9670		500	CHN	KUN	China Radio Int. (CRI)
		100	D	LAM	BBG - VO America (VOA)
		300	G	SKN	Radio Canada Int. (RCI)
		300	J	YAM	Radio Japan (NHK World)
		250	PAK	ISL	Radio Pakistan
		250	PHL	PUG	Radio Veritas Asia
		150	SVK	RSO	Miraya FM
		250	TJK	DSB	BBG - R. Free Asia (RFA)
9675		500	ARS	RIY	BSKSA
		10	B	dom	R. Canção Nova, Cachoeira Paulista
		150	CHN	KUN	China Radio Int. (CRI)
		100	CHN	dom	CNR1 VO China, Beijing
		100	CTR	CRI	R. Exterior de España (REE)
		500	IRN	SIR	VO the Islamic Rep. of Iran
		5	PRU	dom	Pacífico R, Lima
	+	250	RUS	MSK	Voice of Russia (VOR)
9677		5	AZE	dom	Ädälän Säsi Radiosu, Stepanakert
9680		100	ALS	NLS	KNLS International
		250	CLN	IRA	BBG - RFE/RL
		100	D	LAM	BBG - RFE/RL
		125	D	NAU	BBG - VO America (VOA)
		250	INS	dom	RRI, Jakarta
		250	MDG	MDC	BBG - VO America (VOA)
		100	TWN	HUW	Radio Taiwan International
		100	USA	YFR	Family Radio Worldwide
		100	USA	YFR	Radio Taiwan International
9685	†	8	B	dom	R. Gazeta, São Paulo
		150	CHN	BJI	China Radio Int. (CRI)
		500	CHN	KAS	China Radio Int. (CRI)
		500	CHN	KUN	China Radio Int. (CRI)
		500	CHN	URU	China Radio Int. (CRI)
		500	CHN	XIA	China Radio Int. (CRI)
		50	CHN	dom	CNR5 VO Zhonghua, Beijing
		100	D	WER	TWR Europe
		250	IRN	AHW	VO the Islamic Rep. of Iran
		100	SNG	SNG	BBC World Service
		100	TWN	KOU	Radio Taiwan International
		250	UAE	DHA	Family Radio Worldwide
9690		150	CHN	KUN	China Radio Int. (CRI)
		100	CHN	dom	CNR8 VO Minorities, Beijing
		100	D	LAM	BBG - VO America (VOA)
		350	E	NOB	China Radio Int. (CRI)
		250	E	NOB	R. Exterior de España (REE)
		250	G	WOF	Deutsche Welle
		500	IND	BGL	All India Radio (AIR)
		500	IRN	KAM	VO the Islamic Rep. of Iran
		100	LTU	SIT	BBG - R. Free Asia (RFA)
		250	NIG	IKO	Voice of Nigeria
		300	ROU	GAL	Radio Romania International
9690		300	ROU	TIG	Radio Romania International
		500	UAE	DHA	BBG - R. Free Asia (RFA)
		100	USA	YFR	Family Radio Worldwide
9695	†	8	B	dom	R. Rio Mar, Manaus
		500	CHN	JIN	China Radio Int. (CRI)
		500	CHN	KUN	China Radio Int. (CRI)
		250	CVA	SMG	Vatican Radio
		100	D	LAM	BBG - RFE/RL
		500	D	WER	Family Radio Worldwide
		300	J	YAM	Radio Japan (NHK World)
		250	RUS	SAM	Voice of Russia (VOR)
9700	†	50	BUL	SOF	Radio Bulgaria
		500	CHN	KAS	China Radio Int. (CRI)
		500	TUR	EMR	Voice of Turkey (VOT)

kHz	N	kW	Ctry	Site	Station, location
9705		500	CHN	KAS	China Radio Int. (CRI)
		15	CHN	dom	VO Pujiang, Shanghai
		50	CHN	dom	Xinjiang, Ürümqi
		100	ETH	dom	R. Ethiopia, Geja
		250	KOR	KIM	KBS World Radio
		100	MRA	SAI	BBG - VO America (VOA)
		40	NGR	dom	ORTN La Voix du Sahel, Niamey
		100	USA	YFR	Family Radio Worldwide
9710		100	AUS	SHP	Radio Australia
		500	CHN	KAS	China Radio Int. (CRI)
		100	CHN	dom	CNR1 VO China, Shijiazhuang
	±	10	ERI	dom	VO the Broad Masses 2nd prgr, Asmara
		500	IRN	KAM	VO the Islamic Rep. of Iran
9715		250	ARS	RIY	BSKSA
		100	D	BIB	BBG - RFE/RL
		300	G	WOF	Deutsche Welle
		500	IRN	KAM	VO the Islamic Rep. of Iran
		100	MRA	SAI	BBG - RFE/RL
		250	POR	SIN	Deutsche Welle
		250	THA	UDO	BBG - VO America (VOA)
		100	USA	YFR	Family Radio Worldwide
9720		500	CHN	URU	China Radio Int. (CRI)
		500	CHN	XIA	China Radio Int. (CRI)
		150	CHN	dom	CNR2 China Business R, Xi'an
		100	GUM	SDA	AWR Guam (KSDA)
		250	PHL	PHT	R. Netherlands Worldwide
		250	PHL	PUG	Radio Veritas Asia
		1	PRU	dom	R. Victoria, Lima
		250	RRW	KIG	Deutsche Welle
		250	RUS	NVS	Family Radio Worldwide
		120	RUS	KLG	Voice of Russia (VOR)
		250	THA	UDO	R. Thailand World Service
9725		250	PHL	PHT	BBG - VO America (VOA)
		50	PHL	PHT	BBG - VO America (VOA)
		250	THA	UDO	R. Thailand World Service
		250	TUN	SFA	Radio Tunisienne
		500	UAE	DHA	BBG - R. Free Asia (RFA)
		250	USA	GRV	BBG - VO America (VOA)
9730		100	CHN	KAS	China Radio Int. (CRI)
		100	CHN	KUN	China Radio Int. (CRI)
		500	CHN	BEI	China Radio Int. (CRI)
		500	CHN	KAS	China Radio Int. (CRI)
		200	KRE	KUJ	Voice of Korea
		100	PHL	BOC	FEBC Philippines
		250	SNG	SNG	BBC World Service
		100	VTN	VNI	Voice of Vietnam (VOV)
9731		50	BRM	dom	Myanma R, Yangon
9735		250	RRW	KIG	Deutsche Welle
		250	TWN	TNN	Radio Taiwan International
9740		500	CHN	KAS	China Radio Int. (CRI)
		100	CHN	KAS	Radio Canada Int. (RCI)
		500	IRN	KAM	VO the Islamic Rep. of Iran
		250	SNG	SNG	BBC World Service
9743	†	10	INS	dom	RRI, Sorong
9745		10	BHR	ABH	Radio Bahrain
		500	CHN	URU	China Radio Int. (CRI)
		250	CLN	IRA	BBG - VO America (VOA)
		250	EGY	ABZ	Radio Cairo
9750		50	CHN	dom	Nei Menggu, Hohhot
		500	F	ISS	R. Netherlands Worldwide
		500	G	RMP	BBC World Service
		300	J	YAM	Radio Japan (NHK World)
9750		300	KWT	KBD	Radio Kuwait
		100	MLA	KAJ	Voice of Malaysia
9755		250	CAN	SAC	Radio Canada Int. (RCI)
		100	CHN	dom	CNR2 China Business R, Xi'an
		100	CVA	SMG	Vatican Radio
		500	CVA	SMG	Vatican Radio
		250	RRW	KIG	Deutsche Welle
		250	THA	UDO	BBG - VO America (VOA)
9760		500	CHN	KUN	China Radio Int. (CRI)
		250	CYP	CYP	Cyprus Broadcasting Corp.
		100	D	LAM	BBG - R. Farda
		100	D	LAM	BBG - RFE/RL
		100	D	BIB	BBG - VO America (VOA)
		100	F	ISS	Radio Réveil

kHz	N	kW	Ctry	Site	Station, location
	+	90	G	WOF	KBS World Radio
	+	100	G	WOF	Radio Japan (NHK World)
		50	J	dom	R. Nikkei 2, Tokyo
		250	PHL	PHT	BBG - VO America (VOA)
		250	THA	UDO	BBG - VO America (VOA)
9765		100	CHN	NNN	China Radio Int. (CRI)
		500	CHN	KUN	China Radio Int. (CRI)
		500	CHN	XIA	China Radio Int. (CRI)
		100	CTR	CRI	R. Exterior de España (REE)
		500	F	ISS	Radio France Int. (RFI)
		100	NZL	RAN	R. New Zealand Int. (RNZI)
		300	ROU	GAL	Radio Romania International
9770		250	AFS	MEY	Family Radio Worldwide
		300	AUT	MOS	Adventist World R. (AWR)
		500	CHN	KAS	China Radio Int. (CRI)
		100	CHN	KAS	Radio Canada Int. (RCI)
		10	CLN	EKA	Sri Lanka Broadcasting Corp.
		250	CUB	HAB	Radio Habana Cuba
		250	D	WER	BBG - VOA Ashna Radio
		500	F	ISS	Radio Japan (NHK World)
		100	KOR	KIM	KBS World Radio
		100	LTU	SIT	VO the Islamic Rep. of Iran
9774		10	TWN	dom	Fu Hsing Broadcasting Station, Kuanyin
9775		150	CHN	dom	CNR2 China Business R, Beijing
		100	STP	SAO	BBG - VO America (VOA)
9780		250	AFS	MEY	BBG - VO America (VOA)
		100	BOT	BOT	BBG - VO America (VOA)
		50	CHL	SGO	CVC La Voz
		15	CHN	dom	Qinghai PBS, Xining
		250	CVA	SMG	BBG - Afia Darfur Radio
		100	D	BIB	BBG - RFE/RL
	+	100	E	NOB	R. Exterior de España (REE)
		250	MRA	TIN	BBG - R. Free Asia (RFA)
		100	STP	SAO	BBG - VO America (VOA)
		100	TWN	TNN	Furustato/Ilbon-E Baram
		100	TWN	KOU	Radio Taiwan International
	†	50	YEM	SAN	Republic of Yemen Radio
9785		250	CAN	SAC	Radio Canada Int. (RCI)
		150	CHN	KUN	China Radio Int. (CRI)
		500	CHN	JIN	China Radio Int. (CRI)
		100	CHN	dom	CNR8 VO Minorities, Beijing
		250	CLN	TRM	Deutsche Welle
		100	D	LAM	BBG - R. Farda
		250	D	NAU	BBG - VO America (VOA)
		250	PHL	PHT	BBG - VO America (VOA)
		500	TUR	EMR	Voice of Turkey (VOT)
9790		250	CUB	HAB	China Radio Int. (CRI)
		100	CZE	LIT	Radio Prague
		250	D	WER	Radio Japan (NHK World)
		500	F	ISS	Radio France Int. (RFI)
		100	GUM	SDA	AWR Guam (KSDA)
		500	IRN	KAM	VO the Islamic Rep. of Iran
		100	MRA	SAI	BBG - R. Free Asia (RFA)
		300	ROU	TIG	Radio Romania International
		250	THA	UDO	BBG - RFE/RL
9795		250	CAN	SAC	Radio Japan (NHK World)
		500	CHN	URU	China Radio Int. (CRI)
		500	F	ISS	Radio France Int. (RFI)
		250	MRA	TIN	R. Netherlands Worldwide
		100	PHL	BOC	FEBC Philippines
		50	PHL	IBA	FEBC Philippines
		300	POR	LIS	RDP Internacional
9795		250	SNG	SNG	R. Netherlands Worldwide
9800	+	50	BUL	SOF	Radio Bulgaria
		250	CAN	SAC	Radio Canada Int. (RCI)
	+	70	CAN	SAC	Radio Canada Int. (RCI)
	+	70	CAN	SAC	Vatican Radio
		500	CHN	KAS	China Radio Int. (CRI)
		500	D	NAU	Family Radio Worldwide
		500	D	WER	Sawtu Linjiila
		500	F	ISS	Radio France Int. (RFI)
		100	F	MCO	TWR Europe
		100	GUM	SDA	AWR Guam (KSDA)
		250	POR	SIN	Deutsche Welle
		250	RRW	KIG	Deutsche Welle
		250	RUS	IRK	Voice of Russia (VOR)

kHz	N	kW	Ctry	Site	Station, location
9805		300	AUT	MOS	Adventist World R. (AWR)
		100	D	WER	Adventist World R. (AWR)
		250	D	WER	BBG - Afia Darfur Radio
		100	D	LAM	BBG - RFE/RL
		500	F	ISS	Radio France Int. (RFI)
		100	KOR	KIM	KBS World Radio
	+	300	ROU	TIG	Radio Romania International
		100	STP	SAO	BBG - Afia Darfur Radio
9810		250	ATN	BON	R. Netherlands Worldwide
		100	CHN	dom	CNR1 VO China, Nanning
		100	CHN	dom	CNR2 China Business R, Xi'an
		250	CYP	CYP	BBC World Service
		300	G	SKN	BBC World Service
		100	GUM	SDA	AWR Guam (KSDA)
		250	IND	DEL	All India Radio (AIR)
		100	MRA	SAI	BBG - VO America (VOA)
9815		100	BOT	BOT	BBG - VO America (VOA)
		250	CYP	CYP	BBC World Service
		250	D	WER	BBG - Afia Darfur Radio
		100	MRA	SAI	BBG - R. Free Asia (RFA)
		300	POR	LIS	RDP Internacional
	+	80	POR	SIN	RDP Internacional
9820		10	B	dom	R. Nove de Julho, São Paulo
	†	15	CHN	NNN	Beibu Bay Radio
		150	CHN	dom	CNR2 China Business R, Xi'an
		250	CUB	HAB	Radio Habana Cuba
		250	IND	dom	All India R, Panaji
		500	IRN	KAM	VO the Islamic Rep. of Iran
		500	TJK	DSB	Voice of Russia (VOR)
		250	TUR	CAK	Voice of Turkey (VOT)
		250	UAE	DHA	Athmeeya Yatra Radio
9825		100	CAN	SAC	BBG - R. Martí
		100	CHN	KAS	China Radio Int. (CRI)
		125	D	WER	BBG - VO America (VOA)
		250	D	WER	BBG - VO America (VOA)
		250	GUF	GUF	Radio France Int. (RFI)
		300	J	YAM	Radio Japan (NHK World)
		250	MRA	TIN	BBG - R. Free Asia (RFA)
		100	MRA	SAI	BBG - VO America (VOA)
		250	PHL	PHT	BBG - VO America (VOA)
		500	USA	GRV	BBG - R. Martí
9830		300	AUT	MOS	Adventist World R. (AWR)
		100	CHN	dom	CNR1 VO China, Beijing
		250	CVA	SMG	Radio Canada Int. (RCI)
		500	D	WER	R. Netherlands Worldwide
		500	IRN	KAM	VO the Islamic Rep. of Iran
	†	500	JOR	AKA	Radio Jordan
9835		500	AFS	MEY	Radio France Int. (RFI)
		50	CHL	SGO	HCJB La Voz de los Andes
		100	D	LAM	BBG - R. Free Asia (RFA)
		100	D	LAM	BBG - RFE/RL
		50	IND	dom	All India R, Delhi
		250	IND	DEL	All India Radio (AIR)
		300	J	YAM	Radio Japan (NHK World)
9840		300	AUT	MOS	Radio Ö1 International
		250	D	WER	Family Radio Worldwide
		500	F	ISS	Radio Taiwan International
		300	J	YAM	Radio Japan (NHK World)
		250	RUS	MSK	Radio Rossii
		250	RUS	MSK	Voice of Russia (VOR)
		250	RUS	PPK	Voice of Russia (VOR)
		250	TUR	CAK	Voice of Turkey (VOT)
9840		250	UAE	DHA	Sudan Radio Service
		250	USA	HRI	WHRI - World Harvest R. Int.
		100	VTN	VNI	Voice of Vietnam (VOV)
9845		100	CHN	dom	CNR1 VO China, Beijing
		250	CLN	IRA	BBG - Afia Darfur Radio
		250	PHL	PHT	BBG - VO America (VOA)
9850		15	CHN	dom	Qinghai, Xining
		250	CVA	SMG	Vatican Radio
		100	D	LAM	BBG - R. Farda
		250	D	WER	BBG - R. Farda
		100	D	LAM	BBG - RFE/RL
		250	D	WER	Family Radio Worldwide
		500	D	WER	Radio Japan (NHK World)
		125	G	WOF	Polish Radio (Ext Sce)

kHz	N	kW	Ctry	Site	Station, location
		500	IRN	SIR	VO the Islamic Rep. of Iran
		200	KRE	KUJ	Voice of Korea
		100	RUS	ARM	Family Radio Worldwide
		250	UAE	DHA	FEBA Radio
		50	VTN	dom	VO Vietnam Minorities prgr, Xuan Mai
9855		500	CHN	BEI	China Radio Int. (CRI)
		100	CZE	LIT	Radio Prague
		250	EGY	ABZ	Radio Cairo
		250	MRA	TIN	BBG - VO America (VOA)
		250	POR	SIN	Deutsche Welle
		100	POR	LIS	RDP Internacional
		250	RUS	VLD	Voice of Russia (VOR)
		250	UAE	DHA	Family Radio Worldwide
9860		100	AFS	MEY	BBC World Service
		500	CHN	BEI	China Radio Int. (CRI)
		500	CHN	JIN	China Radio Int. (CRI)
		500	CHN	KAS	China Radio Int. (CRI)
		100	CHN	dom	CNR1 VO China, Beijing
		250	CVA	SMG	BBG - VO America (VOA)
		500	D	WER	The Overcomer Ministry
		250	KWT	KWT	BBG - VO America (VOA)
		300	POR	LIS	RDP Internacional
		250	RUS	SAM	Na volne Tatarstana
9865		250	ATN	BON	R. Netherlands Worldwide
		500	CHN	KUN	China Radio Int. (CRI)
		500	IRN	KAM	VO the Islamic Rep. of Iran
		250	PHL	PUG	Radio Veritas Asia
		250	RUS	SAM	Voice of Russia (VOR)
		500	RUS	SAM	Voice of Russia (VOR)
		250	SNG	SNG	Deutsche Welle
		250	USA	HRI	Deutsche Welle
9870		500	ARS	RIY	BSKSA
		500	CHN	XIA	China Radio Int. (CRI)
		500	IND	dom	All India R, Bengaluru (Vividh Bharati)
	+	50	NZL	RAN	R. New Zealand Int. (RNZI)
9875		250	GUF	GUF	Voice of Russia (VOR)
		300	J	YAM	Radio Japan (NHK World)
		100	LTU	SIT	BBG - R. Free Asia (RFA)
		80	PLW	HBN	BBG - R. Free Asia (RFA)
		100	ROU	TIG	Radio Romania International
		250	THA	UDO	BBG - VO America (VOA)
		50	VTN	dom	VO Vietnam 2nd prgr, Hanoi
9880		100	CHN	KUN	China Radio Int. (CRI)
		150	CHN	BEI	China Radio Int. (CRI)
		150	CHN	BJI	China Radio Int. (CRI)
		100	CHN	KUN	Radio Canada Int. (RCI)
		100	CZE	LIT	Radio Prague
		250	RUS	PPK	Voice of Russia (VOR)
9885		100	AFS	MEY	BBG - VO America (VOA)
		500	ARS	RIY	BSKSA
		100	BOT	BOT	BBG - VO America (VOA)
		500	G	RMP	Family Radio Worldwide
		100	IRN	KAM	VO the Islamic Rep. of Iran
		500	IRN	SIR	VO the Islamic Rep. of Iran
		100	STP	SAO	BBG - VO America (VOA)
		250	UAE	DHA	Family Radio Worldwide
		250	USA	GRV	BBG - VO America (VOA)
9890		100	CHN	dom	CNR1 VO China, Lingshi
		100	CHN	dom	CNR8 VO Minorities, Lingshi
	+	50	NZL	RAN	R. New Zealand Int. (RNZI)
9895		100	ALB	SHI	RadioTirana
9895		250	CLN	TRM	R. Netherlands Worldwide
		250	CVA	SMG	R. Netherlands Worldwide
		500	D	NAU	R. Netherlands Worldwide
		500	D	WER	R. Netherlands Worldwide
		500	IRN	ZAH	VO the Islamic Rep. of Iran
		250	MDA	KCH	R. Netherlands Worldwide
		250	UAE	DHA	Family Radio Worldwide
		250	USA	HRI	WHRI - World Harvest R. Int.
9900		300	ARM	ERV	FEBA Radio
		100	CHN	dom	CNR1 VO China, Beijing
		250	EGY	ABZ	Radio Cairo
		500	IRN	KAM	VO the Islamic Rep. of Iran
		250	RUS	SAM	Voice of Russia (VOR)
9905		250	IND	ALG	All India Radio (AIR)

kHz	N	kW	Ctry	Site	Station, location
		500	IRN	SIR	VO the Islamic Rep. of Iran
		250	MRA	TIN	BBG - R. Free Asia (RFA)
		80	PLW	HBN	BBG - R. Free Asia (RFA)
9910		100	GUM	TWR	TWR Asia (KTWR)
		250	IND	ALG	All India Radio (AIR)
		250	IND	DEL	All India Radio (AIR)
9915		250	ASC	ASC	BBC World Service
		300	AUT	MOS	BBC World Service
		250	CYP	CYP	BBC World Service
		300	CYP	CYP	BBC World Service
		250	EGY	ABS	Radio Cairo
		500	G	RMP	BBC World Service
		500	IRN	SIR	VO the Islamic Rep. of Iran
9920		100	GUM	SDA	AWR Guam (KSDA)
		100	GUM	TWR	TWR Asia (KTWR)
		100	PHL	BOC	FEBC Philippines
		100	TWN	TAI	Family Radio Worldwide
9925		100	D	WER	Bible Voice Broadcasting (BVB)
9930		250	CLN	IRA	BBG - VO America (VOA)
		100	PLW	HBN	Radio Hoa-Mai
		100	PLW	HBN	T8WH - World Harvest Radio Int
		100	USA	YFR	Family Radio Worldwide
9935		100	GRC	AVL	RS Makedonias (ERT3)
9940		100	IRN	KAM	VO the Islamic Rep. of Iran
		100	PHL	BOC	FEBC Philippines
9945		250	CLN	IRA	BBG - VO America (VOA)
		100	PLW	HBN	IBRA Radio
		100	TWN	TAI	Family Radio Worldwide
9950		250	IND	DEL	All India Radio (AIR)
	+	50	IND	DEL	All India Radio (AIR)
		100	PLW	HBN	Furustato/Ilbon-E Baram
		100	TWN	TAI	Furustato/Ilbon-E Baram
9955		100	PLW	HBN	T8WH - World Harvest Radio Int
		250	TWN	TNN	Family Radio Worldwide
		100	TWN	RMI	Radio France Int. (RFI)
		50	USA	RMI	WRMI - R. Miami Int.
9960		500	AFS	MEY	Bar-Kulan Radio
		250	MRA	TIN	BBG - VO America (VOA)
		100	TWN	TAI	Family Radio Worldwide
9965	†	-	CTR	—	Radio República
		500	ARM	ERV	Voice of Russia (VOR)
		100	PLW	HBN	Furustato/Ilbon-E Baram
		100	PLW	HBN	Radio Australia
		100	PLW	HBN	T8WH - World Harvest Radio Int
9975		100	GUM	TWR	TWR Asia (KTWR)
		200	KRE	KUJ	Voice of Korea
		250	KWT	KWT	BBG - VOA Ashna Radio
		100	PLW	HBN	Furustato/Ilbon-E Baram
		50	USA	VOH	KVOH - La Voz de Restauración
		100	UZB	TAC	CVC The Voice Asia
9980		100	USA	WCR	WWCR
9985		250	ISR	ISR	Kol Israel
		100	USA	YFR	Family Radio Worldwide
		100	USA	WCR	WWCR
9990		250	EGY	ABS	Radio Cairo
		200	KRE	KUJ	Voice of Korea
		100	MRA	SAI	BBG - R. Free Asia (RFA)
		250	THA	UDO	BBG - R. Free Afghanistan
	±	100	TJK	DSB	Sound of Hope Radio Int.
		100	USA	TWW	WTWW
9996		5	RUS	STF	RWM, Moscow
10000		2	ARG	STF	LOL SHN, Buenos Aires
10000			B	STF	Observatório Nacional, Rio de Janeiro
		20	CHN	STF	BPM, Kinshan
		10	HWA	STF	WWVH, Kauai
		10	USA	STF	WWV, Ft. Collins
10500	±	1	TWN	TAI	Sound of Hope Radio Int.
10970	±	1	TWN	TAI	Sound of Hope Radio Int.
11092		1	SHN	dom	R. St. Helena
11100	±	1	TWN	TAI	Sound of Hope Radio Int.
11250	±	1	TWN	TAI	Sound of Hope Radio Int.
11510		250	EGY	ABZ	Radio Cairo
		250	PAK	ISL	Radio Pakistan
11515		300	ARM	ERV	Democratic VO Burma
		250	MDG	MDC	Radio Dabanga
11520		100	TWN	PAO	Family Radio Worldwide

kHz	N	kW	Ctry	Site	Station, location
		100	TWN	PAO	Radio Taiwan International
		250	USA	EWN	WEWN - EWTN Shortwave
11530		300	UKR	SMF	Dengê Mezopotamya
		100	USA	YFR	Family Radio Worldwide
11535		200	KRE	KUJ	Voice of Korea
		100	TWN	TAI	Family Radio Worldwide
11545		200	KRE	KUJ	Voice of Korea
11550		100	TWN	TAI	Family Radio Worldwide
		250	TWN	TNN	Radio Australia
		100	TWN	TNN	Radio Taiwan International
		250	TWN	TNN	Radio Taiwan International
		250	USA	EWN	WEWN - EWTN Shortwave
11560		300	ARM	ERV	Radio Free Chosun
		250	PHL	PHT	BBG - VO America (VOA)
		100	TWN	HUW	Family Radio Worldwide
11565		100	TWN	TAI	Family Radio Worldwide
		100	USA	YFR	Family Radio Worldwide
		250	USA	HRI	WHRI - World Harvest R. Int.
11570		100	PAK	ISL	Radio Pakistan
		100	TWN	HUW	Family Radio Worldwide
11575		250	PAK	ISL	Radio Pakistan
11580		100	MRA	FBS	FEBC Saipan (KFBS)
		250	PAK	ISL	Radio Pakistan
		100	USA	YFR	Family Radio Worldwide
11585		250	IND	DEL	All India Radio (AIR)
		250	KWT	KWT	BBG - R. Free Asia (RFA)
11590		250	EGY	ABZ	Radio Cairo
		100	GUM	TWR	TWR Asia (KTWR)
		250	KWT	KWT	BBG - R. Free Asia (RFA)
11595		250	ISR	ISR	Kol Israel
11600		150	CHN	BJI	China Radio Int. (CRI)
		100	CZE	LIT	Radio Prague
		500	IRN	KAM	VO the Islamic Rep. of Iran
		250	RUS	MSK	Voice of Russia (VOR)
		150	SVK	RSO	Radio Slovakia Int.
11605		100	AFS	MEY	Radio France Int. (RFI)
		300	G	SKN	BBG - RFE/RL
		250	GUF	GUF	Voice of Russia (VOR)
		250	MDG	MDC	Deutsche Welle
		250	RRW	KIG	Deutsche Welle
		250	TWN	TAI	BBG - R. Free Asia (RFA)
		250	TWN	TNN	Radio Taiwan International
11610		150	CHN	KUN	China Radio Int. (CRI)
		500	CHN	URU	China Radio Int. (CRI)
		150	CHN	dom	CNR2 China Business R, Beijing
		500	D	WER	Family Radio Worldwide
		50	MDG	MDC	Radio VOP
		250	RUS	SAM	Na volne Tatarstana
		150	SVK	RSO	Radio Slovakia Int.
11615		250	AFS	MEY	R. Netherlands Worldwide
		250	ASC	ASC	Family Radio Worldwide
		250	CVA	SMG	BBG - Afia Darfur Radio
		500	D	WER	Radio Dabanga
		250	MDG	MDC	R. Netherlands Worldwide
11620		500	CHN	BEI	China Radio Int. (CRI)
		50	CHN	dom	CNR5 VO Zhonghua, Beijing
		250	CYP	CYP	BBC World Service
		100	GUM	TWR	TWR Asia (KTWR)
		250	IND	dom	All India R, Delhi
		250	IND	ALG	All India Radio (AIR)
		250	IND	DEL	All India Radio (AIR)
		500	IND	BGL	All India Radio (AIR)
11620		500	IND	DEL	All India Radio (AIR)
		500	TUR	EMR	Voice of Turkey (VOT)
11625		100	CVA	SMG	Vatican Radio
		250	CVA	SMG	Vatican Radio
		250	E	NOB	R. Exterior de España (REE)
		100	TWN	PAO	Radio Taiwan International
11630		100	CHN	dom	CNR1 VO China, Lingshi
		100	CHN	dom	CNR8 VO Minorities, Lingshi
		250	KWT	KBD	Radio Kuwait
		1000	TJK	DSB	Voice of Russia (VOR)
		100	TWN	PAO	Family Radio Worldwide
11635		100	ALB	SHI	RadioTirana
		100	MRA	SAI	BBG - VO America (VOA)
		300	POR	LIS	RDP Internacional

kHz	N	kW	Ctry	Site	Station, location
	+	250	RUS	MSK	Voice of Russia (VOR)
		250	THA	UDO	BBG - VO America (VOA)
		100	TWN	PAO	Radio Taiwan International
11640		100	CHN	KAS	China Radio Int. (CRI)
		100	CHN	KUN	China Radio Int. (CRI)
		150	CHN	KUN	China Radio Int. (CRI)
		500	CHN	BEI	China Radio Int. (CRI)
		500	CHN	JIN	China Radio Int. (CRI)
		500	CHN	XIA	China Radio Int. (CRI)
		500	IRN	KAM	VO the Islamic Rep. of Iran
		100	MLI	BKO	China Radio Int. (CRI)
		100	TWN	KOU	Radio Taiwan International
11645		250	CLN	TRM	Deutsche Welle
		250	D	WER	Athmeeya Yatra Radio
		100	GRC	AVL	Voice of Greece (ERA5)
		250	IND	DEL	All India Radio (AIR)
		250	RRW	KIG	Deutsche Welle
11650		100	AUS	SHP	Radio Australia
		100	CHN	KUN	China Radio Int. (CRI)
		150	CHN	KUN	China Radio Int. (CRI)
		500	CHN	BEI	China Radio Int. (CRI)
		500	CHN	URU	China Radio Int. (CRI)
		100	MRA	FBS	FEBC Saipan (KFBS)
11655		250	CAN	SAC	Radio Japan (NHK World)
		100	GUM	SDA	AWR Guam (KSDA)
		500	IRN	KAM	VO the Islamic Rep. of Iran
		250	MDG	MDC	R. Netherlands Worldwide
		250	PHL	PHT	BBG - VO America (VOA)
		300	POR	LIS	RDP Internacional
		100	TWN	TNN	Radio Taiwan International
		300	UKR	LVI	Voice of Russia (VOR)
11660		100	AUS	SHP	Radio Australia
		150	CHN	dom	CNR2 China Business R, Xi'an
		500	TJK	DSB	Voice of Russia (VOR)
11665		250	ASC	ASC	Deutsche Welle
		250	ASC	ASC	Family Radio Worldwide
		500	CHN	URU	China Radio Int. (CRI)
		100	D	WER	Family Radio Worldwide
		500	G	RMP	Family Radio Worldwide
		300	J	YAM	Radio Japan (NHK World)
		300	POR	LIS	RDP Internacional
		100	TWN	TNN	Radio France Int. (RFI)
		300	TWN	TSH	Radio Taiwan International
		100	USA	YFR	Family Radio Worldwide
11670		150	CHN	dom	CNR2 China Business R, Beijing
		100	CUB	HAB	R. Nacional de Venezuela
		500	F	ISS	Radio France Int. (RFI)
		100	IRN	KAM	VO the Islamic Rep. of Iran
		500	IRN	SIR	VO the Islamic Rep. of Iran
11675		100	BOT	BOT	BBG - VO America (VOA)
		250	CLN	IRA	BBG - VO America (VOA)
		250	CLN	IRA	BBG - VOA R. Aap Ki Dunyaa
		250	D	WER	Adventist World R. (AWR)
		250	D	WER	BBG - VO America (VOA)
	+	50	NZL	RAN	R. New Zealand Int. (RNZI)
11680		100	CHN	NNN	China Radio Int. (CRI)
		100	CUB	HAB	R. Nacional de Venezuela
		300	CYP	CYP	BBC World Service
		250	E	NOB	R. Exterior de España (REE)
		250	G	RMP	BBC World Service
		100	GUM	TWR	TWR Asia (KTWR)
			KRE	dom	KCBS, Kanggye
11680		250	TUR	CAK	Voice of Turkey (VOT)
11685		100	CHN	dom	CNR11, Xi'an
		100	GUM	SDA	AWR Guam (KSDA)
		500	IRN	KAM	VO the Islamic Rep. of Iran
		100	SNG	SNG	BBC World Service
11690		250	AFS	MEY	Radio Okapi
		500	CHN	XIA	China Radio Int. (CRI)
		100	CUB	HAB	R. Nacional de Venezuela
		250	CUB	HAB	Radio Habana Cuba
		100	D	LAM	BBG - R. Farda
		100	D	WER	Family Radio Worldwide
		500	F	ISS	Radio France Int. (RFI)
		0.5	FIN	VIR	Scandinavian Weekend R.
		100	GUM	SDA	AWR Guam (KSDA)

kHz	N	kW	Ctry	Site	Station, location
		250	RRW	KIG	Deutsche Welle
11695		100	AUS	SHP	Radio Australia
		500	CHN	BEI	China Radio Int. (CRI)
		250	CLN	TRM	Deutsche Welle
		100	D	WER	TWR Europe
		100	GUM	SDA	AWR Guam (KSDA)
		500	IRN	SIR	VO the Islamic Rep. of Iran
		250	PHL	PHT	BBG - VO America (VOA)
		50	PHL	PHT	BBG - VO America (VOA)
		500	UAE	DHA	BBG - R. Free Asia (RFA)
11700		300	BUL	PLD	Radio Bulgaria
		500	CHN	KUN	China Radio Int. (CRI)
		500	F	ISS	Radio France Int. (RFI)
		100	GUM	SDA	AWR Guam (KSDA)
		100	USA	YFR	Family Radio Worldwide
11705		100	CUB	HAB	R. Nacional de Venezuela
		100	D	BIB	BBG - VO America (VOA)
		250	D	NAU	BBG - VO America (VOA)
		500	F	ISS	Radio France Int. (RFI)
		100	STP	SAO	BBG - VO America (VOA)
		250	THA	UDO	BBG - VO America (VOA)
11710		100	ARG	BUE	Rdif. Argentina al Exterior
		500	CHN	URU	China Radio Int. (CRI)
		100	CHN	dom	CNR1 VO China, Beijing
		50	IND	dom	All India R, Delhi
		50	IND	DEL	All India Radio (AIR)
		200	KRE	KUJ	Voice of Korea
		250	PHL	PUG	Radio Veritas Asia
		300	ROU	TIG	Radio Romania International
		300	TWN	TNN	Radio Taiwan International
11715		100	CVA	SMG	Vatican Radio
		250	CVA	SMG	Vatican Radio
		100	D	LAM	BBG - RFE/RL
		250	IND	DEL	All India Radio (AIR)
		250	IND	PAN	All India Radio (AIR)
		500	IRN	SIR	VO the Islamic Rep. of Iran
		300	J	YAM	Radio Japan (NHK World)
		250	TWN	TNN	Radio Taiwan International
		50	USA	JES	KJES Radio
11720		250	AFS	MEY	BBC World Service
		100	CHN	dom	CNR1 VO China, Shijiazhuang
		100	CHN	dom	CNR8 VO Minorities, Shijiazhuang
		0.5	FIN	VIR	Scandinavian Weekend R.
		500	G	RMP	Deutsche Welle
		250	PHL	PHT	BBG - VO America (VOA)
		250	TJK	DSB	BBG - VO America (VOA)
		100	USA	YFR	Family Radio Worldwide
		100	VTN	dom	VO Vietnam 1st prgr, Hanoi
11725		150	ALB	CER	China Radio Int. (CRI)
		10	B	dom	R. Marumby, Curitiba
		250	D	NAU	Adventist World R. (AWR)
		500	F	ISS	Radio France Int. (RFI)
		100	NZL	RAN	R. New Zealand Int. (RNZI)
		250	RRW	KIG	Deutsche Welle
		100	USA	YFR	Family Radio Worldwide
11730		500	CHN	KUN	China Radio Int. (CRI)
		250	CUB	HAB	Radio Habana Cuba
		250	CYP	CYP	BBC World Service
		100	D	LAM	BBG - RFE/RL
		500	G	RMP	BBC World Service
		100	GUM	SDA	AWR Guam (KSDA)
		250	IND	DEL	All India Radio (AIR)
11730		250	KWT	KWT	BBG - RFE/RL
		250	PHL	PHT	Radio Pilipinas Overseas
		250	PHL	PUG	Radio Veritas Asia
		300	ROU	GAL	Radio Romania International
		300	ROU	TIG	Radio Romania International
		250	THA	UDO	R. Thailand World Service
		100	USA	YFR	Family Radio Worldwide
11735		50	B	dom	R. Transmundial, Santa Maria
		500	IND	BGL	All India Radio (AIR)
		200	KRE	KUJ	Voice of Korea
		500	TUR	EMR	Voice of Turkey (VOT)
	‡	50	TZA	dom	VO Tanzania Zanzibar, Dole
11740		100	AFS	MEY	IBRA Radio
		50	CHN	dom	CNR2 China Business R, Xi'an

kHz	N	kW	Ctry	Site	Station, location
		100	CVA	SMG	Vatican Radio
		250	CVA	SMG	Vatican Radio
		250	EGY	ABS	Radio Cairo
		300	G	WOF	Radio Damal
		250	IND	dom	All India R, Panaji
		250	IND	PAN	All India Radio (AIR)
		250	KWT	KWT	BBG - R. Free Asia (RFA)
		250	SNG	SNG	Radio Japan (NHK World)
		100	STP	SAO	BBG - VO America (VOA)
		250	THA	NAK	BBC World Service
		250	UAE	DHA	Family Radio Worldwide
		100	USA	YFR	Family Radio Worldwide
11745		250	TWN	TNN	Radio Australia
11750		250	AFS	MEY	Adventist World R. (AWR)
		100	AFS	MEY	BBG - VO America (VOA)
		150	ALB	CER	China Radio Int. (CRI)
		50	AUS	KNX	HCJB Global Voice Australia
		10	B	dom	Voz Missionária, Camboriú
		100	CHN	dom	CNR1 VO China, Shijiazhuang
		250	CLN	TRM	Sri Lanka Broadcasting Corp.
		100	D	BIB	BBG - R. Farda
		250	EGY	ABZ	Radio Cairo
		250	PHL	PHT	BBG - VO America (VOA)
		200	SNG	SNG	BBC World Service
		250	SNG	SNG	BBC World Service
		100	STP	SAO	BBG - VO America (VOA)
11755		250	AFS	MEY	Adventist World R. (AWR)
		250	E	NOB	R. Exterior de España (REE)
11760		500	CHN	KUN	China Radio Int. (CRI)
		100	CHN	dom	CNR1 VO China, Shijiazhuang
		250	CUB	HAB	Radio Habana Cuba
		250	CYP	CYP	BBC World Service
		100	D	WER	Adventist World R. (AWR)
		500	D	WER	VO Omoro Liberation Front
		300	J	YAM	Radio Japan (NHK World)
		250	OMA	SLA	BBC World Service
		300	TWN	TSH	Radio Australia
		300	TWN	TSH	Sound of Hope Radio Int.
11765		20	B	dom	Super R. Deus é Amor, Curitiba
		100	D	LAM	BBG - VO America (VOA)
		250	E	NOB	R. Exterior de España (REE)
		100	TWN	TNN	Radio Taiwan International
		100	TWN	TSH	Sound of Hope Radio Int.
11770		250	AFS	MEY	BBC World Service
		100	AFS	MEY	Sudan Radio Service
		250	ASC	ASC	BBC World Service
		100	CHN	KAS	China Radio Int. (CRI)
		500	CHN	BEI	China Radio Int. (CRI)
		50	CHN	dom	Xinjiang, Ürümqi
		250	CUB	HAB	Radio Habana Cuba
		100	GUM	SDA	AWR Guam (KSDA)
		250	NIG	IKO	Voice of Nigeria
11775		100	AIA	AIA	University Network (DGS)
		300	AUT	MOS	Adventist World R. (AWR)
		100	D	BIB	BBG - VO America (VOA)
		125	D	NAU	EPDP Radio/VO Eritrean People
		250	IND	PAN	All India Radio (AIR)
		500	IRN	KAM	VO the Islamic Rep. of Iran
		250	MRA	TIN	BBG - R. Free Asia (RFA)
11780		250	B	dom	R. Nacional da Amazônia, Brasília
		100	CHN	KUN	China Radio Int. (CRI)
		500	CHN	JIN	China Radio Int. (CRI)
11780		50	CHN	dom	CNR8 VO Minorities, Beijing
		250	CLN	IRA	BBG - VO America (VOA)
		250	E	NOB	R. Exterior de España (REE)
		100	UZB	TAC	Radio Japan (NHK World)
11785		100	AFS	MEY	Sudan Radio Service
		150	ALB	CER	China Radio Int. (CRI)
		250	ARS	RIY	BSKSA
		100	AUT	MOS	Polish Radio (Ext Sce)
‡		8	B	dom	R. Guaíba, Porto Alegre
		100	CHN	KAS	China Radio Int. (CRI)
		500	CHN	KAS	China Radio Int. (CRI)
		300	G	SKN	IBRA Radio
		250	KOR	KIM	Radio Canada Int. (RCI)
		250	SEY	SEY	BBC World Service

kHz	N	kW	Ctry	Site	Station, location
11790		100	AFS	MEY	Radio France Int. (RFI)
		100	CHN	KAS	China Radio Int. (CRI)
		500	CHN	XIA	China Radio Int. (CRI)
		250	D	WER	BBG - RFE/RL
		100	MRA	SAI	BBG - R. Free Asia (RFA)
		300	ROU	TIG	Radio Romania International
11795		250	CAN	SAC	KBS World Radio
		100	CHN	KAS	China Radio Int. (CRI)
		250	CLN	IRA	BBG - R. Free Asia (RFA)
		250	D	WER	Adventist World R. (AWR)
		300	G	WOF	BBC World Service
		250	KWT	KWT	BBG - R. Free Asia (RFA)
		250	MRA	TIN	BBG - R. Free Asia (RFA)
		250	THA	UDO	BBG - RFE/RL
		500	TUR	EMR	Voice of Turkey (VOT)
11800		100	AFS	MEY	Adventist World R. (AWR)
		250	ASC	ASC	BBC World Service
		150	CHN	dom	CNR2 China Business R, Beijing
11805†		10	B	dom	Super R. Deus é Amor, Rio de Janeiro
		300	CYP	CYP	BBC World Service
		100	D	BIB	BBG - RFE/RL
		100	D	BIB	BBG - VO America (VOA)
		250	PHL	PHT	BBG - VO America (VOA)
		250	UAE	DHA	Sudan Radio Service
		100	UZB	TAC	CVC The Voice Asia
11810		250	ASC	ASC	BBC World Service
		50	CHN	dom	CNR8 VO Minorities, Beijing
		100	F	ISS	Voice of Oromo Liberation
		250	KOR	KIM	KBS World Radio
11815		8	B	dom	R. Brasil Central, Goiânia
		50	CHN	dom	CNR8 VO Minorities, Beijing
		100	CTR	CRI	R. Exterior de España (REE)
		300	J	YAM	Radio Japan (NHK World)
		250	TUR	CAK	Voice of Turkey (VOT)
11820		500	ARS	RIY	BSKSA
		500	CHN	XIA	China Radio Int. (CRI)
		300	CYP	CYP	BBC World Service
		100	D	LAM	BBG - VO America (VOA)
		500	D	WER	Family Radio Worldwide
		100	PHL	BOC	FEBC Philippines
11825		100	GUM	SDA	AWR Guam (KSDA)
		250	MDG	MDC	BBG - VO America (VOA)
		300	ROU	TIG	Radio Romania International
		100	USA	YFR	Family Radio Worldwide
11830		250	AFS	MEY	Adventist World R. (AWR)
		250	AFS	MEY	Radio France Int. (RFI)
		10	B	dom	R. Daquí, Goiânia
		250	D	NAU	Voice of Ethiopian Unity
		500	F	ISS	Radio France Int. (RFI)
		500	G	RMP	BBC World Service
		50	IND	dom	All India R, Delhi
		250	RUS	PPK	Deutsche Welle
		100	USA	YFR	Family Radio Worldwide
11835		150	CHN	dom	CNR2 China Business R, Xi'an
		100	PHL	BOC	FEBC Philippines
		250	TUR	CAK	Voice of Turkey (VOT)
11840		250	CLN	IRA	BBG - VOA Ashna Radio
		100	D	BIB	BBG - VO America (VOA)
		250	D	NAU	Ev. Missions-Gemeinden
		100	GUM	TWR	TWR Asia (KTWR)
		250	IND	DEL	All India Radio (AIR)
		250	PHL	PHT	BBG - VO America (VOA)
11845		250	CAN	SAC	Radio Canada Int. (RCI)
		500	CHN	XIA	China Radio Int. (CRI)
		250	CVA	SMG	Radio Canada Int. (RCI)
		300	G	SKN	Radio Canada Int. (RCI)
11850		250	CVA	SMG	Vatican Radio
		500	F	ISS	Radio Taiwan International
		100	GUM	SDA	AWR Guam (KSDA)
		50	IND	DEL	All India Radio (AIR)
		250	MRA	TIN	BBG - R. Free Asia (RFA)
		100	MRA	FBS	FEBC Saipan (KFBS)
		250	PHL	PHT	BBG - VO America (VOA)
		250	PHL	PUG	Radio Veritas Asia
		100	SNG	SNG	BBC World Service
11855		150	ALB	CER	China Radio Int. (CRI)

kHz	N	kW	Ctry	Site	Station, location
	†	50	ARS	JED	BSKSA
		1	B	dom	R. Aparecida, Aparecida
		100	D	BIB	BBG - VO America (VOA)
		100	GUM	SDA	AWR Guam (KSDA)
		100	SNG	SNG	Deutsche Welle
		250	THA	UDO	BBG - VO America (VOA)
		100	USA	YFR	Family Radio Worldwide
11860		250	ASC	ASC	BBC World Service
		500	CHN	KUN	China Radio Int. (CRI)
		100	CHN	dom	Xizang, Lhasa
		250	CYP	CYP	BBC World Service
		500	F	ISS	Radio France Int. (RFI)
		125	G	WOF	Polish Radio (Ext Sce)
		250	GUF	GUF	BBC World Service
		500	IRN	KAM	VO the Islamic Rep. of Iran
		500	LBY	SAB	Voice of Africa
		250	SEY	SEY	BBC World Service
		250	SNG	SNG	Radio Japan (NHK World)
		100	UZB	TAC	Radio Japan (NHK World)
11865		250	CAN	SAC	Radio Canada Int. (RCI)
		500	G	RMP	Deutsche Welle
		250	POR	SIN	Deutsche Welle
		250	RRW	KIG	Deutsche Welle
		100	TWN	PAO	Family Radio Worldwide
11870		100	CHN	KAS	China Radio Int. (CRI)
		300	G	WOF	BBG - RFE/RL
		100	GUM	SDA	AWR Guam (KSDA)
		100	GUM	TWR	TWR Asia (KTWR)
		250	PHL	PUG	Radio Veritas Asia
		300	ROU	GAL	Radio Romania International
		300	ROU	TIG	Radio Romania International
		250	THA	UDO	R. Thailand World Service
		250	USA	EWN	WEWN - EWTN Shortwave
11875		250	ASC	ASC	Family Radio Worldwide
		500	CHN	BEI	China Radio Int. (CRI)
		500	CHN	KUN	China Radio Int. (CRI)
		100	D	WER	Bible Voice Broadcasting (BVB)
		500	F	ISS	Radio Taiwan International
		500	G	RMP	Cotton Tree News
		250	RRW	KIG	FEBA Radio
		100	TWN	TNN	Radio France Int. (RFI)
		250	TWN	TNN	Radio Taiwan International
		250	UAE	DHA	Deutsche Welle
11880		100	AUS	SHP	Radio Australia
		100	CHN	KAS	China Radio Int. (CRI)
		100	PAK	ISL	Radio Pakistan
		250	PHL	PHT	Radio Pilipinas Overseas
		500	UAE	DHA	BBG - R. Free Asia (RFA)
11885		250	CAN	SAC	China Radio Int. (CRI)
		500	CHN	XIA	China Radio Int. (CRI)
		50	CHN	dom	Xinjiang, Ürümqi
		250	CVA	SMG	BBG - VO America (VOA)
		500	D	NAU	Family Radio Worldwide
	+	100	MLA	KAJ	Voice of Malaysia
		300	POR	LIS	RDP Internacional
		100	TWN	HUW	Radio Taiwan International
		100	USA	YFR	Family Radio Worldwide
		100	USA	YFR	Radio Taiwan International
11890		500	ASC	ASC	BBC World Service
		300	G	SKN	BBC World Service
		250	PHL	PHT	Radio Pilipinas Overseas
		250	SNG	SNG	BBC World Service
11895†		10	B	dom	R. Boa Vontade, Porto Alegre
		250	CLN	IRA	BBG - VOA Deewa Radio
		250	D	WER	Adventist World R. (AWR)
		250	D	WER	Bible Voice Broadcasting (BVB)
		100	E	NOB	R. Exterior de España (REE)
		100	GUM	SDA	AWR Guam (KSDA)
		300	ROU	GAL	Radio Romania International
		250	THA	NAK	BBC World Service
		100	TWN	TAI	Family Radio Worldwide
11900+		50	BUL	SOF	Radio Bulgaria
		150	CHN	BEI	China Radio Int. (CRI)
		500	CHN	KUN	China Radio Int. (CRI)
		100	MRA	SAI	BBG - R. Free Asia (RFA)
		250	MRA	TIN	BBG - R. Free Asia (RFA)

kHz	N	kW	Ctry	Site	Station, location
11905		100	AFS	MEY	BBG - VO America (VOA)
		50	CHN	dom	CNR6 VO Shenzhou, Beijing
		250	CLN	TRM	Sri Lanka Broadcasting Corp.
		35	CLN	EKA	Sri Lanka Broadcasting Corp.
		250	D	WER	BBG - VO America (VOA)
		500	F	ISS	Radio Japan (NHK World)
		125	G	WOF	Polish Radio (Ext Sce)
		300	ROU	GAL	Radio Romania International
		250	UAE	DHA	BBG - VO America (VOA)
11910		500	CHN	BEI	R. Exterior de España (REE)
		100	CVA	SMG	Vatican Radio
		300	J	YAM	Radio Japan (NHK World)
		200	KRE	KUJ	Voice of Korea
11915		500	ARS	RIY	BSKSA
		10	B	dom	R. Gaúcha, Porto Alegre
		100	CHN	dom	CNR2 China Business R, Xi'an
		250	D	WER	Bible Voice Broadcasting (BVB)
		250	SNG	SNG	BBC World Service
		100	TWN	TNN	Radio Taiwan International
		250	TWN	TNN	Radio Taiwan International
11920		150	ALB	CER	China Radio Int. (CRI)
		50	CHL	SGO	HCJB La Voz de los Andes
		250	PHL	PHT	BBG - VO America (VOA)
		250	SNG	SNG	BBC World Service
11925		250	AFS	MEY	Adventist World R. (AWR)
		250	AFS	MEY	BBC World Service
	†	10	B	dom	R. Bandeirantes, São Paulo
		100	CHN	dom	CNR1 VO China, Lingshi
		500	IRN	SIR	VO the Islamic Rep. of Iran
		250	PHL	PHT	BBG - VO America (VOA)
		250	TUR	CAK	Voice of Turkey (VOT)
11930		500	ARS	RIY	BSKSA
		250	PHL	PHT	BBG - VO America (VOA)
		250	USA	GRV	BBG - R. Martí
11935		500	ARS	RIY	BSKSA
		250	ATN	BON	Radio Japan (NHK World)
		100	CHN	KUN	China Radio Int. (CRI)
		50	CHN	dom	CNR5 VO Zhonghua, Beijing
		250	CVA	SMG	R. Netherlands Worldwide
		250	CVA	SMG	Radio Canada Int. (RCI)
		500	D	NAU	Family Radio Worldwide
		250	G	WOF	Radio Canada Int. (RCI)
		100	GUM	SDA	AWR Guam (KSDA)
		100	IND	MUM	All India Radio (AIR)
		250	IRN	AHW	VO the Islamic Rep. of Iran
		250	PHL	PUG	Radio Veritas Asia
11940		250	E	NOB	R. Exterior de España (REE)
		100	GUM	SDA	AWR Guam (KSDA)
		300	ROU	GAL	Radio Romania International
11945		100	AUS	SHP	Radio Australia
		150	CHN	KUN	China Radio Int. (CRI)
		250	CLN	TRM	Deutsche Welle
		500	F	ISS	Radio Japan (NHK World)
		250	PHL	PUG	Radio Veritas Asia
		250	TJK	DSB	BBG - R. Free Asia (RFA)
11950		100	CHN	dom	Xizang, Lhasa
		250	CYP	CYP	BBC World Service
11955		500	CHN	KUN	China Radio Int. (CRI)
		250	D	WER	Adventist World R. (AWR)
		500	D	NAU	Family Radio Worldwide
		500	F	ISS	Family Radio Worldwide
		100	GUM	SDA	AWR Guam (KSDA)
11955		250	PHL	PHT	BBG - RFE/RL
		100	SNG	SNG	BBC World Service
		250	TUR	CAK	Voice of Turkey (VOT)
11960		100	CHN	dom	CNR1 VO China, Beijing
		500	JOR	AKA	Radio Jordan
		300	POR	LIS	RDP Internacional
		300	ROU	TIG	Radio Romania International
11965		300	G	SKN	BBC World Service
		100	GUM	SDA	AWR Guam (KSDA)
		500	IRN	SIR	VO the Islamic Rep. of Iran
		500	LBY	SAB	Voice of Africa
		250	MRA	TIN	BBG - R. Free Asia (RFA)
		250	PHL	PHT	BBG - VO America (VOA)
		250	RUS	IRK	TWR India

kHz	N	kW	Ctry	Site	Station, location
		250	THA	UDO	BBG - VO America (VOA)
		250	TUR	CAK	Voice of Turkey (VOT)
11970		250	CAN	SAC	China Radio Int. (CRI)
		250	CYP	CYP	BBC World Service
		300	ROU	TIG	Radio Romania International
		100	USA	YFR	Family Radio Worldwide
11975		500	CHN	KUN	China Radio Int. (CRI)
		500	CHN	URU	Radio Canada Int. (RCI)
		50	CHN	dom	Xinjiang, Ürümqi
		100	D	WER	Adventist World R. (AWR)
		500	F	ISS	Family Radio Worldwide
		100	MLI	BKO	China Radio Int. (CRI)
		100	MRA	SAI	BBG - R. Free Asia (RFA)
11980		100	CHN	KUN	China Radio Int. (CRI)
		150	CHN	KUN	China Radio Int. (CRI)
		250	D	WER	BBG - RFE/RL
		250	TJK	DSB	BBG - R. Free Asia (RFA)
†		0.3	UKR	dom	Dniprovska Khvylia, Zaporizhia
11985		500	ARM	ERV	Voice of Russia (VOR)
		250	ASC	ASC	FEBA Radio
		100	D	LAM	BBG - RFE/RL
		250	IND	DEL	All India Radio (AIR)
		500	IND	BGL	All India Radio (AIR)
		500	TUR	EMR	Voice of Turkey (VOT)
		100	TWN	HUW	Radio Taiwan International
11990		250	CAN	SAC	Radio Canada Int. (RCI)
		100	CHN	NNN	China Radio Int. (CRI)
		500	CHN	XIA	China Radio Int. (CRI)
		500	IRN	SIR	VO the Islamic Rep. of Iran
		100	MRA	FBS	FEBC Saipan (KFBS)
11995		250	D	WER	Family Radio Worldwide
		500	F	ISS	Radio France Int. (RFI)
		500	GUF	GUF	Radio Taiwan International
		100	GUM	SDA	AWR Guam (KSDA)
		250	SNG	SNG	BBC World Service
12000		0.02	PRG	dom	R. Licemil, Asunción
		100	VTN	VNI	Voice of Vietnam (VOV)
12005		250	D	WER	Athmeeya Yatra Radio
		500	G	RMP	BBC World Service
		250	PHL	PHT	BBG - VO America (VOA)
		250	RRW	KIG	Deutsche Welle
		500	TUN	SFA	Radio Tunisienne
12010		250	CUB	HAB	Radio Habana Cuba
		100	D	WER	Adventist World R. (AWR)
		250	MRA	TIN	BBG - R. Free Asia (RFA)
		100	SNG	SNG	Deutsche Welle
		100	STP	SAO	BBG - VO America (VOA)
12015		250	AFS	MEY	Radio France Int. (RFI)
		300	ASC	ASC	BBC World Service
		100	CHN	KAS	China Radio Int. (CRI)
		250	CLN	IRA	BBG - R. Farda
		100	D	LAM	BBG - RFE/RL
		250	D	WER	BBG - RFE/RL
		500	F	ISS	Family Radio Worldwide
		500	F	ISS	Radio France Int. (RFI)
		200	KRE	KUJ	Voice of Korea
		250	KWT	KWT	BBG - RFE/RL
		250	PHL	PHT	Radio Canada Int. (RCI)
		250	THA	UDO	BBG - RFE/RL
		250	UAE	DHA	BBC World Service
		250	UAE	DHA	Family Radio Worldwide
12020		300	POR	LIS	RDP Internacional
12020		250	THA	UDO	BBG - VOA R. Aap Ki Dunyaa
		100	VTN	VNI	Voice of Vietnam (VOV)
12025		100	D	LAM	BBG - RFE/RL
		100	GUM	SDA	AWR Guam (KSDA)
		250	IND	PAN	All India Radio (AIR)
		500	IRN	KAM	VO the Islamic Rep. of Iran
		250	KWT	KWT	BBG - RFE/RL
		250	RRW	KIG	Deutsche Welle
		250	RUS	SAM	Voice of Russia (VOR)
		250	THA	UDO	BBG - RFE/RL
12030		250	E	NOB	R. Exterior de España (REE)
		250	RUS	PPK	Voice of Russia (VOR)
12035		250	CYP	CYP	BBC World Service
		250	D	WER	Bible Voice Broadcasting (BVB)
		250	E	NOB	R. Exterior de España (REE)
		300	G	WOF	Deutsche Welle
		100	GUM	SDA	AWR Guam (KSDA)
		250	RUS	SAM	Voice of Russia (VOR)
		500	TUR	CAK	Voice of Turkey (VOT)
12040		250	CUB	HAB	Radio Habana Cuba
		300	POR	LIS	RDP Internacional
		250	THA	UDO	R. Thailand World Service
12045		100	CHN	dom	CNR1 VO China, Beijing
		500	D	WER	Radio Japan (NHK World)
		250	G	WOF	IBRA Radio
		100	MRA	SAI	BBG - VO America (VOA)
		250	PHL	PHT	BBG - VO America (VOA)
		250	RRW	KIG	Deutsche Welle
		250	SNG	SNG	Radio Japan (NHK World)
		250	UAE	DHA	FEBA Radio
12050		250	USA	EWN	WEWN - EWTN Shortwave
12055		100	CHN	dom	CNR1 VO China, Lingshi
		100	CHN	dom	CNR8 VO Minorities, Lingshi
		100	D	LAM	BBG - VO America (VOA)
		500	F	ISS	Radio Taiwan International
		100	PHL	BOC	FEBC Philippines
		250	RRW	KIG	Deutsche Welle
		250	RUS	MSK	Voice of Russia (VOR)
12060		250	RUS	MSK	Voice of Russia (VOR)
12065		250	CLN	TRM	Deutsche Welle
		500	G	RMP	BBC World Service
		250	PHL	PHT	R. Netherlands Worldwide
12070		500	CHN	XIA	China Radio Int. (CRI)
		250	CLN	TRM	Deutsche Welle
		250	G	WOF	Deutsche Welle
		100	PHL	IBA	FEBC Philippines
		250	RRW	KIG	Deutsche Welle
		500	RUS	MSK	Voice of Russia (VOR)
		100	STP	SAO	BBG - VO America (VOA)
12075		100	GUM	TWR	TWR Asia (KTWR)
		250	RUS	MSK	Radio Rossii
		250	SNG	SNG	R. Netherlands Worldwide
12080		10	AUS	BRN	Radio Australia
+		8	AUS	BRN	Radio Australia
		100	BOT	BOT	BBG - VO America (VOA)
		100	CHN	dom	CNR2 China Business R, Xi'an
		250	MDG	MDC	BBG - VOA Studio 7
		100	MRA	SAI	BBG - R. Free Asia (RFA)
		250	PHL	PHT	R. Netherlands Worldwide
		100	STP	SAO	BBG - VOA Studio 7
		250	STP	SAO	BBG - VOA Studio 7
12085		250	MNG	UBA	Voice of Mongolia
‡		500	SYR	ADR	Radio Damascus
12090		100	MRA	FBS	FEBC Saipan (KFBS)
12095		250	ASC	ASC	BBC World Service
		250	CYP	CYP	BBC World Service
		300	CYP	CYP	BBC World Service
		250	G	RMP	BBC World Service
		250	OMA	SLA	BBC World Service
		100	PHL	BOC	FEBC Philippines
		100	SNG	SNG	BBC World Service
		250	THA	NAK	BBC World Service
12105		250	CLN	IRA	BBG - R. Free Asia (RFA)
		100	GRC	AVL	Voice of Greece (ERA5)
		100	GUM	SDA	AWR Guam (KSDA)
		100	GUM	TWR	TWR Asia (KTWR)
12110		100	CHN	KUN	China Radio Int. (CRI)
12115		250	CLN	IRA	BBG - R. Free Asia (RFA)
12120		250	CLN	IRA	BBG - VO America (VOA)
		250	KWT	KWT	BBG - VO America (VOA)
		250	PHL	PHT	BBG - VO America (VOA)
		100	PHL	BOC	FEBC Philippines
12130		250	CLN	IRA	BBG - R. Free Asia (RFA)
		250	CLN	IRA	BBG - Radio Mashaal
12133		3	USA	KEW	AFRTS (AFN Feeder)
12140		250	KWT	KWT	BBG - R. Free Afghanistan

kHz	N	kW	Ctry	Site	Station, location
		250	KWT	KWT	BBG - VOA Ashna Radio
12150		250	CLN	IRA	BBG - VO America (VOA)
12160		100	USA	WCR	WWCR
12170		150	EGY	ABZ	Radio Cairo
12759		3	BIO	DGA	AFRTS (AFN Feeder)
13100±		1	TWN	TAI	Sound of Hope Radio Int.
13146		10	F	MCO	Radio Monaco
13362		3	GUM	BAR	AFRTS (AFN Feeder)
13570		50	USA	INB	WINB
13580		500	CHN	BEI	China Radio Int. (CRI)
		500	CHN	KUN	China Radio Int. (CRI)
		250	CLN	IRA	BBG - R. Free Asia (RFA)
		250	CLN	IRA	BBG - Radio Mashaal
		250	CLN	IRA	BBG - VO America (VOA)
		100	D	LAM	BBG - VO America (VOA)
		250	EGY	ABZ	Radio Cairo
		250	THA	UDO	BBG - Radio Mashaal
13590		500	CHN	BEI	China Radio Int. (CRI)
	+	90	CLN	TRM	BBC World Service
	+	90	CLN	TRM	Deutsche Welle
		100	PLW	HBN	Radio Australia
		500	UAE	DHA	Radio Dabanga
		100	ZMB	LUS	CVC 1Africa Radio
13600		100	CHN	KAS	China Radio Int. (CRI)
		150	CHN	KUN	China Radio Int. (CRI)
		500	CHN	XIA	China Radio Int. (CRI)
		500	IRN	KAM	VO the Islamic Rep. of Iran
		100	STP	SAO	BBG - VO America (VOA)
13605		250	D	WER	Family Radio Worldwide
		250	IND	ALG	All India Radio (AIR)
		500	IND	BGL	All India Radio (AIR)
13610		500	CHN	XIA	China Radio Int. (CRI)
		100	CHN	dom	CNR1 VO China, Nanning
13615		250	CLN	IRA	BBG - R. Farda
		100	D	LAM	BBG - R. Farda
		100	USA	YFR	Family Radio Worldwide
13620		500	CHN	XIA	China Radio Int. (CRI)
		500	IRN	KAM	VO the Islamic Rep. of Iran
13625		250	CLN	TRM	Deutsche Welle
		250	G	WOF	Deutsche Welle
		500	MRA	TIN	BBG - R. Free Asia (RFA)
		150	SVK	RSO	Radio Slovakia Int.
		500	TUR	CAK	Voice of Turkey (VOT)
13630		100	AUS	SHP	Radio Australia
13635		250	CLN	IRA	BBG - VO America (VOA)
		100	D	LAM	BBG - VO America (VOA)
		250	D	WER	BBG - VO America (VOA)
		250	F	ISS	Bible Voice Broadcasting (BVB)
13640		100	ALB	SHI	RadioTirana
		100	CHN	KAS	China Radio Int. (CRI)
		250	GUF	GUF	Radio France Int. (RFI)
		300	J	YAM	Radio Japan (NHK World)
		250	MRA	TIN	BBG - VO America (VOA)
		500	TUR	CAK	Voice of Turkey (VOT)
13645		100	CHN	KAS	China Radio Int. (CRI)
		500	CHN	XIA	China Radio Int. (CRI)
		100	D	WER	Pan American Broadcasting
		250	F	ISS	Pan American Broadcasting
13650		250	CAN	SAC	Radio Canada Int. (RCI)
		500	CHN	URU	China Radio Int. (CRI)
		250	CUB	HAB	China Radio Int. (CRI)
		500	IRN	SIR	VO the Islamic Rep. of Iran
		300	J	YAM	Radio Japan (NHK World)
		200	KRE	KUJ	Voice of Korea
		500	KWT	KBD	Radio Kuwait
		250	MRA	TIN	BBG - VO America (VOA)
13655		500	CHN	XIA	China Radio Int. (CRI)
		500	D	WER	Family Radio Worldwide
13660		250	CYP	CYP	BBC World Service
		500	D	NAU	Family Radio Worldwide
		500	F	ISS	Family Radio Worldwide
		250	G	RMP	BBC World Service
		300	G	SKN	BBC World Service
		100	NZL	RAN	R. New Zealand Int. (RNZI)

kHz	N	kW	Ctry	Site	Station, location
		250	OMA	SLA	BBC World Service
13665		150	ALB	CER	China Radio Int. (CRI)
13670		150	ALB	CER	China Radio Int. (CRI)
		500	CHN	KAS	China Radio Int. (CRI)
		50	CHN	dom	Xinjiang, Ürümqi
		100	D	WER	Bible Voice Broadcasting (BVB)
		250	MRA	TIN	BBG - R. Free Asia (RFA)
13675		250	CAN	SAC	China Radio Int. (CRI)
		250	THA	NAK	BBC World Service
13680		500	CHN	XIA	China Radio Int. (CRI)
		100	CUB	HAB	R. Nacional de Venezuela
		250	CUB	HAB	Radio Habana Cuba
		250	D	WER	BBG - R. Farda
		500	F	ISS	Radio France Int. (RFI)
		500	IRN	KAM	VO the Islamic Rep. of Iran
		500	IRN	SIR	VO the Islamic Rep. of Iran
	±	1	TWN	TAI	Sound of Hope Radio Int.
13690		100	AUS	SHP	Radio Australia
13695		500	F	ISS	Radio France Int. (RFI)
		500	IND	BGL	All India Radio (AIR)
		100	USA	YFR	Family Radio Worldwide
13700		250	ATN	BON	R. Netherlands Worldwide
		250	CAN	SAC	China Radio Int. (CRI)
		250	CAN	SAC	Radio Canada Int. (RCI)
		100	CHN	dom	CNR1 VO China, Lingshi
		100	CHN	dom	CNR8 VO Minorities, Lingshi
		500	D	NAU	Family Radio Worldwide
13710		500	ARS	RIY	BSKSA
		500	CHN	KAS	China Radio Int. (CRI)
		250	D	WER	BBG - VO America (VOA)
		500	IND	BGL	All India Radio (AIR)
		500	IRN	KAM	VO the Islamic Rep. of Iran
		250	MRA	TIN	BBG - R. Free Asia (RFA)
13715		100	CHN	KAS	China Radio Int. (CRI)
		150	SVK	RSO	Radio Slovakia Int.
		250	USA	GRV	BBG - VO America (VOA)
13720		100	CHN	KAS	China Radio Int. (CRI)
		500	CHN	XIA	China Radio Int. (CRI)
		250	E	NOB	R. Exterior de España (REE)
		300	POR	LIS	RDP Internacional
		250	UAE	DHA	Sudan Radio Service
13725		250	IRN	AHW	VO the Islamic Rep. of Iran
		250	MRA	TIN	BBG - R. Free Asia (RFA)
		100	THA	NAK	BBC World Service
13730		500	CHN	XIA	China Radio Int. (CRI)
	+	50	NZL	RAN	R. New Zealand Int. (RNZI)
13735		100	CHN	KAS	China Radio Int. (CRI)
		250	CLN	TRM	Deutsche Welle
		250	RUS	VLD	Voice of Russia (VOR)
13740		100	BOT	BOT	BBG - VO America (VOA)
		250	CUB	HAB	China Radio Int. (CRI)
		500	D	WER	Radio Dabanga
		500	IRN	SIR	VO the Islamic Rep. of Iran
		250	UAE	DHA	Family Radio Worldwide
13745		250	KWT	KWT	BBG - R. Free Asia (RFA)
		100	PLW	HBN	T8WH - World Harvest Radio Int
		250	THA	UDO	R. Thailand World Service
13750		500	CHN	KUN	China Radio Int. (CRI)
		100	CUB	HAB	R. Nacional de Venezuela
		250	CUB	HAB	Radio Habana Cuba
		500	F	ISS	Radio France Int. (RFI)
		500	G	RMP	BBC World Service
13755		250	CLN	IRA	BBG - RFE/RL
		250	D	NAU	Adventist World R. (AWR)
13760		500	IRN	SIR	VO the Islamic Rep. of Iran
		200	KRE	KUJ	Voice of Korea
13765		100	CVA	SMG	Vatican Radio
		500	CVA	SMG	Vatican Radio
13765		100	GUM	TWR	TWR Asia (KTWR)
		250	MDG	MDC	Vatican Radio
		250	THA	UDO	BBG - VO America (VOA)
13770		500	IND	BGL	All India Radio (AIR)
13775		500	ARS	RIY	BSKSA
		250	CYP	CYP	BBC World Service

kHz	N	kW	Ctry	Site	Station, location
13780		500	CHN	KUN	China Radio Int. (CRI)
		250	CLN	TRM	Deutsche Welle
		250	CUB	HAB	Radio Habana Cuba
		250	G	RMP	Deutsche Welle
		300	G	SKN	Deutsche Welle
		250	POR	SIN	Deutsche Welle
		250	RUS	ARM	Deutsche Welle
		250	UAE	DHA	Deutsche Welle
13790		250	ASC	ASC	BBC World Service
		500	CHN	URU	China Radio Int. (CRI)
		500	IRN	KAM	VO the Islamic Rep. of Iran
13795		500	IND	BGL	All India Radio (AIR)
13800		500	IRN	ZAH	VO the Islamic Rep. of Iran
	±	1	TWN	TAI	Sound of Hope Radio Int.
13810		250	CLN	IRA	BBG - R. Free Asia (RFA)
		300	CYP	CYP	BBC World Service
		100	D	NAU	The Overcomer Ministry
		100	F	ISS	Bible Voice Broadcasting (BVB)
		500	IRN	KAM	VO the Islamic Rep. of Iran
		250	RRW	KIG	Deutsche Welle
13820		250	ASC	ASC	BBC World Service
		500	D	NAU	Family Radio Worldwide
		250	OMA	SLA	BBC World Service
		250	USA	GRV	BBG - R. Martí
13825		250	CYP	CYP	BBC World Service
13830		250	USA	EWN	WEWN - EWTN Shortwave
13835		125	G	WOF	Polish Radio (Ext Sce)
13840		250	MDG	MDC	Radio Japan (NHK World)
13845		250	CYP	CYP	BBC World Service
		100	USA	WCR	WWCR
13850		500	CHN	BEI	China Radio Int. (CRI)
		250	G	RMP	Polish Radio (Ext Sce)
13855		500	CHN	KAS	China Radio Int. (CRI)
13865		300	CYP	CYP	BBC World Service
13970±		1	TWN	TAI	Sound of Hope Radio Int.
14400±		1	TWN	TAI	Sound of Hope Radio Int.
14670		10	CAN	STF	CHU, Ottawa
14700±		1	TWN	TAI	Sound of Hope Radio Int.
14970±		1	TWN	TAI	Sound of Hope Radio Int.
14996		5	RUS	STF	RWM, Moscow
15000		20	CHN	STF	BPM, Kinshan
		10	HWA	STF	WWVH, Kauai
		10	USA	STF	WWV, Ft. Collins
15040		100	EGY	ABZ	Radio Cairo
		100	IND	DEL	All India Radio (AIR)
15050		250	IND	DEL	All India Radio (AIR)
15060		250	EGY	ABZ	Radio Cairo
15065		250	EGY	ABZ	Radio Cairo
15075		250	IND	DEL	All India Radio (AIR)
		500	IND	BGL	All India Radio (AIR)
15080		250	EGY	ABS	Radio Cairo
15085		500	IRN	KAM	VO the Islamic Rep. of Iran
15100		200	KRE	KUJ	Voice of Korea
		100	PAK	ISL	Radio Pakistan
15105		500	AFS	MEY	BBC World Service
		250	ASC	ASC	BBC World Service
		150	RUS	SAM	Na volne Tatarstana
15110		100	CHN	KAS	China Radio Int. (CRI)
		500	CHN	URU	China Radio Int. (CRI)
		250	E	NOB	R. Exterior de España (REE)
15115		250	PHL	PHT	BBG - VO America (VOA)
		100	USA	YFR	Family Radio Worldwide
15120		500	ARS	RIY	BSKSA
		500	CHN	BEI	China Radio Int. (CRI)
		250	CLN	IRA	BBG - R. Free Asia (RFA)
		250	CUB	HAB	Radio Habana Cuba
		250	CUB	HAB	Radio Habana Cuba
		100	MRA	SAI	BBG - R. Free Asia (RFA)
		250	NIG	IKO	Voice of Nigeria
15125		500	CHN	KAS	China Radio Int. (CRI)
15125		100	CTR	CRI	R. Exterior de España (REE)
		100	MLI	BKO	China Radio Int. (CRI)
15130		150	CHN	BEI	China Radio Int. (CRI)
		100	D	BIB	BBG - RFE/RL

kHz	N	kW	Ctry	Site	Station, location
		100	D	LAM	BBG - RFE/RL
		100	USA	YFR	Family Radio Worldwide
15135		100	CHN	KUN	China Radio Int. (CRI)
		500	CHN	KAS	China Radio Int. (CRI)
		500	CHN	KUN	China Radio Int. (CRI)
		50	IND	dom	All India R, Delhi
		250	MRA	TIN	BBG - R. Free Asia (RFA)
15140		100	CHN	KAS	China Radio Int. (CRI)
		100	D	LAM	BBG - R. Free Asia (RFA)
		250	IND	DEL	All India Radio (AIR)
		100	OMA	THU	Radio Sultanate of Oman
	±	1	TWN	TAI	Sound of Hope Radio Int.
15145		100	CHN	KAS	China Radio Int. (CRI)
		500	CHN	XIA	China Radio Int. (CRI)
15150		500	G	RMP	BBC World Service
		500	IRN	SIR	VO the Islamic Rep. of Iran
		250	MRA	TIN	BBG - R. Free Asia (RFA)
		300	ROU	TIG	Radio Romania International
15155		250	AFS	MEY	Adventist World R. (AWR)
		500	AFS	MEY	Adventist World R. (AWR)
		300	ROU	GAL	Radio Romania International
15160		100	AUS	SHP	Radio Australia
		100	CHN	NNN	China Radio Int. (CRI)
		500	CHN	JIN	China Radio Int. (CRI)
		500	F	ISS	Radio France Int. (RFI)
		100	KOR	KIM	KBS World Radio
		250	MRA	TIN	BBG - R. Free Asia (RFA)
		300	ROU	TIG	Radio Romania International
15165		50	PHL	PHT	BBG - VO America (VOA)
15170		250	AFS	MEY	Radio France Int. (RFI)
		500	ARS	RIY	BSKSA
		100	CHN	KAS	China Radio Int. (CRI)
		500	CHN	JIN	China Radio Int. (CRI)
		100	CTR	CRI	R. Exterior de España (REE)
		100	GUM	TWR	TWR Asia (KTWR)
		300	ROU	GAL	Radio Romania International
15175		125	G	WOF	Polish Radio (Ext Sce)
		250	IND	PAN	All India Radio (AIR)
15180		250	CYP	CYP	BBC World Service
		200	KRE	KUJ	Voice of Korea
		100	USA	HRI	WHRI - World Harvest R. Int.
15185		50	IND	dom	All India R, Delhi
		250	IND	DEL	All India Radio (AIR)
		50	PHL	PHT	BBG - VO America (VOA)
15190		5	B	dom	R. Inconfidência, Belo Horizonte
		500	CHN	URU	China Radio Int. (CRI)
		50	GNE	BAT	Radio Africa
		250	PHL	PHT	Radio Pilipinas Overseas
15195		250	ASC	ASC	Family Radio Worldwide
		300	J	YAM	Radio Japan (NHK World)
15200		100	GUM	TWR	TWR Asia (KTWR)
		500	TUR	EMR	Voice of Turkey (VOT)
15205		500	ARS	RIY	BSKSA
		500	CHN	KAS	China Radio Int. (CRI)
	+	90	CLN	TRM	Deutsche Welle
		50	PHL	PHT	BBG - VO America (VOA)
		250	RRW	KIG	Deutsche Welle
		250	UAE	DHA	FEBA Radio
		100	UZB	TAC	Radio Japan (NHK World)
15210		100	CHN	KAS	China Radio Int. (CRI)
		500	CHN	KUN	China Radio Int. (CRI)
		100	USA	YFR	Family Radio Worldwide
15215		500	LBY	SAB	Voice of Africa
		250	UAE	DHA	FEBA Radio
15220		300	AUT	MOS	FEBA Radio
		500	CHN	KAS	China Radio Int. (CRI)
		500	CHN	URU	China Radio Int. (CRI)
		500	IRN	SIR	VO the Islamic Rep. of Iran
		250	MRA	TIN	BBG - R. Free Asia (RFA)
15225		500	ARS	RIY	BSKSA
		250	ATN	BON	BBG - VO America (VOA)
		100	BOT	BOT	BBG - VO America (VOA)
15225		500	CHN	KAS	China Radio Int. (CRI)
		500	F	ISS	Radio Taiwan International

kHz	N	kW	Ctry	Site	Station, location
		250	PHL	PUG	Radio Veritas Asia
15230		100	AUS	SHP	Radio Australia
		250	CAN	SAC	China Radio Int. (CRI)
		250	CLN	IRA	BBG - RFE/RL
15235		250	AFS	MEY	Channel Africa
		500	AFS	MEY	Channel Africa
		250	CVA	SMG	Vatican Radio
		250	IND	PAN	All India Radio (AIR)
		500	IND	BGL	All India Radio (AIR)
		500	IRN	KAM	VO the Islamic Rep. of Iran
		500	IRN	SIR	VO the Islamic Rep. of Iran
15240		250	AFS	MEY	Adventist World R. (AWR)
		100	AUS	SHP	Radio Australia
		250	RUS	PPK	Voice of Russia (VOR)
		250	TWN	TNN	Radio Australia
15245		500	CHN	URU	China Radio Int. (CRI)
		125	G	WOF	Polish Radio (Ext Sce)
		200	KRE	KUJ	Voice of Korea
		500	TUR	EMR	Voice of Turkey (VOT)
15250		250	ARS	RIY	BSKSA
		250	ASC	ASC	FEBA Radio
		100	CHN	KUN	China Radio Int. (CRI)
		100	CUB	HAB	R. Nacional de Venezuela
		250	PHL	PHT	BBG - RFE/RL
		100	USA	YFR	Family Radio Worldwide
15255		250	AFS	MEY	Channel Africa
		100	GUM	SDA	AWR Guam (KSDA)
		300	ROU	GAL	Radio Romania International
		500	UAE	DHA	R. Netherlands Worldwide
15260		100	GUM	SDA	AWR Guam (KSDA)
		50	IND	dom	All India R, Delhi
		50	IND	DEL	All India Radio (AIR)
		500	IRN	SIR	VO the Islamic Rep. of Iran
		300	ROU	GAL	Radio Romania International
15265		100	BOT	BOT	BBG - VO America (VOA)
		100	D	LAM	BBG - RFE/RL
		250	G	SKN	Église du Christ
		250	KWT	KWT	BBG - RFE/RL
		250	TWN	TNN	Radio Taiwan International
15270		500	CHN	KAS	China Radio Int. (CRI)
		150	CHN	dom	CNR2 China Business R, Beijing
		100	TWN	PAO	Radio Taiwan International
15275		250	CLN	TRM	Deutsche Welle
		250	G	WOF	Deutsche Welle
		250	POR	SIN	Deutsche Welle
		250	RRW	KIG	Deutsche Welle
		250	THA	UDO	R. Thailand World Service
15280		100	MRA	SAI	R. Netherlands Worldwide
		250	PHL	PUG	Radio Veritas Asia
15285		500	ARS	RIY	BSKSA
		250	D	WER	Athmeeya Yatra Radio
		100	EGY	ABZ	Radio Cairo
		250	PHL	PHT	Radio Pilipinas Overseas
		100	SNG	SNG	BBC World Service
		250	THA	UDO	BBG - RFE/RL
15290		100	CUB	HAB	R. Nacional de Venezuela
		500	F	ISS	Radio Japan (NHK World)
		500	JOR	AKA	Radio Jordan
		250	PHL	PHT	BBG - VO America (VOA)
		300	ROU	TIG	Radio Romania International
		250	TWN	TNN	Radio Australia
		100	TWN	TNN	Radio Taiwan International
15295		100	MLA	KAJ	Voice of Malaysia
15300		500	F	ISS	Radio France Int. (RFI)
15305		250	CAN	SAC	Radio Canada Int. (RCI)
15310		250	OMA	SLA	BBC World Service
		250	THA	NAK	BBC World Service
15315		250	ATN	BON	R. Netherlands Worldwide
		500	D	WER	Family Radio Worldwide
		500	F	ISS	Radio France Int. (RFI)
15320		100	GUM	SDA	AWR Guam (KSDA)
15320		100	PHL	BOC	FEBC Philippines
		100	TWN	PAO	Radio Taiwan International
15325‡	1		B	dom	R. Gazeta, São Paulo

kHz	N	kW	Ctry	Site	Station, location
		500	D	WER	Family Radio Worldwide
		250	E	NOB	R. Exterior de España (REE)
		300	J	YAM	Radio Japan (NHK World)
15330		100	PHL	BOC	FEBC Philippines
		300	ROU	TIG	Radio Romania International
		250	USA	GRV	BBG - R. Martí
15335		500	CHN	KAS	China Radio Int. (CRI)
		100	SNG	SNG	BBC World Service
		250	THA	UDO	BBG - R. Free Afghanistan
15340		100	AUS	KNX	HCJB Global Voice Australia
		500	CHN	XIA	China Radio Int. (CRI)
15341±		250	MRC	NAD	Radio Marocaine
15345		100	ARG	BUE	Rdif. Argentina al Exterior
		250	MRC	NAD	Radio Marocaine
15350		100	CHN	KAS	China Radio Int. (CRI)
		250	PHL	PUG	Radio Veritas Asia
		500	TUR	EMR	Voice of Turkey (VOT)
		250	TWN	TNN	Radio Australia
15355		100	OMA	THU	Radio Sultanate of Oman
		100	USA	YFR	Family Radio Worldwide
15360		250	CUB	HAB	Radio Habana Cuba
		500	F	ISS	Radio France Int. (RFI)
		250	OMA	SLA	BBC World Service
		100	SWZ	MAN	TWR Africa
		250	THA	NAK	BBC World Service
		500	TUR	CAK	Voice of Turkey (VOT)
15365		250	CAN	SAC	Radio Canada Int. (RCI)
15370		100	CHN	dom	CNR1 VO China, Shijiazhuang
		100	CUB	HAB	R. Nacional de Venezuela
		250	CUB	HAB	Radio Habana Cuba
		100	GUM	SDA	AWR Guam (KSDA)
		300	ROU	TIG	Radio Romania International
15375		250	TJK	DSB	BBG - R. Free Asia (RFA)
		10	TWN	dom	Fu Hsing Broadcasting Station, Kuanyin
		500	UAE	DHA	BBG - R. Free Asia (RFA)
15380		500	ARS	RIY	BSKSA
		100	CHN	dom	CNR1 VO China, Beijing
		100	PHL	BOC	FEBC Philippines
		300	ROU	GAL	Radio Romania International
		300	ROU	TIG	Radio Romania International
15385		250	E	NOB	R. Exterior de España (REE)
		250	PHL	PHT	BBG - VO America (VOA)
		50	USA	JES	KJES Radio
15390		100	CHN	dom	CNR1 VO China, Lingshi
		100	CHN	dom	CNR8 VO Minorities, Lingshi
		250	CUB	HAB	Radio Habana Cuba
		500	IRN	KAM	VO the Islamic Rep. of Iran
15400		250	ASC	ASC	BBC World Service
		100	AUS	KNX	HCJB Global Voice Australia
		500	IRN	SIR	VO the Islamic Rep. of Iran
		100	USA	YFR	Family Radio Worldwide
15410		100	D	BIB	BBG - R. Farda
		300	G	SKN	BBG - R. Farda
		250	RRW	KIG	Deutsche Welle
15415		100	AUS	SHP	Radio Australia
		50	CHN	dom	CNR8 VO Minorities, Beijing
15420		250	AFS	MEY	BBC World Service
		250	CYP	CYP	BBC World Service
		250	POR	SIN	BBC World Service
		250	SEY	SEY	BBC World Service
		50	USA	BCQ	WBCQ
15425		500	CHN	XIA	China Radio Int. (CRI)
15430		300	ROU	GAL	Radio Romania International
		300	ROU	TIG	Radio Romania International
		250	UAE	DHA	Voice of Tibet
15435		500	ARS	RIY	BSKSA
		500	CHN	XIA	China Radio Int. (CRI)
		100	PHL	BOC	FEBC Philippines
		250	PHL	PUG	Radio Veritas Asia
15440		300	AUT	MOS	Adventist World R. (AWR)
15440		500	CHN	KUN	China Radio Int. (CRI)
		250	IRN	AHW	VO the Islamic Rep. of Iran
		250	RRW	KIG	Deutsche Welle
		100	USA	YFR	Family Radio Worldwide

kHz	N	kW	Ctry	Site	Station, location
		100	USA	YFR	Radio Taiwan International
15445		500	CHN	KAS	China Radio Int. (CRI)
		100	TWN	TAI	Adventist World R. (AWR)
15450		100	PHL	BOC	FEBC Philippines
		100	PHL	IBA	FEBC Philippines
		250	PHL	PUG	Radio Veritas Asia
		500	RUS	ARM	Voice of Russia (VOR)
15455		100	PHL	BOC	FEBC Philippines
15460		100	D	LAM	BBG - VO America (VOA)
		500	IRN	KAM	VO the Islamic Rep. of Iran
		250	PHL	PUG	Vatican Radio
		300	ROU	TIG	Radio Romania International
		150	SVK	RSO	Radio Slovakia Int.
15465		100	CHN	KAS	China Radio Int. (CRI)
		300	POR	LIS	RDP Internacional
		100	TWN	PAO	Radio Taiwan International
15476†		2	ATA	dom	R. Nacional Arcangel San Gabriel
15480		100	CHN	dom	CNR1 VO China, Beijing
		500	TUR	CAK	Voice of Turkey (VOT)
15490		500	AFS	MEY	BBC World Service
		500	ARS	RIY	BSKSA
		250	PAK	ISL	Radio Pakistan
		250	SEY	SEY	BBC World Service
15495		250	D	NAU	Adventist World R. (AWR)
		100	GUM	SDA	AWR Guam (KSDA)
15500		150	CHN	dom	CNR2 China Business R, Beijing
15510		250	CYP	CYP	BBC World Service
		250	RUS	NVS	BBC World Service
		250	SEY	SEY	BBC World Service
		250	THA	NAK	BBC World Service
15515		100	AUS	SHP	Radio Australia
		250	GUF	GUF	Radio France Int. (RFI)
		500	IRN	KAM	VO the Islamic Rep. of Iran
		500	KWT	KBD	Radio Kuwait
15520		250	ASC	ASC	Family Radio Worldwide
		250	PHL	PUG	Radio Veritas Asia
		300	POR	LIS	RDP Internacional
		250	UAE	DHA	Family Radio Worldwide
15521		100	TJK	DSB	Voice of Tibet
15523		100	TJK	DSB	Voice of Tibet
15525		100	AUS	KNX	HCJB Global Voice Australia
		500	CHN	URU	China Radio Int. (CRI)
		100	PHL	BOC	FEBC Philippines
		100	UAE	DHA	Eternal Good News
15527		100	TJK	DSB	Voice of Tibet
15530		500	F	ISS	Radio France Int. (RFI)
		250	PHL	PUG	Radio Veritas Asia
15535		250	CLN	IRA	BBG - R. Farda
15540		250	ATN	BON	R. Netherlands Worldwide
‡		500	KWT	KBD	Radio Kuwait
		100	PAK	ISL	Radio Pakistan
		250	PHL	PHT	BBG - VO America (VOA)
		250	THA	UDO	BBG - VO America (VOA)
15542		100	TJK	DSB	Voice of Tibet
15548		100	TJK	DSB	Voice of Tibet
15550		100	CHN	dom	CNR1 VO China, Beijing
		500	IRN	SIR	VO the Islamic Rep. of Iran
		250	MRA	TIN	BBG - R. Free Asia (RFA)
		250	PHL	PHT	BBG - VO America (VOA)
		50	USA	JHR	WJHR Radio International
15555		300	POR	LIS	RDP Internacional
15560		100	AUS	SHP	Radio Australia
		500	IRN	SIR	VO the Islamic Rep. of Iran
		300	POR	LIS	RDP Internacional
15565		500	CHN	XIA	China Radio Int. (CRI)
		100	USA	YFR	Family Radio Worldwide
15568		100	TJK	DSB	Voice of Tibet
15570		100	CHN	dom	CNR11, Xi'an
15572		100	TJK	DSB	Voice of Tibet
15575		250	CYP	CYP	BBC World Service
		300	G	SKN	BBC World Service
15578		100	TJK	DSB	Voice of Tibet
15580		250	AFS	MEY	BBG - VO America (VOA)
		250	ATN	BON	BBG - VO America (VOA)
		100	BOT	BOT	BBG - VO America (VOA)
		100	MRA	FBS	FEBC Saipan (KFBS)
		100	STP	SAO	BBG - VO America (VOA)
		250	USA	GRV	BBG - VO America (VOA)
15582		100	TJK	DSB	Voice of Tibet
15585		250	E	NOB	R. Exterior de España (REE)
15590		250	THA	UDO	BBG - RFE/RL
		250	USA	GRV	BBG - VO America (VOA)
15595		100	CVA	SMG	Vatican Radio
		250	CVA	SMG	Vatican Radio
		500	CVA	SMG	Vatican Radio
		250	MDG	MDC	R. Netherlands Worldwide
15600		100	CHN	KUN	China Radio Int. (CRI)
		250	CLN	TRM	Deutsche Welle
		100	PHL	BOC	FEBC Philippines
15605		500	F	ISS	Radio France Int. (RFI)
15610		250	USA	EWN	WEWN - EWTN Shortwave
15620		300	AUT	MOS	Deutsche Welle
		100	BOT	BOT	BBG - VO America (VOA)
		500	CHN	KAS	China Radio Int. (CRI)
		250	CLN	IRA	BBG - VO America (VOA)
		250	CLN	TRM	Deutsche Welle
		250	D	WER	BBG - VO America (VOA)
		250	MDG	MDC	BBG - VO America (VOA)
		100	PAK	ISL	Radio Pakistan
		100	STP	SAO	BBG - VO America (VOA)
15625		500	CHN	KAS	China Radio Int. (CRI)
15630		100	GRC	AVL	Voice of Greece (ERA5)
15640		250	CLN	TRM	Deutsche Welle
		250	POR	SIN	Deutsche Welle
		250	RRW	KIG	Deutsche Welle
		250	UAE	DHA	Deutsche Welle
		250	USA	HRI	Deutsche Welle
15650		100	GRC	AVL	Voice of Greece (ERA5)
15660		100	GUM	SDA	AWR Guam (KSDA)
		500	LBY	SAB	Voice of Africa
15665		500	CHN	KAS	China Radio Int. (CRI)
		100	MRA	SAI	BBG - R. Free Asia (RFA)
		250	USA	HRI	WHRI - World Harvest R. Int.
15670		100	BOT	BOT	BBG - VO America (VOA)
		50	CHN	dom	CNR8 VO Minorities, Beijing
		100	STP	SAO	BBG - VO America (VOA)
		250	THA	UDO	BBG - VO America (VOA)
15680		100	PLW	HBN	T8WH - World Harvest Radio Int
		100	TWN	TNN	Radio France Int. (RFI)
		250	USA	HRI	WHRI - World Harvest R. Int.
15690		250	CLN	IRA	BBG - R. Farda
		500	F	ISS	Radio Taiwan International
		250	MRA	TIN	BBG - R. Free Asia (RFA)
		100	POR	LIS	RDP Internacional
15700		170	BUL	PLD	Radio Bulgaria
		300	BUL	PLD	Radio Bulgaria
		250	MRA	TIN	BBG - R. Free Asia (RFA)
15710		50	CHN	dom	CNR5 VO Zhonghua, Beijing
		250	EGY	ABS	Radio Cairo
		150	SVK	RSO	IRRS Shortwave
		150	SVK	RSO	Miraya FM
15715		250	CLN	IRA	BBG - Radio Mashaal
		250	D	WER	BBG - Radio Mashaal
15720		100	NZL	RAN	R. New Zealand Int. (RNZI)
	+	50	NZL	RAN	R. New Zealand Int. (RNZI)
15730		100	BOT	BOT	BBG - VO America (VOA)
		100	STP	SAO	BBG - VO America (VOA)
15735+		250	RUS	KNA	Voice of Russia (VOR)
15745		500	AFS	MEY	BBC World Service
		35	CLN	EKA	Sri Lanka Broadcasting Corp.
15750		100	MRA	SAI	R. Netherlands Worldwide
		250	OMA	SLA	BBC World Service
		250	THA	UDO	BBG - Radio Mashaal
15750±		100	TJK	DSB	Sound of Hope Radio Int.
		250	UAE	DHA	Bar-Kulan Radio
		250	UAE	DHA	BBC World Service

kHz	N	kW	Ctry	Site	Station, location
15755+		100	ATN	BON	TDPradio
	+	100	ATN	BON	The Disco Palace
15770		100	CHN	dom	CNR2 China Business R, Lingshi
		250	IND	ALG	All India Radio (AIR)
15775		100	STP	SAO	BBG - VOA Studio 7
15785		10	ISR	LOD	Galei Tzahal
15790		250	CYP	CYP	BBC World Service
		250	OMA	SLA	BBC World Service
15795		500	IND	BGL	All India Radio (AIR)
		100	USA	YFR	Family Radio Worldwide
15820		1	ARG	dom	Argentine Armed Forces, Buenos Aires
15825		100	USA	WCR	WWCR
15850±		1	TWN	TAI	Sound of Hope Radio Int.
15896±		0.1	D	dom	BiteXpress, Erlangen
15970±		1	TWN	TAI	Sound of Hope Radio Int.
16100±		1	TWN	TAI	Sound of Hope Radio Int.
16700±		1	TWN	TAI	Sound of Hope Radio Int.
17260		10	F	MCO	Radio Monaco
17485		500	CHN	URU	China Radio Int. (CRI)
		100	D	WER	The Overcomer Ministry
17490		500	CHN	KAS	China Radio Int. (CRI)
17495		500	CHN	BEI	China Radio Int. (CRI)
17500		100	CHN	KAS	China Radio Int. (CRI)
17505		250	ASC	ASC	Family Radio Worldwide
		500	CHN	KAS	China Radio Int. (CRI)
		500	CHN	XIA	China Radio Int. (CRI)
		250	UAE	DHA	Family Radio Worldwide
		500	UAE	DHA	R. Netherlands Worldwide
17510		100	EGY	ABZ	Radio Cairo
		250	IND	DEL	All India Radio (AIR)
17515		200	TJK	DSB	BBG - R. Free Asia (RFA)
17520		250	CLN	TRM	Deutsche Welle
		500	G	RMP	Deutsche Welle
		250	USA	HRI	WHRI - World Harvest R. Int.
17530		250	CLN	IRA	BBG - R. Free Afghanistan
		250	KWT	KWT	BBG - R. Free Afghanistan
17535		100	USA	YFR	Family Radio Worldwide
17540		100	CHN	KAS	China Radio Int. (CRI)
		500	CHN	BEI	China Radio Int. (CRI)
		300	ROU	TIG	Radio Romania International
		100	USA	YFR	Family Radio Worldwide
		250	USA	HRI	WHRI - World Harvest R. Int.
17545		125	D	WER	Bible Voice Broadcasting (BVB)
		100	USA	YFR	Family Radio Worldwide
17550		100	CHN	dom	CNR1 VO China, Beijing
	‡	500	KWT	KBD	Radio Kuwait
17555		100	USA	YFR	Family Radio Worldwide
17560		500	ARS	RIY	BSKSA
		500	CHN	XIA	China Radio Int. (CRI)
		300	J	YAM	Radio Japan (NHK World)
		250	MDG	MDC	Voice of Tibet
17570		500	ARS	RIY	BSKSA
		500	CHN	KAS	China Radio Int. (CRI)
		500	CHN	URU	China Radio Int. (CRI)
17575		250	F	ISS	Adventist World R. (AWR)
		100	USA	YFR	Family Radio Worldwide
17580		100	CHN	dom	CNR1 VO China, Lingshi
17585		100	PLW	HBN	Radio Australia
17590		100	OMA	THU	Radio Sultanate of Oman
17595		100	CHN	dom	CNR1 VO China, Shijiazhuang
		250	E	NOB	R. Exterior de España (REE)
17605		250	ATN	BON	R. Netherlands Worldwide
		250	ATN	BON	Radio Japan (NHK World)
		300	AUT	MOS	Adventist World R. (AWR)
		100	CHN	dom	CNR1 VO China, Beijing
17610		500	F	ISS	Radio France Int. (RFI)
17615		500	ARS	RIY	BSKSA
		100	MRA	SAI	BBG - R. Free Asia (RFA)
17620		500	F	ISS	Radio France Int. (RFI)
17625		500	ARS	RIY	BSKSA
17625		150	CHN	dom	CNR2 China Business R, Beijing
17630		250	GUF	GUF	Radio France Int. (RFI)
		500	IRN	KAM	VO the Islamic Rep. of Iran

kHz	N	kW	Ctry	Site	Station, location
		100	MLI	BKO	China Radio Int. (CRI)
17635		100	GUM	SDA	AWR Guam (KSDA)
17640		250	ASC	ASC	BBC World Service
		500	CHN	XIA	China Radio Int. (CRI)
		250	CYP	CYP	BBC World Service
		250	SEY	SEY	BBC World Service
17645		250	PHL	PHT	BBG - VO America (VOA)
17650		500	CHN	KAS	China Radio Int. (CRI)
		250	D	WER	TWR India
		100	STP	SAO	BBG - VO America (VOA)
		1000	TJK	DSB	Voice of Russia (VOR)
17650±		1	TWN	TAI	Sound of Hope Radio Int.
17660		250	AFS	MEY	Radio France Int. (RFI)
		500	ARS	RIY	BSKSA
		250	ASC	ASC	Family Radio Worldwide
		500	IRN	SIR	VO the Islamic Rep. of Iran
17665		250	RUS	KNA	Voice of Russia (VOR)
17670		500	CHN	KUN	China Radio Int. (CRI)
		125	G	WOF	Polish Radio (Ext Sce)
		500	IND	BGL	All India Radio (AIR)
		250	MDG	MDC	Adventist World R. (AWR)
17675 +		50	NZL	RAN	R. New Zealand Int. (RNZI)
17680		50	CHL	SGO	CVC La Voz
		100	CHN	KUN	China Radio Int. (CRI)
		150	CHN	KUN	China Radio Int. (CRI)
		500	IRN	SIR	VO the Islamic Rep. of Iran
		250	UAE	DHA	IRIN Radio
17685		250	SNG	SNG	BBC World Service
17690		500	CHN	JIN	China Radio Int. (CRI)
		100	CHN	dom	CNR1 VO China, Nanning
		500	IRN	SIR	VO the Islamic Rep. of Iran
		250	MDG	MDC	Radio Japan (NHK World)
		100	USA	YFR	Family Radio Worldwide
17700		250	ASC	ASC	Sudan Radio Service
		250	CLN	TRM	Deutsche Welle
		250	PAK	ISL	Radio Pakistan
17705		500	ARS	RIY	BSKSA
		100	CUB	HAB	R. Nacional de Venezuela
		500	IND	BGL	All India Radio (AIR)
17710		500	CHN	BEI	China Radio Int. (CRI)
		500	CHN	JIN	China Radio Int. (CRI)
		250	CLN	TRM	Deutsche Welle
17715		100	AUS	SHP	Radio Australia
		100	BOT	BOT	BBG - VO America (VOA)
		250	E	NOB	R. Exterior de España (REE)
		125	G	WOF	Polish Radio (Ext Sce)
		250	KWT	KWT	BBG - R. Free Asia (RFA)
		500	TUR	CAK	Voice of Turkey (VOT)
17720		500	CHN	KAS	China Radio Int. (CRI)
		250	THA	NAK	BBC World Service
17725		500	CHN	XIA	China Radio Int. (CRI)
		250	F	ISS	Radio Y'Abaganda
		500	LBY	SAB	Voice of Africa
17730		500	ARS	RIY	BSKSA
		250	MNG	UBA	BBG - R. Free Asia (RFA)
17735		250	CAN	SAC	China Radio Int. (CRI)
		100	CHN	KUN	China Radio Int. (CRI)
		500	F	ISS	Radio Japan (NHK World)
		250	UAE	DHA	Family Radio Worldwide
17740		500	ARS	RIY	BSKSA
		500	CHN	XIA	China Radio Int. (CRI)
		250	IND	DEL	All India Radio (AIR)
		250	USA	GRV	BBG - VO America (VOA)
17745		300	POR	LIS	RDP Internacional
		250	POR	SIN	Sudan Radio Service
		300	ROU	GAL	Radio Romania International
17750		100	AUS	SHP	Radio Australia
		100	CHN	KAS	China Radio Int. (CRI)
		100	CUB	HAB	R. Nacional de Venezuela
		250	CUB	HAB	Radio Habana Cuba
17750		250	KWT	KWT	BBG - R. Free Asia (RFA)
17755		250	E	NOB	R. Exterior de España (REE)
		500	TUR	EMR	Voice of Turkey (VOT)

SW Stations of the World

FREQUENCY LISTS

kHz	N	kW	Ctry	Site	Station, location
17760		250	SNG	SNG	BBC World Service
		250	THA	NAK	BBC World Service
		100	USA	YFR	Family Radio Worldwide
17765		250	CAN	SAC	Radio Canada Int. (RCI)
		250	CVA	SMG	Vatican Radio
		300	ROU	GAL	Radio Romania International
17770		250	AFS	MEY	Channel Africa
		250	ASC	ASC	Deutsche Welle
		250	E	NOB	R. Exterior de España (REE)
		250	PHL	PHT	Radio Pilipinas Overseas
		250	THA	UDO	BBG - RFE/RL
17775		300	ROU	GAL	Radio Romania International
		300	ROU	TIG	Radio Romania International
		50	USA	VOH	KVOH - La Voz de Restauración
17780		250	ASC	ASC	BBC World Service
		250	CYP	CYP	BBC World Service
		500	G	RMP	BBC World Service
		500	IRN	SIR	VO the Islamic Rep. of Iran
		250	PHL	PHT	BBG - VO America (VOA)
		300	ROU	GAL	Radio Romania International
		300	ROU	TIG	Radio Romania International
17785		500	ARS	RIY	BSKSA
17790		250	CAN	SAC	Radio Canada Int. (RCI)
		500	G	RMP	BBC World Service
		250	MDG	MDC	Democratic VO Burma
		250	OMA	SLA	BBC World Service
		250	THA	NAK	BBC World Service
17795		100	AUS	SHP	Radio Australia
17800		250	CLN	TRM	Deutsche Welle
		500	IND	BGL	All India Radio (AIR)
		250	POR	SIN	Deutsche Welle
		300	ROU	GAL	Radio Romania International
		250	RRW	KIG	Deutsche Welle
17805		500	ARS	RIY	BSKSA
		250	RUS	IRK	Voice of Russia (VOR)
17810		250	EGY	ABZ	Radio Cairo
		250	IND	PAN	All India Radio (AIR)
		500	IRN	SIR	VO the Islamic Rep. of Iran
		300	J	YAM	Radio Japan (NHK World)
		250	UAE	DHA	Family Radio Worldwide
17815		100	D	LAM	BBG - R. Farda
17820		300	POR	LIS	RDP Internacional
		250	THA	NAK	BBC World Service
		250	UAE	DHA	BBC World Service
17825		250	THA	NAK	BBC World Service
		300	THA	NAK	BBC World Service
17830		500	AFS	MEY	BBC World Service
		250	ASC	ASC	BBC World Service
		100	CHN	KAS	China Radio Int. (CRI)
		250	PAK	ISL	Radio Pakistan
		250	PHL	PUG	Radio Veritas Asia
17840		250	CLN	IRA	BBG - R. Farda
		100	D	LAM	BBG - R. Farda
		250	D	WER	BBG - R. Farda
		100	PLW	HBN	Radio Australia
		300	POR	LIS	RDP Internacional
17845		100	CHN	dom	CNR1 VO China, Shijiazhuang
		250	IND	DEL	All India Radio (AIR)
		100	PLW	HBN	Radio Australia
17850		250	CLN	IRA	BBG - VO America (VOA)
		100	CTR	CRI	R. Exterior de España (REE)
		500	F	ISS	Radio France Int. (RFI)
17855		300	AUT	MOS	Radio Ö1 International
		500	CHN	BEI	China Radio Int. (CRI)
17860		100	AFS	MEY	Amateur R. Mirror Int.
+		15	CHL	SGO	CVC La Voz
		500	G	RMP	Deutsche Welle
		100	IND	dom	All India R, Delhi
		100	IND	DEL	All India Radio (AIR)
		250	PHL	PUG	Radio Veritas Asia
17860		100	SNG	SNG	Voice of Croatia
		250	THA	UDO	BBG - VO America (VOA)

kHz	N	kW	Ctry	Site	Station, location
17870		250	CYP	CYP	BBC World Service
		250	EGY	ABZ	Radio Cairo
		300	ROU	GAL	Radio Romania International
+		300	ROU	TIG	Radio Romania International
		300	ROU	TIG	Radio Romania International
17875		500	IND	BGL	All India Radio (AIR)
17880		100	GUM	SDA	AWR Guam (KSDA)
		250	MRA	TIN	BBG - R. Free Asia (RFA)
17885		250	ASC	ASC	BBC World Service
17890		100	CHN	dom	CNR1 VO China, Beijing
17895		500	ARS	RIY	BSKSA
		250	ATN	BON	BBG - VO America (VOA)
		100	BOT	BOT	BBG - VO America (VOA)
		500	IND	BGL	All India Radio (AIR)
		100	STP	SAO	BBG - VO America (VOA)
17920±		1	TWN	TAI	Sound of Hope Radio Int.
18180±		1	TWN	TAI	Sound of Hope Radio Int.
18930		100	USA	YFR	Family Radio Worldwide
18980		100	USA	YFR	Family Radio Worldwide
19010		250	KWT	KWT	BBG - R. Free Afghanistan
20000		2.5	USA	STF	WWV, Ft. Collins
21455		100	USA	YFR	Family Radio Worldwide
21470		250	SEY	SEY	BBC World Service
21490		250	MRA	TIN	BBG - R. Free Asia (RFA)
21505		500	ARS	RIY	BSKSA
21525		250	RUS	SAM	La Voix de Djibouti (LVD)
21540		250	E	NOB	R. Exterior de España (REE)
		500	KWT	KBD	Radio Kuwait
		500	MRA	TIN	BBG - R. Free Asia (RFA)
21550		250	POR	SIN	Deutsche Welle
21555		250	RUS	SAM	Radio Democracy
21560		500	F	ISS	Radio Japan (NHK World)
		250	POR	SIN	Deutsche Welle
21570		250	E	NOB	R. Exterior de España (REE)
		250	PHL	PHT	BBG - VO America (VOA)
21580		500	F	ISS	Radio France Int. (RFI)
		500	MRA	TIN	BBG - VO America (VOA)
21595		250	CYP	CYP	BBC World Service
21600		300	ROU	TIG	Radio Romania International
21610		250	E	NOB	R. Exterior de España (REE)
21620		500	F	ISS	Radio France Int. (RFI)
21630		250	ASC	ASC	BBC World Service
		250	POR	SIN	BBC World Service
		250	USA	HRI	WHRI - World Harvest R. Int.
21640		500	IRN	SIR	VO the Islamic Rep. of Iran
21655		300	POR	LIS	RDP Internacional
21670		500	ARS	RIY	BSKSA
21680		250	CVA	SMG	Vatican Radio
		500	CVA	SMG	Vatican Radio
		100	USA	YFR	Family Radio Worldwide
21690		250	GUF	GUF	Radio France Int. (RFI)
21695		500	LBY	SAB	Voice of Africa
		500	UAE	DHA	BBG - R. Free Asia (RFA)
21715		250	CLN	IRA	BBG - R. Farda
		250	THA	UDO	BBG - R. Farda
21725		100	AUS	SHP	Radio Australia
21745		100	CZE	LIT	Radio Prague
		250	PLW	HBN	Family Radio Worldwide
		100	USA	YFR	Family Radio Worldwide
21780		250	CLN	TRM	Deutsche Welle
		500	G	RMP	Deutsche Welle
21840		250	ASC	ASC	Family Radio Worldwide
25000		0.1	FIN	STF	Ctr for Metrology & Accreditation, Espoo
25740+		0.04	RRW	dom	R. Rwanda, Kigali
25775+		0.1	F	dom	TDF, Rennes
26000+		0.1	D	dom	Campus R, Nürnberg
+		1	G	dom	World Radio Network, Croydon
		0.25	I	dom	R. Maria, Andrate
26010+		0.25	I	dom	R. Maria, Andrate
26012+		0.1	D	dom	Campus R, Nürnberg
26040+		1	B	dom	R. Cultura, São Paulo
26045+		0.04	D	dom	Hannover University

International Broadcasts
in English, French, German, Portuguese and Spanish

English			
0000	**English**	**Area**	**kHz**
0000-0030	R. Thailand WS	NAm	13745udo
0000-0100	BBC World Sce	FE	5970sla
0000-0100	BBC World Sce	SAs	6195sng, 9410nak, 12095sng, 13725nak
0000-0100	BBC World Sce	WAs	7360cyp
0000-0100	CRI	NAm	6005sac
0000-0100	CRI	SAs	6180kas, 7425kas
0000-0100	CRI	EAs	9425bei
0000-0100	Deutsche Welle	EAs,SEA	9445trm, 9785trm, 11855sng
0000-0100	Family Radio	NAm	5950yfr, 6085yfr
0000-0100	Family Radio	CAm,SAm	7360yfr, 11720yfr, 11730yfr
0000-0100	KBS World R.	WEu	1440mrn***
0000-0100	R. Bulgaria	NAm	5900pld, 7400pld
0000-0100	RCI	SEA	9880kun
0000-0100	REE	NAm	5970nob*
0000-0100	R. Habana Cuba	Car	5040hab
0000-0200	BBC World Sce	SEA	15335sng
0000-0200	BBC World Sce	FE	15360nak
0000-0200	CRI	SEA	11650bei, 11885xia
0000-0200	CRI	NAm	6020cer, 9570cer
0000-0200	Family Radio	NAm	15440yfr
0000-0200	R. Australia	Pac	17715shp
0000-0200	WHRI	NAm,Eu	7590hri*
0000-0400	VO Russia (VOR)	NAm,CAm	7250arm
0000-0500	Family Radio	NAm	9505yfr
0000-0500	VO Russia (VOR)	NAm,CAm	6240kch
0000-0600	WEWN - EWTN	ME	11520ewn
0000-0800	R. Australia	Pac	15240shp
0000-1200	WWCR	NAm,Eu	4840wcr**
0000-1200	WWCR	Af	5935wcr**
0000-1300	WHRI	NAm	5875hri
0000-2400	AFRTS (AFN)	Atl	5446kew, 7812kew, 12133kew
0000-2400	BBC World Sce	WEu	648orf
0000-2400	BBC World Sce	CHN	675hkg
0000-2400	R. Bahrain	ME	6010abh
0000-2400	WBCQ	NAm,CAm	5110bcq*, 7415bcq*
0000-2400	WTJC	NAm	9370tjc
0005-0105	RCI	NAm	9755sac
0030-0100	BBG-VOA	SEA,Pac	15185pht
0030-0100	BBG-VOA	EAs,SAs,SEA	15290pht
0030-0100	BBG-VOA	SEA	1575bph, 9715udo, 11695pht, 15205pht
0030-0100	BBG-VOA	ME	1593wrt, 6170kwt
0030-0100	BBG-VOA	SAs,SEA	9325pht
0030-0100	BBG-VOA	EAs	9490pht, 12005pht
0030-0100	BVB	ME	5950wer
0030-0100	R. Thailand WS	NAm	13745udo
0100	**English**		
0100-0130	BBG-VOA	ME	1593kwt
0100-0130	R. Prague	NAm	7410lit
0100-0130	R. Slovakia Int	NAm	6040rso
0100-0130	R. Slovakia Int	SAm	9440rso
0100-0130	VO Vietnam	NAm	6175sac

0100	English	Area	kHz
0100-0200	BBC World Sce	SEA	17685sng
0100-0200	BBC World Sce	SAs	5970sla, 11750sng, 12095nak
0100-0200	BBG-VOA	ME	7325ira
0100-0200	BBG-VOA	SAs	9435udo, 11705udo
0100-0200	CRI	NAm	6005sac, 6080sac, 9580hab
0100-0200	CRI	SAs	6075kas, 6175kas, 9420kas
0100-0200	CRI	Eu	9410kas
0100-0200	Family Radio	CAm,SAm	6100yfr
0100-0200	R.Romania Int	NAm	6145gal, 7325gal
0100-0200	R.Taiwan Int.	SAs	11875tnn
0100-0200	R. Ukraine Int.	NAm	7440lvi
0100-0200	VO Korea	LAm	11735kuj, 13760kuj, 15180kuj
0100-0200	VO Korea	EAs	4405kuj, 7220kuj, 9345kuj, 9730kuj
0100-0300	BBC World Sce	SAs	15310nak
0100-0400	BBC World Sce	CAs	5940cyp
0100-0500	Family Radio	NAm	7455yfr
0100-0500	R. Habana Cuba	NAm	6000hab, 6050hab
0100-0520	BBC World Sce	Eu	198dro
0100-0900	WWCR	NAm,Eu,Af	3215wcr**
0100-1100	WWCR	NAm,Af	5890wcr**
0100-1200	WHRI	NAm	7315hri
0100-1300	WWCR	NAm,Eu	4840wcr*
0100-1300	WWCR	Af	5935wcr*
0105-0205	RCI	NAm	9755sac
0130-0145	R. Tirana	NAm	6130shi
0130-0200	BBG-VOA	ME	1593kwt
0130-0200	BBG-VOA	LAm	5960grv, 7465grv
0130-0200	Int. R. Serbia	NAm,Eu	6190bij
0130-0230	VOIRI	NAm	6120sir**, 7250kam**
0130-0300	SLBC	SAs	6005eka, 9770eka, 15745eka
0140-0200	Vatican Radio	As	5895dsb, 7335smg**, 7335tac*, 9580tac**
0200	**English**		
0200-0230	BBC World Sce	SAs	1413sla
0200-0230	R. Prague	NAm	7410lit
0200-0230	R. Thailand WS	NAm	15275udo
0200-0300	BBC World Sce	SAs	12095sla
0200-0300	BBC World Sce	ME	5875cyp
0200-0300	BBC World Sce	EAf	7445sey
0200-0300	CRI	SAs	13640kas
0200-0300	Deutsche Welle	EAs	15205trm+
0200-0300	Family Radio	CAm,SAm	5930yfr, 6885yfr, 6890yfr
0200-0300	Family Radio	NAm	9525yfr
0200-0300	KBS World R.	SAm	9580kim
0200-0300	R.Taiwan Int.	NAm	5950yfr, 9680yfr
0200-0300	RAE	Am	11710bue
0200-0300	VO Korea	SEA	13650kuj, 15100kuj
0200-0300	VO Korea	EAs	3560kuj
0200-0300	WHRI	NAm,Eu	7385hri*
0200-0330	R. Cairo	NAm	6270abz
0200-0330	R. Pilipinas	ME	11880pht, 15285pht, 17770pht
0200-0400	CRI	SAs	11785kas

0200	English	Area	kHz
0200-0400	TWR Bonaire	CAm,SAm	800twb
0200-0500	R. Australia	Pac	15515shp, 21725shp
0200-0500	WRNO Worldwide	NAm,CAm	7505rno
0200-0730	BBC World Sce	ME	1323cyp
0200-1000	WWCR	NAm,Eu,Af	3215wcr*
0200-1200	WWCR	NAm,Af	5890wcr*
0200-2300	BBC World Sce	ME	1323cyp
0230-0300	R.Taiwan Int.	EAs	1422min
0230-0300	VO Vietnam	NAm	6175sac
0245-0300	HCJB Australia	As,SEA	15400knx
0245-0300	R. Tirana	NAm	6130shi
0250-0310	Vatican Radio	CAm,SAm	7305smg
0250-0320	Vatican Radio	NAm	6040sac
0255-0325	TWR Africa	SAf	3200man
0255-0400	TWR Africa	WAf	1566par
0300	**English**		
0300-0320	Vatican Radio	As	15460pug
0300-0330	BBC World Sce	ME	639zak
0300-0330	Vatican Radio	Af	7360smg, 9660mdc, 9660mdc
0300-0400	BBC World Sce	CAs	12095sla
0300-0400	BBC World Sce	ME	1413sla
0300-0400	BBC World Sce	Eu	5940skn
0300-0400	BBC World Sce	EAf	6100cyp, 9460sey
0300-0400	BBC World Sce	WAf	6145mey
0300-0400	BBC World Sce	SAf	7445asc
0300-0400	Channel Africa	EAf	6120mey
0300-0400	CRI	SAs	15110kas
0300-0400	CRI	NAm	9690nob, 9790hab
0300-0400	Deutsche Welle	SAs	1548trm, 11695trm
0300-0400	Family Radio	CAm,SAm	9930yfr, 9985yfr
0300-0400	R. Bulgaria	NAm	5900pld, 7400pld
0300-0400	R. Sult.of Oman	EAf	13355thu
0300-0400	R.Taiwan Int.	SEA	15320pao
0300-0400	R.Taiwan Int.	NAm	6875yfr
0300-0400	VO Korea	EAs	4405kuj, 7220kuj, 9345kuj, 9730kuj
0300-0400	VO Russia (VOR)	NAm	13735vld
0300-0400	VO Russia (VOR)	NAm,SAm	7440lvi
0300-0400	WHRI	NAm,Eu	7590hri*
0300-0430	BBG-VOA	Af	1530sao, 9885bot
0300-0500	BBG-VOA	Af	15580bot
0300-0500	Channel Africa	SAf	3345mey
0300-0500	CRI	EAs	9460bei, 13620xia, 15120bei
0300-0500	Family Radio	SAf	1197mas
0300-0500	SLBC	SAs	6005eka, 9770eka, 15745eka
0300-0500	VO Russia (VOR)	NAm	12030ppk
0300-0600	BBC World Sce	SAs	15310sla
0300-0600	BBC World Sce	SAf	3255mey
0300-0600	BBC World Sce	CAf	7255asc
0300-0600	BBC World Sce	WAs	9410cyp
0300-0600	BBG-VOA	Af	909bot, 4930bot, 6080sao
0300-0600	CRI	NAm	6190sac
0300-0700	BBC World Sce	SAs	17790nak
0300-2200	BBC World Sce	SAf	6190mey
0330-0400	R. Tirana	NAm	6100shi
0330-0400	VO Vietnam	NAm	6175sac
0330-0600	BBC World Sce	EAf	11860sey
0400	**English**		
0400-0430	RFI	EAf	7315iss, 9805iss
0400-0430	R. Prague	NAm	7345lit
0400-0500	BBC World Sce	EAf	12035cyp
0400-0500	BBC World Sce	WAf	9460mey
0400-0500	BBG-VOA	Af	4960sao
0400-0500	Deutsche Welle	Af	5905sin, 5945sin, 6180kig, 9450kig, 15600trm
0400-0500	Family Radio	NAm	9715yfr
0400-0500	R.Romania Int	NAm	6130tig, 7305tig
0400-0500	R.Romania Int	ME	9690gal, 11895gal
0400-0500	VO Russia (VOR)	NAm	13735vld
0400-0500	VO Turkey	ME	7240cak

0400	English	Area	kHz
0400-0500	VO Turkey	Eu,NAm	9655emr
0400-0500	WHRI	Eu,Af	9640hri
0400-0600	BBC World Sce	CAs	15360sla
0400-0600	CRI	CAs	17725xia, 17855bei
0400-0600	1Africa Radio	Af	9430lus
0400-0600	Family Radio	NAm	5950yfr
0400-0600	VO Russia (VOR)	As	15735kna+
0400-0700	BBC World Sce	WAf	6005asc
0400-0700	Channel Africa	SAf	7230mey
0400-0700	Family Radio	NAm	9680yfr
0400-1000	WHRI	NAm,Eu	7465hri
0400-1500	AFRTS (AFN)	IOc	12759dga
0430-0500	R. Australia	As	15415shp
0430-0500	R. Prague	EAf,ME	9855lit
0430-0500	R. Tirana	NAm	6100shi
0430-0500	TWR Africa	WAf	1566par
0430-0500	TWR Africa	SAf	3200man
0430-0600	BBG-VOA	Af	9885sao
0430-0600	TWR Africa	SAf	4775man
0455-0700	VO Nigeria	Eu	15120iko
0500	**English**		
0500-0530	BBC World Sce	EAf	15420sey
0500-0530	Deutsche Welle	Af	6130sin, 6155sin, 6180kig, 9755kig, 12045kig
0500-0530	RFI	EAf	9805iss, 11995iss
0500-0530	R. Japan	SAs	15205tac
0500-0530	R. Japan	SEA	17810yam
0500-0530	R. Japan	Eu	5975rmp
0500-0530	R. Japan	NAm	6110sac
0500-0530	R. Japan	SAf	9770iss
0500-0530	Vatican Radio	Af	7360smg, 9660mdc, 11625smg
0500-0600	BBC World Sce	EAf	15420sey, 15420sey
0500-0600	BBG-VOA	EAf,ME	15580bot
0500-0600	CRI	NAm	5960sac
0500-0600	CRI	ME,NAf	7220cer
0500-0600	R.Taiwan Int.	NAm	6875yfr
0500-0600	WHRI	EAs,SEA	11565hri
0500-0700	BBC World Sce	WAf	11770mey
0500-0700	BBC World Sce	CEu	1296orf+
0500-0700	BBC World Sce	NAf	17640cyp
0500-0700	BBC World Sce	RUS	5875rmp, 12095cyp
0500-0700	CRI	Af,ME	17505kas
0500-0700	R. Habana Cuba	NAm	6010hab, 6060hab
0500-0700	R. Habana Cuba	Am	6150hab
0500-0700	RNZI	Pac	11725ran*, 13730ran*, +
0500-0700	VO Russia (VOR)	NAm	9840ppk, 9855vld
0500-0800	BBC World Sce	Eu	3955skn
0500-0800	R. Australia	Pac	15160shp
0500-0800	TWR Africa	SAf	4775man
0500-0800	TWR Africa	CAf	9500man
0500-0900	CRI	SAs	11880kas, 15465kas, 17540kas
0500-0900	R. Australia	As	13630shp
0500-1100	CRI	SAs	15350kas
0500-1700	Christian Voice	Af	6065lus
0502-0800	TWR Africa	SAf	6120man
0530-0600	R. Australia	As	15415shp
0530-0600	R. Thailand WS	Eu	11730udo
0530-0600	Sudan R. Sce	SDN	13720dha, 13720dha
0530-0600	TWR Africa	WAf	1566par
0600	**English**		
0600-0620	Vatican Radio	Eu	4005vat
0600-0620	Vatican Radio	Eu	7250smg
0600-0630	Deutsche Welle	WAf	5945wof, 7240sin, 15205kig
0600-0630	Lao National R.	SEA	7145vie‡
0600-0630	R. Australia	As	15290tnn
0600-0630	R. Australia	SEA	15415shp
0600-0630	RFI	EAf	13680iss, 15160iss
0600-0630	RFI	CAf,WAf	7315iss*, 9765iss**
0600-0630	Vatican Radio	Eu	1530smg
0600-0630	Vatican Radio	Eu	5965smg
0600-0700	BBC World Sce	WAs	11760cyp

0600	English	Area	kHz
0600-0700	BBC World Sce	WAf	9410asc, 12015asc
0600-0700	BBG-VOA	EAf,SAf	15580bot
0600-0700	BBG-VOA	Af	909bot, 1530sao, 6080sao, 9885mey
0600-0700	Channel Africa	WAf	15255mey
0600-0700	CRI	NAf	11750cer
0600-0700	CRI	ME	11770kas, 15145kas
0600-0700	CRI	NAm	6115sac
0600-0700	Family Radio	Eu	11530yfr
0600-0700	Family Radio	CAm,SAm	6000yfr
0600-0700	Family Radio	Af	9985yfr
0600-0700	VO Greece	Eu	11645avl*
0600-0700	WHRI	Eu,Af	9615hri
0600-0800	BBC World Sce	EAf	15420sey
0600-0800	CRI	SEA	13645xia, 17710bei
0600-0800	Family Radio	Eu	5745yfr
0600-0800	VO Malaysia	SEA,Pac	6175kaj, 9750kaj, 15295kaj
0600-0900	VO Russia (VOR)	Eu	1323wbr
0600-0900	WEWN - EWTN	Af	11520ewn
0600-1400	BBC World Sce	SAs	15310nak
0600-1600	BBC World Sce	SAf	9860mey
0600-2000	1Africa Radio	Af	13590lus
0630-0700	R. Australia	As	15415shp
0630-0700	R.Romania Int	AUS,NZL	17780gal, 21600tig
0630-0700	R.Romania Int	Eu	6020gal+, 7370tig
0630-0700	Vatican Radio	Af	7360smg, 9660smg, 11625smg
0700	**English**		
0700-0730	RFI	WAf,CAf	11725iss*, 15605iss**
0700-0730	R. Slovakia Int	As,AUS	13715rso, 15460rso
0700-0800	BBC World Sce	WAf	11770asc, 13820asc, 17830mey
0700-0800	BBC World Sce	RUS	5875mos+
0700-0800	CRI	Af,ME	15125kas
0700-0800	Family Radio	CAm,SAm	9495yfr
0700-0800	RNZI	Pac	11675ran*, +
0700-0800	TDPradio	Eu	6015iss+
0700-0900	CRI	Eu	1215fla, 11785cer
0700-0900	Family Radio	Af	11580yfr
0700-0900	R. Australia	Pac	9710shp
0700-0900	VO Russia (VOR)	Eu	11635msk+
0700-1000	BBC World Sce	WAf	15400asc
0700-1000	Family Radio	NAm	6875yfr
0700-1100	RNZI	Pac	9765ran*
0700-1100	VO Russia (VOR)	AUS,NZL	17805irk
0700-1100	WHRI	EAs,SEA	11565hri
0700-1200	Channel Africa	SAf	9625mey
0700-1300	BBC World Sce	WAs	15575cyp
0700-1300	BBC World Sce	SAs	17790sla
0700-1300	CRI	Eu	17490kas
0700-1300	R. Australia	Pac	11945shp
0700-1300	R. Australia	As	9475shp
0700-1400	BBC World Sce	ME	11760sla
0700-1400	Family Radio	NAm	7455yfr
0730-0745	Vatican Radio	Af	15595smg
0730-0745	Vatican Radio	Eu	4005vat, 11740smg
0730-0745	Vatican Radio	Eu	585vat, 1530smg, 1611smg, 5965smg, 7250smg, 9645smg
0730-0800	R. Bulgaria	Eu	5900pld, 7400pld
0730-0930	HCJB Australia	As	11750knx
0745-0850	TWR Europe	Eu	6105nau, 9800mco
0800	**English**		
0800-0830	BVB	Eu	7220wer
0800-0830	R. Prague	NEu	7345lit
0800-0845	BVB	Eu	7220wer
0800-0850	TWR Europe	Eu	6105nau, 9800mco
0800-0900	Am.R.Mirr Int.	EAf	17860mey
0800-0900	Am.R.Mirr Int.	SAf	7205mey
0800-0900	BBC World Sce	Eu	5875wof
0800-0900	BBC World Sce	WEu	9610sin+
0800-0900	CRI	Af,ME	15625kas
0800-0900	KBS World R.	SEA	9570kim
0800-0900	R. Australia	Pac	5995brn

0800	English	Area	kHz
0800-0900	R. Ukraine Int.	Eu	9410smf
0800-0900	TDPradio	Eu	6015iss+
0800-0900	WHRI	EAs,SEA	11565hri
0800-1000	BBC World Sce	CAf	17830asc
0800-1000	CRI	EAs	9415xia
0800-1000	VO Malaysia	SEA,Pac	6175kaj*, 9750kaj*, 15295kaj*
0800-1000	VO Russia (VOR)	As	1251dsb
0800-1100	VO Russia (VOR)	SEA,AUS,NZL	17650dsb
0800-1100	VO Russia (VOR)	AUS,NZL	17665kna
0800-1200	RNZI	Pac	9870ran*, +
0800-1300	BBC World Sce	EAf	17640sey
0800-1300	R. Africa	Af	15190bat*
0800-1400	BBC World Sce	SAf	21470sey
0800-1400	R. Australia	Pac	9580shp
0800-1600	R. Australia	Pac	9590shp
0800-2000	AFRTS (AFN)	Pac	5765bar
0815-0850	TWR Europe	Eu	6105nau, 9800mco
0820-0900	TWR Asia (KTWR)	Pac	15170twr
0830-0910	TWR Asia (KTWR)	Pac	11840twr
0900	**English**		
0900-1000	BBC World Sce	FE	17760nak
0900-1000	CRI	Eu	15270kas, 17570uru
0900-1000	CRI	SAs	17750kas
0900-1000	Deutsche Welle	EAs	17710trm, 21780trm
0900-1000	IRRS Shortwave	Eu,ME,NAf	9510rso
0900-1000	TDPradio	Eu	6015iss+
0900-1000	VO Greece	Eu	15630avl
0900-1000	VO Greece	Eu,As,Am	9420avl
0900-1100	BBC World Sce	SEA	6195nak, 6195sng, 15285sng
0900-1100	CRI	Pac	15210kun, 17690jin
0900-1100	Family Radio	SEA	9465pao
0900-1100	WWCR	NAm,Eu,Af	9985wcr**
0900-1200	BBC World Sce	FE	11895nak
0900-1200	WEWN - EWTN	SEA	9390ewn
0900-1500	VO Nigeria	WAf	9690iko
0900-1600	BBC World Sce	SEA	9740sng
0900-2300	BBC World Sce	ME	1323cyp
1000	**English**		
1000-1030	R. Bulgaria	Eu	11900sof+
1000-1030	R. Japan	SAs	11780tac
1000-1030	R. Japan	SEA	9605yam
1000-1030	R. Japan	Pac	9625yam, 9840yam
1000-1030	R. Prague	WAf	21745lit
1000-1030	VO Vietnam	As	9840vni, 12020vni
1000-1100	All India R.	SAs	1053vat, 7270cni, 15260del
1000-1100	All India R.	As,Pac	13710bgl
1000-1100	All India R.	EAs	15235bgl, 17800bgl
1000-1100	All India R.	Pac	17510del, 17895bgl
1000-1100	BBC World Sce	WAf	15400asc
1000-1100	BBC World Sce	CAf	17830asc
1000-1100	CRI	SAs	15190uru
1000-1100	CRI	EAs	5955xia
1000-1100	CRI	CAs	7215xia, 11640bei
1000-1100	Family Radio	EAs	9460irk, 9460nvs
1000-1100	KNLS Int.	EAs	9615nls
1000-1100	RNW	SEA	9720pht, 12065pht
1000-1100	TDPradio	Eu	6015iss+
1000-1100	VO Indonesia	As,Pac	9526jak*
1000-1100	VO Korea	SEA	6185kuj, 9850kuj
1000-1100	VO Korea	LAm	6285kuj, 9335kuj
1000-1200	CRI	SEA	13590bei, 13720xia
1000-1200	VO Russia (VOR)	SEA	7205tch
1000-1200	WHRI	NAm	7520hri
1000-1200	WWCR	NAm,Eu,Af	9985wcr*
1000-1230	BSKSA	WAf,CAf	15250riy
1000-1300	BBC World Sce	FE	9605nak
1000-1400	Family Radio	NAm	6890yfr
1000-1500	WINB	CAm	9265inb**
1000-2200	University Netw	NAm	11775aia
1030-1100	VO Mongolia	As	12085uba
1030-1130	VOIRI	SAs	15460kam, 17630kam
1030-1130	VOIRI	ME	702kia

	English	Area	kHz
1000	**English**		
1030-1130	VOIRI	WAs	765chb
1030-1300	IRRS Shortwave	Eu,ME,NAf	9510rso
1100	**English**		
1100-0100	WWCR	NAm,Af	9980wcr**
1100-1105	R. Pakistan	Eu	15100isl, 17700isl
1100-1130	BBC World Sce	WAf	15400asc
1100-1130	BBG-VOA	SEA	1575bph
1100-1130	KBS World R.	Eu	9760wof+
1100-1130	R. Japan	WEu	9760wof+
1100-1130	Vatican Radio	Eu	7250smg
1100-1130	VO Vietnam	As	7285han
1100-1200	CRI	NAm	5960sac
1100-1200	CRI	SAs	9570kas, 11795kas, 13645kas
1100-1200	Family Radio	CAm,SAm	6000yfr, 11725yfr, 11830yfr
1100-1200	Family Radio	NAm	6875yfr
1100-1200	Family Radio	EAs	7300irk
1100-1200	R. Australia	Pac	5995brn, 12080brn+
1100-1200	R.Taiwan Int.	SEA	7445pao, 11715tnn
1100-1200	TDPradio	Eu	6015iss+
1100-1230	FEBC Korea	SEA	1188seo
1100-1300	CRI	SAs	11650uru
1100-1300	CRI	SEA	1269xuw
1100-1300	CRI	Eu	13665cer
1100-1300	R. Australia	SEA	9965hbn
1100-1300	RNZI	Pac	13660ran*
1100-1400	R. Australia	Pac	6020shp, 9560shp
1100-1600	BBC World Sce	SEA	6195sng
1100-1600	CRI	EAs	5955bei
1100-1600	WINB	CAm	9265inb*
1100-1800	BBC World Sce	CAf	17830asc
1100-2000	WWCR	NAm,Eu,Af	15825wcr**
1100-2200	WTWW	NAm	9480tww***
1100-2200	WTWW	NAm,Eu	9990tww***
1130-1145	Etern.Good News	SAs	15525dha
1130-1200	BBG-VOA	SEA	1575bph
1130-1200	R. Prague	NEu	9880lit
1130-1200	VO Vietnam	As	9840vni, 12020vni
1200	**English**		
1200-0200	WWCR	NAm,Af	9980wcr*
1200-0600	WBCQ	NAm,CAm	9330bcq*
1200-1215	TWR India	As	882put
1200-1230	AWR	As	15495nau
1200-1230	RFI	EAf	21620iss
1200-1230	R. Japan	NAm	6120sac
1200-1230	R. Japan	Pac	9625yam
1200-1230	R. Japan	SEA	9695yam
1200-1230	R. Japan	Eu	9790wer
1200-1300	BBG-VOA	SEA,Pac	11705udo
1200-1300	BBG-VOA	EAs	1170php, 9640pht, 11750pht
1200-1300	BBG-VOA	EAs,SEA	7575ira
1200-1300	CRI	CAs	11690xia
1200-1300	CRI	SEA	684dof, 1188kun, 9600kun, 9645bei, 9730kun
1200-1300	CRI	SAs	7250kas, 9460kas, 12015kas
1200-1300	CRI	Pac	9760kun
1200-1300	Family Radio	CAm,SAm	11530yfr, 17545yfr
1200-1300	KBS World R.	NAm	9650sac
1200-1300	KNLS Int.	EAs	9615nls
1200-1300	R.Romania Int	Eu	11970tig, 15460tig
1200-1300	R.Romania Int	Af	15430gal, 17765gal
1200-1300	TDPradio	Eu	6015iss+
1200-1300	VO Russia (VOR)	As	7340irk+, 7350tch, 9695sam, 11660dsb
1200-1300	WHRI	NAm	7315hri, 7385hri
1200-1400	CRI	Pac	11760kun
1200-1400	CRI	SEA	1341hdu, 11980kun
1200-1400	CRI	Eu	13790uru
1200-1400	Family Radio	NAm	11970yfr
1200-1400	R. Australia	Pac	5995brn+
1200-1400	WHRI	Eu,Af	15665hri

	English	Area	kHz
1200	**English**		
1200-1600	BBC World Sce	FE	5875nak
1200-1600	WWCR	Af	7490wcr**
1200-1900	WEWN - EWTN	Eu	15610ewn
1200-2100	WWCR	NAm,Eu,Af	15825wcr*
1200-2300	WBCQ	NAm,CAm	15420bcq*
1200-2300	WTWW	NAm	9479tww**
1200-2300	WTWW	NAm,Eu	9990tww**
1200-2300	WWRB	NAm,CAm	9385wrb
1200-2400	WWCR	NAm,Eu	13845wcr**
1215-1300	BBC World Sce	Car	9410hri, 11860guf
1215-1330	R. Cairo	SAs	17870abz
1230-1300	HCJB Australia	EAs	15400knx
1230-1300	R. Thailand WS	As,Pac	9720udo
1230-1300	VO Vietnam	As	9840vni, 12020vni
1300	**English**		
1300-0100	WWCR	Af	13845wcr*
1300-1330	HCJB Australia	EAs	15400knx
1300-1330	R. Japan	SAs	9875yam
1300-1400	BBC World Sce	CAs	11805cyp
1300-1400	BBC World Sce	SAs	1413sla
1300-1400	BBC World Sce	EAf	15420sey
1300-1400	BBC World Sce	Eu	15575skn
1300-1400	BBG-VOA	SEA,Pac	11705udo
1300-1400	BBG-VOA	EAs,SEA	7575pht
1300-1400	BBG-VOA	EAs	9640pht
1300-1400	BBG-VOA	SAs,SEA	9760pht
1300-1400	CRI	Pac	11900kun
1300-1400	CRI	Eu	13670kas
1300-1400	CRI	SAs	7300kas, 9655kas
1300-1400	CRI	NAm	9570hab, 11885sac
1300-1400	CRI	SEA	9730bei
1300-1400	Family Radio	NAm	11830yfr
1300-1400	Family Radio	SEA	6075tch, 9310alm, 11520pao
1300-1400	KBS World R.	SEA	9570kim
1300-1400	Polish Radio	Eu	9460mos, 11860wof
1300-1400	TDPradio	Eu	6015iss+
1300-1400	VO Indonesia	As,Pac	9526jak*
1300-1400	VO Korea	EAs	3560kuj
1300-1400	VO Korea	Eu	7570kuj, 12015kuj
1300-1400	VO Korea	NAm	9335kuj, 11710kuj
1300-1400	VO Tajik	WAs,ME	1143dsb, 7245dsb
1300-1500	BBC World Sce	SAs	9410sla
1300-1500	CRI	NAm	15230sac
1300-1500	CRI	CAs	9765xia
1300-1500	Family Radio	SEA	11560huw
1300-1500	VO Russia (VOR)	SEA	7205tch
1300-1550	RNZI	Pac	5950ran*
1300-1600	CRI	SEA	9870xia
1300-1600	Family Radio	NAm	11855yfr
1300-1600	WWCR	NAm,Eu	7490wcr*
1300-1700	BBC World Sce	SAf	17640asc
1300-2200	WHRI	NAm	9840hri
1300-2200	WTWW	NAm	9480tww*
1300-2200	WTWW	NAm,Eu	9990tww*
1330-1400	AWR Guam	SEA	11935sda
1330-1400	AWR Guam	As	15660sda
1330-1400	Lao National R.	SEA	7145vie‡
1330-1400	VO Vietnam	As	9840vni, 12020vni
1330-1430	VO Turkey	As,Pac	11735emr
1330-1430	VO Turkey	Eu	12035cak
1330-1500	All India R.	SEA	9690bgl, 11620del, 13710bgl
1345-1415	BVB	As	13635iss
1400	**English**		
1400-1425	TWR Asia (KTWR)	SEA	9975twr
1400-1430	FEBA Radio	As	12045dha
1400-1430	PanAm Bc	IND	13645wer
1400-1430	R. Japan	CAf	21560iss
1400-1430	R. Japan	SEA	5955yam
1400-1430	R. Japan	SAs	9875yam
1400-1430	R. Prague	SAs	11600lit
1400-1430	R. Thailand WS	As,Pac	9725udo
1400-1435	TWR Asia (KTWR)	SEA	9975twr
1400-1500	BBC World Sce	WAs	11760cyp

1400	English	Area	kHz
1400-1500	BBC World Sce	CAs	9915cyp
1400-1500	BBG-VOA	Af	15580mey, 17650sao, 17715bot
1400-1500	BBG-VOA	SAs	7575udo, 12150ira
1400-1500	BBG-VOA	SAs,SEA	9760udo
1400-1500	BVB	As	13635iss
1400-1500	CRI	ME	11665uru
1400-1500	CRI	NAm	13675sac
1400-1500	CRI	SAs	7300uru, 9460uru
1400-1500	CRI	Eu	9700kas, 9795uru
1400-1500	Family Radio	NAm	13695yfr
1400-1500	Family Radio	EAs	5995ppk, 5995ppk, 6115ppk, 6115ppk
1400-1500	Family Radio	SEA	6070tch
1400-1500	Family Radio	IND	9485irk
1400-1500	RNW	SAs	12080pht
1400-1500	R. Sult.of Oman	Eu,ME	15140thu
1400-1500	Disco Palace	Eu	6015iss+
1400-1500	Overcomer Min.	Eu,NAf	13810nau
1400-1500	VO Russia (VOR)	As	11660dsb, 12055msk
1400-1500	WHRI	Eu,Af	17540hri
1400-1530	BBG-VOA	Af	6080bot
1400-1600	BBC World Sce	SAs	5845nak, 13590trm+
1400-1600	CRI	SEA	1188kun
1400-1600	CRI	NAm	13740hab
1400-1600	CRI	WAf,CAf	17630bko
1400-1600	RNW	SAs	15595mdc
1400-1600	VO Africa	CAf,EAf	17725sab, 21695sab
1400-1700	BBC World Sce	EAf	15420cyp
1400-1700	BBC World Sce	SAs	5975sng
1400-1700	BBG-VOA	Af	4930bot
1400-1700	Family Radio	NAm	11565yfr, 17760yfr
1400-1700	Family Radio	EAs,SEA	1557kou
1400-1700	R. Australia	Pac	7240shp
1400-1800	R. Australia	Pac	5995shp
1400-2000	R. Australia	Pac	6080shp
1400-2200	WJHR	NAm	15550jhr*
1405-1430	BVB	EAs	9345alm
1415-1430	PanAm Bc	IND	13645wer
1415-1430	TWR India	As	882put
1425-1455	TWR Africa	SAf	6025man
1430-1445	PanAm Bc	IND	13645iss
1430-1500	BVB	As	13635iss
1430-1700	R. Australia	As	11660shp
1430-1900	R. Australia	As	9475shp
1445-1500	TWR Europe	CAs	1467bis
1445-1530	HCJB Australia	As	15340knx
1500	**English**		
1500-0400	AFRTS (AFN)	IOc	4319dga
1500-1515	BVB	As	12035wer
1500-1515	TWR India	As	882put
1500-1530	AWR Guam	IND	12025sda
1500-1530	BBC World Sce	EAf	9410sey, 11860sey, 15105mey
1500-1530	VO Vietnam	As	7285han, 9840vni, 12020vni
1500-1600	BBC World Sce	SAs	7395sla
1500-1600	BBC World Sce	WAs	9485cyp
1500-1600	BBG-VOA	ME	11765lam, 12055lam
1500-1600	BBG-VOA	Af	15580sao, 17715bot, 17895sao
1500-1600	BBG-VOA	EAs	6140udo, 9760pht, 9945ira
1500-1600	BBG-VOA	SAs,SEA	7520pht, 7520pht, 9930ira
1500-1600	BBG-VOA	SAs	7575udo, 12150ira
1500-1600	Channel Africa	SAf	9625mey
1500-1600	CRI	NAf,ME	6095kas, 9720uru
1500-1600	CRI	SEA	7325bei
1500-1600	CRI	SAs	7405uru, 9785jin
1500-1600	Family Radio	CAm,SAm	15210yfr
1500-1600	Family Radio	NAm	15795yfr
1500-1600	Family Radio	IND	6280tsh, 9495dha, 12015dha, 21840asc

1500	English	Area	kHz
1500-1600	KNLS Int.	RUS	9655nls
1500-1600	RCI	SAs	9635xia, 11975uru
1500-1600	RNW	SAs	12080pht
1500-1600	Overcomer Min.	Af	17485wer
1500-1600	VO Indonesia	As,Pac	9526jak*
1500-1600	VO Korea	EAs	3560kuj
1500-1600	VO Korea	Eu	7570kuj, 12015kuj
1500-1600	VO Korea	NAm	9335kuj, 11710kuj
1500-1600	VO Nigeria	WAf	15120iko
1500-1600	VO Russia (VOR)	As	1251dsb
1500-1600	VO Russia (VOR)	Eu	5905msk+, 9675msk+
1500-1600	VO Russia (VOR)	SEA	7260vld, 9660xia
1500-1600	WHRI	Af	21630hri
1500-1700	BBC World Sce	Eu,ME	11830rmp
1500-1700	CRI	Eu	9435kas, 9525kas
1500-1800	CRI	SAs	1323uru
1500-2000	VO Russia (VOR)	As,ME	4975dsb
1500-2100	BBC World Sce	WAf	15400asc
1500-2100	WINB	CAm	13570inb**
1500-2300	WHRI	NAm	15180hri
1515-1545	BVB	As	13670wer
1525-1555	TWR Africa	SAf	6025man
1530-1550	Vatican Radio	As	7585dsb, 11850smg, 13765smg
1530-1600	AWR	As	11675wer
1530-1600	BVB	As	13670wer
1530-1600	R. Tirana	NAm	13640shi
1530-1600	Vatican Radio	Af	13765smg
1530-1600	Vatican Radio	As	7585dsb, 11850smg
1530-1600	VO Mongolia	As	9665uba
1530-1630	VOIRI	SAs	9915sir, 11655kam
1530-1700	BBC World Sce	EAf	9410sey, 11860sey, 15105mey
1530-1700	BBG-VOA	Af	6080mey
1550-1650	RNZI	Pac	5950ran*+, 7440ran*
1600	**English**		
1600-1610	R. Pakistan	ME,Af	7510isl, 11575isl
1600-1630	AWR Guam	IND	9585sda, 11690sda
1600-1630	VO Vietnam	ME	7220vni, 9550vni
1600-1630	VO Vietnam	Eu	7280vni, 9730vni
1600-1700	BBC World Sce	SAs	7355sla
1600-1700	BBG-VOA	EAs	1170php
1600-1700	BBG-VOA	Af	909bot, 1530sao, 9395ira, 13600sao, 15460lam, 15580sao, 17895bot
1600-1700	CRI	SEA	6060kun
1600-1700	CRI	ME	6100kas, 9600jin
1600-1700	CRI	SAs	7235kas
1600-1700	CRI	NAf,ME	7420uru
1600-1700	Deutsche Welle	SAs	1548trm, 5845nak+, 5965trm, 9560trm, 13590trm+
1600-1700	Family Radio	IND	11740dha
1600-1700	Family Radio	RUS	1503msk
1600-1700	Family Radio	Af	17540yfr, 17690yfr
1600-1700	Family Radio	CAm,SAm	6085yfr
1600-1700	KBS World R.	Eu	9515kim
1600-1700	KBS World R.	SEA	9640kim
1600-1700	R. Ethiopia	EAf,ME	7165gjw†, 9560gjw±
1600-1700	RFI	Af	15605iss
1600-1700	R.Taiwan Int.	SAs	11550tnn, 12055iss
1600-1700	VO Korea	ME,NAf	9990kuj, 11545kuj
1600-1700	VO Russia (VOR)	Eu	6130msk
1600-1700	VO Russia (VOR)	SEA	7330ppk
1600-1700	VO Russia (VOR)	As	972dsb, 7305nvs, 11630dsb
1600-1800	BBC World Sce	WAf	13790asc
1600-1800	BBC World Sce	SAs	9740sng
1600-1800	CRI	SEA	1080xuw
1600-1800	CRI	Eu	7255kas
1600-1800	CRI	SAf	7435jin, 9570bei
1600-1800	Family Radio	Eu	18980yfr
1600-1800	R. Cairo	CAf,SAf	12170abz

1600	English	Area	kHz
1600-1800	VO Russia (VOR)	ME,Af	9470msk
1600-1900	Family Radio	SAf	1197mas
1600-1900	VO Russia (VOR)	SEA	9880ppk
1600-1900	WHRI	Af	21630hri
1600-2000	Family Radio	NAm	13695yfr
1600-2000	R. Australia	Pac	9710shp
1600-2000	WWCR	Af	12160wcr**
1600-2100	WWCR	NAm,Eu	12160wcr*
1600-2200	BBC World Sce	SAf	3255mey
1600-2200	WINB	CAm	13570inb*
1605-1805	RCI	NAm	9610sac, 9800sac+
1615-1630	TWR India	As	882put
1615-1645	TWR India	As	882put
1615-1700	BBC World Sce	EAf	9410sey, 11860sey, 15105mey
1630-1700	AWR Guam	IND	9790sda
1630-1700	BBC World Sce	EAf	9410sey
1630-1700	BBG-VOA	Af	9785nau*, 11905wer*, 13635wer*
1630-1915	BVB	ME	9460nau
1645-1700	BVB	ME	9460nau
1645-1700	TWR India	As	882put
1645-1715	BVB	ME	9460nau
1645-1715	TWR India	As	882put
1645-1720	BVB	ME	9460nau
1645-1745	BVB	ME	9460nau
1645-1830	BVB	ME	9460nau
1650-1750	RNZI	Pac	9765ran*, 9890ran*+
1700	**English**		
1700-1715	TWR India	As	882put
1700-1730	BBG-VOA	EAf	15580grv
1700-1730	R. Prague	NEu	5930lit
1700-1730	TWR India	As	882put
1700-1746	BBC World Sce	EAf	9410sey, 11860sey
1700-1800	BBC World Sce	CAs	1251dsb
1700-1800	BBC World Sce	SAs	5975nak
1700-1800	BBG-VOA	ME	13635lam
1700-1800	BBG-VOA	Af	6080sao, 17895bon
1700-1800	Channel Africa	WAf	15235mey
1700-1800	CRI	SEA	6090kun, 7420kun
1700-1800	CRI	SAs	6140kas, 7410kas, 7425kas
1700-1800	CRI	ME	6165bei
1700-1800	CRI	Eu	7205bei
1700-1800	Deutsche Welle	SAs	1548trm+, 5845nak+
1700-1800	Family Radio	NAm	15795yfr
1700-1800	Family Radio	Af	7230mdc, 7385mdc, 21680yfr
1700-1800	R.Taiwan Int.	Eu	11850iss*, 15690iss**
1700-1800	VO Russia (VOR)	As	1269xuw
1700-1830	BBC World Sce	SAs	5975sla
1700-1900	BBC World Sce	EAf	15420mey
1700-1900	CRI	Eu	6100bei
1700-1900	VO Russia (VOR)	As	1251dsb, 7240ppk
1700-1900	VO Russia (VOR)	SEA	7330ppk
1700-2000	Family Radio	EAs	1359fan
1700-2000	Family Radio	SEA	1359fan
1700-2000	R. Australia	Pac	9580shp
1700-2000	TWR Africa	SAf	3200man
1700-2030	TWR Africa	SAf	3200man
1700-2100	BBC World Sce	NAf	12095cyp
1700-2100	R. Australia	Pac	11880shp
1700-2105	TWR Africa	SAf	1170man
1700-2200	Christian Voice	Af	4965lus
1700-2200	Family Radio	NAm	17555yfr
1700-2300	R. Africa	Af	15190bat*
1715-1730	Vatican Radio	Eu	585vat, 4005vat, 5885smg, 7250smg, 9645smg
1715-1730	Vatican Radio	NAf	7290smg
1720-1740	BBG-VOA Studio7	ZWE	909bot, 4930bot, 12080mdc, 15775sao
1730-1800	BBC World Sce	SAs	1413sla
1730-1800	BBG-VOA Studio7	ZWE	909bot, 4930bot,

1700	English	Area	kHz
			12080mdc, 15775sao
1730-1800	BBG-VOA	EAf	15580bot
1730-1800	R. Slovakia Int	WEu	5915rso, 6010rso
1730-1800	Sudan R. Sce	SDN	9590dha, 9840dha
1730-1800	Vatican Radio	Af	9755smg, 11625smg, 13765smg
1740-1825	TWR Africa	WAf	1566par
1745-1945	All India R.	Eu	6280bgl, 7550bgl, 9950del+, 11620bgl
1745-1945	All India R.	EAf	7400del, 9415del, 11935mum
1745-1945	All India R.	NAf,WAf	7410del, 9445bgl, 13605bgl
1750-1850	RNZI	Pac	11675ran*, +
1750-2150	RNZI	Pac	11725ran*
1800	**English**		
1800-1815	BVB	ME	7365wer
1800-1830	AWR	SAf	3215mey, 3345mey
1800-1830	BBC World Sce	WAs	7260cyp
1800-1830	BBC World Sce	SAs	7355nak
1800-1830	BBG-VOA	Af	13635ira, 15580bot
1800-1830	BBG-VOA	SAf	4930bot
1800-1830	BBG-VOA	Af	909bot
1800-1830	BVB	ME	6110wer
1800-1830	R. Prague	NEu	5930lit
1800-1830	R.Romania Int	Eu	6020kvi+
1800-1830	VO Vietnam	Eu	5955mos
1800-1900	BVB	ME	6110wer
1800-1900	CRI	Eu	7405bei
1800-1900	Family Radio	Af	6045mey, 7240skn, 9895dha, 11665wer
1800-1900	KBS World R.	Eu	7275kim
1800-1900	Polish Radio	Eu	5895kvi+, 9650dha
1800-1900	RCI	NAf,EAf	11845smg
1800-1900	RCI	CAf,NAf	15365sac
1800-1900	RCI	NAf,WAf	17790sac
1800-1900	RCI	EAf,ME	9740kas
1800-1900	RNW	SAf	6020mdc
1800-1900	R.Romania Int	Eu	6065tig+, 7415tig
1800-1900	R.Taiwan Int.	Eu	3965iss
1800-1900	RAE	Eu	15345bue
1800-1900	Rep of Yemen R.	ME	9780sant
1800-1900	TWR Africa	SAf	9500man
1800-1900	VO Korea	EAs	3560kuj
1800-1900	VO Korea	Eu	7570kuj, 12015kuj
1800-1900	VO Nigeria	Eu,Af	15120iko
1800-1900	VO Russia (VOR)	Af	7270erv
1800-1900	VO Russia (VOR)	ME	7305msk
1800-2000	BBC World Sce	RUS	5875rmp, 7225cyp
1800-2000	BBC World Sce	WAs	5945cyp
1800-2000	BBC World Sce	CAs	5955sla
1800-2000	BBG-VOA	Af	6080sao
1800-2000	Family Radio	Af	7395mdc
1800-2000	R. Australia	Pac	7240shp
1800-2100	BBC World Sce	WAf	11810asc
1800-2100	BBC World Sce	NAf	9615skn
1800-2100	R. Kuwait	Eu,NAm	15540kbd‡
1800-2100	RNW	CAf,WAf	11655mdc
1800-2200	Family Radio	Af	15115yfr
1800-2200	Family Radio	NAm	17535yfr
1800-2200	Family Radio	Eu	6915yfr
1800-2200	VO Russia (VOR)	Eu	7330msk
1830-1900	AWR	EAf	11830mey
1830-1900	BBG-VOA	Af	13635ira
1830-1900	BBG-VOA	Af	909bot
1830-1900	BBG-VOA	SAf	909bot
1830-1900	R. Bulgaria	Eu	6200pld, 7400pld, 9700sof+
1830-1900	R. PMR	Eu	6240kch
1830-1930	BBG-VOA	Af	15580bot
1830-2100	BBC World Sce	EAf	6005sey, 9410sey
1830-2100	BBG-VOA	Af	4930bot
1840-1855	TWR Europe	ME	864erv
1850-1950	RNZI	Pac	15720ran*, +

1900	English	Area	kHz
1900-1930	Deutsche Welle	Af	9735kig, 13780trm, 15275sin
1900-1930	VO Vietnam	Eu	7280vni, 9730vni
1900-2000	BBG-VOA	ME	7480ira, 9585bib
1900-2000	BVB	Eu	6030wer
1900-2000	Family Radio	Eu	15565yfr
1900-2000	Family Radio	Af	3230mey, 6020mdc, 9480yfr, 9705yfr, 9885dha
1900-2000	Family Radio	CAm,SAm	6085yfr
1900-2000	IRRS Shortwave	Eu,ME,NAf	6090rso
1900-2000	REE	Af	9605nob
1900-2000	REE	Eu	9665nob
1900-2000	RNW	WAf	11615mey
1900-2000	RNW	EAf,ME	9895trm
1900-2000	R. Thailand WS	Eu	7570udo
1900-2000	Overcomer Min.	Eu	5945wer
1900-2000	Overcomer Min.	Eu,NAf	6065mos, 9860wer
1900-2000	VO Indonesia	Eu,NAf,ME	9526jak*
1900-2000	VO Korea	SAf	7210kuj, 11910kuj
1900-2000	VO Korea	ME,NAf	9975kuj, 11535kuj
1900-2000	VO Nigeria	WAf	7255iko
1900-2000	VO Russia (VOR)	Af,Eu	12060msk
1900-2000	VO Russia (VOR)	ME	5985msk
1900-2000	WHRI	Af	17520hri
1900-2030	BBG-VOA	Af	4940sao
1900-2030	R. Cairo	WAf	11510abz
1900-2100	BBG-VOA	Af	909bot
1900-2100	CRI	Af,ME	7295kas, 9440kun
1900-2100	RNW	CAf,EAf	7425mdc
1900-2200	Family Radio	EAs	1557kou
1900-2200	Family Radio	Af	9480nau
1900-2200	R. Australia	As	9500shp
1900-2400	WEWN - EWTN	Af	15610ewn
1915-1945	BVB	Eu	6030wer
1930-2000	BBG-VOA	Af	15580bon
1930-2000	Int. R. Serbia	Eu	6100bij
1930-2000	R. Slovakia Int	WEu	5915rso, 7345rso
1930-2030	VOIRI	SAf	11695sir, 11860kam
1930-2030	VOIRI	Eu	6010kam, 6115sit, 7320sir
1930-2030	VO Turkey	Eu	6050cak
1945-2000	R. Tirana	NAm	11635shi
1945-2000	R. Tirana	Eu	7465shi
1950-2050	RNZI	Pac	17675ran*+
2000	**English**		
2000-0800	AFRTS (AFN)	Pac	13362bar
2000-2030	Vatican Radio	Af	7365smg, 9755smg, 11625smg
2000-2100	BBG-VOA	Af	1530sao, 15580bon
2000-2100	BBG-VOA	ME	9420bib, 9490lam
2000-2100	CRI	CAf,SAf	5985bei
2000-2100	Deutsche Welle	WAf	9690wof
2000-2100	Deutsche Welle	Af	9735kig, 13780trm, 15275trm
2000-2100	Family Radio	Af	11615asc, 15520asc
2000-2100	Family Radio	EAs	1503fan
2000-2100	Family Radio	SEA	1503fan
2000-2100	Family Radio	RUS	1503msk
2000-2100	Family Radio	CAm,SAm	17575yfr
2000-2100	Family Radio	Eu	5745yfr, 9850arm
2000-2100	R. Australia	Pac	6080shp, 7240shp, 12080brn
2000-2100	R. Habana Cuba	Am	11760hab
2000-2100	RNW	WAf	5935sin
2000-2100	R. Ukraine Int.	Eu	6030khr
2000-2100	Disco Palace	NAm	15755bon+
2000-2100	WHRI	NAm,Eu	7570hri*
2000-2100	WHRI	NAm,Eu	9895hri
2000-2130	CRI	CAf,SAf	11640bko
2000-2200	BBG-VOA	Af	6080sao
2000-2200	CRI	Eu	5960cer, 7285cer, 7415kas, 9600kas
2000-2200	1Africa Radio	Af	9505lus

2000	English	Area	kHz
2000-2200	Family Radio	Af	15195asc
2000-2200	Family Radio	Eu	7510alm
2000-2200	R. Australia	Pac	11650shp, 11660shp
2000-2200	WHRI	Af	15665hri
2000-2200	WWCR	NAm	7465wcr**
2000-2200	WWCR	Af	9350wcr**
2000-2300	Family Radio	SAf	1197mas
2000-2400	VO Russia (VOR)	Eu	1215klg
2005-2105	Am.R.Mirr Int.	SAf	3215mey
2030-0030	BBG-VOA Ashna R	AFG	1296kab*, 7405kwt*
2030-2045	R. Thailand WS	Eu	9535udo
2030-2100	BBG-VOA	Af	4940sao
2030-2100	R. PMR	Eu	6240kch
2030-2100	VO Vietnam	ME	7220vni, 9550vni
2030-2100	VO Vietnam	Eu	7280vni, 9730vni
2045-2130	Vatican Radio	Eu	9800sac+
2045-2230	All India R.	Eu,Pac	11620bgl
2045-2230	All India R.	Pac	11715pan
2045-2230	All India R.	Eu	6280bgl, 7550del, 9445bgl, 9950del+
2050-2120	Vatican Radio	Eu	1530smg, 4005vat, 5885smg
2050-2120	Vatican Radio	Af	7250smg
2050-2150	RNZI	Pac	15720ran*+
2100	**English**		
2100-0300	WINB	CAm	9265inb**
2100-2130	AWR	WAf	9830mos
2100-2130	R. Prague	NEu	5930lit
2100-2130	R. Tirana	Eu	7530shi
2100-2130	R. Tirana	NAm	9895shi
2100-2200	BBC World Sce	SAf	5910sey
2100-2200	BBC World Sce	SEA	6195nak
2100-2200	BBC World Sce	WAf	7465mey
2100-2200	BBG-VOA	Af	1530sao, 15580bot
2100-2200	CRI	SAf	7205xia
2100-2200	CRI	CAf,SAf	7405bei
2100-2200	Deutsche Welle	WAf	7280sin, 9545trm, 11865kig, 13780trm
2100-2200	R. Australia	As,Pac	11695shp
2100-2200	R. Australia	Pac	9660brn, 12080brn
2100-2200	R .Damascus	Eu,NAm,Pac	9330adr, 12085adr‡
2100-2200	TDPradio	NAm	15755bon+
2100-2200	VO Korea	EAs	3560kuj
2100-2200	VO Korea	Eu	7570kuj, 12015kuj
2100-2200	VO Russia (VOR)	Eu	7290msk
2100-2200	WHRI	NAm,Eu	9490hri**
2100-2300	BBC World Sce	WAf	12095asc
2100-2300	BBC World Sce	FE	5965sla
2100-2300	BBC World Sce	NAf	9915asc
2100-2300	R. Australia	Pac	13630shp, 15515shp
2100-2300	R. Stn Belarus	Eu	1170sas, 6155mns, 7360mns, 7390mns
2100-2300	WHRI	NAm,Eu	7555hri*
2100-2300	WWCR	NAm	7465wcr*
2100-2300	WWCR	NAm,Eu	9350wcr*
2100-2400	BBC World Sce	SEA	3915sng
2100-2400	BBC World Sce	FE	5875nak
2100-2400	Family Radio	NAm	5950yfr
2115-2245	R. Cairo	Eu	6270abz
2120-2300	R. Stn Belarus	Eu	1170sas, 6155mns, 7360mns, 7390mns
2130-2200	R.Romania Int	Eu	6030gal, 7380gal
2130-2200	R.Romania Int	NAm	6115tig, 7310tig
2130-2230	VO Turkey	As,Pac	9610cak
2150-0500	RNZI	Pac	15720ran*, 17675ran*+
2200	**English**		
2200-0100	WWCR	NAm,Eu,Af	7465wcr**
2200-0100	WWRB	NAm	3215wrb
2200-0200	BBC World Sce	SEA	9740sng
2200-0400	WINB	CAm	9265inb*
2200-0400	WWRB	NAf	5745wrb
2200-1000	University Netw	NAm	6090aia
2200-1100	WTWW	NAm,Eu	5080tww***, 5755tww***
2200-1300	WTWW	NAm,Eu	5080tww*, 5755tww*

2200	English	Area	kHz
2200-1300	WWRB	NAm,CAm	5050wrb
2200-2230	Int. R. Serbia	Eu	6100bij
2200-2230	KBS World R.	Eu	3955skn
2200-2300	Angolan Nat R.	Af	945mul, 7217mul*
2200-2300	BBC World Sce	WAf	5910mey
2200-2300	BBC World Sce	FE	6135vld
2200-2300	BBG-VOA	SAs,SEA	5835pht
2200-2300	BBG-VOA	EAs	7220pht, 7570pht
2200-2300	BBG-VOA	CAs	7425kwt
2200-2300	BBG-VOA	SEA,Pac	9490sai
2200-2300	CRI	EAs	5915bei
2200-2300	Family Radio	Af	17690yfr
2200-2300	R. Australia	Pac	9660brn*, 12080brn*
2200-2300	R. Bulgaria	Eu	6200pld, 7400pld
2200-2300	RCI	NAm	9800sac+
2200-2300	REE	Eu	6125nob
2200-2300	VO Russia (VOR)	Eu	7300msk
2200-2300	WHRI	Af	9615hri
2200-2330	R. Australia	As	15240tnn
2200-2400	BBC World Sce	SEA	6195sng
2200-2400	CRI	Eu	1440mrn
2200-2400	Family Radio	NAm	11740yfr, 15440yfr
2200-2400	R. Australia	SEA	13590hbn
2200-2400	R. Australia	Pac	15230shp, 15560shp
2200-2400	WHRI	NAm,Eu	9505hri****
2200-2400	WWCR	Af	5070wcr**
2230-2300	AWR Guam	SEA	15320sda
2230-2300	BBG-VOA	EAs	5850pht, 7230udo, 9570ira
2230-2300	R. PMR	Eu	6240kch
2230-2300	R. Prague	CAf,WAf	7355lit
2230-2400	BBG-VOA	SEA	1575bph
2245-0045	All India R.	EAs	11645del, 13605bgl
2245-0045	All India R.	SEA	6055del, 7305bgl
2300	English		
2300-0030	BBG-VOA	ME	1593kwt
2300-0030	R. Cairo	NAm	11590abz
2300-0100	CRI	SEA	11790xia
2300-0100	WHRI	NAm	9470hri
2300-0100	WWCR	Af	5070wcr*
2300-0200	CRI	Eu	7350kas
2300-0200	R. Australia	Pac	17795shp
2300-0200	WWCR	NAm,Eu,Af	7490wcr**, 7645wcr*
2300-0700	R. Australia	Pac	13690shp
2300-0800	R. Australia	Pac	9660brn
2300-0900	R. Australia	Pac	12080brn
2300-1200	WTWW	NAm,Eu	5080tww**, 5755tww**
2300-1200	WWRB	NAm	3185wrb
2300-2315	TWR Europe	Eu	1467rou
2300-2345	TWR Europe	Eu	1467rou
2300-2345	Vatican Radio	NAm	7370smg+
2300-2400	BBC World Sce	SEA	11955sng
2300-2400	BBC World Sce	FE	6135sla, 7385nak
2300-2400	BBG-VOA	SEA	5830pht
2300-2400	BBG-VOA	EAs	6180udo, 7220pht, 7460pht, 7570pht, 11655pht, 11840pht
2300-2400	BBG-VOA	SAs,SEA	7480pht
2300-2400	BBG-VOA	SEA,Pac	9490sai
2300-2400	CRI	SAs	5915kas, 7415kas
2300-2400	CRI	CAm	5990hab
2300-2400	CRI	NAm	6040sac, 11970sac
2300-2400	CRI	EAs	6145bei
2300-2400	CRI	SEA	9610kun
2300-2400	Family Radio	CAm,SAm	9430yfr, 15400yfr
2300-2400	R.Romania Int	Af	5915tig, 7300tig
2300-2400	R.Romania Int	Eu	6015gal, 7220gal
2300-2400	R. Ukraine Int.	NAm	7440lvi
2300-2400	VO Russia (VOR)	NAm,CAm	7250arm
2300-2400	VO Turkey	Eu,NAm	5960emr
2300-2400	WHRI	NAm,Eu	7335hri***
2330-0400	R. Australia	As	15415shp
2330-0700	R. Australia	As	17750shp
2330-2400	BBC World Sce	FE	6170kim

2300	English	Area	kHz
2330-2400	BBG-VOA	SEA	13640tin
2330-2400	R. Prague	NAm	5930lit
2330-2400	Vatican Radio	Eu	1611smg+
2330-2400	VO Vietnam	As	9840vni, 12020vni
1800	English (Colloquial)		
1800-1830	AWR	EAf	9515mos
1100	English (Mass)		
1130-1300	Vatican Radio	Af	15595smg, 17765smg
0300	English/Arabic		
0300-0600	IRRS Shortwave	Af	7385rso
1400	English/Arabic		
1400-1700	IRRS Shortwave	Af	15710rso
0600	English/Finnish		
0600-0900	Scan.Weekend R.Eu		5980vir
0800	English/Finnish		
0800-1400	Scan.Weekend R.Eu		11720vir
0900	English/Finnish		
0900-1500	Scan.Weekend R.Eu		6170vir
1400	English/Finnish		
1400-1700	Scan.Weekend R.Eu		11690vir
1500	English/Finnish		
1500-1900	Scan.Weekend R.Eu		5980vir
1700	English/Finnish		
1700-1900	Scan.Weekend R.Eu		11720vir
1900	English/Finnish		
1900-2200	Scan.Weekend R.Eu		6170vir, 11690vir
2200	English/Finnish		
2200-0600	Scan.Weekend R.Eu		6170vir
2200-2200	Scan.Weekend R.Eu		1602vir
2200-2300	Scan.Weekend R.Eu		11720vir
2300	English/Finnish		
2300-0800	Scan.Weekend R.Eu		11690vir
1500	English/Hindi		
1530-1630	SLBC	SAs	11905trm
1900	English/Irish Gaelic		
1930-2030	RTE R.Worldwide	Af	6225mey
1400	English/Korean/Japanese/Mandarin		
1400-1430	Shiokaze	KRE	5985yam
2000	English/Korean/Japanese/Mandarin		
2030-2100	Shiokaze	KRE	5955yam
0400	English/Ndebele/Shona		
0430-1530	VO Zimbabwe	Af	5975gwe†
1500	English/Ndebele/Shona		
1530-0430	VO Zimbabwe	Af	4828gwe†
1700	English/Ndebele/Shona		
1700-1900	SW Radio Africa	ZWE	4880mey
1755-1855	Zimbabwe C.R.	ZWE	4895mey
1800	English/Ndebele/Shona		
1800-1830	BBG-VOA Studio7	ZWE	909bot, 4930bot, 12080sao, 15775sao
1830-1900	BBG-VOA Studio7	ZWE	909bot, 15775sao
0200	English/Russian		
0200-2100	BBC World Sce	RUS	666ekb, 1260msk, 1260spb
0000	English/Russian/Ukrainian		
0000-2400	BBC World Sce	UKR	612khr
0400	English/Russian/Ukrainian		
0400-2000	BBC World Sce	UKR	594kyv
1600	English/Russian/Ukrainian		
1600-2100	BBC World Sce	UKR	612khr
0200	English/Spanish		
0200-0330	KJES Radio	NAm	7555jes
1400	English/Spanish		
1400-1700	KJES Radio	NAm,CAm	11715jes
1900	English/Spanish		
1900-2100	KJES Radio	CAm	15385jes

FRENCH

0000	French		
0000-0100	Family Radio	CAm	15400yfr
0000-2400	RTL Radio	F	236bdw*
0100	French		
0100-0200	VO Korea	SEA	13650kuj, 15100kuj

0100	French	Area	kHz
0100-0200	VO Korea	EAs	3560kuj
0130-0200	R. Habana Cuba	Car	5040hab
0200	**French**		
0200-0230	R. Slovakia Int	NAm	6040rso
0200-0230	R. Slovakia Int	SAm	9440rso
0200-0300	R. Bulgaria	NAm	5900pld, 7400pld
0200-0300	R.Romania Int	NAm	5975gal, 7325gal
0230-0250	Vatican Radio	NAm	6040bon, 6040sac
0230-0250	Vatican Radio	CAm,SAm	7305smg
0230-0300	Vatican Radio	Af	7360smg
0300	**French**		
0300-0400	RAE	Am	11710bue
0300-0400	VO Korea	LAm	11735kuj, 13760kuj, 15180kuj
0400	**French**		
0400-0500	RFI	EAf	7215iss
0400-0700	RFI	Af	9790iss
0400-2315	RTBF Int.	WEu	621wav
0430-0500	AWR	NAf	6045mos
0430-0500	BBC World Sce	EAf	15490sey
0430-0500	BBC World Sce	WAf	6135asc
0430-0500	BBC World Sce	CAf	7415asc
0430-0500	Vatican Radio	Af	7360smg, 9660mdc, 9660mdc
0500	**French**		
0500-0600	Family Radio	Eu,Af	9985yfr, 11530yfr
0500-0600	RFI	CAf	7340iss*, 11700iss**
0500-0700	RFI	CAf	11605mey
0500-2300	Africa No.1	Af	9580gab
0500-2315	RTBF Int.	WEu	621wav
0500-2400	RTBF Int.	WEu	621wav
0530-0600	BBG-VOA	Af	1530sao
0530-0600	Lao National R.	SEA	7145vie‡
0530-0600	R. Japan	CAf,WAf	9850wer, 13840mdc
0530-0630	BBG-VOA	Af	4960sao, 6020bot, 7265mdc, 9480bot, 9505sao
0540-0600	Vatican Radio	Eu	4005vat
0540-0600	Vatican Radio	Eu	5965smg, 7250smg
0600	**French**		
0600-0628	Vatican Radio	Af	7360smg, 9660smg, 11625smg
0600-0630	BBC World Sce	NAf	6055skn, 7350skn
0600-0630	BBC World Sce	WAf	6135asc, 7305asc
0600-0630	R.Romania Int	Eu	6100gal+, 7370gal
0600-0630	R.Romania Int	Af	9690tig, 11790tig
0600-0700	Family Radio	Eu,Af	7520yfr, 11580yfr
0600-0700	RFI	CAf	11700iss*, 15300iss**
0600-0700	RFI	NAf,WAf	5925iss*, 7340iss**
0600-0800	CRI	Eu	15220uru
0630-0730	VOIRI	Eu,NAf	13600kam, 15560sir
0700	**French**		
0700-0730	BBC World Sce	WAf	12095asc
0700-0730	BBC World Sce	CAf	15490mey
0700-0730	R. Bulgaria	Eu	5900pld, 7400pld
0700-0730	R. Prague	WEu	5930lit
0700-0800	RFI	CAf	15170mey
0700-0800	RFI	NAf,WAf	9790iss*, 11700iss
0700-0800	VO Greece	Eu	11645avl**
0700-0800	VO Nigeria	Eu	15120iko
0700-1000	RFI	NAf,WAf	13695iss
0700-1100	RFI	CAf	17850iss
0700-1800	RFI	NAf,CAf	15300iss
0715-0730	Vatican Radio	Af	15595smg
0715-0730	Vatican Radio	Eu	4005vat, 11740smg
0715-0730	Vatican Radio	Eu	585vat, 1530smg, 1611smg, 5965smg, 7250smg, 9645smg
0800	**French**		
0800-0830	AWR	NAf	12010wer
0800-0900	Family Radio	Eu,Af	9985yfr
0800-1000	BSKSA	WAf	17785riy
0800-1300	CRI	Eu	702cdm
0800-1600	RFI	WAf	17620iss

0800	French	Area	kHz
0830-0900	R. Prague	SEu,WAf	11600lit
1000	**French**		
1000-1100	Family Radio	CAm	9680yfr, 11740yfr
1000-1100	R.Romania Int	Eu	15260gal, 17870gal
1100	**French**		
1100-1130	RFI	SEA	15680tnn
1100-1200	R.Romania Int	Eu	15150tig, 15255gal, 17800gal
1100-1200	R.Romania Int	Af	17870tig
1100-1200	VO Korea	SEA	6185kuj, 9850kuj
1100-1200	VO Korea	LAm	6285kuj, 9335kuj
1130-1200	RFI	NAm	13640guf
1130-1200	RFI	CAm	17610iss
1130-1200	RFI	Atl	6175iss
1200	**French**		
1200-1203	R. Monaco	Atl,Med	4368mco*, 8728mco*, 13146mco*, 17260mco*
1200-1215	Vatican Radio	Eu	585vat, 1611smg+
1200-1215	Vatican Radio	Eu	5965smg
1200-1230	BBC World Sce	NAf	15420sin
1200-1230	BBC World Sce	WAf	17780asc
1200-1230	BBC World Sce	CAf	21630asc
1200-1230	VO Vietnam	As	7285han
1200-1300	Deutsche Welle	Af	15275kig, 15440kig, 17520rmp, 17800kig, 21550sin
1200-1300	Family Radio	NAm	13695yfr
1200-1300	RFI	CAf	17660mey
1200-1330	RFI	CAf	21580iss
1200-1400	CRI	Eu	15205kas
1230-1300	R. Japan	WAf	17690mdc
1300	**French**		
1300-1330	Lao National R.	SEA	7145vie‡
1300-1400	CRI	Eu	13710kas
1300-1400	Family Radio	CAm	11740yfr
1300-1400	RFI	SEA	684dof
1400	**French**		
1400-1500	VO Korea	EAs	3560kuj
1400-1500	VO Korea	Eu	7570kuj, 12015kuj
1400-1500	VO Korea	NAm	9335kuj, 11710kuj
1400-1600	CRI	WAf,Eu	11920cer, 13670cer
1400-1800	BSKSA	WAf	17660riy
1430-1500	R. Prague	SEu,WAf	11600lit
1440-1525	TWR Africa	SAf	9635man
1500	**French**		
1500-1530	Église d.Christ	NAf	15265skn
1600	**French**		
1600-1700	Channel Africa	WAf	15235mey
1600-1700	Family Radio	NAm	11855yfr
1600-1700	RFI	SEA	1296kun
1600-1700	RFI	CAf,WAf	17850iss
1600-1700	VO Africa	NAf,CAf	15660sab, 17725sab
1600-1700	VO Korea	EAs	3560kuj
1600-1700	VO Korea	Eu	7570kuj, 12015kuj
1600-1700	VO Korea	NAm	9335kuj, 11710kuj
1600-1800	CRI	Eu	7350kas
1630-1700	VO Vietnam	ME	7220vni, 9550vni
1700	**French**		
1700-1715	Vatican Radio	Eu	585vat, 4005vat, 5885smg, 7250smg, 9645smg, 7290smg
1700-1715	Vatican Radio	NAf	7290smg
1700-1728	Vatican Radio	Af	11625smg, 13765smg
1700-1800	Deutsche Welle	Af	9535kig, 12035wof, 13625trm, 15275wof, 17800sin
1700-1800	Family Radio	Eu,Af	6225mey, 15115yfr
1700-1800	R. Ethiopia	ERI	7165gjw†
1700-1800	R. Ethiopia	EAf,ME	9560gjw±
1700-1800	RFI	CAf,WAf	11995iss*
1700-1800	RFI	NAf,WAf	13695iss
1700-1800	R.Romania Int	Eu	7370tig, 9690tig
1700-1800	VO Africa	NAf,CAf	11965sab, 15215sab
1700-1800	VO Russia (VOR)	Eu	6145klg+, 9675msk+

1700	French	Area	kHz
1700-1800	VO Russia (VOR)	Eu,Af	7330msk
1700-1900	VO Russia (VOR)	Af	7295tch
1700-2000	RFI	CAf	11705iss
1700-2000	RFI	WAf	21690guf
1700-2100	VO Russia (VOR)	Af	11985erv
1700-2200	VO Russia (VOR)	Af,Eu	11600msk
1700-2200	VO Russia (VOR)	Eu,Af	6130msk
1730-1800	R. Prague	WEu	5930lit
1800	**French**		
1800-1830	BBC World Sce	CAf	11860asc
1800-1830	BBC World Sce	WAf	17640asc, 17885asc
1800-1830	BBC World Sce	SAf	7465mey
1800-1830	BBC World Sce	NAf	9605rmp
1800-1830	R. Bulgaria	Eu	6200pld, 7400pld, 9700sof+
1800-1830	R. Slovakia Int	WEu	5915rso, 6055rso
1800-1900	Family Radio	Eu,Af	15565yfr, 17690yfr
1800-1900	REE	Eu	9665nob
1800-1900	RFI	NAf,WAf	11995iss
1800-1900	RFI	WAf,CAf	9790iss*, 15300iss**
1800-1900	R.Romania Int	Eu	7350gal
1800-1900	VO Korea	SAf	7210kuj, 11910kuj
1800-1900	VO Korea	ME,NAf	9975kuj, 11535kuj
1800-1900	VO Russia (VOR)	Eu	12060msk
1800-2000	CRI	Eu	5970cer, 7360cer
1800-2000	CRI	NAf,WAf	6055cer, 7385cer
1800-2000	RFI	NAf,WAf	13695iss**
1800-2300	CRI	Eu	702cdm
1805-2005	RCI	NAm	9610sac, 9800sac+
1830-1845	FEBA Radio	CAf,WAf	15250asc
1830-1845	R. Réveil	CAf	9760iss
1830-1900	BBG-VOA	Af	1530sao, 15225bon, 15620bot
1830-1900	R. Tirana	Eu	7465shi
1830-1900	VO Vietnam	Eu	7280vni, 9730vni
1830-1930	Family Radio	Eu,Af	17660asc
1830-1930	VOIRI	WAf	11775kam
1830-1930	VOIRI	Eu	5980kam, 6115sit, 7380sir
1830-2030	CRI	WAf	9645kun
1830-2230	CRI	WAf	7350uru
1900	**French**		
1900-1930	R. PMR	Eu	6240kch
1900-2000	BBG-VOA	Af	1530sao, 15225bon
1900-2000	Family Radio	Eu,Af	9695wer, 15795yfr, 17690yfr
1900-2000	RCI	Af	9670skn, 9770kas, 11845skn, 13650sac, 15365sac, 17790sac
1900-2000	R .Damascus	Eu,NAm	9330adr, 12085adr‡
1900-2000	REE	ME	12030nob
1900-2000	REE	Af	9590nob
1900-2000	RFI	Eu	3965iss+
1900-2000	RFI	WAf	6175iss*
1900-2000	R.Taiwan Int.	Eu	3985skn
1900-2000	R.Taiwan Int.	Af	9365iss*, 11875iss**
1900-2100	RFI	WAf,CAf	9790iss
1900-2200	VO Russia (VOR)	Eu	6120klg
1930-1945	BVB	WAf	9510wer
1930-2000	AWR	CAf	9625mos
1930-2000	R. Habana Cuba	Eu	11770hab
1930-2000	R. Prague	WEu,SEu	5930lit
1930-2000	VO Vietnam	Eu	5955mos, 7280vni, 9730vni
1935-1950	TWR Africa	CAf	9525man
1945-2030	All India R.	NAf,WAf	6180bgl, 7410del, 13605bgl
1950-2020	TWR Africa	CAf	9525man
2000	**French**		
2000-2030	AWR	CAf	11755mey
2000-2030	AWR	NAf	9805wer
2000-2030	BBG-VOA	Af	9780sao, 9815bot, 12080bot, 15225bon, 15620sao
2000-2030	R. Tirana	Eu	7465shi

2000	French	Area	kHz
2000-2100	Family Radio	Eu,Af	7590alm, 9595wer
2000-2100	REE	Af	9570nob
2000-2100	REE	ME	9605nob
2000-2100	R.Romania Int	Eu	7380gal
2000-2100	RAE	Eu	15345bue
2000-2100	VO Indonesia	Eu,NAf,ME	9526jak*
2000-2100	VO Korea	EAs	3560kuj
2000-2100	VO Korea	Eu	7570kuj, 12015kuj
2000-2100	VO Nigeria	WAf	7255iko
2000-2115	R. Cairo	Eu	6270abz
2000-2200	RFI	WAf	6175iss
2000-2200	RFI	CAf,WAf	7205iss
2000-2200	VO Russia (VOR)	Af,Eu	12060msk
2000-2200	VO Russia (VOR)	Eu	1323wbr, 6105klg+
2030-2050	Vatican Radio	Eu	1530smg, 4005vat, 5885smg
2030-2050	Vatican Radio	Af	7250smg, 11625smg
2030-2100	AWR	WAf	9805mos
2030-2100	BBG-VOA	Af	6040sao, 9775sao, 9815bot, 12080bot, 15225bot
2030-2100	R. Slovakia Int	WEu	5915rso, 7345rso
2030-2100	TWR Europe	NAf	1467rou
2030-2100	Vatican Radio	Af	7365smg, 9755smg
2030-2130	CRI	Eu	7215sam
2030-2130	VO Turkey	Eu	5970emr
2030-2130	VO Turkey	Af	6050emr
2030-2230	CRI	Eu	6115bei
2030-2230	R. Cairo	WAf	9280abs
2040-2100	R. Stn Belarus	Eu	1170sas, 6155mns, 7360mns, 7390mns
2040-2245	TWR Africa	WAf	1566par
2045-2115	BVB	Af	6145nau
2100	**French**		
2100-2130	BBG-VOA	Af	9435skn, 9680mdc, 9780sao, 9815bot, 11760hab
2100-2130	R. Habana Cuba	Am	11760hab
2100-2130	R. PMR	Eu	6240kch
2100-2130	R.Romania Int	Eu	6030gal+, 7370gal+
2100-2130	VO Vietnam	ME	7220vni, 9550vni
2100-2130	VO Vietnam	Eu	7280vni, 9730vni
2100-2200	Angolan Nat R.	Af	945mul, 7217mul*
2100-2200	CRI	Eu	1440mrn
2100-2200	Family Radio	CAm	15755yfr
2100-2200	Family Radio	Eu,Af	7305wer
2100-2200	KBS World R.	Eu	3955skn
2100-2200	R. Bulgaria	Eu	6200pld, 7400pld
2100-2200	RCI	NAf	11845sac
2100-2200	RCI	WAf	15365sac
2105-2205	RCI	NAm	6100sac
2130-2200	Int. R. Serbia	Eu	6100bij
2130-2230	CRI	WAf	11975bko
2200	**French**		
2200-1800	RFI	Eu	3965iss+
2200-2300	Family Radio	Eu,Af	9355yfr
2230-2300	R. Habana Cuba	SAm	15370hab
2300	**French**		
2300-2330	RCI	FE	6160kim
2300-2330	R. PMR	Eu	6240kch
2300-2330	R. Prague	NAm	5930lit
2300-2400	Family Radio	NAm	6085yfr
2300-2400	REE	Eu	5970nob
2300-2400	REE	NAm	6055nob
1200	**French/Somali/Various**		
1200-1300	LV Djibouti	DJI	21525sam
0400	**French/Various**		
0400-0500	R. Okapi	COD	11690mey

GERMAN

0000	German	Area	kHz
0000-0100	R. Ö1 Int.	NAm,CAm	7325mos
0000-0200	Deutsche Welle	SAs	1548trm, 6225kch*, 7285trm, 7395rmp**

0000	German	Area	kHz
0000-0200	Deutsche Welle	CAm	9655kig, 11665asc, 12025kig
0000-0800	Deutsche Welle	Eu	6075sin
0100	**German**		
0100-0130	R. Ö1 Int.	SAm	9840mos
0200	**German**		
0200-0400	Deutsche Welle	Eu,ME	6075rmp
0400	**German**		
0400-0430	TWR Africa	SAf	3200man, 4775man
0400-0500	Deutsche Welle	Eu	13780dha**
0400-0500	TWR Africa	SAf	3200man, 4775man
0400-0600	Deutsche Welle	Af	6075wof, 13780arm*, 17800trm
0415-0430	Miss.Heukelbach	Eu	1440mrn
0415-0430	R. Freundes-D.	Eu	1440mrn
0415-0430	Stimme Des Tr.	Eu	1440mrn
0430-0445	R. Freundes-D.	Eu	1440mrn
0445-0500	Miss.Heukelbach	Eu	1440mrn
0500	**German**		
0500-0515	Miss.Heukelbach	Eu	1440mrn
0500-0600	Deutsche Welle	Eu	6075skn
0500-0600	Family Radio	Eu	7730yfr
0500-0800	Deutsche Welle	Eu	3995skn
0520-0540	Vatican Radio	Eu	4005vat, 5965smg
0600	**German**		
0600-0700	Deutsche Welle	NAf	7410wof
0600-0710	R. Ö1 Int.	Eu	6155mos
0600-0715	R. Ö1 Int.	Eu	6155mos
0600-0800	CRI	Eu	15245uru, 17720kas
0600-0800	Deutsche Welle	SAf	12005kig
0600-0800	Deutsche Welle	WAf	15275kig
0600-0800	Deutsche Welle	Eu	6075wof
0600-0800	Deutsche Welle	Eu,Af	9545wof***
0615-0630	Miss.Heukelbach	Eu	1440mrn
0630-0700	R. Bulgaria	Eu	5900pld, 7400pld
0645-0700	RTL Radio	D	6095jun+
0700	**German**		
0700-0730	KBS World R.	WEu	1440mrn+, 6095jun+
0700-0730	R.Romania Int	Eu	7210tig, 9450tig+
0700-0800	Family Radio	Eu	11530yfr
0730-0800	R. Prague	WEu	5930lit
0730-0800	RTL Radio	D	1440mrn+, 6095jun+
0730-0830	VOIRI	Eu	15085kam, 17690sir
0800	**German**		
0800-0830	R. Slovakia Int	WEu	5915rso, 6055rso
0800-1000	Deutsche Welle	AUS,NZL	9450bon
0800-1000	Deutsche Welle	SEu,ME	9545skn
0800-1100	Deutsche Welle	SEu,ME	13780skn
0800-1200	Deutsche Welle	SEA,AUS	17520trm
0800-1200	Deutsche Welle	Eu	6075skn
0800-1800	R. 700	Eu	6005kll
0900	**German**		
0900-0930	VO Greece	Eu	11645avl
1000	**German**		
1000-1100	Chr.Sc.Sentinel	Eu	6055wer
1000-1100	Deutsche Welle	Car,SAm	9865hri
1000-1100	Hamburger LR	Eu	6045wer
1000-1100	MV Baltic Radio	Eu	6140wer
1000-1100	VO Russia (VOR)	Eu	9720klg, 11655lvi
1000-1200	Deutsche Welle	SEA	21780trm
1000-1200	Deutsche Welle	CAm,NAm	5905bon
1000-1300	VO Russia (VOR)	Eu	630klu, 693bln, 1323wbr, 1431dsd
1100	**German**		
1100-1130	R. Prague	CEu	7345lit
1100-1200	Deutsche Welle	SAm	17770asc
1100-1300	VO Russia (VOR)	Eu	7325klg+
1100-1400	Deutsche Welle	SEu,ME	13780sin
1130-1200	Ev.Miss.Gemeind	Eu	6055wer
1200	**German**		
1200-1215	Stimme Des Tr.	Eu	6055wer
1200-1300	R. Santec	Eu	1323wbr
1200-1400	Deutsche Welle	SAs	15640trm
1200-1400	Deutsche Welle	SEA	21780rmp
1200-1430	Deutsche Welle	SAs	1548trm

1200	German	Area	kHz
1200-2000	Deutsche Welle	Eu	6075wof
1230-1300	Polish Radio	Eu	9470wof, 9850wof
1230-1330	VO Turkey	Eu	17755emr
1300	**German**		
1300-1330	R. Ö1 Int.	AUS,NZL	17855mos
1300-1330	R. Prague	CEu	6055lit
1300-1400	R.Romania Int	Eu	11970tig, 15460tig
1300-1500	RTR R.Traumland	Eu	6180wer
1400	**German**		
1400-1600	Deutsche Welle	SEu,ME	13780trm, 15275kig, 17800sin
1430-1500	R. Slovakia Int	WEu	6055rso, 7345rso
1500	**German**		
1500-1515	Vatican Radio	Eu	5885smg, 6060smg+, 7250smg, 9645smg
1600	**German**		
1600-1630	R. Prague	WEu	5930lit
1600-1700	Deutsche Welle	Af	12070kig
1600-1700	R. Santec	Eu	630klu, 693bln, 1431dsd
1600-1700	VO Korea	EAs	4405kuj
1600-1700	VO Korea	Eu	6285kuj, 9325kuj
1600-1700	VO Russia (VOR)	Eu	9675msk+
1600-1800	CRI	Eu	5970cer, 7380cer
1600-1800	Deutsche Welle	Af	12055kig
1600-1800	Deutsche Welle	SEu,ME	13780rmp
1600-1800	Deutsche Welle	EAf	9545wof
1600-1800	VO Russia (VOR)	Eu	630klu, 1431dsd, 7220sam
1600-2000	VO Russia (VOR)	Eu	693bln, 1215klg, 1323wbr
1600-2200	Deutsche Welle	Eu	3995skn
1630-1700	Polish Radio	Eu	6100wof
1700	**German**		
1700-1730	R.Romania Int	Eu	5875kvi+
1700-1730	R. Slovakia Int	WEu	5915rso, 6010rso
1700-1800	Deutsche Welle	Af	12070trm
1700-1800	Family Radio	Eu	17760yfr
1700-1800	RAE	Eu	15345bue
1730-1800	R. Bulgaria	Eu	6200pld, 7400pld, 9700sof+
1730-1830	VOIRI	Eu	3955sit, 6205sir, 7380kam
1800	**German**		
1800-1900	Deutsche Welle	Af	13780trm
1800-1900	Family Radio	Eu	7490erv, 15795yfr, 21455yfr
1800-1900	R .Damascus	Eu	9330adr, 12085adr‡
1800-1900	R. Santec	Eu	1215klg, 1323wbr, 7310msk
1800-1900	R. Ukraine Int.	Eu	6030khr
1800-1900	VO Indonesia	Eu,NAf,ME	9526jak*
1800-1900	VO Russia (VOR)	Eu	1143klg, 5940sam
1800-2000	CRI	Eu	6160xia, 7395kas, 9615uru
1800-2000	Deutsche Welle	Af	9545rmp, 11725kig, 12070wof, 15640sin
1800-2000	VO Korea	EAs	4405kuj
1800-2000	VO Korea	Eu	6285kuj, 9325kuj
1800-2000	VO Russia (VOR)	Eu	7310sam
1815-1830	Stimme Des Tr.	Eu	1440mrn*
1830-1845	R. Freundes-D.	Eu	1440mrn
1830-1930	VO Turkey	Eu	7205cak
1845-1900	Lutherische Std	Eu	1215klg, 1323wbr, 7310msk
1845-1900	Miss.Heukelbach	Eu	1440mrn
1900	**German**		
1900-1930	R. Slovakia Int	WEu	5915rso, 7345rso
1900-1930	R. Tirana	Eu	1458fla
1900-2000	Family Radio	Eu	7490erv
1900-2000	R. Cairo	Eu	6270abz
1900-2000	R.Romania Int	Eu	7370tig, 9805tig+
1900-2000	R.Taiwan Int.	Eu	3955skn
1900-2000	VO Russia (VOR)	Eu	630klu, 1431dsd
1900-2040	R. Stn Belarus	Eu	1170sas, 6155mns, 7360mns, 7390mns

1900	German	Area	kHz
1900-2100	CRI	Eu	1440mrn
1900-2100	R. Stn Belarus	Eu	1170sas, 6155mns, 7360mns, 7390mns
1920-1940	Vatican Radio	Eu	4005vat, 5885smg, 7250smg
1930-2000	R. PMR	Eu	6240kch
1945-2000	Miss.Heukelbach	Eu	630klu, 693bln, 1215klg, 1323wbr, 1431dsd, 7310msk
2000	**German**		
2000-2015	R. Thailand WS	Eu	9535udo
2000-2100	Deutsche Welle	Eu	12070kig
2000-2100	Family Radio	Eu	11565yfr
2000-2100	KBS World R.	Eu	3955skn
2000-2100	R. Bulgaria	Eu	6200pld, 7400pld
2000-2200	Deutsche Welle	AUS,SEA	9510trm, 11605kig
2000-2400	Deutsche Welle	Eu	6075sin
2030-2100	Polish Radio	Eu	3975skn+, 6000dha
2030-2100	R. Tirana	Eu	7465shi
2030-2100	VO Vietnam	Eu	3985skn
2100	**German**		
2100-2130	Int. R. Serbia	Eu	6100bij
2100-2130	VO Vietnam	Eu	3985skn
2100-2200	R.Taiwan Int.	Eu	3965iss
2100-2200	R. Ukraine Int.	Eu	6140khr
2100-2200	RAE	Eu	15345bue
2100-2300	CRI	Eu	963por
2130-2200	R. PMR	Eu	6240kch
2200	**German**		
2200-2400	Deutsche Welle	SAm	11865sin, 15640hri
2200-2400	Deutsche Welle	SEA	11875dha
2200-2400	Deutsche Welle	CAm	12025kig
2300	**German**		
2300-2400	Deutsche Welle	SEA	6050trm
2310-2330	Vatican Radio	Eu	1611smg+
2330-2400	R. PMR	Eu	6240kch
1700	**German (High)**		
1700-1730	HCJB Equador	Eu	3955sit
2300	**German (High)**		
2330-2400	HCJB Equador	SAm	9835sgo
1600	**German (Low)**		
1630-1700	HCJB Equador	Eu	3955sit
2300	**German (Low)**		
2300-2330	HCJB Equador	SAm	9835sgo

PORTUGUESE

0000	Portuguese		
0000-0100	CRI	SAm	9435kas
0000-0100	Family Radio	B	9430yfr, 9690yfr, 11885yfr
0000-0100	RAE	Am	11710bue
0000-0100	VO Russia (VOR)	SAm	6135dsb, 9865sam, 9965erv
0000-0300	RDP Int.	NAm	9455lis
0000-0300	RDP Int.	SAm	9855lis, 11655lis
0005-0105	RCI	NAm	9755sac
0030-0057	Vatican Radio	CAm,SAm	7305smg
0030-0057	Vatican Radio	SAm	9610sac
0030-0100	Vatican Radio	Eu	1260vat
0100	**Portuguese**		
0100-0200	Family Radio	B	7520yfr, 9930yfr, 11825yfr
0200	**Portuguese**		
0200-0300	Family Radio	B	7520yfr
0230-0300	R. Japan	SAm	6145sgo
0300	**Portuguese**		
0300-0400	Family Radio	B	7520yfr, 7730yfr
0400	**Portuguese**		
0400-0500	Family Radio	Af	11580yfr
0430-0530	BBC World Sce	SAf	3380mey, 6145mey, 7305asc
0500	**Portuguese**		
0530-0600	Deutsche Welle	Af	9800kig, 12045kig

0500	Portuguese	Area	kHz
0530-0600	Vatican Radio	Af	7360smg, 9660smg, 11625smg
0600	**Portuguese**		
0600-0700	RFI	CAf	11830mey
0600-0700	RDP Int.	Eu	7345lis
0700	**Portuguese**		
0700-0800	Family Radio	Eu	9355yfr
0700-1300	RDP Int.	Eu	9815lis
0745-0800	TWR Bonaire	CAm,SAm	800twb
0745-0900	RDP Int.	Eu	7360sin
0800	**Portuguese**		
0800-1000	Family Radio	B	9680yfr
0800-1055	RDP Int.	WAf,SAm	15555lis
0800-1100	Family Radio	B	6105yfr, 9605yfr
0800-1100	RDP Int.	Af	15520lis, 17745lis, 17840lis
0800-1200	RDP Int.	Eu	12020lis
0845-0900	TWR Bonaire	CAm,SAm	800twb
0900	**Portuguese**		
0900-0930	Vatican Radio	Eu	1260vat
0900-1100	Family Radio	B	9575yfr
0930-1000	R. Japan	SAm	6145sgo
0930-1100	RDP Int.	Eu	9815sin+
1000	**Portuguese**		
1000-1030	BBG-VOA	SAf	11825mdc, 17850ira
1000-1030	Vatican Radio	Eu	1260vat
1000-1030	Vatican Radio	SAm	21680smg
1000-1100	Family Radio	B	6105yfr
1100	**Portuguese**		
1100-1200	RAE	Am	6060bue, 11710bue
1100-1300	RDP Int.	Af	17745lis
1100-1300	RDP Int.	WAf,SAm	21655lis
1100-1700	RDP Int.	WAf,SAm	21655lis
1200	**Portuguese**		
1200-1300	Family Radio	B	11830yfr
1200-1500	RDP Int.	Eu	11885lis
1300	**Portuguese**		
1300-1400	Family Radio	B	11530yfr
1300-1700	RDP Int.	NAm	15560lis
1300-1700	RDP Int.	NAm	15560lis*
1355-1425	TWR Africa	SAf	7315man
1400	**Portuguese**		
1400-1500	Channel Africa	SAf	9625mey
1400-1500	Family Radio	B	15210yfr
1400-1600	RDP Int.	ME,SAs	15690lis
1415-1430	Vatican Radio	Eu	1260vat, 7250smg
1415-1430	Vatican Radio	Af	9645smg
1425-1455	TWR Africa	SAf	7315man
1500	**Portuguese**		
1500-1600	Family Radio	B	15355yfr
1500-1600	Vatican Radio	Eu	1260vat
1500-1700	RDP Int.	Eu	11635lis
1500-1700	RDP Int.	Af	15520lis
1600	**Portuguese**		
1600-1630	Vatican Radio	Eu	1260vat
1630-1645	TWR Africa	SAf	4760man
1700	**Portuguese**		
1700-1730	BBG-VOA	Af	15670sao
1700-1800	BBG-VOA	Af	1530sao, 9395bot, 17740grv
1700-1800	Family Radio	Af	17690yfr
1700-1800	RFI	Af	12015iss*, 15530iss
1700-1800	RDP Int.	Eu	9860lis
1700-1900	RDP Int.	NAm	17820lis
1700-1900	RDP Int.	NAm	17820lis*
1700-2000	Family Radio	B	17575yfr
1700-2000	RDP Int.	Af	13720lis
1700-2000	RDP Int.	WAf,SAm	15465lis
1700-2100	RDP Int.	Af	13720lis
1700-2100	RDP Int.	WAf,SAm	15465lis
1730-1800	BBG-VOA	Af	15670sao
1800	**Portuguese**		
1800-1830	BBG-VOA	Af	1530sao, 15670bot, 17740grv
1800-1858	Vatican Radio	Af	9755smg, 11625smg, 13765smg

1800	Portuguese	Area	kHz
1800-2000	RDP Int.	Eu	9795lis
1830-1900	REE	SAm	15125cri, 17715nob
1830-1900	REE	LAm	17595nob
1830-1900	REE	Af	17755nob
1830-1900	REE	NAm	17850cri
1830-1900	REE	Eu	7275nob
1830-1900	REE	CAm	9765cri
1850-1905	TWR Africa	SAf	6130man
1900	**Portuguese**		
1900-2000	CRI	CAf,SAf	5985bei, 7365bei, 7405xia, 9535bji
1900-2000	CRI	Eu	7335jin, 9730kas
1900-2000	Family Radio	Af	3955mey, 6100mey
1900-2100	RDP Int.	NAm	12040lis
1900-2400	RDP Int.	NAm	12040lis*
1905-1920	TWR Africa	SAf	6130man
1920-1950	TWR Africa	SAf	6130man
1930-2000	CRI	CAf,SAf	11640bko
1930-2000	Deutsche Welle	Af	9735kig, 13780trm, 15275sin
2000	**Portuguese**		
2000-2030	R. Habana Cuba	Eu	11770hab
2000-2100	RCI	SAm	15305sac, 17765sac
2000-2300	RDP Int.	Eu	9795lis*
2000-2400	RDP Int.	Af	11665lis*
2000-2400	RDP Int.	WAf,SAm	11960lis*
2005-2020	TWR Africa	SAf	6130man
2030-2100	BBC World Sce	SAf	3380mey, 6135mey, 7260mey, 11860asc
2030-2100	BBC World Sce	WAf	5875rmp, 7415skn
2100	**Portuguese**		
2100-2145	Family Radio	Eu	11565yfr
2100-2200	Family Radio	Af	17690yfr
2100-2200	RCI	SAm	15305sac, 17765sac
2100-2200	REE	SAm	11680nob
2100-2400	RDP Int.	NAm	12040lis*
2200	**Portuguese**		
2200-2300	CRI	Eu	6175cer, 7260uru
2200-2300	CRI	SAm	9410kas, 9685kas
2200-2300	RCI	SAm	11990sac, 15305sac
2200-2300	VO Russia (VOR)	Eu	5920dsb, 5940sam, 6090arm, 6120klg, 7340nvs
2200-2400	Family Radio	B	7360guf, 9690yfr, 17575yfr
2215-2330	R. Cairo	SAm	9900abz
2300	**Portuguese**		
2300-0045	HCJB Equador	SAm	11920sgo
2300-0100	CRI	SAm	6100bei
2300-2400	CRI	SAm	13650hab
2300-2400	Family Radio	B	11720yfr
2300-2400	R. Habana Cuba	SAm	15390hab
2300-2400	RDP Int.	Eu	7285lis*
2300-2400	TWR Bonaire	CAm,SAm	800twb
2300-2400	VO Russia (VOR)	SAm	9965erv, 11605guf
2330-2400	R. Habana Cuba	SAm	15370hab

SPANISH

0000	Spanish		
0000-0030	R. Prague	SAm	9790lit
0000-0100	BBG-R.Martí	CUB	9825grv
0000-0100	BBG-VOA	LAm	5890grv, 9725grv, 9885grv
0000-0100	CRI	SAm	15120hab
0000-0100	CRI	CAm	5990hab
0000-0100	Family Radio	CAm,SAm	5985guf, 13615yfr
0000-0100	Family Radio	CAm,NAm	9715yfr, 11855yfr
0000-0100	R. Bulgaria	SAm	6200pld, 7300pld
0000-0100	RCI	SAm	11990sac, 13700sac
0000-0100	R.Romania Int	SAm,CAm	7315gal, 9525gal
0000-0100	R.Romania Int	SAm	9665tig, 11960tig
0000-0100	VO Korea	LAm	11735kuj, 13760kuj, 15180kuj

0000	Spanish	Area	kHz
0000-0200	Family Radio	CAm,SAm	5985yfr
0000-0200	REE	NAm	9630cri+
0000-0200	TWR Bonaire	CAm,SAm	800twb
0000-0300	R. República	CUB	9490sac
0000-0400	BBG-R.Martí	CUB	7365grv
0000-0400	REE	SAm	9765cri
0000-0400	RNW	LAm	6165bon
0000-0500	Family Radio	CAm,SAm	9355yfr
0000-0500	R. Habana Cuba	Am	11760hab
0000-0500	R. Habana Cuba	NAm	6060hab
0000-0500	R. Habana Cuba	SAm	9770hab, 15390hab
0000-0500	WEWN - EWTN	CAm,NAm	5810ewn
0000-1000	WEWN - EWTN	SAm	11870ewn
0000-1100	R. Habana Cuba	Car	6120hab
0000-2400	BBG-R.Martí	CUB	1180mth
0030-0230	VOIRI	LAm	7240kam
0030-0330	VOIRI	LAm	6110kam
0045-0200	R. Cairo	NAm	6270abz
0045-0200	R. Cairo	LAm	9900abz, 9915abs
0100	**Spanish**		
0100-0130	RFI	CAm	5995guf
0100-0145	Vatican Radio	Eu	1260vat
0100-0145	Vatican Radio	CAm,SAm	7305smg
0100-0145	Vatican Radio	SAm	9610bon, 9610sac, 11910smg
0100-0200	BBG-VOA	LAm	5890grv, 9725grv, 9885grv
0100-0200	Family Radio	CAm,NAm	5950yfr, 6890yfr, 9525yfr
0100-0200	Family Radio	CAm,SAm	7570yfr, 11885yfr
0100-0200	KBS World R.	SAm	11810kim
0100-0200	RCI	CAm,NAm	6100sac
0100-0200	VO Russia (VOR)	SAm	6065spb
0100-0300	BBG-R.Martí	CUB	9825sac
0100-0300	CRI	SAm	9590kas, 9710kas
0100-0300	VO Russia (VOR)	SAm	9865sam, 9875guf
0100-0500	VO Russia (VOR)	CAm	9965erv
0100-0600	REE	NAm	6055nob
0100-0600	VO Russia (VOR)	SAm	6135dsb, 7210msk
0100-0600	VO Russia (VOR)	CSAm	7280arm
0100-0800	KVOH	CAm	9975voh
0130-0200	R. Prague	CAm,NAm	7410lit
0145-0230	Vatican Radio	CAm,SAm	7305smg
0145-0230	Vatican Radio	SAm	9610bon, 9610sac, 11910smg
0200	**Spanish**		
0200-0230	KBS World R.	NAm	9560sac
0200-0300	Family Radio	CAm,NAm	9930yfr
0200-0300	Family Radio	CAm,SAm	9985yfr, 11825yfr
0200-0300	R. Bulgaria	CAm	6200pld
0200-0300	R. Bulgaria	SAm	6200pld, 7300pld
0200-0300	RCI	CAm,NAm	9800sac
0200-0300	R.Taiwan Int.	SAm	7570yfr, 11995guf
0200-0300	VO Korea	LAm	11735kuj, 13760kuj, 15180kuj
0200-0300	VO Turkey	SAm	9410emr
0200-0300	VO Turkey	Am	9650cak
0200-0600	REE	CAm	3350cri
0200-0600	REE	NAm	9675cri
0200-0600	VO Russia (VOR)	SAm	9475dsb
0200-1100	R. Habana Cuba	Car	5040hab
0230-0300	R. Slovakia Int	SAm	6080rso, 9440rso
0300	**Spanish**		
0300-0330	R. Prague	SAm	7345lit
0300-0330	VO Vietnam	NAm	6175sac
0300-0400	BBG-R.Martí	CUB	7405grv
0300-0400	CRI	SAm	9665bra
0300-0400	Family Radio	CAm,NAm	6890yfr, 9525yfr, 9680yfr
0300-0400	R.Romania Int	SAm	7325gal, 9635tig, 9765gal, 11825tig
0300-0400	VO Russia (VOR)	SAm	6065spb
0300-0500	Family Radio	CAm,SAm	5985yfr
0300-0600	VO Russia (VOR)	CAm	7335guf
0305-0405	RCI	NAm	9755sac
0320-0400	Vatican Radio	NAm	6040sac

0300	Spanish	Area	kHz
0320-0400	Vatican Radio	CAm,SAm	7305smg
0400	**Spanish**		
0400-0430	R. Japan	SAm	6195bon
0400-0430	VO Vietnam	NAm	6175sac
0400-0500	BBG-R.Martí	CUB	6030grv, 7365grv
0400-0500	Family Radio	CAm,SAm	7730yfr, 9985yfr
0400-0500	Family Radio	CAm,NAm	9930yfr
0400-0500	R.Taiwan Int.	CAm	6890yfr
0400-0700	BBG-R.Martí	CUB	7405grv
0400-0800	REE	SAm	5965cri
0500	**Spanish**		
0500-0530	R. Japan	CAm	6195bon
0500-0600	Family Radio	CAm,NAm	5745yfr
0500-0600	Family Radio	CAm,SAm	6000yfr
0500-0600	Family Radio	Eu	9355yfr
0500-0600	REE	Eu	12035nob
0500-0700	Family Radio	CAm,NAm	9495yfr
0500-0700	REE	ME	11895nob
0500-0900	REE	Eu	9780nob+
0500-1000	BBG-R.Martí	CUB	6030grv
0500-1300	Family Radio	CAm,NAm	9715yfr
0500-1300	WEWN - EWTN	NAm	7555ewn
0530-0630	VOIRI	Eu	13710kam, 15400sir
0600	**Spanish**		
0600-0700	Family Radio	CAm,NAm	5950yfr
0600-0700	KBS World R.	Eu	6045sac
0600-0700	R.Taiwan Int.	CAm	6875yfr
0600-0800	CRI	Eu	15135kas
0600-0900	REE	Eu	12035nob
0700	**Spanish**		
0700-0730	R. Bulgaria	Eu	6200pld, 7300pld
0700-0800	Family Radio	Eu	7520yfr
0700-0800	Family Radio	CAm,NAm	9680yfr
0700-0900	REE	Pac	17770nob
0700-1000	BBG-R.Martí	CUB	5980grv
0700-1000	R. Habana Cuba	SAm	6060hab
0700-1100	Family Radio	CAm,SAm	6000yfr
0700-1100	R. Habana Cuba	NAm	6050hab
0700-1300	R. Habana Cuba	Am	6150hab
0800	**Spanish**		
0800-0900	VO Greece	Eu	11645avl
0800-1000	Family Radio	CAm,SAm	5745yfr, 11740yfr
0800-1000	Family Radio	CAm,NAm	9495yfr
0800-1300	REE	Eu	13720nob
0800-1400	Family Radio	CAm,SAm	9555yfr
0900	**Spanish**		
0900-0930	R. Prague	SEu	11600lit
0900-1000	Family Radio	CAm,NAm	6890yfr
0900-1215	TWR Bonaire	CAm,SAm	800twb
0900-1500	REE	Af	21540nob
0900-1700	REE	Eu	15585nob
0900-1700	REE	ME	21610nob
1000	**Spanish**		
1000-1030	RFI	CAm	7375guf, 9825guf
1000-1030	R. Japan	CAm	6120sac
1000-1030	R. Japan	SAm	6195bon
1000-1200	BBG-R.Martí	CUB	6030grv
1000-1300	BBG-R.Martí	CUB	5980grv
1000-1600	Family Radio	CAm,SAm	6085yfr
1000-1700	WEWN - EWTN	SAm	12050ewn
1100	**Spanish**		
1100-0500	R. Habana Cuba	Am	6140hab
1100-1130	RNW	CAm,NAm	6165bon
1100-1200	CVC La Voz	CAm,SAm	9780sgo
1100-1200	Family Radio	CAm,SAm	9575yfr
1100-1200	KBS World R.	SAm	11795sac
1100-1300	Family Radio	CAm,SAm	11740yfr
1100-1300	Family Radio	CAm,NAm	9605yfr
1100-1300	R. Habana Cuba	NAm	6000hab, 6095hab
1100-1400	BBG-R.Martí	CUB	5745grv
1100-1500	R. Habana Cuba	Am	11760hab
1100-1500	R. Habana Cuba	SAm	15120hab, 15360hab
1100-2000	R. Habana Cuba	Am	11760hab
1100-2200	CVC La Voz	SAm	9635sgo
1100-2300	R. Habana Cuba	SAm	15390hab

1100	Spanish	Area	kHz
1100-2400	R. Habana Cuba	CAm	12040hab
1130-1200	Vatican Radio	Eu	1260vat
1130-1200	Vatican Radio	SAm	21680smg
1130-1230	RNW	SAm	6165bon
1200	**Spanish**		
1200-0100	CVC La Voz	SAm	17680sgo
1200-1215	BBC World Sce	Car	9410hri, 11860guf
1200-1230	RFI	CAm	15515guf
1200-1230	RNW	SAm	9810bon
1200-1400	BBG-R.Martí	CUB	7405grv
1200-1400	REE	SEA	11910bei
1200-1400	RAE	Am	11710bue
1200-1500	REE	SAm	11815cri
1200-1500	REE	NAm	15170cri
1200-1500	REE	CAm	9765cri
1200-1500	R. República	CUB	9965-t,*
1200-1600	Family Radio	CAm,NAm	11725yfr
1200-1600	Family Radio	CAm,SAm	13615yfr
1200-2300	REE	SAm	15125cri
1200-2300	REE	CAm	9765cri
1230-1300	BBG-VOA	LAm	9885grv, 13715grv, 15590grv
1300	**Spanish**		
1300-1400	BBG-VOA	LAm	9885grv, 13715grv, 15590grv
1300-1400	Family Radio	CAm,NAm	15355yfr
1300-1400	REE	Eu	13720nob
1300-1500	KVOH	CAm	9975voh
1300-1500	R. Habana Cuba	NAm	13680hab, 13780hab
1300-1700	REE	SAm	21570nob
1300-1800	WEWN - EWTN	CAm,NAm	11550ewn
1300-1830	REE	LAm	17595nob
1300-1900	REE	Am	17595nob
1300-2000	Family Radio	CAm,NAm	15130yfr
1300-2200	BBG-R.Martí	CUB	11930grv
1300-2330	R. Habana Cuba	CAm	11730hab
1305-1405	RCI	NAm	7325sac
1400	**Spanish**		
1400-1415	Vatican Radio	Eu	1260vat, 7250smg
1400-1415	Vatican Radio	Af	9645smg
1400-1500	Family Radio	CAm,NAm	11830yfr
1400-1500	Family Radio	CAm,SAm	15355yfr
1400-1600	Family Radio	CAm,NAm	11740yfr, 17555yfr
1400-1800	R. Habana Cuba	Car	11690hab**
1400-1800	R. Habana Cuba	CAm	13680hab**
1400-1800	R. Habana Cuba	NAm	13750hab**
1400-1800	R. Habana Cuba	SAm	15370hab**, 17750hab**
1400-1800	R. N. Venezuela	Am	11690hab*, 13680hab*, 13750hab*, 15370hab*, 17750hab*
1400-2000	BBG-R.Martí	CUB	15330grv
1400-2200	BBG-R.Martí	CUB	13820grv
1500	**Spanish**		
1500-0100	KVOH	CAm	17775voh
1500-1530	R. Prague	SEu	11600lit
1500-1600	Family Radio	CAm,NAm	13695yfr
1500-1700	Vatican Radio	Eu	1260vat
1500-1700	REE	Af	15385nob
1500-2000	R. Habana Cuba	Car	11690hab
1500-2200	REE	Af	17755nob
1500-2300	REE	NAm	17850cri
1530-1600	R. Slovakia Int	WEu	9445rso, 11600rso
1530-2000	R. Habana Cuba	Am	11760hab
1600	**Spanish**		
1600-1700	Family Radio	Eu	18930yfr
1600-1800	TWR Bonaire	CAm,SAm	800twb
1600-2300	REE	SAm	15125cri
1600-2300	REE	NAm	17850cri
1600-2300	REE	CAm	9765cri
1700	**Spanish**		
1700-1800	Family Radio	CAm,NAm	17535yfr
1700-1800	VO Indonesia	Eu,NAf,ME	9526jak*
1700-1830	REE	SAm	17715nob
1700-1830	REE	Af	17755nob

1700	Spanish	Area	kHz
1700-1830	REE	Eu	7275nob
1700-1900	Family Radio	CAm,SAm	6085yfr
1700-2200	REE	Af	17755nob
1700-2200	REE	Eu	9665nob
1700-2300	REE	Eu	7275nob
1700-2400	WEWN - EWTN	SAm	13830ewn
1730-1800	R. Bulgaria	Eu	5900pld, 9400pld
1730-1800	Vatican Radio	Eu	1260vat
1730-1830	VO Turkey	Eu	9495cak
1800	**Spanish**		
1800-0300	RAE	Am	6060bue*, 15345bue*
1800-1830	BVB	WAf	7425wer
1800-1830	REE	SAm	15125cri
1800-1830	REE	NAm	17850cri
1800-1830	REE	CAm	9765cri
1800-1830	VO Vietnam	Eu	7280vni, 9730vni
1800-1900	Family Radio	Eu	6120nau, 18930yfr
1800-2000	CVC La Voz	SAm	17860sgo+
1800-2400	WEWN - EWTN	CAm,NAm	12050ewn
1900	**Spanish**		
1900-1930	R. Prague	SEu	5930lit
1900-1930	Vatican Radio	Af	9755smg, 11625smg
1900-2000	REE	SAm	15125cri
1900-2000	REE	NAm	17850cri
1900-2000	REE	CAm	9765cri
1900-2000	VO Korea	EAs	3560kuj
1900-2000	VO Korea	Eu	7570kuj, 12015kuj
1900-2300	REE	LAm	11940nob
1900-2300	REE	NAm	15110nob
1900-2300	REE	Eu	7275nob
2000	**Spanish**		
2000-0230	RAE	Eu	15345bue*
2000-0230	RAE	Am	6060bue*
2000-2030	Int. R. Serbia	Eu	6100bij
2000-2030	R. Prague	SEu	5930lit
2000-2100	Family Radio	CAm,NAm	13695yfr
2000-2100	R.Romania Int	Af	7430tig
2000-2100	R.Romania Int	Eu	9620tig
2000-2100	R.Taiwan Int.	Eu	3965iss
2000-2200	BBG-R.Martí	CUB	9565sac
2000-2300	Family Radio	CAm,SAm	5985yfr
2000-2400	Family Radio	CAm,NAm	11855yfr, 15130yfr
2030-2130	VOIRI	Eu	5950kam, 6055sit, 7200sir
2030-2400	R. Habana Cuba	Car	9820hab
2100	**Spanish**		
2100-0400	R. República	CUB	5954-†,*
2100-2120	R. Stn Belarus	Eu	1170sas, 6155mns, 7360mns, 7390mns
2100-2130	RFI	CAm	17630guf
2100-2130	R. Slovakia Int	SAm	11610rso
2100-2130	R. Slovakia Int	WEu	9460rso
2100-2200	Family Radio	Eu	9355yfr
2100-2200	R. Bulgaria	Eu	5900pld, 7300pld
2100-2200	VO Russia (VOR)	Eu	5920dsb, 5940sam, 6090arm, 7340nvs
2100-2300	CRI	Eu	6020szg, 9640kas
2100-2300	Family Radio	CAm,SAm	11700yfr
2100-2300	TWR Bonaire	CAm,SAm	800twb
2100-2400	R. Habana Cuba	Eu	11770hab
2100-2400	R. Habana Cuba	Car	5040hab
2120-2140	Vatican Radio	Eu	1611smg+, 4005vat, 5885smg
2120-2140	Vatican Radio	Af	7250smg
2130-2200	R. Prague	SEu,SAm	5930lit
2200	**Spanish**		
2200-0100	CVC La Voz	SAm	9635sgo
2200-0400	BBG-R.Martí	CUB	6030grv
2200-2300	CRI	LAm	13700sac
2200-2300	CRI	SAm	9490bei
2200-2300	Family Radio	CAm,SAm	9465yfr, 11580yfr, 11665yfr
2200-2300	R .Damascus	Eu,LAm	9330adr, 12085adr‡
2200-2300	REE	Af	11625nob
2200-2300	REE	Af	7265nob

2200	Spanish	Area	kHz
2200-2300	R.Romania Int	SAm	7380tig, 9790tig
2200-2300	VO Korea	EAs	3560kuj
2200-2300	VO Korea	Eu	7570kuj, 12015kuj
2200-2400	BBG-R.Martí	CUB	7405grv, 9565grv
2200-2400	CRI	Eu	7210cer, 7250uru
2200-2400	RAE	Eu	15345bue
2200-2400	RAE	Am	6060bue, 11710bue
2300	**Spanish**		
2300-0100	CRI	SAm	9590kas, 9800kas
2300-0200	REE	SAm	11680nob
2300-0500	REE	LAm	6125nob
2300-0500	REE	NAm,CAm	9535nob
2300-0500	REE	SAm	9620nob
2300-0500	R. Habana Cuba	SAm	12010hab
2300-2330	R. Nac.RASD	NAf	1550rbn*, 6297rbn±
2300-2400	CRI	Eu	6175cer
2300-2400	Family Radio	CAm,SAm	5985yfr, 9355yfr, 9495yfr, 13615yfr
2300-2400	RCI	CAm,SAm	9785sac, 11990sac
2300-2400	R.Taiwan Int.	SAm	11885yfr
2305-2400	RCI	NAm	6100sac
2330-0100	R. Habana Cuba	NAm	6000hab*, 9640hab*
0300	**Spanish/English**		
0300-0400	WHRI	SAm	7385hri
0400	**Spanish/English**		
0400-0500	WHRI	SAm	7385hri
0500	**Spanish/English**		
0500-0600	WHRI	SAm	7385hri
0600	**Spanish/English**		
0600-0700	WHRI	SAm	7385hri
0700	**Spanish/English**		
0700-0800	WHRI	SAm	7385hri
0800	**Spanish/English**		
0800-0900	WHRI	SAm	7385hri
0900	**Spanish/English**		
0900-1200	WHRI	SAm	7385hri
1000	**Spanish/English**		
1000-1100	R. N. Venezuela	Am	6180hab
1100	**Spanish/English**		
1100-1200	R. N. Venezuela	Am	6060hab
1200	**Spanish/English**		
1200-1300	R. N. Venezuela	Am	11705hab
1200-1300	WHRI	CAm	9410hri
1300	**Spanish/English**		
1300-1400	WHRI	SAm	9495hri
1400	**Spanish/English**		
1400-1900	WHRI	SAm	15665hri
1400-1900	WHRI	SAm	9495hri
1500	**Spanish/English**		
1500-1600	R. N. Venezuela	Am	11680hab
1900	**Spanish/English**		
1900-2000	R. N. Venezuela	Am	15290hab
1900-2300	WHRI	SAm	9595hri
2000	**Spanish/English**		
2000-2100	R. N. Venezuela	Am	17705hab
2200	**Spanish/English**		
2200-2300	R. N. Venezuela	Am	11670hab
2300	**Spanish/English**		
2300-0200	WHRI	SAm	7385hri
2300-1300	WHRI	SAm	5920hri
2300-2400	R. N. Venezuela	Am	13680hab, 15250hab

NB: not all transmissions are daily, please check main schedules under the appropriate country for full details

For *, ** and *** please see Notes under the Country entry for that station in the International Radio section

Key: + = DRM broadcast, † = irregular; ‡ = inactive at time of publication.

© WRTH Publications Ltd, November 2010

DRM International Broadcasts

Time	Language	Area	Station	kHz
0000				
0000-0030	Italian	Eu	Vatican Radio	1611smg
0000-0200	Spanish	NAm	REE	9630cri
0200				
0200-0300	English	EAs	Deutsche Welle	15205trm
0200-0400	Russian	As	VO Russia (VOR)	15735kna
0400				
0400-0600	English	As	VO Russia (VOR)	15735kna
0500				
0500-0800	Bulgarian	Eu	R. Bulgaria	9400sof
0500-0700	English	CEu	BBC World Sce	1296orf
0500-0700	English	Pac	RNZI	13730ran
0500-0530	Mandarin	EAs	R.Romania Int	17870tig
0530-0600	Russian	RUS	R.Romania Int	6175tig
0500-0900	Spanish	Eu	REE	9780nob
0600				
0630-0700	English	Eu	R.Romania Int	6020gal
0600-0630	French	Eu	R.Romania Int	6100gal
0645-0700	German	D	RTL Radio	6095jun
0700				
0700-1000	Bulgarian	Eu	R. Bulgaria	11900sof
0700-0800	English	RUS	BBC World Sce	5875mos
0700-0800	English	Pac	RNZI	11675ran
0700-0800	English	Eu	TDPradio	6015iss
0700-0900	English	Eu	VO Russia (VOR)	11635msk
0700-0730	German	WEu	KBS World R.	1440mrn
0730-0800	German	D	RTL Radio	6095jun
0730-0800	German	D	RTL Radio	1440mrn
0700-0730	German	WEu	KBS World R.	6095jun
0700-0730	German	Eu	R.Romania Int	9450tig
0800				
0800-0900	English	WEu	BBC World Sce	9610sin
0800-0900	English	Eu	TDPradio	6015iss
0800-1200	English	Pac	RNZI	9870ran
0900				
0900-1000	English	Eu	TDPradio	6015iss
0915-1015	Italian	Eu	Vatican Radio	1611smg
0915-1015	Papal Aud	Eu	Vatican Radio	1611smg
0930-1100	Portuguese	Eu	RDP Int.	9815sin
0900-1100	Russian	Eu	VO Russia (VOR)	7325klg
1000				
1000-1300	Bulgarian	Eu	R. Bulgaria	11900sof
1000-1200	Chinese	SEA,Pac	VO Malaysia	11885kaj
1000-1200	Chinese	SEA,Pac	VO Malaysia	7235kaj
1000-1030	English	Eu	R. Bulgaria	11900sof
1000-1100	English	Eu	TDPradio	6015iss
1100				
1100-1130	English	Eu	KBS World R.	9760wof
1100-1200	English	Eu	TDPradio	6015iss
1100-1200	English	Pac	R. Australia	12080brn
1100-1130	English	WEu	R. Japan	9760wof
1100-1300	German	Eu	VO Russia (VOR)	7325klg
1100-1130	Latin	Eu	Vatican Radio	1611smg
1130-1200	Russian	WEu	R. Japan	9760wof
1200				
1200-1300	English	Eu	TDPradio	6015iss
1200-1400	English	Pac	R. Australia	5995brn
1200-1300	English	As	VO Russia (VOR)	7340irk
1200-1215	French	Eu	Vatican Radio	1611smg
1300				
1300-1400	English	Eu	TDPradio	6015iss
1300-1400	Hindi	As	VO Russia (VOR)	7340irk
1300-1330	Italian	Eu	Vatican Radio	1611smg
1300-1500	Russian	Eu	VO Russia (VOR)	9675msk
1300-1600	Russian	Eu	VO Russia (VOR)	7325msk
1400				
1400-1600	English	SAs	BBC World Sce	13590trm
1400-1500	English	Eu	Disco Palace	6015iss
1400-1500	Hindi	SEA	BBC/DW Joint	5845nak
1400-1500	Hindi	SEA	BBC/DW Joint	13590trm
1400-1500	Russian	Eu	VO Russia (VOR)	5905msk
1400-1500	Russian	Eu	VO Russia (VOR)	9675msk
1400-1500	Urdu	As	VO Russia (VOR)	7340irk

Time	Language	Area	Station	kHz
1500				
1550-1650	English	Pac	RNZI	5950ran
1500-1600	English	Eu	VO Russia (VOR)	9675msk
1500-1600	English	Eu	VO Russia (VOR)	5905msk
1500-1600	English	SEA	BBC/DW Joint	5845nak
1500-1600	English	SEA	BBC/DW Joint	13590trm
1500-1515	German	Eu	Vatican Radio	6060smg
1500-1600	Hindi	As	VO Russia (VOR)	7340irk
1515-1530	Polish	Eu	Vatican Radio	6060smg
1600				
1600-1700	English	SAs	Deutsche Welle	13590trm
1600-1700	English	SAs	Deutsche Welle	5845nak
1600-1658	English	SAs	BBC/DW Joint	13590trm
1600-1759	English	SEA	BBC/DW Joint	5845nak
1605-1805	English	NAm	RCI	9800sac
1650-1750	English	Pac	RNZI	9890ran
1600-1700	German	Eu	VO Russia (VOR)	9675msk
1600-1700	Russian	RUS	R.Romania Int	6030tig
1600-1700	Russian	Eu	VO Russia (VOR)	7340arm
1630-1700	Russian	Eu	R. Bulgaria	9800sof
1600-1800	Serbian	Eu	VO Russia (VOR)	6040klg
1700				
1750-1850	English	Pac	RNZI	11675ran
1745-1945	English	Eu	All India R.	9950del
1700-1800	English	SAs	Deutsche Welle	1548trm
1700-1800	French	Eu	VO Russia (VOR)	6145klg
1700-1800	French	Eu	VO Russia (VOR)	9675msk
1730-1800	German	Eu	R. Bulgaria	9700sof
1700-1730	German	Eu	R.Romania Int	5875kvi
1800				
1800-1900	English	Eu	R.Romania Int	6065tig
1850-1950	English	Pac	RNZI	15720ran
1830-1900	English	Eu	R. Bulgaria	9700sof
1800-1830	English	Eu	R.Romania Int	6020kvi
1800-1900	English	Eu	Polish Radio	5895kvi
1800-1830	French	Eu	R. Bulgaria	9700sof
1805-2005	French	NAm	RCI	9800sac
1800-1900	Italian	Eu	VO Russia (VOR)	6040klg
1800-1900	Italian	Eu	VO Russia (VOR)	6145klg
1800-2000	Spanish	SAm	CVC La Voz	17860sgo
1900				
1900-2000	Bulgarian	Eu	R. Bulgaria	9700sof
1950-2050	English	Pac	RNZI	17675ran
1900-2000	French	Eu	RFI	3965iss
1900-2000	German	Eu	R.Romania Int	9805tig
1945-2045	Hindi	Eu	All India R.	9950del
1900-1930	Italian	Eu	R.Romania Int	6180tig
2000				
2045-2130	English	Eu	Vatican Radio	9800sac
2050-2150	English	Pac	RNZI	15720ran
2045-2230	English	Eu	All India R.	9950del
2000-2100	English	NAm	Disco Palace	15755bon
2000-2200	French	Eu	VO Russia (VOR)	6105klg
2030-2100	German	Eu	Polish Radio	3975skn
2100				
2140-2200	Arabic	Eu	Vatican Radio	1611smg
2150-0500	English	Pac	RNZI	17675ran
2130-2200	English	Eu	R. Romania Int.	6030gal
2100-2200	English	NAm	TDPradio	15755bon
2100-2130	French	Eu	R.Romania Int	7370gal
2100-2130	French	Eu	R.Romania Int	6030gal
2120-2140	Spanish	Eu	Vatican Radio	1611smg
2200				
2200-2300	English	NAm	RCI	9800sac
2200-1800	French	Eu	RFI	3965iss
2200-2220	Italian	Eu	Vatican Radio	1611smg
2245-2310	Italian	Eu	Vatican Radio	1611smg
2220-2245	Latin	Eu	Vatican Radio	1611smg
2300				
2330-2400	English	Eu	Vatican Radio	1611smg
2300-2345	English	NAm	Vatican Radio	7370smg
2310-2330	English	Eu	Vatican Radio	1611smg

NB: Not all transmissions are daily
© WRTH Publications Ltd November 2010

TERRESTRIAL TELEVISION

Section Contents

Initial entries for each letter,
see Main Index for full details

Features & Reviews

National Radio

International Radio

Frequency Lists

Terrestrial Television
National Radio via DTT

Reference

CHARACTERISTICS OF ANALOGUE TELEVISION SYSTEMS
(Recommendation ITU-R BT.470-6, Revision 2005)

System	Number of lines	Channel width MHz.	Vision band-width MHz.	Vision/Sound separation MHz.	Vestigial side-band MHz.	Vision mod.	Sound mod.
B	625	7	5	+5.5	0.75	Neg.	FM
B1	625	8	5	+5.5	0.75	Neg.	FM
D	625	8	6	+6.5	0.75	Neg.	FM
D1	625	8	5	+6.5	0.75	Neg.	FM
G	625	8	5	+5.5	0.75	Neg.	FM
H	625	8	5	+5.5	1.25	Neg.	FM
I	625	8	5.5	+5.996	1.25	Neg.	FM
I1	625	8	5.5	+5.996	1.25	Neg.	FM
K	625	8	6	+6.5	0.75	Neg.	FM
K1	625	8	5	+6.5	0.75	Neg.	FM
L	625	8	6	+6.5*	1.25	Pos.	AM
M	525	6	4.2	+4.5	0.75	Neg.	FM
N	625	6	4.2	+4.5	0.75	Neg.	FM

*) On VHF channels L2, L3, L4 (France): -6.5 MHz.

CHANNEL INFORMATION
(Vision carrier frequencies in MHz)

VHF Channels
(in brackets: DTT centre carrier frequencies)

"A" Channels
(Americas, parts of Asia & Pacific) (DTT: ATSC, 6MHz)

A2 = 55.25 (55.31)	A6 = 83.25 (83.31)	A10 = 193.25 (193.31)
A3 = 61.75 (61.31)	A7 = 175.25 (175.31)	A11 = 199.25 (199.31)
A4 = 67.25 (67.31)	A8 = 181.25 (181.31)	A12 = 205.25 (205.31)
A5 = 77.25 (77.31)	A9 = 187.25 (187.31)	A13 = 211.25 (211.31)

"E" Channels
(Most of Europe, Africa, parts of Asia & Pacific)

E2 = 48.25	E6 = 182.25	E10 = 210.25
E3 = 55.25	E7 = 189.25	E11 = 217.25
E4 = 62.25	E9 = 203.25	E12 = 224.25
E5 = 175.25	E9 = 203.25	

"R" Channels
(Parts of Europe, Russia, parts of Asia)

R1 = 49.75	R5 = 93.25	R9 = 199.25
R2 = 59.25	R6 = 175.25	R10 = 207.25
R3 = 77.25	R7 = 183.25	R11 = 215.25
R4 = 85.25	R8 = 191.25	R12 = 223.25

Specific national parameters:

Ireland

A = 45.75	E = 183.25	H = 207.25
B = 53.75	F = 191.25	I = 215.25
C = 61.75	G = 199.25	J = 223.25
D = 175.25		

Italy

A = 53.75	E = 183.75*	H = 210.25
B = 62.25	F = 192.25*	H1 = 217.25
C = 82.25	G = 201.25*	H2 = 218.75
D = 175.25		

(* replaced in 2009 by "E" channels E6, E7, E9)

Morocco

M4 = 163.25	M7 = 187.25	M9 = 203.25
M5 = 171.25	M8 = 195.25	M10 = 211.25
M6 = 179.25		

South Africa & Namibia

SA4 = 175.25	SA7 = 199.25	SA10 = 223.25
SA5 = 183.25	SA8 = 207.25	SA11 = 231.25
SA6 = 191.25	SA9 = 215.25	SA13 = 247.43

China (P.R.)

C1 = 49.75	C5 = 85.25	C9 = 192.25
C2 = 57.75	C6 = 168.25	C10 = 200.25
C3 = 65.75	C7 = 176.25	C11 = 208.25
C4 = 77.25	C8 = 184.25	C12 = 216.25

Japan

J1 = 91.25	J5 = 177.25	J9 = 199.25
J2 = 97.25	J6 = 183.25	J10 = 205.25
J3 = 103.25	J7 = 189.25	J11 = 211.25
J4 = 171.25	J8 = 193.25	J12 = 217.25

Australia & parts of Pacific

AU0 = 46.25	AU5 = 102.25	AU9 = 196.25
AU1 = 57.25	AU5A = 138.25	AU9A = 203.25
AU2 = 64.25	AU6 = 175.25	AU10 = 209.25
AU3 = 86.25	AU7 = 182.25	AU11 = 216.25
AU4 = 95.25	AU8 = 189.25	AU12 = 224.25

New Zealand & parts of Pacific

NZ1 = 45.25	NZ5 = 182.25	NZ9 = 210.25
NZ2 = 55.25	NZ6 = 189.25	NZ10 = 217.25
NZ3 = 62.25	NZ7 = 196.25	NZ11 = 224.25
NZ4 = 175.25	NZ8 = 203.25	

French Overseas Territories

K4 = 175.25	K6 = 191.25	K8 = 207.25
K5 = 183.25	K7 = 199.25	K9 = 215.25

UHF Channels

"A" Channels
(Americas, parts of Asia & Pacific)

14 = 471.25	33 = 585.25	52 = 699.25
15 = 477.25	34 = 591.25	53 = 705.25
16 = 483.25	35 = 597.25	54 = 711.25
17 = 489.25	36 = 603.25	55 = 717.25
18 = 495.25	37 = 609.25	56 = 723.25
19 = 501.25	38 = 615.25	57 = 729.25
20 = 507.25	39 = 621.25	58 = 735.25
21 = 513.25	40 = 627.25	59 = 741.25
22 = 519.25	41 = 633.25	60 = 747.25
23 = 525.25	42 = 639.25	61 = 753.25
24 = 531.25	43 = 645.25	62 = 759.25
25 = 537.25	44 = 651.25	63 = 765.25
26 = 543.25	45 = 657.25	64 = 771.25
27 = 549.25	46 = 663.25	65 = 777.25
28 = 555.25	47 = 669.25	66 = 783.25
29 = 561.25	48 = 675.25	67 = 789.25
30 = 567.25	49 = 681.25	68 = 795.25
31 = 573.25	50 = 687.25	69 = 801.25
32 = 579.25	51 = 693.25	

"E" / "K" / "L" / "R" Channels
(Europe, Russia, Africa, parts of Asia & Pacific) (DTT: DVB-T, 8MHz)

21 = 471.25 (474.00)	38 = 607.25 (610.00)	55 = 743.25 (746.00)
22 = 479.25 (482.00)	39 = 615.25 (618.00)	56 = 751.25 (754.00)
23 = 487.25 (490.00)	40 = 623.25 (626.00)	57 = 759.25 (762.00)
24 = 495.25 (498.00)	41 = 631.25 (634.00)	58 = 767.25 (770.00)
25 = 503.25 (506.00)	42 = 639.25 (642.00)	59 = 775.25 (778.00)
26 = 511.25 (514.00)	43 = 647.25 (650.00)	60 = 783.25 (786.00)
27 = 519.25 (522.00)	44 = 655.25 (658.00)	61 = 791.25 (794.00)
28 = 527.25 (530.00)	45 = 663.25 (666.00)	62 = 799.25 (802.00)
29 = 535.25 (538.00)	46 = 671.25 (674.00)	63 = 807.25 (810.00)
30 = 543.25 (546.00)	47 = 679.25 (682.00)	64 = 815.25 (818.00)
31 = 551.25 (554.00)	48 = 687.25 (690.00)	65 = 823.25 (826.00)
32 = 559.25 (562.00)	49 = 695.25 (698.00)	66 = 831.25 (834.00)
33 = 567.25 (570.00)	50 = 703.25 (706.00)	67 = 839.25 (842.00)
34 = 575.25 (578.00)	51 = 711.25 (714.00)	68 = 847.25 (850.00)
35 = 583.25 (586.00)	52 = 719.25 (722.00)	69 = 855.25 (858.00)
36 = 591.25 (594.00)	53 = 727.25 (730.00)	
37 = 599.25 (602.00)	54 = 735.25 (738.00)	

"J" Channels
(Japan)

13 = 471.25	30 = 573.25	47 = 675.25
14 = 477.25	31 = 579.25	48 = 681.25
15 = 483.25	32 = 585.25	49 = 687.25
16 = 489.25	33 = 591.25	50 = 693.25
17 = 495.25	34 = 597.25	51 = 699.25
18 = 501.25	35 = 603.25	52 = 705.25
19 = 507.25	36 = 609.25	53 = 711.25
20 = 513.25	37 = 615.25	54 = 717.25
21 = 519.25	38 = 621.25	55 = 723.25
22 = 525.25	39 = 627.25	56 = 729.25
23 = 531.25	40 = 633.25	57 = 735.25
24 = 537.25	41 = 639.25	58 = 741.25
25 = 543.25	42 = 645.25	59 = 747.25
26 = 549.25	43 = 651.25	60 = 753.25
27 = 555.25	44 = 657.25	61 = 759.25
28 = 561.25	45 = 663.25	62 = 765.25
29 = 567.25	46 = 669.25	

Australia

28 = 527.25	42 = 625.25	56 = 723.25
29 = 534.25	43 = 632.25	57 = 730.25
30 = 541.25	44 = 639.25	58 = 737.25
31 = 548.25	45 = 646.25	59 = 744.25
32 = 555.25	46 = 653.25	60 = 751.25
33 = 562.25	47 = 660.25	61 = 758.25
34 = 569.25	48 = 667.25	62 = 765.25
35 = 576.25	49 = 674.25	63 = 772.25
36 = 583.25	50 = 681.25	64 = 779.25
37 = 590.25	51 = 688.25	65 = 786.25
38 = 597.25	52 = 695.25	66 = 793.25
39 = 604.25	53 = 702.25	67 = 800.25
40 = 611.25	54 = 709.25	68 = 807.25
41 = 618.25	55 = 716.25	69 = 814.25

China (P.R.)

13 = 471.25	21 = 534.25	29 = 637.25
14 = 479.25	22 = 543.25	30 = 645.25
15 = 487.25	23 = 551.25	31 = 653.25
16 = 495.25	24 = 559.25	32 = 661.25
17 = 503.25	25 = 605.25	33 = 669.25
18 = 511.25	26 = 613.25	34 = 677.25
19 = 519.25	27 = 621.25	36 = 693.25
20 = 527.25	28 = 629.25	

DIGITAL TERRESTRIAL TELEVISION (DTT) SYSTEMS

ATSC
ATSC (Advanced Television Systems Committee) is the digital television standard developed in the USA. ATSC transmits with MPEG-2 video- and audio compression. It produces wide screen 16:9 images up to 1920×1080 pixels in size; up to six standard TV channels can be broadcast from a single TV transmitter using an existing 6MHz channel. ATSC uses the Dolby Digital AC-3 format to provide 5.1-channel surround sound. Numerous auxiliary services can also be provided, incl. radio programmes.

DVB-T
DVB-T (Digital Video Broadcasting - Terrestrial) is the standard of the European DVB consortium for the transmission of digital terrestrial television. This system transmits a compressed digital audio/video stream, using OFDM modulation with concatenated channel coding (COFDM). The source coding methods are MPEG-2 and MPEG-4. The modulation method in DVB-T is COFDM with either 64 or 16 state Quadrature Amplitude Modulation (QAM). 16 and 64QAM constellations can be combined in a single multiplex, providing a controllable degradation for more important programme streams. Several radio programmes can be transmitted as well.

DTMB
DTMB (Digital Terrestrial Multimedia Broadcast) is the DTT standard of the P.R. China. It is understood to be a fusion of TDS-OFDM (Time Domain Synchronous OFDM), developed by the Beijing Tsinghua university and based on the standard used by multi-carriers (similar to DVB-T), and ADTB-T (Advanced Digital Television Broadcast - Terrestrial) developed by the Shanghai Jiatong University.

ISDB-T / SBTVD
ISDB-T (Integrated Services Digital Broadcasting - Terrestrial) was developed in Japan and works similar to the European DVB-T. It uses COFDM modulation with PSK/QAM. The compression system is MPEG-2. ISDB-T can also transmit radio programmes in addition to TV channels. SBTVD (Sistema Brasileiro de Televisão Digital) is a version of ISDB-T developed in Brazil, with MPEG4 compression.

T-DMB
T-DMB (Terrestrial Digital Multimedia Broadcast) is the DTT standard developed in the Republic of Korea.

INTRODUCTION

The TV section contains information about terrestrial TV stations and radio prgrs via DTT, in a compact format. If applicable, each country entry is devided into subsections: "National Stations", "Local Stations", "Regional Stations", "Foreign TV Relays", "Foreign Military Stations". The subsection "DTT Transmitters" contains details of DTT operators and DTT transmitter networks. Contact info for domestic prgrs included in the DTT multiplexes are found in the subsections mentioned above.
Keys: "Systems": # = txs to be phased out, † = txs being phased out, analogue tx details no longer listed ⇩= analogue shutdown date; [A], [E], [K], [K1], [AU], [NZ], [SA] refer to the channel characteristics as shown in the "Channel Information" table. Tx networks for national stations are listed either with main txs (power limit applied) or with key tx(s). Local Stations (if included) are listed in full; if no tx location is given, the site refers to the city of the station's headquarters. (-) = tx details not received at editorial deadline.

AFGHANISTAN

Systems: PAL-B/G [E]. BFBS-TV: DVB-T (MPEG2)

National Stations
AFGHANISTAN NATIONAL TV (ANTV) (Gov) ✉ P.O.Box 544, Kabul ☎ +93 20 2103200 🖷 +93 20 2101086 **E:** info@nrta.org.af **Web:** www.nrta.org.af **L.P:** Pres: Najib Roshan **Txs:** Kabul ch11 (2kW) & relay txs. – **AFGHAN TV (Comm)** ✉ Kabul **L.P:** Dir: Ahmed Shah Afghanzai. **Txs:** Kabul ch24 & relay txs. – **ARIANA TELEVISION NETWORK (ATN) (Comm)** ✉ Darlaman Street, Kabul ☎ +93 70 151515 **E:** marketing@arianatelevision.com **W:** www.arianatelevision.com **Txs:** Kabul ch4 & relay txs. – **AYNA TV (Comm)** ✉ Kabul **Txs:** Kabul ch2 (5kW) & relay txs. – **TOLO TV (Comm)** ✉ P.O.Box 225, Kabul **E:** info@tolo.tv **W:** www.tolo.tv **L.P:** Dir: Saad Mohseni **Txs:** Kabul ch9 & relay txs.

Local Stations
Bamdad TV (Comm): Kabul; ch12. **City TV (Comm):** Karti 3, Kabul; ch49 (1kW). **Emroz TV (Comm):** Kabul; ch(-) (3kW). **Noorin TV (Comm):** Kabul; ch(-). **Noor TV:** Kabul; ch(-) (5kW). **Rai Farda:** Kabul; ch(-) (2kW). **Saba TV:** Kabul; ch(-). **Shamshad TV (Comm):** P.O.Box 150, Kabul; ch6. **Tamadun TV:** Kabul; ch(-) (2kW).**Talim wa Tarbia:** Kabul; ch(-).
NB. Only active stns in Kabul shown. Local stns (esp. in the provinces) are often operating irreg. and may be off the air at any time.

Foreign Military Station
BFBS-TV (British Mil) ✉ Chalfont Grove, Narcot Lane, Chalfont St Peter, Buckinghamshire, SL9 8TN, United Kingdom. **Mux 1+2:** BFBS1, BFBS2, BFBS3 Kids, BFBS4, BFBS1 One Day Later, BFBS+, Sky Sports 1, Sky Sports 2 ⌘ BFBS Radio 1+2. **Txs:** Kandahar, Camp Bastion, Lashkar Gar, Gereshk; channels: (-).

ALASKA (USA)

System: ATSC

Local Stations*
KAKM (Pub) 3877 University Dr, Anchorage, AK 99508-4676. °PBS. Tx: Anchorage ch8 (50kW). **KATN (Comm):** 516 2nd Ave Ste 400, Fairbanks, AK 99701-4729. °ABC. Tx: ch18 (16kW). **KDMD (Comm):** 1310 E 66th Ave, Anchorage, AK 99518-1915. °ION, Telemundo, HSN. Tx: ch32 (50kW). Mux: KDMD, Telemundo. **KFXF (Comm):** 3650 Braddock St Ste 2, Fairbanks, AK 99701-7617. °Fox. Tx: ch22 (11kW). **KIMO (Comm):** 2700 E Tudor Rd, Anchorage, AK 99507-1136. °ABC, CW. Tx: ch12 (41kW). Mux: KIMO-DT (ABC), KWBX-DT (CW). **KJNP-TV (Rlg):** 2501 Mission Rd, North Pole, AK 99705-6361. °TBN. Tx: ch20 (15kW). **KJUD (Comm)** 175 S Franklin St, Juneau, AK 99801-1384. °Fox, CW. Tx: ch11 (0.14kW). Mux: KJUD, CW. **KTBY (Comm):** 440 E Benson Blvd Ste 1, Anchorage, AK 99503-4121. °Fox. Tx: ch20 (234.4kW). **KTNL-TV (Comm):** 520 Lake St, Sitka, AK 99835-7403. °CBS. Tx: ch2 (1kW). **KTOO-TV (Pub):** 360 Egan Dr, Juneau, AK 99801-1748. °PBS. Tx: ch10 (1kW). **KTUU-TV (Comm):** 701 E Tudor Rd Ste 220, Anchorage, AK 99503-7488. °NBC. Tx: ch10 (21kW). **KTVA (Comm):** 1007 W 32nd Ave, Anchorage, AK 99503-3728. °CBS. Tx: ch28 (28.9kW). **KTVF (Comm):** 3528 International Way, Fairbanks, AK 99701-7382. °NBC. Tx: ch12 (12kW). **KUAC-TV (Pub):** 312 Tanana Dr, Fairbanks, AK 99775-2004. °PBS. Tx: ch24 (69kW). **KYES-TV (Comm)** 3700 Woodland Dr Ste 800, Anchorage, AK 99517-2588. °MyNetworkTV. Tx: ch6 (45kW). Mux: KYES-DT, audio channels, Wealth TV. **KYUK-TV (Pub)** 640 Radio St, Bethel, AK 99559. °PBS. Tx: ch3 (4.68kW).
*) Full power licenses (lp licenses not listed); °) Network affiliation.

ALBANIA

System: PAL-B/G [E]; DVB-T (MPEG2)

National Stations
RADIOTELEVISIONI SHQIPTAR (RTSH) (Pub) ✉ Rr. "Ismail Qemali" 11, Tirana ☎ +355 4 2256059 🖷 +355 4 2227745 **W:** www.rtsh.al **L.P:** DG: Artur Zheji; DirTV: Agron Cobani. **Chs:** TVSH1, TVSH2. **Txs: TVSH1:** Dajt ch4 (10kW) & ch57 (5kW), Gllava ch9 (2kW), Cervenake ch11 (5kW), Mide ch12 (5kW), Durres ch25 (1kW) & txs below 1kW; **TVSH2:** Dajt ch11 (1kW). – **TOP CHANNEL (Comm)** ✉ Bul. "Deshmoret e Kombit, Qendra Nderbombetare e Kultures", Tirana ☎ +355 4 2253177 🖷 +355 4 2253178 **E:** info@top-channel.tv **W:** www.top-channel.tv. **Txs:** Tirana ch36 & network. – **TV KLAN (Comm)** ✉ Rr. "Aleksander Moisiu" 97, Tirana ☎ +355 4 2347805 🖷 +355 4 2347808 **E:** info@tvklan.tv **W:** www.rtvklan.com **L.P:** Dir: Pandi Laço **Txs:** Tirana ch28 & network.

Local Stations not shown.

Foreign TV Relay
TV5 Monde Europe (France): Tirana ch43.

DTT Transmitters (under construction)
Operator: DigitALB ✉ Rr. "Themistokli Germenji" 10, Tirana ☎+355 4 2255813 🖷 +355 4 2274831 **E:** info@digitalb.tv **W:** www.digitalb.tv **Mux 1-5** (☉): multiprgr **Txs: Mux 1:** ch62 (SFN), **Mux 2:** ch64 (SFN), **Mux 3:** ch67 (SFN), **Mux 4:** ch69 (SFN). **Mux 5 (DVB-H):** ch38 (Tirana). – **Operator:** Tring Digital ✉ Kompleksi Don Bosko, Kulla 2, kati II, Tirana ☎+355 4 4800008 🖷 +355 4 4800001 **E:** info@tring.tv **W:** www.tring.tv **Mux 1+2** (☉): multiprgr **Txs: Mux 1:** ch47 (Dajt) **Mux 2:** ch59 (Dajt).

ALGERIA

Systems: # PAL-B [E] ⇩2014; DVB-T (MPEG2)

ENTREPRISE NATIONALE DE TÉLÉVISION (ENTV) (Gov) ✉ 21, Boulevard des Martyrs, Alger ☎ +213 21 602300 🖷 +213 21 230914 **E:** commerciale@entv.dz **W:** www.entv.dz **L.P:** DG:Habib-Chawki Hamraoui; Dir.Tech: Chihab Benchikh el Hocine **Chs:** La Chaîne 1, Canal Algérie, A3, TV Tamazight, TV Coran, TV6 **Txs:**(-).

DTT Transmitters (under construction)
Operator: Télédiffusion d'Algérienne (TDA) ✉ BP 50, 16340 Alger ☎ +213 21 901717 🖷 +213 21 902424 **E:** info@tda.dz **W:** www.tda.dz **Mux:** La Chaîne 1, Canal Algérie, A3, TV Tamazight, TV Coran, TV6 ⌘ Chaîne I, II, III, R.Algérie Internationale **Txs:** ch41 (Alger/Bordj El Bahri 0.25kW) & network.

ANDORRA

System: DVB-T (MPEG2)

ANDORRA TELEVISIÓ (Pub) ✉ Baixada del Molí 24, AD500 Andorra la Vella ☎ +376 873777 🖷 +376 863242 **E:** rtva@rtva.ad **W:** www.rtvasa.ad **L.P:** DG: Enric Castellet.

DTT Transmitters
Operator: Andorra Telecom ✉ C/ Mossèn LLuís Pujol, numero 8-14, AD500 Santa Coloma ☎ +376 875274 🖷 +376 863667 **W:** www.sta.ad **Mux 1:** BBC World, RTP Intl, Tele 5, Arte, 3/24 **Tx:** ch28 (La Vella) **Mux 2:** CNN Int., Cuatro, C33, La Sexta, NRJ12 **Tx:** ch36 (La Vella). **Mux 3:** Andorra Televisió, La 2, M6, TF1, TV3 **Tx:** ch42 (La Vella) **Mux 4:** Antena 3, France 2, France 3, La 1, Super 3/3XL **Tx:** ch45 (La Vella).

ANGOLA

System: PAL-I [E]

TELEVISÃO PÚBLICA DE ANGOLA (Pub) ✉ CP 2604, Luanda ☎ +244 22 320326 🖷 +244 22 323622 **E:** tpa.informatica@netangola.com **W:** www.tpa.ao **L.P:** DG: Carlos Cunha; TD: Florindo Ramos **Chs:** TPA1, TPA2 **Txs: TPA 1:** Luanda ch9 (13kW) & relay txs. **TPA2:** (-).

Local Station
TV Zimbo: Avenida de Talatona, Luanda Sul; ch45.

Foreign TV Relay
RTP África (Portugal): (-).

ANGUILLA (UK)

System: NTSC-M [A]

KREATIVE COMMUNICATIONS NETWORK (KCN) (Comm)🖳
P.O.Box 154, The Valley ☎ +1 264 4973519 🖷 +1 264 4973367 **E:**
kcn@anguillanet.com **L.P:** Dir: Carlton Pickering **Stns:** ZJF-TV3 ch3
(0.003kW), ZJF-TV9 ch9 (0.03kW).

ANTARCTICA

NB: No terrestrial TV station.

ANTIGUA & BARBUDA

System: NTSC-M [A]

ABS-TV (Gov)🖳 Public Information Division, Galstron's Palace, Old
Parnham Road, St. John's ☎ +1 268 4620010 🖷 +1 268 4624442 **E:**
absradio@caribmail.com **L.P:** SM: Trevor Parker. **Tx:** ch10V (5kW).

ARGENTINA

Systems: # PAL-N [A]; SBTVD

National Stations
CANAL 7 (Comm)🖳Avenida Figueroa Alcorta 2977, 1425 Buenos
Aires ☎+54 11 8026001 **W:** www.canal7argentina.com.ar **Txs:** ch7
(212kW) & relay txs. – **CANAL 9 (Comm)**🖳Av. Dorrego 1708, 1414
Buenos Aires ☎ +54 11 50936838 **W:** www.canal9.com.ar **Txs:**
ch9 (62kW) & relay txs. – **CANAL 11 (TELEVISIÓN FEDERAL S.
A. - TELEFE) (Comm)**🖳 Pavón 2495, 1248 Buenos Aires ☎ +54
11 43080145 🖷 +54 11 4301522 **W:** www.telefe.com.ar. **Txs:** ch11
(180kW) & relay txs. – **CANAL 13 (Comm)**🖳 Lima 1261, Constitucion,
Capital Federal ☎ +54 11 3050013 🖷 +54 11 3318559 **W:** www.
canal13.com.ar **Txs:** ch13 (116kW) & relay txs.

Local Stations not shown.

DTT Transmitters (Trial)
Operator: Antina Argentina 🖳 Avenida Córdoba 3016, Planta Baja,
C1187AAR ☎ +54 11 52586000 **E:** info@antina.com.ar **W:** www.
antina.com.ar **Mux:** multiprgr (❂) **Txs:** (-). – **Operator:** Telefe **Mux:**
multiprgr **Tx:** ch10 (Buenos Aires).

ARMENIA

Systems: # SECAM-D/K [R], PAL-D/K [R] ⇩2015; DVB-T planned

National Stations
ARMENIAN PUBLIC TELEVISION (Pub)🖳 26, G. Hovsepyan
St., Nork 47, 0047 Yerevan ☎ +374 10 569574 🖷 +374 10 562460
E: director@armtv.com **W:** www.armtv.com **L.P:** CEO: Armen
Arzumanyan; Tech.Dir: Alexander Chitchyan **Txs:** Yerevan ch8 (5kW)
& network. – **ALM (Comm)**🖳 59, Komitas ave., 0014 Yerevan ☎
+374 10 230646 🖷 +374 10 231142 **E:** alm@front.ru **L.P:** Pres: Tigran
Karapetyan. **Txs:** Yerevan ch10 (5kW) & network. – **ARMENIA TV
(Comm)**🖳 Yeghvard Highway N1, 0054 Yerevan ☎ +374 10 365161
🖷 +374 10 365161 **E:** info@armeniatv.am **W:** www.armeniatv.
am **Txs:** Yerevan ch25 (1.5kW) & network. – **h2 (Comm)**🖳 3/1,
Quarter # G-3, 0088 Yerevan ☎ +374 10 398831 🖷 +374 10 395640
E: lraber@tv.am **W:** www.tv.am **L.P:** Dir: Samvel Mayrapetyan **Txs:**
Yerevan ch12 (5kW) & network.

Local Stations
Abovyan: 20, Hanrapetutyan St., Abovyan; ch24 (0.1kW). **Achin:** 1,
Tumanyan St., Nor-Hachin; ch33 (0.1kW). **Ankyun Gumarats 3:** 2/26,
Sarahart St., Alaverdy; ch37 (0.1kW). **Anna:** H.62,23 Ogostosi St., Artashat;
ch30. **AR:** 5a, Tumanyan St., 0010 Yerevan; ch3 (5kW). **ArmenAkob TV:**
Radiotun, 5, A. Manukyan St., Yerevan; ch31 (1kW). **Armenia TV:** 1,
Yeghvard Highway, 0045 Yerevan; ch25 (1.5kW). **Armnews:** 2 floor,1
Yeghvard Highway St., Yerevan; ch63 (1kW). **DAR 21:** 4, Hrazdan Canyon
St., Yerevan; ch27 (1.3kW) **Echmiadzin:** 5a, Araratyan St., Echmiadzin;
ch29 (0.1kW). **Fortuna:** 21, Garegin Nzhdeh St., # 32, Stepanavan; ch29
(0.01kW). **Hai TV:** 78, Hanrapetutyan St., # 37, Yerevan; ch45 (1kW).
Hayrenik: 13, Frunze St., Yerevan; ch21 (1kW). **Hrazdan:** Marzpetaran

building, Hrazdan; ch12 (0.1kW). **Ijevan:** 5, Yerevanyan St., Ijevan; ch12
(0.1kW). **Knetron:** 2, Alikhanyan Yeghbayrneri St., Yerevan; ch37 (1.3kW).
Kyavar: 20, Zoravar Andranik St., Gavar; ch27 (0.2kW). **Last:** 38/1a,
Komitas St., Goris; ch38 (0.1kW). **Lori TV:** 10, Batumi St., Vanadzor; ch2
(0.2kW). **Lusalik:** Culture House, Karen Demirchyan Sq., Charentsavan;
ch32 (0.05kW). **Mig TV:** 46a, Abeghyan St., Vanadzor; ch32 (0.1kW). **Mir
TV:** 2 Arshakunyats Avenue, 6th floor, Yerevan; ch61 (1kW). **Narek TV:**
23, Ogostosi St., Abovyan; ch26 (0.15kW). **Nig-Aparan:** 1, Shahumyan,
St., Aparan; ch10 (0.025kW). **Noy Hayastan:** 22, Yerevanyan St.,
hotel "Armavir", Armavir; ch32 (0.01kW). **Paradise:** 18, Abovyan St.,
Yerevan; ch23 (1kW). **Qamut:** 3, Kamo St., 1st floor, Noyemberyan; ch12
(0.015kW). **RTV:** 66, Myasnikyan St., Dilijan; ch2 (0.05kW). **Shant:** Vazgen
Sargsyan St., TV Center, Gyumri; ch2 (0.1kW), Yerevan ch41 (1kW).
Shirak: 248, Abovyan St., Gyumri; ch6. **Shoghakat:** 8, Hayk Hovsepyan
Str., Yerevan; ch35 (1kW). **Sosi:** 8, R. Meliqyan St., Kapan; Kajaran ch22
(0.1kW), Kapan ch30 (0.2kW). **STV:** 15, Shahumyan St., Spitak; ch7
(0.01kW). **STV1:** 6, Sargis Sevanetsi St., Sevan; ch21 (0.1kW) **Syuni:** 5,
Shirvanzade St., Sisian; ch27 (0.1kW). **Tashirk:** 4, Jahukyan St., Tashir;
ch21 (0.035kW). **Tsayg TV:** 248, Abovyan St., 3rd floors, Gyumri; ch6
(0.1kW), ch28 (0.5kW) **TV5:** 1, Yeghvard Highway, Yerevan; ch39 (1kW).
Yerevan TV: 2, Arshakunyats Avenue, Yerevan; ch51 (1kW). **Yerkir
Media:** 94, Charents St., Yerevan; ch56 (2.5kW). **Zangak:** 43, Getapnya
St., Martuni; ch25 (0.025kW).

Foreign TV Relays not shown.

ARUBA (Netherlands)

System: NTSC-M [A]

ATV - ARUBA BROADCASTING CO. (Gov)🖳 P.O.Box 5040,
Oranjestad ☎+297 5838150 🖷 +297 5838110 **W:** www.15atv.com **Tx:**
Oranjestad ch15. – **CARIBBEAN SUPER STATION (CSS) (Comm)**
🖳 Emmastraat 51, Oranjestad. **Tx:** Oranjestad ch24. – **TELE
ARUBA (Comm)**🖳 P.O.Box 392, Oranjestad ☎ + 297 5857302 🖷 +
297 5851683 **W:** www.telearuba.aw **L.P:** GM: Mrs. Jane Lampkin. **Tx:**
Oranjestad ch13 (3kW H).

ASCENSION ISLAND (UK)

System: PAL-I [E]

BFBS-TV (British Mil)🖳 Chalfont Grove, Narcot Lane, Chalfont
St Peter, Buckinghamshire, SL9 8TN, United Kingdom. **Txs:** BFBS1,
BFBS2: Ascension two UHF frqs (low power).

AUSTRALIA

Systems: DVB-T (MPEG2); † PAL-B/G ⇩31 Dec 2013

National Stations
AUSTRALIAN BROADCASTING CORPORATION (ABC) (Pub)🖳
ABC Ultimo Centre, 700 Harris St, Ultimo, NSW 2007 ☎ +61 2
83331500 🖷 +61 2 83335305 **E:** comments@your.abc.net.au **W:**
www.abc.net.au **L.P:** MD: Mark Scott **Chs:** ABC1, ABC2, ABC3, ABC
News 24. – **SPECIAL BROADCASTING SERVICE (SBS) (Pub)**🖳
Locked bag 028, Crows Nest, NSW 1585 ☎ +61 2 94302828 🖷
+61 2 94303700 **E:** comments@sbs.com.au **W:** www.sbs.com.au
– **NATIONAL INDIGIOUS TELEVISION (NITV) (Pub)**🖳 5 Parsons
Street, Alice Springs, NT 0870 ☎ +61 8 89534763 🖷 +61 8 89534764
E: admin@nitv.org.au **W:** nitv.org.au.

Regional Stations
IMPARJA TELEVISION (Pub)🖳 P.O.Box 2924, Alice Springs, NT
0871 ☎ +61 89 523744, 🖷 +61 89 531014 **W:** www.imparja.com.
au – **NETWORK TEN (Comm)**🖳 P.O. Box 10, Lane Cove, NSW
2066 ☎ +61 2 8870222 **W:** www.ten.com.au; tencorporate.com.au
(corporate) – **NINE NETWORK (Comm)**🖳 P.O.Box 27, Willoughby,
NSW 2068 ☎ +61 2 99069999 🖷 +61 2 99582279 **W:** channelnine.
ninemsn.com.au – **PRIME TELEVISION (Comm)**🖳 PO Box 878,
Dickson, ACT 2602 ☎+61 2 62423700 🖷+61 2 62423764 **W:** www.
primetv.com.au; www.primemedia.com.au (corporate) – **SEVEN
NETWORK (Comm)**🖳 Television Centre, Mobbs Lane, Epping,
NSW 2121 ☎+61 2 8587777 🖷 +61 2 8587888 **W:** au.tv.yahoo.com;
www.sevencorporate.com.au (corporate) – **SOUTHERN CROSS
TELEVISION (Comm)**🖳 70 Park Street, South Melbourne, VIC
3205 ☎ +61 3 92432100 🖷 +61 3 96825158 **W:** www.sctv.com.au
– **WIN TELEVISION (Comm)**🖳 Television Ave, Mt St Thomas,
Wollongong, NSW 2500 ☎ +61 2 42234199 🖷 +61 2 42273682 **W:**
www.wintv.com.au.

Local Stations not shown.

DTT Transmitters
Operator: ABC **Mux:** ABC1, ABC2, ABC3, ABC News 24 ✳ ABC DigMusic, ABC Jazz – **Operator:** Seven Network **Mux:** Seven Digital, 7Two, 7mate – **Operator:** Nine Network **Mux:** Nine Digital, GEM, GO! – **Operator:** Network Ten **Mux:** One HD, One Digital, Ten Digital, Eleven – **Operator:** SBS **Mux:** SBS One, SBS One HD, SBS Two ✳ SBS Radio 1, 2.

Location	ABC	7N	9N	N10	SBS	kW
Sydney	12	6	8	11	34	50

+ nationwide tx networks

AUSTRIA

Systems: DVB-T (MPEG2)

National Stations
ÖSTERREICHISCHER RUNDFUNK (ORF) (Pub)⌧ Würzburgasse 30, 1136 Wien ☎ +43 1 878780 **E:** presse@orf.at **W:** www.orf.at **LP:** DG: Dr.Alexander Wrabetz **Chs:** ORF1, ORF2 incl. reg. stns: a) ORF Burgenland (Buchgraben 51, 7000 Eisenstadt), b) ORF Kärnten (Sponheimer Straße 13, 9020 Klagenfurt), c) ORF Niederösterreich (Radioplatz 1, 3109 St.Pölten), d) ORF Oberösterreich (Europaplatz 3, 4010 Linz), e) ORF Salzburg (Nonntaler Hauptstraße 49d, 5020 Salzburg), f) ORF Steiermark (Marburgerstr. 20, 8042 Graz), g) ORF Tirol (Rennweg 14, 6010 Innsbruck), h) ORF Vorarlberg (Höchstraße 38, 6850 Dornbirn), i) ORF Wien (Argentinierstr. 30a, 1040 Wien). – **ATV (Comm)**⌧ Aspernbrückengasse 2, 1020 Wien ☎ +43 1 213640 ▤ +43 1 21364999 **E:** atv@atv.at **W:** atv.at **LP:** CEO: Franz Prenner. – **PULS 4 (Comm)**⌧ Mariahilferstr. 2/10/16, 1070 Wien ☎ +43 1 999880 ▤ +43 1 999888888 **E:** post@pulstv.at **W:** www.pulstv.at **LP:** CEO: Martin Blank. – **SERVUS TV (Comm)**⌧ Ludwig-Bieringer-Platz 1, 5073 Wals-Himmelreich ▤ +43 662 84224428181 **E:** office@servustv.at **W:** www.servustv.com **LP:** CEO: Rudolf Theierl.

Local Stations not shown (via Mux 3 txs).

DTT Transmitters
Operator Mux 1+2: Österreichische Sender GmbH & Co KG (ORS) ⌧ Würzburggasse 309, 1136 Wien ☎ +43 1 8704012680 ▤ +43 1 874012773 **E:** office@ors.at **W:** www.ors.at **Mux 1:** ORF1, ORF2 incl. reg. prgrs, ATV **Mux 2:** Puls 4, ORF Sport Plus, 3sat, ServusTV. – **Operator Mux 3:** Regional operators (not shown). – **Operator Mux 4:** Media Broadcast GmbH, Joseph-Schumpeter-Allee 17, D-53227 Bonn, Germany **E:** info@media-broadcast.com **W:** www.media-broadcast.com **Mux (DVB-H):** ORF1, ORF 2, ATV, Puls 4, Pro7 Österreich, RTL, Sat1 Österreich, Vox, Laola1.tv, LaLaTV, Red Bull TV, RTL 2, N24, Super RTL, KroneTV ✳ Öi, Hitradio Ö3, FM4, Kronehit, Lounge FM

Location	M1	M2	M4	kW
Bregenz (Pfänder)	24	21	31	2x56/32
Bruck a.d.M. (Mugel)	41	25	-	45/56
Freilassing (Högl)°)	47V			20
Graz (Schöckl)	26	23	39V	2x48/20
Innsbruck (Patscherkofel)	23	27	37	2x63/25
Klagenfurt (Dobratsch)	24	30	54	80/76/10
Linz (Lichtenberg)	43	37	-	50
Mattersberg (Heuberg)	52V			20
Rechnitz (Hirschenstein)	43	23	36V	25/10/20
Salzburg (Gaisberg)	32	29	47V	2x40/13
St.Pölten (Jauerling)	31	21	-	50
Schärding (Schardenberg)	43			16
Schladming (Hauser Kaibling)	40		-	20
Viktring (Stifterkogel)	24	30	46	2x16/14
Waidhofen/Ybbs (Sonntagberg)	43			12
Weitra (Wachberg)	31			14
Wien (Kahlenberg)	24	34	36	2x63/16

+ sites with txs below 10kW. °) Tx located in Germany

AZERBAIJAN

Systems: # PAL-D/K [R] ⇩2012; DVB-T (MPEG4)

National Stations
AZƏRBAYCAN TELEVIZIYA VƏ RADIO VERISLƏRI (Gov)⌧ Mehdi Hüseyn St. 1, AZ 1011 Baki ☎ +994 12 4984720 ▤ +994 12 4972020 **E:** info@aztv.az **W:** www.aztv.az **LP:** Dir: Nizami Xudiyev. **Chs:** AzTV, Idman Azərbaycan, Mədəniyyət (F.pl) **Txs: AzTV:** Baki ch3 (23kW) & network, **Idman Azərbaycan:** Baki ch10 (5kW) & network. – **ICTIMAI TELEVIZIYA (ITV) (Pub)**⌧ Serifzade St. 241, AZ 1012 Baki ☎ +994 12 4335525 ▤ +994 12 4302958 **E:** info@itv.az **W:** www.itv.az **LP:** Dir: Ismail Omarov. **Ch:** 1-ci kanal **Txs:** Baki ch33 (5kW) & network. – **ANS-**

TV (Comm) ⌧ Matbuat ave. 28/11, Baki ☎ +994 12 4977267 ▤ +994 12 4989498 **E:** ans@ans.az **W:** www.ans.az **Txs:** Baki ch31 & network. – **AZAD AZƏRBAYCAN TV (ATV) (Comm)** ⌧ A.Abbaszadä St. 8, AZ 1073 Baki ☎ +994 12 4974621 ▤ +994 12 4932522 **E:** atv@azadazerbaijan.com **Txs:** Baki ch35 & network. – **LIDER TV (Comm)** ⌧ Ä.Äläkbärov St. 83/23, AZ 1141 Baki ☎ +994 12 4978899 ▤ +994 12 4978898 **E:** mail@lidermedia.az **W:** www.lidertv.com **Txs:** Baki ch25 & network. – **SPACE TV (Comm)** ⌧ Hüseyn Cavid ave. 8, AZ 1073 Baki ☎ +994 12 4921256 ▤ +994 12 4927665 **E:** info@spacetv.az **W:** www.spacetv.az **Txs:** Baki ch27 & network. – **XƏZƏR TV (Comm)** ⌧ Atatürk ave. 28, AZ 1000 Baki ☎ +994 12 5621647 ▤ +994 12 5621623 **E:** info@xazar.tv **W:** www.xazar.tv **Txs:** (-).

Local Stations
Alternativ TV: Tagi Arani 6, AZ 2000 Gäncä; ch44. **Aygün TV:** Azadliq St 1, AZ 6200 Zaqatala; ch12. **Cänub TV:** Azad Mirzäyev St. 70, AZ 4200 Länkaran; ch29. **Dünya TV:** 10-su mikrorayon, AZ 5000 Sumqayit; ch42. **Kanal-S MMC:** Q.Qarayev St. 11, Säki; ch53. **Käpäz TV:** Äli Näzmi St. 4, AZ 2000 Gäncä; ch46. **Mingäcevir TV:** S.Vurgun St. 19, AZ 4500 Mingäcevir; ch12. **Qütb TV:** H.Aliyev pr. 156, AZ 4000 Quba; ch27. **Simurq M TV:** S. Mämädov St. 1, AZ 6000 Tovuz; ch45. **RTV:** H.Z.Tagiyev St. 10, Xaçmaz; ch43. **Xäyal TV:** F.Xan 133, AZ 4000 Quba; ch39. **Yevlax TV:** Zärdabi St. 1/6, Yevlax; ch57.

Foreign TV Relays not shown.

DTT Transmitters (under construction)
Operator: RITN Teleradio IB ⌧ A.Abbaszadä 2, AZ 1073 Baki ☎ +994 12 4988066 ▤ +994 12 4988397 **E:** teleradio@azerin.com **W:** www.teleradio.rabita.az **Mux 1:** AzTV, Idman Azärbaycan, Mädäniyyät (F.pl), 1-ci kanal **Mux 2:** multiprgr.

Location	M1	M2	kW
Baki	37	48	0.8/1.2

+ nationwide network (txs being installed at all main sites)

AZORES (Portugal)

Systems: # PAL-B/G [E] ⇩22 Mar 2012; DVB-T (MPEG4)

RTP AÇORES (Pub)⌧ Rua Ernesto do Canto 40, 9500-312 Ponta Delgada ☎ +351 296202700 ▤ +351 296202771 **E:** rtpa@rtp.pt **LP:** Dir: Pedro Bicudo. **Chs:** RTP Açores; RTP1 relay (from mainland) **Txs: RTP Açores:** Pico da Barrosa ch7 (150kW), Santa Bárbara ch9 (115kW) + txs below 10kW; **RTP1:** Santa Bárbara ch21 (470kW), Pico Alto ch23 (1383kW), Pico de Barrosa ch31 (500kW), Cabeço Gordo ch35 (33.6kW), Pico do Geraldo ch36 (11.6kW) + txs below 10kW.

DTT Transmitters (txs under construction)
Operator: Portugal Telecom **Mux 1:** RTP Açores, RTP1, RTP2, SIC, TVI **Txs:** ch47 (São Jorge), ch50 (Pico), ch61 (S.Miguel, Graciosa), ch64 (Faial), ch67 (Terceira, S.Maria, Flores, Carvo). **Mux 2 (✪):** tbd **Txs:** ch48 (São Jorge), ch58 (Pico), ch62 (S.Miguel, Graciosa), ch65 (Faial), ch68 (Terceira, S.Maria, Flores, Carvo). **Mux 3 (✪):** tbd **Txs:** ch49 (São Jorge), ch58 (Pico), ch63 (S.Miguel, Graciosa), ch66 (Faial), ch69 (Terceira, S.Maria, Flores, Corvo).

BAHAMAS

System: NTSC-M [A]

ZNS TV (Pub)⌧ P.O.Box N-1347, Nassau ☎ +1 242 3224623 ▤ +1 242 3223924 **E:** info@znsbahamas.com **W:** www.znsbahamas.com **Chs:** ZNS-TV1, ZNS-TV2 **Txs: ZNS-TV1:** Nassau ch13 (50kW); **ZNS-TV2:** (-).

BAHRAIN

System: PAL-B/G [E]; DVB-T planned

BAHRAIN TELEVISION (BTV) (Gov)⌧ P.O.Box 1075, Bahrain ☎ +973 17686000 ▤ +973 17681544 **E:** marketing@bahraintv.com **W:** www.bahraintv.com **LP:** Dir: Dr. H. Al-Umran **Chs:** BTV (Arabic), Channel 55 (English) **Txs: BTV:** ch4 (5kW), ch44* (500kW), **Channel 55:** ch55 (0.03kW).*) tx to change to DTT.

Foreign TV Relay
BBC World News (UK): ch57 (1kW)

BANGLADESH

System: PAL-B [E]

BANGLADESH TELEVISION (BTV) (Pub)⬚ TV Bhaban, Rampura, Dhaka 1219 ☎ +880 2 8618606 📠 +880 2 8312927 **E:** info@btv.com.bd **W:** www.btv.com.bd **L.P:** DG: Jafar Ahmed Chowdhury. **Txs:** Dhaka ch6 (60kW) & netw. – **EKUSHEY TELEVISION (ETV) (Comm)**⬚ Jahangir Tower, 10, Karwan Bazar, Dhaka 1215 ☎ +880 2 8126535 📠 +880 2 8121270 **E:** info@ekushey-tv.com **W:** www.ekushey-tv.com **Txs:** (-).

BARBADOS

System: NTSC-M [A]

CARIBBEAN BROADCASTING CORP. (CBC-TV) (Gov)⬚ P.O. Box 900, Pine Hill, Bridgetown ☎ +1 246 4675400 📠 +1 246 4294795 **E:** sales@cbcbarbados.bb **W:** www.cbc.bb **L.P:** Sen.Chmn: John Williams **Txs:** Bridgetown ch8 (60kW).

BELARUS

Systems: # SECAM-D/K [R] ⇩2015; DVB-T (MPEG4)

National Stations
BELARUSKAJE TELEBACANNE (BT) (Gov)⬚ Makaionka St. 9, 220807 Minsk ☎ +375 17 2634301 📠 +375 17 2648182 **E:** pr@tvr.by **W:** www.tvr.by **Chs:** Peršy Kanal (PK), Lad, Regional prgrs. Also provides relays of NTV, RTR, Mir (Russia). – **OBSHCHENATSIONALNOYE TELEVIDENIYE (ONT) (Gov)**⬚ Kamunistycny St. 6, 220029 Minsk ☎ +375 17 2170424 **E:** w@ont.by **W:** www.ont.by **L.P:** Pres: Grigoriy L. Kisel. – **STOLICHNOYE TV (STV) (Gov)**⬚ Kamunistycny St. 6, 220029 Minsk ☎ +375 17 2906272 📠 +375 17 2906432 **E:** reklama@ctv.by **W:** www.ctv.by.

Location	PK	Lad	ONT	STV	RTR	NTV	kW
Asipovicy	33	-	-	51	38	-	2/5/1
Babrujsk	12V	25	40	27V	9V	37V	2.5/2x20/5/0.5/5
Brahin	27	37	39	21	-	41V	20/5/20/2x5
Braslaú	23V	37V	12	3	-	6	2x5/2x1/0.1
Brest	7	40	30	47	47	59	5/2x20/2x5
Drahicyn	27	23	-	31	29	-	1/5/2/5
Heraniony	7V	32	29	9V	-	-	2x5/20/1
Homiel	10	24	3	38	8	30	4x5/0.1/5
Hrodna	3	51	34	56	11V	26V	5/2x20/10/0.1/2
Kapyl	7V	47	40V	52	-	36	0.1/4x5
Kastjukovicy	9V	56	27	39V	11V	44	5/0.5/2x5/0.1/2
Mahilioú	4	21	7	51V	10	32	20/20/25/5/0.1/20
Miadziel	8V	51V	41	34	10V	49V	2.5/3x5/0.1/1
Minsk	1	27	6	42	3	37	25/20/25/3x5
Pinsk	4	24	21	33	9	50	2x5/2x20/0.1/5
Salihorsk	11V	53	23	28	2V	-	2.5/1/2/5/0.5
Slonim	10	26	12	8	-	-	40/20/30/0.3
Smarhon	36	22	24	40	-	42	2x5/2x2/1
Smiatanicy	5	41	8	38V	31	-	25/20/25/5/2
Svislac	27	44	39	1	-	-	2x5/2x0.1
Ušacy	9	28	11	36	5	-	25/20/25/5/0.1
Viciebsk	2	31	12V	41V	39V	26	5/20/2x5/0.1/0.3
Vorša	-	36	-	3V	34	38V	5/0.1/5/1
Zlobin	22	34	6	52	-	-	2x20/0.1/5

+ sites with txs below 5kW.

Local Stations (all Comm)
Bug TV: Brest; ch9 (0.1kW). **Inteks:** Baranavicy; ch23. **Nireja:** Homiel; ch35 (1kW). **Njuans TV:** Zlobin; ch29. **MPKET:** Kobrin; ch25. **Ranak:** Svetlahorsk; ch29. **Skif TV:** Viciebsk; ch48 (0.1kW), Palack ch21 (.1kW), Vorša ch34 (0.1kW). **Soltek:** Salihorsk; ch7. **Televid:** Recica; ch12. **TV2:** Mahilioú; Palykovicy ch2 (0.1kW). **Varjag:** Pinsk; ch7 (0.2kW). **Vesta:** Babrujsk; ch9. **8-iy kanal:** Minsk; ch8. **12-iy kanal:** Barysaú; ch12 (0.5kW).

Foreign TV Relays
NTV (Russia), RTR (Russia): see main tx table. **Mir (Russia):** Minsk ch35 (5kW).

DTT Transmitters
Operator: Ministry of Communications **Mux:** Peršy Kanal (incl. reg prgrs), Lad, ONT, NTV-Belarus, RTR-Belarus, STV, Mir, 8-iy kanal ⌘ Belaruskaje R. 1, 2, R. Stalica, Radyus FM

Location	Ch	kW	Location	Ch	kW
Svislac	21	2	Miadziel	44	2
Vorša	25	1	St. Darohi	44	2
Kapyl	34	2	Trokeniki	44	0.2
Salihorsk	34	2	Smetanicy	46	1
Baranavicy	38	2	Babrujsk	47	2
Berazino	41	2	Minsk	48	1
Krupski	41	1	Heraniony	49	2

Location	Ch	kW	Location	Ch	kW
Luki	41	1	Mahiliou	49	1
Slonim	41	2	Mscislaúl	49	2
Hrodna	42	1	Kastjukovicy	50	2
Brahin	43	2	Homiel	51	1
Braslaú	43	1	Rakitnica	51	2
Bycycha	43	2	Pinsk	56	2
Sarkaúscyna	43	0.2	Drahicyn	57	2
Viciebsk	43	0.2	Zlobin	57	2
Zascobye	43	2	Osveja	61	1

Local operator: Kosmos TV **Mux 1+2(♻):** multiprgr **Txs:** ch32 (Minsk), ch57 (Minsk).

NB. Relays of TV prgrs (analogue/DTT) from Russia are subject to cancellation.

BELGIUM

Systems: DVB-T (MPEG2); DVB-T2 (MPEG4) planned

Flanders

National Station
VLAAMSE RADIO EN TELEVISIEOMROEP (VRT) (Pub)⬚ A. Reyerslaan 52, 1043 Brussel ☎ +32 2 7413111 📠 +32 2 7349351 **E:** info@vrt.be **W:** www.vrt.be **L.P:** Dir TV: Aimé van Hecke **Chs:** (in Flemish) Eén, Canvas/Ketnet.

DTT Transmitters
Operator: Norkring België ⬚ Jules Bordetlaan 160, 1140 Evere ☎ +32 2 3639900 📠 +32 2 7454537 **E:** info@norkring.be **W:** www.norkring.be **Mux:** Eén, Canvas/Ketnet ⌘ VRT Radio 1, 2, Klara, Klara continuo, Studio Brussel, MNM, Klara Continuo, NMN Hits, Sporza, Nieuws+.

Location	Ch	kW	Location	Ch	kW
Brussel	22V	20	Veltem	22V	20
Egem	22	20	Antwerpen	25V	10
Gent	22	7	Genk	25	20
Sint-Peters-Leeuw	22	20	Schoten	25	20

Wallonia

National Stations
RADIO TÉLÉVISION BELGE DE LA COMMUNAUTÉ FRANÇAISE (RTBF) (Pub)⬚ Cité Reyers - Local 11M31, Boulevard A. Reyers 52, 1044 Bruxelles ☎ +32 2 7372111 📠 +32 2 7374210 **W:** www.rtbf.be **L.P:** Dir TV: M. Alain Gerlache **Chs:** (in French) La une, La deux, La troix. – **BELGISCHER RUNDFUNK (BRF) (Pub)**⬚ Kehrweg 11, 4700 Eupen ☎ +32 87 591111 📠 +32 87 591199 **E:** info@brf.be **W:** www.brf.be **L.P:** Dir: Arthur Spoden. Ch: (News-magazine in German) Blickpunkt.

DTT Transmitters
Operator: RTBF **Mux:** La une, La deux, La troix, Euronews (via Liège tx: Euronews/BRF Blickpunkt) ⌘ RTBF La Première, Vivacité, Musiq3, Classic 21, Pure FM, BRF.

Location	Ch	kW	Location	Ch	kW
Liège	45	100	Profondville	56	50
Anderlues	56	80	Tournai	56V	40
Bruxelles	56	2	Wavre	56	50
Léglise	56	100	Marche-en-Famene	56	12.5
Namur	56	5			

Local Operator: Be TV ⬚ chaussée de Louvain 656, 1030 Bruxelles ☎ +32 2 7304050 📠 +32 2 7300379 **E:** abonnesweb@betv.be **W:** www.betv.be **Mux (♻):** Be 1, Be Ciné, Be Séries **Tx:** ch55V (Bruxelles 2kW). – **Local Operator:** Télé Bruxelles (TLB) ⬚ rue Gabrielle Petit 32/34, 1080 Bruxelles ☎ +32 2 4212121 📠 +32 2 4212122 **E:** contact@telebruxelles.net **W:** www.telebruxelles.net **Mux:** TLB **Tx:** ch60 (Bruxelles 0.5kW).

BELIZE

System: NTSC-M [A]. BFBS-TV: DVB-T (MPEG2) [E]

BAYMEN BROADCASTING NETWORK (Comm) ⬚ 27 Baymen Ave., Belize City ☎ + 501 2244400 📠 +501 2231243 **Tx:** ch9. – **CHANNEL 5 (Comm)**⬚ P.O.Box 679, Belize City ☎ +501 2277781 📠 +501 2274936 **E:** gbtz@btl.net **W:** www.channel5belize.com **Tx:** ch5. – **TBN (Rlg)** ⬚ Belize City **Tx:** ch13. – **TROPICAL VISION (Comm)**⬚ P.O.Box 89, Belize City ☎ +501 2277246 📠 +501 2275040 **E:** tvseven@btl.net **Tx:** ch7; ch11 (Belize Family Channel).

Foreign Military Station
BFBS-TV (British Mil) ⬚ BFBS Belize, Airport Camp, BFPO 12, United

Kingdom. **Mux(✪):** BFBS1, BFBS2, BFBS3 Kids, BFBS4, Sky Sports 1, Sky Sport 2, Sky News ⌘ BFBS Radio 1+2, BFBS Belize **Tx:** (-).

BENIN

System: SECAM-K1 [E]

National Station
ORTB - TÉLÉVISION NATIONALE (Gov) ▭ BP 366, Cotonou ☎+229 21301096 ▤ +229 21301437 **E:** ortb@intnet.bj **W:** www.ortb.net **LP:** Dir: Pierette Amoussou **Txs:** Cotonou ch4 (20kW) & relay txs.

Local Stations
Canal 3: 02 BP 371, Cotonou; ch42. **Carrefour TV:** 01BP 440 Bohicon; ch67. **Golfe TV:** 06 BP 1624, Cotonou; ch38. **Imalè Africa:** Puerto-Novo; ch68. **LC2:** 05 BP 427, Cotonou; ch44.

BERMUDA (UK)

System: NTSC-M [A]; DVB-T

BERMUDA BROADCASTING CO. LTD. (Gov) ▭ P.O.Box HM452, Hamilton ☎ +1 441 2952828 ▤ +1 441 2954282 **E:** zbmzfb@bermudabroadcasting.com **W:** www.bermudabroadcasting.com **Chs:** ZFB-TV (ABC affiliate), ZBM-TV (CBS affiliate) **Txs:** ZFB-TV: Hamilton ch7 (32.5kW); ZBM-TV: Hamilton ch9 (17.5kW) – **VSB-TV (Comm)** ▭ P.O.Box HM 1450, Hamilton HM FX ☎ +1 441 2761111 ▤ +1 441 2923375 **W:** www.vsb.bm **Tx:** Hamilton ch11 (NBC affiliate, also relays BBC World News).

DTT Transmitters
Operator: The World in Wireless Ltd. **W:** www.wow.bm **Mux:** multiprgr **Tx:** (-).

BHUTAN

System: PAL-B/G [E]

BHUTAN BROADCASTING SERVICE (BBS) (Pub) ▭ P.O.Box 101, Thimphu ☎ +975 2 323580 ▤ +975 2 323 073 **E:** md@bbs.com.bt **W:** www.bbs.com.bt **LP:** MD: Mingbo Dukpa **Txs:** Thimpu ch5 (1kW) & relay txs.

BOLIVIA

System: # NTSC-M [A]; SBTVD planned

National Stations
TELEVISIÓN BOLIVIANA (TVB) (Gov) ▭ Av. Camacho 1485, Ed. La Urbana, La Paz ☎ +591 2 2203404 ▤ +591 2 2203015 **W:** www.television-boliviana.tv.bo **Txs:** La Paz ch7 & relay txs. – **ASOCIACIÓN TELEVISIÓN BOLIVIANO (ATB RED NACIONAL) (Comm)** ▭ Av. Argentina 2057, La Paz ☎ +591 2 2229922 ▤ +591 2 227 935 **E:** atbcbb@atb.com.bo **W:** www.atb.com.bo **Tx:** La Paz ch9 & relay txs. – **BOLIVISION (Comm)** ▭ Av Santa Cruz esq, Tres pasos al frente, Santa Cruz ☎ +591 3 3524544 ▤ +591 3 3530707 **E:** bolivision@cotas.com.bo **W:** www.bolivision.net **Tx:** La Paz ch5 & relay txs. – **RED UNO DE BOLIVIA (Comm)** ▭ Romecin Campos 592, Sopocachi, 14976 La Paz ☎ +591 2 2421111 ▤ +591 2 2415101 **E:** notivision@reduno.com.bo **W:** www.reduno.com.bo **Tx:** La Paz ch11 & relay txs. – **TELEVISIÓN UNIVERSITARIA (Educ)** ▭ Av. 6 de Agosto No. 2170, 13383 La Paz ☎ +591 2 359297 ▤ +591 2 359491 **E:** canal13@umsa.bo **W:** www.umsa.bo. **Txs:** La Paz ch13 (10kW) & relay txs.

Local Stations not shown.

BONAIRE (Netherlands)

NB: No terrestrial TV station.

BOSNIA & HERZEGOVINA

Systems: # PAL-B/G [E]; DVB-T (MPEG4) planned

National (Federal) Station
RADIO TELEVIZIJA BOSNE I HERCEGOVINE (BHRT) (Pub) ▭ Bulevar Meše Selimovica 12, 71000 Sarajevo ☎ +387 33 455124 ▤ +387 33 461523 **E:** sptrgov@bhrt.ba **W:** www.bhrt.ba **LP:** DG: Mehmed Agovic **Ch:** BHT. **Txs:** Tuzla (Ilincica) ch7, Sarajevo (Hum) ch8, Cazin (V. Gomila) ch10, Capljina (C.Brdo) ch11, Travnik (Vlašic) ch11, Zenica (Lisac) ch12, Gorazde (H. Brdo) ch21, Bihac (Brekovica)

ch23, Mostar (Fortica) ch26, Banja Luka (Kozara) ch27, Trebinje (Leotar) ch37, Konjic (Lisin) ch47 & repeaters.

Federacija Bosna i Hercegovina
National Station
FTV (Pub) ▭ Bulevar Meše Selimovica 12, 71000 Sarajevo ☎ +387 33 461539 ▤ +387 33 461539 **E:** press@rtvfbih.ba **W:** www.rtvbih.ba **LP:** Dir TV: Vladimir Bilic. **Txs:** Mostar (Fortica) ch7, Bihac (Brekovica) ch21, Travnik (Vlašic) ch29, Zenica (Lisac) ch39, Gorazde (H. Brdo) ch43, Cazin (V. Gomila) ch47, Tuzla (Ilincica) ch51, Sarajevo (Hum) ch52, Konjic (Lisin) ch54 & repeaters.

Local Stations not shown

Republika Srpska
National Station
RADIO TELEVIZIJA REPUBLIKE SRPSKE (RTRS) (Pub) ▭ ul. Kralja Petra I Karadordevica 129, 78000 Banja Luka ☎ +387 51 301660 **E:** tv@rtrs.tv **W:** www.rtrs.tv **LP:** DG: Dragan Davidovic. **Txs:** Udrigovo ch5, Kmur ch6, Kozara ch6, Banja Luka I ch10, Leotar I ch10, Duge Njive ch12, Trebevic I ch12, Banja Luka II ch31, Trebevic II ch33, Leotar II ch37.

Local Stations not shown

BOTSWANA

System: # PAL-I [E]; DVB-T planned

National Stations
BOTSWANA TV (BTV) (Gov) ▭ P.O.Box 060, Gaborone ☎ +267 3658000 ▤ +267 3900051 **E:** marketing@btv.gov.bw **W:** www.btv.gov.bw **LP:** Dir: Habuji Sosome. **Txs:** (°). – **E-BOTSWANA (Comm)** ▭ P.O.Box 921, Gaborone **LP:** SM: David Coles; ch23 & relay stns.

BRAZIL

Systems: # PAL-M [A]; SBTVD

National Stations
TV BRAZIL (Pub) ▭ An. Gomes Freire 474, Centro, 20231-010 Rio de Janeiro, RJ ☎ +55 21 21176208 **E:** contacto@tvbrasil.org.br **W:** www.tvbrasil.org.br **Txs:** Rio de Janeiro ch2 & relay stns. – **CENTRAL NACIONAL DE TELEVISÃO (CNT) (Comm)** ▭ Rua Francisco Caron 29, Pilarzinho, 82120-200 Curitiba, PR ☎ +55 41 3383377 ▤ +55 41 3384878 **E:** cnt@cnt.com.br **W:** www.cnt.com.br **Txs:** São Paulo ch26 & relay stns. – **REDE BRASIL DE TELEVISÃO (RBTV) (Comm)** ▭ Alameda dos Uapés, 313 - Saúde, 04067-030 São Paulo, SP **W:** www.rbtv.com.br **Txs:** São Paulo ch59 & relay stns. – **REDE CULTURA (Comm)** ▭ Rua Vladimir Herzog 75, Agua Branca, SP 05036-900 São Paulo ☎ +55 11 38743122 ▤ +55 11 36112014 **E:** dirprog@tvcultura.com.br **W:** www.tvcultura.com.br **LP:** Pres: Antonio Carlos Caruso-Ronca. **Txs:** São Paulo ch2 & relays. – **REDE GLOBO (Comm)** ▭ Rua Lopes Quintas 303, Jardim Botanico, 22460-010 Rio de Janeiro, RJ ☎ +55 21 25402000 ▤ +55 21 22942092 **E:** wm@redeglobo.com.br **W:** www.redeglobo.com.br **Txs:** São Paulo ch5 & relay stns – **REDE RECORD (Comm)** ▭ Rua da Várzea 240, 01140-080 São Paulo, SP ☎ +55 11 36604761 ▤ +55 11 36604756 **E:** tvrecord@rederecord.com.br **W:** www.rederecord.com.br **Txs:** São Paulo ch7 & relay stns. – **SISTEMA BRASILEIRO DE TELEVISÃO (SBT) (Comm)** ▭ Av. das Comunicações 4, Vila Jaraguá, 06278-905 Osasco, SP ☎ +55 11 70873000 ▤ +55 11 70873509 **E:** marketing@sbt.com.br **W:** www.sbt.com.br **Txs:** São Paulo ch4 & relay stns.

Local Stations not shown.

DTT Transmitters
Nationwide & local multiprgr DTT networks under construction.

BRITISH INDIAN OCEAN TERRITORY

NB: No terrestrial TV stations.

BRUNEI

Systems: # PAL-B [E] ⇩2014; DVB-T

RADIO TELEVISYEN BRUNEI (RTB) (Gov) ▭ Bandar Seri Begawan, BS8610, Negara ☎ +673 2243111 ▤ +673 2220884 **E:** rtb-dir@rtb.gov.bn **W:** www.rtb.gov.bn **Chs:** RTB1, RTB2, RTB3 HD, RTB4

International, RTB5 **Txs: RTB1:** Bt. Subok ch5 (10kW H), Bt. Andulau ch8 (20kW H); **RTB2:** Subok ch10, Kuala Belait ch11.

DTT Transmitters (Trial)
Operator: RTB **Mux:** RTB1, RTB2, RTB3 HD, RTB4 International, RTB5 **Txs:** ch(-) (Bt. Subok 5kW).

BULGARIA

Systems: # PAL-D/K [R] ↓31 Dec 2012; DVB-T (MPEG2)

National Stations
BALGARSKA NATSIONALNA TELEVIZIYA (BNT) (Pub) ⌨ ul. San Stefano 29, 1504 Sofiya ☎ +359 2 9661149 🖷 +359 2 9634045 **E:** press@bnt.bg **W:** www.bnt.bg **LP:** DG: Ulyana Pramova; TechDir: Dobri Mihaylov **Chs:** BNT1 & regional studios. – **bTV (Comm)** ⌨ Natsionalen Dvorets na Kulturata, 1463 Sofiya ☎ +359 2 9176800 🖷 +359 2 9521483 **E:** pr@btv.bg **W:** www.btv.bg – **NOVA TELEVIZIYA (Comm)** ⌨ bul. N. Vabtsarov 55, Expo 2000 P.K., 1507 Sofiya ☎ +359 2 9151200 **E:** office@ntv.bg **W:** www.ntv.bg

Location	BNT1	bTV	Nova TV	kW
Belogradchik	12	30	46	10/20/10
Burgas	7	26	38	20/40/40
Dobrich	12	28	51	5/10/10
Goce Delchev	9	25	42	10/20/20
Kardzhali	9	34	53	10/20/20
Kyustendil	10	32	51	5/20/20
Montana	9	26	-	10/5
Ruse	-	-	44	10
Sliven	12	31	48	10/20/20
Smolyan	6	38	-	5/10
Sofiya	7	29	36	10
Shumen	5	39	-	50/40
Varna	9	33	50	5/5/10
Vrh Botev	11	25	-	20/40

+ sites with txs below 5kW.
M-SAT (Comm) ⌨ ul. Stefan Karadzha 2, 1000 Sofiya ☎ +359 2 9836666 **E:** pr@sofia.m-sat.bg **W:** www.m-sat.bg **Txs:** network. – **PRO.BG (Comm)** ⌨ bul. Cherin vrah 47, 1407 Sofiya ☎ +359 2 9460208 🖷 +359 2 9460208 **E:** office@probg.bg **W:** www.probg.bg **Txs:** Sofiya ch55 (1kW) & network. – **TV7 (Comm)** ⌨ bul. Dzheyms Baucher 100, Sofiya ☎ +359 2 8162740 🖷 +359 2 8162714 **E:** tv7@tv7.bg **W:** www.tv7.bg **Txs:** Sofiya ch53 (1kW) & network.

Local Stations
Balkan Balgarska Televiziya (BBT): NDK, vhod A3, zala 5, 1414 Sofiya; ch53 (1kW & via local DTT mux. **Kanal 0:** ul. Odrin 15, etazh 2, 1800 Burgas; ch28 (0.1kW). **Rekording Haskovo:** bul. Hakovski 9, 6300 Haskovo; ch51 (0.5kW). **SKAT+:** 8200 Pomorie; ch41. **The Voice:** ul. Srebrena 21, 1407 Sofiya. Tx: via Sofiya local DTT mux. **TV Cherno More:** kv. Chayka, do blok 23, 9005 Varna; Shabla ch31 (0.25kW), Beloslav ch48 (0.01kW), Varna ch48 (0.1kW). **TV Dobrudzha:** ul. Dimitrar Petkov, Kompleks "Vista M", etazh 4, 9300 Dobrich; ch48 (0.1kW). **TV Karnobat:** ul. Asparuh 5, Hotel Karnobat, etazh 8, 8400 Karnobat; ch52 (0.01kW). **TV Rodopi:** Kompleks Vazrozhdenci, blok 14, etaz 15, 6600 Kardzhali; ch21 (0.03kW). **TV Sopot:** ul. Osmi mart 10, 4330 Sopot; ch35. **TV Shumen:** ul. Sedinenie 105, 9700 Shumen; Divdyadovo ch47, Shumen ch52 (0.1kW).

DTT Transmitters (under construction)
Operator: Hannu Pro Bulgaria 🖂 bul. Shipchenski prohod 63, 1574 Sofiya ☎ +359 2 8707350 🖷 +359 2 8077114 **E:** mail@hannu-pro.com **W:** www.hannu-pro.com **Mux 1-4:** tbd **Txs:** 4 national networks planned. – **Operator:** New-Tek Ltd. ⌨ y.k Lulin, bl. 517, Ent. E, ap.167, 1359 Sofiya ☎ +359 2 9625286 🖷 +359 2 8687110 **E:** newtek@mail.orbitel.bg **Mux:** BNT1, bTV, Nova TV, TV7, BBT, The Voice ⌘ Horizont, Hristo Botev ch64 (Sofiya, two txs: Kopitoto 0.8kW, bul. Peyo Yavorov 0.3kW).

BURKINA FASO

System: SECAM-K1 [E]

National Station
TÉLÉVISION NATIONALE DU BURKINA (TNB) (Gov) ⌨ 01 BP 2530, Ouagadougou 01 ☎ +226 50318353 🖷 +226 50318393 **E:** television@rtb.bf **W:** www.tnb.bf **Txs:** Ouagadougou ch6 (10kW H) & relay txs.

Local Stations
Canal Viim Koeega (Rlg): 01 BP 108, Ouagadougou 01; ch24. **Canal 3 (Comm):** 11 BP 340, Ouagadougou 11; ch43.

BURUNDI

System: SECAM-K1 [E]

TÉLÉVISION NATIONALE DU BURUNDI (Gov) ⌨ BP 1900, Bujumbura ☎ +257 22224760 🖷 +257 22244877 **E:** rtnb@cbinf.com **Tx:** Bujumbara ch25 (4kW) & relay txs.

CAMBODIA

System: # PAL-B/G [E] ↓2015; DVB-T planned

National Stations
NATIONAL TV OF CAMBODIA (TVK) (Gov) ⌨ 62 Preah Monivong Boulevard, Sangkat Sras Chork, Khan Daun Penh, Phnon Penh 12202 ☎ +855 23 724149 🖷 +855 23 426407 **E:** tvk@camnet.gov.kh **W:** www.tvk.gov.kh **LP:** SM: Tan Yan; TD: Uy Thuon **Txs:** Phnom Penh ch7 (10kW) & relay txs. – **BAYON TV (Comm)** ⌨ National Road No 1, Boeung Snoa, Chbar Ampeou, Phnom Penh 12357 ☎ +855 23 363695 🖷 +855 23 726619 **E:** bayontv@camnet.gov.kh **W:** www.bayontv.com.kh **Txs:** Phnom Penh ch27 (250kW) & relay txs. – **CAMBODIAN TV NETWORK (CTN) (Comm)** ⌨ National Highway 5, Phum Krol Ko, Sangkat Kilomet 6, Khan Russei Keo, Phnom Penh 12104 ☎ +855 12 800800 🖷 +855 12 801801 **E:** wmaster@ctncambodia.com **W:** www.ctncambodia.com **Chs:** CTN, MyTV **Txs: CTN:** Phnom Penh ch21 & relays txs; **MyTV:** Phnom Phenh ch29.

Local Stations
Apsara TV (Comm): 69, Rue 57, Sangat Beung Keng Kang 1, Khan Chamcarmon, Phnom Penh; ch11 (10kW). **CTV9 (Comm):** 18 rue 562, Toul Kok, Phnom Penh 12151; ch9 (10kW). **TV3 (Phonm Penh Municipality TV):** 2 Bvd Confederation de la Russie (Rue 112), Sangat Monorom, Khan 7 Makra, Phnom Penh; ch5 (10kW). **TV Fark (Royal Cambodian Armed Forces):** rue 169, Borei Keila, Phnom Penh 12253; ch5 (10kW).

Foreign TV Relays
TV5 Monde Asie (France): Phnom Penh ch23; **VTV1 (Vietnam):** Phnom Penh ch25.

CAMEROON

System: PAL-B/G [E]

National Stations
CAMEROON RADIO AND TELEVISION (CRTV) (Gov) ⌨ BP 1634, Yaoundé ☎ +237 2214088 🖷 +237 2204340 **E:** crtv@mail.ditfo.cm **W:** www.crtv.cm **LP:** DG: Amadou Vamoulké **Txs:** Yaoundé ch5 (150kW) & relay txs. – **SPECTRUM TELEVISION (Comm)** ⌨ BP 4883, Douala ☎ +237 3433045 🖷 +237 3433048 **E:** spectrum1@camnet.cm **W:** www.stvgroup.com **Txs:** Douala ch2 (5kW) & relay txs.

Local Stations (all Comm)
Equinoxe Télévision (E.TV): Yaoundé; tx: (-). **RTV Lumière:** Yaoundé; tx: (-). **TV Max:** BP 4527, Douala; tx: (-).

CANADA

Systems: ATSC; † NTSC-M [A] ↓31 Aug 2011 (exc. small markets)

National Networks
CANADIAN BROADCASTING CORP. (CBC) (Pub) ⌨ 181 Queen St, Box 3220 Stn C, Ottawa ON K1Y 1E4 ☎ +1 613 2886000 **W:** www.cbc.ca **Chs:** National English and French networks. **English Network:** ⌨ 250 Front St W, Box 500 Stn A, Toronto ON M5W 1E6 ☎ +1 416 2053311 **W:** www.cbc.ca **Stations:** CBAT Fredericton NB ᵃch4 (54.2kW), CBCT Charlottetown PE ᵃch13 (178kW), CBET Windsor ON ᵃch9 (80.7kW), CBHT Halifax NS ᵃch3 (56kW), CBIT Sydney NS ᵃch5 (54kW), CBKST Saskatoon SK ᵃch11 (325kW), CBKT Regina SK ᵃch9 (140kW), CBLT Toronto ON ch20 (38kW) & ᵃch5 (84kW), CBMT Montréal PQ ch20 (107kW) & ᵃch6 (100kW), CBNT St. John's NF ᵃch8 (196kW), CBOT Ottawa ON ch25 (165kW) & ᵃch4 (100kW), CBRT Calgary AB ᵃch9 (178kW), CBUT Vancouver BC ch58 (30.5kW) & ᵃch2 (50kW), CBWT Winnipeg MB ᵃch6 (100kW), CBXT Edmonton AB ᵃch5 (318kW), and relay txs. NB: Stns identify as 'CBC'. **French Network:** ⌨ 1400 René-Lévesque Boul. E, Box 6000, Montréal PQ H3C 3A8 ☎ +1 514 5976000 **W:** www.radio-canada.ca **LP:** Exec. VP, Télévision de Radio-Canada: Sylvain Lafrance. **Stations:** CBAFT Moncton NB ᵃch11 (163kW), CBFT Montréal PQ ch19 (250kW) & ᵃch2 (100kW), CBKFT Regina SK ᵃch13 (103kW), CBOFT Ottawa ON ch66 (22kW) & ᵃch9 (252kW), CBVT Québec PQ ch12 (2.45kW) & ᵃch11 (128.8kW), CBUFT

Vancouver BC ᵃch26 (34.8kW), CBWFT Winnipeg MB ᵃch3 (59kW), CBXFT Edmonton AB ᵃch11 (90kW), and relay txs. NB: Stns identify as 'Radio-Canada'. – **CTV INC. (Comm)** (Div. of CTVglobemedia) ✉ 9 Channel Nine Court, Scarborough ON M1S 4B5 ☎ +1 416 3325000 📄 +1 416 3325283 **W:** www.ctv.ca **Stations:** CCCF Montréal QC ᵃch12 (325kW), CFCN Calgary AB ch36 (204kW) & ᵃch4 (100kW), CFCN-5 Lethbridge AB ᵃch13 (139kW), CFQC Saskatoon SK ᵃch8 (325kW), CFPL London ON ᵃch10 (325kW), CFRN Edmonton AB ᵃch3 (609kW), CFTO Toronto ON ᵃch9 (325kW), CHBX Sault Ste. Marie ON ᵃch2 (100kW), CHRO Ottawa/Pembroke ON ᵃch5 (100kW), CHWI Windsor ON ᵃch16 (492kW), CICC Yorkton SK ᵃch10 (56kW), CICI Sudbury ON ᵃch5 (100kW), CIPA Prince Albert SK ᵃch9 (325kW), CITO Timmins ON ᵃch3 (100kW), CIVI Victoria BC ᵃch53 (23kW), CIVT Vancouver BC ch33 (0.9kW) (F.pl ch32) & ᵃch32 (2000kW), CJCB Sydney NS ᵃch4 (180kW), CJCH Halifax NS ᵃch5 (100kW), CJOH Ottawa ON ᵃch13 (325kW), CKCK Regina SK ᵃch2 (100kW), CKCO Kitchener ON ᵃch13 (325kW), CKCW Moncton NB ᵃch2 (100kW), CKLT St. John NB ᵃch9 (325kW), CKNY North Bay ON ᵃch10 (132.6kW), CKVR Barrie ON ᵃch3 (100kW), CKY Winnipeg MB ᵃch7 (325kW), and smaller txs. NB: Stns identify as 'A' or 'CTV'. – **GLOBAL TV (Comm)** (Div. of Canwest MediaWorks, Inc.)✉ 3100 CanWest Global Place, 201 Portage Ave, Winnipeg MB R3B 3L7 ☎ +1 204 9562025 📄 +1 204 9479841 **W:** www.globaltv.com **Stations:** CFRE Regina SK ᵃch11 (325kW), CFSK Saskatoon SK ᵃch4 (100kW), CHAN Vancouver BC ch22 (8.7kW) & ᵃch8 (250kW), CHBC Kelowna BC ᵃch2 (3.7kW), CICT Calgary AB ch41 (830kW) & ᵃch2 (100kW), CIHF Halifax/Dartmouth NS ᵃch8 (20kW), CIHF-2 St. John NB ᵃch12 (35.5kW), CIII Toronto ON ch65 (3kW) (F.pl ch41) & ᵃch41 (732kW), CISA Lethbridge AB ᵃch7 (325kW), CITV Edmonton AB ᵃch13 (325kW), CKMI Québec PQ ᵃch20 (86.2kW), CKMI-1 Montréal PQ ᵃch46 (33kW), CKND Winnipeg MB ch28 (f.pl) & ᵃch9 (325kW), and relay txs. NB: Stns identify as 'Global'.

Major Regional Networks
SOCIÉTÉ DE TÉLÉDIFFUSION DU QUEBEC (TÉLÉ-QUÉBEC) (Pub) ✉ 1000 rue Fullum, Montréal PQ H2K 3L7 ☎ +1 514 5212424 📄 +1 514 5255511 **W:** www.telequebec.tv **Stations (French):** CIVM Montréal QC ᵃch17 (889.5kW) & ᵃch27 (15kW) (F.pl: ch26) & relay txs (QC only). – **TV ONTARIO (Pub)** ✉ Box 200 Stn Q, Toronto ON M4T 2T1 ☎ +1 416 4842600 📄 +1 416 4847771 **W:** www.tvo.org **Stations:** CICA Toronto ON ᵃch19 (1080kW) & relay txs (ON only). NB: Stns identify as 'TVO'. – **CHANNEL ZERO INC. (Comm)** ✉Box 6143 Stn A, Toronto ON M5W 1P6 ☎ +1 416 4921595 📄 +1 416 4929539 **W:** www.tvchannelzero.com **Stations:** CHCH Hamilton ON ch18 (60kW) & ᵃch11 (325kW, CJNT Montréal PQ ᵃch62 (11kW) & ch49 (F.pl) – **CORUS ENTERTAINMENT INC. (Comm)** ✉ 630-3rd Ave SW Suite 501, Calgary AB T2P 4L4 ☎+1 403 4444244 📄 +1 403 4444242 **W:** www.corusent.com **Stations:** CKWS (CBC affiliate) Kingston ON ᵃch11 (325kW), CHEX (CBC affiliate) Peterborough ON ᵃch12 (185kW) & relay txs (ON only). – **GROUPE TVA (Comm)** (Div. of Québecor Media) ✉ 1600 boul. de Maisonneuve Est, Montréal PQ H2L 4P2☎ +1 514 5269251 **W:** tva.canoe.com. **Stations (English):** Sun TV (CKXT) Toronto ON ch66 (3kW) (F.pl ch40) & ᵃch52 (30kW); **Stations (French):** CFCM Québec QC ᵃch4 (100kW), CFEM Rouyn-Noranda QC ᵃch13 (346kW), CFER Rimouski QC ᵃch11 (325kW), CFTM Montréal QC ᵃch10 (325kW), CHAU Carleton QC ᵃch5 (81.7kW), CHEM Trois-Rivières QC ᵃch8 (325kW), CHLT Sherbrooke QC ᵃch7 (300kW), CHOT Hull QC ᵃch40 (684kW), CIMT Rivière du Loup QC ᵃch9 (275.4kW), CJPM Chicoutimi QC ᵃch6 (100kW), and relay txs. – **JIM PATTISON BROADCAST GROUP (Comm)** ✉ 460 Pemberton Terrace, Kamloops BC V2C 1T5 ☎ +1 250 3723322 📄 +1 250 3740445 **W:** www.jpbroadcast.com **Stations:** CKPG Prince George BC ch2, CFJC Kamloops BC ch4, CHAT Medicine Hat AB ᵃch6 (58kW) & relay txs (AB & BC only). – **NTV (Comm)** ✉ Box 2020, St. John's NL, A1C 5S2 ☎ +1 709 7225015 📄 +1 709 7265107 **W:** www.ntv.ca **Stations:** CJON St. John's NL ᵃch6 (356kW) & relay txs (NL only). – **ROGERS MEDIA TV (Comm)** ✉ 333 Bloor St. E, 7th flr, Toronto ON M4W 1G9 ☎ +1 416 9358200 **W:** www.rogers.com **Stations:** CFMT Toronto ON ch64 (15kW) & ᵃch47 (1138kW), CHMI Portage LaPrairie/Winnipeg MB ᵃch13 (325kW), CHNM-TV Vancouver BC ch20 (1.5kW) & ᵃch42 (130kW), CITY Toronto ON ch53 (2kW) & ᵃch57 (310kW), CJMT Toronto ON ch44 (15kW) & ᵃch69 (500kW), CJCO-TV Calgary AB ᵃch38 (228kW), CJEO-TV Edmonton AB ᵃch56 (580kW) & ch44 (F.pl), CKAL Calgary AB ᵃch5 (79.4kW), CKEM Edmonton AB ᵃch51 (704kW), CKVU Vancouver ch47 (3.8kW) (F.pl ch10) & ᵃch10 (325kW), and relay txs. NB: Stns identify as 'Citytv' or 'OMNI'. – **V (Comm)** ✉ 612, rue St-Jacques bureau 100, Montréal PQ H3C 5R1 ☎ +1 514 3906035 **W:** vtele.ca **Stations (French):** CFJP Montréal QC ch42 (27kW) (F.pl ch35) & ᵃch35 (697kW) and relay txs (QC only).

Other Regional Networks & Local Stations not shown.

NB: ᵃ= analogue channels. After analogue shutdown date, most txs will continue on the listed ᵃchannels except change to digital mode. Txs on channels above ch51 are required to move to lower channels in ATSC.

CANARY ISLANDS (Spain)

Systems: DVB-T (MPEG2)

National (Regional) Stations
TELEVISIÓN ESPAÑOLA EN CANARIAS (TVE) (Pub) ✉ Plazoleta de Milton 1, 35005 Las Palmas de Gran Canaria ☎ +34 928 293096. – **RADIOTELEVISIÓN CANARIA (RTCV) (Pub)** ✉ Mariucha 2, 35012 Las Palmas de Gran Canaria ☎ +34 928 280188 **W:** www.rtvc.es. **Chs:** TV Canaria, TV Canaria Dos. – **ANTENA 3 TELEVISIÓN (Comm)** ✉Alcalde José Ramírez Bethencourt 25, 35004 Las Palmas de Gran Canaria ☎ +34 928 297300. – **POPULAR TV CANARIAS (Rlg)** ✉Las Palmas de Gran Canaria.

Local Stations not shown.

DTT Transmitters
Operator: TVE **Mux:** La 1, La 2, 24 Horas, Clan ⌘ RNE1, RNE Clásica, RNE3 **Txs:** ch60 (SFN). – **Operator:** Antena 3 **Mux:** Antena 3 **Mux:** Antena 3, Neox, Nova, Gol Televisión✪ ⌘ Onda Cero, Europa FM, Onda Melodía. **Txs:** ch69 (SFN). – **Operator:** Gestevisión Telecinco **Mux:** Telecinco, LaSiete, FDF, Disney Channel. **Txs:** ch68 (SFN). – **Operator:** Sogecable **Mux:** Cuatro, CNN+, Canal+ Dos✪, Canal Club, La Sexta ⌘ SER, 40 Principales, Cadena Dial **Txs:** ch67 (SFN). – **Operator:** Veo TV **Mux:** Veo7, Tienda en Veo, Intereconomía, Teledeporte ⌘ R.Intereconomía, R.Marca, esRadio, Vaughan R. **Txs:** ch66 (SFN). – **Operator:** n/a **Mux:** TV Canaria, TV Canaria Dos, Antena 3 Canarias, Popular TV **Txs:** ch59 (SFN). – **Operator:** n/a **Mux:** Teidevision Canal 6, El Día TV **Txs:** ch56 (SFN). – **Operator:** n/a **Mux:** Canal 7, Mirame TV, Canal 4 **Tx:** ch24 (Santa Cruz de Tenerife). – **Operator:** n/a **Mux:** Canal 11, La Provincia **Tx:** ch33 (La Palma). – **Operator:** n/a **Mux:** Telelinea, RTV Islas Canarias, Canal Ocho **Tx:** ch63 (La Palma). – **Operator:** n/a **Mux:** Canal 7, Canal Ocho, Canal 4 **Txs:** ch30 (Orotava), ch38 (Arona). – **Operator:** n/a **Mux:** RTV Islas Canarias, Estudios Opalo S.L. **Txs:** ch21 (Gomera), ch34 (Hierro). – **Operator:** n/a **Mux:** Canal Ocho, La Opinión de Tenerife **Txs:** ch32 (Hierro), ch62 (Gomera). – **Operator:** n/a **Mux:** Canal 7 Gran Canaria, TIC Canal 8, Canal 4 **Tx:** ch44 (Las Palmas). – **Operator:** n/a **Mux:** Localia, Nueve TV **Txs:** ch43 (Fuerteventura), ch52 (Las Palmas, Mogán, Telde). – **Operator:** n/a **Mux:** Lancelot TV, Enjoy, Canal (L) **Tx:** ch21 (Lanzarote). – **Operator:** n/a **Mux:** RTI, Canal 13, Canal 7 Gran Canaria **Tx:** ch58 (Mogán). – **Operator:** n/a **Mux:** Est Canal, Canal 7 Gran Canaria, TIC Canal 8, Canal 4 **Tx:** ch63 (Telde).

CAPE VERDE

Systems: # SECAM-K1 [E]; DVB-T (MPEG2)

TELEVISÃO NACIONAL DE CABO VERDE (TNCV) (Pub) ✉ Achada Santo Antão, Praia, Ilha de Santiago ☎ +238 614080 **E:** rtc@cvtelecom.cv **W:** www.rtc.cv **L.P:** Dir: Maria Rosario da Luz **Txs:** Praia ch10 (0.5kW H) & relay txs.

Foreign TV Relay
RTP África (Portugal): (-).

DTT Transmitters
Operator: CV Telecom **W:** www.nave.cv/cvtelecom **Mux:** Televisão de Cabo Verde, SIC Noticias, TV Record, Rai Uno, BBC World, TV5, TV Galicia, Infinito, Fox Life, Fashion TV, Euronews, Eurosport, Extreme Sport, TVE Internacional, CNBC, MCM, RTP-África, Lusomundo Premium, Lusomundo Gallery, Playboy, SportTv, SIC, RTP1, RTP2,TVI. **Txs:** (-)

CAYMAN ISLANDS (UK)

System: NTSC-M [A]

CAYMAN INTERNATIONAL TV NETWORK (CITN) (Comm) ✉ P.O. Box 30563 SMB, Grand Cayman ☎ +1 345 9452739 📄 +1 345 9490021 **E:** citn@cayman27.com.ky **W:** www.cayman27.com.ky **Tx:** ch27. – **CAYMAN TELEVISION SERVICE (CTS) (Comm)** ✉ Grand Cayman. **Tx:** ch24. – **CAYMAN ADVENTIST TELEVISION NETWORK (CATN) (Rlg)** ✉ Grand Cayman. **Tx:** ch30. – **CAYMAN CHRISTIAN TELEVISION (CCT) (Rlg)** ✉ Grand Cayman. **Tx:** ch21. Rel. TBN (USA).

CENTRAL AFRICAN REPUBLIC

System: SECAM-K1 [E]
TELÉVISION CENTRAFRICAINE (TVCA) (Gov) ✉ BP 940, Bangui

☎ +236 501412 🗎 +236 615985 **L.P:** Pres: Michel Bata **Txs:** Bangui ch10 (2kW H) & relay txs.

CHAD

System: SECAM-K1 [E]

TÉLÉ TCHAD (Gov) 🖃 BP 5123, N'Djamena ☎ +235 522923 🗎 +235 525163 **E:** tele.tchad@intnet.td **L.P:** Dir: Moussadr Doumngor **Txs:** N'Djamena ch7V (50kW) & relay txs.

CHILE

System: NTSC-M [A]; SBTVD planned

National Stations
TVN CHILE (Gov) 🖃 Bellavista 0990, Providencia, Santiago ☎ +56 +56 2 7077130 🗎 +56 2 7077750 **E:** tvngprog@tvn.cl **W:** www.tvchile. cl. **Txs:** Valparaíso ch12 & network. – **CANAL 13 (Rlg, Comm)** 🖃 Inés Matte Urrejola 0825, Providencia, Santiago ☎ +56 2 6302356 🗎 +56 2 6302341 **E:** mailbag@canal13.cl **W:** cable.canal13.cl **Txs:** Concepción ch5, Valparaíso ch8, Santiago ch13. – **CHILEVISIÓN (Comm)** 🖃 Inés Matte Urrejola 0825, Providencia, Santiago ☎ +56 2 4615460 🗎 +56 2 4615456 **E:** rcarmi@chilevision.cl **W:** www.chilevision.cl **Txs:** ch11 (60kW) & relays. – **MEGA (Comm)** 🖃 Av. Vicuña Mackenna 1348, Santiago ☎ +56 2 8108000 🗎 +56 2 5518916 **E:** mega@mega.cl **W:** www.mega.cl **Txs:** Santiago ch8 & relays. – **RED TV (Comm)** 🖃 Av. Manquehue Sur 1201, Las Condes, Santiago ☎ +56 2 3854000 🗎 +56 2 3854060 **E:** rrpp@redtv.cl **W:** www.redtv. cl **Txs:** Santiago ch4 & relays. – **TELE CANAL (Comm)** 🖃 Nueva Tajamar 481, Oficina 201, Torre Central, Las Condes, Santiago ☎ +56 2 4115600 🗎 +56 2 4115608 **E:** telecanal@telecanal.cl **W:** www.telecanal.cl **Tx:** Santiago ch2 & relays. – **UCV TV (Rlg, Educ)** 🖃 Agua Santa Alta 2455, Viña del Mar ☎ +56 32 611934 🗎 +56 32 610505 **E:** tv@ucv.cl **W:** www.ucvtv.cl **Txs:** Valparaíso ch4, Santiago ch5, Puerto Montt ch7, La Serena ch8.

Local Stations not shown.

CHINA (People's Rep. of)

Systems: # PAL-D ⇩2015/2018; DTMB

National Stations
CHINA CENTRAL TELEVISION (CCTV) (Gov) 🖃 11 Fuxing Lu, Haidian Qu, Beijing 100859 ☎ +86 10 68500114 🗎 +86 10 68508743 **W:** www.cctv.com **L.P:** Dir: Jiao Li. **Chs (terr.): CCTV1:** Beijing ch2 & network **CCTV2:** Beijing ch8 & network. **Children's Channel:** Beijing ch15. **Music Channel:** Beijing ch33. – **CHINA EDUCATION TELEVISION (CETV) Gov)** 🖃 160 Fuxingmennei Dajie, Xicheng Qu, Beijing 100031 ☎ +86 10 66419055. 🗎 +86 10 66084298 **W:** www. cetv.edu.cn.

Regional Stations (all Gov)
Anhui TV: 355 Tongcheng Nanlu, Hefei, Anhui 230011. **W:** www.ahtv. cn **Beijing TV:** 98 Jianguo Lu, Chaoyang QAu, Beijing 100022. **W:** www. btv.com.cn **Chongqing TV:** 68 Yuzhou Lu, Chongqing 400041. **W:** www. cqnews.net **Fujian TV:** 2 Gutian Lu, Fuzhou, Fujian 350001. **W:** www.fjtv. net **Gansu TV:** 561 Zhangsutan, Chengguan Qu, Lanzhou, Gansu 730010. **W:** www.gstv.com.cn **Guangdong TV:** 331 Huangshi Donglu, Guangzhou, Guangdong 510066. **W:** www.gdtv.com.cn **Guangdong Southern TV:** 331 Huangshi Donglu, Guangzhou, Guangdong 510066. **W:** www.tvscn. com **Guangxi TV:** 73 Minzu Dadao, Nanning, Guangxi 530022. **W:** www. gxtv.com.cn **Guizhou TV:** 261 Qingyun Lu, Guiyang, Guizhou 550002. **W:** www.gzstv.com **Hainan TV:** 61 Nansha Lu, Haikou, Hainan 570206. **W:** www.bluehn.com **Hebei TV:** 100 Jianhua Nandajie, Shijiazhuang, Hebei 050031. **W:** www.hebtv.com **Henan TV:** 18 Zhenghua Lu, Zhengzhou, Henan 450008. **W:** www.hntv.ha.cn **Heilongjiang TV:** 181 Zhongshan Lu, Harbin, Heilongjiang 150001. **W:** www.hljtv.com **Hubei TV:** Zijin Cun, Liangdao Jie,Wuchang Qu, Wuhan, Hubei 430071. **W:** www.hbtv.com.cn **Hunan TV:** Liuyang He Daqiao Dong, Changsha, Hunan 410003. **W:** www. hunantv.com **Inner Mongolia TV:** 55 Xinhua Dajie, Hohhot, Nei Menggu 010058. **W:** www.nmtv.cn **Jilin TV:** 2066 Weixing Lu, Changchun, Jilin 130051. **W:** www.jilintv.cn **Jiangsu TV:** 48 Xi Citang Xiang, Zhongshan Donglu, Nanjing, Jiangsu 210002. **W:** www.jsbc.com **Jiangxi TV:** 207 Hongdu Zhong Dadao, Nanchang, Jiangxi 330046. **W:** www.jxgdw.com **Liaoning TV:** 79 Wenhua Lu, Heping Qu, Shenyang, Liaoning 110003. **W:** www.lntv.cn **Ningxia TV:** 66 Beijing Zhonglu, Jinfeng Qu, Yinchuan, Ningxia 75000135. **W:** www.nxtv.cn **Qinghai TV:** 6 Kunlun Lu, Xining,

Qinghai 810001. **W:** www.qhstv.com **Shandong TV:** 81 Jingshi Lu, Jinan, Shandong 250001. **W:** www.sdtv.cn **Shanxi TV:** 318 Yingze Dajie, Taiyuan, Shanxi 030001. **W:** www.sxrtv.com **Shaanxi TV:** 336 Chang'an Nanlu, Xi'an, Shaanxi 710061. **W:** www.sxtvs.com **Shanghai TV:** 651 Nanjing Xilu, Shanghai 200041. **W:** www.smg.cn **Sichuan TV:** 40 Dongsheng Jie, Chengdu, Sichuan 610015. **W:** www.sctv.com **Tianjin TV:** 143 Weijin Lu, Heping Qu, Tianjin 300071. **W:** www.tjtv.com.cn **Tibet TV:** 149 Beijing Zhonglu, Lhasa, Xizang 850000. **W:** www.tibetinfor. com/tibetzt/xztv **Xinjiang TV:** 84 Tuanjie Lu, Urumqi, Xinjiang 830044. **W:** www.xjtvs.com.cn **Yunnan TV:** 182 Renmin Xilu, Kunming, Yunnan 650031. **W:** www.yntv.com **Zhejiang TV:** 111 Moganshan Lu, Hangzhou, Zhejiang 310005. **W:** www.cztv.com.cn.

DTT Transmitters (under construction)
Operator: SARFT **Mux:** CCTV chs **Txs:** nationwide network & regional services. – **Operator:** SARS **Mux:** CCTV chs **Txs:** Beijing. – **Operator:** China DTV Media **Mux:** tbd **Txs:** major Chinese towns. – **Operator:** Shanghai TV **Mux:** tbd **Txs:** major Chinese towns. – **Operator:** China Cable Network Company **Mux:** tbd **Txs:** major Chinese towns. – **Operator:** CHC Home Cinema **Mux:** tbd **Txs:** major Chinese towns. – **Operator:** "DTV alliance" **Mux:** tbd **Txs:** major Chinese towns.

CHRISTMAS ISLAND (Australia)

NB: No terrestrial TV station.

COCOS (Keeling) ISLANDS (Australia)

NB: No terrestrial TV station.

COLOMBIA

System: # NTSC-M [A]; DVB-T

National Stations
RADIO TELEVISIÓN NACIONAL DE COLOMBIA (RTVC) (Pub) 🖃 Avenida El Dorado No. 46 - 76, Bogotá ☎ +57 1 5978000 🗎 +57 1 2222765 **E:** info@rtvc.gov.co **W:** www.rtvc.gov.co **Chs:** Canal Uno, Canal Institucional, Señal Colombia. **Txs: Canal Uno:** Bogotá ch8 & relay txs, **Canal Institucional:** Bogotá ch11 & relay txs, **Señal Colombia/Canal 13:** Bogotá ch13 & relay txs. – **CARACOL TELEVISIÓN (Comm)** 🖃 Calle 103 #69 B 43, Bogota ☎ +57 1 6430430 🗎 +57 1 6430444 **E:** serviciocliente@caracol.com.co **W:** www.caracoltv.com **Txs:** Bogotá ch5 & relay txs. – **RCN TV (Comm)** 🖃 Av Americas 65-82, Bogotá ☎ +57 1 4269292 🗎 +57 1 4140412 **E:** canalrcn@canalrcn.com.co **W:** www.canalrcn.com **Txs:** Bogotá ch4 & relay txs

Local Stations not shown.

DTT Transmitters
Location	RTVC*	Caracol	RCN
Bogotá	16	14	15

+ nationwide netw. under construction
***)** **Mux:** Canal Uno, Canal Institucional, Señal Columbia/Canal 13
Local muxes not shown.

COMOROS

System: SECAM-K1 [E]

National Station
TÉLÉVISION NATIONALE DES COMORES (Pub) 🖃 BP 250, Moroni ☎ +269 7744045 🗎 +269 7731079. **Txs:** (-).

Local Stations (all Comm)
Djabal TV: Iconi; tx: (-). **TV-Sha:** Moroni; tx: (-). **MTV:** Moroni; tx: (-). **Radio Télévision Anjouanaise (RTA):** Mbouyoujou-Ouani, Ile Autonome d'Anjouan; tx: (-).

CONGO (Dem. Rep. of)

System: SECAM-K1 [E]

National Station
RADIOTÉLÉVISION NATIONALE CONGOLAISE (RTNC) (Pub) 🖃 BP 3164, Kinshasa-Gombe ☎ +243 1 5260601 🗎 +243 1 5220655 **E:** ica@ic.cd **L.P:** Pres: M. Atufuka Mbunze **Chs:** RTNC1, RTNC2. **Txs:** **RTNC1:** Kananga ch4 (2kW), Kamina ch4 (2kW), Kinshasa ch5 (27kW),

Kolwezi ch5 (1kW), Mbuji Mayi ch6 (2kW). **RTNC2:** Kinshasa ch37.

Local Stations not shown.

CONGO (Rep. of)

System: SECAM-K1 [E]

TÉLÉ CONGO (Gov) ✉ CP 975, Brazzaville ☎ +242 2810116 🖨 +242 2814128 **LP:** DG: Waméné Ekiaye-Akoly **Txs:** (pol.H): Loubomo ch6 (1kW), Brazzaville ch8 (9kW), Pointe Noire ch10 (4.7kW) & relay txs.

COOK ISLANDS

System: PAL-B [NZ]

COOK ISLANDS TELEVISION (CITV) (Comm) ✉ P.O.Box 126, Avarua, Rarotonga ☎ +682 29460 🖨 +682 21907 **E:** watchus@citv. co.ck **W:** www.citv.co.ck **LP:** Dir: Shona Pitt **Txs:** Airport ch4, Works Depot, TV studio, Mauke ch5, Matavera ch6, Titikaveka ch7, Aitutaki & Rarotonga ch9, Tu Papa ch10, Hospital & Ngatangiia ch11.

COSTA RICA

System: # NTSC-M [A]; SBTVD planned

National Stations
CANAL 13 (Pub) ✉ Apt 7-1908-1000, San José ☎ +506 22313333 🖨 +506 22200072 **E:** canal13@sinart.go.cr **W:** www.sinart.go.cr **LP:** DG: Dr. Ch. Zelaya Goodman **Txs:** San José ch13 & relay txs. – **REPRETEL (Comm)** ✉ Apartado 2860, 1000 San José ☎ +506 22906665 🖨 +506 22324203 **E:** info@repretel.com **W:** www.repretel.com **Chs:** Canal 4, 6, 11. **Txs:** San José ch4 & ch6 & ch11. – **TELETICA (Comm)** ✉ Sabana Oeste, San José ☎ +506 22101201 🖨 +506 22321107 **E:** info@ teletica.com **W:** www.teletica.com **Tx:** San José ch7.

Local Stations not shown.

CROATIA

Systems: DVB-T (MPEG2, MPEG4)

National Stations
HRVATSKA TELEVIZIJA (HRT) (Pub) ✉ Prisavlje 3, 10000 Zagreb. ☎ + 385 1 6163366 🖨 + 385 1 6163392 **W:** www.hrt.hr **LP:** GM: Ivan Parac. **Chs:** HRT1, HRT2, HRT HD – **KAPITAL NETWORK (Comm)** ✉ Rapska 37, 10000 Zagreb ☎ +385 1 7779096 🖨 +385 1 7775506 **E:** redakcija@kapital.tv **W:** www.kapital.tv – **NOVA TV (Comm)** ✉ Remetinecka cesta 139, 10000 Zagreb ☎ +385 1 6008300 🖨 +385 1 6008333 **E:** novatv@novatv.hr **W:** www.novatv.hr – **RTL TELEVIZIJA (Comm)** ✉ Krapinska 45, 10000 Zagreb ☎ +385 1 3660000 🖨 +385 1 3660609 **E:** rtl@rtl.hr **W:** www.rtl.hr **LP:** Chmn: Christoph Mainusch.

Local Stations (via Mux 4, DTT regions: a-i)
Gradska TV: Molatska bb, 23000 Zadar (g) **Kanal RI:** Trg rijecke rezolucije 3, 51000 Rijeka (e). **Nezavisna Istarska Televizija (NIT):** Trg pod lipom 1, 52000 Pazin (e). **RI-TV:** Uzarska 17/3, 51000 Rijeka (e). **Slavonskobrodska Televizija:** Mile Budaka 1/IV, 35 000 Slavonski Brod (b). **TV Cakovec:** Kralja Tomislava 6, 40000 Cakovec (c). **TV Jadran:** Split (h). **TV Nova:** M. Laginje 5, 52100 Pula (e). **TV Slavonije i Baranje:** Hrvatske republike 20, 31000 Osijek (a). **Varazdinska Televizja (VTV):** Kralja P. Kresimira IV 6a, 42000 Varazdin (c). **Vinkovacka Televizija (VKTV):** Trg dr.F.Tudmana 2, Vinkovci (a).

DTT Transmitters (under construction) (MPEG2 exc. *=MPEG4)
Operator: OIV ✉ ul. grada Vukovara 269d, 10000 Zagreb ☎ +385 1 6186000 🖨 +385 1 6186100 **E:** dvbt@oiv.hr **W:** www.oiv.hr **Mux 1:** HRT1, HRT2, RTL Televizija[1], Nova TV[1] **Mux 2:** Nova TV, RTL Televizija, Kapital Network, HRT1[1], HRT2[1] **Mux 3*:** HRT HD **Mux 4:** Local stns. [1] temporary allocations

Location	M1	M2	M3	M4°
Belje	38	44	-	-
Biokovo	33	53	-	-
Borinci	38	44	-	21a
Brac	33	-	-	53h
Celevac	51	59	-	31g
Drenovci	38	-	-	-
Gruda	51	59	-	-
Ivanščica	44V	48V	-	36Vc
Kalnik	44	48	-	-
Krk	28	53	-	-
Labinštica	33	53	-	34h
Lastovo	33	-	-	-
Licka Plješevica	30	54	-	-
Mali Losinj	28	53	-	-
Mirkovica	30	54	-	-
Moslavacka Gora	23	39	-	-
Pag	51	59	-	-
Papuk	23	39	-	-
Petrova Gora	25	48	-	-
Promina	51	59	-	-
Psunj	23	39	-	-
Pula	28	53	-	-
Razromir	28	53	-	-
Rota	33	59	-	-
Sljeme	25	48	-	42d
Srd	51	59	-	28i
Stipanov Gric	30	54	-	-
Sveta Gera	25	48	-	-
Sveta Nedjelja	25	48	56	-
Sibenik	51	59	-	31g
Ucka	28	53	-	29e
Ugljan	51	59	-	31g
Uljenje	51	59	-	-
Zagreb (HRT HQ)	25	48	56	-

+ repeaters. °) DTT regions a-i, see under "Local Stations".

CUBA

System: NTSC-M [A]; DVB-T planned

National Stations
INSTITUTO CUBANO DE RADIO Y TELEVISIÓN (ICRT) (Gov) ✉ Television Nacional, Calle M No. 313, Vedado, La Habana 10400 ☎ +53 7 8309705 🖨 +53 7 8309705 **W:** www.tvcubana.icrt.cu **Chs:** Canal Educativo, Canal Educativo 2, Cubavisión, Tele Rebelde. **Txs:** **Canal Educativo:** La Habana ch4 & network; **Canal Educativo 2:** La Habana ch15 & network; **Cubavisión:** La Habana ch6 & network; **Tele Rebelde:** La Habana ch2 & network.

Local Stations not shown.

CURAÇAO (Netherlands)

System: NTSC-M [A]

TELECURAÇAO (Gov) ✉ P.O.Box 415, Willemstad ☎ +599 9 4611288 🖨 +599 9 4614138 **E:** info@telecuracao.com **W:** www. telecuracao.com **LP:** GM: Norbert Hendrikse; TD: J. Rufina. **Txs** ch8 (20kW H).

CYPRUS

Systems: # PAL-B/G [E] ⬇2012; DVB-T (MPEG2, MPEG4)

National Stations
CYPRUS BROADCASTING CORP. (CYBC) (Pub) ✉ P.O.Box 24824, 1397 Lefkosia ☎ +357 22862000 🖨 +357 223175 85 **E:** rik@cybc.com. cy **W:** www.cybc.com.cy **LP:** DG: Themis Themistocleous **Chs:** RIK1, RIK 2 **Txs: RIK1:** Troodos ch6V (40kW), Kalokhorio ch35 (48kW), Athalassa ch38 (10kW), Limassol ch38 (9kW), Tsadha ch38 (8kW) & txs below 5kW; **RIK2:** Limassol ch21 (9kW), Tsadha ch21 (8kW), Athalassa ch22 (10kW), Troodos ch31 (100kW) & txs below 5kW. – **ALFA TV (Comm)** ✉ P.O.Box 26811, 1648 Lefkosia ☎ +357 22763000 🖨 +357 22760001 **E:** info@alphatv.com.cy **W:** www.alfatv.com.cy **LP:** Pres/DG: Socratis Hassikos **Txs:** DTT ch56 (Lefkosia) & analogue netw.. – **ANTENNA TV (ANT1) (Comm)** ✉ P.O.Box 20923, 1655 Lefkosia ☎+357 22200200 🖨 +357 22200210 **E:** info@antenna.com.cy **W:** www.antenna.com.cy **LP:** GM: Stelios Malekos **Txs:** Lefkosia ch65 & netw. – **CNC PLUS TV (Comm)** ✉ 8 Neas Engomis Str., 2409 Lefkosia ☎ +357 22600600 🖨 +357 22600512 **E:** petrospashias@plustv.com.cy **W:** www.plustv.com.cy **LP:** GM: Costakis Constantinou **Txs:** Lefkosia ch48 & netw.. – **LUMIERE TV (LTV) (Comm)** ✿ ✉ P.O.Box 25614, 1311 Lefkosia ☎ +357 22357272 🖨 +357 22354622 **E:** administration@ltv.com.cy **W:** www.lumieretv.com.cy **LP:** MD: George Xinaris **Txs:** DTT ch63 (Lefkosia) & analogue netw. – **MEGA TV (Comm)** ✉ P.O.Box 27400, 1644 Lefkosia ☎ +357 22477777 🖨 +357 22355138 **E:** newsdpt@megatv.com.cy **W:** www.megatv.com **LP:** GM: George Chouliaras **Txs:** Troodos ch24 & netw. – **SIGMA TV (Comm)** ✉ P.O.Box 21836, 1513 Lefkosia ☎ +357 22580100 🖨 +357 22580252 **E:** info@sigmatv.com **W:** www.sigma.com.cy **LP:** GM: Antis Hadjicostis **Txs:** Lefkosia ch37 & netw.

Local Stations
Capital TV: P.O.Box 55633, 3781 Limassol; ch54 & repeaters. **Extra TV:** P.O.Box 53665, 3317 Limassol; ch59. **Fred TV:** P.O.Box 24729, 1303 Lefkosia; ch23. **Magic TV:** Griva Digeni 16, 8047 Pafos; ch30. **Nimonia TV (NTV):** P.O.Box 55086, 3820 Limassol; ch32. **Omega TV:** P.O.Box 6068, 8100 Pafos; ch27. **Pafos TV:** P.O.Box 62552, 8028 Pafos; ch40. **VOX TV:** P.O.Box 40372, 6304 Larnaca; ch44.

Foreign TV Relays
ERT World (Greece): Troodos ch28 (100kW), Athalassa ch45 (10kW), Limassol ch45 (9kW), Tsadha ch45 (8kW) & txs below 5kW.

DTT Transmitters (MPEG4)
Operator: CYBC **Mux:** RIK1, RIK2, RIK HD, ERT World, Euronews ⌘ RIK1-4 **Txs:** ch33 (Troodos, Vavatsinia).

Northern Cyprus

National Station
BAYRAK RADIO TELEVISYON KURUMU (BRTK) (Gov) ⌨ Dr. Fazil Küçük Bulvari, BRT Sitesi, Lefkosa, via Mersin 10, Turkey ☎ +90 392 2254577 🖷 +90 392 2254577 **E:** info@brtk.net **W:** www.brtk.net **L.P:** Pres: Ahmet Okan **Chs:** Bayrak-Haber TV, Bayrak-Aile TV. **Txs: Bayrak-Haber TV:** Sinandagi ch21 (15kW), Selvilitepe ch44 (390kW); **Bayrak-Aile TV:** Sinandagi ch50 (15kW).

Local Stations not shown.

Foreign TV Relays
TRT (Turkey): TRT1: Sinandagi ch8 (100kW), Selvilitepe ch11 (15kW); TRT2: Selvilitepe ch41 (200kW).

DTT Transmitters (MPEG4 exc. *=MPEG2)
Operator: BRT **Mux:** Bayrak-Haber TV*, TRT1, TRT2, Genç TV, Ada TV, ART, Kanal T, NTV ⌘ Bayrak R*., Bayrak FM* **Txs:** ch33 (Selvilitepe 1.25kW).

AKROTIRI & DHEKELIA (UK)

System: DVB-T (MPEG2)

BFBS-TV Relay (British Mil) ⌨ BFPO 57, Dhekelia Mil 381 ☎ +357 24748518 **Mux(✪):** BFBS1, BFBS2, BFBS 3 Kids, BFBS4 ⌘ BFBS Radio 1+2, BFBS Cyprus **Txs:** (-).

CZECH REPUBLIC

Systems: DVB-T (MPEG2, MPEG4); † PAL-D/K [R] ⇩11 Nov 2011

National Stations
CESKA TELEVIZE (CT) (Pub) ⌨ Kavcí Hory, 140 70 Praha 4 ☎ +420 261131111 🖷 +420 261212891 **E:** info@ceskatelevize.cz **W:** www.ceskatelevize.cz **L.P:** DG: Jiri Janacek **Chs:** CT1, CT2, Reg., CT24, CT4 – **PRIMA TV (Comm)** ⌨ Na Zertvách 24, 180 00 Praha 8 ☎ +420 266700111 🖷 +420 266700201 **E:** informace@iprima.cz **W:** www.iprima.cz **L.P:** Dir: Marek Singer **Chs:** TV Prima, R1 – **TV NOVA (Comm)** ⌨ Krízeneckého nám. 5, 152 52 Praha 5 ☎ +420 233100111 🖷 +420 242424525 **E:** info@nova.cz **W:** www.nova.cz. **L.P:** DG: Petr Dvorák. – **PUBLIC TV (Comm)** ⌨ Václavské nám. 802/56, 110 00 Praha 1 ☎ +420 224032924 🖷 +420 224032922 **E:** info@publiccom. cz **W:** www.publictv.cz – **TV OCKO (Comm)** ⌨ Vrchlického 29, 150 00 Praha ☎ +420 257222263 🖷 +420 257222094 **E:** ocko@ocko.tv **W:** www.ocko.tv – **Z1 (Comm)** ⌨ Jankovcova 53, 170 00 Praha ☎ +420 234602800 **E:** recepce@z1tv.cz **W:** www.z1tv.cz.

Regional & Local Stations not shown.

DTT Transmitters (under construction) (MPEG2 exc. where stated)
Operator Mux 1+2: Ceské Radiokomunikace ⌨ U Nákladového nádrazi 3144, 130 00 Praha 3 ☎ +420 242411111 🖷 +420 242417595 **E:** cra@cra.cz **W:** www.cra.cz **Mux 1:** CT1, CT2, CT24, CT4. **Mux 2:** TV Nova, Nova Cinema, Prima TV, R1, TV Cool, TV Barrandov ⌘ CR 1-4, CR D-dur, CR Leonardo, CR R Cesko. – **Operator Mux 3:** Czech Digital Group ⌨ Na Zertvách 24/132, 180 00 Praha 8 ☎+420 2284827163 🖷 +420 266312531 **E:** mail@digitv.cz **W:** www.digitv.cz **Mux:** Public TV, Z1 ⌘ R.Proglas. – **Operator Mux 4:** Telefónica O2 Czech Republic, a.s. ⌨ Olšanská 5, 130 00 Praha 3 **W:** www.cz.o2.com **Mux (MPEG4):** CT1 HD, TV Nova HD, O2 Info, TV Ócko.

Location	M1	M2	M2	M3	kW
Brno (Kojál)	29	40	59	64	100
Brno (Hády)	25	40	59	64	10
Brno (mesto)	25V	40V	59V	64V	10
C.Budejovice (Klet')	49	50	22	-	100

Location	M1	M2	M2	M3	kW
Cheb (Zelená hora)	36	66	-	-	20
Chomutov (Jedlová hora)	33	58	-	-	32
Domazlice (Vraní vrch)	34	48	-	-	10
Frydek (Lysá hora)	54	-	-	-	25
Jáchymov (Klínovec)	36	66	-	-	32/50
Jeseník (Praded)	36	-	-	-	100
Jihlava (Javorice)	33	35	30	-	100
Kraslice (Snezná)	36	-	-	-	10
Liberec (Ješted)	43	-	-	-	50
Mariánské Lázne	36	-	-	-	10
Mikolov (Devín)	29	40	-	-	25
Nemanice	34	-	-	-	10
Nýrsko	34	-	-	-	10
Ostrava (Hladnov)	54	39	32	-	2x10/2
Ostrava (Hoštálkovice)	54	-	-	63	16
Pardubice	32	39	-	-	25
Plzen (Krašov)	34	48	52	63	100
Praha (Cukrák)	53	41	59	44	100
Praha (Zizkov)	53V	41V	59V	64V	32
Sušice (Svatobor)	49	48	52	-	100/71/100
Trinec (Javorová)	-	57	-	-	100
Trutnov (Cerná hora)	40	61	60	-	100
Ústí n.L. (Buková hora)	33	58	55	62	100
Valašské Klobouky	33	-	-	-	25
Vimperk	49	50	-	-	20
Votice (Mezivrata)	53	41	-	-	32
Zlín (Tlustá hora)	33	-	42	100	100

+ sites with txs below 10kW.

Regional/Local Operators not shown.

DENMARK

Systems: DVB-T (MPEG2, MPEG4)

National Stations
DR (Pub) ⌨ TV Byen, Emil Holms Kanal 20, 0999 København C ☎ +45 35203040 🖷 +45 35202644 **E:** presse@dr.dk **W:** www.dr.dk **L.P:** DG: Kenneth Plummer; Dir. DR Media (TV, Radio, New Media): Mikael Kamber **Chs:** DR1, DR2, DR Update, DR Ramasjang, DR K. – **TV 2 DANMARK (Pub)** ⌨ Rugaardsvej 25, 5100 Odense C ☎ +45 65919191 🖷 +45 65913322 **E:** tv2@tv2.dk **W:** www.tv2.dk **L.P:** MD: Merete Eldrup. **Chs (terr.):** TV 2 incl. reg. prgrs (**W:** www.tv2regionerne. dk): a) TV 2/Bornholm (Brovangen 1, 3720 Aakirkeby), b) TV 2/Fyn (Olfert Fischers Vej 31, 5220 Odense SØ), c) TV 2/Lorry (Allégade 7-9, 2000 Frederiksberg), d) TV/Midt-Vest (Søvej 2, 7500 Holstebro), e) TV 2/Nord (Søparken 4, 9440 Abybro), f) TV Syd (El-vej 2 B, Seest, 6000 Kolding), g) TV2 Øst (Kildemarksvej 7, 4760 Vordingborg), h) TV 2 / Østjylland (Skejbyparken 1, 8200 Århus N).

Local Stations not shown (via Mux 1).

DTT Transmitters (MPEG4 exc. where indicated)
Operator Mux 1+2: DIGI-TV I/S ⌨ Banestrøget 21, 2630 Taastrup **E:** info@digi-tv.dk **W:** www.digi-tv.dk **Mux 1 (MPEG2):** DR1, DR2, TV 2 Danmark, TV2 Danmark Regional stns / indep. local stations. **Mux 2:** DR Update, DR Ramasjang, DR K, Folketinget, DR HD. – **Operator Mux 3-6:** Boxer TV A/S ⌨ Admiralgade 24, 1066 København K **E:** info@boxertv.dk **W:** www.boxertv.dk **Mux 3(✪):** Animal Planet, Canal 9, CNN, Discovery T & L, Discovery Science, Disney Channel, Disney XD, History Channel, Travel Channel, The Voice **Mux 4(✪):** Discovery, Kanal 4, Kanal 5, MTV, TV 2 Charlie, TV 2 Zulu, TV 2 NEWS, TV 2 Film, TV 2 Sport, 6'eren **Mux 5(✪):** Body in Balance/C+ Sport 1, Cartoon Network, C+ First, C+ Hits Sport WE, Nickelodeon/Star!, VH-1, ARD, TV4 Sverige, TV2 Norge. – **Operator Mux 6-8(✪):** tbd.

Location	M1*	M2	M3	M4	M5	M6¹	M7¹	M8¹	kW⁰
Hadsten	26hV	44	69V	56V	55V	36V	24V	5	50
Hedensted	30f	44	33	46	55	36	68	7	50
Jyderup	58g	51	42	31	60	23	65	6	50
København (1)	53c	51	54	31	60	23	67	6	50
København (2)	53c	51	54	31	60	23	67	6	10
Nakskov	58g	34	43	38	66	48	63	6	16
Nibe	29e	57	50	37	35	39	63	5	50
Rø	59a	56	51	32	65	39	62	9	25
Svendborg	25b	49	27	22	43	41	61	7	25
Thisted	31dV	42V	21V	43V	22V	49V	62V	10	25
Tolne	29e	57	50	37	35	39	63	5	10
Tommerup	25b	49	27	22	43	41	61	7	50
Varde	30f	54	33	46	53	28	68	7	50
Viborg	40d	59	66	56	52	45	24	10	50
Videbæk	40d	59	66	22	55	28	34	10	50
Vordingborg	58g	34	42	38	66	48	63	6	50

Location	M1*	M2	M3	M4	M5	M6[1]	M7[1]	M8[1]	kW°
Åbenrå	37f	50	32	22	64	41	67	7	50
Århus	26hV	44V	69V	56V	55V	36V	24V	5	1

+ sites with txs below 1kW. (1) Søborg (2) Gladsaxe °) Power refers to muxes 1-7 *) incl. TV 2 reg prgrs (a-h), MPEG4 from 2012 [1]) F.pl.

DJIBOUTI

System: SECAM-K1 [E]

RADIO TÉLÉVISION DE DJIBOUTI (Gov) ☐ BP 97, Djibouti ☎ +253 352294 🖷 +253 356502 **E:** rtd@intnet.dj **W:** www.rtd.dj **Tx:** Djibouti ch7 (10kW H) & relay txs.

DOMINICA

NB: No terrestrial TV stations.

DOMINICAN REPUBLIC

System: # NTSC-M [A] ⇩Sep 2015; ATSC planned

National Station
CORPORACIÓN ESTATAL DE RADIO Y TELEVISIÓN (CERTV) (Pub) ☐ Av. Dr. Tejeda Florentino 8, Sto. Domingo ☎ +1 829 6891220 🖷+1 829 6886208 **W:** www.certvdominicana.com **LP:** Dir: George Rodriguez. **Txs:** Sto. Domingo ch4 & relay txs.

Local Stations
Antena Latina: Av. Independencia, Sto. Domingo; ch7. **Cadena de Noticias:** C/ Dr Defilló 4, Sto. Domingo; ch37. **Canal 25:** Av. General Lopez, Santiago; ch25. **Canal 27:** Av. Luperon, Sto. Domingo; ch27. **Color Vision:** Corporación Dominicana de Radio & TV, Av. Emilio Morel, Sto. Domingo; ch9. **Digital 15:** Av. San Martin, Sto. Domingo; ch15. **Digital Vision:** Av. Constitución 101, San Cristobal; ch63. **Mango TV:** Av. 27 de Febrero 308, Sto. Domingo; ch59. **Medios Educativa/Canal del Sol:** C/ Cub Scout 19, Sto. Domingo; ch6, ch65. **Radioemisoras Unidas:** Av. Tirandentes 35, Sto. Domingo; ch27. **Supercanal 33:** C/ Rafael A Sanchz, Sto. Domingo; ch33. **Teleamerica:** Av. Abraham Lincoln 1015, Sto. Domingo; ch47. **Teleantillas:** Autopista Duarte Km.7½, Sto. Domingo; ch2. **Telecentro/Transmisiones y Proyecciones:** Av. Pasteur 204, Sto. Domingo; ch13, ch31. **Telecoral:** Av. Independencia 59, Sto. Domingo; ch39. **Telefuturo:** Av. 27 de Febrero 371, Sto. Domingo; ch23. **Telemicro:** Av. San Martin, Sto. Domingo; ch6, ch35. **Telesistema:** Av. 27 de Febrero 52, Sto. Domingo; ch11. **Teleuniverso:** Av. Las Carreras 1, Santiago; ch29. **Televida:** Expreso V Centenario, Sto. Domingo; ch41.

EASTER ISLAND (Chile)

System: NTSC-M [A]

TV RAPA NUI ☐ Hanga Roa, Isla de Pascua. **Tx:** ch13.

ECUADOR

System: NTSC-M [A]

National Stations
ECUADOR TV (Pub) ☐ Quito.**W:** www.ecuadortv.ec. **Txs:** (-). – **CANAL 1 (Comm)** ☐ Av. del Bosque Mz 112, Ciudadela Kennedy Norte, Guayaquil ☎ +593 4 2680200 🖷 +593 4 2680185 **E:** relad_sa@canal1tv.com **W:** www.canal1tv.com. **Txs:** Guayaquil ch12 & relay txs. – **ECUAVISA (Comm)** ☐ Bosmediano 447, José Carb, Quito 1 ☎ +593 2 2995300 🖷 +593 2 2445488 **E:** wmaster@ecuavisa.com **W:** www. ecuavisa.com. **Txs:** Quito ch8 & relay txs. – **GAMAVISIÓN (Comm)** ☐ Av. Eloy Alfaro 5400 y Rio Coca, Quito ☎ +593 2 2262222 🖷 +593 2 2262284 **E:** gamavision@gamavision.com **W:** www.gamavision.com **Txs:** (-). – **TC TELEVISIÓN (Comm)** ☐ Av. de Las Américas y Av. Constitución, Guayaquil ☎ +593 4 2293211 **E:** wmaster@tctv.com.ec **W:** www.tctelevision.com **Txs:** (-). – **TELEAMAZONAS (Comm)** ☐ Av. A. Granda C. 529 y Av. Brasil, Quito ☎ +593 2 2430350 🖷 +593 2 2441620 **E:** contactenos@teleamazonas.com **W:** www.teleamazonas.com. **Txs:** Quito ch4 & relay txs.– **RED TELESISTEMA (Comm)** ☐ 2do Pasaje 32 N.O. y calle 18H N.O., Lomas de Prosperina, Quito ☎ +593 4 2274444 (Guayaquil) 🖷 +593 4 2640540 (Guayaquil) **E:** rts@rts.com.ec **W:** www. rts.com.ec. **Txs:** Quito ch11 & relay txs.

Local Stations not shown.

EGYPT

System: PAL-B/G [E]

EGYPTIAN RADIO AND TV UNION (ERTU) (Gov) ☐ TV Bldg, Cornish El-Nil, Cairo 11511 ☎ +20 2 25757155 🖷 +20 2 25746989 **E:** hamdyemara@egyptradio.tv **W:** www.ertu.gov.eg. **LP:** Head of TV: Nadio Haleem **Chs:** Prgr 1-8, Nile TV, Nile News, Nile Drama, Nile Culture, Nile Live, Nile Family, Nile Sports

Location	P1	P2	kW	Location	P1	P2	kW
Abu Znima	26	29	15.5	Kom Ombo	10	7	40
Alamain	46	48	126	Kuntella	44	-	15.5
Alexandria	6	11	110	Luxor	11	7	19
Assiut	10	6	60	Mahalla	8	10	1600
Asswan	5	9	67	Matruh	10	8	39.2
Baris	7V	5V	10	Nag Hamadi	5	8	17
Barnis	24	29	830	Negela	22	25	5
Bawiti	10	8	22.4	Nuweiba	26V	35V	25.7
Beni Suef	11	7	110	Port Said	5V	7V	10
Cairo	5	9	200	Qena	9	6	30
Dahab	6V	8V	9.3	Quseir	7	5	50
Dumyat	58	61	15.2	Rafah	48	45	350
El Arish	6V	10V	182	Ras El Hekma	32	35	31.6
El Dakhla	8	6	23	Ras Gharib	9	11	66
El Farafra	5V	7V	10	Ras Sedr	58	61	66.2
El Hammam	39	42	69.2	Safaga	11V	9V	50
El Kharga	10V	8V	40	Salum	9	11	6
El Minya	8	5	165	Sharm El Sheikh	27	33	8.9
El Tur	10	8	33	Sidi Barani	49	52	11
Haleyeb	9V	11V	31.6	Siwa	6V	8V	10
Hassana	34	-	35.5	Sohag	7	11	52
Hurghada	5V	7V	89	Suez	7	5	20
Idfu	8	11	165	Taba	32	37	25.7
Ismailia	11V	9V	260	Wadi El Natron	44	41	36.4
Isna	6	9	18				

+ sites with txs below 5kW. P=Prgr.

Prgr 3: Cairo ch7 (200kW); **Prgr 4:** Negila ch28 (74kW), Ismailia ch33V (79.4kW), Port Said ch42V (20.3kW), Zagazig ch52 (158kW); Suez ch30 (20.3kW); **Prgr 5:** Matruh ch5 (11kW), Siwa ch7V (11kW), Negila North ch28 (74kW), Alexandria ch36 (678kW), Ras Hekma ch39 (31.6kW), Hammam ch51 (69kW), Sidi Barani ch55 (11kW); **Prgr 6:** Mahalla ch49 (321kW), Natron ch59 (36kW); **Prgr 7:** Beni Ali ch22V (13.2kW), El Minya ch39V (56kW), Assiut ch48 (117kW), Beni Suef ch51V (43kW), Fayoum ch55V (107kW); **Prgr 8:** Asswan ch21 (67.6kW), Luxor ch22 (69.2kW), Sohag ch27 (340kW), Kom Ombo ch29 (33.9kW), Qena ch30 (85.1kW), Nag Hamady ch32 (77.6kW), Idfu ch40 (66kW), Isna ch49 (70.9kW); **Nile TV:** Qena ch27 (36kW), Cairo ch46 (95kW); **Nile News:** Cairo ch26 (282kW), Alexandria ch24 (110kW); **Nile Cinema:** Sohag ch23 (340kW), Alexandria ch21 (74kW), Asswan ch24 (67.6kW), Luxor ch25 (69.2kW), Cairo ch28 (282kW), Ismailia ch31 (20.7kW), Tanta ch56 (36kW); **Nile Drama:** Cairo ch34 (282kW); **Nile Culture:** Cairo ch43 (91kW); **Nile Live:** Cairo ch22 (282kW); **Nile Family:** Cairo ch40 (89kW); **Nile Sports:** Cairo ch38 (316kW); **Nile Comedy:** Cairo ch30 (282kW) (Txs below 5kW not mentioned; Pol=H exc. where indicated.)

EL SALVADOR

Systems: NTSC-M [A]; ATSC planned

National Stations
TELECORPORACIÓN SALVADOREÑA (TCS) ☐ Alameda Manuel Enrique Araújo, Edifício Canales 2, 4 y 6, San Salvador ☎ +503 22092000 🖷 +503 22092065 **E:** canal2@tcs246.com **W:** www.tcs246. com **Chs:** Canal Dos, Canal Cuatro, Canal Seis. **Txs:** S.Salvador YSR-TV (Canal Dos) ch2 (100kW), YSU-TV (Canal Cuatro) ch4 (75kW), YSLA-TV (Canal Seis) ch6 (150kW). – **TELEVISION CULTURAL EDUCATIVA** ☐ Ap. Postal 4, Santa Tecla ☎ +503 2280499 **Tx:** S.Salvador ch8 (109kW). – **TELEVISION CULTURAL EDUCATIVA CANAL 10** ☐ Ap. Postal No. 104, Neuva San Salvador. ☎ +503 2280499 🖷 +503 2280973 **E:** tydiez@es.com.sv **Tx:** S.Salvador ch10 (109kW). – **TV DOCE** ☐ Boulevard Santa Elena Sur #12, Antiguo Custatlan, La Libertad, San Salvador ☎ +503 25101212 🖷 +503 25101222 **E:** canal12@canal12.com.sv **W:** www.canal12.com.sv. **Tx:** S.Salvador ch12.

Local Stations not shown.

EQUATORIAL GUINEA

System: PAL-B/G [E]

TELEVISIÓN NACIONAL (Gov) ✉ Malabo. **L.P:** Dir: Antonio Nkulu Oye. **Txs:** Santa Isabel ch2 (50kW H) & relay txs.

ERITREA

System: PAL-B/G [E]

ERITREA TELEVISION (ERI-TV) (Gov) ✉ Asmara ☎ +291 1 116033 🖷 +291 1 124847 **E:** shabait@shabait.com **W:** www.shabait.com/eritv.htm **Chs:** ERI-TV, ERI-TV2 **Txs:** **ERI-TV:** (pol.H) Asmara ch5 (5kW), Assab ch11 (5kW) & relay txs. **ERI-TV 2:** (-).

ESTONIA

Systems: DVB-T (MPEG4)

EESTI TELEVISIOON (ETV) (Pub) ✉ Faehlmanni 12, 15029 Tallinn ☎ +372 6284133 🖷 +372 6284155 **E:** etv@etv.ee **W:** www.etv.ee **L.P:** MD: Ainar Ruussaar. **Chs:** ETV, ETV2 – **KANAL 2 (Comm) / KANAL 11 (Comm)** ✉ Maakri 23a, 10145 Tallinn ☎ +372 6662450 🖷 +372 6662451 **E:** info@kanal2.ee / info@kanal11.ee **W:** www.kanal2.ee / www.kanal11.ee **L.P:** Dir: Urmas Oru (Kanal 2) – **TV3 VIASAT (Comm) / TV6 VIASAT (Comm)** ✉ Peterburi tee 81, 11415 Tallinn ☎ +372 6220200 🖷 +372 6220201 **E:** tv3@tv3.ee / tv6@tv6.ee **W:** www.tv3.ee / www.tv6.ee **L.P:** MD: Toomas Vara (TV3).

DTT Transmitters
Operator Mux 1: Levira AS ✉ Kloostrimetsa tee 58 A, 15026 Tallinn ☎ +372 6804000 🖷 +372 6804001 **E:** levira@levira.ee **W:** www.levira.ee **Mux:** ETV, ETV2, Kanal 2, Kanal 11, TV3. **Operator Mux 2+3:** Starman AS ✉ Akadeemia tee 28, 12618 Tallinn ☎ +372 6779977 🖷 +372 6779907 **E:** pressiinfo@starman.ee **W:** www.starman.ee **Mux 2 (❂):** TV6, TV3+, Sony TV, Discovery, Discovery Investigation, Universal, Animal Planet, Cartoon Network/TCM, Eurosport, Fashion TV, PBK Estonia, Ren-TV Estonia **Mux 3 (❂):** TV14, Seitse, Fox Life, Fox Crime, Showtime, National Geographic, Nickelodeon, Eurosport 2, MTV Europe, Euronews, CNN, Hustler TV.

Location	M1	M2	M3	kW
Ellamaa	45	-	-	7.8
Koeru	57	60	63	5/1.7/1.8
Kohtla-Nõmme	33	48	58	12/2x2.3
Pehka	45	-	-	5
Pärnu	53	56	62	12.7/2x2.3
Tallinn (TV-tower)	45	59	64	18.8/2x14.7
Valgjärve	47	52	61	14/2x7

+ sites with txs below 5kW

ETHIOPIA

System: PAL-B/G [E]

National Station
ETHIOPIAN TELEVISION (ETV) (Gov) ✉ P.O.Box 5544, Addis Ababa ☎ +251 11 5505483 🖷 +251 11 5505174 **E:** etv2@ethionet.et **W:** www.erta.gov.et **L.P:** DG: Berhan Haile **Txs:** (pol.H) Shashemene ch5 (1kW), Debrebirhan ch6 (1kW), Debre Markos ch6 (1kW), Goba ch6 (1kW), Addis Ababa ch7 (5kW), Harar ch7 (1kW), Mekele ch7 (1kW), Gambella ch8 (1kW), Araminch ch9 (1kW), Assaita ch9 (1kW), Axum ch9 (1kW), Dessie ch9 (1kW), Godie ch9 (1kW), Assosa ch11 (1kW), Nazereth ch11 (1kW) & txs below 1kW.

Local Station
Dire TV: Dire Dawa; ch9.

FALKLAND ISLANDS (UK)

System: DVB-T (MPEG4)

BFBS-TV (British Mil) ✉ BFBS Falkland Islands, Mt. Pleasant, BFPO 655 ❂ +500 32179 🖷 +500 32193 **Mux 1+2:** BFBS1, BFBS2, BFBS3 Kids, BFBS4, BFBS1 One Day Later, BFBS+, Sky Sports 1, Sky Sports 2 ✺ BFBS Radio 1+2, BFBS Falkland Islands **Txs:** (-).

FAROE ISLANDS (Denmark)

System: # PAL-B/G [E]; DVB-T (MPEG2)

KRINGVARP FØROYA - SJÓNVARPIÐ (Pub) ✉ P.O.Box 1299, 110 Tórshavn ☎ +298 347500 🖷 +298 347501 **E:** kringvarp@kringvarp.fo

W: www.kringvarp.fo **L.P:** GM: Annika M. Jacobsen.**Tx:** Húsareyn ch6 (10kW). – **RÁS 1 (Comm)** ✉ P.O.Box 3128, 110 Tórshavn. Via DTT mux.

DTT Transmitters
Operator: P/F Televarpið ✉ P.O.Box 27, 110 Tórshavn ☎ +298 340340 🖷 +298 340341 **E:** televarp@televarp.fo **W:** www.televarp.fo **Mux 1 (❂):** 3+, National Geographic/CNBC, TV Danmark, BBC Prime, BBC World, Kanal 5 **Mux 2 (partly❂):** SvF, DR1, DR2, TV3, Discovery Danmark, NRK1 **Mux 3 (❂):** Canal+ Danmark, Canal+ Film 1, Canal+ Film 2, Canal+ Sport, Cartoon Network/TCM, Jetix/Hallmark **Mux 4 (❂):** Viasat Sport, Eurosport, Animal Planet, God Channel/Rás 1, MTV Europe, VH1.

Location	M1	M2	M3	M4	kW
Brúnaskarð	57	59	65	67	0.25
Knúkur	53	55	61	63	0.04
Stongin	46	48	50	52	0.04
Støðlafjall	52	50	48	46	0.04
Varða	57	59	65	67	0.25
Velbastaður	60	66	68	58	0.04

+ repeaters.

FIJI

System: NTSC-M [A]

FIJI TELEVISION (Comm) ✉ P.O.Box 2442, Suva ☎ +679 3305100 🖷 +679 3305077 **E:** fijitv@is.com.fj **W:** www.fijitv.com.fj **Chs:** Fiji 1; Sky Fiji (❂) – **PACIFIC BROADCASTING SERVICE (SBS) (Comm)** ✉ P.O.Box 18629, Suva ☎ +679 3372333 🖷 +679 3372336 **E:** info@pacificbroadcastingservices.com **W:** www.pacificbroadcastingservices.com.

FINLAND

System: DVB-T (MPEG2), DVB-T2 (MPEG4)

National Stations
YLEISRADIO OY (Pub) ✉ P.O.Box 66, 00024 Yleisradio ☎ +358 9 14801 🖷 +358 9 14805148 **W:** www.yle.fi **L.P:** CEO: Mikael Jungner. **Chs:** TV1; TV2 (✉/P.O.Box 196, 33101 Tampere); YLE Teema; FST5 in Swedish (✉P.O.Box 83, 00024 Yleisradio). – **CANAL DIGITAL (Comm)** ✉ P.O.Box 2, 00381 Helsinki ☎ +358 9 54264200 🖷 +358 9 54264270 **E:** asiakaspalvelu@canaldigital.fi **W:** www.canaldigital.fi – **MTV3 (Comm)** ✉ 00033 MTV3 ☎ +358 10 300300 🖷 +358 10 3005164 **W:** www.mtv3.fi **L.P:** Pres/CEO: Pekka Karhuvaara. **Chs:** MTV3, MTV3 Max, JIM, Sub, SubJuniori, Leffa – **NELONEN (Comm)** ✉ P.O.Box 350, 00151 Helsinki ☎ +358 9 4545414 **W:** www.nelonen.fi **L.P:** Pres: Juha-Pekka Louhelainen.

DTT Transmitters (MPEG2 exc. where stated)
Operator Mux 1-5: Digita Oy ✉ P.O.Box 135, 00521 Helsinki ☎ +358 20 411711 **E:** info@digita.fi **W:** www.digita.fi **Mux 1:** YLE TV1, YLE TV2, YLE Teema, YLE FST5, SVT World❂. ✺ Ylen Klassinen, R.Peili, YLE Mondo **Mux 2:** MTV3, MTV3 Max❂, Nelonen, JIM, Sub, Liv, SubJuniori/Leffa❂. **Mux 3 (❂exc.*):** Urheilukanava*, The Voice/TV Viisi*, Klubi.tv*, Iskelmä TV/Harju & Pöntinen*, Canal+ First, Canal+ Hits, Canal+ Sport 1, Canal+ Sport 2, Disney Channel, Nelonen Sport Pro, Digi-Viihde, URHOtv **Mux 4 (DVB-H❂):** The Voice TV MTV3 ✺ The Voice, R.Iskelmä, R.Nova (txs not shown) **Mux 5 (❂exc.*):** Liv*, SuomiTV*, Discovery Channel, MTV3 Fakta, MTV, Nickelodeon, KinoTV. – **Operator Mux 6+7:** Anvia Oy ✉ P.O.Box 59, 65101 Vaasa ☎ +358 6 4114111 🖷 +358 6 3170146 **W:** www.anvia.fi **Mux 6 (DVB-T2/MPEG4) (❂):** Discovery Channel HD, Animal Channel HD, Family Channel HD, Canal+ Film HD, MTVn Full HD, Nelonen Sport Pro HD, MTV 1000 HD. **Mux 7 (DVB-T2/MPEG4):** tbd.

Location	M1	M2	M3	M5	M6°	M7°	kW
Anjalankoski	22	27	53	56	-	-	50
Espoo	32	44	46	53	-	-	50
Eurajoki	38	45	52	55	-	-	50
Fiskars	32	44	46	58	-	-	10
Haapavesi	34	42	53	57	-	-	50
Iisalmi	26	38	-	-	-	-	50
Inari	48	25	-	-	-	-	50
Joutseno	47	35	57	32	-	-	50
Jyväskylä	30	60	55	41	-	-	50
Karigasniemi	50	49	-	-	-	-	50
Kerimäki	30	37	33	58	-	-	50
Kiihtelysvaara	26	59	-	-	-	-	30
Koli	25	40	47	51	-	-	60
Kruunupyy	27	22	41	44	-	-	50
Kuopio	24	31	39	52	-	-	50

Location	M1	M2	M3	M5	M6°	M7°	kW
Kuttanen	53	58	-	-	-	-	30
Lahti	33	47	57	51	-	-	50
Lapua	38	37	55	48	-	-	50
Mikkeli	29	43	59	38	-	-	50
Oulu	41	51	54	37	-	-	50
Pernaja	23	50	-	39	-	-	10
Pihtipudas	50	45	58	-	-	-	80
Posio	31	39	-	-	-	-	50
Pyhätunturi	60	41	-	-	-	-	50
Pyhävuori	28	41	-	35	-	-	50
Rovaniemi	43	46	-	53	-	-	50
Ruka	33	48	59	-	-	-	50
Taivalkoski	32	38	-	-	-	-	50
Tammela	22	27	50	43	-	-	60
Tampere	34	23	58	59	-	-	50
Tervola	40	42	-	44	-	-	50
Turku	51	54	57	60	-	-	50
Utsjoki	44	51	-	-	-	-	30
Vaasa	38	37	-	57	-	-	50
Vuokatti	30	52	55	59	-	-	80
Vuotso	31	50	-	-	-	-	30
Ylläs	30	36	-	-	-	-	50
Ähtäri	52	44	-	-	-	-	50

+ sites with txs below 10kW. °) start 2011

FRANCE

Systems: DVB-T (MPEG2, MPEG4); † SECAM-L [L] ⇩ 30 Nov 2011

National Stations

FRANCE TÉLÉVISIONS (Pub) ⌨ HQ: 7 esplanade Henri de France, 75907 Paris Cedex 15 ☎ +33 156226000 **W:** www.francetelevisions. fr **LP:** Pres: Rémy Pflimlin. **Chs: France 2** (www.france2.fr), **France 3** (www.france3.fr) & regional stations, **France 4** (www.france4.fr), **France 5** (www.france5.fr), **France Ô** (for French overseas territories; www.rfo.fr) ⌨ (exc. France 5) 35/37 rue Danton, 92240 Malakoff ☎ +33 1 55227100 ⌨ France 5: 10 rue Horace Vernet, 92785 Issy les Moulineaux Cedex 9 ☎ +33 156229191 ⌨ +33 141080222. — **ARTE (Pub)** ⌨ 8 rue Marceau, 92785 Issy-les-Moulineaux Cedex 9 ☎ +33 155007777 ⌨ +33 155007700 **W:** www.arte.tv **LP:** Pres: Jérôme Clément. — **LA CHAÎNE PARLEMENTAIRE - ASSEMBLÉE NATIONALE (LCP) (Pub)** ⌨ 106 rue de l'Université, 75007 Paris ☎ +33 140639050 ⌨ +33 140639019 **W:** www.lcpan.fr **LP:** DG: Gérard Leclerc. — **PUBLIC SÉNAT (Pub)** ⌨ 92 boulevard Raspail, 75006 Paris ☎ +33 142344400 ⌨ +33 142344450 **W:** www.publicsenat.fr **LP:** DG: Gilles Leclerc. — **BFM TV (Comm)** ⌨ 12 rue d'Oradour-sur-Glane, 75040 Paris Cedex 15 ☎ +33 171191360 ⌨ +33 171191369 **W:** www. bfmtv.fr **LP:** Pres: Alain Weill. — **CANAL PLUS (Comm)** ⌨ 1 place du spectacle, 92863 Issy les Moulineaux Cedex 9 ☎ +33 171353535 ⌨ +33 171351050 **W:** www.canalplus.fr **LP:** Pres.: Bertrand Meheut. **Chs (DTT):** Canal+, Canal+ Cinéma, Canal+ Sport. — **DIRECT 8 / DIRECT STAR (Comm)** ⌨ 31/32 quai de Dion Bouton, 92811 Puteaux Cedex ☎ +33 146964888 ⌨ +33 146964097 **W:** www.direct8.fr / www.directstar.fr **LP:** DG: Yannick Bolloré. — **EUROSPORT FRANCE (Comm)** ⌨ 3, rue Gaston et René Caudron, 92798 Issy-les-Moulineaux Cedex 9 ☎ +33 140938000 **W:** www.eurosport.fr **LP:** DG: Laurent-Eric Le Lay. — **GULLI (Comm)** ⌨ 28 rue François 1er, 75008 Paris ☎ +33 156365555 ⌨ +33 156365559 **W:** www.gulli.fr **LP:** DG: Emmanuelle Guilbart. — **I>TÉLÉ (Comm)** ⌨ 6 allée de la 2ème DB, 75015 Paris ☎ +33 153915000 ⌨ +33 153915123 **W:** www.itele.fr **LP:** DG: Pierre Fraidenraich. — **LCI (Comm)** ⌨ 54 avenue de la Voie Lactée, 92656 Boulogne-Billancourt Cedex ☎ +33 141412345 **W:** www.lci.fr **LP:** DG: Eric Revel. — **M6 (Comm)** ⌨ 89/91 avenue Charles de Gaulle, 92575 Neuilly sur Seine Cedex ☎ +33 141926666 ⌨ +33 141926610 **W:** www.m6.fr **LP:** Pres: Nicolas de Tavernost. — **NRJ 12 (Comm)** ⌨ 46/50 av. Théophile Gautier, 75016 Paris. ☎ +33 140713929 **W:** www. nrj12.fr **LP:** DG: Gérard-Brice Viret. — **PARIS PREMIÈRE (Comm)** ⌨ 89 avenue Charles de Gaulle, 92575 Neuilly sur Seine Cedex ☎ +33 141925700 ⌨ +33 141925703 **W:** www.paris-premiere.fr **LP:** Pres: Karine Blouët. — **PLANÈTE (Comm)** ⌨ 1 place du Spectacle, 92823 Issy les Moulineaux Cedex 9 ☎ +33 171353535 **W:** www.planete.com **LP:** DG: Rodolphe Belmer. — **TF1 / NT1 (Comm)** ⌨ 1 quai du Point du Jour, 92656 Boulogne-Billancourt, Cedex ☎ +33 141411234 ⌨ +33 141412793 **W:** www.tf1.fr / www.nt1.fr **LP:** Pres/DG: Nonce Paolini. — **TF6 (Comm)** ⌨ 120 avenue Charles de Gaulle, 92200 Neuilly sur Seine Cedex ☎ +33 155626666 **W:** www.tf6.fr **LP:** Pres: Laurent Solly. — **TMC (Comm)** ⌨ 6 bis quai Antoine 1er, 98000 Monaco ☎ +377 93151415 ⌨ +377 93151436 **W:** www.tmc.tv **LP:** DG: Jean-Claude Riey. — **TPS STAR (Comm)** ⌨ 1 place du Spectacle, 92863 Issy les Moulineaux Cedex 9 ☎ +33 171353535 **W:** www.tpsstar.fr **LP:** Pres: Bertrand Méheut. — **W9 (Comm)** ⌨ 89/91 avenue Charles de Gaulle,

92575 Neuilly sur Seine Cedex ☎ +33 141926421 ⌨ +33 141926459 **W:** www.w9.fr **LP:** Pres: Jérôme Lefébure.

Local Stations

Pol=H exc. where stated. Only DTT txs shown, high-power DTT txs are generally txs of the national Mux 1 network, cf. main DTT tx table. (*) via Local Paris Mux: ch23 (Paris 5kW, Coulommiers 0.006kW).

Alsace 20: 17 rue Nuée Bleue, 67000 Strasbourg; ch46 (Wissembourg 0.2kW), ch48 (Strasbourg 50kW), ch53 (Mulhouse 100kW) + 28 txs below 0.2kW. **BDM TV:** 50 rue de Clignancourt, 75018 Paris; (*). **BIP TV:** Rue des Noues Chaudes, 36100 Issoudun; ch39 (Argenton sur Creuse 5.7kW), Issoudun ch30 (Issoudun 0.075kW). **Canal Cholet:** La Novathèque, boulevard Pierre Lecoq, 49300 Cholet; ch56 (Nantes 30kW, Les Sables d'Olonne 0.1kW) + txs under 0.1kW. **Canal 15 Vendée:** 8 place Napoléon Galerie Bonaparte, 85000 La Roche sur Yon; ch56 (Nantes 30kW, Les Sables d'Olonne 0.1kW) + 10 txs below 0.1kW. **Canal 32:** 7 rue Raymond Aron, 10120 Saint André les Vergers; ch23 (Troyes 2kW). **CINAPS TV:** 17 rue des Tiphoines, 91240 Saint Michel sur Orge; (*). **Clermont 1ère:** 40 rue Morel Ladeuil, 63006 Clermont-Ferrand Cedex 01; ch39 (Montluçon 0.1kW), ch50 (Clermont-Ferrand 18kW) + 2 txs below 1kW. **Demain IDF:** 1 rue Patry, 92220 Bagneux; (*). **Grand Lille TV:** 101 boulevard Descat, 59200 Tourcoing; ch36 (Lille 1kW). **IDF1:** 7 rue des Bretons, 93210 La Plaine Saint Denis; (*). **LCM:** 37/41 rue Guibal, 13003 Marseille Cedex 3; ch45 (Roquevaire 0.01kW), ch62 (Marseille/Grande Étoile 44kW, Marseille/Pomègues 3kW) + 2 txs below 0.1kW. **LM TV Sarthe:** 21/25 rue Pasteur, 72015 Le Mans Cedex 2; ch26 (Le Mans 79.5kW) + 6 txs below 0.1kW. **Mirabelle TV:** 2 rue Saint Vincent, 57141 Woippy; ch33 (Metz 31.4kW, Verdun 6kW, Longwy 3kW, Forbach 0.7kW), ch41 (Sarrebourg 3.3kW) + 13 txs below 0.1kW. **Nantes 7:** 42 rue de la Tour d'Auvergne, 44200 Nantes; ch47 (Nantes 98kW, Saint-Nazaire 0.27kW) + 2 txs below 0.1kW. **Normandie TV:** 3 square du Théâtre, 14200 Hérouville Saint Clair; ch25 (Caen 100kW, Mortain 0.4kW, Caen (town) 0.3kW) + 17 txs below 0.1kW. **NRJ Paris:** 40/50 avenue Théophile Gautier, 75016 Paris; (*). **Orléans TV:** 12 rue André Dessaux, 45400 Fleury les Aubrais; ch46 (Orléans 2kW). **Télé 102:** 8 rue de l'Hôtel de Ville, 85103 Les Sables d'Olonne; ch64 (Les Sables d'Olonne/Vairé) ch64 0.1kW, Les Sables d'Olonne/Les Roses 0.002kW). **Télé Bocal:** 12 villa Ribérolle, 75020 Paris; (*). **Télé Miroir:** 240 rue Le Corbusier, 30000 Nîmes, ch63 (Nîmes 0.015kW). **TéléGrenoble:** 109 rue Hilaire de Chardonnet, 38100 Grenoble; ch32 (Voiron 0.013kW), ch37 (Grenoble 0.06kW). **Télénantes:** 42 rue de la Tour d'Auvergne, 44262 Nantes Cedex; ch47 (Nantes 98kW, Saint-Nazaire 0.27kW). **TéléPaese:** Cedrajo, 20220 Santa Reparata di Balagna; DTT ch tbd. **Territorial TV:** rue Louis Lepitre, 52200 Langres; ch49 (Bar le Duc 4.6kW) + 5 txs below 0.1kW. **TLM (Télé Lyon Métropole):** 4 rue Montrochet, 69002 Lyon Cedex 07; ch49 (Lyon/Taluyers 0.3kW), ch56 (Lyon/Fourvière 1kW). **TLP (Télé Locale Provence):** Place de l'Europe, 04280 Céreste; ch46 (Manosque 0.025kW), ch56 (Oraison 0.04kW) + 8 txs below 0.02kW. **TL7 Horizon Numérique:** 2 bis rue Joseph Cugnot 42160 Andrézieux Bouthéon; ch50 (Saint-Étienne 0.56kW) + 8 txs below 0.1kW. **TV Rennes 35:** 19 rue de la Quintaine, 35000 Rennes; ch21 (Rennes 80kW) + 22 txs below 0.1kW. **TV Tours:** 232 avenue de Grammont, 37019 Tours Cedex 1; ch34 (Tours 28kW) + 8 txs below 0.1kW. **TV Vendée:** ZI le Séjour, 85170 Dompierre sur Yon; ch56 (Nantes 30kW), ch56 (Les Sables d'Olonne 0.1kW) + 10 txs below 0.1kW. **TVPI:** route de Bayonne, 64210 Bidart; Bayonne ch46 (Lyon/Taluyers 0.3kW) ch56... **TV7 Bordeaux:** 73 av. Thiers, 33000 Bordeaux; ch23 (Bordeaux/Bouliac 33kW) + 2 txs below 0.1kW. **TV8 Mont-Blanc:** route des Pontets, 74320 Sevrier Cedex; ch35 (Cluses 0.25kW), ch57 (Annemasse 0.12kW), ch62 (Chambéry/Les Monts 0.48kW, Montmélian 0.14kW). **TV77:** 15 rue André François Poncet, 77160 Provins; ch55 (Meaux 0.04kW). **Ty Télé:** 8 Auguste Nayel, 56100 Lorient; ch57 (Vannes 20kW, Lorient 0.15kW) + 7 txs below 0.1kW. **Villages TV:** 44 route de la Torchaise, 86580 Vouneuil sous Biard; ch50 (Poitiers 0.5kW) + 1 tx below 0.1kW. **VOO TV:** 20 rue des Ardennes, 21000 Dijon; ch53 (Dijon 0.1kW) + 1 tx below 0.1kW. **Vosges Télévision Images Plus:** 2 rue de la Chipotte, BP 267, 88007 Epinal; ch56 (Vittel 3 kW, Epinal 1kW) + 31 txs below 0.1kW. **Wéo La Télé Nord Pas de Calais:** 17 place Mendès France, 59000 Lille; ch23 (Lille 20kW), ch44 (Maubeuge 0.165kW), ch52 (Valenciennes 1kW). **7L TV:** Avenue de la Pompignane, 34170 Castelnau Le Lez; Montpellier ch49 (Montpellier 35kW).

DTT Transmitters (MPEG2, exc.✿ and HD channels: MPEG4)
Operator Mux 1: Société de gestion du réseau R1 ⌨ 7 esplanade Henri-de-France, 75015 Paris ☎ +33 156224302 ⌨ +33 156225818 **Mux:** France 2, France 3 + reg stns, France 5, France Ô, LCP/Public Sénat, local stations. – **Operator Mux 2:** Nouvelles télévisions numériques ⌨ 28 rue François 1er, 75008 Paris ☎ +33 147232301 ⌨ +33 147232309 **Mux:** BFM TV, Direct 8, Direct Star, France 4, Gulli, i>TELE. – **Operator Mux 3:** Compagnie du numérique hertzien 1 ⌨ place du Spectacle, 92863 Issy-les-Moulineaux ☎ +33 171350130 ⌨ +33 171350626 **Mux** (✿): Canal+ HD**, Canal+ Cinéma, Canal+ Sport**, Planète, TPS Star*. – **Operator Mux 4:** Multi 4 ⌨ 89 avenue

Charles de Gaulle, 92200 Neuilly-sur-Seine ☎ +33 141926140 🖷 +33 141925954 **Mux:** Arte HD, M6, Paris Première✪*, W9. – **Operator Mux 5:** MR5 🖃 1 quai du Point-du-Jour, 92100 Boulogne-Billancourt ☎ +33 141411234 🖷 +33 141413046 **Mux:** TF1 HD, France 2 HD, M6 HD. – **Operator Mux 6:** SMR6 🖃 1 quai du Point-du-Jour, 92100 Boulogne-Billancourt ☎ +33 141411234 🖷 +33 141413046 **Mux:** Arte, Eurosport France✪, LCI✪, NRJ 12, NT1, TF1, TF6✪, TMC.
*) incl. unencrypted sequences. **) in MPEG2-SD during unencrypted sequences

Location	M1	M2	M3	M4	M5	M6	kW
Abbeville	29	25	33	39	37	58	15/4x30/11
Ajaccio	29	26	42	38	59	53	2x24/6.5/2x16/24
Amiens	36	50	43	46	-	40	10
Auxerre	64	61	33	30	-	32	2x11/3x10
Bayonne	65	42	51	49	62	57	4x12/10/12
Bergerac	33	42	45	39	-	30	10
Besançon (Lomont)	22	48	54	37	-	67	9.3/2x11.1/8.5/9.6
Besançon (Montfaucon)	22	48	54	37	-	67	14
Bordeaux	23	59	62	44	41	30	3x33/2x4/33
Bourges	35	24	63	27	60	32	4x23/20/23
Brest	43	58	35	39	30	34	191/144/107/140/118
Caen	25	34	22	29	23	28	100/25/100/2x25/100
Charleville-Mézières	32	47	24	22	-	35	4x22/5
Chartres	41	38	49	31	-	56	10
Cherbourg-Octeville	35	34	59	37	57	32	4x10/7.7/10
Clermont-Ferrand	50	31	37	53	-	32	18
Dijon	25	50	63	66	-	28	20
Hirson	62	59	60	63	-	35	22
Laval	33	58	43	57	60	51	9/14/4x10
Le Mans	26	23	22	31	37	36	79/32/79/2x32/79
Lille	23	48	26	30	31	35	3x20/4.8/2x20
Limoges	48	-	26	55	-	47	25
Lyon	45	36	39	54	42	47	4x43/30/43
Marseille	62	59	28	25	22	30	4x44/32/21
Metz	33	56	61	58	-	36	31.4/3x30.2/31.4
Montpellier	49	55	51	52	-	37	35
Mulhouse	27	55	54	37	-	21	100
Nantes	28	44	62	25	27	24	28/38/39/32/2x36
Niort	37	59	62	54	27	24	4x31/27/31
Paris	35	21	27	24	29	32	41/3x23/20.9/23
Reims	32	31	37	34	45	42	16/31/18/31/2x30
Rennes	21	40	27	49	37	24	80/17/2x80/8/80
Rouen	29	21	27	24	37	32	4x10.5/8/10.5
Saint-Raphaël	23	26	34	53	-	42	33
Strasbourg	48	47	43	22	-	51	50/2x20/10/21
Toulouse	54	48	50	22	26	36	13
Tours	34	38	63	48	37	51	4x28/20/28
Troyes	25	22	41	36	-	28	31
Vannes	57	25	53	48	22	50	20

+ sites with txs below 10kW.

FRENCH GUIANA

System: # SECAM-K1 [K/E]; DVB-T

RFO GUYANE (Pub) 🖃 Avenue le grand boulevard, BP 7013, Cayenne Cedex ☎ +594 594299900 🖷 +594 594302649 **W:** guyane.rfo.fr **LP:** Dir: Henri Neron. TD: Daniel Beugin. **Chs:** Télé Pays, Tempo. **Txs: Télé Pays:** Cayenne (Montagne du Tigre) ch4 (1kW), Iracoubo ch5 (2kW), Kourou ch6 (1.6kW), Mana ch7 (1.3kW), Cayenne (Ville-C.A.T.) ch22 (3.55kW), Ouanary ch48 (4kW) & txs below 1kW; **Tempo:** Iracoubo ch23 (25kW), Cayenne (Ville-C.A.T.) ch25 (3.55kW), Cayenne (Montagne du Tigre) ch33 (14kW), Mana ch40 (17kW), Kourou ch41 (33kW), Ouanary ch51 (4kW) & txs below 1kW. – **ANTENNE CRÉOLE GUYANE (ACG) (Comm)** 🖃 31 avenue Louis Pasteur, 97300 Cayenne ☎ +594 594288288 🖷 +594 594291308 **E:** acg@acg.gf **W:** www.acg.gf **Txs:** Cayenne (Montagne du Tigre) ch36 (14kW), Cayenne (Monte Montabo) ch39 (3kW), Mana ch43 (3kW), Kourou ch44 (33kW) & txs below 1kW. – **CANAL GUYANE (Comm)** ✪ 14, Lotissement Marengo, ZI de Collery, 97300 Cayenne ☎ +594 594295455 🖷 +594 594305335 **E:** gaumont@canalguyane.gf **W:** www.canalguyane.fr **Txs:** (-).

DTT Transmitters
Operator: France Télévisions **Mux:** RFO Guyane, France 2-5, France Ô, Arte, LCP, France 24, Gulli **Txs:** ch(-).

FRENCH POLYNESIA

Systems: # SECAM-K1 [K/E]; DVB-T

RFO POLYNÉSIE (Pub) 🖃 BP 60125, 98702 Faa'a, Polynésie Française ☎ +689 689861616 🖷 +689 689861611 **E:** rfopolyfr@mail.pf **W:** polynesie.rfo.fr **LP:** Dir: Claude Ruben. **Chs:** Télé Pays,

Tempo. **Txs: Télé Pays:** Hiva Ao ch4 (4kW), Bora Bora ch7V (2.4kW), Mahaena ch7 (1kW); **Tempo:** Bora Bora (Pahonu) ch33V (20kW), Bora Bora (Putaiamo) ch33 (5kW) & txs below 1kW. – **TAHITI NUI TV (Comm)** 🖃 BP 348, 98713 Papeete, Polynésie Française ☎ +689 689473636 🖷 +689 689532721 **E:** tntv@tntv.pf **W:** www.tntv.pf **Txs:** Taravao ch26 (3kW), Uturoa Tapioi ch26 (2.2kW), Mahena ch34 (5kW), Mont-Marau ch43 (55kW), Papareva ch51 (6kW) & txs below 1kW.

DTT Transmitters
Operator: France Télévisions **Mux:** RFO Polynésie, France 2-5, France Ô, Arte, LCP, France 24, Gulli **Txs:** ch(-).

FRENCH SO. & ANTARTIC LANDS

NB: No terrestrial TV station.

GABON

System: SECAM-K1 [E]

RADIODIFFUSION-TÉLÉVISION GABONAISE (RTG) (Gov) 🖃 BP 10150, Libreville ☎ +241 732152 🖷 +241 732153 **LP:** DG (TV): Jules César Lekogho. **Chs:** RTG1, RTG2 **Txs: RTG1:** Libreville ch5 (2kW H) & relay txs. **RTG2:** Libreville ch8 (2kW H). – **TV+ CHAÎNE 3 (Comm)** 🖃 BP 8344, Libreville ☎ +241 775740 🖷 +241 729204 **Txs:** Libreville ch(-), Franceville ch41.

GALAPAGOS ISLANDS (Ecuador)

System: NTSC-M [A]

TELEGALAPAGOS (Rlg) 🖃 Misión Franciscana, Puerto Baquerizo Moreno, Isla San Cristobal, Galapagos, Ecuador ☎ +593 5 2520144 **Tx:** ch13.

GAMBIA

System: PAL-I [E]

GAMBIA TELEVISION (Gov) 🖃 P.O.Box 2380, Serrekunda ☎ +220 374251 🖷 +220 374242 **E:** grts@gamtel.gm **W:** www.grts.gm **LP:** DG: Bora Mboge **Txs:** Banjul ch11V (100kW) & relay txs.

GEORGIA

Systems: SECAM-D/K [R], PAL-D/K [R]

National Stations
GEORGIAN PUBLIC BROADCASTING (Pub) 🖃 M. Kostava Street 68, Tbilisi ☎ +995 32 362294 🖷 +995 32 368665 **E:** info@gpb.ge **W:** www.gpb.ge **LP:** Chmn (TV): Levan Tarkhnishvili **Chs:** Public TV 1, Channel 2. **Txs: Public TV1:** Tbilisi ch6 (490kW) & network; **Channel 2:** Tbilisi ch4 (228kW) & network. – **RUSTAVI 2 (Comm)** 🖃 Vazha Pshavela Avenue 4, Tbilisi ☎ +995 32 201111 🖷 +995 32 200012 **E:** tv@rustavi2.com.ge **W:** www.rustavi2.com.ge **LP:** DG: Koba Davarashvili. **Txs:** Tbilisi ch9 (478kW) & network.

Local Stations
ATV12: M. Mashtots St. 56, Akhalkalaki; ch12. **Kvemo Kartlis Tele-Radio Kompania:** Megobroba Ave. 32, 9th Fl, Rustavi; ch8. **Kvemo Kartlis TRC-s Marneulis Philiali:** Maisi St. 40, Marneuli. Tx: ch43. **L-TV:** Zakatala St. 29, Lagodekhi; ch27. **Mega TV:** Solomon Meore St. 1, Khoni. Tx: ch2. **TV Argo:** Argonavtebi St. 1, Zestaphoni; ch9. **TV Borjomi:** Rustaveli Sq. 1, 9th Fl, Borjomi; ch7. **TV Dia:** Tamar Mepe St. 1, 4th Fl, Khashuri; ch11. **TV Edelweiss:** Rustaveli St. 20, Tskaltubo; ch3 (shares tx with TV Imervizia). **TV Egrisi:** Akhalgazrdobis Kheivani St. 9, 2nd Fl, Senaki; ch1. **TV Evrika:** Tsinamdzghvrishvili St. 95, Tbilisi; ch25. **TV Guria:** Aghmashenebeli St. 120, Ozurgeti; ch21 (shares tx with TV Madi). **TV Gurjaani:** Ninoshvili St. 14, Gurjaani; ch11. **TV Imedi:** Lubliana St. 5, Tbilisi; ch7. **TV Imervizia:** Ninoshvili St. 7, Tchiatura; ch3 (shares tx with TV Edelweiss). **TV Jikha:** Gamsakhurdia St. 9, Tsalenjikha; ch5. **TV Kartli:** Chavchavadze St. 51, Gori; ch10. **TV Kavkazia:** Rustaveli St. 14, Tbilisi; ch28. **TV Kolkheti:** Chavchavadze St. 4, Chkhorotsku; ch4. **TV Kutaisi:** Gelati St. 1, Kutaisi; ch1. **TV Lomsia:** Didmamishvili St. 3, Akhaltsikhe; ch2. **TV Madi:** Ramishvili St. 120, Ozurgeti; ch21 (shares tx with TV Guria). **TV Metormete Arkhi:** Svanuri Koshki, Davit Aghmashenebeli St., Bolnisi; ch12. **TV Metskhre Arkhi:** Tamar Mepe St. 6, Akhaltsikhe; ch9. **TV Mze:** Kostava St. 75b, Tbilisi;

ch45. **TV Odishi:** Aghmashenebeli St. 17, Zugdidi; ch39. **TV Parvana:** Freedom St. 39, Ninotsminda; ch12. **TV Rioni:** Tamar Mepe St. 14, Kutaisi; ch31. **TV Sameba:** 19 Kakhetis Gzatketsili St. 19, Sagarejo; tx (-) **TV Samegrelo:** Rustaveli St. 90, Zugdidi; ch21, ch30. **TV Stereo One:** Aleksidze St. 1, 2nd Fl, Tbilisi; ch40. **TV Tanamgzavri:** Sanapiro St. 1, Telavi; ch4. **TV Trialeti:** Chavchavadze St. 45, Gori; ch1. **TV Tvali:** David Aghmashenebeli St. 15, Sagarejo; ch34. **TV Zari:** Javakhishvili St. 8, Samtredia; ch4. **TV 202:** Tsereteli St. 144, Tbilisi; ch12. **TV-1:** Leselidze St. 44, Akhmeta; ch10. **V Channel:** Sulkhan-Saba Orbeliani St. 115, Bolnisi; ch5. **9th Talgha:** Rekvava St. 22, Poti; ch35.

GERMANY

Systems: DVB-T (MPEG2, MPEG4)

National Stations
ARBEITSGEMEINSCHAFT DER ÖFFENTLICH-RECHTLICHEN RUNDFUNKANSTALTEN DEUTSCHLANDS (ARD) (Pub) ☐ Arnulfstrasse 42, 80335 München ☎ +49 89 590001 🖳 +49 89 59003249 **L.P:** Chmn: Monika Piel. **W:** www.daserste.de **NB.** ARD is the head organisation for the regional public service broadcasters: **Bayerischer Rundfunk (BR):** Rundfunkplatz 1, 80335 München ☎ +49 89 59002433 🖳 +49 89 59003199. **Hessischer Rundfunk (HR):** Bertramstrasse 8, 60320 Frankfurt ☎ +49 69 1551 🖳 +49 69 1552900. **Mitteldeutscher Rundfunk (MDR):** Kantstrasse 71-73, 04275 Leipzig ☎ +49 341 22760 🖳 +49 341 5663544. **Norddeutscher Rundfunk (NDR):** Rothenbaumchaussee 132, 20149 Hamburg ☎ +49 40 4131 🖳 +49 40 447602. **Rundfunk Berlin-Brandenburg (RRB)** Masurenallee 8-14, 14057 Berlin ☎ +49 30 9799330141 🖳 +49 30 9799330149. **Radio Bremen Fernsehen (RB),** Diepenau 10, 28195 Bremen ☎ +49 421 2460 🖳 +49 421 2462010. **Saarländischer Rundfunk (SR)** Funkhaus Halberg, 66100 Saarbrücken ☎ +49 681 6020 🖳 +49 681 6023874. **Südwestrundfunk (SWR),** Neckarstrasse 230, 70190 Stuttgart ☎ +49 711 92910001 🖳 +49 711 9291010. **Westdeutscher Rundfunk (WDR),** Appellhoffplatz 1, 50667 Köln ☎ +49 221 2202100 🖳 +49 221 22085724. **Chs:** Das Erste, regional stns. – **ZWEITES DEUTSCHES FERNSEHEN (ZDF) (Pub)** ☐ Postfach 4040, 55030 Mainz ☎ +49 6131 701 🖳 +49 6131 702157 **E:** info@zdf.de **W:** www.zdf.de **L.P:** CEO: Markus Schächter; TD: Albert Ziemer. **Chs:** ZDF, ZDFdoku, ZDFinfo – **PRO7SAT1 MEDIEN AG (Comm)** ☐ Medienallee 7, 85774 Unterföhring ☎ +49 89 950710 🖳 +49 89 950711227 **E:** stefanie.prinz@pro7sat1.de **W:** www.pro7sat1.de **L.P:** CEO: Thomas Ebeling **Chs:** Pro7, Sat1, Kabel 1, N24 – **RTL TELEVISION GMBH (Comm)** ☐ Aachener Str. 1044, 50858 Köln ☎ +49 221 4560 🖳 +49 221 4561690 **E:** ukomm@rtl.de **W:** www.rtl.de; www.rtl-television.de **L.P:** CEO: Anke Schäferkordt **Chs:** RTL Television, RTL2, SuperRTL, Vox, n-tv.

Local Stations not shown.

DTT Transmitters (MPEG2 unless indicated otherwise)
Operator Mux 1: ARD **Mux:** Das Erste, other public channels. – **Operator Mux 2:** ARD Regional **Mux:** public regional stns. – **Operator Mux 3:** ZDF **Mux:** ZDF, 3Sat, KiKa/ZDFneo, ZDF Infokanal. – **Operator Mux 4:** ProSiebenSat.1 **Mux:** Sat1 (incl. reg. prgrs), Pro7, Kabel1, N24. – **Operator Mux 5:** RTL Group **Mux:** RTL (incl. reg. prgrs), RTL2, SuperRTL, Vox.
NB: The content of the muxes is licensed individually by each federal state and varies accordingly.

RE	Location	M1	M2	M3	M4	M5	kW
BB	Cottbus (Calau)	53	57	36	-	-	100
BB	Frankfurt/O.	53V	57V	33V	-	-	50
BE	Berlin (Mitte)	27	7	33	44	25	120/10/50/120/20
BE	Berlin (Charlottenburg)	27	7	-	-	-	10
BE	Berlin (Wannsee)	27	7V	33	44	25	50/5/3x50
BW	Aalen	59	50	23	-	-	50
BW	Baden-Baden	60	49	33	-	-	50
BW	Bad Mergentheim	26	50	23	-	-	5/10/5
BW	Brandenkopf	52	39	33	-	-	50
BW	Donaueschingen	54	41	22	-	-	50
BW	Freiburg	52	39	33	-	-	50
BW	Heidelberg	60	49	21	-	-	50
BW	Hochrhein	52	39	33	-	-	50
BW	Pforzheim	60	49	33	-	-	50
BW	Ravensburg	43	40	22	-	-	50
BW	Raichberg	43	40	22	-	-	50
BW	Stuttgart	43	40	22	-	-	50
BW	Ulm	43	40	22	-	-	50
BW	Waldenburg	26	50	23	-	-	50
BY	Augsburg	36	25	44	-	-	100
BY	Bamberg	29	40	34	-	-	50

RE	Location	M1	M2	M3	M4	M5	kW
BY	Brotjacklriegel	40V	27V	33V	-	-	25/100/50
BY	Büttelberg	55	47	-	-	-	50
BY	Dillberg	55V	47V	34V	-	-	25/2x50
BY	Gelbelsee	36	25	44	-	-	50
BY	Grünten	45	46	28	-	-	50
BY	Hirschau	29	28	23	-	-	50
BY	Hohe Linie	42V	28V	53V	-	-	25/2x50
BY	Hoher Bogen	42V	28V	33V	-	-	25/100/50
BY	Hohenpeißenberg	47	53	28	-	-	2x100/50
BY	Landshut	40	27	33	-	-	20
BY	München	54V	56V	35V	48V	34V	20/5x100
BY	Nürnberg	55V	47V	34V	52V	66V	25/2x50/2x20
BY	Ochsenkopf	29	40	23	-	-	50
BY	Pfaffenberg	36	46	25	-	-	2x100/50
BY	Pfaffenhofen	36	25	44	-	-	50
BY	Pfarrkirchen	40	27	33	-	-	50
BY	Pfänder[o]	45	-	-	-	-	10
BY	Rhön	36	46	25	-	-	2x100/50
BY	Wassertrüdingen	55	47	44	-	-	2x100/50
BY	Wendelstein	54V	56V	35V	48V	34V	25/4x100
BY	Würzburg	36V	46	25	-	-	25/2x50
HB	Bremerhaven	22	29	32	49	42	3/10/3x5
HB	Bremen (Walle)	22	29	32	49	42	40/50/32/2x50
HE	Angelburg	32	24	52	-	-	50
HE	Frankfurt (Ginnheim)	57V	8V	22V	54V	34V	50/20/3x50
HE	Großer Feldberg	57V	8V	22V	54V	34V	50/10/3x50
HE	Habichtswald	32	55	42	-	-	50
HE	Heidelstein	43	39	25	-	-	50
HE	Hohe Wurzel	57V	8V	22V	54V	34V	100/20/2x100
HE	Hoher Meißner	32	55	42	-	-	50
HE	Rimberg	32	39	22	-	-	50
HE	Würzberg	31	53	21	-	-	50
HH	Hamburg	33	54	23	30	40	3x50/2x100
HH	Hamburg (Rahlstedt)	33	54	23	30	40	20/25/3x20
HH	Hamburg (Moorfleet)	-	54V	-	-	-	25
MV	Garz	29	40	-	-	-	20
MV	Helpterberg	22	23	-	-	-	20
MV	Marlow	29V	46V	-	-	-	20
MV	Schwerin	26	53	-	-	-	50
NI	Aurich	48	43	35	-	-	2x50/20
NI	Braunschweig	47V	36V	23V	44V	24V	10/8/5/5/5
NI	Braunschweig (Broitzem)	47V	36V	23V	44V	24V	2x10/5/5/5
NI	Cuxhaven	26	29	31	-	-	2x10/5
NI	Dannenberg	43	58	-	-	-	10/2
NI	Göttingen	21	42	-	-	-	50
NI	Göttingen (Hetjersh.)	59V	21V	-	-	-	40/25
NI	Hannover (Buchholz)	47	36	23	44	24	10/4x20
NI	Hildesheim	47V	36V	23V	-	-	10/8/5
NI	Lingen	41	37	59	-	-	20
NI	Osnabrück	41	37	59	-	-	50
NI	Rosengarten	33	56	23	-	-	5/20/5
NI	Stadthagen	47	36	-	-	-	10
NI	Steinkimmen	55	29	32	49	42	100/4x5
NI	Torfhaus	59V	46V	-	-	-	32
NI	Uelzen	43	58	27	-	-	20/2x50
NI	Visselhövede	43V	58V	27V	-	-	20
NW	Aachen	50V	37V	26V	-	-	2x10/5
NW	Aachen (Stolberg)	50V	37V	26V	-	-	2x50/20
NW	Bielefeld	26	31	33	-	-	20
NW	Bonn	65V	49V	26V	53V	29V	3x50/2x20
NW	Dortmund	48V	25V	35V	55V	29V	50
NW	Düsseldorf	48V	46V	35V	55V	29V	50
NW	Düsseldorf	46V	-	-	-	-	50
NW	Essen	48V	57V	35V	55V	29V	50
NW	Hochsauerland	60	27	30	-	-	50
NW	Köln	65V	49V	26V	53V	29V	3x50/2x20
NW	Langenberg	48V	46V	35V	55V	29V	50
NW	Langenberg	46V	25V	-	-	-	50
NW	Münster	21V	45V	59V	-	-	50
NW	Nordhelle	60V	27V	30V	-	-	100/2x50
NW	Siegen	60	27	30	-	-	20
NW	Teutoburger Wald	26V	31	33	-	-	50
NW	Wesel	48V	46V	35V	29V	-	50
NW	Wesel	46V	-	-	-	-	50
NW	Wuppertal	48	22	35	-	-	20
RP	Ahrweiler	56	33	24	-	-	10
RP	Bad Marienberg	56	33	28	-	-	20
RP	Donnersberg	57	44	30	-	-	50
RP	Eifel	46	48	30	-	-	50
RP	Haardtkopf	46	48	30	-	-	50
RP	Kaiserslautern	57	44	30	-	-	50

RE	Location	M1	M2	M3	M4	M5	kW
RP	Kettrichshof	60	44	30	-	-	50
RP	Koblenz	56	33	28	-	-	50
RP	Saarburg	46	48	30	-	-	50
SH	Bredstedt	26	24	31	-	-	25
SH	Bungsberg	47V	39V	21V	-	-	50
SH	Flensburg	47V	39V	21V	-	-	50
SH	Heide	32	24	31	-	-	32/2x20
SH	Kiel	47	39	21	35	45	2x50/20
SH	Lübeck (Berkenthin)	33	28	23	30	40	20
SH	Lübeck (Stockelsdorf)	33V	28V	23V	30V	40V	20
SH	Mölln	-	28	-	-	-	10
SH	Neumünster	-	28	-	-	-	50
SH	Wedel	-	28	-	-	-	25
SL	Göttelborner Höhe	42V	44V	30V	-	-	50
SL	Schocksberg	42V	-	30V	-	-	20
SL	Spiesen	42V	-	-	-	-	25
ST	Brocken	29V	34V	30V	-	-	50
ST	Dequede	41V	34V	31V	-	-	50
ST	Magdeburg	29V	34V	30V	-	-	2x100/2x50
ST	Halle	24V	35V	22V	-	-	50
ST	Wittenberg	24V	38V	30V	-	-	50
SN	Chemnitz (Geyer)	25V	32V	22V	-	-	50
SN	Chemnitz (Reichenhain)	25V	32V	22V	-	-	2x20/5
SN	Dresden (Wachwitz)	39V	29V	36V	-	-	100
SN	Leipzig	24V	37V	22V	-	-	100
SN	Löbau	39V	27V	36V	-	-	50
SN	Schöneck	25V	32V	22V	-	-	50
TH	Erfurt	21V	27V	50V	-	-	50
TH	Gera	25V	27V	50V	-	-	50
TH	Inselsberg	53V	48V	50V	-	-	50
TH	Jena	21V	27V	50V	-	-	2x10/5
TH	Sonneberg	21V	27V	50V	-	-	2x10/20
TH	Weimar	21V	27V	50V	-	-	25

+ sites with txs below 10kW. °) tx located in Austria

RE) Region codes (federal states): BB=Brandenburg, BE=Berlin, BW=Baden-Württemberg, BY=Bayern, HB=Bremen, HE=Hessen, HH=Hamburg, MV=Mecklenburg-Vorpommern, NI=Niedersachsen, NW=Nordrhein-Westfalen, RP=Rheinland-Pfalz, SH=Schleswig-Holstein, SL=Saarland, ST=Sachsen-Anhalt, SN=Sachsen, TH=Thüringen.

Regional & Local Operators not shown.

GHANA

System: # PAL-B/G [E]; DVB-T (MPEG4)

GHANA BROADCASTING CORP. (Gov) ☐ P.O. Box 1633, Accra ☎ +233 21 221161 🖷 +233 21 773240 **E:** gtv@ncs.com.gh **W:** www.gbcghana.com **LP:** DG: Yaw Owusu-Addo. **Txs:** (-) – **E.TV GHANA (Comm)** ☐ P.O.Box CT 5976, Accra ☎ +233 21 912071 **E:** info@etvghana.com **W:** www.etvghana.com **Txs:** (-) – **METRO TV (Semi-Gov, Comm)** 59 Josiah Tongogara Street, Labone, Accra ☎ +233 21 765701 🖷 +233-21-765703 **E:** admin@metroworld.tv **W:** www.metroworld.tv **Txs:** (-). – **TV3 (Comm)** ☐ Box M83, Accra ☎ +233 21 763458 🖷 +233 21 763450 **E:** info@tv3.com.gh **W:** www.tv3.com.gh **LP:** CEO: Syed Ahmmad Zaidi. **Txs:** (-). – **TV AFRICA (Comm)** ☐ P.O.Box 7151, Accra-North. 🖷 +233 21 224323 🖷 +233 21 223320 **Txs:** (-). – **VIASAT 1 (Comm)** ☐ House 25/8 Abafun Crescent, North Labone, Accra ☎ +233 21 760516 **Tx:** ch30.

DTT Transmitters
Operator: Next Generation Broadcasting (Smart TV) ☐ 3 Asafuatse Road Afua Street, Airport Residential Area, Airport West, Accra ☎ +233 21 766636 **E:** contact@ngbroadcasting.com **W:** www.ngbroadcasting.com **Mux** (☼ **exc.***): TV3*, Viasat 1*, GTV*, TV Africa*, NET2,* BBC World News*, God TV*, FOX Entertainment, Showtime, Hi Nolly, Homebase, Setanta Africa, BBC World News, GOD TV, Kiss, KidsCo **Txs:** (-) – **Operator:** Skyy Digital ☐ Skyy Digital House, 1920 West Fijai, Takoradi ☎ +233 31 25288 **E:** info@myskyyonline.com **W:** www.myskyyonline.com **Mux** (☼): Skyy One, Music World, Channel D, Sports24, Cinimax, Heaven, Planet Kidz, Fiesta, Skyy World, e.TV Ghana ✻ Skyy Power FM, Citi FM **Txs:** (-).

GIBRALTAR (UK)

System: PAL-B/G [E]

GBC TELEVISION ☐ Broadcasting House, 18 So. Barrack Rd,

Gibraltar ☎ +350 20079760 🖷 +350 20078673 **E:** info@gbc.gi **W:** www.gbc.gi **LP:** Chmn: Charles Menes **Txs:** ch6 (0.4kW H) & lp repeaters on ch12, ch53, ch56.

GREECE

Systems: # PAL-B/G [E] ⇩2015; DVB-T (MPEG2, MPEG4)

National Stations
ELLINIKI TILEORASSI (ET) (Pub) ☐ 432 Messogion Ave., 15342 Agia Paraskevi ☎ +30 2106075704 🖷 +30 2106075714 **E:** nkarra@ert.gr **W:** www.ert.gr **LP:** Chmn: Christos Panagopoulos. **Chs:** ET1, NET, ET3. **ET3** ☐ Aggelaki 16, 54621 Thessaloniki ☎ +30 2310299400 🖷 +30 2310299750 **E:** pr@ert3.gr **W:** www.ert3.gr **LP:** DG: Giorgos Frastanlis.

Location	ET1	NET	ET3	kW
Agrinio (Akarnanika)	46	43	23	1.6/1000/-
Alexandroupoli (Plaka)	33	30	27	15/170/-
Athens (Ymittos)	21	5	31	15/30/900
Athens (Ymittos)	-	44	-	-
Athens (Parnitha)	11	34	12	30/450/-
Athens (Parnitha)	56	-	52	-
Chania (Skloka)	7	27	47	10/250/-
Chania (Skloka)	65	-	-	-
Evros (Pythio)	30	35	42	18/-/-
Ioannina (Liggiades)	10	25	21	10/250/-
Irakleio (Achentrias)	10	44	31	10/-/-
Kalamata (Likodimo)	6	32	30	10/-/-
Kastoria (Vitsi)	7	49	39	250/-/-
Kavala (Paggaio)	7	59	64	30/650/-
Kavala (Thasos I.)	39	23	26	550/1000/-
Kefallonia I. (Ainos)	8V	57	29	30/430/-
Kefallonia I. (Ainos)	60	-	-	-
Kerkira Isl. (Pantokrator)	9V	50	53	10/530/-
Korinthos (Gerania)	9	5	63	10/1000/-
Korinthos (Gerania)	-	51	-	-
Lesvos I. (Olympos)	9	48	45	10/200/-
Patra (Panachaiko)	38	22	35	-
Rodos I. (Prof. Ilias)	9V	42	39	10/500/-
Thessaloniki (Chortiatis)	5	30	23	30/1000/27
Thessaloniki (Chortiatis)	37	-	-	-
Thira I. (Profitis Ilias)	8V	29	-	30/500
Tripoli (Tsemperou)	10	48	56	30/500/-
Volos (Pilio)	6	41	44	30/750/-
Volos (Pilio)	62	-	-	-

+ sites with txs below 10kW.
VOULI TV (Pub) ☐ Palaia Anaktora, Syntagma Sq., 10021 Athens ☎ +30 2103733820 **E:** kanali@parliament.gr **W:** www.parliament.gr/video-audio **LP:** Dir: Konstantin Alavanos **Txs:** Athens (Ymittos) ch50 & netw. – **ALPHA TV (Comm)** ☐ 40,2km. Attikis Odou, SEA Mesogion, Ktirio 6, 19002 Paiania ☎ +30 2122124000 🖷 +30 2122124356 **E:** pr@alphatv.gr **W:** www.alphatv.gr **Txs:** Athens (Ymittos) ch24 & netw. – **ALTER CHANNEL (Comm)** ☐ Ag Paraskevi 36-38, 12132 Peristeri ☎ +30 2105707000 🖷 +30 2105707078 **E:** info@alter.gr **Txs:** Athens (Ymittos) ch27 (5kW) & netw. – **ANTENNA TV (ANT1) (Comm)** ☐10-12 Leof. Kifisias 10-12, 15125 Marousi ☎ +30 2106886100 🖷: +30 2106890304 **E:** pr@antenna.gr **W:** tv.antenna.gr **Txs:** Athens (Ymittos) ch38 & netw. – **m. (MAKEDONIA TV) (Comm)** ☐ 26hs Oktovriou 90, 54627 Thessaloniki ☎ +30 2310504300 🖷 +30 2310504344 **E:** info@maketv.gr **W:** www.maketv.gr **Txs:** Athens (Ymittos) ch26 & netw. – **MAD TV (Comm)** ☐ Eth. Antistaseos 253 & E. Kotsopoulou, 15331 Pallini ☎ +30 2106665669 🖷 +30 2106665812 **E:** info@mad.tv **W:** www.mad.tv **Txs:** Athens (Mt.Ymittos) ch68 & netw. – **MEGA CHANNEL (Comm)** ☐ Roussou 4/Leof. Mesogion, 11526 Ambelokipoi ☎ +30 2106903000 🖷 +30 2106983600 **E:** publ_rel@megatv.com **W:** www.megatv.com **Txs:** Athens (Ymittos) ch25 & network. – **MTV GREECE (Comm)** ☐ Leof. Kifisias 304, 15232 Halandri ☎ +30 2106835056 **W:** www.mtv.gr **Txs:** Athens (Ymittos) ch59 & netw. – **NOVA CINEMA / NOVA SPORTS (Comm)** ☼ ☐ Proektasi Odou Manis, Kantza 15351 Pallini ☎ +30 2106602000 🖷 +30 2106722961 **W:** www.novacinema.gr; www.novasports.gr **Txs:** Nova Cinema: Athens (Ymittos) ch53 & netw.; Nova Sports: Athens (Ymittos) ch58 & netw. – **SKAI TV (Comm)** ☐ Eth. Makariou/Falireos 2, 18547 Neo Faliro ☎ +30 2104800170 🖷 +30 2104800120 **E:** technicaltv@skai.gr **W:** www.skai.gr **LP:** CEO: John Alafouzos. **Txs:** Athens (Ymittos) ch49 (10kW) & netw. – **STAR CHANNEL (Comm)** ☐ Viltanioti 36, 14564 Kato Kifisia ☎ +30 2111891000 🖷 +30 2111892000 **E:** info@star.gr **W:** www.star.gr **Txs:** Athens (Ymittos) ch29 & netw. – **TILEASTY (Comm)** ☐ Praxitelous 58, 17674 Kalithea ☎ +30 2109407000 🖷 +30 2109407024 **E:** info@radioasty.gr **W:** www.radioasty.gr **Txs:** Athens

(Ymittos) ch32 & netw. – **902 TV (Comm)** ⌧ Leof. Irakliou 145, 14231 Nea Ionia ☎ +30 2102592902 ▤ +30 2102592532 **E:** mailbox@902.gr **W:** www.902.gr **Txs:** Athens (Ymittos) ch61 & netw.

Regional & Local Stations (ca 140) not shown.

DTT Transmitters (under construction) (MPEG4 exc. *=MPEG2)
Operator Mux 1+2: ERT Digital **Mux 1*:** Cine+, Prisma+, Sport+, RIK SAT **Mux 2*:** ET1, NET, ET3, Vouli TV – **Operator Mux 3-6:** Digea ⌧ Artemidos 3, 151 24 Athens ☎ +30 2106838700 ▤ +30 2106823205 **E:** info@digea.gr **W:** www.digea.gr **Mux 3:** Alpha TV, Alter Channel, ANT1 **Mux 4:** Mega Channel, Skai TV, Star Channel, Makedonia TV, 902 TV **Mux 5:** Attica TV, Nickelodeon, Extra Channel 3, High TV **Mux 6:** Mad TV, MTV Greece, Sport TV, TV 0-6. – **Operator Mux 7+8:** Digital Union ⌧ Ginosati 88, 1st Fl., 144 52 Metamorfosi Athens ☎ +40 2102850880 **E:** info@digital-union.eu **W:** www.digital-union.eu **Mux 7*:** Blue Sky TV **Mux 8*:** Kanali 9, Time Channel, TV Thessaloniki, XTV.

Location	M1	M2	M3	M4	M5	M6	M7	M8
Agrino (Akarnanika)	28	-	-	-	-	-	-	-
Alexandroupoli (Plaka)	41	-	-	-	-	-	-	-
Athens (Aigina)	48	-	46	47	54	63	65	-
Athens (Panitha)	48	-	-	-	-	-	-	-
Athens (Ymittos)	48	-	46	47	-	-	-	-
Chania (Malaxa)	32	-	-	-	-	-	-	-
Drama (Korivolos)	37	-	-	-	-	-	-	-
Ioannina (Liggiades)	49	-	-	-	-	-	-	-
Irakleio (Rogdia)	42	-	-	-	-	-	-	-
Kalamata (Lykodimo)	43	-	-	-	-	-	-	-
Kastoria (Vitsi)	23	-	-	-	-	-	-	-
Kavala (Thasos)	47	-	-	-	-	-	-	-
Korinthos (Osios Potapios)	25	-	-	-	-	-	-	-
Korinthos (Panagia Koryfis)	-	37	43	-	-	-	-	-
Livadia (Chlomos)	22	-	-	-	-	-	-	-
Patra (Panachaiko)	42	-	-	-	-	-	-	-
Mytilini (Profitis Ilias)	31	-	-	-	-	-	-	-
Thessaloniki (Chortiati)	56	26	29	25	-	-	-	-
Thessaloniki (Filippio)	-	-	29	25	-	-	-	59
Tripoli (Doliana)	21	-	-	-	-	-	-	-
Volos (Pilio)	53	-	-	-	-	-	-	-

GREENLAND (Denmark)

Systems: DVB-T (MPEG2); † PAL-B/G

National Station
KNR-TV (Pub) ⌧ P.O.Box 1007, 3900 Nuuk ☎ +299 361500 ▤ +299 325042 **W:** www.knr.gl. **E:** knr@knr.gl **LP:** Chmn (KNR): Peter Jensen.

Local Stations not shown.

DTT Transmitters
Operator: TELE Greenland **Mux:** multiprgr **Tx:** nationwide network under construction. – **Operator:** Nuuk TV **Mux 1-5(☉)** multiprgr **Txs:** (all Nuuk) ch23, ch26, ch29, ch32, ch35. – **Operator:** Sisimiut TV **Mux(☉):** multiprgr **Txs:** (-) – **Operator:** Arctic TV ⌧ P.O.Box 420, 3952 Ilulissat **Mux(☉):** multiprgr **Txs:** (-)

GRENADA

System: NTSC-M [A]

GBN-TV (Gov) ⌧ P.O.Box 535, St. George's ☎ +1 473 4445521 ▤ +1 473 4445054 **E:** gbn@caribsurf.com **W:** www.klassicgrenada.com **LP:** GM: Richard Purcell Sen.Eng: Kennedy Bowen **Txs:** North&East ch7 (4kW), Saint George's ch11 (5kW).

GUADELOUPE (France)

System: # SECAM-K1 [K/E]; DVB-T

RFO GUADELOUPE (Pub) ⌧ BP 180, 97122 Baie-Mahault ☎ +590 590939696 ▤ +590 590939682 **W:** guadeloupe.rfo.fr **LP:** Dir: R.Surjus. **Txs:** Basse-Terre ch5 (2kW) & lp repeaters. Also on St Martin ch7 (0.1kW). – **ARCHIPEL 4 (Comm)** ⌧ Résidence Les Palmiers, Gabarre 2, 97110 Pointe-a-Pitre ☎ +590 590836350 **Txs:** Morne a Louis ch53 (1.3kW). – **CANAL ANTILLES (Comm)** ☉ ⌧ 2 lot. Les Jardins de Houelbourg, 97122 Baie Mahault ☎ +590 590268179 **Txs:** Basse-Terre ch42 (60kW), Morne a Louis ch58 (1.3kW) & lp repeaters under 1kW. – **CANAL 10 (Comm)** ⌧ Bd Marquisart De Houelbourg, BP 416, 97122 Baie Mahault. **W:** www.canal10.com. **Txs:** (-). – **CARRIB'IN TV (Comm)** ⌧ BP 658 Gustavia Cedex, 97099 Saint-Barthélemy ☎ +590 590874362 ▤ +590 590510787 **Txs:** (-). – **ÉCLAIR TV (Comm)** ⌧ Basse-Terre Télévision, Pintade, 97100 Basse-Terre ☎ +590 590811064 ▤ +590 590992510. **Txs:** (-). – **L'A1 GUADELOUPE (Comm)** ⌧ 20 rue Henri Becquerel, 97122 Baie Mahault ☎ +590 590380606 ▤ +590 590380607. **Txs:** (-).

DTT Transmitters
Operator: France Télévisions **Mux:** RFO Guadeloupe, France 2-5, France Ô, Arte, LCP, France 24, Gulli **Txs:** ch(-).

GUAM (USA)

Systems: ATSC

Local Stations*
KGTF (Pub): P.O.Box 21449, GMF, Agana, GU 96921. °PBS. Tx: Agana ch5 (8.26kW). **KTGM (Comm):** 692 N Marine Dr, Tamuning, GU 96913-4454. °ABC. Tx: Tamuning ch17 (2kW). **KUAM-TV (Comm):** 600 Harmon Loop Rd, #102, Dededo, GU 96912-2536. °NBC. Tx: Agana ch2 (0.035kW).
*) Full power licenses (lp licenses not listed). °) Network affiliation.

GUATEMALA

System: NTSC-M [A]; ATSC

CORPORACIÓN ESTATAL DE RADIO Y TELEVISIÓN (CERTV) (Gov) ⌧ 30, Av. 3-40, Zona 11, 01011 Guatemala ☎ +502 25945320 **Chs:** Canal 19, El Súper Canal, Televisiete **W:** www.canal3.com.gt; www.canal7.com.gt **LP:** DG: Hector Olivo. **Txs:** Canal 19: ch19; El Súper Canal: ch3 (240kW); Televisiete: ch7 (180kW). – **CANAL 63 (Rlg)** ⌧ Guatemala. **Tx:** ch63. – **CANAL 65 (Rlg)** ⌧ Guatemala. **Tx:** ch65. Rel. ETWN (USA). – **EL CANAL DE LA ESPERANZA (Rlg)** ⌧ Carretera Vieja a Antigua 2 Calle 23-70, Zona 1 de Mixco, Guatemala ☎ +502 24213434 **E:** canal27@motivacioncristiana.org **W:** www.motivacioncristiana.org/canal27.htm **Tx:** ch27. – **ENLACE (Rlg)** ⌧ Guatemala. **Tx:** ch21. Rel. TBN (USA). – **LATITUD TV (Comm)** ⌧ 12 Avenida 1-96, Zona 2 de Mixco, Colonia Alvarado, Guatemala ☎ +502 24111140 ▤ +502 24111200 **E:** info@latitud.tv **W:** www.latitud.tv **Tx:** ch31. – **TELEONCE (Comm), TRECEVISIÓN (Comm)** ⌧ 20, Calle 5-02, Zona 10, 01010 Guatemala ☎ +502 23682532 ▤ +502 23682221 **E:** jcof@canalonce.tv, jcof@canaltrece.tv **Txs:** TeleOnce: ch11 (316kW); Trecevisión: ch13 (25kW). – **TV USAC (Educ)** ⌧ Guatemala **W:** www.usac.edu.gt **Tx:** ch33. – **TVQ (LA IMAGEN DEL PODER) (Comm)** ⌧ Guatemala.**Tx:** ch58.

GUINEA

System: PAL-K1 [E]

TÉLÉVISION NATIONALE DE GUINÉE (Gov) ⌧ BP 391, Conakry. ☎ +224 30452786 ▤ +224 30451408 **LP:** DG: Aissatou Bella Diallo **Txs:** Kindia ch4 (0.2kW), Conakry ch5 (1kW), Faranah ch5 (0.5kW), Labé ch7 (8kW), Mamou/Mali ch9 (0.2kW), Kankan ch9 (1kW).

GUINEA-BISSAU

System: PAL-B/G [E]

TELEVISÃO NACIONAL DA GUINÉ-BISSAU (Gov) ⌧ CP 178, Bissau ☎ +245 3221924 **LP:** DG: Francelino da Cunha. **Txs:** Nhacra ch7 (200kW) & relay txs.

Foreign TV Relay
RTP África (Portugal): (-).

GUYANA

System: NTSC-M [A]

National Station
NATIONAL COMMUNICATIONS NETWORK (NCN) (Gov) ⌧ Homestretch Ave, D'Urban Park, Georgetown ☎ +592 2271566 ▤ +592 2262253 **E:** gmgtv@sdnp.org.gy **W:** www.ncnguyana.com **LP:** CEO: Mohammed Sattaur **Tx:** ch10.

Local Stations not shown.

HAITI

System: NTSC-M [A]

National Station
TÉLÉVISION NATIONALE D'HAITI (TNH) (Pub) ☒ BP 13400, Delmas 33, Port-au-Prince ☎ +509 2460200 🗎 +509 2463889 **E:** info@tnh.ht **W:** www.tnh.ht **LP:** DG: Pradel Henriquez **Txs:** Port-au-Prince ch8 (0.3kW), ch10 (5kW), Cap. Haïtien ch12.

Local Stations not shown.

HAWAII (USA)

Systems: ATSC

Local Stations*
KAAH-TV (Rlg): 1152 Smith St, Honolulu, HI 96817-5101.°TBN. Tx: ch27 (262kW). **KAII-TV (Comm):** satellite of KHON-TV. Tx: Wailuku 36 (50kW). Mux: Fox, CW. **KALO (Rlg):** 875 Waimanu St, Ste 110, Honolulu, HI 96813-5271. °ETV. Tx: ch10 (21kW). **KBFD (Comm):** 1188 Bishop St PH-1, Honolulu, HI 96813-3300. Tx: ch33 (108kW). **KFVE (Comm):** 150-B Puuhale Rd, Honolulu, HI 96819-2233. Tx: ch23 (50kW). °MNT. **KGMB (Comm):** 1534 Kapiolani Blvd, Honolulu, HI 96814-3715. Tx: ch22 (1000kW). °CBS. **KGMD-TV (Comm):** satellite of KGMB. Tx: Hilo ch8 (3.2kW). **KGMV (Comm):** satellite of KGMB. Tx: Wailuku ch24 (72.4kW). **KHAW-TV (Comm):** satellite of KHON-TV. Tx: Hilo ch21 (50kW). Mux: Fox, CW. **KHBC-TV (Comm):** satellite of KHNL. Tx: Hilo ch22 (8kW). **KHET (Pub):** 2350 Dole St, Honolulu, HI 96822-2410. Tx: ch18 (9.5kW). °PBS. **KHNL (Comm):** 150-B Puuhale Rd, Honolulu, HI 96819-2233. Tx: ch35 (5.9kW). °NBC. **KHON-TV (Comm):** 88 Piikoi St, Honolulu, HI 96814-4245. °Fox, CW. Tx: ch8 (7.2kW). Mux: Fox, CW. **KHVO (Comm):** satellite of KITV. Tx: Hilo ch18 (50kW). **KIKU (Comm):** 737 Bishop St Ste 1430, Honolulu, HI 96813-3204. Tx: ch19 (60.7kW). **KITV (Comm):** 801 S King St, Honolulu, HI 96813-3013. Tx: ch40 (85kW). °ABC. **KKAI (Rlg):** 875 Waimanu St, Ste 110, Honolulu, HI 96813-5271. °Faith TV. Tx: Kailua ch15 (19kW). **KLEI (Comm):** satellite of KPXO. Tx: Kailua-Kona ch25 (700kW). **KMAU (Comm):** satellite of KITV. Tx: Wailuku ch29 (51.2kW). **KMEB (Pub):** satellite of KHET. Tx: Wailuku ch30 (50kW). **KOGG (Comm):** satellite of KHNL. Tx: Wailuku ch16 (50kW). **KPXO (Comm):** 875 Waimanu St Ste 630, Honolulu, HI 96813-5267. Tx: Kane'ohe ch41 (34kW). °ION. **KUPU (Rlg):** 1188 Bishop St Ste 502, Honolulu, HI 96813-3302. Tx: Waimanalo ch23 (19kW). **KWBN (Rlg):** 3901 S Hwy 121 S, Bedford, TX 76021-2066. °ETV. Tx: ch43 (6.46kW). **KWHE (Rlg):** 1188 Bishop St Ste 502, Honolulu, HI 96813-3302. °LeSEA. Tx: ch31 (20.1kW). **KWHH (Rlg):** satellite of KWHE. Tx: Hilo ch23 (14.9kW). **KWHM (Rlg):** rel. KWHE. Tx: Wailuku 45 (87kW).
*) Full power license (lp licenses not listed); °) Network affilation
NB: Txsites are Honolulu unless mentioned otherwise.

HONDURAS

System: NTSC-M [A]

HRJS-TV (Comm) ☒ Apt. Postal 120, San Pedro Sula. **Txs:** ch2, ch9 & relays. – **HRNQ-TV (Comm)** ☒ Casilla 3424, Tegucigalpa. **Tx:** ch13. – **HRGJ-TV (Comm)** ☒ Apt. Postal 882, Barrio Rio Piedras ☎ +504 5505009 🗎 +504 5531810 **W:** www.noti6.com Tx: ch6. – **TELESISTEMA HONDUREÑO (Comm)** ☒ Boulevard Suyapa, Tegucigalpa ☎ +504 2327835 🗎 +504 2320097 **W:** www.televicentro.hn **Chs:** Canal Cinco, Canal 3, Telecadenas Canales 7 y 4. **Txs:** ch3 (Canal 3), ch5 (Canal Cinco), ch4, ch7.

Local Stations not shown.

HONG KONG (China, SAR)

Systems: # PAL-I [E] ⬇2012; DTMB

ASIA TELEVISION LTD. (ATV) (Comm) ☒ 81 Broadcast Drive, Kowloon, Hong Kong ☎ +852 29928888 🗎 +852 23380438 **E:** atv@hkatv.com **W:** www.hkatv.com **LP:** CEO: Mr. Mark Lee **Chs:** ATV Home (Chinese), ATV World (English) **Txs:** Temple Hill ch23 (10kW) (ATV Home) & ch27 (10kW) (ATV World), and lp txs. – **TELEVISION BROADCASTS LTD. (TVB) (Gov)** ☒ TV City, 77 Chun Choi Street, Tseung Kwan O Industrial Estate, Kowloon Hong Kong ☎ +852 23359123 🗎 +852 23581300 **E:** tvbpr@tvb.com.hk **W:** www.tvb.com **LP:** MD: Mr. Louis Page. **Chs:** TVB Jade, TVB Pearl **Txs:** Temple Hill ch21 (10kW) (TVB Jade) & ch25 (10kW) (TVB Pearl), and lp txs.

– **RADIO TELEVISION HONG KONG (RTHK) (Pub)** ☒ 1A Broadcast Drive, Kowloon, Hong Kong. Uses airtime on ATV and TVB.

DTT Transmitters (under construction)
Operator: ATV **Txs:** (-). – **Operator:** TVB **Txs:** (-).

HUNGARY

Systems: DVB-T (MPEG2, MPEG4); † PAL-D/K [R] ⬇31 Dec 2011

National Stations
MAGYAR TELEVÍZIÓ (MTV) (Pub) ☒ Szabadság tér 17, 1810 Budapest 5 ☎ +36 1 3734303 🗎 +36 1 3734133 **E:** info@mtv.hu **W:** www.mtv.hu **LP:** Pres: Rudi Zoltan **Chs:** m1, m2, reg. prgrs: a) Budapesti stúdió, b) Debreceni stúdió, c) Miskolci stúdió, d) Pécsi stúdió, e) Soproni stúdió, f) Szegedi stúdió – **DUNA TELEVÍZIÓ (Pub)** ☒ Mészáros u. 48-54, 1016 Budapest ☎ +36 1 4891200 🗎 +36 1 4891366 **E:** info@dunatv.hu **W:** www.dunatv.hu **LP:** DG: László Cselényi **Chs (terr.):** Duna I, Duna II. – **ATV (Comm)** ☒ Kőrösi Csoma Sándor u. 31, 1102 Budapest ☎ +36 1 8770800 **E:** info@atv.hu **W:** www.atv.hu – **HÍR TV (Comm)** ☒ Szentendrei út 89-93, 1033 Budapest ☎ +36 1 4304000 🗎 +36 1 4304099 **E:** info@hirtv.net **W:** www.hirtv.hu – **RTL KLUB (Comm)** ☒ Fehérvári út 84, 1117 Budapest ☎ +36 1 3828282 🗎 +36 1 3828283 **E:** rtlklub@rtlklub.hu **W:** www.rtlklub.hu **LP:** Dirk Gerkens. – **TV2 (Comm)** ☒ Róna u. 174, 1145 Budapest ☎ +36 1 4676400 🗎 +36 1 4676500 **E:** info@tv2.hu **W:** www.tv2.hu **LP:** CEO: Gábor Kereszty.

Local Stations (all analogue)
Balaton TV: Kele u. 32, 8600 Siófok; Balatonlelle ch41. **Cegléd Városi Televízió:** Teleki u. 12, 2700 Cegléd; ch48. **Csaba TV:** Teleki u. 5, 5600 Békéscsaba; ch47. **Csele TV:** Szegedi út 8, 7714 Mohács; ch25. **Csepp TV:** Simon Bolivár sétány 4-8, 1214 Budapest; ch11. **Csongrád TV:** József A. u. 16, 6640 Csongrád; ch38. **Debrecen Televízió:** Petőfi tér 10, 4025 Debrecen; ch50. **Duna-Híd Televízió:** Kodály Zoltán u. 2, 2400 Dunaújváros; ch33. **Egri Városi Televízió:** Törvényház út 15, 3300 Eger; ch50. **Elektro-Szignál Televízió:** Kossuth L. u. 21, 2600 Vác; Vác ch22, Fót ch48. **Fehérvár Televízió:** Jókai u. 15, 8000 Székesfehérvár; ch48. **Fönix TV:** Gyenes u. 8, 1032 Budapest; ch31. **Gyula Televízió:** Gyár u. 3, 5700 Gyula; ch22. **Hajdúszoboszló VTV:** Szilfakalja u. 7, 4200 Hajdúszoboszló; ch39. **Halas TV:** Bethlen Gábor tér 1, 6400 Kiskunhalas; ch31. **Kalocsa Városi Televízió:** Szt. István király út 2-4, 6300 Kalocsa; ch34. **Kanizsa TV:** Sugár u. 8, 8800 Nagykanizsa; ch55 **Kapos Televízió:** Kossuth L. u. 6, 7400 Kaposvár; ch42. **Kecskeméti Televízió:** Szabadság tér 2, 6000 Kecskemét; ch31. **Keszthely TV:** Pf. 93, 8360 Keszthely; Balatonmáriafürdő ch34. **Kölcsey TV:** Pf.205, 4400 Nyíregyháza; ch52. **Kölcsey Televízió Mátészalka:** Pf.205, 4400 Nyíregyháza; ch39. **KÖR TV:** Szent István út 4, 2300 Ráckeve; ch50. **Körzeti Televízió Esztergom:** Rákóczi tér 4. IV/2, 2500 Esztergom; ch57. **fix.TV:** Dónáti u. 36, 1015 Budapest; ch26. **Makó Városi TV:** Széchenyi tér 6, 6900 Makó; ch51. **Miskolc Városi Televízió:** Kis-Hunyad u. 9, 3525 Miskolc; ch55. **Nyíregyházi Városi Televízió:** Szabadság tér 9, 4400 Nyíregyháza; ch39. **Ozdi Városi Televízió:** Brassoi ut 2, 3600 Özd; ch51. **Pápa Városi Televízió:** Deák Ferenc u. 1, 8500 Pápa; ch46. **Pécs TV:** Lyceum u. 7, 7621 Pécs; ch36 & ch46. **Salgótarjáni Városi Televízió:** Fötér 5, 3100 Salgótarján; ch29. **Sirály TV:** Kele u. 32, 8600 Siófok; ch32. **Sükösd Televízió:** Deák F. u. 123, 6346 Sükösd; ch37. **Szamos Televízió:** Kossuth tér 12, 4900 Fehérgyarmat; ch38. **Szolnok Televízió:** Szigligeti u. 1, 5000 Szolnok; ch59. **Szombathelyi Televízió:** Géfin Gyula u. 22, 9700 Szombathely; ch38. **Telepaks TV:** Dózsa Gy. u. 51-53, 7030 Paks; ch31. **Tapolca Városi Televízió:** Hösök tere 15, 8300 Tapolca; ch43. **Városi Televízió Gyöngyös:** Eszperantó út 6/a, 3200 Gyöngyös; ch27. **Városi Televízió Százhalombatta:** Május 1. tér 3, 2440 Százhalombatta; ch48. **Várpalotai Városi Televízió:** Erdödi Pálffy Tamás u. 19, 8100 Várpalota; ch43. **Vásárhelyi Televízió:** Ady Endre u. 14, 6800 Hódmezövásárhely; ch34. **Veszprém Televízió:** József Attila út 38, 8200 Veszprém; ch42. **VTV Hajdúböszörmény:** Bocskai tér 4, 4221 Hajdúböszörmény; ch35. **Zalaegerszegi Televízió (ZTV):** Kossuth L. u.45-49, 8900 Zalaegerszeg; ch51. **Zemplén TV:** Színház köz 4, 3980 Sátoraljaújhely; ch50.

DTT Transmitters (MPEG4 exc.*=MPEG2)
Operator: Antenna Hungária ☒ Petzváí József u. 31-33, 1119 Budapest ☎ +36 1 2036060 🗎 +36 1 4642525 **E:** antennadigital@ahrt.hu **W:** www.ahrt.hu; www.antennadigital.hu **Mux 1:** m1 HD, m2 HD, reg prgrs, RTL Klub, TV2, ATV✪, hírTV✪, Euronews✖ Kossuth R., Petőfi R., Bartók R. **Mux 2 (DVB-H)*:** tbd (F.pl) **Txs:** ch38V (Budapest/Széchenyi-hegy 26.3kW) **Mux 3:** Duna HD, Duna A.

Location	M1*	M3	kW	Location	M1*	M3	kW
Budapest	55a	62	50	Pécs	52d	67	50
Csávoly	27f	68	50	Sopron	42e	68	50
Csengöd	27f	68	50	Szeged	60f	65	25
Györ	42e	61	25	Szentes	60f	65	50
Kabhegy	64e	61	50	Tokaj	68c	63	50

Location	M1°	M3	kW	Location	M1°	M3	kW
Kékes	44c	69	50	Úzd	52d	67	50
Komádi	51b	62	50	Vasvár	58e	68	50
Nagykanizsa	55d	65	50				

°) incl. reg prgrs (a-f), see above
NB. Txs will be on lower power than shown until analogue switch-off.

ICELAND

Systems: # PAL-B/G [E]; DVB-T (MPEG2)

National Stations
RÚV - SJÓNVARPIÐ (Pub) 🖃 Efstaleiti 1, 150 Reykjavík ☎ +354 5153000 🖷 +354 5153010 **E:** istv@ruv.is **W:** www.ruv.is **LP:** Dir: Páll Magnússon. **Chs:** RÚV, RÚV+ **Txs:** RÚV: Reykjavík ch10 (2kW) & netw. – **365 FJÖLMIÐLAR (Comm)** 🖃 Skaftahlíð 24, 105 Reykjavík ☎ +354 5156000 **E:** 365@365.is **W:** www.365.is **LP:** CEO: Ari Edwald **Chs:** Stöð 2 (partly❂), Stöð 2 Bíó (❂), Stöð 2 Sport (❂) **Txs:** **Stöð 2:** Reykjavík ch6 (10kW) & netw.; **Stöð 2 Bíó:** Reykjavík ch49 (2kW); **Stöð 2 Sport:** Reykjavík ch12 (1kW) & netw. – **SKJÁRINN (Comm)** ❂ 🖃 Skipholti 31, 105 Reykjavík ☎ +354 5956000 **E:** info@skjarinn.is. **Txs: Skjár Einn:** Reykjavík ch55 (1kW). **LP:** CEO: Sigríður Margrét Oddsdóttir.

Local Stations
Extra: 600 Akureyri. **Tx:** Vaðlaheiði ch53 (0.1kW) **Omega (Rlg):** Grensásvegi 8 108 Reykjavík; Bláfjöll ch45 (0.1kW), Reykjavík ch45 (0.002kW) & ch51 (1kW), Mosfellsbær ch57 (0.1kW). **Sjónvarp Flensborg;** Brekkugötu 17-19, 220 Hafnarfjörður; ch58 (0.01kW). **Sjónvarp Ólafsvík:** 355 Ólafsvík; ch63 (0.01kW).

DTT Transmitters
Operator: Og Fjarskipti (Vodafone Digital Ísland) 🖃 Skútuvogi 2, 104 Reykjavík **E:** vodafone@vodafone.is **W:** www.vodafone.is **Mux 1(❂):** RÚV, RÚV+, Stöð 2, Stöð 2 Bíó, Stöð 2 Sport, Stöð Sport 2, Stöð 2 Extra, Stöð 2+, Skjár Einn **Mux 2(❂):** Discovery Channel, E! Entertainment, Sky News, DR1, Cartoon Network, ÍNN, Skjár Golf, Blue Hustler.

Location	M1	M2	kW
Reykjavík	27	28	0.3

+ nationwide tx network

INDIA

Systems: # PAL-B/G [E] ⇩2017; DVB-T (MPEG2), DVB-T2 (MPEG4)

DOORDARSHAN (DD) (Pub) 🖃 Doordarshan Bhawan, Copernicus Marg, New Delhi-110001 ☎ 1+91 11 23386055 🖷 +91 11 23385843 **E:** dddirect@dd.nic.in **W:** www.ddindia.gov.in **LP:** DG: Aruna Sharma, CE: R.R.Prasad **Chs (terr.):** DD-National, DD-News (**W:** www.ddnews.gov.in), DD-Sports, DD-Barathi, DD-Regional Channels **Txs: DD-1:** New Delhi ch5 (20kW) & network; **DD-News:** New Delhi ch7 (20kW) & network.

DTT Transmitters (under construction)
Operator: Doordardshan **Mux 1:** DD-National, DD-News, DD-Sports, DD-Barathi, DD-Regional **Mux 2 (DVB-T2/MPEG4):** (tbd) **Mux 3 (DVB-H):** 16 prgrs.

Location	M1	M2	M3	kW
New Delhi	(*)	29	26	6/50

+ nationwide netw. under construction, mainly Mux 1 (ca. 230 HP sites) *) F.pl

INDONESIA

Systems: # PAL-B/G [E] ⇩2018; DVB-T (MPEG2)

National Stations
TELEVISI REPUBLIK INDONESIA (TVRI) (Gov) 🖃 Jalan Gerbang Pemuda, Senayan, Jakarta 10270 ☎ +62 21 3846740 🖷 +62 21 5737152 **E:** wmaster@tvri.co.id **W:** www.tvri.co.id **LP:** MD: Azis Husein; TD: Djoko Widayat. **Chs:** TVRI1, TVRI2. **Txs:** TVRI1: Jakarta ch6 (5kW) & network; TVRI2: Jakarta ch9. – **ANTEVE (PT CAKRAWALA ANDALAS TELEVISI) (Comm)** 🖃 Mulia Center Building, 19th Floor, Jl. HR Rasuna Said Kav. X-6 No.8, Jakarta 12940 ☎ +62 21 5222084 🖷 +62 62 215222087 **E:** sales@anteve.co.id **W:** www.an.tv **LP:** GM: Dennis M. Cabalfin **Txs:** Jakarta ch47 (40kW) & network. – **GLOBAL TV (PT GLOBAL INFORMASI BERMUTU) (Comm)** 🖃 Jl Jend. Ahmad Yani 31, Jakarta 13230☎ +62 21 5360601 🖷 +62 21 5360602 **E:** globaltv@globaltv.co.id **W:** www. globaltv.co.id **Txs:** Jakarta ch51 & network. – **INDOSIAR (PT.**

INDOSIAR VISUAL MANDIRI) (Comm) 🖃 Jl. Damai No 11, Daan Mogot, Jakarta 11510 ☎ +62 21 5672222 🖷 +62 21 5652221 **E:** program@indosiar.com **W:** www.indosiar.com **Txs:** Jakarta ch41 & network. – **METRO TV (PT MEDIA TELEVISI INDONESIA)** 🖃 Jl. Pilar Mas Raya Kav. A-D., Kedoya, Kebon Jeruk, Jakarta 11520 ☎ +62 21 58300077 🖷 +62 21 5816365 **E:** info@metrotvnews.com **W:** www.metrotvnews.com **Txs:** Jakarta ch57 & network. – **RCTI (PT RAJAWALI CITRA TELEVISI INDONESIA) (Comm)** 🖃 Jl. Raya Perjuangan No. 3, kb. Jeruk, Jakarta 11000 ☎ +62 21 5303540 🖷 +62 21 5493852 **E:** pr@rcti.tv **W:** www.rcti.tv **LP:** Pres Dir: Muchamad Ralie Siregar; TM: Doopy Irwan **Txs:** Jakarta ch43 & network. – **SCTV (PT SURYA CITRA TELEVISI) (Comm)** 🖃 Graha SCTV 2nd floor, Jl. Gatot Subroto Kav 21, Jakarta 12930 ☎ +62 21 5225555 🖷 +62 21 5224777 **E:** pr@sctv.co.id **W:** www.sctv.co.id **LP.:** Dir. Op.: Lanny Ratulangi **Txs:** Jakarta ch45 (120kW) & network. – **TPI (PT CIPTA TELEVISI PENDIDIKAN INDONESIA) (Comm)** 🖃 Jalan Pintu II - Taman Mini Indonesia Indah, Pondok Gede, Jakarta Timur 13810 ☎ +62 21 8412473 🖷 +62 21 8412471 **E:** wmaster@tpi.tv **W:** www.tpi.tv **LP:** GM: Syamsudin C. Haesy **Txs:** Jakarta ch37 (80kW) & network. – **TRANS TV (PT TELEVISI TRANSFORMASI INDONESIA) (Comm)** 🖃 Jl. Kapten Tendean Kav. 12-14A, Jakarta 12790 ☎ +62 21 7944240 🖷 +62 21 7992600 **E:** wmaster@transtv.co.id **W:** www. transtv.co.id **Txs:** Jakarta ch29 (80kW) & network. – **TRANS 7 (PT DUTA VISUAL NUSANTARA TIVI TUJUH) (Comm)** 🖃 Menara Bank Mega Lt. 20, Jl. Kapt. P. Tendean Kav.12-14A, Jakarta 12790 ☎ +62 21 79177000 🖷 +62 21 79184684 **E:** info@trans7.co.id **W:** www.trans7.co.id **Txs:** Jakarta ch49 (60kW) & network. – **TVONE (PT LATIVI MEDIA KARYA (Comm)** 🖃 Kawasan Industri Pulo Gadung, Jl Rawa Teratai II No 2, Jakarta Timur 13260 ☎ +62 21 4613545 🖷 +62 21 4616255 **E:** info@tvone.co.id **W:** www.tvone.co.id **Txs:** Jakarta ch53 & network.

Local Stations not shown.

DTT Transmitters (Trial)
Operator: TVRI/PT Telekom **Mux:** TVRI, SCTV, Trans TV, Trans 7, ANTV, TVOne, Metro TV **Tx:** Jakarta ch(-)

IRAN

Systems: # SECAM-B/G [E]; DVB-T (MPEG4)

ISLAMIC REPUBLIC OF IRAN BROADCASTING (IRIB) (Gov) 🖃 P.O.Box 19395 3333, 19395 Tehran ☎ +98 21 22041093 🖷 +98 21 22014802 **E:** tv@irib.ir **W:** www.irib.com/tv **LP:** Pres: Seyed Ezzatollah **Chs:** IRIB TV1, IRIB TV2, IRIB TV3, IRIB TV4, IRIB TV5, IRINN, regional stations. **Txs:** (-).

DTT Transmitters (under construction)
Operator: IRIB **E:** digitaltv@irib.ir **W:** digitaltv.irib.ir **Mux:** IRIB TV1, IRIB TV2, IRIB TV3, IRIB TV4, IRIB TV5, IRINN **Tx:** ch37 (Tehran). Nationwide netw. planned.

IRAQ

Systems: PAL-B/G [E]

National Stations
AL-IRAQIYA TV (Pub) 🖃 Salhiya, Baghdad ☎ +964 1 8844412 🖷 +964 1 5410480 **W:** www.imn.iq **Chs:** Al-Iraqiya TV, Al-Iraqiya Sport Channel, Al-Iraqiya Atyaf. **Txs: Al-Iraqiya TV:** Najaf area ch5 & ch11, Sinjar ch5, Kirkuk ch6, Tikrit ch6, Amara ch7, Nasirya ch7, Samawa area ch8 & ch10, Ali Al-Gharbi ch9, Baghdad ch9, Basra area ch9 & ch25, Falluja ch9, Karbala area ch5 & ch8, Kut ch9, Mosul ch9, Dewanya ch12, Baquba ch23, Suq Ashiukh ch26, Beiji ch27, Nasir ch33, Babylon area ch38 & ch41, Samarra ch39; **Al-Iraqiya Sport Channel:** Nasirya ch5, Baghdad ch7, Mosul ch7, Nasir ch9, Suq Ashiukh ch10, Amara ch11, Ali Al-Gharbi ch12, Basra ch52; **Al-Iraqiya Atyaf:** Baghdad ch37 – **AL-FORAT (Comm)** 🖃 Baghdad. **E:** info@alforattv.com **W:** www.alforattv.net **Txs:** Baghdad ch26, Najaf ch28, Basra ch46, Kerbala ch57. – **AL-SHARQIYA (Comm)** 🖃 Baghdad. **E:** alsharqiya@alsharqiya.com **W:** www.alsharqiyatv.com **Txs:** Mosul ch42, Basra ch44, Baghdad ch48.

Local Stations
Al-Furatain: Hilla; ch36. **Al-Ghadeer:** Najaf; ch25. **Al-Huda:** Karbala; ch23. **Al-Hurriya:** Baghdad; ch37. **Al-Masar:** Baghdad; ch46. **Al-Mashriq:** Baghdad; ch54. – **Al-Merbad:** Basra; txs: Amara ch42, Nasiriya ch46, Basra ch52. **Al-Nahrain:** Kut; ch6. **Al-Nakheel:** Basra; ch57. **Al-Rasheed:** Baghdad; ch34. **Al-Salam:** Baghdad; ch5. **Al-Sumaryia:** Qadisya; ch56. **Al-Yaum:** Baghdad; ch26. **Ashur:**

Baghdad; ch44. **Kurdistan TV:** ch33. **Kurdsat TV**: Sulaimaniyah ch30.
Nahrian: Baghdad; ch41. **TV Baghdad:** Baghdad; ch51.
Foreign TV Relays
Hurra (USA): Basra ch3, Najaf ch3, Ramadi ch3, Baghdad ch12,
Mosul ch12; **MBC (Saudi Arabia):** Bhaghdad ch61.

Foreign Military Stations
AFN (USA): Balad ch31.

IRELAND

Systems: # PAL-I [E] ⬇2012; DVB-T (MPEG4)

RTÉ TELEVISION (Pub) 🖃 Donnybrook, Dublin 4 ☎ +353 1 208
3111 🖷 +353 1 2082772 **E:** www.rte.ie **W:** www.rte.ie **LP:** DG: Cathal
Goan **Chs:** RTÉ1, RTÉ2, TG4 (Irish-language service, **W:** www.tg4.ie
– **TV3 (Comm)** 🖃 Westgate Business Park, Ballymount, Dublin 24
☎ +353 1 4193333 🖷 +353 1 4193300 **E:** info@tv3.ie **W:** www.tv3.ie
L.P: MD/CEO: Rick Hetherington.

Location	RTE1	RTE2	TV3	TG4	kW
Cairn Hill	40	43	46	50	800
Clermont Carn	52V	56V	66V	68V	300
Holywell Hill	23	26	29	33	20
Kippure	E	H	62	59	2x160/2x500
Maghera	EV	HV	66V	68V	2x200/2x500
Mount Leinster	FV	IV	26V	23V	2x230/2x500
Mullaghanish	DV	GV	27V	31V	2x220/2x500
Spur Hill	53	57	60	63	10
Three Rock	29	33	35	55	25
Truskmore	I	G	60	63	2x280/2x500
+ repeaters.					

DTT Transmitters (under construction)
Operator: RTÉ Transmission Network Ltd **Mux 1:** RTÉ1, RTÉ2, TV3,
TG4. **Mux 2:** tbd

Location	M1	M2	*kW	Location	M1	M2	*kW
Cairn Hill	41	44	-	Mount Leinster	39	42	-
Clermont Carn	53	57	-	Mullaghanish	21	24	-
Dungarvan	55	59	-	Spur Hill	45	49	-
Holywell Hill	25	30	-	Three Rock	54	58	-
Kippure	54	58	-	Truskmore	52	56	-
Maghera	55	48	-	Woodcock Hill	41	44	-
+ repeaters. *) Power varies acc. to stage of DTT transition.							

ISRAEL

Systems: DVB-T (MPEG4)

ISRAEL BROADCASTING AUTHORITY (IBA) (Pub) Chs: Channel 1
(Mabat), Channel 33. **Channel 1:** 🖃 P.O.Box 7139, Jerusalem 91071
☎ +972 2 5301333 🖷 +972 2 6291862 **E:** mabat@iba.org.il **W:** mabat.
iba.org.il **Txs:** Jerusalem ch24 & network. **Channel 33:** 🖃 P.O.Box
13172, Jerusalem 91131 ☎ +972 2 5013800 **E:** arutz33@iba.org.
il. – **THE SECOND AUTHORITY FOR TELEVISION AND RADIO**
(Pub) 🖃 P.O.Box 34465, Jerusalem **Chs:** Channel 2, Channel 10
Channel 2: 🖃 P.O.Box 34122, Jerusalem 95464 ☎ +972 2 6556222
🖷 +972 2 6556286 **E:** rashut@rashut2.org.il **W:** www.rashut2.org.il.
NB: Channel 2 is time-shared by the commercial companies Reshet
(P.O.Box 5577, Herzliya), Keshet (12 Raul Vallenberg St., Tel Aviv
61580), and Israel Educational Television (14 Klauzner St., Ramat
Aviv, Tel Aviv). **Channel 10:** 🖃 53 Derech Hashalom St., Givatayim
53454 ☎ +972 3 7331000 🖷 +972 3 7331040 **W:** www.10.tv **LP:**
DG: Nir Lempert. – **THE KNESSET CHANNEL (Gov)** 🖃 Kiryat Ben-
Gurion, Jerusalem 91950 ☎ +972 2 6541636 **E:** feedback@knesset.
gov.il **W:** www.knesset.gov.il.

DTT Transmitters
Operator: Bezeq 🖃 P.O.Box 1088, Jerusalem 91010 ☎ +972 36
264562 🖷 +972 36 264559 **W:** www.bezeq.com **Mux:** IBA Channel 1,
IBA Channel 33, Channel 2, Channel 10, Knesset Channel **Txs:** ch26
(Tel El Ful, Sha'ar Hanegev, Jerusalem-Shalom Hotel, Mirbatz Mt.),
ch29 (Eilat, Zefat, Carmel Radar, Bney Yehuda, Manara).

WEST BANK & GAZA (Palestinian Authority)

System: PAL-B/G [E]

National Station
PALESTINIAN BROADCASTING CORP. TV (Gov) 🖃 P.O.Box
984, Ramallah Albereih, West Bank ☎ +970 2 2959894 🖷 +970 2
295989 **E:** pbc@palnet.com **W:** www.pbc.gov.ps **LP:** DG: Radwan

Abu Ayyash **Txs:** Nablus ch5, Khan Yunis ch21, Ariha (Jericho) ch21,
Kasser-Elhakim (Gaza) ch23, Ramallah ch25, Halhul ch30, Jenin ch31,
Betjala ch34.

Local Stations not shown.

ITALY

Systems: DVB-T (MPEG2); † PAL-B/G [E] ⬇31 Dec 2012

National Stations
RADIOTELEVISIONE ITALIANA (RAI) (Pub) 🖃 Direzione Centrale
TV, Viale Mazzini 14, 00195 Roma ☎ +39 06 36864046 🖷 +39 06
36226422 **E:** portale@rai.it **W:** www.rai.it **LP:** DG: Mauro Masi **Chs:**
RaiUno, RaiDue, RaiTre, Rai 4, RaiNews24, Raisport Sportpiù, Rai
Storia, Rai Gulp – **LA 7 (Comm)** 🖃 Via della Pineta Sacchetti 229,
00166 Roma (RM) ☎ +39 06 35584 🖷 +39 06 355 84257 **E:** la7@
la7.tv **W:** www.la7.tv – **MEDIASET (Comm)** 🖃 Viale Europa 48,
Palazzo dei Cigni, 20093 Cologno Monzese (MI) ☎ +39 02 21021 🖷
+39 02 85414283 **W:** www.mediaset.it **Chs:** Italia 1, Rete 4, Canale
5. – **MTV ITALIA (Comm)** 🖃 Corso Europa 7, 20122 Milano (MI).
☎ +39 02 7621171 🖷 +39 02 7621227 **E:** segreteria@mtvne.com **W:**
www.mtv.it – **SAT2000 (Comm)** 🖃 Via Aurelia 786, 00165 Roma
(RM) ☎ +39 06 665081 🖷 +39 06 66508581 **W:** www.sat2000.it **E:**
sat2000@sat2000.it.

Local Stations not shown.

DTT Transmitters
Operator: RAIWAY **Mux 1:** Rai 1, Rai 2, Rai 3, RAI News, France 2,
TVB3, TVS ⌘ RAI R.1, 2, 3, FD Leggera. **Txs:** Roma (Monte Cavo) ch11
& netw. **Mux 2:** Rai Sport 1, Rai Sport 2 ⌘ FD Auditorium, FD Leggera
Txs: Roma (Monte Cavo) ch30 & netw. – **Mux 3:** Rai 4, Rai Gulp,
Rai Movie, Rai Premium, Rai YoYo **Txs:** Roma (Monte Cavo) ch26 &
netw. **Mux 4:** Rai Extra, Rai HD, Rai Storia **Txs:** Roma (Monte Cavo)
ch40 & netw. – **Operator:** Dfree **E:** tecnici@dfree.tv **W:** www.dfree.
tv **Mux (◎):** Disney Channel, Joi, Joi + 1, Mya, Cinema, Emotion,
Energy, Steel, Studio Universal. **Txs:** Roma (Monte Cavo) ch50 &
netw. – **Operator:** Mediaset **Mux 1(◎):** Cartoon Network, Playhouse
Disney, Disney Channel +1, Hiro, Moto GP, Calcio HD, Calcio 1-6,
Extra 1, Extra 2, Steel +1 **Txs:** Roma (Monte Cavo) ch52 & netw. **Mux**
2: Canale 5, Rete 4, Italia 1, Boing, Boing +1, Iris, Class News, BBC
World, Coming Soon. **Txs:** Roma (Monte Cavo) ch36 & netw. **Mux 3**
(DVB-H◎): multiprgr **Txs:** Roma (Monte Cavo) ch38 & netw. **Mux 4:**
Canale 5, Italia 1, Rete 4, Cinema HD◎ **Txs:** Roma (Monte Cavo) ch49
& netw. **Mux 5:** Canale 5+1, Italia 1+1, La5, Mya +1◎, Calcio HD 2◎
Txs: Roma (Monte Cavo) ch56 & netw. **Mux 6:** Canale 5 HD, Italia 1
HD, Rete 4 HD **Txs:** Roma (Monte Cavo) ch58 & netw. – **Operator:**
TIMB1 **Mux:** La5, La7, MTV Italia, Cartello K2, Entertainment Fact,
Mya +1, Palermo Channel, QVC, Real Time, Sport Italia, Sport Italia 2,
Sport Italia 24 **Txs:** Roma (Monte Cavo) ch47 & netw.

Other operators: Regional and operators not shown.

South Tyrol

RUNDFUNK-ANSTALT SÜDTIROL (RAS) (Pub) 🖃 Europallee
164A, 39100 Bozen ☎ +39 0471 546666 🖷 +39 0471 200378 **E:** info@
ras.bz.it **W:** www.ras.bz.it **LP:** Pres: Helmuth Hendrich

DTT Transmitters
Operator: RAS **Mux 1:** ORF1, ORF2, Das Erste, ZDF **Mux 2:** SF1, SF
ZWEI, Bayerisches Fernsehen, KiKa **Txs: Mux 1:** ch34 (SFN), **Mux**
2: ch51 (SFN).

IVORY COAST

System: SECAM-K1 [E]

TÉLÉVISION IVOIRIENNE (Gov) 🖃 08 BP 883, Abidjan 08 ☎ +225
22449039 🖷 +225 22447339 **E:** dgrti@aviso.ci **W:** www.rtici.ci **L.P:**
DG: Gnonzie Ouattara **Chs:** La Première, TV2. **Txs: La Première:**
Koun ch4 (10kW), Tiémé ch4 (10kW), Séguéla ch5 (10kW), Digo ch5
(2kW), Dimbroko ch6 (2kW), Touba ch6 (1kW), Man ch7 (10kW),
Dabakala ch7 (2kW), Abidjan ch8 (10kW), Niangbo ch8 (10kW),
Niangué ch8 (10kW), Bouaflé ch9 (10kW) & repeaters. **TV2:** (-).

JAMAICA

System: NTSC-M [A]

PUBLIC BROADCASTING CORP. OF JAMAICA (Pub) 🖃 5-9

South Odeon Avenue, Kingston. – **CVM TELEVISION LTD (Comm)** ⌨ 69 Constant Spring Rd, Kingston 10 ☎ +1 876 9319400 🖷 +1 876 9311573 **E:** customerservice@cvmtv.com **W:** www.cvmtv.com **LP:** Chmn: Neville Blythe **Txs:** Marley Hill ch4, Coopers Hill ch9, Ochos Rios ch10, Montego Bay ch11, Cabbage Hill ch12, Port Antonio ch13. – **LOVE TV (Rlg)** ⌨ 12 Carlton Cresent, Kingston ☎ +1 876 9689596 🖷 +1 876 9685379 **W:** www.love101.org **Txs:** Montego Bay ch2, Ochos Rios ch3, Kingston ch6, Huntley ch8, Kingston ch17. – **TELEVISION JAMAICA LTD (TVJ) (Comm)** ⌨ P.O.Box 100, Kingston 10. ☎ +1 876 9265620 🖷 +1 876 9291029 **E:** tvjadmin@ cwjamaica.com **W:** www.televisionjamaica.com **LP:** Chmn: Milton Samuda **Txs:** ZQI-TV: Coopers Hill ch7, Port Antonio ch8, Yallahs ch9, Montego Bay ch9, Kingston ch11, Oracabessa ch12, Spur Tree ch13.

JAPAN

Systems: ISDB-T; † NTSC-M [A] ⬇24 Jul 2011. AFN: # NTSC-M [A]

National Stations
NIPPON HOSO KYOKAI (NHK) (Pub) ⌨ 2-1, Jinnan 2-chome, Shibuya-ku, Tokyo 150-8001 ☎ +81 3 34651111 **W:** www.nhk. or.jp **Chs:** NHK General, NHK Educational – **ALL-NIPPON NEWS NETWORK (ANN) (Comm)** ⌨ 9-1, Roppongi 6-chome, Minato-ku, Tokyo 106-8001 ☎ +81 3 64061111 **W:** www.tv-asahi.co.jp – **FUJI TELEVISION NETWORK (FTN) (Comm)** ⌨ 4-8, Daiba 2-chome, Minato-ku, Tokyo 137-8088 ☎ +81 3 55008888 **W:** www.fujitv.co.jp – **JAPAN NEWS NETWORK (JNN) (Comm)** ⌨ 3-6, Akasaka 5-chome, Minato-ku, Tokyo 107-8006. ☎ +81 3 37461111 **W:** www.tbs. co.jp – **NIPPON NEWS NETWORK (NNN) (Comm)** ⌨ 6-1, Higashi Shimbashi 1-chome, Minato-ku, Tokyo 105-7444 ☎ +81 3 62154444 **W:** www.ntv.co.jp – **TV TOKYO NETWORK (TXN) (Comm)** ⌨ 3-12, Toranomon 4-chome, Minato-ku, Tokyo 105-8012 ☎ +81 3 54707777 **W:** www.tv-tokyo.co.jp.
NB. Commercial stns are relayed nationwide via local affiliates.

Local Stations not shown.

DTT Transmitters (National Stations)

Location	NHK[1]	NHK[2]	ANN	FTN	JNN	NNN	TXN
Tokyo (Tokyo Tower)	1	2	24	21	22	25	23

+ nationwide tx network [1] NHK General [2] NHK Educational

Foreign Military Station
American Forces Network Japan (U.S. Mil) ⌨ (Misawa) OLAA, AFPBS, APO San Francisco 96519 ⌨ (Okinawa) Det 2, AFPBS, APO San Francisco 96239 **Txs:** Okinawa ch8 (40kW), Misawa ch66 (1kW), Iwakuni ch66 (0.4kW).

JORDAN

System: PAL-B/G [E]

JORDAN TELEVISION (JTV) (Pub) ⌨ P.O.Box 1041, 1118 Amman ☎ +962 6 4749 171 🖷 +962 6 4778 578 **E:** jtv@jrtv.gov.jo **W:** www. jrtv.com **LP:** DG: Mohamed Najib Sarayra; Tech Dir (TV): Ghazi Gammo **Chs:** JTV1, JTV2 **Txs** (pol.H)**: JTV1:** Suweilih ch3 (104kW), Aqaba ch5 (5kW), Ras Munif-Ajlun ch9 (500kW), Deir Alla ch26 (6kW) & repeaters. **JTV2:** Aqaba ch5 (5kW), Suweilih ch6 (108kW), Ras Munif-Ajlun ch11 (500kW), Deir Alla ch29 (6kW) & repeaters.

KAZAKHSTAN

Systems: SECAM-D/K [R], PAL-D/K [R]

National Stations
QAZAQ TELEVIZIYASY (Gov) ⌨ Jeltoqsan 177, 050013 Almati ☎ +7 727 2635579 🖷 +7 727 2631207 **W:** www.kazakstan.kz **Txs:** Almati ch5 & network. – **XABAR (Gov)** ⌨ Respwblïk alana 13, 050013 Almati ☎ +7 727 2625091 🖷 +7 727 2696505 **W:** www.khabar.kz **Chs:** Xabar, Yel Arna. **Txs: Xabar:** Almati ch3 & network.**Yel Arna:** (-). – **KTK (Comm)** ⌨ Respwblïk alana 13, 050013 Almati ☎ +7 727 2583657 🖷 +7 727 2583693 **E:** ktk@ktk.caravan.kz **W:** www.ktk. kz **LP:** Dir: Andrey Osadtsyuk. **Txs:** Almati ch12 & network. – **NTK (Comm)** ⌨ Respwblïk alana 13, 050013 Almati ☎ +7 727 2634255 🖷 +7 727 2582467 **E:** ntk@ntk.kz **W:** www.ntk.kz **LP:** GD: Anuar Salkimbayev. **Txs:** Almati ch7 & network.

Local Stations, Foreign TV Relays not shown.

KENYA

System: # PAL-B/G [E] ⬇2012; DVB-T (MPEG4)

National Stations
KENYA BROADCASTING CORP. (Gov) ⌨ P.O.Box 30456, Harry Thuku Road, 00100 Nairobi ☎ +254 20 334567 🖷 +254 20 220675 **E:** kbctv@swiftkenya.com **W:** www.kbc.co.ke **LP:** Chmn: James Kangwana **Chs:** Channel One, Channel 2, Metro TV **Txs: Channel One:** Timboroa ch2 (10kW), Limuru ch2 (10kW), Nakuru ch6 (10kW), Mazeras ch6 (10kW), Kisii ch8 (2kW), Webuye ch9 (2kW), Nyeri ch10 (10kW), Nyambene ch10 (5kW); **Channel 2** (☉)**:** n/a; **Metro TV:** Nairobi ch31. – **CITIZEN TV (Comm)** ⌨ P.O.Box 7468, 00300 Nairobi ☎ +254 20 2721415 🖷 +254 20 2724220 **E:** news@royalmedia.co.ke **W:** www.citizentv.co.ke **Txs:** Nairobi ch39, Nyeri ch50 **LP:** Owner: Samuel Macharia. – **FAMILY TV (Rlg)** ⌨ P.O.Box 2330 KNH, Nairobi ☎ +254 20 4200000 🖷 +254 20 4200100 **E:** info@familykenya.com **W:** www.familykenya.com **Txs:** Nairobi ch9, Eldoret ch44, Nakuru ch45, Mombasa ch46. – **KENYA TELEVISION NETWORK (KTN) (Comm)** ⌨ P.O.Box 56985, 00100 Nairobi ☎ +254 20 227122 🖷 +254 20 214467 **E:** admin@ktnkenya.com **W:** www.ktnkenya.tv **LP:** Chmn: Mwakio Sio. **Txs:** Mombasa ch12, Nyeri ch54, Eldoret ch57, Nakuru ch58, Nairobi ch59, Kisumu ch61.– **K24 (Comm)** ⌨ 3rd Floor, Longonot Place, Kijabe St., 00100 Nairobi. ☎ +254 21 248000 **W:** www.k24.co.ke **Tx:** Nairobi ch24. – **NATION TV (NTV) (Comm)** ⌨ P.O.Box 49010, Nairobi 00100 GPO ☎ +254 20 3208000 **E:** views@ nation.co.ke **W:** www.nationfm.co.ke **Txs:** Kisumu ch33, Mazeras ch40, Nairobi ch42, Eldoret ch47, Nyeri ch51. – **OXYGEN TV (Comm)** ⌨ P.O.Box 48445, 00100 Nairobi. **LP:** MD: Kass Khimji. – **STELLA VISION (STV) (Comm)** ⌨ P.O. Box 20190, Nairobi ☎ +254 20 2712982 🖷 +254 20 2713146 **Txs:** Kisumu ch30, Eldoret ch37, Nyeri ch43, Nakuru ch54, Mombasa ch55, Machakos ch55, Nairobi ch56.

Local Stations not shown.

DTT Transmitters (network under construction)
Operator: Signet Ltd **Mux:** KBC, NTV, KTN, CNBC Africa, K24, STV, EATN, EATV, Oxygen TV, Family TV, God TV, Kiss TV, Classic TV, Citizen TV, GBS ⌘ KBC English Service, KBC Idha FM, Metro FM, Coro FM **Txs:** (-).

KIRIBATI

NB: No terrestrial TV station.

KOREA, North (D.P.R. of Korea)

Systems: PAL-D/K [R]

KOREAN CENTRAL TV BROADCASTING STATION (Gov)
⌨ Chonsung-dong, Moranbong District, Pyongyang ☎ +850 2 816 035 🖷 +850 2 812100 **LP:** Chmn: Cha, Sung Su; Dir: Chun, Li-Ji; Head of Tech: Chol, Li-Yong. **Txs:** Sangmasan ch1 (10kW), Chayubong ch2 (30kW), Suryongsan ch2 (30kW), Pegebong ch3 (30kW), Hamhung ch3 (70kW), Wonsan ch4 (10kW), Songjinsan ch4 (20kW), Jajiryong ch5 (30kW), Peakam ch5 (10kW), Sambongsan ch5 (10kW), Kangryong ch5 (30kW), Kumgungsan ch5 (30kW), Chongjin ch6 (70kW), Hyangsan ch6 (10kW), Sepo ch6 (70kW), Sinuiju ch6 (70kW), Sariwon ch7 (30kW), Chayubong ch8 (30kW), Haksongsan ch8 (20kW), Kanggye ch8 (70kW), Jaedoksan ch9 (30kW), Unjubong ch9 (70kW), Wangjesan ch9 (30kW), Sepo ch9 (70kW), Sinyang ch9 (30kW), Wonsan ch10 (70kW), Haeju ch11 (70kW), Sambongsan ch11 (10kW), Jonchon ch11 (10kW), Songsan ch12 (10kW), Jajiryong ch12 (30kW), Chongjin ch12 (70kW), Haksongsan ch12 (20kW), Misan ch12 (70kW), Pyongyang ch12 (700kW), Rimbong ch12 (10kW), Robaeksan ch12 (30kW), Tokusan ch12 (20kW) & relay txs below 10kW. – **KOREAN EDUCATIONAL & CULTURAL TELEVISION (Gov)** ⌨ Pyongyang. **Txs:** Kaesong ch8 (30kW), Pyongyang ch9 (140kW). – **MANSUDAE TELEVISION (Gov)** ⌨ Mansudae, Pyongyang. **Tx:** Pyongyang ch5 (350kW).

KOREA, South (Rep. of Korea)

Systems: T-DMB; † NTSC-M [A] ⬇31 Dec 2012. AFN: # NTSC-M [A]

KOREAN BROADCASTING SYSTEM (KBS) (Pub) ⌨ 18 Yeouido-dong Yeoungdeungpo-gu, Seoul 150-790 ☎ +82 2 7812001 🖷 +82 2 7812099 **W:** www.kbs.co.kr **LP:** Pres/CEO: Kim In-Kyu; Exec. MD (Content) Gil Hwan-Young; Exec.MD (New Media & Technics): Kim Seon-Kwon. **Chs:** KBS 1TV, KBS 2TV. **Txs: KBC 1TV** : HLKA-

TV Seoul ªch5 & netw. **KBC 2TV:** HLSA-TV Seoul ªch2 & netw. – **KOREA EDUCATIONAL BROADCASTING SYSTEM (EBS) (Pub)** ☜ 92-6, Umyeon-dong, Seocho-gu. Seoul 137-791 ☎ +82 2 5211586 ▤ +82 2 5210241 **W:** www.ebs.co.kr **Txs:** Seoul ch9 & netw. – **MUNHWA BROADCASTING CORP. (MBC) (Comm)** ☜ 31 Yeouido-dong Yeoungdeungpo-gu, Seoul 150-728 ☎ +82 2 7892851 ▤ +82 2 7823094 **W:** www.imbc.com **Txs:** Seoul ªch11 & netw. – **SEOUL BROADCASTING SYSTEM (SBS) (Comm)** ☜ 920 Mok-dong, Yangcheon-gu, Seoul 158-725 ☎ +82 2 20610006 ▤ +82 2 21133169 **W:** www.sbs.co.kr **Txs:** HLSQ-TV Seoul ªch6. **Affiliates:** CJB (Cheongju), GTB (Chuncheon), JIBS (Jeju), JTV (Jeonju), KBC (Gwangju), KNN (Busan), TBC (Daegu), TJB (Daejeon), UBC (Ulsan). – **OBS KYEONGIN TV (OBS) (Comm)** ☜ 202-7, Ojeong-dong, Ojeong-gu, Bucheon-si, Gyeonggi-do 421-814 ☎ +82 32 6705000 ▤ +82 32 6712069 **W:** www.obs.co.kr **Station:** HLDO-TV. **Txs:** Gwanggyo-san ªch4, Gyeyang-san ªch21. ª) analogue txs

DTT Transmitters (National Stations)

Location	KBS 1TV	KBS 2TV	EBS	MBC	SBS	OBS
Seoul	10	8	9	(*)	(*)	(*)

+ nationwide tx network (*) planned

Foreign Military Station
American Forces Network Korea (U.S. Mil) ☜ Unit #15324, APO AP 96205-0097, USA ☎ +82 2 79146495 **W:** www.afnkorea.net **Txs:** Chinhae ch2, Camp Walker ch12, Pajuri/Munsan ch19, Seoul ch34, Camp Page ch46, Camp Casey/Osan Air Base/Kunsan Air Base/Camp Carroll ch49, Camp Long/Camp Humphreys ch58.

System: PAL-B/G [E]

National Stations
RADIO TELEVIZIONE I KOSOVËS (RTK) (Pub) ☜ Rr. Xhe Prishtina 12, 10000 Prishtinë ☎ +381 38 230102 ▤ +381 38 235336 **E:** post@rtklive.com **W:** www.rtklive.com **LP:** Dir TV: Liridon Cahani **Txs:** Crnusha ch7, Zatriq ch9, Maja e Gjelbërt ch12, Prishtinë ch23. – **RTV21 (Comm)** ☜ Pallati i Mediave, Aneks II, 10000 Prishtinë ☎ +381 38 241526 ▤ +381 38 241526 **E:** lajmet@rtv21.tv **W:** www.rtv21.tv **Txs:** Prishtinë ch37 & network. – **TV KOHAVISION (Comm)** ☜ Rr. Nene Tereza, 10000 Prishtinë ☎ +381 38 248014 ▤ +381 38 248015 **E:** kohavision@koha.net **W:** www.kohavision.net **Txs:** (-).

Local Stations
TV Besa: Rr. Kater Kullat n.n., 20000 Prizereni; ch30. **TV Dukagjini:** Rr. Fehmi Agani 16, Pejë; ch36. **TV Festina:** Rr. Deshmoret e Kombit n.n., Ferizaj; ch40. **TV Herc:** 73000 Shterpcë; ch35. **TV Iliria:** Rr. Hoxhë Jonuzi n.n., 61000 Viti; ch28. **TV Liria:** Rr. Reçak n.n., Ferizaj; ch29. **TV Men:** Rr. Nene Tereza 52, Gjilan; ch47. **TV Mir:** Rr. Vojske Jugoslavije n.n., Leposaviq; ch23. **TV Mitrovica:** Mitrovicë; ch42. **TV Most:** Rr. Nemanjica 14, Zveqan; ch61. **TV Opinion:** Rr. Asdreni 1, 20000 Prizereni; Zym ch28. **TV Prizren:** Rr. Papa Gjon Pali II 1A, Prizereni; ch60. **TV Puls:** Shillovë; Gjilan ch36. **TV Syri Vision:** Rr. Sadik Pozhegu 28, Gjakovë; ch33. **TV Tema:** Rr. Sadik Bega n.n., Ferizaj; ch50. **TV Vali:** Pasjak; Gjilan ch39. **TV Zoom:** Kuvcë e Epërme; ch43. **TV Yeni Donem:** Rr. Gjeravica 13A, 20000 Prizereni; ch53. **TV 3K:** 38217 Soqanicë; ch52.

System: PAL-B/G [E]

KUWAIT TELEVISION (Gov) ☜ P.O. Box 621, 13007 Safat ☎ +965 22415300 ▤ +965 22454233 **W:** www.moinfo.kw.gov **Chs:** KTV1, KTV2, KTV3, KTV4 **Txs: KTV1:** Failaka ch8 (950kW), Failaka ch24 (4500kW), Moi ch26 (25kW), Rawdatein ch38 (4700kW), Ahmadi ch59 (100kW). **KTV2:** Failaka ch10 (1000kW), Moi ch28 (6kW), Failaka ch39 (4800kW), Dibdibah ch65 (400kW). **KTV3:** Mutlaa ch5 (455kW), Mutlaa ch47 (2200kW). **KTV4:** Mutlaa ch12 (600kW), Mutlaa ch45 (2200kW).

Foreign TV Relays
Al Masriyah (Egypt): Kuwait ch5; **MBC (Saudi Arabia):** Kuwait ch11.

System: SECAM-D/K [R], PAL-D/K [R]

KYRGYZ TELEVISION (Gov) ☜ blvd. Jash Gvardiya 59, 720010 Bishkek ☎ +996 312 253404 ▤ +996 312 651064 **E:** ntrk@ntrk.kg **W:** www.ntrk.kg **Txs:** Bishkek ch1 (5kW) & netw. – **ELTR (Pub)** ☜ blvd. Erkindik 122, 720040 Bishkek ☎ +996 312 906144 **LP:** Dir: Shayyrbek Abdrakhmanov **E:** eltr@ktnet.kg **Txs:** Bishkek ch41 & netw.

Local Stations, Foreign TV Relays not shown.

System: # PAL-B/G [E] ⇩2015; DVB-T planned

National Station
TELEVISION NATIONALE LAO (TVNL) (Gov) ☜ BP 5635, Vientiane ☎ + 856 21 710643 ▤ +856 21 710182 **E:** lntv@gov.la **W:** www.lntv.gov.la **LP:** DG: Bouasone Phongphavanh. **Txs:** Vientiane ch9 (5kW) & relay txs.

Local Stations
Attapeu TV: Ban Veuankham, Samakhisai, Attapeu; ch7 (0.1 kW). **Bokeo TV:** Hoay Xai, Bokeo Prov.; ch5 (0.3kW). **Borikhamsai TV:** Ban Hongxai, Paksan, Borikhamsai Prov.; Paksan ch8 (2 kW). **Champassak TV:** Ban Souansawan, Pakse, Champassak Province; Pakse ch11 (2 kW). **Houa Phan TV:** Ban That Muang, Sam Neua, Houa Phan Prov.; Sam Neua ch8 (0.5 kW). **Khammouane TV:** Ban Pakdong, Thakhek, Khammouame Prov.; Thakhek ch6 (2 kW). **Lao Television Channel 3:** PO Box 860, Thatluang Road, Ban Nongbone, Vientiane; ch5 (10 kW). **Luang Namtha TV:** Ban Saysomboun, Luang Namtha Province; Luang Namtha ch9 (0.01 kW). **Luang Prabang TV:** Ban Naxang, Luang Prabang; Luang Prabang ch9 (0.3 kW). **Oudomxai TV:** Muang Xay, Oudomxai Prov.; ch8 (0.1 kW). **Phongsali TV:** Ban Phonekeo, Phongsali Prov.; Phongsali ch7 (0.3 kW). **Saiyabouli TV:** Ban Longpo, Saiyabouli Prov.; Saiyabouli ch8 (2 kW). **Saravane TV:** Ban Laksong, Saravane Prov.; Saravane ch5 (2 kW). **Savannakhet TV:** Ban Dongdangdouane, Khantabouli, Savannakhet; ch12 (2 kW). **Sekong TV:** Ban Phiamai, Lamam, Sekong Prov.; Sekong ch9 (0.3 kW). **Xieng Khouang TV:** Ban Phonekham, Muang Pek, Xieng Khouang Prov.; Xieng Khouang ch8 (2 kW).

Foreign TV Relays
VTV4 (Vietnam): Vientiane ch11 (20kW).

Systems: DVB-T (MPEG4); Local stns: # PAL-D/K [R]

National Stations
LATVIJAS TELEVIZIJA (LTV) (Pub) ☜ Zakusalas krastmala 3, 1509 Riga ☎ +371 67200314 ▤ +371 67200025 **E:** ltv@ltv.lv **W:** www.ltv.lv **LP:** DG: Edgars Kots **Chs:** LTV1, LTV7. – **LATVIJAS NEATKARIGA TELEVIZIJA (LNT) (Comm)** ☜ Elijas iela 17, 1050 Riga ☎ +371 67070200 ▤ +371 67821128 **E:** lnt@lnt.lv **W:** www.lnt.lv **LP:** DG: Andrejs Ekis. – **TV3 VIASAT / TV6 VIASAT (Comm)** ☜ Maskavas iela 322, 1063 Riga ☎ +371 67629366 ▤ +371 67600599 **E:** tv3@tv3.lv / tv6@tv6.lv **W:** www.tv3.lv / www.tv6.lv **LP:** CEO: Baiba Zuzena. – **TV5 (Comm)** ☜ Elijas iela 7, 1050 Riga ☎ +371 7503924 ▤ +371 7503925 **E:** info@tv5.lv **W:** www.tv5.lv **LP:** GD: Michael Sheitelman.

Local Stations (all Comm, all analogue)
Aizputes TV: Jelgavas iela 26, 3456 Aizpute; ch37 (0.05kW). **Dagdas TV:** Alejas iela 29, 5674 Dagda; ch34 (0.8kW). **Gimenes TV:** Lašu iela 5-77, 2010 Jurmala; ch41 (10kW). **Gulbenes TV:** Udensvada iela 2a, 4401 Gulbene; ch11 (0.1kW). **Latgales Regionala Televizija:** 18. Novembra iela 16, 4601 Rezekne; ch10 (0.2kW). **Limbazu TV:** Cesu iela 22, 4001 Limbazi; ch37 (1kW). **Livanu TV:** Rigas iela 112, 5316 Livani; ch12 (0.4kW). **Ogres Televizija:** Rigas iela 98a, 5001 Ogre; ch2 (0.1kW). **Rezeknes TV:** 18.novembra iela 16, 4601 Rezekne; ch10 (0.2kW). **Rujienas TV:** Raina iela 3, 4240 Rujiena; ch7 (0.08kW). **Selijas NTV 6:** Smilšu iela 2, 5237 Viesite; ch6 (1.6kW). **Skrundas TV:** Liela iela 1a, 3326 Skrunda; ch9 (0.1kW). **Smiltenes TV:** Dakteru iela 73, 4729 Smiltene; ch7 (0.15kW). **Talsu TV:** Krišjana Valdemara iela 17a, 3201 Talsi; ch23 (0.8kW). **TV Dzintare:** Graudu 23, 3401 Liepaja; ch12 (1kW). **TV Jelgava:** Aviacijas iela 18, 3001 Jelgava; ch33 (0.85kW). **TV Miljons:** Rigas iela 26, 5403 Daugavpils; ch25 (4kW). **TV Spektrs:** Nakotnes iela 1, 2152 Malpils; ch12 (0.15kW). **TV Vidzeme:** Raunas iela 9a, 4101 Cesis; ch7 (0.06kW). **TV Vilani:** Rezeknes iela 6a, 4650 Vilani; ch2 (0.01kW). **Valmieras TV:** Rigas iela 4, 4201 Valmiera; Valmiera ch9 (0.16kW), Cesis ch46 (1.25kW), Madona ch56 (1.25). **Ventspils TV:** Gertrudes iela 13, 3601 Ventspils; ch45 (0.7kW). **Vidusdaugavas Televizija:** Brivibas iela 2d, 5200 Jekabpils; ch34 (0.4kW). **Zemgales Novadu Televizija:** Krišjana Valdemara iela 7, 3701 Dobele; ch8 (0.35kW).

DTT Transmitters

Operator: Lattelcom SIA ▣ Dzirnavu iela 105, 1011 Riga **E:** lattelecom@lattelecom.lv **W:** www.lattelecom.lv **Mux 1:** LTV1, LTV7, LNT, TV3, TV5○ **Mux 2 (○):** TV3+, TV6, National Geographic Channel, Diva Universal, Cartoon Network/TCM, PBK Latvia, Ren-TV Baltic, RTR Planeta **Mux 3 (○):** oe, LMK, Euronews, Discovery Channel, Animal Planet, Fox Crime, MTV Europe, Eurosport 2, NTV Mir **Mux 4 (○):** Disney Channel, SciFi, Lox Life, ID Investigation Discovery, Disney XD, Silver Baltic, BBC Entertainment **Mux 5 (○):** Discovery HD, Showcase HD.

Location	M1	M2	M3	M4	M5
Aluksne	41	58	22	69	-
Cesvaine	41	58	22	69	-
Daugavpils	47	51	64	27	-
Kuldiga	30	40	47	52	-
Liepaja	21	61	62	69	-
Rezekne	44	50	62	27	-
Riga	43	45	48	66	59
Valmiera	21	51	54	50	-
Ventspils	30	40	47	52	-

+ repeaters.

LEBANON

Systems: SECAM-B/G [E]; DVB-T planned

TÉLÉ-LIBAN (Gov) ▣ BP 115054, Hazmieh, Beirut ☎ +961 1 792000 ▤ +961 1 786921 **E:** tl@tele-liban.com **LP:** DG: Jean-Claude Boulos. **Txs:** Beirut ch5 (50kW) & network. – **AL-MANAR TV** ▣ BP 354/25, Beirut ☎ +961 3 217405 ▤ +961 1 555953 **E:** info@manartv.com **W:** www.manartv.com **Txs:** ch28, ch37, ch46, ch52. – **FUTURE TELEVISION (Comm)** ▣ BP 13-6052, Beirut ☎ +961 1 355355 ▤ +961 1 753232 **E:** future@future.com.lb **W:** www.futuretvnetwork.com **Txs:** ch28, ch37, ch46, ch52. – **LEBANESE BROADCASTING CORP. (LBC) (Comm)** ▣ BP 165853, Zouk 111, Beirut ☎ +961 9 850850 ▤ +961 9 850916 **E:** lbcsat@lbcsat.com.lb **W:** www.lbcgroup.tv **LP:** SM: Pierre Al Daher. TD: Nasim Boustany. **Txs:** (pol.H) ch5 (35kW), ch9 (35kW), ch10 (35kW), ch12 (60kW), ch33 (325kW). – **MTV (Comm)** ▣ Beirut **W:** www.mtv.com.lb **Txs:** (-). – **NATIONAL BROADCASTING NETWORK (NBN) (Comm)** ▣ BP 13-6633 Chouran, Beirut ☎ +961 1 841020 ▤ +961 1 841029 **E:** info@nbn.com.lb **W:** www.nbn.com.lb **Txs:** (-). – **NEW TV (Comm)** ▣ BP 110, 5958 Beirut ☎ +961 1 303300 ▤ +961 1 818389 **E:** info@newtvsat.com **W:** www.newtvsat. **Txs:** (-).

LESOTHO

System: # PAL-I [E] ✃2015 ; DVB-T planned

LESOTHO TELEVISION (Gov) ▣ P.O.Box 552, Maseru 100 ☎ +266 22323561 ▤ +266 22310003 **LP:** Dir: Dada Moqasa; CE: Motlatsi Monyane. **Txs:** Katse (Terata), Semongkong (Thaba-Nts'o), Hilton Hill (Maseru) ch25; Quthing, Hilton Hill (Maseru) ch29; Mokhotlong ch36; Masite ch39; Berea Plateau ch47; Mohale's Hoek, Thabe-Putsoa ch51; Thaba-Tseke ch54; Leribe, Souru (Qacha's Nek) ch58; Baking, Mafeteng ch62.

Foreign TV Relay
TBN (USA): Berea Plateau ch21, Leribe ch41.

LIBERIA

System: PAL-B [E]

CLAR TV (Comm) ▣ Ashmun & Mechlin Street, Monrovia ☎ +231 6522511 **E:** royalcomlr@yahoo.com **Tx:** Monrovia ch5. – **DC TV (Comm)** ▣ P.O.Box 1312, Monrovia. **Tx:** Monrovia ch11. – **POWER TV (Comm)** ▣ Broad & Gurley Street, Monrovia ☎ +231 6514343 **Tx:** Monrovia ch9. – **REAL TV (Comm)** ▣ Monrovia. **Tx:** Monrovia ch3.

LIBYA

System: PAL-B/G [E]; DVB-T planned

LIBYAN JAMAHIRIYA BROADCASTING CORP. (LJBC) (Gov) ▣ P.O.Box 80237, Tripoli ☎ +218 21 3402153 ▤ +218 21 3403458 **W:** www.ljbc.net **LP:** Secr Gen: Abdallah Mansour; TD: Ammar El Mahjoub. **Txs** (pol.H exc. where stated)**:** Benghazi ch5 (10kW V), Tripoli ch6 (20kW), Derna ch6V (5kW), Elmarj ch7V (5kW), Houn ch7

(1kW), Sirte ch7 (1kW), Khoms ch8 (5kW), Tobruk ch8 (5kW), Yevren ch9 (20kW), El Beida ch9 (5kW), Misurata ch10 (5kW), Egdabia ch11 (15kW) & repeaters.

LIECHTENSTEIN

NB: No terrestrial TV station.

LITHUANIA

Systems: DVB-T (MPEG2, MPEG4); † PAL-D/K [R] ✃29 Oct 2012

National Stations

LIETUVOS TELEVIZIJA (LTV) (Pub) ▣ Konarskio g. 49, 03123 Vilnius ☎ +370 5 2363100 ▤ +370 5 2363208 **E:** lrt@lrt.lt **W:** www.lrt.lt **LP:** Dir: Rimvydas Paleckis. **Chs:** LTV, LTV2 – **BALTIJOS TELEVIZIJA (BTV) (Comm)** ▣ Laisves pr. 60, 05120 Vilnius. ☎ +370 5 2780805 ▤ +370 5 2428907 **E:** info@btv.lt **W:** www.btv.lt – **LIETUVOS NEPRIKLAUSIMAS KANALAS (LNK) (Comm)** ▣ Šeškines g. 20, 07156 Vilnius ☎ +370 5 2431058 ▤ +370 5 2123924 **E:** lnk@lnk.lt **W:** www.lnk.lt **Chs:** LNK, Info TV, Liuks! **LIETUVOS RYTAS TV (Comm)** ▣ Gedimino pr. 12 A, 01103 Vilnius ☎ +370 5 2743718 ▤ +370 5 2657338 **E:** tv@lrytas.lt **W:** www.lietuvosrytas.lt – **TV1 (Comm)** ▣ Šeškines g. 20, 07156 Vilnius ☎ +370 5 2615287 ▤ +370 5 2123924 **E:** info@tv1.lt **W:** www.tv1.lt – **TV3 VIASAT / TV6 VIASAT (Comm)** ▣ Kalvariju g. 143, 08221 Vilnius ☎ +370 5 2030101 ▤ +370 5 2030103 **E:** info@@tv3.lt **W:** www.tv3.lt / www.6tv.lt **Chs:** TV3, TV6.

Local Stations (all Comm)

Aidas: Birutes g. 42, 21117 Trakai; Trakai ch6 (0.5kW), Papliauškai ch8 (0.2kW). **Alytaus regionine televizija:** Naujoji g. 124, 62001 Alytus; ch35 (0.4kW). **Aukštaitijos krašto televizija:** Pazagieniu g. 29, Pazagieniai, 35185 Panevežio raj.; ch56 (0.7kW). **Aukštaitijos savaite:** Aušros g. 50, 28001 Utena. Txs: via LTV1 txs Utena ch6, Ignalina ch28, Visaginas ch30. **Balticum TV:** Taikos pr. 101, 94198 Klaipeda; ch12 (2kW). **Dzukijos TV:** Pramones g. 4, 66181 Druskininkai; ch6 (0.2kW). **LN Televizija:** Plateliu g. 17, Plunge; ch32 (0.3kW). **Kedainiu krašto televizija:** Basanaviciaus g. 36, 57288 Kedainiai; ch35 (2kW). **KTV plius:** Nemuno g. 79, 37355 Panevezys; Pazagieniai ch47 (0.5kW). **Kupiškenu studija:** Krantines g. 2-28, 40126 Kupiškis; via LTV1 tx Viešintos ch12. **Marijampoles televizija:** Gedimino g. 11, 68307 Marijampole; ch6 (0.1kW). **PAN TV:** Respublikos g. 19-8, 35185 Panevezys; Pazagieniai ch8 (0.2kW). **PTV:** Peršekininku kaimas, Miroslavo sen., 64262 Alytaus r.; Peršekininkai ch12 (0.03kW). **Pukas-TV:** Ringuvos g. 61, 45242 Kaunas; Juragiai ch52 (1kW). **Raseiniu krašto televizija:** Vaižganto g. 20, 60130 Raseiniai; ch31 (0.15kW). **Roventos TV:** Draugystes g. 18, 89168 Mazeikiai; ch37 (0.5kW). **Siauliu televizija:** Liejyklos g. 10, 78147 Šiauliai; ch55 (5kW). **S Plius:** Tilzes g. 74, 78140 Šiauliai; Bubiai ch38 (1kW). **Televizija „Pajurio švyturys":** Medvalakio g. 27, Palanga; ch6 (0.4kW). **TV11:** Konarskio g. 49, 02123 Vilnius; ch11. **Vakaru Lietuvos televizija:** Zveju g. 2, 91248 Klaipeda; ch10 (1kW). **Ventos regionine televizija:** Ventos g. 32a, Venta, 85316 Akmenes r.; Venta ch9 (0.1kW). **Vidiškiu televizija:** Ignalinos g. 1A, Vidiškiu, 30234 Ignalinos r.; Vidiškes ch12 (0.1kW).

DTT Transmitters (under construction)

Operator: Teo LT ▣ Savanoriu pr. 28, 03501 Vilnius ☎ +370 5 2621511 ▤ +370 5 2126665 **E:** info@teo.lt **W:** www.teo.lt **Mux 1:** LTV, LTV2, LNK, TV3, TV1, BTV, TV6, Info TV, Liuks! Showtime○ **Mux 2 (○):** BBC World News*, CNN International, MTV, Cartoon Channel, RTL, NBA TV, RTR Planeta, Ren-TV Baltics, RTVi, Balticum Auksinis **Mux 3 (○exc.*):** Lietuvos rytas TV*, Animal Planet, Discovery Channel, Discovery World Discovery Science, Eurosport, National Geographic, Travel Channel, Diva Universal, Sport 1 **Mux 4 (○exc.*):** Balticum TV*, TV Polonia*, VH1, BBC Entertainment, Nickeloeon, Eurosport 2, Fashion TV, Multimania, Discovery Travel & Living, Euronews **Mux HD (○):** Discovery Channel HD, Eurosport HD, National Geographic Channel HD **Tx:** Vilnius ch37.

Location	M1	M2	M3	M4	kW
Bubiai	57	63	36	53	2x1.3/2x1.2
Girulai	60	38	36	53	2x0.8/2x1.2
Juragai	44	33	59	60	2x1.3/1.2/1.3
Viešintos	58	62	36	-	2x1/0.8
Vilnius	57	64	50	60	2x1.3/2x1.6

+ sites with txs below 1kW

Operator: Balticum TV ▣ Taikos pr. 101, 94198 Klaipeda ☎ +370 46 390709 ▤ +370 46 342815 **E:** balticum-tv.lt **W:** www.balticum-tv.lt **Mux (○) (MPEG2):** Balticum Auksinis, FOX Crime, FOX Life, National Geographic Wild, NTV Mir, TV1000 East, Viasat History,

Viasat Explorer, Zone Reality **Tx:** Vilnus ch53 (0.6kW).

LORD HOWE ISLAND (Australia)

NB: No terrestrial TV station.

LUXEMBOURG

System: DVB-T (MPEG2)

RTL MULTI MEDIA (Comm) ✉ 45, blvd Pierre Frieden, 1543 Luxembourg ☎ +352 421421 📠 +352 421422760 **W:** www.rtl.lu **L.P:** Chmn: Jacques Santer. **NB:** RTL Multi Media is the company holding for several radio & TV enterprises, a.o.: **RTL Télé Lëtzebuerg** (in Luxembourgish); **RTL9** (in French, for viewers in France); **RTL TVi** (in French, for viewers in Belgium); **RTL4, RTL5, RTL7, RTL8** (in Dutch, for viewers in the Netherlands); **Club RTL, Plug TV** (in Flemish, for viewers in Belgium); **RTL Television** (in German, for viewers in Germany).

DTT Transmitters
Operator: RTL Multi Media **Mux 1:** RTL TVi, Club RTL, Plug TV, RTL4, RTL5, RTL7 **Tx:** ch24 (Dudelange 40kW) **Mux 2:** RTL Télé Letzebuerg, den 2ten RTL **Tx:** ch27 (Dudelange 40kW) **Mux 3:** M6, RTL8 **Tx:** ch7 (Dudelange).

MACAU (China, SAR)

Systems: # PAL-I [E]; DTMB

TELEDIFUSÃO DE MACAU S.A.R.L. (TDM) (Gov) ✉ CP 446, Macau ☎ +853 28520206 📠 +853 28520208 **E:** tdmadm@tdm.com.mo **W:** www.tdm.com.mo **L.P:** CEO: Manuel Maria Dos Santos Gonçalves **Txs:** TV1 (Portuguese): Monte da Guia ch30 (0.2kW) & 1 repeater; TV2 (Chinese): Monte da Guia ch32 (0.2kW) & 1 repeater.

DTT Transmitters (Trial)
Operator: TDM **Mux:** TDM, Canal Macau, Macau HD. **Tx:** (-).

MACEDONIA

Systems: # PAL-B/G [E] ⇩May 2012; DVB-T (MPEG4)

National Stations
MAKEDONSKA TELEVIZIJA (MTV) (Pub) ✉ ul. Dolno Nerezi bb, 1000 Skopje ☎ +389 2 258230 📠 +389 2 3112578 **W:** www.mtv.com.mk **L.P:** DirTV: Jane Teoharevski **Chs:** MTV1, MTV2, Sobranski kanal.

Location	MTV1	MTV2	SK	kW
Belasica	7	35	54	6/-/-
Boskija	8	57	54	10/20/2
Crn Vrv	6	30	40	100/-/10
Golak	5	38	44	10/-/-
Mali Vlaj	9	44	50	10/-/-
Ohrid	6	21	-	-/4
Pelister	4	29	33	30/-/10
Popova Šapka	10	38	34	10/-/-
Stogovo	6	31	37	10/-/-
Turtel	11	22	32	50/1000/-
Vodno	9	26	36	5/2/20

+ sites with txs below 1kW.
A1 TELEVIZIJA (Comm) ✉ ul. Pero Nakov bb, 1000 Skopje ☎ +389 2 2550301.📠 +389 2 2550330 **E:** a1tv@a1.com.mk **W:** www.a1.com.mk **L.P:** GM: Gordana Stoshich **Txs:** Vodno (Skopje) ch47 & network. –
ALSAT-M (Comm) ✉ ul. Krste Misirkov 7, DTC Mavrovka lom. C kat 9, 1000 Skopje ☎ +389 2 3290364 📠 +389 2 3290365 **W:** www.alsat.tv **Txs:** Vodno (Skopje) ch45 & network. NB: relays Alsat-M (Albania). – **KANAL 5 (Comm)** ✉ ul. Skupi bb, 1000 Skopje ☎ +389 2 3091551 📠 +389 2 3091560 **E:** kanal5@kanal5.com.mk **W:** kanal5.com.mk **Txs:** Vodno (Skopje) ch50 & network. – **TV SITEL (Comm)** ✉ ul. Gradski stadion bb, 1000 Skopje ☎ +389 2 3116566 📠 +389 2 3229799 **E:** marketing@sitel.com.mk **W:** www.sitel.com.mk **Txs:** Vodno (Skopje) ch57 & network. – **TV TELMA (Comm)** ✉ ul. Nikola Parapunov bb, 1000 Skopje ☎ +389 2 3076677 📠 +389 2 3070040 **E:** telma@unet.com.mk **W:** telma.com.mk.**Txs:** Vodno (Skopje) ch43 & network.

Local Stations
TV Alfa: Gradski stadion, 1000 Skopje; Raštak ch37, Sredno Vodno ch63. **TV Amazon:** ul. Gemdziska 135-b, 1000 Skopje; Raštak ch58, Sredno Vodno ch65 (time shared with TV Edo/TV KRT). **TV Anisa:** 6530 Plasnica; ch35. **TV Art:** ul. Ohridska 18, 1200 Tetovo; Kaleto ch36. **TV Art Kanal:** ul. Kuzman Šapkarev 12, 6330 Struga; ch22. **TV Boem:** ul. Cvetan Jakovlevski 3, 6250 Kicevo; ch42. **TV BTR Nacional:** ul. 376 br.108, 1000 Skopje; Sredno Vofno ch21. **TV Dalga-KRT:** ul. Partizanska 1, 1321 Kumanovo; Stari Lozja ch34. **TV Due:** ul. Mara Ugrinova bb, 1230 Gostivar; ch59. **TV Edo:** ul. 103 br. 50, 1000 Skopje; Raštak ch58, Sredno Vodno ch65 (time shared with TV Amazon/TV KRT). **TV EMI:** ul. 1-vi maj 34, 2420 Radoviš; ch48. **TV Era:** ul. Dzon Kennedi 9-a, 1010 Skopje-Cair; Sredno Vodno ch29, Gazi Baba ch53. **TV Gurra:** ul. Maršal Tito 172, 6250 Kicevo; Kaleto ch45. **TV Hana:** ul. Sima Pogacarevic bb, 1300 Kumanovo; Stari Lozja ch28. **TV Intel:** ul. Gorgi Trajcev 54, 2400 Strumica; ch24. **TV Iris:** ul. Goce Delcev 44, 2000 Štip; ch30. **TV Kaltrina:** ul. Boro Šain 19-A, 6330 Struga; ch38. **TV Kanal Festa:** ul. Kiro Antevski 15, 1300 Kumanovo; ch38. **TV Kanal 21:** ul. Alekso Deminevski-Bauman bb, 1400 Veles; Bašino ch21. **TV Kiss:** ul. Blagoja Toska 10, 1200 Tetovo; ch31. **TV Ko-Bra:** bul. Aleksandar Makedonski bb, 2420 Radoviš; ch43. **TV Koha:** ul. GTC Trijada 2-ri kat, lok.77, 1200 Tetovo; ch45. **TV KTV-41:** ul. Ploštad Maršal Titobb, 1430 Kavadarci; ch25. **TV KRT:** ul. Anton Popov 103-b, 1000 Skopje-Gazi Baba; Raštak ch58, Sredno Vodno ch65 (time shared with TV Amazon/TV Edo) **TV Medi:** ul. Jorgo Kostovski bb, 7000 Bitola; Brusnik ch51. **TV Menada:** ul. Maršal Tito 36, 1200 Tetovo; Kaleto ch29. **TV Moris:** ul. Pariska komuna 34, 6000 Ohrid; Kaleto ch52. **TV MS (MakSpot):** ul. Nikola Rusinski bb, 1000 Skopje; Raštak ch35, Sredno Vodno ch55. **TV MTM:** ul. Blagoja Stefkovski 40, 1000 Skopje; Gazi Baba ch27, Sredno Vodno ch28. **TV Nova:** ul. Ilindenska bb, 1300 Kumanovo; Stari Lozja ch25. **TV NTV:** ul. Kej Maršal Tito bb, 6000 Ohrid; Gorenicka cuka ch32. **TV Orbis:** ul. Rieka 7a, 7000 Bitola; Kopanki ch39, Pirava ch48. **TV Protel:** ul. Jordan Stojanov 2, 2210 Probištip; ch34. **TV SkajNet:** ul. Vasil Glavinov 16-6, 1000 Skopje; Gazi Baba ch48, Sredno Vodno ch61. **TV Star:** ul. Banco Prke bb, 2000 Štip; ch27. **TV Super Skaj:** ul. 169 br.162, 1201 Mala Recica; ch43.**TV Svet:** ul. Makedonska 29, 2220 Sveti Nikole; ch27. **TV Šutel:** ul. Ce Gevara 37, 1000 Skopje; Sredno Vodno ch59. **TV Tera:** ul. Milton Manaki 21, 7000 Bitola; Pirava ch32, Brusnik ch37. **TV Uskana:** bul. Osloboduvanje 33, 6250 Kicevo; Kicevo ch35, Cocon ch49. **TV Vis:** ul. Blagoj Jankov-Muceto 54, 2400 Strumica; Bair ch28. **TV VTV:** ul. Bojmija 10, 2460 Valandovo; ch47. **TV Zdravkin:** ul. Dimce Mircev 1, 1400 Veles; ch27. **TV Zupa:** 1258 Mal. Papradnik; ch29 & ch55. **TVM:** ul.Nada Fileva 40, 6000 Ohrid; Kaleto ch28.

DTT Transmitters (under construction)
Operator: Cosmofon ✉ Bul. Kuzman Josifovski Pitu 15, 1000 Skopje ☎ +389 2 441000 📠 +389 2 441122 **W:** www.cosmofon.com.mk **Mux 1:** MTV1, MTV2, Sobranski kanal, MKTV Sat **Tx:** ch39 (Skopje 0.1kW) & network under construction. **Mux 2+3 (✿):** multiprgr. **Txs:** (-) Network under construction.

MADAGASCAR

System: SECAM-K1 [E]

National Station
TÉLÉVISION MALAGASY (TVM) (Pub) ✉ BP 271, 101 Antananarivo ☎ +261 20 2221784 📠 +261 20 2232815 **E:** tvm@dts.mg **W:** www.tvm.mg **Txs:** Antananarivo ch5 (1kW H) & relay txs.

Local Stations
CANAL F+ Fianarantsoa: Près Lot 408 IB Andohan'Ivory, 301 Fianarantsoa; ch28. **MATV:** BP 1414, 101 Antananarivo; ch23, ch27. **M3TV:** 29 rue de la Libération, Mahajanga; ch55. **OTV:** BP 4100, 101 Antananarivo; ch39, ch43. **R.T.T. Toamasina:** BP 12170, 101 Antananarivo; ch34. **RTVA:** BP 258, 110 Antsirabe; ch32. **RTV Analamanga:** BP 12170, 101 Antananarivo; ch30, ch33, ch47. **RTV Kalizy:** Zone Zital Ankorondrano, Antananarivo; ch30. **RTV Record:** BP 7522, 101 Antananarivo; ch51. **RTV Soafia:** Zorozoroana Ambalakosoa; ch47. **RTV Soavinandriana:** FKT Ambatombositra Soavinandriana Ambany, Soavinandriana; ch38. **TV Plus Madagascar:** Lot IVW 3D Anosizato - Est, 101 Antananarivo; ch31. **TV Ny Antsika Antalaha:** BP 101, 206 Antalaha; ch21. **TV Soa Menabe:** BP 405 619 Morondava; ch32. **TV Viva:** Parcelle 34; Zone Tana Water Front, 101 Antananarivo; ch53, ch57.

MADEIRA (Portugal)

System: # PAL-B/G [E] ⇩22 Mar 2012; DVB-T (MPEG4)

RTP MADEIRA (Pub) ✉ Rua Caminho de Santo António 145, 9020-002 Funchal ☎ +351 291709100 📠 +351 291741859 **E:** leonel.freitas@rtp.pt **L.P:** Dir: Leonel de Freitas. **Chs:** RTP1 (relay from mainland); RTP Madeira. **Txs:** RTP1: Funchal (Pico do Silva) ch40 (74.6kW) & relay txs. **RTP Madeira:** Funchal (Pico do Silva) ch5 (9.5kW) & relay txs.

DTT Transmitters (txs under construction)
Operator: Portugal Telecom **Mux 1:** RTP1, RTP2, RTP Madeira, RTP Açores, SIC, TVI **Txs:** ch67 (SFN). **Mux 2 (✪):** tbd **Txs:** ch63 (SFN). **Mux 3 (✪):** tbd **Txs:** ch69 (SFN).

MALAWI

System: PAL-I [E]

TELEVISION MALAWI (TVM) (Pub) ✉ Private bag 268, Blantyre ☎ +265 1675033 🖷 +265 1672627 **E:** tvmalawi@malawi.net **L.P:** DG: Rodrick Mulonya. **Txs:** (-). – **AFJ TV (Rlg)** ✉ Blantyre. **Txs:** (-). – **TBN MALAWI (Rlg)** ✉ Blantyre. **Txs:** Lilongwe ch32, Blantyre ch60.

MALAYSIA

Systems: # PAL-B/G [E]; DVB-T (MPEG4)

RADIO TELEVISYEN MALAYSIA (RTM) (Gov) ✉ Dept. of Broadc, Angkasapuri, Kuala Lumpur 50614 ☎ +60 3 22825333 🖷 +60 3 22825103 **E:** programtv@rtm.net.my **W:** www.rtm.net.my **L.P:** DG: Barbara E. Edmonds **Chs:** TV1 (mainly in Malay) iTV2 (in Chinese, English, Malay and Tamil), regional stns. **Txs: TV1:** Kuala Lumpur ch50 & netw.; **iTV:** Kuala Lumpur ch53 & netw.. – **METROPOLITAN TV (8TV)** ✉ Metropolitan TV sdn bhd, Sri Pentas, 3 Persiaran Bandar Utama, 47800 Petaling Jaya, Selangor Darul Ehsan ☎ +60 3 77288282 🖷 +60 3 77268282 **E:** info@8tv.com.my **W:** www.8tv.com. my. **Txs:** Kuala Lumpur ch58 & netw. NB. 8TV is affiliated to TV3. – **NAT SEVEN (NTV7) (Comm)** ✉ 7, Jalan Jurubina U1/18, Hicom-Glenmarie Industrial Park, 40000 Shah Alam, Selangor Darul Ehsan ☎ +60 3 55691777 🖷 +60 3 55692515 **E:** feedback@ntv7.com.my **W:** www.ntv7.com.my. **Txs:** Kuala Lumpur ch7 & netw. – **SYSTEM TV MALAYSIA BERHAD (TV3)** ✉ Sri Pentas (Ground Floor, South Wing) No. 3, Persiaran Banjar Utama, 47800 Petaling Jaya Selangor Darul Ehsan ☎ +60 3 7166333 🖷 +60 3 77278455 **E:** query@tv3.com. my **W:** www.tv3.com.my **L.P:** MD: Hisham Abdul Rahman. **Txs:** Kuala Lumpur ch12 & netw. – **TV9 (Comm)** ✉ Lot 31, Jalan Pelukis U1/46, Temasya Industrial Park, 40150 Shah Alam, Selangor Darul Ehsan ☎ +60 3 55685999 **W:** www.tv9.com.my. **Txs:** Kuala Lumpur ch33 & netw.

DTT Transmitters (under construction)
Operator: MiTV ✉ 3rd Floor, KL Plaza, 179 Jalan Bukit Bintang, 55100 Kuala Lumpur **W:** www.mitv.com.my **Mux (✪):** multiprgr. **Txs:** (-).

MALDIVES

System: PAL-B [E]

TV MALDIVES (Gov) ✉ Buruzu Magu, 20-04 Malé ☎ +960 3323105 🖷 +960 3325083 **E:** admin@tvm.gov.mv **W:** www.tvm.gov. mv **Chs:** TVM, TVM Plus. **Txs: TVM:** Malé ch7 (1kW H); **TVM Plus ✪:** (-).

MALI

System: SECAM-B/G [E]

RADIODIFFUSION TÉLÉVISION DU MALI (Pub) ✉ BP 171, Bamako ☎ +223 20212019 🖷 +223 20214205 **E:** ortm@afribone. net.ml **W:** www.ortm.net **L.P:** DG: Sidiki Konate **Txs:** Bamako ch5 (10kW) & relay txs.

MALTA

Systems: DVB-T (MPEG2)

TELEVISION MALTA (TVM) (Pub) ✉ PBS, 75, Triq San Luqa, Gwardamangia, Pieta', PTA 1022 ☎ +356 21225051 🖷 +356 21244601 **E:** info@pbs.com.mt **W:** www.pbs.com.mt **L.P:** Chmn: Clare Thake Vassallo. – **EDUCATION 22 (E22) (Gov)** ✉ Maria Regina School, Mile End Road, Hamrun, HMR 1716 ☎ +356 21239274 🖷 +356 21240701 **E:** info@e22.com.mt **W:** www.e22.com.mt **L.P:** Chmn: Stephen Azzopardi. – **CALYPSO MUSIC TV (Comm)** ✉ 28, New Street in Valletta Road, Luqa, LQA 6000 ☎ +356 21578022 🖷 +356 21578026 **E:** info@calypsoradio.com **W:** www.calypsoradio.com **L.P:** Dirs: Frank Camilleri, Alfred Cacciatolo – **NET TV (Comm)** ✉ Media. Link Communications, Dar Centrali, Triq Herbert Ganado, Pieta', PTA 1450 ☎ +356 21243641 🖷 +356 21242886 **E:** news@media.link.

com.mt **W:** www.nettv.com.mt **L.P:** Chmn: Joe Saliba. – **ONE TV (Comm)** ✉ ONE Productions Ltd., A28b, Qasam Industrijali, Marsa, MRS 3000 ☎ +356 25682568 🖷 +356 21248249 **E:** sales@one.com. mt **W:** www.one.com.mt **L.P:** Chmn: Jason Micallef. – **SMASH TV (Comm)** ✉ 2, Triq Tax-Xewk, Paola, PLA 1341 ☎ +356 21697829 🖷 +356 21697830 **E:** smash@vol.net.mt **W:** www.smashmalta.com **L.P:** Head: Jesmond Saliba.

DTT Transmitters
Operator: PBS **Mux:** TVM, E22, Net TV, One TV, Smash TV **Txs:** ch5, ch66 (SFN). – **Operator:** GO ✉ Gnien Spencer, Marsa, MRS 1990 ☎ +356 21212121 🖷 +356 21248925 **E:** info@go.com.mt **W:** www. go.com.mt **Muxes (✪):** multiprgr (incl. Calypso Music TV) **Txs:** ch26, ch28, ch31, ch38, ch45, ch55, ch56, ch58, ch60, ch64, ch66 (MFN, txs in Gharghur, S.Juliens, Delimara, Kercem, Zebbug & others).

MARSHALL ISLANDS (USA associated)

NB: No terrestrial TV station.

MARTINIQUE (France)

System: # SECAM-K1 [K/E]; DVB-T

RFO MARTINIQUE (Pub) ✉ BP 662, 97263 Fort de France Cedex ☎ +596 596595200 **W:** martinique.rfo.fr **L.P:** Dir: Fred Jouhoud. **Txs:** Fort de France ch4 (1kW) & relay txs. – **ATV (ANTILLES TÉLÉVISION) (Comm)** ✉ 28 rue Arawaks, 97200 Fort de France ☎ +596 596754444 🖷 +596 596755565 **W:** www.antillestelevision.com **Txs:** Riviere Pilote ch34 (0.19kW), La Trinité ch39 (7kW), Fort de France ch44 (8 kW), La Morne Rouge ch52 (1.4 kW). – **CANAL ANTILLES (Comm) ✪** ✉ Centre Commercial La Galléria, 97232 Le Lamentin ☎ +596 596505787 **Txs:** La Trinite chE25 (7 kW), Port de France ch29 (8kW), Saint Pierre ch34 (0.6kW), Le Morne Rouge ch46 (1.4kW), Riviere Pilote ch50 (0.19 kW). – **KMT (KANAL MARTINIQUE TÉLÉVISION) (Comm)** ✉ Voie n° 7, Renéville, 97200 Fort de France ☎ +596 596718604 🖷 +596 596636485. **Txs:** (-).

DTT Transmitters
Operator: France Télévisions **Mux:** RFO Martinique, France 2-5, France Ô, Arte, LCP, France 24, Gulli **Txs:** ch(-).

MAURITANIA

System: SECAM-B [E]

TÉLÉVISION DE MAURITANIE (Pub) ✉ BP 5522, Nouakchott ☎ +222 5258017 🖷 +222 5254069 **E:** tvm@mauritania.mr **W:** www.tvm. mr **L.P:** DG: Ould Ebnou Abdlem **Chs:** TVM, TVM Plus **Txs: TVM:** Nouakchott ch5 (2kW H) & relay txs; **TVM Plus:** (-).

MAURITIUS

Systems: # SECAM-B/G [E]; DVB-T (MPEG2)

MAURITIUS BROADCASTING CORP. (Pub) ✉ BP 48, Curepipe ☎ +230 6755001 🖷 +230 6757332 **E:** dirgen@mbc.intnet.mu **W:** mbc. intnet.mu **L.P:** DG: Trilock Dwarka. **Chs:** MBC1, MBC2, MBC3; relays of Canal+ (France) & RTL9 (Luxembourg); Rodriges TV. **Txs:** (pol.H) **MBC1:** Malherbes ch4 (10kW) & translators; **MBC2:** Malherbes ch5 (0.1kW) & translators; **MBC3:** Malherbes ch27 (23kW) & translators; **Canal+ ✪:** Malherbes ch21 (23kW) & translators; **RTL9 ✪:** Malherbes ch24 (23kW) & translators; **Rodriges TV:** Mont Malartic ch11 (0.2kW) & translators.

DTT Transmitters
Operator: Multi Carrier (Mauritius Ltd) ✉ Clement Charoux Street, Malherbes, Curepipe ☎ +230 6746547 **E:** mcml@multi-carrier.net **W:** www.multi-carrier.net **Mux:** MBC1, BBC World News, TV5 Monde Afrique, CCTV 9, DD India/DD Bharati/Aastha International, B4U Music/MTV India. **Txs:** ch28 (Malherbes), ch30 (Signal Mt.).

MAYOTTE (France)

System: # SECAM-K1 [K/E]; DVB-T

RFO MAYOTTE (Pub) ✉ BP 103, F-97610 Pamandzi, Ile de Mayotte ☎ +262 269601017 🖷 +262 269601852 **W:** mayotte.rfo.fr **L.P:** Dir:

Robert Xavie; TechDir: Serge Sulpice-Timothee. **Ch:** Télé Pays. **Txs:** (pol.H) Lima Combani ch4 (0.4kW), Mamoudzou ch7 (0.07kW), La Vigie ch9 (0.25kW), Kani-Keli (Vatounkaridi) ch30 (0.031kW), Sada ch40 (0.155kW), Kani-Keli (Choungi) ch49 (0.043kW), Koungou ch49 (0.085kW).

DTT Transmitters
Operator: France Télévisions **Mux:** RFO Mayotte, France 2-5, France Ô, Arte, LCP, France 24, Gulli **Txs:** ch(-).

MEXICO

Systems: # NTSC-M [A] ↧2015; ATSC

National Stations [a]) analogue txs
TELEVISIÓN AZTECA SA de CV (Comm) ▣ Periférico Sur 4121, Col. Fuentes del Pedregal, Mexico, DF 14140 ☎ +52 55 30991313 🖫 +52 55 30991418 **E:** webtv@tvazteca.com.mx **W:** www.tvazteca.com.mx **L.P:** CEO: Pedro Padilla Longoria. **Chs:** Azteca 7, Azteca Trece. **Txs:** **Azteca 7:** XHIMT-TV Mexico City [a]ch7 (267kW)/ch24 & relay txs; **Azteca Trece:** XHDF-TV Mexico City [a]ch13 (320kW)/ch25 & relay txs. – **TELEVISA SA de CV (Comm)** ▣ 2000 Avenida Vasco De Quiroga Santa Fe, Mexico, DF 01210 ☎ +52 55 52612000 🖫 +52 55 52612494 **W:** www.televisa.com.mx **L.P:** CEO: Emilio Azcarraga Jean. **Chs:** Canal 2, Canal 5, Galavisión, Foro TV. **Txs: Canal 2:** XEW-TV Mexico City [a]ch2 (64kW)/ch48 & relay txs; **Canal 5:** XHGC-TV Mexico City [a]ch5 (54kW)/ch50 & relay txs; **Foro TV:** XHTV Mexico City [a]ch4 (64kW)/ch49 & relay txs; **Galavisión:** XEQ-TV Mesico City [a]ch9 (325kW)/ch44 & relay txs.

Regional Networks, Local Stations not shown.

MICRONESIA (USA associated)

System: NTSC-M [A]

KPON-TV (Comm)▣ Central Micronesia Communications, P.O.Box 460, Colonia, Pohnpei, FM 96941. **L.P:** Pres: Bernard Hegenberger. DirTech: David Cliffe. **Tx:** Pohnpei ch7 (1kW). – **TTTK (Comm)** ▣ Chuuk, FM 96942. **Tx:** Moen ch7 (0.1kW). – **WAAB-TV (Gov)** ▣ Department of Youth and Civic Affairs, P.O.Box 30, Colonia, Yap, FS 96943 ☎ +1 691 3502502 **Tx:** ch7 (1kW).

MOLDOVA

Systems: # SECAM-D/K [R], PAL-D/K [R] ↧2013; DVB-T (MPEG2, MPEG4)

National Stations
TELERADIO-MOLDOVA (Pub) ▣ str. Hâncesti nr. 64, 2018 Chisinau ☎ +373 22 723380 🖫 +373 22 723329 **E:** tvdir@trm.md **W:** www.trm. md **L.P:** CEO (TV) Raileanu Adela **Ch:** Moldova1. **Txs:** Straseni ch3 (40kW) & network. – **EURO-TV (Comm)** ▣ str. Columna nr. 106, 2012 Chisinau ☎ +373 22 221149 🖫 +373 22 221147 **E:** eurotv@ yahoo.com **W:** www.eutv.md **Txs:** Straseni ch23 (3kW) & network. – **MUZ-TV (Comm)** ▣ str. Ismail nr. 88/1, 2001 Chisinau ☎ +373 22 207906 🖫 +373 22 270755 **E:** info@muztv.md **W:** www.muztv.md **Txs:** Chisinau ch53 (0.1kW) & network. – **NIT (Comm)** ▣ sos. Hâncesti nr. 59/1, 2028 Chisinau ☎ +373 22 737970 🖫 +373 22 278901 **E:** info@ nit.mdl.net **W:** www.nit.md **Txs:** Chisinau ch49 (1kW) & network. – **N 4 (Comm)** ▣ str. Miorita, nr. 3/5, 2028 Chisinau ☎ +373 22 924275 **Txs:** Chisinau ch51 (0.3kW) & network. – **PRIME (Comm)** ▣str. Banulescu-Bodoni nr. 57/1, 2005 Chisinau ☎ +373 22 244746 🖫 +373 22 244746 **E:** info@prime.md **W:** www.prime.md **Txs:** via Chisinau DTT Mux 2 & analogue netw.– **TV DIXI (Comm)** ▣ str. Sciusev nr. 93, 2012 Chisinau ☎+373 22 233505 🖫: +373 22 237588 **E:** www. info@dixi.md **W:** www.dixi.md **Txs:** via Chisinau (Mega TV tx) ch26 (0.5kW) & own netw.

Local & Regional Stations
Albasat: str. Suveranitatii nr. 1, 6401 Nisporeni; ch8. **ART-TV:** str. Eminescu nr.37, 3736 Straseni; ch7. **AVM:** str. Puskin nr. 16, 4601 Edinet; ch25. **BAS TV:** str. K. Marx nr. 67, 7401 Basarabeasca; ch26. **BTB:** sos. Hîncesti nr. 59/1, 2028 Chisinau; ch49. **Canal-X:** bd. Independentei nr. 48, 4701 Briceni; ch3. **Catrin TV:** str. 31 August nr. 20B, 3100 Balti; ch26. **Drochia TV:** str. Sorocii, nr. 44, ap. 7, 5201 Drochia; ch28. **Elita:** str. 1 mai nr. 2, 5400 Rezina; ch21. **Euronova:** str. Suveranitatii nr.1, 6401 Nisporeni; Ungheni ch39. **FLOR-TV:** str. Stefan cel Mare nr.30 A, 5003 Floresti; ch38. **Impuls TV:** str. 31 august nr.1, of. 304, 7201 Soldanesti; ch7. **Media TV:** str. Stefan cel

Mare nr. 14, 4101 Cimislia; ch43. **Mega TV:** str. Sciusev, nr. 93, 2012 Chisinau; Chisinau ch26 (0.5kW), Singerei ch53, Falesti ch49. **Noroc TV:** bd. Negruzzi 6s, 2001 Chisinau; via Chisinau DTT mux 2. **NTS:** Taraclia; ch41. **SOR-TV:** str. Banulescu-Bodoni nr. 2, 3000 Soroca; ch43. **Studio-L:** bd. M. Eminescu nr. 23, 4301 Causeni; ch35. **ProTV:** bd. Stefan cel Mare nr. 162, 2004 Chisinau; Chisinau ch37 (1.5kW) (+ DTT Mux 2), Balti ch58. **STV-41:** str. Pacii nr. 16, 7401 Taraclia; ch41. **Teleradio Balti:** str. Bucovinei nr. 101a, 3101 Balti; ch21. **TV Prim:** str. Suveranitatii nr. 5, of. 94, 4901 Glodeni; ch35. **TVardita:** str. Frunze 1, 6119 Tvardita; ch48. **TV6 Balti:** str. 31 August nr. 20-B, 3121 Balti; ch26. **TV7:** str. Alecu Russo nr. 1, of. 21, 2068 Chisinau; ch43 (0.5kW) & Chisinau DTT Mux 2. **TVC 21:** str. Alecu Russo nr. 1, 2012 Chisinau; via Chisinau DTT Mux 2.

Gagauzia (autonomous region):
Aiïn-Aciïc: str. Cikalov, nr. 59/7, 6101 Ceadîr-Lunga; ch37. **Bizim Aidïnïc:** str. Tretiacov nr. 4, of. 36, 3801 Comrat; ch38. **Eni Ai:** str. Tretiacov nr. 4, 3801 Comrat; ch23. **TV Gagauzia:** str. Lenin nr.134, 3802 Comrat; Vulcanesti ch24, Comrat ch36, Copceac ch47, Ceadîr-Lunga ch49. **TVK-24:** str. Lenin 124, 6118 Copceac; ch24.

Foreign TV Relays
Prime (Pervyy kanal, Russia), 2 Plus (Antena 1, Romania)

Location	P	2P	kW	Location	P	2P	kW
Cahul	31	1	20/5	Mindrestii Noi	-	41	20
Causeni	28	-	20	Straseni	30	11	20/40
Cimislia	2	-	5	Trifesti	27	25	20/5
Edinet	31	-	20	Ungheni	29	-	20

+ sites with txs below 5kW.
TV5 Monde (France): Chisinau ch8 (0.15kW).

DTT Transmitters (under construction)
Operator: Radiocomunicatii ▣ str. Drumul Viilor nr. 28/2, 2021 Chisinau ☎ +373 22 733914 🖫 +373 22 733874 **E:** crtvr@cni.md **Mux 1:** Moldova 1, Prime (Pervyy kanal), 2 Plus (Antena 1), nit (TVCi), Muz-TV, AltTV (B1 TV). **Tx:** ch61 (Chisinau 10kW). **Mux 2 (MPEG4):** PRO-TV Chisinau, TV7 (NTV), N4, TVC 21 (Realitatea TV), Noroc TV, VDT (TV3) **Tx:** ch58 (Chisinau 10kW).

Transnistria

TV PMR (Gov) ▣ ul. Yunosti 1, 3300 Tiraspol ☎ +373 533 25708 **W:** www.tv-pmr.com **Txs:** Maiac ch5 (1kW), Dnestrovsc ch6, Grigoriopol ch6, Ribnita ch8, Bender & Camenca ch10, Dubasari ch12, Tiraspol ch31, Ribnita ch35. – **TSV (Comm)** ▣ul. Karl Libknekhta 1/2, 3300 Tiraspol ☎ +373 533 63632 🖫 +373 533 63632 **E:** inform@ tsv-tv.idknet.com **Txs:** Ribnita ch44, Camenca & Dubasari ch45, Maiac ch47, Dnestrovsc ch49, Tiraspol ch54.

Local Station
BTV (TV Bender): Bender; ch26 (0.3kW).

DTT Transmitters (Trial)
Operator: n/a **Mux:** TV PMR, Muz-TV, RTR-Planeta, Vesti, RTR Planeta-Sport, Pervyy Kanal. **Tx:** ch63 (Slobozia).

MONACO

Systems: DVB-T (MPEG2)

TÉLÉ MONTE CARLO (TMC) (Comm) ▣ 6 Quai Antoine 1er, 98000 Monaco☎ +377 93151415 🖫 +377 92165481 **E:** tmc@ fr.multithematiques.com **W:** www.tmc.tv **L.P:** Pres: M. Jean Pastorelli.

DTT Transmitters
Operator: Monaco Telecom ▣ 25 Boulevard de Suisse, 98000 Monaco ☎ +377 99663497 **E:** a.segala@monaco-telecom.mc **W:** www.monaco.mc **Mux:** TMC, Monaco Info, TF1, France 2-5, M6, Arte, Canal+ (unencrypted sequences), BFM TV, I>télé, Euronews, RAI 1-3, Canale 5, CNBC **Tx:** ch10 (Mont Agel).

MONGOLIA

System: PAL-D/K [R]; DVB-T planned

National Station
MONGOLÍN ÚNDESNIY TELEVIZ (Pub) ▣ Huvsgalín zam 3, Ulaanbaatar 11 ☎ +976 11 323801 🖫 +976 11 327234 **E:** mrtv@magic-net.mn **W:** www.mnb.mn **L.P:** Dir: Ts. Enkhbat. **Txs:** (-).

Local Stations not shown.

MONTENEGRO

Systems: # PAL-B/G [E] ⇩31 Dec 2012; DVB-T (MPEG4)

National Stations
TELEVIZIJA CRNE GORE (TVCG) (Pub) ✉ Cetinjski put bb, 81000 Podgorica ☎ +382 20 225999 🖃 +382 20 225930 **E:** kontakt@rtcg.me **W:** www.rtcg.me **LP:** Dir TV: Veljo Jaukovic **Chs:** TVSG1, TVSG2

Location	P1	P2	kW	Location	P1	P2	kW
Bjelasica	12	37	50/100	Sjenica	23	29	15
Lovcen	8	31	100/800	Sjudjina Glava	6	24	5
Luština	26	33	10/9	Tvrdaš	9	31	1.5
Mozura	33	43	10	Volujica	12	24	1/10
Obrov	6	44	0.3/10				

+ sites with txs below 1kW. P1=TVSG1, P2=TVSG2
NTV MONTENA (Comm) ✉ Djoka Miraševica 61, 81000 Podgorica ☎ +382 20 266543 **E:** montena@montena.me **W:** www.montena.me **LP:** Dir: Djuro Vucinic. **Txs:** Podgorica ch27 & network.

Local Stations not shown.

DTT Transmitters (under construction)
Operator: Radio-difuzni centar d.o.o. ✉ Bulevar Svetlog Petra Cetinjskog 130/V, 81000 Podgorica ☎ +382 20 408000 🖃 +382 20 408005 **E:** rdc@rdc.co.me **W:** www.rdc.co.me **Mux:** TVCG1, TVCG2 **Tx:** ch67 (Podgorica). National network in preparation.

MONTSERRAT (UK)

System: NTSC-M [A]

ANTILLES TV LTD (Comm) ✉ P.O. Box 342, Plymouth, Montserrat. ☎ +1 664 4912226 🖃 +1 664 4914511 **LP:** GM: K. Osborne; TD: Z.A. Joseph. **Tx:** Chance Pic ch7 (48kW). – **ZJB-TV (Comm)** ✉ Plymouth. **Tx:** Chance Pic ch13 (rel. ABS-TV, Antigua & Barbuda).

MOROCCO

System: # SECAM-B/G [M] ⇩2015; DVB-T (MPEG2)

SOCIÉTÉ NATIONALE DE RADIODIFFUSION ET DE TÉLÉVISION (SNRT) (Gov) ✉ BP 1042, Rabat ☎ +212 37700319 🖃 +212 37722047 **W:** www.snrt.ma **Chs:** Al Aoula, Arriadia, Arradia 2, Arrabia, Assadissa, Aflam TV, Tamazight TV **LP:** DG: Faiçal Laraichi.

Location	ch	kW	Location	ch	kW
Zerhoun	4	120	Tan Tan	8	11
Zaio	4	9	Safi	8	20
Oujda	5	316	Tazerkount	8	90
Boukhouali	5	150	Touzarine	9	9
Tanger	5	20	Tiguelamine	9	9
Sidi Bounouara	5	11	S.Bounoara	9	11
Oukaimeden	6	18	Biougra	9	4
Azougar	6	9	Casablanca	10	180
Dakhla	6	11	Hafa Safa	10	9
Figuig	6	9	Ourzazate	10	267
Rabat	7	180	Bouarfa	10	267
Izeft	7	14	Essaouira	11	20

Regional stn: Lâayoune TV. Tx: chE4 Lâayoune (We. Sahara), 316kW.
TÉLÉVISION 2M (Semi-Gov, Comm) ❂ ✉ Km 7,3 route de Rabat Ain Sebaa, Casablanca 20250 ☎ +212 22354444 🖃 +212 22343390 **E:** 2m@tv2m.co.ma **W:** www.2m.tv **LP:** MD: Tawfik Bennani-Smires. **Txs:** Rabat ch21 & netw.

DTT Transmitters (under construction)
Operator: SNRT **Mux 1:** Al Aoula, 2M, Arriadia, Arradia 2, Arrabia, Assadissa, Aflam TV **Mux 2:** Aflam TV, Tamazight TV

Location	M1	M2
Rabat	30	23

+ nationwide network (txs being installed at all main sites)

CEUTA & MELILLA (Spain)

Systems: DVB-T (MPEG2)

Local Stations
Ceuta Televisión (Comm): Ceuta. **Radio Televisión de Ceuta (RTVCE) (Pub):** Alcalde Sanchez Prados nº 5, 51001 Ceuta. **Melilla Televisión (Comm):** Los Castaños, Urb Los Balandros nº 33, Aguadulce, 04720 Roquetas de Mar. **TV Melilla (TVM) (Comm):**

Miguel Zazo, 31, 2º 52004 Pontevedra.

DTT Transmitters
Operator: TVE **Mux:** La 1, La 2, 24 Horas, Clan ✂ RNE1, RNE Clásica, RNE3 **Txs:** ch52 (Ceuta), ch64 (Melilla). – **Operator:** Antena 3 **Mux:** Antena 3 **Mux:** Antena 3, Neox, Nova, Gol Televisión❂ ✂ Onda Cero, Europa FM, Onda Melodí **Txs:** ch69 (Ceuta, Melilla). – **Operator:** Gestevisión Telecinco **Mux:** Telecinco, La Siete, FDF, Disney Channel.. **Txs:** ch68 (SFN). – **Operator:** Sogecable **Mux:** Cuatro, CNN+, Canal+ Dos❂, Canal Club, La Sexta ✂ SER, 40 Principales, Cadena Dial **Txs:** ch67 (SFN). – **Operator:** Veo TV **Mux:** Veo7, Tienda en Veo, Intereconomía, Teledeporte ✂ R.Intereconomía, R.Marca, esRadio, Vaughan R. **Txs:** ch66 (SFN). – **Operator:** Melilla Tele **Mux:** TV Melilla, Popular TV **Txs:** ch61 (Melilla). – **Operator:** n/a **Mux:** Radio TV de Ceuta, Ceuta TV, Canal Sur **Txs:** ch62 (Ceuta).

MOZAMBIQUE

System: PAL-G [E]

TELEVISÃO DE MOÇAMBIQUE (TVM) (Gov) ✉ CP 2675, Maputo Av 25 de Setembro 154, 2675 Maputo ☎ +258 21 308117 🖃 +258 21 308122 **E:** tvm@tvm.co.mz **W:** www.tvm.co.mz **LP:** Chmn/CEO: Marcos Vetrano **Txs:** Maputo ch33 (1kW) & relay txs. – **RÁDIO E TELEVISÃO KLINT (RTK) (Comm)** ✉ Av. Julius Nyerere 390, Maputo ☎ +258 21 491744 🖃 +258 21 491745 **LP:** Chmn: Carlos Pereira Klint **Txs:** (-). – **SOICOS TELEVISÃO (STV) (Comm)** ✉ Av. Vladimir Lenine, 2015 Maputo ☎ +258 21 418002 🖃 +258 21 415335 **E:** stv@soico.co.mz **W:** www.stv.co.mz **LP:** DG: Daniel David **Txs:** (-). – **TELEVISÃO MIRAMAR (Comm)** ✉ Rua Pereira Lago, 221 - 11o. Andar, Maputo. ☎🖃 +258 21 486813 **E:** contacto@miramar.co.mz **W:** www.redemiramar.co.mz **Txs:** (-).

Foreign TV Relay
RTP África (Portugal): (-).

MYANMAR

Systems: # PAL-B/G [E]; DVB-T (MPEG2)

MYANMA TV (Gov) ✉ 426, Pyay Rd., Kamayut Tsp., Yangon ☎ +95 1 535553 🖃 +95 1 534211 **E:** mrtv@mptmail.net.mm **Chs:** MRTV1, MRTV2, MRTV3, MRTV4 **Txs:** **MRTV1:** Yangon ch6 (17kW H) & relay txs. – **MYAWADY TV (Gov)** ✉ Pyay Rd., Mingalardon Tsp., Yangon ☎ +95 1 600294 **Chs:** MWD1, MWD2, MWD3 **Txs:** (-).

DTT Transmitters
Operator: MRTV **Muxes:** multiprgr **Txs:** ch(-). National network planned. – **Operator:** Myawady TV **Mux:** MWD1, MWD2, MWD3 **Tx:** ch(-) (Yangon/Tarmwe 0.4kW). National network planned.

NAMIBIA

Systems: # PAL-I [SA] ⇩2013; DVB-T (MPEG2) [SA]

NAMIBIAN BROADCASTING CORP. (Gov) ✉ P.O.Box 321, Windhoek 9000 ☎ +264 61 2913111 🖃 +264 61 216209 **E:** pr@nbc.com.na **W:** www.nbc.com.na **LP:** DG: Bob Kandetu. **Txs:** Rundu ch4, Keetmanshoop ch4, Paresis ch5, Windhoek ch6, Erongo ch7 & repeaters. – **ONE AFRICA TV (Comm)** ✉ Storch House, Storch Street, Windhoek ☎ +264 61 253190 🖃 +264 61 220410 **E:** paul@mac.com.na **W:** www.oneafrica.tv **LP:** MD: Waldheim Shiluwa. **Txs:** Windhoek ch48 & netw. – **TBN NAMIBIA (Rlg)** ✉ P.O.Box 1587, Swakopmund ☎ +264 64 401100 🖃 +264 64 403752 **E:** comments@tbnnamibia.tv **W:** www.tbnnamibia.tv **Txs:** Walvis Bay ch37, Windhoek ch40, Rehoboth ch40, Okahandja ch40, Swakopmund ch65.

DTT Transmitters
Operator: MultiChoice Namibia Pty. ✉ P.O.Box 2662, Windhoek ☎ +264 61 2705111 🖃 +264 61 2705247 **Mux (❂):** M-Net, SuperSport 1, SABC Africa, Discovery, Channel O. **Tx:** Windhoek ch13 (0.06kW).

NAURU

System: PAL-B [E]

NAURU TELEVISION (Gov) ✉ Govnt Offices, Yaren District, Rep. of Nauru, Central Pacific. ☎ +674 4443113 🖃 +674 444 3153 **Txs:** one 10W, two 100W txs (channels n/a).

NEPAL

System: PAL-B/G [E]

NEPAL TELEVISION CORP. (Gov) ⬛ P.O.Box 3826, Singha Durba, Kathmandu ☎ +977 1 4228447 🖷 +977 1 4228312 **E:** neptv@ccsl.com **W:** www.explorenepal.com/ntv **L.P:** GM: Durga Nath Sharma; Chief Eng: Deepak Mani Dhital **Txs:** Chamere Danda ch5 (1kW), Namje ch5 (2kW), Phulchoki ch5 (5kW), Jaleswor ch11 (2kW), Ilam ch12 (5kW) + txs below 1kW. – **KANTIPUR TELEVISION (Comm)** ⬛ P.O.Box 1122, New Baneshwor, Kathmandu ☎ +977 1 4480100 **E:** kanti@kantipur.com.np **W:** www.kantipurtv.com **Tx:** Kathmandu ch23.

NETHERLANDS

System: DVB-T (MPEG2)

National Stations
NEDERLANDSE PUBLIEKE OMROEP (NPO) (Pub) ⬛ P.O.Box 26444, 1202 JJ Hilversum ☎ +31 35 6779222 🖷 +31 35 6774188 **E:** voorlichting@omroep.nl **W:** www.omroep.nl **L.P:** Chmn: Henk Hagoort. **Chs:** Nederland 1, Nederland 2, Nederland 3. Prgrs for the NPO are provided by **Nederlandse Omroep Stichting (NOS):** Sumatralaan 45, 1217 GP Hilversum; **NTR:** P.O.Box 29000, 1202 MA Hilversum; and the following major broadcasting organizations: **AVRO (Algemene Vereniging Radio Omroep):** P.O.Box 2, 1200 JA Hilversum; **BNN (Bart's Neverending Network):** P.O.Box 646, 1200 AP Hilversum; **EO (Evangelische Omroep):** P.O.Box 21000, 1202 BB Hilversum; **KRO (Katholieke Radio Omroep):** P.O.Box 23000, 1202 EA Hilversum; **NCRV (Nederlandse Christelijke Radio Vereniging):** P.O.Box 25000, 1202 HB Hilversum; **LLINK:** Lloydstraat 5, 3024 EA Rotterdam; **MAX:** P.O.Box 554, 2700 AN Zoetermeer; **PowNed (Publieke Omroep Weldenkend Nederland En Dergelijke):** P.O.Box 37743, 1030 BG Amsterdam; **TROS (Televisie en Radio Omroep Stichting):** P.O.Box 28450, 1202 LL Hilversum; **VARA (Vereniging Arbeiders Radio Amateurs):** P.O.Box 175, 1200 AD Hilversum; **VPRO (Vrijzinnig Protestantse Radio Omroep):** P.O.Box 11, 1200 JC Hilversum; **WNL (Wakker Nederland):** P.O.Box 376, 1000 EB Amsterdam. In addition, a number of smaller broadcasting organisations is contributing with prgr production. – **RTL.NL (Comm)** ⬛ P.O.Box 20, 1200 AA Hilversum ☎ +31 35 6718711 🖷 +31 35 6236892 **E:** info@rtl.nl **W:** www.rtl.nl **Chs:** RTL4, RTL5, RTL7, RTL8 – **SBS BROADCASTING NEDERLAND (Comm)** ⬛ P.O.Box 18179, 1001 ZB Amsterdam ☎ +31 20 8007000 🖷 +31 20 8007001 **E:** info@sbs.nl **W:** www.sbs.nl **Chs:** NET5, SBS6, Veronica TV. – **TMF (Comm)** ⬛ P.O.Box 999, 1400 AZ Bussum ☎ +31 35 6996666 🖷 +31 35 6947775 **E:** tmfinteractive@tmf.nl **W:** www.tmf.nl.

Regional Stations (Pub) (via DTT Mux 1)
a) L1TV: P.O.Box 31, 6200 AA Maastricht; **b) Omroep Brabant TV:** Postbus 108, 5600 AC Eindhoven; **c) Omroep Fryslân TV:** P.O.Box 7600, 8903 JP Leeuwarden; **d) Omroep Zeeland TV:** P.O.Box 1090, 4388 ZH Oost-Souburg; **e) Regio TV Utrecht:** P.O.Box 9043, 3506 GA Utrecht; **f) TV Drenthe:** P.O.Box 999, 9400 AZ Assen; **g) TV Flevoland:** P.O.Box 567, 8200 AN Lelystad; **h) TV Gelderland:** P.O.Box 747, 6800 AS Arnhem; **i) TV Noord:** P.O.Box 30101, 9701 BH Groningen; **j) TV Noord-Holland:** P.O.Box 9823, 1006 AM Amsterdam; **k) TV Oost:** P.O.Box 1000, 7550 BA Hengelo; **l) TV Rijmond:** P.O.Box 350, 3000 AJ Rotterdam; **m) TV West:** P.O.Box 24012, 2490 AA Den Haag.

Local Stations
AT5: P.O.Box 3976, 1001 AT Amsterdam; via DTT Mux 3. **Haarlem 105:** P.O.Box 3355, 2001 DJ Haarlem; ch46V (0.025kW).

DTT Transmitters
Operator: KPN Digitenne ⬛ P.O.Box 396, 1200 AJ Hilversum **E:** klantenservice@digitenne.nl **W:** www.digitenne.nl **Mux 1:** Nederland 1, Nederland 2, Nederland 3, Public Regional Stations (a-m) ✳ R.1, 2, 3FM, 4, 5, 6, FunX, Public reg. stns **Mux 2 (◐):** NET5, RTL4, RTL5, RTL7/RTLZ, SBS6 ✳ Classic FM, Q-Music, R.Veronica, R.10 Gold, Sky R. 101 FM, Slam!FM, 100%NL **Mux 3 (◐):** Animal Planet, BBC Entertainment, CNN, MTV, NGC, Eredivisie Live 1, Eredivisie Live 2/Cartoon Netw./AT5, TMF ✳ Arrow Classic Rock, BNR Nieuwsradio, R.538 **Mux 4 (◐):** Eén, Canvas/Ketnet, Discovery Channel, Eurosport, Nickelodeon/Comedy Central, Private Sprice, RTL8, Veronica TV/Disney HD. **Mux 5:** DVB-H mux (8 prgrs).

Location	M1*	M2	M3	M4	M5	kW
Alkmar	39j	45	34	35	27	40/4x20
Alphen a.d.R.	52m	64	57	24	27	10/15/3x10
Amsterdam (RAI)	39j	64	57	24	27	10

Location	M1*	M2	M3	M4	M5	kW
Apeldoorn	42h	65	58	66	28	20
Arnhem	42h	65	58	66	28	10/2x40/2x20
Breda	30b	60	31	32	27	2x20/3x15
Den Bosch	30b	60	31	56	33	5/10/5/10/5
Den Haag	52m	64	57	24	27	10
Den Haag (Zichtenb.)	52m	64	57	24	27	10
Deventer	22k	65	23	47		10
Doetinchem	42h	65	58	66	28	2x20/40/2x20
Eindhoven Oost	30b	60	31	56	33	10/3x15
Eindhoven West	30b	60	31	56	33	3x10/5
Enschede	22k	65	23	47	28	20
Goes	54d	48	29	32	35	10
Groningen	66i	30	54	33	25	20
Haarlem	39j	64	57	24		15
Heerlen	54a	34	51	64	27	3x40/20
Helmond	30b	60	31	56		20
Hengelo	22k	65	23	47	28	40/20/40/2x20
Hilversum	39j	64	57	24		15
Ijsselstein	30b	64	57	24		3x15/10
Krimpen a.d.IJ.	21l	64	57	24	27	10
Leeuwarden	32c	55	34	21	44	20
Lelystad	26g	65	23	47	44	2x20/10/2x20
Leusden	50e	64	57	24		3x10/5
Loon op Zand	30b	60	31	56		15
Maarssen	50e	64	57	24		20
Maastricht	54a	34	51	64	27	3x10/15/10
Nijmegen	42h	60	31	56	33	10/20/10/20/15
Oegstgeest	52m	64	57	24	27	10
Oss	42h	60	31	56	33	20/4x10
Oss	30b	-	-	-		10
Roermond	54a	34	24	64		20
Rosendaal	30b	48	29	32		20/3x10
Rotterdam (Waalhaven)	21l	64	57	24	27	10
Sittard	54a	34	24	64	27	20
Sliedrecht	21l	64	57	24		10
Smilde	60f	30	54	33	25	3x40/30/40
Utrecht	50e	64	57	24		10/5/2x4
Veenendal	50e	65	58	66	28	20
Veenendal	42h	-	-	-		20
Venlo	54a	34	31	56	27	20/40/20/40/20
Zoetermeer	52m	64	57	24		12
Zwolle	22k	65	23	47	28	20

+ sites with txs below 10kW. Pol=V. *) incl. regional stns (see above).

NEW CALEDONIA (France)

System: # SECAM-K1 [E]; DVB-T

RFO NOUVELLE CALEDONIE (Pub) ⬛ BP G3 Mont Coffin, 98848 Nouméa Cedex ☎ +687 687274327 🖷 +687 687281252 **W:** nouvel-lecaledonie.rfo.fr **L.P:** Dir: Alain Le Garrec. **Txs:** (pol.H) Mont Coffyn ch43 (11kW), Dumbea/Mont Khogi ch25 (9kW), Bouloupari ch54 (50kW) & repeaters. – **CANAL CALEDONIE (Comm)** ◐ ⬛ 8 rue de Verneilh, Noumea. **Txs:** Noumea (Mt Koghi) ch25 (9kW), Noumea (Mt Coffyn) ch43 (11kW), Noumea (Town) ch33 (0.06kW).

DTT Transmitters
Operator: France Télévisions **Mux:** RFO Nouvelle Caledonie, France 2-5, France Ô, Arte, LCP, France 24, Gulli **Txs:** ch(-).

NEW ZEALAND

Systems: DVB-T (MPEG4); † PAL-B/G [B=NZ;G=E] ⬇2013

National Stations
TELEVISION NEW ZEALAND (TVNZ) (Pub) ⬛ P.O.Box 3819, Auckland ☎ +64 9 9167000 🖷 +64 9 9167934 **W:** tvnz.co.nz **Chs:** TV One, TV2. – **PRIME TELEVISION (Comm)** ⬛ 1 John Glenn Ave., North Harbour, Auckland ☎ +64 9 4140700 🖷 +64 9 4140701 **E:** info@primetv.co.nz **W:** primetv.co.nz – **TAB TRACKSIDE (Comm)** ⬛ P.O. Box 388-99, Wellington Mail Centre, Wellington ☎ +64 4 576 6999 🖷 +64 4 576 6996 **E:** corporate@tab.co.nz **W:** www.tab.co.nz – **TV3 (Comm)** ⬛ P.O. Box 5185, Auckland ☎ +64 9 779730 🖷 +64 9 3667029 **W:** www.tv3.co.nz

Local Stations not shown.

DTT Transmitters (under construction)
Operator Mux 1: TVNZ **Mux:** TV One, TV2, TVNZ 6, TVNZ 7, TVNZ

Sport Extra – **Operator Mux 2:** MediaWorks ⬛ P.O.Box 92624, Symonds Street, Auckland ☎ +69 9 3779730 ▤ +69 9 3665999 **W:** www.mediaworks.co.nz **Mux:** TV3, TV3+, C4 – **Operator Mux 3:** Kordia ⬛ P.O.Box 2495, Auckland ☎ +64 9 9166400 ▤ +64 9 9166403 **W:** www.kordiasolutions.com **Mux:** Parliment TV, Maori TV, CTV8, Prime.

Location	M1	M2	M3
Auckland (Waiatarua)	29	33	45
+ nationwide tx network			

NICARAGUA

System: NTSC-M [A]; SBTVD planned

CANAL 6 ⬛ 3 1/2 Carretera Sur Contig o Shell, Managua☎ +505 22660118▤ +505 22666522 **Tx:** ch6 (25kW). – **CANAL 10** ⬛ Mansión, Teodolinda, 2c, al Ote ☎ +505 22665021 **Tx:** ch10. – **NICAVISIÓN** ⬛ Apdo 2766, Managua ☎ +505 22660691 ▤ +505 22661424 **Tx:** ch12. – **NUEVA IMAGEN SA** ⬛ Del Montoya, 1c al Sur, 1c al Este, Managua ☎ +505 22663420 ▤ +505 22663467 **Tx:** ch4. – **TELENICA** ⬛ Apdo Postal 3611, Mansión Teodolinda 1c al Sur, y ½ Abajo ☎ +505 22665021 ▤ +505 22665024 **Tx:** ch8. – **TELESAT** ⬛ Carretera a Masaya Km. 4½, Motorama ½c al Su, Managua ☎ +505 22670170 **Tx:** ch23. – **TELEVICENTRO SA** ⬛ Apdo Postal 688, Managua☎ +505 22682222 **E:** canal2@ibw.com.ni **Tx:** ch2 (25kW). – **TELEVISIÓN CRISTIANA DE NICARAGUA** ⬛ De la Casa de Obrero, 5c. al Sur y 2c. Arriba ☎ +505 22668688 ▤ +505 22683132 **Tx:** ch21.

NIGER

System: SECAM-K1 [E]

TÉLÉ-SAHEL (Gov) ⬛ BP 309, Niamey ☎ +227 20723686 ▤ +227 20723153 **E:** ortny@intnet.net **W:** www.ortn-niger.com **LP:** Dir TV: Moussa Abdou Saley. **Chs:** Télé Sahel, Tal TV. **Txs:** (pol.H exc. where stated) Télé Sahel: Agadez ch4 (10kW), Dosso ch4 (10kW), Zinder ch5 (10kW), Arlit ch6 (1kW), Maradi ch7V (10kW), Dogondoutchi ch7 (1kW), Gaya ch8V (1kW), Niamey ch9 (10kW), Konni ch9 (10kW), Diffa ch9 (10kW) & repeaters. **Tal TV:** (-). – **TELESTAR (Comm)** ⬦ ⬛ Taiwo Street 17, Niamey. **Txs:** (-). – **TÉNÉRÉ TV (Comm)** ⬛ BP 13600, Niamey ☎ +227 20736576 ▤ +227 20737775 **E:** tenerefm@ intnet.net **Txs:** (-).

NIGERIA

System: PAL-B/G [E]

NIGERIAN TELEVISION AUTHORITY (NTA) (Pub) ⬛ P.M.B 113, Garki-Abuja ☎ +234 9 2346907 ▤ +234 9 2345914 **LP:** Chmn: Alhaji Ibrahim Buba **Chs:** NTA, NTA2, reg. stns. **Txs:** NTA: ch10 (100kW) & relay txs; **NTA 2:** ch5 (100kW) & relay txs. – **DBN TELEVISION (Comm)** ⬛ P.O.Box 51162, Ikoyi, Lagos ☎ +234 1 2690051 ▤ +234 1 2693888 **E:** info@dbninternational.net **LP:** CEO: Osa Sonny Adun. **Txs:** ch32 & relay txs. – **LAGOS TV (Comm)** ⬛ Lagos. **Tx:** Lagos ch8. – **MINAJ BROADCAST INTERNATIONAL (Comm)** ⬛ P.O.Box 3975, Mushin, Lagos ☎ +234 1 4529203 ▤ +234 1 4528500 **E:** minaj@minaj-hq.com. **W:** www.minaj.com **LP:** Exec. Chairman: Mike Ajegbo **Txs:** ch41, ch43.

NIUE

System: PAL-B [NZ]

TV NIUE ⬛ P.O.Box 68, Alofi ☎ +683 4026 ▤ +683 4217 **E:** gm.bcn@ mail.gov.nu **LP:** GM: Patrick Lino **Txs:** Makefu ch4 (0.01kW), Alofi ch6 (0.75kW), Mutulau ch8 (0.04kW).

NORFOLK ISLAND (Australia)

System: PAL-B [AU]

NORFOLK ISLAND TELEVISION SCE. (Gov) ⬛ New Cascade Rd, Norfolk Island 2899, Australia ☎ +672 22137 ▤ +672 23298 **Chs:** rel. ABC, SBS & Central 7 TV from Australia. **Txs:** Mt. Pitt ch7V (0.02kW); ch10 (local).

NO. MARIANA IS (USA associated)

System: ATSC

WSZE-TV (Comm) ⬛ Saipan. **Tx:** ch10 (0.5kW). **Mux:** NBC; CBS / KUAM-TV (Guam).

NORWAY

Systems: DVB-T (MPEG4)

National Stations
NORSK RIKSKRINGKASTING (NRK) (Pub) ⬛ 0340 Oslo ☎ +47 23047000 ▤ +47 23047799 **E:** info@nrk.no **W:** www.nrk.no **LP:** CEO: Hans-Tore Bjerkaas. **Chs:** NRK1, NRK2, NRK3, regional stns – **CANAL DIGITAL (Comm)** ⬛ 4896 Grimstad ☎ +47 81559600 ▤ +47 22939305 **E:** kundeservice@canaldigital.no **W:** www.canaldigital. no – **TV2 (Comm)** ⬛ Postboks 7222, 5002 Bergen ☎ +47 55908070 ▤ +47 55908090 **E:** info@tv2.no **W:** www.tv2.no **L.P:** MD: Kåre Valebrokk. – **TV Norge (Comm)** ⬛ Postboks 11 Sentrum, 0101 Oslo ☎ +47 21022000 ▤ +47 22051000 **E:** tvnorge@tvnorge.no **W:** www. tvnorge.no **L.P:** MD: Morten Aass.

Local Stations (all Comm) (via DTT Mux 3)
a) BTV: P.O.Box 7240, 5020 Bergen; **b) TV Aust-Agder (TV-A):** P.O.Box 349, 4801 Arendal; **c) TV-Adressa:** Industriveien 13, 7003 Trondheim; **d) TV Budstikka*:** P.O.Box 133, 1376 Billingstad; **e) TV Drammen*:** P.O.Box 7033, 3007 Drammen; **f) TV Follo*:** Idrettsveien 11, 1400 Ski. **g) TV Haugaland:** P.O.Box 408, 5501 Haugesund; **h) TV Hålogaland:** P.O.Box 85, 9481 Harstad; **i) TV Innlandet:** P.O.Box 94, 2801 Gjøvik; **j) TV Nord:** P.O.Box 1193 Sentrum, 9504 Alta; **k) TV Nordland:** P.O.Box 564, 8601 Mo i Rana; **l) TV Nord-Trøndelag:** Drivhuset, Skippergata 11 d, 7725 Steinkjer; **m) TV Nordvest*:** P.O.Box 471, 6501 Kristiansund. **n) TV Oslo:** P.O.Box 11, Sentrum, 0101 Oslo;**o) TV Romerike:** Roseveien 1, 2007 Kjeller; **p) TV Sunnmøre:** Kirkegata 10, 6004 Ålesund; **q) TV Sør:** P.O.Box 342, 4663 Kristiansand; **r) TV Telemark:** P.O.Box 2833, 3702 Skien; **s) TV Tromsø:** P.O.Box 815, 9258 Tromsø; **t) TV Vest:** Auglendsmyrå 6, 4016 Stavanger; **u) TV Vestfold*:** P.O.Box 2003, 3103 Tønsberg. **v) TV Østfold:** P.O.Box 48, 1701 Sarpsborg.
NB: all stns relay TV Norge outside own prgrs, except (*).

DTT Transmitters
Operator: Norges televisjon AS ⬛ P.O.Box 313, 0511 Oslo ☎ +47 22883700 ▤ +47 22883781 **E:** info@ntv.no **W:** www.ntv.no **Mux (✪exc.*):** NRK1 (incl. reg prgrs)*, NRK2*, NRK3*, NRK Super*, TV3, Viasat 4, Disney Channel. **Mux 2 (✪exc.*):** TV2*, TV2 Filmkanalen, TV2 Nyhetskanalen, TV2 Sport, TV2 Zebra, Animal Planet, The Voice TV. **Mux 3 (✪exc.*):** TVNorge, FEM, National Geographic, BBC World News, Canal+ First, Canal+ Hits, Canal+ Sport 1, Åpen kanal*, Local Stations*. **Local Mux Oslo:** TV Oslo, TV Follo og TV Budstikka **Tx:** Oslo ch46.

Location	M1	M2	M3	kW
Bagn	32	39	42	50
Bergen	33	49	39	50
Bjerkreim	23	26	30	50
Bokn	36	54	57	50
Bremager	25	28	31	50
Gamlemsveten	37	38	54	50
Gausta	25	27	35	10
Greipstad	51	54	47	50
Grong	21	31	35	50
Gulen	37	42	26	50
Hadsel	45	48	58	50
Halden	38	42	62	60
Hammerfest	33	37	48	50
Hemnes	42	45	48	50
Hovdefjell	41	52	48	40
Jetta	45	48	58	50
Kautokeino	46	56	59	50
Kistefjell	26	46	43	50
Kongsberg	60	66	51	50
Kongsvinger	24	48	55	50
Kopparen	26	40	45	50
Lyngdal	25	53	47	50
Lønahorgi	31	41	44	50
Melhus	55	28	25	50
Mosvik	44	47	46	50
Narvik	21	27	37	50
Nordfjordeid	40	44	33	10

Location	M1	M2	M3	kW
Nordhue	33	43	56	50
Nordkapp	30	40	43	50
Oslo	52	58	61	50
Reinsfjell	39	42	35	50
Salten	50	43	60	50
Skien	60	66	54	50
Sogndal	21	24	34	50
Steigen	31	41	44	50
Stord	55	58	60	50
Trolltind	27	39	42	50
Tron	26	34	49	50
Varanger	28	33	50	50
Vega	25	32	37	50

+ sites with txs below 10kW.

OMAN

System: PAL-B/G [E]

SULTANATE OF OMAN TELEVISION (Gov) ✉ P.O.Box 600, 113 Muscat, Oman ☎ +968 24 603888 🖷 +968 24 604629 **E:** tvradio@ omantel.net.om **W:** www.oman-tv.gov.om **LP:** DG: Salum Habsi; DG (Engineering): Moh'd Salim Al Marhouby. **Txs:** Bahlah ch5 (4kW), Shinas ch5 (7kW), Sur ch7 (15kW), Thamret ch8V (100kW), Al-Amirat ch10, Nizwa ch10 (100kW), Quriat ch11 (0.3kW), Saham ch11 (200kW), Al-Berami ch12, Ibra ch12 (6kW), Maserah ch24, Dhank ch25 (2kW), Haima ch28, Adam ch40, Madha ch48, Barka ch51, Ibri ch55, Jabal Qahwi ch60 & low power txs.

PAKISTAN

System: PAL-B/G [E]

PAKISTAN TELEVISION CORP. LTD (PTV) (Gov) ✉ P.O.Box 1221, Islamabad 44000 ☎ +92 51 9208651 🖷 +92 51 9203406 **E:** ptvhq@ hotmail.com **W:** www.ptv.com.pk **LP:** Chmn: Shahid Rafi. **Chs:** PTV1, PTV2, regional stns. **Txs: PTV1:** Islamabad ch6 (50kW) & network; **PTV2:** (-) – **ATV (Comm)** ✉ 11 -F, Model Town, Lahore ☎ +92 42 5853669 🖷 +92 42 5853668 **E:** info@atv.com.pk **W:** www.atv.com.pk **LP:** Chmn: Abdul Jabbar. **Txs:** (-).

PALAU (USA associated)

NB: No terrestrial TV station.

PANAMA

System: NTSC-M [A]; DVB-T planned

National Stations
TELEVISORA NACIONAL (CANAL 2) (Pub) ✉ Apt. 0819-07129, El Dorado, Panamá ☎ +507 2793700 🖷 +507 2362987 **E:** tvn@tvn-2.com **W:** www.tvn-2.com **LP:** Pres: Stanley Motta; DG: Pedro Diaz **Tx:** ch2 (18kW). – **CADENA MILENIUM RCM (CANAL 21) (Comm)** ✉ Via Espana Sector de Carrasquilla, Apdo. postal 87-1989, Zona 7, Panamá **E:** ventas@rcmtv.tv **W:** www.rcmtv.tv **Tx:** ch21 (20kW). – **TELEMETRO (CANAL 13) (Comm)** ✉ Ave 12 de Octubre, Apartado 1-1425, Panama 8, Panamá ☎ +507 2106845 🖷 +507 2106929 **W:** www.telemetro.com.pa **LP:** Pres: Fernando Eleta Almarán **Tx:** ch13 (30kW). – **RPC TELEVISION (CANAL 4) (Comm)** ✉ Ave 12 de Octubre, Apartado 1-1425, Panamá 8 ☎ +507 2104104 **W:** www.rpctv. com **Tx:** ch4 (30kW). – **FETV (CANAL 5) (Comm)** ✉ Ave Ricardo J. Alfaro Contiguo al Gimnasio de la USMA, Apdo.6-7295, El Dorado, Panamá ☎ +507 2308000 🖷 +507 2301955 **W:** www.fetv.org **LP:** DG: Manuel Santiago Blanquer i Planells **Tx:** ch5 (30kW). – **RADIO Y TELEVISIÓN EDUCATIVA (CANAL ONCE) (Educ)** ✉ Curundú. Diagonal al Ministerio de Obras Publicas, Estafeta Universitaria, Universidad de Panamá ☎ +507 2328100 🖷 +507 2327466 **E:** rtvel@ ancon.up.ac.pa **LP** DG: Carlos Aguilar **Txs:** Prov. Panamá, Colón, Prov. Centrales, Sant. de Veraguas y Darien: 10 kW.

Local Stations
Canal +23 (Comm): Plaza Hispanidad, Ave. 12 de Octubre, (Ap. 6A-9292 El Dorado), Panamá; ch23 (30kW). **Canal 29 (Comm):** Avenida Ricardo J. Alfaro, Sun Tower Mall, Piso 2, Apdo. postal 1465, Balboa, Ancón Panamá; ch29. **Hosanna Vision (Canal 37) (Comm):** Ave. Martin Sosa, Ed. Hosanna Vision (A.P. Hosanna, El Dorado 6-7981), Panamá; ch37 (30kW).

PAPUA NEW GUINEA

System: PAL-B/G [E]

EMTV (Comm) ✉ P.O.Box 443, Boroko NCD ☎ +675 3257322 🖷 +675 3254450 **E:** emtv@datec.com.pg **W:** www.emtv.com.pg **LP:** CEO: Ken Clark. **Txs:** Burns Peak ch9 (1.1kW), Air Niugini Hill ch31 (0.17kW), Garden City ch68 (0.02kW) (all Port Moresby area).

PARAGUAY

System: # PAL-N [A]; SBTVD planned

National Stations
LATELE (Comm) ✉ Av. Eusebio Ayala No. 2995, Esq. Pasaje Tembetary, Asunción ☎ +595 21 622253 **E:** info@latele.com.py **W:** www.atele.com.py **Txs:** Asunción ch11 (40kW) & relays. – **PARAVISION (Comm)** ✉ Av. Mariscal López esq. Bélgica, Asunción ☎ +595 21 664380 **E:** info@paravision.com.py **W:** www.paravision. com.py **Txs:** Asunción ch5 (20kW) & relays. – **RED GUARANI (Comm)** ✉ Gral. Santos 1024 c/Concordia, Asunción ☎ +595 21 205444 **E:** info@redguarani.com.py **W:** www.redguarani.com.py **Txs:** Asunción ch2 (20kW) & relays. – **RED PRIVADA DE COMUNICACIÓN (RPC) (Comm)** ✉ Calles Comendador Nicolás Bó y Guaranies, Lambaré, Asunción ☎ +595 21 332823 🖷 +595 21 331695 **E:** commercial@rpc. com.py **W:** www.rpc.com.py **Txs:** Asunción ch13 (40kW) & relays. – **SISTEMA NACIONAL DE TELEVISIÓN (SNT) (Comm)** ✉ Av. Carlos Antonio Lopez 572, Asunción ☎ +595 21 424222 🖷 +595 21 480230 **E:** snt@snt.com.py **W:** www.snt.com.py **Txs:** Asunción ch9 (40kW) & relays. – **TELEFUTURO (Comm)** ✉ Andrade c/ O'Higgins, Villa Morra, Asunción ☎ +595 21 608756 **W:** www.telefuturo.com.py **Txs:** Asunción ch4 (60kW) & relays.

Local Stations not shown.

PERU

System: # NTSC-M [A] ⇩2020; SBTVD

National Stations
TELEVISIÓN NACIONAL DEL PERÚ (TNP) (Gov) ✉ Av. Jose Galvez 1040, Santa Beatriz, Líma ☎ +51 1 6190707 🖷 +51 1 6190711 **E:** comereirtp@wayna.rcp.net.pe **W:** www.tnp.com.pe **LP:** Dir: Juan Carlos Vicente **Txs:** ch7 (10kW) & relays. – **AMÉRICA TELEVISION (Comm)** ✉ Montero Rosas 1099, Santa Beatriz, Líma ☎ +51 1 2657361 🖷 +51 1 2656976 **E:** americanoticias@americatv.com.pe **W:** www.americatv. com.pe **LP:** Pres: Jose Antonio Miroquesada **Txs:** Piura ch2 (2kW), Chiclayo ch4 (5kW), Huancayo ch4 (2kW), Trujillo ch6 (2kW), Tacna ch9 (2kW), & repeaters. – **ANDINA DE RADIODIFUSIÓN (ATV) (Comm)** ✉ Arequipa 3570, San Isidro, Apartado 270077, Líma ☎ +51 1 2212261 🖷 +51 1 4217263 **E:** andinatelevision@atv.com.pe **W:** atv.com.pe **Txs:** ch3 (20kW), ch4 (20kW), ch6 (2kW), ch8 (5kW), ch9 (315kW) & relays. – **FRECUENCIA LATINA (Comm)** ✉ Av. San Felipe 968, Jesús Mariá, Líma 11 ☎ +51 1 4707272 🖷 +51 1 4712688 **E:** flatina@frecuencialatina. com.pe **W:** www.frecuencialatina.com.pe **Txs:** ch2 (22.5kW), ch13 & relays. – **PANAMERICANA (Comm)** ✉ Av. Arequipa 1110, Líma ☎ +51 1 4113200 🖷 +51 1 4113309 **E:** pantel@pantel.com.pe **W:** www. pantel.com.pe **Txs:** ch5 (290kW) & relays. – **RBC TELEVISIÓN (Comm)** ✉ Manco Capac 333, La Victoria, Líma ☎ +51 1 4337674 🖷 +51 1 4331237 **Txs:** ch11 (30kW) & relays.

Local Stations not shown.

PHILIPPINES

Systems: # NTSC-M [A] ⇩31 Dec 2015; ISDB-T

National Stations
INTERCONTINENTAL BROADCASTING CORP. (IBC) (Gov) ✉ Broadcast City, Capitol Hills, Diliman, Quezon City ☎ +63 2 9318781 🖷 +63 2 9324611 **W:** www.ibc.com.ph **Txs:** DZTV-TV Manila ch13 (50kW) & relay stns. – **NATIONAL BROADCASTING NETWORK (NBN) (Gov)** ✉ Broadcast Complex, Visayas Ave, Quezon City 1100 ☎ +63 2 9206521 🖷 +63 2 9204342 **W:** www.nbni.tv **LP:** Chwmn: Ms Mia A. Concio; GM: Ramon S. Diez; CE: Antonio M Leduna. **Txs:** DGWT-TV Manila ch4 (50kW) & relay stns. – **ABS-CBN BROADCASTING CORP. (Comm)** ✉ Eugenio Lopez Jnr St, Quezon C. ☎ +63 2 4111166 🖷 +63 2 4152272 **W:** www.abs-cbn.com **LP:** Chmn: Eugenio Lopez III. **Txs:** DWAC-TV Manila ch23 (1125kW) & relay stns. – **ACQ KINGDOM BROADCASTING NETWORK (ACQ-KBN)**

(Rlg) Suite 3102 31/F Jollibee Plaza, F. Ortigas Jr. Road, Ortigas Center, Pasig City, 1600 ☎ +63 2 6830772 🖷 +63 2 6830775 **E:** sonshine@sonshinemedia.com **W:** www.sonshinetv.com **Txs:** DWBP-TV Manila ch39 (50kW) & relay stns. – **ASSOCIATED BROADCASTING CO. (ABC) (Comm)** AMPC Bldg., 136 Amorsolo cor. Gamboa Sts., Legaspi Village, Makati City ☎ +63 2 8923801 🖷 +63 2 8154314 **W:** www.abc.com.ph **Txs:** DWET-TV Manila ch5 (55kW) & relay stns. – **GATEWAY UHF BROADCASTING (3ABN) (Rlg)** Manila. **W:** www.3abn.org **Txs:** DWVN-TV Manila ch45 (5kW) & relay stns. – **PROGRESSIVE BROADCASTING CORP. (UNTV) (Pub)** Manila. **W:** www.untvw.com **Txs:** DWAO-TV Manila ch37 (2058kW) & relay stns. – **RADIO MINDANAO NETWORK (Comm)** 4F State Condominium I, Salcedo St., Legaspi Village, Makati City ☎ +63 2 8120540 🖷 +63 2 8163680 **W:** www.rmn.com.ph **L.P:** Chmn: Henry R. Canoy **Txs:** DWKC-TV Manila ch31 (50kW) & relay stns. – **RADIO PHILIPPINES NETWORK (RPN) (Comm)** Broadcast City, Capitol Hills, Quezon City ☎ +63 2 9315080 🖷 +63 2 9321470 **W:** rpn9.com **L.P:** Chmn: Cerge M. Remonde **Txs:** DZKB-TV Manila ch9 (50kW) & relay stns. – **RAJAH BROADCASTING NETWORK (Comm)** 3/F Save-A-Lot Mall, 2284 Pasong Tamo Ext., Makati City ☎ +63 2 8933404 🖷 +63 2 8932360 **E:** rjofc@compass.com.ph **W:** www. rjplanet.com **L.P:** Pres: Ramon Jacinto **Txs:** DZRJ-TV Manila ch29 (1354kW) & relay stns. – **REPUBLIC BROADCASTING SYSTEM (GMA) (Comm)** EDSA, Diliman, Quezon City, Metro Manila ☎ +63 2 9285041 🖷 +63 2 9285041 **W:** www.igma.tv **Txs:** DZBB-TV Manila ch7 (100kW) & relay stns. – **SOUTHERN BROADCASTING NETWORK (SBN) (Comm)** Suite 2901 Jollibee Plaza, Emerald Ave., Ortigas Center, Pasig City ☎ +63 2 6363286 🖷 +63 2 6363288 **E:** genceo@sbnphilippines.net **L.P:** Pres/CEO: Teofilo A. Henson **Txs:** DWCP-TV Manila ch21 (40kW) & relay stns. – **STUDIO 23 (Comm)** 3rd/F, Main Building, ABS-CBN Broadcasting Center, Mo. Ignacia Street cor. Sgt Esguerra, Quezon City, Philippines 1103 ☎ +63 2 4152272 **E:** studio23@abs-cbn.com **W:** www.studio23.tv **Txs:** DWAC-TV Manila ch23 (1126kW) & relay stns. Affiliated with ABS-CBN. – **ZOE BROADCASTING NETWORK (Rlg)** Manila.**W:** www.universitv.net **Txs:** DZOZ-TV Manila ch33 & relay stns.

Local Stations not shown.

DTT Transmitters (under construction)
Operator: GEM-TV **Mux:** GEM-TV, Net 25. **Tx:** ch49 (Manila).

POLAND

Systems: # PAL-D/K [R] ↧31 Jul 2013; DVB-T (MPEG2, MPEG4)

National Stations
TELEWIZJA POLSKA S.A. (TVP) (Pub) ul. Woronicza 17, 00-999 Warszawa ☎ +48 225478000 🖷 +48 225478000 **E:** tvp@tvp. pl **W:** www.tvp.com.pl **L.P:** DG: Romuald Orzel **Chs (terr.):** TVP1, TVP2, TVP Info/TVP Regionalna **Txs: TVP1:** Warszawa (Raszyn) ch11 (250kW) & netw. **TVP2:** Warszawa (Raszyn) ch27 (800kW) & netw. – **CZWÓRKA (Comm)** ul. Gen. Okulickiego 6, 05-500 Piaseczno ☎ +48 227569711 🖷 +48 227503090 **E:** sekretariat@tv4. pl **W:** www.tv4.pl **Txs:** Warszawa (Raszyn) ch33 (20kW) & netw. – **TVN (Comm)** ul. Wiertnicza 166, 02-952 Warszawa ☎ +48 228566060 🖷 +48 228566666 **E:** tzn@tvn.pl **W:** www.tvn.pl **Txs:** Warszawa (Raszyn) ch35 (100kW) & netw. – **TV POLSAT (Comm)** ul. Ostrobramska 77, 04-175 Warszawa ☎ +48 225145533 🖷 +48 225145550 **E:** poczta@polsat.com.pl **W:** www.polsat.com.pl **Txs:** Warszawa (Raszyn) ch44 (316kW) & netw. – **TV PULS (Rlg)** ul. Al. Stanów Zjednoczonych 53, 04-028 Warszawa ☎ +48 223233301 🖷 +48 223233302 **E:** info@pulstv.pl **W:** www.pulstv.pl **Txs:** Warszawa (Raszyn) ch41 (50kW) & netw.

Local Stations (all Comm)
NTL Radomsko*: ul. 11-go Listopada 2, 97-500 Radomsko; ch9 (1kW). Rel. TVN. **Telewizja Luzyce*:** ul. Bracka 12,59-800 Luban; ch51 (5kW). **Telewizja Odra**:** ul. Muchoborska 6, 54-424 Wroclaw; Lubin ch22 (1kW), Opole ch30 (0.2kW), Wroclaw ch31 (1kW), Swidnica ch39 (1kW), Gorzów Wlkp. ch40 (1kW), Jelenia Góra ch43 (1kW), Zielona Góra ch51 (1kW), Glogów ch56 (1kW), Legnica ch57 (1kW). **Telewizja TVT*:** ul. Rynek 1, 44-200 Rybnik; Rybnik ch22 (0.2kW), Zory ch30 (0.2kW). *) Rel. TVN + Loc **) Rel. Czwórka + Loc

DTT Transmitters (under construction)
Operator: TP EmiTel ul.Wadowicka 8W, 30-415 Kraków ☎ +48 122637355 🖷 +48 122637611 **E:** sekretariat@emitel.pl **W:** www. emitel.pl **Mux 1:** Temporary allocations: TVP1, TVP2, TVP INFO/TVP Regionalna; Comm. stns tbd **Mux 2:** Polsat, TVN, Czwórka, TV Puls

Mux 3: TVP1, TVP2, TVP INFO/TVP Regionalna.

Location	M1	M2	M3	kW
Biala Podlaska	-	-	60V	5
Chelm (Kumowa Dolina)	-	-	25V	1
Chodziez	-	-	51V	1
Ciechanów	-	-	35V	1
Czestochowa (Bleszno)	-	-	41	2
Czluchów	-	-	54	50
Elblag (Jagodnik)	-	-	45	10
Gizycko (Milki)	-	-	50V	20
Goldap (Rokna Góra)	-	-	50	1
Gorzów Wielkopolski (Janice)	-	-	21V	5
Ilawa (Kisielice)	-	-	46	100
Katowice (Kosztowy)	-	-	51V	50
Kudowa-Zdrój (G.Parkowa)	-	-	25V	1
Lobez (Toporzyk)	-	-	38V	5
Lódz (Zygry)	-	-	47V	50
Olsztyn (Pieczewo)	-	-	44V	100
Opole (Chrzelice)	-	-	43V	100
Plock (Rachocin)	-	-	35	20
Poznan (Srem)	-	39	28	100/20
Radom (Wacyn)	-	-	40V	3
Ryki	-	-	32V	40
Siedlce (Losice)	-	-	56	5
Suwalki	-	-	24V	1
Szczecin (Kolowo)	-	-	49	10
Walcz (Rusinowo)	-	-	51	20
Warszawa (Raszyn)	58	48	55	2x100/50
Wloclawek	-	-	35	5
Wroclaw (Zórawina)	-	-	33V	10
Zagan (Wichów)	-	41	-	50
Zielona Góra	-	46	-	80

Temporary/Local muxes (MPEG2) not shown.

PORTUGAL

Systems: # PAL-B/G [E] ↧26 Apr 2012; DVB-T (MPEG4)

RÁDIO E TELEVISÃO DE PORTUGAL, SGPS, S.A. (RTP) (Pub) Av. Marechal Gomes da Costa 37, 1849-030 Lisboa ☎ +351 217947000 🖷 +351 217947570 **E:** rtp@rtp.pt **W:** www.rtp.pt **L.P:** Pres: Guilherme Costa; Dir RTP África: Nuno Santos. **Chs (terr.):** RTP1, RTP2; RTP Açores and RTP Madeira (see Azores, Madeira), RTP África (see Angola, Cape Verde, Guinea-Bissau, Mozambique, São Tomé & Príncipe). – **SOCIEDADE INDEPENDENTE DE COMUNICAÇÃO, S.A. (SIC) (Comm)** Estrada da Outurela 119, 2794-052 Carnaxide ☎ +351 214179550 🖷 +351 214173118 **E:** contacto@siconline.pt **W:** www.sic.pt **L.P:** Chmn: Francisco Pinto Balsemão. – **TELEVISÃO INDEPENDENTE, S.A. (TVI) (Comm)** R. Mário Castelhano, 40, Queluz de Baixo, 2749-502 Barcarena ☎ +351 214347500 🖷 +351 214355076 **E:** relacoes.exteriores@iol.pt **W:** www.tvi.iol.pt **L.P:** Chmn: Miguel Pais do Amaral.

Location	RTP1	RTP2	SIC	TVI	kW
Bornes	7	25	22	28	12/3x200
Bragança (Nogueira)	10	33	30	52	0.65/2/2x10
Faro (São Miguel)	6	31	34	37	20/3x250
Fóia	8	47	50	57	20/3x550
Gardunha	8	34	37	31	3/3x20
Leiranco	8	34	56	31	4/3x40
Leiria	-	-	-	42	100
Lisboa (Monsanto)	7	25	28	22	100/3x450
Lousã	3	26	29	32	60/3x540
Marão	6	36	38	24	40/3x300
Marofa	5	48	51	54	16/3x300
Mendro	6	34	37	30	30/3x560
Montejunto	6	46	49	52	22/3x200
Mosteiro	10	21	27	24	1/3x10
Muro	2V	27	24	30	67/3x500
Palmela	35	9	39	32	128/20/2x128
Portalegre (São Mamede)	54	57	51	48	2x60/2x100
Porto (Monte da Virgem)	9	41	52	44	100
Santiago do Cacém	-	-	-	45	100
São Macário	10	50V	57	47	10/3x75
Valença	7	46	43	49	7/3x70

+ sites with txs below 10kW.

DTT Transmitters (txs under construction)
Operator: Portugal Telecom Av. Casal Ribeiro, nº 14 1º, 1000-092 Lisboa ☎ +351 213308100 🖷 +351 213308160 **E:** contact@ptcontact. pt **W:** www.telecom.pt **Mux 1:** RTP1, RTP2, RTP Açores, RTP Madeira, SIC, TVI **Mux 2-6 (☉):** tbd (F.pl).

Location	M1	M2	M3	M4	M5	M6
SFN	67	69	60	65*	66*	68*

*) parts of the country

PUERTO RICO (USA Commonwealth)

Systems: ATSC

Local Stations*
WAPA-TV (Comm): Carr 19 Kilometro 0.5, Guaynabo, PR 00966. Txs: ch27 (1000kW). Mux: WAPA-TV, El Canal del Tiempo. **WCCV-TV (Rlg):** Carr No 2 K92.6, Camuy, PR 00627-2348. Tx: Arecibo ch46 (50kW). **WDWL (Rlg):** Ave Sabana Seca Section 5, Toa Baja, PR 00949. °TBN. Tx: Bayamon ch30 (50kW). **WECN (Rlg):** Carr 167 KM 18.9, Bayamon, PR 00957. Tx: Naranjito ch18 (23kW). **WELU (Rlg):** Carr #2 Km162.8, Hormigueros, PR 00660. Tx: ch34 (250kW). **WIDP (Rlg):** Loma Verde San Jose 1820, Rio Piedras, PR 00926. Tx: ch45 (50.1kW). **WIMN-CA (Rlg):** PO Box 1350, Hatillo, PR 00659. Tx: Arecibo ch20 (0.035kW). **WIPM-TV (Pub):** satellite of WIPR-TV. Tx: Mayagüez ch35 (620kW). **WIPR-TV (Pub):** 570 Ave. Hostos U.Baldrich, Hato Rey, PR 00918. °PBS. Tx: ch43 (790kW). **WIRS (Comm):** satellite of WJPX. Tx: Jauco ch41 (185kW). **WJPX (Comm):** Carr 19 Kilometro 0.5, Guaynabo, PR 00966. °CaribeVisión. Tx: ch21 (1000kW). **WJWN-TV (Comm):** satellite of WJPX. Tx: San Sebastán ch39 (700kW). **WKAQ-TV (Comm):** 383 Roosevelt Ave, Hato Rey, PR 00919. °Telemundo Tx: ch28 (924kW). **WKPV (Comm):** satellite of WJPX. Tx: Ponce ch19 (700kW). **WLII (Comm):** Calle Carazo 62, Guaynabo, PR 00969. °Univision. Tx: Caguas ch56 (71kW). **WMEI (Comm):** 1095 Avenida Wilson, Edificio Puerta del Condado, Suite 2, San Juan, PR 00907. Tx: Arecibo ch14 (50kW). **WMTJ (Pub):** Isodoro Colon Estatal176, San Juan, PR 00928-1345. °PBS. Tx: ch16 (140kW). **WNJX-TV (Comm):** satellite of WAPA-TV. Tx: Mayagüez ch23 (400kW). Mux: WAPA-TV, El Canal del Tiempo. **WOLE-TV (Comm):** Carr 111 Bario Palmar, Aguadilla, PR 00603-5125. Repeater for WKAQ-TV. Tx: Aguadilla ch69 (120kW). **WORA-TV (Comm):** satellite of WLII. Tx: Mayagüez ch29 (650kW). **WORO-TV (Rlg):** Ave Iturreguy/ Baldorioti, Carolina, PR 00902. Tx: ch33 (6kW). **WQHA (Rlg):** satellite of WUJA. Tx: Aguada ch50 (50kW). **WQQZ-CA (Comm):** satellite of WMEI. Tx: Ponce ch33 (3kW). **WQTO (Pub):** satellite of WMTJ. Tx: Ponce ch25 (200kW). **WRFB (Comm):** #21Clle B Sabana Abajo Ind, Carolina, PR 00982. Tx: ch51 (16kW). **WRUA (Rlg):** satellite of WECN. Tx: Fajardo ch33 (37kW). **WSJN-CA (Rlg):** Carr 861 KM 4.4, Toa Alta, PR 00953. Tx: ch15 (38.8kW). **WSJU-TV (Comm):** 1508 Calle Bori Urb Antonsant, San Juan, PR 00927-6116. Tx: ch31 (66kW). **WSTE (Comm):** Calle Carazo 64, Guaynabo, PR 00969. Tx: Ponce ch8 (50kW). Boosters in San Juan (WSTE1), Mayagüez (WSTE2), Arecibo (WSTE3). **WSUR-TV (Comm):** satellite of WLII. Tx: Ponce 43 (68kW). **WTCV (Comm):** Calle Bori # 1554, San Juan, PR 00927-6113. Tx: ch32 (50kW). **WTIN (Comm):** satellite of WAPA-TV. Tx: Ponce ch15 (380.2kW). Mux: WAPA-TV, El Canal del Tiempo. **WUJA (Rlg):** Calle B #24 Urb Ind, Sabana Abajo Carolina, PR 00984-4039. Tx: Caguas ch48 (2.5kW). **WVEO (Comm):** satellite of WTCV. Tx: Aguadilla ch17 (42kW). **WVOZ-TV (Comm):** satellite of WTCV. Tx: Ponce ch47 (50.1kW). **WVSN (Rlg):** Satellite of WCCV-TV. Tx: Humaco ch49 (46kW).

*) Full power licenses (lp licenses not shown). °) Network affiliation. Tx sites are San Juan, unless indicated otherwise.

QATAR

System: PAL-B/G [E]

QATAR TELEVISION (Gov) ✉ P.O.Box 1944, Doha ☎ +974 4894444 🖷 +974 4864611 **E:** info@qatar-tv.net **W:** qtv.iscool.net **Chs:** Prgr 1 (Arabic), Prgr 2 (English). **Txs: Prgr 1:** Jumaliyah ch11 (200kW) & netw. **Prgr 2:** Jumaliyah ch52 (400kW) & netw.

RÉUNION (France)

System: # SECAM-K1 [K/E]; DVB-T

RFO RÉUNION (Pub) ✉ 1 rue Jean-Chatel, 97716 Saint Denis ☎ +262 262406767 🖷 +262 262406771 **W:** reunion.rfo.fr **Chs:** Télé Pays, Tempo **Txs:** P. Textor ch9 (0.5kW) & low power repeaters. – **ANTENNE RÉUNION (Comm)** ✉ BP 80 001, 97801 Saint-Denis Cedex 009 ☎ +262 262482828 🖷 +262 262482829 **W:** www. antennereunion.fr **LP:** DG: Christophe Ducasse **Txs:** Saint Benoit ch26 (2.4kW), Saint Denis ch33 (2kW), Saint Leu ch36 (1kW), Sainte Suzanne ch42 (2kW), Saint Joseph ch55 (1.7kW), Piton Textor ch56 (2kW), Le Port ch61 (2kW). – **CANAL RÉUNION (Comm)** ✪ 6 rue René Demarne, Technopole, 97490 Sainte Clotilde ☎ +262 262979898 🖷 +262 262291709 **E:** contact@

canalreunion.net **W:** www.canal-reunion.com **Txs:** Saint Denis ch25 (2kW), Saint Pierre ch26 (2kW), Sainte Suzanne ch39 (2kW), Saint Joseph ch52 (1.7kW), Piton Textor ch53 (2kW), Le Port ch54 (9kW). – **TÉLÉ KRÉOL (Comm)** ✉ 16 rue du Fangourin, 97460 Savannah ☎ +262 262452017 **W:** www.telekreol.net **Tx:** ch59.

DTT Transmitters
Operator: France Télévisions **Mux:** RFO Réunion, France 2- 5, France Ô, Arte, LCP, France 24, Gulli **Txs:** ch(-).

ROMANIA

Systems: # PAL-D/G [R/E] ⇓31 Dec 2015; DVB-T (MPEG2; MPEG4 pl.)

National Stations
TELEVIZIUNEA ROMÂNA (TVR) (Pub)✉ Calea Dorobantilor nr. 191, sector 1, Bucuresti ☎ +40 21 3199112 🖷 +40 21 3199264 **E:** office@tvr.ro **W:** www.tvr.ro **LP:** DG: Dumitru Popa. **Chs:** TVR1, TVR2, Regional stns.

Location	TVR1	TVR2	REG	kW
Balota	51	-	-	5
Bucuresti	34	51	57	5
Cerbu Novaci	8	-	-	20
Comanesti	23	40	-	5
Costila	6	22	-	10/20
Cozia	44	27	-	2/10
Craiova	-	-	50	10
Dobrogea Sud	46	-	-	5
Feleac	11	21	42	10
Gheorghieni	32	-	-	5
Heniu	59	21	-	10
Magura Boiu	27	-	-	5
Magura Odobesti	9	38	-	2/5
Mogosa	10	37	-	5
Paltinis	7	-	-	5
Pietrarie	-	-	59	10
Saveni	22	-	-	5
Semenic	58	-	-	20
Siria	52	-	-	5
Terghirghiol	8	42	-	5/10
Urseni	9	21	53	10/40/10
Vacareni	43	26	-	5

+ sites with txs below 5kW.

ANTENA 1 (Comm) ✉ Bd. Ficusului 44A, sector 1, Bucuresti ☎ +40 21 2303202 🖷 +40 21 2327707 **E:** office@antena1.ro **W:** www.antena1.ro **LP:** DG: F. Bratescu. **Txs:** Bucuresti ch57 (1kW) & network. – **PRO TV (Comm)** ✉ Bd. Pache Protopopescu 109, Bucuresti ☎ +40 21 2501430 🖷 +40 21 3124218 **E:** doinita@protv.ro **W:** www.protv.ro. **Txs:** Bucuresti ch31 (1kW) & network. – **REALITATEA TV (Comm)** ✉ Sos. Dudesti Pantelimon 1-3, 033091 Bucuresti ☎ 402 1 3160019 🖷 402 1 3160019 **E:** office@realitatea.net **W:** www.realitatea.net **LP:** DG: Sorin Enache. **Txs:** Bucuresti ch42 (1kW) & network.

Local & Regional Stations
1 TV Bacau: Bacau; ch30. **7 Est:** Iasi; ch54. **Activ TV:** Bucuresti; ch49. **Activ TV Turnu Magurelesc:** Turnu Magurele; ch45. **Actual TV:** Târgu Neamt; ch44. **Alpha TV:** Ploesti; Ploiesti ch26, Câmpulung ch26, Pitesti ch33, Costesti ch40, Curtea De Arges ch40, Focsani ch49, Cluj-Napoca ch50, Târgu Jiu ch50. **Ardealeanul TV:** Abrud; ch45. **AS-TV:** Bistrita; ch56. **Banat TV:** Resita; Resita ch28, Moldova-Noua ch44. **Bucovina TV:** Suceava; ch28. **City TV:** Suceava; ch45. **Club TV:** Zalau; ch50. **Columna Wyl TV:** Târgoviste; ch47. **CTV:** Constanta; ch40. **Dâmbovita TV:** Târgoviste; ch25. **Datina TV:** Drobeta-Turnu Severin; ch22. **Deva TV:** Deva; ch37. **Digital 3:** Odorheiu Secuiesc; ch39. **Diplomatic TV Focsani:** Focsani; ch53. **Etno TV:** Sos. Dudesti-Pantelimon 1-3, Sector 3, 033091 Bucuresti; Mizil ch37 (0.1kW), Câmpulung-Moldovenesc ch41, Calafat ch47, Bacau ch51, Turnu Magurele ch52. **Fagaras TV:** Fagaras; ch28. **Fény TV:** 535500 Gheorgheni; Gheorgheni ch28, Odorheiu Secuiesc ch29, Miercurea Ciuc ch42, Tâgu Mures ch46. **Focus TV:** Buzau; Negresti-Oas ch47, Cehu Silvaniei ch51, Buzau ch57. **Global TV:** Tâgu Secuiesc ch25, Lupeni ch26, Vulcan ch33, Bacau ch44, Caransebes ch46, Orastie ch51. **Gorj TV:** Târgu Jiu; ch48. **M Plus TV:** Roman; ch52. **Media TV:** Bârlad; ch56. **Minisat Telecom:** Gaesti; ch35. **Mix TV:** Str. Oltet nr. 10-11, 500152 Brasov; ch37. **Mix 2:** Brasov; ch31. **Neptun TV:** Bul. Mamaia nr. 296, 900581 Constanta; ch36. **Nova TV:** Str. Republicii nr. 62, 500030 Brasov; ch24. **Nova TV Fagaras:** Str. Unirii nr. 2, Fagaras; ch41. **Obiectiv TV:** Cluj-Napoca; ch38. **Oltenia:** Craiova; ch57. **One TV:** Txs: Hunedoara ch24, Cluj-Napoca ch30, Sebes ch33, Deva ch34, Alba Iulia ch41, Onesti ch44, Petrosani ch47. **Orizont**

TV: Varatec; ch44. **Panoramic TV:** Bd. Mihail Kog\lniceanu nr. 19, bl. C5, et. 11, ap. 1A, 500173 Brasov; ch43. **Prima TV:** Slatina; Piatra Neamt ch42, Comanesti ch57. **Radioteleviziunea Severis:** Drobeta-Turnu Severin; ch53. **Regio TV:** Sfantu Gheorghe; ch38. **RTT:** Brasov; ch27. **RTT Fagaras:** Fagaras; ch30. **RTV-1:** Tãgu Bujor; ch29. **Studioul de Televiziune:** Vaslui; ch36. **Sud Est TV:** Slobozia; ch29. **Sun TV:** Miercurea Ciuc; ch44. **Szatmar TV:** Satu Mare; ch52. **Târgoviste TV:** Târgoviste; ch37. **Tele 7 Muscel:** Câmpulung; ch32. **Tele 'M (Bârlad):** Bârlad; ch42. **Tele 'M (Botosani):** Botosani; ch35. **Tele 'M (Iasi):** Iasi; ch28 (rel. Prima-TV). **Tele 'M (Piatra Neamt):** Piatra Neamt; ch22. **Tele 'M (Suceava):** Suceava; ch51. **Tele 'M (Neamt Piatra Neamt):** ch28. **Teleuniversitatea TV:** Cluj-Napoca; ch26. **Teleuniversitatea TV:** Timisoara; ch31. **Teleuniversitatea TV:** Craiova; ch36. **Televiziunea Arad:** Arad; ch29. **Televiziunea Eveniment:** Sibiu; ch41. **Ten TV:** Lugoj; ch37. **Tansilvania TV:** Oradea; ch56. **TV Baile Herculane:** Baile Herculane; ch11. **TV Bacau:** Bacau; ch27. **TV Bistrita:** Bistrita; ch31. **TV BIT:** Iasi; ch38. **TV Buzau:** Buzau; ch25. **TV Cristal:** Petrosani; ch51. **TV Etalon:** Râmnicu Vâlcea; ch37. **TV Europa Nova (Cluj-Napoca):** Cluj-Napoca; ch26. **TV Europa Nova (Timisoara):** Timisoara; ch31. **TV Focus:** Sighetu Marmatiei; ch46. **TV Galati:** Galati; Tãrgu Bujor ch42, Tecuci ch47, Galati ch49. **TV Impact:** Baia Mare; Sighetu Marmatiei ch24, Baia Mare ch54. **TV Info Transilvania:** Sinaia; ch25. **TV Modreni Sat:** Valea Salciei; ch46. **TV Parâng:** Petrosani; ch22. **TV Piatra Neamt:** Piatra Neamt; ch35. **TV Resita:** Resita; ch26. **TV Sibiu:** Sibiu; ch9. **TV Sigma:** Bucuresti; ch25. **TV Sinaia:** Sinaia. Tx: ch57. **TV Total:** Vaslui. Tx: ch55. **TV Fagaras.** Tx: ch45. **Unu TV:** Piatra Neamt; ch45. **Vâlcea 1:** Calea lui Traian, nr. 115, Complex Sarguinta, et. 4, Râmnicu Vâlcea; ch21 (Vâlcea 1), Horezu ch36 (Vâlcea 2), Dragasani ch48 (Vâlcea 3). **Wyl TV:** Str. Torcatori nr. 2B, 100275 Ploiesti; Ploiesti ch32, Pacureti ch33.

DTT Transmitters (Trial)
Operator: Radiocom ✉ Bd. Libertatii 14, sector 5, 050706 Bucuresti ☎ +40 31 5003007 📠 +40 21 3149798 **W:** www.radiocom.ro **Mux 1:** TVR1, TVR2, TVR HD **Tx:** ch54 (Bucuresti 1.5kW, Paltinis 1.3kW) **Mux 2:** TVR3, TVR Cultural **Tx:** ch47 (Paltinis), ch59 (Bucuresti). – **Operator:** Pro TV **Mux:** Pro TV HD **Tx:** ch30 (Bucuresti 1kW).

RUSSIA

Systems: # SECAM-D/K [R] ⇩2015; DVB-T (MPEG2, MPEG4)

National Stations
NTV (Gov) ✉ 127000 Moskva, ul. Ak. Korolyova 12 ☎ +7 495 2177895 📠 +7 495 2175103 **E:** ntv@ntv.ru **W:** www.ntv.ru **LP:** Dir: Nikolay Senkevich. **Txs:** Moskva ch8 (40kW) & network. – **PERVYY KANAL (Gov)** ✉ 127000 Moskva, ul. Ak. Korolyova 12 ☎ +7 495 2179838 📠 +7 495 2151976. **E:** dip@1tv.ru **W:** www.1tv.ru **LP:** Dir: Konstantin Ernst. **Txs:** Moskva ch1 (40kW) & network. – **ROSSIYA 24 (Gov)** ✉ 15162 Moskva, ul. Shabolovka 37 **W:** www.vesti.ru **LP:** G. **Txs:** via national DTT mux. – **TELEKANAL KULTURA (Gov)** ✉ 123995 Moskva, ul. Malaya Nikitskaya 24 ☎ +7 495 2900421 📠 +7 495 2900421 **E:** kultura@tvkultura.ru **W:** www.tvkultura.ru **LP:** DG: Tatyana Paykhova. **Txs:** Moskva ch33 (20kW) & network. – **TELEKANAL ROSSIYA (Gov)** ✉ 115162 Moskva, ul. Shabolovka 37 ☎ +7 495 2348600 📠 +7 495 2142347 **E:** info@rutv.ru **W:** www.rutv.ru **LP:** Chmn: Oleg Dobrodeyev. **Txs:** Moskva ch11 (60kW) & network. – **TELEKANAL SPORT (Gov)** ✉ 115162 Moskva, ul. Shabolovka 37 ☎ +7 495 9540461 **W:** www.sport-tv.ru **Txs:** Moskva ch6 (1kW) & network. – **TV TSENTR (Gov)** ✉ 115184 Moskva, ul. Bolshaya Tatarskaya 33-1 ☎ +7 495 9593900 **E:** press@tvc.ru **W:** www.tvc.ru **LP:** DG: Oleg M. Poptsov. **Txs:** Moskva ch3 (40kW) & network. – **TV ZVEZDA (Gov)** ✉ 129110 Moskva, Suvorovskaya pl. 2 ☎ +7 495 6316883 **E:** info@tvzvezda.ru **W:** www.tvzvezda.ru **Txs:** Moskva ch57 (5kW) & network. – **5 KANAL (Gov)** ✉ 197376 St.Peterburg, ul. Chapygina 6 ☎ +7 812 3351560 📠 +7 812 2343846 **E:** trk@spbtv.ru **W:** www.5-tv.ru **LP:** GM: Vladimir V.Troyepolskiy. **Txs:** St.Peterburg ch3 (50kW) & network. – **DTV (Comm)** ✉ 129226 Moskva, ul. Dokukina 16 ☎ +7 495 7893818 📠 +7 495 7893824 **E:** sales@dtv.ru **W:** www.dtv.ru **LP:** DG: Mart Luik. **Txs:** Moskva ch23 (10kW) & network. – **REN-TV (Comm)** ✉ 119847 Moskva, Zubovskiy bul. 17-1 ☎ +7 495 2465933. 📠 +7 495 2460655. **E:** site@ren-tv.com **W:** www.ren-tv.com **LP:** GD: Boris I. Mints. **Txs:** Moskva ch49 (20kW) & network. – **STS (Comm)** ✉ 123298 Moskva, ul. 3-ya Khoroshevskaya 12 ☎ +7 495 7974126 📠 +7 495 7974101 **E:** ctc@ctc-tv.ru **W:** www.ctc-tv.ru **LP:** DG: Roman E. Petrenko. **Txs:** Moskva ch27 (50kW) & network. – **TNT (Comm)** ✉ 123298 Moskva, ul. Trifonovskaya 57-3 ☎ +7 495 2178188 📠 +7 495 7481490 **E:** info@tnt-tv.ru **W:** www.tnt-tv.ru **LP:** GM: Andrey V. Skutin. **Txs:** Moskva ch35 (5kW) & network. – **TV3 (Comm)** ✉ 125195 Moskva, ul. Ak. Korolyova 4-4 ☎ +7 495 9374039 **W:** www.tv3russia.ru **LP:** GM: Aleksandr Karpov. **Txs:** Moskva ch46 (5kW) & network.

Regional Stations (Gov)
GTRK "Adygeya": 385000 Maykop, ul. Zhukovskogo 24. **GTRK "Alaniya":** 362007 Vladikavkaz, Osetinskaya gorka 2. **GTRK "Altay":** 656045 Barnaul, Zmeinogorskiy trakt 27a. **GTRK "Amur":** 675000 Blagoveshchensk, per. Svyatitelya Innokentiya 15. **GTRK "Bashkortostan":** 450076 Ufa, ul. Gafuri 9/1. **GTRK "Belgorod":** 308000 Belgorod, pr. Slavy 60. **GTRK "Bira":** 679016 Birobidzhan, ul. Oktyabrskaya 15. **GTRK "Bryansk":** 241033 Bryansk, ul. Stanke Dimitrova 77. **GTRK "Buryatiya":** 670000 Ulan-Ude, ul. Erbanova 7. **GTRK "Chita":** 672090 Chita, ul. Kostyushko-Grigorovicha 27. **GTRK "Chukotka":** 686710 Anadyr, ul. Lenina 18. **GTRK "Chuvashiya":** 428003 Cheboksary, ul. Nikolayeva 4. **GTRK "Dagestan":** 367032 Makhachkala, ul. Magomeda Gadzhieva 182. **GTRK "Dalnevostochnaya":** 682632 Khabarovsk, ul. Lenina 4. **GTRK "Don-TR":** 344101 Rostov-na-Donu, ul. 1-ya Barrikadnaya 18. **GTRK "Gornyy Altay":** 659700 Gorno-Altaysk, ul. Choros-Gurkina 38. **GTRK "Ingushetiya":** 366720 Nazran, pr. Bazorkina 72. **GTRK "Irkutsk":** 664003 Irkutsk, ul. Gorkogo 15. **GTRK "Irtysh":** 644050 Omsk, pr. Mira 2. **GTRK "Ivteleradio":** 153647 Ivanovo, ul. Teatralnaya 11. **GTRK "Kabardino-Balkariya":** 360000 Nalchik, pr. Lenina 3. **GTRK "Kaliningrad":** 236016 Kaliningrad, ul. Klinicheskaya 19. **GTRK "Kalmykiya":** 358000 Elista, ul. M. Gorkogo 34a. **GTRK "Kaluga":** 248021 Kaluga, Pole Svobody 40a. **GTRK "Kamchatka":** 683000 Petropavlovsk-Kamchatskiy, ul. Sovetskaya 62. **GTRK "Karachayevo-Cherkesiya":** 357100 Cherkessk, ul. Krasnoarmeyskaya 51. **GTRK "Kareliya":** 185630 Petrozavodsk, ul. Pirogova 2. **GTRK "Khakasiya":** 662000 Abakan, ul. Vyatkina 12. **GTRK "Komi Gor":** 167610 Syktyvkar, Oktyabrskiy pr. 164. **GTRK "Kostroma":** 156005 Kostroma, ul. Nikitskaya 10. **GTRK "Krasnoyarsk":** 660028 Krasnoyarsk, ul. Mechnikova 44a. **GTRK "Kuban":** 350038 Krasnodar, ul. Rashpilevskaya 106. **GTRK "Kurgan":** 640018 Kurgan, ul. Sovetskaya 105. **GTRK "Kursk":** 305016 Kursk, ul. Sovetskaya 32. **GTRK "Kuzbass":** 650099 Kemorovo, ul. Krasnoarmeyskaya 137a. **GTRK "Lipetsk":** 398050 Lipetsk, pl. Plekhanova 1. **GTRK "Lotos":** 414000 Astrakhan, ul. Molodoy Gvardii 17. **GTRK "Magadan":** 685024 Magadan, ul. Kommuny 8/12. **GTRK "Mariy-El":** 424014 Yoshkar-Ola, ul. Osipenko 50. **GTRK "Mordoviya":** 430000 Saransk, ul. Dokuchayeva 29. **GTRK "Murman":** 183032 Murmansk, per. Rusanova 7. **GTRK "Nizhniy Novgorod":** 603600 Nizhniy Novgorod, ul. Belinskogo 9a. **GTRK "Norilsk":** 663300 Norilsk, nab. Urvantseva 10. **GTRK "Oka":** 390006 Ryazan, ul. Skomoroshinskaya 20. **GTRK "Orenburg":** 460024 Orenburg, per. Televizionnyy 3. **GTRK "Oryol":** 302028 Oryol, ul. 7 Noyabrya 43. **GTRK "Penza":** 440602 Penza, ul. Lermontova 39. **GTRK "Perm":** 614070 Perm, ul. Tekhnicheskaya 7. **GTRK "Pomorye":** 163061 Arkhangelsk, ul. Popova 2. **GTRK "Pskov":** 180000 Pskov, ul. Nekrasova 50. **GTRK "Region-Tyumen":** 625013 Tyumen, ul. Permyakova 6. **GTRK "Sakha (NVK "Sakha"):** 677007 Yakutsk, ul. Ordzhonikidze 48. **GTRK "Sakhalin":** 693000 Yuzhno-Sakhalinsk, ul. Komsomolskaya 209. **GTRK "Samara":** 443011 Samara, ul. Sovetskoy Armii 205. **GTRK "Sankt-Peterburg":** 197022 St.Peterburg, nab. reki Karpovki 43. **GTRK "Saratov":** 410004 Saratov, 2-ya Sadovaya ul. 7. **GTRK "Slaviya":** 173620 Velikiy Novgorod, ul. B.Moskovskaya 106. **GTRK "Smolensk":** 214025 Smolensk, ul. Nakhimova 1. **GTRK "Stavropolye":** 355000 Stavropol, ul. Artema 35a. **GTRK "Tambov":** 392720 Tambov, ul. Michurinskaya 8a. **GTRK "Tatarstan":** 420015 Kazan, ul. M. Gorkogo 15. **GTRK "Tomsk":** 634050 Tomsk, ul. Pushkina 19. **GTRK "Tula":** 300600 Tula, Staronikitskaya ul. 1. **GTRK "Tver":** 170000 Tver, ul. Vagzhanova 9. **GTRK "Tyva":** 667003 Kyzyl, ul. Gornaya 31. **GTRK "Udmurtiya":** 426014 Izhevsk, ul. Komunarov 216. **GTRK "Ural":** 620026 Yekaterinburg, ul. Lunacharskogo 212. **GTRK "Vaynakh":** 364000 Groznyy, ul. B.Khmelnitskogo 147, korpus 5. **GTRK "Vladimir":** 600000 Vladimir, ul. Bol. Moskovskaya 62. **GTRK "Vladivostok":** 690091 Vladivostok, ul. Uborevicha 20a. **GTRK "Volga":** 432030 Ulyanovsk, ul. Simbirskaya 5. **GTRK "Volgograd-TRV":** 400066 Volgograd, ul. Mira 9. **GTRK "Vologda":** 160000 Vologda, ul. Predtecheskaya 2. **GTRK "Voronezh":** 394625 Voronezh, ul. Karl Marksa 114. **GTRK "Vyatka":** 610002 Kirov, ul. Uritskogo 34. **GTRK "Yamal":** 626060 Salekhard, ul. Lambinykh 3. **GTRK "Yaroslaviya":** 150014 Yaroslavl, ul. Bogdanovicha 20. **GTRK "Yugoriya":** 626200 Khanty-Mansiysk, ul. Mira 7. **GTRK "Yuzhnyy Ural":** 454000 Chelyabinsk, ul. Ordzhonikidze 54b. **GTRK "Komsomolsk-na-Amure":** 681000 Komsomolsk-na-Amure, ul. Molododvargeyskaya 7. **Territorialnoye otdeleniye GTRK "Krasnoyarsk":** 663370 Tura, ul. 50 let Oktyabrya 28. **Territorialnoye otdeleniye GTRK "Kuban":** 354000 Sochi, ul. Teatralnaya 11a. **Territorialnoye otdeleniye GTRK "Norilsk":** 647000 Dudinka, ul. Gorgogo 15. **Territorialnoye otdeleniye GTRK "Pomorye":** 164700 Naryan-Mar, ul. Smidovicha 19. **TRK "RTV-Podmoskovye":** 123007 Moskva, ul. 1-ya Magistralnaya 14.

Other Regional & Local Stations not shown.

DTT Transmitters (under construction)
Operator: RTRS ✉ Moskva, ul. Ak. Korolyova 13 ☎ +7 495 6480111 📠

+7 495 6480111 **E:** glavred@rtrn.ru **W:** www.rtrs.ru **Mux:** Pervyy kanal, Rossiya 1 Rossiya 2, Rossiya 24, Kultura, NTV, 5-kanal ⌘ R.Rossii, R. Mayak, Vesti FM. **Txs:** nationwide network under construction. **Regional/Local operators** not shown.

RWANDA

System: # PAL-B/G [E] ⇩2015; DVB-T

TÉLÉVISION RWANDAISE (TVR) (Gov)⊠ BP 83, Kigali ☎ +250 577519 🖳 +250 577520 **E:** telerwa@rwanda1.com **LP:** Dir: Rodgers Kayihura. **Txs:** (-). – **TELE 10 (Comm)** ⊠ BP 4307, Kigali ☎ +250 512022 🖳 +250 512024 **E:** tele10@rwanda1.com **W:** www.tele-10.com **LP:** CEO: Eugene Nyagahene. **Txs:** (-).

DTT Transmitters (Trial)
Operator: n/a **Mux:** TVR, CNN, TV5 **Tx:** Kigali ch(-). National network planned.

SABA (Netherlands)

NB: No terrestrial TV station.

SAMOA

System: PAL-B [E]

TELEVISE SAMOA (Gov) ⊠ P.O.Box 3691, Apia ☎ +685 626641 🖳 +685 624789 **E:** ceotvsamoa@samoa.ws **LP:** CEO: Faiesea Lei Sam Matafeo **Txs:** Mount Aflau ch4 (0.05kW), Api Park ch5 (0.005kW), Mount Fiamoe ch6 (0.01kW), Mount Vaea ch8 (0.05kW), Faleasiu ch10 (0.01kW), Apia ch11 (0.01kW). – **LAU TV (Comm)** ⊠ Apia. **LP:** CEO: Tuiasau Uelese Petaia. – **TV3 (Comm)** ⊠ Apia. **LP:** CEO: Atanoa Herbert Crichton. **Txs:** (-) – **VAIALA BEACH TELEVISION (VBTV) (Comm)** ⊠ Apia. **LP:** SM: Muaausa Shane Rivers **Txs:** ch31, ch35.

Foreign TV Relay
CCTV-9 (China): (-).

SAMOA (AMERICAN) (USA)

System: ATSC; lp stns only: # NTSC-M [A]

KVZK-TV (Gov) ⊠ P.O.Box 3511, Pago Pago AS 96799, USA ☎ + 1 684 6334191 🖳 +1 684 6331044 **E:** kvzk-tv@samoatelco.com **LP:** Dir: Vaoita Sava; Dir.Tech: Robert Blauvelt **Tx:** Pago Pago ch5. Mux: KVZK, KGMB/CBS, PBS, BBC World.
NB: Full power license (lp licenses not listed)

SAN MARINO

System: DVB-T (MPEG2)

SAN MARINO RTV (Pub) ⊠ Viale J.F.Kennedy 13, San Marino A-3, SM-43031 ☎ +378 0549 882000 🖳 +378 0549 882850 **E:** amminist-razione@sanmarinortv.sm **W:** www.sanmarinortv.sm **LP:** Head of TV: Michele Mangiafico **Tx:** San Marino ch51 (10kW).

SÃO TOMÉ & PRÍNCIPE

System: PAL-B [E]

TELEVISÃO DE SÃO TOMÉ (TVS) (Pub) ⊠ CP 420, São Tomé ☎ +239 221041 🖳 +239 221942 **E:** tvs@cstome.net **LP:** DG: Víctor Correia **Txs:** ch5, ch7, ch11.

Foreign TV Relay
RTP África (Portugal): (-).

SAUDI ARABIA

Systems: # SECAM-B/G [E], PAL-B/G [E]; DVB-T (MPEG2)

SAUDI ARABIAN TELEVISION (Gov) ⊠ P.O.Box 570, Riyadh 11421 ☎ +966 1 4014440 🖳 +966 1 4044192 **Chs:** Saudi TV 1, Saudi TV2, Al-Riyadh, Al-Ekhbariya. **Txs:** TV1: Riyadh ch5 (11kW) & netw., **TV2:** Riyadh ch7 (50kW) & netw., **Al-Ekhbariya:** Riyadh ch30 (500kW) & netw.

DTT Transmitters (under construction)
Operator: Ministry of Culture and Information **Mux:** Saudi TV1, Saudi TV2, Arriyadiya, Al-Ekhbariya ⌘ General prgr, R.Qu'ran, Second prgr, European prgr **Txs:** (-).

SENEGAL

System: SECAM-K1 [E]

RADIODIFFUSION TÉLÉVISION SÉNÉGALAISE (Gov) ⊠ BP 1765, Dakar ☎ +221 33 8217801 🖳 +221 33 8223490 **E:** rts@rts.sn **W:** www. rts.sn **LP:** DG: Guila Thiam; Dir TV: Babacar Diagne;Tech Dir: Seydou Diallo. **Txs:** (pol.H) Dakar ch7 (10kW), Thiès ch8 (10kW), Tambacounda ch10 (10kW), Ziguinchor ch10 (10kW), Louga ch11 (10kW) & repeaters. – **TÉLÉ FUTURS MÉDIAS (TFM) (Comm)** ⊠ BP 17795, Dakar ☎ +221 8491644 **E:** info@futursmedias.net **LP:** DG: Youssou Ndour **Tx:** Dakar ch42; network planned, pending national licence.

SERBIA

Systems: # PAL-B/G [E] ⇩4 Apr 2012; DVB-T2 (MPEG4)

National Stations
RADIOTELEVIZIJA SRBIJE (RTS) (Pub) ⊠ Takovska 10, 11000 Beograd ☎ +381 11 3212000 🖳 +381 11 3212211 **E:** rtstv@rts.rs **W:** www.rts.rs **LP:** DG: Aleksandar Tijanic; TD: Radisa Petrovic **Chs:** RTS1, RTS2, RTS Digital, RTS HD

Location	P1	P2	kW	Location	P1	P2	kW
Avala	6	22	100/500	Jastrebac	5	27	100/500
Besna Kobila	8	49	10/50	Maljen	9	26	5/250
Deli Jovan	6	23	10/50	Ovcar	8	42	50/400
Cer	7	37	3/300	Tornik	7	53	1/50
Crni Vrh	11	35	35/500	Tupiznica	10	25	35/500
Kapaonik	3	41	50/400				

+ sites with txs below 1kW. P1=RTS1, P2=RTS2
B92 (Comm) ⊠ Bulevar AVNOJ-a 64, 11000 Beograd ☎ +381 11 3012000 🖳 +381 11 3012001 **E:** tvpitanja@b92.net **W:** www.b92. net/tv. **Txs:** Beograd ch57 & network. – **FOX TELEVIZIJA (Comm)** ⊠ Lepenicka 7, 11000 Beograd ☎ +381 11 2091000. **E:** info@foxtv. co.rs **W:** www.foxtv.co.rs **Txs:** Beograd ch12 & network. – **HAPPY TV (Comm)** ⊠ Admirala Geprata 14, 11000 Beograd ☎ +381 11 3613495 🖳 +381 11 3613497 **E:** office@happytv.tv **W:** www.happytv.tv (Shares network with Košava). **Txs:** Beograd ch64 & network. – **KOŠAVA (Comm)** ⊠ Masarikova 5/X, 11000 Beograd ☎ +381 11 3061491 🖳 +381 11 3612135 **E:** office@kanal1.co.yu **W:** www.kosava.co.rs (Shares network with Happy TV). – **PINK TV (Comm)** ⊠ Neznanog Junaka 1, 11000 Beograd ☎ +381 11 3063400 🖳 +381 11 2636862 **E:** marketing@rtvpink.com **W:** www.rtvpink.com **Txs:** Beograd ch45 & network. – **TV AVALA (Comm)** ⊠ Bulevar Vojvode Mišica 39a, 11000 Beograd ☎ +381 11 3644172 🖳 +381 11 3644153 **E:** office@tv-avala. com **W:** www.tv-avala.com **Txs:** Beograd ch28 & network.

Local Stations not shown.

DTT Transmitters (under construction; MPEG2 exc.*=MPEG4)
Operator: RTS **Mux:** RTS1, RTS2, RTS Digital, RTS HD*⌘ R.Beograd 1, 2, 202 **Txs:** ch27 (Avala 0.4kW), txs being installed at all main sites.

Vojvodina

RADIOTELEVIZIJA VOJVODINE (RTV) (Pub) ⊠ Sutjeska ulica 1, 21000 Novi Sad ☎ +381 21 422829 🖳 +381 21 420139 **E:** jelena. kazic@rtv.rs **W:** www.rtv.rs. **Chs:** RTV1, RTV2.

Location	RTS1	RTS2	RTV1	RTV2	kW
Crveni Cot	10	24	41	48	100/500/2x100
Subotica	5	-	43	40	5/2x100
Vršac	11	-	56	62	5/2x100

Local Stations not shown.

DTT Transmitters (under construction; MPEG2 exc.*=MPEG4)
Operator: RTS **Mux:** RTS1, RTS2, RTS Digital, RTS HD* ⌘ R.Beograd 1, 2, 202 **Tx:** ch31 (Crveni Cot 0.2kW).

SEYCHELLES

System: PAL-B [E]

SEYCHELLES BROADCASTING CORP. (SBC-TV) (Pub) ⊠ P.O.Box 321, Hermitage, Mahé ☎ +248 289600 🖳 +248 225641 **E:** sbcradtv@seychelles.sc **W:** www.sbc.sc **LP:** MD: Ibrahim Afif **Txs:** La Misère ch5 (1kW), St. Louis ch7 (6kW) & repeaters.

SIERRA LEONE

System: PAL-G [E]

SIERRA LEONE BROADCASTING SERVICE (SLBS-TV) (Gov) ☑
New England Ville, Freetown ☎ +232 22 240123 🖷 +232 22 240922
E: slbs@sierratel.sl **LP:** Dir: Gina Banda-Thomas **Txs:** Freetown ch24;
other towns: ch29, ch31.

SINGAPORE

Systems: # PAL-B/G [E]; DVB-T (MPEG4)

MEDIACORP TV HOLDINGS PTE. LTD (Comm) ☑ Caldecott
Broadcast Centre, Andrew Rd. Singapore, 299939 ☎ +65 63333888
🖷 +65 62538119 **W:** mediacorptv.com; www.corporate.mediacorp.
sg (corporate) **LP:** CEO: Lucas Chow. **Txs:** Channel 5: Bukit Batok ch5
(120kW); Channel 8: Bukit Batok ch8 (120kW); Suria: ch12 (120kW);
Vasantham: ch24; Channel U: ch28; okto: ch30; NewsAsia: ch32.

DTT Transmitters
Operator: Mediacorp **Mux 1:** Channel 5, Channel 8, NewsAsia **Txs:**
ch37 (SFN). **Mux 2:** Channel 5 (HD5) **Txs:** ch38 (Singapore).

SLOVAKIA

Systems: DVB-T (MPEG2); DVB-T2 (MPEG4); † PAL-B/G [E] ⇩2012

National Stations
SLOVENSKÁ TELEVÍZIA (Pub) ☑ Mlynská Dolina 28, 845 45
Bratislava ☎ +421 2 60611111 **W:** www.stv.sk **LP:** DG: Jozef
Darmo. **Chs:** STV1, STV2, STV3, reg prgrs. – **MARKÍZA TV (Comm)**
☑ P.O.Box 7, 843 56 Bratislava 48 ☎ +421 2 68274111 🖷 +421 2
65956824 **E:** generalnyriaditel.pr@markiza.sk **W:** www.markiza.sk
LP: CEO: Pavol Rusko. – **TV JOJ (Comm)** ☑ P.O.Box 33, 830 07
Bratislava 37 ☎ +421 2 59888111 🖷 +421 2 59888112 **E:** joj@joj.
sk **W:** www.joj.sk. – **TA3 (Comm)** ☑ P.O.Box 31, 820 15 Bratislava
215 ☎ +421 2 48203511 🖷 +421 2 48203549 **E:** ta3@ta3.com **W:**
www.ta3.com.

Local Stations
Bardejovská TV: Radnicné nám. 16, 08501 Bardejov; ch24 (0.025kW).
Kysucké televízne vysielanie: Podvysoká 370, 023 57 Podvysoká;
Cadka ch55 (0.03kW). Rel. TV JOJ. **MTT Trnava:** Hlavná 1, 917 01
Trnava; ch53 (0.2kW). **MsTV Komárno:** P.O.Box 136, 945 01 Komárno;
ch53 (0.2kW). **RTV Banská Bystrica:** Skuteckého 3, 974 00 Banská
Bystrica; ch51 (0.5kW).**Šturovská TV:** Petöfiho 9, 943 01 Šturovo; ch34
(0.01kW). **Teleprior:** Slovenskej jednoty 8, 040 01 Košice; ch33 (0.2kW).
TV B52: Stanicná 1329, 093 01 Vranov n. Topľou; ch49 (0.1kW). Rel.
TV JOJ. **TV Liptov:** Šturova 1989/41, 031 42 Liptovský Mikuláš; ch60
(0.1kW). **TV Naša:** Hutnóicka 1, 040 01 Košice; ch50 (1kW). **TV Myjava:**
Parizánska 290/17, 907 01 Myjava; ch30 (0.01kW). **TV Nové Zámky:**
Radnicná 3, 940 01 Nové Zámky; ch52 (0.2kW). **TV Pezinok:** Holubyho
42, 902 01 Pezinok; ch55 (0.1kW). **TV Poprad:** Podtatranská 1/149,
058 01 Poprad; ch52 (0.2kW). **TV Prievidza:** Hviezdoslavova 3, 971
01 Prievidza; ch46 (0.05kW). **TV Púchov:** Nám. slobody 1400, 020 01
Púchov; ch40 (0.05kW). **TV Reduta:** Radicné nám. 4, 052 01 Spišská
Nová Ves; ch26 (0.25kW). **TV Sen:** Nám. oslobodenia 11, 905 01 Senica;
Skalica ch32 (0.05kW), Senica ch48 (0.03kW). **TV Trencín:** Mierové
nám. 22, 911 01 Trencín; Trencín ch60 (0.3kW), Púchov ch60 (0.04kW).
Zilinská TV: Horný val 3, 010 01 Zilina; ch55 (0.2kW).

DTT Transmitters (MPEG2)
Operator: Towercom, a.s. ☑ Cesta na Kamzík 14, 831 01 Bratislava
☎ +421 2 49220111 🖷 +421 2 44461042 **E:** info@towercom.sk **W:**
www.towercom.sk **Mux 1:** STV1, STV2, STV3, JOJ, JOJ Plus, TV
Markíza **Mux 2:** tbd **Mux 3:** tbd.

Location	M1	M2	M3	kW
Banská Bystrica	65	51	33	25
Banská Štiavnica	21	-	-	12.6
Borský Mikuláš	66	-	-	21
Bratislava	66	56	27	50
Košice (Heringeš)	64	59	25	20
Košice (Makovica)	64	59	25	20
Lucenec	65	-	-	15.6
Námestovo	68	59	26	15.5
Nitra	21	-	48	39
Nové Mesto n.V.	69	-	-	50
Roznava	61	-	54	14.3
Rozomberok	68	-	-	10

Location	M1	M2	M3	kW
Snina	64	-	-	10
Stará Ľubovna	66	-	-	10
Štúrovo	21	-	-	14
Trencín	69	52	57	45.5
Zilina	68	52	32	19.9

+ sites with txs below 10kW. Pol=V.
NB: Subject to changes while DTT transition is progressing.

SLOVENIA

Systems: DVB-T (MPEG4); † PAL-B/G [E] ⇩2011

National Stations
TELEVIZIJA SLOVENIJA (TVS) (Pub) ☑ Kolodvorska 2-4, 1000
Ljubljana ☎ +38 61 1311333 🖷 +38 61 1319171 **E:** info@rtvslo.si **W:**
www.rtvslo.si **LP:** DG: Zarko Petan **Chs:** TV SLO1, TV SLO2, TV SLO3,
TV Koper/Capodistria, Tele M. – **KANAL A (Comm)** ☑ Kranjceva 26,
1113 Ljubljana ☎ +386 1 5893 200 🖷 +386 1 5893200 **E:** television@
kanal-a.si **W:** info@kanal-a.si **LP:** Pres: Douglas Fulton. – **POP TV
(Comm)** ☑ Kranjceva 26, 1113 Ljubljana ☎ +386 1 5893200 🖷 +386 1
5893200 **E:** info@pop-tv.si **W:** www.pop-tv.si **LP:** GD: Marjan Jurenec.
– **TV3 (Comm)** ☑ Vojkova 58, 1000 Ljubljana ☎ +386 1 2807800
🖷 +386 1 2807840 **E:** info@www.tv3.si **W:** www.tv3.si. – **PINK SI
(Comm)** ☑ Ljubljana **E:** info@tvpink.si **W:** www.tvpink.si.

Local Stations
ATV Signal Litija: Pokopališka pot 8, 1270 Litija; Jevnica 2 ch22, Litija
ch23, Sava ch26, Jevnica 1 ch53. **EPTV:** Tivolska 50, 1000 Ljubljana;
Krvavec ch34. **Kanal 10:** Kocljeva ulica 9, 9000 Murska Sobota;
Pecarovci ch38. **Loka TV:** Kapucinski trg 7, 4220 Škofja Loka; Lubnik
ch51. **MOJ TV:** Mariborska cesta 65 A, 2352 Selnica ob Dravi; ch48.
RTS: Meljska cesta 34, 2000 Maribor; Meljski Hrib ch44, Stoperce-
Jelovice ch58. **Sponka.tv:** Liminjanska 96, 6320 Portoroz; Ankaran ch30,
Tinjan ch52. **Studio AS:** Slovenska 52, 9000 Murska Sobota; ch54. **TV
Celje:** Mariborska 86, 3000 Celje; Malic Laško ch40, Celje ch45. **TV
Pika:** Mala vas 23b, 1000 Ljubljana; ch60. **TV Primorka:** Polje 5, 5290
Šempeter pri Gorici; Tolmin ch56, Trstelj ch67. **Vaš Kanal:** Podbevškova
ulica 12, 8000 Novo Mesto; Zuzemberk ch40, Trdinov Vrh ch41. **Vaša
Televizija:** Žarova cesta 10, 3320 Velenje; Malic Laško ch27, Mozirje
ch46, Plešivec ch52. **Vitel:** Gregorciceva 13, 5294 Dornberk; ch9.

DTT Transmitters
Operator Mux 1: TVS **Mux:** TV SLO1, TV SLO2, TV SLO3, Tele M,
Pop TV, Kanal A, TV3, local stns – **Operator Mux 2:** Norkring **Mux:**
TV3, Pink SI.

Location	M1	M2*	kW	Location	M1	M2*	kW
Beli Kriz	51	66	5	Pecarovci	66	-	5
Boc	66	67	5	Plešivec	66	-	25
Kambreško	31	-	5	Pohorje	66	67	100
Krim	37	64	5	Skalnica	51	66	5
Krvavec	37	64	25	Slavnik	51	-	5
Kuk	31	66	5	Tinjan	67	-	5
Kum	45	64	5	Trdinov vrh	45	64	20
Nanos	51	66	200	Trstelj	31	66	5

+ sites with txs below 5kW *) under construction

SOLOMON ISLANDS

NB: No terrestrial TV station.

SOMALIA

System: PAL-B/G [E]

HORNAFRIK TV (Comm) ☑ Mogadishu **E:** info@hornafrik.com **W:**
www.hornafrik.com. **Txs:** (-). – **SHABELLE TV (Comm)** ☑ Mogadishu
☎ +252 5 933111 **E:** info@shabelle.net **W:** www.shabelle.net **Txs:** (-).
– **SOMALI BROADCASTING CORP. (SBC) (Comm)** ☑ SBC Building,
Airport Road, Bossaso ☎ +252 5 824600 **E:** sbc@allsbc.com **W:** www.
allsbc.com **Txs:** (-). – **SOMALILAND NATIONAL TV (SLNTV) (Comm)**
☑ Hargeisa. **E:** slntv@slntv.net **W:** www.slntv.net **Txs:** (-). – **SOMALI
TELEMEDIA NETWORK (STN) (Comm)** ☑ STN Building, Bakaaraha
Market, Mogadishu ☎ +252 5 933300 **Txs:** (-).

SOUTH AFRICA

Systems: # PAL-I [SA] ⇩ 2013; DVB-T (MPEG2), DVB-T2 (MPEG4)
planned [SA]

SOUTH AFRICAN BROADCASTING CORP. (SABC) (Pub) ⌨
Private Bag XI, Auckland Park 2006 ☎ +27 11 7149111 ▤ +27
11 7143106 **E:** info@sabc.co.za **W:** www.sabc.co.za **LP:** Chmn:
Dr Vincent Maphai. **Chs:** SABC1, SABC2, SABC3, Bop TV. – **E.TV
(Comm)** ⌨ Block B, Longkloof Studios, Darters Road, Gardens,
Cape Town 8001 ☎ +27 21 4814500 ▤ +27 21 4814510 **E:** info@
e.tv **W:** www.etv.co.za **LP:** Chief Exec. Marcel Golding. – **M-NET
(Comm)** ✪ ⌨ P.O.Box 4950, Randburg 2125 ☎ +27 11 2893000 ▤
+27 11 7875763 **E:** inquiries@mnet.co.za **W:** www.mnet.co.za **LP:**
PD: Sheryl Raine.

DTT Transmitters (DVB-T/MPEG2, under construction)
Operator: Sentech **W:** www.sentech.co.za **Txs:** SABC1, SABC2,
SABC3, e.tv **Txs:** ch65 (Johannesburg) & netw. – **Operator:**
Multichoice **W:** www.multichoice.co.za **Mux (✪):** multiprgr (Orbicom/
M-Net) **Txs:** ch62V (Johannesburg, Kyalami, Helderkruin).
NB: Nationwide networks with 2 multiplexes under construction.

Systems: DVB-T (MPEG2, MPEG4)

National Stations
TELEVISION ESPAÑOLA (TVE) (Pub) ⌨ Prado del Rey, 28223
Pozuelo de Alarcon (Madrid) ☎ +34 91 5817000 ▤ +34 91 5815476 **E:**
direccion.comunicacion@rtve.es **W:** www.rtve.es **LP:** Chmn: Javier
Pons Tubio. **Chs:** TVE1, TVE2, 24 horas, Clan, Teledeporte – **ANTENA
3 (Comm)** ⌨ Carretera San Sebastian de los Reyes, 28700 Madrid
☎ +34 1 6320500 ▤ +34 1 6327144 **E:** antena3tv@antena3tv.es **W:**
www. antena3tv.com **LP:** Pres: Javier de Godó. – **CUATRO (Comm)**
⌨ Avenida de los Artesanos 6, 28760 Madrid ☎ +34 91 7367000 **E:**
internet@cuatro.com **W:** www.cuatro.com **LP:** – **LA SEXTA (Comm)**
C/ Virgilio, 2 Edificio 4, 28223 Pozuelo de Alarcón ☎ +34 91 8382966
E: rr.hh@lasexta.com **W:** www.lasexta.com **LP:** Pres: Emilio Aragón
Chs: La Sexta, TeleHit – **TELECINCO (Comm)** ⌨ Ctra de Irún, km
11,700, 28049 Madrid ☎ +34 902 155555 **E:** inversores@telecinco.
es **W:** www.telecinco.es.

Regional Stations (Pub)
Andalucía TV: Apartado de Correos 132, 41920 San Juan de
Aznalfarache (Sevilla). **W:** www.canalsur.es. **Aragón Televisión:**
Coso 33, Planta 3ª, 50003 Zaragoza. **W:** www.cartv.es. **Castilla-
La Mancha Televisió:** C/ Río Alberche, s/n Polígono Santa Mª
de Benquerencia, 45007 Toledo **W:** www.rtvcm.es. **Euskal Irrati
Telebista:** Barria Lurreta, 48200 Durango (Vizcaya). **W:** www.
eitb.com. **IB3 TV:** c/ Madalena, 21 Polígon Son Bugadelles, 07180
Santa Ponça **W:** www.ib3.es. **Televisión Castilla y León:** c/Manuel
Canesi Acevedo 1, 47016 Valladoid **W:** www.tvcyl.es. **Televisió de
Catalunya:** Jacint Verdaguer, s/n 08970- Sant Joan Despí, Catalunya.
W: www.tv3.cat. **Televisión de Galicia:** Apt. 707, San Marcos
(Santiago de Compostela). **W:** www.crtvg.es. **Televisión de Madrid:**
Paseo del Príncipe 3, 28223 Pozuelo de Alarcón (Madrid). **W:** www.
telemadrid.es. **Televisión del Principado de Asturias:** Camino
de las Clarisas, 263, 33203 Gijón **W:** www.rtpa.net. **Televisió
Valenciana:** Polígon Accés Ademús s/n; 46100 Burjassot, València.
W: www.rtvv.es.**7 Región de Murcia:** Plaza San Agustín 5, 30005
Murcia. **W:** www.rtrm.es

Local Stations not shown.

DTT Transmitters (MPEG2 exc. *=MPEG4)
Operator: TVE **Mux:** La 1, La 2, 24 Horas, Clan ⌘ RNE1, RNE Clásica,
RNE3 **Txs:** ch66 (SFN), ch57-65 (MFN). **Mux HD*:** TVE HD, Teledeporte
⌘ RNE Clásica, RNE3 **Txs:** ch55 (Madrid) & netw. – **Operator:**
Antena 3 **Mux:** Antena 3, Neox, Nova, Gol Televisión✪ ⌘ Onda
Cero, Europa FM, Onda Melodía **Txs:** ch69 (SFN). **Mux HD*:** Antena
3 HD, Nitro, Marca TV, 13TV **Txs:** ch49 (Madrid) & netw. – **Operator:**
Gestevisión Telecinco **Mux:** Telecinco, La Siete, FDF, Disney Channel.
Txs: ch68 (SFN). **Mux HD*:** Telecinco HD **Txs:** ch33 (Madrid) & netw.
– **Operator:** Sogecable ⌨ Avenida de los Artesanos, 6, 28760 Madrid
☎ +34 91 7367000 ▤ +34 91 7368911 **E:** info@sogetel.com **W:**
www.sogecable.es **Mux:** Cuatro, CNN+, Canal+ Dos✪, Canal Club,
La Sexta ⌘ SER, 40 Principales, Cadena Dial **Txs:** ch67 (SFN). **Mux
HD*:** LaSexta 2, LaSexta 3, LaSexta HD **Txs:** ch33 (Madrid) & netw.
– **Operator:** Veo TV ⌨ Pº Castellana 40, 28046 Madrid ☎ +34 91
4316666 ▤ +34 91 4312999 **E:** rrhh@veo.es **W:** www.veo.es **Mux:**
Veo7, Tienda en Veo, Intereconomía, Teledeporte ⌘ R.Intereconomía,
R.Marca, esRadio, Vaughan R. **Txs:** ch66 (SFN).
NB. HD muxes are currently only available in selected areas.

Regional/Local operators not shown.

System: # PAL-B/G [E]; DVB-T

National Stations
SRI LANKA RUPAVAHINI CORP. (SLRC) (Pub) ⌨ P.O. Box 2204,
Colombo 7 ☎ +94 11 2697491 ▤ +94 11 2695488 **LP:** CEO/DG:
Sisira Kothalawala **E:** dg@rupavahini.lk **W:** www.rupavahini.lk **Chs:**
Rupavahini, Channel Eye/Nethra TV. **Txs: Rupavahini:** Pidurutalagala
ch5 (20kW), Kokavil ch8 (20 kW) & relay txs; **Channel Eye/Nethra
TV:** Pidurutalagala ch7 (20kW) & network. – **ASIA BROADCASTING
CORP. (PVT) LTD. (HIRU TV) (Comm)** 35th Floor, East Tower, World
Trade Centre, Colombo 1 ☎ +94 11 2346888 ▤ +94 11 2346880 **Txs:**
Kandy & Rathnapura ch21, Kurunegala & Nayabadda ch22, Badulla &
Hunnasgiriya ch23, Gammaduwa ch26, Jaffna & Kikiliyamana ch38,
Kokawil & Magalkanda ch45, Colombo ch56. – **INDEPENDENT
TELEVISION NETWORK (ITN) (Comm)** ⌨ Wickramasinghepura,
Battaramulla ☎ +94 11 28684231 ▤ +94 11 2864595 **E:** itnadm@slt.
lk **W:** www.itn.lk **LP:** Chmn: Newton Gunaratne **Txs:** Deniyaya ch9
(20kW), Colombo ch12 (100kW), Yatiyantota ch12 (100kW), Nayabedde
ch12 (3kW). – **MTV CHANNEL (PVT) LTD. (Comm)** ⌨ 7, Braybrook
Pl., Colombo 2 ☎ +94 11 4792600 ▤ +94 11 2447308 **E:** info@media.
maharaja.lk **LP:** DG/GM: Ms Nedra Weerasinghe; TechDir: Tharake
Mohotti. **Chs:** MTV (English), Shakhti TV (Tamil), Sirasa TV (Singalese).
Txs: Channel One/Shakhti: Colombo ch25 (5kW) & network; **Sirasa:**
Colombo ch23 (5kW) & network. – **SWARNAVAHINI (Comm)** (EAP
Networks (PVT) Ltd.) ⌨ 676 Galle Rd, Colombo 3 ☎ +94 11 2599642
▤ +94 11 2503788 **E:** eapnet@slt.lk **W:** www.swarnavahini.lk **LP:**
Chwmn: Mrs. Soma Edirisinghe. **Txs:** Colombo ch34 (5kW) & network.
– **TELSHAN NETWORK (PVT) LTD. (TNL) (Comm)** ⌨ Innagale
Estate Dampe-Piliyandala ☎ +94 11 2501681 ▤ +94 11 2575436 **LP:**
Chmn/MD: Shantilal Nilkant Wickremesinghe **Txs:** Piliyandala ch3
(20kW), Polgahaweda ch3 (1kW), Nuweraeliya ch4 (40kW), Colombo
ch21 (22kW), Hantana (Kandy) ch21 (22kW), Piliyandala ch26 (22kW),
Ratnapura ch26 (1kW). – **TV LANKA (Comm)** ⌨ 68 Attidiya Road,
Ratmalana ☎ +94 11 4213771 ▤ +94 11 4213980 **Txs:** Colombo &
Matale & Matara ch48, Badulla & Kandy & Vauniya ch53.

Local Stations
Art TV: 451 Kandy Road, Kelaniya; Colombo ch28, Kandy ch52. **Derana
TV:** 1072/1 5th Lane, Kotte Road, Rajagiriya; Matale ch28, Matara ch31,
Badulla ch32, Nuwara Eliya ch36, Colombo ch37, Kalutara ch56. **Extra
Terrestrial Vision:** 31 Shady Grove Avenue, Colombo 08; Colombo
ch35, Kalutara ch40. **Max TV:** 221 Stanley Thilakaratne Mawatha,
Nugedoga; Colombo ch30, Ratnapura ch32, Karagahatenna ch46,
Nayabedda ch47, Hunnasgiriya & Kandy ch56. **TV2:** Media House,
594/1 Galle Road, Colombo 03; Karagahatenna ch42, Colombo ch53.

DTT Transmitters (Trial)
Operator: Dialog Telekom PLC ⌨ 475, Union Place, Colombo 02 ☎
+94 11 2678700 **W:** www.dialog.lk **Mux:** Rupavahini Eye/Nethra TV,
ITN, Derana TV, Swarnavahini, The Buddhist, Al Jazeera & others
Txs: ch(-) (Colombo).

NB: No terrestrial TV station.

NB: No terrestrial TV station.

System: PAL-I [E]

NB: Cable & Wireless St. Helena is distributing satellite TV prgrs in
St. Helena on three UHF channels: Channel A (Supersport), Channel B
(MNET), Channel C (BBC World/Discovery Channel).

System: NTSC-M [A]

ZIZ TELEVISION (Gov) ⌨ P.O.Box 331, Basseterre, St. Kitts ☎
+1 869 4652621 ▤ +1 869 4652159 **E:** zbc@caribsurf.com **W:** www.
zizonline.com **LP:** GM: Winston McMahon **Txs:** (pol.H) Basseterre
ch2 (0.2kW), Bayfords (St. Kitts) ch5 (20kW), Brimstone ch9 (0.15kW),
Ottleys Mount ch11 (0.04kW), Nevis ch13 (0.01kW).

ST LUCIA

System: NTSC-M [A]

NATIONAL TELEVISION NETWORK (NTN) (Gov) ☐ Greaham Louisy Administrative Building, The Waterfront, Castries, Saint Lucia ☎ +1 758 4682116 ☐ +1 758 4531614 **E:** ntn@candw.lc **Txs:** (-). – **HELEN TELEVISION (HTS) (Comm)** ☐ P.O. Box 621, The Morne, Castries ☎ +1 758 4524982 ☐ +1 758 4531737 **E:** news@htsstlucia.com **W:** www.htsstlucia.com **LP:** MD: Linford Fevrier; CE: Stephenson Anius. **Txs:** Castries ch4 (20kW H) & ch5 (20kW H).

ST MAARTEN (Netherlands)

System: NTSC-M [A]

LEEWARD BROADCASTING CORP. (LBC) ☐ P.O.Box 375, Philipsburg. **Tx:** ch7 (5kW).

ST MARTIN (France)

System: SECAM-K1 [K/E]

MSRTV 6 (Comm) ☐ #238 Rue De Hollande Agrement, F-97150 St Martin ☎ +590 590876150 **E:** info@msrtv6.com **W:** www.msrtv6.com. **Txs:** (-).

ST PIERRE & MIQUELON (France)

System: # SECAM-K1 [K/E]; DVB-T

RFO ST PIERRE ET MIQUELON (Pub) ☐ BP 4227, F-97500 St. Pierre et Miquelon ☎ +508 411111 ☐ +508 412219 **W:** saintpier-remiquelon.rfo.fr **LP:** Dir: Joseph Eden. **Ch:** Télé Pays **Txs:** (pol.H) St Pierre (Cap a l'Aigle) ch4, Miquelon ch6 (0.2kW), St Pierre (Phare de Galantry) ch39 (0.1kW).

DTT Transmitters
Operator: France Télévisions **Mux:** RFO St Pierre et Miquelon, France 2-5, France Ô, Arte, LCP, France 24, Gulli **Txs:** ch(-).

ST VINCENT & THE GRENADINES

System: NTSC-M [A]

SVG-TV (Gov) ☐ P.O.Box 705, Kingstown, St. Vincent ☎ +1 784 457111 ☐ +1 784 4562759 **E:** svgbc@vincysurf.com **W:** www.svgbc.com **LP:** Managing Dir: R. Paul MacLeish. **Txs:** (pol.H) Dorsetshirehill ch9 (0.4kW), Layouhill ch7 (0.04kW), Maroonhill ch7 (0.04kW), Belleislehill ch11 (0.06kW), Mustique ch11 (0.06kW), Bequia ch13 (0.06kW).

Foreign TV Relay
TBN (USA): Kingstown ch4.

SUDAN

System: PAL-B/G [E]

SUDAN TELEVISION (Gov) ☐ P.O.Box 1094, Omdurman ☎ +249 11 557398 ☐ +249 11 553538 **E:** sudantvlive@sudanmail.net **W:** www.sudantv.tv **LP:** GM: Amin Hasan Umar **Txs:** Omdurman ch5 (5kW H), Gezira ch7 (10kW), Atbara ch9 (0.5kW) & relay txs. – **JUBA TELEVISION (Gov)** ☐ Juba **LP:** Dir: Obeid Lado Kundu. Txs: (-).

SURINAME

System: NTSC-M [A]

National Stations
SURINAAMSE TELEVISIE STICHTING (STVS) (Gov) ☐ P.O.Box 535, Paramaribo ☎ +597 473032 ☐ +597 477216 **E:** stvs@sr.net **Txs:** Wageningen ch7 (0.10kW), Paramaribo ch8 (1kW), Moengo ch9 (0.01kW), Caranis ch10 (0.10kW), Nickerie ch11 (1kW). – **ALGEMENE TELEVISIE VERZORGING (ATV) (Comm)** ☐ P.O.Box 1839, Paramaribo ☎ +597 404661 ☐ +597 402660 **E:** cooman@sr.net **W:** www.atv.sr **Txs:** Borokopondo ch2 (1kW), Wageningen ch6 (0.25kW), Moengo ch7, Paramaribo ch12 (0.4kW), Nickerie ch13 (0.5kW).

Local Stations (all Comm)
Ampies Broadcasting Corp: P.O.Box 885, Paramaribo; ch4 (1kW). **Garuda TV:** Goudstraat 20, Paramaribo; ch23. **Radika TV:** P.O.Box 1083, Paramaribo; ch14 (1kW). **Rapar Broadcasting Network:** P.O.Box 975, Paramaribo. Tx: ch5 (2kW). **Rasonic:** Bataviastraat 2, Nickerie; ch7 (1kW). **TV Apinti:** P.O.Box 595, Paramaribo; ch10 (1kW). **TV Sookha:** Batavaiastraat 25, Nickerie; tx: (-).

SWAZILAND

System: PAL-B/G [E]

SWAZI TV (Gov) ☐ P.O. Box A146, Mbabane ☎ +268 4043036 ☐ +268 4042093 **E:** nomcebo@swazitv.co.sz **W:** www.swazitv.co.sz **LP:** CEO: Vukani Maziya **Chs:** 2 national networks **Txs:** (pol. H exc. where stated) **Prgr 1:** Bulembu ch5 (1.5kW), Ntondozi ch25 (15kW), Mbabane ch27V (1kW) & relay txs. **Prgr 2:** Unknown loc. ch4, unknown loc. ch9, Ntondozi ch21 (50kW).

SWEDEN

System: DVB-T (MPEG2, MPEG4), DVB-T2 (MPEG4)

National Stations
SVERIGES TELEVISION AB (SVT) (Pub) ☐ Oxenstiernsgatan 26-34, 105 10 Stockholm ☎ +46 8 7840000 ☐ +46 8 7841500 **E:** info@svt.se **W:** www.svt.se **LP:** CEO: Eva Hamilton. **Chs:** SVT1, SVT2, SVT24, regional stns. – **CANAL DIGITAL (Comm)** ☐ Tegeluddsvägen 7, 115 80 Stockholm ☎ +46 8 7722700 ☐ +46 8 7722555 **E:** kundservice@canaldigital.se **W:** www.canaldigital.se. – **TV4 AB (Comm)** ☐ Tegeluddsvägen 3-5, 115 79 Stockholm ☎ +46 8 4594000 ☐ +46 8 4594444 **E:** info@tv4.se **W:** www.tv4.se **LP:** CEO: Jan Scherman – **VIASAT AB (Comm)** ☐ P.O.Box 17115, 104 62 Stockholm ☎ +46 8 56241060 ☐ +46 8 56202330 **E:** info@viasat.se **W:** www.viasat.se.

Local Stations not shown.

DTT Transmitters (MPEG2 exc. where stated)
Operator: Boxer TV Access AB ☐ Esplanaden 3c, 3 tr, 172 67 Sundbyberg ☎ +46 8 58789900 ☐ +46 8 58789999 **E:** kundtjanst@boxer.se **W:** www.boxer.se **Mux 1:** SVT1 (incl. reg. prgrs), SVT2 (incl. reg. prgrs), Kunskapskanalen, SVTB/SVT24 **Mux 2:** TV4 (incl. reg. prgrs), TV4 Plus, TV400, TV4 Fakta, TV4 Film, CNN International, TV6. **Mux 3** (✪): TV3, Kanal 5, TV8, Disney Channel/VH1, Canal+ First, Canal+ Hits, SF-kanalen/Canal+ Sport 1. **Mux 4** (✪): Eurosport, Discovery Channel, Nickelodeon/Comedy Central, Kanal 9, MTV, Animal Planet, TV10 **Mux 5** (**MPEG4**) (✪exc.*): Canal 7, Silver, TCM, Axess TV*, TV4 Sport, Local stns*, TV Finland (Mälardalen only)*, BBC World News, Discovery Science/Discovery Travel & Living, Disney XD/Showtime, Silver, Star!/7 **Mux 6** (**DVB-T2/MPEG4**) (✪): SVT1 HD, SVT2 HD, Canal+ HD Mix, Kanal 5 HD, MTVN HD, National Geographic Channel HD, TV3 HD, TV4 HD, Viasat Sport HD, Canal+ Series, Cartoon Network, Eurosport 2, Viasat Football, 24.UNT **Mux 7** (**DVB-T2/MPEG4**) (✪): F.pl

Location	M1	M2	M3	M4	M5	M6*	M7°	kW
Arvidsjaur (Julträsk)	21	24	30	34	42	57	51	50
Bollnäs	29	49	34	39	23	53	6	50
Borlänge (Idkerberget)	47	52	43	41	54	60	28	50
Borås (Dalsjöfors)	44	54	29	42	55	36	41	50
Bäckefors	26	22	35	25	56	49	7	50
Emmaboda (Bälshult)	31	28	46	21	53	47	8	50
Filipstad (Klockarhöjden)	33	23	30	42	40	27	59	50
Finnveden	26	56	52	60	48	58	7	50
Gällivare	33	26	40	28	46	43	22	50
Gävle (Skogmur)	27	24	32	30	46	50	9	50
Göteborg (Brudaremossen)	30	27	46	40	59	33	9	50
Halmstad (Oskarström)	21	28	38	45	47	32	7	10
Helsingborg (Olympia)	33	43	41	25	30	30	10	10
Hudiksvall (Forsa)	31	44	34	39	23	53	60	50
Hörby (Sallerup)	33	43	41	25	61	30	10	50
Jönköping (Bondberget)	31	28	35	33	51	26	6	50
Kalix	35	29	60	55	58	58	27	50
Karlshamn	27	24	42	55	26	30	8	50
Karlskrona (Vämö)	27	24	42	55	26	30	8	50
Karlstad (Sörmon)	43	46	30	42	40	27	59	50
Kiruna (Kirunavaara)	39	35	32	49	42	29	44	50
Kisa	29	55	50	56	59	49	6	50
Lycksele (Knaften)	45	53	22	28	48	58	38	50
Malmö (Jägersro)	33	43	41	25	27	27	10	50
Mora (Eldris)	22	25	35	42	44	51	38	50
Motala (Ervasteby)	27	40	21	42	52	39	53	50
Norrköping (Krokek)	36	46	60	28	54	32	5	50

Location	M1	M2	M3	M4	M5	M6*	M7°	kW¹
Nässjö	22	23	35	33	51	25	6	50
Pajala	34	23	31	37	54	47	51	50
Skellefteå	23	26	49	43	59	46	6	50
Skövde	37	24	32	34	57	69	47	50
Sollefteå (Multrå)	46	24	31	26	44	49	59	50
Stockholm (Nacka)	23	42	56	50	55	59	53	50
Storuman	23	43	36	46	56	60	49	50
Sundsvall (S Stadsberget)	47	27	30	43	56	50	58	50
Sunne (Blåbärskullen)	36	39	50	53	47	60	7	50
Sveg (Brickan)	21	24	46	41	36	59	9	50
Trollhättan	23	43	31	25	56	53	9	50
Tåsjö	37	40	51	41	50	57	30	50
Uddevalla (Herrestad)	23	43	31	25	56	53	9	50
Uppsala (Vedyxa)	40	21	43	49	33	58	52	50
Varberg (Grimeton)	21	28	38	45	47	32	7	10
Visby (Follingbo)	41	44	48	37	58	51	9	50
Vislanda (Nydala)	40	49	34	37	39	57	8	50
Vännäs (Granlundsberget)	47	50	56	36	52	60	8	50
Västervik (Fårhult)	26	34	24	30	40	43	57	50
Västerås (Lillhärad)	37	31	22	34	38	51	57	50
Västerås (Lillhärad)	44	-	-	-	-	-	-	50
Ånge (Snöberg)	42	37	57	28	55	52	22	50
Älvsbyn	36	39	47	32	38	52	56	50
Örebro (Lockhyttan)	35	29	25	49	55	58	48	50
Örnsköldsvik (Ås)	23	21	34	25	42	39	29	50
Östersund (Brattåsen)	27	45	58	53	54	56	48	50
Östhammar (Valö)	40	21	43	26	48	58	6	50
Överkalix	45	48	60	55	50	58	27	50

+ sites with txs below 10kW. *) under construction °) F.pl ¹) Power refers to muxes 1-6

System: DVB-T (MPEG2)

SCHWEIZERISCHE RUNDFUNK- UND FERNSEHGESELLSCHAFT (SRG SSR idée suisse) (Pub) 🖃 Giacomettistrasse 3, 3000 Bern 15 ☎ +41 31 3509710 🖹 +41 31 3509709 **E:** info@srgssrideesuisse.ch **W:** www.srgssrideesuisse.ch **L.P:** Pres: Jean-Bernard Münch.
NB: SRG SSR idée suisse is the holding institution for the regional TV enterprises SF, TSR, TSI and TVR.
Schweizer Fernsehen (SF): Fernsehstrasse 1-4, 8052 Zürich ☎ +41 44 3056611 🖹 +41 44 3055001 **E:** sf@sf.tv **W:** www.sf.tv **L.P:** Dir: Ingrid Deltenre **Chs:** SF1, SF ZWEI (in German and Rumansh), SF Info.
Télé Suisse Romande (TSR): 20 Quai Ernest Ansermet, 1205 Genève ☎ +41 22 7082020 🖹 +41 22 7089800 **E:** dir@tsr.ch **W:** www.tsr.ch **L.P:** DG: Gilles Marchand **Chs:** TSR1, TSR2 (in French). **Televisione Svizzera di lingua Italiana (TSI):** Casella postale, 6949 Comano ☎ +41 91 8035111 🖹 +41 91 803 9314 **E:** info@rtsi.ch **W:** www.rtsi.ch **L.P:** Dir: Prof. Remigio Ratti **Chs:** LA 1, LA 2 (in Italian). **Televisiun Rumantscha (TVR):** Via da Masans 2, Plazza dal teater, 7002 Cuira ☎ +41 81 2557575 🖹 +41 81 2557500 **E:** info@rtr.ch **W:** www.rtr.ch **L.P:** Dir: Bernard Cathomas **Ch:** in Rumansh via SF1.

DTT Transmitters
Operator: SRG SSR idée suisse **Mux 1:** SF1, SF ZWEI, SF Info, TSR1, TSI1 **Mux 2:** TSR1, TSR2, SF1, TSI1 **Mux 3:** LA 1, LA 2, SF1, TSR1.

Location	M1	M2	M3	kW	Location	M1	M2	M3	kW
Bantiger	48V	51V	-	10	M.Ceneri	-	-	49	34
Brueelberg	34V	-	-	10	Mt.Pelerin	-	-	47V	16
Buclards	-	-	34V	10	Pizzo Matro	-	29	-	14
Champ Lequet	-	56V	-	14	Rigi Kulm	32V	-	-	16
Chasseral	62V	56V	-	10	Säntis	34	-	-	42
Gebidem	45V	-	-	10	Tremblex	-	-	56V	10
Gibloux	-	56V	-	11	Tüllingen*	31V	-	-	20
Hoher Kasten	34V	-	-	21	Uetliberg	32	-	-	45

+ sites with txs below 10kW. *) Tx located in Germany
Other operators: Local/regional operators not shown.

System: PAL-B/G [E]

TÉLÉVISION ARABE SYRIENNE (Gov) 🖃 pl. Ommayad, Damas ☎ +963 11 2720700 🖹 +963 11 2234930 **L.P:** Dir TV: Fouad Sherbaji **E:** contact@rtv.gov.sy **W:** www.rtv.gov.sy **Txs:** (pol.H) **Prgr 1:** Abou-Kmal ch3 (200kW), Nabi-Saleh ch3 (100kW), Hassakeh ch4 (200kW), Aleppo ch4 (200kW), Damas ch4 (100kW), Deir-Al Zoor ch6 (100kW), Saroukhieh ch6 (10kW), Soueida ch7 (350kW), Al Soweida ch7 (200kW), Homs ch7 (200kW), Aein-Al-Arab ch7 (100kW), Tabqua ch8 (100kW), Kaldoun ch8 (10kW), Slenfeh ch9 (200kW), Salhieh ch11 (30kW), Afrien ch11 (10kW), Palmyra ch11 (10kW) & txs below

10kW; **Prgr 2:** Al-Malkeih ch12 (200kW), Lattakia ch26 (60kW) & txs below 10kW.

Systems: # NTSC [A] ⇓2012; DVB-T (MPEG2, MPEG4)

PUBLIC TELEVISION SERVICE (PTS) (Pub) 🖃 50, Lane 75, Kang-ning Rd., Sec-3, Taipei 114 ☎ 886 2 26339122 🖹 +886 2 26338124 **E:** pub@mail.pts.org.tw **W:** www.pts.org.tw – **CHINA TELEVISION CO. (CTV) (Comm)** 🖃 120, Chung-Yang Rd, Nankang District, Taipei ☎ +886 2 27838308 🖹 +886 2 2782 6007 **E:** pubr@mail.chinatv.com.tw **W:** www.ctv.com.tw **L.P:** Pres: Hu Ping Chung. – **CHINESE TELEVISION SYSTEM (CTS) (Comm)** 🖃 100, Kuang Fu South Rd, Taipei ☎ +886 2 27510321 🖹 +886 2 27775414 **E:** wwwpub@mail.cts.com.tw **W:** www.cts.com.tw **L.P:** Chmn: Chien-Chiu Yee. – **FORMOSA TELEVISION (FTV) (Comm)** 🖃 24F, No.366, Po Ai 1st Road San Min Dist, Kaohsiung City 807 ☎ +886 2 25702570 🖹 +886 2 25796633 **E:** service@ftv.com.tw **W:** www.ftv.com.tw – **TAIWAN TELEVISION ENTERPRISE (TTV) (Comm)** 🖃 10, Pa Te Rd, Section 3, Taipei 10560 ☎ +886 2 25781515 🖹 +886 2 25799625 **E:** ref@email.ttv.com.tw **W:** www.ttv.com.tw **L.P:** Chmn: Ching-Teh Hsu.

DTT Transmitters
Operator: n/a **Mux 1:** tbd **Txs:** ch25, ch31, ch33, ch49. **Mux 2:** tbd. **Txs:** ch27, ch29. – **Operator:** n/a **Mux:** n/a **Tx:** ch36 (Taipei 3.4kW) DVB-H trial.

Systems: SECAM-D/K [R], PAL-D/K [R]

National Stations
TAJIK TELEVISION (TVT) (Gov) 🖃 Bekhzod St. 7a, 734013 Dushanbe ☎ +992 37 2224357 🖹 +992 37 2213459 **E:** administrator@tvt.tj **W:** www.tvt.tj. **Txs:** (-). – **TV SAFINA (Gov)** 🖃 Bukhoro St. 43, 734025 Dushanbe ☎ +992 37 2278029 🖹 +992 37 2277905 **E:** tv_safina@mail.tj **W:** www.safina.tj. **Txs:** (-).

Local Stations, Foreign TV Relays not shown.

System: PAL-B/G [E]

National Stations
TELEVISHENI YA TAIFA (TVT) (Gov) 🖃 P.O.Box 31519, Dar es Salaam ☎ +255 22 2700011 🖹 +255 22 2700468 **E:** tvt-dg@africanon-line.co.tz **Txs:** Dar es Salaam ch5 & relay txs. – **STAR TV (Comm)** 🖃 P.O.Box 1732, Zanzibar **W:** www.startvtz.com **Txs:** Lake Victoria zone ch31, Arusha ch34, Dar es Salaam ch42.

Local Stations not shown.

Systems: # PAL-B/G [E] ⇓2015; DVB-T

MASS COMMUNICATIONS ORGANISATION OF THAILAND (MCOT) (MODERNINE TV) (Gov) 🖃 63/1 Rama IX Road, Huay Khwang, Bangkok 10320 ☎ +66 2 22016000 🖹 +66 2 22451960 **E:** tnanews@mcot.or.th **W:** www.thaitv9.tv **Txs:** Bangkok ch9 (20kW) & network. – **ROYAL ARMY TELEVISION (TV5) (Gov)** 🖃 210 Phaholyothin Rd, Sanam Pao, Bangkok 10400 ☎ +66 2 22710060 🖹 +66 2 22712515 **E:** army@tv5.co.th **W:** www.tv5.co.th **L.P:** DG: Maj. Gen. Vijit Junapart. **Txs:** Bangkok ch5 (20kW) & net-work. – **TELEVISION OF THAILAND (TVT) (Gov)** 🖃 90-91 New Phetchaburi Road, Huay Khwang, Bangkok 10320 ☎ +66 2 3182110 🖹 +66 2 3182991 **E:** tv11@prd.go.th **W:** tv11.prd.go.th **Txs:** Bangkok ch11 (200kW) & relays. – **THAI PUBLIC BROADCASTING SERVICE (TTPBS) (Pub)** 🖃 1010 Shinawatra Tower III, 13 Vibhavadi Rangsit Road, Chatchuchak, Bangkok 10900 ☎ +66 2 27911000 🖹 +66 2 27911010 **E:** webmaster@thaipbs.or.th **W:** www.thaipbs.or.th **Txs:** Bangkok ch29 (1000kW) & network. – **BANGKOK BROADCASTING & TELEVISION (BBTV) (Comm)** 🖃 P.O.Box 4-56, Bangkok 10900 ☎ +66 2 2720010 🖹 +66 2 27202106 **E:** marketing@ch7.com **W:** www.ch7.com. **L.P:** SM: Chatchur Karnasuta; TD: Supoch Sangsayan. **Txs:** Bangkok ch7 (20kW) & network. – **BANGKOK ENTERTAINMENT CO. Ltd. (THAI COLOR CHANNEL 3) (Comm)** 🖃 Floors 7, 15, 16, The Emporium Tower, Sukhumvit Road, Khlong Tan, Khlong Toey,

Bangkok 10110 ☎ +66 2 22623333 🖥 +66 2 22041384 **E:** internet@
tv3.co.th **W:** www.thaitv3.com **LP:** Prgr. Dir: Pravit Maleenont. TD:
Manoontham Thachai. **Txs:** Bangkok ch32 (650kW) & network.

DTT Transmitters (Trial)
Operator: TVT **Mux:** multiprgr **Txs:** ch60 (Bangkok: Jewellery Trade
Centre/Silom Road; Family Complex/Phayathai; The Emporium Tower/
Sukhumvit Road).

TIMOR-LESTE

System: PAL-B [E]

TELEVIZAUN TIMOR LOROSAE (Gov)⊡ Edifício da Rádio e
Televisão, Rua de Caicoli, Dili ☎+670 3321825 **E**: tv@rttl.org **W:** www.
rttl.org **L.P:** Mgr: Antonio Diaz **Txs:** Dili ch7 (1.5 kW), Baucau ch12.

TOGO

System: SECAM-K1 [E]

National Stations
TÉLÉVISION TOGOLAISE (Gov) ⊡ BP 3286, Lomé ☎ +228
2215357 🖥 +228 2215786 **E:** televisiontogolaise@yahoo.fr **W:** www.
tvt.tg **L.P:** Dir: Kouessan Yovodevi **Txs:** Mt. Agou ch7 (10kW H), Lomé
ch9V (10kW), Aledjo-Kadara ch9 (10kW H) & relay txs. – **RTV DELTA
SANTÉ (RTDS) (Comm)** ⊡ BP 202, Aneho, Lomé ☎ +228 3310573
🖥 +228 2221477 **E:** tvdeltasante@wanadoo.fr. **Txs:** (-).

Local Stations
TV2: Hotel du 2 février, Lomé; tx: (-) **TV7:** BP 81104, Lomé; tx: (-).

TOKELAU (New Zealand)

NB: No terrestrial TV station.

TONGA

Systems: NTSC-M [A]; PAL-B [E]

TELEVISION TONGA (Gov) ⊡ P.O.Box 36, Nuku'alofa ☎ +676
23555 🖥 +676 24417 **E:** a3z-mgt@kalianet.to **W:** www.tonga-broad-
casting.com **L.P:** Controler TV: Taulupe Aleamotu'a **Tx:** Fasi-moe-afi
ch n/a (PAL-B). – **A3M-TV7 (Rlg)** ⊡ P.O.Box 91, Nuku'alofa ☎ +676
23314 🖥 +676 23658 **E:** a3mtonga@kalianet.to **Txs:** Nuku'alofa ch7
(0.5kW, NTSC-M); ch9 (PAL-B). – **ASTL-TV3 (Comm)** ⊡ P.O.Box 66,
Nuku'alofa. ☎ +676 22325. 🖥 +676 22811 **L.P:** Pres: Latu Tupouniua
Tx: Nuku'alofa ch3 (0.025kW, NTSC-M).

Foreign TV Relay
CCTV9 (China): (-)

TRINIDAD & TOBAGO

System: NTSC-M [A]; DVB-T (MPEG4)

NCC-TV (Gov) ⊡ 11A Maraval Rd., P.O. Box 665, Port-of-Spain,
Trinidad ☎ +1 868 62241414 🖥 +1 868 6220344 **E:** nbnl@nbn.co.tt
Txs: ch4, ch16. – **ADVANCED COMMUNITY TV STATION (ACTS)**
⊡ San Fernando, Trinidad **Tx:** ch25. – **CCN TV6 (Comm)** ⊡ 35
Independence Sq, Port-of-Spain, Trinidad ☎ +1 868 6278806 🖥 +1
868 6271451 **E:** info@sixpointproductions.com **W:** www.tv6tnt.com
L.P: CEO: Craig Reynald **Txs:** ch6 & ch18 (Trinidad), ch19 (Tobago). – **C
TELEVISION (CNMG) (Comm)** ⊡ 11A Maraval Road, Port-of-Spain,
Trinidad **L.P:** CEO: Dominic Beaubrun **Txs:** ch9, ch13. – **GAYELLE
TV (Comm)** ⊡ 161 Western Main Road, St. James, Port-of-Spain,
Trinidad ☎ +1 868 6227954 🖥 +1 868 6224601 **W:** www.gayelletv.com
L.P: CEO: Christopher Laird **Tx:** ch23. – **WIN TV (Comm)** ⊡ Mulchan
Seuchan & Endeavors Road, Lange Park, Chaguanas, Trinidad ☎ +1 809
6716937 🖥 +1 809 6721059 **E:** feedback@wintvworld.com **W:** www.
wintvworld.com **L.P:** Chmn: Mohan Jaikaran **Tx:** ch37, ch39.

DTT Transmitters (under construction)
Operator: Green Dot Ltd 61 Mucurapo Road, St James, Trinidad
☎ +868 6284388 🖥 +868 6285197 **E:** info@gd.tt **W:** www.gd.tt **Mux**
1-4: multiprgr **Txs:** (-) (SFN)

TRISTAN DA CUNHA (UK)

NB: No terrestrial TV station.

TUNISIA

Systems: DVB-T (MPEG4); † PAL-B [E], SECAM-B/G [E] ↻2015

**ENTREPRISE DE LA RADIODIFFUSION-TÉLÉVISION TUNISI-
ENNE (E.R.T.T.) (Gov)** ⊡ 71 Ave de la Liberté, 1002 Tunis
Belvedere ☎ +216 71287300 🖥 +216 71781058 **E:** info@tunisiatv.
com **W:** www.tunisiatv.com **L.P:** Dir: Abdeh Afidh Hardudm. **Chs:** TV7
(Arabic), Canal 21 (French). – **HANNIBAL TV (Comm)** ⊡ 85 Avenue
du 13 Aout, Choutrana 2, La Soukra 2036 ☎ +216 70944944 🖥 +216
70944411 **E:** info@hannibaltv.com.tn **W:** www.hannibaltv.com.tn.

DTT Transmitters
Operator: Office National de la Télédiffusion (ONT) ⊡ BP 399, Tunis
1080 ☎ +216 71801177 🖥 +216 71781927 **E:** ont@telediffusion.net.
tn **W:** www.telediffusion.net.tn **Mux:** TV7, Canal 21, Hannibal TV, RAI
1 **Tx:** ch36 (Tunis-Boukornine) & network.

TURKEY

Systems: # PAL-B/G [E] ↻2015; DVB-T (MPEG2)

National Stations
TÜRKIYE RADYO TELEVIZYON KURUMU (TRT) (Pub) ⊡ TRT-TV
Department, TRT Sitesi A Blok 427 Oran, 06109 Ankara ☎ +90 312
4901058 🖥 +90 312 4901109 **E:** genel.sekreterlik@trt.net.tr **W:** www.
trt.net.tr. **L.P:** Head of TV Dept: Nurullah Karakas **Chs (terr.):** TRT1,
TRT2, TRT3, TRT4, TRT-GAP, TRT Çocuk

Location	TRT1	TRT2	TRT3	TRT4	kW
Adana	6	22	25	33	30/3x450
Agri	5	32	35	38	30/3x450
Aksehir	9	23	26	29	30/3x450
Amasya	9	22	25	28	100/3x450
Ankara	5	37	34	40	100/3x450
Antalya	5	47	50	53	30/3x450
Bingöl	10	48	51	54	30/3x450
Bursa	6	55	58	45	100/3x450
Çanakkale	8	22	25	43	30/3x450
Cizre	5	21	24	27	30/3x450
Denizli	8	28	32	35	100/3x450
Diyarbakir	9	56	59	49	100/3x450
Edirne	9	28	31	41	30/3x450
Ekisehir	7	22	25	33	100/3x450
Elazig	7	21	24	31	100/3x450
Elbistan	5	34	37	44	30/3x450
Erzincan	8	42	39	45	100/3x450
Erzurum	6	23	26	29	100/3x450
Gaziantep	10	27	30	40	30/3x450
Isparta	6	40	37	43	30/3x450
Istanbul	5	51	48	54	100/3x450
Izmir	10	23	26	29	30/3x450
Izmit	10	21	24	27	30/3x450
Kars	9	21	24	27	100/3x450
Kastamonu	6	48	45	-	30/2x450
Kayseri	8	26	23	29	30/3x450
Kirsehir	7	42	45	49	30/3x450
Konya	10	32	35	52	100/2x450
Mugla	7	55	58	52	30/3x450
Ordu	5	57	60	44	30/3x450
Samsun	7	52	49	55	30/3x450
Silifke	7	39	42	45	30/3x450
Sivas	10	47	50	54	30/3x450
Trabzon	9	28	31	35	30/3x450
Van	7	22	25	28	30/3x450
Zonguldak	9	23	26	29	30/2x450

+ sites with txs below 30kW.
ATV (Comm) ⊡ Gayrettepe Mah. Barbaros Bul. 125, 80700
Balmumcu-Besiktas-Istanbul ☎ +90 212 3543000 🖥 +90 212 3544064
E: wmaster@atv.com.tr **W:** www.atv.com.tr – **CINE 5 (Comm)** ⊡
Büyükdere Cad. 163, 80504 Esentepe-Istanbul ☎ +90 212 3361515
🖥 +90 212 2173052 **E:** cine5@cine5.com.tr **W:** www.cine5.com.
tr – **CNBC-E (Comm)** ⊡ Eskibüyükdere Cad. Uso Center 61,
80660 Maslak-Istanbul ☎ +90 212 3350000 🖥 +90 212 3350087
E: insankaynaklari@cnbce.com **W:** www.cnbce.com – **CNN TÜRK
(Comm)** ⊡ Dogan Medya Center, 34204 Bagcilar-Istanbul ☎ +90
212 4135600 🖥 +90 212 4135850 **E:** info@cnnturk.com.tr **W:** www.
cnnturk.com – **EUROTÜRK (Comm)** ⊡ Meliha Avni Sözen Cad.
17, 34387 Mecidiyeköy-Istanbul ☎ +90 212 355 8500 **E:** euroturk@
kanalturk.com.tr **W:** www.euroturkonline.com – **KANAL D (Comm)**
⊡ Kanal D TV Center Bagcilar, 34204 Istanbul ☎ +90 212 4135111

📶 +90 212 4135400 **E:** info@kanald.com.tr **W:** www.kanald.com.tr – **KANAL 1 (Comm)** ▣ Gazi Osman Pasa Bulvari, Yeni Asir Ishani 3, Kat 8, 35210 Cankaya-Izmir ☎ +90 232 4453900 ▤ +90 232 4453858 **W:** www.kanal1.com.tr – **KANAL 6 (Comm)** ▣ Seyrantepe Nato Cad. No:5, Levent, Istanbul ☎ +90 212 2830316 ▤ +90 212 2830080 **E:** kanal6@kanal6.com.tr **W:** www.kanal6.com.tr – **KANAL 7 (Comm)** ▣ Otakçilar Cad. 60, Eyüp-Istanbul ☎ +90 212 61290907 ▤ +90 212 61227760 **E:** kanal7@kanal7com **W:** www.kanal7.com – **KRAL TV (Comm)** ▣ Basin Ekspres Yolu Star Sokak 2, 34540 Gunesli-Istanbul ☎ +90 212 6979797 ▤ +90 212 6984970 **E:** kraltv@kraltv.com.tr **W:** www.kraltv.com.tr – **MELTEM TV (Comm)** ▣ Mahmutbey Yolu Cad. 5, Sirinevler-Istanbul ☎ +90 212 6542161 ▤ +90 212 6521765 **E:** bilgi@meltemtv.com.tr **W:** www.meltemtv.com.tr – **NTV (Comm)** ▣ Eskibüyükdere Cad. Uso Center 61, 80660 Maslak-Istanbul ☎ +90 212 3350000 ▤ +90 212 3300050 **E:** ntv@ntv.com.tr **W:** www.ntvmsnbc.com – **SHOW (Comm)** ▣ Yapi Kredi Plaza E Blok 1, Levent, 80620 Istanbul ☎ +90 212 3550101 ▤ +90 212 2792575 **E:** info@showtvnet.com **W:** www.showtvnet.com – **STAR (Comm)** ▣ Star TV Dogan TV Center, Bagcilar-Istanbul ☎: +90 212 4135000 ▤ +90 212 4489364 **E:** bizeyazin@startv.com.tr **W:** www.startv.com.tr – **STV (Comm)** ▣ Kisikli Mah. Ferah Cad. Resatbey Sok. 12 , Çamlica-Istanbul ☎ +90 216 3448560 ▤ +90 216 3448568 **E:** haber@stv.com.tr **W:** www.stv.com.tr – **TGRT HABER (Comm)** ▣ 29 Ekim Cad. 23, Yenibosna, Istanbul ☎ +90 212 4545600 ▤ +90 212 4545666 **E:** tgrt@tgrt.com.tr **W:** www.tgrthaber.com.tr – **TV5 (Comm)** ▣ Inonu Mah. Muammer Aksoy Cad. Milsan Tesisleri 38/A, Sefakoy, 34620 Istanbul ☎ +90 212 2885150 ▤ +90 212 4725455 **E:** program@tv5.com.tr **W:** www.tv5.com.tr – **TV8 (Comm)** ▣ Hlamurdere Cad. Yesilcimen Sok. No.5 OTIM, Besiktas, 80820 Istanbul ☎ +90 212 2885152 ▤ +90 212 2880413 **E:** tv8@tv8.com.tr **W:** www.tv8.com.tr **Txs:** Ankara (Polatli) ch21 & netw. – **Yeni TV (Comm)** ▣ Gazi Osman Pasa Bulvari No.5, 35200 Izmir ☎ +90 232 4415000 ▤ +90 232 4883720 **E:** yenitv@yenitv.com.tr **W:** www.yenitv.tr

Local Stations not shown.

DTT Transmitters (Trials)
Operator: TRT **Mux:** TRT1; two other channels for commercial broadcasters (rotating): Kanal 7, TGRT, CNN Türk, NTV, Kanal D, Star, ATV, Show, CNBC-E, Cine 5, Kral TV, STV, Kanal 1, EuroTürk, Meltem TV **Txs:** ch23 (Istanbul, 2 sites 1.5kW), ch31 (Ankara 1.5kW). – **Operator:** Anten A.S. **Mux:** multiprgr **Txs:** 13 major cities (F.pl)

TURKMENISTAN

Systems: SECAM-D/K [R], PAL-D/K [R]

TURKMEN TELEVISION (Gov) ▣ Mollanepes St. 3, 744000 Asgabat ☎ +993 12 351515 ▤ +993 12 356850 **L.P:** Chmn: Murad Orazov. **Chs:** Altyn Asyr Türkmenistan, Miras, Yaslyk, TV4-Türkmenistan; Türkmen Owazy. **Txs:** (-).

TURKS & CAICOS ISLANDS (UK)

NB: No terrestrial TV station.

TUVALU

System: PAL-B [E]

TUVALU MEDIA CORP. (TMC) (Pub) ▣ Private Mail Bag, Funafuti ☎ +688 20139 ▤ +688 20732 **E:** media@tuvalu.tv **L.P:** CEO/GM: Mrs Tia Taui **Tx:** Funafuti ch n/a (0.02kW).

UGANDA

System: PAL-B/G [E]; DVB-T (MPEG4)

National Stations
UGANDA BROADCASTING CORP. (UBC) (Pub) ▣ P.O.Box 4260, Kampala ☎ +256 41 4254468 ▤ +256 41 425688 **E:** leeta@utlonline.co.ug **W:** www.ubc.ug **L.P:** Head TV: Mrs Proscovia Njuki; Head Engineer: Godfrey Lugya **Txs:** (pol.H exc. where stated) Arua ch5 (1kW), Kampala ch5 (100kW), Koloto ch5 (55kW), Fort Portal ch6 (5kW), Lira ch7 (50kW), Kabale ch7 (5kW), Masaka ch8 (50kW), Mbale ch8 (50kW), Jinja ch9 (5kW), Gulu ch9 (50kW), Hoima ch9V (5kW), Mbarara ch10 (50kW), Soroti ch10 (50kW), Masindi ch11 (5kW) & txs below 1kW. – **WBS TELEVISION (Comm)** ▣ P.O.Box 5914, Kampala **E:** email@wbs-tv.com **W:** www.wbs-tv.com **Txs:** Kampala ch25, Mbara ch25,

Masaka ch45, Jinja ch55.

Local Stations
Born Free TV (Comm): Arua; ch23. **East Africa TV (EATV) (Comm):** Kampala; ch31. **Lighthouse TV (Rlg):** Kampala; ch22. **Nation TV (NTV) (Comm):** Kampala; ch54. **NB TV (Comm):** Lira; ch61. **Pulse TV (Comm):** Kampala; ch12. **Top TV (Rlg):** Kampala; ch28.

DTT Transmitters
Operator: Next Generation Broadcasting (Smart TV) ▣ Plot 6-A, Acacia Avenue, Kampala ☎ +256 312 112842 **E:** contact@ngbroadcasting.com **W:** www.ngbroadcasting.com **Mux (❂):** multiprgr **Txs:** (-)

UKRAINE

Systems: # SECAM-D/K [R], PAL-D/K [R]; DVB-T (MPEG2, MPEG4)

National Stations
NATSIONALNA TELEKOMPANIA UKRAINY (Gov) ▣ vul. Melnykova 42, 04119 Kyiv ☎ +380 44 2413909 ▤ +380 44 2468848 **E:** pr@ntu.com.ua **W:** www.1tv.com.ua **L.P:** Dir: Ihor Storozhuk. **Chs:** Pershyi Natsionalnyi, regional stns – **1+1 (Comm)** ▣ vul. Kreschatyk 7/11, 01001 Kyiv ☎ +380 44 4900101 ▤ +380 44 4907097 **E:** feedback@1plus1.net **W:** www.1plus1.tv **L.P:** DG: Oleksandr Rodnianskyi – **INTER (Comm)** ▣ vul. Dmytrivska 30, 01601 Kyiv ☎ +380 44 4906765 ▤ +380 44 4906765 **E:** program@inter.ua **W:** www.inter.ua

Location	PN	1+1	Inter	kW
Andriivka	12	7	21	2x5/-
Bershad	22	34	32	2x20/0.1
Bilopillia	5V	7	24	2x25/20
Buky	5	1V	28	2x5/0.1
Cherkasy	12	3	33	5
Chernihiv	6	11	28	2x5/1
Dnipropetrovsk	5	7	28	25/2x5
Donetsk	4	30	10	50/20/50
Ivano-Frankivsk	7	9	40	5
Izium	11	29	40	2x25/20
Izmayil	27	3	39	1/0.1/-
Kamyanets-Podilsky	-	21	33	20
Kamyanske	4	7	29	2x5/-
Kerch	12	25	30	5/2x1
Kharkiv	3	9	32	2x5/20
Kherson	3	12	26	2x5/0.1
Khmelnitskyi	28	30	33	2x20/0.1
Kholmy	21	33	-	20
Khust	22	34	28	25/20/5
Kirovohrad	6	21	11	5
Komysh-Zoria	8V	3	28	2x5/0.1
Kotovsk	6	39	-	5
Kovel	6	5	26	25/5/-
Kramatorsk	22	34	37	20
Krasnohorivka	8	10	28	25/50/-
Krasnoperekopsk	32	12V	39	25/5/20
Kryvyi Rih	9	1	40	5
Kyiv	2	4	9	50
Luhansk	6	2	3	5
Lviv	8	1	6	25/40/5
Mariupol	5V	25	31	5
Melitopol	11	29	41	5/2x20
Mykolaiv	10	33	2	5
Nikopol	41	3	31	0.1/2.5/1
Novodnistrovsk	-	38	-	5
Odesa	5	9	12	5
Olevsk	34	22	-	20
Pervomaysk	7	29	12	2x5/0.1
Pryluky	34	22	41	25/5/-
Rivne	3	27	39	50/25/20
Rovenky	12	6	29	2x5/0.1
Sevastopol	5	31	11	5/20/5
Shostka	35	38	51	2x5/20
Simferopol	6	3V	1	2x5/25
Sovietske	7	2V	-	5
Starobilsk	8	28	-	5/25
Ternopil	5V	11V	38	2x25/20
Trostianets	1	11V	25	5
Uzhhorod	2	11	33	5
Vasylivka	4	8	28	2x5/0.1
Vinnitsia	8	10	40	40/35/20

Location	PN	1+1	Inter	kW
Zaporizhia	6	12	33	5

+ sites with txs below 5kW.

Local Stations not shown.

DTT Transmitters (MPEG2 exc.*)
Operator: Concern RRT (Kvant Efir) ✉ Kyiv. **Mux:** UR1*, 1+1*, Inter*, 5 kanal, STB, TRK Ukraina*, NTN* ✂ UR1, PUR2, UR3 **Txs:** Kyiv, ch22 (Chernihiv), ch27 (Lviv 1.2kW), ch28 (Zhytomyr, 3 sites), ch35 (Dnipropetrovsk 1.2kW), ch36 (Simferpol 0.6kW), ch39 (Ternopil 1.2kW), ch56 (Uzhhorod), ch58 (Mykolaiv 1.2kW). – **Operator:** Era Production ✉ Kyiv. **Mux:** UT1, TRK Era, TRK Kyiv, K1, Rada **Tx:** ch41 (Kyiv 1.3kW). – **Operator:** Express Inform ✉ Kyiv. **Mux:** 5 Kanal, Tonis, RU Music, O-TV, News One, Menyu TB **Tx:** ch43 (Kyiv 0.5kW). – **Operator:** Ukrainian Digital TV Network (UDTVN) ✉ Kyiv. **Mux:** Inter, Mega, NTN, Kyivska DRTRK, Music Box UA, Humor/Babai, M2, Futbol, Unian **Tx:** ch51 (Kyiv 1kW), ch61 (Uzhhorod). – **Operator:** Gamma Consulting ✉ Kyiv. **Mux:** M1, M2, Gamma, Pershyi Dilovyi, TRK Ukraina **Tx:** ch64 (Kyiv 0.2kW).

UNITED ARAB EMIRATES

System: # PAL-B/G [E]; DVB-T

ABU DHABI TV (Gov) ✉ P.O.Box 637, Abu Dhabi ☎ +971 2 44451111 **E:** adtv@emi.co.ae **W:** www.emi.co.ae **Txs: Prgr 1:** Abu Dhabi ch1 (2kW) & network, **Prgr 2:** Abu Dhabi ch5 & network. – **AJMAN TV (Gov)** ✉ P.O.Box 422, Ajman ☎ +971 6 7465000 ▤ +971 6 7465135 **E:** progajtv@ajmantv.com **W:** www.ajmantv.com **Tx:** Ajman ch26 (100kW) – **DUBAI TV (Gov)** ✉ P.O.Box 1695, Dubai ☎ +971 4 3077245 ▤ +971 4 3374111 **E:** commercial@dubaitv.gov.ae **W:** www.dubaitv.ae **Txs: Prgr 1:** Dubai ch2 (150kW), Zabeel ch10 (455kW), Jebel Hatta ch41 (1600kW); **Prgr 2:** on Zabeel ch33 (1700kW). – **SHARJAH TV (Gov)** ✉ P.O.Box 111, Sharjah ☎ +971 6 5661111 ▤ +971 6 5669999 **E:** admin@ sharjahtv.ae **W:** www.sharjahtv.ae **Txs: Prgr 1** (Arabic): Sharjah ch54 (199kW); **Prgr 2** (English): Jabel ch22 (398kW).

UNITED KINGDOM

Systems: DVB-T (MPEG2), DVB-T2 (MPEG4); † PAL-I [E] ⬇2012

National Stations
BRITISH BROADCASTING CORP. (BBC) (Pub) ✉ BBC Television Centre, 80 Wood Lane, London W12 0TT ☎ +44 20 87438000 ▤ +44 20 87497520 **W:** www.bbc.co.uk/tv **L.P:** Dir.TV: Jana Bennett. **Chs:** BBC1, BBC2, BBC3, BBC4, CBeebies, CBBC, BBC News, regional stns: **a) BBC Channel Islands:** 18-21 Parade Road, St Helier JE2 3PL; **b) BBC East:** The Forum, Millennium Plain, Norwich NR2 1BH; **c) BBC East Midlands:** London Road, Nottingham NG2 4UU; **d) BBC East Yorkshire & Lincolnshire:** Queen's Court, Hull HU1 3RH; **e) BBC London:** Marylebone High St., London W1A 6FL; **f) BBC North East & Cumbria:** Broadcasting Centre, Barrack Rd, Newcastle upon Tyne NE99 2NE; **g) BBC North West:** New Broadcasting House, Oxford Road, Manchester M60 1SJ; **h) BBC Northern Ireland:** Ormeau Avenue, Belfast BT2 8HQ; **i) BBC Scotland:** 40 Pacific Quay, Glasgow G51 1DA; **j) BBC South:** Broadcasting House, 10 Havelock Road, Southampton SO14 7PU; **k) BBC South East:** The Great Hall, Mount Pleasant Road, Tunbridge Wells TN1 1QQ; **l) BBC South West:** Broadcasting House, Seymour Road, Plymouth PL3 5BD; **m) BBC Wales:** Llantrisant Rd, Cardiff CF5 2YQ; **n) BBC West:** Broadcasting House, Whiteladies Road, Bristol BS8 2LR; **o) BBC West Midlands:** Level 7, The Mailbox, Birmingham B1 1RF; **p) BBC Yorkshire:** 2 St Peter's Square, Leeds LS9 8AH. – **INDEPENDENT TELEVISION NETWORK (ITV) (Comm)** ✉ London Television Centre, Upper Ground, London SE1 9LT ☎ +44 20 76201620 **W:** www.itv.com **L.P:** CEO: Adam Crozier. **Chs:** ITV1, ITV2, ITV3, ITV4, regional stations (see also www.itvlocal.com): **a) ITV Anglia:** Anglia House, Norwich NR1 3JG; **b) ITV Border:** The Television Centre, Carlisle CA1 3NT; **c) ITV Central:** Gas Street, Birmingham B1 2JT; **d) ITV Channel Television:** Television Centre, St Helier, Jersey, Channel Islands JE1 3ZD; **e) ITV Granada:** Quay Street, Manchester M60 9EA; **f) ITV London:** 200 Gray's Inn Road, London WC1X 8XZ; **g) ITV Meridian:** Solent Business Park, Whiteley PO15 7PA; **h) ITV Thames Valley:** 9 Windrush Court, Abingdon Business Park, Abingdon OX14 1SA; **i) ITV Tyne Tees:** Television House, The Watermark, Gateshead, Tyne and Wear NE11 9SZ; **j) ITV Wales:** The Television Centre, Culverhouse Cross, Cardiff CF5 6XJ; **k) ITV West:** 470 Bath Road, Bristol BS4 3HG; **l) ITV Westcountry:** Langage Science Park, Western Wood Way, Plymouth PL7 5BQ; **m) ITV Yorkshire:** The Television Centre, 104 Kirkstall Rd, Leeds LS3 1JS; **n1) STV:** Pacific

Quay, Glasgow G51 1PQ; **o) UTV:** Havelock House, Ormeau Road, Belfast BT7 1EB. NB. ITV plans to reduce the regions to 6. – **CHANNEL FOUR TELEVISION CORP. (Comm)** ✉ 124 Horseferry Road, London SW1P 2TX ☎ +44 20 73964444 ▤ +44 20 73068366 **W:** www.channel4. com **L.P:** CEO: David Abraham. – **CHANNEL 5 BROADCASTING LTD (Comm)** ✉ 22 Long Acre, London WC2E 9LY ☎ +44 20 75505555 ▤ +44 20 75505554 **W:** www.five.tv **L.P:** CEO: Jane Lighting. – **BRITISH SKY BROADCASTING LTD (Comm)** ✉ Grant Way, Isleworth, London TW7 5QD ☎ +44 20 77053000 ▤ +44 20 77053453 **W:** www.sky. com **L.P:** CEO: Jeremy Darroch. – **S4C (WELSH FOURTH CHANNEL AUTHORITY) (Pub)** ✉ Parc Ty Glas, Llanishen, Caerdydd/Cardiff CF14 5DU ☎ +44 2920 747444 ▤ +44 2920 754444 **E:** s4c@s4c.co.uk **W:** www.s4c.co.uk **L.P:** CEO: Arwel Ellis Owen (interim).

DTT Transmitters (under construction) (MPEG2 except where indicated)
Operator Mux 1+3: BBC **Mux 1:** BBC1 (incl. reg. prgrs a-p), BBC2, BBC3, CBBC, BBC News **Mux 3:** BBC4, Cbeebies, BBC Parliament, Community, BBCi 301, BBCi 302 ✂ BBC R.1, 1Xtra, R.2, R.3, R.4, R.5Live, R.5 Live Sports Extra, R.6 Music, R.7, BBC Asian Network, BBCWS **Mux 3HD (DVB-T2/MPEG4):** BBC One HD, ITV1 HD, Channel 4 HD, BBC HD – **Operator Mux 2:** Digital 3&4 Ltd. **Mux:** ITV1 (incl. reg. prgrs a-o), ITV2, ITV2+1, Channel 4, FIVE, Channel 4+1, More4, E4 ✂ Heart – **Operator Mux 4:** SDN Ltd. **Mux:** ITV3, QVC, Bid TV, FIVER, FIVE USA, Quest, CNN International, S4C/2, Teachers' TV, TOPUP Stream 1-3✪ ✂ Smash Hits! – **Operator Mux 5+6:** Arqiva **Mux 5:** Sky3, Dave, Dave ja vu, E4+1, price-drop tv, Sky 3+1, Sky News ✂ talkSPORT, Premier Christian R., Absolute R. **Mux 6:** Yesterday, Film4, 4Music, VIVA, Ideal World, ITV4, Gems TV, Lottery Xtra, Russia Today, Al Jazeera English ✂ The Hits R., Kiss R., Heat R., Magic R., Q R., Smooth R., Kerrang! – **Operator Mux 7:** tbd **Mux (DVB-T2/MPEG4):** tbd (F.pl). **NB.** Some regional variations apply.

Location	M1°	M2°	M3	M4	M5	M6	M7²	kW
Angus	60i	53n	57¹	54	58	61	48	3x20/3x10
Beacon Hill	60i	53l	57¹	42	45	51	49	3x20/3x10
Belmont	30d	48m	66	68	60	57	21	10/20/2x10/2x4
Bilsdale	34f	21i	24	31	27	42	24	4.8/4x6/1.6
Black Hill	41i	47n	59¹	44	55	65	51	2x20/10/3x20
Blaenplwyf	27m	24j	21¹	25	22	28	–	3x40/3x10
Bluebell Hill	59f	24g	45	27	42	39	56	3/2x2/3x3
Bressay	21i	24n	31	27	66	68	30	2
Caldbeck	25f	28b	30¹	23	26	29	21	3x100/3x50
Caradon Hill	28l	25l	22¹	21	24	27	30	3x100/3x50
Carmel	60m	53j	57¹	54	58	61	52	3x20/3x10
Chatton	40f	50i	46	43	47	51	–	4x6/2x4
Craigkelly	33i	29n	26¹	23	42	39	52	4x4/2x2
Crystal Palace	25e	22f	31¹	32	34	29³	29	2x20/10x2/10/2x4
Darvel	22i	25n	28	32	30	34	30	4x4/2x2
Divis	29h	33o	23	26	48	34	–	4x2.3/2/1.6
Dover	68f	61g	58	55	57	60	57	1/3x2/1/0.5
Durris	28i	25n	22¹	52	41	51	30	3x100/20/2x5
Emley More	52b	40a	39¹	43	50	49	45	2x10/5/2x10/4
Fremont Point*	41a	44d	–	47	–	–	–	–
Hannington	50j	43h	46	40	44	41	43	4x20/2x10
Heathfield	34h	29g	47	48	54	51	54	2x1.6/4x1
Huntshaw Cross	62l	59l	55¹	48	52	56	51	3x20/3x10
Keelylang Hill	46i	43n	50¹	42	45	49	48	3x20/3x10
Knock More	26i	23n	29¹	53	57	60	56	3x20/3x10
Llanddona	57m	60j	53¹	43	46	50	51	3x20/3x10
Mendip	61n	54k	58¹	62	56	59	55	3x100/3x10
Midhurst	56j	65g	59	62	64	60	46	4x2/2.5/1
Moel-y-Parc	45m	49j	62	51	52	48	30	3x20/3x10
Oxford	34j	68h	52	51	48	29	49	2x10/2x6/2x8
Pontop Pike	48f	55i	63¹	59	62	53	56	10
Presely	43m	46j	50¹	42	45	49	30	3x20/3x10
Redruth	44l	41l	47¹	48	52	51	51	3x20/3x10
Ridge Hill	53o	57c	63	60	42	45	30	2
Rosemarkie	45i	49n	42¹	43	46	50	52	3x20/3x10
Rowridge	34j	32g	28	30	37	33	29	2x20
Rumster Forest	27i	24n	21¹	30	59	62	52	3x20/3x10
Sandy Heath	42b	45a	67	43	40	46	49	20
Selkirk	62i	59n	55¹	57	53	60	56	3x10/3x5
Stockland Hill	26l	23l	25¹	22	28	30	30	3x50/3x10
Sudbury	49b	68a	39	48	54	50	49	7/8.1/5/7.5/1.5/1.1
Sutton Coldfield	41o	44c	51	47	52	55	51	8
Talconeston	63b	60a	58	53	61	64	57	2x10/4x5
The Wrekin	21o	31c	27	24	48	52	51	4x2/2x1
Waltham	49c	23c	33	26	45	53	56	10/3x8/2x5
Wenvoe	41m	44j	47	43	28	34	30	3x100/3x50
Winter Hill	62g	59e	54¹	58	61	55	–	4x100/12.5/1

+ sites with txs below 1kW. °) incl. reg. prgrs *) F.pl (start: 2013) ¹) txs carry mux 3HD ²) F.pl (power not indicated) ³) to move to channel 28 prior to start of mux 7
NB. Powers and/or channels for many txs are subject to change while the transition to DTT is progressing.

UNITED STATES OF AMERICA

Systems: ATSC

Main National Networks
ABC, Inc. (Comm) (Subsidiary of Walt Disney Co.) ☒ 77 W. 66th St., New York, NY 10023-6298 ☎ +1 212 4567777 ▤ +1 212 4566850 **W:** abc.go.com **LP:** Chmn/CEO: Thomas S. Murphy. **O&O Stations**[1]: KABC-TV Los Angeles, CA: ch53 (182kW); KFSN-TV Fresno, CA: ch9 (8.7kW); KGO-TV San Francisco, CA: ch24 (561kW); KTRK-TV Houston, TX: ch32 (797kW); WABC-TV New York, NY: ch45 (219kW); WJRT-TV Flint, MI: ch36 (860kW); WLS-TV Chicago, IL: ch52 (153.6kW); WPVI-TV Philadelphia, PA: ch64 (500kW); WTVD Durham, NC: ch52 (1000kW); WTVG-TV Toledo, OH: ch19 (795kW). **Full-Power Affiliates:** 229. – **CBS BROADCASTING, Inc. (Comm)** (Subsidiary of CBS Corporation) ☒ 51 W 52nd St, New York, NY 10019-6119 ☎ +1 212 9754321 ▤ +1 212 9754516 **E:** marketing@cbs.com **W:** www.cbs.com **LP:** Chmn: Leslie Moonves. **O&O Stations**[1]: KCBS-TV Los Angeles, CA: ch60 (469kW); KCNC-TV Denver, CO: ch35 (1000kW); KDKA-TV Pittsburgh, PA: ch25 (1000kW); KOVR-TV Sacramento, CA: ch25 (760kW); KPIX-TV San Francisco, CA: ch29 (1000kW); KTVT-TV Dallas, TX: ch19 (695kW); KYW-TV Philadelphia, PA: ch26 (770kW); WBBM-TV Chicago, IL: ch3 (4.4kW); WBZ-TV Boston, MA: ch30 (825kW); WCBS-TV New York, NY: ch56 (349kW); WFOR-TV Miami, FL: ch22 (1000kW); WJZ-TV Baltimore, MD: ch38 (1000kW); WCCO-TV Minneapolis, MN: ch32 (1000kW); WKBD-TV Detroit, MI: ch14 (200kW); WWJ-TV Detroit, MI: ch44 (200kW). **Full-Power Affiliates:** 215. – **FOX BROADCASTING CO. (Comm)** (Subsidiary of Fox Entertainment Group, Inc.) ☒ 10201 W. Pico Blvd., Los Angeles, CA 90035 ☎ +1 310 3693716 ▤ +1 310 9693300 **E:** foxnet@delphi.com **W:** www.fox.com **LP:** Chmn: Peter Liguori **O&O Stations**[1]: KDFW Dallas/Fort Worth, TX: ch35 (857kW); KMSP-TV Minneapolis, MN: ch26 (691kW); KRIV Houston, TX: ch27 (500kW); KSAZ-TV Phoenix, AZ: ch10 (1000kW); KTBC Austin, TX: ch7 (30kW); KTTV Los Angeles, CA: ch65 (1000kW); WAGA Atlanta, GA: ch27 (1000kW); WFLD Chicago, IL: ch31 (690kW); WFXT Boston, MA: ch31 (1000kW); WHBQ-TV Memphis, TN: ch53 (1000kW); WJBK Detroit, MI: ch58 (1000kW); WOFL Orlando, FL: ch22 (1000kW); WOGX Ocala/Gainesville, FL: ch31 (500kW); WNYW New York, NY: ch44 (426kW); WTTG Washington, D.C.: ch36 (1000kW); WTVT Tampa Bay, FL: ch12 (17.5kW); WTXF-TV Philadelphia, PA: ch42 (1000kW). **Full-Power Affiliates:** 223. – **NBC UNIVERSAL, Inc. (Comm)** (Subsidiary of General Electric Co.) ☒ 30 Rockefeller Plaza, New York, NY 10112 ☎ +1 212 6644444 ▤ +1 212 212 6644085 **W:** www.nbcuni.com **LP:** CEO: Jeff Zucker. **O&O Stations**[1]: KNBC Los Angeles, CA: ch36 (380kW); KNSD San Diego, CA: ch40 (370kW); KNTV San Francisco, CA: ch12 (103.1kW); KXAS-TV Dallas/Fort Worth, TX: ch41 (891kW); WCAU Philadelphia, PA: ch67 (560kW); WMAQ-TV Chicago, IL: ch29 (350kW); WNBC New York, NY: ch28 (200.1kW); WRC-TV Washington, D.C.: ch48 (813kW); WTVJ Miami, FL: ch31 (1000kW); WVIT Hartford, CT: ch35 (350kW). **Full Power Affiliates:** 226. – **PUBLIC BROADCASTING SERVICE (PBS) (Pub)** ☒ 1320 Braddock Place, Alexandria, VA 22314-1698 ☎ +1 703 7395000 ▤ +1 703 7390775 **W:** www.pbs.org **LP:** Pres/CEO: Ervin S. Duggan. **Member Stations:** 350.
[1] O&O = owned-and-operated.

Other Major Networks
ION TELEVISION (Comm) (Subsidiary of ION Media Networks, Inc.) ☒ 601 Clearwater Park Road, West Palm Beach, FL 33401 ☎ +1 561 6594122 ▤ +1 561 6594252 **W:** www.iontelevision.com **LP:** Pres/CEO (ION Media Networks): Brandon Burgess. **Full-Power Affiliates:** 64. – **MYNETWORKTV, Inc. (Comm)** (Subsidiary of Fox Entertainment Group, Inc.) ☒ 1999 S. Bundy Dr., Los Angeles, CA 90025-5203 ☎ +1 310 5842000 ▤ +1 310 5842288 **W:** www.mynetworktv.com **LP:** Chief Operating Officer: Jack Abernethy. **Full-Power Affiliates:** 162. – **TELEFUTURA NETWORK (Comm)** (Subsidiary of Univision Communications, Inc.; Spanish-language netw.) ☒ 9405 NW 41st Street, Miami, FL 33178-2301 ☎ +1 305 4713900 ▤ +1 305 4714065 **W:** www.telefutura.com **LP:** Executive Vice President and Operating Manager: Alina Falcon. **Full-Power Affiliates:** 43. – **TELEMUNDO NETWORK GROUP, LLC (Comm)** (Subsidiary of Telemundo Holdings, Inc.; Spanish-language netw.) ☒ 2290 West 8th Avenue, Hialeah, FL 33010 ☎ +1 305 8848200 ▤ +1 305 8897950 **W:** www.telemundo.com **LP:** Chief Operating Officer: Alan J. Sokol. **Full-Power Affiliates:** 42. – **THE CW NETWORK, LLC (Comm)** (Subsidiary of WB Communications, Inc. and CBS Corporation) ☒ 4000 Warner Blvd., Burbank, CA 91522 ☎ +1 818 9775000 ▤ +1 818 9778310 **W:** www.cwtv.com **LP:** Chief Operating Officer: John Maatta. **Full-Power Affiliates:** 204. – **UNIVISION NETWORK, LP (Comm)** (Subsidiary of Univision Communications, Inc.; Spanish-language netw.) ☒ 9405 NW 41 St., Miami, FL 33178-2301 ☎ +1

305 4713900 ▤ +1 305 4714065 **W:** www.univision.com **LP:** Executive Vice Pres: Alina Falcon. **Full-Power Affiliates:** 44. – **V-ME (Educ)** (V-me Media Inc.; Spanish-language netw.) ☒ 450 West 33 St., 6th Floor, New York, NY 10016 ☎ +1 212 5608700 ▤ +1 212 5608720 **W:** www.vmetv.com **Full-Power Affiliates:** 45. – **JCTV (Rlg)**(TBN channel)☒ 2442 Michelle Dr, Tustin, CA 92780 ☎ +1 714 8322950 ▤ +1 714 6652191 **W:** www.jctv.org **Full-Power Affiliates:** 33. – **SMILE OF A CHILD (Rlg)** (TBN channel) ☒ P.O.Box 10700, Santa Ana, CA 92711-0700 ☎ +1 714 6652100 ▤ +1 714 7085428 **W:** www.smileofachildtv.org **Full-Power Affiliates:** 33. – **THE WORSHIP NETWORK (Rlg)** (The Christian Network, Inc.) ☒ P.O.Box 428, Safety Harbor, FL 34695 ☎ +1 877 2967744 **W:** www.worship.net **LP:** CEO/Pres (CNI): Dustin Rubeck. **Full-Power Affiliates:** 58. – **TBN ENLACE USA (Rlg)** (Spanish-language TBN channel) ☒ 2823 West Irving Blvd., Irving, TX 75061 ☎ +1 972 3139500 **W:** www.tbnenla-ceusa.tv **Full-Power Affiliates:** 33. – **THE CHURCH CHANNEL (Rlg)** (TBN channel) ☒ 14171 Chambers Rd., Tustin, CA 92780 ☎ +1 714 6652153 **W:** www.churchchannel.tv **Full-Power Affiliates:** 33. – **TRINITY BROADCASTING NETWORK (TBN) (Rlg)** ☒ P. O.Box A, Santa Ana, CA 92711 ☎ +1 714 8322950 **W:** www.tbn.org **LP:** Chmn/Pres: Paul Crouch. **Full-Power Affiliates:** 37.

Other Networks, Local Stations not shown.

URUGUAY

Systems: # PAL-N [A]; DVB-T planned

Key Local Stations
MONTE CARLO TELEVISIÓN (Comm) ☒ Paraguay 2253, 11800 Montevideo ☎ +598 2 9247924 ▤ +598 2 9244444 **E:** c4ventas@netgate.com.uy **W:** www.canal4.com.uy **Tx:** Montevideo ch4 (300kW). – **SAETA TV (Comm)** ☒ Lorenzo Carnelli 1234, 11200 Montevideo ☎ +598 2 4002120 ▤ +598 2 4095812 **E:** canal10@multi.com.uy **W:** www.canal10.com.uy **Tx:** Montevideo ch10 (600kW). – **TELEDOCE (Comm)** ☒ Enriqueta Compte y Rique 1276, 11800 Montevideo ☎ +598 2 2083363 ▤ +598 2 2083555 **E:** teledoce@teledoce.com **W:** www.teledoce.com **Tx:** Montevideo ch12 (600kW). – **TEVEO (TELEVISIÓN NACIONAL) (Comm)** ☒ Bvrd. Artigas 2476,11600 Montevideo ☎ +598 2 4871129 **E:** info@tnu.com.uy **W:** www.tveo.com.uy. **Tx:** Montevideo ch5 (96kW).

Other Local Stations not shown.

UZBEKISTAN

Systems: # SECAM-D/K [R], # PAL-D/K [R]; DVB-T (MPEG4)

National Stations
UZBEK TELEVISION (Gov) ☒ Navoiy Str. 69, 100011 Toshkent ☎+998 71 1141250 ▤ +998 71 1441332 **E:** info@mtrk.uz **W:** www.mtrk.uz **LP:** Chmn: Alisher Djurakulovich Hadjayev. **Chs:** O'zbekiston, Yoshlar, TV Toshkent, regional stns. **Txs: O'zbekiston:** Toshkent ch5 (20kW) & network; **Yoshlar:** Toshkent ch9 (20kW) & network; **TV Toshkent:** Toshkent ch12.

Local Stations, Foreign TV Relays not shown.

DTT Transmitters (under construction)
Operator: Uzdigital TV ☒ Amir Timur St. 109-A, 100084 Toshkent ☎ +998 71 1299000 ▤ +998 71 1505884 **E:** info@uzdtv.uz **W:** www.uzdigitaltv.uz **Mux 1:** O'zbekiston, Yoshlar, TV Toshkent, Sport TV, 1 kanal, NTV, Telenyanya, dom kino, Home, Vremya, MzTV, TNT, Test **Txs:** ch42 (Toshkent 2kW, Samarqand) **Mux 2:** 5 kanal, 7 TV, BBC, DTV, Euronews, Yumor TV, Luxe TV HD, Rossiya 1, Rossiya 2, Rossiya 24, TV3, TV Zvezda **Tx:** ch41 (Toshkent 2kW) **Mux 3:** tbd **Tx:** ch37 (Toshkent).

VANUATU

System: PAL-B [E]

TV BLONG VANUATU (Gov) ☒ P.M.B. 049, Port Vila **W:** www.vbtc.com.vu **LP:** GM: Jonas Cullwick. **Txs:** (-).

Foreign TV Relay
CCTV9 (China): (-).

VATICAN CITY STATE

System: T-DMB

VATICAN TELEVISION CENTER ☒ Via del Pellegrino, I-00120

Vatican City ☎ +39 06 69885467 ▤ +39 06 69885192 **E:** ctv@ctv.va **L.P:** Dir: Fr. Federico Lombardi, S.I. **Tx:** (-).

VENEZUELA

System: # NTSC-M [A]; SBTVD

National Stations
TELEVISORA VENEZOLANA SOCIAL (TVES) (Gov) ▤ Caracas **E:** info@tves.com.ve **W:** www.tves.org.ve **Txs:** Caracas ch2 & relay txs. – **VENEZOLANA DE TELEVISIÓN (VTV) (Gov)** ▤ Ap. 2979, Caracas 1050. ☎ +58 212 2349581 **E:** atencionciudadano@vtv.gov.ve **W:** www.vtv.gov.ve **Txs:** Caracas ch8 (190kW) & relay txs. – **VISIÓN VENEZUELA (VIVE) (Gov)** ▤ Final Av. Panteón, Foro Libertador, Edf. Biblioteca Nacional, AP-4, Altagracia, Caracas ☎ +58 212 5051611 **E:** atencionciudadana@vive.gob.ve **W:** www.vive.gob.ve **Txs:** Caracas ch25 & relay tx. – **GLOBOVISIÓN (Comm)** ▤ av. Los Pinos, cruce con Calle Alameda, Qta. Globovisión, Urb. Alta Florida, Caracas ☎ +58 212 7301134 **E:** info@globovision.com **W:** www.globovision.com. **Txs:** Caracas ch31 & relay txs. – **MERIDIANO TELEVISIÓN (Comm)** ▤ Final av. San Martin con Av. La Paz, Edificio Bloque De Armas, Caracas ☎ +58 212 4064516 ▤ +58 212 4515627 **E:** meridianotv@internet.ve **W:** www. meridiano.com.ve **Txs:** Caracas ch39 & relay txs. – **TELEVEN (Comm)** ▤ Av. Romulo Gallegos con 4ta. transversal de horizonte, Edificio Televen, Caracas 1071 ☎ +58 212 2800151 **E:** webmaster@televen.com **W:** www.televen.com. **Txs:** Caracas ch8 & relay txs. – **VALE TV (Rlg)** ▤ Final Av. La Salle, Quinta ValeTV, Colinas de los Caobos, Caracas ☎ +58 212 7939215 ▤ +58 212 7089743 **E:** webmaster@valetv.com **W:** www.valetv.com **Txs:** Caracas ch5 (210kW) & relay txs. – **VENEVISIÓN (Comm)** ▤ Av. La Salle, Edif, Venevision,Colinas de Los Caobos, Caracas 1050 ☎ +58 212 7089444 **E:** saladeredaccion@venevision.com **W:** www.vene-vision.com. **Txs:** Caracas ch4 (132kW) & relay txs.

Local Stations (Comm. exc. where stated)
Amavision (Rlg): Calle Selesiano, Colegio Pio XI, Puerto Ayacucho, Amazonas; Puerto Ayacucho ch7 (6kW). **Canal 10:** Av. Francisco de Miranda, con Principal de los Ruices, Centro Empresarial Miranda PHD, Caracas; ch10. **La Tele:** Calle Republica Dominicana, Boleita Sur, Caracas**;** ch12. **Canal Metropolitano de Televisión:** Av. Circumvalacion El Sol, Centro Professional Santa Paula, Torre B, Piso 4, Santa Paula, Caracas; ch51. **NCTV:** Urv. La Paz, Avenida 57 y Maracaibo, Maracaibo; ch11 (108kW). **Omnivision:** Calle Milan, Edif. Omnivision, Los Ruices Sur, Caracas. **Puma TV:** Av. Sanatorio del Avila, Boleíta Norte, Caracas 1071; Maracaibo ch53, Caracas ch57. **Televisora Andina de Merida (Rlg):** Av. Bolivar, Calle 23 entre Av. 4-5, Merida 5101; Tachira ch3 (33kW), Merida ch6 (20kW). **Tele Bocono (Cult):** Calle 3, Qta. Caleuche, El Saman. Bocono; Trujillo ch13 (4kW). **Telecaribe:** Centro Banaven (Cubo Negro), Torre C, Piso 1, of C-12, Chuao, Caracas; Anzoategui ch9 (50kW), Nueva Esparta ch12 (30kW). **Telecentro:** Avenide Pedro León Torres, esquina de la calle 47, Edificio Telecentro, Barquisimeto, (3001) Lara; ch11 (100kW). **TV Guyana:** Puerto Ordaz, Bolivar; ch12 (125kW). **Telesol:** Calle Sucre no 15, Cumana, Sucre; ch7 (12kW). **Televisora Regional del Tachira:** Av. Libertador, edif. Servicios Unidos, Piso 3, San Cristobal, Tachira; ch6 (144kW). **Televisora de Oriente (TVO):** Puerto la Cruz, Anzoategui; ch5 (50kW).

VIETNAM

Systems: # PAL-D/K [R] ⇩2015; DVB-T (MPEG2)

National Stations
VIETNAM TELEVISION (Gov) ▤ 43 Nguyen Chi Thanh, Ba Dinh District, Hanoi ☎ +84 8 8224403 ▤ +84 8 8223422 **E:** qhqt-vtv@vtv.org.vn **W:** www.vtv.org.vn **L.P:** DG: Vu Van Hien **Chs (terr.):** VTV1, VTV2, VTV3. **VTV1:** Hanoi ch9 (10kW) & netw. **VTV2:** Hanoi ch11 (10kW) & netw. **VTV3:** Hanoi ch22 (20kW) & netw.

Regional and Local Stations not shown.

DTT Transmitters (under construction)
Operator: VTCi Media ▤ 67B Ham Long Street, Hoan Kiem District, Hanoi ☎ +84 4 9433409 ▤ +84 4 9439867 **E:** tv.vtc@hn.vnn.vn **W:** www.vtci.com.vn **Mux:** VTV1-7, Hanoi TV 1, HTV1 **Txs:** ch26 (Hanoi 1.3kW) & nationwide netw. under construction.

VIRGIN ISLANDS (USA)

Systems: ATSC

Local Stations*
WCVI-TV (Comm): 1 k Little Princess, St. Croix, VI 00820-4027. °CW.

Tx: St. Croix ch23 (0.66kW). **WSVI (Comm):** Sunny Isle Shopping Cente, St Croix, VI 00820-4493. °ABC. Tx: St. Croix ch20 (459kW). **WTJX-TV (Pub):** 58-158A Hay Place Hill S, St. Thomas, VI 20801. °PBS. Tx: Charlotte Amalie ch44 (50kW). **WVIF (Comm):** 4200 United Shipping Plaza #3, St Croix, VI 00820. Tx: Christiansted ch15 (16.2kW).
WVXF (Comm): 8000 Nisky Center, Suite 714, Saint Thomas, VI 00802. °CBS. Tx: Charlotte Amalie ch48 (50kW). **WZVI (Comm):** satellite of WSVI. Tx: Charlotte Amalie ch43 (1.4kW).
*) Full power licenses (lp licenses not listed); °) Network affiliation

VIRGIN ISLANDS (UK)

System: NTSC-M [A]

VIRGIN ISLANDS BROADCASTING LTD (Comm) ▤ P.O.Box 78, Road Town, Tortola ☎ +1 284 4942250 **E:** zbvi@surfbvi.com **W:** www.zbvi.com **L.P:** GM: Harvey Herbert **Tx:** ch5 (30kW).

WALLIS & FUTUNA (France)

System: # SECAM-K1 [K]; DVB-T

RFO WALLIS ET FUNUTA (Pub) ▤ BP 102, Mata Utu, 98600 Uvea, Iles de Wallis-et-Futuna, Pacifique sud (par Nouméa, Nouvelle-Calédonie) ☎ +681 681722020 ▤ +681 681722346 **E:** rfo.wallis@wallis.co.nc **W:** wallisfutuna.rfo.fr **L.P:** TechDir: Taniela Heafala **Ch:** Télé Pays **Txs:** Royaume de Sigave (Nuku) ch5 (0.01kW), Royaume de Sigave (Mont-Utulimu) ch5 (0.019kW), Uvea (Pointe Matalaa) ch6 (0.02kW), Royaume d'Alo ch7 (0.02kW), Royaume de Sigave (Sa'alauniu) ch 7 (0.02kW), Royaume de Sigave (Sausau) ch7 (0.007kW), Royaume de Sigave (Apipi) ch9 (0.05kW), Uvea (Mont Loka) ch9 (0.006kW).

DTT Transmitters
Operator: France Télévisions **Mux 1:** RFO Wallis et Funuta, France 2-5, France Ô, Arte, LCP, France 24, Gulli **Txs:** (-)

YEMEN

System: PAL-B [E]

REPUBLIC OF YEMEN TELEVISION (Gov) Channel 1: ▤ P.O.Box 1140, al-Guraf, Sana'a ☎ +967 1 332001 ▤ +967 1 332086 **Channel 2:** ▤ P.O.Box 1264, Tawahi, Aden ☎ +967 2 202481 ▤ +967 2 221121. **Txs:** (-).

ZAMBIA

System: PAL-B [E]

ZAMBIA NATIONAL BROADCASTING CORP. (Gov) ▤ P.O.Box 50015, Lusaka 10101 ☎ +260 21 1254989 ▤ +260 21 1254317 **E:** znbctv@znbc.co.zm **W:** www.znbc.co.zm **L.P:** DG: Eddy Mupeso; Tech Dir: Edward Mwanza **Txs:** Solwezi ch3 (0.075kW H), Kapiri Mposhi ch6 (200kW H), Pemba ch8 (200kW V), Kasama ch8 (16kW H), Mubwa ch8 (0.1kW H), Kitwe ch9 (200kW H), Lusaka ch10 (200kW H), Senkobo ch10 (200kW V), Chipata ch11 (16kW H). **MUVI TV (Comm)** ▤ P.O.Box 33932, Lusaka ☎ +260 21 1253271 **E:** frontoffice@muvitv.com **W:** www.muvitv.com **Txs:** Lusaka ch(-) 2kW & relay txs.

Local Stations (all Comm)
CBC: Lusaka; tx: Lusaka ch(-) 2kW. **Copperbelt Broadcasting Services:** Ndola, tx: Ndola ch(-) 2kW. **Mobi TV:** Lusaka; tx: Lusaka ch(-) 2kW. **Northrise TV:** Lusaka; tx: Lusaka ch(-).

Foreign TV Relay
TBN (USA): Lusaka ch(-).

ZIMBABWE

System: PAL-B/G [E]; DVB-T planned

ZIMBABWE TELEVISION (ZTV) (Gov) ▤ P.O.Box HG 444, Highlands, Harare ☎ +263 4 498610 ▤ +263 4 498613 **E:** zbc@zbc.co.zw **W:** www.zbc.co.zw **L.P:** CEO: Susan Makore **Chs:** ZTV1, ZTV2 **Txs: ZTV1:** Gweru ch2 (17.6kW H), Bulawayo ch3 (3kW H) & relay txs; **ZTV 2:** (-).

REFERENCE

Section Contents

Features & Reviews

National Radio

International Radio

Frequency Lists

Terrestrial Television

Reference

MAIN COUNTRY INDEX

	Nat	Int	CTB	TV
Afghanistan	66	428		612
Alaska	66	428		612
Albania	67	428		612
Algeria	68	428		612
Andorra	69			612
Angola	69	428		612
Anguilla	70	429		613
Antarctica	70			613
Antigua & Barbuda	70			613
Argentina	71	429		613
Armenia	78	429		613
Aruba	78			613
Ascension Island	78	429		613
Australia	79	429		612
Austria	87	431		614
Azerbaijan	88			614
Azores	89			614
Bahamas	89			614
Bahrain	90	432		614
Bangladesh	90	432		614
Barbados	90			615
Belarus	91	432		615
Belgium	92	433		615
Belize	94			615
Benin	94	433		616
Bermuda	94			616
Bhutan	95			616
Bolivia	95			616
Bonaire	99	433		616
Bosnia & Herzegovina	99	434		616
Botswana	100	434		616
Brazil	100	434		616
British Indian Ocean Territory	130			616
Brunei	130			616
Bulgaria	130	434		617
Burkina Faso	131			617
Burundi	131			617
Cambodia	132			617
Cameroon	133		501	617
Canada	133	435		617
Canary Islands	140			618
Cape Verde	140			618
Cayman Islands	140			618
Central African Rep.	141			618
Chad	141			619
Chile	141	436		619
China	144	436	501	619
Christmas Island	159			619
Cocos Islands	159			619
Colombia	159			619
Comoros	166			619
Congo (Dem. Rep.)	166		502	619
Congo (Rep. of)	167			620
Cook Islands	167			620
Costa Rica	167	440		620
Croatia	168	440		620
Cuba	170	441	502	620
Curaçao	172			620
Cyprus	172	441		620
Czech Republic	173	442		621
Denmark	175			621
Djibouti	176	442	502	622
Dominica	176			622
Dominican Republic	176			622
Easter Island	178			622
Ecuador	178	442		622
Egypt	183	443		622
El Salvador	184			622
Equatorial Guinea	185	443		622
Eritrea	185		502	623
Estonia	185	443		623
Ethiopia	186	443	503	623
Falkland Islands	186			623
Faroe Islands	187			623
Fiji	187			623
Finland	187	444		623
France	189	444		624
French Guiana	197	445		625
French Polynesia	197			625
French So. & Antarctic Lands	198			625
Gabon	198	445		625
Galapagos Islands	199			625
Gambia	199		504	625
Georgia	199			625
Germany	200	445		626
Ghana	214			627
Gibraltar	214			627
Greece	214	447		627
Greenland	216			628
Grenada	217			628
Guadeloupe	217			628
Guam	217	448		628
Guatemala	218			628
Guinea	219			628
Guinea-Bissau	220			628
Guyana	220			628
Haiti	220			629
Hawaii	220			629
Honduras	222			629
Hong Kong	225			629
Hungary	225			629
Iceland	227			630
India	227	449	504	630
Indonesia	232	453		630
Iran	237	454	504	630
Iraq	238		504	630
Ireland	240	455		631
Israel	242	455		631
Italy	243	455		631
Ivory Coast	245			631
Jamaica	245			631
Japan	246	456		632
Jordan	249	457		632
Kazakstan	250	457		632
Kenya	250			632
Kiribati	251			632
Korea (North)	251	457	505	632
Korea (South)	252	458	506	632
Kosovo	256			633
Kuwait	256	459		633
Kyrgyzstan	256	460		633
Laos	257	460	506	633

GEOGRAPHICAL AREA CODES USED IN WRTH
Codes assigned by the International Telecommunications Union ITU (except * = WRTH code)

Code	Country	Code	Country	Code	Country	Code	Country
ABW	Aruba	CYM	Cayman Islands	KWT	Kuwait	RRW	Rwanda
AFG	Afghanistan	CYP	Cyprus	LAO	Laos	RUS	Russia
AFS	South Africa	CZE	Czech Republic	LBN	Lebanon	S	Sweden
AGL	Angola	D	Germany	LBR	Liberia	SCN	St. Kitts & Nevis
AIA	Anguilla	DGA	Diego Garcia	LBY	Libya	SDN	Sudan
ALB	Albania	DJI	Djibouti	LCA	St. Lucia	SEN	Senegal
ALG	Algeria	DMA	Dominica	LHW*	Lord Howe Island	SEY	Seychelles
ALS	Alaska	DNK	Denmark	LIE	Liechtenstein	SHN	St. Helena
AND	Andorra	DOM	Dominican Republic	LSO	Lesotho	SLM	Solomon Islands
AOE	Western Sahara	E	Spain	LTU	Lithuania	SLV	El Salvador
ARG	Argentina	EGY	Egypt	LUX	Luxembourg	SMA	American Samoa
ARM	Armenia	EQA	Ecuador	LVA	Latvia	SMO	Samoa
ARS	Saudi Arabia	ERI	Eritrea	MAC	Macao	SMR	San Marino
ASC	Ascension Island	EST	Estonia	MAF	St. Martin	SNG	Singapore
ATA	Antarctica	ETH	Ethiopia	MAU	Mauritius	SOM	Somalia
ATG	Antigua & Barbuda	F	France	MCO	Monaco	SPM	St. Pierre & Miquelon
ATN	Bonaire	FIN	Finland	MDA	Moldova	SRB	Serbia
ATN	Curaçao	FJI	Fiji	MDG	Madagascar	SRL	Sierra Leone
ATN	Saba	FLK	Falkland Islands	MDR	Madeira	STP	São Tomé & Príncipe
ATN	St. Eustatius	FRO	Faroe Islands	MEX	Mexico	SUI	Switzerland
ATN	St. Maarten	FSA*	French So. & Ant. Lands	MHL	Marshall Islands	SUR	Suriname
AUS	Australia	FSM	Micronesia	MKD	Macedonia	SVK	Slovakia
AUT	Austria	G	United Kingdom	MLA	Malaysia	SVN	Slovenia
AZE	Azerbaijan	GAB	Gabon	MLD	Maldives	SWZ	Swaziland
AZR	Azores	GAL*	Galapagos Islands	MLI	Mali	SYR	Syria
B	Brazil	GEO	Georgia	MLT	Malta	TCA	Turks & Caicos Islands
BAH	Bahamas	GHA	Ghana	MNE	Montenegro	TCD	Chad
BDI	Burundi	GIB	Gibraltar	MNG	Mongolia	TGO	Togo
BEL	Belgium	GLP	Guadeloupe	MOZ	Mozambique	THA	Thailand
BEN	Benin	GMB	Gambia	MRA	Northern Mariana Is	TJK	Tajikistan
BER	Bermuda	GNB	Guinea-Bissau	MRC	Morocco	TKM	Turkmenistan
BFA	Burkina Faso	GNE	Equatorial Guinea	MRT	Martinique	TKL	Tokelau
BGD	Bangladesh	GRC	Greece	MSR	Montserrat	TLS	Timor-Leste
BHR	Bahrain	GRD	Grenada	MTN	Mauritania	TON	Tonga
BIH	Bosnia & Herzegovina	GRL	Greenland	MWI	Malawi	TRC	Tristan da Cunha
BIO	British Indian Ocean	GTM	Guatemala	MYT	Mayotte	TRD	Trinidad & Tobago
	Territory	GUF	French Guiana	NCG	Nicaragua	TUN	Tunisia
BLM	St. Barthélemy	GUI	Guinea	NCL	New Caledonia	TUR	Turkey
BLR	Belarus	GUM	Guam	NFK	Norfolk Island	TUV	Tuvalu
BLZ	Belize	GUY	Guyana	NGR	Niger	TWN*	Taiwan
BOL	Bolivia	HKG	Hong Kong	NIG	Nigeria	TZA	Tanzania
BOT	Botswana	HND	Honduras	NIU	Niue	UAE	United Arab Emirates
BRB	Barbados	HNG	Hungary	NMB	Namibia	UGA	Uganda
BRM	Myanmar	HOL	Netherlands	NOR	Norway	UKR	Ukraine
BRU	Brunei	HRV	Croatia	NPL	Nepal	URG	Uruguay
BTN	Bhutan	HTI	Haiti	NRU	Nauru	USA	United States of America
BUL	Bulgaria	HWA	Hawaii	NZL	New Zealand	UZB	Uzbekistan
CAF	Central African Republic	I	Italy	OCE	French Polynesia	VCT	St. Vincent &
CAN	Canada	ICO	Cocos (Keeling) Islands	OMA	Oman		the Grenadines
CBG	Cambodia	IND	India	PAK	Pakistan	VEN	Venezuela
CHL	Chile	INS	Indonesia	PAQ	Easter Island	VIR	Virgin Islands
CHN	China (People's Rep. of)	IRL	Ireland	PHL	Philippines	VRG	British Virgin Islands
CHR	Christmas Island	IRN	Iran	PLW	Palau	VTN	Vietnam
CKH	Cook Islands	IRQ	Iraq	PNG	Papua New Guinea	VUT	Vanuatu
CLM	Colombia	ISL	Iceland	PNR	Panama	WAL	Wallis & Futuna
CLN	Sri Lanka	ISR	Israel	POL	Poland	XGZ	Gaza Strip[1]
CME	Cameroon	J	Japan	POR	Portugal	XWB	West Bank[1]
CNR	Canary Islands	JMC	Jamaica	PRG	Paraguay	YEM	Yemen
COD	Congo (Dem. Rep. of the)	JOR	Jordan	PRU	Peru	ZMB	Zambia
COG	Congo (Rep. of the)	KAZ	Kazakhstan	PSE*	Palestine[1]	ZWE	Zimbabwe
COM	Comoros	KEN	Kenya	PTR	Puerto Rico		
CPV	Cape Verde	KER	Iles Kerguelen	QAT	Qatar		
CTI	Côte d'Ivoire	KGZ	Kyrgyzstan	REU	Réunion		
CTR	Costa Rica	KIR	Kiribati	RKS*	Kosovo		
CUB	Cuba	KOR	Korea, South	ROD	Rodrigues		
CVA	Vatican City State	KRE	Korea, North	ROU	Romania		

[1] The code "PSE" is used as target designation in the "COTB" section of "International Radio/COTB"; otherwise the codes "XGZ"/"XWB" are used.

ABBREVIATIONS & SYMBOLS USED IN WRTH

✉	=	Address
☎	=	Telephone
🖹	=	Fax
✪	=	Encrypted
⌘	=	Radio via DTT
acc.	=	accepted
Admin.	=	Administration
alt.	=	alternate, alternative
AM	=	Amplitude Modulation
Ann.	=	Announcement
Ap.	=	Apartado
approx.	=	approximate(ly)
Assoc.	=	Association
Asst.	=	Assistant
Ave	=	Avenue, Avenida
B.P.	=	Boîte Postale
B'caster	=	Broadcaster
Bldg	=	Building
Broadc.	=	Broadcast(ing)
BS	=	Broadc. Stn/Sce
C	=	Chinese
C.P.	=	Case/Caixa Postal, Construction Permit
Cad.	=	Cadena
Cas.	=	Casilla
Cd.	=	Ciudad
Ce.	=	Central
CEO	=	Chief Exec. Officer
cf.	=	refer to
Ch.	=	Channel
Chmn.	=	Chairman/Chair
Cl.	=	Club(e)
Clan.	=	Clandestine
Co.	=	Company
Com.	=	Comunicações
comm.	=	commercial
Contr.	=	Controller
Corp.	=	Corporation
Cra.	=	Carrera
Cult.	=	Cultura, Cultural
D	=	Daily
d	=	directional antenna
D.Prgr	=	Daily Programme(s)
DAB	=	Digital Audio Broadc.
DMB	=	Digital Multimedia Broadcasting
Dem.	=	Democratic
Dep.	=	Deputy
Dept.	=	Department
Depto.	=	Departamento
Desp.	=	Despacho
DG	=	Director General
Dif.	=	Difusora, Difusão
Diff.	=	Diffusion
Dir.	=	Director
Div.	=	Division
dom	=	domestic
DRM	=	Digital Radio Mondiale
DSB	=	Double Side Band
DST	=	Daylight Saving Time
DTT	=	Digital Terrestrial TV
DVB	=	Digital Video Broadc.

DX	=	Long Distance (Reception)
E	=	English
E:	=	Email
E.C	=	Electric Current
Ea.	=	East(ern)
Edif.	=	Edificio
Educ.	=	Education(al), Educación
e.g.	=	for example
Em.	=	Emis(s)ora
Eng.	=	Engineer(ing)
ERP	=	Effective Radiated Power
Esq.	=	Esquina
est.	=	estimated
Est.	=	Estado
exc.	=	except
excl.	=	excluding
exec.	=	executive
ext.	=	external
F	=	French
F.Pl.	=	Future Plan(s)
fed.	=	federal
FM	=	Frequency Modulation
Fr.	=	Father
Freq.	=	Frequency
Fri	=	Friday
FS	=	Foreign Service
Ft.	=	Fort
G	=	German
G.C	=	Geographical Coordinates
GD	=	General Director
gen.	=	general
GM	=	General Manager
Gov.	=	Government(al)
Gte.	=	Gerente
H	=	Horizontal Pol.
h(rs)	=	hour(s)
HD	=	High Definition
HQ	=	Headquarters
HS	=	Home Service
I	=	Italian
ID	=	(Station) Identification
i.e.	=	that is
Inc.	=	Incorporated
incl.	=	including
Inf.	=	Information
int.	=	international
IRC	=	Int. Reply Coupon
irr.	=	irregular
IS	=	Interval Signal
I./Is	=	Island/Islands
kHz	=	kiloHertz
L	=	Local
L.P	=	Leading Personnel
L.T	=	Local Time
Langs.	=	Languages
Lp.	=	Low power (transmitter)
LSB	=	Lower Side Band
Ltd	=	Limited
LV	=	La Voz, La Voce

LW	=	Longwave
max.	=	maximum
MD	=	Managing Director
MF	=	Mondays-Fridays
MFN	=	Multiple Freq. Netw.
Mgr	=	Manager
MHz	=	MegaHertz
mil.	=	military
Min.	=	Ministry, Ministerio, Ministério
min(s)	=	minute(s)
Mon	=	Monday
Mpal.	=	Municipal
Mpo.	=	Município
Mt	=	Mount, Mountain
Mux	=	Multiplex
MW	=	Mediumwave
N.	=	News
NB	=	Note (Nota Bene)
n.f.	=	nominal frequency
n/a	=	not available, not applicable
nal.	=	nacional
nat.	=	national
nd	=	nondirectional antenna
NE	=	North East(ern)
Netw.	=	Network
No.	=	North(ern), Number
nom.	=	nominal
Nte	=	Norte
NW	=	North West(ern)
occ.	=	occasional(ly)
Op(s)	=	Operation(s)
Org.	=	Organisation
Ote.	=	Oeste
P	=	Portuguese
P.O.	=	Post Office
P.R.	=	Public Relations, People's Republic
PD	=	Programme Director
pl.	=	planned
Pol.	=	Polarisation
Pop.	=	Population
Pr.	=	Praça
Pr.L	=	Principal Language(s)
Pres.	=	President
Priv.	=	Private
Prgr(s).	=	Programme(s)
Prod.	=	Production
Prov.	=	Province, Provincial
Pt.	=	Point
Pte.	=	Presidente
Pto.	=	Puerto
Pub	=	Public service
Pub(s)	=	Publication(s)
QSL	=	Reception Confirmation
R.	=	Radio, Rádio, Rádió Radyjo, Radyo
r.	=	reported, repeater
Rdif.	=	Radiodifusion
R. Dif.	=	Radio Difusora
Rec.	=	Recording(s)
Reg.	=	Region(al)

Rel.	=	Relay(s), Relations
Rep.	=	Republic
Rev.	=	Reverend
rlg	=	religious
Rp.	=	Return Postage
Rpt.	=	(Reception) Report
S.	=	San(ta), Sán, Santo
s/off	=	sign off
s/on	=	sign on
SAE	=	Self Addressed Envelope
SAR	=	Special Administrative Region
Sat	=	Saturday, satellite
Sce.	=	Service
Sched.	=	Schedule
SE	=	South East(ern)
Secr.	=	Secretary
Sen.	=	Senior
SFN	=	Single Freq. Netw.
Sist.	=	Sistema
SM	=	Station Manager
So.	=	South(ern)
Soc.	=	Sociedad(e)
Sp.	=	Spanish
SS	=	Sat/Sun
SSB	=	Single Side Band
St	=	Saint, Street
Stn	=	Station
Str.	=	Street, Straße
Su.	=	Summer
Sun	=	Sunday
Superv.	=	Supervisor
SW	=	Shortwave South West(ern)
Syst.	=	System
tbd	=	to be defined
TD	=	Technical Director
techn.	=	technical
terr.	=	terrestrial
Thu	=	Thursday
tr(s)	=	transmission(s)
TRP	=	Transmitter Power
Tue	=	Tuesday
tx(s)	=	transmitter(s)
ul.	=	ulitsa, ulica
u.c.	=	under construction
Univ.	=	University
unk.	=	unknown
UHF	=	Ultra High Frequency
USB	=	Upper Side Band
UTC	=	Coordinated Universal Time
V	=	Vertical Pol.
V.	=	Verification
v.	=	varying
VHF	=	Very High Frequency
VO	=	Voice of
W	=	Weekdays (Mon-Sat)
W:	=	Web
We.	=	West(ern)
Wed	=	Wednesday
Wi.	=	Winter
Wrp.	=	Weather Report

TRANSMITTER SITES
Location & Decode Tables

INTERNATIONAL TRANSMITTER SITES

Code	Site	Ctry	Lat	Long	SW	MW
-	unidentified	-	-	-	×	×
ABH	Abu Hayan	BHR	26N02	050E37	✓	×
ABS	Abis	EGY	31N08	030E04	✓	×
ABZ	Abu Zaabal	EGY	30N16	031E22	✓	×
ADR	Adra	SYR	33N33	036E34	✓	×
AHW	Ahwaz, Bandar-e Mahshar	IRN	30N37	049E12	✓	×
AIA	The Valley	AIA	18N13	063W01	✓	✓
AJA	Abuja, Lugbe	NIG	08N58	007E22	✓	×
AKA	Al Karanah	JOR	31N45	036E28	✓	×
ALG	Aligarh	IND	28N00	078E06	✓	×
ALM	Almati	KAZ	43N30	077E00	✓	✓
ARM	Krasnodar, Tbilisskaya	RUS	45N28	040E06	✓	✓
ASC	Ascension, English Bay	ASC	07S54	014W23	✓	×
ASM	Asmara	ERI	15N13	038E52	✓	✓
AVL	Vathy (Avlida municipality)	GRC	38N23	023E36	✓	×
BAR	Barrigada	GUM	13N29	144E50	✓	×
BAT	Bata	GNE	01N50	009E47	✓	×
BCQ	Monticello, ME	USA	46N20	067W49	✓	×
BDW	Beidweiler	LUX	49N44	006E19	×	×
BEI	Beijing, Doudian	CHN	39N38	116E06	✓	×
BEO	Beograd, Stubline	SRB	44N34	020E09	✓	×
BGL	Bengaluru, Doddaballapur	IND	13N15	077E29	✓	×
BIB	Biblis	D	49N41	008E29	✓	×
BIJ	Bijeljina, Jabanuša	BIH	44N42	019E10	✓	×
BIS	Bishkek, Krasnaya Rechka	KGZ	42N53	074E59	✓	✓
BJI	Baoji, Shaanxi, Xinjie	CHN	34N39	106E58	✓	×
BKO	Bamako, Kati	MLI	12N45	008W03	✓	×
BLG	Belogorsk, Konstantinogradovka	RUS	50N31	128E18	×	✓
BLN	Berlin, Zehlendorf	D	52N47	013E23	✓	×
BNB	Bonab	IRN	37N18	046E03	×	×
BNT	Bandar-e Torkaman	IRN	36N54	054E03	×	×
BOC	Bocaue, Bulacan	PHL	14N48	120E55	✓	×
BON	Bonaire, Tolo	ATN	12N13	068W19	✓	×
BOT	Moepeng Hill	BOT	21S57	027E38	✓	×
BPH	Ban Phachi, Rasom	THA	14N24	100E47	×	×
BRA	Brasília, Rodeador Park	B	15S36	048W08	✓	×
BRN	Brandon	AUS	19S31	147E20	✓	×
BUE	Buenos Aires, General Pacheco	ARG	34S26	058W37	✓	×
CAH	Changchun, Jilin	CHN	43N44	125E24	×	×
CAK	Çakirlar	TUR	39N58	032E41	✓	×
CDM	Col de La Madone	F	43N47	007E25	✓	×
CER	Cërrik, Shtermen	ALB	41N00	020E00	✓	×
CGR	Cape Gkreko	CYP	34N58	034E05	×	✓
CHB	Chabahar	IRN	25N29	060E32	×	✓
CHJ	Chongjin	KRE	41N45	129E42	×	✓
CHO	Choybalsan	MNG	48N00	114E26	×	✓
CLS	Caltanissetta	I	37N30	014E04	×	✓
CLZ	Calabozo	VEN	08N55	067W23	✓	×
CNI	Chennai	IND	13N08	080E07	✓	×
CRI	Cariari de Pococí	CTR	10N25	083W43	✓	×
CRN	Chernivtsi	UKR	48N18	025E50	×	✓
CYP	Zygi	CYP	34N43	033E19	✓	✓
DAN	Dangjin	KOR	36N58	126E37	✓	✓
DEA	Deanovec	HRV	45N42	016E28	✓	✓
DEH	Luxi, Dehong, Yunnan	CHN	24N27	098E36	✓	×
DEL	Delhi	IND	28N43	077E12	✓	×
DGA	Diego Garcia	BIO	07S26	072E26	✓	×
DHA	Dhabbaya	UAE	24N10	054E15	✓	×
DJI	Djibouti, Doraleh	DJI	11N34	043E04	✓	×
DKA	Dhaka, Khabirpur	BGD	24N00	090E15	✓	×
DOF	Dongfang, Hainan	CHN	18N53	108E39	✓	×
DRO	Droitwich	G	52N18	002W06	✓	✓
DSB	Dushanbe	TJK	38N29	068E48	✓	×
DSD	Dresden, Wilsdruff	D	51N04	013E30	✓	×
EKA	Colombo, Ekala	CLN	07N06	079E54	✓	×
EKB	Yekaterinburg	RUS	56N53	060E41	×	✓
EMR	Emirler	TUR	39N24	032E51	✓	×
ERV	Gavar, Noratus	ARM	40N25	045E11	✓	×
EWN	Vandiver, AL	USA	33N30	086W29	✓	×
FAN	Fangliao	TWN	22N23	120E34	✓	×
FBS	Saipan, Marpi	MRA	15N16	145E48	✓	×
FLA	Fllaka	ALB	41N22	019E30	×	✓
GAB	Mouanda, Moyabi	GAB	01S41	013E18	✓	×
GAL	Bacau, Galbeni	ROU	46N45	026E51	✓	×
GJW	Geja Jewe	ETH	08N47	038E39	✓	×
GKP	Gorakhpur	IND	26N53	083E28	✓	×
GOY	Goyang	KOR	37N36	126E51	✓	×
GRV	Greenville, NC	USA	35N28	077W12	✓	×
GUF	Cayenne, Montsinéry	GUF	04N54	052W30	✓	×
GUW	Guwahati	IND	26N09	091E39	✓	×
GWE	Gweru, Guinea Fowl	ZWE	19S31	029E56	✓	×
HAB	La Habana	CUB	22N57	082W33	✓	×
HAN	Hanoi, Me Tri	VTN	21N00	105E47	✓	×
HBN	Babeldaob, Medorm	PLW	07N27	134E29	✓	×
HDN	Huadian, Jilin	CHN	43N07	126E31	×	✓
HDU	Huadu, Guangzhou, Guangdong	CHN	23N24	113E14	×	✓
HEI	Shuangyashan, Heilongjiang	CHN	46N43	131E13	×	✓
HJU	Haeju	KRE	41N47	129E50	✓	✓

Code	Site	Ctry	Lat	Long	SW	MW
HKG	Hongkong, Peng Chau	CHN	22N17	114E03	×	✓
HMS	Horns, Saraqeb	SYR	35N52	036E48	×	✓
HNL	Henglin, Changzhou, Jiangsu	CHN	31N42	120E07	×	✓
HRI	Cypress Creek, SC	USA	32N41	081W08	✓	×
HSI	Hsin-feng	TWN	24N56	121E00	×	✓
HUH	Hohhot, Bikeqi, Nei Menggu	CHN	40N48	111E12	✓	×
HUW	Huwei	TWN	23N43	120E25	✓	×
HWA	Hwaseong	KOR	37N13	126E47	✓	×
HWD	Hwadae	KRE	40N51	129E26	✓	✓
IBA	Iba, Zambales	PHL	15N22	119E57	✓	×
IKO	Ikorodu	NIG	06N36	003E30	✓	×
INB	Red Lion, PA	USA	39N54	076W35	✓	×
IRA	Iranawila	CLN	07N30	079E48	✓	×
IRK	Irkutsk, Angarsk	RUS	52N26	103E40	✓	✓
ISK	Yeni Iskele	CYP	35N18	033E55	✓	×
ISL	Islamabad, Rawat	PAK	33N28	073E12	✓	×
ISR	Yavne	ISR	31N54	034E45	✓	×
ISS	Issoudun	F	46N56	001E53	✓	×
IUJ	Yuzhno-Sakhalinsk, Vestochka	RUS	46N50	142E53	×	✓
JAK	Jakarta, Cimanggis	INS	06S24	106E52	✓	×
JAL	Jalandhar	IND	31N09	075E47	×	✓
JED	Jeddah	ARS	21N23	039E25	✓	✓
JEJ	Jeju	KOR	33N29	126E23	×	✓
JES	Vado, NM	USA	32N08	106W36	✓	×
JHR	Milton, FL	USA	30N39	087W05	✓	×
JIN	Jinhua, Lanxi, Zhejiang	CHN	29N07	119E19	✓	×
JUN	Junglinster	LUX	49N43	006E16	✓	✓
KAB	Kabul, Pol-e Charkhi	AFG	34N32	069E20	✓	✓
KAJ	Kajang	MLA	03N01	101E47	✓	×
KAM	Tehran, Kamalabad	IRN	35N50	050E52	✓	×
KAN	Kangnam	KRE	39N05	125E33	×	✓
KAS	Kashi, Saibagh, Xinjiang	CHN	39N21	075E46	✓	✓
KBD	Kuwait, Kabd	KWT	29N09	047E46	✓	×
KCH	Grigoriopol, Maiac	MDA	47N17	029E25	✓	✓
KER	Kerman	IRN	30N17	057E05	×	✓
KEW	Saddlebunch Keys, FL	USA	24N39	081W36	✓	×
KHB	Khabarovsk	RUS	48N35	135E05	✓	×
KHO	Khost, Tani	AFG	33N20	069N56	×	✓
KHR	Kharkiv, Taranivka	UKR	49N38	036E07	✓	✓
KIA	Bandar e-Kiashahr	IRN	37N25	050E01	✓	×
KIG	Kigali, Kinyinya	RRW	01S55	030E07	✓	×
KIM	Gimje	KOR	35N49	126E52	✓	✓
KKT	Chinsurah	IND	23N02	088E21	×	✓
KLG	Kaliningrad, Bolshakovo	RUS	54N55	021E43	✓	✓
KLL	Kall, Krekel	D	50N29	006E31	✓	×
KLU	Königslutter	D	52N17	010E44	×	✓
KNA	Komsomolsk-na-Amure	RUS	50N39	136E55	✓	×
KNG	Kanggye	KRE	40N58	126E36	✓	×
KNX	Kununurra	AUS	15S48	128E41	✓	×
KOU	Kouhu	TWN	23N32	120E10	✓	✓
KUJ	Kujang	KRE	40N05	126E07	✓	×
KUN	Kunming, Anning, Yunnan	CHN	24N53	102E30	✓	✓
KVI	Kvitsøy	NOR	59N04	005E27	✓	✓
KWT	Kuwait, Umm Al-Rimam	KWT	29N31	047E40	✓	✓
KYV	Kyiv, Brovary	UKR	50N30	030E49	✓	✓
LAM	Lampertheim	D	49N36	008E32	✓	×
LIS	São Gabriel	POR	38N47	008W42	✓	×
LIT	Litomyšl	CZE	49N49	016E18	✓	×
LND	London, Crystal Palace	G	51N25	000W05	×	✓
LOD	Lod	ISR	32N00	034N50	✓	×
LUK	Lukang	TWN	24N03	120E25	×	✓
LUS	Lusaka, Makeni Ranch	ZMB	15S32	028E00	✓	×
LVI	Lviv, Krasne	UKR	49N54	024E40	✓	×
MAN	Manzini, Mpangela Ranch	SWZ	26S20	031E36	✓	×
MAS	Maseru, Lancer's Gap	LSO	29S19	027E33	✓	×
MBC	Chuncheon	KOR	37N56	127E43	×	✓
MCO	Fontbonne, Mont Agel	F	43N46	007E26	✓	×
MDC	Talata Volonondry	MDG	18S45	047E37	✓	×
MEA	Metula	ISR	32N30	035E00	×	✓
MEK	Mek'ele	ETH	13N30	039E29	×	✓
MEY	Meyerton, Tygerberg	AFS	26S35	028E08	✓	×
MHD	Mahidasht	IRN	34N16	046E48	×	✓
MHJ	Mahajanga	MDG	15S40	046E21	✓	×
MIL	Milano	I	45N20	009E12	×	✓
MIN	Minhsiung	TWN	23N34	120E26	✓	×
MNS	Minsk, Kalodziscy	BLR	53N58	027E47	✓	×
MOS	Moosbrunn	AUT	48N00	016E28	✓	×
MRN	Marnach	LUX	50N03	006E05	✓	×
MSK	Moskva	RUS	55N45	037E37	✓	✓
MSU	Maseru, Lancers Gap	LSO	29S19	027E33	✓	×
MTH	Marathon Key, FL	USA	24N42	081W05	×	✓
MUL	Luanda, Mulenvos	AGL	08S51	013E19	✓	×
MUM	Mumbai	IND	19N11	072E48	✓	×
NAD	Nador	MRC	35N03	002W55	✓	×
NAK	Nakhon Sawan	THA	15N49	100E04	✓	×
NAP	Napoli	I	41N00	014E19	×	✓
NAU	Nauen	D	52N39	012E55	✓	×
NLS	Anchor Pt, AK	ALS	59N45	151W44	✓	×
NNN	Nanning, Guangxi	CHN	22N48	108E11	✓	×
NOB	Noblejas	E	39N57	003W26	✓	×
NVS	Novosibirsk	RUS	55N55	082E51	✓	×
OMO	Can Tho, Ô Môn	VTN	10N07	105E34	×	✓
ORF	Orfordness	G	52N06	001E34	✓	×
PAN	Panaji	IND	15N21	073E51	✓	×
PAO	Paochung	TWN	23N43	120E18	✓	×
PAR	Parakou	BEN	09N21	002E37	✓	×
PET	Petrich, Novo Konomladi	BUL	41N28	023E20	✓	×
PHP	Poro Point	PHL	16N37	120E17	✓	×
PHT	Tinang	PHL	15N22	120E37	✓	×
PLD	Plovdiv, Padarsko	BUL	42N23	024E52	✓	×
POR	Pori	FIN	61N29	021E34	×	✓
PPK	Petropavlovsk, Yelizovo	RUS	53N11	158E25	✓	×
PUG	Palauig	PHL	15N28	119E55	✓	×
PUT	Puttalam	CLN	07N58	079E48	✓	×
PYO	Pyongyang	KRE	39N03	125E42	✓	×
QSH	Qasr-e Shirin	IRN	34N27	045E37	×	✓
QUI	Quito, Mount Pichincha	EQU	00S10	078W32	✓	×
RAN	Rangitaiki	NZL	38S51	176E26	✓	×
RBN	Rabouni	ALG	27N33	008W06	×	✓
RIY	Riyadh	ARS	24N49	046E52	✓	×

Code	Site	Ctry	Lat	Long	SW	MW
RMI	Hialeah, FL	USA	25N54	080W22	✓	✗
RMP	Rampisham	G	50N48	002W38	✓	✗
RNO	New Orleans, LA	USA	29N50	090W07	✓	✗
ROM	Roma	I	41N55	012E26	✗	✓
ROU	Roumoules	F	43N48	006E10	✗	✓
RSO	Rimavská Sobota	SVK	48N24	020E08	✓	✗
SAB	Sabrata	LBY	32N36	012E21	✓	✗
SAC	Sackville, NB	CAN	45N54	064W19	✓	✗
SAG	Samgo	KRE	38N02	126E32	✗	✓
SAI	Saipan, Agingan Point	MRA	15N07	145E42	✓	✗
SAM	Samara	RUS	53N17	050E14	✓	✗
SAN	Sana'a	YEM	15N23	044E12	✓	✗
SAO	Pinheira	STP	00N18	006E45	✓	✓
SAS	Sasnovy	BLR	53N25	028E31	✗	✓
SDA	Agat, Facpi Point	GUM	13N20	144E39	✓	✗
SEB	Seeb	OMA	23N34	058E14	✓	✗
SEL	Selebi-Phikwe	BOT	21S57	027E36	✗	✓
SEO	Seoul, Incheon	KOR	37N25	126E45	✗	✓
SEP	Sepo	KRE	38N37	127E22	✗	✓
SEY	Mahé	SEY	04S41	055E27	✓	✗
SFA	Sfax, Sidi Mansour	TUN	34N49	010E51	✓	✗
SGO	Santiago, Calera de Tango	CHL	33S38	070W51	✓	✗
SHI	Shijak	ALB	41N20	019E33	✓	✗
SHP	Shepparton	AUS	36S19	145E25	✓	✗
SIN	Sines	POR	37N56	008W46	✓	✗
SIR	Sirjan	IRN	29N36	055E47	✓	✗
SIT	Kaunas, Sitkunai	LTU	55N03	023E49	✓	✓
SKN	Skelton	G	54N44	002W53	✓	✗
SKO	Skopje, Sveti Nikole	MKD	41N47	021E53	✗	✓
SLA	A'Seela	OMA	21N55	059E37	✓	✗
SMF	Mykolaiv, Luch	UKR	46N49	032E13	✗	✓
SMG	Santa Maria di Galeria	CVA	42N03	012E19	✓	✓
SNG	Singapore	SNG	01N25	103E43	✓	✗
SOF	Sofia, Kostinbrod	BUL	42N49	023E11	✓	✗
SPB	St. Peterburg, Krasnyy Bor	RUS	59N40	030E41	✓	✓
SWO	Sangwon	KRE	38N51	126E06	✗	✓
SZG	Shijiazhuang, Nanpozhuang, HB	CHN	38N13	114E06	✓	✗
TAC	Toshkent	UZB	41N13	069E09	✓	✗
TAI	Taiwan (exact site unknown)	TWN	-	-	✓	✓
TAR	Tartus	SYR	34N57	035E53	✗	✓
TCH	Chita, Kruchina	RUS	51N50	113E43	✓	✓
THU	Thumrait	OMA	17N38	053E56	✓	✗
TIG	Bucuresti	ROU	44N45	026E06	✓	✗
TIN	Tinian	MRA	15N03	145E36	✓	✗
TJC	Newport, NC	USA	34N47	076W53	✓	✗
TNN	Tainan	TWN	23N03	120E10	✓	✗
TRI	Tripoli	LBY	32N51	013E04	✗	✓
TRM	Trincomalee	CLN	08N45	081E08	✓	✓
TSH	Tanshui	TWN	25N11	121E25	✓	✗
TTU	Tartu, Kavastu	EST	58N25	027E06	✗	✓
TUA	Kota Kinabalu	MLA	06N10	116E10	✗	✓
TUT	Tuticorin	IND	08N49	078E05	✗	✓
TWB	Bonaire, Belnem	ATN	12N06	068W17	✗	✓
TWR	Merizo	GUM	13N17	144E40	✓	✗
TWW	Lebanon, TN	USA	36N17	086W06	✓	✗
TYB	Tayebad	IRN	34N44	060E48	✗	✓
UBA	Ulaanbaatar, Hönhör	MNG	47N48	107E11	✓	✓
UDO	Udon Thani (Udorn), Ban Dung	THA	17N40	103E12	✓	✗
URU	Ürümqi, Hutubi, Xinjiang	CHN	44N09	086E54	✓	✗
VAT	Vatican City	CVA	41N54	012E27	✓	✓
VDN	Vidin, Vodna	BUL	43N50	022E43	✗	✓
VIE	Vientiane	LAO	18N00	102E38	✗	✗
VIR	Virrat, Liedenpohja	FIN	62N23	023E37	✓	✗
VLD	Vladivostok, Razdolnoye	RUS	43N33	131E55	✓	✗
VLN	Vilnius	LTU	54N42	025E13	✗	✓
VNI	Son Tay	VTN	21N12	105E22	✗	✗
VOH	Rancho Simi, CA	USA	34N15	118W39	✓	✗
WAV	Wavre	BEL	50N45	004E35	✓	✗
WBR	Wachenbrunn	D	50N29	010E33	✓	✗
WCR	Nashville, TN	USA	36N12	086W54	✓	✗
WER	Wertachtal	D	48N05	010E42	✓	✗
WOF	Woofferton	G	52N19	002W43	✓	✗
WOL	Wolvertem	BEL	50N59	004E18	✗	✓
WON	Wonsan	KRE	39N05	127E25	✗	✗
WRB	Manchester, TN	USA	35N37	086W01	✓	✗
XIA	Xi'an, Xianyang, Shaanxi	CHN	34N22	108E37	✓	✗
XUW	Xuanwei, Yunnan	CHN	26N11	104E02	✗	✓
YAM	Koga, Yamata, Ibaraki pref.	J	36N10	139E49	✓	✗
YER	Yerevan, Arinj	ARM	40N14	044E36	✓	✗
YFR	Okeechobee, FL	USA	27N27	080W56	✓	✗
ZAB	Zabol	IRN	31N02	061E33	✗	✓
ZAD	Zadar	HRV	44N14	015E14	✗	✓
ZAH	Zahedan	IRN	29N28	060E52	✓	✗
ZAK	Zakaki (Lady's Mile, Akrotiri SBA)	CYP	34N37	033E00	✗	✓

TARGET AREA CODES

Code	Target Area	Code	Target Area	Code	Target Area	Code	Target Area
Af	Africa	Cau	Caucasia	IOc	Indian Ocean	SAf	Southern Africa
Am	Americas	CEu	Central Europe	LAm	Latin America	SAm	South America
As	Asia	EAf	Eastern Africa	ME	Middle East	SAs	Southern Asia
Atl	Atlantic Ocean	EAs	Eastern Asia	Med	Mediterranean	SEA	South East Asia
CAf	Central Africa	EEu	Eastern Europe	NAf	Northern Africa	SEu	Southern Europe
CAm	Central America	Eu	Europe	NAm	North America	WAf	Western Africa
Car	Caribbean	FE	Far East	NEu	Northern Europe	WAs	West Asia
CAs	Central Asia	FER	Far Eastern Russia	Pac	Pacific Ocean	WEu	Western Europe

DOMESTIC SW TRANSMITTER SITES

Coordinate System: WGS84 (rounded)

NB: For coordinates of sites that are jointly used for National and International/COTB services, see the International Transmitter Sites table

*) Sites not co-located with Int. services: (a) ORTB (b) ZNBC **) utility tx site °) presumed locations

Ctry	Site	Lat	Long	Ctry	Site	Lat	Long	Ctry	Site	Lat	Long	Ctry	Site	Lat	Long
ARG	Malargüe	35S30	069W35	B	Natal	05S47	035W13	BOL	S. Ignacio de Vel.	16S22	060W57	DOM	Santo Domingo	18N30	069W57
ARG	Mendoza	32S50	068W47	B	Óbidos	01S52	055W30	BOL	S. José de Chiqu.	17S53	060W45	EQA	Ibarra	00N21	078W08
ARG	Puerto Iguazú	25S36	054W34	B	Parintins	02S38	056W45	BOL	S. Ana del Yac.	13S45	065W32	EQA	Lago Agrio	00N05	076W52
ATA	Base Esperanza	63S23	056W60	B	Petrolina	09S22	040W30	BOL	Santa Cruz	17S46	063W11	EQA	Otavalo	00N18	078W11
AUS	Alice Springs	23S42	133E53	B	Porto Alegre	30S03	051W10	BOL	Siglo Veinte	18S23	066W38	EQA	Quito	00S11	078W32
AUS	Katherine	14S28	132E16	B	Porto Velho	08S45	063W54	BOL	Sucre	19S02	065W18	EQA	Saquisilí	00S58	078W24
AUS	Sydney	33S53	151E13	B	Ribeirão Preto	21S10	047W44	BOL	Tazna	20S41	066W22	EQA	Saraguro	03S42	079W18
AUS	Tennant Creek	19S40	134E10	B	Rio Branco	09S59	067W49	BOL	Tocla	20S40	065W47	EQA	Tena	01S00	077W48
AZE	Stepanakert	39N49	046E44	B	Rio de Janeiro	22S57	043W13	BOL	Trinidad	14S48	064W48	ERI	Asmara	15N13	038E53
B	Altamira	03S13	052W15	B	Rondonópolis	16S29	054W37	BOL	Tumupasa	14S09	067W55	ETH	Addis Ababa	08N58	038E43
B	Anápolis	16S20	048W58	B	Santa Maria	29S42	053W42	BOL	Tupiza	21S27	065W43	ETH	Geja Dera	08N46	038E40
B	Aparecida	23S00	045W00	B	Santarém	02S26	054W41	BOL	Uncia	18S22	066W37	F	Rennes	48N07	001E40
B	Aquidauana	20S27	055W45	B	São Carlos	22S02	047W53	BOL	Villa Montes	14S18	062W21	FSM	Ninseitamw	06N58	158E12
B	Araguaína	07S16	048W18	B	S. Gabriel da Ca.	00S09	067W03	BOL	Villa Serrano	19S07	064W20	GEO	Sokhumi	43N00	041E04
B	Araraquara	21S47	048W10	B	São Luís	02S34	044W16	BOL	Yura	20S06	066W10	GHA	Accra	05N31	000W10
B	Barra do Garças	15S54	052W15	B	São Paulo	23S33	046W39	BRM	Naypyitaw	19N45	096E11	GNE	Malabo	03N45	008E47
B	Belém	01S27	048W29	B	Sen. Guiomard	10S10	067W50	BRM	Taunggyi	20N49	097E02	GRL	Tasiilaq**	65N36	037W38
B	Belo Horizonte	19S54	043W54	B	Sorocaba	23S29	047W27	BRM	Yangon	16N52	096E09	GTM	Chiquimula	14N48	089W32
B	Boa Vista	02N51	060W43	B	Taubaté	23S00	045W36	BTN	Thimphu	27N29	089E37	GTM	Guatemala City	14N37	090W31
B	Bragança	01S02	046W46	B	Tefé	03S24	064W45	CAF	Bangui	04N20	018E31	GTM	S. Pedro La Lag.	14N46	091W11
B	Brasiléia	11S00	068W44	B	Teresina	05S09	042W46	CAF	Boali	04N52	018E02	GTM	S. Sebastián Coa.	15N30	091W30
B	Cáceres	16S05	057W40	B	Varginha	21S33	045W25	CAN	Calgary	50N54	113W53	GUI	Conakry	09N41	013W32
B	Cachoeira Paul.	22S39	045W01	B	Vitória	20S19	040W21	CAN	St. John's	47N34	052W49	GUY	Georgetown	06N46	058W14
B	Camboriú	26S59	048W38	B	Xapuri	10S40	068W30	CAN	Toronto	43N30	079W38	HND	Comayagüela	14N15	087W20
B	Campina Grande	07S13	035W53	BEN	Parakou* (a)	09N20	002E38	CAN	Vancouver	49N08	123W12	HND	San Luís	15N05	088W23
B	Campinas	22S54	047W06	BFA	Ouagadougou	12N26	001W33	CHL	Putre	18S12	069W35	HND	Tegucigalpa	14N05	087W14
B	Campo Grande	20S24	054W35	BGD	Savar	23N52	090E16	CHL	Temuco	38S41	072W35	I	Andrate	45N31	007W53
B	Campos dos Goit.	21S46	041W21	BLR	Brest	52N08	023E59	CHN	Bayanhot	38N58	105E35	IND	Aizawl	23N43	092E43
B	Coari	04S08	063W07	BLR	Hrodna	53N40	023E52	CHN	Changsha	28N12	112E58	IND	Bhopal	23N15	077E29
B	Congonhas	20S30	043W53	BLR	Mahilioú	53N59	030E21	CHN	Fuzhou	26N06	119E24	IND	Gangtok	27N20	088E40
B	Corumbá	19S01	057W39	BOL	Animas	20S58	066W18	CHN	Gejiu	23N21	103E08	IND	Guwahati	26N09	091E39
B	Cruzeiro do Sul	07S40	072W39	BOL	Aripalca	20S13	065W29	CHN	Guiyang	26N25	106E36	IND	Hyderabad	17N20	078E34
B	Cuiabá	15S32	056W05	BOL	Bermejo	22S10	064W42	CHN	Hailar	49N02	119E45	IND	Imphal	24N37	093E54
B	Curitiba	25S23	049W10	BOL	Camargo	20S38	065W15	CHN	Hezuo	35N06	102E54	IND	Itanagar	27N05	093E35
B	Descalvado	21S53	047W40	BOL	Camíri	20S03	063W32	CHN	Nanjing	32N02	118E44	IND	Jaipur	26N55	075E45
B	Dourados	22S09	054W52	BOL	Caranavi	15S49	067W33	CHN	Shanghai	31N15	121E29	IND	Jammu	32N47	074E49
B	Florianópolis	27S35	048W31	BOL	Cobija	11S01	068W45	CHN	Shangzhi	45N02	128E00	IND	Jeypore	18N55	082E34
B	Foz do Iguaçu	25S34	054W33	BOL	Cochabamba	17S23	066W11	CHN	Wuchang	44N54	127E11	IND	Kohima	25N43	094E02
B	Goiânia	16S43	049W18	BOL	El Alto	16S30	068W12	CHN	Xichang	27N49	102E14	IND	Kolkata	22N22	088E17
B	Gov. Valadares	18S51	041W57	BOL	Guanay	12S31	066W50	CHN	Xining	36N38	101E36	IND	Kurseong	26N55	088E19
B	Guajará Mirim	10S50	065W21	BOL	Guayaramerín	10S51	065W23	CLM	Bogotá	04N38	074W05	IND	Leh	34N07	077E35
B	Guarujá	23S55	046W17	BOL	Huanuni	18S47	066W48	CLM	Puerto Lleras	03N16	073W22	IND	Lucknow	26N53	081E03
B	Guarulhos	23S28	046W32	BOL	Independencia	17S04	066W49	CLM	S. José del Guav.	02N34	072W38	IND	Port Blair	11N37	092E45
B	Humaitá	07S31	063W01	BOL	La Paz	16S30	068W08	CME	Buea	04N10	009E14	IND	Ranchi	23N24	085E14
B	Ibitinga	21S43	048W47	BOL	Mina Bolívar	18S29	066W53	COD	Bukavu	02S30	028E50	IND	Shillong	25N34	091E56
B	Jataí	17S58	051W45	BOL	Montero	17S15	063W15	COD	Bunia	01N32	030E11	IND	Shimla	31N10	077E12
B	Ji-Paraná	10S50	061W58	BOL	Oruro	17S58	067W07	COD	Goma (F.pl)	01S41	029E13	IND	Srinagar	34N02	074E54
B	Limeira	22S34	047W25	BOL	Padilla	19S18	064W20	CTR	San José	09N56	084W05	IND	Thiruvananthap.	08N27	076E56
B	Londrina	23S18	051W13	BOL	Reyes	14S18	067W23	CTR	San Pedro M.d.O.	09N56	084W03	INS	Bajawa	08S46	120E59
B	Macapá	00N04	051W04	BOL	Riberalta	10S59	066W06	D	Berlin	52N30	013E20	INS	Biak	01S10	136E06
B	Manaus	03S04	060W00	BOL	San Borja	14S52	066W53	D	Erlangen	49N58	011E00	INS	Fakfak	02S55	132E18
B	Marília	22S13	049W56	BOL	S. Ignacio de Mo.	14S56	065W38	DJI	Arta	11N35	043E05	INS	Jambi	01S38	103E34

Ctry	Site	Lat	Long	Ctry	Site	Lat	Long	Ctry	Site	Lat	Long	Ctry	Site	Lat	Long
INS	Kendari	03S58	122E34	PAK	Rawalpindi	33N37	072E59	PRU	Cusco	13S32	071W57	PRU	Villa Atalaya	10S44	073W46
INS	Makassar	05S10	119E25	PNG	Alotau	10S18	150E28	PRU	Cutervo	06S23	078W51	PRU	Wanchaq	13S31	071W58
INS	Manokwari	00S52	134E05	PNG	Buka	05S25	154E40	PRU	Huamachuco	07S50	078W01	PRU	Yurimaguas	05S54	076W07
INS	Nabire	03S22	135E29	PNG	Daru	09S05	143E10	PRU	Huamanga	07S10	078W31	RUS	Arkhangelsk	64N21	041E23
INS	Palangkaraya	02S12	113E50	PNG	Kavieng	02S34	150E48	PRU	Huancabamba	05S14	079W24	RUS	Arman	59N42	150E10
INS	Pontianak	00S05	109E16	PNG	Kerema	07S59	145E46	PRU	Huancavelica	12S45	075W03	RUS	Krasnoyarsk	56N02	092E45
INS	Ruteng	08S35	120E28	PNG	Kimbe	05S36	150E10	PRU	Huancayo	12S05	075W12	RUS	Kyzyl	51N41	094E36
INS	Ternate	00N48	127E23	PNG	Kiunga	06S07	141E17	PRU	Huanta	12S54	074W13	RUS	Monchegorsk	68N04	032E58
IRQ	Salah ad-Din°	33N58	044E10	PNG	Kundiawa	06S00	144E57	PRU	Huánuco	09S55	076W11	RUS	Selenginsk	52N02	106E56
IRQ	Sulaimaniyah°	35N34	045E19	PNG	Lae	06S41	146E54	PRU	Huaraz	09S33	077W31	RUS	Yakutsk	62N14	129E49
J	Nagara	35N28	140E12	PNG	Lorengau	02S01	147E15	PRU	Huarmaca	05S34	079W32	SDN	Omdurman	15N35	032E26
J	Nemuro	43N17	145E34	PNG	Madang	05S14	145E45	PRU	Iquitos	03S51	073W13	SDN	Juba	04N51	031E36
KRE	Hamhung	39N55	127E31	PNG	Mendi	06S13	143E39	PRU	Juliaca	15S29	070W09	SHN	Jamestown	15S57	005W33
KRE	Hyesan	41N23	128E10	PNG	Popondetta	08S45	148E15	PRU	Junín	11S11	076W00	SLM	Honiara	09S25	160E03
KRE	Pyongsong	40N14	125E49	PNG	Port Moresby	09S26	147E11	PRU	La Oroya	11S36	075W54	SOM	Hargeysa	09N34	044E04
KRE	Sariwon	38N31	125E46	PNG	Rabaul	04S13	152E07	PRU	Lambayeque	06S36	079W45	SUR	Paramaribo	05N49	055W12
LAO	Sam Neua	20N25	104E04	PNG	Vanimo	02S42	141E18	PRU	Líma	12S06	077W03	TCD	N'Djaména	12N07	015E05
LBR	Monrovia	06N14	010W42	PNG	Wabag	05S28	143E40	PRU	Nueva Cajam.	06S30	077W30	TKM	Asgabat	37N51	058E22
MDG	Ambohidrano	18S47	047E29	PNG	Wewak	03S35	143E40	PRU	Otuzco	07S54	078W35	TWN	Guanyin	25N02	121E06
MEX	Mérida	20N58	089W37	PRG	Asunción	25S16	057W38	PRU	Panao	09S54	075W58	TZA	Dole	06S06	039E15
MEX	México	19N26	099W08	PRG	Villarrica	25S45	056W26	PRU	Paucartambo	10S54	075W51	UGA	Kampala	00N19	032E37
MEX	San Luis Potosí	22N01	100W59	PRG	Ypané	25S27	057W32	PRU	Puerto Maldon.	12S37	069W11	UGA	Mukono	00N21	032E45
MLA	Kuching	01N31	110E19	PRU	Abancay	13S37	072W52	PRU	Quillabamba	12S49	072W41	URG	Artigas	30S25	056W29
MLA	Sibu	02N18	111E49	PRU	Arequipa	16S25	071W32	PRU	Rodrigues de M.	06S19	077W28	URG	Castillos	34S16	053W56
MNG	Altay	46N19	096E15	PRU	Bambamarca	06S43	078W34	PRU	S. Miguel de El F.	05S17	079W28	URG	Montevideo	34S50	056W18
MNG	Mörön	49N37	100E10	PRU	Bolívar	07S16	077W47	PRU	San Andrés	06S13	078W40	URG	Sarandí del Yi	33S22	055W38
MTN	Nouakchott	18N08	015W60	PRU	Callalli	15S51	072W51	PRU	San Ignacio	05S09	079W00	VEN	Pto. Ayacucho	05N35	067W40
MWI	Lilongwe (F.pl)	13S47	033E47	PRU	Celendín	06S53	078W09	PRU	San Miguel	07S03	078W54	VTN	Buôn Mê Thuôt	12N40	108E12
NGR	Niamey	13N32	002E04	PRU	Cerro de Pasco	10S41	076W16	PRU	Santa Cruz	06S40	079W00	VTN	Son La	21N20	103E55
NIG	Enugu	06N27	007E28	PRU	Chachapoyas	06S10	077W50	PRU	Santa Monica	07S10	078W30	VTN	Xuân Mai	20N43	105E33
NIG	Kaduna	10N45	007E33	PRU	Chiclayo	06S47	079W47	PRU	Sicuani	14S15	071W12	VUT	Port-Vila	17S45	168E22
NPL	Khumaltar	27N39	085E20	PRU	Chota	06S21	078W39	PRU	Tacna	18S00	070W13	ZMB	Lusaka* (b)	15S30	028E15
PAK	Karachi (F.pl)	24N52	067E02	PRU	Comas	11S57	077W04	PRU	Tarma	11S28	075W41				
PAK	Peshawar	34N01	071E37	PRU	Cortegana	06S31	078W20	PRU	Trujillo	08S06	079W00				

CLUBS FOR DXERS & INTERNATIONAL LISTENERS

This section lists non-commercial hobby clubs serving international radio enthusiasts. Most clubs are orientated to DXing, the reception of distant radio stations, some are oriented to programme listening. Many clubs produce bulletins on a regular basis. Sample copies of printed periodicals are generally available for return postage (3 or 4 IRCs, or equivalent currency). For officially multilingual countries, the language(s) used in club publications (and/or activities) is indicated when known; for non-English speaking countries also if a publication is partly or entirely in English: EE = English, FF = French, GG = German, II = Italian, JJ = Japanese, SS = Spanish. This list does not include clubs run by commercial publications or by individual broadcasters.

EUROPE

European DX Council (EDXC) (Umbrella organization of DX Clubs in Europe) c/o Tibor Szilagyi, Ringvägen 86, 13731 Västerhaninge, Sweden. General Secretary: Tibor Szilagyi (**E:** tiszi2035@yahoo.com), Vice General Secretary: Ingvar Kohlström (**E:** ingvar.kohlstrom@telia.com) **W:** www.edxc.org; edxcnews.wordpress.com (blog)

AUSTRIA: Austrian DX Board (ADXB-OE) (Club der Freunde elektronischer Medien - Rundfunk global) , Postfach 1000, 1082 Wien. **E:** office@adxb-oe.org **W:** www.adxb-oe.org Pub: Rundbrief quartely. Annual DX camp. Member of AGDX (Germany)

BELGIUM: DX-Antwerp, P.O.Box 2660 Hoboken. (Flemish) **E:** info@dx-antwerp.com **W:** www.dx-antwerp.com – **Belgique Radio-Loisirs**, B.P. 12, 7160 Chapelle-lez-Herlaimont. (FF)

BULGARIA: Association of Balkan Cross-band DXers (ABCDX), c/o Rumen Pankov, P.O.Box 199, 1000 Sofia-C **E:** rumen_pankov@yahoo.co.uk – **Bulgarian DX Club plus Satellite,** c/o Ivan Penev, Simeon Radev 40-V, 1618 Sofia ☎ +359 2 8557143 **E:** ipenev@mail.orbitel.bg **W:** tele-satellite.hit.bg/bdxcs

CZECH REPUBLIC/SLOVAKIA: Czechoslovak DX Club (CSDXC), c/o Václav Dosoudil, Horní 9, 768 21 Kvasice, Czech Republic. **E:** mail@dx.cz **W:** www.dx.cz Pub: RADIO revue

DENMARK: Danish Shortwave Club International (DSWCI), Tavleager 31, 2670 Greve. **E:** kaj.bredahl@mail.dk **W:** www.dswci.org Pub (all EE): Shortwave News, DX-Window (bi-weekly by email), Tropical Bands Monitor (monthly), Domestic Broadcasting Survey (Annual) – **Dansk DX Lytter Klub (DDXLK)**, P.O.Box 112, 8960 Randers SØ. **E:** ddxlk@ddxlk.net **W:** www.ddxlk.net Pub: DX-FOKUS (bimonthly)

FINLAND: Suomen DX-liitto ry (SDXL), P.O.Box 454, 00101 Helsinki (Umbrella organization of Finnish language DX clubs) **E:** toimisto@sdxl.org **W:** www.sdxl.org Pub: Radiomaailma (Finnish/EE) – **Finlands Svenska DX-Förbund rf (FSDXF)**, P.O.Box 9, 68601 Jakobstad (Umbrella organization of Swedish language DX clubs in Finland). **E:** info@fsdxf.org **W:** uk.groups.yahoo.com/group/fsdxf

FRANCE: Amitié Radio, 49-51 rue Marcel Bourdarias, 94140 Alfortville. **E:** amitieradio@libertysurf.fr **W:** amitieradio.monsite-orange.fr Pub: RadioPanorama, A l'écoute du monde – **Monde & Radiodiffusion**, 65 Montée des Princes, 84100 Orange. – **Radio Club de la Poste**, c/o Marcel Lecerf, 13 avenue St Michel, 54220 Malzéville. – **Radio Club du Perche**, 12 rue du Grand Thuret, 72320 Greez sur Roc. – **Radio Club International Ondes Courtes**, 19 lot Saturne, 26120 Malissard. – **Radio DX Club d'Auvergne**, Centre Municipal P. et M. Curie, 2 bis, Rue du Clos Perret, 63100 Clermont-Ferrand. – **Union des Écouteurs Français**, BP 31, 92242 Malakoff Cédéx. **E:** tsfinfo@u-e-f.net **W:** www.u-e-f.net

GERMANY: Arbeitsgemeinschaft DX e.V. (AGDX), Postfach 1214, 61282 Bad Homburg (Umbrella organization for the German DX clubs adxb-DL, Kurzwellenfreunde Sachsen, UKW/TV Arbeitskreis der AGDX, Worldwide DX Club, and for the Austrian DX-Board) **E:** mail@agdx.de **W:** www.agdx.de – **Assoziation Deutschsprachiger Kurzwellenhörer e.V. (ADDX)**, Scharsbergweg 14, 41189 Mönchengladbach. **E:** kurier@addx.de **W:** www.addx.de Pub: Radio-Kurier – **Assoziation Junger DXer e.V. (adxb-DL),** c/o Thomas Schubaur, Neufnachstr. 30, 86850 Fischach. **E:** dl1ts@t-online.de **W:** www.adxb-dl.de – **Deutscher Welt Radio Club e.V. (DWRC),** c/o Bernd Schilling, Hüling 11, 53332 Bornheim. – **Eastside DX (EDX)**, c/o Jens Adolph, Postfach 100137, 04001 Leipzig. **E:** ctu33jens@gmx.net Pub: Eastside Radiogeschichte – **Freundeskreis Berliner Empfangsamateure**, Postfach 200113, 13511 Berlin. **E:** mittelwelle@genion.de – **Hamburger Freunde des Rundfunkfernempfangs**, c/o Dieter Schäfer, Am Sportplatz 18, 24629 Kisdorf. **E:** dl1lad@darc.de – **Kurzwellenfreunde Brand**, c/o Hans-Jürgen Schmelzer, Mitterteicher Str. 15, 95643 Tirschenreuth. **E:** hugotir@t-online.de – **Kurzwellenclub Schwalmtal,** c/o Helmut Reitzer Jr, Willy-Rösler-Str. 41, 41366 Schwalmtal. **E:** dk0kws@qsl.net **W:** www.qsl.net/dk0kws – **Kurzwellenfreunde Rhein-Ruhr,** c/o U. Schnelle, Kurfürstenstr. 37, 45883 Gelsenkirchen. **E:** infohq@kwfr.de **W:** www.kwfr.de – **Kurzwellenfreunde Sachsen (KWFS),** c/o AGDX e.V., Postfach 1214, 61242 Bad Homburg. **E:** dk5tl@qsl.net – **Kurzwellenfreunde Wuppertal (KWFW)**, Postfach 220342, 42373 Wuppertal. – **Oldenburger Kurzwellenfreunde,** c/o Olaf C. Hänßler, Sandweg 98, 26135 Oldenburg. **E:** olaf.haenssler@informatik.uni-oldenburg.de – **Radio Japan Club Brilon (RJCB),** c/o Reinhard Reese, Niederbeckstr. 23, 40472 Düsseldorf. **E:** rreese@gmx.net **W:** radio-japan-club-brilon.gmxhome.de – **Rhein-Main-Radio-Club e.V. (RMRC),** Postfach 700849, 60558 Frankfurt. **E:** mail@rmrc.de **W:** www.rmrc.de. **UKW/TV Arbeitskreis der AGDX**, c/o H.-J. Kuhlo, Wilhelm-Leuschner-Str. 293B, 64347 Griesheim. (RM/TV only) **E:** sekretariat@ukwtv.de **W:** www.ukwtv.de – **Worldwide DX Club (WWDXC)**, Postfach 1214, 61282 Bad Homburg. **E:** mail@wwdxc.de **W:** www.wwdxc.de. Pub: DX-Magazine (EE)

HUNGARY: FM DX Club, **E:** fmdx@ha5kfu.hu **W:** fmdx.ha5kfu.hu – **Hungarian DX Club**, Beke utca 85, 2519 Piliscsév. **E:** tiszi2035@yahoo.com

IRELAND: Irish DX Radio Club, c/o Edward Dunne, 17 Anville Drive, Kilmacud, Stillorgan, Co. Dublin **E:** irishdxclub@live.ie Pub: MediaWatch (by email)

ITALY: Associazione Italiana Radioascolto (A.I.R.), C.P. 1338, 10100 Torino (AD). **E:** info@air-radio.it **W:** www.air-radio.it Pub: Radioarama (monthly) – **BCL Sicilia Club**, c/o Roberto Scaglione, C.P. 119, Succ. 34, 90144 Palermo (PA). **E:** bclsiciliaclub@inwind.it **W:** www.bclnews.it – **Coordinamento del Radioascolto (Co. Rad)**, c/o Dario Monferini (Web-based umbrella organisation of various Italian DX clubs) **E:** info@corad.net **W:** www.corad.net – **FM-DX Italy,** c/o Fabrizio Carnevalini (Web-based, FM-TV DX only) **E:** fabrizio58it@yahoo.it **W:** www.fmdx.altervista.org – **Gruppo d'Ascolto della Marca Trevigiana**, C.P. 3, Succ. 10, 31100 Treviso (TV). **E:** gamt@ntt.it – **Gruppo d'Ascolto Radio dello Stretto**, c/o Giovanni Sergi, Via Sibari 40, 98149 Messina (Camaro Inferiore). **E:** gsergi5050@hotmail.com **W:** www.polistenaweb.it/gars Pub: Radio Notizie (quarterly) – **Gruppo d'Ascolto Radio Televisivo della Sicilia**, c/o Gioacchino Stallone, Via G.Falcone 11, Lotto 27, interno 3, 91025 Marsala (TP). (local activity) – **Gruppo Radio Ascolto Bologna**, c/o Elio Antonucci (Web-based) **E:** radioascolto@elio.org **W:** www.elio.org/radioascolto – **Play-DX**, c/o Dario Monferini, Via Davanzati 8, 20158 Milano (MI). (II/EE/SS) (specialises in difficult DX) **E:** info@playdx.com **W:** www.playdx.com Pub: PLAY-DX (weekly)

NETHERLANDS: Benelux DX Club (BDXC), Rietdekkerstraat 40, 1445 KG Purmerend. **E:** secretaris@bdxc.nl **W:** www.bdxc.nl Pub: BDXC-Bulletin (Dutch/EE)

NORWAY: DX Listeners' Club (DXLC), c/o Jan Alvestad, Vigdelsveien 637B, 4054 Tjelta. **E:** dx-news@dxlc.com **W:** dxlc. wordpress.com Pub: DX-News

RUSSIA: Club of DX-ers, c/o Vadim Alexeew, P.O.Box 65, 125581 Moscow. **E:** gusev@itep.ru **W:** www.radio.hobby.ru – **Irkutsk DX Club:** c/o Feodor Brazhnikov, P.O.Box 3036, 664059 Irkutsk. **E:** brazhnikov@yahoo.com **W:** www.irkutsk.com/radio – **Novosibirsk DX Club**, c/o Igor Yaremenko, Novosibirsk. **E:** dx@ngs.ru **W:** www. novosibdx.info – **Russian DX League**, c/o Anatoly Klepov, ul. Tvardovskogo 23-365, 123458 Moscow. **E:** rusdx@yandex.ru **W:** rusdx.narod.ru – **Sankt-Peterburg DX Club**, c/o Alexey Osipov, P.O.Box 46, 195213 Sankt-Peterburg. **E:** dxspb@vfemail.net – **Tomsk DX Club**, c/o Vladimir Kovalenko, Tomsk. **E:** tomskdx@mail.ru

SPAIN: Asociación DX Barcelona (ADXB), P.O. Box 335, 08080 Barcelona. **E:** info@mundodx.net **W:** www.mundodx.net Pub: Mundo DX - multimedia (SS/Catalan) – **Asociación Española de Radioescucha (AER)**, Apartado 10014, 50080 Zaragoza. **E:** sedano@lander.es **W:** www.aer-dx.org – **S500 DX Club**, c/o Álvaro López Osuma, c/ Santa Micaela, 1 2° Derecha, 18015 Granada. **E:** alvak7@yahoo.es **W:** www.upv.es/~csahuqui/julio/s500. Pub: SW DX newsletter (four-monthly)

SWEDEN: Arctic Radio Club (ARC), c/o Tore Larsson, Frejagatan 14A, 52143 Falköping (MW only) **E:** tore.larsson@beta.telenordia.se Pub: MV-Eko (Swedish/EE) – **Sveriges DX Förbund (SDXF)**, Box 1097, 40523 Göteborg (Umbrella organization). **E:** sdxf@sdxf.se **W:** www.sdxf.se Pub: Eter-aktuellt

SWITZERLAND: Radio- und Fernseh-Club Basel und Umgebung (RFCB), Postfach 354, 4015 Basel. (GG) **E:** hb9b@rfcb. ch **W:** www.rfcb.ch

UNITED KINGDOM: British DX Club (BDXC), 10 Hemdean Hill, Caversham, Reading RG4 7SB. **E:** bdxc@bdxc.org.uk **W:** www.bdxc. org.uk Pub: Communication – **International Shortwave League (ISWL)**, c/o Peter Lewis, 18 Bittaford Wood, Ivybridge, Devon, PL21 0ET. **E:** vfgnsu@yahoo.co.uk **W:** www.iswl.org.uk – **Medium Wave Circle (MWC)**, c/o Herman Boel, Papeveld 3, B-9320 Erembodegem-Aalst, Belgium (LW/MW only) **E:** herman@hermanboel.eu **W:** www. mwcircle.org – **World DX Club (WDXC)**, c/o Arthur Ward, 17 Motspur Drive, Northampton NN2 6LY. **E:** barraclough.mike@gmail. com **W:** www.worlddxclub.org.uk Pub: Contact. For North American branch see USA

IVORY COAST: DX-Ivoire, c/o Jibirila Liasu, B.P. 197, Abidjan 20. (FF)

KENYA: DX Listeners' Club, c/o Oscar Machuki, PO Box 646, Kisii 4-0200. (EE/Swahili) (SW only) ☎ +254 721 534171 **E:** oscarmogire@yahoo.com

NIGERIA: Africa DX Association, c/o Mr. Friday I. Okoloise, NITEL, P.M.B. 23, Lafia, Plateau State. (EE) – **International DX Club**, Emmanuel Ezeani, P.O.Box 1633, Sokoto, Sokoto State. (EE) **E:** idxerclub@yahoo.com; emmanuel_ezeani@yahoo.com

SÃO TOMÉ & PRÍNCIPE: Clube DX-STP, c/o Leal Bouças, Av. 12 de Julho, Vila Maria (C.P. 490), São Tomé. **E:** petterboudx@ hotmail.com

TANZANIA: Kemogemba DX Listeners Club, c/o Ras Franz Manko Ngogo, P.O.Box 71, Tarime, Mara. (EE/Swahili)☎ + 255 755 814704 **E:** kemogemba@yahoo.com

TOGO: Club Inter Amitié Radio, CCF, B.P. 2090, Lomé. (F) – **Groupe Endoc**, B.P. 2667, Lomé. (FF)

TUNISIA: Club des Auditeurs et de l'Amitié, c/o De Riadh Sakka, Route de Gremda Merkez Sahnoun, 3012 Sfax. (FF)

UGANDA: International DX Club of East Africa, c/o Ouma Samuel, P.B.Box 565, Iganga. (EE) ☎ +256 772 444201 **E:** samuel. ouma@talk21.com; 0772444201@mtnconnect.co.ug

BANGLADESH: Aurora Listeners' Club, c/o Miss Kakali Rani, Harida Khalsi-6403, Madhnagar-Natore-6400 – **Basupara DX Listeners Club**, c/o Asfaqul Alam, Basupara, Nandangachi, Rajshahi 6260. **E:** bdxls@uymail.com – **International Radio Listeners Club**, Konabari, P.O.Nilnagor, Gazipur, Dhaka. (EE/Bengali) **E:** irlclub@ hotmail.com **W:**irlclub.googlepages.com – **Online DX Forum**, c/o MD Azizul Alam Al-Amin, Gourhanga, Ghoramara, Rajshahi 6100. **E:** mtech@rajbd.com – **Rose DW Listeners Club**, c/o Ashik Eqbal "Tokon", Luximpur Greater Rd, G.P.O.Box 56, Rajshahi 6000. **E:** rosedwlc@yahoo.com **W:** www.rosedwlc.webs.com. Pub: DX-Net (Bengali)

INDIA: Apollo DX International, c/o Deepak Kumar Das, Dholi Sakra 843105, Dist. Muzaffarpur, Bihar. **E:** deepakdx@rediffmail. com – **Ardic DX Club**, c/o Jaisakthivel, T., 59 Annai Sathya Nagar, Arumbakkam, Chennai 600106, Tamil Nadu. ☎ +91 98413 66086 Pub: DXers Guide (EE quartely), Sarvadesa Vanoli (Tamil monthly). DX Prgr: Vaanoli Ulagam on AIR. **E:** ardicdxclub@yahoo. co.in **W:** dxersguide.blogspot.com – **Chaudhary Srota Sangh**, c/o Santosh Kumar (President), Kharauna Jairam, Kharauna DIH 843113, Dist. Muzaffarpur, Bihar – **Chennai DX Club**, c/o K. Raja, 21 JP Koil St, Old Washermenpet, Chennai 600021, Tamil Nadu. **E:** chennaidxclub@gmail.com – **El Nino Electronics DX Club**, c/o Partha Sarathi Goswami, Kishalay, College Road, Siliguri 734001, Darjeeling, West Bengal. **E:** elnino@dxinginfo.com **W:** dxinginfo.com ☎ +91 94343 27414 – **Foreign Radio Listeners' Club**, c/o Prasenjit Bhakat, 313/8 Ghoradhara, P.O Jhargram 721507, West Bengal. ☎ +91 3221 256084 **E:** frlclub@gmail.com – **Globe Radio DX Club (GRDXC)**, c/o Harjot Singh Brar, P.O.Box 158, Chandigarh 160017, Chandigarh. **E:** grdxc@yahoo.co.in – **International DX Association**, c/o Bedanta Das, 1-No,Galiahati, Near Night School, Barpeta 781301, Assam. ☎ +91 3665 236267 **E:** das884@gmail.com Pub: DX Times (EE) – **Metali Listeners' Club**, c/o Mr Shivendu Paul, 49/36, Dr SG Dhar Lane, P.O. Khagra, Dist. Murshidabad 742103, West Bengal. **E:** metalilistenersclub@gmail.com ☎ +91 94348 58497 – **Minnakkal Kurinji DX Club**, c/o E. Selvaraj, Choolaimedu Street, Minnakkal Post, Dist. Namakkal 637505, Tamil Nadu. **E:** selvarajminnakkal@gmail.com – **Paribar Bandhu SWL Club**, c/o Mr Anand Mohan Bain, UCO Bank, 47/6, Nehru Nagar, P.O. Nehru Nagar, Bhilai, Dist. Durg 490020, Chattisgarh. **E:** anand_mohan10@ yahoo.com ☎ +91 94255 21083, +91 788 4031648. – **Pollachi Radio Club**, c/o Mr. N. Lakshmanan, Sri Mugha Bhavan, 44/77 Lac Colony, Dr Ansari Street, Pollachi 642001, Tamil Nadu. ☎ +91 98650 16402 **E:** pollachidxclub@yahoo.co.in; pollachiradioclub. blogspot.com (blog) – **Span Radio Listeners' Club**, c/o A Ragu, Nandavankula Theru, Vedaraniam 614810, Tamil Nadu. ☎ +91 4369 318808 – **Utkarna Shrota Sangha**, c/o Mr. Rajib Bandopadhyay, Amrita Bhaban, P.O. Makardah 711409, Dist. Howrah, West Bengal ☎ +91 94343 28609, **E:** ussrajib@gmail.com – **World DX Club & Library**, c/o Baidyanath Upadhyaya, At Khairabarigaon, P.O. Khawrang, Udalguri 784509, Darrang, Assam. – **World DXing Club**, c/o Mr Madhab Ch. Sagour, 93/1, Mitrapara Road, P.O. Naihati 743 165, 24 Parganas (North), West Bengal. – **World Radio Club**, c/o Mr. Biswanath Mandal, Chak Harharia, P.O. Islampur 742304, Dist. Murshidabad, West Bengal ☎ +91 3481 236534 **E:** bmandalwrc@rediffmail.com – **Young Stars Radio Club** **(YSRC)**, c/o Mr Hari Madugula, 40 Hastinapura Colony, Raghavendra Residency-FF4, Sainikpuri, Secunderabad 500094, Andhra Pradesh. – **Youth International Radio Listeners' Club**, c/o Mr. Pranab Kumar Roy, Vill + PO, Shyamnagar, 741 155, Via Palashipara, Nadia, West Bengal. ☎ +91 3471 252163 **E:** etherbarta@gmail.com Pub: Etherbarta (Bengali), Radio Monitors' Guide (EE)

INDONESIA: Indonesian DX Club (IDXC), P.O.Box 2001 DPPS, Depok 16432. (EE/Indonesian). **E:** info@idxc.org **W:** www.idxc.org – **MAPEM Club**, c/o M. Jayadi D., 02 Tromolpos Pringgabaya, East Lombok, West Nusa Tenggara 83654. **E:** mapemclub2020@gmail. com **W:** mapem-club.org– **Media Monitoring Club**, c/o Summase A. Sanjaya, P.O.Box 1157 MKS, Makassar 90000. **E:** monitoringclub@ yahoo.co.id **W:** monitoringclub.org

JAPAN: Asian Broadcasting Institute (ABI), P.O.Box 2334, Ginza Branch, Japan Post, Tokyo 100-8698. **E:** info@abiweb.jp **W:** www. abiweb.jp – **Indonesian DX Circle Japan**, c/o Atusnori Ishida, 1-16-201 Teranishi, Saichi-cho, Iwakura-shi, Aichi 42-0036 – **Japan**

BCL Federation, 1-9-23 Tadao, Machida City, Tokyo 194-0035. E: mywave@m2.ocv.ne.jp – **Japanese Association of DXers**, P.O.Box 1766, Tokyo 100-91. (JJ/EE) – **Japan Short Wave Club (JSWC)**, P.O.Box 44, Kamakura 248-8691. (JJ/EE) ☎ 📠 +81 467 43 2167 E: jswchq@live.jp Pub: SW DX Guide (JJ/EE) – **Nagoya DXers Circle (NDXC)**, c/o Shigenori Aoki, 2-51 Kasumori-cho, Nakamura-ku, Nagoya 453-0855. W: www.ndxc.org – **Radio Nuevo Mundo**, c/o Tetsuyu Hirahara, 5-6-6 Nukuikita, Koganei-shi, Tokyo 184-0015

KOREA (SOUTH): Northeast Asian Broadcasting Institute (NEABI), c/o SeKyung Park, #103-302, Geumho Apt, 240-32 Yeomchang-dong, Gangseo-gu, Seoul 157-861. E: neabipress@gmail.com W: www.neabi.com

NEPAL: Friendship Radio Club, c/o Mr Umesh Regmi, Tanki Sinuwari 5, District Morang, Biratnagar. E: friendshipradioclub@yahoo.com – **Listeners' Club of Nepal** (Reg. No.144), P.O.Box 126, Biratnagar-4. – **Small Giant Radio Listener Club** (Reg. No.17), P.O.Box 21110, Kathmandu

PAKISTAN: National Society of Pakistani DXers, E-161/1, Iqbal Park, opposite Adil Hospital, Defence Housing Society Rd, Lahore Cantt. – **Pakistan Aafaqie Lehrain Society (PALS)**, c/o Asrar Chaudhary, Dusehra Ground, Sheikhupura 39350, Punjab. Pub: Newsletter Aafaqie Lehrain (Urdu with EE section) E: pals_swlc@yahoo.com W: www.pals2000.webs.com – **Pakistani Shortwave Listeners' Association**, c/o Muhammad Imran Mehr, 38/2 Habib Colony, Bahawalpur 63108, Punjab. ☎ +92 334 6865847, +92 300 6801719. E: imran.mehr@gmail.com Pub: Radio World (bi-monthly by email) – **Wonderful World of Shortwave (WWSW)**, c/o Baber Shehzad, 43 Habib Colony, Bahawalpur 63108, Punjab. E: baber73@yahoo.com. Pub: News Letter of Pakistani DX-ers (by email)

SRI LANKA: Union of Asian DXers (UADX), c/o Victor Goonetilleke, "Shangri-La" 298 Kolamunne, Piliyandala. E: victor.goonetileke@gmail.com Pub: UADX Asian DX Report (EE)

PACIFIC

AUSTRALIA: Australian Radio DX Club (ARDXC), 29 Milford Road, Peakhurst, NSW 2210. E: dxer1234@gmail.com W: www.ardxc.info – **The Electronic DX Press Radio Monitoring Association (EDXP)**, c/o Bob Padula, 404 Mont Albert Road, Mont Albert, VIC 3127. (Web-based club). E: bobpadula@mydesk.net.au W: edxp.yolasite.com

NEW ZEALAND: New Zealand Radio DX League, P.O.Box 39-596, Howick, Manukau 2145. E: secretary@radiodx.com W: www.radiodx.com Pub: NZ DX Times (monthly)

NORTH AMERICA

CANADA: Canadian International DX Club (CIDX), P.O.Box 67063-Lemoyne, St. Lambert, QC J4R 2T8. E: CIDXclub@yahoo.com W: www.cidx.ca – **Club d'Ondes Cortes du Québec** c/o Dominique Duplessis, 5120 35ème rue, Grand-Mère, QC G9T 3N6. (FF) E: dduplessis@infoteck.qc.ca – **Ontario DX Association (ODXA)**, 3211 Centennial D., Apt. 23, Vernon, BC V1T 2T8. E: odxa@rogers.com W: www.odxa.on.ca – **Vancouver Shortwave Association**, P.O.Box 500, 2245 Eton St., Vancouver, BC V5L 1C9

MEXICO: Audio Pico DX Club, c/o César Granillo, Ap. Postal 309, 94301 Orizaba, Veracruz – **Club DX Miguel Auza**, c/o Luis Antero Aguilar, Ap. Postal 38, 98330 Miguel Auza, Zacatecas – **Consultorio DX**, c/o Miguel Angel Rocha Gámez, Ap. Postal 31, 31820 Ascensión, Chihuahua – **Nayarit DX Club**, Ap. Postal 62, 63001 Tepic, Nayarit. E: naydx@tepic.megared.net.mx W: www.naydx.8m.com – **Sociedad de Ingenieros Radioescuchas**, c/o Rafael Gustavo Grajeda Rosado, Ap. Postal 203, Admon. No.1, 91701 Veracruz, Veracruz. E: rggr681121@hotmail.com

USA: American Shortwave Listeners Club (ASWLC), c/o Stewart MacKenzie, 16182 Ballad Lane, Huntington Beach, CA 92649. E: wdx6aa@earthlink.net – **Association of Clandestine Enthusiasts (A*C*E*)**, c/o Pat Murphy, P.O.Box 12112, Nofolk, VA 23541. E: acehdq@localnet.com W: www.frn.net/ace – **Boston**

Area DXers, c/o Paul Graveline, 9 Stirling St., Andover, MA 01810-1408. E: beaconti@aol.com – **Central Indiana Shortwave Club,** c/o Steve Hammer, 2517 E. DePauw Road, Indianapolis, IN 46227-4404. – **Chicago Area DX Club,** c/o Edward G. Stroh, 53 Arrowhead Dr., Thornton, IL 60476. – **DecaloMania**, c/o Paul Richards, P.O.Box 126, Lincroft, NJ 07738 (Club for collectors of station promo, items and airchecks). E: decalcomania@aol.com W: www.anarc.org/decal Pub: DecaloMania – **Indiana Recording Club**, c/o Bill Davies, 1729 E. 77th St., Indianapolis, IN 46240. (Club for airchecks and recordings of mediumwave stations). – **International Radio Club of America (IRCA)**, P.O.Box 60241, Lafayette, LA 70596. (MW only) E: ircamember@ircaonline.org W: www.ircaonline.org Pub: DX Monitor (by online or hard-copy subscription) – **Longwave Club of America (LWCA)**, 45 Wildflower Road, Levittown, PA 19057. E: billoliver@verizon.net W: www.lwca.org Pub: The Lowdown – **Miami Valley DX Club (MVDXC)**, P.O.Box 292132, Columbus, OH 43229. E: mvdxc@core.com W: www.anarc.org/mvdxc Pub: DX World. – **Michigan Area Radio Enthusiasts**, P.O.Box 300, Manchester, MI 48158. E: mare.radio@gmail.com W: mare.radio.tripod.com – **Minnesota DX Club (MDXC)** c/o James Dale, 16330 Germane Ct W, Rosemount, MN 55068. W: www.frontiernet.net/~jadale Pub: MDXC Newsletter – **National Radio Club (NRC)**, P.O.Box 473251, Aurora, CO 80047-3251. (MW only) E: plsbcbdxer@aol.com W: www.nrcdxas.org Pub: DX News – **North American Shortwave Association (NASWA)**, 45 Wildflower Road, Levittown, PA 19057. E: weoliver@comcast.net W: www.naswa.net Pub: The NASWA Journal (by post) and Flashsheet (by email) – **Pacific Northwest/British Columbia DX Club**, c/o Bruce Portzer, 6546 19th Ave NE, Seattle WA 98115. E: bytheway@atk.com W: www.anarc.org/pnbcdxc – **Puna DX Club,** c/o Jerry Witham, P.O.Box 596, Keaau, HI 96749. E: punadxclub@mailcity.com – **Rocky Mountain Radio Listeners**, c/o Mike Curta, P.O.Box 470776, Aurora, CO 80047-0776. – **Southern California Area DXers**, c/o Bill Fisher Sr., 6398 Pheasant Drive, Buena Park, CA 90620-1356. E: billfishernow@netzero.net – **World DX Club (North America branch)** c/o Richard A. D'Angelo, 2216 Burkey Drive, Wyomissing, PA 19610. E: rdangelo3@aol.com – **Worldwide TV-FM DX Association (WFTDA)**, P.O.Box 501, Somersville, CT 06072 (FM/TV only). W: www.wtfda.org Pub: VHF-UHF Digest.

SOUTH AMERICA

ARGENTINA: Asociación DX del Litoral, c/o Emilio Pedro Povrzenic, 1 de Mayo 1071, 2124 Villa Diego, Santa Fé. E: adx-lboletin@yahoo.com.ar – **Grupo DX Suquía**, c/o Carolina J.G. Vandenberghe, Estafeta Rivera Indarte, C.C. No.26, 5149 Córdoba. E: gdxs@hotmail.com – **Grupo Radioescucha Argentino (GRA)**, c/o Marcelo A. Cornachioni, Alvarez Thomas 248, 1832 Lomas de Zamora, Buenos Aires. E: conexiongra@fullzero.com.ar W: www.conexiongra.com.ar

BRAZIL: DX Clube do Brasil (DXCP), C.P. 1594, 09571-970 São Caetano do Sul (SP). E: dxcb@mandic.com.br W: www.ondascurtas.com – **Santa Rita DX Clube**, C.P. 4, 58300-970 Santa Rita (PB)

CHILE: Club Diexista de Chile, Calle 3 Ponienta 55, Talca. E: chiledxclub@mixmail.com W: www.galeon.com/chiledxclub Pub: Radiograma – **Federación de Clubes de Radioaficionados de Chile (FEDERACHI)**, c/o Héctor Frías Jofre, Dr. Eduardo Cruz Coke 389, 3° piso, (Cas. 260-2) Santiago. E: federachi@federachi.cl W: www.federachi.cl

COLOMBIA: Grupo Internacional de Diexistas y Radioaficionados, c/o Miguel Bayona. E: m-bayona@hotmail.com W: www.colombiadx.tk

URUGUAY: DX Club Montevideo, Calle Batovi 2068, 11800 Montevideo. E: dxclubmontevideo@yahoo.com W: groups.yahoo.com/group/dxclubmontevideo

VENEZUELA: Asociación Dixista de Venezuela, Ap. Postal 65657, Caracas 1066-A. E: marl1@hotmail.com – **Club Diexistas de la Amistad**, c/o Ing. Santiago San Gil Gonzáles, Ap. Postal 202, Barinas 5201-A, Estado Barinas. E: cdxainternacional@hotmail.com W: diexismovenezolano.blogspot.com – **Venezuelan QSL Help**, c/o Winter Monges, Ap. Postal 1.116, Barquisimeto 3001-A, Lara. (SS/EE) E: wintermonges@yahoo.com

STANDARD TIME & FREQUENCY TRANSMISSIONS

What are STFTs?

Standard Time and Frequency Transmissions (STFTs) are transmissions aimed at testing and calibrating radio receivers and synchronizing clocks. When broadcast on shortwave, STFTs usually consist of continuous AM and/or SSB transmissions of 'beeps' or 'pips' every second (often referred to as 'Time Signals'), with the time in UTC (or local time) announced at certain intervals. Some stations broadcast for 24 hours a day while other stations run for up to a few hours daily or on certain days of the week.

Also listed in this chapter is a selection of standard broadcast and VLF utility stations that broadcast regular timechecks in modes other than AM/SSB. The VLF transmitters of the maritime navigation networks Loran-C/Chayka on 100kHz and similar VLF txs primarily used for navigation purposes, except for txs of the "Beta" system of the Russian Navy, are not included.

Using STFTs

STFTs are invaluable aids for the SW radio user. Not only do they allow listeners to synchronise clocks to UTC but they are also a handy tool for checking propagation and reception paths. Their most useful role for serious shortwave listeners, however, is for checking that equipment is performing as it should and to test for receiver frequency calibration errors.

Checking Performance

It is possible to carry out tests on a variety of frequencies ranging from 2500 to 20000kHz. First select an appropriate set of STFTs, perhaps by saving them into a set of memory channels if your radio has the facility. A quick check can be made for the characteristic ticks and pulses to ensure that there is a good reception path and that the STFT is currently active, before moving on to the tests themselves. Don't forget that it is essential to allow the radio to warm up for at least hour before starting these tests.

The object of the exercise is to mix the incoming STFT signal with an internally generated signal from the radio's Beat Frequency Oscillator (BFO) and then tune the radio until the resulting whistle, or heterodyne as it is known, drops down to zero. This process of tuning for 'zero beat' then ensures that the radio is on exactly the same frequency as the transmission – any error shown on the dial will be the receiver error and any drift in tone will be receiver drift.

For most radios it will suffice to select either upper or lower sideband mode with a wide filter setting and use a loudspeaker or pair of headphones with a good low frequency audio performance. Many SW receivers have internal speakers which are not good when it comes to reproducing low notes so are useless for this procedure.

As you carefully tune down and hear the note drop, you should find that the S-meter needle starts to fluctuate. This means that you are very close to the transmission – any error shown on the dial will be indicating the 'phase error' between the STFT and your BFO. When your needle moves at its slowest rate, you have reached zero beat. Make a note of the dial reading as this will show the receiver error.

For those radios with particularly good filters, which may prevent a 'zero beat' approach, a similar technique can be used by switching to CW mode. This method requires the use of an audio digital frequency meter – ask around at your radio club and you will probably find one. First of all, refer to the manual and find the 'CW offset' frequency. This is the audio frequency which the receiver will produce when it is exactly tuned to the carrier of the STFT. Common values are 600Hz to 800Hz. Some radios allow the user to programme the CW offset, so make sure that it has not been changed before you start the tests.

The procedure is essentially the same as has just been described, except that you will be tuning the radio until the DFM reads 600Hz exactly, then the receiver dial will show you any error. By repeating this test on a number of frequencies, you can be confident that your radio is accurate, or at least be aware of any errors or developing problems, and of course these are handy techniques for testing a radio you are considering buying.

STFT Stations, Schedules and Contact Information

ARGENTINA

⌨ Servicio de Hidrografía Naval, Observatorio Naval, Av. España 2099, 1107 Buenos Aires, Argentina ☎ +54 11 43611162 📠 +54 11 43611162 **E**: onba@hidro.gov.ar **W**: www.hidro.gov.ar

Location	Call	kHz	kW	Mode	Schedule
Buenos Aires	LOL	10000	2	AM	1400-1500 (MF)

BELARUS

⌨ 43-y uzel svyazi Voenno-morskogo Flota RF, Vileyka, Belarus.

Location	Call	kHz	kW	Mode	Schedule**
Vileyka	RJH69	*25	300	CW	0706-0747

Key: *) :06-:25 on 25.0, :27-:30 on 25.1, :32-:35 on 25.5, :38-:41 on 23.0, :44-:47 on 20.5kHz; **) 1h earlier in summer (DST in Belarus). NB: Site is operated by Russian Navy; see Russia for HQ.

BRAZIL

⌨ Divisão Serviço da Hora (DSHO), Observatório Nacional, R. Gal. José Cristino 77, São Cristóvão, Rio de Janeiro, CEP 20921-400, Brazil ☎ +55 21 35049100 📠 +55 21 25806041 **E**: dsh@on.br **W**: www.horalegalbrasil.mct.on.br

Location	Call	kHz	kW	Mode	Schedule
Rio de Janeiro	PPE	10000	1	AM/U	24h

CANADA

⌨ Institute for National Measurement Standards, National Research Council of Canada (NRC-INMS), 1200 Montreal Road, Bldg M-36, Ottawa, Ontario, K1A 0R6, Canada ☎ +1 613 9935698 📠 +1 613 9521394 **E**: radio.chu@nrc-cnrc.gc.ca **W**: inms-ienm.nrc-cnrc.gc.ca

Location	Call	kHz	kW	Mode	Schedule
Ottawa	CHU	3330	10	AM/U	24h
	CHU	7850	3	AM/U	24h
	CHU	14670	3	AM/U	24h

CHINA (P.R.)

⌨ National Time Service Center (NTSC), Chinese Acadamy of Sciences, P.O.Box 18, Lintong 710600, Shaanxi, P.R.China ☎ +86 29 83890344 📠 +86 29 83890196 **E**: time@ntsc.ac.cn **W**: time.ntsc.ac.cn

Location	Call	kHz	kW	Mode	Schedule
Shangqiu	BPC	68.5	-	CW	24h
Lintong	BPM	2500	10	AM	0730-0100
(Mt. Li)	BPM	5000	20	AM	24h
	BPM	10000	20	AM	24h
	BPM	15000	20	AM	0100-0900

ECUADOR

⌨ Instituto Oceanográfico de la Armada, Casilla 5940, Guayaquil, Ecuador ☎ +593 4 2481300 📠 +593 4 2485166 **E**: inocar@inocar.mil.ec **W**: www.inocar.mil.ec

Location	Call	kHz	kW	Mode	Schedule
Guayaquil	HD2IOA	3810	1	AM/U	0000-1200

FINLAND

⌨ Centre for Metrology and Accreditation (MIKES), P.O.Box 9, Tekniikantie 1, 02151 Espoo, Finland ☎ +358 10 6054000 📠 +358 10 6054299 **E**: kalevi.kalliomaki@mikes.fi **W**: www.mikes.fi

Location	Call	kHz	kW	Mode	Schedule
Espoo	(none)	25000	0.1	AM	24h

NB: Tx is intended for local service only. Projected to move to 2500kHz, subject to international frequency coordination.

FRANCE

📻 Laboratoire national de métrologie et d'essais - Système de Références Temps-Espace (LNE-SYRTE), 61 avenue de l'Observatoire, 75014 Paris, France ☎+33 140512070 🖩 +33 143255542 **E**: info.syrte@obspm.fr **W**: syrte.obspm.fr

Location	Call	kHz	kW	Mode	Schedule
Allouis	(none)	162	2000*	PSK	24h

Key: *) 1000kW evening/nighttime
NB: Tx is provided by Télédiffusion de France (TDF) and carries the France Inter radio prgr.

GERMANY

📻 Physikalisch-Technische Bundesanstalt (PTB), Bundesallee 100, 38116 Braunschweig, Germany ☎ +49 531 5923006 🖩 +49 531 5923008 **E**: time@ptb.de **W**: www.ptb.de

Location	Call	kHz	kW	Mode	Schedule
Mainflingen	DCF77	77.5	50	CW/PSK	24h

NB: Tx is leased from Media Broadcast.

HAWAII (USA)

📻 NIST radio station WWVH, P.O.Box 417, Kekaha, HI 96752, USA **E**: wwvh@boulder.nist.gov

Location	Call	kHz	kW	Mode	Schedule
Kekaha, HI	WWVH	2500	5	AM	24h
	WWVH	5000	10	AM	24h
	WWVH	10000	10	AM	24h
	WWVH	15000	10	AM	24h

NB: Time announced in female voice. NIST HQ: see USA.

JAPAN

📻 National Institute of Information and CommunicationsTechnology (NICT), Applied Research and Standards Frequency Division, Japan Standard Time Group, 4-2-1, Nukui-Kitamachi, Koganei, Tokyo 184-8795, Japan ☎ +81 42 3277567 🖩 +81 42 3276689 **E**: horonet @nict.go.jp **W**: jjy.nict.go.jp

Location	Call	kHz	kW	Mode	Schedule
Mt. Ohtakadoya	JJY	40	10	CW	24h
Mt. Hagane	JJY	60	10	CW	24h

KOREA, SOUTH

📻 Time & Frequency Laboratory, Korea Research Institute of Standards & Science (KRISS), P.O.Box 102, Yusong, Daejeon 305-600, Rep. of Korea ☎ +82 42 8685114 🖩 +82 42 8611494 **E**: hslee@kriss.re.kr **W**: krissol.kriss.re.kr

Location	Call	kHz	kW	Mode	Schedule
Daejeon	HLA	5000	2	AM	24h

KYRGYZSTAN

📻 Uzel svyazi Voenno-morskogo Flota RF, Chaldovar, Kyrgyzstan.

Location	Call	kHz	kW	Mode	Schedule**
Chaldovar	RJH66	*25	300	CW	0406-0447,1006-1047

Key: *) :06-:25 on 25.0, :27-:30 on 25.1, :32-:35 on 25.5, :38-:41 on 23.0, :44-:47 on 20.5kHz; **) 1h earlier in summer.
NB: Site is operated by Russian Navy; see Russia for HQ.

RUSSIA

📻 Generalniy Shtab Voenno-morskogo Flota RF (Russian Navy), St.Petersburg, Russia **W**: www.navy.ru

Location	Call	kHz	kW	Mode	Schedule°
Arkhangelsk	RJH77	*25	300	CW	0906-0947
Khabarovsk	RAB99	**25	300	CW	0206-0240, 0606-0640
Krasnodar	RJH63	**25	300	+CW	1106-1140
N.Novgorod	RJH90	*25	300	CW	0806-0847

Key: *) :06-:25 on 25.0, :27-:30 on 25.1, :32-:35 on 25.5, :38-:41 on 23.0, :44-:47 on 20.5kHz; **) :06-:20 on 25.0, :21-:23 on 25.1, :24-:26 on 25.5, :27-:31 on 23.0, :32-:40 on 20.5kHz; +) also in FSK mode :36-:40 on 20.5kHz; °) 1h earlier in summer (DST in Russia)

📻 Main Metrological Center of the State Service for Time, Frequency and Earth rotation parameters determination (SSTF), National Research Institute for Physicotechnical and Radio Engineering Measurements (FSUE "VNIIFTRI"), Moscow Region, 141570 Mendeleevo, Russia ☎+7 495 5350836 🖩 +7 495 5350871 **E**: office@vniiftri.ru **W**: www.vniiftri.ru

Location	Call	kHz	kW	Mode	Schedule*
Taldom	RBU	66.66	10	AM	24h
Taldom	RWM	4996	5	CW	24h
Taldom	RWM	9996	5	CW	24h
Taldom	RWM	14996	8	CW	24h

Key: *) 1h earlier in summer (DST in Russia).

📻 East Siberian Branch of FSUE "VNIIFTRI", 664056 Irkutsk, ul. Borodina 57, Russia ☎+7 3952 468303 🖩 +7 3952 463848 **E**: office@niiftri.irk.ru **W**: www.vniiftri-irk.ru

Location	Call	kHz	kW	Mode	Schedule*
Angarsk	RTZ	50	10	CW	2200-2100

Key: *) 1h earlier in summer (DST in Russia).

SPAIN

📻 Real Instituto y Observatorio de la Armada (ROA), Calle Cecilio Pujazón s/n, 11110 San Fernando ☎+34 956545590 🖩 +34 956599366 **W**: www.roa.es

Location	Call	kHz	kW	Mode	Schedule
San Fernando	EBC	4998	10		1030-1055 (MF)
		15006	10		1000-1025 (MF)

SWITZERLAND

📻 Bundesamt für Metrologie und Akkreditierung (METAS), Lindenweg 50, 3003 Bern-Wabern, Switzerland ☎ +41 31 3233111 🖩 +41 31 3233210 **E**: laurent-guy.bernier@metas.ch **W**: www.metas.ch

Location	Call	kHz	kW	Mode	Schedule
Prangins	HBG	75	20	CW	24h

NB: Service is scheduled to cease on 31 December 2011.

UNITED KINGDOM

📻 National Physical Laboratory, Hampton Road, Teddington, Middlesex, TW11 0LW, United Kingdom ☎ +44 20 89773222 🖩 +44 20 86140446 **E**: time@npl.co.uk **W**: www.npl.co.uk

Location	Call	kHz	kW	Mode	Schedule
Anthorn*	MSF	60	15	CW	24h
Droitwich**	(none)	198	400	PSK	24h
Burghead**	(none)	198	50	PSK	24h
Westerglen**	(none)	198	50	PSK	24h

Key: *) Tx is leased from Babcock; **) Txs are provided by Arqiva and carry the BBC Radio 4/BBCWS prgrs.

UNITED STATES OF AMERICA

📻 National Institute of Standards and Technology (NIST), Physics Laboratory, Time & Frequency Division, 325 Broadway, Mailcode 847.00, Boulder, CO 80305-3328, USA ☎ +1 303 4973295 🖩 +1 303 4976461 **W**: www.boulder.nist.gov **W**: www.bldrdoc.gov/timefreq

📻 NIST radio stations WWV/WWVB, 2000 East County Rd. 58, Ft. Collins, CO 80524 **E**: nist.radio@boulder.nist.gov

Location	Call	kHz	kW	Mode	Schedule
Ft. Collins, CO	WWVB	60	50	CW	24h
	WWV	2500	2.5	AM	24h
	WWV	5000	10	AM	24h
	WWV	10000	10	AM	24h
	WWV	15000	10	AM	24h
	WWV	20000	2.5	AM	24h

NB: WWV: Time announced in male voice. See also under Hawaii.

VENEZUELA

📻 Dirección de Hidrografía y Navegación (DHN), Estación Transmisora YVTO, Apartado Postal 6745, Caracas, Venezuela ☎ +58 212 482266 🖩 +58 212 4835879 **E**: dhn@truevision.net **W**: www.dhn.mil.ve

Location	Call	kHz	kW	Mode	Schedule
Caracas	YVTO	5000	2	AM	24h

INTERNATIONAL ORGANISATIONS

ARAB STATES BROADCASTING UNION (ASBU)
✉ CP 250, 1080 Tunis Cedex, Tunisia Street address: Rue 8840, centre urbain nord, Tunis, Tunisia
☎ +216 71849000 📠 +216 71843054
E: asbu@asbu.intl.tn **W:** www.asbu.net
LP: Pres: Riyadh Kamal Najm; Vice Pres: Momtaz al-Skeikh

ASIA-PACIFIC BROADCASTING UNION (ABU)
✉ P.O.Box 1164, Lorong Maarof, 59000 Kuala Lumpur, Malaysia
☎ +60 3 22823592 📠 +60 3 22825292
E: info@abu.org.my **W:** www.abu.org.my
LP: Pres/Exec. Vice Pres: Yoshinori Imai; Secr. General: Javad Mottaghi

ASIA-PACIFIC INSTITUTE FOR BROADCASTING DEVELOPMENT (AIBD)
✉ P.O.Box 1137, Pantai, 59700 Kuala Lumpur, Malaysia. Street address: Angkasapuri, Jalan Pantai Dalam, 50614 Kuala Lumpur, Malaysia
☎ +60 3 22824618 📠 +60 3 22822761
E: info@aibd.org.my **W:** www.aibd.org.my
LP: Dir: Yang Binyuan

ASIA-PACIFIC SATELLITE COMMUNICATIONS COUNCIL (APSCC)
✉ Suite T-1602 Poonglim Iwantplus, 255-1 Seohyun-dong, Bundang-gu, Seongnam, Kyungi-do 463-862, Republic of Korea
☎ +82 31 7836244 📠 +82 31 7836249
E: info@apscc.or.kr **W:** www.apscc.or.kr
LP: Dir Project Development & Coordination: Inho Seo; Dir Membership & External Relation: Chloe Song

ASOCIACIÓN INTERNACIONAL DE RADIODIFUSIÓN (AIR)
✉ Oficina Central, Carlos Quijano 1264, 11100 Montevideo, Uruquay ☎ +598 2 9011319 📠 +598 2 9080458
E: mail@airiab.com **W:** www.airiab.com
LP: Pres: Luís Pardo Sainz; DG: Héctor Oscar Amengual

CARIBBEAN BROADCASTING UNION (CBU)
✉ CBU Secretariat, Suite 1B, Building #6A, Harbour Industrial Estate, Harbour Road, St. Michael, 11145, Barbados
☎ +1 246 4301006 📠 +1 246 2289524
E: patrick.cozier@caribsurf.com **W:** www.caribunion.com
LP: Secr. General: Patrick Cozier

COMMONWEALTH BROADCASTING ORGANISATION (CBA)
✉ Secretariat, 17 Fleet Street, London EC4Y 1AA, United Kingdom
☎ +44 20 75835550 📠 +44 20 75835549
E: cba@cba.org.uk **W:** www.cba.org.uk
LP: Pres/GM: Moneeza Hashimi; Secr. General: Elizabeth Smith

DIGITAL RADIO MONDIALE
✉ DRM Project Office (London), c/o BBC World Service, Bush House, Aldwych, Strand, London WC2B 4PH, United Kingdom
☎ +44 207 5573271📠 +44 207 5570225
E: projectoffice@drm.org **W:** www.drm.org
LP: Chmn (DRM Consortium)/Pres (DRM Association): Ruxandra Obeja; Project Dir: Vineeta Dwivedi

EUROPEAN BROADCASTING UNION (EBU)
✉ L'Ancienne-Route 17A, CH-1218 Grand-Saconnex, Switzerland
☎ +41 22 7172111 📠 +41 22 7474000
E: ebu@ebu.ch **W:** www.ebu.ch
LP: Pres: Jean-Paul Philippot; DG: Ingrid Deltenre

HIGH FREQUENCY CO-ORDINATION CONFERENCE (HFCC)
✉ Vinohradská 12, 12099 Praha 2, Czech Republic
☎ +42 2 22715005 📠 +42 2 22715005
E: info@hfcc.org **W:** www.hfcc.org
LP: Chmn: Oldrich Cip

INTERNATIONAL INSTITUTE OF COMMUNICATIONS (IIC)
✉ Regent House, 24-25 Nutford Place, London W1H 5YN, United Kingdom ☎ +44 20 77237210 📠 +44 20 77236982
E: enquiries@iicom.org **W:** www.iicom.org
LP: Pres: Fabio Colosanti

INTERNATIONAL TELECOMMUNICATIONS UNION (ITU)
✉ Place des Nations, 1211 Genève 20, Switzerland
☎ +41 22 7305111 📠 +41 22 7337256
E: itumail@itu.int **W:** www.itu.int
LP: Secr. General: Hamadoun Touré; Deputy Secr. General: Houlin Zhao; Dir Radiocommunications Bureau: Valery Timofeev

ISLAMIC BROADCASTING UNION (IBU)
✉ P.O.Box 6351, 21442 Jeddah, Saudi Arabia; Street address: 80 Rabita Islamia Street, Mushrafa, Jeddah, Saudi Arabia
☎ +966 2 6722269 📠 +966 2 6722600
E: isboo@isboo.org **W:** www.isboo.org
LP: Dir: Zainal Iberahim; Secr. General: Abdelouahed Belkeziz

NORTH AMERICAN BROADCASTERS ASSOCIATION (NABA)
✉ P.O.Box 500, Station A, Toronto, Ontario, M5W 1E6, Canada
☎ +1 416 5989877 📠 +1 416 5989774
E: contact@nabanet.com **W:** www.nabanet.com
LP: Pres: Leonardo Ramos Mateos; Secr. General: John Harding

SOUTHERN AFRICAN BROADCASTING ASSOCIATION (SABA)
✉ c/o SABC, P.O.Box 8606, Auckland Park, Johannesburg, South Africa
☎ +27 11 8884481 📠 +27 11 8884489
Email: sabahq@saba.co.za **Web:** www.saba.co.za
LP: Pres: Albertus Aochamub; Secr. General: Arlindo Lopez

UNION AFRICAINE DE RADIODIFFUSION (UAR)
✉ BP 3237, Dakar, Senegal. Street address: 101 Rue Carnot, Dakar, Senegal
☎ +221 8215970 📠 +221 8215113
W: www.aub-uar.org
LP: Pres: Julien Pierre Akpaki; CEO: Lawrence Atiase

WORLD ASSOCIATION OF COMMUNITY RADIO BROADCASTERS (AMARC)
✉ International Secretariat, 705 Rue Bourget #100, Montreal, Quebec, H4C 2M6, Canada
☎ +1 514 9820351 📠 +1 514 8497129
E: secretariat@si.amarc.org **W:** www.amarc.org
LP: Pres: Steve Buckley; Secr. General: Marcelo Solervicens

WORLD DMB FORUM
✉ Project Office, 77 Shaftesbury Avenue, London W1D 5DU, United Kingdom
☎ +44 20 72884642 📠 +44 20 72884643
E: info@worlddab.org **W:** www.worlddab.org
LP: Project Dir: Letty Zambrano

SELECTED INTERNET RESOURCES FOR DXERS & RADIO LISTENERS

GENERAL & LINK PAGES:

BCL News: bclnews.it
BCL News (Chinese): www.5bcl.com
Bruce's AM Log: bamlog.com
DX Information Centre: dxinfocentre.com
DXing.com: dxing.com
DXing.info: dxing.info
DX Portal (Russian): dxing.ru
Canada's Original World Band Radio page: dxer.ca
Cumbre DX: www.cumbredx.net
El Mundo de la R. (Spanish): elmundodelaradio.com
Hard-Core-DX: hard-core-dx.com
History of SW: ontheshortwaves.com
Ontario DX Association: odxa.on.ca
Predavatel Europa: predavatel.eu
Radio Monde (French): radiomonde.voila.net
Radio Communication Archive: dokufunk.org
Radio Directory: radiodirectory.com
Radio Intelligencer: radiointel.com
Radio Portal: radio-portal.org
Radio Heritage Foundation: radioheritage.net
World of Radio: worldofradio.com

DATABASES & SCHEDULES:

ADDX Foreign Frequency List:
addx.de/Hfpdat/plaene.php
AM Query: fcc.gov/mb/audio/amq.html
Asian Broadcasting Institute: abiweb.jp
Bi Newsletter: geocities.jp/binewsjp
Broadcasting in South & South East Asia:
asiawaves.net
Combined HF Skeds: hfskeds.com/skeds
DXAsia: dxasia.info
EiBi Shortwave Schedules: eibi.de.vu
Euro-African Medium Wave Guide: emwg.info
FM List: fmlist.org
MW List: mwlist.org
KOJE-KOH-KOMEX: www.tapiokalmi.net/dx/koje
Pacific Asian Log: radioheritage.net/PAL_search.asp
Prime Time Shortwave: primetimeshortwave.com
Shortwave Radio Resource Center: shortwave.hfradio.org

NEWS SOURCES AND BULLETINS:

BC-DX Top News: wwdxc.de/topnews.htm
DX-Listening Digest: worldofradio.com/dxldmid.html
Electronic DX Press Radio Monitoring Association:
edxp.org

Shortwave Bulletin: hard-core-dx.com/swb
Ydun's Mediumwave Info: mediumwave.info

ONLINE RADIO:

mikesradioworld.com – live365.com – multilingualbooks.
com/online-radio.html –publicradiofan.com – radiobeta.
com –radiotower.com – reciva.com – shoutcast.com
– tvradionetwork.com – web-radio.fm – vtuner.com
Web DX-listening: globaltuners.com – onlinereceivers.
net – sdr-radio.com

RECEIVERS, ANTENNAS:

Dave's Radio Receiver page: n9ewo.angelfire.com
RF Circuit Building Blocks Page: qsl.net/wa1ion

STATION IDENTIFICATION:

Interval Signals Online: intervalsignals.net
National Anthems: national-anthems.net –
nationalanthems.info
Interval Signal Database: intervalsignals.org

PROPAGATION:

Solar Terrestrial Activity Report: solen.info/solar
Space Weather Prediction Center: swpc.noaa.gov

MAILING LISTS & IRC:

A-DX Mailingliste (G): ratzer.at/A_DX_Info.php
Hard-Core-DX mailing list:
hard-core-dx.com/mailman/listinfo/hard-core-dx
IRC at irc.starchat.net: #mwdx & #swl
Lists at Mailman.qth.net: AMFMTVDX – SWL
Lists at Yahoogroups.com: ABDX – ASWLC – ausS-
WDXgroup – bangladx – bclnews (It.) – condiglist
(S) – conexiondigital (S) – dexismo (Port.) – DRM-L
– DRM-shortwave-dx – drmna – dxld – dxplorer
– dx_india – dx_sasia – fmdxitalia (I) – harmonics
– LatinMWDX – media_dx (Ru.) – mwdx – mw-br
(Port.) – NORDX (Sw.) – NoticiasDX (S) – ondescourtes
(F) – open_dx (Ru.) – PakistanDXers – playdx2003
(I) – ptdx (Port.) – RadioBroadcasting – radioescutas
(Port.) – RealDX – recradioshortwave – rusdx (Ru.)
– SDRlist – shortwave – ShortwaveBasics – short-
wavedxing – shortwavelistening – ShortWaveRadio
– ShortwaveRadios – shortwave-radio – shortwaves
– shortwavesites – Shortwave-SWL-Antenna –sky-
wavesdx – skywavesmw – soft_radio – SWPirates
– SWR-Wordwide – thebasicsofshortwave – UDXF
– webreceivers
Radio Listening Interest Group in facebook.com

ADVERTISERS' INDEX

AOR	27
Australian Radio DX Club	48
Babcock International	Back cover
BBC World Service	2
British DX Club	46
Committee to Preserve Radio Verifications	49
Continental Electronics	Inside back cover
Cumbre DX	64
Danish SW Club International	46
Deutsche Welle	4
Electronic DX Press	49
Focal Press	35
George Jacobs Associates	48
MonitoR	49
Grove Enterprises	36
National Radio Club	48 & 385
New Zealand Radio DX League	48
Popular Communications	40
PW Publishing	30
Radio Heritage Foundation	64
Sherwood Engineering Inc	49
Studio DX	243
Transradio	Inside front cover
Universal Radio	21
Waters and Stanton	35
WinRadio	6
World DX Club	64

ADVERTISING SALES

Advertising sales manager:

Beth Leinbach
2698 Green Cove Road
Brasstown, NC 28902
USA
Tel/fax: +1 828 389 4007
Email: bleinbach@brmemc.net

Advertising sales rep (Europe):

Enrico Callerio
Media Age srl
via Stefano Jacini, 4
20121 Milano
ITALY
Tel: +39 02 876038
Fax: +39 02 86450149
Email: callerio@monitor-radiotv.com